# WEATHER AMERICA

# WEATHER AMERICA

## The Latest Detailed Climatological Data for Over 4,000 Places — *with rankings*

Alfred N. Garwood
EDITOR

TOUCAN VALLEY PUBLICATIONS

ISBN 1-884925-60-X

Library of Congress Cataloging-in-Publication Data is available.



**Toucan Valley Publications, Inc.**
142 N. Milpitas Blvd., Suite 260
Milpitas, CA 95035

Manufactured in the United States of America
First Edition

# CONTENTS

# INTRODUCTION

***Weather America*** is a new reference source which provides key climatological data for over 4,000 places in the United States. Based on original work with Federal government weather station data tapes, ***Weather America*** is designed to provide a comprehensive single-volume handbook that is both easy to use and up-to-date. It represents a different approach in the way weather data is compiled as well as a step forward in the way such data is presented and published. To fully understand the reasons for this book and what makes it different from other ostensibly similar works, it is necessary to briefly review weather publications already available.

Over the years a good deal of weather information about the United States has been published, both by private sector publishers and by the government itself, but with few exceptions almost all weather data published has been the same. And well might this be expected, as almost all of it originates with the National Oceanic and Atmospheric Administration (NOAA), and more specifically within NOAA, the National Climatic Data Center (NCDC) in Asheville, NC. NCDC organizes almost 10,000 weather stations across the country and, on a regular basis, collects, compiles, edits, and adjusts climatological data, and then publishes it in a number of regular serial publications.

Perhaps the best known and most widely used of these is a series of individual state data pamphlets titled *Climatography of the United States*. These state pamphlets provide monthly station averages of temperature, precipitation, and heating and cooling degree days for a thirty year period (most recently, 1961-1990) in four consecutive tables. In addition to this series, NOAA also regularly publishes a number of other series on storm data, precipitation, world climatic conditions, and the like.

As is true in so many other areas that require vast resources to collect and manage information, the Federal government is the major compiler/publisher of weather data, and often provides the only published source for specific types of information. But as pamphlets and serials are a chore to use (and are not readily available) for the average researcher, a number of private sector publishers have taken existing Federal weather data compilations and re-compiled them into single- or multi-volume reference works. Such efforts have been helpful in gaining a wider distribution of government data, and have served the needs of a number of researchers in different fields. In fact, with the exception of depository libraries (which collect and make available the original government serial publications), the weather data sources found in most public and academic libraries are from such private-sector publishers who have adapted NOAA's publications specifically for library use.

But no matter how valuable such efforts are, the re-collecting and republishing of data does not always meet all needs. And in considering much of the Federal statistical publishing output, it must be remembered that the collection and publication of government data is largely driven by governmental concerns. For example, most of the statistical information the Federal government collects about poverty (and persons in poverty) is collected not for the research use of social scientists (although social scientists do constitute a large group of secondary users of this data), but principally for those who administer

Federal programs relating to poverty. As a result, secondary users have had to make do with what has been produced for others. This paucity of desired information may be due to the fact that the data sought may not have been collected by the government at all, but more frequently it is because a good deal of collected data remains unpublished. This is also true for weather data. And unlike information about poverty where there have been numerous monographs and reports which reaggregate and analyze Federal measures based on the source data itself, most of the weather data that has been published has been a simple rearrangement or (to a more limited extent) a re-editing of previously published Federal data. There are many data users who would like to have weather information that is both more appropriate to their purpose, and available in more accessible form.

This is the origin of the idea for ***Weather America***. As the government collects a lot more weather data than it actually publishes, and as current computer technology enables editors with the equipment and expertise to review all the data collected, it is possible to select and calculate that which is appropriate for general research use without being limited by the previous editorial concerns of NOAA or NCDC. Such an endeavor represents a clear advantage to data users, as earlier private sector publishers have essentially relied on (and thus been limited by) what the government has chosen to publish. The notion here is to go beyond what is offered by the Federal government through its serials, and beyond the efforts of earlier publishers who have relied exclusively on the government's published output, by going directly to the source data itself, analyzing and evaluating it in order to provide a more complete and appropriate compilation for libraries and their clientele.

Specifically, by relying on the primary source itself, there are five things that the editors were able to achieve that make ***Weather America*** different:

1) **Breadth of coverage**. Most weather data books provide information for only a few hundred weather stations. Because climate is highly localized, the more places a publication covers, the more complete the picture presented. ***Weather America*** presents data for 4,158 places.

2) **Depth of data for each place**. In reviewing all the weather data that NCDC collects, thirteen key data items were selected for presentation here. Source data for these thirteen items were then used for computations to provide averages for all twelve months along with a separate annual average, for every station. Most notably, ***Weather America*** includes weather station averages for some temperature and precipitation measures (e.g., snowfall), and heating and cooling degree days, which are not included in the other reference sources, and no other publisher (not even the Federal government itself) computes this data for as many places as has been done here.

3) **Currency.** By using the data tapes themselves, it has been possible to provide the most recent data. As thirty years is considered the standard time period for computation, that is what was employed. Traditionally weather data is collected and presented for every third decade, e.g. 1951-1980, 1961-1990, but because more recent data is available, ***Weather America*** is based on the most recent thirty year period for which complete year data is available, 1965-1994.

4) **Organization and format.** Use of the original tapes freed the editors from the organization, format, and structure of Federal publications. This has allowed for the display of all climatological data about a given station (temperature, precipitation, etc.) in a single station table, each station table being identical format. Not only are the tables concise and easy to use, but the fact that the format is uniform

throughout facilitates comparisons between places anywhere in the nation. Additionally, it has been possible to provide appropriate indexes to the tables, one by county of location, and another by station elevation.

5) **Ranking tables**. Use of the source tapes has allowed for the compilation of ranking tables. Such tables have a clear reference use for those interested in the hottest (or coldest) place in a state, those with the most (or least) snowfall, and the like. In every state section there are four ranking tables, and at the end of the state sections there are four national ranking tables.

There are also a number of other attributes of ***Weather America*** which enhance its usefulness. Each state section opens with a narrative description which provides a lucid general overview of each state's climatic conditions. In addition, each state section contains a map displaying the location of weather stations within the state. In combination with the station data tables, the indexes and map allow multiple access points to weather information in the most commonly sought after ways.

Taken together ***Weather America*'s** innovations and supplemental resources combine to create a source of choice for researchers looking for basic climatic data for places across the United States.

## Inclusion Criteria: How the Data and Stations Were Selected

There were two central goals in the preparation of ***Weather America***. The first was to select those data elements which would have the broadest possible use by the greatest range of potential users. For the approximately 280 First Order stations there is a substantial quantity and variety of climatological data collected, however for the majority of stations the data is more limited. After evaluating the available data set, the editors chose seven temperature measures, four precipitation measures, and heating and cooling degree days -- thirteen key data elements that are widely requested and are believed to be of the greatest general interest.

The second goal was to provide data for as many weather stations as possible. Although there are almost 10,000 such stations, not every station collects data for both precipitation and temperature, and even among those that do, the data is not always complete for the last thirty years. As the editors used a different methodology than that of NCDC to compute data (see below), a formal data sufficiency criteria was devised and applied to the source tapes in order to select stations for inclusion. The basic criteria was that a station must have data for temperature, precipitation, heating and cooling degree days of sufficient quantity to create a meaningful average. More specifically, the definition of sufficiency here has two parts. First, there must be 22 values for a given data element, with the exception of cooling degree days which required only eleven values in order to be considered sufficient (more about this later). Second, seven of the thirteen elements included in the table must pass this sufficiency test. For example, in regard to average maximum temperature (the first element on every data table), a given station needs to have a value for every month of at least 22 of the last thirty years in order to meet the criteria, and, in addition, every station included must have at least seven of the thirteen elements to this minimal level of completeness in order to fulfill the criteria. By using this procedure, 4,158 stations met the requirements for inclusion here.

## Organization

The main body of ***Weather America*** is composed of fifty state sections (weather information for the District of Columbia is interfiled by station name in the Maryland section). Each section begins with a narrative description of the climatic conditions of the state, extracted from the *Climatography of the United States No. 60* series by NOAA.

This is followed by two indexes to the data tables. The first index is by the name of the county in which the station is located (or by the name of the county equivalent, or by the name of the NOAA division in the case of Alaska), and the second index is by station elevation.

Following the indexes is a state map (or maps) with the location of weather stations clearly marked. It should be noted that as the maps are reproduced from official NCDC maps, and as not all NCDC weather stations meet our inclusion criteria, some maps may include a few stations for which there are no tables here.

The map(s) are followed by individual tables for each weather station selected.

The final element of each state section is a collection of four ranking tables which cover the July high temperature, the January low, annual precipitation, and annual snowfall displaying the 25 highest (and lowest) places in every state.

The book concludes with a short section devoted to ranking tables for those same elements covered in the state rankings, but with the 100 highest (and lowest) places nationwide displayed.

## What the Data in the Tables and Rankings Mean

The data and ranking tables have been designed to be self-explanatory; however a short review of some basics is appropriate.

Data values that appear in bold italics indicate a different base for computing the data was used (see Methodology below for a complete explanation).

The tables are organized alphabetically within each state section by station name. The station name is almost always a place name, and is shown here just as it appears in NCDC data. There are just six exceptions, cases where the order of the name was changed slightly as an aid to alphabetization. These five are: South Entr Yosemite was changed to Yosemite South Entr; UCLA was changed to Los Angeles UCLA; Upr San Leandro Fltr was changed to San Leandro Upr Fltr; LSU Ben Hur Farm was changed to Baton Rouge LSU Ben; Nat'l Arboretum DC was changed to Washington Natl Arbo; and in Oregon, Headworks Portland W was changed to Portland Headworks W).

The station name is followed by the county in which the station is located (or by county equivalent name, or by the NOAA division name in Alaska), the elevation of the station (at the beginning of the thirty year period), and the longitude and latitude.

The definitions of the thirteen descriptors of the data elements are also straightforward. The average temperatures (maximum, minimum, and mean) are the average (see Methodology below) of those temperatures for all available values for a given month. For example, for a given station the average maximum temperature for July is the arithmetic average of all available maximum July temperatures for that station. (Maximum means the highest recorded temperature, minimum means the lowest recorded temperature, and mean means an arithmetic average temperature.)

The days for maximum temperature and minimum temperature are likewise the average number of days those criteria were met for all available instances. The symbol ≥ means greater than or equal to, the symbol ≤ means less than or equal to. For example, for a given station, the number of days the maximum temperature was greater than or equal to 90°F in July, is an arithmetic average of the number of days in all the available Julys for that station.

Heating and cooling degree days are based on the median temperature for a given day and its variance from 65°F. For example, for a given station if the day's high temperature was 50°F and the day's low temperature was 30°F, the median (midpoint) temperature was 40°F. 40°F is 25 degrees below 65°F, hence on this day there would be 25 heating degree days. This also applies for cooling degree days. For example, for a given station if the day's high temperature was 80°F and the day's low temperature was 70°F, the median (midpoint) temperature was 75°F. 75°F is 10 degrees above 65°F, hence on this day there would be 10 cooling degree days. All heating and/or cooling degree days in a month are summed giving respective totals for each element for that month. These sums for a given month for a given station over the past thirty years are again summed and then arithmetically averaged.

It should be noted that the heating and cooling degree days do not cancel each other out. It is possible to have both for a given station in the same month.

Precipitation data follows the same computational pattern as heating and cooling degree days. Total precipitation and total snowfall are arithmetic averages of cumulative totals for the month. All available values for the thirty year period for a given month for a given station are summed and then divided by the number of values. The same is true for days of greater than or equal to 0.1" of precipitation, and days of greater than or equal to 1" of snow depth on the ground.

The rankings (both for individual states and the national rankings) are a listing of the 25 (or in the case of national rankings, 100) highest (or lowest) stations for a given item. In the case of a tie, the stations are listed alphabetically. If a tie is the final element in a ranking table, it is possible that other stations with the same value for that element have been left off the table due to lack of space. In cases where a given place has more than one station, the full station name is provided.

Note that all values presented in the tables and the rankings are averages of available data (see Methodology below) for that specific data element for the last thirty years (1965-1994).

## Source of the data

The source of the data in ***Weather America*** is the original NCDC computer tapes of station data (TD 3220 Summary of Month Co-Operative). This data has been collected from the almost 10,000 weather stations around the country.

NCDC has two main classes or types of weather stations, First Order stations staffed by professional meteorologists (there are approximately 280 such stations), and cooperative stations staffed by volunteers. In ***Weather America***, all first order stations are included, plus approximately 3,900 cooperative stations.

There are a number of variables in collecting weather data. However, consistency is key, especially in regard to the time of observation, the instruments used, the location of the instruments (and the station itself) and the like. Those who wish to learn more about basic data collection protocols are referred to NCDC documents concerning these areas.

## Our Methodology

***Weather America*** is based on an arithmetic average of all available data for a specific data element at a given station. For example, the average maximum daily high temperature during July for Alma, Michigan, was abstracted from NCDC source tapes for the thirty Julys, starting in July, 1965 and ending in July, 1994. These thirty figures were then summed and divided by thirty to produce an arithmetic average. As might be expected, there were not thirty values for every data element on every table. For a variety of reasons, NCDC data is sometimes incomplete. Thus the following standards were established.

For those data elements where there were 26-30 values, the data was taken to be essentially complete and an average was computed. For data elements where there were 22-25 values, the data was taken as being partly complete but still valid enough to use to compute an average. Such averages are shown in bold italic type to indicate that there were fewer than 26 values. For the few data elements where there were fewer than 22 values, no average was computed and 'na' appears in the space. If any of the twelve months for a given data element reported a value of 'na', no annual average was computed and the annual average was reported as 'na' as well.

This procedure was followed for twelve of the thirteen data elements. The one exception is cooling degree days. The collection of this data began in 1980 so the following standards were adopted: for those data elements where there were 13-15 values, the data was taken to be essentially complete and an average was computed. For data elements where there were 11-12 values, the data was taken as being partly complete but still valid enough to use to compute an average. Such averages are shown in bold italic type to indicate that there were 11-12 values. For the few data elements where there were fewer than eleven values, no average was computed and 'na' appears in the space. If any of the twelve months for a given data element reported a value of 'na', no annual average was computed and the annual average was reported as 'na' as well.

It is interesting to note how complete the data is. Out of a possible 702,702 values which appear in this book (169 values per table, 4,158 tables in the book), only 29,916 appear in bold italic (4.3%), and 9,059 (1.3%) have an 'na'.

Thus the basic computational methodology of ***Weather America*** is to provide an arithmetic average. Because of this, it needs to be stated that such a pure arithmetic average is somewhat different from the special type of average (called a 'normal') which NCDC procedures produce and which appears in Federal publications.

Perhaps the best outline of the contrasting normalization methodology is found in the following paragraph (which appears as part of an NCDC technical document titled, *CLIM81 1961-1990 NORMALS TD-9641* prepared by Lewis France of NCDC in May, 1992):

> Normals have been defined as the arithmetic mean of a climatological element computed over a long time period. International agreements eventually led to the decision that the appropriate time period would be three consecutive decades (Guttman, 1989). The data record should be consistent (have no changes in location, instruments, observation practices, etc.; these are identified here as "exposure changes") and have no missing values so a normal will reflect the actual average climatic conditions. If any significant exposure changes have occurred, the data record is said to be "inhomogeneous", and the normal may not reflect a true climatic average. Such data need to be adjusted to remove the nonclimatic inhomogeneities. The resulting (adjusted) record is then said to be "homogeneous". If no exposure changes have occurred at a station, the normal is calculated simply by averaging the appropriate 30 values from the 1961-1990 record.

In the main, there are two "inhomogeneities" for which NCDC is correcting with normalization: first, adjusting for variances in time of day of observation (at the so-called First Order stations data is based on midnight to midnight observation times; this practice is not necessarily followed at cooperative stations which are staffed by volunteers); and second, estimating data that is either missing or incongruent.

A long discussion of the normalization process is not required here but a short note concerning comparative results of the two methodologies is appropriate.

With the compilation of ***Weather America***, a concern arose because the normalization process would not be replicated; would our methodology produce strikingly different results than NCDC's? To allay concerns, results of the two processes were compared for the time period normalized results are available (1961-1990). In short, what was found was that the answer to this question is "no." Here is what we did:

In the key area of temperature (which represents seven of the thirteen data elements in each station table) the editors selected a sample of stations where it was believed the possibility of variance was greatest (for example those stations where the July high temperature was quite hot, and those stations where the January temperature was quite cold) along with a sample of control stations selected at random. Arithmetic averages for selected data elements were computed according to our methodology for the years 1961-1990 and compared with the normals published by NCDC. On average, our methodology yielded a difference of a little more than one tenth of one degree. In the course of our sampling, no difference greater than seven-tenths of a degree was noted. These tests were repeated and similar tests were conducted on other data elements with similar results. We were surprised at the congruity of results produced by our method compared with the more elaborate normalization procedure. This is not to reflect on the normalization process, but just to suggest that the results published here might not prove greatly different from normalized averages computed by NCDC methodology for the time period covered.

Never-the-less, users should be aware that because of both the time period covered (1965-1994) and the methodology used, data in ***Weather America*** is not compatible with data from other sources.

## Some potential cautions in using *Weather America*

There are two cautions in using ***Weather America***, both of which apply to statistical reference works of this type in general.

First, users need to be aware of the source of the data. The information here comes from NOAA, and it is the most comprehensive and reliable core data available. Although it is the best, it is not perfect. Most weather stations are staffed by volunteers, times of observation sometimes vary, stations occasionally are moved (especially over a thirty year period), equipment is changed or upgraded, and all of these factors (along with some others) affect the uniformity of the data. Nothing has been done here to correct for these factors. ***Weather America*** is not intended for either climatologists or atmospheric scientists. Users with concerns about data collection and reporting protocols are referred to NCDC technical documentation, and to the original computer tapes as well.

Second, there is the trap of that informal logical fallacy known as "hasty generalization," and its corollaries. This may involve presuming the future will be like the past (specifically, next year will be an average year), or it may involve misunderstanding the limitations of an arithmetic average, but more interestingly, it may involve those mistakes made most innocently by generalizing informally on too broad a basis. As weather is highly localized, the data should be taken in that context. A weather station collects data about climatic conditions at that spot, and that spot may or may not be an effective paradigm for an entire town or area. For example, the weather station in Burlington, Vermont is located at the airport about three miles east of the center of town. Most of Burlington is a lot closer to Lake Champlain, and that should mean to a careful user that there could be a significant difference between the temperature readings gathered at the weather station and readings that might be gathered at another place, say, City Hall downtown. How much would this difference be? How could it be estimated? There are no answers here for these sorts of questions, but it is important for users of this book to raise them for themselves. (It is interesting to note that similar situations abound across the country. For example, compare different readings for the multiple stations in San Francisco, CA or for those around the city of New York.)

Our source of data has been consistent, as has our methodology. The data has been computed and reported consistently as well. As a result, ***Weather America*** will prove valuable to the careful and informed reader.

## A Final Word

It is hoped that ***Weather America*** becomes a standard reference in the field. Questions, comments, and criticisms from users are a vital part of the informed choices that need to be made for upcoming editions. If you have a suggestion or comment, please be assured that it is both appreciated and will be carefully considered. If you find an error here, please let us know so that we may promptly correct it. Our goal is to create an accurate, comprehensive, easy to use source which serves your needs. Your help will enable us to serve you better.

# ALABAMA

PHYSICAL FEATURES. The surface of Alabama rises as a rolling plain from the Gulf of Mexico in the southwest to foothills in the central part of the State. Thence there is a rise to the Appalachian Mountains which extend into the northeastern counties. Ridges from the Appalachians extend southward through the eastern counties, with elevations along these ridges as much as 600 to 800 feet above sea level in the southeast. The general elevation of the high northeastern area is about 800 feet above sea level, but some mountain summits rise to over 2,000 feet, the highest (Mount Cheaha in southwestern Cleburne County) being 2,407 feet.

GENERAL CLIMATE. The climate is temperate, becoming largely subtropical near the coast. The summers are long, hot, and humid, with little day-to-day temperature change. In the northeastern counties, higher altitudes help make the summer nights more comfortable. From late June through middle August, approximately a third of the evenings are made comfortable by local afternoon thundershowers which bring cool breezes over the areas where they occur.

In the coldest months of December, January, and February, there are frequent shifts between mild air, which has been moistened and warmed by the Gulf, and dry, cool continental air. Severely cold weather seldom occurs. Even in the northern third of the State, temperatures of zero or lower are rare and occur only when there is snow on the ground. Since cold air on clear nights collects in low places, there is considerable irregularity in the distribution of the last spring or first fall freezes in all sections

PRECIPITATION. Precipitation is nearly all in the form of rain. Snow falls in the northern counties on an average of about twice each winter. The average fall in that area is only about 3 inches per year, and since this includes unusually heavy snows in a few individual winters, some winters have little or none. From late June through the first half of August, nearly all precipitation is from local thundershowers which occur mostly in the afternoons. During late August and in September, summer conditions of atmospheric temperature and moisture persist, but thundershowers become less frequent. However, late night and early morning thundershowers, characteristic of late summer on the coast, continue in the coastal counties until mid-September. Rains during October are nearly always from showers or thundershowers occurring ahead of temperature drops. Such changes become more frequent and more pronounced as winter approaches. Dry, sunny weather prevails most of the time in September and October, but from August through early October, heavy general rain may occur with a tropical disturbance or hurricane moving inland from the Gulf of Mexico. Since summer rain is heavier near the coast than elsewhere and winter rain is heavier in the north, the middle areas of the State get somewhat less precipitation for the year as a whole than the other areas.

Droughts may occur any time during the growing season from late April through October. Relatively long periods with little or no rain are more likely to occur in late summer and autumn that at any other time, while a secondary maximum of such periods occurs in May and June. Severe local droughts occur nearly every year, but severe statewide droughts are practically unknown.

Rivers in Alabama overflow about once a year on an average. Most floods occur from rains in late winter and early spring, with March the month of greatest flood frequency. The lower Tombigbee overflows most often, and in some stretches may stay over the banks most of the time in wet winter and spring seasons.

STORMS. Nearly all tornadoes occur during the season from November through early May. The greatest frequency is in March and April. The area covered by the average tornado is small. Destructive tropical hurricanes visit the coastal area on an average of about once in 7 years between July and November. Windstorm damage may occur in local thundersqualls any time of the year.

Thunderstorms in the north and central sections occur on an average of 1 day each month in winter, on about 13 days in July, and on about 60 days during the year. Almost all the hail that falls in Alabama occurs in the period from February through May, although in the northern counties there are rare occurrences of damaging hail in June.

Heavy fog occurs mostly in winter. It occurs on an average of 5 days per year in Birmingham, 8 days per year at

Montgomery, and 31 days per year at Mobile, near the coast.

WINDS. In winter, winds from a northerly direction are most frequent. In summer, the wind is quite variable, but most often comes from southerly directions.

## COUNTY INDEX

***Baldwin County***
BAY MINETTE 3 NNW
FAIRHOPE 2 NE
ROBERTSDALE 5 NE

***Barbour County***
CLAYTON

***Blount County***
ONEONTA

***Bullock County***
UNION SPRINGS 9 S

***Butler County***
GREENVILLE

***Calhoun County***
ANNISTON-CALHOUN CO

***Chilton County***
CLANTON
THORSBY EXP STATION

***Clarke County***
THOMASVILLE
WHATLEY

***Clay County***
ASHLAND 3 ENE

***Cleburne County***
HEFLIN

***Coffee County***
ENTERPRISE 5 NNW

***Colbert County***
MUSCLE SHOALS REG AP

***Conecuh County***
EVERGREEN

***Coosa County***
ROCKFORD 3 ESE

***Covington County***
ANDALUSIA 3 W

***Crenshaw County***
HIGHLAND HOME

***Cullman County***
SAINT BERNARD

***Dallas County***
MARION JUNCTION 2 NE
SELMA

***DeKalb County***
SAND MOUNTAIN SUBSTN
VALLEY HEAD

***Escambia County***
BREWTON 3 SSE

***Etowah County***
GADSDEN

***Franklin County***
RUSSELLVILLE 2

***Greene County***
GAINESVILLE LOCK

***Hale County***
GREENSBORO

***Henry County***
HEADLAND

***Jackson County***
SCOTTSBORO

***Jefferson County***
BIRMINGHAM MUNI AP

***Lamar County***
VERNON 2 N

***Lawrence County***
MOULTON 2

***Lee County***
OPELIKA

***Limestone County***
BELLE MINA 2 N

***Madison County***
HUNTSVILLE MADISON

***Marengo County***
DEMOPOLIS L & D

***Marion County***
HAMILTON 3 S

***Marshall County***
GUNTERSVILLE

***Mobile County***
CODEN
MOBILE REGIONAL AP

***Monroe County***
FRISCO CITY 3 SSW

***Montgomery County***
MONTGOMERY DANNELLY

***Pickens County***
ALICEVILLE

***Pike County***
TROY

***Sumter County***
LIVINGSTON 2 SW

***Talladega County***
CHILDERSBURG WTR PLT
SYLACAUGA 4 NE
TALLADEGA

***Tuscaloosa County***
BANKHEAD L & D
TUSCALOOSA MUNI AP
TUSCALOOSA OLIVER DM

***Walker County***
JASPER

***Washington County***
CHATOM

***Winston County***
HALEYVILLE 2 ENE

## ELEVATION INDEX

| FEET | STATION NAME |
|---|---|
| 10 | CODEN |
| 23 | FAIRHOPE 2 NE |
| 85 | BREWTON 3 SSE |
| 102 | DEMOPOLIS L & D |
| 131 | GAINESVILLE LOCK |
| 151 | SELMA |
| 151 | TUSCALOOSA OLIVER DM |
| 161 | LIVINGSTON 2 SW |
| 171 | WHATLEY |
| 175 | ROBERTSDALE 5 NE |
| 180 | VERNON 2 N |
| 187 | TUSCALOOSA MUNI AP |
| 190 | ALICEVILLE |
| 200 | MARION JUNCTION 2 NE |
| 200 | MONTGOMERY DANNELLY |
| 217 | MOBILE REGIONAL AP |
| 220 | GREENSBORO |
| 269 | BAY MINETTE 3 NNW |
| 280 | BANKHEAD L & D |
| 282 | FRISCO CITY 3 SSW |
| 285 | CHATOM |
| 341 | EVERGREEN |
| 351 | ANDALUSIA 3 W |
| 370 | HEADLAND |
| 390 | THOMASVILLE |
| 418 | CHILDERSBURG WTR PLT |
| 440 | GREENVILLE |
| 459 | UNION SPRINGS 9 S |
| 469 | ENTERPRISE 5 NNW |
| 489 | SYLACAUGA 4 NE |
| 522 | HAMILTON 3 S |
| 531 | JASPER |
| 540 | MUSCLE SHOALS REG AP |
| 551 | TALLADEGA |
| 571 | GADSDEN |
| 591 | CLANTON |
| 591 | CLAYTON |
| 591 | TROY |
| 594 | HIGHLAND HOME |
| 600 | BELLE MINA 2 N |
| 604 | ANNISTON-CALHOUN CO |
| 623 | HUNTSVILLE MADISON |
| 630 | BIRMINGHAM MUNI AP |
| 640 | GUNTERSVILLE |
| 645 | MOULTON 2 |
| 650 | ROCKFORD 3 ESE |
| 659 | SCOTTSBORO |
| 680 | THORSBY EXP STATION |
| 741 | OPELIKA |
| 800 | SAINT BERNARD |
| 850 | HEFLIN |
| 860 | ONEONTA |
| 869 | RUSSELLVILLE 2 |
| 951 | HALEYVILLE 2 ENE |
| 990 | ASHLAND 3 ENE |
| 1089 | VALLEY HEAD |
| 1201 | SAND MOUNTAIN SUBSTN |

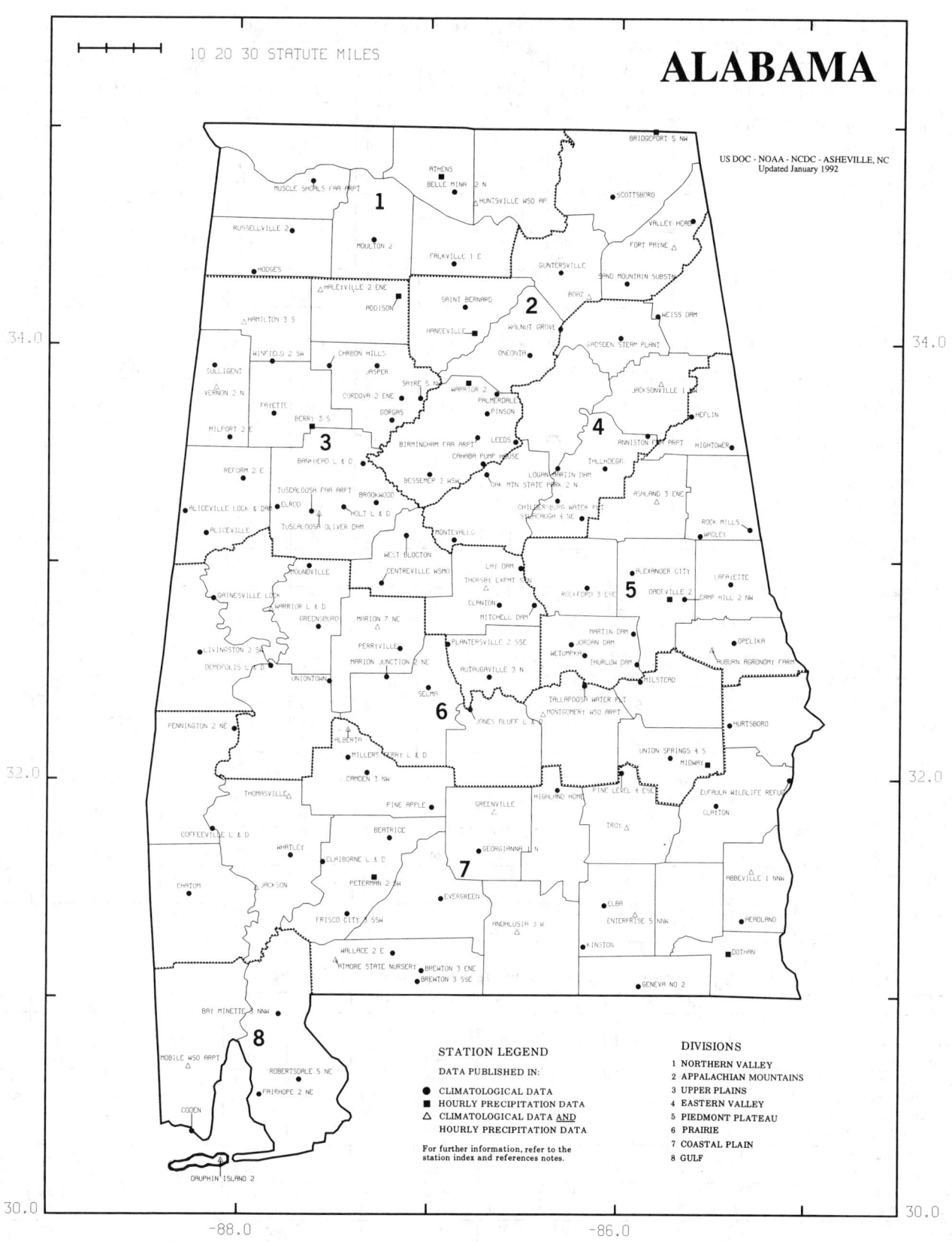

# ALABAMA

10 20 30 STATUTE MILES

US DOC - NOAA - NCDC - ASHEVILLE, NC
Updated January 1992

STATION LEGEND

DATA PUBLISHED IN:

- ● CLIMATOLOGICAL DATA
- ■ HOURLY PRECIPITATION DATA
- △ CLIMATOLOGICAL DATA AND HOURLY PRECIPITATION DATA

For further information, refer to the station index and references notes.

DIVISIONS

1. NORTHERN VALLEY
2. APPALACHIAN MOUNTAINS
3. UPPER PLAINS
4. EASTERN VALLEY
5. PIEDMONT PLATEAU
6. PRAIRIE
7. COASTAL PLAIN
8. GULF

## ALICEVILLE *Pickens County* ELEVATION 190 ft LAT/LONG 33° 8 ' N / 88° 8 ' W

| | JAN | FEB | MAR | APR | MAY | JUN | JUL | AUG | SEP | OCT | NOV | DEC | YEAR |
|---|---|---|---|---|---|---|---|---|---|---|---|---|---|
| Maximum Temp °F | 53.2 | 57.7 | 66.8 | 75.5 | 81.9 | 89.3 | 91.4 | 90.6 | 86.2 | 76.0 | 65.3 | 57.1 | 74.3 |
| Minimum Temp °F | 31.4 | 33.7 | 41.1 | 49.1 | 57.6 | 65.9 | ***69.9*** | 68.1 | 62.4 | 49.4 | 40.0 | 34.0 | 50.2 |
| Mean Temp °F | 42.3 | 45.7 | 54.0 | 62.3 | 69.7 | 77.6 | ***80.7*** | 79.4 | 74.3 | 62.8 | 52.7 | 45.6 | 62.3 |
| Days Max Temp ≥ 90 °F | 0 | 0 | 0 | 0 | 3 | 16 | 20 | 20 | 9 | 1 | 0 | 0 | 69 |
| Days Max Temp ≤ 32 °F | 1 | 0 | 0 | 0 | 0 | 0 | 0 | 0 | 0 | 0 | 0 | 0 | 1 |
| Days Min Temp ≤ 32 °F | 18 | 14 | ***7*** | 1 | 0 | 0 | 0 | 0 | 0 | 1 | 8 | 17 | 66 |
| Days Min Temp ≤ 0 °F | 0 | 0 | 0 | 0 | 0 | 0 | 0 | 0 | 0 | 0 | 0 | 0 | 0 |
| Heating Degree Days | 698 | 540 | 348 | 133 | 25 | 0 | 0 | 0 | 9 | 136 | 372 | 599 | 2860 |
| Cooling Degree Days | 0 | 4 | 15 | 52 | 173 | 380 | 489 | 462 | ***297*** | 74 | 13 | 4 | 1963 |
| Total Precipitation (") | 5.14 | 5.35 | 6.06 | 6.02 | 4.38 | 3.48 | 4.69 | 3.42 | 4.07 | 3.46 | 4.53 | 5.46 | 56.06 |
| Days ≥ 0.1" Precip | 7 | ***6*** | ***7*** | ***6*** | 6 | 5 | ***6*** | 5 | 5 | 5 | ***6*** | ***7*** | 71 |
| Total Snowfall (") | 0.5 | 0.0 | 0.1 | 0.0 | 0.0 | 0.0 | 0.0 | 0.0 | 0.0 | 0.0 | 0.0 | 0.0 | 0.6 |
| Days ≥ 1" Snow Depth | ***0*** | 0 | 0 | 0 | 0 | 0 | 0 | 0 | 0 | 0 | 0 | 0 | 0 |

## ANDALUSIA 3 W *Covington County* ELEVATION 351 ft LAT/LONG 31° 19 ' N / 86° 30 ' W

| | JAN | FEB | MAR | APR | MAY | JUN | JUL | AUG | SEP | OCT | NOV | DEC | YEAR |
|---|---|---|---|---|---|---|---|---|---|---|---|---|---|
| Maximum Temp °F | 59.1 | 63.0 | 71.3 | 79.2 | 84.6 | 90.1 | 91.5 | 90.6 | 87.0 | 78.4 | 69.5 | 62.6 | 77.2 |
| Minimum Temp °F | 33.7 | 35.2 | 42.0 | 48.6 | 56.3 | 64.1 | 67.3 | 66.9 | 62.3 | 49.7 | 41.1 | 36.0 | 50.3 |
| Mean Temp °F | 46.6 | 49.1 | 56.7 | 63.9 | 70.5 | 77.2 | 79.5 | 78.8 | 74.7 | 63.9 | 55.0 | 49.5 | 63.8 |
| Days Max Temp ≥ 90 °F | 0 | 0 | 0 | 1 | 5 | 19 | 23 | 21 | 12 | 1 | 0 | 0 | 82 |
| Days Max Temp ≤ 32 °F | 0 | 0 | 0 | 0 | 0 | 0 | 0 | 0 | 0 | 0 | 0 | 0 | 0 |
| Days Min Temp ≤ 32 °F | 16 | 14 | 7 | 1 | 0 | 0 | 0 | 0 | 0 | 1 | 8 | 14 | 61 |
| Days Min Temp ≤ 0 °F | 0 | 0 | 0 | 0 | 0 | 0 | 0 | 0 | 0 | 0 | 0 | 0 | 0 |
| Heating Degree Days | 565 | 448 | 276 | 105 | 15 | 0 | 0 | 0 | 7 | 110 | 311 | 482 | 2319 |
| Cooling Degree Days | 1 | 7 | 27 | 65 | 189 | 379 | 468 | 450 | 310 | 94 | 22 | 9 | 2021 |
| Total Precipitation (") | 5.53 | 5.52 | 6.73 | 3.91 | 4.88 | 5.31 | 6.26 | 5.21 | 4.24 | 3.16 | 4.24 | 5.10 | 60.09 |
| Days ≥ 0.1" Precip | 8 | 7 | 8 | 5 | 7 | 8 | 11 | 9 | 6 | 4 | 6 | 7 | 86 |
| Total Snowfall (") | 0.0 | 0.0 | 0.2 | 0.0 | 0.0 | 0.0 | 0.0 | 0.0 | 0.0 | 0.0 | 0.0 | 0.0 | 0.2 |
| Days ≥ 1" Snow Depth | 0 | 0 | 0 | 0 | 0 | 0 | 0 | 0 | 0 | 0 | 0 | 0 | 0 |

## ANNISTON-CALHOUN CO *Calhoun County* ELEVATION 604 ft LAT/LONG 33° 35 ' N / 85° 51 ' W

| | JAN | FEB | MAR | APR | MAY | JUN | JUL | AUG | SEP | OCT | NOV | DEC | YEAR |
|---|---|---|---|---|---|---|---|---|---|---|---|---|---|
| Maximum Temp °F | 52.5 | 57.2 | 66.2 | 75.1 | 81.0 | 87.7 | 90.2 | 89.3 | 83.9 | 74.5 | 64.9 | 56.7 | 73.3 |
| Minimum Temp °F | 32.4 | 34.8 | 42.1 | 49.6 | 57.6 | 65.4 | 69.4 | 68.6 | 62.8 | 50.3 | 41.2 | 35.4 | 50.8 |
| Mean Temp °F | 42.5 | 46.1 | 54.2 | 62.4 | 69.3 | 76.6 | 79.8 | 79.0 | 73.4 | 62.4 | 53.1 | 46.1 | 62.1 |
| Days Max Temp ≥ 90 °F | 0 | 0 | 0 | 0 | 2 | 12 | 18 | 16 | 6 | 0 | 0 | 0 | 54 |
| Days Max Temp ≤ 32 °F | 1 | 0 | 0 | 0 | 0 | 0 | 0 | 0 | 0 | 0 | 0 | 0 | 1 |
| Days Min Temp ≤ 32 °F | 16 | 13 | 5 | 1 | 0 | 0 | 0 | 0 | 0 | 1 | 7 | 14 | 57 |
| Days Min Temp ≤ 0 °F | 0 | 0 | 0 | 0 | 0 | 0 | 0 | 0 | 0 | 0 | 0 | 0 | 0 |
| Heating Degree Days | 692 | 530 | 342 | 130 | 26 | 1 | 0 | 0 | 11 | 133 | 360 | 583 | 2808 |
| Cooling Degree Days | 0 | 1 | 15 | 50 | 167 | 368 | 479 | 453 | 268 | 66 | 11 | 3 | 1881 |
| Total Precipitation (") | 5.03 | 4.71 | 6.25 | 4.87 | 4.52 | 3.95 | 4.48 | 4.07 | 3.49 | 2.75 | 3.93 | 4.49 | 52.54 |
| Days ≥ 0.1" Precip | 8 | 7 | 8 | 7 | 7 | 6 | 8 | 6 | 5 | 4 | 6 | 7 | 79 |
| Total Snowfall (") | 0.6 | 0.2 | 0.4 | 0.1 | 0.0 | 0.0 | 0.0 | 0.0 | 0.0 | 0.0 | 0.0 | 0.0 | 1.3 |
| Days ≥ 1" Snow Depth | 0 | 0 | 0 | 0 | 0 | 0 | 0 | 0 | 0 | 0 | 0 | 0 | 0 |

## ASHLAND 3 ENE *Clay County* ELEVATION 990 ft LAT/LONG 33° 17 ' N / 85° 48 ' W

| | JAN | FEB | MAR | APR | MAY | JUN | JUL | AUG | SEP | OCT | NOV | DEC | YEAR |
|---|---|---|---|---|---|---|---|---|---|---|---|---|---|
| Maximum Temp °F | 51.9 | 56.6 | 64.8 | 73.7 | 79.2 | 85.7 | 88.2 | 87.1 | 82.5 | 73.3 | 64.2 | 56.1 | 71.9 |
| Minimum Temp °F | 29.7 | 31.9 | 39.0 | 46.5 | 54.5 | 62.3 | 66.6 | 65.5 | 59.5 | 47.3 | 39.2 | 33.0 | 47.9 |
| Mean Temp °F | 40.8 | 44.3 | 51.9 | 60.2 | 66.9 | 74.0 | 77.4 | 76.4 | 71.0 | 60.3 | 51.7 | 44.6 | 60.0 |
| Days Max Temp ≥ 90 °F | 0 | 0 | 0 | 0 | 0 | 6 | 12 | 9 | 3 | 0 | 0 | 0 | 30 |
| Days Max Temp ≤ 32 °F | 2 | 0 | 0 | 0 | 0 | 0 | 0 | 0 | 0 | 0 | 0 | 1 | 3 |
| Days Min Temp ≤ 32 °F | 20 | 16 | 9 | 2 | 0 | 0 | 0 | 0 | 0 | 1 | 9 | 17 | 74 |
| Days Min Temp ≤ 0 °F | 0 | 0 | 0 | 0 | 0 | 0 | 0 | 0 | 0 | 0 | 0 | 0 | 0 |
| Heating Degree Days | 743 | 579 | 405 | 171 | 46 | 2 | 0 | 0 | 19 | 172 | 396 | 628 | 3161 |
| Cooling Degree Days | 0 | 0 | 7 | 23 | 109 | 294 | 408 | 374 | 202 | 36 | 3 | 1 | 1457 |
| Total Precipitation (") | 5.72 | 5.43 | 7.06 | 4.81 | 5.02 | 4.19 | 5.39 | 4.12 | 3.81 | 3.69 | 4.39 | 5.08 | 58.71 |
| Days ≥ 0.1" Precip | 9 | 7 | 8 | 6 | 7 | 6 | 8 | 7 | 6 | 5 | 6 | 7 | 82 |
| Total Snowfall (") | 0.8 | 0.4 | 0.6 | 0.2 | 0.0 | 0.0 | 0.0 | 0.0 | 0.0 | 0.0 | 0.0 | 0.0 | 2.0 |
| Days ≥ 1" Snow Depth | 1 | 0 | 0 | 0 | 0 | 0 | 0 | 0 | 0 | 0 | 0 | 0 | 1 |

## BANKHEAD L & D *Tuscaloosa County* ELEVATION 280 ft LAT/LONG 33° 27 ' N / 87° 21 ' W

| | JAN | FEB | MAR | APR | MAY | JUN | JUL | AUG | SEP | OCT | NOV | DEC | YEAR |
|---|---|---|---|---|---|---|---|---|---|---|---|---|---|
| Maximum Temp °F | 52.2 | 57.2 | 66.1 | 75.1 | 81.8 | 89.0 | 91.6 | 90.7 | 85.5 | 75.6 | 65.6 | 56.7 | 73.9 |
| Minimum Temp °F | 30.8 | 32.8 | 39.8 | 47.9 | 56.3 | 64.0 | 68.3 | 67.2 | 61.8 | 49.5 | 40.8 | 34.3 | 49.5 |
| Mean Temp °F | 41.6 | 45.0 | 53.0 | 61.5 | 69.1 | 76.6 | 80.0 | 79.0 | 73.7 | 62.6 | 53.2 | 45.5 | 61.7 |
| Days Max Temp ≥ 90 °F | 0 | 0 | 0 | 0 | 3 | 16 | 21 | 19 | 9 | 0 | 0 | 0 | 68 |
| Days Max Temp ≤ 32 °F | 1 | 0 | 0 | 0 | 0 | 0 | 0 | 0 | 0 | 0 | 0 | 1 | 2 |
| Days Min Temp ≤ 32 °F | 18 | 15 | 8 | 1 | 0 | 0 | 0 | 0 | 0 | 1 | 7 | 15 | 65 |
| Days Min Temp ≤ 0 °F | 0 | 0 | 0 | 0 | 0 | 0 | 0 | 0 | 0 | 0 | 0 | 0 | 0 |
| Heating Degree Days | 720 | 558 | 375 | 149 | 27 | 1 | 0 | 0 | 11 | 131 | 355 | 598 | 2925 |
| Cooling Degree Days | 0 | 1 | 10 | 47 | 156 | 366 | 487 | 460 | 282 | 63 | 8 | 1 | 1881 |
| Total Precipitation (") | 5.70 | 5.01 | 6.40 | 4.90 | 5.05 | 4.40 | 5.09 | 4.13 | 3.91 | 3.88 | 4.45 | 5.60 | 58.52 |
| Days ≥ 0.1" Precip | 8 | 7 | 8 | 6 | 7 | 6 | 8 | 7 | 6 | 5 | 7 | 7 | 82 |
| Total Snowfall (") | 0.2 | 0.0 | 0.0 | 0.0 | 0.0 | 0.0 | 0.0 | 0.0 | 0.0 | 0.0 | 0.0 | 0.0 | 0.2 |
| Days ≥ 1" Snow Depth | *0* | 0 | 0 | 0 | 0 | 0 | 0 | 0 | 0 | 0 | 0 | 0 | 0 |

## BAY MINETTE 3 NNW *Baldwin County* ELEVATION 269 ft LAT/LONG 30° 53 ' N / 87° 47 ' W

| | JAN | FEB | MAR | APR | MAY | JUN | JUL | AUG | SEP | OCT | NOV | DEC | YEAR |
|---|---|---|---|---|---|---|---|---|---|---|---|---|---|
| Maximum Temp °F | 60.5 | 64.8 | 71.9 | 79.1 | 84.4 | 89.5 | 90.6 | 89.9 | 86.8 | 79.1 | 70.2 | 63.5 | 77.5 |
| Minimum Temp °F | 39.9 | 41.9 | 48.5 | 55.6 | 62.7 | 68.8 | 71.2 | 70.7 | 66.9 | 56.6 | 48.5 | 42.8 | 56.2 |
| Mean Temp °F | 50.2 | 53.4 | 60.2 | 67.4 | 73.5 | 79.2 | 80.9 | 80.3 | 76.9 | 67.9 | 59.4 | 53.2 | 66.9 |
| Days Max Temp ≥ 90 °F | 0 | 0 | 0 | 0 | 4 | 17 | 21 | 19 | 11 | 1 | 0 | 0 | 73 |
| Days Max Temp ≤ 32 °F | 0 | 0 | 0 | 0 | 0 | 0 | 0 | 0 | 0 | 0 | 0 | 0 | 0 |
| Days Min Temp ≤ 32 °F | 9 | 6 | 2 | 0 | 0 | 0 | 0 | 0 | 0 | 0 | 2 | 7 | 26 |
| Days Min Temp ≤ 0 °F | 0 | 0 | 0 | 0 | 0 | 0 | 0 | 0 | 0 | 0 | 0 | 0 | 0 |
| Heating Degree Days | 460 | 331 | 184 | 46 | 2 | 0 | 0 | 0 | 3 | 48 | 203 | 379 | 1656 |
| Cooling Degree Days | 5 | 12 | 44 | 109 | 270 | 433 | 501 | 487 | 357 | 140 | 45 | 18 | 2421 |
| Total Precipitation (") | 5.82 | 5.25 | 6.29 | 4.32 | 5.92 | 5.51 | 7.59 | 6.55 | 5.31 | 3.12 | 4.61 | 5.49 | 65.78 |
| Days ≥ 0.1" Precip | 8 | 6 | 7 | 5 | 6 | 8 | 11 | 10 | 7 | 4 | 6 | 6 | 84 |
| Total Snowfall (") | 0.1 | 0.2 | 0.1 | 0.0 | 0.0 | 0.0 | 0.0 | 0.0 | 0.0 | 0.0 | 0.0 | 0.0 | 0.4 |
| Days ≥ 1" Snow Depth | 0 | 0 | 0 | 0 | 0 | 0 | 0 | 0 | 0 | 0 | 0 | 0 | 0 |

## BELLE MINA 2 N *Limestone County* ELEVATION 600 ft LAT/LONG 34° 42 ' N / 86° 53 ' W

| | JAN | FEB | MAR | APR | MAY | JUN | JUL | AUG | SEP | OCT | NOV | DEC | YEAR |
|---|---|---|---|---|---|---|---|---|---|---|---|---|---|
| Maximum Temp °F | 48.8 | 53.8 | 63.1 | 72.9 | 79.6 | 86.8 | 89.6 | 89.1 | 83.3 | 73.5 | 62.8 | 52.9 | 71.4 |
| Minimum Temp °F | 29.1 | 31.6 | 39.4 | 48.0 | 56.1 | 63.8 | 67.7 | 65.8 | 59.7 | 47.2 | 39.0 | 32.2 | 48.3 |
| Mean Temp °F | 39.0 | 42.7 | 51.3 | 60.5 | 67.9 | 75.3 | 78.7 | 77.4 | 71.6 | 60.4 | 50.9 | 42.6 | 59.9 |
| Days Max Temp ≥ 90 °F | 0 | 0 | 0 | 0 | 1 | 10 | 17 | 15 | 5 | 0 | 0 | 0 | 48 |
| Days Max Temp ≤ 32 °F | 3 | 1 | 0 | 0 | 0 | 0 | 0 | 0 | 0 | 0 | 0 | 1 | 5 |
| Days Min Temp ≤ 32 °F | 20 | 16 | 9 | 1 | 0 | 0 | 0 | 0 | 0 | 1 | 9 | 17 | 73 |
| Days Min Temp ≤ 0 °F | 0 | 0 | 0 | 0 | 0 | 0 | 0 | 0 | 0 | 0 | 0 | 0 | 0 |
| Heating Degree Days | 800 | 622 | 426 | 175 | 42 | 1 | 0 | 0 | 20 | 179 | 419 | 689 | 3373 |
| Cooling Degree Days | 0 | 0 | 7 | 39 | 140 | 330 | 450 | 409 | 224 | 43 | 5 | 1 | 1648 |
| Total Precipitation (") | 4.94 | 4.66 | 6.17 | 4.46 | 5.02 | 4.21 | 4.57 | 3.58 | 3.82 | 3.38 | 4.70 | 5.76 | 55.27 |
| Days ≥ 0.1" Precip | 8 | 7 | 8 | 7 | 7 | 6 | 8 | 6 | 6 | 5 | 7 | 8 | 83 |
| Total Snowfall (") | 1.2 | 0.9 | 0.5 | 0.0 | 0.0 | 0.0 | 0.0 | 0.0 | 0.0 | 0.0 | 0.1 | 0.3 | 3.0 |
| Days ≥ 1" Snow Depth | *0* | 0 | 0 | 0 | 0 | 0 | 0 | 0 | 0 | 0 | 0 | 0 | 0 |

## BIRMINGHAM MUNI AP *Jefferson County* ELEVATION 630 ft LAT/LONG 33° 34 ' N / 86° 45 ' W

| | JAN | FEB | MAR | APR | MAY | JUN | JUL | AUG | SEP | OCT | NOV | DEC | YEAR |
|---|---|---|---|---|---|---|---|---|---|---|---|---|---|
| Maximum Temp °F | 52.1 | 57.2 | 66.1 | 74.7 | 80.7 | 87.5 | 90.1 | 89.2 | 84.0 | 74.4 | 64.5 | 56.2 | 73.1 |
| Minimum Temp °F | 32.1 | 34.7 | 42.2 | 49.4 | 57.5 | 65.4 | 69.8 | 68.9 | 63.0 | 50.6 | 41.5 | 35.5 | 50.9 |
| Mean Temp °F | 42.1 | 46.0 | 54.1 | 62.1 | 69.1 | 76.5 | 80.0 | 79.1 | 73.5 | 62.5 | 53.0 | 45.9 | 62.0 |
| Days Max Temp ≥ 90 °F | 0 | 0 | 0 | 0 | 1 | 11 | 18 | 15 | 6 | 0 | 0 | 0 | 51 |
| Days Max Temp ≤ 32 °F | 1 | 0 | 0 | 0 | 0 | 0 | 0 | 0 | 0 | 0 | 0 | 1 | 2 |
| Days Min Temp ≤ 32 °F | 17 | 13 | 6 | 1 | 0 | 0 | 0 | 0 | 0 | 0 | 7 | 14 | 58 |
| Days Min Temp ≤ 0 °F | 0 | 0 | 0 | 0 | 0 | 0 | 0 | 0 | 0 | 0 | 0 | 0 | 0 |
| Heating Degree Days | 703 | 533 | 346 | 139 | 30 | 1 | 0 | 0 | 11 | 135 | 362 | 589 | 2849 |
| Cooling Degree Days | 0 | 2 | 17 | 53 | 171 | 369 | 491 | 468 | 284 | 75 | 11 | 3 | 1944 |
| Total Precipitation (") | 4.93 | 4.25 | 5.88 | 4.64 | 5.09 | 3.69 | 5.10 | 3.84 | 4.18 | 2.94 | 4.33 | 4.77 | 53.64 |
| Days ≥ 0.1" Precip | 8 | 7 | 8 | 7 | 7 | 7 | 8 | 6 | 6 | 4 | 7 | 7 | 82 |
| Total Snowfall (") | 0.8 | 0.1 | 0.6 | 0.2 | 0.0 | 0.0 | 0.0 | 0.0 | 0.0 | 0.0 | 0.0 | 0.0 | 1.7 |
| Days ≥ 1" Snow Depth | 0 | 0 | 0 | 0 | 0 | 0 | 0 | 0 | 0 | 0 | 0 | 0 | 0 |

## BREWTON 3 SSE *Escambia County* ELEVATION 85 ft LAT/LONG 31° 4 ' N / 87° 3 ' W

| | JAN | FEB | MAR | APR | MAY | JUN | JUL | AUG | SEP | OCT | NOV | DEC | YEAR |
|---|---|---|---|---|---|---|---|---|---|---|---|---|---|
| Maximum Temp °F | 61.1 | 65.3 | 73.1 | 80.3 | 85.8 | 90.9 | 92.1 | 91.1 | 87.5 | 79.3 | 70.4 | 64.0 | 78.4 |
| Minimum Temp °F | 36.0 | 37.8 | 44.3 | 50.4 | 57.6 | 64.7 | 68.2 | 67.6 | 62.9 | 50.4 | 42.4 | 38.3 | 51.7 |
| Mean Temp °F | 48.6 | 51.6 | 58.7 | 65.4 | 71.7 | 77.8 | 80.2 | 79.4 | 75.2 | 64.9 | 56.4 | 51.2 | 65.1 |
| Days Max Temp ≥ 90 °F | 0 | 0 | 0 | 2 | 8 | 19 | 23 | 21 | 13 | 1 | 0 | 0 | 87 |
| Days Max Temp ≤ 32 °F | 0 | 0 | 0 | 0 | 0 | 0 | 0 | 0 | 0 | 0 | 0 | 0 | 0 |
| Days Min Temp ≤ 32 °F | 14 | 11 | 5 | 1 | 0 | 0 | 0 | 0 | 0 | 2 | 8 | 12 | 53 |
| Days Min Temp ≤ 0 °F | 0 | 0 | 0 | 0 | 0 | 0 | 0 | 0 | 0 | 0 | 0 | 0 | 0 |
| Heating Degree Days | 510 | 381 | 225 | 79 | 10 | 0 | 0 | 0 | 6 | 97 | 276 | 436 | 2020 |
| Cooling Degree Days | 4 | 7 | 30 | 75 | 201 | 378 | 471 | 453 | 310 | 106 | 29 | 14 | 2078 |
| Total Precipitation (") | 5.83 | 5.91 | 6.58 | 3.86 | 5.48 | 6.01 | 7.23 | 6.07 | 4.81 | 3.29 | 4.67 | 5.52 | 65.26 |
| Days ≥ 0.1" Precip | 8 | 7 | 7 | 5 | 7 | 8 | 11 | 10 | 7 | 4 | 6 | 7 | 87 |
| Total Snowfall (") | 0.0 | 0.3 | 0.0 | 0.0 | 0.0 | 0.0 | 0.0 | 0.0 | 0.0 | 0.0 | 0.0 | 0.0 | 0.3 |
| Days ≥ 1" Snow Depth | 0 | 0 | 0 | 0 | 0 | 0 | 0 | 0 | 0 | 0 | 0 | 0 | 0 |

## CHATOM *Washington County* ELEVATION 285 ft LAT/LONG 31° 28 ' N / 88° 15 ' W

| | JAN | FEB | MAR | APR | MAY | JUN | JUL | AUG | SEP | OCT | NOV | DEC | YEAR |
|---|---|---|---|---|---|---|---|---|---|---|---|---|---|
| Maximum Temp °F | 59.2 | 64.3 | 72.1 | 79.7 | 85.1 | 90.2 | 92.1 | 91.2 | 87.7 | 79.6 | 70.1 | 63.2 | 77.9 |
| Minimum Temp °F | 35.5 | 37.9 | 44.7 | 52.0 | 59.4 | 65.6 | 68.4 | 67.5 | 63.1 | 52.2 | 43.2 | 38.4 | 52.3 |
| Mean Temp °F | 47.4 | 51.2 | 58.4 | 65.9 | 72.4 | 78.0 | 80.3 | 79.5 | 75.4 | 66.0 | 56.7 | 50.8 | 65.2 |
| Days Max Temp ≥ 90 °F | 0 | 0 | 0 | 0 | 6 | 18 | 24 | 23 | 13 | 1 | 0 | 0 | 85 |
| Days Max Temp ≤ 32 °F | 0 | 0 | 0 | 0 | 0 | 0 | 0 | 0 | 0 | 0 | 0 | 0 | 0 |
| Days Min Temp ≤ 32 °F | 13 | 9 | 4 | 1 | 0 | 0 | 0 | 0 | 0 | 0 | 5 | 11 | 43 |
| Days Min Temp ≤ 0 °F | 0 | 0 | 0 | 0 | 0 | 0 | 0 | 0 | 0 | 0 | 0 | 0 | 0 |
| Heating Degree Days | 544 | 388 | 230 | 68 | 6 | 0 | 0 | 0 | 6 | 72 | 267 | 445 | 2026 |
| Cooling Degree Days | 2 | *6* | 26 | *84* | *230* | *393* | *490* | 457 | *318* | *110* | 24 | 9 | 2149 |
| Total Precipitation (") | 6.04 | 6.09 | 6.53 | 4.79 | 5.64 | 4.04 | 5.99 | 4.77 | 4.44 | 3.29 | 4.69 | 5.81 | 62.12 |
| Days ≥ 0.1" Precip | 7 | 6 | 7 | 5 | 6 | 6 | 8 | 7 | 6 | 4 | 6 | 7 | 75 |
| Total Snowfall (") | 0.0 | 0.1 | 0.2 | 0.0 | 0.0 | 0.0 | 0.0 | 0.0 | 0.0 | 0.0 | 0.0 | 0.1 | 0.4 |
| Days ≥ 1" Snow Depth | 0 | 0 | 0 | 0 | 0 | 0 | 0 | 0 | 0 | 0 | 0 | 0 | 0 |

## CHILDERSBURG WTR PLT *Talladega County* ELEVATION 418 ft LAT/LONG 33° 17 ' N / 86° 20 ' W

| | JAN | FEB | MAR | APR | MAY | JUN | JUL | AUG | SEP | OCT | NOV | DEC | YEAR |
|---|---|---|---|---|---|---|---|---|---|---|---|---|---|
| Maximum Temp °F | 55.7 | 60.8 | 69.8 | 77.8 | 83.1 | 89.6 | 91.8 | 90.9 | 86.5 | 77.4 | 67.8 | 59.6 | 75.9 |
| Minimum Temp °F | 32.4 | 34.6 | 42.0 | 48.9 | 56.6 | 64.2 | 68.4 | 67.2 | 61.4 | 48.8 | 40.5 | 35.4 | 50.0 |
| Mean Temp °F | 44.1 | 47.7 | 55.9 | 63.4 | 69.9 | 76.9 | 80.1 | 79.1 | 73.9 | 63.1 | 54.2 | 47.6 | 63.0 |
| Days Max Temp ≥ 90 °F | 0 | 0 | 0 | 0 | 4 | 16 | 23 | 20 | 10 | 1 | 0 | 0 | 74 |
| Days Max Temp ≤ 32 °F | 1 | 0 | 0 | 0 | 0 | 0 | 0 | 0 | 0 | 0 | 0 | 0 | 1 |
| Days Min Temp ≤ 32 °F | 16 | 13 | 7 | 2 | 0 | 0 | 0 | 0 | 0 | 2 | 9 | 14 | 63 |
| Days Min Temp ≤ 0 °F | 0 | 0 | 0 | 0 | 0 | 0 | 0 | 0 | 0 | 0 | 0 | 0 | 0 |
| Heating Degree Days | 642 | 485 | 298 | 112 | 22 | 1 | 0 | 0 | 9 | 120 | 332 | 539 | 2560 |
| Cooling Degree Days | 1 | 4 | 22 | 63 | 183 | 382 | 491 | 457 | 283 | 77 | 16 | 5 | 1984 |
| Total Precipitation (") | 5.76 | 5.10 | 6.61 | 5.10 | 4.80 | 3.96 | 4.94 | 4.10 | 3.98 | 3.14 | 4.00 | 5.18 | 56.67 |
| Days ≥ 0.1" Precip | 9 | 7 | 8 | 7 | 7 | 7 | 8 | 6 | 5 | 5 | 6 | 7 | 82 |
| Total Snowfall (") | 0.3 | 0.0 | 0.4 | 0.0 | 0.0 | 0.0 | 0.0 | 0.0 | 0.0 | 0.0 | 0.0 | 0.0 | 0.7 |
| Days ≥ 1" Snow Depth | 0 | 0 | 0 | 0 | 0 | 0 | 0 | 0 | 0 | 0 | 0 | 0 | 0 |

## CLANTON *Chilton County* ELEVATION 591 ft LAT/LONG 32° 51 ' N / 86° 38 ' W

| | JAN | FEB | MAR | APR | MAY | JUN | JUL | AUG | SEP | OCT | NOV | DEC | YEAR |
|---|---|---|---|---|---|---|---|---|---|---|---|---|---|
| Maximum Temp °F | 53.1 | 57.9 | 66.6 | 75.7 | 81.5 | 88.1 | 90.5 | 89.4 | 85.4 | 76.1 | 66.2 | 57.7 | 74.0 |
| Minimum Temp °F | 31.0 | 33.3 | 41.2 | 49.2 | 56.9 | 64.5 | 68.5 | 67.2 | 61.6 | 49.5 | 40.2 | 34.6 | 49.8 |
| Mean Temp °F | 42.1 | 45.6 | 53.9 | 62.4 | 69.2 | 76.3 | 79.5 | 78.3 | 73.5 | 62.8 | 53.2 | 46.2 | 61.9 |
| Days Max Temp ≥ 90 °F | 0 | 0 | 0 | 0 | 2 | 12 | 19 | 16 | 8 | 1 | 0 | 0 | 58 |
| Days Max Temp ≤ 32 °F | 1 | 0 | 0 | 0 | 0 | 0 | 0 | 0 | 0 | 0 | 0 | 0 | 1 |
| Days Min Temp ≤ 32 °F | 18 | 15 | 7 | 1 | 0 | 0 | 0 | 0 | 0 | 1 | 8 | 15 | 65 |
| Days Min Temp ≤ 0 °F | 0 | 0 | 0 | 0 | 0 | 0 | 0 | 0 | 0 | 0 | 0 | 0 | 0 |
| Heating Degree Days | 704 | 542 | 350 | 127 | 26 | 1 | 0 | 0 | 11 | 126 | 357 | 578 | 2822 |
| Cooling Degree Days | *1* | *0* | *14* | *49* | *164* | *361* | *471* | *427* | *272* | *77* | *12* | *2* | 1850 |
| Total Precipitation (") | 5.36 | 5.19 | 6.69 | 5.61 | 4.69 | 3.89 | 5.75 | 4.72 | 4.32 | 3.24 | 4.74 | 5.40 | 59.60 |
| Days ≥ 0.1" Precip | 8 | 7 | 8 | 6 | 7 | 6 | 8 | 7 | 6 | 4 | 6 | 7 | 80 |
| Total Snowfall (") | 0.4 | 0.1 | 0.6 | 0.0 | 0.0 | 0.0 | 0.0 | 0.0 | 0.0 | 0.0 | 0.0 | 0.0 | 1.1 |
| Days ≥ 1" Snow Depth | *0* | 0 | *0* | 0 | 0 | 0 | 0 | 0 | 0 | 0 | 0 | 0 | 0 |

## CLAYTON *Barbour County* ELEVATION 591 ft LAT/LONG 31° 52 ' N / 85° 27 ' W

| | JAN | FEB | MAR | APR | MAY | JUN | JUL | AUG | SEP | OCT | NOV | DEC | YEAR |
|---|---|---|---|---|---|---|---|---|---|---|---|---|---|
| Maximum Temp °F | 56.3 | 60.3 | 68.7 | 76.5 | 81.7 | 87.5 | 89.4 | 88.8 | 85.5 | 76.9 | 68.2 | 60.5 | 75.0 |
| Minimum Temp °F | 35.6 | 37.7 | 45.3 | 52.7 | 60.5 | 66.8 | 69.2 | 68.9 | 64.7 | 53.9 | 45.4 | 38.7 | 53.3 |
| Mean Temp °F | 46.0 | 49.0 | 57.0 | 64.6 | 71.1 | 77.1 | 79.4 | 78.9 | 75.1 | 65.4 | 56.8 | 49.6 | 64.2 |
| Days Max Temp ≥ 90 °F | 0 | 0 | 0 | 0 | 1 | 10 | 16 | 14 | 8 | 1 | 0 | 0 | 50 |
| Days Max Temp ≤ 32 °F | 1 | 0 | 0 | 0 | 0 | 0 | 0 | 0 | 0 | 0 | 0 | 0 | 1 |
| Days Min Temp ≤ 32 °F | 12 | 9 | 3 | 0 | 0 | 0 | 0 | 0 | 0 | 0 | 3 | 9 | 36 |
| Days Min Temp ≤ 0 °F | 0 | 0 | 0 | 0 | 0 | 0 | 0 | 0 | 0 | 0 | 0 | 0 | 0 |
| Heating Degree Days | 584 | 449 | 267 | 91 | 9 | 0 | 0 | 0 | 4 | 81 | 260 | 476 | 2221 |
| Cooling Degree Days | 1 | 6 | 29 | 76 | 212 | 390 | 469 | 455 | 315 | 103 | 25 | 5 | 2086 |
| Total Precipitation (") | 4.82 | 4.59 | 6.47 | 3.42 | 4.22 | 4.56 | 5.47 | 4.22 | 3.28 | 2.60 | 3.79 | 4.46 | 51.90 |
| Days ≥ 0.1" Precip | 7 | *6* | 7 | 5 | 6 | 6 | 8 | 7 | 5 | 3 | 5 | 6 | 71 |
| Total Snowfall (") | 0.0 | 0.5 | 0.0 | 0.0 | 0.0 | 0.0 | 0.0 | 0.0 | 0.0 | 0.0 | 0.0 | 0.2 | 0.7 |
| Days ≥ 1" Snow Depth | 0 | 0 | 0 | 0 | 0 | 0 | 0 | 0 | 0 | 0 | 0 | 0 | 0 |

## CODEN *Mobile County* ELEVATION 10 ft LAT/LONG 30° 23 ' N / 88° 13 ' W

| | JAN | FEB | MAR | APR | MAY | JUN | JUL | AUG | SEP | OCT | NOV | DEC | YEAR |
|---|---|---|---|---|---|---|---|---|---|---|---|---|---|
| Maximum Temp °F | *60.9* | *63.5* | *69.8* | *77.2* | *82.7* | *88.6* | *90.2* | *89.5* | *86.9* | *79.9* | *70.8* | *64.4* | 77.0 |
| Minimum Temp °F | *38.7* | *41.1* | *47.7* | *56.3* | *62.6* | *68.8* | *71.4* | *71.0* | *66.6* | na | *46.3* | *42.1* | na |
| Mean Temp °F | *49.5* | *52.3* | *58.7* | *66.8* | *72.6* | *78.7* | *80.8* | *80.3* | *76.7* | na | *58.4* | *53.3* | na |
| Days Max Temp ≥ 90 °F | 0 | *0* | *0* | *0* | *1* | *12* | *20* | *17* | *10* | *1* | *0* | 0 | 61 |
| Days Max Temp ≤ 32 °F | 0 | *0* | *0* | *0* | *0* | *0* | *0* | *0* | *0* | *0* | *0* | 0 | 0 |
| Days Min Temp ≤ 32 °F | *9* | *7* | *2* | *0* | *0* | *0* | *0* | *0* | *0* | *0* | *3* | *8* | 29 |
| Days Min Temp ≤ 0 °F | 0 | *0* | *0* | *0* | *0* | *0* | *0* | *0* | *0* | *0* | *0* | 0 | 0 |
| Heating Degree Days | *476* | *356* | *212* | *52* | *4* | *0* | *0* | *0* | *3* | na | *221* | *372* | na |
| Cooling Degree Days | *1* | *3* | na | na | na | na | na | na | na | *135* | *28* | *13* | na |
| Total Precipitation (") | *5.95* | *6.39* | *6.30* | *4.01* | *5.14* | *4.44* | *6.99* | *7.62* | *5.23* | *3.66* | *4.07* | *4.82* | 64.62 |
| Days ≥ 0.1" Precip | *7* | *6* | na | na | *5* | *6* | *9* | *9* | *6* | *3* | *5* | *6* | na |
| Total Snowfall (") | *0.0* | *0.1* | *0.0* | *0.0* | *0.0* | *0.0* | *0.0* | *0.0* | *0.0* | 0.0 | 0.0 | 0.0 | 0.1 |
| Days ≥ 1" Snow Depth | *0* | *0* | *0* | *0* | *0* | *0* | *0* | *0* | *0* | *0* | *0* | 0 | 0 |

## DEMOPOLIS L & D *Marengo County* ELEVATION 102 ft LAT/LONG 32° 31 ' N / 87° 50 ' W

| | JAN | FEB | MAR | APR | MAY | JUN | JUL | AUG | SEP | OCT | NOV | DEC | YEAR |
|---|---|---|---|---|---|---|---|---|---|---|---|---|---|
| Maximum Temp °F | 54.7 | 59.1 | 67.7 | 76.0 | 82.1 | 89.1 | 91.4 | 90.6 | 86.0 | 76.6 | 67.2 | 58.9 | 75.0 |
| Minimum Temp °F | 32.8 | 35.3 | 42.7 | 50.4 | 58.3 | 65.6 | 69.0 | 68.1 | 62.9 | 50.1 | 41.1 | 35.3 | 51.0 |
| Mean Temp °F | 43.7 | 47.2 | 55.3 | 63.1 | 70.2 | 77.4 | 80.2 | 79.4 | 74.5 | 63.4 | 54.1 | 47.2 | 63.0 |
| Days Max Temp ≥ 90 °F | 0 | 0 | 0 | 0 | 3 | 15 | 21 | 20 | 10 | 1 | 0 | 0 | 70 |
| Days Max Temp ≤ 32 °F | 1 | 0 | 0 | 0 | 0 | 0 | 0 | 0 | 0 | 0 | 0 | 0 | 1 |
| Days Min Temp ≤ 32 °F | 16 | 12 | 5 | 1 | 0 | 0 | 0 | 0 | 0 | 1 | 7 | 15 | 57 |
| Days Min Temp ≤ 0 °F | 0 | 0 | 0 | 0 | 0 | 0 | 0 | 0 | 0 | 0 | 0 | 0 | 0 |
| Heating Degree Days | 654 | 499 | 316 | 117 | 19 | 0 | 0 | 0 | 9 | 122 | 335 | 551 | 2622 |
| Cooling Degree Days | 2 | 5 | 20 | 62 | 194 | 389 | 490 | 466 | 303 | 82 | 16 | 6 | 2035 |
| Total Precipitation (") | 5.60 | 4.87 | 6.15 | 4.98 | 4.66 | 3.27 | 4.75 | 4.13 | 3.83 | 3.37 | 4.31 | 4.98 | 54.90 |
| Days ≥ 0.1" Precip | 8 | 7 | 7 | 6 | 7 | 6 | 8 | 6 | 6 | 4 | 6 | 7 | 78 |
| Total Snowfall (") | 0.1 | 0.0 | 0.1 | 0.0 | 0.0 | 0.0 | 0.0 | 0.0 | 0.0 | 0.0 | 0.0 | 0.0 | 0.2 |
| Days ≥ 1" Snow Depth | 0 | 0 | 0 | 0 | 0 | 0 | 0 | 0 | 0 | 0 | 0 | 0 | 0 |

## ENTERPRISE 5 NNW *Coffee County* ELEVATION 469 ft LAT/LONG 31° 23 ' N / 85° 54 ' W

| | JAN | FEB | MAR | APR | MAY | JUN | JUL | AUG | SEP | OCT | NOV | DEC | YEAR |
|---|---|---|---|---|---|---|---|---|---|---|---|---|---|
| Maximum Temp °F | 57.3 | *61.0* | 69.6 | *76.9* | 82.6 | 88.6 | 90.3 | *89.2* | 86.4 | 77.7 | *68.7* | 61.1 | 75.8 |
| Minimum Temp °F | 37.4 | 39.4 | 46.7 | *53.6* | 60.9 | 67.9 | 70.2 | 69.7 | 65.7 | 55.2 | *46.8* | 40.2 | 54.5 |
| Mean Temp °F | 47.4 | *50.2* | 58.1 | *65.3* | 71.8 | 78.2 | 80.3 | *79.5* | 76.1 | 66.5 | *57.8* | 50.7 | 65.2 |
| Days Max Temp ≥ 90 °F | 0 | 0 | 0 | 0 | 3 | *13* | *19* | *15* | *9* | 1 | 0 | 0 | 60 |
| Days Max Temp ≤ 32 °F | 0 | 0 | 0 | 0 | 0 | 0 | 0 | 0 | 0 | 0 | 0 | 0 | 0 |
| Days Min Temp ≤ 32 °F | *11* | *8* | 2 | 0 | 0 | 0 | 0 | 0 | 0 | 0 | *2* | *8* | 31 |
| Days Min Temp ≤ 0 °F | 0 | 0 | 0 | 0 | 0 | 0 | 0 | 0 | 0 | 0 | 0 | 0 | 0 |
| Heating Degree Days | 543 | *417* | 235 | *80* | 9 | 0 | 0 | *0* | 3 | 68 | *239* | 445 | 2039 |
| Cooling Degree Days | 1 | 8 | 31 | 85 | 232 | 415 | 488 | 461 | 339 | 123 | 36 | 7 | 2226 |
| Total Precipitation (") | *5.95* | 4.98 | 6.34 | *3.62* | 4.94 | 5.05 | 6.19 | 4.07 | 3.27 | 2.70 | 4.06 | 4.86 | 56.03 |
| Days ≥ 0.1" Precip | 7 | *7* | *7* | *5* | *6* | 7 | 9 | 7 | 5 | 4 | *5* | 6 | 75 |
| Total Snowfall (") | *0.2* | 0.3 | 0.0 | 0.0 | 0.0 | 0.0 | 0.0 | 0.0 | 0.0 | 0.0 | 0.0 | 0.0 | 0.5 |
| Days ≥ 1" Snow Depth | *0* | *0* | 0 | 0 | 0 | 0 | 0 | 0 | 0 | 0 | 0 | 0 | 0 |

## EVERGREEN *Conecuh County* ELEVATION 341 ft LAT/LONG 31° 26 ' N / 86° 58 ' W

| | JAN | FEB | MAR | APR | MAY | JUN | JUL | AUG | SEP | OCT | NOV | DEC | YEAR |
|---|---|---|---|---|---|---|---|---|---|---|---|---|---|
| Maximum Temp °F | 58.0 | 62.4 | 70.4 | 77.6 | 83.4 | 89.0 | 90.8 | 90.1 | 86.7 | 78.5 | 69.2 | 61.4 | 76.5 |
| Minimum Temp °F | 35.0 | 37.0 | 44.1 | 51.0 | 59.5 | 66.8 | 69.8 | 69.3 | 64.7 | 52.7 | 43.6 | 37.8 | 52.6 |
| Mean Temp °F | 46.5 | 49.7 | 57.2 | 64.4 | 71.5 | 77.9 | 80.4 | 79.7 | 75.7 | 65.7 | 56.6 | 49.6 | 64.6 |
| Days Max Temp ≥ 90 °F | 0 | 0 | 0 | 0 | 4 | 14 | 20 | 18 | 11 | 1 | 0 | 0 | 68 |
| Days Max Temp ≤ 32 °F | 0 | 0 | 0 | 0 | 0 | 0 | 0 | 0 | 0 | 0 | 0 | 0 | 0 |
| Days Min Temp ≤ 32 °F | 15 | 11 | 4 | 0 | 0 | 0 | 0 | 0 | 0 | 0 | 5 | ***12*** | 47 |
| Days Min Temp ≤ 0 °F | 0 | 0 | 0 | 0 | 0 | 0 | 0 | 0 | 0 | 0 | 0 | 0 | 0 |
| Heating Degree Days | 570 | 431 | 263 | 94 | 12 | 0 | 0 | 0 | 5 | 82 | 270 | 480 | 2207 |
| Cooling Degree Days | 2 | 7 | 27 | 72 | 210 | 395 | 476 | 461 | 320 | 106 | 25 | 9 | 2110 |
| Total Precipitation (") | 6.10 | 5.82 | 7.02 | 4.64 | 5.41 | 5.36 | 7.08 | 4.56 | 4.27 | 2.69 | 4.33 | 5.57 | 62.85 |
| Days ≥ 0.1" Precip | 8 | 7 | 8 | 6 | 7 | 8 | 10 | 8 | 6 | 4 | 6 | 6 | 84 |
| Total Snowfall (") | 0.2 | 0.0 | 0.2 | 0.0 | 0.0 | 0.0 | 0.0 | 0.0 | 0.0 | 0.0 | 0.0 | 0.0 | 0.4 |
| Days ≥ 1" Snow Depth | 0 | 0 | 0 | 0 | 0 | 0 | 0 | 0 | 0 | 0 | 0 | 0 | 0 |

## FAIRHOPE 2 NE *Baldwin County* ELEVATION 23 ft LAT/LONG 30° 33 ' N / 87° 53 ' W

| | JAN | FEB | MAR | APR | MAY | JUN | JUL | AUG | SEP | OCT | NOV | DEC | YEAR |
|---|---|---|---|---|---|---|---|---|---|---|---|---|---|
| Maximum Temp °F | 60.2 | 63.6 | 70.2 | 77.4 | 83.3 | 88.6 | 90.0 | 89.5 | 86.6 | 79.0 | 70.2 | 63.3 | 76.8 |
| Minimum Temp °F | 40.2 | 42.3 | 49.3 | 56.3 | 63.1 | 69.9 | 72.4 | 71.7 | 67.7 | 57.0 | 48.9 | 43.0 | 56.8 |
| Mean Temp °F | 50.2 | 53.0 | 59.8 | 66.9 | 73.2 | 79.3 | 81.2 | 80.6 | 77.2 | 68.0 | 59.6 | 53.1 | 66.8 |
| Days Max Temp ≥ 90 °F | 0 | 0 | 0 | 0 | 1 | 13 | 19 | 18 | 9 | 1 | 0 | 0 | 61 |
| Days Max Temp ≤ 32 °F | 0 | 0 | 0 | 0 | 0 | 0 | 0 | 0 | 0 | 0 | 0 | 0 | 0 |
| Days Min Temp ≤ 32 °F | 8 | 5 | 1 | 0 | 0 | 0 | 0 | 0 | 0 | 0 | 2 | 5 | 21 |
| Days Min Temp ≤ 0 °F | 0 | 0 | 0 | 0 | 0 | 0 | 0 | 0 | 0 | 0 | 0 | 0 | 0 |
| Heating Degree Days | 459 | 343 | 192 | 50 | 2 | 0 | 0 | 0 | 2 | 47 | 199 | 379 | 1673 |
| Cooling Degree Days | 4 | 9 | 33 | 91 | 255 | 432 | 516 | 502 | 372 | 147 | 47 | 16 | 2424 |
| Total Precipitation (") | 5.54 | 5.76 | 6.26 | 3.96 | 5.78 | 6.18 | 7.39 | 6.99 | 5.56 | 3.44 | 4.30 | 4.62 | 65.78 |
| Days ≥ 0.1" Precip | 7 | 6 | 7 | 5 | 6 | 7 | 10 | 10 | 7 | 4 | 6 | 6 | 81 |
| Total Snowfall (") | 0.0 | 0.0 | 0.0 | 0.0 | 0.0 | 0.0 | 0.0 | 0.0 | 0.0 | 0.0 | 0.0 | 0.0 | 0.0 |
| Days ≥ 1" Snow Depth | 0 | 0 | 0 | 0 | 0 | 0 | 0 | 0 | 0 | 0 | 0 | 0 | 0 |

## FRISCO CITY 3 SSW *Monroe County* ELEVATION 282 ft LAT/LONG 31° 23 ' N / 87° 26 ' W

| | JAN | FEB | MAR | APR | MAY | JUN | JUL | AUG | SEP | OCT | NOV | DEC | YEAR |
|---|---|---|---|---|---|---|---|---|---|---|---|---|---|
| Maximum Temp °F | ***58.3*** | ***62.1*** | 70.1 | 77.8 | 83.7 | 88.9 | ***90.8*** | 89.7 | 86.7 | 77.8 | ***69.0*** | 61.6 | 76.4 |
| Minimum Temp °F | ***36.6*** | ***38.3*** | 45.9 | 53.0 | 60.4 | 67.0 | 69.4 | 68.7 | 64.6 | 53.5 | 45.0 | 39.5 | 53.5 |
| Mean Temp °F | ***47.5*** | ***50.3*** | 58.0 | 65.4 | 72.2 | 78.0 | ***80.1*** | 79.2 | 75.6 | 65.7 | ***57.1*** | 50.6 | 65.0 |
| Days Max Temp ≥ 90 °F | 0 | 0 | 0 | 0 | 4 | 15 | 20 | 17 | 10 | 1 | 0 | 0 | 67 |
| Days Max Temp ≤ 32 °F | 0 | 0 | 0 | 0 | 0 | 0 | 0 | 0 | 0 | 0 | 0 | 0 | 0 |
| Days Min Temp ≤ 32 °F | 11 | 9 | 3 | 0 | 0 | 0 | 0 | 0 | 0 | 0 | ***5*** | ***9*** | 37 |
| Days Min Temp ≤ 0 °F | 0 | 0 | 0 | 0 | 0 | 0 | 0 | 0 | 0 | 0 | 0 | 0 | 0 |
| Heating Degree Days | ***540*** | ***414*** | 242 | 77 | 7 | 0 | ***0*** | 0 | 5 | 80 | ***259*** | 450 | 2074 |
| Cooling Degree Days | ***3*** | ***6*** | ***36*** | 83 | 242 | 400 | ***476*** | 442 | 328 | 112 | ***32*** | 11 | 2171 |
| Total Precipitation (") | 5.50 | 5.52 | 6.32 | 4.28 | 6.05 | 5.03 | 5.84 | 4.94 | 4.42 | 3.01 | 4.35 | 5.31 | 60.57 |
| Days ≥ 0.1" Precip | 7 | ***7*** | 7 | 5 | 6 | 7 | 8 | 8 | 6 | 4 | 5 | 6 | 76 |
| Total Snowfall (") | ***0.3*** | 0.2 | 0.2 | 0.0 | 0.0 | 0.0 | 0.0 | 0.0 | 0.0 | 0.0 | 0.0 | 0.2 | 0.9 |
| Days ≥ 1" Snow Depth | ***0*** | ***0*** | 0 | 0 | 0 | 0 | 0 | 0 | 0 | 0 | 0 | 0 | 0 |

## GADSDEN *Etowah County* ELEVATION 571 ft LAT/LONG 34° 1 ' N / 86° 0 ' W

| | JAN | FEB | MAR | APR | MAY | JUN | JUL | AUG | SEP | OCT | NOV | DEC | YEAR |
|---|---|---|---|---|---|---|---|---|---|---|---|---|---|
| Maximum Temp °F | 50.0 | 55.0 | 65.1 | 74.2 | 80.4 | 87.0 | 89.9 | 89.0 | 83.4 | ***73.5*** | 63.6 | 55.0 | 72.2 |
| Minimum Temp °F | 29.9 | 32.5 | 40.1 | 48.7 | 56.8 | 64.8 | 69.2 | 68.0 | 61.9 | ***49.2*** | 40.1 | 33.5 | 49.6 |
| Mean Temp °F | 40.0 | 43.8 | 52.5 | 61.5 | 68.6 | 76.0 | 79.6 | 78.5 | 72.7 | ***61.4*** | 51.9 | 44.2 | 60.9 |
| Days Max Temp ≥ 90 °F | 0 | 0 | 0 | 0 | 2 | 10 | 18 | 14 | 6 | 0 | 0 | 0 | 50 |
| Days Max Temp ≤ 32 °F | 2 | 0 | 0 | 0 | 0 | 0 | 0 | 0 | 0 | 0 | 0 | 0 | 2 |
| Days Min Temp ≤ 32 °F | 19 | 15 | 7 | 1 | 0 | 0 | 0 | 0 | 0 | ***0*** | 7 | 16 | 65 |
| Days Min Temp ≤ 0 °F | 0 | 0 | 0 | 0 | 0 | 0 | 0 | 0 | 0 | 0 | 0 | 0 | 0 |
| Heating Degree Days | 769 | 595 | 387 | 149 | 31 | 1 | 0 | 0 | 13 | ***154*** | 392 | 638 | 3129 |
| Cooling Degree Days | 0 | 1 | 9 | 49 | 156 | 356 | 485 | 449 | 258 | ***59*** | 6 | ***2*** | 1830 |
| Total Precipitation (") | 5.24 | 4.74 | 6.40 | 5.36 | 5.06 | 3.74 | 4.84 | 3.80 | 3.73 | 3.06 | 4.37 | 4.90 | 55.24 |
| Days ≥ 0.1" Precip | 8 | ***6*** | 8 | 7 | 7 | 6 | 7 | 6 | 6 | ***5*** | 6 | 7 | 79 |
| Total Snowfall (") | ***0.0*** | 0.1 | 0.0 | 0.0 | 0.0 | 0.0 | 0.0 | 0.0 | 0.0 | 0.0 | 0.0 | 0.0 | 0.1 |
| Days ≥ 1" Snow Depth | ***0*** | 0 | 0 | 0 | 0 | 0 | 0 | 0 | 0 | 0 | 0 | 0 | 0 |

## GAINESVILLE LOCK *Greene County* ELEVATION 131 ft LAT/LONG 32° 49 ' N / 88° 9 ' W

| | JAN | FEB | MAR | APR | MAY | JUN | JUL | AUG | SEP | OCT | NOV | DEC | YEAR |
|---|---|---|---|---|---|---|---|---|---|---|---|---|---|
| Maximum Temp °F | 53.3 | 58.3 | 67.0 | 75.6 | 81.6 | 88.4 | 91.1 | 90.3 | 85.6 | 76.0 | 66.2 | 57.4 | 74.2 |
| Minimum Temp °F | 32.2 | 34.7 | 42.7 | 50.5 | 58.7 | 66.3 | 70.1 | 69.2 | 63.3 | 50.2 | 41.1 | 35.1 | 51.2 |
| Mean Temp °F | 42.8 | 46.5 | 54.9 | 63.2 | 70.1 | 77.4 | 80.6 | 79.8 | 74.5 | 63.1 | 53.7 | 46.3 | 62.7 |
| Days Max Temp ≥ 90 °F | 0 | 0 | 0 | 0 | 2 | 13 | 21 | 19 | 9 | 0 | 0 | 0 | 64 |
| Days Max Temp ≤ 32 °F | 1 | 0 | 0 | 0 | 0 | 0 | 0 | 0 | 0 | 0 | 0 | 0 | 1 |
| Days Min Temp ≤ 32 °F | 17 | 13 | 4 | 0 | 0 | 0 | 0 | 0 | 0 | 0 | 7 | 15 | 56 |
| Days Min Temp ≤ 0 °F | 0 | 0 | 0 | 0 | 0 | 0 | 0 | 0 | 0 | 0 | 0 | 0 | 0 |
| Heating Degree Days | 682 | 518 | 325 | 117 | 22 | 1 | 0 | 0 | 9 | 123 | 345 | 576 | 2718 |
| Cooling Degree Days | 0 | 3 | 17 | 58 | 188 | 386 | 503 | 485 | 303 | 73 | 12 | 4 | 2032 |
| Total Precipitation (") | 5.25 | 5.05 | 6.18 | 5.24 | 4.63 | 3.79 | 4.13 | 3.26 | 3.52 | 3.12 | 4.18 | 5.14 | 53.49 |
| Days ≥ 0.1" Precip | 8 | 7 | 7 | 6 | 7 | 5 | 7 | 6 | 5 | 4 | 6 | 7 | 75 |
| Total Snowfall (") | 0.4 | 0.1 | 0.1 | 0.0 | 0.0 | 0.0 | 0.0 | 0.0 | 0.0 | 0.0 | 0.0 | 0.0 | 0.6 |
| Days ≥ 1" Snow Depth | 0 | 0 | 0 | 0 | 0 | 0 | 0 | 0 | 0 | 0 | 0 | 0 | 0 |

## GREENSBORO *Hale County* ELEVATION 220 ft LAT/LONG 32° 42 ' N / 87° 36 ' W

| | JAN | FEB | MAR | APR | MAY | JUN | JUL | AUG | SEP | OCT | NOV | DEC | YEAR |
|---|---|---|---|---|---|---|---|---|---|---|---|---|---|
| Maximum Temp °F | 55.5 | 60.6 | 69.0 | 77.8 | 83.8 | 90.4 | 92.4 | 91.6 | 86.9 | 77.5 | 67.5 | 59.3 | 76.0 |
| Minimum Temp °F | 34.2 | 36.8 | 43.4 | 50.9 | 59.1 | 66.6 | 70.0 | 69.3 | 64.3 | 52.2 | 43.1 | 37.3 | 52.3 |
| Mean Temp °F | 44.9 | 48.7 | 56.3 | 64.4 | 71.5 | 78.5 | 81.2 | 80.5 | 75.6 | 64.9 | 55.3 | 48.3 | 64.2 |
| Days Max Temp ≥ 90 °F | 0 | 0 | 0 | 0 | 5 | 19 | 24 | 22 | 11 | 1 | 0 | 0 | 82 |
| Days Max Temp ≤ 32 °F | 1 | 0 | 0 | 0 | 0 | 0 | 0 | 0 | 0 | 0 | 0 | 0 | 1 |
| Days Min Temp ≤ 32 °F | 14 | 10 | 5 | 1 | 0 | 0 | 0 | 0 | 0 | 0 | 5 | 12 | 47 |
| Days Min Temp ≤ 0 °F | 0 | 0 | 0 | 0 | 0 | 0 | 0 | 0 | 0 | 0 | 0 | 0 | 0 |
| Heating Degree Days | 618 | 456 | 285 | 91 | 10 | 0 | 0 | 0 | 6 | 90 | 300 | 516 | 2372 |
| Cooling Degree Days | 0 | 4 | 23 | 76 | 225 | 432 | 530 | 510 | 338 | 99 | 18 | 6 | 2261 |
| Total Precipitation (") | 5.80 | 5.22 | 6.39 | 5.33 | 4.62 | 3.62 | 5.13 | 3.84 | 3.84 | 3.12 | 4.13 | 5.23 | 56.27 |
| Days ≥ 0.1" Precip | 8 | 6 | 7 | 6 | 7 | 6 | 8 | 6 | 6 | 4 | 6 | 7 | 77 |
| Total Snowfall (") | 0.2 | 0.1 | 0.1 | 0.1 | 0.0 | 0.0 | 0.0 | 0.0 | 0.0 | 0.0 | 0.0 | 0.0 | 0.5 |
| Days ≥ 1" Snow Depth | 0 | 0 | 0 | 0 | 0 | 0 | 0 | 0 | 0 | 0 | 0 | 0 | 0 |

## GREENVILLE *Butler County* ELEVATION 440 ft LAT/LONG 31° 50 ' N / 86° 38 ' W

| | JAN | FEB | MAR | APR | MAY | JUN | JUL | AUG | SEP | OCT | NOV | DEC | YEAR |
|---|---|---|---|---|---|---|---|---|---|---|---|---|---|
| Maximum Temp °F | 58.4 | 62.3 | 70.6 | 78.5 | 84.2 | ***89.6*** | 91.6 | 90.8 | 87.2 | 78.6 | 69.8 | 62.0 | 77.0 |
| Minimum Temp °F | 36.5 | 38.0 | 44.8 | 52.2 | 59.6 | 66.6 | 69.6 | 68.9 | 64.8 | 53.7 | 44.9 | 38.7 | 53.2 |
| Mean Temp °F | 47.7 | 50.2 | 57.8 | 65.5 | 71.9 | ***78.1*** | 80.6 | 79.8 | 76.0 | 66.2 | 57.4 | 50.4 | 65.1 |
| Days Max Temp ≥ 90 °F | 0 | 0 | 0 | 1 | 5 | ***17*** | 23 | 20 | 13 | 1 | 0 | 0 | 80 |
| Days Max Temp ≤ 32 °F | 0 | 0 | 0 | 0 | 0 | 0 | 0 | 0 | 0 | 0 | 0 | 0 | 0 |
| Days Min Temp ≤ 32 °F | 11 | ***10*** | 4 | 0 | 0 | 0 | 0 | 0 | 0 | 0 | 4 | 10 | 39 |
| Days Min Temp ≤ 0 °F | 0 | 0 | 0 | 0 | 0 | 0 | 0 | 0 | 0 | 0 | 0 | 0 | 0 |
| Heating Degree Days | 533 | 418 | 246 | 75 | 9 | 0 | 0 | 0 | 3 | 71 | 250 | 456 | 2061 |
| Cooling Degree Days | 2 | 8 | 34 | 92 | 244 | 423 | 510 | 490 | 352 | 135 | 37 | 11 | 2338 |
| Total Precipitation (") | 5.73 | 5.49 | 6.24 | 3.97 | 4.31 | 4.78 | 5.45 | 4.79 | 3.88 | 2.47 | 4.15 | 5.00 | 56.26 |
| Days ≥ 0.1" Precip | 8 | ***7*** | 7 | ***5*** | 6 | 7 | 9 | 7 | 5 | 4 | 6 | 7 | 78 |
| Total Snowfall (") | 0.0 | 0.4 | 0.1 | 0.0 | 0.0 | 0.0 | 0.0 | 0.0 | 0.0 | 0.0 | 0.0 | 0.1 | 0.6 |
| Days ≥ 1" Snow Depth | 0 | 0 | 0 | 0 | 0 | 0 | 0 | 0 | 0 | 0 | 0 | 0 | 0 |

## GUNTERSVILLE *Marshall County* ELEVATION 640 ft LAT/LONG 34° 21 ' N / 86° 17 ' W

| | JAN | FEB | MAR | APR | MAY | JUN | JUL | AUG | SEP | OCT | NOV | DEC | YEAR |
|---|---|---|---|---|---|---|---|---|---|---|---|---|---|
| Maximum Temp °F | 51.4 | 57.0 | 65.9 | 75.0 | 81.2 | 87.9 | 90.7 | 90.0 | ***85.0*** | ***74.6*** | 64.6 | 55.4 | 73.2 |
| Minimum Temp °F | 30.7 | 33.2 | 41.1 | 48.9 | ***57.2*** | 64.7 | 68.3 | 67.2 | ***61.9*** | ***49.9*** | 41.1 | 34.4 | 49.9 |
| Mean Temp °F | 41.1 | 45.1 | 53.5 | 62.0 | ***69.2*** | 76.4 | 79.6 | 78.7 | ***73.4*** | ***62.4*** | 52.9 | 46.6 | 61.7 |
| Days Max Temp ≥ 90 °F | 0 | 0 | 0 | 0 | 2 | 11 | 19 | 16 | ***7*** | ***0*** | 0 | 0 | 55 |
| Days Max Temp ≤ 32 °F | 1 | 0 | 0 | 0 | 0 | 0 | 0 | 0 | ***0*** | 0 | 0 | ***0*** | 1 |
| Days Min Temp ≤ 32 °F | 18 | ***15*** | 7 | 1 | ***0*** | 0 | 0 | 0 | ***0*** | 0 | ***5*** | 14 | 60 |
| Days Min Temp ≤ 0 °F | 0 | 0 | 0 | 0 | 0 | 0 | 0 | 0 | ***0*** | 0 | 0 | 0 | 0 |
| Heating Degree Days | 734 | ***552*** | 360 | 137 | ***24*** | 1 | 0 | 0 | ***9*** | ***127*** | 363 | 615 | 2922 |
| Cooling Degree Days | 0 | ***1*** | 11 | ***44*** | ***173*** | 369 | ***485*** | 447 | 265 | 54 | 8 | 1 | 1858 |
| Total Precipitation (") | 4.59 | 4.84 | 5.96 | 4.75 | 4.96 | 3.42 | 4.61 | 3.32 | 4.21 | 3.16 | 3.94 | 4.98 | 52.74 |
| Days ≥ 0.1" Precip | 7 | ***6*** | 8 | 6 | ***7*** | 5 | 7 | 6 | na | ***5*** | ***6*** | ***7*** | na |
| Total Snowfall (") | ***0.7*** | ***0.2*** | 0.3 | 0.0 | 0.0 | 0.0 | 0.0 | 0.0 | 0.0 | 0.0 | ***0.0*** | 0.0 | 1.2 |
| Days ≥ 1" Snow Depth | ***0*** | ***0*** | 0 | 0 | 0 | 0 | 0 | 0 | ***0*** | 0 | 0 | ***0*** | 0 |

## HALEYVILLE 2 ENE *Winston County* ELEVATION 951 ft LAT/LONG 34° 14 ' N / 87° 37 ' W

| | JAN | FEB | MAR | APR | MAY | JUN | JUL | AUG | SEP | OCT | NOV | DEC | YEAR |
|---|---|---|---|---|---|---|---|---|---|---|---|---|---|
| Maximum Temp °F | 48.6 | 53.6 | 62.8 | 72.6 | 78.7 | 85.7 | 88.7 | 88.0 | 82.7 | 72.6 | 61.8 | 52.9 | 70.7 |
| Minimum Temp °F | 28.7 | 31.1 | 39.8 | 48.0 | 55.9 | 63.3 | 67.1 | 65.9 | 60.3 | 48.2 | 39.6 | 32.5 | 48.4 |
| Mean Temp °F | 38.7 | 42.4 | 51.3 | 60.3 | 67.3 | 74.5 | 77.9 | 77.0 | 71.5 | 60.4 | 50.7 | 42.7 | 59.6 |
| Days Max Temp ≥ 90 °F | 0 | 0 | 0 | 0 | 1 | 6 | 15 | 13 | 5 | 0 | 0 | 0 | 40 |
| Days Max Temp ≤ 32 °F | 3 | 1 | 0 | 0 | 0 | 0 | 0 | 0 | 0 | 0 | 0 | 1 | 5 |
| Days Min Temp ≤ 32 °F | 20 | 16 | 9 | 2 | 0 | 0 | 0 | 0 | 0 | 1 | 8 | 16 | 72 |
| Days Min Temp ≤ 0 °F | 0 | 0 | 0 | 0 | 0 | 0 | 0 | 0 | 0 | 0 | 0 | 0 | 0 |
| Heating Degree Days | 809 | 634 | 425 | 177 | 46 | 2 | 0 | 0 | 21 | 177 | 426 | 685 | 3402 |
| Cooling Degree Days | 0 | 0 | 8 | 40 | 126 | 303 | 427 | 404 | 235 | 48 | 5 | 1 | 1597 |
| Total Precipitation (") | 5.23 | 5.31 | 6.70 | 5.28 | 5.90 | 4.01 | 4.86 | 4.12 | 4.74 | 3.65 | 5.18 | 6.33 | 61.31 |
| Days ≥ 0.1" Precip | 7 | 7 | 8 | 7 | 7 | 7 | 8 | 6 | 6 | 5 | 7 | 7 | 82 |
| Total Snowfall (") | 1.0 | 0.4 | 0.5 | 0.1 | 0.0 | 0.0 | 0.0 | 0.0 | 0.0 | 0.0 | 0.2 | 0.2 | 2.4 |
| Days ≥ 1" Snow Depth | 0 | 0 | 0 | 0 | 0 | 0 | 0 | 0 | 0 | 0 | 0 | 0 | 0 |

## HAMILTON 3 S *Marion County* ELEVATION 522 ft LAT/LONG 34° 6 ' N / 87° 59 ' W

| | JAN | FEB | MAR | APR | MAY | JUN | JUL | AUG | SEP | OCT | NOV | DEC | YEAR |
|---|---|---|---|---|---|---|---|---|---|---|---|---|---|
| Maximum Temp °F | 50.7 | 55.8 | 65.2 | 74.9 | 80.9 | 87.8 | 90.9 | 90.4 | 85.1 | 75.1 | 64.7 | 55.2 | 73.1 |
| Minimum Temp °F | 25.3 | 27.3 | 34.9 | 42.8 | 52.0 | 60.4 | 65.4 | 63.9 | 57.6 | 43.6 | 34.9 | 28.9 | 44.8 |
| Mean Temp °F | 38.1 | 41.6 | 50.1 | 58.9 | 66.5 | 74.1 | 78.2 | 77.2 | 71.4 | 59.3 | 49.8 | 42.1 | 58.9 |
| Days Max Temp ≥ 90 °F | 0 | 0 | 0 | 0 | 2 | 13 | 21 | 19 | 9 | 0 | 0 | 0 | 64 |
| Days Max Temp ≤ 32 °F | 2 | 1 | 0 | 0 | 0 | 0 | 0 | 0 | 0 | 0 | 0 | 1 | 4 |
| Days Min Temp ≤ 32 °F | 23 | 20 | 14 | 6 | 0 | 0 | 0 | 0 | 0 | 5 | 15 | 20 | 103 |
| Days Min Temp ≤ 0 °F | 0 | 0 | 0 | 0 | 0 | 0 | 0 | 0 | 0 | 0 | 0 | 0 | 0 |
| Heating Degree Days | 829 | 656 | 462 | 209 | 59 | 4 | 0 | 0 | 24 | 205 | 452 | 704 | 3604 |
| Cooling Degree Days | 0 | 1 | 7 | 30 | 119 | 310 | 442 | 413 | 232 | 44 | 4 | 1 | 1603 |
| Total Precipitation (") | 5.06 | 5.19 | 6.28 | 5.55 | 5.97 | 4.08 | 4.66 | 4.04 | 4.51 | 3.55 | 4.87 | 6.04 | 59.80 |
| Days ≥ 0.1" Precip | 7 | 7 | 7 | 7 | 7 | 6 | 7 | 6 | 6 | 5 | 6 | 8 | 79 |
| Total Snowfall (") | 1.0 | 0.6 | 0.4 | 0.0 | 0.0 | 0.0 | 0.0 | 0.0 | 0.0 | 0.0 | 0.0 | 0.1 | 2.1 |
| Days ≥ 1" Snow Depth | 1 | 0 | 0 | 0 | 0 | 0 | 0 | 0 | 0 | 0 | 0 | 0 | 1 |

## HEADLAND *Henry County* ELEVATION 370 ft LAT/LONG 31° 21 ' N / 85° 20 ' W

| | JAN | FEB | MAR | APR | MAY | JUN | JUL | AUG | SEP | OCT | NOV | DEC | YEAR |
|---|---|---|---|---|---|---|---|---|---|---|---|---|---|
| Maximum Temp °F | 58.0 | 61.8 | 70.1 | 78.5 | 84.5 | 89.8 | 91.5 | 90.7 | 87.4 | 78.7 | 69.4 | 61.8 | 76.9 |
| Minimum Temp °F | 36.1 | 38.3 | 45.8 | 53.3 | 60.8 | 67.2 | 69.5 | 68.5 | 64.1 | 52.7 | 45.1 | 38.8 | 53.4 |
| Mean Temp °F | 47.1 | 50.1 | 58.0 | 65.9 | 72.8 | 78.6 | 80.5 | 79.6 | 75.8 | 65.7 | 57.3 | 50.3 | 65.1 |
| Days Max Temp ≥ 90 °F | 0 | 0 | 0 | 1 | 5 | 18 | 23 | 21 | 13 | 1 | 0 | 0 | 82 |
| Days Max Temp ≤ 32 °F | 0 | 0 | 0 | 0 | 0 | 0 | 0 | 0 | 0 | 0 | 0 | 0 | 0 |
| Days Min Temp ≤ 32 °F | 13 | 9 | 2 | 0 | 0 | 0 | 0 | 0 | 0 | 0 | 3 | 9 | 36 |
| Days Min Temp ≤ 0 °F | 0 | 0 | 0 | 0 | 0 | 0 | 0 | 0 | 0 | 0 | 0 | 0 | 0 |
| Heating Degree Days | 551 | 420 | 239 | 72 | 6 | 0 | 0 | 0 | 4 | 77 | 251 | 453 | 2073 |
| Cooling Degree Days | 1 | 8 | 29 | 89 | 245 | 422 | 492 | 461 | 321 | 104 | 29 | 7 | 2208 |
| Total Precipitation (") | 6.24 | 5.20 | 6.13 | 3.70 | 4.65 | 4.99 | 6.30 | 4.43 | 3.49 | 2.92 | 3.87 | 4.44 | 56.36 |
| Days ≥ 0.1" Precip | 8 | 7 | 7 | 5 | 6 | 8 | 10 | 8 | 5 | 4 | 5 | 6 | 79 |
| Total Snowfall (") | 0.1 | 0.1 | 0.0 | 0.0 | 0.0 | 0.0 | 0.0 | 0.0 | 0.0 | 0.0 | 0.0 | 0.0 | 0.2 |
| Days ≥ 1" Snow Depth | 0 | 0 | 0 | 0 | 0 | 0 | 0 | 0 | 0 | 0 | 0 | 0 | 0 |

## HEFLIN *Cleburne County* ELEVATION 850 ft LAT/LONG 33° 39 ' N / 85° 36 ' W

| | JAN | FEB | MAR | APR | MAY | JUN | JUL | AUG | SEP | OCT | NOV | DEC | YEAR |
|---|---|---|---|---|---|---|---|---|---|---|---|---|---|
| Maximum Temp °F | 50.5 | 55.6 | 64.6 | 73.8 | 79.9 | 86.9 | 90.0 | 89.2 | 83.8 | 74.0 | 64.1 | 54.8 | 72.3 |
| Minimum Temp °F | 27.4 | 29.6 | 36.6 | 44.1 | 52.3 | 61.1 | 66.0 | 64.8 | 58.6 | 44.5 | 36.3 | 30.8 | 46.0 |
| Mean Temp °F | 39.0 | 42.6 | 50.6 | 59.0 | 66.1 | 74.0 | 78.0 | 77.0 | 71.2 | 59.2 | 50.2 | 42.8 | 59.1 |
| Days Max Temp ≥ 90 °F | 0 | 0 | 0 | 0 | 1 | 9 | 18 | 16 | 6 | 0 | 0 | 0 | 50 |
| Days Max Temp ≤ 32 °F | 2 | 1 | 0 | 0 | 0 | 0 | 0 | 0 | 0 | 0 | 0 | 1 | 4 |
| Days Min Temp ≤ 32 °F | 21 | 18 | 11 | 3 | 0 | 0 | 0 | 0 | 0 | 4 | 12 | *18* | 87 |
| Days Min Temp ≤ 0 °F | 0 | 0 | 0 | 0 | 0 | 0 | 0 | 0 | 0 | 0 | 0 | 0 | 0 |
| Heating Degree Days | 801 | 625 | 443 | 200 | 60 | 4 | 0 | 0 | 21 | 202 | 441 | 683 | 3480 |
| Cooling Degree Days | 0 | 0 | 7 | 24 | 105 | 301 | 429 | 396 | 221 | *33* | 3 | *1* | 1520 |
| Total Precipitation (") | 5.55 | 5.26 | 6.97 | 5.19 | 5.33 | 4.41 | 4.80 | 3.49 | 4.12 | 3.16 | 4.42 | 5.06 | 57.76 |
| Days ≥ 0.1" Precip | na | *6* | *8* | 6 | 7 | 6 | 8 | 6 | *5* | *4* | 6 | *7* | na |
| Total Snowfall (") | 1.0 | *0.2* | 0.6 | 0.2 | 0.0 | 0.0 | 0.0 | 0.0 | 0.0 | 0.0 | 0.0 | 0.1 | 2.1 |
| Days ≥ 1" Snow Depth | *0* | *0* | 0 | 0 | 0 | 0 | 0 | 0 | 0 | 0 | 0 | 0 | 0 |

## HIGHLAND HOME *Crenshaw County* ELEVATION 594 ft LAT/LONG 31° 57 ' N / 86° 19 ' W

| | JAN | FEB | MAR | APR | MAY | JUN | JUL | AUG | SEP | OCT | NOV | DEC | YEAR |
|---|---|---|---|---|---|---|---|---|---|---|---|---|---|
| Maximum Temp °F | 55.9 | 60.4 | 68.9 | 76.9 | 83.0 | 88.8 | 90.9 | 90.0 | 86.7 | 77.8 | 68.2 | 60.0 | 75.6 |
| Minimum Temp °F | 34.2 | 36.7 | 44.0 | 51.3 | 58.9 | 65.8 | 68.9 | 68.1 | 63.6 | 52.2 | 43.9 | 37.4 | 52.1 |
| Mean Temp °F | 45.1 | 48.6 | 56.5 | 64.1 | 70.9 | 77.3 | 79.9 | 79.1 | 75.2 | 65.0 | 56.1 | 48.7 | 63.9 |
| Days Max Temp ≥ 90 °F | 0 | 0 | 0 | 0 | 3 | 15 | 20 | 19 | 12 | 1 | 0 | 0 | 70 |
| Days Max Temp ≤ 32 °F | 1 | 0 | 0 | 0 | 0 | 0 | 0 | 0 | 0 | 0 | 0 | 0 | 1 |
| Days Min Temp ≤ 32 °F | 14 | 11 | 4 | 0 | 0 | 0 | 0 | 0 | 0 | 0 | 4 | 11 | 44 |
| Days Min Temp ≤ 0 °F | 0 | 0 | 0 | 0 | 0 | 0 | 0 | 0 | 0 | 0 | 0 | 0 | 0 |
| Heating Degree Days | 612 | 459 | 282 | 101 | 14 | 0 | 0 | 0 | 6 | 90 | 281 | 504 | 2349 |
| Cooling Degree Days | 1 | 4 | 24 | 68 | 199 | 385 | 468 | 445 | 311 | 98 | 20 | 6 | 2029 |
| Total Precipitation (") | 5.15 | 5.42 | 6.89 | 3.98 | 4.62 | 4.49 | 4.88 | 4.43 | 3.75 | 2.68 | 4.20 | 4.79 | 55.28 |
| Days ≥ 0.1" Precip | 8 | 7 | 8 | 6 | 6 | 7 | 9 | 7 | 5 | 4 | 5 | 7 | 79 |
| Total Snowfall (") | 0.2 | 0.7 | 0.3 | 0.0 | 0.0 | 0.0 | 0.0 | 0.0 | 0.0 | 0.0 | 0.1 | 0.0 | 1.3 |
| Days ≥ 1" Snow Depth | 0 | 0 | 0 | 0 | 0 | 0 | 0 | 0 | 0 | 0 | 0 | 0 | 0 |

## HUNTSVILLE MADISON *Madison County* ELEVATION 623 ft LAT/LONG 34° 39 ' N / 86° 46 ' W

| | JAN | FEB | MAR | APR | MAY | JUN | JUL | AUG | SEP | OCT | NOV | DEC | YEAR |
|---|---|---|---|---|---|---|---|---|---|---|---|---|---|
| Maximum Temp °F | 48.6 | 53.6 | 63.0 | 72.8 | 79.2 | 86.8 | 89.5 | 88.6 | 82.7 | 72.8 | 62.2 | 53.0 | 71.1 |
| Minimum Temp °F | 29.9 | 33.1 | 40.9 | 49.3 | 57.4 | 65.3 | 69.3 | 67.9 | 61.7 | 49.4 | 40.4 | 33.8 | 49.9 |
| Mean Temp °F | 39.3 | 43.4 | 52.0 | 61.1 | 68.3 | 76.1 | 79.4 | 78.3 | 72.2 | 61.1 | 51.3 | 43.4 | 60.5 |
| Days Max Temp ≥ 90 °F | 0 | 0 | 0 | 0 | 1 | 9 | 16 | 14 | 4 | 0 | 0 | 0 | 44 |
| Days Max Temp ≤ 32 °F | 3 | 1 | 0 | 0 | 0 | 0 | 0 | 0 | 0 | 0 | 0 | 1 | 5 |
| Days Min Temp ≤ 32 °F | 19 | 14 | 7 | 1 | 0 | 0 | 0 | 0 | 0 | 0 | 8 | 16 | 65 |
| Days Min Temp ≤ 0 °F | 0 | 0 | 0 | 0 | 0 | 0 | 0 | 0 | 0 | 0 | 0 | 0 | 0 |
| Heating Degree Days | 791 | 605 | 406 | 163 | 38 | 1 | 0 | 0 | 17 | 161 | 408 | 663 | 3253 |
| Cooling Degree Days | 0 | 0 | 9 | 43 | 151 | 348 | 468 | 433 | 241 | 49 | 6 | 2 | 1750 |
| Total Precipitation (") | 5.09 | 4.91 | 6.48 | 4.63 | 5.54 | 4.26 | 4.75 | 3.71 | 4.29 | 3.39 | 4.83 | 5.88 | 57.76 |
| Days ≥ 0.1" Precip | 8 | 7 | 8 | 7 | 7 | 6 | 7 | 6 | 6 | 5 | 7 | 8 | 82 |
| Total Snowfall (") | 1.6 | 0.7 | 0.5 | 0.0 | 0.0 | 0.0 | 0.0 | 0.0 | 0.0 | 0.0 | 0.2 | 0.2 | 3.2 |
| Days ≥ 1" Snow Depth | 1 | 0 | 0 | 0 | 0 | 0 | 0 | 0 | 0 | 0 | 0 | 0 | 1 |

## JASPER *Walker County* ELEVATION 531 ft LAT/LONG 33° 54 ' N / 87° 16 ' W

| | JAN | FEB | MAR | APR | MAY | JUN | JUL | AUG | SEP | OCT | NOV | DEC | YEAR |
|---|---|---|---|---|---|---|---|---|---|---|---|---|---|
| Maximum Temp °F | 50.8 | 55.9 | 66.0 | 74.9 | 79.8 | 86.9 | 89.8 | 89.2 | 83.8 | 73.6 | 63.7 | ***55.6*** | 72.5 |
| Minimum Temp °F | 28.0 | 30.4 | 37.5 | 45.7 | 53.9 | 62.3 | 66.9 | 65.7 | 59.5 | 46.4 | 37.2 | ***31.9*** | 47.1 |
| Mean Temp °F | 39.4 | 43.1 | 51.8 | 60.3 | 66.9 | 74.6 | 78.3 | 77.5 | 71.7 | 60.0 | 50.4 | ***43.8*** | 59.8 |
| Days Max Temp ≥ 90 °F | 0 | 0 | 0 | 0 | 1 | 10 | 17 | 15 | 6 | 0 | 0 | ***0*** | 49 |
| Days Max Temp ≤ 32 °F | 2 | 1 | 0 | 0 | 0 | 0 | 0 | 0 | 0 | 0 | 0 | ***0*** | 3 |
| Days Min Temp ≤ 32 °F | 21 | 18 | 11 | 3 | 0 | 0 | 0 | 0 | 0 | 2 | 11 | ***18*** | 84 |
| Days Min Temp ≤ 0 °F | 0 | 0 | 0 | 0 | 0 | 0 | 0 | 0 | 0 | 0 | 0 | ***0*** | 0 |
| Heating Degree Days | 786 | 611 | 412 | 172 | 50 | 2 | 0 | 0 | 18 | 183 | 434 | ***652*** | 3320 |
| Cooling Degree Days | ***0*** | ***1*** | ***9*** | ***33*** | ***120*** | 310 | ***439*** | ***414*** | ***231*** | ***46*** | ***4*** | na | na |
| Total Precipitation (") | 5.39 | 5.05 | 6.47 | 5.15 | 5.43 | 4.14 | 4.91 | 3.59 | 4.32 | 3.64 | 4.11 | 6.19 | 58.39 |
| Days ≥ 0.1" Precip | 8 | 7 | 8 | 7 | 8 | 7 | 8 | 6 | 6 | 5 | 7 | 8 | 85 |
| Total Snowfall (") | 0.8 | 0.2 | 0.4 | 0.0 | 0.0 | 0.0 | 0.0 | 0.0 | 0.0 | 0.0 | 0.0 | 0.0 | 1.4 |
| Days ≥ 1" Snow Depth | 1 | 0 | 0 | 0 | 0 | 0 | 0 | 0 | 0 | 0 | 0 | 0 | 1 |

## LIVINGSTON 2 SW *Sumter County* ELEVATION 161 ft LAT/LONG 32° 35 ' N / 88° 11 ' W

| | JAN | FEB | MAR | APR | MAY | JUN | JUL | AUG | SEP | OCT | NOV | DEC | YEAR |
|---|---|---|---|---|---|---|---|---|---|---|---|---|---|
| Maximum Temp °F | 55.5 | 60.9 | 68.5 | 76.7 | 83.2 | 90.0 | 92.3 | 91.2 | 87.2 | 77.8 | 68.6 | 59.4 | 75.9 |
| Minimum Temp °F | 31.9 | 34.9 | 42.1 | 49.3 | 57.9 | 65.3 | 68.8 | 67.6 | 62.2 | ***49.6*** | 40.5 | 34.6 | 50.4 |
| Mean Temp °F | 43.8 | 47.9 | 55.4 | 63.0 | 70.6 | 77.7 | 80.6 | 79.4 | 74.7 | ***63.7*** | 54.6 | 47.0 | 63.2 |
| Days Max Temp ≥ 90 °F | 0 | 0 | 0 | 0 | 4 | 17 | 23 | 22 | 12 | 1 | 0 | 0 | 79 |
| Days Max Temp ≤ 32 °F | 1 | 0 | 0 | 0 | 0 | 0 | 0 | 0 | 0 | 0 | 0 | 0 | 1 |
| Days Min Temp ≤ 32 °F | 17 | 12 | 5 | 1 | 0 | 0 | 0 | 0 | 0 | 1 | 7 | 14 | 57 |
| Days Min Temp ≤ 0 °F | 0 | 0 | 0 | 0 | 0 | 0 | 0 | 0 | 0 | 0 | 0 | 0 | 0 |
| Heating Degree Days | 653 | 478 | 310 | 118 | 15 | 0 | 0 | 0 | 8 | ***115*** | 319 | 556 | 2572 |
| Cooling Degree Days | 1 | 5 | 19 | 61 | 195 | 390 | 494 | 464 | 302 | 84 | 15 | 5 | 2035 |
| Total Precipitation (") | 5.38 | 5.01 | 6.40 | 5.23 | 4.74 | 3.84 | 5.55 | 3.76 | 3.32 | 3.55 | 4.34 | 5.51 | 56.63 |
| Days ≥ 0.1" Precip | ***7*** | 6 | 7 | 5 | 6 | 6 | 8 | 5 | 5 | 4 | 5 | 7 | 71 |
| Total Snowfall (") | 0.3 | 0.0 | 0.2 | 0.1 | 0.0 | 0.0 | 0.0 | 0.0 | 0.0 | 0.0 | 0.0 | 0.0 | 0.6 |
| Days ≥ 1" Snow Depth | ***0*** | 0 | 0 | 0 | 0 | 0 | 0 | 0 | 0 | 0 | 0 | 0 | 0 |

## MARION JUNCTION 2 NE *Dallas County* ELEVATION 200 ft LAT/LONG 32° 28 ' N / 87° 13 ' W

| | JAN | FEB | MAR | APR | MAY | JUN | JUL | AUG | SEP | OCT | NOV | DEC | YEAR |
|---|---|---|---|---|---|---|---|---|---|---|---|---|---|
| Maximum Temp °F | 54.4 | 59.0 | 67.5 | 75.7 | 81.9 | 88.7 | 90.9 | 89.9 | 85.7 | 76.3 | 66.8 | 58.4 | 74.6 |
| Minimum Temp °F | 32.8 | 35.4 | 42.9 | 50.5 | 58.4 | 65.9 | 69.2 | 68.3 | 62.7 | 50.3 | 41.2 | 35.5 | 51.1 |
| Mean Temp °F | 43.7 | 47.2 | 55.2 | 63.1 | 70.2 | 77.3 | 80.1 | 79.1 | 74.2 | 63.3 | 54.0 | 46.9 | 62.9 |
| Days Max Temp ≥ 90 °F | 0 | 0 | 0 | 0 | 3 | 14 | 21 | 18 | 9 | 1 | 0 | 0 | 66 |
| Days Max Temp ≤ 32 °F | 1 | 0 | 0 | 0 | 0 | 0 | 0 | 0 | 0 | 0 | 0 | 0 | 1 |
| Days Min Temp ≤ 32 °F | 16 | 12 | 5 | 0 | 0 | 0 | 0 | 0 | 0 | 1 | 7 | 14 | 55 |
| Days Min Temp ≤ 0 °F | 0 | 0 | 0 | 0 | 0 | 0 | 0 | 0 | 0 | 0 | 0 | 0 | 0 |
| Heating Degree Days | 656 | 498 | 315 | 117 | 19 | 1 | 0 | 0 | 10 | 120 | 337 | 558 | 2631 |
| Cooling Degree Days | 1 | 3 | 17 | 55 | 181 | 382 | 478 | 449 | 285 | 77 | 15 | 4 | 1947 |
| Total Precipitation (") | 5.35 | 4.78 | 6.18 | 4.83 | 4.19 | 3.86 | 5.06 | 3.85 | 3.48 | 2.86 | 4.10 | 5.31 | 53.85 |
| Days ≥ 0.1" Precip | 8 | 7 | 8 | 6 | 7 | 6 | 8 | 6 | 6 | 4 | 6 | 7 | 79 |
| Total Snowfall (") | 0.0 | 0.1 | 0.0 | 0.0 | 0.0 | 0.0 | 0.0 | 0.0 | 0.0 | 0.0 | 0.0 | 0.0 | 0.1 |
| Days ≥ 1" Snow Depth | 0 | 0 | 0 | 0 | 0 | 0 | 0 | 0 | 0 | 0 | 0 | 0 | 0 |

## MOBILE REGIONAL AP *Mobile County* ELEVATION 217 ft LAT/LONG 30° 41 ' N / 88° 14 ' W

| | JAN | FEB | MAR | APR | MAY | JUN | JUL | AUG | SEP | OCT | NOV | DEC | YEAR |
|---|---|---|---|---|---|---|---|---|---|---|---|---|---|
| Maximum Temp °F | 60.1 | 63.8 | 70.8 | 78.3 | 84.1 | 89.9 | 91.2 | 90.3 | 86.8 | 79.2 | 70.2 | 63.3 | 77.3 |
| Minimum Temp °F | 40.5 | 42.8 | 49.8 | 57.0 | 64.2 | 70.8 | 73.2 | 72.8 | 68.8 | 57.6 | 48.9 | 43.4 | 57.5 |
| Mean Temp °F | 50.3 | 53.3 | 60.4 | 67.7 | 74.2 | 80.4 | 82.2 | 81.6 | 77.8 | 68.4 | 59.6 | 53.4 | 67.4 |
| Days Max Temp ≥ 90 °F | 0 | 0 | 0 | 0 | 3 | 17 | 22 | 20 | 10 | 1 | 0 | 0 | 73 |
| Days Max Temp ≤ 32 °F | 0 | 0 | 0 | 0 | 0 | 0 | 0 | 0 | 0 | 0 | 0 | 0 | 0 |
| Days Min Temp ≤ 32 °F | 8 | 5 | 1 | 0 | 0 | 0 | 0 | 0 | 0 | 0 | 1 | 5 | 20 |
| Days Min Temp ≤ 0 °F | 0 | 0 | 0 | 0 | 0 | 0 | 0 | 0 | 0 | 0 | 0 | 0 | 0 |
| Heating Degree Days | 458 | 334 | 182 | 45 | 2 | 0 | 0 | 0 | 2 | 45 | 200 | 372 | 1640 |
| Cooling Degree Days | 4 | 10 | 42 | 105 | 274 | 451 | 529 | 516 | 378 | 149 | 45 | 16 | 2519 |
| Total Precipitation (") | 5.27 | 5.42 | 6.42 | 4.43 | 6.24 | 4.67 | 7.02 | 6.78 | 5.78 | 3.16 | 4.40 | 4.94 | 64.53 |
| Days ≥ 0.1" Precip | 7 | 7 | 7 | 5 | 6 | 7 | 10 | 10 | 7 | 4 | 5 | 7 | 82 |
| Total Snowfall (") | 0.1 | 0.2 | 0.1 | 0.0 | 0.0 | 0.0 | 0.0 | 0.0 | 0.0 | 0.0 | 0.0 | 0.0 | 0.4 |
| Days ≥ 1" Snow Depth | 0 | 0 | 0 | 0 | 0 | 0 | 0 | 0 | 0 | 0 | 0 | 0 | 0 |

## MONTGOMERY DANNELLY *Montgomery County* ELEVATION 200 ft LAT/LONG 32° 18 ' N / 86° 24 ' W

| | JAN | FEB | MAR | APR | MAY | JUN | JUL | AUG | SEP | OCT | NOV | DEC | YEAR |
|---|---|---|---|---|---|---|---|---|---|---|---|---|---|
| Maximum Temp °F | 56.5 | 60.8 | 69.0 | 77.0 | 83.1 | 89.6 | 91.5 | 90.8 | 87.0 | 78.0 | 68.5 | 60.4 | 76.0 |
| Minimum Temp °F | 35.8 | 38.5 | 45.5 | 52.8 | 60.7 | 67.9 | 71.6 | 70.8 | 65.8 | 53.2 | 44.2 | 38.6 | 53.8 |
| Mean Temp °F | 46.2 | 49.6 | 57.3 | 64.9 | 71.9 | 78.8 | 81.6 | 80.8 | 76.4 | 65.6 | 56.4 | 49.5 | 64.9 |
| Days Max Temp ≥ 90 °F | 0 | 0 | 0 | 0 | 3 | 17 | 22 | 21 | 12 | 1 | 0 | 0 | 76 |
| Days Max Temp ≤ 32 °F | 0 | 0 | 0 | 0 | 0 | 0 | 0 | 0 | 0 | 0 | 0 | 0 | 0 |
| Days Min Temp ≤ 32 °F | 13 | 9 | 3 | 0 | 0 | 0 | 0 | 0 | 0 | 0 | 4 | 10 | 39 |
| Days Min Temp ≤ 0 °F | 0 | 0 | 0 | 0 | 0 | 0 | 0 | 0 | 0 | 0 | 0 | 0 | 0 |
| Heating Degree Days | 579 | 431 | 257 | 82 | 8 | 0 | 0 | 0 | 4 | 80 | 275 | 481 | 2197 |
| Cooling Degree Days | 1 | 5 | 25 | 78 | 233 | 432 | 533 | 520 | 359 | 115 | 28 | 9 | 2338 |
| Total Precipitation (") | 4.82 | 5.41 | 6.24 | 4.08 | 4.22 | 4.03 | 5.50 | 3.77 | 4.02 | 2.67 | 4.41 | 5.16 | 54.33 |
| Days ≥ 0.1" Precip | 8 | 7 | 8 | 5 | 6 | 6 | 9 | 6 | 5 | 4 | 6 | 7 | 77 |
| Total Snowfall (") | 0.2 | 0.1 | 0.1 | 0.0 | 0.0 | 0.0 | 0.0 | 0.0 | 0.0 | 0.0 | 0.0 | 0.0 | 0.4 |
| Days ≥ 1" Snow Depth | 0 | 0 | 0 | 0 | 0 | 0 | 0 | 0 | 0 | 0 | 0 | 0 | 0 |

## MOULTON 2 *Lawrence County* ELEVATION 645 ft LAT/LONG 34° 29 ' N / 87° 18 ' W

| | JAN | FEB | MAR | APR | MAY | JUN | JUL | AUG | SEP | OCT | NOV | DEC | YEAR |
|---|---|---|---|---|---|---|---|---|---|---|---|---|---|
| Maximum Temp °F | 50.4 | 55.3 | 64.7 | 74.5 | 80.6 | 87.5 | 90.3 | 89.9 | 84.4 | 74.3 | 63.6 | 54.7 | 72.5 |
| Minimum Temp °F | 30.5 | 33.1 | 41.1 | 49.1 | 57.0 | 64.7 | 68.7 | 67.1 | 61.3 | 48.9 | 40.7 | 34.4 | 49.7 |
| Mean Temp °F | 40.4 | 44.2 | 52.9 | 61.8 | 68.8 | 76.1 | 79.5 | 78.5 | 72.8 | 61.6 | 52.2 | 44.6 | 61.1 |
| Days Max Temp ≥ 90 °F | 0 | 0 | 0 | 0 | 2 | 11 | 19 | 17 | 7 | 0 | 0 | 0 | 56 |
| Days Max Temp ≤ 32 °F | 2 | 1 | 0 | 0 | 0 | 0 | 0 | 0 | 0 | 0 | 0 | 1 | 4 |
| Days Min Temp ≤ 32 °F | 18 | 15 | 8 | 1 | 0 | 0 | 0 | 0 | 0 | 1 | 8 | 15 | 66 |
| Days Min Temp ≤ 0 °F | 0 | 0 | 0 | 0 | 0 | 0 | 0 | 0 | 0 | 0 | 0 | 0 | 0 |
| Heating Degree Days | 755 | 581 | 379 | 147 | 32 | 1 | 0 | 0 | 14 | 152 | 385 | 627 | 3073 |
| Cooling Degree Days | 0 | 1 | 14 | 54 | 160 | 348 | 469 | 437 | 256 | 58 | 7 | 1 | 1805 |
| Total Precipitation (") | 5.05 | 4.90 | 6.41 | 4.82 | 5.32 | 3.98 | 4.49 | 3.48 | 4.20 | 3.59 | 4.88 | 6.01 | 57.13 |
| Days ≥ 0.1" Precip | 8 | 8 | 9 | 7 | 8 | 7 | 8 | 6 | 6 | 5 | 7 | 8 | 87 |
| Total Snowfall (") | 1.8 | 0.9 | 0.6 | 0.0 | 0.0 | 0.0 | 0.0 | 0.0 | 0.0 | 0.0 | 0.2 | 0.3 | 3.8 |
| Days ≥ 1" Snow Depth | 1 | 0 | 0 | 0 | 0 | 0 | 0 | 0 | 0 | 0 | 0 | 0 | 1 |

## MUSCLE SHOALS REG AP *Colbert County* ELEVATION 540 ft LAT/LONG 34° 45 ' N / 87° 37 ' W

| | JAN | FEB | MAR | APR | MAY | JUN | JUL | AUG | SEP | OCT | NOV | DEC | YEAR |
|---|---|---|---|---|---|---|---|---|---|---|---|---|---|
| Maximum Temp °F | 48.6 | 53.8 | 63.3 | 73.0 | 79.8 | 87.6 | 90.6 | 89.5 | 83.5 | 73.2 | 62.3 | 53.0 | 71.5 |
| Minimum Temp °F | 30.0 | 33.1 | 41.1 | 49.3 | 57.3 | 65.3 | 69.5 | 67.9 | 61.8 | 49.1 | 40.6 | 34.0 | 49.9 |
| Mean Temp °F | 39.3 | 43.5 | 52.2 | 61.2 | 68.6 | 76.5 | 80.1 | 78.7 | 72.7 | 61.2 | 51.5 | 43.5 | 60.8 |
| Days Max Temp ≥ 90 °F | 0 | 0 | 0 | 0 | 2 | 12 | 19 | 16 | 6 | 0 | 0 | 0 | 55 |
| Days Max Temp ≤ 32 °F | 3 | 1 | 0 | 0 | 0 | 0 | 0 | 0 | 0 | 0 | 0 | 1 | 5 |
| Days Min Temp ≤ 32 °F | 20 | 15 | 6 | 1 | 0 | 0 | 0 | 0 | 0 | 0 | 7 | 15 | 64 |
| Days Min Temp ≤ 0 °F | 0 | 0 | 0 | 0 | 0 | 0 | 0 | 0 | 0 | 0 | 0 | 0 | 0 |
| Heating Degree Days | 789 | 602 | 400 | 160 | 36 | 0 | 0 | 0 | 17 | 160 | 404 | 659 | 3227 |
| Cooling Degree Days | 0 | 1 | 10 | 46 | 157 | 369 | 499 | 463 | 258 | 51 | 6 | 1 | 1861 |
| Total Precipitation (") | 4.50 | 4.53 | 5.94 | 4.50 | 5.53 | 4.37 | 4.65 | 3.43 | 4.39 | 3.30 | 4.62 | 5.87 | 55.63 |
| Days ≥ 0.1" Precip | 7 | 7 | 8 | 7 | 7 | 7 | 7 | 5 | 6 | 5 | 7 | 8 | 81 |
| Total Snowfall (") | 1.3 | 1.0 | 0.4 | 0.0 | 0.0 | 0.0 | 0.0 | 0.0 | 0.0 | 0.0 | 0.0 | 0.4 | 3.1 |
| Days ≥ 1" Snow Depth | 1 | 1 | 0 | 0 | 0 | 0 | 0 | 0 | 0 | 0 | 0 | 0 | 2 |

## ONEONTA *Blount County* ELEVATION 860 ft LAT/LONG 33° 57 ' N / 86° 29 ' W

| | JAN | FEB | MAR | APR | MAY | JUN | JUL | AUG | SEP | OCT | NOV | DEC | YEAR |
|---|---|---|---|---|---|---|---|---|---|---|---|---|---|
| Maximum Temp °F | 49.8 | 54.6 | 63.8 | 72.8 | 79.2 | 86.3 | 89.4 | 88.7 | 83.3 | 73.3 | 63.4 | 54.3 | 71.6 |
| Minimum Temp °F | 28.8 | 31.2 | 38.8 | 46.6 | 54.8 | 62.9 | 67.3 | 66.2 | 60.1 | 46.9 | 38.8 | 32.2 | 47.9 |
| Mean Temp °F | 39.4 | 42.9 | 51.3 | 59.7 | 67.1 | 74.6 | 78.4 | 77.4 | 71.8 | 60.1 | 51.1 | 43.3 | 59.8 |
| Days Max Temp ≥ 90 °F | 0 | 0 | 0 | 0 | 1 | 9 | 17 | 15 | 6 | 0 | 0 | 0 | 48 |
| Days Max Temp ≤ 32 °F | 2 | 1 | 0 | 0 | 0 | 0 | 0 | 0 | 0 | 0 | 0 | 1 | 4 |
| Days Min Temp ≤ 32 °F | 20 | 17 | 10 | 2 | 0 | 0 | 0 | 0 | 0 | 2 | 10 | 18 | 79 |
| Days Min Temp ≤ 0 °F | 0 | 0 | 0 | 0 | 0 | 0 | 0 | 0 | 0 | 0 | 0 | 0 | 0 |
| Heating Degree Days | 788 | 619 | 425 | 189 | 52 | 3 | 0 | 0 | 20 | 184 | 417 | 669 | 3366 |
| Cooling Degree Days | 0 | 1 | 9 | 34 | 128 | 314 | 439 | 409 | 229 | 44 | 6 | 2 | 1615 |
| Total Precipitation (") | 5.46 | 4.97 | 6.21 | 5.44 | 5.08 | 3.94 | 5.39 | 3.62 | 4.42 | 3.28 | 4.40 | 5.15 | 57.36 |
| Days ≥ 0.1" Precip | 8 | 7 | 8 | 7 | 7 | 6 | 7 | 6 | 6 | 5 | 7 | 7 | 81 |
| Total Snowfall (") | 1.1 | 0.4 | 0.8 | 0.0 | 0.0 | 0.0 | 0.0 | 0.0 | 0.0 | 0.0 | 0.2 | 0.1 | 2.6 |
| Days ≥ 1" Snow Depth | 1 | 0 | 0 | 0 | 0 | 0 | 0 | 0 | 0 | 0 | 0 | 0 | 1 |

## OPELIKA *Lee County* ELEVATION 741 ft LAT/LONG 32° 38 ' N / 85° 23 ' W

| | JAN | FEB | MAR | APR | MAY | JUN | JUL | AUG | SEP | OCT | NOV | DEC | YEAR |
|---|---|---|---|---|---|---|---|---|---|---|---|---|---|
| Maximum Temp °F | 55.0 | 58.8 | 67.6 | 75.8 | 81.6 | 88.0 | 90.1 | 89.2 | *85.0* | 75.8 | 67.0 | 58.5 | 74.4 |
| Minimum Temp °F | 32.0 | 33.4 | 40.2 | 47.6 | 55.4 | 63.5 | 67.8 | 67.2 | *62.5* | 49.7 | 41.0 | 34.9 | 49.6 |
| Mean Temp °F | 43.5 | 46.1 | 53.9 | 61.7 | 68.5 | 75.8 | 78.9 | 78.2 | *73.8* | 62.7 | 54.0 | 46.7 | 62.0 |
| Days Max Temp ≥ 90 °F | 0 | 0 | 0 | 0 | 2 | 12 | 17 | 15 | *7* | 0 | 0 | 0 | 53 |
| Days Max Temp ≤ 32 °F | 1 | 0 | 0 | 0 | 0 | 0 | 0 | 0 | *0* | 0 | 0 | 0 | 1 |
| Days Min Temp ≤ 32 °F | 17 | 14 | 7 | 1 | 0 | 0 | 0 | 0 | *0* | 1 | 7 | *14* | 61 |
| Days Min Temp ≤ 0 °F | 0 | 0 | 0 | 0 | 0 | 0 | 0 | 0 | *0* | 0 | 0 | 0 | 0 |
| Heating Degree Days | 661 | 530 | 351 | 139 | 30 | 1 | 0 | 0 | *8* | 125 | 334 | 563 | 2742 |
| Cooling Degree Days | 0 | 1 | *13* | 36 | 146 | 347 | 450 | 436 | na | *72* | 13 | 3 | na |
| Total Precipitation (") | 5.28 | 5.09 | 6.85 | 4.72 | 4.09 | 3.96 | 5.92 | 4.06 | 3.86 | 3.24 | 4.20 | 5.34 | 56.61 |
| Days ≥ 0.1" Precip | 9 | *7* | *8* | 6 | 6 | *6* | 9 | 6 | *6* | 4 | 6 | *7* | 80 |
| Total Snowfall (") | 0.2 | 0.3 | 0.1 | 0.0 | 0.0 | 0.0 | 0.0 | 0.0 | 0.0 | 0.0 | 0.0 | 0.0 | 0.6 |
| Days ≥ 1" Snow Depth | 0 | 0 | 0 | 0 | 0 | 0 | 0 | 0 | *0* | 0 | 0 | 0 | 0 |

## ROBERTSDALE 5 NE *Baldwin County* ELEVATION 175 ft LAT/LONG 30° 37 ' N / 87° 40 ' W

| | JAN | FEB | MAR | APR | MAY | JUN | JUL | AUG | SEP | OCT | NOV | DEC | YEAR |
|---|---|---|---|---|---|---|---|---|---|---|---|---|---|
| Maximum Temp °F | 60.6 | 64.0 | 70.8 | 77.8 | 83.9 | 89.3 | 90.9 | 90.0 | 87.1 | 79.7 | 71.1 | 63.8 | 77.4 |
| Minimum Temp °F | 38.4 | 40.6 | 47.4 | 54.3 | 61.4 | 68.2 | 71.0 | 70.3 | 66.1 | 54.8 | 46.7 | 40.8 | 55.0 |
| Mean Temp °F | 49.5 | 52.3 | 59.1 | 66.0 | 72.7 | 78.8 | 81.0 | 80.2 | 76.6 | 67.2 | 58.9 | 52.4 | 66.2 |
| Days Max Temp ≥ 90 °F | 0 | 0 | 0 | 0 | 3 | 15 | 21 | 19 | 10 | 1 | 0 | 0 | 69 |
| Days Max Temp ≤ 32 °F | 0 | 0 | 0 | 0 | 0 | 0 | 0 | 0 | 0 | 0 | 0 | 0 | 0 |
| Days Min Temp ≤ 32 °F | 11 | 7 | 2 | 0 | 0 | 0 | 0 | 0 | 0 | 0 | 3 | 8 | 31 |
| Days Min Temp ≤ 0 °F | 0 | 0 | 0 | 0 | 0 | 0 | 0 | 0 | 0 | 0 | 0 | 0 | 0 |
| Heating Degree Days | 481 | 362 | 211 | 64 | 5 | 0 | 0 | 0 | 3 | 59 | 215 | 399 | 1799 |
| Cooling Degree Days | 5 | 10 | 33 | 80 | 237 | 416 | 506 | 486 | 350 | 130 | 39 | 13 | 2305 |
| Total Precipitation (") | 5.70 | 5.46 | 6.52 | 3.80 | 5.38 | 5.56 | 7.79 | 7.50 | 5.75 | 3.88 | 4.60 | 4.21 | 66.15 |
| Days ≥ 0.1" Precip | 7 | 6 | 7 | 5 | 6 | 8 | 10 | 11 | 7 | 4 | 5 | 6 | 82 |
| Total Snowfall (") | 0.0 | 0.1 | 0.1 | 0.0 | 0.0 | 0.0 | 0.0 | 0.0 | 0.0 | 0.0 | 0.0 | 0.0 | 0.2 |
| Days ≥ 1" Snow Depth | 0 | 0 | 0 | 0 | 0 | 0 | 0 | 0 | 0 | 0 | 0 | 0 | 0 |

## ROCKFORD 3 ESE *Coosa County* ELEVATION 650 ft LAT/LONG 32° 53' N / 86° 13' W

| | JAN | FEB | MAR | APR | MAY | JUN | JUL | AUG | SEP | OCT | NOV | DEC | YEAR |
|---|---|---|---|---|---|---|---|---|---|---|---|---|---|
| Maximum Temp °F | 53.4 | 58.1 | 67.2 | 75.4 | 80.9 | 87.4 | 89.3 | 88.5 | 84.2 | 75.4 | 65.3 | 56.8 | 73.5 |
| Minimum Temp °F | 31.7 | 34.0 | 41.6 | 48.7 | 56.1 | 62.9 | 66.7 | *65.8* | 61.0 | 49.6 | 41.3 | 34.6 | 49.5 |
| Mean Temp °F | 42.6 | 46.1 | 54.4 | 62.0 | 68.5 | 75.2 | 78.0 | *77.1* | 72.6 | 62.5 | 53.3 | 45.7 | 61.5 |
| Days Max Temp ≥ 90 °F | 0 | 0 | 0 | 0 | 1 | *11* | *15* | *13* | 6 | 0 | 0 | 0 | 46 |
| Days Max Temp ≤ 32 °F | 1 | 0 | 0 | 0 | 0 | 0 | 0 | 0 | 0 | 0 | 0 | 0 | 1 |
| Days Min Temp ≤ 32 °F | 17 | 13 | 7 | 1 | 0 | 0 | 0 | 0 | 0 | 1 | 7 | *15* | 61 |
| Days Min Temp ≤ 0 °F | 0 | 0 | 0 | 0 | 0 | 0 | 0 | 0 | 0 | 0 | 0 | 0 | 0 |
| Heating Degree Days | 689 | 529 | 334 | 132 | 27 | 1 | 0 | *0* | 11 | 126 | 352 | 593 | 2794 |
| Cooling Degree Days | *0* | *1* | *13* | *38* | 146 | 338 | *434* | *409* | 261 | *66* | *8* | *2* | 1716 |
| Total Precipitation (") | 5.70 | 5.49 | 6.45 | 4.79 | 4.41 | 3.81 | 6.05 | 4.45 | 4.05 | 3.02 | 4.11 | 5.17 | 57.50 |
| Days ≥ 0.1" Precip | 8 | 7 | 8 | 6 | 6 | 6 | 9 | 7 | 5 | 4 | 6 | *7* | 79 |
| Total Snowfall (") | 0.4 | 0.1 | *0.2* | 0.1 | 0.0 | 0.0 | 0.0 | 0.0 | 0.0 | 0.0 | 0.0 | 0.0 | 0.8 |
| Days ≥ 1" Snow Depth | *0* | 0 | 0 | 0 | 0 | 0 | 0 | 0 | 0 | 0 | *0* | 0 | 0 |

## RUSSELLVILLE 2 *Franklin County* ELEVATION 869 ft LAT/LONG 34° 30' N / 87° 44' W

| | JAN | FEB | MAR | APR | MAY | JUN | JUL | AUG | SEP | OCT | NOV | DEC | YEAR |
|---|---|---|---|---|---|---|---|---|---|---|---|---|---|
| Maximum Temp °F | 48.5 | 52.8 | 62.4 | 72.0 | 78.6 | 85.8 | 89.1 | 88.4 | 82.9 | 72.7 | 62.2 | 52.8 | 70.7 |
| Minimum Temp °F | 26.6 | 28.6 | 36.5 | 44.3 | 52.4 | 60.4 | 65.1 | 63.5 | 57.4 | 44.4 | 36.5 | 30.1 | 45.5 |
| Mean Temp °F | 37.7 | 40.8 | 49.5 | 58.2 | 65.5 | 73.1 | 77.1 | 75.9 | 70.2 | 58.6 | 49.4 | 41.5 | 58.1 |
| Days Max Temp ≥ 90 °F | 0 | 0 | 0 | 0 | 1 | 7 | 15 | 13 | 5 | 0 | 0 | 0 | 41 |
| Days Max Temp ≤ 32 °F | 3 | 1 | 0 | 0 | 0 | 0 | 0 | 0 | 0 | 0 | 0 | 1 | 5 |
| Days Min Temp ≤ 32 °F | 22 | 19 | 12 | 4 | 0 | 0 | 0 | 0 | 0 | 4 | 12 | 19 | 92 |
| Days Min Temp ≤ 0 °F | 0 | 0 | 0 | 0 | 0 | 0 | 0 | 0 | 0 | 0 | 0 | 0 | 0 |
| Heating Degree Days | 842 | 679 | 480 | 225 | 73 | 6 | 0 | 0 | 30 | 222 | 464 | 721 | 3742 |
| Cooling Degree Days | 0 | 1 | 6 | 22 | 96 | 270 | 396 | 367 | 195 | 33 | 3 | 0 | 1389 |
| Total Precipitation (") | 4.84 | 4.94 | 6.31 | 5.08 | 6.12 | 3.86 | 4.66 | 3.41 | 4.39 | 3.58 | 4.46 | 5.71 | 57.36 |
| Days ≥ 0.1" Precip | 7 | 7 | 8 | 7 | 7 | 6 | 7 | 6 | 6 | 5 | 6 | 7 | 79 |
| Total Snowfall (") | 0.9 | 0.3 | 0.2 | 0.0 | 0.0 | 0.0 | 0.0 | 0.0 | 0.0 | 0.0 | 0.1 | 0.2 | 1.7 |
| Days ≥ 1" Snow Depth | *0* | 0 | 0 | 0 | 0 | 0 | 0 | 0 | 0 | 0 | 0 | 0 | 0 |

## SAINT BERNARD *Cullman County* ELEVATION 800 ft LAT/LONG 34° 10' N / 86° 49' W

| | JAN | FEB | MAR | APR | MAY | JUN | JUL | AUG | SEP | OCT | NOV | DEC | YEAR |
|---|---|---|---|---|---|---|---|---|---|---|---|---|---|
| Maximum Temp °F | 49.8 | 54.7 | 64.3 | 73.6 | 79.8 | 86.5 | 89.7 | 89.0 | 83.6 | 73.6 | 63.4 | 54.1 | 71.8 |
| Minimum Temp °F | 27.5 | 30.1 | 37.9 | 46.0 | 54.0 | 61.6 | 65.5 | 64.1 | 58.1 | 45.4 | 37.3 | 30.8 | 46.5 |
| Mean Temp °F | 38.7 | 42.4 | 51.1 | 59.8 | 66.9 | 74.1 | 77.7 | 76.6 | 70.9 | 59.6 | 50.4 | 42.5 | 59.2 |
| Days Max Temp ≥ 90 °F | 0 | 0 | 0 | 0 | 1 | 9 | 16 | 14 | 6 | 0 | 0 | 0 | 46 |
| Days Max Temp ≤ 32 °F | 2 | 1 | 0 | 0 | 0 | 0 | 0 | 0 | 0 | 0 | 0 | 1 | 4 |
| Days Min Temp ≤ 32 °F | 21 | 18 | 10 | 3 | 0 | 0 | 0 | 0 | 0 | 3 | 11 | 18 | 84 |
| Days Min Temp ≤ 0 °F | 0 | 0 | 0 | 0 | 0 | 0 | 0 | 0 | 0 | 0 | 0 | 0 | 0 |
| Heating Degree Days | 810 | 632 | 430 | 186 | 49 | 3 | 0 | 0 | 22 | 193 | 435 | 692 | 3452 |
| Cooling Degree Days | 0 | 1 | 8 | 34 | 117 | 289 | 414 | 377 | 200 | 34 | 3 | 1 | 1478 |
| Total Precipitation (") | 5.41 | 5.42 | 6.33 | 5.13 | 5.23 | 4.07 | 4.86 | 3.64 | 5.09 | 3.50 | 4.40 | 5.87 | 58.95 |
| Days ≥ 0.1" Precip | 7 | 7 | *8* | 6 | 7 | 7 | 7 | 6 | 6 | 5 | 6 | 8 | 80 |
| Total Snowfall (") | 0.9 | 0.3 | 0.5 | 0.1 | 0.0 | 0.0 | 0.0 | 0.0 | 0.0 | 0.0 | 0.1 | 0.0 | 1.9 |
| Days ≥ 1" Snow Depth | *0* | 0 | 0 | 0 | 0 | 0 | 0 | 0 | 0 | 0 | 0 | 0 | 0 |

## SAND MOUNTAIN SUBSTN *DeKalb County* ELEVATION 1201 ft LAT/LONG 34° 17' N / 85° 58' W

| | JAN | FEB | MAR | APR | MAY | JUN | JUL | AUG | SEP | OCT | NOV | DEC | YEAR |
|---|---|---|---|---|---|---|---|---|---|---|---|---|---|
| Maximum Temp °F | 47.9 | 52.8 | 61.9 | 71.1 | 77.3 | 84.3 | 87.4 | 86.9 | 81.5 | 71.6 | 61.4 | 52.2 | 69.7 |
| Minimum Temp °F | 28.5 | 31.3 | 39.6 | 47.5 | 55.4 | 62.9 | 66.2 | 64.8 | 59.0 | 47.1 | 39.2 | 32.0 | 47.8 |
| Mean Temp °F | 38.3 | 42.1 | 50.8 | 59.4 | 66.4 | 73.6 | 76.8 | 75.9 | 70.3 | 59.4 | 50.4 | 42.1 | 58.8 |
| Days Max Temp ≥ 90 °F | 0 | 0 | 0 | 0 | 0 | 5 | 11 | 9 | 3 | 0 | 0 | 0 | 28 |
| Days Max Temp ≤ 32 °F | 3 | 1 | 0 | 0 | 0 | 0 | 0 | 0 | 0 | 0 | 0 | 1 | 5 |
| Days Min Temp ≤ 32 °F | 20 | 16 | 9 | 2 | 0 | 0 | 0 | 0 | 0 | 2 | 9 | 17 | 75 |
| Days Min Temp ≤ 0 °F | 1 | 0 | 0 | 0 | 0 | 0 | 0 | 0 | 0 | 0 | 0 | 0 | 1 |
| Heating Degree Days | 822 | 642 | 439 | 197 | 59 | 4 | 0 | 0 | 26 | 197 | 436 | 702 | 3524 |
| Cooling Degree Days | 0 | 0 | 5 | 28 | 103 | 276 | 391 | 360 | 191 | 33 | 4 | 0 | 1391 |
| Total Precipitation (") | 5.06 | 5.11 | 5.93 | 4.70 | 4.82 | 3.70 | 4.65 | 3.44 | 4.64 | 3.15 | 4.28 | 5.33 | 54.81 |
| Days ≥ 0.1" Precip | 8 | 7 | 8 | 7 | 7 | 6 | 7 | 6 | 6 | 5 | 7 | 7 | 81 |
| Total Snowfall (") | 1.1 | 0.5 | 0.2 | 0.0 | 0.0 | 0.0 | 0.0 | 0.0 | 0.0 | 0.0 | 0.0 | 0.1 | 1.9 |
| Days ≥ 1" Snow Depth | 0 | 0 | 0 | 0 | 0 | 0 | 0 | 0 | 0 | 0 | 0 | 0 | 0 |

## SCOTTSBORO *Jackson County* ELEVATION 659 ft LAT/LONG 34° 41 ' N / 86° 2 ' W

| | JAN | FEB | MAR | APR | MAY | JUN | JUL | AUG | SEP | OCT | NOV | DEC | YEAR |
|---|---|---|---|---|---|---|---|---|---|---|---|---|---|
| Maximum Temp °F | 49.1 | 54.1 | 63.1 | 72.6 | 80.0 | 87.0 | 90.3 | 89.8 | 84.2 | 73.5 | 63.3 | 53.9 | 71.7 |
| Minimum Temp °F | 27.4 | 30.3 | 37.8 | 45.7 | 54.2 | 62.2 | 66.7 | 64.8 | 58.8 | 45.3 | 36.8 | 30.7 | 46.7 |
| Mean Temp °F | 38.3 | 42.2 | 50.5 | 59.2 | 67.1 | 74.6 | 78.5 | 77.3 | 71.5 | 59.4 | 50.1 | 42.4 | 59.3 |
| Days Max Temp ≥ 90 °F | 0 | 0 | 0 | 0 | 2 | 9 | 18 | 16 | 7 | 0 | 0 | 0 | 52 |
| Days Max Temp ≤ 32 °F | 2 | 1 | 0 | 0 | 0 | 0 | 0 | 0 | 0 | 0 | 0 | 1 | 4 |
| Days Min Temp ≤ 32 °F | 20 | 18 | 10 | 2 | 0 | 0 | 0 | 0 | 0 | 3 | 11 | 18 | 82 |
| Days Min Temp ≤ 0 °F | 0 | 0 | 0 | 0 | 0 | 0 | 0 | 0 | 0 | 0 | 0 | 0 | 0 |
| Heating Degree Days | 821 | 629 | 448 | 199 | 51 | 2 | 0 | 0 | 22 | 196 | 445 | 696 | 3509 |
| Cooling Degree Days | 0 | 1 | 9 | 28 | 132 | 316 | 446 | 411 | 223 | 31 | 4 | 1 | 1602 |
| Total Precipitation (") | 4.94 | 5.06 | 6.71 | 4.50 | 4.67 | 4.36 | 4.33 | 3.66 | 4.93 | 3.32 | 4.54 | 5.91 | 56.93 |
| Days ≥ 0.1" Precip | 8 | 7 | 9 | 7 | 7 | 7 | 7 | 6 | 7 | 5 | 6 | 8 | 84 |
| Total Snowfall (") | 0.8 | 0.1 | 0.5 | 0.0 | 0.0 | 0.0 | 0.0 | 0.0 | 0.0 | 0.0 | 0.0 | 0.0 | 1.4 |
| Days ≥ 1" Snow Depth | 0 | 0 | 0 | 0 | 0 | 0 | 0 | 0 | 0 | 0 | 0 | 0 | 0 |

## SELMA *Dallas County* ELEVATION 151 ft LAT/LONG 32° 25 ' N / 87° 0 ' W

| | JAN | FEB | MAR | APR | MAY | JUN | JUL | AUG | SEP | OCT | NOV | DEC | YEAR |
|---|---|---|---|---|---|---|---|---|---|---|---|---|---|
| Maximum Temp °F | 57.2 | 61.8 | 70.3 | 78.1 | 84.3 | 90.5 | 92.1 | 91.5 | 87.5 | 78.4 | 68.7 | 60.9 | 76.8 |
| Minimum Temp °F | 37.2 | 39.5 | 46.4 | 53.3 | 61.2 | 68.3 | 71.6 | 70.7 | 65.7 | 54.0 | 44.9 | 39.0 | 54.3 |
| Mean Temp °F | 47.3 | 50.7 | 58.4 | 65.7 | 72.8 | 79.4 | 81.9 | 81.1 | 76.6 | 66.2 | 56.8 | 50.0 | 65.6 |
| Days Max Temp ≥ 90 °F | 0 | 0 | 0 | 1 | 5 | 18 | 24 | 22 | 13 | 1 | 0 | 0 | 84 |
| Days Max Temp ≤ 32 °F | 0 | 0 | 0 | 0 | 0 | 0 | 0 | 0 | 0 | 0 | 0 | 0 | 0 |
| Days Min Temp ≤ 32 °F | 11 | 7 | 2 | 0 | 0 | 0 | 0 | 0 | 0 | 0 | 3 | 10 | 33 |
| Days Min Temp ≤ 0 °F | 0 | 0 | 0 | 0 | 0 | 0 | 0 | 0 | 0 | 0 | 0 | 0 | 0 |
| Heating Degree Days | 547 | 401 | 233 | 72 | 6 | 0 | 0 | 0 | 3 | 69 | 260 | 469 | 2060 |
| Cooling Degree Days | 2 | 6 | 28 | 79 | 238 | 444 | 533 | 509 | 351 | 110 | 21 | 4 | 2325 |
| Total Precipitation (") | 5.20 | 5.12 | 6.44 | 4.24 | 3.89 | 3.85 | 4.59 | 4.01 | 3.58 | 2.78 | 3.96 | 5.07 | 52.73 |
| Days ≥ 0.1" Precip | 8 | 7 | 7 | 5 | 6 | 6 | 8 | 6 | 5 | 4 | 6 | 7 | 75 |
| Total Snowfall (") | 0.3 | 0.0 | 0.0 | 0.1 | 0.0 | 0.0 | 0.0 | 0.0 | 0.0 | 0.0 | 0.0 | 0.0 | 0.4 |
| Days ≥ 1" Snow Depth | 0 | 0 | 0 | 0 | 0 | 0 | 0 | 0 | 0 | 0 | 0 | 0 | 0 |

## SYLACAUGA 4 NE *Talladega County* ELEVATION 489 ft LAT/LONG 33° 12 ' N / 86° 12 ' W

| | JAN | FEB | MAR | APR | MAY | JUN | JUL | AUG | SEP | OCT | NOV | DEC | YEAR |
|---|---|---|---|---|---|---|---|---|---|---|---|---|---|
| Maximum Temp °F | 55.2 | 59.4 | 67.8 | 75.9 | 82.3 | 88.2 | 90.5 | 90.1 | 85.2 | 76.5 | 66.7 | 58.1 | 74.7 |
| Minimum Temp °F | 31.7 | 33.7 | 40.6 | 47.2 | 55.2 | 62.2 | 65.9 | 64.8 | 59.1 | 47.5 | 38.6 | 33.8 | 48.4 |
| Mean Temp °F | 43.5 | 46.6 | 54.2 | 61.6 | 68.9 | 75.2 | 78.3 | 77.5 | 72.2 | 62.0 | 52.7 | 46.0 | 61.6 |
| Days Max Temp ≥ 90 °F | 0 | 0 | 0 | 0 | 2 | 12 | 19 | 18 | 8 | 1 | 0 | 0 | 60 |
| Days Max Temp ≤ 32 °F | 1 | 0 | 0 | 0 | 0 | 0 | 0 | 0 | 0 | 0 | 0 | 0 | 1 |
| Days Min Temp ≤ 32 °F | 18 | 14 | 8 | 2 | 0 | 0 | 0 | 0 | 0 | 3 | 10 | 15 | 70 |
| Days Min Temp ≤ 0 °F | 0 | 0 | 0 | 0 | 0 | 0 | 0 | 0 | 0 | 0 | 0 | 0 | 0 |
| Heating Degree Days | 661 | 515 | 343 | 151 | 30 | 2 | 0 | 0 | 16 | 143 | 373 | 585 | 2819 |
| Cooling Degree Days | 0 | 1 | 14 | 43 | 150 | 325 | 428 | 397 | 230 | 56 | 9 | 3 | 1656 |
| Total Precipitation (") | 5.30 | 5.02 | 5.76 | 4.57 | 4.16 | 4.11 | 5.25 | 4.43 | 3.93 | 3.09 | 4.45 | 5.44 | 55.51 |
| Days ≥ 0.1" Precip | 8 | 7 | 8 | 6 | 7 | 7 | 8 | 7 | 5 | 5 | 6 | 7 | 81 |
| Total Snowfall (") | 0.3 | 0.1 | 0.4 | 0.1 | 0.0 | 0.0 | 0.0 | 0.0 | 0.0 | 0.0 | 0.0 | 0.0 | 0.9 |
| Days ≥ 1" Snow Depth | 0 | 0 | 0 | 0 | 0 | 0 | 0 | 0 | 0 | 0 | 0 | 0 | 0 |

## TALLADEGA *Talladega County* ELEVATION 551 ft LAT/LONG 33° 26 ' N / 86° 6 ' W

| | JAN | FEB | MAR | APR | MAY | JUN | JUL | AUG | SEP | OCT | NOV | DEC | YEAR |
|---|---|---|---|---|---|---|---|---|---|---|---|---|---|
| Maximum Temp °F | 54.1 | 59.1 | 68.2 | 76.3 | 81.8 | 88.4 | 90.6 | 89.7 | 84.7 | 75.9 | 66.3 | 58.1 | 74.4 |
| Minimum Temp °F | 31.8 | 34.1 | 41.1 | 48.0 | 55.3 | 62.4 | 67.1 | 65.9 | 60.4 | 48.2 | 40.0 | 34.8 | 49.1 |
| Mean Temp °F | 42.9 | 46.6 | 54.7 | 62.2 | 68.6 | 75.3 | 78.9 | 77.8 | 72.6 | 62.1 | 53.2 | 46.5 | 61.8 |
| Days Max Temp ≥ 90 °F | 0 | 0 | 0 | 0 | 2 | 12 | 19 | 16 | 6 | 0 | 0 | 0 | 55 |
| Days Max Temp ≤ 32 °F | 1 | 0 | 0 | 0 | 0 | 0 | 0 | 0 | 0 | 0 | 0 | 0 | 1 |
| Days Min Temp ≤ 32 °F | 17 | 13 | 7 | 2 | 0 | 0 | 0 | 0 | 0 | 2 | 8 | 13 | 62 |
| Days Min Temp ≤ 0 °F | 0 | 0 | 0 | 0 | 0 | 0 | 0 | 0 | 0 | 0 | 0 | 0 | 0 |
| Heating Degree Days | 674 | 516 | 327 | 131 | 29 | 1 | 0 | 0 | 11 | 138 | 358 | 568 | 2753 |
| Cooling Degree Days | 0 | 3 | 17 | 48 | 158 | 327 | 471 | 432 | 253 | 72 | 11 | 3 | 1795 |
| Total Precipitation (") | 5.73 | 4.94 | 6.75 | 4.63 | 4.95 | 4.33 | 4.72 | 3.92 | 3.80 | 3.10 | 4.38 | 4.43 | 55.68 |
| Days ≥ 0.1" Precip | 8 | 7 | 7 | 6 | 7 | 6 | 8 | 7 | 5 | 4 | 6 | 7 | 78 |
| Total Snowfall (") | 0.5 | 0.0 | 0.6 | 0.1 | 0.0 | 0.0 | 0.0 | 0.0 | 0.0 | 0.0 | 0.0 | 0.0 | 1.2 |
| Days ≥ 1" Snow Depth | 0 | 0 | 0 | 0 | 0 | 0 | 0 | 0 | 0 | 0 | 0 | 0 | 0 |

## THOMASVILLE *Clarke County* ELEVATION 390 ft LAT/LONG 31° 55' N / 87° 45' W

| | JAN | FEB | MAR | APR | MAY | JUN | JUL | AUG | SEP | OCT | NOV | DEC | YEAR |
|---|---|---|---|---|---|---|---|---|---|---|---|---|---|
| Maximum Temp °F | 56.6 | 61.2 | 69.6 | 77.4 | 82.8 | 89.3 | 91.1 | 90.5 | 86.7 | 77.8 | 68.2 | 60.2 | 76.0 |
| Minimum Temp °F | 33.9 | 36.3 | 43.8 | 51.6 | 59.3 | 66.7 | 69.7 | 68.8 | 63.9 | 52.0 | 43.2 | 36.9 | 52.2 |
| Mean Temp °F | 45.3 | 48.8 | 56.7 | 64.5 | 71.1 | 78.0 | 80.4 | 79.7 | 75.3 | 64.9 | 55.7 | 48.5 | 64.1 |
| Days Max Temp ≥ 90 °F | 0 | 0 | 0 | 0 | 3 | 16 | 22 | 20 | 11 | 1 | 0 | 0 | 73 |
| Days Max Temp ≤ 32 °F | 0 | 0 | 0 | 0 | 0 | 0 | 0 | 0 | 0 | 0 | 0 | 0 | 0 |
| Days Min Temp ≤ 32 °F | 15 | 11 | 4 | 0 | 0 | 0 | 0 | 0 | 0 | 0 | 5 | 13 | 48 |
| Days Min Temp ≤ 0 °F | 0 | 0 | 0 | 0 | 0 | 0 | 0 | 0 | 0 | 0 | 0 | 0 | 0 |
| Heating Degree Days | 606 | 457 | 276 | 94 | 13 | 0 | 0 | 0 | 6 | 94 | 294 | 511 | 2351 |
| Cooling Degree Days | 1 | 6 | 24 | 75 | 210 | 405 | 495 | 482 | 329 | 105 | 24 | 7 | 2163 |
| Total Precipitation (") | 5.66 | 5.32 | 6.78 | 4.40 | 4.93 | 4.23 | 6.16 | 4.07 | 3.98 | 2.78 | 4.82 | 5.56 | 58.69 |
| Days ≥ 0.1" Precip | 8 | 7 | 7 | 6 | 7 | 7 | 9 | 7 | 6 | 4 | 6 | 7 | 81 |
| Total Snowfall (") | 0.1 | 0.1 | 0.4 | 0.0 | 0.0 | 0.0 | 0.0 | 0.0 | 0.0 | 0.0 | 0.0 | 0.1 | 0.7 |
| Days ≥ 1" Snow Depth | 0 | 0 | 0 | 0 | 0 | 0 | 0 | 0 | 0 | 0 | 0 | 0 | 0 |

## THORSBY EXP STATION *Chilton County* ELEVATION 680 ft LAT/LONG 32° 53' N / 86° 42' W

| | JAN | FEB | MAR | APR | MAY | JUN | JUL | AUG | SEP | OCT | NOV | DEC | YEAR |
|---|---|---|---|---|---|---|---|---|---|---|---|---|---|
| Maximum Temp °F | 53.1 | 58.2 | 66.8 | 75.4 | 81.3 | 88.0 | 89.9 | 88.8 | 84.3 | 75.3 | 65.3 | 57.0 | 73.6 |
| Minimum Temp °F | 32.7 | 35.2 | 42.9 | 50.2 | 58.1 | 65.2 | 68.3 | 67.1 | 62.2 | 50.6 | 42.0 | 36.1 | 50.9 |
| Mean Temp °F | 42.9 | 46.7 | 54.9 | 62.8 | 69.8 | 76.7 | 79.1 | 78.0 | 73.3 | 63.0 | 53.7 | 46.6 | 62.3 |
| Days Max Temp ≥ 90 °F | 0 | 0 | 0 | 0 | 2 | 12 | 17 | 14 | 6 | 0 | 0 | 0 | 51 |
| Days Max Temp ≤ 32 °F | 1 | 0 | 0 | 0 | 0 | 0 | 0 | 0 | 0 | 0 | 0 | 0 | 1 |
| Days Min Temp ≤ 32 °F | 16 | 12 | 5 | 1 | 0 | 0 | 0 | 0 | 0 | 0 | 6 | 13 | 53 |
| Days Min Temp ≤ 0 °F | 0 | 0 | 0 | 0 | 0 | 0 | 0 | 0 | 0 | 0 | 0 | 0 | 0 |
| Heating Degree Days | 678 | 512 | 321 | 118 | 19 | 1 | 0 | 0 | 9 | 115 | 342 | 568 | 2683 |
| Cooling Degree Days | 0 | 2 | 12 | 43 | 162 | 364 | 459 | 428 | 264 | 61 | 9 | 3 | 1807 |
| Total Precipitation (") | 5.88 | 5.16 | 6.50 | 4.92 | 4.24 | 3.99 | 5.31 | 4.21 | 4.05 | 3.00 | 4.27 | 5.03 | 56.56 |
| Days ≥ 0.1" Precip | 9 | 7 | 8 | 6 | 7 | 6 | 8 | 6 | 6 | 4 | 6 | 7 | 80 |
| Total Snowfall (") | 0.1 | 0.1 | 0.5 | 0.1 | 0.0 | 0.0 | 0.0 | 0.0 | 0.0 | 0.0 | 0.0 | 0.0 | 0.8 |
| Days ≥ 1" Snow Depth | *0* | 0 | 0 | 0 | 0 | 0 | 0 | 0 | 0 | 0 | 0 | 0 | 0 |

## TROY *Pike County* ELEVATION 591 ft LAT/LONG 31° 49' N / 85° 58' W

| | JAN | FEB | MAR | APR | MAY | JUN | JUL | AUG | SEP | OCT | NOV | DEC | YEAR |
|---|---|---|---|---|---|---|---|---|---|---|---|---|---|
| Maximum Temp °F | 57.5 | 61.5 | 69.9 | 77.5 | 83.1 | 88.8 | 90.3 | 89.8 | 86.3 | 77.5 | 68.3 | 60.6 | 75.9 |
| Minimum Temp °F | 36.6 | 38.6 | 45.4 | 52.4 | 59.8 | 66.7 | 69.7 | 69.3 | 65.2 | 54.2 | 45.2 | 39.3 | 53.5 |
| Mean Temp °F | 47.1 | 50.0 | 57.6 | 65.0 | 71.5 | 77.8 | 80.0 | 79.6 | 75.8 | 65.8 | 56.8 | 50.0 | 64.7 |
| Days Max Temp ≥ 90 °F | 0 | 0 | 0 | 0 | 3 | 14 | 19 | 17 | 9 | 1 | 0 | 0 | 63 |
| Days Max Temp ≤ 32 °F | 0 | 0 | 0 | 0 | 0 | 0 | 0 | 0 | 0 | 0 | 0 | 0 | 0 |
| Days Min Temp ≤ 32 °F | 12 | 9 | 3 | 0 | 0 | 0 | 0 | 0 | 0 | 0 | 4 | 9 | 37 |
| Days Min Temp ≤ 0 °F | 0 | 0 | 0 | 0 | 0 | 0 | 0 | 0 | 0 | 0 | 0 | 0 | 0 |
| Heating Degree Days | 550 | 420 | 248 | 82 | 10 | 0 | 0 | 0 | 4 | 76 | 262 | 465 | 2117 |
| Cooling Degree Days | 1 | 5 | 23 | 68 | 200 | 394 | 486 | 473 | 326 | 101 | 21 | 4 | 2102 |
| Total Precipitation (") | 4.87 | 4.91 | 6.30 | 3.72 | 3.88 | 4.64 | 5.92 | 3.82 | 3.30 | 2.64 | 4.15 | 4.64 | 52.79 |
| Days ≥ 0.1" Precip | 7 | 7 | 7 | 5 | 6 | 7 | 9 | 6 | 5 | 4 | 5 | 6 | 74 |
| Total Snowfall (") | 0.1 | 0.4 | 0.1 | 0.0 | 0.0 | 0.0 | 0.0 | 0.0 | 0.0 | 0.0 | 0.0 | 0.1 | 0.7 |
| Days ≥ 1" Snow Depth | 0 | 0 | 0 | 0 | 0 | 0 | 0 | 0 | 0 | 0 | 0 | 0 | 0 |

## TUSCALOOSA MUNI AP *Tuscaloosa County* ELEVATION 187 ft LAT/LONG 33° 14' N / 87° 37' W

| | JAN | FEB | MAR | APR | MAY | JUN | JUL | AUG | SEP | OCT | NOV | DEC | YEAR |
|---|---|---|---|---|---|---|---|---|---|---|---|---|---|
| Maximum Temp °F | 53.5 | 58.6 | 67.7 | 76.2 | 82.3 | 89.0 | 91.4 | 90.6 | 86.0 | 76.6 | 66.0 | 57.7 | 74.6 |
| Minimum Temp °F | 33.1 | 35.8 | 43.4 | 50.9 | 59.4 | 67.0 | 71.3 | 70.3 | 64.5 | 51.2 | 41.9 | 36.5 | 52.1 |
| Mean Temp °F | 43.3 | 47.3 | 55.6 | 63.5 | 70.9 | 78.0 | 81.4 | 80.5 | 75.3 | 64.0 | 54.0 | 47.1 | 63.4 |
| Days Max Temp ≥ 90 °F | 0 | 0 | 0 | 0 | 4 | 16 | 21 | 20 | 10 | 1 | 0 | 0 | 72 |
| Days Max Temp ≤ 32 °F | 1 | 0 | 0 | 0 | 0 | 0 | 0 | 0 | 0 | 0 | 0 | 0 | 1 |
| Days Min Temp ≤ 32 °F | 16 | 12 | 5 | 0 | 0 | 0 | 0 | 0 | 0 | 0 | 7 | 13 | 53 |
| Days Min Temp ≤ 0 °F | 0 | 0 | 0 | 0 | 0 | 0 | 0 | 0 | 0 | 0 | 0 | 0 | 0 |
| Heating Degree Days | 666 | 497 | 306 | 108 | 17 | 0 | 0 | 0 | 7 | 108 | 337 | 551 | 2597 |
| Cooling Degree Days | 0 | 3 | 20 | 64 | 211 | 405 | 530 | 508 | 331 | 89 | 18 | 6 | 2185 |
| Total Precipitation (") | 5.15 | 4.87 | 6.00 | 5.25 | 4.61 | 3.86 | 5.26 | 4.10 | 3.54 | 3.17 | 4.22 | 4.91 | 54.94 |
| Days ≥ 0.1" Precip | 8 | 7 | 7 | 6 | 7 | 6 | 8 | 6 | 5 | 4 | 6 | 7 | 77 |
| Total Snowfall (") | 0.3 | 0.1 | 0.2 | 0.0 | 0.0 | 0.0 | 0.0 | 0.0 | 0.0 | 0.0 | 0.0 | 0.0 | 0.6 |
| Days ≥ 1" Snow Depth | 0 | 0 | 0 | 0 | 0 | 0 | 0 | 0 | 0 | 0 | 0 | 0 | 0 |

## TUSCALOOSA OLIVER DM *Tuscaloosa County* ELEVATION 151 ft LAT/LONG 33° 13 ' N / 87° 35 ' W

| | JAN | FEB | MAR | APR | MAY | JUN | JUL | AUG | SEP | OCT | NOV | DEC | YEAR |
|---|---|---|---|---|---|---|---|---|---|---|---|---|---|
| Maximum Temp °F | 52.8 | 57.6 | 66.6 | 75.8 | na | 88.2 | 91.4 | na | 85.8 | 76.3 | 65.7 | 57.5 | na |
| Minimum Temp °F | 30.7 | 33.5 | 41.4 | 49.3 | 57.6 | 65.3 | 68.9 | 68.0 | 62.6 | 50.6 | 41.1 | 34.7 | 50.3 |
| Mean Temp °F | 41.8 | 45.6 | 54.1 | 62.6 | na | 76.7 | 80.2 | na | 74.2 | 63.5 | 53.4 | 46.1 | na |
| Days Max Temp ≥ 90 °F | 0 | 0 | 0 | 0 | 3 | 13 | na | na | 9 | 1 | 0 | 0 | na |
| Days Max Temp ≤ 32 °F | 1 | 0 | 0 | 0 | 0 | 0 | 0 | 0 | 0 | 0 | 0 | 0 | 1 |
| Days Min Temp ≤ 32 °F | 19 | 14 | 6 | 0 | 0 | 0 | 0 | 0 | 0 | 0 | 6 | 15 | 60 |
| Days Min Temp ≤ 0 °F | 0 | 0 | 0 | 0 | 0 | 0 | 0 | 0 | 0 | 0 | 0 | 0 | 0 |
| Heating Degree Days | 714 | 542 | 348 | 128 | na | 1 | 0 | 0 | 10 | 117 | 352 | 580 | na |
| Cooling Degree Days | na | na | na | na | na | na | na | na | na | na | na | na | na |
| Total Precipitation (") | 5.38 | 5.03 | 6.29 | 5.34 | 4.78 | 3.69 | 4.61 | 3.75 | 3.28 | 3.67 | 4.02 | 5.07 | 54.91 |
| Days ≥ 0.1" Precip | 8 | 7 | 7 | 6 | 7 | 6 | 7 | 6 | 5 | 4 | 6 | 7 | 76 |
| Total Snowfall (") | 0.0 | 0.0 | 0.0 | 0.0 | 0.0 | 0.0 | 0.0 | 0.0 | 0.0 | 0.0 | 0.0 | 0.0 | 0.0 |
| Days ≥ 1" Snow Depth | 0 | 0 | 0 | 0 | 0 | 0 | 0 | 0 | 0 | 0 | 0 | 0 | 0 |

## UNION SPRINGS 9 S *Bullock County* ELEVATION 459 ft LAT/LONG 32° 9 ' N / 85° 43 ' W

| | JAN | FEB | MAR | APR | MAY | JUN | JUL | AUG | SEP | OCT | NOV | DEC | YEAR |
|---|---|---|---|---|---|---|---|---|---|---|---|---|---|
| Maximum Temp °F | 55.7 | 60.0 | 68.7 | 76.9 | 82.8 | 89.1 | 91.0 | 90.1 | 86.3 | 77.1 | 68.0 | 59.7 | 75.5 |
| Minimum Temp °F | 33.7 | 35.9 | 43.1 | 50.2 | 58.0 | 66.1 | 69.4 | 68.5 | 63.8 | 51.8 | 43.1 | 36.3 | 51.7 |
| Mean Temp °F | 44.7 | 48.0 | 55.9 | 63.5 | 70.5 | 77.6 | 80.2 | 79.3 | 75.1 | 64.5 | 55.6 | 48.1 | 63.6 |
| Days Max Temp ≥ 90 °F | 0 | 0 | 0 | 0 | 4 | 16 | 21 | 19 | 11 | 1 | 0 | 0 | 72 |
| Days Max Temp ≤ 32 °F | 1 | 0 | 0 | 0 | 0 | 0 | 0 | 0 | 0 | 0 | 0 | 0 | 1 |
| Days Min Temp ≤ 32 °F | 16 | 12 | 5 | 0 | 0 | 0 | 0 | 0 | 0 | 0 | 6 | 13 | 52 |
| Days Min Temp ≤ 0 °F | 0 | 0 | 0 | 0 | 0 | 0 | 0 | 0 | 0 | 0 | 0 | 0 | 0 |
| Heating Degree Days | 624 | 478 | 296 | 108 | 16 | 0 | 0 | 0 | 6 | 99 | 294 | 524 | 2445 |
| Cooling Degree Days | 1 | 5 | 20 | 58 | 186 | 397 | 491 | 462 | 308 | 91 | 23 | 4 | 2046 |
| Total Precipitation (") | 4.99 | 4.74 | 6.44 | 3.90 | 4.15 | 5.25 | 5.47 | 4.18 | 3.69 | 2.73 | 4.20 | 4.78 | 54.52 |
| Days ≥ 0.1" Precip | 8 | 7 | 7 | 5 | 6 | 6 | 9 | 7 | 6 | 4 | 5 | 7 | 77 |
| Total Snowfall (") | 0.3 | 0.5 | 0.2 | 0.0 | 0.0 | 0.0 | 0.0 | 0.0 | 0.0 | 0.0 | 0.1 | 0.1 | 1.2 |
| Days ≥ 1" Snow Depth | 0 | 0 | 0 | 0 | 0 | 0 | 0 | 0 | 0 | 0 | 0 | 0 | 0 |

## VALLEY HEAD *DeKalb County* ELEVATION 1089 ft LAT/LONG 34° 33 ' N / 85° 37 ' W

| | JAN | FEB | MAR | APR | MAY | JUN | JUL | AUG | SEP | OCT | NOV | DEC | YEAR |
|---|---|---|---|---|---|---|---|---|---|---|---|---|---|
| Maximum Temp °F | 47.3 | 52.0 | 61.0 | 70.6 | 77.1 | 84.1 | 87.3 | 86.7 | 81.4 | 71.5 | 61.5 | 51.8 | 69.4 |
| Minimum Temp °F | 25.3 | 27.1 | 34.6 | 42.5 | 51.1 | 59.6 | 64.1 | 63.1 | 57.0 | 43.5 | 35.1 | 28.5 | 44.3 |
| Mean Temp °F | 36.3 | 39.6 | 47.8 | 56.6 | 64.1 | 71.9 | 75.7 | 74.9 | 69.2 | 57.5 | 48.3 | 40.1 | 56.8 |
| Days Max Temp ≥ 90 °F | 0 | 0 | 0 | 0 | 0 | 4 | 10 | 8 | 3 | 0 | 0 | 0 | 25 |
| Days Max Temp ≤ 32 °F | 3 | 1 | 0 | 0 | 0 | 0 | 0 | 0 | 0 | 0 | 0 | 1 | 5 |
| Days Min Temp ≤ 32 °F | 22 | 20 | 14 | 6 | 0 | 0 | 0 | 0 | 0 | 5 | 14 | 21 | 102 |
| Days Min Temp ≤ 0 °F | 1 | 0 | 0 | 0 | 0 | 0 | 0 | 0 | 0 | 0 | 0 | 0 | 1 |
| Heating Degree Days | 882 | 712 | 527 | 264 | 95 | 10 | 1 | 0 | 39 | 247 | 496 | 764 | 4037 |
| Cooling Degree Days | 0 | 0 | 2 | 11 | 76 | 234 | 362 | 329 | 166 | 22 | 2 | 0 | 1204 |
| Total Precipitation (") | 5.42 | 5.44 | 6.43 | 4.73 | 4.86 | 3.97 | 5.40 | 3.75 | 4.24 | 3.24 | 4.63 | 5.39 | 57.50 |
| Days ≥ 0.1" Precip | 9 | 7 | 8 | 7 | 8 | 7 | 8 | 7 | 6 | 5 | 7 | 8 | 87 |
| Total Snowfall (") | 2.7 | 1.6 | 1.1 | 0.4 | 0.0 | 0.0 | 0.0 | 0.0 | 0.0 | 0.0 | 0.2 | 0.3 | 6.3 |
| Days ≥ 1" Snow Depth | 2 | 1 | 0 | 0 | 0 | 0 | 0 | 0 | 0 | 0 | 0 | 0 | 3 |

## VERNON 2 N *Lamar County* ELEVATION 180 ft LAT/LONG 33° 45 ' N / 88° 7 ' W

| | JAN | FEB | MAR | APR | MAY | JUN | JUL | AUG | SEP | OCT | NOV | DEC | YEAR |
|---|---|---|---|---|---|---|---|---|---|---|---|---|---|
| Maximum Temp °F | 51.6 | 57.1 | 66.5 | 75.8 | 81.8 | 88.7 | 91.5 | 90.7 | 85.2 | 75.2 | 64.5 | 55.7 | 73.7 |
| Minimum Temp °F | 28.3 | 30.9 | 38.6 | 46.4 | 54.7 | 62.9 | 67.1 | 65.5 | 59.6 | 46.3 | 38.2 | 31.6 | 47.5 |
| Mean Temp °F | 40.0 | 44.0 | 52.6 | 61.1 | 68.3 | 75.8 | 79.3 | 78.2 | 72.4 | 60.7 | 51.4 | 43.7 | 60.6 |
| Days Max Temp ≥ 90 °F | 0 | 0 | 0 | 0 | 3 | 14 | 22 | 20 | 9 | 0 | 0 | 0 | 68 |
| Days Max Temp ≤ 32 °F | 1 | 1 | 0 | 0 | 0 | 0 | 0 | 0 | 0 | 0 | 0 | 1 | 3 |
| Days Min Temp ≤ 32 °F | 21 | 17 | 10 | 3 | 0 | 0 | 0 | 0 | 0 | 4 | 11 | 18 | 84 |
| Days Min Temp ≤ 0 °F | 0 | 0 | 0 | 0 | 0 | 0 | 0 | 0 | 0 | 0 | 0 | 0 | 0 |
| Heating Degree Days | 768 | 587 | 389 | 161 | 40 | 2 | 0 | 0 | 18 | 176 | 409 | 656 | 3206 |
| Cooling Degree Days | 0 | 2 | 7 | 33 | 127 | 316 | 440 | 409 | 229 | 47 | 7 | 2 | 1619 |
| Total Precipitation (") | 5.54 | 5.28 | 6.58 | 5.40 | 5.61 | 3.98 | 4.73 | 3.61 | 4.09 | 3.50 | 5.10 | 5.98 | 59.40 |
| Days ≥ 0.1" Precip | 8 | 7 | 8 | 6 | 7 | 6 | 7 | 6 | 6 | 5 | 7 | 8 | 81 |
| Total Snowfall (") | 0.6 | 0.1 | 0.3 | 0.0 | 0.0 | 0.0 | 0.0 | 0.0 | 0.0 | 0.0 | 0.1 | 0.0 | 1.1 |
| Days ≥ 1" Snow Depth | 0 | 0 | 0 | 0 | 0 | 0 | 0 | 0 | 0 | 0 | 0 | 0 | 0 |

## WHATLEY *Clarke County* ELEVATION 171 ft LAT/LONG 31° 39 ' N / 87° 43 ' W

| | JAN | FEB | MAR | APR | MAY | JUN | JUL | AUG | SEP | OCT | NOV | DEC | YEAR |
|---|---|---|---|---|---|---|---|---|---|---|---|---|---|
| Maximum Temp °F | 59.1 | 64.0 | 71.7 | 79.0 | 84.4 | 90.5 | 92.3 | 91.5 | 87.8 | 79.8 | 70.3 | 62.8 | 77.8 |
| Minimum Temp °F | 33.9 | 36.1 | 42.4 | 49.5 | 56.9 | 64.7 | 67.9 | 67.4 | 62.4 | 50.0 | 41.1 | 36.2 | 50.7 |
| Mean Temp °F | 46.6 | 50.1 | 57.1 | 64.2 | 70.6 | 77.7 | 80.0 | 79.4 | 75.1 | 64.9 | 55.8 | 49.6 | 64.3 |
| Days Max Temp ≥ 90 °F | 0 | 0 | 0 | 1 | 5 | 19 | 23 | 22 | 14 | 2 | 0 | 0 | 86 |
| Days Max Temp ≤ 32 °F | 0 | 0 | 0 | 0 | 0 | 0 | 0 | 0 | 0 | 0 | 0 | 0 | 0 |
| Days Min Temp ≤ 32 °F | 15 | 12 | 5 | 1 | 0 | 0 | 0 | 0 | 0 | 1 | 7 | 13 | 54 |
| Days Min Temp ≤ 0 °F | 0 | 0 | 0 | 0 | 0 | 0 | 0 | 0 | 0 | 0 | 0 | 0 | 0 |
| Heating Degree Days | 580 | 420 | 263 | 91 | 14 | 0 | 0 | 0 | 6 | 90 | 288 | 480 | 2232 |
| Cooling Degree Days | 2 | 5 | 21 | 67 | 194 | 371 | 457 | 455 | 307 | 97 | 19 | 10 | 2005 |
| Total Precipitation (") | 6.00 | 5.86 | 7.46 | 4.73 | 5.24 | 4.93 | 5.71 | 4.15 | 4.02 | 3.01 | 4.57 | 5.92 | 61.60 |
| Days ≥ 0.1" Precip | 8 | 7 | 6 | 5 | 6 | 6 | 9 | 7 | 5 | 4 | 6 | 7 | 76 |
| Total Snowfall (") | 0.2 | 0.3 | 0.2 | 0.1 | 0.0 | 0.0 | 0.0 | 0.0 | 0.0 | 0.0 | 0.0 | 0.1 | 0.9 |
| Days ≥ 1" Snow Depth | 0 | 0 | 0 | 0 | 0 | 0 | 0 | 0 | 0 | 0 | 0 | 0 | 0 |

## JANUARY MINIMUM TEMPERATURE °F

### LOWEST

| | | |
|---|---|---|
| 1 | Hamilton | 25.3 |
| | Valley Head | 25.3 |
| 3 | Russellville | 26.6 |
| 4 | Heflin | 27.4 |
| | Scottsboro | 27.4 |
| 6 | St. Bernard | 27.5 |
| 7 | Jasper | 28.0 |
| 8 | Vernon | 28.3 |
| 9 | Sand Mountain | 28.5 |
| 10 | Haleyville | 28.7 |
| 11 | Oneonta | 28.8 |
| 12 | Belle Mina | 29.1 |
| 13 | Ashland | 29.7 |
| 14 | Gadsden | 29.9 |
| | Huntsville | 29.9 |
| 16 | Muscle Shoals | 30.0 |
| 17 | Moulton | 30.5 |
| 18 | Guntersville | 30.7 |
| | Tuscaloosa-Oliver | 30.7 |
| 20 | Bankhead L & D | 30.8 |
| 21 | Clanton | 31.0 |
| 22 | Aliceville | 31.4 |
| 23 | Rockford | 31.7 |
| | Sylacauga | 31.7 |
| 25 | Talladega | 31.8 |

### HIGHEST

| | | |
|---|---|---|
| 1 | Mobile | 40.5 |
| 2 | Fairhope | 40.2 |
| 3 | Bay Minette | 39.9 |
| 4 | Coden | 38.7 |
| 5 | Robertsdale | 38.4 |
| 6 | Enterprise | 37.4 |
| 7 | Selma | 37.2 |
| 8 | Frisco City | 36.6 |
| | Troy | 36.6 |
| 10 | Greenville | 36.5 |
| 11 | Headland | 36.1 |
| 12 | Brewton | 36.0 |
| 13 | Montgomery | 35.8 |
| 14 | Clayton | 35.6 |
| 15 | Chatom | 35.5 |
| 16 | Evergreen | 35.0 |
| 17 | Greensboro | 34.2 |
| | Highland Home | 34.2 |
| 19 | Thomasville | 33.9 |
| | Whatley | 33.9 |
| 21 | Andalusia | 33.7 |
| | Union Springs | 33.7 |
| 23 | Tuscaloosa-Muni | 33.1 |
| 24 | Demopolis L & D | 32.8 |
| | Marion Junction | 32.8 |

## JULY MAXIMUM TEMPERATURE °F

### HIGHEST

| | | |
|---|---|---|
| 1 | Greensboro | 92.4 |
| 2 | Livingston | 92.3 |
| | Whatley | 92.3 |
| 4 | Brewton | 92.1 |
| | Chatom | 92.1 |
| | Selma | 92.1 |
| 7 | Childersburg | 91.8 |
| 8 | Bankhead L & D | 91.6 |
| | Greenville | 91.6 |
| 10 | Andalusia | 91.5 |
| | Headland | 91.5 |
| | Montgomery | 91.5 |
| | Vernon | 91.5 |
| 14 | Aliceville | 91.4 |
| | Demopolis L & D | 91.4 |
| | Tuscaloosa-Muni | 91.4 |
| | Tuscaloosa-Oliver | 91.4 |
| 18 | Mobile | 91.2 |
| 19 | Gainesville Lock | 91.1 |
| | Thomasville | 91.1 |
| 21 | Union Springs | 91.0 |
| 22 | Hamilton | 90.9 |
| | Highland Home | 90.9 |
| | Marion Junction | 90.9 |
| | Robertsdale | 90.9 |

### LOWEST

| | | |
|---|---|---|
| 1 | Valley Head | 87.3 |
| 2 | Sand Mountain | 87.4 |
| 3 | Ashland | 88.2 |
| 4 | Haleyville | 88.7 |
| 5 | Russellville | 89.1 |
| 6 | Rockford | 89.3 |
| 7 | Clayton | 89.4 |
| | Oneonta | 89.4 |
| 9 | Huntsville | 89.5 |
| 10 | Belle Mina | 89.6 |
| 11 | St. Bernard | 89.7 |
| 12 | Jasper | 89.8 |
| 13 | Gadsden | 89.9 |
| | Thorsby | 89.9 |
| 15 | Fairhope | 90.0 |
| | Heflin | 90.0 |
| 17 | Birmingham | 90.1 |
| | Opelika | 90.1 |
| 19 | Anniston | 90.2 |
| | Coden | 90.2 |
| 21 | Enterprise | 90.3 |
| | Moulton | 90.3 |
| | Scottsboro | 90.3 |
| | Troy | 90.3 |
| 25 | Clanton | 90.5 |

## ANNUAL PRECIPITATION (")

### HIGHEST

| | | |
|---|---|---|
| 1 | Robertsdale | 66.15 |
| 2 | Bay Minette | 65.78 |
| | Fairhope | 65.78 |
| 4 | Brewton | 65.26 |
| 5 | Coden | 64.62 |
| 6 | Mobile | 64.53 |
| 7 | Evergreen | 62.85 |
| 8 | Chatom | 62.12 |
| 9 | Whatley | 61.60 |
| 10 | Haleyville | 61.31 |
| 11 | Frisco City | 60.57 |
| 12 | Andalusia | 60.09 |
| 13 | Hamilton | 59.80 |
| 14 | Clanton | 59.60 |
| 15 | Vernon | 59.40 |
| 16 | St. Bernard | 58.95 |
| 17 | Ashland | 58.71 |
| 18 | Thomasville | 58.69 |
| 19 | Bankhead L & D | 58.52 |
| 20 | Jasper | 58.39 |
| 21 | Heflin | 57.76 |
| | Huntsville | 57.76 |
| 23 | Rockford | 57.50 |
| | Valley Head | 57.50 |
| 25 | Oneonta | 57.36 |

### LOWEST

| | | |
|---|---|---|
| 1 | Clayton | 51.90 |
| 2 | Anniston | 52.54 |
| 3 | Selma | 52.73 |
| 4 | Guntersville | 52.74 |
| 5 | Troy | 52.79 |
| 6 | Gainesville Lock | 53.49 |
| 7 | Birmingham | 53.64 |
| 8 | Marion Junction | 53.85 |
| 9 | Montgomery | 54.33 |
| 10 | Union Springs | 54.52 |
| 11 | Sand Mountain | 54.81 |
| 12 | Demopolis L & D | 54.90 |
| 13 | Tuscaloosa-Oliver | 54.91 |
| 14 | Tuscaloosa-Muni | 54.94 |
| 15 | Gadsden | 55.24 |
| 16 | Belle Mina | 55.27 |
| 17 | Highland Home | 55.28 |
| 18 | Sylacauga | 55.51 |
| 19 | Muscle Shoals | 55.63 |
| 20 | Talladega | 55.68 |
| 21 | Enterprise | 56.03 |
| 22 | Aliceville | 56.06 |
| 23 | Greenville | 56.26 |
| 24 | Greensboro | 56.27 |
| 25 | Headland | 56.36 |

## ANNUAL SNOWFALL (")

### HIGHEST

| | | |
|---|---|---|
| 1 | Valley Head | 6.3 |
| 2 | Moulton | 3.8 |
| 3 | Huntsville | 3.2 |
| 4 | Muscle Shoals | 3.1 |
| 5 | Belle Mina | 3.0 |
| 6 | Oneonta | 2.6 |
| 7 | Haleyville | 2.4 |
| 8 | Hamilton | 2.1 |
| | Heflin | 2.1 |
| 10 | Ashland | 2.0 |
| 11 | Sand Mountain | 1.9 |
| | St. Bernard | 1.9 |
| 13 | Birmingham | 1.7 |
| | Russellville | 1.7 |
| 15 | Jasper | 1.4 |
| | Scottsboro | 1.4 |
| 17 | Anniston | 1.3 |
| | Highland Home | 1.3 |
| 19 | Guntersville | 1.2 |
| | Talladega | 1.2 |
| | Union Springs | 1.2 |
| 22 | Clanton | 1.1 |
| | Vernon | 1.1 |
| 24 | Frisco City | 0.9 |
| | Sylacauga | 0.9 |

### LOWEST

| | | |
|---|---|---|
| 1 | Fairhope | 0.0 |
| | Tuscaloosa-Oliver | 0.0 |
| 3 | Coden | 0.1 |
| | Gadsden | 0.1 |
| | Marion Junction | 0.1 |
| 6 | Andalusia | 0.2 |
| | Bankhead L & D | 0.2 |
| | Demopolis L & D | 0.2 |
| | Headland | 0.2 |
| | Robertsdale | 0.2 |
| 11 | Brewton | 0.3 |
| 12 | Bay Minette | 0.4 |
| | Chatom | 0.4 |
| | Evergreen | 0.4 |
| | Mobile | 0.4 |
| | Montgomery | 0.4 |
| | Selma | 0.4 |
| 18 | Enterprise | 0.5 |
| | Greensboro | 0.5 |
| 20 | Aliceville | 0.6 |
| | Gainesville Lock | 0.6 |
| | Greenville | 0.6 |
| | Livingston | 0.6 |
| | Opelika | 0.6 |
| | Tuscaloosa-Muni | 0.6 |

# ALASKA

PHYSICAL FEATURES. Alaska is the westernmost extension of the North American continent. Its east-west span covers a distance of 2,000 miles, and from north to south a distance of 1,100 miles. The state's coastline, 33,000 miles in length, is 50% longer than that of the conterminous U.S. In addition to the Aleutian Islands, hundreds of other islands are found along the northern coast of the Gulf of Alaska, the Alaska Peninsula, and the Bering Sea Coast. Alaska contains 375 million acres of land, and over 3 million lakes.

There are 12 major rivers plus three major tributaries of the Yukon, all of which drain two-thirds of the State. Four rivers, the Yukon, Stikine, Alek, and Taku, can be classed as major international rivers.

The two longest mountain ranges are the Brooks Range which separates the Arctic region from the interior, and the Alaska-Aleutian Range, which extends westward along the Alaska Peninsula and the Aleutian Islands, and northward about 200 miles from the Peninsula, then eastward to Canada. Other shorter but important ranges are the Chugach Mountains which form a rim to the central north Gulf of Alaska, and the Wrangell Mountains lying to the northeast of the Chugach Range and south of the Alaska Range. Both of these shorter ranges merge with the St. Elias Mountains, extending southeastward through Canada and across southeastern Alaska as the Coast Range. Numerous peaks in excess of 10,000 feet are found in all but the Brooks Range. The highest peak (20,320 feet above sea level) in the North American continent, Mt. McKinley, is found in Alaska, and several others tower above 16,000 feet.

Permafrost is a major factor in the geography of Alaska. It is defined as a layer of soil at variable depths beneath the surface of the earth in which the temperature has been below freezing continuously from a few to several thousands of years. It exists where summer heating fails to penetrate to the base of the layer of frozen ground. Permafrost covers most of the northern third of the State. Discontinuous or isolated patches also exist over the central portion in an overall area covering nearly a third of the State. No permafrost exists in the south-central and southern coastal portions, including southeastern Alaska, the Alaska Peninsula, and the Aleutian chain.

GENERAL CLIMATE. The geographical features already mentioned have a significant effect on Alaska's climate, which falls into four major zones. The climate zones are: (1) a Maritime Zone which includes southeastern Alaska, the South Coast, and southwestern islands, (2) a transition zone between marine and continental influences (this zone is difficult to define but generally comprises a very narrow band along the southern portion of the Copper River and the northern extreme of the South Coast--specifically the Chugach Mountains, Cook Inlet, Bristol Bay, and the coastal regions of the West-Central Division), (3) a continental zone made up of the remainders of the Copper River and West-Central Divisions, and the Interior Basin, and (4) an Arctic zone.

PRECIPITATION. In the maritime zone a coastal mountain range coupled with plentiful moisture produces annual precipitation amounts up to 200 inches in the southeastern panhandle, and up to 150 inches along the northern coast of the Gulf of Alaska. Amounts taper to near 60 inches on the southern side of the Alaska Range in the Peninsula and Aleutian Island sections. Precipitation amounts decrease rapidly to the north, with an average of 12 inches in the continental zone and less than 6 inches in the Arctic Region.

Snowfall makes up a large portion of the total annual precipitation.. Total snow depths on the ground are controlled by the temperature of an area. Fortunately, most of the areas of heavy snow have relatively mild temperatures which prevent total depths from becoming excessive.

TEMPERATURE. Mean annual temperatures in Alaska range from the low 40's under the maritime influence in the south to a chilly 10 degrees along the Arctic Slope north of the Brooks Mountain Range. The greatest seasonal temperature contrast between seasons is found in the central and eastern portion of the Continental Interior. In this area summer heating produces average maximum temperatures in the upper 70's with extreme readings in the 90's. In winter the lack of sunshine permits radiation to lower temperatures to the minus 50's and occasionally colder for two or three weeks at a time. Average winter minimums in this area are 20 to 30 degrees below zero. Elsewhere in the State, temperature contrasts are much more moderate. In the maritime zone the summer to winter range of average temperatures is from near 60 to the 20's. In the transition zone, temperatures range from the low 60's to near zero, except for the colder northern coastal region of the West-Central Division, where the range is from the mid 50's to near 10 below zero. The Arctic slope has a range extending from the upper 40's to 20 below zero.

Winter temperatures play a principal role in the flow of most of Alaska's rivers. Usually beginning in late October and extending into May (and sometimes early June for the northernmost streams), thick layers of ice form, permitting passage with all types of heavy equipment. Several rivers cease to flow completely during the coldest months.

WIND. A normal storm track along the Aleutian Island chain, the Alaska Peninsula and all of the coastal area of the Gulf of Alaska exposes these parts of the State to a large majority of the storms crossing the north Pacific, resulting in a variety of wind problems. Direct exposure to the wind of the storms themselves results in the frequent occurrence of winds in excess of 50 m.p.h. during all but the summer months, and on occasion even then for the land areas along the storm track. Wind velocities approaching 100 m.p.h. are not common but do occur, usually associated with mountainous terrain and narrow passes.

An occasional storm will either develop in or move into the Bering Sea, then move north or northeastward, creating strong winds along the western coastal area. Because of the quite low flat ground in many places along the coast, these winds will cause flooding during the time the winds are blowing onshore. Winter storms moving eastward across the southern Arctic Ocean cause winds of 50 m.p.h. or higher along the Arctic Coast. Except for local strong wind conditions, winds are generally light in the interior sections.

Strong winds, or in fact any wind occurring in the areas of extreme winter cold, create a definite hazard to personnel exposed for even brief periods of time. For example, (using a wind chill chart developed by the U.S. Army) a temperature of a minus 13 and an accompanying wind of 15 m.p.h. results in an equivalent temperature of a minus 49. That is, the chilling would be equivalent to the conditions that would be experienced with a temperature of minus 49 degrees and no wind. If the temperature is a minus 49 and the winds only 10 m.p.h., the resulting equivalent temperature is a minus 81.

## COUNTY INDEX

### *Arctic Drainage Division*
BARROW W POST-W ROGE
KOTZEBUE RALPH WEIN

### *Bristol Bay Division*
DILLINGHAM ARPT
ILIAMNA ARPT
INTRICATE BAY
KING SALMON AP
PORT ALSWORTH

### *Cook Inlet Division*
ANCHORAGE ELMENDORF
ANCHORAGE INTL AP
EKLUTNA PROJECT
HOMER AP
KASILOF 3 NW
KENAI MUNICIPAL AP
MATANUSKA AES
PALMER IAS
PUNTILLA
TALKEETNA STATE AP
WASILLA 3 S

### *Copper River Division*
GLENNALLEN KCAM
GULKANA INTERMEDIATE
SNOWSHOE LAKE
TONSINA

### *Interior Basin Division*
BETTLES FIELD
BIG DELTA ALLEN AAF
CLEARWATER
COLLEGE OBSERVATORY
EAGLE
EIELSON FIELD
FAIRBANKS INTL AP
MC GRATH AP
MCKINLEY PARK
NORTHWAY AIRPORT
TANANA CALHOUN ARPT
TOK

### *South Coast Division*
COOPER LAKE PROJECT
CORDOVA
KITOI BAY
SEWARD
VALDEZ WSO
YAKUTAT STATE AP

### *Southeastern Division*
ANNETTE ISLAND AP
AUKE BAY
BEAVER FALLS
JUNEAU AP
KETCHIKAN
LITTLE PORT WALTER
SITKA JAPONSKI AP
WRANGELL AIRPORT

### *Southwestern Islands Division*
COLD BAY AP
SHEMYA USAF BASE
ST PAUL ISLAND AP

### *West Central Division*
BETHEL AIRPORT
NOME MUNICIPAL AP
WALES

## ELEVATION INDEX

| FEET | STATION NAME |
|---|---|
| 10 | LITTLE PORT WALTER |
| 13 | NOME MUNICIPAL AP |
| 15 | SITKA JAPONSKI AP |
| 16 | BETHEL AIRPORT |
| 16 | WALES |
| 20 | KETCHIKAN |
| 20 | KITOI BAY |
| 20 | KOTZEBUE RALPH WEIN |
| 23 | JUNEAU AP |
| 30 | BARROW W POST-W ROGE |
| 30 | DILLINGHAM ARPT |
| 30 | ST PAUL ISLAND AP |
| 30 | YAKUTAT STATE AP |
| 38 | EKLUTNA PROJECT |
| 39 | AUKE BAY |
| 39 | BEAVER FALLS |
| 43 | CORDOVA |
| 44 | WRANGELL AIRPORT |
| 49 | KING SALMON AP |
| 59 | KASILOF 3 NW |
| 72 | HOMER AP |
| 75 | SEWARD |
| 87 | VALDEZ WSO |
| 92 | KENAI MUNICIPAL AP |
| 92 | SHEMYA USAF BASE |
| 98 | COLD BAY AP |
| 110 | ANNETTE ISLAND AP |
| 112 | ANCHORAGE INTL AP |
| 151 | ILIAMNA ARPT |
| 151 | MATANUSKA AES |
| 171 | INTRICATE BAY |
| 194 | ANCHORAGE ELMENDORF |
| 230 | PORT ALSWORTH |
| 239 | TANANA CALHOUN ARPT |
| 302 | PALMER IAS |
| 331 | WASILLA 3 S |
| 341 | MC GRATH AP |
| 351 | TALKEETNA STATE AP |
| 443 | FAIRBANKS INTL AP |
| 449 | COOPER LAKE PROJECT |
| 547 | EIELSON FIELD |
| 620 | COLLEGE OBSERVATORY |
| 673 | BETTLES FIELD |
| 807 | EAGLE |
| 1102 | CLEARWATER |
| 1276 | BIG DELTA ALLEN AAF |
| 1460 | GLENNALLEN KCAM |
| 1503 | TONSINA |
| 1578 | GULKANA INTERMEDIATE |
| 1631 | TOK |
| 1719 | NORTHWAY AIRPORT |
| 1730 | MCKINLEY PARK |
| 1837 | PUNTILLA |
| 2297 | SNOWSHOE LAKE |

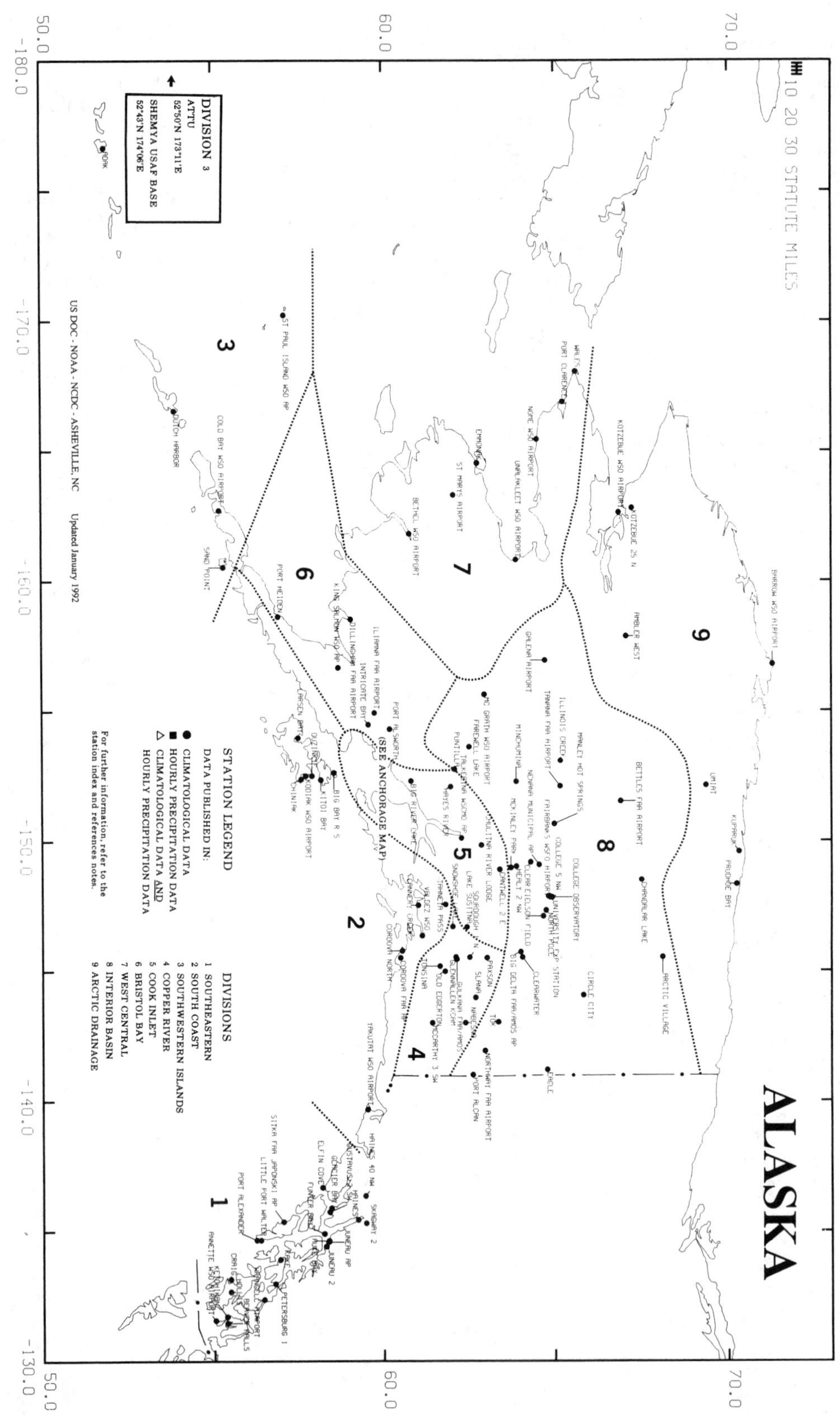
ALASKA
10 20 30 STATUTE MILES
DIVISION 3
ATTU
52°50'N 173°11'E
SHEMYA USAF BASE
52°43'N 174°06'E
US DOC - NOAA - NCDC - ASHEVILLE, NC Updated January 1992
STATION LEGEND
DATA PUBLISHED IN:
● CLIMATOLOGICAL DATA
■ HOURLY PRECIPITATION DATA
△ CLIMATOLOGICAL DATA AND HOURLY PRECIPITATION DATA
For further information, refer to the station index and references notes.
(SEE ANCHORAGE MAP)
DIVISIONS
1 SOUTHEASTERN
2 SOUTH COAST
3 SOUTHWESTERN ISLANDS
4 COPPER RIVER
5 COOK INLET
6 BRISTOL BAY
7 WEST CENTRAL
8 INTERIOR BASIN
9 ARCTIC DRAINAGE

# ANCHORAGE

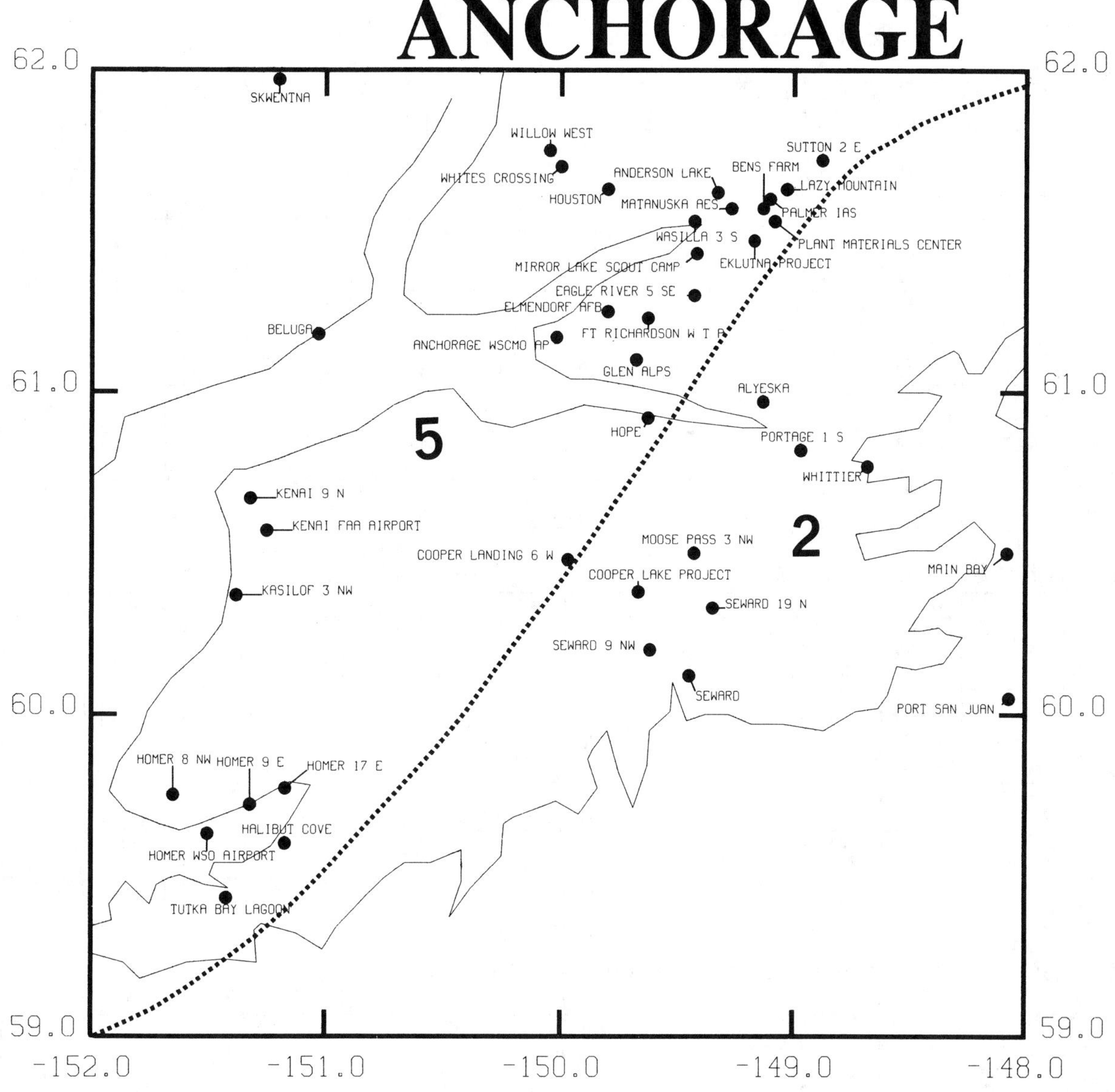

62.0
61.0
60.0
59.0
-152.0
-151.0
-150.0
-149.0
-148.0
SKWENTNA
WILLOW WEST
WHITES CROSSING
HOUSTON
ANDERSON LAKE
MATANUSKA AES
BENS FARM
SUTTON 2 E
LAZY MOUNTAIN
PALMER IAS
PLANT MATERIALS CENTER
WASILLA 3 S
EKLUTNA PROJECT
MIRROR LAKE SCOUT CAMP
EAGLE RIVER 5 SE
ELMENDORF AFB
BELUGA
ANCHORAGE WSCMO AP
FT RICHARDSON W T P
GLEN ALPS
ALYESKA
HOPE
PORTAGE 1 S
WHITTIER
5
2
KENAI 9 N
KENAI FAA AIRPORT
COOPER LANDING 6 W
MOOSE PASS 3 NW
COOPER LAKE PROJECT
MAIN BAY
KASILOF 3 NW
SEWARD 19 N
SEWARD 9 NW
SEWARD
PORT SAN JUAN
HOMER 8 NW
HOMER 9 E
HOMER 17 E
HALIBUT COVE
HOMER WSO AIRPORT
TUTKA BAY LAGOON

## ANCHORAGE ELMENDORF *Cook Inlet Division* ELEVATION 194 ft LAT/LONG 61° 15 ' N / 149° 48 ' W

| | JAN | FEB | MAR | APR | MAY | JUN | JUL | AUG | SEP | OCT | NOV | DEC | YEAR |
|---|---|---|---|---|---|---|---|---|---|---|---|---|---|
| Maximum Temp °F | 19.0 | 23.8 | 32.8 | 43.4 | ***54.3*** | 61.4 | 64.7 | 63.1 | ***54.5*** | 38.6 | 27.1 | 21.8 | 42.0 |
| Minimum Temp °F | 5.4 | 8.7 | 17.7 | 28.9 | ***39.4*** | 47.8 | 52.2 | 49.9 | ***41.6*** | 27.4 | 15.3 | 9.2 | 28.6 |
| Mean Temp °F | 12.3 | 16.3 | 25.3 | 36.2 | ***46.9*** | 54.6 | 58.5 | 56.5 | ***48.0*** | 32.9 | 21.2 | 15.5 | 35.4 |
| Days Max Temp ≥ 90 °F | na | na | na | na | na | na | na | na | na | na | na | na | na |
| Days Max Temp ≤ 32 °F | 26 | 21 | 13 | 2 | ***0*** | 0 | 0 | 0 | ***0*** | 7 | 21 | 25 | 115 |
| Days Min Temp ≤ 32 °F | 31 | 28 | 29 | 21 | ***2*** | 0 | 0 | 0 | ***3*** | 21 | 28 | 31 | 194 |
| Days Min Temp ≤ 0 °F | 12 | 9 | 3 | 0 | 0 | 0 | 0 | 0 | ***0*** | 0 | 4 | 8 | 36 |
| Heating Degree Days | 1632 | 1371 | 1224 | 859 | ***555*** | 304 | 198 | 256 | ***503*** | 985 | 1307 | 1529 | 10723 |
| Cooling Degree Days | 0 | ***0*** | 0 | 0 | 0 | 0 | 3 | 2 | ***0*** | 0 | 0 | 0 | 5 |
| Total Precipitation (") | 0.72 | 0.79 | 0.64 | 0.50 | 0.66 | 1.00 | 1.80 | 2.25 | 2.49 | 1.81 | 1.11 | 1.19 | 14.96 |
| Days ≥ 0.1" Precip | 2 | 3 | 2 | 2 | ***2*** | 3 | 5 | 6 | ***7*** | 5 | 4 | 4 | 45 |
| Total Snowfall (") | 10.9 | 11.2 | 8.8 | 4.1 | ***0.1*** | 0.0 | 0.0 | 0.0 | ***0.2*** | 9.2 | 13.8 | 17.5 | 75.8 |
| Days ≥ 1" Snow Depth | ***31*** | ***28*** | ***29*** | 12 | ***0*** | 0 | 0 | 0 | ***0*** | 8 | ***22*** | 30 | 160 |

## ANCHORAGE INTL AP *Cook Inlet Division* ELEVATION 112 ft LAT/LONG 61° 10 ' N / 149° 59 ' W

| | JAN | FEB | MAR | APR | MAY | JUN | JUL | AUG | SEP | OCT | NOV | DEC | YEAR |
|---|---|---|---|---|---|---|---|---|---|---|---|---|---|
| Maximum Temp °F | 21.5 | 25.5 | 33.8 | 43.4 | 54.6 | 62.0 | 65.3 | 63.2 | 54.9 | 40.3 | 27.7 | 23.3 | 43.0 |
| Minimum Temp °F | 8.9 | 11.2 | 19.1 | 29.0 | 39.2 | 47.4 | 51.9 | 49.7 | 41.4 | 28.7 | 15.9 | 11.1 | 29.5 |
| Mean Temp °F | 15.1 | 18.4 | 26.5 | 36.2 | 46.9 | 54.7 | 58.6 | 56.5 | 48.2 | 34.6 | 21.6 | 17.3 | 36.2 |
| Days Max Temp ≥ 90 °F | na | na | na | na | na | na | na | na | na | na | na | na | na |
| Days Max Temp ≤ 32 °F | 25 | 20 | 11 | 2 | 0 | 0 | 0 | 0 | 0 | 5 | 21 | 24 | 108 |
| Days Min Temp ≤ 32 °F | 31 | 27 | 28 | 21 | 3 | 0 | 0 | 0 | 3 | 20 | 28 | 30 | 191 |
| Days Min Temp ≤ 0 °F | 10 | 7 | 2 | 0 | 0 | 0 | 0 | 0 | 0 | 0 | 3 | 7 | 29 |
| Heating Degree Days | 1544 | 1310 | 1187 | 857 | 553 | 302 | 193 | 258 | 498 | 937 | 1288 | 1476 | 10403 |
| Cooling Degree Days | 0 | 0 | 0 | 0 | 0 | 0 | 2 | 0 | 0 | 0 | 0 | 0 | 2 |
| Total Precipitation (") | 0.74 | 0.76 | 0.66 | 0.56 | 0.70 | 0.96 | 1.64 | 2.51 | 2.82 | 2.01 | 1.21 | 1.13 | 15.70 |
| Days ≥ 0.1" Precip | 2 | 3 | 2 | 2 | 2 | 3 | 5 | 7 | 8 | 6 | 4 | 4 | 48 |
| Total Snowfall (") | 9.3 | 10.8 | 9.3 | 4.6 | 0.1 | 0.0 | 0.0 | 0.0 | 0.4 | 8.0 | 11.6 | 14.9 | 69.0 |
| Days ≥ 1" Snow Depth | 29 | 26 | 28 | 12 | 0 | 0 | 0 | 0 | 0 | 7 | 21 | 28 | 151 |

## ANNETTE ISLAND AP *Southeastern Division* ELEVATION 110 ft LAT/LONG 55° 2 ' N / 131° 34 ' W

| | JAN | FEB | MAR | APR | MAY | JUN | JUL | AUG | SEP | OCT | NOV | DEC | YEAR |
|---|---|---|---|---|---|---|---|---|---|---|---|---|---|
| Maximum Temp °F | 39.0 | 41.8 | 45.0 | 49.9 | 55.9 | 60.8 | 64.5 | 65.1 | 60.1 | 51.8 | 44.3 | 40.5 | 51.6 |
| Minimum Temp °F | 29.7 | 32.0 | 34.1 | 37.3 | 42.7 | 48.0 | 52.1 | 52.4 | 48.1 | 41.8 | 35.0 | 31.8 | 40.4 |
| Mean Temp °F | 34.4 | 36.9 | 39.7 | 43.6 | 49.3 | 54.5 | 58.3 | 58.7 | 54.1 | 46.8 | 39.6 | 36.2 | 46.0 |
| Days Max Temp ≥ 90 °F | na | na | na | na | na | na | na | na | na | na | na | na | na |
| Days Max Temp ≤ 32 °F | 5 | 2 | 0 | 0 | 0 | 0 | 0 | 0 | 0 | 0 | 1 | 4 | 12 |
| Days Min Temp ≤ 32 °F | 18 | 14 | 11 | 4 | 0 | 0 | 0 | 0 | 0 | 2 | 10 | 14 | 73 |
| Days Min Temp ≤ 0 °F | 0 | 0 | 0 | 0 | 0 | 0 | 0 | 0 | 0 | 0 | 0 | 0 | 0 |
| Heating Degree Days | 943 | 787 | 782 | 635 | 479 | 309 | 207 | 194 | 320 | 557 | 754 | 886 | 6853 |
| Cooling Degree Days | 0 | 0 | 0 | 0 | 1 | 2 | 6 | 9 | 1 | 0 | 0 | 0 | 19 |
| Total Precipitation (") | 10.01 | 8.65 | 7.89 | 7.52 | 6.39 | 4.43 | 4.36 | 5.93 | 9.40 | 14.77 | 12.22 | 11.32 | 102.89 |
| Days ≥ 0.1" Precip | 16 | 15 | 15 | 14 | 12 | 9 | 9 | 10 | 14 | 19 | 18 | 17 | 168 |
| Total Snowfall (") | 13.1 | 11.7 | 7.2 | 2.7 | 0.0 | 0.0 | 0.0 | 0.0 | 0.0 | 0.2 | 4.0 | 9.8 | 48.7 |
| Days ≥ 1" Snow Depth | 9 | 6 | 2 | 0 | 0 | 0 | 0 | 0 | 0 | 0 | 2 | 5 | 24 |

## AUKE BAY *Southeastern Division* ELEVATION 39 ft LAT/LONG 58° 23 ' N / 134° 38 ' W

| | JAN | FEB | MAR | APR | MAY | JUN | JUL | AUG | SEP | OCT | NOV | DEC | YEAR |
|---|---|---|---|---|---|---|---|---|---|---|---|---|---|
| Maximum Temp °F | 29.8 | 34.8 | 40.7 | 48.8 | 56.4 | 63.0 | 65.4 | 64.3 | 56.5 | 47.1 | 36.9 | 32.4 | 48.0 |
| Minimum Temp °F | 20.9 | 24.1 | 28.5 | 33.2 | 40.0 | 46.3 | 49.8 | 49.1 | 44.7 | 38.3 | 29.3 | 24.9 | 35.8 |
| Mean Temp °F | 25.4 | 29.4 | 34.6 | 41.0 | 48.2 | 54.7 | 57.6 | 56.7 | 50.6 | 42.7 | 33.1 | 28.7 | 41.9 |
| Days Max Temp ≥ 90 °F | na | na | na | na | na | na | na | na | na | na | na | na | na |
| Days Max Temp ≤ 32 °F | 16 | 8 | 2 | 0 | 0 | 0 | 0 | 0 | 0 | 0 | 6 | 12 | 44 |
| Days Min Temp ≤ 32 °F | 26 | 22 | 21 | 13 | 2 | 0 | 0 | 0 | 0 | 5 | 18 | 24 | 131 |
| Days Min Temp ≤ 0 °F | 3 | 1 | 0 | 0 | 0 | 0 | 0 | 0 | 0 | 0 | 0 | 1 | 5 |
| Heating Degree Days | 1222 | 998 | 935 | 714 | 513 | 303 | 227 | 251 | 424 | 684 | 949 | 1118 | 8338 |
| Cooling Degree Days | 0 | 0 | 0 | 0 | 0 | 2 | 4 | 2 | 0 | 0 | 0 | 0 | 8 |
| Total Precipitation (") | 5.19 | 3.78 | 3.42 | 2.84 | 4.10 | 4.05 | 5.05 | 6.65 | 8.72 | 8.97 | 5.68 | 4.90 | 63.35 |
| Days ≥ 0.1" Precip | 12 | 10 | 12 | 9 | 12 | 10 | 11 | 13 | 16 | 19 | 13 | 14 | 151 |
| Total Snowfall (") | 29.8 | 18.3 | 11.1 | 1.2 | 0.0 | 0.0 | 0.0 | 0.0 | 0.0 | 0.7 | 13.8 | 21.4 | 96.3 |
| Days ≥ 1" Snow Depth | 25 | 22 | 19 | 6 | 0 | 0 | 0 | 0 | 0 | 0 | 11 | 20 | 103 |

## BARROW W POST-W ROGE *Arctic Drainage Division* ELEVATION 30 ft LAT/LONG 71° 18 ' N / 156° 47 ' W

| | JAN | FEB | MAR | APR | MAY | JUN | JUL | AUG | SEP | OCT | NOV | DEC | YEAR |
|---|---|---|---|---|---|---|---|---|---|---|---|---|---|
| Maximum Temp °F | -7.8 | -11.2 | -8.3 | 5.2 | 24.4 | 38.6 | 45.5 | 42.8 | 33.7 | 18.4 | 3.2 | -5.1 | 14.9 |
| Minimum Temp °F | -19.7 | -23.0 | -20.7 | -8.2 | 14.8 | 29.8 | 33.8 | 33.3 | 26.7 | 8.8 | -7.8 | -17.1 | 4.2 |
| Mean Temp °F | -13.8 | -17.1 | -14.5 | -1.6 | 19.6 | 34.3 | 39.7 | 38.1 | 30.3 | 13.6 | -2.3 | -11.1 | 9.6 |
| Days Max Temp ≥ 90 °F | na | na | na | na | na | na | na | na | na | na | na | na | na |
| Days Max Temp ≤ 32 °F | 31 | 28 | 31 | 29 | 26 | 4 | 0 | 2 | 14 | 29 | 30 | 31 | 255 |
| Days Min Temp ≤ 32 °F | 31 | 28 | 31 | 30 | 31 | 23 | 13 | 16 | 25 | 31 | 30 | 31 | 320 |
| Days Min Temp ≤ 0 °F | 29 | 27 | 30 | 23 | 3 | 0 | 0 | 0 | 0 | 8 | 22 | 29 | 171 |
| Heating Degree Days | 2445 | 2324 | 2471 | 1991 | 1400 | 916 | 778 | 828 | 1036 | 1589 | 2019 | 2364 | 20161 |
| Cooling Degree Days | 0 | 0 | 0 | 0 | 0 | 0 | 0 | 0 | 0 | 0 | 0 | 0 | 0 |
| Total Precipitation (") | 0.13 | 0.12 | 0.11 | 0.15 | 0.14 | 0.29 | 0.84 | 0.89 | 0.64 | 0.41 | 0.25 | 0.17 | 4.14 |
| Days ≥ 0.1" Precip | 0 | 0 | 0 | 0 | 0 | 1 | 3 | 3 | 2 | 1 | 1 | 0 | 11 |
| Total Snowfall (") | 1.9 | 1.7 | 1.8 | 2.1 | 1.9 | 0.8 | 0.2 | 0.9 | 4.7 | 6.3 | 3.5 | 2.4 | 28.2 |
| Days ≥ 1" Snow Depth | 31 | 28 | 31 | 30 | 29 | 4 | 0 | 0 | 8 | 29 | 30 | 31 | 251 |

## BEAVER FALLS *Southeastern Division* ELEVATION 39 ft LAT/LONG 55° 23 ' N / 131° 28 ' W

| | JAN | FEB | MAR | APR | MAY | JUN | JUL | AUG | SEP | OCT | NOV | DEC | YEAR |
|---|---|---|---|---|---|---|---|---|---|---|---|---|---|
| Maximum Temp °F | 36.3 | 39.8 | 43.3 | 48.9 | 55.1 | 61.2 | 64.3 | 64.5 | 59.2 | 50.3 | 42.3 | 38.1 | 50.3 |
| Minimum Temp °F | 27.1 | 29.5 | 31.9 | 35.3 | 40.8 | 46.9 | 51.3 | 51.6 | 47.5 | 40.4 | 33.3 | 29.5 | 38.8 |
| Mean Temp °F | 31.7 | 34.7 | 37.6 | 42.0 | 48.0 | 54.1 | 57.8 | 57.9 | 53.4 | 45.3 | 37.8 | 33.8 | 44.5 |
| Days Max Temp ≥ 90 °F | na | na | na | na | na | na | na | na | na | na | na | na | na |
| Days Max Temp ≤ 32 °F | 7 | 3 | 0 | 0 | 0 | 0 | 0 | 0 | 0 | 0 | 1 | 4 | 15 |
| Days Min Temp ≤ 32 °F | 22 | 18 | 16 | 7 | 1 | 0 | 0 | 0 | 0 | 2 | 12 | 19 | 97 |
| Days Min Temp ≤ 0 °F | 0 | 0 | 0 | 0 | 0 | 0 | 0 | 0 | 0 | 0 | 0 | 0 | 0 |
| Heating Degree Days | 1024 | 850 | 843 | 682 | 522 | 324 | 223 | 215 | 341 | 604 | 810 | 959 | 7397 |
| Cooling Degree Days | 0 | *0* | 0 | 0 | 0 | *2* | *6* | *6* | *0* | *0* | *0* | *0* | 14 |
| Total Precipitation (") | 14.93 | 11.67 | 11.37 | 9.72 | 8.02 | 6.02 | 5.46 | 9.24 | 15.37 | 22.17 | 17.86 | 16.56 | 148.39 |
| Days ≥ 0.1" Precip | 17 | 15 | 16 | 15 | 13 | 11 | 10 | 12 | 15 | 21 | 19 | 19 | 183 |
| Total Snowfall (") | 21.8 | 20.7 | 7.9 | 1.9 | 0.1 | 0.0 | 0.0 | 0.0 | 0.0 | 0.3 | *5.8* | 16.3 | 74.8 |
| Days ≥ 1" Snow Depth | 18 | 16 | 12 | 3 | 0 | 0 | 0 | 0 | 0 | 0 | *5* | 13 | 67 |

## BETHEL AIRPORT *West Central Division* ELEVATION 16 ft LAT/LONG 60° 47 ' N / 161° 43 ' W

| | JAN | FEB | MAR | APR | MAY | JUN | JUL | AUG | SEP | OCT | NOV | DEC | YEAR |
|---|---|---|---|---|---|---|---|---|---|---|---|---|---|
| Maximum Temp °F | 12.5 | 13.1 | 21.6 | 32.5 | 49.0 | 59.3 | 62.5 | 59.7 | 51.8 | 35.3 | 23.5 | 15.6 | 36.4 |
| Minimum Temp °F | 0.2 | -0.3 | 6.3 | 17.1 | 32.1 | 42.7 | 47.9 | 46.6 | 38.1 | 24.2 | 11.7 | 3.0 | 22.5 |
| Mean Temp °F | 6.4 | 6.4 | 13.9 | 24.8 | 40.6 | 50.9 | 55.3 | 53.2 | 45.0 | 29.8 | 17.6 | 9.3 | 29.4 |
| Days Max Temp ≥ 90 °F | na | na | na | na | na | na | na | na | na | na | na | na | na |
| Days Max Temp ≤ 32 °F | 25 | 23 | 22 | 12 | 1 | 0 | 0 | 0 | 0 | 11 | 21 | 24 | 139 |
| Days Min Temp ≤ 32 °F | 30 | 28 | 31 | 28 | 16 | 1 | 0 | 0 | 6 | 25 | 29 | 30 | 224 |
| Days Min Temp ≤ 0 °F | 16 | 15 | 12 | 5 | 0 | 0 | 0 | 0 | 0 | 1 | 6 | 15 | 70 |
| Heating Degree Days | 1817 | 1655 | 1579 | 1199 | 749 | 413 | 297 | 360 | 594 | 1084 | 1417 | 1724 | 12888 |
| Cooling Degree Days | 0 | 0 | 0 | 0 | 0 | 0 | 2 | 0 | 0 | 0 | 0 | 0 | 2 |
| Total Precipitation (") | 0.60 | 0.45 | 0.67 | 0.66 | 0.74 | 1.46 | 1.95 | 2.93 | 2.07 | 1.49 | 1.33 | 1.16 | 15.51 |
| Days ≥ 0.1" Precip | 2 | 1 | 2 | 2 | 2 | 5 | 6 | 8 | 6 | 5 | 4 | 4 | 47 |
| Total Snowfall (") | 6.2 | 5.3 | 8.0 | 4.6 | 1.7 | 0.1 | 0.0 | 0.0 | 0.2 | 3.6 | 9.9 | 10.3 | 49.9 |
| Days ≥ 1" Snow Depth | 30 | 26 | 28 | 20 | 3 | 0 | 0 | 0 | 0 | 5 | 21 | 28 | 161 |

## BETTLES FIELD *Interior Basin Division* ELEVATION 673 ft LAT/LONG 66° 55 ' N / 151° 31 ' W

| | JAN | FEB | MAR | APR | MAY | JUN | JUL | AUG | SEP | OCT | NOV | DEC | YEAR |
|---|---|---|---|---|---|---|---|---|---|---|---|---|---|
| Maximum Temp °F | -4.6 | 0.2 | 15.6 | 32.6 | 53.4 | 68.0 | 70.2 | 62.7 | 48.6 | 24.8 | 5.3 | -0.8 | 31.3 |
| Minimum Temp °F | -20.1 | -18.7 | -7.0 | 10.5 | 33.9 | 47.3 | 49.6 | 43.6 | 32.0 | 11.8 | -8.9 | -15.7 | 13.2 |
| Mean Temp °F | -12.4 | -9.3 | 4.3 | 21.6 | 43.9 | 57.7 | 59.9 | 53.2 | 40.3 | 18.3 | -1.8 | -8.3 | 22.3 |
| Days Max Temp ≥ 90 °F | na | na | na | na | na | na | na | na | na | na | na | na | na |
| Days Max Temp ≤ 32 °F | 30 | 28 | 28 | 14 | 0 | 0 | 0 | 0 | 1 | 23 | 29 | 31 | 184 |
| Days Min Temp ≤ 32 °F | 31 | 28 | 31 | 29 | 13 | 0 | 0 | 2 | 15 | 30 | 30 | 31 | 240 |
| Days Min Temp ≤ 0 °F | 25 | 23 | 21 | 8 | 0 | 0 | 0 | 0 | 0 | 7 | 21 | 24 | 129 |
| Heating Degree Days | 2402 | 2100 | 1879 | 1295 | 654 | 230 | 174 | 365 | 733 | 1443 | 2005 | 2275 | 15555 |
| Cooling Degree Days | 0 | 0 | 0 | 0 | 1 | 20 | 24 | 3 | 0 | 0 | 0 | 0 | 48 |
| Total Precipitation (") | 0.78 | 0.57 | 0.69 | 0.52 | 0.70 | 1.49 | 1.93 | 2.43 | 1.68 | 1.08 | 1.07 | 0.99 | 13.93 |
| Days ≥ 0.1" Precip | 2 | 2 | 2 | 2 | 2 | 5 | 6 | 7 | 5 | 4 | 3 | 3 | 43 |
| Total Snowfall (") | 13.4 | 9.0 | 11.3 | 7.3 | 0.8 | 0.0 | 0.0 | 0.1 | 2.3 | 12.6 | 16.0 | 17.7 | 90.5 |
| Days ≥ 1" Snow Depth | 31 | 28 | 31 | 30 | 11 | 0 | 0 | 0 | 2 | 23 | 30 | 31 | 217 |

## BIG DELTA ALLEN AAF *Interior Basin Division* ELEVATION 1276 ft LAT/LONG 64° 0 ' N / 145° 44 ' W

| | JAN | FEB | MAR | APR | MAY | JUN | JUL | AUG | SEP | OCT | NOV | DEC | YEAR |
|---|---|---|---|---|---|---|---|---|---|---|---|---|---|
| Maximum Temp °F | 3.2 | 10.2 | 25.8 | 41.6 | 57.3 | 66.9 | 70.2 | 65.0 | 53.1 | 31.4 | 13.6 | 6.9 | 37.1 |
| Minimum Temp °F | -11.1 | -7.8 | 3.9 | 21.2 | 37.3 | 47.3 | 50.8 | 46.0 | 35.2 | 17.3 | -1.2 | -7.7 | 19.3 |
| Mean Temp °F | -4.0 | 1.1 | 14.9 | 31.4 | 47.5 | 57.1 | 60.5 | 55.5 | 44.2 | 24.4 | 6.2 | -0.5 | 28.2 |
| Days Max Temp ≥ 90 °F | na | na | na | na | na | na | na | na | na | na | na | na | na |
| Days Max Temp ≤ 32 °F | 28 | 24 | 19 | 6 | 0 | 0 | 0 | 0 | 1 | 17 | 26 | 28 | 149 |
| Days Min Temp ≤ 32 °F | 30 | 27 | 29 | 25 | 7 | 0 | 0 | 1 | 10 | 27 | 29 | 30 | 215 |
| Days Min Temp ≤ 0 °F | 20 | 17 | 13 | 2 | 0 | 0 | 0 | 0 | 0 | 4 | 15 | 20 | 91 |
| Heating Degree Days | 2139 | 1806 | 1548 | 1000 | 541 | 237 | 148 | 295 | 616 | 1253 | 1763 | 2028 | 13374 |
| Cooling Degree Days | 0 | 0 | 0 | 0 | *1* | 7 | 16 | 10 | 0 | 0 | 0 | 0 | 34 |
| Total Precipitation (") | 0.33 | 0.29 | 0.23 | 0.21 | 0.91 | 2.41 | 2.75 | 1.86 | 1.04 | 0.74 | 0.63 | 0.36 | 11.76 |
| Days ≥ 0.1" Precip | 1 | 1 | 1 | 1 | 3 | 7 | 7 | 6 | 3 | 3 | 2 | 1 | 36 |
| Total Snowfall (") | 5.3 | 4.8 | 3.8 | 2.8 | 0.9 | 0.0 | 0.0 | 0.0 | 2.2 | 11.3 | 10.9 | 6.4 | 48.4 |
| Days ≥ 1" Snow Depth | 30 | 27 | 28 | 19 | 2 | 0 | 0 | 0 | 2 | 18 | 27 | 29 | 182 |

## CLEARWATER *Interior Basin Division* ELEVATION 1102 ft LAT/LONG 64° 3 ' N / 145° 31 ' W

| | JAN | FEB | MAR | APR | MAY | JUN | JUL | AUG | SEP | OCT | NOV | DEC | YEAR |
|---|---|---|---|---|---|---|---|---|---|---|---|---|---|
| Maximum Temp °F | 0.1 | 9.1 | 27.2 | 43.4 | 60.2 | 69.7 | 72.5 | 67.4 | 54.3 | 32.0 | 12.1 | 3.9 | 37.7 |
| Minimum Temp °F | -18.5 | -14.0 | -2.4 | 16.5 | 31.5 | 41.7 | 44.9 | 39.9 | 29.3 | 13.0 | -6.5 | -14.2 | 13.4 |
| Mean Temp °F | -9.3 | -2.4 | 12.4 | 30.0 | 45.8 | 55.7 | 58.7 | 53.7 | 41.8 | 22.5 | 2.8 | -5.2 | 25.5 |
| Days Max Temp ≥ 90 °F | na | na | na | na | na | na | na | na | na | na | na | na | na |
| Days Max Temp ≤ 32 °F | 30 | 26 | 18 | 5 | 0 | 0 | 0 | 0 | 1 | *16* | 28 | 30 | 154 |
| Days Min Temp ≤ 32 °F | 31 | 28 | 31 | 28 | 18 | 3 | 0 | 6 | 19 | 29 | 30 | 31 | 254 |
| Days Min Temp ≤ 0 °F | 25 | 20 | 17 | 4 | 0 | 0 | 0 | 0 | 0 | *6* | 19 | 24 | 115 |
| Heating Degree Days | 2306 | 1906 | 1627 | 1045 | 588 | 276 | 193 | 348 | 689 | 1311 | 1865 | 2176 | 14330 |
| Cooling Degree Days | 0 | 0 | 0 | 0 | 0 | 5 | 7 | 4 | 0 | *0* | 0 | 0 | 16 |
| Total Precipitation (") | 0.91 | 0.69 | 0.63 | 0.48 | 1.00 | 2.48 | 2.85 | 1.88 | 1.40 | 1.41 | 1.23 | 0.82 | 15.78 |
| Days ≥ 0.1" Precip | 3 | 2 | 2 | 2 | 3 | 7 | 7 | 6 | 5 | 5 | 5 | 3 | 50 |
| Total Snowfall (") | 7.5 | 6.8 | 5.6 | 3.5 | 0.5 | 0.0 | 0.0 | 0.0 | 1.8 | *11.6* | 12.0 | 8.5 | 57.8 |
| Days ≥ 1" Snow Depth | 30 | 28 | 30 | 22 | 2 | 0 | 0 | 0 | 2 | *19* | 29 | 30 | 192 |

## COLD BAY AP *Southwestern Islands Division* ELEVATION 98 ft LAT/LONG 55° 12 ' N / 162° 43 ' W

| | JAN | FEB | MAR | APR | MAY | JUN | JUL | AUG | SEP | OCT | NOV | DEC | YEAR |
|---|---|---|---|---|---|---|---|---|---|---|---|---|---|
| Maximum Temp °F | 33.1 | 32.1 | 35.0 | 38.3 | 44.6 | 50.5 | 55.2 | 56.0 | 52.4 | 44.6 | 39.3 | 35.6 | 43.1 |
| Minimum Temp °F | 24.1 | 22.7 | 25.1 | 28.9 | 34.8 | 41.1 | 46.1 | 47.4 | 43.1 | 35.0 | 30.0 | 26.8 | 33.8 |
| Mean Temp °F | 28.6 | 27.4 | 30.0 | 33.6 | 39.7 | 45.8 | 50.7 | 51.7 | 47.7 | 39.8 | 34.6 | 31.2 | 38.4 |
| Days Max Temp ≥ 90 °F | na | na | na | na | na | na | na | na | na | na | na | na | na |
| Days Max Temp ≤ 32 °F | 11 | 11 | 9 | 5 | 1 | 0 | 0 | 0 | 0 | 1 | 4 | 8 | 50 |
| Days Min Temp ≤ 32 °F | 24 | 23 | 24 | 20 | 9 | 0 | 0 | 0 | 1 | 9 | 18 | 22 | 150 |
| Days Min Temp ≤ 0 °F | 1 | 0 | 1 | 0 | 0 | 0 | 0 | 0 | 0 | 0 | 0 | 0 | 2 |
| Heating Degree Days | 1121 | 1056 | 1077 | 935 | 777 | 569 | 437 | 405 | 511 | 774 | 902 | 1041 | 9605 |
| Cooling Degree Days | 0 | 0 | 0 | 0 | 0 | 0 | 0 | 0 | 0 | 0 | 0 | 0 | 0 |
| Total Precipitation (") | 2.91 | 2.38 | 2.28 | 2.08 | 2.42 | 2.42 | 2.44 | 3.38 | 4.59 | 4.49 | 4.38 | 4.05 | 37.82 |
| Days ≥ 0.1" Precip | 9 | 7 | 7 | 6 | 7 | 7 | 7 | 9 | 11 | 11 | 11 | 11 | 103 |
| Total Snowfall (") | 12.1 | 12.9 | 11.7 | 6.3 | 2.1 | 0.0 | 0.0 | 0.0 | 0.0 | 3.6 | 8.7 | 11.7 | 69.1 |
| Days ≥ 1" Snow Depth | 15 | 15 | 13 | 5 | 1 | 0 | 0 | 0 | 0 | 2 | 8 | 14 | 73 |

## COLLEGE OBSERVATORY *Interior Basin Division* ELEVATION 620 ft LAT/LONG 64° 52 ' N / 147° 50 ' W

| | JAN | FEB | MAR | APR | MAY | JUN | JUL | AUG | SEP | OCT | NOV | DEC | YEAR |
|---|---|---|---|---|---|---|---|---|---|---|---|---|---|
| Maximum Temp °F | 1.4 | 8.1 | 25.8 | 42.1 | 59.9 | 70.4 | 72.7 | 66.9 | 54.5 | 32.1 | 12.2 | 4.7 | 37.6 |
| Minimum Temp °F | -12.9 | -10.4 | 3.0 | 20.1 | 36.0 | 47.1 | 50.3 | 45.3 | 34.0 | 16.7 | -2.2 | -9.7 | 18.1 |
| Mean Temp °F | -5.8 | -1.2 | 14.3 | 31.1 | 48.0 | 58.7 | 61.5 | 56.1 | 44.3 | 24.5 | 5.0 | -2.5 | 27.8 |
| Days Max Temp ≥ 90 °F | na | na | na | na | na | na | na | na | na | na | na | na | na |
| Days Max Temp ≤ 32 °F | 29 | 25 | 20 | 6 | 0 | 0 | 0 | 0 | 1 | 16 | 28 | 29 | 154 |
| Days Min Temp ≤ 32 °F | 31 | 28 | 31 | 28 | 9 | 0 | 0 | 2 | 12 | 29 | 30 | 31 | 231 |
| Days Min Temp ≤ 0 °F | 22 | 20 | 13 | 2 | 0 | 0 | 0 | 0 | 0 | 3 | 16 | 22 | 98 |
| Heating Degree Days | 2194 | 1868 | 1566 | 1012 | 522 | 199 | 130 | 278 | 615 | 1251 | 1798 | 2093 | 13526 |
| Cooling Degree Days | 0 | 0 | 0 | 0 | 1 | 22 | 32 | 8 | 0 | 0 | 0 | 0 | 63 |
| Total Precipitation (") | 0.59 | 0.43 | 0.43 | 0.28 | 0.59 | 1.65 | 1.88 | 1.89 | 1.14 | 0.95 | 1.01 | 0.96 | 11.80 |
| Days ≥ 0.1" Precip | 2 | 1 | 1 | 1 | 2 | 5 | 5 | 5 | 4 | 3 | 3 | 3 | 35 |
| Total Snowfall (") | 9.6 | 7.1 | 6.4 | 3.3 | 0.5 | 0.0 | 0.0 | 0.0 | 1.8 | 12.0 | 16.1 | 14.1 | 70.9 |
| Days ≥ 1" Snow Depth | 31 | 28 | 31 | 26 | 3 | 0 | 0 | 0 | 1 | 19 | 30 | 31 | 200 |

## COOPER LAKE PROJECT *South Coast Division* ELEVATION 449 ft LAT/LONG 60° 22 ' N / 149° 40 ' W

| | JAN | FEB | MAR | APR | MAY | JUN | JUL | AUG | SEP | OCT | NOV | DEC | YEAR |
|---|---|---|---|---|---|---|---|---|---|---|---|---|---|
| Maximum Temp °F | 26.3 | 28.7 | 35.9 | 44.2 | 53.4 | 61.2 | 65.7 | 63.8 | 53.8 | 41.9 | 33.3 | 29.1 | 44.8 |
| Minimum Temp °F | 14.0 | 14.0 | 19.7 | 27.9 | 35.8 | 42.7 | 47.9 | 47.6 | 41.0 | 31.4 | 23.1 | 18.3 | 30.3 |
| Mean Temp °F | 20.2 | 21.4 | 27.7 | 36.0 | 44.7 | 52.0 | 56.8 | 55.7 | 47.4 | 36.7 | 28.2 | 23.7 | 37.5 |
| Days Max Temp ≥ 90 °F | na | na | na | na | na | na | na | na | na | na | na | na | na |
| Days Max Temp ≤ 32 °F | 17 | 14 | 8 | 1 | 0 | 0 | 0 | 0 | 0 | 3 | 11 | 16 | 70 |
| Days Min Temp ≤ 32 °F | 27 | 25 | 26 | 22 | 6 | 0 | 0 | 0 | 2 | 16 | 24 | 27 | 175 |
| Days Min Temp ≤ 0 °F | 7 | 6 | 3 | 0 | 0 | 0 | 0 | 0 | 0 | 0 | 0 | 3 | 19 |
| Heating Degree Days | 1385 | 1226 | 1148 | 862 | 624 | 384 | 248 | 281 | 521 | 871 | 1097 | 1273 | 9920 |
| Cooling Degree Days | 0 | 0 | 0 | 0 | 0 | 0 | 2 | 0 | 0 | 0 | 0 | 0 | 2 |
| Total Precipitation (") | 2.61 | 2.37 | 1.47 | 1.12 | 0.98 | 1.06 | 1.37 | 3.00 | 5.53 | 5.34 | 3.39 | 3.87 | 32.11 |
| Days ≥ 0.1" Precip | 5 | 5 | 4 | 3 | 3 | 3 | 4 | *6* | 8 | 9 | 7 | 8 | 65 |
| Total Snowfall (") | 12.7 | 16.4 | 12.2 | 3.4 | 0.1 | 0.0 | 0.0 | 0.0 | 0.1 | 5.1 | 14.6 | 22.3 | 86.9 |
| Days ≥ 1" Snow Depth | 28 | 24 | 28 | 16 | 1 | 0 | 0 | 0 | 0 | 4 | 17 | 25 | 143 |

## CORDOVA *South Coast Division* ELEVATION 43 ft LAT/LONG 60° 30 ' N / 145° 30 ' W

| | JAN | FEB | MAR | APR | MAY | JUN | JUL | AUG | SEP | OCT | NOV | DEC | YEAR |
|---|---|---|---|---|---|---|---|---|---|---|---|---|---|
| Maximum Temp °F | 30.5 | 34.1 | 38.6 | 44.9 | 52.2 | 58.0 | 61.7 | 61.6 | 56.1 | 46.3 | 36.6 | 32.6 | 46.1 |
| Minimum Temp °F | 15.6 | 18.3 | 23.0 | 29.2 | 36.7 | 43.3 | 47.2 | 46.0 | 40.2 | 32.8 | 22.9 | 19.0 | 31.2 |
| Mean Temp °F | 23.0 | 26.2 | 30.8 | 37.1 | 44.5 | 50.7 | 54.5 | 53.8 | 48.2 | 39.5 | 29.8 | 25.8 | 38.7 |
| Days Max Temp ≥ 90 °F | na | na | na | na | na | na | na | na | na | na | na | na | na |
| Days Max Temp ≤ 32 °F | 15 | 10 | 4 | 1 | 0 | 0 | 0 | 0 | 0 | 1 | 8 | 12 | 51 |
| Days Min Temp ≤ 32 °F | 25 | 22 | 24 | 20 | 7 | 0 | 0 | 0 | 5 | 14 | 22 | 25 | 164 |
| Days Min Temp ≤ 0 °F | 7 | 5 | 2 | 0 | 0 | 0 | 0 | 0 | 0 | 0 | 2 | 4 | 20 |
| Heating Degree Days | 1295 | 1088 | 1053 | 832 | 629 | 422 | 320 | 339 | 498 | 782 | 1049 | 1209 | 9516 |
| Cooling Degree Days | 0 | 0 | 0 | 0 | 0 | 0 | 0 | 0 | 0 | 0 | 0 | 0 | 0 |
| Total Precipitation (") | 6.89 | 6.72 | 6.82 | 5.92 | 6.46 | 5.64 | 5.65 | 9.64 | 14.64 | 13.33 | 8.00 | 9.85 | 99.56 |
| Days ≥ 0.1" Precip | 12 | 11 | 12 | 12 | 14 | 13 | 12 | 14 | 17 | 18 | 12 | 14 | 161 |
| Total Snowfall (") | 21.2 | 19.3 | 24.9 | 11.8 | 1.1 | 0.0 | 0.0 | 0.0 | 0.0 | 3.1 | 12.5 | 26.6 | 120.5 |
| Days ≥ 1" Snow Depth | 24 | 23 | 24 | 14 | 1 | 0 | 0 | 0 | 0 | 1 | 12 | 23 | 122 |

## DILLINGHAM ARPT *Bristol Bay Division* ELEVATION 30 ft LAT/LONG 59° 3 ' N / 158° 27 ' W

| | JAN | FEB | MAR | APR | MAY | JUN | JUL | AUG | SEP | OCT | NOV | DEC | YEAR |
|---|---|---|---|---|---|---|---|---|---|---|---|---|---|
| Maximum Temp °F | 22.1 | 21.5 | 29.4 | 37.2 | 49.3 | *57.5* | *61.8* | *60.1* | *53.2* | *39.1* | *28.5* | *22.2* | 40.2 |
| Minimum Temp °F | 10.6 | 8.4 | 16.1 | 24.3 | 34.9 | *43.0* | *48.2* | *47.7* | *41.1* | *27.0* | *17.1* | *10.6* | 27.4 |
| Mean Temp °F | 16.4 | 15.0 | 22.8 | 30.8 | 42.2 | *50.3* | *55.0* | *53.9* | *47.1* | *33.1* | *22.8* | *16.4* | 33.8 |
| Days Max Temp ≥ 90 °F | na | na | na | na | na | na | na | na | na | na | na | na | na |
| Days Max Temp ≤ 32 °F | 22 | 20 | 16 | 5 | 0 | 0 | *0* | *0* | *0* | *6* | *17* | *22* | 108 |
| Days Min Temp ≤ 32 °F | 29 | 27 | 29 | 25 | 9 | 1 | *0* | *0* | *4* | *20* | *27* | *29* | 200 |
| Days Min Temp ≤ 0 °F | 8 | 10 | 4 | 1 | 0 | 0 | *0* | *0* | *0* | *0* | *3* | *9* | 35 |
| Heating Degree Days | 1503 | 1409 | 1302 | 1020 | 701 | *435* | *305* | *338* | *529* | *980* | *1259* | *1503* | 11284 |
| Cooling Degree Days | 0 | 0 | 0 | 0 | 0 | 0 | 1 | 0 | 0 | 0 | 0 | 0 | 1 |
| Total Precipitation (") | *1.77* | *1.28* | *1.75* | *1.05* | *1.42* | *1.94* | *2.93* | *3.87* | *3.73* | *2.35* | *2.39* | *2.05* | 26.53 |
| Days ≥ 0.1" Precip | *5* | *4* | *5* | *4* | *4* | *6* | *7* | *10* | *9* | *6* | *6* | *7* | 73 |
| Total Snowfall (") | *19.4* | *12.4* | *17.5* | *6.5* | *0.3* | 0.0 | *0.0* | *0.0* | *0.0* | na | na | *20.1* | na |
| Days ≥ 1" Snow Depth | *30* | *27* | *28* | *19* | *3* | 0 | *0* | *0* | *0* | na | na | na | na |

## EAGLE *Interior Basin Division* ELEVATION 807 ft LAT/LONG 64° 46 ' N / 141° 12 ' W

| | JAN | FEB | MAR | APR | MAY | JUN | JUL | AUG | SEP | OCT | NOV | DEC | YEAR |
|---|---|---|---|---|---|---|---|---|---|---|---|---|---|
| Maximum Temp °F | -3.2 | *2.4* | 24.0 | *42.0* | *59.0* | *70.0* | 73.2 | *67.0* | *54.0* | *31.7* | 10.6 | *2.9* | 36.1 |
| Minimum Temp °F | -21.0 | *-20.5* | -5.6 | *14.5* | *31.8* | *43.3* | 46.8 | *40.9* | *30.0* | *14.6* | -6.2 | *-14.3* | 12.9 |
| Mean Temp °F | -12.1 | *-9.1* | 9.2 | *28.3* | *45.4* | *56.7* | 60.0 | *54.0* | *42.0* | *23.1* | 2.2 | *-5.7* | 24.5 |
| Days Max Temp ≥ 90 °F | na | na | na | na | na | na | na | na | na | na | na | na | na |
| Days Max Temp ≤ 32 °F | 29 | *26* | 21 | *6* | *0* | *0* | 0 | *0* | *1* | *16* | 28 | *29* | 156 |
| Days Min Temp ≤ 32 °F | 31 | *28* | 31 | *29* | *17* | *1* | 0 | *5* | *19* | *30* | 30 | *31* | 252 |
| Days Min Temp ≤ 0 °F | 25 | *23* | 20 | *5* | *0* | *0* | 0 | *0* | *0* | *5* | 19 | *23* | 120 |
| Heating Degree Days | 2394 | *2093* | 1727 | *1096* | *601* | *252* | 164 | *340* | *682* | *1292* | 1884 | *2193* | 14718 |
| Cooling Degree Days | 0 | 0 | 0 | 0 | 1 | 10 | 16 | 7 | 0 | 0 | 0 | 0 | 34 |
| Total Precipitation (") | 0.49 | *0.44* | 0.33 | *0.33* | *1.03* | *1.67* | 2.08 | *1.88* | *1.12* | 0.96 | 0.79 | *0.70* | 11.82 |
| Days ≥ 0.1" Precip | *2* | *2* | 1 | *1* | *4* | *6* | 7 | *6* | *4* | *4* | 3 | *2* | 42 |
| Total Snowfall (") | 7.8 | *7.2* | 5.4 | *4.3* | *0.8* | *0.0* | 0.0 | *0.0* | *1.6* | *10.8* | 12.6 | *11.9* | 62.4 |
| Days ≥ 1" Snow Depth | *31* | *28* | *31* | na | *2* | *0* | 0 | *0* | *2* | *18* | 29 | *31* | na |

## EIELSON FIELD *Interior Basin Division* ELEVATION 547 ft LAT/LONG 64° 40 ' N / 147° 6 ' W

| | JAN | FEB | MAR | APR | MAY | JUN | JUL | AUG | SEP | OCT | NOV | DEC | YEAR |
|---|---|---|---|---|---|---|---|---|---|---|---|---|---|
| Maximum Temp °F | -2.6 | 5.1 | 23.9 | 41.2 | 58.6 | 69.0 | 70.9 | 65.7 | 54.1 | 30.9 | 10.6 | 1.6 | 35.8 |
| Minimum Temp °F | -18.7 | -15.6 | -0.8 | 20.1 | 37.4 | 48.5 | 51.5 | 46.5 | 34.9 | 16.5 | -4.9 | -14.3 | 16.8 |
| Mean Temp °F | -10.5 | -5.2 | 11.5 | 30.7 | 48.0 | 58.7 | 61.2 | 56.1 | 44.6 | 23.7 | 2.8 | -6.3 | 26.3 |
| Days Max Temp ≥ 90 °F | na | na | na | na | na | na | na | na | na | na | na | na | na |
| Days Max Temp ≤ 32 °F | 29 | 26 | 22 | 7 | 0 | 0 | 0 | 0 | 1 | 17 | 27 | 29 | 158 |
| Days Min Temp ≤ 32 °F | 31 | 28 | 31 | 27 | 7 | 0 | 0 | 1 | 11 | 28 | 30 | 31 | 225 |
| Days Min Temp ≤ 0 °F | 26 | 22 | 16 | 3 | 0 | 0 | 0 | 0 | 0 | 4 | 19 | 25 | 115 |
| Heating Degree Days | 2350 | 1986 | 1654 | 1023 | 519 | 198 | 129 | 275 | 607 | 1272 | 1863 | 2214 | 14090 |
| Cooling Degree Days | 0 | 0 | 0 | 0 | 1 | 20 | 25 | 7 | 0 | 0 | 0 | 0 | 53 |
| Total Precipitation (") | 0.50 | 0.33 | 0.37 | 0.27 | 0.68 | 1.67 | 2.23 | 2.16 | 1.21 | 0.90 | 0.82 | 0.66 | 11.80 |
| Days ≥ 0.1" Precip | 1 | 1 | 1 | 1 | 2 | 5 | 7 | 6 | 4 | 3 | 3 | 2 | 36 |
| Total Snowfall (") | 11.2 | 7.6 | 7.4 | 4.3 | 1.2 | 0.0 | 0.0 | 0.0 | 2.5 | 13.3 | 19.1 | 14.6 | 81.2 |
| Days ≥ 1" Snow Depth | 28 | 26 | 30 | 16 | 1 | 0 | 0 | 0 | 1 | 17 | 29 | 30 | 178 |

## EKLUTNA PROJECT *Cook Inlet Division* ELEVATION 38 ft LAT/LONG 61° 28 ' N / 149° 10 ' W

| | JAN | FEB | MAR | APR | MAY | JUN | JUL | AUG | SEP | OCT | NOV | DEC | YEAR |
|---|---|---|---|---|---|---|---|---|---|---|---|---|---|
| Maximum Temp °F | 17.4 | 24.9 | 35.2 | 47.0 | 59.2 | 66.6 | 69.1 | 66.1 | 56.1 | 40.1 | 26.3 | 18.8 | 43.9 |
| Minimum Temp °F | 0.6 | 6.9 | 13.7 | 24.9 | 35.1 | 43.5 | 47.0 | 44.0 | 36.8 | 24.8 | 11.0 | 2.8 | 24.3 |
| Mean Temp °F | 9.4 | 15.9 | 24.5 | 36.0 | 47.2 | 55.1 | 58.1 | 55.1 | 46.5 | 32.5 | 18.7 | 10.9 | 34.2 |
| Days Max Temp ≥ 90 °F | na | na | na | na | na | na | na | na | na | na | na | na | na |
| Days Max Temp ≤ 32 °F | 22 | 17 | 9 | 1 | 0 | 0 | 0 | 0 | 0 | 6 | 19 | 22 | 96 |
| Days Min Temp ≤ 32 °F | 28 | 25 | 27 | 24 | 9 | 0 | 0 | 1 | 8 | 22 | 27 | 27 | 198 |
| Days Min Temp ≤ 0 °F | 14 | 10 | 6 | 0 | 0 | 0 | 0 | 0 | 0 | 1 | 7 | 11 | 49 |
| Heating Degree Days | 1735 | 1382 | 1249 | 865 | 546 | 292 | 209 | 301 | 549 | 998 | 1383 | 1692 | 11201 |
| Cooling Degree Days | na | na | na | na | na | na | 1 | 0 | 0 | 0 | na | na | na |
| Total Precipitation (") | 0.91 | 0.90 | 0.80 | 0.66 | 0.82 | 1.76 | 2.55 | 2.35 | 2.76 | 1.70 | 1.50 | 1.57 | 18.28 |
| Days ≥ 0.1" Precip | 3 | 3 | 2 | 2 | 2 | 5 | 7 | 6 | 6 | 5 | 4 | 4 | 49 |
| Total Snowfall (") | 7.8 | 10.0 | 8.6 | 2.8 | 0.0 | 0.0 | 0.0 | 0.0 | 0.0 | 5.8 | 10.0 | 15.6 | 60.6 |
| Days ≥ 1" Snow Depth | 28 | 25 | 28 | 21 | 1 | 0 | 0 | 0 | 0 | 7 | 22 | 28 | 160 |

## FAIRBANKS INTL AP *Interior Basin Division* ELEVATION 443 ft LAT/LONG 64° 50 ' N / 147° 43 ' W

| | JAN | FEB | MAR | APR | MAY | JUN | JUL | AUG | SEP | OCT | NOV | DEC | YEAR |
|---|---|---|---|---|---|---|---|---|---|---|---|---|---|
| Maximum Temp °F | -1.3 | 6.8 | 25.0 | 42.3 | 60.0 | 70.7 | 72.9 | 66.6 | 54.5 | 31.7 | 11.7 | 3.0 | 37.0 |
| Minimum Temp °F | -18.2 | -15.1 | -0.3 | 21.1 | 38.3 | 49.7 | 52.9 | 47.1 | 35.7 | 17.4 | -4.6 | -13.7 | 17.5 |
| Mean Temp °F | -9.8 | -4.2 | 12.4 | 31.8 | 49.2 | 60.3 | 62.9 | 56.9 | 45.1 | 24.6 | 3.6 | -5.4 | 27.3 |
| Days Max Temp ≥ 90 °F | na | na | na | na | na | na | na | na | na | na | na | na | na |
| Days Max Temp ≤ 32 °F | 30 | 26 | 21 | 6 | 0 | 0 | 0 | 0 | 0 | 16 | 28 | 30 | 157 |
| Days Min Temp ≤ 32 °F | 31 | 28 | 31 | 26 | 6 | 0 | 0 | 1 | 9 | 29 | 30 | 31 | 222 |
| Days Min Temp ≤ 0 °F | 26 | 22 | 16 | 2 | 0 | 0 | 0 | 0 | 0 | 4 | 18 | 24 | 112 |
| Heating Degree Days | 2320 | 1953 | 1627 | 990 | 484 | 161 | 99 | 254 | 589 | 1246 | 1841 | 2182 | 13746 |
| Cooling Degree Days | 0 | 0 | 0 | 0 | 1 | 30 | 42 | 8 | 0 | 0 | 0 | 0 | 81 |
| Total Precipitation (") | 0.53 | 0.38 | 0.36 | 0.28 | 0.62 | 1.37 | 1.71 | 1.69 | 1.00 | 0.89 | 0.88 | 0.90 | 10.61 |
| Days ≥ 0.1" Precip | 2 | 1 | 1 | 1 | 2 | 4 | 5 | 5 | 3 | 3 | 3 | 2 | 32 |
| Total Snowfall (") | 10.5 | 7.9 | 6.3 | 3.3 | 0.8 | 0.0 | 0.0 | 0.0 | 2.0 | 12.0 | 16.6 | 16.0 | 75.4 |
| Days ≥ 1" Snow Depth | 31 | 28 | 31 | 22 | 1 | 0 | 0 | 0 | 1 | 17 | 30 | 31 | 192 |

## GLENNALLEN KCAM *Copper River Division* ELEVATION 1460 ft LAT/LONG 62° 7 ' N / 145° 32 ' W

| | JAN | FEB | MAR | APR | MAY | JUN | JUL | AUG | SEP | OCT | NOV | DEC | YEAR |
|---|---|---|---|---|---|---|---|---|---|---|---|---|---|
| Maximum Temp °F | 5.0 | 15.0 | 30.5 | 43.9 | 57.3 | 67.0 | 70.7 | 66.7 | 55.2 | 36.5 | 15.2 | 6.1 | 39.1 |
| Minimum Temp °F | -17.0 | -11.3 | -0.2 | 17.1 | 29.4 | 38.7 | 42.9 | 38.2 | 29.6 | 16.0 | -5.5 | -14.2 | 13.6 |
| Mean Temp °F | -5.7 | 1.8 | 15.3 | 30.5 | 43.4 | 52.9 | 56.8 | 52.6 | 42.4 | 26.2 | 4.9 | -4.1 | 26.4 |
| Days Max Temp ≥ 90 °F | na | na | na | na | na | na | na | na | na | na | na | na | na |
| Days Max Temp ≤ 32 °F | 28 | 23 | 15 | 2 | 0 | 0 | 0 | 0 | 0 | 10 | 25 | 28 | 131 |
| Days Min Temp ≤ 32 °F | 30 | 28 | 30 | 28 | 21 | 6 | 2 | 9 | 18 | 28 | 29 | 31 | 260 |
| Days Min Temp ≤ 0 °F | 24 | 20 | 16 | 3 | 0 | 0 | 0 | 0 | 0 | 5 | 18 | 24 | 110 |
| Heating Degree Days | 2183 | 1779 | 1537 | 1027 | 663 | 358 | 250 | 379 | 670 | 1194 | 1803 | 2136 | 13979 |
| Cooling Degree Days | 0 | 0 | 0 | 0 | 0 | 1 | 4 | 2 | 0 | 0 | 0 | 0 | 7 |
| Total Precipitation (") | 0.51 | 0.53 | 0.39 | 0.22 | 0.55 | 1.52 | 1.65 | 1.68 | 1.20 | 1.00 | 0.73 | 1.27 | 11.25 |
| Days ≥ 0.1" Precip | 2 | 2 | 1 | 1 | 2 | 5 | 5 | 4 | 4 | 3 | 2 | na | na |
| Total Snowfall (") | 7.1 | 7.1 | 4.2 | 2.7 | 0.4 | 0.0 | 0.0 | 0.0 | 0.5 | 8.0 | na | 12.4 | na |
| Days ≥ 1" Snow Depth | 30 | 27 | 29 | 16 | 1 | 0 | 0 | 0 | 0 | 11 | 27 | na | na |

## GULKANA INTERMEDIATE *Copper River Division* ELEVATION 1578 ft LAT/LONG 62° 9' N / 145° 27' W

| | JAN | FEB | MAR | APR | MAY | JUN | JUL | AUG | SEP | OCT | NOV | DEC | YEAR |
|---|---|---|---|---|---|---|---|---|---|---|---|---|---|
| Maximum Temp °F | 2.7 | 13.2 | 28.9 | 42.1 | 55.2 | 64.7 | 68.5 | 64.9 | 53.7 | 34.9 | 13.6 | 5.7 | 37.3 |
| Minimum Temp °F | -13.5 | -7.5 | 4.4 | 20.2 | 33.0 | 42.3 | 46.4 | 42.3 | 33.3 | 19.1 | -1.4 | -10.0 | 17.4 |
| Mean Temp °F | -5.4 | 2.8 | 16.7 | 31.2 | 44.1 | 53.5 | 57.5 | 53.6 | 43.5 | 27.0 | 6.3 | -2.1 | 27.4 |
| Days Max Temp ≥ 90 °F | na | na | na | na | na | na | na | na | na | na | na | na | na |
| Days Max Temp ≤ 32 °F | 29 | 24 | 17 | 3 | 0 | 0 | 0 | 0 | 1 | 12 | 27 | 28 | 141 |
| Days Min Temp ≤ 32 °F | 31 | 28 | 31 | 28 | 15 | 1 | 0 | 4 | 14 | 26 | 29 | 31 | 238 |
| Days Min Temp ≤ 0 °F | 23 | 18 | 13 | 2 | 0 | 0 | 0 | 0 | 0 | 4 | 17 | 23 | 100 |
| Heating Degree Days | 2184 | 1755 | 1492 | 1009 | 641 | 338 | 229 | 348 | 637 | 1170 | 1760 | 2082 | 13645 |
| Cooling Degree Days | 0 | 0 | 0 | 0 | 0 | 1 | 4 | 2 | 0 | 0 | 0 | 0 | 7 |
| Total Precipitation (") | 0.45 | 0.53 | 0.34 | 0.24 | 0.62 | 1.54 | 1.79 | 1.55 | 1.40 | 0.96 | 0.72 | 0.90 | 11.04 |
| Days ≥ 0.1" Precip | 2 | 2 | 1 | 1 | 2 | 5 | 5 | 5 | 5 | 3 | 2 | 2 | 35 |
| Total Snowfall (") | 6.9 | 7.7 | 5.0 | 3.3 | 0.5 | 0.0 | 0.0 | 0.0 | 1.3 | 8.2 | 9.1 | 10.5 | 52.5 |
| Days ≥ 1" Snow Depth | 31 | 28 | 31 | 20 | 1 | 0 | 0 | 0 | 1 | 14 | 27 | 30 | 183 |

## HOMER AP *Cook Inlet Division* ELEVATION 72 ft LAT/LONG 59° 38' N / 151° 30' W

| | JAN | FEB | MAR | APR | MAY | JUN | JUL | AUG | SEP | OCT | NOV | DEC | YEAR |
|---|---|---|---|---|---|---|---|---|---|---|---|---|---|
| Maximum Temp °F | 28.6 | 31.0 | 36.5 | 42.8 | 50.2 | 56.6 | 60.6 | 60.6 | 54.6 | 43.9 | 35.0 | 30.9 | 44.3 |
| Minimum Temp °F | 17.1 | 17.9 | 22.8 | 29.0 | 36.4 | 42.6 | 46.7 | 46.4 | 40.7 | 31.3 | 23.3 | 19.4 | 31.1 |
| Mean Temp °F | 22.9 | 24.5 | 29.7 | 35.9 | 43.3 | 49.6 | 53.7 | 53.5 | 47.7 | 37.7 | 29.1 | 25.2 | 37.7 |
| Days Max Temp ≥ 90 °F | na | na | na | na | na | na | na | na | na | na | na | na | na |
| Days Max Temp ≤ 32 °F | 17 | 14 | 7 | 1 | 0 | 0 | 0 | 0 | 0 | 2 | 10 | 15 | 66 |
| Days Min Temp ≤ 32 °F | 27 | 25 | 26 | 21 | 8 | 0 | 0 | 0 | 4 | 17 | 25 | 27 | 180 |
| Days Min Temp ≤ 0 °F | 5 | 2 | 1 | 0 | 0 | 0 | 0 | 0 | 0 | 0 | 0 | 2 | 10 |
| Heating Degree Days | 1300 | 1138 | 1088 | 865 | 666 | 455 | 343 | 348 | 513 | 841 | 1067 | 1228 | 9852 |
| Cooling Degree Days | 0 | 0 | 0 | 0 | 0 | 0 | 0 | 0 | 0 | 0 | 0 | 0 | 0 |
| Total Precipitation (") | 2.41 | 2.03 | 1.89 | 1.21 | 1.12 | 0.94 | 1.43 | 2.29 | 3.29 | 3.03 | 2.73 | 3.10 | 25.47 |
| Days ≥ 0.1" Precip | 6 | 6 | 5 | 3 | 4 | 3 | 5 | 6 | 9 | 8 | 7 | 8 | 70 |
| Total Snowfall (") | 9.1 | 12.7 | 9.9 | 3.4 | 0.4 | 0.0 | 0.0 | 0.0 | 0.0 | 2.5 | 8.2 | 14.1 | 60.3 |
| Days ≥ 1" Snow Depth | 22 | 19 | 18 | 7 | 0 | 0 | 0 | 0 | 0 | 2 | 10 | 21 | 99 |

## ILIAMNA ARPT *Bristol Bay Division* ELEVATION 151 ft LAT/LONG 59° 45' N / 154° 55' W

| | JAN | FEB | MAR | APR | MAY | JUN | JUL | AUG | SEP | OCT | NOV | DEC | YEAR |
|---|---|---|---|---|---|---|---|---|---|---|---|---|---|
| Maximum Temp °F | 23.2 | 23.8 | 31.8 | *39.3* | *50.6* | *58.6* | *63.0* | 61.0 | *54.1* | 40.1 | 30.8 | *24.7* | 41.8 |
| Minimum Temp °F | 11.3 | 10.0 | 18.0 | *25.8* | *36.0* | *43.8* | *49.1* | 48.2 | *42.0* | 28.8 | 19.1 | *12.5* | 28.7 |
| Mean Temp °F | 17.5 | 16.9 | 24.9 | *32.5* | *43.3* | *51.3* | *56.1* | 54.7 | *48.0* | 34.4 | 25.1 | *18.6* | 35.3 |
| Days Max Temp ≥ 90 °F | na | na | na | na | na | na | na | na | na | na | na | na | na |
| Days Max Temp ≤ 32 °F | *17* | *16* | 12 | 5 | *0* | 0 | 0 | 0 | 0 | *6* | *14* | *17* | 87 |
| Days Min Temp ≤ 32 °F | *27* | *25* | 26 | 22 | *7* | 0 | 0 | 0 | 3 | *18* | *24* | *26* | 178 |
| Days Min Temp ≤ 0 °F | *8* | *9* | *3* | 1 | 0 | 0 | 0 | 0 | 0 | 0 | 2 | *9* | 32 |
| Heating Degree Days | 1467 | 1356 | 1237 | *967* | *665* | *404* | *271* | 315 | *502* | 940 | 1191 | *1434* | 10749 |
| Cooling Degree Days | 0 | 0 | 0 | 0 | 0 | 0 | 0 | 0 | 0 | 0 | 0 | 0 | 0 |
| Total Precipitation (") | 1.37 | 1.03 | *1.26* | *1.08* | *1.37* | *1.62* | *2.61* | 4.41 | 4.28 | *3.24* | 2.46 | *1.76* | 26.49 |
| Days ≥ 0.1" Precip | 4 | *4* | *4* | *4* | *4* | *4* | *7* | 9 | *9* | *8* | 6 | *5* | 68 |
| Total Snowfall (") | *11.9* | 8.2 | 9.8 | *5.0* | *0.9* | 0.0 | 0.0 | 0.0 | 0.0 | 2.2 | 10.5 | *11.1* | 59.6 |
| Days ≥ 1" Snow Depth | 25 | 23 | 23 | 15 | 3 | 0 | 0 | 0 | 0 | 3 | *16* | *23* | 131 |

## INTRICATE BAY *Bristol Bay Division* ELEVATION 171 ft LAT/LONG 59° 34' N / 154° 28' W

| | JAN | FEB | MAR | APR | MAY | JUN | JUL | AUG | SEP | OCT | NOV | DEC | YEAR |
|---|---|---|---|---|---|---|---|---|---|---|---|---|---|
| Maximum Temp °F | 24.8 | 25.4 | 33.4 | 41.1 | 52.5 | 60.8 | 65.1 | 63.0 | 55.3 | 42.3 | 33.1 | 27.1 | 43.7 |
| Minimum Temp °F | 8.2 | 7.1 | 14.8 | 23.2 | 33.2 | 41.0 | 46.0 | 45.1 | 38.7 | 27.6 | 18.5 | 11.6 | 26.3 |
| Mean Temp °F | 16.5 | 16.3 | 24.1 | 32.2 | 42.9 | 50.9 | 55.6 | 54.1 | 47.0 | 35.0 | 25.8 | 19.3 | 35.0 |
| Days Max Temp ≥ 90 °F | na | na | na | na | na | na | na | na | na | na | na | na | na |
| Days Max Temp ≤ 32 °F | 17 | 16 | 10 | 4 | 0 | 0 | 0 | 0 | 0 | 3 | 12 | 16 | 78 |
| Days Min Temp ≤ 32 °F | 28 | 25 | 27 | 24 | 13 | 3 | 0 | 1 | 7 | 20 | 26 | 28 | 202 |
| Days Min Temp ≤ 0 °F | 11 | 11 | 7 | 2 | 0 | 0 | 0 | 0 | 0 | 0 | 3 | 10 | 44 |
| Heating Degree Days | 1499 | 1374 | 1261 | 977 | 680 | 415 | 286 | 330 | 532 | 923 | 1169 | 1412 | 10858 |
| Cooling Degree Days | 0 | 0 | 0 | 0 | 0 | 0 | 0 | 0 | 0 | 0 | 0 | 0 | 0 |
| Total Precipitation (") | 2.68 | 2.39 | 2.22 | 2.17 | 2.50 | 1.81 | 2.25 | 3.98 | 4.37 | 4.01 | 3.39 | 3.33 | 35.10 |
| Days ≥ 0.1" Precip | 6 | 6 | 5 | 6 | 7 | 6 | 6 | 9 | 10 | 8 | 8 | 8 | 85 |
| Total Snowfall (") | 11.2 | 11.7 | 10.2 | 7.0 | 2.0 | 0.0 | 0.0 | 0.0 | 0.0 | 3.5 | 11.1 | 15.2 | 71.9 |
| Days ≥ 1" Snow Depth | 25 | 21 | 20 | 11 | 1 | 0 | 0 | 0 | 0 | 3 | 15 | 25 | 121 |

## JUNEAU AP *Southeastern Division* ELEVATION 23 ft LAT/LONG 58° 22 ' N / 134° 35 ' W

| | JAN | FEB | MAR | APR | MAY | JUN | JUL | AUG | SEP | OCT | NOV | DEC | YEAR |
|---|---|---|---|---|---|---|---|---|---|---|---|---|---|
| Maximum Temp °F | 29.4 | 33.8 | 39.2 | 47.8 | 55.2 | 61.5 | 64.0 | 63.0 | 55.7 | 47.0 | 37.3 | 32.2 | 47.2 |
| Minimum Temp °F | 18.9 | 22.4 | 27.4 | 32.7 | 39.5 | 45.3 | 48.5 | 47.6 | 43.3 | 37.3 | 28.1 | 23.4 | 34.5 |
| Mean Temp °F | 24.2 | 28.1 | 33.3 | 40.3 | 47.4 | 53.5 | 56.2 | 55.3 | 49.5 | 42.1 | 32.7 | 27.8 | 40.9 |
| Days Max Temp ≥ 90 °F | na | na | na | na | na | na | na | na | na | na | na | na | na |
| Days Max Temp ≤ 32 °F | 16 | 9 | 3 | 0 | 0 | 0 | 0 | 0 | 0 | 0 | 6 | 12 | 46 |
| Days Min Temp ≤ 32 °F | 25 | 22 | 21 | 13 | 3 | 0 | 0 | 0 | 1 | 7 | 19 | 24 | 135 |
| Days Min Temp ≤ 0 °F | 5 | 2 | 0 | 0 | 0 | 0 | 0 | 0 | 0 | 0 | 0 | 1 | 8 |
| Heating Degree Days | 1259 | 1035 | 975 | 734 | 541 | 340 | 266 | 294 | 458 | 702 | 961 | 1145 | 8710 |
| Cooling Degree Days | 0 | 0 | 0 | 0 | 0 | 1 | 2 | 0 | 0 | 0 | 0 | 0 | 3 |
| Total Precipitation (") | 4.82 | 3.92 | 3.45 | 2.67 | 3.61 | 3.06 | 3.95 | 5.27 | 7.28 | 7.72 | 5.65 | 4.68 | 56.08 |
| Days ≥ 0.1" Precip | 12 | 10 | 11 | 8 | 10 | 9 | 10 | 12 | 15 | 17 | 13 | 13 | 140 |
| Total Snowfall (") | 32.0 | 20.1 | 12.4 | 1.4 | 0.0 | 0.0 | 0.0 | 0.0 | 0.0 | 1.1 | 14.8 | 21.7 | 103.5 |
| Days ≥ 1" Snow Depth | 22 | 18 | 13 | 2 | 0 | 0 | 0 | 0 | 0 | 0 | 9 | 17 | 81 |

## KASILOF 3 NW *Cook Inlet Division* ELEVATION 59 ft LAT/LONG 60° 19 ' N / 151° 17 ' W

| | JAN | FEB | MAR | APR | MAY | JUN | JUL | AUG | SEP | OCT | NOV | DEC | YEAR |
|---|---|---|---|---|---|---|---|---|---|---|---|---|---|
| Maximum Temp °F | 22.5 | 26.8 | 34.3 | 42.8 | 52.5 | 59.3 | 62.7 | 61.9 | 56.0 | 43.2 | 30.1 | 23.9 | 43.0 |
| Minimum Temp °F | 6.2 | 7.0 | 13.6 | 24.0 | 31.7 | 39.6 | 44.5 | 42.8 | 35.2 | 25.1 | 13.7 | 8.8 | 24.3 |
| Mean Temp °F | 14.2 | 17.0 | 24.0 | 33.4 | 42.1 | 49.5 | 53.7 | 52.4 | 45.6 | 34.2 | 21.9 | 16.4 | 33.7 |
| Days Max Temp ≥ 90 °F | na | na | na | na | na | na | na | na | na | na | na | na | na |
| Days Max Temp ≤ 32 °F | 22 | ***18*** | ***11*** | 1 | 0 | 0 | 0 | 0 | 0 | 3 | 17 | 22 | 94 |
| Days Min Temp ≤ 32 °F | 30 | 27 | ***29*** | ***27*** | ***17*** | 4 | 0 | 2 | 10 | ***24*** | 28 | 30 | 228 |
| Days Min Temp ≤ 0 °F | 11 | ***9*** | ***6*** | 0 | 0 | 0 | 0 | 0 | 0 | 1 | 6 | ***9*** | 42 |
| Heating Degree Days | 1570 | 1354 | 1265 | 940 | 703 | 460 | 346 | 385 | 575 | 947 | 1285 | 1502 | 11332 |
| Cooling Degree Days | 0 | 0 | 0 | 0 | 0 | 0 | 0 | 0 | 0 | 0 | 0 | 0 | 0 |
| Total Precipitation (") | 1.02 | 1.03 | 0.87 | 0.59 | 0.69 | 0.98 | 1.49 | 2.55 | 3.01 | 2.25 | 1.65 | 1.66 | 17.79 |
| Days ≥ 0.1" Precip | 3 | ***4*** | ***3*** | ***2*** | 2 | 3 | 4 | ***7*** | 8 | ***6*** | 5 | 5 | 52 |
| Total Snowfall (") | 7.3 | ***8.2*** | ***6.1*** | ***2.7*** | 0.1 | 0.0 | 0.0 | 0.0 | 0.0 | ***3.2*** | ***10.7*** | 11.2 | 49.5 |
| Days ≥ 1" Snow Depth | ***26*** | ***26*** | ***26*** | ***16*** | 0 | 0 | 0 | 0 | 0 | ***3*** | ***18*** | 26 | 141 |

## KENAI MUNICIPAL AP *Cook Inlet Division* ELEVATION 92 ft LAT/LONG 60° 34 ' N / 151° 15 ' W

| | JAN | FEB | MAR | APR | MAY | JUN | JUL | AUG | SEP | OCT | NOV | DEC | YEAR |
|---|---|---|---|---|---|---|---|---|---|---|---|---|---|
| Maximum Temp °F | 20.8 | 25.6 | 33.1 | 42.1 | 52.4 | 58.1 | 61.9 | 61.5 | 54.8 | 41.5 | 29.1 | 23.1 | 42.0 |
| Minimum Temp °F | 4.3 | 6.5 | 14.6 | 26.1 | 35.9 | 42.9 | 47.6 | 46.0 | 38.8 | 27.6 | 14.1 | 7.8 | 26.0 |
| Mean Temp °F | 12.5 | 16.1 | 23.9 | 34.2 | 44.1 | 50.5 | 54.8 | 53.8 | 46.8 | 34.6 | 21.6 | 15.4 | 34.0 |
| Days Max Temp ≥ 90 °F | na | na | na | na | na | na | na | na | na | na | na | na | na |
| Days Max Temp ≤ 32 °F | 24 | 20 | 13 | 3 | 0 | 0 | 0 | 0 | 0 | 3 | 18 | 23 | 104 |
| Days Min Temp ≤ 32 °F | 30 | 27 | 29 | 24 | 8 | 0 | 0 | 0 | 6 | 20 | 27 | 30 | 201 |
| Days Min Temp ≤ 0 °F | 13 | 11 | 6 | 1 | 0 | 0 | 0 | 0 | 0 | 0 | 6 | 11 | 48 |
| Heating Degree Days | 1626 | 1376 | 1268 | 919 | 639 | 427 | 312 | 342 | 540 | 937 | 1296 | 1533 | 11215 |
| Cooling Degree Days | 0 | 0 | 0 | 0 | 0 | 0 | 0 | 0 | 0 | 0 | 0 | 0 | 0 |
| Total Precipitation (") | 1.01 | 0.95 | 0.89 | 0.71 | 1.04 | 1.03 | 1.58 | 2.55 | 3.20 | 2.47 | 1.70 | 1.44 | 18.57 |
| Days ≥ 0.1" Precip | 3 | 3 | 3 | 2 | 3 | 3 | 5 | 7 | 9 | 7 | 5 | 5 | 55 |
| Total Snowfall (") | 8.8 | 9.3 | 7.8 | 3.1 | 0.5 | 0.0 | 0.0 | 0.0 | 0.2 | 4.8 | 11.0 | 13.3 | 58.8 |
| Days ≥ 1" Snow Depth | 29 | 26 | 26 | 13 | 0 | 0 | 0 | 0 | 0 | 4 | 20 | 27 | 145 |

## KETCHIKAN *Southeastern Division* ELEVATION 20 ft LAT/LONG 55° 21 ' N / 131° 39 ' W

| | JAN | FEB | MAR | APR | MAY | JUN | JUL | AUG | SEP | OCT | NOV | DEC | YEAR |
|---|---|---|---|---|---|---|---|---|---|---|---|---|---|
| Maximum Temp °F | 38.5 | ***42.0*** | ***44.7*** | 50.2 | ***56.4*** | 61.6 | 65.0 | 65.5 | 59.8 | 51.8 | 44.8 | 40.6 | 51.7 |
| Minimum Temp °F | 27.5 | ***31.1*** | ***32.8*** | 36.2 | ***41.2*** | ***46.6*** | 50.9 | 51.7 | 46.9 | 40.6 | 34.3 | 30.7 | 39.2 |
| Mean Temp °F | 33.1 | ***36.6*** | ***38.7*** | 43.3 | ***48.8*** | 54.2 | 58.0 | 58.6 | 53.4 | 46.2 | 39.6 | 35.7 | 45.5 |
| Days Max Temp ≥ 90 °F | na | na | na | na | na | na | na | na | na | na | na | na | na |
| Days Max Temp ≤ 32 °F | 6 | ***2*** | ***0*** | 0 | ***0*** | 0 | 0 | 0 | 0 | 0 | 1 | 3 | 12 |
| Days Min Temp ≤ 32 °F | 20 | ***15*** | ***15*** | 6 | ***0*** | 0 | 0 | 0 | 0 | 3 | 11 | ***17*** | 87 |
| Days Min Temp ≤ 0 °F | 0 | ***0*** | ***0*** | 0 | ***0*** | 0 | 0 | 0 | 0 | 0 | 0 | 0 | 0 |
| Heating Degree Days | 984 | ***797*** | ***809*** | 646 | ***495*** | 320 | 215 | 198 | 341 | 576 | 757 | 903 | 7041 |
| Cooling Degree Days | ***0*** | ***0*** | na | 0 | ***0*** | ***2*** | ***4*** | 7 | 0 | 0 | ***0*** | ***0*** | na |
| Total Precipitation (") | 12.95 | ***12.65*** | ***10.27*** | 10.51 | 9.53 | 7.13 | 7.07 | 9.48 | 12.81 | 21.28 | 16.77 | 14.30 | 144.75 |
| Days ≥ 0.1" Precip | ***16*** | ***15*** | ***15*** | 14 | ***13*** | ***11*** | 11 | 11 | 14 | 20 | 19 | ***17*** | 176 |
| Total Snowfall (") | na | na | na | na | ***0.0*** | 0.0 | ***0.0*** | 0.0 | 0.0 | ***0.1*** | na | na | na |
| Days ≥ 1" Snow Depth | na | na | na | na | ***0*** | ***0*** | ***0*** | ***0*** | ***0*** | ***0*** | na | na | na |

## KING SALMON AP *Bristol Bay Division* ELEVATION 49 ft LAT/LONG 58° 41 ' N / 156° 39 ' W

| | JAN | FEB | MAR | APR | MAY | JUN | JUL | AUG | SEP | OCT | NOV | DEC | YEAR |
|---|---|---|---|---|---|---|---|---|---|---|---|---|---|
| Maximum Temp °F | 22.1 | 22.7 | 31.6 | 40.0 | 51.4 | 58.9 | 63.2 | 61.7 | 54.6 | 40.0 | 30.3 | 24.4 | 41.7 |
| Minimum Temp °F | 7.7 | 6.6 | 15.4 | 24.2 | 34.2 | 41.6 | 46.6 | 46.6 | 39.6 | 25.3 | 15.6 | 9.1 | 26.0 |
| Mean Temp °F | 15.0 | 14.6 | 23.5 | 32.2 | 42.8 | 50.3 | 54.9 | 54.1 | 47.1 | 32.8 | 22.9 | 16.9 | 33.9 |
| Days Max Temp ≥ 90 °F | na | na | na | na | na | na | na | na | na | na | na | na | na |
| Days Max Temp ≤ 32 °F | 19 | 17 | 12 | 5 | 0 | 0 | 0 | 0 | 0 | 6 | 15 | 17 | 91 |
| Days Min Temp ≤ 32 °F | 28 | 26 | 27 | 25 | 11 | 0 | 0 | 0 | 6 | 21 | 25 | 28 | 197 |
| Days Min Temp ≤ 0 °F | 11 | 12 | 6 | 1 | 0 | 0 | 0 | 0 | 0 | 1 | 5 | 11 | 47 |
| Heating Degree Days | 1548 | 1420 | 1280 | 979 | 680 | 435 | 305 | 329 | 529 | 995 | 1256 | 1491 | 11247 |
| Cooling Degree Days | 0 | 0 | 0 | 0 | 0 | 0 | 0 | 0 | 0 | 0 | 0 | 0 | 0 |
| Total Precipitation (") | 1.05 | 0.78 | 1.03 | 0.99 | 1.29 | 1.63 | 2.22 | 2.94 | 2.68 | 2.03 | 1.63 | 1.48 | 19.75 |
| Days ≥ 0.1" Precip | 4 | 2 | 3 | 4 | 5 | 6 | 7 | 9 | 8 | 6 | 5 | 5 | 64 |
| Total Snowfall (") | 9.4 | 6.4 | 7.1 | 4.9 | 1.1 | 0.0 | 0.0 | 0.0 | 0.0 | 3.0 | 7.0 | 9.4 | 48.3 |
| Days ≥ 1" Snow Depth | 20 | 17 | 16 | 6 | 0 | 0 | 0 | 0 | 0 | 3 | 12 | 20 | 94 |

## KITOI BAY *South Coast Division* ELEVATION 20 ft LAT/LONG 58° 11 ' N / 152° 21 ' W

| | JAN | FEB | MAR | APR | MAY | JUN | JUL | AUG | SEP | OCT | NOV | DEC | YEAR |
|---|---|---|---|---|---|---|---|---|---|---|---|---|---|
| Maximum Temp °F | 33.4 | 34.1 | 38.3 | 42.6 | 48.6 | 54.8 | 60.5 | 61.4 | 55.3 | 45.3 | 37.4 | 34.1 | 45.5 |
| Minimum Temp °F | 23.8 | 23.4 | 25.9 | 29.9 | 36.1 | 42.7 | 47.6 | 47.7 | 42.7 | 33.3 | 27.8 | 24.1 | 33.8 |
| Mean Temp °F | 28.6 | 28.8 | 32.1 | 36.3 | 42.4 | 48.8 | 54.1 | 54.6 | 49.0 | 39.3 | 32.6 | 29.1 | 39.6 |
| Days Max Temp ≥ 90 °F | na | na | na | na | na | na | na | na | na | na | na | na | na |
| Days Max Temp ≤ 32 °F | 10 | 8 | 4 | 1 | 0 | 0 | 0 | 0 | 0 | 1 | 5 | 11 | 40 |
| Days Min Temp ≤ 32 °F | 25 | 23 | 25 | 20 | 7 | 0 | 0 | 0 | 1 | 15 | 23 | 26 | 165 |
| Days Min Temp ≤ 0 °F | 1 | 1 | 0 | 0 | 0 | 0 | 0 | 0 | 0 | 0 | 0 | 0 | 2 |
| Heating Degree Days | 1122 | 1017 | 1013 | 854 | 694 | 481 | 332 | 315 | 472 | 789 | 964 | 1107 | 9160 |
| Cooling Degree Days | 0 | 0 | 0 | 0 | 0 | 0 | 1 | 0 | 0 | 0 | 0 | 0 | 1 |
| Total Precipitation (") | 6.28 | 4.99 | 4.66 | 4.92 | 5.66 | 4.57 | 3.64 | 5.54 | 6.73 | 6.59 | 5.69 | 6.23 | 65.50 |
| Days ≥ 0.1" Precip | 13 | 11 | 11 | 12 | 14 | 10 | 8 | 10 | 13 | 12 | 11 | 13 | 138 |
| Total Snowfall (") | 15.1 | 15.6 | 10.0 | 4.2 | 0.4 | 0.0 | 0.0 | 0.0 | 0.0 | 1.2 | 5.8 | 15.2 | 67.5 |
| Days ≥ 1" Snow Depth | 19 | 18 | 15 | 7 | 1 | 0 | 0 | 0 | 0 | 1 | 6 | 16 | 83 |

## KOTZEBUE RALPH WEIN *Arctic Drainage Division* ELEVATION 20 ft LAT/LONG 66° 52 ' N / 162° 38 ' W

| | JAN | FEB | MAR | APR | MAY | JUN | JUL | AUG | SEP | OCT | NOV | DEC | YEAR |
|---|---|---|---|---|---|---|---|---|---|---|---|---|---|
| Maximum Temp °F | 5.0 | 3.0 | 8.7 | 20.5 | 37.7 | 50.3 | 59.7 | 56.7 | 46.5 | 27.7 | 13.5 | 6.2 | 28.0 |
| Minimum Temp °F | -8.2 | -11.2 | -7.7 | 2.9 | 24.5 | 38.3 | 49.2 | 47.2 | 36.7 | 18.5 | 2.7 | -6.7 | 15.5 |
| Mean Temp °F | -1.6 | -4.1 | 0.5 | 11.7 | 31.2 | 44.3 | 54.5 | 52.0 | 41.6 | 23.1 | 8.1 | -0.2 | 21.8 |
| Days Max Temp ≥ 90 °F | na | na | na | na | na | na | na | na | na | na | na | na | na |
| Days Max Temp ≤ 32 °F | 30 | 27 | 30 | 24 | 8 | 0 | 0 | 0 | 1 | 20 | 28 | 30 | 198 |
| Days Min Temp ≤ 32 °F | 31 | 28 | 31 | 30 | 25 | 6 | 0 | 0 | 7 | 28 | 30 | 31 | 247 |
| Days Min Temp ≤ 0 °F | 21 | 21 | 22 | 14 | 1 | 0 | 0 | 0 | 0 | 2 | 13 | 21 | 115 |
| Heating Degree Days | 2065 | 1955 | 1999 | 1595 | 1042 | 614 | 323 | 398 | 695 | 1291 | 1703 | 2023 | 15703 |
| Cooling Degree Days | 0 | 0 | 0 | 0 | 0 | 3 | 4 | 0 | 0 | 0 | 0 | 0 | 7 |
| Total Precipitation (") | 0.45 | 0.34 | 0.42 | 0.35 | 0.28 | 0.49 | 1.47 | 1.83 | 1.50 | 0.87 | 0.68 | 0.56 | 9.24 |
| Days ≥ 0.1" Precip | 2 | 1 | 1 | 1 | 1 | 2 | 5 | 6 | 5 | 3 | 2 | 2 | 31 |
| Total Snowfall (") | 6.6 | 4.7 | 6.1 | 5.0 | 1.4 | 0.0 | 0.0 | 0.0 | 1.0 | 7.1 | 9.2 | 8.9 | 50.0 |
| Days ≥ 1" Snow Depth | 31 | 28 | 31 | 30 | 21 | 1 | 0 | 0 | 1 | 16 | 28 | 31 | 218 |

## LITTLE PORT WALTER *Southeastern Division* ELEVATION 10 ft LAT/LONG 56° 23 ' N / 134° 39 ' W

| | JAN | FEB | MAR | APR | MAY | JUN | JUL | AUG | SEP | OCT | NOV | DEC | YEAR |
|---|---|---|---|---|---|---|---|---|---|---|---|---|---|
| Maximum Temp °F | 36.5 | 38.9 | 42.3 | 46.8 | 52.6 | 58.3 | 61.9 | 62.1 | 56.7 | 49.5 | 42.2 | 38.6 | 48.9 |
| Minimum Temp °F | 28.3 | 29.7 | 31.6 | 34.2 | 38.9 | 44.4 | 48.6 | 48.9 | 45.4 | 39.9 | 33.6 | 30.5 | 37.8 |
| Mean Temp °F | 32.4 | 34.3 | 36.9 | 40.6 | 45.8 | 51.4 | 55.3 | 55.5 | 51.1 | 44.7 | 37.9 | 34.6 | 43.4 |
| Days Max Temp ≥ 90 °F | na | na | na | na | na | na | na | na | na | na | na | na | na |
| Days Max Temp ≤ 32 °F | 7 | 4 | 1 | 0 | 0 | 0 | 0 | 0 | 0 | 0 | 2 | 4 | 18 |
| Days Min Temp ≤ 32 °F | 22 | 19 | 19 | 10 | 2 | 0 | 0 | 0 | 0 | 3 | 13 | 19 | 107 |
| Days Min Temp ≤ 0 °F | 0 | 0 | 0 | 0 | 0 | 0 | 0 | 0 | 0 | 0 | 0 | 0 | 0 |
| Heating Degree Days | 1003 | 861 | 865 | 727 | 589 | 403 | 294 | 287 | 410 | 623 | 805 | 937 | 7804 |
| Cooling Degree Days | 0 | 0 | 0 | 0 | 0 | 0 | 0 | 1 | 0 | 0 | 0 | 0 | 1 |
| Total Precipitation (") | 23.88 | 19.40 | 17.80 | 14.34 | 13.37 | 7.94 | 8.14 | 12.08 | 22.71 | 34.35 | 27.84 | 23.77 | 225.62 |
| Days ≥ 0.1" Precip | 20 | 17 | 18 | 16 | 14 | 10 | 10 | 12 | 17 | 23 | 21 | 21 | 199 |
| Total Snowfall (") | 32.4 | 26.3 | 14.7 | 2.3 | 0.2 | 0.0 | 0.0 | 0.0 | 0.0 | 0.3 | ***11.7*** | ***24.5*** | 112.4 |
| Days ≥ 1" Snow Depth | 23 | 20 | 18 | 9 | 1 | 0 | 0 | 0 | 0 | 0 | 8 | 18 | 97 |

## MATANUSKA AES *Cook Inlet Division* ELEVATION 151 ft LAT/LONG 61° 34 ' N / 149° 16 ' W

| | JAN | FEB | MAR | APR | MAY | JUN | JUL | AUG | SEP | OCT | NOV | DEC | YEAR |
|---|---|---|---|---|---|---|---|---|---|---|---|---|---|
| Maximum Temp °F | 21.7 | 26.8 | 36.0 | 46.4 | 57.7 | 64.5 | 67.3 | 65.1 | 56.3 | 41.8 | 28.3 | 23.4 | 44.6 |
| Minimum Temp °F | 5.0 | 9.0 | 17.7 | 27.2 | 36.4 | 44.4 | 48.5 | 46.3 | 38.5 | 25.7 | 12.1 | 7.1 | 26.5 |
| Mean Temp °F | 13.4 | 18.0 | 26.9 | 36.8 | 47.1 | 54.5 | 57.9 | 55.7 | 47.4 | 33.8 | 20.2 | 15.2 | 35.6 |
| Days Max Temp ≥ 90 °F | na | na | na | na | na | na | na | na | na | na | na | na | na |
| Days Max Temp ≤ 32 °F | 22 | 18 | 9 | 1 | 0 | 0 | 0 | 0 | 0 | 5 | 19 | 22 | 96 |
| Days Min Temp ≤ 32 °F | 30 | 27 | 29 | 23 | 8 | 0 | 0 | 0 | 6 | 23 | 29 | 30 | 205 |
| Days Min Temp ≤ 0 °F | 13 | 9 | 3 | 0 | 0 | 0 | 0 | 0 | 0 | 1 | 7 | 11 | 44 |
| Heating Degree Days | 1597 | 1325 | 1175 | 839 | 550 | 309 | 214 | 280 | 519 | 962 | 1337 | 1538 | 10645 |
| Cooling Degree Days | 0 | 0 | 0 | 0 | 0 | 1 | 2 | 0 | 0 | 0 | 0 | 0 | 3 |
| Total Precipitation (") | 0.70 | 0.66 | 0.53 | 0.48 | 0.74 | 1.39 | 2.26 | 2.10 | 2.48 | 1.48 | 1.02 | 1.23 | 15.07 |
| Days ≥ 0.1" Precip | 2 | 2 | 2 | 2 | 2 | 5 | 7 | 7 | 7 | 5 | 4 | 4 | 49 |
| Total Snowfall (") | 7.6 | 8.6 | 5.7 | 2.0 | 0.1 | 0.0 | 0.0 | 0.0 | 0.0 | 5.2 | 9.3 | 11.8 | 50.3 |
| Days ≥ 1" Snow Depth | na | na | ***13*** | 3 | 0 | 0 | 0 | 0 | 0 | 6 | ***17*** | ***21*** | na |

## MC GRATH AP *Interior Basin Division* ELEVATION 341 ft LAT/LONG 62° 58 ' N / 155° 37 ' W

| | JAN | FEB | MAR | APR | MAY | JUN | JUL | AUG | SEP | OCT | NOV | DEC | YEAR |
|---|---|---|---|---|---|---|---|---|---|---|---|---|---|
| Maximum Temp °F | 0.8 | 8.8 | 24.2 | 38.2 | 55.4 | 66.2 | 68.6 | 63.6 | 52.5 | 31.2 | 13.3 | 4.0 | 35.6 |
| Minimum Temp °F | -17.3 | -14.2 | -1.7 | 16.7 | 35.0 | 45.3 | 49.3 | 45.3 | 35.3 | 18.1 | -2.8 | -13.5 | 16.3 |
| Mean Temp °F | -8.3 | -2.7 | 11.3 | 27.5 | 45.2 | 55.8 | 59.0 | 54.5 | 43.9 | 24.7 | 5.0 | -4.7 | 25.9 |
| Days Max Temp ≥ 90 °F | na | na | na | na | na | na | na | na | na | na | na | na | na |
| Days Max Temp ≤ 32 °F | 29 | 26 | 21 | 8 | 0 | 0 | 0 | 0 | 0 | 16 | 27 | 29 | 156 |
| Days Min Temp ≤ 32 °F | 31 | 28 | 31 | 28 | 11 | 0 | 0 | 1 | 10 | 28 | 30 | 31 | 229 |
| Days Min Temp ≤ 0 °F | 23 | 21 | 17 | 5 | 0 | 0 | 0 | 0 | 0 | 4 | 18 | 23 | 111 |
| Heating Degree Days | 2272 | 1913 | 1662 | 1120 | 609 | 273 | 191 | 321 | 625 | 1244 | 1789 | 2162 | 14181 |
| Cooling Degree Days | 0 | 0 | 0 | 0 | 0 | 4 | 15 | 1 | 0 | 0 | 0 | 0 | 20 |
| Total Precipitation (") | 0.92 | 0.68 | 0.90 | 0.70 | 0.91 | 1.47 | 2.11 | 2.58 | 2.14 | 1.46 | 1.53 | 1.51 | 16.91 |
| Days ≥ 0.1" Precip | 3 | 2 | 3 | 2 | 3 | 5 | 6 | 8 | 6 | 4 | 4 | 4 | 50 |
| Total Snowfall (") | 14.3 | 11.3 | 12.6 | 6.5 | 1.1 | 0.0 | 0.0 | 0.0 | 1.3 | 11.5 | 20.7 | 21.9 | 101.2 |
| Days ≥ 1" Snow Depth | 31 | 28 | 31 | 28 | 4 | 0 | 0 | 0 | 1 | 16 | 29 | 31 | 199 |

## MCKINLEY PARK *Interior Basin Division* ELEVATION 1730 ft LAT/LONG 63° 44 ' N / 148° 55 ' W

| | JAN | FEB | MAR | APR | MAY | JUN | JUL | AUG | SEP | OCT | NOV | DEC | YEAR |
|---|---|---|---|---|---|---|---|---|---|---|---|---|---|
| Maximum Temp °F | 9.3 | 13.9 | 26.3 | 38.6 | ***53.5*** | 64.0 | 66.9 | 61.9 | ***50.5*** | 31.8 | 17.4 | 12.6 | 37.2 |
| Minimum Temp °F | -8.3 | -6.8 | 1.6 | 14.7 | ***28.8*** | 38.6 | 42.4 | 38.9 | ***29.8*** | 13.1 | -0.2 | -4.9 | 15.6 |
| Mean Temp °F | 0.5 | 3.5 | 14.0 | 26.7 | ***41.2*** | 51.4 | 54.6 | 50.4 | ***40.1*** | 22.5 | 8.6 | 3.9 | 26.5 |
| Days Max Temp ≥ 90 °F | na | na | na | na | na | na | na | na | na | na | na | na | na |
| Days Max Temp ≤ 32 °F | 27 | 23 | 19 | 7 | ***0*** | 0 | 0 | 0 | 1 | 15 | 25 | 27 | 144 |
| Days Min Temp ≤ 32 °F | 31 | 28 | 31 | 29 | ***23*** | 4 | 1 | 5 | ***18*** | 29 | 30 | 31 | 260 |
| Days Min Temp ≤ 0 °F | 20 | 18 | 15 | 4 | ***0*** | 0 | 0 | 0 | 0 | 6 | 15 | 19 | 97 |
| Heating Degree Days | 1999 | 1735 | 1575 | 1143 | ***732*** | 404 | 314 | 446 | ***739*** | 1312 | 1689 | 1893 | 13981 |
| Cooling Degree Days | 0 | 0 | 0 | 0 | na | ***2*** | ***2*** | 1 | 0 | ***0*** | 0 | 0 | na |
| Total Precipitation (") | 0.76 | 0.52 | 0.50 | 0.33 | ***0.84*** | 2.40 | 3.25 | 2.41 | 1.65 | 1.12 | 1.08 | 1.01 | 15.87 |
| Days ≥ 0.1" Precip | 3 | 2 | 1 | 1 | ***2*** | ***6*** | 8 | 7 | 5 | ***3*** | 3 | 3 | 44 |
| Total Snowfall (") | 11.5 | 9.0 | 8.6 | 4.5 | ***3.5*** | 0.3 | 0.0 | 0.1 | 5.0 | 14.4 | 17.9 | 15.7 | 90.5 |
| Days ≥ 1" Snow Depth | 31 | 28 | 31 | 28 | ***6*** | 0 | ***0*** | 0 | ***3*** | 20 | 30 | 31 | 208 |

## NOME MUNICIPAL AP *West Central Division* ELEVATION 13 ft LAT/LONG 64° 30 ' N / 165° 26 ' W

| | JAN | FEB | MAR | APR | MAY | JUN | JUL | AUG | SEP | OCT | NOV | DEC | YEAR |
|---|---|---|---|---|---|---|---|---|---|---|---|---|---|
| Maximum Temp °F | 14.0 | 12.7 | 17.6 | 26.1 | 42.4 | 53.8 | 58.3 | 56.2 | 48.6 | 33.8 | 22.9 | 15.6 | 33.5 |
| Minimum Temp °F | -1.8 | -3.9 | 0.2 | 10.8 | 29.5 | 39.3 | 45.5 | 44.1 | 35.9 | 22.3 | 9.4 | 0.0 | 19.3 |
| Mean Temp °F | 6.0 | 4.4 | 8.9 | 18.3 | 36.0 | 46.6 | 51.9 | 50.2 | 42.3 | 28.1 | 16.2 | 7.7 | 26.4 |
| Days Max Temp ≥ 90 °F | na | na | na | na | na | na | na | na | na | na | na | na | na |
| Days Max Temp ≤ 32 °F | 28 | 25 | 27 | 20 | 5 | 0 | 0 | 0 | 0 | 13 | 24 | 27 | 169 |
| Days Min Temp ≤ 32 °F | 31 | 28 | 31 | 29 | 19 | 3 | 0 | 2 | 10 | 25 | 29 | 31 | 238 |
| Days Min Temp ≤ 0 °F | 16 | 17 | 15 | 8 | 0 | 0 | 0 | 0 | 0 | 1 | 8 | 16 | 81 |
| Heating Degree Days | 1821 | 1711 | 1735 | 1391 | 893 | 546 | 401 | 454 | 676 | 1138 | 1460 | 1770 | 13996 |
| Cooling Degree Days | 0 | 0 | 0 | 0 | 0 | 1 | 3 | 0 | 0 | 0 | 0 | 0 | 4 |
| Total Precipitation (") | 0.74 | 0.64 | 0.60 | 0.60 | 0.58 | 1.09 | 2.21 | 2.97 | 2.31 | 1.50 | 1.18 | 0.93 | 15.35 |
| Days ≥ 0.1" Precip | 2 | 2 | 2 | 2 | 2 | 3 | 6 | 8 | 7 | 5 | 4 | 3 | 46 |
| Total Snowfall (") | 8.2 | 6.6 | 6.8 | 6.2 | 2.4 | 0.2 | 0.0 | 0.0 | 0.4 | 4.8 | 12.7 | 10.6 | 58.9 |
| Days ≥ 1" Snow Depth | 30 | 28 | 31 | 28 | 12 | 0 | 0 | 0 | 0 | 6 | 23 | 30 | 188 |

## NORTHWAY AIRPORT *Interior Basin Division* ELEVATION 1719 ft LAT/LONG 62° 57' N / 141° 56' W

| | JAN | FEB | MAR | APR | MAY | JUN | JUL | AUG | SEP | OCT | NOV | DEC | YEAR |
|---|---|---|---|---|---|---|---|---|---|---|---|---|---|
| Maximum Temp °F | -10.6 | 1.4 | 24.8 | 42.2 | 56.7 | 66.6 | 69.8 | 65.2 | 52.2 | 29.6 | 5.5 | -6.1 | 33.1 |
| Minimum Temp °F | -27.3 | -21.8 | -6.9 | 15.3 | 32.9 | 44.0 | 48.3 | 42.7 | 30.8 | 13.1 | -11.6 | -22.1 | 11.5 |
| Mean Temp °F | -18.9 | -10.2 | 9.0 | 28.8 | 44.8 | 55.3 | 59.1 | 54.0 | 41.5 | 21.4 | -2.9 | -14.1 | 22.3 |
| Days Max Temp ≥ 90 °F | na | na | na | na | na | na | na | na | na | na | na | na | na |
| Days Max Temp ≤ 32 °F | 31 | 28 | 22 | 5 | 0 | 0 | 0 | 0 | 1 | 19 | 30 | 31 | 167 |
| Days Min Temp ≤ 32 °F | 31 | 28 | 31 | 29 | 16 | 0 | 0 | 2 | 18 | 31 | 30 | 31 | 247 |
| Days Min Temp ≤ 0 °F | 29 | 26 | 20 | 3 | 0 | 0 | 0 | 0 | 0 | 5 | 23 | 29 | 135 |
| Heating Degree Days | 2609 | 2124 | 1733 | 1080 | 619 | 287 | 184 | 338 | 697 | 1346 | 2038 | 2456 | 15511 |
| Cooling Degree Days | 0 | 0 | 0 | 0 | 0 | 4 | 7 | 4 | 0 | 0 | 0 | 0 | 15 |
| Total Precipitation (") | 0.26 | 0.26 | 0.17 | 0.19 | 0.97 | 1.69 | 2.31 | 1.32 | 1.01 | 0.51 | 0.36 | 0.25 | 9.30 |
| Days ≥ 0.1" Precip | 1 | 1 | 1 | 1 | 3 | 5 | 6 | 4 | 3 | 2 | 1 | 1 | 29 |
| Total Snowfall (") | 5.4 | 4.6 | 3.2 | 1.9 | 0.7 | 0.0 | 0.0 | 0.3 | 1.6 | 7.0 | 7.1 | 5.5 | 37.3 |
| Days ≥ 1" Snow Depth | 31 | 28 | 31 | 24 | 1 | 0 | 0 | 0 | 1 | 18 | 30 | 31 | 195 |

## PALMER IAS *Cook Inlet Division* ELEVATION 302 ft LAT/LONG 61° 36' N / 149° 8' W

| | JAN | FEB | MAR | APR | MAY | JUN | JUL | AUG | SEP | OCT | NOV | DEC | YEAR |
|---|---|---|---|---|---|---|---|---|---|---|---|---|---|
| Maximum Temp °F | 21.2 | 25.9 | 35.3 | 45.5 | 57.1 | 64.0 | 66.6 | 64.3 | 56.1 | 41.4 | 27.9 | 23.1 | 44.0 |
| Minimum Temp °F | 5.5 | 8.5 | 17.2 | 27.5 | 37.1 | 45.1 | 48.7 | 46.7 | 39.0 | 26.1 | 12.5 | 7.9 | 26.8 |
| Mean Temp °F | 13.3 | 17.2 | 26.3 | 36.5 | 47.1 | 54.5 | 57.7 | 55.5 | 47.6 | 33.7 | 20.2 | 15.5 | 35.4 |
| Days Max Temp ≥ 90 °F | na | na | na | na | na | na | na | na | na | na | na | na | na |
| Days Max Temp ≤ 32 °F | 23 | 18 | 10 | 2 | 0 | 0 | 0 | 0 | 0 | 5 | 19 | 23 | 100 |
| Days Min Temp ≤ 32 °F | 30 | 27 | 29 | 23 | 6 | 0 | 0 | 0 | 4 | 22 | 28 | 30 | 199 |
| Days Min Temp ≤ 0 °F | 13 | 9 | 3 | 0 | 0 | 0 | 0 | 0 | 0 | 0 | 6 | 10 | 41 |
| Heating Degree Days | 1599 | 1345 | 1193 | 848 | 547 | 308 | 222 | 288 | 517 | 961 | 1338 | 1529 | 10695 |
| Cooling Degree Days | 0 | 0 | 0 | 0 | 0 | 0 | 1 | 0 | 0 | 0 | 0 | 0 | 1 |
| Total Precipitation (") | 0.82 | 0.78 | 0.64 | 0.55 | 0.68 | 1.32 | 2.06 | 2.08 | 2.60 | 1.64 | 1.05 | 1.27 | 15.49 |
| Days ≥ 0.1" Precip | 2 | 2 | 2 | 2 | 2 | 4 | 6 | 6 | 7 | 5 | 4 | 4 | 46 |
| Total Snowfall (") | 8.7 | 10.5 | 8.3 | 3.6 | 0.1 | 0.0 | 0.0 | 0.0 | 0.0 | ***6.8*** | 9.1 | 14.2 | 61.3 |
| Days ≥ 1" Snow Depth | ***23*** | ***22*** | 19 | 5 | 0 | 0 | 0 | 0 | 0 | 8 | 20 | ***26*** | 123 |

## PORT ALSWORTH *Bristol Bay Division* ELEVATION 230 ft LAT/LONG 60° 12' N / 154° 18' W

| | JAN | FEB | MAR | APR | MAY | JUN | JUL | AUG | SEP | OCT | NOV | DEC | YEAR |
|---|---|---|---|---|---|---|---|---|---|---|---|---|---|
| Maximum Temp °F | 22.8 | 25.5 | 34.2 | 43.1 | 55.1 | 63.7 | 67.6 | 65.1 | 56.1 | 41.3 | 31.3 | 24.7 | 44.2 |
| Minimum Temp °F | 4.4 | 4.5 | 12.3 | 23.2 | 33.2 | 41.0 | 45.5 | 44.8 | 37.9 | 26.4 | 16.5 | 8.4 | 24.8 |
| Mean Temp °F | 13.9 | 15.0 | 23.2 | 33.2 | 44.2 | 52.4 | 56.6 | 55.0 | 47.1 | 33.9 | 23.9 | 16.5 | 34.6 |
| Days Max Temp ≥ 90 °F | na | na | na | na | na | na | na | na | na | na | na | na | na |
| Days Max Temp ≤ 32 °F | 18 | 16 | 10 | 3 | 0 | 0 | 0 | 0 | 0 | 5 | 14 | 18 | 84 |
| Days Min Temp ≤ 32 °F | 28 | 25 | 27 | 24 | 14 | 2 | 0 | 1 | 7 | 21 | 26 | 27 | 202 |
| Days Min Temp ≤ 0 °F | 13 | 13 | 8 | 2 | 0 | 0 | 0 | 0 | 0 | 0 | 4 | 12 | 52 |
| Heating Degree Days | 1582 | 1410 | 1290 | 947 | 639 | 372 | 256 | 304 | 531 | 958 | 1226 | 1501 | 11016 |
| Cooling Degree Days | 0 | 0 | 0 | 0 | 0 | 0 | 1 | 1 | 0 | 0 | 0 | 0 | 2 |
| Total Precipitation (") | 0.74 | 0.52 | 0.72 | 0.54 | 0.53 | 1.10 | 1.56 | 2.40 | 2.14 | 1.56 | 1.46 | 1.11 | 14.38 |
| Days ≥ 0.1" Precip | 3 | 2 | 3 | 2 | 2 | 3 | 5 | 6 | 6 | 5 | 4 | 3 | 44 |
| Total Snowfall (") | na | na | ***10.8*** | ***4.0*** | 0.3 | 0.0 | 0.0 | 0.0 | 0.0 | ***3.7*** | na | ***18.8*** | na |
| Days ≥ 1" Snow Depth | 22 | 18 | 19 | 8 | 0 | 0 | 0 | 0 | 0 | 4 | 18 | 23 | 112 |

## PUNTILLA *Cook Inlet Division* ELEVATION 1837 ft LAT/LONG 62° 6' N / 152° 45' W

| | JAN | FEB | MAR | APR | MAY | JUN | JUL | AUG | SEP | OCT | NOV | DEC | YEAR |
|---|---|---|---|---|---|---|---|---|---|---|---|---|---|
| Maximum Temp °F | 12.4 | 17.0 | 27.2 | 37.0 | 49.2 | 60.5 | 63.6 | 59.9 | 49.8 | ***33.7*** | 19.7 | ***15.1*** | 37.1 |
| Minimum Temp °F | -6.0 | -3.0 | 4.0 | 14.5 | 28.5 | 37.6 | 42.9 | 40.0 | 32.6 | ***16.6*** | 0.8 | ***-2.8*** | 17.1 |
| Mean Temp °F | 3.2 | 7.0 | ***15.6*** | 25.8 | 38.9 | 49.0 | 53.3 | 50.0 | 41.2 | ***25.1*** | 10.3 | ***6.0*** | 27.1 |
| Days Max Temp ≥ 90 °F | na | na | na | na | na | na | na | na | na | na | na | na | na |
| Days Max Temp ≤ 32 °F | 29 | 24 | ***20*** | 7 | 0 | 0 | 0 | 0 | 0 | ***13*** | 27 | ***29*** | 149 |
| Days Min Temp ≤ 32 °F | 31 | 28 | 31 | 29 | 23 | 6 | 1 | 4 | 14 | ***28*** | 30 | 31 | 256 |
| Days Min Temp ≤ 0 °F | 20 | 16 | ***13*** | 5 | 0 | 0 | 0 | 0 | 0 | ***4*** | 15 | ***18*** | 91 |
| Heating Degree Days | 1914 | 1636 | 1525 | 1169 | 802 | 472 | 356 | 460 | 706 | ***1230*** | 1637 | ***1825*** | 13732 |
| Cooling Degree Days | 0 | 0 | 0 | 0 | 0 | 0 | 0 | 0 | 0 | 0 | 0 | 0 | 0 |
| Total Precipitation (") | 1.19 | 1.70 | ***1.28*** | ***0.57*** | 0.61 | 1.84 | 2.20 | 2.56 | 3.36 | ***1.82*** | 1.26 | ***1.58*** | 19.97 |
| Days ≥ 0.1" Precip | 4 | 4 | ***3*** | 3 | 2 | 4 | 6 | 6 | 6 | na | ***4*** | ***6*** | na |
| Total Snowfall (") | 16.4 | 12.3 | ***12.3*** | ***5.3*** | 1.5 | 0.0 | 0.0 | 0.0 | 0.6 | ***9.4*** | 15.2 | ***18.3*** | 91.3 |
| Days ≥ 1" Snow Depth | 29 | 27 | ***31*** | 30 | 14 | 0 | 0 | 0 | 0 | ***17*** | ***29*** | ***31*** | 208 |

## SEWARD *South Coast Division* ELEVATION 75 ft LAT/LONG 60° 7 ' N / 149° 27 ' W

| | JAN | FEB | MAR | APR | MAY | JUN | JUL | AUG | SEP | OCT | NOV | DEC | YEAR |
|---|---|---|---|---|---|---|---|---|---|---|---|---|---|
| Maximum Temp °F | 30.0 | 32.1 | 37.5 | 44.1 | 51.4 | 57.6 | 62.0 | 61.4 | 54.8 | 44.1 | 35.9 | 31.9 | 45.2 |
| Minimum Temp °F | 20.7 | 21.4 | 26.3 | 32.0 | 38.9 | 45.2 | 49.9 | 49.3 | 43.6 | 34.3 | 26.4 | 22.5 | 34.2 |
| Mean Temp °F | 25.4 | 26.8 | 31.9 | 38.1 | 45.2 | 51.4 | 56.0 | 55.4 | 49.2 | 39.2 | 31.2 | 27.2 | 39.7 |
| Days Max Temp ≥ 90 °F | na | na | na | na | na | na | na | na | na | na | na | na | na |
| Days Max Temp ≤ 32 °F | 14 | 12 | 5 | 1 | 0 | 0 | 0 | 0 | 0 | 1 | 8 | 13 | 54 |
| Days Min Temp ≤ 32 °F | 25 | 23 | 24 | 14 | 1 | 0 | 0 | 0 | 1 | 11 | 23 | 26 | 148 |
| Days Min Temp ≤ 0 °F | 2 | 1 | 0 | 0 | 0 | 0 | 0 | 0 | 0 | 0 | 0 | 1 | 4 |
| Heating Degree Days | 1221 | 1073 | 1019 | 802 | 607 | 401 | 276 | 295 | 467 | 792 | 1008 | 1164 | 9125 |
| Cooling Degree Days | 0 | 0 | 0 | 0 | 0 | 0 | 3 | 3 | 0 | 0 | 0 | 0 | 6 |
| Total Precipitation (") | 6.61 | 5.40 | 4.15 | 4.14 | 4.43 | 2.43 | 2.11 | 5.60 | 10.17 | 10.23 | 6.79 | 7.39 | 69.45 |
| Days ≥ 0.1" Precip | 10 | 9 | 9 | 8 | 9 | 6 | 6 | 10 | 13 | 13 | 9 | 11 | 113 |
| Total Snowfall (") | 12.4 | 17.6 | 11.2 | 5.1 | 0.3 | 0.0 | 0.0 | 0.0 | 0.0 | 2.2 | 8.6 | 18.7 | 76.1 |
| Days ≥ 1" Snow Depth | 21 | 18 | 17 | 7 | 0 | 0 | 0 | 0 | 0 | 1 | 8 | 19 | 91 |

## SHEMYA USAF BASE *Southwestern Islands Division* ELEVATION 92 ft LAT/LONG 52° 43 ' N / 174° 6 ' E

| | JAN | FEB | MAR | APR | MAY | JUN | JUL | AUG | SEP | OCT | NOV | DEC | YEAR |
|---|---|---|---|---|---|---|---|---|---|---|---|---|---|
| Maximum Temp °F | 33.4 | 33.4 | 34.6 | 37.8 | 41.4 | 45.0 | 49.2 | 51.9 | 50.9 | 45.6 | 39.5 | 35.3 | 41.5 |
| Minimum Temp °F | 28.0 | 28.3 | 28.9 | 32.1 | 35.9 | 40.1 | 44.4 | 47.1 | 45.1 | 39.4 | 33.4 | 29.6 | 36.0 |
| Mean Temp °F | 30.7 | 30.8 | 31.8 | 35.0 | 38.7 | 42.5 | 46.8 | 49.5 | 48.0 | 42.5 | 36.5 | 32.5 | 38.8 |
| Days Max Temp ≥ 90 °F | na | na | na | na | na | na | na | na | na | na | na | na | na |
| Days Max Temp ≤ 32 °F | 11 | 10 | 7 | 1 | 0 | 0 | 0 | 0 | 0 | 0 | 1 | 7 | 37 |
| Days Min Temp ≤ 32 °F | 26 | 23 | 24 | 15 | 3 | 0 | 0 | 0 | 0 | 1 | 11 | 22 | 125 |
| Days Min Temp ≤ 0 °F | 0 | 0 | 0 | 0 | 0 | 0 | 0 | 0 | 0 | 0 | 0 | 0 | 0 |
| Heating Degree Days | 1055 | 959 | 1022 | 893 | 809 | 667 | 556 | 473 | 502 | 690 | 849 | 1002 | 9477 |
| Cooling Degree Days | 0 | 0 | 0 | 0 | 0 | 0 | 0 | 0 | 0 | 0 | 0 | 0 | 0 |
| Total Precipitation (") | 2.60 | 1.96 | 2.03 | 1.77 | 1.66 | 1.78 | 2.81 | 3.66 | 2.95 | 3.62 | 4.05 | 3.13 | 32.02 |
| Days ≥ 0.1" Precip | 8 | 6 | 6 | 5 | 5 | 5 | 7 | 8 | 7 | 9 | 11 | 10 | 87 |
| Total Snowfall (") | 17.2 | 13.1 | 12.2 | 5.2 | 1.1 | 0.0 | 0.0 | 0.0 | 0.0 | 0.5 | 8.3 | 17.7 | 75.3 |
| Days ≥ 1" Snow Depth | 20 | 20 | 15 | 4 | 0 | 0 | 0 | 0 | 0 | 0 | 5 | 14 | 78 |

## SITKA JAPONSKI AP *Southeastern Division* ELEVATION 15 ft LAT/LONG 57° 4 ' N / 135° 21 ' W

| | JAN | FEB | MAR | APR | MAY | JUN | JUL | AUG | SEP | OCT | NOV | DEC | YEAR |
|---|---|---|---|---|---|---|---|---|---|---|---|---|---|
| Maximum Temp °F | 38.1 | 40.6 | 43.6 | 48.1 | 53.2 | 57.6 | 61.0 | 62.1 | 58.0 | 50.5 | 43.3 | 39.8 | 49.7 |
| Minimum Temp °F | 29.9 | 31.2 | 33.3 | 36.4 | 41.5 | 46.9 | 51.2 | 51.9 | 47.9 | 41.6 | 34.9 | 32.1 | 39.9 |
| Mean Temp °F | 34.1 | 36.0 | 38.4 | 42.3 | 47.3 | 52.3 | 56.2 | 57.1 | 52.9 | 46.0 | 39.1 | 35.9 | 44.8 |
| Days Max Temp ≥ 90 °F | na | na | na | na | na | na | na | na | na | na | na | na | na |
| Days Max Temp ≤ 32 °F | 6 | 3 | 1 | 0 | 0 | 0 | 0 | 0 | 0 | 0 | 1 | 4 | 15 |
| Days Min Temp ≤ 32 °F | 16 | 15 | 12 | 5 | 0 | 0 | 0 | 0 | 0 | 2 | 9 | 13 | 72 |
| Days Min Temp ≤ 0 °F | 0 | 0 | 0 | 0 | 0 | 0 | 0 | 0 | 0 | 0 | 0 | 0 | 0 |
| Heating Degree Days | 952 | 814 | 819 | 675 | 541 | 376 | 268 | 239 | 355 | 580 | 769 | 894 | 7282 |
| Cooling Degree Days | 0 | 0 | 0 | 0 | 0 | 0 | 0 | 1 | 0 | 0 | 0 | 0 | 1 |
| Total Precipitation (") | 7.79 | 5.97 | 5.73 | 4.56 | 4.63 | 3.35 | 3.90 | 6.21 | 11.38 | 13.81 | 9.34 | 8.24 | 84.91 |
| Days ≥ 0.1" Precip | 15 | 12 | 13 | 12 | 11 | 9 | 10 | 12 | 16 | 20 | 16 | 16 | 162 |
| Total Snowfall (") | 11.0 | 8.1 | 5.8 | 1.4 | 0.1 | 0.0 | 0.0 | 0.0 | 0.0 | 0.3 | 4.1 | 7.9 | 38.7 |
| Days ≥ 1" Snow Depth | 11 | 9 | 6 | 0 | 0 | 0 | 0 | 0 | 0 | 0 | 4 | 8 | 38 |

## SNOWSHOE LAKE *Copper River Division* ELEVATION 2297 ft LAT/LONG 62° 2 ' N / 146° 40 ' W

| | JAN | FEB | MAR | APR | MAY | JUN | JUL | AUG | SEP | OCT | NOV | DEC | YEAR |
|---|---|---|---|---|---|---|---|---|---|---|---|---|---|
| Maximum Temp °F | 2.1 | 13.7 | 28.0 | 40.2 | 52.5 | 62.6 | 66.8 | 63.4 | 51.7 | 33.5 | 13.9 | 5.1 | 36.1 |
| Minimum Temp °F | -18.8 | -14.0 | -5.4 | 10.7 | 27.2 | 36.3 | 40.3 | 36.2 | 27.4 | 12.7 | -6.1 | -13.7 | 11.1 |
| Mean Temp °F | -8.4 | 0.0 | 11.2 | 25.5 | 39.8 | 49.5 | 53.6 | 49.8 | 39.6 | 23.2 | 3.9 | -4.2 | 23.6 |
| Days Max Temp ≥ 90 °F | na | na | na | na | na | na | na | na | na | na | na | na | na |
| Days Max Temp ≤ 32 °F | 29 | 26 | 18 | 4 | 0 | 0 | 0 | 0 | 1 | 14 | 28 | 29 | 149 |
| Days Min Temp ≤ 32 °F | 30 | 28 | 30 | 30 | 25 | 8 | 3 | 10 | 21 | 30 | 30 | 30 | 275 |
| Days Min Temp ≤ 0 °F | 26 | 22 | 19 | 6 | 0 | 0 | 0 | 0 | 0 | 5 | 21 | 24 | 123 |
| Heating Degree Days | 2277 | 1839 | 1663 | 1178 | 775 | 459 | 346 | 465 | 756 | 1290 | 1829 | 2147 | 15024 |
| Cooling Degree Days | 0 | 0 | 0 | 0 | 0 | 0 | 0 | 0 | 0 | 0 | 0 | 0 | 0 |
| Total Precipitation (") | 0.42 | 0.57 | 0.41 | 0.23 | 0.75 | 2.10 | 2.32 | 1.48 | 1.08 | 0.95 | 0.81 | 0.83 | 11.95 |
| Days ≥ 0.1" Precip | 1 | 2 | 1 | 1 | 2 | 6 | 7 | 5 | 3 | 3 | 3 | 2 | 36 |
| Total Snowfall (") | 6.4 | 7.1 | 5.1 | 3.1 | 1.5 | 0.1 | 0.0 | 0.1 | 1.6 | 10.3 | 9.7 | 10.3 | 55.3 |
| Days ≥ 1" Snow Depth | 31 | 28 | 30 | 29 | 6 | 0 | 0 | 0 | 1 | 19 | 30 | 31 | 205 |

## ST PAUL ISLAND AP *Southwestern Islands Division* ELEVATION 30 ft LAT/LONG 57° 9 ' N / 170° 13 ' W

| | JAN | FEB | MAR | APR | MAY | JUN | JUL | AUG | SEP | OCT | NOV | DEC | YEAR |
|---|---|---|---|---|---|---|---|---|---|---|---|---|---|
| Maximum Temp °F | 30.5 | 27.0 | 28.8 | 32.6 | 39.4 | 46.0 | 50.0 | 51.4 | 49.0 | 42.2 | 37.2 | 33.3 | 38.9 |
| Minimum Temp °F | 22.2 | 18.2 | 19.2 | 23.9 | 31.1 | 37.2 | 42.7 | 44.7 | 40.4 | 34.0 | 29.1 | 24.9 | 30.6 |
| Mean Temp °F | 26.4 | 22.5 | 24.0 | 28.3 | 35.2 | 41.6 | 46.4 | 48.1 | 44.7 | 38.1 | 33.2 | 29.1 | 34.8 |
| Days Max Temp ≥ 90 °F | na | na | na | na | na | na | na | na | na | na | na | na | na |
| Days Max Temp ≤ 32 °F | 15 | 16 | 17 | 12 | 3 | 0 | 0 | 0 | 0 | 1 | 5 | 11 | 80 |
| Days Min Temp ≤ 32 °F | 25 | 25 | 28 | 26 | 18 | 4 | 0 | 0 | 3 | 12 | 19 | 24 | 184 |
| Days Min Temp ≤ 0 °F | 1 | 3 | 2 | 0 | 0 | 0 | 0 | 0 | 0 | 0 | 0 | 0 | 6 |
| Heating Degree Days | 1190 | 1196 | 1264 | 1095 | 915 | 694 | 569 | 518 | 602 | 826 | 946 | 1105 | 10920 |
| Cooling Degree Days | 0 | 0 | 0 | 0 | 0 | 0 | 0 | 0 | 0 | 0 | 0 | 0 | 0 |
| Total Precipitation (") | 1.68 | 1.32 | 1.19 | 1.17 | 1.21 | 1.26 | 1.89 | 2.91 | 2.72 | 2.71 | 2.87 | 2.21 | 23.14 |
| Days ≥ 0.1" Precip | 6 | 5 | 4 | 4 | 4 | 4 | 6 | 8 | 8 | 8 | 9 | 8 | 74 |
| Total Snowfall (") | 10.0 | 9.4 | 10.2 | 6.3 | 2.3 | 0.1 | 0.0 | 0.0 | 0.1 | 2.8 | 7.4 | 9.3 | 57.9 |
| Days ≥ 1" Snow Depth | 19 | 21 | 25 | 19 | 7 | 1 | 0 | 0 | 0 | 2 | 9 | 18 | 121 |

## TALKEETNA STATE AP *Cook Inlet Division* ELEVATION 351 ft LAT/LONG 62° 18 ' N / 150° 6 ' W

| | JAN | FEB | MAR | APR | MAY | JUN | JUL | AUG | SEP | OCT | NOV | DEC | YEAR |
|---|---|---|---|---|---|---|---|---|---|---|---|---|---|
| Maximum Temp °F | 19.0 | 25.2 | 34.0 | 44.0 | 56.5 | 65.0 | 67.8 | 64.5 | 55.1 | 39.1 | 25.4 | 20.4 | 43.0 |
| Minimum Temp °F | 0.8 | 4.2 | 11.3 | 22.9 | 34.4 | 44.7 | 49.2 | 46.1 | 36.5 | 23.3 | 8.7 | 3.6 | 23.8 |
| Mean Temp °F | 9.9 | 14.7 | 22.7 | 33.5 | 45.5 | 54.8 | 58.5 | 55.3 | 45.9 | 31.3 | 17.1 | 12.1 | 33.4 |
| Days Max Temp ≥ 90 °F | na | na | na | na | na | na | na | na | na | na | na | na | na |
| Days Max Temp ≤ 32 °F | 26 | 20 | 12 | 2 | 0 | 0 | 0 | 0 | 0 | 6 | 22 | 26 | 114 |
| Days Min Temp ≤ 32 °F | 31 | 28 | 30 | 28 | 13 | 0 | 0 | 1 | 9 | 25 | 29 | 31 | 225 |
| Days Min Temp ≤ 0 °F | 15 | 12 | 8 | 1 | 0 | 0 | 0 | 0 | 0 | 1 | 9 | 13 | 59 |
| Heating Degree Days | 1706 | 1415 | 1305 | 939 | 599 | 299 | 198 | 294 | 567 | 1039 | 1432 | 1638 | 11431 |
| Cooling Degree Days | 0 | 0 | 0 | 0 | 0 | 2 | 4 | 1 | 0 | 0 | 0 | 0 | 7 |
| Total Precipitation (") | 1.35 | 1.31 | 1.36 | 1.42 | 1.66 | 2.44 | 3.48 | 4.54 | 4.32 | 2.99 | 1.93 | 2.04 | 28.84 |
| Days ≥ 0.1" Precip | 4 | 4 | 4 | 4 | 6 | 6 | 9 | 10 | 9 | 8 | 5 | 6 | 75 |
| Total Snowfall (") | 20.5 | 18.0 | 18.9 | 10.3 | 0.8 | 0.0 | 0.0 | 0.0 | 0.4 | 12.6 | 22.5 | 27.8 | 131.8 |
| Days ≥ 1" Snow Depth | 31 | 28 | 31 | 28 | 5 | 0 | 0 | 0 | 0 | 12 | 28 | 31 | 194 |

## TANANA CALHOUN ARPT *Interior Basin Division* ELEVATION 239 ft LAT/LONG 65° 10 ' N / 152° 6 ' W

| | JAN | FEB | MAR | APR | MAY | JUN | JUL | AUG | SEP | OCT | NOV | DEC | YEAR |
|---|---|---|---|---|---|---|---|---|---|---|---|---|---|
| Maximum Temp °F | -2.8 | 3.0 | 20.1 | 37.9 | 57.4 | 69.6 | 71.4 | 64.9 | 51.4 | 28.9 | 8.6 | 0.6 | 34.3 |
| Minimum Temp °F | -18.1 | -16.3 | -3.6 | 14.7 | 33.9 | 45.7 | 49.0 | 44.2 | 33.4 | 15.3 | -5.5 | -14.4 | 14.9 |
| Mean Temp °F | -10.5 | -6.7 | 8.3 | 26.3 | 45.7 | 57.7 | 60.2 | 54.6 | 42.4 | 22.1 | 1.6 | -6.9 | 24.6 |
| Days Max Temp ≥ 90 °F | na | na | na | na | na | na | na | na | na | na | na | na | na |
| Days Max Temp ≤ 32 °F | 31 | 28 | 26 | 9 | 0 | 0 | 0 | 0 | 1 | 19 | 29 | 31 | 174 |
| Days Min Temp ≤ 32 °F | 31 | 28 | 31 | 28 | 13 | 0 | 0 | 2 | 13 | 29 | 30 | 31 | 236 |
| Days Min Temp ≤ 0 °F | 25 | 22 | 18 | 6 | 0 | 0 | 0 | 0 | 0 | 5 | 19 | 23 | 118 |
| Heating Degree Days | 2344 | 2025 | 1756 | 1154 | 592 | 226 | 159 | 319 | 671 | 1323 | 1901 | 2231 | 14701 |
| Cooling Degree Days | 0 | 0 | 0 | 0 | 0 | 16 | 22 | 4 | 0 | 0 | 0 | 0 | 42 |
| Total Precipitation (") | 0.53 | 0.47 | 0.58 | 0.38 | 0.53 | 1.38 | 2.00 | 2.47 | 1.56 | 0.89 | 0.79 | 0.84 | 12.42 |
| Days ≥ 0.1" Precip | 2 | 2 | 2 | 1 | 2 | 4 | 6 | 7 | 5 | 3 | 2 | 3 | 39 |
| Total Snowfall (") | 5.8 | 5.3 | 6.3 | 2.8 | 0.4 | 0.0 | 0.0 | 0.0 | 0.7 | 6.3 | 8.8 | 9.9 | 46.3 |
| Days ≥ 1" Snow Depth | 31 | 28 | 31 | 27 | 4 | 0 | 0 | 0 | 0 | 19 | 30 | 31 | 201 |

## TOK *Interior Basin Division* ELEVATION 1631 ft LAT/LONG 63° 20 ' N / 143° 2 ' W

| | JAN | FEB | MAR | APR | MAY | JUN | JUL | AUG | SEP | OCT | NOV | DEC | YEAR |
|---|---|---|---|---|---|---|---|---|---|---|---|---|---|
| Maximum Temp °F | -7.1 | 4.9 | 27.7 | 43.7 | 59.9 | ***70.2*** | ***73.1*** | 68.5 | 54.2 | 31.3 | 8.2 | -2.5 | 36.0 |
| Minimum Temp °F | -25.7 | -20.1 | -3.7 | 15.7 | 28.9 | ***39.3*** | ***43.5*** | 37.8 | 28.8 | 12.5 | -11.2 | -20.8 | 10.4 |
| Mean Temp °F | -16.4 | -7.6 | 12.0 | 29.7 | 44.5 | ***54.8*** | ***58.3*** | 53.2 | 41.5 | 22.0 | -1.5 | -11.7 | 23.2 |
| Days Max Temp ≥ 90 °F | na | na | na | na | na | na | na | na | na | na | na | na | na |
| Days Max Temp ≤ 32 °F | 31 | 28 | 19 | ***3*** | 0 | ***0*** | ***0*** | 0 | 0 | ***16*** | 30 | 31 | 158 |
| Days Min Temp ≤ 32 °F | 31 | 28 | 30 | 29 | ***22*** | na | ***2*** | ***8*** | ***20*** | 30 | 30 | 31 | na |
| Days Min Temp ≤ 0 °F | 28 | 24 | 18 | ***3*** | 0 | ***0*** | ***0*** | 0 | 0 | 5 | 25 | ***28*** | 131 |
| Heating Degree Days | 2528 | 2053 | 1638 | 1054 | 630 | ***304*** | ***209*** | 364 | 696 | 1329 | 1995 | 2382 | 15182 |
| Cooling Degree Days | 0 | 0 | 0 | 0 | 0 | ***4*** | 11 | 8 | 0 | 0 | 0 | 0 | 23 |
| Total Precipitation (") | ***0.37*** | 0.19 | ***0.12*** | ***0.15*** | ***0.50*** | na | na | na | ***0.68*** | ***0.58*** | 0.54 | ***0.22*** | na |
| Days ≥ 0.1" Precip | 1 | ***1*** | ***0*** | ***0*** | ***1*** | na | na | na | ***2*** | ***2*** | ***2*** | ***1*** | na |
| Total Snowfall (") | 4.9 | ***2.7*** | 2.1 | ***2.0*** | ***0.6*** | ***0.0*** | ***0.0*** | 0.4 | ***1.9*** | ***7.0*** | 6.4 | 3.9 | 31.9 |
| Days ≥ 1" Snow Depth | 31 | ***28*** | 31 | ***24*** | ***3*** | ***0*** | ***0*** | 0 | ***2*** | 19 | 30 | 31 | 199 |

### TONSINA *Copper River Division* ELEVATION 1503 ft LAT/LONG 61° 39 ' N / 145° 10 ' W

| | JAN | FEB | MAR | APR | MAY | JUN | JUL | AUG | SEP | OCT | NOV | DEC | YEAR |
|---|---|---|---|---|---|---|---|---|---|---|---|---|---|
| Maximum Temp °F | 4.3 | 14.8 | 30.6 | 43.4 | 55.7 | 65.7 | 69.5 | 65.6 | 54.4 | 36.3 | 15.5 | 6.4 | 38.5 |
| Minimum Temp °F | -14.4 | -8.5 | 1.8 | 17.7 | 30.0 | 38.5 | 42.8 | 38.7 | 29.6 | 17.8 | -2.4 | -10.9 | 15.1 |
| Mean Temp °F | -5.0 | 3.1 | 16.1 | 30.6 | 42.9 | 52.1 | 56.2 | 52.2 | 42.0 | 27.1 | 6.6 | -2.3 | 26.8 |
| Days Max Temp ≥ 90 °F | na | na | na | na | na | na | na | na | na | na | na | na | na |
| Days Max Temp ≤ 32 °F | 29 | 23 | 15 | 2 | 0 | 0 | 0 | 0 | 0 | 10 | 27 | 29 | 135 |
| Days Min Temp ≤ 32 °F | 31 | 28 | 31 | 29 | 22 | 5 | 1 | 7 | 20 | 29 | 30 | 31 | 264 |
| Days Min Temp ≤ 0 °F | 24 | 19 | 14 | 2 | 0 | 0 | 0 | 0 | 0 | 3 | 17 | 23 | 102 |
| Heating Degree Days | 2172 | 1748 | 1510 | 1026 | 678 | 379 | 269 | 391 | 682 | 1167 | 1751 | 2086 | 13859 |
| Cooling Degree Days | 0 | 0 | 0 | 0 | 0 | 0 | 1 | 0 | 0 | 0 | 0 | 0 | 1 |
| Total Precipitation (") | 0.80 | 0.94 | 0.43 | 0.32 | 0.52 | 1.35 | 1.73 | 1.22 | 1.27 | 1.39 | 1.41 | 1.16 | 12.54 |
| Days ≥ 0.1" Precip | 3 | 3 | 2 | 1 | 2 | 4 | 5 | 4 | 4 | 4 | 4 | 4 | 40 |
| Total Snowfall (") | 9.0 | 8.7 | 5.3 | *3.2* | 0.4 | 0.0 | 0.0 | 0.0 | 0.8 | 9.8 | 14.3 | 12.7 | 64.2 |
| Days ≥ 1" Snow Depth | 31 | 28 | 31 | 28 | 4 | 0 | 0 | 0 | 1 | 16 | 29 | 31 | 199 |

### VALDEZ WSO *South Coast Division* ELEVATION 87 ft LAT/LONG 61° 8 ' N / 146° 21 ' W

| | JAN | FEB | MAR | APR | MAY | JUN | JUL | AUG | SEP | OCT | NOV | DEC | YEAR |
|---|---|---|---|---|---|---|---|---|---|---|---|---|---|
| Maximum Temp °F | *27.8* | *30.0* | *36.8* | 44.9 | *52.7* | *59.5* | *62.5* | 60.7 | *53.5* | *43.2* | *32.5* | *28.7* | 44.4 |
| Minimum Temp °F | *17.6* | *18.6* | *24.1* | 30.4 | *38.4* | *45.0* | 48.0 | 46.6 | *40.7* | *33.1* | *22.9* | *19.2* | 32.1 |
| Mean Temp °F | *22.8* | *24.3* | *30.6* | 37.7 | *45.7* | *52.3* | 55.3 | 53.7 | *47.1* | *38.2* | *27.7* | *24.0* | 38.3 |
| Days Max Temp ≥ 90 °F | na | na | na | na | na | na | na | na | na | na | na | na | na |
| Days Max Temp ≤ 32 °F | *19* | *14* | *6* | 1 | *0* | *0* | 0 | 0 | *0* | *1* | *13* | *19* | 73 |
| Days Min Temp ≤ 32 °F | *30* | *27* | *29* | 18 | *2* | *0* | 0 | 0 | *1* | *13* | *28* | *30* | 178 |
| Days Min Temp ≤ 0 °F | *3* | *1* | *0* | 0 | *0* | *0* | 0 | 0 | *0* | *0* | *0* | *1* | 5 |
| Heating Degree Days | *1303* | *1143* | *1063* | 813 | *593* | *375* | 298 | 345 | *529* | *824* | *1111* | *1265* | 9662 |
| Cooling Degree Days | 0 | 0 | 0 | 0 | 0 | 0 | 2 | 0 | 0 | 0 | 0 | 0 | 2 |
| Total Precipitation (") | *5.96* | *5.33* | *4.99* | *3.15* | *3.04* | *2.76* | 3.59 | 6.37 | *9.33* | *8.29* | 6.02 | *7.87* | 66.70 |
| Days ≥ 0.1" Precip | *11* | *9* | *10* | *8* | *9* | *7* | 9 | 11 | *15* | *13* | *10* | *13* | 125 |
| Total Snowfall (") | *57.1* | *50.4* | *54.9* | 18.9 | *0.6* | 0.0 | 0.0 | 0.0 | *0.2* | *9.5* | *42.3* | *76.0* | 309.9 |
| Days ≥ 1" Snow Depth | *31* | *28* | *31* | 28 | *5* | *0* | 0 | 0 | *0* | *4* | *25* | *30* | 182 |

### WALES *West Central Division* ELEVATION 16 ft LAT/LONG 65° 37 ' N / 168° 3 ' W

| | JAN | FEB | MAR | APR | MAY | JUN | JUL | AUG | SEP | OCT | NOV | DEC | YEAR |
|---|---|---|---|---|---|---|---|---|---|---|---|---|---|
| Maximum Temp °F | 7.9 | 3.1 | 5.5 | 15.6 | 31.8 | 43.6 | 51.8 | 51.0 | 44.0 | 32.7 | 21.8 | 10.4 | 26.6 |
| Minimum Temp °F | -6.2 | -10.5 | -8.3 | 2.5 | 22.2 | 33.5 | 42.6 | 42.9 | 36.5 | 24.6 | 10.5 | -2.4 | 15.7 |
| Mean Temp °F | 0.9 | -3.8 | -1.4 | 9.1 | 27.0 | 38.6 | 47.2 | 46.9 | 40.3 | 28.7 | 16.2 | 4.0 | 21.1 |
| Days Max Temp ≥ 90 °F | na | na | na | na | na | na | na | na | na | na | na | na | na |
| Days Max Temp ≤ 32 °F | 29 | 26 | 29 | 26 | 15 | 1 | 0 | 0 | 0 | 14 | 25 | 28 | 193 |
| Days Min Temp ≤ 32 °F | 31 | 28 | 31 | 30 | 29 | 14 | 0 | 0 | 7 | 27 | 29 | 31 | 257 |
| Days Min Temp ≤ 0 °F | 21 | 22 | 24 | 14 | 1 | 0 | 0 | 0 | 0 | 0 | 6 | *19* | 107 |
| Heating Degree Days | 1989 | 1944 | 2061 | 1677 | 1169 | 787 | 544 | 552 | 734 | 1119 | 1459 | 1891 | 15926 |
| Cooling Degree Days | 0 | 0 | 0 | 0 | 0 | 0 | 0 | 0 | 0 | 0 | 0 | 0 | 0 |
| Total Precipitation (") | 0.38 | 0.39 | 0.54 | 0.30 | 0.61 | 0.74 | 1.54 | 2.48 | 1.90 | 1.43 | 0.78 | 0.54 | 11.63 |
| Days ≥ 0.1" Precip | 1 | *2* | 1 | 1 | 2 | 3 | 5 | 8 | 6 | 5 | 3 | 2 | 39 |
| Total Snowfall (") | 3.9 | 4.0 | 5.3 | 2.8 | 2.6 | 0.2 | 0.2 | 0.0 | 1.5 | *7.3* | *7.6* | *5.2* | 40.6 |
| Days ≥ 1" Snow Depth | 30 | 27 | 31 | 30 | 29 | 5 | 0 | 0 | 1 | 11 | 25 | 29 | 218 |

### WASILLA 3 S *Cook Inlet Division* ELEVATION 331 ft LAT/LONG 61° 35 ' N / 149° 27 ' W

| | JAN | FEB | MAR | APR | MAY | JUN | JUL | AUG | SEP | OCT | NOV | DEC | YEAR |
|---|---|---|---|---|---|---|---|---|---|---|---|---|---|
| Maximum Temp °F | 22.3 | 27.6 | 37.0 | 47.8 | 59.7 | 66.5 | 69.6 | 67.2 | 58.5 | 42.7 | 28.8 | 24.1 | 46.0 |
| Minimum Temp °F | 6.3 | 10.3 | 18.5 | 27.4 | 36.0 | 43.9 | 48.5 | 46.1 | 38.7 | 26.8 | 13.3 | 8.5 | 27.0 |
| Mean Temp °F | 14.3 | 19.0 | 27.8 | 37.6 | 47.9 | 55.2 | 59.1 | 56.7 | 48.6 | 34.7 | 21.1 | 16.3 | 36.5 |
| Days Max Temp ≥ 90 °F | na | na | na | na | na | na | na | na | na | na | na | na | na |
| Days Max Temp ≤ 32 °F | 22 | 17 | 8 | 1 | 0 | 0 | 0 | 0 | 0 | 3 | 18 | 21 | 90 |
| Days Min Temp ≤ 32 °F | 29 | 26 | 28 | 23 | 9 | 0 | 0 | 0 | 7 | 21 | 28 | 30 | 201 |
| Days Min Temp ≤ 0 °F | 11 | 8 | 3 | 0 | 0 | 0 | 0 | 0 | 0 | 0 | 6 | 9 | 37 |
| Heating Degree Days | 1567 | 1295 | 1146 | 815 | 524 | 288 | 180 | 251 | 486 | 931 | 1311 | 1506 | 10300 |
| Cooling Degree Days | 0 | 0 | 0 | 0 | 0 | 0 | 2 | 0 | 0 | 0 | 0 | 0 | 2 |
| Total Precipitation (") | 0.66 | 0.82 | 0.62 | 0.50 | 0.81 | 1.56 | 2.30 | 2.51 | 2.87 | 1.99 | 1.24 | 1.16 | 17.04 |
| Days ≥ 0.1" Precip | 3 | 3 | 2 | 2 | 3 | 5 | 7 | 7 | 7 | 5 | 4 | 4 | 52 |
| Total Snowfall (") | 8.5 | 8.2 | 6.2 | *2.6* | 0.2 | 0.0 | 0.0 | 0.0 | 0.0 | 6.1 | 9.4 | 14.1 | 55.3 |
| Days ≥ 1" Snow Depth | na | na | *12* | *3* | 0 | 0 | 0 | 0 | 0 | *4* | na | na | na |

## WRANGELL AIRPORT *Southeastern Division* ELEVATION 44 ft LAT/LONG 56° 29' N / 132° 22' W

| | JAN | FEB | MAR | APR | MAY | JUN | JUL | AUG | SEP | OCT | NOV | DEC | YEAR |
|---|---|---|---|---|---|---|---|---|---|---|---|---|---|
| Maximum Temp °F | 33.8 | 37.8 | 42.9 | 49.3 | 56.1 | 62.1 | 64.5 | 64.0 | 57.8 | 49.4 | 40.7 | 36.1 | 49.5 |
| Minimum Temp °F | 23.8 | 27.1 | 30.9 | 34.8 | 40.1 | 45.2 | 48.7 | 48.9 | 44.9 | 38.4 | 31.2 | 26.8 | 36.7 |
| Mean Temp °F | 28.8 | 32.5 | 36.9 | 42.0 | 48.2 | 53.7 | 56.6 | 56.4 | 51.4 | 43.9 | 36.0 | 31.5 | 43.2 |
| Days Max Temp ≥ 90 °F | na | na | na | na | na | na | na | na | na | na | na | na | na |
| Days Max Temp ≤ 32 °F | 11 | 6 | 1 | 0 | 0 | 0 | 0 | 0 | 0 | 0 | 4 | 7 | 29 |
| Days Min Temp ≤ 32 °F | 23 | 19 | 16 | 8 | 1 | 0 | 0 | 0 | 0 | 3 | 14 | *20* | 104 |
| Days Min Temp ≤ 0 °F | 1 | 0 | 0 | 0 | 0 | 0 | 0 | 0 | 0 | 0 | 0 | 0 | 1 |
| Heating Degree Days | 1115 | 912 | 864 | 683 | 516 | 335 | 254 | 258 | 403 | 646 | 863 | 1032 | 7881 |
| Cooling Degree Days | 0 | 0 | 0 | 0 | 0 | 2 | 1 | 1 | 0 | 0 | 0 | 0 | 4 |
| Total Precipitation (") | 7.18 | 5.55 | 5.01 | 4.54 | 4.71 | 3.90 | 4.47 | 5.51 | 9.47 | 12.61 | 8.61 | 7.69 | 79.25 |
| Days ≥ 0.1" Precip | 13 | 12 | 12 | 12 | 13 | 10 | 11 | 11 | 15 | *19* | 16 | *15* | 159 |
| Total Snowfall (") | 16.1 | 8.2 | 3.8 | 0.7 | 0.0 | 0.0 | 0.0 | 0.0 | 0.0 | 0.0 | 3.9 | 10.9 | 43.6 |
| Days ≥ 1" Snow Depth | 16 | 13 | 5 | 0 | 0 | 0 | 0 | 0 | 0 | 0 | 4 | 11 | 49 |

## YAKUTAT STATE AP *South Coast Division* ELEVATION 30 ft LAT/LONG 59° 31' N / 139° 40' W

| | JAN | FEB | MAR | APR | MAY | JUN | JUL | AUG | SEP | OCT | NOV | DEC | YEAR |
|---|---|---|---|---|---|---|---|---|---|---|---|---|---|
| Maximum Temp °F | 31.2 | 34.8 | 38.7 | 44.1 | 50.1 | 56.0 | 59.7 | 59.8 | 55.2 | 46.9 | 37.7 | 33.5 | 45.6 |
| Minimum Temp °F | 18.1 | 20.4 | 24.1 | 29.4 | 36.8 | 43.4 | 47.9 | 46.5 | 40.8 | 34.6 | 25.5 | 21.7 | 32.4 |
| Mean Temp °F | 24.6 | 27.7 | 31.4 | 36.7 | 43.5 | 49.7 | 53.8 | 53.2 | 48.0 | 40.8 | 31.6 | 27.6 | 39.1 |
| Days Max Temp ≥ 90 °F | na | na | na | na | na | na | na | na | na | na | na | na | na |
| Days Max Temp ≤ 32 °F | 15 | 9 | 3 | 0 | 0 | 0 | 0 | 0 | 0 | 0 | 6 | 12 | 45 |
| Days Min Temp ≤ 32 °F | 25 | 23 | 24 | 21 | 7 | 0 | 0 | 0 | 5 | 11 | 22 | 24 | 162 |
| Days Min Temp ≤ 0 °F | 5 | 3 | 1 | 0 | 0 | 0 | 0 | 0 | 0 | 0 | 1 | 2 | 12 |
| Heating Degree Days | 1246 | 1048 | 1034 | 841 | 661 | 452 | 340 | 360 | 502 | 743 | 994 | 1154 | 9375 |
| Cooling Degree Days | 0 | 0 | 0 | 0 | 0 | 0 | 0 | 0 | 0 | 0 | 0 | 0 | 0 |
| Total Precipitation (") | 12.58 | 10.31 | 11.97 | 9.90 | 9.93 | 6.85 | 7.76 | 12.28 | 20.02 | 23.65 | 15.03 | 15.15 | 155.43 |
| Days ≥ 0.1" Precip | 16 | 14 | 16 | 14 | 15 | 11 | 12 | 14 | 18 | 22 | 18 | 18 | 188 |
| Total Snowfall (") | 40.0 | 36.7 | 34.7 | 17.6 | 1.5 | 0.0 | 0.0 | 0.0 | 0.0 | 5.9 | 22.0 | 35.0 | 193.4 |
| Days ≥ 1" Snow Depth | 25 | 23 | 24 | 16 | 2 | 0 | 0 | 0 | 0 | 2 | 13 | 24 | 129 |

## JANUARY MINIMUM TEMPERATURE °F

### LOWEST

| Rank | Station | Value |
|---|---|---|
| 1 | Northway | -27.3 |
| 2 | Tok | -25.7 |
| 3 | Eagle | -21.0 |
| 4 | Bettles | -20.1 |
| 5 | Barrow | -19.7 |
| 6 | Snowshoe Lake | -18.8 |
| 7 | Eielson | -18.7 |
| 8 | Clearwater | -18.5 |
| 9 | Fairbanks | -18.2 |
| 10 | Tanana | -18.1 |
| 11 | McGrath | -17.3 |
| 12 | Glennallen | -17.0 |
| 13 | Tonsina | -14.4 |
| 14 | Gulkana | -13.5 |
| 15 | College | -12.9 |
| 16 | Big Delta | -11.1 |
| 17 | McKinley | -8.3 |
| 18 | Kotzebue | -8.2 |
| 19 | Wales | -6.2 |
| 20 | Puntilla | -6.0 |
| 21 | Nome | -1.8 |
| 22 | Bethel | 0.2 |
| 23 | Eklutna | 0.6 |
| 24 | Talkeetna | 0.8 |
| 25 | Kenai | 4.3 |

### HIGHEST

| Rank | Station | Value |
|---|---|---|
| 1 | Sitka | 29.9 |
| 2 | Annette Island | 29.7 |
| 3 | Little Port Walter | 28.3 |
| 4 | Shemya | 28.0 |
| 5 | Ketchikan | 27.5 |
| 6 | Beaver Falls | 27.1 |
| 7 | Cold Bay | 24.1 |
| 8 | Kitoi Bay | 23.8 |
| | Wrangell | 23.8 |
| 10 | St. Paul Island | 22.2 |
| 11 | Auke Bay | 20.9 |
| 12 | Seward | 20.7 |
| 13 | Juneau | 18.9 |
| 14 | Yakutat | 18.1 |
| 15 | Valdez | 17.6 |
| 16 | Homer | 17.1 |
| 17 | Cordova | 15.6 |
| 18 | Cooper Lake | 14.0 |
| 19 | Iliamna | 11.3 |
| 20 | Dillingham | 10.6 |
| 21 | Anchorage-Intl | 8.9 |
| 22 | Intricate Bay | 8.2 |
| 23 | King Salmon | 7.7 |
| 24 | Wasilla | 6.3 |
| 25 | Kasilof | 6.2 |

## JULY MAXIMUM TEMPERATURE °F

### HIGHEST

| Rank | Station | Value |
|---|---|---|
| 1 | Eagle | 73.2 |
| 2 | Tok | 73.1 |
| 3 | Fairbanks | 72.9 |
| 4 | College | 72.7 |
| 5 | Clearwater | 72.5 |
| 6 | Tanana | 71.4 |
| 7 | Eielson | 70.9 |
| 8 | Glennallen | 70.7 |
| 9 | Bettles | 70.2 |
| | Big Delta | 70.2 |
| 11 | Northway | 69.8 |
| 12 | Wasilla | 69.6 |
| 13 | Tonsina | 69.5 |
| 14 | Eklutna | 69.1 |
| 15 | McGrath | 68.6 |
| 16 | Gulkana | 68.5 |
| 17 | Talkeetna | 67.8 |
| 18 | Port Alsworth | 67.6 |
| 19 | Matanuska | 67.3 |
| 20 | McKinley | 66.9 |
| 21 | Snowshoe Lake | 66.8 |
| 22 | Palmer | 66.6 |
| 23 | Cooper Lake | 65.7 |
| 24 | Auke Bay | 65.4 |
| 25 | Anchorage-Intl | 65.3 |

### LOWEST

| Rank | Station | Value |
|---|---|---|
| 1 | Barrow | 45.5 |
| 2 | Shemya | 49.2 |
| 3 | St. Paul Island | 50.0 |
| 4 | Wales | 51.8 |
| 5 | Cold Bay | 55.2 |
| 6 | Nome | 58.3 |
| 7 | Kotzebue | 59.7 |
| | Yakutat | 59.7 |
| 9 | Kitoi Bay | 60.5 |
| 10 | Homer | 60.6 |
| 11 | Sitka | 61.0 |
| 12 | Cordova | 61.7 |
| 13 | Dillingham | 61.8 |
| 14 | Kenai | 61.9 |
| | Little Port Walter | 61.9 |
| 16 | Seward | 62.0 |
| 17 | Bethel | 62.5 |
| | Valdez | 62.5 |
| 19 | Kasilof | 62.7 |
| 20 | Iliamna | 63.0 |
| 21 | King Salmon | 63.2 |
| 22 | Puntilla | 63.6 |
| 23 | Juneau | 64.0 |
| 24 | Beaver Falls | 64.3 |
| 25 | Annette Island | 64.5 |

## ANNUAL PRECIPITATION (")

### HIGHEST

| Rank | Station | Value |
|---|---|---|
| 1 | Little Port Walter | 225.62 |
| 2 | Yakutat | 155.43 |
| 3 | Beaver Falls | 148.39 |
| 4 | Ketchikan | 144.75 |
| 5 | Annette Island | 102.89 |
| 6 | Cordova | 99.56 |
| 7 | Sitka | 84.91 |
| 8 | Wrangell | 79.25 |
| 9 | Seward | 69.45 |
| 10 | Valdez | 66.70 |
| 11 | Kitoi Bay | 65.50 |
| 12 | Auke Bay | 63.35 |
| 13 | Juneau | 56.08 |
| 14 | Cold Bay | 37.82 |
| 15 | Intricate Bay | 35.10 |
| 16 | Cooper Lake | 32.11 |
| 17 | Shemya | 32.02 |
| 18 | Talkeetna | 28.84 |
| 19 | Dillingham | 26.53 |
| 20 | Iliamna | 26.49 |
| 21 | Homer | 25.47 |
| 22 | St. Paul Island | 23.14 |
| 23 | Puntilla | 19.97 |
| 24 | King Salmon | 19.75 |
| 25 | Kenai | 18.57 |

### LOWEST

| Rank | Station | Value |
|---|---|---|
| 1 | Barrow | 4.14 |
| 2 | Kotzebue | 9.24 |
| 3 | Northway | 9.30 |
| 4 | Fairbanks | 10.61 |
| 5 | Gulkana | 11.04 |
| 6 | Glennallen | 11.25 |
| 7 | Wales | 11.63 |
| 8 | Big Delta | 11.76 |
| 9 | College | 11.80 |
| | Eielson | 11.80 |
| 11 | Eagle | 11.82 |
| 12 | Snowshoe Lake | 11.95 |
| 13 | Tanana | 12.42 |
| 14 | Tonsina | 12.54 |
| 15 | Bettles | 13.93 |
| 16 | Port Alsworth | 14.38 |
| 17 | Anchrge-Elmndrf | 14.96 |
| 18 | Matanuska | 15.07 |
| 19 | Nome | 15.35 |
| 20 | Palmer | 15.49 |
| 21 | Bethel | 15.51 |
| 22 | Anchorage-Intl | 15.70 |
| 23 | Clearwater | 15.78 |
| 24 | McKinley | 15.87 |
| 25 | McGrath | 16.91 |

## ANNUAL SNOWFALL (")

### HIGHEST

| Rank | Station | Value |
|---|---|---|
| 1 | Valdez | 309.9 |
| 2 | Yakutat | 193.4 |
| 3 | Talkeetna | 131.8 |
| 4 | Cordova | 120.5 |
| 5 | Little Port Walter | 112.4 |
| 6 | Juneau | 103.5 |
| 7 | McGrath | 101.2 |
| 8 | Auke Bay | 96.3 |
| 9 | Puntilla | 91.3 |
| 10 | Bettles | 90.5 |
| | McKinley | 90.5 |
| 12 | Cooper Lake | 86.9 |
| 13 | Eielson | 81.2 |
| 14 | Seward | 76.1 |
| 15 | Anchrge-Elmndrf | 75.8 |
| 16 | Fairbanks | 75.4 |
| 17 | Shemya | 75.3 |
| 18 | Beaver Falls | 74.8 |
| 19 | Intricate Bay | 71.9 |
| 20 | College | 70.9 |
| 21 | Cold Bay | 69.1 |
| 22 | Anchorage-Intl | 69.0 |
| 23 | Kitoi Bay | 67.5 |
| 24 | Tonsina | 64.2 |
| 25 | Eagle | 62.4 |

### LOWEST

| Rank | Station | Value |
|---|---|---|
| 1 | Barrow | 28.2 |
| 2 | Tok | 31.9 |
| 3 | Northway | 37.3 |
| 4 | Sitka | 38.7 |
| 5 | Wales | 40.6 |
| 6 | Wrangell | 43.6 |
| 7 | Tanana | 46.3 |
| 8 | King Salmon | 48.3 |
| 9 | Big Delta | 48.4 |
| 10 | Annette Island | 48.7 |
| 11 | Kasilof | 49.5 |
| 12 | Bethel | 49.9 |
| 13 | Kotzebue | 50.0 |
| 14 | Matanuska | 50.3 |
| 15 | Gulkana | 52.5 |
| 16 | Snowshoe Lake | 55.3 |
| | Wasilla | 55.3 |
| 18 | Clearwater | 57.8 |
| 19 | St. Paul Island | 57.9 |
| 20 | Kenai | 58.8 |
| 21 | Nome | 58.9 |
| 22 | Iliamna | 59.6 |
| 23 | Homer | 60.3 |
| 24 | Eklutna | 60.6 |
| 25 | Palmer | 61.3 |

# ARIZONA

PHYSICAL FEATURES. Arizona covers 113,909 square miles, with about 350 square miles of water surface. The State can be divided into three main topographical areas: (1) the northeastern portion is a high plateau averaging between 5,000 and 7,000 feet in elevation; (2) running diagonally from the southeastern to the northwestern corners of the State is a mountainous region with maximum elevations between about 9,000 and 12,000 feet above mean sea level; (3) the southwestern third of the State is made up of low mountain ranges and desert valleys. From the White Mountain area across the Mogollon Rim to the San Francisco Peaks is an unbroken stand of Ponderosa pine. The Kaibab plateau north of the Grand Canyon continues this timbered strip into southern Utah. The highest point in the State is Humphreys Peak with an elevation of 12,611 feet, located just northwest of Flagstaff. Baldy Peak, in the White Mountains of eastern Arizona, is the second highest in the State with an elevation of 11,490 feet.

The higher elevations of the State, running diagonally from the southeast to the northwest, average between 25 and 30 inches of precipitation (rain plus melted snow) annually, while the desert southwest has averages as low as 3 or 4 inches per year. The desert valleys of southwestern Arizona are an extension of the Sonora Desert of Mexico, with elevations as low as about 100 feet above sea level in the Lower Colorado River Valley. The plateau country in the northeastern corner of the State receives approximately 10 inches of precipitation annually. Higher ridges here are covered with junipers and pinon trees.

Nearly the entire State is in the Colorado River drainage basin emptying into the Gulf of California. The world-famed Grand Canyon lies within the State, extending from the junction of the Little Colorado with the main stream southwestward for approximately 217 miles. The Grand Canyon varies in width from 4 to 18 miles, and depths from the rim to the river bed range from 2,700 to as much as 5,700 feet. This is an outstanding example of arid or semiarid land erosion by a major river whose source is in a more rainy area.

GENERAL CLIMATE. Cold air masses from Canada sometimes penetrate into the State bringing temperatures well below zero in the high plateau and mountainous regions of central and northern Arizona. Great extremes occur between day and night temperatures throughout Arizona. The daily range between maximum and minimum temperatures sometimes runs as high as 50° to 60° during drier portions of the year. During winter months, daytime temperatures may average 70°, with night temperatures often falling to freezing or slightly below in lower desert valleys. In the summer the pine-clad forests in the central part of the State may have afternoon temperatures of 80°, while night temperatures drop to 35° to 40°.

PRECIPITATION. Precipitation throughout Arizona is governed to a great extent by elevation and the season of the year. From November through March, storm systems from the Pacific Ocean cross the State. These winter storms occur frequently in the higher mountains of the central and northern parts of the State and sometimes bring with them heavy snows. Snow accumulation may reach depths of 100 inches of more during the winter. The gradual melting of this snow during the spring serves to maintain a supply of water in the main rivers of the State. Reservoirs on these streams supply water to the desert areas in the lower Salt River Valley and the lower Gila River Valley areas.

Summer rainfall begins early in July and usually extends to mid-September. Moisture-bearing winds sweep into Arizona from the southeast, with their source region in the Gulf of Mexico. Summer rains occur in the form of thundershowers which are caused, to a great extent, by excessive heating of the ground and the lifting of moisture-laden air along main mountain ranges. Thus, the heaviest thundershowers are usually found in mountainous regions of the central and southeastern portions of Arizona. These thunderstorms are often accompanied by strong winds and brief periods of blowing dust prior to the onset of rain. Hail occurs rather infrequently.

The average number of days with measurable precipitation per year varies from 72 in the Flagstaff region to 34 at Phoenix, 50 in Tucson, 53 at Winslow, and 15 at Yuma. A large portion of Arizona is included in the semiarid region of the United States. Long periods often occur with little or no precipitation. The air is generally dry and clear, with relatively low humidity and a high percentage of sunshine. April, May and June are the months with the greatest number of clear days, while July and August, as well as December, January and February have the cloudiest weather and lowest percent of sunshine. Humidities, while low when compared to most other states, are higher throughout much of Arizona during July and August, which is the thunderstorm season. Annual average humidity values, based on four readings per day, show Flagstaff with 55%, Phoenix 38%, Tucson 38%, Winslow 46% and Yuma 33%. Yearly averages of percent of possible sunshine show Phoenix with 86%, Tucson 86%, and Yuma 92%. Due to high temperatures, the dryness of the air, and the high percentage of sunshine, evaporation rates in Arizona are high. Mean annual lake evaporation varies from about 80 inches in the southwestern corner of the State to about 50 inches in the northeastern corner. Phoenix averages about 72 inches and Tucson 70 inches per year.

The length of the growing season (period between freezes) varies tremendously over Arizona, averaging less than 3 months in some of the elevated areas of northern and eastern portions of the State. On the other hand, lower desert valleys sometimes have 2 or 3 years in succession without freezes.

Flood conditions occur infrequently, although heavy thundershowers during July and August at times cause flash floods that do considerable local damage. Floods on main rivers are mostly limited to the upper basins. Heaviest runoff usually occurs in connection with the arrival of tropical air over Arizona, which had its origin in hurricanes that dissipated in or off the west coast of Mexico. Heavy rains associated with these systems usually occur during August or September. High winds accompanying heavy thunderstorms during July and August sometimes reach peak gusts of about 100 miles per hour in local areas. Tornado funnels have been reported in Arizona.

## COUNTY INDEX

### *Apache County*
ALPINE
GANADO
GREER
LUKACHUKAI
MC NARY 2 N
SAINT JOHNS
SANDERS
SPRINGERVILLE
TEEC NOS POS
WINDOW ROCK 4 SW

### *Cochise County*
BISBEE-DOUGLAS INTL
BOWIE
CHIRICAHUA NATL MONU
CORONADO N M HDQTRS
DOUGLAS
PORTAL 4 SW
TOMBSTONE
WILLCOX

### *Coconino County*
CHEVELON R S
FLAGSTAFF PULLIAM AP
FORT VALLEY
LEES FERRY
PAGE
SEDONA R S
WAHWEAP
WILLIAMS
WUPATKI NATL MONUMNT

### *Gila County*
MIAMI
PAYSON
PLEASANT VALLEY R S
ROOSEVELT 1 WNW
SAN CARLOS RESERVOIR

### *Graham County*
BLACK RIVER PUMPS
SAFFORD AGRI CENTER

### *Greenlee County*
CLIFTON
DUNCAN

### *La Paz County*
BOUSE
PARKER 6 NE

### *Maricopa County*
AGUILA
BARTLETT DAM
BUCKEYE
CHANDLER HEIGHTS
GILA BEND
LAVEEN 3 SSE
LITCHFIELD PARK
MESA
MORMON FLAT
PHOENIX CITY
PHOENIX SKY HARBOR
SOUTH PHOENIX
STEWART MOUNTAIN
TEMPE A S U
WICKENBURG
YOUNGTOWN

### *Mohave County*
COLORADO CITY
PIPE SPRINGS N M
WIKIEUP
YUCCA 1 NNE

### *Navajo County*
BETATAKIN
HEBER R S
HOLBROOK
KEAMS CANYON
PETRIFIED FOREST N P
SHOW LOW AP
SNOWFLAKE
SNOWFLAKE 15 W
WHITERIVER 1 SW
WINSLOW MUNICIPAL AP

### *Pima County*
AJO
ANVIL RANCH
KITT PEAK
ORGAN PIPE CACTUS NM
SANTA RITA EXP RANGE
SASABE
TUCSON CP AVE EXP FM
TUCSON INTL ARPT
TUCSON MAGNETIC OBSY
TUCSON U OF ARIZONA

### *Pinal County*
CASA GRANDE
CASA GRANDE NATL MON
ELOY 4 NE
FLORENCE
MARICOPA 4 N
ORACLE 2 SE
SACATON
SUPERIOR

### *Santa Cruz County*
CANELO 1 NW
NOGALES 6 N
TUMACACORI NATL MONM

### *Yavapai County*
BEAVER CREEK R S
CASTLE HOT SPRINGS
CHILDS
CHINO VALLEY
CORDES
HILLSIDE 4 NNE
JEROME
MONTEZUMA CASTLE NM
PRESCOTT
SELIGMAN
WALNUT CREEK

### *Yuma County*
KOFA MINE
YUMA CITRUS STATION
YUMA INTL AP
YUMA PROVING GROUND

## ELEVATION INDEX

| FEET | STATION NAME |
|---|---|
| 191 | YUMA CITRUS STATION |
| 207 | YUMA INTL AP |
| 324 | YUMA PROVING GROUND |
| 410 | PARKER 6 NE |
| 741 | GILA BEND |
| 932 | BOUSE |
| 981 | BUCKEYE |
| 1030 | LITCHFIELD PARK |
| 1106 | PHOENIX CITY |
| 1112 | LAVEEN 3 SSE |
| 1112 | PHOENIX SKY HARBOR |
| 1142 | YOUNGTOWN |
| 1152 | SOUTH PHOENIX |
| 1170 | TEMPE A S U |
| 1171 | MARICOPA 4 N |
| 1235 | MESA |
| 1280 | SACATON |
| 1391 | CASA GRANDE |
| 1421 | CASA GRANDE NATL MON |
| 1422 | STEWART MOUNTAIN |
| 1450 | CHANDLER HEIGHTS |
| 1503 | FLORENCE |
| 1545 | ELOY 4 NE |
| 1650 | BARTLETT DAM |

| FEET | STATION NAME |
|---|---|
| 1660 | ORGAN PIPE CACTUS NM |
| 1762 | AJO |
| 1781 | KOFA MINE |
| 1811 | YUCCA 1 NNE |
| 1821 | MORMON FLAT |
| 1972 | CASTLE HOT SPRINGS |
| 2070 | WICKENBURG |
| 2133 | WIKIEUP |
| 2182 | AGUILA |
| 2205 | ROOSEVELT 1 WNW |
| 2330 | TUCSON CP AVE EXP FM |
| 2421 | TUCSON U OF ARIZONA |
| 2526 | TUCSON MAGNETIC OBSY |
| 2532 | SAN CARLOS RESERVOIR |
| 2559 | TUCSON INTL ARPT |
| 2650 | CHILDS |
| 2753 | ANVIL RANCH |
| 2954 | SAFFORD AGRI CENTER |
| 3002 | SUPERIOR |
| 3143 | LEES FERRY |
| 3182 | MONTEZUMA CASTLE NM |
| 3202 | HILLSIDE 4 NNE |
| 3267 | TUMACACORI NATL MONM |
| 3471 | CLIFTON |
| 3563 | NOGALES 6 N |
| 3590 | SASABE |
| 3602 | MIAMI |
| 3642 | DUNCAN |
| 3720 | WAHWEAP |
| 3763 | BOWIE |
| 3771 | CORDES |
| 3832 | BEAVER CREEK R S |
| 4104 | BISBEE-DOUGLAS INTL |
| 4144 | DOUGLAS |
| 4203 | WILLCOX |
| 4222 | SEDONA R S |
| 4272 | PAGE |
| 4304 | SANTA RITA EXP RANGE |
| 4373 | ORACLE 2 SE |
| 4544 | TOMBSTONE |
| 4672 | CHINO VALLEY |
| 4885 | WINSLOW MUNICIPAL AP |
| 4900 | WUPATKI NATL MONUMNT |
| 4920 | PIPE SPRINGS N M |
| 5003 | CANELO 1 NW |
| 5007 | PAYSON |
| 5013 | COLORADO CITY |
| 5052 | PLEASANT VALLEY R S |
| 5072 | HOLBROOK |
| 5120 | WHITERIVER 1 SW |

| FEET | STATION NAME |
|---|---|
| 5164 | WALNUT CREEK |
| 5174 | TEEC NOS POS |
| 5205 | PRESCOTT |
| 5223 | SELIGMAN |
| 5243 | CORONADO N M HDQTRS |
| 5253 | JEROME |
| 5300 | CHIRICAHUA NATL MONU |
| 5394 | PORTAL 4 SW |
| 5463 | PETRIFIED FOREST N P |
| 5642 | SNOWFLAKE |
| 5735 | SAINT JOHNS |
| 5853 | SANDERS |
| 6040 | BLACK RIVER PUMPS |
| 6080 | SNOWFLAKE 15 W |
| 6214 | KEAMS CANYON |
| 6355 | GANADO |
| 6411 | SHOW LOW AP |
| 6483 | LUKACHUKAI |
| 6506 | HEBER R S |
| 6750 | WILLIAMS |
| 6844 | KITT PEAK |
| 6900 | WINDOW ROCK 4 SW |
| 7006 | FLAGSTAFF PULLIAM AP |
| 7014 | CHEVELON R S |
| 7037 | SPRINGERVILLE |
| 7205 | BETATAKIN |
| 7316 | MC NARY 2 N |
| 7356 | FORT VALLEY |
| 8005 | ALPINE |
| 8405 | GREER |

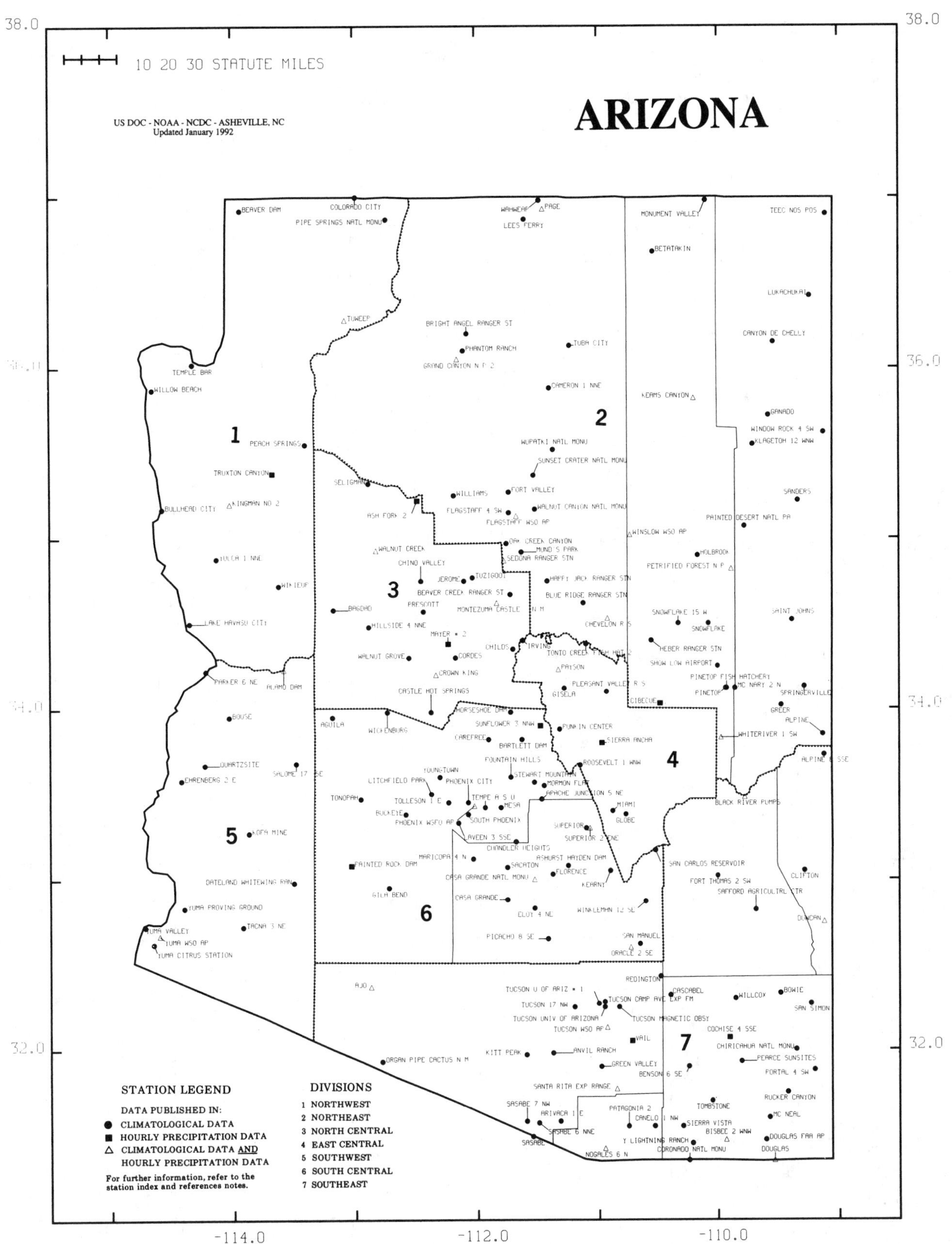
10 20 30 STATUTE MILES
ARIZONA
US DOC - NOAA - NCDC - ASHEVILLE, NC
Updated January 1992
38.0
36.0
34.0
32.0
-114.0
-112.0
-110.0
STATION LEGEND
DATA PUBLISHED IN:
● CLIMATOLOGICAL DATA
■ HOURLY PRECIPITATION DATA
△ CLIMATOLOGICAL DATA AND HOURLY PRECIPITATION DATA
For further information, refer to the station index and references notes.
DIVISIONS
1 NORTHWEST
2 NORTHEAST
3 NORTH CENTRAL
4 EAST CENTRAL
5 SOUTHWEST
6 SOUTH CENTRAL
7 SOUTHEAST

## AGUILA *Maricopa County* ELEVATION 2182 ft LAT/LONG 33° 57 ' N / 113° 11 ' W

| | JAN | FEB | MAR | APR | MAY | JUN | JUL | AUG | SEP | OCT | NOV | DEC | YEAR |
|---|---|---|---|---|---|---|---|---|---|---|---|---|---|
| Maximum Temp °F | 62.9 | 65.6 | 70.3 | 79.5 | 88.7 | 98.7 | 101.5 | 99.6 | 94.1 | 84.2 | 71.1 | 62.5 | 81.6 |
| Minimum Temp °F | 34.1 | 36.4 | 39.7 | 43.8 | 52.4 | 60.8 | 69.7 | 69.0 | 61.8 | 50.3 | 38.7 | 32.9 | 49.1 |
| Mean Temp °F | 48.6 | 51.0 | 55.0 | 61.7 | 70.6 | 79.8 | 85.6 | 84.3 | 77.9 | 67.3 | 54.9 | 47.8 | 65.4 |
| Days Max Temp ≥ 90 °F | 0 | 0 | 0 | 4 | 15 | 27 | 30 | 30 | 24 | 9 | 0 | 0 | 139 |
| Days Max Temp ≤ 32 °F | 0 | 0 | 0 | 0 | 0 | 0 | 0 | 0 | 0 | 0 | 0 | 0 | 0 |
| Days Min Temp ≤ 32 °F | na | 9 | 3 | 1 | 0 | 0 | 0 | 0 | 0 | 0 | 5 | 16 | na |
| Days Min Temp ≤ 0 °F | 0 | 0 | 0 | 0 | 0 | 0 | 0 | 0 | 0 | 0 | 0 | 0 | 0 |
| Heating Degree Days | 503 | 389 | 307 | 132 | 25 | 1 | 0 | 0 | 2 | 50 | 299 | 528 | 2236 |
| Cooling Degree Days | 0 | 1 | 5 | 56 | 202 | 461 | 641 | 616 | 392 | 124 | 3 | 0 | 2501 |
| Total Precipitation (") | 1.07 | 1.15 | 1.22 | 0.23 | 0.16 | 0.06 | 0.87 | 1.28 | 0.78 | 0.57 | 0.80 | 1.00 | 9.19 |
| Days ≥ 0.1" Precip | na | 2 | 3 | 1 | 1 | 0 | 2 | 2 | 2 | 1 | 2 | 2 | na |
| Total Snowfall (") | 0.0 | 0.3 | 0.0 | 0.0 | 0.0 | 0.0 | 0.0 | 0.0 | 0.0 | 0.0 | 0.0 | 0.3 | 0.6 |
| Days ≥ 1" Snow Depth | na | 0 | 0 | 0 | 0 | 0 | 0 | 0 | 0 | 0 | 0 | 0 | na |

## AJO *Pima County* ELEVATION 1762 ft LAT/LONG 32° 22 ' N / 112° 52 ' W

| | JAN | FEB | MAR | APR | MAY | JUN | JUL | AUG | SEP | OCT | NOV | DEC | YEAR |
|---|---|---|---|---|---|---|---|---|---|---|---|---|---|
| Maximum Temp °F | 64.5 | 69.1 | 73.8 | 81.3 | 89.5 | 99.5 | 102.1 | 100.7 | 96.5 | 86.5 | na | na | na |
| Minimum Temp °F | 43.8 | 47.3 | 51.2 | 56.8 | 64.3 | 73.6 | 78.7 | 77.1 | 72.6 | 63.1 | na | na | na |
| Mean Temp °F | 54.2 | 58.2 | 62.5 | 69.1 | 76.9 | 86.6 | 90.4 | 88.9 | 84.5 | 74.8 | na | na | na |
| Days Max Temp ≥ 90 °F | 0 | 0 | 1 | 6 | 15 | 26 | 28 | 28 | 23 | 12 | 1 | 0 | 140 |
| Days Max Temp ≤ 32 °F | 0 | 0 | 0 | 0 | 0 | 0 | 0 | 0 | 0 | 0 | 0 | 0 | 0 |
| Days Min Temp ≤ 32 °F | 2 | 1 | 0 | 0 | 0 | 0 | 0 | 0 | 0 | 0 | 0 | 1 | 4 |
| Days Min Temp ≤ 0 °F | 0 | 0 | 0 | 0 | 0 | 0 | 0 | 0 | 0 | 0 | 0 | 0 | 0 |
| Heating Degree Days | 331 | 204 | 136 | 46 | 6 | 0 | 0 | 0 | 0 | 15 | na | na | na |
| Cooling Degree Days | na | 25 | 70 | 214 | na | 674 | na | 755 | 605 | 343 | na | na | na |
| Total Precipitation (") | 0.69 | 0.73 | 0.82 | 0.22 | 0.16 | 0.06 | 0.99 | 1.79 | 0.92 | 0.58 | 0.67 | 1.11 | 8.74 |
| Days ≥ 0.1" Precip | 1 | 2 | 2 | 1 | 0 | 0 | 2 | 3 | 2 | 1 | 1 | 2 | 17 |
| Total Snowfall (") | 0.0 | 0.0 | 0.0 | 0.0 | 0.0 | 0.0 | 0.0 | 0.0 | 0.0 | 0.0 | 0.0 | 0.1 | 0.1 |
| Days ≥ 1" Snow Depth | 0 | 0 | 0 | 0 | 0 | 0 | 0 | 0 | 0 | 0 | 0 | 0 | 0 |

## ALPINE *Apache County* ELEVATION 8005 ft LAT/LONG 33° 51 ' N / 109° 8 ' W

| | JAN | FEB | MAR | APR | MAY | JUN | JUL | AUG | SEP | OCT | NOV | DEC | YEAR |
|---|---|---|---|---|---|---|---|---|---|---|---|---|---|
| Maximum Temp °F | 45.0 | 46.9 | 51.3 | 59.0 | 66.5 | 76.6 | 77.6 | 74.4 | 70.3 | 62.5 | 52.4 | 46.3 | 60.7 |
| Minimum Temp °F | 13.0 | 15.1 | 20.3 | 24.1 | 29.2 | 36.2 | 44.6 | 43.4 | 36.9 | 27.6 | 19.3 | 14.4 | 27.0 |
| Mean Temp °F | 29.0 | 31.0 | 35.8 | 41.6 | 47.9 | 56.4 | 61.1 | 59.0 | 53.6 | 45.1 | 35.8 | 30.5 | 43.9 |
| Days Max Temp ≥ 90 °F | 0 | 0 | 0 | 0 | 0 | 0 | 0 | 0 | 0 | 0 | 0 | 0 | 0 |
| Days Max Temp ≤ 32 °F | 2 | 1 | 0 | 0 | 0 | 0 | 0 | 0 | 0 | 0 | 1 | 2 | 6 |
| Days Min Temp ≤ 32 °F | 31 | 27 | 30 | 27 | 21 | 9 | 0 | 0 | 7 | na | na | 30 | na |
| Days Min Temp ≤ 0 °F | 3 | 2 | 0 | 0 | 0 | 0 | 0 | 0 | 0 | 0 | 0 | 2 | 7 |
| Heating Degree Days | 1104 | 947 | 894 | 697 | 523 | 254 | 122 | 181 | 337 | 611 | 869 | 1064 | 7603 |
| Cooling Degree Days | 0 | 0 | 0 | 0 | 0 | 3 | 3 | 0 | 0 | 0 | 0 | 0 | 6 |
| Total Precipitation (") | 1.33 | 1.31 | 1.28 | 0.62 | 0.78 | 0.78 | 3.53 | 4.19 | 2.48 | 1.98 | 1.50 | 1.77 | 21.55 |
| Days ≥ 0.1" Precip | 3 | 3 | 4 | 2 | 2 | 2 | 10 | 10 | 5 | 4 | 3 | 4 | 52 |
| Total Snowfall (") | na | 6.7 | na | 1.3 | 0.4 | 0.0 | 0.0 | 0.0 | 0.0 | 0.4 | 3.3 | 10.9 | na |
| Days ≥ 1" Snow Depth | na | 5 | na | 0 | 0 | 0 | 0 | 0 | 0 | 0 | 1 | na | na |

## ANVIL RANCH *Pima County* ELEVATION 2753 ft LAT/LONG 31° 58 ' N / 111° 23 ' W

| | JAN | FEB | MAR | APR | MAY | JUN | JUL | AUG | SEP | OCT | NOV | DEC | YEAR |
|---|---|---|---|---|---|---|---|---|---|---|---|---|---|
| Maximum Temp °F | 65.6 | 69.6 | 74.2 | 82.1 | 90.0 | 99.8 | 99.4 | 97.2 | 94.1 | 85.0 | 74.0 | 65.8 | 83.1 |
| Minimum Temp °F | 32.8 | 35.5 | 39.5 | 44.4 | 52.1 | 61.9 | 70.0 | 68.1 | 61.3 | 49.7 | 38.7 | 33.3 | 48.9 |
| Mean Temp °F | 49.2 | 52.5 | 56.9 | 63.3 | 71.1 | 80.9 | 84.7 | 82.7 | 77.7 | 67.4 | 56.3 | 49.7 | 66.0 |
| Days Max Temp ≥ 90 °F | 0 | 0 | 1 | 5 | 17 | 29 | 30 | 29 | 25 | 9 | 0 | 0 | 145 |
| Days Max Temp ≤ 32 °F | 0 | 0 | 0 | 0 | 0 | 0 | 0 | 0 | 0 | 0 | 0 | 0 | 0 |
| Days Min Temp ≤ 32 °F | 15 | 10 | 4 | 1 | 0 | 0 | 0 | 0 | 0 | 0 | 6 | na | na |
| Days Min Temp ≤ 0 °F | 0 | 0 | 0 | 0 | 0 | 0 | 0 | 0 | 0 | 0 | 0 | 0 | 0 |
| Heating Degree Days | 484 | 346 | 253 | 101 | 17 | 0 | 0 | 0 | 1 | 46 | 258 | 466 | 1972 |
| Cooling Degree Days | 0 | 1 | 7 | 71 | 216 | 493 | 605 | 542 | 381 | 123 | 7 | 0 | 2446 |
| Total Precipitation (") | 0.88 | 0.78 | 0.79 | 0.18 | 0.25 | 0.31 | 2.63 | 2.59 | 1.47 | 1.00 | 0.58 | 1.35 | 12.81 |
| Days ≥ 0.1" Precip | 2 | 2 | 2 | 1 | 1 | 1 | 6 | 5 | 3 | 2 | 2 | 3 | 30 |
| Total Snowfall (") | 0.0 | 0.0 | 0.0 | 0.0 | 0.0 | 0.0 | 0.0 | 0.0 | 0.0 | 0.0 | 0.0 | 0.2 | 0.2 |
| Days ≥ 1" Snow Depth | 0 | 0 | 0 | 0 | 0 | 0 | 0 | 0 | 0 | 0 | 0 | 0 | 0 |

## BARTLETT DAM *Maricopa County* ELEVATION 1650 ft LAT/LONG 33° 49 ' N / 111° 38 ' W

| | JAN | FEB | MAR | APR | MAY | JUN | JUL | AUG | SEP | OCT | NOV | DEC | YEAR |
|---|---|---|---|---|---|---|---|---|---|---|---|---|---|
| Maximum Temp °F | 65.3 | 69.4 | 73.4 | 82.1 | 91.4 | 101.7 | 104.7 | 102.8 | 97.4 | 86.9 | 74.2 | 65.4 | 84.6 |
| Minimum Temp °F | 40.4 | 42.7 | 45.7 | 51.8 | 59.5 | 68.1 | 74.7 | 74.3 | 69.1 | 59.1 | 47.6 | 41.4 | 56.2 |
| Mean Temp °F | 52.9 | 56.1 | 59.6 | 67.0 | 75.5 | 84.9 | 89.7 | 88.6 | 83.3 | 73.0 | 60.7 | 53.5 | 70.4 |
| Days Max Temp ≥ 90 °F | 0 | 0 | 1 | 7 | 20 | 29 | 31 | 31 | 27 | 13 | 1 | 0 | 160 |
| Days Max Temp ≤ 32 °F | 0 | 0 | 0 | 0 | 0 | 0 | 0 | 0 | 0 | 0 | 0 | 0 | 0 |
| Days Min Temp ≤ 32 °F | 3 | 2 | 1 | 0 | 0 | 0 | 0 | 0 | 0 | 0 | 1 | 2 | 9 |
| Days Min Temp ≤ 0 °F | 0 | 0 | 0 | 0 | 0 | 0 | 0 | 0 | 0 | 0 | 0 | 0 | 0 |
| Heating Degree Days | 369 | 252 | 195 | 64 | 7 | 0 | 0 | 0 | 0 | 20 | 155 | 353 | 1415 |
| Cooling Degree Days | 0 | 8 | 32 | 160 | 346 | 610 | 763 | 733 | 554 | 280 | 40 | 2 | 3528 |
| Total Precipitation (") | 1.75 | 1.65 | 2.24 | 0.59 | 0.36 | 0.28 | 1.23 | 1.75 | 1.36 | 1.20 | 1.44 | 2.02 | 15.87 |
| Days ≥ 0.1" Precip | 4 | 3 | 4 | 1 | 1 | 0 | 3 | 4 | 2 | 2 | 3 | 3 | 30 |
| Total Snowfall (") | 0.0 | 0.0 | 0.0 | 0.0 | 0.0 | 0.0 | 0.0 | 0.0 | 0.0 | 0.0 | 0.0 | 0.0 | 0.0 |
| Days ≥ 1" Snow Depth | 0 | 0 | 0 | 0 | 0 | 0 | 0 | 0 | 0 | 0 | 0 | 0 | 0 |

## BEAVER CREEK R S *Yavapai County* ELEVATION 3832 ft LAT/LONG 34° 40 ' N / 111° 43 ' W

| | JAN | FEB | MAR | APR | MAY | JUN | JUL | AUG | SEP | OCT | NOV | DEC | YEAR |
|---|---|---|---|---|---|---|---|---|---|---|---|---|---|
| Maximum Temp °F | 57.6 | 61.8 | 65.9 | 74.2 | 82.7 | 93.1 | 96.3 | 93.8 | 88.5 | 78.0 | 66.1 | ***56.8*** | 76.2 |
| Minimum Temp °F | 29.7 | 32.8 | 37.3 | 43.6 | 51.3 | 60.4 | 66.2 | 64.2 | 57.8 | 46.6 | 35.1 | 29.5 | 46.2 |
| Mean Temp °F | 43.7 | 47.3 | 51.5 | 58.9 | 67.0 | 76.8 | 81.2 | 79.0 | 73.1 | 62.3 | 50.6 | ***43.2*** | 61.2 |
| Days Max Temp ≥ 90 °F | 0 | 0 | 0 | 1 | 6 | 22 | 28 | 25 | 13 | 3 | 0 | 0 | 98 |
| Days Max Temp ≤ 32 °F | 0 | 0 | 0 | 0 | 0 | 0 | 0 | 0 | 0 | 0 | 0 | 0 | 0 |
| Days Min Temp ≤ 32 °F | 20 | 12 | 7 | 2 | 0 | 0 | 0 | 0 | 0 | 1 | 11 | 21 | 74 |
| Days Min Temp ≤ 0 °F | 0 | 0 | 0 | 0 | 0 | 0 | 0 | 0 | 0 | 0 | 0 | 0 | 0 |
| Heating Degree Days | 653 | 492 | 412 | 198 | 55 | 4 | 0 | 0 | 8 | 128 | 427 | ***669*** | 3046 |
| Cooling Degree Days | 0 | 0 | 1 | 36 | 142 | 390 | 507 | 450 | 264 | 50 | 0 | 0 | 1840 |
| Total Precipitation (") | 1.46 | 1.58 | 2.03 | 0.92 | 0.49 | 0.20 | 1.63 | 2.08 | 1.60 | 1.37 | 1.40 | 1.59 | 16.35 |
| Days ≥ 0.1" Precip | 3 | 3 | 4 | 2 | 2 | 1 | 3 | 4 | 3 | 3 | 2 | 3 | 33 |
| Total Snowfall (") | 0.7 | ***0.5*** | 0.2 | 0.4 | 0.0 | 0.0 | 0.0 | 0.0 | 0.0 | 0.0 | 0.0 | 0.8 | 2.6 |
| Days ≥ 1" Snow Depth | 0 | ***0*** | 0 | 0 | 0 | 0 | 0 | 0 | 0 | 0 | 0 | ***0*** | 0 |

## BETATAKIN *Navajo County* ELEVATION 7205 ft LAT/LONG 36° 41 ' N / 110° 32 ' W

| | JAN | FEB | MAR | APR | MAY | JUN | JUL | AUG | SEP | OCT | NOV | DEC | YEAR |
|---|---|---|---|---|---|---|---|---|---|---|---|---|---|
| Maximum Temp °F | 39.1 | 43.0 | 49.6 | 59.5 | 69.4 | 80.5 | 84.7 | 82.1 | 74.6 | 62.9 | 48.2 | 39.7 | 61.1 |
| Minimum Temp °F | 20.1 | 23.2 | 27.8 | 33.9 | 42.6 | 52.6 | 58.2 | 56.7 | 50.2 | 39.9 | 28.6 | 21.4 | 37.9 |
| Mean Temp °F | 29.6 | 33.1 | 38.8 | 46.7 | 56.0 | 66.6 | 71.5 | 69.4 | 62.5 | 51.4 | 38.5 | 30.6 | 49.6 |
| Days Max Temp ≥ 90 °F | 0 | 0 | 0 | 0 | 0 | 3 | 6 | 2 | 0 | 0 | 0 | 0 | 11 |
| Days Max Temp ≤ 32 °F | 6 | 3 | 1 | 0 | 0 | 0 | 0 | 0 | 0 | 0 | 2 | 6 | 18 |
| Days Min Temp ≤ 32 °F | 30 | 26 | 23 | 13 | 4 | 0 | 0 | 0 | 0 | 6 | 20 | 29 | 151 |
| Days Min Temp ≤ 0 °F | 0 | 0 | 0 | 0 | 0 | 0 | 0 | 0 | 0 | 0 | 0 | 0 | 0 |
| Heating Degree Days | 1091 | 893 | 807 | 542 | 279 | 59 | 2 | 11 | 107 | 414 | 790 | 1060 | 6055 |
| Cooling Degree Days | 0 | 0 | 0 | 0 | 7 | 115 | 201 | 146 | 36 | 0 | 0 | 0 | 505 |
| Total Precipitation (") | 1.02 | 1.01 | 1.13 | 0.73 | 0.63 | 0.38 | 1.47 | 1.67 | 1.16 | 1.30 | 1.23 | 1.20 | 12.93 |
| Days ≥ 0.1" Precip | 3 | 3 | 4 | 2 | 2 | 1 | 4 | 4 | 3 | 3 | 3 | 3 | 35 |
| Total Snowfall (") | 12.0 | 10.2 | 7.3 | 4.2 | 0.9 | 0.0 | 0.0 | 0.0 | 0.0 | 0.7 | 6.6 | 14.5 | 56.4 |
| Days ≥ 1" Snow Depth | ***18*** | ***14*** | 5 | ***1*** | 0 | 0 | 0 | 0 | 0 | 0 | ***4*** | ***13*** | 55 |

## BISBEE-DOUGLAS INTL *Cochise County* ELEVATION 4104 ft LAT/LONG 31° 28 ' N / 109° 36 ' W

| | JAN | FEB | MAR | APR | MAY | JUN | JUL | AUG | SEP | OCT | NOV | DEC | YEAR |
|---|---|---|---|---|---|---|---|---|---|---|---|---|---|
| Maximum Temp °F | 61.0 | 64.9 | 70.1 | 77.9 | 85.1 | 94.2 | 92.8 | 90.4 | 87.2 | 79.4 | 68.9 | 61.4 | 77.8 |
| Minimum Temp °F | 29.2 | 31.9 | 36.4 | 42.2 | 50.3 | 59.5 | 65.1 | 63.5 | 58.4 | 46.8 | 35.5 | 29.6 | 45.7 |
| Mean Temp °F | 45.1 | 48.4 | 53.3 | 60.1 | 67.7 | 76.9 | 78.9 | 77.0 | 72.8 | 63.1 | 52.2 | 45.6 | 61.8 |
| Days Max Temp ≥ 90 °F | 0 | 0 | 0 | 1 | 8 | 24 | 23 | 19 | 11 | 2 | 0 | 0 | 88 |
| Days Max Temp ≤ 32 °F | 0 | 0 | 0 | 0 | 0 | 0 | 0 | 0 | 0 | 0 | 0 | 0 | 0 |
| Days Min Temp ≤ 32 °F | 21 | 15 | 9 | 3 | 0 | 0 | 0 | 0 | 0 | 1 | 11 | 21 | 81 |
| Days Min Temp ≤ 0 °F | 0 | 0 | 0 | 0 | 0 | 0 | 0 | 0 | 0 | 0 | 0 | 0 | 0 |
| Heating Degree Days | 609 | 462 | 357 | 160 | 31 | 1 | 0 | 0 | 3 | 98 | 377 | 596 | 2694 |
| Cooling Degree Days | 0 | 0 | 1 | 26 | 138 | 382 | 445 | 389 | 253 | 47 | 0 | 0 | 1681 |
| Total Precipitation (") | 0.83 | 0.61 | 0.49 | 0.21 | 0.35 | 0.53 | 3.23 | 3.06 | 1.64 | 1.06 | 0.70 | 1.22 | 13.93 |
| Days ≥ 0.1" Precip | 2 | 2 | 2 | 1 | 1 | 1 | 7 | 7 | 3 | 2 | 2 | 3 | 33 |
| Total Snowfall (") | 0.2 | 0.2 | 0.1 | 0.0 | 0.0 | 0.0 | 0.0 | 0.0 | 0.0 | 0.0 | 0.1 | 0.5 | 1.1 |
| Days ≥ 1" Snow Depth | 0 | 0 | 0 | 0 | 0 | 0 | 0 | 0 | 0 | 0 | 0 | 0 | 0 |

## BLACK RIVER PUMPS *Graham County* ELEVATION 6040 ft LAT/LONG 33° 29 ' N / 109° 46 ' W

| | JAN | FEB | MAR | APR | MAY | JUN | JUL | AUG | SEP | OCT | NOV | DEC | YEAR |
|---|---|---|---|---|---|---|---|---|---|---|---|---|---|
| Maximum Temp °F | 48.6 | 52.0 | 56.7 | 65.8 | 74.7 | 85.3 | 87.0 | 83.8 | 78.8 | 69.3 | 57.4 | 49.6 | 67.4 |
| Minimum Temp °F | 20.7 | 23.6 | 27.1 | 32.2 | 40.2 | 49.3 | 56.1 | 55.2 | 48.9 | 38.0 | 27.0 | 21.3 | 36.6 |
| Mean Temp °F | 34.7 | 37.8 | 42.0 | 49.1 | 57.5 | 67.3 | 71.6 | 69.5 | 63.9 | 53.7 | 42.2 | 35.5 | 52.1 |
| Days Max Temp ≥ 90 °F | 0 | 0 | 0 | 0 | 0 | 9 | 11 | 4 | 0 | 0 | 0 | 0 | 24 |
| Days Max Temp ≤ 32 °F | 1 | 0 | 0 | 0 | 0 | 0 | 0 | 0 | 0 | 0 | 0 | 1 | 2 |
| Days Min Temp ≤ 32 °F | 29 | 26 | 26 | 16 | 4 | 0 | 0 | 0 | 0 | 7 | 24 | 29 | 161 |
| Days Min Temp ≤ 0 °F | 1 | 0 | 0 | 0 | 0 | 0 | 0 | 0 | 0 | 0 | 0 | 1 | 2 |
| Heating Degree Days | 933 | 761 | 708 | 471 | 235 | 38 | 1 | 5 | 71 | 346 | 676 | 910 | 5155 |
| Cooling Degree Days | 0 | 0 | 0 | 0 | 9 | 128 | 207 | 159 | 45 | 1 | 0 | 0 | 549 |
| Total Precipitation (") | 1.67 | 1.56 | 1.69 | 0.62 | 0.66 | 0.68 | 3.29 | 3.21 | 1.88 | 1.60 | 1.28 | 1.95 | 20.09 |
| Days ≥ 0.1" Precip | 4 | 5 | 5 | 2 | 2 | 2 | 8 | 8 | 4 | 3 | 3 | 5 | 51 |
| Total Snowfall (") | 2.0 | *1.3* | *0.9* | 0.1 | 0.0 | 0.0 | 0.0 | 0.0 | 0.0 | 0.0 | 0.4 | 3.3 | 8.0 |
| Days ≥ 1" Snow Depth | *0* | *0* | *0* | 0 | 0 | 0 | 0 | 0 | 0 | 0 | 0 | *0* | 0 |

## BOUSE *La Paz County* ELEVATION 932 ft LAT/LONG 33° 57 ' N / 114° 1 ' W

| | JAN | FEB | MAR | APR | MAY | JUN | JUL | AUG | SEP | OCT | NOV | DEC | YEAR |
|---|---|---|---|---|---|---|---|---|---|---|---|---|---|
| Maximum Temp °F | 65.2 | 71.2 | 76.6 | 85.3 | 94.2 | 104.1 | 107.6 | 105.5 | 99.6 | 88.0 | 74.1 | 64.7 | 86.3 |
| Minimum Temp °F | 34.9 | 38.9 | 44.0 | 49.8 | 58.8 | 67.4 | 76.6 | 75.2 | 66.4 | 53.7 | 41.4 | 34.7 | 53.5 |
| Mean Temp °F | 50.1 | 55.1 | 60.3 | 67.6 | 76.5 | 85.8 | 92.1 | 90.4 | 83.0 | 70.9 | 57.7 | 49.8 | 69.9 |
| Days Max Temp ≥ 90 °F | 0 | 0 | 2 | 10 | 24 | 29 | 31 | 31 | 28 | 15 | 1 | 0 | 171 |
| Days Max Temp ≤ 32 °F | 0 | 0 | 0 | 0 | 0 | 0 | 0 | 0 | 0 | 0 | 0 | 0 | 0 |
| Days Min Temp ≤ 32 °F | 12 | 6 | 2 | 0 | 0 | 0 | 0 | 0 | 0 | 0 | 4 | 12 | 36 |
| Days Min Temp ≤ 0 °F | 0 | 0 | 0 | 0 | 0 | 0 | 0 | 0 | 0 | 0 | 0 | 0 | 0 |
| Heating Degree Days | 456 | 278 | 165 | 45 | 4 | 0 | 0 | 0 | 0 | 26 | 222 | 465 | 1661 |
| Cooling Degree Days | 0 | 5 | 28 | 166 | 387 | 655 | 844 | 802 | 556 | 217 | 12 | 0 | 3672 |
| Total Precipitation (") | 0.70 | 0.66 | 0.72 | 0.20 | 0.13 | 0.05 | 0.60 | 0.99 | 0.48 | 0.45 | 0.46 | 0.59 | 6.03 |
| Days ≥ 0.1" Precip | 2 | 2 | 2 | 1 | 0 | 0 | 1 | 2 | 1 | 1 | 1 | 2 | 15 |
| Total Snowfall (") | 0.0 | 0.0 | 0.0 | 0.0 | 0.0 | 0.0 | 0.0 | 0.0 | 0.0 | 0.0 | 0.0 | 0.0 | 0.0 |
| Days ≥ 1" Snow Depth | 0 | 0 | 0 | 0 | 0 | 0 | 0 | 0 | 0 | 0 | 0 | 0 | 0 |

## BOWIE *Cochise County* ELEVATION 3763 ft LAT/LONG 32° 20 ' N / 109° 29 ' W

| | JAN | FEB | MAR | APR | MAY | JUN | JUL | AUG | SEP | OCT | NOV | DEC | YEAR |
|---|---|---|---|---|---|---|---|---|---|---|---|---|---|
| Maximum Temp °F | 60.0 | 65.3 | 71.3 | 79.9 | *87.5* | 97.4 | 97.1 | 94.3 | 90.0 | 79.8 | 68.2 | 60.2 | 79.3 |
| Minimum Temp °F | 31.4 | 34.3 | 38.7 | 44.4 | *52.5* | 61.8 | 67.6 | 65.7 | 59.6 | 48.5 | 37.0 | 31.5 | 47.8 |
| Mean Temp °F | 45.7 | 49.8 | 55.0 | 62.2 | *70.0* | 79.6 | 82.4 | 80.0 | 74.8 | 64.2 | 52.6 | 45.9 | 63.5 |
| Days Max Temp ≥ 90 °F | 0 | 0 | 0 | 2 | *12* | 27 | 29 | 26 | 18 | 3 | 0 | 0 | 117 |
| Days Max Temp ≤ 32 °F | 0 | 0 | 0 | 0 | 0 | 0 | 0 | 0 | 0 | 0 | 0 | 0 | 0 |
| Days Min Temp ≤ 32 °F | 18 | 11 | 6 | 2 | *0* | 0 | 0 | 0 | 0 | 1 | 9 | 19 | 66 |
| Days Min Temp ≤ 0 °F | 0 | 0 | 0 | 0 | 0 | 0 | 0 | 0 | 0 | 0 | 0 | 0 | 0 |
| Heating Degree Days | 592 | 422 | 307 | 118 | *17* | 0 | 0 | 0 | 2 | 85 | 366 | 585 | 2494 |
| Cooling Degree Days | 0 | 0 | 4 | 50 | 193 | 457 | 543 | 470 | 308 | 67 | 1 | 0 | 2093 |
| Total Precipitation (") | 1.08 | 0.83 | 0.66 | 0.30 | 0.48 | 0.34 | 2.01 | 2.19 | 1.15 | 1.40 | 0.75 | 1.33 | 12.52 |
| Days ≥ 0.1" Precip | 3 | 2 | 2 | 1 | *1* | 1 | 5 | 5 | 3 | 3 | 2 | *3* | 31 |
| Total Snowfall (") | 0.6 | *0.5* | 0.3 | 0.1 | *0.0* | 0.0 | 0.0 | 0.0 | 0.0 | 0.0 | 0.3 | 0.8 | 2.6 |
| Days ≥ 1" Snow Depth | 0 | *0* | *0* | *0* | *0* | 0 | 0 | 0 | 0 | 0 | *0* | *0* | 0 |

## BUCKEYE *Maricopa County* ELEVATION 981 ft LAT/LONG 33° 22 ' N / 112° 35 ' W

| | JAN | FEB | MAR | APR | MAY | JUN | JUL | AUG | SEP | OCT | NOV | DEC | YEAR |
|---|---|---|---|---|---|---|---|---|---|---|---|---|---|
| Maximum Temp °F | 68.2 | 73.4 | 78.8 | 87.7 | 96.4 | 106.2 | 109.0 | 106.8 | 101.0 | 90.3 | 76.9 | 67.8 | 88.5 |
| Minimum Temp °F | 37.2 | 40.5 | 44.9 | 50.2 | 58.3 | 66.4 | 75.9 | 74.7 | 66.8 | 54.2 | 43.0 | 37.2 | 54.1 |
| Mean Temp °F | 52.7 | 57.0 | 61.8 | 69.0 | 77.4 | 86.3 | 92.4 | 90.8 | 83.9 | 72.3 | 60.0 | 52.5 | 71.3 |
| Days Max Temp ≥ 90 °F | 0 | 0 | 3 | 13 | 26 | 30 | 31 | 31 | 29 | 18 | 1 | 0 | 182 |
| Days Max Temp ≤ 32 °F | 0 | 0 | 0 | 0 | 0 | 0 | 0 | 0 | 0 | 0 | 0 | 0 | 0 |
| Days Min Temp ≤ 32 °F | 8 | 3 | 1 | 0 | 0 | 0 | 0 | 0 | 0 | 0 | 1 | 8 | 21 |
| Days Min Temp ≤ 0 °F | 0 | 0 | 0 | 0 | 0 | 0 | 0 | 0 | 0 | 0 | 0 | 0 | 0 |
| Heating Degree Days | 374 | 225 | 129 | 30 | 2 | 0 | 0 | 0 | 0 | 16 | 164 | 380 | 1320 |
| Cooling Degree Days | 0 | 6 | 39 | 197 | 414 | 665 | 843 | 815 | 577 | 257 | 22 | 1 | 3836 |
| Total Precipitation (") | 0.82 | 0.87 | 0.98 | 0.25 | 0.17 | 0.03 | 0.64 | 1.12 | 0.97 | 0.56 | 0.75 | 1.14 | 8.30 |
| Days ≥ 0.1" Precip | 2 | 2 | 2 | 1 | 0 | 0 | 1 | 2 | 1 | 1 | 1 | 2 | 15 |
| Total Snowfall (") | 0.0 | 0.0 | 0.0 | 0.0 | 0.0 | 0.0 | 0.0 | 0.0 | 0.0 | 0.0 | 0.0 | 0.0 | 0.0 |
| Days ≥ 1" Snow Depth | 0 | 0 | 0 | 0 | 0 | 0 | 0 | 0 | 0 | 0 | 0 | 0 | 0 |

## CANELO 1 NW *Santa Cruz County* ELEVATION 5003 ft LAT/LONG 31° 33 ' N / 110° 31 ' W

| | JAN | FEB | MAR | APR | MAY | JUN | JUL | AUG | SEP | OCT | NOV | DEC | YEAR |
|---|---|---|---|---|---|---|---|---|---|---|---|---|---|
| Maximum Temp °F | 58.0 | 61.2 | 65.6 | 73.3 | 80.8 | 90.3 | 88.6 | 85.9 | 83.2 | 75.6 | 65.4 | 58.2 | 73.8 |
| Minimum Temp °F | 27.7 | 29.2 | 32.4 | 37.2 | 44.2 | 53.1 | 60.3 | 58.7 | 53.0 | 42.8 | 33.1 | 28.4 | 41.7 |
| Mean Temp °F | 42.9 | 45.2 | 49.1 | 55.3 | 62.6 | 71.8 | 74.5 | 72.3 | 68.1 | 59.2 | 49.3 | 43.3 | 57.8 |
| Days Max Temp ≥ 90 °F | 0 | 0 | 0 | 0 | 3 | 17 | 14 | 6 | 2 | 0 | 0 | 0 | 42 |
| Days Max Temp ≤ 32 °F | 0 | 0 | 0 | 0 | 0 | 0 | 0 | 0 | 0 | 0 | 0 | 0 | 0 |
| Days Min Temp ≤ 32 °F | 23 | 19 | 16 | 7 | 1 | 0 | 0 | 0 | 0 | 3 | 14 | 22 | 105 |
| Days Min Temp ≤ 0 °F | 0 | 0 | 0 | 0 | 0 | 0 | 0 | 0 | 0 | 0 | 0 | 0 | 0 |
| Heating Degree Days | 680 | 553 | 488 | 286 | 107 | 6 | 0 | 0 | 15 | 185 | 465 | 664 | 3449 |
| Cooling Degree Days | 0 | 0 | 0 | 4 | 45 | 233 | 304 | 238 | 120 | 11 | 0 | 0 | 955 |
| Total Precipitation (") | 1.39 | 1.10 | 1.10 | 0.44 | 0.25 | 0.48 | 3.94 | 3.84 | 1.87 | 1.17 | 0.91 | 1.79 | 18.28 |
| Days ≥ 0.1" Precip | 3 | 3 | 3 | 1 | 1 | 1 | 9 | 8 | 4 | 2 | 2 | 3 | 40 |
| Total Snowfall (") | 0.0 | 0.2 | 0.3 | 0.1 | 0.0 | 0.0 | 0.0 | 0.0 | 0.0 | 0.0 | 0.0 | ***0.8*** | 1.4 |
| Days ≥ 1" Snow Depth | 0 | 0 | 0 | 0 | 0 | 0 | 0 | 0 | 0 | 0 | 0 | ***0*** | 0 |

## CASA GRANDE *Pinal County* ELEVATION 1391 ft LAT/LONG 32° 53 ' N / 111° 45 ' W

| | JAN | FEB | MAR | APR | MAY | JUN | JUL | AUG | SEP | OCT | NOV | DEC | YEAR |
|---|---|---|---|---|---|---|---|---|---|---|---|---|---|
| Maximum Temp °F | 66.9 | 71.4 | 76.8 | 86.0 | 94.9 | 104.5 | 105.8 | 103.5 | 98.8 | 88.4 | 75.6 | ***66.1*** | 86.6 |
| Minimum Temp °F | 36.9 | 39.6 | 44.1 | 49.6 | 58.2 | 66.8 | 75.6 | 74.0 | 66.7 | 54.3 | 42.6 | 36.7 | 53.8 |
| Mean Temp °F | 52.0 | 55.5 | 60.4 | 67.8 | 76.6 | 85.7 | 90.7 | 88.8 | 82.8 | 71.4 | 59.2 | ***51.4*** | 70.2 |
| Days Max Temp ≥ 90 °F | 0 | 0 | 2 | 11 | 24 | 30 | 31 | 31 | 28 | 15 | 1 | 0 | 173 |
| Days Max Temp ≤ 32 °F | 0 | 0 | 0 | 0 | 0 | 0 | 0 | 0 | 0 | 0 | 0 | 0 | 0 |
| Days Min Temp ≤ 32 °F | 9 | 4 | 1 | 0 | 0 | 0 | 0 | 0 | 0 | 0 | 2 | 8 | 24 |
| Days Min Temp ≤ 0 °F | 0 | 0 | 0 | 0 | 0 | 0 | 0 | 0 | 0 | 0 | 0 | 0 | 0 |
| Heating Degree Days | 397 | 263 | 163 | 43 | 3 | 0 | 0 | 0 | 0 | 23 | 183 | ***417*** | 1492 |
| Cooling Degree Days | 0 | 5 | 29 | 170 | 389 | 645 | 792 | 749 | 548 | 235 | 18 | 0 | 3580 |
| Total Precipitation (") | 0.79 | 0.82 | 1.02 | 0.27 | 0.19 | 0.13 | 0.87 | 1.83 | 0.87 | 0.80 | 0.81 | 1.30 | 9.70 |
| Days ≥ 0.1" Precip | 2 | 2 | 2 | 1 | 1 | 1 | 2 | 3 | 2 | 2 | 2 | 3 | 23 |
| Total Snowfall (") | 0.0 | 0.0 | 0.0 | 0.0 | 0.0 | 0.0 | 0.0 | 0.0 | 0.0 | 0.0 | 0.0 | 0.0 | 0.0 |
| Days ≥ 1" Snow Depth | 0 | 0 | 0 | 0 | 0 | 0 | 0 | 0 | 0 | 0 | 0 | 0 | 0 |

## CASA GRANDE NATL MON *Pinal County* ELEVATION 1421 ft LAT/LONG 33° 0 ' N / 111° 32 ' W

| | JAN | FEB | MAR | APR | MAY | JUN | JUL | AUG | SEP | OCT | NOV | DEC | YEAR |
|---|---|---|---|---|---|---|---|---|---|---|---|---|---|
| Maximum Temp °F | 67.3 | 72.1 | 77.3 | 86.7 | 95.9 | 105.5 | 107.1 | 105.0 | 100.2 | 89.7 | 76.2 | 67.0 | 87.5 |
| Minimum Temp °F | 35.3 | 37.7 | 41.7 | 46.8 | 55.3 | 64.2 | 73.9 | 72.6 | 65.4 | 52.9 | 40.9 | 35.3 | 51.8 |
| Mean Temp °F | 51.3 | 55.0 | 59.5 | 66.8 | 75.6 | 84.9 | 90.5 | 88.8 | 82.8 | 71.3 | 58.6 | 51.2 | 69.7 |
| Days Max Temp ≥ 90 °F | 0 | 0 | 2 | 12 | 25 | 30 | 31 | 31 | 29 | 17 | 1 | 0 | 178 |
| Days Max Temp ≤ 32 °F | 0 | 0 | 0 | 0 | 0 | 0 | 0 | 0 | 0 | 0 | 0 | 0 | 0 |
| Days Min Temp ≤ 32 °F | 13 | 7 | 3 | 0 | 0 | 0 | 0 | 0 | 0 | 0 | 4 | 12 | 39 |
| Days Min Temp ≤ 0 °F | 0 | 0 | 0 | 0 | 0 | 0 | 0 | 0 | 0 | 0 | 0 | 0 | 0 |
| Heating Degree Days | 418 | 280 | 185 | 55 | 5 | 0 | 0 | 0 | 0 | 25 | 203 | 421 | 1592 |
| Cooling Degree Days | 0 | 4 | 27 | 161 | 376 | 630 | 793 | 759 | 550 | 237 | 17 | 0 | 3554 |
| Total Precipitation (") | 0.87 | 0.92 | 1.22 | 0.34 | 0.21 | 0.11 | 0.88 | 1.18 | 0.78 | 0.92 | 0.85 | 1.48 | 9.76 |
| Days ≥ 0.1" Precip | 2 | 2 | 3 | 1 | 1 | 0 | 2 | 2 | 2 | 2 | 2 | 3 | 22 |
| Total Snowfall (") | 0.0 | 0.0 | 0.0 | 0.0 | 0.0 | 0.0 | 0.0 | 0.0 | 0.0 | 0.0 | 0.0 | 0.0 | 0.0 |
| Days ≥ 1" Snow Depth | 0 | 0 | 0 | 0 | 0 | 0 | 0 | 0 | 0 | 0 | 0 | 0 | 0 |

## CASTLE HOT SPRINGS *Yavapai County* ELEVATION 1972 ft LAT/LONG 33° 59 ' N / 112° 22 ' W

| | JAN | FEB | MAR | APR | MAY | JUN | JUL | AUG | SEP | OCT | NOV | DEC | YEAR |
|---|---|---|---|---|---|---|---|---|---|---|---|---|---|
| Maximum Temp °F | 65.5 | 68.8 | 72.3 | 80.1 | 89.4 | 99.6 | 102.7 | 100.6 | 95.3 | 85.6 | 73.7 | 65.7 | 83.3 |
| Minimum Temp °F | 39.9 | 42.7 | 45.6 | 51.0 | 58.7 | 68.0 | 74.9 | 73.4 | 68.0 | 57.8 | 46.6 | 40.5 | 55.6 |
| Mean Temp °F | 52.8 | 55.8 | 59.0 | 65.6 | 74.1 | 83.8 | 88.8 | 87.0 | 81.7 | 71.7 | 60.2 | 53.1 | 69.5 |
| Days Max Temp ≥ 90 °F | 0 | 0 | 0 | 4 | 16 | 28 | 31 | 30 | 26 | 10 | 0 | 0 | 145 |
| Days Max Temp ≤ 32 °F | 0 | 0 | 0 | 0 | 0 | 0 | 0 | 0 | 0 | 0 | 0 | 0 | 0 |
| Days Min Temp ≤ 32 °F | 4 | 1 | 1 | 0 | 0 | 0 | 0 | 0 | 0 | 0 | 1 | 3 | 10 |
| Days Min Temp ≤ 0 °F | 0 | 0 | 0 | 0 | 0 | 0 | 0 | 0 | 0 | 0 | 0 | 0 | 0 |
| Heating Degree Days | 373 | 258 | 206 | 73 | 8 | 0 | 0 | 0 | 0 | 20 | 165 | 361 | 1464 |
| Cooling Degree Days | 0 | 8 | 26 | 128 | 293 | 564 | 715 | 676 | 496 | 233 | 27 | 0 | 3166 |
| Total Precipitation (") | 1.92 | 1.95 | 2.23 | 0.69 | 0.34 | 0.17 | 1.38 | 2.00 | 1.48 | 0.88 | 1.51 | 2.04 | 16.59 |
| Days ≥ 0.1" Precip | 3 | 3 | 3 | 2 | 1 | 1 | 3 | 4 | 2 | 2 | 2 | 3 | 29 |
| Total Snowfall (") | 0.0 | 0.2 | 0.0 | 0.0 | 0.0 | 0.0 | 0.0 | 0.0 | 0.0 | 0.0 | 0.0 | 0.0 | 0.2 |
| Days ≥ 1" Snow Depth | 0 | 0 | 0 | 0 | 0 | 0 | 0 | 0 | 0 | 0 | 0 | 0 | 0 |

## CHANDLER HEIGHTS *Maricopa County* ELEVATION 1450 ft LAT/LONG 33° 13' N / 111° 41' W

| | JAN | FEB | MAR | APR | MAY | JUN | JUL | AUG | SEP | OCT | NOV | DEC | YEAR |
|---|---|---|---|---|---|---|---|---|---|---|---|---|---|
| Maximum Temp °F | 65.0 | 69.7 | 74.9 | 83.9 | 93.0 | 102.0 | 104.0 | 101.7 | 97.3 | 87.0 | 73.7 | 64.5 | 84.7 |
| Minimum Temp °F | 38.3 | 41.6 | 45.1 | 50.8 | 59.4 | 68.3 | 75.4 | 73.7 | 67.8 | 56.5 | 44.9 | 38.8 | 55.0 |
| Mean Temp °F | 51.7 | 55.7 | 60.0 | 67.4 | 76.2 | 85.2 | 89.7 | 87.8 | 82.6 | 71.8 | 59.3 | 51.7 | 69.9 |
| Days Max Temp ≥ 90 °F | 0 | 0 | 1 | 8 | 22 | 29 | 31 | 30 | 27 | 13 | 0 | 0 | 161 |
| Days Max Temp ≤ 32 °F | 0 | 0 | 0 | 0 | 0 | 0 | 0 | 0 | 0 | 0 | 0 | 0 | 0 |
| Days Min Temp ≤ 32 °F | 6 | 3 | 1 | 0 | 0 | 0 | 0 | 0 | 0 | 0 | 1 | 6 | 17 |
| Days Min Temp ≤ 0 °F | 0 | 0 | 0 | 0 | 0 | 0 | 0 | 0 | 0 | 0 | 0 | 0 | 0 |
| Heating Degree Days | 405 | 262 | 177 | 55 | 5 | 0 | 0 | 0 | 0 | 24 | 189 | 407 | 1524 |
| Cooling Degree Days | 0 | 5 | 25 | 168 | 387 | 648 | 782 | 738 | 556 | 253 | 30 | 1 | 3593 |
| Total Precipitation (") | 0.96 | 1.03 | 1.25 | 0.33 | 0.21 | 0.05 | 0.80 | 1.23 | 0.86 | 0.80 | 0.95 | 1.36 | 9.83 |
| Days ≥ 0.1" Precip | 3 | 2 | 3 | 1 | 1 | 0 | 2 | 2 | 2 | 2 | 2 | 3 | 23 |
| Total Snowfall (") | 0.0 | 0.0 | 0.0 | 0.0 | 0.0 | 0.0 | 0.0 | 0.0 | 0.0 | 0.0 | 0.0 | 0.0 | 0.0 |
| Days ≥ 1" Snow Depth | 0 | 0 | 0 | 0 | 0 | 0 | 0 | 0 | 0 | 0 | 0 | 0 | 0 |

## CHEVELON R S *Coconino County* ELEVATION 7014 ft LAT/LONG 34° 32' N / 110° 55' W

| | JAN | FEB | MAR | APR | MAY | JUN | JUL | AUG | SEP | OCT | NOV | DEC | YEAR |
|---|---|---|---|---|---|---|---|---|---|---|---|---|---|
| Maximum Temp °F | 42.4 | 45.9 | 50.6 | 59.2 | 67.8 | 78.8 | 82.0 | 79.2 | 74.0 | 63.5 | na | *43.8* | na |
| Minimum Temp °F | 18.8 | 22.1 | 26.5 | 32.2 | 39.2 | 47.9 | 55.6 | 53.6 | 46.7 | 36.2 | *26.6* | *19.9* | 35.4 |
| Mean Temp °F | 30.6 | 34.0 | 38.6 | 45.7 | 53.5 | 63.4 | 68.8 | 66.4 | 60.4 | 49.9 | na | *31.9* | na |
| Days Max Temp ≥ 90 °F | 0 | 0 | 0 | 0 | 0 | 2 | 3 | 1 | 0 | 0 | 0 | 0 | 6 |
| Days Max Temp ≤ 32 °F | 4 | 2 | 1 | 0 | 0 | 0 | 0 | 0 | 0 | 0 | *1* | 3 | 11 |
| Days Min Temp ≤ 32 °F | 29 | 26 | 25 | 15 | 6 | 1 | 0 | 0 | 1 | 9 | na | 28 | na |
| Days Min Temp ≤ 0 °F | 1 | 0 | 0 | 0 | 0 | 0 | 0 | 0 | 0 | 0 | 0 | 1 | 2 |
| Heating Degree Days | 1058 | 866 | 813 | 572 | 352 | 100 | 11 | 30 | 148 | 461 | na | *1021* | na |
| Cooling Degree Days | 0 | 0 | 0 | 0 | 5 | 60 | 125 | 71 | 14 | 0 | 0 | 0 | 275 |
| Total Precipitation (") | 1.48 | 1.31 | 1.69 | 0.74 | 0.90 | 0.42 | 2.92 | 3.19 | 1.60 | 1.44 | 1.79 | 1.93 | 19.41 |
| Days ≥ 0.1" Precip | 4 | 3 | 4 | 2 | 2 | 1 | 7 | 7 | 3 | 3 | *3* | 4 | 43 |
| Total Snowfall (") | *10.8* | 8.5 | *9.2* | 4.2 | 0.2 | 0.0 | 0.0 | 0.0 | 0.1 | 0.5 | *4.3* | *11.0* | 48.8 |
| Days ≥ 1" Snow Depth | na | na | *2* | 1 | 0 | 0 | 0 | 0 | 0 | 0 | *1* | 6 | na |

## CHILDS *Yavapai County* ELEVATION 2650 ft LAT/LONG 34° 21' N / 111° 42' W

| | JAN | FEB | MAR | APR | MAY | JUN | JUL | AUG | SEP | OCT | NOV | DEC | YEAR |
|---|---|---|---|---|---|---|---|---|---|---|---|---|---|
| Maximum Temp °F | 60.2 | 65.5 | 69.8 | 78.6 | 87.8 | 98.6 | 102.1 | 99.3 | 94.0 | 83.6 | 70.4 | 59.8 | 80.8 |
| Minimum Temp °F | 32.3 | 34.9 | 38.6 | 43.2 | 50.9 | 59.0 | 67.9 | 66.3 | 59.3 | 48.3 | 37.5 | 32.5 | 47.6 |
| Mean Temp °F | 46.3 | 50.3 | 54.3 | 60.9 | 69.4 | 78.8 | 85.0 | 82.8 | 76.7 | 66.0 | 54.0 | 46.2 | 64.2 |
| Days Max Temp ≥ 90 °F | 0 | 0 | 0 | 3 | 15 | 26 | 30 | 29 | 24 | 9 | 0 | 0 | 136 |
| Days Max Temp ≤ 32 °F | 0 | 0 | 0 | 0 | 0 | 0 | 0 | 0 | 0 | 0 | 0 | 0 | 0 |
| Days Min Temp ≤ 32 °F | 16 | 9 | 4 | 1 | 0 | 0 | 0 | 0 | 0 | 0 | 6 | 16 | 52 |
| Days Min Temp ≤ 0 °F | 0 | 0 | 0 | 0 | 0 | 0 | 0 | 0 | 0 | 0 | 0 | 0 | 0 |
| Heating Degree Days | 572 | 409 | 332 | 150 | 34 | 1 | 0 | 0 | 3 | 71 | 327 | 576 | 2475 |
| Cooling Degree Days | 0 | 0 | 5 | 49 | 177 | 433 | 612 | 564 | 360 | 110 | 5 | 0 | 2315 |
| Total Precipitation (") | 2.13 | 2.01 | 2.28 | 0.88 | 0.51 | 0.28 | 1.95 | 2.58 | 1.69 | 1.42 | 1.72 | 2.60 | 20.05 |
| Days ≥ 0.1" Precip | 4 | 4 | 4 | 2 | 2 | 1 | 5 | 6 | 3 | 2 | 3 | 5 | 41 |
| Total Snowfall (") | 0.1 | 0.6 | 0.1 | 0.0 | 0.0 | 0.0 | 0.0 | 0.0 | 0.0 | 0.0 | 0.2 | 0.2 | 1.2 |
| Days ≥ 1" Snow Depth | 0 | 0 | 0 | 0 | 0 | 0 | 0 | 0 | 0 | 0 | 0 | 0 | 0 |

## CHINO VALLEY *Yavapai County* ELEVATION 4672 ft LAT/LONG 34° 46' N / 112° 27' W

| | JAN | FEB | MAR | APR | MAY | JUN | JUL | AUG | SEP | OCT | NOV | DEC | YEAR |
|---|---|---|---|---|---|---|---|---|---|---|---|---|---|
| Maximum Temp °F | na | *56.8* | *62.3* | 69.4 | *77.5* | *88.5* | 91.5 | 89.2 | *84.5* | *74.3* | na | na | na |
| Minimum Temp °F | na | *25.0* | *29.3* | 34.3 | *42.0* | 50.6 | 59.3 | *57.4* | *49.9* | *38.7* | na | na | na |
| Mean Temp °F | na | *40.9* | *45.8* | 51.8 | *59.8* | *69.5* | 75.4 | *73.3* | *67.2* | *56.5* | na | na | na |
| Days Max Temp ≥ 90 °F | 0 | 0 | 0 | 0 | 1 | 13 | 19 | 15 | 6 | 1 | 0 | 0 | 55 |
| Days Max Temp ≤ 32 °F | 0 | 0 | 0 | 0 | 0 | 0 | 0 | 0 | 0 | 0 | 0 | 1 | 1 |
| Days Min Temp ≤ 32 °F | 25 | 22 | 20 | 11 | 2 | 0 | 0 | 0 | 0 | 5 | 19 | 25 | 129 |
| Days Min Temp ≤ 0 °F | 0 | 0 | 0 | 0 | 0 | 0 | 0 | 0 | 0 | 0 | 0 | 0 | 0 |
| Heating Degree Days | na | *674* | *588* | 389 | *176* | *23* | 0 | *1* | *36* | *265* | na | na | na |
| Cooling Degree Days | na | na | na | 1 | na | *180* | *324* | na | na | na | na | na | na |
| Total Precipitation (") | 1.01 | 1.31 | 1.26 | 0.55 | 0.50 | 0.37 | 1.86 | 2.10 | 1.22 | 0.94 | 0.84 | 1.14 | 13.10 |
| Days ≥ 0.1" Precip | 2 | 3 | 3 | 1 | 1 | 1 | 4 | 4 | 2 | 2 | 1 | 3 | 27 |
| Total Snowfall (") | 1.2 | 1.2 | 1.0 | 0.3 | 0.0 | 0.0 | 0.0 | 0.0 | 0.0 | 0.0 | 0.2 | 1.7 | 5.6 |
| Days ≥ 1" Snow Depth | *2* | *0* | 0 | 0 | 0 | 0 | 0 | 0 | 0 | 0 | 0 | *1* | 3 |

## CHIRICAHUA NATL MONU *Cochise County* ELEVATION 5300 ft LAT/LONG 32° 0 ' N / 109° 21 ' W

| | JAN | FEB | MAR | APR | MAY | JUN | JUL | AUG | SEP | OCT | NOV | DEC | YEAR |
|---|---|---|---|---|---|---|---|---|---|---|---|---|---|
| Maximum Temp °F | 56.5 | 59.3 | 64.9 | 72.7 | 80.4 | 89.7 | 88.6 | 85.6 | 82.6 | 74.7 | 64.2 | 57.2 | 73.0 |
| Minimum Temp °F | 30.2 | 31.3 | 34.4 | 39.6 | 46.8 | 55.7 | 60.1 | 59.1 | 54.9 | 45.7 | 35.9 | 30.3 | 43.7 |
| Mean Temp °F | 43.4 | 45.3 | 49.7 | 56.2 | 63.6 | 72.7 | 74.4 | 72.4 | 68.8 | 60.2 | 50.1 | 43.7 | 58.4 |
| Days Max Temp ≥ 90 °F | 0 | 0 | 0 | 0 | 2 | 15 | 14 | 6 | 3 | 0 | 0 | 0 | 40 |
| Days Max Temp ≤ 32 °F | 0 | 0 | 0 | 0 | 0 | 0 | 0 | 0 | 0 | 0 | 0 | 0 | 0 |
| Days Min Temp ≤ 32 °F | 20 | 16 | 13 | 5 | 1 | 0 | 0 | 0 | 0 | 2 | 9 | 19 | 85 |
| Days Min Temp ≤ 0 °F | 0 | 0 | 0 | 0 | 0 | 0 | 0 | 0 | 0 | 0 | 0 | 0 | 0 |
| Heating Degree Days | 664 | 551 | 468 | 263 | 90 | 5 | 0 | 0 | 17 | 164 | 441 | 651 | 3314 |
| Cooling Degree Days | 0 | 0 | 0 | 8 | 58 | 259 | 298 | 236 | 147 | 22 | 0 | 0 | 1028 |
| Total Precipitation (") | 1.54 | 1.26 | 1.28 | 0.48 | 0.44 | 0.65 | 4.01 | 3.74 | 2.05 | 1.41 | 1.21 | 2.17 | 20.24 |
| Days ≥ 0.1" Precip | 4 | 3 | 3 | 1 | 1 | 2 | 8 | 8 | 4 | 3 | 2 | 4 | 43 |
| Total Snowfall (") | 1.6 | 1.5 | 1.7 | 0.5 | 0.0 | 0.0 | 0.0 | 0.0 | 0.0 | 0.0 | 0.6 | 3.8 | 9.7 |
| Days ≥ 1" Snow Depth | 0 | 1 | 0 | 0 | 0 | 0 | 0 | 0 | 0 | 0 | 0 | 1 | 2 |

## CLIFTON *Greenlee County* ELEVATION 3471 ft LAT/LONG 33° 3 ' N / 109° 17 ' W

| | JAN | FEB | MAR | APR | MAY | JUN | JUL | AUG | SEP | OCT | NOV | DEC | YEAR |
|---|---|---|---|---|---|---|---|---|---|---|---|---|---|
| Maximum Temp °F | 60.5 | 65.6 | 71.6 | 80.0 | 88.4 | 98.4 | 99.6 | 97.0 | 91.7 | 81.6 | 68.9 | 59.9 | 80.3 |
| Minimum Temp °F | 31.9 | 36.0 | 41.2 | 47.6 | 55.7 | 64.9 | 70.2 | 69.0 | 63.2 | 52.1 | 39.5 | 32.4 | 50.3 |
| Mean Temp °F | 46.3 | 50.8 | 56.4 | 63.8 | 72.1 | 81.7 | 84.9 | 83.1 | 77.5 | 66.8 | 54.2 | 46.1 | 65.3 |
| Days Max Temp ≥ 90 °F | 0 | 0 | 1 | 2 | 14 | 28 | 30 | 29 | 21 | 4 | 0 | 0 | 129 |
| Days Max Temp ≤ 32 °F | 0 | 0 | 0 | 0 | 0 | 0 | 0 | 0 | 0 | 0 | 0 | 0 | 0 |
| Days Min Temp ≤ 32 °F | 17 | 9 | 3 | 0 | 0 | 0 | 0 | 0 | 0 | 0 | 4 | 17 | 50 |
| Days Min Temp ≤ 0 °F | 0 | 0 | 0 | 0 | 0 | 0 | 0 | 0 | 0 | 0 | 0 | 0 | 0 |
| Heating Degree Days | 575 | 394 | 269 | 87 | 12 | 0 | 0 | 0 | 0 | 52 | 318 | 578 | 2285 |
| Cooling Degree Days | 0 | 0 | 7 | 73 | 251 | 518 | 613 | 567 | 392 | 113 | 1 | 0 | 2535 |
| Total Precipitation (") | 1.08 | 1.01 | 0.87 | 0.29 | 0.44 | 0.33 | 2.06 | 2.34 | 1.64 | 1.06 | 0.86 | 1.72 | 13.70 |
| Days ≥ 0.1" Precip | 3 | 3 | 3 | 1 | 2 | 1 | 5 | 6 | 4 | 3 | 2 | 4 | 37 |
| Total Snowfall (") | 0.0 | 0.0 | 0.0 | 0.0 | 0.0 | 0.0 | 0.0 | 0.0 | 0.0 | 0.0 | 0.0 | 0.1 | 0.1 |
| Days ≥ 1" Snow Depth | 0 | 0 | 0 | 0 | 0 | 0 | 0 | 0 | 0 | 0 | 0 | 0 | 0 |

## COLORADO CITY *Mohave County* ELEVATION 5013 ft LAT/LONG 37° 0 ' N / 112° 59 ' W

| | JAN | FEB | MAR | APR | MAY | JUN | JUL | AUG | SEP | OCT | NOV | DEC | YEAR |
|---|---|---|---|---|---|---|---|---|---|---|---|---|---|
| Maximum Temp °F | 47.3 | 52.2 | 57.9 | 66.5 | 76.3 | 87.5 | 92.3 | 89.9 | 82.7 | 71.6 | 57.3 | 48.5 | 69.2 |
| Minimum Temp °F | 23.1 | 27.9 | 32.1 | 37.5 | 45.7 | 54.2 | 60.4 | 59.5 | 52.5 | 41.6 | 31.1 | 23.6 | 40.8 |
| Mean Temp °F | 35.2 | 40.1 | 45.0 | 52.1 | 61.0 | 70.9 | 76.4 | 74.7 | 67.6 | 56.7 | 44.2 | 36.1 | 55.0 |
| Days Max Temp ≥ 90 °F | 0 | 0 | 0 | 0 | 1 | 13 | 23 | 18 | 5 | 0 | 0 | 0 | 60 |
| Days Max Temp ≤ 32 °F | 1 | 0 | 0 | 0 | 0 | 0 | 0 | 0 | 0 | 0 | 0 | 1 | 2 |
| Days Min Temp ≤ 32 °F | 27 | 20 | 16 | 7 | 1 | 0 | 0 | 0 | 0 | 4 | 17 | 26 | 118 |
| Days Min Temp ≤ 0 °F | 0 | 0 | 0 | 0 | 0 | 0 | 0 | 0 | 0 | 0 | 0 | 1 | 1 |
| Heating Degree Days | 916 | 697 | 612 | 384 | 152 | 20 | 0 | 1 | 40 | 263 | 615 | 889 | 4589 |
| Cooling Degree Days | 0 | 0 | 0 | 5 | 40 | 221 | 363 | 316 | 133 | 12 | 0 | 0 | 1090 |
| Total Precipitation (") | 1.32 | 1.34 | 1.56 | 0.78 | 0.71 | 0.37 | 1.21 | 1.53 | 0.94 | 0.86 | 1.11 | 0.97 | 12.70 |
| Days ≥ 0.1" Precip | 4 | 4 | 4 | 2 | 2 | 1 | 3 | 4 | 2 | 2 | 3 | 3 | 34 |
| Total Snowfall (") | 5.3 | 4.4 | 3.6 | 1.4 | 0.0 | 0.0 | 0.0 | 0.0 | 0.0 | 0.4 | 2.7 | 4.0 | 21.8 |
| Days ≥ 1" Snow Depth | 7 | 3 | 1 | 0 | 0 | 0 | 0 | 0 | 0 | 0 | 2 | 4 | 17 |

## CORDES *Yavapai County* ELEVATION 3771 ft LAT/LONG 34° 18 ' N / 112° 10 ' W

| | JAN | FEB | MAR | APR | MAY | JUN | JUL | AUG | SEP | OCT | NOV | DEC | YEAR |
|---|---|---|---|---|---|---|---|---|---|---|---|---|---|
| Maximum Temp °F | 57.4 | 60.6 | 64.4 | 72.6 | 81.4 | 92.0 | 95.4 | 93.1 | 87.6 | 77.3 | 65.3 | 57.4 | 75.4 |
| Minimum Temp °F | 32.5 | 34.4 | 36.9 | 41.3 | 48.4 | 57.1 | 65.0 | 64.1 | 57.8 | 48.4 | 38.3 | 32.6 | 46.4 |
| Mean Temp °F | 45.0 | 47.5 | 50.7 | 56.9 | 64.9 | 74.5 | 80.2 | 78.6 | 72.7 | 62.9 | 51.8 | 45.0 | 60.9 |
| Days Max Temp ≥ 90 °F | 0 | 0 | 0 | 0 | 5 | 19 | 27 | 24 | 12 | 3 | 0 | 0 | 90 |
| Days Max Temp ≤ 32 °F | 0 | 0 | 0 | 0 | 0 | 0 | 0 | 0 | 0 | 0 | 0 | 0 | 0 |
| Days Min Temp ≤ 32 °F | 15 | 11 | 7 | 3 | 0 | 0 | 0 | 0 | 0 | 0 | 6 | 15 | 57 |
| Days Min Temp ≤ 0 °F | 0 | 0 | 0 | 0 | 0 | 0 | 0 | 0 | 0 | 0 | 0 | 0 | 0 |
| Heating Degree Days | 614 | 487 | 438 | 247 | 79 | 6 | 0 | 0 | 9 | 116 | 389 | 612 | 2997 |
| Cooling Degree Days | 0 | 0 | 1 | 20 | 87 | 311 | 470 | 434 | 252 | 57 | 1 | 0 | 1633 |
| Total Precipitation (") | 1.66 | 1.77 | 1.67 | 0.64 | 0.47 | 0.23 | 1.74 | 2.12 | 1.67 | 1.08 | 1.39 | 1.83 | 16.27 |
| Days ≥ 0.1" Precip | 3 | 3 | 4 | 2 | 1 | 1 | 4 | 4 | 3 | 2 | 2 | 4 | 33 |
| Total Snowfall (") | 0.1 | 0.5 | 0.4 | 0.1 | 0.0 | 0.0 | 0.0 | 0.0 | 0.0 | 0.0 | 0.3 | 0.9 | 2.3 |
| Days ≥ 1" Snow Depth | 0 | 0 | 0 | 0 | 0 | 0 | 0 | 0 | 0 | 0 | 0 | 1 | 1 |

## CORONADO N M HDQTRS *Cochise County* ELEVATION 5243 ft LAT/LONG 31° 21' N / 110° 15' W

| | JAN | FEB | MAR | APR | MAY | JUN | JUL | AUG | SEP | OCT | NOV | DEC | YEAR |
|---|---|---|---|---|---|---|---|---|---|---|---|---|---|
| Maximum Temp °F | 58.3 | 62.1 | 67.0 | 74.8 | 82.0 | 91.6 | 89.9 | 87.1 | 84.3 | 76.4 | 66.1 | 58.8 | 74.9 |
| Minimum Temp °F | 32.4 | 34.1 | 37.2 | 43.2 | 50.3 | 59.1 | 62.0 | 60.2 | 56.7 | 48.2 | 38.5 | 33.0 | 46.2 |
| Mean Temp °F | 45.3 | 48.1 | 52.1 | 59.0 | 66.2 | 75.3 | 76.0 | 73.7 | 70.5 | 62.3 | 52.3 | 45.9 | 60.6 |
| Days Max Temp ≥ 90 °F | 0 | 0 | 0 | 0 | 4 | 19 | 17 | 9 | 5 | 1 | 0 | 0 | 55 |
| Days Max Temp ≤ 32 °F | 0 | 0 | 0 | 0 | 0 | 0 | 0 | 0 | 0 | 0 | 0 | 0 | 0 |
| Days Min Temp ≤ 32 °F | 16 | 12 | 9 | 3 | 0 | 0 | 0 | 0 | 0 | 1 | 7 | 15 | 63 |
| Days Min Temp ≤ 0 °F | 0 | 0 | 0 | 0 | 0 | 0 | 0 | 0 | 0 | 0 | 0 | 0 | 0 |
| Heating Degree Days | 603 | 470 | 393 | 187 | 51 | 2 | 0 | 0 | 7 | 116 | 374 | 584 | 2787 |
| Cooling Degree Days | 0 | 0 | 1 | 18 | 97 | 323 | 342 | 266 | 179 | 29 | 0 | 0 | 1255 |
| Total Precipitation (") | 1.95 | 1.47 | 1.32 | 0.45 | 0.31 | 0.56 | 4.68 | 3.63 | 2.04 | 1.46 | 1.07 | 2.41 | 21.35 |
| Days ≥ 0.1" Precip | 4 | 3 | 3 | 1 | 1 | 2 | 10 | 8 | 4 | 2 | 2 | 4 | 44 |
| Total Snowfall (") | 0.7 | 1.6 | 1.7 | 0.2 | 0.0 | 0.0 | 0.0 | 0.0 | 0.0 | 0.0 | 0.3 | 2.6 | 7.1 |
| Days ≥ 1" Snow Depth | ***1*** | ***0*** | ***0*** | 0 | 0 | 0 | 0 | 0 | 0 | 0 | 0 | 0 | 1 |

## DOUGLAS *Cochise County* ELEVATION 4144 ft LAT/LONG 31° 21' N / 109° 33' W

| | JAN | FEB | MAR | APR | MAY | JUN | JUL | AUG | SEP | OCT | NOV | DEC | YEAR |
|---|---|---|---|---|---|---|---|---|---|---|---|---|---|
| Maximum Temp °F | 62.3 | 66.7 | 71.9 | 79.3 | 86.4 | 95.3 | 93.9 | 91.4 | 88.6 | 80.8 | 70.1 | 62.6 | 79.1 |
| Minimum Temp °F | 28.9 | 31.6 | 36.0 | 41.5 | 49.2 | 57.8 | 64.2 | 62.6 | 56.8 | 45.3 | 34.3 | 28.6 | 44.7 |
| Mean Temp °F | 45.6 | 49.2 | 54.0 | 60.4 | 67.8 | 76.6 | 79.1 | 77.0 | 72.7 | 63.1 | 52.2 | 45.6 | 61.9 |
| Days Max Temp ≥ 90 °F | 0 | 0 | 0 | 1 | 10 | 26 | 25 | 21 | 14 | 3 | 0 | 0 | 100 |
| Days Max Temp ≤ 32 °F | 0 | 0 | 0 | 0 | 0 | 0 | 0 | 0 | 0 | 0 | 0 | 0 | 0 |
| Days Min Temp ≤ 32 °F | 22 | 16 | 10 | 3 | 0 | 0 | 0 | 0 | 0 | 2 | 13 | 22 | 88 |
| Days Min Temp ≤ 0 °F | 0 | 0 | 0 | 0 | 0 | 0 | 0 | 0 | 0 | 0 | 0 | 0 | 0 |
| Heating Degree Days | 594 | 440 | 337 | 151 | 30 | 0 | 0 | 0 | 2 | 98 | 377 | 595 | 2624 |
| Cooling Degree Days | 0 | 0 | 1 | 25 | 134 | 366 | 441 | 380 | 251 | 43 | 1 | 0 | 1642 |
| Total Precipitation (") | 0.85 | 0.66 | 0.54 | 0.28 | 0.34 | 0.55 | 3.77 | 3.34 | 1.63 | 1.07 | 0.69 | 1.23 | 14.95 |
| Days ≥ 0.1" Precip | 2 | 2 | 1 | 1 | 1 | 1 | 8 | 7 | 3 | 2 | 2 | 3 | 33 |
| Total Snowfall (") | 0.0 | 0.0 | 0.0 | 0.0 | 0.0 | 0.0 | 0.0 | 0.0 | 0.0 | 0.0 | 0.0 | 0.0 | 0.0 |
| Days ≥ 1" Snow Depth | 0 | 0 | 0 | 0 | 0 | 0 | 0 | 0 | 0 | 0 | 0 | 0 | 0 |

## DUNCAN *Greenlee County* ELEVATION 3642 ft LAT/LONG 32° 44' N / 109° 6' W

| | JAN | FEB | MAR | APR | MAY | JUN | JUL | AUG | SEP | OCT | NOV | DEC | YEAR |
|---|---|---|---|---|---|---|---|---|---|---|---|---|---|
| Maximum Temp °F | 59.0 | 64.1 | 70.2 | 78.8 | 86.8 | 96.0 | 96.4 | 93.7 | 88.8 | 79.6 | 67.6 | 58.8 | 78.3 |
| Minimum Temp °F | 23.9 | 26.4 | 31.2 | 35.9 | 44.3 | 53.5 | 63.4 | 61.9 | 53.8 | 40.6 | 28.7 | 23.3 | 40.6 |
| Mean Temp °F | 41.5 | 45.3 | 50.7 | 57.4 | 65.6 | 74.8 | 79.9 | 77.8 | 71.4 | 60.1 | 48.2 | 41.1 | 59.5 |
| Days Max Temp ≥ 90 °F | 0 | 0 | 0 | 2 | 11 | 26 | 28 | 26 | 15 | 3 | 0 | 0 | 111 |
| Days Max Temp ≤ 32 °F | 0 | 0 | 0 | 0 | 0 | 0 | 0 | 0 | 0 | 0 | 0 | 0 | 0 |
| Days Min Temp ≤ 32 °F | 25 | 22 | 18 | 10 | 1 | 0 | 0 | 0 | 0 | 5 | 21 | 26 | 128 |
| Days Min Temp ≤ 0 °F | 0 | 0 | 0 | 0 | 0 | 0 | 0 | 0 | 0 | 0 | 0 | 0 | 0 |
| Heating Degree Days | 722 | 551 | 436 | 229 | 55 | 1 | 0 | 0 | 8 | 167 | 497 | 735 | 3401 |
| Cooling Degree Days | 0 | 0 | 0 | 12 | 88 | 332 | 477 | 425 | 221 | 21 | 0 | 0 | 1576 |
| Total Precipitation (") | 0.93 | 0.92 | 0.64 | 0.22 | 0.37 | 0.34 | 2.39 | 2.07 | 1.11 | 1.13 | 0.72 | 1.43 | 12.27 |
| Days ≥ 0.1" Precip | 3 | 3 | 2 | 1 | 1 | 1 | 5 | 5 | 3 | 2 | 2 | 3 | 31 |
| Total Snowfall (") | 0.1 | ***0.1*** | 0.0 | 0.0 | 0.0 | 0.0 | 0.0 | 0.0 | 0.0 | 0.0 | 0.0 | ***0.3*** | 0.5 |
| Days ≥ 1" Snow Depth | 0 | ***0*** | 0 | 0 | 0 | 0 | 0 | 0 | 0 | 0 | 0 | ***0*** | 0 |

## ELOY 4 NE *Pinal County* ELEVATION 1545 ft LAT/LONG 32° 50' N / 111° 32' W

| | JAN | FEB | MAR | APR | MAY | JUN | JUL | AUG | SEP | OCT | NOV | DEC | YEAR |
|---|---|---|---|---|---|---|---|---|---|---|---|---|---|
| Maximum Temp °F | 67.1 | 71.5 | 76.2 | 84.9 | 93.4 | ***103.0*** | 104.5 | ***102.3*** | ***98.0*** | 88.5 | 75.9 | 66.9 | 86.0 |
| Minimum Temp °F | 36.6 | 40.1 | 44.8 | 50.8 | 59.2 | ***67.6*** | 74.2 | ***73.3*** | ***66.4*** | 55.1 | 43.4 | 37.1 | 54.1 |
| Mean Temp °F | 51.9 | 55.8 | 60.6 | 67.9 | 76.3 | ***85.4*** | 89.4 | ***87.8*** | ***82.2*** | 71.8 | 59.7 | 52.0 | 70.1 |
| Days Max Temp ≥ 90 °F | 0 | 0 | 2 | 9 | 23 | 29 | 31 | ***31*** | ***28*** | 14 | 1 | 0 | 168 |
| Days Max Temp ≤ 32 °F | 0 | 0 | 0 | 0 | 0 | 0 | 0 | 0 | 0 | 0 | 0 | 0 | 0 |
| Days Min Temp ≤ 32 °F | 9 | 3 | 1 | 0 | 0 | 0 | 0 | 0 | 0 | 0 | 3 | 8 | 24 |
| Days Min Temp ≤ 0 °F | 0 | 0 | 0 | 0 | 0 | 0 | 0 | 0 | 0 | 0 | 0 | 0 | 0 |
| Heating Degree Days | 400 | 257 | 159 | 46 | 5 | 0 | 0 | ***0*** | ***0*** | 20 | 184 | 396 | 1467 |
| Cooling Degree Days | 0 | 5 | 25 | 151 | 364 | 624 | 751 | 709 | 520 | 221 | 17 | 0 | 3387 |
| Total Precipitation (") | 0.97 | 0.96 | 1.20 | 0.31 | 0.27 | 0.13 | 1.19 | 1.54 | 0.96 | 1.05 | 0.92 | 1.51 | 11.01 |
| Days ≥ 0.1" Precip | 2 | 2 | 3 | 1 | 1 | 0 | ***3*** | ***3*** | ***2*** | 2 | 2 | 3 | 24 |
| Total Snowfall (") | 0.0 | 0.0 | 0.0 | 0.0 | 0.0 | 0.0 | 0.0 | 0.0 | 0.0 | 0.0 | 0.0 | 0.0 | 0.0 |
| Days ≥ 1" Snow Depth | 0 | 0 | 0 | 0 | 0 | 0 | 0 | 0 | 0 | 0 | 0 | 0 | 0 |

## FLAGSTAFF PULLIAM AP *Coconino County* ELEVATION 7006 ft LAT/LONG 35° 8 ' N / 111° 40 ' W

| | JAN | FEB | MAR | APR | MAY | JUN | JUL | AUG | SEP | OCT | NOV | DEC | YEAR |
|---|---|---|---|---|---|---|---|---|---|---|---|---|---|
| Maximum Temp °F | 42.5 | 45.3 | 49.7 | 58.3 | 67.6 | 78.3 | 82.0 | 79.6 | 73.7 | 63.3 | 50.7 | 43.0 | 61.2 |
| Minimum Temp °F | 15.9 | 18.2 | 22.3 | 27.0 | 33.8 | 41.4 | 50.2 | 48.9 | 41.2 | 31.1 | 22.4 | 16.1 | 30.7 |
| Mean Temp °F | 29.2 | 31.8 | 36.0 | 42.7 | 50.7 | 59.8 | 66.2 | 64.3 | 57.5 | 47.2 | 36.6 | 29.6 | 46.0 |
| Days Max Temp ≥ 90 °F | 0 | 0 | 0 | 0 | 0 | 2 | 2 | 1 | 0 | 0 | 0 | 0 | 5 |
| Days Max Temp ≤ 32 °F | 4 | 2 | 1 | 0 | 0 | 0 | 0 | 0 | 0 | 0 | 1 | 5 | 13 |
| Days Min Temp ≤ 32 °F | 30 | 28 | 30 | 25 | 13 | 3 | 0 | 0 | 3 | 18 | 28 | 30 | 208 |
| Days Min Temp ≤ 0 °F | 3 | 1 | 0 | 0 | 0 | 0 | 0 | 0 | 0 | 0 | 0 | 2 | 6 |
| Heating Degree Days | 1103 | 931 | 890 | 663 | 435 | 172 | 26 | 55 | 222 | 545 | 845 | 1090 | 6977 |
| Cooling Degree Days | 0 | 0 | 0 | 0 | 0 | 26 | 69 | 46 | 3 | 0 | 0 | 0 | 144 |
| Total Precipitation (") | 2.30 | 2.51 | 2.77 | 1.44 | 0.84 | 0.43 | 2.63 | 2.93 | 1.99 | 1.77 | 2.10 | 2.56 | 24.27 |
| Days ≥ 0.1" Precip | 5 | 5 | 6 | 3 | 2 | 1 | 6 | 7 | 4 | 3 | 4 | 5 | 51 |
| Total Snowfall (") | 21.4 | 20.7 | 26.1 | 11.9 | 1.5 | 0.0 | 0.0 | 0.0 | 0.1 | 2.5 | 11.9 | 18.6 | 114.7 |
| Days ≥ 1" Snow Depth | 18 | 15 | 10 | 4 | 0 | *0* | 0 | *0* | *0* | *1* | 4 | 13 | 65 |

## FLORENCE *Pinal County* ELEVATION 1503 ft LAT/LONG 33° 2 ' N / 111° 23 ' W

| | JAN | FEB | MAR | APR | MAY | JUN | JUL | AUG | SEP | OCT | NOV | DEC | YEAR |
|---|---|---|---|---|---|---|---|---|---|---|---|---|---|
| Maximum Temp °F | 66.4 | 71.1 | 75.8 | 85.0 | 94.1 | 103.9 | 105.3 | 103.5 | 98.9 | 88.4 | 75.2 | 66.4 | 86.2 |
| Minimum Temp °F | 38.0 | 40.7 | 44.0 | 49.2 | 57.3 | 66.0 | 74.9 | 73.9 | 67.4 | 55.6 | 44.2 | 38.4 | 54.1 |
| Mean Temp °F | 52.2 | 55.9 | 59.9 | 67.1 | 75.7 | 85.0 | 90.2 | 88.7 | 83.2 | 72.0 | 59.7 | 52.4 | 70.2 |
| Days Max Temp ≥ 90 °F | 0 | 0 | 2 | 10 | 23 | 29 | 31 | 31 | 28 | 15 | 1 | 0 | 170 |
| Days Max Temp ≤ 32 °F | 0 | 0 | 0 | 0 | 0 | 0 | 0 | 0 | 0 | 0 | 0 | 0 | 0 |
| Days Min Temp ≤ 32 °F | 7 | 4 | 1 | 0 | 0 | 0 | 0 | 0 | 0 | 0 | 2 | 7 | 21 |
| Days Min Temp ≤ 0 °F | 0 | 0 | 0 | 0 | 0 | 0 | 0 | 0 | 0 | 0 | 0 | 0 | 0 |
| Heating Degree Days | 389 | 254 | 179 | 52 | 4 | 0 | 0 | 0 | 0 | 21 | 177 | 384 | 1460 |
| Cooling Degree Days | 1 | 7 | 31 | 160 | 362 | 620 | 761 | 736 | 545 | 253 | 22 | 1 | 3499 |
| Total Precipitation (") | 1.05 | 1.03 | 1.30 | 0.45 | 0.28 | 0.17 | 1.05 | 1.11 | 0.93 | 0.93 | 0.88 | 1.70 | 10.88 |
| Days ≥ 0.1" Precip | 3 | 3 | 3 | 1 | 1 | 0 | 2 | 2 | 2 | 2 | 2 | 3 | 24 |
| Total Snowfall (") | 0.0 | 0.0 | 0.0 | 0.0 | 0.0 | 0.0 | 0.0 | 0.0 | 0.0 | 0.0 | 0.0 | 0.0 | 0.0 |
| Days ≥ 1" Snow Depth | 0 | 0 | 0 | 0 | 0 | 0 | 0 | 0 | 0 | 0 | 0 | 0 | 0 |

## FORT VALLEY *Coconino County* ELEVATION 7356 ft LAT/LONG 35° 16 ' N / 111° 44 ' W

| | JAN | FEB | MAR | APR | MAY | JUN | JUL | AUG | SEP | OCT | NOV | DEC | YEAR |
|---|---|---|---|---|---|---|---|---|---|---|---|---|---|
| Maximum Temp °F | 42.8 | 45.1 | 49.2 | 57.8 | 67.3 | 78.3 | 80.9 | 78.5 | 73.1 | 63.6 | 51.3 | 43.2 | 60.9 |
| Minimum Temp °F | 10.0 | 13.1 | 17.6 | 22.0 | 27.4 | 33.7 | 43.6 | 42.9 | 35.5 | 25.8 | 17.5 | 10.5 | 25.0 |
| Mean Temp °F | 26.4 | 29.1 | 33.4 | 39.9 | 47.4 | 56.0 | 62.2 | 60.7 | 54.3 | 44.7 | 34.5 | 26.9 | 43.0 |
| Days Max Temp ≥ 90 °F | 0 | 0 | 0 | 0 | 0 | 2 | 2 | 1 | 0 | 0 | 0 | 0 | 5 |
| Days Max Temp ≤ 32 °F | 4 | 2 | 1 | 0 | 0 | 0 | 0 | 0 | 0 | 0 | 1 | 4 | 12 |
| Days Min Temp ≤ 32 °F | 31 | 28 | 30 | 29 | 25 | 14 | 2 | 2 | 10 | 26 | 29 | 31 | 257 |
| Days Min Temp ≤ 0 °F | 7 | 3 | 1 | 0 | 0 | 0 | 0 | 0 | 0 | 0 | 1 | 5 | 17 |
| Heating Degree Days | 1190 | 1006 | 972 | 745 | 539 | 270 | 98 | 134 | 314 | 621 | 909 | 1175 | 7973 |
| Cooling Degree Days | 0 | 0 | 0 | 0 | 0 | 8 | 17 | 10 | 0 | 0 | 0 | 0 | 35 |
| Total Precipitation (") | 2.10 | 2.13 | 2.65 | 1.34 | 0.92 | 0.51 | 2.96 | 2.98 | 1.90 | 1.56 | 1.78 | 2.13 | 22.96 |
| Days ≥ 0.1" Precip | 4 | 4 | 5 | 3 | 3 | 2 | 7 | 6 | 4 | 3 | 3 | 4 | 48 |
| Total Snowfall (") | 19.7 | 18.1 | 22.2 | 9.3 | 0.5 | 0.0 | 0.0 | 0.0 | 0.0 | 1.6 | 9.4 | 17.5 | 98.3 |
| Days ≥ 1" Snow Depth | *23* | *19* | *17* | 4 | 0 | 0 | 0 | 0 | 0 | 0 | *6* | *16* | 85 |

## GANADO *Apache County* ELEVATION 6355 ft LAT/LONG 35° 43 ' N / 109° 33 ' W

| | JAN | FEB | MAR | APR | MAY | JUN | JUL | AUG | SEP | OCT | NOV | DEC | YEAR |
|---|---|---|---|---|---|---|---|---|---|---|---|---|---|
| Maximum Temp °F | 41.3 | *46.3* | 53.9 | 62.3 | *72.0* | 82.9 | *87.0* | *84.3* | *77.3* | *66.1* | *52.5* | 42.9 | 64.1 |
| Minimum Temp °F | 14.8 | *19.7* | 25.5 | 30.5 | *38.4* | 47.0 | *55.1* | *54.3* | *46.3* | *35.3* | *24.3* | 16.4 | 34.0 |
| Mean Temp °F | 28.1 | *33.0* | 39.7 | 46.4 | *55.3* | 64.9 | *71.1* | *69.4* | *61.9* | *50.7* | *38.4* | 29.7 | 49.1 |
| Days Max Temp ≥ 90 °F | 1 | 0 | 0 | 0 | 0 | 5 | *10* | 5 | 0 | 0 | 0 | 0 | 21 |
| Days Max Temp ≤ 32 °F | 5 | 2 | 1 | 0 | 0 | 0 | 0 | 0 | 0 | 0 | 1 | 5 | 14 |
| Days Min Temp ≤ 32 °F | 30 | 26 | 26 | 18 | 5 | 0 | 0 | 0 | 1 | 10 | 25 | 29 | 170 |
| Days Min Temp ≤ 0 °F | 3 | 1 | 0 | 0 | 0 | 0 | 0 | 0 | 0 | 0 | 0 | 2 | 6 |
| Heating Degree Days | 1141 | *896* | 777 | 551 | *295* | 68 | *2* | *9* | *118* | *437* | *789* | 1098 | 6181 |
| Cooling Degree Days | *0* | na | *0* | 0 | na | *79* | *188* | *168* | na | *0* | *0* | *0* | na |
| Total Precipitation (") | 0.77 | 0.77 | 1.00 | 0.70 | 0.57 | 0.32 | 1.53 | 1.70 | 1.11 | 1.15 | 0.97 | 1.06 | 11.65 |
| Days ≥ 0.1" Precip | *2* | *2* | 3 | 2 | 2 | 1 | 4 | 4 | 3 | 3 | 2 | 3 | 31 |
| Total Snowfall (") | *5.1* | *5.9* | *4.0* | *1.1* | 0.0 | 0.0 | 0.0 | 0.0 | 0.0 | 0.4 | *2.6* | *7.4* | 26.5 |
| Days ≥ 1" Snow Depth | na | na | na | *1* | 0 | 0 | 0 | 0 | 0 | 0 | na | na | na |

## GILA BEND *Maricopa County* ELEVATION 741 ft LAT/LONG 32° 57 ' N / 112° 43 ' W

| | JAN | FEB | MAR | APR | MAY | JUN | JUL | AUG | SEP | OCT | NOV | DEC | YEAR |
|---|---|---|---|---|---|---|---|---|---|---|---|---|---|
| Maximum Temp °F | 68.6 | 73.7 | 78.9 | 87.6 | 96.6 | 106.2 | 109.0 | 106.9 | 102.2 | 91.1 | 78.0 | 68.6 | 88.9 |
| Minimum Temp °F | 39.9 | 43.0 | 47.5 | 53.3 | 61.5 | 70.2 | 79.2 | 78.1 | 71.1 | 58.6 | 46.9 | 40.3 | 57.5 |
| Mean Temp °F | 54.3 | 58.4 | 63.3 | 70.5 | 79.1 | 88.2 | 94.1 | 92.6 | 86.6 | 74.9 | 62.5 | 54.5 | 73.3 |
| Days Max Temp ≥ 90 °F | 0 | 0 | 3 | 14 | 26 | 30 | 31 | 31 | 29 | 20 | 2 | 0 | 186 |
| Days Max Temp ≤ 32 °F | 0 | 0 | 0 | 0 | 0 | 0 | 0 | 0 | 0 | 0 | 0 | 0 | 0 |
| Days Min Temp ≤ 32 °F | 4 | 2 | 0 | 0 | 0 | 0 | 0 | 0 | 0 | 0 | 1 | 3 | 10 |
| Days Min Temp ≤ 0 °F | 0 | 0 | 0 | 0 | 0 | 0 | 0 | 0 | 0 | 0 | 0 | 0 | 0 |
| Heating Degree Days | 327 | 191 | 105 | 25 | 1 | 0 | 0 | 0 | 0 | 12 | 118 | 320 | 1099 |
| Cooling Degree Days | 2 | 17 | 64 | 246 | 470 | 728 | 902 | 878 | 675 | 348 | 52 | 1 | 4383 |
| Total Precipitation (") | 0.66 | 0.80 | 0.71 | 0.21 | 0.18 | 0.03 | 0.65 | 1.12 | 0.64 | 0.43 | 0.69 | 1.02 | 7.14 |
| Days ≥ 0.1" Precip | 2 | 2 | 2 | 1 | 0 | 0 | 2 | 3 | 1 | 1 | 2 | 2 | 18 |
| Total Snowfall (") | 0.0 | 0.0 | 0.0 | 0.0 | 0.0 | 0.0 | 0.0 | 0.0 | 0.0 | 0.0 | 0.0 | 0.1 | 0.1 |
| Days ≥ 1" Snow Depth | 0 | 0 | 0 | 0 | 0 | 0 | 0 | 0 | 0 | 0 | 0 | 0 | 0 |

## GREER *Apache County* ELEVATION 8405 ft LAT/LONG 34° 2 ' N / 109° 28 ' W

| | JAN | FEB | MAR | APR | MAY | JUN | JUL | AUG | SEP | OCT | NOV | DEC | YEAR |
|---|---|---|---|---|---|---|---|---|---|---|---|---|---|
| Maximum Temp °F | 41.4 | 43.3 | 47.7 | 56.4 | 64.9 | 74.6 | 75.6 | 72.4 | 67.8 | 59.7 | 49.6 | 42.4 | 58.0 |
| Minimum Temp °F | 15.0 | 17.0 | 21.0 | 26.3 | 32.8 | 40.6 | 47.2 | 46.0 | 40.4 | 30.9 | 22.2 | 16.6 | 29.7 |
| Mean Temp °F | 28.4 | 30.2 | 34.3 | 41.4 | 48.8 | 57.6 | 61.4 | 59.2 | 54.2 | 45.3 | 35.9 | 29.5 | 43.8 |
| Days Max Temp ≥ 90 °F | 0 | 0 | 0 | 0 | 0 | 0 | 0 | 0 | 0 | 0 | 0 | 0 | 0 |
| Days Max Temp ≤ 32 °F | 4 | 3 | 1 | 0 | 0 | 0 | 0 | 0 | 0 | 0 | 1 | *4* | 13 |
| Days Min Temp ≤ 32 °F | 31 | 28 | 30 | 25 | 15 | 3 | 0 | 0 | 2 | 18 | 28 | 31 | 211 |
| Days Min Temp ≤ 0 °F | 3 | 1 | 0 | 0 | 0 | 0 | 0 | 0 | 0 | 0 | 0 | 2 | 6 |
| Heating Degree Days | 1130 | 976 | 944 | 702 | 495 | 219 | 110 | 173 | 318 | 603 | 865 | 1092 | 7627 |
| Cooling Degree Days | 0 | 0 | 0 | 0 | 0 | 7 | 5 | 1 | 0 | 0 | 0 | 0 | 13 |
| Total Precipitation (") | 1.55 | 1.38 | 1.73 | 0.81 | 0.95 | 1.01 | 4.10 | 4.55 | 2.55 | 1.93 | 1.60 | 2.03 | 24.19 |
| Days ≥ 0.1" Precip | 4 | 4 | 5 | 3 | 3 | 3 | 11 | 11 | 6 | 4 | 4 | 4 | 62 |
| Total Snowfall (") | na | *19.9* | na | 6.7 | 0.5 | 0.1 | 0.0 | 0.0 | 0.0 | 1.5 | *11.2* | *28.4* | na |
| Days ≥ 1" Snow Depth | *23* | *18* | na | *2* | 0 | 0 | 0 | 0 | 0 | 1 | 6 | *19* | na |

## HEBER R S *Navajo County* ELEVATION 6506 ft LAT/LONG 34° 25 ' N / 110° 34 ' W

| | JAN | FEB | MAR | APR | MAY | JUN | JUL | AUG | SEP | OCT | NOV | DEC | YEAR |
|---|---|---|---|---|---|---|---|---|---|---|---|---|---|
| Maximum Temp °F | na | 50.5 | 55.3 | 63.5 | 72.2 | 82.7 | 84.7 | 81.9 | 76.8 | 67.0 | 55.0 | 46.9 | na |
| Minimum Temp °F | na | 19.7 | 24.0 | 28.3 | 35.2 | 43.1 | 52.1 | 51.2 | 44.1 | 32.9 | 23.5 | 17.5 | na |
| Mean Temp °F | na | 35.2 | 39.7 | 45.9 | 53.7 | 62.9 | 68.4 | 66.6 | 60.5 | 50.0 | 39.3 | 32.2 | na |
| Days Max Temp ≥ 90 °F | 0 | 0 | 0 | 0 | 0 | 4 | 6 | 2 | 0 | 0 | 0 | 0 | 12 |
| Days Max Temp ≤ 32 °F | 2 | 1 | 0 | 0 | 0 | 0 | 0 | 0 | 0 | 0 | 0 | 2 | 5 |
| Days Min Temp ≤ 32 °F | 25 | 24 | 26 | 21 | 11 | 2 | 0 | 0 | 1 | 15 | 26 | 28 | 179 |
| Days Min Temp ≤ 0 °F | 2 | 1 | 0 | 0 | 0 | 0 | 0 | 0 | 0 | 0 | 0 | 2 | 5 |
| Heating Degree Days | na | 834 | 779 | 566 | 343 | 104 | 9 | 22 | 144 | 459 | 766 | 1010 | na |
| Cooling Degree Days | na | *0* | 0 | 0 | 1 | 52 | 119 | 80 | 14 | 0 | 0 | 0 | na |
| Total Precipitation (") | 1.42 | 1.46 | 1.58 | 0.63 | 0.82 | 0.40 | 2.99 | 3.01 | 1.76 | 1.54 | 1.52 | 2.05 | 19.18 |
| Days ≥ 0.1" Precip | 3 | 3 | 4 | 2 | 2 | 1 | 6 | 7 | 4 | 3 | 3 | 4 | 42 |
| Total Snowfall (") | *8.7* | *9.2* | 5.9 | 2.6 | 0.0 | 0.0 | 0.0 | 0.0 | 0.0 | 0.0 | 3.6 | 14.1 | 44.1 |
| Days ≥ 1" Snow Depth | *10* | *5* | 2 | 0 | 0 | 0 | 0 | 0 | 0 | 0 | 1 | 7 | 25 |

## HILLSIDE 4 NNE *Yavapai County* ELEVATION 3202 ft LAT/LONG 34° 28 ' N / 112° 52 ' W

| | JAN | FEB | MAR | APR | MAY | JUN | JUL | AUG | SEP | OCT | NOV | DEC | YEAR |
|---|---|---|---|---|---|---|---|---|---|---|---|---|---|
| Maximum Temp °F | 59.3 | 62.7 | 66.1 | 74.1 | 83.3 | 93.7 | 97.1 | 94.7 | 89.8 | 79.1 | 67.1 | 59.5 | 77.2 |
| Minimum Temp °F | 25.6 | 27.8 | 31.0 | 34.4 | 41.3 | 48.5 | 60.3 | 60.0 | 51.6 | 39.6 | 29.5 | 25.3 | 39.6 |
| Mean Temp °F | 42.4 | 45.3 | 48.6 | 54.3 | 62.3 | 71.1 | 78.7 | 77.4 | 70.7 | 59.3 | 48.3 | 42.5 | 58.4 |
| Days Max Temp ≥ 90 °F | 0 | 0 | 0 | 1 | 6 | 22 | 28 | 25 | *16* | 3 | 0 | 0 | 101 |
| Days Max Temp ≤ 32 °F | 0 | 0 | 0 | 0 | 0 | 0 | 0 | 0 | 0 | 0 | 0 | 0 | 0 |
| Days Min Temp ≤ 32 °F | 26 | 21 | 20 | 12 | 3 | 0 | 0 | 0 | 0 | 5 | 21 | 26 | 134 |
| Days Min Temp ≤ 0 °F | 0 | 0 | 0 | 0 | 0 | 0 | 0 | 0 | 0 | 0 | 0 | 0 | 0 |
| Heating Degree Days | 692 | 550 | 501 | 317 | 117 | 15 | 0 | 0 | 16 | 185 | 494 | 692 | 3579 |
| Cooling Degree Days | 0 | 0 | 0 | 3 | 38 | 192 | 394 | 375 | 175 | 15 | 0 | 0 | 1192 |
| Total Precipitation (") | 1.83 | 1.80 | 2.10 | 0.76 | 0.42 | 0.23 | 1.46 | 2.50 | 1.25 | 1.05 | 1.44 | 1.67 | 16.51 |
| Days ≥ 0.1" Precip | 4 | 4 | 4 | 2 | 1 | 1 | 4 | 5 | 3 | 2 | 2 | 4 | 36 |
| Total Snowfall (") | 0.3 | 0.7 | 0.4 | 0.1 | 0.0 | 0.0 | 0.0 | 0.0 | 0.0 | 0.0 | 0.0 | 0.3 | 1.8 |
| Days ≥ 1" Snow Depth | 0 | 0 | 0 | 0 | 0 | 0 | 0 | 0 | 0 | 0 | 0 | 0 | 0 |

## HOLBROOK *Navajo County* ELEVATION 5072 ft LAT/LONG 34° 54 ' N / 110° 10 ' W

| | JAN | FEB | MAR | APR | MAY | JUN | JUL | AUG | SEP | OCT | NOV | DEC | YEAR |
|---|---|---|---|---|---|---|---|---|---|---|---|---|---|
| Maximum Temp °F | 47.4 | 55.1 | 62.3 | 71.2 | 79.9 | 90.6 | 94.0 | 91.4 | 85.0 | 73.6 | 60.0 | 48.7 | 71.6 |
| Minimum Temp °F | 18.1 | 22.6 | 28.0 | 34.1 | 41.4 | 50.3 | 59.1 | 57.9 | 49.4 | 36.9 | 25.6 | 18.6 | 36.8 |
| Mean Temp °F | 32.8 | 38.9 | 45.2 | 52.7 | 60.7 | 70.4 | 76.6 | 74.6 | 67.2 | 55.3 | 42.8 | 33.7 | 54.2 |
| Days Max Temp ≥ 90 °F | 0 | 0 | 0 | 0 | 3 | 18 | 24 | *21* | 7 | 0 | 0 | 0 | 73 |
| Days Max Temp ≤ 32 °F | 3 | 0 | 0 | 0 | 0 | 0 | 0 | 0 | 0 | 0 | 0 | 2 | 5 |
| Days Min Temp ≤ 32 °F | 29 | 26 | 23 | 13 | 3 | 0 | 0 | 0 | 0 | *8* | 25 | 30 | 157 |
| Days Min Temp ≤ 0 °F | 1 | 0 | 0 | 0 | 0 | 0 | 0 | 0 | 0 | 0 | 0 | 1 | 2 |
| Heating Degree Days | 993 | 731 | 609 | 366 | 154 | 18 | 0 | 1 | 37 | 298 | 659 | 965 | 4831 |
| Cooling Degree Days | 0 | 0 | 0 | 4 | 37 | 211 | 369 | 324 | 118 | 5 | 0 | 0 | 1068 |
| Total Precipitation (") | 0.66 | 0.60 | 0.69 | 0.33 | 0.41 | 0.18 | 1.23 | 1.53 | 1.02 | 0.88 | 0.66 | 0.77 | 8.96 |
| Days ≥ 0.1" Precip | 2 | 2 | 2 | 1 | 1 | 1 | 4 | 4 | 2 | 2 | 2 | 2 | 25 |
| Total Snowfall (") | *2.0* | na | 0.5 | 0.3 | 0.0 | 0.0 | 0.0 | 0.0 | 0.0 | 0.0 | 0.9 | *2.7* | na |
| Days ≥ 1" Snow Depth | na | na | 0 | 0 | 0 | 0 | 0 | 0 | 0 | 0 | 0 | *2* | na |

## JEROME *Yavapai County* ELEVATION 5253 ft LAT/LONG 34° 45 ' N / 112° 7 ' W

| | JAN | FEB | MAR | APR | MAY | JUN | JUL | AUG | SEP | OCT | NOV | DEC | YEAR |
|---|---|---|---|---|---|---|---|---|---|---|---|---|---|
| Maximum Temp °F | 49.8 | 53.7 | 58.7 | 67.1 | 76.1 | 86.4 | 90.1 | 87.1 | 81.7 | 71.1 | 58.4 | 50.5 | 69.2 |
| Minimum Temp °F | 33.2 | 35.3 | 38.5 | 44.7 | 52.7 | 62.6 | 66.8 | 65.2 | 60.2 | 50.5 | 40.0 | 33.6 | 48.6 |
| Mean Temp °F | 41.5 | 44.4 | 48.6 | 55.9 | 64.4 | 74.6 | 78.6 | 76.2 | 71.0 | 60.7 | 49.2 | 42.1 | 58.9 |
| Days Max Temp ≥ 90 °F | 0 | 0 | 0 | 0 | 1 | 11 | 17 | 11 | 3 | 0 | 0 | 0 | 43 |
| Days Max Temp ≤ 32 °F | 0 | 0 | 0 | 0 | 0 | 0 | 0 | 0 | 0 | 0 | 0 | 0 | 0 |
| Days Min Temp ≤ 32 °F | 13 | 9 | 7 | 3 | 0 | 0 | 0 | 0 | 0 | 0 | 5 | 13 | 50 |
| Days Min Temp ≤ 0 °F | 0 | 0 | 0 | 0 | 0 | 0 | 0 | 0 | 0 | 0 | 0 | 0 | 0 |
| Heating Degree Days | 720 | 576 | 501 | 280 | 96 | 7 | 0 | 1 | 15 | 169 | 471 | 704 | 3540 |
| Cooling Degree Days | 0 | 0 | 1 | 25 | 92 | 328 | 439 | 366 | 207 | 44 | 0 | 0 | 1502 |
| Total Precipitation (") | 1.72 | 2.07 | 2.25 | 1.06 | 0.78 | 0.39 | 2.76 | 3.25 | 1.50 | 1.37 | 1.64 | 1.54 | 20.33 |
| Days ≥ 0.1" Precip | 4 | 4 | 5 | 3 | 2 | 1 | 5 | 6 | 3 | 2 | 3 | *3* | 41 |
| Total Snowfall (") | 3.1 | 2.3 | 2.7 | 0.6 | 0.0 | 0.0 | 0.0 | 0.0 | 0.0 | 0.0 | 0.5 | 3.6 | 12.8 |
| Days ≥ 1" Snow Depth | *1* | *0* | *0* | 0 | 0 | 0 | 0 | 0 | 0 | 0 | 0 | na | na |

## KEAMS CANYON *Navajo County* ELEVATION 6214 ft LAT/LONG 35° 49 ' N / 110° 12 ' W

| | JAN | FEB | MAR | APR | MAY | JUN | JUL | AUG | SEP | OCT | NOV | DEC | YEAR |
|---|---|---|---|---|---|---|---|---|---|---|---|---|---|
| Maximum Temp °F | 43.4 | 49.1 | 55.8 | 64.7 | 72.8 | 84.9 | *88.7* | 85.9 | 79.1 | 68.3 | *54.7* | 44.3 | 66.0 |
| Minimum Temp °F | 16.1 | 20.7 | 25.1 | 30.5 | 37.8 | 46.2 | *54.7* | 54.4 | 46.5 | 34.7 | *25.2* | 16.6 | 34.0 |
| Mean Temp °F | 29.9 | 34.9 | 40.5 | 47.7 | 55.3 | 65.6 | *71.7* | 70.2 | 62.8 | 51.6 | *40.0* | 30.5 | 50.1 |
| Days Max Temp ≥ 90 °F | 0 | 0 | 0 | 0 | 0 | 7 | *14* | 6 | 1 | 0 | 0 | 0 | 28 |
| Days Max Temp ≤ 32 °F | 3 | 1 | 0 | 0 | 0 | 0 | 0 | 0 | 0 | 0 | 0 | 2 | 6 |
| Days Min Temp ≤ 32 °F | 30 | 26 | 27 | 18 | *6* | 0 | 0 | 0 | 1 | 10 | *24* | 29 | 171 |
| Days Min Temp ≤ 0 °F | 2 | 1 | 0 | 0 | 0 | 0 | 0 | 0 | 0 | 0 | 0 | 1 | 4 |
| Heating Degree Days | 1082 | 844 | 754 | 514 | 294 | *59* | *1* | 4 | 100 | 413 | *743* | 1064 | 5872 |
| Cooling Degree Days | 0 | 0 | 0 | 0 | 4 | *90* | *215* | 163 | 41 | *0* | *0* | 0 | 513 |
| Total Precipitation (") | 0.76 | 0.88 | 1.02 | 0.58 | 0.53 | 0.28 | 1.40 | 1.52 | 0.88 | 1.10 | 0.79 | 1.10 | 10.84 |
| Days ≥ 0.1" Precip | 2 | 2 | 3 | 2 | 2 | 1 | 3 | 3 | 2 | 2 | 2 | 2 | 26 |
| Total Snowfall (") | na | na | *1.4* | 0.4 | 0.0 | 0.0 | 0.0 | 0.0 | 0.0 | 0.0 | *0.5* | na | na |
| Days ≥ 1" Snow Depth | na | *2* | *1* | 0 | 0 | 0 | 0 | 0 | 0 | 0 | *1* | na | na |

## KITT PEAK *Pima County* ELEVATION 6844 ft LAT/LONG 31° 58 ' N / 111° 36 ' W

| | JAN | FEB | MAR | APR | MAY | JUN | JUL | AUG | SEP | OCT | NOV | DEC | YEAR |
|---|---|---|---|---|---|---|---|---|---|---|---|---|---|
| Maximum Temp °F | 48.9 | 50.5 | 53.7 | 61.6 | 69.8 | 79.4 | 80.2 | 77.6 | 74.6 | 66.5 | 56.3 | 49.7 | 64.1 |
| Minimum Temp °F | 32.7 | 33.3 | 35.4 | 41.1 | 48.3 | 58.0 | 60.4 | 59.2 | 56.2 | 48.0 | 38.9 | 33.5 | 45.4 |
| Mean Temp °F | 40.8 | 41.9 | 44.6 | 51.3 | 59.1 | 68.7 | 70.3 | 68.4 | 65.4 | 57.3 | 47.7 | 41.6 | 54.8 |
| Days Max Temp ≥ 90 °F | 0 | 0 | 0 | 0 | 0 | 2 | 2 | 0 | 0 | 0 | 0 | 0 | 4 |
| Days Max Temp ≤ 32 °F | 1 | 1 | 1 | 0 | 0 | 0 | 0 | 0 | 0 | 0 | 0 | 2 | 5 |
| Days Min Temp ≤ 32 °F | 14 | 12 | 12 | 6 | 1 | 0 | 0 | 0 | 0 | 1 | 7 | 13 | 66 |
| Days Min Temp ≤ 0 °F | 0 | 0 | 0 | 0 | 0 | 0 | 0 | 0 | 0 | 0 | 0 | 0 | 0 |
| Heating Degree Days | 743 | 644 | 627 | 407 | 202 | 36 | 6 | 14 | 53 | 248 | 514 | 718 | 4212 |
| Cooling Degree Days | 0 | 0 | 0 | 6 | 30 | 174 | 191 | 136 | 80 | 17 | 0 | 0 | 634 |
| Total Precipitation (") | 2.03 | 1.86 | 2.04 | 0.61 | 0.51 | 0.46 | 4.58 | 4.82 | 2.34 | 1.73 | 1.33 | 3.17 | 25.48 |
| Days ≥ 0.1" Precip | 4 | 3 | 3 | 1 | 1 | 1 | 8 | 8 | 4 | 3 | 2 | 4 | 42 |
| Total Snowfall (") | 3.6 | 5.6 | 4.3 | 1.1 | 0.1 | 0.0 | 0.0 | 0.0 | 0.0 | 0.1 | 1.2 | 4.6 | 20.6 |
| Days ≥ 1" Snow Depth | 4 | 4 | 3 | 1 | 0 | 0 | 0 | 0 | 0 | 0 | 0 | 3 | 15 |

## KOFA MINE *Yuma County* ELEVATION 1781 ft LAT/LONG 33° 16 ' N / 113° 52 ' W

| | JAN | FEB | MAR | APR | MAY | JUN | JUL | AUG | SEP | OCT | NOV | DEC | YEAR |
|---|---|---|---|---|---|---|---|---|---|---|---|---|---|
| Maximum Temp °F | 65.8 | 69.8 | 74.3 | 82.3 | 90.1 | ***99.8*** | 103.3 | 101.9 | 96.2 | 85.8 | 73.4 | ***64.9*** | 84.0 |
| Minimum Temp °F | 45.6 | 48.9 | 51.6 | 57.3 | 64.5 | ***73.3*** | 79.1 | 78.0 | 72.8 | 63.5 | 52.7 | ***45.2*** | 61.0 |
| Mean Temp °F | 55.8 | 59.4 | 63.0 | 69.8 | 77.3 | ***86.6*** | 91.2 | 90.0 | 84.5 | 74.7 | 63.1 | ***55.2*** | 72.6 |
| Days Max Temp ≥ 90 °F | 0 | 0 | 1 | 7 | 18 | ***28*** | 31 | 30 | 26 | 11 | 0 | ***0*** | 152 |
| Days Max Temp ≤ 32 °F | 0 | 0 | 0 | 0 | 0 | ***0*** | 0 | 0 | 0 | 0 | 0 | ***0*** | 0 |
| Days Min Temp ≤ 32 °F | 1 | 0 | 0 | 0 | 0 | ***0*** | 0 | 0 | 0 | 0 | 0 | ***1*** | 2 |
| Days Min Temp ≤ 0 °F | 0 | 0 | 0 | 0 | 0 | 0 | 0 | 0 | 0 | 0 | 0 | 0 | 0 |
| Heating Degree Days | 286 | 176 | 128 | 42 | 5 | 0 | 0 | 0 | 0 | 13 | 113 | ***301*** | 1064 |
| Cooling Degree Days | 5 | 27 | 66 | 224 | 406 | 682 | 819 | 787 | 607 | 335 | 60 | 3 | 4021 |
| Total Precipitation (") | 0.88 | 0.66 | 0.80 | 0.28 | 0.17 | 0.02 | 0.73 | 1.03 | 1.21 | 0.59 | 0.52 | 0.76 | 7.65 |
| Days ≥ 0.1" Precip | ***2*** | 2 | 2 | 1 | 0 | 0 | 1 | 2 | 2 | 1 | 1 | 2 | 16 |
| Total Snowfall (") | 0.0 | 0.0 | 0.0 | 0.0 | 0.0 | 0.0 | 0.0 | 0.0 | 0.0 | 0.0 | 0.0 | 0.0 | 0.0 |
| Days ≥ 1" Snow Depth | 0 | 0 | 0 | 0 | 0 | 0 | 0 | 0 | 0 | 0 | 0 | 0 | 0 |

## LAVEEN 3 SSE *Maricopa County* ELEVATION 1112 ft LAT/LONG 33° 20 ' N / 112° 9 ' W

| | JAN | FEB | MAR | APR | MAY | JUN | JUL | AUG | SEP | OCT | NOV | DEC | YEAR |
|---|---|---|---|---|---|---|---|---|---|---|---|---|---|
| Maximum Temp °F | 66.4 | 71.4 | 75.9 | 84.9 | 93.6 | 103.0 | 104.9 | 103.0 | 98.0 | 88.0 | 75.0 | 66.2 | 85.9 |
| Minimum Temp °F | 38.6 | 41.9 | 45.5 | 51.5 | 60.0 | 69.4 | 77.1 | 75.4 | 68.3 | 56.2 | 44.9 | 38.7 | 55.6 |
| Mean Temp °F | 52.5 | 56.7 | 60.8 | 68.2 | 76.9 | 86.3 | 91.0 | 89.2 | 83.2 | 72.2 | 60.0 | 52.5 | 70.8 |
| Days Max Temp ≥ 90 °F | 0 | 0 | 2 | 10 | 23 | 29 | 31 | 31 | 27 | 15 | 1 | 0 | 169 |
| Days Max Temp ≤ 32 °F | 0 | 0 | 0 | 0 | 0 | 0 | 0 | 0 | 0 | 0 | 0 | 0 | 0 |
| Days Min Temp ≤ 32 °F | 6 | 2 | 0 | 0 | 0 | 0 | 0 | 0 | 0 | 0 | 1 | 5 | 14 |
| Days Min Temp ≤ 0 °F | 0 | 0 | 0 | 0 | 0 | 0 | 0 | 0 | 0 | 0 | 0 | 0 | 0 |
| Heating Degree Days | 381 | 234 | 158 | 42 | 3 | 0 | 0 | 0 | 0 | 19 | 167 | 383 | 1387 |
| Cooling Degree Days | 0 | 8 | 36 | 182 | 391 | 651 | 791 | 755 | 551 | 255 | 23 | 0 | 3643 |
| Total Precipitation (") | 0.81 | 0.84 | 1.06 | 0.24 | 0.18 | 0.19 | 1.04 | 0.98 | 0.83 | 0.78 | 0.77 | 1.16 | 8.88 |
| Days ≥ 0.1" Precip | 2 | 2 | 3 | 1 | 1 | 0 | 2 | 2 | 2 | 2 | 2 | 2 | 21 |
| Total Snowfall (") | 0.0 | 0.0 | 0.0 | 0.0 | 0.0 | 0.0 | 0.0 | 0.0 | 0.0 | 0.0 | 0.0 | 0.0 | 0.0 |
| Days ≥ 1" Snow Depth | 0 | 0 | 0 | 0 | 0 | 0 | 0 | 0 | 0 | 0 | 0 | 0 | 0 |

## LEES FERRY *Coconino County* ELEVATION 3143 ft LAT/LONG 36° 52 ' N / 111° 35 ' W

| | JAN | FEB | MAR | APR | MAY | JUN | JUL | AUG | SEP | OCT | NOV | DEC | YEAR |
|---|---|---|---|---|---|---|---|---|---|---|---|---|---|
| Maximum Temp °F | 48.6 | 56.9 | 66.9 | 77.5 | 87.6 | 99.0 | 103.1 | 100.1 | 91.5 | 77.1 | 60.8 | 48.4 | 76.5 |
| Minimum Temp °F | 28.0 | 33.1 | 40.0 | 47.0 | 56.5 | 65.6 | 72.2 | 70.0 | 61.2 | 48.6 | 37.0 | 28.3 | 49.0 |
| Mean Temp °F | 38.3 | 45.0 | 53.5 | 62.3 | 72.1 | 82.3 | 87.7 | 85.1 | 76.4 | 62.9 | 48.9 | 38.5 | 62.8 |
| Days Max Temp ≥ 90 °F | 0 | 0 | 0 | 3 | 14 | 26 | 30 | 29 | 19 | 2 | 0 | 0 | 123 |
| Days Max Temp ≤ 32 °F | 1 | 0 | 0 | 0 | 0 | 0 | 0 | 0 | 0 | 0 | 0 | 0 | 1 |
| Days Min Temp ≤ 32 °F | 23 | 13 | 3 | 0 | 0 | 0 | 0 | 0 | 0 | 0 | ***8*** | 23 | 70 |
| Days Min Temp ≤ 0 °F | 0 | 0 | 0 | 0 | 0 | 0 | 0 | 0 | 0 | 0 | 0 | 0 | 0 |
| Heating Degree Days | 820 | 556 | 353 | 128 | 20 | 0 | 0 | 0 | 4 | 119 | 476 | 815 | 3291 |
| Cooling Degree Days | 0 | 0 | 2 | 67 | ***248*** | 550 | 711 | 626 | 342 | 60 | 0 | 0 | 2606 |
| Total Precipitation (") | 0.50 | 0.46 | 0.55 | 0.44 | 0.43 | 0.17 | 0.92 | 0.93 | 0.50 | 0.66 | 0.52 | 0.45 | 6.53 |
| Days ≥ 0.1" Precip | 1 | 2 | 2 | 1 | 2 | 0 | 3 | 2 | 2 | 2 | 2 | 1 | 20 |
| Total Snowfall (") | 0.3 | 0.2 | 0.0 | 0.0 | 0.0 | 0.0 | 0.0 | 0.0 | 0.0 | 0.0 | 0.1 | 1.0 | 1.6 |
| Days ≥ 1" Snow Depth | 0 | 0 | 0 | 0 | 0 | 0 | 0 | 0 | 0 | 0 | 0 | 1 | 1 |

## LITCHFIELD PARK *Maricopa County* ELEVATION 1030 ft LAT/LONG 33° 30 ' N / 112° 22 ' W

| | JAN | FEB | MAR | APR | MAY | JUN | JUL | AUG | SEP | OCT | NOV | DEC | YEAR |
|---|---|---|---|---|---|---|---|---|---|---|---|---|---|
| Maximum Temp °F | 67.6 | 72.6 | 77.5 | 86.8 | 95.9 | 105.4 | 107.8 | 106.0 | 100.7 | 90.3 | 76.6 | 67.2 | 87.9 |
| Minimum Temp °F | 37.4 | 40.7 | 44.9 | 50.3 | 58.5 | 67.1 | 75.4 | 74.0 | 66.4 | 54.3 | 43.2 | 37.1 | 54.1 |
| Mean Temp °F | 52.5 | 56.7 | 61.3 | 68.5 | 77.2 | 86.3 | 91.6 | 90.0 | 83.6 | 72.3 | 59.9 | 52.2 | 71.0 |
| Days Max Temp ≥ 90 °F | 0 | 0 | 2 | 12 | 25 | 30 | 30 | 31 | 29 | 18 | 1 | 0 | 178 |
| Days Max Temp ≤ 32 °F | 0 | 0 | 0 | 0 | 0 | 0 | 0 | 0 | 0 | 0 | 0 | 0 | 0 |
| Days Min Temp ≤ 32 °F | 7 | 3 | 1 | 0 | 0 | 0 | 0 | 0 | 0 | 0 | 1 | 8 | 20 |
| Days Min Temp ≤ 0 °F | 0 | 0 | 0 | 0 | 0 | 0 | 0 | 0 | 0 | 0 | 0 | 0 | 0 |
| Heating Degree Days | 380 | 232 | 145 | 39 | 3 | 0 | 0 | 0 | 0 | 16 | 164 | 391 | 1370 |
| Cooling Degree Days | 0 | 6 | 36 | ***201*** | 419 | 675 | 820 | 794 | 575 | 267 | 19 | 0 | 3812 |
| Total Precipitation (") | 1.01 | 1.07 | 1.07 | 0.31 | 0.14 | 0.06 | 0.67 | 0.88 | 1.05 | 0.64 | 0.78 | 1.28 | 8.96 |
| Days ≥ 0.1" Precip | 2 | 2 | 2 | 1 | 1 | 0 | 2 | 2 | 2 | 1 | 2 | 2 | 19 |
| Total Snowfall (") | 0.0 | 0.0 | 0.0 | 0.0 | 0.0 | 0.0 | 0.0 | 0.0 | 0.0 | 0.0 | 0.0 | 0.0 | 0.0 |
| Days ≥ 1" Snow Depth | 0 | 0 | 0 | 0 | 0 | 0 | 0 | 0 | 0 | 0 | 0 | 0 | 0 |

## LUKACHUKAI *Apache County* ELEVATION 6483 ft LAT/LONG 36° 25' N / 109° 14' W

| | JAN | FEB | MAR | APR | MAY | JUN | JUL | AUG | SEP | OCT | NOV | DEC | YEAR |
|---|---|---|---|---|---|---|---|---|---|---|---|---|---|
| Maximum Temp °F | 40.1 | 45.4 | *51.9* | 61.1 | 70.6 | 81.6 | *85.9* | 83.2 | 76.1 | 64.7 | 50.2 | 40.9 | 62.6 |
| Minimum Temp °F | 16.1 | 21.8 | 27.5 | 33.6 | 42.0 | 51.2 | 57.8 | 56.3 | 48.8 | 37.6 | 27.0 | 18.2 | 36.5 |
| Mean Temp °F | 28.1 | 33.6 | *39.8* | 47.3 | 56.3 | 66.4 | *71.8* | 69.8 | 62.5 | 51.2 | 38.6 | 29.6 | 49.6 |
| Days Max Temp ≥ 90 °F | 0 | 0 | 0 | 0 | 0 | 4 | *9* | 4 | 0 | 0 | 0 | 0 | 17 |
| Days Max Temp ≤ 32 °F | *6* | 2 | 0 | 0 | 0 | 0 | 0 | 0 | 0 | 0 | 1 | 5 | 14 |
| Days Min Temp ≤ 32 °F | 30 | 25 | 22 | 12 | 3 | 0 | 0 | 0 | 1 | 8 | 22 | 30 | 153 |
| Days Min Temp ≤ 0 °F | *2* | 1 | 0 | 0 | 0 | 0 | 0 | 0 | 0 | 0 | 0 | 1 | 4 |
| Heating Degree Days | 1137 | 879 | *774* | 525 | 270 | 59 | 4 | 11 | 111 | 422 | 784 | 1091 | 6067 |
| Cooling Degree Days | 0 | 0 | 0 | 1 | 10 | 116 | 235 | 180 | 46 | 0 | 0 | 0 | 588 |
| Total Precipitation (") | *0.92* | 0.45 | *0.72* | *0.52* | 0.49 | 0.30 | 1.28 | 1.42 | 1.15 | 1.07 | 0.80 | *0.94* | 10.06 |
| Days ≥ 0.1" Precip | *2* | *1* | na | 1 | 1 | 1 | 4 | 4 | 3 | 2 | *2* | *2* | na |
| Total Snowfall (") | na | na | na | *0.8* | 0.0 | 0.0 | 0.0 | 0.0 | 0.0 | 0.1 | *1.1* | na | na |
| Days ≥ 1" Snow Depth | na | na | na | *0* | 0 | 0 | 0 | 0 | 0 | 0 | na | na | na |

## MARICOPA 4 N *Pinal County* ELEVATION 1171 ft LAT/LONG 33° 7' N / 112° 2' W

| | JAN | FEB | MAR | APR | MAY | JUN | JUL | AUG | SEP | OCT | NOV | DEC | YEAR |
|---|---|---|---|---|---|---|---|---|---|---|---|---|---|
| Maximum Temp °F | 65.8 | 70.8 | 76.3 | 85.7 | 94.6 | 104.8 | 106.8 | 104.4 | 99.3 | 88.1 | 74.5 | 65.3 | 86.4 |
| Minimum Temp °F | 34.5 | 38.0 | 42.7 | 48.1 | 56.9 | 66.2 | 75.7 | 73.9 | 65.6 | 52.6 | 40.4 | 34.3 | 52.4 |
| Mean Temp °F | 50.1 | 54.4 | 59.5 | 66.9 | 75.8 | 85.5 | 91.2 | 89.2 | 82.5 | 70.4 | 57.5 | 49.9 | 69.4 |
| Days Max Temp ≥ 90 °F | 0 | 0 | 2 | 11 | 24 | 30 | 31 | 31 | 28 | 15 | 0 | 0 | 172 |
| Days Max Temp ≤ 32 °F | 0 | 0 | 0 | 0 | 0 | 0 | 0 | 0 | 0 | 0 | 0 | 0 | 0 |
| Days Min Temp ≤ 32 °F | 14 | 7 | 2 | 0 | 0 | 0 | 0 | 0 | 0 | 0 | 4 | 13 | 40 |
| Days Min Temp ≤ 0 °F | 0 | 0 | 0 | 0 | 0 | 0 | 0 | 0 | 0 | 0 | 0 | 0 | 0 |
| Heating Degree Days | 454 | 294 | 186 | 52 | 4 | 0 | 0 | 0 | 0 | 28 | 227 | 462 | 1707 |
| Cooling Degree Days | 0 | 2 | 22 | 151 | 376 | 649 | 811 | 773 | 543 | 215 | 10 | 0 | 3552 |
| Total Precipitation (") | 0.79 | 0.86 | 1.00 | 0.24 | 0.16 | 0.16 | 0.98 | 0.97 | 0.72 | 0.65 | 0.76 | 1.19 | 8.48 |
| Days ≥ 0.1" Precip | 2 | 2 | 2 | 1 | 0 | 0 | 2 | 2 | 2 | 1 | 2 | 3 | 19 |
| Total Snowfall (") | 0.0 | 0.0 | 0.0 | 0.0 | 0.0 | 0.0 | 0.0 | 0.0 | 0.0 | 0.0 | 0.0 | 0.0 | 0.0 |
| Days ≥ 1" Snow Depth | 0 | 0 | 0 | 0 | 0 | 0 | 0 | 0 | 0 | 0 | 0 | 0 | 0 |

## MC NARY 2 N *Apache County* ELEVATION 7316 ft LAT/LONG 34° 4' N / 109° 51' W

| | JAN | FEB | MAR | APR | MAY | JUN | JUL | AUG | SEP | OCT | NOV | DEC | YEAR |
|---|---|---|---|---|---|---|---|---|---|---|---|---|---|
| Maximum Temp °F | 44.8 | 47.0 | 51.4 | 60.3 | 69.0 | 79.4 | 81.2 | 78.3 | 74.0 | 65.6 | 53.5 | 46.2 | 62.6 |
| Minimum Temp °F | 17.6 | 19.7 | 23.6 | 28.6 | 35.1 | 42.7 | 49.5 | 48.3 | 42.5 | 33.3 | 24.6 | 18.7 | 32.0 |
| Mean Temp °F | 31.2 | 33.3 | 37.5 | 44.5 | 52.1 | 61.1 | 65.4 | 63.3 | 58.3 | 49.4 | 39.0 | 32.5 | 47.3 |
| Days Max Temp ≥ 90 °F | 0 | 0 | 0 | 0 | 0 | 2 | 2 | 0 | 0 | 0 | 0 | 0 | 4 |
| Days Max Temp ≤ 32 °F | 3 | 2 | 1 | 0 | 0 | 0 | 0 | 0 | 0 | 0 | 1 | 3 | 10 |
| Days Min Temp ≤ 32 °F | 30 | 27 | 29 | 21 | 9 | 1 | 0 | 0 | 1 | 13 | 26 | 29 | 186 |
| Days Min Temp ≤ 0 °F | 1 | 1 | 0 | 0 | 0 | 0 | 0 | 0 | 0 | 0 | 0 | 1 | 3 |
| Heating Degree Days | 1041 | 886 | 846 | 608 | 393 | 137 | 35 | 68 | 198 | 476 | 772 | 1002 | 6462 |
| Cooling Degree Days | 0 | 0 | 0 | 0 | 0 | 28 | 50 | 22 | 3 | 0 | 0 | 0 | 103 |
| Total Precipitation (") | 2.85 | 2.42 | 3.28 | 1.23 | 0.94 | 0.79 | 3.50 | 3.84 | 2.42 | 2.31 | 2.45 | 3.24 | 29.27 |
| Days ≥ 0.1" Precip | 6 | 5 | 7 | 3 | 3 | 2 | 8 | 10 | 5 | 4 | 4 | 5 | 62 |
| Total Snowfall (") | 19.1 | 17.5 | 20.1 | *7.1* | 0.9 | 0.0 | 0.0 | 0.0 | 0.0 | 2.6 | 7.8 | *19.5* | 94.6 |
| Days ≥ 1" Snow Depth | 22 | 17 | 13 | *4* | 0 | 0 | 0 | 0 | 0 | 1 | *6* | 15 | 78 |

## MESA *Maricopa County* ELEVATION 1235 ft LAT/LONG 33° 25' N / 111° 48' W

| | JAN | FEB | MAR | APR | MAY | JUN | JUL | AUG | SEP | OCT | NOV | DEC | YEAR |
|---|---|---|---|---|---|---|---|---|---|---|---|---|---|
| Maximum Temp °F | *65.8* | *70.0* | *74.5* | *82.6* | *92.4* | *102.1* | *104.9* | *103.1* | *97.8* | *87.1* | *74.7* | *66.1* | 85.1 |
| Minimum Temp °F | *39.1* | *42.1* | *46.1* | *51.6* | *59.8* | *68.4* | *76.7* | *75.3* | *68.7* | *57.0* | *46.0* | *39.2* | 55.8 |
| Mean Temp °F | *52.4* | *56.1* | *60.3* | *67.1* | *76.1* | *85.3* | *90.8* | *89.2* | *83.3* | *72.1* | *60.4* | *52.7* | 70.5 |
| Days Max Temp ≥ 90 °F | *0* | *0* | *1* | *7* | *21* | *29* | *31* | *30* | *27* | *14* | *0* | *0* | 160 |
| Days Max Temp ≤ 32 °F | *0* | *0* | *0* | *0* | *0* | *0* | *0* | *0* | *0* | *0* | *0* | *0* | 0 |
| Days Min Temp ≤ 32 °F | *4* | *2* | *0* | *0* | *0* | *0* | *0* | *0* | *0* | *0* | *0* | *4* | 10 |
| Days Min Temp ≤ 0 °F | *0* | *0* | *0* | *0* | *0* | *0* | *0* | *0* | *0* | *0* | *0* | *0* | 0 |
| Heating Degree Days | *383* | *249* | *170* | *57* | *6* | *0* | *0* | *0* | *0* | *20* | *157* | *375* | 1417 |
| Cooling Degree Days | na | na | na | na | na | na | na | na | na | na | na | na | na |
| Total Precipitation (") | 1.03 | 0.92 | 1.20 | 0.32 | 0.19 | 0.07 | 0.92 | 1.05 | 0.89 | 0.73 | 0.86 | 1.27 | 9.45 |
| Days ≥ 0.1" Precip | 3 | 2 | 3 | 1 | 1 | 0 | 2 | 3 | 2 | 2 | 2 | 3 | 24 |
| Total Snowfall (") | 0.0 | 0.0 | 0.0 | 0.0 | 0.0 | 0.0 | 0.0 | 0.0 | 0.0 | 0.0 | 0.0 | 0.0 | 0.0 |
| Days ≥ 1" Snow Depth | 0 | 0 | 0 | 0 | 0 | 0 | 0 | 0 | 0 | 0 | 0 | 0 | 0 |

## MIAMI *Gila County* ELEVATION 3602 ft LAT/LONG 33° 24' N / 110° 53' W

| | JAN | FEB | MAR | APR | MAY | JUN | JUL | AUG | SEP | OCT | NOV | DEC | YEAR |
|---|---|---|---|---|---|---|---|---|---|---|---|---|---|
| Maximum Temp °F | 56.5 | 60.5 | 65.4 | 74.1 | 83.1 | 93.4 | 96.2 | 93.8 | 88.6 | 78.3 | 65.5 | 56.5 | 76.0 |
| Minimum Temp °F | 32.9 | 35.5 | 40.0 | 46.7 | 55.1 | 63.8 | 70.0 | 68.0 | 62.1 | 51.0 | 39.7 | 33.2 | 49.8 |
| Mean Temp °F | 44.7 | 48.0 | 52.7 | 60.4 | 69.1 | 78.6 | 83.1 | 80.9 | 75.3 | 64.7 | 52.6 | 44.9 | 62.9 |
| Days Max Temp ≥ 90 °F | 0 | 0 | 0 | 1 | 6 | 22 | 28 | 25 | 14 | 2 | 0 | 0 | 98 |
| Days Max Temp ≤ 32 °F | 0 | 0 | 0 | 0 | 0 | 0 | 0 | 0 | 0 | 0 | 0 | 0 | 0 |
| Days Min Temp ≤ 32 °F | 16 | 8 | 3 | 1 | 0 | 0 | 0 | 0 | 0 | 0 | 4 | 14 | 46 |
| Days Min Temp ≤ 0 °F | 0 | 0 | 0 | 0 | 0 | 0 | 0 | 0 | 0 | 0 | 0 | 0 | 0 |
| Heating Degree Days | 621 | 474 | 378 | 166 | 36 | 1 | 0 | 0 | 4 | 90 | 366 | 616 | 2752 |
| Cooling Degree Days | 0 | 0 | 3 | 53 | 183 | 442 | 562 | 504 | 332 | 90 | 2 | 0 | 2171 |
| Total Precipitation (") | 2.06 | 1.93 | 2.26 | 0.58 | 0.49 | 0.28 | 2.15 | 3.02 | 1.66 | 1.62 | 1.72 | 2.79 | 20.56 |
| Days ≥ 0.1" Precip | 4 | 4 | 4 | 2 | 1 | 1 | 5 | 6 | 3 | 3 | 3 | 4 | 40 |
| Total Snowfall (") | 0.0 | 0.6 | 0.1 | 0.1 | 0.0 | 0.0 | 0.0 | 0.0 | 0.0 | 0.0 | 0.1 | 1.5 | 2.4 |
| Days ≥ 1" Snow Depth | 0 | 0 | 0 | 0 | 0 | 0 | 0 | 0 | 0 | 0 | 0 | 0 | 0 |

## MONTEZUMA CASTLE NM *Yavapai County* ELEVATION 3182 ft LAT/LONG 34° 37' N / 111° 50' W

| | JAN | FEB | MAR | APR | MAY | JUN | JUL | AUG | SEP | OCT | NOV | DEC | YEAR |
|---|---|---|---|---|---|---|---|---|---|---|---|---|---|
| Maximum Temp °F | 60.1 | 65.3 | 70.2 | 78.7 | 87.4 | 97.7 | 100.9 | 98.2 | 92.6 | 82.1 | 69.0 | 59.1 | 80.1 |
| Minimum Temp °F | 26.2 | 29.4 | 34.5 | 39.1 | 46.3 | 54.0 | 63.6 | 62.3 | 54.5 | 42.9 | 32.1 | 26.1 | 42.6 |
| Mean Temp °F | 43.1 | 47.4 | 52.3 | 58.9 | 66.9 | 75.9 | 82.3 | 80.3 | 73.6 | 62.5 | 50.6 | 42.6 | 61.4 |
| Days Max Temp ≥ 90 °F | 0 | 0 | 0 | 3 | 13 | 26 | 30 | 29 | 21 | 6 | 0 | 0 | 128 |
| Days Max Temp ≤ 32 °F | 0 | 0 | 0 | 0 | 0 | 0 | 0 | 0 | 0 | 0 | 0 | 0 | 0 |
| Days Min Temp ≤ 32 °F | 25 | 19 | 11 | 5 | 1 | 0 | 0 | 0 | 0 | 3 | 17 | 26 | 107 |
| Days Min Temp ≤ 0 °F | 0 | 0 | 0 | 0 | 0 | 0 | 0 | 0 | 0 | 0 | 0 | 0 | 0 |
| Heating Degree Days | 671 | 491 | 387 | 193 | 47 | 2 | 0 | 0 | 5 | 115 | 426 | 688 | 3025 |
| Cooling Degree Days | 0 | 0 | 1 | 28 | 119 | 348 | 534 | 486 | 274 | 47 | 0 | 0 | 1837 |
| Total Precipitation (") | 1.28 | 1.33 | 1.43 | 0.68 | 0.39 | 0.30 | 1.49 | 2.09 | 1.52 | 1.16 | 1.17 | 1.39 | 14.23 |
| Days ≥ 0.1" Precip | 3 | 3 | 4 | 2 | 1 | 1 | 4 | 5 | 3 | 3 | 2 | 3 | 34 |
| Total Snowfall (") | 0.2 | 0.5 | 0.2 | 0.0 | 0.0 | 0.0 | 0.0 | 0.0 | 0.0 | 0.0 | 0.0 | 1.8 | 2.7 |
| Days ≥ 1" Snow Depth | 0 | 0 | 0 | 0 | 0 | 0 | 0 | 0 | 0 | 0 | 0 | 1 | 1 |

## MORMON FLAT *Maricopa County* ELEVATION 1821 ft LAT/LONG 33° 33' N / 111° 27' W

| | JAN | FEB | MAR | APR | MAY | JUN | JUL | AUG | SEP | OCT | NOV | DEC | YEAR |
|---|---|---|---|---|---|---|---|---|---|---|---|---|---|
| Maximum Temp °F | 63.8 | 68.3 | 72.3 | 81.9 | 91.1 | 101.6 | 103.4 | 101.8 | 96.5 | 86.0 | 73.0 | 63.6 | 83.6 |
| Minimum Temp °F | 42.3 | 43.9 | 46.6 | 52.9 | 61.9 | 71.0 | 77.3 | 75.8 | 70.7 | 60.8 | 49.3 | 42.4 | 57.9 |
| Mean Temp °F | 53.0 | 56.1 | 59.5 | 67.4 | 76.5 | 86.3 | 90.4 | 88.8 | 83.7 | 73.5 | 61.2 | 53.0 | 70.8 |
| Days Max Temp ≥ 90 °F | 0 | 0 | 1 | 6 | 19 | 28 | 29 | 29 | 26 | 11 | 0 | 0 | 149 |
| Days Max Temp ≤ 32 °F | 0 | 0 | 0 | 0 | 0 | 0 | 0 | 0 | 0 | 0 | 0 | 0 | 0 |
| Days Min Temp ≤ 32 °F | 2 | 1 | 0 | 0 | 0 | 0 | 0 | 0 | 0 | 0 | 0 | 2 | 5 |
| Days Min Temp ≤ 0 °F | 0 | 0 | 0 | 0 | 0 | 0 | 0 | 0 | 0 | 0 | 0 | 0 | 0 |
| Heating Degree Days | 367 | 251 | 195 | 60 | 7 | 0 | 0 | 0 | 0 | 19 | 146 | ***365*** | 1410 |
| Cooling Degree Days | 1 | 8 | 35 | 166 | 380 | 648 | 778 | ***745*** | 570 | 285 | 43 | ***2*** | 3661 |
| Total Precipitation (") | 1.63 | 1.44 | 2.02 | 0.51 | 0.44 | 0.11 | 1.53 | 1.85 | 1.30 | 1.31 | 1.32 | 1.93 | 15.39 |
| Days ≥ 0.1" Precip | 3 | 3 | 4 | 1 | 1 | 0 | 3 | 4 | 2 | 2 | 2 | 3 | 28 |
| Total Snowfall (") | 0.0 | 0.0 | 0.0 | 0.0 | 0.0 | 0.0 | 0.0 | 0.0 | 0.0 | 0.0 | 0.0 | 0.0 | 0.0 |
| Days ≥ 1" Snow Depth | 0 | 0 | 0 | 0 | 0 | 0 | 0 | 0 | 0 | 0 | 0 | 0 | 0 |

## NOGALES 6 N *Santa Cruz County* ELEVATION 3563 ft LAT/LONG 31° 25' N / 111° 57' W

| | JAN | FEB | MAR | APR | MAY | JUN | JUL | AUG | SEP | OCT | NOV | DEC | YEAR |
|---|---|---|---|---|---|---|---|---|---|---|---|---|---|
| Maximum Temp °F | 63.6 | 66.5 | 70.7 | 78.2 | 85.7 | 95.4 | 93.9 | 92.0 | 89.7 | 82.1 | 71.5 | 64.4 | 79.5 |
| Minimum Temp °F | 27.3 | 29.6 | 33.7 | 38.3 | 44.8 | 54.0 | 63.4 | 62.5 | 55.1 | 43.5 | 32.9 | 27.8 | 42.7 |
| Mean Temp °F | 45.5 | 48.1 | 52.2 | 58.3 | 65.4 | 74.7 | 78.7 | 77.1 | 72.4 | 62.8 | 52.2 | 46.1 | 61.1 |
| Days Max Temp ≥ 90 °F | 0 | 0 | 0 | 1 | 9 | 25 | 24 | 23 | 17 | 5 | 0 | 0 | 104 |
| Days Max Temp ≤ 32 °F | 0 | 0 | 0 | 0 | 0 | 0 | 0 | 0 | 0 | 0 | 0 | 0 | 0 |
| Days Min Temp ≤ 32 °F | 24 | 19 | 14 | 6 | 1 | 0 | 0 | 0 | 0 | 2 | 15 | 24 | 105 |
| Days Min Temp ≤ 0 °F | 0 | 0 | 0 | 0 | 0 | 0 | 0 | 0 | 0 | 0 | 0 | 0 | 0 |
| Heating Degree Days | 598 | 471 | 391 | 208 | 64 | 4 | 0 | 0 | 4 | 107 | 377 | 579 | 2803 |
| Cooling Degree Days | 0 | 0 | 1 | 15 | 86 | 311 | 443 | 403 | 237 | 40 | 0 | 0 | 1536 |
| Total Precipitation (") | 1.32 | 1.03 | 1.01 | 0.43 | 0.33 | 0.46 | 4.73 | 3.98 | 1.88 | 1.50 | 0.73 | 1.88 | 19.28 |
| Days ≥ 0.1" Precip | 3 | 2 | 2 | 1 | 1 | 1 | 9 | 8 | 4 | 2 | 2 | 4 | 39 |
| Total Snowfall (") | 0.0 | 0.0 | 0.0 | 0.0 | 0.0 | 0.0 | 0.0 | 0.0 | 0.0 | 0.0 | 0.0 | 0.3 | 0.3 |
| Days ≥ 1" Snow Depth | 0 | 0 | 0 | 0 | 0 | 0 | 0 | 0 | 0 | 0 | 0 | 0 | 0 |

## ORACLE 2 SE *Pinal County* ELEVATION 4373 ft LAT/LONG 32° 34 ' N / 110° 43 ' W

| | JAN | FEB | MAR | APR | MAY | JUN | JUL | AUG | SEP | OCT | NOV | DEC | YEAR |
|---|---|---|---|---|---|---|---|---|---|---|---|---|---|
| Maximum Temp °F | 56.8 | 60.3 | 64.6 | 72.7 | 81.7 | 91.7 | 92.1 | 89.4 | 85.6 | 76.7 | 65.0 | 57.0 | 74.5 |
| Minimum Temp °F | 34.9 | 36.8 | 39.8 | 45.6 | 53.8 | 63.4 | 66.7 | 65.3 | 60.9 | 51.2 | 41.0 | 35.2 | 49.6 |
| Mean Temp °F | 45.9 | 48.6 | 52.2 | 59.2 | 67.8 | 77.6 | 79.4 | 77.4 | 73.3 | 64.0 | 53.0 | 46.2 | 62.0 |
| Days Max Temp ≥ 90 °F | 0 | 0 | 0 | 0 | 4 | 20 | 21 | 17 | 7 | 1 | 0 | 0 | 70 |
| Days Max Temp ≤ 32 °F | 0 | 0 | 0 | 0 | 0 | 0 | 0 | 0 | 0 | 0 | 0 | 0 | 0 |
| Days Min Temp ≤ 32 °F | 11 | 8 | 6 | 1 | 0 | 0 | 0 | 0 | 0 | 0 | 4 | 11 | 41 |
| Days Min Temp ≤ 0 °F | 0 | 0 | 0 | 0 | 0 | 0 | 0 | 0 | 0 | 0 | 0 | 0 | 0 |
| Heating Degree Days | 587 | 457 | 393 | 190 | 44 | 2 | 0 | 0 | 4 | 97 | 355 | 578 | 2707 |
| Cooling Degree Days | 0 | 0 | 1 | 29 | 131 | 396 | 447 | 376 | 251 | 58 | 1 | 0 | 1690 |
| Total Precipitation (") | 2.42 | 2.32 | 2.62 | 0.94 | 0.57 | 0.31 | 3.00 | 3.87 | 1.87 | 1.83 | 1.85 | 2.90 | 24.50 |
| Days ≥ 0.1" Precip | 4 | 4 | 5 | 2 | 2 | 1 | 6 | 7 | 4 | 3 | 3 | 5 | 46 |
| Total Snowfall (") | 2.3 | 2.7 | *2.0* | 0.5 | 0.0 | 0.0 | 0.0 | 0.0 | 0.0 | 0.0 | 0.5 | 3.6 | 11.6 |
| Days ≥ 1" Snow Depth | *0* | *0* | *0* | 0 | 0 | 0 | 0 | 0 | 0 | 0 | 0 | na | na |

## ORGAN PIPE CACTUS NM *Pima County* ELEVATION 1660 ft LAT/LONG 31° 56 ' N / 112° 47 ' W

| | JAN | FEB | MAR | APR | MAY | JUN | JUL | AUG | SEP | OCT | NOV | DEC | YEAR |
|---|---|---|---|---|---|---|---|---|---|---|---|---|---|
| Maximum Temp °F | 68.3 | 72.3 | 76.5 | 84.3 | 91.8 | 100.8 | 103.2 | 101.6 | 97.7 | 88.3 | 76.3 | 68.2 | 85.8 |
| Minimum Temp °F | 38.9 | 41.0 | 44.0 | 49.0 | 56.1 | 64.5 | 73.6 | 72.3 | 66.5 | 55.8 | 44.9 | 38.9 | 53.8 |
| Mean Temp °F | 53.6 | 56.6 | 60.3 | 66.7 | 74.0 | 82.7 | 88.4 | 87.0 | 82.1 | 72.0 | 60.6 | 53.6 | 69.8 |
| Days Max Temp ≥ 90 °F | 0 | 0 | 2 | 8 | 20 | 28 | 31 | 31 | 27 | 15 | 1 | 0 | 163 |
| Days Max Temp ≤ 32 °F | 0 | 0 | 0 | 0 | 0 | 0 | 0 | 0 | 0 | 0 | 0 | 0 | 0 |
| Days Min Temp ≤ 32 °F | 5 | 2 | 1 | 0 | 0 | 0 | 0 | 0 | 0 | 0 | 1 | 5 | 14 |
| Days Min Temp ≤ 0 °F | 0 | 0 | 0 | 0 | 0 | 0 | 0 | 0 | 0 | 0 | 0 | 0 | 0 |
| Heating Degree Days | 347 | 235 | 169 | 57 | 7 | 0 | 0 | 0 | 0 | 16 | 147 | 348 | 1326 |
| Cooling Degree Days | 0 | 9 | 33 | 152 | 318 | 567 | 735 | 701 | 533 | 258 | 25 | 0 | 3331 |
| Total Precipitation (") | 0.89 | 0.77 | 0.94 | 0.18 | 0.12 | 0.03 | 1.45 | 2.11 | 0.98 | 0.74 | 0.68 | 1.39 | 10.28 |
| Days ≥ 0.1" Precip | 2 | 2 | 2 | 1 | 0 | 0 | 3 | 4 | 2 | 1 | 2 | 3 | 22 |
| Total Snowfall (") | 0.0 | 0.0 | 0.0 | 0.0 | 0.0 | 0.0 | 0.0 | 0.0 | 0.0 | 0.0 | 0.0 | 0.0 | 0.0 |
| Days ≥ 1" Snow Depth | 0 | 0 | 0 | 0 | 0 | 0 | 0 | 0 | 0 | 0 | 0 | 0 | 0 |

## PAGE *Coconino County* ELEVATION 4272 ft LAT/LONG 36° 56 ' N / 111° 27 ' W

| | JAN | FEB | MAR | APR | MAY | JUN | JUL | AUG | SEP | OCT | NOV | DEC | YEAR |
|---|---|---|---|---|---|---|---|---|---|---|---|---|---|
| Maximum Temp °F | 42.5 | 50.1 | 59.6 | 69.1 | 79.0 | 91.0 | 95.9 | 93.0 | 84.4 | 70.7 | 54.7 | 43.5 | 69.5 |
| Minimum Temp °F | 25.2 | 30.5 | 37.5 | 44.4 | 53.1 | 63.1 | 69.1 | 67.1 | 58.6 | 47.0 | 35.3 | 26.2 | 46.4 |
| Mean Temp °F | 33.9 | 40.3 | 48.5 | 56.8 | 66.1 | 77.0 | 82.5 | 80.1 | 71.5 | 58.9 | 45.0 | 34.9 | 58.0 |
| Days Max Temp ≥ 90 °F | 0 | 0 | 0 | 0 | 3 | 19 | 28 | 24 | 8 | 0 | 0 | 0 | 82 |
| Days Max Temp ≤ 32 °F | 3 | 1 | 0 | 0 | 0 | 0 | 0 | 0 | 0 | 0 | 0 | 3 | 7 |
| Days Min Temp ≤ 32 °F | 27 | 17 | 6 | 1 | 0 | 0 | 0 | 0 | 0 | 1 | 10 | 26 | 88 |
| Days Min Temp ≤ 0 °F | 0 | 0 | 0 | 0 | 0 | 0 | 0 | 0 | 0 | 0 | 0 | 0 | 0 |
| Heating Degree Days | 958 | 690 | 504 | 256 | 73 | 4 | 0 | 0 | 18 | 208 | 592 | 926 | 4229 |
| Cooling Degree Days | 0 | 0 | 0 | 26 | 111 | 399 | 546 | 473 | 229 | 26 | 0 | 0 | 1810 |
| Total Precipitation (") | 0.60 | 0.56 | 0.66 | 0.42 | 0.49 | 0.18 | 0.56 | 0.62 | 0.54 | 0.78 | 0.58 | 0.55 | 6.54 |
| Days ≥ 0.1" Precip | 2 | 2 | 2 | 1 | 1 | 1 | 2 | 2 | 2 | 2 | 2 | 1 | 20 |
| Total Snowfall (") | *1.9* | *1.0* | 0.1 | 0.0 | 0.0 | 0.0 | 0.0 | 0.0 | 0.0 | 0.0 | 0.6 | *1.8* | 5.4 |
| Days ≥ 1" Snow Depth | *2* | *0* | 0 | 0 | 0 | 0 | 0 | 0 | 0 | 0 | 0 | na | na |

## PARKER 6 NE *La Paz County* ELEVATION 410 ft LAT/LONG 34° 13 ' N / 114° 13 ' W

| | JAN | FEB | MAR | APR | MAY | JUN | JUL | AUG | SEP | OCT | NOV | DEC | YEAR |
|---|---|---|---|---|---|---|---|---|---|---|---|---|---|
| Maximum Temp °F | 66.9 | 72.6 | 78.3 | 86.6 | 95.3 | 104.3 | 108.4 | 106.4 | 101.0 | 89.9 | 76.1 | 66.5 | 87.7 |
| Minimum Temp °F | 39.7 | 43.9 | 48.6 | 54.3 | 63.2 | 71.2 | 78.9 | 78.3 | 70.9 | 58.9 | 47.0 | 39.7 | 57.9 |
| Mean Temp °F | 53.3 | 58.2 | 63.5 | 70.5 | 79.3 | 87.8 | 93.7 | 92.4 | 86.0 | 74.4 | 61.6 | 53.1 | 72.8 |
| Days Max Temp ≥ 90 °F | 0 | 0 | 3 | 12 | 25 | 28 | 31 | 31 | 29 | 18 | 1 | 0 | 178 |
| Days Max Temp ≤ 32 °F | 0 | 0 | 0 | 0 | 0 | 0 | 0 | 0 | 0 | 0 | 0 | 0 | 0 |
| Days Min Temp ≤ 32 °F | 4 | 1 | 0 | 0 | 0 | 0 | 0 | 0 | 0 | 0 | 0 | 4 | 9 |
| Days Min Temp ≤ 0 °F | 0 | 0 | 0 | 0 | 0 | 0 | 0 | 0 | 0 | 0 | 0 | 0 | 0 |
| Heating Degree Days | 356 | 193 | 96 | 21 | 1 | 0 | 0 | 0 | 0 | 8 | 131 | 361 | 1167 |
| Cooling Degree Days | 0 | 11 | 54 | 230 | 458 | 688 | 868 | 852 | 631 | 314 | 38 | 0 | 4144 |
| Total Precipitation (") | 0.76 | 0.59 | 0.72 | 0.17 | 0.09 | 0.02 | 0.32 | 0.62 | 0.49 | 0.37 | 0.43 | 0.67 | 5.25 |
| Days ≥ 0.1" Precip | 2 | 1 | 2 | 1 | 0 | 0 | 1 | 1 | 1 | 1 | 1 | 2 | 13 |
| Total Snowfall (") | 0.0 | 0.0 | 0.0 | 0.0 | 0.0 | 0.0 | 0.0 | 0.0 | 0.0 | 0.0 | 0.0 | 0.0 | 0.0 |
| Days ≥ 1" Snow Depth | 0 | 0 | 0 | 0 | 0 | 0 | 0 | 0 | 0 | 0 | 0 | 0 | 0 |

## PAYSON *Gila County* ELEVATION 5007 ft LAT/LONG 34° 14' N / 111° 20' W

| | JAN | FEB | MAR | APR | MAY | JUN | JUL | AUG | SEP | OCT | NOV | DEC | YEAR |
|---|---|---|---|---|---|---|---|---|---|---|---|---|---|
| Maximum Temp °F | 53.5 | 57.4 | 61.8 | 70.4 | 78.8 | 89.2 | 92.1 | 89.6 | 84.3 | 74.2 | 61.7 | 53.6 | 72.2 |
| Minimum Temp °F | 24.3 | 26.3 | 29.7 | 34.3 | 41.3 | 49.1 | 58.0 | 57.0 | 49.6 | 39.8 | 29.6 | 24.2 | 38.6 |
| Mean Temp °F | 38.9 | 41.9 | 45.8 | 52.4 | 60.1 | 69.2 | 75.1 | 73.3 | 67.0 | 57.0 | 45.7 | 38.9 | 55.4 |
| Days Max Temp ≥ 90 °F | 0 | 0 | 0 | 0 | 2 | 14 | 22 | 17 | 6 | 1 | 0 | 0 | 62 |
| Days Max Temp ≤ 32 °F | 0 | 0 | 0 | 0 | 0 | 0 | 0 | 0 | 0 | 0 | 0 | 0 | 0 |
| Days Min Temp ≤ 32 °F | 27 | 24 | 22 | 13 | 3 | 0 | 0 | 0 | 0 | 4 | 21 | 28 | 142 |
| Days Min Temp ≤ 0 °F | 0 | 0 | 0 | 0 | 0 | 0 | 0 | 0 | 0 | 0 | 0 | 0 | 0 |
| Heating Degree Days | 801 | 645 | 589 | 373 | 164 | 21 | 0 | 0 | 35 | 246 | 573 | 802 | 4249 |
| Cooling Degree Days | 0 | 0 | 0 | 1 | 21 | 162 | 311 | 278 | 114 | 7 | 0 | 0 | 894 |
| Total Precipitation (") | 2.32 | 2.10 | 2.69 | 1.05 | 0.76 | 0.40 | 2.46 | 2.99 | 1.88 | 1.74 | 1.87 | 2.44 | 22.70 |
| Days ≥ 0.1" Precip | 5 | 4 | 5 | 2 | 2 | 1 | 6 | 6 | 4 | 3 | 3 | 5 | 46 |
| Total Snowfall (") | 4.8 | 5.2 | 5.4 | 1.6 | 0.0 | 0.0 | 0.0 | 0.0 | 0.0 | 0.0 | 2.4 | 7.1 | 26.5 |
| Days ≥ 1" Snow Depth | 3 | 2 | 1 | 0 | 0 | 0 | 0 | 0 | 0 | 0 | 0 | 3 | 9 |

## PETRIFIED FOREST N P *Navajo County* ELEVATION 5463 ft LAT/LONG 34° 48' N / 109° 52' W

| | JAN | FEB | MAR | APR | MAY | JUN | JUL | AUG | SEP | OCT | NOV | DEC | YEAR |
|---|---|---|---|---|---|---|---|---|---|---|---|---|---|
| Maximum Temp °F | 47.0 | 54.3 | 61.2 | 70.0 | 78.4 | 89.1 | 92.1 | 89.3 | 83.0 | 71.8 | 58.5 | 48.1 | 70.2 |
| Minimum Temp °F | 20.6 | 24.3 | 29.0 | 35.0 | 42.9 | 51.8 | 59.8 | 58.7 | 51.6 | 39.6 | 28.5 | 21.1 | 38.6 |
| Mean Temp °F | 33.8 | 39.3 | 45.1 | 52.5 | 60.7 | 70.5 | 76.0 | 74.0 | 67.3 | 55.7 | 43.5 | 34.7 | 54.4 |
| Days Max Temp ≥ 90 °F | 0 | 0 | 0 | 0 | 2 | 15 | 22 | 16 | 3 | 0 | 0 | 0 | 58 |
| Days Max Temp ≤ 32 °F | 2 | 0 | 0 | 0 | 0 | 0 | 0 | 0 | 0 | 0 | 0 | 2 | 4 |
| Days Min Temp ≤ 32 °F | 28 | 24 | 22 | 11 | 2 | 0 | 0 | 0 | 0 | 5 | 21 | 28 | 141 |
| Days Min Temp ≤ 0 °F | 1 | 0 | 0 | 0 | 0 | 0 | 0 | 0 | 0 | 0 | 0 | 1 | 2 |
| Heating Degree Days | 959 | 718 | 609 | 368 | 151 | 13 | 0 | 1 | 31 | 284 | 638 | 934 | 4706 |
| Cooling Degree Days | 0 | 0 | 0 | 2 | 30 | 203 | 342 | 292 | 109 | 3 | 0 | 0 | 981 |
| Total Precipitation (") | 0.65 | 0.57 | 0.76 | 0.41 | 0.56 | 0.32 | 1.41 | 1.72 | 1.28 | 1.13 | 0.76 | 0.85 | 10.42 |
| Days ≥ 0.1" Precip | 2 | 2 | 3 | 1 | 1 | 1 | 4 | 4 | 3 | 3 | 2 | 3 | 29 |
| Total Snowfall (") | 2.6 | 1.6 | 0.9 | 0.8 | 0.0 | 0.0 | 0.0 | 0.0 | 0.0 | 0.0 | 1.0 | 3.1 | 10.0 |
| Days ≥ 1" Snow Depth | na | 1 | 0 | 0 | 0 | 0 | 0 | 0 | 0 | 0 | 0 | ***3*** | na |

## PHOENIX CITY *Maricopa County* ELEVATION 1106 ft LAT/LONG 33° 28' N / 112° 4' W

| | JAN | FEB | MAR | APR | MAY | JUN | JUL | AUG | SEP | OCT | NOV | DEC | YEAR |
|---|---|---|---|---|---|---|---|---|---|---|---|---|---|
| Maximum Temp °F | 67.5 | 72.1 | ***76.6*** | 85.6 | 94.0 | 103.7 | 105.7 | ***104.5*** | ***99.5*** | ***89.6*** | ***76.4*** | 67.7 | 86.9 |
| Minimum Temp °F | 43.9 | 47.1 | ***51.4*** | 58.2 | 66.4 | 75.2 | 81.1 | ***80.2*** | ***74.2*** | ***63.3*** | ***50.4*** | 43.9 | 61.3 |
| Mean Temp °F | 55.7 | 59.6 | ***64.0*** | 71.9 | 80.2 | 89.5 | 93.4 | ***92.4*** | ***86.9*** | ***76.5*** | ***63.4*** | 55.8 | 74.1 |
| Days Max Temp ≥ 90 °F | 0 | 0 | ***2*** | 10 | 23 | 29 | 31 | ***31*** | 28 | ***17*** | 1 | 0 | 172 |
| Days Max Temp ≤ 32 °F | 0 | 0 | ***0*** | 0 | 0 | 0 | 0 | ***0*** | 0 | ***0*** | 0 | 0 | 0 |
| Days Min Temp ≤ 32 °F | 1 | 0 | ***0*** | ***0*** | 0 | 0 | 0 | ***0*** | 0 | ***0*** | 0 | 1 | 2 |
| Days Min Temp ≤ 0 °F | 0 | 0 | ***0*** | 0 | 0 | 0 | 0 | ***0*** | 0 | ***0*** | 0 | 0 | 0 |
| Heating Degree Days | 283 | 161 | ***92*** | 23 | 1 | 0 | 0 | ***0*** | ***0*** | ***6*** | ***102*** | 281 | 949 |
| Cooling Degree Days | 3 | 24 | 86 | 288 | 513 | 772 | 898 | 868 | 677 | 380 | 67 | 2 | 4578 |
| Total Precipitation (") | ***0.99*** | 0.86 | ***1.22*** | 0.26 | 0.20 | 0.09 | 0.86 | ***0.94*** | ***0.72*** | ***0.75*** | ***0.79*** | 1.30 | 8.98 |
| Days ≥ 0.1" Precip | ***2*** | 2 | ***3*** | 1 | 1 | 0 | 2 | ***2*** | 1 | ***1*** | 2 | 2 | 19 |
| Total Snowfall (") | 0.0 | ***0.0*** | ***0.0*** | 0.0 | 0.0 | 0.0 | 0.0 | ***0.0*** | 0.0 | ***0.0*** | ***0.0*** | 0.0 | 0.0 |
| Days ≥ 1" Snow Depth | 0 | ***0*** | ***0*** | 0 | 0 | 0 | 0 | ***0*** | 0 | ***0*** | 0 | 0 | 0 |

## PHOENIX SKY HARBOR *Maricopa County* ELEVATION 1112 ft LAT/LONG 33° 26' N / 112° 2' W

| | JAN | FEB | MAR | APR | MAY | JUN | JUL | AUG | SEP | OCT | NOV | DEC | YEAR |
|---|---|---|---|---|---|---|---|---|---|---|---|---|---|
| Maximum Temp °F | 66.4 | 71.0 | 75.8 | 84.8 | 93.8 | 103.6 | 105.8 | 104.0 | 98.8 | 88.2 | 74.9 | 66.2 | 86.1 |
| Minimum Temp °F | 42.4 | 45.9 | 50.1 | 56.6 | 65.3 | 74.3 | 81.6 | 80.3 | 73.7 | 62.0 | 49.5 | 42.8 | 60.4 |
| Mean Temp °F | 54.5 | 58.5 | 63.0 | 70.7 | 79.6 | 89.0 | 93.7 | 92.2 | 86.3 | 75.1 | 62.2 | 54.5 | 73.3 |
| Days Max Temp ≥ 90 °F | 0 | 0 | 2 | 10 | 23 | 29 | 31 | 31 | 28 | 15 | 1 | 0 | 170 |
| Days Max Temp ≤ 32 °F | 0 | 0 | 0 | 0 | 0 | 0 | 0 | 0 | 0 | 0 | 0 | 0 | 0 |
| Days Min Temp ≤ 32 °F | 3 | 1 | 0 | 0 | 0 | 0 | 0 | 0 | 0 | 0 | 0 | 2 | 6 |
| Days Min Temp ≤ 0 °F | 0 | 0 | 0 | 0 | 0 | 0 | 0 | 0 | 0 | 0 | 0 | 0 | 0 |
| Heating Degree Days | 321 | 191 | 116 | 29 | 1 | 0 | 0 | 0 | 0 | 10 | 123 | 320 | 1111 |
| Cooling Degree Days | 2 | 23 | 80 | 282 | 521 | 786 | 914 | 882 | 684 | 373 | 61 | 1 | 4609 |
| Total Precipitation (") | 0.85 | 0.74 | 1.07 | 0.23 | 0.17 | 0.13 | 0.90 | 0.82 | 0.87 | 0.70 | 0.78 | 1.18 | 8.44 |
| Days ≥ 0.1" Precip | 2 | 2 | 3 | 1 | 1 | 0 | 2 | 2 | 2 | 1 | 2 | 2 | 20 |
| Total Snowfall (") | 0.0 | 0.0 | 0.0 | 0.0 | 0.0 | 0.0 | 0.0 | 0.0 | 0.0 | 0.0 | 0.0 | 0.0 | 0.0 |
| Days ≥ 1" Snow Depth | 0 | 0 | 0 | 0 | 0 | 0 | 0 | 0 | 0 | 0 | 0 | 0 | 0 |

## PIPE SPRINGS N M *Mohave County* ELEVATION 4920 ft LAT/LONG 36° 52 ' N / 112° 44 ' W

| | JAN | FEB | MAR | APR | MAY | JUN | JUL | AUG | SEP | OCT | NOV | DEC | YEAR |
|---|---|---|---|---|---|---|---|---|---|---|---|---|---|
| Maximum Temp °F | 47.5 | 53.6 | 60.7 | 69.0 | 78.8 | 89.6 | 94.4 | 91.6 | 84.8 | 73.2 | 58.2 | 48.4 | 70.8 |
| Minimum Temp °F | 21.2 | 25.3 | 29.9 | 34.8 | 42.8 | 50.9 | 58.5 | 57.1 | 49.2 | 38.4 | 28.5 | 22.1 | 38.2 |
| Mean Temp °F | 34.4 | 39.6 | 45.3 | 51.9 | 60.8 | 70.3 | 76.5 | 74.4 | 67.0 | 55.8 | 43.4 | 35.3 | 54.6 |
| Days Max Temp ≥ 90 °F | 0 | 0 | 0 | 0 | 2 | 17 | 26 | 21 | 7 | 0 | 0 | 0 | 73 |
| Days Max Temp ≤ 32 °F | 1 | 1 | 0 | 0 | 0 | 0 | 0 | 0 | 0 | 0 | 0 | 1 | 3 |
| Days Min Temp ≤ 32 °F | 28 | 24 | 20 | 11 | 2 | 0 | 0 | 0 | 0 | 6 | 21 | 28 | 140 |
| Days Min Temp ≤ 0 °F | 1 | 0 | 0 | 0 | 0 | 0 | 0 | 0 | 0 | 0 | 0 | 1 | 2 |
| Heating Degree Days | 943 | 711 | 602 | 387 | 152 | 17 | 0 | 1 | 41 | 283 | 640 | 915 | 4692 |
| Cooling Degree Days | 0 | 0 | 0 | 3 | 34 | 200 | 366 | 313 | 111 | 6 | 0 | 0 | 1033 |
| Total Precipitation (") | 1.21 | 1.14 | 1.20 | 0.65 | 0.59 | 0.36 | 0.90 | 1.41 | 0.82 | 0.83 | 0.93 | 0.86 | 10.90 |
| Days ≥ 0.1" Precip | 3 | 3 | 4 | 2 | 2 | 1 | 2 | 3 | 2 | 2 | 2 | 3 | 29 |
| Total Snowfall (") | 3.3 | 1.6 | 1.3 | 0.2 | 0.0 | 0.0 | 0.0 | 0.0 | 0.0 | 0.0 | 1.5 | 2.2 | 10.1 |
| Days ≥ 1" Snow Depth | 3 | *1* | *0* | 0 | 0 | 0 | 0 | 0 | 0 | 0 | 0 | *2* | 6 |

## PLEASANT VALLEY R S *Gila County* ELEVATION 5052 ft LAT/LONG 34° 6 ' N / 110° 56 ' W

| | JAN | FEB | MAR | APR | MAY | JUN | JUL | AUG | SEP | OCT | NOV | DEC | YEAR |
|---|---|---|---|---|---|---|---|---|---|---|---|---|---|
| Maximum Temp °F | 54.1 | 57.5 | 61.2 | 69.1 | 77.2 | 87.3 | 90.1 | 87.5 | 83.1 | 74.3 | 62.3 | 54.5 | 71.5 |
| Minimum Temp °F | 20.8 | 23.2 | 26.4 | 29.9 | 36.2 | 43.7 | 54.4 | 53.8 | 45.9 | 35.0 | 25.6 | 20.7 | 34.6 |
| Mean Temp °F | 37.5 | 40.4 | 43.8 | 49.5 | 56.7 | 65.6 | 72.3 | 70.7 | 64.5 | 54.6 | 44.0 | 37.6 | 53.1 |
| Days Max Temp ≥ 90 °F | 0 | 0 | 0 | 0 | 1 | 12 | 18 | 11 | 3 | 0 | 0 | 0 | 45 |
| Days Max Temp ≤ 32 °F | 0 | 0 | 0 | 0 | 0 | 0 | 0 | 0 | 0 | 0 | 0 | 0 | 0 |
| Days Min Temp ≤ 32 °F | 28 | 25 | 26 | 20 | 9 | 1 | 0 | 0 | 1 | 11 | 26 | 28 | 175 |
| Days Min Temp ≤ 0 °F | 1 | 0 | 0 | 0 | 0 | 0 | 0 | 0 | 0 | 0 | 0 | 1 | 2 |
| Heating Degree Days | 845 | 688 | 650 | 459 | 253 | 60 | 2 | 4 | 65 | 316 | 625 | 842 | 4809 |
| Cooling Degree Days | 0 | 0 | 0 | 0 | 4 | 91 | 230 | 205 | 65 | 1 | 0 | 0 | 596 |
| Total Precipitation (") | 2.23 | 1.87 | 2.60 | 0.86 | 0.79 | 0.39 | 2.61 | 3.10 | 1.90 | 1.66 | 1.81 | 2.31 | 22.13 |
| Days ≥ 0.1" Precip | 4 | 4 | 6 | 2 | 2 | 1 | 7 | 6 | 3 | 3 | 3 | 5 | 46 |
| Total Snowfall (") | na | na | na | 0.4 | 0.0 | 0.0 | 0.0 | 0.0 | 0.0 | 0.0 | *1.3* | na | na |
| Days ≥ 1" Snow Depth | na | na | *0* | 0 | 0 | 0 | 0 | 0 | 0 | 0 | *0* | na | na |

## PORTAL 4 SW *Cochise County* ELEVATION 5394 ft LAT/LONG 31° 53 ' N / 109° 12 ' W

| | JAN | FEB | MAR | APR | MAY | JUN | JUL | AUG | SEP | OCT | NOV | DEC | YEAR |
|---|---|---|---|---|---|---|---|---|---|---|---|---|---|
| Maximum Temp °F | 53.1 | 57.7 | 63.6 | 71.3 | 78.5 | 87.5 | 86.3 | 83.2 | 79.3 | 71.8 | 61.2 | 53.3 | 70.6 |
| Minimum Temp °F | 22.9 | 24.7 | 29.1 | 33.8 | 40.2 | 48.3 | 55.6 | 54.1 | 47.9 | 37.7 | 28.5 | 23.7 | 37.2 |
| Mean Temp °F | 38.0 | 41.2 | 46.3 | 52.6 | 59.4 | 67.9 | 71.0 | 68.7 | 63.6 | 54.8 | 44.9 | 38.5 | 53.9 |
| Days Max Temp ≥ 90 °F | 0 | 0 | 0 | 0 | 1 | 11 | 9 | 2 | 0 | 0 | 0 | 0 | 23 |
| Days Max Temp ≤ 32 °F | 0 | 0 | 0 | 0 | 0 | 0 | 0 | 0 | 0 | 0 | 0 | 0 | 0 |
| Days Min Temp ≤ 32 °F | 27 | 24 | 22 | 14 | 4 | 0 | 0 | 0 | 0 | 8 | 23 | 26 | 148 |
| Days Min Temp ≤ 0 °F | 0 | 0 | 0 | 0 | 0 | 0 | 0 | 0 | 0 | 0 | 0 | 0 | 0 |
| Heating Degree Days | 829 | 665 | 572 | 367 | 176 | 21 | 0 | 4 | 64 | 311 | 597 | 815 | 4421 |
| Cooling Degree Days | 0 | 0 | 0 | 0 | 10 | 124 | 187 | 129 | 30 | 0 | 0 | 0 | 480 |
| Total Precipitation (") | 1.43 | 1.26 | 1.04 | 0.55 | 0.51 | 0.85 | 4.62 | 3.88 | 2.43 | 1.61 | 1.39 | 2.35 | 21.92 |
| Days ≥ 0.1" Precip | 4 | 3 | 3 | 1 | 1 | 2 | 9 | 8 | 5 | 3 | 3 | 4 | 46 |
| Total Snowfall (") | 2.2 | 1.0 | 1.1 | 0.3 | 0.0 | 0.0 | 0.0 | 0.0 | 0.0 | 0.0 | 0.6 | 2.3 | 7.5 |
| Days ≥ 1" Snow Depth | *2* | *1* | 0 | 0 | 0 | 0 | 0 | 0 | 0 | 0 | 0 | 1 | 4 |

## PRESCOTT *Yavapai County* ELEVATION 5205 ft LAT/LONG 34° 34 ' N / 112° 26 ' W

| | JAN | FEB | MAR | APR | MAY | JUN | JUL | AUG | SEP | OCT | NOV | DEC | YEAR |
|---|---|---|---|---|---|---|---|---|---|---|---|---|---|
| Maximum Temp °F | 50.7 | 54.2 | 58.0 | 65.7 | 73.9 | 84.5 | 88.0 | 85.2 | 80.4 | 71.2 | 59.4 | 50.9 | 68.5 |
| Minimum Temp °F | 22.5 | 24.9 | 29.3 | 34.9 | 42.1 | 50.5 | 58.2 | 56.6 | 49.5 | 38.7 | 28.7 | 22.8 | 38.2 |
| Mean Temp °F | 36.6 | 39.6 | 43.7 | 50.4 | 58.1 | 67.5 | 73.1 | 70.9 | 65.0 | 55.0 | 44.1 | 36.9 | 53.4 |
| Days Max Temp ≥ 90 °F | 0 | 0 | 0 | 0 | 0 | 8 | 13 | 6 | 2 | 0 | 0 | 0 | 29 |
| Days Max Temp ≤ 32 °F | 1 | 0 | 0 | 0 | 0 | 0 | 0 | 0 | 0 | 0 | 0 | 1 | 2 |
| Days Min Temp ≤ 32 °F | 29 | 25 | 21 | 11 | 3 | 0 | 0 | 0 | 0 | 6 | 22 | 28 | 145 |
| Days Min Temp ≤ 0 °F | 0 | 0 | 0 | 0 | 0 | 0 | 0 | 0 | 0 | 0 | 0 | 0 | 0 |
| Heating Degree Days | 873 | 712 | 653 | 433 | 223 | 45 | 1 | 5 | 62 | 308 | 622 | 864 | 4801 |
| Cooling Degree Days | 0 | 0 | 0 | 1 | 23 | 147 | 270 | 216 | 74 | 5 | 0 | 0 | 736 |
| Total Precipitation (") | 1.63 | 1.79 | 1.87 | 0.82 | 0.66 | 0.42 | 3.10 | 3.38 | 1.95 | 1.18 | 1.54 | 1.81 | 20.15 |
| Days ≥ 0.1" Precip | 4 | 3 | 4 | 2 | 2 | 1 | 6 | 6 | 3 | 2 | 3 | 4 | 40 |
| Total Snowfall (") | 3.6 | 4.9 | 6.4 | 2.2 | 0.0 | 0.0 | 0.0 | 0.0 | 0.0 | 0.2 | 1.5 | 5.1 | 23.9 |
| Days ≥ 1" Snow Depth | 4 | 2 | 2 | 1 | 0 | 0 | 0 | 0 | 0 | 0 | 1 | 4 | 14 |

## ROOSEVELT 1 WNW *Gila County* ELEVATION 2205 ft LAT/LONG 33° 40 ' N / 111° 9 ' W

| | JAN | FEB | MAR | APR | MAY | JUN | JUL | AUG | SEP | OCT | NOV | DEC | YEAR |
|---|---|---|---|---|---|---|---|---|---|---|---|---|---|
| Maximum Temp °F | 58.6 | 64.2 | 69.9 | 79.3 | 88.6 | 98.8 | 101.8 | 99.2 | 94.0 | 82.6 | 68.7 | 58.7 | 80.4 |
| Minimum Temp °F | 37.6 | 40.2 | 44.6 | 51.2 | 59.8 | 68.8 | 74.4 | 73.0 | 67.4 | 56.8 | 45.4 | 38.3 | 54.8 |
| Mean Temp °F | 48.1 | 52.2 | 57.3 | 65.3 | 74.2 | 83.8 | 88.1 | 86.2 | 80.7 | 69.7 | 57.1 | 48.5 | 67.6 |
| Days Max Temp ≥ 90 °F | 0 | 0 | 0 | 3 | 15 | 27 | 31 | 29 | 24 | 6 | 0 | 0 | 135 |
| Days Max Temp ≤ 32 °F | 0 | 0 | 0 | 0 | 0 | 0 | 0 | 0 | 0 | 0 | 0 | 0 | 0 |
| Days Min Temp ≤ 32 °F | 4 | 2 | 1 | 0 | 0 | 0 | 0 | 0 | 0 | 0 | 1 | 4 | 12 |
| Days Min Temp ≤ 0 °F | 0 | 0 | 0 | 0 | 0 | 0 | 0 | 0 | 0 | 0 | 0 | 0 | 0 |
| Heating Degree Days | 516 | 356 | 246 | 76 | 10 | 0 | 0 | 0 | 0 | 29 | 237 | 503 | 1973 |
| Cooling Degree Days | 0 | 0 | 10 | 119 | 318 | 593 | 725 | 680 | 498 | 189 | 7 | 0 | 3139 |
| Total Precipitation (") | 2.11 | 1.83 | 2.30 | 0.53 | 0.49 | 0.15 | 1.34 | 1.67 | 1.28 | 1.36 | 1.68 | 2.39 | 17.13 |
| Days ≥ 0.1" Precip | 4 | 4 | 4 | 2 | 1 | 1 | 3 | 4 | 3 | 2 | 3 | 4 | 35 |
| Total Snowfall (") | 0.0 | 0.1 | 0.0 | 0.0 | 0.0 | 0.0 | 0.0 | 0.0 | 0.0 | 0.0 | 0.0 | 0.0 | 0.1 |
| Days ≥ 1" Snow Depth | 0 | 0 | 0 | 0 | 0 | 0 | 0 | 0 | 0 | 0 | 0 | 0 | 0 |

## SACATON *Pinal County* ELEVATION 1280 ft LAT/LONG 33° 5 ' N / 111° 45 ' W

| | JAN | FEB | MAR | APR | MAY | JUN | JUL | AUG | SEP | OCT | NOV | DEC | YEAR |
|---|---|---|---|---|---|---|---|---|---|---|---|---|---|
| Maximum Temp °F | 66.3 | 71.4 | 76.5 | 86.1 | 94.8 | 104.6 | 106.5 | 104.7 | ***100.3*** | ***89.2*** | 77.0 | 66.4 | 87.0 |
| Minimum Temp °F | 34.9 | 38.2 | 42.4 | 47.8 | 56.3 | 66.0 | 75.0 | 73.1 | ***65.5*** | ***53.0*** | 41.0 | 34.2 | 52.3 |
| Mean Temp °F | 50.6 | 54.9 | 59.5 | 67.0 | 75.6 | 85.3 | 90.8 | 88.9 | ***82.9*** | ***71.1*** | 59.1 | 50.3 | 69.7 |
| Days Max Temp ≥ 90 °F | 0 | 0 | 2 | 12 | 24 | 29 | 31 | 31 | ***28*** | ***16*** | 2 | 0 | 175 |
| Days Max Temp ≤ 32 °F | 0 | 0 | 0 | 0 | 0 | 0 | 0 | 0 | ***0*** | ***0*** | 0 | 0 | 0 |
| Days Min Temp ≤ 32 °F | 13 | 6 | 2 | 0 | 0 | 0 | 0 | 0 | ***0*** | ***0*** | ***4*** | 14 | 39 |
| Days Min Temp ≤ 0 °F | 0 | 0 | 0 | 0 | 0 | 0 | 0 | 0 | ***0*** | ***0*** | 0 | 0 | 0 |
| Heating Degree Days | 439 | 283 | 191 | 59 | 8 | 0 | 0 | 0 | ***0*** | ***31*** | 193 | 448 | 1652 |
| Cooling Degree Days | 0 | 4 | 28 | 162 | 362 | 647 | 798 | 754 | na | ***242*** | ***29*** | ***0*** | na |
| Total Precipitation (") | 0.90 | 0.86 | 1.11 | 0.31 | 0.21 | 0.06 | 1.01 | 1.33 | 0.71 | 0.73 | 0.81 | 1.22 | 9.26 |
| Days ≥ 0.1" Precip | 3 | 2 | 2 | 1 | 1 | 0 | 2 | 2 | ***2*** | 2 | ***2*** | 3 | 22 |
| Total Snowfall (") | 0.0 | 0.0 | 0.0 | 0.0 | 0.0 | 0.0 | 0.0 | 0.0 | 0.0 | 0.0 | 0.0 | 0.0 | 0.0 |
| Days ≥ 1" Snow Depth | 0 | 0 | 0 | 0 | 0 | 0 | 0 | 0 | 0 | 0 | 0 | 0 | 0 |

## SAFFORD AGRI CENTER *Graham County* ELEVATION 2954 ft LAT/LONG 32° 49 ' N / 109° 41 ' W

| | JAN | FEB | MAR | APR | MAY | JUN | JUL | AUG | SEP | OCT | NOV | DEC | YEAR |
|---|---|---|---|---|---|---|---|---|---|---|---|---|---|
| Maximum Temp °F | 59.3 | 64.3 | 70.2 | 79.1 | 87.8 | 97.6 | 97.9 | 95.4 | 90.9 | 81.4 | 68.7 | 59.3 | 79.3 |
| Minimum Temp °F | 29.0 | 32.4 | 37.4 | 42.9 | 51.0 | 60.4 | 67.9 | 66.2 | 59.1 | 47.1 | 35.4 | 29.2 | 46.5 |
| Mean Temp °F | 44.2 | 48.4 | 53.8 | 61.0 | 69.4 | 79.0 | 83.0 | 80.8 | 75.0 | 64.3 | 52.1 | 44.3 | 62.9 |
| Days Max Temp ≥ 90 °F | 0 | 0 | 0 | 2 | 14 | 27 | 29 | 28 | 21 | 5 | 0 | 0 | 126 |
| Days Max Temp ≤ 32 °F | 0 | 0 | 0 | 0 | 0 | 0 | 0 | 0 | 0 | 0 | 0 | 0 | 0 |
| Days Min Temp ≤ 32 °F | 23 | 15 | 7 | 2 | 0 | 0 | 0 | 0 | 0 | 1 | 10 | 23 | 81 |
| Days Min Temp ≤ 0 °F | 0 | 0 | 0 | 0 | 0 | 0 | 0 | 0 | 0 | 0 | 0 | 0 | 0 |
| Heating Degree Days | 638 | 463 | 343 | 149 | 28 | 0 | 0 | 0 | 2 | 89 | 382 | 636 | 2730 |
| Cooling Degree Days | 0 | 0 | 4 | 52 | 195 | 451 | 557 | 503 | 322 | 74 | 1 | 0 | 2159 |
| Total Precipitation (") | 0.73 | 0.75 | 0.62 | 0.21 | 0.27 | 0.27 | 1.66 | 1.64 | 1.16 | 0.96 | 0.55 | 1.14 | 9.96 |
| Days ≥ 0.1" Precip | 2 | 2 | 2 | 1 | 1 | 1 | 4 | 4 | 3 | 2 | 2 | 3 | 27 |
| Total Snowfall (") | 0.2 | 0.4 | 0.1 | 0.1 | 0.0 | 0.0 | 0.0 | 0.0 | 0.0 | 0.0 | 0.0 | 0.2 | 1.0 |
| Days ≥ 1" Snow Depth | 0 | 0 | 0 | 0 | 0 | 0 | 0 | 0 | 0 | 0 | 0 | 0 | 0 |

## SAINT JOHNS *Apache County* ELEVATION 5735 ft LAT/LONG 34° 30 ' N / 109° 22 ' W

| | JAN | FEB | MAR | APR | MAY | JUN | JUL | AUG | SEP | OCT | NOV | DEC | YEAR |
|---|---|---|---|---|---|---|---|---|---|---|---|---|---|
| Maximum Temp °F | 48.2 | 54.8 | 61.1 | 70.0 | 78.1 | 87.8 | 89.9 | 87.2 | 81.8 | 71.7 | 59.1 | 48.9 | 69.9 |
| Minimum Temp °F | 19.1 | 22.8 | 28.7 | 34.0 | 42.4 | 50.9 | 58.0 | 56.2 | 48.8 | 37.1 | 26.2 | 19.0 | 36.9 |
| Mean Temp °F | 33.6 | 38.8 | 44.9 | 52.0 | 60.3 | 69.4 | 73.9 | 71.7 | 65.3 | 54.5 | 42.6 | 34.0 | 53.4 |
| Days Max Temp ≥ 90 °F | 0 | 0 | 0 | 0 | 1 | 13 | 17 | 11 | 2 | 0 | 0 | 0 | 44 |
| Days Max Temp ≤ 32 °F | 2 | 0 | 0 | 0 | 0 | 0 | 0 | 0 | 0 | 0 | 0 | 2 | 4 |
| Days Min Temp ≤ 32 °F | 29 | 25 | 22 | 13 | 2 | 0 | 0 | 0 | 0 | 8 | 24 | 29 | 152 |
| Days Min Temp ≤ 0 °F | 1 | 0 | 0 | 0 | 0 | 0 | 0 | 0 | 0 | 0 | 0 | 1 | 2 |
| Heating Degree Days | 965 | 732 | 616 | 382 | 160 | 16 | 0 | 1 | 47 | 321 | 664 | 956 | 4860 |
| Cooling Degree Days | 0 | 0 | 0 | 1 | 20 | 166 | 271 | 211 | 60 | 1 | 0 | 0 | 730 |
| Total Precipitation (") | 0.75 | 0.56 | 0.81 | 0.46 | 0.52 | 0.53 | 1.85 | 2.33 | 1.28 | 1.05 | 0.63 | 0.80 | 11.57 |
| Days ≥ 0.1" Precip | 2 | 2 | 2 | 1 | 2 | 1 | 6 | 6 | 3 | 3 | 2 | 2 | 32 |
| Total Snowfall (") | 3.6 | 2.3 | 3.1 | 1.0 | 0.0 | 0.0 | 0.0 | 0.0 | 0.0 | 0.3 | 1.3 | 4.6 | 16.2 |
| Days ≥ 1" Snow Depth | ***2*** | ***0*** | na | 0 | 0 | 0 | 0 | 0 | 0 | 0 | 0 | na | na |

## SAN CARLOS RESERVOIR *Gila County* ELEVATION 2532 ft LAT/LONG 33° 10 ' N / 110° 31 ' W

| | JAN | FEB | MAR | APR | MAY | JUN | JUL | AUG | SEP | OCT | NOV | DEC | YEAR |
|---|---|---|---|---|---|---|---|---|---|---|---|---|---|
| Maximum Temp °F | 58.4 | 63.7 | 69.3 | 78.5 | 87.7 | 98.1 | 99.8 | 97.5 | 92.6 | 82.0 | 68.6 | 58.4 | 79.6 |
| Minimum Temp °F | 33.4 | 37.0 | 41.3 | 47.7 | 56.6 | 65.9 | 73.0 | 70.9 | 64.3 | 52.9 | 40.6 | 33.9 | 51.5 |
| Mean Temp °F | 45.9 | 50.4 | 55.3 | 63.1 | 72.2 | 82.0 | 86.4 | 84.2 | 78.4 | 67.5 | 54.6 | 46.2 | 65.5 |
| Days Max Temp ≥ 90 °F | 0 | 0 | 0 | 2 | 14 | 27 | 30 | 29 | 22 | 5 | 0 | 0 | 129 |
| Days Max Temp ≤ 32 °F | 0 | 0 | 0 | 0 | 0 | 0 | 0 | 0 | 0 | 0 | 0 | 0 | 0 |
| Days Min Temp ≤ 32 °F | 14 | 6 | 2 | 0 | 0 | 0 | 0 | 0 | 0 | 0 | 3 | 12 | 37 |
| Days Min Temp ≤ 0 °F | 0 | 0 | 0 | 0 | 0 | 0 | 0 | 0 | 0 | 0 | 0 | 0 | 0 |
| Heating Degree Days | 584 | 407 | 299 | 111 | 18 | 1 | 0 | 0 | 1 | 51 | 308 | 577 | 2357 |
| Cooling Degree Days | 0 | 0 | 7 | 84 | 267 | 552 | 673 | 613 | 427 | 139 | 4 | 0 | 2766 |
| Total Precipitation (") | 1.83 | 1.61 | 1.84 | 0.45 | 0.44 | 0.21 | 1.55 | 2.31 | 1.43 | 1.32 | 1.34 | 2.25 | 16.58 |
| Days ≥ 0.1" Precip | 4 | 4 | 4 | 1 | 1 | 1 | 4 | 5 | 3 | 2 | 3 | 4 | 36 |
| Total Snowfall (") | 0.0 | 0.3 | 0.0 | 0.0 | 0.0 | 0.0 | 0.0 | 0.0 | 0.0 | 0.0 | 0.0 | 0.0 | 0.3 |
| Days ≥ 1" Snow Depth | 0 | 0 | 0 | 0 | 0 | 0 | 0 | 0 | 0 | 0 | 0 | 0 | 0 |

## SANDERS *Apache County* ELEVATION 5853 ft LAT/LONG 35° 13 ' N / 109° 20 ' W

| | JAN | FEB | MAR | APR | MAY | JUN | JUL | AUG | SEP | OCT | NOV | DEC | YEAR |
|---|---|---|---|---|---|---|---|---|---|---|---|---|---|
| Maximum Temp °F | 45.9 | 52.1 | 58.7 | 67.4 | 75.8 | 87.2 | 90.8 | 88.0 | 81.4 | 70.3 | 57.4 | 47.3 | 68.5 |
| Minimum Temp °F | 17.4 | 22.4 | 25.8 | 30.6 | 38.5 | 46.8 | 55.5 | 55.2 | 47.6 | 36.6 | 25.7 | 18.7 | 35.1 |
| Mean Temp °F | 31.7 | 37.3 | 42.1 | 49.0 | 57.2 | 67.0 | 73.2 | 71.6 | 64.5 | 53.5 | 41.6 | 33.0 | 51.8 |
| Days Max Temp ≥ 90 °F | 0 | 0 | 0 | 0 | 0 | 12 | 20 | 13 | 2 | 0 | 0 | 0 | 47 |
| Days Max Temp ≤ 32 °F | 2 | 1 | 0 | 0 | 0 | 0 | 0 | 0 | 0 | 0 | 0 | 2 | 5 |
| Days Min Temp ≤ 32 °F | 30 | 26 | 26 | 19 | 5 | 0 | 0 | 0 | 0 | 8 | 24 | 29 | 167 |
| Days Min Temp ≤ 0 °F | 2 | 0 | 0 | 0 | 0 | 0 | 0 | 0 | 0 | 0 | 0 | 1 | 3 |
| Heating Degree Days | 1026 | 776 | 701 | 474 | 241 | 41 | 1 | 3 | 67 | 352 | 696 | 984 | 5362 |
| Cooling Degree Days | 0 | 0 | 0 | 0 | 6 | 118 | 257 | 237 | 59 | 1 | 0 | 0 | 678 |
| Total Precipitation (") | 0.96 | 1.09 | 1.15 | 0.55 | 0.70 | 0.36 | 1.34 | 2.01 | 1.10 | 1.22 | 1.08 | 1.06 | 12.62 |
| Days ≥ 0.1" Precip | 3 | 3 | 3 | 2 | 2 | 1 | 4 | 4 | 3 | 3 | 3 | 3 | 34 |
| Total Snowfall (") | na | na | na | na | 0.0 | 0.0 | 0.0 | 0.0 | 0.0 | 0.0 | ***1.0*** | na | na |
| Days ≥ 1" Snow Depth | na | na | na | ***0*** | 0 | 0 | 0 | 0 | 0 | 0 | 0 | na | na |

## SANTA RITA EXP RANGE *Pima County* ELEVATION 4304 ft LAT/LONG 31° 46 ' N / 110° 51 ' W

| | JAN | FEB | MAR | APR | MAY | JUN | JUL | AUG | SEP | OCT | NOV | DEC | YEAR |
|---|---|---|---|---|---|---|---|---|---|---|---|---|---|
| Maximum Temp °F | 60.2 | 64.0 | 68.0 | 75.7 | 83.4 | 93.2 | 91.8 | 88.9 | 86.0 | 78.7 | 67.6 | 60.6 | 76.5 |
| Minimum Temp °F | 36.4 | 38.6 | 42.1 | 47.7 | 55.1 | 63.8 | 66.2 | 64.6 | 61.4 | 53.8 | 42.2 | ***36.8*** | 50.7 |
| Mean Temp °F | 48.3 | 51.4 | 55.1 | 61.7 | 69.3 | 78.6 | 79.0 | 76.8 | 73.7 | 66.3 | 54.9 | ***48.6*** | 63.6 |
| Days Max Temp ≥ 90 °F | 0 | 0 | 0 | 0 | 6 | 21 | 20 | 15 | 7 | 2 | 0 | 0 | 71 |
| Days Max Temp ≤ 32 °F | 0 | 0 | 0 | 0 | 0 | 0 | 0 | 0 | 0 | 0 | 0 | 0 | 0 |
| Days Min Temp ≤ 32 °F | 9 | 6 | 3 | 1 | 0 | 0 | 0 | 0 | 0 | 0 | 4 | na | na |
| Days Min Temp ≤ 0 °F | 0 | 0 | 0 | 0 | 0 | 0 | 0 | 0 | 0 | 0 | 0 | 0 | 0 |
| Heating Degree Days | 510 | 380 | 312 | 137 | 31 | 1 | 0 | 0 | 2 | 65 | 300 | ***500*** | 2238 |
| Cooling Degree Days | 0 | 2 | 10 | 62 | 184 | 441 | 446 | 370 | 282 | ***110*** | 4 | 0 | 1911 |
| Total Precipitation (") | 1.77 | 1.59 | 1.78 | 0.61 | 0.29 | 0.64 | 5.06 | 4.25 | 2.47 | 1.76 | 1.28 | 2.35 | 23.85 |
| Days ≥ 0.1" Precip | 4 | 3 | 3 | 1 | 1 | 1 | 9 | 8 | 4 | 2 | 2 | 4 | 42 |
| Total Snowfall (") | 0.4 | ***1.0*** | 0.6 | 0.2 | 0.0 | 0.0 | 0.0 | 0.0 | 0.0 | 0.0 | 0.1 | 1.5 | 3.8 |
| Days ≥ 1" Snow Depth | 0 | ***0*** | 0 | 0 | 0 | 0 | 0 | 0 | 0 | 0 | 0 | ***0*** | 0 |

## SASABE *Pima County* ELEVATION 3590 ft LAT/LONG 31° 29 ' N / 111° 33 ' W

| | JAN | FEB | MAR | APR | MAY | JUN | JUL | AUG | SEP | OCT | NOV | DEC | YEAR |
|---|---|---|---|---|---|---|---|---|---|---|---|---|---|
| Maximum Temp °F | 63.1 | 66.0 | 69.3 | 77.3 | 84.7 | 94.4 | 94.4 | 91.9 | 89.3 | 81.5 | 70.4 | 62.9 | 78.8 |
| Minimum Temp °F | 35.5 | 37.4 | 40.1 | 44.9 | 50.8 | 60.8 | 66.5 | 65.1 | 60.3 | 51.0 | 41.3 | 35.9 | 49.1 |
| Mean Temp °F | 49.4 | 51.8 | 54.7 | 61.1 | 67.8 | 77.7 | 80.5 | 78.5 | 74.8 | 66.3 | 55.9 | 49.4 | 64.0 |
| Days Max Temp ≥ 90 °F | 0 | 0 | 0 | 1 | 8 | 24 | 25 | 22 | 16 | 4 | 0 | 0 | 100 |
| Days Max Temp ≤ 32 °F | 0 | 0 | 0 | 0 | 0 | 0 | 0 | 0 | 0 | 0 | 0 | 0 | 0 |
| Days Min Temp ≤ 32 °F | 10 | 7 | 4 | 1 | 0 | 0 | 0 | 0 | 0 | 0 | 3 | 8 | 33 |
| Days Min Temp ≤ 0 °F | 0 | 0 | 0 | 0 | 0 | 0 | 0 | 0 | 0 | 0 | 0 | 0 | 0 |
| Heating Degree Days | 478 | 368 | 317 | 141 | 41 | 1 | 0 | 0 | 1 | 54 | 272 | 476 | 2149 |
| Cooling Degree Days | 0 | 1 | 3 | 43 | 141 | 393 | 481 | 421 | 297 | 94 | 3 | 0 | 1877 |
| Total Precipitation (") | 1.37 | 1.46 | 1.29 | 0.39 | 0.23 | 0.29 | 3.81 | 3.30 | 1.80 | 1.27 | 0.95 | 2.15 | 18.31 |
| Days ≥ 0.1" Precip | 3 | 3 | 3 | 1 | 1 | 1 | 7 | 7 | 3 | 2 | 2 | 3 | 36 |
| Total Snowfall (") | 0.1 | 0.3 | 0.0 | 0.0 | 0.0 | 0.0 | 0.0 | 0.0 | 0.0 | 0.0 | 0.1 | 0.6 | 1.1 |
| Days ≥ 1" Snow Depth | 0 | 0 | 0 | 0 | 0 | 0 | 0 | 0 | 0 | 0 | 0 | 0 | 0 |

## SEDONA R S *Coconino County* ELEVATION 4222 ft LAT/LONG 34° 52' N / 111° 46' W

| | JAN | FEB | MAR | APR | MAY | JUN | JUL | AUG | SEP | OCT | NOV | DEC | YEAR |
|---|---|---|---|---|---|---|---|---|---|---|---|---|---|
| Maximum Temp °F | 56.0 | 60.0 | 64.6 | 73.1 | 82.2 | 93.0 | 96.5 | 94.1 | 88.1 | 77.6 | 64.2 | 55.9 | 75.4 |
| Minimum Temp °F | 30.1 | 32.4 | 35.7 | 41.0 | 48.0 | 56.6 | 63.6 | 62.5 | 56.4 | 47.0 | 35.9 | 30.0 | 44.9 |
| Mean Temp °F | 43.1 | 46.3 | 50.1 | 57.1 | 65.1 | 74.8 | 80.1 | 78.3 | 72.3 | 62.3 | 50.1 | 43.0 | 60.2 |
| Days Max Temp ≥ 90 °F | 0 | 0 | 0 | 0 | 5 | 21 | 28 | 25 | 13 | 2 | 0 | 0 | 94 |
| Days Max Temp ≤ 32 °F | 0 | 0 | 0 | 0 | 0 | 0 | 0 | 0 | 0 | 0 | 0 | 0 | 0 |
| Days Min Temp ≤ 32 °F | 20 | 14 | 10 | 4 | 1 | 0 | 0 | 0 | 0 | 1 | 9 | 20 | 79 |
| Days Min Temp ≤ 0 °F | 0 | 0 | 0 | 0 | 0 | 0 | 0 | 0 | 0 | 0 | 0 | 0 | 0 |
| Heating Degree Days | 672 | 522 | 456 | 248 | 79 | 6 | 0 | 0 | 10 | 125 | 441 | 676 | 3235 |
| Cooling Degree Days | 0 | 0 | 1 | 29 | 116 | 360 | 497 | 445 | 262 | 60 | 0 | 0 | 1770 |
| Total Precipitation (") | 2.02 | 2.02 | 2.45 | 1.19 | 0.75 | 0.37 | 1.69 | 2.04 | 1.91 | 1.54 | 1.69 | 1.96 | 19.63 |
| Days ≥ 0.1" Precip | 4 | 4 | 5 | 2 | 2 | 1 | 4 | 5 | 3 | 3 | 3 | 4 | 40 |
| Total Snowfall (") | *0.4* | *0.4* | 0.5 | 0.0 | 0.0 | 0.0 | 0.0 | 0.0 | 0.0 | 0.0 | 0.2 | 1.8 | 3.3 |
| Days ≥ 1" Snow Depth | *0* | *0* | 0 | 0 | 0 | 0 | 0 | 0 | 0 | 0 | 0 | 0 | 0 |

## SELIGMAN *Yavapai County* ELEVATION 5223 ft LAT/LONG 35° 19' N / 112° 53' W

| | JAN | FEB | MAR | APR | MAY | JUN | JUL | AUG | SEP | OCT | NOV | DEC | YEAR |
|---|---|---|---|---|---|---|---|---|---|---|---|---|---|
| Maximum Temp °F | 50.9 | 55.1 | 60.2 | 67.8 | 76.6 | 86.8 | 90.5 | 88.1 | 82.9 | 73.1 | 60.6 | 51.8 | 70.4 |
| Minimum Temp °F | 21.9 | 23.8 | 27.3 | 31.4 | 38.6 | 46.6 | 54.8 | 53.9 | 46.9 | 36.4 | 27.3 | 21.4 | 35.9 |
| Mean Temp °F | 36.4 | 39.5 | 43.8 | 49.6 | 57.5 | 66.7 | 72.7 | 71.0 | 64.9 | 54.8 | 43.9 | 36.6 | 53.1 |
| Days Max Temp ≥ 90 °F | 0 | 0 | 0 | 0 | 1 | 11 | 19 | 13 | 4 | 0 | 0 | 0 | 48 |
| Days Max Temp ≤ 32 °F | 1 | 0 | 0 | 0 | 0 | 0 | 0 | 0 | 0 | 0 | 0 | 1 | 2 |
| Days Min Temp ≤ 32 °F | 29 | 25 | 25 | 16 | 5 | 0 | 0 | 0 | 0 | *8* | 24 | 28 | 160 |
| Days Min Temp ≤ 0 °F | 0 | 0 | 0 | 0 | 0 | 0 | 0 | 0 | 0 | 0 | 0 | 1 | 1 |
| Heating Degree Days | 879 | 714 | 651 | 455 | 233 | 45 | 1 | 4 | 59 | 311 | 625 | 872 | 4849 |
| Cooling Degree Days | 0 | 0 | 0 | 1 | 13 | 117 | 255 | 211 | 64 | 2 | 0 | 0 | 663 |
| Total Precipitation (") | 1.14 | 1.06 | 1.43 | 0.51 | 0.58 | 0.38 | 1.98 | 2.11 | 1.10 | 0.84 | 0.93 | 1.18 | 13.24 |
| Days ≥ 0.1" Precip | 3 | 3 | 4 | 2 | 2 | 1 | 5 | 5 | 2 | 2 | 2 | 3 | 34 |
| Total Snowfall (") | 1.9 | *1.2* | *1.1* | 0.1 | 0.0 | 0.0 | 0.0 | 0.0 | 0.0 | 0.1 | 0.7 | 3.9 | 9.0 |
| Days ≥ 1" Snow Depth | 2 | 1 | *0* | 0 | 0 | 0 | 0 | 0 | 0 | 0 | 0 | *2* | 5 |

## SHOW LOW AP *Navajo County* ELEVATION 6411 ft LAT/LONG 34° 15' N / 110° 0' W

| | JAN | FEB | MAR | APR | MAY | JUN | JUL | AUG | SEP | OCT | NOV | DEC | YEAR |
|---|---|---|---|---|---|---|---|---|---|---|---|---|---|
| Maximum Temp °F | 44.5 | 49.8 | 55.4 | 64.0 | 72.7 | 83.4 | 85.6 | 82.6 | 77.6 | 67.1 | 54.8 | 46.0 | 65.3 |
| Minimum Temp °F | 18.7 | 22.9 | 28.0 | 33.4 | 40.7 | 49.7 | 56.8 | 55.1 | 48.9 | 37.4 | 26.9 | 20.4 | 36.6 |
| Mean Temp °F | 31.6 | 36.4 | 41.7 | 48.7 | 56.8 | 66.6 | 71.2 | 68.9 | 63.3 | 52.3 | 40.9 | 33.2 | 51.0 |
| Days Max Temp ≥ 90 °F | 0 | 0 | 0 | 0 | 0 | 5 | 8 | 2 | 0 | 0 | 0 | 0 | 15 |
| Days Max Temp ≤ 32 °F | 3 | 1 | 0 | 0 | 0 | 0 | 0 | 0 | 0 | 0 | 0 | 3 | 7 |
| Days Min Temp ≤ 32 °F | 29 | 25 | 23 | 13 | 4 | 0 | 0 | 0 | 0 | 8 | 22 | 28 | 152 |
| Days Min Temp ≤ 0 °F | 2 | 0 | 0 | 0 | 0 | 0 | 0 | 0 | 0 | 0 | 0 | 1 | 3 |
| Heating Degree Days | 1028 | 801 | 713 | 481 | 255 | 47 | 2 | 8 | 84 | 388 | 716 | 978 | 5501 |
| Cooling Degree Days | 0 | 0 | 0 | 0 | 11 | 122 | 209 | 149 | 42 | 0 | 0 | 0 | 533 |
| Total Precipitation (") | 1.32 | 1.37 | 1.46 | 0.61 | 0.77 | 0.45 | 2.38 | 3.08 | 1.66 | 1.69 | 1.55 | 2.11 | 18.45 |
| Days ≥ 0.1" Precip | 3 | 3 | 4 | 2 | 2 | 1 | 6 | 7 | 4 | 3 | 3 | 4 | 42 |
| Total Snowfall (") | *6.9* | 5.4 | *6.5* | 1.6 | 0.0 | 0.0 | 0.0 | 0.0 | 0.0 | 0.6 | 2.7 | 6.7 | 30.4 |
| Days ≥ 1" Snow Depth | na | na | *2* | 0 | 0 | 0 | 0 | 0 | 0 | 0 | *1* | na | na |

## SNOWFLAKE *Navajo County* ELEVATION 5642 ft LAT/LONG 34° 30' N / 110° 5' W

| | JAN | FEB | MAR | APR | MAY | JUN | JUL | AUG | SEP | OCT | NOV | DEC | YEAR |
|---|---|---|---|---|---|---|---|---|---|---|---|---|---|
| Maximum Temp °F | 48.2 | 54.6 | 60.2 | 68.4 | 76.7 | 87.0 | 89.8 | 87.2 | 81.8 | 71.3 | 58.4 | 49.0 | 69.4 |
| Minimum Temp °F | 18.8 | 21.9 | 26.7 | 31.8 | 39.2 | 47.1 | 56.3 | 54.8 | 47.0 | 35.5 | 25.7 | 19.5 | 35.4 |
| Mean Temp °F | 33.5 | 38.3 | 43.5 | 50.1 | 58.0 | 67.1 | 73.1 | 71.0 | 64.4 | 53.4 | 42.1 | 34.3 | 52.4 |
| Days Max Temp ≥ 90 °F | 0 | 0 | 0 | 0 | 1 | 11 | 17 | 11 | 3 | 0 | 0 | 0 | 43 |
| Days Max Temp ≤ 32 °F | 2 | 0 | 0 | 0 | 0 | 0 | 0 | 0 | 0 | 0 | 0 | 2 | 4 |
| Days Min Temp ≤ 32 °F | 28 | 25 | 25 | 16 | 5 | 0 | 0 | 0 | 1 | 11 | 24 | 28 | 163 |
| Days Min Temp ≤ 0 °F | 1 | 0 | 0 | 0 | 0 | 0 | 0 | 0 | 0 | 0 | 0 | 1 | 2 |
| Heating Degree Days | 969 | 748 | 661 | 440 | 217 | 38 | 0 | 3 | 66 | 352 | 680 | 946 | 5120 |
| Cooling Degree Days | 0 | 0 | 0 | 0 | 12 | 124 | 261 | 208 | 61 | 0 | 0 | 0 | 666 |
| Total Precipitation (") | 0.71 | 0.67 | 1.00 | 0.41 | 0.71 | 0.32 | 2.15 | 2.66 | 1.43 | 1.22 | 0.93 | 1.08 | 13.29 |
| Days ≥ 0.1" Precip | 2 | 2 | 3 | 2 | 2 | 1 | 6 | 6 | 3 | 3 | 2 | 3 | 35 |
| Total Snowfall (") | 3.9 | 2.8 | 2.2 | 0.9 | 0.3 | 0.0 | 0.0 | 0.0 | 0.0 | 0.4 | 2.0 | 5.3 | 17.8 |
| Days ≥ 1" Snow Depth | 4 | 1 | 0 | 0 | 0 | 0 | 0 | 0 | 0 | 0 | 1 | 4 | 10 |

## SNOWFLAKE 15 W *Navajo County* ELEVATION 6080 ft LAT/LONG 34° 30 ' N / 110° 20 ' W

| | JAN | FEB | MAR | APR | MAY | JUN | JUL | AUG | SEP | OCT | NOV | DEC | YEAR |
|---|---|---|---|---|---|---|---|---|---|---|---|---|---|
| Maximum Temp °F | 44.3 | 50.1 | 56.2 | 64.5 | 73.0 | 83.7 | 87.0 | 84.3 | 78.6 | 68.0 | 55.4 | 45.7 | 65.9 |
| Minimum Temp °F | 19.4 | 23.1 | 28.1 | 33.5 | 41.0 | 50.4 | 57.9 | 56.5 | 49.6 | 38.6 | 27.8 | 20.4 | 37.2 |
| Mean Temp °F | 31.9 | 36.6 | 42.2 | 49.0 | 57.0 | 67.1 | 72.5 | 70.4 | 64.1 | 53.4 | 41.6 | 33.1 | 51.6 |
| Days Max Temp ≥ 90 °F | 0 | 0 | 0 | 0 | 0 | 7 | 11 | 5 | 1 | 0 | 0 | 0 | 24 |
| Days Max Temp ≤ 32 °F | 4 | 1 | 0 | 0 | 0 | 0 | 0 | 0 | 0 | 0 | 0 | 3 | 8 |
| Days Min Temp ≤ 32 °F | 29 | 26 | 22 | 14 | 4 | 0 | 0 | 0 | 0 | 7 | 21 | 28 | 151 |
| Days Min Temp ≤ 0 °F | 1 | 0 | 0 | 0 | 0 | 0 | 0 | 0 | 0 | 0 | 0 | 1 | 2 |
| Heating Degree Days | 1021 | 795 | 701 | 474 | 252 | 49 | 1 | 7 | 77 | 356 | 694 | 983 | 5410 |
| Cooling Degree Days | 0 | 0 | 0 | 1 | 14 | 117 | 234 | 184 | 57 | 1 | 0 | 0 | 608 |
| Total Precipitation (") | 0.68 | 0.59 | 0.86 | 0.38 | 0.63 | 0.38 | 2.06 | 2.53 | 1.38 | 1.12 | 0.89 | 0.95 | 12.45 |
| Days ≥ 0.1" Precip | 2 | 2 | 3 | 1 | 2 | 1 | 5 | 6 | 3 | 3 | 2 | 3 | 33 |
| Total Snowfall (") | ***3.9*** | 2.5 | 2.6 | 0.5 | 0.0 | 0.0 | 0.0 | 0.0 | 0.0 | 0.1 | 1.7 | ***5.2*** | 16.5 |
| Days ≥ 1" Snow Depth | na | ***2*** | 1 | 0 | 0 | 0 | 0 | 0 | 0 | 0 | 1 | ***5*** | na |

## SOUTH PHOENIX *Maricopa County* ELEVATION 1152 ft LAT/LONG 33° 23 ' N / 112° 4 ' W

| | JAN | FEB | MAR | APR | MAY | JUN | JUL | AUG | SEP | OCT | NOV | DEC | YEAR |
|---|---|---|---|---|---|---|---|---|---|---|---|---|---|
| Maximum Temp °F | 66.1 | 71.3 | 76.4 | 84.4 | 91.9 | 100.3 | 102.2 | 100.8 | 96.6 | 86.5 | 74.1 | 65.7 | 84.7 |
| Minimum Temp °F | 38.5 | 41.4 | 44.8 | 49.5 | 56.8 | 64.9 | 73.5 | 72.5 | 65.7 | 54.8 | 44.3 | 38.7 | 53.8 |
| Mean Temp °F | 52.3 | 56.4 | 60.6 | 67.0 | 74.4 | 82.6 | 87.9 | 86.8 | 81.3 | 70.7 | 59.2 | 52.2 | 69.3 |
| Days Max Temp ≥ 90 °F | 0 | 0 | 2 | 8 | 21 | 29 | 31 | 31 | 26 | 11 | 0 | 0 | 159 |
| Days Max Temp ≤ 32 °F | 0 | 0 | 0 | 0 | 0 | 0 | 0 | 0 | 0 | 0 | 0 | 0 | 0 |
| Days Min Temp ≤ 32 °F | 6 | 3 | 1 | 0 | 0 | 0 | 0 | 0 | 0 | 0 | 1 | 5 | 16 |
| Days Min Temp ≤ 0 °F | 0 | 0 | 0 | 0 | 0 | 0 | 0 | 0 | 0 | 0 | 0 | 0 | 0 |
| Heating Degree Days | 386 | 240 | 156 | 45 | 5 | 0 | 0 | 0 | 0 | 21 | 181 | 390 | 1424 |
| Cooling Degree Days | 0 | 5 | 28 | 152 | 335 | 561 | 718 | 703 | 507 | 218 | 14 | 0 | 3241 |
| Total Precipitation (") | 0.95 | 0.88 | 1.21 | 0.27 | 0.22 | 0.17 | 0.87 | 1.15 | 0.87 | 0.74 | 0.81 | 1.20 | 9.34 |
| Days ≥ 0.1" Precip | 2 | 2 | 3 | 1 | 1 | 0 | 2 | 3 | 2 | 2 | 2 | 2 | 22 |
| Total Snowfall (") | 0.0 | 0.0 | 0.0 | 0.0 | 0.0 | 0.0 | 0.0 | 0.0 | 0.0 | 0.0 | 0.0 | 0.0 | 0.0 |
| Days ≥ 1" Snow Depth | 0 | 0 | 0 | 0 | 0 | 0 | 0 | 0 | 0 | 0 | 0 | 0 | 0 |

## SPRINGERVILLE *Apache County* ELEVATION 7037 ft LAT/LONG 34° 8 ' N / 109° 18 ' W

| | JAN | FEB | MAR | APR | MAY | JUN | JUL | AUG | SEP | OCT | NOV | DEC | YEAR |
|---|---|---|---|---|---|---|---|---|---|---|---|---|---|
| Maximum Temp °F | 47.6 | 50.8 | 56.2 | 64.0 | 71.1 | 80.1 | 81.9 | 79.6 | 75.2 | 67.0 | 56.4 | 48.6 | 64.9 |
| Minimum Temp °F | 15.9 | 18.6 | 22.7 | 27.5 | 35.1 | 42.9 | 50.9 | 49.4 | 42.3 | 30.6 | 21.5 | 16.0 | 31.1 |
| Mean Temp °F | 31.8 | 34.7 | 39.5 | 45.8 | 53.1 | 61.5 | 66.4 | 64.5 | 58.8 | 48.9 | 39.0 | 32.3 | 48.0 |
| Days Max Temp ≥ 90 °F | 0 | 0 | 0 | 0 | 0 | 1 | 2 | 0 | 0 | 0 | 0 | 0 | 3 |
| Days Max Temp ≤ 32 °F | 1 | 1 | 0 | 0 | 0 | 0 | 0 | 0 | 0 | 0 | 0 | 1 | 3 |
| Days Min Temp ≤ 32 °F | 29 | 26 | 27 | 21 | 10 | 2 | 0 | 0 | 3 | 19 | 26 | 29 | 192 |
| Days Min Temp ≤ 0 °F | 2 | 1 | 0 | 0 | 0 | 0 | 0 | 0 | 0 | 0 | 0 | 2 | 5 |
| Heating Degree Days | 1023 | 848 | 784 | 571 | 362 | 125 | 18 | 46 | 185 | 493 | 773 | 1006 | 6234 |
| Cooling Degree Days | 0 | 0 | 0 | 0 | 1 | 31 | 68 | 39 | 5 | 0 | 0 | 0 | 144 |
| Total Precipitation (") | 0.46 | 0.49 | 0.47 | 0.31 | 0.48 | 0.58 | 2.58 | 3.18 | 1.55 | 0.93 | 0.54 | 0.54 | 12.11 |
| Days ≥ 0.1" Precip | 1 | 2 | 1 | 1 | 2 | 2 | 7 | 8 | 4 | 3 | 2 | 2 | 35 |
| Total Snowfall (") | 2.4 | ***1.7*** | 1.6 | 0.6 | 0.1 | 0.0 | 0.0 | 0.0 | 0.0 | 0.7 | ***1.4*** | ***3.1*** | 11.6 |
| Days ≥ 1" Snow Depth | 2 | ***0*** | ***0*** | 0 | 0 | 0 | 0 | 0 | 0 | 0 | ***0*** | ***2*** | 4 |

## STEWART MOUNTAIN *Maricopa County* ELEVATION 1422 ft LAT/LONG 33° 34 ' N / 111° 32 ' W

| | JAN | FEB | MAR | APR | MAY | JUN | JUL | AUG | SEP | OCT | NOV | DEC | YEAR |
|---|---|---|---|---|---|---|---|---|---|---|---|---|---|
| Maximum Temp °F | 65.4 | 69.3 | 74.0 | 82.4 | 91.3 | 101.2 | 103.6 | 102.2 | 97.4 | 87.2 | 74.4 | 65.6 | 84.5 |
| Minimum Temp °F | 37.2 | 39.4 | 43.0 | 49.2 | 57.5 | 66.1 | 74.1 | 72.7 | 66.2 | 54.0 | 42.5 | 37.0 | 53.2 |
| Mean Temp °F | 51.2 | 54.4 | 58.5 | 65.8 | 74.4 | 83.7 | 88.9 | 87.5 | 81.8 | 70.6 | 58.3 | 51.4 | 68.9 |
| Days Max Temp ≥ 90 °F | 0 | 0 | 1 | 7 | 20 | 28 | 31 | 30 | 27 | 13 | 0 | 0 | 157 |
| Days Max Temp ≤ 32 °F | 0 | 0 | 0 | 0 | 0 | 0 | 0 | 0 | 0 | 0 | 0 | 0 | 0 |
| Days Min Temp ≤ 32 °F | 8 | 4 | 1 | 0 | 0 | 0 | 0 | 0 | 0 | 0 | 3 | 8 | 24 |
| Days Min Temp ≤ 0 °F | 0 | 0 | 0 | 0 | 0 | 0 | 0 | 0 | 0 | 0 | 0 | 0 | 0 |
| Heating Degree Days | 421 | 294 | 213 | 70 | 9 | 0 | 0 | 0 | 0 | 28 | 209 | 416 | 1660 |
| Cooling Degree Days | 0 | 2 | 18 | 127 | 313 | 573 | 722 | 690 | 496 | 204 | 12 | 0 | 3157 |
| Total Precipitation (") | 1.63 | 1.42 | 1.87 | 0.50 | 0.34 | 0.13 | 1.25 | 1.64 | 1.20 | 1.11 | 1.37 | 1.85 | 14.31 |
| Days ≥ 0.1" Precip | 4 | 3 | 4 | 1 | 1 | 0 | 2 | 3 | 2 | 2 | 2 | 3 | 27 |
| Total Snowfall (") | 0.0 | 0.0 | 0.0 | 0.0 | 0.0 | 0.0 | 0.0 | 0.0 | 0.0 | 0.0 | 0.0 | 0.0 | 0.0 |
| Days ≥ 1" Snow Depth | 0 | 0 | 0 | 0 | 0 | 0 | 0 | 0 | 0 | 0 | 0 | 0 | 0 |

## SUPERIOR *Pinal County* ELEVATION 3002 ft LAT/LONG 33° 18 ' N / 111° 6 ' W

| | JAN | FEB | MAR | APR | MAY | JUN | JUL | AUG | SEP | OCT | NOV | DEC | YEAR |
|---|---|---|---|---|---|---|---|---|---|---|---|---|---|
| Maximum Temp °F | 60.7 | 63.9 | 68.2 | 76.1 | 85.4 | 95.5 | 97.5 | 95.8 | 91.7 | 82.2 | 69.4 | 61.0 | 79.0 |
| Minimum Temp °F | 42.3 | 45.0 | 47.7 | 53.7 | 61.8 | 71.6 | 75.2 | 74.0 | 70.3 | 61.3 | 50.5 | 43.2 | 58.1 |
| Mean Temp °F | 51.5 | 54.5 | 58.0 | 64.9 | 73.6 | 83.6 | 86.4 | 84.9 | 81.1 | 71.7 | 60.0 | 52.1 | 68.5 |
| Days Max Temp ≥ 90 °F | 0 | 0 | 0 | 1 | 10 | 25 | 29 | 27 | 20 | 6 | 0 | 0 | 118 |
| Days Max Temp ≤ 32 °F | 0 | 0 | 0 | 0 | 0 | 0 | 0 | 0 | 0 | 0 | 0 | 0 | 0 |
| Days Min Temp ≤ 32 °F | 2 | 1 | 1 | 0 | 0 | 0 | 0 | 0 | 0 | 0 | 0 | 2 | 6 |
| Days Min Temp ≤ 0 °F | 0 | 0 | 0 | 0 | 0 | 0 | 0 | 0 | 0 | 0 | 0 | 0 | 0 |
| Heating Degree Days | 411 | 296 | 239 | 91 | 14 | 0 | 0 | 0 | 1 | 29 | 177 | 393 | 1651 |
| Cooling Degree Days | 1 | 9 | 27 | 120 | 306 | 573 | 652 | 614 | 489 | 243 | 29 | 1 | 3064 |
| Total Precipitation (") | 2.37 | 2.12 | 2.52 | 0.70 | 0.48 | 0.28 | 1.87 | 2.76 | 1.83 | 1.32 | 1.86 | 2.77 | 20.88 |
| Days ≥ 0.1" Precip | 4 | 4 | 5 | 2 | 1 | 1 | 4 | 5 | 3 | 2 | 3 | 4 | 38 |
| Total Snowfall (") | 0.0 | 0.3 | 0.3 | 0.1 | 0.0 | 0.0 | 0.0 | 0.0 | 0.0 | 0.0 | 0.0 | 0.5 | 1.2 |
| Days ≥ 1" Snow Depth | 0 | 0 | 0 | 0 | 0 | 0 | 0 | 0 | 0 | 0 | 0 | 0 | 0 |

## TEEC NOS POS *Apache County* ELEVATION 5174 ft LAT/LONG 36° 54 ' N / 109° 6 ' W

| | JAN | FEB | MAR | APR | MAY | JUN | JUL | AUG | SEP | OCT | NOV | DEC | YEAR |
|---|---|---|---|---|---|---|---|---|---|---|---|---|---|
| Maximum Temp °F | 40.9 | 48.7 | 57.6 | 68.1 | 77.6 | 88.7 | 92.7 | 90.1 | 81.7 | 69.4 | 54.2 | 42.7 | 67.7 |
| Minimum Temp °F | 19.0 | 25.2 | 32.4 | 39.6 | 49.6 | 58.8 | 65.2 | 62.9 | 54.1 | 41.2 | 30.5 | 20.4 | 41.6 |
| Mean Temp °F | 30.0 | 37.0 | 45.0 | 53.9 | 63.7 | 73.8 | 79.0 | 76.5 | 67.9 | 55.3 | 42.4 | 31.5 | 54.7 |
| Days Max Temp ≥ 90 °F | 0 | 0 | 0 | 0 | 2 | 15 | *24* | 18 | 3 | 0 | 0 | 0 | 62 |
| Days Max Temp ≤ 32 °F | 6 | 1 | 0 | 0 | 0 | 0 | 0 | 0 | 0 | 0 | 0 | 4 | 11 |
| Days Min Temp ≤ 32 °F | 29 | 23 | 16 | 5 | 0 | 0 | 0 | 0 | 0 | 4 | 18 | 29 | 124 |
| Days Min Temp ≤ 0 °F | 1 | 0 | 0 | 0 | 0 | 0 | 0 | 0 | 0 | 0 | 0 | 1 | 2 |
| Heating Degree Days | 1079 | 784 | 614 | 334 | 111 | 11 | 0 | 1 | 39 | 298 | 672 | 1030 | 4973 |
| Cooling Degree Days | 0 | 0 | 0 | 8 | 56 | 272 | 424 | 351 | 129 | 4 | 0 | 0 | 1244 |
| Total Precipitation (") | 0.64 | 0.48 | 0.73 | 0.45 | 0.57 | 0.32 | 0.98 | 1.01 | 0.80 | 0.80 | 0.61 | 0.71 | 8.10 |
| Days ≥ 0.1" Precip | 3 | 2 | 2 | 2 | 2 | 1 | 3 | 3 | 2 | 2 | 2 | 2 | 26 |
| Total Snowfall (") | *2.8* | *0.9* | *0.2* | 0.0 | 0.0 | 0.0 | 0.0 | 0.0 | 0.0 | 0.0 | 0.7 | *1.7* | 6.3 |
| Days ≥ 1" Snow Depth | *3* | *0* | 0 | 0 | 0 | 0 | 0 | 0 | 0 | 0 | 0 | na | na |

## TEMPE A S U *Maricopa County* ELEVATION 1170 ft LAT/LONG 33° 25 ' N / 111° 56 ' W

| | JAN | FEB | MAR | APR | MAY | JUN | JUL | AUG | SEP | OCT | NOV | DEC | YEAR |
|---|---|---|---|---|---|---|---|---|---|---|---|---|---|
| Maximum Temp °F | 67.1 | 71.9 | 76.9 | 85.7 | 94.3 | 103.6 | 105.8 | 104.0 | 99.1 | 88.9 | 75.7 | 66.9 | 86.7 |
| Minimum Temp °F | 38.0 | 40.6 | 44.8 | 50.2 | 57.9 | 65.7 | 74.9 | 73.6 | 66.7 | 55.5 | 44.1 | 38.1 | 54.2 |
| Mean Temp °F | 52.6 | 56.3 | 60.9 | 68.0 | 76.2 | 84.7 | 90.4 | 88.8 | 82.9 | 72.2 | 59.9 | 52.5 | 70.5 |
| Days Max Temp ≥ 90 °F | 0 | 0 | 2 | 11 | 24 | 29 | 31 | 31 | 28 | 16 | 1 | 0 | 173 |
| Days Max Temp ≤ 32 °F | 0 | 0 | 0 | 0 | 0 | 0 | 0 | 0 | 0 | 0 | 0 | 0 | 0 |
| Days Min Temp ≤ 32 °F | 7 | 4 | 1 | 0 | 0 | 0 | 0 | 0 | 0 | 0 | 1 | 7 | 20 |
| Days Min Temp ≤ 0 °F | 0 | 0 | 0 | 0 | 0 | 0 | 0 | 0 | 0 | 0 | 0 | 0 | 0 |
| Heating Degree Days | 378 | 245 | 156 | 47 | 5 | 0 | 0 | 0 | 0 | 21 | 169 | 381 | 1402 |
| Cooling Degree Days | 1 | 11 | 47 | 205 | 416 | 654 | 807 | 778 | 584 | 301 | 36 | 0 | 3840 |
| Total Precipitation (") | 1.02 | 0.98 | 1.21 | 0.29 | 0.23 | 0.11 | 0.85 | 1.07 | 1.04 | 0.80 | 0.83 | 1.41 | 9.84 |
| Days ≥ 0.1" Precip | 3 | 2 | 3 | 1 | 1 | 0 | 2 | 2 | 2 | 2 | 2 | 3 | 23 |
| Total Snowfall (") | 0.0 | 0.0 | 0.0 | 0.0 | 0.0 | 0.0 | 0.0 | 0.0 | 0.0 | 0.0 | 0.0 | 0.0 | 0.0 |
| Days ≥ 1" Snow Depth | 0 | 0 | 0 | 0 | 0 | 0 | 0 | 0 | 0 | 0 | 0 | 0 | 0 |

## TOMBSTONE *Cochise County* ELEVATION 4544 ft LAT/LONG 31° 43 ' N / 110° 4 ' W

| | JAN | FEB | MAR | APR | MAY | JUN | JUL | AUG | SEP | OCT | NOV | DEC | YEAR |
|---|---|---|---|---|---|---|---|---|---|---|---|---|---|
| Maximum Temp °F | 59.1 | 63.2 | 68.2 | 76.6 | 84.5 | 94.3 | 93.0 | 90.4 | 86.9 | 78.2 | 67.5 | 59.5 | 76.8 |
| Minimum Temp °F | 35.7 | 37.8 | 41.3 | 46.9 | 54.1 | 62.6 | 66.2 | 64.9 | 61.0 | 52.0 | 42.5 | 36.8 | 50.2 |
| Mean Temp °F | 47.4 | 50.6 | 54.8 | 61.8 | 69.4 | 78.5 | 79.6 | 77.7 | 74.0 | 65.1 | 55.0 | 48.2 | 63.5 |
| Days Max Temp ≥ 90 °F | 0 | 0 | 0 | 1 | 7 | 24 | 23 | 19 | 11 | 2 | 0 | 0 | 87 |
| Days Max Temp ≤ 32 °F | 0 | 0 | 0 | 0 | 0 | 0 | 0 | 0 | 0 | 0 | 0 | 0 | 0 |
| Days Min Temp ≤ 32 °F | 10 | 6 | 3 | 1 | 0 | 0 | 0 | 0 | 0 | 0 | 3 | 8 | 31 |
| Days Min Temp ≤ 0 °F | 0 | 0 | 0 | 0 | 0 | 0 | 0 | 0 | 0 | 0 | 0 | 0 | 0 |
| Heating Degree Days | 538 | 401 | 317 | 130 | 25 | 0 | 0 | 0 | 2 | 70 | 295 | 514 | 2292 |
| Cooling Degree Days | 0 | 0 | 6 | 53 | 183 | 426 | 455 | 389 | 285 | 77 | 2 | 0 | 1876 |
| Total Precipitation (") | 1.05 | 0.71 | 0.72 | 0.25 | 0.26 | 0.44 | 3.03 | 2.96 | 1.56 | 1.07 | 0.62 | 1.15 | 13.82 |
| Days ≥ 0.1" Precip | 3 | 2 | 2 | 1 | 1 | 1 | 8 | 7 | 4 | 2 | 2 | 3 | 36 |
| Total Snowfall (") | 0.0 | *0.0* | 0.0 | 0.0 | 0.0 | 0.0 | 0.0 | 0.0 | 0.0 | 0.0 | 0.0 | 0.3 | 0.3 |
| Days ≥ 1" Snow Depth | 0 | *0* | 0 | 0 | 0 | 0 | 0 | 0 | 0 | 0 | 0 | 0 | 0 |

## TUCSON CP AVE EXP FM *Pima County* ELEVATION 2330 ft LAT/LONG 32° 17 ' N / 110° 57 ' W

| | JAN | FEB | MAR | APR | MAY | JUN | JUL | AUG | SEP | OCT | NOV | DEC | YEAR |
|---|---|---|---|---|---|---|---|---|---|---|---|---|---|
| Maximum Temp °F | 66.0 | 69.6 | 74.2 | 82.5 | 90.8 | 100.2 | 100.5 | 98.5 | 95.2 | 85.8 | 74.1 | 66.1 | 83.6 |
| Minimum Temp °F | 34.0 | 36.5 | 40.6 | 45.6 | 53.9 | 63.3 | 71.4 | 70.1 | 64.1 | 51.6 | 39.7 | 34.4 | 50.4 |
| Mean Temp °F | 50.0 | 53.1 | 57.4 | 64.1 | 72.3 | 81.7 | 86.0 | 84.3 | 79.6 | 68.7 | 57.0 | 50.3 | 67.0 |
| Days Max Temp ≥ 90 °F | 0 | 0 | 1 | 5 | 19 | 29 | 30 | 30 | 26 | 10 | 0 | 0 | 150 |
| Days Max Temp ≤ 32 °F | 0 | 0 | 0 | 0 | 0 | 0 | 0 | 0 | 0 | 0 | 0 | 0 | 0 |
| Days Min Temp ≤ 32 °F | 14 | 9 | 3 | 1 | 0 | 0 | 0 | 0 | 0 | 0 | 5 | 13 | 45 |
| Days Min Temp ≤ 0 °F | 0 | 0 | 0 | 0 | 0 | 0 | 0 | 0 | 0 | 0 | 0 | 0 | 0 |
| Heating Degree Days | 457 | 330 | 237 | 86 | 11 | 0 | 0 | 0 | 0 | 37 | 241 | 450 | 1849 |
| Cooling Degree Days | 0 | 1 | 8 | 88 | 274 | 538 | 653 | 615 | 460 | 165 | 8 | 0 | 2810 |
| Total Precipitation (") | 1.13 | 1.02 | 1.04 | 0.36 | 0.23 | 0.19 | 1.79 | 2.15 | 1.09 | 1.16 | 0.81 | 1.58 | 12.55 |
| Days ≥ 0.1" Precip | 3 | 2 | 3 | 1 | 1 | 0 | 4 | 5 | 2 | 2 | 2 | 3 | 28 |
| Total Snowfall (") | 0.2 | 0.1 | 0.0 | 0.0 | 0.0 | 0.0 | 0.0 | 0.0 | 0.0 | 0.0 | 0.0 | 0.3 | 0.6 |
| Days ≥ 1" Snow Depth | 0 | 0 | 0 | 0 | 0 | 0 | 0 | 0 | 0 | 0 | 0 | 0 | 0 |

## TUCSON INTL ARPT *Pima County* ELEVATION 2559 ft LAT/LONG 32° 8 ' N / 110° 57 ' W

| | JAN | FEB | MAR | APR | MAY | JUN | JUL | AUG | SEP | OCT | NOV | DEC | YEAR |
|---|---|---|---|---|---|---|---|---|---|---|---|---|---|
| Maximum Temp °F | 64.7 | 68.4 | 72.9 | 81.3 | 89.8 | 99.7 | 99.3 | 97.2 | 93.5 | 84.1 | 72.6 | 64.8 | 82.4 |
| Minimum Temp °F | 39.0 | 41.5 | 44.9 | 50.4 | 58.3 | 67.8 | 73.4 | 72.2 | 67.3 | 56.5 | 45.4 | 39.3 | 54.7 |
| Mean Temp °F | 51.9 | 55.0 | 58.9 | 65.9 | 74.1 | 83.8 | 86.4 | 84.8 | 80.4 | 70.3 | 59.0 | 52.1 | 68.6 |
| Days Max Temp ≥ 90 °F | 0 | 0 | 1 | 4 | 17 | 28 | 29 | 29 | 23 | 8 | 0 | 0 | 139 |
| Days Max Temp ≤ 32 °F | 0 | 0 | 0 | 0 | 0 | 0 | 0 | 0 | 0 | 0 | 0 | 0 | 0 |
| Days Min Temp ≤ 32 °F | 5 | 3 | 1 | 0 | 0 | 0 | 0 | 0 | 0 | 0 | 1 | 5 | 15 |
| Days Min Temp ≤ 0 °F | 0 | 0 | 0 | 0 | 0 | 0 | 0 | 0 | 0 | 0 | 0 | 0 | 0 |
| Heating Degree Days | 401 | 280 | 205 | 71 | 8 | 0 | 0 | 0 | 0 | 31 | 192 | 394 | 1582 |
| Cooling Degree Days | 0 | 5 | 27 | 141 | 333 | 608 | 690 | 640 | 492 | 217 | 19 | 1 | 3173 |
| Total Precipitation (") | 1.00 | 0.83 | 0.84 | 0.27 | 0.26 | 0.20 | 2.11 | 2.21 | 1.49 | 1.06 | 0.69 | 1.25 | 12.21 |
| Days ≥ 0.1" Precip | 2 | 2 | 3 | 1 | 1 | 1 | 5 | 5 | 3 | 2 | 2 | 3 | 30 |
| Total Snowfall (") | 0.3 | 0.3 | 0.2 | 0.1 | 0.0 | 0.0 | 0.0 | 0.0 | 0.0 | 0.0 | 0.0 | 0.4 | 1.3 |
| Days ≥ 1" Snow Depth | 0 | 0 | 0 | 0 | 0 | 0 | 0 | 0 | 0 | 0 | 0 | 0 | 0 |

## TUCSON MAGNETIC OBSY *Pima County* ELEVATION 2526 ft LAT/LONG 32° 15 ' N / 110° 50 ' W

| | JAN | FEB | MAR | APR | MAY | JUN | JUL | AUG | SEP | OCT | NOV | DEC | YEAR |
|---|---|---|---|---|---|---|---|---|---|---|---|---|---|
| Maximum Temp °F | 65.2 | 68.7 | 72.9 | 81.6 | 90.6 | 100.5 | 100.7 | 98.6 | 95.2 | 85.7 | 74.0 | 65.7 | 83.3 |
| Minimum Temp °F | 34.9 | 37.6 | 40.8 | 45.7 | 53.8 | 63.3 | 71.7 | 70.3 | 64.4 | 52.4 | 40.9 | 35.1 | 50.9 |
| Mean Temp °F | 50.1 | 53.2 | 56.9 | 63.6 | 72.2 | 81.9 | 86.3 | 84.5 | 79.8 | 69.1 | 57.5 | 50.4 | 67.1 |
| Days Max Temp ≥ 90 °F | 0 | 0 | 1 | 5 | 19 | 29 | 30 | 29 | 25 | 11 | 0 | 0 | 149 |
| Days Max Temp ≤ 32 °F | 0 | 0 | 0 | 0 | 0 | 0 | 0 | 0 | 0 | 0 | 0 | 0 | 0 |
| Days Min Temp ≤ 32 °F | 13 | 7 | 3 | 1 | 0 | 0 | 0 | 0 | 0 | 0 | 4 | 13 | 41 |
| Days Min Temp ≤ 0 °F | 0 | 0 | 0 | 0 | 0 | 0 | 0 | 0 | 0 | 0 | 0 | 0 | 0 |
| Heating Degree Days | 456 | 329 | 257 | 102 | 17 | 0 | 0 | 0 | 1 | 39 | 230 | 446 | 1877 |
| Cooling Degree Days | 0 | 1 | 14 | 94 | 274 | 538 | 665 | 611 | 454 | 179 | 12 | 0 | 2842 |
| Total Precipitation (") | 1.33 | 1.12 | 1.29 | 0.35 | 0.31 | 0.27 | 1.98 | 2.43 | 1.34 | 1.19 | 0.89 | 1.61 | 14.11 |
| Days ≥ 0.1" Precip | 4 | 3 | 3 | 1 | 1 | 1 | 5 | 5 | 3 | 2 | 2 | 4 | 34 |
| Total Snowfall (") | 0.1 | 0.2 | 0.0 | 0.1 | 0.0 | 0.0 | 0.0 | 0.0 | 0.0 | 0.0 | 0.0 | 0.3 | 0.7 |
| Days ≥ 1" Snow Depth | 0 | 0 | 0 | 0 | 0 | 0 | 0 | 0 | 0 | 0 | 0 | 0 | 0 |

## TUCSON U OF ARIZONA *Pima County* ELEVATION 2421 ft LAT/LONG 32° 14 ' N / 110° 57 ' W

| | JAN | FEB | MAR | APR | MAY | JUN | JUL | AUG | SEP | OCT | NOV | DEC | YEAR |
|---|---|---|---|---|---|---|---|---|---|---|---|---|---|
| Maximum Temp °F | 66.3 | 70.3 | 74.7 | 82.9 | 91.0 | 100.6 | 100.7 | 99.0 | 95.2 | 85.9 | 74.4 | 66.5 | 84.0 |
| Minimum Temp °F | 40.8 | 43.6 | 47.3 | 53.3 | 61.2 | 70.5 | 75.8 | 74.5 | 69.7 | 58.5 | 47.9 | 41.6 | 57.1 |
| Mean Temp °F | 53.6 | 57.0 | 61.0 | 68.1 | 76.1 | 85.5 | 88.3 | 86.7 | 82.5 | 72.3 | 60.9 | 54.1 | 70.5 |
| Days Max Temp ≥ 90 °F | 0 | 0 | 1 | 7 | 19 | 29 | 30 | 30 | 25 | 11 | 0 | 0 | 152 |
| Days Max Temp ≤ 32 °F | 0 | 0 | 0 | 0 | 0 | 0 | 0 | 0 | 0 | 0 | 0 | 0 | 0 |
| Days Min Temp ≤ 32 °F | 4 | 2 | 0 | 0 | 0 | 0 | 0 | 0 | 0 | 0 | 0 | 3 | 9 |
| Days Min Temp ≤ 0 °F | 0 | 0 | 0 | 0 | 0 | 0 | 0 | 0 | 0 | 0 | 0 | 0 | 0 |
| Heating Degree Days | 348 | 225 | 157 | 46 | 5 | 0 | 0 | 0 | 0 | 18 | 148 | 333 | 1280 |
| Cooling Degree Days | 2 | 9 | 45 | 189 | 393 | 662 | 734 | 688 | 553 | 270 | 35 | 2 | 3582 |
| Total Precipitation (") | 1.09 | 0.95 | 0.94 | 0.35 | 0.23 | 0.25 | 2.08 | 2.13 | 1.17 | 1.09 | 0.74 | 1.42 | 12.44 |
| Days ≥ 0.1" Precip | 3 | 2 | 3 | 1 | 1 | 1 | 5 | 5 | 3 | 2 | 2 | 3 | 31 |
| Total Snowfall (") | 0.2 | 0.1 | 0.2 | 0.0 | 0.0 | 0.0 | 0.0 | 0.0 | 0.0 | 0.0 | 0.0 | 0.1 | 0.6 |
| Days ≥ 1" Snow Depth | 0 | 0 | 0 | 0 | 0 | 0 | 0 | 0 | 0 | 0 | 0 | 0 | 0 |

## TUMACACORI NATL MONM *Santa Cruz County* ELEVATION 3267 ft LAT/LONG 31° 34 ' N / 111° 3 ' W

| | JAN | FEB | MAR | APR | MAY | JUN | JUL | AUG | SEP | OCT | NOV | DEC | YEAR |
|---|---|---|---|---|---|---|---|---|---|---|---|---|---|
| Maximum Temp °F | 66.0 | 69.2 | 73.3 | 81.2 | 88.9 | 98.7 | 96.8 | 94.1 | 91.5 | 83.8 | 73.3 | 66.2 | 81.9 |
| Minimum Temp °F | 32.0 | 33.7 | 37.3 | 41.5 | 48.0 | 57.7 | 65.8 | 64.3 | 58.2 | 47.2 | 37.3 | 32.9 | 46.3 |
| Mean Temp °F | 49.0 | 51.5 | 55.4 | 61.4 | 68.4 | 78.2 | 81.4 | 79.2 | 74.9 | 65.5 | 55.3 | 49.6 | 64.2 |
| Days Max Temp ≥ 90 °F | 0 | 0 | 1 | 4 | 14 | 28 | 27 | 26 | 20 | 7 | 0 | 0 | 127 |
| Days Max Temp ≤ 32 °F | 0 | 0 | 0 | 0 | 0 | 0 | 0 | 0 | 0 | 0 | 0 | 0 | 0 |
| Days Min Temp ≤ 32 °F | 17 | 12 | 7 | 2 | 0 | 0 | 0 | 0 | 0 | 1 | 7 | 15 | 61 |
| Days Min Temp ≤ 0 °F | 0 | 0 | 0 | 0 | 0 | 0 | 0 | 0 | 0 | 0 | 0 | 0 | 0 |
| Heating Degree Days | 489 | 375 | 294 | 132 | 29 | 1 | 0 | 0 | 1 | 60 | 285 | 471 | 2137 |
| Cooling Degree Days | 0 | 0 | 5 | 44 | 149 | 413 | 512 | 445 | 295 | 79 | 3 | 0 | 1945 |
| Total Precipitation (") | 1.25 | 1.02 | 0.98 | 0.31 | 0.21 | 0.37 | 4.15 | 3.77 | 1.74 | 1.11 | 0.75 | 1.67 | 17.33 |
| Days ≥ 0.1" Precip | 3 | 2 | 2 | 1 | 1 | 1 | 8 | 7 | 4 | 2 | 2 | 3 | 36 |
| Total Snowfall (") | 0.1 | 0.2 | 0.0 | 0.0 | 0.0 | 0.0 | 0.0 | 0.0 | 0.0 | 0.0 | 0.0 | 0.5 | 0.8 |
| Days ≥ 1" Snow Depth | 0 | 0 | 0 | 0 | 0 | 0 | 0 | 0 | 0 | 0 | 0 | 0 | 0 |

## WAHWEAP *Coconino County* ELEVATION 3720 ft LAT/LONG 36° 59 ' N / 111° 29 ' W

| | JAN | FEB | MAR | APR | MAY | JUN | JUL | AUG | SEP | OCT | NOV | DEC | YEAR |
|---|---|---|---|---|---|---|---|---|---|---|---|---|---|
| Maximum Temp °F | 46.3 | 53.5 | 63.1 | 72.9 | 83.0 | 93.5 | 97.9 | 95.4 | 87.4 | 73.3 | 58.0 | 47.2 | 72.6 |
| Minimum Temp °F | 26.4 | 31.8 | 38.1 | 44.6 | 54.5 | 63.8 | 70.3 | 68.9 | 60.3 | 49.0 | 36.7 | 27.8 | 47.7 |
| Mean Temp °F | 36.4 | 42.7 | 50.6 | 58.8 | 68.8 | 78.6 | 84.2 | 82.2 | 73.9 | 61.2 | 47.4 | 37.5 | 60.2 |
| Days Max Temp ≥ 90 °F | 0 | 0 | 0 | 0 | 7 | 22 | 28 | 27 | 12 | 0 | 0 | 0 | 96 |
| Days Max Temp ≤ 32 °F | 1 | 0 | 0 | 0 | 0 | 0 | 0 | 0 | 0 | 0 | 0 | 1 | 2 |
| Days Min Temp ≤ 32 °F | 25 | 15 | 5 | 1 | 0 | 0 | 0 | 0 | 0 | 1 | 8 | 25 | 80 |
| Days Min Temp ≤ 0 °F | 0 | 0 | 0 | 0 | 0 | 0 | 0 | 0 | 0 | 0 | 0 | 0 | 0 |
| Heating Degree Days | 880 | 623 | 440 | 208 | 43 | 1 | 0 | 0 | 7 | 151 | 523 | 846 | 3722 |
| Cooling Degree Days | 0 | 0 | 0 | 29 | 162 | 418 | 578 | 520 | 273 | 35 | 0 | 0 | 2015 |
| Total Precipitation (") | 0.59 | 0.60 | 0.66 | 0.37 | 0.42 | 0.23 | 0.64 | 0.74 | 0.51 | 0.67 | 0.67 | 0.47 | 6.57 |
| Days ≥ 0.1" Precip | 2 | 2 | 2 | 1 | 1 | 1 | 2 | 2 | 1 | 2 | 2 | 1 | 19 |
| Total Snowfall (") | 0.1 | ***0.3*** | 0.0 | 0.0 | 0.0 | 0.0 | 0.0 | 0.0 | 0.0 | 0.0 | 0.0 | ***0.0*** | 0.4 |
| Days ≥ 1" Snow Depth | 1 | ***0*** | 0 | 0 | 0 | 0 | 0 | 0 | 0 | 0 | 0 | ***1*** | 2 |

## WALNUT CREEK *Yavapai County* ELEVATION 5164 ft LAT/LONG 34° 56 ' N / 112° 51 ' W

| | JAN | FEB | MAR | APR | MAY | JUN | JUL | AUG | SEP | OCT | NOV | DEC | YEAR |
|---|---|---|---|---|---|---|---|---|---|---|---|---|---|
| Maximum Temp °F | 50.9 | 55.7 | 60.9 | 69.3 | 77.2 | 87.2 | 89.8 | 87.4 | 82.6 | 72.7 | 59.9 | 50.6 | 70.4 |
| Minimum Temp °F | 21.1 | 23.2 | 26.6 | 30.1 | 37.0 | 44.1 | 53.6 | 52.5 | 44.6 | 33.7 | 25.2 | 19.8 | 34.3 |
| Mean Temp °F | 36.1 | 39.4 | 43.7 | 49.7 | 57.1 | 65.6 | 71.8 | 70.0 | 63.6 | 53.3 | 42.6 | 35.2 | 52.3 |
| Days Max Temp ≥ 90 °F | 0 | 0 | 0 | 0 | 1 | 12 | 17 | 11 | 4 | 0 | 0 | 0 | 45 |
| Days Max Temp ≤ 32 °F | 0 | 0 | 0 | 0 | 0 | 0 | 0 | 0 | 0 | 0 | 0 | 1 | 1 |
| Days Min Temp ≤ 32 °F | 29 | 25 | 26 | 19 | 8 | 1 | 0 | 0 | 2 | 14 | 26 | 28 | 178 |
| Days Min Temp ≤ 0 °F | 1 | 0 | 0 | 0 | 0 | 0 | 0 | 0 | 0 | 0 | 0 | 1 | 2 |
| Heating Degree Days | 890 | 715 | 652 | 451 | 243 | 60 | 3 | 7 | 83 | 358 | 667 | 917 | 5046 |
| Cooling Degree Days | 0 | 0 | 0 | 0 | 8 | 90 | 220 | 186 | 56 | 1 | 0 | 0 | 561 |
| Total Precipitation (") | 1.53 | 1.85 | 1.73 | 0.60 | 0.61 | 0.28 | 2.31 | 2.45 | 1.23 | 1.06 | 1.30 | 1.49 | 16.44 |
| Days ≥ 0.1" Precip | 4 | 4 | 4 | 2 | 2 | 1 | 5 | 6 | 3 | 2 | 2 | 3 | 38 |
| Total Snowfall (") | 0.6 | ***0.8*** | 0.7 | 0.0 | 0.0 | 0.0 | 0.0 | 0.0 | 0.0 | 0.0 | 0.0 | 0.6 | 2.7 |
| Days ≥ 1" Snow Depth | 0 | 0 | 0 | 0 | 0 | 0 | 0 | 0 | 0 | 0 | 0 | 0 | 0 |

## WHITERIVER 1 SW *Navajo County* ELEVATION 5120 ft LAT/LONG 33° 50 ' N / 109° 58 ' W

| | JAN | FEB | MAR | APR | MAY | JUN | JUL | AUG | SEP | OCT | NOV | DEC | YEAR |
|---|---|---|---|---|---|---|---|---|---|---|---|---|---|
| Maximum Temp °F | ***52.9*** | ***55.8*** | ***60.0*** | 67.8 | 76.3 | 87.0 | 89.3 | 86.7 | 82.2 | ***72.5*** | ***61.7*** | na | na |
| Minimum Temp °F | ***23.3*** | ***25.7*** | ***30.7*** | 35.6 | 42.8 | 51.6 | 58.5 | 56.9 | 51.1 | ***40.6*** | ***30.2*** | na | na |
| Mean Temp °F | ***38.2*** | ***40.8*** | ***45.4*** | 51.8 | 59.6 | 69.3 | 73.9 | 71.8 | 66.7 | ***56.6*** | ***46.0*** | na | na |
| Days Max Temp ≥ 90 °F | 0 | 0 | 0 | 0 | 1 | 11 | 16 | 8 | 3 | 0 | 0 | 0 | 39 |
| Days Max Temp ≤ 32 °F | 0 | 0 | 0 | 0 | 0 | 0 | 0 | 0 | 0 | 0 | 0 | 0 | 0 |
| Days Min Temp ≤ 32 °F | 25 | 22 | 19 | 9 | 1 | 0 | 0 | 0 | 0 | 4 | 17 | 24 | 121 |
| Days Min Temp ≤ 0 °F | 0 | 0 | 0 | 0 | 0 | 0 | 0 | 0 | 0 | 0 | 0 | 0 | 0 |
| Heating Degree Days | ***825*** | ***677*** | ***601*** | 391 | 179 | 21 | 0 | 2 | 35 | ***260*** | ***562*** | na | na |
| Cooling Degree Days | na | na | na | 1 | ***19*** | 163 | 269 | 215 | ***91*** | na | na | na | na |
| Total Precipitation (") | 1.75 | 1.50 | 2.01 | 0.83 | 0.71 | 0.52 | 2.48 | 3.61 | 1.78 | 1.65 | 1.50 | 1.79 | 20.13 |
| Days ≥ 0.1" Precip | 4 | 3 | 4 | 2 | 2 | 1 | 6 | 7 | 4 | 3 | 3 | 3 | 42 |
| Total Snowfall (") | 3.8 | 2.1 | 4.1 | 1.5 | 0.1 | 0.0 | 0.0 | 0.0 | 0.0 | 0.1 | 1.3 | ***2.9*** | 15.9 |
| Days ≥ 1" Snow Depth | 2 | 1 | 2 | 0 | 0 | 0 | 0 | 0 | 0 | 0 | 1 | 2 | 8 |

## WICKENBURG *Maricopa County* ELEVATION 2070 ft LAT/LONG 33° 58 ' N / 112° 44 ' W

| | JAN | FEB | MAR | APR | MAY | JUN | JUL | AUG | SEP | OCT | NOV | DEC | YEAR |
|---|---|---|---|---|---|---|---|---|---|---|---|---|---|
| Maximum Temp °F | 64.7 | 68.7 | 73.3 | 82.2 | 90.6 | 100.8 | 104.0 | 101.6 | 96.0 | 85.7 | 73.4 | 64.9 | 83.8 |
| Minimum Temp °F | 31.8 | 34.9 | 38.7 | 43.2 | 50.6 | 58.8 | 68.5 | 67.5 | 59.3 | 47.8 | 37.1 | 31.5 | 47.5 |
| Mean Temp °F | 48.3 | 51.8 | 56.0 | 62.7 | 70.6 | 79.9 | 86.3 | 84.6 | 77.7 | 66.8 | 55.2 | 48.3 | 65.7 |
| Days Max Temp ≥ 90 °F | 0 | 0 | 1 | 6 | 19 | 28 | 31 | 31 | 26 | 10 | 0 | 0 | 152 |
| Days Max Temp ≤ 32 °F | 0 | 0 | 0 | 0 | 0 | 0 | 0 | 0 | 0 | 0 | 0 | 0 | 0 |
| Days Min Temp ≤ 32 °F | 18 | 10 | 5 | 1 | 0 | 0 | 0 | 0 | 0 | 0 | 7 | 19 | 60 |
| Days Min Temp ≤ 0 °F | 0 | 0 | 0 | 0 | 0 | 0 | 0 | 0 | 0 | 0 | 0 | 0 | 0 |
| Heating Degree Days | 512 | 366 | 280 | 110 | 23 | 0 | 0 | 0 | 1 | 54 | 288 | 510 | 2144 |
| Cooling Degree Days | 0 | 0 | 5 | 61 | 193 | 436 | 634 | 616 | 394 | 119 | 2 | 0 | 2460 |
| Total Precipitation (") | 1.36 | 1.38 | 1.73 | 0.44 | 0.30 | 0.14 | 1.27 | 1.78 | 1.36 | 0.56 | 1.03 | 1.34 | 12.69 |
| Days ≥ 0.1" Precip | 3 | 2 | 3 | 1 | 1 | 0 | 3 | 3 | 2 | 1 | 2 | 3 | 24 |
| Total Snowfall (") | 0.0 | 0.2 | 0.0 | 0.0 | 0.0 | 0.0 | 0.0 | 0.0 | 0.0 | 0.0 | 0.0 | 0.0 | 0.2 |
| Days ≥ 1" Snow Depth | 0 | 0 | 0 | 0 | 0 | 0 | 0 | 0 | 0 | 0 | 0 | 0 | 0 |

## WIKIEUP *Mohave County* ELEVATION 2133 ft LAT/LONG 34° 44 ' N / 113° 37 ' W

| | JAN | FEB | MAR | APR | MAY | JUN | JUL | AUG | SEP | OCT | NOV | DEC | YEAR |
|---|---|---|---|---|---|---|---|---|---|---|---|---|---|
| Maximum Temp °F | 64.2 | 68.2 | 72.6 | 81.6 | 90.3 | 100.3 | 104.6 | 102.6 | 96.7 | 87.1 | 73.0 | 64.6 | 83.8 |
| Minimum Temp °F | 33.7 | 35.8 | 38.8 | 43.5 | 51.6 | 59.1 | 68.4 | 67.3 | 59.9 | 49.0 | 38.6 | 32.6 | 48.2 |
| Mean Temp °F | 49.0 | 52.0 | 55.7 | 62.6 | 71.0 | 79.6 | 86.6 | 85.0 | 78.3 | 68.1 | 55.8 | 48.6 | 66.0 |
| Days Max Temp ≥ 90 °F | 0 | 0 | 1 | 7 | 18 | 27 | 31 | 30 | 26 | 12 | 0 | 0 | 152 |
| Days Max Temp ≤ 32 °F | 0 | 0 | 0 | 0 | 0 | 0 | 0 | 0 | 0 | 0 | 0 | 0 | 0 |
| Days Min Temp ≤ 32 °F | 14 | 9 | 4 | 1 | 0 | 0 | 0 | 0 | 0 | 0 | 6 | 16 | 50 |
| Days Min Temp ≤ 0 °F | 0 | 0 | 0 | 0 | 0 | 0 | 0 | 0 | 0 | 0 | 0 | 0 | 0 |
| Heating Degree Days | 491 | 360 | 286 | 119 | 26 | 1 | 0 | 0 | 3 | 43 | 274 | 501 | 2104 |
| Cooling Degree Days | 0 | 1 | 5 | 66 | 221 | 450 | 668 | 635 | 403 | 132 | 5 | 0 | 2586 |
| Total Precipitation (") | 1.63 | 1.57 | 1.51 | 0.47 | 0.28 | 0.07 | 0.82 | 1.26 | 0.95 | 0.58 | 0.96 | 1.26 | 11.36 |
| Days ≥ 0.1" Precip | 3 | 3 | 3 | 1 | 1 | 0 | 2 | 3 | 2 | 1 | 2 | 3 | 24 |
| Total Snowfall (") | 0.1 | 0.0 | 0.0 | 0.0 | 0.0 | 0.0 | 0.0 | 0.0 | 0.0 | 0.0 | 0.0 | 0.0 | 0.1 |
| Days ≥ 1" Snow Depth | 0 | 0 | 0 | 0 | 0 | 0 | 0 | 0 | 0 | 0 | 0 | 0 | 0 |

## WILLCOX *Cochise County* ELEVATION 4203 ft LAT/LONG 32° 15 ' N / 109° 50 ' W

| | JAN | FEB | MAR | APR | MAY | JUN | JUL | AUG | SEP | OCT | NOV | DEC | YEAR |
|---|---|---|---|---|---|---|---|---|---|---|---|---|---|
| Maximum Temp °F | 59.4 | 64.1 | 70.0 | 78.1 | 85.9 | 95.1 | 94.9 | 92.2 | 88.1 | 79.1 | 67.8 | 59.5 | 77.9 |
| Minimum Temp °F | 27.0 | 28.9 | 32.7 | 37.3 | 44.9 | 54.1 | 63.2 | 61.9 | 54.5 | 42.3 | 31.6 | 26.8 | 42.1 |
| Mean Temp °F | 43.2 | 46.5 | 51.4 | 57.7 | 65.4 | 74.6 | 79.1 | 77.1 | 71.3 | 60.7 | 49.7 | 43.2 | 60.0 |
| Days Max Temp ≥ 90 °F | 0 | 0 | 0 | 1 | 9 | 25 | 27 | 23 | 14 | 2 | 0 | 0 | 101 |
| Days Max Temp ≤ 32 °F | 0 | 0 | 0 | 0 | 0 | 0 | 0 | 0 | 0 | 0 | 0 | 0 | 0 |
| Days Min Temp ≤ 32 °F | 24 | 20 | 15 | 7 | 1 | 0 | 0 | 0 | 0 | 4 | 17 | 24 | 112 |
| Days Min Temp ≤ 0 °F | 0 | 0 | 0 | 0 | 0 | 0 | 0 | 0 | 0 | 0 | 0 | 0 | 0 |
| Heating Degree Days | 669 | 515 | 416 | 221 | 59 | 3 | 0 | 0 | 7 | 151 | 452 | 670 | 3163 |
| Cooling Degree Days | 0 | 0 | 0 | 15 | 102 | 327 | 448 | 395 | 217 | 24 | 0 | 0 | 1528 |
| Total Precipitation (") | 1.10 | 0.87 | 0.70 | 0.26 | 0.35 | 0.32 | 2.50 | 2.75 | 1.36 | 1.11 | 0.67 | 1.56 | 13.55 |
| Days ≥ 0.1" Precip | 3 | 2 | 2 | 1 | 1 | 1 | 6 | 6 | 3 | 2 | 2 | 3 | 32 |
| Total Snowfall (") | 0.9 | 1.0 | 0.6 | 0.2 | 0.0 | 0.0 | 0.0 | 0.0 | 0.0 | 0.0 | 0.1 | 1.5 | 4.3 |
| Days ≥ 1" Snow Depth | 0 | 0 | 0 | 0 | 0 | 0 | 0 | 0 | 0 | 0 | 0 | 1 | 1 |

## WILLIAMS *Coconino County* ELEVATION 6750 ft LAT/LONG 35° 15 ' N / 112° 11 ' W

| | JAN | FEB | MAR | APR | MAY | JUN | JUL | AUG | SEP | OCT | NOV | DEC | YEAR |
|---|---|---|---|---|---|---|---|---|---|---|---|---|---|
| Maximum Temp °F | 45.8 | 48.2 | 51.9 | 60.1 | 68.6 | 79.2 | 82.5 | 80.1 | 74.7 | 65.0 | 53.2 | 46.4 | 63.0 |
| Minimum Temp °F | 22.1 | 23.8 | 27.0 | 32.2 | 39.7 | 48.4 | 54.3 | 53.0 | 47.2 | 37.7 | 28.2 | 22.8 | 36.4 |
| Mean Temp °F | 34.0 | 36.0 | 39.5 | 46.2 | 54.2 | 63.8 | 68.4 | 66.6 | 61.0 | 51.4 | 40.7 | 34.6 | 49.7 |
| Days Max Temp ≥ 90 °F | 0 | 0 | 0 | 0 | 0 | 2 | 2 | 1 | 0 | 0 | 0 | 0 | 5 |
| Days Max Temp ≤ 32 °F | 2 | 1 | 1 | 0 | 0 | 0 | 0 | 0 | 0 | 0 | 1 | 2 | 7 |
| Days Min Temp ≤ 32 °F | 28 | 25 | 24 | 15 | 5 | 0 | 0 | 0 | 1 | 7 | 21 | 28 | 154 |
| Days Min Temp ≤ 0 °F | 0 | 0 | 0 | 0 | 0 | 0 | 0 | 0 | 0 | 0 | 0 | 0 | 0 |
| Heating Degree Days | 955 | 812 | 784 | 558 | 331 | 90 | 9 | 24 | 133 | 415 | 722 | 934 | 5767 |
| Cooling Degree Days | 0 | 0 | 0 | 0 | 5 | 68 | 126 | 83 | 18 | 0 | 0 | 0 | 300 |
| Total Precipitation (") | 2.00 | 2.10 | 2.37 | 1.04 | 0.84 | 0.48 | 2.93 | 2.86 | 1.60 | 1.59 | 2.02 | 2.17 | 22.00 |
| Days ≥ 0.1" Precip | 4 | 4 | 6 | 3 | 3 | 1 | 7 | 7 | 4 | 3 | 3 | 5 | 50 |
| Total Snowfall (") | 16.1 | 14.8 | 18.1 | 7.4 | 0.6 | 0.0 | 0.0 | 0.0 | 0.0 | 0.6 | 6.9 | 13.3 | 77.8 |
| Days ≥ 1" Snow Depth | 11 | 8 | na | 2 | 0 | 0 | 0 | 0 | 0 | 0 | 2 | 8 | na |

## WINDOW ROCK 4 SW *Apache County* ELEVATION 6900 ft LAT/LONG 35° 37 ' N / 109° 7 ' W

| | JAN | FEB | MAR | APR | MAY | JUN | JUL | AUG | SEP | OCT | NOV | DEC | YEAR |
|---|---|---|---|---|---|---|---|---|---|---|---|---|---|
| Maximum Temp °F | 42.5 | 46.6 | 52.8 | 62.1 | 70.9 | 81.5 | 84.4 | 82.1 | 75.7 | 65.6 | 52.4 | 43.8 | 63.4 |
| Minimum Temp °F | 14.2 | 18.9 | 24.3 | 29.8 | 37.6 | 47.2 | 54.6 | 52.7 | 43.8 | 32.9 | 23.4 | 15.6 | 32.9 |
| Mean Temp °F | 28.4 | 32.8 | 38.6 | 46.0 | 54.3 | 64.3 | 69.5 | 67.4 | 59.8 | 49.3 | 37.9 | 29.7 | 48.2 |
| Days Max Temp ≥ 90 °F | 0 | 0 | 0 | 0 | 0 | 4 | 5 | *1* | 0 | 0 | 0 | 0 | 10 |
| Days Max Temp ≤ 32 °F | 4 | 2 | 1 | 0 | 0 | 0 | 0 | 0 | 0 | 0 | 1 | 4 | 12 |
| Days Min Temp ≤ 32 °F | 30 | 28 | 28 | 19 | 7 | 0 | 0 | 0 | 1 | 15 | 26 | 30 | 184 |
| Days Min Temp ≤ 0 °F | 3 | 1 | 0 | 0 | 0 | 0 | 0 | 0 | 0 | 0 | 0 | 2 | 6 |
| Heating Degree Days | 1128 | 903 | 813 | 564 | 328 | 84 | 10 | 16 | 162 | 480 | 805 | 1086 | 6379 |
| Cooling Degree Days | 0 | 0 | 0 | 0 | 4 | 86 | 166 | 108 | 17 | 0 | 0 | 0 | 381 |
| Total Precipitation (") | 0.75 | 0.67 | 0.98 | 0.57 | 0.49 | 0.44 | 1.83 | 1.83 | 1.15 | 1.15 | 1.05 | 0.96 | 11.87 |
| Days ≥ 0.1" Precip | 3 | 2 | 3 | 2 | 1 | 1 | 5 | 5 | 3 | 3 | 3 | 3 | 34 |
| Total Snowfall (") | *4.7* | *5.1* | *4.2* | 2.2 | 0.1 | 0.0 | 0.0 | 0.0 | 0.0 | 0.4 | *1.9* | na | na |
| Days ≥ 1" Snow Depth | na | na | 2 | 1 | 0 | 0 | 0 | 0 | 0 | 0 | *1* | na | na |

## WINSLOW MUNICIPAL AP *Navajo County* ELEVATION 4885 ft LAT/LONG 35° 1 ' N / 110° 44 ' W

| | JAN | FEB | MAR | APR | MAY | JUN | JUL | AUG | SEP | OCT | NOV | DEC | YEAR |
|---|---|---|---|---|---|---|---|---|---|---|---|---|---|
| Maximum Temp °F | 45.2 | 53.4 | 60.8 | 69.8 | 78.8 | 89.8 | 92.8 | 90.0 | 83.5 | 71.5 | 57.7 | 45.8 | 69.9 |
| Minimum Temp °F | 20.1 | 25.0 | 30.6 | 36.8 | 45.1 | 54.0 | 62.3 | 60.9 | 52.6 | 40.1 | 28.9 | 20.6 | 39.8 |
| Mean Temp °F | 32.7 | 39.2 | 45.8 | 53.3 | 62.0 | 71.9 | 77.6 | 75.5 | 68.1 | 55.8 | 43.3 | 33.2 | 54.9 |
| Days Max Temp ≥ 90 °F | 0 | 0 | 0 | 0 | 2 | 16 | 23 | 18 | 5 | 0 | 0 | 0 | 64 |
| Days Max Temp ≤ 32 °F | 4 | 0 | 0 | 0 | 0 | 0 | 0 | 0 | 0 | 0 | 0 | 3 | 7 |
| Days Min Temp ≤ 32 °F | 28 | 23 | 18 | 8 | 1 | 0 | 0 | 0 | 0 | 5 | 21 | 28 | 132 |
| Days Min Temp ≤ 0 °F | 2 | 0 | 0 | 0 | 0 | 0 | 0 | 0 | 0 | 0 | 0 | 1 | 3 |
| Heating Degree Days | 995 | 721 | 590 | 346 | 131 | 10 | 0 | 1 | 31 | 283 | 643 | 978 | 4729 |
| Cooling Degree Days | 0 | 0 | 0 | 4 | 45 | 226 | 370 | 317 | 121 | 3 | 0 | 0 | 1086 |
| Total Precipitation (") | 0.49 | 0.50 | 0.59 | 0.27 | 0.41 | 0.34 | 1.25 | 1.25 | 0.83 | 0.86 | 0.57 | 0.73 | 8.09 |
| Days ≥ 0.1" Precip | 2 | 2 | 2 | 1 | 1 | 1 | 3 | 3 | 2 | 2 | 2 | 2 | 23 |
| Total Snowfall (") | 2.5 | 2.0 | 2.2 | 0.5 | 0.0 | 0.0 | 0.0 | 0.0 | 0.0 | 0.2 | 0.8 | 4.0 | 12.2 |
| Days ≥ 1" Snow Depth | 4 | 1 | 1 | 0 | 0 | 0 | 0 | 0 | 0 | 0 | 0 | 4 | 10 |

## WUPATKI NATL MONUMNT *Coconino County* ELEVATION 4900 ft LAT/LONG 35° 31 ' N / 111° 22 ' W

| | JAN | FEB | MAR | APR | MAY | JUN | JUL | AUG | SEP | OCT | NOV | DEC | YEAR |
|---|---|---|---|---|---|---|---|---|---|---|---|---|---|
| Maximum Temp °F | 46.9 | 55.3 | 63.1 | 72.0 | 81.0 | 91.9 | 95.3 | 92.2 | 85.5 | 73.5 | 58.8 | 46.9 | 71.9 |
| Minimum Temp °F | 24.3 | 29.2 | 35.1 | 41.5 | 49.6 | 59.0 | 64.8 | 62.4 | 55.8 | 44.9 | 33.5 | 24.6 | 43.7 |
| Mean Temp °F | 35.7 | 42.3 | 49.1 | 56.8 | 65.3 | 75.5 | 80.1 | 77.3 | 70.6 | 59.2 | 46.2 | 35.8 | 57.8 |
| Days Max Temp ≥ 90 °F | 0 | 0 | 0 | 0 | 4 | 20 | 27 | 22 | 8 | 0 | 0 | 0 | 81 |
| Days Max Temp ≤ 32 °F | 2 | 0 | 0 | 0 | 0 | 0 | 0 | 0 | 0 | 0 | 0 | 3 | 5 |
| Days Min Temp ≤ 32 °F | 27 | 19 | 10 | 3 | 0 | 0 | 0 | 0 | 0 | 1 | 13 | 27 | 100 |
| Days Min Temp ≤ 0 °F | 0 | 0 | 0 | 0 | 0 | 0 | 0 | 0 | 0 | 0 | 0 | 0 | 0 |
| Heating Degree Days | 903 | 635 | 484 | 252 | 80 | 6 | 0 | 2 | 18 | 198 | 558 | 899 | 4035 |
| Cooling Degree Days | 0 | 0 | 0 | 19 | 95 | 336 | 471 | 389 | 190 | 22 | 0 | 0 | 1522 |
| Total Precipitation (") | 0.53 | 0.51 | 0.74 | 0.34 | 0.42 | 0.29 | 1.34 | 1.62 | 0.81 | 0.76 | 0.63 | 0.66 | 8.65 |
| Days ≥ 0.1" Precip | 2 | 1 | 2 | 1 | 1 | 1 | 4 | 4 | 2 | 2 | 1 | 2 | 23 |
| Total Snowfall (") | *1.2* | 1.2 | 1.3 | 0.1 | 0.0 | 0.0 | 0.0 | 0.0 | 0.0 | 0.0 | 0.4 | 2.9 | 7.1 |
| Days ≥ 1" Snow Depth | na | 0 | 0 | 0 | 0 | 0 | 0 | 0 | 0 | 0 | 0 | *1* | na |

## YOUNGTOWN *Maricopa County* ELEVATION 1142 ft LAT/LONG 33° 36 ' N / 112° 18 ' W

| | JAN | FEB | MAR | APR | MAY | JUN | JUL | AUG | SEP | OCT | NOV | DEC | YEAR |
|---|---|---|---|---|---|---|---|---|---|---|---|---|---|
| Maximum Temp °F | 66.7 | 71.5 | 76.4 | 85.2 | 93.9 | 103.1 | 105.4 | 103.8 | 98.6 | 88.1 | 75.0 | 66.3 | 86.2 |
| Minimum Temp °F | 39.2 | 42.5 | 46.4 | 52.1 | 60.7 | 69.3 | 77.2 | 75.7 | 68.5 | 56.8 | 45.3 | 38.9 | 56.0 |
| Mean Temp °F | 53.0 | 57.0 | 61.4 | 68.7 | 77.3 | 86.2 | 91.4 | 89.8 | 83.6 | 72.5 | 60.2 | 52.7 | 71.2 |
| Days Max Temp ≥ 90 °F | 0 | 0 | 2 | 9 | 23 | 29 | 31 | 31 | 28 | 14 | 0 | 0 | 167 |
| Days Max Temp ≤ 32 °F | 0 | 0 | 0 | 0 | 0 | 0 | 0 | 0 | 0 | 0 | 0 | 0 | 0 |
| Days Min Temp ≤ 32 °F | 5 | 2 | 0 | 0 | 0 | 0 | 0 | 0 | 0 | 0 | 0 | 5 | 12 |
| Days Min Temp ≤ 0 °F | 0 | 0 | 0 | 0 | 0 | 0 | 0 | 0 | 0 | 0 | 0 | 0 | 0 |
| Heating Degree Days | 366 | 223 | 142 | 36 | 2 | 0 | 0 | 0 | 0 | 15 | 159 | 376 | 1319 |
| Cooling Degree Days | 0 | 9 | 41 | 204 | 428 | 683 | 834 | 805 | 587 | 281 | 25 | 0 | 3897 |
| Total Precipitation (") | 1.08 | 1.02 | 1.17 | 0.26 | 0.16 | 0.08 | 0.80 | 1.13 | 0.90 | 0.65 | 0.79 | 1.38 | 9.42 |
| Days ≥ 0.1" Precip | 2 | 2 | 3 | 1 | 1 | 0 | 2 | 2 | 2 | 1 | 2 | 3 | 21 |
| Total Snowfall (") | 0.0 | 0.0 | 0.0 | 0.0 | 0.0 | 0.0 | 0.0 | 0.0 | 0.0 | 0.0 | 0.0 | 0.0 | 0.0 |
| Days ≥ 1" Snow Depth | 0 | 0 | 0 | 0 | 0 | 0 | 0 | 0 | 0 | 0 | 0 | 0 | 0 |

## YUCCA 1 NNE *Mohave County* ELEVATION 1811 ft LAT/LONG 34° 52 ' N / 114° 9 ' W

| | JAN | FEB | MAR | APR | MAY | JUN | JUL | AUG | SEP | OCT | NOV | DEC | YEAR |
|---|---|---|---|---|---|---|---|---|---|---|---|---|---|
| Maximum Temp °F | 59.3 | 64.4 | 69.2 | 77.5 | 86.9 | 97.2 | 101.6 | 99.8 | 93.8 | 82.2 | 68.2 | 58.9 | 79.9 |
| Minimum Temp °F | 36.8 | 39.8 | 43.3 | 48.9 | 58.3 | 67.2 | 75.8 | 74.2 | 65.9 | 54.4 | 43.4 | 36.6 | 53.7 |
| Mean Temp °F | 48.1 | 52.2 | 56.3 | 63.3 | 72.6 | 82.2 | 88.7 | 87.0 | 79.9 | 68.3 | 55.8 | 47.8 | 66.8 |
| Days Max Temp ≥ 90 °F | 0 | 0 | 0 | 3 | 13 | 26 | 30 | 30 | 22 | 7 | 0 | 0 | 131 |
| Days Max Temp ≤ 32 °F | 0 | 0 | 0 | 0 | 0 | 0 | 0 | 0 | 0 | 0 | 0 | 0 | 0 |
| Days Min Temp ≤ 32 °F | 8 | 4 | 2 | 0 | 0 | 0 | 0 | 0 | 0 | 0 | 2 | 8 | 24 |
| Days Min Temp ≤ 0 °F | 0 | 0 | 0 | 0 | 0 | 0 | 0 | 0 | 0 | 0 | 0 | 0 | 0 |
| Heating Degree Days | 518 | 357 | 275 | 113 | 20 | 0 | 0 | 0 | 2 | 48 | 275 | 526 | 2134 |
| Cooling Degree Days | 0 | 2 | 13 | 96 | 277 | 557 | 757 | 715 | 472 | 165 | 9 | 0 | 3063 |
| Total Precipitation (") | 1.03 | 0.88 | 1.34 | 0.41 | 0.20 | 0.10 | 0.76 | 0.87 | 0.67 | 0.48 | 0.61 | 0.82 | 8.17 |
| Days ≥ 0.1" Precip | 2 | 2 | 3 | 1 | 1 | 0 | 2 | 2 | 1 | 1 | 1 | 2 | 18 |
| Total Snowfall (") | 0.1 | 0.1 | 0.0 | 0.0 | 0.0 | 0.0 | 0.0 | 0.0 | 0.0 | 0.0 | 0.0 | 0.0 | 0.2 |
| Days ≥ 1" Snow Depth | 0 | 0 | 0 | 0 | 0 | 0 | 0 | 0 | 0 | 0 | 0 | 0 | 0 |

## YUMA CITRUS STATION *Yuma County* ELEVATION 191 ft LAT/LONG 32° 37 ' N / 114° 39 ' W

| | JAN | FEB | MAR | APR | MAY | JUN | JUL | AUG | SEP | OCT | NOV | DEC | YEAR |
|---|---|---|---|---|---|---|---|---|---|---|---|---|---|
| Maximum Temp °F | 67.7 | 73.0 | 78.4 | 85.5 | 93.3 | 102.3 | 105.6 | 104.8 | 100.2 | 89.7 | 76.6 | 67.4 | 87.0 |
| Minimum Temp °F | 39.5 | 41.5 | 46.0 | 50.8 | 57.9 | 65.8 | 74.8 | 74.7 | 68.3 | 56.5 | 45.3 | 39.6 | 55.1 |
| Mean Temp °F | 53.6 | 57.3 | 62.3 | 68.2 | 75.7 | 84.1 | 90.2 | 89.8 | 84.2 | 73.1 | 61.0 | 53.5 | 71.1 |
| Days Max Temp ≥ 90 °F | 0 | 0 | 3 | 10 | 22 | 29 | 31 | 31 | 28 | 17 | 1 | 0 | 172 |
| Days Max Temp ≤ 32 °F | 0 | 0 | 0 | 0 | 0 | 0 | 0 | 0 | 0 | 0 | 0 | 0 | 0 |
| Days Min Temp ≤ 32 °F | 4 | 2 | 0 | 0 | 0 | 0 | 0 | 0 | 0 | 0 | 0 | 4 | 10 |
| Days Min Temp ≤ 0 °F | 0 | 0 | 0 | 0 | 0 | 0 | 0 | 0 | 0 | 0 | 0 | 0 | 0 |
| Heating Degree Days | 346 | 217 | 122 | 41 | 5 | 0 | 0 | 0 | 0 | 13 | 141 | 350 | 1235 |
| Cooling Degree Days | 0 | 7 | 46 | 176 | 363 | 600 | 792 | 794 | 602 | 289 | 28 | 0 | 3697 |
| Total Precipitation (") | 0.49 | 0.36 | 0.38 | 0.14 | 0.05 | 0.03 | 0.30 | 0.67 | 0.35 | 0.30 | 0.28 | 0.55 | 3.90 |
| Days ≥ 0.1" Precip | 1 | 1 | 1 | 0 | 0 | 0 | 0 | 1 | 1 | 1 | 1 | 1 | 8 |
| Total Snowfall (") | 0.0 | 0.0 | 0.0 | 0.0 | 0.0 | 0.0 | 0.0 | 0.0 | 0.0 | 0.0 | 0.0 | 0.0 | 0.0 |
| Days ≥ 1" Snow Depth | 0 | 0 | 0 | 0 | 0 | 0 | 0 | 0 | 0 | 0 | 0 | 0 | 0 |

## YUMA INTL AP *Yuma County* ELEVATION 207 ft LAT/LONG 32° 40 ' N / 114° 36 ' W

| | JAN | FEB | MAR | APR | MAY | JUN | JUL | AUG | SEP | OCT | NOV | DEC | YEAR |
|---|---|---|---|---|---|---|---|---|---|---|---|---|---|
| Maximum Temp °F | 68.5 | 74.0 | 78.8 | 86.1 | 94.1 | 103.4 | 106.6 | 105.4 | 100.4 | 90.0 | 76.9 | 67.9 | 87.7 |
| Minimum Temp °F | 44.9 | 47.4 | 51.6 | 57.0 | 64.4 | 72.4 | 80.5 | 80.0 | 73.7 | 62.6 | 51.2 | 44.7 | 60.9 |
| Mean Temp °F | 56.7 | 60.7 | 65.2 | 71.6 | 79.2 | 87.9 | 93.6 | 92.8 | 87.0 | 76.3 | 64.1 | 56.3 | 74.3 |
| Days Max Temp ≥ 90 °F | 0 | 0 | 3 | 12 | 23 | 29 | 31 | 31 | 28 | 18 | 1 | 0 | 176 |
| Days Max Temp ≤ 32 °F | 0 | 0 | 0 | 0 | 0 | 0 | 0 | 0 | 0 | 0 | 0 | 0 | 0 |
| Days Min Temp ≤ 32 °F | 1 | 0 | 0 | 0 | 0 | 0 | 0 | 0 | 0 | 0 | 0 | 1 | 2 |
| Days Min Temp ≤ 0 °F | 0 | 0 | 0 | 0 | 0 | 0 | 0 | 0 | 0 | 0 | 0 | 0 | 0 |
| Heating Degree Days | 252 | 136 | 69 | 18 | 1 | 0 | 0 | 0 | 0 | 5 | 82 | 265 | 828 |
| Cooling Degree Days | 3 | 34 | 100 | 283 | 486 | 731 | 905 | 892 | 699 | 400 | 74 | 2 | 4609 |
| Total Precipitation (") | 0.40 | 0.29 | 0.29 | 0.12 | 0.06 | 0.02 | 0.26 | 0.63 | 0.30 | 0.25 | 0.26 | 0.50 | 3.38 |
| Days ≥ 0.1" Precip | 1 | 1 | 1 | 0 | 0 | 0 | 0 | 1 | 1 | 0 | 1 | 1 | 7 |
| Total Snowfall (") | 0.0 | 0.0 | 0.0 | 0.0 | 0.0 | 0.0 | 0.0 | 0.0 | 0.0 | 0.0 | 0.0 | 0.0 | 0.0 |
| Days ≥ 1" Snow Depth | 0 | 0 | 0 | 0 | 0 | 0 | 0 | 0 | 0 | 0 | 0 | 0 | 0 |

## YUMA PROVING GROUND *Yuma County* ELEVATION 324 ft LAT/LONG 32° 50 ' N / 114° 24 ' W

| | JAN | FEB | MAR | APR | MAY | JUN | JUL | AUG | SEP | OCT | NOV | DEC | YEAR |
|---|---|---|---|---|---|---|---|---|---|---|---|---|---|
| Maximum Temp °F | 67.9 | 73.2 | 78.0 | 85.4 | 93.3 | 102.9 | 106.1 | 104.9 | 100.0 | 89.5 | 76.4 | 67.3 | 87.1 |
| Minimum Temp °F | 43.2 | 46.8 | 51.3 | 56.9 | 64.2 | 72.4 | 80.3 | 80.0 | 73.3 | 61.8 | 49.7 | 42.7 | 60.2 |
| Mean Temp °F | 55.5 | 60.0 | 64.7 | 71.2 | 78.8 | 87.7 | 93.2 | 92.5 | 86.7 | 75.7 | 63.1 | 55.1 | 73.7 |
| Days Max Temp ≥ 90 °F | 0 | 0 | 3 | 10 | 22 | 29 | 31 | 31 | 28 | 17 | 1 | 0 | 172 |
| Days Max Temp ≤ 32 °F | 0 | 0 | 0 | 0 | 0 | 0 | 0 | 0 | 0 | 0 | 0 | 0 | 0 |
| Days Min Temp ≤ 32 °F | 2 | 0 | 0 | 0 | 0 | 0 | 0 | 0 | 0 | 0 | 0 | 1 | 3 |
| Days Min Temp ≤ 0 °F | 0 | 0 | 0 | 0 | 0 | 0 | 0 | 0 | 0 | 0 | 0 | 0 | 0 |
| Heating Degree Days | 288 | 149 | 79 | 20 | 1 | 0 | 0 | 0 | 0 | 6 | 101 | 302 | 946 |
| Cooling Degree Days | 0 | 22 | 78 | 255 | 450 | 708 | 877 | 870 | 666 | 357 | 51 | 1 | 4335 |
| Total Precipitation (") | 0.53 | 0.39 | 0.42 | 0.17 | 0.04 | 0.05 | 0.22 | 0.66 | 0.46 | 0.35 | 0.35 | 0.46 | 4.10 |
| Days ≥ 0.1" Precip | 1 | 1 | 1 | 0 | 0 | 0 | 1 | 1 | 1 | 0 | 1 | 1 | 8 |
| Total Snowfall (") | 0.0 | 0.0 | 0.0 | 0.0 | 0.0 | 0.0 | 0.0 | 0.0 | 0.0 | 0.0 | 0.0 | 0.0 | 0.0 |
| Days ≥ 1" Snow Depth | 0 | 0 | 0 | 0 | 0 | 0 | 0 | 0 | 0 | 0 | 0 | 0 | 0 |

## JANUARY MINIMUM TEMPERATURE °F

**LOWEST**

| | | |
|---|---|---|
| 1 | Fort Valley | 10.0 |
| 2 | Alpine | 13.0 |
| 3 | Window Rock | 14.2 |
| 4 | Ganado | 14.8 |
| 5 | Greer | 15.0 |
| 6 | Flagstaff | 15.9 |
| | Springerville | 15.9 |
| 8 | Keams Canyon | 16.1 |
| | Lukachukai | 16.1 |
| 10 | Sanders | 17.4 |
| 11 | McNary | 17.6 |
| 12 | Holbrook | 18.1 |
| 13 | Show Low | 18.7 |
| 14 | Chevelon | 18.8 |
| | Snowflake | 18.8 |
| 16 | Teec Nos Pos | 19.0 |
| 17 | St. Johns | 19.1 |
| 18 | Snowflake-15 W | 19.4 |
| 19 | Betatakin | 20.1 |
| | Winslow | 20.1 |
| 21 | Petrified Forest | 20.6 |
| 22 | Black River | 20.7 |
| 23 | Pleasant Valley | 20.8 |
| 24 | Walnut Creek | 21.1 |
| 25 | Pipe Springs | 21.2 |

**HIGHEST**

| | | |
|---|---|---|
| 1 | Kofa | 45.6 |
| 2 | Yuma-Intl | 44.9 |
| 3 | Phoenix | 43.9 |
| 4 | Ajo | 43.8 |
| 5 | Yuma-Proving | 43.2 |
| 6 | Phoenx-Sky Harbr | 42.4 |
| 7 | Mormon Flat | 42.3 |
| | Superior | 42.3 |
| 9 | Tucson-U of A | 40.8 |
| 10 | Bartlett Dam | 40.4 |
| 11 | Castle Hot Springs | 39.9 |
| | Gila Bend | 39.9 |
| 13 | Parker | 39.7 |
| 14 | Yuma-Citrus | 39.5 |
| 15 | Youngtown | 39.2 |
| 16 | Mesa | 39.1 |
| 17 | Tucson-Intl | 39.0 |
| 18 | Organ Pipe Cactus | 38.9 |
| 19 | Laveen | 38.6 |
| 20 | South Phoenix | 38.5 |
| 21 | Chandler Heights | 38.3 |
| 22 | Florence | 38.0 |
| | Tempe | 38.0 |
| 24 | Roosevelt | 37.6 |
| 25 | Litchfield Park | 37.4 |

## JULY MAXIMUM TEMPERATURE °F

**HIGHEST**

| | | |
|---|---|---|
| 1 | Buckeye | 109.0 |
| | Gila Bend | 109.0 |
| 3 | Parker | 108.4 |
| 4 | Litchfield Park | 107.8 |
| 5 | Bouse | 107.6 |
| 6 | Casa Grande-NM | 107.1 |
| 7 | Maricopa | 106.8 |
| 8 | Yuma-Intl | 106.6 |
| 9 | Sacaton | 106.5 |
| 10 | Yuma-Proving | 106.1 |
| 11 | Casa Grande | 105.8 |
| | Phoenx-Sky Harbr | 105.8 |
| | Tempe | 105.8 |
| 14 | Phoenix | 105.7 |
| 15 | Yuma-Citrus | 105.6 |
| 16 | Youngtown | 105.4 |
| 17 | Florence | 105.3 |
| 18 | Laveen | 104.9 |
| | Mesa | 104.9 |
| 20 | Bartlett Dam | 104.7 |
| 21 | Wikieup | 104.6 |
| 22 | Eloy | 104.5 |
| 23 | Chandler Heights | 104.0 |
| | Wickenburg | 104.0 |
| 25 | Stewart Mountain | 103.6 |

**LOWEST**

| | | |
|---|---|---|
| 1 | Greer | 75.6 |
| 2 | Alpine | 77.6 |
| 3 | Kitt Peak | 80.2 |
| 4 | Fort Valley | 80.9 |
| 5 | McNary | 81.2 |
| 6 | Springerville | 81.9 |
| 7 | Chevelon | 82.0 |
| | Flagstaff | 82.0 |
| 9 | Williams | 82.5 |
| 10 | Window Rock | 84.4 |
| 11 | Betatakin | 84.7 |
| | Heber | 84.7 |
| 13 | Show Low | 85.6 |
| 14 | Lukachukai | 85.9 |
| 15 | Portal | 86.3 |
| 16 | Black River | 87.0 |
| | Ganado | 87.0 |
| | Snowflake-15 W | 87.0 |
| 19 | Prescott | 88.0 |
| 20 | Canelo | 88.6 |
| | Chiricahua | 88.6 |
| 22 | Keams Canyon | 88.7 |
| 23 | Whiteriver | 89.3 |
| 24 | Snowflake | 89.8 |
| | Walnut Creek | 89.8 |

## ANNUAL PRECIPITATION (")

**HIGHEST**

| | | |
|---|---|---|
| 1 | McNary | 29.27 |
| 2 | Kitt Peak | 25.48 |
| 3 | Oracle | 24.50 |
| 4 | Flagstaff | 24.27 |
| 5 | Greer | 24.19 |
| 6 | Santa Rita | 23.85 |
| 7 | Fort Valley | 22.96 |
| 8 | Payson | 22.70 |
| 9 | Pleasant Valley | 22.13 |
| 10 | Williams | 22.00 |
| 11 | Portal | 21.92 |
| 12 | Alpine | 21.55 |
| 13 | Coronado | 21.35 |
| 14 | Superior | 20.88 |
| 15 | Miami | 20.56 |
| 16 | Jerome | 20.33 |
| 17 | Chiricahua | 20.24 |
| 18 | Prescott | 20.15 |
| 19 | Whiteriver | 20.13 |
| 20 | Black River | 20.09 |
| 21 | Childs | 20.05 |
| 22 | Sedona | 19.63 |
| 23 | Chevelon | 19.41 |
| 24 | Nogales | 19.28 |
| 25 | Heber | 19.18 |

**LOWEST**

| | | |
|---|---|---|
| 1 | Yuma-Intl | 3.38 |
| 2 | Yuma-Citrus | 3.90 |
| 3 | Yuma-Proving | 4.10 |
| 4 | Parker | 5.25 |
| 5 | Bouse | 6.03 |
| 6 | Lees Ferry | 6.53 |
| 7 | Page | 6.54 |
| 8 | Wahweap | 6.57 |
| 9 | Gila Bend | 7.14 |
| 10 | Kofa | 7.65 |
| 11 | Winslow | 8.09 |
| 12 | Teec Nos Pos | 8.10 |
| 13 | Yucca | 8.17 |
| 14 | Buckeye | 8.30 |
| 15 | Phoenx-Sky Harbr | 8.44 |
| 16 | Maricopa | 8.48 |
| 17 | Wupatki | 8.65 |
| 18 | Ajo | 8.74 |
| 19 | Laveen | 8.88 |
| 20 | Holbrook | 8.96 |
| | Litchfield Park | 8.96 |
| 22 | Phoenix | 8.98 |
| 23 | Aguila | 9.19 |
| 24 | Sacaton | 9.26 |
| 25 | South Phoenix | 9.34 |

## ANNUAL SNOWFALL (")

**HIGHEST**

| | | |
|---|---|---|
| 1 | Flagstaff | 114.7 |
| 2 | Fort Valley | 98.3 |
| 3 | McNary | 94.6 |
| 4 | Williams | 77.8 |
| 5 | Betatakin | 56.4 |
| 6 | Chevelon | 48.8 |
| 7 | Heber | 44.1 |
| 8 | Show Low | 30.4 |
| 9 | Ganado | 26.5 |
| | Payson | 26.5 |
| 11 | Prescott | 23.9 |
| 12 | Colorado City | 21.8 |
| 13 | Kitt Peak | 20.6 |
| 14 | Snowflake | 17.8 |
| 15 | Snowflake-15 W | 16.5 |
| 16 | St. Johns | 16.2 |
| 17 | Whiteriver | 15.9 |
| 18 | Jerome | 12.8 |
| 19 | Winslow | 12.2 |
| 20 | Oracle | 11.6 |
| | Springerville | 11.6 |
| 22 | Pipe Springs | 10.1 |
| 23 | Petrified Forest | 10.0 |
| 24 | Chiricahua | 9.7 |
| 25 | Seligman | 9.0 |

**LOWEST***

| | | |
|---|---|---|
| 1 | Ajo | 0.1 |
| | Clifton | 0.1 |
| | Gila Bend | 0.1 |
| | Roosevelt | 0.1 |
| | Wikieup | 0.1 |
| 6 | Anvil | 0.2 |
| | Castle Hot Springs | 0.2 |
| | Wickenburg | 0.2 |
| | Yucca | 0.2 |
| 10 | Nogales | 0.3 |
| | San Carlos Rsrvr | 0.3 |
| | Tombstone | 0.3 |
| 13 | Wahweap | 0.4 |
| 14 | Duncan | 0.5 |
| 15 | Aguila | 0.6 |
| | Tucson-CP Ave | 0.6 |
| | Tucson-U of A | 0.6 |
| 18 | Tucson-Magnetic | 0.7 |
| 19 | Tumacacori | 0.8 |
| 20 | Safford | 1.0 |
| 21 | Bisbee | 1.1 |
| | Sasabe | 1.1 |
| 23 | Childs | 1.2 |
| | Superior | 1.2 |
| 25 | Tucson-Intl | 1.3 |

* Does not include stations which receive no snowfall.

# ARKANSAS

PHYSICAL FEATURES. Arkansas is divided geographically into two principal divisions on the basis of topography, and to a lesser extent, climate. The dividing line between these two sections cuts diagonally across the State from the northeast to the southwest. West and north of this line are the interior highlands; to the east and south are the lowlands.

Much of western and northern Arkansas is hilly and mountainous. In the southern part, or that portion south of the Arkansas River, are the Ouachita Mountains made up of a number of narrow east-west ridges separated by rather narrow valleys. Some of these ridges reach elevations of 2,500 feet or more. The Arkansas valley, between the Ozark and Ouachita highlands, is an area of fairly low relief with a few isolated ridges and mountains. One of these mountains in the Arkansas valley, Mt. Magazine, with an elevation of 2,823 feet above sea level, is the highest point in the State.

The Ozark Mountains and particularly that portion known as the Boston Mountains are the largest and most massive in Arkansas. It is this topographical feature of the State that has the most noticeable effect upon Arkansas weather.

GENERAL CLIMATE. Climatic differences between the two areas are not as great as the local differences between mountain and valley stations in the highlands. Generally, the climate of western and northern Arkansas is a little cooler and there are greater temperature extremes; humidities are lower and there is less cloudiness.

Average maximum or minimum temperatures show little variation over the State. Winter temperatures vary more noticeably from northwest to southeast than is the case in the summer. Maximum temperatures exceed 100° at times during July and August, particularly in the valley stations in the highlands. The winters are short, but cold periods of brief duration do occur. In the northern part of the State, zero temperatures are of occasional occurrence in January and February and zero has been recorded to the southern border.

PRECIPITATION. Precipitation in Arkansas is predominantly of the shower type except for occasional periods of general rain during the late fall, winter, and early spring. The average number of days with measurable precipitation averages around 100 per year.

Rainfall is normally abundant and well distributed throughout the year. However, extended rain-free periods, as well as flooding local storms, are by no means unusual.

Annual precipitation amounts display both local orographic influence and geographic location with the State. Just by virtue of being closer to the Gulf of Mexico moisture source, the southeast counties receive, on the average, 5 to 6 inches more rainfall per year than the northwest counties. However, noticeable exceptions to this are a number of Ozark and Ouachita Mountain stations where the year's totals average 55 to nearly 56 inches.

Winter and spring are the wettest times of the year. December and January are the wet months on the average in the southern counties and March through May is the wet period in the north. The fall of the year is uniformly the dry time of the year when monthly precipitation totals average 2 to 3 inches.

The State is subject to heavy local rains which frequently give storm totals of from 5 to 10 inches. Floods are frequent along the White, Black, and Ouachita Rivers. Disastrous floods are of rare occurrence.

Most of the State's precipitation falls as rain. Snow does occur, principally in the northwest. The average annual totals range from a little over a foot on the ground in the higher Ozark elevations in the northwest to 1 to 2 inches in the delta flat lands of the extreme southeast counties. Snowfall in these southern and eastern lowlands is generally light and remains on the ground only briefly.

Despite the generally abundant rainfall, short periods of dry weather are frequent over small areas of the State. Occasionally severe droughts of longer duration and involving large areas do occur. Severe droughts covering the greater part of the State occur infrequently.

OTHER CLIMATIC ELEMENTS. The long growing season, averaging from 180 days in the northwest up to more than 230 days in the principal cotton producing areas, favors agricultural activity. In addition to adequate moisture conditions, the eastern and southern Arkansas areas have dry, sunny weather during the early fall. Extended warm and humid summer periods are common.

An average of 17 tornadoes per year have been observed in Arkansas. The severe thunderstorms and tornadoes occur most frequently in the period March through May. With the advent of summer heat in June, the tornado occurrence falls off sharply.

## COUNTY INDEX

***Randolph County***
POCAHONTAS 1

***Saline County***
ALUM FORK
BENTON

***Scott County***
WALDRON

***Searcy County***
GILBERT

***Sebastian County***
FT SMITH MUNICIPL AP

***Sevier County***
DEQUEEN

***Sharp County***
EVENING SHADE 1 NNE

***Stone County***
MOUNTAIN VIEW

***Union County***
EL DORADO GOODWIN FD

***Washington County***
FAYETTEVILLE EXP STN

***White County***
SEARCY

***Yell County***
BLUE MOUNTAIN DAM
DARDANELLE

## ELEVATION INDEX

| FEET | STATION NAME |
|---|---|
| 121 | CAMDEN 1 |
| 121 | PORTLAND |
| 141 | EUDORA |
| 143 | DERMOTT 3 NE |
| 150 | ROHWER 2 NNE |
| 161 | DUMAS |
| 171 | CLARENDON |
| 171 | SPARKMAN |
| 180 | ARKANSAS POST |
| 180 | CROSSETT 2 SSE |
| 198 | STUTTGART 9 ESE |
| 200 | ARKADELPHIA 2 N |
| 200 | DES ARC |
| 200 | ST CHARLES |
| 210 | BRINKLEY |
| 210 | WARREN 2 WSW |
| 220 | NEWPORT |
| 220 | PINE BLUFF |
| 220 | WEST MEMPHIS |
| 221 | BEEDEVILLE 4 NE |
| 230 | KEO |
| 230 | MARIANNA 2 S |
| 232 | KEISER |
| 239 | HELENA |
| 249 | BLYTHEVILLE |
| 249 | MONTICELLO 3 SW |
| 249 | SEARCY |
| 253 | EL DORADO GOODWIN FD |
| 259 | ALICIA |
| 259 | FORDYCE |
| 259 | WYNNE |
| 261 | LEOLA |
| 276 | LITTLE ROCK ADAMS FD |
| 277 | BATESVILLE L & D 1 |
| 285 | BENTON |
| 289 | CABOT 4 SW |
| 289 | CORNING |
| 310 | CONWAY |
| 312 | MALVERN |
| 322 | MAGNOLIA 3 N |
| 322 | PRESCOTT |
| 331 | DARDANELLE |
| 331 | MORRILTON |
| 331 | POCAHONTAS 1 |
| 351 | JONESBORO 4 N |
| 361 | GREENBRIER |
| 381 | CALICO ROCK 2 WSW |
| 381 | HOPE 3 NE |
| 396 | OZARK |
| 430 | BLAKELY MOUNTAIN DAM |
| 430 | DEQUEEN |
| 447 | FT SMITH MUNICIPL AP |
| 459 | BLUE MOUNTAIN DAM |
| 469 | NIMROD DAM |
| 489 | EVENING SHADE 1 NNE |
| 502 | SUBIACO |
| 527 | GREERS FERRY DAM |
| 550 | NASHVILLE |
| 571 | BATESVILLE LIVESTOCK |
| 600 | GILBERT |
| 600 | MAMMOTH SPRING |
| 659 | MOUNT IDA 3 SE |
| 675 | WALDRON |
| 712 | HOT SPRINGS 1 NNE |
| 712 | LEAD HILL |
| 761 | ALUM FORK |
| 771 | MOUNTAIN VIEW |
| 801 | MOUNTAIN HOME 1 NNW |
| 1211 | MENA |
| 1220 | BENTONVILLE 4 S |
| 1250 | GRAVETTE |
| 1270 | FAYETTEVILLE EXP STN |
| 1375 | HARRISON FAA AP |
| 1470 | EUREKA SPRINGS 3 WNW |

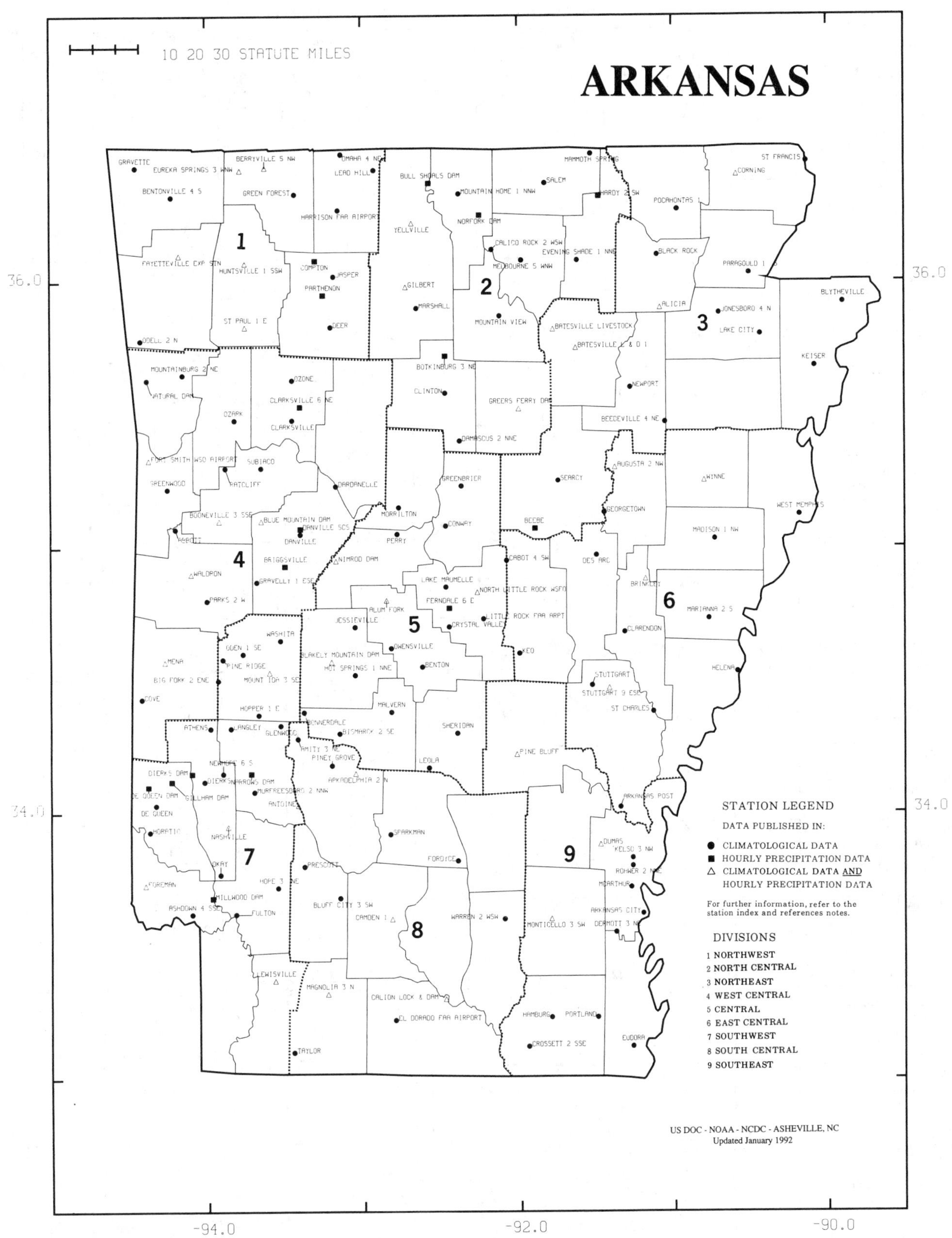
10 20 30 STATUTE MILES
ARKANSAS
36.0
34.0
-94.0
-92.0
-90.0
STATION LEGEND
DATA PUBLISHED IN:
● CLIMATOLOGICAL DATA
■ HOURLY PRECIPITATION DATA
△ CLIMATOLOGICAL DATA AND HOURLY PRECIPITATION DATA
For further information, refer to the station index and references notes.
DIVISIONS
1 NORTHWEST
2 NORTH CENTRAL
3 NORTHEAST
4 WEST CENTRAL
5 CENTRAL
6 EAST CENTRAL
7 SOUTHWEST
8 SOUTH CENTRAL
9 SOUTHEAST
US DOC - NOAA - NCDC - ASHEVILLE, NC
Updated January 1992

## ALICIA *Lawrence County* ELEVATION 259 ft LAT/LONG 35° 54 ' N / 91° 5 ' W

| | JAN | FEB | MAR | APR | MAY | JUN | JUL | AUG | SEP | OCT | NOV | DEC | YEAR |
|---|---|---|---|---|---|---|---|---|---|---|---|---|---|
| Maximum Temp °F | 46.4 | 51.9 | 62.2 | 73.6 | 81.3 | 89.9 | 93.6 | 91.8 | 84.5 | 74.8 | 61.1 | 50.3 | 71.8 |
| Minimum Temp °F | 27.5 | 31.2 | 40.1 | 49.3 | 57.5 | 66.2 | 69.6 | 67.0 | 60.0 | 48.0 | 39.3 | 31.4 | 48.9 |
| Mean Temp °F | 37.0 | 41.5 | 51.2 | 61.5 | 69.4 | 78.0 | 81.6 | 79.4 | 72.3 | 61.4 | 50.2 | 40.9 | 60.4 |
| Days Max Temp ≥ 90 °F | 0 | 0 | 0 | 0 | 3 | 17 | 24 | 20 | 8 | 1 | 0 | 0 | 73 |
| Days Max Temp ≤ 32 °F | 4 | 1 | 0 | 0 | 0 | 0 | 0 | 0 | 0 | 0 | 0 | 1 | 6 |
| Days Min Temp ≤ 32 °F | 22 | 16 | 7 | 1 | 0 | 0 | 0 | 0 | 0 | 1 | 8 | 17 | 72 |
| Days Min Temp ≤ 0 °F | 0 | 0 | 0 | 0 | 0 | 0 | 0 | 0 | 0 | 0 | 0 | 0 | 0 |
| Heating Degree Days | 864 | 655 | 431 | 153 | 29 | 0 | 0 | 0 | 21 | 159 | 443 | 741 | 3496 |
| Cooling Degree Days | 0 | 0 | 7 | 52 | 178 | 411 | 555 | 493 | 275 | 55 | 5 | 1 | 2032 |
| Total Precipitation (") | 3.49 | 3.45 | 4.61 | 4.85 | 4.90 | 3.35 | 3.03 | 3.05 | 4.16 | 3.55 | 4.92 | 4.76 | 48.12 |
| Days ≥ 0.1" Precip | 5 | 5 | 7 | 7 | 7 | 5 | 5 | 5 | 5 | 5 | 6 | 6 | 68 |
| Total Snowfall (") | *2.2* | *1.7* | 0.8 | 0.0 | 0.0 | 0.0 | 0.0 | 0.0 | 0.0 | 0.0 | 0.0 | 0.6 | 5.3 |
| Days ≥ 1" Snow Depth | na | *0* | 0 | 0 | 0 | 0 | 0 | 0 | 0 | 0 | 0 | 0 | na |

## ALUM FORK *Saline County* ELEVATION 761 ft LAT/LONG 34° 48 ' N / 92° 52 ' W

| | JAN | FEB | MAR | APR | MAY | JUN | JUL | AUG | SEP | OCT | NOV | DEC | YEAR |
|---|---|---|---|---|---|---|---|---|---|---|---|---|---|
| Maximum Temp °F | 50.7 | 56.4 | 65.5 | 75.2 | 80.7 | 88.0 | 92.6 | 91.0 | 84.6 | 75.5 | 63.1 | 54.0 | 73.1 |
| Minimum Temp °F | 29.8 | 32.6 | 41.1 | 50.0 | 57.6 | 65.4 | 69.3 | 67.7 | 62.0 | 50.9 | 41.4 | 33.6 | 50.1 |
| Mean Temp °F | 40.2 | 44.5 | 53.3 | 62.6 | 69.2 | 76.7 | 80.9 | 79.4 | 73.3 | 63.3 | 52.3 | 43.8 | 61.6 |
| Days Max Temp ≥ 90 °F | 0 | 0 | 0 | 0 | 1 | 12 | 23 | 20 | 8 | 1 | 0 | 0 | 65 |
| Days Max Temp ≤ 32 °F | 2 | 1 | 0 | 0 | 0 | 0 | 0 | 0 | 0 | 0 | 0 | 1 | 4 |
| Days Min Temp ≤ 32 °F | 20 | 14 | 6 | 1 | 0 | 0 | 0 | 0 | 0 | 0 | 6 | 15 | 62 |
| Days Min Temp ≤ 0 °F | 0 | 0 | 0 | 0 | 0 | 0 | 0 | 0 | 0 | 0 | 0 | 0 | 0 |
| Heating Degree Days | 762 | 572 | 367 | 126 | 25 | 0 | 0 | 0 | 14 | 120 | 381 | 651 | 3018 |
| Cooling Degree Days | 0 | 0 | 12 | 60 | 159 | 369 | 507 | 469 | 279 | 71 | 8 | 1 | 1935 |
| Total Precipitation (") | 3.66 | 3.86 | 5.15 | 5.41 | 6.06 | 4.20 | 3.83 | 3.38 | 3.97 | 4.55 | 5.55 | 5.47 | 55.09 |
| Days ≥ 0.1" Precip | 6 | 6 | 7 | 7 | 8 | 6 | 6 | 5 | 5 | 5 | 6 | 7 | 74 |
| Total Snowfall (") | 1.6 | 0.9 | 0.2 | 0.0 | 0.0 | 0.0 | 0.0 | 0.0 | 0.0 | 0.0 | 0.1 | 0.3 | 3.1 |
| Days ≥ 1" Snow Depth | *1* | 1 | 0 | 0 | 0 | 0 | 0 | 0 | 0 | 0 | 0 | 0 | 2 |

## ARKADELPHIA 2 N *Clark County* ELEVATION 200 ft LAT/LONG 34° 7 ' N / 93° 3 ' W

| | JAN | FEB | MAR | APR | MAY | JUN | JUL | AUG | SEP | OCT | NOV | DEC | YEAR |
|---|---|---|---|---|---|---|---|---|---|---|---|---|---|
| Maximum Temp °F | 51.8 | 57.8 | 67.1 | 76.2 | 81.8 | 88.9 | 92.6 | 91.8 | 85.4 | 75.7 | 63.8 | 54.7 | 74.0 |
| Minimum Temp °F | 30.4 | 33.8 | 41.7 | 50.0 | 58.0 | 65.9 | 69.6 | 68.1 | 61.9 | 49.2 | 40.7 | 33.4 | 50.2 |
| Mean Temp °F | 41.1 | 45.8 | 54.4 | 63.1 | 69.9 | 77.4 | 81.1 | 80.0 | 73.6 | 62.5 | 52.3 | 44.1 | 62.1 |
| Days Max Temp ≥ 90 °F | 0 | 0 | 0 | 0 | 2 | 15 | 24 | 22 | 9 | 1 | 0 | 0 | 73 |
| Days Max Temp ≤ 32 °F | 1 | 0 | 0 | 0 | 0 | 0 | 0 | 0 | 0 | 0 | 0 | 1 | 2 |
| Days Min Temp ≤ 32 °F | 19 | 14 | 6 | 1 | 0 | 0 | 0 | 0 | 0 | 1 | 7 | 15 | 63 |
| Days Min Temp ≤ 0 °F | 0 | 0 | 0 | 0 | 0 | 0 | 0 | 0 | 0 | 0 | 0 | 0 | 0 |
| Heating Degree Days | 735 | 536 | 337 | 114 | 19 | 0 | 0 | 0 | 13 | 133 | 383 | 643 | 2913 |
| Cooling Degree Days | 0 | 1 | 15 | 61 | 183 | 391 | 514 | 489 | 289 | 61 | 11 | 2 | 2017 |
| Total Precipitation (") | 3.76 | 3.67 | 4.85 | 4.76 | 6.26 | 4.07 | 4.17 | 3.00 | 3.51 | 4.50 | 5.51 | 4.96 | 53.02 |
| Days ≥ 0.1" Precip | 6 | 6 | 6 | 6 | 8 | 6 | 6 | 4 | 5 | 5 | 6 | 6 | 70 |
| Total Snowfall (") | 1.2 | 1.0 | 0.1 | 0.0 | 0.0 | 0.0 | 0.0 | 0.0 | 0.0 | 0.0 | 0.1 | 0.5 | 2.9 |
| Days ≥ 1" Snow Depth | 1 | 1 | 0 | 0 | 0 | 0 | 0 | 0 | 0 | 0 | 0 | 1 | 3 |

## ARKANSAS POST *Arkansas County* ELEVATION 180 ft LAT/LONG 34° 2 ' N / 91° 21 ' W

| | JAN | FEB | MAR | APR | MAY | JUN | JUL | AUG | SEP | OCT | NOV | DEC | YEAR |
|---|---|---|---|---|---|---|---|---|---|---|---|---|---|
| Maximum Temp °F | 50.6 | 56.5 | 66.0 | 75.4 | 82.4 | 89.6 | 92.7 | 91.0 | 85.6 | 76.4 | 63.9 | 54.5 | 73.7 |
| Minimum Temp °F | 31.9 | 35.5 | 44.1 | 53.1 | 60.6 | 68.0 | 71.1 | 69.2 | 63.4 | 51.9 | 43.5 | 36.1 | 52.4 |
| Mean Temp °F | 41.3 | 46.1 | 55.1 | 64.3 | 71.5 | 78.8 | 81.9 | 80.2 | 74.5 | 64.2 | 53.7 | 45.3 | 63.1 |
| Days Max Temp ≥ 90 °F | 0 | 0 | 0 | 0 | 3 | 16 | 24 | 20 | 9 | 1 | 0 | 0 | 73 |
| Days Max Temp ≤ 32 °F | 2 | 1 | 0 | 0 | 0 | 0 | 0 | 0 | 0 | 0 | 0 | 1 | 4 |
| Days Min Temp ≤ 32 °F | 17 | 12 | 4 | 0 | 0 | 0 | 0 | 0 | 0 | 0 | 5 | 12 | 50 |
| Days Min Temp ≤ 0 °F | 0 | 0 | 0 | 0 | 0 | 0 | 0 | 0 | 0 | 0 | 0 | 0 | 0 |
| Heating Degree Days | 730 | 530 | 318 | 101 | 14 | 0 | 0 | 0 | 10 | 107 | 344 | 605 | 2759 |
| Cooling Degree Days | 0 | 2 | 16 | 82 | 226 | 436 | 547 | 504 | 319 | 86 | 13 | 3 | 2234 |
| Total Precipitation (") | 4.41 | 4.71 | 5.59 | 4.67 | 5.28 | 3.71 | 3.67 | 3.10 | 3.58 | 3.76 | 4.91 | 5.08 | 52.47 |
| Days ≥ 0.1" Precip | 7 | 6 | 7 | 6 | 7 | 5 | 5 | 4 | 5 | 4 | 6 | 7 | 69 |
| Total Snowfall (") | 1.8 | 0.8 | 0.3 | 0.0 | 0.0 | 0.0 | 0.0 | 0.0 | 0.0 | 0.0 | 0.0 | 0.3 | 3.2 |
| Days ≥ 1" Snow Depth | 1 | 1 | 0 | 0 | 0 | 0 | 0 | 0 | 0 | 0 | 0 | 0 | 2 |

## BATESVILLE L & D 1 *Independence County* ELEVATION 277 ft LAT/LONG 35° 45 ' N / 91° 38 ' W

| | JAN | FEB | MAR | APR | MAY | JUN | JUL | AUG | SEP | OCT | NOV | DEC | YEAR |
|---|---|---|---|---|---|---|---|---|---|---|---|---|---|
| Maximum Temp °F | 47.0 | 52.7 | 62.5 | 73.0 | 80.6 | 88.7 | 93.4 | 91.4 | 83.9 | 73.3 | 61.1 | 51.1 | 71.6 |
| Minimum Temp °F | 26.2 | 29.7 | 39.1 | 48.1 | 56.1 | 64.2 | 68.6 | 66.3 | 60.0 | 47.7 | 38.9 | 30.7 | 48.0 |
| Mean Temp °F | 36.6 | 41.2 | 50.8 | 60.5 | 68.4 | 76.5 | 81.0 | 78.9 | 72.0 | 60.5 | 50.0 | 40.9 | 59.8 |
| Days Max Temp ≥ 90 °F | 0 | 0 | 0 | 0 | 3 | 15 | 24 | 20 | 7 | 1 | 0 | 0 | 70 |
| Days Max Temp ≤ 32 °F | 4 | 1 | 0 | 0 | 0 | 0 | 0 | 0 | 0 | 0 | 0 | 1 | 6 |
| Days Min Temp ≤ 32 °F | 23 | 18 | 9 | 2 | 0 | 0 | 0 | 0 | 0 | 2 | 9 | 19 | 82 |
| Days Min Temp ≤ 0 °F | 1 | 0 | 0 | 0 | 0 | 0 | 0 | 0 | 0 | 0 | 0 | 0 | 1 |
| Heating Degree Days | 872 | 666 | 443 | 176 | 40 | 1 | 0 | 0 | 25 | 183 | 449 | 740 | 3595 |
| Cooling Degree Days | 0 | 0 | 8 | 44 | 156 | 363 | 520 | 464 | 250 | 46 | 6 | 1 | 1858 |
| Total Precipitation (") | 3.18 | 3.12 | 4.50 | 5.01 | 5.07 | 3.29 | 3.42 | 3.35 | 4.31 | 4.03 | 4.90 | 4.61 | 48.79 |
| Days ≥ 0.1" Precip | 6 | 5 | 7 | 7 | 8 | 6 | 5 | 6 | 5 | 5 | 6 | 7 | 73 |
| Total Snowfall (") | 2.8 | 3.2 | 0.7 | 0.0 | 0.0 | 0.0 | 0.0 | 0.0 | 0.0 | 0.0 | 0.6 | 1.2 | 8.5 |
| Days ≥ 1" Snow Depth | 3 | 2 | 0 | 0 | 0 | 0 | 0 | 0 | 0 | 0 | 0 | 1 | 6 |

## BATESVILLE LIVESTOCK *Independence County* ELEVATION 571 ft LAT/LONG 35° 49 ' N / 91° 47 ' W

| | JAN | FEB | MAR | APR | MAY | JUN | JUL | AUG | SEP | OCT | NOV | DEC | YEAR |
|---|---|---|---|---|---|---|---|---|---|---|---|---|---|
| Maximum Temp °F | 47.8 | 53.6 | 63.1 | 73.6 | 79.6 | 87.5 | 92.4 | 91.0 | 84.1 | 74.4 | 61.7 | 51.4 | 71.7 |
| Minimum Temp °F | 26.7 | 30.5 | 39.3 | 48.3 | 55.4 | 63.7 | 68.2 | 65.9 | 59.6 | 48.1 | 39.8 | 30.9 | 48.0 |
| Mean Temp °F | 37.3 | 42.1 | 51.3 | 61.0 | 67.6 | 75.6 | 80.3 | 78.5 | 71.8 | 61.3 | 50.8 | 41.2 | 59.9 |
| Days Max Temp ≥ 90 °F | 0 | 0 | 0 | 0 | 1 | 11 | 22 | 19 | 8 | 1 | 0 | 0 | 62 |
| Days Max Temp ≤ 32 °F | 4 | 1 | 0 | 0 | 0 | 0 | 0 | 0 | 0 | 0 | 0 | 1 | 6 |
| Days Min Temp ≤ 32 °F | 22 | 17 | 9 | 2 | 0 | 0 | 0 | 0 | 0 | 2 | 8 | 18 | 78 |
| Days Min Temp ≤ 0 °F | 1 | 0 | 0 | 0 | 0 | 0 | 0 | 0 | 0 | 0 | 0 | 0 | 1 |
| Heating Degree Days | 852 | 640 | 431 | 164 | 44 | 2 | 0 | 0 | 23 | 163 | 425 | 733 | 3477 |
| Cooling Degree Days | 0 | 0 | 10 | 55 | 143 | 349 | 509 | 463 | 259 | 59 | 7 | 1 | 1855 |
| Total Precipitation (") | 2.91 | 3.24 | 4.51 | 4.76 | 4.80 | 3.45 | 3.13 | 3.29 | 4.45 | 3.84 | 4.92 | 4.35 | 47.65 |
| Days ≥ 0.1" Precip | 5 | 5 | 6 | 7 | 8 | 5 | 5 | 5 | 5 | 5 | 6 | 6 | 68 |
| Total Snowfall (") | 2.7 | 2.6 | 0.7 | 0.0 | 0.0 | 0.0 | 0.0 | 0.0 | 0.0 | 0.1 | 1.0 | 1.1 | 8.2 |
| Days ≥ 1" Snow Depth | 4 | 3 | 0 | 0 | 0 | 0 | 0 | 0 | 0 | 0 | 0 | 1 | 8 |

## BEEDEVILLE 4 NE *Jackson County* ELEVATION 221 ft LAT/LONG 35° 26 ' N / 91° 6 ' W

| | JAN | FEB | MAR | APR | MAY | JUN | JUL | AUG | SEP | OCT | NOV | DEC | YEAR |
|---|---|---|---|---|---|---|---|---|---|---|---|---|---|
| Maximum Temp °F | 46.1 | 52.0 | 62.2 | 73.2 | 80.8 | 89.2 | 93.4 | 91.5 | 84.4 | 74.5 | 61.4 | 51.0 | 71.6 |
| Minimum Temp °F | 28.5 | 32.3 | 41.0 | 50.2 | 58.5 | 66.6 | 70.4 | 67.3 | 60.4 | 48.7 | 40.3 | 32.5 | 49.7 |
| Mean Temp °F | 37.3 | 42.2 | 51.6 | 61.7 | 69.7 | 77.9 | 81.9 | 79.4 | 72.5 | 61.6 | 50.9 | 41.8 | 60.7 |
| Days Max Temp ≥ 90 °F | 0 | 0 | 0 | 0 | 3 | 16 | 24 | 21 | 8 | 1 | 0 | 0 | 73 |
| Days Max Temp ≤ 32 °F | 4 | 1 | 0 | 0 | 0 | 0 | 0 | 0 | 0 | 0 | 0 | 1 | 6 |
| Days Min Temp ≤ 32 °F | 21 | 15 | 6 | 1 | 0 | 0 | 0 | 0 | 0 | 1 | 7 | 16 | 67 |
| Days Min Temp ≤ 0 °F | 0 | 0 | 0 | 0 | 0 | 0 | 0 | 0 | 0 | 0 | 0 | 0 | 0 |
| Heating Degree Days | 852 | 638 | 417 | 147 | 27 | 0 | 0 | 0 | 21 | 155 | 424 | 714 | 3395 |
| Cooling Degree Days | 0 | 0 | 8 | 52 | 179 | 403 | 550 | 464 | 253 | 52 | 6 | 1 | 1968 |
| Total Precipitation (") | 3.88 | 3.57 | 4.56 | 5.13 | 5.21 | 3.85 | 2.85 | 2.73 | 4.20 | 3.41 | 5.03 | 5.05 | 49.47 |
| Days ≥ 0.1" Precip | 5 | 5 | 6 | 6 | 6 | 5 | 4 | 4 | 5 | 4 | 6 | 5 | 61 |
| Total Snowfall (") | 1.9 | 1.7 | *0.2* | 0.0 | 0.0 | 0.0 | 0.0 | 0.0 | 0.0 | 0.0 | 0.3 | 0.2 | 4.3 |
| Days ≥ 1" Snow Depth | na | *0* | *0* | 0 | 0 | 0 | 0 | 0 | 0 | 0 | 0 | 0 | na |

## BENTON *Saline County* ELEVATION 285 ft LAT/LONG 34° 33 ' N / 92° 37 ' W

| | JAN | FEB | MAR | APR | MAY | JUN | JUL | AUG | SEP | OCT | NOV | DEC | YEAR |
|---|---|---|---|---|---|---|---|---|---|---|---|---|---|
| Maximum Temp °F | 52.2 | 58.4 | 67.0 | 75.9 | 81.7 | 88.5 | 92.8 | 91.6 | 85.3 | 76.2 | 64.3 | 55.4 | 74.1 |
| Minimum Temp °F | 28.8 | 32.4 | 41.3 | 49.4 | 56.9 | 64.6 | 68.4 | 66.3 | 60.2 | 48.2 | 39.7 | 32.3 | 49.0 |
| Mean Temp °F | 40.5 | 45.4 | 54.2 | 62.7 | 69.3 | 76.6 | 80.6 | 79.0 | 72.8 | 62.2 | 52.0 | 43.9 | 61.6 |
| Days Max Temp ≥ 90 °F | 0 | 0 | 0 | 0 | 2 | 13 | 23 | 21 | 9 | 1 | 0 | 0 | 69 |
| Days Max Temp ≤ 32 °F | 2 | 1 | 0 | 0 | 0 | 0 | 0 | 0 | 0 | 0 | 0 | 1 | 4 |
| Days Min Temp ≤ 32 °F | 20 | 15 | 7 | 1 | 0 | 0 | 0 | 0 | 0 | 1 | 9 | 17 | 70 |
| Days Min Temp ≤ 0 °F | 0 | 0 | 0 | 0 | 0 | 0 | 0 | 0 | 0 | 0 | 0 | 0 | 0 |
| Heating Degree Days | 754 | 548 | 346 | 124 | 25 | 0 | 0 | 0 | 16 | 142 | 391 | 649 | 2995 |
| Cooling Degree Days | 0 | 2 | 16 | 60 | 176 | 375 | 512 | 468 | 277 | 64 | 9 | 2 | 1961 |
| Total Precipitation (") | 3.58 | 3.37 | 4.82 | 5.51 | 5.56 | 4.17 | 4.06 | 3.51 | 4.45 | 4.38 | 5.56 | 5.21 | 54.18 |
| Days ≥ 0.1" Precip | 6 | 5 | 6 | 6 | 7 | 6 | 6 | 4 | 5 | 5 | 6 | 6 | 68 |
| Total Snowfall (") | *2.1* | 1.7 | 0.2 | 0.0 | 0.0 | 0.0 | 0.0 | 0.0 | 0.0 | 0.0 | 0.3 | 0.2 | 4.5 |
| Days ≥ 1" Snow Depth | *1* | 1 | 0 | 0 | 0 | 0 | 0 | 0 | 0 | 0 | 0 | 0 | 2 |

## BENTONVILLE 4 S *Benton County* ELEVATION 1220 ft LAT/LONG 36° 19 ' N / 94° 13 ' W

| | JAN | FEB | MAR | APR | MAY | JUN | JUL | AUG | SEP | OCT | NOV | DEC | YEAR |
|---|---|---|---|---|---|---|---|---|---|---|---|---|---|
| Maximum Temp °F | 44.0 | 49.1 | 58.7 | 69.4 | 75.8 | 83.6 | 89.0 | 88.2 | 80.7 | 70.6 | 57.7 | 48.1 | 67.9 |
| Minimum Temp °F | 22.2 | 25.9 | 35.1 | 44.7 | 53.0 | 61.7 | 66.1 | 64.1 | 57.1 | 45.2 | 35.3 | 26.8 | 44.8 |
| Mean Temp °F | 33.1 | 37.5 | 46.9 | 57.1 | 64.4 | 72.7 | 77.6 | 76.2 | 68.9 | 57.9 | 46.5 | 37.4 | 56.4 |
| Days Max Temp ≥ 90 °F | 0 | 0 | 0 | 0 | 0 | 3 | 15 | 13 | 4 | 0 | 0 | 0 | 35 |
| Days Max Temp ≤ 32 °F | 6 | 3 | 1 | 0 | 0 | 0 | 0 | 0 | 0 | 0 | 0 | 3 | 13 |
| Days Min Temp ≤ 32 °F | 26 | 21 | 13 | 3 | 0 | 0 | 0 | 0 | 0 | 3 | 12 | 23 | 101 |
| Days Min Temp ≤ 0 °F | 2 | 1 | 0 | 0 | 0 | 0 | 0 | 0 | 0 | 0 | 0 | 1 | 4 |
| Heating Degree Days | 982 | 769 | 557 | 255 | 90 | 9 | 1 | 2 | 51 | 241 | 550 | 847 | 4354 |
| Cooling Degree Days | 0 | 0 | 3 | 21 | 73 | 251 | 399 | 367 | 176 | 23 | 2 | 0 | 1315 |
| Total Precipitation (") | 2.16 | 2.50 | 4.16 | 4.55 | 5.16 | 4.91 | 3.18 | 3.47 | 4.62 | 3.91 | 4.40 | 3.59 | 46.61 |
| Days ≥ 0.1" Precip | 4 | 5 | 7 | 7 | 8 | 7 | 5 | 5 | 7 | 6 | 6 | 5 | 72 |
| Total Snowfall (") | ***2.6*** | 3.3 | 2.3 | 0.0 | 0.0 | 0.0 | 0.0 | 0.0 | 0.0 | 0.0 | 0.7 | 1.3 | 10.2 |
| Days ≥ 1" Snow Depth | 3 | 3 | 1 | 0 | 0 | 0 | 0 | 0 | 0 | 0 | 0 | 2 | 9 |

## BLAKELY MOUNTAIN DAM *Garland County* ELEVATION 430 ft LAT/LONG 34° 34 ' N / 93° 12 ' W

| | JAN | FEB | MAR | APR | MAY | JUN | JUL | AUG | SEP | OCT | NOV | DEC | YEAR |
|---|---|---|---|---|---|---|---|---|---|---|---|---|---|
| Maximum Temp °F | ***49.3*** | 54.3 | 63.6 | 73.8 | ***79.4*** | 87.8 | 92.5 | 91.0 | 84.3 | 74.4 | 62.3 | 53.9 | 72.2 |
| Minimum Temp °F | 26.2 | 28.5 | 36.8 | 45.6 | ***53.5*** | 62.3 | 66.3 | 64.5 | 58.7 | 46.4 | ***37.4*** | 30.2 | 46.4 |
| Mean Temp °F | ***37.6*** | 41.4 | 50.2 | 59.7 | ***66.5*** | 74.9 | 79.4 | 77.8 | 71.6 | 60.4 | ***49.8*** | 42.2 | 59.3 |
| Days Max Temp ≥ 90 °F | 0 | 0 | 0 | 0 | 1 | 12 | 21 | 19 | 7 | 1 | 0 | 0 | 61 |
| Days Max Temp ≤ 32 °F | 2 | 1 | 0 | 0 | 0 | 0 | 0 | 0 | 0 | 0 | 0 | 0 | 3 |
| Days Min Temp ≤ 32 °F | 23 | 19 | 10 | 3 | 0 | 0 | 0 | 0 | 0 | 1 | 10 | 20 | 86 |
| Days Min Temp ≤ 0 °F | 0 | 0 | 0 | 0 | 0 | 0 | 0 | 0 | 0 | 0 | 0 | 0 | 0 |
| Heating Degree Days | ***841*** | 660 | 458 | 185 | ***54*** | 2 | 0 | 0 | 23 | 177 | ***453*** | 702 | 3555 |
| Cooling Degree Days | na | ***0*** | ***5*** | 30 | na | 318 | ***463*** | ***420*** | ***234*** | ***42*** | na | ***1*** | na |
| Total Precipitation (") | 3.51 | 3.82 | 5.06 | 5.38 | 6.06 | 4.42 | 4.66 | 3.52 | 3.87 | 4.86 | 6.08 | 5.33 | 56.57 |
| Days ≥ 0.1" Precip | 6 | 6 | 6 | 6 | 8 | 6 | 5 | 5 | 5 | 5 | 6 | 6 | 70 |
| Total Snowfall (") | ***1.0*** | 1.2 | 0.1 | 0.0 | 0.0 | 0.0 | 0.0 | 0.0 | 0.0 | 0.0 | 0.1 | 0.1 | 2.5 |
| Days ≥ 1" Snow Depth | na | ***0*** | 0 | 0 | 0 | 0 | 0 | 0 | 0 | 0 | 0 | 0 | na |

## BLUE MOUNTAIN DAM *Yell County* ELEVATION 459 ft LAT/LONG 35° 6 ' N / 93° 39 ' W

| | JAN | FEB | MAR | APR | MAY | JUN | JUL | AUG | SEP | OCT | NOV | DEC | YEAR |
|---|---|---|---|---|---|---|---|---|---|---|---|---|---|
| Maximum Temp °F | 47.3 | 53.4 | 63.3 | 73.2 | 79.4 | 87.8 | 92.9 | 91.9 | 84.2 | 73.4 | 60.4 | 51.0 | 71.5 |
| Minimum Temp °F | 26.7 | 30.3 | 39.7 | 48.4 | 56.4 | 64.4 | 68.2 | 66.7 | 60.5 | 48.5 | 38.4 | 30.3 | 48.2 |
| Mean Temp °F | 37.0 | 41.9 | 51.5 | 60.8 | 67.9 | 76.1 | 80.6 | 79.3 | 72.4 | 61.0 | 49.4 | 40.7 | 59.9 |
| Days Max Temp ≥ 90 °F | 0 | 0 | 0 | 0 | 2 | 13 | 23 | 21 | 9 | 1 | 0 | 0 | 69 |
| Days Max Temp ≤ 32 °F | 3 | 1 | 0 | 0 | 0 | 0 | 0 | 0 | 0 | 0 | 0 | 1 | 5 |
| Days Min Temp ≤ 32 °F | 23 | 17 | 7 | 1 | 0 | 0 | 0 | 0 | 0 | 1 | 9 | 19 | 77 |
| Days Min Temp ≤ 0 °F | 0 | 0 | 0 | 0 | 0 | 0 | 0 | 0 | 0 | 0 | 0 | 0 | 0 |
| Heating Degree Days | 860 | 647 | 422 | 169 | 43 | 2 | 0 | 0 | 24 | 173 | 467 | 747 | 3554 |
| Cooling Degree Days | 0 | 0 | 10 | 49 | 139 | 348 | 502 | 465 | 260 | 50 | 8 | 1 | 1832 |
| Total Precipitation (") | 2.76 | 2.94 | 4.59 | 4.38 | 6.07 | 3.82 | 3.71 | 2.77 | 3.68 | 4.12 | 4.68 | 4.39 | 47.91 |
| Days ≥ 0.1" Precip | 5 | 5 | 7 | 6 | 8 | 6 | 5 | 5 | 6 | 6 | 6 | 6 | 71 |
| Total Snowfall (") | 2.3 | 1.3 | 0.3 | 0.0 | 0.0 | 0.0 | 0.0 | 0.0 | 0.0 | 0.0 | 0.3 | 0.7 | 4.9 |
| Days ≥ 1" Snow Depth | 2 | 2 | 0 | 0 | 0 | 0 | 0 | 0 | 0 | 0 | 0 | 1 | 5 |

## BLYTHEVILLE *Mississippi County* ELEVATION 249 ft LAT/LONG 35° 56 ' N / 89° 55 ' W

| | JAN | FEB | MAR | APR | MAY | JUN | JUL | AUG | SEP | OCT | NOV | DEC | YEAR |
|---|---|---|---|---|---|---|---|---|---|---|---|---|---|
| Maximum Temp °F | 45.1 | 50.0 | 60.6 | 72.0 | 81.2 | 89.6 | 92.8 | 90.1 | 83.9 | 73.7 | 60.9 | 49.9 | 70.8 |
| Minimum Temp °F | 28.2 | 31.8 | 41.1 | 51.2 | 60.0 | 68.2 | 72.0 | 69.2 | 62.7 | 50.8 | 42.1 | 33.2 | 50.9 |
| Mean Temp °F | 36.7 | 40.9 | 50.8 | 61.6 | 70.6 | 78.9 | 82.4 | 79.7 | 73.3 | 62.3 | 51.5 | 41.6 | 60.9 |
| Days Max Temp ≥ 90 °F | 0 | 0 | 0 | 0 | ***4*** | 16 | 23 | 17 | 7 | 1 | 0 | 0 | 68 |
| Days Max Temp ≤ 32 °F | 5 | 3 | 0 | 0 | 0 | 0 | 0 | 0 | 0 | 0 | 0 | 2 | 10 |
| Days Min Temp ≤ 32 °F | 20 | 15 | 6 | 0 | 0 | 0 | 0 | 0 | 0 | 0 | 5 | 15 | 61 |
| Days Min Temp ≤ 0 °F | 0 | 0 | 0 | 0 | 0 | 0 | 0 | 0 | 0 | 0 | 0 | 0 | 0 |
| Heating Degree Days | 869 | 673 | 444 | 157 | 25 | 0 | 0 | 0 | 17 | 147 | 407 | 719 | 3458 |
| Cooling Degree Days | 0 | 0 | 10 | 60 | 192 | 427 | 555 | 477 | 272 | 64 | 8 | 1 | 2066 |
| Total Precipitation (") | 3.44 | 3.93 | 4.62 | 5.19 | 4.97 | 4.39 | 3.83 | 3.12 | 3.81 | 3.42 | 4.67 | 5.09 | 50.48 |
| Days ≥ 0.1" Precip | 6 | 6 | 8 | 7 | 7 | 6 | 5 | 4 | 5 | 5 | 6 | 7 | 72 |
| Total Snowfall (") | na | na | ***0.8*** | 0.0 | 0.0 | 0.0 | 0.0 | 0.0 | 0.0 | 0.0 | 0.1 | 0.7 | na |
| Days ≥ 1" Snow Depth | na | ***0*** | ***0*** | 0 | 0 | 0 | 0 | 0 | 0 | 0 | 0 | ***0*** | na |

## BRINKLEY *Monroe County* ELEVATION 210 ft LAT/LONG 34° 54' N / 91° 11' W

| | JAN | FEB | MAR | APR | MAY | JUN | JUL | AUG | SEP | OCT | NOV | DEC | YEAR |
|---|---|---|---|---|---|---|---|---|---|---|---|---|---|
| Maximum Temp °F | 47.1 | 52.2 | 61.6 | 72.2 | 80.0 | 88.4 | 91.7 | 90.1 | 83.9 | 74.2 | 61.5 | 51.3 | 71.2 |
| Minimum Temp °F | 28.6 | 32.0 | 40.5 | 49.9 | 58.6 | 66.7 | 70.3 | 67.7 | 60.9 | 48.2 | 39.7 | 32.6 | 49.6 |
| Mean Temp °F | 37.9 | 42.2 | 51.1 | 61.1 | 69.3 | 77.6 | 81.1 | 79.0 | 72.4 | 61.3 | 50.9 | 42.0 | 60.5 |
| Days Max Temp ≥ 90 °F | 0 | 0 | 0 | 0 | 2 | 14 | 21 | 18 | 8 | 1 | 0 | 0 | 64 |
| Days Max Temp ≤ 32 °F | 4 | 2 | 0 | 0 | 0 | 0 | 0 | 0 | 0 | 0 | 0 | 2 | 8 |
| Days Min Temp ≤ 32 °F | 21 | 15 | 6 | 1 | 0 | 0 | 0 | 0 | 0 | 1 | 7 | 16 | 67 |
| Days Min Temp ≤ 0 °F | 0 | 0 | 0 | 0 | 0 | 0 | 0 | 0 | 0 | 0 | 0 | 0 | 0 |
| Heating Degree Days | 835 | 639 | 434 | 164 | 33 | 1 | 0 | 0 | 23 | 166 | 422 | 708 | 3425 |
| Cooling Degree Days | 0 | 0 | 9 | 51 | 175 | 397 | 517 | 461 | 266 | 58 | 9 | 1 | 1944 |
| Total Precipitation (") | 3.60 | 4.04 | 4.81 | 5.14 | 5.78 | 3.61 | 3.49 | 3.24 | 3.35 | 3.49 | 4.57 | 4.78 | 49.90 |
| Days ≥ 0.1" Precip | 6 | 6 | 7 | 7 | 7 | 5 | 5 | 5 | 5 | 5 | 6 | 6 | 70 |
| Total Snowfall (") | 2.2 | 1.7 | 0.6 | 0.0 | 0.0 | 0.0 | 0.0 | 0.0 | 0.0 | 0.0 | 0.1 | 0.2 | 4.8 |
| Days ≥ 1" Snow Depth | 2 | 1 | 0 | 0 | 0 | 0 | 0 | 0 | 0 | 0 | 0 | 0 | 3 |

## CABOT 4 SW *Pulaski County* ELEVATION 289 ft LAT/LONG 34° 57' N / 92° 5' W

| | JAN | FEB | MAR | APR | MAY | JUN | JUL | AUG | SEP | OCT | NOV | DEC | YEAR |
|---|---|---|---|---|---|---|---|---|---|---|---|---|---|
| Maximum Temp °F | 48.6 | 54.4 | 64.3 | 73.3 | 79.7 | 87.6 | 91.5 | 89.9 | 83.7 | 74.0 | 61.9 | 52.4 | 71.8 |
| Minimum Temp °F | 28.8 | 32.7 | 41.8 | 49.9 | 57.9 | 65.8 | 69.3 | 67.4 | 61.5 | 49.2 | 40.4 | 32.8 | 49.8 |
| Mean Temp °F | 38.7 | 43.6 | 53.1 | 61.6 | 68.8 | 76.7 | 80.4 | 78.7 | 72.6 | 61.6 | 51.2 | 42.6 | 60.8 |
| Days Max Temp ≥ 90 °F | 0 | 0 | 0 | 0 | 1 | 11 | 20 | 17 | 6 | 1 | 0 | 0 | 56 |
| Days Max Temp ≤ 32 °F | 3 | 1 | 0 | 0 | 0 | 0 | 0 | 0 | 0 | 0 | 0 | 1 | 5 |
| Days Min Temp ≤ 32 °F | 21 | 15 | 6 | 1 | 0 | 0 | 0 | 0 | 0 | 1 | 8 | 16 | 68 |
| Days Min Temp ≤ 0 °F | 0 | 0 | 0 | 0 | 0 | 0 | 0 | 0 | 0 | 0 | 0 | 0 | 0 |
| Heating Degree Days | 809 | 599 | 376 | 146 | 30 | 1 | 0 | 0 | 18 | 153 | 415 | 689 | 3236 |
| Cooling Degree Days | 0 | 1 | 12 | 49 | 157 | 366 | 487 | 445 | 262 | 52 | 7 | 2 | 1840 |
| Total Precipitation (") | 3.53 | 3.38 | 4.72 | 5.07 | 5.58 | 3.66 | 3.24 | 3.32 | 3.70 | 4.00 | 5.60 | 5.04 | 50.84 |
| Days ≥ 0.1" Precip | 6 | 6 | 7 | 7 | 7 | 6 | 5 | 5 | 5 | 5 | 6 | 6 | 71 |
| Total Snowfall (") | 2.1 | 1.7 | 0.3 | 0.0 | 0.0 | 0.0 | 0.0 | 0.0 | 0.0 | 0.0 | 0.4 | 0.4 | 4.9 |
| Days ≥ 1" Snow Depth | 2 | 1 | 0 | 0 | 0 | 0 | 0 | 0 | 0 | 0 | 0 | 1 | 4 |

## CALICO ROCK 2 WSW *Izard County* ELEVATION 381 ft LAT/LONG 36° 8' N / 92° 8' W

| | JAN | FEB | MAR | APR | MAY | JUN | JUL | AUG | SEP | OCT | NOV | DEC | YEAR |
|---|---|---|---|---|---|---|---|---|---|---|---|---|---|
| Maximum Temp °F | 47.5 | 52.7 | 62.9 | 73.7 | 80.2 | 88.1 | 93.1 | 91.2 | 83.7 | 73.6 | 60.7 | 51.2 | 71.6 |
| Minimum Temp °F | 23.6 | 26.5 | 35.0 | 44.0 | 51.9 | 60.1 | 64.1 | 62.4 | 56.2 | 44.0 | 35.9 | 28.0 | 44.3 |
| Mean Temp °F | 35.5 | 39.6 | 48.9 | 58.8 | 66.1 | 74.1 | 78.6 | 76.8 | 70.0 | 58.8 | 48.3 | 39.6 | 57.9 |
| Days Max Temp ≥ 90 °F | 0 | 0 | 0 | 1 | 2 | 13 | 23 | 19 | 7 | 1 | 0 | 0 | 66 |
| Days Max Temp ≤ 32 °F | 4 | 2 | 0 | 0 | 0 | 0 | 0 | 0 | 0 | 0 | 0 | 1 | 7 |
| Days Min Temp ≤ 32 °F | 25 | 20 | 14 | 5 | 0 | 0 | 0 | 0 | 0 | 4 | 13 | 21 | 102 |
| Days Min Temp ≤ 0 °F | 1 | 0 | 0 | 0 | 0 | 0 | 0 | 0 | 0 | 0 | 0 | 0 | 1 |
| Heating Degree Days | 907 | 711 | 499 | 214 | 64 | 3 | 0 | 1 | 36 | 217 | 497 | 781 | 3930 |
| Cooling Degree Days | 0 | 0 | 6 | 32 | 106 | 284 | 431 | 391 | 197 | 35 | 3 | 1 | 1486 |
| Total Precipitation (") | 2.72 | 3.11 | 4.40 | 4.37 | 4.58 | 3.77 | 3.22 | 3.25 | 4.30 | 3.67 | 4.62 | 4.30 | 46.31 |
| Days ≥ 0.1" Precip | 5 | 5 | 7 | 7 | 8 | 6 | 6 | 5 | 6 | 5 | 6 | 6 | 72 |
| Total Snowfall (") | *2.1* | na | 0.7 | 0.0 | 0.0 | 0.0 | 0.0 | 0.0 | 0.0 | 0.0 | 0.6 | 0.5 | na |
| Days ≥ 1" Snow Depth | *1* | *1* | *0* | 0 | 0 | 0 | 0 | 0 | 0 | 0 | 0 | 1 | 3 |

## CAMDEN 1 *Ouachita County* ELEVATION 121 ft LAT/LONG 33° 36' N / 92° 49' W

| | JAN | FEB | MAR | APR | MAY | JUN | JUL | AUG | SEP | OCT | NOV | DEC | YEAR |
|---|---|---|---|---|---|---|---|---|---|---|---|---|---|
| Maximum Temp °F | 52.6 | 58.0 | 67.2 | 76.1 | 81.7 | 88.9 | 92.4 | 91.6 | 85.5 | 76.0 | 65.3 | 56.3 | 74.3 |
| Minimum Temp °F | 30.2 | 33.3 | 41.4 | 50.2 | 58.3 | 66.3 | 70.0 | 68.2 | 61.8 | 49.4 | 40.2 | 33.7 | 50.3 |
| Mean Temp °F | 41.4 | 45.7 | 54.3 | 63.2 | 70.0 | 77.7 | 81.2 | 79.9 | 73.6 | 62.7 | 52.8 | 45.1 | 62.3 |
| Days Max Temp ≥ 90 °F | 0 | 0 | 0 | 0 | 2 | 15 | 23 | 21 | 10 | 1 | 0 | 0 | 72 |
| Days Max Temp ≤ 32 °F | 2 | 1 | 0 | 0 | 0 | 0 | 0 | 0 | 0 | 0 | 0 | 1 | 4 |
| Days Min Temp ≤ 32 °F | 20 | 14 | 6 | 1 | 0 | 0 | 0 | 0 | 0 | 1 | 8 | 16 | 66 |
| Days Min Temp ≤ 0 °F | 0 | 0 | 0 | 0 | 0 | 0 | 0 | 0 | 0 | 0 | 0 | 0 | 0 |
| Heating Degree Days | 726 | 542 | 341 | 119 | 23 | 0 | 0 | 0 | 15 | 136 | 372 | 615 | 2889 |
| Cooling Degree Days | 0 | 2 | 14 | 61 | 178 | 387 | 509 | 478 | 281 | 66 | 11 | 2 | 1989 |
| Total Precipitation (") | 4.19 | 4.52 | 4.95 | 4.50 | 5.31 | 4.51 | 4.17 | 3.02 | 3.74 | 4.47 | 5.05 | 5.05 | 53.48 |
| Days ≥ 0.1" Precip | 7 | 6 | 7 | 6 | 8 | 6 | 6 | 4 | 5 | 5 | 7 | 7 | 74 |
| Total Snowfall (") | 0.7 | 0.9 | 0.2 | 0.0 | 0.0 | 0.0 | 0.0 | 0.0 | 0.0 | 0.0 | 0.0 | 0.4 | 2.2 |
| Days ≥ 1" Snow Depth | 0 | 1 | 0 | 0 | 0 | 0 | 0 | 0 | 0 | 0 | 0 | 0 | 1 |

## CLARENDON *Monroe County* ELEVATION 171 ft LAT/LONG 34° 41 ' N / 91° 18 ' W

| | JAN | FEB | MAR | APR | MAY | JUN | JUL | AUG | SEP | OCT | NOV | DEC | YEAR |
|---|---|---|---|---|---|---|---|---|---|---|---|---|---|
| Maximum Temp °F | 49.2 | 55.0 | 64.0 | 73.3 | 81.1 | 88.6 | 91.6 | 89.9 | 83.9 | 75.7 | 63.2 | 53.0 | 72.4 |
| Minimum Temp °F | 30.9 | 34.5 | 43.1 | 51.1 | 58.8 | 66.5 | 69.7 | 67.3 | 60.4 | 49.0 | 41.5 | 34.3 | 50.6 |
| Mean Temp °F | 40.1 | 44.8 | 53.6 | 62.2 | 70.0 | 77.6 | 80.7 | 78.7 | 72.2 | 62.4 | 52.4 | 43.6 | 61.5 |
| Days Max Temp ≥ 90 °F | 0 | 0 | 0 | 0 | 2 | 15 | 21 | 17 | 7 | 1 | 0 | 0 | 63 |
| Days Max Temp ≤ 32 °F | 2 | 1 | 0 | 0 | 0 | 0 | 0 | 0 | 0 | 0 | 0 | 1 | 4 |
| Days Min Temp ≤ 32 °F | 19 | 12 | 4 | 1 | 0 | 0 | 0 | 0 | 0 | 1 | 6 | 14 | 57 |
| Days Min Temp ≤ 0 °F | 0 | 0 | 0 | 0 | 0 | 0 | 0 | 0 | 0 | 0 | 0 | 0 | 0 |
| Heating Degree Days | 765 | 564 | 362 | 140 | 22 | 0 | 0 | 0 | 21 | 142 | 381 | 657 | 3054 |
| Cooling Degree Days | 0 | 0 | 11 | 56 | 175 | 397 | 509 | 456 | 246 | 59 | 8 | 1 | 1918 |
| Total Precipitation (") | 3.36 | 4.07 | 5.14 | 5.03 | 5.19 | 3.40 | 3.49 | 3.13 | 3.50 | 3.91 | 4.97 | 5.44 | 50.63 |
| Days ≥ 0.1" Precip | 6 | 6 | 7 | 6 | 7 | 6 | 5 | 4 | 5 | 5 | 6 | 7 | 70 |
| Total Snowfall (") | 1.6 | 0.5 | 0.3 | 0.0 | 0.0 | 0.0 | 0.0 | 0.0 | 0.0 | 0.0 | 0.0 | 0.0 | 2.4 |
| Days ≥ 1" Snow Depth | na | na | 0 | 0 | 0 | 0 | 0 | 0 | 0 | 0 | 0 | 0 | na |

## CONWAY *Faulkner County* ELEVATION 310 ft LAT/LONG 35° 5 ' N / 92° 28 ' W

| | JAN | FEB | MAR | APR | MAY | JUN | JUL | AUG | SEP | OCT | NOV | DEC | YEAR |
|---|---|---|---|---|---|---|---|---|---|---|---|---|---|
| Maximum Temp °F | 49.6 | 55.9 | 65.7 | 75.3 | 81.2 | 88.8 | 93.1 | 92.1 | 85.4 | 75.5 | 63.0 | 53.4 | 73.3 |
| Minimum Temp °F | 28.8 | 32.1 | 41.3 | 50.3 | 57.9 | 66.2 | 70.0 | 68.2 | 62.0 | 49.4 | 40.5 | 32.8 | 50.0 |
| Mean Temp °F | 39.2 | 44.0 | 53.5 | 62.8 | 69.6 | 77.5 | 81.6 | 80.2 | 73.7 | 62.5 | 51.8 | 43.1 | 61.6 |
| Days Max Temp ≥ 90 °F | 0 | 0 | 0 | 0 | 2 | 15 | 24 | 21 | 9 | 1 | 0 | 0 | 72 |
| Days Max Temp ≤ 32 °F | 3 | 1 | 0 | 0 | 0 | 0 | 0 | 0 | 0 | 0 | 0 | 1 | 5 |
| Days Min Temp ≤ 32 °F | 21 | 16 | 6 | 1 | 0 | 0 | 0 | 0 | 0 | 1 | 8 | 16 | 69 |
| Days Min Temp ≤ 0 °F | 0 | 0 | 0 | 0 | 0 | 0 | 0 | 0 | 0 | 0 | 0 | 0 | 0 |
| Heating Degree Days | 794 | 587 | 369 | 126 | 25 | 0 | 0 | 0 | 16 | 139 | 398 | 673 | 3127 |
| Cooling Degree Days | 0 | 0 | 17 | 67 | 178 | 402 | 539 | 502 | 298 | 63 | 9 | 2 | 2077 |
| Total Precipitation (") | 3.43 | 3.23 | 4.41 | 4.78 | 5.18 | 4.45 | 3.05 | 2.93 | 4.17 | 4.13 | 4.84 | 5.14 | 49.74 |
| Days ≥ 0.1" Precip | 6 | 5 | 6 | 7 | 7 | 6 | 5 | 5 | 6 | 5 | 6 | 7 | 71 |
| Total Snowfall (") | 2.3 | 1.9 | 0.5 | 0.0 | 0.0 | 0.0 | 0.0 | 0.0 | 0.0 | 0.0 | 0.3 | 0.4 | 5.4 |
| Days ≥ 1" Snow Depth | 3 | 1 | 0 | 0 | 0 | 0 | 0 | 0 | 0 | 0 | 0 | 0 | 4 |

## CORNING *Clay County* ELEVATION 289 ft LAT/LONG 36° 24 ' N / 90° 35 ' W

| | JAN | FEB | MAR | APR | MAY | JUN | JUL | AUG | SEP | OCT | NOV | DEC | YEAR |
|---|---|---|---|---|---|---|---|---|---|---|---|---|---|
| Maximum Temp °F | 44.7 | 50.5 | 61.1 | 72.4 | 80.4 | 88.5 | 92.3 | 90.1 | 83.6 | 73.2 | 60.1 | 48.9 | 70.5 |
| Minimum Temp °F | 26.1 | 30.1 | 39.6 | 49.0 | 57.3 | 65.8 | 69.7 | 66.7 | 59.7 | 47.6 | 39.5 | 30.7 | 48.5 |
| Mean Temp °F | 35.4 | 40.3 | 50.4 | 60.8 | 68.9 | 77.2 | 81.0 | 78.4 | 71.7 | 60.4 | 49.8 | 39.8 | 59.5 |
| Days Max Temp ≥ 90 °F | 0 | 0 | 0 | 0 | 2 | 14 | 22 | 17 | 7 | 0 | 0 | 0 | 62 |
| Days Max Temp ≤ 32 °F | 5 | 2 | 0 | 0 | 0 | 0 | 0 | 0 | 0 | 0 | 0 | 2 | 9 |
| Days Min Temp ≤ 32 °F | 23 | 17 | 9 | 1 | 0 | 0 | 0 | 0 | 0 | 1 | 8 | 18 | 77 |
| Days Min Temp ≤ 0 °F | 1 | 0 | 0 | 0 | 0 | 0 | 0 | 0 | 0 | 0 | 0 | 0 | 1 |
| Heating Degree Days | 911 | 690 | 456 | 172 | 36 | 1 | 0 | 0 | 26 | 183 | 454 | 774 | 3703 |
| Cooling Degree Days | 0 | 0 | 10 | 51 | 167 | 387 | 529 | 454 | 252 | 47 | 5 | 1 | 1903 |
| Total Precipitation (") | 3.56 | 3.67 | 4.81 | 4.57 | 5.23 | 3.23 | 3.49 | 3.24 | 3.98 | 3.35 | 4.86 | 4.75 | 48.74 |
| Days ≥ 0.1" Precip | 5 | 6 | 7 | 7 | 7 | 5 | 5 | 5 | 5 | 5 | 6 | 7 | 70 |
| Total Snowfall (") | 4.2 | 3.8 | 1.4 | 0.0 | 0.0 | 0.0 | 0.0 | 0.0 | 0.0 | 0.0 | 0.7 | 1.5 | 11.6 |
| Days ≥ 1" Snow Depth | na | 1 | 0 | 0 | 0 | 0 | 0 | 0 | 0 | 0 | 0 | 1 | na |

## CROSSETT 2 SSE *Ashley County* ELEVATION 180 ft LAT/LONG 33° 7 ' N / 91° 57 ' W

| | JAN | FEB | MAR | APR | MAY | JUN | JUL | AUG | SEP | OCT | NOV | DEC | YEAR |
|---|---|---|---|---|---|---|---|---|---|---|---|---|---|
| Maximum Temp °F | 52.9 | 58.4 | 67.2 | 76.4 | 82.2 | 89.0 | 92.2 | 91.7 | 86.2 | 76.9 | 65.5 | 56.5 | 74.6 |
| Minimum Temp °F | 30.3 | 33.3 | 40.9 | 50.1 | 56.9 | 64.8 | 68.2 | 66.7 | 60.5 | 47.9 | 40.5 | 33.7 | 49.5 |
| Mean Temp °F | 41.6 | 45.9 | 54.1 | 63.3 | 69.5 | 76.9 | 80.2 | 79.2 | 73.3 | 62.4 | 53.0 | 45.1 | 62.0 |
| Days Max Temp ≥ 90 °F | 0 | 0 | 0 | 0 | 3 | 16 | 24 | 22 | 12 | 1 | 0 | 0 | 78 |
| Days Max Temp ≤ 32 °F | 1 | 0 | 0 | 0 | 0 | 0 | 0 | 0 | 0 | 0 | 0 | 1 | 2 |
| Days Min Temp ≤ 32 °F | 19 | 15 | 7 | 1 | 0 | 0 | 0 | 0 | 0 | 2 | 8 | 15 | 67 |
| Days Min Temp ≤ 0 °F | 0 | 0 | 0 | 0 | 0 | 0 | 0 | 0 | 0 | 0 | 0 | 0 | 0 |
| Heating Degree Days | 719 | 534 | 348 | 116 | 26 | 1 | 0 | 0 | 17 | 143 | 366 | 612 | 2882 |
| Cooling Degree Days | 0 | 1 | 13 | 58 | 150 | 360 | 465 | 444 | 255 | 59 | 11 | 2 | 1818 |
| Total Precipitation (") | 5.14 | 5.71 | 5.59 | 5.36 | 6.05 | 4.42 | 4.40 | 3.43 | 3.40 | 4.41 | 4.87 | 5.32 | 58.10 |
| Days ≥ 0.1" Precip | 8 | 6 | 7 | 6 | 7 | 6 | 7 | 5 | 5 | 5 | 6 | 8 | 76 |
| Total Snowfall (") | 0.6 | 0.3 | 0.2 | 0.0 | 0.0 | 0.0 | 0.0 | 0.0 | 0.0 | 0.0 | 0.0 | 0.2 | 1.3 |
| Days ≥ 1" Snow Depth | 0 | 0 | 0 | 0 | 0 | 0 | 0 | 0 | 0 | 0 | 0 | 0 | 0 |

## DARDANELLE *Yell County* ELEVATION 331 ft LAT/LONG 35° 13 ' N / 93° 9 ' W

| | JAN | FEB | MAR | APR | MAY | JUN | JUL | AUG | SEP | OCT | NOV | DEC | YEAR |
|---|---|---|---|---|---|---|---|---|---|---|---|---|---|
| Maximum Temp °F | 49.2 | 55.6 | 65.1 | 75.2 | 81.3 | 88.9 | 93.0 | 91.6 | 85.0 | 75.5 | 62.3 | 52.3 | 72.9 |
| Minimum Temp °F | 28.0 | 31.9 | 40.5 | 49.4 | 57.4 | 65.4 | 69.3 | 67.5 | 61.5 | 49.3 | 39.8 | 31.7 | 49.3 |
| Mean Temp °F | 38.6 | 43.8 | 52.8 | 62.3 | 69.4 | 77.2 | 81.2 | 79.6 | 73.2 | 62.4 | 51.1 | 42.0 | 61.1 |
| Days Max Temp ≥ 90 °F | 0 | 0 | 0 | 0 | 2 | 15 | 24 | 20 | 9 | 1 | 0 | 0 | 71 |
| Days Max Temp ≤ 32 °F | 2 | 1 | 0 | 0 | 0 | 0 | 0 | 0 | 0 | 0 | 0 | 1 | 4 |
| Days Min Temp ≤ 32 °F | 21 | 15 | 7 | 1 | 0 | 0 | 0 | 0 | 0 | 1 | 8 | 18 | 71 |
| Days Min Temp ≤ 0 °F | 0 | 0 | 0 | 0 | 0 | 0 | 0 | 0 | 0 | 0 | 0 | 0 | 0 |
| Heating Degree Days | 811 | 591 | 380 | 131 | 25 | 0 | 0 | 0 | 15 | 134 | 417 | 707 | 3211 |
| Cooling Degree Days | 0 | 0 | 11 | 56 | 167 | 383 | 515 | 482 | 276 | 60 | 7 | 1 | 1958 |
| Total Precipitation (") | 3.09 | 3.34 | 4.77 | 4.51 | 5.51 | 4.18 | 3.35 | 3.12 | 3.94 | 4.20 | 4.91 | 4.36 | 49.28 |
| Days ≥ 0.1" Precip | 5 | 5 | 7 | 6 | 8 | 6 | 5 | 5 | 6 | 5 | 6 | 6 | 70 |
| Total Snowfall (") | 2.2 | 1.2 | 0.6 | 0.0 | 0.0 | 0.0 | 0.0 | 0.0 | 0.0 | 0.0 | 0.4 | 0.7 | 5.1 |
| Days ≥ 1" Snow Depth | 3 | 1 | 0 | 0 | 0 | 0 | 0 | 0 | 0 | 0 | 0 | 1 | 5 |

## DEQUEEN *Sevier County* ELEVATION 430 ft LAT/LONG 34° 2 ' N / 94° 21 ' W

| | JAN | FEB | MAR | APR | MAY | JUN | JUL | AUG | SEP | OCT | NOV | DEC | YEAR |
|---|---|---|---|---|---|---|---|---|---|---|---|---|---|
| Maximum Temp °F | 52.6 | 58.3 | 67.3 | 75.9 | 81.8 | 88.9 | 93.1 | 92.9 | 86.3 | 76.7 | 64.7 | 55.7 | 74.5 |
| Minimum Temp °F | 29.2 | 32.4 | 40.6 | 49.2 | 57.9 | 65.7 | 69.1 | 68.0 | 61.6 | 49.4 | 39.7 | 32.6 | 49.6 |
| Mean Temp °F | 40.9 | 45.4 | 54.0 | 62.6 | 69.9 | 77.3 | 81.1 | 80.5 | 74.0 | 63.1 | 52.2 | 44.2 | 62.1 |
| Days Max Temp ≥ 90 °F | 0 | 0 | 0 | 0 | 2 | 15 | 24 | 23 | 11 | 1 | 0 | 0 | 76 |
| Days Max Temp ≤ 32 °F | 1 | 0 | 0 | 0 | 0 | 0 | 0 | 0 | 0 | 0 | 0 | 1 | 2 |
| Days Min Temp ≤ 32 °F | 21 | 16 | 7 | 1 | 0 | 0 | 0 | 0 | 0 | 1 | 9 | 17 | 72 |
| Days Min Temp ≤ 0 °F | 0 | 0 | 0 | 0 | 0 | 0 | 0 | 0 | 0 | 0 | 0 | 0 | 0 |
| Heating Degree Days | 739 | 548 | 350 | 126 | 22 | 1 | 0 | 0 | 14 | 127 | 389 | 641 | 2957 |
| Cooling Degree Days | 0 | 1 | 11 | 49 | 168 | 376 | 507 | 510 | 304 | 76 | 12 | 2 | 2016 |
| Total Precipitation (") | 3.53 | 3.75 | 5.36 | 4.87 | 7.03 | 4.49 | 4.08 | 2.50 | 4.43 | 5.23 | 5.08 | 5.20 | 55.55 |
| Days ≥ 0.1" Precip | 6 | 6 | 7 | 7 | 8 | 6 | 6 | 4 | 6 | 5 | 6 | 6 | 73 |
| Total Snowfall (") | na | 0.2 | 0.0 | 0.0 | 0.0 | 0.0 | 0.0 | 0.0 | 0.0 | 0.0 | 0.0 | 0.1 | na |
| Days ≥ 1" Snow Depth | na | 0 | 0 | 0 | 0 | 0 | 0 | 0 | 0 | 0 | 0 | 0 | na |

## DERMOTT 3 NE *Chicot County* ELEVATION 143 ft LAT/LONG 33° 33 ' N / 91° 23 ' W

| | JAN | FEB | MAR | APR | MAY | JUN | JUL | AUG | SEP | OCT | NOV | DEC | YEAR |
|---|---|---|---|---|---|---|---|---|---|---|---|---|---|
| Maximum Temp °F | 50.7 | 56.3 | 65.5 | 75.2 | 82.9 | 90.4 | 93.2 | 91.8 | 86.5 | 77.1 | 65.2 | 55.2 | 74.2 |
| Minimum Temp °F | 31.7 | 35.1 | 43.6 | 51.9 | 59.9 | 67.8 | 70.8 | 68.4 | 62.0 | 49.7 | 42.4 | 35.3 | 51.6 |
| Mean Temp °F | 41.2 | 45.7 | 54.6 | 63.6 | 71.5 | 79.1 | 82.0 | 80.2 | 74.3 | 63.4 | 53.8 | 45.3 | 62.9 |
| Days Max Temp ≥ 90 °F | 0 | 0 | 0 | 1 | 5 | 19 | 24 | 21 | 12 | 2 | 0 | 0 | 84 |
| Days Max Temp ≤ 32 °F | 3 | 1 | 0 | 0 | 0 | 0 | 0 | 0 | 0 | 0 | 0 | 1 | 5 |
| Days Min Temp ≤ 32 °F | 18 | 12 | 4 | 0 | 0 | 0 | 0 | 0 | 0 | 1 | 6 | 13 | 54 |
| Days Min Temp ≤ 0 °F | 0 | 0 | 0 | 0 | 0 | 0 | 0 | 0 | 0 | 0 | 0 | 0 | 0 |
| Heating Degree Days | 731 | 539 | 335 | 118 | 20 | 0 | 0 | 0 | 14 | 129 | 345 | 607 | 2838 |
| Cooling Degree Days | 0 | 2 | 20 | 85 | 235 | 449 | 557 | 509 | 320 | 90 | 19 | 4 | 2290 |
| Total Precipitation (") | 4.68 | 5.41 | 5.30 | 4.87 | 4.89 | 4.12 | 4.30 | 2.90 | 3.20 | 3.94 | 5.00 | 5.56 | 54.17 |
| Days ≥ 0.1" Precip | 8 | 6 | 7 | 6 | 7 | 5 | 6 | 5 | 5 | 5 | 6 | 7 | 73 |
| Total Snowfall (") | 1.6 | 0.6 | 0.4 | 0.0 | 0.0 | 0.0 | 0.0 | 0.0 | 0.0 | 0.0 | 0.0 | 0.2 | 2.8 |
| Days ≥ 1" Snow Depth | 1 | 0 | 0 | 0 | 0 | 0 | 0 | 0 | 0 | 0 | 0 | 0 | 1 |

## DES ARC *Prairie County* ELEVATION 200 ft LAT/LONG 34° 58 ' N / 91° 30 ' W

| | JAN | FEB | MAR | APR | MAY | JUN | JUL | AUG | SEP | OCT | NOV | DEC | YEAR |
|---|---|---|---|---|---|---|---|---|---|---|---|---|---|
| Maximum Temp °F | 48.2 | 54.0 | 63.6 | 74.4 | 81.2 | 89.4 | 92.9 | 90.9 | 84.4 | 75.4 | 62.7 | 52.8 | 72.5 |
| Minimum Temp °F | 30.6 | 33.9 | 42.9 | 52.6 | 60.2 | 68.2 | 71.5 | 68.6 | 61.7 | 50.9 | 42.7 | 34.5 | 51.5 |
| Mean Temp °F | 39.4 | 44.0 | 53.3 | 63.5 | 70.8 | 78.8 | 82.2 | 79.7 | 73.1 | 63.1 | 52.7 | 43.7 | 62.0 |
| Days Max Temp ≥ 90 °F | 0 | 0 | 0 | 0 | 3 | 16 | 24 | 20 | 8 | 1 | 0 | 0 | 72 |
| Days Max Temp ≤ 32 °F | 3 | 1 | 0 | 0 | 0 | 0 | 0 | 0 | 0 | 0 | 0 | 1 | 5 |
| Days Min Temp ≤ 32 °F | 19 | 12 | 4 | 0 | 0 | 0 | 0 | 0 | 0 | 0 | 5 | 13 | 53 |
| Days Min Temp ≤ 0 °F | 0 | 0 | 0 | 0 | 0 | 0 | 0 | 0 | 0 | 0 | 0 | 0 | 0 |
| Heating Degree Days | 787 | 587 | 371 | 115 | 20 | 0 | 0 | 0 | 17 | 128 | 375 | 655 | 3055 |
| Cooling Degree Days | 0 | 0 | 13 | 70 | 197 | 434 | 555 | 483 | 266 | 63 | 11 | 1 | 2093 |
| Total Precipitation (") | 3.85 | 4.10 | 5.13 | 5.63 | 5.15 | 3.57 | 3.64 | 3.20 | 3.45 | 3.62 | 5.02 | 4.61 | 50.97 |
| Days ≥ 0.1" Precip | 6 | 6 | 7 | 7 | 7 | 6 | 5 | 4 | 5 | 5 | 6 | 6 | 70 |
| Total Snowfall (") | 1.5 | 1.4 | 0.5 | 0.0 | 0.0 | 0.0 | 0.0 | 0.0 | 0.0 | 0.0 | 0.1 | 0.1 | 3.6 |
| Days ≥ 1" Snow Depth | na | 0 | 0 | 0 | 0 | 0 | 0 | 0 | 0 | 0 | 0 | 0 | na |

## DUMAS *Desha County* ELEVATION 161 ft LAT/LONG 33° 53 ' N / 91° 30 ' W

| | JAN | FEB | MAR | APR | MAY | JUN | JUL | AUG | SEP | OCT | NOV | DEC | YEAR |
|---|---|---|---|---|---|---|---|---|---|---|---|---|---|
| Maximum Temp °F | 51.5 | 57.5 | 66.8 | 76.4 | 83.0 | 90.2 | 92.7 | 91.2 | 85.4 | 75.8 | 64.2 | 55.4 | 74.2 |
| Minimum Temp °F | 32.6 | 36.2 | 44.2 | 53.0 | 61.1 | 68.5 | 71.3 | 69.5 | 63.5 | 52.1 | 43.6 | 36.5 | 52.7 |
| Mean Temp °F | 42.1 | 46.9 | 55.5 | 64.7 | 72.1 | 79.4 | 82.1 | 80.4 | 74.5 | 64.0 | 53.9 | 46.0 | 63.5 |
| Days Max Temp ≥ 90 °F | 0 | 0 | 0 | 1 | 4 | 18 | 24 | 21 | 9 | 1 | 0 | 0 | 78 |
| Days Max Temp ≤ 32 °F | 2 | 1 | 0 | 0 | 0 | 0 | 0 | 0 | 0 | 0 | 0 | 1 | 4 |
| Days Min Temp ≤ 32 °F | 16 | 11 | 3 | 0 | 0 | 0 | 0 | 0 | 0 | 0 | 4 | 12 | 46 |
| Days Min Temp ≤ 0 °F | 0 | 0 | 0 | 0 | 0 | 0 | 0 | 0 | 0 | 0 | 0 | 0 | 0 |
| Heating Degree Days | 704 | 506 | 307 | 95 | 10 | 0 | 0 | 0 | 10 | 107 | 338 | 586 | 2663 |
| Cooling Degree Days | 0 | 3 | 20 | 93 | 241 | 448 | 542 | 500 | 307 | 79 | 14 | 3 | 2250 |
| Total Precipitation (") | 4.37 | 4.65 | 5.48 | 4.72 | 4.89 | 3.54 | 4.18 | 3.22 | 3.63 | 4.35 | 4.94 | 4.93 | 52.90 |
| Days ≥ 0.1" Precip | 6 | 6 | 7 | 6 | 7 | 5 | 6 | 4 | 5 | 5 | 6 | 7 | 70 |
| Total Snowfall (") | 1.8 | 0.5 | 0.5 | 0.0 | 0.0 | 0.0 | 0.0 | 0.0 | 0.0 | 0.0 | 0.0 | 0.2 | 3.0 |
| Days ≥ 1" Snow Depth | 2 | 0 | 0 | 0 | 0 | 0 | 0 | 0 | 0 | 0 | 0 | 0 | 2 |

## EL DORADO GOODWIN FD *Union County* ELEVATION 253 ft LAT/LONG 33° 13 ' N / 92° 48 ' W

| | JAN | FEB | MAR | APR | MAY | JUN | JUL | AUG | SEP | OCT | NOV | DEC | YEAR |
|---|---|---|---|---|---|---|---|---|---|---|---|---|---|
| Maximum Temp °F | 53.4 | 58.7 | 67.7 | 76.5 | 82.2 | 89.1 | 92.1 | 91.6 | 85.9 | 76.3 | 65.4 | 57.1 | 74.7 |
| Minimum Temp °F | 32.4 | 35.1 | 43.1 | 51.6 | 59.5 | 67.2 | 70.9 | 69.4 | 63.4 | 51.2 | 42.2 | 35.5 | 51.8 |
| Mean Temp °F | 42.9 | 46.9 | 55.4 | 64.0 | 70.9 | 78.2 | 81.6 | 80.5 | 74.7 | 63.8 | 53.8 | 46.3 | 63.2 |
| Days Max Temp ≥ 90 °F | 0 | 0 | 0 | 0 | 2 | 15 | 23 | 22 | 11 | 1 | 0 | 0 | 74 |
| Days Max Temp ≤ 32 °F | 1 | 1 | 0 | 0 | 0 | 0 | 0 | 0 | 0 | 0 | 0 | 1 | 3 |
| Days Min Temp ≤ 32 °F | 17 | 13 | 5 | 1 | 0 | 0 | 0 | 0 | 0 | 0 | 6 | 14 | 56 |
| Days Min Temp ≤ 0 °F | 0 | 0 | 0 | 0 | 0 | 0 | 0 | 0 | 0 | 0 | 0 | 0 | 0 |
| Heating Degree Days | 681 | 507 | 312 | 103 | 17 | 0 | 0 | 0 | 11 | 116 | 347 | 576 | 2670 |
| Cooling Degree Days | 1 | 5 | 21 | 81 | 214 | 414 | 534 | 516 | 321 | 88 | 21 | 5 | 2221 |
| Total Precipitation (") | 4.61 | 4.56 | 5.01 | 4.45 | 5.62 | 4.88 | 3.99 | 3.05 | 3.32 | 4.40 | 4.69 | 4.73 | 53.31 |
| Days ≥ 0.1" Precip | 7 | 6 | 7 | 6 | 8 | 6 | 6 | 5 | 4 | 5 | 5 | 7 | 72 |
| Total Snowfall (") | 0.7 | 0.4 | 0.4 | 0.0 | 0.0 | 0.0 | 0.0 | 0.0 | 0.0 | 0.0 | 0.0 | 0.1 | 1.6 |
| Days ≥ 1" Snow Depth | 1 | 0 | 0 | 0 | 0 | 0 | 0 | 0 | 0 | 0 | 0 | 0 | 1 |

## EUDORA *Chicot County* ELEVATION 141 ft LAT/LONG 33° 7 ' N / 91° 16 ' W

| | JAN | FEB | MAR | APR | MAY | JUN | JUL | AUG | SEP | OCT | NOV | DEC | YEAR |
|---|---|---|---|---|---|---|---|---|---|---|---|---|---|
| Maximum Temp °F | 51.7 | 56.8 | 65.9 | 75.2 | 82.6 | 89.9 | 92.4 | 91.4 | 86.3 | 77.1 | 65.7 | 56.1 | 74.3 |
| Minimum Temp °F | 33.0 | 36.3 | 44.3 | 53.0 | 61.3 | 68.8 | 71.5 | 69.9 | 63.8 | 52.0 | 43.4 | 36.6 | 52.8 |
| Mean Temp °F | 42.4 | 46.6 | 55.1 | 64.1 | 71.9 | 79.4 | 82.0 | 80.7 | 75.1 | 64.6 | 54.6 | 46.3 | 63.6 |
| Days Max Temp ≥ 90 °F | 0 | 0 | 0 | 0 | 4 | 18 | 24 | 22 | 12 | 1 | 0 | 0 | 81 |
| Days Max Temp ≤ 32 °F | 2 | 1 | 0 | 0 | 0 | 0 | 0 | 0 | 0 | 0 | 0 | 1 | 4 |
| Days Min Temp ≤ 32 °F | 16 | 11 | 3 | 0 | 0 | 0 | 0 | 0 | 0 | 0 | 4 | 11 | 45 |
| Days Min Temp ≤ 0 °F | 0 | 0 | 0 | 0 | 0 | 0 | 0 | 0 | 0 | 0 | 0 | 0 | 0 |
| Heating Degree Days | 696 | 515 | 319 | 106 | 15 | 0 | 0 | 0 | 10 | 106 | 325 | 575 | 2667 |
| Cooling Degree Days | 0 | 3 | 18 | 77 | 232 | 446 | 542 | 508 | 324 | 101 | 21 | 5 | 2277 |
| Total Precipitation (") | 4.76 | 5.28 | 5.37 | 5.70 | 5.42 | 3.91 | 4.04 | 3.21 | 2.77 | 4.00 | 5.14 | 5.83 | 55.43 |
| Days ≥ 0.1" Precip | 7 | 6 | 7 | 6 | 7 | 6 | 6 | 5 | 4 | 5 | 6 | 7 | 72 |
| Total Snowfall (") | ***0.0*** | 0.0 | 0.0 | 0.0 | 0.0 | 0.0 | 0.0 | 0.0 | 0.0 | 0.0 | 0.0 | 0.1 | 0.1 |
| Days ≥ 1" Snow Depth | ***0*** | 0 | 0 | 0 | 0 | 0 | 0 | 0 | 0 | 0 | 0 | 0 | 0 |

## EUREKA SPRINGS 3 WNW *Carroll County* ELEVATION 1470 ft LAT/LONG 36° 24 ' N / 93° 45 ' W

| | JAN | FEB | MAR | APR | MAY | JUN | JUL | AUG | SEP | OCT | NOV | DEC | YEAR |
|---|---|---|---|---|---|---|---|---|---|---|---|---|---|
| Maximum Temp °F | 45.6 | 51.0 | 61.5 | 72.4 | 77.8 | 85.1 | 90.6 | 89.5 | 81.2 | ***71.8*** | 58.8 | 49.2 | 69.5 |
| Minimum Temp °F | 25.9 | 29.8 | 38.9 | 48.3 | 55.2 | 63.4 | 68.0 | 65.9 | 59.4 | ***49.1*** | 39.1 | 30.0 | 47.8 |
| Mean Temp °F | 35.8 | 40.4 | 50.2 | 60.4 | 66.5 | 74.3 | 79.3 | 77.7 | 70.3 | ***60.5*** | 48.9 | 39.6 | 58.7 |
| Days Max Temp ≥ 90 °F | 0 | 0 | 0 | 0 | 1 | 7 | 19 | 17 | 4 | 0 | 0 | 0 | 48 |
| Days Max Temp ≤ 32 °F | 5 | 3 | 0 | 0 | 0 | 0 | 0 | 0 | 0 | 0 | 0 | 3 | 11 |
| Days Min Temp ≤ 32 °F | 22 | 17 | 10 | 2 | 0 | 0 | 0 | 0 | 0 | 1 | 9 | 18 | 79 |
| Days Min Temp ≤ 0 °F | 1 | 0 | 0 | 0 | 0 | 0 | 0 | 0 | 0 | 0 | 0 | 0 | 1 |
| Heating Degree Days | 899 | 688 | 463 | 183 | 57 | 4 | 0 | 1 | 36 | ***182*** | 480 | 780 | 3773 |
| Cooling Degree Days | 0 | 1 | 12 | 59 | 118 | 306 | 475 | 434 | 221 | na | 6 | 0 | na |
| Total Precipitation (") | 2.22 | 2.78 | 4.19 | 4.62 | 4.84 | 4.33 | 3.31 | 3.59 | 4.34 | 3.96 | 4.44 | 3.54 | 46.16 |
| Days ≥ 0.1" Precip | 4 | 4 | 7 | 7 | 8 | 7 | 4 | 5 | 6 | 5 | 5 | 5 | 67 |
| Total Snowfall (") | 3.6 | ***3.6*** | ***2.6*** | 0.4 | 0.0 | 0.0 | 0.0 | 0.0 | 0.0 | 0.1 | 0.9 | 1.8 | 13.0 |
| Days ≥ 1" Snow Depth | 6 | 4 | ***1*** | 0 | 0 | 0 | 0 | 0 | 0 | 0 | 0 | 2 | 13 |

## EVENING SHADE 1 NNE *Sharp County* ELEVATION 489 ft LAT/LONG 36° 5 ' N / 91° 37 ' W

| | JAN | FEB | MAR | APR | MAY | JUN | JUL | AUG | SEP | OCT | NOV | DEC | YEAR |
|---|---|---|---|---|---|---|---|---|---|---|---|---|---|
| Maximum Temp °F | 48.4 | 53.5 | 63.0 | 73.8 | 80.1 | 88.0 | 93.1 | 91.3 | 84.3 | 74.6 | 62.2 | 51.6 | 72.0 |
| Minimum Temp °F | 23.9 | 27.1 | 35.9 | 45.0 | 52.7 | 61.3 | 65.8 | 63.5 | 57.3 | 44.0 | 36.1 | 27.8 | 45.0 |
| Mean Temp °F | 36.1 | 40.3 | 49.5 | 59.4 | 66.4 | 74.6 | 79.5 | 77.4 | 70.8 | 59.4 | 49.0 | 39.7 | 58.5 |
| Days Max Temp ≥ 90 °F | 0 | 0 | 0 | 0 | 2 | 13 | 24 | 20 | 7 | 1 | 0 | 0 | 67 |
| Days Max Temp ≤ 32 °F | 3 | 1 | 0 | 0 | 0 | 0 | 0 | 0 | 0 | 0 | 0 | 1 | 5 |
| Days Min Temp ≤ 32 °F | 25 | 19 | 12 | 4 | 0 | 0 | 0 | 0 | 0 | 5 | 12 | 20 | 97 |
| Days Min Temp ≤ 0 °F | 1 | 0 | 0 | 0 | 0 | 0 | 0 | 0 | 0 | 0 | 0 | 0 | 1 |
| Heating Degree Days | 889 | 690 | 481 | 198 | 56 | 2 | 0 | 0 | 31 | 203 | 475 | 777 | 3802 |
| Cooling Degree Days | 0 | 0 | 5 | 38 | 113 | 316 | 476 | 422 | 223 | 35 | 3 | 1 | 1632 |
| Total Precipitation (") | 3.20 | 3.41 | 4.37 | 4.66 | 4.72 | 3.56 | 3.42 | 3.50 | 4.36 | 3.71 | 5.11 | 4.59 | 48.61 |
| Days ≥ 0.1" Precip | 5 | 6 | 7 | 7 | 8 | 6 | 5 | 5 | 6 | 5 | 6 | 7 | 73 |
| Total Snowfall (") | *3.3* | 3.7 | 1.8 | 0.0 | 0.0 | 0.0 | 0.0 | 0.0 | 0.0 | 0.1 | 0.5 | 1.7 | 11.1 |
| Days ≥ 1" Snow Depth | na | na | *0* | 0 | 0 | 0 | 0 | 0 | 0 | 0 | 0 | *0* | na |

## FAYETTEVILLE EXP STN *Washington County* ELEVATION 1270 ft LAT/LONG 36° 6 ' N / 94° 10 ' W

| | JAN | FEB | MAR | APR | MAY | JUN | JUL | AUG | SEP | OCT | NOV | DEC | YEAR |
|---|---|---|---|---|---|---|---|---|---|---|---|---|---|
| Maximum Temp °F | 45.4 | 50.5 | 59.5 | 70.0 | 76.3 | 84.1 | 89.5 | 88.9 | 81.1 | 70.9 | 58.6 | 49.5 | 68.7 |
| Minimum Temp °F | 24.5 | 28.4 | 37.8 | 47.3 | 55.0 | 63.7 | 68.4 | 66.5 | 59.3 | 47.4 | 37.9 | 29.0 | 47.1 |
| Mean Temp °F | 35.0 | 39.5 | 48.7 | 58.7 | 65.7 | 74.0 | 79.0 | 77.7 | 70.2 | 59.2 | 48.3 | 39.3 | 57.9 |
| Days Max Temp ≥ 90 °F | 0 | 0 | 0 | 0 | 0 | 4 | 16 | 14 | 4 | 0 | 0 | 0 | 38 |
| Days Max Temp ≤ 32 °F | 5 | 3 | 1 | 0 | 0 | 0 | 0 | 0 | 0 | 0 | 0 | 2 | 11 |
| Days Min Temp ≤ 32 °F | 24 | 18 | 11 | 2 | 0 | 0 | 0 | 0 | 0 | 2 | 10 | 20 | 87 |
| Days Min Temp ≤ 0 °F | 1 | 1 | 0 | 0 | 0 | 0 | 0 | 0 | 0 | 0 | 0 | 1 | 3 |
| Heating Degree Days | 924 | 716 | 509 | 221 | 76 | 6 | 0 | 1 | 41 | 214 | 500 | 791 | 3999 |
| Cooling Degree Days | 0 | 0 | 6 | 35 | 95 | 284 | 446 | 417 | 209 | 33 | 4 | 0 | 1529 |
| Total Precipitation (") | 2.00 | 2.55 | 3.99 | 4.56 | 4.73 | 4.88 | 2.96 | 3.19 | 4.54 | 4.02 | 4.32 | 3.35 | 45.09 |
| Days ≥ 0.1" Precip | 4 | 4 | 6 | 7 | 8 | 7 | 5 | 5 | 6 | 5 | 5 | 5 | 67 |
| Total Snowfall (") | 2.0 | 2.8 | 1.1 | 0.0 | 0.0 | 0.0 | 0.0 | 0.0 | 0.0 | 0.0 | 0.7 | 1.3 | 7.9 |
| Days ≥ 1" Snow Depth | 2 | 2 | 0 | 0 | 0 | 0 | 0 | 0 | 0 | 0 | 0 | 1 | 5 |

## FORDYCE *Dallas County* ELEVATION 259 ft LAT/LONG 33° 49 ' N / 92° 25 ' W

| | JAN | FEB | MAR | APR | MAY | JUN | JUL | AUG | SEP | OCT | NOV | DEC | YEAR |
|---|---|---|---|---|---|---|---|---|---|---|---|---|---|
| Maximum Temp °F | 52.9 | 57.9 | 66.9 | 75.9 | 81.9 | *88.2* | 92.1 | 91.8 | 85.3 | 75.5 | *64.3* | 55.3 | 74.0 |
| Minimum Temp °F | 30.8 | 33.9 | 41.8 | 50.6 | 57.8 | *65.6* | 69.2 | 67.6 | 61.4 | 50.1 | *40.7* | 33.6 | 50.3 |
| Mean Temp °F | 41.9 | 45.9 | 54.4 | 63.3 | 70.0 | *76.9* | 80.6 | 79.7 | 73.4 | 62.9 | *52.5* | 44.5 | 62.2 |
| Days Max Temp ≥ 90 °F | 0 | 0 | 0 | 1 | 3 | *14* | 23 | 22 | 10 | 1 | *0* | 0 | 74 |
| Days Max Temp ≤ 32 °F | 1 | *0* | 0 | 0 | 0 | *0* | 0 | 0 | 0 | 0 | *0* | 1 | 2 |
| Days Min Temp ≤ 32 °F | 19 | *13* | 6 | 1 | 0 | *0* | 0 | 0 | 0 | 1 | *7* | 15 | 62 |
| Days Min Temp ≤ 0 °F | 0 | 0 | 0 | 0 | 0 | *0* | 0 | 0 | 0 | 0 | *0* | 0 | 0 |
| Heating Degree Days | 711 | 532 | 339 | 116 | 26 | *1* | 0 | 0 | 17 | 134 | *377* | 632 | 2885 |
| Cooling Degree Days | 0 | 0 | 12 | 56 | 175 | *341* | 479 | 471 | 267 | 66 | 9 | 2 | 1878 |
| Total Precipitation (") | 4.22 | 4.63 | 5.03 | 4.56 | 5.24 | *4.09* | 4.11 | 2.99 | 3.26 | 4.79 | 5.07 | 5.54 | 53.53 |
| Days ≥ 0.1" Precip | 7 | 6 | 7 | 6 | 7 | *6* | 6 | 4 | 5 | 6 | *7* | 7 | 74 |
| Total Snowfall (") | *1.1* | *0.5* | 0.3 | 0.0 | 0.0 | *0.0* | 0.0 | 0.0 | 0.0 | 0.0 | *0.0* | 0.2 | 2.1 |
| Days ≥ 1" Snow Depth | *1* | *0* | 0 | 0 | 0 | 0 | 0 | 0 | 0 | 0 | *0* | 0 | 1 |

## FT SMITH MUNICIPL AP *Sebastian County* ELEVATION 447 ft LAT/LONG 35° 20 ' N / 94° 22 ' W

| | JAN | FEB | MAR | APR | MAY | JUN | JUL | AUG | SEP | OCT | NOV | DEC | YEAR |
|---|---|---|---|---|---|---|---|---|---|---|---|---|---|
| Maximum Temp °F | 47.9 | 53.7 | 63.9 | 73.9 | 80.0 | 87.9 | 92.9 | 92.1 | 84.8 | 74.9 | 61.9 | 51.6 | 72.1 |
| Minimum Temp °F | 26.5 | 30.6 | 39.5 | 48.8 | 57.7 | 66.4 | 70.4 | 68.9 | 61.8 | 48.9 | 38.6 | 30.3 | 49.0 |
| Mean Temp °F | 37.2 | 42.2 | 51.7 | 61.4 | 68.9 | 77.2 | 81.7 | 80.5 | 73.3 | 61.9 | 50.3 | 41.0 | 60.6 |
| Days Max Temp ≥ 90 °F | 0 | 0 | 0 | 1 | 2 | 13 | 23 | 21 | 10 | 1 | 0 | 0 | 71 |
| Days Max Temp ≤ 32 °F | 3 | 1 | 0 | 0 | 0 | 0 | 0 | 0 | 0 | 0 | 0 | 1 | 5 |
| Days Min Temp ≤ 32 °F | 23 | 17 | 7 | 1 | 0 | 0 | 0 | 0 | 0 | 1 | 9 | 19 | 77 |
| Days Min Temp ≤ 0 °F | 0 | 0 | 0 | 0 | 0 | 0 | 0 | 0 | 0 | 0 | 0 | 0 | 0 |
| Heating Degree Days | 854 | 638 | 414 | 153 | 32 | 1 | 0 | 0 | 20 | 150 | 441 | 739 | 3442 |
| Cooling Degree Days | 0 | 0 | 10 | 52 | 164 | 394 | 540 | 510 | 287 | 63 | 7 | 1 | 2028 |
| Total Precipitation (") | 2.11 | 2.60 | 3.80 | 4.23 | 5.41 | 3.64 | 2.92 | 2.77 | 3.40 | 4.03 | 4.48 | 3.31 | 42.70 |
| Days ≥ 0.1" Precip | 4 | 4 | 6 | 7 | 8 | 6 | 5 | 5 | 6 | 5 | 5 | 5 | 66 |
| Total Snowfall (") | 2.9 | 1.8 | 0.9 | 0.0 | 0.0 | 0.0 | 0.0 | 0.0 | 0.0 | 0.0 | 0.5 | 1.0 | 7.1 |
| Days ≥ 1" Snow Depth | 3 | 2 | 0 | 0 | 0 | 0 | 0 | 0 | 0 | 0 | 0 | 1 | 6 |

## GILBERT *Searcy County* ELEVATION 600 ft LAT/LONG 35° 59 ' N / 92° 43 ' W

| | JAN | FEB | MAR | APR | MAY | JUN | JUL | AUG | SEP | OCT | NOV | DEC | YEAR |
|---|---|---|---|---|---|---|---|---|---|---|---|---|---|
| Maximum Temp °F | 49.6 | 54.5 | 64.2 | 74.2 | 79.9 | 87.1 | 92.2 | 91.1 | 84.2 | 74.8 | 62.5 | 52.9 | 72.3 |
| Minimum Temp °F | 23.0 | 26.2 | 35.3 | 44.2 | 52.4 | 60.6 | 64.3 | 62.4 | 56.2 | 43.3 | 34.7 | 26.8 | 44.1 |
| Mean Temp °F | 36.3 | 40.4 | 49.8 | 59.2 | 66.2 | 73.9 | 78.3 | 76.8 | 70.2 | 59.1 | 48.6 | 39.9 | 58.2 |
| Days Max Temp ≥ 90 °F | 0 | 0 | 0 | 1 | 1 | 11 | 22 | 19 | 8 | 1 | 0 | 0 | 63 |
| Days Max Temp ≤ 32 °F | 3 | 1 | 0 | 0 | 0 | 0 | 0 | 0 | 0 | 0 | 0 | 1 | 5 |
| Days Min Temp ≤ 32 °F | 25 | 20 | 14 | 5 | 0 | 0 | 0 | 0 | 0 | 6 | 14 | 21 | 105 |
| Days Min Temp ≤ 0 °F | 1 | 0 | 0 | 0 | 0 | 0 | 0 | 0 | 0 | 0 | 0 | 1 | 2 |
| Heating Degree Days | 881 | 688 | 474 | 204 | 61 | 4 | 0 | 1 | 34 | 210 | 489 | 774 | 3820 |
| Cooling Degree Days | 0 | 0 | 8 | 31 | 104 | 284 | 426 | 391 | 204 | 34 | 4 | 1 | 1487 |
| Total Precipitation (") | 2.55 | 3.00 | 3.90 | 4.35 | 5.00 | 3.85 | 2.71 | 3.03 | 4.06 | 3.69 | 4.75 | 3.93 | 44.82 |
| Days ≥ 0.1" Precip | 5 | 5 | 7 | 7 | 8 | 6 | 5 | 5 | 6 | 5 | 5 | 6 | 70 |
| Total Snowfall (") | *2.1* | 1.9 | 1.5 | 0.1 | 0.0 | 0.0 | 0.0 | 0.0 | 0.0 | 0.0 | 0.9 | 1.6 | 8.1 |
| Days ≥ 1" Snow Depth | 3 | *1* | 0 | 0 | 0 | 0 | 0 | 0 | 0 | 0 | 0 | 1 | 5 |

## GRAVETTE *Benton County* ELEVATION 1250 ft LAT/LONG 36° 24 ' N / 94° 28 ' W

| | JAN | FEB | MAR | APR | MAY | JUN | JUL | AUG | SEP | OCT | NOV | DEC | YEAR |
|---|---|---|---|---|---|---|---|---|---|---|---|---|---|
| Maximum Temp °F | 45.6 | 51.1 | 61.1 | 71.4 | 77.6 | 85.3 | 90.9 | 90.1 | 82.2 | 71.9 | 58.9 | 49.6 | 69.6 |
| Minimum Temp °F | 24.2 | 28.3 | 37.5 | 47.0 | 54.3 | 62.6 | 66.4 | 64.8 | 58.0 | 47.3 | 37.5 | 28.5 | 46.4 |
| Mean Temp °F | 34.9 | 39.7 | 49.4 | 59.2 | 66.0 | 74.0 | 78.7 | 77.5 | 70.1 | 59.6 | 48.2 | 39.1 | 58.0 |
| Days Max Temp ≥ 90 °F | 0 | 0 | 0 | 0 | 0 | 6 | 19 | 18 | 6 | 0 | 0 | 0 | 49 |
| Days Max Temp ≤ 32 °F | 5 | 2 | 0 | 0 | 0 | 0 | 0 | 0 | 0 | 0 | 0 | 2 | 9 |
| Days Min Temp ≤ 32 °F | 24 | 18 | 11 | 3 | 0 | 0 | 0 | 0 | 0 | 2 | 10 | 20 | 88 |
| Days Min Temp ≤ 0 °F | 1 | 1 | 0 | 0 | 0 | 0 | 0 | 0 | 0 | 0 | 0 | 1 | 3 |
| Heating Degree Days | 926 | 707 | 486 | 204 | 67 | 5 | 1 | 1 | 41 | 200 | 501 | 797 | 3936 |
| Cooling Degree Days | 0 | 0 | 7 | 40 | 108 | 294 | 440 | 412 | 208 | 37 | 4 | 0 | 1550 |
| Total Precipitation (") | 2.20 | 2.39 | 4.06 | 4.85 | 5.10 | 5.04 | 2.99 | 3.36 | 4.67 | 4.01 | 4.43 | 3.53 | 46.63 |
| Days ≥ 0.1" Precip | 4 | 4 | 7 | 7 | 8 | 6 | 5 | 5 | 6 | 5 | 6 | 5 | 68 |
| Total Snowfall (") | *4.3* | 4.5 | 3.2 | 0.0 | 0.0 | 0.0 | 0.0 | 0.0 | 0.0 | 0.1 | 0.7 | 2.7 | 15.5 |
| Days ≥ 1" Snow Depth | 5 | 3 | 1 | 0 | 0 | 0 | 0 | 0 | 0 | 0 | 0 | 2 | 11 |

## GREENBRIER *Faulkner County* ELEVATION 361 ft LAT/LONG 35° 14 ' N / 92° 23 ' W

| | JAN | FEB | MAR | APR | MAY | JUN | JUL | AUG | SEP | OCT | NOV | DEC | YEAR |
|---|---|---|---|---|---|---|---|---|---|---|---|---|---|
| Maximum Temp °F | *49.3* | *54.8* | *63.8* | 73.4 | *79.8* | *87.5* | 92.0 | *91.1* | *84.0* | 74.2 | 61.6 | 51.7 | 71.9 |
| Minimum Temp °F | *27.0* | *30.6* | *39.0* | 47.7 | *55.6* | *63.5* | 67.5 | *65.6* | *59.3* | 47.0 | 37.8 | 30.2 | 47.6 |
| Mean Temp °F | *38.2* | *42.7* | *51.4* | 60.6 | *67.7* | *75.6* | 79.8 | *78.4* | *71.7* | 60.7 | 49.7 | 41.0 | 59.8 |
| Days Max Temp ≥ 90 °F | *0* | *0* | *0* | 0 | *1* | *11* | 22 | *20* | 7 | 1 | 0 | 0 | 62 |
| Days Max Temp ≤ 32 °F | *3* | *1* | *0* | 0 | *0* | *0* | 0 | *0* | 0 | 0 | 0 | 1 | 5 |
| Days Min Temp ≤ 32 °F | *23* | *17* | *9* | 2 | *0* | *0* | 0 | *0* | 0 | 2 | 10 | 19 | 82 |
| Days Min Temp ≤ 0 °F | *0* | *0* | *0* | 0 | *0* | *0* | 0 | *0* | 0 | 0 | 0 | 0 | 0 |
| Heating Degree Days | *824* | *622* | *424* | 168 | *40* | *1* | 0 | *1* | *26* | 176 | 457 | 739 | 3478 |
| Cooling Degree Days | 0 | 0 | 8 | 38 | 135 | 328 | 465 | 424 | 236 | 44 | 7 | 1 | 1686 |
| Total Precipitation (") | *3.46* | *3.55* | *4.56* | 5.12 | *5.33* | *4.37* | 3.29 | *2.72* | *4.10* | 4.49 | 4.98 | 4.89 | 50.86 |
| Days ≥ 0.1" Precip | *6* | *6* | *6* | 7 | *7* | *6* | 5 | *5* | 5 | 5 | 6 | 6 | 70 |
| Total Snowfall (") | *2.2* | *1.2* | *1.0* | 0.0 | *0.0* | *0.0* | 0.0 | *0.0* | 0.0 | 0.0 | 0.4 | 0.7 | 5.5 |
| Days ≥ 1" Snow Depth | *2* | *1* | *0* | 0 | *0* | *0* | 0 | *0* | 0 | 0 | 0 | 1 | 4 |

## GREERS FERRY DAM *Cleburne County* ELEVATION 527 ft LAT/LONG 35° 31 ' N / 92° 0 ' W

| | JAN | FEB | MAR | APR | MAY | JUN | JUL | AUG | SEP | OCT | NOV | DEC | YEAR |
|---|---|---|---|---|---|---|---|---|---|---|---|---|---|
| Maximum Temp °F | 46.7 | 51.9 | 61.3 | 72.0 | 78.7 | 86.7 | 91.5 | 90.2 | 83.2 | 73.3 | 60.8 | 50.9 | 70.6 |
| Minimum Temp °F | 24.8 | 27.7 | 36.8 | 46.3 | 54.5 | 63.7 | 68.0 | 65.7 | 59.0 | 46.6 | 37.6 | 29.3 | 46.7 |
| Mean Temp °F | 35.8 | 39.6 | 49.1 | 59.2 | 66.6 | 75.2 | 79.8 | 78.0 | 71.2 | 60.0 | 49.2 | 40.1 | 58.7 |
| Days Max Temp ≥ 90 °F | 0 | 0 | 0 | 0 | 1 | 10 | 21 | 17 | 6 | 0 | 0 | 0 | 55 |
| Days Max Temp ≤ 32 °F | 4 | 2 | 0 | 0 | 0 | 0 | 0 | 0 | 0 | 0 | 0 | 2 | 8 |
| Days Min Temp ≤ 32 °F | 25 | 19 | 11 | 2 | 0 | 0 | 0 | 0 | 0 | 1 | 10 | 20 | 88 |
| Days Min Temp ≤ 0 °F | 0 | 0 | 0 | 0 | 0 | 0 | 0 | 0 | 0 | 0 | 0 | 0 | 0 |
| Heating Degree Days | 898 | 710 | 493 | 201 | 59 | 2 | 0 | 0 | 27 | 189 | 470 | 766 | 3815 |
| Cooling Degree Days | 0 | 0 | 6 | 33 | 118 | 332 | 489 | 441 | 239 | 43 | 5 | 1 | 1707 |
| Total Precipitation (") | 3.41 | 3.55 | 4.91 | 5.26 | 5.24 | 3.77 | 3.75 | 3.48 | 4.54 | 4.31 | 5.48 | 4.81 | 52.51 |
| Days ≥ 0.1" Precip | 5 | 6 | 7 | 7 | 8 | 6 | 5 | 6 | 5 | 5 | 6 | 7 | 73 |
| Total Snowfall (") | na | *1.5* | 0.0 | 0.0 | 0.0 | 0.0 | 0.0 | 0.0 | 0.0 | 0.0 | 0.1 | 0.1 | na |
| Days ≥ 1" Snow Depth | na | na | 0 | 0 | 0 | 0 | 0 | 0 | 0 | 0 | 0 | *0* | na |

## HARRISON FAA AP *Boone County* ELEVATION 1375 ft LAT/LONG 36° 16 ' N / 93° 9 ' W

| | JAN | FEB | MAR | APR | MAY | JUN | JUL | AUG | SEP | OCT | NOV | DEC | YEAR |
|---|---|---|---|---|---|---|---|---|---|---|---|---|---|
| Maximum Temp °F | 44.4 | 49.4 | 59.1 | 69.6 | 76.0 | 83.7 | 88.8 | 87.3 | 79.7 | 70.1 | 57.3 | 48.2 | 67.8 |
| Minimum Temp °F | 25.0 | 28.6 | 37.8 | 47.5 | 55.1 | 63.4 | 68.1 | 65.8 | 58.9 | 48.1 | 38.2 | 29.4 | 47.2 |
| Mean Temp °F | 34.7 | 39.0 | 48.5 | 58.6 | 65.6 | 73.6 | 78.5 | 76.6 | 69.3 | 59.1 | 47.8 | 38.8 | 57.5 |
| Days Max Temp ≥ 90 °F | 0 | 0 | 0 | 0 | 0 | 4 | 15 | 12 | 3 | 0 | 0 | 0 | 34 |
| Days Max Temp ≤ 32 °F | 6 | 4 | 1 | 0 | 0 | 0 | 0 | 0 | 0 | 0 | 0 | 3 | 14 |
| Days Min Temp ≤ 32 °F | 23 | 18 | 11 | 2 | 0 | 0 | 0 | 0 | 0 | 1 | 9 | 20 | 84 |
| Days Min Temp ≤ 0 °F | 1 | 0 | 0 | 0 | 0 | 0 | 0 | 0 | 0 | 0 | 0 | 0 | 1 |
| Heating Degree Days | 932 | 727 | 513 | 223 | 73 | 4 | 0 | 1 | 43 | 212 | 512 | 805 | 4045 |
| Cooling Degree Days | 0 | 0 | 7 | 42 | 101 | 280 | 442 | 392 | 191 | 36 | 4 | 0 | 1495 |
| Total Precipitation (") | 2.41 | 2.86 | 4.29 | 4.28 | 5.00 | 4.57 | 3.01 | 3.52 | 4.29 | 4.06 | 4.67 | 3.87 | 46.83 |
| Days ≥ 0.1" Precip | 4 | 5 | 7 | 7 | 8 | 7 | 5 | 5 | 6 | 6 | 6 | 5 | 71 |
| Total Snowfall (") | 3.6 | 4.3 | 2.9 | 0.6 | 0.0 | 0.0 | 0.0 | 0.0 | 0.0 | 0.1 | 0.9 | 2.6 | 15.0 |
| Days ≥ 1" Snow Depth | 5 | 4 | 1 | 0 | 0 | 0 | 0 | 0 | 0 | 0 | 1 | 2 | 13 |

## HELENA *Phillips County* ELEVATION 239 ft LAT/LONG 34° 33 ' N / 90° 38 ' W

| | JAN | FEB | MAR | APR | MAY | JUN | JUL | AUG | SEP | OCT | NOV | DEC | YEAR |
|---|---|---|---|---|---|---|---|---|---|---|---|---|---|
| Maximum Temp °F | 47.2 | 52.8 | 62.2 | 73.1 | 80.4 | 88.4 | 91.5 | 89.8 | 83.8 | 74.2 | 62.1 | 52.2 | 71.5 |
| Minimum Temp °F | 28.4 | 32.3 | 41.1 | 50.7 | 59.3 | 67.0 | 70.8 | 68.7 | 61.7 | 49.5 | 40.5 | 33.0 | 50.3 |
| Mean Temp °F | 37.9 | 42.6 | 51.7 | 61.9 | 69.8 | 77.7 | 81.2 | 79.3 | 72.7 | 61.8 | 51.3 | 42.6 | 60.9 |
| Days Max Temp ≥ 90 °F | 0 | 0 | 0 | 0 | 2 | 15 | 22 | 17 | 7 | 1 | 0 | 0 | 64 |
| Days Max Temp ≤ 32 °F | 4 | 2 | 0 | 0 | 0 | 0 | 0 | 0 | 0 | 0 | 0 | 1 | 7 |
| Days Min Temp ≤ 32 °F | 21 | 15 | 6 | 1 | 0 | 0 | 0 | 0 | 0 | 1 | 6 | 15 | 65 |
| Days Min Temp ≤ 0 °F | 0 | 0 | 0 | 0 | 0 | 0 | 0 | 0 | 0 | 0 | 0 | 0 | 0 |
| Heating Degree Days | 835 | 626 | 416 | 148 | 30 | 1 | 0 | 0 | 20 | 152 | 411 | 688 | 3327 |
| Cooling Degree Days | 0 | 0 | 12 | 70 | 206 | 430 | 547 | 496 | 295 | 67 | 10 | 1 | 2134 |
| Total Precipitation (") | 4.21 | 4.61 | 4.77 | 5.32 | 5.82 | 4.61 | 3.84 | 3.26 | 3.64 | 3.80 | 5.28 | 5.81 | 54.97 |
| Days ≥ 0.1" Precip | 7 | 7 | 7 | 7 | 7 | 6 | 5 | 5 | 5 | 5 | 7 | 7 | 75 |
| Total Snowfall (") | ***0.8*** | ***0.3*** | 0.1 | 0.0 | 0.0 | 0.0 | 0.0 | 0.0 | 0.0 | 0.0 | 0.0 | 0.0 | 1.2 |
| Days ≥ 1" Snow Depth | na | ***0*** | 0 | 0 | 0 | 0 | 0 | 0 | 0 | 0 | 0 | 0 | na |

## HOPE 3 NE *Hempstead County* ELEVATION 381 ft LAT/LONG 33° 43 ' N / 93° 33 ' W

| | JAN | FEB | MAR | APR | MAY | JUN | JUL | AUG | SEP | OCT | NOV | DEC | YEAR |
|---|---|---|---|---|---|---|---|---|---|---|---|---|---|
| Maximum Temp °F | 50.7 | 56.3 | 65.2 | 74.5 | 80.7 | 88.1 | 92.0 | 91.5 | 84.9 | 75.6 | 63.9 | 54.9 | 73.2 |
| Minimum Temp °F | 29.0 | 32.2 | 40.2 | 49.0 | 57.4 | 65.4 | 69.0 | 67.4 | 61.0 | 48.9 | 39.7 | 32.5 | 49.3 |
| Mean Temp °F | 39.9 | 44.3 | 52.7 | 61.8 | 69.1 | 76.8 | 80.5 | 79.5 | 73.0 | 62.3 | 51.8 | 43.7 | 61.3 |
| Days Max Temp ≥ 90 °F | 0 | 0 | 0 | 0 | 1 | 13 | 23 | 21 | 9 | 1 | 0 | 0 | 68 |
| Days Max Temp ≤ 32 °F | 2 | 1 | 0 | 0 | 0 | 0 | 0 | 0 | 0 | 0 | 0 | 1 | 4 |
| Days Min Temp ≤ 32 °F | 21 | 16 | 8 | 1 | 0 | 0 | 0 | 0 | 0 | 1 | 8 | 18 | 73 |
| Days Min Temp ≤ 0 °F | 0 | 0 | 0 | 0 | 0 | 0 | 0 | 0 | 0 | 0 | 0 | 0 | 0 |
| Heating Degree Days | 773 | 580 | 386 | 143 | 31 | 1 | 0 | 0 | 19 | 145 | 399 | 654 | 3131 |
| Cooling Degree Days | 0 | 1 | 11 | 49 | 160 | 361 | 491 | 473 | 274 | 67 | 12 | 2 | 1901 |
| Total Precipitation (") | 3.72 | 4.14 | 5.01 | 5.20 | 5.64 | 4.44 | 4.01 | 3.97 | 4.06 | 4.36 | 5.42 | 5.27 | 55.24 |
| Days ≥ 0.1" Precip | 6 | 6 | 7 | 7 | 7 | 7 | 5 | 5 | 5 | 5 | 6 | 7 | 73 |
| Total Snowfall (") | ***1.2*** | 1.0 | 0.1 | 0.0 | 0.0 | 0.0 | 0.0 | 0.0 | 0.0 | 0.0 | 0.0 | 0.4 | 2.7 |
| Days ≥ 1" Snow Depth | na | ***0*** | 0 | 0 | 0 | 0 | 0 | 0 | 0 | 0 | 0 | 0 | na |

## HOT SPRINGS 1 NNE *Garland County* ELEVATION 712 ft LAT/LONG 34° 31 ' N / 93° 3 ' W

| | JAN | FEB | MAR | APR | MAY | JUN | JUL | AUG | SEP | OCT | NOV | DEC | YEAR |
|---|---|---|---|---|---|---|---|---|---|---|---|---|---|
| Maximum Temp °F | 50.4 | 56.2 | 65.2 | 74.9 | 81.2 | 89.0 | 93.8 | 92.7 | 85.9 | 75.9 | 63.0 | 53.8 | 73.5 |
| Minimum Temp °F | 29.6 | 32.7 | 41.2 | 50.2 | 57.8 | 65.8 | 70.1 | 68.2 | 61.9 | 50.8 | 41.2 | 33.3 | 50.2 |
| Mean Temp °F | 40.0 | 44.5 | 53.2 | 62.6 | 69.5 | 77.4 | 82.0 | 80.5 | 73.9 | 63.4 | 52.1 | 43.6 | 61.9 |
| Days Max Temp ≥ 90 °F | 0 | 0 | 0 | 1 | 3 | 15 | 24 | 22 | 11 | 2 | 0 | 0 | 78 |
| Days Max Temp ≤ 32 °F | 2 | 1 | 0 | 0 | 0 | 0 | 0 | 0 | 0 | 0 | 0 | 1 | 4 |
| Days Min Temp ≤ 32 °F | 20 | 14 | 6 | 1 | 0 | 0 | 0 | 0 | 0 | 0 | 6 | 15 | 62 |
| Days Min Temp ≤ 0 °F | 0 | 0 | 0 | 0 | 0 | 0 | 0 | 0 | 0 | 0 | 0 | 0 | 0 |
| Heating Degree Days | 767 | 574 | 373 | 132 | 28 | 0 | 0 | 0 | 15 | 124 | 388 | 658 | 3059 |
| Cooling Degree Days | 0 | 1 | 14 | 62 | 172 | 391 | 548 | 512 | 302 | 73 | 8 | 1 | 2084 |
| Total Precipitation (") | 3.61 | 3.92 | 5.08 | 5.44 | 6.73 | 4.69 | 4.61 | 3.53 | 4.06 | 4.81 | 5.89 | 5.36 | 57.73 |
| Days ≥ 0.1" Precip | 6 | 6 | 7 | 7 | 8 | 6 | 6 | 5 | 6 | 6 | 6 | 7 | 76 |
| Total Snowfall (") | ***1.7*** | 0.6 | 0.3 | 0.0 | 0.0 | 0.0 | 0.0 | 0.0 | 0.0 | 0.0 | 0.2 | 0.2 | 3.0 |
| Days ≥ 1" Snow Depth | ***0*** | ***0*** | 0 | 0 | 0 | 0 | 0 | 0 | 0 | 0 | 0 | 0 | 0 |

## JONESBORO 4 N *Craighead County* ELEVATION 351 ft LAT/LONG 35° 50 ' N / 90° 42 ' W

| | JAN | FEB | MAR | APR | MAY | JUN | JUL | AUG | SEP | OCT | NOV | DEC | YEAR |
|---|---|---|---|---|---|---|---|---|---|---|---|---|---|
| Maximum Temp °F | 44.9 | 50.1 | 60.8 | 72.1 | 79.9 | 88.2 | 91.9 | 89.7 | 83.1 | 73.3 | 59.9 | 49.3 | 70.3 |
| Minimum Temp °F | 27.6 | 31.6 | 40.8 | 50.5 | 58.6 | 67.2 | 71.1 | 68.6 | 61.5 | 49.5 | 40.7 | 32.3 | 50.0 |
| Mean Temp °F | 36.3 | 40.9 | 50.8 | 61.3 | 69.3 | 77.7 | 81.5 | 79.2 | 72.3 | 61.5 | 50.3 | 40.8 | 60.2 |
| Days Max Temp ≥ 90 °F | 0 | 0 | 0 | 0 | 2 | 14 | 22 | 16 | 6 | 0 | 0 | 0 | 60 |
| Days Max Temp ≤ 32 °F | 5 | 3 | 0 | 0 | 0 | 0 | 0 | 0 | 0 | 0 | 0 | 2 | 10 |
| Days Min Temp ≤ 32 °F | 21 | 15 | 7 | 1 | 0 | 0 | 0 | 0 | 0 | 0 | 7 | 16 | 67 |
| Days Min Temp ≤ 0 °F | 0 | 0 | 0 | 0 | 0 | 0 | 0 | 0 | 0 | 0 | 0 | 0 | 0 |
| Heating Degree Days | 883 | 675 | 440 | 160 | 33 | 0 | 0 | 0 | 21 | 162 | 439 | 743 | 3556 |
| Cooling Degree Days | 0 | 0 | 8 | 49 | 164 | 393 | 530 | 466 | 257 | 54 | 5 | 1 | 1927 |
| Total Precipitation (") | 3.59 | 3.63 | 4.33 | 5.36 | 4.89 | 3.33 | 2.81 | 2.94 | 3.64 | 3.77 | 4.70 | 4.60 | 47.59 |
| Days ≥ 0.1" Precip | 6 | 5 | 7 | 7 | 7 | 5 | 5 | 4 | 5 | 5 | 6 | 7 | 69 |
| Total Snowfall (") | *3.1* | 2.2 | 1.5 | 0.0 | 0.0 | 0.0 | 0.0 | 0.0 | 0.0 | 0.0 | 0.2 | 0.7 | 7.7 |
| Days ≥ 1" Snow Depth | *1* | *1* | 0 | 0 | 0 | 0 | 0 | 0 | 0 | 0 | 0 | 0 | 2 |

## KEISER *Mississippi County* ELEVATION 232 ft LAT/LONG 35° 41 ' N / 90° 5 ' W

| | JAN | FEB | MAR | APR | MAY | JUN | JUL | AUG | SEP | OCT | NOV | DEC | YEAR |
|---|---|---|---|---|---|---|---|---|---|---|---|---|---|
| Maximum Temp °F | 44.3 | 49.0 | 59.6 | 70.9 | 79.7 | 88.5 | 91.3 | 88.9 | 82.7 | 72.9 | 59.9 | 49.3 | 69.8 |
| Minimum Temp °F | 27.1 | 30.5 | 39.6 | 49.6 | 58.2 | 66.7 | 70.1 | 67.1 | 59.9 | 47.3 | 39.5 | 31.4 | 48.9 |
| Mean Temp °F | 35.7 | 39.8 | 49.6 | 60.3 | 69.0 | 77.6 | 80.7 | 78.0 | 71.3 | 60.1 | 49.8 | 40.4 | 59.4 |
| Days Max Temp ≥ 90 °F | 0 | 0 | 0 | 0 | 3 | 15 | 21 | 14 | 6 | 0 | 0 | 0 | 59 |
| Days Max Temp ≤ 32 °F | 5 | 3 | 0 | 0 | 0 | 0 | 0 | 0 | 0 | 0 | 0 | 2 | 10 |
| Days Min Temp ≤ 32 °F | 22 | 16 | 8 | 1 | 0 | 0 | 0 | 0 | 0 | 1 | 8 | 18 | 74 |
| Days Min Temp ≤ 0 °F | 0 | 0 | 0 | 0 | 0 | 0 | 0 | 0 | 0 | 0 | 0 | 0 | 0 |
| Heating Degree Days | 900 | 705 | 477 | 186 | 40 | 1 | 0 | 0 | 29 | 193 | 455 | 757 | 3743 |
| Cooling Degree Days | 0 | 0 | 8 | 47 | 167 | 396 | 509 | 431 | 232 | 47 | 5 | 1 | 1843 |
| Total Precipitation (") | 3.58 | 3.69 | 4.65 | 5.03 | 5.56 | 3.89 | 3.52 | 2.96 | 4.07 | 3.20 | 4.57 | 4.97 | 49.69 |
| Days ≥ 0.1" Precip | 7 | 6 | 8 | 7 | 8 | 6 | 6 | 5 | 5 | 5 | 6 | 7 | 76 |
| Total Snowfall (") | *1.1* | *0.8* | 0.7 | 0.0 | 0.0 | 0.0 | 0.0 | 0.0 | 0.0 | 0.0 | 0.1 | 0.3 | 3.0 |
| Days ≥ 1" Snow Depth | na | *0* | 0 | 0 | 0 | 0 | 0 | 0 | 0 | 0 | 0 | 0 | na |

## KEO *Lonoke County* ELEVATION 230 ft LAT/LONG 34° 36 ' N / 92° 0 ' W

| | JAN | FEB | MAR | APR | MAY | JUN | JUL | AUG | SEP | OCT | NOV | DEC | YEAR |
|---|---|---|---|---|---|---|---|---|---|---|---|---|---|
| Maximum Temp °F | 48.5 | 54.3 | 64.1 | 74.0 | 80.6 | 87.9 | 90.8 | 89.1 | 83.2 | 74.2 | 62.0 | 52.7 | 71.8 |
| Minimum Temp °F | 30.7 | 34.7 | 42.9 | 51.9 | 60.0 | 67.6 | 70.7 | 68.7 | 62.2 | 50.9 | 42.2 | 34.7 | 51.4 |
| Mean Temp °F | 39.6 | 44.5 | 53.5 | 63.0 | 70.3 | 77.8 | 80.8 | 78.9 | 72.7 | 62.6 | 52.2 | 43.8 | 61.6 |
| Days Max Temp ≥ 90 °F | 0 | 0 | 0 | 0 | 1 | 12 | 20 | 16 | 6 | 0 | 0 | 0 | 55 |
| Days Max Temp ≤ 32 °F | 3 | 1 | 0 | 0 | 0 | 0 | 0 | 0 | 0 | 0 | 0 | 1 | 5 |
| Days Min Temp ≤ 32 °F | 19 | 12 | 5 | 0 | 0 | 0 | 0 | 0 | 0 | 0 | 6 | 14 | 56 |
| Days Min Temp ≤ 0 °F | 0 | 0 | 0 | 0 | 0 | 0 | 0 | 0 | 0 | 0 | 0 | 0 | 0 |
| Heating Degree Days | 780 | 573 | 364 | 124 | 20 | 0 | 0 | 0 | 17 | 134 | 387 | 653 | 3052 |
| Cooling Degree Days | 0 | 1 | 13 | 68 | 197 | 411 | 519 | 470 | 276 | 66 | 11 | 2 | 2034 |
| Total Precipitation (") | 3.50 | 3.72 | 4.65 | 5.07 | 5.15 | 3.72 | 3.64 | 2.46 | 3.22 | 3.86 | 4.55 | 4.87 | 48.41 |
| Days ≥ 0.1" Precip | 6 | 6 | 7 | 7 | 7 | 5 | 5 | 4 | 5 | 5 | 6 | 7 | 70 |
| Total Snowfall (") | 2.5 | 1.4 | 0.3 | 0.0 | 0.0 | 0.0 | 0.0 | 0.0 | 0.0 | 0.0 | 0.2 | 0.1 | 4.5 |
| Days ≥ 1" Snow Depth | 3 | 1 | 0 | 0 | 0 | 0 | 0 | 0 | 0 | 0 | 0 | 1 | 5 |

## LEAD HILL *Boone County* ELEVATION 712 ft LAT/LONG 36° 24 ' N / 92° 54 ' W

| | JAN | FEB | MAR | APR | MAY | JUN | JUL | AUG | SEP | OCT | NOV | DEC | YEAR |
|---|---|---|---|---|---|---|---|---|---|---|---|---|---|
| Maximum Temp °F | 47.6 | 52.7 | 62.9 | 72.7 | 78.9 | 86.5 | 92.3 | 90.7 | 82.9 | 73.4 | 60.9 | 50.8 | 71.0 |
| Minimum Temp °F | 24.1 | 27.6 | 36.1 | 45.3 | 54.1 | 62.4 | 67.1 | 64.8 | 57.8 | 46.2 | 36.8 | 28.4 | 45.9 |
| Mean Temp °F | 35.9 | 40.2 | 49.7 | 58.9 | 66.6 | 74.5 | 79.8 | 77.7 | 70.4 | 59.9 | 48.9 | 39.7 | 58.5 |
| Days Max Temp ≥ 90 °F | 0 | 0 | 0 | 0 | 1 | 10 | 22 | 19 | 6 | 1 | 0 | 0 | 59 |
| Days Max Temp ≤ 32 °F | 4 | 2 | 0 | 0 | 0 | 0 | 0 | 0 | 0 | 0 | 0 | 2 | 8 |
| Days Min Temp ≤ 32 °F | 24 | 19 | 13 | 3 | 0 | 0 | 0 | 0 | 0 | 3 | 11 | 20 | 93 |
| Days Min Temp ≤ 0 °F | 1 | 0 | 0 | 0 | 0 | 0 | 0 | 0 | 0 | 0 | 0 | 0 | 1 |
| Heating Degree Days | 897 | 697 | 477 | 211 | 59 | 4 | 0 | 1 | 35 | 195 | 481 | 779 | 3836 |
| Cooling Degree Days | 0 | 0 | 6 | 30 | 103 | 284 | 446 | 397 | 191 | 37 | 4 | 1 | 1499 |
| Total Precipitation (") | 2.30 | 2.79 | 4.07 | 4.23 | 4.71 | 4.53 | 3.07 | 3.23 | 4.01 | 3.69 | 4.65 | 3.63 | 44.91 |
| Days ≥ 0.1" Precip | 4 | 5 | 6 | 7 | 8 | 7 | 5 | 6 | 6 | 5 | 6 | 6 | 71 |
| Total Snowfall (") | na | *2.5* | *1.7* | 0.0 | 0.0 | 0.0 | 0.0 | 0.0 | 0.0 | 0.0 | 0.2 | *1.1* | na |
| Days ≥ 1" Snow Depth | na | na | *0* | 0 | 0 | 0 | 0 | 0 | 0 | 0 | 0 | *0* | na |

## LEOLA *Grant County* ELEVATION 261 ft LAT/LONG 34° 10 ' N / 92° 35 ' W

| | JAN | FEB | MAR | APR | MAY | JUN | JUL | AUG | SEP | OCT | NOV | DEC | YEAR |
|---|---|---|---|---|---|---|---|---|---|---|---|---|---|
| Maximum Temp °F | 51.6 | 57.5 | 66.5 | 75.7 | 81.4 | 88.6 | 92.2 | *91.1* | *84.9* | *75.1* | 63.6 | 55.3 | 73.6 |
| Minimum Temp °F | 30.0 | 33.3 | 41.3 | 49.8 | 57.5 | 65.4 | 69.3 | *67.2* | *61.2* | *48.3* | 40.2 | 33.6 | 49.8 |
| Mean Temp °F | 40.8 | 45.5 | 53.9 | 62.8 | 69.5 | 77.0 | 80.7 | *79.2* | *73.1* | *61.7* | 51.9 | 44.4 | 61.7 |
| Days Max Temp ≥ 90 °F | 0 | 0 | 0 | 0 | 2 | 15 | 22 | 19 | *8* | *1* | 0 | 0 | 67 |
| Days Max Temp ≤ 32 °F | 2 | 1 | 0 | 0 | 0 | 0 | 0 | 0 | *0* | *0* | 0 | 1 | 4 |
| Days Min Temp ≤ 32 °F | 20 | 15 | 7 | 1 | 0 | 0 | 0 | 0 | *0* | *2* | 9 | 16 | 70 |
| Days Min Temp ≤ 0 °F | 0 | 0 | 0 | 0 | 0 | 0 | 0 | 0 | 0 | *0* | 0 | 0 | 0 |
| Heating Degree Days | 745 | 547 | 354 | 127 | 28 | 0 | 0 | *0* | *17* | *159* | 396 | 633 | 3006 |
| Cooling Degree Days | 0 | 2 | 17 | 63 | 173 | 376 | 499 | na | na | na | 11 | 3 | na |
| Total Precipitation (") | 4.00 | 4.07 | 4.90 | 4.96 | 5.82 | 3.94 | 4.15 | *3.51* | *4.02* | *4.47* | 4.95 | 5.33 | 54.12 |
| Days ≥ 0.1" Precip | 6 | 6 | 7 | 6 | 8 | 6 | 6 | 5 | *5* | *5* | 6 | 7 | 73 |
| Total Snowfall (") | 1.7 | 1.3 | 0.3 | 0.0 | 0.0 | 0.0 | 0.0 | 0.0 | 0.0 | *0.0* | 0.2 | 0.4 | 3.9 |
| Days ≥ 1" Snow Depth | 2 | 1 | 0 | 0 | 0 | 0 | 0 | 0 | 0 | *0* | 0 | 1 | 4 |

## LITTLE ROCK ADAMS FD *Pulaski County* ELEVATION 276 ft LAT/LONG 34° 44 ' N / 92° 14 ' W

| | JAN | FEB | MAR | APR | MAY | JUN | JUL | AUG | SEP | OCT | NOV | DEC | YEAR |
|---|---|---|---|---|---|---|---|---|---|---|---|---|---|
| Maximum Temp °F | 48.9 | 54.2 | 63.8 | 73.4 | 80.8 | 89.2 | 92.4 | 91.2 | 84.5 | 74.5 | 62.3 | 52.9 | 72.3 |
| Minimum Temp °F | 30.1 | 33.6 | 42.4 | 50.8 | 59.1 | 67.7 | 71.6 | 69.9 | 63.4 | 51.1 | 41.6 | 34.0 | 51.3 |
| Mean Temp °F | 39.5 | 43.9 | 53.1 | 62.1 | 70.0 | 78.5 | 82.0 | 80.5 | 74.0 | 62.8 | 52.0 | 43.5 | 61.8 |
| Days Max Temp ≥ 90 °F | 0 | 0 | 0 | 0 | 3 | 16 | 22 | 19 | 9 | 1 | 0 | 0 | 70 |
| Days Max Temp ≤ 32 °F | 3 | 1 | 0 | 0 | 0 | 0 | 0 | 0 | 0 | 0 | 0 | 1 | 5 |
| Days Min Temp ≤ 32 °F | 19 | 13 | 5 | 1 | 0 | 0 | 0 | 0 | 0 | 0 | 5 | 14 | 57 |
| Days Min Temp ≤ 0 °F | 0 | 0 | 0 | 0 | 0 | 0 | 0 | 0 | 0 | 0 | 0 | 0 | 0 |
| Heating Degree Days | 783 | 589 | 375 | 137 | 24 | 0 | 0 | 0 | 13 | 130 | 393 | 662 | 3106 |
| Cooling Degree Days | 0 | 1 | 13 | 56 | 189 | 424 | 556 | 515 | 306 | 71 | 10 | 3 | 2144 |
| Total Precipitation (") | 3.72 | 3.42 | 4.69 | 5.63 | 5.35 | 3.76 | 3.57 | 3.45 | 3.95 | 4.11 | 5.41 | 4.89 | 51.95 |
| Days ≥ 0.1" Precip | 6 | 5 | 7 | 7 | 8 | 6 | 6 | 5 | 6 | 5 | 6 | 7 | 74 |
| Total Snowfall (") | 2.3 | 1.7 | 0.7 | 0.0 | 0.0 | 0.0 | 0.0 | 0.0 | 0.0 | 0.0 | 0.3 | 0.1 | 5.1 |
| Days ≥ 1" Snow Depth | 2 | 1 | 0 | 0 | 0 | 0 | 0 | 0 | 0 | 0 | 0 | 0 | 3 |

## MAGNOLIA 3 N *Columbia County* ELEVATION 322 ft LAT/LONG 33° 17 ' N / 93° 14 ' W

| | JAN | FEB | MAR | APR | MAY | JUN | JUL | AUG | SEP | OCT | NOV | DEC | YEAR |
|---|---|---|---|---|---|---|---|---|---|---|---|---|---|
| Maximum Temp °F | 53.9 | 59.8 | 68.4 | 76.5 | 82.0 | 88.7 | 92.1 | 91.8 | 85.7 | 76.5 | 65.6 | 57.2 | 74.9 |
| Minimum Temp °F | 32.4 | 35.4 | 43.2 | 51.2 | 58.6 | 66.0 | 69.2 | 67.7 | 62.1 | 50.8 | 42.1 | 35.4 | 51.2 |
| Mean Temp °F | 43.2 | 47.6 | 55.9 | 63.9 | 70.3 | 77.4 | 80.7 | 79.7 | 73.9 | 63.7 | 53.9 | 46.4 | 63.1 |
| Days Max Temp ≥ 90 °F | 0 | 0 | 0 | 0 | 2 | 14 | 23 | 22 | 10 | 1 | 0 | 0 | 72 |
| Days Max Temp ≤ 32 °F | 1 | 0 | 0 | 0 | 0 | 0 | 0 | 0 | 0 | 0 | 0 | 1 | 2 |
| Days Min Temp ≤ 32 °F | 17 | 13 | 6 | 1 | 0 | 0 | 0 | 0 | 0 | 1 | 7 | 14 | 59 |
| Days Min Temp ≤ 0 °F | 0 | 0 | 0 | 0 | 0 | 0 | 0 | 0 | 0 | 0 | 0 | 0 | 0 |
| Heating Degree Days | 672 | 488 | 300 | 106 | 19 | 1 | 0 | 0 | 14 | 120 | 345 | 575 | 2640 |
| Cooling Degree Days | 0 | 5 | 21 | 76 | 199 | 405 | 509 | 498 | 309 | 89 | 20 | 4 | 2135 |
| Total Precipitation (") | 4.01 | 4.25 | 5.02 | 4.53 | 5.11 | 4.56 | 3.48 | 3.56 | 3.34 | 4.10 | 5.04 | 5.05 | 52.05 |
| Days ≥ 0.1" Precip | 6 | 6 | 7 | 6 | 7 | 6 | 6 | 5 | 5 | 5 | 6 | 7 | 72 |
| Total Snowfall (") | 0.9 | 0.9 | 0.1 | 0.0 | 0.0 | 0.0 | 0.0 | 0.0 | 0.0 | 0.0 | 0.0 | 0.4 | 2.3 |
| Days ≥ 1" Snow Depth | *0* | 0 | 0 | 0 | 0 | 0 | 0 | 0 | 0 | 0 | 0 | 0 | 0 |

## MALVERN *Hot Spring County* ELEVATION 312 ft LAT/LONG 34° 22 ' N / 92° 49 ' W

| | JAN | FEB | MAR | APR | MAY | JUN | JUL | AUG | SEP | OCT | NOV | DEC | YEAR |
|---|---|---|---|---|---|---|---|---|---|---|---|---|---|
| Maximum Temp °F | 51.9 | 58.2 | 67.5 | 76.8 | 82.2 | 89.2 | 93.0 | 91.3 | 84.5 | 75.0 | 63.6 | 54.8 | 74.0 |
| Minimum Temp °F | 28.9 | 32.0 | 40.1 | 48.6 | 56.4 | 64.0 | 67.6 | 66.1 | 60.5 | 48.4 | 39.6 | 32.5 | 48.7 |
| Mean Temp °F | 40.4 | 45.1 | 53.8 | 62.7 | 69.3 | 76.7 | 80.3 | 78.7 | 72.5 | 61.7 | 51.6 | 43.7 | 61.4 |
| Days Max Temp ≥ 90 °F | 0 | 0 | 0 | 1 | 3 | 15 | 24 | 20 | 7 | 0 | 0 | 0 | 70 |
| Days Max Temp ≤ 32 °F | 2 | 1 | 0 | 0 | 0 | 0 | 0 | 0 | 0 | 0 | 0 | 1 | 4 |
| Days Min Temp ≤ 32 °F | 21 | 16 | 8 | 2 | 0 | 0 | 0 | 0 | 0 | 1 | 9 | 17 | 74 |
| Days Min Temp ≤ 0 °F | 0 | 0 | 0 | 0 | 0 | 0 | 0 | 0 | 0 | 0 | 0 | 0 | 0 |
| Heating Degree Days | 757 | 555 | 353 | 124 | 25 | 0 | 0 | 0 | 17 | 149 | 402 | 656 | 3038 |
| Cooling Degree Days | 0 | 1 | 13 | 57 | 168 | 376 | 499 | 461 | 266 | 56 | 9 | 2 | 1908 |
| Total Precipitation (") | 3.74 | 3.79 | 5.14 | 5.39 | 6.14 | 4.45 | 4.59 | 3.42 | 3.87 | 4.68 | 5.53 | 5.48 | 56.22 |
| Days ≥ 0.1" Precip | 6 | 6 | 7 | 6 | 8 | 6 | 6 | 5 | 6 | 6 | 6 | 7 | 75 |
| Total Snowfall (") | 1.3 | 1.5 | 0.3 | 0.0 | 0.0 | 0.0 | 0.0 | 0.0 | 0.0 | 0.0 | 0.2 | 0.2 | 3.5 |
| Days ≥ 1" Snow Depth | 2 | 1 | 0 | 0 | 0 | 0 | 0 | 0 | 0 | 0 | 0 | 1 | 4 |

## MAMMOTH SPRING *Fulton County* ELEVATION 600 ft LAT/LONG 36° 29' N / 91° 32' W

| | JAN | FEB | MAR | APR | MAY | JUN | JUL | AUG | SEP | OCT | NOV | DEC | YEAR |
|---|---|---|---|---|---|---|---|---|---|---|---|---|---|
| Maximum Temp °F | 46.0 | 51.7 | 62.3 | 73.0 | 79.2 | 86.4 | 91.6 | 89.9 | 82.6 | 73.3 | 60.3 | 50.1 | 70.5 |
| Minimum Temp °F | 22.6 | 26.3 | 35.8 | 45.3 | 53.1 | 61.5 | 65.5 | 63.7 | 56.8 | 44.6 | 35.4 | 27.0 | 44.8 |
| Mean Temp °F | 34.3 | 39.0 | 49.1 | 59.2 | 66.1 | 74.0 | 78.6 | 76.8 | 69.8 | 58.9 | 47.9 | 38.6 | 57.7 |
| Days Max Temp ≥ 90 °F | 0 | 0 | 0 | 1 | 1 | 9 | 20 | 17 | 5 | 1 | 0 | 0 | 54 |
| Days Max Temp ≤ 32 °F | 5 | 2 | 0 | 0 | 0 | 0 | 0 | 0 | 0 | 0 | 0 | 2 | 9 |
| Days Min Temp ≤ 32 °F | 25 | 20 | 13 | 4 | 0 | 0 | 0 | 0 | 0 | 4 | 13 | 22 | 101 |
| Days Min Temp ≤ 0 °F | 1 | 0 | 0 | 0 | 0 | 0 | 0 | 0 | 0 | 0 | 0 | 1 | 2 |
| Heating Degree Days | 945 | 726 | 494 | 206 | 65 | 3 | 0 | 1 | 40 | 216 | 511 | 812 | 4019 |
| Cooling Degree Days | 0 | 0 | 6 | 35 | 110 | 283 | 441 | 393 | 193 | 34 | 4 | 0 | 1499 |
| Total Precipitation (") | 2.86 | 2.87 | 3.97 | 4.50 | 4.03 | 3.72 | 3.35 | 3.48 | 4.28 | 3.60 | 4.96 | 4.36 | 45.98 |
| Days ≥ 0.1" Precip | *5* | 5 | 7 | 7 | 7 | 6 | 6 | 5 | 5 | 5 | 6 | 6 | 70 |
| Total Snowfall (") | *2.3* | *2.2* | 0.8 | 0.1 | 0.0 | 0.0 | 0.0 | 0.0 | 0.0 | 0.0 | 0.9 | 0.9 | 7.2 |
| Days ≥ 1" Snow Depth | na | *1* | *0* | 0 | 0 | 0 | 0 | 0 | 0 | 0 | 0 | 1 | na |

## MARIANNA 2 S *Lee County* ELEVATION 230 ft LAT/LONG 34° 44' N / 90° 49' W

| | JAN | FEB | MAR | APR | MAY | JUN | JUL | AUG | SEP | OCT | NOV | DEC | YEAR |
|---|---|---|---|---|---|---|---|---|---|---|---|---|---|
| Maximum Temp °F | 47.9 | 53.6 | 63.0 | 73.5 | 81.2 | 89.2 | 92.0 | 90.2 | 84.6 | 75.0 | 62.4 | 52.5 | 72.1 |
| Minimum Temp °F | 29.2 | 33.3 | 41.9 | 51.1 | 59.4 | 67.2 | 70.2 | 67.6 | 61.2 | 49.6 | 41.2 | 33.4 | 50.4 |
| Mean Temp °F | 38.6 | 43.5 | 52.4 | 62.3 | 70.3 | 78.2 | 81.1 | 78.9 | 72.9 | 62.3 | 51.8 | 43.0 | 61.3 |
| Days Max Temp ≥ 90 °F | 0 | 0 | 0 | 0 | 3 | 16 | 22 | 19 | 8 | 1 | 0 | 0 | 69 |
| Days Max Temp ≤ 32 °F | 4 | 1 | 0 | 0 | 0 | 0 | 0 | 0 | 0 | 0 | 0 | 1 | 6 |
| Days Min Temp ≤ 32 °F | 21 | 14 | 6 | 1 | 0 | 0 | 0 | 0 | 0 | 1 | 6 | 15 | 64 |
| Days Min Temp ≤ 0 °F | 0 | 0 | 0 | 0 | 0 | 0 | 0 | 0 | 0 | 0 | 0 | 0 | 0 |
| Heating Degree Days | 813 | 602 | 393 | 139 | 24 | 0 | 0 | 0 | 19 | 146 | 397 | 677 | 3210 |
| Cooling Degree Days | 0 | 0 | 10 | 54 | 186 | 408 | 515 | 452 | 264 | 64 | 9 | 2 | 1964 |
| Total Precipitation (") | 3.91 | 4.03 | 5.42 | 5.32 | 5.65 | 4.15 | 3.83 | 3.17 | 3.48 | 3.80 | 5.05 | 5.53 | 53.34 |
| Days ≥ 0.1" Precip | 6 | 6 | 7 | 7 | 7 | 6 | 5 | 5 | 5 | 5 | 6 | 7 | 72 |
| Total Snowfall (") | *1.6* | 0.7 | 0.7 | 0.0 | 0.0 | 0.0 | 0.0 | 0.0 | 0.0 | 0.0 | 0.1 | 0.1 | 3.2 |
| Days ≥ 1" Snow Depth | *2* | 1 | 0 | 0 | 0 | 0 | 0 | 0 | 0 | 0 | 0 | 0 | 3 |

## MENA *Polk County* ELEVATION 1211 ft LAT/LONG 34° 35' N / 94° 15' W

| | JAN | FEB | MAR | APR | MAY | JUN | JUL | AUG | SEP | OCT | NOV | DEC | YEAR |
|---|---|---|---|---|---|---|---|---|---|---|---|---|---|
| Maximum Temp °F | 48.9 | 54.3 | 63.0 | 72.1 | 77.9 | 85.3 | 89.8 | 89.3 | 82.5 | 72.7 | 60.8 | 51.9 | 70.7 |
| Minimum Temp °F | 27.6 | 31.0 | 39.4 | 48.1 | 56.0 | 63.9 | 67.3 | 65.8 | 59.8 | 48.7 | 39.3 | 31.2 | 48.2 |
| Mean Temp °F | 38.3 | 42.7 | 51.2 | 60.1 | 67.0 | 74.6 | 78.6 | 77.6 | 71.2 | 60.7 | 50.1 | 41.6 | 59.5 |
| Days Max Temp ≥ 90 °F | 0 | 0 | 0 | 0 | 0 | 6 | 17 | 17 | 5 | 0 | 0 | 0 | 45 |
| Days Max Temp ≤ 32 °F | 3 | 1 | 0 | 0 | 0 | 0 | 0 | 0 | 0 | 0 | 0 | 1 | 5 |
| Days Min Temp ≤ 32 °F | 22 | 16 | 9 | 1 | 0 | 0 | 0 | 0 | 0 | 1 | 9 | 18 | 76 |
| Days Min Temp ≤ 0 °F | 0 | 0 | 0 | 0 | 0 | 0 | 0 | 0 | 0 | 0 | 0 | 0 | 0 |
| Heating Degree Days | 821 | 623 | 428 | 176 | 49 | 3 | 0 | 1 | 27 | 172 | 445 | 721 | 3466 |
| Cooling Degree Days | 0 | 0 | 5 | 29 | 100 | 290 | 414 | 394 | 212 | 40 | 5 | 0 | 1489 |
| Total Precipitation (") | 3.19 | 3.67 | 5.10 | 5.54 | 6.74 | 4.72 | 5.20 | 2.48 | 5.08 | 5.54 | 5.33 | 4.85 | 57.44 |
| Days ≥ 0.1" Precip | 5 | 6 | 7 | 7 | 9 | 7 | 6 | 5 | 6 | 6 | 6 | 6 | 76 |
| Total Snowfall (") | *1.5* | *1.7* | 0.2 | 0.0 | 0.0 | 0.0 | 0.0 | 0.0 | 0.0 | 0.0 | 0.2 | 0.2 | 3.8 |
| Days ≥ 1" Snow Depth | 2 | *1* | 0 | 0 | 0 | 0 | 0 | 0 | 0 | 0 | 0 | 0 | 3 |

## MONTICELLO 3 SW *Drew County* ELEVATION 249 ft LAT/LONG 33° 35' N / 91° 50' W

| | JAN | FEB | MAR | APR | MAY | JUN | JUL | AUG | SEP | OCT | NOV | DEC | YEAR |
|---|---|---|---|---|---|---|---|---|---|---|---|---|---|
| Maximum Temp °F | 50.9 | 56.3 | 65.2 | 74.1 | 81.1 | 88.2 | 91.7 | 90.8 | 85.3 | 75.5 | 64.2 | 55.4 | 73.2 |
| Minimum Temp °F | 30.3 | 33.5 | 42.3 | 50.1 | 58.3 | 66.1 | 69.2 | 67.5 | 61.8 | 49.7 | 42.0 | 34.2 | 50.4 |
| Mean Temp °F | 40.7 | 44.9 | 53.8 | 62.1 | 69.7 | 77.1 | 80.5 | 79.2 | 73.6 | 62.6 | 53.3 | 44.8 | 61.9 |
| Days Max Temp ≥ 90 °F | 0 | 0 | 0 | 0 | 1 | 13 | 22 | 20 | 10 | 1 | 0 | 0 | 67 |
| Days Max Temp ≤ 32 °F | 2 | 1 | 0 | 0 | 0 | 0 | 0 | 0 | 0 | 0 | 0 | 1 | 4 |
| Days Min Temp ≤ 32 °F | 19 | 14 | 5 | 1 | 0 | 0 | 0 | 0 | 0 | 1 | 6 | 15 | 61 |
| Days Min Temp ≤ 0 °F | 0 | 0 | 0 | 0 | 0 | 0 | 0 | 0 | 0 | 0 | 0 | 0 | 0 |
| Heating Degree Days | 749 | 562 | 359 | 144 | 29 | 1 | 0 | 0 | 17 | 141 | 360 | 621 | 2983 |
| Cooling Degree Days | 0 | 2 | 17 | 63 | 180 | 380 | 489 | 461 | 288 | 73 | 14 | 3 | 1970 |
| Total Precipitation (") | 4.70 | 4.99 | 5.75 | 5.03 | 4.91 | 4.09 | 4.42 | 3.61 | 3.36 | 4.80 | 5.03 | 5.06 | 55.75 |
| Days ≥ 0.1" Precip | 7 | 6 | 7 | 6 | 7 | 6 | 6 | 5 | 5 | 5 | 6 | 7 | 73 |
| Total Snowfall (") | 0.9 | 0.3 | 0.5 | 0.0 | 0.0 | 0.0 | 0.0 | 0.0 | 0.0 | 0.0 | 0.0 | 0.0 | 1.7 |
| Days ≥ 1" Snow Depth | *0* | *0* | 0 | 0 | 0 | 0 | 0 | 0 | 0 | 0 | 0 | 0 | 0 |

## MORRILTON *Conway County* ELEVATION 331 ft LAT/LONG 35° 8 ' N / 92° 44 ' W

| | JAN | FEB | MAR | APR | MAY | JUN | JUL | AUG | SEP | OCT | NOV | DEC | YEAR |
|---|---|---|---|---|---|---|---|---|---|---|---|---|---|
| Maximum Temp °F | 50.1 | 56.2 | 65.2 | 75.6 | 81.9 | 89.6 | 94.1 | 92.3 | 85.8 | 76.0 | 62.5 | 53.4 | 73.6 |
| Minimum Temp °F | 29.0 | 32.4 | 40.9 | 50.2 | 58.0 | 65.8 | 69.4 | 66.8 | 61.1 | 49.2 | 39.7 | 32.4 | 49.6 |
| Mean Temp °F | 39.6 | 44.3 | 53.1 | 62.9 | 70.0 | 77.7 | 81.8 | 79.6 | 73.5 | 62.6 | 51.1 | 42.9 | 61.6 |
| Days Max Temp ≥ 90 °F | 0 | 0 | 0 | 1 | 3 | 17 | 25 | 22 | 10 | 1 | 0 | 0 | 79 |
| Days Max Temp ≤ 32 °F | 2 | 1 | 0 | 0 | 0 | 0 | 0 | 0 | 0 | 0 | 0 | 1 | 4 |
| Days Min Temp ≤ 32 °F | 21 | 15 | 6 | 1 | 0 | 0 | 0 | 0 | 0 | 1 | 8 | 17 | 69 |
| Days Min Temp ≤ 0 °F | 0 | 0 | 0 | 0 | 0 | 0 | 0 | 0 | 0 | 0 | 0 | 0 | 0 |
| Heating Degree Days | 782 | 576 | 375 | 121 | 24 | 0 | 0 | 0 | 15 | 133 | 416 | 679 | 3121 |
| Cooling Degree Days | 0 | na | 7 | 53 | na | na | 539 | na | na | na | na | 1 | na |
| Total Precipitation (") | 3.13 | 3.18 | 4.34 | 4.35 | 5.25 | 4.22 | 2.98 | 2.91 | 3.92 | 4.50 | 5.07 | 4.73 | 48.58 |
| Days ≥ 0.1" Precip | 6 | 5 | 7 | 6 | 7 | 6 | 5 | 4 | 5 | 5 | 6 | 6 | 68 |
| Total Snowfall (") | 1.8 | 1.2 | 0.8 | 0.0 | 0.0 | 0.0 | 0.0 | 0.0 | 0.0 | 0.0 | 0.2 | 0.5 | 4.5 |
| Days ≥ 1" Snow Depth | na | 1 | 0 | 0 | 0 | 0 | 0 | 0 | 0 | 0 | 0 | 0 | na |

## MOUNT IDA 3 SE *Montgomery County* ELEVATION 659 ft LAT/LONG 34° 33 ' N / 93° 38 ' W

| | JAN | FEB | MAR | APR | MAY | JUN | JUL | AUG | SEP | OCT | NOV | DEC | YEAR |
|---|---|---|---|---|---|---|---|---|---|---|---|---|---|
| Maximum Temp °F | 49.6 | 54.8 | 63.5 | 73.3 | 79.2 | 87.0 | 92.0 | 91.0 | 83.9 | 73.6 | 62.4 | 53.1 | 72.0 |
| Minimum Temp °F | 25.2 | 28.2 | 36.9 | 45.5 | 53.6 | 62.2 | 66.4 | 64.4 | 57.8 | 45.0 | 36.1 | 28.7 | 45.8 |
| Mean Temp °F | 37.4 | 41.5 | 50.2 | 59.4 | 66.5 | 74.6 | 79.2 | 77.7 | 70.9 | 59.3 | 49.3 | 41.0 | 58.9 |
| Days Max Temp ≥ 90 °F | 0 | 0 | 0 | 0 | 1 | 11 | 22 | 20 | 7 | 1 | 0 | 0 | 62 |
| Days Max Temp ≤ 32 °F | 3 | 1 | 0 | 0 | 0 | 0 | 0 | 0 | 0 | 0 | 0 | 1 | 5 |
| Days Min Temp ≤ 32 °F | 25 | 20 | 12 | 4 | 0 | 0 | 0 | 0 | 0 | 4 | 14 | 21 | 100 |
| Days Min Temp ≤ 0 °F | 0 | 0 | 0 | 0 | 0 | 0 | 0 | 0 | 0 | 0 | 0 | 0 | 0 |
| Heating Degree Days | 849 | 657 | 459 | 200 | 61 | 4 | 0 | 1 | 33 | 208 | 471 | 740 | 3683 |
| Cooling Degree Days | 0 | 0 | 8 | 34 | 105 | 305 | 453 | 418 | 219 | 37 | 7 | 1 | 1587 |
| Total Precipitation (") | 3.49 | 3.94 | 5.37 | 5.28 | 6.71 | 4.76 | 4.45 | 3.04 | 4.81 | 5.14 | 5.63 | 5.51 | 58.13 |
| Days ≥ 0.1" Precip | 6 | 6 | 7 | 7 | 8 | 6 | 7 | 5 | 6 | 6 | 6 | 7 | 77 |
| Total Snowfall (") | 2.3 | 1.9 | 0.4 | 0.1 | 0.0 | 0.0 | 0.0 | 0.0 | 0.0 | 0.0 | 0.2 | 0.4 | 5.3 |
| Days ≥ 1" Snow Depth | 3 | 2 | 0 | 0 | 0 | 0 | 0 | 0 | 0 | 0 | 0 | 1 | 6 |

## MOUNTAIN HOME 1 NNW *Baxter County* ELEVATION 801 ft LAT/LONG 36° 20 ' N / 92° 23 ' W

| | JAN | FEB | MAR | APR | MAY | JUN | JUL | AUG | SEP | OCT | NOV | DEC | YEAR |
|---|---|---|---|---|---|---|---|---|---|---|---|---|---|
| Maximum Temp °F | 45.2 | 50.7 | 60.6 | 71.3 | 77.8 | 85.4 | 90.8 | 89.4 | 81.9 | 72.4 | 58.8 | 48.9 | 69.4 |
| Minimum Temp °F | 23.8 | 27.4 | 36.6 | 46.3 | 54.6 | 62.8 | 67.4 | 65.3 | 58.4 | 46.5 | 37.0 | 28.0 | 46.2 |
| Mean Temp °F | 34.5 | 39.1 | 48.6 | 58.8 | 66.3 | 74.1 | 79.1 | 77.4 | 70.2 | 59.5 | 47.9 | 38.5 | 57.8 |
| Days Max Temp ≥ 90 °F | 0 | 0 | 0 | 0 | 1 | 7 | 19 | 16 | 5 | 0 | 0 | 0 | 48 |
| Days Max Temp ≤ 32 °F | 6 | 3 | 0 | 0 | 0 | 0 | 0 | 0 | 0 | 0 | 0 | 3 | 12 |
| Days Min Temp ≤ 32 °F | 25 | 19 | 12 | 2 | 0 | 0 | 0 | 0 | 0 | 2 | 10 | 21 | 91 |
| Days Min Temp ≤ 0 °F | 1 | 0 | 0 | 0 | 0 | 0 | 0 | 0 | 0 | 0 | 0 | 0 | 1 |
| Heating Degree Days | 938 | 725 | 508 | 216 | 61 | 3 | 1 | 1 | 34 | 202 | 507 | 815 | 4011 |
| Cooling Degree Days | 0 | 0 | 6 | 38 | 105 | 293 | 454 | 413 | 201 | 33 | 2 | 0 | 1545 |
| Total Precipitation (") | 2.69 | 3.04 | 4.20 | 4.45 | 4.75 | 4.12 | 2.68 | 2.94 | 4.39 | 3.50 | 4.99 | 4.02 | 45.77 |
| Days ≥ 0.1" Precip | 5 | 5 | 7 | 7 | 8 | 6 | 5 | 5 | 6 | 5 | 6 | 6 | 71 |
| Total Snowfall (") | 3.8 | 3.6 | 2.1 | 0.1 | 0.0 | 0.0 | 0.0 | 0.0 | 0.0 | 0.1 | 0.7 | 2.1 | 12.5 |
| Days ≥ 1" Snow Depth | na | na | 0 | 0 | 0 | 0 | 0 | 0 | 0 | 0 | 0 | 0 | na |

## MOUNTAIN VIEW *Stone County* ELEVATION 771 ft LAT/LONG 35° 52 ' N / 92° 8 ' W

| | JAN | FEB | MAR | APR | MAY | JUN | JUL | AUG | SEP | OCT | NOV | DEC | YEAR |
|---|---|---|---|---|---|---|---|---|---|---|---|---|---|
| Maximum Temp °F | 46.0 | 51.4 | 61.0 | 71.8 | 78.5 | 86.5 | 91.5 | 89.9 | 83.0 | 73.3 | 60.7 | 50.2 | 70.3 |
| Minimum Temp °F | 24.1 | 27.9 | 36.9 | 46.5 | 54.2 | 63.0 | 67.6 | 65.2 | 58.2 | 46.0 | 37.7 | 28.9 | 46.4 |
| Mean Temp °F | 35.1 | 39.7 | 49.0 | 59.1 | 66.4 | 74.8 | 79.6 | 77.6 | 70.6 | 59.7 | 49.2 | 39.6 | 58.4 |
| Days Max Temp ≥ 90 °F | 0 | 0 | 0 | 0 | 1 | 10 | 21 | 17 | 7 | 1 | 0 | 0 | 57 |
| Days Max Temp ≤ 32 °F | 5 | 2 | 0 | 0 | 0 | 0 | 0 | 0 | 0 | 0 | 0 | 2 | 9 |
| Days Min Temp ≤ 32 °F | 25 | 19 | 13 | 3 | 0 | 0 | 0 | 0 | 0 | 3 | 11 | 20 | 94 |
| Days Min Temp ≤ 0 °F | 1 | 0 | 0 | 0 | 0 | 0 | 0 | 0 | 0 | 0 | 0 | 0 | 1 |
| Heating Degree Days | 920 | 708 | 499 | 210 | 65 | 4 | 0 | 1 | 36 | 207 | 473 | 781 | 3904 |
| Cooling Degree Days | 0 | 0 | 8 | 43 | 119 | 314 | 476 | 420 | 224 | 44 | 7 | 0 | 1655 |
| Total Precipitation (") | 3.08 | 3.24 | 4.76 | 5.01 | 5.10 | 3.69 | 3.43 | 3.60 | 4.92 | 4.08 | 5.03 | 4.78 | 50.72 |
| Days ≥ 0.1" Precip | 5 | 5 | 7 | 7 | 8 | 6 | 5 | 5 | 6 | 5 | 6 | 6 | 71 |
| Total Snowfall (") | 1.5 | 1.9 | 0.7 | 0.0 | 0.0 | 0.0 | 0.0 | 0.0 | 0.0 | 0.0 | 1.0 | 0.7 | 5.8 |
| Days ≥ 1" Snow Depth | 5 | 3 | 1 | 0 | 0 | 0 | 0 | 0 | 0 | 0 | 0 | 2 | 11 |

## NASHVILLE *Howard County* ELEVATION 550 ft LAT/LONG 34° 0 ' N / 93° 56 ' W

| | JAN | FEB | MAR | APR | MAY | JUN | JUL | AUG | SEP | OCT | NOV | DEC | YEAR |
|---|---|---|---|---|---|---|---|---|---|---|---|---|---|
| Maximum Temp °F | 50.4 | 55.5 | 64.1 | 73.4 | 79.8 | 87.2 | 91.4 | 91.1 | 84.5 | 75.0 | 63.2 | 54.2 | 72.5 |
| Minimum Temp °F | 29.9 | 33.1 | 41.0 | 49.6 | 57.8 | 65.7 | 69.3 | 68.3 | 62.7 | 51.1 | 41.3 | 33.7 | 50.3 |
| Mean Temp °F | 40.2 | 44.4 | 52.6 | 61.5 | 68.8 | 76.5 | 80.4 | 79.8 | 73.6 | 63.0 | 52.3 | 44.0 | 61.4 |
| Days Max Temp ≥ 90 °F | 0 | 0 | 0 | 0 | 1 | 10 | 21 | 20 | 8 | 1 | 0 | 0 | 61 |
| Days Max Temp ≤ 32 °F | 2 | 1 | 0 | 0 | 0 | 0 | 0 | 0 | 0 | 0 | 0 | 1 | 4 |
| Days Min Temp ≤ 32 °F | 20 | 14 | 6 | 1 | 0 | 0 | 0 | 0 | 0 | 0 | 7 | 15 | 63 |
| Days Min Temp ≤ 0 °F | 0 | 0 | 0 | 0 | 0 | 0 | 0 | 0 | 0 | 0 | 0 | 0 | 0 |
| Heating Degree Days | 762 | 577 | 387 | 146 | 30 | 1 | 0 | 0 | 15 | 129 | 385 | 647 | 3079 |
| Cooling Degree Days | 0 | 1 | 11 | 50 | 161 | 367 | 491 | 484 | 290 | 77 | 13 | 2 | 1947 |
| Total Precipitation (") | 3.28 | 3.96 | 4.97 | 4.95 | 5.87 | 4.85 | 4.35 | 3.47 | 3.86 | 5.06 | 5.11 | 5.27 | 55.00 |
| Days ≥ 0.1" Precip | 6 | 6 | 7 | 7 | 8 | 6 | 5 | 5 | 5 | 5 | 6 | 7 | 73 |
| Total Snowfall (") | 1.1 | 1.4 | 0.2 | 0.0 | 0.0 | 0.0 | 0.0 | 0.0 | 0.0 | 0.0 | 0.1 | 0.4 | 3.2 |
| Days ≥ 1" Snow Depth | 2 | 1 | 0 | 0 | 0 | 0 | 0 | 0 | 0 | 0 | 0 | 0 | 3 |

## NEWPORT *Jackson County* ELEVATION 220 ft LAT/LONG 35° 36 ' N / 91° 17 ' W

| | JAN | FEB | MAR | APR | MAY | JUN | JUL | AUG | SEP | OCT | NOV | DEC | YEAR |
|---|---|---|---|---|---|---|---|---|---|---|---|---|---|
| Maximum Temp °F | 45.1 | 50.7 | 60.8 | 71.6 | 79.5 | 87.9 | 91.7 | 89.7 | 83.0 | 73.1 | 60.4 | 49.5 | 70.3 |
| Minimum Temp °F | 27.3 | 31.5 | 41.4 | 51.1 | 59.4 | 67.3 | 71.0 | 68.4 | 61.1 | 48.8 | 40.5 | 32.0 | 50.0 |
| Mean Temp °F | 36.2 | 41.2 | 51.1 | 61.4 | 69.5 | 77.6 | 81.4 | 79.1 | 72.1 | 61.0 | 50.5 | 40.7 | 60.2 |
| Days Max Temp ≥ 90 °F | 0 | 0 | 0 | 0 | 2 | 13 | 22 | 16 | 6 | 1 | 0 | 0 | 60 |
| Days Max Temp ≤ 32 °F | 5 | 2 | 0 | 0 | 0 | 0 | 0 | 0 | 0 | 0 | 0 | 2 | 9 |
| Days Min Temp ≤ 32 °F | 21 | 15 | 6 | 0 | 0 | 0 | 0 | 0 | 0 | 0 | 7 | 17 | 66 |
| Days Min Temp ≤ 0 °F | 0 | 0 | 0 | 0 | 0 | 0 | 0 | 0 | 0 | 0 | 0 | 0 | 0 |
| Heating Degree Days | 885 | 667 | 432 | 157 | 31 | 0 | 0 | 0 | 22 | 171 | 435 | 745 | 3545 |
| Cooling Degree Days | 0 | 0 | 8 | 53 | 176 | 400 | 544 | 486 | 263 | 52 | 5 | 1 | 1988 |
| Total Precipitation (") | 3.49 | 3.40 | 4.75 | 4.97 | 4.89 | 3.82 | 3.45 | 3.62 | 4.11 | 3.64 | 5.06 | 4.90 | 50.10 |
| Days ≥ 0.1" Precip | 6 | 6 | 7 | 7 | 7 | 5 | 6 | 5 | 5 | 5 | 6 | 7 | 72 |
| Total Snowfall (") | *2.8* | 1.9 | 0.4 | 0.0 | 0.0 | 0.0 | 0.0 | 0.0 | 0.0 | 0.0 | 0.3 | 0.5 | 5.9 |
| Days ≥ 1" Snow Depth | na | na | *0* | 0 | 0 | 0 | 0 | 0 | 0 | 0 | 0 | 0 | na |

## NIMROD DAM *Perry County* ELEVATION 469 ft LAT/LONG 34° 57 ' N / 93° 10 ' W

| | JAN | FEB | MAR | APR | MAY | JUN | JUL | AUG | SEP | OCT | NOV | DEC | YEAR |
|---|---|---|---|---|---|---|---|---|---|---|---|---|---|
| Maximum Temp °F | 47.6 | 53.4 | 62.3 | 72.9 | 79.0 | 86.8 | 91.8 | 90.5 | 83.1 | 73.5 | 61.5 | 51.6 | 71.2 |
| Minimum Temp °F | 25.4 | 28.8 | 37.8 | 47.1 | 55.1 | 63.2 | 67.4 | 65.4 | 59.3 | 47.1 | 37.3 | 29.4 | 46.9 |
| Mean Temp °F | 36.5 | 41.1 | 50.0 | 60.0 | 67.1 | 75.0 | 79.6 | 78.0 | 71.3 | 60.3 | 49.4 | 40.6 | 59.1 |
| Days Max Temp ≥ 90 °F | 0 | 0 | 0 | 1 | 1 | 10 | 21 | 18 | 7 | 1 | 0 | 0 | 59 |
| Days Max Temp ≤ 32 °F | 4 | 2 | 0 | 0 | 0 | 0 | 0 | 0 | 0 | 0 | 0 | 1 | 7 |
| Days Min Temp ≤ 32 °F | 24 | 19 | 10 | 1 | 0 | 0 | 0 | 0 | 0 | 2 | 11 | 21 | 88 |
| Days Min Temp ≤ 0 °F | 0 | 0 | 0 | 0 | 0 | 0 | 0 | 0 | 0 | 0 | 0 | 0 | 0 |
| Heating Degree Days | 876 | 667 | 464 | 183 | 51 | 2 | 0 | 1 | 27 | 181 | 466 | 752 | 3670 |
| Cooling Degree Days | 0 | 0 | 8 | 39 | 125 | 319 | 469 | 432 | 233 | 45 | 7 | 1 | 1678 |
| Total Precipitation (") | 3.16 | 3.49 | 4.81 | 4.66 | 5.70 | 3.83 | 3.37 | 3.04 | 3.75 | 4.13 | 4.34 | 4.94 | 49.22 |
| Days ≥ 0.1" Precip | 6 | 6 | 7 | 6 | 8 | 6 | 5 | 5 | 5 | 5 | 5 | 6 | 70 |
| Total Snowfall (") | *1.0* | na | 0.3 | 0.0 | 0.0 | 0.0 | 0.0 | 0.0 | 0.0 | 0.0 | 0.2 | 0.5 | na |
| Days ≥ 1" Snow Depth | *1* | na | 0 | 0 | 0 | 0 | 0 | 0 | 0 | 0 | 0 | 0 | na |

## OZARK *Franklin County* ELEVATION 396 ft LAT/LONG 35° 29 ' N / 93° 50 ' W

| | JAN | FEB | MAR | APR | MAY | JUN | JUL | AUG | SEP | OCT | NOV | DEC | YEAR |
|---|---|---|---|---|---|---|---|---|---|---|---|---|---|
| Maximum Temp °F | 48.8 | 54.4 | 63.6 | 73.3 | 79.9 | 87.4 | 92.8 | 91.7 | 84.7 | 74.5 | 61.9 | 52.5 | 72.1 |
| Minimum Temp °F | 27.0 | 30.6 | 39.3 | 48.4 | 56.9 | 65.1 | 69.1 | 67.6 | 61.5 | 48.9 | 39.2 | 30.9 | 48.7 |
| Mean Temp °F | 37.9 | 42.6 | 51.5 | 60.9 | 68.4 | 76.3 | 81.0 | 79.7 | 73.1 | 61.7 | 50.6 | 41.7 | 60.5 |
| Days Max Temp ≥ 90 °F | 0 | 0 | 0 | 0 | 1 | 12 | 23 | 21 | 8 | 1 | 0 | 0 | 66 |
| Days Max Temp ≤ 32 °F | 3 | 1 | 0 | 0 | 0 | 0 | 0 | 0 | 0 | 0 | 0 | 1 | 5 |
| Days Min Temp ≤ 32 °F | 22 | 17 | 8 | 2 | 0 | 0 | 0 | 0 | 0 | 1 | 8 | 19 | 77 |
| Days Min Temp ≤ 0 °F | 0 | 0 | 0 | 0 | 0 | 0 | 0 | 0 | 0 | 0 | 0 | 0 | 0 |
| Heating Degree Days | 834 | 627 | 421 | 163 | 36 | 1 | 0 | 1 | 19 | 152 | 430 | 716 | 3400 |
| Cooling Degree Days | 0 | 0 | 6 | 36 | 132 | 334 | *494* | 458 | 258 | 43 | *4* | 0 | 1765 |
| Total Precipitation (") | 2.71 | 2.94 | 4.29 | 3.80 | 5.36 | 3.85 | 3.25 | 2.63 | 4.12 | 4.14 | 4.37 | 4.03 | 45.49 |
| Days ≥ 0.1" Precip | 5 | 5 | 6 | 7 | 8 | 7 | 5 | 5 | 6 | 5 | 5 | 6 | 70 |
| Total Snowfall (") | 2.2 | 1.2 | 0.5 | 0.0 | 0.0 | 0.0 | 0.0 | 0.0 | 0.0 | 0.0 | 0.3 | 0.7 | 4.9 |
| Days ≥ 1" Snow Depth | 2 | 1 | 0 | 0 | 0 | 0 | 0 | 0 | 0 | 0 | 0 | 0 | 3 |

## PINE BLUFF *Jefferson County* ELEVATION 220 ft LAT/LONG 34° 12' N / 92° 0' W

| | JAN | FEB | MAR | APR | MAY | JUN | JUL | AUG | SEP | OCT | NOV | DEC | YEAR |
|---|---|---|---|---|---|---|---|---|---|---|---|---|---|
| Maximum Temp °F | 50.4 | 56.3 | 65.5 | 75.5 | 81.9 | 89.0 | 92.5 | 91.1 | 85.3 | 76.0 | 64.4 | 54.9 | 73.6 |
| Minimum Temp °F | 31.2 | 34.7 | 43.0 | 52.1 | 60.1 | 68.1 | 71.4 | 69.7 | 62.9 | 51.4 | 42.2 | 34.7 | 51.8 |
| Mean Temp °F | 40.8 | 45.5 | 54.3 | 63.8 | 71.0 | 78.5 | 82.0 | 80.4 | 74.1 | 63.7 | 53.3 | 44.8 | 62.7 |
| Days Max Temp ≥ 90 °F | 0 | 0 | 0 | 0 | 3 | 15 | 23 | 20 | 9 | 1 | 0 | 0 | 71 |
| Days Max Temp ≤ 32 °F | 2 | 1 | 0 | 0 | 0 | 0 | 0 | 0 | 0 | 0 | 0 | 1 | 4 |
| Days Min Temp ≤ 32 °F | 18 | 13 | 4 | 0 | 0 | 0 | 0 | 0 | 0 | 0 | 6 | 14 | 55 |
| Days Min Temp ≤ 0 °F | 0 | 0 | 0 | 0 | 0 | 0 | 0 | 0 | 0 | 0 | 0 | 0 | 0 |
| Heating Degree Days | 745 | 545 | 348 | 113 | 20 | 0 | 0 | 0 | 13 | 120 | 358 | 622 | 2884 |
| Cooling Degree Days | 0 | 1 | 14 | 69 | 201 | *413* | 531 | 495 | 292 | *81* | 14 | 3 | 2114 |
| Total Precipitation (") | 4.13 | 4.38 | 4.96 | 5.32 | 5.51 | 3.53 | 4.18 | 4.02 | 3.65 | 4.61 | 4.29 | 5.31 | 53.89 |
| Days ≥ 0.1" Precip | 7 | 6 | 7 | 6 | 7 | 6 | 6 | 5 | 5 | 5 | 6 | 7 | 73 |
| Total Snowfall (") | 1.2 | *0.9* | 0.3 | 0.0 | 0.0 | 0.0 | 0.0 | 0.0 | 0.0 | 0.0 | 0.0 | 0.2 | 2.6 |
| Days ≥ 1" Snow Depth | *0* | *0* | 0 | 0 | 0 | 0 | 0 | 0 | 0 | 0 | 0 | 0 | 0 |

## POCAHONTAS 1 *Randolph County* ELEVATION 331 ft LAT/LONG 36° 16' N / 90° 59' W

| | JAN | FEB | MAR | APR | MAY | JUN | JUL | AUG | SEP | OCT | NOV | DEC | YEAR |
|---|---|---|---|---|---|---|---|---|---|---|---|---|---|
| Maximum Temp °F | 45.1 | 50.8 | 61.4 | 72.6 | 80.2 | 88.3 | 92.3 | 89.8 | 82.8 | 72.7 | 59.6 | 49.2 | 70.4 |
| Minimum Temp °F | 25.9 | 29.6 | 38.8 | 48.2 | 56.3 | 65.2 | 69.4 | 66.8 | 60.2 | 47.9 | 39.0 | 30.3 | 48.1 |
| Mean Temp °F | 35.5 | 40.2 | 50.2 | 60.4 | 68.3 | 76.8 | 80.9 | 78.3 | 71.5 | 60.3 | 49.3 | 39.8 | 59.3 |
| Days Max Temp ≥ 90 °F | 0 | 0 | 0 | 0 | 2 | 14 | 22 | 16 | 5 | 0 | 0 | 0 | 59 |
| Days Max Temp ≤ 32 °F | 5 | 2 | 0 | 0 | 0 | 0 | 0 | 0 | 0 | 0 | 0 | 2 | 9 |
| Days Min Temp ≤ 32 °F | 23 | 17 | 9 | 1 | 0 | 0 | 0 | 0 | 0 | 1 | 9 | 18 | 78 |
| Days Min Temp ≤ 0 °F | 1 | 0 | 0 | 0 | 0 | 0 | 0 | 0 | 0 | 0 | 0 | 0 | 1 |
| Heating Degree Days | 907 | 693 | 459 | 174 | 38 | 0 | 0 | 0 | 24 | 181 | 467 | 776 | 3719 |
| Cooling Degree Days | 0 | 0 | 7 | 45 | 153 | 369 | 518 | 450 | 232 | 41 | 3 | 0 | 1818 |
| Total Precipitation (") | 3.83 | 3.71 | 5.13 | 4.68 | 4.98 | 3.30 | 3.55 | 3.48 | 4.09 | 3.50 | 5.26 | 4.88 | 50.39 |
| Days ≥ 0.1" Precip | 6 | 5 | 7 | 8 | 7 | 5 | 5 | 5 | 5 | 5 | 7 | 7 | 72 |
| Total Snowfall (") | 4.2 | 3.6 | 1.4 | 0.0 | 0.0 | 0.0 | 0.0 | 0.0 | 0.0 | 0.1 | 0.7 | 1.6 | 11.6 |
| Days ≥ 1" Snow Depth | *3* | *1* | 0 | 0 | 0 | 0 | 0 | 0 | 0 | 0 | 0 | 0 | 4 |

## PORTLAND *Ashley County* ELEVATION 121 ft LAT/LONG 33° 14' N / 91° 30' W

| | JAN | FEB | MAR | APR | MAY | JUN | JUL | AUG | SEP | OCT | NOV | DEC | YEAR |
|---|---|---|---|---|---|---|---|---|---|---|---|---|---|
| Maximum Temp °F | *52.4* | *58.7* | *68.1* | 76.0 | 82.9 | 89.8 | 92.6 | 91.7 | 86.5 | 77.7 | 66.6 | 56.9 | 75.0 |
| Minimum Temp °F | *31.9* | *36.2* | *44.4* | 52.4 | 60.6 | 68.1 | 71.1 | 69.0 | 62.9 | 51.2 | 43.6 | 35.9 | 52.3 |
| Mean Temp °F | *42.2* | *47.5* | *56.3* | 64.2 | 71.8 | 79.0 | 81.9 | 80.4 | 74.8 | 64.5 | 55.2 | 46.4 | 63.7 |
| Days Max Temp ≥ 90 °F | *0* | *0* | *0* | 0 | 4 | 17 | 24 | 22 | 12 | 1 | 0 | 0 | 80 |
| Days Max Temp ≤ 32 °F | *2* | *0* | *0* | 0 | 0 | 0 | 0 | 0 | 0 | 0 | 0 | 1 | 3 |
| Days Min Temp ≤ 32 °F | *17* | *11* | *4* | 0 | 0 | 0 | 0 | 0 | 0 | 0 | 5 | 12 | 49 |
| Days Min Temp ≤ 0 °F | 0 | *0* | *0* | 0 | 0 | 0 | 0 | 0 | 0 | 0 | 0 | 0 | 0 |
| Heating Degree Days | *703* | *490* | *284* | 99 | 12 | 0 | 0 | 0 | 11 | 100 | 309 | 573 | 2581 |
| Cooling Degree Days | 0 | 2 | 20 | 92 | 245 | 452 | 552 | 517 | 330 | 100 | 21 | 3 | 2334 |
| Total Precipitation (") | *4.98* | *5.81* | 5.18 | 5.50 | 5.67 | 4.43 | 4.15 | 2.92 | 2.89 | 4.14 | 5.18 | 5.61 | 56.46 |
| Days ≥ 0.1" Precip | *7* | *6* | *6* | 6 | *8* | 6 | 5 | 5 | 4 | 5 | *6* | 7 | 71 |
| Total Snowfall (") | na | *0.3* | 0.3 | 0.0 | 0.0 | 0.0 | 0.0 | 0.0 | 0.0 | 0.0 | 0.0 | 0.1 | na |
| Days ≥ 1" Snow Depth | na | *0* | *0* | 0 | 0 | 0 | 0 | 0 | 0 | 0 | 0 | *0* | na |

## PRESCOTT *Nevada County* ELEVATION 322 ft LAT/LONG 33° 48' N / 93° 23' W

| | JAN | FEB | MAR | APR | MAY | JUN | JUL | AUG | SEP | OCT | NOV | DEC | YEAR |
|---|---|---|---|---|---|---|---|---|---|---|---|---|---|
| Maximum Temp °F | 52.4 | 58.9 | 68.1 | 76.9 | 82.6 | 89.3 | 92.8 | 92.3 | 85.9 | 76.3 | 64.0 | 55.4 | 74.6 |
| Minimum Temp °F | 32.0 | 35.4 | 43.2 | 51.8 | 59.5 | 67.2 | 70.5 | 69.3 | 63.4 | 52.1 | 42.5 | 35.2 | 51.8 |
| Mean Temp °F | 42.2 | 47.2 | 55.7 | 64.3 | 71.1 | 78.2 | 81.7 | 80.9 | 74.7 | 64.2 | 53.3 | 45.3 | 63.2 |
| Days Max Temp ≥ 90 °F | 0 | 0 | 0 | 0 | 3 | 16 | 24 | 22 | 11 | 1 | 0 | 0 | 77 |
| Days Max Temp ≤ 32 °F | 1 | 0 | 0 | 0 | 0 | 0 | 0 | 0 | 0 | 0 | 0 | 1 | 2 |
| Days Min Temp ≤ 32 °F | 17 | 12 | 4 | 0 | 0 | 0 | 0 | 0 | 0 | 0 | 5 | 14 | 52 |
| Days Min Temp ≤ 0 °F | 0 | 0 | 0 | 0 | 0 | 0 | 0 | 0 | 0 | 0 | 0 | 0 | 0 |
| Heating Degree Days | 701 | 498 | 301 | 92 | 12 | 0 | 0 | 0 | 11 | 105 | 357 | 606 | 2683 |
| Cooling Degree Days | 0 | 3 | 16 | 73 | 200 | 406 | 518 | 507 | 308 | 82 | 12 | 2 | 2127 |
| Total Precipitation (") | 4.05 | 4.21 | 5.01 | 5.30 | 5.61 | 4.34 | 4.37 | 3.51 | 4.22 | 4.88 | 5.47 | 5.67 | 56.64 |
| Days ≥ 0.1" Precip | 7 | 6 | 7 | 6 | 8 | 6 | 6 | 5 | 5 | 6 | 6 | 7 | 75 |
| Total Snowfall (") | 2.2 | 1.7 | 0.3 | 0.0 | 0.0 | 0.0 | 0.0 | 0.0 | 0.0 | 0.0 | 0.0 | 0.7 | 4.9 |
| Days ≥ 1" Snow Depth | 2 | 1 | 0 | 0 | 0 | 0 | 0 | 0 | 0 | 0 | 0 | 1 | 4 |

## ROHWER 2 NNE *Desha County* ELEVATION 150 ft LAT/LONG 33° 48' N / 91° 16' W

| | JAN | FEB | MAR | APR | MAY | JUN | JUL | AUG | SEP | OCT | NOV | DEC | YEAR |
|---|---|---|---|---|---|---|---|---|---|---|---|---|---|
| Maximum Temp °F | 49.8 | 54.9 | 64.1 | 73.9 | 81.3 | 89.2 | 92.0 | 90.6 | 85.1 | 75.6 | 64.0 | 54.5 | 72.9 |
| Minimum Temp °F | 31.1 | 34.4 | 42.2 | 51.6 | 60.0 | 67.7 | 70.7 | 68.2 | 61.5 | 49.3 | 41.4 | 34.5 | 51.0 |
| Mean Temp °F | 40.4 | 44.7 | 53.1 | 62.7 | 70.7 | 78.5 | 81.4 | 79.5 | 73.3 | 62.5 | 52.7 | 44.6 | 62.0 |
| Days Max Temp ≥ 90 °F | 0 | 0 | 0 | 0 | 3 | 16 | 22 | 19 | 10 | 1 | 0 | 0 | 71 |
| Days Max Temp ≤ 32 °F | 3 | 1 | 0 | 0 | 0 | 0 | 0 | 0 | 0 | 0 | 0 | 1 | 5 |
| Days Min Temp ≤ 32 °F | 18 | 13 | 5 | 0 | 0 | 0 | 0 | 0 | 0 | 1 | 6 | 15 | 58 |
| Days Min Temp ≤ 0 °F | 0 | 0 | 0 | 0 | 0 | 0 | 0 | 0 | 0 | 0 | 0 | 0 | 0 |
| Heating Degree Days | 755 | 568 | 375 | 131 | 22 | 0 | 0 | 0 | 17 | 143 | 373 | 628 | 3012 |
| Cooling Degree Days | 0 | 1 | 12 | 68 | 211 | 431 | 530 | 479 | 284 | 75 | 12 | 3 | 2106 |
| Total Precipitation (") | 4.58 | 4.72 | 5.06 | 4.70 | 4.91 | 3.45 | 3.76 | 2.86 | 3.25 | 3.51 | 5.05 | 5.78 | 51.63 |
| Days ≥ 0.1" Precip | 7 | 6 | 7 | 6 | 7 | 5 | 6 | 5 | 5 | 5 | 6 | 7 | 72 |
| Total Snowfall (") | 0.9 | 0.1 | 0.4 | 0.0 | 0.0 | 0.0 | 0.0 | 0.0 | 0.0 | 0.0 | 0.0 | 0.0 | 1.4 |
| Days ≥ 1" Snow Depth | 0 | 0 | 0 | 0 | 0 | 0 | 0 | 0 | 0 | 0 | 0 | 0 | 0 |

## SEARCY *White County* ELEVATION 249 ft LAT/LONG 35° 15' N / 91° 44' W

| | JAN | FEB | MAR | APR | MAY | JUN | JUL | AUG | SEP | OCT | NOV | DEC | YEAR |
|---|---|---|---|---|---|---|---|---|---|---|---|---|---|
| Maximum Temp °F | 49.2 | 54.7 | 64.6 | 74.9 | 81.7 | 89.6 | 93.3 | 91.7 | 85.3 | 75.6 | 62.4 | 52.6 | 73.0 |
| Minimum Temp °F | 28.6 | 32.0 | 40.8 | 49.8 | 57.9 | 66.1 | 69.9 | 67.8 | 61.5 | 48.9 | 40.3 | 32.4 | 49.7 |
| Mean Temp °F | 38.9 | 43.4 | 52.7 | 62.4 | 69.8 | 77.9 | 81.7 | 79.7 | 73.4 | 62.3 | 51.4 | 42.5 | 61.3 |
| Days Max Temp ≥ 90 °F | 0 | 0 | 0 | 0 | 3 | 16 | 24 | 21 | 9 | 1 | 0 | 0 | 74 |
| Days Max Temp ≤ 32 °F | 3 | 1 | 0 | 0 | 0 | 0 | 0 | 0 | 0 | 0 | 0 | 1 | 5 |
| Days Min Temp ≤ 32 °F | 21 | 15 | 7 | 1 | 0 | 0 | 0 | 0 | 0 | 1 | 8 | 17 | 70 |
| Days Min Temp ≤ 0 °F | 0 | 0 | 0 | 0 | 0 | 0 | 0 | 0 | 0 | 0 | 0 | 0 | 0 |
| Heating Degree Days | 802 | 604 | 386 | 135 | 25 | 0 | 0 | 0 | 15 | 140 | 409 | 690 | 3206 |
| Cooling Degree Days | 0 | 0 | 12 | 64 | 191 | 416 | 547 | 498 | 295 | 64 | 9 | 1 | 2097 |
| Total Precipitation (") | 3.90 | 3.52 | 5.16 | 5.13 | 5.67 | 3.62 | 3.78 | 3.68 | 3.82 | 3.98 | 5.37 | 5.26 | 52.89 |
| Days ≥ 0.1" Precip | 6 | 6 | 7 | 7 | 8 | 6 | 6 | 5 | 5 | 5 | 6 | 7 | 74 |
| Total Snowfall (") | ***1.9*** | 1.9 | 0.3 | 0.0 | 0.0 | 0.0 | 0.0 | 0.0 | 0.0 | 0.0 | 0.2 | 0.2 | 4.5 |
| Days ≥ 1" Snow Depth | na | 1 | 0 | 0 | 0 | 0 | 0 | 0 | 0 | 0 | 0 | 0 | na |

## SPARKMAN *Dallas County* ELEVATION 171 ft LAT/LONG 33° 55' N / 92° 51' W

| | JAN | FEB | MAR | APR | MAY | JUN | JUL | AUG | SEP | OCT | NOV | DEC | YEAR |
|---|---|---|---|---|---|---|---|---|---|---|---|---|---|
| Maximum Temp °F | 52.1 | 57.2 | 67.0 | 75.7 | 81.7 | 89.2 | 93.2 | 92.0 | 84.9 | 75.2 | 64.4 | 55.1 | 74.0 |
| Minimum Temp °F | 29.7 | 32.4 | 40.5 | ***48.9*** | 57.3 | 64.9 | 69.0 | 67.6 | 61.0 | 48.4 | 39.6 | 32.5 | 49.3 |
| Mean Temp °F | 40.9 | 44.8 | 53.8 | ***62.4*** | 69.6 | 77.1 | 81.2 | 79.7 | 73.0 | 61.8 | 52.0 | 43.8 | 61.7 |
| Days Max Temp ≥ 90 °F | 0 | 0 | 0 | 0 | 2 | 15 | 25 | 22 | ***8*** | 1 | 0 | 0 | 73 |
| Days Max Temp ≤ 32 °F | 2 | 1 | 0 | 0 | 0 | 0 | 0 | 0 | 0 | 0 | 0 | 1 | 4 |
| Days Min Temp ≤ 32 °F | 20 | 16 | 7 | 1 | 0 | 0 | 0 | 0 | 0 | 1 | ***8*** | ***15*** | 68 |
| Days Min Temp ≤ 0 °F | 0 | 0 | 0 | 0 | 0 | 0 | 0 | 0 | 0 | 0 | 0 | 0 | 0 |
| Heating Degree Days | 743 | 566 | 351 | ***130*** | 24 | 1 | 0 | 0 | 16 | 150 | 391 | 652 | 3024 |
| Cooling Degree Days | 0 | 1 | 6 | ***42*** | 152 | 381 | ***531*** | 490 | 282 | 56 | 8 | 1 | 1950 |
| Total Precipitation (") | 3.80 | 3.69 | 4.79 | 4.63 | 5.20 | 3.99 | 3.55 | 3.04 | 3.20 | 4.41 | 5.04 | 4.92 | 50.26 |
| Days ≥ 0.1" Precip | 5 | 4 | 6 | 5 | 6 | 5 | 5 | 5 | 5 | 5 | 5 | 5 | 61 |
| Total Snowfall (") | 0.4 | ***0.5*** | 0.0 | 0.0 | 0.0 | 0.0 | 0.0 | 0.0 | 0.0 | 0.0 | 0.0 | 0.0 | 0.9 |
| Days ≥ 1" Snow Depth | 0 | ***0*** | 0 | 0 | 0 | 0 | 0 | 0 | 0 | 0 | 0 | 0 | 0 |

## ST CHARLES *Arkansas County* ELEVATION 200 ft LAT/LONG 34° 23' N / 91° 8' W

| | JAN | FEB | MAR | APR | MAY | JUN | JUL | AUG | SEP | OCT | NOV | DEC | YEAR |
|---|---|---|---|---|---|---|---|---|---|---|---|---|---|
| Maximum Temp °F | 48.6 | 53.7 | 63.3 | 73.6 | 80.6 | 88.7 | 92.6 | 91.2 | 85.0 | 75.5 | 63.5 | 53.4 | 72.5 |
| Minimum Temp °F | 29.6 | 33.1 | 42.1 | 50.9 | 59.0 | 67.0 | 70.7 | 68.7 | 62.1 | 49.9 | 40.7 | 33.8 | 50.6 |
| Mean Temp °F | 39.1 | 43.4 | 52.7 | 62.3 | 69.8 | 78.0 | 81.8 | 80.0 | 73.6 | 62.7 | 52.1 | 43.7 | 61.6 |
| Days Max Temp ≥ 90 °F | 0 | 0 | 0 | 0 | 2 | 15 | 21 | 19 | 10 | 1 | 0 | 0 | 68 |
| Days Max Temp ≤ 32 °F | 3 | 2 | 0 | 0 | 0 | 0 | 0 | 0 | 0 | 0 | 0 | 1 | 6 |
| Days Min Temp ≤ 32 °F | 20 | 14 | 5 | 0 | 0 | 0 | 0 | 0 | 0 | 0 | 5 | 14 | 58 |
| Days Min Temp ≤ 0 °F | 0 | 0 | 0 | 0 | 0 | 0 | 0 | 0 | 0 | 0 | 0 | 0 | 0 |
| Heating Degree Days | 796 | 609 | 385 | 139 | 26 | 0 | 0 | 0 | 17 | 137 | 390 | 656 | 3155 |
| Cooling Degree Days | 0 | 0 | 12 | 62 | 180 | ***417*** | ***555*** | 510 | 298 | 75 | 9 | 2 | 2120 |
| Total Precipitation (") | 3.94 | 4.22 | 5.48 | 5.30 | 5.18 | 3.87 | 3.66 | 2.68 | 3.37 | 3.91 | 5.30 | 5.12 | 52.03 |
| Days ≥ 0.1" Precip | 6 | 5 | 7 | 6 | 7 | 5 | 5 | 4 | 5 | 5 | 6 | 6 | 67 |
| Total Snowfall (") | ***1.5*** | 0.6 | 0.7 | 0.0 | 0.0 | 0.0 | 0.0 | 0.0 | 0.0 | 0.0 | 0.0 | 0.2 | 3.0 |
| Days ≥ 1" Snow Depth | ***1*** | ***0*** | 0 | 0 | 0 | 0 | 0 | 0 | 0 | 0 | 0 | 0 | 1 |

## STUTTGART 9 ESE *Arkansas County* ELEVATION 198 ft LAT/LONG 34° 28' N / 91° 25' W

| | JAN | FEB | MAR | APR | MAY | JUN | JUL | AUG | SEP | OCT | NOV | DEC | YEAR |
|---|---|---|---|---|---|---|---|---|---|---|---|---|---|
| Maximum Temp °F | 47.3 | 52.5 | 62.0 | 72.4 | 79.8 | 87.8 | 91.1 | 89.6 | 83.7 | 74.2 | 62.0 | 52.2 | 71.2 |
| Minimum Temp °F | 29.8 | 33.5 | 42.3 | 51.7 | 60.6 | 68.6 | 71.8 | 69.3 | 62.4 | 50.0 | 41.8 | 34.0 | 51.3 |
| Mean Temp °F | 38.6 | 43.0 | 52.2 | 62.1 | 70.3 | 78.2 | 81.5 | 79.5 | 73.1 | 62.1 | 51.9 | 43.1 | 61.3 |
| Days Max Temp ≥ 90 °F | 0 | 0 | 0 | 0 | 2 | 13 | 21 | 17 | 7 | 0 | 0 | 0 | 60 |
| Days Max Temp ≤ 32 °F | 4 | 2 | 0 | 0 | 0 | 0 | 0 | 0 | 0 | 0 | 0 | 1 | 7 |
| Days Min Temp ≤ 32 °F | 20 | 14 | 4 | 0 | 0 | 0 | 0 | 0 | 0 | 0 | 5 | 15 | 58 |
| Days Min Temp ≤ 0 °F | 0 | 0 | 0 | 0 | 0 | 0 | 0 | 0 | 0 | 0 | 0 | 0 | 0 |
| Heating Degree Days | 812 | 614 | 400 | 143 | 25 | 0 | 0 | 0 | 19 | 149 | 393 | 672 | 3227 |
| Cooling Degree Days | 0 | 0 | 10 | 58 | 192 | 417 | 533 | 479 | 284 | 68 | 8 | 1 | 2050 |
| Total Precipitation (") | 3.58 | 3.81 | 4.79 | 4.91 | 5.17 | 3.78 | 3.32 | 2.72 | 3.19 | 3.68 | 4.76 | 5.10 | 48.81 |
| Days ≥ 0.1" Precip | 6 | 5 | 7 | 6 | 7 | 6 | 5 | 4 | 5 | 5 | 6 | 7 | 69 |
| Total Snowfall (") | *1.1* | 0.5 | 0.4 | 0.0 | 0.0 | 0.0 | 0.0 | 0.0 | 0.0 | 0.0 | 0.0 | 0.0 | 2.0 |
| Days ≥ 1" Snow Depth | *2* | 0 | 0 | 0 | 0 | 0 | 0 | 0 | 0 | 0 | 0 | 0 | 2 |

## SUBIACO *Logan County* ELEVATION 502 ft LAT/LONG 35° 18' N / 93° 39' W

| | JAN | FEB | MAR | APR | MAY | JUN | JUL | AUG | SEP | OCT | NOV | DEC | YEAR |
|---|---|---|---|---|---|---|---|---|---|---|---|---|---|
| Maximum Temp °F | 48.9 | 55.1 | 64.9 | 74.9 | 80.7 | 88.7 | 93.4 | 91.8 | 84.6 | 74.5 | 62.1 | 52.3 | 72.7 |
| Minimum Temp °F | 28.8 | 32.6 | 41.4 | 50.5 | 58.1 | 65.9 | 69.8 | 68.0 | 61.9 | 50.6 | 40.9 | 32.6 | 50.1 |
| Mean Temp °F | 38.9 | 43.9 | 53.2 | 62.7 | 69.4 | 77.3 | 81.6 | 79.9 | 73.3 | 62.6 | 51.5 | 42.5 | 61.4 |
| Days Max Temp ≥ 90 °F | 0 | 0 | 0 | 1 | 3 | 15 | 24 | 21 | 9 | 1 | 0 | 0 | 74 |
| Days Max Temp ≤ 32 °F | 3 | 1 | 0 | 0 | 0 | 0 | 0 | 0 | 0 | 0 | 0 | 1 | 5 |
| Days Min Temp ≤ 32 °F | 20 | 14 | 6 | 1 | 0 | 0 | 0 | 0 | 0 | 1 | 6 | 16 | 64 |
| Days Min Temp ≤ 0 °F | 0 | 0 | 0 | 0 | 0 | 0 | 0 | 0 | 0 | 0 | 0 | 0 | 0 |
| Heating Degree Days | 803 | 590 | 373 | 129 | 25 | 1 | 0 | 0 | 16 | 135 | 405 | 693 | 3170 |
| Cooling Degree Days | 0 | 0 | 13 | 59 | 157 | 375 | 523 | 479 | 271 | 61 | 7 | 2 | 1947 |
| Total Precipitation (") | 2.61 | 3.11 | 3.98 | 4.31 | 5.40 | 4.11 | 3.28 | 3.25 | 3.95 | 3.93 | 4.48 | 4.03 | 46.44 |
| Days ≥ 0.1" Precip | 5 | 5 | 6 | 7 | 8 | 6 | 5 | 5 | 6 | 5 | 6 | 6 | 70 |
| Total Snowfall (") | 2.2 | 1.3 | 0.3 | 0.0 | 0.0 | 0.0 | 0.0 | 0.0 | 0.0 | 0.0 | 0.4 | 0.8 | 5.0 |
| Days ≥ 1" Snow Depth | 3 | 2 | 0 | 0 | 0 | 0 | 0 | 0 | 0 | 0 | 0 | 1 | 6 |

## WALDRON *Scott County* ELEVATION 675 ft LAT/LONG 34° 54' N / 94° 6' W

| | JAN | FEB | MAR | APR | MAY | JUN | JUL | AUG | SEP | OCT | NOV | DEC | YEAR |
|---|---|---|---|---|---|---|---|---|---|---|---|---|---|
| Maximum Temp °F | 51.8 | 57.5 | 66.5 | 76.2 | 82.0 | 89.4 | 94.1 | 93.0 | 86.0 | 76.4 | 63.9 | 54.9 | 74.3 |
| Minimum Temp °F | 27.0 | 30.8 | 39.6 | 48.5 | 56.4 | 64.3 | 67.6 | 65.9 | 59.3 | 47.5 | 38.2 | 30.6 | 48.0 |
| Mean Temp °F | 39.4 | 44.2 | 53.1 | 62.4 | 69.2 | 76.9 | 80.9 | 79.5 | 72.6 | 62.0 | 51.1 | 42.8 | 61.2 |
| Days Max Temp ≥ 90 °F | 0 | 0 | 0 | 0 | 3 | 16 | 25 | 23 | 10 | 2 | 0 | 0 | 79 |
| Days Max Temp ≤ 32 °F | 2 | 1 | 0 | 0 | 0 | 0 | 0 | 0 | 0 | 0 | 0 | 1 | 4 |
| Days Min Temp ≤ 32 °F | 22 | 17 | 9 | 2 | 0 | 0 | 0 | 0 | 0 | 2 | 10 | 18 | 80 |
| Days Min Temp ≤ 0 °F | 0 | 0 | 0 | 0 | 0 | 0 | 0 | 0 | 0 | 0 | 0 | 0 | 0 |
| Heating Degree Days | 786 | 582 | 376 | 136 | 30 | 1 | 0 | 0 | 23 | 149 | 418 | 684 | 3185 |
| Cooling Degree Days | 0 | 1 | 12 | 56 | 159 | 365 | 497 | 463 | 255 | 59 | 8 | 2 | 1877 |
| Total Precipitation (") | 2.77 | 2.99 | 4.26 | 4.96 | 6.25 | 4.39 | 3.92 | 2.89 | 4.07 | 4.28 | 4.39 | 4.37 | 49.54 |
| Days ≥ 0.1" Precip | 5 | 5 | 7 | 7 | 8 | 6 | 5 | 5 | 6 | 5 | 5 | 6 | 70 |
| Total Snowfall (") | 3.4 | 2.3 | 0.4 | 0.0 | 0.0 | 0.0 | 0.0 | 0.0 | 0.0 | 0.0 | 0.3 | 0.9 | 7.3 |
| Days ≥ 1" Snow Depth | 2 | 1 | 0 | 0 | 0 | 0 | 0 | 0 | 0 | 0 | 0 | 0 | 3 |

## WARREN 2 WSW *Bradley County* ELEVATION 210 ft LAT/LONG 33° 36' N / 92° 4' W

| | JAN | FEB | MAR | APR | MAY | JUN | JUL | AUG | SEP | OCT | NOV | DEC | YEAR |
|---|---|---|---|---|---|---|---|---|---|---|---|---|---|
| Maximum Temp °F | 51.7 | 56.8 | 65.7 | 74.9 | 80.8 | 88.4 | 92.0 | 91.4 | 85.8 | 75.5 | 64.6 | 55.9 | 73.6 |
| Minimum Temp °F | 30.2 | 33.5 | 41.3 | 50.4 | 57.7 | 66.1 | 69.3 | 67.3 | 61.6 | 49.1 | 40.8 | 34.0 | 50.1 |
| Mean Temp °F | 41.0 | 45.2 | 53.5 | 62.7 | 69.3 | 77.3 | 80.7 | 79.4 | 73.7 | 62.3 | 52.7 | 45.0 | 61.9 |
| Days Max Temp ≥ 90 °F | 0 | 0 | 0 | 0 | 2 | 14 | 22 | 21 | 10 | 1 | 0 | 0 | 70 |
| Days Max Temp ≤ 32 °F | 2 | 1 | 0 | 0 | 0 | 0 | 0 | 0 | 0 | 0 | 0 | 1 | 4 |
| Days Min Temp ≤ 32 °F | 20 | 14 | 6 | 0 | 0 | 0 | 0 | 0 | 0 | 1 | 7 | 16 | 64 |
| Days Min Temp ≤ 0 °F | 0 | 0 | 0 | 0 | 0 | 0 | 0 | 0 | 0 | 0 | 0 | 0 | 0 |
| Heating Degree Days | 740 | 554 | 362 | 127 | 29 | 1 | 0 | 0 | 15 | 144 | 373 | 616 | 2961 |
| Cooling Degree Days | 0 | 1 | 13 | 58 | 163 | 373 | 496 | 463 | 278 | 67 | 13 | 3 | 1928 |
| Total Precipitation (") | 4.46 | 4.89 | 5.88 | 5.07 | 4.86 | 3.60 | 4.32 | 3.08 | 3.81 | 4.93 | 4.78 | 5.42 | 55.10 |
| Days ≥ 0.1" Precip | 7 | 6 | 8 | 6 | 7 | 5 | 6 | 4 | 5 | 5 | 6 | 7 | 72 |
| Total Snowfall (") | 1.3 | *0.3* | 0.5 | 0.0 | 0.0 | 0.0 | 0.0 | 0.0 | 0.0 | 0.0 | 0.1 | 0.3 | 2.5 |
| Days ≥ 1" Snow Depth | 1 | *0* | 0 | 0 | 0 | 0 | 0 | 0 | 0 | 0 | 0 | 0 | 1 |

## WEST MEMPHIS *Crittenden County* ELEVATION 220 ft LAT/LONG 35° 8 ' N / 90° 11 ' W

| | JAN | FEB | MAR | APR | MAY | JUN | JUL | AUG | SEP | OCT | NOV | DEC | YEAR |
|---|---|---|---|---|---|---|---|---|---|---|---|---|---|
| Maximum Temp °F | 47.6 | 53.4 | 62.7 | 73.3 | 81.1 | 88.8 | 91.6 | 89.6 | 83.9 | 74.8 | 62.1 | 52.5 | 71.8 |
| Minimum Temp °F | 29.6 | 33.6 | 41.9 | 51.2 | 59.5 | 68.0 | 71.6 | 69.2 | 62.8 | 51.0 | 42.3 | 34.5 | 51.3 |
| Mean Temp °F | 38.6 | 43.5 | 52.3 | 62.3 | 70.3 | 78.5 | 81.6 | 79.5 | 73.4 | 62.8 | 52.2 | 43.5 | 61.5 |
| Days Max Temp ≥ 90 °F | 0 | 0 | 0 | 0 | 3 | 15 | 22 | 17 | 6 | 0 | 0 | 0 | 63 |
| Days Max Temp ≤ 32 °F | 3 | 1 | 0 | 0 | 0 | 0 | 0 | 0 | 0 | 0 | 0 | 1 | 5 |
| Days Min Temp ≤ 32 °F | 19 | 13 | 6 | 1 | 0 | 0 | 0 | 0 | 0 | 0 | 6 | 13 | 58 |
| Days Min Temp ≤ 0 °F | 0 | 0 | 0 | 0 | 0 | 0 | 0 | 0 | 0 | 0 | 0 | 0 | 0 |
| Heating Degree Days | 812 | 599 | 395 | 139 | 23 | 0 | 0 | 0 | 18 | 135 | 387 | 661 | 3169 |
| Cooling Degree Days | 0 | 0 | 10 | *59* | 188 | 413 | 529 | 470 | 286 | 69 | 12 | 2 | 2038 |
| Total Precipitation (") | 3.77 | 4.02 | 5.24 | 5.64 | 4.98 | 4.05 | 3.11 | 3.37 | 3.58 | 3.24 | 5.19 | 5.73 | 51.92 |
| Days ≥ 0.1" Precip | 7 | 7 | 8 | 8 | 7 | 6 | 5 | 5 | 6 | 5 | 7 | 7 | 78 |
| Total Snowfall (") | *1.7* | *0.7* | 0.6 | 0.0 | 0.0 | 0.0 | 0.0 | 0.0 | 0.0 | 0.0 | 0.1 | 0.0 | 3.1 |
| Days ≥ 1" Snow Depth | *0* | *0* | 0 | 0 | 0 | 0 | 0 | 0 | 0 | 0 | 0 | 0 | 0 |

## WYNNE *Cross County* ELEVATION 259 ft LAT/LONG 35° 13 ' N / 90° 46 ' W

| | JAN | FEB | MAR | APR | MAY | JUN | JUL | AUG | SEP | OCT | NOV | DEC | YEAR |
|---|---|---|---|---|---|---|---|---|---|---|---|---|---|
| Maximum Temp °F | 46.8 | 52.9 | 63.1 | 73.6 | 81.3 | 89.3 | 92.2 | 90.1 | 84.2 | 75.3 | 62.0 | 51.9 | 71.9 |
| Minimum Temp °F | 28.8 | 33.0 | 41.8 | 50.6 | 58.8 | 66.8 | 70.5 | 67.4 | 61.1 | 49.7 | 41.2 | 33.3 | 50.3 |
| Mean Temp °F | 37.8 | 43.0 | 52.5 | 62.1 | 70.1 | 78.1 | 81.4 | 78.8 | 72.6 | 62.5 | 51.6 | 42.6 | 61.1 |
| Days Max Temp ≥ 90 °F | 0 | 0 | 0 | 0 | 2 | 16 | 23 | 18 | 7 | 1 | 0 | 0 | 67 |
| Days Max Temp ≤ 32 °F | 4 | 1 | 0 | 0 | 0 | 0 | 0 | 0 | 0 | 0 | 0 | 1 | 6 |
| Days Min Temp ≤ 32 °F | 21 | 14 | 6 | 1 | 0 | 0 | 0 | 0 | 0 | 1 | 6 | 15 | 64 |
| Days Min Temp ≤ 0 °F | 0 | 0 | 0 | 0 | 0 | 0 | 0 | 0 | 0 | 0 | 0 | 0 | 0 |
| Heating Degree Days | 836 | 615 | 394 | 144 | 27 | 0 | 0 | 0 | 21 | 143 | 403 | 688 | 3271 |
| Cooling Degree Days | 0 | 0 | 12 | 55 | 181 | 418 | 517 | 439 | 255 | 66 | 10 | 2 | 1955 |
| Total Precipitation (") | 3.47 | 3.53 | 4.72 | 5.78 | 4.99 | 3.51 | 2.89 | 2.73 | 3.96 | 3.51 | 4.87 | 5.21 | 49.17 |
| Days ≥ 0.1" Precip | 6 | 6 | 7 | 7 | 7 | 5 | 4 | 4 | 5 | 5 | 6 | 7 | 69 |
| Total Snowfall (") | 2.8 | 1.8 | 0.2 | 0.0 | 0.0 | 0.0 | 0.0 | 0.0 | 0.0 | 0.0 | 0.0 | 0.2 | 5.0 |
| Days ≥ 1" Snow Depth | 2 | 1 | 0 | 0 | 0 | 0 | 0 | 0 | 0 | 0 | 0 | 0 | 3 |

## JANUARY MINIMUM TEMPERATURE °F

### LOWEST

| | | |
|---|---|---|
| 1 | Bentonville | 22.2 |
| 2 | Mammoth Spring | 22.6 |
| 3 | Gilbert | 23.0 |
| 4 | Calico Rock | 23.6 |
| 5 | Mountain Home | 23.8 |
| 6 | Evening Shade | 23.9 |
| 7 | Lead Hill | 24.1 |
| | Mountain View | 24.1 |
| 9 | Gravette | 24.2 |
| 10 | Fayetteville | 24.5 |
| 11 | Greers Ferry Dam | 24.8 |
| 12 | Harrison | 25.0 |
| 13 | Mount Ida | 25.2 |
| 14 | Nimrod Dam | 25.4 |
| 15 | Eureka Springs | 25.9 |
| | Pocahontas | 25.9 |
| 17 | Corning | 26.1 |
| 18 | Batesville L & D | 26.2 |
| | Blakely Mntn Dm | 26.2 |
| 20 | Ft. Smith | 26.5 |
| 21 | Batesville | 26.7 |
| | Blue Mntain Dm | 26.7 |
| 23 | Greenbrier | 27.0 |
| | Ozark | 27.0 |
| | Waldron | 27.0 |

### HIGHEST

| | | |
|---|---|---|
| 1 | Eudora | 33.0 |
| 2 | Dumas | 32.6 |
| 3 | El Dorado | 32.4 |
| | Magnolia | 32.4 |
| 5 | Prescott | 32.0 |
| 6 | Arkansas Post | 31.9 |
| | Portland | 31.9 |
| 8 | Dermott | 31.7 |
| 9 | Pine Bluff | 31.2 |
| 10 | Rohwer | 31.1 |
| 11 | Clarendon | 30.9 |
| 12 | Fordyce | 30.8 |
| 13 | Keo | 30.7 |
| 14 | Des Arc | 30.6 |
| 15 | Arkadelphia | 30.4 |
| 16 | Crossett | 30.3 |
| | Monticello | 30.3 |
| 18 | Camden | 30.2 |
| | Warren | 30.2 |
| 20 | Little Rock | 30.1 |
| 21 | Leola | 30.0 |
| 22 | Nashville | 29.9 |
| 23 | Alum Fork | 29.8 |
| | Stuttgart | 29.8 |
| 25 | Sparkman | 29.7 |

## JULY MAXIMUM TEMPERATURE °F

### HIGHEST

| | | |
|---|---|---|
| 1 | Morrilton | 94.1 |
| | Waldron | 94.1 |
| 3 | Hot Springs | 93.8 |
| 4 | Alicia | 93.6 |
| 5 | Batesville L & D | 93.4 |
| | Beedeville | 93.4 |
| | Subiaco | 93.4 |
| 8 | Searcy | 93.3 |
| 9 | Dermott | 93.2 |
| | Sparkman | 93.2 |
| 11 | Calico Rock | 93.1 |
| | Conway | 93.1 |
| | DeQueen | 93.1 |
| | Evening Shade | 93.1 |
| 15 | Dardanelle | 93.0 |
| | Malvern | 93.0 |
| 17 | Blue Mntain Dm | 92.9 |
| | Des Arc | 92.9 |
| | Ft. Smith | 92.9 |
| 20 | Benton | 92.8 |
| | Blytheville | 92.8 |
| | Ozark | 92.8 |
| | Prescott | 92.8 |
| 24 | Arkansas Post | 92.7 |
| | Dumas | 92.7 |

### LOWEST

| | | |
|---|---|---|
| 1 | Harrison | 88.8 |
| 2 | Bentonville | 89.0 |
| 3 | Fayetteville | 89.5 |
| 4 | Mena | 89.8 |
| 5 | Eureka Springs | 90.6 |
| 6 | Keo | 90.8 |
| | Mountain Home | 90.8 |
| 8 | Gravette | 90.9 |
| 9 | Stuttgart | 91.1 |
| 10 | Keiser | 91.3 |
| 11 | Nashville | 91.4 |
| 12 | Cabot | 91.5 |
| | Greers Ferry Dam | 91.5 |
| | Helena | 91.5 |
| | Mountain View | 91.5 |
| 16 | Clarendon | 91.6 |
| | Mammoth Spring | 91.6 |
| | West Memphis | 91.6 |
| 19 | Brinkley | 91.7 |
| | Monticello | 91.7 |
| | Newport | 91.7 |
| 22 | Nimrod Dam | 91.8 |
| 23 | Jonesboro | 91.9 |
| 24 | Greenbrier | 92.0 |
| | Hope | 92.0 |

## ANNUAL PRECIPITATION (")

### HIGHEST

| | | |
|---|---|---|
| 1 | Mount Ida | 58.13 |
| 2 | Crossett | 58.10 |
| 3 | Hot Springs | 57.73 |
| 4 | Mena | 57.44 |
| 5 | Prescott | 56.64 |
| 6 | Blakely Mntn Dm | 56.57 |
| 7 | Portland | 56.46 |
| 8 | Malvern | 56.22 |
| 9 | Monticello | 55.75 |
| 10 | DeQueen | 55.55 |
| 11 | Eudora | 55.43 |
| 12 | Hope | 55.24 |
| 13 | Warren | 55.10 |
| 14 | Alum Fork | 55.09 |
| 15 | Nashville | 55.00 |
| 16 | Helena | 54.97 |
| 17 | Benton | 54.18 |
| 18 | Dermott | 54.17 |
| 19 | Leola | 54.12 |
| 20 | Pine Bluff | 53.89 |
| 21 | Fordyce | 53.53 |
| 22 | Camden | 53.48 |
| 23 | Marianna | 53.34 |
| 24 | El Dorado | 53.31 |
| 25 | Arkadelphia | 53.02 |

### LOWEST

| | | |
|---|---|---|
| 1 | Ft. Smith | 42.70 |
| 2 | Gilbert | 44.82 |
| 3 | Lead Hill | 44.91 |
| 4 | Fayetteville | 45.09 |
| 5 | Ozark | 45.49 |
| 6 | Mountain Home | 45.77 |
| 7 | Mammoth Spring | 45.98 |
| 8 | Eureka Springs | 46.16 |
| 9 | Calico Rock | 46.31 |
| 10 | Subiaco | 46.44 |
| 11 | Bentonville | 46.61 |
| 12 | Gravette | 46.63 |
| 13 | Harrison | 46.83 |
| 14 | Jonesboro | 47.59 |
| 15 | Batesville | 47.65 |
| 16 | Blue Mntain Dm | 47.91 |
| 17 | Alicia | 48.12 |
| 18 | Keo | 48.41 |
| 19 | Morrilton | 48.58 |
| 20 | Evening Shade | 48.61 |
| 21 | Corning | 48.74 |
| 22 | Batesville L & D | 48.79 |
| 23 | Stuttgart | 48.81 |
| 24 | Wynne | 49.17 |
| 25 | Nimrod Dam | 49.22 |

## ANNUAL SNOWFALL (")

### HIGHEST

| | | |
|---|---|---|
| 1 | Gravette | 15.5 |
| 2 | Harrison | 15.0 |
| 3 | Eureka Springs | 13.0 |
| 4 | Mountain Home | 12.5 |
| 5 | Corning | 11.6 |
| | Pocahontas | 11.6 |
| 7 | Evening Shade | 11.1 |
| 8 | Bentonville | 10.2 |
| 9 | Batesville L & D | 8.5 |
| 10 | Batesville | 8.2 |
| 11 | Gilbert | 8.1 |
| 12 | Fayetteville | 7.9 |
| 13 | Jonesboro | 7.7 |
| 14 | Waldron | 7.3 |
| 15 | Mammoth Spring | 7.2 |
| 16 | Ft. Smith | 7.1 |
| 17 | Newport | 5.9 |
| 18 | Mountain View | 5.8 |
| 19 | Greenbrier | 5.5 |
| 20 | Conway | 5.4 |
| 21 | Alicia | 5.3 |
| | Mount Ida | 5.3 |
| 23 | Dardanelle | 5.1 |
| | Little Rock | 5.1 |
| 25 | Subiaco | 5.0 |

### LOWEST

| | | |
|---|---|---|
| 1 | Eudora | 0.1 |
| 2 | Sparkman | 0.9 |
| 3 | Helena | 1.2 |
| 4 | Crossett | 1.3 |
| 5 | Rohwer | 1.4 |
| 6 | El Dorado | 1.6 |
| 7 | Monticello | 1.7 |
| 8 | Stuttgart | 2.0 |
| 9 | Fordyce | 2.1 |
| 10 | Camden | 2.2 |
| 11 | Magnolia | 2.3 |
| 12 | Clarendon | 2.4 |
| 13 | Blakely Mntn Dm | 2.5 |
| | Warren | 2.5 |
| 15 | Pine Bluff | 2.6 |
| 16 | Hope | 2.7 |
| 17 | Dermott | 2.8 |
| 18 | Arkadelphia | 2.9 |
| 19 | Dumas | 3.0 |
| | Hot Springs | 3.0 |
| | Keiser | 3.0 |
| | St. Charles | 3.0 |
| 23 | Alum Fork | 3.1 |
| | West Memphis | 3.1 |
| 25 | Arkansas Post | 3.2 |

# CALIFORNIA

PHYSICAL FEATURES. The State of California extends along the shore of the Pacific Ocean between latitudes 32½° N. and 42° N. Its more than 1,340 miles of coastline constitute nearly three-fourths of the Pacific coastline of the conterminous United States. The total land area amounts to 158,693 square miles. With its major axis oriented in a northwest-southeast direction, the State is 800 miles in length. Its greatest east-west dimension at a given latitude is about 360 miles though its average width is only 250 miles. However, it spreads over more than 10° of longitude, a distance of 550 miles.

The topography of the State is varied. Included are Death Valley, the lowest point in the U.S., with an elevation of 276 feet below sea level, and less than 85 miles away, Mt. Whitney, the highest peak in the conterminous states, reaching to 14,495 feet above sea level. These wide ranges of altitude and latitude are responsible in part for the variety of climates and vegetation found in various areas of the State. Another significant factor is the continuous interaction of maritime air masses with those of continental origin. The combination of these influences results in pronounced climatic changes with short distances.

The Coast Range parallels the coastline from the Oregon border to just north of the Los Angeles Basin. It is generally no more than 50 miles from the coast to the crest of the range. The principal break in the Coast Range is at San Francisco Bay where a sea level opening permits an abundant inflow of marine air to the interior of the State under certain circulation patterns. In the northern end of the State, the Coast Range merges with the Cascade Range, farther inland, to create an extensive area of rugged terrain more than 200 miles in width. The Cascades, in turn, extend southeastward until they merge into the Sierra Nevada. The Sierra Nevada, like the Coast Range, lies parallel to the coast, but the crest over most of its length is about 150 miles inland. Thus, between the two ranges there is a broad, flat valley averaging 45 miles or more in width. In length the valley extends nearly 500 miles.

Both the extreme northeastern portion of California and the desert area of southern California east of the mountains lie within the Great Basin. The Great Basin extends from Utah to the Sierra Nevada and has no surface drainage to the ocean. It is an area of climatological extremes.

GENERAL CLIMATE. Along the western side of the Coast Range the climate is dominated by the Pacific Ocean. Warm winters, cool summers, small daily and seasonal temperature ranges, and high relative humidities are characteristic of this area. With increasing distance from the ocean the maritime influence decreases. Areas that are well protected from the ocean experience a more continental type of climate with warmer summers, colder winters, greater daily and seasonal temperature ranges, and generally lower relative humidities. Many parts of the State lie within a transitional zone, where conditions range between these two climatic extremes. Summer is a dry period over most of the State. With the northward migration of the semi-permanent Pacific high during summer, most storm tracks are deflected far to the north. In winter, the Pacific high decreases in intensity and drops further south, permitting storms to move into and across the State, producing widespread rain at low elevations and snow at high elevations.

The easternmost mountain chains form a barrier that protects much of California from the extremely cold air of the Great Basin in winter. The ranges of mountains to the west offer some protection to the interior from the strong flow of air off the Pacific Ocean. As a result, precipitation is heavy on the coastward or western side of both the Coast Range and the Sierra Nevada, and lighter on the eastern slopes. Temperature tends toward uniformity from day to day and from season to season on the ocean side of the Coast Range and in coastal valleys. East of the Sierra Nevada temperature patterns are continental in character with wide excursions from high readings to low. Between the two mountain chains and over much of the desert area the temperature regime is intermediate between the maritime and the continental models. Hot summers are the rule while winters are moderate to cold. In the basins and valleys adjoining the coast, climate is subject to wide variations within short distances as a result of the influence of topography on the circulation of marine air. The Los Angeles Basin and the San Francisco Bay area offer many varieties of climate within a few miles.

A dominating factor in the weather of California is the semi-permanent high pressure area of the north Pacific Ocean.

This pressure center moves northward in summer, holding storm tracks well to the north, and as a result California receives little or no precipitation from this source during that period. In winter, the Pacific high retreats southward permitting storm centers to swing into and across California. These are the storms that bring widespread, moderate precipitation to the State. When changes in the circulation pattern permit storm centers to approach the California coast from a southwesterly direction, copious amounts of moisture are carried by the northeastward streaming air. This results in heavy rains and often produces widespread flooding.

There is another California weather characteristic that results from the location of the Pacific high. The steady flow of air from the northwest during the summer helps to drive the California Current of the Pacific Ocean as it sweeps southward almost parallel to the California coastline. However, since the mean drift is slightly offshore there is a band of upwelling immediately off the coast as water from deeper layers is drawn into the surface circulation. The water from below the surface is colder than the surface water, and as a result there is a semi-permanent band of cold water just offshore. The temperature of water reaching the surface from deeper levels ranges from about 49° in winter to 55° in late summer along the northern California coast, and from 57° to 65° on the southern California coast. Comparatively warm, moist Pacific air masses drifting over this band of cold water form a bank of fog which is swept inland by the prevailing northwest winds out of the high pressure center. In general, heat is added to the air as it moves inland during these summer months, and the fog quickly lifts to form a deck of low clouds that extend inland only a short distance before evaporating completely. Characteristically this deck of clouds extends inland further during the night and then recedes to the vicinity of the coast during the day.

PRECIPITATION. In the northern part of the State the months of heaviest precipitation are October through April. The rainy season becomes shorter in the southern part of the State, November through March marking the wet period here. During the rest of the year precipitation is infrequent and usually light. In the north and over the central and northern mountains there are usually from 60 to 100 days of precipitation per year, while in the southern desert there may be as few as 10 days. It is apparent, therefore, that the rainy season is made up of periods of stormy weather alternating with longer periods of pleasant weather. A typical winter storm situation brings intermittent rain over a period of from 2 to 5 days, followed by 7 to 14 days of dry weather.

SNOWFALL. Snow has been reported at one time or another in nearly every part of California but it is very infrequent west of the Sierra Nevada except at high elevations of the Coast Range and the Cascades. In the Sierra Nevada, snow in moderate amounts is reported nearly every winter at elevations as low as 2,000 feet. Amounts and intensities increase with elevation to around 7,000 or 8,000 feet. Above 4,000 feet elevation snow remains on the ground for appreciable lengths of time each winter.

RELATIVE HUMIDITY. In general, relative humidities are moderate to high along the coast throughout the year. Inland humidities are high during the winter and low during the summer. Since the ocean is the source of the cool, humid, maritime air of summer, it follows that with increasing distance from the ocean, relative humidity tends to decrease. Where mountain barriers prevent the free flow of marine air inland, humidities decrease rapidly. Where openings in these barriers permit a significant influx of cool, moist air it mixes with the drier inland air, resulting in a more gradual decrease of moisture. This pattern is characteristic of most coastal valleys.

STORMS. Thunderstorms may occur in California at any time of the year. Near the coast and over the Central Valley there appears to be no prevailing season. The storms are usually light and infrequent. Over the interior mountain areas storms are more intense, and they may become unusually strong on occasion at intermediate and high elevations of the Sierra Nevada. Many California thunderstorms produce so little precipitation that range and forest fires often result from the lightning strikes. Heavy precipitation occasionally results. Some flash flooding has been reported as a result of thundershowers. Hail diameters from ¼ to ½-inch are sometimes reported. Serious hail damage is infrequent. Tornadoes have been reported in California but with a frequency of only 1 or 2 per year. They are generally not severe, in many cases amounting to little more than a local whirlwind.

DROUGHT. Drought must be evaluated on a different basis than in other parts of the country. Typically there are extended periods every summer with little or no precipitation. This is the normal and expected condition. A deficiency of precipitation becomes significant in the State when the normal winter water supply fails to materialize.

# COUNTY INDEX

### *Monterey County*
KING CITY
MONTEREY
PRIEST VALLEY
SALINAS 3 E
SALINAS MUNICIPAL AP

### *Napa County*
ANGWIN PAC UNION COL
NAPA STATE HOSPITAL
SAINT HELENA

### *Nevada County*
BOCA
BOWMAN DAM
DONNER MEMORIAL ST P
LAKE SPAULDING
NEVADA CITY
TRUCKEE RANGER STN

### *Orange County*
LAGUNA BEACH
NEWPORT BEACH HARBOR
SANTA ANA FIRE STN
TUSTIN IRVINE RANCH

### *Placer County*
AUBURN
BLUE CANYON
COLFAX
TAHOE CITY

### *Plumas County*
CANYON DAM
CHESTER
PORTOLA

### *Riverside County*
BEAUMONT 1 E
BLYTHE
BLYTHE RIVERSIDE AP
EAGLE MOUNTAIN
ELSINORE
HAYFIELD PUMP PLANT
IDYLLWILD FIRE DEPT
INDIO FIRE STATION
MECCA FIRE STATION
PALM SPRINGS
PALM SPRINGS THERMAL
RIVERSIDE CITRUS EXP
RIVERSIDE FIRE STN 3

### *Sacramento County*
SACRAMENTO EXEC ARPT
SACRAMENTO WSO CITY

### *San Benito County*
PINNACLES NM

### *San Bernardino County*
BIG BEAR LAKE
DAGGETT SAN BERN AP
IRON MOUNTAIN
LAKE ARROWHEAD
MITCHELL CAVERNS
MOUNTAIN PASS
NEEDLES AIRPORT
PARKER RESERVOIR
REDLANDS
S BERNARDINO CO HOSP
TRONA
TWENTYNINE PALMS
VICTORVILLE PUMP PLT

### *San Diego County*
BORREGO DESERT PARK
CAMPO
CHULA VISTA
CUYAMACA
EL CAPITAN DAM
HENSHAW DAM
LA MESA
OCEANSIDE HARBOR
PALOMAR MTN OBS
SAN DIEGO LINDBERGH
VISTA 2 NNE

### *San Francisco County*
SAN FRANCISCO
SAN FRANCISCO OCEANS

### *San Joaquin County*
LODI
STOCKTON FIRE STN #
STOCKTON METRO ARPT
TRACY CARBONA

### *San Luis Obispo County*
MORRO BAY FIRE DEPT
PASO ROBLES
PASO ROBLES MUNI AP
PISMO BEACH
SAN LUIS OBISPO POLY

### *San Mateo County*
HALF MOON BAY
REDWOOD CITY
SAN FRANCISCO INT AP
SAN GREGORIO 2 SE

### *Santa Barbara County*
CACHUMA LAKE
LOMPOC
SANTA BARBARA
SANTA BARBARA MUNI
SANTA MARIA PBLC AP
TWITCHELL DAM

### *Santa Clara County*
GILROY
LOS GATOS
MOUNT HAMILTON
PALO ALTO
SAN JOSE

### *Santa Cruz County*
SANTA CRUZ
WATSONVILLE WATERWOR

### *Shasta County*
HAT CREEK PH 1
MANZANITA LAKE
SHASTA DAM
WHISKEYTOWN RESERVOI

### *Sierra County*
SIERRA CITY
SIERRAVILLE RANGER S

### *Siskiyou County*
CALLAHAN
CECILVILLE 1 SE
HAPPY CAMP RANGER ST
LAVA BEDS NATL MONUM
MC CLOUD
MOUNT SHASTA
TULELAKE
YREKA

### *Solano County*
FAIRFIELD
VACAVILLE

### *Sonoma County*
CLOVERDALE
FORT ROSS
GRATON
HEALDSBURG
PETALUMA FIRE STA 2
SANTA ROSA
SONOMA

### *Stanislaus County*
MODESTO
NEWMAN

### *Tehama County*

MINERAL
RED BLUFF MUNI AP

### *Trinity County*

BIG BAR RANGER STN

### *Tulare County*

ASH MOUNTAIN
GRANT GROVE
LEMON COVE
LINDSAY
PORTERVILLE
THREE RIVERS ED PH 1
VISALIA

### *Tuolumne County*

CHERRY VALLEY DAM
HETCH HETCHY
SONORA RS

### *Ventura County*

OJAI
OXNARD
SANTA PAULA

### *Yolo County*

DAVIS 2 WSW EXP FARM
WINTERS
WOODLAND 1 WNW

### *Yuba County*

MARYSVILLE
STRAWBERRY VALLEY

## ELEVATION INDEX

| FEET | STATION NAME |
|---|---|
| -182 | MECCA FIRE STATION |
| -171 | DEATH VALLEY |
| -121 | BRAWLEY 2 SW |
| -115 | PALM SPRINGS THERMAL |
| -64 | IMPERIAL |
| -49 | EL CENTRO 2 SSW |
| -21 | INDIO FIRE STATION |
| 7 | SAN FRANCISCO INT AP |
| 9 | CHULA VISTA |
| 10 | KENTFIELD |
| 10 | NEWARK |
| 10 | NEWPORT BEACH HARBOR |
| 10 | OCEANSIDE HARBOR |
| 10 | PETALUMA FIRE STA 2 |
| 10 | STOCKTON FIRE STN # |
| 14 | SANTA MONICA PIER |
| 16 | SANTA BARBARA MUNI |
| 20 | FAIRFIELD |
| 20 | PALO ALTO |
| 20 | SAN FRANCISCO OCEANS |
| 23 | SACRAMENTO EXEC ARPT |
| 25 | KLAMATH |
| 25 | LONG BEACH DAUGHERTY |
| 26 | STOCKTON METRO ARPT |
| 30 | PISMO BEACH |
| 30 | SAN DIEGO LINDBERGH |
| 30 | SAN RAFAEL CIVIC CEN |
| 31 | REDWOOD CITY |
| 39 | HALF MOON BAY |
| 40 | LODI |
| 43 | EUREKA WSO CITY |
| 49 | DAVIS 2 WSW EXP FARM |
| 49 | OXNARD |
| 52 | SALINAS MUNICIPAL AP |
| 55 | RICHMOND |
| 59 | ANTIOCH PUMP PLANT 3 |
| 59 | COLUSA 2 SSW |
| 59 | LAGUNA BEACH |
| 59 | WOODLAND 1 WNW |
| 60 | NAPA STATE HOSPITAL |
| 61 | TRACY PUMPING PLANT |
| 66 | SACRAMENTO WSO CITY |
| 69 | CRESCENT CITY 1 N |
| 69 | MARYSVILLE |
| 69 | SONOMA |
| 79 | CULVER CITY |
| 79 | SALINAS 3 E |
| 90 | NEWMAN |
| 91 | MODESTO |
| 95 | SAN JOSE |
| 95 | TORRANCE MUNICIPAL A |
| 102 | HEALDSBURG |
| 102 | LOMPOC |
| 102 | WATSONVILLE WATERWOR |
| 112 | FORT ROSS |
| 118 | TUSTIN IRVINE RANCH |
| 120 | FORT BRAGG 5 N |
| 121 | MORRO BAY FIRE DEPT |
| 121 | SANTA BARBARA |
| 125 | LOS ANGELES INTL AP |
| 131 | LOS BANOS |
| 131 | SANTA ANA FIRE STN |
| 131 | SANTA CRUZ |
| 131 | WINTERS |
| 139 | SCOTIA |
| 140 | TRACY CARBONA |
| 153 | MERCED MUNICIPAL AP |
| 161 | ORICK PRAIRIE CRK PK |
| 167 | SANTA ROSA |
| 180 | VACAVILLE |
| 194 | GILROY |
| 200 | CORCORAN IRRIG DIST |
| 200 | SAN LUIS OBISPO POLY |
| 205 | CHICO UNIVERSITY FAR |
| 207 | SAN FRANCISCO |
| 210 | GRATON |
| 233 | WILLOWS 6 W |
| 237 | SANTA PAULA |
| 239 | SANTA MARIA PBLC AP |
| 249 | HANFORD 1 S |
| 254 | ORLAND |
| 259 | MONTEREY |
| 259 | SAINT HELENA |
| 259 | SAN LUIS DAM |
| 268 | BLYTHE |
| 269 | BUTTONWILLOW |
| 270 | LOS ANGELES CVC CNTR |
| 279 | FIVE POINTS 5 SSW |
| 302 | BERKELEY |
| 302 | MADERA |
| 320 | KING CITY |
| 322 | CLOVERDALE |
| 331 | WASCO |
| 338 | FRESNO AIR TERMINAL |
| 349 | RED BLUFF MUNI AP |
| 351 | VISALIA |
| 361 | SAN GREGORIO 2 SE |
| 390 | BLYTHE RIVERSIDE AP |
| 393 | PORTERVILLE |
| 394 | SAN LEANDRO UPR FLTR |
| 400 | ORLEANS |
| 400 | SAN GABRIEL FIRE DEP |
| 410 | FRIANT GOVERNMENT CP |
| 420 | LINDSAY |
| 425 | PALM SPRINGS |
| 430 | LOS GATOS |
| 440 | LOS ANGELES U C L A |
| 479 | LIVERMORE |
| 485 | GOLD ROCK RANCH |
| 492 | BAKERSFIELD MEADOWS |
| 500 | RICHARDSON GROVE S P |
| 513 | LEMON COVE |

| FEET | STATION NAME |
|---|---|
| 581 | LA MESA |
| 582 | TWITCHELL DAM |
| 600 | VISTA 2 NNE |
| 610 | EL CAPITAN DAM |
| 623 | UKIAH |
| 655 | BURBANK VALLEY PUMP |
| 658 | CAMP PARDEE |
| 663 | COALINGA |
| 700 | PASO ROBLES |
| 702 | ELECTRA POWER HOUSE |
| 738 | PARKER RESERVOIR |
| 740 | POMONA CAL POLY |
| 751 | OJAI |
| 781 | CACHUMA LAKE |
| 790 | CANOGA PARK PIERCE C |
| 791 | STONY GORGE RESERVOI |
| 807 | PASO ROBLES MUNI AP |
| 850 | BORREGO DESERT PARK |
| 860 | PASADENA |
| 902 | RIVERSIDE FIRE STN 3 |
| 919 | NEEDLES AIRPORT |
| 942 | IRON MOUNTAIN |
| 973 | EAGLE MOUNTAIN |
| 1010 | POTTER VALLEY P H |
| 1047 | RIVERSIDE CITRUS EXP |
| 1075 | SHASTA DAM |
| 1089 | HAPPY CAMP RANGER ST |
| 1132 | S BERNARDINO CO HOSP |
| 1191 | THREE RIVERS ED PH 1 |
| 1211 | EAST PARK RESERVOIR |
| 1270 | BIG BAR RANGER STN |
| 1302 | ELSINORE |
| 1307 | PINNACLES NM |
| 1312 | WHISKEYTOWN RESERVOI |
| 1342 | LAKEPORT |
| 1349 | CLEARLAKE 4 SE |
| 1352 | REDLANDS |
| 1352 | WILLITS 1 NE |
| 1362 | AUBURN |
| 1370 | HAYFIELD PUMP PLANT |
| 1391 | COVELO |
| 1425 | TEJON RANCHO |
| 1690 | ASH MOUNTAIN |
| 1703 | TRONA |
| 1713 | PARADISE |
| 1752 | BALCH POWER HOUSE |
| 1821 | ANGWIN PAC UNION COL |
| 1831 | SONORA RS |
| 1850 | PLACERVILLE |
| 1929 | DAGGETT SAN BERN AP |

| FEET | STATION NAME |
|---|---|
| 1990 | TWENTYNINE PALMS |
| 1991 | AUBERRY 1 NW |
| 2170 | MT DIABLO JUNCTION |
| 2251 | PRIEST VALLEY |
| 2362 | TIGER CREEK PH |
| 2421 | COLFAX |
| 2440 | INYOKERN |
| 2592 | BEAUMONT 1 E |
| 2592 | CAMPO |
| 2602 | NEVADA CITY |
| 2631 | YREKA |
| 2661 | PALMDALE |
| 2700 | HENSHAW DAM |
| 2703 | KERN RIVER PH 3 |
| 2720 | DE SABLA |
| 2735 | MOJAVE |
| 2841 | VICTORVILLE PUMP PLT |
| 2960 | CECILVILLE 1 SE |
| 3022 | HAT CREEK PH 1 |
| 3060 | FAIRMONT |
| 3133 | GLENNVILLE |
| 3143 | CALLAHAN |
| 3251 | MC CLOUD |
| 3570 | RANDSBURG |
| 3590 | MOUNT SHASTA |
| 3700 | SALT SPRINGS PH |
| 3783 | STRAWBERRY VALLEY |
| 3832 | HAIWEE |
| 3870 | HETCH HETCHY |
| 3914 | INDEPENDENCE |
| 3983 | TEHACHAPI |
| 3993 | YOSEMITE PARK HDQTRS |
| 4042 | TULELAKE |
| 4147 | BISHOP AP |
| 4160 | SUSANVILLE MUNI AP |
| 4193 | ADIN RANGER STN |
| 4203 | SIERRA CITY |
| 4213 | MOUNT HAMILTON |
| 4314 | MITCHELL CAVERNS |
| 4393 | DOYLE 4 SSE |
| 4462 | ALTURAS |
| 4505 | FORT BIDWELL |
| 4524 | SANDBERG WSMO |
| 4534 | CHESTER |
| 4560 | CANYON DAM |
| 4652 | CEDARVILLE |
| 4672 | CUYAMACA |
| 4705 | CALAVERAS BIG TREES |
| 4744 | MOUNTAIN PASS |
| 4770 | LAVA BEDS NATL MONUM |

| FEET | STATION NAME |
|---|---|
| 4774 | CHERRY VALLEY DAM |
| 4833 | PORTOLA |
| 4954 | MINERAL |
| 4984 | SIERRAVILLE RANGER S |
| 5144 | YOSEMITE SOUTH ENTR |
| 5164 | LAKE SPAULDING |
| 5233 | DEEP SPRINGS COLLEGE |
| 5233 | LAKE ARROWHEAD |
| 5280 | BLUE CANYON |
| 5292 | JESS VALLEY |
| 5305 | TERMO 1 E |
| 5354 | BOWMAN DAM |
| 5403 | IDYLLWILD FIRE DEPT |
| 5584 | BOCA |
| 5604 | PALOMAR MTN OBS |
| 5709 | MT WILSON NO 2 |
| 5853 | MANZANITA LAKE |
| 5945 | DONNER MEMORIAL ST P |
| 5984 | TRUCKEE RANGER STN |
| 6234 | TAHOE CITY |
| 6375 | BRIDGEPORT |
| 6585 | GRANT GROVE |
| 6755 | BIG BEAR LAKE |
| 7020 | HUNTINGTON LAKE |
| 7835 | TWIN LAKES |
| 8369 | BODIE |

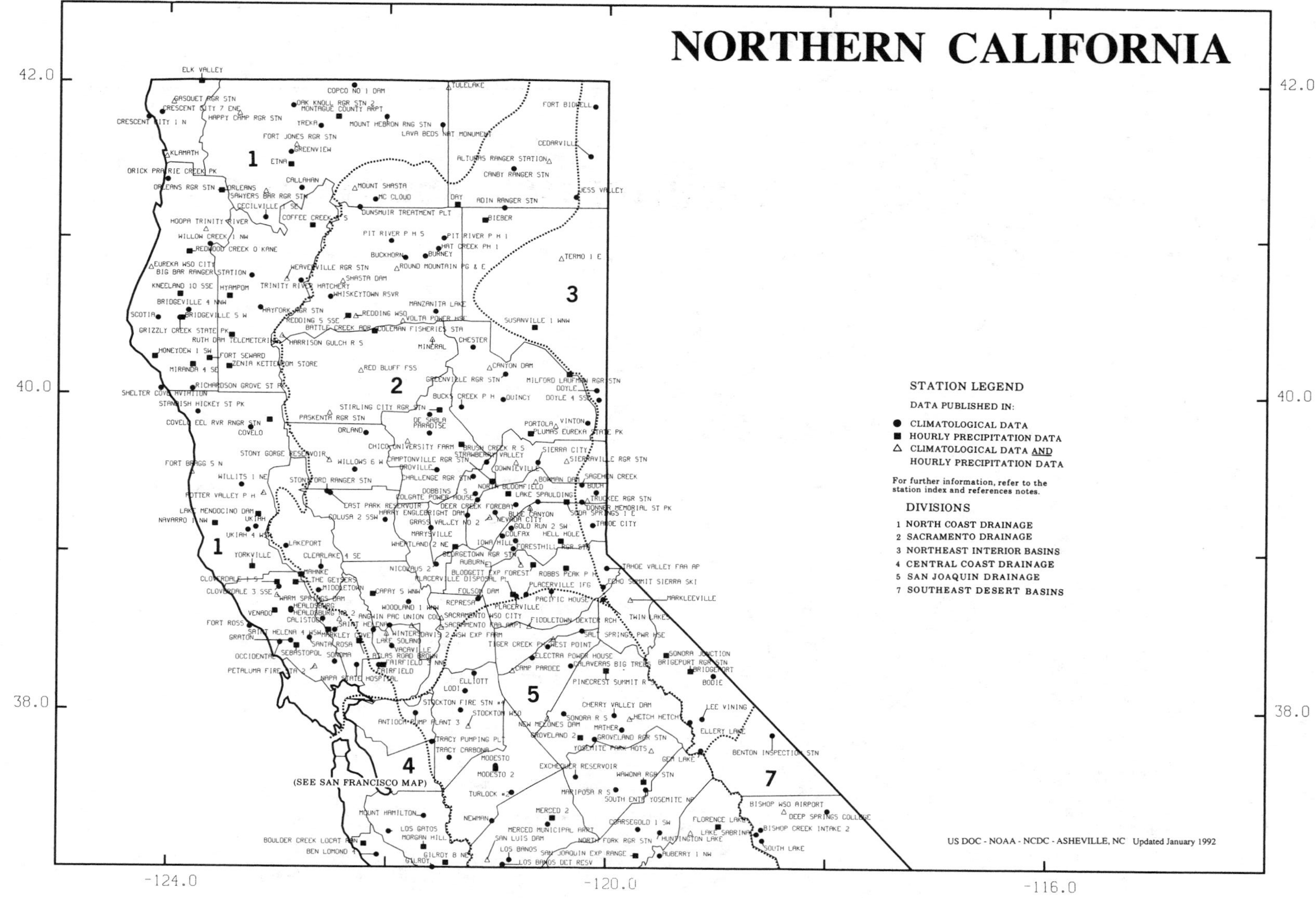
NORTHERN CALIFORNIA
(SEE SAN FRANCISCO MAP)
STATION LEGEND
DATA PUBLISHED IN:
● CLIMATOLOGICAL DATA
■ HOURLY PRECIPITATION DATA
△ CLIMATOLOGICAL DATA AND HOURLY PRECIPITATION DATA
For further information, refer to the station index and references notes.
DIVISIONS
1 NORTH COAST DRAINAGE
2 SACRAMENTO DRAINAGE
3 NORTHEAST INTERIOR BASINS
4 CENTRAL COAST DRAINAGE
5 SAN JOAQUIN DRAINAGE
7 SOUTHEAST DESERT BASINS
US DOC - NOAA - NCDC - ASHEVILLE, NC Updated January 1992

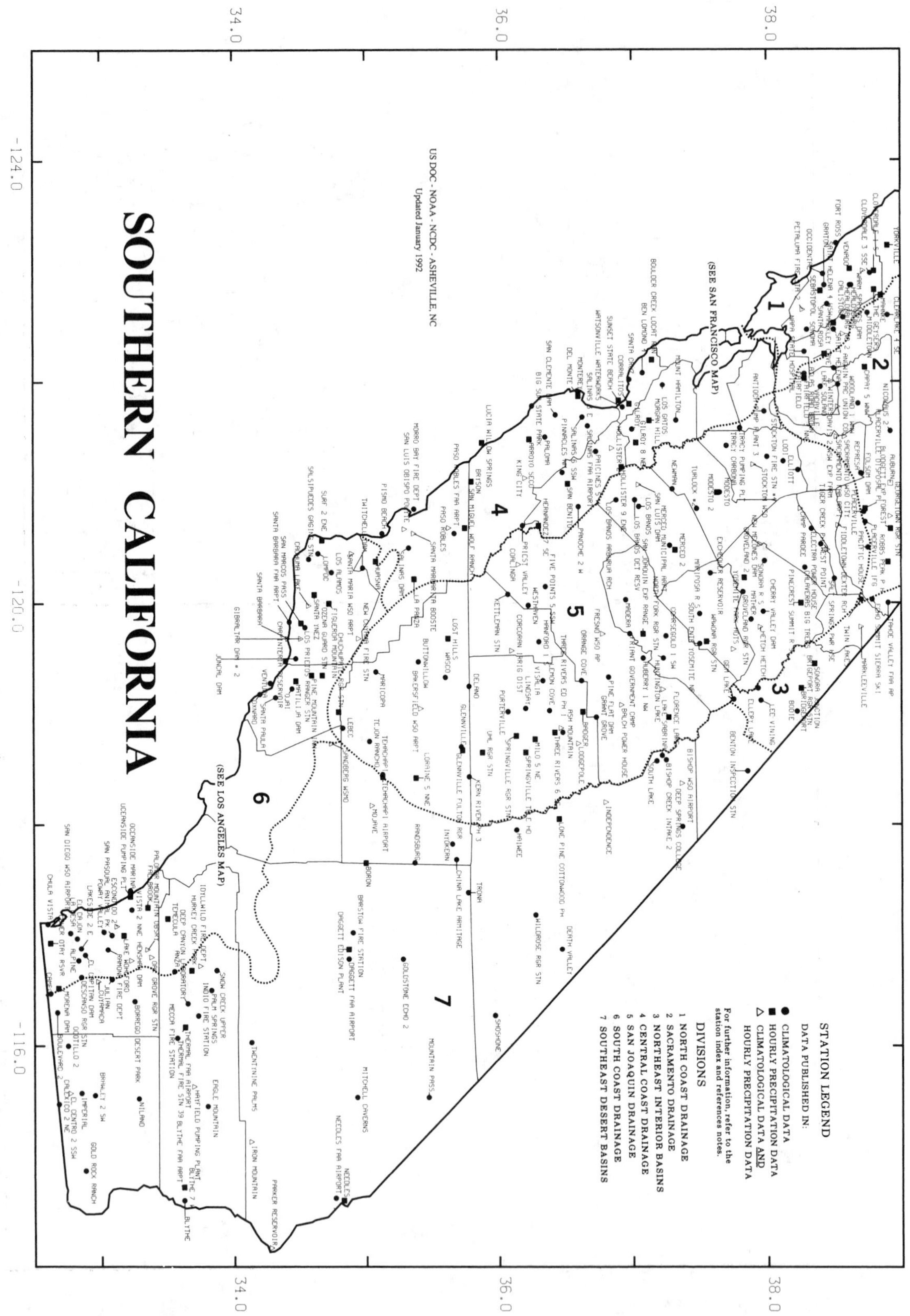
SOUTHERN CALIFORNIA
US DOC - NOAA - NCDC - ASHEVILLE, NC
Updated January 1992
(SEE SAN FRANCISCO MAP)
(SEE LOS ANGELES MAP)
STATION LEGEND
DATA PUBLISHED IN:
● CLIMATOLOGICAL DATA
■ HOURLY PRECIPITATION DATA
△ CLIMATOLOGICAL DATA AND HOURLY PRECIPITATION DATA
For further information, refer to the station index and references notes.
DIVISIONS
1 NORTH COAST DRAINAGE
2 SACRAMENTO DRAINAGE
3 NORTHEAST INTERIOR BASINS
4 CENTRAL COAST DRAINAGE
5 SAN JOAQUIN DRAINAGE
6 SOUTH COAST DRAINAGE
7 SOUTHEAST DESERT BASINS

# LOS ANGELES

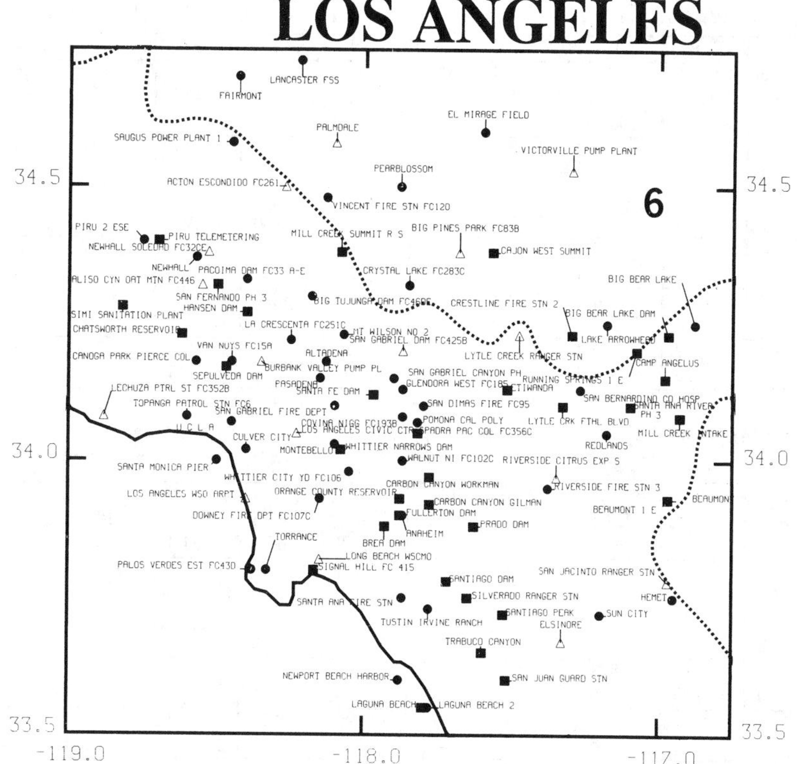


# SAN FRANCISCO

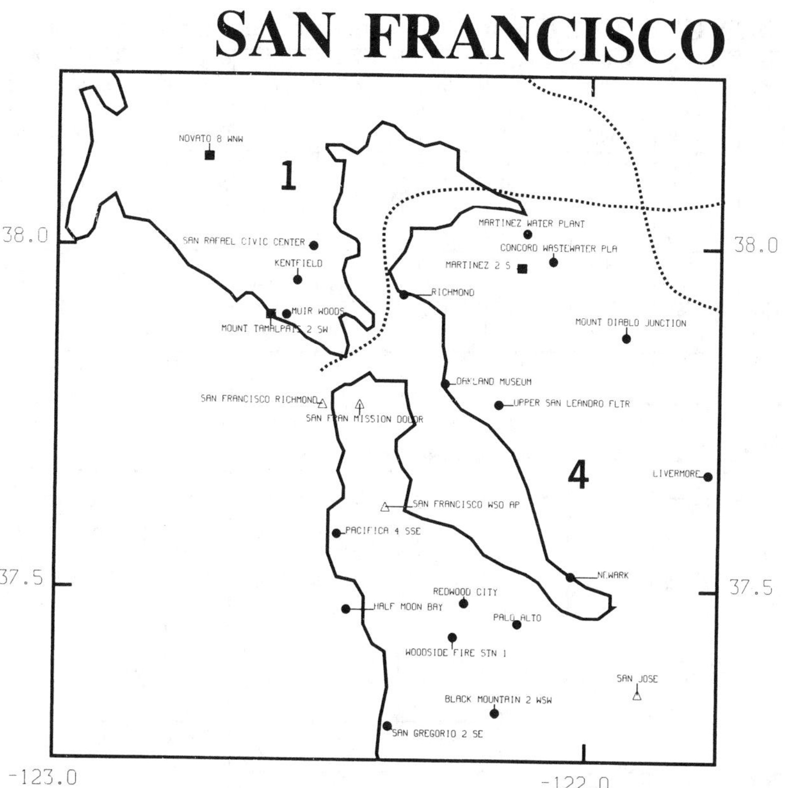

## ADIN RANGER STN *Modoc County* ELEVATION 4193 ft LAT/LONG 41° 12 ' N / 120° 57 ' W

| | JAN | FEB | MAR | APR | MAY | JUN | JUL | AUG | SEP | OCT | NOV | DEC | YEAR |
|---|---|---|---|---|---|---|---|---|---|---|---|---|---|
| Maximum Temp °F | na | *47.5* | *52.8* | 58.9 | *68.7* | 76.5 | *85.8* | 84.3 | *78.2* | *66.4* | na | na | na |
| Minimum Temp °F | na | 23.1 | 26.9 | 30.3 | *36.1* | 42.6 | *47.6* | 46.3 | *40.2* | *32.8* | na | na | na |
| Mean Temp °F | na | *35.7* | *39.8* | 44.6 | *52.5* | 59.5 | *66.8* | 65.3 | *59.4* | *49.6* | na | na | na |
| Days Max Temp ≥ 90 °F | 0 | 0 | 0 | 0 | 0 | 2 | 8 | 8 | 3 | 0 | 0 | 0 | 21 |
| Days Max Temp ≤ 32 °F | 3 | 1 | 0 | 0 | 0 | 0 | 0 | 0 | 0 | 0 | 1 | 4 | 9 |
| Days Min Temp ≤ 32 °F | 24 | 22 | 23 | *17* | *7* | 2 | 0 | 0 | 4 | *12* | 21 | *25* | 157 |
| Days Min Temp ≤ 0 °F | 1 | 0 | 0 | 0 | 0 | 0 | 0 | 0 | 0 | 0 | 0 | 1 | 2 |
| Heating Degree Days | na | *822* | *777* | 605 | *385* | 185 | *43* | 62 | *184* | *472* | na | na | na |
| Cooling Degree Days | na | na | na | 0 | na | *22* | na | 72 | na | na | na | na | na |
| Total Precipitation (") | 2.08 | 1.57 | 1.74 | 1.22 | 1.39 | 0.93 | 0.28 | 0.51 | 0.76 | 1.14 | 1.96 | 1.85 | 15.43 |
| Days ≥ 0.1" Precip | 4 | 4 | 4 | 3 | 3 | 2 | 1 | 1 | 2 | 3 | 5 | 5 | 37 |
| Total Snowfall (") | *11.4* | *7.2* | *7.7* | 3.1 | 1.1 | 0.1 | 0.0 | 0.0 | 0.1 | 0.5 | *4.9* | *11.4* | 47.5 |
| Days ≥ 1" Snow Depth | na | na | na | *0* | 0 | 0 | 0 | 0 | 0 | 0 | na | na | na |

## ALTURAS *Modoc County* ELEVATION 4462 ft LAT/LONG 41° 30 ' N / 120° 32 ' W

| | JAN | FEB | MAR | APR | MAY | JUN | JUL | AUG | SEP | OCT | NOV | DEC | YEAR |
|---|---|---|---|---|---|---|---|---|---|---|---|---|---|
| Maximum Temp °F | 41.7 | 46.6 | 51.8 | 58.3 | 68.2 | 77.3 | 87.1 | 86.5 | 78.3 | 66.7 | 50.0 | 41.8 | 62.9 |
| Minimum Temp °F | 16.0 | 20.0 | 24.5 | 27.7 | 33.9 | 40.2 | 43.5 | 42.1 | 35.4 | 27.6 | 22.5 | 16.0 | 29.1 |
| Mean Temp °F | 28.9 | 33.3 | 38.2 | 43.0 | 51.1 | 58.8 | 65.3 | 64.3 | 56.9 | 47.2 | 36.3 | 28.9 | 46.0 |
| Days Max Temp ≥ 90 °F | 0 | 0 | 0 | 0 | 1 | 4 | 14 | 13 | 4 | 0 | 0 | 0 | 36 |
| Days Max Temp ≤ 32 °F | 4 | 1 | 0 | 0 | 0 | 0 | 0 | 0 | 0 | 0 | 1 | 4 | 10 |
| Days Min Temp ≤ 32 °F | 29 | 26 | 26 | 23 | 13 | 4 | 1 | 1 | 10 | 25 | 26 | 29 | 213 |
| Days Min Temp ≤ 0 °F | 4 | 1 | 0 | 0 | 0 | 0 | 0 | 0 | 0 | 0 | 0 | 3 | 8 |
| Heating Degree Days | 1112 | 888 | 823 | 653 | 427 | 206 | 69 | 82 | 246 | 545 | 854 | 1111 | 7016 |
| Cooling Degree Days | 0 | 0 | 0 | 0 | 3 | 29 | 84 | 62 | 10 | 0 | 0 | 0 | 188 |
| Total Precipitation (") | 1.39 | 1.13 | 1.42 | 1.08 | 1.13 | 0.90 | 0.23 | 0.47 | 0.55 | 0.71 | 1.59 | 1.30 | 11.90 |
| Days ≥ 0.1" Precip | 4 | 3 | 4 | 4 | 3 | 2 | 1 | 1 | 2 | 2 | 5 | 4 | 35 |
| Total Snowfall (") | 7.9 | 5.4 | 4.8 | 2.7 | 0.9 | 0.0 | 0.0 | 0.0 | 0.0 | 0.2 | 4.1 | 7.4 | 33.4 |
| Days ≥ 1" Snow Depth | na | *5* | na | 1 | 0 | 0 | 0 | 0 | 0 | 0 | *4* | na | na |

## ANGWIN PAC UNION COL *Napa County* ELEVATION 1821 ft LAT/LONG 38° 35 ' N / 122° 26 ' W

| | JAN | FEB | MAR | APR | MAY | JUN | JUL | AUG | SEP | OCT | NOV | DEC | YEAR |
|---|---|---|---|---|---|---|---|---|---|---|---|---|---|
| Maximum Temp °F | 52.5 | 55.0 | 58.3 | 64.6 | 73.7 | 81.0 | 86.9 | 85.9 | 81.0 | 72.1 | 58.9 | 51.4 | 68.4 |
| Minimum Temp °F | 37.6 | 38.6 | 39.7 | 41.9 | 46.5 | 50.6 | 54.3 | 53.7 | 52.8 | 49.2 | 41.9 | 36.9 | 45.3 |
| Mean Temp °F | 45.1 | 46.8 | 49.0 | 53.3 | 60.1 | 65.8 | 70.6 | 69.8 | 66.9 | 60.7 | 50.4 | 44.1 | 56.9 |
| Days Max Temp ≥ 90 °F | 0 | 0 | 0 | 0 | 1 | 6 | 11 | 11 | 5 | 1 | 0 | 0 | 35 |
| Days Max Temp ≤ 32 °F | 0 | 0 | 0 | 0 | 0 | 0 | 0 | 0 | 0 | 0 | 0 | 0 | 0 |
| Days Min Temp ≤ 32 °F | 7 | 5 | 4 | 2 | 0 | 0 | 0 | 0 | 0 | 0 | 2 | *8* | 28 |
| Days Min Temp ≤ 0 °F | 0 | 0 | 0 | 0 | 0 | 0 | 0 | 0 | 0 | 0 | 0 | 0 | 0 |
| Heating Degree Days | 611 | 506 | 490 | 353 | 191 | 78 | 17 | 22 | 58 | 179 | 435 | 644 | 3584 |
| Cooling Degree Days | 0 | 0 | 1 | 10 | 39 | 100 | 189 | 160 | 128 | 51 | 0 | 0 | 678 |
| Total Precipitation (") | 8.81 | 6.93 | 5.75 | 2.28 | 0.66 | 0.33 | 0.05 | 0.16 | 0.57 | 2.21 | 5.85 | 6.78 | 40.38 |
| Days ≥ 0.1" Precip | 9 | 8 | 8 | 5 | 2 | 1 | 0 | 1 | 1 | 3 | 7 | 8 | 53 |
| Total Snowfall (") | 1.5 | 0.1 | 0.4 | 0.0 | 0.0 | 0.0 | 0.0 | 0.0 | 0.0 | 0.0 | 0.0 | 0.3 | 2.3 |
| Days ≥ 1" Snow Depth | 1 | 0 | 0 | 0 | 0 | 0 | 0 | 0 | 0 | 0 | 0 | 0 | 1 |

## ANTIOCH PUMP PLANT 3 *Contra Costa County* ELEVATION 59 ft LAT/LONG 37° 59 ' N / 121° 44 ' W

| | JAN | FEB | MAR | APR | MAY | JUN | JUL | AUG | SEP | OCT | NOV | DEC | YEAR |
|---|---|---|---|---|---|---|---|---|---|---|---|---|---|
| Maximum Temp °F | 53.3 | 60.1 | 65.1 | 71.4 | 79.4 | 85.5 | 90.6 | 89.8 | 86.0 | 77.7 | 64.0 | 53.9 | 73.1 |
| Minimum Temp °F | 36.6 | 40.3 | 43.6 | 46.4 | 51.3 | 56.0 | 57.3 | 56.7 | 55.1 | 49.9 | 43.1 | 36.5 | 47.7 |
| Mean Temp °F | 45.0 | 50.2 | 54.4 | 58.9 | 65.4 | 70.7 | 74.0 | 73.3 | 70.6 | 63.8 | 53.5 | 45.3 | 60.4 |
| Days Max Temp ≥ 90 °F | 0 | 0 | 0 | 0 | 4 | 9 | 18 | 16 | 10 | 2 | 0 | 0 | 59 |
| Days Max Temp ≤ 32 °F | 0 | 0 | 0 | 0 | 0 | 0 | 0 | 0 | 0 | 0 | 0 | 0 | 0 |
| Days Min Temp ≤ 32 °F | 9 | 2 | 1 | 0 | 0 | 0 | 0 | 0 | 0 | 0 | 2 | 9 | 23 |
| Days Min Temp ≤ 0 °F | 0 | 0 | 0 | 0 | 0 | 0 | 0 | 0 | 0 | 0 | 0 | 0 | 0 |
| Heating Degree Days | 614 | 410 | 322 | 193 | 71 | 16 | 2 | 1 | 9 | 89 | 338 | 603 | 2668 |
| Cooling Degree Days | 0 | 0 | 1 | 23 | 105 | 219 | 317 | 279 | 204 | 76 | 1 | 0 | 1225 |
| Total Precipitation (") | 2.61 | 2.15 | 2.04 | 0.78 | 0.31 | 0.08 | 0.04 | 0.06 | 0.22 | 0.71 | 1.87 | 1.91 | 12.78 |
| Days ≥ 0.1" Precip | 6 | 5 | 6 | 3 | 1 | 0 | 0 | 0 | 1 | 2 | 4 | 5 | 33 |
| Total Snowfall (") | 0.0 | 0.0 | 0.0 | 0.0 | 0.0 | 0.0 | 0.0 | 0.0 | 0.0 | 0.0 | 0.0 | 0.0 | 0.0 |
| Days ≥ 1" Snow Depth | 0 | 0 | 0 | 0 | 0 | 0 | 0 | 0 | 0 | 0 | 0 | 0 | 0 |

## ASH MOUNTAIN *Tulare County* ELEVATION 1690 ft LAT/LONG 36° 29' N / 118° 50' W

| | JAN | FEB | MAR | APR | MAY | JUN | JUL | AUG | SEP | OCT | NOV | DEC | YEAR |
|---|---|---|---|---|---|---|---|---|---|---|---|---|---|
| Maximum Temp °F | 57.7 | 61.6 | 64.3 | 69.9 | 80.0 | 89.5 | 97.0 | 96.1 | 90.1 | 80.0 | 66.0 | 57.8 | 75.8 |
| Minimum Temp °F | 36.5 | 39.7 | 42.3 | 46.1 | 53.3 | 61.1 | 67.8 | 67.2 | 60.8 | 52.5 | 42.9 | 36.7 | 50.6 |
| Mean Temp °F | 47.1 | 50.6 | 53.3 | 58.0 | 66.6 | 75.4 | 82.4 | 81.7 | 75.5 | 66.3 | 54.5 | 47.3 | 63.2 |
| Days Max Temp ≥ 90 °F | 0 | 0 | 0 | 0 | 6 | 16 | 28 | 26 | 17 | 5 | 0 | 0 | 98 |
| Days Max Temp ≤ 32 °F | 0 | 0 | 0 | 0 | 0 | 0 | 0 | 0 | 0 | 0 | 0 | 0 | 0 |
| Days Min Temp ≤ 32 °F | 9 | 3 | 1 | 1 | 0 | 0 | 0 | 0 | 0 | 0 | 1 | 8 | 23 |
| Days Min Temp ≤ 0 °F | 0 | 0 | 0 | 0 | 0 | 0 | 0 | 0 | 0 | 0 | 0 | 0 | 0 |
| Heating Degree Days | 547 | 399 | 357 | 228 | 82 | 11 | 0 | 0 | 8 | 76 | 317 | 543 | 2568 |
| Cooling Degree Days | 0 | 0 | 2 | 33 | 141 | 336 | 547 | 533 | 340 | 136 | 6 | 0 | 2074 |
| Total Precipitation (") | 4.75 | 4.19 | 4.62 | 2.36 | 0.67 | 0.30 | 0.09 | 0.15 | 0.70 | 1.34 | 3.07 | 3.78 | 26.02 |
| Days ≥ 0.1" Precip | 6 | 6 | 7 | 4 | 2 | 1 | 0 | 0 | 1 | 2 | 5 | 5 | 39 |
| Total Snowfall (") | 0.6 | 0.1 | 0.2 | 0.2 | 0.0 | 0.0 | 0.0 | 0.0 | 0.0 | 0.0 | 0.6 | 0.3 | 2.0 |
| Days ≥ 1" Snow Depth | 0 | 0 | 0 | 0 | 0 | 0 | 0 | 0 | 0 | 0 | 0 | 0 | 0 |

## AUBERRY 1 NW *Fresno County* ELEVATION 1991 ft LAT/LONG 37° 5' N / 119° 29' W

| | JAN | FEB | MAR | APR | MAY | JUN | JUL | AUG | SEP | OCT | NOV | DEC | YEAR |
|---|---|---|---|---|---|---|---|---|---|---|---|---|---|
| Maximum Temp °F | 55.2 | 58.2 | 60.6 | 67.4 | 77.9 | 87.1 | 93.7 | 92.3 | 86.1 | 76.1 | 62.1 | 55.0 | 72.6 |
| Minimum Temp °F | 35.6 | 38.4 | 40.7 | 44.5 | 51.6 | 59.4 | 67.0 | 65.7 | 60.4 | 51.9 | 41.4 | 35.8 | 49.4 |
| Mean Temp °F | 45.4 | 48.3 | 50.7 | 56.0 | 64.8 | 73.3 | 80.3 | 79.0 | 73.3 | 64.0 | 51.8 | 45.4 | 61.0 |
| Days Max Temp ≥ 90 °F | 0 | 0 | 0 | 0 | 3 | 13 | 25 | 22 | 11 | 2 | 0 | 0 | 76 |
| Days Max Temp ≤ 32 °F | 0 | 0 | 0 | 0 | 0 | 0 | 0 | 0 | 0 | 0 | 0 | 0 | 0 |
| Days Min Temp ≤ 32 °F | 10 | 6 | 3 | 1 | 0 | 0 | 0 | 0 | 0 | 0 | 3 | 10 | 33 |
| Days Min Temp ≤ 0 °F | 0 | 0 | 0 | 0 | 0 | 0 | 0 | 0 | 0 | 0 | 0 | 0 | 0 |
| Heating Degree Days | 599 | 464 | 439 | 280 | 106 | 21 | 0 | 1 | 16 | 110 | 393 | 601 | 3030 |
| Cooling Degree Days | 0 | 0 | 1 | 23 | 106 | 276 | 479 | 457 | 284 | 96 | 2 | 0 | 1724 |
| Total Precipitation (") | 4.50 | 4.34 | 4.46 | 2.08 | 0.64 | 0.18 | 0.08 | 0.05 | 0.52 | 1.33 | 3.06 | 3.82 | 25.06 |
| Days ≥ 0.1" Precip | 6 | 6 | 7 | 3 | 1 | 0 | 0 | 0 | 1 | 2 | 4 | 6 | 36 |
| Total Snowfall (") | 0.5 | 0.1 | 0.9 | 0.3 | 0.0 | 0.0 | 0.0 | 0.0 | 0.0 | 0.0 | 0.0 | 0.6 | 2.4 |
| Days ≥ 1" Snow Depth | 0 | 0 | 0 | 0 | 0 | 0 | 0 | 0 | 0 | 0 | 0 | 1 | 1 |

## AUBURN *Placer County* ELEVATION 1362 ft LAT/LONG 38° 53' N / 121° 4' W

| | JAN | FEB | MAR | APR | MAY | JUN | JUL | AUG | SEP | OCT | NOV | DEC | YEAR |
|---|---|---|---|---|---|---|---|---|---|---|---|---|---|
| Maximum Temp °F | 54.1 | 58.6 | 61.8 | 67.6 | 76.8 | 84.8 | 91.9 | 91.1 | 85.6 | 76.0 | 61.8 | 54.0 | 72.0 |
| Minimum Temp °F | 36.2 | 39.4 | 41.5 | 44.8 | 50.5 | 56.8 | 62.2 | 61.3 | 57.6 | 51.1 | 42.3 | 36.4 | 48.3 |
| Mean Temp °F | 45.1 | 48.9 | 51.7 | 56.2 | 63.7 | 70.8 | 77.1 | 76.2 | 71.6 | 63.6 | 52.1 | 45.2 | 60.2 |
| Days Max Temp ≥ 90 °F | 0 | 0 | 0 | 0 | 3 | 10 | 21 | 19 | 11 | 2 | 0 | 0 | 66 |
| Days Max Temp ≤ 32 °F | 0 | 0 | 0 | 0 | 0 | 0 | 0 | 0 | 0 | 0 | 0 | 0 | 0 |
| Days Min Temp ≤ 32 °F | 9 | 3 | 1 | 0 | 0 | 0 | 0 | 0 | 0 | 0 | 1 | 9 | 23 |
| Days Min Temp ≤ 0 °F | 0 | 0 | 0 | 0 | 0 | 0 | 0 | 0 | 0 | 0 | 0 | 0 | 0 |
| Heating Degree Days | 610 | 447 | 406 | 268 | 117 | 27 | 2 | 2 | 19 | 113 | 384 | 607 | 3002 |
| Cooling Degree Days | 0 | 0 | 1 | 15 | 83 | 204 | 383 | 350 | 223 | 83 | 1 | 0 | 1343 |
| Total Precipitation (") | 6.21 | 5.31 | 5.94 | 2.50 | 0.80 | 0.36 | 0.14 | 0.16 | 0.71 | 1.92 | 5.46 | 5.56 | 35.07 |
| Days ≥ 0.1" Precip | 8 | 8 | 9 | 5 | 2 | 1 | 0 | 0 | 1 | 3 | 7 | 8 | 52 |
| Total Snowfall (") | 0.5 | 0.2 | 0.3 | 0.3 | 0.0 | 0.0 | 0.0 | 0.0 | 0.0 | 0.0 | 0.2 | 0.2 | 1.7 |
| Days ≥ 1" Snow Depth | 0 | 0 | 0 | 0 | 0 | 0 | 0 | 0 | 0 | 0 | 0 | 0 | 0 |

## BAKERSFIELD MEADOWS *Kern County* ELEVATION 492 ft LAT/LONG 35° 25' N / 119° 3' W

| | JAN | FEB | MAR | APR | MAY | JUN | JUL | AUG | SEP | OCT | NOV | DEC | YEAR |
|---|---|---|---|---|---|---|---|---|---|---|---|---|---|
| Maximum Temp °F | 57.1 | 64.3 | 69.0 | 75.9 | 85.0 | 92.5 | 98.6 | 96.9 | 90.8 | 81.2 | 67.1 | 56.8 | 77.9 |
| Minimum Temp °F | 39.1 | 42.9 | 46.6 | 50.1 | 57.6 | 64.4 | 70.1 | 69.0 | 64.2 | 55.3 | 45.0 | 38.4 | 53.6 |
| Mean Temp °F | 48.1 | 53.6 | 57.8 | 63.1 | 71.4 | 78.5 | 84.4 | 83.0 | 77.5 | 68.3 | 56.1 | 47.6 | 65.8 |
| Days Max Temp ≥ 90 °F | 0 | 0 | 0 | 2 | 11 | 20 | 28 | 26 | 18 | 6 | 0 | 0 | 111 |
| Days Max Temp ≤ 32 °F | 0 | 0 | 0 | 0 | 0 | 0 | 0 | 0 | 0 | 0 | 0 | 0 | 0 |
| Days Min Temp ≤ 32 °F | 5 | 1 | 0 | 0 | 0 | 0 | 0 | 0 | 0 | 0 | 0 | 5 | 11 |
| Days Min Temp ≤ 0 °F | 0 | 0 | 0 | 0 | 0 | 0 | 0 | 0 | 0 | 0 | 0 | 0 | 0 |
| Heating Degree Days | 516 | 316 | 224 | 113 | 25 | 1 | 0 | 0 | 2 | 41 | 266 | 532 | 2036 |
| Cooling Degree Days | 0 | 1 | 5 | 68 | 203 | 383 | 576 | 539 | 364 | 144 | 3 | 0 | 2286 |
| Total Precipitation (") | 0.99 | 1.03 | 1.25 | 0.55 | 0.19 | 0.10 | 0.01 | 0.08 | 0.14 | 0.28 | 0.69 | 0.76 | 6.07 |
| Days ≥ 0.1" Precip | 3 | 3 | 4 | 2 | 1 | 0 | 0 | 0 | 0 | 1 | 2 | 2 | 18 |
| Total Snowfall (") | 0.0 | 0.0 | 0.1 | 0.0 | 0.0 | 0.0 | 0.0 | 0.0 | 0.0 | 0.0 | 0.0 | 0.0 | 0.1 |
| Days ≥ 1" Snow Depth | 0 | 0 | 0 | 0 | 0 | 0 | 0 | 0 | 0 | 0 | 0 | 0 | 0 |

## BALCH POWER HOUSE *Fresno County* ELEVATION 1752 ft LAT/LONG 36° 54 ' N / 119° 6 ' W

| | JAN | FEB | MAR | APR | MAY | JUN | JUL | AUG | SEP | OCT | NOV | DEC | YEAR |
|---|---|---|---|---|---|---|---|---|---|---|---|---|---|
| Maximum Temp °F | 52.5 | 58.5 | 63.1 | 69.3 | 77.8 | 85.8 | 93.1 | 92.9 | 87.7 | 77.2 | 60.3 | 52.1 | 72.5 |
| Minimum Temp °F | 37.4 | 39.8 | 42.3 | 45.8 | 52.6 | 59.7 | 66.9 | 67.1 | 61.5 | 53.3 | 43.7 | 37.7 | 50.7 |
| Mean Temp °F | 45.0 | 49.2 | 52.7 | 57.6 | 65.2 | 72.8 | 80.0 | 80.1 | 74.6 | 65.3 | 52.0 | 44.9 | 61.6 |
| Days Max Temp ≥ 90 °F | 0 | 0 | 0 | 0 | 3 | 11 | 24 | 23 | 14 | 3 | 0 | 0 | 78 |
| Days Max Temp ≤ 32 °F | 0 | 0 | 0 | 0 | 0 | 0 | 0 | 0 | 0 | 0 | 0 | 0 | 0 |
| Days Min Temp ≤ 32 °F | 6 | 2 | 1 | 0 | 0 | 0 | 0 | 0 | 0 | 0 | 1 | 5 | 15 |
| Days Min Temp ≤ 0 °F | 0 | 0 | 0 | 0 | 0 | 0 | 0 | 0 | 0 | 0 | 0 | 0 | 0 |
| Heating Degree Days | 614 | 441 | 374 | 236 | 84 | 13 | 0 | 0 | 9 | 83 | 384 | 615 | 2853 |
| Cooling Degree Days | 0 | 0 | 0 | 24 | 94 | 252 | 469 | 478 | 317 | 103 | 1 | 0 | 1738 |
| Total Precipitation (") | 5.88 | 5.08 | 5.08 | 2.69 | 0.76 | 0.31 | 0.17 | 0.06 | 0.85 | 1.56 | 3.57 | 4.42 | 30.43 |
| Days ≥ 0.1" Precip | 6 | 6 | 8 | 5 | 2 | 1 | 0 | 0 | 1 | 2 | 5 | 6 | 42 |
| Total Snowfall (") | 0.1 | 0.0 | 0.0 | 0.0 | 0.0 | 0.0 | 0.0 | 0.0 | 0.0 | 0.0 | 0.0 | 0.2 | 0.3 |
| Days ≥ 1" Snow Depth | 0 | 1 | 0 | 0 | 0 | 0 | 0 | 0 | 0 | 0 | 0 | 0 | 1 |

## BEAUMONT 1 E *Riverside County* ELEVATION 2592 ft LAT/LONG 33° 56 ' N / 116° 56 ' W

| | JAN | FEB | MAR | APR | MAY | JUN | JUL | AUG | SEP | OCT | NOV | DEC | YEAR |
|---|---|---|---|---|---|---|---|---|---|---|---|---|---|
| Maximum Temp °F | 61.4 | 64.5 | 66.7 | 72.9 | 80.3 | 88.7 | 95.7 | 95.5 | 89.9 | 80.8 | 69.0 | 61.3 | 77.2 |
| Minimum Temp °F | 39.5 | 39.8 | 40.9 | 43.3 | 48.7 | 53.5 | 59.3 | 59.8 | 56.6 | 50.1 | 43.3 | 39.3 | 47.8 |
| Mean Temp °F | 50.5 | 52.2 | 53.8 | 58.1 | 64.5 | 71.1 | 77.5 | 77.7 | 73.3 | 65.5 | 56.2 | 50.3 | 62.6 |
| Days Max Temp ≥ 90 °F | 0 | 0 | 0 | 1 | 5 | 15 | 27 | 26 | 17 | 5 | 0 | 0 | 96 |
| Days Max Temp ≤ 32 °F | 0 | 0 | 0 | 0 | 0 | 0 | 0 | 0 | 0 | 0 | 0 | 0 | 0 |
| Days Min Temp ≤ 32 °F | 5 | 4 | 3 | 1 | 0 | 0 | 0 | 0 | 0 | 0 | 2 | 5 | 20 |
| Days Min Temp ≤ 0 °F | 0 | 0 | 0 | 0 | 0 | 0 | 0 | 0 | 0 | 0 | 0 | 0 | 0 |
| Heating Degree Days | 444 | 356 | 344 | 216 | 85 | 19 | 0 | 0 | 9 | 69 | 265 | 448 | 2255 |
| Cooling Degree Days | 0 | 1 | 3 | 27 | 81 | 226 | 392 | 408 | 278 | 94 | 8 | 1 | 1519 |
| Total Precipitation (") | 3.84 | 3.61 | 3.62 | 1.29 | 0.61 | 0.18 | 0.28 | 0.33 | 0.53 | 0.64 | 1.90 | 2.44 | 19.27 |
| Days ≥ 0.1" Precip | 5 | 4 | 6 | 3 | 1 | 0 | 1 | 1 | 1 | 2 | 3 | 4 | 31 |
| Total Snowfall (") | 0.5 | 0.3 | 0.0 | 0.0 | 0.0 | 0.0 | 0.0 | 0.0 | 0.0 | 0.0 | 0.1 | 0.4 | 1.3 |
| Days ≥ 1" Snow Depth | 0 | 0 | 0 | 0 | 0 | 0 | 0 | 0 | 0 | 0 | 0 | 0 | 0 |

## BERKELEY *Alameda County* ELEVATION 302 ft LAT/LONG 37° 52 ' N / 122° 16 ' W

| | JAN | FEB | MAR | APR | MAY | JUN | JUL | AUG | SEP | OCT | NOV | DEC | YEAR |
|---|---|---|---|---|---|---|---|---|---|---|---|---|---|
| Maximum Temp °F | 56.6 | 59.6 | 61.2 | 63.4 | 66.4 | ***68.9*** | ***69.9*** | 70.0 | 71.7 | 69.7 | ***62.4*** | 56.4 | 64.7 |
| Minimum Temp °F | 43.7 | 46.1 | 46.9 | 48.1 | 50.6 | ***53.2*** | ***54.3*** | ***55.2*** | ***55.8*** | 53.2 | ***48.5*** | 43.7 | 49.9 |
| Mean Temp °F | 50.2 | 52.9 | 54.1 | 55.8 | 58.5 | ***61.1*** | ***62.1*** | ***62.7*** | ***63.8*** | 61.5 | ***55.5*** | 50.0 | 57.4 |
| Days Max Temp ≥ 90 °F | 0 | 0 | 0 | 0 | 0 | 1 | 0 | 0 | 1 | 0 | ***0*** | 0 | 2 |
| Days Max Temp ≤ 32 °F | 0 | 0 | 0 | 0 | 0 | 0 | 0 | 0 | 0 | 0 | ***0*** | 0 | 0 |
| Days Min Temp ≤ 32 °F | 0 | 0 | 0 | 0 | 0 | 0 | 0 | 0 | 0 | 0 | ***0*** | 1 | 1 |
| Days Min Temp ≤ 0 °F | 0 | 0 | 0 | 0 | 0 | 0 | 0 | 0 | 0 | 0 | ***0*** | 0 | 0 |
| Heating Degree Days | 454 | 336 | 333 | 274 | 206 | ***135*** | ***102*** | ***81*** | ***73*** | 128 | ***281*** | 457 | 2860 |
| Cooling Degree Days | ***0*** | ***0*** | ***0*** | ***10*** | ***8*** | ***22*** | ***21*** | ***15*** | na | ***27*** | na | ***0*** | na |
| Total Precipitation (") | 5.30 | 3.95 | 3.61 | 1.81 | 0.37 | 0.14 | 0.08 | 0.11 | 0.39 | 1.31 | ***3.91*** | 3.73 | 24.71 |
| Days ≥ 0.1" Precip | 7 | 7 | ***6*** | 4 | 1 | 0 | 0 | 0 | 1 | 2 | ***6*** | 6 | 40 |
| Total Snowfall (") | 0.0 | 0.0 | 0.0 | 0.0 | 0.0 | 0.0 | 0.0 | 0.0 | 0.0 | 0.0 | ***0.0*** | 0.0 | 0.0 |
| Days ≥ 1" Snow Depth | 0 | 0 | ***0*** | 0 | 0 | 0 | 0 | 0 | 0 | 0 | ***0*** | 0 | 0 |

## BIG BAR RANGER STN *Trinity County* ELEVATION 1270 ft LAT/LONG 40° 45 ' N / 123° 15 ' W

| | JAN | FEB | MAR | APR | MAY | JUN | JUL | AUG | SEP | OCT | NOV | DEC | YEAR |
|---|---|---|---|---|---|---|---|---|---|---|---|---|---|
| Maximum Temp °F | na | na | na | na | ***78.4*** | ***87.1*** | 95.5 | ***94.9*** | ***88.2*** | ***74.8*** | na | na | na |
| Minimum Temp °F | na | na | na | na | ***43.2*** | ***49.7*** | 52.8 | ***52.5*** | ***47.8*** | ***41.0*** | na | na | na |
| Mean Temp °F | na | na | na | na | ***60.8*** | ***68.4*** | 74.2 | ***73.7*** | ***68.1*** | ***57.9*** | na | na | na |
| Days Max Temp ≥ 90 °F | 0 | 0 | 0 | 1 | 5 | 13 | 22 | 22 | 15 | 3 | 0 | 0 | 81 |
| Days Max Temp ≤ 32 °F | 0 | 0 | 0 | 0 | 0 | 0 | 0 | 0 | 0 | 0 | 0 | 0 | 0 |
| Days Min Temp ≤ 32 °F | ***12*** | ***10*** | ***7*** | ***3*** | 1 | 0 | 0 | 0 | 0 | 2 | 7 | ***14*** | 56 |
| Days Min Temp ≤ 0 °F | 0 | 0 | 0 | 0 | 0 | 0 | 0 | 0 | 0 | 0 | 0 | 0 | 0 |
| Heating Degree Days | na | na | na | na | ***169*** | ***45*** | 5 | ***8*** | ***42*** | na | na | na | na |
| Cooling Degree Days | na | na | na | na | na | na | 284 | na | na | na | na | na | na |
| Total Precipitation (") | 7.06 | 5.28 | 4.88 | 1.94 | 1.11 | 0.46 | 0.18 | 0.46 | 0.89 | 2.21 | 5.81 | 6.79 | 37.07 |
| Days ≥ 0.1" Precip | na | ***5*** | ***7*** | ***4*** | 3 | 1 | 0 | 1 | 2 | 3 | ***5*** | na | na |
| Total Snowfall (") | ***2.3*** | 0.3 | 0.2 | 0.0 | 0.0 | 0.0 | 0.0 | 0.0 | 0.0 | 0.0 | 0.4 | 1.4 | 4.6 |
| Days ≥ 1" Snow Depth | ***1*** | 0 | 0 | 0 | 0 | 0 | 0 | 0 | 0 | 0 | 0 | 1 | 2 |

## BIG BEAR LAKE *San Bernardino County* ELEVATION 6755 ft LAT/LONG 34° 15' N / 116° 55' W

| | JAN | FEB | MAR | APR | MAY | JUN | JUL | AUG | SEP | OCT | NOV | DEC | YEAR |
|---|---|---|---|---|---|---|---|---|---|---|---|---|---|
| Maximum Temp °F | 47.2 | 47.9 | 50.5 | 57.3 | 65.9 | 75.3 | 80.3 | 78.8 | 73.3 | 64.9 | 54.0 | 47.0 | 61.9 |
| Minimum Temp °F | 19.7 | 21.5 | 24.1 | 27.9 | 34.2 | 41.0 | 47.5 | 46.6 | 40.6 | 32.2 | 25.4 | 19.9 | 31.7 |
| Mean Temp °F | 33.4 | 34.7 | 37.3 | 42.6 | 50.1 | 58.2 | 63.9 | 62.7 | 57.0 | 48.5 | 39.7 | 33.4 | 46.8 |
| Days Max Temp ≥ 90 °F | 0 | 0 | 0 | 0 | 0 | 0 | 1 | 0 | 0 | 0 | 0 | 0 | 1 |
| Days Max Temp ≤ 32 °F | 2 | 1 | 1 | 0 | 0 | 0 | 0 | 0 | 0 | 0 | 0 | 2 | 6 |
| Days Min Temp ≤ 32 °F | 29 | 27 | 29 | 24 | 13 | 2 | 0 | 0 | 3 | 17 | 26 | 30 | 200 |
| Days Min Temp ≤ 0 °F | 1 | 0 | 0 | 0 | 0 | 0 | 0 | 0 | 0 | 0 | 0 | 1 | 2 |
| Heating Degree Days | 971 | 850 | 851 | 665 | 455 | 206 | 62 | 90 | 237 | 503 | 751 | 971 | 6612 |
| Cooling Degree Days | 0 | 0 | 0 | 0 | 0 | 9 | 31 | 23 | 1 | 0 | 0 | 0 | 64 |
| Total Precipitation (") | 4.57 | 4.31 | 3.52 | 1.39 | 0.51 | 0.16 | 0.86 | 1.02 | 0.50 | 0.71 | 2.35 | 3.77 | 23.67 |
| Days ≥ 0.1" Precip | 5 | 4 | 5 | 2 | 1 | 0 | 2 | 2 | 1 | 2 | 3 | 4 | 31 |
| Total Snowfall (") | *12.8* | *15.6* | *16.6* | 2.0 | 0.5 | 0.0 | 0.0 | 0.0 | 0.1 | 0.2 | 6.5 | *11.3* | 65.6 |
| Days ≥ 1" Snow Depth | na | na | na | *2* | 0 | 0 | 0 | 0 | 0 | 0 | *2* | na | na |

## BISHOP AP *Inyo County* ELEVATION 4147 ft LAT/LONG 37° 22' N / 118° 25' W

| | JAN | FEB | MAR | APR | MAY | JUN | JUL | AUG | SEP | OCT | NOV | DEC | YEAR |
|---|---|---|---|---|---|---|---|---|---|---|---|---|---|
| Maximum Temp °F | 53.1 | 58.0 | 63.6 | 71.2 | 80.9 | 90.7 | 97.2 | 95.1 | 86.9 | 76.0 | 62.1 | 53.2 | 74.0 |
| Minimum Temp °F | 22.1 | 26.0 | 30.7 | 35.9 | 44.0 | 51.0 | 56.2 | 54.3 | 46.8 | 37.3 | 27.5 | 21.5 | 37.8 |
| Mean Temp °F | 37.6 | 42.0 | 47.2 | 53.6 | 62.5 | 70.9 | 76.7 | 74.7 | 66.9 | 56.7 | 44.9 | 37.4 | 55.9 |
| Days Max Temp ≥ 90 °F | 0 | 0 | 0 | 0 | 5 | 18 | 29 | 26 | 13 | 1 | 0 | 0 | 92 |
| Days Max Temp ≤ 32 °F | 1 | 0 | 0 | 0 | 0 | 0 | 0 | 0 | 0 | 0 | 0 | 1 | 2 |
| Days Min Temp ≤ 32 °F | 29 | 24 | 19 | 8 | 1 | 0 | 0 | 0 | 0 | 7 | 23 | 29 | 140 |
| Days Min Temp ≤ 0 °F | 0 | 0 | 0 | 0 | 0 | 0 | 0 | 0 | 0 | 0 | 0 | 0 | 0 |
| Heating Degree Days | 842 | 642 | 545 | 338 | 125 | 17 | 1 | 1 | 45 | 260 | 598 | 850 | 4264 |
| Cooling Degree Days | 0 | 0 | 0 | 5 | 52 | 202 | 362 | 304 | 109 | 10 | 0 | 0 | 1044 |
| Total Precipitation (") | 1.05 | 0.94 | 0.53 | 0.24 | 0.23 | 0.16 | 0.22 | 0.15 | 0.27 | 0.16 | 0.55 | 0.88 | 5.38 |
| Days ≥ 0.1" Precip | 2 | 2 | 1 | 1 | 1 | 0 | 1 | 1 | 1 | 1 | 1 | 2 | 14 |
| Total Snowfall (") | 4.0 | 2.3 | 0.6 | 0.2 | 0.0 | 0.0 | 0.0 | 0.0 | 0.0 | 0.1 | 0.3 | 1.6 | 9.1 |
| Days ≥ 1" Snow Depth | 3 | 2 | 1 | 0 | 0 | 0 | 0 | 0 | 0 | 0 | 0 | 2 | 8 |

## BLUE CANYON *Placer County* ELEVATION 5280 ft LAT/LONG 39° 17' N / 120° 42' W

| | JAN | FEB | MAR | APR | MAY | JUN | JUL | AUG | SEP | OCT | NOV | DEC | YEAR |
|---|---|---|---|---|---|---|---|---|---|---|---|---|---|
| Maximum Temp °F | *44.7* | *45.5* | *46.3* | *51.7* | *61.9* | *70.1* | *77.3* | *77.2* | *71.5* | *62.5* | *49.8* | *44.7* | 58.6 |
| Minimum Temp °F | *32.1* | *32.2* | *32.5* | *35.8* | *44.2* | *51.9* | *58.7* | *58.1* | *52.9* | *45.7* | *36.4* | *32.1* | 42.7 |
| Mean Temp °F | *38.4* | *38.9* | *39.4* | *43.8* | *53.1* | *61.1* | *68.0* | *67.6* | *62.2* | *54.1* | *43.2* | *38.4* | 50.7 |
| Days Max Temp ≥ 90 °F | *0* | *0* | *0* | *0* | *0* | *0* | *1* | *1* | *0* | *0* | *0* | *0* | 2 |
| Days Max Temp ≤ 32 °F | *4* | *3* | *2* | *1* | *0* | *0* | *0* | *0* | *0* | *0* | *1* | *4* | 15 |
| Days Min Temp ≤ 32 °F | *17* | *16* | *17* | *12* | *4* | *0* | *0* | *0* | *0* | *2* | *10* | *16* | 94 |
| Days Min Temp ≤ 0 °F | *0* | *0* | *0* | *0* | *0* | *0* | *0* | *0* | *0* | *0* | *0* | *0* | 0 |
| Heating Degree Days | *817* | *731* | *787* | *629* | *374* | *168* | *44* | *56* | *143* | *345* | *649* | *816* | 5559 |
| Cooling Degree Days | na | na | na | na | na | na | na | na | na | na | na | na | na |
| Total Precipitation (") | *12.62* | *10.03* | *9.17* | *5.09* | *2.04* | *0.75* | *0.41* | *0.65* | *1.45* | *3.66* | *10.42* | *11.08* | 67.37 |
| Days ≥ 0.1" Precip | *10* | *9* | *10* | *8* | *4* | *2* | *0* | *1* | *2* | *5* | *9* | *10* | 70 |
| Total Snowfall (") | *43.3* | *44.1* | *44.5* | *26.0* | *5.4* | *0.5* | *0.0* | *0.0* | *0.3* | *3.5* | *27.5* | *41.8* | 236.9 |
| Days ≥ 1" Snow Depth | *26* | *24* | *25* | *15* | *4* | *0* | *0* | *0* | *0* | *1* | *10* | *21* | 126 |

## BLYTHE *Riverside County* ELEVATION 268 ft LAT/LONG 33° 37' N / 114° 36' W

| | JAN | FEB | MAR | APR | MAY | JUN | JUL | AUG | SEP | OCT | NOV | DEC | YEAR |
|---|---|---|---|---|---|---|---|---|---|---|---|---|---|
| Maximum Temp °F | 67.0 | 72.9 | 78.6 | 86.7 | 95.4 | 104.2 | 108.2 | 106.5 | 100.9 | 89.6 | 75.4 | 66.0 | 87.6 |
| Minimum Temp °F | 38.9 | 42.8 | 47.4 | 53.0 | 61.0 | 68.3 | 76.6 | 75.5 | 67.9 | 56.4 | 44.9 | 38.5 | 55.9 |
| Mean Temp °F | 53.0 | 57.8 | 63.0 | 69.9 | 78.2 | 86.3 | 92.4 | 91.1 | 84.4 | 73.0 | 60.2 | 52.3 | 71.8 |
| Days Max Temp ≥ 90 °F | 0 | 0 | 3 | 12 | 24 | 29 | 31 | 31 | 28 | 17 | 0 | 0 | 175 |
| Days Max Temp ≤ 32 °F | 0 | 0 | 0 | 0 | 0 | 0 | 0 | 0 | 0 | 0 | 0 | 0 | 0 |
| Days Min Temp ≤ 32 °F | 6 | 2 | 0 | 0 | 0 | 0 | 0 | 0 | 0 | 0 | 0 | 4 | 12 |
| Days Min Temp ≤ 0 °F | 0 | 0 | 0 | 0 | 0 | 0 | 0 | 0 | 0 | 0 | 0 | 0 | 0 |
| Heating Degree Days | 367 | 202 | 105 | 24 | 1 | 0 | 0 | 0 | 0 | 10 | 158 | 388 | 1255 |
| Cooling Degree Days | 0 | 7 | 55 | 219 | 449 | 661 | *845* | 821 | 593 | 275 | 21 | 0 | 3946 |
| Total Precipitation (") | 0.48 | 0.44 | 0.37 | 0.13 | 0.06 | 0.03 | 0.16 | 0.74 | 0.39 | 0.29 | 0.32 | 0.57 | 3.98 |
| Days ≥ 0.1" Precip | 1 | 1 | 1 | 0 | 0 | 0 | 0 | 1 | 1 | 1 | 1 | 1 | 8 |
| Total Snowfall (") | 0.0 | 0.0 | 0.0 | 0.0 | 0.0 | 0.0 | 0.0 | 0.0 | 0.0 | 0.0 | 0.0 | 0.0 | 0.0 |
| Days ≥ 1" Snow Depth | 0 | 0 | 0 | 0 | 0 | 0 | 0 | 0 | 0 | 0 | 0 | 0 | 0 |

## BLYTHE RIVERSIDE AP *Riverside County* ELEVATION 390 ft LAT/LONG 33° 37 ' N / 114° 43 ' W

| | JAN | FEB | MAR | APR | MAY | JUN | JUL | AUG | SEP | OCT | NOV | DEC | YEAR |
|---|---|---|---|---|---|---|---|---|---|---|---|---|---|
| Maximum Temp °F | 66.4 | 72.1 | 78.1 | 86.2 | 94.9 | 104.9 | 108.4 | 106.5 | 100.5 | 89.1 | 74.9 | 65.7 | 87.3 |
| Minimum Temp °F | 41.5 | 45.5 | 50.3 | 56.2 | 64.1 | 72.7 | 80.8 | 79.6 | 72.3 | 60.3 | 48.1 | 41.0 | 59.4 |
| Mean Temp °F | 54.0 | 58.8 | 64.2 | 71.2 | 79.5 | 88.8 | 94.6 | 93.1 | 86.4 | 74.8 | 61.5 | 53.4 | 73.4 |
| Days Max Temp ≥ 90 °F | 0 | 0 | 3 | 11 | 24 | 29 | 31 | 31 | 28 | 16 | 1 | 0 | 174 |
| Days Max Temp ≤ 32 °F | 0 | 0 | 0 | 0 | 0 | 0 | 0 | 0 | 0 | 0 | 0 | 0 | 0 |
| Days Min Temp ≤ 32 °F | 3 | 1 | 0 | 0 | 0 | 0 | 0 | 0 | 0 | 0 | 0 | 2 | 6 |
| Days Min Temp ≤ 0 °F | 0 | 0 | 0 | 0 | 0 | 0 | 0 | 0 | 0 | 0 | 0 | 0 | 0 |
| Heating Degree Days | 336 | 179 | 83 | 21 | 1 | 0 | 0 | 0 | 0 | 10 | 132 | 354 | 1116 |
| Cooling Degree Days | 1 | 16 | 73 | 261 | 468 | 727 | 903 | 876 | 652 | 332 | 33 | 1 | 4343 |
| Total Precipitation (") | 0.45 | 0.45 | 0.42 | 0.21 | 0.02 | 0.01 | 0.32 | 0.65 | 0.42 | 0.28 | 0.25 | 0.54 | 4.02 |
| Days ≥ 0.1" Precip | 1 | 1 | 1 | 0 | 0 | 0 | 1 | 1 | 1 | 1 | 1 | 1 | 9 |
| Total Snowfall (") | 0.0 | 0.0 | 0.0 | 0.0 | 0.0 | 0.0 | 0.0 | 0.0 | 0.0 | 0.0 | 0.0 | 0.0 | 0.0 |
| Days ≥ 1" Snow Depth | 0 | 0 | 0 | 0 | 0 | 0 | 0 | 0 | 0 | 0 | 0 | 0 | 0 |

## BOCA *Nevada County* ELEVATION 5584 ft LAT/LONG 39° 23 ' N / 120° 6 ' W

| | JAN | FEB | MAR | APR | MAY | JUN | JUL | AUG | SEP | OCT | NOV | DEC | YEAR |
|---|---|---|---|---|---|---|---|---|---|---|---|---|---|
| Maximum Temp °F | 41.6 | 45.3 | 49.8 | 56.8 | 66.3 | 75.0 | 83.4 | 82.9 | 75.9 | 65.9 | 51.1 | 42.1 | 61.3 |
| Minimum Temp °F | 9.4 | 12.7 | 19.0 | 23.4 | 29.5 | 34.2 | 37.6 | 36.0 | 30.5 | 24.6 | 20.0 | 12.0 | 24.1 |
| Mean Temp °F | 25.5 | 29.0 | 34.4 | 40.1 | 47.9 | 54.6 | 60.5 | 59.5 | 53.3 | 45.3 | 35.6 | 27.1 | 42.7 |
| Days Max Temp ≥ 90 °F | 0 | 0 | 0 | 0 | 0 | 1 | 6 | 5 | 1 | 0 | 0 | 0 | 13 |
| Days Max Temp ≤ 32 °F | 4 | 2 | 1 | 0 | 0 | 0 | 0 | 0 | 0 | 0 | 1 | 4 | 12 |
| Days Min Temp ≤ 32 °F | 30 | 28 | 30 | 28 | 22 | 12 | 7 | 9 | 19 | 28 | 28 | 30 | 271 |
| Days Min Temp ≤ 0 °F | 8 | 4 | 1 | 0 | 0 | 0 | 0 | 0 | 0 | 0 | 0 | 5 | 18 |
| Heating Degree Days | 1218 | 1009 | 941 | 741 | 524 | 308 | 151 | 176 | 346 | 605 | 876 | 1169 | 8064 |
| Cooling Degree Days | 0 | 0 | 0 | 0 | 0 | 4 | 17 | 10 | 0 | 0 | 0 | 0 | 31 |
| Total Precipitation (") | 3.76 | 3.17 | 2.83 | 1.09 | 0.90 | 0.75 | 0.59 | 0.59 | 0.87 | 1.39 | 2.99 | 3.23 | 22.16 |
| Days ≥ 0.1" Precip | 6 | 5 | 6 | 4 | 3 | 3 | 1 | 2 | 2 | 3 | 5 | 6 | 46 |
| Total Snowfall (") | 20.1 | 19.9 | 16.9 | 5.7 | 1.5 | 0.1 | 0.0 | 0.0 | 0.2 | 1.2 | 10.3 | 20.4 | 96.3 |
| Days ≥ 1" Snow Depth | 24 | 22 | 16 | 4 | 1 | 0 | 0 | 0 | 0 | 1 | 7 | 20 | 95 |

## BODIE *Mono County* ELEVATION 8369 ft LAT/LONG 38° 13 ' N / 119° 1 ' W

| | JAN | FEB | MAR | APR | MAY | JUN | JUL | AUG | SEP | OCT | NOV | DEC | YEAR |
|---|---|---|---|---|---|---|---|---|---|---|---|---|---|
| Maximum Temp °F | 39.8 | 41.3 | 43.4 | 49.8 | 59.6 | 68.9 | 76.4 | 75.7 | 68.9 | 59.6 | 47.8 | 39.9 | 55.9 |
| Minimum Temp °F | 5.0 | 7.1 | 12.4 | 17.5 | 24.7 | 31.0 | 35.2 | 34.0 | 27.0 | 19.6 | 12.8 | 5.4 | 19.3 |
| Mean Temp °F | 22.4 | 24.2 | 27.9 | 33.7 | 42.2 | 50.0 | 55.9 | 54.9 | 48.0 | 39.6 | 30.3 | 22.7 | 37.7 |
| Days Max Temp ≥ 90 °F | 0 | 0 | 0 | 0 | 0 | 0 | 0 | 0 | 0 | 0 | 0 | 0 | 0 |
| Days Max Temp ≤ 32 °F | 7 | 5 | 4 | 2 | 0 | 0 | 0 | 0 | 0 | 0 | 2 | 7 | 27 |
| Days Min Temp ≤ 32 °F | 31 | 28 | 31 | 30 | 27 | 18 | 11 | 13 | 23 | 29 | 29 | 31 | 301 |
| Days Min Temp ≤ 0 °F | 11 | 7 | 4 | 1 | 0 | 0 | 0 | 0 | 0 | 0 | 4 | 10 | 37 |
| Heating Degree Days | 1313 | 1144 | 1144 | 933 | 700 | 444 | 279 | 307 | 503 | 781 | 1033 | 1306 | 9887 |
| Cooling Degree Days | 0 | 0 | 0 | 0 | 0 | 0 | 1 | 0 | 0 | 0 | 0 | 0 | 1 |
| Total Precipitation (") | 1.85 | 1.84 | 1.57 | 0.90 | 0.73 | 0.73 | 1.05 | 0.66 | 0.62 | 0.61 | 1.42 | 1.57 | 13.55 |
| Days ≥ 0.1" Precip | 4 | 4 | 4 | 3 | 2 | 2 | 3 | 2 | 2 | 2 | *3* | *4* | 35 |
| Total Snowfall (") | *19.0* | *20.6* | 16.8 | 7.5 | 4.4 | 0.7 | 0.0 | 0.1 | 0.9 | *3.2* | *13.4* | *20.1* | 106.7 |
| Days ≥ 1" Snow Depth | *27* | *25* | *25* | 12 | 3 | 0 | 0 | 0 | 0 | *1* | *11* | *24* | 128 |

## BORREGO DESERT PARK *San Diego County* ELEVATION 850 ft LAT/LONG 33° 16 ' N / 116° 25 ' W

| | JAN | FEB | MAR | APR | MAY | JUN | JUL | AUG | SEP | OCT | NOV | DEC | YEAR |
|---|---|---|---|---|---|---|---|---|---|---|---|---|---|
| Maximum Temp °F | 68.9 | 73.0 | 77.3 | 84.9 | 92.8 | 103.0 | 107.1 | 105.6 | 100.6 | 89.8 | 77.6 | 68.9 | 87.5 |
| Minimum Temp °F | 43.2 | 46.6 | 49.6 | 53.5 | 60.0 | 68.2 | 74.9 | 74.5 | 69.2 | 60.2 | 49.8 | 43.0 | 57.7 |
| Mean Temp °F | 56.1 | 59.8 | 63.5 | 69.2 | 76.4 | 85.6 | 91.0 | 90.1 | 84.9 | 75.0 | 63.7 | 55.9 | 72.6 |
| Days Max Temp ≥ 90 °F | 0 | 1 | 3 | 10 | 22 | 28 | 31 | 31 | 28 | 17 | 2 | 0 | 173 |
| Days Max Temp ≤ 32 °F | 0 | 0 | 0 | 0 | 0 | 0 | 0 | 0 | 0 | 0 | 0 | 0 | 0 |
| Days Min Temp ≤ 32 °F | 2 | 0 | 0 | 0 | 0 | 0 | 0 | 0 | 0 | 0 | 0 | 1 | 3 |
| Days Min Temp ≤ 0 °F | 0 | 0 | 0 | 0 | 0 | 0 | 0 | 0 | 0 | 0 | 0 | 0 | 0 |
| Heating Degree Days | 272 | 165 | 110 | 40 | 9 | 0 | 0 | 0 | 0 | 10 | 101 | 277 | 984 |
| Cooling Degree Days | 3 | 26 | 66 | 211 | 374 | 620 | 804 | 794 | 605 | 340 | 67 | 5 | 3915 |
| Total Precipitation (") | 1.33 | 1.27 | 0.90 | 0.23 | 0.10 | 0.02 | 0.36 | 0.66 | 0.41 | 0.24 | 0.65 | 0.94 | 7.11 |
| Days ≥ 0.1" Precip | 3 | 2 | 2 | 1 | 0 | 0 | 1 | 1 | 1 | 1 | 1 | 2 | 15 |
| Total Snowfall (") | 0.0 | 0.0 | 0.0 | 0.0 | 0.0 | 0.0 | 0.0 | 0.0 | 0.0 | 0.0 | 0.0 | 0.2 | 0.2 |
| Days ≥ 1" Snow Depth | 0 | 0 | 0 | 0 | 0 | 0 | 0 | 0 | 0 | 0 | 0 | 0 | 0 |

## BOWMAN DAM *Nevada County* ELEVATION 5354 ft LAT/LONG 39° 27 ' N / 120° 39 ' W

| | JAN | FEB | MAR | APR | MAY | JUN | JUL | AUG | SEP | OCT | NOV | DEC | YEAR |
|---|---|---|---|---|---|---|---|---|---|---|---|---|---|
| Maximum Temp °F | 45.7 | 46.4 | 48.5 | 54.0 | 63.1 | 72.1 | 79.5 | 79.2 | 73.8 | 64.1 | 50.9 | 44.9 | 60.2 |
| Minimum Temp °F | 27.2 | 26.7 | 28.4 | 31.6 | 38.7 | 46.3 | 52.5 | 52.4 | 48.4 | 41.0 | 32.3 | 27.0 | 37.7 |
| Mean Temp °F | 36.5 | 36.6 | 38.5 | 42.8 | 50.9 | 59.2 | 66.0 | 65.9 | 61.1 | 52.6 | 41.6 | 36.0 | 49.0 |
| Days Max Temp ≥ 90 °F | 0 | 0 | 0 | 0 | 0 | 0 | 1 | 1 | 0 | 0 | 0 | 0 | 2 |
| Days Max Temp ≤ 32 °F | 2 | 2 | 1 | 0 | 0 | 0 | 0 | 0 | 0 | 0 | 1 | 3 | 9 |
| Days Min Temp ≤ 32 °F | 24 | 23 | 24 | 16 | 6 | 1 | 0 | 0 | 0 | 4 | 15 | 24 | 137 |
| Days Min Temp ≤ 0 °F | 0 | 0 | 0 | 0 | 0 | 0 | 0 | 0 | 0 | 0 | 0 | 0 | 0 |
| Heating Degree Days | 878 | 797 | 816 | 659 | 430 | 194 | 55 | 62 | 147 | 385 | 695 | 893 | 6011 |
| Cooling Degree Days | 0 | 0 | 0 | *0* | 2 | 26 | 82 | *89* | 39 | 8 | 0 | 0 | 246 |
| Total Precipitation (") | 11.23 | 9.39 | 9.22 | 4.21 | 2.41 | 1.26 | 0.37 | 0.83 | 1.33 | 4.25 | 9.88 | 10.02 | 64.40 |
| Days ≥ 0.1" Precip | 9 | 9 | 10 | 7 | 5 | 3 | 0 | 1 | 2 | 4 | 8 | 9 | 67 |
| Total Snowfall (") | 44.2 | 49.8 | 48.7 | 23.7 | 5.6 | 0.4 | 0.0 | 0.0 | 0.3 | 2.2 | 25.1 | 45.6 | 245.6 |
| Days ≥ 1" Snow Depth | 27 | 26 | 28 | 16 | 4 | 0 | 0 | 0 | 0 | 1 | 10 | 23 | 135 |

## BRAWLEY 2 SW *Imperial County* ELEVATION -121 ft LAT/LONG 32° 59 ' N / 115° 32 ' W

| | JAN | FEB | MAR | APR | MAY | JUN | JUL | AUG | SEP | OCT | NOV | DEC | YEAR |
|---|---|---|---|---|---|---|---|---|---|---|---|---|---|
| Maximum Temp °F | 69.3 | 73.8 | 78.5 | 85.2 | 93.2 | 102.6 | 106.5 | 105.4 | 101.0 | 90.9 | 78.3 | 69.2 | 87.8 |
| Minimum Temp °F | 39.2 | 42.9 | 47.2 | 52.0 | 58.7 | 65.8 | 74.2 | 74.8 | 68.5 | 57.6 | 46.0 | 39.0 | 55.5 |
| Mean Temp °F | 54.3 | 58.4 | 62.9 | 68.6 | 76.0 | 84.3 | 90.4 | 90.1 | 84.8 | 74.3 | 62.2 | 54.1 | 71.7 |
| Days Max Temp ≥ 90 °F | 0 | 0 | 3 | 10 | 22 | 29 | 31 | 31 | 28 | 19 | 2 | 0 | 175 |
| Days Max Temp ≤ 32 °F | 0 | 0 | 0 | 0 | 0 | 0 | 0 | 0 | 0 | 0 | 0 | 0 | 0 |
| Days Min Temp ≤ 32 °F | 5 | 1 | 0 | 0 | 0 | 0 | 0 | 0 | 0 | 0 | 0 | 4 | 10 |
| Days Min Temp ≤ 0 °F | 0 | 0 | 0 | 0 | 0 | 0 | 0 | 0 | 0 | 0 | 0 | 0 | 0 |
| Heating Degree Days | 325 | 189 | 106 | 32 | 3 | 0 | 0 | 0 | 0 | 8 | 118 | 331 | 1112 |
| Cooling Degree Days | 1 | 11 | 53 | 187 | 367 | 600 | 793 | 802 | 619 | 326 | 42 | 0 | 3801 |
| Total Precipitation (") | 0.51 | 0.40 | 0.44 | 0.10 | 0.05 | 0.01 | 0.10 | 0.47 | 0.36 | 0.28 | 0.27 | 0.46 | 3.45 |
| Days ≥ 0.1" Precip | 1 | 1 | 1 | 0 | 0 | 0 | 0 | 1 | 1 | 1 | 1 | 1 | 8 |
| Total Snowfall (") | 0.0 | 0.0 | 0.0 | 0.0 | 0.0 | 0.0 | 0.0 | 0.0 | 0.0 | 0.0 | 0.0 | 0.0 | 0.0 |
| Days ≥ 1" Snow Depth | 0 | 0 | 0 | 0 | 0 | 0 | 0 | 0 | 0 | 0 | 0 | 0 | 0 |

## BRIDGEPORT *Mono County* ELEVATION 6375 ft LAT/LONG 38° 16 ' N / 119° 14 ' W

| | JAN | FEB | MAR | APR | MAY | JUN | JUL | AUG | SEP | OCT | NOV | DEC | YEAR |
|---|---|---|---|---|---|---|---|---|---|---|---|---|---|
| Maximum Temp °F | 42.2 | 45.7 | 51.6 | 59.0 | 67.5 | 76.0 | 83.4 | 82.2 | 75.7 | 66.9 | 53.3 | 43.7 | 62.3 |
| Minimum Temp °F | 8.1 | 11.5 | 18.3 | 22.3 | 29.1 | 35.7 | 39.9 | 38.3 | 30.4 | 22.0 | 16.4 | 9.4 | 23.5 |
| Mean Temp °F | 25.2 | 28.7 | 35.0 | 40.7 | 48.3 | 55.9 | 61.7 | 60.3 | 53.0 | 44.5 | 34.9 | 26.6 | 42.9 |
| Days Max Temp ≥ 90 °F | 0 | 0 | 0 | 0 | 0 | 0 | 3 | 2 | 0 | 0 | 0 | 0 | 5 |
| Days Max Temp ≤ 32 °F | 5 | 2 | 1 | 0 | 0 | 0 | 0 | 0 | 0 | 0 | 1 | 5 | 14 |
| Days Min Temp ≤ 32 °F | 31 | 28 | 29 | 27 | 21 | 9 | 4 | 6 | 19 | 28 | 28 | 30 | 260 |
| Days Min Temp ≤ 0 °F | 8 | 5 | 1 | 0 | 0 | 0 | 0 | 0 | 0 | 0 | 2 | 7 | 23 |
| Heating Degree Days | 1230 | 1019 | 923 | 723 | 510 | 270 | 113 | 148 | 353 | 629 | 897 | 1185 | 8000 |
| Cooling Degree Days | 0 | 0 | 0 | 0 | 0 | 5 | 28 | 16 | 0 | 0 | 0 | 0 | 49 |
| Total Precipitation (") | 1.45 | 1.33 | 0.90 | 0.44 | 0.50 | 0.63 | 0.46 | 0.51 | 0.55 | 0.36 | 1.07 | 1.14 | 9.34 |
| Days ≥ 0.1" Precip | 3 | 3 | 2 | 1 | 2 | 2 | 1 | 2 | 1 | 1 | 2 | 3 | 23 |
| Total Snowfall (") | 8.9 | 12.0 | 6.2 | 2.5 | 1.4 | 0.0 | 0.0 | 0.0 | 0.3 | 0.7 | 4.4 | *8.3* | 44.7 |
| Days ≥ 1" Snow Depth | *15* | 13 | 6 | 1 | 0 | 0 | 0 | 0 | 0 | 0 | 3 | *11* | 49 |

## BURBANK VALLEY PUMP *Los Angeles County* ELEVATION 655 ft LAT/LONG 34° 11 ' N / 118° 21 ' W

| | JAN | FEB | MAR | APR | MAY | JUN | JUL | AUG | SEP | OCT | NOV | DEC | YEAR |
|---|---|---|---|---|---|---|---|---|---|---|---|---|---|
| Maximum Temp °F | 67.8 | 70.2 | 71.3 | 75.5 | 78.6 | 83.9 | 90.1 | 90.7 | 88.0 | 82.4 | 73.5 | 67.4 | 78.3 |
| Minimum Temp °F | 41.2 | 43.5 | 45.7 | 48.6 | 53.6 | 57.6 | 61.3 | 61.7 | 59.4 | 53.4 | 45.6 | 40.7 | 51.0 |
| Mean Temp °F | 54.5 | 56.9 | 58.5 | 62.1 | 66.1 | 70.8 | 75.7 | 76.2 | 73.8 | 67.9 | 59.6 | 54.1 | 64.7 |
| Days Max Temp ≥ 90 °F | 0 | 0 | 1 | 2 | 3 | 7 | 17 | 17 | 13 | 7 | 1 | 0 | 68 |
| Days Max Temp ≤ 32 °F | 0 | 0 | 0 | 0 | 0 | 0 | 0 | 0 | 0 | 0 | 0 | 0 | 0 |
| Days Min Temp ≤ 32 °F | 2 | 1 | 1 | 0 | 0 | 0 | 0 | 0 | 0 | 0 | 0 | 2 | 6 |
| Days Min Temp ≤ 0 °F | 0 | 0 | 0 | 0 | 0 | 0 | 0 | 0 | 0 | 0 | 0 | 0 | 0 |
| Heating Degree Days | 320 | 227 | 205 | 121 | 47 | 8 | 0 | 0 | 2 | 29 | 172 | 332 | 1463 |
| Cooling Degree Days | 2 | 8 | 15 | 60 | 96 | 201 | 343 | 364 | 291 | 138 | 14 | 1 | 1533 |
| Total Precipitation (") | 3.37 | 3.70 | 3.55 | 1.09 | 0.22 | 0.06 | 0.02 | 0.18 | 0.34 | 0.51 | 1.91 | 2.36 | 17.31 |
| Days ≥ 0.1" Precip | 4 | 4 | 4 | 2 | 0 | 0 | 0 | 0 | 1 | 1 | 2 | 3 | 21 |
| Total Snowfall (") | 0.0 | 0.0 | 0.0 | 0.0 | 0.0 | 0.0 | 0.0 | 0.0 | 0.0 | 0.0 | 0.0 | 0.0 | 0.0 |
| Days ≥ 1" Snow Depth | 0 | 0 | 0 | 0 | 0 | 0 | 0 | 0 | 0 | 0 | 0 | 0 | 0 |

## BUTTONWILLOW *Kern County* ELEVATION 269 ft LAT/LONG 35° 24 ' N / 119° 28 ' W

| | JAN | FEB | MAR | APR | MAY | JUN | JUL | AUG | SEP | OCT | NOV | DEC | YEAR |
|---|---|---|---|---|---|---|---|---|---|---|---|---|---|
| Maximum Temp °F | 55.5 | 63.5 | 68.8 | 75.8 | 85.2 | 92.3 | 97.4 | 96.1 | 90.8 | 81.8 | 67.1 | 55.9 | 77.5 |
| Minimum Temp °F | 34.7 | 38.9 | 43.2 | 46.6 | 53.6 | 59.9 | 64.8 | 63.2 | 57.8 | 48.5 | 39.1 | 32.9 | 48.6 |
| Mean Temp °F | 45.2 | 51.2 | 56.1 | 61.2 | 69.4 | 76.1 | 81.1 | 79.7 | 74.4 | 65.2 | 53.1 | 44.4 | 63.1 |
| Days Max Temp ≥ 90 °F | 0 | 0 | 0 | 2 | 10 | 19 | 28 | 26 | 18 | 6 | 0 | 0 | 109 |
| Days Max Temp ≤ 32 °F | 0 | 0 | 0 | 0 | 0 | 0 | 0 | 0 | 0 | 0 | 0 | 0 | 0 |
| Days Min Temp ≤ 32 °F | 12 | 5 | 1 | 0 | 0 | 0 | 0 | 0 | 0 | 0 | 6 | 15 | 39 |
| Days Min Temp ≤ 0 °F | 0 | 0 | 0 | 0 | 0 | 0 | 0 | 0 | 0 | 0 | 0 | 0 | 0 |
| Heating Degree Days | 608 | 383 | 274 | 143 | 32 | 2 | 0 | 0 | 4 | 75 | 351 | 630 | 2502 |
| Cooling Degree Days | 0 | 0 | 4 | 47 | 178 | 340 | 496 | 459 | 292 | 93 | 1 | 0 | 1910 |
| Total Precipitation (") | 1.04 | 1.01 | 1.18 | 0.54 | 0.15 | 0.06 | 0.03 | 0.04 | 0.21 | 0.24 | 0.65 | 0.66 | 5.81 |
| Days ≥ 0.1" Precip | 3 | 3 | 3 | 2 | 0 | 0 | 0 | 0 | 1 | 1 | 2 | 2 | 17 |
| Total Snowfall (") | 0.0 | 0.0 | 0.0 | 0.0 | 0.0 | 0.0 | 0.0 | 0.0 | 0.0 | 0.0 | 0.0 | 0.0 | 0.0 |
| Days ≥ 1" Snow Depth | 0 | 0 | 0 | 0 | 0 | 0 | 0 | 0 | 0 | 0 | 0 | 0 | 0 |

## CACHUMA LAKE *Santa Barbara County* ELEVATION 781 ft LAT/LONG 34° 35 ' N / 119° 59 ' W

| | JAN | FEB | MAR | APR | MAY | JUN | JUL | AUG | SEP | OCT | NOV | DEC | YEAR |
|---|---|---|---|---|---|---|---|---|---|---|---|---|---|
| Maximum Temp °F | 65.3 | 66.8 | 68.0 | 72.6 | 77.4 | 83.7 | 89.6 | 90.6 | 87.0 | 82.0 | 72.7 | 65.5 | 76.8 |
| Minimum Temp °F | 39.0 | 40.7 | 42.3 | 43.6 | 47.3 | 49.9 | 52.3 | 53.0 | 51.9 | 48.6 | 43.2 | 38.5 | 45.9 |
| Mean Temp °F | 52.2 | 53.8 | 55.2 | 58.1 | 62.4 | 66.8 | 71.0 | 71.8 | 69.5 | 65.3 | 58.0 | 52.1 | 61.4 |
| Days Max Temp ≥ 90 °F | 0 | 0 | 0 | 1 | 3 | 8 | 16 | 17 | 11 | 7 | 1 | 0 | 64 |
| Days Max Temp ≤ 32 °F | 0 | 0 | 0 | 0 | 0 | 0 | 0 | 0 | 0 | 0 | 0 | 0 | 0 |
| Days Min Temp ≤ 32 °F | 5 | 2 | 1 | 0 | 0 | 0 | 0 | 0 | 0 | 0 | 1 | 5 | 14 |
| Days Min Temp ≤ 0 °F | 0 | 0 | 0 | 0 | 0 | 0 | 0 | 0 | 0 | 0 | 0 | 0 | 0 |
| Heating Degree Days | 391 | 312 | 300 | 214 | 115 | 32 | 3 | 2 | 12 | 65 | 218 | 395 | 2059 |
| Cooling Degree Days | 0 | 2 | 4 | 22 | 44 | 107 | 207 | 233 | 154 | 88 | 14 | 1 | 876 |
| Total Precipitation (") | 3.92 | 4.53 | 4.37 | 1.34 | 0.22 | 0.03 | 0.01 | 0.03 | 0.29 | 0.58 | 2.09 | 2.99 | 20.40 |
| Days ≥ 0.1" Precip | 5 | 5 | 5 | 2 | 1 | 0 | 0 | 0 | 1 | 1 | 3 | 4 | 27 |
| Total Snowfall (") | 0.0 | 0.0 | 0.0 | 0.0 | 0.0 | 0.0 | 0.0 | 0.0 | 0.0 | 0.0 | 0.0 | 0.0 | 0.0 |
| Days ≥ 1" Snow Depth | 0 | 0 | 0 | 0 | 0 | 0 | 0 | 0 | 0 | 0 | 0 | 0 | 0 |

## CALAVERAS BIG TREES *Calaveras County* ELEVATION 4705 ft LAT/LONG 38° 17 ' N / 120° 19 ' W

| | JAN | FEB | MAR | APR | MAY | JUN | JUL | AUG | SEP | OCT | NOV | DEC | YEAR |
|---|---|---|---|---|---|---|---|---|---|---|---|---|---|
| Maximum Temp °F | 44.9 | 46.8 | 49.3 | 55.4 | 65.5 | 74.8 | 82.2 | 81.7 | 74.9 | 64.7 | 51.5 | 44.4 | 61.3 |
| Minimum Temp °F | 27.2 | 27.8 | 29.4 | 32.6 | 38.9 | 46.0 | 51.5 | 50.9 | 46.3 | 39.4 | 31.8 | 27.2 | 37.4 |
| Mean Temp °F | 36.1 | 37.3 | 39.4 | 44.0 | 52.3 | 60.4 | 66.9 | 66.3 | 60.7 | 52.1 | 41.7 | 35.8 | 49.4 |
| Days Max Temp ≥ 90 °F | 0 | 0 | 0 | 0 | 0 | 1 | 3 | 3 | 1 | 0 | 0 | 0 | 8 |
| Days Max Temp ≤ 32 °F | 2 | 1 | 1 | 0 | 0 | 0 | 0 | 0 | 0 | 0 | 0 | 3 | 7 |
| Days Min Temp ≤ 32 °F | 26 | 23 | 23 | 16 | 6 | 1 | 0 | 0 | 1 | 5 | 18 | 25 | 144 |
| Days Min Temp ≤ 0 °F | 0 | 0 | 0 | 0 | 0 | 0 | 0 | 0 | 0 | 0 | 0 | 0 | 0 |
| Heating Degree Days | 890 | 775 | 788 | 623 | 392 | 165 | 41 | 49 | 157 | 398 | 693 | 898 | 5869 |
| Cooling Degree Days | 0 | 0 | 0 | 0 | 5 | 33 | 104 | 99 | 34 | 7 | 0 | 0 | 282 |
| Total Precipitation (") | 9.84 | 8.53 | 8.59 | 4.31 | 1.38 | 0.66 | 0.28 | 0.32 | 0.99 | 3.18 | 7.38 | 8.15 | 53.61 |
| Days ≥ 0.1" Precip | 9 | 8 | 10 | 6 | 3 | 1 | 0 | 0 | 1 | 3 | 7 | 8 | 56 |
| Total Snowfall (") | 25.9 | 24.2 | 27.2 | 15.7 | 1.0 | 0.1 | 0.0 | 0.0 | 0.0 | 0.5 | 7.9 | 21.4 | 123.9 |
| Days ≥ 1" Snow Depth | *14* | *15* | *15* | na | 1 | 0 | 0 | 0 | 0 | 0 | *3* | *10* | na |

## CALLAHAN *Siskiyou County* ELEVATION 3143 ft LAT/LONG 41° 18 ' N / 122° 48 ' W

| | JAN | FEB | MAR | APR | MAY | JUN | JUL | AUG | SEP | OCT | NOV | DEC | YEAR |
|---|---|---|---|---|---|---|---|---|---|---|---|---|---|
| Maximum Temp °F | 44.2 | 50.8 | 56.4 | 62.2 | 71.9 | 79.5 | 87.1 | 86.2 | 79.4 | 67.8 | 51.0 | 43.5 | 65.0 |
| Minimum Temp °F | 25.2 | 28.0 | 30.3 | 33.2 | 38.4 | 44.5 | 49.2 | 48.1 | 42.3 | 35.8 | 30.0 | 25.5 | 35.9 |
| Mean Temp °F | 34.8 | 39.5 | 43.4 | 47.7 | 55.2 | 62.0 | 68.1 | 67.2 | 60.9 | 51.8 | 40.5 | 34.5 | 50.5 |
| Days Max Temp ≥ 90 °F | 0 | 0 | 0 | 0 | 1 | 4 | 13 | 11 | 3 | 0 | 0 | 0 | 32 |
| Days Max Temp ≤ 32 °F | 2 | 0 | 0 | 0 | 0 | 0 | 0 | 0 | 0 | 0 | 0 | 2 | 4 |
| Days Min Temp ≤ 32 °F | 26 | 21 | 20 | 15 | 6 | 1 | 0 | 0 | 2 | 9 | 21 | 25 | 146 |
| Days Min Temp ≤ 0 °F | 0 | 0 | 0 | 0 | 0 | 0 | 0 | 0 | 0 | 0 | 0 | 0 | 0 |
| Heating Degree Days | 931 | 715 | 663 | 512 | 305 | 128 | 31 | 37 | 142 | 402 | 728 | 939 | 5533 |
| Cooling Degree Days | 0 | 0 | 0 | 0 | 7 | 39 | 118 | 101 | 24 | 1 | 0 | 0 | 290 |
| Total Precipitation (") | 3.32 | 2.55 | 2.15 | 1.33 | 0.94 | 0.85 | 0.33 | 0.37 | 0.61 | 1.34 | 3.05 | 3.44 | 20.28 |
| Days ≥ 0.1" Precip | 6 | 5 | 5 | 4 | 3 | 2 | 1 | 1 | 1 | 3 | 7 | 7 | 45 |
| Total Snowfall (") | *3.8* | *1.7* | 1.7 | 0.2 | 0.1 | 0.0 | 0.0 | 0.0 | 0.0 | 0.1 | 0.9 | *2.6* | 11.1 |
| Days ≥ 1" Snow Depth | na | *1* | 0 | 0 | 0 | 0 | 0 | 0 | 0 | 0 | 0 | *2* | na |

## CAMP PARDEE *Calaveras County* ELEVATION 658 ft LAT/LONG 38° 15' N / 120° 51' W

| | JAN | FEB | MAR | APR | MAY | JUN | JUL | AUG | SEP | OCT | NOV | DEC | YEAR |
|---|---|---|---|---|---|---|---|---|---|---|---|---|---|
| Maximum Temp °F | 53.4 | 59.3 | 63.4 | 70.3 | 80.5 | 88.8 | 95.3 | 93.7 | 88.3 | 78.1 | 63.4 | 53.7 | 74.0 |
| Minimum Temp °F | 38.2 | 41.3 | 43.7 | 46.2 | 51.3 | 56.6 | 61.5 | 60.8 | 58.0 | 52.5 | 44.8 | 38.4 | 49.4 |
| Mean Temp °F | 45.8 | 50.4 | 53.6 | 58.3 | 65.9 | 72.7 | 78.4 | 77.3 | 73.2 | 65.3 | 54.1 | 46.1 | 61.8 |
| Days Max Temp ≥ 90 °F | 0 | 0 | 0 | 0 | 6 | 14 | 25 | 23 | 14 | 3 | 0 | 0 | 85 |
| Days Max Temp ≤ 32 °F | 0 | 0 | 0 | 0 | 0 | 0 | 0 | 0 | 0 | 0 | 0 | 0 | 0 |
| Days Min Temp ≤ 32 °F | 5 | 1 | 0 | 0 | 0 | 0 | 0 | 0 | 0 | 0 | 0 | 5 | 11 |
| Days Min Temp ≤ 0 °F | 0 | 0 | 0 | 0 | 0 | 0 | 0 | 0 | 0 | 0 | 0 | 0 | 0 |
| Heating Degree Days | 589 | 408 | 349 | 214 | 75 | 13 | 0 | 1 | 9 | 78 | 323 | 581 | 2640 |
| Cooling Degree Days | 0 | 0 | 1 | 24 | 111 | 257 | 422 | 387 | 268 | 107 | 2 | 0 | 1579 |
| Total Precipitation (") | 3.82 | 3.36 | 3.83 | 1.89 | 0.47 | 0.23 | 0.10 | 0.14 | 0.51 | 1.22 | 3.19 | 3.21 | 21.97 |
| Days ≥ 0.1" Precip | 7 | 6 | 7 | 4 | 1 | 1 | 0 | 0 | 1 | 2 | 5 | 7 | 41 |
| Total Snowfall (") | 0.0 | 0.0 | 0.0 | 0.0 | 0.0 | 0.0 | 0.0 | 0.0 | 0.0 | 0.0 | 0.0 | 0.0 | 0.0 |
| Days ≥ 1" Snow Depth | 0 | 0 | 0 | 0 | 0 | 0 | 0 | 0 | 0 | 0 | 0 | 0 | 0 |

## CAMPO *San Diego County* ELEVATION 2592 ft LAT/LONG 32° 37' N / 116° 28' W

| | JAN | FEB | MAR | APR | MAY | JUN | JUL | AUG | SEP | OCT | NOV | DEC | YEAR |
|---|---|---|---|---|---|---|---|---|---|---|---|---|---|
| Maximum Temp °F | 61.8 | 64.0 | 65.7 | 71.4 | 77.7 | 86.8 | 93.3 | 93.3 | 88.2 | 79.3 | 68.7 | 62.1 | 76.0 |
| Minimum Temp °F | 33.9 | 33.9 | 35.6 | 36.9 | 40.9 | 44.9 | 51.8 | 52.8 | 48.6 | 42.1 | 36.3 | 32.6 | 40.9 |
| Mean Temp °F | 47.9 | 49.0 | 50.7 | 54.2 | 59.3 | 65.9 | 72.6 | 73.0 | 68.4 | 60.8 | 52.5 | 47.4 | 58.5 |
| Days Max Temp ≥ 90 °F | 0 | 0 | 0 | 1 | 3 | 13 | 24 | 24 | 15 | 5 | 0 | 0 | 85 |
| Days Max Temp ≤ 32 °F | 0 | 0 | 0 | 0 | 0 | 0 | 0 | 0 | 0 | 0 | 0 | 0 | 0 |
| Days Min Temp ≤ 32 °F | 15 | 13 | 10 | 7 | 3 | 0 | 0 | 0 | 0 | 2 | 10 | 17 | 77 |
| Days Min Temp ≤ 0 °F | 0 | 0 | 0 | 0 | 0 | 0 | 0 | 0 | 0 | 0 | 0 | 0 | 0 |
| Heating Degree Days | 523 | 445 | 439 | 320 | 187 | 62 | 5 | 8 | 37 | 155 | 368 | 540 | 3089 |
| Cooling Degree Days | 0 | 0 | 0 | 3 | 24 | 106 | 231 | 265 | 153 | 32 | 1 | 0 | 815 |
| Total Precipitation (") | 3.13 | 2.62 | 2.66 | 1.06 | 0.23 | 0.07 | 0.37 | 0.63 | 0.36 | 0.59 | 1.68 | 2.02 | 15.42 |
| Days ≥ 0.1" Precip | 5 | 5 | 5 | 3 | 1 | 0 | 1 | 1 | 1 | 1 | 3 | 4 | 30 |
| Total Snowfall (") | 0.0 | 0.0 | 0.0 | 0.0 | 0.0 | 0.0 | 0.0 | 0.0 | 0.0 | 0.0 | 0.0 | 0.1 | 0.1 |
| Days ≥ 1" Snow Depth | 0 | 0 | 0 | 0 | 0 | 0 | 0 | 0 | 0 | 0 | 0 | 0 | 0 |

## CANOGA PARK PIERCE C *Los Angeles County* ELEVATION 790 ft LAT/LONG 34° 11' N / 118° 34' W

| | JAN | FEB | MAR | APR | MAY | JUN | JUL | AUG | SEP | OCT | NOV | DEC | YEAR |
|---|---|---|---|---|---|---|---|---|---|---|---|---|---|
| Maximum Temp °F | 68.7 | 70.8 | 72.4 | 77.4 | 81.5 | 88.4 | 95.2 | 95.7 | 91.5 | 84.6 | 74.9 | 68.5 | 80.8 |
| Minimum Temp °F | 39.6 | 41.0 | 42.6 | 44.9 | 49.7 | 53.8 | 57.6 | 58.3 | 55.6 | 49.8 | 42.9 | 38.6 | 47.9 |
| Mean Temp °F | 54.2 | 55.9 | 57.5 | 61.2 | 65.7 | 71.1 | 76.4 | 77.0 | 73.6 | 67.2 | 58.9 | 53.6 | 64.4 |
| Days Max Temp ≥ 90 °F | 0 | 0 | 1 | 3 | 7 | 14 | 24 | 25 | 17 | 9 | 2 | 0 | 102 |
| Days Max Temp ≤ 32 °F | 0 | 0 | 0 | 0 | 0 | 0 | 0 | 0 | 0 | 0 | 0 | 0 | 0 |
| Days Min Temp ≤ 32 °F | 5 | 2 | 1 | 0 | 0 | 0 | 0 | 0 | 0 | 0 | 1 | 5 | 14 |
| Days Min Temp ≤ 0 °F | 0 | 0 | 0 | 0 | 0 | 0 | 0 | 0 | 0 | 0 | 0 | 0 | 0 |
| Heating Degree Days | 331 | 253 | 234 | 142 | 64 | 13 | 0 | 0 | 4 | 37 | 190 | 350 | 1618 |
| Cooling Degree Days | 2 | 4 | 10 | 51 | 99 | 221 | 366 | 395 | 282 | 124 | 14 | 1 | 1569 |
| Total Precipitation (") | 3.67 | 3.89 | 3.44 | 0.97 | 0.19 | 0.03 | 0.01 | 0.17 | 0.25 | 0.58 | 2.25 | 2.55 | 18.00 |
| Days ≥ 0.1" Precip | 4 | 4 | 4 | 2 | 0 | 0 | 0 | 0 | 1 | 1 | 3 | 3 | 22 |
| Total Snowfall (") | 0.0 | 0.0 | 0.0 | 0.0 | 0.0 | 0.0 | 0.0 | 0.0 | 0.0 | 0.0 | 0.0 | 0.0 | 0.0 |
| Days ≥ 1" Snow Depth | 0 | 0 | 0 | 0 | 0 | 0 | 0 | 0 | 0 | 0 | 0 | 0 | 0 |

## CANYON DAM *Plumas County* ELEVATION 4560 ft LAT/LONG 40° 10' N / 121° 5' W

| | JAN | FEB | MAR | APR | MAY | JUN | JUL | AUG | SEP | OCT | NOV | DEC | YEAR |
|---|---|---|---|---|---|---|---|---|---|---|---|---|---|
| Maximum Temp °F | 39.5 | 43.7 | 49.1 | 57.3 | 67.8 | 76.0 | 84.4 | 83.6 | 76.3 | 64.1 | 48.1 | 39.8 | 60.8 |
| Minimum Temp °F | 21.6 | 23.8 | 26.8 | 30.0 | 36.4 | 42.5 | 47.3 | 46.2 | 40.9 | 34.3 | 28.3 | 22.7 | 33.4 |
| Mean Temp °F | 30.6 | 33.8 | 38.0 | 43.7 | 52.1 | 59.3 | 65.9 | 64.9 | 58.6 | 49.2 | 38.2 | 31.3 | 47.1 |
| Days Max Temp ≥ 90 °F | 0 | 0 | 0 | 0 | 0 | 1 | 7 | 6 | 1 | 0 | 0 | 0 | 15 |
| Days Max Temp ≤ 32 °F | 3 | 1 | 0 | 0 | 0 | 0 | 0 | 0 | 0 | 0 | 0 | 3 | 7 |
| Days Min Temp ≤ 32 °F | 29 | 26 | 27 | 21 | 8 | 1 | 0 | 0 | 2 | 12 | 24 | 30 | 180 |
| Days Min Temp ≤ 0 °F | 0 | 0 | 0 | 0 | 0 | 0 | 0 | 0 | 0 | 0 | 0 | 0 | 0 |
| Heating Degree Days | 1060 | 875 | 831 | 634 | 395 | 186 | 51 | 65 | 196 | 483 | 797 | 1038 | 6611 |
| Cooling Degree Days | 0 | 0 | 0 | 0 | 3 | 20 | 83 | 62 | 16 | 0 | 0 | 0 | 184 |
| Total Precipitation (") | 7.02 | 5.84 | 5.45 | 2.34 | 1.24 | 0.70 | 0.19 | 0.46 | 0.86 | 2.12 | 5.20 | 5.78 | 37.20 |
| Days ≥ 0.1" Precip | 8 | 8 | 9 | 5 | 3 | 2 | 0 | 1 | 2 | 3 | 8 | 7 | 56 |
| Total Snowfall (") | 28.9 | 22.7 | 20.2 | 7.4 | 0.7 | 0.2 | 0.0 | 0.0 | 0.1 | 0.5 | 11.1 | 26.0 | 117.8 |
| Days ≥ 1" Snow Depth | 29 | 26 | 22 | 8 | 1 | 0 | 0 | 0 | 0 | 0 | 7 | 22 | 115 |

## CECILVILLE 1 SE *Siskiyou County* ELEVATION 2960 ft LAT/LONG 41° 8 ' N / 123° 8 ' W

| | JAN | FEB | MAR | APR | MAY | JUN | JUL | AUG | SEP | OCT | NOV | DEC | YEAR |
|---|---|---|---|---|---|---|---|---|---|---|---|---|---|
| Maximum Temp °F | 47.3 | 52.8 | 57.5 | 63.8 | 73.4 | 81.6 | 89.7 | 89.9 | 82.3 | 71.1 | 54.2 | 45.3 | 67.4 |
| Minimum Temp °F | 27.6 | 29.5 | 31.4 | 33.8 | 39.5 | 45.5 | 49.5 | 48.4 | 43.6 | 37.1 | 31.2 | 27.0 | 37.0 |
| Mean Temp °F | 37.4 | 41.3 | 44.4 | 48.8 | 56.5 | 63.6 | 69.6 | 69.2 | 63.0 | 54.1 | 42.7 | 36.2 | 52.2 |
| Days Max Temp ≥ 90 °F | 0 | 0 | 0 | 0 | 1 | 7 | 17 | 19 | na | 1 | 0 | 0 | na |
| Days Max Temp ≤ 32 °F | 0 | 0 | 0 | 0 | 0 | 0 | 0 | 0 | 0 | 0 | 0 | 1 | 1 |
| Days Min Temp ≤ 32 °F | 24 | 20 | 18 | 13 | 4 | 0 | 0 | 0 | 1 | 6 | 19 | 25 | 130 |
| Days Min Temp ≤ 0 °F | 0 | 0 | 0 | 0 | 0 | 0 | 0 | 0 | 0 | 0 | 0 | 0 | 0 |
| Heating Degree Days | 848 | 663 | 634 | 480 | 270 | 107 | 27 | 27 | 112 | 337 | 661 | 886 | 5052 |
| Cooling Degree Days | 0 | 0 | 0 | 2 | 20 | 88 | 207 | 193 | 75 | 12 | 0 | 0 | 597 |
| Total Precipitation (") | 6.40 | 4.60 | 4.28 | 2.04 | 1.44 | 0.77 | 0.35 | 0.67 | 0.97 | 2.35 | 4.89 | 5.97 | 34.73 |
| Days ≥ 0.1" Precip | 9 | 8 | 8 | 5 | 4 | 2 | 1 | 2 | 2 | 4 | 8 | na | na |
| Total Snowfall (") | 6.2 | 5.7 | 1.8 | 1.3 | 0.0 | 0.0 | 0.0 | 0.0 | 0.0 | 0.0 | 1.6 | 8.4 | 25.0 |
| Days ≥ 1" Snow Depth | 5 | 3 | 1 | 0 | 0 | 0 | 0 | 0 | 0 | 0 | 1 | 6 | 16 |

## CEDARVILLE *Modoc County* ELEVATION 4652 ft LAT/LONG 41° 32 ' N / 120° 10 ' W

| | JAN | FEB | MAR | APR | MAY | JUN | JUL | AUG | SEP | OCT | NOV | DEC | YEAR |
|---|---|---|---|---|---|---|---|---|---|---|---|---|---|
| Maximum Temp °F | 39.2 | 44.3 | 50.1 | 57.0 | 66.9 | 76.2 | 86.0 | 85.0 | 76.9 | 64.8 | 48.2 | 39.9 | 61.2 |
| Minimum Temp °F | 19.8 | 23.9 | 28.7 | 33.3 | 40.2 | 47.6 | 54.1 | 52.3 | 43.5 | 34.6 | 26.5 | 20.0 | 35.4 |
| Mean Temp °F | 29.5 | 34.1 | 39.5 | 45.2 | 53.6 | 61.9 | 70.1 | 68.7 | 60.3 | 49.7 | 37.4 | 29.9 | 48.3 |
| Days Max Temp ≥ 90 °F | 0 | 0 | 0 | 0 | 0 | 3 | 11 | 10 | 2 | 0 | 0 | 0 | 26 |
| Days Max Temp ≤ 32 °F | 7 | 2 | 1 | 0 | 0 | 0 | 0 | 0 | 0 | 0 | 2 | 6 | 18 |
| Days Min Temp ≤ 32 °F | 27 | 24 | 22 | 14 | 4 | 0 | 0 | 0 | 2 | 12 | 23 | 28 | 156 |
| Days Min Temp ≤ 0 °F | 1 | 1 | 0 | 0 | 0 | 0 | 0 | 0 | 0 | 0 | 0 | 2 | 4 |
| Heating Degree Days | 1093 | 866 | 785 | 588 | 357 | 149 | 28 | 41 | 172 | 467 | 822 | 1080 | 6448 |
| Cooling Degree Days | 0 | 0 | 0 | 0 | 11 | 62 | 173 | 152 | 35 | 1 | 0 | 0 | 434 |
| Total Precipitation (") | 1.65 | 1.30 | 1.47 | 1.09 | 1.03 | 0.69 | 0.30 | 0.44 | 0.57 | 0.92 | 1.81 | 1.53 | 12.80 |
| Days ≥ 0.1" Precip | 4 | 4 | 4 | 4 | 3 | 2 | 1 | 1 | 2 | 3 | 6 | 5 | 39 |
| Total Snowfall (") | 8.1 | 4.7 | 4.2 | 2.2 | 0.6 | 0.0 | 0.0 | 0.0 | 0.0 | 0.5 | 2.4 | 6.5 | 29.2 |
| Days ≥ 1" Snow Depth | 11 | 4 | 3 | 1 | 0 | 0 | 0 | 0 | 0 | 0 | 2 | 6 | 27 |

## CHERRY VALLEY DAM *Tuolumne County* ELEVATION 4774 ft LAT/LONG 37° 58 ' N / 119° 55 ' W

| | JAN | FEB | MAR | APR | MAY | JUN | JUL | AUG | SEP | OCT | NOV | DEC | YEAR |
|---|---|---|---|---|---|---|---|---|---|---|---|---|---|
| Maximum Temp °F | 49.0 | 51.2 | 53.7 | 60.2 | 69.5 | 78.3 | 86.6 | 86.2 | 79.2 | 68.7 | 55.4 | 48.3 | 65.5 |
| Minimum Temp °F | 28.1 | 28.4 | 30.2 | 34.3 | 41.9 | 48.9 | 55.3 | 54.8 | 49.4 | 41.7 | 33.4 | 28.2 | 39.6 |
| Mean Temp °F | 38.6 | 39.8 | 42.0 | 47.3 | 55.7 | 63.6 | 71.0 | 70.5 | 64.3 | 55.2 | 44.4 | 38.2 | 52.6 |
| Days Max Temp ≥ 90 °F | 0 | 0 | 0 | 0 | 0 | 3 | 11 | 11 | 3 | 0 | 0 | 0 | 28 |
| Days Max Temp ≤ 32 °F | 1 | 1 | 0 | 0 | 0 | 0 | 0 | 0 | 0 | 0 | 0 | 1 | 3 |
| Days Min Temp ≤ 32 °F | 22 | 21 | 20 | 12 | 3 | 0 | 0 | 0 | 0 | 3 | 14 | 22 | 117 |
| Days Min Temp ≤ 0 °F | 0 | 0 | 0 | 0 | 0 | 0 | 0 | 0 | 0 | 0 | 0 | 0 | 0 |
| Heating Degree Days | 812 | 698 | 707 | 526 | 294 | 108 | 13 | 19 | 94 | 310 | 612 | 822 | 5015 |
| Cooling Degree Days | 0 | 0 | 0 | 1 | 15 | 81 | 207 | 206 | 82 | 16 | 0 | 0 | 608 |
| Total Precipitation (") | 7.93 | 7.12 | 6.95 | 3.62 | 1.14 | 0.73 | 0.18 | 0.27 | 0.96 | 2.73 | 6.70 | 6.79 | 45.12 |
| Days ≥ 0.1" Precip | 7 | 6 | 8 | 5 | 2 | 2 | 0 | 1 | 2 | 3 | 6 | 6 | 48 |
| Total Snowfall (") | 23.6 | 19.9 | 26.8 | 12.5 | 0.7 | 0.1 | 0.0 | 0.0 | 0.0 | 0.3 | 8.5 | 20.9 | 113.3 |
| Days ≥ 1" Snow Depth | 18 | 16 | 11 | 5 | 0 | 0 | 0 | 0 | 0 | 0 | 5 | 13 | 68 |

## CHESTER *Plumas County* ELEVATION 4534 ft LAT/LONG 40° 18 ' N / 121° 13 ' W

| | JAN | FEB | MAR | APR | MAY | JUN | JUL | AUG | SEP | OCT | NOV | DEC | YEAR |
|---|---|---|---|---|---|---|---|---|---|---|---|---|---|
| Maximum Temp °F | 42.0 | 46.5 | 51.4 | 58.7 | 68.7 | 76.9 | 85.0 | 84.1 | 78.1 | 66.8 | 50.3 | 41.8 | 62.5 |
| Minimum Temp °F | 18.8 | 21.7 | 25.3 | 28.5 | 34.2 | 40.4 | 44.1 | 42.5 | 37.4 | 31.1 | 25.7 | 19.9 | 30.8 |
| Mean Temp °F | 30.4 | 34.2 | 38.4 | 43.6 | 51.5 | 58.7 | 64.6 | 63.3 | 57.8 | 49.0 | 38.0 | 30.9 | 46.7 |
| Days Max Temp ≥ 90 °F | 0 | 0 | 0 | 0 | 0 | 2 | 7 | 7 | 2 | 0 | 0 | 0 | 18 |
| Days Max Temp ≤ 32 °F | 2 | 1 | 0 | 0 | 0 | 0 | 0 | 0 | 0 | 0 | 0 | 3 | 6 |
| Days Min Temp ≤ 32 °F | 30 | 27 | 28 | 24 | 13 | 3 | 0 | 1 | 5 | 20 | 26 | 29 | 206 |
| Days Min Temp ≤ 0 °F | 1 | 1 | 0 | 0 | 0 | 0 | 0 | 0 | 0 | 0 | 0 | 1 | 3 |
| Heating Degree Days | 1064 | 864 | 818 | 635 | 413 | 198 | 66 | 87 | 217 | 490 | 802 | 1052 | 6706 |
| Cooling Degree Days | 0 | 0 | 0 | 0 | 2 | 20 | 68 | 44 | 9 | 0 | 0 | 0 | 143 |
| Total Precipitation (") | 5.86 | 4.99 | 4.46 | 1.91 | 1.18 | 0.80 | 0.30 | 0.47 | 0.80 | 1.93 | 4.46 | 4.98 | 32.14 |
| Days ≥ 0.1" Precip | 8 | 8 | 8 | 4 | 3 | 2 | 1 | 1 | 2 | 4 | 7 | 7 | 55 |
| Total Snowfall (") | 34.3 | 25.5 | 19.8 | 7.0 | 0.6 | 0.2 | 0.0 | 0.0 | 0.1 | 0.5 | 12.8 | 30.9 | 131.7 |
| Days ≥ 1" Snow Depth | 27 | 23 | 16 | 4 | 0 | 0 | 0 | 0 | 0 | 0 | 5 | 20 | 95 |

### CHICO UNIVERSITY FAR *Butte County* ELEVATION 205 ft LAT/LONG 39° 42 ' N / 121° 49 ' W

| | JAN | FEB | MAR | APR | MAY | JUN | JUL | AUG | SEP | OCT | NOV | DEC | YEAR |
|---|---|---|---|---|---|---|---|---|---|---|---|---|---|
| Maximum Temp °F | 54.3 | 60.7 | 65.2 | 72.1 | 81.7 | 89.1 | 94.5 | 93.5 | 89.1 | ***78.7*** | 63.7 | 54.3 | 74.7 |
| Minimum Temp °F | 34.9 | 38.6 | 41.4 | 44.4 | 51.3 | 57.2 | 60.9 | 58.9 | 54.6 | 47.2 | 40.3 | 35.2 | 47.1 |
| Mean Temp °F | 44.6 | 49.7 | 53.3 | 58.2 | 66.5 | 73.1 | 77.8 | 76.2 | 71.9 | ***62.9*** | 52.1 | 44.8 | 60.9 |
| Days Max Temp ≥ 90 °F | 0 | 0 | 0 | 0 | 7 | 14 | 24 | 23 | 16 | 4 | 0 | 0 | 88 |
| Days Max Temp ≤ 32 °F | 0 | 0 | 0 | 0 | 0 | 0 | 0 | 0 | 0 | 0 | 0 | 0 | 0 |
| Days Min Temp ≤ 32 °F | 13 | 6 | 3 | 0 | 0 | 0 | 0 | 0 | 0 | 0 | 4 | 11 | 37 |
| Days Min Temp ≤ 0 °F | 0 | 0 | 0 | 0 | 0 | 0 | 0 | 0 | 0 | 0 | 0 | 0 | 0 |
| Heating Degree Days | 625 | 425 | 356 | 211 | 61 | 9 | 0 | 1 | 12 | ***109*** | 382 | 621 | 2812 |
| Cooling Degree Days | ***0*** | 0 | 0 | 20 | 112 | 254 | ***406*** | 337 | 226 | ***51*** | 1 | ***0*** | 1407 |
| Total Precipitation (") | 5.27 | 3.99 | 3.97 | 1.58 | 0.58 | 0.45 | 0.05 | 0.18 | 0.55 | 1.22 | 3.63 | 4.01 | 25.48 |
| Days ≥ 0.1" Precip | ***6*** | 6 | 7 | 3 | 2 | 1 | 0 | 0 | 1 | 2 | ***5*** | ***5*** | 38 |
| Total Snowfall (") | 0.1 | 0.0 | 0.0 | 0.0 | 0.0 | 0.0 | 0.0 | 0.0 | 0.0 | 0.0 | 0.0 | 0.1 | 0.2 |
| Days ≥ 1" Snow Depth | 0 | 0 | 0 | 0 | 0 | 0 | 0 | 0 | 0 | 0 | 0 | 0 | 0 |

### CHULA VISTA *San Diego County* ELEVATION 9 ft LAT/LONG 32° 36 ' N / 117° 6 ' W

| | JAN | FEB | MAR | APR | MAY | JUN | JUL | AUG | SEP | OCT | NOV | DEC | YEAR |
|---|---|---|---|---|---|---|---|---|---|---|---|---|---|
| Maximum Temp °F | 65.3 | 65.6 | 65.3 | 67.1 | 67.6 | 69.8 | 73.3 | 75.4 | 75.5 | 73.5 | 69.1 | 65.3 | 69.4 |
| Minimum Temp °F | 45.2 | 46.8 | 49.4 | 52.3 | 56.6 | 59.8 | 63.6 | 65.2 | 62.9 | 57.2 | 49.9 | 45.1 | 54.5 |
| Mean Temp °F | 55.3 | 56.2 | 57.4 | 59.7 | 62.1 | 64.8 | 68.5 | 70.3 | 69.2 | 65.4 | 59.6 | 55.2 | 62.0 |
| Days Max Temp ≥ 90 °F | 0 | 0 | 0 | 0 | 0 | 0 | 0 | 0 | 1 | 1 | 0 | 0 | 2 |
| Days Max Temp ≤ 32 °F | 0 | 0 | 0 | 0 | 0 | 0 | 0 | 0 | 0 | 0 | 0 | 0 | 0 |
| Days Min Temp ≤ 32 °F | 0 | 0 | 0 | 0 | 0 | 0 | 0 | 0 | 0 | 0 | 0 | 0 | 0 |
| Days Min Temp ≤ 0 °F | 0 | 0 | 0 | 0 | 0 | 0 | 0 | 0 | 0 | 0 | 0 | 0 | 0 |
| Heating Degree Days | 296 | 244 | 235 | 161 | 96 | 38 | 4 | 1 | 5 | 42 | 165 | 297 | 1584 |
| Cooling Degree Days | 1 | 4 | 5 | 16 | 23 | 63 | 169 | 216 | 175 | 87 | 9 | 1 | 769 |
| Total Precipitation (") | 1.78 | 1.69 | 2.08 | 0.78 | 0.13 | 0.07 | 0.03 | 0.07 | 0.16 | 0.37 | 1.55 | 1.50 | 10.21 |
| Days ≥ 0.1" Precip | 4 | 4 | 4 | 2 | 0 | 0 | 0 | 0 | 0 | 1 | 3 | 3 | 21 |
| Total Snowfall (") | 0.0 | 0.0 | 0.0 | 0.0 | 0.0 | 0.0 | 0.0 | 0.0 | 0.0 | 0.0 | 0.0 | 0.1 | 0.1 |
| Days ≥ 1" Snow Depth | 0 | 0 | 0 | 0 | 0 | 0 | 0 | 0 | 0 | 0 | 0 | 0 | 0 |

### CLEARLAKE 4 SE *Lake County* ELEVATION 1349 ft LAT/LONG 38° 54 ' N / 122° 36 ' W

| | JAN | FEB | MAR | APR | MAY | JUN | JUL | AUG | SEP | OCT | NOV | DEC | YEAR |
|---|---|---|---|---|---|---|---|---|---|---|---|---|---|
| Maximum Temp °F | 55.0 | 58.4 | 61.7 | 68.0 | 77.4 | 85.7 | 93.4 | 92.7 | 86.6 | 76.2 | 62.1 | 54.3 | 72.6 |
| Minimum Temp °F | 29.7 | 32.4 | 35.4 | 38.2 | 44.2 | 50.5 | 53.8 | 52.4 | 47.3 | 40.7 | 34.5 | 30.3 | 40.8 |
| Mean Temp °F | 42.4 | 45.4 | 48.5 | 53.2 | 60.8 | 68.1 | 73.6 | 72.6 | 67.0 | 58.5 | 48.3 | 42.3 | 56.7 |
| Days Max Temp ≥ 90 °F | 0 | 0 | 0 | 0 | 4 | 11 | 23 | 21 | 13 | 3 | 0 | 0 | 75 |
| Days Max Temp ≤ 32 °F | 0 | 0 | 0 | 0 | 0 | 0 | 0 | 0 | 0 | 0 | 0 | 0 | 0 |
| Days Min Temp ≤ 32 °F | 22 | 15 | 10 | 5 | 1 | 0 | 0 | 0 | 0 | 3 | 13 | 21 | 90 |
| Days Min Temp ≤ 0 °F | 0 | 0 | 0 | 0 | 0 | 0 | 0 | 0 | 0 | 0 | 0 | 0 | 0 |
| Heating Degree Days | 693 | 546 | 503 | 351 | 161 | 40 | 6 | 6 | 46 | 211 | 494 | 696 | 3753 |
| Cooling Degree Days | 0 | 0 | 0 | 3 | 35 | 130 | 270 | 234 | 115 | 23 | 0 | 0 | 810 |
| Total Precipitation (") | 5.51 | 4.06 | 3.78 | 1.39 | 0.55 | 0.23 | 0.06 | 0.15 | 0.44 | 1.39 | 3.56 | 4.57 | 25.69 |
| Days ≥ 0.1" Precip | 7 | 7 | 7 | 4 | 2 | 1 | 0 | 0 | 1 | 3 | 6 | 7 | 45 |
| Total Snowfall (") | 1.9 | 0.1 | 0.3 | 0.0 | 0.0 | 0.0 | 0.0 | 0.0 | 0.0 | 0.0 | 0.0 | 0.1 | 2.4 |
| Days ≥ 1" Snow Depth | ***1*** | 0 | 0 | 0 | 0 | 0 | 0 | 0 | 0 | 0 | 0 | ***0*** | 1 |

### CLOVERDALE *Sonoma County* ELEVATION 322 ft LAT/LONG 38° 46 ' N / 122° 59 ' W

| | JAN | FEB | MAR | APR | MAY | JUN | JUL | AUG | SEP | OCT | NOV | DEC | YEAR |
|---|---|---|---|---|---|---|---|---|---|---|---|---|---|
| Maximum Temp °F | 57.4 | 61.5 | 64.9 | 70.4 | 78.7 | ***85.5*** | 91.1 | 90.6 | 86.4 | 77.9 | 64.7 | 56.8 | 73.8 |
| Minimum Temp °F | 37.3 | 40.1 | 42.1 | 44.5 | 48.8 | ***53.1*** | 54.0 | 54.2 | 52.6 | 48.7 | 42.7 | 37.3 | 46.3 |
| Mean Temp °F | 47.4 | 50.8 | 53.5 | 57.5 | 63.8 | ***69.3*** | 72.5 | 72.4 | 69.5 | 63.2 | 53.7 | 47.1 | 60.1 |
| Days Max Temp ≥ 90 °F | 0 | 0 | 0 | 1 | 5 | ***10*** | 18 | 18 | 11 | 4 | 0 | 0 | 67 |
| Days Max Temp ≤ 32 °F | 0 | 0 | 0 | 0 | 0 | ***0*** | 0 | 0 | 0 | 0 | 0 | 0 | 0 |
| Days Min Temp ≤ 32 °F | 7 | 2 | 1 | 0 | 0 | ***0*** | 0 | 0 | 0 | 0 | 1 | 7 | 18 |
| Days Min Temp ≤ 0 °F | 0 | 0 | 0 | 0 | 0 | ***0*** | 0 | 0 | 0 | 0 | 0 | 0 | 0 |
| Heating Degree Days | 540 | 394 | 350 | 235 | 101 | ***25*** | 3 | 3 | 20 | 107 | 337 | 549 | 2664 |
| Cooling Degree Days | 0 | 0 | 1 | 24 | 78 | na | ***278*** | ***280*** | 181 | ***62*** | 2 | 0 | na |
| Total Precipitation (") | 9.26 | 6.78 | 5.82 | 2.57 | 0.72 | ***0.18*** | 0.07 | 0.19 | 0.61 | 2.24 | 6.46 | 7.64 | 42.54 |
| Days ≥ 0.1" Precip | 9 | 8 | 8 | 5 | 2 | ***0*** | 0 | 1 | 1 | 3 | 8 | 9 | 54 |
| Total Snowfall (") | 0.2 | 0.0 | 0.1 | 0.0 | 0.0 | ***0.0*** | 0.0 | 0.0 | 0.0 | 0.0 | 0.0 | 0.0 | 0.3 |
| Days ≥ 1" Snow Depth | 0 | 0 | 0 | 0 | 0 | ***0*** | 0 | 0 | 0 | 0 | 0 | 0 | 0 |

## COALINGA *Fresno County* ELEVATION 663 ft LAT/LONG 36° 9 ' N / 120° 21 ' W

| | JAN | FEB | MAR | APR | MAY | JUN | JUL | AUG | SEP | OCT | NOV | DEC | YEAR |
|---|---|---|---|---|---|---|---|---|---|---|---|---|---|
| Maximum Temp °F | 57.5 | 64.3 | 69.4 | 76.5 | 86.3 | 93.4 | 99.0 | 97.5 | 92.0 | 82.4 | 67.4 | 57.7 | 78.6 |
| Minimum Temp °F | 36.1 | 39.4 | 42.3 | 45.7 | 52.5 | 59.2 | 64.8 | 63.4 | 58.3 | 49.9 | 40.7 | 34.9 | 48.9 |
| Mean Temp °F | 46.8 | 51.9 | 55.9 | 61.2 | 69.4 | 76.3 | 81.9 | 80.5 | 75.2 | 66.1 | 54.0 | 46.4 | 63.8 |
| Days Max Temp ≥ 90 °F | 0 | 0 | 0 | 3 | 12 | 22 | 29 | 28 | 20 | 7 | 0 | 0 | 121 |
| Days Max Temp ≤ 32 °F | 0 | 0 | 0 | 0 | 0 | 0 | 0 | 0 | 0 | 0 | 0 | 0 | 0 |
| Days Min Temp ≤ 32 °F | 10 | 4 | 1 | 0 | 0 | 0 | 0 | 0 | 0 | 0 | 4 | 12 | 31 |
| Days Min Temp ≤ 0 °F | 0 | 0 | 0 | 0 | 0 | 0 | 0 | 0 | 0 | 0 | 0 | 0 | 0 |
| Heating Degree Days | 557 | 363 | 280 | 150 | 36 | 3 | 0 | 0 | 4 | 62 | 324 | 570 | 2349 |
| Cooling Degree Days | 0 | 0 | 5 | 58 | 195 | 372 | 550 | 501 | 341 | 123 | 1 | 0 | 2146 |
| Total Precipitation (") | 1.61 | 1.65 | 1.35 | 0.55 | 0.17 | 0.04 | 0.02 | 0.03 | 0.34 | 0.35 | 0.92 | 1.19 | 8.22 |
| Days ≥ 0.1" Precip | 3 | 4 | 3 | 2 | 1 | 0 | 0 | 0 | 1 | 1 | 3 | 3 | 21 |
| Total Snowfall (") | 0.0 | 0.0 | 0.0 | 0.0 | 0.0 | 0.0 | 0.0 | 0.0 | 0.0 | 0.0 | 0.0 | 0.0 | 0.0 |
| Days ≥ 1" Snow Depth | 0 | 0 | 0 | 0 | 0 | 0 | 0 | 0 | 0 | 0 | 0 | 0 | 0 |

## COLFAX *Placer County* ELEVATION 2421 ft LAT/LONG 39° 6 ' N / 120° 58 ' W

| | JAN | FEB | MAR | APR | MAY | JUN | JUL | AUG | SEP | OCT | NOV | DEC | YEAR |
|---|---|---|---|---|---|---|---|---|---|---|---|---|---|
| Maximum Temp °F | 55.6 | 57.8 | 60.2 | 66.2 | 75.1 | 83.8 | 91.1 | 90.1 | 85.1 | 75.1 | 60.8 | 54.6 | 71.3 |
| Minimum Temp °F | 34.8 | 36.9 | 39.1 | 42.0 | 48.4 | 55.9 | 61.8 | 60.0 | 55.4 | 47.8 | 39.1 | 34.0 | 46.3 |
| Mean Temp °F | 45.2 | 47.4 | 49.7 | 54.1 | 61.8 | 69.9 | 76.5 | 75.1 | 70.3 | 61.5 | 49.9 | 44.3 | 58.8 |
| Days Max Temp ≥ 90 °F | 0 | 0 | 0 | 0 | 1 | 9 | 20 | 18 | 10 | 2 | 0 | 0 | 60 |
| Days Max Temp ≤ 32 °F | 0 | 0 | 0 | 0 | 0 | 0 | 0 | 0 | 0 | 0 | 0 | 0 | 0 |
| Days Min Temp ≤ 32 °F | 12 | 7 | 5 | 2 | 0 | 0 | 0 | 0 | 0 | 0 | 5 | 12 | 43 |
| Days Min Temp ≤ 0 °F | 0 | 0 | 0 | 0 | 0 | 0 | 0 | 0 | 0 | 0 | 0 | 0 | 0 |
| Heating Degree Days | 607 | 491 | 470 | 327 | 153 | 39 | 3 | 5 | 33 | 154 | 445 | 635 | 3362 |
| Cooling Degree Days | 0 | 0 | 0 | 10 | 56 | 179 | 339 | 315 | 188 | 53 | 0 | 0 | 1140 |
| Total Precipitation (") | 8.26 | 7.16 | 7.34 | 3.42 | 1.15 | 0.63 | 0.20 | 0.27 | 1.00 | 2.57 | 7.10 | 7.36 | 46.46 |
| Days ≥ 0.1" Precip | 7 | 7 | *9* | 6 | 2 | 1 | 0 | 1 | 1 | 3 | 7 | 7 | 51 |
| Total Snowfall (") | 3.8 | 3.3 | 2.1 | 0.7 | 0.0 | 0.0 | 0.0 | 0.0 | 0.0 | 0.0 | 0.4 | 1.8 | 12.1 |
| Days ≥ 1" Snow Depth | *1* | *0* | *0* | 0 | 0 | 0 | 0 | 0 | 0 | 0 | 0 | *0* | 1 |

## COLUSA 2 SSW *Colusa County* ELEVATION 59 ft LAT/LONG 39° 13 ' N / 122° 0 ' W

| | JAN | FEB | MAR | APR | MAY | JUN | JUL | AUG | SEP | OCT | NOV | DEC | YEAR |
|---|---|---|---|---|---|---|---|---|---|---|---|---|---|
| Maximum Temp °F | 53.9 | 60.8 | 66.1 | 73.4 | 82.7 | 89.8 | 95.0 | 93.7 | 89.1 | 79.1 | 63.4 | 53.8 | 75.1 |
| Minimum Temp °F | 36.7 | 40.0 | 42.7 | 44.9 | 52.2 | 56.7 | 58.8 | 57.2 | 53.9 | 47.8 | 40.9 | 36.2 | 47.3 |
| Mean Temp °F | 45.3 | 50.4 | 54.4 | 59.2 | 67.5 | 73.3 | 77.0 | 75.5 | 71.5 | 63.5 | 52.2 | 45.0 | 61.2 |
| Days Max Temp ≥ 90 °F | 0 | 0 | 0 | 1 | 8 | 16 | 25 | 23 | 15 | 4 | 0 | 0 | 92 |
| Days Max Temp ≤ 32 °F | 0 | 0 | 0 | 0 | 0 | 0 | 0 | 0 | 0 | 0 | 0 | 0 | 0 |
| Days Min Temp ≤ 32 °F | 9 | 3 | 1 | 0 | 0 | 0 | 0 | 0 | 0 | 0 | 3 | 10 | 26 |
| Days Min Temp ≤ 0 °F | 0 | 0 | 0 | 0 | 0 | 0 | 0 | 0 | 0 | 0 | 0 | 0 | 0 |
| Heating Degree Days | 603 | 406 | 322 | 200 | 49 | 6 | 0 | 0 | 10 | 97 | 377 | 613 | 2683 |
| Cooling Degree Days | 0 | 0 | 1 | 23 | 132 | 259 | 372 | 312 | 207 | 66 | 1 | 0 | 1373 |
| Total Precipitation (") | 3.37 | 2.76 | 2.45 | 0.82 | 0.38 | 0.25 | 0.04 | 0.07 | 0.31 | 0.85 | 2.45 | 2.56 | 16.31 |
| Days ≥ 0.1" Precip | 6 | 6 | 5 | 2 | 1 | 1 | 0 | 0 | 1 | 2 | 5 | 6 | 35 |
| Total Snowfall (") | 0.0 | 0.0 | 0.0 | 0.0 | 0.0 | 0.0 | 0.0 | 0.0 | 0.0 | 0.0 | 0.0 | 0.1 | 0.1 |
| Days ≥ 1" Snow Depth | 0 | 0 | 0 | 0 | 0 | 0 | 0 | 0 | 0 | 0 | 0 | 0 | 0 |

## CORCORAN IRRIG DIST *Kings County* ELEVATION 200 ft LAT/LONG 36° 6 ' N / 119° 34 ' W

| | JAN | FEB | MAR | APR | MAY | JUN | JUL | AUG | SEP | OCT | NOV | DEC | YEAR |
|---|---|---|---|---|---|---|---|---|---|---|---|---|---|
| Maximum Temp °F | 54.3 | 62.4 | 68.2 | 76.1 | 85.9 | 93.1 | 98.4 | 96.6 | 90.9 | 81.2 | 65.8 | 54.2 | 77.3 |
| Minimum Temp °F | 36.4 | 39.6 | 42.9 | 46.0 | 52.2 | 58.2 | 62.4 | 61.5 | 56.9 | 49.2 | 40.2 | 35.1 | 48.4 |
| Mean Temp °F | 45.4 | 51.0 | 55.6 | 61.1 | 69.1 | 75.7 | 80.4 | 79.1 | 73.9 | 65.2 | 53.0 | 44.7 | 62.9 |
| Days Max Temp ≥ 90 °F | 0 | 0 | 0 | 2 | 11 | 21 | 29 | 27 | 18 | 5 | 0 | 0 | 113 |
| Days Max Temp ≤ 32 °F | 0 | 0 | 0 | 0 | 0 | 0 | 0 | 0 | 0 | 0 | 0 | 0 | 0 |
| Days Min Temp ≤ 32 °F | 9 | 4 | 1 | 0 | 0 | 0 | 0 | 0 | 0 | 0 | 4 | 11 | 29 |
| Days Min Temp ≤ 0 °F | 0 | 0 | 0 | 0 | 0 | 0 | 0 | 0 | 0 | 0 | 0 | 0 | 0 |
| Heating Degree Days | 602 | 388 | 287 | 146 | 35 | 3 | 0 | 0 | 4 | 72 | 353 | 624 | 2514 |
| Cooling Degree Days | 0 | 0 | 2 | 47 | 168 | 331 | 472 | 430 | 276 | 95 | 0 | 0 | 1821 |
| Total Precipitation (") | 1.41 | 1.44 | 1.23 | 0.57 | 0.16 | 0.05 | 0.01 | 0.01 | 0.25 | 0.30 | 0.81 | 1.05 | 7.29 |
| Days ≥ 0.1" Precip | 4 | 4 | 4 | 2 | 1 | 0 | 0 | 0 | 1 | 1 | 2 | 3 | 22 |
| Total Snowfall (") | 0.0 | 0.0 | 0.0 | 0.0 | 0.0 | 0.0 | 0.0 | 0.0 | 0.0 | 0.0 | 0.0 | 0.0 | 0.0 |
| Days ≥ 1" Snow Depth | 0 | 0 | 0 | 0 | 0 | 0 | 0 | 0 | 0 | 0 | 0 | 0 | 0 |

## COVELO *Mendocino County* ELEVATION 1391 ft LAT/LONG 39° 47' N / 123° 15' W

| | JAN | FEB | MAR | APR | MAY | JUN | JUL | AUG | SEP | OCT | NOV | DEC | YEAR |
|---|---|---|---|---|---|---|---|---|---|---|---|---|---|
| Maximum Temp °F | 53.3 | 58.1 | 62.2 | 68.2 | 77.0 | 85.1 | 92.9 | 92.5 | 87.3 | 75.8 | 59.7 | 51.9 | 72.0 |
| Minimum Temp °F | 29.8 | 32.7 | 35.3 | 37.0 | 41.4 | 47.2 | 51.1 | 49.9 | 45.0 | 38.7 | 34.2 | 29.8 | 39.3 |
| Mean Temp °F | 41.6 | 45.4 | 48.8 | 52.6 | 59.3 | 66.2 | 72.0 | 71.2 | 66.2 | 57.3 | 47.0 | 40.9 | 55.7 |
| Days Max Temp ≥ 90 °F | 0 | 0 | 0 | 0 | 4 | 10 | 22 | 22 | ***14*** | 3 | 0 | 0 | 75 |
| Days Max Temp ≤ 32 °F | 0 | 0 | 0 | 0 | 0 | 0 | 0 | 0 | 0 | 0 | 0 | 0 | 0 |
| Days Min Temp ≤ 32 °F | 21 | 15 | 11 | 7 | 1 | 0 | 0 | 0 | 0 | 5 | 15 | ***20*** | 95 |
| Days Min Temp ≤ 0 °F | 0 | 0 | 0 | 0 | 0 | 0 | 0 | 0 | 0 | 0 | 0 | 0 | 0 |
| Heating Degree Days | 720 | 546 | 494 | 366 | 194 | 58 | 9 | 9 | 52 | 245 | 535 | 741 | 3969 |
| Cooling Degree Days | 0 | 0 | 0 | 3 | 26 | 107 | 254 | 224 | 110 | 19 | 0 | 0 | 743 |
| Total Precipitation (") | 7.61 | 6.38 | 5.82 | 2.27 | 1.08 | 0.32 | 0.06 | 0.38 | 0.68 | 2.35 | 6.13 | 7.19 | 40.27 |
| Days ≥ 0.1" Precip | 9 | 9 | 9 | 5 | 3 | 1 | 0 | 1 | 1 | 4 | 8 | ***9*** | 59 |
| Total Snowfall (") | ***1.3*** | 0.3 | 0.7 | 0.0 | 0.0 | 0.0 | 0.0 | 0.0 | 0.0 | 0.0 | 0.1 | ***0.4*** | 2.8 |
| Days ≥ 1" Snow Depth | ***0*** | 0 | 0 | 0 | 0 | 0 | 0 | 0 | 0 | 0 | 0 | ***0*** | 0 |

## CRESCENT CITY 1 N *Del Norte County* ELEVATION 69 ft LAT/LONG 41° 44' N / 124° 12' W

| | JAN | FEB | MAR | APR | MAY | JUN | JUL | AUG | SEP | OCT | NOV | DEC | YEAR |
|---|---|---|---|---|---|---|---|---|---|---|---|---|---|
| Maximum Temp °F | 55.3 | 56.2 | 56.8 | 58.1 | 60.9 | 63.7 | 65.5 | 66.1 | 66.6 | 64.0 | 58.2 | ***55.5*** | 60.6 |
| Minimum Temp °F | 40.5 | 41.6 | 42.5 | 43.2 | 46.1 | 49.5 | 51.2 | 52.1 | 50.3 | 47.2 | 44.0 | ***40.4*** | 45.7 |
| Mean Temp °F | 47.9 | 49.0 | 49.7 | 50.7 | 53.5 | 56.6 | 58.4 | 59.1 | 58.5 | 55.6 | 51.1 | ***47.9*** | 53.2 |
| Days Max Temp ≥ 90 °F | 0 | 0 | 0 | 0 | 0 | 0 | 0 | 0 | 0 | 0 | 0 | 0 | 0 |
| Days Max Temp ≤ 32 °F | 0 | 0 | 0 | 0 | 0 | 0 | 0 | 0 | 0 | 0 | 0 | 0 | 0 |
| Days Min Temp ≤ 32 °F | ***2*** | 2 | 1 | 0 | 0 | 0 | 0 | 0 | 0 | 0 | 1 | 3 | 9 |
| Days Min Temp ≤ 0 °F | 0 | 0 | 0 | 0 | 0 | 0 | 0 | 0 | 0 | 0 | 0 | 0 | 0 |
| Heating Degree Days | 519 | 446 | 469 | 423 | 348 | 250 | 200 | 176 | ***193*** | 285 | 410 | ***526*** | 4245 |
| Cooling Degree Days | 0 | 0 | 0 | 0 | 1 | 1 | 1 | 1 | 3 | 4 | 0 | 0 | 11 |
| Total Precipitation (") | 10.09 | 8.18 | 8.66 | 4.95 | 2.78 | 1.43 | 0.41 | 0.87 | 1.67 | 4.38 | 9.72 | 10.93 | 64.07 |
| Days ≥ 0.1" Precip | na | na | na | na | ***4*** | 2 | 1 | 1 | ***2*** | na | na | na | na |
| Total Snowfall (") | 0.2 | 0.0 | 0.0 | 0.0 | 0.0 | 0.0 | 0.0 | 0.0 | 0.0 | 0.0 | 0.0 | 0.0 | 0.2 |
| Days ≥ 1" Snow Depth | 0 | 0 | 0 | 0 | 0 | 0 | 0 | 0 | 0 | 0 | 0 | 0 | 0 |

## CULVER CITY *Los Angeles County* ELEVATION 79 ft LAT/LONG 34° 1' N / 118° 23' W

| | JAN | FEB | MAR | APR | MAY | JUN | JUL | AUG | SEP | OCT | NOV | DEC | YEAR |
|---|---|---|---|---|---|---|---|---|---|---|---|---|---|
| Maximum Temp °F | 67.3 | 68.7 | ***69.1*** | 72.1 | 73.2 | 76.5 | 80.0 | 81.4 | 80.4 | ***77.7*** | 71.6 | ***67.0*** | 73.8 |
| Minimum Temp °F | 46.8 | 48.2 | ***50.3*** | 52.5 | 55.5 | 59.0 | 62.1 | 63.3 | 61.9 | ***57.7*** | 51.7 | ***47.1*** | 54.7 |
| Mean Temp °F | 57.0 | 58.5 | ***59.7*** | 62.3 | 64.4 | 67.8 | 71.1 | 72.4 | 71.2 | ***67.7*** | 61.7 | ***57.0*** | 64.2 |
| Days Max Temp ≥ 90 °F | 0 | 0 | 0 | 0 | 0 | 1 | 1 | 2 | 2 | 2 | 0 | 0 | 8 |
| Days Max Temp ≤ 32 °F | 0 | 0 | 0 | 0 | 0 | 0 | 0 | 0 | 0 | 0 | 0 | 0 | 0 |
| Days Min Temp ≤ 32 °F | 0 | 0 | 0 | 0 | 0 | 0 | 0 | 0 | 0 | 0 | 0 | 0 | 0 |
| Days Min Temp ≤ 0 °F | 0 | 0 | 0 | 0 | 0 | 0 | 0 | 0 | 0 | 0 | 0 | 0 | 0 |
| Heating Degree Days | 246 | 185 | ***168*** | 102 | 53 | 12 | 1 | 0 | 1 | ***15*** | 116 | ***246*** | 1145 |
| Cooling Degree Days | ***2*** | ***8*** | na | 49 | ***55*** | ***136*** | 235 | ***267*** | 215 | na | 15 | ***2*** | na |
| Total Precipitation (") | 3.20 | 2.61 | ***2.51*** | 0.71 | 0.14 | 0.01 | 0.02 | 0.09 | 0.09 | 0.26 | 1.69 | 2.23 | 13.56 |
| Days ≥ 0.1" Precip | ***4*** | ***3*** | ***3*** | 1 | 0 | 0 | 0 | 0 | 0 | 0 | 2 | ***3*** | 16 |
| Total Snowfall (") | 0.0 | 0.0 | 0.0 | 0.0 | 0.0 | 0.0 | 0.0 | 0.0 | 0.0 | 0.0 | 0.0 | 0.0 | 0.0 |
| Days ≥ 1" Snow Depth | 0 | 0 | 0 | 0 | 0 | 0 | 0 | 0 | 0 | 0 | 0 | ***0*** | 0 |

## CUYAMACA *San Diego County* ELEVATION 4672 ft LAT/LONG 32° 59' N / 116° 35' W

| | JAN | FEB | MAR | APR | MAY | JUN | JUL | AUG | SEP | OCT | NOV | DEC | YEAR |
|---|---|---|---|---|---|---|---|---|---|---|---|---|---|
| Maximum Temp °F | 50.9 | 53.1 | 54.7 | 60.4 | 67.5 | 76.8 | 83.8 | 83.5 | 78.8 | 69.5 | 58.6 | 51.4 | 65.8 |
| Minimum Temp °F | 30.1 | 30.9 | 33.3 | 36.4 | 41.2 | 48.8 | 55.1 | 54.5 | 48.8 | 40.0 | 34.2 | 29.2 | 40.2 |
| Mean Temp °F | 40.5 | 42.0 | 44.0 | 48.4 | 54.3 | 62.8 | 69.5 | 69.0 | 63.8 | 54.8 | 46.4 | 40.3 | 53.0 |
| Days Max Temp ≥ 90 °F | 0 | 0 | 0 | 0 | 0 | 2 | 4 | 4 | 2 | 0 | 0 | 0 | 12 |
| Days Max Temp ≤ 32 °F | 1 | 0 | 0 | 0 | 0 | 0 | 0 | 0 | 0 | 0 | 0 | 1 | 2 |
| Days Min Temp ≤ 32 °F | 20 | 17 | 14 | 8 | 2 | 0 | 0 | 0 | 0 | 4 | 13 | 21 | 99 |
| Days Min Temp ≤ 0 °F | 0 | 0 | 0 | 0 | 0 | 0 | 0 | 0 | 0 | 0 | 0 | 0 | 0 |
| Heating Degree Days | 750 | 642 | 644 | 491 | 330 | 120 | 19 | 25 | 94 | 315 | 550 | 758 | 4738 |
| Cooling Degree Days | 0 | 0 | 0 | 0 | 5 | 55 | 139 | 152 | 65 | 6 | 0 | 0 | 422 |
| Total Precipitation (") | 6.12 | 5.93 | 6.63 | 2.92 | 0.91 | 0.18 | 0.49 | 0.90 | 0.96 | 1.55 | 4.13 | 5.71 | 36.43 |
| Days ≥ 0.1" Precip | 6 | 6 | 7 | 5 | 2 | 1 | 1 | 2 | 2 | 2 | 4 | 6 | 44 |
| Total Snowfall (") | 5.8 | 6.7 | 10.6 | 2.5 | 0.0 | 0.0 | 0.0 | 0.0 | 0.0 | 0.2 | 1.4 | 5.1 | 32.3 |
| Days ≥ 1" Snow Depth | 5 | 3 | 4 | 1 | 0 | 0 | 0 | 0 | 0 | 0 | 0 | 4 | 17 |

## DAGGETT SAN BERN AP *San Bernardino County* ELEVATION 1929 ft LAT/LONG 34° 52' N / 116° 47' W

| | JAN | FEB | MAR | APR | MAY | JUN | JUL | AUG | SEP | OCT | NOV | DEC | YEAR |
|---|---|---|---|---|---|---|---|---|---|---|---|---|---|
| Maximum Temp °F | 60.4 | 65.7 | 70.5 | 78.3 | 87.8 | 98.1 | 103.9 | 101.9 | 94.2 | 83.0 | 69.2 | 60.1 | 81.1 |
| Minimum Temp °F | 36.8 | 41.3 | 46.4 | 51.6 | 59.8 | 67.6 | 74.0 | 72.8 | 66.1 | 55.5 | 44.0 | 36.3 | 54.4 |
| Mean Temp °F | 48.6 | 53.6 | 58.5 | 65.0 | 73.8 | 82.9 | 89.0 | 87.4 | 80.1 | 69.3 | 56.6 | 48.2 | 67.8 |
| Days Max Temp ≥ 90 °F | 0 | 0 | 0 | 4 | 15 | 26 | 31 | 30 | 23 | 7 | 0 | 0 | 136 |
| Days Max Temp ≤ 32 °F | 0 | 0 | 0 | 0 | 0 | 0 | 0 | 0 | 0 | 0 | 0 | 0 | 0 |
| Days Min Temp ≤ 32 °F | 8 | 3 | 0 | 0 | 0 | 0 | 0 | 0 | 0 | 0 | 2 | 9 | 22 |
| Days Min Temp ≤ 0 °F | 0 | 0 | 0 | 0 | 0 | 0 | 0 | 0 | 0 | 0 | 0 | 0 | 0 |
| Heating Degree Days | 502 | 319 | 215 | 85 | 15 | 0 | 0 | 0 | 1 | 36 | 255 | 513 | 1941 |
| Cooling Degree Days | 0 | 4 | 18 | 110 | 291 | 547 | 735 | 698 | 469 | 186 | 13 | 0 | 3071 |
| Total Precipitation (") | 0.63 | 0.52 | 0.59 | 0.22 | 0.09 | 0.11 | 0.44 | 0.44 | 0.28 | 0.17 | 0.24 | 0.55 | 4.28 |
| Days ≥ 0.1" Precip | 2 | 2 | 2 | 1 | 0 | 0 | 1 | 1 | 1 | 1 | 1 | 2 | 14 |
| Total Snowfall (") | 0.0 | 0.1 | 0.0 | 0.0 | 0.0 | 0.0 | 0.0 | 0.0 | 0.0 | 0.0 | 0.0 | 0.5 | 0.6 |
| Days ≥ 1" Snow Depth | 0 | 0 | 0 | 0 | 0 | 0 | 0 | 0 | 0 | 0 | 0 | 0 | 0 |

## DAVIS 2 WSW EXP FARM *Yolo County* ELEVATION 49 ft LAT/LONG 38° 32' N / 121° 45' W

| | JAN | FEB | MAR | APR | MAY | JUN | JUL | AUG | SEP | OCT | NOV | DEC | YEAR |
|---|---|---|---|---|---|---|---|---|---|---|---|---|---|
| Maximum Temp °F | 53.0 | 59.9 | 64.6 | 71.5 | 80.8 | 88.2 | 93.0 | 91.8 | 87.8 | 78.8 | 63.6 | 53.1 | 73.8 |
| Minimum Temp °F | 36.6 | 39.5 | 42.1 | 44.6 | 49.8 | 54.4 | 55.6 | 54.8 | 53.0 | 47.8 | 41.0 | 36.2 | 46.3 |
| Mean Temp °F | 44.8 | 49.7 | 53.4 | 58.1 | 65.3 | 71.3 | 74.3 | 73.3 | 70.4 | 63.3 | 52.3 | 44.7 | 60.1 |
| Days Max Temp ≥ 90 °F | 0 | 0 | 0 | 0 | 6 | 13 | 22 | 20 | 13 | 4 | 0 | 0 | 78 |
| Days Max Temp ≤ 32 °F | 0 | 0 | 0 | 0 | 0 | 0 | 0 | 0 | 0 | 0 | 0 | 0 | 0 |
| Days Min Temp ≤ 32 °F | 9 | 4 | 1 | 0 | 0 | 0 | 0 | 0 | 0 | 0 | 3 | 9 | 26 |
| Days Min Temp ≤ 0 °F | 0 | 0 | 0 | 0 | 0 | 0 | 0 | 0 | 0 | 0 | 0 | 0 | 0 |
| Heating Degree Days | 618 | 425 | 353 | 214 | 77 | 15 | 1 | 1 | 13 | 97 | 374 | 623 | 2811 |
| Cooling Degree Days | 0 | 0 | 1 | 18 | 93 | 215 | 308 | 266 | 189 | 62 | 1 | 0 | 1153 |
| Total Precipitation (") | 3.99 | 3.15 | 2.83 | 1.03 | 0.32 | 0.20 | 0.03 | 0.06 | 0.28 | 0.82 | 2.73 | 2.94 | 18.38 |
| Days ≥ 0.1" Precip | 6 | 6 | 6 | 3 | 1 | 1 | 0 | 0 | 1 | 2 | 5 | 6 | 37 |
| Total Snowfall (") | 0.1 | 0.0 | 0.0 | 0.0 | 0.0 | 0.0 | 0.0 | 0.0 | 0.0 | 0.0 | 0.0 | 0.0 | 0.1 |
| Days ≥ 1" Snow Depth | 0 | 0 | 0 | 0 | 0 | 0 | 0 | 0 | 0 | 0 | 0 | 0 | 0 |

## DE SABLA *Butte County* ELEVATION 2720 ft LAT/LONG 39° 52' N / 121° 37' W

| | JAN | FEB | MAR | APR | MAY | JUN | JUL | AUG | SEP | OCT | NOV | DEC | YEAR |
|---|---|---|---|---|---|---|---|---|---|---|---|---|---|
| Maximum Temp °F | 51.8 | 54.7 | 57.7 | 63.9 | 73.7 | 81.9 | 89.0 | 88.6 | 83.1 | 72.5 | 57.5 | 50.8 | 68.8 |
| Minimum Temp °F | 31.2 | 33.0 | 34.8 | 37.9 | 44.0 | 50.1 | 54.1 | 53.1 | 49.4 | 43.1 | 35.6 | 30.9 | 41.4 |
| Mean Temp °F | 41.5 | 43.9 | 46.3 | 51.0 | 58.9 | 66.0 | 71.6 | 70.9 | 66.3 | 57.8 | 46.6 | 40.9 | 55.1 |
| Days Max Temp ≥ 90 °F | 0 | 0 | 0 | 0 | 1 | 7 | 16 | 16 | 8 | 2 | 0 | 0 | 50 |
| Days Max Temp ≤ 32 °F | 0 | 0 | 0 | 0 | 0 | 0 | 0 | 0 | 0 | 0 | 0 | 0 | 0 |
| Days Min Temp ≤ 32 °F | 19 | 14 | 12 | 7 | 1 | 0 | 0 | 0 | 0 | 2 | 10 | 19 | 84 |
| Days Min Temp ≤ 0 °F | 0 | 0 | 0 | 0 | 0 | 0 | 0 | 0 | 0 | 0 | 0 | 0 | 0 |
| Heating Degree Days | 720 | 589 | 574 | 417 | 212 | 71 | 11 | 15 | 64 | 241 | 547 | 740 | 4201 |
| Cooling Degree Days | 0 | 0 | 0 | 4 | 24 | 104 | 208 | 186 | 102 | 27 | 0 | 0 | 655 |
| Total Precipitation (") | 12.03 | 9.84 | 9.71 | 4.51 | 1.73 | 0.80 | 0.16 | 0.41 | 1.35 | 3.54 | 9.45 | 10.20 | 63.73 |
| Days ≥ 0.1" Precip | 10 | *9* | 10 | 6 | 3 | 2 | 0 | 1 | 2 | 4 | 8 | 9 | 64 |
| Total Snowfall (") | 9.1 | 2.5 | 5.4 | 1.8 | 0.0 | 0.0 | 0.0 | 0.0 | 0.0 | 0.0 | 0.9 | 3.6 | 23.3 |
| Days ≥ 1" Snow Depth | *2* | *1* | 1 | 1 | 0 | 0 | 0 | 0 | 0 | 0 | 0 | *1* | 6 |

## DEATH VALLEY *Inyo County* ELEVATION -171 ft LAT/LONG 36° 28' N / 116° 52' W

| | JAN | FEB | MAR | APR | MAY | JUN | JUL | AUG | SEP | OCT | NOV | DEC | YEAR |
|---|---|---|---|---|---|---|---|---|---|---|---|---|---|
| Maximum Temp °F | 65.5 | 73.3 | 80.8 | 89.5 | 99.4 | 108.9 | 114.9 | 113.1 | 105.4 | 92.5 | 76.1 | 64.6 | 90.3 |
| Minimum Temp °F | 39.1 | 46.0 | 54.1 | 61.7 | 71.6 | 80.6 | 87.0 | 85.2 | 75.3 | 61.5 | 47.9 | 38.2 | 62.4 |
| Mean Temp °F | 52.3 | 59.7 | 67.5 | 75.6 | 85.5 | 94.8 | 100.9 | 99.2 | 90.4 | 77.0 | 62.0 | 51.4 | 76.4 |
| Days Max Temp ≥ 90 °F | 0 | 0 | 4 | 16 | 26 | 30 | 31 | 31 | 29 | 20 | 1 | 0 | 188 |
| Days Max Temp ≤ 32 °F | 0 | 0 | 0 | 0 | 0 | 0 | 0 | 0 | 0 | 0 | 0 | 0 | 0 |
| Days Min Temp ≤ 32 °F | 4 | 1 | 0 | 0 | 0 | 0 | 0 | 0 | 0 | 0 | 0 | 6 | 11 |
| Days Min Temp ≤ 0 °F | 0 | 0 | 0 | 0 | 0 | 0 | 0 | 0 | 0 | 0 | 0 | 0 | 0 |
| Heating Degree Days | 388 | 162 | 47 | 8 | 0 | 0 | 0 | 0 | 0 | 4 | 131 | 415 | 1155 |
| Cooling Degree Days | 1 | 23 | 130 | 366 | 637 | 890 | 1097 | 1056 | 759 | 376 | 50 | 1 | 5386 |
| Total Precipitation (") | 0.27 | 0.46 | 0.38 | 0.15 | 0.08 | 0.04 | 0.13 | 0.15 | 0.13 | 0.11 | 0.21 | 0.19 | 2.30 |
| Days ≥ 0.1" Precip | 1 | 1 | 1 | 0 | 0 | 0 | 0 | 0 | 0 | 0 | 1 | 1 | 5 |
| Total Snowfall (") | 0.0 | 0.0 | 0.0 | 0.0 | 0.0 | 0.0 | 0.0 | 0.0 | 0.0 | 0.0 | 0.0 | 0.0 | 0.0 |
| Days ≥ 1" Snow Depth | 0 | 0 | 0 | 0 | 0 | 0 | 0 | 0 | 0 | 0 | 0 | 0 | 0 |

## DEEP SPRINGS COLLEGE *Inyo County* ELEVATION 5233 ft LAT/LONG 37° 22 ' N / 117° 59 ' W

| | JAN | FEB | MAR | APR | MAY | JUN | JUL | AUG | SEP | OCT | NOV | DEC | YEAR |
|---|---|---|---|---|---|---|---|---|---|---|---|---|---|
| Maximum Temp °F | *47.1* | *51.4* | 57.1 | 65.6 | *75.2* | na | *91.3* | *88.7* | 80.6 | *69.6* | *55.6* | *46.5* | na |
| Minimum Temp °F | *18.1* | *24.3* | 29.5 | 35.4 | *43.9* | na | *58.8* | *56.0* | 48.4 | *37.2* | *25.9* | *18.0* | na |
| Mean Temp °F | *32.8* | *37.8* | 43.4 | 50.6 | *59.5* | na | *75.0* | *72.3* | 64.5 | *53.4* | *40.8* | *32.3* | na |
| Days Max Temp ≥ 90 °F | *0* | 0 | 0 | 0 | *1* | na | *19* | *14* | *2* | 0 | 0 | 0 | na |
| Days Max Temp ≤ 32 °F | *2* | 0 | 0 | 0 | *0* | *0* | 0 | 0 | 0 | 0 | 0 | 2 | 4 |
| Days Min Temp ≤ 32 °F | *30* | *25* | 20 | *9* | *2* | *0* | 0 | 0 | 0 | *7* | *25* | *28* | 146 |
| Days Min Temp ≤ 0 °F | *0* | 0 | 0 | 0 | *0* | *0* | 0 | 0 | 0 | 0 | 0 | 1 | 1 |
| Heating Degree Days | *989* | *760* | 664 | 426 | *186* | na | *0* | *4* | 75 | *357* | *720* | *1008* | na |
| Cooling Degree Days | *0* | *0* | 0 | 1 | *27* | na | na | *243* | 74 | *4* | 0 | *0* | na |
| Total Precipitation (") | *0.64* | *0.84* | 0.71 | 0.53 | *0.44* | *0.28* | 0.68 | 0.60 | 0.50 | 0.23 | 0.50 | 0.61 | 6.56 |
| Days ≥ 0.1" Precip | *2* | *2* | 2 | 1 | *1* | *1* | 1 | 1 | 1 | 1 | 1 | 1 | 15 |
| Total Snowfall (") | *4.2* | *1.9* | 0.7 | 0.6 | *0.0* | *0.0* | 0.0 | 0.0 | 0.0 | 0.1 | *0.6* | *1.7* | 9.8 |
| Days ≥ 1" Snow Depth | *4* | *2* | 1 | 0 | 0 | *0* | 0 | 0 | 0 | 0 | 0 | *2* | 9 |

## DONNER MEMORIAL ST P *Nevada County* ELEVATION 5945 ft LAT/LONG 39° 19 ' N / 120° 14 ' W

| | JAN | FEB | MAR | APR | MAY | JUN | JUL | AUG | SEP | OCT | NOV | DEC | YEAR |
|---|---|---|---|---|---|---|---|---|---|---|---|---|---|
| Maximum Temp °F | 40.2 | 43.4 | 46.7 | 53.7 | 63.5 | 72.6 | 80.8 | 79.9 | 73.0 | 62.7 | 48.1 | 39.9 | 58.7 |
| Minimum Temp °F | 13.0 | 15.2 | 20.6 | 25.1 | 31.0 | 36.7 | 40.7 | 39.9 | 34.0 | 27.3 | 21.4 | 14.5 | 26.6 |
| Mean Temp °F | 26.6 | 29.3 | 33.7 | 39.4 | 47.2 | 54.6 | 60.8 | 59.9 | 53.6 | 45.0 | 34.7 | 27.2 | 42.7 |
| Days Max Temp ≥ 90 °F | 0 | 0 | 0 | 0 | 0 | 0 | 2 | 2 | 0 | 0 | 0 | 0 | 4 |
| Days Max Temp ≤ 32 °F | 5 | 3 | 2 | 0 | 0 | 0 | 0 | 0 | 0 | 0 | 2 | 6 | 18 |
| Days Min Temp ≤ 32 °F | 31 | 28 | 30 | 28 | 20 | 7 | 2 | 2 | 12 | 26 | 28 | 31 | 245 |
| Days Min Temp ≤ 0 °F | 3 | 2 | 1 | 0 | 0 | 0 | 0 | 0 | 0 | 0 | 0 | 2 | 8 |
| Heating Degree Days | 1183 | 1001 | 965 | 761 | 544 | 308 | 143 | 164 | 337 | 612 | 900 | 1166 | 8084 |
| Cooling Degree Days | 0 | 0 | 0 | 0 | 0 | 5 | 20 | 16 | 1 | 0 | 0 | 0 | 42 |
| Total Precipitation (") | 6.61 | 5.80 | 5.45 | 2.37 | 1.14 | 0.92 | 0.45 | 0.67 | 0.92 | 2.10 | 5.62 | 5.76 | 37.81 |
| Days ≥ 0.1" Precip | 8 | 8 | 8 | 5 | 3 | 3 | 1 | 2 | 2 | 3 | 7 | 7 | 57 |
| Total Snowfall (") | 37.4 | 37.5 | 33.9 | 14.3 | 3.8 | 0.3 | 0.0 | 0.0 | 0.3 | 2.0 | *19.0* | 35.2 | 183.7 |
| Days ≥ 1" Snow Depth | 31 | 28 | 29 | 14 | 4 | 0 | 0 | 0 | 0 | 1 | *11* | 24 | 142 |

## DOYLE 4 SSE *Lassen County* ELEVATION 4393 ft LAT/LONG 39° 57 ' N / 120° 5 ' W

| | JAN | FEB | MAR | APR | MAY | JUN | JUL | AUG | SEP | OCT | NOV | DEC | YEAR |
|---|---|---|---|---|---|---|---|---|---|---|---|---|---|
| Maximum Temp °F | 42.2 | 48.2 | 54.5 | 61.4 | 71.3 | 80.4 | 89.2 | 87.9 | 79.7 | 67.5 | 51.7 | 42.2 | 64.7 |
| Minimum Temp °F | 21.9 | 25.8 | 29.4 | 32.4 | 38.6 | 44.5 | 49.5 | 48.1 | 42.1 | 34.1 | 27.4 | 21.5 | 34.6 |
| Mean Temp °F | 32.1 | 37.1 | 42.0 | 46.9 | 55.0 | 62.5 | 69.3 | 68.0 | 61.1 | 50.8 | 39.6 | 31.9 | 49.7 |
| Days Max Temp ≥ 90 °F | 0 | 0 | 0 | 0 | 1 | 5 | 17 | 14 | 4 | 0 | 0 | 0 | 41 |
| Days Max Temp ≤ 32 °F | 4 | 1 | 0 | 0 | 0 | 0 | 0 | 0 | 0 | 0 | 1 | 4 | 10 |
| Days Min Temp ≤ 32 °F | 27 | 22 | 21 | 16 | 6 | 1 | 0 | 0 | 2 | 13 | 21 | 27 | 156 |
| Days Min Temp ≤ 0 °F | 1 | 0 | 0 | 0 | 0 | 0 | 0 | 0 | 0 | 0 | 0 | 1 | 2 |
| Heating Degree Days | 1014 | 783 | 707 | 537 | 311 | 122 | 23 | 32 | 145 | 434 | 756 | 1020 | 5884 |
| Cooling Degree Days | 0 | 0 | 0 | 0 | 7 | 48 | 147 | 124 | 33 | 0 | 0 | 0 | 359 |
| Total Precipitation (") | 2.76 | 2.18 | 1.93 | 0.82 | 1.04 | 0.76 | 0.50 | 0.47 | 0.65 | 1.08 | 2.64 | 2.35 | 17.18 |
| Days ≥ 0.1" Precip | 4 | 4 | 4 | 2 | 3 | 2 | 1 | 1 | 2 | 2 | 5 | 4 | 34 |
| Total Snowfall (") | 8.7 | 5.4 | 5.1 | 2.1 | 1.7 | 0.0 | 0.0 | 0.0 | 0.1 | 0.3 | 2.8 | 7.6 | 33.8 |
| Days ≥ 1" Snow Depth | 10 | 4 | 1 | 0 | 0 | 0 | 0 | 0 | 0 | 0 | 2 | 7 | 24 |

## EAGLE MOUNTAIN *Riverside County* ELEVATION 973 ft LAT/LONG 33° 48 ' N / 115° 27 ' W

| | JAN | FEB | MAR | APR | MAY | JUN | JUL | AUG | SEP | OCT | NOV | DEC | YEAR |
|---|---|---|---|---|---|---|---|---|---|---|---|---|---|
| Maximum Temp °F | 64.8 | 69.6 | 74.9 | 82.3 | 90.6 | 100.2 | 104.3 | 103.1 | 97.4 | 86.5 | 73.5 | 64.8 | 84.3 |
| Minimum Temp °F | 44.9 | 48.9 | 53.2 | 59.7 | 67.9 | 76.9 | 82.4 | 80.9 | 74.9 | 64.3 | 53.0 | 45.2 | 62.7 |
| Mean Temp °F | 54.9 | 59.3 | 64.1 | 71.0 | 79.3 | 88.6 | 93.3 | 92.1 | 86.2 | 75.4 | 63.3 | 55.0 | 73.5 |
| Days Max Temp ≥ 90 °F | 0 | 0 | 1 | 6 | 19 | 27 | 31 | 31 | 26 | 11 | 0 | 0 | 152 |
| Days Max Temp ≤ 32 °F | 0 | 0 | 0 | 0 | 0 | 0 | 0 | 0 | 0 | 0 | 0 | 0 | 0 |
| Days Min Temp ≤ 32 °F | 1 | 0 | 0 | 0 | 0 | 0 | 0 | 0 | 0 | 0 | 0 | 1 | 2 |
| Days Min Temp ≤ 0 °F | 0 | 0 | 0 | 0 | 0 | 0 | 0 | 0 | 0 | 0 | 0 | 0 | 0 |
| Heating Degree Days | 309 | 174 | 96 | 28 | 2 | 0 | 0 | 0 | 0 | 9 | 105 | 305 | 1028 |
| Cooling Degree Days | 0 | 21 | 74 | 255 | 462 | 725 | 872 | 846 | 643 | 336 | 55 | 2 | 4291 |
| Total Precipitation (") | 0.55 | 0.45 | 0.50 | 0.10 | 0.08 | 0.06 | 0.48 | 0.78 | 0.44 | 0.26 | 0.24 | 0.52 | 4.46 |
| Days ≥ 0.1" Precip | 1 | 1 | 1 | 0 | 0 | 0 | 0 | 1 | 1 | 1 | 1 | 1 | 8 |
| Total Snowfall (") | 0.0 | 0.0 | 0.0 | 0.0 | 0.0 | 0.0 | 0.0 | 0.0 | 0.0 | 0.0 | 0.0 | 0.0 | 0.0 |
| Days ≥ 1" Snow Depth | 0 | 0 | 0 | 0 | 0 | 0 | 0 | 0 | 0 | 0 | 0 | 0 | 0 |

## EAST PARK RESERVOIR *Colusa County* ELEVATION 1211 ft LAT/LONG 39° 22 ' N / 122° 31 ' W

| | JAN | FEB | MAR | APR | MAY | JUN | JUL | AUG | SEP | OCT | NOV | DEC | YEAR |
|---|---|---|---|---|---|---|---|---|---|---|---|---|---|
| Maximum Temp °F | 55.7 | 59.0 | 62.1 | 68.2 | 78.1 | 86.6 | 93.3 | 92.2 | 87.0 | 77.4 | 63.4 | 55.8 | 73.2 |
| Minimum Temp °F | 31.2 | 34.5 | 37.4 | 40.0 | 46.6 | 53.7 | 59.0 | 57.2 | 52.4 | 44.5 | 36.9 | 31.3 | 43.7 |
| Mean Temp °F | 43.5 | 46.8 | 49.8 | 54.1 | 62.4 | 70.2 | 76.2 | 74.7 | 69.8 | 61.0 | 50.2 | 43.6 | 58.5 |
| Days Max Temp ≥ 90 °F | 0 | 0 | 0 | 0 | 4 | 12 | 23 | 22 | 13 | 3 | 0 | 0 | 77 |
| Days Max Temp ≤ 32 °F | 0 | 0 | 0 | 0 | 0 | 0 | 0 | 0 | 0 | 0 | 0 | 0 | 0 |
| Days Min Temp ≤ 32 °F | 20 | 11 | 7 | 2 | 0 | 0 | 0 | 0 | 0 | 1 | 9 | 19 | 69 |
| Days Min Temp ≤ 0 °F | 0 | 0 | 0 | 0 | 0 | 0 | 0 | 0 | 0 | 0 | 0 | 0 | 0 |
| Heating Degree Days | 659 | 509 | 465 | 322 | 134 | 30 | 2 | 4 | 26 | 157 | 439 | 657 | 3404 |
| Cooling Degree Days | 0 | 0 | 0 | 3 | 51 | 178 | 355 | 313 | 184 | 50 | 0 | 0 | 1134 |
| Total Precipitation (") | 4.52 | 3.66 | 2.72 | 1.06 | 0.35 | 0.30 | 0.05 | 0.13 | 0.31 | 0.98 | 2.88 | 3.57 | 20.53 |
| Days ≥ 0.1" Precip | 7 | 6 | 6 | 3 | 1 | 1 | 0 | 0 | 1 | 2 | 5 | 6 | 38 |
| Total Snowfall (") | 2.2 | 0.4 | 0.4 | 0.1 | 0.0 | 0.0 | 0.0 | 0.0 | 0.0 | 0.0 | 0.4 | 0.6 | 4.1 |
| Days ≥ 1" Snow Depth | 1 | 0 | 0 | 0 | 0 | 0 | 0 | 0 | 0 | 0 | 0 | 0 | 1 |

## EL CAPITAN DAM *San Diego County* ELEVATION 610 ft LAT/LONG 32° 53 ' N / 116° 49 ' W

| | JAN | FEB | MAR | APR | MAY | JUN | JUL | AUG | SEP | OCT | NOV | DEC | YEAR |
|---|---|---|---|---|---|---|---|---|---|---|---|---|---|
| Maximum Temp °F | 69.1 | 70.4 | 70.7 | 75.3 | 79.2 | 86.2 | 92.7 | 93.5 | 90.2 | 84.1 | 76.2 | 69.4 | 79.8 |
| Minimum Temp °F | 41.5 | 43.0 | 44.8 | 47.3 | 51.1 | 54.4 | 57.9 | 59.7 | 57.9 | 52.8 | 46.1 | 41.6 | 49.8 |
| Mean Temp °F | 55.3 | 56.7 | 57.8 | 61.3 | 65.1 | 70.3 | 75.4 | 76.6 | 74.1 | 68.5 | 61.2 | 55.6 | 64.8 |
| Days Max Temp ≥ 90 °F | 0 | 0 | 0 | 2 | 4 | 11 | 22 | 23 | 16 | 9 | 2 | 0 | 89 |
| Days Max Temp ≤ 32 °F | 0 | 0 | 0 | 0 | 0 | 0 | 0 | 0 | 0 | 0 | 0 | 0 | 0 |
| Days Min Temp ≤ 32 °F | 2 | 0 | 0 | 0 | 0 | 0 | 0 | 0 | 0 | 0 | 0 | 1 | 3 |
| Days Min Temp ≤ 0 °F | 0 | 0 | 0 | 0 | 0 | 0 | 0 | 0 | 0 | 0 | 0 | 0 | 0 |
| Heating Degree Days | 297 | 233 | 228 | 137 | 64 | 12 | 0 | 0 | 3 | 28 | 139 | 289 | 1430 |
| Cooling Degree Days | 2 | 8 | 12 | 48 | 83 | 184 | 332 | 373 | ***297*** | 147 | 29 | 4 | 1519 |
| Total Precipitation (") | 2.97 | 2.89 | 3.35 | 1.30 | 0.38 | 0.15 | 0.10 | 0.14 | 0.33 | 0.80 | 1.90 | 2.55 | 16.86 |
| Days ≥ 0.1" Precip | 5 | 5 | 6 | 3 | 1 | 0 | 0 | 0 | 1 | 2 | 3 | 4 | 30 |
| Total Snowfall (") | 0.0 | 0.0 | 0.0 | 0.0 | 0.0 | 0.0 | 0.0 | 0.0 | 0.0 | 0.0 | 0.0 | 0.0 | 0.0 |
| Days ≥ 1" Snow Depth | 0 | 0 | 0 | 0 | 0 | 0 | 0 | 0 | 0 | 0 | 0 | 0 | 0 |

## EL CENTRO 2 SSW *Imperial County* ELEVATION -49 ft LAT/LONG 32° 47 ' N / 115° 34 ' W

| | JAN | FEB | MAR | APR | MAY | JUN | JUL | AUG | SEP | OCT | NOV | DEC | YEAR |
|---|---|---|---|---|---|---|---|---|---|---|---|---|---|
| Maximum Temp °F | 69.8 | 74.4 | 79.1 | 85.9 | 94.3 | 103.5 | 107.2 | 106.0 | 101.2 | 91.1 | 78.0 | 69.2 | 88.3 |
| Minimum Temp °F | 40.0 | 43.6 | 47.6 | 52.2 | 59.3 | 67.2 | 75.3 | 75.8 | 69.3 | 58.3 | 46.8 | 39.7 | 56.3 |
| Mean Temp °F | 55.0 | 59.0 | 63.4 | 69.1 | 76.8 | 85.4 | 91.3 | 90.9 | 85.3 | 74.7 | 62.4 | 54.5 | 72.3 |
| Days Max Temp ≥ 90 °F | 0 | 0 | 3 | 11 | 24 | 29 | 31 | 31 | 28 | 19 | 1 | 0 | 177 |
| Days Max Temp ≤ 32 °F | 0 | 0 | 0 | 0 | 0 | 0 | 0 | 0 | 0 | 0 | 0 | 0 | 0 |
| Days Min Temp ≤ 32 °F | 4 | 1 | 0 | 0 | 0 | 0 | 0 | 0 | 0 | 0 | 0 | 3 | 8 |
| Days Min Temp ≤ 0 °F | 0 | 0 | 0 | 0 | 0 | 0 | 0 | 0 | 0 | 0 | 0 | 0 | 0 |
| Heating Degree Days | 305 | 172 | 98 | 30 | 2 | 0 | 0 | 0 | 0 | 7 | 109 | 320 | 1043 |
| Cooling Degree Days | 0 | 15 | 65 | 206 | 395 | 637 | 822 | 817 | 632 | 335 | 44 | 0 | 3968 |
| Total Precipitation (") | 0.48 | 0.36 | 0.32 | 0.06 | 0.03 | 0.01 | 0.09 | 0.35 | 0.30 | 0.36 | 0.29 | 0.48 | 3.13 |
| Days ≥ 0.1" Precip | 1 | 1 | 1 | 0 | 0 | 0 | 0 | 1 | 0 | 1 | 1 | 1 | 7 |
| Total Snowfall (") | 0.0 | 0.0 | 0.0 | 0.0 | 0.0 | 0.0 | 0.0 | 0.0 | 0.0 | 0.0 | 0.0 | 0.0 | 0.0 |
| Days ≥ 1" Snow Depth | 0 | 0 | 0 | 0 | 0 | 0 | 0 | 0 | 0 | 0 | 0 | 0 | 0 |

## ELECTRA POWER HOUSE *Amador County* ELEVATION 702 ft LAT/LONG 38° 20 ' N / 120° 40 ' W

| | JAN | FEB | MAR | APR | MAY | JUN | JUL | AUG | SEP | OCT | NOV | DEC | YEAR |
|---|---|---|---|---|---|---|---|---|---|---|---|---|---|
| Maximum Temp °F | 57.3 | 62.3 | 65.6 | 71.6 | 81.5 | 89.8 | 96.4 | 95.3 | 89.9 | 79.9 | 65.4 | 56.6 | 76.0 |
| Minimum Temp °F | 33.9 | 36.6 | 39.4 | 41.8 | 46.9 | 51.9 | 56.1 | 55.1 | 51.8 | 45.9 | 39.4 | 34.0 | 44.4 |
| Mean Temp °F | 45.6 | 49.5 | 52.6 | 56.7 | 64.3 | 70.9 | 76.3 | 75.2 | 70.9 | 62.9 | 52.4 | 45.3 | 60.2 |
| Days Max Temp ≥ 90 °F | 0 | 0 | 0 | 0 | 7 | 15 | 27 | 25 | 17 | 5 | 0 | 0 | 96 |
| Days Max Temp ≤ 32 °F | 0 | 0 | 0 | 0 | 0 | 0 | 0 | 0 | 0 | 0 | 0 | 0 | 0 |
| Days Min Temp ≤ 32 °F | 15 | 8 | 4 | 1 | 0 | 0 | 0 | 0 | 0 | 0 | 5 | 15 | 48 |
| Days Min Temp ≤ 0 °F | 0 | 0 | 0 | 0 | 0 | 0 | 0 | 0 | 0 | 0 | 0 | 0 | 0 |
| Heating Degree Days | 594 | 432 | 377 | 249 | 90 | 15 | 1 | 1 | 14 | 107 | 370 | 603 | 2853 |
| Cooling Degree Days | 0 | 0 | 0 | 9 | 68 | 187 | 337 | 307 | 188 | 54 | 0 | 0 | 1150 |
| Total Precipitation (") | 5.35 | 4.58 | 5.19 | 2.56 | 0.68 | 0.31 | 0.18 | 0.17 | 0.63 | 1.76 | 4.28 | 4.30 | 29.99 |
| Days ≥ 0.1" Precip | 7 | 7 | 8 | 4 | 2 | 1 | 0 | 0 | 1 | 3 | 6 | 6 | 45 |
| Total Snowfall (") | 0.0 | 0.1 | 0.3 | 0.2 | 0.0 | 0.0 | 0.0 | 0.0 | 0.0 | 0.0 | 0.0 | 0.1 | 0.7 |
| Days ≥ 1" Snow Depth | 0 | 0 | 0 | 0 | 0 | 0 | 0 | 0 | 0 | 0 | 0 | 0 | 0 |

## ELSINORE *Riverside County* ELEVATION 1302 ft LAT/LONG 33° 40' N / 117° 21' W

| | JAN | FEB | MAR | APR | MAY | JUN | JUL | AUG | SEP | OCT | NOV | DEC | YEAR |
|---|---|---|---|---|---|---|---|---|---|---|---|---|---|
| Maximum Temp °F | *65.5* | 68.2 | 70.6 | 76.6 | *82.3* | 90.7 | 98.1 | *98.1* | *92.5* | 83.6 | 73.0 | 65.8 | 80.4 |
| Minimum Temp °F | *37.1* | 39.5 | 42.4 | 45.2 | 50.7 | 56.0 | 60.7 | *61.9* | *57.7* | 50.3 | 42.3 | 36.6 | 48.4 |
| Mean Temp °F | *51.3* | 53.9 | 56.5 | 61.0 | *66.5* | 73.5 | 79.5 | *80.0* | *75.2* | 66.9 | 57.7 | 51.2 | 64.4 |
| Days Max Temp ≥ 90 °F | 0 | 0 | 0 | 3 | *8* | 17 | 28 | *28* | *19* | 9 | 1 | 0 | 113 |
| Days Max Temp ≤ 32 °F | 0 | 0 | 0 | 0 | 0 | 0 | 0 | 0 | 0 | 0 | 0 | 0 | 0 |
| Days Min Temp ≤ 32 °F | 8 | 4 | 2 | 0 | 0 | 0 | 0 | 0 | 0 | 0 | 2 | 8 | 24 |
| Days Min Temp ≤ 0 °F | 0 | 0 | 0 | 0 | 0 | 0 | 0 | 0 | 0 | 0 | 0 | 0 | 0 |
| Heating Degree Days | *417* | 308 | 260 | 146 | *51* | 8 | 0 | *0* | *3* | 45 | 221 | 420 | 1879 |
| Cooling Degree Days | *0* | 1 | 4 | 47 | *130* | 284 | 463 | 487 | 336 | 124 | 6 | 1 | 1883 |
| Total Precipitation (") | 2.75 | 2.71 | 2.39 | 0.65 | 0.19 | 0.02 | 0.10 | 0.12 | 0.29 | 0.34 | 1.21 | 2.07 | 12.84 |
| Days ≥ 0.1" Precip | 4 | 4 | 4 | 2 | 0 | 0 | 0 | 0 | 0 | 1 | 2 | 4 | 21 |
| Total Snowfall (") | 0.0 | 0.0 | 0.0 | 0.0 | 0.0 | 0.0 | 0.0 | 0.0 | 0.0 | 0.0 | 0.0 | 0.0 | 0.0 |
| Days ≥ 1" Snow Depth | 0 | 0 | 0 | 0 | 0 | 0 | 0 | 0 | 0 | 0 | 0 | 0 | 0 |

## EUREKA WSO CITY *Humboldt County* ELEVATION 43 ft LAT/LONG 40° 48' N / 124° 10' W

| | JAN | FEB | MAR | APR | MAY | JUN | JUL | AUG | SEP | OCT | NOV | DEC | YEAR |
|---|---|---|---|---|---|---|---|---|---|---|---|---|---|
| Maximum Temp °F | 54.7 | 55.8 | 56.0 | 56.6 | 58.5 | 60.7 | 62.1 | 63.1 | 63.1 | 61.2 | 58.1 | 54.8 | 58.7 |
| Minimum Temp °F | 41.6 | 43.0 | 43.7 | 44.8 | 47.8 | 50.7 | 52.4 | 53.3 | 51.6 | 48.7 | 44.9 | 41.7 | 47.0 |
| Mean Temp °F | 48.2 | 49.4 | 49.9 | 50.7 | 53.2 | 55.7 | 57.3 | 58.2 | 57.4 | 54.9 | 51.5 | 48.3 | 52.9 |
| Days Max Temp ≥ 90 °F | 0 | 0 | 0 | 0 | 0 | 0 | 0 | 0 | 0 | 0 | 0 | 0 | 0 |
| Days Max Temp ≤ 32 °F | 0 | 0 | 0 | 0 | 0 | 0 | 0 | 0 | 0 | 0 | 0 | 0 | 0 |
| Days Min Temp ≤ 32 °F | 1 | 1 | 0 | 0 | 0 | 0 | 0 | 0 | 0 | 0 | 1 | 2 | 5 |
| Days Min Temp ≤ 0 °F | 0 | 0 | 0 | 0 | 0 | 0 | 0 | 0 | 0 | 0 | 0 | 0 | 0 |
| Heating Degree Days | 515 | 434 | 462 | 421 | 359 | 272 | 232 | 203 | 222 | 306 | 398 | 512 | 4336 |
| Cooling Degree Days | 0 | 0 | 0 | 0 | 0 | 0 | 0 | 1 | 1 | 2 | 0 | 0 | 4 |
| Total Precipitation (") | 5.91 | 4.73 | 5.09 | 2.77 | 1.43 | 0.57 | 0.16 | 0.43 | 0.84 | 2.28 | 5.85 | 6.23 | 36.29 |
| Days ≥ 0.1" Precip | 10 | 9 | 11 | 7 | 4 | 2 | 0 | 1 | 2 | 5 | 10 | 12 | 73 |
| Total Snowfall (") | 0.1 | 0.2 | 0.0 | 0.0 | 0.0 | 0.0 | 0.0 | 0.0 | 0.0 | 0.0 | 0.0 | 0.1 | 0.4 |
| Days ≥ 1" Snow Depth | 0 | 0 | 0 | 0 | 0 | 0 | 0 | 0 | 0 | 0 | 0 | 0 | 0 |

## FAIRFIELD *Solano County* ELEVATION 20 ft LAT/LONG 38° 15' N / 122° 3' W

| | JAN | FEB | MAR | APR | MAY | JUN | JUL | AUG | SEP | OCT | NOV | DEC | YEAR |
|---|---|---|---|---|---|---|---|---|---|---|---|---|---|
| Maximum Temp °F | 55.0 | 61.6 | 65.7 | 71.1 | 78.4 | 84.4 | 88.6 | 88.4 | 86.2 | 78.4 | 64.9 | 55.0 | 73.1 |
| Minimum Temp °F | 36.8 | 40.4 | 43.7 | 45.9 | 50.1 | 53.8 | 55.8 | 55.9 | 54.3 | 49.6 | 42.5 | 36.6 | 47.1 |
| Mean Temp °F | 45.9 | 51.0 | 54.7 | 58.5 | 64.3 | 69.1 | 72.3 | 72.1 | 70.3 | 64.0 | 53.7 | 45.8 | 60.1 |
| Days Max Temp ≥ 90 °F | 0 | 0 | 0 | 0 | 4 | 8 | 14 | 13 | 10 | 3 | 0 | 0 | 52 |
| Days Max Temp ≤ 32 °F | 0 | 0 | 0 | 0 | 0 | 0 | 0 | 0 | 0 | 0 | 0 | 0 | 0 |
| Days Min Temp ≤ 32 °F | 8 | 2 | 1 | 0 | 0 | 0 | 0 | 0 | 0 | 0 | 2 | 8 | 21 |
| Days Min Temp ≤ 0 °F | 0 | 0 | 0 | 0 | 0 | 0 | 0 | 0 | 0 | 0 | 0 | 0 | 0 |
| Heating Degree Days | 585 | 389 | 313 | 202 | 86 | 22 | 3 | 3 | 11 | 83 | 333 | 588 | 2618 |
| Cooling Degree Days | 0 | 1 | 1 | 20 | 81 | 176 | 269 | 241 | 186 | 72 | 1 | 0 | 1048 |
| Total Precipitation (") | 4.91 | 3.69 | 3.31 | 1.16 | 0.37 | 0.15 | 0.03 | 0.09 | 0.29 | 1.16 | 3.18 | 3.58 | 21.92 |
| Days ≥ 0.1" Precip | 7 | 6 | 7 | 3 | 1 | 0 | 0 | 0 | 1 | 2 | 5 | 6 | 38 |
| Total Snowfall (") | 0.0 | 0.0 | 0.0 | 0.0 | 0.0 | 0.0 | 0.0 | 0.0 | 0.0 | 0.0 | 0.0 | 0.1 | 0.1 |
| Days ≥ 1" Snow Depth | 0 | 0 | 0 | 0 | 0 | 0 | 0 | 0 | 0 | 0 | 0 | 0 | 0 |

## FAIRMONT *Los Angeles County* ELEVATION 3060 ft LAT/LONG 34° 42' N / 118° 26' W

| | JAN | FEB | MAR | APR | MAY | JUN | JUL | AUG | SEP | OCT | NOV | DEC | YEAR |
|---|---|---|---|---|---|---|---|---|---|---|---|---|---|
| Maximum Temp °F | 53.3 | 56.6 | 59.9 | 65.6 | 74.0 | 82.7 | 89.7 | 89.8 | 84.6 | 74.3 | 61.4 | 53.1 | 70.4 |
| Minimum Temp °F | 35.9 | 38.2 | 40.8 | 44.6 | 51.3 | 59.3 | 66.1 | 65.8 | 60.3 | 51.8 | 42.5 | 35.9 | 49.4 |
| Mean Temp °F | 44.7 | 47.4 | 50.4 | 55.1 | 62.6 | 71.0 | 77.9 | 77.8 | 72.5 | 63.1 | 52.0 | 44.5 | 59.9 |
| Days Max Temp ≥ 90 °F | 0 | 0 | 0 | 0 | 2 | 9 | 17 | 18 | 10 | 1 | 0 | 0 | 57 |
| Days Max Temp ≤ 32 °F | 0 | 0 | 0 | 0 | 0 | 0 | 0 | 0 | 0 | 0 | 0 | 0 | 0 |
| Days Min Temp ≤ 32 °F | 9 | 5 | 3 | 1 | 0 | 0 | 0 | 0 | 0 | 0 | 2 | 9 | 29 |
| Days Min Temp ≤ 0 °F | 0 | 0 | 0 | 0 | 0 | 0 | 0 | 0 | 0 | 0 | 0 | 0 | 0 |
| Heating Degree Days | 624 | 489 | 447 | 309 | 154 | 44 | 2 | 3 | 26 | 129 | 386 | 627 | 3240 |
| Cooling Degree Days | 0 | 0 | 1 | 26 | 84 | 233 | 386 | 403 | 261 | 71 | 3 | 0 | 1468 |
| Total Precipitation (") | 3.40 | 3.59 | 2.75 | 1.12 | 0.26 | 0.07 | 0.06 | 0.16 | 0.31 | 0.40 | 2.24 | 2.53 | 16.89 |
| Days ≥ 0.1" Precip | 4 | 4 | 4 | 2 | 1 | 0 | 0 | 0 | 1 | 1 | 2 | 4 | 23 |
| Total Snowfall (") | 2.1 | 1.1 | 0.3 | 0.0 | 0.0 | 0.0 | 0.0 | 0.0 | 0.0 | 0.0 | 0.0 | 3.5 | 7.0 |
| Days ≥ 1" Snow Depth | 0 | 0 | 0 | 0 | 0 | 0 | 0 | 0 | 0 | 0 | 0 | 1 | 1 |

## FIVE POINTS 5 SSW *Fresno County* ELEVATION 279 ft LAT/LONG 36° 21 ' N / 120° 9 ' W

| | JAN | FEB | MAR | APR | MAY | JUN | JUL | AUG | SEP | OCT | NOV | DEC | YEAR |
|---|---|---|---|---|---|---|---|---|---|---|---|---|---|
| Maximum Temp °F | 54.8 | 62.9 | ***67.9*** | ***75.3*** | 84.7 | 91.4 | ***96.0*** | 93.9 | ***88.7*** | 79.1 | 64.9 | ***54.7*** | 76.2 |
| Minimum Temp °F | 37.1 | 39.8 | ***42.4*** | ***45.3*** | 51.5 | 57.5 | ***62.4*** | 61.8 | ***58.2*** | 50.3 | 41.5 | ***35.7*** | 48.6 |
| Mean Temp °F | 46.0 | 51.3 | ***55.2*** | ***60.3*** | 68.2 | 74.4 | ***79.2*** | 77.9 | ***73.5*** | 64.7 | 53.3 | ***45.3*** | 62.4 |
| Days Max Temp ≥ 90 °F | 0 | 0 | 0 | 1 | ***9*** | ***18*** | 25 | 23 | 14 | 3 | 0 | 0 | 93 |
| Days Max Temp ≤ 32 °F | 0 | 0 | 0 | 0 | 0 | 0 | 0 | 0 | 0 | 0 | 0 | 0 | 0 |
| Days Min Temp ≤ 32 °F | ***8*** | 3 | 1 | 0 | 0 | 0 | 0 | 0 | 0 | 0 | 3 | ***9*** | 24 |
| Days Min Temp ≤ 0 °F | 0 | 0 | 0 | 0 | 0 | 0 | 0 | 0 | 0 | 0 | 0 | 0 | 0 |
| Heating Degree Days | ***580*** | 378 | ***297*** | ***173*** | ***46*** | 7 | ***0*** | 1 | ***5*** | ***73*** | 347 | na | na |
| Cooling Degree Days | 0 | ***0*** | 7 | ***34*** | ***151*** | 289 | na | 390 | ***257*** | 91 | 0 | na | na |
| Total Precipitation (") | 1.43 | 1.33 | 1.23 | 0.45 | 0.20 | 0.14 | 0.03 | 0.03 | 0.27 | 0.35 | 0.75 | 0.86 | 7.07 |
| Days ≥ 0.1" Precip | ***3*** | ***3*** | 3 | 1 | 1 | 0 | 0 | 0 | 1 | 1 | 2 | 2 | 17 |
| Total Snowfall (") | 0.0 | 0.0 | 0.0 | 0.0 | 0.0 | 0.0 | 0.0 | 0.0 | 0.0 | 0.0 | 0.0 | 0.1 | 0.1 |
| Days ≥ 1" Snow Depth | 0 | 0 | 0 | 0 | 0 | 0 | 0 | 0 | 0 | 0 | 0 | 0 | 0 |

## FORT BIDWELL *Modoc County* ELEVATION 4505 ft LAT/LONG 41° 51 ' N / 120° 8 ' W

| | JAN | FEB | MAR | APR | MAY | JUN | JUL | AUG | SEP | OCT | NOV | DEC | YEAR |
|---|---|---|---|---|---|---|---|---|---|---|---|---|---|
| Maximum Temp °F | 39.7 | 45.6 | 52.5 | 59.8 | 69.3 | 77.0 | 85.6 | 84.8 | 77.5 | 66.1 | 49.2 | 40.0 | 62.3 |
| Minimum Temp °F | 19.6 | 23.7 | 27.9 | 31.7 | 38.0 | 44.1 | 49.1 | 47.8 | 40.8 | 33.4 | 26.4 | 20.0 | 33.5 |
| Mean Temp °F | 29.7 | 34.7 | 40.2 | 45.8 | 53.7 | 60.6 | 67.3 | 66.3 | 59.2 | 49.8 | 37.9 | 30.0 | 47.9 |
| Days Max Temp ≥ 90 °F | 0 | 0 | 0 | 0 | 0 | 2 | 10 | 9 | 2 | 0 | 0 | 0 | 23 |
| Days Max Temp ≤ 32 °F | 6 | 2 | 0 | 0 | 0 | 0 | 0 | 0 | 0 | 0 | 1 | 6 | 15 |
| Days Min Temp ≤ 32 °F | 27 | 24 | 23 | 16 | 7 | 1 | 0 | 0 | 3 | 14 | 23 | 28 | 166 |
| Days Min Temp ≤ 0 °F | 2 | 1 | 0 | 0 | 0 | 0 | 0 | 0 | 0 | 0 | 0 | 2 | 5 |
| Heating Degree Days | 1087 | 850 | 761 | 570 | 350 | 156 | 36 | 48 | 185 | 465 | 808 | 1079 | 6395 |
| Cooling Degree Days | 0 | 0 | 0 | 0 | 9 | 34 | 122 | 103 | 24 | 0 | 0 | 0 | 292 |
| Total Precipitation (") | 2.31 | 1.99 | 1.92 | 1.37 | 1.18 | 0.93 | 0.35 | 0.54 | 0.66 | 1.08 | 2.50 | 2.37 | 17.20 |
| Days ≥ 0.1" Precip | 7 | 6 | 6 | 5 | 4 | 3 | 1 | 1 | 2 | 3 | 7 | 8 | 53 |
| Total Snowfall (") | 11.8 | 9.4 | 6.3 | 3.6 | 0.9 | 0.0 | 0.0 | 0.0 | 0.0 | 1.0 | 6.5 | 13.3 | 52.8 |
| Days ≥ 1" Snow Depth | 13 | 7 | 2 | 0 | 0 | 0 | 0 | 0 | 0 | 0 | 4 | 12 | 38 |

## FORT BRAGG 5 N *Mendocino County* ELEVATION 120 ft LAT/LONG 39° 30 ' N / 123° 47 ' W

| | JAN | FEB | MAR | APR | MAY | JUN | JUL | AUG | SEP | OCT | NOV | DEC | YEAR |
|---|---|---|---|---|---|---|---|---|---|---|---|---|---|
| Maximum Temp °F | 55.3 | 56.7 | 58.1 | 59.6 | 61.6 | 64.0 | 65.5 | 65.7 | 66.2 | 63.8 | 59.3 | 55.1 | 60.9 |
| Minimum Temp °F | 39.5 | 40.7 | 42.2 | 43.0 | 45.4 | 48.2 | 49.4 | 50.0 | 49.2 | 46.6 | 43.0 | 39.6 | 44.7 |
| Mean Temp °F | 47.4 | 48.7 | 50.1 | 51.3 | 53.6 | 56.1 | 57.5 | 57.9 | 57.7 | 55.2 | 51.1 | 47.4 | 52.8 |
| Days Max Temp ≥ 90 °F | 0 | 0 | 0 | 0 | 0 | 0 | 0 | 0 | 0 | 0 | 0 | 0 | 0 |
| Days Max Temp ≤ 32 °F | 0 | 0 | 0 | 0 | 0 | 0 | 0 | 0 | 0 | 0 | 0 | 0 | 0 |
| Days Min Temp ≤ 32 °F | 4 | 2 | 1 | 0 | 0 | 0 | 0 | 0 | 0 | 0 | 1 | 4 | 12 |
| Days Min Temp ≤ 0 °F | 0 | 0 | 0 | 0 | 0 | 0 | 0 | 0 | 0 | 0 | 0 | 0 | 0 |
| Heating Degree Days | 537 | 453 | 454 | 403 | 350 | 261 | 227 | 215 | 214 | 298 | 410 | 538 | 4360 |
| Cooling Degree Days | 0 | 0 | 0 | ***0*** | 0 | 1 | 1 | 1 | 1 | 2 | 0 | 0 | 6 |
| Total Precipitation (") | 6.64 | 5.91 | 5.65 | 2.69 | 1.25 | 0.34 | 0.14 | 0.37 | 0.76 | 2.53 | 5.65 | 6.68 | 38.61 |
| Days ≥ 0.1" Precip | 10 | 10 | 10 | 6 | 3 | 1 | 0 | 1 | 1 | 4 | 8 | 10 | 64 |
| Total Snowfall (") | 0.0 | 0.0 | 0.0 | 0.0 | 0.0 | 0.0 | 0.0 | 0.0 | 0.0 | 0.0 | 0.0 | 0.0 | 0.0 |
| Days ≥ 1" Snow Depth | 0 | 0 | 0 | 0 | 0 | 0 | 0 | 0 | 0 | 0 | 0 | 0 | 0 |

## FORT ROSS *Sonoma County* ELEVATION 112 ft LAT/LONG 38° 31 ' N / 123° 15 ' W

| | JAN | FEB | MAR | APR | MAY | JUN | JUL | AUG | SEP | OCT | NOV | DEC | YEAR |
|---|---|---|---|---|---|---|---|---|---|---|---|---|---|
| Maximum Temp °F | 56.9 | 58.3 | 59.0 | 60.9 | 62.6 | 65.1 | 66.1 | 67.0 | 68.0 | 66.1 | 61.4 | 57.1 | 62.4 |
| Minimum Temp °F | 40.5 | 42.0 | 42.3 | 41.8 | 43.6 | 46.4 | 47.4 | 48.4 | 48.1 | 46.2 | 43.4 | 40.1 | 44.2 |
| Mean Temp °F | 48.8 | 50.1 | 50.7 | 51.4 | 53.1 | 55.8 | 56.7 | 57.8 | 58.1 | 56.2 | 52.4 | 48.6 | 53.3 |
| Days Max Temp ≥ 90 °F | 0 | 0 | 0 | 0 | 0 | 0 | 0 | 0 | 0 | 0 | 0 | 0 | 0 |
| Days Max Temp ≤ 32 °F | 0 | 0 | 0 | 0 | 0 | 0 | 0 | 0 | 0 | 0 | 0 | 0 | 0 |
| Days Min Temp ≤ 32 °F | 2 | 1 | 0 | 1 | 0 | 0 | 0 | 0 | 0 | 0 | 1 | 3 | 8 |
| Days Min Temp ≤ 0 °F | 0 | 0 | 0 | 0 | 0 | 0 | 0 | 0 | 0 | 0 | 0 | 0 | 0 |
| Heating Degree Days | 495 | 413 | 437 | 402 | 363 | 271 | 249 | 218 | 205 | 271 | 371 | 501 | 4196 |
| Cooling Degree Days | 0 | 0 | 0 | 0 | 0 | 2 | 0 | 2 | 2 | 2 | 0 | 0 | 8 |
| Total Precipitation (") | 7.37 | 5.55 | 5.34 | 2.32 | 0.72 | 0.37 | 0.15 | 0.28 | 0.65 | 2.32 | 5.65 | 6.16 | 36.88 |
| Days ≥ 0.1" Precip | 9 | 8 | 9 | 5 | 2 | 1 | 0 | 1 | 1 | 4 | 7 | 8 | 55 |
| Total Snowfall (") | 0.0 | 0.0 | 0.0 | 0.0 | 0.0 | 0.0 | 0.0 | 0.0 | 0.0 | 0.0 | 0.0 | 0.0 | 0.0 |
| Days ≥ 1" Snow Depth | 0 | 0 | 0 | 0 | 0 | 0 | 0 | 0 | 0 | 0 | 0 | 0 | 0 |

## FRESNO AIR TERMINAL *Fresno County* ELEVATION 338 ft LAT/LONG 36° 47 ' N / 119° 42 ' W

| | JAN | FEB | MAR | APR | MAY | JUN | JUL | AUG | SEP | OCT | NOV | DEC | YEAR |
|---|---|---|---|---|---|---|---|---|---|---|---|---|---|
| Maximum Temp °F | 54.2 | 62.0 | 67.1 | 74.9 | 84.6 | 92.4 | 98.3 | 96.4 | 90.5 | 80.0 | 64.7 | 53.9 | 76.6 |
| Minimum Temp °F | 37.6 | 40.7 | 44.3 | 47.8 | 54.4 | 60.7 | 65.6 | 64.1 | 59.4 | 51.1 | 42.1 | 36.6 | 50.4 |
| Mean Temp °F | 45.9 | 51.4 | 55.7 | 61.4 | 69.5 | 76.6 | 82.0 | 80.3 | 74.9 | 65.6 | 53.4 | 45.3 | 63.5 |
| Days Max Temp ≥ 90 °F | 0 | 0 | 0 | 2 | 10 | 20 | 29 | 26 | 17 | 4 | 0 | 0 | 108 |
| Days Max Temp ≤ 32 °F | 0 | 0 | 0 | 0 | 0 | 0 | 0 | 0 | 0 | 0 | 0 | 0 | 0 |
| Days Min Temp ≤ 32 °F | 7 | 3 | 1 | 0 | 0 | 0 | 0 | 0 | 0 | 0 | 2 | 9 | 22 |
| Days Min Temp ≤ 0 °F | 0 | 0 | 0 | 0 | 0 | 0 | 0 | 0 | 0 | 0 | 0 | 0 | 0 |
| Heating Degree Days | 585 | 377 | 283 | 142 | 34 | 3 | 0 | 0 | 3 | 68 | 340 | 604 | 2439 |
| Cooling Degree Days | 0 | 0 | 2 | 53 | 189 | 374 | 559 | 512 | 335 | 111 | 1 | 0 | 2136 |
| Total Precipitation (") | 2.05 | 1.86 | 2.10 | 0.87 | 0.31 | 0.13 | 0.01 | 0.01 | 0.24 | 0.56 | 1.28 | 1.47 | 10.89 |
| Days ≥ 0.1" Precip | 4 | 5 | 5 | 2 | 1 | 0 | 0 | 0 | 1 | 1 | 3 | 4 | 26 |
| Total Snowfall (") | 0.0 | 0.0 | 0.0 | 0.0 | 0.0 | 0.0 | 0.0 | 0.0 | 0.0 | 0.0 | 0.0 | 0.0 | 0.0 |
| Days ≥ 1" Snow Depth | 0 | 0 | 0 | 0 | 0 | 0 | 0 | 0 | 0 | 0 | 0 | 0 | 0 |

## FRIANT GOVERNMENT CP *Fresno County* ELEVATION 410 ft LAT/LONG 36° 59 ' N / 119° 43 ' W

| | JAN | FEB | MAR | APR | MAY | JUN | JUL | AUG | SEP | OCT | NOV | DEC | YEAR |
|---|---|---|---|---|---|---|---|---|---|---|---|---|---|
| Maximum Temp °F | 54.8 | 61.4 | 65.7 | 73.4 | 84.4 | 92.8 | 99.3 | 98.1 | 91.7 | 81.2 | 65.9 | 54.9 | 77.0 |
| Minimum Temp °F | 36.5 | 39.4 | 41.2 | 42.7 | 48.4 | 55.0 | 60.2 | 59.3 | 55.7 | 49.1 | 41.3 | 35.6 | 47.0 |
| Mean Temp °F | 45.7 | 50.4 | 53.5 | 58.1 | 66.4 | 73.9 | 79.8 | 78.7 | 73.7 | 65.2 | 53.6 | 45.3 | 62.0 |
| Days Max Temp ≥ 90 °F | 0 | 0 | 0 | 1 | 10 | 20 | 29 | 27 | 19 | 6 | 0 | 0 | 112 |
| Days Max Temp ≤ 32 °F | 0 | 0 | 0 | 0 | 0 | 0 | 0 | 0 | 0 | 0 | 0 | 0 | 0 |
| Days Min Temp ≤ 32 °F | 9 | 4 | 2 | 1 | 0 | 0 | 0 | 0 | 0 | 0 | 2 | 10 | 28 |
| Days Min Temp ≤ 0 °F | 0 | 0 | 0 | 0 | 0 | 0 | 0 | 0 | 0 | 0 | 0 | 0 | 0 |
| Heating Degree Days | 592 | 406 | 351 | 220 | 70 | 8 | 0 | 0 | 8 | 79 | 337 | 604 | 2675 |
| Cooling Degree Days | 0 | 0 | 1 | 25 | 122 | 282 | 458 | 427 | 279 | 101 | 1 | 0 | 1696 |
| Total Precipitation (") | 2.43 | 2.35 | 2.69 | 1.22 | 0.34 | 0.13 | 0.02 | 0.02 | 0.25 | 0.73 | 1.80 | 2.06 | 14.04 |
| Days ≥ 0.1" Precip | 5 | 5 | 6 | 3 | 1 | 0 | 0 | 0 | 1 | 1 | 4 | 4 | 30 |
| Total Snowfall (") | 0.0 | 0.0 | 0.0 | 0.0 | 0.0 | 0.0 | 0.0 | 0.0 | 0.0 | 0.0 | 0.0 | 0.0 | 0.0 |
| Days ≥ 1" Snow Depth | 0 | 0 | 0 | 0 | 0 | 0 | 0 | 0 | 0 | 0 | 0 | 0 | 0 |

## GILROY *Santa Clara County* ELEVATION 194 ft LAT/LONG 37° 0 ' N / 121° 34 ' W

| | JAN | FEB | MAR | APR | MAY | JUN | JUL | AUG | SEP | OCT | NOV | DEC | YEAR |
|---|---|---|---|---|---|---|---|---|---|---|---|---|---|
| Maximum Temp °F | 59.5 | 63.6 | 66.7 | 72.2 | 78.2 | 83.5 | 88.0 | 87.7 | 85.5 | 78.7 | 66.9 | 59.3 | 74.1 |
| Minimum Temp °F | 36.3 | 39.5 | 42.5 | 44.0 | 48.0 | 51.8 | 53.8 | 54.2 | 52.6 | 47.8 | 41.3 | 35.9 | 45.6 |
| Mean Temp °F | 47.9 | 51.6 | 54.6 | 58.1 | 63.1 | 67.7 | 71.0 | 71.0 | 69.1 | 63.3 | 54.1 | 47.6 | 59.9 |
| Days Max Temp ≥ 90 °F | 0 | 0 | 0 | 1 | 4 | 6 | 12 | 12 | 9 | 4 | 0 | 0 | 48 |
| Days Max Temp ≤ 32 °F | 0 | 0 | 0 | 0 | 0 | 0 | 0 | 0 | 0 | 0 | 0 | 0 | 0 |
| Days Min Temp ≤ 32 °F | 10 | 4 | 1 | 0 | 0 | 0 | 0 | 0 | 0 | 0 | 3 | 10 | 28 |
| Days Min Temp ≤ 0 °F | 0 | 0 | 0 | 0 | 0 | 0 | 0 | 0 | 0 | 0 | 0 | 0 | 0 |
| Heating Degree Days | 524 | 373 | 316 | 211 | 103 | 29 | 2 | 2 | 15 | 100 | 321 | 531 | 2527 |
| Cooling Degree Days | 0 | *0* | *1* | 19 | 53 | 120 | 200 | 198 | 142 | 62 | 2 | 0 | 797 |
| Total Precipitation (") | 4.03 | 3.16 | 3.68 | 1.33 | 0.26 | 0.12 | 0.07 | 0.07 | 0.34 | 0.86 | 2.72 | 3.22 | 19.86 |
| Days ≥ 0.1" Precip | 6 | 6 | 6 | 3 | 1 | 1 | 0 | 0 | 1 | 2 | 4 | 5 | 35 |
| Total Snowfall (") | 0.0 | 0.0 | 0.0 | 0.0 | 0.0 | 0.0 | 0.0 | 0.0 | 0.0 | 0.0 | 0.0 | 0.0 | 0.0 |
| Days ≥ 1" Snow Depth | 0 | 0 | 0 | 0 | 0 | 0 | 0 | 0 | 0 | 0 | 0 | 0 | 0 |

## GLENNVILLE *Kern County* ELEVATION 3133 ft LAT/LONG 35° 43 ' N / 118° 42 ' W

| | JAN | FEB | MAR | APR | MAY | JUN | JUL | AUG | SEP | OCT | NOV | DEC | YEAR |
|---|---|---|---|---|---|---|---|---|---|---|---|---|---|
| Maximum Temp °F | 57.0 | 58.8 | 59.5 | 65.4 | 74.7 | 83.6 | 90.3 | 89.3 | 83.5 | 74.0 | 62.2 | 56.2 | 71.2 |
| Minimum Temp °F | 28.6 | 30.8 | 33.1 | 35.0 | 39.7 | 45.2 | 50.9 | 50.1 | 45.8 | 38.6 | 32.1 | 27.9 | 38.2 |
| Mean Temp °F | 42.9 | 44.8 | 46.3 | 50.2 | 57.2 | 64.4 | 70.7 | 69.8 | 64.7 | 56.3 | 47.2 | 42.0 | 54.7 |
| Days Max Temp ≥ 90 °F | 0 | 0 | 0 | 0 | 1 | 8 | 18 | 15 | 8 | 1 | 0 | 0 | 51 |
| Days Max Temp ≤ 32 °F | 0 | 0 | 0 | 0 | 0 | 0 | 0 | 0 | 0 | 0 | 0 | 0 | 0 |
| Days Min Temp ≤ 32 °F | 23 | 18 | 15 | 10 | 3 | 1 | 0 | 0 | 1 | 5 | 17 | 25 | 118 |
| Days Min Temp ≤ 0 °F | 0 | 0 | 0 | 0 | 0 | 0 | 0 | 0 | 0 | 0 | 0 | 0 | 0 |
| Heating Degree Days | 680 | 562 | 572 | 437 | 245 | 80 | 9 | 16 | 73 | 271 | 528 | 705 | 4178 |
| Cooling Degree Days | 0 | 0 | 0 | 0 | 12 | 69 | 179 | 168 | 73 | 10 | 0 | 0 | 511 |
| Total Precipitation (") | 3.48 | 2.95 | 3.45 | 1.79 | 0.54 | 0.11 | 0.05 | 0.17 | 0.54 | 0.84 | 2.28 | 2.91 | 19.11 |
| Days ≥ 0.1" Precip | 5 | 5 | 7 | 4 | 1 | 0 | 0 | 0 | 1 | 2 | 4 | 5 | 34 |
| Total Snowfall (") | 2.1 | 1.4 | 2.7 | 0.8 | 0.0 | 0.0 | 0.0 | 0.0 | 0.0 | 0.0 | 1.1 | 1.9 | 10.0 |
| Days ≥ 1" Snow Depth | *0* | *0* | *0* | 0 | 0 | 0 | 0 | 0 | 0 | 0 | 0 | *0* | 0 |

## GOLD ROCK RANCH *Imperial County* ELEVATION 485 ft LAT/LONG 32° 53 ' N / 114° 52 ' W

| | JAN | FEB | MAR | APR | MAY | JUN | JUL | AUG | SEP | OCT | NOV | DEC | YEAR |
|---|---|---|---|---|---|---|---|---|---|---|---|---|---|
| Maximum Temp °F | 67.9 | 73.3 | 77.8 | 86.0 | 94.1 | 103.3 | 106.9 | 105.5 | 100.2 | 89.5 | 76.5 | 67.4 | 87.4 |
| Minimum Temp °F | 45.9 | 48.1 | 50.8 | 56.0 | 63.1 | 72.2 | 79.6 | 79.2 | 73.4 | 62.9 | 52.1 | 45.7 | 60.8 |
| Mean Temp °F | 56.9 | 60.8 | 64.2 | 71.0 | 78.6 | 87.8 | 93.2 | 92.4 | 86.8 | 76.3 | 64.3 | 56.6 | 74.1 |
| Days Max Temp ≥ 90 °F | 0 | 0 | 3 | 11 | 23 | 29 | 30 | 31 | 28 | 17 | 1 | 0 | 173 |
| Days Max Temp ≤ 32 °F | 0 | 0 | 0 | 0 | 0 | 0 | 0 | 0 | 0 | 0 | 0 | 0 | 0 |
| Days Min Temp ≤ 32 °F | 0 | 0 | 0 | 0 | 0 | 0 | 0 | 0 | 0 | 0 | 0 | 1 | 1 |
| Days Min Temp ≤ 0 °F | 0 | 0 | 0 | 0 | 0 | 0 | 0 | 0 | 0 | 0 | 0 | 0 | 0 |
| Heating Degree Days | 250 | 140 | 90 | 23 | 2 | 0 | 0 | 0 | 0 | 5 | 85 | 258 | 853 |
| Cooling Degree Days | 6 | 36 | 83 | 258 | 454 | 722 | 886 | 874 | 672 | 367 | 64 | 4 | 4426 |
| Total Precipitation (") | 0.51 | 0.41 | 0.39 | 0.12 | 0.07 | 0.01 | 0.31 | 0.57 | 0.39 | 0.37 | 0.27 | 0.53 | 3.95 |
| Days ≥ 0.1" Precip | 1 | 1 | 1 | 0 | 0 | 0 | 1 | 1 | 1 | 1 | 1 | 1 | 9 |
| Total Snowfall (") | 0.0 | 0.0 | 0.0 | 0.0 | 0.0 | 0.0 | 0.0 | 0.0 | 0.0 | 0.0 | 0.0 | 0.0 | 0.0 |
| Days ≥ 1" Snow Depth | 0 | 0 | 0 | 0 | 0 | 0 | 0 | 0 | 0 | 0 | 0 | 0 | 0 |

## GRANT GROVE *Tulare County* ELEVATION 6585 ft LAT/LONG 36° 44 ' N / 118° 58 ' W

| | JAN | FEB | MAR | APR | MAY | JUN | JUL | AUG | SEP | OCT | NOV | DEC | YEAR |
|---|---|---|---|---|---|---|---|---|---|---|---|---|---|
| Maximum Temp °F | 43.8 | 43.8 | 44.3 | 48.9 | 57.2 | 67.2 | 74.9 | 74.1 | 68.1 | 59.0 | 48.8 | 43.7 | 56.2 |
| Minimum Temp °F | 26.6 | 26.1 | 26.9 | 30.5 | 37.4 | 45.6 | 51.8 | 51.3 | 46.4 | 39.5 | 31.6 | 26.7 | 36.7 |
| Mean Temp °F | 35.2 | 35.0 | 35.7 | 39.7 | 47.3 | 56.4 | 63.4 | 62.7 | 57.3 | 49.2 | 40.2 | 35.3 | 46.4 |
| Days Max Temp ≥ 90 °F | 0 | 0 | 0 | 0 | 0 | 0 | 0 | 0 | 0 | 0 | 0 | 0 | 0 |
| Days Max Temp ≤ 32 °F | 4 | 4 | 3 | 2 | 0 | 0 | 0 | 0 | 0 | 0 | 2 | 5 | 20 |
| Days Min Temp ≤ 32 °F | 24 | 22 | 24 | 18 | 8 | 1 | 0 | 0 | 1 | 6 | 16 | 22 | 142 |
| Days Min Temp ≤ 0 °F | 0 | 0 | 0 | 0 | 0 | 0 | 0 | 0 | 0 | 0 | 0 | 0 | 0 |
| Heating Degree Days | 918 | 840 | 903 | 752 | 543 | 261 | 88 | 106 | 237 | 483 | 735 | 915 | 6781 |
| Cooling Degree Days | 0 | 0 | 0 | 0 | 1 | 11 | 45 | 41 | 11 | 2 | 0 | 0 | 111 |
| Total Precipitation (") | 7.80 | 6.50 | 7.07 | 3.40 | 1.10 | 0.42 | 0.21 | 0.15 | 1.25 | 2.08 | 4.83 | 6.45 | 41.26 |
| Days ≥ 0.1" Precip | 6 | 7 | 8 | 5 | 3 | 1 | 1 | 1 | 2 | 3 | 6 | 6 | 49 |
| Total Snowfall (") | 31.6 | 33.7 | 43.7 | 22.7 | 3.9 | 0.0 | 0.0 | 0.0 | 0.4 | 1.9 | 17.1 | 29.8 | 184.8 |
| Days ≥ 1" Snow Depth | 27 | 26 | 29 | 19 | 8 | 0 | 0 | 0 | 0 | 1 | 11 | 23 | 144 |

## GRATON *Sonoma County* ELEVATION 210 ft LAT/LONG 38° 26 ' N / 122° 53 ' W

| | JAN | FEB | MAR | APR | MAY | JUN | JUL | AUG | SEP | OCT | NOV | DEC | YEAR |
|---|---|---|---|---|---|---|---|---|---|---|---|---|---|
| Maximum Temp °F | 56.8 | 61.8 | 64.7 | 69.9 | 76.3 | 81.3 | 83.9 | 83.9 | 82.6 | 76.9 | 65.3 | 56.7 | 71.7 |
| Minimum Temp °F | 34.8 | 37.2 | 39.2 | 39.9 | 43.3 | 46.7 | 48.2 | 48.2 | 46.8 | 43.0 | 38.2 | 34.5 | 41.7 |
| Mean Temp °F | 45.8 | 49.5 | 52.0 | 54.9 | 59.8 | 64.0 | 66.1 | 66.1 | 64.7 | 60.0 | 51.8 | 45.6 | 56.7 |
| Days Max Temp ≥ 90 °F | 0 | 0 | 0 | 0 | 2 | 5 | 7 | 7 | 7 | 2 | 0 | 0 | 30 |
| Days Max Temp ≤ 32 °F | 0 | 0 | 0 | 0 | 0 | 0 | 0 | 0 | 0 | 0 | 0 | 0 | 0 |
| Days Min Temp ≤ 32 °F | 14 | 8 | 4 | 2 | 0 | 0 | 0 | 0 | 0 | 1 | 7 | 14 | 50 |
| Days Min Temp ≤ 0 °F | 0 | 0 | 0 | 0 | 0 | 0 | 0 | 0 | 0 | 0 | 0 | 0 | 0 |
| Heating Degree Days | 588 | 431 | 398 | 298 | 171 | 74 | 30 | 28 | 56 | 166 | 390 | 595 | 3225 |
| Cooling Degree Days | 0 | 0 | 0 | 4 | 19 | 56 | 75 | 69 | 51 | 21 | 0 | 0 | 295 |
| Total Precipitation (") | 8.93 | 6.44 | 5.69 | 2.19 | 0.56 | 0.29 | 0.08 | 0.11 | 0.48 | 2.10 | 6.18 | 7.27 | 40.32 |
| Days ≥ 0.1" Precip | 9 | 8 | 8 | 5 | 1 | 1 | 0 | 0 | 1 | 3 | 8 | 8 | 52 |
| Total Snowfall (") | 0.0 | 0.0 | 0.0 | 0.0 | 0.0 | 0.0 | 0.0 | 0.0 | 0.0 | 0.0 | 0.0 | 0.0 | 0.0 |
| Days ≥ 1" Snow Depth | 0 | 0 | 0 | 0 | 0 | 0 | 0 | 0 | 0 | 0 | 0 | 0 | 0 |

## HAIWEE *Inyo County* ELEVATION 3832 ft LAT/LONG 36° 8 ' N / 117° 57 ' W

| | JAN | FEB | MAR | APR | MAY | JUN | JUL | AUG | SEP | OCT | NOV | DEC | YEAR |
|---|---|---|---|---|---|---|---|---|---|---|---|---|---|
| Maximum Temp °F | 51.7 | 57.0 | 62.4 | 69.5 | 79.4 | 89.2 | 95.3 | 93.6 | 86.2 | 75.7 | 61.6 | 51.9 | 72.8 |
| Minimum Temp °F | 28.7 | 32.0 | 36.7 | 42.2 | 50.6 | 58.9 | 64.7 | 63.4 | 56.4 | 46.6 | 35.8 | 28.6 | 45.4 |
| Mean Temp °F | 40.2 | 44.5 | 49.6 | 55.9 | 65.0 | 74.1 | 80.0 | 78.6 | 71.4 | 61.2 | 48.7 | 40.3 | 59.1 |
| Days Max Temp ≥ 90 °F | 0 | 0 | 0 | 0 | 3 | 16 | 27 | 24 | 11 | 1 | 0 | 0 | 82 |
| Days Max Temp ≤ 32 °F | 0 | 0 | 0 | 0 | 0 | 0 | 0 | 0 | 0 | 0 | 0 | 1 | 1 |
| Days Min Temp ≤ 32 °F | 24 | 15 | 7 | 2 | 0 | 0 | 0 | 0 | 0 | 1 | 9 | 23 | 81 |
| Days Min Temp ≤ 0 °F | 0 | 0 | 0 | 0 | 0 | 0 | 0 | 0 | 0 | 0 | 0 | 0 | 0 |
| Heating Degree Days | 762 | 571 | 472 | 278 | 91 | 10 | 0 | 0 | 17 | 153 | 482 | 760 | 3596 |
| Cooling Degree Days | 0 | 0 | 0 | 19 | 102 | 301 | 468 | 435 | 218 | 46 | 0 | 0 | 1589 |
| Total Precipitation (") | 1.13 | 1.33 | 1.26 | 0.36 | 0.30 | 0.08 | 0.37 | 0.40 | 0.33 | 0.13 | 0.77 | 1.03 | 7.49 |
| Days ≥ 0.1" Precip | 3 | 2 | 3 | 1 | 1 | 0 | 1 | 1 | 1 | 0 | 2 | 2 | 17 |
| Total Snowfall (") | 1.0 | 1.0 | 0.1 | 0.3 | 0.0 | 0.0 | 0.0 | 0.0 | 0.0 | 0.0 | 0.1 | *0.5* | 3.0 |
| Days ≥ 1" Snow Depth | 1 | 1 | 0 | 0 | 0 | 0 | 0 | 0 | 0 | 0 | 0 | 1 | 3 |

## HALF MOON BAY *San Mateo County* ELEVATION 39 ft LAT/LONG 37° 28 ' N / 122° 26 ' W

| | JAN | FEB | MAR | APR | MAY | JUN | JUL | AUG | SEP | OCT | NOV | DEC | YEAR |
|---|---|---|---|---|---|---|---|---|---|---|---|---|---|
| Maximum Temp °F | 58.7 | 59.6 | 59.8 | 60.8 | 61.3 | 62.9 | 64.0 | 65.6 | 67.1 | 66.0 | 62.9 | 58.5 | 62.3 |
| Minimum Temp °F | 42.9 | 43.8 | 44.4 | 44.2 | 46.9 | 49.6 | 51.4 | 52.7 | 51.4 | 48.6 | 45.7 | 42.8 | 47.0 |
| Mean Temp °F | 50.8 | 51.7 | 52.1 | 52.5 | 54.1 | 56.3 | 57.7 | 59.2 | 59.4 | 57.3 | 54.3 | 50.7 | 54.7 |
| Days Max Temp ≥ 90 °F | 0 | 0 | 0 | 0 | 0 | 0 | 0 | 0 | 0 | 0 | 0 | 0 | 0 |
| Days Max Temp ≤ 32 °F | 0 | 0 | 0 | 0 | 0 | 0 | 0 | 0 | 0 | 0 | 0 | 0 | 0 |
| Days Min Temp ≤ 32 °F | 1 | 0 | 0 | 0 | 0 | 0 | 0 | 0 | 0 | 0 | 0 | 1 | 2 |
| Days Min Temp ≤ 0 °F | 0 | 0 | 0 | 0 | 0 | 0 | 0 | 0 | 0 | 0 | 0 | 0 | 0 |
| Heating Degree Days | 434 | 368 | 392 | 369 | 331 | 255 | 219 | 177 | 167 | 236 | 315 | 436 | 3699 |
| Cooling Degree Days | 0 | 0 | 0 | 1 | 0 | 1 | 0 | 2 | 4 | 5 | 0 | 0 | 13 |
| Total Precipitation (") | 5.08 | 4.13 | 4.22 | 1.87 | 0.49 | 0.25 | 0.13 | 0.24 | 0.42 | 1.69 | 3.69 | 4.27 | 26.48 |
| Days ≥ 0.1" Precip | 8 | 7 | 8 | 4 | 1 | 1 | 0 | 1 | 1 | 3 | 6 | 7 | 47 |
| Total Snowfall (") | 0.0 | 0.0 | 0.0 | 0.0 | 0.0 | 0.0 | 0.0 | 0.0 | 0.0 | 0.0 | 0.0 | 0.0 | 0.0 |
| Days ≥ 1" Snow Depth | 0 | 0 | 0 | 0 | 0 | 0 | 0 | 0 | 0 | 0 | 0 | 0 | 0 |

## HANFORD 1 S *Kings County* ELEVATION 249 ft LAT/LONG 36° 19 ' N / 119° 40 ' W

| | JAN | FEB | MAR | APR | MAY | JUN | JUL | AUG | SEP | OCT | NOV | DEC | YEAR |
|---|---|---|---|---|---|---|---|---|---|---|---|---|---|
| Maximum Temp °F | 53.6 | 61.6 | 67.1 | 74.9 | 84.2 | 90.9 | 95.9 | 94.5 | 89.2 | 80.4 | 65.1 | 53.4 | 75.9 |
| Minimum Temp °F | 35.3 | 38.7 | 43.0 | 46.3 | 52.7 | 58.3 | 62.1 | 60.8 | 56.2 | 48.2 | 39.6 | 34.1 | 47.9 |
| Mean Temp °F | 44.5 | 50.2 | 55.0 | 60.6 | 68.5 | 74.7 | 79.0 | 77.7 | 72.7 | 64.3 | 52.6 | 43.8 | 62.0 |
| Days Max Temp ≥ 90 °F | 0 | 0 | 0 | 1 | 9 | 17 | 27 | 25 | 16 | 4 | 0 | 0 | 99 |
| Days Max Temp ≤ 32 °F | 0 | 0 | 0 | 0 | 0 | 0 | 0 | 0 | 0 | 0 | 0 | 0 | 0 |
| Days Min Temp ≤ 32 °F | 11 | 4 | 1 | 0 | 0 | 0 | 0 | 0 | 0 | 0 | 5 | 13 | 34 |
| Days Min Temp ≤ 0 °F | 0 | 0 | 0 | 0 | 0 | 0 | 0 | 0 | 0 | 0 | 0 | 0 | 0 |
| Heating Degree Days | 628 | 411 | 303 | 157 | 38 | 3 | 0 | 0 | 6 | 81 | 372 | 649 | 2648 |
| Cooling Degree Days | 0 | 0 | 2 | 44 | 160 | 313 | 455 | 417 | 258 | 80 | 1 | 0 | 1730 |
| Total Precipitation (") | 1.53 | 1.50 | 1.56 | 0.65 | 0.19 | 0.07 | 0.01 | 0.01 | 0.27 | 0.33 | 0.98 | 1.10 | 8.20 |
| Days ≥ 0.1" Precip | 4 | 4 | 4 | 2 | 0 | 0 | 0 | 0 | 1 | 1 | 3 | 3 | 22 |
| Total Snowfall (") | 0.0 | 0.0 | 0.0 | 0.0 | 0.0 | 0.0 | 0.0 | 0.0 | 0.0 | 0.0 | 0.0 | 0.0 | 0.0 |
| Days ≥ 1" Snow Depth | 0 | 0 | 0 | 0 | 0 | 0 | 0 | 0 | 0 | 0 | 0 | 0 | 0 |

## HAPPY CAMP RANGER ST *Siskiyou County* ELEVATION 1089 ft LAT/LONG 41° 48 ' N / 123° 23 ' W

| | JAN | FEB | MAR | APR | MAY | JUN | JUL | AUG | SEP | OCT | NOV | DEC | YEAR |
|---|---|---|---|---|---|---|---|---|---|---|---|---|---|
| Maximum Temp °F | na | ***54.6*** | 61.7 | ***67.9*** | 78.0 | 85.8 | 94.3 | 93.4 | 86.6 | ***73.6*** | na | na | na |
| Minimum Temp °F | na | ***32.2*** | 34.2 | 36.0 | 41.3 | 47.6 | 51.6 | 50.9 | 45.4 | ***38.9*** | na | na | na |
| Mean Temp °F | na | ***43.3*** | 48.0 | ***52.2*** | 59.7 | 66.6 | 73.0 | 72.2 | 66.0 | ***56.3*** | na | na | na |
| Days Max Temp ≥ 90 °F | 0 | 0 | 0 | 0 | 4 | 12 | 22 | 23 | 13 | 2 | 0 | 0 | 76 |
| Days Max Temp ≤ 32 °F | 0 | 0 | 0 | 0 | 0 | 0 | 0 | 0 | 0 | 0 | 0 | 0 | 0 |
| Days Min Temp ≤ 32 °F | ***17*** | ***14*** | ***11*** | ***7*** | 2 | 0 | 0 | 0 | 0 | 4 | 9 | 15 | 79 |
| Days Min Temp ≤ 0 °F | 0 | 0 | 0 | 0 | 0 | 0 | 0 | 0 | 0 | 0 | 0 | 0 | 0 |
| Heating Degree Days | na | ***613*** | 522 | ***380*** | 191 | 65 | 10 | 11 | 59 | ***272*** | na | na | na |
| Cooling Degree Days | na | na | ***0*** | ***2*** | 30 | 99 | 236 | 223 | 94 | na | na | na | na |
| Total Precipitation (") | 9.64 | 7.16 | 6.37 | 2.69 | 1.33 | 0.61 | 0.28 | 0.66 | 1.23 | 3.07 | 8.17 | 9.52 | 50.73 |
| Days ≥ 0.1" Precip | 8 | ***6*** | ***8*** | 4 | 3 | 2 | 1 | 1 | 2 | 4 | ***7*** | na | na |
| Total Snowfall (") | ***8.9*** | 1.6 | 1.5 | 0.0 | 0.0 | 0.0 | 0.0 | 0.0 | 0.0 | 0.0 | 0.8 | 5.3 | 18.1 |
| Days ≥ 1" Snow Depth | ***4*** | ***0*** | 0 | 0 | 0 | 0 | 0 | 0 | 0 | 0 | 0 | ***2*** | 6 |

## HAT CREEK PH 1 *Shasta County* ELEVATION 3022 ft LAT/LONG 40° 56 ' N / 121° 33 ' W

| | JAN | FEB | MAR | APR | MAY | JUN | JUL | AUG | SEP | OCT | NOV | DEC | YEAR |
|---|---|---|---|---|---|---|---|---|---|---|---|---|---|
| Maximum Temp °F | 47.3 | 52.7 | 57.6 | 64.3 | 74.4 | 82.1 | 90.2 | 89.6 | 83.2 | 71.7 | 55.0 | 46.4 | 67.9 |
| Minimum Temp °F | 21.6 | 25.4 | 29.1 | 32.7 | 38.7 | 44.5 | 47.6 | 45.5 | 39.4 | 32.4 | 27.6 | 22.2 | 33.9 |
| Mean Temp °F | 34.5 | 39.1 | 43.4 | 48.5 | 56.5 | 63.3 | 68.9 | 67.6 | 61.3 | 52.0 | 41.3 | 34.3 | 50.9 |
| Days Max Temp ≥ 90 °F | 0 | 0 | 0 | 0 | 2 | 7 | 18 | 18 | 8 | 1 | 0 | 0 | 54 |
| Days Max Temp ≤ 32 °F | 0 | 0 | 0 | 0 | 0 | 0 | 0 | 0 | 0 | 0 | 0 | 1 | 1 |
| Days Min Temp ≤ 32 °F | 27 | 24 | 22 | 15 | 5 | 1 | 0 | 0 | 4 | 17 | 23 | 27 | 165 |
| Days Min Temp ≤ 0 °F | 1 | 0 | 0 | 0 | 0 | 0 | 0 | 0 | 0 | 0 | 0 | 0 | 1 |
| Heating Degree Days | 939 | 725 | 663 | 488 | 269 | 106 | 25 | 36 | 136 | 397 | 703 | 944 | 5431 |
| Cooling Degree Days | 0 | 0 | 0 | 0 | 18 | 69 | 163 | 131 | 40 | 2 | 0 | 0 | 423 |
| Total Precipitation (") | 3.04 | 2.52 | 2.54 | 1.34 | 1.10 | 0.78 | 0.17 | 0.42 | 0.68 | 1.25 | 2.47 | 2.80 | 19.11 |
| Days ≥ 0.1" Precip | 7 | 6 | 7 | 4 | 3 | 3 | 1 | 1 | 2 | 3 | 6 | 7 | 50 |
| Total Snowfall (") | 8.9 | 3.8 | 2.9 | 1.0 | 0.0 | 0.0 | 0.0 | 0.0 | 0.1 | 0.1 | 1.0 | 7.8 | 25.6 |
| Days ≥ 1" Snow Depth | 10 | 6 | 2 | 0 | 0 | 0 | 0 | 0 | 0 | 0 | 1 | 7 | 26 |

## HAYFIELD PUMP PLANT *Riverside County* ELEVATION 1370 ft LAT/LONG 33° 42 ' N / 115° 38 ' W

| | JAN | FEB | MAR | APR | MAY | JUN | JUL | AUG | SEP | OCT | NOV | DEC | YEAR |
|---|---|---|---|---|---|---|---|---|---|---|---|---|---|
| Maximum Temp °F | 65.1 | 69.5 | 74.2 | 81.4 | 89.5 | 99.2 | 103.8 | 102.4 | 97.0 | 86.6 | 73.9 | 65.5 | 84.0 |
| Minimum Temp °F | 39.6 | 42.8 | 46.4 | 51.6 | 59.1 | 66.8 | 74.8 | 73.7 | 66.4 | 55.6 | 45.6 | 39.1 | 55.1 |
| Mean Temp °F | 52.4 | 56.2 | 60.3 | 66.5 | 74.3 | 83.0 | 89.3 | 88.0 | 81.7 | 71.1 | 59.8 | 52.3 | 69.6 |
| Days Max Temp ≥ 90 °F | 0 | 0 | 1 | 6 | 17 | 27 | 31 | 30 | 26 | 12 | 1 | 0 | 151 |
| Days Max Temp ≤ 32 °F | 0 | 0 | 0 | 0 | 0 | 0 | 0 | 0 | 0 | 0 | 0 | 0 | 0 |
| Days Min Temp ≤ 32 °F | 6 | 2 | 1 | 0 | 0 | 0 | 0 | 0 | 0 | 0 | 1 | 6 | 16 |
| Days Min Temp ≤ 0 °F | 0 | 0 | 0 | 0 | 0 | 0 | 0 | 0 | 0 | 0 | 0 | 0 | 0 |
| Heating Degree Days | 387 | 249 | 171 | 63 | 10 | 0 | 0 | 0 | 0 | 22 | 174 | 388 | 1464 |
| Cooling Degree Days | 0 | 7 | 30 | 140 | 315 | 553 | 745 | 723 | 510 | 216 | 23 | 1 | 3263 |
| Total Precipitation (") | 0.68 | 0.60 | 0.60 | 0.12 | 0.10 | 0.01 | 0.29 | 0.75 | 0.34 | 0.30 | 0.30 | 0.47 | 4.56 |
| Days ≥ 0.1" Precip | 2 | 1 | 1 | 0 | 0 | 0 | 1 | 2 | 1 | 1 | 1 | 2 | 12 |
| Total Snowfall (") | 0.0 | 0.0 | 0.0 | 0.0 | 0.0 | 0.0 | 0.0 | 0.0 | 0.0 | 0.0 | 0.0 | 0.0 | 0.0 |
| Days ≥ 1" Snow Depth | 0 | 0 | 0 | 0 | 0 | 0 | 0 | 0 | 0 | 0 | 0 | 0 | 0 |

## HEALDSBURG *Sonoma County* ELEVATION 102 ft LAT/LONG 38° 37 ' N / 122° 52 ' W

| | JAN | FEB | MAR | APR | MAY | JUN | JUL | AUG | SEP | OCT | NOV | DEC | YEAR |
|---|---|---|---|---|---|---|---|---|---|---|---|---|---|
| Maximum Temp °F | 57.8 | 63.2 | 66.6 | 72.1 | 79.1 | 84.8 | 89.0 | 88.4 | 85.9 | 78.6 | 65.7 | 57.5 | 74.1 |
| Minimum Temp °F | 38.6 | 41.6 | 43.6 | 45.3 | 49.4 | 53.0 | 53.8 | 53.8 | 52.8 | 48.9 | 43.3 | 38.5 | 46.9 |
| Mean Temp °F | 48.2 | 52.4 | 55.1 | 58.7 | 64.3 | 68.9 | 71.4 | 71.1 | 69.4 | 63.8 | 54.5 | 48.0 | 60.5 |
| Days Max Temp ≥ 90 °F | 0 | 0 | 0 | 1 | 5 | 9 | 14 | 14 | 10 | 4 | 0 | 0 | 57 |
| Days Max Temp ≤ 32 °F | 0 | 0 | 0 | 0 | 0 | 0 | 0 | 0 | 0 | 0 | 0 | 0 | 0 |
| Days Min Temp ≤ 32 °F | 5 | 1 | 0 | 0 | 0 | 0 | 0 | 0 | 0 | 0 | 1 | 7 | 14 |
| Days Min Temp ≤ 0 °F | 0 | 0 | 0 | 0 | 0 | 0 | 0 | 0 | 0 | 0 | 0 | 0 | 0 |
| Heating Degree Days | 513 | 349 | 301 | 194 | 81 | 19 | 2 | 2 | 14 | 84 | 309 | 520 | 2388 |
| Cooling Degree Days | 0 | 0 | 2 | 16 | 62 | 143 | 199 | 183 | 134 | 49 | 2 | 0 | 790 |
| Total Precipitation (") | 9.02 | 7.02 | 5.90 | 2.31 | 0.58 | 0.21 | 0.07 | 0.15 | 0.53 | 2.13 | 6.45 | 7.47 | 41.84 |
| Days ≥ 0.1" Precip | 9 | 8 | 8 | 4 | 1 | 0 | 0 | 0 | 1 | 4 | 7 | 8 | 50 |
| Total Snowfall (") | 0.0 | 0.0 | 0.0 | 0.0 | 0.0 | 0.0 | 0.0 | 0.0 | 0.0 | 0.0 | 0.0 | 0.0 | 0.0 |
| Days ≥ 1" Snow Depth | 0 | 0 | 0 | 0 | 0 | 0 | 0 | 0 | 0 | 0 | 0 | 0 | 0 |

## HENSHAW DAM *San Diego County* ELEVATION 2700 ft LAT/LONG 33° 14 ' N / 116° 46 ' W

| | JAN | FEB | MAR | APR | MAY | JUN | JUL | AUG | SEP | OCT | NOV | DEC | YEAR |
|---|---|---|---|---|---|---|---|---|---|---|---|---|---|
| Maximum Temp °F | 60.1 | 62.3 | 63.6 | 68.4 | 74.6 | 83.9 | 91.8 | 92.4 | 87.1 | 78.6 | 67.4 | 60.2 | 74.2 |
| Minimum Temp °F | 29.8 | 31.5 | 34.8 | 37.6 | 42.5 | 46.7 | 52.8 | 53.3 | 47.8 | 39.6 | 32.8 | 28.5 | 39.8 |
| Mean Temp °F | 44.9 | 46.9 | 49.2 | 53.0 | 58.6 | 65.3 | 72.3 | 72.8 | 67.5 | ***59.2*** | 50.1 | 44.4 | 57.0 |
| Days Max Temp ≥ 90 °F | 0 | 0 | 0 | 0 | 2 | 10 | 21 | 22 | 14 | 5 | 0 | 0 | 74 |
| Days Max Temp ≤ 32 °F | 0 | 0 | 0 | 0 | 0 | 0 | 0 | 0 | 0 | 0 | 0 | 0 | 0 |
| Days Min Temp ≤ 32 °F | 22 | 17 | 11 | 4 | 1 | 0 | 0 | 0 | 0 | 3 | 16 | 24 | 98 |
| Days Min Temp ≤ 0 °F | 0 | 0 | 0 | 0 | 0 | 0 | 0 | 0 | 0 | 0 | 0 | 0 | 0 |
| Heating Degree Days | 615 | 503 | 482 | 354 | 207 | 72 | 7 | 7 | 48 | ***189*** | 440 | 633 | 3557 |
| Cooling Degree Days | 0 | 0 | 0 | 1 | 13 | 88 | 222 | 255 | 129 | ***19*** | 0 | 0 | 727 |
| Total Precipitation (") | 5.46 | 5.01 | 5.58 | 1.95 | 0.52 | 0.11 | 0.43 | 0.65 | 0.58 | 0.87 | 2.90 | 4.04 | 28.10 |
| Days ≥ 0.1" Precip | 5 | 5 | 7 | 4 | 2 | 0 | 1 | 1 | 1 | 2 | 3 | 5 | 36 |
| Total Snowfall (") | 0.1 | 0.8 | 0.1 | 0.0 | 0.0 | 0.0 | 0.0 | 0.0 | 0.0 | 0.0 | 0.0 | 0.3 | 1.3 |
| Days ≥ 1" Snow Depth | 0 | 0 | 0 | 0 | 0 | 0 | 0 | 0 | 0 | 0 | 0 | 0 | 0 |

## HETCH HETCHY *Tuolumne County* ELEVATION 3870 ft LAT/LONG 37° 57 ' N / 119° 47 ' W

| | JAN | FEB | MAR | APR | MAY | JUN | JUL | AUG | SEP | OCT | NOV | DEC | YEAR |
|---|---|---|---|---|---|---|---|---|---|---|---|---|---|
| Maximum Temp °F | 48.2 | 53.0 | 56.1 | 61.8 | 70.0 | 77.8 | 84.9 | 84.9 | 79.5 | 70.4 | 56.2 | 47.5 | 65.9 |
| Minimum Temp °F | 29.5 | 30.6 | 33.0 | 37.1 | 43.7 | 50.0 | 55.8 | 55.6 | 50.4 | 42.3 | 34.4 | 29.5 | 41.0 |
| Mean Temp °F | 38.9 | 41.9 | 44.6 | 49.5 | 56.9 | 63.9 | 70.4 | 70.3 | 65.0 | 56.4 | 45.3 | 38.5 | 53.5 |
| Days Max Temp ≥ 90 °F | 0 | 0 | 0 | 0 | 0 | 2 | 6 | 7 | 2 | 0 | 0 | 0 | 17 |
| Days Max Temp ≤ 32 °F | 0 | 0 | 0 | 0 | 0 | 0 | 0 | 0 | 0 | 0 | 0 | 1 | 1 |
| Days Min Temp ≤ 32 °F | 22 | 18 | 15 | 7 | 1 | 0 | 0 | 0 | 0 | 2 | 11 | 22 | 98 |
| Days Min Temp ≤ 0 °F | 0 | 0 | 0 | 0 | 0 | 0 | 0 | 0 | 0 | 0 | 0 | 0 | 0 |
| Heating Degree Days | 802 | 648 | 626 | 460 | 262 | 95 | 12 | 15 | 75 | 275 | 584 | 815 | 4669 |
| Cooling Degree Days | 0 | 0 | 0 | 2 | 15 | 72 | 188 | 192 | 89 | 21 | 0 | 0 | 579 |
| Total Precipitation (") | 5.71 | 5.30 | 5.45 | 3.15 | 1.28 | 0.76 | 0.30 | 0.34 | 0.88 | 2.21 | 4.95 | 4.88 | 35.21 |
| Days ≥ 0.1" Precip | 7 | ***6*** | 9 | 5 | 3 | 2 | 0 | 1 | 2 | 3 | 5 | ***6*** | 49 |
| Total Snowfall (") | 9.8 | 7.7 | 10.3 | 5.9 | 0.1 | 0.0 | 0.0 | 0.0 | 0.0 | 0.0 | ***3.2*** | ***10.1*** | 47.1 |
| Days ≥ 1" Snow Depth | 11 | 7 | 4 | 2 | 0 | 0 | 0 | 0 | 0 | 0 | 3 | ***7*** | 34 |

## HUNTINGTON LAKE *Fresno County* ELEVATION 7020 ft LAT/LONG 37° 14 ' N / 119° 13 ' W

| | JAN | FEB | MAR | APR | MAY | JUN | JUL | AUG | SEP | OCT | NOV | DEC | YEAR |
|---|---|---|---|---|---|---|---|---|---|---|---|---|---|
| Maximum Temp °F | 45.5 | 45.9 | 46.4 | 50.4 | 57.4 | 66.4 | 73.8 | 73.3 | 67.4 | 59.2 | 49.4 | 44.4 | 56.6 |
| Minimum Temp °F | 24.9 | 24.5 | 25.5 | 28.3 | 34.4 | 41.3 | 47.5 | 47.6 | 42.9 | 36.5 | 29.5 | 25.4 | 34.0 |
| Mean Temp °F | 35.3 | 35.2 | 35.9 | 39.4 | 45.9 | 53.9 | 60.7 | 60.4 | 55.2 | 47.9 | 39.5 | 35.0 | 45.4 |
| Days Max Temp ≥ 90 °F | 0 | 0 | 0 | 0 | 0 | 0 | 0 | 0 | 0 | 0 | 0 | 0 | 0 |
| Days Max Temp ≤ 32 °F | 2 | 2 | 2 | 1 | 0 | 0 | 0 | 0 | 0 | 0 | 1 | 4 | 12 |
| Days Min Temp ≤ 32 °F | 26 | 24 | 26 | 22 | 11 | 2 | 0 | 0 | 2 | 7 | 19 | 25 | 164 |
| Days Min Temp ≤ 0 °F | 0 | 0 | 0 | 0 | 0 | 0 | 0 | 0 | 0 | 0 | 0 | 0 | 0 |
| Heating Degree Days | 915 | 833 | 894 | 763 | 585 | 328 | 141 | 150 | 290 | 522 | 759 | 924 | 7104 |
| Cooling Degree Days | 0 | 0 | 0 | 0 | 0 | 2 | 21 | 18 | 3 | 0 | 0 | 0 | 44 |
| Total Precipitation (") | na | na | na | na | na | na | na | na | na | na | na | na | na |
| Days ≥ 0.1" Precip | na | na | na | na | na | na | na | na | na | na | na | na | na |
| Total Snowfall (") | ***31.2*** | na | ***28.0*** | ***22.4*** | ***2.1*** | 0.1 | 0.0 | 0.0 | 0.0 | 1.2 | na | ***28.6*** | na |
| Days ≥ 1" Snow Depth | ***16*** | na | na | ***11*** | ***2*** | ***0*** | ***0*** | 0 | 0 | 0 | na | na | na |

## IDYLLWILD FIRE DEPT *Riverside County* ELEVATION 5403 ft LAT/LONG 33° 45 ' N / 116° 43 ' W

| | JAN | FEB | MAR | APR | MAY | JUN | JUL | AUG | SEP | OCT | NOV | DEC | YEAR |
|---|---|---|---|---|---|---|---|---|---|---|---|---|---|
| Maximum Temp °F | 53.6 | 54.5 | 56.1 | 62.0 | 69.8 | 79.1 | 84.4 | 83.4 | 78.6 | 70.2 | 60.2 | 54.0 | 67.2 |
| Minimum Temp °F | 28.0 | 28.3 | 29.5 | 32.8 | 38.1 | 45.1 | 51.7 | 51.3 | 46.5 | 38.9 | 31.9 | 28.0 | 37.5 |
| Mean Temp °F | 40.9 | 41.4 | 42.8 | 47.4 | 54.0 | 62.1 | 68.1 | 67.4 | 62.6 | 54.6 | 46.1 | 41.0 | 52.4 |
| Days Max Temp ≥ 90 °F | 0 | 0 | 0 | 0 | 0 | 2 | 5 | 4 | 1 | 0 | 0 | 0 | 12 |
| Days Max Temp ≤ 32 °F | 0 | 0 | 0 | 0 | 0 | 0 | 0 | 0 | 0 | 0 | 0 | 0 | 0 |
| Days Min Temp ≤ 32 °F | 23 | 22 | 23 | 15 | 6 | 1 | 0 | 0 | 1 | 6 | 16 | 24 | 137 |
| Days Min Temp ≤ 0 °F | 0 | 0 | 0 | 0 | 0 | 0 | 0 | 0 | 0 | 0 | 0 | 0 | 0 |
| Heating Degree Days | 746 | 660 | 682 | 520 | 339 | 125 | 23 | 31 | 113 | 320 | 563 | 737 | 4859 |
| Cooling Degree Days | 0 | 0 | 0 | 1 | 3 | 44 | 114 | 116 | 49 | 6 | 0 | 0 | 333 |
| Total Precipitation (") | 5.10 | 4.66 | 4.75 | 1.88 | 0.69 | 0.16 | 0.85 | 0.98 | 0.92 | 1.02 | 2.73 | 4.04 | 27.78 |
| Days ≥ 0.1" Precip | 5 | 5 | 7 | 4 | 1 | 0 | 1 | 2 | 1 | 2 | 3 | 5 | 36 |
| Total Snowfall (") | ***9.3*** | 4.8 | 10.2 | ***2.4*** | 0.2 | 0.0 | 0.0 | 0.0 | 0.0 | 0.1 | 2.0 | 6.2 | 35.2 |
| Days ≥ 1" Snow Depth | na | 3 | ***3*** | 1 | 0 | 0 | 0 | 0 | 0 | 0 | 1 | 4 | na |

## IMPERIAL *Imperial County* ELEVATION -64 ft LAT/LONG 32° 51 ' N / 115° 34 ' W

| | JAN | FEB | MAR | APR | MAY | JUN | JUL | AUG | SEP | OCT | NOV | DEC | YEAR |
|---|---|---|---|---|---|---|---|---|---|---|---|---|---|
| Maximum Temp °F | 69.6 | 74.2 | 78.7 | 85.4 | 93.3 | 102.6 | 106.2 | 105.0 | 100.0 | 89.8 | 77.5 | 68.9 | 87.6 |
| Minimum Temp °F | 42.7 | 46.6 | 51.1 | 56.1 | 62.6 | 70.1 | 77.6 | 77.9 | 72.0 | 61.4 | 49.9 | 42.4 | 59.2 |
| Mean Temp °F | 56.2 | 60.4 | 64.9 | 70.8 | 78.0 | 86.4 | 91.9 | 91.5 | 86.0 | 75.6 | 63.7 | 55.7 | 73.4 |
| Days Max Temp ≥ 90 °F | 0 | 0 | 2 | 10 | 22 | 29 | 31 | 31 | 28 | 17 | 1 | 0 | 171 |
| Days Max Temp ≤ 32 °F | 0 | 0 | 0 | 0 | 0 | 0 | 0 | 0 | 0 | 0 | 0 | 0 | 0 |
| Days Min Temp ≤ 32 °F | 1 | 0 | 0 | 0 | 0 | 0 | 0 | 0 | 0 | 0 | 0 | 1 | 2 |
| Days Min Temp ≤ 0 °F | 0 | 0 | 0 | 0 | 0 | 0 | 0 | 0 | 0 | 0 | 0 | 0 | 0 |
| Heating Degree Days | 268 | 140 | 66 | 17 | 1 | 0 | 0 | 0 | 0 | 5 | 89 | 282 | 868 |
| Cooling Degree Days | 2 | 20 | 72 | 234 | 415 | 654 | 831 | 829 | 648 | 348 | 52 | 1 | 4106 |
| Total Precipitation (") | 0.47 | 0.36 | 0.37 | 0.07 | 0.04 | 0.00 | 0.13 | 0.32 | 0.33 | 0.28 | 0.29 | 0.47 | 3.13 |
| Days ≥ 0.1" Precip | 1 | 1 | 1 | 0 | 0 | 0 | 0 | 1 | 1 | 1 | 1 | 1 | 8 |
| Total Snowfall (") | 0.0 | 0.0 | 0.0 | 0.0 | 0.0 | 0.0 | 0.0 | 0.0 | 0.0 | 0.0 | 0.0 | 0.0 | 0.0 |
| Days ≥ 1" Snow Depth | 0 | 0 | 0 | 0 | 0 | 0 | 0 | 0 | 0 | 0 | 0 | 0 | 0 |

## INDEPENDENCE *Inyo County* ELEVATION 3914 ft LAT/LONG 36° 48 ' N / 118° 12 ' W

| | JAN | FEB | MAR | APR | MAY | JUN | JUL | AUG | SEP | OCT | NOV | DEC | YEAR |
|---|---|---|---|---|---|---|---|---|---|---|---|---|---|
| Maximum Temp °F | ***54.4*** | 59.4 | 65.7 | 73.3 | 82.3 | 91.7 | 98.0 | 95.8 | 88.8 | 77.5 | ***63.5*** | 54.8 | 75.4 |
| Minimum Temp °F | ***27.8*** | 31.4 | 36.8 | 42.8 | 51.0 | 59.0 | 64.6 | 62.4 | 55.6 | 45.2 | ***34.3*** | 27.5 | 44.9 |
| Mean Temp °F | ***41.1*** | 45.4 | 51.3 | 58.1 | 66.7 | 75.4 | 81.3 | 79.1 | 72.2 | 61.4 | ***48.9*** | 41.2 | 60.2 |
| Days Max Temp ≥ 90 °F | 0 | 0 | 0 | 1 | ***7*** | ***20*** | 29 | 26 | ***15*** | 3 | 0 | 0 | 101 |
| Days Max Temp ≤ 32 °F | 0 | 0 | 0 | 0 | 0 | 0 | 0 | 0 | 0 | 0 | 0 | 0 | 0 |
| Days Min Temp ≤ 32 °F | 25 | ***16*** | ***7*** | 2 | 0 | 0 | 0 | 0 | 0 | 1 | ***12*** | ***24*** | 87 |
| Days Min Temp ≤ 0 °F | 0 | 0 | 0 | 0 | 0 | 0 | 0 | 0 | 0 | 0 | 0 | 0 | 0 |
| Heating Degree Days | ***734*** | 538 | 419 | 228 | 75 | 10 | 0 | 1 | 14 | 150 | ***478*** | 732 | 3379 |
| Cooling Degree Days | 0 | 1 | 4 | 40 | 151 | 364 | 531 | 476 | 268 | 58 | 2 | 0 | 1895 |
| Total Precipitation (") | 1.18 | 0.96 | 0.62 | 0.23 | 0.19 | 0.10 | 0.15 | 0.13 | 0.23 | 0.16 | 0.62 | 0.86 | 5.43 |
| Days ≥ 0.1" Precip | 2 | 2 | 1 | 1 | 1 | 0 | 0 | 0 | 1 | 1 | 1 | 1 | 11 |
| Total Snowfall (") | ***1.8*** | ***0.2*** | 0.0 | 0.0 | 0.0 | 0.0 | 0.0 | 0.0 | 0.0 | 0.0 | 0.0 | ***0.5*** | 2.5 |
| Days ≥ 1" Snow Depth | ***0*** | ***0*** | ***0*** | 0 | 0 | 0 | 0 | 0 | 0 | 0 | 0 | ***0*** | 0 |

## INDIO FIRE STATION *Riverside County* ELEVATION -21 ft LAT/LONG 33° 44 ' N / 116° 16 ' W

| | JAN | FEB | MAR | APR | MAY | JUN | JUL | AUG | SEP | OCT | NOV | DEC | YEAR |
|---|---|---|---|---|---|---|---|---|---|---|---|---|---|
| Maximum Temp °F | 70.9 | 75.7 | 79.7 | 86.3 | 93.4 | 102.5 | 106.7 | ***105.6*** | 101.0 | 91.9 | 79.7 | 71.3 | 88.7 |
| Minimum Temp °F | 39.8 | 45.2 | 51.2 | 57.6 | 64.5 | 72.2 | 77.9 | ***77.3*** | 70.6 | 60.1 | 47.2 | 39.3 | 58.6 |
| Mean Temp °F | 55.4 | 60.5 | 65.5 | 72.0 | 79.0 | 87.4 | 92.3 | ***91.5*** | 85.8 | 76.0 | 63.5 | 55.3 | 73.7 |
| Days Max Temp ≥ 90 °F | 0 | 1 | 4 | 12 | 23 | 28 | 31 | ***31*** | 28 | 20 | 3 | 0 | 181 |
| Days Max Temp ≤ 32 °F | 0 | 0 | 0 | 0 | 0 | 0 | 0 | ***0*** | 0 | 0 | 0 | 0 | 0 |
| Days Min Temp ≤ 32 °F | 5 | 1 | 0 | 0 | 0 | 0 | 0 | ***0*** | 0 | 0 | 0 | 4 | 10 |
| Days Min Temp ≤ 0 °F | 0 | 0 | 0 | 0 | 0 | 0 | 0 | ***0*** | 0 | 0 | 0 | 0 | 0 |
| Heating Degree Days | 294 | 148 | 73 | 19 | 1 | 0 | 0 | ***0*** | 0 | 6 | 98 | 295 | 934 |
| Cooling Degree Days | ***3*** | ***36*** | ***102*** | ***291*** | ***455*** | ***707*** | ***854*** | na | ***634*** | ***377*** | ***60*** | ***3*** | na |
| Total Precipitation (") | ***0.74*** | 0.66 | 0.52 | 0.08 | 0.09 | 0.01 | 0.18 | ***0.33*** | 0.27 | 0.16 | 0.31 | 0.33 | 3.68 |
| Days ≥ 0.1" Precip | ***2*** | 1 | 1 | 0 | 0 | 0 | 0 | ***0*** | 1 | 0 | 1 | 1 | 7 |
| Total Snowfall (") | 0.0 | 0.0 | 0.0 | 0.0 | 0.0 | 0.0 | 0.0 | ***0.0*** | 0.0 | 0.0 | 0.0 | 0.0 | 0.0 |
| Days ≥ 1" Snow Depth | 0 | 0 | 0 | 0 | 0 | 0 | 0 | ***0*** | 0 | 0 | 0 | 0 | 0 |

## INYOKERN *Kern County* ELEVATION 2440 ft LAT/LONG 35° 39 ' N / 117° 49 ' W

| | JAN | FEB | MAR | APR | MAY | JUN | JUL | AUG | SEP | OCT | NOV | DEC | YEAR |
|---|---|---|---|---|---|---|---|---|---|---|---|---|---|
| Maximum Temp °F | 59.8 | 65.3 | 70.7 | 77.8 | 87.2 | 96.9 | 102.6 | 101.0 | 93.5 | 83.1 | 68.7 | 59.2 | 80.5 |
| Minimum Temp °F | 30.2 | 34.4 | 38.7 | 43.5 | 52.6 | 60.1 | 65.8 | 64.6 | 57.6 | 47.9 | 37.0 | 29.7 | 46.8 |
| Mean Temp °F | 45.0 | 49.9 | 54.8 | 60.7 | 70.0 | 78.5 | 84.3 | 82.8 | 75.6 | 65.5 | 52.9 | 44.5 | 63.7 |
| Days Max Temp ≥ 90 °F | 0 | 0 | 0 | 3 | 14 | 26 | 31 | 30 | 22 | 7 | 0 | 0 | 133 |
| Days Max Temp ≤ 32 °F | 0 | 0 | 0 | 0 | 0 | 0 | 0 | 0 | 0 | 0 | 0 | 0 | 0 |
| Days Min Temp ≤ 32 °F | 20 | 12 | 5 | 2 | 0 | 0 | 0 | 0 | 0 | 1 | 9 | 22 | 71 |
| Days Min Temp ≤ 0 °F | 0 | 0 | 0 | 0 | 0 | 0 | 0 | 0 | 0 | 0 | 0 | 0 | 0 |
| Heating Degree Days | 612 | 422 | 315 | 158 | 35 | 2 | 0 | 0 | 4 | 74 | 360 | 629 | 2611 |
| Cooling Degree Days | 0 | 1 | 7 | 55 | 204 | 427 | 595 | 572 | 346 | 109 | 4 | 0 | 2320 |
| Total Precipitation (") | 0.78 | 1.13 | 0.79 | 0.22 | 0.09 | 0.02 | 0.22 | 0.35 | 0.22 | 0.07 | 0.48 | 0.59 | 4.96 |
| Days ≥ 0.1" Precip | 2 | 2 | 2 | 1 | 0 | 0 | 1 | 0 | 1 | 0 | 1 | 1 | 11 |
| Total Snowfall (") | 0.3 | 0.0 | 0.1 | 0.0 | 0.0 | 0.0 | 0.0 | 0.0 | 0.0 | 0.0 | 0.0 | 0.3 | 0.7 |
| Days ≥ 1" Snow Depth | 0 | 0 | 0 | 0 | 0 | 0 | 0 | 0 | 0 | 0 | 0 | 0 | 0 |

## IRON MOUNTAIN *San Bernardino County* ELEVATION 942 ft LAT/LONG 34° 9 ' N / 115° 7 ' W

| | JAN | FEB | MAR | APR | MAY | JUN | JUL | AUG | SEP | OCT | NOV | DEC | YEAR |
|---|---|---|---|---|---|---|---|---|---|---|---|---|---|
| Maximum Temp °F | 64.8 | 70.5 | 76.3 | 84.3 | 93.4 | 103.0 | 107.9 | 106.1 | 99.9 | 88.4 | 74.4 | 64.7 | 86.1 |
| Minimum Temp °F | 43.0 | 46.8 | 51.3 | 57.6 | 66.1 | 75.3 | 81.0 | 79.4 | 72.6 | 61.1 | 50.1 | 42.7 | 60.6 |
| Mean Temp °F | 53.9 | 58.7 | 63.8 | 71.0 | 79.8 | 89.2 | 94.5 | 92.8 | 86.2 | 74.8 | 62.3 | 53.7 | 73.4 |
| Days Max Temp ≥ 90 °F | 0 | 0 | 2 | 9 | 22 | 29 | 31 | 31 | 27 | 15 | 1 | 0 | 167 |
| Days Max Temp ≤ 32 °F | 0 | 0 | 0 | 0 | 0 | 0 | 0 | 0 | 0 | 0 | 0 | 0 | 0 |
| Days Min Temp ≤ 32 °F | 2 | 1 | 0 | 0 | 0 | 0 | 0 | 0 | 0 | 0 | 0 | 1 | 4 |
| Days Min Temp ≤ 0 °F | 0 | 0 | 0 | 0 | 0 | 0 | 0 | 0 | 0 | 0 | 0 | 0 | 0 |
| Heating Degree Days | 339 | 187 | 101 | 27 | 1 | 0 | 0 | 0 | 0 | 11 | 126 | 344 | 1136 |
| Cooling Degree Days | 1 | 19 | 74 | 256 | 480 | 739 | 907 | 867 | 648 | 326 | 54 | 2 | 4373 |
| Total Precipitation (") | 0.57 | 0.41 | 0.51 | 0.14 | 0.10 | 0.05 | 0.31 | 0.42 | 0.23 | 0.41 | 0.26 | 0.50 | 3.91 |
| Days ≥ 0.1" Precip | 2 | 1 | 1 | 0 | 0 | 0 | 1 | 1 | 1 | 1 | 1 | 1 | 10 |
| Total Snowfall (") | 0.0 | 0.0 | 0.0 | 0.0 | 0.0 | 0.0 | 0.0 | 0.0 | 0.0 | 0.0 | 0.0 | 0.0 | 0.0 |
| Days ≥ 1" Snow Depth | 0 | 0 | 0 | 0 | 0 | 0 | 0 | 0 | 0 | 0 | 0 | 0 | 0 |

## JESS VALLEY *Modoc County* ELEVATION 5292 ft LAT/LONG 41° 16 ' N / 120° 18 ' W

| | JAN | FEB | MAR | APR | MAY | JUN | JUL | AUG | SEP | OCT | NOV | DEC | YEAR |
|---|---|---|---|---|---|---|---|---|---|---|---|---|---|
| Maximum Temp °F | 42.0 | 44.8 | 48.2 | 55.0 | 65.1 | 73.6 | 83.2 | 82.0 | 74.9 | 63.8 | 48.4 | 41.2 | 60.2 |
| Minimum Temp °F | 18.9 | 21.7 | 24.7 | 27.8 | 34.3 | 40.9 | 46.1 | 45.2 | 39.0 | 32.2 | 24.7 | 18.8 | 31.2 |
| Mean Temp °F | 30.4 | 33.1 | 36.5 | 41.5 | 49.8 | 57.3 | 64.7 | 63.6 | 57.0 | 48.0 | 36.6 | 29.8 | 45.7 |
| Days Max Temp ≥ 90 °F | 0 | 0 | 0 | 0 | 0 | 1 | 6 | 5 | 0 | 0 | 0 | 0 | 12 |
| Days Max Temp ≤ 32 °F | 4 | 2 | 1 | 0 | 0 | 0 | 0 | 0 | 0 | 0 | 2 | 6 | 15 |
| Days Min Temp ≤ 32 °F | 29 | 25 | 27 | 23 | 13 | 3 | 0 | 0 | 5 | 15 | 25 | 29 | 194 |
| Days Min Temp ≤ 0 °F | 2 | 1 | 0 | 0 | 0 | 0 | 0 | 0 | 0 | 0 | 0 | 2 | 5 |
| Heating Degree Days | 1068 | 892 | 877 | 699 | 464 | 239 | 70 | 92 | 241 | 521 | 846 | 1085 | 7094 |
| Cooling Degree Days | 0 | 0 | 0 | 0 | 2 | 14 | 59 | 49 | 10 | 0 | 0 | 0 | 134 |
| Total Precipitation (") | 1.81 | 1.55 | 2.03 | 1.87 | 2.06 | 1.45 | 0.48 | 0.68 | 0.86 | 1.25 | 2.30 | 1.89 | 18.23 |
| Days ≥ 0.1" Precip | 6 | 5 | 7 | 6 | 5 | 4 | 2 | 2 | 2 | 4 | 7 | 7 | 57 |
| Total Snowfall (") | 12.5 | 9.7 | 12.0 | 8.8 | 5.0 | 0.4 | 0.1 | 0.0 | 0.6 | 1.8 | 10.4 | 13.7 | 75.0 |
| Days ≥ 1" Snow Depth | 20 | 13 | 8 | 1 | 0 | 0 | 0 | 0 | 0 | 1 | 6 | 15 | 64 |

## KENTFIELD *Marin County* ELEVATION 10 ft LAT/LONG 37° 57 ' N / 122° 33 ' W

| | JAN | FEB | MAR | APR | MAY | JUN | JUL | AUG | SEP | OCT | NOV | DEC | YEAR |
|---|---|---|---|---|---|---|---|---|---|---|---|---|---|
| Maximum Temp °F | 55.7 | 60.9 | 64.8 | 69.8 | 75.9 | 81.2 | 84.4 | 83.9 | 81.9 | 75.1 | 63.0 | 55.4 | 71.0 |
| Minimum Temp °F | 39.6 | 42.3 | 43.5 | 44.6 | 47.5 | 50.6 | 51.7 | 52.0 | 51.4 | 49.0 | 44.6 | 40.0 | 46.4 |
| Mean Temp °F | 47.7 | 51.6 | 54.2 | 57.2 | 61.7 | 65.9 | 68.1 | 68.0 | 66.7 | 62.0 | 53.8 | 47.7 | 58.7 |
| Days Max Temp ≥ 90 °F | 0 | 0 | 0 | 0 | 2 | 5 | 6 | 6 | 5 | 1 | 0 | 0 | 25 |
| Days Max Temp ≤ 32 °F | 0 | 0 | 0 | 0 | 0 | 0 | 0 | 0 | 0 | 0 | 0 | 0 | 0 |
| Days Min Temp ≤ 32 °F | 3 | 1 | 0 | 0 | 0 | 0 | 0 | 0 | 0 | 0 | 0 | 3 | 7 |
| Days Min Temp ≤ 0 °F | 0 | 0 | 0 | 0 | 0 | 0 | 0 | 0 | 0 | 0 | 0 | 0 | 0 |
| Heating Degree Days | 531 | 373 | 328 | 231 | 123 | 43 | 11 | 11 | 29 | 114 | 330 | 530 | 2654 |
| Cooling Degree Days | 0 | 0 | 0 | 9 | 29 | 86 | 128 | 117 | 87 | 37 | 1 | 0 | 494 |
| Total Precipitation (") | 10.46 | 7.92 | 6.94 | 2.63 | 0.77 | 0.35 | 0.12 | 0.13 | 0.49 | 2.55 | 8.00 | 8.50 | 48.86 |
| Days ≥ 0.1" Precip | 8 | 8 | 8 | 5 | 1 | 1 | 0 | 0 | 1 | 3 | 7 | 8 | 50 |
| Total Snowfall (") | 0.0 | 0.0 | 0.0 | 0.0 | 0.0 | 0.0 | 0.0 | 0.0 | 0.0 | 0.0 | 0.0 | 0.0 | 0.0 |
| Days ≥ 1" Snow Depth | 0 | 0 | 0 | 0 | 0 | 0 | 0 | 0 | 0 | 0 | 0 | 0 | 0 |

## KERN RIVER PH 3 *Kern County* ELEVATION 2703 ft LAT/LONG 35° 47 ' N / 118° 26 ' W

| | JAN | FEB | MAR | APR | MAY | JUN | JUL | AUG | SEP | OCT | NOV | DEC | YEAR |
|---|---|---|---|---|---|---|---|---|---|---|---|---|---|
| Maximum Temp °F | 59.8 | 63.2 | 66.4 | 73.1 | 81.6 | 90.6 | 97.3 | 96.4 | 90.5 | 80.1 | 67.0 | 59.6 | 77.1 |
| Minimum Temp °F | 33.2 | 36.0 | 39.0 | 43.9 | 51.5 | 59.1 | 65.1 | 63.9 | 58.3 | 48.6 | 38.5 | 32.7 | 47.5 |
| Mean Temp °F | 46.5 | 49.6 | 52.7 | 58.5 | 66.6 | 74.9 | 81.2 | 80.2 | 74.4 | 64.4 | 52.7 | 46.2 | 62.3 |
| Days Max Temp ≥ 90 °F | 0 | 0 | 0 | 1 | 6 | 18 | 29 | 27 | 19 | 5 | 0 | 0 | 105 |
| Days Max Temp ≤ 32 °F | 0 | 0 | 0 | 0 | 0 | 0 | 0 | 0 | 0 | 0 | 0 | 0 | 0 |
| Days Min Temp ≤ 32 °F | 16 | 10 | 5 | 1 | 0 | 0 | 0 | 0 | 0 | 0 | 7 | 16 | 55 |
| Days Min Temp ≤ 0 °F | 0 | 0 | 0 | 0 | 0 | 0 | 0 | 0 | 0 | 0 | 0 | 0 | 0 |
| Heating Degree Days | 568 | 430 | 377 | 216 | 71 | 8 | 0 | 0 | 8 | 98 | 366 | 578 | 2720 |
| Cooling Degree Days | 0 | 0 | 2 | 38 | 129 | 318 | 512 | 493 | 305 | 92 | 2 | 0 | 1891 |
| Total Precipitation (") | 2.76 | 2.50 | 2.28 | 0.81 | 0.22 | 0.13 | 0.16 | 0.20 | 0.43 | 0.36 | 1.49 | 1.83 | 13.17 |
| Days ≥ 0.1" Precip | 4 | 5 | 5 | 2 | 1 | 0 | 0 | 1 | 1 | 1 | 3 | 4 | 27 |
| Total Snowfall (") | 0.0 | 0.1 | 0.0 | 0.1 | 0.0 | 0.0 | 0.0 | 0.0 | 0.0 | 0.0 | 0.0 | 0.6 | 0.8 |
| Days ≥ 1" Snow Depth | 0 | 0 | 0 | 0 | 0 | 0 | 0 | 0 | 0 | 0 | 0 | 0 | 0 |

## KING CITY *Monterey County* ELEVATION 320 ft LAT/LONG 36° 12 ' N / 121° 8 ' W

| | JAN | FEB | MAR | APR | MAY | JUN | JUL | AUG | SEP | OCT | NOV | DEC | YEAR |
|---|---|---|---|---|---|---|---|---|---|---|---|---|---|
| Maximum Temp °F | 63.0 | 66.5 | 68.9 | 74.0 | 78.5 | 82.6 | 84.9 | 85.0 | 84.7 | 79.8 | 68.8 | 62.4 | 74.9 |
| Minimum Temp °F | 34.9 | 38.1 | 40.7 | 41.5 | 45.4 | 49.1 | 51.3 | 51.6 | 49.4 | 44.9 | 38.9 | 34.3 | 43.3 |
| Mean Temp °F | 48.9 | 52.3 | 54.9 | 57.8 | 62.0 | 65.9 | 68.1 | 68.3 | 67.1 | 62.4 | 53.9 | 48.4 | 59.2 |
| Days Max Temp ≥ 90 °F | 0 | 0 | 0 | 1 | 4 | 5 | 6 | 6 | 7 | 5 | 0 | 0 | 34 |
| Days Max Temp ≤ 32 °F | 0 | 0 | 0 | 0 | 0 | 0 | 0 | 0 | 0 | 0 | 0 | 0 | 0 |
| Days Min Temp ≤ 32 °F | 13 | 6 | 3 | 1 | 0 | 0 | 0 | 0 | 0 | 1 | 7 | 14 | 45 |
| Days Min Temp ≤ 0 °F | 0 | 0 | 0 | 0 | 0 | 0 | 0 | 0 | 0 | 0 | 0 | 0 | 0 |
| Heating Degree Days | 487 | 347 | 307 | 221 | 120 | 38 | 9 | 7 | 25 | 112 | 327 | 508 | 2508 |
| Cooling Degree Days | 0 | 0 | 1 | 15 | 32 | 73 | 115 | 123 | 97 | 43 | 1 | 0 | 500 |
| Total Precipitation (") | 2.26 | 2.25 | 2.27 | 0.79 | 0.15 | 0.04 | 0.01 | 0.04 | 0.28 | 0.43 | 1.46 | 1.69 | 11.67 |
| Days ≥ 0.1" Precip | 4 | 4 | 5 | 2 | 0 | 0 | 0 | 0 | 0 | 1 | 3 | 4 | 23 |
| Total Snowfall (") | 0.0 | 0.0 | 0.0 | 0.0 | 0.0 | 0.0 | 0.0 | 0.0 | 0.0 | 0.0 | 0.0 | 0.0 | 0.0 |
| Days ≥ 1" Snow Depth | 0 | 0 | 0 | 0 | 0 | 0 | 0 | 0 | 0 | 0 | 0 | 0 | 0 |

## KLAMATH *Del Norte County* ELEVATION 25 ft LAT/LONG 41° 31 ' N / 124° 2 ' W

| | JAN | FEB | MAR | APR | MAY | JUN | JUL | AUG | SEP | OCT | NOV | DEC | YEAR |
|---|---|---|---|---|---|---|---|---|---|---|---|---|---|
| Maximum Temp °F | 55.1 | 56.5 | 57.3 | 58.8 | 62.1 | 64.8 | 66.6 | 66.8 | 67.1 | 64.4 | 58.6 | 54.6 | 61.1 |
| Minimum Temp °F | 38.2 | 39.8 | 40.7 | 42.1 | 45.7 | 49.6 | 52.0 | 52.7 | 50.1 | 46.3 | 42.2 | 38.2 | 44.8 |
| Mean Temp °F | 46.7 | 48.2 | 49.1 | 50.5 | 53.9 | 57.2 | 59.3 | 59.8 | 58.7 | 55.4 | 50.4 | 46.4 | 53.0 |
| Days Max Temp ≥ 90 °F | 0 | 0 | 0 | 0 | 0 | 0 | 0 | 0 | 0 | 0 | 0 | 0 | 0 |
| Days Max Temp ≤ 32 °F | 0 | 0 | 0 | 0 | 0 | 0 | 0 | 0 | 0 | 0 | 0 | 0 | 0 |
| Days Min Temp ≤ 32 °F | 7 | 4 | 2 | 1 | 0 | 0 | 0 | 0 | 0 | 0 | 2 | 6 | 22 |
| Days Min Temp ≤ 0 °F | 0 | 0 | 0 | 0 | 0 | 0 | 0 | 0 | 0 | 0 | 0 | 0 | 0 |
| Heating Degree Days | 560 | 468 | 487 | 429 | 337 | 229 | 169 | 157 | 187 | 293 | 431 | 569 | 4316 |
| Cooling Degree Days | 0 | 0 | 0 | 0 | 1 | 2 | 1 | 2 | 3 | 2 | 0 | 0 | 11 |
| Total Precipitation (") | 12.28 | 9.77 | 10.41 | 5.72 | 3.30 | 1.50 | 0.39 | 0.95 | 1.91 | 5.05 | 11.97 | 13.08 | 76.33 |
| Days ≥ 0.1" Precip | 13 | 12 | 13 | 9 | 5 | 3 | 1 | 2 | 3 | 6 | 13 | 14 | 94 |
| Total Snowfall (") | 0.5 | 0.7 | 0.0 | 0.0 | 0.0 | 0.0 | 0.0 | 0.0 | 0.0 | 0.0 | 0.0 | 0.3 | 1.5 |
| Days ≥ 1" Snow Depth | 0 | 0 | 0 | 0 | 0 | 0 | 0 | 0 | 0 | 0 | 0 | 0 | 0 |

## LA MESA *San Diego County* ELEVATION 581 ft LAT/LONG 32° 46' N / 117° 1' W

| | JAN | FEB | MAR | APR | MAY | JUN | JUL | AUG | SEP | OCT | NOV | DEC | YEAR |
|---|---|---|---|---|---|---|---|---|---|---|---|---|---|
| Maximum Temp °F | 68.5 | 69.7 | 69.7 | 72.6 | 74.1 | 78.3 | 83.5 | 85.2 | 83.8 | 79.7 | 73.4 | 68.4 | 75.6 |
| Minimum Temp °F | 44.9 | 46.2 | 48.1 | 50.9 | 54.8 | 58.2 | 62.0 | 63.5 | 61.5 | 56.3 | 49.3 | 44.9 | 53.4 |
| Mean Temp °F | 56.8 | 58.0 | 58.9 | 61.8 | 64.4 | 68.3 | 72.8 | 74.4 | 72.7 | 68.1 | 61.4 | 56.7 | 64.5 |
| Days Max Temp ≥ 90 °F | 0 | 0 | 0 | 1 | 1 | 3 | 4 | 7 | 7 | 4 | 0 | 0 | 27 |
| Days Max Temp ≤ 32 °F | 0 | 0 | 0 | 0 | 0 | 0 | 0 | 0 | 0 | 0 | 0 | 0 | 0 |
| Days Min Temp ≤ 32 °F | 0 | 0 | 0 | 0 | 0 | 0 | 0 | 0 | 0 | 0 | 0 | 0 | 0 |
| Days Min Temp ≤ 0 °F | 0 | 0 | 0 | 0 | 0 | 0 | 0 | 0 | 0 | 0 | 0 | 0 | 0 |
| Heating Degree Days | 256 | 199 | 191 | 117 | 59 | 15 | 0 | 0 | 2 | 20 | 125 | 256 | 1240 |
| Cooling Degree Days | 3 | 10 | 13 | 42 | 57 | 138 | 271 | 320 | 260 | 130 | 22 | 2 | 1268 |
| Total Precipitation (") | 2.66 | 2.11 | 2.87 | 1.08 | 0.27 | 0.09 | 0.07 | 0.11 | 0.26 | 0.53 | 1.67 | 1.93 | 13.65 |
| Days ≥ 0.1" Precip | 4 | 4 | 5 | 2 | 1 | 0 | 0 | 0 | 1 | 1 | 3 | 4 | 25 |
| Total Snowfall (") | 0.0 | 0.0 | 0.0 | 0.0 | 0.0 | 0.0 | 0.0 | 0.0 | 0.0 | 0.0 | 0.0 | 0.0 | 0.0 |
| Days ≥ 1" Snow Depth | 0 | 0 | 0 | 0 | 0 | 0 | 0 | 0 | 0 | 0 | 0 | 0 | 0 |

## LAGUNA BEACH *Orange County* ELEVATION 59 ft LAT/LONG 33° 32' N / 117° 47' W

| | JAN | FEB | MAR | APR | MAY | JUN | JUL | AUG | SEP | OCT | NOV | DEC | YEAR |
|---|---|---|---|---|---|---|---|---|---|---|---|---|---|
| Maximum Temp °F | 66.1 | 67.1 | 67.0 | 69.5 | 70.2 | 73.0 | 76.4 | 78.1 | 77.8 | 75.7 | 71.0 | 66.3 | 71.5 |
| Minimum Temp °F | 41.5 | 43.2 | 44.7 | 46.9 | 52.2 | 55.3 | 58.8 | 59.2 | 57.7 | 52.7 | 46.2 | 41.4 | 50.0 |
| Mean Temp °F | 53.8 | 55.2 | 55.9 | 58.2 | 61.1 | 64.2 | 67.6 | 68.8 | 67.8 | 64.3 | 58.6 | 53.8 | 60.8 |
| Days Max Temp ≥ 90 °F | 0 | 0 | 0 | 0 | 0 | 0 | 0 | 1 | 1 | 1 | 0 | 0 | 3 |
| Days Max Temp ≤ 32 °F | 0 | 0 | 0 | 0 | 0 | 0 | 0 | 0 | 0 | 0 | 0 | 0 | 0 |
| Days Min Temp ≤ 32 °F | 1 | 1 | 0 | 0 | 0 | 0 | 0 | 0 | 0 | 0 | 0 | 1 | 3 |
| Days Min Temp ≤ 0 °F | 0 | 0 | 0 | 0 | 0 | 0 | 0 | 0 | 0 | 0 | 0 | 0 | 0 |
| Heating Degree Days | 341 | 273 | 279 | 202 | 125 | 52 | 14 | 9 | 18 | 64 | 192 | 342 | 1911 |
| Cooling Degree Days | *2* | 4 | 3 | 9 | 15 | 46 | 136 | 150 | 123 | *58* | 7 | 0 | 553 |
| Total Precipitation (") | 2.53 | 2.49 | 2.49 | 0.90 | 0.23 | 0.12 | 0.05 | 0.11 | 0.34 | 0.35 | 1.55 | 1.90 | 13.06 |
| Days ≥ 0.1" Precip | 4 | 4 | 4 | 2 | 1 | 0 | 0 | 0 | 1 | 1 | 3 | 3 | 23 |
| Total Snowfall (") | 0.0 | 0.0 | 0.0 | 0.0 | 0.0 | 0.0 | 0.0 | 0.0 | 0.0 | 0.0 | 0.0 | 0.0 | 0.0 |
| Days ≥ 1" Snow Depth | 0 | 0 | 0 | 0 | 0 | 0 | 0 | 0 | 0 | 0 | 0 | 0 | 0 |

## LAKE ARROWHEAD *San Bernardino County* ELEVATION 5233 ft LAT/LONG 34° 14' N / 117° 11' W

| | JAN | FEB | MAR | APR | MAY | JUN | JUL | AUG | SEP | OCT | NOV | DEC | YEAR |
|---|---|---|---|---|---|---|---|---|---|---|---|---|---|
| Maximum Temp °F | 45.5 | 48.1 | 52.8 | 59.6 | 67.3 | 76.1 | 81.3 | 81.0 | 75.8 | 65.0 | 52.5 | 45.4 | 62.5 |
| Minimum Temp °F | 29.0 | 30.1 | 31.7 | 34.9 | 41.3 | 48.9 | 56.1 | 55.9 | 50.5 | 42.1 | 34.1 | 29.1 | 40.3 |
| Mean Temp °F | 37.3 | 39.1 | 42.3 | 47.3 | 54.3 | 62.5 | 68.7 | 68.5 | 63.2 | 53.5 | 43.3 | 37.3 | 51.4 |
| Days Max Temp ≥ 90 °F | 0 | 0 | 0 | 0 | 0 | 1 | 2 | 2 | 0 | 0 | 0 | 0 | 5 |
| Days Max Temp ≤ 32 °F | 1 | 1 | 0 | 0 | 0 | 0 | 0 | 0 | 0 | 0 | 0 | 2 | 4 |
| Days Min Temp ≤ 32 °F | 22 | 20 | 17 | 11 | 3 | 0 | 0 | 0 | 0 | 2 | 13 | 22 | 110 |
| Days Min Temp ≤ 0 °F | 0 | 0 | 0 | 0 | 0 | 0 | 0 | 0 | 0 | 0 | 0 | 0 | 0 |
| Heating Degree Days | 853 | 724 | 698 | 525 | 328 | 121 | 19 | 25 | 102 | 352 | 644 | 852 | 5243 |
| Cooling Degree Days | 0 | 0 | 0 | 0 | 4 | 53 | 128 | 140 | 58 | 5 | 0 | 0 | 388 |
| Total Precipitation (") | 8.55 | 8.44 | 7.70 | 3.11 | 1.07 | 0.20 | 0.18 | 0.38 | 1.10 | 1.85 | 4.95 | 6.56 | 44.09 |
| Days ≥ 0.1" Precip | 6 | 5 | 6 | 4 | 2 | 0 | 0 | 1 | 1 | 2 | 4 | 5 | 36 |
| Total Snowfall (") | 8.9 | *10.3* | *10.5* | 6.9 | 0.4 | 0.0 | 0.0 | 0.0 | 0.0 | 0.4 | 1.9 | 7.2 | 46.5 |
| Days ≥ 1" Snow Depth | *6* | na | *4* | 3 | 0 | 0 | 0 | 0 | 0 | 0 | 1 | 4 | na |

## LAKE SPAULDING *Nevada County* ELEVATION 5164 ft LAT/LONG 39° 19' N / 120° 38' W

| | JAN | FEB | MAR | APR | MAY | JUN | JUL | AUG | SEP | OCT | NOV | DEC | YEAR |
|---|---|---|---|---|---|---|---|---|---|---|---|---|---|
| Maximum Temp °F | 45.7 | 48.1 | 50.7 | 56.5 | 66.1 | 75.0 | 82.2 | 81.3 | 75.8 | 65.6 | 50.9 | 44.7 | 61.9 |
| Minimum Temp °F | 23.9 | 24.7 | 26.6 | 29.2 | 35.7 | 42.2 | 46.3 | 45.7 | 42.0 | 35.9 | 28.6 | 23.6 | 33.7 |
| Mean Temp °F | 34.8 | 36.4 | 38.7 | 42.9 | 50.9 | 58.7 | 64.3 | 63.5 | 58.9 | 50.8 | 39.8 | 34.1 | 47.8 |
| Days Max Temp ≥ 90 °F | 0 | 0 | 0 | 0 | 0 | 1 | 3 | 3 | 1 | 0 | 0 | 0 | 8 |
| Days Max Temp ≤ 32 °F | 2 | 1 | 1 | 0 | 0 | 0 | 0 | 0 | 0 | 0 | 0 | 3 | 7 |
| Days Min Temp ≤ 32 °F | 28 | 26 | 28 | 22 | 9 | 1 | 0 | 0 | 2 | 9 | 23 | 29 | 177 |
| Days Min Temp ≤ 0 °F | 0 | 0 | 0 | 0 | 0 | 0 | 0 | 0 | 0 | 0 | 0 | 0 | 0 |
| Heating Degree Days | 930 | 802 | 809 | 657 | 431 | 201 | 69 | 83 | 188 | 436 | 749 | 950 | 6305 |
| Cooling Degree Days | 0 | 0 | 0 | 0 | 2 | 16 | 54 | 47 | 14 | 3 | 0 | 0 | 136 |
| Total Precipitation (") | 12.23 | 10.74 | 10.47 | 5.19 | 2.48 | 1.13 | 0.36 | 0.68 | 1.60 | 4.03 | 10.03 | 10.94 | 69.88 |
| Days ≥ 0.1" Precip | 10 | 10 | 11 | 8 | 4 | 2 | 0 | 1 | 2 | 4 | 9 | 9 | 70 |
| Total Snowfall (") | 49.3 | 50.2 | 47.3 | 26.3 | 4.6 | 0.5 | 0.0 | 0.0 | 0.0 | 2.2 | 24.7 | 47.9 | 253.0 |
| Days ≥ 1" Snow Depth | 28 | 27 | 29 | 18 | 4 | 0 | 0 | 0 | 0 | 1 | 10 | 24 | 141 |

## LAKEPORT *Lake County* ELEVATION 1342 ft LAT/LONG 39° 3 ' N / 122° 55 ' W

| | JAN | FEB | MAR | APR | MAY | JUN | JUL | AUG | SEP | OCT | NOV | DEC | YEAR |
|---|---|---|---|---|---|---|---|---|---|---|---|---|---|
| Maximum Temp °F | ***54.6*** | ***58.4*** | 63.4 | 69.0 | ***79.0*** | 87.1 | 94.6 | ***93.9*** | 87.1 | ***76.0*** | ***60.8*** | ***53.9*** | 73.1 |
| Minimum Temp °F | ***31.6*** | ***34.4*** | 36.5 | 38.6 | ***43.4*** | 50.0 | 54.0 | ***53.1*** | 48.0 | ***42.5*** | ***37.0*** | ***32.5*** | 41.8 |
| Mean Temp °F | ***43.1*** | ***46.5*** | 50.0 | 53.8 | ***61.2*** | 68.6 | 74.3 | ***73.5*** | 67.6 | ***59.3*** | ***48.9*** | ***43.2*** | 57.5 |
| Days Max Temp ≥ 90 °F | 0 | 0 | 0 | 0 | 4 | 12 | 23 | 22 | 13 | 2 | 0 | 0 | 76 |
| Days Max Temp ≤ 32 °F | 0 | 0 | 0 | 0 | 0 | 0 | 0 | 0 | 0 | 0 | 0 | 0 | 0 |
| Days Min Temp ≤ 32 °F | 18 | 11 | 9 | 4 | 0 | 0 | 0 | 0 | 1 | ***1*** | 9 | 16 | 69 |
| Days Min Temp ≤ 0 °F | 0 | 0 | 0 | 0 | 0 | 0 | 0 | 0 | 0 | ***0*** | 0 | 0 | 0 |
| Heating Degree Days | ***671*** | ***517*** | 459 | 331 | ***152*** | 38 | 4 | ***5*** | 45 | ***192*** | ***475*** | ***669*** | 3558 |
| Cooling Degree Days | na | na | 0 | 4 | na | 146 | ***276*** | na | ***128*** | na | na | na | na |
| Total Precipitation (") | 6.41 | 4.98 | 3.92 | 1.70 | 0.52 | 0.23 | 0.05 | 0.14 | 0.48 | 1.53 | 4.59 | 5.33 | 29.88 |
| Days ≥ 0.1" Precip | 8 | 7 | 6 | 4 | 1 | 0 | 0 | 0 | 1 | 3 | 7 | 8 | 45 |
| Total Snowfall (") | na | 0.0 | 0.2 | 0.0 | 0.0 | 0.0 | 0.0 | 0.0 | 0.0 | 0.0 | 0.0 | 0.2 | na |
| Days ≥ 1" Snow Depth | 0 | 0 | 0 | 0 | 0 | 0 | 0 | 0 | 0 | 0 | 0 | 0 | 0 |

## LAVA BEDS NATL MONUM *Siskiyou County* ELEVATION 4770 ft LAT/LONG 41° 44 ' N / 121° 31 ' W

| | JAN | FEB | MAR | APR | MAY | JUN | JUL | AUG | SEP | OCT | NOV | DEC | YEAR |
|---|---|---|---|---|---|---|---|---|---|---|---|---|---|
| Maximum Temp °F | 40.4 | 44.8 | 49.2 | 55.9 | 65.9 | 73.8 | 83.3 | 83.1 | 75.3 | ***63.2*** | 48.1 | 40.1 | 60.3 |
| Minimum Temp °F | 22.0 | 25.1 | 27.6 | 31.4 | 38.4 | 44.4 | 50.9 | 50.7 | 44.3 | ***36.2*** | 28.2 | 22.2 | 35.1 |
| Mean Temp °F | 31.2 | 35.0 | 38.4 | 43.7 | 52.2 | 59.1 | 67.2 | 66.9 | 59.8 | ***49.8*** | 38.2 | 31.2 | 47.7 |
| Days Max Temp ≥ 90 °F | 0 | 0 | 0 | 0 | 0 | 1 | 7 | 7 | 1 | 0 | 0 | 0 | 16 |
| Days Max Temp ≤ 32 °F | 5 | 2 | 1 | 0 | 0 | 0 | 0 | 0 | 0 | 0 | 1 | 5 | 14 |
| Days Min Temp ≤ 32 °F | 28 | 24 | 24 | 16 | 8 | 2 | 0 | 0 | 2 | ***10*** | 22 | 28 | 164 |
| Days Min Temp ≤ 0 °F | 0 | 0 | 0 | 0 | 0 | 0 | 0 | 0 | 0 | 0 | 0 | 1 | 1 |
| Heating Degree Days | 1040 | 841 | 817 | 633 | 399 | 208 | 57 | 62 | 184 | ***468*** | 799 | 1042 | 6550 |
| Cooling Degree Days | 0 | 0 | 0 | 0 | 12 | 40 | 130 | 136 | 43 | ***3*** | 0 | 0 | 364 |
| Total Precipitation (") | 1.66 | 1.55 | 1.76 | 1.07 | 1.25 | 1.05 | 0.47 | 0.67 | 0.50 | 1.19 | 1.67 | 1.87 | 14.71 |
| Days ≥ 0.1" Precip | 4 | 4 | 5 | 4 | 4 | 3 | 1 | 2 | 2 | 3 | 4 | 4 | 40 |
| Total Snowfall (") | 10.3 | 7.9 | 7.9 | 3.1 | 1.3 | 0.2 | 0.0 | 0.0 | 0.0 | 0.5 | 4.5 | 10.1 | 45.8 |
| Days ≥ 1" Snow Depth | 16 | ***10*** | ***6*** | 2 | 1 | 0 | 0 | 0 | 0 | 0 | ***5*** | ***13*** | 53 |

## LEMON COVE *Tulare County* ELEVATION 513 ft LAT/LONG 36° 23 ' N / 119° 2 ' W

| | JAN | FEB | MAR | APR | MAY | JUN | JUL | AUG | SEP | OCT | NOV | DEC | YEAR |
|---|---|---|---|---|---|---|---|---|---|---|---|---|---|
| Maximum Temp °F | 55.7 | 62.8 | 67.7 | 74.6 | 84.0 | 91.6 | 97.2 | 95.7 | 90.0 | 80.4 | 66.1 | 55.8 | 76.8 |
| Minimum Temp °F | 36.9 | 40.8 | 44.3 | 47.1 | 52.8 | 58.3 | 63.0 | 61.8 | 57.6 | 50.4 | 41.7 | 36.2 | 49.2 |
| Mean Temp °F | 46.3 | 51.8 | 56.1 | 60.8 | 68.4 | 75.0 | 80.1 | 78.7 | 73.8 | 65.4 | 53.9 | 46.0 | 63.0 |
| Days Max Temp ≥ 90 °F | 0 | 0 | 0 | 2 | 9 | 19 | 29 | 26 | 17 | 4 | 0 | 0 | 106 |
| Days Max Temp ≤ 32 °F | 0 | 0 | 0 | 0 | 0 | 0 | 0 | 0 | 0 | 0 | 0 | 0 | 0 |
| Days Min Temp ≤ 32 °F | 8 | 2 | 1 | 0 | 0 | 0 | 0 | 0 | 0 | 0 | 2 | 9 | 22 |
| Days Min Temp ≤ 0 °F | 0 | 0 | 0 | 0 | 0 | 0 | 0 | 0 | 0 | 0 | 0 | 0 | 0 |
| Heating Degree Days | 572 | 367 | 273 | 157 | 46 | 5 | 0 | 0 | 7 | 72 | 328 | 582 | 2409 |
| Cooling Degree Days | 0 | 0 | 3 | 50 | 159 | 314 | 472 | 439 | 287 | 105 | 1 | 0 | 1830 |
| Total Precipitation (") | 2.54 | 2.29 | 2.64 | 1.16 | 0.40 | 0.18 | 0.02 | 0.04 | 0.38 | 0.72 | 1.68 | 1.98 | 14.03 |
| Days ≥ 0.1" Precip | 5 | 5 | 5 | 3 | 1 | 0 | 0 | 0 | 1 | 1 | 4 | 4 | 29 |
| Total Snowfall (") | 0.0 | 0.0 | 0.0 | 0.0 | 0.0 | 0.0 | 0.0 | 0.0 | 0.0 | 0.0 | 0.0 | 0.0 | 0.0 |
| Days ≥ 1" Snow Depth | 0 | 0 | 0 | 0 | 0 | 0 | 0 | 0 | 0 | 0 | 0 | 0 | 0 |

## LINDSAY *Tulare County* ELEVATION 420 ft LAT/LONG 36° 11 ' N / 119° 3 ' W

| | JAN | FEB | MAR | APR | MAY | JUN | JUL | AUG | SEP | OCT | NOV | DEC | YEAR |
|---|---|---|---|---|---|---|---|---|---|---|---|---|---|
| Maximum Temp °F | 56.9 | 64.1 | 69.1 | 76.4 | 85.5 | 92.4 | 97.8 | 96.2 | 90.6 | 80.7 | 66.5 | 56.4 | 77.7 |
| Minimum Temp °F | 35.9 | 38.5 | 41.7 | 44.8 | 50.9 | 56.5 | 61.4 | 59.8 | 54.9 | 47.2 | 39.6 | 34.7 | 47.2 |
| Mean Temp °F | 46.4 | 51.3 | 55.4 | 60.6 | 68.2 | 74.5 | 79.6 | 78.0 | 72.8 | 64.0 | 53.1 | 45.5 | 62.5 |
| Days Max Temp ≥ 90 °F | 0 | 0 | 0 | 2 | 11 | 20 | 29 | 27 | 18 | 4 | 0 | 0 | 111 |
| Days Max Temp ≤ 32 °F | 0 | 0 | 0 | 0 | 0 | 0 | 0 | 0 | 0 | 0 | 0 | 0 | 0 |
| Days Min Temp ≤ 32 °F | 11 | 5 | 2 | 0 | 0 | 0 | 0 | 0 | 0 | 0 | 4 | 12 | 34 |
| Days Min Temp ≤ 0 °F | 0 | 0 | 0 | 0 | 0 | 0 | 0 | 0 | 0 | 0 | 0 | 0 | 0 |
| Heating Degree Days | 569 | 380 | 292 | 155 | 43 | 4 | 0 | 0 | 7 | 85 | 352 | 596 | 2483 |
| Cooling Degree Days | 0 | 0 | 1 | 39 | 146 | 290 | 448 | 410 | 250 | 65 | 0 | 0 | 1649 |
| Total Precipitation (") | 2.16 | 2.04 | 2.29 | 1.03 | 0.32 | 0.11 | 0.01 | 0.02 | 0.37 | 0.60 | 1.50 | 1.64 | 12.09 |
| Days ≥ 0.1" Precip | 4 | 5 | 6 | 3 | 1 | 0 | 0 | 0 | 1 | 2 | 3 | 4 | 29 |
| Total Snowfall (") | 0.0 | 0.0 | 0.0 | 0.0 | 0.0 | 0.0 | 0.0 | 0.0 | 0.0 | 0.0 | 0.0 | 0.0 | 0.0 |
| Days ≥ 1" Snow Depth | 0 | 0 | 0 | 0 | 0 | 0 | 0 | 0 | 0 | 0 | 0 | 0 | 0 |

## LIVERMORE *Alameda County* ELEVATION 479 ft LAT/LONG 37° 41' N / 121° 46' W

| | JAN | FEB | MAR | APR | MAY | JUN | JUL | AUG | SEP | OCT | NOV | DEC | YEAR |
|---|---|---|---|---|---|---|---|---|---|---|---|---|---|
| Maximum Temp °F | 57.0 | 61.8 | 65.4 | 70.9 | 77.7 | 83.9 | 89.6 | 89.2 | 86.4 | 78.4 | 65.4 | 56.6 | 73.5 |
| Minimum Temp °F | 36.1 | 39.3 | 41.4 | 43.0 | 47.5 | 51.8 | 54.1 | 54.0 | 52.3 | 47.6 | 41.1 | 36.1 | 45.4 |
| Mean Temp °F | 46.6 | 50.6 | 53.4 | 57.0 | 62.6 | 67.8 | 71.8 | 71.7 | 69.4 | 63.1 | 53.3 | 46.4 | 59.5 |
| Days Max Temp ≥ 90 °F | 0 | 0 | 0 | 1 | 4 | 9 | 16 | 16 | 11 | 4 | 0 | 0 | 61 |
| Days Max Temp ≤ 32 °F | 0 | 0 | 0 | 0 | 0 | 0 | 0 | 0 | 0 | 0 | 0 | 0 | 0 |
| Days Min Temp ≤ 32 °F | 11 | 5 | 2 | 1 | 0 | 0 | 0 | 0 | 0 | 0 | 3 | 10 | 32 |
| Days Min Temp ≤ 0 °F | 0 | 0 | 0 | 0 | 0 | 0 | 0 | 0 | 0 | 0 | 0 | 0 | 0 |
| Heating Degree Days | 565 | 402 | 352 | 241 | 116 | 35 | 5 | 3 | 15 | 99 | 345 | 569 | 2747 |
| Cooling Degree Days | 0 | 0 | 0 | 12 | 48 | 134 | 234 | 222 | 161 | 54 | 1 | 0 | 866 |
| Total Precipitation (") | 2.79 | 2.40 | 2.34 | 1.04 | 0.29 | 0.08 | 0.04 | 0.07 | 0.22 | 0.80 | 2.04 | 2.19 | 14.30 |
| Days ≥ 0.1" Precip | 5 | 6 | 6 | 3 | 1 | 0 | 0 | 0 | 1 | 2 | 4 | 5 | 33 |
| Total Snowfall (") | 0.0 | 0.0 | 0.0 | 0.0 | 0.0 | 0.0 | 0.0 | 0.0 | 0.0 | 0.0 | 0.0 | 0.0 | 0.0 |
| Days ≥ 1" Snow Depth | 0 | 0 | 0 | 0 | 0 | 0 | 0 | 0 | 0 | 0 | 0 | 0 | 0 |

## LODI *San Joaquin County* ELEVATION 40 ft LAT/LONG 38° 7' N / 121° 17' W

| | JAN | FEB | MAR | APR | MAY | JUN | JUL | AUG | SEP | OCT | NOV | DEC | YEAR |
|---|---|---|---|---|---|---|---|---|---|---|---|---|---|
| Maximum Temp °F | 54.2 | 61.6 | 66.5 | 73.1 | 81.5 | 87.1 | 91.2 | 90.3 | 87.0 | 78.6 | 64.1 | 54.1 | 74.1 |
| Minimum Temp °F | 36.6 | 39.5 | 42.2 | 44.7 | 49.6 | 53.7 | 56.3 | 55.8 | 53.1 | 47.2 | 40.7 | 35.9 | 46.3 |
| Mean Temp °F | 45.4 | 50.8 | 54.4 | 58.9 | 65.6 | 70.4 | 73.7 | 73.1 | 70.1 | 62.9 | 52.4 | 45.0 | 60.2 |
| Days Max Temp ≥ 90 °F | 0 | 0 | 0 | 1 | 6 | 12 | 20 | 17 | 12 | 3 | 0 | 0 | 71 |
| Days Max Temp ≤ 32 °F | 0 | 0 | 0 | 0 | 0 | 0 | 0 | 0 | 0 | 0 | 0 | 0 | 0 |
| Days Min Temp ≤ 32 °F | 10 | 4 | 1 | 0 | 0 | 0 | 0 | 0 | 0 | 0 | 4 | 11 | 30 |
| Days Min Temp ≤ 0 °F | 0 | 0 | 0 | 0 | 0 | 0 | 0 | 0 | 0 | 0 | 0 | 0 | 0 |
| Heating Degree Days | 599 | 394 | 323 | 189 | 66 | 12 | 1 | 1 | 12 | 102 | 371 | 613 | 2683 |
| Cooling Degree Days | 0 | 0 | 0 | 15 | 82 | 166 | 255 | 237 | 156 | 46 | 0 | 0 | 957 |
| Total Precipitation (") | 3.25 | 2.86 | 2.96 | 1.17 | 0.30 | 0.12 | 0.08 | 0.07 | 0.35 | 0.94 | 2.61 | 2.71 | 17.42 |
| Days ≥ 0.1" Precip | 6 | 6 | 7 | 3 | 1 | 0 | 0 | 0 | 1 | 2 | 5 | 6 | 37 |
| Total Snowfall (") | 0.0 | 0.0 | 0.0 | 0.0 | 0.0 | 0.0 | 0.0 | 0.0 | 0.0 | 0.0 | 0.0 | 0.0 | 0.0 |
| Days ≥ 1" Snow Depth | 0 | 0 | 0 | 0 | 0 | 0 | 0 | 0 | 0 | 0 | 0 | 0 | 0 |

## LOMPOC *Santa Barbara County* ELEVATION 102 ft LAT/LONG 34° 38' N / 120° 28' W

| | JAN | FEB | MAR | APR | MAY | JUN | JUL | AUG | SEP | OCT | NOV | DEC | YEAR |
|---|---|---|---|---|---|---|---|---|---|---|---|---|---|
| Maximum Temp °F | 65.7 | 66.6 | 67.0 | 69.4 | 69.8 | 72.2 | 73.7 | 74.9 | 76.2 | 75.6 | 70.3 | 65.6 | 70.6 |
| Minimum Temp °F | 40.7 | 42.6 | 44.2 | 45.3 | 48.5 | 51.3 | 53.4 | 54.2 | 53.1 | 49.4 | 44.0 | 39.7 | 47.2 |
| Mean Temp °F | 53.2 | 54.7 | 55.6 | 57.4 | 59.2 | 61.8 | 63.6 | 64.6 | 64.7 | 62.6 | 57.2 | 52.7 | 58.9 |
| Days Max Temp ≥ 90 °F | 0 | 0 | 0 | 1 | 0 | 1 | 0 | 0 | 1 | 1 | 0 | 0 | 4 |
| Days Max Temp ≤ 32 °F | 0 | 0 | 0 | 0 | 0 | 0 | 0 | 0 | 0 | 0 | 0 | 0 | 0 |
| Days Min Temp ≤ 32 °F | 3 | 1 | 1 | 0 | 0 | 0 | 0 | 0 | 0 | 0 | 1 | 4 | 10 |
| Days Min Temp ≤ 0 °F | 0 | 0 | 0 | 0 | 0 | 0 | 0 | 0 | 0 | 0 | 0 | 0 | 0 |
| Heating Degree Days | 358 | 286 | 285 | 229 | 179 | 104 | 63 | 39 | 45 | 99 | 233 | 376 | 2296 |
| Cooling Degree Days | 0 | 1 | 2 | 12 | 8 | 19 | 45 | 55 | 54 | 40 | 6 | 1 | 243 |
| Total Precipitation (") | 2.82 | 2.98 | 2.95 | 1.04 | 0.21 | 0.02 | 0.01 | 0.04 | 0.23 | 0.43 | 1.58 | 2.22 | 14.53 |
| Days ≥ 0.1" Precip | 5 | 5 | 5 | 3 | 1 | 0 | 0 | 0 | 0 | 1 | 3 | 4 | 27 |
| Total Snowfall (") | 0.0 | 0.0 | 0.0 | 0.0 | 0.0 | 0.0 | 0.0 | 0.0 | 0.0 | 0.0 | 0.0 | 0.0 | 0.0 |
| Days ≥ 1" Snow Depth | 0 | 0 | 0 | 0 | 0 | 0 | 0 | 0 | 0 | 0 | 0 | 0 | 0 |

## LONG BEACH DAUGHERTY *Los Angeles County* ELEVATION 25 ft LAT/LONG 33° 49' N / 118° 9' W

| | JAN | FEB | MAR | APR | MAY | JUN | JUL | AUG | SEP | OCT | NOV | DEC | YEAR |
|---|---|---|---|---|---|---|---|---|---|---|---|---|---|
| Maximum Temp °F | 66.9 | 67.8 | 68.3 | 71.9 | 73.7 | 77.6 | 82.7 | 84.3 | 82.5 | 78.6 | 72.6 | 67.3 | 74.5 |
| Minimum Temp °F | 45.3 | 47.3 | 49.8 | 52.4 | 56.9 | 60.3 | 63.9 | 65.2 | 63.1 | 58.0 | 50.4 | 45.0 | 54.8 |
| Mean Temp °F | 56.1 | 57.6 | 59.1 | 62.2 | 65.3 | 69.0 | 73.3 | 74.8 | 72.8 | 68.3 | 61.6 | 56.2 | 64.7 |
| Days Max Temp ≥ 90 °F | 0 | 0 | 0 | 1 | 1 | 2 | 4 | 6 | 6 | 3 | 1 | 0 | 24 |
| Days Max Temp ≤ 32 °F | 0 | 0 | 0 | 0 | 0 | 0 | 0 | 0 | 0 | 0 | 0 | 0 | 0 |
| Days Min Temp ≤ 32 °F | 0 | 0 | 0 | 0 | 0 | 0 | 0 | 0 | 0 | 0 | 0 | 0 | 0 |
| Days Min Temp ≤ 0 °F | 0 | 0 | 0 | 0 | 0 | 0 | 0 | 0 | 0 | 0 | 0 | 0 | 0 |
| Heating Degree Days | 273 | 206 | 186 | 103 | 38 | 6 | 0 | 0 | 1 | 14 | 120 | 269 | 1216 |
| Cooling Degree Days | 2 | 6 | 10 | 38 | 56 | 136 | 269 | 308 | 246 | 124 | 17 | 2 | 1214 |
| Total Precipitation (") | 2.73 | 2.68 | 2.22 | 0.65 | 0.17 | 0.04 | 0.02 | 0.11 | 0.25 | 0.28 | 1.51 | 1.87 | 12.53 |
| Days ≥ 0.1" Precip | 4 | 4 | 4 | 2 | 0 | 0 | 0 | 0 | 1 | 1 | 2 | 3 | 21 |
| Total Snowfall (") | 0.0 | 0.0 | 0.0 | 0.0 | 0.0 | 0.0 | 0.0 | 0.0 | 0.0 | 0.0 | 0.0 | 0.0 | 0.0 |
| Days ≥ 1" Snow Depth | 0 | 0 | 0 | 0 | 0 | 0 | 0 | 0 | 0 | 0 | 0 | 0 | 0 |

## LOS ANGELES CVC CNTR *Los Angeles County* ELEVATION 270 ft LAT/LONG 34° 3 ' N / 118° 14 ' W

| | JAN | FEB | MAR | APR | MAY | JUN | JUL | AUG | SEP | OCT | NOV | DEC | YEAR |
|---|---|---|---|---|---|---|---|---|---|---|---|---|---|
| Maximum Temp °F | 68.1 | 69.5 | 69.7 | 72.6 | 74.3 | 78.8 | 83.8 | 84.7 | 83.0 | 79.2 | 73.1 | 68.1 | 75.4 |
| Minimum Temp °F | 49.0 | 50.7 | 52.3 | 54.8 | 58.3 | 61.8 | 65.1 | 66.3 | 65.0 | 60.7 | 53.8 | 49.1 | 57.2 |
| Mean Temp °F | 58.6 | 60.1 | 61.0 | 63.7 | 66.3 | 70.3 | 74.5 | 75.5 | 74.0 | 70.0 | 63.5 | 58.6 | 66.3 |
| Days Max Temp ≥ 90 °F | 0 | 0 | 0 | 1 | 1 | 2 | 4 | 6 | 6 | 4 | 1 | 0 | 25 |
| Days Max Temp ≤ 32 °F | 0 | 0 | 0 | 0 | 0 | 0 | 0 | 0 | 0 | 0 | 0 | 0 | 0 |
| Days Min Temp ≤ 32 °F | 0 | 0 | 0 | 0 | 0 | 0 | 0 | 0 | 0 | 0 | 0 | 0 | 0 |
| Days Min Temp ≤ 0 °F | 0 | 0 | 0 | 0 | 0 | 0 | 0 | 0 | 0 | 0 | 0 | 0 | 0 |
| Heating Degree Days | 208 | 153 | 142 | 86 | 36 | 6 | 0 | 0 | 0 | 9 | 86 | 205 | 931 |
| Cooling Degree Days | 18 | 32 | 34 | 78 | 102 | 196 | 319 | 345 | 300 | 191 | 48 | 16 | 1679 |
| Total Precipitation (") | 3.23 | 3.28 | 2.99 | 0.98 | 0.21 | 0.04 | 0.01 | 0.14 | 0.41 | 0.32 | 1.83 | 2.24 | 15.68 |
| Days ≥ 0.1" Precip | 4 | 4 | 5 | 2 | 0 | 0 | 0 | 0 | 1 | 1 | 3 | 3 | 23 |
| Total Snowfall (") | na | na | na | *0.0* | na | na | na | na | na | na | na | na | na |
| Days ≥ 1" Snow Depth | na | na | na | *0* | na | na | na | na | na | na | na | na | na |

## LOS ANGELES INTL AP *Los Angeles County* ELEVATION 125 ft LAT/LONG 33° 56 ' N / 118° 23 ' W

| | JAN | FEB | MAR | APR | MAY | JUN | JUL | AUG | SEP | OCT | NOV | DEC | YEAR |
|---|---|---|---|---|---|---|---|---|---|---|---|---|---|
| Maximum Temp °F | 65.5 | 65.9 | 65.3 | 67.8 | 69.3 | 72.2 | 75.1 | 76.7 | 76.2 | 74.6 | 70.6 | 66.1 | 70.4 |
| Minimum Temp °F | 48.2 | 49.7 | 51.3 | 53.4 | 56.8 | 60.0 | 63.1 | 64.5 | 63.4 | 59.5 | 53.1 | 48.3 | 55.9 |
| Mean Temp °F | 56.9 | 57.8 | 58.3 | 60.6 | 63.0 | 66.1 | 69.1 | 70.6 | 69.8 | 67.1 | 61.9 | 57.2 | 63.2 |
| Days Max Temp ≥ 90 °F | 0 | 0 | 0 | 0 | 0 | 0 | 0 | 0 | 2 | 2 | 0 | 0 | 4 |
| Days Max Temp ≤ 32 °F | 0 | 0 | 0 | 0 | 0 | 0 | 0 | 0 | 0 | 0 | 0 | 0 | 0 |
| Days Min Temp ≤ 32 °F | 0 | 0 | 0 | 0 | 0 | 0 | 0 | 0 | 0 | 0 | 0 | 0 | 0 |
| Days Min Temp ≤ 0 °F | 0 | 0 | 0 | 0 | 0 | 0 | 0 | 0 | 0 | 0 | 0 | 0 | 0 |
| Heating Degree Days | 250 | 202 | 206 | 139 | 74 | 18 | 1 | 0 | 2 | 18 | 111 | 238 | 1259 |
| Cooling Degree Days | 6 | 8 | 8 | 22 | 18 | 62 | 147 | 180 | 152 | 92 | 22 | 5 | 722 |
| Total Precipitation (") | 2.66 | 2.56 | 2.20 | 0.69 | 0.14 | 0.04 | 0.03 | 0.14 | 0.27 | 0.34 | 1.63 | 1.86 | 12.56 |
| Days ≥ 0.1" Precip | 4 | 4 | 4 | 2 | 0 | 0 | 0 | 0 | 0 | 1 | 2 | 3 | 20 |
| Total Snowfall (") | 0.0 | 0.0 | 0.0 | 0.0 | 0.0 | 0.0 | 0.0 | 0.0 | 0.0 | 0.0 | 0.0 | 0.0 | 0.0 |
| Days ≥ 1" Snow Depth | 0 | 0 | 0 | 0 | 0 | 0 | 0 | 0 | 0 | 0 | 0 | 0 | 0 |

## LOS ANGELES U C L A *Los Angeles County* ELEVATION 440 ft LAT/LONG 34° 4 ' N / 118° 27 ' W

| | JAN | FEB | MAR | APR | MAY | JUN | JUL | AUG | SEP | OCT | NOV | DEC | YEAR |
|---|---|---|---|---|---|---|---|---|---|---|---|---|---|
| Maximum Temp °F | 66.3 | 67.0 | 66.7 | 68.9 | 69.5 | 72.9 | 76.9 | 78.1 | 77.8 | 75.4 | 70.6 | 66.3 | 71.4 |
| Minimum Temp °F | 50.3 | 50.6 | 50.7 | 52.6 | 55.1 | 58.0 | 61.1 | 62.3 | 61.6 | 58.8 | 54.3 | 50.5 | 55.5 |
| Mean Temp °F | 58.3 | 58.8 | 58.8 | 60.8 | 62.3 | 65.5 | 69.0 | 70.2 | 69.7 | 67.1 | 62.5 | 58.4 | 63.5 |
| Days Max Temp ≥ 90 °F | 0 | 0 | 0 | 0 | 0 | 1 | 1 | 1 | 3 | 2 | 0 | 0 | 8 |
| Days Max Temp ≤ 32 °F | 0 | 0 | 0 | 0 | 0 | 0 | 0 | 0 | 0 | 0 | 0 | 0 | 0 |
| Days Min Temp ≤ 32 °F | 0 | 0 | 0 | 0 | 0 | 0 | 0 | 0 | 0 | 0 | 0 | 0 | 0 |
| Days Min Temp ≤ 0 °F | 0 | 0 | 0 | 0 | 0 | 0 | 0 | 0 | 0 | 0 | 0 | 0 | 0 |
| Heating Degree Days | 222 | 187 | 202 | 149 | 103 | 37 | 3 | 2 | 7 | 29 | 113 | 218 | 1272 |
| Cooling Degree Days | 21 | 23 | 16 | 39 | 21 | 61 | 136 | 171 | 155 | 101 | 40 | 18 | 802 |
| Total Precipitation (") | 3.79 | 4.13 | 3.25 | 0.92 | 0.22 | 0.07 | 0.03 | 0.17 | 0.34 | 0.47 | 2.16 | 2.59 | 18.14 |
| Days ≥ 0.1" Precip | 4 | 4 | 5 | 2 | 0 | 0 | 0 | 0 | 1 | 1 | 3 | 3 | 23 |
| Total Snowfall (") | 0.0 | 0.0 | 0.0 | 0.0 | 0.0 | 0.0 | 0.0 | 0.0 | 0.0 | 0.0 | 0.0 | 0.0 | 0.0 |
| Days ≥ 1" Snow Depth | 0 | 0 | 0 | 0 | 0 | 0 | 0 | 0 | 0 | 0 | 0 | 0 | 0 |

## LOS BANOS *Merced County* ELEVATION 131 ft LAT/LONG 37° 3 ' N / 120° 51 ' W

| | JAN | FEB | MAR | APR | MAY | JUN | JUL | AUG | SEP | OCT | NOV | DEC | YEAR |
|---|---|---|---|---|---|---|---|---|---|---|---|---|---|
| Maximum Temp °F | 54.9 | 62.7 | 68.0 | 74.4 | 82.6 | 89.6 | 95.4 | 94.0 | 89.3 | 80.0 | 65.4 | 54.7 | 75.9 |
| Minimum Temp °F | 36.3 | 40.0 | 43.4 | 46.2 | 52.1 | 56.9 | 60.6 | 59.7 | 56.6 | 50.2 | 41.7 | 35.5 | 48.3 |
| Mean Temp °F | 45.6 | 51.4 | 55.7 | 60.4 | 67.4 | 73.3 | 78.0 | 76.9 | 73.0 | 65.1 | 53.6 | 45.1 | 62.1 |
| Days Max Temp ≥ 90 °F | 0 | 0 | 0 | 1 | 7 | 15 | 26 | 24 | 16 | 4 | 0 | 0 | 93 |
| Days Max Temp ≤ 32 °F | 0 | 0 | 0 | 0 | 0 | 0 | 0 | 0 | 0 | 0 | 0 | 0 | 0 |
| Days Min Temp ≤ 32 °F | 10 | 3 | 1 | 0 | 0 | 0 | 0 | 0 | 0 | 0 | 3 | 11 | 28 |
| Days Min Temp ≤ 0 °F | 0 | 0 | 0 | 0 | 0 | 0 | 0 | 0 | 0 | 0 | 0 | 0 | 0 |
| Heating Degree Days | 594 | 378 | 282 | 159 | 45 | 6 | 0 | 0 | 5 | 68 | 337 | 610 | 2484 |
| Cooling Degree Days | 0 | 0 | 1 | 36 | 131 | 262 | 411 | 376 | 259 | 93 | 1 | 0 | 1570 |
| Total Precipitation (") | 1.65 | 1.67 | 1.49 | 0.68 | 0.26 | 0.05 | 0.04 | 0.06 | 0.33 | 0.48 | 1.32 | 1.27 | 9.30 |
| Days ≥ 0.1" Precip | 5 | 4 | 4 | 2 | 1 | 0 | 0 | 0 | 1 | 2 | 3 | 4 | 26 |
| Total Snowfall (") | 0.0 | 0.0 | 0.0 | 0.0 | 0.0 | 0.0 | 0.0 | 0.0 | 0.0 | 0.0 | 0.0 | 0.0 | 0.0 |
| Days ≥ 1" Snow Depth | 0 | 0 | 0 | 0 | 0 | 0 | 0 | 0 | 0 | 0 | 0 | 0 | 0 |

## LOS GATOS *Santa Clara County* ELEVATION 430 ft LAT/LONG 37° 13 ' N / 121° 59 ' W

| | JAN | FEB | MAR | APR | MAY | JUN | JUL | AUG | SEP | OCT | NOV | DEC | YEAR |
|---|---|---|---|---|---|---|---|---|---|---|---|---|---|
| Maximum Temp °F | 58.5 | 62.6 | 65.5 | 70.4 | 76.4 | 81.7 | 85.6 | 85.1 | 82.8 | 75.9 | 65.0 | 57.9 | 72.3 |
| Minimum Temp °F | 37.8 | 40.4 | 42.4 | 43.4 | 47.2 | 51.0 | 54.0 | 54.1 | 52.7 | 48.3 | 42.2 | 37.5 | 45.9 |
| Mean Temp °F | 48.2 | 51.5 | 54.0 | 56.9 | 61.8 | 66.4 | 69.8 | 69.6 | 67.8 | 62.1 | 53.6 | 47.7 | 59.1 |
| Days Max Temp ≥ 90 °F | 0 | 0 | 0 | 0 | 2 | 6 | 8 | 8 | 6 | 1 | 0 | 0 | 31 |
| Days Max Temp ≤ 32 °F | 0 | 0 | 0 | 0 | 0 | 0 | 0 | 0 | 0 | 0 | 0 | 0 | 0 |
| Days Min Temp ≤ 32 °F | 8 | 3 | 1 | 0 | 0 | 0 | 0 | 0 | 0 | 0 | 1 | 8 | 21 |
| Days Min Temp ≤ 0 °F | 0 | 0 | 0 | 0 | 0 | 0 | 0 | 0 | 0 | 0 | 0 | 0 | 0 |
| Heating Degree Days | 515 | 375 | 336 | 242 | 126 | 41 | 7 | 4 | 18 | 113 | 336 | 528 | 2641 |
| Cooling Degree Days | 0 | 0 | 0 | 10 | 34 | 96 | 171 | 153 | 105 | 34 | 0 | 0 | 603 |
| Total Precipitation (") | 4.95 | 4.24 | 3.76 | 1.37 | 0.26 | 0.07 | 0.05 | 0.07 | 0.25 | 0.93 | 2.82 | 3.55 | 22.32 |
| Days ≥ 0.1" Precip | 6 | 6 | 6 | 3 | 1 | 0 | 0 | 0 | 1 | 2 | 5 | 6 | 36 |
| Total Snowfall (") | 0.0 | 0.1 | 0.0 | 0.0 | 0.0 | 0.0 | 0.0 | 0.0 | 0.0 | 0.0 | 0.0 | 0.0 | 0.1 |
| Days ≥ 1" Snow Depth | 0 | 0 | 0 | 0 | 0 | 0 | 0 | 0 | 0 | 0 | 0 | 0 | 0 |

## MADERA *Madera County* ELEVATION 302 ft LAT/LONG 36° 58 ' N / 120° 1 ' W

| | JAN | FEB | MAR | APR | MAY | JUN | JUL | AUG | SEP | OCT | NOV | DEC | YEAR |
|---|---|---|---|---|---|---|---|---|---|---|---|---|---|
| Maximum Temp °F | 53.9 | 61.4 | 66.8 | 74.1 | 83.9 | 91.1 | 97.2 | 95.7 | 90.1 | 80.3 | 65.1 | 53.9 | 76.1 |
| Minimum Temp °F | 35.9 | 38.4 | 42.7 | 45.8 | 51.8 | 57.6 | 61.9 | 60.6 | 55.8 | 48.0 | 40.1 | 34.7 | 47.8 |
| Mean Temp °F | 44.9 | 49.9 | 54.8 | 59.9 | 67.9 | 74.4 | 79.6 | 78.2 | 73.0 | 64.2 | 52.6 | 44.3 | 62.0 |
| Days Max Temp ≥ 90 °F | 0 | 0 | 0 | 1 | 9 | 18 | 27 | 25 | 17 | 4 | 0 | 0 | 101 |
| Days Max Temp ≤ 32 °F | 0 | 0 | 0 | 0 | 0 | 0 | 0 | 0 | 0 | 0 | 0 | 0 | 0 |
| Days Min Temp ≤ 32 °F | 10 | 4 | 1 | 0 | 0 | 0 | 0 | 0 | 0 | 0 | 4 | 12 | 31 |
| Days Min Temp ≤ 0 °F | 0 | 0 | 0 | 0 | 0 | 0 | 0 | 0 | 0 | 0 | 0 | 0 | 0 |
| Heating Degree Days | 615 | 418 | 313 | 171 | 45 | 6 | 0 | 0 | 7 | 88 | 365 | 634 | 2662 |
| Cooling Degree Days | 0 | 0 | 1 | 34 | 142 | 293 | 452 | 421 | 255 | 69 | 0 | 0 | 1667 |
| Total Precipitation (") | 1.94 | 1.77 | 2.10 | 1.11 | 0.36 | 0.06 | 0.01 | 0.02 | 0.20 | 0.64 | 1.46 | 1.59 | 11.26 |
| Days ≥ 0.1" Precip | 5 | 5 | 5 | 3 | 1 | 0 | 0 | 0 | 0 | 2 | 3 | 4 | 28 |
| Total Snowfall (") | 0.0 | 0.0 | 0.0 | 0.0 | 0.0 | 0.0 | 0.0 | 0.0 | 0.0 | 0.0 | 0.0 | 0.0 | 0.0 |
| Days ≥ 1" Snow Depth | 0 | 0 | 0 | 0 | 0 | 0 | 0 | 0 | 0 | 0 | 0 | 0 | 0 |

## MANZANITA LAKE *Shasta County* ELEVATION 5853 ft LAT/LONG 40° 32 ' N / 121° 34 ' W

| | JAN | FEB | MAR | APR | MAY | JUN | JUL | AUG | SEP | OCT | NOV | DEC | YEAR |
|---|---|---|---|---|---|---|---|---|---|---|---|---|---|
| Maximum Temp °F | 42.5 | 43.5 | 45.7 | 51.4 | 61.5 | 69.8 | 78.2 | 77.2 | 71.3 | 60.9 | 47.1 | 41.7 | 57.6 |
| Minimum Temp °F | 20.4 | 20.8 | 23.5 | 27.3 | 34.3 | 40.8 | 45.0 | 44.1 | 39.8 | 33.5 | 26.3 | 20.9 | 31.4 |
| Mean Temp °F | 31.4 | 32.2 | 34.6 | 39.4 | 47.9 | 55.3 | 61.7 | 60.7 | 55.5 | 47.2 | 36.7 | 31.4 | 44.5 |
| Days Max Temp ≥ 90 °F | 0 | 0 | 0 | 0 | 0 | 0 | 1 | 1 | 0 | 0 | 0 | 0 | 2 |
| Days Max Temp ≤ 32 °F | 4 | 3 | 2 | 0 | 0 | 0 | 0 | 0 | 0 | 0 | 2 | 5 | 16 |
| Days Min Temp ≤ 32 °F | 29 | 27 | 29 | 23 | 13 | 3 | 1 | 1 | 4 | 14 | 24 | 29 | 197 |
| Days Min Temp ≤ 0 °F | 1 | 1 | 0 | 0 | 0 | 0 | 0 | 0 | 0 | 0 | 0 | 1 | 3 |
| Heating Degree Days | 1033 | 920 | 936 | 762 | 523 | 292 | 131 | 154 | 282 | 544 | 842 | 1036 | 7455 |
| Cooling Degree Days | 0 | 0 | 0 | 0 | 1 | 6 | 31 | 23 | 7 | 1 | 0 | 0 | 69 |
| Total Precipitation (") | 5.84 | 4.92 | 5.45 | 3.13 | 2.27 | 1.57 | 0.47 | 0.91 | 1.46 | 2.97 | 5.70 | 5.27 | 39.96 |
| Days ≥ 0.1" Precip | 9 | 8 | 10 | 7 | 5 | 3 | 1 | 2 | 2 | 4 | 9 | 9 | 69 |
| Total Snowfall (") | 32.8 | 30.6 | 34.7 | 20.1 | 6.6 | 1.3 | 0.0 | 0.0 | 0.5 | 2.7 | 19.1 | 33.0 | 181.4 |
| Days ≥ 1" Snow Depth | 29 | 26 | 28 | 15 | 5 | 0 | 0 | 0 | 0 | 2 | 12 | 26 | 143 |

## MARYSVILLE *Yuba County* ELEVATION 69 ft LAT/LONG 39° 8 ' N / 121° 36 ' W

| | JAN | FEB | MAR | APR | MAY | JUN | JUL | AUG | SEP | OCT | NOV | DEC | YEAR |
|---|---|---|---|---|---|---|---|---|---|---|---|---|---|
| Maximum Temp °F | 54.2 | 61.6 | 66.9 | 73.4 | 82.9 | 90.0 | 95.8 | 94.3 | 89.2 | 79.5 | 64.3 | 54.3 | 75.5 |
| Minimum Temp °F | 37.5 | 41.5 | 44.7 | 47.8 | 54.2 | 59.1 | 62.0 | 60.6 | 57.2 | 51.1 | 43.4 | 37.7 | 49.7 |
| Mean Temp °F | 45.9 | 51.6 | 55.8 | 60.6 | 68.6 | 74.6 | 78.9 | 77.5 | 73.3 | 65.3 | 53.8 | 46.0 | 62.7 |
| Days Max Temp ≥ 90 °F | 0 | 0 | 0 | 1 | 8 | 16 | 25 | 23 | 16 | 4 | 0 | 0 | 93 |
| Days Max Temp ≤ 32 °F | 0 | 0 | 0 | 0 | 0 | 0 | 0 | 0 | 0 | 0 | 0 | 0 | 0 |
| Days Min Temp ≤ 32 °F | 6 | 2 | 0 | 0 | 0 | 0 | 0 | 0 | 0 | 0 | 1 | 7 | 16 |
| Days Min Temp ≤ 0 °F | 0 | 0 | 0 | 0 | 0 | 0 | 0 | 0 | 0 | 0 | 0 | 0 | 0 |
| Heating Degree Days | 586 | 373 | 281 | 154 | 39 | 6 | 0 | 0 | 6 | 68 | 329 | 582 | 2424 |
| Cooling Degree Days | 0 | 0 | 2 | 39 | 158 | 307 | 447 | 401 | 268 | 97 | 1 | 0 | 1720 |
| Total Precipitation (") | 4.22 | 3.41 | 3.30 | 1.43 | 0.43 | 0.23 | 0.07 | 0.10 | 0.37 | 1.13 | 3.14 | 3.65 | 21.48 |
| Days ≥ 0.1" Precip | 7 | 6 | 6 | 3 | 1 | 1 | 0 | 0 | 1 | 2 | 6 | 7 | 40 |
| Total Snowfall (") | 0.0 | 0.0 | 0.0 | 0.0 | 0.0 | 0.0 | 0.0 | 0.0 | 0.0 | 0.0 | 0.0 | 0.0 | 0.0 |
| Days ≥ 1" Snow Depth | 0 | 0 | 0 | 0 | 0 | 0 | 0 | 0 | 0 | 0 | 0 | 0 | 0 |

## MC CLOUD *Siskiyou County* ELEVATION 3251 ft LAT/LONG 41° 15 ' N / 122° 8 ' W

| | JAN | FEB | MAR | APR | MAY | JUN | JUL | AUG | SEP | OCT | NOV | DEC | YEAR |
|---|---|---|---|---|---|---|---|---|---|---|---|---|---|
| Maximum Temp °F | 47.1 | 49.9 | 53.9 | 60.4 | 71.1 | 79.0 | 87.4 | 86.7 | 80.5 | 69.4 | 53.5 | 46.7 | 65.5 |
| Minimum Temp °F | 23.6 | 26.0 | 28.4 | 31.8 | 37.7 | 44.6 | 48.4 | 46.6 | 41.0 | 34.3 | 28.4 | 24.1 | 34.6 |
| Mean Temp °F | 35.3 | 38.0 | 41.2 | 46.1 | 54.4 | 61.9 | 67.9 | 66.7 | 60.8 | 51.9 | 41.0 | 35.4 | 50.0 |
| Days Max Temp ≥ 90 °F | 0 | 0 | 0 | 0 | 1 | 5 | 13 | 13 | 6 | 1 | 0 | 0 | 39 |
| Days Max Temp ≤ 32 °F | 1 | 0 | 0 | 0 | 0 | 0 | 0 | 0 | 0 | 0 | 0 | 1 | 2 |
| Days Min Temp ≤ 32 °F | 28 | 24 | 24 | 17 | 7 | 1 | 0 | 0 | 2 | 12 | 23 | 28 | 166 |
| Days Min Temp ≤ 0 °F | 0 | 0 | 0 | 0 | 0 | 0 | 0 | 0 | 0 | 0 | 0 | 0 | 0 |
| Heating Degree Days | 912 | 757 | 730 | 560 | 332 | 141 | 38 | 49 | 153 | 403 | 714 | 909 | 5698 |
| Cooling Degree Days | 0 | 0 | 0 | 0 | 15 | 59 | 143 | 111 | 39 | 6 | 0 | 0 | 373 |
| Total Precipitation (") | 8.49 | 6.88 | 7.18 | 3.17 | 1.89 | 0.86 | 0.23 | 0.45 | 1.04 | 2.82 | 6.71 | 7.73 | 47.45 |
| Days ≥ 0.1" Precip | 9 | 8 | 10 | 6 | 4 | 2 | 1 | 1 | 2 | 4 | 8 | 9 | 64 |
| Total Snowfall (") | 22.4 | 13.9 | 11.4 | 3.3 | 0.1 | 0.0 | 0.0 | 0.0 | 0.0 | 0.2 | 6.4 | 22.1 | 79.8 |
| Days ≥ 1" Snow Depth | 17 | 12 | 6 | 2 | 0 | 0 | 0 | 0 | 0 | 0 | 4 | 12 | 53 |

## MECCA FIRE STATION *Riverside County* ELEVATION -182 ft LAT/LONG 33° 33 ' N / 116° 2 ' W

| | JAN | FEB | MAR | APR | MAY | JUN | JUL | AUG | SEP | OCT | NOV | DEC | YEAR |
|---|---|---|---|---|---|---|---|---|---|---|---|---|---|
| Maximum Temp °F | 70.2 | 75.7 | 80.6 | 88.3 | 95.9 | 104.1 | 107.9 | 106.6 | 101.9 | 91.9 | 78.9 | 69.8 | 89.3 |
| Minimum Temp °F | 38.2 | 42.6 | 48.2 | 53.5 | 60.8 | 67.2 | 73.9 | 73.9 | 67.3 | 56.4 | 44.5 | 36.9 | 55.3 |
| Mean Temp °F | 54.2 | 59.2 | 64.5 | 70.9 | 78.3 | 85.7 | 90.9 | 90.3 | 84.6 | 74.1 | 61.7 | 53.4 | 72.3 |
| Days Max Temp ≥ 90 °F | 0 | 1 | 4 | 14 | 25 | 29 | 31 | 31 | 29 | 20 | 2 | 0 | 186 |
| Days Max Temp ≤ 32 °F | 0 | 0 | 0 | 0 | 0 | 0 | 0 | 0 | 0 | 0 | 0 | 0 | 0 |
| Days Min Temp ≤ 32 °F | 7 | 2 | 0 | 0 | 0 | 0 | 0 | 0 | 0 | 0 | 1 | 7 | 17 |
| Days Min Temp ≤ 0 °F | 0 | 0 | 0 | 0 | 0 | 0 | 0 | 0 | 0 | 0 | 0 | 0 | 0 |
| Heating Degree Days | 329 | 172 | 77 | 18 | 1 | 0 | 0 | 0 | 0 | 9 | 127 | 355 | 1088 |
| Cooling Degree Days | 1 | 17 | 66 | 246 | 442 | 648 | 812 | 804 | 613 | 321 | *34* | 1 | 4005 |
| Total Precipitation (") | 0.60 | 0.56 | 0.44 | 0.09 | 0.04 | 0.01 | 0.10 | 0.28 | 0.35 | 0.27 | 0.31 | 0.40 | 3.45 |
| Days ≥ 0.1" Precip | 1 | 1 | 1 | 0 | 0 | 0 | 0 | 1 | 1 | 1 | 1 | 1 | 8 |
| Total Snowfall (") | 0.0 | 0.0 | 0.0 | 0.0 | 0.0 | 0.0 | 0.0 | 0.0 | 0.0 | 0.0 | 0.0 | 0.0 | 0.0 |
| Days ≥ 1" Snow Depth | 0 | 0 | 0 | 0 | 0 | 0 | 0 | 0 | 0 | 0 | 0 | 0 | 0 |

## MERCED MUNICIPAL AP *Merced County* ELEVATION 153 ft LAT/LONG 37° 17 ' N / 120° 31 ' W

| | JAN | FEB | MAR | APR | MAY | JUN | JUL | AUG | SEP | OCT | NOV | DEC | YEAR |
|---|---|---|---|---|---|---|---|---|---|---|---|---|---|
| Maximum Temp °F | 54.9 | 62.6 | 67.8 | 75.0 | 84.2 | 91.3 | 96.8 | 95.2 | 90.4 | 81.1 | 65.6 | 54.7 | 76.6 |
| Minimum Temp °F | 36.3 | 38.8 | 42.0 | 44.5 | 50.7 | 56.2 | 60.5 | 58.9 | 55.0 | 47.6 | 40.3 | 35.1 | 47.2 |
| Mean Temp °F | 45.6 | 50.7 | 54.9 | 59.8 | 67.5 | 73.8 | 78.7 | 77.1 | 72.8 | 64.3 | 53.0 | 44.9 | 61.9 |
| Days Max Temp ≥ 90 °F | 0 | 0 | 0 | 1 | 8 | 18 | 27 | 26 | 18 | 5 | 0 | 0 | 103 |
| Days Max Temp ≤ 32 °F | 0 | 0 | 0 | 0 | 0 | 0 | 0 | 0 | 0 | 0 | 0 | 0 | 0 |
| Days Min Temp ≤ 32 °F | 9 | 5 | 1 | 0 | 0 | 0 | 0 | 0 | 0 | 0 | 4 | 12 | 31 |
| Days Min Temp ≤ 0 °F | 0 | 0 | 0 | 0 | 0 | 0 | 0 | 0 | 0 | 0 | 0 | 0 | 0 |
| Heating Degree Days | 594 | 397 | 308 | 172 | 49 | 5 | 0 | 0 | 5 | 80 | 354 | 616 | 2580 |
| Cooling Degree Days | 0 | 0 | 0 | 28 | *124* | 277 | 433 | 389 | 247 | 73 | 0 | 0 | 1571 |
| Total Precipitation (") | 2.19 | 2.15 | 2.06 | 1.00 | 0.34 | 0.06 | 0.03 | 0.04 | 0.19 | 0.64 | 1.71 | 1.69 | 12.10 |
| Days ≥ 0.1" Precip | 5 | 5 | 5 | 3 | 1 | 0 | 0 | 0 | 1 | 2 | 4 | 4 | 30 |
| Total Snowfall (") | 0.0 | 0.0 | 0.0 | 0.0 | 0.0 | 0.0 | 0.0 | 0.0 | 0.0 | 0.0 | 0.0 | 0.0 | 0.0 |
| Days ≥ 1" Snow Depth | 0 | 0 | 0 | 0 | 0 | 0 | 0 | 0 | 0 | 0 | 0 | 0 | 0 |

## MINERAL *Tehama County* ELEVATION 4954 ft LAT/LONG 40° 21 ' N / 121° 34 ' W

| | JAN | FEB | MAR | APR | MAY | JUN | JUL | AUG | SEP | OCT | NOV | DEC | YEAR |
|---|---|---|---|---|---|---|---|---|---|---|---|---|---|
| Maximum Temp °F | 41.6 | 43.6 | 46.5 | 52.9 | 63.2 | 71.5 | 80.4 | 80.1 | 73.3 | 62.3 | 47.3 | 41.0 | 58.6 |
| Minimum Temp °F | 21.1 | 22.8 | 25.5 | 28.1 | 33.3 | 39.1 | 42.6 | 41.5 | 37.2 | 31.6 | 26.5 | 21.8 | 30.9 |
| Mean Temp °F | 31.3 | 33.2 | 36.0 | 40.5 | 48.2 | 55.3 | 61.5 | 60.8 | 55.2 | 47.0 | 36.9 | 31.5 | 44.8 |
| Days Max Temp ≥ 90 °F | 0 | 0 | 0 | 0 | 0 | 0 | 3 | 3 | 1 | 0 | 0 | 0 | 7 |
| Days Max Temp ≤ 32 °F | 4 | 2 | 1 | 0 | 0 | 0 | 0 | 0 | 0 | 0 | 1 | 5 | 13 |
| Days Min Temp ≤ 32 °F | 29 | 27 | 28 | 24 | 15 | 4 | 1 | 1 | 6 | 18 | 26 | 29 | 208 |
| Days Min Temp ≤ 0 °F | 0 | 0 | 0 | 0 | 0 | 0 | 0 | 0 | 0 | 0 | 0 | 0 | 0 |
| Heating Degree Days | 1036 | 891 | 891 | 727 | 513 | 291 | 130 | 149 | 290 | 551 | 838 | 1033 | 7340 |
| Cooling Degree Days | 0 | 0 | 0 | 0 | 0 | 5 | 26 | 17 | 4 | 0 | 0 | 0 | 52 |
| Total Precipitation (") | 9.31 | 7.45 | 7.83 | 3.72 | 2.22 | 1.38 | 0.33 | 0.70 | 1.31 | 3.81 | 7.91 | 7.96 | 53.93 |
| Days ≥ 0.1" Precip | 10 | 9 | 10 | 7 | 5 | 3 | 1 | 1 | 2 | 4 | 9 | 10 | 71 |
| Total Snowfall (") | 31.7 | 27.8 | 31.9 | 13.4 | 2.4 | 0.2 | 0.0 | 0.0 | 0.1 | 1.4 | 12.0 | 31.6 | 152.5 |
| Days ≥ 1" Snow Depth | 27 | 24 | 23 | *11* | 2 | 0 | 0 | 0 | 0 | 1 | 8 | 21 | 117 |

## MITCHELL CAVERNS *San Bernardino County* ELEVATION 4314 ft LAT/LONG 34° 56 ' N / 115° 32 ' W

| | JAN | FEB | MAR | APR | MAY | JUN | JUL | AUG | SEP | OCT | NOV | DEC | YEAR |
|---|---|---|---|---|---|---|---|---|---|---|---|---|---|
| Maximum Temp °F | 54.0 | 57.0 | 60.9 | 69.3 | 78.5 | 88.5 | 93.5 | 91.2 | 84.9 | 74.5 | 61.9 | 54.2 | 72.4 |
| Minimum Temp °F | 37.6 | 39.9 | 42.4 | 47.9 | 56.1 | 65.5 | 70.6 | 69.1 | 63.4 | 54.5 | 44.0 | 37.6 | 52.4 |
| Mean Temp °F | 45.9 | 48.4 | 51.5 | 58.6 | 67.3 | 77.0 | 82.0 | 80.2 | 74.1 | 64.5 | 53.0 | 46.0 | 62.4 |
| Days Max Temp ≥ 90 °F | 0 | 0 | 0 | 0 | 2 | 14 | 24 | 20 | 8 | 1 | 0 | 0 | 69 |
| Days Max Temp ≤ 32 °F | 0 | 0 | 0 | 0 | 0 | 0 | 0 | 0 | 0 | 0 | 0 | 0 | 0 |
| Days Min Temp ≤ 32 °F | 7 | 4 | 2 | 1 | 0 | 0 | 0 | 0 | 0 | 0 | 3 | 8 | 25 |
| Days Min Temp ≤ 0 °F | 0 | 0 | 0 | 0 | 0 | 0 | 0 | 0 | 0 | 0 | 0 | 0 | 0 |
| Heating Degree Days | 585 | 462 | 415 | 222 | 67 | 5 | 0 | 0 | 11 | 105 | 360 | 584 | 2816 |
| Cooling Degree Days | 0 | 1 | 3 | 45 | 136 | 372 | 495 | 453 | 274 | 79 | 4 | 0 | 1862 |
| Total Precipitation (") | 1.39 | 1.55 | 1.63 | 0.53 | 0.29 | 0.14 | 0.85 | 1.88 | 0.64 | 0.65 | 0.80 | 1.15 | 11.50 |
| Days ≥ 0.1" Precip | 3 | 3 | 3 | 1 | 1 | 0 | 1 | 2 | 1 | 1 | 2 | 2 | 20 |
| Total Snowfall (") | 0.5 | 0.4 | 0.7 | 0.3 | 0.0 | 0.0 | 0.0 | 0.0 | 0.0 | 0.0 | 0.1 | 0.7 | 2.7 |
| Days ≥ 1" Snow Depth | 0 | 0 | 0 | 0 | 0 | 0 | 0 | 0 | 0 | 0 | 0 | *0* | 0 |

## MODESTO *Stanislaus County* ELEVATION 91 ft LAT/LONG 37° 39 ' N / 121° 0 ' W

| | JAN | FEB | MAR | APR | MAY | JUN | JUL | AUG | SEP | OCT | NOV | DEC | YEAR |
|---|---|---|---|---|---|---|---|---|---|---|---|---|---|
| Maximum Temp °F | 54.0 | 62.0 | 67.4 | 73.8 | 82.0 | 88.6 | 93.9 | 92.3 | 87.8 | 78.6 | 64.0 | 53.6 | 74.8 |
| Minimum Temp °F | 38.1 | 41.6 | 44.4 | 47.3 | 52.2 | 57.2 | 60.5 | 60.0 | 57.0 | 50.8 | 43.1 | 37.6 | 49.2 |
| Mean Temp °F | 46.1 | 51.8 | 55.9 | 60.5 | 67.2 | 72.9 | 77.2 | 76.1 | 72.4 | 64.7 | 53.6 | 45.6 | 62.0 |
| Days Max Temp ≥ 90 °F | 0 | 0 | 0 | 1 | 7 | 14 | 24 | 20 | 13 | 3 | 0 | 0 | 82 |
| Days Max Temp ≤ 32 °F | 0 | 0 | 0 | 0 | 0 | 0 | 0 | 0 | 0 | 0 | 0 | 0 | 0 |
| Days Min Temp ≤ 32 °F | 5 | 2 | 0 | 0 | 0 | 0 | 0 | 0 | 0 | 0 | 1 | 7 | 15 |
| Days Min Temp ≤ 0 °F | 0 | 0 | 0 | 0 | 0 | 0 | 0 | 0 | 0 | 0 | 0 | 0 | 0 |
| Heating Degree Days | 579 | 366 | 277 | 153 | 49 | 8 | 0 | 0 | 5 | 75 | 337 | 594 | 2443 |
| Cooling Degree Days | 0 | 0 | 2 | 38 | 132 | 267 | 397 | 362 | 247 | 89 | 1 | 0 | 1535 |
| Total Precipitation (") | 2.23 | 1.96 | 2.26 | 0.93 | 0.27 | 0.09 | 0.05 | 0.09 | 0.28 | 0.69 | 1.77 | 1.72 | 12.34 |
| Days ≥ 0.1" Precip | 5 | 5 | 6 | 3 | 1 | 0 | 0 | 0 | 1 | 2 | 4 | 4 | 31 |
| Total Snowfall (") | 0.0 | 0.0 | 0.0 | 0.0 | 0.0 | 0.0 | 0.0 | 0.0 | 0.0 | 0.0 | 0.0 | 0.0 | 0.0 |
| Days ≥ 1" Snow Depth | 0 | 0 | 0 | 0 | 0 | 0 | 0 | 0 | 0 | 0 | 0 | 0 | 0 |

## MOJAVE *Kern County* ELEVATION 2735 ft LAT/LONG 35° 3 ' N / 118° 10 ' W

| | JAN | FEB | MAR | APR | MAY | JUN | JUL | AUG | SEP | OCT | NOV | DEC | YEAR |
|---|---|---|---|---|---|---|---|---|---|---|---|---|---|
| Maximum Temp °F | 57.3 | 61.3 | 65.0 | 71.2 | 80.7 | 89.8 | 96.7 | 95.7 | 89.1 | 78.5 | 65.1 | 56.8 | 75.6 |
| Minimum Temp °F | 32.3 | 36.0 | 40.3 | 45.7 | 54.5 | 62.9 | 68.6 | 67.1 | 59.8 | 49.3 | 39.4 | 32.1 | 49.0 |
| Mean Temp °F | 44.9 | 48.7 | 52.6 | 58.5 | 67.7 | 76.4 | 82.6 | 81.3 | 74.5 | 63.9 | 52.3 | 44.5 | 62.3 |
| Days Max Temp ≥ 90 °F | 0 | 0 | 0 | 1 | 5 | 17 | 28 | 27 | 16 | 3 | 0 | 0 | 97 |
| Days Max Temp ≤ 32 °F | 0 | 0 | 0 | 0 | 0 | 0 | 0 | 0 | 0 | 0 | 0 | 0 | 0 |
| Days Min Temp ≤ 32 °F | 16 | 9 | *4* | 1 | 0 | 0 | 0 | 0 | 0 | 0 | 5 | 17 | 52 |
| Days Min Temp ≤ 0 °F | 0 | 0 | 0 | 0 | 0 | 0 | 0 | 0 | 0 | 0 | 0 | 0 | 0 |
| Heating Degree Days | 615 | 452 | 378 | 219 | 68 | 7 | 0 | 0 | 10 | 102 | 376 | 630 | 2857 |
| Cooling Degree Days | 0 | 0 | 3 | 48 | 163 | 367 | 548 | 527 | 315 | 88 | 2 | 0 | 2061 |
| Total Precipitation (") | 1.27 | 1.38 | 1.08 | 0.32 | 0.11 | 0.05 | 0.16 | 0.17 | 0.25 | 0.26 | 0.78 | 0.91 | 6.74 |
| Days ≥ 0.1" Precip | 3 | 2 | 2 | 1 | 0 | 0 | 0 | 0 | 1 | 1 | 2 | 2 | 14 |
| Total Snowfall (") | 0.3 | 0.1 | 0.1 | 0.0 | 0.0 | 0.0 | 0.0 | 0.0 | 0.0 | 0.0 | 0.0 | 0.2 | 0.7 |
| Days ≥ 1" Snow Depth | 0 | 0 | 0 | 0 | 0 | 0 | 0 | 0 | 0 | 0 | 0 | 0 | 0 |

## MONTEREY *Monterey County* ELEVATION 259 ft LAT/LONG 36° 36 ' N / 121° 55 ' W

| | JAN | FEB | MAR | APR | MAY | JUN | JUL | AUG | SEP | OCT | NOV | DEC | YEAR |
|---|---|---|---|---|---|---|---|---|---|---|---|---|---|
| Maximum Temp °F | 59.9 | 61.5 | 61.9 | 63.7 | 64.4 | 66.8 | 68.1 | 69.5 | 71.8 | 70.4 | 64.8 | 59.7 | 65.2 |
| Minimum Temp °F | 43.5 | 44.8 | 45.5 | 46.0 | 48.0 | 50.4 | 52.2 | 53.2 | 53.3 | 51.4 | 47.3 | 43.4 | 48.2 |
| Mean Temp °F | 51.7 | 53.2 | 53.7 | 54.8 | 56.2 | 58.6 | 60.2 | 61.4 | 62.6 | 60.9 | 56.1 | 51.6 | 56.8 |
| Days Max Temp ≥ 90 °F | 0 | 0 | 0 | 0 | 0 | 0 | 0 | 0 | 1 | 1 | 0 | 0 | 2 |
| Days Max Temp ≤ 32 °F | 0 | 0 | 0 | 0 | 0 | 0 | 0 | 0 | 0 | 0 | 0 | 0 | 0 |
| Days Min Temp ≤ 32 °F | 0 | 0 | 0 | 0 | 0 | 0 | 0 | 0 | 0 | 0 | 0 | 1 | 1 |
| Days Min Temp ≤ 0 °F | 0 | 0 | 0 | 0 | 0 | 0 | 0 | 0 | 0 | 0 | 0 | 0 | 0 |
| Heating Degree Days | 404 | 328 | 343 | 303 | 269 | 193 | 152 | 115 | 93 | 146 | 266 | 409 | 3021 |
| Cooling Degree Days | 0 | 0 | 0 | 7 | 2 | 8 | 12 | 10 | 20 | 27 | 3 | 0 | 89 |
| Total Precipitation (") | 3.64 | 3.04 | 3.35 | 1.60 | 0.35 | 0.21 | 0.09 | 0.11 | 0.27 | 0.90 | 2.67 | 2.97 | 19.20 |
| Days ≥ 0.1" Precip | 7 | 7 | 7 | 4 | 1 | 1 | 0 | 0 | 1 | 2 | 5 | 6 | 41 |
| Total Snowfall (") | 0.0 | 0.0 | 0.0 | 0.0 | 0.0 | 0.0 | 0.0 | 0.0 | 0.0 | 0.0 | 0.0 | 0.0 | 0.0 |
| Days ≥ 1" Snow Depth | 0 | 0 | 0 | 0 | 0 | 0 | 0 | 0 | 0 | 0 | 0 | 0 | 0 |

## MORRO BAY FIRE DEPT *San Luis Obispo County* ELEVATION 121 ft LAT/LONG 35° 22 ' N / 120° 51 ' W

| | JAN | FEB | MAR | APR | MAY | JUN | JUL | AUG | SEP | OCT | NOV | DEC | YEAR |
|---|---|---|---|---|---|---|---|---|---|---|---|---|---|
| Maximum Temp °F | 62.3 | 63.0 | 63.1 | 63.9 | 63.1 | 64.6 | 65.6 | 66.6 | 68.6 | 69.6 | 66.9 | 62.5 | 65.0 |
| Minimum Temp °F | 41.9 | 43.2 | 44.0 | 44.7 | 47.1 | 50.0 | 51.8 | 52.8 | 52.1 | 50.0 | 46.1 | 41.4 | 47.1 |
| Mean Temp °F | 52.2 | 53.1 | 53.6 | 54.4 | 55.1 | 57.3 | 58.7 | 59.7 | 60.4 | 59.8 | 56.5 | 52.0 | 56.1 |
| Days Max Temp ≥ 90 °F | 0 | 0 | 0 | 0 | 0 | 0 | 0 | 0 | 1 | 1 | 0 | 0 | 2 |
| Days Max Temp ≤ 32 °F | 0 | 0 | 0 | 0 | 0 | 0 | 0 | 0 | 0 | 0 | 0 | 0 | 0 |
| Days Min Temp ≤ 32 °F | 2 | 1 | 0 | 0 | 0 | 0 | 0 | 0 | 0 | 0 | 0 | 2 | 5 |
| Days Min Temp ≤ 0 °F | 0 | 0 | 0 | 0 | 0 | 0 | 0 | 0 | 0 | 0 | 0 | 0 | 0 |
| Heating Degree Days | 392 | 330 | 348 | 316 | 302 | 231 | 189 | 161 | 151 | 172 | 253 | 397 | 3242 |
| Cooling Degree Days | 0 | 0 | 1 | 5 | 1 | 3 | 3 | 5 | 18 | 13 | 4 | 0 | 53 |
| Total Precipitation (") | 3.11 | 2.96 | 3.07 | 1.25 | 0.20 | 0.04 | 0.05 | 0.09 | 0.41 | 0.69 | 1.76 | 2.59 | 16.22 |
| Days ≥ 0.1" Precip | 6 | 6 | 6 | 3 | 1 | 0 | 0 | 0 | 1 | 2 | 4 | 5 | 34 |
| Total Snowfall (") | 0.0 | 0.0 | 0.0 | 0.0 | 0.0 | 0.0 | 0.0 | 0.0 | 0.0 | 0.0 | 0.0 | 0.0 | 0.0 |
| Days ≥ 1" Snow Depth | 0 | 0 | 0 | 0 | 0 | 0 | 0 | 0 | 0 | 0 | 0 | 0 | 0 |

## MOUNT HAMILTON *Santa Clara County* ELEVATION 4213 ft LAT/LONG 37° 20 ' N / 121° 39 ' W

| | JAN | FEB | MAR | APR | MAY | JUN | JUL | AUG | SEP | OCT | NOV | DEC | YEAR |
|---|---|---|---|---|---|---|---|---|---|---|---|---|---|
| Maximum Temp °F | 49.6 | 50.0 | 50.4 | 55.7 | 64.5 | 72.5 | 78.9 | 78.3 | 73.9 | 66.0 | 54.4 | 49.1 | 61.9 |
| Minimum Temp °F | 37.3 | 37.1 | 36.9 | 39.9 | 47.0 | 55.0 | 63.2 | 62.8 | 57.6 | 50.8 | 41.4 | 36.7 | 47.1 |
| Mean Temp °F | 43.5 | 43.5 | 43.7 | 47.8 | 55.8 | 63.8 | 71.0 | 70.6 | 65.7 | 58.4 | 47.9 | 42.9 | 54.6 |
| Days Max Temp ≥ 90 °F | 0 | 0 | 0 | 0 | 0 | 0 | 2 | 2 | 1 | 0 | 0 | 0 | 5 |
| Days Max Temp ≤ 32 °F | 1 | 0 | 0 | 0 | 0 | 0 | 0 | 0 | 0 | 0 | 0 | 1 | 2 |
| Days Min Temp ≤ 32 °F | 10 | 9 | 9 | 8 | 2 | 0 | 0 | 0 | 0 | 1 | 5 | ***10*** | 54 |
| Days Min Temp ≤ 0 °F | 0 | 0 | 0 | 0 | 0 | 0 | 0 | 0 | 0 | 0 | 0 | 0 | 0 |
| Heating Degree Days | 660 | 600 | 653 | 512 | 308 | 131 | 31 | 37 | 96 | 242 | 507 | 676 | 4453 |
| Cooling Degree Days | 0 | 0 | 0 | 5 | 25 | 94 | 215 | 205 | 128 | 45 | 1 | 0 | 718 |
| Total Precipitation (") | 3.55 | 3.30 | 3.40 | 1.80 | 0.58 | 0.17 | 0.05 | 0.08 | 0.42 | 1.32 | 3.27 | 3.09 | 21.03 |
| Days ≥ 0.1" Precip | 7 | 7 | 8 | 4 | 2 | 1 | 0 | 0 | 1 | 2 | 6 | ***6*** | 44 |
| Total Snowfall (") | ***4.6*** | 2.9 | 4.2 | 1.7 | 0.0 | 0.0 | 0.0 | 0.0 | 0.0 | 0.0 | ***0.2*** | ***1.5*** | 15.1 |
| Days ≥ 1" Snow Depth | 2 | 1 | 2 | 1 | 0 | 0 | 0 | 0 | 0 | 0 | ***0*** | 2 | 8 |

## MOUNT SHASTA *Siskiyou County* ELEVATION 3590 ft LAT/LONG 41° 19 ' N / 122° 19 ' W

| | JAN | FEB | MAR | APR | MAY | JUN | JUL | AUG | SEP | OCT | NOV | DEC | YEAR |
|---|---|---|---|---|---|---|---|---|---|---|---|---|---|
| Maximum Temp °F | 43.3 | 47.7 | 52.0 | 58.4 | 68.3 | 76.1 | 84.8 | 83.5 | 76.6 | 65.3 | 50.1 | 43.0 | 62.4 |
| Minimum Temp °F | 25.4 | 28.1 | 30.2 | 33.3 | 39.6 | 46.1 | 50.2 | 48.7 | 43.5 | 36.9 | 30.4 | 25.7 | 36.5 |
| Mean Temp °F | 34.4 | 38.0 | 41.1 | 45.9 | 54.0 | 61.1 | 67.5 | 66.1 | 60.1 | 51.1 | 40.3 | 34.4 | 49.5 |
| Days Max Temp ≥ 90 °F | 0 | 0 | 0 | 0 | 0 | 2 | 9 | 7 | 2 | 0 | 0 | 0 | 20 |
| Days Max Temp ≤ 32 °F | 2 | 0 | 0 | 0 | 0 | 0 | 0 | 0 | 0 | 0 | 0 | 2 | 4 |
| Days Min Temp ≤ 32 °F | 26 | 21 | 21 | 15 | 4 | 0 | 0 | 0 | 1 | 7 | 19 | 26 | 140 |
| Days Min Temp ≤ 0 °F | 0 | 0 | 0 | 0 | 0 | 0 | 0 | 0 | 0 | 0 | 0 | 0 | 0 |
| Heating Degree Days | 942 | 757 | 733 | 566 | 341 | 152 | 37 | 54 | 168 | 424 | 736 | 942 | 5852 |
| Cooling Degree Days | ***0*** | 0 | 0 | 0 | 6 | 37 | 111 | 79 | 26 | ***2*** | ***0*** | ***0*** | 261 |
| Total Precipitation (") | 6.54 | 4.92 | 5.15 | 2.60 | 1.53 | 0.85 | 0.29 | 0.47 | 0.81 | 2.11 | 5.24 | 5.81 | 36.32 |
| Days ≥ 0.1" Precip | 8 | 7 | 9 | 5 | 4 | 3 | 1 | 1 | 2 | 4 | 7 | 8 | 59 |
| Total Snowfall (") | 25.2 | 16.4 | 13.1 | 6.1 | 0.6 | 0.0 | 0.0 | 0.0 | 0.0 | 0.4 | 10.4 | 26.6 | 98.8 |
| Days ≥ 1" Snow Depth | 15 | 10 | 5 | 2 | 0 | 0 | 0 | 0 | 0 | 0 | 4 | 12 | 48 |

## MOUNTAIN PASS *San Bernardino County* ELEVATION 4744 ft LAT/LONG 35° 28 ' N / 115° 32 ' W

| | JAN | FEB | MAR | APR | MAY | JUN | JUL | AUG | SEP | OCT | NOV | DEC | YEAR |
|---|---|---|---|---|---|---|---|---|---|---|---|---|---|
| Maximum Temp °F | ***50.4*** | ***53.5*** | 59.1 | 66.5 | ***76.1*** | na | na | na | na | na | na | ***50.7*** | na |
| Minimum Temp °F | ***29.3*** | ***31.8*** | 35.8 | 41.0 | ***49.0*** | na | na | na | na | na | na | ***29.0*** | na |
| Mean Temp °F | ***39.9*** | ***42.8*** | 47.5 | 53.8 | ***62.5*** | na | na | na | na | na | na | ***39.9*** | na |
| Days Max Temp ≥ 90 °F | 0 | 0 | 0 | 0 | 2 | ***12*** | ***20*** | ***18*** | ***8*** | 1 | 0 | 0 | 61 |
| Days Max Temp ≤ 32 °F | 1 | 0 | 0 | 0 | 0 | 0 | 0 | 0 | 0 | 0 | 0 | 1 | 2 |
| Days Min Temp ≤ 32 °F | 19 | 14 | 10 | 4 | 1 | 0 | 0 | 0 | 0 | 1 | 8 | 19 | 76 |
| Days Min Temp ≤ 0 °F | 0 | 0 | 0 | 0 | 0 | 0 | 0 | 0 | 0 | 0 | 0 | 0 | 0 |
| Heating Degree Days | ***772*** | ***621*** | 537 | 340 | ***148*** | na | na | na | na | na | na | ***771*** | na |
| Cooling Degree Days | na | na | 0 | 18 | na | na | na | na | na | na | na | na | na |
| Total Precipitation (") | 1.01 | 0.92 | 1.07 | 0.44 | 0.28 | 0.27 | 1.04 | 1.37 | 0.50 | 0.40 | 0.71 | 0.82 | 8.83 |
| Days ≥ 0.1" Precip | 2 | 2 | 3 | 1 | 1 | 0 | 2 | 2 | ***1*** | 1 | 1 | 2 | 18 |
| Total Snowfall (") | 2.4 | 2.7 | ***1.9*** | 0.6 | 0.0 | 0.0 | 0.0 | 0.0 | 0.0 | 0.1 | ***0.4*** | 1.7 | 9.8 |
| Days ≥ 1" Snow Depth | 2 | ***1*** | ***1*** | 0 | 0 | 0 | 0 | 0 | 0 | 0 | 1 | 1 | 6 |

## MT DIABLO JUNCTION *Contra Costa County* ELEVATION 2170 ft LAT/LONG 37° 52 ' N / 121° 56 ' W

| | JAN | FEB | MAR | APR | MAY | JUN | JUL | AUG | SEP | OCT | NOV | DEC | YEAR |
|---|---|---|---|---|---|---|---|---|---|---|---|---|---|
| Maximum Temp °F | 55.6 | 57.6 | 59.3 | 64.6 | 71.9 | 79.2 | 86.4 | 86.3 | 82.7 | 74.6 | 61.9 | 55.6 | 69.6 |
| Minimum Temp °F | 39.4 | 40.7 | 40.9 | 43.1 | 47.6 | 53.1 | 60.1 | 60.0 | 57.3 | 52.0 | 44.2 | 39.6 | 48.2 |
| Mean Temp °F | 47.5 | 49.2 | 50.1 | 53.9 | 59.8 | 66.1 | 73.3 | 73.2 | 70.0 | 63.4 | 53.1 | 47.6 | 58.9 |
| Days Max Temp ≥ 90 °F | 0 | 0 | 0 | 0 | 1 | 6 | 12 | 12 | 7 | 3 | 0 | 0 | 41 |
| Days Max Temp ≤ 32 °F | 0 | 0 | 0 | 0 | 0 | 0 | 0 | 0 | 0 | 0 | 0 | 0 | 0 |
| Days Min Temp ≤ 32 °F | 4 | 3 | 2 | 1 | 0 | 0 | 0 | 0 | 0 | 0 | 1 | 4 | 15 |
| Days Min Temp ≤ 0 °F | 0 | 0 | 0 | 0 | 0 | 0 | 0 | 0 | 0 | 0 | 0 | 0 | 0 |
| Heating Degree Days | 535 | 441 | 456 | 336 | 211 | 93 | 21 | 20 | 39 | 131 | 357 | 533 | 3173 |
| Cooling Degree Days | 0 | 0 | 0 | 10 | 45 | 114 | 247 | 240 | 179 | 76 | 3 | 0 | 914 |
| Total Precipitation (") | 4.53 | 3.78 | 3.71 | 1.52 | 0.47 | 0.13 | 0.05 | 0.07 | 0.35 | 1.22 | 3.53 | 3.60 | 22.96 |
| Days ≥ 0.1" Precip | 7 | 7 | 7 | 4 | 1 | 0 | 0 | 0 | 1 | 2 | 6 | 6 | 41 |
| Total Snowfall (") | 0.7 | 0.0 | 0.4 | 0.7 | 0.0 | 0.0 | 0.0 | 0.0 | 0.0 | 0.0 | 0.1 | 0.1 | 2.0 |
| Days ≥ 1" Snow Depth | 0 | 0 | 0 | 0 | 0 | 0 | 0 | 0 | 0 | 0 | 0 | 0 | 0 |

## MT WILSON NO 2 *Los Angeles County* ELEVATION 5709 ft LAT/LONG 34° 14 ' N / 118° 4 ' W

| | JAN | FEB | MAR | APR | MAY | JUN | JUL | AUG | SEP | OCT | NOV | DEC | YEAR |
|---|---|---|---|---|---|---|---|---|---|---|---|---|---|
| Maximum Temp °F | 52.7 | 53.1 | 54.1 | 59.2 | 67.0 | 76.0 | 81.3 | 80.7 | 76.4 | 68.4 | 58.5 | 52.7 | 65.0 |
| Minimum Temp °F | 37.1 | 37.0 | 37.1 | 40.6 | 47.8 | 56.6 | 62.7 | 62.0 | 57.8 | 51.0 | 42.3 | 37.1 | 47.4 |
| Mean Temp °F | 44.9 | 45.1 | 45.6 | 50.1 | 57.4 | 66.3 | 72.0 | 71.4 | 67.2 | 59.6 | 50.4 | 44.9 | 56.2 |
| Days Max Temp ≥ 90 °F | 0 | 0 | 0 | 0 | 0 | 1 | 2 | 2 | 1 | 0 | 0 | 0 | 6 |
| Days Max Temp ≤ 32 °F | 1 | 0 | 0 | 0 | 0 | 0 | 0 | 0 | 0 | 0 | 0 | 1 | 2 |
| Days Min Temp ≤ 32 °F | 9 | 9 | 11 | 7 | 2 | 0 | 0 | 0 | 0 | 1 | 4 | 10 | 53 |
| Days Min Temp ≤ 0 °F | 0 | 0 | 0 | 0 | 0 | 0 | 0 | 0 | 0 | 0 | 0 | 0 | 0 |
| Heating Degree Days | 615 | 557 | 594 | 447 | 261 | 84 | 10 | 16 | 62 | 208 | 436 | 616 | 3906 |
| Cooling Degree Days | 0 | 0 | 0 | 10 | 31 | 136 | 244 | 229 | 150 | 49 | 3 | 0 | 852 |
| Total Precipitation (") | 8.02 | 8.51 | 7.97 | 2.68 | 0.63 | 0.13 | 0.11 | 0.30 | 1.04 | 1.46 | 4.96 | 5.57 | 41.38 |
| Days ≥ 0.1" Precip | 5 | 5 | 6 | 3 | 1 | 0 | 0 | 1 | 1 | 2 | 4 | 5 | 33 |
| Total Snowfall (") | na | na | na | na | 0.3 | 0.0 | 0.0 | 0.0 | 0.0 | 0.0 | ***0.8*** | na | na |
| Days ≥ 1" Snow Depth | na | 4 | ***6*** | ***3*** | 0 | 0 | 0 | 0 | 0 | 0 | 1 | ***4*** | na |

## NAPA STATE HOSPITAL *Napa County* ELEVATION 60 ft LAT/LONG 38° 17 ' N / 122° 16 ' W

| | JAN | FEB | MAR | APR | MAY | JUN | JUL | AUG | SEP | OCT | NOV | DEC | YEAR |
|---|---|---|---|---|---|---|---|---|---|---|---|---|---|
| Maximum Temp °F | 57.4 | 62.7 | 65.9 | 70.7 | 76.1 | 80.3 | 82.5 | 82.5 | 82.4 | 77.2 | 65.2 | 57.2 | 71.7 |
| Minimum Temp °F | 38.3 | 41.3 | 42.8 | 44.1 | 48.3 | 52.3 | 54.1 | 54.1 | 52.7 | 48.8 | 42.8 | 38.3 | 46.5 |
| Mean Temp °F | 47.9 | 52.0 | 54.4 | 57.4 | 62.2 | 66.3 | 68.3 | 68.3 | 67.6 | 63.1 | 54.0 | 47.7 | 59.1 |
| Days Max Temp ≥ 90 °F | 0 | 0 | 0 | 0 | 3 | 5 | 5 | 5 | 6 | 2 | 0 | 0 | 26 |
| Days Max Temp ≤ 32 °F | 0 | 0 | 0 | 0 | 0 | 0 | 0 | 0 | 0 | 0 | 0 | 0 | 0 |
| Days Min Temp ≤ 32 °F | 7 | 2 | 1 | 0 | 0 | 0 | 0 | 0 | 0 | 0 | 1 | 7 | 18 |
| Days Min Temp ≤ 0 °F | 0 | 0 | 0 | 0 | 0 | 0 | 0 | 0 | 0 | 0 | 0 | 0 | 0 |
| Heating Degree Days | 524 | 361 | 323 | 229 | 120 | 42 | 10 | 9 | 21 | 95 | 323 | 528 | 2585 |
| Cooling Degree Days | 0 | 1 | 0 | 12 | 39 | 97 | 128 | 121 | 99 | 44 | 1 | 0 | 542 |
| Total Precipitation (") | 5.40 | 4.11 | 3.76 | 1.44 | 0.40 | 0.20 | 0.05 | 0.13 | 0.40 | 1.44 | 3.86 | 4.01 | 25.20 |
| Days ≥ 0.1" Precip | 7 | 7 | 7 | 4 | 1 | 0 | 0 | 0 | 1 | 2 | 6 | 7 | 42 |
| Total Snowfall (") | 0.0 | 0.0 | 0.0 | 0.0 | 0.0 | 0.0 | 0.0 | 0.0 | 0.0 | 0.0 | 0.0 | 0.0 | 0.0 |
| Days ≥ 1" Snow Depth | 0 | 0 | 0 | 0 | 0 | 0 | 0 | 0 | 0 | 0 | 0 | 0 | 0 |

## NEEDLES AIRPORT *San Bernardino County* ELEVATION 919 ft LAT/LONG 34° 46 ' N / 114° 37 ' W

| | JAN | FEB | MAR | APR | MAY | JUN | JUL | AUG | SEP | OCT | NOV | DEC | YEAR |
|---|---|---|---|---|---|---|---|---|---|---|---|---|---|
| Maximum Temp °F | 63.9 | 70.1 | 76.3 | 84.3 | 93.7 | 103.9 | 108.4 | 106.3 | 99.9 | 87.7 | 73.2 | 63.5 | 85.9 |
| Minimum Temp °F | 41.8 | 45.7 | 50.4 | 56.9 | 66.5 | 75.8 | 83.2 | 81.2 | 73.5 | 61.0 | 49.3 | 41.8 | 60.6 |
| Mean Temp °F | 52.9 | 58.0 | 63.4 | 70.6 | 80.1 | 89.9 | 95.9 | 93.8 | 86.7 | 74.4 | 61.3 | 52.7 | 73.3 |
| Days Max Temp ≥ 90 °F | 0 | 0 | 2 | 9 | 22 | 29 | 31 | 30 | 27 | 14 | 0 | 0 | 164 |
| Days Max Temp ≤ 32 °F | 0 | 0 | 0 | 0 | 0 | 0 | 0 | 0 | 0 | 0 | 0 | 0 | 0 |
| Days Min Temp ≤ 32 °F | 2 | 0 | 0 | 0 | 0 | 0 | 0 | 0 | 0 | 0 | 0 | 2 | 4 |
| Days Min Temp ≤ 0 °F | 0 | 0 | 0 | 0 | 0 | 0 | 0 | 0 | 0 | 0 | 0 | 0 | 0 |
| Heating Degree Days | 369 | 203 | 109 | 29 | 1 | 0 | 0 | 0 | 0 | 12 | 140 | 375 | 1238 |
| Cooling Degree Days | 0 | 13 | 70 | 248 | 485 | 762 | 956 | 906 | 670 | 321 | 42 | 1 | 4474 |
| Total Precipitation (") | 0.67 | 0.56 | 0.70 | 0.30 | 0.11 | 0.03 | 0.36 | 0.75 | 0.51 | 0.30 | 0.42 | 0.59 | 5.30 |
| Days ≥ 0.1" Precip | 2 | 1 | 2 | 1 | 0 | 0 | 1 | 1 | 1 | 1 | 1 | 1 | 12 |
| Total Snowfall (") | 0.0 | 0.0 | 0.0 | 0.0 | 0.0 | 0.0 | 0.0 | 0.0 | 0.0 | 0.0 | 0.0 | 0.0 | 0.0 |
| Days ≥ 1" Snow Depth | 0 | 0 | 0 | 0 | 0 | 0 | 0 | 0 | 0 | 0 | 0 | 0 | 0 |

## NEVADA CITY *Nevada County* ELEVATION 2602 ft LAT/LONG 39° 16' N / 121° 2' W

| | JAN | FEB | MAR | APR | MAY | JUN | JUL | AUG | SEP | OCT | NOV | DEC | YEAR |
|---|---|---|---|---|---|---|---|---|---|---|---|---|---|
| Maximum Temp °F | 50.5 | 53.2 | 56.1 | 62.3 | 71.6 | 79.9 | 87.3 | 86.3 | 80.0 | 69.9 | 56.3 | 49.6 | 66.9 |
| Minimum Temp °F | 30.9 | 32.3 | 34.5 | 37.6 | 43.5 | 50.0 | 54.5 | 53.3 | 48.7 | 42.4 | 35.4 | 31.0 | 41.2 |
| Mean Temp °F | 40.7 | 42.8 | 45.3 | 50.0 | 57.6 | 65.0 | 71.0 | 69.8 | 64.4 | 56.2 | 45.8 | 40.3 | 54.1 |
| Days Max Temp ≥ 90 °F | 0 | 0 | 0 | 0 | 1 | 5 | 13 | 12 | 4 | 1 | 0 | 0 | 36 |
| Days Max Temp ≤ 32 °F | 0 | 0 | 0 | 0 | 0 | 0 | 0 | 0 | 0 | 0 | 0 | 0 | 0 |
| Days Min Temp ≤ 32 °F | 20 | 16 | 13 | 8 | 1 | 0 | 0 | 0 | 0 | 3 | 11 | 19 | 91 |
| Days Min Temp ≤ 0 °F | 0 | 0 | 0 | 0 | 0 | 0 | 0 | 0 | 0 | 0 | 0 | 0 | 0 |
| Heating Degree Days | 745 | 619 | 603 | 446 | 243 | 84 | 13 | 21 | 90 | 287 | 568 | 758 | 4477 |
| Cooling Degree Days | 0 | 0 | 0 | 3 | 28 | 103 | 229 | 214 | 114 | 37 | 0 | 0 | 728 |
| Total Precipitation (") | 10.40 | 9.05 | 8.75 | 4.01 | 1.48 | 0.62 | 0.18 | 0.31 | 1.10 | 2.78 | 8.33 | 8.86 | 55.87 |
| Days ≥ 0.1" Precip | 9 | 9 | 10 | 6 | 3 | 2 | 0 | 1 | 2 | 4 | 8 | 9 | 63 |
| Total Snowfall (") | 4.7 | 3.3 | 3.6 | 0.4 | 0.0 | 0.0 | 0.0 | 0.0 | 0.0 | 0.0 | 0.9 | 4.3 | 17.2 |
| Days ≥ 1" Snow Depth | *4* | 3 | 1 | 0 | 0 | 0 | 0 | 0 | 0 | 0 | 0 | 3 | 11 |

## NEWARK *Alameda County* ELEVATION 10 ft LAT/LONG 37° 31' N / 122° 2' W

| | JAN | FEB | MAR | APR | MAY | JUN | JUL | AUG | SEP | OCT | NOV | DEC | YEAR |
|---|---|---|---|---|---|---|---|---|---|---|---|---|---|
| Maximum Temp °F | 57.0 | 60.7 | 63.4 | 66.4 | 70.1 | 73.8 | 76.2 | 76.6 | 76.6 | 72.6 | 64.1 | 57.0 | 67.9 |
| Minimum Temp °F | 41.3 | 44.6 | 47.2 | 49.3 | 52.7 | 55.9 | 57.3 | 58.0 | 57.2 | 53.6 | 47.1 | 41.3 | 50.5 |
| Mean Temp °F | 49.2 | 52.7 | 55.3 | 57.9 | 61.4 | 64.9 | 66.8 | 67.3 | 66.9 | 63.2 | 55.6 | 49.2 | 59.2 |
| Days Max Temp ≥ 90 °F | 0 | 0 | 0 | 0 | 1 | 1 | 1 | 1 | 2 | 0 | 0 | 0 | 6 |
| Days Max Temp ≤ 32 °F | 0 | 0 | 0 | 0 | 0 | 0 | 0 | 0 | 0 | 0 | 0 | 0 | 0 |
| Days Min Temp ≤ 32 °F | 1 | 0 | 0 | 0 | 0 | 0 | 0 | 0 | 0 | 0 | 0 | 2 | 3 |
| Days Min Temp ≤ 0 °F | 0 | 0 | 0 | 0 | 0 | 0 | 0 | 0 | 0 | 0 | 0 | 0 | 0 |
| Heating Degree Days | 484 | 341 | 295 | 217 | 133 | 56 | 21 | 11 | 23 | 86 | 277 | 483 | 2427 |
| Cooling Degree Days | 0 | 0 | 1 | 17 | 30 | 67 | 95 | 96 | 85 | 42 | 1 | 0 | 434 |
| Total Precipitation (") | 2.70 | 2.30 | 2.23 | 1.00 | 0.26 | 0.09 | 0.04 | 0.08 | 0.19 | 0.81 | 2.06 | 2.12 | 13.88 |
| Days ≥ 0.1" Precip | 5 | *5* | *5* | 3 | 1 | 0 | 0 | 0 | 1 | 2 | *4* | 5 | 31 |
| Total Snowfall (") | 0.0 | 0.0 | 0.0 | 0.0 | 0.0 | 0.0 | 0.0 | 0.0 | 0.0 | 0.0 | 0.0 | 0.0 | 0.0 |
| Days ≥ 1" Snow Depth | 0 | 0 | 0 | 0 | 0 | 0 | 0 | 0 | 0 | 0 | 0 | 0 | 0 |

## NEWMAN *Stanislaus County* ELEVATION 90 ft LAT/LONG 37° 18' N / 121° 2' W

| | JAN | FEB | MAR | APR | MAY | JUN | JUL | AUG | SEP | OCT | NOV | DEC | YEAR |
|---|---|---|---|---|---|---|---|---|---|---|---|---|---|
| Maximum Temp °F | 55.0 | 62.8 | 68.3 | 75.4 | *84.7* | *91.5* | 96.3 | 94.2 | 89.6 | 80.8 | 65.6 | 55.1 | 76.6 |
| Minimum Temp °F | 36.0 | 39.2 | 42.0 | 44.4 | *50.3* | *55.4* | 58.9 | 58.1 | 54.9 | 48.6 | 40.5 | 35.2 | 47.0 |
| Mean Temp °F | 45.5 | 50.9 | 55.2 | 59.9 | *67.5* | *73.5* | 77.6 | 76.2 | 72.3 | 64.7 | 53.1 | 45.2 | 61.8 |
| Days Max Temp ≥ 90 °F | 0 | 0 | 0 | 2 | *10* | *18* | 27 | 25 | 15 | 4 | 0 | 0 | 101 |
| Days Max Temp ≤ 32 °F | 0 | 0 | 0 | 0 | *0* | 0 | 0 | 0 | 0 | 0 | 0 | 0 | 0 |
| Days Min Temp ≤ 32 °F | 10 | 4 | 1 | 0 | *0* | 0 | 0 | 0 | 0 | 0 | 3 | 11 | 29 |
| Days Min Temp ≤ 0 °F | 0 | 0 | 0 | 0 | *0* | 0 | 0 | 0 | 0 | 0 | 0 | 0 | 0 |
| Heating Degree Days | 596 | 390 | 299 | 171 | *47* | *6* | 0 | 0 | 5 | 74 | 352 | 607 | 2547 |
| Cooling Degree Days | 0 | 0 | 1 | 36 | na | *287* | *414* | 373 | 246 | 88 | 0 | 0 | na |
| Total Precipitation (") | 2.17 | 1.86 | 1.76 | 0.72 | 0.19 | 0.05 | 0.03 | 0.03 | 0.30 | 0.54 | 1.62 | 1.55 | 10.82 |
| Days ≥ 0.1" Precip | 5 | 5 | 5 | 2 | *1* | 0 | 0 | 0 | 1 | 2 | 4 | 4 | 29 |
| Total Snowfall (") | 0.0 | 0.0 | 0.0 | 0.0 | 0.0 | 0.0 | 0.0 | 0.0 | 0.0 | 0.0 | 0.0 | 0.0 | 0.0 |
| Days ≥ 1" Snow Depth | 0 | 0 | 0 | 0 | 0 | 0 | 0 | 0 | 0 | 0 | 0 | 0 | 0 |

## NEWPORT BEACH HARBOR *Orange County* ELEVATION 10 ft LAT/LONG 33° 36' N / 117° 53' W

| | JAN | FEB | MAR | APR | MAY | JUN | JUL | AUG | SEP | OCT | NOV | DEC | YEAR |
|---|---|---|---|---|---|---|---|---|---|---|---|---|---|
| Maximum Temp °F | 63.3 | 63.6 | 63.4 | 65.2 | 66.2 | 68.6 | 71.5 | 73.1 | 73.0 | 71.5 | 67.5 | 63.7 | 67.6 |
| Minimum Temp °F | 47.3 | 48.8 | 50.5 | 52.8 | 56.5 | 59.6 | 62.7 | 64.2 | 62.6 | 58.4 | 52.1 | 47.4 | 55.2 |
| Mean Temp °F | 55.4 | 56.2 | 57.0 | 59.0 | 61.4 | 64.1 | 67.1 | 68.6 | 67.8 | 64.9 | 59.8 | 55.6 | 61.4 |
| Days Max Temp ≥ 90 °F | 0 | 0 | 0 | 0 | 0 | 0 | 0 | 0 | 1 | 0 | 0 | 0 | 1 |
| Days Max Temp ≤ 32 °F | 0 | 0 | 0 | 0 | 0 | 0 | 0 | 0 | 0 | 0 | 0 | 0 | 0 |
| Days Min Temp ≤ 32 °F | 0 | 0 | 0 | 0 | 0 | 0 | 0 | 0 | 0 | 0 | 0 | 0 | 0 |
| Days Min Temp ≤ 0 °F | 0 | 0 | 0 | 0 | 0 | 0 | 0 | 0 | 0 | 0 | 0 | 0 | 0 |
| Heating Degree Days | 295 | 244 | 243 | 178 | 112 | 45 | 7 | 2 | 7 | 42 | 156 | 287 | 1618 |
| Cooling Degree Days | 3 | 4 | 1 | 7 | 7 | 29 | 92 | 122 | 103 | 55 | 9 | 1 | 433 |
| Total Precipitation (") | 2.37 | 2.29 | 2.15 | 0.88 | 0.16 | 0.06 | 0.02 | 0.08 | 0.30 | 0.20 | 1.37 | 1.68 | 11.56 |
| Days ≥ 0.1" Precip | 4 | 4 | 4 | 2 | 0 | 0 | 0 | 0 | 1 | 1 | 3 | 3 | 22 |
| Total Snowfall (") | 0.0 | 0.0 | 0.0 | 0.0 | 0.0 | 0.0 | 0.0 | 0.0 | 0.0 | 0.0 | 0.0 | 0.0 | 0.0 |
| Days ≥ 1" Snow Depth | 0 | 0 | 0 | 0 | 0 | 0 | 0 | 0 | 0 | 0 | 0 | 0 | 0 |

## OCEANSIDE HARBOR *San Diego County* ELEVATION 10 ft LAT/LONG 33° 13 ' N / 117° 24 ' W

| | JAN | FEB | MAR | APR | MAY | JUN | JUL | AUG | SEP | OCT | NOV | DEC | YEAR |
|---|---|---|---|---|---|---|---|---|---|---|---|---|---|
| Maximum Temp °F | 64.5 | 64.6 | 64.8 | 66.2 | 67.6 | 69.8 | 73.4 | 75.1 | 74.8 | 72.8 | 68.9 | 65.0 | 69.0 |
| Minimum Temp °F | 44.2 | 45.4 | 47.6 | 50.2 | 54.6 | 58.4 | 62.0 | 63.4 | 60.9 | 55.8 | 48.7 | 44.2 | 53.0 |
| Mean Temp °F | 54.4 | 55.1 | 56.2 | 58.2 | 61.1 | 64.1 | 67.8 | 69.3 | 67.9 | 64.3 | 58.8 | 54.7 | 61.0 |
| Days Max Temp ≥ 90 °F | 0 | 0 | 0 | 0 | 0 | 0 | 0 | 0 | 0 | 1 | 0 | 0 | 1 |
| Days Max Temp ≤ 32 °F | 0 | 0 | 0 | 0 | 0 | 0 | 0 | 0 | 0 | 0 | 0 | 0 | 0 |
| Days Min Temp ≤ 32 °F | 0 | 0 | 0 | 0 | 0 | 0 | 0 | 0 | 0 | 0 | 0 | 0 | 0 |
| Days Min Temp ≤ 0 °F | 0 | 0 | 0 | 0 | 0 | 0 | 0 | 0 | 0 | 0 | 0 | 0 | 0 |
| Heating Degree Days | 323 | 275 | 268 | 198 | 118 | 48 | 7 | 4 | 11 | 56 | 188 | 315 | 1811 |
| Cooling Degree Days | 0 | 1 | 0 | 3 | 4 | 22 | 91 | 117 | 82 | 33 | 4 | 0 | 357 |
| Total Precipitation (") | 2.29 | 2.08 | 2.13 | 0.91 | 0.20 | 0.09 | 0.03 | 0.13 | 0.26 | 0.35 | 1.24 | 1.67 | 11.38 |
| Days ≥ 0.1" Precip | 4 | 4 | 4 | 2 | 1 | 0 | 0 | 0 | 1 | 1 | 2 | 4 | 23 |
| Total Snowfall (") | 0.0 | 0.0 | 0.0 | 0.0 | 0.0 | 0.0 | 0.0 | 0.0 | 0.0 | 0.0 | 0.0 | 0.0 | 0.0 |
| Days ≥ 1" Snow Depth | 0 | 0 | 0 | 0 | 0 | 0 | 0 | 0 | 0 | 0 | 0 | 0 | 0 |

## OJAI *Ventura County* ELEVATION 751 ft LAT/LONG 34° 27 ' N / 119° 15 ' W

| | JAN | FEB | MAR | APR | MAY | JUN | JUL | AUG | SEP | OCT | NOV | DEC | YEAR |
|---|---|---|---|---|---|---|---|---|---|---|---|---|---|
| Maximum Temp °F | 67.0 | 68.5 | 69.6 | 74.0 | 76.9 | 82.6 | 88.8 | 89.7 | 86.6 | 81.4 | 73.1 | 67.2 | 77.1 |
| Minimum Temp °F | 36.8 | 39.0 | 41.2 | 43.6 | 48.1 | 51.7 | 55.3 | 55.9 | 53.5 | 47.9 | 41.1 | 36.3 | 45.9 |
| Mean Temp °F | 51.9 | 53.8 | 55.4 | 58.8 | 62.5 | 67.2 | 72.1 | 72.8 | 70.1 | 64.7 | 57.1 | 51.8 | 61.5 |
| Days Max Temp ≥ 90 °F | 0 | 0 | 0 | 2 | 3 | 7 | 15 | 15 | 12 | 6 | 1 | 0 | 61 |
| Days Max Temp ≤ 32 °F | 0 | 0 | 0 | 0 | 0 | 0 | 0 | 0 | 0 | 0 | 0 | 0 | 0 |
| Days Min Temp ≤ 32 °F | 9 | 3 | 2 | 0 | 0 | 0 | 0 | 0 | 0 | 0 | 2 | 8 | 24 |
| Days Min Temp ≤ 0 °F | 0 | 0 | 0 | 0 | 0 | 0 | 0 | 0 | 0 | 0 | 0 | 0 | 0 |
| Heating Degree Days | 398 | 311 | 293 | 193 | 111 | 37 | 1 | 1 | 13 | 73 | 237 | 404 | 2072 |
| Cooling Degree Days | 0 | 1 | 2 | 20 | 38 | 106 | 211 | 233 | 178 | 66 | 6 | 0 | 861 |
| Total Precipitation (") | 4.69 | 4.91 | 4.09 | 1.20 | 0.27 | 0.05 | 0.04 | 0.08 | 0.43 | 0.50 | 2.37 | 3.16 | 21.79 |
| Days ≥ 0.1" Precip | 4 | 4 | 5 | 2 | 1 | 0 | 0 | 0 | 1 | 1 | 3 | 3 | 24 |
| Total Snowfall (") | 0.0 | 0.0 | 0.0 | 0.0 | 0.0 | 0.0 | 0.0 | 0.0 | 0.0 | 0.0 | 0.0 | 0.0 | 0.0 |
| Days ≥ 1" Snow Depth | 0 | 0 | 0 | 0 | 0 | 0 | 0 | 0 | 0 | 0 | 0 | 0 | 0 |

## ORICK PRAIRIE CRK PK *Humboldt County* ELEVATION 161 ft LAT/LONG 41° 20 ' N / 124° 2 ' W

| | JAN | FEB | MAR | APR | MAY | JUN | JUL | AUG | SEP | OCT | NOV | DEC | YEAR |
|---|---|---|---|---|---|---|---|---|---|---|---|---|---|
| Maximum Temp °F | 53.0 | 55.7 | 57.9 | 59.6 | 63.1 | 65.9 | 68.7 | 69.8 | 71.1 | 66.2 | 57.5 | 51.8 | 61.7 |
| Minimum Temp °F | 36.9 | 38.3 | 38.6 | 39.1 | 42.0 | 46.0 | 48.2 | 48.7 | 45.7 | 42.5 | 39.8 | 36.8 | 41.9 |
| Mean Temp °F | 45.0 | 47.0 | 48.3 | 49.4 | 52.6 | 56.0 | 58.4 | 59.3 | 58.5 | 54.3 | 48.6 | 44.2 | 51.8 |
| Days Max Temp ≥ 90 °F | 0 | 0 | 0 | 0 | 0 | 0 | 0 | 0 | 0 | 0 | 0 | 0 | 0 |
| Days Max Temp ≤ 32 °F | 0 | 0 | 0 | 0 | 0 | 0 | 0 | 0 | 0 | 0 | 0 | 0 | 0 |
| Days Min Temp ≤ 32 °F | 9 | 6 | 6 | 4 | 2 | 0 | 0 | 0 | 0 | 1 | 5 | 10 | 43 |
| Days Min Temp ≤ 0 °F | 0 | 0 | 0 | 0 | 0 | 0 | 0 | 0 | 0 | 0 | 0 | 0 | 0 |
| Heating Degree Days | 614 | 502 | 511 | 462 | 379 | 266 | 197 | 172 | 193 | 325 | 484 | 636 | 4741 |
| Cooling Degree Days | 0 | 0 | 0 | 0 | 1 | 2 | 2 | *1* | 3 | 1 | *0* | 0 | 10 |
| Total Precipitation (") | 9.98 | 8.62 | 8.76 | 4.98 | 2.80 | 1.20 | 0.33 | 0.70 | 1.57 | 4.05 | 10.40 | 10.92 | 64.31 |
| Days ≥ 0.1" Precip | 12 | 12 | 12 | 9 | 5 | 3 | 1 | 2 | 3 | 6 | *11* | 13 | 89 |
| Total Snowfall (") | 0.3 | 0.0 | 0.0 | 0.0 | 0.0 | 0.0 | 0.0 | 0.0 | 0.0 | 0.0 | 0.0 | 0.2 | 0.5 |
| Days ≥ 1" Snow Depth | 0 | 0 | 0 | 0 | 0 | 0 | 0 | 0 | 0 | 0 | 0 | 0 | 0 |

## ORLAND *Glenn County* ELEVATION 254 ft LAT/LONG 39° 45 ' N / 122° 12 ' W

| | JAN | FEB | MAR | APR | MAY | JUN | JUL | AUG | SEP | OCT | NOV | DEC | YEAR |
|---|---|---|---|---|---|---|---|---|---|---|---|---|---|
| Maximum Temp °F | 54.7 | 60.7 | 65.3 | 72.3 | 82.0 | 89.3 | 94.9 | 93.5 | 89.2 | 79.2 | 63.8 | 54.3 | 74.9 |
| Minimum Temp °F | 35.6 | 39.2 | 42.5 | 45.8 | 53.0 | 59.0 | 62.0 | 60.0 | 56.3 | 49.1 | 41.3 | 35.6 | 48.3 |
| Mean Temp °F | 45.2 | 50.0 | 53.9 | 59.0 | 67.5 | 74.1 | 78.5 | 76.8 | 72.8 | 64.2 | 52.6 | 45.0 | 61.6 |
| Days Max Temp ≥ 90 °F | 0 | 0 | 0 | 1 | 7 | 15 | 24 | 22 | 16 | 5 | 0 | 0 | 90 |
| Days Max Temp ≤ 32 °F | 0 | 0 | 0 | 0 | 0 | 0 | 0 | 0 | 0 | 0 | 0 | 0 | 0 |
| Days Min Temp ≤ 32 °F | 11 | 4 | 1 | 0 | 0 | 0 | 0 | 0 | 0 | 0 | 2 | 10 | 28 |
| Days Min Temp ≤ 0 °F | 0 | 0 | 0 | 0 | 0 | 0 | 0 | 0 | 0 | 0 | 0 | 0 | 0 |
| Heating Degree Days | 608 | 417 | 339 | 195 | 55 | 9 | 0 | 1 | 9 | 92 | 369 | 614 | 2708 |
| Cooling Degree Days | 0 | 0 | 1 | 28 | 136 | 284 | 421 | 361 | 247 | 79 | 2 | 0 | 1559 |
| Total Precipitation (") | 4.04 | 3.27 | 3.15 | 1.10 | 0.63 | 0.39 | 0.07 | 0.22 | 0.38 | 1.06 | 3.02 | 3.29 | 20.62 |
| Days ≥ 0.1" Precip | 6 | 6 | 6 | 3 | 2 | 1 | 0 | 0 | 1 | 2 | 5 | 6 | 38 |
| Total Snowfall (") | 0.2 | 0.0 | 0.1 | 0.0 | 0.0 | 0.0 | 0.0 | 0.0 | 0.0 | 0.0 | 0.0 | 0.1 | 0.4 |
| Days ≥ 1" Snow Depth | 0 | 0 | 0 | 0 | 0 | 0 | 0 | 0 | 0 | 0 | 0 | 0 | 0 |

## ORLEANS *Humboldt County* ELEVATION 400 ft LAT/LONG 41° 18 ' N / 123° 32 ' W

| | JAN | FEB | MAR | APR | MAY | JUN | JUL | AUG | SEP | OCT | NOV | DEC | YEAR |
|---|---|---|---|---|---|---|---|---|---|---|---|---|---|
| Maximum Temp °F | 51.7 | 57.5 | 63.8 | 69.9 | 77.8 | 85.0 | 92.6 | 92.2 | 87.4 | 74.6 | 57.8 | 50.7 | 71.8 |
| Minimum Temp °F | 34.5 | 37.1 | 39.3 | 40.9 | 45.2 | 50.1 | 53.8 | 53.4 | 48.9 | 43.9 | 40.2 | 34.9 | 43.5 |
| Mean Temp °F | 43.1 | 47.3 | 51.6 | 55.4 | 61.5 | 67.6 | 73.2 | 72.8 | 68.1 | 59.3 | 49.0 | 42.8 | 57.6 |
| Days Max Temp ≥ 90 °F | 0 | 0 | 0 | 1 | 4 | 11 | 20 | 21 | 14 | 2 | 0 | 0 | 73 |
| Days Max Temp ≤ 32 °F | 0 | 0 | 0 | 0 | 0 | 0 | 0 | 0 | 0 | 0 | 0 | 0 | 0 |
| Days Min Temp ≤ 32 °F | 12 | 6 | 4 | 1 | 0 | 0 | 0 | 0 | 0 | 1 | 3 | 11 | 38 |
| Days Min Temp ≤ 0 °F | 0 | 0 | 0 | 0 | 0 | 0 | 0 | 0 | 0 | 0 | 0 | 0 | 0 |
| Heating Degree Days | 671 | 493 | 410 | 285 | 140 | 42 | 4 | 4 | 30 | 186 | 472 | 681 | 3418 |
| Cooling Degree Days | 0 | 0 | 0 | 8 | 42 | 119 | 254 | 245 | 124 | 20 | 0 | 0 | 812 |
| Total Precipitation (") | 8.71 | 6.94 | 6.62 | 3.07 | 1.90 | 0.68 | 0.19 | 0.61 | 1.17 | 3.63 | 8.27 | 9.14 | 50.93 |
| Days ≥ 0.1" Precip | 10 | 10 | 10 | 6 | 4 | 2 | 0 | 1 | 2 | 5 | 10 | 11 | 71 |
| Total Snowfall (") | 1.3 | 0.3 | 0.3 | 0.0 | 0.0 | 0.0 | 0.0 | 0.0 | 0.0 | 0.0 | 0.3 | 1.7 | 3.9 |
| Days ≥ 1" Snow Depth | 0 | 0 | 0 | 0 | 0 | 0 | 0 | 0 | 0 | 0 | 0 | 1 | 1 |

## OXNARD *Ventura County* ELEVATION 49 ft LAT/LONG 34° 11 ' N / 119° 10 ' W

| | JAN | FEB | MAR | APR | MAY | JUN | JUL | AUG | SEP | OCT | NOV | DEC | YEAR |
|---|---|---|---|---|---|---|---|---|---|---|---|---|---|
| Maximum Temp °F | 66.2 | 66.5 | 66.5 | 68.5 | 69.0 | 71.7 | 74.4 | 75.8 | 75.4 | 74.4 | 70.7 | 66.4 | 70.5 |
| Minimum Temp °F | 44.9 | 45.8 | 47.3 | 49.0 | 52.5 | 55.7 | 58.4 | 59.5 | 58.0 | 54.0 | 48.9 | 44.5 | 51.5 |
| Mean Temp °F | 55.6 | 56.2 | 56.9 | 58.8 | 60.8 | 63.7 | 66.4 | 67.7 | 66.7 | 64.2 | 59.8 | 55.5 | 61.0 |
| Days Max Temp ≥ 90 °F | 0 | 0 | 0 | 0 | 0 | 0 | 0 | 0 | 1 | 1 | 0 | 0 | 2 |
| Days Max Temp ≤ 32 °F | 0 | 0 | 0 | 0 | 0 | 0 | 0 | 0 | 0 | 0 | 0 | 0 | 0 |
| Days Min Temp ≤ 32 °F | 0 | 0 | 0 | 0 | 0 | 0 | 0 | 0 | 0 | 0 | 0 | 0 | 0 |
| Days Min Temp ≤ 0 °F | 0 | 0 | 0 | 0 | 0 | 0 | 0 | 0 | 0 | 0 | 0 | 0 | 0 |
| Heating Degree Days | 292 | 248 | 249 | 189 | 133 | 56 | 16 | 6 | 18 | 62 | 167 | 292 | 1728 |
| Cooling Degree Days | 8 | 8 | 6 | 13 | 5 | 27 | 77 | 96 | 70 | 46 | 16 | 6 | 378 |
| Total Precipitation (") | 3.16 | 3.24 | 2.84 | 0.86 | 0.11 | 0.03 | 0.02 | 0.08 | 0.37 | 0.34 | 1.87 | 2.10 | 15.02 |
| Days ≥ 0.1" Precip | 4 | 4 | 4 | 2 | 0 | 0 | 0 | 0 | 1 | 1 | 3 | 3 | 22 |
| Total Snowfall (") | 0.0 | 0.0 | 0.0 | 0.0 | 0.0 | 0.0 | 0.0 | 0.0 | 0.0 | 0.0 | 0.0 | 0.0 | 0.0 |
| Days ≥ 1" Snow Depth | 0 | 0 | 0 | 0 | 0 | 0 | 0 | 0 | 0 | 0 | 0 | 0 | 0 |

## PALM SPRINGS *Riverside County* ELEVATION 425 ft LAT/LONG 33° 50 ' N / 116° 30 ' W

| | JAN | FEB | MAR | APR | MAY | JUN | JUL | AUG | SEP | OCT | NOV | DEC | YEAR |
|---|---|---|---|---|---|---|---|---|---|---|---|---|---|
| Maximum Temp °F | 70.3 | 75.2 | 79.7 | 87.1 | 94.8 | 103.8 | 108.5 | 106.8 | 101.3 | 91.3 | 78.3 | 69.5 | 88.9 |
| Minimum Temp °F | 43.0 | 46.3 | 49.7 | 54.8 | 61.9 | 68.5 | 75.4 | 75.1 | 69.0 | 60.1 | 49.2 | 42.3 | 57.9 |
| Mean Temp °F | 56.7 | 60.8 | 64.8 | 71.0 | 78.4 | 86.2 | 92.0 | 91.0 | 85.2 | 75.7 | 63.8 | 55.9 | 73.5 |
| Days Max Temp ≥ 90 °F | 0 | 1 | 4 | 13 | 23 | 29 | 31 | 31 | 28 | 19 | 2 | 0 | 181 |
| Days Max Temp ≤ 32 °F | 0 | 0 | 0 | 0 | 0 | 0 | 0 | 0 | 0 | 0 | 0 | 0 | 0 |
| Days Min Temp ≤ 32 °F | 2 | 0 | 0 | 0 | 0 | 0 | 0 | 0 | 0 | 0 | 0 | 2 | 4 |
| Days Min Temp ≤ 0 °F | 0 | 0 | 0 | 0 | 0 | 0 | 0 | 0 | 0 | 0 | 0 | 0 | 0 |
| Heating Degree Days | 260 | 142 | 84 | 27 | 1 | 0 | 0 | 0 | 0 | 6 | 93 | 279 | 892 |
| Cooling Degree Days | 8 | 34 | 87 | 268 | 450 | 669 | 840 | 825 | 625 | 367 | 67 | 6 | 4246 |
| Total Precipitation (") | 1.22 | 1.04 | 0.65 | 0.12 | 0.09 | 0.05 | 0.20 | 0.42 | 0.35 | 0.12 | 0.60 | 0.78 | 5.64 |
| Days ≥ 0.1" Precip | 2 | 2 | 2 | 0 | 0 | 0 | 0 | 1 | 1 | 0 | 1 | 2 | 11 |
| Total Snowfall (") | 0.1 | 0.0 | 0.0 | 0.0 | 0.0 | 0.0 | 0.0 | 0.0 | 0.0 | 0.0 | 0.0 | 0.0 | 0.1 |
| Days ≥ 1" Snow Depth | 0 | 0 | 0 | 0 | 0 | 0 | 0 | 0 | 0 | 0 | 0 | 0 | 0 |

## PALM SPRINGS THERMAL *Riverside County* ELEVATION -115 ft LAT/LONG 33° 38 ' N / 116° 10 ' W

| | JAN | FEB | MAR | APR | MAY | JUN | JUL | AUG | SEP | OCT | NOV | DEC | YEAR |
|---|---|---|---|---|---|---|---|---|---|---|---|---|---|
| Maximum Temp °F | 70.4 | 74.8 | 79.5 | 86.3 | 93.6 | 102.4 | 106.2 | 104.8 | 100.1 | 90.4 | 78.2 | 69.7 | 88.0 |
| Minimum Temp °F | 38.8 | 43.1 | 49.2 | 55.1 | 62.7 | 69.3 | 75.7 | 75.1 | 68.4 | 57.2 | 44.8 | 37.7 | 56.4 |
| Mean Temp °F | 54.6 | 59.0 | 64.3 | 70.7 | 78.2 | 85.9 | 91.0 | 90.0 | 84.3 | 73.8 | 61.5 | 53.7 | 72.2 |
| Days Max Temp ≥ 90 °F | 0 | 1 | 3 | 11 | 22 | 28 | 31 | 31 | 28 | 18 | 2 | 0 | 175 |
| Days Max Temp ≤ 32 °F | 0 | 0 | 0 | 0 | 0 | 0 | 0 | 0 | 0 | 0 | 0 | 0 | 0 |
| Days Min Temp ≤ 32 °F | 7 | 2 | 0 | 0 | 0 | 0 | 0 | 0 | 0 | 0 | 1 | 7 | 17 |
| Days Min Temp ≤ 0 °F | 0 | 0 | 0 | 0 | 0 | 0 | 0 | 0 | 0 | 0 | 0 | 0 | 0 |
| Heating Degree Days | 317 | 178 | 83 | 20 | 1 | 0 | 0 | 0 | 0 | 9 | 134 | 344 | 1086 |
| Cooling Degree Days | 2 | 20 | 71 | 236 | 431 | 648 | 799 | 777 | 583 | 297 | 35 | 1 | 3900 |
| Total Precipitation (") | 0.65 | 0.59 | 0.45 | 0.06 | 0.07 | 0.02 | 0.19 | 0.38 | 0.32 | 0.18 | 0.31 | 0.38 | 3.60 |
| Days ≥ 0.1" Precip | 2 | 1 | 1 | 0 | 0 | 0 | 0 | 1 | 1 | 1 | 1 | 1 | 9 |
| Total Snowfall (") | 0.0 | 0.0 | 0.0 | 0.0 | 0.0 | 0.0 | 0.0 | 0.0 | 0.0 | 0.0 | 0.0 | 0.0 | 0.0 |
| Days ≥ 1" Snow Depth | 0 | 0 | 0 | 0 | 0 | 0 | 0 | 0 | 0 | 0 | 0 | 0 | 0 |

## PALMDALE *Los Angeles County* ELEVATION 2661 ft LAT/LONG 34° 35 ' N / 118° 7 ' W

| | JAN | FEB | MAR | APR | MAY | JUN | JUL | AUG | SEP | OCT | NOV | DEC | YEAR |
|---|---|---|---|---|---|---|---|---|---|---|---|---|---|
| Maximum Temp °F | 58.2 | 62.9 | 66.8 | 73.6 | 82.1 | 90.8 | 97.0 | 96.4 | 90.3 | 79.9 | 66.5 | 57.5 | 76.8 |
| Minimum Temp °F | 32.2 | 35.6 | 39.0 | 43.1 | 50.6 | 57.7 | 64.2 | 63.5 | 56.9 | 47.6 | 37.8 | 31.4 | 46.6 |
| Mean Temp °F | 45.3 | 49.3 | 52.9 | 58.4 | 66.4 | 74.3 | 80.6 | 79.9 | 73.6 | 63.7 | 52.2 | 44.5 | 61.8 |
| Days Max Temp ≥ 90 °F | 0 | 0 | 0 | 2 | 7 | 18 | 28 | 27 | 18 | 4 | 0 | 0 | 104 |
| Days Max Temp ≤ 32 °F | 0 | 0 | 0 | 0 | 0 | 0 | 0 | 0 | 0 | 0 | 0 | 0 | 0 |
| Days Min Temp ≤ 32 °F | 16 | 10 | 5 | 1 | 0 | 0 | 0 | 0 | 0 | 1 | 8 | 18 | 59 |
| Days Min Temp ≤ 0 °F | 0 | 0 | 0 | 0 | 0 | 0 | 0 | 0 | 0 | 0 | 0 | 0 | 0 |
| Heating Degree Days | 605 | 437 | 369 | 215 | 71 | 11 | 0 | 0 | 10 | 103 | 380 | 629 | 2830 |
| Cooling Degree Days | 0 | 0 | 1 | 38 | 128 | 313 | 494 | 479 | 289 | 78 | 2 | 0 | 1822 |
| Total Precipitation (") | 1.48 | 1.52 | 1.28 | 0.39 | 0.14 | 0.07 | 0.06 | 0.18 | 0.20 | 0.20 | 0.86 | 1.17 | 7.55 |
| Days ≥ 0.1" Precip | 3 | 3 | 3 | 1 | 0 | 0 | 0 | 0 | 0 | 1 | 2 | 2 | 15 |
| Total Snowfall (") | 0.7 | 0.0 | 0.0 | 0.0 | 0.0 | 0.0 | 0.0 | 0.0 | 0.0 | 0.0 | 0.0 | 0.2 | 0.9 |
| Days ≥ 1" Snow Depth | 0 | 0 | 0 | 0 | 0 | 0 | 0 | 0 | 0 | 0 | 0 | 0 | 0 |

## PALO ALTO *Santa Clara County* ELEVATION 20 ft LAT/LONG 37° 27 ' N / 122° 8 ' W

| | JAN | FEB | MAR | APR | MAY | JUN | JUL | AUG | SEP | OCT | NOV | DEC | YEAR |
|---|---|---|---|---|---|---|---|---|---|---|---|---|---|
| Maximum Temp °F | 57.3 | 61.2 | 64.0 | 68.2 | 73.2 | 77.0 | 78.2 | 78.2 | 77.4 | 72.7 | 64.0 | 57.4 | 69.1 |
| Minimum Temp °F | 37.7 | 41.0 | 43.5 | 44.5 | 48.3 | 52.3 | 54.4 | 54.7 | 52.5 | 47.9 | 42.1 | 37.4 | 46.4 |
| Mean Temp °F | 47.6 | 51.1 | 53.8 | 56.4 | 60.7 | 64.7 | 66.3 | 66.5 | 65.0 | 60.3 | 53.1 | 47.4 | 57.7 |
| Days Max Temp ≥ 90 °F | 0 | 0 | 0 | 0 | 1 | 2 | 2 | 1 | 2 | 0 | 0 | 0 | 8 |
| Days Max Temp ≤ 32 °F | 0 | 0 | 0 | 0 | 0 | 0 | 0 | 0 | 0 | 0 | 0 | 0 | 0 |
| Days Min Temp ≤ 32 °F | 7 | 2 | 0 | 0 | 0 | 0 | 0 | 0 | 0 | 0 | 1 | 7 | 17 |
| Days Min Temp ≤ 0 °F | 0 | 0 | 0 | 0 | 0 | 0 | 0 | 0 | 0 | 0 | 0 | 0 | 0 |
| Heating Degree Days | 533 | 385 | 340 | 256 | 150 | 61 | 24 | 17 | 46 | 153 | 350 | 541 | 2856 |
| Cooling Degree Days | 0 | 0 | 0 | 7 | 22 | 54 | 70 | 58 | 44 | 17 | 0 | 0 | 272 |
| Total Precipitation (") | 3.11 | 2.59 | 2.49 | 1.02 | 0.19 | 0.06 | 0.04 | 0.06 | 0.19 | 0.79 | 1.99 | 2.54 | 15.07 |
| Days ≥ 0.1" Precip | 6 | 6 | 6 | 3 | 1 | 0 | 0 | 0 | 1 | 2 | 4 | 6 | 35 |
| Total Snowfall (") | 0.0 | 0.0 | 0.0 | 0.0 | 0.0 | 0.0 | 0.0 | 0.0 | 0.0 | 0.0 | 0.0 | 0.0 | 0.0 |
| Days ≥ 1" Snow Depth | 0 | 0 | 0 | 0 | 0 | 0 | 0 | 0 | 0 | 0 | 0 | 0 | 0 |

## PALOMAR MTN OBS *San Diego County* ELEVATION 5604 ft LAT/LONG 33° 21 ' N / 116° 51 ' W

| | JAN | FEB | MAR | APR | MAY | JUN | JUL | AUG | SEP | OCT | NOV | DEC | YEAR |
|---|---|---|---|---|---|---|---|---|---|---|---|---|---|
| Maximum Temp °F | 52.0 | 52.9 | 55.6 | 61.9 | 69.3 | 79.0 | 84.7 | 83.9 | 79.2 | 69.2 | 58.4 | 51.5 | 66.5 |
| Minimum Temp °F | 33.5 | 34.1 | 35.0 | 38.9 | 45.1 | 54.2 | 60.3 | 60.9 | 55.5 | 47.5 | 38.9 | 33.6 | 44.8 |
| Mean Temp °F | 42.8 | 43.6 | 45.4 | 50.4 | 57.2 | 66.6 | 72.7 | 72.5 | 67.3 | 58.4 | 48.7 | 42.6 | 55.7 |
| Days Max Temp ≥ 90 °F | 0 | 0 | 0 | 0 | 0 | 2 | 5 | 4 | 2 | 0 | 0 | 0 | 13 |
| Days Max Temp ≤ 32 °F | 1 | 0 | 0 | 0 | 0 | 0 | 0 | 0 | 0 | 0 | 0 | 1 | 2 |
| Days Min Temp ≤ 32 °F | 14 | 13 | 13 | 9 | 3 | 0 | 0 | 0 | 0 | 1 | 7 | 13 | 73 |
| Days Min Temp ≤ 0 °F | 0 | 0 | 0 | 0 | 0 | 0 | 0 | 0 | 0 | 0 | 0 | 0 | 0 |
| Heating Degree Days | 682 | 599 | 601 | 434 | 256 | 72 | 7 | 9 | 53 | 230 | 483 | 689 | 4115 |
| Cooling Degree Days | 0 | 0 | 0 | 5 | 20 | 127 | 238 | 239 | 131 | 32 | 0 | 0 | 792 |
| Total Precipitation (") | 5.76 | 5.72 | 5.81 | 2.09 | 0.54 | 0.15 | 0.47 | 0.98 | 0.61 | 0.94 | 3.32 | 4.68 | 31.07 |
| Days ≥ 0.1" Precip | 5 | 4 | 5 | 3 | 1 | 0 | 1 | 1 | 1 | 2 | 3 | 4 | 30 |
| Total Snowfall (") | 5.1 | 4.0 | 11.5 | 4.3 | 0.3 | 0.0 | 0.0 | 0.0 | 0.0 | 0.1 | 2.0 | 6.4 | 33.7 |
| Days ≥ 1" Snow Depth | 2 | 2 | 4 | 2 | 0 | 0 | 0 | 0 | 0 | 0 | 1 | 4 | 15 |

## PARADISE *Butte County* ELEVATION 1713 ft LAT/LONG 39° 45 ' N / 121° 37 ' W

| | JAN | FEB | MAR | APR | MAY | JUN | JUL | AUG | SEP | OCT | NOV | DEC | YEAR |
|---|---|---|---|---|---|---|---|---|---|---|---|---|---|
| Maximum Temp °F | 53.8 | 56.7 | 59.8 | 66.0 | 75.6 | 84.0 | 90.9 | 89.8 | 84.4 | 74.0 | 59.9 | 53.1 | 70.7 |
| Minimum Temp °F | 37.6 | 40.2 | 42.1 | 45.8 | 52.2 | 58.7 | 64.1 | 63.1 | 59.3 | 52.3 | 43.2 | 37.4 | 49.7 |
| Mean Temp °F | 45.7 | 48.5 | 50.9 | 55.9 | 63.9 | 71.3 | 77.5 | 76.4 | 71.9 | 63.2 | 51.5 | 45.3 | 60.2 |
| Days Max Temp ≥ 90 °F | 0 | 0 | 0 | 0 | 2 | 9 | 20 | 18 | 9 | 2 | 0 | 0 | 60 |
| Days Max Temp ≤ 32 °F | 0 | 0 | 0 | 0 | 0 | 0 | 0 | 0 | 0 | 0 | 0 | 0 | 0 |
| Days Min Temp ≤ 32 °F | 6 | 3 | 2 | 0 | 0 | 0 | 0 | 0 | 0 | 0 | 1 | 7 | 19 |
| Days Min Temp ≤ 0 °F | 0 | 0 | 0 | 0 | 0 | 0 | 0 | 0 | 0 | 0 | 0 | 0 | 0 |
| Heating Degree Days | 591 | 460 | 430 | 283 | 120 | 31 | 3 | 4 | 23 | 127 | 401 | 605 | 3078 |
| Cooling Degree Days | 0 | 0 | 0 | 25 | 96 | 237 | 411 | 384 | 255 | 89 | 1 | 0 | 1498 |
| Total Precipitation (") | 10.45 | 8.10 | 8.38 | 3.73 | 1.21 | 0.59 | 0.12 | 0.31 | 1.06 | 2.65 | 7.99 | 8.69 | 53.28 |
| Days ≥ 0.1" Precip | 9 | 9 | 9 | 6 | 3 | 1 | 0 | 1 | 2 | 3 | 8 | 9 | 60 |
| Total Snowfall (") | 1.9 | 0.5 | 0.6 | 0.0 | 0.0 | 0.0 | 0.0 | 0.0 | 0.0 | 0.0 | 0.0 | 0.8 | 3.8 |
| Days ≥ 1" Snow Depth | 1 | 0 | 0 | 0 | 0 | 0 | 0 | 0 | 0 | 0 | 0 | 0 | 1 |

## PARKER RESERVOIR *San Bernardino County* ELEVATION 738 ft LAT/LONG 34° 17' N / 114° 10' W

| | JAN | FEB | MAR | APR | MAY | JUN | JUL | AUG | SEP | OCT | NOV | DEC | YEAR |
|---|---|---|---|---|---|---|---|---|---|---|---|---|---|
| Maximum Temp °F | 64.3 | 70.2 | 75.9 | 84.1 | 93.1 | 103.1 | 107.4 | 105.8 | 100.1 | 88.3 | 74.0 | 64.3 | 85.9 |
| Minimum Temp °F | 42.8 | 47.1 | 52.4 | 59.1 | 67.7 | 76.9 | 83.0 | 81.5 | 75.2 | 63.6 | 51.1 | 43.0 | 62.0 |
| Mean Temp °F | 53.6 | 58.7 | 64.2 | 71.6 | 80.4 | 90.0 | 95.2 | 93.7 | 87.7 | 76.0 | 62.6 | 53.6 | 73.9 |
| Days Max Temp ≥ 90 °F | 0 | 0 | 2 | 9 | 22 | 29 | 31 | 31 | 27 | 15 | 1 | 0 | 167 |
| Days Max Temp ≤ 32 °F | 0 | 0 | 0 | 0 | 0 | 0 | 0 | 0 | 0 | 0 | 0 | 0 | 0 |
| Days Min Temp ≤ 32 °F | 1 | 0 | 0 | 0 | 0 | 0 | 0 | 0 | 0 | 0 | 0 | 1 | 2 |
| Days Min Temp ≤ 0 °F | 0 | 0 | 0 | 0 | 0 | 0 | 0 | 0 | 0 | 0 | 0 | 0 | 0 |
| Heating Degree Days | 348 | 187 | 99 | 24 | 1 | 0 | 0 | 0 | 0 | 9 | 122 | 345 | 1135 |
| Cooling Degree Days | 0 | 17 | 77 | 267 | 496 | 771 | 928 | 891 | 675 | 348 | 51 | 0 | 4521 |
| Total Precipitation (") | 0.94 | 0.78 | 0.87 | 0.22 | 0.12 | 0.05 | 0.35 | 0.65 | 0.51 | 0.46 | 0.52 | 0.70 | 6.17 |
| Days ≥ 0.1" Precip | 2 | 2 | 2 | 1 | 0 | 0 | 1 | 1 | 1 | 1 | 1 | 2 | 14 |
| Total Snowfall (") | 0.0 | 0.0 | 0.0 | 0.0 | 0.0 | 0.0 | 0.0 | 0.0 | 0.0 | 0.0 | 0.0 | 0.0 | 0.0 |
| Days ≥ 1" Snow Depth | 0 | 0 | 0 | 0 | 0 | 0 | 0 | 0 | 0 | 0 | 0 | 0 | 0 |

## PASADENA *Los Angeles County* ELEVATION 860 ft LAT/LONG 34° 9' N / 118° 9' W

| | JAN | FEB | MAR | APR | MAY | JUN | JUL | AUG | SEP | OCT | NOV | DEC | YEAR |
|---|---|---|---|---|---|---|---|---|---|---|---|---|---|
| Maximum Temp °F | 68.0 | 70.2 | 70.9 | 75.0 | 77.7 | 82.9 | 88.9 | 90.0 | 87.9 | 82.4 | 73.7 | 67.6 | 77.9 |
| Minimum Temp °F | 43.8 | 45.5 | 47.0 | 49.6 | 53.3 | 57.2 | 61.1 | 62.0 | 60.2 | 55.1 | 48.3 | 43.7 | 52.2 |
| Mean Temp °F | 55.9 | 57.9 | 59.0 | 62.4 | 65.6 | 70.1 | 75.0 | 76.0 | 74.1 | 68.8 | 61.0 | 55.7 | 65.1 |
| Days Max Temp ≥ 90 °F | 0 | 0 | 0 | 2 | 3 | 6 | 14 | 16 | 13 | 7 | 1 | 0 | 62 |
| Days Max Temp ≤ 32 °F | 0 | 0 | 0 | 0 | 0 | 0 | 0 | 0 | 0 | 0 | 0 | 0 | 0 |
| Days Min Temp ≤ 32 °F | 0 | 0 | 0 | 0 | 0 | 0 | 0 | 0 | 0 | 0 | 0 | 1 | 1 |
| Days Min Temp ≤ 0 °F | 0 | 0 | 0 | 0 | 0 | 0 | 0 | 0 | 0 | 0 | 0 | 0 | 0 |
| Heating Degree Days | 280 | 205 | 194 | 115 | 56 | 13 | 0 | 0 | 1 | 22 | 140 | 285 | 1311 |
| Cooling Degree Days | 5 | 14 | 17 | 65 | 88 | 190 | 327 | 360 | 298 | 155 | 26 | 2 | 1547 |
| Total Precipitation (") | 4.21 | 4.34 | 4.13 | 1.34 | 0.32 | 0.11 | 0.05 | 0.20 | 0.51 | 0.58 | 2.40 | 2.72 | 20.91 |
| Days ≥ 0.1" Precip | 5 | 4 | 5 | 2 | 1 | 0 | 0 | 0 | 1 | 1 | 3 | 3 | 25 |
| Total Snowfall (") | 0.0 | 0.0 | 0.0 | 0.0 | 0.0 | 0.0 | 0.0 | 0.0 | 0.0 | 0.0 | 0.0 | 0.0 | 0.0 |
| Days ≥ 1" Snow Depth | 0 | 0 | 0 | 0 | 0 | 0 | 0 | 0 | 0 | 0 | 0 | 0 | 0 |

## PASO ROBLES *San Luis Obispo County* ELEVATION 700 ft LAT/LONG 35° 38' N / 120° 41' W

| | JAN | FEB | MAR | APR | MAY | JUN | JUL | AUG | SEP | OCT | NOV | DEC | YEAR |
|---|---|---|---|---|---|---|---|---|---|---|---|---|---|
| Maximum Temp °F | 61.9 | 65.5 | 68.2 | 74.1 | 81.1 | 87.3 | 92.7 | 93.0 | 89.2 | 81.7 | 68.8 | 61.5 | 77.1 |
| Minimum Temp °F | 33.2 | 36.6 | 39.4 | 40.2 | 44.0 | 48.1 | 51.1 | 50.8 | 47.9 | 42.2 | 36.3 | 31.3 | 41.8 |
| Mean Temp °F | 47.6 | 51.1 | 53.8 | 57.2 | 62.6 | 67.7 | 71.9 | 71.9 | 68.6 | 62.0 | 52.6 | 46.4 | 59.5 |
| Days Max Temp ≥ 90 °F | 0 | 0 | 0 | 1 | 7 | 13 | 20 | 22 | 16 | 7 | 0 | 0 | 86 |
| Days Max Temp ≤ 32 °F | 0 | 0 | 0 | 0 | 0 | 0 | 0 | 0 | 0 | 0 | 0 | 0 | 0 |
| Days Min Temp ≤ 32 °F | 16 | 9 | 5 | 2 | 0 | 0 | 0 | 0 | 0 | 2 | 11 | 19 | 64 |
| Days Min Temp ≤ 0 °F | 0 | 0 | 0 | 0 | 0 | 0 | 0 | 0 | 0 | 0 | 0 | 0 | 0 |
| Heating Degree Days | 533 | 387 | 341 | 234 | 115 | 34 | 6 | 4 | 25 | 123 | 367 | 569 | 2738 |
| Cooling Degree Days | 0 | 0 | 0 | 10 | 53 | 131 | 228 | 231 | 149 | 43 | 0 | 0 | 845 |
| Total Precipitation (") | 3.13 | 2.92 | 2.47 | 0.93 | 0.15 | 0.03 | 0.03 | 0.06 | 0.35 | 0.55 | 1.54 | 2.25 | 14.41 |
| Days ≥ 0.1" Precip | 5 | 5 | 5 | 2 | 0 | 0 | 0 | 0 | 1 | 1 | 4 | 4 | 27 |
| Total Snowfall (") | 0.0 | 0.0 | 0.0 | 0.0 | 0.0 | 0.0 | 0.0 | 0.0 | 0.0 | 0.0 | 0.0 | 0.1 | 0.1 |
| Days ≥ 1" Snow Depth | 0 | 0 | 0 | 0 | 0 | 0 | 0 | 0 | 0 | 0 | 0 | 0 | 0 |

## PASO ROBLES MUNI AP *San Luis Obispo County* ELEVATION 807 ft LAT/LONG 35° 40' N / 120° 38' W

| | JAN | FEB | MAR | APR | MAY | JUN | JUL | AUG | SEP | OCT | NOV | DEC | YEAR |
|---|---|---|---|---|---|---|---|---|---|---|---|---|---|
| Maximum Temp °F | 59.6 | 63.2 | 66.0 | 72.6 | 80.5 | 87.8 | 93.6 | 93.2 | 88.2 | 79.6 | 66.9 | 59.4 | 75.9 |
| Minimum Temp °F | 34.1 | 37.2 | 39.7 | 40.7 | 45.5 | 50.3 | 53.8 | 53.6 | 50.8 | 44.6 | 37.7 | 32.7 | 43.4 |
| Mean Temp °F | 46.9 | 50.2 | 52.9 | 56.7 | 63.0 | 69.1 | 73.8 | 73.5 | 69.5 | 62.2 | 52.3 | 46.1 | 59.7 |
| Days Max Temp ≥ 90 °F | 0 | 0 | 0 | 1 | 7 | 14 | 22 | 22 | 15 | 6 | 0 | 0 | 87 |
| Days Max Temp ≤ 32 °F | 0 | 0 | 0 | 0 | 0 | 0 | 0 | 0 | 0 | 0 | 0 | 0 | 0 |
| Days Min Temp ≤ 32 °F | 15 | 8 | 4 | 2 | 0 | 0 | 0 | 0 | 0 | 1 | 8 | 17 | 55 |
| Days Min Temp ≤ 0 °F | 0 | 0 | 0 | 0 | 0 | 0 | 0 | 0 | 0 | 0 | 0 | 0 | 0 |
| Heating Degree Days | 555 | 411 | 369 | 250 | 110 | 24 | 3 | 4 | 22 | 123 | 376 | 581 | 2828 |
| Cooling Degree Days | 0 | 0 | 0 | 10 | 57 | 154 | 273 | 265 | 168 | 50 | 0 | 0 | 977 |
| Total Precipitation (") | 2.74 | 2.63 | 2.31 | 0.86 | 0.13 | 0.02 | 0.03 | 0.06 | 0.37 | 0.49 | 1.37 | 1.99 | 13.00 |
| Days ≥ 0.1" Precip | 5 | 5 | 4 | 2 | 0 | 0 | 0 | 0 | 1 | 1 | 3 | 4 | 25 |
| Total Snowfall (") | 0.0 | 0.0 | 0.0 | 0.0 | 0.0 | 0.0 | 0.0 | 0.0 | 0.0 | 0.0 | 0.0 | 0.1 | 0.1 |
| Days ≥ 1" Snow Depth | 0 | 0 | 0 | 0 | 0 | 0 | 0 | 0 | 0 | 0 | 0 | 0 | 0 |

## PETALUMA FIRE STA 2 *Sonoma County* ELEVATION 10 ft LAT/LONG 38° 14 ' N / 122° 38 ' W

| | JAN | FEB | MAR | APR | MAY | JUN | JUL | AUG | SEP | OCT | NOV | DEC | YEAR |
|---|---|---|---|---|---|---|---|---|---|---|---|---|---|
| Maximum Temp °F | 57.0 | 61.9 | 64.5 | 68.6 | 73.6 | 78.4 | 82.6 | 82.5 | 81.7 | 76.2 | 64.9 | 56.7 | 70.7 |
| Minimum Temp °F | 37.4 | 40.0 | 42.1 | 43.3 | 46.6 | 50.4 | 51.7 | 52.1 | 51.3 | 47.1 | 41.6 | 37.3 | 45.1 |
| Mean Temp °F | 47.2 | 51.0 | 53.3 | 56.0 | 60.1 | 64.4 | 67.2 | 67.4 | 66.5 | 61.7 | 53.3 | 47.0 | 57.9 |
| Days Max Temp ≥ 90 °F | 0 | 0 | 0 | 0 | 1 | 3 | 5 | 6 | 6 | 2 | 0 | 0 | 23 |
| Days Max Temp ≤ 32 °F | 0 | 0 | 0 | 0 | 0 | 0 | 0 | 0 | 0 | 0 | 0 | 0 | 0 |
| Days Min Temp ≤ 32 °F | 8 | 3 | 1 | 0 | 0 | 0 | 0 | 0 | 0 | 0 | 3 | 9 | 24 |
| Days Min Temp ≤ 0 °F | 0 | 0 | 0 | 0 | 0 | 0 | 0 | 0 | 0 | 0 | 0 | 0 | 0 |
| Heating Degree Days | 545 | 388 | 356 | 268 | 162 | 63 | 16 | 12 | 28 | 120 | 345 | 551 | 2854 |
| Cooling Degree Days | 0 | 0 | 0 | 5 | 18 | 54 | 95 | 85 | 67 | 24 | 1 | 0 | 349 |
| Total Precipitation (") | 5.39 | 4.28 | 3.55 | 1.41 | 0.32 | 0.22 | 0.05 | 0.09 | 0.28 | 1.41 | 3.73 | 4.20 | 24.93 |
| Days ≥ 0.1" Precip | 7 | 7 | 7 | 3 | 1 | 0 | 0 | 0 | 1 | 2 | 6 | 7 | 41 |
| Total Snowfall (") | 0.0 | 0.0 | 0.0 | 0.0 | 0.0 | 0.0 | 0.0 | 0.0 | 0.0 | 0.0 | 0.0 | 0.0 | 0.0 |
| Days ≥ 1" Snow Depth | 0 | 0 | 0 | 0 | 0 | 0 | 0 | 0 | 0 | 0 | 0 | 0 | 0 |

## PINNACLES NM *San Benito County* ELEVATION 1307 ft LAT/LONG 36° 29 ' N / 121° 11 ' W

| | JAN | FEB | MAR | APR | MAY | JUN | JUL | AUG | SEP | OCT | NOV | DEC | YEAR |
|---|---|---|---|---|---|---|---|---|---|---|---|---|---|
| Maximum Temp °F | 61.8 | 64.0 | 65.7 | 71.2 | 79.6 | 87.8 | 94.7 | 94.4 | 89.8 | 81.3 | 68.8 | 61.4 | 76.7 |
| Minimum Temp °F | 32.8 | 35.1 | 37.2 | 38.7 | 42.5 | 46.8 | 50.5 | 50.2 | 48.0 | 42.6 | 36.8 | 32.2 | 41.1 |
| Mean Temp °F | 47.3 | 49.5 | 51.5 | 55.0 | 61.1 | 67.3 | 72.6 | 72.4 | 69.0 | 62.0 | 52.8 | 46.8 | 58.9 |
| Days Max Temp ≥ 90 °F | 0 | 0 | 0 | 1 | 6 | 14 | 24 | 23 | 16 | 7 | 0 | 0 | 91 |
| Days Max Temp ≤ 32 °F | 0 | 0 | 0 | 0 | 0 | 0 | 0 | 0 | 0 | 0 | 0 | 0 | 0 |
| Days Min Temp ≤ 32 °F | 17 | 10 | 7 | 3 | 1 | 0 | 0 | 0 | 0 | 1 | 8 | 18 | 65 |
| Days Min Temp ≤ 0 °F | 0 | 0 | 0 | 0 | 0 | 0 | 0 | 0 | 0 | 0 | 0 | 0 | 0 |
| Heating Degree Days | 542 | 430 | 413 | 301 | 158 | 45 | 7 | 5 | 30 | 137 | 361 | 557 | 2986 |
| Cooling Degree Days | 0 | 0 | 0 | 11 | 39 | 116 | 232 | 222 | 144 | 51 | 1 | 0 | 816 |
| Total Precipitation (") | 2.99 | 3.06 | 3.24 | 1.29 | 0.31 | 0.04 | 0.04 | 0.09 | 0.33 | 0.80 | 2.05 | 2.46 | 16.70 |
| Days ≥ 0.1" Precip | 6 | 5 | 6 | 3 | 1 | 0 | 0 | 0 | 1 | 2 | 4 | 5 | 33 |
| Total Snowfall (") | 0.2 | 0.0 | 0.0 | 0.0 | 0.0 | 0.0 | 0.0 | 0.0 | 0.0 | 0.0 | 0.0 | 0.1 | 0.3 |
| Days ≥ 1" Snow Depth | 0 | 0 | 0 | 0 | 0 | 0 | 0 | 0 | 0 | 0 | 0 | 0 | 0 |

## PISMO BEACH *San Luis Obispo County* ELEVATION 30 ft LAT/LONG 35° 8 ' N / 120° 38 ' W

| | JAN | FEB | MAR | APR | MAY | JUN | JUL | AUG | SEP | OCT | NOV | DEC | YEAR |
|---|---|---|---|---|---|---|---|---|---|---|---|---|---|
| Maximum Temp °F | 64.2 | 65.4 | 66.4 | 68.9 | 69.2 | 70.9 | 70.9 | 72.1 | 73.0 | 72.7 | 69.0 | 64.3 | 68.9 |
| Minimum Temp °F | 41.9 | 43.5 | 44.3 | 45.3 | 47.1 | 50.3 | 52.5 | 53.2 | 52.8 | 50.0 | 46.1 | 41.8 | 47.4 |
| Mean Temp °F | 53.1 | 54.5 | 55.3 | 57.1 | 58.2 | 60.6 | 61.7 | 62.7 | 62.9 | 61.4 | 57.6 | 53.1 | 58.2 |
| Days Max Temp ≥ 90 °F | 0 | 0 | 0 | 0 | 0 | 1 | 1 | 0 | 1 | 1 | 0 | 0 | 4 |
| Days Max Temp ≤ 32 °F | 0 | 0 | 0 | 0 | 0 | 0 | 0 | 0 | 0 | 0 | 0 | 0 | 0 |
| Days Min Temp ≤ 32 °F | 2 | 0 | 0 | 0 | 0 | 0 | 0 | 0 | 0 | 0 | 0 | 1 | 3 |
| Days Min Temp ≤ 0 °F | 0 | 0 | 0 | 0 | 0 | 0 | 0 | 0 | 0 | 0 | 0 | 0 | 0 |
| Heating Degree Days | 363 | 292 | 294 | 236 | 210 | 139 | 108 | 83 | 84 | 125 | 220 | 364 | 2518 |
| Cooling Degree Days | *0* | 1 | 1 | 10 | 4 | 9 | 16 | 14 | 24 | 15 | 5 | 0 | 99 |
| Total Precipitation (") | 3.61 | 3.21 | 3.18 | 1.19 | 0.23 | 0.03 | 0.03 | 0.03 | 0.36 | 0.73 | 1.97 | 2.82 | 17.39 |
| Days ≥ 0.1" Precip | 6 | 5 | 5 | 2 | 0 | 0 | 0 | 0 | 1 | 1 | 4 | 5 | 29 |
| Total Snowfall (") | 0.0 | 0.0 | 0.0 | 0.0 | 0.0 | 0.0 | 0.0 | 0.0 | 0.0 | 0.0 | 0.0 | 0.0 | 0.0 |
| Days ≥ 1" Snow Depth | 0 | 0 | 0 | 0 | 0 | 0 | 0 | 0 | 0 | 0 | 0 | 0 | 0 |

## PLACERVILLE *El Dorado County* ELEVATION 1850 ft LAT/LONG 38° 43 ' N / 120° 49 ' W

| | JAN | FEB | MAR | APR | MAY | JUN | JUL | AUG | SEP | OCT | NOV | DEC | YEAR |
|---|---|---|---|---|---|---|---|---|---|---|---|---|---|
| Maximum Temp °F | 53.9 | 57.5 | 60.3 | 66.0 | 75.1 | 83.2 | 91.2 | 90.7 | 85.4 | 74.6 | 60.8 | 53.8 | 71.0 |
| Minimum Temp °F | 31.7 | 34.5 | 37.3 | 40.2 | 45.7 | 51.5 | 56.7 | 56.1 | 52.0 | 44.6 | 37.0 | 32.2 | 43.3 |
| Mean Temp °F | 42.8 | 46.0 | 48.8 | 53.1 | 60.4 | 67.4 | 74.0 | 73.4 | 68.8 | 59.6 | 48.9 | 43.0 | 57.2 |
| Days Max Temp ≥ 90 °F | 0 | 0 | 0 | 0 | 2 | 8 | 20 | 19 | 10 | 2 | 0 | 0 | 61 |
| Days Max Temp ≤ 32 °F | 0 | 0 | 0 | 0 | 0 | 0 | 0 | 0 | 0 | 0 | 0 | 0 | 0 |
| Days Min Temp ≤ 32 °F | 18 | 11 | 7 | 3 | 0 | 0 | 0 | 0 | 0 | 1 | 7 | 18 | 65 |
| Days Min Temp ≤ 0 °F | 0 | 0 | 0 | 0 | 0 | 0 | 0 | 0 | 0 | 0 | 0 | 0 | 0 |
| Heating Degree Days | 681 | 529 | 495 | 353 | 177 | 56 | 9 | 7 | 39 | 198 | 476 | 674 | 3694 |
| Cooling Degree Days | 0 | 0 | 0 | 8 | 49 | 149 | ***319*** | 298 | 188 | 56 | 0 | 0 | 1067 |
| Total Precipitation (") | 6.81 | 5.51 | 5.73 | 2.86 | 0.89 | 0.41 | 0.21 | 0.20 | 0.84 | 2.00 | 5.49 | 5.78 | 36.73 |
| Days ≥ 0.1" Precip | 8 | 8 | 9 | 5 | 2 | 1 | 0 | 0 | 1 | 3 | 7 | 7 | 51 |
| Total Snowfall (") | ***0.8*** | 0.0 | 0.3 | 0.4 | 0.0 | 0.0 | 0.0 | 0.0 | 0.0 | 0.0 | 0.0 | 0.3 | 1.8 |
| Days ≥ 1" Snow Depth | ***0*** | 0 | 0 | 0 | 0 | 0 | 0 | 0 | 0 | 0 | 0 | ***0*** | 0 |

## POMONA CAL POLY *Los Angeles County* ELEVATION 740 ft LAT/LONG 34° 4 ' N / 117° 49 ' W

| | JAN | FEB | MAR | APR | MAY | JUN | JUL | AUG | SEP | OCT | NOV | DEC | YEAR |
|---|---|---|---|---|---|---|---|---|---|---|---|---|---|
| Maximum Temp °F | 67.8 | 69.6 | 69.8 | 74.3 | 77.3 | 82.7 | 89.5 | 89.8 | 87.5 | 81.5 | 73.6 | 68.1 | 77.6 |
| Minimum Temp °F | 41.0 | 42.9 | 44.6 | 46.9 | 51.4 | 54.7 | 59.0 | 59.8 | 58.2 | 52.8 | 45.9 | 40.7 | 49.8 |
| Mean Temp °F | 54.4 | 56.3 | 57.2 | 60.6 | 64.4 | 68.8 | 74.3 | 74.9 | 72.9 | 67.2 | 59.8 | 54.4 | 63.8 |
| Days Max Temp ≥ 90 °F | 0 | 0 | 0 | 1 | 3 | 6 | 16 | 16 | 13 | 7 | 1 | 0 | 63 |
| Days Max Temp ≤ 32 °F | 0 | 0 | 0 | 0 | 0 | 0 | 0 | 0 | 0 | 0 | 0 | 0 | 0 |
| Days Min Temp ≤ 32 °F | 3 | 1 | 0 | 0 | 0 | 0 | 0 | 0 | 0 | 0 | 0 | 3 | 7 |
| Days Min Temp ≤ 0 °F | 0 | 0 | 0 | 0 | 0 | 0 | 0 | 0 | 0 | 0 | 0 | 0 | 0 |
| Heating Degree Days | 325 | 246 | 243 | 151 | 76 | 21 | 0 | 0 | 4 | 39 | 169 | 323 | 1597 |
| Cooling Degree Days | 3 | 9 | 11 | 37 | 64 | 129 | 288 | 296 | 252 | 115 | 15 | 1 | 1220 |
| Total Precipitation (") | 3.87 | 3.72 | 3.40 | 1.04 | 0.23 | 0.05 | 0.02 | 0.13 | 0.32 | 0.58 | 1.83 | 2.45 | 17.64 |
| Days ≥ 0.1" Precip | 5 | 4 | 4 | 2 | 1 | 0 | 0 | 0 | 1 | 1 | 2 | 3 | 23 |
| Total Snowfall (") | 0.0 | 0.0 | 0.0 | 0.0 | 0.0 | 0.0 | 0.0 | 0.0 | 0.0 | 0.0 | 0.0 | 0.0 | 0.0 |
| Days ≥ 1" Snow Depth | 0 | 0 | 0 | 0 | 0 | 0 | 0 | 0 | 0 | 0 | 0 | 0 | 0 |

## PORTERVILLE *Tulare County* ELEVATION 393 ft LAT/LONG 36° 4 ' N / 119° 1 ' W

| | JAN | FEB | MAR | APR | MAY | JUN | JUL | AUG | SEP | OCT | NOV | DEC | YEAR |
|---|---|---|---|---|---|---|---|---|---|---|---|---|---|
| Maximum Temp °F | 56.9 | 64.2 | 69.3 | 76.2 | 85.5 | 92.8 | 98.2 | 96.9 | 91.6 | 82.2 | 67.7 | 57.0 | 78.2 |
| Minimum Temp °F | 36.8 | 40.4 | 44.3 | 47.7 | 54.1 | 60.1 | 65.3 | 63.9 | 58.6 | 50.8 | 41.9 | 36.0 | 50.0 |
| Mean Temp °F | 46.8 | 52.3 | 56.8 | 62.0 | 69.8 | 76.5 | 81.8 | 80.4 | 75.1 | 66.5 | 54.8 | 46.5 | 64.1 |
| Days Max Temp ≥ 90 °F | 0 | 0 | 0 | 3 | 11 | 20 | 29 | 27 | 19 | 6 | 0 | 0 | 115 |
| Days Max Temp ≤ 32 °F | 0 | 0 | 0 | 0 | 0 | 0 | 0 | 0 | 0 | 0 | 0 | 0 | 0 |
| Days Min Temp ≤ 32 °F | 8 | 2 | 0 | 0 | 0 | 0 | 0 | 0 | 0 | 0 | 2 | 9 | 21 |
| Days Min Temp ≤ 0 °F | 0 | 0 | 0 | 0 | 0 | 0 | 0 | 0 | 0 | 0 | 0 | 0 | 0 |
| Heating Degree Days | 556 | 352 | 252 | 132 | 34 | 3 | 0 | 0 | 4 | 55 | 302 | 566 | 2256 |
| Cooling Degree Days | 0 | 1 | 5 | 64 | 190 | 359 | 531 | 496 | 331 | 123 | 2 | 0 | 2102 |
| Total Precipitation (") | 1.96 | 1.80 | 2.11 | 0.99 | 0.35 | 0.08 | 0.02 | 0.02 | 0.37 | 0.60 | 1.32 | 1.58 | 11.20 |
| Days ≥ 0.1" Precip | 4 | 4 | 5 | 3 | 1 | 0 | 0 | 0 | 1 | 1 | 3 | 4 | 26 |
| Total Snowfall (") | 0.0 | 0.0 | 0.0 | 0.0 | 0.0 | 0.0 | 0.0 | 0.0 | 0.0 | 0.0 | 0.0 | 0.0 | 0.0 |
| Days ≥ 1" Snow Depth | 0 | 0 | 0 | 0 | 0 | 0 | 0 | 0 | 0 | 0 | 0 | 0 | 0 |

## PORTOLA *Plumas County* ELEVATION 4833 ft LAT/LONG 39° 48 ' N / 120° 28 ' W

| | JAN | FEB | MAR | APR | MAY | JUN | JUL | AUG | SEP | OCT | NOV | DEC | YEAR |
|---|---|---|---|---|---|---|---|---|---|---|---|---|---|
| Maximum Temp °F | 41.8 | 46.4 | 51.7 | 58.3 | 68.3 | 77.2 | 85.7 | 84.6 | 77.7 | ***67.3*** | ***50.8*** | ***42.8*** | 62.7 |
| Minimum Temp °F | 18.2 | 20.8 | 25.3 | 28.0 | 32.8 | 38.1 | 41.5 | 40.5 | 35.5 | ***29.8*** | ***24.4*** | ***19.0*** | 29.5 |
| Mean Temp °F | 30.0 | 33.6 | 38.5 | 43.2 | 50.6 | 57.7 | 63.6 | 62.6 | 56.6 | ***48.5*** | ***37.7*** | ***30.9*** | 46.1 |
| Days Max Temp ≥ 90 °F | 0 | 0 | 0 | 0 | 0 | 2 | 9 | 6 | 2 | 0 | 0 | 0 | 19 |
| Days Max Temp ≤ 32 °F | 3 | 1 | 0 | 0 | 0 | 0 | 0 | 0 | 0 | 0 | 1 | ***3*** | 8 |
| Days Min Temp ≤ 32 °F | 29 | ***26*** | ***27*** | 24 | ***15*** | ***5*** | 2 | 2 | 8 | ***22*** | ***26*** | ***29*** | 215 |
| Days Min Temp ≤ 0 °F | 1 | 1 | 0 | 0 | 0 | 0 | 0 | 0 | 0 | 0 | 0 | 1 | 3 |
| Heating Degree Days | 1079 | 879 | ***809*** | 647 | 442 | 223 | 84 | 105 | 248 | ***504*** | ***813*** | ***1050*** | 6883 |
| Cooling Degree Days | 0 | 0 | 0 | 0 | 1 | 11 | 56 | 35 | 5 | 0 | 0 | 0 | 108 |
| Total Precipitation (") | 3.84 | 3.28 | 3.32 | 1.13 | 0.95 | 0.69 | 0.38 | 0.53 | 0.71 | 1.12 | 2.79 | 3.36 | 22.10 |
| Days ≥ 0.1" Precip | 6 | ***5*** | 6 | 3 | 3 | 2 | 1 | 1 | 2 | 3 | 5 | ***5*** | 42 |
| Total Snowfall (") | 16.1 | ***13.5*** | 8.9 | 4.1 | 0.9 | 0.0 | 0.0 | 0.0 | 0.1 | 0.1 | 5.1 | ***16.9*** | 65.7 |
| Days ≥ 1" Snow Depth | ***19*** | 16 | 4 | 1 | 0 | 0 | 0 | 0 | 0 | 0 | 3 | ***11*** | 54 |

## POTTER VALLEY P H *Mendocino County* ELEVATION 1010 ft LAT/LONG 39° 22 ' N / 123° 8 ' W

| | JAN | FEB | MAR | APR | MAY | JUN | JUL | AUG | SEP | OCT | NOV | DEC | YEAR |
|---|---|---|---|---|---|---|---|---|---|---|---|---|---|
| Maximum Temp °F | 56.8 | 60.7 | 63.6 | 68.9 | 77.6 | 85.1 | 92.7 | 91.6 | 87.5 | 77.6 | 62.7 | 55.8 | 73.4 |
| Minimum Temp °F | 33.6 | 36.5 | 38.1 | 39.4 | 43.9 | 49.2 | 53.0 | 52.1 | 48.6 | 43.0 | 37.4 | 33.7 | 42.4 |
| Mean Temp °F | 45.3 | 48.6 | 50.9 | 54.2 | 60.8 | 67.2 | 72.9 | 71.9 | 68.1 | 60.3 | 50.0 | 44.7 | 57.9 |
| Days Max Temp ≥ 90 °F | 0 | 0 | 0 | 0 | 5 | 10 | 21 | 20 | 13 | 4 | 0 | 0 | 73 |
| Days Max Temp ≤ 32 °F | 0 | 0 | 0 | 0 | 0 | 0 | 0 | 0 | 0 | 0 | 0 | 0 | 0 |
| Days Min Temp ≤ 32 °F | 15 | 8 | 5 | 2 | 0 | 0 | 0 | 0 | 0 | 1 | ***8*** | 15 | 54 |
| Days Min Temp ≤ 0 °F | 0 | 0 | 0 | 0 | 0 | 0 | 0 | 0 | 0 | 0 | 0 | 0 | 0 |
| Heating Degree Days | 604 | 456 | 433 | 320 | 165 | 51 | 7 | 7 | 35 | 171 | 442 | 622 | 3313 |
| Cooling Degree Days | 0 | 0 | 0 | 3 | 32 | 106 | 240 | 208 | 119 | 31 | 0 | 0 | 739 |
| Total Precipitation (") | 8.90 | 6.85 | 6.22 | 2.70 | 1.19 | 0.32 | 0.07 | 0.26 | 0.78 | 2.61 | 6.71 | 7.48 | 44.09 |
| Days ≥ 0.1" Precip | 9 | 9 | 9 | 5 | 2 | 1 | 0 | 1 | 1 | 4 | 8 | 9 | 58 |
| Total Snowfall (") | 0.4 | 0.0 | 0.0 | 0.0 | 0.0 | 0.0 | 0.0 | 0.0 | 0.0 | 0.0 | 0.0 | 0.1 | 0.5 |
| Days ≥ 1" Snow Depth | 0 | 0 | 0 | 0 | 0 | 0 | 0 | 0 | 0 | 0 | 0 | 0 | 0 |

## PRIEST VALLEY *Monterey County* ELEVATION 2251 ft LAT/LONG 36° 12 ' N / 120° 42 ' W

| | JAN | FEB | MAR | APR | MAY | JUN | JUL | AUG | SEP | OCT | NOV | DEC | YEAR |
|---|---|---|---|---|---|---|---|---|---|---|---|---|---|
| Maximum Temp °F | 57.5 | 59.8 | 62.3 | 68.6 | 77.9 | 86.9 | 93.9 | 93.1 | 87.3 | 77.6 | 64.0 | 57.3 | 73.9 |
| Minimum Temp °F | 29.1 | 31.3 | 33.3 | 34.2 | 39.1 | 44.6 | 49.6 | 48.9 | 44.9 | 38.1 | 32.0 | 28.0 | 37.8 |
| Mean Temp °F | 43.3 | 45.6 | 47.8 | 51.4 | 58.5 | 65.7 | 71.7 | 71.0 | 66.1 | 57.9 | 48.0 | 42.7 | 55.8 |
| Days Max Temp ≥ 90 °F | 0 | 0 | 0 | 0 | 4 | 12 | 24 | 23 | 14 | 4 | 0 | 0 | 81 |
| Days Max Temp ≤ 32 °F | 0 | 0 | 0 | 0 | 0 | 0 | 0 | 0 | 0 | 0 | 0 | 0 | 0 |
| Days Min Temp ≤ 32 °F | 22 | 18 | 15 | 12 | 4 | 0 | 0 | 0 | 1 | 6 | 18 | 24 | 120 |
| Days Min Temp ≤ 0 °F | 0 | 0 | 0 | 0 | 0 | 0 | 0 | 0 | 0 | 0 | 0 | 0 | 0 |
| Heating Degree Days | 665 | 542 | 526 | 401 | 213 | 62 | 8 | 10 | 55 | 229 | 503 | 685 | 3899 |
| Cooling Degree Days | 0 | 0 | 0 | 1 | 19 | 92 | 222 | 202 | 100 | 20 | 0 | 0 | 656 |
| Total Precipitation (") | 4.18 | 3.63 | 3.71 | 1.49 | 0.29 | 0.08 | 0.08 | 0.08 | 0.48 | 0.97 | 2.49 | 3.22 | 20.70 |
| Days ≥ 0.1" Precip | 6 | 6 | 6 | 3 | 1 | 0 | 0 | 0 | 1 | 2 | 4 | 5 | 34 |
| Total Snowfall (") | 0.8 | 0.2 | 0.3 | 0.1 | 0.0 | 0.0 | 0.0 | 0.0 | 0.0 | 0.0 | 0.1 | 0.4 | 1.9 |
| Days ≥ 1" Snow Depth | 0 | 0 | 0 | 0 | 0 | 0 | 0 | 0 | 0 | 0 | 0 | 0 | 0 |

## RANDSBURG *Kern County* ELEVATION 3570 ft LAT/LONG 35° 22 ' N / 117° 39 ' W

| | JAN | FEB | MAR | APR | MAY | JUN | JUL | AUG | SEP | OCT | NOV | DEC | YEAR |
|---|---|---|---|---|---|---|---|---|---|---|---|---|---|
| Maximum Temp °F | 53.9 | 58.5 | 63.5 | 71.4 | 81.7 | 91.5 | 98.0 | 96.2 | 88.3 | 76.4 | 62.2 | 53.5 | 74.6 |
| Minimum Temp °F | 36.1 | 38.9 | 41.1 | 45.9 | 54.0 | 62.2 | 68.0 | 66.8 | 61.4 | 52.7 | 42.5 | 35.9 | 50.5 |
| Mean Temp °F | 45.0 | 48.7 | 52.3 | 58.7 | 67.8 | 76.9 | 83.1 | 81.5 | 74.9 | 64.6 | 52.3 | 44.6 | 62.5 |
| Days Max Temp ≥ 90 °F | 0 | 0 | 0 | 0 | 6 | 19 | 29 | 27 | 15 | 2 | 0 | 0 | 98 |
| Days Max Temp ≤ 32 °F | 0 | 0 | 0 | 0 | 0 | 0 | 0 | 0 | 0 | 0 | 0 | 0 | 0 |
| Days Min Temp ≤ 32 °F | 9 | 4 | 3 | 1 | 0 | 0 | 0 | 0 | 0 | 0 | 2 | 9 | 28 |
| Days Min Temp ≤ 0 °F | 0 | 0 | 0 | 0 | 0 | 0 | 0 | 0 | 0 | 0 | 0 | 0 | 0 |
| Heating Degree Days | 613 | 452 | 389 | 215 | 65 | 7 | 0 | 0 | 9 | 99 | 376 | 624 | 2849 |
| Cooling Degree Days | 0 | 0 | 2 | 46 | 156 | 369 | 557 | 516 | 314 | 92 | 3 | 0 | 2055 |
| Total Precipitation (") | 1.28 | 1.37 | 1.20 | 0.38 | 0.10 | 0.05 | 0.15 | 0.24 | 0.18 | 0.31 | 0.66 | 0.89 | 6.81 |
| Days ≥ 0.1" Precip | 2 | 2 | 3 | 1 | 0 | 0 | 0 | 1 | 1 | 1 | 1 | 2 | 14 |
| Total Snowfall (") | 1.1 | 0.4 | 0.5 | 0.1 | 0.0 | 0.0 | 0.0 | 0.0 | 0.0 | 0.0 | 0.1 | 1.2 | 3.4 |
| Days ≥ 1" Snow Depth | 0 | 0 | 0 | 0 | 0 | 0 | 0 | 0 | 0 | 0 | 0 | 1 | 1 |

## RED BLUFF MUNI AP *Tehama County* ELEVATION 349 ft LAT/LONG 40° 9 ' N / 122° 15 ' W

| | JAN | FEB | MAR | APR | MAY | JUN | JUL | AUG | SEP | OCT | NOV | DEC | YEAR |
|---|---|---|---|---|---|---|---|---|---|---|---|---|---|
| Maximum Temp °F | 55.3 | 60.6 | 65.0 | 71.6 | 82.5 | 90.7 | 97.8 | 96.2 | 90.5 | 79.2 | 63.0 | 54.6 | 75.6 |
| Minimum Temp °F | 37.1 | 40.6 | 43.7 | 46.9 | 54.3 | 61.6 | 65.7 | 63.8 | 59.4 | 51.3 | 42.7 | 37.4 | 50.4 |
| Mean Temp °F | 46.3 | 50.6 | 54.4 | 59.3 | 68.4 | 76.2 | 81.8 | 80.0 | 75.0 | 65.3 | 52.9 | 46.1 | 63.0 |
| Days Max Temp ≥ 90 °F | 0 | 0 | 0 | 1 | 8 | 17 | 27 | 25 | 18 | 5 | 0 | 0 | 101 |
| Days Max Temp ≤ 32 °F | 0 | 0 | 0 | 0 | 0 | 0 | 0 | 0 | 0 | 0 | 0 | 0 | 0 |
| Days Min Temp ≤ 32 °F | 8 | 3 | 1 | 0 | 0 | 0 | 0 | 0 | 0 | 0 | 2 | 7 | 21 |
| Days Min Temp ≤ 0 °F | 0 | 0 | 0 | 0 | 0 | 0 | 0 | 0 | 0 | 0 | 0 | 0 | 0 |
| Heating Degree Days | 574 | 400 | 324 | 189 | 48 | 6 | 0 | 0 | 7 | 80 | 360 | 580 | 2568 |
| Cooling Degree Days | 0 | 1 | 2 | 34 | 156 | 341 | 518 | 458 | 301 | 96 | 1 | 0 | 1908 |
| Total Precipitation (") | 4.31 | 3.66 | 3.07 | 1.35 | 0.86 | 0.42 | 0.07 | 0.22 | 0.65 | 1.26 | 3.34 | 3.79 | 23.00 |
| Days ≥ 0.1" Precip | 7 | 6 | 6 | 3 | 2 | 1 | 0 | 1 | 1 | 2 | 6 | 7 | 42 |
| Total Snowfall (") | 0.9 | 0.0 | 0.3 | 0.0 | 0.0 | 0.0 | 0.0 | 0.0 | 0.0 | 0.0 | 0.1 | 1.1 | 2.4 |
| Days ≥ 1" Snow Depth | 0 | 0 | 0 | 0 | 0 | 0 | 0 | 0 | 0 | 0 | 0 | 0 | 0 |

## REDLANDS *San Bernardino County* ELEVATION 1352 ft LAT/LONG 34° 3 ' N / 117° 11 ' W

| | JAN | FEB | MAR | APR | MAY | JUN | JUL | AUG | SEP | OCT | NOV | DEC | YEAR |
|---|---|---|---|---|---|---|---|---|---|---|---|---|---|
| Maximum Temp °F | 66.2 | 68.6 | 69.5 | 75.1 | 79.8 | 87.8 | 95.1 | 95.2 | 90.3 | 82.2 | 72.8 | 66.2 | 79.1 |
| Minimum Temp °F | 40.0 | 42.0 | 44.2 | 47.1 | 52.1 | 56.4 | 61.3 | 61.9 | 58.6 | 52.3 | 44.1 | 39.4 | 49.9 |
| Mean Temp °F | 53.1 | 55.3 | 56.9 | 61.1 | 66.0 | 72.1 | 78.2 | 78.6 | 74.5 | 67.3 | 58.5 | 52.8 | 64.5 |
| Days Max Temp ≥ 90 °F | 0 | 0 | 0 | 3 | 6 | 14 | 26 | 26 | 17 | 8 | 1 | 0 | 101 |
| Days Max Temp ≤ 32 °F | 0 | 0 | 0 | 0 | 0 | 0 | 0 | 0 | 0 | 0 | 0 | 0 | 0 |
| Days Min Temp ≤ 32 °F | 4 | 1 | 0 | 0 | 0 | 0 | 0 | 0 | 0 | 0 | 1 | 3 | 9 |
| Days Min Temp ≤ 0 °F | 0 | 0 | 0 | 0 | 0 | 0 | 0 | 0 | 0 | 0 | 0 | 0 | 0 |
| Heating Degree Days | 363 | 273 | 255 | 147 | 63 | 12 | 0 | 0 | 3 | 46 | 202 | 370 | 1734 |
| Cooling Degree Days | 1 | 4 | 10 | 57 | 110 | 251 | 415 | 435 | 317 | 125 | 15 | 0 | 1740 |
| Total Precipitation (") | 2.78 | 2.78 | 2.51 | 0.96 | 0.38 | 0.10 | 0.11 | 0.25 | 0.39 | 0.46 | 1.31 | 1.93 | 13.96 |
| Days ≥ 0.1" Precip | 5 | 4 | 5 | 2 | 1 | 0 | 0 | 0 | 1 | 1 | 3 | 4 | 26 |
| Total Snowfall (") | 0.0 | 0.0 | 0.0 | 0.0 | 0.0 | 0.0 | 0.0 | 0.0 | 0.0 | 0.0 | 0.0 | 0.0 | 0.0 |
| Days ≥ 1" Snow Depth | 0 | 0 | 0 | 0 | 0 | 0 | 0 | 0 | 0 | 0 | 0 | 0 | 0 |

## REDWOOD CITY *San Mateo County* ELEVATION 31 ft LAT/LONG 37° 29 ' N / 122° 14 ' W

| | JAN | FEB | MAR | APR | MAY | JUN | JUL | AUG | SEP | OCT | NOV | DEC | YEAR |
|---|---|---|---|---|---|---|---|---|---|---|---|---|---|
| Maximum Temp °F | 58.5 | 62.6 | 65.5 | 70.6 | 75.8 | 80.5 | 83.1 | 82.9 | 81.2 | 75.3 | 65.2 | 58.2 | 71.6 |
| Minimum Temp °F | 39.0 | 41.9 | 44.0 | 45.1 | 48.4 | 52.3 | 54.5 | 54.6 | 52.9 | 48.9 | 43.5 | 39.0 | 47.0 |
| Mean Temp °F | 48.8 | 52.3 | 54.8 | 57.9 | 62.2 | 66.4 | 68.8 | 68.8 | 67.1 | 62.1 | 54.4 | 48.7 | 59.4 |
| Days Max Temp ≥ 90 °F | 0 | 0 | 0 | 0 | 2 | 5 | 5 | 5 | 5 | 1 | 0 | 0 | 23 |
| Days Max Temp ≤ 32 °F | 0 | 0 | 0 | 0 | 0 | 0 | 0 | 0 | 0 | 0 | 0 | 0 | 0 |
| Days Min Temp ≤ 32 °F | 5 | 1 | 0 | 0 | 0 | 0 | 0 | 0 | 0 | 0 | 1 | 5 | 12 |
| Days Min Temp ≤ 0 °F | 0 | 0 | 0 | 0 | 0 | 0 | 0 | 0 | 0 | 0 | 0 | 0 | 0 |
| Heating Degree Days | 495 | 354 | 311 | 215 | 118 | 40 | 9 | 4 | 21 | 108 | 312 | 500 | 2487 |
| Cooling Degree Days | 0 | 0 | 0 | 10 | 35 | 98 | 142 | 130 | 89 | 35 | 0 | 0 | 539 |
| Total Precipitation (") | 4.25 | 3.49 | 3.08 | 1.22 | 0.25 | 0.07 | 0.04 | 0.08 | 0.20 | 0.98 | 2.74 | 3.30 | 19.70 |
| Days ≥ 0.1" Precip | 6 | 6 | 6 | 4 | 1 | 0 | 0 | 0 | 1 | 2 | 5 | 6 | 37 |
| Total Snowfall (") | 0.0 | 0.0 | 0.0 | 0.0 | 0.0 | 0.0 | 0.0 | 0.0 | 0.0 | 0.0 | 0.0 | 0.0 | 0.0 |
| Days ≥ 1" Snow Depth | 0 | 0 | 0 | 0 | 0 | 0 | 0 | 0 | 0 | 0 | 0 | 0 | 0 |

## RICHARDSON GROVE S P *Humboldt County* ELEVATION 500 ft LAT/LONG 40° 2 ' N / 123° 47 ' W

| | JAN | FEB | MAR | APR | MAY | JUN | JUL | AUG | SEP | OCT | NOV | DEC | YEAR |
|---|---|---|---|---|---|---|---|---|---|---|---|---|---|
| Maximum Temp °F | 49.9 | 54.7 | 59.6 | 64.6 | 72.0 | 78.8 | 86.4 | 87.0 | 83.2 | 70.6 | 55.7 | 49.2 | 67.6 |
| Minimum Temp °F | 36.8 | 37.9 | 39.6 | 41.1 | 45.4 | 50.0 | 53.0 | 52.8 | 49.3 | 44.8 | 40.3 | 36.7 | 44.0 |
| Mean Temp °F | 43.4 | 46.3 | 49.6 | 52.9 | 58.7 | 64.4 | 69.7 | 70.0 | 66.3 | 57.7 | 48.0 | 43.0 | 55.8 |
| Days Max Temp ≥ 90 °F | 0 | 0 | 0 | 0 | 2 | 4 | 12 | 12 | 8 | 1 | 0 | 0 | 39 |
| Days Max Temp ≤ 32 °F | 0 | 0 | 0 | 0 | 0 | 0 | 0 | 0 | 0 | 0 | 0 | 0 | 0 |
| Days Min Temp ≤ 32 °F | 8 | 5 | 2 | 0 | 0 | 0 | 0 | 0 | 0 | 0 | 3 | 9 | 27 |
| Days Min Temp ≤ 0 °F | 0 | 0 | 0 | 0 | 0 | 0 | 0 | 0 | 0 | 0 | 0 | 0 | 0 |
| Heating Degree Days | 664 | 521 | 470 | 359 | 206 | 76 | 14 | 11 | 48 | 229 | 502 | 676 | 3776 |
| Cooling Degree Days | 0 | 0 | 0 | 1 | 18 | 58 | 150 | 159 | 84 | 13 | 0 | 0 | 483 |
| Total Precipitation (") | 12.64 | 9.91 | 9.29 | 4.33 | 1.72 | 0.57 | 0.09 | 0.47 | 1.28 | 3.92 | 10.02 | 12.14 | 66.38 |
| Days ≥ 0.1" Precip | 11 | 10 | 11 | 7 | 3 | 1 | 0 | 1 | 2 | 5 | 10 | 12 | 73 |
| Total Snowfall (") | 0.2 | 0.0 | 0.1 | 0.0 | 0.0 | 0.0 | 0.0 | 0.0 | 0.0 | 0.0 | 0.0 | 0.1 | 0.4 |
| Days ≥ 1" Snow Depth | 0 | 0 | 0 | 0 | 0 | 0 | 0 | 0 | 0 | 0 | 0 | 0 | 0 |

## RICHMOND *Contra Costa County* ELEVATION 55 ft LAT/LONG 37° 56 ' N / 122° 21 ' W

| | JAN | FEB | MAR | APR | MAY | JUN | JUL | AUG | SEP | OCT | NOV | DEC | YEAR |
|---|---|---|---|---|---|---|---|---|---|---|---|---|---|
| Maximum Temp °F | 57.8 | 61.9 | 64.1 | 67.1 | 69.2 | 71.2 | 71.0 | 71.4 | 74.1 | 72.5 | 64.7 | 57.7 | 66.9 |
| Minimum Temp °F | 42.1 | 45.1 | 46.9 | 48.6 | 51.6 | 54.6 | 55.7 | 56.3 | 56.3 | 53.1 | 47.7 | 42.4 | 50.0 |
| Mean Temp °F | 50.0 | 53.5 | 55.6 | 57.9 | 60.4 | 62.9 | 63.4 | 63.9 | 65.3 | 62.8 | 56.2 | 50.1 | 58.5 |
| Days Max Temp ≥ 90 °F | 0 | 0 | 0 | 0 | 0 | 1 | 0 | 0 | 2 | 1 | 0 | 0 | 4 |
| Days Max Temp ≤ 32 °F | 0 | 0 | 0 | 0 | 0 | 0 | 0 | 0 | 0 | 0 | 0 | 0 | 0 |
| Days Min Temp ≤ 32 °F | 1 | 0 | 0 | 0 | 0 | 0 | 0 | 0 | 0 | 0 | 0 | 1 | 2 |
| Days Min Temp ≤ 0 °F | 0 | 0 | 0 | 0 | 0 | 0 | 0 | 0 | 0 | 0 | 0 | 0 | 0 |
| Heating Degree Days | 460 | 317 | 287 | 214 | 149 | 87 | 71 | 55 | 41 | 91 | 258 | 456 | 2486 |
| Cooling Degree Days | 0 | 0 | 1 | 10 | 13 | 37 | 37 | 26 | 48 | 33 | 1 | 0 | 206 |
| Total Precipitation (") | 4.67 | 3.60 | 3.32 | 1.42 | 0.26 | 0.17 | 0.07 | 0.07 | 0.27 | 1.18 | 3.57 | 3.54 | 22.14 |
| Days ≥ 0.1" Precip | 7 | 6 | 7 | 4 | 1 | 0 | 0 | 0 | 1 | 2 | 6 | 6 | 40 |
| Total Snowfall (") | 0.0 | 0.0 | 0.0 | 0.0 | 0.0 | 0.0 | 0.0 | 0.0 | 0.0 | 0.0 | 0.0 | 0.0 | 0.0 |
| Days ≥ 1" Snow Depth | 0 | 0 | 0 | 0 | 0 | 0 | 0 | 0 | 0 | 0 | 0 | 0 | 0 |

## RIVERSIDE CITRUS EXP *Riverside County* ELEVATION 1047 ft LAT/LONG 33° 58 ' N / 117° 20 ' W

| | JAN | FEB | MAR | APR | MAY | JUN | JUL | AUG | SEP | OCT | NOV | DEC | YEAR |
|---|---|---|---|---|---|---|---|---|---|---|---|---|---|
| Maximum Temp °F | 65.9 | 68.1 | 69.2 | 74.7 | 79.2 | 86.3 | 93.1 | 93.5 | 89.3 | 81.8 | 72.5 | 66.1 | 78.3 |
| Minimum Temp °F | 41.7 | 43.2 | 45.3 | 48.0 | 52.7 | 56.5 | 60.7 | 61.4 | 58.5 | 52.4 | 45.3 | 40.9 | 50.5 |
| Mean Temp °F | 53.8 | 55.7 | 57.3 | 61.4 | 66.0 | 71.4 | 76.9 | 77.5 | 73.9 | 67.1 | 58.9 | 53.6 | 64.5 |
| Days Max Temp ≥ 90 °F | 0 | 0 | 0 | 2 | 4 | 12 | 23 | 24 | 15 | 7 | 1 | 0 | 88 |
| Days Max Temp ≤ 32 °F | 0 | 0 | 0 | 0 | 0 | 0 | 0 | 0 | 0 | 0 | 0 | 0 | 0 |
| Days Min Temp ≤ 32 °F | 2 | 1 | 0 | 0 | 0 | 0 | 0 | 0 | 0 | 0 | 0 | 2 | 5 |
| Days Min Temp ≤ 0 °F | 0 | 0 | 0 | 0 | 0 | 0 | 0 | 0 | 0 | 0 | 0 | 0 | 0 |
| Heating Degree Days | 341 | 261 | 240 | 136 | 54 | 12 | 0 | 0 | 5 | 44 | 192 | 348 | 1633 |
| Cooling Degree Days | 0 | 5 | 8 | 47 | 96 | 226 | 381 | 404 | 294 | 119 | 12 | 1 | 1593 |
| Total Precipitation (") | 2.19 | 2.11 | 2.12 | 0.79 | 0.20 | 0.05 | 0.06 | 0.19 | 0.26 | 0.27 | 1.10 | 1.46 | 10.80 |
| Days ≥ 0.1" Precip | 4 | 4 | 4 | 2 | 1 | 0 | 0 | 0 | 1 | 1 | 2 | 3 | 22 |
| Total Snowfall (") | 0.0 | 0.0 | 0.0 | 0.0 | 0.0 | 0.0 | 0.0 | 0.0 | 0.0 | 0.0 | 0.0 | 0.0 | 0.0 |
| Days ≥ 1" Snow Depth | 0 | 0 | 0 | 0 | 0 | 0 | 0 | 0 | 0 | 0 | 0 | 0 | 0 |

## RIVERSIDE FIRE STN 3 *Riverside County* ELEVATION 902 ft LAT/LONG 33° 58 ' N / 117° 21 ' W

| | JAN | FEB | MAR | APR | MAY | JUN | JUL | AUG | SEP | OCT | NOV | DEC | YEAR |
|---|---|---|---|---|---|---|---|---|---|---|---|---|---|
| Maximum Temp °F | 67.5 | 70.1 | 71.4 | 76.6 | 81.3 | 88.0 | 94.3 | 94.6 | 90.2 | 83.0 | 73.6 | 67.3 | 79.8 |
| Minimum Temp °F | 40.9 | 42.7 | 45.2 | 48.1 | 53.2 | 57.4 | 61.9 | 62.6 | 59.3 | 52.6 | 44.8 | 40.1 | 50.7 |
| Mean Temp °F | 54.3 | 56.4 | 58.3 | 62.3 | 67.3 | 72.7 | 78.1 | 78.6 | 74.8 | 67.8 | 59.2 | 53.7 | 65.3 |
| Days Max Temp ≥ 90 °F | 0 | 0 | 1 | 3 | 5 | 13 | 25 | 25 | 16 | 8 | 1 | 0 | 97 |
| Days Max Temp ≤ 32 °F | 0 | 0 | 0 | 0 | 0 | 0 | 0 | 0 | 0 | 0 | 0 | 0 | 0 |
| Days Min Temp ≤ 32 °F | 3 | 1 | 0 | 0 | 0 | 0 | 0 | 0 | 0 | 0 | 0 | 3 | 7 |
| Days Min Temp ≤ 0 °F | 0 | 0 | 0 | 0 | 0 | 0 | 0 | 0 | 0 | 0 | 0 | 0 | 0 |
| Heating Degree Days | 327 | 238 | 209 | 112 | 34 | 5 | 0 | 0 | 2 | 31 | 179 | 344 | 1481 |
| Cooling Degree Days | 2 | 5 | 12 | 61 | 126 | 266 | 428 | 453 | 331 | 145 | 14 | 1 | 1844 |
| Total Precipitation (") | 2.12 | 2.11 | 2.07 | 0.69 | 0.16 | 0.05 | 0.04 | 0.19 | 0.25 | 0.29 | 1.02 | 1.35 | 10.34 |
| Days ≥ 0.1" Precip | 4 | 4 | 4 | 2 | 1 | 0 | 0 | 0 | 0 | 1 | 2 | 3 | 21 |
| Total Snowfall (") | 0.0 | 0.0 | 0.0 | 0.0 | 0.0 | 0.0 | 0.0 | 0.0 | 0.0 | 0.0 | 0.0 | 0.0 | 0.0 |
| Days ≥ 1" Snow Depth | 0 | 0 | 0 | 0 | 0 | 0 | 0 | 0 | 0 | 0 | 0 | 0 | 0 |

## S BERNARDINO CO HOSP *San Bernardino County* ELEVATION 1132 ft LAT/LONG 34° 8 ' N / 117° 16 ' W

| | JAN | FEB | MAR | APR | MAY | JUN | JUL | AUG | SEP | OCT | NOV | DEC | YEAR |
|---|---|---|---|---|---|---|---|---|---|---|---|---|---|
| Maximum Temp °F | 66.7 | 68.8 | 70.0 | 75.5 | 80.9 | 88.8 | 96.0 | 96.0 | 91.1 | 82.7 | 72.9 | 66.5 | 79.7 |
| Minimum Temp °F | 40.9 | 43.0 | 45.2 | 48.3 | 53.1 | 57.5 | 62.3 | 63.1 | 60.0 | 53.1 | 45.3 | 40.4 | 51.0 |
| Mean Temp °F | 53.8 | 55.9 | 57.6 | 61.9 | 67.0 | 73.2 | 79.2 | 79.5 | 75.5 | 67.9 | 59.1 | 53.4 | 65.3 |
| Days Max Temp ≥ 90 °F | 0 | 0 | 0 | 2 | 6 | 15 | 26 | 26 | 17 | 8 | 1 | 0 | 101 |
| Days Max Temp ≤ 32 °F | 0 | 0 | 0 | 0 | 0 | 0 | 0 | 0 | 0 | 0 | 0 | 0 | 0 |
| Days Min Temp ≤ 32 °F | 3 | 1 | 0 | 0 | 0 | 0 | 0 | 0 | 0 | 0 | 0 | 3 | 7 |
| Days Min Temp ≤ 0 °F | 0 | 0 | 0 | 0 | 0 | 0 | 0 | 0 | 0 | 0 | 0 | 0 | 0 |
| Heating Degree Days | 340 | 255 | 230 | 124 | 45 | 7 | 0 | 0 | 2 | 33 | 188 | 353 | 1577 |
| Cooling Degree Days | 1 | 5 | 8 | 57 | 123 | 271 | 439 | 459 | 334 | 133 | 12 | 1 | 1843 |
| Total Precipitation (") | 3.32 | 3.21 | 3.14 | 1.15 | 0.35 | 0.05 | 0.05 | 0.24 | 0.39 | 0.65 | 1.61 | 2.35 | 16.51 |
| Days ≥ 0.1" Precip | 5 | 4 | 5 | 3 | 1 | 0 | 0 | 0 | 1 | 1 | 3 | 4 | 27 |
| Total Snowfall (") | 0.0 | 0.0 | 0.0 | 0.0 | 0.0 | 0.0 | 0.0 | 0.0 | 0.0 | 0.0 | 0.0 | 0.0 | 0.0 |
| Days ≥ 1" Snow Depth | 0 | 0 | 0 | 0 | 0 | 0 | 0 | 0 | 0 | 0 | 0 | 0 | 0 |

## SACRAMENTO EXEC ARPT *Sacramento County* ELEVATION 23 ft LAT/LONG 38° 31 ' N / 121° 30 ' W

| | JAN | FEB | MAR | APR | MAY | JUN | JUL | AUG | SEP | OCT | NOV | DEC | YEAR |
|---|---|---|---|---|---|---|---|---|---|---|---|---|---|
| Maximum Temp °F | 53.2 | 60.1 | 64.4 | 71.1 | 80.7 | 87.6 | 92.9 | 91.9 | 87.6 | 78.3 | 63.4 | 52.8 | 73.7 |
| Minimum Temp °F | 38.1 | 41.4 | 43.9 | 45.8 | 50.7 | 55.3 | 58.2 | 58.0 | 55.8 | 50.5 | 43.1 | 37.5 | 48.2 |
| Mean Temp °F | 45.6 | 50.8 | 54.1 | 58.5 | 65.7 | 71.5 | 75.6 | 75.0 | 71.7 | 64.4 | 53.3 | 45.2 | 61.0 |
| Days Max Temp ≥ 90 °F | 0 | 0 | 0 | 0 | 6 | 12 | 22 | 20 | 13 | 3 | 0 | 0 | 76 |
| Days Max Temp ≤ 32 °F | 0 | 0 | 0 | 0 | 0 | 0 | 0 | 0 | 0 | 0 | 0 | 0 | 0 |
| Days Min Temp ≤ 32 °F | 6 | 2 | 0 | 0 | 0 | 0 | 0 | 0 | 0 | 0 | 1 | 7 | 16 |
| Days Min Temp ≤ 0 °F | 0 | 0 | 0 | 0 | 0 | 0 | 0 | 0 | 0 | 0 | 0 | 0 | 0 |
| Heating Degree Days | 593 | 394 | 330 | 201 | 67 | 10 | 1 | 0 | 8 | 74 | 346 | 607 | 2631 |
| Cooling Degree Days | 0 | 0 | 1 | 19 | 90 | 210 | 331 | 303 | 207 | 72 | 0 | 0 | 1233 |
| Total Precipitation (") | 3.71 | 3.02 | 2.61 | 1.13 | 0.34 | 0.17 | 0.05 | 0.07 | 0.36 | 0.84 | 2.48 | 2.51 | 17.29 |
| Days ≥ 0.1" Precip | 6 | 6 | 6 | 3 | 1 | 0 | 0 | 0 | 1 | 2 | 5 | 5 | 35 |
| Total Snowfall (") | 0.0 | 0.1 | 0.0 | 0.0 | 0.0 | 0.0 | 0.0 | 0.0 | 0.0 | 0.0 | 0.0 | 0.0 | 0.1 |
| Days ≥ 1" Snow Depth | 0 | 0 | 0 | 0 | 0 | 0 | 0 | 0 | 0 | 0 | 0 | 0 | 0 |

## SACRAMENTO WSO CITY *Sacramento County* ELEVATION 66 ft LAT/LONG 38° 35 ' N / 121° 30 ' W

| | JAN | FEB | MAR | APR | MAY | JUN | JUL | AUG | SEP | OCT | NOV | DEC | YEAR |
|---|---|---|---|---|---|---|---|---|---|---|---|---|---|
| Maximum Temp °F | 54.6 | 62.0 | 66.7 | 73.2 | 81.9 | 88.3 | 93.7 | 92.4 | 88.4 | 79.1 | 64.0 | 54.1 | 74.9 |
| Minimum Temp °F | 40.7 | 44.2 | 46.7 | 49.2 | 54.0 | 58.2 | 60.7 | 60.5 | 58.8 | 53.4 | 46.0 | 40.1 | 51.0 |
| Mean Temp °F | 47.7 | 53.1 | 56.7 | 61.2 | 68.0 | 73.3 | 77.2 | 76.5 | 73.6 | 66.3 | 55.0 | 47.1 | 63.0 |
| Days Max Temp ≥ 90 °F | 0 | 0 | 0 | 1 | 8 | 13 | 23 | 21 | 14 | 4 | 0 | 0 | 84 |
| Days Max Temp ≤ 32 °F | 0 | 0 | 0 | 0 | 0 | 0 | 0 | 0 | 0 | 0 | 0 | 0 | 0 |
| Days Min Temp ≤ 32 °F | 2 | 0 | 0 | 0 | 0 | 0 | 0 | 0 | 0 | 0 | 0 | 3 | 5 |
| Days Min Temp ≤ 0 °F | 0 | 0 | 0 | 0 | 0 | 0 | 0 | 0 | 0 | 0 | 0 | 0 | 0 |
| Heating Degree Days | 530 | 330 | 253 | 140 | 44 | 7 | 0 | 0 | 5 | 55 | 295 | 547 | 2206 |
| Cooling Degree Days | 0 | 0 | 3 | 43 | 145 | 271 | 394 | 363 | 274 | 115 | 2 | 0 | 1610 |
| Total Precipitation (") | 3.96 | 3.24 | 2.96 | 1.19 | 0.36 | 0.17 | 0.05 | 0.06 | 0.35 | 0.91 | 2.88 | 2.91 | 19.04 |
| Days ≥ 0.1" Precip | 6 | 6 | 6 | 3 | 1 | 1 | 0 | 0 | 1 | 2 | 5 | 6 | 37 |
| Total Snowfall (") | na | na | na | na | na | na | na | na | na | na | na | na | na |
| Days ≥ 1" Snow Depth | na | na | na | na | na | na | na | na | na | na | na | na | na |

## SAINT HELENA *Napa County* ELEVATION 259 ft LAT/LONG 38° 30 ' N / 122° 28 ' W

| | JAN | FEB | MAR | APR | MAY | JUN | JUL | AUG | SEP | OCT | NOV | DEC | YEAR |
|---|---|---|---|---|---|---|---|---|---|---|---|---|---|
| Maximum Temp °F | *57.2* | 61.8 | 65.0 | 71.5 | 79.4 | 84.7 | *89.4* | 89.0 | 85.6 | 77.6 | 65.3 | 57.3 | 73.6 |
| Minimum Temp °F | *35.8* | 38.9 | 40.9 | 43.0 | 47.4 | 51.5 | *53.0* | 53.1 | 50.6 | 46.4 | 40.8 | 35.8 | 44.8 |
| Mean Temp °F | *46.5* | 50.4 | 53.0 | 57.3 | 63.4 | 68.1 | *71.2* | 71.1 | 68.1 | 62.0 | 53.0 | 46.6 | 59.2 |
| Days Max Temp ≥ 90 °F | *0* | 0 | 0 | 1 | 5 | 9 | *14* | 14 | 10 | 3 | 0 | 0 | 56 |
| Days Max Temp ≤ 32 °F | *0* | 0 | 0 | 0 | 0 | 0 | *0* | 0 | 0 | 0 | 0 | 0 | 0 |
| Days Min Temp ≤ 32 °F | *12* | 5 | 2 | 1 | 0 | 0 | *0* | 0 | 0 | 0 | 3 | 11 | 34 |
| Days Min Temp ≤ 0 °F | *0* | 0 | 0 | 0 | 0 | 0 | *0* | 0 | 0 | 0 | 0 | 0 | 0 |
| Heating Degree Days | *566* | 405 | 367 | 236 | 99 | 26 | *3* | 4 | 23 | 118 | 351 | 565 | 2763 |
| Cooling Degree Days | na | 0 | *0* | 19 | *72* | *134* | *225* | *219* | 131 | 39 | 1 | 0 | na |
| Total Precipitation (") | 7.96 | 6.12 | 5.06 | 1.74 | 0.34 | 0.23 | 0.05 | 0.13 | 0.44 | 1.84 | 5.03 | 5.97 | 34.91 |
| Days ≥ 0.1" Precip | *8* | 8 | 8 | 4 | *1* | 0 | 0 | 1 | 1 | 3 | 6 | 7 | 47 |
| Total Snowfall (") | *0.2* | 0.0 | 0.1 | 0.0 | 0.0 | 0.0 | *0.0* | 0.0 | 0.0 | 0.0 | 0.0 | 0.0 | 0.3 |
| Days ≥ 1" Snow Depth | *0* | 0 | 0 | 0 | 0 | 0 | *0* | 0 | 0 | 0 | 0 | 0 | 0 |

## SALINAS 3 E *Monterey County* ELEVATION 79 ft LAT/LONG 36° 40 ' N / 121° 37 ' W

| | JAN | FEB | MAR | APR | MAY | JUN | JUL | AUG | SEP | OCT | NOV | DEC | YEAR |
|---|---|---|---|---|---|---|---|---|---|---|---|---|---|
| Maximum Temp °F | 62.4 | 64.4 | 65.0 | 66.9 | 68.2 | *70.2* | 71.1 | 72.6 | 74.7 | 73.6 | 67.3 | 62.2 | 68.2 |
| Minimum Temp °F | 40.5 | 42.6 | 44.2 | 45.4 | 49.2 | *51.8* | 54.1 | 54.8 | 53.5 | 49.6 | 44.4 | 40.3 | 47.5 |
| Mean Temp °F | 51.4 | 53.5 | 54.7 | 56.2 | 58.7 | *61.0* | 62.6 | 63.8 | 64.1 | 61.6 | 55.9 | 51.3 | 57.9 |
| Days Max Temp ≥ 90 °F | 0 | 0 | 0 | 0 | 0 | 1 | 1 | 0 | 2 | 2 | 0 | 0 | 6 |
| Days Max Temp ≤ 32 °F | 0 | 0 | 0 | 0 | 0 | 0 | 0 | 0 | 0 | 0 | 0 | 0 | 0 |
| Days Min Temp ≤ 32 °F | 4 | 1 | 1 | 0 | 0 | 0 | 0 | 0 | 0 | 0 | 0 | 3 | 9 |
| Days Min Temp ≤ 0 °F | 0 | 0 | 0 | 0 | 0 | 0 | 0 | 0 | 0 | 0 | 0 | 0 | 0 |
| Heating Degree Days | 414 | 318 | 315 | 263 | 196 | *125* | 82 | 53 | 57 | 124 | 269 | 418 | 2634 |
| Cooling Degree Days | 0 | 0 | 1 | 7 | 6 | 10 | 19 | 28 | 37 | 30 | 3 | 0 | 141 |
| Total Precipitation (") | 2.72 | 2.43 | 2.60 | 1.12 | 0.21 | 0.09 | 0.05 | 0.07 | 0.18 | 0.67 | 1.93 | 2.28 | 14.35 |
| Days ≥ 0.1" Precip | 4 | 5 | *5* | 3 | 1 | 0 | 0 | 0 | 0 | 1 | 4 | 4 | 27 |
| Total Snowfall (") | 0.0 | 0.0 | 0.0 | 0.0 | 0.0 | 0.0 | 0.0 | 0.0 | 0.0 | 0.0 | 0.0 | 0.0 | 0.0 |
| Days ≥ 1" Snow Depth | 0 | 0 | 0 | 0 | 0 | 0 | 0 | 0 | 0 | 0 | 0 | 0 | 0 |

## SALINAS MUNICIPAL AP *Monterey County* ELEVATION 52 ft LAT/LONG 36° 42 ' N / 120° 40 ' W

| | JAN | FEB | MAR | APR | MAY | JUN | JUL | AUG | SEP | OCT | NOV | DEC | YEAR |
|---|---|---|---|---|---|---|---|---|---|---|---|---|---|
| Maximum Temp °F | 60.9 | 62.6 | 63.6 | 66.4 | 67.7 | 69.9 | 71.0 | 72.1 | 74.6 | 73.1 | 66.5 | 60.7 | 67.4 |
| Minimum Temp °F | 40.4 | 42.4 | 44.0 | 45.4 | 49.4 | 52.3 | 54.1 | 54.8 | 53.9 | 49.9 | 44.1 | 39.8 | 47.5 |
| Mean Temp °F | 50.7 | 52.5 | 53.8 | 55.9 | 58.6 | 61.1 | 62.6 | 63.4 | 64.2 | 61.5 | 55.3 | 50.3 | 57.5 |
| Days Max Temp ≥ 90 °F | 0 | 0 | 0 | 0 | 0 | 1 | 0 | 0 | 2 | 2 | 0 | 0 | 5 |
| Days Max Temp ≤ 32 °F | 0 | 0 | 0 | 0 | 0 | 0 | 0 | 0 | 0 | 0 | 0 | 0 | 0 |
| Days Min Temp ≤ 32 °F | 3 | 1 | 0 | 0 | 0 | 0 | 0 | 0 | 0 | 0 | 0 | 3 | 7 |
| Days Min Temp ≤ 0 °F | 0 | 0 | 0 | 0 | 0 | 0 | 0 | 0 | 0 | 0 | 0 | 0 | 0 |
| Heating Degree Days | 436 | 345 | 341 | 272 | 199 | 123 | 87 | 62 | 56 | 128 | 286 | 448 | 2783 |
| Cooling Degree Days | 0 | 0 | 1 | 8 | 7 | 14 | 25 | 23 | 37 | 30 | 2 | 0 | 147 |
| Total Precipitation (") | 2.35 | 2.07 | 2.20 | 0.98 | 0.20 | 0.08 | 0.05 | 0.06 | 0.22 | 0.54 | 1.68 | 2.01 | 12.44 |
| Days ≥ 0.1" Precip | 5 | 5 | 5 | 3 | 1 | 0 | 0 | 0 | 1 | 1 | 4 | 5 | 30 |
| Total Snowfall (") | 0.0 | 0.0 | 0.0 | 0.0 | 0.0 | 0.0 | 0.0 | 0.0 | 0.0 | 0.0 | 0.0 | 0.0 | 0.0 |
| Days ≥ 1" Snow Depth | 0 | 0 | 0 | 0 | 0 | 0 | 0 | 0 | 0 | 0 | 0 | 0 | 0 |

## SALT SPRINGS PH *Amador County* ELEVATION 3700 ft LAT/LONG 38° 30 ' N / 120° 13 ' W

| | JAN | FEB | MAR | APR | MAY | JUN | JUL | AUG | SEP | OCT | NOV | DEC | YEAR |
|---|---|---|---|---|---|---|---|---|---|---|---|---|---|
| Maximum Temp °F | 53.0 | 56.0 | 58.9 | 63.9 | 72.4 | 79.9 | 87.1 | 87.4 | 82.2 | 73.1 | 59.8 | 52.2 | 68.8 |
| Minimum Temp °F | 34.2 | 35.0 | 36.2 | 39.7 | 46.5 | 53.1 | 59.9 | 59.5 | 55.1 | 47.9 | 39.2 | 34.2 | 45.0 |
| Mean Temp °F | 43.6 | 45.6 | 47.5 | 51.7 | 59.5 | 66.5 | 73.5 | 73.5 | 68.7 | 60.5 | 49.5 | 43.2 | 56.9 |
| Days Max Temp ≥ 90 °F | 0 | 0 | 0 | 0 | 0 | 4 | 12 | 12 | 6 | 1 | 0 | 0 | 35 |
| Days Max Temp ≤ 32 °F | 0 | 0 | 0 | 0 | 0 | 0 | 0 | 0 | 0 | 0 | 0 | 0 | 0 |
| Days Min Temp ≤ 32 °F | 13 | 11 | 9 | 5 | 1 | 0 | 0 | 0 | 0 | 1 | 5 | 12 | 57 |
| Days Min Temp ≤ 0 °F | 0 | 0 | 0 | 0 | 0 | 0 | 0 | 0 | 0 | 0 | 0 | 0 | 0 |
| Heating Degree Days | 657 | 542 | 535 | 398 | 202 | 66 | 7 | 9 | 47 | 179 | 459 | 668 | 3769 |
| Cooling Degree Days | 0 | 0 | 0 | 11 | 36 | 125 | 266 | 288 | 175 | 49 | 1 | 0 | 951 |
| Total Precipitation (") | 7.81 | 6.87 | 6.93 | 3.70 | 1.53 | 0.75 | 0.32 | 0.42 | 1.25 | 2.71 | 6.64 | 6.48 | 45.41 |
| Days ≥ 0.1" Precip | 8 | 8 | 9 | 7 | 4 | 2 | 1 | 1 | 2 | 4 | 7 | 7 | 60 |
| Total Snowfall (") | 16.1 | 13.3 | 15.7 | 9.7 | 0.2 | 0.0 | 0.0 | 0.0 | 0.0 | 0.4 | 3.3 | 12.8 | 71.5 |
| Days ≥ 1" Snow Depth | 8 | 4 | 3 | 2 | 0 | 0 | 0 | 0 | 0 | 0 | 1 | 6 | 24 |

## SAN DIEGO LINDBERGH *San Diego County* ELEVATION 30 ft LAT/LONG 32° 44 ' N / 117° 10 ' W

| | JAN | FEB | MAR | APR | MAY | JUN | JUL | AUG | SEP | OCT | NOV | DEC | YEAR |
|---|---|---|---|---|---|---|---|---|---|---|---|---|---|
| Maximum Temp °F | 65.9 | 66.4 | 66.4 | 68.6 | 69.3 | 72.1 | 76.1 | 77.7 | 77.1 | 74.5 | 70.1 | 66.0 | 70.9 |
| Minimum Temp °F | 49.0 | 50.8 | 53.2 | 56.1 | 59.4 | 62.3 | 65.9 | 67.5 | 65.8 | 61.1 | 53.8 | 48.7 | 57.8 |
| Mean Temp °F | 57.5 | 58.6 | 59.8 | 62.3 | 64.4 | 67.2 | 71.0 | 72.6 | 71.5 | 67.8 | 62.0 | 57.4 | 64.3 |
| Days Max Temp ≥ 90 °F | 0 | 0 | 0 | 0 | 0 | 1 | 0 | 0 | 1 | 1 | 0 | 0 | 3 |
| Days Max Temp ≤ 32 °F | 0 | 0 | 0 | 0 | 0 | 0 | 0 | 0 | 0 | 0 | 0 | 0 | 0 |
| Days Min Temp ≤ 32 °F | 0 | 0 | 0 | 0 | 0 | 0 | 0 | 0 | 0 | 0 | 0 | 0 | 0 |
| Days Min Temp ≤ 0 °F | 0 | 0 | 0 | 0 | 0 | 0 | 0 | 0 | 0 | 0 | 0 | 0 | 0 |
| Heating Degree Days | 229 | 177 | 158 | 89 | 43 | 8 | 0 | 0 | 0 | 10 | 100 | 231 | 1045 |
| Cooling Degree Days | 2 | 7 | 5 | 26 | 38 | 96 | 210 | 251 | 207 | 106 | 15 | 1 | 964 |
| Total Precipitation (") | 2.04 | 1.81 | 2.19 | 0.80 | 0.17 | 0.07 | 0.03 | 0.10 | 0.19 | 0.40 | 1.37 | 1.67 | 10.84 |
| Days ≥ 0.1" Precip | 4 | 4 | 4 | 2 | 0 | 0 | 0 | 0 | 0 | 1 | 3 | 4 | 22 |
| Total Snowfall (") | 0.0 | 0.0 | 0.0 | 0.0 | 0.0 | 0.0 | 0.0 | 0.0 | 0.0 | 0.0 | 0.0 | 0.0 | 0.0 |
| Days ≥ 1" Snow Depth | 0 | 0 | 0 | 0 | 0 | 0 | 0 | 0 | 0 | 0 | 0 | 0 | 0 |

## SAN FRANCISCO *San Francisco County* ELEVATION 207 ft LAT/LONG 37° 48 ' N / 122° 24 ' W

| | JAN | FEB | MAR | APR | MAY | JUN | JUL | AUG | SEP | OCT | NOV | DEC | YEAR |
|---|---|---|---|---|---|---|---|---|---|---|---|---|---|
| Maximum Temp °F | 57.1 | 60.5 | 61.5 | 63.3 | 64.4 | 66.3 | 66.6 | 67.7 | 70.4 | 69.9 | 63.5 | 57.1 | 64.0 |
| Minimum Temp °F | 46.1 | 48.5 | 49.3 | 49.9 | 51.1 | 53.1 | 54.0 | 55.3 | 56.1 | 54.8 | 51.2 | 46.5 | 51.3 |
| Mean Temp °F | 51.6 | 54.5 | 55.4 | 56.6 | 57.8 | 59.8 | 60.3 | 61.5 | 63.3 | 62.4 | 57.4 | 51.9 | 57.7 |
| Days Max Temp ≥ 90 °F | 0 | 0 | 0 | 0 | 0 | 1 | 0 | 0 | 1 | 1 | 0 | 0 | 3 |
| Days Max Temp ≤ 32 °F | 0 | 0 | 0 | 0 | 0 | 0 | 0 | 0 | 0 | 0 | 0 | 0 | 0 |
| Days Min Temp ≤ 32 °F | 0 | 0 | 0 | 0 | 0 | 0 | 0 | 0 | 0 | 0 | 0 | 0 | 0 |
| Days Min Temp ≤ 0 °F | 0 | 0 | 0 | 0 | 0 | 0 | 0 | 0 | 0 | 0 | 0 | 0 | 0 |
| Heating Degree Days | 408 | 290 | 290 | 252 | 226 | 168 | 151 | 116 | 83 | 109 | 227 | 400 | 2720 |
| Cooling Degree Days | 0 | 1 | 2 | 12 | 11 | 23 | 22 | 24 | 38 | 40 | 4 | 0 | 177 |
| Total Precipitation (") | 4.21 | 3.26 | 3.16 | 1.25 | 0.28 | 0.16 | 0.04 | 0.08 | 0.25 | 1.11 | 3.24 | 3.17 | 20.21 |
| Days ≥ 0.1" Precip | 7 | 7 | 6 | 3 | 1 | 0 | 0 | 0 | 1 | 2 | 6 | 7 | 40 |
| Total Snowfall (") | na | na | na | na | na | na | na | na | na | na | na | na | na |
| Days ≥ 1" Snow Depth | na | na | na | na | na | na | na | na | na | na | na | na | na |

## SAN FRANCISCO INT AP *San Mateo County* ELEVATION 7 ft LAT/LONG 37° 37 ' N / 122° 23 ' W

| | JAN | FEB | MAR | APR | MAY | JUN | JUL | AUG | SEP | OCT | NOV | DEC | YEAR |
|---|---|---|---|---|---|---|---|---|---|---|---|---|---|
| Maximum Temp °F | 55.9 | 59.4 | 61.3 | 64.1 | 67.3 | 70.3 | 71.8 | 72.4 | 73.5 | 70.5 | 62.7 | 55.9 | 65.4 |
| Minimum Temp °F | 42.0 | 44.9 | 46.4 | 47.6 | 50.0 | 52.7 | 54.1 | 55.1 | 54.8 | 52.1 | 47.2 | 42.4 | 49.1 |
| Mean Temp °F | 49.0 | 52.2 | 53.9 | 55.9 | 58.7 | 61.5 | 63.0 | 63.8 | 64.2 | 61.3 | 54.9 | 49.2 | 57.3 |
| Days Max Temp ≥ 90 °F | 0 | 0 | 0 | 0 | 0 | 1 | 1 | 0 | 1 | 0 | 0 | 0 | 3 |
| Days Max Temp ≤ 32 °F | 0 | 0 | 0 | 0 | 0 | 0 | 0 | 0 | 0 | 0 | 0 | 0 | 0 |
| Days Min Temp ≤ 32 °F | 1 | 0 | 0 | 0 | 0 | 0 | 0 | 0 | 0 | 0 | 0 | 1 | 2 |
| Days Min Temp ≤ 0 °F | 0 | 0 | 0 | 0 | 0 | 0 | 0 | 0 | 0 | 0 | 0 | 0 | 0 |
| Heating Degree Days | 490 | 355 | 338 | 271 | 199 | 117 | 79 | 55 | 57 | 128 | 295 | 483 | 2867 |
| Cooling Degree Days | 0 | 0 | 0 | 6 | 10 | 22 | 28 | 30 | 37 | 25 | 0 | 0 | 158 |
| Total Precipitation (") | 4.44 | 3.44 | 3.07 | 1.27 | 0.22 | 0.11 | 0.03 | 0.06 | 0.19 | 1.00 | 2.73 | 3.18 | 19.74 |
| Days ≥ 0.1" Precip | 7 | 7 | 6 | 3 | 1 | 0 | 0 | 0 | 1 | 2 | 5 | 6 | 38 |
| Total Snowfall (") | 0.0 | 0.0 | 0.0 | 0.0 | 0.0 | 0.0 | 0.0 | 0.0 | 0.0 | 0.0 | 0.0 | 0.0 | 0.0 |
| Days ≥ 1" Snow Depth | 0 | 0 | 0 | 0 | 0 | 0 | 0 | 0 | 0 | 0 | 0 | 0 | 0 |

## SAN FRANCISCO OCEANS *San Francisco County* ELEVATION 20 ft LAT/LONG 37° 46 ' N / 122° 30 ' W

| | JAN | FEB | MAR | APR | MAY | JUN | JUL | AUG | SEP | OCT | NOV | DEC | YEAR |
|---|---|---|---|---|---|---|---|---|---|---|---|---|---|
| Maximum Temp °F | ***57.6*** | 60.0 | 60.5 | 61.2 | 61.7 | 63.2 | 63.9 | 64.8 | 66.8 | ***66.6*** | 62.6 | 57.7 | 62.2 |
| Minimum Temp °F | ***43.5*** | 45.7 | 46.6 | 47.2 | 49.2 | 51.3 | 53.4 | ***54.4*** | 54.2 | ***52.1*** | 47.9 | 43.9 | 49.1 |
| Mean Temp °F | ***50.6*** | 52.9 | 53.6 | 54.2 | 55.5 | 57.3 | 58.7 | ***59.6*** | 60.5 | ***59.4*** | 55.3 | 50.8 | 55.7 |
| Days Max Temp ≥ 90 °F | 0 | 0 | 0 | 0 | 0 | 0 | 0 | 0 | 0 | 0 | 0 | 0 | 0 |
| Days Max Temp ≤ 32 °F | 0 | 0 | 0 | 0 | 0 | 0 | 0 | 0 | 0 | 0 | 0 | 0 | 0 |
| Days Min Temp ≤ 32 °F | 0 | 0 | 0 | 0 | 0 | 0 | 0 | 0 | 0 | 0 | 0 | 1 | 1 |
| Days Min Temp ≤ 0 °F | 0 | 0 | 0 | 0 | 0 | 0 | 0 | 0 | 0 | 0 | 0 | 0 | 0 |
| Heating Degree Days | ***440*** | 336 | 347 | 318 | 290 | 226 | 190 | ***163*** | 138 | ***182*** | 287 | 433 | 3350 |
| Cooling Degree Days | na | ***0*** | 0 | 3 | ***0*** | ***0*** | ***2*** | na | ***6*** | na | ***3*** | ***0*** | na |
| Total Precipitation (") | ***4.03*** | ***3.09*** | ***3.10*** | ***1.27*** | 0.31 | 0.16 | 0.03 | ***0.13*** | 0.18 | ***1.00*** | ***2.89*** | ***3.30*** | 19.49 |
| Days ≥ 0.1" Precip | ***6*** | ***6*** | ***6*** | ***3*** | 1 | 0 | 0 | ***0*** | 1 | ***2*** | ***5*** | ***7*** | 37 |
| Total Snowfall (") | 0.0 | 0.0 | 0.0 | 0.0 | 0.0 | 0.0 | 0.0 | 0.0 | 0.0 | ***0.0*** | 0.0 | 0.0 | 0.0 |
| Days ≥ 1" Snow Depth | 0 | 0 | 0 | 0 | 0 | 0 | 0 | 0 | 0 | ***0*** | 0 | 0 | 0 |

## SAN GABRIEL FIRE DEP *Los Angeles County* ELEVATION 400 ft LAT/LONG 34° 6 ' N / 118° 6 ' W

| | JAN | FEB | MAR | APR | MAY | JUN | JUL | AUG | SEP | OCT | NOV | DEC | YEAR |
|---|---|---|---|---|---|---|---|---|---|---|---|---|---|
| Maximum Temp °F | 69.7 | 71.5 | 72.1 | 76.0 | 78.3 | 83.5 | 89.0 | 90.2 | 88.0 | 83.0 | 75.1 | 69.6 | 78.8 |
| Minimum Temp °F | 41.9 | 44.1 | 46.7 | 49.3 | 54.0 | 57.9 | 61.7 | 62.6 | 60.1 | 54.3 | 46.5 | 41.5 | 51.7 |
| Mean Temp °F | 55.9 | 57.8 | 59.4 | 62.7 | 66.2 | 70.7 | 75.4 | 76.4 | 74.1 | 68.7 | 60.8 | 55.6 | 65.3 |
| Days Max Temp ≥ 90 °F | 0 | 0 | 1 | 2 | 3 | 6 | 14 | 17 | 13 | 7 | 1 | 0 | 64 |
| Days Max Temp ≤ 32 °F | 0 | 0 | 0 | 0 | 0 | 0 | 0 | 0 | 0 | 0 | 0 | 0 | 0 |
| Days Min Temp ≤ 32 °F | 2 | 0 | 0 | 0 | 0 | 0 | 0 | 0 | 0 | 0 | 0 | 2 | 4 |
| Days Min Temp ≤ 0 °F | 0 | 0 | 0 | 0 | 0 | 0 | 0 | 0 | 0 | 0 | 0 | 0 | 0 |
| Heating Degree Days | 279 | 203 | 180 | 104 | 43 | 8 | 0 | 0 | 1 | 19 | 139 | 286 | 1262 |
| Cooling Degree Days | 3 | 10 | 18 | 62 | 102 | 212 | 347 | 381 | 306 | 162 | 19 | 1 | 1623 |
| Total Precipitation (") | 4.07 | 4.14 | 3.65 | 1.13 | 0.23 | 0.08 | 0.04 | 0.11 | 0.49 | 0.50 | 2.02 | 2.43 | 18.89 |
| Days ≥ 0.1" Precip | 4 | 4 | 5 | 2 | 0 | 0 | 0 | 0 | 1 | 1 | 3 | 3 | 23 |
| Total Snowfall (") | 0.0 | 0.0 | 0.0 | 0.0 | 0.0 | 0.0 | 0.0 | 0.0 | 0.0 | 0.0 | 0.0 | 0.0 | 0.0 |
| Days ≥ 1" Snow Depth | 0 | 0 | 0 | 0 | 0 | 0 | 0 | 0 | 0 | 0 | 0 | 0 | 0 |

## SAN GREGORIO 2 SE *San Mateo County* ELEVATION 361 ft LAT/LONG 37° 18 ' N / 122° 20 ' W

| | JAN | FEB | MAR | APR | MAY | JUN | JUL | AUG | SEP | OCT | NOV | DEC | YEAR |
|---|---|---|---|---|---|---|---|---|---|---|---|---|---|
| Maximum Temp °F | 58.7 | 59.9 | 60.2 | 62.3 | 64.1 | 67.2 | 69.4 | 70.1 | 71.2 | 68.4 | 62.8 | 58.3 | 64.4 |
| Minimum Temp °F | 40.0 | 41.5 | 42.4 | 42.5 | 45.7 | 48.4 | 50.3 | 50.8 | 49.1 | 45.6 | 42.1 | 39.4 | 44.8 |
| Mean Temp °F | 49.4 | 50.7 | 51.3 | 52.4 | 54.9 | 57.8 | 59.9 | 60.5 | 60.2 | 57.0 | 52.5 | 48.9 | 54.6 |
| Days Max Temp ≥ 90 °F | 0 | 0 | 0 | 0 | 0 | 0 | 0 | 0 | 1 | 0 | 0 | 0 | 1 |
| Days Max Temp ≤ 32 °F | 0 | 0 | 0 | 0 | 0 | 0 | 0 | 0 | 0 | 0 | 0 | 0 | 0 |
| Days Min Temp ≤ 32 °F | 4 | 2 | 1 | 1 | 0 | 0 | 0 | 0 | 0 | 0 | 1 | 4 | 13 |
| Days Min Temp ≤ 0 °F | 0 | 0 | 0 | 0 | 0 | 0 | 0 | 0 | 0 | 0 | 0 | 0 | 0 |
| Heating Degree Days | 477 | 397 | 417 | 372 | 306 | 212 | 156 | 136 | 146 | 246 | 370 | 492 | 3727 |
| Cooling Degree Days | 0 | 0 | 0 | 2 | 0 | 3 | 6 | 4 | 6 | 6 | 0 | 0 | 27 |
| Total Precipitation (") | 5.37 | 4.44 | 4.51 | 2.04 | 0.59 | 0.30 | 0.14 | 0.21 | 0.40 | 1.59 | 4.08 | 4.64 | 28.31 |
| Days ≥ 0.1" Precip | 8 | 8 | 8 | 5 | 2 | 1 | 0 | 1 | 1 | 3 | 6 | 8 | 51 |
| Total Snowfall (") | 0.0 | 0.0 | 0.0 | 0.0 | 0.0 | 0.0 | 0.0 | 0.0 | 0.0 | 0.0 | 0.0 | 0.0 | 0.0 |
| Days ≥ 1" Snow Depth | 0 | 0 | 0 | 0 | 0 | 0 | 0 | 0 | 0 | 0 | 0 | 0 | 0 |

## SAN JOSE *Santa Clara County* ELEVATION 95 ft LAT/LONG 37° 20 ' N / 121° 53 ' W

| | JAN | FEB | MAR | APR | MAY | JUN | JUL | AUG | SEP | OCT | NOV | DEC | YEAR |
|---|---|---|---|---|---|---|---|---|---|---|---|---|---|
| Maximum Temp °F | 58.0 | 62.3 | 65.7 | 70.1 | 75.3 | 79.7 | 82.5 | 82.1 | 80.6 | 74.7 | 64.3 | 57.4 | 71.1 |
| Minimum Temp °F | 40.6 | 43.9 | 45.9 | 47.4 | 51.2 | 54.8 | 56.9 | 57.1 | 56.0 | 51.8 | 45.3 | 40.5 | 49.3 |
| Mean Temp °F | 49.3 | 53.1 | 55.8 | 58.8 | 63.3 | 67.3 | 69.7 | 69.6 | 68.3 | 63.3 | 54.9 | 48.9 | 60.2 |
| Days Max Temp ≥ 90 °F | 0 | 0 | 0 | 0 | 2 | 4 | 4 | 4 | 4 | 1 | 0 | 0 | 19 |
| Days Max Temp ≤ 32 °F | 0 | 0 | 0 | 0 | 0 | 0 | 0 | 0 | 0 | 0 | 0 | 0 | 0 |
| Days Min Temp ≤ 32 °F | 3 | 1 | 0 | 0 | 0 | 0 | 0 | 0 | 0 | 0 | 0 | 4 | 8 |
| Days Min Temp ≤ 0 °F | 0 | 0 | 0 | 0 | 0 | 0 | 0 | 0 | 0 | 0 | 0 | 0 | 0 |
| Heating Degree Days | 479 | 329 | 279 | 192 | 94 | 27 | 4 | 2 | 10 | 85 | 298 | 490 | 2289 |
| Cooling Degree Days | 0 | 0 | 1 | 18 | 48 | 115 | 175 | 166 | 124 | 53 | 1 | 0 | 701 |
| Total Precipitation (") | 2.80 | 2.42 | 2.48 | 1.12 | 0.27 | 0.06 | 0.06 | 0.12 | 0.23 | 0.76 | 2.00 | 2.07 | 14.39 |
| Days ≥ 0.1" Precip | 6 | 6 | 6 | 3 | 1 | 0 | 0 | 0 | 1 | 2 | 5 | 5 | 35 |
| Total Snowfall (") | 0.0 | 0.0 | 0.0 | 0.0 | 0.0 | 0.0 | 0.0 | 0.0 | 0.0 | 0.0 | 0.0 | 0.0 | 0.0 |
| Days ≥ 1" Snow Depth | 0 | 0 | 0 | 0 | 0 | 0 | 0 | 0 | 0 | 0 | 0 | 0 | 0 |

## SAN LEANDRO UPR FLTR *Alameda County* ELEVATION 394 ft LAT/LONG 37° 46 ' N / 122° 10 ' W

| | JAN | FEB | MAR | APR | MAY | JUN | JUL | AUG | SEP | OCT | NOV | DEC | YEAR |
|---|---|---|---|---|---|---|---|---|---|---|---|---|---|
| Maximum Temp °F | 57.3 | *60.9* | *62.7* | *66.1* | *69.2* | *72.2* | *75.0* | *75.5* | *76.1* | *72.7* | *64.3* | *57.8* | 67.5 |
| Minimum Temp °F | 40.5 | 42.3 | *43.4* | *44.4* | *47.6* | *51.2* | *52.8* | *53.8* | *53.7* | *50.7* | *45.5* | *41.0* | 47.2 |
| Mean Temp °F | 48.9 | *51.7* | *53.1* | *55.3* | *58.5* | *61.7* | *64.0* | *64.7* | *64.9* | *61.7* | *54.9* | *49.4* | 57.4 |
| Days Max Temp ≥ 90 °F | 0 | 0 | *0* | *0* | *1* | *2* | *2* | *1* | *2* | *1* | *0* | *0* | 9 |
| Days Max Temp ≤ 32 °F | 0 | 0 | *0* | *0* | *0* | *0* | *0* | *0* | *0* | *0* | *0* | *0* | 0 |
| Days Min Temp ≤ 32 °F | 2 | 1 | *0* | *0* | *0* | *0* | *0* | *0* | *0* | *0* | *0* | *2* | 5 |
| Days Min Temp ≤ 0 °F | 0 | 0 | *0* | *0* | *0* | *0* | *0* | *0* | *0* | *0* | *0* | *0* | 0 |
| Heating Degree Days | 492 | *371* | *363* | *293* | *211* | *122* | *69* | *46* | *55* | *131* | *297* | *476* | 2926 |
| Cooling Degree Days | *0* | *1* | na | na | na | na | na | na | na | na | na | na | na |
| Total Precipitation (") | 4.93 | 3.94 | 4.20 | 1.76 | 0.42 | 0.16 | 0.06 | 0.08 | 0.34 | 1.51 | 3.84 | 3.97 | 25.21 |
| Days ≥ 0.1" Precip | 7 | 7 | 8 | 4 | 1 | 0 | 0 | 0 | 1 | 2 | 6 | 7 | 43 |
| Total Snowfall (") | 0.0 | 0.0 | 0.0 | 0.0 | 0.0 | 0.0 | 0.0 | 0.0 | 0.0 | 0.0 | 0.0 | 0.0 | 0.0 |
| Days ≥ 1" Snow Depth | 0 | 0 | 0 | 0 | 0 | 0 | 0 | 0 | 0 | 0 | 0 | 0 | 0 |

## SAN LUIS DAM *Merced County* ELEVATION 259 ft LAT/LONG 37° 3 ' N / 121° 4 ' W

| | JAN | FEB | MAR | APR | MAY | JUN | JUL | AUG | SEP | OCT | NOV | DEC | YEAR |
|---|---|---|---|---|---|---|---|---|---|---|---|---|---|
| Maximum Temp °F | 53.6 | 60.1 | 65.2 | 71.1 | 78.7 | 85.3 | 91.6 | 90.6 | 86.3 | 77.7 | 64.2 | 54.0 | 73.2 |
| Minimum Temp °F | 37.4 | 41.6 | 46.0 | 49.3 | 54.0 | 58.8 | 63.8 | 63.6 | 60.3 | 53.8 | 45.1 | 37.5 | 50.9 |
| Mean Temp °F | 45.5 | 50.9 | 55.6 | 60.2 | 66.4 | 72.1 | 77.7 | 77.1 | 73.3 | 65.8 | 54.7 | 45.8 | 62.1 |
| Days Max Temp ≥ 90 °F | 0 | 0 | 0 | 0 | 5 | 10 | 20 | 18 | 11 | 2 | 0 | 0 | 66 |
| Days Max Temp ≤ 32 °F | 0 | 0 | 0 | 0 | 0 | 0 | 0 | 0 | 0 | 0 | 0 | 0 | 0 |
| Days Min Temp ≤ 32 °F | 7 | 1 | 0 | 0 | 0 | 0 | 0 | 0 | 0 | 0 | 1 | 6 | 15 |
| Days Min Temp ≤ 0 °F | 0 | 0 | 0 | 0 | 0 | 0 | 0 | 0 | 0 | 0 | 0 | 0 | 0 |
| Heating Degree Days | 597 | 392 | 285 | 165 | 66 | 16 | 1 | 1 | 4 | 59 | 306 | 589 | 2481 |
| Cooling Degree Days | 0 | 0 | 1 | 36 | 118 | 240 | 398 | 380 | 264 | 99 | 2 | 0 | 1538 |
| Total Precipitation (") | 1.87 | 1.78 | 1.53 | 0.63 | 0.25 | 0.06 | 0.03 | 0.10 | 0.26 | 0.45 | 1.41 | 1.40 | 9.77 |
| Days ≥ 0.1" Precip | 5 | 4 | 4 | 2 | 1 | 0 | 0 | 0 | 1 | 1 | 3 | 4 | 25 |
| Total Snowfall (") | 0.0 | 0.0 | 0.0 | 0.0 | 0.0 | 0.0 | 0.0 | 0.0 | 0.0 | 0.0 | 0.0 | 0.0 | 0.0 |
| Days ≥ 1" Snow Depth | 0 | 0 | 0 | 0 | 0 | 0 | 0 | 0 | 0 | 0 | 0 | 0 | 0 |

## SAN LUIS OBISPO POLY *San Luis Obispo County* ELEVATION 200 ft LAT/LONG 35° 17 ' N / 120° 40 ' W

| | JAN | FEB | MAR | APR | MAY | JUN | JUL | AUG | SEP | OCT | NOV | DEC | YEAR |
|---|---|---|---|---|---|---|---|---|---|---|---|---|---|
| Maximum Temp °F | 63.5 | 65.0 | 65.4 | 68.3 | 70.7 | 74.9 | 78.4 | 79.5 | 79.2 | 76.5 | 70.0 | 64.1 | 71.3 |
| Minimum Temp °F | 41.6 | 43.1 | 43.9 | 45.0 | 47.1 | 50.3 | 52.4 | 52.9 | 52.6 | 49.8 | 45.5 | 41.0 | 47.1 |
| Mean Temp °F | 52.6 | 54.1 | 54.7 | 56.7 | 58.9 | 62.6 | 65.4 | 66.2 | 66.0 | 63.2 | 57.8 | 52.6 | 59.2 |
| Days Max Temp ≥ 90 °F | 0 | 0 | 0 | 1 | 1 | 2 | 2 | 2 | 4 | 3 | 0 | 0 | 15 |
| Days Max Temp ≤ 32 °F | 0 | 0 | 0 | 0 | 0 | 0 | 0 | 0 | 0 | 0 | 0 | 0 | 0 |
| Days Min Temp ≤ 32 °F | 2 | 0 | 0 | 0 | 0 | 0 | 0 | 0 | 0 | 0 | 0 | 2 | 4 |
| Days Min Temp ≤ 0 °F | 0 | 0 | 0 | 0 | 0 | 0 | 0 | 0 | 0 | 0 | 0 | 0 | 0 |
| Heating Degree Days | 379 | 303 | 315 | 254 | 200 | 100 | 36 | 22 | 39 | 98 | 221 | 380 | 2347 |
| Cooling Degree Days | 0 | 2 | 2 | 18 | 14 | 41 | 64 | 75 | ***81*** | 50 | 11 | 1 | 359 |
| Total Precipitation (") | 5.29 | 4.73 | 4.13 | 1.50 | 0.33 | 0.06 | 0.05 | 0.09 | 0.54 | 0.96 | 2.44 | 3.84 | 23.96 |
| Days ≥ 0.1" Precip | 6 | 6 | 6 | 3 | 1 | 0 | 0 | 0 | 1 | 2 | 4 | 5 | 34 |
| Total Snowfall (") | 0.0 | 0.0 | 0.0 | 0.0 | 0.0 | 0.0 | 0.0 | 0.0 | 0.0 | 0.0 | 0.0 | 0.0 | 0.0 |
| Days ≥ 1" Snow Depth | 0 | 0 | 0 | 0 | 0 | 0 | 0 | 0 | 0 | 0 | 0 | 0 | 0 |

## SAN RAFAEL CIVIC CEN *Marin County* ELEVATION 30 ft LAT/LONG 37° 58 ' N / 122° 32 ' W

| | JAN | FEB | MAR | APR | MAY | JUN | JUL | AUG | SEP | OCT | NOV | DEC | YEAR |
|---|---|---|---|---|---|---|---|---|---|---|---|---|---|
| Maximum Temp °F | ***57.2*** | 62.0 | 65.3 | 69.5 | 73.9 | 78.1 | 81.6 | 81.7 | 80.7 | 75.8 | 64.8 | 56.6 | 70.6 |
| Minimum Temp °F | ***41.1*** | 43.7 | 45.2 | ***46.7*** | 49.6 | 52.8 | 54.4 | 54.7 | 53.9 | 50.9 | 46.2 | 41.1 | 48.4 |
| Mean Temp °F | ***49.1*** | 52.9 | 55.3 | ***58.2*** | 61.8 | 65.4 | 68.0 | 68.2 | 67.3 | 63.3 | 55.5 | 48.9 | 59.5 |
| Days Max Temp ≥ 90 °F | ***0*** | 0 | 0 | 0 | 2 | 3 | 4 | 4 | 5 | 2 | 0 | 0 | 20 |
| Days Max Temp ≤ 32 °F | ***0*** | 0 | 0 | 0 | 0 | 0 | 0 | 0 | 0 | 0 | 0 | 0 | 0 |
| Days Min Temp ≤ 32 °F | ***1*** | 0 | 0 | 0 | 0 | 0 | 0 | 0 | 0 | 0 | 0 | 2 | 3 |
| Days Min Temp ≤ 0 °F | ***0*** | 0 | 0 | 0 | 0 | 0 | 0 | 0 | 0 | 0 | 0 | 0 | 0 |
| Heating Degree Days | ***485*** | 336 | 296 | ***205*** | 132 | 51 | 13 | 8 | 26 | 87 | 278 | 492 | 2409 |
| Cooling Degree Days | na | ***1*** | ***1*** | ***11*** | ***29*** | ***65*** | ***113*** | ***105*** | ***82*** | ***38*** | ***2*** | ***0*** | na |
| Total Precipitation (") | ***8.26*** | 6.24 | ***4.40*** | 1.63 | 0.38 | 0.18 | 0.07 | 0.10 | 0.33 | ***1.73*** | 4.76 | 5.90 | 33.98 |
| Days ≥ 0.1" Precip | ***8*** | 7 | ***7*** | 4 | 1 | 0 | 0 | 0 | 1 | ***2*** | 6 | ***6*** | 42 |
| Total Snowfall (") | ***0.0*** | 0.0 | ***0.0*** | 0.0 | 0.0 | 0.0 | 0.0 | 0.0 | 0.0 | 0.0 | 0.0 | 0.0 | 0.0 |
| Days ≥ 1" Snow Depth | ***0*** | 0 | 0 | 0 | 0 | 0 | 0 | 0 | 0 | 0 | 0 | 0 | 0 |

## SANDBERG WSMO *Los Angeles County* ELEVATION 4524 ft LAT/LONG 34° 45 ' N / 118° 44 ' W

| | JAN | FEB | MAR | APR | MAY | JUN | JUL | AUG | SEP | OCT | NOV | DEC | YEAR |
|---|---|---|---|---|---|---|---|---|---|---|---|---|---|
| Maximum Temp °F | 48.4 | 50.4 | 53.1 | 58.4 | 67.0 | 77.1 | 84.3 | 84.0 | 78.2 | 68.3 | 55.5 | ***47.0*** | 64.3 |
| Minimum Temp °F | 36.0 | 37.2 | 37.5 | 40.8 | 47.2 | 55.4 | 62.6 | 62.7 | 58.5 | 51.2 | 41.5 | ***34.6*** | 47.1 |
| Mean Temp °F | 42.2 | 43.8 | 45.3 | 49.6 | 57.2 | 66.3 | 73.4 | 73.4 | 68.3 | 59.8 | 48.5 | ***40.8*** | 55.7 |
| Days Max Temp ≥ 90 °F | 0 | 0 | 0 | 0 | 0 | 3 | 7 | 7 | 2 | 0 | 0 | 0 | 19 |
| Days Max Temp ≤ 32 °F | 1 | 1 | 0 | 0 | 0 | 0 | 0 | 0 | 0 | 0 | 0 | 2 | 4 |
| Days Min Temp ≤ 32 °F | 10 | 8 | 9 | 5 | 1 | 0 | 0 | 0 | 0 | 0 | 4 | 11 | 48 |
| Days Min Temp ≤ 0 °F | 0 | 0 | 0 | 0 | 0 | 0 | 0 | 0 | 0 | 0 | 0 | 0 | 0 |
| Heating Degree Days | 699 | 590 | 603 | 460 | 270 | 90 | 9 | 17 | 65 | 211 | 488 | ***742*** | 4244 |
| Cooling Degree Days | 0 | 0 | 0 | 9 | ***36*** | 148 | 281 | 304 | 196 | 69 | 2 | ***0*** | 1045 |
| Total Precipitation (") | 2.36 | 3.05 | 2.02 | 0.80 | 0.24 | 0.07 | 0.04 | 0.09 | 0.29 | 0.27 | 1.77 | 2.31 | 13.31 |
| Days ≥ 0.1" Precip | 3 | 3 | 3 | 2 | 1 | 0 | 0 | 0 | 1 | 1 | 2 | 3 | 19 |
| Total Snowfall (") | 3.8 | 1.8 | 2.9 | 2.9 | 0.3 | 0.0 | 0.0 | 0.0 | 0.0 | 0.0 | 0.3 | 3.8 | 15.8 |
| Days ≥ 1" Snow Depth | 3 | 1 | 1 | 1 | 0 | 0 | 0 | 0 | 0 | 0 | 0 | 2 | 8 |

## SANTA ANA FIRE STN *Orange County* ELEVATION 131 ft LAT/LONG 33° 46' N / 117° 51' W

| | JAN | FEB | MAR | APR | MAY | JUN | JUL | AUG | SEP | OCT | NOV | DEC | YEAR |
|---|---|---|---|---|---|---|---|---|---|---|---|---|---|
| Maximum Temp °F | 69.1 | 70.0 | 70.2 | 73.0 | 74.4 | 77.8 | 82.4 | 84.0 | 83.3 | 79.7 | 73.6 | 68.5 | 75.5 |
| Minimum Temp °F | 46.1 | 47.5 | 49.4 | 52.0 | 56.1 | 59.6 | 63.0 | 64.4 | 62.4 | 57.7 | 50.6 | 45.6 | 54.5 |
| Mean Temp °F | 57.6 | 58.8 | 59.8 | 62.5 | 65.3 | 68.7 | 72.7 | 74.2 | 72.9 | 68.7 | 62.1 | 57.1 | 65.0 |
| Days Max Temp ≥ 90 °F | 0 | 0 | 0 | 1 | 1 | 2 | 3 | 5 | 5 | 3 | 0 | 0 | 20 |
| Days Max Temp ≤ 32 °F | 0 | 0 | 0 | 0 | 0 | 0 | 0 | 0 | 0 | 0 | 0 | 0 | 0 |
| Days Min Temp ≤ 32 °F | 0 | 0 | 0 | 0 | 0 | 0 | 0 | 0 | 0 | 0 | 0 | 0 | 0 |
| Days Min Temp ≤ 0 °F | 0 | 0 | 0 | 0 | 0 | 0 | 0 | 0 | 0 | 0 | 0 | 0 | 0 |
| Heating Degree Days | 229 | 177 | 166 | 97 | 41 | 7 | 0 | 0 | 0 | 13 | 103 | 244 | 1077 |
| Cooling Degree Days | 7 | 12 | 14 | 44 | 63 | 134 | 257 | 297 | 247 | 144 | 24 | 5 | 1248 |
| Total Precipitation (") | 2.87 | 2.65 | 2.65 | 0.80 | 0.15 | 0.08 | 0.02 | 0.12 | 0.31 | 0.28 | 1.56 | 1.89 | 13.38 |
| Days ≥ 0.1" Precip | 5 | 4 | 5 | 2 | 0 | 0 | 0 | 0 | 1 | 1 | 2 | 3 | 23 |
| Total Snowfall (") | 0.0 | 0.0 | 0.0 | 0.0 | 0.0 | 0.0 | 0.0 | 0.0 | 0.0 | 0.0 | 0.0 | 0.0 | 0.0 |
| Days ≥ 1" Snow Depth | 0 | 0 | 0 | 0 | 0 | 0 | 0 | 0 | 0 | 0 | 0 | 0 | 0 |

## SANTA BARBARA *Santa Barbara County* ELEVATION 121 ft LAT/LONG 34° 25' N / 119° 43' W

| | JAN | FEB | MAR | APR | MAY | JUN | JUL | AUG | SEP | OCT | NOV | DEC | YEAR |
|---|---|---|---|---|---|---|---|---|---|---|---|---|---|
| Maximum Temp °F | 64.8 | 65.4 | 66.2 | 69.0 | 69.4 | 72.1 | 75.1 | 76.9 | 75.4 | 73.8 | 69.2 | ***65.0*** | 70.2 |
| Minimum Temp °F | 43.6 | 45.4 | 47.6 | 49.5 | 52.4 | 55.4 | 58.0 | 59.3 | 58.0 | 54.2 | 48.5 | ***43.5*** | 51.3 |
| Mean Temp °F | 54.3 | 55.5 | 56.9 | 59.3 | 60.9 | 63.7 | 66.6 | 68.1 | 66.7 | 64.1 | 58.8 | ***54.2*** | 60.8 |
| Days Max Temp ≥ 90 °F | 0 | 0 | 0 | 0 | 0 | 1 | 0 | 0 | 1 | 1 | 0 | ***0*** | 3 |
| Days Max Temp ≤ 32 °F | 0 | 0 | 0 | 0 | 0 | 0 | 0 | 0 | 0 | 0 | 0 | 0 | 0 |
| Days Min Temp ≤ 32 °F | 0 | 0 | 0 | 0 | 0 | 0 | 0 | 0 | 0 | 0 | 0 | ***1*** | 1 |
| Days Min Temp ≤ 0 °F | 0 | 0 | 0 | 0 | 0 | 0 | 0 | 0 | 0 | 0 | 0 | 0 | 0 |
| Heating Degree Days | 327 | 264 | 247 | 177 | 132 | 60 | 15 | 6 | 19 | 60 | 185 | ***329*** | 1821 |
| Cooling Degree Days | ***1*** | ***2*** | 5 | ***17*** | 14 | 33 | 84 | 109 | 78 | 40 | 4 | ***0*** | 387 |
| Total Precipitation (") | 3.55 | 3.85 | 2.99 | 0.95 | 0.21 | 0.06 | 0.01 | 0.03 | 0.28 | 0.33 | 1.90 | 2.56 | 16.72 |
| Days ≥ 0.1" Precip | 4 | 5 | 4 | 2 | 0 | 0 | 0 | 0 | 0 | 1 | 3 | ***4*** | 23 |
| Total Snowfall (") | 0.0 | 0.0 | 0.0 | 0.0 | 0.0 | 0.0 | 0.0 | 0.0 | 0.0 | 0.0 | 0.0 | 0.0 | 0.0 |
| Days ≥ 1" Snow Depth | 0 | 0 | 0 | 0 | 0 | 0 | 0 | 0 | 0 | 0 | 0 | ***0*** | 0 |

## SANTA BARBARA MUNI *Santa Barbara County* ELEVATION 16 ft LAT/LONG 34° 26' N / 119° 50' W

| | JAN | FEB | MAR | APR | MAY | JUN | JUL | AUG | SEP | OCT | NOV | DEC | YEAR |
|---|---|---|---|---|---|---|---|---|---|---|---|---|---|
| Maximum Temp °F | 63.9 | 64.7 | 65.4 | 67.9 | 69.0 | 71.8 | 74.1 | 75.6 | 75.2 | 73.3 | 69.1 | 64.5 | 69.5 |
| Minimum Temp °F | 40.5 | 43.3 | 45.6 | 47.3 | 50.5 | 53.8 | 57.1 | 58.3 | 56.4 | 51.4 | 44.5 | 39.8 | 49.0 |
| Mean Temp °F | 52.2 | 54.0 | 55.6 | 57.6 | 59.7 | 62.8 | 65.6 | 67.0 | 65.8 | 62.4 | 56.8 | 52.2 | 59.3 |
| Days Max Temp ≥ 90 °F | 0 | 0 | 0 | 0 | 0 | 0 | 0 | 0 | 1 | 1 | 0 | 0 | 2 |
| Days Max Temp ≤ 32 °F | 0 | 0 | 0 | 0 | 0 | 0 | 0 | 0 | 0 | 0 | 0 | 0 | 0 |
| Days Min Temp ≤ 32 °F | 3 | 1 | 0 | 0 | 0 | 0 | 0 | 0 | 0 | 0 | 0 | 2 | 6 |
| Days Min Temp ≤ 0 °F | 0 | 0 | 0 | 0 | 0 | 0 | 0 | 0 | 0 | 0 | 0 | 0 | 0 |
| Heating Degree Days | 390 | 304 | 288 | 220 | 164 | 79 | 23 | 12 | 30 | 96 | 241 | 390 | 2237 |
| Cooling Degree Days | 1 | 0 | 3 | 7 | 8 | 23 | 61 | 78 | 61 | 21 | 2 | 0 | 265 |
| Total Precipitation (") | 3.43 | 3.72 | 3.25 | 0.99 | 0.16 | 0.05 | 0.03 | 0.11 | 0.42 | 0.48 | 1.87 | 2.42 | 16.93 |
| Days ≥ 0.1" Precip | 4 | 5 | 5 | 2 | 0 | 0 | 0 | 0 | 1 | 1 | 3 | 4 | 25 |
| Total Snowfall (") | 0.0 | 0.0 | 0.0 | 0.0 | 0.0 | 0.0 | 0.0 | 0.0 | 0.0 | 0.0 | 0.0 | 0.0 | 0.0 |
| Days ≥ 1" Snow Depth | 0 | 0 | 0 | 0 | 0 | 0 | 0 | 0 | 0 | 0 | 0 | 0 | 0 |

## SANTA CRUZ *Santa Cruz County* ELEVATION 131 ft LAT/LONG 36° 58' N / 122° 1' W

| | JAN | FEB | MAR | APR | MAY | JUN | JUL | AUG | SEP | OCT | NOV | DEC | YEAR |
|---|---|---|---|---|---|---|---|---|---|---|---|---|---|
| Maximum Temp °F | 60.7 | 62.9 | 64.6 | 68.1 | 71.1 | 74.1 | 74.9 | 75.6 | 76.1 | 73.2 | 65.5 | 60.1 | 68.9 |
| Minimum Temp °F | 38.9 | 41.1 | 42.5 | 43.3 | 46.5 | 49.8 | 51.8 | 52.3 | 51.2 | 47.7 | 42.8 | 38.7 | 45.6 |
| Mean Temp °F | 49.9 | 52.0 | 53.6 | 55.7 | 58.8 | 62.0 | 63.4 | 64.0 | 63.7 | 60.4 | 54.2 | 49.4 | 57.3 |
| Days Max Temp ≥ 90 °F | 0 | 0 | 0 | 0 | 1 | 1 | 1 | 1 | 2 | 2 | 0 | 0 | 8 |
| Days Max Temp ≤ 32 °F | 0 | 0 | 0 | 0 | 0 | 0 | 0 | 0 | 0 | 0 | 0 | 0 | 0 |
| Days Min Temp ≤ 32 °F | 5 | 2 | 1 | 0 | 0 | 0 | 0 | 0 | 0 | 0 | 1 | 4 | 13 |
| Days Min Temp ≤ 0 °F | 0 | 0 | 0 | 0 | 0 | 0 | 0 | 0 | 0 | 0 | 0 | 0 | 0 |
| Heating Degree Days | 463 | 359 | 348 | 274 | 194 | 102 | 65 | 52 | 65 | 152 | 318 | 476 | 2868 |
| Cooling Degree Days | 0 | 0 | 0 | 5 | 7 | 20 | 28 | 30 | 29 | 17 | 0 | 0 | 136 |
| Total Precipitation (") | 5.67 | 5.28 | 4.70 | 2.02 | 0.43 | 0.18 | 0.15 | 0.11 | 0.40 | 1.30 | 4.20 | 4.53 | 28.97 |
| Days ≥ 0.1" Precip | 8 | 7 | 7 | 4 | 1 | 0 | 0 | 0 | 1 | 2 | 6 | 7 | 43 |
| Total Snowfall (") | 0.0 | 0.0 | 0.0 | 0.0 | 0.0 | 0.0 | 0.0 | 0.0 | 0.0 | 0.0 | 0.0 | 0.0 | 0.0 |
| Days ≥ 1" Snow Depth | 0 | 0 | 0 | 0 | 0 | 0 | 0 | 0 | 0 | 0 | 0 | 0 | 0 |

## SANTA MARIA PBLC AP *Santa Barbara County* ELEVATION 239 ft LAT/LONG 34° 54 ' N / 120° 27 ' W

| | JAN | FEB | MAR | APR | MAY | JUN | JUL | AUG | SEP | OCT | NOV | DEC | YEAR |
|---|---|---|---|---|---|---|---|---|---|---|---|---|---|
| Maximum Temp °F | 63.9 | 64.8 | 64.6 | 67.2 | 68.5 | 71.2 | 73.3 | 74.1 | 74.7 | 74.1 | 69.1 | 64.2 | 69.1 |
| Minimum Temp °F | 38.8 | 40.7 | 42.2 | 43.1 | 46.8 | 50.4 | 53.4 | 54.1 | 52.6 | 48.1 | 42.2 | 37.8 | 45.9 |
| Mean Temp °F | 51.4 | 52.8 | 53.4 | 55.2 | 57.7 | 60.8 | 63.4 | 64.1 | 63.7 | 61.1 | 55.7 | 51.1 | 57.5 |
| Days Max Temp ≥ 90 °F | 0 | 0 | 0 | 0 | 0 | 1 | 0 | 0 | 2 | 1 | 0 | 0 | 4 |
| Days Max Temp ≤ 32 °F | 0 | 0 | 0 | 0 | 0 | 0 | 0 | 0 | 0 | 0 | 0 | 0 | 0 |
| Days Min Temp ≤ 32 °F | 6 | 3 | 1 | 0 | 0 | 0 | 0 | 0 | 0 | 0 | 2 | 6 | 18 |
| Days Min Temp ≤ 0 °F | 0 | 0 | 0 | 0 | 0 | 0 | 0 | 0 | 0 | 0 | 0 | 0 | 0 |
| Heating Degree Days | 415 | 338 | 353 | 292 | 225 | 129 | 66 | 46 | 65 | 134 | 277 | 425 | 2765 |
| Cooling Degree Days | 0 | 1 | 1 | 6 | 3 | 10 | 33 | 33 | 38 | 23 | 3 | 0 | 151 |
| Total Precipitation (") | 2.33 | 2.70 | 2.59 | 0.93 | 0.19 | 0.03 | 0.02 | 0.05 | 0.29 | 0.43 | 1.35 | 1.96 | 12.87 |
| Days ≥ 0.1" Precip | 5 | 5 | 5 | 2 | 1 | 0 | 0 | 0 | 0 | 1 | 3 | 4 | 26 |
| Total Snowfall (") | 0.0 | 0.0 | 0.0 | 0.0 | 0.0 | 0.0 | 0.0 | 0.0 | 0.0 | 0.0 | 0.0 | 0.0 | 0.0 |
| Days ≥ 1" Snow Depth | 0 | 0 | 0 | 0 | 0 | 0 | 0 | 0 | 0 | 0 | 0 | 0 | 0 |

## SANTA MONICA PIER *Los Angeles County* ELEVATION 14 ft LAT/LONG 34° 1 ' N / 118° 30 ' W

| | JAN | FEB | MAR | APR | MAY | JUN | JUL | AUG | SEP | OCT | NOV | DEC | YEAR |
|---|---|---|---|---|---|---|---|---|---|---|---|---|---|
| Maximum Temp °F | 64.7 | 64.1 | 63.1 | 64.1 | 64.2 | 66.7 | 69.4 | 71.0 | 71.1 | ***70.8*** | 68.4 | 65.0 | 66.9 |
| Minimum Temp °F | 49.9 | 50.7 | 51.6 | 53.5 | 56.0 | 59.0 | 61.8 | 63.2 | 62.2 | ***59.4*** | 54.5 | 50.0 | 56.0 |
| Mean Temp °F | 57.3 | 57.4 | 57.4 | 58.8 | 60.1 | 62.9 | 65.6 | 67.2 | 66.7 | ***65.1*** | 61.5 | 57.6 | 61.5 |
| Days Max Temp ≥ 90 °F | 0 | 0 | 0 | 0 | 0 | 0 | 0 | 0 | 0 | 0 | 0 | 0 | 0 |
| Days Max Temp ≤ 32 °F | 0 | 0 | 0 | 0 | 0 | 0 | 0 | 0 | 0 | 0 | 0 | 0 | 0 |
| Days Min Temp ≤ 32 °F | 0 | 0 | 0 | 0 | 0 | 0 | 0 | 0 | 0 | 0 | 0 | 0 | 0 |
| Days Min Temp ≤ 0 °F | 0 | 0 | 0 | 0 | 0 | 0 | 0 | 0 | 0 | 0 | 0 | 0 | 0 |
| Heating Degree Days | 236 | 213 | 232 | 186 | 149 | 74 | 22 | 10 | 18 | ***41*** | 120 | 230 | 1531 |
| Cooling Degree Days | 7 | 9 | 4 | 11 | 5 | 23 | 66 | 90 | 77 | 52 | 17 | 4 | 365 |
| Total Precipitation (") | 2.70 | 2.86 | 2.24 | 0.64 | 0.14 | 0.02 | 0.03 | 0.13 | 0.20 | 0.27 | 1.67 | 1.99 | 12.89 |
| Days ≥ 0.1" Precip | 4 | 4 | 3 | 2 | 0 | 0 | 0 | 0 | 1 | 1 | 2 | 3 | 20 |
| Total Snowfall (") | 0.0 | 0.0 | 0.0 | 0.0 | 0.0 | 0.0 | 0.0 | 0.0 | 0.0 | 0.0 | 0.0 | 0.0 | 0.0 |
| Days ≥ 1" Snow Depth | 0 | 0 | 0 | 0 | 0 | 0 | 0 | 0 | 0 | 0 | ***0*** | ***0*** | 0 |

## SANTA PAULA *Ventura County* ELEVATION 237 ft LAT/LONG 34° 19 ' N / 119° 9 ' W

| | JAN | FEB | MAR | APR | MAY | JUN | JUL | AUG | SEP | OCT | NOV | DEC | YEAR |
|---|---|---|---|---|---|---|---|---|---|---|---|---|---|
| Maximum Temp °F | 68.1 | 69.0 | 69.8 | 73.1 | 74.2 | ***77.3*** | ***80.7*** | 81.6 | 80.8 | 78.3 | 72.8 | 67.7 | 74.5 |
| Minimum Temp °F | 41.6 | 42.4 | 43.6 | 45.5 | 49.3 | 52.4 | ***55.0*** | 55.9 | 54.4 | 50.0 | 44.1 | 40.6 | 47.9 |
| Mean Temp °F | ***54.7*** | 55.7 | 56.7 | 59.3 | 61.7 | ***64.9*** | ***67.9*** | 68.7 | 67.7 | 64.3 | 58.5 | 54.2 | 61.2 |
| Days Max Temp ≥ 90 °F | 0 | 0 | 0 | 1 | 1 | 1 | 2 | 3 | 4 | 3 | 1 | 0 | 16 |
| Days Max Temp ≤ 32 °F | 0 | 0 | 0 | 0 | 0 | 0 | 0 | 0 | 0 | 0 | 0 | 0 | 0 |
| Days Min Temp ≤ 32 °F | 3 | 1 | 0 | 0 | 0 | 0 | 0 | 0 | 0 | 0 | 1 | 3 | 8 |
| Days Min Temp ≤ 0 °F | 0 | 0 | 0 | 0 | 0 | 0 | 0 | 0 | 0 | 0 | 0 | 0 | 0 |
| Heating Degree Days | ***319*** | 262 | 254 | 177 | 116 | ***44*** | 10 | 9 | 19 | 70 | 203 | 331 | 1814 |
| Cooling Degree Days | 5 | 9 | 5 | 20 | 20 | ***46*** | ***129*** | 134 | 107 | 55 | 11 | 2 | 543 |
| Total Precipitation (") | 3.96 | 3.94 | 3.27 | 0.93 | 0.16 | 0.03 | 0.01 | 0.09 | 0.33 | 0.44 | 2.25 | 2.78 | 18.19 |
| Days ≥ 0.1" Precip | 4 | 4 | 4 | 2 | 0 | 0 | 0 | 0 | 0 | 1 | 3 | 3 | 21 |
| Total Snowfall (") | 0.0 | 0.0 | 0.0 | 0.0 | 0.0 | 0.0 | 0.0 | 0.0 | 0.0 | 0.0 | 0.0 | 0.0 | 0.0 |
| Days ≥ 1" Snow Depth | 0 | 0 | 0 | 0 | 0 | 0 | 0 | 0 | 0 | 0 | 0 | 0 | 0 |

## SANTA ROSA *Sonoma County* ELEVATION 167 ft LAT/LONG 38° 27 ' N / 122° 42 ' W

| | JAN | FEB | MAR | APR | MAY | JUN | JUL | AUG | SEP | OCT | NOV | DEC | YEAR |
|---|---|---|---|---|---|---|---|---|---|---|---|---|---|
| Maximum Temp °F | 57.6 | 62.3 | 64.9 | 69.4 | 74.9 | 79.8 | 82.8 | 82.8 | 82.4 | 77.2 | 65.3 | 57.4 | 71.4 |
| Minimum Temp °F | 37.4 | 40.4 | 42.1 | 43.3 | 47.0 | 50.8 | 51.7 | 52.1 | 51.2 | 47.4 | 42.1 | 37.6 | 45.3 |
| Mean Temp °F | 47.5 | 51.4 | 53.5 | 56.4 | 60.9 | 65.3 | 67.3 | 67.5 | 66.8 | 62.3 | 53.7 | 47.5 | 58.3 |
| Days Max Temp ≥ 90 °F | 0 | 0 | 0 | 0 | 2 | 5 | 6 | 6 | 6 | 3 | 0 | 0 | 28 |
| Days Max Temp ≤ 32 °F | 0 | 0 | 0 | 0 | 0 | 0 | 0 | 0 | 0 | 0 | 0 | 0 | 0 |
| Days Min Temp ≤ 32 °F | 8 | 3 | 2 | 0 | 0 | 0 | 0 | 0 | 0 | 0 | 2 | 8 | 23 |
| Days Min Temp ≤ 0 °F | 0 | 0 | 0 | 0 | 0 | 0 | 0 | 0 | 0 | 0 | 0 | 0 | 0 |
| Heating Degree Days | 535 | 378 | 350 | 257 | 145 | 58 | 22 | 20 | 31 | 111 | 333 | 536 | 2776 |
| Cooling Degree Days | 0 | 0 | 1 | 8 | 24 | 73 | 89 | 84 | 78 | 41 | 1 | 0 | 399 |
| Total Precipitation (") | 6.33 | 4.85 | 4.43 | 1.74 | 0.49 | 0.23 | 0.06 | 0.15 | 0.47 | 1.81 | 4.46 | 5.12 | 30.14 |
| Days ≥ 0.1" Precip | 8 | 7 | 8 | 4 | 1 | 1 | 0 | 0 | 1 | 3 | 7 | 7 | 47 |
| Total Snowfall (") | 0.0 | 0.0 | 0.0 | 0.0 | 0.0 | 0.0 | 0.0 | 0.0 | 0.0 | 0.0 | 0.0 | 0.0 | 0.0 |
| Days ≥ 1" Snow Depth | 0 | 0 | 0 | 0 | 0 | 0 | 0 | 0 | 0 | 0 | 0 | 0 | 0 |

## SCOTIA *Humboldt County* ELEVATION 139 ft LAT/LONG 40° 29' N / 124° 6' W

| | JAN | FEB | MAR | APR | MAY | JUN | JUL | AUG | SEP | OCT | NOV | DEC | YEAR |
|---|---|---|---|---|---|---|---|---|---|---|---|---|---|
| Maximum Temp °F | 55.2 | 57.2 | 58.2 | 59.6 | 62.6 | 65.9 | 68.6 | 70.0 | 70.4 | 66.8 | 59.6 | 54.8 | 62.4 |
| Minimum Temp °F | 39.2 | 40.6 | 41.9 | 43.3 | 46.7 | 50.2 | 52.0 | 52.7 | 50.3 | 46.9 | 43.0 | 39.1 | 45.5 |
| Mean Temp °F | 47.2 | 48.9 | 50.1 | 51.5 | 54.7 | 58.0 | 60.4 | 61.4 | 60.4 | 56.9 | 51.3 | 47.0 | 54.0 |
| Days Max Temp ≥ 90 °F | 0 | 0 | 0 | 0 | 0 | 0 | 0 | 0 | 0 | 0 | 0 | 0 | 0 |
| Days Max Temp ≤ 32 °F | 0 | 0 | 0 | 0 | 0 | 0 | 0 | 0 | 0 | 0 | 0 | 0 | 0 |
| Days Min Temp ≤ 32 °F | 4 | 2 | 1 | 0 | 0 | 0 | 0 | 0 | 0 | 0 | 1 | 5 | 13 |
| Days Min Temp ≤ 0 °F | 0 | 0 | 0 | 0 | 0 | 0 | 0 | 0 | 0 | 0 | 0 | 0 | 0 |
| Heating Degree Days | 545 | 447 | 456 | 399 | 315 | 206 | 141 | 113 | 140 | 249 | 404 | 552 | 3967 |
| Cooling Degree Days | 0 | 0 | 0 | 1 | 3 | 2 | 4 | 6 | 5 | 6 | 0 | 0 | 27 |
| Total Precipitation (") | 8.38 | 6.75 | 6.91 | 3.28 | 1.45 | 0.49 | 0.10 | 0.39 | 0.85 | 2.72 | 7.18 | 8.64 | 47.14 |
| Days ≥ 0.1" Precip | 11 | 10 | 11 | 7 | 4 | 1 | 0 | 1 | 2 | 5 | 11 | 12 | 75 |
| Total Snowfall (") | 0.1 | 0.3 | 0.0 | 0.0 | 0.0 | 0.0 | 0.0 | 0.0 | 0.0 | 0.0 | 0.0 | 0.3 | 0.7 |
| Days ≥ 1" Snow Depth | 0 | 0 | 0 | 0 | 0 | 0 | 0 | 0 | 0 | 0 | 0 | 0 | 0 |

## SHASTA DAM *Shasta County* ELEVATION 1075 ft LAT/LONG 40° 43' N / 122° 25' W

| | JAN | FEB | MAR | APR | MAY | JUN | JUL | AUG | SEP | OCT | NOV | DEC | YEAR |
|---|---|---|---|---|---|---|---|---|---|---|---|---|---|
| Maximum Temp °F | 52.9 | 57.0 | 61.3 | 67.8 | 78.2 | 86.5 | 94.8 | 93.3 | 87.0 | 75.4 | 60.2 | 52.5 | 72.2 |
| Minimum Temp °F | 38.9 | 40.9 | 43.2 | 47.3 | 55.0 | 62.4 | 68.1 | 66.5 | 61.8 | 54.5 | 45.3 | 39.1 | 51.9 |
| Mean Temp °F | 45.9 | 49.0 | 52.3 | 57.6 | 66.7 | 74.5 | 81.4 | 80.0 | 74.4 | 65.0 | 52.8 | 45.8 | 62.1 |
| Days Max Temp ≥ 90 °F | 0 | 0 | 0 | 0 | 4 | 12 | 24 | 23 | 13 | 3 | 0 | 0 | 79 |
| Days Max Temp ≤ 32 °F | 0 | 0 | 0 | 0 | 0 | 0 | 0 | 0 | 0 | 0 | 0 | 0 | 0 |
| Days Min Temp ≤ 32 °F | 4 | 1 | 1 | 0 | 0 | 0 | 0 | 0 | 0 | 0 | 0 | 4 | 10 |
| Days Min Temp ≤ 0 °F | 0 | 0 | 0 | 0 | 0 | 0 | 0 | 0 | 0 | 0 | 0 | 0 | 0 |
| Heating Degree Days | 585 | 446 | 389 | 238 | 77 | 15 | 1 | 2 | 14 | 98 | 363 | 587 | 2815 |
| Cooling Degree Days | 0 | 0 | 1 | 32 | 133 | 298 | 512 | 474 | 312 | 115 | 2 | 0 | 1879 |
| Total Precipitation (") | 11.25 | 8.89 | 9.61 | 4.20 | 2.19 | 1.05 | 0.25 | 0.54 | 1.44 | 2.91 | 8.65 | 9.44 | 60.42 |
| Days ≥ 0.1" Precip | 9 | 9 | 10 | 6 | 3 | 2 | 0 | 1 | 2 | 3 | 9 | 9 | 63 |
| Total Snowfall (") | 0.7 | 0.3 | 0.6 | 0.0 | 0.0 | 0.0 | 0.0 | 0.0 | 0.0 | 0.0 | 0.4 | 1.5 | 3.5 |
| Days ≥ 1" Snow Depth | 0 | 0 | 0 | 0 | 0 | 0 | 0 | 0 | 0 | 0 | 0 | 0 | 0 |

## SIERRA CITY *Sierra County* ELEVATION 4203 ft LAT/LONG 39° 34' N / 120° 38' W

| | JAN | FEB | MAR | APR | MAY | JUN | JUL | AUG | SEP | OCT | NOV | DEC | YEAR |
|---|---|---|---|---|---|---|---|---|---|---|---|---|---|
| Maximum Temp °F | 48.4 | 51.4 | 54.5 | 60.4 | 70.5 | 78.8 | 86.3 | 85.9 | 79.9 | 69.9 | 54.7 | 47.3 | 65.7 |
| Minimum Temp °F | 28.4 | 29.1 | 31.2 | 34.3 | 40.6 | 47.1 | 52.3 | 51.6 | 47.2 | 40.6 | 32.9 | 28.4 | 38.6 |
| Mean Temp °F | 38.5 | 40.2 | 42.9 | 47.4 | 55.6 | 63.0 | 69.3 | 68.8 | 63.6 | 55.3 | 43.8 | 37.9 | 52.2 |
| Days Max Temp ≥ 90 °F | 0 | 0 | 0 | 0 | 0 | 3 | 10 | 10 | 4 | 0 | 0 | 0 | 27 |
| Days Max Temp ≤ 32 °F | 1 | 1 | 0 | 0 | 0 | 0 | 0 | 0 | 0 | 0 | 0 | 1 | 3 |
| Days Min Temp ≤ 32 °F | 24 | 21 | 19 | 12 | 4 | 0 | 0 | 0 | 0 | 3 | 16 | 24 | 123 |
| Days Min Temp ≤ 0 °F | 0 | 0 | 0 | 0 | 0 | 0 | 0 | 0 | 0 | 0 | 0 | 0 | 0 |
| Heating Degree Days | 816 | 692 | 679 | 523 | 297 | 118 | 25 | 31 | 104 | 305 | 629 | 833 | 5052 |
| Cooling Degree Days | ***0*** | ***0*** | ***0*** | ***1*** | 15 | 72 | 175 | 169 | 76 | ***15*** | 0 | 0 | 523 |
| Total Precipitation (") | 11.44 | 9.90 | 9.75 | 4.35 | 2.41 | 1.05 | 0.38 | 0.83 | 1.68 | 4.14 | 9.78 | 10.06 | 65.77 |
| Days ≥ 0.1" Precip | 9 | ***8*** | 10 | 7 | 5 | 3 | 1 | 1 | 3 | 4 | 8 | 9 | 68 |
| Total Snowfall (") | na | na | na | na | 0.4 | 0.0 | 0.0 | 0.0 | 0.0 | ***0.9*** | ***8.5*** | na | na |
| Days ≥ 1" Snow Depth | 14 | ***12*** | 8 | ***3*** | 0 | 0 | 0 | 0 | 0 | 0 | 4 | 10 | 51 |

## SIERRAVILLE RANGER S *Sierra County* ELEVATION 4984 ft LAT/LONG 39° 35' N / 120° 22' W

| | JAN | FEB | MAR | APR | MAY | JUN | JUL | AUG | SEP | OCT | NOV | DEC | YEAR |
|---|---|---|---|---|---|---|---|---|---|---|---|---|---|
| Maximum Temp °F | na | ***47.8*** | ***51.4*** | ***58.2*** | 67.8 | 76.0 | 84.0 | 83.8 | 77.3 | 67.7 | ***52.7*** | na | na |
| Minimum Temp °F | na | ***20.0*** | ***25.2*** | ***28.5*** | 35.5 | 41.5 | 44.6 | 42.9 | 36.8 | 29.7 | ***24.0*** | na | na |
| Mean Temp °F | na | ***33.9*** | ***38.4*** | ***43.2*** | 51.5 | 58.8 | 64.3 | 63.4 | 57.1 | 48.7 | ***38.4*** | na | na |
| Days Max Temp ≥ 90 °F | 0 | 0 | 0 | 0 | 0 | 1 | 7 | 6 | 1 | 0 | 0 | 0 | 15 |
| Days Max Temp ≤ 32 °F | 3 | 1 | 0 | 0 | 0 | 0 | 0 | 0 | 0 | 0 | 1 | 3 | 8 |
| Days Min Temp ≤ 32 °F | 24 | 22 | 24 | ***21*** | ***10*** | 2 | 0 | 1 | 7 | 21 | 23 | 26 | 181 |
| Days Min Temp ≤ 0 °F | 3 | 1 | 0 | 0 | 0 | 0 | 0 | 0 | 0 | 0 | 0 | 2 | 6 |
| Heating Degree Days | na | na | na | ***650*** | ***411*** | 201 | 77 | 91 | 239 | 497 | ***792*** | na | na |
| Cooling Degree Days | na | na | na | na | ***2*** | 16 | 56 | 47 | 8 | 0 | na | na | na |
| Total Precipitation (") | 4.64 | 3.72 | 3.41 | 1.21 | 0.90 | 0.60 | 0.28 | 0.50 | 0.76 | 1.65 | 4.12 | 3.98 | 25.77 |
| Days ≥ 0.1" Precip | 5 | 4 | 5 | 3 | 2 | 2 | 1 | 1 | 2 | 2 | 5 | 5 | 37 |
| Total Snowfall (") | 14.9 | 13.9 | 13.3 | 5.0 | 1.5 | 0.0 | 0.0 | 0.0 | 0.0 | 0.6 | 6.6 | 14.7 | 70.5 |
| Days ≥ 1" Snow Depth | 15 | 11 | ***6*** | 2 | 0 | 0 | 0 | 0 | 0 | 0 | 4 | 10 | 48 |

## SONOMA *Sonoma County* ELEVATION 69 ft LAT/LONG 38° 17 ' N / 122° 27 ' W

| | JAN | FEB | MAR | APR | MAY | JUN | JUL | AUG | SEP | OCT | NOV | DEC | YEAR |
|---|---|---|---|---|---|---|---|---|---|---|---|---|---|
| Maximum Temp °F | 58.0 | 63.7 | 67.0 | 71.7 | 78.6 | 85.0 | 89.8 | 89.5 | 87.4 | 80.1 | 66.2 | 57.8 | 74.6 |
| Minimum Temp °F | 36.4 | 39.1 | 41.0 | 41.8 | 45.6 | 49.7 | 50.9 | 50.8 | 49.3 | 45.5 | 40.4 | 36.1 | 43.9 |
| Mean Temp °F | 47.2 | 51.4 | 54.0 | 56.8 | 62.1 | 67.5 | 70.4 | 70.2 | 68.4 | 62.8 | 53.3 | 47.0 | 59.3 |
| Days Max Temp ≥ 90 °F | 0 | 0 | 0 | 0 | 4 | 9 | 15 | 15 | 12 | 5 | 0 | 0 | 60 |
| Days Max Temp ≤ 32 °F | 0 | 0 | 0 | 0 | 0 | 0 | 0 | 0 | 0 | 0 | 0 | 0 | 0 |
| Days Min Temp ≤ 32 °F | 11 | 5 | 3 | 1 | 0 | 0 | 0 | 0 | 0 | 0 | 4 | 11 | 35 |
| Days Min Temp ≤ 0 °F | 0 | 0 | 0 | 0 | 0 | 0 | 0 | 0 | 0 | 0 | 0 | 0 | 0 |
| Heating Degree Days | 545 | 378 | 333 | 245 | 121 | 33 | 7 | 5 | 19 | 100 | 344 | 551 | 2681 |
| Cooling Degree Days | 0 | 1 | 0 | 9 | 36 | 116 | 182 | 170 | 122 | 43 | 1 | 0 | 680 |
| Total Precipitation (") | 6.38 | 5.09 | 4.26 | 1.57 | 0.46 | 0.23 | 0.05 | 0.12 | 0.31 | 1.61 | 4.53 | 4.63 | 29.24 |
| Days ≥ 0.1" Precip | 7 | 7 | 7 | 4 | 1 | 1 | 0 | 0 | 1 | 3 | 7 | 7 | 45 |
| Total Snowfall (") | 0.0 | 0.0 | 0.0 | 0.0 | 0.0 | 0.0 | 0.0 | 0.0 | 0.0 | 0.0 | 0.0 | 0.0 | 0.0 |
| Days ≥ 1" Snow Depth | 0 | 0 | 0 | 0 | 0 | 0 | 0 | 0 | 0 | 0 | 0 | 0 | 0 |

## SONORA RS *Tuolumne County* ELEVATION 1831 ft LAT/LONG 37° 59 ' N / 120° 23 ' W

| | JAN | FEB | MAR | APR | MAY | JUN | JUL | AUG | SEP | OCT | NOV | DEC | YEAR |
|---|---|---|---|---|---|---|---|---|---|---|---|---|---|
| Maximum Temp °F | na | na | ***61.9*** | ***67.2*** | 77.9 | 87.1 | 94.8 | 93.8 | 87.3 | 76.8 | ***62.6*** | na | na |
| Minimum Temp °F | na | na | 37.0 | ***39.4*** | 45.3 | 51.1 | 56.6 | 55.5 | 50.6 | 42.7 | ***35.9*** | na | na |
| Mean Temp °F | na | na | ***49.6*** | ***53.2*** | 61.6 | 69.1 | 75.7 | 74.6 | 69.0 | 59.8 | ***49.3*** | na | na |
| Days Max Temp ≥ 90 °F | 0 | 0 | 0 | 0 | 4 | 13 | 25 | 23 | 14 | 3 | 0 | 0 | 82 |
| Days Max Temp ≤ 32 °F | 0 | 0 | 0 | 0 | 0 | 0 | 0 | 0 | 0 | 0 | 0 | 0 | 0 |
| Days Min Temp ≤ 32 °F | 17 | 12 | 7 | 3 | 0 | 0 | 0 | 0 | 0 | 1 | 9 | ***17*** | 66 |
| Days Min Temp ≤ 0 °F | 0 | 0 | 0 | 0 | 0 | 0 | 0 | 0 | 0 | 0 | 0 | 0 | 0 |
| Heating Degree Days | na | na | ***468*** | na | 152 | 38 | 3 | 4 | 34 | 182 | ***464*** | na | na |
| Cooling Degree Days | na | na | na | na | 51 | 164 | 328 | 305 | 158 | 30 | ***0*** | na | na |
| Total Precipitation (") | 5.84 | 4.99 | 5.04 | 2.65 | 0.75 | 0.23 | 0.07 | 0.19 | 0.55 | 1.83 | 4.46 | 4.48 | 31.08 |
| Days ≥ 0.1" Precip | 6 | 6 | 6 | 4 | 2 | 1 | 0 | 0 | 1 | 3 | 5 | 6 | 40 |
| Total Snowfall (") | na | 0.2 | 0.2 | 0.0 | 0.0 | 0.0 | 0.0 | 0.0 | 0.0 | 0.0 | 0.0 | ***0.1*** | na |
| Days ≥ 1" Snow Depth | ***0*** | 0 | 0 | 0 | 0 | 0 | 0 | 0 | 0 | 0 | 0 | ***0*** | 0 |

## STOCKTON FIRE STN # *San Joaquin County* ELEVATION 10 ft LAT/LONG 37° 58 ' N / 121° 18 ' W

| | JAN | FEB | MAR | APR | MAY | JUN | JUL | AUG | SEP | OCT | NOV | DEC | YEAR |
|---|---|---|---|---|---|---|---|---|---|---|---|---|---|
| Maximum Temp °F | 54.3 | 61.7 | 66.6 | 72.9 | 81.2 | 87.4 | 92.1 | 91.3 | 87.7 | 79.1 | 64.9 | 54.7 | 74.5 |
| Minimum Temp °F | 35.7 | 38.9 | 42.0 | 44.7 | 49.4 | 54.1 | 56.2 | 55.5 | 52.6 | 47.0 | 40.4 | 35.0 | 46.0 |
| Mean Temp °F | 45.0 | 50.3 | 54.3 | 58.8 | 65.3 | 70.8 | 74.2 | 73.5 | 70.2 | 63.1 | 52.7 | 44.9 | 60.3 |
| Days Max Temp ≥ 90 °F | 0 | 0 | 0 | 1 | 6 | 12 | 21 | 19 | 13 | 4 | 0 | 0 | 76 |
| Days Max Temp ≤ 32 °F | 0 | 0 | 0 | 0 | 0 | 0 | 0 | 0 | 0 | 0 | 0 | 0 | 0 |
| Days Min Temp ≤ 32 °F | 10 | 4 | 2 | 0 | 0 | 0 | 0 | 0 | 0 | 0 | 3 | 10 | 29 |
| Days Min Temp ≤ 0 °F | 0 | 0 | 0 | 0 | 0 | 0 | 0 | 0 | 0 | 0 | 0 | 0 | 0 |
| Heating Degree Days | 613 | 407 | 325 | 194 | 76 | 16 | 1 | 2 | 16 | 105 | 363 | 616 | 2734 |
| Cooling Degree Days | 0 | 0 | 1 | 19 | 94 | 194 | 281 | 250 | 170 | 57 | 0 | 0 | 1066 |
| Total Precipitation (") | 2.99 | 2.49 | 2.52 | 1.17 | 0.33 | 0.09 | 0.05 | 0.06 | 0.34 | 0.86 | 2.18 | 2.37 | 15.45 |
| Days ≥ 0.1" Precip | 6 | 5 | 6 | 3 | 1 | 0 | 0 | 0 | 1 | 2 | 4 | 5 | 33 |
| Total Snowfall (") | 0.0 | 0.0 | 0.0 | 0.0 | 0.0 | 0.0 | 0.0 | 0.0 | 0.0 | 0.0 | 0.0 | 0.0 | 0.0 |
| Days ≥ 1" Snow Depth | 0 | 0 | 0 | 0 | 0 | 0 | 0 | 0 | 0 | 0 | 0 | 0 | 0 |

## STOCKTON METRO ARPT *San Joaquin County* ELEVATION 26 ft LAT/LONG 37° 54 ' N / 121° 15 ' W

| | JAN | FEB | MAR | APR | MAY | JUN | JUL | AUG | SEP | OCT | NOV | DEC | YEAR |
|---|---|---|---|---|---|---|---|---|---|---|---|---|---|
| Maximum Temp °F | 53.2 | 60.8 | 65.9 | 72.8 | 81.8 | 88.8 | 94.1 | 92.8 | 88.2 | 78.7 | 63.8 | 53.0 | 74.5 |
| Minimum Temp °F | 37.5 | 40.7 | 43.4 | 46.3 | 52.3 | 57.7 | 61.2 | 60.8 | 57.9 | 51.0 | 42.7 | 36.9 | 49.0 |
| Mean Temp °F | 45.4 | 50.8 | 54.7 | 59.6 | 67.1 | 73.3 | 77.7 | 76.8 | 73.1 | 64.9 | 53.3 | 45.0 | 61.8 |
| Days Max Temp ≥ 90 °F | 0 | 0 | 0 | 1 | 7 | 14 | 23 | 21 | 14 | 3 | 0 | 0 | 83 |
| Days Max Temp ≤ 32 °F | 0 | 0 | 0 | 0 | 0 | 0 | 0 | 0 | 0 | 0 | 0 | 0 | 0 |
| Days Min Temp ≤ 32 °F | 8 | 3 | 1 | 0 | 0 | 0 | 0 | 0 | 0 | 0 | 2 | 9 | 23 |
| Days Min Temp ≤ 0 °F | 0 | 0 | 0 | 0 | 0 | 0 | 0 | 0 | 0 | 0 | 0 | 0 | 0 |
| Heating Degree Days | 602 | 395 | 314 | 175 | 48 | 6 | 0 | 0 | 4 | 71 | 346 | 613 | 2574 |
| Cooling Degree Days | 0 | 0 | 0 | 23 | 118 | 260 | 398 | 363 | 248 | 78 | 0 | 0 | 1488 |
| Total Precipitation (") | 2.76 | 2.14 | 2.30 | 1.03 | 0.29 | 0.08 | 0.06 | 0.06 | 0.34 | 0.75 | 2.01 | 2.05 | 13.87 |
| Days ≥ 0.1" Precip | 6 | 5 | 6 | 3 | 1 | 0 | 0 | 0 | 1 | 2 | 4 | 5 | 33 |
| Total Snowfall (") | 0.0 | 0.0 | 0.0 | 0.0 | 0.0 | 0.0 | 0.0 | 0.0 | 0.0 | 0.0 | 0.0 | 0.0 | 0.0 |
| Days ≥ 1" Snow Depth | 0 | 0 | 0 | 0 | 0 | 0 | 0 | 0 | 0 | 0 | 0 | 0 | 0 |

## STONY GORGE RESERVOI *Glenn County* ELEVATION 791 ft LAT/LONG 39° 35' N / 122° 32' W

| | JAN | FEB | MAR | APR | MAY | JUN | JUL | AUG | SEP | OCT | NOV | DEC | YEAR |
|---|---|---|---|---|---|---|---|---|---|---|---|---|---|
| Maximum Temp °F | 55.5 | 59.6 | 63.2 | 69.8 | 80.6 | 89.0 | 95.7 | 94.4 | 89.3 | 78.9 | 63.5 | 55.3 | 74.6 |
| Minimum Temp °F | 33.1 | 36.1 | 38.8 | 41.6 | 48.6 | 55.9 | 60.6 | 59.2 | 54.2 | 46.6 | 37.8 | 32.6 | 45.4 |
| Mean Temp °F | 44.3 | 47.9 | 51.1 | 55.7 | 64.6 | 72.4 | 78.2 | 76.8 | 71.8 | 62.8 | 50.7 | 44.0 | 60.0 |
| Days Max Temp ≥ 90 °F | 0 | 0 | 0 | 0 | 6 | 15 | 25 | 24 | 16 | 5 | 0 | 0 | 91 |
| Days Max Temp ≤ 32 °F | 0 | 0 | 0 | 0 | 0 | 0 | 0 | 0 | 0 | 0 | 0 | 0 | 0 |
| Days Min Temp ≤ 32 °F | 16 | 8 | 4 | 1 | 0 | 0 | 0 | 0 | 0 | 0 | 7 | 16 | 52 |
| Days Min Temp ≤ 0 °F | 0 | 0 | 0 | 0 | 0 | 0 | 0 | 0 | 0 | 0 | 0 | 0 | 0 |
| Heating Degree Days | 634 | 476 | 425 | 279 | 97 | 16 | 1 | 2 | 16 | 122 | 423 | 643 | 3134 |
| Cooling Degree Days | 0 | 0 | *0* | 11 | 89 | 233 | 411 | 370 | 227 | 71 | 0 | 0 | 1412 |
| Total Precipitation (") | 4.39 | 3.47 | 2.87 | 1.16 | 0.54 | 0.43 | 0.11 | 0.29 | 0.29 | 1.04 | 2.60 | 3.43 | 20.62 |
| Days ≥ 0.1" Precip | 7 | 6 | 6 | 3 | 2 | 1 | 0 | 1 | 1 | 2 | 5 | 6 | 40 |
| Total Snowfall (") | 0.9 | 0.2 | 0.1 | 0.0 | 0.0 | 0.0 | 0.0 | 0.0 | 0.0 | 0.0 | 0.0 | 0.4 | 1.6 |
| Days ≥ 1" Snow Depth | 0 | 0 | 0 | 0 | 0 | 0 | 0 | 0 | 0 | 0 | 0 | 0 | 0 |

## STRAWBERRY VALLEY *Yuba County* ELEVATION 3783 ft LAT/LONG 39° 34' N / 121° 6' W

| | JAN | FEB | MAR | APR | MAY | JUN | JUL | AUG | SEP | OCT | NOV | DEC | YEAR |
|---|---|---|---|---|---|---|---|---|---|---|---|---|---|
| Maximum Temp °F | 49.4 | 50.8 | 53.4 | 59.5 | 68.7 | 76.2 | 83.4 | 82.9 | 78.0 | 68.5 | 54.6 | 48.9 | 64.5 |
| Minimum Temp °F | 29.0 | 29.9 | 31.4 | 34.1 | 40.2 | 46.4 | 50.7 | 49.9 | 46.4 | 40.5 | 33.4 | 29.0 | 38.4 |
| Mean Temp °F | 39.2 | 40.4 | 42.4 | 46.8 | 54.5 | 61.2 | 67.0 | 66.4 | 62.2 | 54.5 | 44.0 | 39.0 | 51.5 |
| Days Max Temp ≥ 90 °F | 0 | 0 | 0 | 0 | 0 | 2 | 5 | 5 | 2 | 0 | 0 | 0 | 14 |
| Days Max Temp ≤ 32 °F | 1 | 1 | 0 | 0 | 0 | 0 | 0 | 0 | 0 | 0 | 0 | 1 | 3 |
| Days Min Temp ≤ 32 °F | 23 | 20 | 19 | 13 | 4 | 1 | 0 | 0 | 0 | 3 | 14 | 23 | 120 |
| Days Min Temp ≤ 0 °F | 0 | 0 | 0 | 0 | 0 | 0 | 0 | 0 | 0 | 0 | 0 | 0 | 0 |
| Heating Degree Days | 792 | 689 | 693 | 539 | 327 | 147 | 37 | 46 | 121 | 328 | 623 | 800 | 5142 |
| Cooling Degree Days | 0 | 0 | 0 | 1 | 7 | 39 | 102 | 92 | 46 | 17 | 0 | 0 | 304 |
| Total Precipitation (") | 14.74 | 12.04 | 11.85 | 5.46 | 2.26 | 0.97 | 0.27 | 0.43 | 1.55 | 4.27 | 11.31 | 12.52 | 77.67 |
| Days ≥ 0.1" Precip | 10 | 10 | 11 | 7 | 4 | 2 | 0 | 1 | 2 | 4 | 9 | 10 | 70 |
| Total Snowfall (") | 25.8 | 21.7 | 24.6 | 11.0 | 0.7 | 0.0 | 0.0 | 0.0 | 0.0 | 0.2 | 6.7 | 22.9 | 113.6 |
| Days ≥ 1" Snow Depth | 15 | 13 | 9 | 4 | 0 | 0 | 0 | 0 | 0 | 0 | 3 | 11 | 55 |

## SUSANVILLE MUNI AP *Lassen County* ELEVATION 4160 ft LAT/LONG 40° 23' N / 120° 34' W

| | JAN | FEB | MAR | APR | MAY | JUN | JUL | AUG | SEP | OCT | NOV | DEC | YEAR |
|---|---|---|---|---|---|---|---|---|---|---|---|---|---|
| Maximum Temp °F | 40.2 | 46.4 | 53.9 | 61.3 | 71.6 | 80.7 | 89.3 | 88.0 | 79.0 | 66.2 | 51.0 | 41.3 | 64.1 |
| Minimum Temp °F | 19.5 | 23.6 | 28.4 | 32.0 | 38.4 | 45.1 | 49.3 | 47.7 | 40.8 | 33.1 | 26.5 | 20.1 | 33.7 |
| Mean Temp °F | 29.9 | 35.0 | 41.2 | 46.7 | 55.0 | 62.9 | 69.4 | 67.8 | 59.9 | 49.6 | 38.8 | 30.7 | 48.9 |
| Days Max Temp ≥ 90 °F | 0 | 0 | 0 | 0 | 1 | 6 | 17 | 15 | 3 | 0 | 0 | 0 | 42 |
| Days Max Temp ≤ 32 °F | 6 | 2 | 0 | 0 | 0 | 0 | 0 | 0 | 0 | 0 | 0 | 5 | 13 |
| Days Min Temp ≤ 32 °F | 28 | 24 | 22 | 16 | 6 | 1 | 0 | 0 | 3 | *15* | 22 | 28 | 165 |
| Days Min Temp ≤ 0 °F | 2 | 0 | 0 | 0 | 0 | 0 | 0 | 0 | 0 | 0 | 0 | 1 | 3 |
| Heating Degree Days | 1081 | 840 | 731 | 544 | 312 | 117 | 26 | 38 | 171 | 470 | 780 | 1057 | 6167 |
| Cooling Degree Days | 0 | 0 | 0 | 0 | 10 | 61 | 148 | 129 | 27 | *0* | *0* | *0* | 375 |
| Total Precipitation (") | 2.78 | 1.80 | 1.48 | 0.56 | 0.79 | 0.62 | 0.29 | 0.26 | 0.45 | 0.99 | 1.99 | 2.20 | 14.21 |
| Days ≥ 0.1" Precip | 6 | 4 | 4 | 2 | 2 | 2 | 1 | 1 | 1 | 2 | 4 | *5* | 34 |
| Total Snowfall (") | na | na | *3.8* | *0.4* | *0.1* | 0.0 | 0.0 | 0.0 | 0.1 | *0.1* | na | na | na |
| Days ≥ 1" Snow Depth | na | *4* | *1* | *0* | *0* | 0 | 0 | 0 | 0 | *0* | na | na | na |

## TAHOE CITY *Placer County* ELEVATION 6234 ft LAT/LONG 39° 10' N / 120° 9' W

| | JAN | FEB | MAR | APR | MAY | JUN | JUL | AUG | SEP | OCT | NOV | DEC | YEAR |
|---|---|---|---|---|---|---|---|---|---|---|---|---|---|
| Maximum Temp °F | 40.1 | 41.8 | 44.6 | 50.9 | 60.5 | 69.1 | 77.5 | 77.0 | 69.8 | 59.6 | 47.4 | 40.8 | 56.6 |
| Minimum Temp °F | 19.3 | 20.5 | 23.6 | 27.0 | 33.1 | 39.3 | 44.5 | 44.6 | 39.2 | 32.2 | 25.7 | 19.8 | 30.7 |
| Mean Temp °F | 29.7 | 31.2 | 34.1 | 38.9 | 46.8 | 54.2 | 61.1 | 60.8 | 54.5 | 45.9 | 36.6 | 30.4 | 43.7 |
| Days Max Temp ≥ 90 °F | 0 | 0 | 0 | 0 | 0 | 0 | 0 | 0 | 0 | 0 | 0 | 0 | 0 |
| Days Max Temp ≤ 32 °F | 4 | 3 | 2 | 0 | 0 | 0 | 0 | 0 | 0 | 0 | 1 | 4 | 14 |
| Days Min Temp ≤ 32 °F | 30 | 28 | 29 | 26 | 14 | 3 | 0 | 0 | 3 | 16 | 26 | 30 | 205 |
| Days Min Temp ≤ 0 °F | 0 | 0 | 0 | 0 | 0 | 0 | 0 | 0 | 0 | 0 | 0 | 0 | 0 |
| Heating Degree Days | 1087 | 949 | 951 | 774 | 557 | 319 | 130 | 134 | 309 | 585 | 845 | 1067 | 7707 |
| Cooling Degree Days | 0 | 0 | 0 | 0 | 0 | 3 | 16 | 14 | 1 | 0 | 0 | 0 | 34 |
| Total Precipitation (") | 5.58 | 4.71 | 4.09 | 1.66 | 0.88 | 0.80 | 0.34 | 0.55 | 0.87 | 1.93 | 4.53 | 4.88 | 30.82 |
| Days ≥ 0.1" Precip | 7 | 7 | 7 | 5 | 3 | 3 | 1 | 1 | 2 | 3 | 7 | 7 | 53 |
| Total Snowfall (") | 35.8 | 33.0 | 32.9 | 13.5 | 2.6 | 0.2 | 0.0 | 0.0 | 0.4 | 2.3 | 19.6 | 32.6 | 172.9 |
| Days ≥ 1" Snow Depth | 29 | 26 | 27 | 12 | 3 | 0 | 0 | 0 | 0 | 1 | 11 | 24 | 133 |

## TEHACHAPI *Kern County* ELEVATION 3983 ft LAT/LONG 35° 8 ' N / 118° 28 ' W

| | JAN | FEB | MAR | APR | MAY | JUN | JUL | AUG | SEP | OCT | NOV | DEC | YEAR |
|---|---|---|---|---|---|---|---|---|---|---|---|---|---|
| Maximum Temp °F | 52.2 | 54.6 | 56.5 | 62.3 | 71.3 | 79.5 | 86.4 | 85.8 | 80.1 | 70.9 | 58.5 | 51.4 | 67.5 |
| Minimum Temp °F | 30.9 | 32.1 | 34.5 | 37.4 | 44.5 | 51.9 | 58.4 | 56.5 | 50.1 | 41.8 | 35.1 | 30.2 | 42.0 |
| Mean Temp °F | 41.6 | 43.4 | 45.5 | 49.9 | 57.9 | 65.7 | 72.4 | 71.2 | 65.2 | 56.4 | 46.8 | 40.8 | 54.7 |
| Days Max Temp ≥ 90 °F | 0 | 0 | 0 | 0 | 1 | 4 | 9 | 9 | 4 | 0 | 0 | 0 | 27 |
| Days Max Temp ≤ 32 °F | 0 | 0 | 0 | 0 | 0 | 0 | 0 | 0 | 0 | 0 | 0 | 1 | 1 |
| Days Min Temp ≤ 32 °F | 18 | 15 | 12 | 7 | 1 | 0 | 0 | 0 | 0 | 2 | 11 | 19 | 85 |
| Days Min Temp ≤ 0 °F | 0 | 0 | 0 | 0 | 0 | 0 | 0 | 0 | 0 | 0 | 0 | 0 | 0 |
| Heating Degree Days | 719 | 604 | 597 | 448 | 233 | 71 | 6 | 10 | 69 | 269 | 539 | 743 | 4308 |
| Cooling Degree Days | 0 | 0 | 0 | 1 | 21 | 98 | 238 | 207 | 86 | 15 | 0 | 0 | 666 |
| Total Precipitation (") | 1.79 | 1.77 | 2.15 | 0.85 | 0.41 | 0.12 | 0.11 | 0.32 | 0.29 | 0.46 | 1.50 | 1.70 | 11.47 |
| Days ≥ 0.1" Precip | 4 | 4 | 5 | 2 | 1 | 0 | 0 | 1 | 1 | 1 | 3 | 4 | 26 |
| Total Snowfall (") | 3.2 | 2.0 | 4.5 | 2.5 | 0.1 | 0.0 | 0.0 | 0.0 | 0.0 | 0.0 | 1.3 | 3.2 | 16.8 |
| Days ≥ 1" Snow Depth | *1* | *0* | *0* | 1 | 0 | 0 | 0 | 0 | 0 | 0 | 0 | *1* | 3 |

## TEJON RANCHO *Kern County* ELEVATION 1425 ft LAT/LONG 35° 2 ' N / 118° 45 ' W

| | JAN | FEB | MAR | APR | MAY | JUN | JUL | AUG | SEP | OCT | NOV | DEC | YEAR |
|---|---|---|---|---|---|---|---|---|---|---|---|---|---|
| Maximum Temp °F | 57.2 | 62.3 | 66.7 | 73.6 | 82.9 | 90.8 | 96.6 | 94.9 | 89.6 | 80.1 | 66.3 | 57.1 | 76.5 |
| Minimum Temp °F | 35.0 | 39.4 | 42.8 | 46.6 | 54.0 | 60.8 | 66.8 | 64.9 | 60.2 | 52.4 | 41.7 | 35.0 | 50.0 |
| Mean Temp °F | 46.1 | 50.9 | 54.8 | 60.1 | 68.5 | 75.8 | 81.8 | 79.9 | 74.9 | 66.2 | 54.0 | 46.1 | 63.3 |
| Days Max Temp ≥ 90 °F | 0 | 0 | 0 | 1 | 8 | 18 | 28 | 26 | 17 | 4 | 0 | 0 | 102 |
| Days Max Temp ≤ 32 °F | 0 | 0 | 0 | 0 | 0 | 0 | 0 | 0 | 0 | 0 | 0 | 0 | 0 |
| Days Min Temp ≤ 32 °F | 12 | 3 | 1 | 0 | 0 | 0 | 0 | 0 | 0 | 0 | 2 | 11 | 29 |
| Days Min Temp ≤ 0 °F | 0 | 0 | 0 | 0 | 0 | 0 | 0 | 0 | 0 | 0 | 0 | 0 | 0 |
| Heating Degree Days | 578 | 392 | 315 | 177 | 53 | 6 | 0 | 0 | 6 | 70 | 327 | 581 | 2505 |
| Cooling Degree Days | 0 | 0 | 2 | 44 | 160 | 333 | 523 | 477 | 318 | 128 | 2 | 0 | 1987 |
| Total Precipitation (") | 1.85 | 1.49 | 2.39 | 1.24 | 0.43 | 0.12 | 0.05 | 0.11 | 0.32 | 0.55 | 1.61 | 1.42 | 11.58 |
| Days ≥ 0.1" Precip | 4 | 4 | 5 | 3 | 1 | 0 | 0 | 0 | 1 | 1 | 3 | 4 | 26 |
| Total Snowfall (") | 0.0 | 0.0 | 0.0 | 0.0 | 0.0 | 0.0 | 0.0 | 0.0 | 0.0 | 0.0 | 0.0 | 0.0 | 0.0 |
| Days ≥ 1" Snow Depth | 0 | 0 | 0 | 0 | 0 | 0 | 0 | 0 | 0 | 0 | 0 | 0 | 0 |

## TERMO 1 E *Lassen County* ELEVATION 5305 ft LAT/LONG 40° 52 ' N / 120° 27 ' W

| | JAN | FEB | MAR | APR | MAY | JUN | JUL | AUG | SEP | OCT | NOV | DEC | YEAR |
|---|---|---|---|---|---|---|---|---|---|---|---|---|---|
| Maximum Temp °F | 38.4 | 43.1 | 48.9 | 56.5 | 66.4 | 75.3 | 85.1 | 83.9 | 75.6 | 63.9 | 47.7 | 38.9 | 60.3 |
| Minimum Temp °F | 13.8 | 19.0 | 23.8 | 26.1 | 32.4 | 38.3 | 42.8 | 41.0 | 34.3 | 26.7 | 20.8 | 14.2 | 27.8 |
| Mean Temp °F | 26.1 | 31.1 | 36.3 | 41.4 | 49.4 | 56.9 | 63.9 | 62.5 | 55.0 | 45.3 | 34.3 | 26.6 | 44.1 |
| Days Max Temp ≥ 90 °F | 0 | 0 | 0 | 0 | 0 | 2 | 9 | 7 | 1 | 0 | 0 | 0 | 19 |
| Days Max Temp ≤ 32 °F | 7 | 3 | 1 | 0 | 0 | 0 | 0 | 0 | 0 | 0 | 2 | 7 | 20 |
| Days Min Temp ≤ 32 °F | 29 | 26 | 27 | 25 | 16 | 6 | 2 | 3 | 12 | 24 | 26 | 30 | 226 |
| Days Min Temp ≤ 0 °F | 6 | 2 | 1 | 0 | 0 | 0 | 0 | 0 | 0 | 0 | 1 | 5 | 15 |
| Heating Degree Days | 1198 | 951 | 881 | 702 | 477 | 249 | 86 | 114 | 296 | 604 | 914 | 1183 | 7655 |
| Cooling Degree Days | 0 | 0 | 0 | 0 | 1 | 11 | 55 | 38 | 3 | 0 | 0 | 0 | 108 |
| Total Precipitation (") | 1.12 | 0.99 | 1.05 | 0.76 | 1.05 | 0.90 | 0.36 | 0.37 | 0.55 | 0.81 | 1.31 | 1.25 | 10.52 |
| Days ≥ 0.1" Precip | 3 | 4 | 3 | 2 | 3 | 2 | 1 | 1 | 2 | 2 | 4 | 4 | 31 |
| Total Snowfall (") | na | na | na | *2.9* | 2.1 | 0.4 | 0.1 | 0.0 | 0.4 | 1.5 | *6.2* | na | na |
| Days ≥ 1" Snow Depth | 21 | *13* | *4* | *0* | 0 | 0 | 0 | 0 | 0 | 0 | *6* | 16 | 60 |

## THREE RIVERS ED PH 1 *Tulare County* ELEVATION 1191 ft LAT/LONG 36° 28 ' N / 118° 51 ' W

| | JAN | FEB | MAR | APR | MAY | JUN | JUL | AUG | SEP | OCT | NOV | DEC | YEAR |
|---|---|---|---|---|---|---|---|---|---|---|---|---|---|
| Maximum Temp °F | *58.9* | *64.6* | *67.9* | *75.6* | *85.0* | *93.7* | *99.3* | *97.8* | *92.3* | *82.0* | *66.7* | *58.8* | 78.6 |
| Minimum Temp °F | *34.5* | *37.9* | *41.3* | *44.8* | *51.3* | *58.4* | *64.8* | *63.8* | *58.2* | *49.3* | *39.4* | *34.2* | 48.2 |
| Mean Temp °F | *46.7* | *51.3* | *54.6* | *60.2* | *68.1* | *76.1* | *82.1* | *80.8* | *75.2* | *65.6* | *53.1* | *46.6* | 63.4 |
| Days Max Temp ≥ 90 °F | *0* | *0* | *0* | *2* | *11* | *22* | *30* | *28* | *20* | *7* | *0* | *0* | 120 |
| Days Max Temp ≤ 32 °F | *0* | *0* | *0* | *0* | *0* | *0* | *0* | *0* | *0* | *0* | *0* | *0* | 0 |
| Days Min Temp ≤ 32 °F | na | *5* | *2* | *1* | *0* | *0* | *0* | *0* | *0* | *0* | *4* | *13* | na |
| Days Min Temp ≤ 0 °F | *0* | *0* | *0* | *0* | *0* | *0* | *0* | *0* | *0* | *0* | *0* | *0* | 0 |
| Heating Degree Days | *561* | *381* | *318* | *167* | *50* | *5* | *0* | *0* | *6* | *72* | *353* | *565* | 2478 |
| Cooling Degree Days | 0 | 0 | 1 | 37 | 149 | 336 | 532 | 505 | 320 | 97 | 1 | 0 | 1978 |
| Total Precipitation (") | *4.10* | *4.13* | *4.97* | *1.68* | *0.62* | *0.25* | *0.11* | *0.08* | *0.78* | *1.18* | *2.79* | *3.02* | 23.71 |
| Days ≥ 0.1" Precip | *5* | *6* | *7* | *3* | *2* | *0* | *0* | *0* | *1* | *2* | *4* | *5* | 35 |
| Total Snowfall (") | *0.0* | *0.0* | *0.0* | *0.0* | *0.0* | *0.0* | *0.0* | *0.0* | *0.0* | *0.0* | *0.0* | *0.0* | 0.0 |
| Days ≥ 1" Snow Depth | *0* | *0* | *0* | *0* | *0* | *0* | *0* | *0* | *0* | *0* | *0* | *0* | 0 |

## TIGER CREEK PH *Amador County* ELEVATION 2362 ft LAT/LONG 38° 27 ' N / 120° 29 ' W

| | JAN | FEB | MAR | APR | MAY | JUN | JUL | AUG | SEP | OCT | NOV | DEC | YEAR |
|---|---|---|---|---|---|---|---|---|---|---|---|---|---|
| Maximum Temp °F | 51.4 | 57.8 | 60.6 | 66.5 | 75.4 | 83.7 | 91.0 | 90.5 | 85.2 | 75.4 | 59.4 | 48.9 | 70.5 |
| Minimum Temp °F | 32.8 | 34.3 | 36.3 | 39.1 | 45.0 | 50.7 | 56.0 | 55.9 | 51.8 | 44.8 | 37.3 | 32.6 | 43.1 |
| Mean Temp °F | 42.1 | 46.1 | 48.5 | 52.9 | 60.3 | 67.2 | 73.6 | 73.2 | 68.5 | 60.1 | 48.3 | 40.8 | 56.8 |
| Days Max Temp ≥ 90 °F | 0 | 0 | 0 | 0 | 2 | 8 | 20 | 19 | 10 | 2 | 0 | 0 | 61 |
| Days Max Temp ≤ 32 °F | 0 | 0 | 0 | 0 | 0 | 0 | 0 | 0 | 0 | 0 | 0 | 0 | 0 |
| Days Min Temp ≤ 32 °F | 16 | 11 | 8 | 4 | 0 | 0 | 0 | 0 | 0 | 1 | 7 | 16 | 63 |
| Days Min Temp ≤ 0 °F | 0 | 0 | 0 | 0 | 0 | 0 | 0 | 0 | 0 | 0 | 0 | 0 | 0 |
| Heating Degree Days | 703 | 527 | 505 | 360 | 174 | 50 | 6 | 6 | 37 | 176 | 494 | 743 | 3781 |
| Cooling Degree Days | 0 | 0 | 0 | 4 | 31 | 117 | 252 | 257 | 145 | 35 | 0 | 0 | 841 |
| Total Precipitation (") | 8.20 | 6.86 | 7.29 | 3.73 | 1.21 | 0.47 | 0.21 | 0.28 | 0.95 | 2.61 | 6.63 | 7.00 | 45.44 |
| Days ≥ 0.1" Precip | 8 | 8 | 9 | 6 | 3 | 1 | 0 | 0 | 2 | 3 | 8 | 8 | 56 |
| Total Snowfall (") | 4.3 | 3.6 | 4.2 | 4.0 | 0.0 | 0.0 | 0.0 | 0.0 | 0.0 | 0.0 | 0.5 | 5.6 | 22.2 |
| Days ≥ 1" Snow Depth | 1 | 1 | 1 | 1 | 0 | 0 | 0 | 0 | 0 | 0 | 0 | 2 | 6 |

## TORRANCE MUNICIPAL A *Los Angeles County* ELEVATION 95 ft LAT/LONG 33° 48 ' N / 118° 20 ' W

| | JAN | FEB | MAR | APR | MAY | JUN | JUL | AUG | SEP | OCT | NOV | DEC | YEAR |
|---|---|---|---|---|---|---|---|---|---|---|---|---|---|
| Maximum Temp °F | 66.8 | 67.7 | 67.9 | 70.8 | 72.1 | 75.1 | 78.7 | 80.1 | 79.0 | 76.7 | 71.6 | 66.9 | 72.8 |
| Minimum Temp °F | 45.7 | 46.9 | 48.4 | 50.8 | 54.4 | 57.5 | 60.9 | 62.2 | 60.7 | 56.6 | 50.4 | 45.7 | 53.4 |
| Mean Temp °F | 56.3 | 57.3 | 58.2 | 60.8 | 63.3 | 66.3 | 69.8 | 71.2 | 69.9 | 66.7 | 61.0 | 56.3 | 63.1 |
| Days Max Temp ≥ 90 °F | 0 | 0 | 0 | 1 | 0 | 1 | 1 | 2 | 2 | 2 | 0 | 0 | 9 |
| Days Max Temp ≤ 32 °F | 0 | 0 | 0 | 0 | 0 | 0 | 0 | 0 | 0 | 0 | 0 | 0 | 0 |
| Days Min Temp ≤ 32 °F | 0 | 0 | 0 | 0 | 0 | 0 | 0 | 0 | 0 | 0 | 0 | 0 | 0 |
| Days Min Temp ≤ 0 °F | 0 | 0 | 0 | 0 | 0 | 0 | 0 | 0 | 0 | 0 | 0 | 0 | 0 |
| Heating Degree Days | 267 | 215 | 211 | 138 | 75 | 21 | 2 | 0 | 2 | 26 | 133 | 264 | 1354 |
| Cooling Degree Days | 3 | 7 | 8 | 28 | 29 | 77 | 176 | 201 | 160 | 93 | 17 | 2 | 801 |
| Total Precipitation (") | 3.33 | 2.82 | 2.48 | 0.83 | 0.19 | 0.04 | 0.02 | 0.13 | 0.23 | 0.37 | 1.84 | 2.21 | 14.49 |
| Days ≥ 0.1" Precip | 5 | 4 | 4 | 2 | 0 | 0 | 0 | 0 | 1 | 1 | 3 | 3 | 23 |
| Total Snowfall (") | 0.0 | 0.0 | 0.0 | 0.0 | 0.0 | 0.0 | 0.0 | 0.0 | 0.0 | 0.0 | 0.0 | 0.0 | 0.0 |
| Days ≥ 1" Snow Depth | 0 | 0 | 0 | 0 | 0 | 0 | 0 | 0 | 0 | 0 | 0 | 0 | 0 |

## TRACY CARBONA *San Joaquin County* ELEVATION 140 ft LAT/LONG 37° 42 ' N / 121° 25 ' W

| | JAN | FEB | MAR | APR | MAY | JUN | JUL | AUG | SEP | OCT | NOV | DEC | YEAR |
|---|---|---|---|---|---|---|---|---|---|---|---|---|---|
| Maximum Temp °F | ***54.4*** | 61.3 | 66.8 | 72.9 | 81.5 | 88.3 | 93.7 | 92.3 | 88.0 | 79.1 | ***64.8*** | ***54.6*** | 74.8 |
| Minimum Temp °F | ***36.4*** | 39.9 | 43.2 | 45.6 | 50.2 | 55.0 | 57.0 | 55.8 | 54.1 | 49.0 | ***42.1*** | ***36.3*** | 47.1 |
| Mean Temp °F | ***45.5*** | 50.6 | 55.0 | 59.3 | 65.9 | 71.7 | 75.4 | 74.1 | 71.0 | 64.0 | ***53.4*** | ***45.3*** | 60.9 |
| Days Max Temp ≥ 90 °F | 0 | 0 | 0 | 1 | 6 | 14 | 24 | 21 | 13 | 3 | 0 | 0 | 82 |
| Days Max Temp ≤ 32 °F | 0 | 0 | 0 | 0 | 0 | 0 | 0 | 0 | 0 | 0 | 0 | 0 | 0 |
| Days Min Temp ≤ 32 °F | 9 | 2 | 0 | 0 | 0 | 0 | 0 | 0 | 0 | 0 | 1 | 8 | 20 |
| Days Min Temp ≤ 0 °F | 0 | 0 | 0 | 0 | 0 | 0 | 0 | 0 | 0 | 0 | 0 | 0 | 0 |
| Heating Degree Days | ***599*** | 400 | 303 | 185 | 67 | 13 | 1 | 2 | 8 | 82 | ***341*** | ***602*** | 2603 |
| Cooling Degree Days | na | 0 | 1 | 21 | 105 | 217 | 328 | 271 | 188 | 63 | na | na | na |
| Total Precipitation (") | 1.89 | 1.61 | 1.65 | 0.80 | 0.42 | 0.08 | 0.05 | 0.15 | 0.27 | 0.56 | 1.41 | 1.48 | 10.37 |
| Days ≥ 0.1" Precip | 5 | 4 | 5 | 2 | 1 | 0 | 0 | 0 | 1 | 2 | 3 | 4 | 27 |
| Total Snowfall (") | 0.0 | 0.0 | 0.0 | 0.0 | 0.0 | 0.0 | 0.0 | 0.0 | 0.0 | 0.0 | 0.0 | 0.0 | 0.0 |
| Days ≥ 1" Snow Depth | 0 | 0 | 0 | 0 | 0 | 0 | 0 | 0 | 0 | 0 | 0 | 0 | 0 |

## TRACY PUMPING PLANT *Alameda County* ELEVATION 61 ft LAT/LONG 37° 48 ' N / 121° 35 ' W

| | JAN | FEB | MAR | APR | MAY | JUN | JUL | AUG | SEP | OCT | NOV | DEC | YEAR |
|---|---|---|---|---|---|---|---|---|---|---|---|---|---|
| Maximum Temp °F | ***54.3*** | 61.3 | 66.1 | 72.3 | 80.5 | 86.9 | 92.4 | 91.9 | 87.4 | 78.4 | ***64.4*** | ***54.4*** | 74.2 |
| Minimum Temp °F | ***37.6*** | 41.0 | 44.5 | 47.2 | 52.9 | 57.0 | 60.0 | 60.0 | 57.7 | 51.6 | ***43.5*** | 37.3 | 49.2 |
| Mean Temp °F | ***45.9*** | 51.1 | 55.4 | 59.8 | 66.7 | 72.0 | 76.2 | 76.0 | 72.5 | 65.1 | ***54.0*** | ***45.9*** | 61.7 |
| Days Max Temp ≥ 90 °F | 0 | 0 | 0 | 1 | 6 | 12 | 21 | 19 | 12 | 3 | 0 | 0 | 74 |
| Days Max Temp ≤ 32 °F | 0 | 0 | 0 | 0 | 0 | 0 | 0 | 0 | 0 | 0 | 0 | 0 | 0 |
| Days Min Temp ≤ 32 °F | 7 | 2 | 0 | 0 | 0 | 0 | 0 | 0 | 0 | 0 | 1 | 8 | 18 |
| Days Min Temp ≤ 0 °F | 0 | 0 | 0 | 0 | 0 | 0 | 0 | 0 | 0 | 0 | 0 | 0 | 0 |
| Heating Degree Days | ***584*** | 384 | 292 | 171 | 60 | 14 | 1 | 0 | 5 | 64 | ***325*** | ***585*** | 2485 |
| Cooling Degree Days | na | ***0*** | 0 | 26 | ***125*** | ***238*** | 358 | 351 | ***240*** | ***85*** | na | na | na |
| Total Precipitation (") | 2.46 | 2.00 | 1.82 | 0.78 | 0.26 | 0.09 | 0.05 | 0.09 | 0.25 | 0.59 | 1.78 | 1.76 | 11.93 |
| Days ≥ 0.1" Precip | 5 | 5 | 5 | 2 | 1 | 0 | 0 | 0 | 0 | 1 | 4 | 5 | 28 |
| Total Snowfall (") | 0.0 | 0.0 | 0.0 | 0.0 | 0.0 | 0.0 | 0.0 | 0.0 | 0.0 | 0.0 | 0.0 | 0.0 | 0.0 |
| Days ≥ 1" Snow Depth | 0 | 0 | 0 | 0 | 0 | 0 | 0 | 0 | 0 | 0 | 0 | 0 | 0 |

## TRONA *San Bernardino County* ELEVATION 1703 ft LAT/LONG 35° 47 ' N / 117° 23 ' W

| | JAN | FEB | MAR | APR | MAY | JUN | JUL | AUG | SEP | OCT | NOV | DEC | YEAR |
|---|---|---|---|---|---|---|---|---|---|---|---|---|---|
| Maximum Temp °F | 58.5 | 65.4 | 71.0 | 78.9 | 88.7 | 98.8 | 105.3 | 103.1 | 95.2 | 84.0 | 69.7 | 58.1 | 81.4 |
| Minimum Temp °F | 33.0 | 38.1 | 44.1 | 50.2 | 59.1 | 67.4 | 74.1 | 73.0 | 65.6 | 53.5 | 41.4 | 32.2 | 52.6 |
| Mean Temp °F | 45.8 | 52.0 | 57.6 | 64.6 | 73.9 | 83.1 | 89.8 | 88.1 | 80.4 | 68.8 | 55.6 | 45.1 | 67.1 |
| Days Max Temp ≥ 90 °F | 0 | 0 | 0 | 4 | 17 | 26 | 31 | 30 | 23 | 8 | 0 | 0 | 139 |
| Days Max Temp ≤ 32 °F | 0 | 0 | 0 | 0 | 0 | 0 | 0 | 0 | 0 | 0 | 0 | 0 | 0 |
| Days Min Temp ≤ 32 °F | 15 | 6 | 1 | 0 | 0 | 0 | 0 | 0 | 0 | 0 | 3 | 17 | 42 |
| Days Min Temp ≤ 0 °F | 0 | 0 | 0 | 0 | 0 | 0 | 0 | 0 | 0 | 0 | 0 | 0 | 0 |
| Heating Degree Days | 590 | 362 | 234 | 94 | 16 | 1 | 0 | 0 | 1 | 39 | 284 | 612 | 2233 |
| Cooling Degree Days | 0 | 4 | 19 | 126 | 338 | 604 | 810 | 755 | 507 | 188 | *13* | 0 | 3364 |
| Total Precipitation (") | 0.83 | 0.82 | 0.62 | 0.17 | 0.10 | 0.13 | 0.23 | 0.34 | 0.20 | 0.12 | 0.43 | 0.49 | 4.48 |
| Days ≥ 0.1" Precip | 2 | 2 | 2 | 1 | 0 | 0 | 0 | 1 | 0 | 0 | 1 | 1 | 10 |
| Total Snowfall (") | 0.3 | 0.0 | 0.0 | 0.0 | 0.0 | 0.0 | 0.0 | 0.0 | 0.0 | 0.0 | 0.0 | 0.0 | 0.3 |
| Days ≥ 1" Snow Depth | 0 | 0 | 0 | 0 | 0 | 0 | 0 | 0 | 0 | 0 | 0 | 0 | 0 |

## TRUCKEE RANGER STN *Nevada County* ELEVATION 5984 ft LAT/LONG 39° 20 ' N / 120° 11 ' W

| | JAN | FEB | MAR | APR | MAY | JUN | JUL | AUG | SEP | OCT | NOV | DEC | YEAR |
|---|---|---|---|---|---|---|---|---|---|---|---|---|---|
| Maximum Temp °F | 39.6 | 43.2 | 47.1 | 53.7 | 63.7 | 72.8 | 81.7 | 80.7 | 74.0 | 63.6 | 48.3 | 40.0 | 59.0 |
| Minimum Temp °F | 14.2 | 16.4 | 21.8 | 25.7 | 31.5 | 37.2 | 41.8 | 40.9 | 35.5 | 28.5 | 22.1 | 15.2 | 27.6 |
| Mean Temp °F | 26.9 | 29.9 | 34.5 | 39.7 | 47.6 | 55.0 | 61.7 | 60.8 | 54.8 | 46.1 | 35.2 | 27.6 | 43.3 |
| Days Max Temp ≥ 90 °F | 0 | 0 | 0 | 0 | 0 | 0 | 3 | 3 | 1 | 0 | 0 | 0 | 7 |
| Days Max Temp ≤ 32 °F | 6 | 3 | 2 | 0 | 0 | 0 | 0 | 0 | 0 | 0 | 2 | 6 | 19 |
| Days Min Temp ≤ 32 °F | 31 | 27 | 29 | 27 | 19 | 7 | 2 | 2 | 9 | 23 | 28 | 30 | 234 |
| Days Min Temp ≤ 0 °F | 3 | 2 | 0 | 0 | 0 | 0 | 0 | 0 | 0 | 0 | 0 | 3 | 8 |
| Heating Degree Days | 1174 | 985 | 937 | 751 | 532 | 298 | 124 | 145 | 302 | 579 | 887 | 1152 | 7866 |
| Cooling Degree Days | 0 | 0 | 0 | 0 | 0 | 7 | 33 | *30* | 3 | 0 | 0 | 0 | 73 |
| Total Precipitation (") | 5.41 | 4.53 | 4.24 | 1.82 | 1.04 | 0.74 | 0.45 | 0.65 | 0.90 | 1.72 | 4.26 | 4.77 | 30.53 |
| Days ≥ 0.1" Precip | 8 | 7 | 7 | 5 | 3 | 2 | 1 | 2 | 2 | 3 | 6 | 7 | 53 |
| Total Snowfall (") | 41.6 | 40.1 | 36.3 | 14.7 | 3.9 | 0.5 | 0.0 | 0.0 | 0.6 | 3.1 | 21.7 | 38.6 | 201.1 |
| Days ≥ 1" Snow Depth | 30 | 27 | 26 | 11 | 3 | 0 | 0 | 0 | 0 | 2 | 12 | 26 | 137 |

## TULELAKE *Siskiyou County* ELEVATION 4042 ft LAT/LONG 41° 58 ' N / 121° 28 ' W

| | JAN | FEB | MAR | APR | MAY | JUN | JUL | AUG | SEP | OCT | NOV | DEC | YEAR |
|---|---|---|---|---|---|---|---|---|---|---|---|---|---|
| Maximum Temp °F | 39.7 | 46.6 | 52.3 | 59.3 | 69.4 | 76.6 | 84.5 | 83.9 | 77.3 | 66.3 | 48.5 | 39.3 | 62.0 |
| Minimum Temp °F | 18.9 | 23.1 | 25.6 | 28.9 | 36.2 | 42.6 | 45.7 | 43.6 | 37.0 | 30.1 | 24.6 | 18.8 | 31.3 |
| Mean Temp °F | 29.3 | 34.9 | 38.9 | 44.1 | 52.8 | 59.6 | 65.1 | 63.8 | 57.2 | 48.2 | 36.6 | 29.1 | 46.6 |
| Days Max Temp ≥ 90 °F | 0 | 0 | 0 | 0 | 0 | 2 | 8 | 7 | 2 | 0 | 0 | 0 | 19 |
| Days Max Temp ≤ 32 °F | 5 | 1 | 0 | 0 | 0 | 0 | 0 | 0 | 0 | 0 | 1 | 6 | 13 |
| Days Min Temp ≤ 32 °F | 28 | 25 | 25 | 21 | 10 | 2 | 0 | 0 | 7 | 20 | 25 | 28 | 191 |
| Days Min Temp ≤ 0 °F | 2 | 1 | 0 | 0 | 0 | 0 | 0 | 0 | 0 | 0 | 0 | 2 | 5 |
| Heating Degree Days | 1098 | 843 | 798 | 619 | 374 | 179 | 65 | 84 | 235 | 513 | 846 | 1107 | 6761 |
| Cooling Degree Days | 0 | 0 | 0 | *0* | 7 | 26 | 63 | 43 | 9 | 0 | 0 | 0 | 148 |
| Total Precipitation (") | 1.19 | 0.96 | 1.10 | 0.84 | 0.84 | 0.84 | 0.32 | 0.61 | 0.49 | 0.88 | 1.39 | 1.30 | 10.76 |
| Days ≥ 0.1" Precip | 4 | 3 | 3 | 3 | 3 | 2 | 1 | 1 | 1 | 3 | 4 | 4 | 32 |
| Total Snowfall (") | *5.6* | 3.1 | 3.2 | na | 0.0 | 0.0 | 0.0 | 0.0 | 0.0 | 0.1 | 3.0 | *5.8* | na |
| Days ≥ 1" Snow Depth | 10 | 4 | 2 | *0* | 0 | 0 | 0 | 0 | 0 | 0 | 3 | 10 | 29 |

## TUSTIN IRVINE RANCH *Orange County* ELEVATION 118 ft LAT/LONG 33° 44 ' N / 117° 47 ' W

| | JAN | FEB | MAR | APR | MAY | JUN | JUL | AUG | SEP | OCT | NOV | DEC | YEAR |
|---|---|---|---|---|---|---|---|---|---|---|---|---|---|
| Maximum Temp °F | 67.3 | 68.7 | 69.1 | 72.9 | 74.7 | 78.7 | 83.5 | 85.1 | 83.5 | 79.3 | 72.7 | 67.4 | 75.2 |
| Minimum Temp °F | 41.1 | 42.7 | 45.1 | 48.0 | 52.7 | 56.4 | 59.9 | 60.3 | 58.2 | 52.6 | 45.1 | 40.1 | 50.2 |
| Mean Temp °F | 54.2 | 55.7 | 57.1 | 60.5 | 63.7 | 67.6 | 71.7 | 72.8 | 70.9 | 66.0 | 59.0 | 53.8 | 62.7 |
| Days Max Temp ≥ 90 °F | 0 | 0 | 0 | 1 | 1 | 2 | 4 | 6 | 7 | 3 | 0 | 0 | 24 |
| Days Max Temp ≤ 32 °F | 0 | 0 | 0 | 0 | 0 | 0 | 0 | 0 | 0 | 0 | 0 | 0 | 0 |
| Days Min Temp ≤ 32 °F | 2 | 1 | 0 | 0 | 0 | 0 | 0 | 0 | 0 | 0 | 0 | 2 | 5 |
| Days Min Temp ≤ 0 °F | 0 | 0 | 0 | 0 | 0 | 0 | 0 | 0 | 0 | 0 | 0 | 0 | 0 |
| Heating Degree Days | 330 | 257 | 242 | 146 | 72 | 18 | 0 | 1 | 4 | 40 | 182 | 340 | 1632 |
| Cooling Degree Days | 3 | 4 | 6 | 28 | *43* | 111 | 242 | *268* | 193 | 86 | 6 | 1 | 991 |
| Total Precipitation (") | 2.62 | 2.63 | 2.49 | 0.87 | 0.16 | 0.07 | 0.02 | 0.14 | 0.31 | 0.33 | 1.49 | 1.91 | 13.04 |
| Days ≥ 0.1" Precip | 4 | 4 | 4 | 2 | 0 | 0 | 0 | 0 | 1 | 1 | 2 | 4 | 22 |
| Total Snowfall (") | 0.0 | 0.0 | 0.0 | 0.0 | 0.0 | 0.0 | 0.0 | 0.0 | 0.0 | 0.0 | 0.0 | 0.0 | 0.0 |
| Days ≥ 1" Snow Depth | 0 | 0 | 0 | 0 | 0 | 0 | 0 | 0 | 0 | 0 | 0 | 0 | 0 |

## TWENTYNINE PALMS *San Bernardino County* ELEVATION 1990 ft LAT/LONG 34° 8 ' N / 116° 3 ' W

| | JAN | FEB | MAR | APR | MAY | JUN | JUL | AUG | SEP | OCT | NOV | DEC | YEAR |
|---|---|---|---|---|---|---|---|---|---|---|---|---|---|
| Maximum Temp °F | 63.3 | 68.3 | 74.2 | 82.0 | 91.0 | 100.9 | 105.4 | 103.1 | 96.7 | 85.5 | 71.3 | 62.4 | 83.7 |
| Minimum Temp °F | 35.7 | 38.9 | 43.2 | 48.6 | 56.7 | 64.4 | 70.9 | 69.9 | 63.0 | 52.2 | 41.3 | 34.8 | 51.6 |
| Mean Temp °F | 49.5 | 53.6 | 58.7 | 65.3 | 73.9 | 82.7 | 88.2 | 86.5 | 79.8 | 68.9 | 56.3 | 48.6 | 67.7 |
| Days Max Temp ≥ 90 °F | 0 | 0 | 1 | 6 | 19 | 28 | 31 | 30 | 26 | 10 | 0 | 0 | 151 |
| Days Max Temp ≤ 32 °F | 0 | 0 | 0 | 0 | 0 | 0 | 0 | 0 | 0 | 0 | 0 | 0 | 0 |
| Days Min Temp ≤ 32 °F | 10 | 4 | 2 | 0 | 0 | 0 | 0 | 0 | 0 | 0 | 3 | 11 | 30 |
| Days Min Temp ≤ 0 °F | 0 | 0 | 0 | 0 | 0 | 0 | 0 | 0 | 0 | 0 | 0 | 0 | 0 |
| Heating Degree Days | 473 | 316 | 207 | 79 | 12 | 0 | 0 | 0 | 1 | 38 | 260 | 500 | 1886 |
| Cooling Degree Days | 0 | 3 | 21 | 125 | 308 | 561 | 718 | 680 | 461 | 170 | 7 | 0 | 3054 |
| Total Precipitation (") | 0.48 | 0.46 | 0.50 | 0.10 | 0.13 | 0.01 | 0.67 | 0.78 | 0.44 | 0.15 | 0.24 | 0.51 | 4.47 |
| Days ≥ 0.1" Precip | 2 | 1 | 1 | 0 | 0 | 0 | 1 | 2 | 1 | 1 | 1 | 1 | 11 |
| Total Snowfall (") | 0.0 | 0.0 | 0.0 | 0.0 | 0.0 | 0.0 | 0.0 | 0.0 | 0.0 | 0.0 | 0.0 | 0.4 | 0.4 |
| Days ≥ 1" Snow Depth | 0 | 0 | 0 | 0 | 0 | 0 | 0 | 0 | 0 | 0 | 0 | 0 | 0 |

## TWIN LAKES *Alpine County* ELEVATION 7835 ft LAT/LONG 38° 42 ' N / 120° 3 ' W

| | JAN | FEB | MAR | APR | MAY | JUN | JUL | AUG | SEP | OCT | NOV | DEC | YEAR |
|---|---|---|---|---|---|---|---|---|---|---|---|---|---|
| Maximum Temp °F | 38.8 | 39.5 | 41.2 | 46.5 | 54.1 | 62.7 | 70.8 | ***70.3*** | 64.0 | 55.4 | 43.8 | 38.6 | 52.1 |
| Minimum Temp °F | 17.7 | 17.5 | 19.5 | 23.4 | 30.4 | 37.8 | 43.9 | 44.1 | 38.9 | 31.8 | 23.4 | 18.1 | 28.9 |
| Mean Temp °F | 28.3 | 28.5 | 30.4 | 35.0 | 42.3 | 50.3 | 57.3 | ***57.2*** | 51.5 | 43.6 | 33.6 | 28.3 | 40.5 |
| Days Max Temp ≥ 90 °F | 0 | 0 | 0 | 0 | 0 | 0 | 0 | 0 | 0 | 0 | 0 | 0 | 0 |
| Days Max Temp ≤ 32 °F | 9 | 7 | 5 | 2 | 1 | 0 | 0 | 0 | 0 | 1 | 4 | 8 | 37 |
| Days Min Temp ≤ 32 °F | 31 | 27 | 30 | 27 | 19 | 7 | 1 | 1 | 4 | 16 | 25 | 30 | 218 |
| Days Min Temp ≤ 0 °F | 1 | 1 | 1 | 0 | 0 | 0 | 0 | 0 | 0 | 0 | 0 | 1 | 4 |
| Heating Degree Days | 1132 | 1025 | 1067 | 901 | 698 | 436 | 233 | ***240*** | 400 | 658 | 934 | 1132 | 8856 |
| Cooling Degree Days | 0 | 0 | 0 | 0 | 0 | 1 | ***4*** | ***4*** | ***0*** | 0 | 0 | 0 | 9 |
| Total Precipitation (") | 7.96 | 6.74 | 6.41 | 3.71 | 1.76 | 1.15 | 0.65 | 0.96 | 1.28 | 2.66 | 6.73 | 7.09 | 47.10 |
| Days ≥ 0.1" Precip | 9 | 9 | 10 | 8 | 4 | 3 | 2 | 1 | 2 | 4 | 7 | 8 | 67 |
| Total Snowfall (") | ***65.5*** | na | ***74.9*** | ***30.5*** | 9.3 | 1.7 | 0.0 | 0.2 | 1.3 | 10.7 | ***48.4*** | ***62.8*** | na |
| Days ≥ 1" Snow Depth | na | na | na | na | na | 2 | 0 | 0 | 0 | ***2*** | na | na | na |

## TWITCHELL DAM *Santa Barbara County* ELEVATION 582 ft LAT/LONG 34° 59 ' N / 120° 19 ' W

| | JAN | FEB | MAR | APR | MAY | JUN | JUL | AUG | SEP | OCT | NOV | DEC | YEAR |
|---|---|---|---|---|---|---|---|---|---|---|---|---|---|
| Maximum Temp °F | 65.4 | 66.4 | 67.0 | 70.7 | 72.8 | 77.2 | 79.9 | 80.8 | 80.6 | 78.5 | 71.6 | 65.9 | 73.1 |
| Minimum Temp °F | 40.1 | 42.4 | 43.5 | 44.5 | 47.0 | 50.0 | 51.9 | 52.1 | 51.5 | 48.4 | 44.2 | 39.4 | 46.3 |
| Mean Temp °F | 52.8 | 54.4 | 55.3 | 57.6 | 59.9 | 63.6 | 65.9 | 66.5 | 66.1 | 63.5 | 57.9 | 52.7 | 59.7 |
| Days Max Temp ≥ 90 °F | 0 | 0 | 0 | 1 | 1 | 2 | 2 | 3 | 4 | 4 | 0 | 0 | 17 |
| Days Max Temp ≤ 32 °F | 0 | 0 | 0 | 0 | 0 | 0 | 0 | 0 | 0 | 0 | 0 | 0 | 0 |
| Days Min Temp ≤ 32 °F | 4 | 1 | 1 | 0 | 0 | 0 | 0 | 0 | 0 | 0 | 1 | 4 | 11 |
| Days Min Temp ≤ 0 °F | 0 | 0 | 0 | 0 | 0 | 0 | 0 | 0 | 0 | 0 | 0 | 0 | 0 |
| Heating Degree Days | 373 | 295 | 298 | 228 | 173 | 81 | 29 | 21 | 40 | 95 | 215 | 377 | 2225 |
| Cooling Degree Days | 0 | 3 | 5 | 22 | ***19*** | 51 | 74 | 71 | 89 | 57 | 11 | 1 | 403 |
| Total Precipitation (") | 3.30 | 3.25 | 3.40 | 1.35 | 0.23 | 0.06 | 0.05 | 0.05 | 0.42 | 0.75 | 1.85 | 2.56 | 17.27 |
| Days ≥ 0.1" Precip | 5 | 5 | 5 | 3 | 1 | 0 | 0 | 0 | 1 | 2 | 4 | 4 | 30 |
| Total Snowfall (") | 0.0 | 0.0 | 0.0 | 0.0 | 0.0 | 0.0 | 0.0 | 0.0 | 0.0 | 0.0 | 0.0 | 0.0 | 0.0 |
| Days ≥ 1" Snow Depth | 0 | 0 | 0 | 0 | 0 | 0 | 0 | 0 | 0 | 0 | 0 | 0 | 0 |

## UKIAH *Mendocino County* ELEVATION 623 ft LAT/LONG 39° 9 ' N / 123° 12 ' W

| | JAN | FEB | MAR | APR | MAY | JUN | JUL | AUG | SEP | OCT | NOV | DEC | YEAR |
|---|---|---|---|---|---|---|---|---|---|---|---|---|---|
| Maximum Temp °F | 57.6 | 61.4 | 64.8 | 69.7 | 77.6 | 84.5 | 92.0 | 91.6 | 87.4 | 77.7 | 63.4 | 56.4 | 73.7 |
| Minimum Temp °F | 36.2 | 38.9 | 40.4 | 42.0 | 47.0 | 52.2 | 55.5 | 54.7 | 51.3 | 45.8 | 40.3 | 36.0 | 45.0 |
| Mean Temp °F | 46.9 | 50.2 | 52.6 | 55.8 | 62.3 | 68.4 | 73.8 | 73.2 | 69.4 | 61.8 | 51.9 | 46.2 | 59.4 |
| Days Max Temp ≥ 90 °F | 0 | 0 | 0 | 1 | 5 | 10 | 19 | 19 | 13 | 4 | 0 | 0 | 71 |
| Days Max Temp ≤ 32 °F | 0 | 0 | 0 | 0 | 0 | 0 | 0 | 0 | 0 | 0 | 0 | 0 | 0 |
| Days Min Temp ≤ 32 °F | 11 | 6 | 3 | 1 | 0 | 0 | 0 | 0 | 0 | 1 | 4 | 12 | 38 |
| Days Min Temp ≤ 0 °F | 0 | 0 | 0 | 0 | 0 | 0 | 0 | 0 | 0 | 0 | 0 | 0 | 0 |
| Heating Degree Days | 554 | 413 | 377 | 275 | 128 | 34 | 4 | 3 | 21 | 133 | 388 | 576 | 2906 |
| Cooling Degree Days | 0 | 0 | 0 | 11 | 49 | 139 | 280 | 257 | 158 | 46 | 0 | 0 | 940 |
| Total Precipitation (") | 7.96 | 6.04 | 5.23 | 2.22 | 0.72 | 0.26 | 0.06 | 0.18 | 0.63 | 2.00 | 5.65 | 6.58 | 37.53 |
| Days ≥ 0.1" Precip | 9 | 9 | 8 | 5 | 2 | 1 | 0 | 0 | 1 | 4 | 8 | 9 | 56 |
| Total Snowfall (") | 0.0 | 0.0 | 0.1 | 0.0 | 0.0 | 0.0 | 0.0 | 0.0 | 0.0 | 0.0 | 0.0 | 0.0 | 0.1 |
| Days ≥ 1" Snow Depth | 0 | 0 | 0 | 0 | 0 | 0 | 0 | 0 | 0 | 0 | 0 | 0 | 0 |

## VACAVILLE *Solano County* ELEVATION 180 ft LAT/LONG 38° 22 ' N / 122° 0 ' W

| | JAN | FEB | MAR | APR | MAY | JUN | JUL | AUG | SEP | OCT | NOV | DEC | YEAR |
|---|---|---|---|---|---|---|---|---|---|---|---|---|---|
| Maximum Temp °F | 54.7 | 61.6 | 66.2 | 72.8 | 81.7 | 88.7 | 94.2 | 93.3 | 89.0 | 79.5 | 64.5 | 54.7 | 75.1 |
| Minimum Temp °F | 36.5 | 39.6 | 42.2 | 44.1 | 49.2 | 53.8 | 56.4 | 55.6 | 53.3 | 48.5 | 41.6 | 36.5 | 46.4 |
| Mean Temp °F | 45.6 | 50.6 | 54.2 | 58.4 | 65.5 | 71.3 | 75.3 | 74.5 | 71.2 | 64.0 | 53.1 | 45.6 | 60.8 |
| Days Max Temp ≥ 90 °F | 0 | 0 | 0 | 1 | 7 | 14 | 24 | 22 | 15 | 4 | 0 | 0 | 87 |
| Days Max Temp ≤ 32 °F | 0 | 0 | 0 | 0 | 0 | 0 | 0 | 0 | 0 | 0 | 0 | 0 | 0 |
| Days Min Temp ≤ 32 °F | 10 | 4 | 1 | 1 | 0 | 0 | 0 | 0 | 0 | 0 | 2 | 10 | 28 |
| Days Min Temp ≤ 0 °F | 0 | 0 | 0 | 0 | 0 | 0 | 0 | 0 | 0 | 0 | 0 | 0 | 0 |
| Heating Degree Days | 594 | 399 | 328 | 207 | 69 | 11 | 1 | 0 | 9 | 85 | 352 | 594 | 2649 |
| Cooling Degree Days | 0 | 0 | 1 | 22 | 94 | 214 | 344 | 304 | 211 | 79 | 2 | 0 | 1271 |
| Total Precipitation (") | 5.53 | 4.12 | 3.48 | 1.31 | 0.30 | 0.12 | 0.05 | 0.04 | 0.33 | 1.10 | 3.49 | 3.93 | 23.80 |
| Days ≥ 0.1" Precip | 7 | 6 | 6 | 3 | 1 | 0 | 0 | 0 | 1 | 2 | 5 | 6 | 37 |
| Total Snowfall (") | 0.1 | 0.0 | 0.0 | 0.0 | 0.0 | 0.0 | 0.0 | 0.0 | 0.0 | 0.0 | 0.0 | 0.1 | 0.2 |
| Days ≥ 1" Snow Depth | 0 | 0 | 0 | 0 | 0 | 0 | 0 | 0 | 0 | 0 | 0 | 0 | 0 |

## VICTORVILLE PUMP PLT *San Bernardino County* ELEVATION 2841 ft LAT/LONG 34° 32 ' N / 117° 18 ' W

| | JAN | FEB | MAR | APR | MAY | JUN | JUL | AUG | SEP | OCT | NOV | DEC | YEAR |
|---|---|---|---|---|---|---|---|---|---|---|---|---|---|
| Maximum Temp °F | 58.3 | 62.3 | 66.2 | 73.2 | 81.9 | 91.3 | 97.2 | 96.4 | 90.0 | 79.8 | 67.1 | 58.2 | 76.8 |
| Minimum Temp °F | 30.4 | 33.9 | 37.4 | 41.6 | 48.3 | 54.7 | 61.0 | 60.8 | 54.8 | 45.0 | 35.7 | 29.7 | 44.4 |
| Mean Temp °F | 44.4 | 48.2 | 51.9 | 57.4 | 65.1 | 73.0 | 79.1 | 78.6 | 72.4 | 62.4 | 51.5 | 44.0 | 60.7 |
| Days Max Temp ≥ 90 °F | 0 | 0 | 0 | 1 | 7 | 19 | 28 | 27 | 18 | 4 | 0 | 0 | 104 |
| Days Max Temp ≤ 32 °F | 0 | 0 | 0 | 0 | 0 | 0 | 0 | 0 | 0 | 0 | 0 | 0 | 0 |
| Days Min Temp ≤ 32 °F | 20 | 12 | 6 | 2 | 0 | 0 | 0 | 0 | 0 | 1 | 10 | 21 | 72 |
| Days Min Temp ≤ 0 °F | 0 | 0 | 0 | 0 | 0 | 0 | 0 | 0 | 0 | 0 | 0 | 0 | 0 |
| Heating Degree Days | 633 | 468 | 401 | 238 | 87 | 14 | 1 | 1 | 17 | 128 | 401 | 644 | 3033 |
| Cooling Degree Days | 0 | 0 | 2 | 28 | 101 | 266 | 433 | 435 | 256 | 60 | 1 | 0 | 1582 |
| Total Precipitation (") | 1.04 | 1.05 | 1.10 | 0.36 | 0.20 | 0.08 | 0.16 | 0.27 | 0.26 | 0.25 | 0.54 | 0.87 | 6.18 |
| Days ≥ 0.1" Precip | 3 | 2 | 2 | 1 | 0 | 0 | 0 | 1 | 1 | 1 | 1 | 2 | 14 |
| Total Snowfall (") | 0.6 | 0.0 | 0.0 | 0.0 | 0.0 | 0.0 | 0.0 | 0.0 | 0.0 | 0.0 | 0.0 | 0.1 | 0.7 |
| Days ≥ 1" Snow Depth | 0 | 0 | 0 | 0 | 0 | 0 | 0 | 0 | 0 | 0 | 0 | 0 | 0 |

## VISALIA *Tulare County* ELEVATION 351 ft LAT/LONG 36° 20 ' N / 119° 18 ' W

| | JAN | FEB | MAR | APR | MAY | JUN | JUL | AUG | SEP | OCT | NOV | DEC | YEAR |
|---|---|---|---|---|---|---|---|---|---|---|---|---|---|
| Maximum Temp °F | 53.7 | 61.4 | 66.9 | 73.7 | 82.8 | 90.1 | 95.3 | 93.7 | 88.1 | 79.1 | 64.7 | 53.8 | 75.3 |
| Minimum Temp °F | 37.5 | 41.3 | 45.3 | 48.6 | 54.6 | 60.5 | 65.1 | 63.6 | 59.2 | 51.9 | 43.0 | 36.7 | 50.6 |
| Mean Temp °F | 45.7 | 51.4 | 56.2 | 61.2 | 68.7 | 75.3 | 80.2 | 78.7 | 73.7 | 65.5 | 53.9 | 45.2 | 63.0 |
| Days Max Temp ≥ 90 °F | 0 | 0 | 0 | 1 | 7 | 16 | 26 | 24 | 14 | 3 | 0 | 0 | 91 |
| Days Max Temp ≤ 32 °F | 0 | 0 | 0 | 0 | 0 | 0 | 0 | 0 | 0 | 0 | 0 | 0 | 0 |
| Days Min Temp ≤ 32 °F | 6 | 1 | 0 | 0 | 0 | 0 | 0 | 0 | 0 | 0 | 1 | 7 | 15 |
| Days Min Temp ≤ 0 °F | 0 | 0 | 0 | 0 | 0 | 0 | 0 | 0 | 0 | 0 | 0 | 0 | 0 |
| Heating Degree Days | 593 | 378 | 270 | 145 | 38 | 3 | 0 | 0 | 4 | 65 | 329 | 605 | 2430 |
| Cooling Degree Days | 0 | 0 | 2 | 46 | 158 | 318 | 470 | 428 | 271 | 92 | 1 | 0 | 1786 |
| Total Precipitation (") | 1.85 | 1.79 | 1.83 | 0.85 | 0.27 | 0.08 | 0.01 | 0.01 | 0.22 | 0.57 | 1.22 | 1.58 | 10.28 |
| Days ≥ 0.1" Precip | 4 | 4 | 5 | 2 | 1 | 0 | 0 | 0 | 1 | 1 | 3 | 4 | 25 |
| Total Snowfall (") | 0.0 | 0.0 | 0.0 | 0.0 | 0.0 | 0.0 | 0.0 | 0.0 | 0.0 | 0.0 | 0.0 | 0.0 | 0.0 |
| Days ≥ 1" Snow Depth | 0 | 0 | 0 | 0 | 0 | 0 | 0 | 0 | 0 | 0 | 0 | 0 | 0 |

## VISTA 2 NNE *San Diego County* ELEVATION 600 ft LAT/LONG 33° 13 ' N / 117° 14 ' W

| | JAN | FEB | MAR | APR | MAY | JUN | JUL | AUG | SEP | OCT | NOV | DEC | YEAR |
|---|---|---|---|---|---|---|---|---|---|---|---|---|---|
| Maximum Temp °F | 67.2 | 67.8 | 67.7 | 70.8 | 72.5 | 76.7 | 81.5 | 83.0 | 81.9 | 78.2 | 72.6 | 67.4 | 73.9 |
| Minimum Temp °F | 43.7 | 44.6 | 46.0 | 48.6 | 53.0 | 56.4 | 60.1 | 61.6 | 59.6 | 54.5 | 47.9 | 43.6 | 51.6 |
| Mean Temp °F | 55.5 | 56.3 | 56.9 | 59.7 | 62.8 | 66.6 | 70.8 | 72.3 | 70.8 | 66.3 | 60.3 | 55.5 | 62.8 |
| Days Max Temp ≥ 90 °F | 0 | 0 | 0 | 0 | 1 | 2 | 2 | 4 | 5 | 3 | 0 | 0 | 17 |
| Days Max Temp ≤ 32 °F | 0 | 0 | 0 | 0 | 0 | 0 | 0 | 0 | 0 | 0 | 0 | 0 | 0 |
| Days Min Temp ≤ 32 °F | 1 | 0 | 0 | 0 | 0 | 0 | 0 | 0 | 0 | 0 | 0 | 1 | 2 |
| Days Min Temp ≤ 0 °F | 0 | 0 | 0 | 0 | 0 | 0 | 0 | 0 | 0 | 0 | 0 | 0 | 0 |
| Heating Degree Days | 294 | 246 | 249 | 166 | 90 | 26 | 1 | 0 | 3 | 34 | 152 | 290 | 1551 |
| Cooling Degree Days | 5 | 9 | 7 | 25 | 37 | 106 | 225 | 268 | 214 | 102 | 19 | 4 | 1021 |
| Total Precipitation (") | 2.83 | 2.44 | 2.70 | 1.05 | 0.25 | 0.12 | 0.03 | 0.11 | 0.27 | 0.41 | 1.57 | 2.01 | 13.79 |
| Days ≥ 0.1" Precip | 5 | 4 | 5 | 3 | 1 | 0 | 0 | 0 | 1 | 1 | 3 | 4 | 27 |
| Total Snowfall (") | 0.0 | 0.0 | 0.0 | 0.0 | 0.0 | 0.0 | 0.0 | 0.0 | 0.0 | 0.0 | 0.0 | 0.1 | 0.1 |
| Days ≥ 1" Snow Depth | 0 | 0 | 0 | 0 | 0 | 0 | 0 | 0 | 0 | 0 | 0 | 0 | 0 |

## WASCO *Kern County* ELEVATION 331 ft LAT/LONG 35° 36' N / 119° 20' W

| | JAN | FEB | MAR | APR | MAY | JUN | JUL | AUG | SEP | OCT | NOV | DEC | YEAR |
|---|---|---|---|---|---|---|---|---|---|---|---|---|---|
| Maximum Temp °F | 55.8 | 63.8 | 69.6 | 77.1 | 86.4 | 93.6 | 98.9 | 97.3 | 91.4 | 81.8 | 66.7 | 55.7 | 78.2 |
| Minimum Temp °F | 36.3 | 40.1 | 44.3 | 48.1 | 54.6 | 60.7 | 65.4 | 63.9 | 59.0 | 50.2 | 40.9 | 35.0 | 49.9 |
| Mean Temp °F | 46.1 | 52.0 | 57.0 | 62.6 | 70.5 | 77.2 | 82.2 | 80.6 | 75.2 | 65.9 | 53.9 | 45.3 | 64.0 |
| Days Max Temp ≥ 90 °F | 0 | 0 | 0 | 3 | 12 | 21 | 29 | 28 | 19 | 5 | 0 | 0 | 117 |
| Days Max Temp ≤ 32 °F | 0 | 0 | 0 | 0 | 0 | 0 | 0 | 0 | 0 | 0 | 0 | 0 | 0 |
| Days Min Temp ≤ 32 °F | 9 | 3 | 0 | 0 | 0 | 0 | 0 | 0 | 0 | 0 | 3 | 11 | 26 |
| Days Min Temp ≤ 0 °F | 0 | 0 | 0 | 0 | 0 | 0 | 0 | 0 | 0 | 0 | 0 | 0 | 0 |
| Heating Degree Days | 579 | 361 | 245 | 116 | 25 | 2 | 0 | 0 | 3 | 61 | 329 | 602 | 2323 |
| Cooling Degree Days | 0 | 0 | 4 | 68 | 207 | 370 | 529 | 486 | 321 | 104 | 0 | 0 | 2089 |
| Total Precipitation (") | 1.22 | 1.17 | 1.42 | 0.68 | 0.17 | 0.07 | 0.02 | 0.03 | 0.17 | 0.31 | 0.71 | 0.90 | 6.87 |
| Days ≥ 0.1" Precip | 4 | 3 | 4 | 2 | 0 | 0 | 0 | 0 | 1 | 1 | 2 | 3 | 20 |
| Total Snowfall (") | 0.0 | 0.0 | 0.0 | 0.0 | 0.0 | 0.0 | 0.0 | 0.0 | 0.0 | 0.0 | 0.0 | 0.0 | 0.0 |
| Days ≥ 1" Snow Depth | 0 | 0 | 0 | 0 | 0 | 0 | 0 | 0 | 0 | 0 | 0 | 0 | 0 |

## WATSONVILLE WATERWOR *Santa Cruz County* ELEVATION 102 ft LAT/LONG 36° 56' N / 121° 46' W

| | JAN | FEB | MAR | APR | MAY | JUN | JUL | AUG | SEP | OCT | NOV | DEC | YEAR |
|---|---|---|---|---|---|---|---|---|---|---|---|---|---|
| Maximum Temp °F | 60.5 | 62.5 | 63.7 | 66.7 | 68.3 | 70.4 | 71.2 | 72.0 | 73.4 | 72.0 | 66.0 | 60.2 | 67.2 |
| Minimum Temp °F | 38.5 | 41.2 | 43.3 | 44.6 | 47.8 | 50.8 | 52.7 | 53.3 | 52.0 | 47.9 | 42.6 | 38.1 | 46.1 |
| Mean Temp °F | 49.5 | 51.9 | 53.5 | 55.7 | 58.0 | 60.6 | 62.0 | 62.7 | 62.7 | 60.0 | 54.3 | 49.2 | 56.7 |
| Days Max Temp ≥ 90 °F | 0 | 0 | 0 | 0 | 0 | 1 | 0 | 0 | 1 | 1 | 0 | 0 | 3 |
| Days Max Temp ≤ 32 °F | 0 | 0 | 0 | 0 | 0 | 0 | 0 | 0 | 0 | 0 | 0 | 0 | 0 |
| Days Min Temp ≤ 32 °F | 5 | 1 | 0 | 0 | 0 | 0 | 0 | 0 | 0 | 0 | 1 | 5 | 12 |
| Days Min Temp ≤ 0 °F | 0 | 0 | 0 | 0 | 0 | 0 | 0 | 0 | 0 | 0 | 0 | 0 | 0 |
| Heating Degree Days | 473 | 366 | 351 | 277 | 216 | 139 | 102 | 81 | 92 | 168 | 316 | 484 | 3065 |
| Cooling Degree Days | 0 | 0 | 0 | 8 | 8 | 19 | 24 | 23 | 29 | 19 | 1 | 0 | 131 |
| Total Precipitation (") | 4.17 | 3.60 | 3.86 | 1.69 | 0.32 | 0.13 | 0.09 | 0.07 | 0.29 | 0.98 | 3.12 | 3.56 | 21.88 |
| Days ≥ 0.1" Precip | 6 | 6 | 7 | 3 | 1 | 0 | 0 | 0 | 1 | 2 | 5 | 6 | 37 |
| Total Snowfall (") | 0.0 | 0.0 | 0.0 | 0.0 | 0.0 | 0.0 | 0.0 | 0.0 | 0.0 | 0.0 | 0.0 | 0.0 | 0.0 |
| Days ≥ 1" Snow Depth | 0 | 0 | 0 | 0 | 0 | 0 | 0 | 0 | 0 | 0 | 0 | 0 | 0 |

## WHISKEYTOWN RESERVOI *Shasta County* ELEVATION 1312 ft LAT/LONG 40° 37' N / 122° 32' W

| | JAN | FEB | MAR | APR | MAY | JUN | JUL | AUG | SEP | OCT | NOV | DEC | YEAR |
|---|---|---|---|---|---|---|---|---|---|---|---|---|---|
| Maximum Temp °F | 53.7 | 57.4 | 61.5 | 67.7 | 77.8 | 86.4 | 95.2 | 94.8 | 88.2 | 76.4 | 60.4 | 53.1 | 72.7 |
| Minimum Temp °F | 35.7 | 37.9 | 40.6 | 44.5 | 52.0 | 58.6 | 63.5 | 62.0 | 57.3 | 49.9 | 41.2 | 35.7 | 48.2 |
| Mean Temp °F | 44.7 | 47.6 | 51.0 | 56.1 | 64.9 | 72.5 | 79.4 | 78.4 | 72.7 | 63.2 | 50.8 | 44.4 | 60.5 |
| Days Max Temp ≥ 90 °F | 0 | 0 | 0 | 0 | 4 | 12 | 25 | 24 | 15 | 4 | 0 | 0 | 84 |
| Days Max Temp ≤ 32 °F | 0 | 0 | 0 | 0 | 0 | 0 | 0 | 0 | 0 | 0 | 0 | 0 | 0 |
| Days Min Temp ≤ 32 °F | 9 | 5 | 2 | 1 | 0 | 0 | 0 | 0 | 0 | 0 | 2 | 8 | 27 |
| Days Min Temp ≤ 0 °F | 0 | 0 | 0 | 0 | 0 | 0 | 0 | 0 | 0 | 0 | 0 | 0 | 0 |
| Heating Degree Days | 621 | 484 | 427 | 273 | 97 | 21 | 1 | 2 | 21 | 130 | 421 | 632 | 3130 |
| Cooling Degree Days | 0 | 0 | 0 | 17 | 99 | 245 | 444 | 419 | 256 | 86 | 1 | 0 | 1567 |
| Total Precipitation (") | 10.98 | 8.85 | 9.95 | 4.21 | 2.08 | 0.85 | 0.39 | 0.37 | 1.39 | 2.85 | 8.49 | 9.32 | 59.73 |
| Days ≥ 0.1" Precip | 9 | 9 | 9 | 6 | 3 | 2 | 0 | 1 | 2 | 3 | 8 | 9 | 61 |
| Total Snowfall (") | 2.7 | 0.4 | 0.0 | 0.1 | 0.0 | 0.0 | 0.0 | 0.0 | 0.0 | 0.0 | 0.6 | 0.9 | 4.7 |
| Days ≥ 1" Snow Depth | 1 | 0 | 0 | 0 | 0 | 0 | 0 | 0 | 0 | 0 | 0 | 1 | 2 |

## WILLITS 1 NE *Mendocino County* ELEVATION 1352 ft LAT/LONG 39° 25' N / 123° 21' W

| | JAN | FEB | MAR | APR | MAY | JUN | JUL | AUG | SEP | OCT | NOV | DEC | YEAR |
|---|---|---|---|---|---|---|---|---|---|---|---|---|---|
| Maximum Temp °F | 55.5 | 58.5 | 60.9 | 65.0 | 72.8 | 78.9 | 85.7 | 85.5 | 83.0 | 74.7 | 61.2 | 54.3 | 69.7 |
| Minimum Temp °F | 33.1 | 35.1 | 36.9 | 37.2 | 40.8 | 44.6 | 47.3 | 46.1 | 43.0 | 39.2 | 35.5 | 32.5 | 39.3 |
| Mean Temp °F | 44.3 | 46.8 | 48.9 | 51.1 | 56.8 | 61.8 | 66.5 | 65.8 | 63.0 | 56.9 | 48.4 | 43.4 | 54.5 |
| Days Max Temp ≥ 90 °F | 0 | 0 | 0 | 0 | 1 | 5 | 11 | 10 | 8 | 2 | 0 | 0 | 37 |
| Days Max Temp ≤ 32 °F | 0 | 0 | 0 | 0 | 0 | 0 | 0 | 0 | 0 | 0 | 0 | 0 | 0 |
| Days Min Temp ≤ 32 °F | 16 | 11 | 8 | 6 | 1 | 0 | 0 | 0 | 1 | 4 | 12 | 16 | 75 |
| Days Min Temp ≤ 0 °F | 0 | 0 | 0 | 0 | 0 | 0 | 0 | 0 | 0 | 0 | 0 | 0 | 0 |
| Heating Degree Days | 634 | 507 | 491 | 410 | 257 | 128 | 43 | 47 | 93 | 248 | 492 | 662 | 4012 |
| Cooling Degree Days | 0 | 0 | 0 | 1 | 7 | 33 | 88 | 62 | 34 | 7 | 0 | 0 | 232 |
| Total Precipitation (") | 9.97 | 7.77 | 6.97 | 2.98 | 1.18 | 0.29 | 0.07 | 0.26 | 0.87 | 2.74 | 7.06 | 8.42 | 48.58 |
| Days ≥ 0.1" Precip | 10 | 9 | 10 | 6 | 2 | 1 | 0 | 1 | 2 | 4 | 9 | 11 | 65 |
| Total Snowfall (") | 2.1 | 0.8 | 0.9 | 0.0 | 0.0 | 0.0 | 0.0 | 0.0 | 0.0 | 0.0 | 0.0 | 0.6 | 4.4 |
| Days ≥ 1" Snow Depth | 0 | 0 | 0 | 0 | 0 | 0 | 0 | 0 | 0 | 0 | 0 | 0 | 0 |

## WILLOWS 6 W *Glenn County* ELEVATION 233 ft LAT/LONG 39° 31' N / 122° 18' W

| | JAN | FEB | MAR | APR | MAY | JUN | JUL | AUG | SEP | OCT | NOV | DEC | YEAR |
|---|---|---|---|---|---|---|---|---|---|---|---|---|---|
| Maximum Temp °F | 55.1 | 61.1 | 65.7 | 72.6 | 81.8 | 88.7 | 93.8 | 92.4 | 88.9 | 79.5 | 64.5 | 55.1 | 74.9 |
| Minimum Temp °F | 35.6 | 38.6 | 41.3 | 44.3 | 51.5 | 57.5 | 60.5 | 58.6 | 55.8 | 49.1 | 40.9 | 35.4 | 47.4 |
| Mean Temp °F | 45.4 | 49.9 | 53.5 | 58.5 | 66.7 | 73.2 | 77.2 | 75.6 | 72.4 | 64.3 | 52.7 | 45.3 | 61.2 |
| Days Max Temp ≥ 90 °F | 0 | 0 | 0 | 0 | 7 | 14 | 23 | 21 | 15 | 4 | 0 | 0 | 84 |
| Days Max Temp ≤ 32 °F | 0 | 0 | 0 | 0 | 0 | 0 | 0 | 0 | 0 | 0 | 0 | 0 | 0 |
| Days Min Temp ≤ 32 °F | 11 | 5 | 2 | 0 | 0 | 0 | 0 | 0 | 0 | 0 | 2 | 11 | 31 |
| Days Min Temp ≤ 0 °F | 0 | 0 | 0 | 0 | 0 | 0 | 0 | 0 | 0 | 0 | 0 | 0 | 0 |
| Heating Degree Days | 601 | 420 | 349 | 207 | 64 | 12 | 0 | 1 | 9 | 91 | 363 | 604 | 2721 |
| Cooling Degree Days | 0 | 0 | 1 | 23 | 113 | 254 | 370 | 318 | 237 | 89 | 2 | 0 | 1407 |
| Total Precipitation (") | 3.57 | 3.08 | 2.54 | 1.02 | 0.47 | 0.29 | 0.05 | 0.14 | 0.32 | 0.89 | 2.66 | 2.83 | 17.86 |
| Days ≥ 0.1" Precip | 6 | 5 | 5 | 3 | 1 | 1 | 0 | 0 | 1 | 2 | 5 | 5 | 34 |
| Total Snowfall (") | 0.3 | 0.0 | 0.0 | 0.0 | 0.0 | 0.0 | 0.0 | 0.0 | 0.0 | 0.0 | 0.0 | 0.3 | 0.6 |
| Days ≥ 1" Snow Depth | 0 | 0 | 0 | 0 | 0 | 0 | 0 | 0 | 0 | 0 | 0 | 0 | 0 |

## WINTERS *Yolo County* ELEVATION 131 ft LAT/LONG 38° 31' N / 121° 58' W

| | JAN | FEB | MAR | APR | MAY | JUN | JUL | AUG | SEP | OCT | NOV | DEC | YEAR |
|---|---|---|---|---|---|---|---|---|---|---|---|---|---|
| Maximum Temp °F | 55.1 | 61.9 | 66.8 | 74.1 | 83.5 | 91.0 | 96.5 | 95.0 | 90.3 | 80.8 | 65.6 | 55.2 | 76.3 |
| Minimum Temp °F | 37.0 | 40.8 | 44.1 | 47.4 | 53.3 | 58.0 | 59.8 | 59.0 | 56.7 | 50.5 | 42.6 | 36.8 | 48.8 |
| Mean Temp °F | 46.1 | 51.4 | 55.5 | 60.8 | 68.4 | 74.5 | 78.2 | 77.0 | 73.5 | 65.7 | 54.1 | 46.0 | 62.6 |
| Days Max Temp ≥ 90 °F | 0 | 0 | 0 | 1 | 9 | 18 | 27 | 25 | 18 | 5 | 0 | 0 | 103 |
| Days Max Temp ≤ 32 °F | 0 | 0 | 0 | 0 | 0 | 0 | 0 | 0 | 0 | 0 | 0 | 0 | 0 |
| Days Min Temp ≤ 32 °F | 8 | 2 | 0 | 0 | 0 | 0 | 0 | 0 | 0 | 0 | 1 | 8 | 19 |
| Days Min Temp ≤ 0 °F | 0 | 0 | 0 | 0 | 0 | 0 | 0 | 0 | 0 | 0 | 0 | 0 | 0 |
| Heating Degree Days | 579 | 377 | 290 | 154 | 42 | 6 | 0 | 0 | 5 | 63 | 321 | 580 | 2417 |
| Cooling Degree Days | 0 | 1 | 3 | 45 | 161 | 308 | 424 | 381 | 276 | 107 | 4 | 0 | 1710 |
| Total Precipitation (") | 5.09 | 3.87 | 3.29 | 1.10 | 0.35 | 0.15 | 0.03 | 0.06 | 0.25 | 0.91 | 3.19 | 3.60 | 21.89 |
| Days ≥ 0.1" Precip | 7 | 6 | 6 | 3 | 1 | 1 | 0 | 0 | 1 | 2 | 5 | 6 | 38 |
| Total Snowfall (") | 0.1 | 0.0 | 0.0 | 0.0 | 0.0 | 0.0 | 0.0 | 0.0 | 0.0 | 0.0 | 0.0 | 0.0 | 0.1 |
| Days ≥ 1" Snow Depth | 0 | 0 | 0 | 0 | 0 | 0 | 0 | 0 | 0 | 0 | 0 | 0 | 0 |

## WOODLAND 1 WNW *Yolo County* ELEVATION 59 ft LAT/LONG 38° 41' N / 121° 46' W

| | JAN | FEB | MAR | APR | MAY | JUN | JUL | AUG | SEP | OCT | NOV | DEC | YEAR |
|---|---|---|---|---|---|---|---|---|---|---|---|---|---|
| Maximum Temp °F | 53.3 | 60.1 | 65.7 | 73.0 | 83.1 | 90.7 | 96.1 | 94.6 | 90.0 | 79.4 | 63.5 | 53.1 | 75.2 |
| Minimum Temp °F | 37.1 | 40.4 | 43.4 | 46.3 | 51.6 | 56.3 | 58.0 | 57.4 | 55.3 | 49.5 | 42.4 | 36.4 | 47.8 |
| Mean Temp °F | 45.2 | 50.3 | 54.6 | 59.7 | 67.4 | 73.5 | 77.0 | 76.0 | 72.7 | 64.5 | 53.0 | 44.8 | 61.6 |
| Days Max Temp ≥ 90 °F | 0 | 0 | 0 | 1 | 8 | 17 | 26 | 24 | 17 | 4 | 0 | 0 | 97 |
| Days Max Temp ≤ 32 °F | 0 | 0 | 0 | 0 | 0 | 0 | 0 | 0 | 0 | 0 | 0 | 0 | 0 |
| Days Min Temp ≤ 32 °F | 8 | 2 | 1 | 0 | 0 | 0 | 0 | 0 | 0 | 0 | 1 | 9 | 21 |
| Days Min Temp ≤ 0 °F | 0 | 0 | 0 | 0 | 0 | 0 | 0 | 0 | 0 | 0 | 0 | 0 | 0 |
| Heating Degree Days | 605 | 408 | 317 | 176 | 48 | 6 | 0 | 0 | 5 | 79 | 355 | 619 | 2618 |
| Cooling Degree Days | 0 | 0 | 1 | ***34*** | ***132*** | 277 | 389 | 351 | ***269*** | ***94*** | 1 | 0 | 1548 |
| Total Precipitation (") | 4.22 | 3.67 | 3.09 | 1.19 | 0.35 | 0.18 | 0.04 | 0.07 | 0.40 | 1.04 | 2.87 | 3.11 | 20.23 |
| Days ≥ 0.1" Precip | 7 | 6 | 6 | 3 | 1 | 1 | 0 | 0 | 1 | 2 | 5 | 6 | 38 |
| Total Snowfall (") | 0.1 | 0.0 | 0.0 | 0.0 | 0.0 | 0.0 | 0.0 | 0.0 | 0.0 | 0.0 | 0.0 | 0.0 | 0.1 |
| Days ≥ 1" Snow Depth | 0 | 0 | 0 | 0 | 0 | 0 | 0 | 0 | 0 | 0 | 0 | 0 | 0 |

## YOSEMITE PARK HDQTRS *Mariposa County* ELEVATION 3993 ft LAT/LONG 37° 45' N / 119° 35' W

| | JAN | FEB | MAR | APR | MAY | JUN | JUL | AUG | SEP | OCT | NOV | DEC | YEAR |
|---|---|---|---|---|---|---|---|---|---|---|---|---|---|
| Maximum Temp °F | 48.7 | 54.8 | 59.0 | 65.6 | 73.7 | 82.0 | 89.7 | 90.0 | 84.4 | 74.3 | 58.0 | 47.8 | 69.0 |
| Minimum Temp °F | 26.5 | 28.6 | 31.5 | 35.9 | 42.2 | 48.3 | 53.9 | 53.0 | 47.5 | 39.3 | 31.0 | 26.0 | 38.6 |
| Mean Temp °F | 37.6 | 41.7 | 45.2 | 50.8 | 58.0 | 65.1 | 71.8 | 71.5 | 66.0 | 56.8 | 44.5 | 36.9 | 53.8 |
| Days Max Temp ≥ 90 °F | 0 | 0 | 0 | 0 | 1 | 5 | 17 | 18 | 9 | 2 | 0 | 0 | 52 |
| Days Max Temp ≤ 32 °F | 0 | 0 | 0 | 0 | 0 | 0 | 0 | 0 | 0 | 0 | 0 | 1 | 1 |
| Days Min Temp ≤ 32 °F | 27 | 23 | 19 | 10 | 2 | 0 | 0 | 0 | 0 | 6 | 19 | 27 | 133 |
| Days Min Temp ≤ 0 °F | 0 | 0 | 0 | 0 | 0 | 0 | 0 | 0 | 0 | 0 | 0 | 0 | 0 |
| Heating Degree Days | 842 | 650 | 606 | 420 | 225 | 74 | 8 | 11 | 64 | 266 | 608 | 865 | 4639 |
| Cooling Degree Days | 0 | 0 | 0 | 0 | 15 | 97 | 251 | 241 | 110 | 24 | 0 | 0 | 738 |
| Total Precipitation (") | 6.18 | 6.04 | 5.43 | 2.76 | 1.20 | 0.73 | 0.53 | 0.27 | 0.88 | 2.24 | 5.16 | 5.50 | 36.92 |
| Days ≥ 0.1" Precip | 7 | 7 | 8 | 5 | 3 | 2 | 1 | 1 | 2 | 3 | 6 | 7 | 52 |
| Total Snowfall (") | 16.9 | ***11.6*** | 9.2 | 5.5 | 0.1 | 0.0 | 0.0 | 0.0 | 0.0 | 0.0 | 3.9 | 10.1 | 57.3 |
| Days ≥ 1" Snow Depth | ***11*** | ***8*** | 5 | 2 | 0 | 0 | 0 | 0 | 0 | 0 | 1 | 8 | 35 |

## YOSEMITE SOUTH ENTR *Mariposa County* ELEVATION 5144 ft LAT/LONG 37° 30 ' N / 119° 38 ' W

| | JAN | FEB | MAR | APR | MAY | JUN | JUL | AUG | SEP | OCT | NOV | DEC | YEAR |
|---|---|---|---|---|---|---|---|---|---|---|---|---|---|
| Maximum Temp °F | 46.3 | 48.0 | 49.2 | 55.5 | 64.8 | 73.2 | 80.7 | 79.9 | 74.0 | 65.0 | 52.6 | 47.0 | 61.4 |
| Minimum Temp °F | 26.1 | 26.6 | 28.1 | 31.2 | 37.7 | 44.5 | 49.9 | 49.5 | 45.0 | 37.7 | 30.4 | 25.9 | 36.1 |
| Mean Temp °F | 36.2 | 37.3 | 38.7 | 43.3 | 51.3 | 58.9 | 65.4 | 64.7 | 59.5 | 51.4 | 41.5 | 36.5 | 48.7 |
| Days Max Temp ≥ 90 °F | 0 | 0 | 0 | 0 | 0 | 1 | 3 | 2 | 0 | 0 | 0 | 0 | 6 |
| Days Max Temp ≤ 32 °F | 1 | 1 | 1 | 0 | 0 | 0 | 0 | 0 | 0 | 0 | 1 | 2 | 6 |
| Days Min Temp ≤ 32 °F | 27 | 25 | 26 | 18 | 7 | 1 | 0 | 0 | 1 | 7 | 20 | 27 | 159 |
| Days Min Temp ≤ 0 °F | 0 | 0 | 0 | 0 | 0 | 0 | 0 | 0 | 0 | 0 | 0 | 0 | 0 |
| Heating Degree Days | 885 | 775 | 808 | 644 | 418 | 199 | 62 | 70 | 175 | 418 | 698 | 878 | 6030 |
| Cooling Degree Days | 0 | 0 | 0 | 0 | 1 | 14 | 57 | 50 | 12 | 3 | 0 | 0 | 137 |
| Total Precipitation (") | 8.19 | 7.35 | 7.20 | 3.49 | 1.19 | 0.68 | 0.17 | 0.11 | 0.85 | 2.29 | 5.41 | 5.77 | 42.70 |
| Days ≥ 0.1" Precip | 7 | *6* | 8 | 5 | 3 | 1 | 0 | 0 | 2 | 3 | 6 | 5 | 46 |
| Total Snowfall (") | na | na | na | *8.6* | 0.7 | 0.0 | 0.0 | 0.0 | 0.0 | 0.4 | *7.4* | na | na |
| Days ≥ 1" Snow Depth | *17* | na | na | *6* | 0 | 0 | 0 | 0 | 0 | 0 | *4* | na | na |

## YREKA *Siskiyou County* ELEVATION 2631 ft LAT/LONG 41° 43 ' N / 122° 38 ' W

| | JAN | FEB | MAR | APR | MAY | JUN | JUL | AUG | SEP | OCT | NOV | DEC | YEAR |
|---|---|---|---|---|---|---|---|---|---|---|---|---|---|
| Maximum Temp °F | 44.5 | 51.2 | 56.8 | 62.7 | 73.2 | 81.1 | 90.5 | 89.8 | 81.8 | 69.7 | 52.5 | 43.9 | 66.5 |
| Minimum Temp °F | 24.2 | 26.9 | 30.5 | 34.2 | 40.4 | 47.3 | 52.2 | 51.6 | 44.8 | 36.2 | 29.4 | 24.4 | 36.8 |
| Mean Temp °F | 34.3 | 39.0 | 43.7 | 48.5 | 56.8 | 64.2 | 71.4 | 70.7 | 63.3 | 53.0 | 41.0 | 34.2 | 51.7 |
| Days Max Temp ≥ 90 °F | 0 | 0 | 0 | 0 | 2 | 7 | 18 | 19 | 7 | 1 | 0 | 0 | 54 |
| Days Max Temp ≤ 32 °F | 1 | 0 | 0 | 0 | 0 | 0 | 0 | 0 | 0 | 0 | 0 | 2 | 3 |
| Days Min Temp ≤ 32 °F | 27 | 23 | 20 | 13 | 3 | 0 | 0 | 0 | 0 | 8 | 21 | 26 | 141 |
| Days Min Temp ≤ 0 °F | 0 | 0 | 0 | 0 | 0 | 0 | 0 | 0 | 0 | 0 | 0 | 0 | 0 |
| Heating Degree Days | 943 | 727 | 653 | 490 | 268 | 107 | 18 | 21 | 107 | 369 | 713 | 949 | 5365 |
| Cooling Degree Days | 0 | 0 | 0 | *2* | 22 | 79 | 203 | 195 | 60 | 5 | 0 | 0 | 566 |
| Total Precipitation (") | 3.06 | 1.78 | 1.83 | 1.01 | 0.94 | 0.96 | 0.45 | 0.62 | 0.80 | 1.04 | 2.76 | 3.31 | 18.56 |
| Days ≥ 0.1" Precip | 6 | 5 | 4 | 3 | 3 | 2 | 1 | 2 | 2 | 3 | 6 | 7 | 44 |
| Total Snowfall (") | 5.9 | 2.3 | 2.2 | 0.6 | 0.0 | 0.0 | 0.0 | 0.0 | 0.0 | 0.1 | 1.7 | 5.3 | 18.1 |
| Days ≥ 1" Snow Depth | 6 | 2 | 1 | 0 | 0 | 0 | 0 | 0 | 0 | 0 | 1 | 4 | 14 |

## JANUARY MINIMUM TEMPERATURE °F

### LOWEST

| Rank | Station | Value |
|---|---|---|
| 1 | Bodie | 5.0 |
| 2 | Bridgeport | 8.1 |
| 3 | Boca | 9.4 |
| 4 | Donner Memorial | 13.0 |
| 5 | Termo | 13.8 |
| 6 | Truckee | 14.2 |
| 7 | Alturas | 16.0 |
| 8 | Twin Lakes | 17.7 |
| 9 | Deep Springs | 18.1 |
| 10 | Portola | 18.2 |
| 11 | Chester | 18.8 |
| 12 | Jess Valley | 18.9 |
|  | Tulelake | 18.9 |
| 14 | Tahoe City | 19.3 |
| 15 | Susanville | 19.5 |
| 16 | Fort Bidwell | 19.6 |
| 17 | Big Bear Lake | 19.7 |
| 18 | Cedarville | 19.8 |
| 19 | Manzanita Lake | 20.4 |
| 20 | Mineral | 21.1 |
| 21 | Canyon Dam | 21.6 |
|  | Hat Creek | 21.6 |
| 23 | Doyle | 21.9 |
| 24 | Lava Beds | 22.0 |
| 25 | Bishop | 22.1 |

### HIGHEST

| Rank | Station | Value |
|---|---|---|
| 1 | Ls Angeles-UCLA | 50.3 |
| 2 | Santa Monica | 49.9 |
| 3 | Los Angeles-CC | 49.0 |
|  | San Diego | 49.0 |
| 5 | Los Angeles-Intl | 48.2 |
| 6 | Newport Beach | 47.3 |
| 7 | Culver City | 46.8 |
| 8 | San Francisco | 46.1 |
|  | Santa Ana | 46.1 |
| 10 | Gold Rock | 45.9 |
| 11 | Torrance | 45.7 |
| 12 | Long Beach | 45.3 |
| 13 | Chula Vista | 45.2 |
| 14 | Eagle Mountain | 44.9 |
|  | La Mesa | 44.9 |
|  | Oxnard | 44.9 |
| 17 | Oceanside | 44.2 |
| 18 | Pasadena | 43.8 |
| 19 | Berkeley | 43.7 |
|  | Vista | 43.7 |
| 21 | Santa Barbara | 43.6 |
| 22 | Monterey | 43.5 |
|  | San Frncsco-Ocns | 43.5 |
| 24 | Borrego Desert | 43.2 |
| 25 | Iron Mountain | 43.0 |

## JULY MAXIMUM TEMPERATURE °F

### HIGHEST

| Rank | Station | Value |
|---|---|---|
| 1 | Death Valley | 114.9 |
| 2 | Palm Springs | 108.5 |
| 3 | Blythe-Riverside | 108.4 |
|  | Needles | 108.4 |
| 5 | Blythe | 108.2 |
| 6 | Iron Mountain | 107.9 |
|  | Mecca | 107.9 |
| 8 | Parker Reservoir | 107.4 |
| 9 | El Centro | 107.2 |
| 10 | Borrego Desert | 107.1 |
| 11 | Gold Rock | 106.9 |
| 12 | Indio | 106.7 |
| 13 | Brawley | 106.5 |
| 14 | Imperial | 106.2 |
|  | Palm Sprgs-Thrml | 106.2 |
| 16 | Twentynine Palms | 105.4 |
| 17 | Trona | 105.3 |
| 18 | Eagle Mountain | 104.3 |
| 19 | Daggett | 103.9 |
| 20 | Hayfield | 103.8 |
| 21 | Inyokern | 102.6 |
| 22 | Friant | 99.3 |
|  | Three Rivers | 99.3 |
| 24 | Coalinga | 99.0 |
| 25 | Wasco | 98.9 |

### LOWEST

| Rank | Station | Value |
|---|---|---|
| 1 | Eureka | 62.1 |
| 2 | San Frncsco-Ocns | 63.9 |
| 3 | Half Moon Bay | 64.0 |
| 4 | Crescent City | 65.5 |
|  | Fort Bragg | 65.5 |
| 6 | Morro Bay | 65.6 |
| 7 | Fort Ross | 66.1 |
| 8 | Klamath | 66.6 |
|  | San Francisco | 66.6 |
| 10 | Monterey | 68.1 |
| 11 | Scotia | 68.6 |
| 12 | Orick Prairie Crk | 68.7 |
| 13 | San Gregorio | 69.4 |
|  | Santa Monica | 69.4 |
| 15 | Berkeley | 69.9 |
| 16 | Twin Lakes | 70.8 |
| 17 | Pismo Beach | 70.9 |
| 18 | Richmond | 71.0 |
|  | Salinas-Muni | 71.0 |
| 20 | Salinas-3 E | 71.1 |
| 21 | Watsonville | 71.2 |
| 22 | Newport Beach | 71.5 |
| 23 | San Francisco-Int | 71.8 |
| 24 | Chula Vista | 73.3 |
|  | Santa Maria | 73.3 |

## ANNUAL PRECIPITATION (")

### HIGHEST

| Rank | Station | Value |
|---|---|---|
| 1 | Strawberry Valley | 77.67 |
| 2 | Klamath | 76.33 |
| 3 | Lake Spaulding | 69.88 |
| 4 | Blue Canyon | 67.37 |
| 5 | Richardson Grove | 66.38 |
| 6 | Sierra City | 65.77 |
| 7 | Bowman Dam | 64.40 |
| 8 | Orick Prairie Crk | 64.31 |
| 9 | Crescent City | 64.07 |
| 10 | De Sabla | 63.73 |
| 11 | Shasta Dam | 60.42 |
| 12 | Whiskeytown rsrv | 59.73 |
| 13 | Nevada City | 55.87 |
| 14 | Mineral | 53.93 |
| 15 | Calaveras | 53.61 |
| 16 | Paradise | 53.28 |
| 17 | Orleans | 50.93 |
| 18 | Happy Camp | 50.73 |
| 19 | Kentfield | 48.86 |
| 20 | Willits | 48.58 |
| 21 | McCloud | 47.45 |
| 22 | Scotia | 47.14 |
| 23 | Twin Lakes | 47.10 |
| 24 | Colfax | 46.46 |
| 25 | Tiger Creek | 45.44 |

### LOWEST

| Rank | Station | Value |
|---|---|---|
| 1 | Death Valley | 2.30 |
| 2 | El Centro | 3.13 |
|  | Imperial | 3.13 |
| 4 | Brawley | 3.45 |
|  | Mecca | 3.45 |
| 6 | Palm Sprgs-Thrml | 3.60 |
| 7 | Indio | 3.68 |
| 8 | Iron Mountain | 3.91 |
| 9 | Gold Rock | 3.95 |
| 10 | Blythe | 3.98 |
| 11 | Blythe-Riverside | 4.02 |
| 12 | Daggett | 4.28 |
| 13 | Eagle Mountain | 4.46 |
| 14 | Twentynine Palms | 4.47 |
| 15 | Trona | 4.48 |
| 16 | Hayfield | 4.56 |
| 17 | Inyokern | 4.96 |
| 18 | Needles | 5.30 |
| 19 | Bishop | 5.38 |
| 20 | Independence | 5.43 |
| 21 | Palm Springs | 5.64 |
| 22 | Buttonwillow | 5.81 |
| 23 | Bakersfield | 6.07 |
| 24 | Parker Reservoir | 6.17 |
| 25 | Victorville | 6.18 |

## ANNUAL SNOWFALL (")

### HIGHEST

| Rank | Station | Value |
|---|---|---|
| 1 | Lake Spaulding | 253.0 |
| 2 | Bowman Dam | 245.6 |
| 3 | Blue Canyon | 236.9 |
| 4 | Truckee | 201.1 |
| 5 | Grant Grove | 184.8 |
| 6 | Donner Memorial | 183.7 |
| 7 | Manzanita Lake | 181.4 |
| 8 | Tahoe City | 172.9 |
| 9 | Mineral | 152.5 |
| 10 | Chester | 131.7 |
| 11 | Calaveras | 123.9 |
| 12 | Canyon Dam | 117.8 |
| 13 | Strawberry Valley | 113.6 |
| 14 | Cherry Valley Dm | 113.3 |
| 15 | Bodie | 106.7 |
| 16 | Mount Shasta | 98.8 |
| 17 | Boca | 96.3 |
| 18 | McCloud | 79.8 |
| 19 | Jess Valley | 75.0 |
| 20 | Salt Springs | 71.5 |
| 21 | Sierraville | 70.5 |
| 22 | Portola | 65.7 |
| 23 | Big Bear Lake | 65.6 |
| 24 | Yosemite-Hdqtrs | 57.3 |
| 25 | Fort Bidwell | 52.8 |

### LOWEST*

| Rank | Station | Value |
|---|---|---|
| 1 | Bakersfield | 0.1 |
|  | Campo | 0.1 |
|  | Chula Vista | 0.1 |
|  | Colusa | 0.1 |
|  | Davis | 0.1 |
|  | Fairfield | 0.1 |
|  | Five Points | 0.1 |
|  | Los Gatos | 0.1 |
|  | Palm Springs | 0.1 |
|  | Paso Robles | 0.1 |
|  | Paso Robles-Muni | 0.1 |
|  | Sacramento-Exec | 0.1 |
|  | Ukiah | 0.1 |
|  | Vista | 0.1 |
|  | Winters | 0.1 |
|  | Woodland | 0.1 |
| 17 | Borrego Desert | 0.2 |
|  | Chico | 0.2 |
|  | Crescent City | 0.2 |
|  | Vacaville | 0.2 |
| 21 | Balch | 0.3 |
|  | Cloverdale | 0.3 |
|  | Pinnacles | 0.3 |
|  | St. Helena | 0.3 |
|  | Trona | 0.3 |

* Does not include stations which receive no snowfall.

# COLORADO

PHYSICAL FEATURES. Colorado lies astride the highest mountains of the Continental Divide. Nearly rectangular, its north and south boundaries are the 41° and 37° N. parallels, and the east and west boundaries are the 102° and 109° W. meridians. It is eighth in size among the 50 states, with an area of 104,247 square miles. Although primarily a mountain state, nearly 40 percent of its area is taken up by the eastern high plains.

The principal features of Colorado geography are its inland continental location in the middle latitudes, and the mountains and ranges extending north and south approximately through the middle of the State. With an average altitude about 6,800 feet above sea level, Colorado is the highest state in the Union. Roughly three-quarters of the Nation's land above 10,000 feet altitude lies within its borders. The State has 54 mountains 14,000 feet or higher, and about 830 mountains between 11,000 and 14,000 feet in elevation.

Emerging gradually from the plains of Kansas and Nebraska, the high plains of Colorado slope gently upward for a distance of some 200 miles from the eastern border to the base of the foothills of the Rocky Mountains. The eastern portion of the State is generally level to rolling prairie broken by occasional hills and bluffs. The northern part of the plains area slopes to the northeast and the southern part to the southeast, divided by higher country and hills extending eastward from the mountains near the center of the State. Elevations along the eastern border range from about 3,350 feet at the lowest point in the State (where the Arkansas River crosses the border) to near 4,000 feet.

At elevations between 5,000 and 6,000 feet the plains give way abruptly to foothills with elevations of 7,000 to 9,000 feet. Backing the foothills are the mountain ranges above 9,000 feet with the higher peaks over 14,000 feet. West of these "front ranges" are additional ranges, generally extending north and south, but with many spurs and extensions in other directions. These ranges enclose numerous high mountain peaks and valleys. Farther westward the mountains give way to rugged plateau country in the form of high mesas (some more than 10,000 feet in elevation) which extends to the western border of the State. This land is often cut by rugged canyons, the work of the many streams fed by accumulations of winter snow.

All rivers in Colorado rise within its borders and flow outward, with the exception of the Green River which flows diagonally across the extreme northwestern corner of the State. Four of the Nation's major rivers have their source in Colorado: the Colorado, the Rio Grande, the Arkansas, and the Platte.

GENERAL CLIMATE. Most of Colorado has a cool and invigorating climate that could be termed a highland or mountain climate of a continental location. During summer there are hot days in the plains, but these are often relieved by afternoon thundershowers. Mountain regions are nearly always cool. Humidity is generally quite low; this favors rapid evaporation and a relatively comfortable feeling even on hot days. The thin atmosphere allows greater penetration of solar radiation and results in pleasant daytime conditions even during the winter. This is why skiers at high elevations are often pictured in very light clothing, although surrounded by heavy snow.

The climates of local areas are profoundly affected by differences in elevation, and to a lesser degree, by the orientation of mountain ranges and valleys with respect to general air movements. While temperature decreases, and precipitation generally increases with altitude, these patterns are modified by the orientation of mountain slopes with respect to the prevailing winds and by the effect of topographical features in creating local air movements.

As a result of the State's distance from major sources of moisture (the Pacific Ocean and the Gulf of Mexico), precipitation is generally light in the lower elevations. Prevailing air currents reach Colorado from westerly directions. Eastward-moving storms originating in the Pacific Ocean lose much of their moisture in passage over mountain ranges to the west; a large part of the remaining moisture falls as rain or snow on the mountaintops and westward-facing slopes. Eastern slope areas receive relatively small amounts of precipitation from these storms.

Storms moving from the north usually carry little moisture. The frequency of such storms increases during the fall and winter months, and decreases rapidly in the spring. The accompanying outbreaks of polar air are responsible for the sudden drops in temperature often experienced in the plains section. Occasionally these outbreaks are attended by strong northerly winds which come in contact with moist air from the south; the interaction of these air masses causes a heavy fall of snow and the most severe of all weather conditions of the high plains, the blizzard. This cold air is frequently too shallow to cross the mountains to the western portion of the State so while the plains are in the grip of a very severe storm, the weather in the mountains and western valleys may be mild.

Occasionally, when the plains are covered with a shallow layer of cold air, strong westerly winds aloft work their way to the surface. Warmed by rapid descent from higher levels, these winds bring large and sudden temperature rises. This phenomenon is called the "chinook" of the high plains and temperature rises of 25° - 35° within a short time are not uncommon. Chinook winds greatly moderate average winter temperatures in areas near enough to the mountains to experience them frequently.

Warm, moist air from the south moves into Colorado most frequently in the spring. As this air is carried northward and westward to higher elevations, the heaviest and most general rainfalls of the year occur over the eastern portions of the State. Frequent showers and thunderstorms continue well into the summer.

CLIMATE OF THE EASTERN PLAINS. The climate of the plains is comparatively uniform from place to place, with characteristic features of low relative humidity, abundant sunshine, light rainfall, moderate to high wind movement, and a large daily range in temperature. Because of the very low relative humidity, hot days cause less discomfort than in more humid areas. Summer precipitation in the plains is largely from thunderstorm activity and is sometimes extremely heavy. Strong winds occur frequently in winter and spring. These winds tend to dry out soils, which are not well supplied with moisture because of the low annual precipitation. During periods of drought such winds give rise to the duststorms which are especially characteristic of the southeastern plains.

At the western edge of the plains and near the foothills of the mountains, there are a number of significant changes in climate as compared to the plains proper. Average wind movement is less. Temperature changes from day to day are not as great; summer temperatures are lower, and winter temperatures are higher. Precipitation, which decreases gradually from the eastern border to a minimum near the mountains, increases rapidly with the increasing elevation of the foothills and proximity to higher ranges.

CLIMATE OF WESTERN COLORADO. The rugged topography of western Colorado causes large variations in climate within short distances, and few climatic generalizations apply to the whole area. Snow-covered mountain peaks and valleys often have very cold nighttime temperatures in winter, when skies are clear and the air is still -- occasionally to 50° F below zero. Summer in the mountains is a cool and refreshing season. At typical mountain stations the average July temperature is in the neighborhood of 60° F. The highest temperatures are usually in the 70's and 80's, but may reach 95° F. Above 7,000 feet, the nights are quite cool throughout the summer, while bright sunshine makes the days comfortably warm.

Precipitation west of the Continental Divide is more evenly distributed throughout the year than in the eastern plains. For most of western Colorado, the greatest monthly precipitation occurs in the winter months, while June is the driest month. In contrast, June is one of the wetter months in most of the eastern portion of the State.

STORMS. Thunderstorms are quite prevalent in the eastern plains and along the eastern slopes of the mountains during the spring and summer. These often become quite severe, and the frequency of hail damage to crops in northeastern Colorado is quite high. Tornadoes almost never occur in the mountains, and are relatively infrequent over the eastern plains. A spring flood potential results from the melting of snow. In years when snow cover is heavy, or when there is a sudden warming in the spring at high elevations, there may be extensive flooding. Heavy thunderstorms in the eastern foothills and plains occasionally cause damaging flash floods. Similar flash floods occur on the western slopes but with somewhat lower frequency.

## COUNTY INDEX

### *Adams County*
BYERS 5 ENE

### *Alamosa County*
ALAMOSA BERGMAN FLD
GREAT SAND DUNES NAT

### *Arapahoe County*
CHERRY CREEK DAM

### *Baca County*
SPRINGFIELD 7 WSW
WALSH 1 W

### *Bent County*
LAS ANIMAS

### *Boulder County*
BOULDER
LONGMONT 2 ESE

### *Chaffee County*
BUENA VISTA

### *Cheyenne County*
CHEYENNE WELLS
KIT CARSON

### *Conejos County*
MANASSA

### *Costilla County*
BLANCA

### *Custer County*
WESTCLIFFE

### *Delta County*
CEDAREDGE
DELTA
PAONIA 1 SW

### *Denver County*
DENVER STAPLETON AP

### *Dolores County*
NORTHDALE
RICO

### *Douglas County*
CASTLE ROCK

### *Eagle County*
EAGLE COUNTY AP

### *Elbert County*
PARKER 6 E

### *El Paso County*
COLORADO SPRGS MUNI
RUSH
RUXTON PARK

### *Fremont County*
CANON CITY

### *Garfield County*
ALTENBERN
GLENWOOD SPRINGS # 2
RIFLE

### *Grand County*
GRAND LAKE 1 NW
GRAND LAKE 6 SSW

### *Gunnison County*
COCHETOPA CREEK
CRESTED BUTTE
GUNNISON 3 SW
TAYLOR PARK

### *Hinsdale County*
LAKE CITY

### *Huerfano County*
WALSENBURG

### *Jackson County*
SPICER
WALDEN

### *Jefferson County*
CHEESMAN
EVERGREEN
KASSLER
LAKEWOOD

### *Kiowa County*
EADS

### *Kit Carson County*
BURLINGTON
FLAGLER 2 NW
STRATTON

### *Lake County*
CLIMAX
SUGARLOAF RESERVOIR

### *La Plata County*
FORT LEWIS
IGNACIO 1 N
VALLECITO DAM

### *Larimer County*
ESTES PARK
FORT COLLINS
WATERDALE

### *Las Animas County*
TRINIDAD
TRINIDAD LAS ANIMAS

### *Logan County*
STERLING

### *Mesa County*
COLLBRAN 1 W
COLORADO NATL MONUME
FRUITA 1 W
GATEWAY 1 SE
GRAND JUNCTION 6 ESE
GRAND JUNCTION WLKR
PALISADE

### *Mineral County*
HERMIT 7 ESE
WOLF CREEK PASS 1 E

### *Moffat County*
DINOSAUR NATL MON

### *Montezuma County*
CORTEZ
MESA VERDE NATL PARK
YELLOW JACKET 2 W

### *Montrose County*
CIMARRON
MONTROSE 2
URAVAN

### *Morgan County*
FORT MORGAN 2 S

### *Otero County*
LA JUNTA 4 NNE
ROCKY FORD 2 SE

### *Ouray County*
OURAY

### *Park County*
ANTERO RESERVOIR
BAILEY
GRANT
LAKE GEORGE 8 SW

### *Phillips County*
HOLYOKE

### *Pitkin County*
MEREDITH

### *Prowers County*
HOLLY
LAMAR

### *Pueblo County*
PUEBLO MEMORIAL AP
TACONY 10 SE

### *Rio Blanco County*
RANGELY 1 E

### *Rio Grande County*
CENTER 4 SSW
DEL NORTE
MONTE VISTA

### *Routt County*
HAYDEN
STEAMBOAT SPRINGS
YAMPA

### *Saguache County*
SAGUACHE

### *San Juan County*
SILVERTON

### *San Miguel County*
NORWOOD
TELLURIDE 4 WNW

### *Sedgwick County*
JULESBURG
SEDGWICK 5 S

### *Summit County*
DILLON 1 E
GREEN MOUNTAIN DAM

### *Washington County*
AKRON WASHINGTON CO

### *Weld County*
GREELEY UNC
NEW RAYMER

### *Yuma County*
BONNY LAKE
WRAY
YUMA

## ELEVATION INDEX

| FEET | STATION NAME |
|---|---|
| 3390 | HOLLY |
| 3471 | JULESBURG |
| 3514 | WRAY |
| 3642 | LAMAR |
| 3652 | BONNY LAKE |
| 3730 | HOLYOKE |
| 3891 | LAS ANIMAS |
| 3944 | STERLING |
| 3969 | WALSH 1 W |
| 3989 | SEDGWICK 5 S |
| 4134 | YUMA |
| 4173 | BURLINGTON |
| 4173 | ROCKY FORD 2 SE |
| 4196 | LA JUNTA 4 NNE |
| 4202 | KIT CARSON |
| 4250 | CHEYENNE WELLS |
| 4262 | EADS |
| 4320 | FORT MORGAN 2 S |
| 4403 | STRATTON |
| 4524 | FRUITA 1 W |
| 4564 | GATEWAY 1 SE |
| 4580 | SPRINGFIELD 7 WSW |
| 4587 | AKRON WASHINGTON CO |
| 4646 | PUEBLO MEMORIAL AP |
| 4652 | GREELEY UNC |
| 4724 | PALISADE |
| 4759 | GRAND JUNCTION 6 ESE |
| 4849 | GRAND JUNCTION WLKR |
| 4954 | LONGMONT 2 ESE |
| 4954 | NEW RAYMER |

| FEET | STATION NAME |
|---|---|
| 4984 | FLAGLER 2 NW |
| 5003 | TACONY 10 SE |
| 5007 | FORT COLLINS |
| 5023 | URAVAN |
| 5100 | BYERS 5 ENE |
| 5125 | DELTA |
| 5203 | WATERDALE |
| 5282 | RANGELY 1 E |
| 5298 | DENVER STAPLETON AP |
| 5299 | RIFLE |
| 5355 | CANON CITY |
| 5411 | BOULDER |
| 5501 | KASSLER |
| 5636 | LAKEWOOD |
| 5647 | CHERRY CREEK DAM |
| 5685 | PAONIA 1 SW |
| 5690 | ALTENBERN |
| 5738 | TRINIDAD LAS ANIMAS |
| 5784 | COLORADO NATL MONUME |
| 5822 | GLENWOOD SPRINGS # 2 |
| 5833 | MONTROSE 2 |
| 5921 | DINOSAUR NATL MON |
| 6020 | RUSH |
| 6145 | COLLBRAN 1 W |
| 6184 | CEDAREDGE |
| 6184 | CORTEZ |
| 6224 | WALSENBURG |
| 6253 | CASTLE ROCK |
| 6306 | PARKER 6 E |
| 6306 | TRINIDAD |
| 6375 | HAYDEN |
| 6424 | IGNACIO 1 N |
| 6483 | NORTHDALE |
| 6608 | EAGLE COUNTY AP |
| 6769 | STEAMBOAT SPRINGS |
| 6863 | YELLOW JACKET 2 W |
| 6893 | CHEESMAN |
| 6965 | MESA VERDE NATL PARK |
| 7005 | EVERGREEN |
| 7019 | NORWOOD |
| 7244 | CIMARRON |
| 7539 | ALAMOSA BERGMAN FLD |
| 7615 | FORT LEWIS |
| 7650 | VALLECITO DAM |
| 7664 | GUNNISON 3 SW |
| 7674 | CENTER 4 SSW |
| 7674 | MONTE VISTA |
| 7707 | MANASSA |
| 7707 | SAGUACHE |
| 7726 | OURAY |

| FEET | STATION NAME |
|---|---|
| 7736 | BAILEY |
| 7746 | BLANCA |
| 7756 | ESTES PARK |
| 7766 | GREEN MOUNTAIN DAM |
| 7805 | MEREDITH |
| 7859 | WESTCLIFFE |
| 7884 | DEL NORTE |
| 7891 | YAMPA |
| 7985 | BUENA VISTA |
| 8000 | COCHETOPA CREEK |
| 8107 | WALDEN |
| 8119 | GREAT SAND DUNES NAT |
| 8307 | SPICER |
| 8386 | GRAND LAKE 1 NW |
| 8386 | GRAND LAKE 6 SSW |
| 8507 | LAKE GEORGE 8 SW |
| 8694 | GRANT |
| 8756 | TELLURIDE 4 WNW |
| 8825 | RICO |
| 8875 | CRESTED BUTTE |
| 8894 | LAKE CITY |
| 8899 | DILLON 1 E |
| 8937 | ANTERO RESERVOIR |
| 9006 | HERMIT 7 ESE |
| 9049 | RUXTON PARK |
| 9199 | COLORADO SPRGS MUNI |
| 9216 | TAYLOR PARK |
| 9426 | SILVERTON |
| 10003 | SUGARLOAF RESERVOIR |
| 10641 | WOLF CREEK PASS 1 E |
| 11529 | CLIMAX |

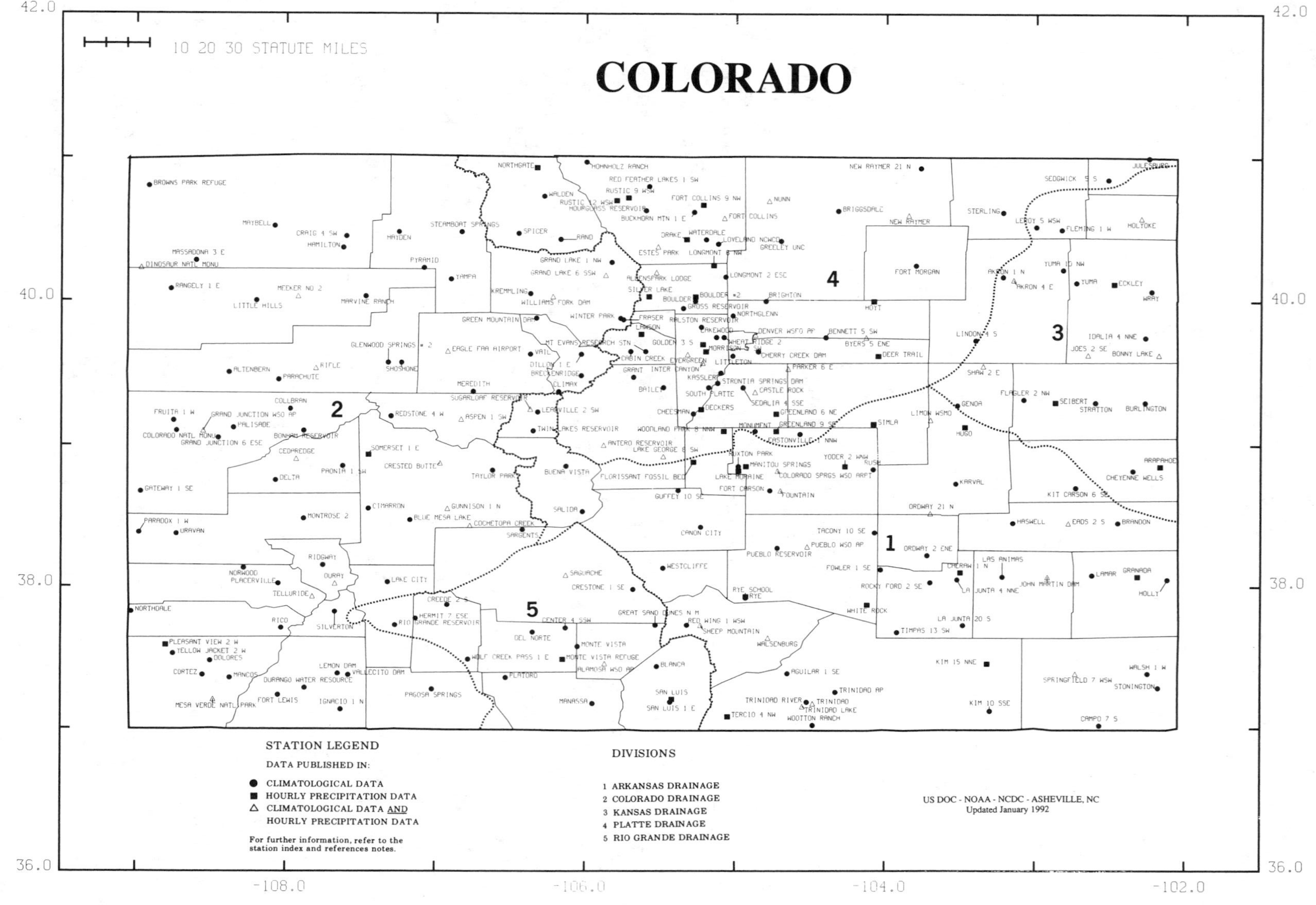
COLORADO
10 20 30 STATUTE MILES
STATION LEGEND
DATA PUBLISHED IN:
● CLIMATOLOGICAL DATA
■ HOURLY PRECIPITATION DATA
△ CLIMATOLOGICAL DATA AND HOURLY PRECIPITATION DATA
For further information, refer to the station index and references notes.
DIVISIONS
1 ARKANSAS DRAINAGE
2 COLORADO DRAINAGE
3 KANSAS DRAINAGE
4 PLATTE DRAINAGE
5 RIO GRANDE DRAINAGE
US DOC - NOAA - NCDC - ASHEVILLE, NC
Updated January 1992

## AKRON WASHINGTON CO *Washington County* ELEVATION 4587 ft LAT/LONG 40° 7 ' N / 103° 10 ' W

| | JAN | FEB | MAR | APR | MAY | JUN | JUL | AUG | SEP | OCT | NOV | DEC | YEAR |
|---|---|---|---|---|---|---|---|---|---|---|---|---|---|
| Maximum Temp °F | 38.1 | 42.9 | 50.6 | 60.5 | 70.0 | 81.1 | 87.4 | 85.3 | 76.3 | 64.1 | 48.5 | 39.9 | 62.1 |
| Minimum Temp °F | 14.5 | 18.4 | 24.6 | 33.3 | 43.3 | 53.0 | 58.9 | 57.1 | 48.0 | 36.2 | 24.1 | 16.3 | 35.6 |
| Mean Temp °F | 26.4 | 30.7 | 37.6 | 46.9 | 56.6 | 67.1 | 73.2 | 71.2 | 62.2 | 50.2 | 36.3 | 28.1 | 48.9 |
| Days Max Temp ≥ 90 °F | 0 | 0 | 0 | 0 | 1 | 6 | 13 | 11 | 3 | 0 | 0 | 0 | 34 |
| Days Max Temp ≤ 32 °F | 10 | 6 | 4 | 1 | 0 | 0 | 0 | 0 | 0 | 0 | 4 | 8 | 33 |
| Days Min Temp ≤ 32 °F | 30 | 27 | 26 | 14 | 3 | 0 | 0 | 0 | 1 | 10 | 24 | 30 | 165 |
| Days Min Temp ≤ 0 °F | 5 | 2 | 1 | 0 | 0 | 0 | 0 | 0 | 0 | 0 | 0 | 3 | 11 |
| Heating Degree Days | 1191 | 963 | 843 | 538 | 270 | 61 | 8 | 17 | 148 | 454 | 853 | 1136 | 6482 |
| Cooling Degree Days | 0 | 0 | 0 | 3 | 18 | 140 | 267 | 227 | 78 | 3 | 0 | 0 | 736 |
| Total Precipitation (") | 0.32 | 0.34 | 0.99 | 1.46 | 3.26 | 2.58 | 2.87 | 1.88 | 0.86 | 0.90 | 0.62 | 0.41 | 16.49 |
| Days ≥ 0.1" Precip | 1 | 1 | 3 | 4 | 6 | 6 | 6 | 4 | 3 | 2 | 2 | 1 | 39 |
| Total Snowfall (") | 5.6 | 4.2 | 9.4 | 4.4 | 0.7 | 0.0 | 0.0 | 0.0 | 0.5 | 3.6 | 6.7 | 6.0 | 41.1 |
| Days ≥ 1" Snow Depth | 16 | 10 | 7 | 3 | 0 | 0 | 0 | 0 | 0 | 1 | 6 | 12 | 55 |

## ALAMOSA BERGMAN FLD *Alamosa County* ELEVATION 7539 ft LAT/LONG 37° 26 ' N / 105° 51 ' W

| | JAN | FEB | MAR | APR | MAY | JUN | JUL | AUG | SEP | OCT | NOV | DEC | YEAR |
|---|---|---|---|---|---|---|---|---|---|---|---|---|---|
| Maximum Temp °F | 32.7 | 39.5 | 49.1 | 58.8 | 67.9 | 77.9 | 81.9 | 79.1 | 72.8 | 61.9 | 46.5 | 35.2 | 58.6 |
| Minimum Temp °F | -4.0 | 4.5 | 16.5 | 23.6 | 33.0 | 41.0 | 47.4 | 45.4 | 36.5 | 24.4 | 11.7 | -0.2 | 23.3 |
| Mean Temp °F | 14.4 | 22.0 | 32.8 | 41.2 | 50.5 | 59.5 | 64.7 | 62.3 | 54.7 | 43.2 | 29.1 | 17.5 | 41.0 |
| Days Max Temp ≥ 90 °F | 0 | 0 | 0 | 0 | 0 | 0 | 1 | 0 | 0 | 0 | 0 | 0 | 1 |
| Days Max Temp ≤ 32 °F | 15 | 6 | 1 | 0 | 0 | 0 | 0 | 0 | 0 | 0 | 3 | 11 | 36 |
| Days Min Temp ≤ 32 °F | 31 | 28 | 31 | 27 | 14 | 2 | 0 | 0 | 8 | 27 | 30 | 31 | 229 |
| Days Min Temp ≤ 0 °F | 19 | 10 | 1 | 0 | 0 | 0 | 0 | 0 | 0 | 0 | 4 | 16 | 50 |
| Heating Degree Days | 1564 | 1208 | 992 | 707 | 444 | 167 | 38 | 90 | 304 | 670 | 1070 | 1467 | 8721 |
| Cooling Degree Days | 0 | 0 | 0 | 0 | 0 | 11 | 33 | 14 | 1 | 0 | 0 | 0 | 59 |
| Total Precipitation (") | 0.25 | 0.25 | 0.47 | 0.47 | 0.74 | 0.66 | 1.15 | 1.28 | 0.87 | 0.73 | 0.46 | 0.42 | 7.75 |
| Days ≥ 0.1" Precip | 1 | 1 | 1 | 1 | 2 | 2 | 3 | 4 | 3 | 2 | 1 | 1 | 22 |
| Total Snowfall (") | 4.4 | 4.0 | 6.5 | 3.7 | 1.9 | 0.0 | 0.0 | 0.0 | 0.1 | 4.1 | 4.3 | 6.7 | 35.7 |
| Days ≥ 1" Snow Depth | 18 | 9 | 4 | 1 | 0 | 0 | 0 | 0 | 0 | 1 | 5 | 13 | 51 |

## ALTENBERN *Garfield County* ELEVATION 5690 ft LAT/LONG 39° 30 ' N / 108° 23 ' W

| | JAN | FEB | MAR | APR | MAY | JUN | JUL | AUG | SEP | OCT | NOV | DEC | YEAR |
|---|---|---|---|---|---|---|---|---|---|---|---|---|---|
| Maximum Temp °F | 35.5 | 42.5 | 52.2 | 62.0 | 71.7 | 82.7 | 88.3 | 85.0 | 76.9 | 64.5 | 48.4 | 37.3 | 62.2 |
| Minimum Temp °F | 9.3 | 15.8 | 23.8 | 29.8 | 37.8 | 44.0 | 50.3 | 48.8 | 41.1 | 31.3 | 21.1 | 11.6 | 30.4 |
| Mean Temp °F | 22.4 | 29.2 | 38.0 | 45.9 | 54.8 | 63.4 | 69.3 | 67.0 | 59.0 | 47.9 | 34.8 | 24.5 | 46.3 |
| Days Max Temp ≥ 90 °F | 0 | 0 | 0 | 0 | 0 | 6 | 14 | 7 | 1 | 0 | 0 | 0 | 28 |
| Days Max Temp ≤ 32 °F | 11 | 2 | 0 | 0 | 0 | 0 | 0 | 0 | 0 | 0 | 2 | 8 | 23 |
| Days Min Temp ≤ 32 °F | 31 | 28 | 28 | 20 | 6 | 1 | 0 | 0 | 3 | 18 | 28 | 31 | 194 |
| Days Min Temp ≤ 0 °F | 7 | 2 | 0 | 0 | 0 | 0 | 0 | 0 | 0 | 0 | 0 | 5 | 14 |
| Heating Degree Days | 1313 | 1005 | 829 | 566 | 312 | 93 | 7 | 25 | 184 | 522 | 900 | 1250 | 7006 |
| Cooling Degree Days | 0 | 0 | 0 | 0 | 2 | 51 | 136 | 94 | 13 | 0 | 0 | 0 | 296 |
| Total Precipitation (") | 1.12 | 1.11 | 1.62 | 1.35 | 1.66 | 1.03 | 1.40 | 1.45 | 1.39 | 1.75 | 1.42 | 1.24 | 16.54 |
| Days ≥ 0.1" Precip | 4 | 4 | 5 | 5 | 5 | 3 | 4 | 4 | 4 | 5 | 4 | 4 | 51 |
| Total Snowfall (") | 14.9 | 10.0 | 7.7 | 2.7 | 0.6 | 0.1 | 0.0 | 0.0 | 0.3 | 1.3 | 7.6 | 15.0 | 60.2 |
| Days ≥ 1" Snow Depth | 23 | 15 | 5 | 0 | 0 | 0 | 0 | 0 | 0 | 0 | 4 | 17 | 64 |

## ANTERO RESERVOIR *Park County* ELEVATION 8937 ft LAT/LONG 39° 0 ' N / 105° 53 ' W

| | JAN | FEB | MAR | APR | MAY | JUN | JUL | AUG | SEP | OCT | NOV | DEC | YEAR |
|---|---|---|---|---|---|---|---|---|---|---|---|---|---|
| Maximum Temp °F | 31.9 | 35.0 | 40.8 | 48.6 | 59.2 | 70.0 | 75.5 | 72.8 | 66.4 | 55.8 | 40.8 | 32.5 | 52.4 |
| Minimum Temp °F | -3.6 | -1.1 | 8.9 | 18.1 | 27.3 | 34.3 | 39.9 | 38.9 | 30.7 | 19.6 | 8.2 | -1.8 | 18.3 |
| Mean Temp °F | 14.2 | 17.0 | 24.9 | 33.4 | 43.3 | 52.2 | 57.7 | 55.9 | 48.6 | 37.7 | 24.5 | 15.4 | 35.4 |
| Days Max Temp ≥ 90 °F | 0 | 0 | 0 | 0 | 0 | 0 | 0 | 0 | 0 | 0 | 0 | 0 | 0 |
| Days Max Temp ≤ 32 °F | 15 | 10 | 6 | 2 | 0 | 0 | 0 | 0 | 0 | 1 | 7 | 14 | 55 |
| Days Min Temp ≤ 32 °F | 31 | 28 | 31 | 29 | 25 | 11 | 2 | 3 | 18 | 29 | 30 | 31 | 268 |
| Days Min Temp ≤ 0 °F | 19 | 15 | 7 | 2 | 0 | 0 | 0 | 0 | 0 | 1 | 8 | 17 | 69 |
| Heating Degree Days | 1570 | 1350 | 1238 | 941 | 667 | 379 | 219 | 277 | 485 | 839 | 1208 | 1533 | 10706 |
| Cooling Degree Days | 0 | 0 | 0 | 0 | 0 | 0 | 1 | 0 | 0 | 0 | 0 | 0 | 1 |
| Total Precipitation (") | 0.18 | 0.24 | 0.49 | 0.63 | 0.97 | 1.15 | 2.04 | 2.13 | 0.94 | 0.68 | 0.37 | 0.30 | 10.12 |
| Days ≥ 0.1" Precip | 1 | 1 | 2 | 2 | 3 | 3 | 6 | 5 | 3 | 2 | 1 | 1 | 30 |
| Total Snowfall (") | 4.0 | 5.4 | 8.4 | 7.2 | 2.7 | 0.6 | 0.0 | 0.0 | 1.5 | 5.5 | 6.3 | 7.3 | 48.9 |
| Days ≥ 1" Snow Depth | 22 | 19 | 15 | 7 | 1 | 0 | 0 | 0 | 0 | 4 | 13 | 21 | 102 |

## BAILEY *Park County* ELEVATION 7736 ft LAT/LONG 39° 24 ' N / 105° 29 ' W

| | JAN | FEB | MAR | APR | MAY | JUN | JUL | AUG | SEP | OCT | NOV | DEC | YEAR |
|---|---|---|---|---|---|---|---|---|---|---|---|---|---|
| Maximum Temp °F | 39.3 | 42.0 | 46.9 | 54.3 | 63.1 | 73.9 | 78.6 | 76.2 | 69.7 | 59.7 | 46.4 | 39.2 | 57.4 |
| Minimum Temp °F | 8.3 | 10.0 | 16.2 | 22.6 | 30.5 | 37.4 | 43.3 | 41.8 | 33.6 | 24.0 | 15.9 | 9.2 | 24.4 |
| Mean Temp °F | 23.8 | 26.1 | 31.6 | 38.5 | 46.8 | 55.7 | 61.0 | 59.0 | 51.7 | 41.9 | 31.2 | 24.2 | 41.0 |
| Days Max Temp ≥ 90 °F | 0 | 0 | 0 | 0 | 0 | 0 | 1 | 0 | 0 | 0 | 0 | 0 | 1 |
| Days Max Temp ≤ 32 °F | 8 | 4 | 3 | 1 | 0 | 0 | 0 | 0 | 0 | 1 | 4 | *8* | 29 |
| Days Min Temp ≤ 32 °F | 30 | 28 | 30 | 28 | 20 | 5 | 0 | 1 | 12 | 28 | 29 | 31 | 242 |
| Days Min Temp ≤ 0 °F | 8 | 5 | 3 | 0 | 0 | 0 | 0 | 0 | 0 | 0 | 2 | 6 | 24 |
| Heating Degree Days | 1270 | 1091 | 1029 | 788 | 556 | 276 | 127 | 181 | 392 | 709 | 1008 | 1257 | 8684 |
| Cooling Degree Days | 0 | 0 | 0 | 0 | 0 | 2 | 6 | 2 | 0 | 0 | 0 | 0 | 10 |
| Total Precipitation (") | 0.35 | 0.54 | 1.20 | 1.58 | 2.04 | 1.71 | 2.52 | 2.53 | 1.24 | 1.39 | 0.83 | 0.59 | 16.52 |
| Days ≥ 0.1" Precip | 1 | 2 | 4 | 4 | 5 | 5 | 7 | 7 | 3 | 3 | 3 | 2 | 46 |
| Total Snowfall (") | 6.6 | 9.6 | *17.6* | 13.3 | 2.9 | 0.3 | 0.0 | 0.0 | 1.3 | 9.4 | 11.9 | 10.5 | 83.4 |
| Days ≥ 1" Snow Depth | na | na | na | na | *1* | 0 | 0 | 0 | 0 | na | na | na | na |

## BLANCA *Costilla County* ELEVATION 7746 ft LAT/LONG 37° 26 ' N / 105° 31 ' W

| | JAN | FEB | MAR | APR | MAY | JUN | JUL | AUG | SEP | OCT | NOV | DEC | YEAR |
|---|---|---|---|---|---|---|---|---|---|---|---|---|---|
| Maximum Temp °F | 34.0 | 39.8 | 48.6 | 58.0 | 67.2 | 77.4 | 81.8 | 78.8 | 72.5 | 61.8 | 47.0 | 36.1 | 58.6 |
| Minimum Temp °F | 0.5 | 7.7 | 17.7 | 24.9 | 33.6 | 41.7 | 47.6 | 45.4 | 37.0 | 25.9 | 14.3 | 3.2 | 25.0 |
| Mean Temp °F | 17.3 | 23.8 | 33.2 | 41.5 | 50.4 | 59.6 | 64.7 | 62.1 | 54.8 | 44.1 | 30.8 | 19.7 | 41.8 |
| Days Max Temp ≥ 90 °F | 0 | 0 | 0 | 0 | 0 | 0 | 1 | 0 | 0 | 0 | 0 | 0 | 1 |
| Days Max Temp ≤ 32 °F | 13 | 6 | 1 | 0 | 0 | 0 | 0 | 0 | 0 | 0 | 3 | 10 | 33 |
| Days Min Temp ≤ 32 °F | 31 | 28 | 30 | 25 | 13 | 2 | 0 | 0 | 7 | 25 | 29 | 31 | 221 |
| Days Min Temp ≤ 0 °F | 15 | 7 | 1 | 0 | 0 | 0 | 0 | 0 | 0 | 0 | 3 | 12 | 38 |
| Heating Degree Days | 1474 | 1158 | 980 | 698 | 446 | 166 | 43 | 97 | 299 | 642 | 1020 | 1398 | 8421 |
| Cooling Degree Days | 0 | 0 | 0 | 0 | 0 | 16 | 45 | 20 | 2 | 0 | 0 | 0 | 83 |
| Total Precipitation (") | 0.28 | 0.30 | 0.54 | 0.57 | 0.85 | 0.83 | 1.43 | 1.59 | 1.03 | 0.71 | 0.47 | 0.41 | 9.01 |
| Days ≥ 0.1" Precip | 1 | 1 | 2 | 2 | 2 | 3 | 4 | 5 | 3 | 2 | 1 | 2 | 28 |
| Total Snowfall (") | 4.3 | 3.7 | 5.4 | 3.3 | 0.4 | 0.0 | 0.0 | 0.0 | 0.0 | 2.0 | 3.8 | *5.3* | 28.2 |
| Days ≥ 1" Snow Depth | *11* | na | *4* | *1* | 0 | 0 | 0 | 0 | 0 | *1* | *4* | *10* | na |

## BONNY LAKE *Yuma County* ELEVATION 3652 ft LAT/LONG 39° 37 ' N / 102° 11 ' W

| | JAN | FEB | MAR | APR | MAY | JUN | JUL | AUG | SEP | OCT | NOV | DEC | YEAR |
|---|---|---|---|---|---|---|---|---|---|---|---|---|---|
| Maximum Temp °F | *40.2* | 45.8 | *53.1* | *63.3* | 72.7 | *83.5* | 90.0 | *87.8* | *78.9* | *67.8* | *52.3* | *43.1* | 64.9 |
| Minimum Temp °F | *13.0* | 17.3 | *24.0* | *33.8* | 43.7 | *53.2* | 59.7 | *57.6* | *47.8* | 35.2 | *23.7* | *15.8* | 35.4 |
| Mean Temp °F | *26.6* | 31.7 | *38.6* | *48.6* | 58.3 | *68.4* | 74.9 | *72.7* | *63.4* | *51.5* | *38.0* | *29.5* | 50.2 |
| Days Max Temp ≥ 90 °F | 0 | 0 | 0 | 0 | 1 | 7 | 17 | 15 | 5 | 0 | 0 | 0 | 45 |
| Days Max Temp ≤ 32 °F | 8 | 5 | 3 | 0 | 0 | 0 | 0 | 0 | 0 | 0 | 2 | 6 | 24 |
| Days Min Temp ≤ 32 °F | 29 | 27 | 26 | 12 | 2 | 0 | 0 | 0 | 1 | 10 | 24 | 27 | 158 |
| Days Min Temp ≤ 0 °F | 4 | 2 | 1 | 0 | 0 | 0 | 0 | 0 | 0 | 0 | 0 | 2 | 9 |
| Heating Degree Days | *1183* | 934 | *813* | *487* | 224 | *44* | 4 | *11* | *125* | *416* | *803* | *1095* | 6139 |
| Cooling Degree Days | na | 0 | na | *2* | *23* | na | *299* | na | *99* | na | na | na | na |
| Total Precipitation (") | 0.45 | 0.41 | 1.17 | 1.69 | 3.37 | 2.81 | 2.71 | 2.21 | 1.37 | 1.05 | 0.78 | 0.38 | 18.40 |
| Days ≥ 0.1" Precip | 1 | 1 | 3 | 4 | 6 | 5 | 5 | 4 | 3 | 2 | 2 | 1 | 37 |
| Total Snowfall (") | na | na | na | na | 0.1 | 0.0 | 0.0 | 0.0 | 0.1 | 1.6 | na | na | na |
| Days ≥ 1" Snow Depth | na | na | na | *1* | 0 | 0 | 0 | 0 | 0 | *0* | na | *8* | na |

## BOULDER *Boulder County* ELEVATION 5411 ft LAT/LONG 40° 2 ' N / 105° 16 ' W

| | JAN | FEB | MAR | APR | MAY | JUN | JUL | AUG | SEP | OCT | NOV | DEC | YEAR |
|---|---|---|---|---|---|---|---|---|---|---|---|---|---|
| Maximum Temp °F | 45.1 | 48.2 | 54.7 | 62.6 | 71.6 | 81.5 | 86.9 | 84.9 | 76.8 | 66.3 | 52.0 | 45.9 | 64.7 |
| Minimum Temp °F | 19.8 | 22.9 | 28.4 | 35.5 | 43.9 | 52.3 | 57.7 | 56.4 | 47.7 | 38.0 | 27.7 | 21.8 | 37.7 |
| Mean Temp °F | 32.5 | 35.6 | 41.6 | 49.1 | 57.8 | 67.0 | 72.4 | 70.7 | 62.3 | 52.2 | 39.9 | 33.9 | 51.3 |
| Days Max Temp ≥ 90 °F | 0 | 0 | 0 | 0 | 0 | 6 | 12 | 7 | 2 | 0 | 0 | 0 | 27 |
| Days Max Temp ≤ 32 °F | 5 | 3 | 1 | 0 | 0 | 0 | 0 | 0 | 0 | 0 | 2 | 4 | 15 |
| Days Min Temp ≤ 32 °F | 27 | 23 | 21 | 11 | 1 | 0 | 0 | 0 | 1 | 7 | 20 | 26 | 137 |
| Days Min Temp ≤ 0 °F | 2 | 1 | 0 | 0 | 0 | 0 | 0 | 0 | 0 | 0 | 0 | 1 | 4 |
| Heating Degree Days | 1002 | 825 | 720 | 473 | 236 | 54 | 6 | 10 | 129 | 394 | 747 | 957 | 5553 |
| Cooling Degree Days | 0 | 0 | 0 | 3 | 14 | 120 | 219 | 192 | 51 | 3 | 0 | 0 | 602 |
| Total Precipitation (") | 0.58 | 0.73 | 1.66 | 2.36 | 2.95 | 2.17 | 1.93 | 1.45 | 1.72 | 1.38 | 1.37 | 0.77 | 19.07 |
| Days ≥ 0.1" Precip | 2 | 2 | 4 | 5 | 6 | 4 | 5 | 4 | 4 | 3 | 3 | 3 | 45 |
| Total Snowfall (") | 9.5 | 10.6 | 15.8 | 9.9 | 1.6 | 0.0 | 0.0 | 0.0 | 1.6 | 5.5 | 14.8 | 12.6 | 81.9 |
| Days ≥ 1" Snow Depth | 12 | 9 | 6 | 3 | 0 | 0 | 0 | 0 | 0 | 2 | *8* | *12* | 52 |

## BUENA VISTA *Chaffee County* ELEVATION 7985 ft LAT/LONG 38° 50 ' N / 106° 8 ' W

| | JAN | FEB | MAR | APR | MAY | JUN | JUL | AUG | SEP | OCT | NOV | DEC | YEAR |
|---|---|---|---|---|---|---|---|---|---|---|---|---|---|
| Maximum Temp °F | 40.4 | 42.9 | 48.7 | 56.4 | 65.5 | 76.4 | 81.4 | 78.6 | 72.2 | 61.8 | 47.8 | 40.2 | 59.4 |
| Minimum Temp °F | 10.8 | 13.6 | 20.3 | 26.3 | 34.2 | 42.0 | 47.4 | 45.9 | 37.7 | 28.0 | 18.7 | 10.9 | 28.0 |
| Mean Temp °F | 25.8 | 28.3 | 34.5 | 41.3 | 49.9 | 59.2 | 64.4 | 62.2 | 54.9 | 44.9 | 33.3 | 25.6 | 43.7 |
| Days Max Temp ≥ 90 °F | 0 | 0 | 0 | 0 | 0 | 0 | 1 | 0 | 0 | 0 | 0 | 0 | 1 |
| Days Max Temp ≤ 32 °F | 6 | 3 | 2 | 0 | 0 | 0 | 0 | 0 | 0 | 0 | 2 | *7* | 20 |
| Days Min Temp ≤ 32 °F | 31 | 28 | 30 | 25 | 11 | 1 | 0 | 0 | 5 | *24* | 28 | 31 | 214 |
| Days Min Temp ≤ 0 °F | 4 | 2 | 0 | 0 | 0 | 0 | 0 | 0 | 0 | 0 | 1 | 4 | 11 |
| Heating Degree Days | 1208 | 1027 | 937 | 703 | 461 | 176 | 48 | 96 | 296 | 618 | 945 | 1214 | 7729 |
| Cooling Degree Days | 0 | 0 | 0 | 0 | 0 | 11 | 34 | 15 | 1 | 0 | 0 | 0 | 61 |
| Total Precipitation (") | 0.26 | 0.38 | 0.66 | 0.81 | 1.06 | 0.91 | 1.61 | 1.82 | 1.05 | 0.83 | 0.51 | 0.44 | 10.34 |
| Days ≥ 0.1" Precip | 1 | 1 | 3 | 3 | 3 | 3 | 5 | 5 | 3 | 2 | 2 | 2 | 33 |
| Total Snowfall (") | *3.8* | *4.1* | *6.4* | *3.7* | 1.7 | 0.3 | 0.0 | 0.0 | 0.4 | 1.9 | *3.5* | *4.7* | 30.5 |
| Days ≥ 1" Snow Depth | na | na | na | *1* | 0 | 0 | 0 | 0 | 0 | *1* | na | na | na |

## BURLINGTON *Kit Carson County* ELEVATION 4173 ft LAT/LONG 39° 18 ' N / 102° 16 ' W

| | JAN | FEB | MAR | APR | MAY | JUN | JUL | AUG | SEP | OCT | NOV | DEC | YEAR |
|---|---|---|---|---|---|---|---|---|---|---|---|---|---|
| Maximum Temp °F | 42.0 | 46.7 | 54.8 | 64.6 | 73.2 | 84.0 | 89.3 | 86.6 | 78.9 | 67.0 | 51.3 | 43.1 | 65.1 |
| Minimum Temp °F | 17.2 | 20.6 | 26.3 | 35.7 | 45.1 | 55.1 | 60.6 | 59.0 | 50.0 | 38.2 | 26.4 | 19.1 | 37.8 |
| Mean Temp °F | 29.6 | 33.7 | 40.6 | 50.2 | 59.2 | 69.6 | 75.0 | 72.8 | 64.4 | 52.6 | 38.9 | 31.1 | 51.5 |
| Days Max Temp ≥ 90 °F | 0 | 0 | 0 | 0 | 1 | 8 | *16* | 13 | 4 | 0 | 0 | 0 | 42 |
| Days Max Temp ≤ 32 °F | 7 | 4 | 2 | 0 | 0 | 0 | 0 | 0 | 0 | 0 | 3 | 6 | 22 |
| Days Min Temp ≤ 32 °F | 29 | 26 | 22 | 10 | 2 | 0 | 0 | 0 | 1 | 7 | 22 | 28 | 147 |
| Days Min Temp ≤ 0 °F | 3 | 1 | 0 | 0 | 0 | 0 | 0 | 0 | 0 | 0 | 0 | 1 | 5 |
| Heating Degree Days | 1090 | 877 | 753 | 443 | 207 | 34 | 3 | 10 | 102 | 384 | 776 | 1044 | 5723 |
| Cooling Degree Days | 0 | 0 | 0 | 5 | 27 | 182 | 312 | 274 | 102 | 4 | 0 | 0 | 906 |
| Total Precipitation (") | 0.28 | 0.32 | 0.91 | 1.22 | 2.68 | 2.54 | 2.51 | 2.18 | 1.11 | 0.97 | 0.54 | 0.36 | 15.62 |
| Days ≥ 0.1" Precip | 1 | 1 | 2 | 3 | 5 | 5 | 5 | 4 | 3 | 2 | 1 | 1 | 33 |
| Total Snowfall (") | *5.2* | *3.6* | na | 3.7 | 0.2 | 0.0 | 0.0 | 0.0 | 0.0 | *0.9* | na | *4.0* | na |
| Days ≥ 1" Snow Depth | na | na | na | *0* | *0* | 0 | 0 | 0 | 0 | *0* | na | na | na |

## BYERS 5 ENE *Adams County* ELEVATION 5100 ft LAT/LONG 39° 45 ' N / 104° 8 ' W

| | JAN | FEB | MAR | APR | MAY | JUN | JUL | AUG | SEP | OCT | NOV | DEC | YEAR |
|---|---|---|---|---|---|---|---|---|---|---|---|---|---|
| Maximum Temp °F | 41.5 | 45.9 | 53.6 | 63.3 | 72.3 | 83.3 | 89.4 | 87.2 | 78.8 | 66.9 | 51.3 | 42.8 | 64.7 |
| Minimum Temp °F | 11.5 | 16.3 | 23.2 | 31.4 | 40.9 | 49.9 | 55.6 | 53.9 | 44.8 | 32.9 | 20.8 | 13.0 | 32.9 |
| Mean Temp °F | 26.5 | 31.1 | 38.4 | 47.4 | 56.6 | 66.6 | 72.5 | 70.6 | 61.8 | 50.0 | 36.1 | 27.9 | 48.8 |
| Days Max Temp ≥ 90 °F | 0 | 0 | 0 | 0 | 1 | 8 | 17 | 13 | 4 | 0 | 0 | 0 | 43 |
| Days Max Temp ≤ 32 °F | 7 | 4 | 2 | 0 | 0 | 0 | 0 | 0 | 0 | 0 | 3 | 6 | 22 |
| Days Min Temp ≤ 32 °F | 31 | 28 | 28 | 17 | 4 | 0 | 0 | 0 | 2 | 15 | 27 | 30 | 182 |
| Days Min Temp ≤ 0 °F | 6 | 3 | 1 | 0 | 0 | 0 | 0 | 0 | 0 | 0 | 1 | 4 | 15 |
| Heating Degree Days | 1185 | 951 | 818 | 522 | 266 | 62 | 5 | 10 | 140 | 461 | 860 | 1143 | 6423 |
| Cooling Degree Days | 0 | 0 | 0 | 1 | 16 | 129 | 248 | 206 | 59 | 1 | 0 | 0 | 660 |
| Total Precipitation (") | 0.44 | 0.34 | 1.05 | 1.39 | 2.73 | 2.15 | 2.36 | 1.76 | 1.24 | 0.90 | 0.78 | 0.44 | 15.58 |
| Days ≥ 0.1" Precip | 2 | 1 | 3 | 4 | 5 | 4 | 5 | 4 | 3 | 2 | 2 | 1 | 36 |
| Total Snowfall (") | *7.2* | 4.6 | *10.7* | 4.5 | 0.3 | 0.0 | 0.0 | 0.0 | 0.8 | 3.7 | *7.3* | *5.9* | 45.0 |
| Days ≥ 1" Snow Depth | 16 | 12 | 6 | 2 | 0 | 0 | 0 | 0 | 0 | 1 | 7 | 13 | 57 |

## CANON CITY *Fremont County* ELEVATION 5355 ft LAT/LONG 38° 25 ' N / 105° 13 ' W

| | JAN | FEB | MAR | APR | MAY | JUN | JUL | AUG | SEP | OCT | NOV | DEC | YEAR |
|---|---|---|---|---|---|---|---|---|---|---|---|---|---|
| Maximum Temp °F | 49.3 | 51.9 | 57.0 | 65.0 | 73.0 | 83.1 | 88.5 | 85.8 | 78.5 | 69.1 | 56.3 | 49.2 | 67.2 |
| Minimum Temp °F | 21.5 | 23.6 | 29.2 | 37.2 | 45.2 | 54.2 | 60.4 | 58.6 | 49.4 | 39.6 | 28.7 | 22.3 | 39.2 |
| Mean Temp °F | 35.4 | 37.8 | 43.2 | 51.1 | 59.1 | 68.7 | 74.5 | 72.3 | 63.9 | 54.4 | 42.7 | 35.8 | 53.2 |
| Days Max Temp ≥ 90 °F | 0 | 0 | 0 | 0 | 1 | 7 | 15 | 9 | 2 | 0 | 0 | 0 | 34 |
| Days Max Temp ≤ 32 °F | 3 | 2 | 1 | 0 | 0 | 0 | 0 | 0 | 0 | 0 | 1 | 4 | 11 |
| Days Min Temp ≤ 32 °F | 26 | 23 | 20 | 8 | 1 | 0 | 0 | 0 | 1 | *6* | *19* | 25 | 129 |
| Days Min Temp ≤ 0 °F | 2 | 1 | 0 | 0 | 0 | 0 | 0 | 0 | 0 | 0 | *0* | 1 | 4 |
| Heating Degree Days | 910 | 763 | 669 | 413 | 201 | 43 | 3 | 10 | 102 | 329 | 662 | 900 | 5005 |
| Cooling Degree Days | 0 | 0 | 0 | 5 | 20 | 135 | 282 | 227 | 68 | 4 | 0 | 0 | 741 |
| Total Precipitation (") | 0.41 | 0.39 | 0.94 | 1.23 | 1.55 | 1.31 | 1.91 | 2.07 | 1.17 | 0.79 | 0.78 | 0.59 | 13.14 |
| Days ≥ 0.1" Precip | 1 | 1 | 3 | 3 | 4 | 4 | 5 | 5 | 3 | 2 | *2* | 2 | 35 |
| Total Snowfall (") | *5.3* | *5.9* | na | 3.6 | 0.5 | 0.0 | 0.0 | 0.0 | 0.1 | *2.4* | na | *8.5* | na |
| Days ≥ 1" Snow Depth | na | na | na | *1* | 0 | 0 | 0 | 0 | 0 | *0* | na | na | na |

## CASTLE ROCK *Douglas County* ELEVATION 6253 ft LAT/LONG 39° 22' N / 104° 52' W

| | JAN | FEB | MAR | APR | MAY | JUN | JUL | AUG | SEP | OCT | NOV | DEC | YEAR |
|---|---|---|---|---|---|---|---|---|---|---|---|---|---|
| Maximum Temp °F | 45.0 | 47.1 | 52.4 | 60.3 | 69.2 | 79.7 | 85.6 | 83.4 | 76.5 | 66.0 | 52.8 | 45.3 | 63.6 |
| Minimum Temp °F | 12.8 | 15.8 | 22.0 | 29.9 | 38.9 | 47.6 | 53.1 | 50.9 | 42.2 | 31.4 | 20.7 | 13.7 | 31.6 |
| Mean Temp °F | 29.0 | 31.5 | 37.1 | 45.1 | 54.1 | 63.7 | 69.4 | 67.2 | 59.3 | 48.7 | 36.8 | 29.5 | 47.6 |
| Days Max Temp ≥ 90 °F | 0 | 0 | 0 | 0 | 0 | 4 | 8 | 4 | 1 | 0 | 0 | 0 | 17 |
| Days Max Temp ≤ 32 °F | 5 | 3 | 2 | 0 | 0 | 0 | 0 | 0 | 0 | 0 | 2 | 5 | 17 |
| Days Min Temp ≤ 32 °F | 30 | 27 | 27 | 18 | 6 | 0 | 0 | 0 | *3* | 17 | 27 | 30 | 185 |
| Days Min Temp ≤ 0 °F | 5 | 2 | 1 | 0 | 0 | 0 | 0 | 0 | 0 | 0 | 1 | 4 | 13 |
| Heating Degree Days | 1111 | 939 | 856 | 589 | 333 | 95 | 12 | 26 | 185 | 494 | 839 | *1095* | 6574 |
| Cooling Degree Days | 0 | *0* | 0 | 1 | 5 | 72 | 148 | *109* | *24* | *0* | 0 | 0 | 359 |
| Total Precipitation (") | 0.56 | 0.55 | 1.45 | 1.55 | 2.52 | 2.17 | 2.53 | 1.95 | 1.22 | 1.28 | 0.96 | 0.69 | 17.43 |
| Days ≥ 0.1" Precip | *2* | 2 | 3 | 4 | 5 | 5 | 6 | 5 | 3 | 3 | *3* | *2* | 43 |
| Total Snowfall (") | *7.5* | *8.0* | *14.3* | *11.0* | 1.5 | 0.0 | 0.0 | 0.0 | 0.7 | 6.2 | na | *12.1* | na |
| Days ≥ 1" Snow Depth | na | na | na | na | 0 | 0 | 0 | 0 | 0 | na | na | na | na |

## CEDAREDGE *Delta County* ELEVATION 6184 ft LAT/LONG 38° 54' N / 107° 56' W

| | JAN | FEB | MAR | APR | MAY | JUN | JUL | AUG | SEP | OCT | NOV | DEC | YEAR |
|---|---|---|---|---|---|---|---|---|---|---|---|---|---|
| Maximum Temp °F | 38.6 | 44.4 | 53.2 | 62.1 | 72.2 | 82.9 | 88.1 | 85.7 | 77.5 | 65.2 | 49.4 | 39.6 | 63.2 |
| Minimum Temp °F | 14.9 | 20.2 | 26.8 | 33.0 | 40.8 | 49.3 | 55.6 | 53.8 | 45.8 | 35.5 | 25.3 | 16.9 | 34.8 |
| Mean Temp °F | 26.8 | 32.3 | 40.1 | 47.6 | 56.5 | 66.1 | 71.9 | 69.8 | 61.7 | 50.4 | 37.4 | 28.3 | 49.1 |
| Days Max Temp ≥ 90 °F | 0 | 0 | 0 | 0 | 0 | 6 | 13 | 8 | 1 | 0 | 0 | 0 | 28 |
| Days Max Temp ≤ 32 °F | 7 | 2 | 0 | 0 | 0 | 0 | 0 | 0 | 0 | 0 | 1 | 6 | 16 |
| Days Min Temp ≤ 32 °F | 30 | 27 | 25 | 14 | 4 | 0 | 0 | 0 | 1 | 10 | 24 | 30 | 165 |
| Days Min Temp ≤ 0 °F | 3 | 1 | 0 | 0 | 0 | 0 | 0 | 0 | 0 | 0 | 0 | 1 | 5 |
| Heating Degree Days | 1178 | 916 | 765 | 517 | 262 | 61 | 2 | 10 | 125 | 446 | 822 | 1131 | 6235 |
| Cooling Degree Days | 0 | 0 | 0 | 0 | 7 | 116 | 224 | 176 | 35 | 0 | 0 | 0 | 558 |
| Total Precipitation (") | 1.02 | 0.86 | 1.28 | 0.90 | 1.16 | 0.75 | 0.97 | 1.18 | 1.23 | 1.54 | 1.24 | 1.08 | 13.21 |
| Days ≥ 0.1" Precip | 3 | 3 | 4 | 3 | 4 | 2 | 3 | 4 | 3 | 4 | 4 | 4 | 41 |
| Total Snowfall (") | 12.9 | 8.6 | 8.4 | 1.5 | 0.5 | 0.0 | 0.0 | 0.0 | 0.0 | 1.9 | 5.2 | 10.6 | 49.6 |
| Days ≥ 1" Snow Depth | 17 | 9 | 1 | 0 | 0 | 0 | 0 | 0 | 0 | 0 | 2 | 11 | 40 |

## CENTER 4 SSW *Rio Grande County* ELEVATION 7674 ft LAT/LONG 37° 45' N / 106° 7' W

| | JAN | FEB | MAR | APR | MAY | JUN | JUL | AUG | SEP | OCT | NOV | DEC | YEAR |
|---|---|---|---|---|---|---|---|---|---|---|---|---|---|
| Maximum Temp °F | 31.5 | 38.5 | 48.2 | 58.3 | 66.5 | 75.3 | 79.1 | 77.2 | 71.9 | 61.5 | 45.7 | 34.1 | 57.3 |
| Minimum Temp °F | -3.6 | 3.7 | 16.8 | 24.5 | 33.3 | 39.7 | 45.2 | 43.2 | 36.0 | 25.7 | 13.0 | 0.7 | 23.2 |
| Mean Temp °F | 14.0 | 21.1 | 32.5 | 41.4 | 50.0 | 57.5 | 62.2 | 60.2 | 54.0 | 43.6 | 29.4 | 17.4 | 40.3 |
| Days Max Temp ≥ 90 °F | 0 | 0 | 0 | 0 | 0 | 0 | 0 | 0 | 0 | 0 | 0 | 0 | 0 |
| Days Max Temp ≤ 32 °F | 16 | 8 | 1 | 0 | 0 | 0 | 0 | 0 | 0 | 0 | 4 | 14 | 43 |
| Days Min Temp ≤ 32 °F | 31 | 28 | 31 | 27 | 13 | 2 | 0 | 0 | 7 | 26 | 30 | 31 | 226 |
| Days Min Temp ≤ 0 °F | 19 | 11 | 1 | 0 | 0 | 0 | 0 | 0 | 0 | 0 | 3 | 13 | 47 |
| Heating Degree Days | 1577 | 1234 | 999 | 701 | 460 | 219 | 91 | 145 | 324 | 656 | 1062 | 1469 | 8937 |
| Cooling Degree Days | 0 | 0 | 0 | 0 | 0 | 3 | 8 | 3 | 0 | 0 | 0 | 0 | 14 |
| Total Precipitation (") | 0.16 | 0.19 | 0.35 | 0.35 | 0.64 | 0.71 | 1.17 | 1.39 | 0.95 | 0.56 | 0.44 | 0.35 | 7.26 |
| Days ≥ 0.1" Precip | 0 | 1 | 1 | 1 | 2 | 2 | 3 | 4 | 3 | 2 | 1 | 1 | 21 |
| Total Snowfall (") | *2.9* | na | *4.8* | 1.2 | 0.4 | 0.0 | 0.0 | 0.0 | 0.0 | 1.6 | na | *6.2* | na |
| Days ≥ 1" Snow Depth | na | na | na | *0* | 0 | 0 | 0 | 0 | 0 | 0 | na | na | na |

## CHEESMAN *Jefferson County* ELEVATION 6893 ft LAT/LONG 39° 13' N / 105° 17' W

| | JAN | FEB | MAR | APR | MAY | JUN | JUL | AUG | SEP | OCT | NOV | DEC | YEAR |
|---|---|---|---|---|---|---|---|---|---|---|---|---|---|
| Maximum Temp °F | 45.5 | 47.7 | 51.5 | 58.5 | 67.3 | 78.3 | 83.5 | 81.3 | 74.8 | 65.2 | 52.1 | 45.7 | 62.6 |
| Minimum Temp °F | 8.2 | 9.6 | 17.4 | 25.2 | 33.0 | 41.4 | 46.2 | 44.7 | 37.0 | 26.4 | 17.3 | 10.1 | 26.4 |
| Mean Temp °F | 26.9 | 28.7 | 34.5 | 41.9 | 50.2 | 59.9 | 64.9 | 63.0 | 55.9 | 45.8 | 34.7 | 28.0 | 44.5 |
| Days Max Temp ≥ 90 °F | 0 | 0 | 0 | 0 | 0 | 2 | 5 | 2 | 0 | 0 | 0 | 0 | 9 |
| Days Max Temp ≤ 32 °F | 4 | 2 | 2 | 0 | 0 | 0 | 0 | 0 | 0 | 0 | 2 | 4 | 14 |
| Days Min Temp ≤ 32 °F | 30 | 28 | 30 | 25 | 14 | 2 | 0 | 0 | 8 | 25 | 29 | 30 | 221 |
| Days Min Temp ≤ 0 °F | 8 | 6 | 2 | 0 | 0 | 0 | 0 | 0 | 0 | 0 | 2 | 6 | 24 |
| Heating Degree Days | 1175 | 1020 | 938 | 688 | 452 | 163 | 46 | 83 | 268 | 588 | 902 | 1141 | 7464 |
| Cooling Degree Days | 0 | 0 | 0 | 0 | 1 | 19 | 52 | 29 | 2 | 0 | 0 | 0 | 103 |
| Total Precipitation (") | 0.38 | 0.58 | 1.40 | 1.60 | 2.03 | 1.96 | 2.80 | 2.61 | 1.18 | 1.30 | 0.88 | 0.70 | 17.42 |
| Days ≥ 0.1" Precip | 1 | 2 | 4 | 4 | 5 | 5 | 7 | 7 | 3 | 3 | 2 | 2 | 45 |
| Total Snowfall (") | 5.7 | 7.7 | *14.3* | *9.4* | 1.7 | 0.0 | 0.0 | 0.0 | 0.9 | 4.4 | *11.1* | 10.1 | 65.3 |
| Days ≥ 1" Snow Depth | 12 | 8 | 8 | 3 | 1 | 0 | 0 | 0 | 0 | 2 | 7 | 11 | 52 |

## CHERRY CREEK DAM *Arapahoe County* ELEVATION 5647 ft LAT/LONG 39° 39 ' N / 104° 51 ' W

| | JAN | FEB | MAR | APR | MAY | JUN | JUL | AUG | SEP | OCT | NOV | DEC | YEAR |
|---|---|---|---|---|---|---|---|---|---|---|---|---|---|
| Maximum Temp °F | 45.1 | 48.2 | 54.4 | 62.1 | 71.9 | 82.7 | 89.6 | 87.0 | 78.5 | 67.1 | ***53.4*** | 46.8 | 65.6 |
| Minimum Temp °F | 14.3 | 17.6 | 23.8 | 31.4 | 40.0 | 48.8 | 54.6 | 52.9 | 44.2 | 33.3 | ***22.9*** | 16.1 | 33.3 |
| Mean Temp °F | 29.7 | 32.9 | 39.1 | 46.8 | 56.0 | 65.8 | 72.2 | 70.0 | 61.4 | 50.2 | ***38.2*** | 31.5 | 49.5 |
| Days Max Temp ≥ 90 °F | 0 | 0 | 0 | 0 | 1 | 8 | 17 | 12 | 4 | 0 | 0 | 0 | 42 |
| Days Max Temp ≤ 32 °F | 5 | 3 | 2 | 0 | 0 | 0 | 0 | 0 | 0 | 1 | 2 | 4 | 17 |
| Days Min Temp ≤ 32 °F | 30 | 27 | 25 | 17 | 4 | 0 | 0 | 0 | 2 | 13 | ***25*** | 29 | 172 |
| Days Min Temp ≤ 0 °F | 4 | 2 | 1 | 0 | 0 | 0 | 0 | 0 | 0 | 0 | 1 | 3 | 11 |
| Heating Degree Days | 1088 | 899 | 795 | 541 | 285 | 74 | 8 | 15 | 151 | 453 | ***797*** | 1032 | 6138 |
| Cooling Degree Days | 0 | 0 | 0 | 1 | 13 | 120 | 244 | 191 | 59 | 2 | 0 | 0 | 630 |
| Total Precipitation (") | 0.52 | 0.51 | 1.43 | 1.79 | 2.66 | 2.28 | 2.48 | 1.93 | 1.36 | 1.26 | 1.24 | 0.55 | 18.01 |
| Days ≥ 0.1" Precip | 2 | 2 | ***4*** | 4 | 6 | 5 | 5 | 4 | 3 | 3 | ***3*** | 2 | 43 |
| Total Snowfall (") | ***7.3*** | ***7.5*** | ***10.7*** | 6.8 | 0.8 | 0.0 | 0.0 | 0.0 | 1.5 | 3.9 | ***10.0*** | ***7.4*** | 55.9 |
| Days ≥ 1" Snow Depth | 15 | 10 | ***5*** | 3 | 0 | 0 | 0 | 0 | 0 | ***2*** | ***8*** | ***14*** | 57 |

## CHEYENNE WELLS *Cheyenne County* ELEVATION 4250 ft LAT/LONG 38° 49 ' N / 102° 21 ' W

| | JAN | FEB | MAR | APR | MAY | JUN | JUL | AUG | SEP | OCT | NOV | DEC | YEAR |
|---|---|---|---|---|---|---|---|---|---|---|---|---|---|
| Maximum Temp °F | 42.7 | 47.7 | 55.7 | 65.8 | 74.2 | 85.2 | 90.8 | 88.0 | 80.2 | 69.0 | 52.9 | 43.9 | 66.3 |
| Minimum Temp °F | 15.7 | 19.2 | 25.9 | 34.8 | 44.7 | 53.8 | 59.1 | 57.6 | 48.7 | 37.1 | 25.0 | 17.3 | 36.6 |
| Mean Temp °F | 29.4 | 33.4 | 40.8 | 50.3 | 59.5 | 69.5 | 74.9 | 72.8 | 64.5 | 53.1 | 38.9 | 30.7 | 51.5 |
| Days Max Temp ≥ 90 °F | 0 | 0 | 0 | 0 | 1 | 10 | 19 | 15 | 6 | 1 | 0 | 0 | 52 |
| Days Max Temp ≤ 32 °F | 7 | 4 | 2 | 0 | 0 | 0 | 0 | 0 | 0 | 0 | 2 | 5 | 20 |
| Days Min Temp ≤ 32 °F | 31 | 27 | 25 | 11 | 2 | 0 | 0 | 0 | 1 | 9 | 24 | 30 | 160 |
| Days Min Temp ≤ 0 °F | 3 | 2 | 0 | 0 | 0 | 0 | 0 | 0 | 0 | 0 | 0 | 2 | 7 |
| Heating Degree Days | 1098 | 884 | 743 | 437 | 196 | 33 | 2 | 7 | 101 | 369 | 775 | 1058 | 5703 |
| Cooling Degree Days | 0 | 0 | 0 | 6 | 28 | 179 | 316 | 277 | 105 | 7 | 0 | 0 | 918 |
| Total Precipitation (") | 0.23 | 0.28 | 0.78 | 1.05 | 2.76 | 2.48 | 2.62 | 2.33 | 1.65 | 0.94 | 0.57 | 0.26 | 15.95 |
| Days ≥ 0.1" Precip | 1 | 1 | 3 | 3 | 6 | 5 | 5 | 4 | 3 | 2 | 2 | 1 | 36 |
| Total Snowfall (") | 3.7 | 2.8 | 4.4 | 2.7 | 0.8 | 0.0 | 0.0 | 0.0 | 0.1 | 2.1 | 3.3 | 3.7 | 23.6 |
| Days ≥ 1" Snow Depth | 8 | 4 | 3 | 1 | 0 | 0 | 0 | 0 | 0 | 1 | 3 | 6 | 26 |

## CIMARRON *Montrose County* ELEVATION 7244 ft LAT/LONG 38° 24 ' N / 107° 31 ' W

| | JAN | FEB | MAR | APR | MAY | JUN | JUL | AUG | SEP | OCT | NOV | DEC | YEAR |
|---|---|---|---|---|---|---|---|---|---|---|---|---|---|
| Maximum Temp °F | 32.7 | 37.6 | 47.0 | 57.2 | 67.7 | 79.6 | 84.4 | 82.6 | 74.7 | 63.3 | 46.3 | 34.8 | 59.0 |
| Minimum Temp °F | 0.0 | 5.4 | 17.4 | 24.0 | 31.0 | 36.6 | 43.7 | 43.0 | 33.9 | 24.2 | 15.6 | 4.0 | 23.2 |
| Mean Temp °F | 16.4 | 21.5 | 32.2 | 40.6 | 49.4 | 58.1 | 64.1 | 62.8 | 54.3 | 43.8 | 31.0 | 19.4 | 41.1 |
| Days Max Temp ≥ 90 °F | 0 | 0 | 0 | 0 | 0 | 1 | 4 | 2 | 0 | 0 | 0 | 0 | 7 |
| Days Max Temp ≤ 32 °F | 15 | 8 | 1 | 0 | 0 | 0 | 0 | 0 | 0 | 0 | 3 | 12 | 39 |
| Days Min Temp ≤ 32 °F | 31 | 28 | 30 | 27 | 20 | 7 | 1 | 2 | 14 | 27 | 28 | 31 | 246 |
| Days Min Temp ≤ 0 °F | 17 | 10 | 2 | 0 | 0 | 0 | 0 | 0 | 0 | 0 | 2 | 13 | 44 |
| Heating Degree Days | 1502 | 1223 | 1009 | 728 | 477 | 205 | 51 | 82 | 315 | 652 | 1014 | 1407 | 8665 |
| Cooling Degree Days | 0 | 0 | 0 | 0 | 0 | 7 | 31 | 25 | 1 | 0 | 0 | 0 | 64 |
| Total Precipitation (") | 0.88 | 0.75 | 1.02 | 0.97 | 1.11 | 0.82 | 1.43 | 1.40 | 1.30 | 1.43 | 1.23 | 0.72 | 13.06 |
| Days ≥ 0.1" Precip | 3 | 3 | 3 | 3 | 4 | 3 | 4 | 4 | 4 | 4 | 4 | 3 | 42 |
| Total Snowfall (") | 13.8 | ***11.6*** | ***11.8*** | ***4.5*** | 0.3 | 0.0 | 0.0 | 0.0 | 0.0 | na | na | ***13.3*** | na |
| Days ≥ 1" Snow Depth | na | na | na | na | 0 | 0 | 0 | 0 | 0 | ***0*** | na | na | na |

## CLIMAX *Lake County* ELEVATION 11529 ft LAT/LONG 39° 23 ' N / 106° 12 ' W

| | JAN | FEB | MAR | APR | MAY | JUN | JUL | AUG | SEP | OCT | NOV | DEC | YEAR |
|---|---|---|---|---|---|---|---|---|---|---|---|---|---|
| Maximum Temp °F | 24.9 | 27.6 | 32.3 | 38.6 | 47.3 | 58.0 | 64.5 | 62.4 | 55.9 | 45.1 | 32.1 | 26.2 | 42.9 |
| Minimum Temp °F | 1.5 | 2.6 | 6.5 | 13.3 | 23.6 | 32.9 | 38.7 | 37.4 | 30.7 | 21.0 | 9.0 | 3.0 | 18.3 |
| Mean Temp °F | 13.2 | 15.1 | 19.5 | 26.0 | 35.4 | 45.6 | 51.6 | 49.9 | 43.3 | 33.1 | 20.6 | 14.6 | 30.7 |
| Days Max Temp ≥ 90 °F | 0 | 0 | 0 | 0 | 0 | 0 | 0 | 0 | 0 | 0 | 0 | 0 | 0 |
| Days Max Temp ≤ 32 °F | 24 | 19 | 15 | 8 | 1 | 0 | 0 | 0 | 0 | 3 | 15 | 22 | 107 |
| Days Min Temp ≤ 32 °F | 31 | 28 | 31 | 30 | 29 | 14 | 2 | 3 | 17 | 30 | 30 | 31 | 276 |
| Days Min Temp ≤ 0 °F | 13 | 11 | 8 | 3 | 0 | 0 | 0 | 0 | 0 | 1 | 6 | 12 | 54 |
| Heating Degree Days | 1601 | 1402 | 1405 | 1165 | 910 | 576 | 410 | 461 | 643 | 982 | 1327 | 1556 | 12438 |
| Cooling Degree Days | 0 | 0 | 0 | 0 | 0 | 0 | 0 | 0 | 0 | 0 | 0 | 0 | 0 |
| Total Precipitation (") | 1.78 | 1.67 | 2.27 | 2.21 | 1.91 | 1.39 | 2.26 | 2.12 | 1.46 | 1.43 | 1.98 | 1.86 | 22.34 |
| Days ≥ 0.1" Precip | 6 | 6 | 8 | 8 | 6 | 5 | 7 | 7 | 5 | 5 | 7 | 6 | 76 |
| Total Snowfall (") | 34.4 | 32.2 | 41.0 | 36.4 | 19.5 | 5.3 | 0.1 | 0.1 | 4.9 | 18.9 | 35.7 | 35.3 | 263.8 |
| Days ≥ 1" Snow Depth | 31 | 28 | 31 | 30 | 26 | 3 | 0 | 0 | 1 | 10 | 26 | 31 | 217 |

## COCHETOPA CREEK *Gunnison County* ELEVATION 8000 ft LAT/LONG 38° 26' N / 106° 46' W

| | JAN | FEB | MAR | APR | MAY | JUN | JUL | AUG | SEP | OCT | NOV | DEC | YEAR |
|---|---|---|---|---|---|---|---|---|---|---|---|---|---|
| Maximum Temp °F | 27.3 | 33.0 | 43.2 | 54.1 | 65.0 | 75.3 | 80.5 | 78.6 | 71.5 | 60.2 | 43.9 | 31.3 | 55.3 |
| Minimum Temp °F | -6.1 | -1.0 | 12.3 | 20.8 | 28.2 | 35.1 | 41.9 | 40.8 | 32.0 | 21.1 | 11.1 | -1.3 | 19.6 |
| Mean Temp °F | 10.6 | 16.0 | 27.8 | 37.4 | 46.6 | 55.2 | 61.2 | 59.7 | 51.8 | 40.7 | 27.5 | 15.0 | 37.5 |
| Days Max Temp ≥ 90 °F | 0 | 0 | 0 | 0 | 0 | 0 | 1 | 0 | 0 | 0 | 0 | 0 | 1 |
| Days Max Temp ≤ 32 °F | 21 | 13 | 4 | 0 | 0 | 0 | 0 | 0 | 0 | 0 | 4 | 16 | 58 |
| Days Min Temp ≤ 32 °F | 31 | 28 | 31 | 29 | 24 | 10 | 1 | 2 | 16 | 29 | 30 | 31 | 262 |
| Days Min Temp ≤ 0 °F | 22 | 15 | 5 | 0 | 0 | 0 | 0 | 0 | 0 | 0 | 4 | 18 | 64 |
| Heating Degree Days | 1683 | 1379 | 1147 | 820 | 562 | 288 | 118 | 162 | 390 | 748 | 1120 | 1544 | 9961 |
| Cooling Degree Days | 0 | 0 | 0 | 0 | 0 | 2 | 12 | 8 | 1 | 0 | 0 | 0 | 23 |
| Total Precipitation (") | 0.68 | 0.65 | 0.79 | 0.81 | 0.86 | 0.78 | 1.69 | 1.88 | 1.14 | 0.88 | 0.74 | 0.81 | 11.71 |
| Days ≥ 0.1" Precip | 2 | 2 | 4 | 3 | 3 | 3 | 5 | 6 | 3 | 3 | 3 | 3 | 40 |
| Total Snowfall (") | 9.6 | 8.7 | 8.0 | 3.9 | 0.8 | 0.0 | 0.0 | 0.0 | 0.0 | 2.2 | 7.5 | 10.6 | 51.3 |
| Days ≥ 1" Snow Depth | 17 | 17 | 16 | 3 | 1 | 0 | 0 | 0 | 0 | 1 | 7 | 15 | 77 |

## COLLBRAN 1 W *Mesa County* ELEVATION 6145 ft LAT/LONG 39° 14' N / 107° 58' W

| | JAN | FEB | MAR | APR | MAY | JUN | JUL | AUG | SEP | OCT | NOV | DEC | YEAR |
|---|---|---|---|---|---|---|---|---|---|---|---|---|---|
| Maximum Temp °F | 36.5 | 42.7 | 50.9 | 60.5 | 70.5 | 81.0 | 87.0 | 84.8 | 76.4 | 64.8 | 48.2 | 38.1 | 61.8 |
| Minimum Temp °F | 7.2 | 13.9 | 22.9 | 30.0 | 37.9 | 45.5 | 51.2 | 49.3 | 41.0 | 30.8 | 20.6 | 11.2 | 30.1 |
| Mean Temp °F | 21.9 | 28.3 | 36.9 | 45.2 | 54.2 | 63.3 | 69.1 | 67.1 | 58.7 | 47.8 | 34.4 | 24.7 | 46.0 |
| Days Max Temp ≥ 90 °F | 0 | 0 | 0 | 0 | 0 | 4 | 11 | 6 | 1 | 0 | 0 | 0 | 22 |
| Days Max Temp ≤ 32 °F | 10 | 3 | 1 | 0 | 0 | 0 | 0 | 0 | 0 | 0 | 2 | 7 | 23 |
| Days Min Temp ≤ 32 °F | 31 | 28 | 28 | 18 | 7 | 1 | 0 | 0 | 4 | 19 | 28 | 31 | 195 |
| Days Min Temp ≤ 0 °F | 9 | 3 | 1 | 0 | 0 | 0 | 0 | 0 | 0 | 0 | 0 | 4 | 17 |
| Heating Degree Days | 1331 | 1032 | 865 | 587 | 329 | 98 | 10 | 30 | 196 | 525 | 911 | 1243 | 7157 |
| Cooling Degree Days | 0 | 0 | 0 | 0 | 2 | 60 | 151 | 116 | 16 | 0 | 0 | 0 | 345 |
| Total Precipitation (") | 0.77 | 0.77 | 1.60 | 1.33 | 1.39 | 0.79 | 1.07 | 1.11 | 1.19 | 1.57 | 1.29 | 1.06 | 13.94 |
| Days ≥ 0.1" Precip | 3 | 3 | 5 | 4 | 4 | 2 | 4 | 3 | 4 | 4 | 4 | 3 | 43 |
| Total Snowfall (") | 13.3 | 8.4 | 10.0 | na | 1.0 | 0.0 | 0.0 | 0.0 | 0.0 | 2.0 | na | 13.5 | na |
| Days ≥ 1" Snow Depth | na | 13 | na | na | 0 | 0 | 0 | 0 | 0 | 1 | na | na | na |

## COLORADO NATL MONUME *Mesa County* ELEVATION 5784 ft LAT/LONG 39° 6' N / 108° 44' W

| | JAN | FEB | MAR | APR | MAY | JUN | JUL | AUG | SEP | OCT | NOV | DEC | YEAR |
|---|---|---|---|---|---|---|---|---|---|---|---|---|---|
| Maximum Temp °F | 36.4 | 43.6 | 53.3 | 63.5 | 73.9 | 85.3 | 91.1 | 88.7 | 79.5 | 65.3 | 48.5 | 38.2 | 63.9 |
| Minimum Temp °F | 17.6 | 23.2 | 30.0 | 36.9 | 46.4 | 55.9 | 62.1 | 59.7 | 51.5 | 40.6 | 28.8 | 20.1 | 39.4 |
| Mean Temp °F | 27.0 | 33.4 | 41.7 | 50.3 | 60.1 | 70.7 | 76.6 | 74.2 | 65.5 | 53.0 | 38.7 | 29.2 | 51.7 |
| Days Max Temp ≥ 90 °F | 0 | 0 | 0 | 0 | 0 | 10 | 21 | 15 | 2 | 0 | 0 | 0 | 48 |
| Days Max Temp ≤ 32 °F | 10 | 2 | 0 | 0 | 0 | 0 | 0 | 0 | 0 | 0 | 1 | 7 | 20 |
| Days Min Temp ≤ 32 °F | 30 | 25 | 20 | 9 | 2 | 0 | 0 | 0 | 0 | 5 | 20 | 30 | 141 |
| Days Min Temp ≤ 0 °F | 1 | 0 | 0 | 0 | 0 | 0 | 0 | 0 | 0 | 0 | 0 | 0 | 1 |
| Heating Degree Days | 1170 | 885 | 716 | 439 | 181 | 30 | 1 | 4 | 71 | 369 | 783 | 1104 | 5753 |
| Cooling Degree Days | 0 | 0 | 0 | 5 | 34 | 215 | 354 | 286 | 94 | 2 | 0 | 0 | 990 |
| Total Precipitation (") | 0.76 | 0.64 | 1.11 | 0.91 | 1.12 | 0.68 | 0.91 | 1.13 | 0.86 | 1.34 | 0.97 | 0.96 | 11.39 |
| Days ≥ 0.1" Precip | 3 | 2 | 4 | 3 | 3 | 2 | 3 | 3 | 3 | 3 | 3 | 3 | 35 |
| Total Snowfall (") | 7.7 | 5.3 | 4.2 | 1.3 | 0.2 | 0.0 | 0.0 | 0.0 | 0.2 | 1.2 | 3.4 | 8.3 | 31.8 |
| Days ≥ 1" Snow Depth | 19 | 10 | 3 | 0 | 0 | 0 | 0 | 0 | 0 | 1 | 2 | 12 | 47 |

## COLORADO SPRGS MUNI *El Paso County* ELEVATION 9199 ft LAT/LONG 38° 49' N / 104° 42' W

| | JAN | FEB | MAR | APR | MAY | JUN | JUL | AUG | SEP | OCT | NOV | DEC | YEAR |
|---|---|---|---|---|---|---|---|---|---|---|---|---|---|
| Maximum Temp °F | 41.6 | 44.8 | 50.7 | 59.9 | 68.4 | 79.2 | 84.4 | 81.3 | 73.9 | 63.0 | 49.9 | 42.3 | 61.6 |
| Minimum Temp °F | 16.8 | 19.6 | 25.2 | 33.3 | 42.3 | 51.4 | 57.0 | 55.5 | 47.2 | 36.3 | 24.7 | 17.8 | 35.6 |
| Mean Temp °F | 29.3 | 32.2 | 38.0 | 46.6 | 55.3 | 65.3 | 70.7 | 68.4 | 60.6 | 49.7 | 37.3 | 30.1 | 48.6 |
| Days Max Temp ≥ 90 °F | 0 | 0 | 0 | 0 | 0 | 4 | 8 | 3 | 1 | 0 | 0 | 0 | 16 |
| Days Max Temp ≤ 32 °F | 7 | 5 | 3 | 1 | 0 | 0 | 0 | 0 | 0 | 0 | 3 | 7 | 26 |
| Days Min Temp ≤ 32 °F | 30 | 27 | 25 | 13 | 2 | 0 | 0 | 0 | 1 | 9 | 24 | 29 | 160 |
| Days Min Temp ≤ 0 °F | 2 | 1 | 0 | 0 | 0 | 0 | 0 | 0 | 0 | 0 | 0 | 2 | 5 |
| Heating Degree Days | 1101 | 918 | 830 | 545 | 298 | 73 | 9 | 22 | 159 | 469 | 823 | 1076 | 6323 |
| Cooling Degree Days | 0 | 0 | 0 | 1 | 6 | 96 | 194 | 138 | 35 | 1 | 0 | 0 | 471 |
| Total Precipitation (") | 0.28 | 0.37 | 0.95 | 1.33 | 2.28 | 2.36 | 2.93 | 3.22 | 1.27 | 0.92 | 0.54 | 0.45 | 16.90 |
| Days ≥ 0.1" Precip | 1 | 1 | 3 | 4 | 5 | 4 | 7 | 6 | 4 | 3 | 2 | 1 | 41 |
| Total Snowfall (") | 5.3 | 5.5 | 9.5 | 5.5 | 1.4 | 0.0 | 0.0 | 0.0 | 0.5 | 3.6 | 6.1 | 7.1 | 44.5 |
| Days ≥ 1" Snow Depth | 8 | 5 | 5 | 2 | 0 | 0 | 0 | 0 | 0 | 1 | 4 | 9 | 34 |

## CORTEZ *Montezuma County* ELEVATION 6184 ft LAT/LONG 37° 21 ' N / 108° 34 ' W

| | JAN | FEB | MAR | APR | MAY | JUN | JUL | AUG | SEP | OCT | NOV | DEC | YEAR |
|---|---|---|---|---|---|---|---|---|---|---|---|---|---|
| Maximum Temp °F | 39.7 | 44.8 | 52.6 | 61.6 | 71.5 | 82.5 | 87.8 | 85.3 | 77.6 | 65.8 | 51.2 | 41.3 | 63.5 |
| Minimum Temp °F | 11.9 | 17.7 | 24.8 | 29.9 | 38.0 | 46.0 | ***54.0*** | 52.8 | 44.0 | 33.4 | 23.2 | 14.7 | 32.5 |
| Mean Temp °F | 25.8 | 31.3 | 38.7 | 45.7 | 54.8 | 64.2 | ***70.9*** | 69.1 | 60.9 | 49.6 | 37.3 | 28.1 | 48.0 |
| Days Max Temp ≥ 90 °F | 0 | 0 | 0 | 0 | 0 | 5 | 12 | 7 | 0 | 0 | 0 | 0 | 24 |
| Days Max Temp ≤ 32 °F | 5 | 2 | 0 | 0 | 0 | 0 | 0 | 0 | 0 | 0 | 1 | 4 | 12 |
| Days Min Temp ≤ 32 °F | 30 | 27 | 27 | 20 | 6 | 0 | 0 | 0 | 1 | ***14*** | 26 | 30 | 181 |
| Days Min Temp ≤ 0 °F | 5 | 2 | 0 | 0 | 0 | 0 | 0 | 0 | 0 | 0 | 0 | 2 | 9 |
| Heating Degree Days | 1207 | 945 | 807 | 571 | 313 | 80 | ***3*** | 11 | 139 | 471 | 826 | 1138 | 6511 |
| Cooling Degree Days | 0 | 0 | 0 | 0 | 3 | 63 | ***169*** | 141 | 21 | 0 | 0 | 0 | 397 |
| Total Precipitation (") | 0.94 | 0.94 | 1.30 | 0.87 | 1.02 | 0.48 | 1.23 | 1.43 | 1.26 | 1.55 | 1.19 | 1.19 | 13.40 |
| Days ≥ 0.1" Precip | 4 | 3 | 4 | 3 | 3 | 2 | 4 | 4 | 3 | 4 | 3 | 4 | 41 |
| Total Snowfall (") | na | na | na | na | 0.0 | 0.0 | 0.0 | 0.0 | 0.1 | ***0.2*** | na | na | na |
| Days ≥ 1" Snow Depth | na | na | na | ***0*** | 0 | 0 | 0 | 0 | 0 | ***0*** | na | na | na |

## CRESTED BUTTE *Gunnison County* ELEVATION 8875 ft LAT/LONG 38° 52 ' N / 106° 58 ' W

| | JAN | FEB | MAR | APR | MAY | JUN | JUL | AUG | SEP | OCT | NOV | DEC | YEAR |
|---|---|---|---|---|---|---|---|---|---|---|---|---|---|
| Maximum Temp °F | 26.9 | 31.3 | 38.6 | 46.8 | 58.1 | 69.1 | 74.6 | 73.0 | 65.4 | 54.6 | 38.7 | 28.5 | 50.5 |
| Minimum Temp °F | -5.9 | -2.3 | 7.6 | 17.7 | 27.9 | 33.4 | 38.7 | 37.8 | 30.6 | 22.0 | 9.5 | -3.0 | 17.8 |
| Mean Temp °F | 10.5 | 14.5 | 23.1 | 32.2 | 43.0 | 51.3 | 56.7 | 55.4 | 48.0 | 38.3 | 24.1 | 12.8 | 34.2 |
| Days Max Temp ≥ 90 °F | 0 | 0 | 0 | 0 | 0 | 0 | 0 | 0 | 0 | 0 | 0 | 0 | 0 |
| Days Max Temp ≤ 32 °F | 24 | 15 | 5 | 1 | 0 | 0 | 0 | 0 | 0 | 1 | 8 | 21 | 75 |
| Days Min Temp ≤ 32 °F | 31 | 28 | 31 | 29 | 25 | 13 | 3 | 5 | 19 | 28 | 30 | 31 | 273 |
| Days Min Temp ≤ 0 °F | 21 | 16 | 9 | 2 | 0 | 0 | 0 | 0 | 0 | 0 | 7 | 20 | 75 |
| Heating Degree Days | 1686 | 1421 | 1291 | 977 | 675 | 404 | 252 | 290 | 503 | 821 | 1220 | 1614 | 11154 |
| Cooling Degree Days | 0 | 0 | 0 | 0 | 0 | 0 | 0 | 0 | 0 | 0 | 0 | 0 | 0 |
| Total Precipitation (") | 2.37 | 2.19 | 2.32 | 1.68 | 1.47 | 1.27 | 1.96 | 2.01 | 2.02 | 1.72 | 2.18 | 2.45 | 23.64 |
| Days ≥ 0.1" Precip | 7 | 7 | 7 | 6 | 5 | 4 | 6 | 6 | 6 | 5 | 6 | 7 | 72 |
| Total Snowfall (") | 40.6 | 38.5 | 35.3 | 18.7 | 6.8 | 0.9 | 0.0 | 0.0 | 0.9 | 9.9 | 32.2 | 40.3 | 224.1 |
| Days ≥ 1" Snow Depth | 31 | 28 | 30 | 20 | 3 | 0 | 0 | 0 | 0 | 5 | 22 | 29 | 168 |

## DEL NORTE *Rio Grande County* ELEVATION 7884 ft LAT/LONG 37° 40 ' N / 106° 21 ' W

| | JAN | FEB | MAR | APR | MAY | JUN | JUL | AUG | SEP | OCT | NOV | DEC | YEAR |
|---|---|---|---|---|---|---|---|---|---|---|---|---|---|
| Maximum Temp °F | 34.3 | 40.1 | 49.4 | 58.9 | 67.2 | 75.4 | 78.3 | 76.2 | 70.7 | 62.0 | 47.2 | 36.6 | 58.0 |
| Minimum Temp °F | 5.9 | 11.7 | 20.7 | 27.4 | 35.4 | 42.6 | 48.5 | 47.1 | 39.9 | 30.6 | 18.8 | 8.7 | 28.1 |
| Mean Temp °F | 20.1 | 25.9 | 35.1 | 43.1 | 51.3 | 59.0 | 63.4 | 61.7 | 55.3 | 46.3 | 33.1 | 22.7 | 43.1 |
| Days Max Temp ≥ 90 °F | 0 | 0 | 0 | 0 | 0 | 0 | 0 | 0 | 0 | 0 | 0 | 0 | 0 |
| Days Max Temp ≤ 32 °F | 13 | 6 | 1 | 0 | 0 | 0 | 0 | 0 | 0 | 0 | 3 | 10 | 33 |
| Days Min Temp ≤ 32 °F | 31 | 28 | 30 | 23 | 9 | 1 | 0 | 0 | 2 | 19 | 29 | 31 | 203 |
| Days Min Temp ≤ 0 °F | 9 | 4 | 1 | 0 | 0 | 0 | 0 | 0 | 0 | 0 | 1 | 6 | 21 |
| Heating Degree Days | 1384 | 1098 | 921 | 649 | 417 | 180 | 63 | 106 | 285 | 573 | 951 | 1305 | 7932 |
| Cooling Degree Days | 0 | 0 | 0 | 0 | 0 | 10 | 27 | 13 | 0 | 0 | 0 | 0 | 50 |
| Total Precipitation (") | 0.32 | 0.36 | 0.75 | 0.63 | 0.96 | 0.78 | 1.65 | 1.91 | 1.14 | 0.82 | 0.61 | 0.61 | 10.54 |
| Days ≥ 0.1" Precip | 1 | 1 | 2 | 2 | 3 | 3 | 5 | 6 | 4 | 3 | 2 | 2 | 34 |
| Total Snowfall (") | 5.6 | 6.2 | 8.9 | 4.4 | 1.6 | 0.0 | 0.0 | 0.0 | 0.0 | 3.9 | 6.1 | 9.8 | 46.5 |
| Days ≥ 1" Snow Depth | na | na | na | na | 0 | 0 | 0 | 0 | 0 | 1 | na | na | na |

## DELTA *Delta County* ELEVATION 5125 ft LAT/LONG 38° 45 ' N / 108° 4 ' W

| | JAN | FEB | MAR | APR | MAY | JUN | JUL | AUG | SEP | OCT | NOV | DEC | YEAR |
|---|---|---|---|---|---|---|---|---|---|---|---|---|---|
| Maximum Temp °F | 39.4 | 48.1 | 58.0 | 67.4 | 76.6 | ***87.9*** | ***92.2*** | ***89.5*** | 81.5 | ***69.1*** | ***52.8*** | 42.1 | 67.1 |
| Minimum Temp °F | 13.1 | 19.0 | 26.5 | 33.4 | 41.7 | ***48.8*** | ***54.7*** | ***52.6*** | 44.0 | ***33.8*** | ***24.1*** | 15.5 | 33.9 |
| Mean Temp °F | 26.3 | 33.6 | 42.2 | 50.4 | 59.2 | ***68.5*** | ***73.4*** | ***71.1*** | 62.7 | ***51.5*** | ***38.5*** | 28.8 | 50.5 |
| Days Max Temp ≥ 90 °F | 0 | 0 | 0 | 0 | 1 | 13 | ***22*** | ***17*** | 4 | 0 | 0 | 0 | 57 |
| Days Max Temp ≤ 32 °F | 8 | 2 | 0 | 0 | 0 | 0 | ***0*** | ***0*** | 0 | ***0*** | 0 | 5 | 15 |
| Days Min Temp ≤ 32 °F | 30 | 26 | 24 | 14 | 3 | 0 | ***0*** | ***0*** | 2 | ***14*** | 25 | 30 | 168 |
| Days Min Temp ≤ 0 °F | 4 | 1 | 0 | 0 | 0 | 0 | ***0*** | ***0*** | 0 | ***0*** | 0 | 1 | 6 |
| Heating Degree Days | 1193 | 881 | 698 | 432 | 188 | ***27*** | ***1*** | ***8*** | 105 | ***414*** | ***790*** | 1115 | 5852 |
| Cooling Degree Days | 0 | 0 | 0 | 0 | 14 | ***158*** | 265 | ***210*** | 53 | 1 | ***0*** | 0 | 701 |
| Total Precipitation (") | 0.39 | ***0.36*** | 0.60 | 0.45 | 0.67 | ***0.47*** | ***0.70*** | ***0.86*** | 0.92 | ***1.06*** | 0.67 | 0.48 | 7.63 |
| Days ≥ 0.1" Precip | ***1*** | ***1*** | 2 | 1 | 2 | 1 | ***3*** | ***3*** | 3 | ***3*** | 2 | ***2*** | 24 |
| Total Snowfall (") | na | na | na | 0.1 | 0.0 | 0.0 | ***0.0*** | ***0.0*** | 0.0 | ***0.2*** | na | ***2.7*** | na |
| Days ≥ 1" Snow Depth | na | na | na | 0 | 0 | 0 | ***0*** | ***0*** | 0 | ***0*** | na | na | na |

## DENVER STAPLETON AP *Denver County* ELEVATION 5298 ft LAT/LONG 39° 46 ' N / 104° 53 ' W

| | JAN | FEB | MAR | APR | MAY | JUN | JUL | AUG | SEP | OCT | NOV | DEC | YEAR |
|---|---|---|---|---|---|---|---|---|---|---|---|---|---|
| Maximum Temp °F | 43.8 | 47.0 | 53.3 | 61.9 | 70.9 | 81.7 | 87.9 | 85.7 | 77.1 | 65.8 | 51.8 | 44.6 | 64.3 |
| Minimum Temp °F | 16.9 | 20.7 | 26.7 | 34.9 | 44.0 | 52.9 | 58.7 | 57.1 | 47.9 | 36.4 | 25.1 | 17.7 | 36.6 |
| Mean Temp °F | 30.4 | 33.9 | 40.0 | 48.4 | 57.5 | 67.3 | 73.3 | 71.5 | 62.5 | 51.1 | 38.5 | 31.2 | 50.5 |
| Days Max Temp ≥ 90 °F | 0 | 0 | 0 | 0 | 0 | 7 | 14 | 10 | 2 | 0 | 0 | 0 | 33 |
| Days Max Temp ≤ 32 °F | 6 | 4 | 3 | 0 | 0 | 0 | 0 | 0 | 0 | 0 | 2 | 5 | 20 |
| Days Min Temp ≤ 32 °F | 30 | 26 | 24 | 11 | 2 | 0 | 0 | 0 | 1 | 9 | 24 | 29 | 156 |
| Days Min Temp ≤ 0 °F | 3 | 2 | 0 | 0 | 0 | 0 | 0 | 0 | 0 | 0 | 0 | 3 | 8 |
| Heating Degree Days | 1067 | 872 | 768 | 493 | 243 | 54 | 5 | 9 | 129 | 426 | 789 | 1042 | 5897 |
| Cooling Degree Days | 0 | 0 | 0 | 3 | 21 | 153 | 280 | 233 | 72 | 1 | 0 | 0 | 763 |
| Total Precipitation (") | 0.51 | 0.54 | 1.29 | 1.81 | 2.32 | 1.79 | 2.05 | 1.61 | 1.14 | 1.10 | 1.00 | 0.65 | 15.81 |
| Days ≥ 0.1" Precip | 1 | 2 | 4 | 4 | 5 | 4 | 4 | 4 | 3 | 3 | 3 | 2 | 39 |
| Total Snowfall (") | 7.7 | 6.9 | 11.9 | 7.9 | 1.5 | 0.0 | 0.0 | 0.0 | 1.7 | 4.5 | 10.6 | 8.9 | 61.6 |
| Days ≥ 1" Snow Depth | 13 | 8 | 6 | 3 | 0 | 0 | 0 | 0 | 0 | 2 | 8 | 11 | 51 |

## DILLON 1 E *Summit County* ELEVATION 8899 ft LAT/LONG 39° 36 ' N / 106° 3 ' W

| | JAN | FEB | MAR | APR | MAY | JUN | JUL | AUG | SEP | OCT | NOV | DEC | YEAR |
|---|---|---|---|---|---|---|---|---|---|---|---|---|---|
| Maximum Temp °F | 31.2 | 33.9 | 38.6 | 46.9 | 57.0 | 67.7 | 73.6 | 71.7 | 64.5 | 54.2 | 39.6 | 32.3 | 50.9 |
| Minimum Temp °F | 1.2 | 3.4 | 10.8 | 18.5 | 26.9 | 33.3 | 38.4 | 37.4 | 30.5 | 21.8 | 12.4 | 4.1 | 19.9 |
| Mean Temp °F | 16.3 | 18.7 | 24.7 | 32.7 | 42.0 | 50.5 | 56.1 | 54.6 | 47.5 | 38.0 | 26.0 | 18.2 | 35.4 |
| Days Max Temp ≥ 90 °F | 0 | 0 | 0 | 0 | 0 | 0 | 0 | 0 | 0 | 0 | 0 | 0 | 0 |
| Days Max Temp ≤ 32 °F | 16 | 11 | 8 | 3 | 0 | 0 | 0 | 0 | 0 | 1 | 8 | 15 | 62 |
| Days Min Temp ≤ 32 °F | 31 | 28 | 31 | 30 | 27 | 13 | 2 | 5 | 19 | 30 | 30 | 31 | 277 |
| Days Min Temp ≤ 0 °F | 15 | 11 | 5 | 1 | 0 | 0 | 0 | 0 | 0 | 0 | 3 | 11 | 46 |
| Heating Degree Days | 1505 | 1301 | 1242 | 962 | 708 | 428 | 271 | 317 | 517 | 829 | 1162 | 1444 | 10686 |
| Cooling Degree Days | 0 | 0 | 0 | 0 | 0 | 0 | 0 | 0 | 0 | 0 | 0 | 0 | 0 |
| Total Precipitation (") | 0.74 | 0.80 | 1.15 | 1.08 | 1.33 | 1.22 | 1.75 | 1.66 | 1.35 | 0.82 | 0.87 | 0.81 | 13.58 |
| Days ≥ 0.1" Precip | 2 | 3 | 4 | 4 | 4 | 4 | 6 | 6 | 4 | 3 | 3 | 3 | 46 |
| Total Snowfall (") | 14.9 | 15.2 | 20.1 | 15.7 | 7.9 | 1.4 | 0.0 | 0.0 | 1.8 | 7.5 | 15.7 | 15.8 | 116.0 |
| Days ≥ 1" Snow Depth | 29 | 26 | 23 | 11 | 3 | 0 | 0 | 0 | 1 | 4 | 16 | 26 | 139 |

## DINOSAUR NATL MON *Moffat County* ELEVATION 5921 ft LAT/LONG 40° 14 ' N / 108° 58 ' W

| | JAN | FEB | MAR | APR | MAY | JUN | JUL | AUG | SEP | OCT | NOV | DEC | YEAR |
|---|---|---|---|---|---|---|---|---|---|---|---|---|---|
| Maximum Temp °F | 31.7 | 38.3 | 49.7 | 60.9 | 71.8 | 83.4 | 90.1 | 87.9 | 77.9 | 63.4 | 45.4 | 33.7 | 61.2 |
| Minimum Temp °F | 8.5 | 13.5 | 24.2 | 31.3 | 40.1 | 48.5 | 55.7 | 53.9 | 44.8 | 34.4 | 22.7 | 11.8 | 32.4 |
| Mean Temp °F | 20.1 | 25.9 | 37.0 | 46.1 | 56.0 | 66.0 | 73.0 | 70.9 | 61.3 | 48.9 | 34.1 | 22.8 | 46.8 |
| Days Max Temp ≥ 90 °F | 0 | 0 | 0 | 0 | 0 | 8 | 17 | 13 | 2 | 0 | 0 | 0 | 40 |
| Days Max Temp ≤ 32 °F | 16 | 7 | 1 | 0 | 0 | 0 | 0 | 0 | 0 | 0 | 3 | 13 | 40 |
| Days Min Temp ≤ 32 °F | 31 | 28 | 27 | 17 | 5 | 1 | 0 | 0 | 2 | 12 | 27 | 31 | 181 |
| Days Min Temp ≤ 0 °F | 7 | 3 | 0 | 0 | 0 | 0 | 0 | 0 | 0 | 0 | 0 | 5 | 15 |
| Heating Degree Days | 1385 | 1096 | 862 | 560 | 279 | 73 | 2 | 10 | 140 | 492 | 921 | 1301 | 7121 |
| Cooling Degree Days | 0 | 0 | 0 | 0 | 8 | 127 | 265 | 215 | 42 | 0 | 0 | 0 | 657 |
| Total Precipitation (") | 0.64 | 0.54 | 1.03 | 1.06 | 1.33 | 1.11 | 1.01 | 0.82 | 1.09 | 1.48 | 0.81 | 0.71 | 11.63 |
| Days ≥ 0.1" Precip | 2 | 2 | 3 | 3 | 4 | 3 | 3 | 3 | 3 | 4 | 3 | 2 | 35 |
| Total Snowfall (") | 10.7 | 7.4 | 7.7 | 4.4 | 0.9 | 0.3 | 0.0 | 0.0 | 0.3 | 2.0 | 5.6 | 9.7 | 49.0 |
| Days ≥ 1" Snow Depth | 26 | 20 | 8 | 1 | 0 | 0 | 0 | 0 | 0 | 0 | 6 | 19 | 80 |

## EADS *Kiowa County* ELEVATION 4262 ft LAT/LONG 38° 29 ' N / 102° 46 ' W

| | JAN | FEB | MAR | APR | MAY | JUN | JUL | AUG | SEP | OCT | NOV | DEC | YEAR |
|---|---|---|---|---|---|---|---|---|---|---|---|---|---|
| Maximum Temp °F | 42.7 | 49.2 | 58.0 | 67.7 | 76.2 | 87.4 | 92.2 | 89.2 | 81.6 | 70.2 | 54.4 | ***45.2*** | 67.8 |
| Minimum Temp °F | 12.9 | 18.1 | 26.1 | 34.9 | 44.5 | 54.3 | 60.4 | 57.6 | 48.6 | 35.5 | 23.4 | ***15.2*** | 36.0 |
| Mean Temp °F | 27.9 | 33.7 | 42.1 | 51.3 | 60.4 | 70.9 | 76.3 | 73.5 | 65.1 | 53.2 | 38.9 | ***30.2*** | 52.0 |
| Days Max Temp ≥ 90 °F | 0 | 0 | 0 | 0 | 2 | 13 | 21 | 17 | 7 | 1 | 0 | ***0*** | 61 |
| Days Max Temp ≤ 32 °F | 8 | 3 | 1 | 0 | 0 | 0 | 0 | 0 | 0 | 0 | 2 | ***5*** | 19 |
| Days Min Temp ≤ 32 °F | 30 | 27 | 24 | 11 | 2 | 0 | 0 | 0 | 1 | 11 | 26 | ***31*** | 163 |
| Days Min Temp ≤ 0 °F | 4 | 1 | 0 | 0 | 0 | 0 | 0 | 0 | 0 | 0 | 0 | ***3*** | 8 |
| Heating Degree Days | 1145 | 879 | 704 | 409 | 176 | 25 | 2 | 6 | 90 | 363 | 775 | ***1072*** | 5646 |
| Cooling Degree Days | ***0*** | ***0*** | ***0*** | ***4*** | ***38*** | 209 | 351 | 275 | 104 | ***5*** | 0 | na | na |
| Total Precipitation (") | 0.29 | 0.31 | 0.92 | 1.13 | 2.33 | 2.00 | 2.75 | 2.11 | 1.33 | 0.92 | 0.63 | ***0.38*** | 15.10 |
| Days ≥ 0.1" Precip | 1 | 1 | ***3*** | 3 | 5 | 4 | 5 | 4 | 2 | 2 | 2 | ***1*** | 33 |
| Total Snowfall (") | 5.2 | ***2.4*** | ***5.4*** | ***1.8*** | 0.0 | 0.0 | 0.0 | 0.0 | 0.2 | ***0.9*** | ***4.3*** | ***4.2*** | 24.4 |
| Days ≥ 1" Snow Depth | na | na | na | ***0*** | 0 | 0 | 0 | 0 | 0 | ***0*** | na | na | na |

## EAGLE COUNTY AP *Eagle County* ELEVATION 6608 ft LAT/LONG 39° 40 ' N / 106° 50 ' W

| | JAN | FEB | MAR | APR | MAY | JUN | JUL | AUG | SEP | OCT | NOV | DEC | YEAR |
|---|---|---|---|---|---|---|---|---|---|---|---|---|---|
| Maximum Temp °F | 34.3 | 40.6 | 49.2 | 58.5 | 69.0 | 80.0 | 85.5 | 83.2 | 75.1 | 63.0 | 46.1 | 34.8 | 59.9 |
| Minimum Temp °F | 4.6 | 10.5 | 20.6 | 26.3 | 34.0 | 40.5 | 47.0 | 45.3 | 36.9 | 26.6 | 17.1 | 6.1 | 26.3 |
| Mean Temp °F | 19.6 | 25.6 | 34.9 | 42.4 | 51.5 | 60.3 | 66.2 | 64.3 | 56.0 | 44.8 | 31.6 | 20.5 | 43.1 |
| Days Max Temp ≥ 90 °F | 0 | 0 | 0 | 0 | 0 | 3 | 7 | 4 | 1 | 0 | 0 | 0 | 15 |
| Days Max Temp ≤ 32 °F | 12 | 5 | 1 | 0 | 0 | 0 | 0 | 0 | 0 | 0 | 3 | 12 | 33 |
| Days Min Temp ≤ 32 °F | 30 | 28 | 30 | 25 | 12 | 2 | 0 | 0 | 8 | 25 | 29 | 30 | 219 |
| Days Min Temp ≤ 0 °F | 11 | 6 | 1 | 0 | 0 | 0 | 0 | 0 | 0 | 0 | 1 | 9 | 28 |
| Heating Degree Days | 1402 | 1107 | 925 | 671 | 411 | 151 | 23 | 55 | 266 | 619 | 994 | 1375 | 7999 |
| Cooling Degree Days | 0 | 0 | 0 | 0 | 0 | 20 | 67 | 44 | 4 | 0 | 0 | 0 | 135 |
| Total Precipitation (") | 0.74 | 0.58 | 0.76 | 0.73 | 0.83 | 0.86 | 1.34 | 0.89 | 1.10 | 1.12 | 0.72 | 0.92 | 10.59 |
| Days ≥ 0.1" Precip | 2 | 2 | 3 | 2 | 3 | 3 | 4 | 3 | 3 | 3 | 3 | 3 | 34 |
| Total Snowfall (") | 8.6 | 5.7 | 5.9 | 3.3 | 1.5 | 0.1 | 0.0 | 0.0 | 0.4 | 3.1 | 6.0 | 10.6 | 45.2 |
| Days ≥ 1" Snow Depth | 23 | 19 | 8 | 1 | 0 | 0 | 0 | 0 | 0 | 1 | 6 | 17 | 75 |

## ESTES PARK *Larimer County* ELEVATION 7756 ft LAT/LONG 40° 23 ' N / 105° 31 ' W

| | JAN | FEB | MAR | APR | MAY | JUN | JUL | AUG | SEP | OCT | NOV | DEC | YEAR |
|---|---|---|---|---|---|---|---|---|---|---|---|---|---|
| Maximum Temp °F | 38.7 | 40.8 | 45.7 | 53.6 | 62.3 | 72.4 | ***78.3*** | 76.3 | 69.3 | 59.2 | 45.6 | 39.2 | 56.8 |
| Minimum Temp °F | 16.9 | 17.2 | 21.6 | 27.1 | 34.4 | 41.3 | ***46.6*** | 44.7 | 38.0 | 30.2 | 22.7 | 16.5 | 29.8 |
| Mean Temp °F | 27.8 | 29.0 | 33.7 | 40.4 | 48.4 | 56.9 | ***62.5*** | 60.5 | 53.7 | 44.7 | 34.2 | 27.9 | 43.3 |
| Days Max Temp ≥ 90 °F | 0 | 0 | 0 | 0 | 0 | 0 | ***0*** | 0 | 0 | 0 | 0 | 0 | 0 |
| Days Max Temp ≤ 32 °F | 7 | 5 | 3 | 1 | 0 | 0 | ***0*** | 0 | 0 | 0 | 3 | 8 | 27 |
| Days Min Temp ≤ 32 °F | 28 | 25 | 27 | 22 | 12 | 1 | ***0*** | 0 | 7 | 19 | 24 | 28 | 193 |
| Days Min Temp ≤ 0 °F | 4 | 3 | 2 | 0 | 0 | 0 | ***0*** | 0 | 0 | 0 | 1 | ***3*** | 13 |
| Heating Degree Days | 1145 | 1008 | 965 | 731 | 509 | 241 | ***92*** | 141 | 334 | 621 | 918 | 1143 | 7848 |
| Cooling Degree Days | 0 | 0 | 0 | 0 | 0 | 6 | ***26*** | ***10*** | 2 | 0 | 0 | ***0*** | 44 |
| Total Precipitation (") | 0.31 | 0.48 | 0.90 | 1.27 | 2.04 | 1.69 | 2.23 | 1.89 | 1.21 | 0.93 | 0.64 | 0.41 | 14.00 |
| Days ≥ 0.1" Precip | 1 | 2 | 2 | 3 | 4 | 4 | ***6*** | 5 | 3 | 3 | 2 | 1 | 36 |
| Total Snowfall (") | na | na | na | na | ***0.5*** | 0.1 | ***0.0*** | 0.0 | 0.6 | na | na | na | na |
| Days ≥ 1" Snow Depth | na | na | na | na | 0 | 0 | ***0*** | 0 | 0 | ***0*** | na | na | na |

## EVERGREEN *Jefferson County* ELEVATION 7005 ft LAT/LONG 39° 38 ' N / 105° 19 ' W

| | JAN | FEB | MAR | APR | MAY | JUN | JUL | AUG | SEP | OCT | NOV | DEC | YEAR |
|---|---|---|---|---|---|---|---|---|---|---|---|---|---|
| Maximum Temp °F | 44.2 | 45.5 | 49.4 | 56.4 | 64.3 | 75.0 | 80.5 | 78.9 | 71.8 | 61.7 | 49.9 | 44.9 | 60.2 |
| Minimum Temp °F | 10.1 | 12.2 | 18.1 | 25.6 | 33.5 | 41.0 | 46.3 | 44.8 | 36.7 | 26.6 | 17.6 | 10.7 | 26.9 |
| Mean Temp °F | 27.2 | 28.8 | 33.8 | 41.1 | 48.9 | 58.1 | 63.4 | 61.9 | 54.2 | 44.2 | 33.8 | 27.8 | 43.6 |
| Days Max Temp ≥ 90 °F | 0 | 0 | 0 | 0 | 0 | 1 | 1 | 1 | 0 | 0 | 0 | 0 | 3 |
| Days Max Temp ≤ 32 °F | 5 | 4 | 3 | 1 | 0 | 0 | 0 | 0 | 0 | 1 | 3 | 5 | 22 |
| Days Min Temp ≤ 32 °F | 31 | 28 | 30 | 26 | 13 | 1 | 0 | 0 | 7 | 26 | 29 | 31 | 222 |
| Days Min Temp ≤ 0 °F | 6 | 4 | 2 | 0 | 0 | 0 | 0 | 0 | 0 | 0 | 1 | 5 | 18 |
| Heating Degree Days | 1165 | 1013 | 961 | 712 | 492 | 209 | 69 | 104 | 317 | 638 | 929 | 1146 | 7755 |
| Cooling Degree Days | 0 | 0 | 0 | 0 | 0 | 10 | 30 | 17 | 1 | 0 | 0 | 0 | 58 |
| Total Precipitation (") | 0.49 | 0.69 | 1.63 | 2.10 | 2.78 | 2.16 | 2.53 | 2.22 | 1.40 | 1.44 | 1.09 | 0.75 | 19.28 |
| Days ≥ 0.1" Precip | 2 | 3 | 4 | 4 | 6 | 5 | 7 | 6 | 4 | 3 | 3 | 2 | 49 |
| Total Snowfall (") | 7.9 | 9.3 | 18.7 | 13.8 | 3.6 | 0.2 | 0.0 | 0.0 | 1.7 | 8.4 | 14.7 | 9.3 | 87.6 |
| Days ≥ 1" Snow Depth | na | na | na | na | ***0*** | 0 | 0 | 0 | 0 | na | na | na | na |

## FLAGLER 2 NW *Kit Carson County* ELEVATION 4984 ft LAT/LONG 39° 19 ' N / 103° 5 ' W

| | JAN | FEB | MAR | APR | MAY | JUN | JUL | AUG | SEP | OCT | NOV | DEC | YEAR |
|---|---|---|---|---|---|---|---|---|---|---|---|---|---|
| Maximum Temp °F | 41.7 | 45.7 | 53.4 | 63.0 | 71.4 | 82.5 | 88.7 | 86.1 | 78.3 | 66.9 | 50.9 | 43.0 | 64.3 |
| Minimum Temp °F | 12.9 | 16.6 | 23.0 | 31.4 | 41.5 | 50.8 | 57.1 | 55.2 | 46.2 | 33.9 | 21.7 | 14.8 | 33.8 |
| Mean Temp °F | 27.3 | 31.2 | 38.2 | 47.2 | 56.5 | 66.7 | 72.9 | 70.7 | 62.3 | 50.4 | 36.4 | 29.0 | 49.1 |
| Days Max Temp ≥ 90 °F | 0 | 0 | 0 | 0 | 0 | 7 | 15 | 11 | 3 | 0 | 0 | 0 | 36 |
| Days Max Temp ≤ 32 °F | 7 | 5 | 3 | 0 | 0 | 0 | 0 | 0 | 0 | 0 | 3 | 6 | 24 |
| Days Min Temp ≤ 32 °F | 31 | 28 | 28 | 17 | 4 | 0 | 0 | 0 | 2 | 13 | 27 | 30 | 180 |
| Days Min Temp ≤ 0 °F | 5 | 2 | 1 | 0 | 0 | 0 | 0 | 0 | 0 | 0 | 1 | 3 | 12 |
| Heating Degree Days | 1161 | 949 | 823 | 527 | 269 | 57 | 6 | 15 | 132 | 447 | 852 | 1110 | 6348 |
| Cooling Degree Days | 0 | 0 | 0 | 1 | 9 | 108 | 240 | 200 | 61 | 1 | 0 | 0 | 620 |
| Total Precipitation (") | 0.33 | 0.33 | 0.92 | 1.32 | 2.59 | 2.58 | 2.56 | 2.15 | 1.16 | 0.84 | 0.68 | 0.35 | 15.81 |
| Days ≥ 0.1" Precip | 1 | 1 | 3 | 3 | 6 | 5 | 5 | 4 | 3 | 2 | 2 | 1 | 36 |
| Total Snowfall (") | 4.8 | 3.7 | 7.0 | 4.8 | 0.9 | 0.0 | 0.0 | 0.0 | 0.3 | 3.1 | 6.6 | 5.2 | 36.4 |
| Days ≥ 1" Snow Depth | ***6*** | 4 | 4 | 2 | 0 | 0 | 0 | 0 | 0 | ***1*** | 4 | 4 | 25 |

## FORT COLLINS *Larimer County* ELEVATION 5007 ft LAT/LONG 40° 35 ' N / 105° 5 ' W

| | JAN | FEB | MAR | APR | MAY | JUN | JUL | AUG | SEP | OCT | NOV | DEC | YEAR |
|---|---|---|---|---|---|---|---|---|---|---|---|---|---|
| Maximum Temp °F | 41.8 | 45.7 | 52.7 | 61.3 | 70.0 | 79.9 | 85.2 | 83.0 | 74.9 | 63.9 | 49.9 | 42.2 | 62.5 |
| Minimum Temp °F | 15.1 | 19.4 | 25.9 | 34.2 | 43.3 | 51.8 | 57.3 | 55.2 | 46.0 | 35.2 | 24.2 | 16.5 | 35.3 |
| Mean Temp °F | 28.5 | 32.6 | 39.3 | 47.7 | 56.6 | 65.9 | 71.3 | 69.1 | 60.5 | 49.6 | 37.1 | 29.4 | 49.0 |
| Days Max Temp ≥ 90 °F | 0 | 0 | 0 | 0 | 0 | 4 | 8 | 4 | 1 | 0 | 0 | 0 | 17 |
| Days Max Temp ≤ 32 °F | 7 | 4 | 2 | 0 | 0 | 0 | 0 | 0 | 0 | 0 | 2 | 6 | 21 |
| Days Min Temp ≤ 32 °F | 30 | 26 | 25 | 12 | 2 | 0 | 0 | 0 | 1 | 10 | 25 | 30 | 161 |
| Days Min Temp ≤ 0 °F | 3 | 2 | 0 | 0 | 0 | 0 | 0 | 0 | 0 | 0 | 0 | 3 | 8 |
| Heating Degree Days | 1124 | 909 | 789 | 511 | 259 | 62 | 7 | 13 | 155 | 471 | 830 | 1097 | 6227 |
| Cooling Degree Days | 0 | 0 | 0 | 1 | 8 | 102 | 204 | 158 | 31 | 0 | 0 | 0 | 504 |
| Total Precipitation (") | 0.40 | 0.37 | 1.37 | 1.83 | 2.50 | 2.08 | 1.81 | 1.31 | 1.20 | 1.03 | 0.81 | 0.51 | 15.22 |
| Days ≥ 0.1" Precip | 1 | 1 | 3 | 4 | 5 | 4 | 4 | 3 | 3 | 3 | 2 | 2 | 35 |
| Total Snowfall (") | 7.8 | 6.6 | 11.9 | 6.5 | 1.5 | 0.0 | 0.0 | 0.0 | 0.9 | 3.7 | 9.5 | 8.8 | 57.2 |
| Days ≥ 1" Snow Depth | 14 | 8 | 6 | 2 | 0 | 0 | 0 | 0 | 0 | 1 | 6 | 11 | 48 |

## FORT LEWIS *La Plata County* ELEVATION 7615 ft LAT/LONG 37° 14 ' N / 108° 3 ' W

| | JAN | FEB | MAR | APR | MAY | JUN | JUL | AUG | SEP | OCT | NOV | DEC | YEAR |
|---|---|---|---|---|---|---|---|---|---|---|---|---|---|
| Maximum Temp °F | 36.1 | 40.3 | 46.3 | 56.0 | 65.7 | 76.3 | 80.9 | 78.3 | 71.3 | 60.7 | 46.1 | 37.6 | 58.0 |
| Minimum Temp °F | 8.7 | 12.6 | 19.8 | 26.1 | 33.3 | 40.8 | 48.7 | 47.2 | 39.6 | 29.8 | 19.3 | 11.3 | 28.1 |
| Mean Temp °F | 22.4 | 26.5 | 33.1 | 41.0 | 49.5 | 58.6 | 64.8 | 62.8 | 55.6 | 45.3 | 32.7 | 24.5 | 43.1 |
| Days Max Temp ≥ 90 °F | 0 | 0 | 0 | 0 | 0 | 0 | 1 | 0 | 0 | 0 | 0 | 0 | 1 |
| Days Max Temp ≤ 32 °F | 10 | 4 | 1 | 0 | 0 | 0 | 0 | 0 | 0 | 0 | 2 | 8 | 25 |
| Days Min Temp ≤ 32 °F | 31 | 28 | 30 | 24 | 13 | 3 | 0 | 0 | 4 | 19 | 28 | 31 | 211 |
| Days Min Temp ≤ 0 °F | 7 | 3 | 1 | 0 | 0 | 0 | 0 | 0 | 0 | 0 | 1 | 5 | 17 |
| Heating Degree Days | 1314 | 1081 | 983 | 712 | 474 | 197 | 43 | 81 | 278 | 605 | 963 | 1250 | 7981 |
| Cooling Degree Days | 0 | 0 | 0 | 0 | 0 | 13 | 42 | 24 | 1 | 0 | 0 | 0 | 80 |
| Total Precipitation (") | 1.59 | 1.42 | 1.70 | 1.03 | 1.27 | 0.69 | 2.14 | 2.29 | 1.95 | 1.91 | 1.70 | 1.54 | 19.23 |
| Days ≥ 0.1" Precip | 4 | 4 | 5 | 3 | 3 | 2 | 6 | 6 | 4 | 4 | 4 | 4 | 49 |
| Total Snowfall (") | ***16.7*** | ***14.5*** | na | ***3.1*** | ***0.0*** | 0.0 | 0.0 | 0.0 | 0.0 | 1.5 | ***10.2*** | ***20.0*** | na |
| Days ≥ 1" Snow Depth | ***27*** | 22 | ***20*** | ***4*** | 0 | 0 | 0 | 0 | 0 | 1 | na | ***23*** | na |

## FORT MORGAN 2 S *Morgan County* ELEVATION 4320 ft LAT/LONG 40° 15 ' N / 103° 48 ' W

| | JAN | FEB | MAR | APR | MAY | JUN | JUL | AUG | SEP | OCT | NOV | DEC | YEAR |
|---|---|---|---|---|---|---|---|---|---|---|---|---|---|
| Maximum Temp °F | 38.4 | 44.7 | 53.2 | 62.9 | 72.3 | 83.1 | 89.8 | 87.4 | 78.6 | 66.9 | 50.7 | 40.7 | 64.1 |
| Minimum Temp °F | 9.9 | 16.1 | 24.7 | 34.4 | 45.0 | 54.5 | 59.9 | 57.3 | 46.9 | 33.9 | 22.3 | 12.2 | 34.8 |
| Mean Temp °F | 24.2 | 30.4 | 39.0 | 48.7 | 58.6 | 68.8 | 74.9 | 72.4 | 62.8 | 50.5 | 36.5 | 26.5 | 49.4 |
| Days Max Temp ≥ 90 °F | 0 | 0 | 0 | 0 | 1 | 9 | 17 | 14 | 5 | 0 | 0 | 0 | 46 |
| Days Max Temp ≤ 32 °F | 9 | 5 | 3 | 0 | 0 | 0 | 0 | 0 | 0 | 0 | 3 | 8 | 28 |
| Days Min Temp ≤ 32 °F | 31 | 28 | 26 | 12 | 1 | 0 | 0 | 0 | 1 | 13 | 27 | 31 | 170 |
| Days Min Temp ≤ 0 °F | 6 | 3 | 1 | 0 | 0 | 0 | 0 | 0 | 0 | 0 | 1 | 4 | 15 |
| Heating Degree Days | 1259 | 970 | 800 | 485 | 220 | 44 | 5 | 10 | 130 | 446 | 848 | 1187 | 6404 |
| Cooling Degree Days | 0 | 0 | 0 | 2 | 32 | 173 | 314 | 260 | 80 | 2 | 0 | 0 | 863 |
| Total Precipitation (") | 0.20 | 0.15 | 0.72 | 1.24 | 2.41 | 2.10 | 1.79 | 1.54 | 1.06 | 0.83 | 0.47 | 0.29 | 12.80 |
| Days ≥ 0.1" Precip | 1 | 1 | 2 | 3 | 5 | 5 | 4 | 3 | 3 | 2 | 2 | 1 | 32 |
| Total Snowfall (") | ***5.0*** | 2.4 | ***5.9*** | 2.7 | 0.4 | 0.0 | 0.0 | 0.0 | 0.2 | 1.3 | 4.0 | 4.8 | 26.7 |
| Days ≥ 1" Snow Depth | na | na | na | na | 0 | 0 | 0 | 0 | 0 | 0 | na | na | na |

## FRUITA 1 W *Mesa County* ELEVATION 4524 ft LAT/LONG 39° 9 ' N / 108° 44 ' W

| | JAN | FEB | MAR | APR | MAY | JUN | JUL | AUG | SEP | OCT | NOV | DEC | YEAR |
|---|---|---|---|---|---|---|---|---|---|---|---|---|---|
| Maximum Temp °F | 36.6 | 45.7 | 56.8 | 66.4 | 76.3 | 87.3 | 92.6 | 90.0 | 81.9 | 68.5 | 52.3 | 40.1 | 66.2 |
| Minimum Temp °F | 11.3 | 18.4 | 26.8 | 33.6 | 42.8 | 49.9 | 56.7 | 54.9 | 44.7 | 33.4 | 23.4 | 14.7 | 34.2 |
| Mean Temp °F | 24.0 | 32.1 | 41.9 | 50.0 | 59.6 | 68.6 | 74.7 | 72.5 | 63.3 | 51.0 | 37.9 | 27.4 | 50.3 |
| Days Max Temp ≥ 90 °F | 0 | 0 | 0 | 0 | 1 | 13 | 23 | 18 | 4 | 0 | 0 | 0 | 59 |
| Days Max Temp ≤ 32 °F | 10 | 2 | 0 | 0 | 0 | 0 | 0 | 0 | 0 | 0 | 0 | 6 | 18 |
| Days Min Temp ≤ 32 °F | 30 | 26 | 23 | 14 | 2 | 0 | 0 | 0 | 1 | 15 | 26 | 29 | 166 |
| Days Min Temp ≤ 0 °F | 6 | 1 | 0 | 0 | 0 | 0 | 0 | 0 | 0 | 0 | 0 | 2 | 9 |
| Heating Degree Days | 1265 | 924 | 711 | 442 | 181 | 33 | 1 | 3 | 94 | 427 | 806 | 1159 | 6046 |
| Cooling Degree Days | 0 | ***0*** | 0 | 1 | 22 | ***156*** | 300 | 241 | 52 | 0 | 0 | 0 | 772 |
| Total Precipitation (") | 0.65 | 0.51 | 0.91 | 0.71 | 1.03 | 0.53 | 0.78 | 0.79 | 0.73 | 1.04 | 0.77 | 0.70 | 9.15 |
| Days ≥ 0.1" Precip | 2 | 2 | 3 | 2 | 3 | 1 | 2 | 2 | 2 | 3 | 2 | 2 | 26 |
| Total Snowfall (") | 5.2 | ***1.9*** | 1.3 | 0.2 | 0.0 | 0.0 | 0.0 | 0.0 | 0.0 | 0.3 | ***1.0*** | 3.9 | 13.8 |
| Days ≥ 1" Snow Depth | ***11*** | ***4*** | 1 | 0 | 0 | 0 | 0 | 0 | 0 | 0 | ***1*** | 6 | 23 |

## GATEWAY 1 SE *Mesa County* ELEVATION 4564 ft LAT/LONG 38° 40' N / 108° 59' W

| | JAN | FEB | MAR | APR | MAY | JUN | JUL | AUG | SEP | OCT | NOV | DEC | YEAR |
|---|---|---|---|---|---|---|---|---|---|---|---|---|---|
| Maximum Temp °F | 40.8 | 49.3 | 58.5 | 67.7 | 76.6 | 86.8 | *91.5* | na | *81.9* | *69.2* | 55.0 | 43.3 | na |
| Minimum Temp °F | 16.7 | 23.7 | 31.5 | 37.8 | 46.2 | 54.7 | *61.0* | na | *50.2* | *38.0* | 28.9 | 20.0 | na |
| Mean Temp °F | 28.8 | 36.5 | 45.0 | 52.8 | 61.4 | 70.8 | *76.3* | na | *66.1* | *53.6* | 41.9 | 31.7 | na |
| Days Max Temp ≥ 90 °F | 0 | 0 | 0 | 0 | 1 | 13 | *19* | 15 | 4 | 0 | 0 | 0 | 52 |
| Days Max Temp ≤ 32 °F | 6 | 1 | 0 | 0 | 0 | 0 | 0 | 0 | 0 | 0 | 0 | 3 | 10 |
| Days Min Temp ≤ 32 °F | 29 | 24 | 17 | *8* | 1 | 0 | 0 | 0 | 0 | *6* | 20 | 29 | 134 |
| Days Min Temp ≤ 0 °F | 3 | 1 | 0 | 0 | 0 | 0 | 0 | 0 | 0 | 0 | 0 | 1 | 5 |
| Heating Degree Days | 1116 | 799 | 612 | 366 | 144 | 24 | *0* | na | *58* | *351* | 683 | 1027 | na |
| Cooling Degree Days | 0 | 0 | 0 | 9 | 48 | 248 | na | na | na | na | 0 | 0 | na |
| Total Precipitation (") | 0.72 | 0.70 | 1.16 | 0.96 | 1.13 | 0.62 | 1.15 | 1.17 | 0.87 | 1.25 | 1.05 | 0.86 | 11.64 |
| Days ≥ 0.1" Precip | 3 | 3 | 4 | 3 | 3 | 2 | 3 | 3 | 2 | 4 | 3 | 3 | 36 |
| Total Snowfall (") | 6.0 | 1.5 | 2.5 | 0.9 | 0.1 | 0.0 | 0.0 | 0.0 | 0.0 | 0.1 | *1.2* | 5.8 | 18.1 |
| Days ≥ 1" Snow Depth | *7* | 2 | 0 | 0 | 0 | 0 | 0 | 0 | 0 | 0 | *1* | *4* | 14 |

## GLENWOOD SPRINGS # 2 *Garfield County* ELEVATION 5822 ft LAT/LONG 39° 34' N / 107° 20' W

| | JAN | FEB | MAR | APR | MAY | JUN | JUL | AUG | SEP | OCT | NOV | DEC | YEAR |
|---|---|---|---|---|---|---|---|---|---|---|---|---|---|
| Maximum Temp °F | 36.2 | 43.1 | 52.1 | 61.4 | *71.3* | 82.5 | 88.3 | 86.3 | 77.4 | 65.2 | 48.1 | 37.1 | 62.4 |
| Minimum Temp °F | 11.7 | 17.0 | 24.6 | 30.3 | *37.9* | 44.6 | 51.3 | 50.0 | 41.5 | 31.5 | 22.1 | 13.4 | 31.3 |
| Mean Temp °F | 24.0 | 30.0 | 38.4 | 45.9 | *54.6* | 63.6 | 69.8 | 68.2 | 59.5 | 48.3 | 35.1 | 25.3 | 46.9 |
| Days Max Temp ≥ 90 °F | 0 | 0 | 0 | 0 | 0 | 6 | 14 | 10 | 1 | 0 | 0 | 0 | 31 |
| Days Max Temp ≤ 32 °F | 10 | 3 | 1 | 0 | *0* | 0 | 0 | 0 | 0 | 0 | 2 | 8 | 24 |
| Days Min Temp ≤ 32 °F | 31 | 27 | 27 | 18 | *5* | 0 | 0 | 0 | 3 | 17 | 27 | 31 | 186 |
| Days Min Temp ≤ 0 °F | 5 | 1 | 0 | 0 | 0 | 0 | 0 | 0 | 0 | 0 | 0 | 3 | 9 |
| Heating Degree Days | 1265 | 980 | 820 | 567 | *316* | 90 | 5 | 20 | 174 | 510 | 889 | 1224 | 6860 |
| Cooling Degree Days | 0 | 0 | 0 | 0 | *2* | 68 | 170 | 147 | 22 | 0 | 0 | 0 | 409 |
| Total Precipitation (") | 1.36 | 1.13 | 1.42 | 1.46 | 1.69 | 1.26 | 1.32 | 1.26 | 1.61 | 1.84 | 1.28 | 1.43 | 17.06 |
| Days ≥ 0.1" Precip | 4 | 4 | 5 | 5 | *4* | 3 | 4 | 4 | 4 | 5 | 4 | 4 | 50 |
| Total Snowfall (") | *15.4* | *10.3* | *5.6* | na | *0.5* | 0.0 | 0.0 | 0.0 | 0.0 | 0.8 | *6.7* | *16.1* | na |
| Days ≥ 1" Snow Depth | na | na | na | na | *0* | 0 | 0 | 0 | 0 | 0 | na | na | na |

## GRAND JUNCTION 6 ESE *Mesa County* ELEVATION 4759 ft LAT/LONG 39° 3' N / 108° 27' W

| | JAN | FEB | MAR | APR | MAY | JUN | JUL | AUG | SEP | OCT | NOV | DEC | YEAR |
|---|---|---|---|---|---|---|---|---|---|---|---|---|---|
| Maximum Temp °F | 37.0 | 45.5 | 56.1 | 65.3 | 75.2 | 86.4 | 92.0 | 89.5 | 80.7 | 67.3 | 51.4 | 39.8 | 65.5 |
| Minimum Temp °F | 15.7 | 23.1 | 32.0 | 39.1 | 47.8 | 56.8 | 62.9 | 61.1 | 52.1 | 40.4 | 28.9 | 19.3 | 39.9 |
| Mean Temp °F | 26.4 | 34.3 | 44.1 | 52.2 | 61.5 | 71.6 | 77.5 | 75.3 | 66.4 | 53.9 | 40.2 | 29.6 | 52.8 |
| Days Max Temp ≥ 90 °F | 0 | 0 | 0 | 0 | 1 | 12 | 23 | 18 | 4 | 0 | 0 | 0 | 58 |
| Days Max Temp ≤ 32 °F | 10 | 2 | 0 | 0 | 0 | 0 | 0 | 0 | 0 | 0 | 1 | 6 | 19 |
| Days Min Temp ≤ 32 °F | 30 | 25 | 16 | 7 | 1 | 0 | 0 | 0 | 0 | 5 | 20 | 30 | 134 |
| Days Min Temp ≤ 0 °F | 3 | 1 | 0 | 0 | 0 | 0 | 0 | 0 | 0 | 0 | 0 | 1 | 5 |
| Heating Degree Days | 1191 | 861 | 641 | 383 | 153 | 25 | 0 | 2 | 62 | 343 | 738 | 1091 | 5490 |
| Cooling Degree Days | 0 | 0 | 0 | 8 | 55 | 249 | 395 | 338 | 119 | 3 | 0 | 0 | 1167 |
| Total Precipitation (") | 0.53 | 0.48 | 0.94 | 0.73 | 1.04 | 0.50 | 0.72 | 0.76 | 0.74 | 1.03 | 0.79 | 0.64 | 8.90 |
| Days ≥ 0.1" Precip | 2 | 2 | 3 | 2 | 3 | 1 | 2 | 3 | 3 | 3 | 3 | 2 | 29 |
| Total Snowfall (") | na | *2.0* | 1.7 | 0.1 | 0.1 | 0.0 | 0.0 | 0.0 | 0.0 | 0.4 | *1.3* | 4.2 | na |
| Days ≥ 1" Snow Depth | na | na | *1* | 0 | 0 | 0 | 0 | 0 | 0 | 0 | na | na | na |

## GRAND JUNCTION WLKR *Mesa County* ELEVATION 4849 ft LAT/LONG 39° 7' N / 108° 32' W

| | JAN | FEB | MAR | APR | MAY | JUN | JUL | AUG | SEP | OCT | NOV | DEC | YEAR |
|---|---|---|---|---|---|---|---|---|---|---|---|---|---|
| Maximum Temp °F | 35.7 | 45.0 | 55.9 | 65.6 | 75.9 | 87.8 | 93.3 | 90.4 | 81.2 | 67.2 | 50.6 | 38.5 | 65.6 |
| Minimum Temp °F | 15.1 | 23.1 | 31.7 | 38.7 | 47.9 | 57.2 | 63.9 | 62.1 | 52.8 | 41.1 | 28.7 | 18.6 | 40.1 |
| Mean Temp °F | 25.4 | 34.1 | 43.8 | 52.2 | 62.0 | 72.5 | 78.6 | 76.3 | 67.0 | 54.2 | 39.7 | 28.6 | 52.9 |
| Days Max Temp ≥ 90 °F | 0 | 0 | 0 | 0 | 1 | 15 | 24 | 20 | 4 | 0 | 0 | 0 | 64 |
| Days Max Temp ≤ 32 °F | 12 | 2 | 0 | 0 | 0 | 0 | 0 | 0 | 0 | 0 | 1 | 7 | 22 |
| Days Min Temp ≤ 32 °F | 30 | 25 | 16 | 6 | 0 | 0 | 0 | 0 | 0 | 3 | 21 | 30 | 131 |
| Days Min Temp ≤ 0 °F | 4 | 1 | 0 | 0 | 0 | 0 | 0 | 0 | 0 | 0 | 0 | 1 | 6 |
| Heating Degree Days | 1221 | 866 | 650 | 382 | 141 | 18 | 0 | 1 | 55 | 333 | 753 | 1121 | 5541 |
| Cooling Degree Days | 0 | 0 | 0 | 5 | 52 | 265 | 414 | 357 | 123 | 4 | 0 | 0 | 1220 |
| Total Precipitation (") | 0.59 | 0.50 | 0.95 | 0.78 | 0.95 | 0.48 | 0.65 | 0.76 | 0.82 | 1.08 | 0.72 | 0.62 | 8.90 |
| Days ≥ 0.1" Precip | 2 | 1 | 3 | 3 | 3 | 1 | 2 | 2 | 2 | 3 | 2 | 2 | 26 |
| Total Snowfall (") | 6.5 | 3.6 | 3.1 | 1.2 | 0.2 | 0.0 | 0.0 | 0.0 | 0.1 | 0.8 | 2.3 | 5.6 | 23.4 |
| Days ≥ 1" Snow Depth | 17 | 8 | 1 | 0 | 0 | 0 | 0 | 0 | 0 | 0 | 1 | 8 | 35 |

## GRAND LAKE 1 NW *Grand County* ELEVATION 8386 ft LAT/LONG 40° 16 ' N / 105° 50 ' W

| | JAN | FEB | MAR | APR | MAY | JUN | JUL | AUG | SEP | OCT | NOV | DEC | YEAR |
|---|---|---|---|---|---|---|---|---|---|---|---|---|---|
| Maximum Temp °F | 31.2 | 35.0 | 40.9 | 49.0 | 59.0 | 70.1 | 75.3 | 73.4 | 66.6 | 55.5 | 39.4 | 31.9 | 52.3 |
| Minimum Temp °F | 2.6 | 4.0 | 11.0 | 18.8 | 27.0 | 33.3 | 38.4 | 36.8 | 30.1 | 22.3 | 12.0 | 3.8 | 20.0 |
| Mean Temp °F | 16.9 | 19.5 | 26.0 | 33.9 | 43.0 | 51.7 | 56.9 | 55.1 | 48.4 | 38.9 | 25.7 | 17.8 | 36.2 |
| Days Max Temp ≥ 90 °F | 0 | 0 | 0 | 0 | 0 | 0 | 0 | 0 | 0 | 0 | 0 | 0 | 0 |
| Days Max Temp ≤ 32 °F | 16 | 10 | 5 | 2 | 0 | 0 | 0 | 0 | 0 | 1 | 8 | 15 | 57 |
| Days Min Temp ≤ 32 °F | 31 | 28 | 31 | 30 | 27 | 13 | 3 | 5 | 20 | 29 | 30 | 31 | 278 |
| Days Min Temp ≤ 0 °F | 13 | 11 | 5 | 1 | 0 | 0 | 0 | 0 | 0 | 0 | 5 | 12 | 47 |
| Heating Degree Days | 1486 | 1277 | 1201 | 926 | 673 | 391 | 245 | 300 | 493 | 800 | 1172 | 1454 | 10418 |
| Cooling Degree Days | 0 | 0 | 0 | 0 | 0 | 0 | 1 | 0 | 0 | 0 | 0 | 0 | 1 |
| Total Precipitation (") | 1.64 | 1.50 | 1.55 | 1.91 | 1.92 | 1.55 | 2.07 | 2.07 | 1.66 | 1.59 | 1.42 | 1.56 | 20.44 |
| Days ≥ 0.1" Precip | 6 | 5 | 6 | 6 | 6 | 5 | 6 | 6 | 5 | 5 | 5 | 5 | 66 |
| Total Snowfall (") | 26.9 | 22.1 | 17.9 | 16.6 | 5.0 | 0.7 | 0.0 | 0.0 | 0.6 | ***6.6*** | 21.7 | 24.7 | 142.8 |
| Days ≥ 1" Snow Depth | 30 | 28 | 30 | 21 | 3 | 0 | 0 | 0 | 0 | ***3*** | 18 | 30 | 163 |

## GRAND LAKE 6 SSW *Grand County* ELEVATION 8386 ft LAT/LONG 40° 15 ' N / 105° 51 ' W

| | JAN | FEB | MAR | APR | MAY | JUN | JUL | AUG | SEP | OCT | NOV | DEC | YEAR |
|---|---|---|---|---|---|---|---|---|---|---|---|---|---|
| Maximum Temp °F | 26.7 | 30.7 | 38.1 | 47.4 | 58.5 | 68.7 | 74.0 | 72.4 | 65.2 | 54.2 | 38.5 | 28.7 | 50.3 |
| Minimum Temp °F | 0.9 | 1.9 | 10.7 | 19.9 | 29.3 | 36.2 | 41.7 | 40.6 | 33.6 | 25.1 | 15.8 | 5.3 | 21.8 |
| Mean Temp °F | 13.8 | 16.3 | 24.4 | 33.7 | 43.9 | 52.5 | 57.9 | 56.5 | 49.5 | 39.7 | 27.2 | 17.0 | 36.0 |
| Days Max Temp ≥ 90 °F | 0 | 0 | 0 | 0 | 0 | 0 | 0 | 0 | 0 | 0 | 0 | 0 | 0 |
| Days Max Temp ≤ 32 °F | 23 | 16 | 7 | 2 | 0 | 0 | 0 | 0 | 0 | 1 | 8 | 20 | 77 |
| Days Min Temp ≤ 32 °F | 31 | 28 | 31 | 29 | 23 | 7 | 0 | 2 | 13 | 27 | 29 | 31 | 251 |
| Days Min Temp ≤ 0 °F | 15 | 13 | 6 | 1 | 0 | 0 | 0 | 0 | 0 | 0 | 2 | 11 | 48 |
| Heating Degree Days | 1583 | 1368 | 1251 | 933 | 647 | 369 | 214 | 256 | 460 | 779 | 1127 | 1482 | 10469 |
| Cooling Degree Days | 0 | 0 | 0 | 0 | 0 | 0 | 0 | 0 | 0 | 0 | 0 | 0 | 0 |
| Total Precipitation (") | 0.97 | 0.80 | 0.95 | 1.23 | 1.44 | 1.23 | 1.52 | 1.57 | 1.14 | 1.09 | 0.98 | 0.85 | 13.77 |
| Days ≥ 0.1" Precip | 3 | 3 | 4 | 5 | 5 | 4 | 5 | 6 | 4 | 4 | 3 | 3 | 49 |
| Total Snowfall (") | na | ***13.7*** | ***14.2*** | 10.8 | 2.4 | 0.6 | 0.0 | 0.0 | 0.6 | ***3.9*** | na | ***14.2*** | na |
| Days ≥ 1" Snow Depth | ***29*** | ***26*** | ***28*** | 16 | 2 | 0 | 0 | 0 | 0 | ***2*** | na | ***22*** | na |

## GRANT *Park County* ELEVATION 8694 ft LAT/LONG 39° 28 ' N / 105° 41 ' W

| | JAN | FEB | MAR | APR | MAY | JUN | JUL | AUG | SEP | OCT | NOV | DEC | YEAR |
|---|---|---|---|---|---|---|---|---|---|---|---|---|---|
| Maximum Temp °F | 33.3 | 37.3 | 43.0 | 51.0 | 60.8 | 70.8 | 75.3 | 72.7 | 65.8 | 56.0 | 40.5 | 33.1 | 53.3 |
| Minimum Temp °F | 8.6 | 10.0 | 15.4 | 22.1 | 29.9 | 36.6 | 42.0 | 41.0 | 33.7 | 25.6 | 15.8 | 9.9 | 24.2 |
| Mean Temp °F | 21.0 | 23.6 | 29.2 | 36.5 | 45.4 | 53.8 | 58.7 | 56.8 | 49.8 | 40.8 | 28.2 | 21.5 | 38.8 |
| Days Max Temp ≥ 90 °F | 0 | 0 | 0 | 0 | 0 | 0 | 0 | 0 | 0 | 0 | 0 | 0 | 0 |
| Days Max Temp ≤ 32 °F | 13 | 8 | 5 | 1 | 0 | 0 | 0 | 0 | 0 | 1 | 6 | 14 | 48 |
| Days Min Temp ≤ 32 °F | 31 | 28 | 31 | 29 | 21 | 6 | 0 | 1 | 11 | 28 | 30 | 31 | 247 |
| Days Min Temp ≤ 0 °F | 7 | 5 | 2 | 1 | 0 | 0 | 0 | 0 | 0 | 0 | 2 | 6 | 23 |
| Heating Degree Days | 1358 | 1161 | 1103 | 847 | 602 | 330 | 189 | 246 | 450 | 742 | 1098 | 1342 | 9468 |
| Cooling Degree Days | 0 | 0 | 0 | 0 | 0 | 0 | 1 | 0 | 0 | 0 | 0 | 0 | 1 |
| Total Precipitation (") | 0.44 | 0.52 | 1.06 | 1.36 | 1.67 | 1.64 | 2.35 | 2.45 | 1.35 | 1.18 | 0.86 | 0.70 | 15.58 |
| Days ≥ 0.1" Precip | 2 | 2 | 4 | 4 | 5 | 5 | 7 | 7 | 4 | 3 | 3 | 2 | 48 |
| Total Snowfall (") | 8.6 | 10.8 | 18.0 | 16.0 | 7.0 | 1.3 | 0.0 | 0.0 | 2.6 | 10.2 | 13.3 | 12.5 | 100.3 |
| Days ≥ 1" Snow Depth | 29 | 27 | 28 | 14 | 3 | 0 | 0 | 0 | 1 | 6 | 22 | 28 | 158 |

## GREAT SAND DUNES NAT *Alamosa County* ELEVATION 8119 ft LAT/LONG 37° 43 ' N / 105° 32 ' W

| | JAN | FEB | MAR | APR | MAY | JUN | JUL | AUG | SEP | OCT | NOV | DEC | YEAR |
|---|---|---|---|---|---|---|---|---|---|---|---|---|---|
| Maximum Temp °F | 34.3 | 38.7 | 47.0 | 56.7 | 65.9 | 76.3 | 80.5 | 77.6 | 70.9 | 60.1 | 45.3 | 36.1 | 57.5 |
| Minimum Temp °F | 8.2 | 12.9 | 21.0 | 27.6 | 36.2 | 44.7 | 50.2 | 48.2 | 41.2 | 31.2 | 19.7 | 10.4 | 29.3 |
| Mean Temp °F | 21.3 | 25.8 | 34.0 | 42.2 | 51.1 | 60.5 | 65.4 | 62.9 | 56.1 | 45.7 | 32.5 | 23.4 | 43.4 |
| Days Max Temp ≥ 90 °F | 0 | 0 | 0 | 0 | 0 | 0 | 1 | 0 | 0 | 0 | 0 | 0 | 1 |
| Days Max Temp ≤ 32 °F | 13 | 6 | 2 | 0 | 0 | 0 | 0 | 0 | 0 | 0 | 4 | 11 | 36 |
| Days Min Temp ≤ 32 °F | 31 | 28 | 29 | 22 | 8 | 1 | 0 | 0 | 2 | 17 | 29 | 31 | 198 |
| Days Min Temp ≤ 0 °F | 7 | 3 | 0 | 0 | 0 | 0 | 0 | 0 | 0 | 0 | 1 | 5 | 16 |
| Heating Degree Days | 1349 | 1098 | 955 | 678 | 427 | 149 | 38 | 80 | 263 | 592 | 969 | 1284 | 7882 |
| Cooling Degree Days | 0 | 0 | 0 | 0 | 1 | 28 | 66 | 25 | 4 | 0 | 0 | 0 | 124 |
| Total Precipitation (") | 0.37 | 0.36 | 0.81 | 0.69 | 1.11 | 0.98 | 1.82 | 2.15 | 1.28 | 0.90 | 0.49 | 0.46 | 11.42 |
| Days ≥ 0.1" Precip | 1 | 1 | 3 | 2 | 3 | 3 | 5 | 6 | 4 | 2 | 1 | 2 | 33 |
| Total Snowfall (") | ***6.9*** | ***5.1*** | 8.4 | ***3.7*** | ***1.1*** | 0.0 | 0.0 | 0.0 | 0.1 | 3.6 | ***4.3*** | ***7.5*** | 40.7 |
| Days ≥ 1" Snow Depth | ***16*** | ***12*** | na | ***2*** | 0 | 0 | 0 | 0 | 0 | 1 | ***6*** | na | na |

## GREELEY UNC *Weld County* ELEVATION 4652 ft LAT/LONG 40° 25' N / 104° 42' W

| | JAN | FEB | MAR | APR | MAY | JUN | JUL | AUG | SEP | OCT | NOV | DEC | YEAR |
|---|---|---|---|---|---|---|---|---|---|---|---|---|---|
| Maximum Temp °F | 40.4 | 46.5 | 54.9 | 63.9 | 72.7 | 83.3 | 88.7 | 86.6 | 78.6 | 66.1 | 50.0 | 41.3 | 64.4 |
| Minimum Temp °F | 13.9 | 19.3 | 26.4 | 34.5 | 43.8 | 52.7 | 57.6 | 55.5 | 46.0 | 34.9 | 23.8 | 15.6 | 35.3 |
| Mean Temp °F | 27.2 | 32.9 | 40.7 | 49.2 | 58.3 | 68.1 | 73.2 | 71.1 | 62.4 | 50.5 | 36.9 | 28.5 | 49.9 |
| Days Max Temp ≥ 90 °F | 0 | 0 | 0 | 0 | 1 | 9 | 15 | 12 | 3 | 0 | 0 | 0 | 40 |
| Days Max Temp ≤ 32 °F | 8 | 4 | 2 | 0 | 0 | 0 | 0 | 0 | 0 | 0 | 3 | 7 | 24 |
| Days Min Temp ≤ 32 °F | 31 | 27 | 25 | 12 | 1 | 0 | 0 | 0 | 1 | 11 | 26 | 30 | 164 |
| Days Min Temp ≤ 0 °F | 4 | 1 | 0 | 0 | 0 | 0 | 0 | 0 | 0 | 0 | 0 | 3 | 8 |
| Heating Degree Days | 1166 | 901 | 746 | 468 | 219 | 43 | 4 | 8 | 120 | 442 | 836 | 1126 | 6079 |
| Cooling Degree Days | 0 | 0 | 0 | 1 | 19 | 145 | 260 | 210 | 53 | 0 | 0 | 0 | 688 |
| Total Precipitation (") | 0.50 | 0.38 | 1.20 | 1.68 | 2.53 | 1.87 | 1.41 | 1.14 | 1.07 | 0.99 | 0.90 | 0.49 | 14.16 |
| Days ≥ 0.1" Precip | 2 | 1 | 3 | 4 | 6 | 4 | 3 | 3 | 3 | 3 | 3 | 2 | 37 |
| Total Snowfall (") | 6.0 | 4.3 | 7.6 | *6.0* | 0.9 | 0.0 | 0.0 | 0.0 | 0.7 | *3.3* | 8.8 | *6.3* | 43.9 |
| Days ≥ 1" Snow Depth | 16 | *8* | *4* | *2* | 0 | 0 | 0 | 0 | 0 | 1 | *7* | *14* | 52 |

## GREEN MOUNTAIN DAM *Summit County* ELEVATION 7766 ft LAT/LONG 39° 53' N / 106° 20' W

| | JAN | FEB | MAR | APR | MAY | JUN | JUL | AUG | SEP | OCT | NOV | DEC | YEAR |
|---|---|---|---|---|---|---|---|---|---|---|---|---|---|
| Maximum Temp °F | 29.1 | 33.6 | 42.2 | 52.4 | 62.8 | 73.3 | 78.5 | 76.7 | 69.0 | 57.3 | 40.8 | 30.5 | 53.9 |
| Minimum Temp °F | 4.6 | 5.9 | 15.1 | 24.5 | 32.9 | 39.4 | 44.5 | 43.2 | 36.2 | 27.2 | 17.5 | 7.5 | 24.9 |
| Mean Temp °F | 16.9 | 19.7 | 28.7 | 38.5 | 47.9 | 56.4 | 61.5 | 60.0 | 52.6 | 42.2 | 29.2 | 19.0 | 39.4 |
| Days Max Temp ≥ 90 °F | 0 | 0 | 0 | 0 | 0 | 0 | 0 | 0 | 0 | 0 | 0 | 0 | 0 |
| Days Max Temp ≤ 32 °F | 19 | 11 | 5 | 1 | 0 | 0 | 0 | 0 | 0 | 0 | 6 | 18 | 60 |
| Days Min Temp ≤ 32 °F | 31 | 28 | 31 | 26 | 14 | 3 | 1 | 1 | 8 | 25 | 29 | 31 | 228 |
| Days Min Temp ≤ 0 °F | 12 | 10 | 4 | 0 | 0 | 0 | 0 | 0 | 0 | 0 | 2 | 9 | 37 |
| Heating Degree Days | 1485 | 1272 | 1119 | 790 | 523 | 253 | 112 | 155 | 366 | 699 | 1069 | 1418 | 9261 |
| Cooling Degree Days | 0 | 0 | 0 | 0 | 0 | 2 | 7 | 5 | 0 | 0 | 0 | 0 | 14 |
| Total Precipitation (") | 0.90 | 0.90 | 1.34 | 1.25 | 1.53 | 1.26 | 1.53 | 1.38 | 1.30 | 1.21 | 1.12 | 1.05 | 14.77 |
| Days ≥ 0.1" Precip | 4 | 3 | 5 | 4 | 5 | 4 | 5 | 5 | 4 | 4 | 4 | 4 | 51 |
| Total Snowfall (") | *12.8* | *12.8* | *14.7* | *8.0* | 2.5 | 0.3 | 0.0 | 0.0 | 0.4 | 3.5 | *11.7* | *12.9* | 79.6 |
| Days ≥ 1" Snow Depth | 30 | 27 | 23 | 6 | 1 | 0 | 0 | 0 | 0 | 2 | 11 | 26 | 126 |

## GUNNISON 3 SW *Gunnison County* ELEVATION 7664 ft LAT/LONG 38° 32' N / 106° 56' W

| | JAN | FEB | MAR | APR | MAY | JUN | JUL | AUG | SEP | OCT | NOV | DEC | YEAR |
|---|---|---|---|---|---|---|---|---|---|---|---|---|---|
| Maximum Temp °F | 25.3 | 31.0 | 42.8 | 55.6 | 65.7 | 75.6 | 80.4 | 78.3 | 71.5 | 60.6 | 43.6 | 29.7 | 55.0 |
| Minimum Temp °F | -7.0 | -1.7 | 12.8 | 21.2 | 29.0 | 35.6 | 42.4 | 40.3 | 31.9 | 21.6 | 11.6 | -1.5 | 19.7 |
| Mean Temp °F | 9.2 | 14.7 | 27.8 | 38.4 | 47.4 | 55.6 | 61.4 | 59.3 | 51.8 | 41.1 | 27.6 | 14.1 | 37.4 |
| Days Max Temp ≥ 90 °F | 0 | 0 | 0 | 0 | 0 | 0 | 1 | 0 | 0 | 0 | 0 | 0 | 1 |
| Days Max Temp ≤ 32 °F | 22 | 14 | 4 | 0 | 0 | 0 | 0 | 0 | 0 | 0 | 4 | 19 | 63 |
| Days Min Temp ≤ 32 °F | 30 | 28 | 30 | 29 | 22 | 9 | 0 | 3 | 15 | 28 | 29 | 31 | 254 |
| Days Min Temp ≤ 0 °F | 22 | 15 | 4 | 0 | 0 | 0 | 0 | 0 | 0 | 0 | 4 | 18 | 63 |
| Heating Degree Days | 1728 | 1414 | 1145 | 792 | 540 | 275 | 110 | 172 | 389 | 733 | 1114 | 1571 | 9983 |
| Cooling Degree Days | 0 | 0 | 0 | *0* | *0* | *2* | *10* | *4* | *0* | 0 | 0 | 0 | 16 |
| Total Precipitation (") | 0.71 | 0.57 | 0.49 | 0.49 | 0.70 | 0.62 | 1.45 | 1.60 | 1.06 | 0.79 | 0.60 | 0.72 | 9.80 |
| Days ≥ 0.1" Precip | 2 | 2 | 2 | 2 | 2 | 2 | 5 | 5 | 3 | 2 | 2 | 2 | 31 |
| Total Snowfall (") | 11.0 | *7.2* | *6.4* | *3.0* | *0.6* | 0.0 | 0.0 | 0.0 | 0.1 | 2.1 | 5.9 | 10.3 | 46.6 |
| Days ≥ 1" Snow Depth | *26* | *21* | *9* | na | 0 | 0 | 0 | 0 | 0 | *1* | *5* | *17* | na |

## HAYDEN *Routt County* ELEVATION 6375 ft LAT/LONG 40° 29' N / 107° 15' W

| | JAN | FEB | MAR | APR | MAY | JUN | JUL | AUG | SEP | OCT | NOV | DEC | YEAR |
|---|---|---|---|---|---|---|---|---|---|---|---|---|---|
| Maximum Temp °F | 29.4 | 34.1 | 43.7 | 57.0 | 67.9 | 78.4 | 84.8 | 82.7 | 73.9 | 61.2 | 43.7 | 31.3 | 57.3 |
| Minimum Temp °F | 5.9 | 9.1 | 19.0 | 28.1 | 36.1 | 42.8 | 48.7 | 47.2 | 39.0 | 29.0 | 19.1 | 8.3 | 27.7 |
| Mean Temp °F | 17.7 | 21.6 | 31.4 | 42.6 | 52.0 | 60.6 | 66.8 | 65.0 | 56.5 | 45.1 | 31.4 | 19.9 | 42.6 |
| Days Max Temp ≥ 90 °F | 0 | 0 | 0 | 0 | 0 | 2 | 5 | 3 | 0 | 0 | 0 | 0 | 10 |
| Days Max Temp ≤ 32 °F | 19 | 11 | 4 | 0 | 0 | 0 | 0 | 0 | 0 | 0 | 5 | 17 | 56 |
| Days Min Temp ≤ 32 °F | 31 | 28 | 30 | 22 | 8 | 1 | 0 | 0 | 6 | 22 | 28 | 31 | 207 |
| Days Min Temp ≤ 0 °F | 10 | 7 | 2 | 0 | 0 | 0 | 0 | 0 | 0 | 0 | 1 | 8 | 28 |
| Heating Degree Days | 1461 | 1219 | 1035 | 666 | 396 | 147 | 21 | 47 | 255 | 609 | 1001 | 1392 | 8249 |
| Cooling Degree Days | 0 | 0 | 0 | 0 | 0 | 27 | 86 | 65 | 5 | 0 | 0 | 0 | 183 |
| Total Precipitation (") | 1.44 | 1.12 | 1.27 | 1.38 | 1.46 | 1.26 | 1.53 | 1.36 | 1.25 | 1.70 | 1.51 | 1.56 | 16.84 |
| Days ≥ 0.1" Precip | 5 | 4 | 5 | 5 | 5 | 4 | 4 | 4 | 4 | 4 | 5 | 5 | 54 |
| Total Snowfall (") | 26.7 | 19.0 | 15.1 | 8.2 | 1.4 | 0.3 | 0.0 | 0.0 | 0.7 | 5.4 | 17.2 | 25.9 | 119.9 |
| Days ≥ 1" Snow Depth | 30 | 27 | 19 | *2* | 0 | 0 | 0 | 0 | 0 | 2 | 12 | 26 | 118 |

## HERMIT 7 ESE *Mineral County* ELEVATION 9006 ft LAT/LONG 37° 45' N / 107° 7' W

| | JAN | FEB | MAR | APR | MAY | JUN | JUL | AUG | SEP | OCT | NOV | DEC | YEAR |
|---|---|---|---|---|---|---|---|---|---|---|---|---|---|
| Maximum Temp °F | 27.4 | 31.3 | 35.7 | 44.9 | 58.9 | 70.3 | 74.8 | 72.3 | 66.8 | 57.7 | 42.3 | 29.9 | 51.0 |
| Minimum Temp °F | -8.2 | -4.4 | 4.1 | 14.1 | 24.1 | 30.0 | 37.1 | 36.6 | 28.7 | 19.1 | 6.6 | -5.5 | 15.2 |
| Mean Temp °F | 9.6 | 13.4 | 19.9 | 29.6 | 41.5 | 50.2 | 55.9 | 54.5 | 47.7 | 38.4 | 24.5 | 12.2 | 33.1 |
| Days Max Temp ≥ 90 °F | 0 | 0 | 0 | 0 | 0 | 0 | 0 | 0 | 0 | 0 | 0 | 0 | 0 |
| Days Max Temp ≤ 32 °F | 22 | 17 | 12 | 2 | 0 | 0 | 0 | 0 | 0 | 0 | 6 | 21 | 80 |
| Days Min Temp ≤ 32 °F | 31 | 28 | 31 | 30 | 28 | 22 | 6 | 8 | 21 | 30 | 30 | 31 | 296 |
| Days Min Temp ≤ 0 °F | 24 | 19 | 11 | 2 | 0 | 0 | 0 | 0 | 0 | 0 | 8 | 23 | 87 |
| Heating Degree Days | 1713 | 1451 | 1390 | 1056 | 721 | 439 | 275 | 320 | 511 | 817 | 1210 | 1632 | 11535 |
| Cooling Degree Days | 0 | 0 | 0 | 0 | 0 | 0 | 0 | 0 | 0 | 0 | 0 | 0 | 0 |
| Total Precipitation (") | 0.80 | 0.77 | 1.12 | 1.18 | 1.13 | 0.78 | 2.31 | 2.39 | 1.54 | 1.43 | 1.13 | 1.22 | 15.80 |
| Days ≥ 0.1" Precip | 2 | 2 | 3 | 3 | 4 | 3 | 8 | 7 | 4 | 3 | 3 | 3 | 45 |
| Total Snowfall (") | 10.6 | 10.9 | 15.0 | *10.4* | 1.7 | 0.0 | 0.0 | 0.0 | 0.2 | 4.4 | *11.0* | 15.2 | 79.4 |
| Days ≥ 1" Snow Depth | na | na | na | na | 0 | 0 | 0 | 0 | 0 | *1* | na | na | na |

## HOLLY *Prowers County* ELEVATION 3390 ft LAT/LONG 38° 3' N / 102° 7' W

| | JAN | FEB | MAR | APR | MAY | JUN | JUL | AUG | SEP | OCT | NOV | DEC | YEAR |
|---|---|---|---|---|---|---|---|---|---|---|---|---|---|
| Maximum Temp °F | 43.6 | 49.5 | 58.8 | 69.5 | 77.7 | 88.8 | 93.9 | 91.1 | 82.3 | 71.6 | 55.5 | 45.7 | 69.0 |
| Minimum Temp °F | 13.0 | 17.3 | 25.5 | 36.3 | 46.2 | 56.5 | 62.2 | 59.4 | 49.3 | 36.0 | 23.3 | 15.4 | 36.7 |
| Mean Temp °F | 28.4 | 33.4 | 42.2 | 52.9 | 62.0 | 72.7 | 78.0 | 75.3 | 65.9 | 53.8 | 39.4 | 30.6 | 52.9 |
| Days Max Temp ≥ 90 °F | 0 | 0 | 0 | 1 | 4 | 15 | 23 | 19 | 9 | 1 | 0 | 0 | 72 |
| Days Max Temp ≤ 32 °F | 7 | 4 | 2 | 0 | 0 | 0 | 0 | 0 | 0 | 0 | 2 | 5 | 20 |
| Days Min Temp ≤ 32 °F | 31 | 27 | 25 | 10 | 1 | 0 | 0 | 0 | 1 | 9 | 26 | 31 | 161 |
| Days Min Temp ≤ 0 °F | 4 | 2 | 0 | 0 | 0 | 0 | 0 | 0 | 0 | 0 | 0 | 2 | 8 |
| Heating Degree Days | 1130 | 886 | 702 | 366 | 145 | 17 | 2 | 4 | 84 | 348 | 761 | 1062 | 5507 |
| Cooling Degree Days | 0 | 0 | 0 | 13 | 55 | 259 | 409 | 342 | 126 | 8 | 0 | 0 | 1212 |
| Total Precipitation (") | 0.30 | 0.37 | 0.86 | 1.17 | 2.40 | 3.06 | 2.42 | 2.47 | 1.47 | 0.89 | 0.64 | 0.30 | 16.35 |
| Days ≥ 0.1" Precip | 1 | 1 | 2 | 3 | 5 | 5 | 5 | 4 | 3 | 2 | 2 | 1 | 34 |
| Total Snowfall (") | 5.3 | 3.9 | *5.1* | 1.8 | 0.1 | 0.0 | 0.0 | 0.0 | 0.0 | 0.3 | 3.7 | 4.4 | 24.6 |
| Days ≥ 1" Snow Depth | *9* | *5* | *2* | 0 | 0 | 0 | 0 | 0 | 0 | 0 | 2 | *5* | 23 |

## HOLYOKE *Phillips County* ELEVATION 3730 ft LAT/LONG 40° 35' N / 102° 18' W

| | JAN | FEB | MAR | APR | MAY | JUN | JUL | AUG | SEP | OCT | NOV | DEC | YEAR |
|---|---|---|---|---|---|---|---|---|---|---|---|---|---|
| Maximum Temp °F | 40.1 | 45.7 | 53.7 | 64.2 | 73.4 | 83.4 | 88.8 | 86.9 | 78.0 | 66.5 | 51.2 | 42.4 | 64.5 |
| Minimum Temp °F | 13.5 | 17.7 | 24.7 | 34.2 | 44.4 | 53.9 | 59.1 | 57.0 | 47.1 | 35.3 | 23.8 | 15.7 | 35.5 |
| Mean Temp °F | 26.8 | 31.7 | 39.2 | 49.2 | 58.9 | 68.7 | 74.0 | 72.0 | 62.6 | 50.9 | 37.5 | 29.1 | 50.1 |
| Days Max Temp ≥ 90 °F | 0 | 0 | 0 | 0 | 2 | 9 | 16 | 13 | 4 | 0 | 0 | 0 | 44 |
| Days Max Temp ≤ 32 °F | 8 | 5 | 2 | 0 | 0 | 0 | 0 | 0 | 0 | 0 | 3 | 7 | 25 |
| Days Min Temp ≤ 32 °F | 30 | 28 | 26 | 13 | 2 | 0 | 0 | 0 | 1 | 11 | 26 | 30 | 167 |
| Days Min Temp ≤ 0 °F | 5 | 2 | 1 | 0 | 0 | 0 | 0 | 0 | 0 | 0 | 0 | 3 | 11 |
| Heating Degree Days | 1177 | 933 | 793 | 470 | 213 | 39 | 5 | 11 | 129 | 432 | 817 | 1107 | 6126 |
| Cooling Degree Days | 0 | 0 | 0 | 4 | 30 | 152 | 270 | 233 | 65 | 2 | 0 | 0 | 756 |
| Total Precipitation (") | 0.52 | 0.45 | 1.23 | 1.64 | 3.36 | 3.08 | 2.75 | 2.03 | 1.08 | 0.84 | 0.68 | 0.42 | 18.08 |
| Days ≥ 0.1" Precip | 2 | 1 | 3 | 4 | 6 | 6 | 5 | 4 | 3 | 2 | 2 | 1 | 39 |
| Total Snowfall (") | 6.8 | 4.5 | *7.9* | 2.2 | 0.2 | 0.0 | 0.0 | 0.0 | 0.3 | 1.9 | 5.5 | 5.3 | 34.6 |
| Days ≥ 1" Snow Depth | 11 | 8 | 4 | 1 | 0 | 0 | 0 | 0 | 0 | 1 | 4 | 7 | 36 |

## IGNACIO 1 N *La Plata County* ELEVATION 6424 ft LAT/LONG 37° 8' N / 107° 38' W

| | JAN | FEB | MAR | APR | MAY | JUN | JUL | AUG | SEP | OCT | NOV | DEC | YEAR |
|---|---|---|---|---|---|---|---|---|---|---|---|---|---|
| Maximum Temp °F | *38.1* | 44.6 | 52.3 | 61.6 | 71.6 | 82.6 | *87.3* | 84.6 | *76.3* | 65.2 | *50.9* | *40.8* | 63.0 |
| Minimum Temp °F | *6.9* | 13.0 | 20.9 | 26.5 | 34.0 | 41.1 | *49.4* | 47.8 | *39.8* | *30.0* | *20.4* | *10.5* | 28.4 |
| Mean Temp °F | *22.5* | 28.8 | 36.6 | 44.1 | 52.8 | 61.9 | *68.4* | 66.2 | *58.1* | *47.5* | *35.8* | *25.7* | 45.7 |
| Days Max Temp ≥ 90 °F | 0 | 0 | 0 | 0 | 0 | 5 | 11 | 6 | 1 | 0 | 0 | 0 | 23 |
| Days Max Temp ≤ 32 °F | 7 | 2 | 0 | 0 | 0 | 0 | 0 | 0 | 0 | 0 | *1* | *5* | 15 |
| Days Min Temp ≤ 32 °F | *31* | 27 | 30 | 25 | *12* | 2 | 0 | 0 | *4* | *21* | *27* | 31 | 210 |
| Days Min Temp ≤ 0 °F | 8 | 3 | 0 | 0 | 0 | 0 | 0 | 0 | 0 | 0 | 0 | *5* | 16 |
| Heating Degree Days | *1310* | 1013 | 872 | 620 | 371 | 120 | *12* | 35 | *210* | *534* | *870* | *1210* | 7177 |
| Cooling Degree Days | 0 | 0 | 0 | 0 | 0 | 43 | *140* | *95* | *12* | 0 | *0* | 0 | 290 |
| Total Precipitation (") | 1.33 | 1.04 | 1.25 | 0.76 | 0.91 | 0.53 | 1.30 | 1.67 | 1.63 | 1.37 | 1.25 | 1.34 | 14.38 |
| Days ≥ 0.1" Precip | 3 | 3 | 3 | 2 | 3 | 2 | 3 | 5 | 4 | 4 | 4 | *3* | 39 |
| Total Snowfall (") | na | na | na | na | *0.0* | 0.0 | 0.0 | 0.0 | 0.0 | 0.4 | na | na | na |
| Days ≥ 1" Snow Depth | na | na | na | na | 0 | 0 | 0 | 0 | 0 | 0 | na | na | na |

## JULESBURG *Sedgwick County* ELEVATION 3471 ft LAT/LONG 41° 0 ' N / 102° 15 ' W

| | JAN | FEB | MAR | APR | MAY | JUN | JUL | AUG | SEP | OCT | NOV | DEC | YEAR |
|---|---|---|---|---|---|---|---|---|---|---|---|---|---|
| Maximum Temp °F | 40.1 | 46.2 | 54.0 | 63.9 | 73.5 | 84.2 | ***90.8*** | 88.8 | 79.8 | ***68.3*** | ***51.7*** | ***41.8*** | 65.3 |
| Minimum Temp °F | 13.0 | 17.6 | 24.4 | 34.4 | 45.1 | 54.6 | ***60.9*** | 58.3 | 47.4 | ***34.7*** | 23.2 | ***14.4*** | 35.7 |
| Mean Temp °F | 26.6 | 31.9 | 39.2 | 49.2 | 59.3 | 69.4 | ***75.9*** | 73.6 | 63.6 | ***51.6*** | ***37.4*** | ***28.2*** | 50.5 |
| Days Max Temp ≥ 90 °F | 0 | 0 | 0 | 0 | 2 | ***9*** | ***19*** | 16 | 6 | ***0*** | ***0*** | ***0*** | 52 |
| Days Max Temp ≤ 32 °F | 8 | 5 | 2 | 0 | 0 | 0 | ***0*** | 0 | 0 | ***0*** | ***3*** | ***7*** | 25 |
| Days Min Temp ≤ 32 °F | 31 | 27 | 26 | 13 | 2 | 0 | ***0*** | 0 | 1 | ***12*** | 26 | ***30*** | 168 |
| Days Min Temp ≤ 0 °F | 5 | 2 | 1 | 0 | 0 | 0 | ***0*** | 0 | 0 | ***0*** | 0 | ***3*** | 11 |
| Heating Degree Days | 1185 | 929 | 792 | 471 | 205 | 35 | ***3*** | 6 | 120 | ***413*** | ***820*** | ***1135*** | 6114 |
| Cooling Degree Days | 0 | 0 | 0 | 6 | 30 | ***174*** | ***348*** | ***292*** | ***84*** | ***1*** | ***0*** | ***0*** | 935 |
| Total Precipitation (") | 0.43 | 0.33 | 1.16 | 1.52 | 3.34 | 2.92 | ***2.57*** | 2.03 | 1.17 | 0.81 | 0.55 | 0.33 | 17.16 |
| Days ≥ 0.1" Precip | ***1*** | 1 | ***3*** | 4 | 6 | 6 | ***5*** | 4 | 3 | ***2*** | ***2*** | ***1*** | 38 |
| Total Snowfall (") | na | na | na | ***2.1*** | 0.0 | 0.0 | ***0.0*** | 0.0 | 0.2 | ***0.9*** | na | na | na |
| Days ≥ 1" Snow Depth | na | na | na | ***1*** | 0 | 0 | ***0*** | 0 | 0 | ***0*** | na | na | na |

## KASSLER *Jefferson County* ELEVATION 5501 ft LAT/LONG 39° 30 ' N / 105° 6 ' W

| | JAN | FEB | MAR | APR | MAY | JUN | JUL | AUG | SEP | OCT | NOV | DEC | YEAR |
|---|---|---|---|---|---|---|---|---|---|---|---|---|---|
| Maximum Temp °F | 46.0 | 48.4 | 53.9 | 62.1 | 70.7 | 81.3 | 87.1 | 85.0 | 77.3 | 66.4 | 53.3 | 46.4 | 64.8 |
| Minimum Temp °F | 15.9 | 19.2 | 25.4 | 33.3 | 42.7 | 51.9 | 58.1 | 56.5 | 47.8 | 36.0 | 24.3 | 16.5 | 35.6 |
| Mean Temp °F | 30.8 | 33.8 | 39.7 | 47.7 | 56.7 | 66.6 | 72.6 | 70.8 | 62.5 | 51.1 | 38.9 | 31.5 | 50.2 |
| Days Max Temp ≥ 90 °F | 0 | 0 | 0 | 0 | 0 | 6 | 12 | 8 | 2 | 0 | 0 | 0 | 28 |
| Days Max Temp ≤ 32 °F | 5 | 3 | 2 | 0 | 0 | 0 | 0 | 0 | 0 | 0 | 2 | 5 | 17 |
| Days Min Temp ≤ 32 °F | 28 | 26 | 24 | 14 | 3 | 0 | 0 | 0 | 1 | 11 | 23 | 28 | 158 |
| Days Min Temp ≤ 0 °F | 4 | 2 | 1 | 0 | 0 | 0 | 0 | 0 | 0 | 0 | 1 | 4 | 12 |
| Heating Degree Days | 1053 | 875 | 778 | 512 | 265 | 64 | 7 | 13 | 133 | 426 | 778 | 1033 | 5937 |
| Cooling Degree Days | 0 | 0 | 0 | 2 | 14 | 118 | 237 | 196 | 68 | 2 | 0 | 0 | 637 |
| Total Precipitation (") | 0.55 | 0.65 | 1.66 | 2.08 | 2.95 | 1.88 | 1.75 | 1.58 | 1.42 | 1.51 | 1.36 | 0.78 | 18.17 |
| Days ≥ 0.1" Precip | 2 | 2 | 5 | 5 | 6 | 4 | 4 | 4 | 3 | 3 | 3 | 2 | 43 |
| Total Snowfall (") | 10.2 | 9.7 | 15.7 | 9.4 | 0.2 | 0.0 | 0.0 | 0.0 | 0.9 | 5.3 | 13.6 | ***13.2*** | 78.2 |
| Days ≥ 1" Snow Depth | 12 | 8 | 7 | 3 | 0 | 0 | 0 | 0 | 0 | 2 | 8 | 13 | 53 |

## KIT CARSON *Cheyenne County* ELEVATION 4202 ft LAT/LONG 38° 42 ' N / 102° 44 ' W

| | JAN | FEB | MAR | APR | MAY | JUN | JUL | AUG | SEP | OCT | NOV | DEC | YEAR |
|---|---|---|---|---|---|---|---|---|---|---|---|---|---|
| Maximum Temp °F | 42.2 | 47.4 | 56.5 | 66.4 | 75.5 | 86.3 | 91.8 | 89.7 | 80.6 | 68.8 | 53.2 | 43.9 | 66.9 |
| Minimum Temp °F | 10.8 | 15.4 | 23.7 | 32.9 | 43.5 | 53.6 | 59.2 | 57.0 | 47.7 | 33.9 | 21.6 | 12.9 | 34.4 |
| Mean Temp °F | 26.5 | 31.4 | 40.0 | 49.7 | ***59.6*** | ***70.0*** | 75.6 | 73.5 | 64.3 | 51.4 | 37.6 | 28.4 | 50.7 |
| Days Max Temp ≥ 90 °F | 0 | 0 | 0 | 0 | ***2*** | 12 | 20 | 18 | 7 | 0 | 0 | 0 | 59 |
| Days Max Temp ≤ 32 °F | 7 | 4 | 2 | 0 | 0 | 0 | 0 | 0 | 0 | 0 | 2 | 6 | 21 |
| Days Min Temp ≤ 32 °F | 31 | 28 | 27 | 14 | 2 | 0 | 0 | 0 | 1 | 14 | 27 | 31 | 175 |
| Days Min Temp ≤ 0 °F | 5 | 2 | 1 | 0 | 0 | 0 | 0 | 0 | 0 | 0 | 0 | 3 | 11 |
| Heating Degree Days | 1186 | 942 | 768 | 453 | ***194*** | ***29*** | 2 | 6 | 105 | 416 | 817 | 1129 | 6047 |
| Cooling Degree Days | 0 | 0 | 0 | 3 | ***29*** | ***185*** | ***327*** | ***281*** | ***95*** | 1 | 0 | 0 | 921 |
| Total Precipitation (") | 0.25 | 0.26 | 0.80 | 0.89 | 2.52 | 2.10 | 2.59 | 2.30 | 1.32 | 0.88 | 0.50 | 0.30 | 14.71 |
| Days ≥ 0.1" Precip | 1 | 1 | 2 | 2 | 5 | 4 | 5 | 4 | 3 | 2 | ***2*** | 1 | 32 |
| Total Snowfall (") | ***3.4*** | ***2.6*** | ***3.0*** | 1.6 | 0.4 | 0.0 | 0.0 | 0.0 | 0.2 | 1.0 | na | 3.4 | na |
| Days ≥ 1" Snow Depth | na | na | na | ***0*** | 0 | 0 | 0 | 0 | 0 | 0 | na | na | na |

## LA JUNTA 4 NNE *Otero County* ELEVATION 4196 ft LAT/LONG 38° 3 ' N / 103° 31 ' W

| | JAN | FEB | MAR | APR | MAY | JUN | JUL | AUG | SEP | OCT | NOV | DEC | YEAR |
|---|---|---|---|---|---|---|---|---|---|---|---|---|---|
| Maximum Temp °F | 43.4 | 49.4 | 58.7 | 69.2 | 77.8 | 89.6 | 94.3 | 91.2 | 82.8 | 70.6 | 54.9 | 44.9 | 68.9 |
| Minimum Temp °F | 15.8 | 20.6 | 28.2 | 37.8 | 47.8 | 57.8 | 63.4 | 61.0 | 51.7 | 38.6 | 25.4 | 17.1 | 38.8 |
| Mean Temp °F | 29.7 | 35.0 | 43.5 | 53.5 | 62.9 | 73.7 | 78.9 | 76.1 | 67.3 | 54.6 | 40.2 | 31.0 | 53.9 |
| Days Max Temp ≥ 90 °F | 0 | 0 | 0 | 1 | 4 | 16 | 24 | 20 | 9 | 1 | 0 | 0 | 75 |
| Days Max Temp ≤ 32 °F | 8 | 4 | 2 | 0 | 0 | 0 | 0 | 0 | 0 | 0 | 2 | 6 | 22 |
| Days Min Temp ≤ 32 °F | 30 | 26 | 22 | 7 | 0 | 0 | 0 | 0 | 0 | 6 | 24 | 30 | 145 |
| Days Min Temp ≤ 0 °F | 3 | 2 | 0 | 0 | 0 | 0 | 0 | 0 | 0 | 0 | 0 | 2 | 7 |
| Heating Degree Days | 1085 | 840 | 660 | 348 | 128 | 13 | 1 | 3 | 68 | 325 | 738 | 1048 | 5257 |
| Cooling Degree Days | 0 | 0 | 1 | 12 | 70 | 288 | 436 | 375 | 158 | 9 | 0 | 0 | 1349 |
| Total Precipitation (") | 0.30 | 0.33 | 0.74 | 1.07 | 1.65 | 1.37 | 2.18 | 1.51 | 0.93 | 0.69 | 0.47 | 0.30 | 11.54 |
| Days ≥ 0.1" Precip | 1 | 1 | 2 | 3 | 4 | 3 | 5 | 3 | 2 | 2 | 1 | 1 | 28 |
| Total Snowfall (") | 4.4 | 4.3 | 6.7 | 2.6 | 0.5 | 0.0 | 0.0 | 0.0 | 0.1 | 1.5 | 3.6 | 4.4 | 28.1 |
| Days ≥ 1" Snow Depth | 10 | 6 | 3 | 1 | 0 | 0 | 0 | 0 | 0 | 0 | 3 | 8 | 31 |

## LAKE CITY *Hinsdale County* ELEVATION 8894 ft LAT/LONG 38° 3 ' N / 107° 19 ' W

| | JAN | FEB | MAR | APR | MAY | JUN | JUL | AUG | SEP | OCT | NOV | DEC | YEAR |
|---|---|---|---|---|---|---|---|---|---|---|---|---|---|
| Maximum Temp °F | 34.6 | 38.7 | 44.8 | 53.3 | 62.3 | 73.0 | 77.0 | 74.4 | 69.2 | 60.1 | 44.7 | 35.1 | 55.6 |
| Minimum Temp °F | -3.6 | 1.5 | 12.1 | 21.4 | 30.2 | 37.6 | 43.7 | 42.3 | 34.7 | 24.7 | 11.9 | 0.3 | 21.4 |
| Mean Temp °F | 15.5 | 20.1 | 28.4 | 37.4 | 46.3 | 55.4 | 60.4 | 58.3 | 52.0 | 42.4 | 28.4 | 17.7 | 38.5 |
| Days Max Temp ≥ 90 °F | 0 | 0 | 0 | 0 | 0 | 0 | 0 | 0 | 0 | 0 | 0 | 0 | 0 |
| Days Max Temp ≤ 32 °F | 12 | 6 | 2 | 0 | 0 | 0 | 0 | 0 | 0 | 0 | 3 | 11 | 34 |
| Days Min Temp ≤ 32 °F | 31 | 28 | 31 | 28 | 20 | 6 | 0 | 1 | 11 | 27 | 29 | 31 | 243 |
| Days Min Temp ≤ 0 °F | 21 | 14 | 4 | 0 | 0 | 0 | 0 | 0 | 0 | 0 | 4 | 16 | 59 |
| Heating Degree Days | 1529 | 1262 | 1127 | 821 | 574 | 285 | 142 | 201 | 385 | 694 | 1093 | 1459 | 9572 |
| Cooling Degree Days | 0 | 0 | 0 | 0 | 0 | 3 | 6 | 2 | 0 | 0 | 0 | 0 | 11 |
| Total Precipitation (") | 0.71 | 0.70 | 0.91 | 1.01 | 1.06 | 0.77 | 1.97 | 2.17 | 1.33 | 1.14 | 1.04 | 0.98 | 13.79 |
| Days ≥ 0.1" Precip | 3 | 2 | 3 | 3 | 3 | 3 | 6 | 7 | 4 | 4 | 4 | 3 | 45 |
| Total Snowfall (") | 11.4 | 9.7 | 14.3 | 10.5 | 3.8 | 0.3 | 0.0 | 0.0 | 0.1 | 5.1 | 14.1 | 15.4 | 84.7 |
| Days ≥ 1" Snow Depth | na | *14* | na | na | *1* | 0 | 0 | 0 | 0 | *1* | na | *15* | na |

## LAKE GEORGE 8 SW *Park County* ELEVATION 8507 ft LAT/LONG 38° 55 ' N / 105° 29 ' W

| | JAN | FEB | MAR | APR | MAY | JUN | JUL | AUG | SEP | OCT | NOV | DEC | YEAR |
|---|---|---|---|---|---|---|---|---|---|---|---|---|---|
| Maximum Temp °F | 31.3 | 35.1 | 41.8 | 50.4 | 60.0 | 70.7 | 75.7 | 73.2 | 66.9 | 56.7 | 41.4 | 32.4 | 53.0 |
| Minimum Temp °F | -1.5 | 1.8 | 12.5 | 22.1 | 31.7 | 39.6 | 45.7 | 44.5 | 36.3 | 25.9 | 14.3 | 2.1 | 22.9 |
| Mean Temp °F | 14.9 | 18.5 | 27.2 | 36.3 | 45.9 | 55.2 | 60.7 | 58.9 | 51.6 | 41.3 | 27.9 | 17.3 | 38.0 |
| Days Max Temp ≥ 90 °F | 0 | 0 | 0 | 0 | 0 | 0 | 0 | 0 | 0 | 0 | 0 | 0 | 0 |
| Days Max Temp ≤ 32 °F | 15 | 10 | 6 | 2 | 0 | 0 | 0 | 0 | 0 | 1 | 6 | 14 | 54 |
| Days Min Temp ≤ 32 °F | 31 | 28 | 31 | 28 | 17 | 3 | 0 | 0 | 7 | 27 | 30 | 31 | 233 |
| Days Min Temp ≤ 0 °F | 17 | 13 | 5 | 1 | 0 | 0 | 0 | 0 | 0 | 0 | 3 | 13 | 52 |
| Heating Degree Days | 1547 | 1308 | 1166 | 855 | 585 | 288 | 130 | 184 | 394 | 727 | 1105 | 1474 | 9763 |
| Cooling Degree Days | 0 | 0 | 0 | 0 | 0 | 1 | 5 | 1 | 0 | 0 | 0 | 0 | 7 |
| Total Precipitation (") | 0.28 | 0.31 | 0.80 | 0.91 | 1.44 | 1.39 | 2.51 | 2.64 | 1.10 | 0.84 | 0.48 | 0.44 | 13.14 |
| Days ≥ 0.1" Precip | 1 | 1 | 3 | 3 | 4 | 4 | 7 | 7 | 3 | 2 | 1 | 2 | 38 |
| Total Snowfall (") | 5.5 | 6.0 | 13.9 | 12.6 | 3.2 | 0.3 | 0.0 | 0.0 | 1.2 | 6.3 | 7.0 | 8.9 | 64.9 |
| Days ≥ 1" Snow Depth | 20 | 16 | 14 | 7 | 1 | 0 | 0 | 0 | 0 | 3 | 10 | 19 | 90 |

## LAKEWOOD *Jefferson County* ELEVATION 5636 ft LAT/LONG 39° 45 ' N / 105° 8 ' W

| | JAN | FEB | MAR | APR | MAY | JUN | JUL | AUG | SEP | OCT | NOV | DEC | YEAR |
|---|---|---|---|---|---|---|---|---|---|---|---|---|---|
| Maximum Temp °F | 43.9 | 46.3 | 52.0 | 59.8 | 68.8 | 79.4 | 85.4 | 83.6 | 75.4 | 64.6 | 51.2 | 44.9 | 62.9 |
| Minimum Temp °F | 18.2 | 20.9 | 26.7 | 33.8 | 42.8 | 52.0 | 57.9 | 56.2 | 47.3 | 36.5 | 25.9 | 19.3 | 36.5 |
| Mean Temp °F | 31.1 | 33.6 | 39.3 | 46.9 | 55.8 | 65.7 | 71.7 | 69.9 | 61.4 | 50.6 | 38.6 | 32.1 | 49.7 |
| Days Max Temp ≥ 90 °F | 0 | 0 | 0 | 0 | 0 | 4 | 9 | 5 | 1 | 0 | 0 | 0 | 19 |
| Days Max Temp ≤ 32 °F | 6 | 4 | 2 | 0 | 0 | 0 | 0 | 0 | 0 | 0 | 3 | 5 | 20 |
| Days Min Temp ≤ 32 °F | 28 | 25 | 23 | 13 | 3 | 0 | 0 | 0 | 1 | 9 | 23 | 28 | 153 |
| Days Min Temp ≤ 0 °F | 3 | 1 | 0 | 0 | 0 | 0 | 0 | 0 | 0 | 0 | 0 | 2 | 6 |
| Heating Degree Days | 1046 | 879 | 789 | 538 | 289 | 78 | 10 | 15 | 151 | 442 | 785 | 1014 | 6036 |
| Cooling Degree Days | 0 | 0 | 0 | 2 | 12 | 112 | 222 | 179 | 53 | 1 | 0 | 0 | 581 |
| Total Precipitation (") | 0.45 | 0.48 | 1.36 | 1.73 | 2.55 | 2.18 | 1.90 | 1.58 | 1.32 | 1.13 | 1.12 | 0.58 | 16.38 |
| Days ≥ 0.1" Precip | 2 | 2 | 4 | 4 | 5 | 4 | 5 | 3 | 3 | 3 | 3 | 2 | 40 |
| Total Snowfall (") | 7.4 | 7.9 | 10.8 | 9.2 | 1.2 | 0.0 | 0.0 | 0.0 | 0.8 | 4.3 | 11.1 | 8.8 | 61.5 |
| Days ≥ 1" Snow Depth | 12 | 9 | 7 | 4 | 1 | 0 | 0 | 0 | 0 | 2 | 9 | 12 | 56 |

## LAMAR *Prowers County* ELEVATION 3642 ft LAT/LONG 38° 4 ' N / 102° 37 ' W

| | JAN | FEB | MAR | APR | MAY | JUN | JUL | AUG | SEP | OCT | NOV | DEC | YEAR |
|---|---|---|---|---|---|---|---|---|---|---|---|---|---|
| Maximum Temp °F | 43.9 | 50.1 | 59.5 | 70.1 | 77.9 | 88.5 | 93.4 | 90.7 | 82.3 | 70.8 | 55.2 | 45.5 | 69.0 |
| Minimum Temp °F | 14.5 | 19.6 | 27.5 | 37.3 | 47.3 | 57.1 | 62.6 | 60.5 | 51.1 | 37.2 | 24.4 | 16.2 | 37.9 |
| Mean Temp °F | 29.2 | 34.9 | 43.5 | 53.7 | 62.6 | 72.8 | 78.0 | 75.6 | 66.7 | 54.0 | 39.8 | 30.9 | 53.5 |
| Days Max Temp ≥ 90 °F | 0 | 0 | 0 | 1 | 4 | 15 | 23 | 19 | 9 | 1 | 0 | 0 | 72 |
| Days Max Temp ≤ 32 °F | 7 | 3 | 1 | 0 | 0 | 0 | 0 | 0 | 0 | 0 | 1 | 5 | 17 |
| Days Min Temp ≤ 32 °F | 31 | 27 | 22 | 9 | 1 | 0 | 0 | 0 | 0 | 8 | 25 | 31 | 154 |
| Days Min Temp ≤ 0 °F | 3 | 2 | 0 | 0 | 0 | 0 | 0 | 0 | 0 | 0 | 0 | 2 | 7 |
| Heating Degree Days | 1101 | 843 | 660 | 344 | 135 | 15 | 1 | 4 | 74 | 341 | 748 | 1051 | 5317 |
| Cooling Degree Days | 0 | 0 | 1 | 12 | 65 | 252 | 397 | 344 | 137 | 6 | 0 | 0 | 1214 |
| Total Precipitation (") | 0.42 | 0.43 | 0.93 | 1.24 | 2.42 | 2.33 | 2.20 | 2.18 | 1.34 | 0.77 | 0.67 | 0.40 | 15.33 |
| Days ≥ 0.1" Precip | 2 | 1 | 2 | 3 | 5 | 5 | 5 | 4 | 3 | 2 | 2 | 1 | 35 |
| Total Snowfall (") | 5.5 | 3.7 | 5.8 | 2.0 | 0.0 | 0.0 | 0.0 | 0.0 | 0.1 | 1.3 | 4.8 | 4.7 | 27.9 |
| Days ≥ 1" Snow Depth | 8 | 4 | 2 | 1 | 0 | 0 | 0 | 0 | 0 | 0 | 3 | 6 | 24 |

## LAS ANIMAS *Bent County* ELEVATION 3891 ft LAT/LONG 38° 5 ' N / 103° 13 ' W

| | JAN | FEB | MAR | APR | MAY | JUN | JUL | AUG | SEP | OCT | NOV | DEC | YEAR |
|---|---|---|---|---|---|---|---|---|---|---|---|---|---|
| Maximum Temp °F | 45.4 | 52.0 | 62.0 | 71.5 | 79.7 | 90.8 | 95.4 | 92.6 | 84.3 | 72.8 | 56.9 | 47.1 | 70.9 |
| Minimum Temp °F | 14.0 | 19.3 | 27.5 | 36.7 | 46.9 | 56.5 | 62.1 | 59.9 | 50.6 | 37.1 | 24.3 | 15.8 | 37.6 |
| Mean Temp °F | 29.7 | 35.7 | 44.7 | 54.1 | 63.3 | 73.7 | 78.8 | 76.3 | 67.5 | 55.1 | 40.7 | 31.4 | 54.3 |
| Days Max Temp ≥ 90 °F | 0 | 0 | 0 | 1 | 5 | 18 | 25 | 22 | 10 | 1 | 0 | 0 | 82 |
| Days Max Temp ≤ 32 °F | 6 | 3 | 1 | 0 | 0 | 0 | 0 | 0 | 0 | 0 | 1 | 5 | 16 |
| Days Min Temp ≤ 32 °F | 31 | 27 | 22 | 9 | 1 | 0 | 0 | 0 | 0 | 9 | 25 | 30 | 154 |
| Days Min Temp ≤ 0 °F | 4 | 2 | 0 | 0 | 0 | 0 | 0 | 0 | 0 | 0 | 1 | 2 | 9 |
| Heating Degree Days | 1087 | 820 | 621 | 330 | 117 | 11 | 1 | 2 | 57 | 311 | 724 | 1036 | 5117 |
| Cooling Degree Days | 0 | 0 | 1 | 14 | 79 | 284 | 432 | 374 | 146 | 7 | 0 | 0 | 1337 |
| Total Precipitation (") | 0.30 | 0.38 | 0.71 | 1.03 | 1.89 | 1.79 | 2.11 | 1.43 | 1.17 | 0.73 | 0.53 | 0.28 | 12.35 |
| Days ≥ 0.1" Precip | 1 | 1 | 2 | 3 | 4 | 4 | 4 | 3 | 3 | 2 | 1 | 1 | 29 |
| Total Snowfall (") | 4.0 | 3.0 | 4.3 | 1.2 | 0.1 | 0.0 | 0.0 | 0.0 | 0.2 | 0.7 | 3.8 | 3.3 | 20.6 |
| Days ≥ 1" Snow Depth | 8 | 4 | 2 | 0 | 0 | 0 | 0 | 0 | 0 | 0 | 3 | 5 | 22 |

## LONGMONT 2 ESE *Boulder County* ELEVATION 4954 ft LAT/LONG 40° 10 ' N / 105° 4 ' W

| | JAN | FEB | MAR | APR | MAY | JUN | JUL | AUG | SEP | OCT | NOV | DEC | YEAR |
|---|---|---|---|---|---|---|---|---|---|---|---|---|---|
| Maximum Temp °F | 41.9 | 45.7 | 53.3 | 62.5 | 71.8 | 82.3 | 88.5 | 86.3 | 77.7 | 66.2 | 51.1 | 43.1 | 64.2 |
| Minimum Temp °F | 12.3 | 16.9 | 24.3 | 32.7 | 42.4 | 50.8 | 55.6 | 53.7 | 44.2 | 32.8 | 22.0 | 13.7 | 33.4 |
| Mean Temp °F | 27.1 | 31.3 | 38.8 | 47.6 | 57.1 | 66.6 | 72.1 | 70.0 | 60.9 | 49.5 | 36.5 | 28.4 | 48.8 |
| Days Max Temp ≥ 90 °F | 0 | 0 | 0 | 0 | 1 | 8 | 15 | 11 | 3 | 0 | 0 | 0 | 38 |
| Days Max Temp ≤ 32 °F | 8 | 5 | 2 | 0 | 0 | 0 | 0 | 0 | 0 | 0 | 3 | 6 | 24 |
| Days Min Temp ≤ 32 °F | 30 | 27 | 27 | 14 | 2 | 0 | 0 | 0 | 2 | 14 | 28 | 31 | 175 |
| Days Min Temp ≤ 0 °F | 5 | 2 | 1 | 0 | 0 | 0 | 0 | 0 | 0 | 0 | 1 | 4 | 13 |
| Heating Degree Days | 1168 | 943 | 804 | 516 | 252 | 61 | 8 | 15 | 158 | 474 | 847 | 1127 | 6373 |
| Cooling Degree Days | 0 | 0 | 0 | 1 | 15 | 122 | 231 | 188 | 47 | 1 | 0 | 0 | 605 |
| Total Precipitation (") | 0.38 | 0.38 | 1.15 | 1.83 | 2.25 | 1.80 | 1.14 | 1.37 | 1.30 | 0.90 | 0.79 | 0.58 | 13.87 |
| Days ≥ 0.1" Precip | 1 | 2 | 3 | 4 | 5 | 4 | 4 | 3 | 3 | 2 | 2 | 2 | 35 |
| Total Snowfall (") | 5.4 | 4.1 | 6.0 | 4.2 | 0.5 | 0.0 | 0.0 | 0.0 | 0.5 | 1.6 | 5.9 | 7.2 | 35.4 |
| Days ≥ 1" Snow Depth | *8* | na | *4* | *1* | 0 | 0 | 0 | 0 | 0 | 1 | 3 | na | na |

## MANASSA *Conejos County* ELEVATION 7707 ft LAT/LONG 37° 11 ' N / 105° 57 ' W

| | JAN | FEB | MAR | APR | MAY | JUN | JUL | AUG | SEP | OCT | NOV | DEC | YEAR |
|---|---|---|---|---|---|---|---|---|---|---|---|---|---|
| Maximum Temp °F | 34.2 | 40.7 | 50.4 | 59.7 | 68.3 | 77.3 | 81.1 | 78.7 | 73.0 | 63.1 | 48.2 | 37.1 | 59.3 |
| Minimum Temp °F | 0.1 | 7.5 | 18.1 | 24.3 | 32.8 | 41.1 | 46.5 | 44.8 | 37.1 | 26.3 | 14.3 | 3.7 | 24.7 |
| Mean Temp °F | 17.2 | 24.1 | 34.3 | 42.0 | 50.6 | 59.2 | 63.8 | 61.8 | 55.1 | 44.7 | 31.2 | 20.4 | 42.0 |
| Days Max Temp ≥ 90 °F | 0 | 0 | 0 | 0 | 0 | 0 | 1 | 0 | 0 | 0 | 0 | 0 | 1 |
| Days Max Temp ≤ 32 °F | *13* | 5 | 1 | 0 | 0 | 0 | 0 | 0 | 0 | 0 | 3 | *9* | 31 |
| Days Min Temp ≤ 32 °F | 31 | 28 | 30 | 25 | *13* | *2* | 0 | 0 | *7* | 24 | 28 | 31 | 219 |
| Days Min Temp ≤ 0 °F | *16* | *7* | 1 | 0 | 0 | 0 | 0 | 0 | 0 | 0 | *3* | na | na |
| Heating Degree Days | 1478 | 1149 | 944 | 685 | 440 | 175 | 56 | 105 | 291 | 622 | 1007 | 1375 | 8327 |
| Cooling Degree Days | 0 | 0 | 0 | 0 | 0 | 9 | 24 | 9 | 1 | 0 | 0 | 0 | 43 |
| Total Precipitation (") | 0.28 | 0.27 | 0.43 | 0.40 | 0.79 | 0.55 | 1.28 | 1.57 | 0.99 | 0.69 | 0.50 | 0.39 | 8.14 |
| Days ≥ 0.1" Precip | 1 | 1 | 1 | 2 | 3 | 2 | 4 | 4 | 3 | 2 | 2 | 2 | 27 |
| Total Snowfall (") | na | na | *4.9* | *2.0* | 0.9 | 0.0 | 0.0 | 0.0 | 0.0 | 2.9 | *3.8* | *6.6* | na |
| Days ≥ 1" Snow Depth | na | na | na | *0* | 0 | 0 | 0 | 0 | 0 | *1* | na | na | na |

## MEREDITH *Pitkin County* ELEVATION 7805 ft LAT/LONG 39° 22 ' N / 106° 45 ' W

| | JAN | FEB | MAR | APR | MAY | JUN | JUL | AUG | SEP | OCT | NOV | DEC | YEAR |
|---|---|---|---|---|---|---|---|---|---|---|---|---|---|
| Maximum Temp °F | *33.4* | 36.7 | 41.8 | 50.7 | 62.7 | 73.3 | 79.9 | 78.1 | 70.6 | *59.0* | 42.5 | *35.0* | 55.3 |
| Minimum Temp °F | *2.1* | 4.3 | 13.0 | 20.7 | 29.1 | 34.7 | 40.5 | 39.2 | 32.3 | *23.5* | 13.9 | *5.5* | 21.6 |
| Mean Temp °F | *17.7* | 20.6 | 27.4 | 35.7 | 45.9 | 54.0 | 60.2 | 58.7 | 51.5 | *41.3* | 28.2 | *20.2* | 38.4 |
| Days Max Temp ≥ 90 °F | 0 | 0 | 0 | 0 | 0 | 0 | 1 | 0 | 0 | 0 | 0 | 0 | 1 |
| Days Max Temp ≤ 32 °F | *14* | *8* | *4* | 1 | 0 | 0 | 0 | 0 | 0 | 0 | *6* | *11* | 44 |
| Days Min Temp ≤ 32 °F | 31 | 27 | 31 | *29* | *23* | 11 | 2 | 3 | *16* | *28* | *29* | 30 | 260 |
| Days Min Temp ≤ 0 °F | *14* | *10* | 3 | 0 | 0 | 0 | 0 | 0 | 0 | 0 | *3* | *10* | 40 |
| Heating Degree Days | *1460* | 1248 | *1163* | 860 | 585 | 325 | 146 | 192 | 399 | *729* | *1095* | *1381* | 9583 |
| Cooling Degree Days | 0 | 0 | 0 | 0 | 0 | 1 | 7 | 4 | 0 | 0 | 0 | 0 | 12 |
| Total Precipitation (") | 1.25 | 0.97 | 1.28 | 1.15 | 1.46 | 1.39 | 1.69 | 1.71 | 1.57 | 1.46 | 1.24 | 1.30 | 16.47 |
| Days ≥ 0.1" Precip | 4 | 4 | 4 | 4 | 4 | 3 | 5 | 5 | 3 | *4* | *4* | *3* | 47 |
| Total Snowfall (") | 20.4 | 17.7 | *16.8* | *7.7* | 1.3 | 0.2 | 0.0 | 0.0 | 0.0 | 1.7 | *11.8* | *17.9* | 95.5 |
| Days ≥ 1" Snow Depth | *27* | 24 | *25* | *11* | 1 | 0 | 0 | 0 | 0 | *1* | na | na | na |

## MESA VERDE NATL PARK *Montezuma County* ELEVATION 6965 ft LAT/LONG 37° 12 ' N / 108° 30 ' W

| | JAN | FEB | MAR | APR | MAY | JUN | JUL | AUG | SEP | OCT | NOV | DEC | YEAR |
|---|---|---|---|---|---|---|---|---|---|---|---|---|---|
| Maximum Temp °F | 39.0 | 43.3 | 49.7 | 59.1 | 69.3 | 80.6 | 85.8 | 83.3 | 75.5 | 63.6 | 48.7 | 39.8 | 61.5 |
| Minimum Temp °F | 17.8 | 21.7 | 26.9 | 32.8 | 41.1 | 50.2 | 56.4 | 54.7 | 47.8 | 37.7 | 27.1 | 19.3 | 36.1 |
| Mean Temp °F | 28.4 | 32.5 | 38.3 | 46.0 | 55.3 | 65.4 | 71.1 | 69.0 | 61.7 | 50.7 | 37.9 | 29.6 | 48.8 |
| Days Max Temp ≥ 90 °F | 0 | 0 | 0 | 0 | 0 | 3 | 7 | 3 | 0 | 0 | 0 | 0 | 13 |
| Days Max Temp ≤ 32 °F | 7 | 3 | 1 | 0 | 0 | 0 | 0 | 0 | 0 | 0 | 2 | 6 | 19 |
| Days Min Temp ≤ 32 °F | 30 | 27 | 25 | 15 | 4 | 0 | 0 | 0 | 0 | 7 | 22 | 30 | 160 |
| Days Min Temp ≤ 0 °F | 1 | 0 | 0 | 0 | 0 | 0 | 0 | 0 | 0 | 0 | 0 | 1 | 2 |
| Heating Degree Days | 1128 | 909 | 819 | 563 | 302 | 73 | 4 | 14 | 126 | 438 | 807 | 1091 | 6274 |
| Cooling Degree Days | 0 | 0 | 0 | 0 | 5 | 81 | 166 | 125 | 24 | 0 | 0 | 0 | 401 |
| Total Precipitation (") | 1.77 | 1.54 | 1.83 | 1.10 | 1.26 | 0.63 | 1.87 | 1.96 | 1.55 | 1.80 | 1.71 | 1.73 | 18.75 |
| Days ≥ 0.1" Precip | 5 | 4 | 5 | 3 | 4 | 2 | 5 | 5 | 4 | 4 | 4 | 4 | 49 |
| Total Snowfall (") | 19.6 | 15.4 | 15.7 | 5.8 | 1.0 | 0.0 | 0.0 | 0.0 | 0.1 | 2.2 | 9.6 | 17.4 | 86.8 |
| Days ≥ 1" Snow Depth | 27 | 23 | 14 | 2 | 0 | 0 | 0 | 0 | 0 | 1 | 9 | 22 | 98 |

## MONTE VISTA *Rio Grande County* ELEVATION 7674 ft LAT/LONG 37° 34 ' N / 106° 9 ' W

| | JAN | FEB | MAR | APR | MAY | JUN | JUL | AUG | SEP | OCT | NOV | DEC | YEAR |
|---|---|---|---|---|---|---|---|---|---|---|---|---|---|
| Maximum Temp °F | 32.7 | 38.6 | ***48.6*** | 58.5 | 67.7 | 76.7 | ***80.3*** | 78.2 | 72.2 | 61.9 | 46.2 | ***35.0*** | 58.1 |
| Minimum Temp °F | 0.3 | 6.3 | ***17.2*** | 24.7 | 34.1 | 40.9 | 47.2 | 45.6 | 37.2 | 26.8 | 14.0 | ***3.6*** | 24.8 |
| Mean Temp °F | 16.5 | 22.5 | ***32.9*** | 41.6 | 50.9 | 58.9 | ***63.8*** | 61.9 | 54.7 | 44.4 | 30.1 | ***19.3*** | 41.5 |
| Days Max Temp ≥ 90 °F | 0 | 0 | 0 | 0 | 0 | 0 | 0 | 0 | 0 | 0 | 0 | 0 | 0 |
| Days Max Temp ≤ 32 °F | ***16*** | ***8*** | ***1*** | 1 | 0 | 0 | 0 | 0 | 0 | 0 | 3 | 12 | 41 |
| Days Min Temp ≤ 32 °F | 31 | 28 | ***30*** | ***26*** | ***11*** | 1 | 0 | ***0*** | 6 | 25 | 30 | 31 | 219 |
| Days Min Temp ≤ 0 °F | ***16*** | ***9*** | ***1*** | 0 | 0 | 0 | 0 | 0 | 0 | 0 | ***2*** | ***11*** | 39 |
| Heating Degree Days | 1497 | 1195 | ***988*** | 695 | 430 | 185 | ***59*** | 102 | ***303*** | 633 | ***1039*** | ***1409*** | 8535 |
| Cooling Degree Days | 0 | 0 | ***0*** | ***0*** | ***0*** | 4 | ***10*** | ***5*** | 0 | 0 | 0 | 0 | 19 |
| Total Precipitation (") | 0.22 | 0.28 | 0.48 | 0.39 | 0.68 | 0.54 | 1.33 | 1.49 | 0.99 | 0.57 | 0.50 | 0.40 | 7.87 |
| Days ≥ 0.1" Precip | 1 | 1 | ***2*** | 1 | 2 | 2 | 4 | 5 | 3 | 2 | 2 | 1 | 26 |
| Total Snowfall (") | ***3.0*** | 4.3 | ***5.4*** | ***2.4*** | 0.5 | 0.0 | 0.0 | 0.0 | 0.0 | 2.1 | 4.1 | 6.0 | 27.8 |
| Days ≥ 1" Snow Depth | ***18*** | ***12*** | ***3*** | ***1*** | 0 | 0 | 0 | 0 | 0 | 0 | ***5*** | ***14*** | 53 |

## MONTROSE 2 *Montrose County* ELEVATION 5833 ft LAT/LONG 38° 29 ' N / 107° 53 ' W

| | JAN | FEB | MAR | APR | MAY | JUN | JUL | AUG | SEP | OCT | NOV | DEC | YEAR |
|---|---|---|---|---|---|---|---|---|---|---|---|---|---|
| Maximum Temp °F | 37.3 | 44.0 | 53.3 | 62.1 | 71.8 | 82.5 | 87.8 | 85.5 | 77.4 | 65.1 | 49.3 | 39.0 | 62.9 |
| Minimum Temp °F | 13.1 | 19.1 | 27.1 | 33.9 | 42.3 | 50.7 | 56.5 | 54.5 | 45.7 | 35.1 | 24.7 | 15.7 | 34.9 |
| Mean Temp °F | 25.2 | 31.5 | 40.2 | 48.0 | 57.1 | 66.6 | 72.2 | 70.0 | 61.6 | 50.1 | 37.0 | 27.3 | 48.9 |
| Days Max Temp ≥ 90 °F | 0 | 0 | 0 | 0 | 0 | 5 | 13 | 7 | 1 | 0 | 0 | 0 | 26 |
| Days Max Temp ≤ 32 °F | 10 | 2 | 0 | 0 | 0 | 0 | 0 | 0 | 0 | 0 | 2 | 7 | 21 |
| Days Min Temp ≤ 32 °F | 31 | 27 | 24 | 13 | 2 | 0 | 0 | 0 | 1 | 11 | 25 | 30 | 164 |
| Days Min Temp ≤ 0 °F | 4 | 1 | 0 | 0 | 0 | 0 | 0 | 0 | 0 | 0 | 0 | 1 | 6 |
| Heating Degree Days | 1226 | 938 | 761 | 503 | 250 | 58 | 2 | 11 | 130 | 455 | 833 | 1160 | 6327 |
| Cooling Degree Days | 0 | 0 | 0 | 0 | 14 | 129 | 231 | 186 | 38 | 0 | 0 | 0 | 598 |
| Total Precipitation (") | 0.50 | 0.40 | 0.66 | 0.74 | 0.93 | 0.62 | 1.00 | 1.10 | 1.05 | 1.10 | 0.87 | 0.62 | 9.59 |
| Days ≥ 0.1" Precip | 2 | 2 | 2 | 2 | 3 | 2 | 3 | 3 | 3 | 3 | 3 | 2 | 30 |
| Total Snowfall (") | na | na | na | ***0.7*** | 0.0 | 0.0 | 0.0 | 0.0 | 0.0 | 0.5 | na | na | na |
| Days ≥ 1" Snow Depth | na | na | na | ***0*** | 0 | 0 | 0 | 0 | 0 | 0 | na | na | na |

## NEW RAYMER *Weld County* ELEVATION 4954 ft LAT/LONG 40° 35 ' N / 103° 50 ' W

| | JAN | FEB | MAR | APR | MAY | JUN | JUL | AUG | SEP | OCT | NOV | DEC | YEAR |
|---|---|---|---|---|---|---|---|---|---|---|---|---|---|
| Maximum Temp °F | 39.6 | 45.7 | 53.7 | 62.9 | 72.4 | 82.7 | 89.7 | 87.6 | 79.4 | 66.7 | 49.1 | 40.7 | 64.2 |
| Minimum Temp °F | 11.4 | 16.0 | 22.6 | 30.3 | 40.5 | 49.7 | 55.7 | 53.8 | 44.5 | 32.6 | 20.5 | 12.9 | 32.5 |
| Mean Temp °F | 25.6 | 30.9 | 38.2 | 46.6 | 56.5 | 66.2 | 72.7 | 70.8 | 62.0 | 49.7 | 34.8 | 26.9 | 48.4 |
| Days Max Temp ≥ 90 °F | 0 | 0 | 0 | 0 | 1 | 7 | 17 | 13 | 4 | 0 | 0 | 0 | 42 |
| Days Max Temp ≤ 32 °F | 8 | 4 | 2 | 0 | 0 | 0 | 0 | 0 | 0 | 0 | 4 | 7 | 25 |
| Days Min Temp ≤ 32 °F | 31 | 28 | 28 | ***18*** | 4 | 0 | 0 | 0 | 2 | 15 | 27 | 31 | 184 |
| Days Min Temp ≤ 0 °F | 6 | 2 | 1 | 0 | 0 | 0 | 0 | 0 | 0 | 0 | 1 | 4 | 14 |
| Heating Degree Days | 1215 | 958 | 824 | 545 | 268 | 64 | 6 | 11 | 134 | 470 | 898 | 1174 | 6567 |
| Cooling Degree Days | 0 | 0 | 0 | 1 | 12 | 114 | 242 | 196 | 50 | 1 | 0 | 0 | 616 |
| Total Precipitation (") | 0.25 | 0.16 | 0.83 | 1.36 | 2.44 | 2.44 | 2.15 | 1.93 | 1.05 | 0.82 | 0.46 | 0.24 | 14.13 |
| Days ≥ 0.1" Precip | 1 | 0 | 2 | 3 | 6 | 5 | 5 | 4 | 3 | 2 | 2 | 1 | 34 |
| Total Snowfall (") | 5.4 | 3.2 | 7.2 | 5.2 | 0.6 | 0.0 | 0.0 | 0.0 | 0.5 | 2.7 | 6.1 | 5.3 | 36.2 |
| Days ≥ 1" Snow Depth | 15 | 7 | 4 | 2 | 0 | 0 | 0 | 0 | 0 | 1 | 6 | 11 | 46 |

## NORTHDALE *Dolores County* ELEVATION 6483 ft LAT/LONG 37° 49 ' N / 109° 2 ' W

| | JAN | FEB | MAR | APR | MAY | JUN | JUL | AUG | SEP | OCT | NOV | DEC | YEAR |
|---|---|---|---|---|---|---|---|---|---|---|---|---|---|
| Maximum Temp °F | 36.4 | 41.0 | 49.7 | 59.1 | 69.4 | 80.9 | 86.1 | 83.6 | 75.8 | 63.5 | 48.1 | 38.0 | 61.0 |
| Minimum Temp °F | 8.6 | 13.7 | 23.0 | 27.9 | 35.1 | 42.6 | 50.5 | 49.6 | 41.1 | 30.7 | 20.8 | 11.8 | 29.6 |
| Mean Temp °F | 22.5 | 27.4 | 36.3 | 43.5 | 52.3 | 61.8 | 68.4 | 66.6 | 58.5 | 47.1 | 34.5 | 24.9 | 45.3 |
| Days Max Temp ≥ 90 °F | 0 | 0 | 0 | 0 | 0 | 4 | 9 | 4 | 0 | 0 | 0 | 0 | 17 |
| Days Max Temp ≤ 32 °F | 10 | 4 | 1 | 0 | 0 | 0 | 0 | 0 | 0 | 0 | 2 | 8 | 25 |
| Days Min Temp ≤ 32 °F | 31 | 28 | 28 | 23 | 11 | 2 | 0 | 0 | 3 | 20 | 28 | 30 | 204 |
| Days Min Temp ≤ 0 °F | 9 | 4 | 0 | 0 | 0 | 0 | 0 | 0 | 0 | 0 | 1 | 5 | 19 |
| Heating Degree Days | 1310 | 1054 | 881 | 638 | 388 | 127 | 10 | 28 | 199 | 547 | 909 | 1237 | 7328 |
| Cooling Degree Days | 0 | 0 | 0 | 0 | 1 | 47 | 123 | 95 | 10 | 0 | 0 | 0 | 276 |
| Total Precipitation (") | 0.78 | 0.77 | 0.89 | 0.77 | 0.93 | 0.46 | 1.38 | 1.34 | 1.35 | 1.70 | 1.11 | 1.07 | 12.55 |
| Days ≥ 0.1" Precip | 3 | 2 | 3 | 3 | 3 | 1 | 4 | 4 | 3 | 4 | 3 | 3 | 36 |
| Total Snowfall (") | na | na | na | *1.7* | 0.0 | 0.0 | 0.0 | 0.0 | 0.0 | 1.1 | *2.9* | na | na |
| Days ≥ 1" Snow Depth | na | na | na | *1* | 0 | 0 | 0 | 0 | 0 | 0 | na | na | na |

## NORWOOD *San Miguel County* ELEVATION 7019 ft LAT/LONG 38° 8 ' N / 108° 17 ' W

| | JAN | FEB | MAR | APR | MAY | JUN | JUL | AUG | SEP | OCT | NOV | DEC | YEAR |
|---|---|---|---|---|---|---|---|---|---|---|---|---|---|
| Maximum Temp °F | 37.5 | 41.8 | 48.9 | 57.7 | 67.3 | 78.3 | 83.2 | 80.9 | 73.1 | 61.2 | 47.3 | 38.4 | 59.6 |
| Minimum Temp °F | 8.8 | 14.4 | 22.2 | 27.3 | 35.2 | 43.0 | 49.4 | 48.2 | 41.2 | 31.0 | 20.3 | 11.2 | 29.4 |
| Mean Temp °F | 23.2 | 28.1 | 35.6 | 42.6 | 51.3 | 60.7 | 66.3 | 64.6 | 57.1 | 46.1 | 33.8 | 24.8 | 44.5 |
| Days Max Temp ≥ 90 °F | 0 | 0 | 0 | 0 | 0 | 1 | 3 | 1 | 0 | 0 | 0 | 0 | 5 |
| Days Max Temp ≤ 32 °F | 9 | 3 | 1 | 0 | 0 | 0 | 0 | 0 | 0 | 0 | 2 | 6 | 21 |
| Days Min Temp ≤ 32 °F | 31 | 28 | 28 | 21 | 10 | 2 | 0 | 0 | *3* | 16 | 28 | 31 | 198 |
| Days Min Temp ≤ 0 °F | 7 | 3 | 1 | 0 | 0 | 0 | 0 | 0 | 0 | 0 | 1 | 5 | 17 |
| Heating Degree Days | 1289 | 1034 | 906 | 666 | 419 | 147 | 23 | 50 | 234 | 581 | 928 | 1239 | 7516 |
| Cooling Degree Days | 0 | 0 | 0 | 0 | 0 | 31 | 78 | 50 | 5 | 0 | 0 | 0 | 164 |
| Total Precipitation (") | 0.99 | 0.75 | 1.13 | 1.09 | 1.23 | 0.91 | 1.86 | 1.78 | 1.63 | 1.59 | 1.33 | 0.98 | 15.27 |
| Days ≥ 0.1" Precip | 3 | 3 | 4 | 4 | 4 | 3 | 5 | 5 | 4 | 4 | 4 | *4* | 47 |
| Total Snowfall (") | 13.4 | 10.4 | *10.2* | *6.6* | 0.9 | 0.0 | 0.0 | 0.0 | 0.0 | 3.6 | *7.9* | *12.6* | 65.6 |
| Days ≥ 1" Snow Depth | na | na | na | na | 0 | 0 | 0 | 0 | 0 | *0* | na | na | na |

## OURAY *Ouray County* ELEVATION 7726 ft LAT/LONG 38° 1 ' N / 107° 41 ' W

| | JAN | FEB | MAR | APR | MAY | JUN | JUL | AUG | SEP | OCT | NOV | DEC | YEAR |
|---|---|---|---|---|---|---|---|---|---|---|---|---|---|
| Maximum Temp °F | 36.8 | 39.1 | 44.7 | 53.6 | 63.1 | 73.5 | 77.8 | 76.0 | 69.5 | 58.9 | 44.3 | 36.9 | 56.2 |
| Minimum Temp °F | 14.2 | 16.8 | 22.4 | 29.3 | 37.2 | 44.8 | 50.9 | 50.0 | 43.1 | 33.5 | 23.0 | 15.6 | 31.7 |
| Mean Temp °F | 25.5 | 28.0 | 33.6 | 41.5 | 50.2 | 59.2 | 64.4 | 63.0 | 56.4 | 46.2 | 33.6 | 26.3 | 44.0 |
| Days Max Temp ≥ 90 °F | 0 | 0 | 0 | 0 | 0 | 0 | 0 | 0 | 0 | 0 | 0 | 0 | 0 |
| Days Max Temp ≤ 32 °F | 9 | 6 | 3 | 1 | 0 | 0 | 0 | 0 | 0 | 0 | 4 | 9 | 32 |
| Days Min Temp ≤ 32 °F | 31 | 28 | 28 | 19 | 7 | 1 | 0 | 0 | 2 | 12 | 26 | 30 | 184 |
| Days Min Temp ≤ 0 °F | 2 | 1 | 0 | 0 | 0 | 0 | 0 | 0 | 0 | 0 | 0 | 2 | 5 |
| Heating Degree Days | 1219 | 1039 | 967 | 699 | 452 | 182 | 49 | 78 | 255 | 575 | 934 | 1194 | 7643 |
| Cooling Degree Days | 0 | 0 | 0 | 0 | 0 | 18 | 41 | 24 | 2 | 0 | 0 | 0 | 85 |
| Total Precipitation (") | 1.59 | 1.78 | 2.35 | 2.02 | 1.93 | 1.31 | 2.16 | 2.32 | 2.01 | 2.26 | 2.29 | 1.90 | 23.92 |
| Days ≥ 0.1" Precip | 6 | 6 | 7 | 6 | 5 | 4 | 7 | 8 | 6 | 6 | 6 | 6 | 73 |
| Total Snowfall (") | 22.9 | 22.8 | 25.3 | 14.2 | 4.3 | 0.4 | 0.0 | 0.0 | 0.0 | 7.6 | 21.7 | 23.7 | 142.9 |
| Days ≥ 1" Snow Depth | 30 | 27 | 27 | 11 | 2 | 0 | 0 | 0 | 0 | 4 | 16 | 29 | 146 |

## PALISADE *Mesa County* ELEVATION 4724 ft LAT/LONG 39° 6 ' N / 108° 21 ' W

| | JAN | FEB | MAR | APR | MAY | JUN | JUL | AUG | SEP | OCT | NOV | DEC | YEAR |
|---|---|---|---|---|---|---|---|---|---|---|---|---|---|
| Maximum Temp °F | 39.0 | 47.3 | 57.3 | 66.8 | 77.0 | 88.6 | 94.8 | 92.1 | 83.0 | 69.6 | 53.0 | 41.5 | 67.5 |
| Minimum Temp °F | 17.8 | 24.8 | 33.2 | 40.4 | 49.0 | 57.9 | 64.1 | 62.1 | 53.0 | 41.8 | 30.5 | 20.8 | 41.3 |
| Mean Temp °F | 28.4 | 36.0 | 45.3 | 53.6 | 63.1 | 73.3 | 79.5 | 77.1 | 68.0 | 55.7 | 41.8 | 31.2 | 54.4 |
| Days Max Temp ≥ 90 °F | 0 | 0 | 0 | 0 | 2 | 15 | 26 | 21 | 6 | 0 | 0 | 0 | 70 |
| Days Max Temp ≤ 32 °F | 8 | 1 | 0 | 1 | 0 | 0 | 0 | 0 | 0 | 0 | 0 | 4 | 14 |
| Days Min Temp ≤ 32 °F | 30 | 23 | 13 | 4 | 0 | 0 | 0 | 0 | 0 | 3 | 18 | 30 | 121 |
| Days Min Temp ≤ 0 °F | 1 | 0 | 0 | 0 | 0 | 0 | 0 | 0 | 0 | 0 | 0 | 0 | 1 |
| Heating Degree Days | 1126 | 810 | 605 | 342 | 125 | 17 | 0 | 0 | 46 | 292 | 689 | 1042 | 5094 |
| Cooling Degree Days | 0 | 0 | 0 | 12 | 73 | 295 | 457 | 388 | 146 | 10 | 0 | 0 | 1381 |
| Total Precipitation (") | 0.56 | 0.60 | 1.13 | 1.01 | 1.10 | 0.73 | 0.73 | 0.85 | 0.96 | 1.26 | 0.96 | 0.71 | 10.60 |
| Days ≥ 0.1" Precip | 2 | 2 | 4 | 3 | 3 | 2 | 2 | 3 | 3 | 3 | 3 | 2 | 32 |
| Total Snowfall (") | na | na | *1.1* | 0.2 | 0.1 | 0.0 | 0.0 | 0.0 | 0.0 | 0.1 | 1.6 | *4.1* | na |
| Days ≥ 1" Snow Depth | na | na | *1* | 0 | 0 | 0 | 0 | 0 | 0 | 0 | *1* | na | na |

## PAONIA 1 SW *Delta County* ELEVATION 5685 ft LAT/LONG 38° 52 ' N / 107° 35 ' W

| | JAN | FEB | MAR | APR | MAY | JUN | JUL | AUG | SEP | OCT | NOV | DEC | YEAR |
|---|---|---|---|---|---|---|---|---|---|---|---|---|---|
| Maximum Temp °F | 38.8 | 45.0 | 53.9 | 63.1 | 73.3 | 83.8 | 89.6 | 87.4 | 78.8 | 67.0 | 51.2 | 41.2 | 64.4 |
| Minimum Temp °F | 12.8 | 19.1 | 27.1 | 33.4 | 41.3 | 49.5 | 56.0 | 54.5 | 46.5 | 36.4 | 25.8 | 16.1 | 34.9 |
| Mean Temp °F | 25.8 | 32.1 | 40.5 | 48.3 | 57.3 | 66.7 | 72.8 | 71.0 | 62.7 | 51.7 | 38.5 | 28.9 | 49.7 |
| Days Max Temp ≥ 90 °F | 0 | 0 | 0 | 0 | 0 | 7 | 16 | 12 | 2 | 0 | 0 | 0 | 37 |
| Days Max Temp ≤ 32 °F | 7 | 2 | 0 | 0 | 0 | 0 | 0 | 0 | 0 | 0 | 1 | 4 | 14 |
| Days Min Temp ≤ 32 °F | 31 | 27 | 23 | 13 | 3 | 0 | 0 | 0 | 1 | 8 | 23 | 30 | 159 |
| Days Min Temp ≤ 0 °F | 5 | 1 | 0 | 0 | 0 | 0 | 0 | 0 | 0 | 0 | 0 | 2 | 8 |
| Heating Degree Days | 1207 | 923 | 754 | 496 | 241 | 53 | 1 | 6 | 107 | 405 | 788 | 1113 | 6094 |
| Cooling Degree Days | 0 | 0 | 0 | 0 | 12 | 135 | 271 | 229 | 55 | 0 | 0 | 0 | 702 |
| Total Precipitation (") | 1.12 | 1.04 | 1.46 | 1.16 | 1.38 | 0.83 | 1.11 | 1.17 | 1.35 | 1.70 | 1.43 | 1.37 | 15.12 |
| Days ≥ 0.1" Precip | 4 | 4 | 5 | 4 | 4 | 3 | 3 | 4 | 4 | 4 | 5 | 5 | 49 |
| Total Snowfall (") | 13.1 | 8.6 | ***7.0*** | 2.5 | 0.3 | 0.0 | 0.0 | 0.0 | 0.0 | 0.9 | 5.0 | ***13.1*** | 50.5 |
| Days ≥ 1" Snow Depth | 20 | 12 | ***3*** | 0 | 0 | 0 | 0 | 0 | 0 | 0 | 3 | 13 | 51 |

## PARKER 6 E *Elbert County* ELEVATION 6306 ft LAT/LONG 39° 31 ' N / 104° 39 ' W

| | JAN | FEB | MAR | APR | MAY | JUN | JUL | AUG | SEP | OCT | NOV | DEC | YEAR |
|---|---|---|---|---|---|---|---|---|---|---|---|---|---|
| Maximum Temp °F | 43.6 | 46.0 | 51.5 | 60.4 | 69.0 | 79.4 | 85.2 | 83.2 | 75.9 | 65.7 | 51.2 | 44.4 | 63.0 |
| Minimum Temp °F | 16.3 | 18.8 | 23.9 | 31.7 | 40.4 | 49.7 | 56.1 | 54.4 | 46.3 | 35.5 | 24.2 | 17.5 | 34.6 |
| Mean Temp °F | 30.0 | 32.5 | 37.7 | 46.1 | 54.7 | 64.6 | 70.7 | 68.8 | 61.1 | 50.6 | 37.7 | 30.9 | 48.8 |
| Days Max Temp ≥ 90 °F | 0 | 0 | 0 | 0 | 0 | 3 | 8 | 4 | 1 | 0 | 0 | 0 | 16 |
| Days Max Temp ≤ 32 °F | 6 | 4 | 2 | 0 | 0 | 0 | 0 | 0 | 0 | 0 | 3 | 5 | 20 |
| Days Min Temp ≤ 32 °F | 29 | 26 | 25 | 17 | 5 | 0 | 0 | 0 | 2 | 11 | 23 | 28 | 166 |
| Days Min Temp ≤ 0 °F | 4 | 2 | 1 | 0 | 0 | 0 | 0 | 0 | 0 | 0 | 1 | 3 | 11 |
| Heating Degree Days | 1079 | 912 | 830 | 562 | 318 | 88 | 10 | 17 | 153 | 441 | 812 | 1049 | 6271 |
| Cooling Degree Days | 0 | 0 | 0 | 1 | 6 | 97 | 204 | 164 | 53 | 3 | 0 | 0 | 528 |
| Total Precipitation (") | 0.29 | 0.31 | 0.94 | 1.41 | 2.46 | 2.05 | 2.48 | 1.94 | 1.15 | 0.85 | 0.80 | 0.34 | 15.02 |
| Days ≥ 0.1" Precip | 1 | 1 | 3 | 4 | 6 | 5 | 6 | 5 | 3 | 2 | 2 | 1 | 39 |
| Total Snowfall (") | 5.5 | 5.3 | 9.5 | 9.5 | 2.3 | 0.2 | 0.0 | 0.0 | 1.5 | 4.7 | 9.5 | 6.2 | 54.2 |
| Days ≥ 1" Snow Depth | 15 | 10 | 6 | 3 | 0 | 0 | 0 | 0 | 0 | 2 | 8 | 12 | 56 |

## PUEBLO MEMORIAL AP *Pueblo County* ELEVATION 4646 ft LAT/LONG 38° 17 ' N / 104° 30 ' W

| | JAN | FEB | MAR | APR | MAY | JUN | JUL | AUG | SEP | OCT | NOV | DEC | YEAR |
|---|---|---|---|---|---|---|---|---|---|---|---|---|---|
| Maximum Temp °F | 45.5 | 51.0 | 58.2 | 67.9 | 76.3 | 87.7 | 92.7 | 89.5 | 81.4 | 69.9 | 55.6 | 46.7 | 68.5 |
| Minimum Temp °F | 14.5 | 19.2 | 26.5 | 35.6 | 45.5 | 54.2 | 60.5 | 58.6 | 49.3 | 36.1 | 23.5 | 15.7 | 36.6 |
| Mean Temp °F | 30.0 | 35.1 | 42.4 | 51.8 | 60.9 | 71.0 | 76.6 | 74.1 | 65.4 | 53.0 | 39.6 | 31.2 | 52.6 |
| Days Max Temp ≥ 90 °F | 0 | 0 | 0 | 0 | 2 | 14 | 22 | 18 | 7 | 0 | 0 | 0 | 63 |
| Days Max Temp ≤ 32 °F | 6 | 3 | 1 | 0 | 0 | 0 | 0 | 0 | 0 | 0 | 2 | 5 | 17 |
| Days Min Temp ≤ 32 °F | 30 | 27 | 24 | 9 | 1 | 0 | 0 | 0 | 1 | 10 | 26 | 30 | 158 |
| Days Min Temp ≤ 0 °F | 4 | 1 | 0 | 0 | 0 | 0 | 0 | 0 | 0 | 0 | 0 | 3 | 8 |
| Heating Degree Days | 1077 | 837 | 695 | 393 | 157 | 20 | 1 | 4 | 79 | 368 | 757 | 1040 | 5428 |
| Cooling Degree Days | 0 | 0 | 0 | 3 | 36 | 203 | 354 | 291 | 95 | 3 | 0 | 0 | 985 |
| Total Precipitation (") | 0.33 | 0.28 | 0.83 | 0.99 | 1.31 | 1.37 | 2.10 | 2.13 | 0.88 | 0.66 | 0.54 | 0.42 | 11.84 |
| Days ≥ 0.1" Precip | 1 | 1 | 3 | 3 | 4 | 3 | 5 | 4 | 2 | 2 | 2 | 1 | 31 |
| Total Snowfall (") | 5.9 | 4.0 | 6.5 | 3.8 | 0.6 | 0.0 | 0.0 | 0.0 | 0.4 | 1.8 | 4.4 | 5.9 | 33.3 |
| Days ≥ 1" Snow Depth | 8 | 3 | 2 | 1 | 0 | 0 | 0 | 0 | 0 | 0 | 3 | 6 | 23 |

## RANGELY 1 E *Rio Blanco County* ELEVATION 5282 ft LAT/LONG 40° 5 ' N / 108° 48 ' W

| | JAN | FEB | MAR | APR | MAY | JUN | JUL | AUG | SEP | OCT | NOV | DEC | YEAR |
|---|---|---|---|---|---|---|---|---|---|---|---|---|---|
| Maximum Temp °F | 30.7 | 38.6 | 51.9 | 63.5 | 73.3 | 85.2 | 91.2 | ***89.1*** | 79.5 | 65.9 | 47.7 | 33.6 | 62.5 |
| Minimum Temp °F | 2.3 | 9.2 | 23.0 | 32.1 | 40.1 | 48.7 | 55.3 | ***52.7*** | 43.2 | 31.3 | 20.0 | 6.3 | 30.4 |
| Mean Temp °F | 16.5 | 23.9 | 37.5 | 47.8 | 56.7 | 67.0 | 73.3 | ***70.9*** | 61.4 | 48.7 | 33.8 | 20.0 | 46.5 |
| Days Max Temp ≥ 90 °F | 0 | 0 | 0 | 0 | 0 | 10 | 20 | ***15*** | 2 | 0 | 0 | 0 | 47 |
| Days Max Temp ≤ 32 °F | 17 | 7 | 1 | 0 | 0 | 0 | 0 | 0 | 0 | 0 | 2 | 14 | 41 |
| Days Min Temp ≤ 32 °F | 31 | 28 | 28 | 16 | 3 | 1 | ***0*** | ***0*** | 3 | 16 | 28 | 30 | 184 |
| Days Min Temp ≤ 0 °F | 14 | 7 | 1 | 0 | 0 | 0 | 0 | 0 | 0 | 0 | 1 | 9 | 32 |
| Heating Degree Days | 1497 | 1154 | 847 | 509 | 258 | 50 | 1 | ***10*** | 137 | 497 | 929 | 1389 | 7278 |
| Cooling Degree Days | 0 | 0 | 0 | 0 | 9 | 141 | 267 | ***226*** | 43 | 0 | 0 | 0 | 686 |
| Total Precipitation (") | 0.52 | 0.44 | 0.98 | 1.15 | 1.09 | 0.83 | 1.05 | 0.87 | 1.09 | 1.30 | 0.68 | 0.57 | 10.57 |
| Days ≥ 0.1" Precip | 2 | 2 | 3 | 4 | 4 | 2 | 3 | ***3*** | 3 | 4 | 2 | 2 | 34 |
| Total Snowfall (") | na | na | na | ***2.0*** | 0.0 | 0.0 | 0.0 | ***0.0*** | 0.0 | ***0.1*** | na | na | na |
| Days ≥ 1" Snow Depth | na | na | na | na | 0 | 0 | 0 | ***0*** | 0 | 0 | na | na | na |

## RICO *Dolores County* ELEVATION 8825 ft LAT/LONG 37° 41 ' N / 108° 2 ' W

| | JAN | FEB | MAR | APR | MAY | JUN | JUL | AUG | SEP | OCT | NOV | DEC | YEAR |
|---|---|---|---|---|---|---|---|---|---|---|---|---|---|
| Maximum Temp °F | 38.4 | 40.2 | 43.5 | 50.8 | 60.5 | 70.6 | 75.1 | 73.0 | 66.4 | 57.7 | 45.2 | 39.1 | 55.0 |
| Minimum Temp °F | 5.1 | 7.1 | 13.3 | 20.7 | 27.7 | 33.6 | 39.8 | 39.0 | 31.9 | 24.5 | 14.6 | 6.8 | 22.0 |
| Mean Temp °F | 21.8 | 23.7 | 28.4 | 35.8 | 44.1 | 52.1 | 57.5 | 56.0 | 49.2 | 41.1 | 29.9 | 23.0 | 38.6 |
| Days Max Temp ≥ 90 °F | 0 | 0 | 0 | 0 | 0 | 0 | 0 | 0 | 0 | 0 | 0 | 0 | 0 |
| Days Max Temp ≤ 32 °F | 9 | 6 | 3 | 1 | 0 | 0 | 0 | 0 | 0 | 0 | 3 | 8 | 30 |
| Days Min Temp ≤ 32 °F | 31 | 28 | 31 | 29 | 26 | 13 | 2 | 3 | 16 | 28 | 30 | 31 | 268 |
| Days Min Temp ≤ 0 °F | 10 | 8 | 4 | 1 | 0 | 0 | 0 | 0 | 0 | 0 | 3 | 9 | 35 |
| Heating Degree Days | 1335 | 1159 | 1127 | 871 | 641 | 381 | 228 | 276 | 468 | 733 | 1047 | 1297 | 9563 |
| Cooling Degree Days | 0 | 0 | 0 | 0 | 0 | 0 | 0 | 0 | 0 | 0 | 0 | 0 | 0 |
| Total Precipitation (") | 2.39 | 2.40 | 2.76 | 1.83 | 1.82 | 1.65 | 3.14 | 3.24 | 2.72 | 2.34 | 2.33 | 2.54 | 29.16 |
| Days ≥ 0.1" Precip | 6 | 6 | 8 | 6 | 5 | 5 | 9 | 9 | 7 | 6 | 6 | 7 | 80 |
| Total Snowfall (") | 28.9 | 31.1 | 33.7 | 19.5 | 5.6 | 0.1 | 0.0 | 0.0 | 0.4 | 8.3 | 24.4 | 28.8 | 180.8 |
| Days ≥ 1" Snow Depth | 29 | 27 | 29 | 18 | 2 | 0 | 0 | 0 | 0 | 4 | 17 | 26 | 152 |

## RIFLE *Garfield County* ELEVATION 5299 ft LAT/LONG 39° 31 ' N / 107° 47 ' W

| | JAN | FEB | MAR | APR | MAY | JUN | JUL | AUG | SEP | OCT | NOV | DEC | YEAR |
|---|---|---|---|---|---|---|---|---|---|---|---|---|---|
| Maximum Temp °F | 36.6 | 44.5 | 54.6 | 63.8 | 73.3 | 83.6 | 89.1 | 87.4 | 78.8 | 66.7 | 50.3 | 38.7 | 64.0 |
| Minimum Temp °F | 9.1 | 16.3 | 24.6 | 30.7 | 38.5 | 45.3 | 52.2 | 50.5 | 41.6 | 31.0 | 22.3 | 12.1 | 31.2 |
| Mean Temp °F | 22.9 | 30.4 | 39.6 | 47.2 | 55.9 | 64.5 | 70.7 | 69.0 | 60.2 | 48.9 | 36.3 | 25.4 | 47.6 |
| Days Max Temp ≥ 90 °F | 0 | 0 | 0 | 0 | 0 | 8 | 16 | 12 | 2 | 0 | 0 | 0 | 38 |
| Days Max Temp ≤ 32 °F | 10 | 2 | 0 | 0 | 0 | 0 | 0 | 0 | 0 | 0 | 1 | 7 | 20 |
| Days Min Temp ≤ 32 °F | 31 | 28 | 27 | 18 | 5 | 0 | 0 | 0 | 3 | 19 | 26 | 30 | 187 |
| Days Min Temp ≤ 0 °F | 7 | 2 | 0 | 0 | 0 | 0 | 0 | 0 | 0 | 0 | 0 | 4 | 13 |
| Heating Degree Days | 1299 | 970 | 779 | 526 | 277 | 71 | 2 | 10 | 155 | 493 | 854 | 1220 | 6656 |
| Cooling Degree Days | 0 | 0 | 0 | 0 | 2 | 75 | 194 | 159 | 24 | 0 | 0 | 0 | 454 |
| Total Precipitation (") | 0.90 | 0.78 | 0.97 | 0.98 | 1.11 | 0.96 | 1.10 | 1.03 | 1.18 | 1.38 | 1.06 | 1.16 | 12.61 |
| Days ≥ 0.1" Precip | 3 | 3 | 4 | 3 | 4 | 3 | 3 | 3 | 4 | 5 | 4 | 4 | 43 |
| Total Snowfall (") | 14.5 | 9.2 | 4.0 | 0.6 | 0.0 | 0.0 | 0.0 | 0.0 | 0.2 | 1.0 | 5.5 | 15.7 | 50.7 |
| Days ≥ 1" Snow Depth | 22 | 13 | 2 | 0 | 0 | 0 | 0 | 0 | 0 | 0 | 2 | 14 | 53 |

## ROCKY FORD 2 SE *Otero County* ELEVATION 4173 ft LAT/LONG 38° 2 ' N / 103° 42 ' W

| | JAN | FEB | MAR | APR | MAY | JUN | JUL | AUG | SEP | OCT | NOV | DEC | YEAR |
|---|---|---|---|---|---|---|---|---|---|---|---|---|---|
| Maximum Temp °F | 46.1 | 52.1 | 60.9 | 70.7 | 78.8 | 89.4 | 93.4 | 90.8 | 83.8 | 72.7 | 56.5 | 47.2 | 70.2 |
| Minimum Temp °F | 13.2 | 18.2 | 26.0 | 35.4 | 45.4 | 54.7 | 59.7 | 57.2 | 48.1 | 35.1 | 22.8 | 15.1 | 35.9 |
| Mean Temp °F | 29.7 | 35.2 | 43.5 | 53.1 | 62.1 | 72.1 | 76.6 | 74.0 | 66.0 | 53.9 | 39.7 | 31.2 | 53.1 |
| Days Max Temp ≥ 90 °F | 0 | 0 | 0 | 0 | 3 | 16 | 24 | 20 | 9 | 1 | 0 | 0 | 73 |
| Days Max Temp ≤ 32 °F | 6 | 3 | 1 | 0 | 0 | 0 | 0 | 0 | 0 | 0 | 1 | 4 | 15 |
| Days Min Temp ≤ 32 °F | 30 | 27 | 24 | 10 | 1 | 0 | 0 | 0 | 1 | 12 | 27 | 30 | 162 |
| Days Min Temp ≤ 0 °F | 4 | 2 | 0 | 0 | 0 | 0 | 0 | 0 | 0 | 0 | 1 | 3 | 10 |
| Heating Degree Days | 1090 | 835 | 660 | 356 | 133 | 12 | 1 | 2 | 66 | 341 | 753 | 1041 | 5290 |
| Cooling Degree Days | 0 | 0 | 0 | 6 | 55 | 240 | 362 | 305 | 111 | 4 | 0 | 0 | 1083 |
| Total Precipitation (") | 0.26 | 0.29 | 0.75 | 1.03 | 1.60 | 1.36 | 2.00 | 1.48 | 0.94 | 0.70 | 0.50 | 0.29 | 11.20 |
| Days ≥ 0.1" Precip | 1 | 1 | 2 | 3 | 4 | 3 | 4 | 4 | 2 | 2 | 2 | 1 | 29 |
| Total Snowfall (") | 4.0 | 3.4 | 4.9 | 2.1 | 0.5 | 0.0 | 0.0 | 0.0 | 0.1 | 1.0 | 3.5 | 4.4 | 23.9 |
| Days ≥ 1" Snow Depth | 7 | 3 | 1 | 0 | 0 | 0 | 0 | 0 | 0 | 0 | 3 | 5 | 19 |

## RUSH *El Paso County* ELEVATION 6020 ft LAT/LONG 38° 50 ' N / 104° 5 ' W

| | JAN | FEB | MAR | APR | MAY | JUN | JUL | AUG | SEP | OCT | NOV | DEC | YEAR |
|---|---|---|---|---|---|---|---|---|---|---|---|---|---|
| Maximum Temp °F | 41.5 | 44.8 | ***50.3*** | 60.3 | 68.5 | 79.0 | 84.6 | 81.6 | 74.6 | 63.6 | 49.9 | ***42.4*** | 61.8 |
| Minimum Temp °F | 12.4 | 15.1 | ***20.8*** | 29.4 | 38.3 | 48.3 | 53.3 | 51.4 | 42.8 | 31.5 | 20.6 | ***13.6*** | 31.5 |
| Mean Temp °F | 27.0 | 30.0 | ***35.6*** | 44.9 | 53.4 | 63.6 | 69.0 | 66.5 | 58.7 | 47.6 | 35.3 | ***28.0*** | 46.6 |
| Days Max Temp ≥ 90 °F | 0 | 0 | ***0*** | 0 | 0 | 3 | 9 | 3 | 1 | 0 | 0 | 0 | 16 |
| Days Max Temp ≤ 32 °F | 8 | 5 | ***3*** | 1 | 0 | 0 | 0 | 0 | 0 | 1 | 4 | 7 | 29 |
| Days Min Temp ≤ 32 °F | 31 | 27 | ***28*** | 19 | 6 | 0 | 0 | 0 | 3 | 17 | 27 | 30 | 188 |
| Days Min Temp ≤ 0 °F | 5 | 3 | ***1*** | 0 | 0 | 0 | 0 | 0 | 0 | 0 | 1 | 3 | 13 |
| Heating Degree Days | 1173 | 982 | ***905*** | 596 | 355 | 100 | 18 | 40 | 201 | 534 | 884 | ***1137*** | 6925 |
| Cooling Degree Days | 0 | 0 | 0 | 0 | 3 | 54 | 125 | 81 | 14 | 0 | 0 | 0 | 277 |
| Total Precipitation (") | 0.25 | 0.19 | ***0.72*** | 1.03 | 2.16 | 2.10 | 2.79 | 2.41 | 1.07 | 0.63 | 0.40 | 0.29 | 14.04 |
| Days ≥ 0.1" Precip | ***1*** | 1 | ***2*** | 3 | 5 | 5 | 6 | 5 | 3 | 2 | 1 | 1 | 35 |
| Total Snowfall (") | na | na | na | na | ***0.2*** | 0.0 | 0.0 | 0.0 | 0.4 | ***1.1*** | na | na | na |
| Days ≥ 1" Snow Depth | na | na | na | ***1*** | ***0*** | 0 | 0 | 0 | 0 | na | na | na | na |

## RUXTON PARK *El Paso County* ELEVATION 9049 ft LAT/LONG 38° 51 ' N / 104° 59 ' W

| | JAN | FEB | MAR | APR | MAY | JUN | JUL | AUG | SEP | OCT | NOV | DEC | YEAR |
|---|---|---|---|---|---|---|---|---|---|---|---|---|---|
| Maximum Temp °F | 33.2 | 34.6 | 38.6 | 46.1 | 55.8 | 66.3 | 70.9 | 68.5 | 62.5 | 52.4 | 39.5 | 34.0 | 50.2 |
| Minimum Temp °F | 6.3 | 6.8 | 12.1 | 19.5 | 27.9 | 35.4 | 40.0 | 38.4 | 31.7 | 23.1 | 13.7 | 7.6 | 21.9 |
| Mean Temp °F | 19.8 | 20.7 | 25.4 | 32.9 | 41.9 | 50.9 | 55.5 | 53.5 | 47.2 | 37.7 | 26.6 | 20.8 | 36.1 |
| Days Max Temp ≥ 90 °F | 0 | 0 | 0 | 0 | 0 | 0 | 0 | 0 | 0 | 0 | 0 | 0 | 0 |
| Days Max Temp ≤ 32 °F | 13 | 11 | 8 | 3 | 1 | 0 | 0 | 0 | 0 | 1 | 8 | 13 | 58 |
| Days Min Temp ≤ 32 °F | 31 | 28 | 31 | 29 | 25 | 10 | 1 | 2 | 15 | 28 | 30 | 31 | 261 |
| Days Min Temp ≤ 0 °F | 9 | 7 | 4 | 1 | 0 | 0 | 0 | 0 | 0 | 0 | 4 | 8 | 33 |
| Heating Degree Days | 1396 | 1244 | 1224 | 958 | 710 | 417 | 288 | 350 | 528 | 838 | 1145 | 1364 | 10462 |
| Cooling Degree Days | 0 | 0 | 0 | 0 | 0 | 0 | 1 | 0 | 0 | 0 | 0 | 0 | 1 |
| Total Precipitation (") | 0.58 | 0.83 | 1.96 | 2.55 | 2.83 | 2.39 | 3.95 | 3.87 | 1.73 | 1.51 | 1.04 | 0.85 | 24.09 |
| Days ≥ 0.1" Precip | 2 | 3 | 6 | 6 | 7 | 6 | 10 | 10 | 5 | 4 | 3 | 3 | 65 |
| Total Snowfall (") | 10.3 | 14.8 | 28.9 | 27.9 | 11.9 | 1.6 | 0.0 | 0.0 | 1.4 | 13.9 | 17.3 | 16.0 | 144.0 |
| Days ≥ 1" Snow Depth | *25* | *23* | *23* | *17* | na | 0 | 0 | 0 | 0 | na | na | na | na |

## SAGUACHE *Saguache County* ELEVATION 7707 ft LAT/LONG 38° 5 ' N / 106° 9 ' W

| | JAN | FEB | MAR | APR | MAY | JUN | JUL | AUG | SEP | OCT | NOV | DEC | YEAR |
|---|---|---|---|---|---|---|---|---|---|---|---|---|---|
| Maximum Temp °F | 33.7 | 39.5 | 48.7 | 58.2 | 66.8 | 76.1 | 79.9 | 77.5 | 71.4 | 61.6 | 46.6 | 35.7 | 58.0 |
| Minimum Temp °F | 1.5 | 8.2 | 18.2 | 24.5 | 32.8 | 40.9 | 46.6 | 44.9 | 36.6 | 26.6 | 14.7 | 4.3 | 25.0 |
| Mean Temp °F | 17.6 | 23.9 | 33.5 | 41.3 | 49.9 | 58.5 | 63.3 | 61.3 | 54.0 | 44.1 | 30.7 | 20.1 | 41.5 |
| Days Max Temp ≥ 90 °F | 0 | 0 | 0 | 0 | 0 | 0 | 0 | 0 | 0 | 0 | 0 | 0 | 0 |
| Days Max Temp ≤ 32 °F | 14 | 7 | 1 | 0 | 0 | 0 | 0 | 0 | 0 | 0 | 3 | 11 | 36 |
| Days Min Temp ≤ 32 °F | 31 | 28 | 31 | 27 | 14 | 2 | 0 | 1 | 7 | 25 | 30 | 31 | 227 |
| Days Min Temp ≤ 0 °F | 14 | 6 | 1 | 0 | 0 | 0 | 0 | 0 | 0 | 0 | 2 | 11 | 34 |
| Heating Degree Days | 1463 | 1155 | 971 | 704 | 463 | 194 | 66 | 119 | 323 | 640 | 1023 | 1386 | 8507 |
| Cooling Degree Days | 0 | 0 | 0 | 0 | 0 | 6 | 18 | 9 | 0 | 0 | 0 | 0 | 33 |
| Total Precipitation (") | 0.28 | 0.20 | 0.38 | 0.53 | 0.78 | 0.61 | 1.57 | 1.69 | 1.01 | 0.66 | 0.46 | 0.36 | 8.53 |
| Days ≥ 0.1" Precip | 1 | 1 | 2 | 2 | 3 | 2 | 5 | 6 | 3 | 2 | 1 | 1 | 29 |
| Total Snowfall (") | 4.3 | 2.7 | 4.1 | *2.2* | 0.3 | 0.0 | 0.0 | 0.0 | 0.1 | 1.3 | 2.2 | *5.6* | 22.8 |
| Days ≥ 1" Snow Depth | na | na | na | 1 | 0 | 0 | 0 | 0 | 0 | 1 | *2* | na | na |

## SEDGWICK 5 S *Sedgwick County* ELEVATION 3989 ft LAT/LONG 40° 51 ' N / 102° 31 ' W

| | JAN | FEB | MAR | APR | MAY | JUN | JUL | AUG | SEP | OCT | NOV | DEC | YEAR |
|---|---|---|---|---|---|---|---|---|---|---|---|---|---|
| Maximum Temp °F | 38.0 | 44.1 | 52.6 | 63.1 | 72.4 | 82.8 | 89.2 | 88.1 | 79.5 | 67.0 | 49.7 | 40.0 | 63.9 |
| Minimum Temp °F | 14.1 | 18.7 | 25.4 | 34.4 | 44.4 | 53.8 | 59.4 | 57.7 | 48.3 | 36.8 | 24.7 | 16.2 | 36.2 |
| Mean Temp °F | 26.1 | 31.4 | 39.0 | 48.8 | 58.4 | 68.4 | 74.3 | 72.9 | 63.9 | 51.9 | 37.2 | 28.1 | 50.0 |
| Days Max Temp ≥ 90 °F | 0 | 0 | 0 | 0 | 1 | 7 | 15 | 14 | 6 | 0 | 0 | 0 | 43 |
| Days Max Temp ≤ 32 °F | 9 | 5 | 3 | 0 | 0 | 0 | 0 | 0 | 0 | 0 | 3 | 8 | 28 |
| Days Min Temp ≤ 32 °F | 30 | 26 | 25 | 12 | 1 | 0 | 0 | 0 | 1 | 9 | 24 | 30 | 158 |
| Days Min Temp ≤ 0 °F | 5 | 2 | 1 | 0 | 0 | 0 | 0 | 0 | 0 | 0 | 0 | 3 | 11 |
| Heating Degree Days | 1200 | 941 | 799 | 483 | 222 | 41 | 4 | 7 | 111 | 403 | 828 | 1138 | 6177 |
| Cooling Degree Days | 0 | 0 | 0 | 4 | 24 | 163 | 298 | 268 | 93 | 4 | 0 | 0 | 854 |
| Total Precipitation (") | 0.48 | 0.54 | 1.23 | 1.66 | 3.43 | 3.07 | 2.57 | 1.72 | 1.20 | 0.93 | 0.75 | 0.50 | 18.08 |
| Days ≥ 0.1" Precip | 2 | 2 | 3 | 4 | 7 | 6 | 6 | 4 | 3 | 3 | 2 | 2 | 44 |
| Total Snowfall (") | 7.8 | 6.5 | 10.0 | 4.7 | 0.2 | 0.0 | 0.0 | 0.0 | 0.3 | 1.8 | 7.1 | 6.8 | 45.2 |
| Days ≥ 1" Snow Depth | 16 | 10 | 6 | 2 | 0 | 0 | 0 | 0 | 0 | 1 | 5 | 13 | 53 |

## SILVERTON *San Juan County* ELEVATION 9426 ft LAT/LONG 37° 48 ' N / 107° 40 ' W

| | JAN | FEB | MAR | APR | MAY | JUN | JUL | AUG | SEP | OCT | NOV | DEC | YEAR |
|---|---|---|---|---|---|---|---|---|---|---|---|---|---|
| Maximum Temp °F | 34.3 | 37.6 | 41.9 | *48.3* | 58.6 | 68.6 | 73.4 | *71.0* | *64.3* | *55.2* | *42.8* | 34.5 | 52.5 |
| Minimum Temp °F | -5.0 | -1.6 | 7.2 | *17.4* | 26.4 | 31.7 | 37.3 | *36.5* | *29.8* | *21.5* | *8.9* | -1.6 | 17.4 |
| Mean Temp °F | 14.7 | 18.1 | 24.6 | *32.9* | 42.5 | 50.2 | 55.4 | *53.8* | *47.1* | *38.4* | *25.9* | 16.5 | 35.0 |
| Days Max Temp ≥ 90 °F | 0 | 0 | 0 | 0 | 0 | 0 | 0 | 0 | *0* | *0* | 0 | 0 | 0 |
| Days Max Temp ≤ 32 °F | 12 | 8 | 4 | 1 | 0 | 0 | 0 | 0 | *0* | *0* | *4* | 11 | 40 |
| Days Min Temp ≤ 32 °F | 31 | 28 | 31 | 30 | *27* | *17* | 5 | *7* | *21* | *29* | *30* | 31 | 287 |
| Days Min Temp ≤ 0 °F | *22* | 17 | 9 | 1 | 0 | 0 | 0 | 0 | *0* | *0* | *7* | 20 | 76 |
| Heating Degree Days | 1553 | 1319 | 1247 | *958* | 690 | 439 | 291 | *341* | *531* | *817* | *1167* | 1497 | 10850 |
| Cooling Degree Days | 0 | 0 | 0 | *0* | 0 | 0 | *0* | *0* | *0* | *0* | 0 | 0 | 0 |
| Total Precipitation (") | 1.54 | 1.83 | 2.16 | 1.57 | 1.65 | 1.31 | 2.92 | 3.10 | 2.72 | 2.23 | 1.83 | 2.13 | 24.99 |
| Days ≥ 0.1" Precip | *5* | 5 | 7 | 5 | 5 | 5 | 9 | *10* | *7* | *6* | 6 | 6 | 76 |
| Total Snowfall (") | *21.9* | 25.7 | 24.8 | 13.4 | *3.7* | 0.1 | 0.0 | 0.0 | *0.7* | *7.3* | 25.8 | 25.0 | 148.4 |
| Days ≥ 1" Snow Depth | *30* | 27 | 28 | *16* | 2 | 0 | 0 | 0 | *0* | *3* | *20* | 29 | 155 |

## SPICER *Jackson County* ELEVATION 8307 ft LAT/LONG 40° 29' N / 106° 25' W

| | JAN | FEB | MAR | APR | MAY | JUN | JUL | AUG | SEP | OCT | NOV | DEC | YEAR |
|---|---|---|---|---|---|---|---|---|---|---|---|---|---|
| Maximum Temp °F | 28.8 | 32.0 | 38.5 | 47.9 | 58.9 | 69.0 | 74.8 | 73.4 | 66.0 | 53.9 | 37.6 | 29.6 | 50.9 |
| Minimum Temp °F | 6.3 | 6.9 | 12.5 | 20.3 | 28.3 | 34.8 | 39.8 | 38.2 | 31.0 | 22.8 | 13.3 | 6.8 | 21.8 |
| Mean Temp °F | 17.5 | 19.5 | 25.5 | 34.2 | 43.6 | 51.9 | 57.3 | 55.8 | 48.5 | 38.4 | 25.4 | 18.2 | 36.3 |
| Days Max Temp ≥ 90 °F | 0 | 0 | 0 | 0 | 0 | 0 | 0 | 0 | 0 | 0 | 0 | 0 | 0 |
| Days Max Temp ≤ 32 °F | 19 | 13 | 7 | 2 | 0 | 0 | 0 | 0 | 0 | 1 | 9 | 18 | 69 |
| Days Min Temp ≤ 32 °F | 31 | 28 | 31 | 28 | 23 | 11 | 3 | 5 | ***16*** | 28 | 30 | 31 | 265 |
| Days Min Temp ≤ 0 °F | 9 | 8 | 4 | 1 | 0 | 0 | 0 | 0 | 0 | 0 | 4 | 8 | 34 |
| Heating Degree Days | 1466 | 1280 | 1218 | 918 | 655 | 385 | 232 | 278 | 488 | 818 | 1181 | 1443 | 10362 |
| Cooling Degree Days | 0 | 0 | 0 | 0 | 0 | 0 | 0 | 0 | 0 | 0 | 0 | 0 | 0 |
| Total Precipitation (") | 0.92 | 0.74 | 0.92 | 1.08 | 1.46 | 1.28 | 1.76 | 1.33 | 1.25 | 1.19 | 1.02 | 0.94 | 13.89 |
| Days ≥ 0.1" Precip | 3 | 3 | 3 | 4 | 5 | 4 | 6 | 4 | 5 | 4 | 4 | 3 | 48 |
| Total Snowfall (") | ***24.0*** | 21.5 | 27.0 | 17.3 | 5.1 | 0.7 | 0.0 | 0.0 | 0.6 | ***6.6*** | 21.9 | ***24.9*** | 149.6 |
| Days ≥ 1" Snow Depth | ***23*** | ***22*** | ***20*** | ***12*** | 1 | 0 | 0 | 0 | 0 | 3 | ***14*** | na | na |

## SPRINGFIELD 7 WSW *Baca County* ELEVATION 4580 ft LAT/LONG 37° 23' N / 102° 42' W

| | JAN | FEB | MAR | APR | MAY | JUN | JUL | AUG | SEP | OCT | NOV | DEC | YEAR |
|---|---|---|---|---|---|---|---|---|---|---|---|---|---|
| Maximum Temp °F | 46.6 | 51.1 | 59.0 | 68.8 | 76.0 | 86.7 | 91.2 | 88.2 | 81.0 | 70.7 | 56.2 | 47.8 | 68.6 |
| Minimum Temp °F | 17.8 | 21.1 | 27.3 | 36.0 | 45.3 | 55.1 | 60.3 | 58.7 | 50.8 | 39.0 | 27.2 | 19.9 | 38.2 |
| Mean Temp °F | 32.2 | 36.1 | 43.2 | 52.4 | 60.7 | 70.9 | 75.8 | 73.5 | 65.9 | 54.9 | 41.7 | 33.9 | 53.4 |
| Days Max Temp ≥ 90 °F | 0 | 0 | 0 | 0 | 2 | 11 | 20 | 14 | 6 | 0 | 0 | 0 | 53 |
| Days Max Temp ≤ 32 °F | 5 | 3 | 2 | 0 | 0 | 0 | 0 | 0 | 0 | 0 | 2 | 4 | 16 |
| Days Min Temp ≤ 32 °F | 30 | 26 | 22 | 10 | 2 | 0 | 0 | 0 | 0 | 6 | 21 | 29 | 146 |
| Days Min Temp ≤ 0 °F | 2 | 1 | 0 | 0 | 0 | 0 | 0 | 0 | 0 | 0 | 0 | 1 | 4 |
| Heating Degree Days | 1009 | 809 | 670 | 377 | 166 | 22 | 1 | 4 | 76 | 317 | 691 | 958 | 5100 |
| Cooling Degree Days | 0 | 0 | 1 | 9 | 50 | 233 | 365 | 308 | 125 | 10 | 0 | 0 | 1101 |
| Total Precipitation (") | 0.41 | 0.45 | 0.99 | 1.46 | 2.77 | 2.21 | 2.48 | 2.24 | 1.38 | 0.83 | 0.78 | 0.38 | 16.38 |
| Days ≥ 0.1" Precip | 1 | 1 | 3 | 3 | 5 | 4 | 5 | 5 | 3 | 2 | 2 | 1 | 35 |
| Total Snowfall (") | 5.4 | 4.0 | 7.4 | 3.4 | 1.2 | 0.0 | 0.0 | 0.0 | 0.2 | 1.5 | 5.0 | 4.8 | 32.9 |
| Days ≥ 1" Snow Depth | 8 | 4 | 3 | 1 | 0 | 0 | 0 | 0 | 0 | 0 | 3 | 5 | 24 |

## STEAMBOAT SPRINGS *Routt County* ELEVATION 6769 ft LAT/LONG 40° 30' N / 106° 50' W

| | JAN | FEB | MAR | APR | MAY | JUN | JUL | AUG | SEP | OCT | NOV | DEC | YEAR |
|---|---|---|---|---|---|---|---|---|---|---|---|---|---|
| Maximum Temp °F | 28.4 | 33.4 | 42.7 | 53.5 | 64.4 | 75.0 | 81.7 | 80.2 | 71.7 | 59.5 | 41.7 | 29.3 | 55.1 |
| Minimum Temp °F | 2.4 | 4.6 | 15.8 | 24.2 | 31.6 | 36.1 | 41.7 | 40.4 | 33.1 | 24.5 | 15.2 | 3.6 | 22.8 |
| Mean Temp °F | 15.4 | 19.0 | 29.3 | 38.9 | 48.0 | 55.6 | 61.7 | 60.3 | 52.4 | 42.0 | 28.5 | 16.5 | 39.0 |
| Days Max Temp ≥ 90 °F | 0 | 0 | 0 | 0 | 0 | 0 | 1 | 1 | 0 | 0 | 0 | 0 | 2 |
| Days Max Temp ≤ 32 °F | 21 | 12 | 3 | 0 | 0 | 0 | 0 | 0 | 0 | 0 | 6 | 20 | 62 |
| Days Min Temp ≤ 32 °F | 31 | 28 | 30 | 27 | 18 | 7 | 1 | 2 | 14 | 27 | 29 | 31 | 245 |
| Days Min Temp ≤ 0 °F | 14 | 11 | 3 | 0 | 0 | 0 | 0 | 0 | 0 | 0 | 3 | 13 | 44 |
| Heating Degree Days | 1531 | 1291 | 1101 | 777 | 520 | 278 | 106 | 145 | 371 | 706 | 1090 | 1498 | 9414 |
| Cooling Degree Days | 0 | 0 | 0 | 0 | 0 | 5 | 15 | 9 | 0 | 0 | 0 | 0 | 29 |
| Total Precipitation (") | 2.44 | 2.03 | 2.04 | 2.20 | 2.16 | 1.53 | 1.59 | 1.52 | 1.46 | 1.91 | 2.27 | 2.51 | 23.66 |
| Days ≥ 0.1" Precip | 8 | 7 | 7 | 7 | 7 | 4 | 5 | 5 | 5 | 5 | 7 | 8 | 75 |
| Total Snowfall (") | 36.2 | 27.8 | 20.7 | 13.2 | 2.7 | 0.3 | 0.0 | 0.0 | 0.7 | 7.6 | 24.7 | 36.3 | 170.2 |
| Days ≥ 1" Snow Depth | 30 | 28 | 24 | 6 | ***0*** | 0 | 0 | 0 | 0 | ***2*** | 16 | 30 | 136 |

## STERLING *Logan County* ELEVATION 3944 ft LAT/LONG 40° 37' N / 103° 12' W

| | JAN | FEB | MAR | APR | MAY | JUN | JUL | AUG | SEP | OCT | NOV | DEC | YEAR |
|---|---|---|---|---|---|---|---|---|---|---|---|---|---|
| Maximum Temp °F | 38.1 | 44.6 | 52.7 | 62.7 | 72.3 | 83.4 | 89.8 | 87.8 | 78.3 | 66.3 | 50.1 | 40.1 | 63.9 |
| Minimum Temp °F | 11.0 | 16.5 | 24.2 | 33.7 | 44.2 | 53.9 | 59.5 | 57.1 | 46.1 | 33.5 | 22.0 | 12.9 | 34.6 |
| Mean Temp °F | 24.5 | 30.6 | 38.5 | 48.2 | 58.3 | 68.7 | 74.6 | 72.5 | 62.3 | 49.9 | 36.1 | 26.5 | 49.2 |
| Days Max Temp ≥ 90 °F | 0 | 0 | 0 | 0 | 1 | 9 | 17 | 14 | 5 | 0 | 0 | 0 | 46 |
| Days Max Temp ≤ 32 °F | 9 | 5 | 3 | 0 | 0 | 0 | 0 | 0 | 0 | 0 | 3 | 8 | 28 |
| Days Min Temp ≤ 32 °F | 31 | 28 | 26 | 13 | 2 | 0 | 0 | 0 | 1 | 13 | 27 | 31 | 172 |
| Days Min Temp ≤ 0 °F | 6 | 3 | 1 | 0 | 0 | 0 | 0 | 0 | 0 | 0 | 1 | 4 | 15 |
| Heating Degree Days | 1247 | 965 | 815 | 497 | 233 | 48 | 5 | 11 | 142 | 461 | 860 | 1186 | 6470 |
| Cooling Degree Days | 0 | 0 | 0 | 2 | 33 | 178 | 311 | 272 | 79 | 1 | 0 | 0 | 876 |
| Total Precipitation (") | 0.37 | 0.27 | 1.06 | 1.33 | 2.90 | 3.05 | 2.29 | 1.91 | 1.07 | 0.93 | 0.58 | 0.34 | 16.10 |
| Days ≥ 0.1" Precip | 1 | 1 | 3 | 3 | 6 | 5 | 5 | 4 | 3 | 2 | 2 | 1 | 36 |
| Total Snowfall (") | ***5.5*** | ***3.8*** | na | ***2.2*** | 0.1 | 0.0 | 0.0 | 0.0 | 0.3 | ***0.4*** | ***5.0*** | ***5.0*** | na |
| Days ≥ 1" Snow Depth | na | na | na | ***1*** | 0 | 0 | 0 | 0 | 0 | ***0*** | na | na | na |

## STRATTON *Kit Carson County* ELEVATION 4403 ft LAT/LONG 39° 18' N / 102° 35' W

| | JAN | FEB | MAR | APR | MAY | JUN | JUL | AUG | SEP | OCT | NOV | DEC | YEAR |
|---|---|---|---|---|---|---|---|---|---|---|---|---|---|
| Maximum Temp °F | 41.8 | 46.7 | 55.5 | 65.6 | 74.1 | 84.5 | 90.0 | 87.8 | 79.4 | 68.5 | 50.9 | 43.3 | 65.7 |
| Minimum Temp °F | 15.3 | 18.5 | 25.5 | 35.0 | 44.7 | 53.7 | 59.8 | 57.8 | 48.6 | 36.9 | 24.8 | 17.3 | 36.5 |
| Mean Temp °F | 28.6 | 32.7 | 40.5 | 50.4 | 59.5 | 69.1 | 74.8 | 72.8 | 64.0 | 52.7 | 37.8 | 30.3 | 51.1 |
| Days Max Temp ≥ 90 °F | 0 | 0 | 0 | 0 | 1 | 10 | 18 | 15 | 5 | 0 | 0 | 0 | 49 |
| Days Max Temp ≤ 32 °F | 8 | 4 | 2 | 0 | 0 | 0 | 0 | 0 | 0 | 0 | 3 | 6 | 23 |
| Days Min Temp ≤ 32 °F | 30 | 27 | 24 | 11 | 2 | 0 | 0 | 0 | 1 | 9 | 24 | 30 | 158 |
| Days Min Temp ≤ 0 °F | 4 | 2 | 1 | 0 | 0 | 0 | 0 | 0 | 0 | 0 | 0 | 2 | 9 |
| Heating Degree Days | 1123 | 905 | 754 | 437 | 197 | 36 | 4 | 10 | 104 | 378 | 810 | 1068 | 5826 |
| Cooling Degree Days | na | 0 | na | na | na | na | na | na | na | na | na | 0 | na |
| Total Precipitation (") | 0.28 | 0.37 | 0.72 | 1.32 | 2.70 | 2.65 | 3.19 | 2.83 | 1.18 | 1.02 | 0.72 | 0.31 | 17.29 |
| Days ≥ 0.1" Precip | 1 | 1 | 2 | 3 | 6 | 5 | 6 | 5 | 2 | 2 | 2 | 1 | 36 |
| Total Snowfall (") | na | na | na | 2.3 | 0.3 | 0.0 | 0.0 | 0.0 | 0.2 | 2.6 | 6.3 | na | na |
| Days ≥ 1" Snow Depth | na | na | na | 1 | 0 | 0 | 0 | 0 | 0 | 1 | na | na | na |

## SUGARLOAF RESERVOIR *Lake County* ELEVATION 10003 ft LAT/LONG 39° 15' N / 106° 22' W

| | JAN | FEB | MAR | APR | MAY | JUN | JUL | AUG | SEP | OCT | NOV | DEC | YEAR |
|---|---|---|---|---|---|---|---|---|---|---|---|---|---|
| Maximum Temp °F | 29.9 | 32.4 | 37.0 | 43.7 | 53.9 | 65.4 | 71.1 | 68.7 | 62.0 | 51.3 | 37.6 | 31.4 | 48.7 |
| Minimum Temp °F | 1.1 | 1.2 | 7.6 | 16.6 | 26.6 | 33.9 | 38.9 | 37.6 | 31.5 | 23.7 | 13.0 | 5.3 | 19.8 |
| Mean Temp °F | 15.6 | 16.8 | 22.3 | 30.2 | 40.3 | 49.7 | 55.0 | 53.2 | 46.7 | 37.5 | 25.3 | 18.4 | 34.3 |
| Days Max Temp ≥ 90 °F | 0 | 0 | 0 | 0 | 0 | 0 | 0 | 0 | 0 | 0 | 0 | 0 | 0 |
| Days Max Temp ≤ 32 °F | 18 | 13 | 8 | 3 | 0 | 0 | 0 | 0 | 0 | 1 | 10 | 16 | 69 |
| Days Min Temp ≤ 32 °F | 31 | 28 | 31 | 30 | 27 | 11 | 2 | 4 | 16 | 29 | 30 | 30 | 269 |
| Days Min Temp ≤ 0 °F | 13 | 12 | 6 | 1 | 0 | 0 | 0 | 0 | 0 | 0 | 3 | 9 | 44 |
| Heating Degree Days | 1527 | 1356 | 1317 | 1038 | 760 | 452 | 302 | 359 | 541 | 845 | 1177 | 1438 | 11112 |
| Cooling Degree Days | 0 | 0 | 0 | 0 | 0 | 0 | 0 | 0 | 0 | 0 | 0 | 0 | 0 |
| Total Precipitation (") | 1.24 | 1.10 | 1.44 | 1.19 | 1.27 | 1.08 | 1.93 | 1.89 | 1.31 | 1.04 | 1.36 | 1.21 | 16.06 |
| Days ≥ 0.1" Precip | 4 | 4 | 4 | 4 | 4 | 4 | 5 | 6 | 4 | 3 | 4 | 3 | 49 |
| Total Snowfall (") | 18.7 | 17.0 | 22.1 | 16.2 | 8.1 | 1.5 | 0.0 | 0.0 | 1.6 | 7.7 | 20.5 | na | na |
| Days ≥ 1" Snow Depth | 28 | 24 | 27 | 21 | 5 | 0 | 0 | 0 | 0 | 3 | 17 | 27 | 152 |

## TACONY 10 SE *Pueblo County* ELEVATION 5003 ft LAT/LONG 38° 25' N / 104° 7' W

| | JAN | FEB | MAR | APR | MAY | JUN | JUL | AUG | SEP | OCT | NOV | DEC | YEAR |
|---|---|---|---|---|---|---|---|---|---|---|---|---|---|
| Maximum Temp °F | 44.1 | 49.7 | 57.7 | 67.6 | 75.6 | 87.4 | 91.6 | 88.5 | 81.1 | 69.4 | 54.2 | 45.8 | 67.7 |
| Minimum Temp °F | 14.0 | 18.5 | 25.2 | 33.9 | 43.8 | 52.9 | 58.6 | 56.5 | 47.7 | 35.8 | 23.5 | 16.1 | 35.5 |
| Mean Temp °F | 29.1 | 34.1 | 41.5 | 50.8 | 59.7 | 70.2 | 75.1 | 72.6 | 64.4 | 52.6 | 38.9 | 31.0 | 51.7 |
| Days Max Temp ≥ 90 °F | 0 | 0 | 0 | 0 | 2 | 13 | 20 | 16 | 6 | 0 | 0 | 0 | 57 |
| Days Max Temp ≤ 32 °F | 6 | 3 | 1 | 0 | 0 | 0 | 0 | 0 | 0 | 0 | 2 | 5 | 17 |
| Days Min Temp ≤ 32 °F | 31 | 27 | 25 | 12 | 2 | 0 | 0 | 0 | 1 | 10 | 26 | 30 | 164 |
| Days Min Temp ≤ 0 °F | 4 | 1 | 0 | 0 | 0 | 0 | 0 | 0 | 0 | 0 | 0 | 2 | 7 |
| Heating Degree Days | 1106 | 866 | 724 | 422 | 183 | 24 | 2 | 5 | 90 | 379 | 777 | 1049 | 5627 |
| Cooling Degree Days | 0 | 0 | 0 | 2 | 26 | 201 | 315 | 255 | 83 | 1 | 0 | 0 | 883 |
| Total Precipitation (") | 0.20 | 0.18 | 0.50 | 0.89 | 1.56 | 1.42 | 2.04 | 2.00 | 0.93 | 0.62 | 0.34 | 0.26 | 10.94 |
| Days ≥ 0.1" Precip | 1 | 1 | 1 | 2 | 4 | 3 | 4 | 5 | 2 | 2 | 1 | 1 | 27 |
| Total Snowfall (") | 4.2 | 3.2 | 5.0 | 3.4 | 0.8 | 0.0 | 0.0 | 0.0 | 0.2 | 1.4 | 4.1 | 5.1 | 27.4 |
| Days ≥ 1" Snow Depth | 8 | 3 | 1 | 1 | 0 | 0 | 0 | 0 | 0 | 0 | 3 | 6 | 22 |

## TAYLOR PARK *Gunnison County* ELEVATION 9216 ft LAT/LONG 38° 49' N / 106° 37' W

| | JAN | FEB | MAR | APR | MAY | JUN | JUL | AUG | SEP | OCT | NOV | DEC | YEAR |
|---|---|---|---|---|---|---|---|---|---|---|---|---|---|
| Maximum Temp °F | 26.3 | 32.0 | 37.9 | 45.2 | 55.5 | 66.9 | 71.4 | 69.1 | 62.6 | 52.6 | 37.7 | 27.1 | 48.7 |
| Minimum Temp °F | -12.7 | -10.8 | -1.2 | 12.0 | 25.0 | 33.0 | 39.6 | 38.5 | 31.0 | 22.0 | 9.0 | -7.5 | 14.8 |
| Mean Temp °F | 6.8 | 10.7 | 18.4 | 28.6 | 40.3 | 50.0 | 55.5 | 53.8 | 46.8 | 37.3 | 23.4 | 9.8 | 31.8 |
| Days Max Temp ≥ 90 °F | 0 | 0 | 0 | 0 | 0 | 0 | 0 | 0 | 0 | 0 | 0 | 0 | 0 |
| Days Max Temp ≤ 32 °F | 23 | 14 | 6 | 2 | 0 | 0 | 0 | 0 | 0 | 1 | 9 | 22 | 77 |
| Days Min Temp ≤ 32 °F | 31 | 28 | 31 | 30 | 28 | 13 | 2 | 3 | 17 | 29 | 30 | 31 | 273 |
| Days Min Temp ≤ 0 °F | 25 | 21 | 17 | 5 | 0 | 0 | 0 | 0 | 0 | 0 | 7 | 23 | 98 |
| Heating Degree Days | 1800 | 1529 | 1438 | 1084 | 759 | 444 | 286 | 341 | 539 | 851 | 1242 | 1706 | 12019 |
| Cooling Degree Days | 0 | 0 | 0 | 0 | 0 | 0 | 0 | 0 | 0 | 0 | 0 | 0 | 0 |
| Total Precipitation (") | 1.22 | 1.23 | 1.46 | 1.19 | 1.34 | 1.12 | 1.78 | 1.73 | 1.56 | 1.30 | 1.38 | 1.44 | 16.75 |
| Days ≥ 0.1" Precip | 5 | 5 | 6 | 5 | 5 | 4 | 7 | 7 | 5 | 5 | 6 | 6 | 66 |
| Total Snowfall (") | na | na | na | na | 1.6 | 0.1 | 0.0 | 0.0 | 0.5 | 4.1 | na | na | na |
| Days ≥ 1" Snow Depth | na | na | na | na | 1 | 0 | 0 | 0 | 0 | 2 | na | na | na |

## TELLURIDE 4 WNW *San Miguel County* ELEVATION 8756 ft LAT/LONG 37° 57' N / 107° 49' W

| | JAN | FEB | MAR | APR | MAY | JUN | JUL | AUG | SEP | OCT | NOV | DEC | YEAR |
|---|---|---|---|---|---|---|---|---|---|---|---|---|---|
| Maximum Temp °F | 37.9 | 40.6 | 44.5 | 52.8 | 62.5 | 73.3 | 77.8 | 75.4 | 69.2 | 59.7 | 45.6 | 38.3 | 56.5 |
| Minimum Temp °F | 6.6 | 9.4 | 16.0 | 23.1 | 30.5 | 36.6 | 42.4 | 41.7 | 35.0 | 26.0 | 16.0 | 8.3 | 24.3 |
| Mean Temp °F | 22.3 | 25.1 | 30.3 | 38.0 | 46.5 | 55.0 | 60.1 | 58.6 | 52.1 | 42.9 | 30.8 | 23.3 | 40.4 |
| Days Max Temp ≥ 90 °F | 0 | 0 | 0 | 0 | 0 | 0 | 0 | 0 | 0 | 0 | 0 | 0 | 0 |
| Days Max Temp ≤ 32 °F | 8 | 5 | 2 | 1 | 0 | 0 | 0 | 0 | 0 | 0 | 3 | 8 | 27 |
| Days Min Temp ≤ 32 °F | 31 | 28 | 31 | 27 | 20 | 7 | 0 | 1 | 10 | 25 | 29 | 31 | 240 |
| Days Min Temp ≤ 0 °F | 9 | 6 | 2 | 0 | 0 | 0 | 0 | 0 | 0 | 0 | 2 | 7 | 26 |
| Heating Degree Days | 1318 | 1121 | 1069 | 804 | 566 | 296 | 145 | 194 | 380 | 679 | 1018 | 1285 | 8875 |
| Cooling Degree Days | 0 | 0 | 0 | 0 | 0 | 1 | 3 | 2 | 0 | 0 | 0 | 0 | 6 |
| Total Precipitation (") | 1.57 | 1.50 | 1.99 | 1.89 | 1.93 | 1.36 | 2.53 | 2.64 | 2.34 | 2.09 | 1.85 | 1.68 | 23.37 |
| Days ≥ 0.1" Precip | 6 | 5 | 6 | 6 | 6 | 4 | 9 | 8 | 7 | 5 | 5 | 6 | 73 |
| Total Snowfall (") | 32.2 | 28.6 | 38.4 | 25.1 | 9.2 | 0.9 | 0.0 | 0.0 | 0.6 | 11.5 | 27.8 | 31.6 | 205.9 |
| Days ≥ 1" Snow Depth | na | *27* | na | *12* | *2* | 0 | 0 | 0 | 0 | na | na | na | na |

## TRINIDAD *Las Animas County* ELEVATION 6306 ft LAT/LONG 37° 10' N / 104° 30' W

| | JAN | FEB | MAR | APR | MAY | JUN | JUL | AUG | SEP | OCT | NOV | DEC | YEAR |
|---|---|---|---|---|---|---|---|---|---|---|---|---|---|
| Maximum Temp °F | 48.2 | 50.9 | 56.9 | 65.5 | 73.2 | 82.9 | 86.3 | 83.6 | 78.1 | 68.8 | 55.7 | 48.4 | 66.5 |
| Minimum Temp °F | 18.9 | 21.5 | 27.6 | 35.0 | 43.5 | 52.6 | 57.5 | 55.7 | 48.4 | 37.4 | 27.0 | 20.3 | 37.1 |
| Mean Temp °F | 33.5 | 36.2 | 42.3 | 50.3 | 58.4 | 67.7 | 71.9 | 69.7 | 63.2 | 53.1 | 41.4 | 34.4 | 51.8 |
| Days Max Temp ≥ 90 °F | 0 | 0 | 0 | 0 | 0 | 6 | 9 | 4 | 1 | 0 | 0 | 0 | 20 |
| Days Max Temp ≤ 32 °F | 4 | 2 | 1 | 0 | 0 | 0 | 0 | 0 | 0 | 0 | 1 | 3 | 11 |
| Days Min Temp ≤ 32 °F | 29 | 26 | 22 | 12 | 3 | 0 | 0 | 0 | 1 | 8 | 22 | 28 | 151 |
| Days Min Temp ≤ 0 °F | 2 | 1 | 0 | 0 | 0 | 0 | 0 | 0 | 0 | 0 | 0 | 1 | 4 |
| Heating Degree Days | 968 | 806 | 698 | 435 | 214 | 39 | 2 | 9 | 97 | 364 | 702 | 943 | 5277 |
| Cooling Degree Days | 0 | 0 | 0 | 1 | 19 | 136 | 226 | 166 | 54 | 1 | 0 | 0 | 603 |
| Total Precipitation (") | 0.43 | 0.50 | 1.05 | 1.12 | 2.03 | 1.72 | 2.75 | 2.78 | 1.30 | 0.90 | 0.92 | 0.51 | 16.01 |
| Days ≥ 0.1" Precip | 2 | 2 | 3 | 3 | 5 | 4 | 6 | 6 | 3 | 2 | 3 | 2 | 41 |
| Total Snowfall (") | *7.4* | *7.8* | *9.9* | *4.1* | 1.1 | 0.0 | 0.0 | 0.0 | 0.1 | *0.9* | na | *7.1* | na |
| Days ≥ 1" Snow Depth | *7* | *4* | *4* | 1 | 0 | 0 | 0 | 0 | 0 | 0 | *4* | na | na |

## TRINIDAD LAS ANIMAS *Las Animas County* ELEVATION 5738 ft LAT/LONG 37° 15' N / 104° 20' W

| | JAN | FEB | MAR | APR | MAY | JUN | JUL | AUG | SEP | OCT | NOV | DEC | YEAR |
|---|---|---|---|---|---|---|---|---|---|---|---|---|---|
| Maximum Temp °F | 46.8 | 50.7 | 57.0 | 65.7 | 73.7 | 84.3 | 88.6 | 85.9 | 78.9 | 68.9 | 55.6 | 47.4 | 67.0 |
| Minimum Temp °F | 16.8 | 19.9 | 26.0 | 34.6 | 43.7 | 53.1 | 58.7 | 57.0 | 48.8 | 37.1 | 25.5 | 17.6 | 36.6 |
| Mean Temp °F | 31.8 | 35.3 | 41.5 | 50.1 | 58.7 | 68.7 | 73.7 | 71.5 | 63.9 | 53.0 | 40.5 | 32.5 | 51.8 |
| Days Max Temp ≥ 90 °F | 0 | 0 | 0 | 0 | 0 | 9 | 15 | 9 | 2 | 0 | 0 | 0 | 35 |
| Days Max Temp ≤ 32 °F | 4 | 3 | 2 | 0 | 0 | 0 | 0 | 0 | 0 | 0 | 1 | 4 | 14 |
| Days Min Temp ≤ 32 °F | 30 | 26 | 24 | 12 | 2 | 0 | 0 | 0 | 1 | 8 | 24 | 30 | 157 |
| Days Min Temp ≤ 0 °F | 3 | 2 | 0 | 0 | 0 | 0 | 0 | 0 | 0 | 0 | 0 | 2 | 7 |
| Heating Degree Days | 1019 | 832 | 721 | 441 | 210 | 35 | 3 | 7 | 98 | 368 | 727 | 1000 | 5461 |
| Cooling Degree Days | 0 | 0 | 0 | 3 | 22 | 159 | 276 | 222 | 76 | 2 | 0 | 0 | 760 |
| Total Precipitation (") | 0.44 | 0.45 | 0.94 | 0.96 | 1.78 | 1.58 | 2.17 | 2.23 | 1.18 | 0.80 | 0.78 | 0.59 | 13.90 |
| Days ≥ 0.1" Precip | 2 | 2 | 3 | 3 | 5 | 4 | 5 | 5 | 3 | 2 | 3 | 2 | 39 |
| Total Snowfall (") | 5.1 | 5.4 | 8.4 | 4.9 | 2.0 | 0.0 | 0.0 | 0.0 | 0.5 | 3.8 | 7.2 | 6.9 | 44.2 |
| Days ≥ 1" Snow Depth | 9 | 5 | 4 | 2 | 0 | 0 | 0 | 0 | 0 | 1 | 5 | 10 | 36 |

## URAVAN *Montrose County* ELEVATION 5023 ft LAT/LONG 38° 22' N / 108° 44' W

| | JAN | FEB | MAR | APR | MAY | JUN | JUL | AUG | SEP | OCT | NOV | DEC | YEAR |
|---|---|---|---|---|---|---|---|---|---|---|---|---|---|
| Maximum Temp °F | 41.2 | 49.2 | 58.5 | 67.7 | 78.0 | 89.3 | 94.7 | 92.1 | 83.8 | 71.0 | 54.3 | 42.8 | 68.6 |
| Minimum Temp °F | 14.9 | 21.9 | 29.7 | 35.8 | 44.6 | 52.6 | 59.3 | 58.1 | 48.4 | 37.1 | 27.0 | 18.2 | 37.3 |
| Mean Temp °F | 28.1 | 35.6 | 44.1 | 51.8 | 61.3 | 71.0 | 77.0 | 75.1 | 66.0 | 54.1 | 40.7 | 30.6 | 53.0 |
| Days Max Temp ≥ 90 °F | 0 | 0 | 0 | 0 | 2 | 17 | 26 | 22 | 7 | 0 | 0 | 0 | 74 |
| Days Max Temp ≤ 32 °F | 4 | 1 | 0 | 0 | 0 | 0 | 0 | 0 | 0 | 0 | 0 | 3 | 8 |
| Days Min Temp ≤ 32 °F | 29 | 25 | 21 | 10 | 1 | 0 | 0 | 0 | 0 | 7 | 24 | 30 | 147 |
| Days Min Temp ≤ 0 °F | 3 | 0 | 0 | 0 | 0 | 0 | 0 | 0 | 0 | 0 | 0 | 1 | 4 |
| Heating Degree Days | 1137 | 822 | 640 | 391 | 144 | 19 | 0 | 1 | 55 | 333 | 723 | 1061 | 5326 |
| Cooling Degree Days | 0 | 0 | 0 | 1 | 37 | 227 | 372 | 330 | 96 | 0 | 0 | 0 | 1063 |
| Total Precipitation (") | 0.91 | 0.70 | 0.94 | 0.95 | 1.08 | 0.48 | 1.26 | 1.40 | 1.23 | 1.61 | 1.10 | 0.97 | 12.63 |
| Days ≥ 0.1" Precip | 3 | 3 | 3 | 3 | 4 | 1 | 4 | 4 | 3 | 4 | 4 | 3 | 39 |
| Total Snowfall (") | na | na | na | *0.3* | 0.0 | 0.0 | 0.0 | 0.0 | 0.0 | 0.2 | na | na | na |
| Days ≥ 1" Snow Depth | na | na | *0* | 0 | 0 | 0 | 0 | 0 | 0 | 0 | na | na | na |

## VALLECITO DAM *La Plata County* ELEVATION 7650 ft LAT/LONG 37° 22 ' N / 107° 35 ' W

| | JAN | FEB | MAR | APR | MAY | JUN | JUL | AUG | SEP | OCT | NOV | DEC | YEAR |
|---|---|---|---|---|---|---|---|---|---|---|---|---|---|
| Maximum Temp °F | 36.4 | 40.3 | 46.3 | 55.5 | 64.1 | 74.7 | 79.5 | 77.0 | 70.5 | 60.3 | 46.5 | 38.2 | 57.4 |
| Minimum Temp °F | 5.5 | 9.1 | 17.5 | 24.8 | 32.2 | 39.4 | 46.5 | 45.2 | 38.2 | 29.3 | 19.6 | 11.1 | 26.5 |
| Mean Temp °F | 21.0 | 24.7 | 31.9 | 40.2 | 48.2 | 57.1 | 63.0 | 61.1 | 54.4 | 44.9 | 33.1 | 24.6 | 42.0 |
| Days Max Temp ≥ 90 °F | 0 | 0 | 0 | 0 | 0 | 0 | 0 | 0 | 0 | 0 | 0 | 0 | 0 |
| Days Max Temp ≤ 32 °F | 10 | 5 | 2 | 0 | 0 | 0 | 0 | 0 | 0 | 0 | 3 | 8 | 28 |
| Days Min Temp ≤ 32 °F | 31 | 28 | 31 | 27 | 16 | 3 | 0 | 0 | 5 | 22 | 29 | 31 | 223 |
| Days Min Temp ≤ 0 °F | 10 | 6 | 1 | 0 | 0 | 0 | 0 | 0 | 0 | 0 | 1 | 5 | 23 |
| Heating Degree Days | 1358 | 1130 | 1019 | 738 | 515 | 236 | 78 | 122 | 312 | 618 | 952 | 1245 | 8323 |
| Cooling Degree Days | 0 | 0 | 0 | 0 | 0 | 5 | 20 | 10 | 1 | 0 | 0 | 0 | 36 |
| Total Precipitation (") | 2.20 | 2.10 | 2.35 | 1.64 | 1.68 | 1.12 | 2.63 | 3.48 | 2.78 | 2.62 | 2.44 | 2.63 | 27.67 |
| Days ≥ 0.1" Precip | 5 | 5 | 6 | 4 | 5 | 3 | 7 | 9 | 6 | 5 | 5 | 5 | 65 |
| Total Snowfall (") | 26.8 | 23.1 | 22.6 | 8.5 | 1.6 | 0.0 | 0.0 | 0.0 | 0.0 | 3.0 | 14.0 | 26.6 | 126.2 |
| Days ≥ 1" Snow Depth | na | na | na | na | 0 | 0 | 0 | 0 | 0 | 0 | *5* | na | na |

## WALDEN *Jackson County* ELEVATION 8107 ft LAT/LONG 40° 44 ' N / 106° 16 ' W

| | JAN | FEB | MAR | APR | MAY | JUN | JUL | AUG | SEP | OCT | NOV | DEC | YEAR |
|---|---|---|---|---|---|---|---|---|---|---|---|---|---|
| Maximum Temp °F | 29.3 | 32.4 | 39.6 | 49.8 | 60.6 | 71.4 | 77.8 | 76.0 | 67.4 | 55.3 | 39.2 | 30.7 | 52.5 |
| Minimum Temp °F | 4.6 | 6.1 | 13.4 | 20.6 | 28.0 | 36.0 | 39.8 | 37.3 | 30.3 | 22.0 | 12.8 | 5.7 | 21.4 |
| Mean Temp °F | 17.0 | 19.3 | 26.5 | 35.2 | 44.3 | 53.7 | 58.8 | 56.7 | 48.9 | 38.7 | 26.0 | 18.2 | 36.9 |
| Days Max Temp ≥ 90 °F | 0 | 0 | 0 | 0 | 0 | 0 | 0 | 0 | 0 | 0 | 0 | 0 | 0 |
| Days Max Temp ≤ 32 °F | 18 | 12 | 7 | 2 | 0 | 0 | 0 | 0 | 0 | 1 | 8 | 17 | 65 |
| Days Min Temp ≤ 32 °F | 31 | 28 | 31 | 29 | 24 | 8 | 2 | 6 | 19 | 28 | 29 | 31 | 266 |
| Days Min Temp ≤ 0 °F | 11 | 9 | 4 | 1 | 0 | 0 | 0 | 0 | 0 | 0 | 5 | 11 | 41 |
| Heating Degree Days | 1484 | 1285 | 1186 | 887 | 634 | 332 | 185 | 252 | 478 | 810 | 1163 | 1444 | 10140 |
| Cooling Degree Days | 0 | 0 | 0 | 0 | 0 | 0 | 2 | 0 | 0 | 0 | 0 | 0 | 2 |
| Total Precipitation (") | 0.53 | 0.55 | 0.79 | 0.95 | 1.28 | 1.05 | 1.27 | 1.04 | 1.12 | 0.96 | 0.80 | 0.61 | 10.95 |
| Days ≥ 0.1" Precip | 2 | 2 | 3 | 3 | 4 | 3 | 4 | 4 | 4 | 3 | 3 | 2 | 37 |
| Total Snowfall (") | 8.0 | 7.6 | 9.6 | 8.2 | 3.6 | 0.8 | 0.0 | 0.0 | 1.1 | 4.6 | 10.4 | 9.0 | 62.9 |
| Days ≥ 1" Snow Depth | 27 | 23 | 11 | 2 | 0 | 0 | 0 | 0 | 0 | 2 | 12 | 23 | 100 |

## WALSENBURG *Huerfano County* ELEVATION 6224 ft LAT/LONG 37° 37 ' N / 104° 48 ' W

| | JAN | FEB | MAR | APR | MAY | JUN | JUL | AUG | SEP | OCT | NOV | DEC | YEAR |
|---|---|---|---|---|---|---|---|---|---|---|---|---|---|
| Maximum Temp °F | 46.9 | 49.9 | 56.1 | 65.1 | 73.4 | 83.6 | 87.7 | 84.5 | 78.5 | 68.9 | 54.9 | 47.3 | 66.4 |
| Minimum Temp °F | 20.4 | 22.1 | 27.0 | 33.4 | 42.0 | 50.4 | 56.4 | 54.8 | 47.3 | 36.9 | 27.6 | 21.5 | 36.7 |
| Mean Temp °F | 33.7 | 36.0 | 41.5 | 49.3 | 57.7 | 67.0 | 72.1 | 69.7 | 62.9 | 52.9 | 41.2 | 34.4 | 51.5 |
| Days Max Temp ≥ 90 °F | 0 | 0 | 0 | 0 | 0 | 6 | 12 | 5 | 1 | 0 | 0 | 0 | 24 |
| Days Max Temp ≤ 32 °F | 3 | 2 | 1 | 0 | 0 | 0 | 0 | 0 | 0 | 0 | 1 | 3 | 10 |
| Days Min Temp ≤ 32 °F | 27 | 23 | 22 | 14 | 3 | 0 | 0 | 0 | 1 | 9 | 20 | 25 | 144 |
| Days Min Temp ≤ 0 °F | 2 | 1 | 0 | 0 | 0 | 0 | 0 | 0 | 0 | 0 | 0 | 2 | 5 |
| Heating Degree Days | 965 | 812 | 721 | 465 | 231 | 43 | 3 | 10 | 107 | 370 | 706 | 941 | 5374 |
| Cooling Degree Days | 0 | 0 | 0 | 1 | 15 | 121 | 235 | 178 | 56 | 2 | 0 | 0 | 608 |
| Total Precipitation (") | 0.64 | 0.85 | 1.62 | 1.68 | 1.91 | 1.35 | 2.38 | 2.26 | 1.06 | 0.99 | 1.24 | 0.92 | 16.90 |
| Days ≥ 0.1" Precip | 2 | 3 | 4 | 3 | 4 | 3 | 5 | 5 | 3 | 3 | 3 | 3 | 41 |
| Total Snowfall (") | 11.3 | 12.6 | 18.0 | 12.3 | 2.9 | 0.1 | 0.0 | 0.0 | 0.7 | 5.9 | 14.5 | 15.6 | 93.9 |
| Days ≥ 1" Snow Depth | *7* | *5* | 4 | 2 | 0 | 0 | 0 | 0 | 0 | 1 | na | *9* | na |

## WALSH 1 W *Baca County* ELEVATION 3969 ft LAT/LONG 37° 23 ' N / 102° 15 ' W

| | JAN | FEB | MAR | APR | MAY | JUN | JUL | AUG | SEP | OCT | NOV | DEC | YEAR |
|---|---|---|---|---|---|---|---|---|---|---|---|---|---|
| Maximum Temp °F | 44.8 | 49.9 | 58.2 | 67.9 | 76.0 | 87.1 | 91.6 | 88.9 | 81.1 | 69.8 | 55.3 | 46.8 | 68.1 |
| Minimum Temp °F | 16.9 | 20.7 | 28.2 | 36.8 | *47.2* | 57.0 | 61.9 | 60.4 | 51.6 | 38.7 | 26.7 | 19.1 | 38.8 |
| Mean Temp °F | 30.9 | 35.3 | 43.2 | 52.4 | *61.6* | 72.1 | 76.8 | 74.7 | 66.4 | 54.3 | 41.0 | 32.8 | 53.5 |
| Days Max Temp ≥ 90 °F | 0 | 0 | 0 | 1 | 2 | 13 | 20 | 16 | 7 | 1 | 0 | 0 | 60 |
| Days Max Temp ≤ 32 °F | *6* | 4 | 2 | 0 | 0 | 0 | 0 | 0 | 0 | 0 | 2 | 5 | 19 |
| Days Min Temp ≤ 32 °F | 30 | 26 | 21 | 10 | *1* | 0 | 0 | 0 | 0 | 7 | 23 | 29 | 147 |
| Days Min Temp ≤ 0 °F | 2 | 1 | 0 | 0 | 0 | 0 | 0 | 0 | 0 | 0 | 0 | 2 | 5 |
| Heating Degree Days | 1052 | 833 | 668 | 381 | *148* | 21 | 1 | 5 | 78 | 337 | 713 | 993 | 5230 |
| Cooling Degree Days | 0 | 0 | 0 | 11 | 52 | 239 | 374 | 324 | 131 | 8 | 0 | 0 | 1139 |
| Total Precipitation (") | 0.38 | 0.40 | 0.94 | 1.24 | 2.74 | 2.41 | 2.98 | 2.43 | 1.53 | 0.82 | 0.68 | 0.34 | 16.89 |
| Days ≥ 0.1" Precip | 1 | 1 | 3 | 3 | 5 | 5 | 5 | 4 | 3 | 2 | 2 | 1 | 35 |
| Total Snowfall (") | 4.8 | 3.4 | *4.5* | *2.1* | 0.4 | 0.0 | 0.0 | 0.0 | 0.1 | 0.9 | 3.3 | *4.1* | 23.6 |
| Days ≥ 1" Snow Depth | *7* | *3* | *2* | *1* | 0 | 0 | 0 | 0 | 0 | *0* | *3* | *4* | 20 |

## WATERDALE *Larimer County* ELEVATION 5203 ft LAT/LONG 40° 25' N / 105° 12' W

| | JAN | FEB | MAR | APR | MAY | JUN | JUL | AUG | SEP | OCT | NOV | DEC | YEAR |
|---|---|---|---|---|---|---|---|---|---|---|---|---|---|
| Maximum Temp °F | 42.6 | 45.7 | 52.3 | 60.5 | 69.8 | 80.0 | 86.3 | 84.1 | 75.3 | 64.1 | 50.5 | 43.6 | 62.9 |
| Minimum Temp °F | 14.1 | 17.5 | 24.2 | 32.1 | 40.9 | 48.8 | 54.4 | 53.1 | 44.5 | 34.1 | 23.5 | 15.6 | 33.6 |
| Mean Temp °F | 28.4 | 31.6 | 38.3 | 46.3 | 55.4 | 64.4 | 70.4 | 68.6 | 59.9 | 49.1 | 37.0 | 29.6 | 48.3 |
| Days Max Temp ≥ 90 °F | 0 | 0 | 0 | 0 | 0 | 5 | 11 | 6 | 1 | 0 | 0 | 0 | 23 |
| Days Max Temp ≤ 32 °F | 7 | 4 | 2 | 0 | 0 | 0 | 0 | 0 | 0 | 0 | 3 | 6 | 22 |
| Days Min Temp ≤ 32 °F | 30 | 27 | 26 | 15 | 3 | 0 | 0 | 0 | 2 | 12 | 26 | 30 | 171 |
| Days Min Temp ≤ 0 °F | 4 | 2 | 1 | 0 | 0 | 0 | 0 | 0 | 0 | 0 | 0 | 3 | 10 |
| Heating Degree Days | 1129 | 936 | 821 | 554 | 298 | 85 | 11 | 20 | 175 | 486 | 833 | 1089 | 6437 |
| Cooling Degree Days | 0 | 0 | 0 | 1 | 8 | 79 | 178 | 144 | 34 | 0 | 0 | 0 | 444 |
| Total Precipitation (") | 0.41 | 0.46 | 1.26 | 1.93 | 2.87 | 1.88 | 1.88 | 1.66 | 1.54 | 1.09 | 0.76 | 0.54 | 16.28 |
| Days ≥ 0.1" Precip | 2 | 2 | 3 | 4 | 6 | 4 | 5 | 4 | 4 | 3 | 2 | 2 | 41 |
| Total Snowfall (") | 6.7 | 5.4 | 8.8 | 5.4 | 0.2 | 0.0 | 0.0 | 0.0 | 0.7 | 3.2 | 7.3 | 9.2 | 46.9 |
| Days ≥ 1" Snow Depth | ***6*** | ***6*** | 4 | 2 | 0 | 0 | 0 | 0 | 0 | 1 | ***3*** | ***5*** | 27 |

## WESTCLIFFE *Custer County* ELEVATION 7859 ft LAT/LONG 38° 8' N / 105° 29' W

| | JAN | FEB | MAR | APR | MAY | JUN | JUL | AUG | SEP | OCT | NOV | DEC | YEAR |
|---|---|---|---|---|---|---|---|---|---|---|---|---|---|
| Maximum Temp °F | 39.3 | 42.2 | 47.7 | 56.3 | 65.4 | 76.0 | 80.7 | 77.8 | 71.4 | 61.5 | 47.8 | 40.5 | 58.9 |
| Minimum Temp °F | 6.3 | 9.3 | 17.7 | 25.2 | 33.1 | 40.1 | 44.8 | 43.6 | 35.9 | 25.7 | 15.9 | 7.7 | 25.4 |
| Mean Temp °F | 22.8 | 25.7 | 32.7 | 40.8 | 49.3 | 58.1 | 62.8 | 60.7 | 53.7 | 43.6 | 31.9 | 24.1 | 42.2 |
| Days Max Temp ≥ 90 °F | 0 | 0 | 0 | 0 | 0 | 0 | 0 | 0 | 0 | 0 | 0 | 0 | 0 |
| Days Max Temp ≤ 32 °F | 7 | 4 | 2 | 0 | 0 | 0 | 0 | 0 | 0 | 0 | 3 | 7 | 23 |
| Days Min Temp ≤ 32 °F | 31 | 28 | 29 | 24 | ***15*** | 3 | 0 | 0 | na | 25 | 28 | 30 | na |
| Days Min Temp ≤ 0 °F | 9 | 7 | 3 | 1 | 0 | 0 | 0 | 0 | 0 | 0 | 3 | 9 | 32 |
| Heating Degree Days | 1302 | 1102 | 992 | 720 | 481 | 207 | 80 | 133 | 332 | 657 | 985 | 1262 | 8253 |
| Cooling Degree Days | 0 | 0 | 0 | 0 | 0 | 9 | 16 | 6 | 0 | 0 | 0 | 0 | 31 |
| Total Precipitation (") | 0.44 | 0.53 | 1.21 | 1.13 | 1.50 | 1.09 | 2.28 | 2.86 | 1.21 | 1.16 | 0.91 | 0.72 | 15.04 |
| Days ≥ 0.1" Precip | 1 | 2 | 3 | 3 | 4 | 4 | 6 | 7 | 3 | 3 | 2 | 2 | 40 |
| Total Snowfall (") | 8.9 | 11.2 | 20.8 | 13.1 | 5.1 | 0.2 | 0.0 | 0.0 | 1.5 | 10.1 | 14.3 | 14.1 | 99.3 |
| Days ≥ 1" Snow Depth | na | na | ***6*** | 2 | 1 | 0 | 0 | 0 | 0 | 2 | ***7*** | ***9*** | na |

## WOLF CREEK PASS 1 E *Mineral County* ELEVATION 10641 ft LAT/LONG 37° 29' N / 106° 47' W

| | JAN | FEB | MAR | APR | MAY | JUN | JUL | AUG | SEP | OCT | NOV | DEC | YEAR |
|---|---|---|---|---|---|---|---|---|---|---|---|---|---|
| Maximum Temp °F | 31.4 | 31.8 | 35.2 | 41.6 | ***51.0*** | 60.8 | 65.9 | 64.0 | 57.1 | 48.2 | ***36.9*** | ***31.1*** | 46.3 |
| Minimum Temp °F | 3.5 | 5.2 | 10.1 | 17.3 | ***27.1*** | 34.5 | ***40.3*** | 38.6 | 33.0 | 23.7 | 11.9 | ***4.7*** | 20.8 |
| Mean Temp °F | 17.5 | 18.5 | 22.7 | 29.5 | ***39.1*** | 47.7 | ***53.1*** | 51.2 | 45.0 | 35.7 | ***24.4*** | ***17.9*** | 33.5 |
| Days Max Temp ≥ 90 °F | 0 | 0 | 0 | 0 | 0 | 0 | 0 | 0 | 0 | 0 | 0 | 0 | 0 |
| Days Max Temp ≤ 32 °F | 16 | 14 | 12 | 5 | 0 | 0 | 0 | 0 | 0 | 2 | 10 | 16 | 75 |
| Days Min Temp ≤ 32 °F | 30 | 28 | 31 | 29 | ***24*** | 10 | 0 | 2 | 13 | 28 | 30 | ***31*** | 256 |
| Days Min Temp ≤ 0 °F | 12 | 8 | 4 | 1 | 0 | 0 | 0 | 0 | 0 | 0 | 3 | 10 | 38 |
| Heating Degree Days | 1466 | 1305 | 1306 | 1058 | ***799*** | 512 | ***363*** | 420 | 593 | 900 | ***1211*** | ***1453*** | 11386 |
| Cooling Degree Days | ***0*** | 0 | 0 | ***0*** | ***0*** | ***0*** | ***0*** | 0 | 0 | 0 | ***0*** | ***0*** | 0 |
| Total Precipitation (") | 3.88 | 3.88 | 4.92 | 3.02 | 2.29 | 1.99 | 3.65 | 4.65 | 4.47 | 4.63 | 4.37 | 4.72 | 46.47 |
| Days ≥ 0.1" Precip | 7 | 8 | 9 | 7 | ***6*** | 5 | 9 | 11 | 7 | 7 | 7 | ***7*** | 90 |
| Total Snowfall (") | ***77.0*** | ***65.9*** | ***75.2*** | ***39.4*** | ***14.6*** | 1.6 | 0.0 | 0.0 | 2.3 | ***25.7*** | ***58.7*** | ***91.7*** | 452.1 |
| Days ≥ 1" Snow Depth | na | na | na | na | na | ***1*** | 0 | 0 | ***1*** | na | na | na | na |

## WRAY *Yuma County* ELEVATION 3514 ft LAT/LONG 40° 4' N / 102° 13' W

| | JAN | FEB | MAR | APR | MAY | JUN | JUL | AUG | SEP | OCT | NOV | DEC | YEAR |
|---|---|---|---|---|---|---|---|---|---|---|---|---|---|
| Maximum Temp °F | 44.5 | 50.1 | 57.6 | 67.5 | ***76.3*** | 86.3 | 91.6 | 90.3 | 82.4 | ***71.2*** | ***54.2*** | ***44.8*** | 68.1 |
| Minimum Temp °F | 12.8 | 17.5 | 24.5 | 34.2 | 44.8 | 54.7 | 60.8 | 58.0 | 47.7 | 34.6 | 22.9 | 15.4 | 35.7 |
| Mean Temp °F | 28.7 | 33.6 | 41.1 | 50.9 | ***60.5*** | 70.6 | 76.3 | 74.1 | 65.0 | ***53.0*** | ***38.6*** | ***29.8*** | 51.9 |
| Days Max Temp ≥ 90 °F | 0 | 0 | 0 | 0 | 3 | 11 | 20 | 19 | 9 | 1 | ***0*** | ***0*** | 63 |
| Days Max Temp ≤ 32 °F | 6 | 3 | 2 | 0 | 0 | 0 | 0 | 0 | 0 | 0 | ***2*** | ***5*** | 18 |
| Days Min Temp ≤ 32 °F | 31 | ***27*** | 25 | 12 | 2 | 0 | 0 | 0 | 1 | 12 | 27 | 30 | 167 |
| Days Min Temp ≤ 0 °F | 5 | 2 | 1 | 0 | 0 | 0 | 0 | 0 | 0 | 0 | 0 | 3 | 11 |
| Heating Degree Days | 1118 | 877 | 734 | 420 | ***179*** | 26 | 3 | 6 | 100 | ***373*** | ***786*** | ***1086*** | 5708 |
| Cooling Degree Days | 0 | ***0*** | ***0*** | ***5*** | ***40*** | ***202*** | ***334*** | ***301*** | ***115*** | ***5*** | na | na | na |
| Total Precipitation (") | 0.56 | 0.38 | 1.13 | 1.67 | 3.08 | 2.90 | 2.90 | 2.01 | 1.13 | 0.98 | 0.73 | 0.42 | 17.89 |
| Days ≥ 0.1" Precip | 2 | 1 | 3 | 4 | 6 | 5 | 5 | 5 | 3 | 2 | 2 | 1 | 39 |
| Total Snowfall (") | 6.4 | 3.4 | 6.7 | 2.7 | 0.1 | 0.0 | 0.0 | 0.0 | 0.1 | 1.9 | ***4.8*** | 4.4 | 30.5 |
| Days ≥ 1" Snow Depth | na | 4 | ***3*** | 1 | 0 | 0 | 0 | 0 | 0 | 1 | ***3*** | na | na |

## YAMPA *Routt County* ELEVATION 7891 ft LAT/LONG 40° 9' N / 106° 54' W

| | JAN | FEB | MAR | APR | MAY | JUN | JUL | AUG | SEP | OCT | NOV | DEC | YEAR |
|---|---|---|---|---|---|---|---|---|---|---|---|---|---|
| Maximum Temp °F | 31.6 | 34.9 | 41.3 | 51.2 | 62.2 | 71.4 | 76.5 | 75.5 | 68.0 | 56.9 | 41.1 | 32.5 | 53.6 |
| Minimum Temp °F | 6.3 | 8.1 | 15.3 | 23.3 | 31.9 | 38.9 | 45.1 | 43.8 | 36.2 | 26.7 | 16.2 | 8.2 | 25.0 |
| Mean Temp °F | 19.0 | 21.5 | 28.3 | 37.3 | 47.1 | 55.2 | 60.8 | 59.7 | 52.1 | 41.9 | 28.7 | 20.4 | 39.3 |
| Days Max Temp ≥ 90 °F | 0 | 0 | 0 | 0 | 0 | 0 | 0 | 0 | 0 | 0 | 0 | 0 | 0 |
| Days Max Temp ≤ 32 °F | 16 | 10 | 5 | 1 | 0 | 0 | 0 | 0 | 0 | 1 | 7 | 15 | 55 |
| Days Min Temp ≤ 32 °F | 31 | 28 | 30 | 26 | 16 | 4 | 0 | 1 | 8 | 24 | 29 | 31 | 228 |
| Days Min Temp ≤ 0 °F | 9 | 6 | 3 | 1 | 0 | 0 | 0 | 0 | 0 | 0 | 3 | 8 | 30 |
| Heating Degree Days | 1419 | 1223 | 1131 | 826 | 549 | 289 | 130 | 159 | 381 | 708 | 1082 | 1374 | 9271 |
| Cooling Degree Days | 0 | 0 | 0 | 0 | 0 | 3 | 8 | 3 | 0 | 0 | 0 | 0 | 14 |
| Total Precipitation (") | 1.09 | 0.86 | 1.21 | 1.18 | 1.46 | 1.44 | 2.11 | 1.68 | 1.33 | 1.34 | 1.22 | 1.15 | 16.07 |
| Days ≥ 0.1" Precip | 4 | 3 | 4 | 4 | 5 | 5 | 6 | 6 | 5 | 4 | 4 | 4 | 54 |
| Total Snowfall (") | 18.2 | 13.9 | 17.2 | 11.3 | 2.7 | 0.2 | 0.0 | 0.1 | 0.8 | 8.8 | 18.8 | 19.6 | 111.6 |
| Days ≥ 1" Snow Depth | 30 | 28 | 27 | 7 | 0 | 0 | 0 | 0 | 0 | 3 | 17 | 29 | 141 |

## YELLOW JACKET 2 W *Montezuma County* ELEVATION 6863 ft LAT/LONG 37° 33' N / 108° 44' W

| | JAN | FEB | MAR | APR | MAY | JUN | JUL | AUG | SEP | OCT | NOV | DEC | YEAR |
|---|---|---|---|---|---|---|---|---|---|---|---|---|---|
| Maximum Temp °F | 37.4 | 42.0 | 48.6 | 58.7 | 68.9 | 80.6 | 86.2 | 83.6 | 75.5 | 63.0 | 48.1 | 38.9 | 61.0 |
| Minimum Temp °F | 14.1 | 18.2 | 24.7 | 31.0 | 39.4 | 47.9 | 54.4 | 53.1 | 46.0 | 36.4 | 25.2 | 17.0 | 33.9 |
| Mean Temp °F | 25.7 | 30.1 | 36.7 | 44.9 | 54.1 | 64.3 | 70.4 | 68.4 | 60.8 | 49.7 | 36.7 | 28.0 | 47.5 |
| Days Max Temp ≥ 90 °F | 0 | 0 | 0 | 0 | 0 | 3 | 7 | 4 | 0 | 0 | 0 | 0 | 14 |
| Days Max Temp ≤ 32 °F | 8 | 4 | 1 | 0 | 0 | 0 | 0 | 0 | 0 | 0 | 2 | 7 | 22 |
| Days Min Temp ≤ 32 °F | 31 | 28 | 26 | 17 | 5 | 0 | 0 | 0 | 1 | 8 | 24 | 30 | 170 |
| Days Min Temp ≤ 0 °F | 3 | 1 | 0 | 0 | 0 | 0 | 0 | 0 | 0 | 0 | 0 | 2 | 6 |
| Heating Degree Days | 1210 | 979 | 871 | 597 | 331 | 83 | 3 | 15 | 144 | 468 | 844 | 1142 | 6687 |
| Cooling Degree Days | 0 | 0 | 0 | 0 | 1 | 78 | 176 | 140 | 25 | 0 | 0 | 0 | 420 |
| Total Precipitation (") | 1.19 | 1.32 | 1.30 | 0.93 | 1.30 | 0.60 | 1.47 | 1.61 | 1.54 | 1.88 | 1.56 | 1.38 | 16.08 |
| Days ≥ 0.1" Precip | 4 | 4 | 4 | 3 | 4 | 2 | 4 | 5 | 4 | 4 | 4 | 4 | 46 |
| Total Snowfall (") | 16.1 | 15.2 | 11.7 | 4.2 | 1.0 | 0.0 | 0.0 | 0.0 | 0.1 | 1.3 | 7.7 | 17.0 | 74.3 |
| Days ≥ 1" Snow Depth | *20* | *17* | na | *0* | 0 | 0 | 0 | 0 | 0 | 0 | *2* | na | na |

## YUMA *Yuma County* ELEVATION 4134 ft LAT/LONG 40° 7' N / 102° 44' W

| | JAN | FEB | MAR | APR | MAY | JUN | JUL | AUG | SEP | OCT | NOV | DEC | YEAR |
|---|---|---|---|---|---|---|---|---|---|---|---|---|---|
| Maximum Temp °F | 41.8 | *46.9* | *55.3* | *65.4* | *73.9* | 84.6 | 89.9 | 87.8 | 79.9 | 67.8 | 51.1 | *43.3* | 65.6 |
| Minimum Temp °F | 14.7 | *18.2* | *25.4* | *34.2* | *44.4* | 53.2 | 58.1 | 56.6 | 47.4 | 35.6 | 23.9 | *16.1* | 35.7 |
| Mean Temp °F | 28.3 | *32.5* | *40.4* | *49.8* | *59.2* | 68.9 | 74.0 | 72.3 | 63.7 | 51.7 | 37.5 | *29.7* | 50.7 |
| Days Max Temp ≥ 90 °F | 0 | *0* | *0* | 0 | 1 | 9 | 18 | 14 | 4 | 0 | 0 | 0 | 46 |
| Days Max Temp ≤ 32 °F | *7* | *4* | *2* | 0 | 0 | 0 | 0 | 0 | 0 | 0 | 3 | *6* | 22 |
| Days Min Temp ≤ 32 °F | 30 | *27* | *25* | *13* | *2* | 0 | 0 | 0 | 1 | *10* | 26 | *30* | 164 |
| Days Min Temp ≤ 0 °F | *4* | *2* | *1* | 0 | *0* | 0 | 0 | 0 | 0 | 0 | 0 | *3* | 10 |
| Heating Degree Days | 1132 | *907* | *756* | *451* | *200* | 34 | 3 | 8 | 109 | 407 | 817 | *1088* | 5912 |
| Cooling Degree Days | 0 | 0 | 0 | 3 | 27 | 158 | 271 | 237 | 73 | 2 | 0 | 0 | 771 |
| Total Precipitation (") | 0.42 | *0.30* | *0.91* | *1.46* | 3.06 | 2.75 | 2.89 | 1.95 | 1.08 | 1.02 | 0.55 | *0.35* | 16.74 |
| Days ≥ 0.1" Precip | 1 | *1* | *3* | 4 | 7 | *6* | 6 | 4 | 3 | 2 | 2 | *1* | 40 |
| Total Snowfall (") | na | na | na | *3.3* | 0.4 | 0.0 | 0.0 | 0.0 | 0.2 | *3.4* | *5.2* | na | na |
| Days ≥ 1" Snow Depth | na | na | na | *1* | 0 | 0 | 0 | 0 | 0 | 1 | na | na | na |

## JANUARY MINIMUM TEMPERATURE °F

### LOWEST

| | | |
|---|---|---|
| 1 | Taylor Park | -12.7 |
| 2 | Hermit | -8.2 |
| 3 | Gunnison | -7.0 |
| 4 | Cochetopa Creek | -6.1 |
| 5 | Crested Butte | -5.9 |
| 6 | Silverton | -5.0 |
| 7 | Alamosa | -4.0 |
| 8 | Antero Reservoir | -3.6 |
| | Center | -3.6 |
| | Lake City | -3.6 |
| 11 | Lake George | -1.5 |
| 12 | Cimarron | 0.0 |
| 13 | Manassa | 0.1 |
| 14 | Monte Vista | 0.3 |
| 15 | Blanca | 0.5 |
| 16 | Grand Lke-6 SSW | 0.9 |
| 17 | Sugarloaf Rervr | 1.1 |
| 18 | Dillon | 1.2 |
| 19 | Climax | 1.5 |
| | Saguache | 1.5 |
| 21 | Meredith | 2.1 |
| 22 | Rangely | 2.3 |
| 23 | Steamboat Springs | 2.4 |
| 24 | Grand Lake-1 NW | 2.6 |
| 25 | Wolf Creek Pass | 3.5 |

### HIGHEST

| | | |
|---|---|---|
| 1 | Canon City | 21.5 |
| 2 | Walsenburg | 20.4 |
| 3 | Boulder | 19.8 |
| 4 | Trinidad | 18.9 |
| 5 | Lakewood | 18.2 |
| 6 | Mesa Verde | 17.8 |
| | Palisade | 17.8 |
| | Springfield | 17.8 |
| 9 | Colorado | 17.6 |
| 10 | Burlington | 17.2 |
| 11 | Denver | 16.9 |
| | Estes Park | 16.9 |
| | Walsh | 16.9 |
| 14 | Colorado Springs | 16.8 |
| | Trinidad-Ls Anms | 16.8 |
| 16 | Gateway | 16.7 |
| 17 | Parker | 16.3 |
| 18 | Kassler | 15.9 |
| 19 | La Junta | 15.8 |
| 20 | Cheyenne Wells | 15.7 |
| | Grnd Jnctn-6 ESE | 15.7 |
| 22 | Stratton | 15.3 |
| 23 | Fort Collins | 15.1 |
| | Grand Jnctn-wlkr | 15.1 |
| 25 | Cedaredge | 14.9 |

## JULY MAXIMUM TEMPERATURE °F

### HIGHEST

| | | |
|---|---|---|
| 1 | Las Animas | 95.4 |
| 2 | Palisade | 94.8 |
| 3 | Uravan | 94.7 |
| 4 | La Junta | 94.3 |
| 5 | Holly | 93.9 |
| 6 | Lamar | 93.4 |
| | Rocky Ford | 93.4 |
| 8 | Grand Jnctn-wlkr | 93.3 |
| 9 | Pueblo | 92.7 |
| 10 | Fruita | 92.6 |
| 11 | Delta | 92.2 |
| | Eads | 92.2 |
| 13 | Grnd Jnctn-6 ESE | 92.0 |
| 14 | Kit Carson | 91.8 |
| 15 | Tacony | 91.6 |
| | Walsh | 91.6 |
| | Wray | 91.6 |
| 18 | Gateway | 91.5 |
| 19 | Rangely | 91.2 |
| | Springfield | 91.2 |
| 21 | Colorado | 91.1 |
| 22 | Cheyenne Wells | 90.8 |
| | Julesburg | 90.8 |
| 24 | Dinosaur | 90.1 |
| 25 | Bonny Lake | 90.0 |

### LOWEST

| | | |
|---|---|---|
| 1 | Climax | 64.5 |
| 2 | Wolf Creek Pass | 65.9 |
| 3 | Ruxton Park | 70.9 |
| 4 | Sugarloaf Rsrvr | 71.1 |
| 5 | Taylor Park | 71.4 |
| 6 | Silverton | 73.4 |
| 7 | Dillon | 73.6 |
| 8 | Grand Lke-6 SSW | 74.0 |
| 9 | Crested Butte | 74.6 |
| 10 | Hermit | 74.8 |
| | Spicer | 74.8 |
| 12 | Rico | 75.1 |
| 13 | Grand Lake-1 NW | 75.3 |
| | Grant | 75.3 |
| 15 | Antero Reservoir | 75.5 |
| 16 | Lake George | 75.7 |
| 17 | Yampa | 76.5 |
| 18 | Lake City | 77.0 |
| 19 | Ouray | 77.8 |
| | Telluride | 77.8 |
| | Walden | 77.8 |
| 22 | Del Norte | 78.3 |
| | Estes Park | 78.3 |
| 24 | Green Mntn Dm | 78.5 |
| 25 | Bailey | 78.6 |

## ANNUAL PRECIPITATION (")

### HIGHEST

| | | |
|---|---|---|
| 1 | Wolf Creek Pass | 46.47 |
| 2 | Rico | 29.16 |
| 3 | Vallecito Dam | 27.67 |
| 4 | Silverton | 24.99 |
| 5 | Ruxton Park | 24.09 |
| 6 | Ouray | 23.92 |
| 7 | Steamboat Springs | 23.66 |
| 8 | Crested Butte | 23.64 |
| 9 | Telluride | 23.37 |
| 10 | Climax | 22.34 |
| 11 | Grand Lake-1 NW | 20.44 |
| 12 | Evergreen | 19.28 |
| 13 | Fort Lewis | 19.23 |
| 14 | Boulder | 19.07 |
| 15 | Mesa Verde | 18.75 |
| 16 | Bonny Lake | 18.40 |
| 17 | Kassler | 18.17 |
| 18 | Holyoke | 18.08 |
| | Sedgwick | 18.08 |
| 20 | Cherry Creek Dm | 18.01 |
| 21 | Wray | 17.89 |
| 22 | Castle Rock | 17.43 |
| 23 | Cheesman | 17.42 |
| 24 | Stratton | 17.29 |
| 25 | Julesburg | 17.16 |

### LOWEST

| | | |
|---|---|---|
| 1 | Center | 7.26 |
| 2 | Delta | 7.63 |
| 3 | Alamosa | 7.75 |
| 4 | Monte Vista | 7.87 |
| 5 | Manassa | 8.14 |
| 6 | Saguache | 8.53 |
| 7 | Grand Jnct-6 ESE | 8.90 |
| | Grand Jnctn-wlkr | 8.90 |
| 9 | Blanca | 9.01 |
| 10 | Fruita | 9.15 |
| 11 | Montrose | 9.59 |
| 12 | Gunnison | 9.80 |
| 13 | Antero Reservoir | 10.12 |
| 14 | Buena Vista | 10.34 |
| 15 | Del Norte | 10.54 |
| 16 | Rangely | 10.57 |
| 17 | Eagle | 10.59 |
| 18 | Palisade | 10.60 |
| 19 | Tacony | 10.94 |
| 20 | Walden | 10.95 |
| 21 | Rocky Ford | 11.20 |
| 22 | Colorado | 11.39 |
| 23 | Great Sand Dunes | 11.42 |
| 24 | La Junta | 11.54 |
| 25 | Dinosaur | 11.63 |

## ANNUAL SNOWFALL (")

### HIGHEST

| | | |
|---|---|---|
| 1 | Wolf Creek Pass | 452.1 |
| 2 | Climax | 263.8 |
| 3 | Crested Butte | 224.1 |
| 4 | Telluride | 205.9 |
| 5 | Rico | 180.8 |
| 6 | Steamboat Springs | 170.2 |
| 7 | Spicer | 149.6 |
| 8 | Silverton | 148.4 |
| 9 | Ruxton Park | 144.0 |
| 10 | Ouray | 142.9 |
| 11 | Grand Lake-1 NW | 142.8 |
| 12 | Vallecito Dam | 126.2 |
| 13 | Hayden | 119.9 |
| 14 | Dillon | 116.0 |
| 15 | Yampa | 111.6 |
| 16 | Grant | 100.3 |
| 17 | Westcliffe | 99.3 |
| 18 | Meredith | 95.5 |
| 19 | Walsenburg | 93.9 |
| 20 | Evergreen | 87.6 |
| 21 | Mesa Verde | 86.8 |
| 22 | Lake City | 84.7 |
| 23 | Bailey | 83.4 |
| 24 | Boulder | 81.9 |
| 25 | Green Mntn Dm | 79.6 |

### LOWEST

| | | |
|---|---|---|
| 1 | Fruita | 13.8 |
| 2 | Gateway | 18.1 |
| 3 | Las Animas | 20.6 |
| 4 | Saguache | 22.8 |
| 5 | Grand Jnctn-wlkr | 23.4 |
| 6 | Cheyenne Wells | 23.6 |
| | Walsh | 23.6 |
| 8 | Rocky Ford | 23.9 |
| 9 | Eads | 24.4 |
| 10 | Holly | 24.6 |
| 11 | Fort Morgan | 26.7 |
| 12 | Tacony | 27.4 |
| 13 | Monte Vista | 27.8 |
| 14 | Lamar | 27.9 |
| 15 | La Junta | 28.1 |
| 16 | Blanca | 28.2 |
| 17 | Buena Vista | 30.5 |
| | Wray | 30.5 |
| 19 | Colorado | 31.8 |
| 20 | Springfield | 32.9 |
| 21 | Pueblo | 33.3 |
| 22 | Holyoke | 34.6 |
| 23 | Longmont | 35.4 |
| 24 | Alamosa | 35.7 |
| 25 | New Raymer | 36.2 |

# CONNECTICUT

PHYSICAL FEATURES. Connecticut occupies the southwestern portion of the region known as New England. The State extends for 90 miles in an east-west direction and 75 miles from north to south. The total area of 5,009 square miles makes Connecticut the third smallest state in the Nation.

The topography of Connecticut is predominantly hilly. The highest terrain is found in the northwest portion of the State, with elevations of 1,000 to 2,000 feet. The southwestern quarter and most of the eastern half have elevations of 300 to 1,000 feet. The State of Connecticut is bisected by the Connecticut River which rises in Canada. Smaller river basins in the State with their headwaters in the southern half of Massachusetts include the Housatonic in the west and the Shetucket, Quinebaug, and Thames in the east. The narrow river valleys and steep hillsides in much of the western highlands make for destructive flash floods during periods of unusually heavy or intense rainfall.

The entire southern border of Connecticut is washed by the waters of Long Island Sound. The coastline of approximately 100 miles is indented by small coves and the mouths of numerous rivers and streams. Beaches are found along the greater length.

GENERAL CLIMATE. The chief characteristics of Connecticut's climate are: (1) equable distribution of precipitation among the four seasons, (2) large ranges of temperature both daily and annually, (3) great differences in the same season or month of different years, and (4) considerable diversity of the weather over short periods of time.

Connecticut lies in the "prevailing westerlies", the belt of generally eastward air movement which encircles the globe in middle latitudes. Embedded in this circulation are extensive masses of air originating in higher and lower latitudes and interacting to produce low-pressure storm systems. A large number of storm centers and air-mass fronts pass near or over Connecticut during a year. Three types of air affect this State: (1) cold, dry air pouring down from subarctic North America, (2) warm, moist air streaming up on a long overland journey from the Gulf of Mexico and subtropical waters of the Atlantic, and (3) cool, damp air moving in from the North Atlantic. Because the flow of air is usually from continental areas, Connecticut is more influenced by the first two types than it is by the third. The procession of contrasting air masses and the relatively frequent passage of storms bring about a roughly twice weekly alternation from fair to cloudy or storm conditions, usually attended by abrupt changes in temperature, moisture, sunshine, and wind direction and speed. There is no regular or persistent rhythm to this sequence; it is sometimes interrupted by periods during which the weather pattern continues much the same for several days, and infrequently for a few weeks.

TEMPERATURE. Despite the small size of Connecticut, there is a difference of about 6° F. in mean annual temperature from north to south. The greater contrast of temperature over the State occurs during the winter season. The number of days with minimum temperatures of zero or below average about 10 per year at the higher elevations, about 5 in the lower uplands and central valley, and 2 or less along the shore of Long Island Sound. Summer temperatures are comparatively uniform over the State. The central valley experiences the greatest number of hot days. Temperatures of 90° F. or higher occur on an average on about 10 days per year. At the higher elevations and near the coast, the average number is approximately 3 days per year. In much of the western and eastern highlands, the occurrence of 90° F. temperatures is a little less frequent than in the central valley.

During the warmest month of the summer the average minimum temperature ranges from about 56° F. in the cool northwestern corner of the State to about 63° F. in the warmer coastal sections. Over most of the State the average July minimum temperature is within a degree or two of 60°F.

The period free from temperatures of 32° F. or lower has an average length of 155 to 170 days over the greater portion of Connecticut. In the northwest as well as in local areas of the western and eastern highlands, the freeze-free season lasts about 125 to 135 days. Along the immediate coast approximately 190 days will elapse between the last spring and first fall freeze.

PRECIPITATION. Precipitation tends to become evenly distributed throughout the year in all parts of Connecticut.

Low-pressure centers and their accompanying air mass fronts are the principal year-round producers of precipitation. Storms moving up the Atlantic coast generally yield the heaviest amounts of rain and snow. In the summer bands and patches of thunderstorms and convective showers add considerable precipitation and make up the difference resulting from decreased activity of low-pressure storm centers. Thunderstorms are of brief duration and often scattered in comparison with the general storms, but they yield the heaviest local rainfall.

Variations in precipitation from month to month are sometimes extreme. A month yielding 5 inches or more may be preceded or followed by one with less than 2 inches of precipitation, in any season. Months with less than 1 inch are known to occur, as well as those with precipitation in excess of 10 inches. Such large fluctuations, however, are not characteristic of the precipitation supply in Connecticut. Consequently, prolonged droughts and widespread floods are infrequent.

While there are no pronounced wet and dry months as in other climates, February and October are relatively dry. The average total precipitation for each of these months is 3 inches or slightly less in comparison with 3.5 to 4 inches in the other 10 months. Measurable precipitation falls on an average of 1 day in 3, with the yearly total approximating 120 days. Periods of 5 days or more of successive daily precipitation occur a few times during most years.

The average annual snowfall increases from the coast to the northwestern corner of the State. Most of the snow falls in January and February, but in the majority of winters substantial amounts fall in December or March storms as well. Except for the northwestern highlands, snowfalls of more than 1 inch are quite rare before mid-November and after April 15. The average number of days per year with snow on the ground similarly shows an increase from the shore to the northwest. During an average winter a measurable snow cover is present most of the time from late December through the early half of March in the greater portion of the State. In the immediate coastal areas a snow cover does not last more than a few days unless a heavy snowstorm is followed by prolonged cold temperatures.

OTHER CLIMATIC ELEMENTS. During the colder months the prevailing wind is northwest to north over Connecticut, while from April through September southwest or south winds predominate. The mean hourly speed ranges from about 7 m.p.h. in the summer and early fall to about 10 m.p.h. in the winter and spring seasons. An important feature of the climate is the sea breeze along the coast. During the summer and late spring this onshore wind blows from cool ocean during the afternoon and penetrates inland from 5 to 10 miles. It occurs often enough to give lower mean summer maximum temperatures in a narrow coastal belt than prevail over interior lowlands.

Thunderstorms occur on an average of 20 to 30 days per year, with the greatest frequency during the summer months and in the afternoon or evening hours. Often these storms are accompanied by destructive hail and/or wind. Aside from infrequent tornadoes and hurricanes, coastal storms or "northeasters" are the most serious weather hazard in Connecticut. They generate very strong winds and heavy rain and produce the greatest snowstorms in the winter. If these storms occur at the time of high tide, heavy water damage results along the shore. In occasional years a tornado or storm with tornadic characteristics strikes some part of the State. The central valley appears to be the most likely to be struck, and the summer months the most likely season. Storms of tropical origin occasionally affect Connecticut during the summer or fall months, as they move on a path well out over the ocean.

The Connecticut River shows an annual rise in early spring as the result of the melting of high elevation snow in northern and central New England. A secondary period of flooding (occasionally of major proportions) is caused by heavy rains which may be associated with hurricanes or storms of tropical origin in late summer or fall, normally the low water season.

The percentage of possible sunshine averages 55 to 60 percent, ranging from 45 percent in the interior during the months of November through January to near 65 percent along the coast in the summer. The average number of clear days per year is between 100 and 125, with the greatest number per month usually occurring in September and October. An average of about 140 cloudy days occur per year. Heavy or dense fog is observed on an average of about 25 days per year in both coastal and inland sections. In the former section, heavy fog is most common during the late winter and spring seasons, while inland the late summer and fall is the period of maximum occurrence. The humidity tends to be lowest in the spring and highest in the late summer and early fall.

## COUNTY INDEX

### *Fairfield County*
BRIDGEPORT SIKORSKY
DANBURY
STAMFORD 5 N

### *Hartford County*
BURLINGTON
HARTFORD BRADLEY AP
HARTFORD BRAINARD FD

### *Litchfield County*
BULLS BRIDGE DAM
FALLS VILLAGE
NORFOLK 2 SW
SHEPAUG DAM
WIGWAM RESERVOIR

### *Middlesex County*
MIDDLETOWN 4 W

### *New Haven County*
MOUNT CARMEL

### *New London County*
GROTON
NORWICH PUB UTIL PLA

### *Tolland County*
MANSFIELD HOLLOW LAK
STORRS

### *Windham County*
WEST THOMPSON LAKE

## ELEVATION INDEX

| FEET | STATION NAME |
|---|---|
| 20 | HARTFORD BRAINARD FD |
| 20 | NORWICH PUB UTIL PLA |
| 26 | BRIDGEPORT SIKORSKY |
| 40 | GROTON |
| 180 | MOUNT CARMEL |
| 190 | STAMFORD 5 N |
| 213 | HARTFORD BRADLEY AP |
| 249 | MANSFIELD HOLLOW LAK |
| 259 | BULLS BRIDGE DAM |
| 360 | WEST THOMPSON LAKE |
| 369 | MIDDLETOWN 4 W |
| 405 | DANBURY |
| 510 | BURLINGTON |
| 581 | FALLS VILLAGE |
| 600 | WIGWAM RESERVOIR |
| 650 | STORRS |
| 840 | SHEPAUG DAM |
| 1342 | NORFOLK 2 SW |

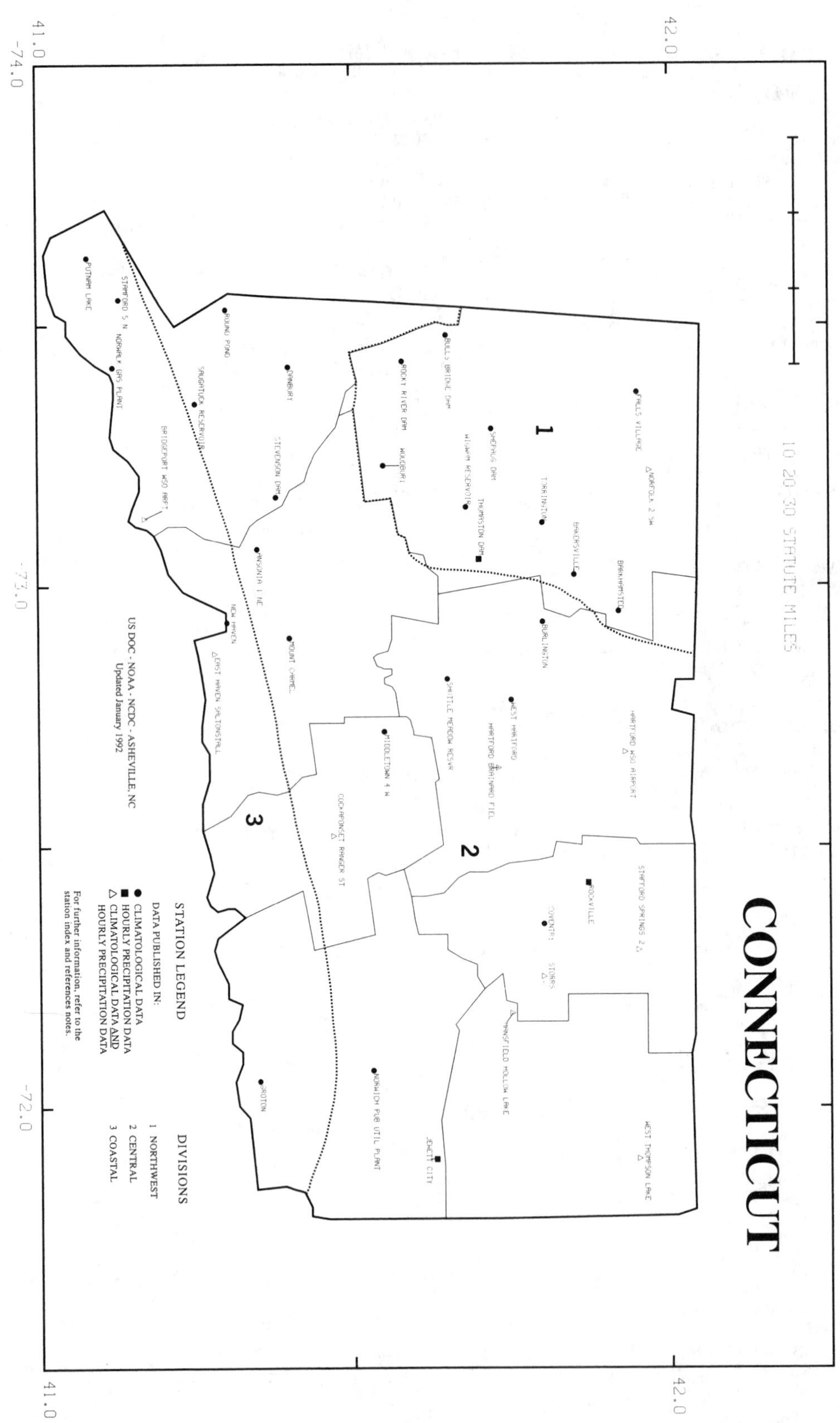
CONNECTICUT
10 20 30 STATUTE MILES
41.0
42.0
-74.0
-73.0
-72.0
1
2
3
US DOC - NOAA - NCDC - ASHEVILLE, NC
Updated January 1992
STATION LEGEND
DATA PUBLISHED IN:
● CLIMATOLOGICAL DATA
■ HOURLY PRECIPITATION DATA
△ CLIMATOLOGICAL DATA AND HOURLY PRECIPITATION DATA
For further information, refer to the station index and references notes.
DIVISIONS
1 NORTHWEST
2 CENTRAL
3 COASTAL

## BRIDGEPORT SIKORSKY *Fairfield County* ELEVATION 26 ft LAT/LONG 41° 10' N / 73° 8' W

| | JAN | FEB | MAR | APR | MAY | JUN | JUL | AUG | SEP | OCT | NOV | DEC | YEAR |
|---|---|---|---|---|---|---|---|---|---|---|---|---|---|
| Maximum Temp °F | 36.2 | 37.9 | 46.2 | 56.9 | 67.2 | 76.3 | 82.1 | 81.1 | 74.0 | 63.5 | 53.0 | 41.9 | 59.7 |
| Minimum Temp °F | 22.2 | 23.5 | 30.9 | 40.2 | 50.3 | 59.4 | 66.0 | 65.4 | 57.6 | 46.8 | 38.2 | 28.2 | 44.1 |
| Mean Temp °F | 29.2 | 30.7 | 38.6 | 48.6 | 58.7 | 67.9 | 74.1 | 73.3 | 65.8 | 55.2 | 45.6 | 35.0 | 51.9 |
| Days Max Temp ≥ 90 °F | 0 | 0 | 0 | 0 | 0 | 1 | 3 | 2 | 0 | 0 | 0 | 0 | 6 |
| Days Max Temp ≤ 32 °F | 10 | 7 | 1 | 0 | 0 | 0 | 0 | 0 | 0 | 0 | 0 | 5 | 23 |
| Days Min Temp ≤ 32 °F | 26 | 23 | 17 | 3 | 0 | 0 | 0 | 0 | 0 | 1 | 7 | 21 | 98 |
| Days Min Temp ≤ 0 °F | 1 | 0 | 0 | 0 | 0 | 0 | 0 | 0 | 0 | 0 | 0 | 0 | 1 |
| Heating Degree Days | 1104 | 961 | 811 | 488 | 208 | 31 | 2 | 4 | 64 | 307 | 575 | 922 | 5477 |
| Cooling Degree Days | 0 | 0 | 0 | 1 | 26 | 134 | 297 | 255 | 87 | 8 | 0 | 0 | 808 |
| Total Precipitation (") | 3.29 | 2.96 | 4.13 | 3.67 | 4.04 | 3.50 | 3.70 | 3.65 | 3.46 | 3.22 | 3.79 | 3.76 | 43.17 |
| Days ≥ 0.1" Precip | 6 | 6 | 7 | 6 | 7 | 6 | 6 | 6 | 6 | 5 | 6 | 7 | 74 |
| Total Snowfall (") | 8.1 | 7.9 | 4.7 | 0.4 | 0.0 | 0.0 | 0.0 | 0.0 | 0.0 | 0.0 | 0.6 | 4.0 | 25.7 |
| Days ≥ 1" Snow Depth | 11 | 8 | 3 | 0 | 0 | 0 | 0 | 0 | 0 | 0 | 0 | 4 | 26 |

## BULLS BRIDGE DAM *Litchfield County* ELEVATION 259 ft LAT/LONG 41° 39' N / 73° 29' W

| | JAN | FEB | MAR | APR | MAY | JUN | JUL | AUG | SEP | OCT | NOV | DEC | YEAR |
|---|---|---|---|---|---|---|---|---|---|---|---|---|---|
| Maximum Temp °F | 34.2 | 37.3 | 47.0 | 59.9 | 71.1 | 79.1 | 83.7 | 81.7 | 73.8 | 62.9 | 50.7 | 39.0 | 60.0 |
| Minimum Temp °F | 13.7 | 16.3 | 25.1 | 34.8 | 45.1 | 53.9 | 59.2 | 58.2 | 50.3 | 38.0 | 30.2 | 20.5 | 37.1 |
| Mean Temp °F | 23.9 | 26.8 | 36.1 | 47.4 | 58.1 | 66.5 | 71.5 | 70.0 | 62.1 | 50.5 | 40.4 | 29.8 | 48.6 |
| Days Max Temp ≥ 90 °F | 0 | 0 | 0 | 0 | 0 | 2 | 5 | 3 | 1 | 0 | 0 | 0 | 11 |
| Days Max Temp ≤ 32 °F | 13 | 9 | 2 | 0 | 0 | 0 | 0 | 0 | 0 | 0 | 0 | 7 | 31 |
| Days Min Temp ≤ 32 °F | 30 | 27 | 25 | 13 | 2 | 0 | 0 | 0 | 0 | 10 | 20 | 28 | 155 |
| Days Min Temp ≤ 0 °F | 5 | 2 | 0 | 0 | 0 | 0 | 0 | 0 | 0 | 0 | 0 | 1 | 8 |
| Heating Degree Days | 1266 | 1071 | 891 | 523 | 231 | 54 | 9 | 18 | 133 | 444 | 730 | 1086 | 6456 |
| Cooling Degree Days | 0 | 0 | 0 | 2 | 30 | 105 | 228 | 172 | 51 | 3 | 0 | 0 | 591 |
| Total Precipitation (") | 3.11 | 3.03 | 3.80 | 4.05 | 4.11 | 3.92 | 4.28 | 4.18 | 3.87 | 3.66 | 4.07 | 3.85 | 45.93 |
| Days ≥ 0.1" Precip | 6 | 6 | 7 | 7 | 8 | 7 | 7 | 7 | 6 | 6 | 7 | 7 | 81 |
| Total Snowfall (") | na | na | na | na | 0.0 | 0.0 | 0.0 | 0.0 | 0.0 | 0.0 | na | na | na |
| Days ≥ 1" Snow Depth | na | na | na | na | 0 | 0 | 0 | 0 | 0 | 0 | na | na | na |

## BURLINGTON *Hartford County* ELEVATION 510 ft LAT/LONG 41° 48' N / 72° 56' W

| | JAN | FEB | MAR | APR | MAY | JUN | JUL | AUG | SEP | OCT | NOV | DEC | YEAR |
|---|---|---|---|---|---|---|---|---|---|---|---|---|---|
| Maximum Temp °F | 33.8 | 36.2 | 45.2 | 57.8 | 69.1 | 77.3 | 82.4 | 80.4 | 72.5 | 61.8 | 50.8 | 38.7 | 58.8 |
| Minimum Temp °F | 13.7 | 15.5 | 24.7 | 35.2 | 45.3 | 54.2 | 59.6 | 58.0 | 49.9 | 38.3 | 30.9 | 20.4 | 37.1 |
| Mean Temp °F | 23.8 | 25.9 | 35.0 | 46.6 | 57.2 | 65.8 | 71.0 | 69.2 | 61.2 | 50.1 | 40.8 | 29.5 | 48.0 |
| Days Max Temp ≥ 90 °F | 0 | 0 | 0 | 0 | 0 | 2 | 4 | 2 | 1 | 0 | 0 | 0 | 9 |
| Days Max Temp ≤ 32 °F | 14 | 10 | 3 | 0 | 0 | 0 | 0 | 0 | 0 | 0 | 1 | 8 | 36 |
| Days Min Temp ≤ 32 °F | 30 | 27 | 26 | 12 | 1 | 0 | 0 | 0 | 0 | 8 | 19 | 28 | 151 |
| Days Min Temp ≤ 0 °F | 4 | 2 | 0 | 0 | 0 | 0 | 0 | 0 | 0 | 0 | 0 | 1 | 7 |
| Heating Degree Days | 1272 | 1098 | 923 | 549 | 258 | 67 | 10 | 25 | 153 | 458 | 718 | 1092 | 6623 |
| Cooling Degree Days | 0 | 0 | 0 | 1 | 25 | 96 | 209 | 158 | 47 | 3 | 0 | 0 | 539 |
| Total Precipitation (") | 3.69 | 3.32 | 4.45 | 4.25 | 4.59 | 4.25 | 3.95 | 4.58 | 4.54 | 3.92 | 4.55 | 4.27 | 50.36 |
| Days ≥ 0.1" Precip | 6 | 6 | 7 | 7 | 8 | 6 | 7 | 7 | 6 | 6 | 7 | 7 | 80 |
| Total Snowfall (") | ***8.4*** | na | ***7.2*** | ***1.4*** | 0.0 | 0.0 | 0.0 | 0.0 | 0.0 | 0.1 | ***1.9*** | na | na |
| Days ≥ 1" Snow Depth | na | na | na | na | ***0*** | ***0*** | 0 | 0 | 0 | ***0*** | na | na | na |

## DANBURY *Fairfield County* ELEVATION 405 ft LAT/LONG 41° 24' N / 73° 25' W

| | JAN | FEB | MAR | APR | MAY | JUN | JUL | AUG | SEP | OCT | NOV | DEC | YEAR |
|---|---|---|---|---|---|---|---|---|---|---|---|---|---|
| Maximum Temp °F | ***34.3*** | 37.7 | 46.8 | ***59.8*** | 71.0 | 79.0 | 83.9 | 81.7 | ***73.6*** | ***62.4*** | ***51.0*** | ***39.6*** | 60.1 |
| Minimum Temp °F | ***17.0*** | 19.1 | 27.5 | ***36.9*** | ***46.8*** | ***55.8*** | ***60.9*** | 60.1 | ***51.9*** | ***40.8*** | ***33.2*** | ***23.3*** | 39.4 |
| Mean Temp °F | ***25.7*** | 28.4 | 37.2 | ***48.4*** | ***58.9*** | ***67.4*** | ***72.4*** | 70.9 | ***62.8*** | ***51.7*** | ***42.1*** | ***31.5*** | 49.8 |
| Days Max Temp ≥ 90 °F | ***0*** | 0 | 0 | 0 | 1 | 2 | 5 | 3 | ***1*** | ***0*** | ***0*** | ***0*** | 12 |
| Days Max Temp ≤ 32 °F | ***13*** | 9 | 2 | 0 | 0 | 0 | 0 | 0 | ***0*** | ***0*** | ***0*** | ***7*** | 31 |
| Days Min Temp ≤ 32 °F | ***29*** | 25 | 22 | 9 | ***1*** | ***0*** | ***0*** | 0 | ***0*** | 7 | ***15*** | ***26*** | 134 |
| Days Min Temp ≤ 0 °F | ***3*** | 1 | 0 | 0 | ***0*** | ***0*** | ***0*** | 0 | ***0*** | ***0*** | ***0*** | ***0*** | 4 |
| Heating Degree Days | ***1213*** | 1026 | 856 | ***496*** | ***215*** | ***46*** | ***5*** | 13 | ***125*** | ***411*** | ***679*** | ***1032*** | 6117 |
| Cooling Degree Days | na | ***0*** | ***0*** | na | na | na | na | ***212*** | na | na | na | na | na |
| Total Precipitation (") | ***3.59*** | 3.25 | 4.41 | ***4.17*** | 4.25 | 4.25 | 4.11 | 4.33 | ***4.66*** | ***3.57*** | ***4.30*** | ***4.39*** | 49.28 |
| Days ≥ 0.1" Precip | ***7*** | 6 | 8 | 7 | ***8*** | 8 | 6 | 6 | ***6*** | ***6*** | ***7*** | ***7*** | 82 |
| Total Snowfall (") | ***10.6*** | 12.2 | 7.9 | ***0.9*** | 0.0 | 0.0 | 0.0 | 0.0 | ***0.0*** | ***0.2*** | ***1.2*** | ***6.9*** | 39.9 |
| Days ≥ 1" Snow Depth | ***17*** | 14 | 7 | 0 | 0 | 0 | 0 | 0 | ***0*** | ***0*** | ***1*** | ***8*** | 47 |

## FALLS VILLAGE *Litchfield County* ELEVATION 581 ft LAT/LONG 41° 57 ' N / 73° 22 ' W

| | JAN | FEB | MAR | APR | MAY | JUN | JUL | AUG | SEP | OCT | NOV | DEC | YEAR |
|---|---|---|---|---|---|---|---|---|---|---|---|---|---|
| Maximum Temp °F | 33.7 | 36.7 | 46.2 | 59.6 | 71.9 | 79.3 | 83.9 | 81.6 | 73.3 | 62.3 | 50.2 | 38.4 | 59.8 |
| Minimum Temp °F | 12.1 | 14.5 | 23.7 | 33.3 | 43.9 | 52.5 | 57.6 | 56.4 | 48.5 | 36.7 | 29.2 | 18.8 | 35.6 |
| Mean Temp °F | 22.9 | 25.6 | 35.0 | 46.5 | 57.9 | 65.9 | 70.8 | 69.0 | 61.0 | 49.5 | 39.7 | 28.6 | 47.7 |
| Days Max Temp ≥ 90 °F | 0 | 0 | 0 | 0 | 0 | 2 | 5 | 2 | 0 | 0 | 0 | 0 | 9 |
| Days Max Temp ≤ 32 °F | 14 | 10 | 2 | 0 | 0 | 0 | 0 | 0 | 0 | 0 | 1 | 8 | 35 |
| Days Min Temp ≤ 32 °F | 30 | 27 | 26 | 15 | 3 | 0 | 0 | 0 | 1 | 12 | 20 | 29 | 163 |
| Days Min Temp ≤ 0 °F | 6 | 4 | 0 | 0 | 0 | 0 | 0 | 0 | 0 | 0 | 0 | 2 | 12 |
| Heating Degree Days | 1297 | 1105 | 924 | 551 | 237 | 59 | 10 | 26 | 159 | 474 | 753 | 1121 | 6716 |
| Cooling Degree Days | 0 | 0 | 0 | 2 | 28 | 97 | 216 | 167 | 49 | 2 | 0 | 0 | 561 |
| Total Precipitation (") | 2.90 | 2.66 | 3.37 | 3.73 | 4.09 | 4.21 | 3.87 | 4.58 | 3.76 | 3.44 | 3.75 | 3.61 | 43.97 |
| Days ≥ 0.1" Precip | 6 | 6 | 7 | 7 | 8 | 7 | 7 | 7 | 6 | 6 | 8 | 7 | 82 |
| Total Snowfall (") | na | na | *7.5* | 1.6 | 0.0 | 0.0 | 0.0 | 0.0 | 0.0 | 0.1 | 1.2 | na | na |
| Days ≥ 1" Snow Depth | na | *13* | na | 1 | 0 | 0 | 0 | 0 | 0 | 0 | *1* | na | na |

## GROTON *New London County* ELEVATION 40 ft LAT/LONG 41° 21 ' N / 72° 3 ' W

| | JAN | FEB | MAR | APR | MAY | JUN | JUL | AUG | SEP | OCT | NOV | DEC | YEAR |
|---|---|---|---|---|---|---|---|---|---|---|---|---|---|
| Maximum Temp °F | 37.1 | 38.7 | 46.6 | 56.5 | 66.3 | 75.0 | 80.9 | 80.0 | 72.9 | 62.7 | 53.0 | 42.5 | 59.4 |
| Minimum Temp °F | 18.8 | 20.6 | 28.4 | 37.4 | 47.0 | 56.1 | 62.5 | 61.8 | 53.8 | 43.0 | 35.0 | 25.1 | 40.8 |
| Mean Temp °F | 28.0 | 29.6 | 37.6 | 47.0 | 56.7 | 65.6 | 71.7 | 70.9 | 63.4 | 52.9 | 44.0 | 33.8 | 50.1 |
| Days Max Temp ≥ 90 °F | 0 | 0 | 0 | 0 | 0 | 1 | 2 | 1 | 0 | 0 | 0 | 0 | 4 |
| Days Max Temp ≤ 32 °F | 10 | 6 | 1 | 0 | 0 | 0 | 0 | 0 | 0 | 0 | 0 | 4 | 21 |
| Days Min Temp ≤ 32 °F | 28 | 25 | 22 | 7 | 0 | 0 | 0 | 0 | 0 | 4 | 13 | 25 | 124 |
| Days Min Temp ≤ 0 °F | 2 | 1 | 0 | 0 | 0 | 0 | 0 | 0 | 0 | 0 | 0 | 0 | 3 |
| Heating Degree Days | 1141 | 991 | 843 | 534 | 263 | 53 | 4 | 10 | 104 | 373 | 623 | 959 | 5898 |
| Cooling Degree Days | 0 | 0 | 0 | 0 | 14 | 85 | 235 | 200 | 68 | 6 | 0 | 0 | 608 |
| Total Precipitation (") | 4.01 | 3.54 | 4.65 | 3.97 | 3.81 | 3.41 | 3.30 | 4.25 | 3.72 | 3.73 | 4.89 | 4.81 | 48.09 |
| Days ≥ 0.1" Precip | 7 | 6 | 8 | 7 | 7 | 6 | 5 | 6 | 6 | 6 | 7 | 8 | 79 |
| Total Snowfall (") | 7.4 | 6.5 | 3.7 | 0.3 | 0.0 | 0.0 | 0.0 | 0.0 | 0.0 | 0.0 | 0.5 | 3.6 | 22.0 |
| Days ≥ 1" Snow Depth | 10 | 7 | 3 | 0 | 0 | 0 | 0 | 0 | 0 | 0 | 0 | 4 | 24 |

## HARTFORD BRADLEY AP *Hartford County* ELEVATION 213 ft LAT/LONG 41° 56 ' N / 72° 41 ' W

| | JAN | FEB | MAR | APR | MAY | JUN | JUL | AUG | SEP | OCT | NOV | DEC | YEAR |
|---|---|---|---|---|---|---|---|---|---|---|---|---|---|
| Maximum Temp °F | 33.4 | 36.8 | 46.8 | 60.2 | 71.8 | 80.1 | 85.1 | 82.8 | 74.6 | 63.3 | 51.0 | 38.5 | 60.4 |
| Minimum Temp °F | 16.3 | 19.1 | 28.1 | 37.9 | 47.9 | 57.0 | 62.5 | 60.8 | 52.0 | 40.6 | 32.8 | 22.2 | 39.8 |
| Mean Temp °F | 24.9 | 27.9 | 37.5 | 49.1 | 59.9 | 68.6 | 73.8 | 71.8 | 63.3 | 52.0 | 41.9 | 30.4 | 50.1 |
| Days Max Temp ≥ 90 °F | 0 | 0 | 0 | 0 | 1 | 4 | 8 | 5 | 1 | 0 | 0 | 0 | 19 |
| Days Max Temp ≤ 32 °F | 14 | 9 | 2 | 0 | 0 | 0 | 0 | 0 | 0 | 0 | 1 | 8 | 34 |
| Days Min Temp ≤ 32 °F | 29 | 26 | 21 | 8 | 1 | 0 | 0 | 0 | 0 | 7 | 16 | 27 | 135 |
| Days Min Temp ≤ 0 °F | 3 | 1 | 0 | 0 | 0 | 0 | 0 | 0 | 0 | 0 | 0 | 1 | 5 |
| Heating Degree Days | 1236 | 1040 | 847 | 477 | 192 | 36 | 3 | 12 | 115 | 403 | 686 | 1066 | 6113 |
| Cooling Degree Days | 0 | 0 | 0 | 4 | 40 | 136 | 281 | 222 | 70 | 7 | 0 | 0 | 760 |
| Total Precipitation (") | 3.37 | 3.10 | 3.88 | 3.87 | 4.28 | 3.86 | 3.45 | 3.84 | 3.89 | 3.63 | 4.11 | 4.05 | 45.33 |
| Days ≥ 0.1" Precip | 6 | 6 | 8 | 7 | 8 | 7 | 6 | 6 | 6 | 6 | 7 | 7 | 80 |
| Total Snowfall (") | 12.8 | 11.6 | 8.6 | 1.3 | 0.0 | 0.0 | 0.0 | 0.0 | 0.0 | 0.1 | 2.2 | 9.8 | 46.4 |
| Days ≥ 1" Snow Depth | 19 | 15 | 7 | 1 | 0 | 0 | 0 | 0 | 0 | 0 | 1 | 9 | 52 |

## HARTFORD BRAINARD FD *Hartford County* ELEVATION 20 ft LAT/LONG 41° 44 ' N / 72° 39 ' W

| | JAN | FEB | MAR | APR | MAY | JUN | JUL | AUG | SEP | OCT | NOV | DEC | YEAR |
|---|---|---|---|---|---|---|---|---|---|---|---|---|---|
| Maximum Temp °F | 34.7 | 37.6 | 46.5 | 58.8 | 70.1 | 78.6 | 83.8 | 82.0 | 74.2 | 62.8 | 51.6 | 39.7 | 60.0 |
| Minimum Temp °F | 16.5 | 19.2 | 27.7 | 37.6 | 47.8 | 57.0 | 62.8 | 61.2 | 52.1 | 40.6 | 33.1 | 22.8 | 39.9 |
| Mean Temp °F | 25.6 | 28.4 | 37.1 | 48.2 | 59.0 | 67.8 | 73.3 | 71.6 | 63.2 | 51.7 | 42.4 | 31.3 | 50.0 |
| Days Max Temp ≥ 90 °F | 0 | 0 | 0 | 0 | 1 | 3 | 5 | 4 | 1 | 0 | 0 | 0 | 14 |
| Days Max Temp ≤ 32 °F | 13 | 9 | 2 | 0 | 0 | 0 | 0 | 0 | 0 | 0 | 1 | 8 | 33 |
| Days Min Temp ≤ 32 °F | 29 | 26 | 22 | 8 | 1 | 0 | 0 | 0 | 0 | 6 | 16 | 27 | 135 |
| Days Min Temp ≤ 0 °F | 3 | 1 | 0 | 0 | 0 | 0 | 0 | 0 | 0 | 0 | 0 | 0 | 4 |
| Heating Degree Days | 1214 | 1027 | 857 | 502 | 211 | 42 | 4 | 11 | 116 | 410 | 673 | 1038 | 6105 |
| Cooling Degree Days | 0 | 0 | 0 | 3 | 37 | 137 | 280 | 223 | 71 | 5 | 0 | 0 | 756 |
| Total Precipitation (") | 3.22 | 2.85 | 3.62 | 3.80 | 4.02 | 3.86 | 3.68 | 3.75 | 3.81 | 3.65 | 3.81 | 3.87 | 43.94 |
| Days ≥ 0.1" Precip | 6 | 5 | 7 | 7 | 7 | 6 | 6 | 5 | 6 | 5 | 7 | 7 | 74 |
| Total Snowfall (") | na | 7.5 | 5.4 | 0.8 | 0.0 | 0.0 | 0.0 | 0.0 | 0.0 | 0.0 | 0.6 | ***6.3*** | na |
| Days ≥ 1" Snow Depth | 13 | 9 | 4 | 0 | 0 | 0 | 0 | 0 | 0 | 0 | 1 | 6 | 33 |

## MANSFIELD HOLLOW LAK *Tolland County* ELEVATION 249 ft LAT/LONG 41° 45 ' N / 72° 11 ' W

| | JAN | FEB | MAR | APR | MAY | JUN | JUL | AUG | SEP | OCT | NOV | DEC | YEAR |
|---|---|---|---|---|---|---|---|---|---|---|---|---|---|
| Maximum Temp °F | 34.8 | 37.1 | 46.4 | 57.9 | 69.3 | 77.6 | 82.9 | 81.3 | 73.4 | 63.1 | 51.8 | 39.9 | 59.6 |
| Minimum Temp °F | 12.6 | 15.1 | 25.0 | 34.6 | 43.9 | 53.2 | 58.9 | 57.3 | 48.3 | 36.6 | 30.2 | 19.6 | 36.3 |
| Mean Temp °F | 23.7 | 26.1 | 35.8 | 46.3 | 56.6 | 65.4 | 70.9 | 69.4 | 60.9 | 49.9 | 41.1 | 29.8 | 48.0 |
| Days Max Temp ≥ 90 °F | 0 | 0 | 0 | 0 | 0 | 2 | 4 | 2 | 1 | 0 | 0 | 0 | 9 |
| Days Max Temp ≤ 32 °F | 12 | 9 | 2 | 0 | 0 | 0 | 0 | 0 | 0 | 0 | 0 | 7 | 30 |
| Days Min Temp ≤ 32 °F | 29 | 25 | 24 | 13 | 3 | 0 | 0 | 0 | 1 | 11 | 19 | 27 | 152 |
| Days Min Temp ≤ 0 °F | 5 | 3 | 0 | 0 | 0 | 0 | 0 | 0 | 0 | 0 | 0 | 1 | 9 |
| Heating Degree Days | 1273 | 1092 | 899 | 556 | 271 | 71 | 12 | 26 | 160 | 464 | 711 | 1085 | 6620 |
| Cooling Degree Days | ***0*** | 0 | ***0*** | 1 | ***18*** | 90 | ***219*** | ***172*** | 44 | ***4*** | ***0*** | ***0*** | 548 |
| Total Precipitation (") | 3.89 | 3.29 | 4.23 | 4.12 | 4.03 | 3.77 | 4.21 | 4.06 | 3.98 | 4.00 | 4.62 | 4.47 | 48.67 |
| Days ≥ 0.1" Precip | 7 | 6 | 7 | 7 | 7 | 7 | 7 | 6 | 6 | 6 | 8 | 8 | 82 |
| Total Snowfall (") | 10.2 | 9.4 | 6.6 | 1.2 | 0.0 | 0.0 | 0.0 | 0.0 | 0.0 | 0.1 | 1.2 | 6.5 | 35.2 |
| Days ≥ 1" Snow Depth | 17 | 16 | 8 | 1 | 0 | 0 | 0 | 0 | 0 | 0 | 1 | 8 | 51 |

## MIDDLETOWN 4 W *Middlesex County* ELEVATION 369 ft LAT/LONG 41° 33 ' N / 72° 43 ' W

| | JAN | FEB | MAR | APR | MAY | JUN | JUL | AUG | SEP | OCT | NOV | DEC | YEAR |
|---|---|---|---|---|---|---|---|---|---|---|---|---|---|
| Maximum Temp °F | 35.3 | 37.9 | 46.7 | 58.7 | 70.3 | 79.2 | 84.0 | 81.3 | 72.7 | 61.7 | 51.3 | 40.2 | 59.9 |
| Minimum Temp °F | 19.4 | 21.2 | 29.2 | 38.6 | 48.2 | 57.3 | 62.6 | 61.5 | 53.6 | 43.2 | 35.3 | 25.4 | 41.3 |
| Mean Temp °F | 27.4 | 29.6 | 38.0 | 48.7 | 59.3 | 68.3 | 73.3 | 71.4 | 63.2 | 52.5 | 43.3 | 32.8 | 50.7 |
| Days Max Temp ≥ 90 °F | 0 | 0 | 0 | 0 | 0 | 2 | 5 | 2 | 0 | 0 | 0 | 0 | 9 |
| Days Max Temp ≤ 32 °F | 12 | 8 | 2 | 0 | 0 | 0 | 0 | 0 | 0 | 0 | 0 | 6 | 28 |
| Days Min Temp ≤ 32 °F | 27 | 24 | 20 | 7 | 0 | 0 | 0 | 0 | 0 | 4 | 12 | 24 | 118 |
| Days Min Temp ≤ 0 °F | 2 | 1 | 0 | 0 | 0 | 0 | 0 | 0 | 0 | 0 | 0 | 0 | 3 |
| Heating Degree Days | 1160 | 994 | 831 | 485 | 202 | 38 | 4 | 15 | 119 | 389 | 645 | 991 | 5873 |
| Cooling Degree Days | 0 | 0 | 0 | 2 | 39 | 157 | 294 | 231 | 74 | 6 | 0 | 0 | 803 |
| Total Precipitation (") | 3.88 | 3.39 | 4.44 | 4.28 | 4.47 | 4.37 | 3.99 | 4.37 | 4.40 | 4.30 | 4.60 | 4.40 | 50.89 |
| Days ≥ 0.1" Precip | 6 | 6 | 7 | 7 | 8 | 7 | 6 | 6 | 6 | 6 | 7 | 7 | 79 |
| Total Snowfall (") | 10.5 | 10.0 | 6.6 | 0.8 | 0.0 | 0.0 | 0.0 | 0.0 | 0.0 | 0.1 | 1.6 | 6.2 | 35.8 |
| Days ≥ 1" Snow Depth | 15 | 14 | 7 | 0 | 0 | 0 | 0 | 0 | 0 | 0 | 1 | ***6*** | 43 |

## MOUNT CARMEL *New Haven County* ELEVATION 180 ft LAT/LONG 41° 24 ' N / 72° 54 ' W

| | JAN | FEB | MAR | APR | MAY | JUN | JUL | AUG | SEP | OCT | NOV | DEC | YEAR |
|---|---|---|---|---|---|---|---|---|---|---|---|---|---|
| Maximum Temp °F | 36.4 | 39.1 | 47.9 | 59.7 | 70.2 | 78.6 | 83.6 | 82.1 | 74.5 | 64.0 | 52.7 | 40.9 | 60.8 |
| Minimum Temp °F | 18.6 | 20.5 | 28.5 | 37.6 | 47.2 | 56.2 | 62.0 | 61.1 | 53.2 | 42.3 | 34.3 | 24.3 | 40.5 |
| Mean Temp °F | 27.5 | 29.8 | 38.3 | 48.7 | 58.7 | 67.4 | 72.9 | 71.6 | 63.9 | 53.2 | 43.5 | 32.6 | 50.7 |
| Days Max Temp ≥ 90 °F | 0 | 0 | 0 | 0 | 1 | 2 | 5 | 3 | 1 | 0 | 0 | 0 | 12 |
| Days Max Temp ≤ 32 °F | 10 | 7 | 1 | 0 | 0 | 0 | 0 | 0 | 0 | 0 | 0 | 6 | 24 |
| Days Min Temp ≤ 32 °F | 28 | 25 | 21 | 7 | 1 | 0 | 0 | 0 | 0 | 5 | 14 | 25 | 126 |
| Days Min Temp ≤ 0 °F | 2 | 1 | 0 | 0 | 0 | 0 | 0 | 0 | 0 | 0 | 0 | 0 | 3 |
| Heating Degree Days | 1155 | 987 | 823 | 486 | 215 | 41 | 3 | 10 | 103 | 365 | 638 | 996 | 5822 |
| Cooling Degree Days | 0 | 0 | 0 | 2 | 35 | 132 | 278 | 231 | 85 | 7 | 0 | 0 | 770 |
| Total Precipitation (") | 3.77 | 3.26 | 4.52 | 4.42 | 4.62 | 4.44 | 4.24 | 4.18 | 4.61 | 4.05 | 4.44 | 4.27 | 50.82 |
| Days ≥ 0.1" Precip | 6 | 6 | 7 | 7 | 7 | 7 | 6 | 6 | 6 | 6 | 7 | 7 | 78 |
| Total Snowfall (") | 9.0 | 8.9 | 5.5 | 0.5 | 0.0 | 0.0 | 0.0 | 0.0 | 0.0 | 0.1 | 1.0 | 5.5 | 30.5 |
| Days ≥ 1" Snow Depth | 16 | 13 | 5 | 0 | 0 | 0 | 0 | 0 | 0 | 0 | 0 | 7 | 41 |

## NORFOLK 2 SW *Litchfield County* ELEVATION 1342 ft LAT/LONG 41° 58 ' N / 73° 13 ' W

| | JAN | FEB | MAR | APR | MAY | JUN | JUL | AUG | SEP | OCT | NOV | DEC | YEAR |
|---|---|---|---|---|---|---|---|---|---|---|---|---|---|
| Maximum Temp °F | 27.6 | 29.9 | 39.6 | 52.7 | 65.0 | 72.8 | 77.6 | 75.6 | 67.5 | 56.1 | 44.4 | 32.7 | 53.5 |
| Minimum Temp °F | 11.0 | 12.3 | 21.3 | 32.5 | 43.7 | 52.6 | 57.9 | 56.4 | 48.4 | 37.4 | 29.3 | 17.8 | 35.1 |
| Mean Temp °F | 19.3 | 21.1 | 30.5 | 42.7 | 54.4 | 62.7 | 67.8 | 66.0 | 58.0 | 46.8 | 36.9 | 25.3 | 44.3 |
| Days Max Temp ≥ 90 °F | 0 | 0 | 0 | 0 | 0 | 0 | 0 | 0 | 0 | 0 | 0 | 0 | 0 |
| Days Max Temp ≤ 32 °F | 21 | 17 | 8 | 1 | 0 | 0 | 0 | 0 | 0 | 0 | 3 | 16 | 66 |
| Days Min Temp ≤ 32 °F | 30 | 27 | 28 | 16 | 3 | 0 | 0 | 0 | 1 | 10 | 21 | 29 | 165 |
| Days Min Temp ≤ 0 °F | 6 | 5 | 1 | 0 | 0 | 0 | 0 | 0 | 0 | 0 | 0 | 2 | 14 |
| Heating Degree Days | 1411 | 1232 | 1062 | 665 | 334 | 115 | 31 | 55 | 227 | 559 | 837 | 1224 | 7752 |
| Cooling Degree Days | 0 | 0 | 0 | 1 | 13 | 52 | 129 | 93 | 23 | 2 | 0 | 0 | 313 |
| Total Precipitation (") | 3.81 | 3.66 | 4.40 | 4.49 | 4.70 | 4.65 | 4.43 | 4.70 | 4.30 | 4.07 | 4.60 | 4.51 | 52.32 |
| Days ≥ 0.1" Precip | 7 | 7 | 8 | 8 | 9 | 8 | 8 | 7 | 7 | 7 | 9 | 8 | 93 |
| Total Snowfall (") | 21.2 | 20.3 | 17.9 | 6.3 | 0.8 | 0.0 | 0.0 | 0.0 | 0.0 | 0.8 | 6.6 | 19.4 | 93.3 |
| Days ≥ 1" Snow Depth | 27 | 26 | 23 | 6 | 0 | 0 | 0 | 0 | 0 | 0 | 5 | 21 | 108 |

## NORWICH PUB UTIL PLA *New London County* ELEVATION 20 ft LAT/LONG 41° 32 ' N / 72° 4 ' W

| | JAN | FEB | MAR | APR | MAY | JUN | JUL | AUG | SEP | OCT | NOV | DEC | YEAR |
|---|---|---|---|---|---|---|---|---|---|---|---|---|---|
| Maximum Temp °F | 37.8 | 39.7 | 48.4 | 59.1 | 70.0 | 78.2 | 83.2 | 81.3 | *74.2* | 64.0 | 53.3 | 42.4 | 61.0 |
| Minimum Temp °F | 17.6 | 19.9 | 28.3 | 37.1 | 47.0 | 55.5 | 62.1 | 60.7 | 52.4 | 40.4 | 32.9 | 23.6 | 39.8 |
| Mean Temp °F | *27.7* | *29.7* | *38.4* | 48.2 | 58.5 | *66.8* | *72.7* | *71.0* | *63.4* | *52.0* | 43.1 | 33.0 | 50.4 |
| Days Max Temp ≥ 90 °F | 0 | 0 | 0 | 0 | 1 | *2* | 5 | *2* | 1 | 0 | 0 | 0 | 11 |
| Days Max Temp ≤ 32 °F | *9* | 6 | 1 | 0 | 0 | 0 | 0 | 0 | 0 | 0 | 0 | 5 | 21 |
| Days Min Temp ≤ 32 °F | 29 | 25 | 20 | 9 | 0 | 0 | 0 | 0 | 0 | 7 | 16 | 26 | 132 |
| Days Min Temp ≤ 0 °F | *2* | 1 | 0 | 0 | 0 | 0 | 0 | 0 | 0 | 0 | 0 | 0 | 3 |
| Heating Degree Days | *1147* | *985* | *818* | 499 | 216 | *48* | *4* | *12* | *109* | *400* | 650 | *983* | 5871 |
| Cooling Degree Days | 0 | 0 | 0 | 2 | 24 | 109 | 261 | 214 | 67 | 6 | 0 | 0 | 683 |
| Total Precipitation (") | 4.04 | 3.73 | 4.81 | 4.33 | 4.27 | 3.28 | 3.69 | 4.43 | 3.99 | 4.07 | 5.09 | 4.99 | 50.72 |
| Days ≥ 0.1" Precip | 6 | 6 | 8 | 7 | 7 | 6 | 6 | 6 | 6 | 6 | 7 | 8 | 79 |
| Total Snowfall (") | na | na | na | *0.3* | 0.0 | 0.0 | 0.0 | 0.0 | 0.0 | 0.0 | *0.2* | na | na |
| Days ≥ 1" Snow Depth | na | na | na | *0* | 0 | 0 | 0 | 0 | 0 | 0 | *0* | na | na |

## SHEPAUG DAM *Litchfield County* ELEVATION 840 ft LAT/LONG 41° 43 ' N / 73° 18 ' W

| | JAN | FEB | MAR | APR | MAY | JUN | JUL | AUG | SEP | OCT | NOV | DEC | YEAR |
|---|---|---|---|---|---|---|---|---|---|---|---|---|---|
| Maximum Temp °F | 33.1 | 35.8 | 45.0 | 57.5 | 68.3 | 75.8 | 80.3 | 78.7 | 71.1 | 61.1 | 50.0 | 38.1 | 57.9 |
| Minimum Temp °F | 13.8 | 15.5 | 24.5 | 34.7 | 45.1 | 54.1 | 59.4 | 58.0 | 50.6 | 39.7 | 31.6 | 20.5 | 37.3 |
| Mean Temp °F | 23.5 | 25.7 | 34.7 | 46.1 | 56.7 | 64.9 | 69.9 | 68.4 | 60.9 | 50.4 | 40.8 | 29.3 | 47.6 |
| Days Max Temp ≥ 90 °F | 0 | 0 | 0 | 0 | 0 | 1 | 2 | 1 | 0 | 0 | 0 | 0 | 4 |
| Days Max Temp ≤ 32 °F | 15 | 10 | 3 | 0 | 0 | 0 | 0 | 0 | 0 | 0 | 1 | 8 | 37 |
| Days Min Temp ≤ 32 °F | 30 | 26 | 26 | 13 | 2 | 0 | 0 | 0 | 0 | 7 | 17 | 28 | 149 |
| Days Min Temp ≤ 0 °F | 5 | 3 | 1 | 0 | 0 | 0 | 0 | 0 | 0 | 0 | 0 | 1 | 10 |
| Heating Degree Days | 1281 | 1103 | 932 | 562 | 268 | 74 | 13 | 28 | 158 | 448 | 718 | 1099 | 6684 |
| Cooling Degree Days | 0 | 0 | 0 | 2 | 16 | 76 | 178 | 135 | 37 | 2 | 0 | 0 | 446 |
| Total Precipitation (") | 3.46 | 3.20 | 4.35 | 4.31 | 4.61 | 4.30 | 4.45 | 4.61 | 4.36 | 4.14 | 4.43 | 4.16 | 50.38 |
| Days ≥ 0.1" Precip | 7 | 7 | 8 | 9 | 9 | 8 | 8 | 7 | 7 | 6 | 8 | 8 | 92 |
| Total Snowfall (") | 14.6 | 12.9 | 10.7 | 2.1 | 0.1 | 0.0 | 0.0 | 0.0 | 0.0 | 0.3 | 2.9 | 10.4 | 54.0 |
| Days ≥ 1" Snow Depth | na | na | na | na | *0* | 0 | 0 | 0 | 0 | 0 | na | na | na |

## STAMFORD 5 N *Fairfield County* ELEVATION 190 ft LAT/LONG 41° 8 ' N / 73° 33 ' W

| | JAN | FEB | MAR | APR | MAY | JUN | JUL | AUG | SEP | OCT | NOV | DEC | YEAR |
|---|---|---|---|---|---|---|---|---|---|---|---|---|---|
| Maximum Temp °F | 37.2 | 40.3 | 49.5 | 61.6 | 72.1 | 80.2 | 84.8 | 82.9 | 75.5 | 64.9 | 53.5 | 41.9 | 62.0 |
| Minimum Temp °F | 17.9 | 19.7 | 28.2 | 37.1 | 46.9 | 55.8 | 61.3 | 60.3 | 52.9 | 41.4 | 33.5 | 24.3 | 39.9 |
| Mean Temp °F | 27.6 | 30.0 | 38.8 | 49.4 | 59.5 | 68.0 | 73.1 | 71.6 | 64.2 | 53.2 | 43.5 | 33.1 | 51.0 |
| Days Max Temp ≥ 90 °F | 0 | 0 | 0 | 0 | 1 | 2 | 6 | 3 | 1 | 0 | 0 | 0 | 13 |
| Days Max Temp ≤ 32 °F | 9 | 5 | 1 | 0 | 0 | 0 | 0 | 0 | 0 | 0 | 0 | 4 | 19 |
| Days Min Temp ≤ 32 °F | 29 | 26 | 21 | 9 | 1 | 0 | 0 | 0 | 0 | 6 | 15 | 25 | 132 |
| Days Min Temp ≤ 0 °F | 2 | 1 | 0 | 0 | 0 | 0 | 0 | 0 | 0 | 0 | 0 | 0 | 3 |
| Heating Degree Days | 1152 | 982 | 804 | 465 | 194 | 34 | 2 | 8 | 91 | 365 | 638 | 980 | 5715 |
| Cooling Degree Days | *0* | 0 | 0 | 2 | 37 | 145 | 289 | 226 | 78 | 6 | 0 | 0 | 783 |
| Total Precipitation (") | 3.91 | 3.38 | 4.65 | 4.33 | 4.83 | 4.14 | 3.97 | 4.29 | 4.42 | 4.12 | 4.58 | 4.38 | 51.00 |
| Days ≥ 0.1" Precip | 7 | 6 | 8 | 7 | 7 | 7 | 6 | 7 | 6 | 6 | 7 | *7* | 81 |
| Total Snowfall (") | *9.5* | 9.7 | 5.6 | 0.6 | 0.0 | 0.0 | 0.0 | 0.0 | 0.0 | 0.0 | 0.7 | 4.6 | 30.7 |
| Days ≥ 1" Snow Depth | na | na | *2* | 0 | 0 | 0 | 0 | 0 | 0 | 0 | 0 | na | na |

## STORRS *Tolland County* ELEVATION 650 ft LAT/LONG 41° 48 ' N / 72° 15 ' W

| | JAN | FEB | MAR | APR | MAY | JUN | JUL | AUG | SEP | OCT | NOV | DEC | YEAR |
|---|---|---|---|---|---|---|---|---|---|---|---|---|---|
| Maximum Temp °F | 32.9 | 35.2 | 44.3 | 56.2 | 67.5 | 75.2 | 79.8 | 78.2 | 70.9 | 60.8 | 49.8 | 37.8 | 57.4 |
| Minimum Temp °F | 16.6 | 18.6 | 27.2 | 36.8 | 46.4 | 55.2 | 60.8 | 59.5 | 51.6 | 41.2 | 33.5 | 22.5 | 39.2 |
| Mean Temp °F | 24.8 | 26.9 | 35.8 | 46.5 | 57.0 | 65.2 | 70.3 | 68.9 | 61.3 | 51.0 | 41.6 | 30.2 | 48.3 |
| Days Max Temp ≥ 90 °F | 0 | 0 | 0 | 0 | 0 | 0 | 1 | 0 | 0 | 0 | 0 | 0 | 1 |
| Days Max Temp ≤ 32 °F | 14 | 11 | 3 | 0 | 0 | 0 | 0 | 0 | 0 | 0 | 1 | 9 | 38 |
| Days Min Temp ≤ 32 °F | 29 | 26 | 23 | 9 | 0 | 0 | 0 | 0 | 0 | 5 | 14 | 27 | 133 |
| Days Min Temp ≤ 0 °F | 3 | 1 | 0 | 0 | 0 | 0 | 0 | 0 | 0 | 0 | 0 | 1 | 5 |
| Heating Degree Days | 1240 | 1070 | 901 | 551 | 259 | 70 | 11 | 24 | 148 | 430 | 693 | 1072 | 6469 |
| Cooling Degree Days | 0 | *0* | 0 | 1 | 19 | 85 | 192 | 149 | 44 | 4 | 0 | 0 | 494 |
| Total Precipitation (") | 3.93 | 3.61 | 4.28 | 4.09 | 4.11 | 4.00 | 4.20 | 4.36 | 4.36 | 4.16 | 4.60 | 4.44 | 50.14 |
| Days ≥ 0.1" Precip | 6 | *6* | 7 | 7 | 7 | 7 | 7 | 7 | 6 | 6 | 7 | 7 | 80 |
| Total Snowfall (") | na | na | *6.1* | 0.8 | 0.1 | 0.0 | 0.0 | 0.0 | 0.0 | 0.2 | *1.4* | na | na |
| Days ≥ 1" Snow Depth | na | na | na | *0* | 0 | 0 | 0 | 0 | 0 | 0 | *1* | na | na |

## WEST THOMPSON LAKE *Windham County* ELEVATION 360 ft LAT/LONG 41° 57 ' N / 71° 54 ' W

| | JAN | FEB | MAR | APR | MAY | JUN | JUL | AUG | SEP | OCT | NOV | DEC | YEAR |
|---|---|---|---|---|---|---|---|---|---|---|---|---|---|
| Maximum Temp °F | *34.5* | *36.9* | 46.0 | 57.8 | 69.2 | 77.1 | 82.5 | 80.6 | 72.8 | 62.4 | 51.1 | 38.9 | 59.2 |
| Minimum Temp °F | 12.0 | *14.3* | 24.3 | 34.1 | 43.9 | 53.1 | 58.7 | 57.4 | 48.3 | 36.3 | 29.6 | 18.6 | 35.9 |
| Mean Temp °F | *23.3* | *25.6* | 35.2 | 46.0 | 56.6 | 65.1 | 70.6 | 69.0 | 60.6 | 49.4 | 40.4 | 28.7 | 47.5 |
| Days Max Temp ≥ 90 °F | 0 | 0 | 0 | 0 | 1 | 2 | 4 | 2 | 1 | 0 | 0 | 0 | 10 |
| Days Max Temp ≤ 32 °F | 12 | 9 | 2 | 0 | 0 | 0 | 0 | 0 | 0 | 0 | 1 | 7 | 31 |
| Days Min Temp ≤ 32 °F | 29 | 25 | 25 | 14 | 3 | 0 | 0 | 0 | 1 | 11 | 19 | 27 | 154 |
| Days Min Temp ≤ 0 °F | 6 | 3 | 0 | 0 | 0 | 0 | 0 | 0 | 0 | 0 | 0 | 2 | 11 |
| Heating Degree Days | *1287* | *1106* | 918 | 567 | 272 | 76 | 14 | 28 | 166 | 480 | 733 | 1121 | 6768 |
| Cooling Degree Days | *0* | *0* | 0 | 1 | 20 | 84 | 195 | 155 | 41 | 3 | *0* | *0* | 499 |
| Total Precipitation (") | 3.96 | 3.19 | 4.31 | 4.18 | 4.11 | 3.99 | 4.11 | 4.54 | 3.94 | 3.97 | 4.77 | 4.42 | 49.49 |
| Days ≥ 0.1" Precip | 6 | 6 | 7 | 7 | 7 | 7 | 7 | 7 | 6 | 5 | 7 | 7 | 79 |
| Total Snowfall (") | *10.3* | *9.6* | *5.5* | 1.1 | 0.0 | 0.0 | 0.0 | 0.0 | 0.0 | 0.1 | 1.8 | *7.1* | 35.5 |
| Days ≥ 1" Snow Depth | *15* | *14* | 7 | 1 | 0 | 0 | 0 | 0 | 0 | 0 | 1 | 8 | 46 |

## WIGWAM RESERVOIR *Litchfield County* ELEVATION 600 ft LAT/LONG 41° 40 ' N / 73° 8 ' W

| | JAN | FEB | MAR | APR | MAY | JUN | JUL | AUG | SEP | OCT | NOV | DEC | YEAR |
|---|---|---|---|---|---|---|---|---|---|---|---|---|---|
| Maximum Temp °F | 34.5 | 37.3 | 46.6 | 59.2 | 70.3 | 78.1 | 83.1 | 81.1 | 73.2 | 62.6 | 51.4 | 39.4 | 59.7 |
| Minimum Temp °F | 13.8 | 15.2 | 24.1 | 33.3 | 43.4 | 52.7 | 57.9 | 56.8 | 49.0 | 37.8 | 30.0 | 20.2 | 36.2 |
| Mean Temp °F | 24.2 | 26.3 | 35.3 | 46.3 | 56.9 | 65.4 | 70.5 | 68.9 | 61.2 | 50.3 | 40.7 | 30.0 | 48.0 |
| Days Max Temp ≥ 90 °F | 0 | 0 | 0 | 0 | 0 | 2 | 4 | 2 | 1 | 0 | 0 | 0 | 9 |
| Days Max Temp ≤ 32 °F | 12 | 8 | 2 | 0 | 0 | 0 | 0 | 0 | 0 | 0 | 0 | 7 | 29 |
| Days Min Temp ≤ 32 °F | 30 | 27 | 26 | 15 | 3 | 0 | 0 | 0 | 1 | 11 | 19 | 28 | 160 |
| Days Min Temp ≤ 0 °F | 4 | 3 | 1 | 0 | 0 | 0 | 0 | 0 | 0 | 0 | 0 | 1 | 9 |
| Heating Degree Days | 1259 | 1087 | 913 | 557 | 263 | 70 | 11 | 25 | 155 | 453 | 721 | 1077 | 6591 |
| Cooling Degree Days | 0 | 0 | 0 | 2 | 19 | 91 | 214 | 162 | 46 | 2 | 0 | 0 | 536 |
| Total Precipitation (") | 3.46 | 3.16 | 4.29 | 4.11 | 4.31 | 4.04 | 4.13 | 4.29 | 4.27 | 4.06 | 4.36 | 4.37 | 48.85 |
| Days ≥ 0.1" Precip | 6 | 6 | 7 | 7 | 8 | 7 | 7 | 6 | 6 | 6 | 7 | 7 | 80 |
| Total Snowfall (") | na | na | na | 0.8 | 0.0 | 0.0 | 0.0 | 0.0 | 0.0 | 0.0 | na | na | na |
| Days ≥ 1" Snow Depth | na | na | na | na | 0 | 0 | 0 | 0 | 0 | 0 | na | na | na |

## JANUARY MINIMUM TEMPERATURE °F

### LOWEST

| | | |
|---|---|---|
| 1 | Norfolk | 11.0 |
| 2 | West Thmpson Lk | 12.0 |
| 3 | Falls Village | 12.1 |
| 4 | Mnsfld Hollow lk | 12.6 |
| 5 | Bulls Bridge Dam | 13.7 |
| | Burlington | 13.7 |
| 7 | Shepaug Dam | 13.8 |
| | Wigwam Rservoir | 13.8 |
| 9 | Hartford-Bradley | 16.3 |
| 10 | Hartford-Brainard | 16.5 |
| 11 | Storrs | 16.6 |
| 12 | Danbury | 17.0 |
| 13 | Norwich | 17.6 |
| 14 | Stamford | 17.9 |
| 15 | Mount Carmel | 18.6 |
| 16 | Groton | 18.8 |
| 17 | Middletown | 19.4 |
| 18 | Bridgeport | 22.2 |

### HIGHEST

| | | |
|---|---|---|
| 1 | Bridgeport | 22.2 |
| 2 | Middletown | 19.4 |
| 3 | Groton | 18.8 |
| 4 | Mount Carmel | 18.6 |
| 5 | Stamford | 17.9 |
| 6 | Norwich | 17.6 |
| 7 | Danbury | 17.0 |
| 8 | Storrs | 16.6 |
| 9 | Hartford-Brainard | 16.5 |
| 10 | Hartford-Bradley | 16.3 |
| 11 | Shepaug Dam | 13.8 |
| | Wigwam Rservoir | 13.8 |
| 13 | Bulls Bridge Dam | 13.7 |
| | Burlington | 13.7 |
| 15 | Mnsfld Hollow lk | 12.6 |
| 16 | Falls Village | 12.1 |
| 17 | West Thmpson Lk | 12.0 |
| 18 | Norfolk | 11.0 |

## JULY MAXIMUM TEMPERATURE °F

### HIGHEST

| | | |
|---|---|---|
| 1 | Hartford-Bradley | 85.1 |
| 2 | Stamford | 84.8 |
| 3 | Middletown | 84.0 |
| 4 | Danbury | 83.9 |
| | Falls Village | 83.9 |
| 6 | Hartford-Brainard | 83.8 |
| 7 | Bulls Bridge Dam | 83.7 |
| 8 | Mount Carmel | 83.6 |
| 9 | Norwich | 83.2 |
| 10 | Wigwam Rservoir | 83.1 |
| 11 | Mnsfld Hollow lk | 82.9 |
| 12 | West Thmpson Lk | 82.5 |
| 13 | Burlington | 82.4 |
| 14 | Bridgeport | 82.1 |
| 15 | Groton | 80.9 |
| 16 | Shepaug Dam | 80.3 |
| 17 | Storrs | 79.8 |
| 18 | Norfolk | 77.6 |

### LOWEST

| | | |
|---|---|---|
| 1 | Norfolk | 77.6 |
| 2 | Storrs | 79.8 |
| 3 | Shepaug Dam | 80.3 |
| 4 | Groton | 80.9 |
| 5 | Bridgeport | 82.1 |
| 6 | Burlington | 82.4 |
| 7 | West Thmpson Lk | 82.5 |
| 8 | Mnsfld Hollow lk | 82.9 |
| 9 | Wigwam Rservoir | 83.1 |
| 10 | Norwich | 83.2 |
| 11 | Mount Carmel | 83.6 |
| 12 | Bulls Bridge Dam | 83.7 |
| 13 | Hartford-Brainard | 83.8 |
| 14 | Danbury | 83.9 |
| | Falls Village | 83.9 |
| 16 | Middletown | 84.0 |
| 17 | Stamford | 84.8 |
| 18 | Hartford-Bradley | 85.1 |

## ANNUAL PRECIPITATION (")

### HIGHEST

| | | |
|---|---|---|
| 1 | Norfolk | 52.32 |
| 2 | Stamford | 51.00 |
| 3 | Middletown | 50.89 |
| 4 | Mount Carmel | 50.82 |
| 5 | Norwich | 50.72 |
| 6 | Shepaug Dam | 50.38 |
| 7 | Burlington | 50.36 |
| 8 | Storrs | 50.14 |
| 9 | West Thmpson Lk | 49.49 |
| 10 | Danbury | 49.28 |
| 11 | Wigwam Rservoir | 48.85 |
| 12 | Mnsfld Hollow lk | 48.67 |
| 13 | Groton | 48.09 |
| 14 | Bulls Bridge Dam | 45.93 |
| 15 | Hartford-Bradley | 45.33 |
| 16 | Falls Village | 43.97 |
| 17 | Hartford-Brainard | 43.94 |
| 18 | Bridgeport | 43.17 |

### LOWEST

| | | |
|---|---|---|
| 1 | Bridgeport | 43.17 |
| 2 | Hartford-Brainard | 43.94 |
| 3 | Falls Village | 43.97 |
| 4 | Hartford-Bradley | 45.33 |
| 5 | Bulls Bridge Dam | 45.93 |
| 6 | Groton | 48.09 |
| 7 | Mnsfld Hollow lk | 48.67 |
| 8 | Wigwam Rservoir | 48.85 |
| 9 | Danbury | 49.28 |
| 10 | West Thmpson Lk | 49.49 |
| 11 | Storrs | 50.14 |
| 12 | Burlington | 50.36 |
| 13 | Shepaug Dam | 50.38 |
| 14 | Norwich | 50.72 |
| 15 | Mount Carmel | 50.82 |
| 16 | Middletown | 50.89 |
| 17 | Stamford | 51.00 |
| 18 | Norfolk | 52.32 |

## ANNUAL SNOWFALL (")

### HIGHEST

| | | |
|---|---|---|
| 1 | Norfolk | 93.3 |
| 2 | Shepaug Dam | 54.0 |
| 3 | Hartford-Bradley | 46.4 |
| 4 | Danbury | 39.9 |
| 5 | Middletown | 35.8 |
| 6 | West Thmpson Lk | 35.5 |
| 7 | Mnsfld Hollow lk | 35.2 |
| 8 | Stamford | 30.7 |
| 9 | Mount Carmel | 30.5 |
| 10 | Bridgeport | 25.7 |
| 11 | Groton | 22.0 |

### LOWEST

| | | |
|---|---|---|
| 1 | Groton | 22.0 |
| 2 | Bridgeport | 25.7 |
| 3 | Mount Carmel | 30.5 |
| 4 | Stamford | 30.7 |
| 5 | Mnsfld Hollow lk | 35.2 |
| 6 | West Thmpson Lk | 35.5 |
| 7 | Middletown | 35.8 |
| 8 | Danbury | 39.9 |
| 9 | Hartford-Bradley | 46.4 |
| 10 | Shepaug Dam | 54.0 |
| 11 | Norfolk | 93.3 |

# DELAWARE

PHYSICAL FEATURES. The State of Delaware is located on the east coast of the United States midway between the north and the south. Delaware lies in a north-south position, spanning a distance of 96 miles. The width increases from 9 miles in the northern portion to 35 miles in the extreme southern portion. The State occupies the eastern and northern portion of the Delmarva Peninsula which is bounded by the Chesapeake Bay on the west and the Delaware Bay and Atlantic Ocean on the east. The total area of Delaware is 2,057 square miles.

Over 95 percent of the land area of the State is more or less flat and without topographic features; however, the extreme northern portion, about 120 square miles, which lies on the Piedmont, is undulating and hilly with elevations rising to 438 feet above mean sea level. This increase in elevation no doubt contributes to a slight decrease in local temperatures under certain circumstances.

GENERAL CLIMATE. Since the flow of the atmosphere in temperate latitudes is from west to east, the distribution of land and water masses, i.e., the expansive North American continent situated immediately to the west, predisposes the Delaware area to a continental type of climate. This type of climate in middle latitudes is marked by well-defined seasons. Winter is the dormant season for plant growth and is one of low temperature rather than drought. In spring and fall the changeableness of the weather is a striking characteristic. It is occasioned by a rapid succession of warm and cold periods associated with storms, which generally move from a westerly direction over the eastern portion of the United States. Summers are warm to hot. The higher atmospheric humidity along the sea coast causes the summer heat to be more oppressive or sultry and the winter cold more raw and penetrating than in drier climates of the interior.

The topography of the eastern United States is characterized by the Appalachian Mountains, which extend along a northeast-southwest axis about 150 miles to the northwest of Delaware. To the west and northwest of Delaware, these mountains range in height from 2,000 to 3,000 feet above mean sea level and contribute to some slight tempering of the cold air masses which move rapidly out of the interior of the continent over the Delaware region in the winter.

A semipermanent high pressure area with a clockwise circulation virtually overspreads the entire Atlantic Ocean at middle latitudes and exerts a pronounced effect on the weather regimes of the east coast. During the winter season the Atlantic High (or Azores High) maintains an average position between latitude 30° N. and 33° N. and longitudes 25° W. and 35° W. and overspreads the eastern portion of the south Atlantic Ocean. As the summer season approaches, the Atlantic High moves westward and slightly northward to a mean position between latitudes 32° N. and 35° N. and longitudes 40° W. and 45° W. During this period it becomes more intense and widespread as the semipermanent low of the north Atlantic Ocean becomes smaller and weaker. In the summer location the Atlantic High dominates the flow of air over the eastern United States much of the time. A persistence of the Atlantic High in a westerly position in the vicinity of Bermuda results in a prolonged flow of moist, warm tropical air over the entire eastern United States. Weather in this type of air mass consists of scattered thunderstorms, considerable daytime cloudiness, and hot, sultry conditions. In the westerly position the High exerts blocking action on Lows which are forced to travel across more northerly latitudes. Persistence of this High over the eastern United States frequently results in drought conditions over the Delaware region, as the dry, subsiding air of the High prevents the formation of precipitation.

WINDS. Prevailing surface winds in northern Delaware blow from the northwesterly quadrant in all months except June, when southerly winds prevail. However, during the periods of May and July through September, winds come from the southwesterly quadrant a high proportion of the time. In southern Delaware surface winds prevail from the southwesterly quadrant from May through September and from the northwesterly quadrant from October through April.

Average wind speeds are higher during the period January through April, largely due to the rapid succession of well-developed storm systems which migrate from a westerly to easterly direction. During this period average wind speeds of about 10 miles per hour prevail. From July through October winds are somewhat lighter, averaging from 7 to 9 miles per hour.

During the fall, winter, and spring seasons, it is not unusual to experience brief windstorms associated with violent, fast-moving cold fronts with gusts from 50 to 60 m.p.h. In the summer, rare occurrences of violent windstorms are associated with severe thunderstorms. From June through October, it is estimated that wind speeds of more than 75 m.p.h. could occur anywhere in Delaware during the rare event of a hurricane traversing or passing very near the State.

Delaware lies in the mean zone of the westerlies in the winter and slightly south of the tracks followed by most of the migrating cyclones in their movement from some point in the United States to the region of semipermanent low pressure in the Iceland or North Atlantic area. Cyclones which have their origin in the south Pacific coastal region, Texas, or the Gulf or South Atlantic States have a greater tendency to follow a track through the Delaware region. Storms of the south Pacific coast, Texas, east Gulf, and sometimes of the south Atlantic bring the heaviest widespread rains to the Delaware area.

TEMPERATURE. The difference in latitude of northern Delaware and southern Delaware contributes in some part to the difference in mean temperature between these two regions of the State. The mean temperature difference of 3° to 4° between northern and southern portions in winter and 1° to 2° in summer is largely but not entirely due to the variation in solar radiational heating. In the extreme northern portion where elevations range from 300 to 400 feet on the higher hills, altitude is a controlling factor, although a small one, and reduces temperatures by approximately 1° on the average as compared to the nearby lower terrain.

In order for ocean currents to have a direct temperature control, the winds must be prevailing onshore. The relatively frequent occurrence of easterly winds associated with cyclonic storms to the southeast brings about advection of air off the mild waters and consequently tends to raise the normal winter temperatures and lower the summer temperatures. Therefore, mean winter temperatures of Delaware are roughly 5° higher than for regions of the continental interior at the same latitude.

The climate of Delaware is humid, temperate, with hot summers and mild winters. The winter climate is intermediate between the cold of the northeast and the mild weather of the south. The average frost penetration ranges from about 5 inches in southern Delaware to about 10 inches in northern Delaware. Summer weather is characterized by considerable warm weather, including at least several hot, humid periods. However, nights are usually quite comfortable. The average length of the growing (frost-free) season ranges from about 175 to 195 days.

PRECIPITATION. The average annual precipitation ranges from 44 inches in northern Delaware to 47 inches in southern Delaware. The monthly distribution is fairly uniform throughout the year, with July and August the months with heaviest amounts. Precipitation in the summer season is less dependable and more variable than in winter. The seasonal increase in evapotranspiration during the summer results in a rapid loss of soil moisture and contributes to the development of drought conditions. Flooding occurs infrequently, and results largely from tides pushed by strong easterly winds. The passage through the area of storms of tropical origin, usually during the late summer or fall, with their high winds and intense rains constitute the most serious flood threat.

The mean snowfall is 18 inches in northern Delaware and 14 inches in southern Delaware. The snow season runs from December through March, with a few light flurries in some years as early as November or late October and as late as early April. Heaviest snowfalls in Delaware generally occur in February and March.

STORMS. Thunderstorms occur at a given station on the average of 30 to 33 days per year. The Atlantic coastal region has fewer thunderstorms than interior portions, on the average. They have been observed in every month of the year; however, July is the month with the greatest frequency of thunderstorms, on the average. Hail is uncommon in Delaware. The frequency of occurrence of tornadoes in Delaware is estimated at about 1 in 2 or 3 years, on the average.

Average relative humidity in Delaware is lowest in winter and early spring, and highest in the late summer and early fall. February and March have average relative humidities of about 60 to 65 percent, whereas August, September, and October have average relative humidities of about 75 to 80 percent.

# COUNTY INDEX

### *Kent County*
DOVER

### *New Castle County*
NEWARK UNIVERSITY FA
WILMINGTN NEW CASTLE
WILMNGTON PORTER RSV

### *Sussex County*
GEORGETOWN 5 SW
LEWES

# ELEVATION INDEX

| FEET | STATION NAME |
|---|---|
| 10 | LEWES |
| 30 | DOVER |
| 49 | GEORGETOWN 5 SW |
| 79 | WILMINGTN NEW CASTLE |
| 112 | NEWARK UNIVERSITY FA |
| 259 | WILMNGTON PORTER RSV |

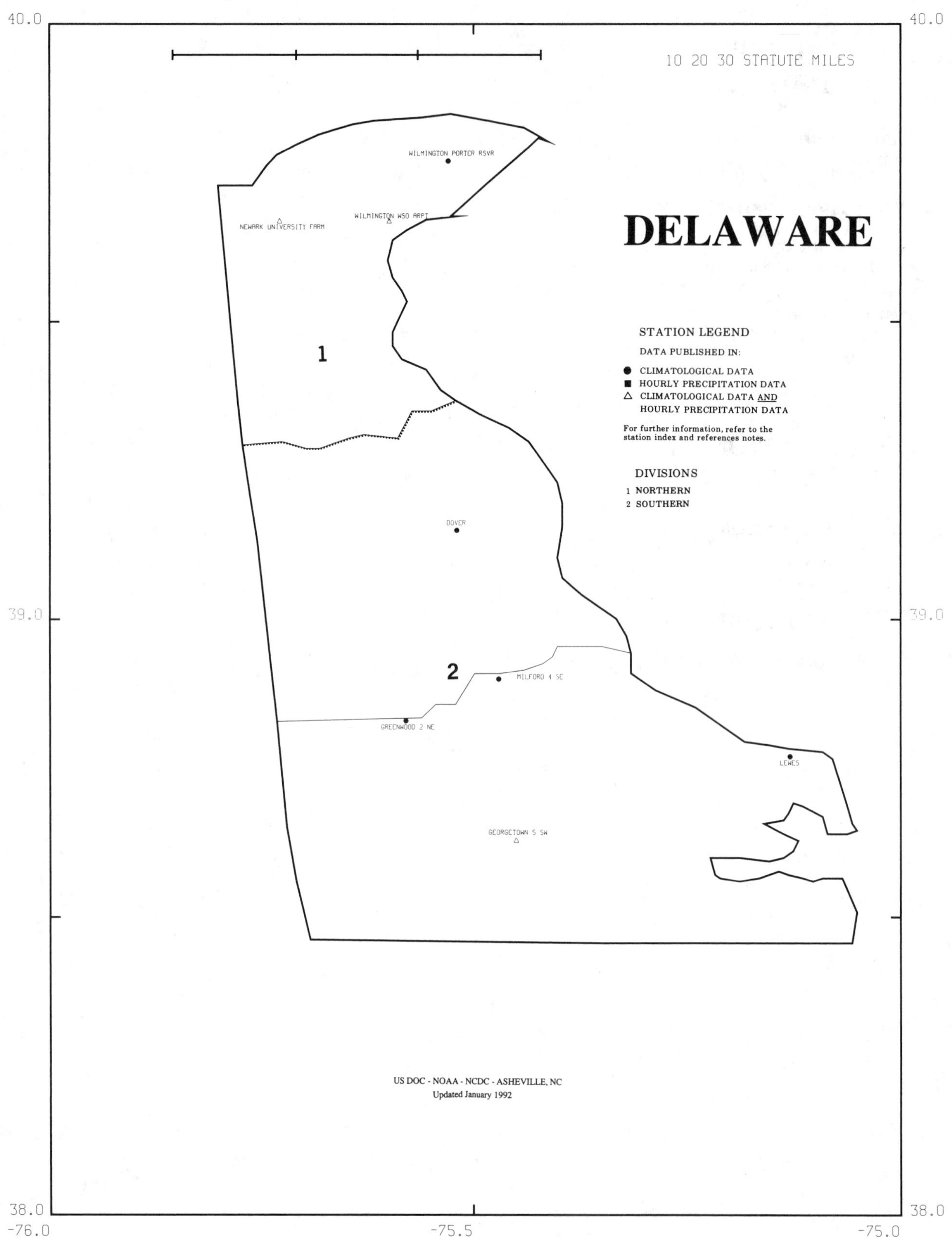
40.0
40.0
10 20 30 STATUTE MILES
WILMINGTON PORTER RSVR
NEWARK UNIVERSITY FARM
WILMINGTON WSO ARPT
DELAWARE
1
STATION LEGEND
DATA PUBLISHED IN:
CLIMATOLOGICAL DATA
HOURLY PRECIPITATION DATA
CLIMATOLOGICAL DATA AND HOURLY PRECIPITATION DATA
For further information, refer to the station index and references notes.
DIVISIONS
1 NORTHERN
2 SOUTHERN
DOVER
39.0
39.0
2
MILFORD 4 SE
GREENWOOD 2 NE
LEWES
GEORGETOWN 5 SW
US DOC - NOAA - NCDC - ASHEVILLE, NC
Updated January 1992
38.0
38.0
-76.0
-75.5
-75.0

## DOVER *Kent County* ELEVATION 30 ft LAT/LONG 39° 9 ' N / 75° 31 ' W

| | JAN | FEB | MAR | APR | MAY | JUN | JUL | AUG | SEP | OCT | NOV | DEC | YEAR |
|---|---|---|---|---|---|---|---|---|---|---|---|---|---|
| Maximum Temp °F | 42.8 | 46.0 | 54.9 | 65.5 | 75.3 | 83.6 | 87.7 | 86.0 | 80.1 | 69.1 | 59.0 | 48.3 | 66.5 |
| Minimum Temp °F | 25.5 | 27.1 | 34.6 | 43.1 | 53.1 | 62.1 | 67.3 | 65.9 | 59.1 | 47.5 | 39.2 | 30.6 | 46.3 |
| Mean Temp °F | 34.2 | 36.6 | 44.8 | 54.3 | 64.2 | 72.9 | 77.5 | 76.0 | 69.6 | 58.3 | 49.1 | 39.5 | 56.4 |
| Days Max Temp ≥ 90 °F | 0 | 0 | 0 | 0 | 2 | 6 | 12 | 9 | 3 | 0 | 0 | 0 | 32 |
| Days Max Temp ≤ 32 °F | 5 | 3 | 0 | 0 | 0 | 0 | 0 | 0 | 0 | 0 | 0 | 2 | 10 |
| Days Min Temp ≤ 32 °F | 23 | 20 | 13 | 2 | 0 | 0 | 0 | 0 | 0 | 2 | 8 | 19 | 87 |
| Days Min Temp ≤ 0 °F | 0 | 0 | 0 | 0 | 0 | 0 | 0 | 0 | 0 | 0 | 0 | 0 | 0 |
| Heating Degree Days | 948 | 796 | 623 | 328 | 102 | 9 | 0 | 1 | 30 | 229 | 474 | 784 | 4324 |
| Cooling Degree Days | 0 | 0 | 3 | 15 | 96 | 264 | 425 | 348 | 177 | 33 | 3 | 0 | 1364 |
| Total Precipitation (") | 3.45 | 2.82 | 4.11 | 3.38 | 4.11 | 3.65 | 4.30 | 4.61 | 4.47 | 3.23 | 2.98 | 3.62 | 44.73 |
| Days ≥ 0.1" Precip | 6 | 6 | 7 | 7 | 8 | 6 | 7 | 7 | 6 | 5 | 6 | 7 | 78 |
| Total Snowfall (") | 5.3 | 6.8 | 1.9 | 0.0 | 0.0 | 0.0 | 0.0 | 0.0 | 0.0 | 0.0 | 0.6 | 2.2 | 16.8 |
| Days ≥ 1" Snow Depth | 4 | 4 | 1 | 0 | 0 | 0 | 0 | 0 | 0 | 0 | 0 | 2 | 11 |

## GEORGETOWN 5 SW *Sussex County* ELEVATION 49 ft LAT/LONG 38° 38 ' N / 75° 28 ' W

| | JAN | FEB | MAR | APR | MAY | JUN | JUL | AUG | SEP | OCT | NOV | DEC | YEAR |
|---|---|---|---|---|---|---|---|---|---|---|---|---|---|
| Maximum Temp °F | 42.6 | 45.1 | 53.9 | 64.2 | 73.5 | 82.2 | 86.7 | 84.9 | 79.0 | 68.0 | 58.3 | 48.0 | 65.5 |
| Minimum Temp °F | 24.0 | 25.8 | 33.2 | 41.2 | 51.4 | 60.4 | 65.4 | 64.0 | 56.6 | 44.9 | 37.0 | 28.8 | 44.4 |
| Mean Temp °F | 33.3 | 35.5 | 43.6 | 52.7 | 62.4 | 71.4 | 76.1 | 74.5 | 67.8 | 56.5 | 47.7 | 38.4 | 55.0 |
| Days Max Temp ≥ 90 °F | 0 | 0 | 0 | 0 | 1 | 5 | 11 | 6 | 2 | 0 | 0 | 0 | 25 |
| Days Max Temp ≤ 32 °F | 6 | 4 | 1 | 0 | 0 | 0 | 0 | 0 | 0 | 0 | 0 | 2 | 13 |
| Days Min Temp ≤ 32 °F | 25 | 21 | 15 | 5 | 0 | 0 | 0 | 0 | 0 | 4 | 10 | 20 | 100 |
| Days Min Temp ≤ 0 °F | 1 | 0 | 0 | 0 | 0 | 0 | 0 | 0 | 0 | 0 | 0 | 0 | 1 |
| Heating Degree Days | 975 | 827 | 659 | 373 | 137 | 20 | 1 | 5 | 51 | 279 | 516 | 817 | 4660 |
| Cooling Degree Days | 0 | 0 | 2 | 11 | 70 | 227 | 374 | 297 | 141 | 25 | 2 | 0 | 1149 |
| Total Precipitation (") | 3.59 | 3.04 | 4.08 | 3.44 | 3.73 | 3.49 | 3.75 | 5.85 | 3.44 | 3.28 | 3.06 | 3.51 | 44.26 |
| Days ≥ 0.1" Precip | 7 | 6 | 7 | 7 | 7 | 6 | 6 | 7 | 5 | 5 | 6 | 6 | 75 |
| Total Snowfall (") | 5.0 | *5.2* | 1.6 | 0.1 | 0.0 | 0.0 | 0.0 | 0.0 | 0.0 | 0.0 | 0.3 | 1.8 | 14.0 |
| Days ≥ 1" Snow Depth | *4* | *4* | 1 | 0 | 0 | 0 | 0 | 0 | 0 | 0 | 0 | 2 | 11 |

## LEWES *Sussex County* ELEVATION 10 ft LAT/LONG 38° 46 ' N / 75° 8 ' W

| | JAN | FEB | MAR | APR | MAY | JUN | JUL | AUG | SEP | OCT | NOV | DEC | YEAR |
|---|---|---|---|---|---|---|---|---|---|---|---|---|---|
| Maximum Temp °F | 43.1 | 45.5 | 53.5 | 63.5 | 72.8 | 81.2 | 84.9 | 83.9 | 78.1 | 67.8 | 58.8 | 48.9 | 65.2 |
| Minimum Temp °F | 26.6 | 28.3 | 35.1 | 43.2 | 53.1 | 62.4 | 67.5 | 66.4 | 60.3 | 48.9 | 40.7 | 31.9 | 47.0 |
| Mean Temp °F | 34.9 | 36.9 | 44.3 | 53.4 | 63.0 | 71.8 | 76.2 | 75.2 | 69.2 | 58.4 | 49.8 | 40.5 | 56.1 |
| Days Max Temp ≥ 90 °F | 0 | 0 | 0 | 0 | 1 | 4 | 8 | 5 | 2 | 0 | 0 | 0 | 20 |
| Days Max Temp ≤ 32 °F | 5 | 3 | 0 | 0 | 0 | 0 | 0 | 0 | 0 | 0 | 0 | 2 | 10 |
| Days Min Temp ≤ 32 °F | 22 | 19 | 12 | 3 | 0 | 0 | 0 | 0 | 0 | 1 | 7 | 17 | 81 |
| Days Min Temp ≤ 0 °F | 0 | 0 | 0 | 0 | 0 | 0 | 0 | 0 | 0 | 0 | 0 | 0 | 0 |
| Heating Degree Days | 928 | 786 | 639 | 355 | 125 | 15 | 1 | 2 | 29 | 225 | 455 | 754 | 4314 |
| Cooling Degree Days | 0 | 0 | 5 | 16 | 90 | 263 | 407 | 343 | 181 | 35 | 6 | 1 | 1347 |
| Total Precipitation (") | 3.69 | 3.19 | 4.16 | 3.45 | 3.76 | 3.33 | 4.13 | 5.90 | 3.18 | 3.09 | 3.17 | 3.81 | 44.86 |
| Days ≥ 0.1" Precip | 7 | 6 | 7 | 6 | 7 | 6 | 7 | 7 | 5 | 5 | 6 | 6 | 75 |
| Total Snowfall (") | 4.5 | 4.0 | 2.0 | 0.0 | 0.0 | 0.0 | 0.0 | 0.0 | 0.0 | 0.0 | 0.4 | 1.7 | 12.6 |
| Days ≥ 1" Snow Depth | 3 | 3 | 1 | 0 | 0 | 0 | 0 | 0 | 0 | 0 | 0 | 1 | 8 |

## NEWARK UNIVERSITY FA *New Castle County* ELEVATION 112 ft LAT/LONG 39° 39 ' N / 75° 46 ' W

| | JAN | FEB | MAR | APR | MAY | JUN | JUL | AUG | SEP | OCT | NOV | DEC | YEAR |
|---|---|---|---|---|---|---|---|---|---|---|---|---|---|
| Maximum Temp °F | 40.3 | 44.2 | 53.6 | 65.0 | 75.3 | 83.5 | 87.3 | 85.7 | 79.3 | 67.7 | 56.7 | 45.6 | 65.4 |
| Minimum Temp °F | 22.8 | 24.6 | 32.1 | 40.8 | 51.0 | 60.0 | 65.3 | 63.8 | 56.6 | 44.6 | 36.5 | 28.0 | 43.8 |
| Mean Temp °F | 31.6 | 34.4 | 42.8 | 52.9 | 63.2 | 71.8 | 76.3 | 74.8 | 68.0 | 56.2 | 46.6 | 36.8 | 54.6 |
| Days Max Temp ≥ 90 °F | 0 | 0 | 0 | 0 | 1 | 6 | 10 | 7 | 2 | 0 | 0 | 0 | 26 |
| Days Max Temp ≤ 32 °F | 6 | 4 | 0 | 0 | 0 | 0 | 0 | 0 | 0 | 0 | 0 | 3 | 13 |
| Days Min Temp ≤ 32 °F | 26 | 22 | 17 | 5 | 0 | 0 | 0 | 0 | 0 | 3 | 10 | 22 | 105 |
| Days Min Temp ≤ 0 °F | 1 | 0 | 0 | 0 | 0 | 0 | 0 | 0 | 0 | 0 | 0 | 0 | 1 |
| Heating Degree Days | 1029 | 856 | 681 | 365 | 119 | 13 | 1 | 2 | 47 | 283 | 546 | 867 | 4809 |
| Cooling Degree Days | 0 | 0 | 1 | 10 | 77 | 234 | 385 | 311 | 143 | 19 | 1 | 0 | 1181 |
| Total Precipitation (") | 2.98 | 2.59 | 3.63 | 3.44 | 4.33 | 3.93 | 4.89 | 4.00 | 3.94 | 3.20 | 3.25 | 3.62 | 43.80 |
| Days ≥ 0.1" Precip | 6 | 5 | 7 | 7 | 7 | 6 | 7 | 6 | 5 | 5 | 6 | 6 | 73 |
| Total Snowfall (") | 5.9 | 4.0 | 1.7 | 0.0 | 0.0 | 0.0 | 0.0 | 0.0 | 0.0 | 0.0 | 0.3 | 1.8 | 13.7 |
| Days ≥ 1" Snow Depth | 7 | 4 | 1 | 0 | 0 | 0 | 0 | 0 | 0 | 0 | 0 | 2 | 14 |

## WILMINGTN NEW CASTLE *New Castle County* ELEVATION 79 ft LAT/LONG 39° 40' N / 75° 36' W

| | JAN | FEB | MAR | APR | MAY | JUN | JUL | AUG | SEP | OCT | NOV | DEC | YEAR |
|---|---|---|---|---|---|---|---|---|---|---|---|---|---|
| Maximum Temp °F | 39.0 | 42.4 | 51.9 | 63.1 | 73.0 | 81.6 | 85.9 | 84.3 | 77.6 | 66.2 | 55.6 | 44.7 | 63.8 |
| Minimum Temp °F | 22.8 | 25.0 | 33.1 | 42.1 | 52.4 | 61.8 | 67.5 | 66.0 | 58.2 | 45.7 | 37.2 | 28.3 | 45.0 |
| Mean Temp °F | 30.9 | 33.7 | 42.5 | 52.6 | 62.7 | 71.8 | 76.7 | 75.2 | 68.0 | 56.0 | 46.4 | 36.5 | 54.4 |
| Days Max Temp ≥ 90 °F | 0 | 0 | 0 | 0 | 1 | 4 | 8 | 5 | 2 | 0 | 0 | 0 | 20 |
| Days Max Temp ≤ 32 °F | 8 | 5 | 1 | 0 | 0 | 0 | 0 | 0 | 0 | 0 | 0 | 3 | 17 |
| Days Min Temp ≤ 32 °F | 26 | 22 | 14 | 3 | 0 | 0 | 0 | 0 | 0 | 2 | 10 | 21 | 98 |
| Days Min Temp ≤ 0 °F | 1 | 0 | 0 | 0 | 0 | 0 | 0 | 0 | 0 | 0 | 0 | 0 | 1 |
| Heating Degree Days | 1049 | 877 | 692 | 373 | 130 | 12 | 1 | 3 | 47 | 290 | 552 | 876 | 4902 |
| Cooling Degree Days | 0 | 0 | 1 | 11 | 76 | 232 | 396 | 317 | 137 | 19 | 0 | 0 | 1189 |
| Total Precipitation (") | 3.08 | 2.81 | 3.70 | 3.33 | 4.12 | 3.54 | 4.47 | 3.55 | 3.62 | 2.91 | 3.12 | 3.55 | 41.80 |
| Days ≥ 0.1" Precip | 6 | 5 | 7 | 6 | 7 | 6 | 6 | 6 | 5 | 5 | 6 | 6 | 71 |
| Total Snowfall (") | 6.9 | 6.2 | 2.5 | 0.3 | 0.0 | 0.0 | 0.0 | 0.0 | 0.0 | 0.1 | 0.6 | 2.7 | 19.3 |
| Days ≥ 1" Snow Depth | 8 | 5 | 2 | 0 | 0 | 0 | 0 | 0 | 0 | 0 | 0 | 2 | 17 |

## WILMNGTON PORTER RSV *New Castle County* ELEVATION 259 ft LAT/LONG 39° 46' N / 75° 32' W

| | JAN | FEB | MAR | APR | MAY | JUN | JUL | AUG | SEP | OCT | NOV | DEC | YEAR |
|---|---|---|---|---|---|---|---|---|---|---|---|---|---|
| Maximum Temp °F | 38.1 | 41.3 | 50.5 | 61.8 | 72.1 | 80.7 | 84.8 | 83.2 | 76.2 | 64.9 | 54.4 | 43.7 | 62.6 |
| Minimum Temp °F | 22.8 | 24.9 | 32.7 | 42.0 | 52.2 | 61.4 | 66.5 | 64.9 | 57.5 | 45.9 | 37.4 | 28.4 | 44.7 |
| Mean Temp °F | 30.5 | 33.2 | 41.6 | 51.9 | 62.2 | 71.1 | 75.6 | 74.1 | 66.9 | 55.4 | 45.9 | 36.1 | 53.7 |
| Days Max Temp ≥ 90 °F | 0 | 0 | 0 | 0 | 1 | 3 | 6 | 4 | 1 | 0 | 0 | 0 | 15 |
| Days Max Temp ≤ 32 °F | 9 | 6 | 1 | 0 | 0 | 0 | 0 | 0 | 0 | 0 | 0 | 4 | 20 |
| Days Min Temp ≤ 32 °F | 26 | 22 | 15 | 3 | 0 | 0 | 0 | 0 | 0 | 1 | 9 | 21 | 97 |
| Days Min Temp ≤ 0 °F | 1 | 0 | 0 | 0 | 0 | 0 | 0 | 0 | 0 | 0 | 0 | 0 | 1 |
| Heating Degree Days | 1064 | 892 | 718 | 393 | 139 | 15 | 1 | 3 | 58 | 304 | 568 | 889 | 5044 |
| Cooling Degree Days | 0 | 0 | 2 | 9 | 74 | 228 | 379 | 304 | 129 | 17 | 1 | 0 | 1143 |
| Total Precipitation (") | 3.64 | 3.01 | 4.12 | 3.96 | 4.62 | 4.07 | 4.78 | 4.04 | 4.28 | 3.34 | 3.66 | 3.96 | 47.48 |
| Days ≥ 0.1" Precip | 7 | 5 | 7 | 7 | 7 | 6 | 6 | 6 | 6 | 5 | 6 | 6 | 74 |
| Total Snowfall (") | 5.9 | 5.8 | 2.0 | 0.1 | 0.0 | 0.0 | 0.0 | 0.0 | 0.0 | 0.0 | 0.6 | 1.9 | 16.3 |
| Days ≥ 1" Snow Depth | 6 | 5 | 1 | 0 | 0 | 0 | 0 | 0 | 0 | 0 | 0 | 2 | 14 |

## JANUARY MINIMUM TEMPERATURE °F

### LOWEST

| | | |
|---|---|---|
| 1 | Newark | 22.8 |
| | Wilmingtn-New C | 22.8 |
| | Wilmington-Prter | 22.8 |
| 4 | Georgetown | 24.0 |
| 5 | Dover | 25.5 |
| 6 | Lewes | 26.6 |

### HIGHEST

| | | |
|---|---|---|
| 1 | Lewes | 26.6 |
| 2 | Dover | 25.5 |
| 3 | Georgetown | 24.0 |
| 4 | Newark | 22.8 |
| | Wilmingtn-New C | 22.8 |
| | Wilmington-Prter | 22.8 |

## JULY MAXIMUM TEMPERATURE °F

### HIGHEST

| | | |
|---|---|---|
| 1 | Dover | 87.7 |
| 2 | Newark | 87.3 |
| 3 | Georgetown | 86.7 |
| 4 | Wilmingtn-New C | 85.9 |
| 5 | Lewes | 84.9 |
| 6 | Wilmington-Prter | 84.8 |

### LOWEST

| | | |
|---|---|---|
| 1 | Wilmington-Prter | 84.8 |
| 2 | Lewes | 84.9 |
| 3 | Wilmingtn-New C | 85.9 |
| 4 | Georgetown | 86.7 |
| 5 | Newark | 87.3 |
| 6 | Dover | 87.7 |

## ANNUAL PRECIPITATION (")

### HIGHEST

| | | |
|---|---|---|
| 1 | Wilmington-Prter | 47.48 |
| 2 | Lewes | 44.86 |
| 3 | Dover | 44.73 |
| 4 | Georgetown | 44.26 |
| 5 | Newark | 43.80 |
| 6 | Wilmingtn-New C | 41.80 |

### LOWEST

| | | |
|---|---|---|
| 1 | Wilmingtn-New C | 41.80 |
| 2 | Newark | 43.80 |
| 3 | Georgetown | 44.26 |
| 4 | Dover | 44.73 |
| 5 | Lewes | 44.86 |
| 6 | Wilmington-Prter | 47.48 |

## ANNUAL SNOWFALL (")

### HIGHEST

| | | |
|---|---|---|
| 1 | Wilmingtn-New C | 19.3 |
| 2 | Dover | 16.8 |
| 3 | Wilmington-Prter | 16.3 |
| 4 | Georgetown | 14.0 |
| 5 | Newark | 13.7 |
| 6 | Lewes | 12.6 |

### LOWEST

| | | |
|---|---|---|
| 1 | Lewes | 12.6 |
| 2 | Newark | 13.7 |
| 3 | Georgetown | 14.0 |
| 4 | Wilmington-Prter | 16.3 |
| 5 | Dover | 16.8 |
| 6 | Wilmingtn-New C | 19.3 |

# FLORIDA

PHYSICAL FEATURES. Florida, situated between latitudes 24° 30' and 31° N, and longitudes 80° and 87° 30' W, is largely a lowland peninsula comprising about 54,100 square miles of land area and is surrounded on three sides by the waters of the Atlantic Ocean and the Gulf of Mexico. Countless shallow lakes, which exist particularly on the peninsula and range in size from small cypress ponds to that of Lake Okeechobee, account for approximately 4,400 square miles of additional water area.

No point in the State is more than 70 miles from salt water, and the highest natural land in the Northwest Division is only 345 feet above sea level. Coastal areas are low and flat and are indented by many small bays or inlets. Many small islands dot the shorelines. The elevation of most of the interior ranges from 50 to 100 feet above sea level, though gentle hills in the interior of the peninsula and across the northern and western portions of the State rise above 200 feet.

A large portion of the southern one-third of the peninsula is the swampland known as the Everglades. An ill-defined divide of low, rolling hills, extending north-to-south near the middle of the peninsula and terminating north of Lake Okeechobee, gives rise to most peninsula streams, chains of lakes, and many springs. Stream gradients are slight and often insufficient to handle the runoff following heavy rainfall. Consequently, there are sizable areas of swamp and marshland near these streams.

GENERAL CLIMATE. Climate is probably Florida's greatest natural resource. General climatic conditions range from a zone of transition between temperate and subtropical conditions in the extreme northern interior portion of the State to the tropical conditions found on the Florida Keys. The chief factors of climatic control are: latitude, proximity to the Atlantic Ocean and Gulf of Mexico, and numerous inland lakes.

Summers throughout the State are long, warm, and relatively humid; winters, although punctuated with periodic invasions of cool to occasionally cold air from the north, are mild because of the southern latitude and relatively warm adjacent ocean waters. The Gulf Stream, which flows around the western tip of Cuba, through the Straits of Florida, and northward along the lower east coast, exerts a warming influence to the southern east coast largely because the predominate wind direction is from the east. Coastal stations throughout the State average slightly warmer in winter and cooler in summer than do inland stations at the same latitude.

Florida enjoys abundant rainfall. Except for the northwestern portion of the State, the average year can be divided into two seasons--the so-called "rainy season" and the long, relatively dry season. On the peninsula, generally more than one-half of the precipitation for an average year can be expected to fall during the 4-month period, June through September. In northwest Florida, there is a secondary rainfall maximum in late winter and in early spring.

The summer heat is tempered by sea breezes along the coast and by frequent afternoon or early evening thunderstorms in all areas. During the warm season, sea breezes are felt almost daily within several miles of the coast and occasionally 20 to 30 miles inland. Thundershowers, which on the average occur about one-half of the days in summer, frequently are accompanied by as much as a rapid 10° to 20° drop in temperature, resulting in comfortable weather for the remainder of the day. Gentle breezes occur almost daily in all areas and serve to mitigate further the oppressiveness that otherwise would accompany the prevailing summer temperature and humidity conditions. Because most of the large-scale wind patterns affecting Florida have passed over water surfaces, hot drying winds seldom occur.

Most of the summer rainfall is derived from "local" showers or thundershowers. Many stations average more than 80 thundershowers per year, and some average more than 100. Showers are often heavy, usually lasting only 1 or 2 hours, and generally occur near the hottest part of the day. The more severe thundershowers are occasionally attended by hail or locally strong winds which may inflict serious local damage to crops and property. Day-long summer rains are usually associated with tropical disturbances and are infrequent. Even in the wet season, the rainfall duration is generally less than 10 percent of the time.

DROUGHTS. Florida is not immune from drought, even though annual rainfall amounts are relatively large. Prolonged periods of deficient rainfall are occasionally experienced even during the time of the expected rainy season. Several such dry periods, in the course of 1 or 2 years, can lead to significantly lowered water tables and lake levels which, in turn, may cause serious water shortages for those communities that depend upon lakes and shallow wells for their water supply. Statewide droughts during summer are rare, but it is not unusual during a drought in one portion of the State for other portions to receive generous rainfall. In a few instances, individual stations have experienced periods of a month or more without rainfall.

SNOW. Snowfall in Florida is unusual, although measurable amounts have fallen in the northern portions at irregular intervals, and a trace of snow has been recorded as far south as Fort Myers.

WIND. Prevailing winds over the southern peninsula are southeast and east. Over the remainder of the State, wind directions are influenced locally by convectional forces inland and by the land-and-sea-breeze-effect near the coast. Consequently, prevailing directions are somewhat erratic, but, in general, follow a pattern from the north in winter and from the south in summer. The windiest months are March and April. High local winds of short duration occur occasionally in connection with thunderstorms in summer and with cold fronts moving across the State in other seasons. Tornadoes, funnel clouds, and waterspouts also occur, averaging 10 to 15 in a year. Tornadoes have occurred in all seasons, but are most frequent in spring; they also occur in connection with tropical storms. Generally, tornado paths in Florida are short. Occasionally, waterspouts come inland, but they usually dissipate soon after reaching land and affect only very small areas.

TROPICAL STORMS. Storms that produce high winds and are often destructive are usually tropical in origin. Florida, jutting out into the ocean between the subtropical Atlantic and the Gulf of Mexico, is the most exposed of all States to these storms. In particular, hurricanes can approach from the Atlantic Ocean to the east, from the Caribbean Sea to the south, and from the Gulf of Mexico to the west.

The vulnerability of the State to tropical storms varies with the progress of the hurricane season. In August and early September, tropical storms normally approach the State from the east or southeast, but as the season progresses into late September and October, the region of maximum hurricane activity (insofar as Florida is concerned) shifts to the western Caribbean. Most of those storms that move into Florida approach the State from the south or southwest, entering the Keys, the Miami area, or along the west coast. Some of the world's heaviest rainfalls have occurred within tropical cyclones. Rainfall over 20 inches in 24 hours is not uncommon. The intensity of the rainfall, however, does not seem to bear any relation to the intensity of the wind circulation.

OTHER CLIMATIC ELEMENTS. The climate of Florida is humid. Inland areas with greater temperature extremes enjoy slightly lower relative humidity, especially during times of hot weather. On the average, variations in relative humidity from one place to another are small; humidities range from about 50 to 65 percent during the afternoon hours to about 85 to 95 percent during the night and early morning hours.

Heavy fogs are usually confined to the night and early morning hours in the late fall, winter, and early spring months. On the average, they occur about 35 to 40 days in a year over the extreme northern portion; about 25 to 30 days in a year over the central portion; and less than 10 days in a year over the extreme southern portion of the State. These fogs usually dissipate or thin soon after sunrise; heavy daytime fog is seldom observed in Florida.

Florida has been nicknamed the Sunshine State. Sunshine measurements made at widely separated stations in the State indicate the sun shines about two-thirds of the possible sunlight hours during the year, ranging from slightly more than 60 percent of possible in December and January to more than 70 percent of possible in April and May. In general, southern Florida enjoys a higher percentage of possible sunshine hours than does northern Florida. The length of day operates to Florida's advantage. In winter, when sunshine is highly valued, the sun can shine longer in Florida than in the more northern latitudes. In summer, the situation reverses itself with longer days returning to the north.

## COUNTY INDEX

*Alachua County*
HIGH SPRINGS

*Baker County*
GLEN ST MARY 1 W

*Brevard County*
MELBOURNE REGIONL AP
TITUSVILLE

*Broward County*
FORT LAUDERDALE
POMPANO BEACH

*Charlotte County*
PUNTA GORDA 4 ESE

*Citrus County*
INVERNESS 3 SE

*Collier County*
EVERGLADES
IMMOKALEE 3 NNW
NAPLES

*Columbia County*
LAKE CITY 2 E

*Dade County*
HIALEAH
MIAMI BEACH
MIAMI INTL ARPT
TAMIAMI TRAIL 40 MI

*DeSoto County*
ARCADIA

*Dixie County*
CROSS CITY 2 WNW
STEINHATCHEE 6 ENE

*Duval County*
JACKSONVILLE BEACH
JACKSONVILLE INTL AP

*Escambia County*
PENSACOLA REGIONL AP

*Franklin County*
APALACHICOLA MUNI AP

*Gadsden County*
QUINCY 3 SSW

*Glades County*
MOORE HAVEN LOCK 1

*Gulf County*
WEWAHITCHKA

*Hamilton County*
JASPER

*Hardee County*
WAUCHULA 2 N

*Hendry County*
CLEWISTON US ENG
DEVILS GARDEN
LA BELLE

*Hernando County*
BROOKSVILLE CHIN HIL

*Highlands County*
AVON PARK 2 W

*Hillsborough County*
PLANT CITY
TAMPA INTL ARPT

*Indian River County*
VERO BEACH 4 W

*Jefferson County*
MONTICELLO 3 W

*Lafayette County*
MAYO

*Lake County*
CLERMONT 7 S
LISBON

*Lee County*
FORT MYERS PAGE FLD

*Leon County*
TALLAHASSEE MUNI AP

*Levy County*
USHER TOWER

*Madison County*
MADISON 4 N

*Manatee County*
BRADENTON 5 ESE
PARRISH

*Marion County*
OCALA

*Martin County*
STUART 1 N

*Monroe County*
FLAMINGO RANGER STN
KEY WEST INTL ARPT
TAVERNIER

*Nassau County*
FERNANDINA BEACH

*Okaloosa County*
NICEVILLE

*Okeechobee County*
FORT DRUM 5 NW
OKEECHOBEE HRCN GATE

*Osceola County*
KISSIMMEE 2

*Palm Beach County*
BELLE GLADE EXP STN
CANAL POINT USDA
W PALM BEACH INTL AP

*Pasco County*
ST LEO

*Pinellas County*
ST PETERSBURG WHITTD
TARPON SPNGS SWG PLT

*Polk County*
BARTOW
LAKE ALFRED EXP STN
LAKELAND
MOUNTAIN LAKE
WINTER HAVEN

***Putnam County***
FEDERAL POINT

***St. Lucie County***
FORT PIERCE

***Santa Rosa County***
MILTON EXPERIMENT ST

***Sarasota County***
MYAKKA RIVER STATE P
VENICE

***Sumter County***
BUSHNELL 2 E

***Suwannee County***
LIVE OAK

***Taylor County***
PERRY

***Volusia County***
DAYTONA BEACH REG AP
DELAND 1 SSE

***Walton County***
DE FUNIAK SPRINGS

***Washington County***
CHIPLEY 3 E

## ELEVATION INDEX

| FEET | STATION NAME |
|---|---|
| 3 | FLAMINGO RANGER STN |
| 4 | KEY WEST INTL ARPT |
| 5 | MIAMI BEACH |
| 6 | EVERGLADES |
| 7 | TAVERNIER |
| 8 | ST PETERSBURG WHITTD |
| 10 | FEDERAL POINT |
| 10 | FORT LAUDERDALE |
| 10 | FORT PIERCE |
| 10 | JACKSONVILLE BEACH |
| 10 | NAPLES |
| 10 | PUNTA GORDA 4 ESE |
| 10 | STUART 1 N |
| 10 | TAMIAMI TRAIL 40 MI |
| 10 | VENICE |
| 12 | HIALEAH |
| 12 | LA BELLE |
| 13 | TITUSVILLE |
| 15 | FORT MYERS PAGE FLD |
| 15 | POMPANO BEACH |
| 16 | BELLE GLADE EXP STN |
| 19 | APALACHICOLA MUNI AP |
| 20 | BRADENTON 5 ESE |
| 20 | CLEWISTON US ENG |
| 20 | DEVILS GARDEN |
| 20 | FERNANDINA BEACH |
| 20 | MOORE HAVEN LOCK 1 |
| 20 | MYAKKA RIVER STATE P |
| 20 | TARPON SPNGS SWG PLT |
| 20 | VERO BEACH 4 W |
| 20 | W PALM BEACH INTL AP |
| 23 | MIAMI INTL ARPT |
| 26 | JACKSONVILLE INTL AP |
| 30 | CANAL POINT USDA |
| 30 | DELAND 1 SSE |
| 30 | OKEECHOBEE HRCN GATE |
| 30 | STEINHATCHEE 6 ENE |
| 30 | USHER TOWER |
| 33 | DAYTONA BEACH REG AP |
| 35 | MELBOURNE REGIONL AP |
| 36 | TAMPA INTL ARPT |
| 39 | IMMOKALEE 3 NNW |
| 39 | PARRISH |
| 46 | PERRY |
| 49 | CROSS CITY 2 WNW |
| 49 | WEWAHITCHKA |
| 55 | TALLAHASSEE MUNI AP |
| 59 | FORT DRUM 5 NW |
| 60 | NICEVILLE |
| 65 | MAYO |
| 69 | ARCADIA |
| 69 | HIGH SPRINGS |
| 69 | INVERNESS 3 SE |
| 69 | KISSIMMEE 2 |
| 69 | LISBON |
| 75 | BUSHNELL 2 E |
| 89 | OCALA |
| 102 | LIVE OAK |
| 112 | PENSACOLA REGIONL AP |
| 120 | PLANT CITY |
| 121 | BARTOW |
| 121 | WAUCHULA 2 N |
| 130 | CHIPLEY 3 E |
| 131 | CLERMONT 7 S |
| 141 | GLEN ST MARY 1 W |
| 147 | JASPER |
| 151 | AVON PARK 2 W |
| 151 | LAKE ALFRED EXP STN |
| 151 | LAKELAND |
| 161 | MOUNTAIN LAKE |
| 180 | MADISON 4 N |
| 180 | WINTER HAVEN |
| 200 | ST LEO |
| 210 | LAKE CITY 2 E |
| 210 | MONTICELLO 3 W |
| 217 | MILTON EXPERIMENT ST |
| 249 | QUINCY 3 SSW |
| 259 | DE FUNIAK SPRINGS |
| 279 | BROOKSVILLE CHIN HIL |

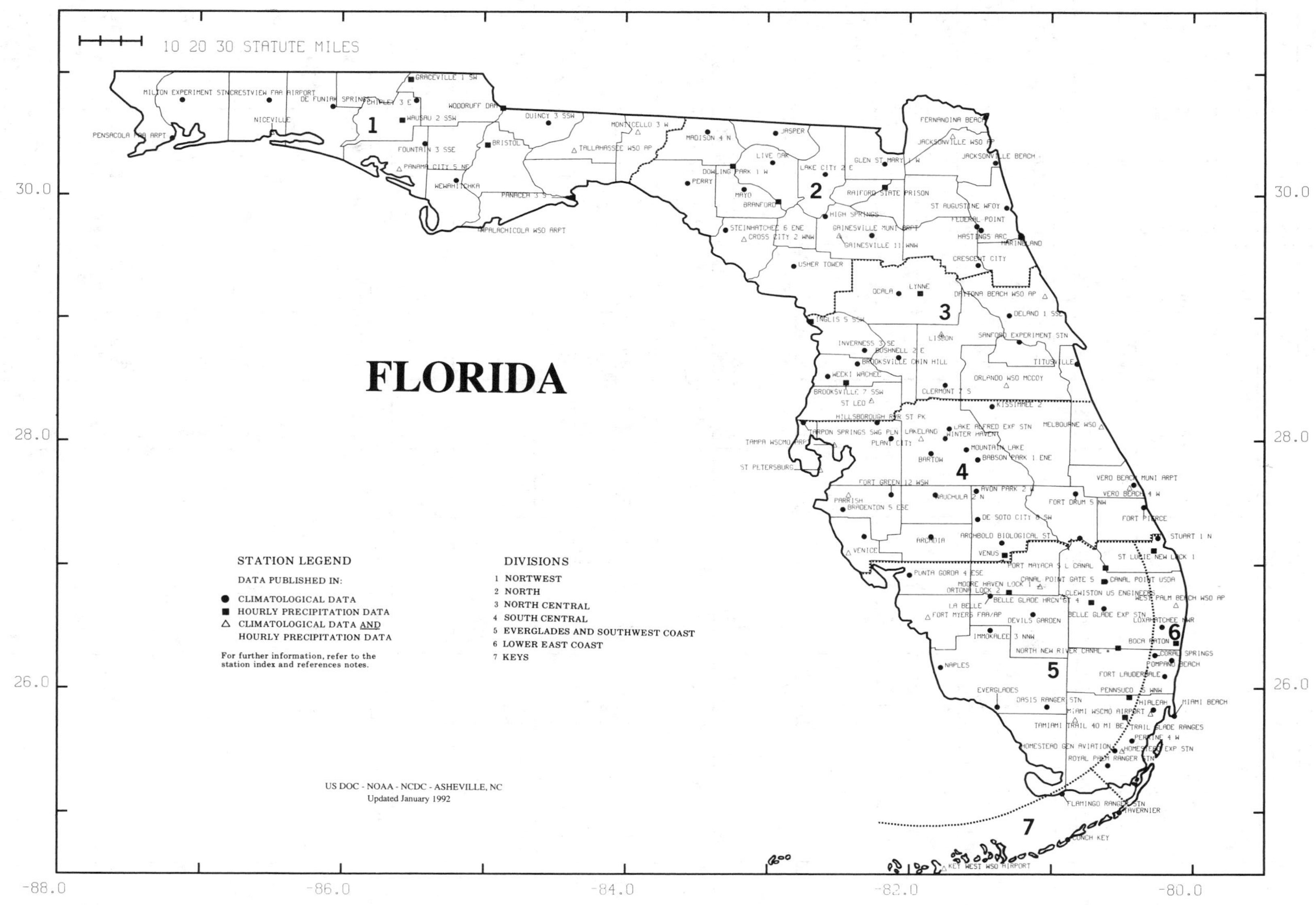
FLORIDA
10 20 30 STATUTE MILES
STATION LEGEND
DATA PUBLISHED IN:
● CLIMATOLOGICAL DATA
■ HOURLY PRECIPITATION DATA
△ CLIMATOLOGICAL DATA AND HOURLY PRECIPITATION DATA
For further information, refer to the station index and references notes.
DIVISIONS
1 NORTWEST
2 NORTH
3 NORTH CENTRAL
4 SOUTH CENTRAL
5 EVERGLADES AND SOUTHWEST COAST
6 LOWER EAST COAST
7 KEYS
US DOC - NOAA - NCDC - ASHEVILLE, NC
Updated January 1992
-88.0
-86.0
-84.0
-82.0
-80.0
26.0
28.0
30.0

## APALACHICOLA MUNI AP *Franklin County* ELEVATION 19 ft LAT/LONG 29° 44 ' N / 85° 2 ' W

| | JAN | FEB | MAR | APR | MAY | JUN | JUL | AUG | SEP | OCT | NOV | DEC | YEAR |
|---|---|---|---|---|---|---|---|---|---|---|---|---|---|
| Maximum Temp °F | 61.2 | 63.4 | 69.0 | 75.9 | 82.3 | 87.7 | 88.9 | 88.6 | 86.2 | 79.2 | 71.0 | 64.4 | 76.5 |
| Minimum Temp °F | 44.3 | 46.2 | 52.4 | 59.0 | 65.8 | 72.0 | 74.1 | 74.1 | 71.2 | 61.5 | 52.9 | 46.8 | 60.0 |
| Mean Temp °F | 52.8 | 54.8 | 60.7 | 67.5 | 74.1 | 79.8 | 81.6 | 81.3 | 78.7 | 70.3 | 62.0 | 55.6 | 68.3 |
| Days Max Temp ≥ 90 °F | 0 | 0 | 0 | 0 | 1 | 9 | 13 | 12 | 5 | 0 | 0 | 0 | 40 |
| Days Max Temp ≤ 32 °F | 0 | 0 | 0 | 0 | 0 | 0 | 0 | 0 | 0 | 0 | 0 | 0 | 0 |
| Days Min Temp ≤ 32 °F | 5 | 2 | 0 | 0 | 0 | 0 | 0 | 0 | 0 | 0 | 0 | 3 | 10 |
| Days Min Temp ≤ 0 °F | 0 | 0 | 0 | 0 | 0 | 0 | 0 | 0 | 0 | 0 | 0 | 0 | 0 |
| Heating Degree Days | 378 | 288 | 159 | 36 | 1 | 0 | 0 | 0 | 0 | 22 | 142 | 297 | 1323 |
| Cooling Degree Days | 6 | 10 | 36 | 100 | 281 | 457 | 528 | 520 | 414 | 195 | 70 | 19 | 2636 |
| Total Precipitation (") | 4.70 | 3.76 | 4.80 | 2.71 | 2.93 | 4.76 | 7.59 | 8.31 | 7.27 | 3.81 | 3.15 | 3.83 | 57.62 |
| Days ≥ 0.1" Precip | 7 | 5 | 6 | 4 | 4 | 6 | 10 | 10 | 8 | 4 | 4 | 5 | 73 |
| Total Snowfall (") | 0.0 | 0.0 | 0.0 | 0.0 | 0.0 | 0.0 | 0.0 | 0.0 | 0.0 | 0.0 | 0.0 | 0.0 | 0.0 |
| Days ≥ 1" Snow Depth | 0 | 0 | 0 | 0 | 0 | 0 | 0 | 0 | 0 | 0 | 0 | 0 | 0 |

## ARCADIA *DeSoto County* ELEVATION 69 ft LAT/LONG 27° 13 ' N / 81° 51 ' W

| | JAN | FEB | MAR | APR | MAY | JUN | JUL | AUG | SEP | OCT | NOV | DEC | YEAR |
|---|---|---|---|---|---|---|---|---|---|---|---|---|---|
| Maximum Temp °F | 74.0 | 75.3 | 80.0 | 84.9 | 89.4 | 91.0 | 91.8 | 91.4 | 89.7 | 85.4 | 79.7 | 75.1 | 84.0 |
| Minimum Temp °F | 49.0 | 49.1 | 53.8 | 57.3 | 63.0 | 68.5 | 70.5 | 70.9 | 70.1 | 64.2 | 56.5 | 51.1 | 60.3 |
| Mean Temp °F | 61.5 | 62.2 | 67.0 | 71.2 | 76.4 | 79.8 | 81.2 | 81.2 | 79.9 | 74.8 | 68.1 | 63.2 | 72.2 |
| Days Max Temp ≥ 90 °F | 0 | 0 | 2 | 6 | 16 | 21 | 26 | 26 | 19 | 5 | 0 | 0 | 121 |
| Days Max Temp ≤ 32 °F | 0 | 0 | 0 | 0 | 0 | 0 | 0 | 0 | 0 | 0 | 0 | 0 | 0 |
| Days Min Temp ≤ 32 °F | 3 | 2 | 0 | 0 | 0 | 0 | 0 | 0 | 0 | 0 | 0 | 2 | 7 |
| Days Min Temp ≤ 0 °F | 0 | 0 | 0 | 0 | 0 | 0 | 0 | 0 | 0 | 0 | 0 | 0 | 0 |
| Heating Degree Days | 160 | 128 | 59 | 11 | 0 | 0 | 0 | 0 | 0 | 3 | 44 | 126 | 531 |
| Cooling Degree Days | 52 | *60* | 120 | 184 | 359 | 461 | 518 | 508 | 452 | 324 | 164 | 78 | 3280 |
| Total Precipitation (") | 1.95 | 2.79 | 3.08 | 1.41 | 3.87 | 8.61 | 7.76 | 7.71 | 6.59 | 3.38 | 2.01 | 1.82 | 50.98 |
| Days ≥ 0.1" Precip | 3 | 4 | 4 | 2 | 5 | 10 | 11 | 11 | 9 | 5 | 3 | 3 | 70 |
| Total Snowfall (") | 0.0 | 0.0 | 0.0 | 0.0 | 0.0 | 0.0 | 0.0 | 0.0 | 0.0 | 0.0 | 0.0 | 0.0 | 0.0 |
| Days ≥ 1" Snow Depth | 0 | 0 | 0 | 0 | 0 | 0 | 0 | 0 | 0 | 0 | 0 | 0 | 0 |

## AVON PARK 2 W *Highlands County* ELEVATION 151 ft LAT/LONG 27° 36 ' N / 81° 30 ' W

| | JAN | FEB | MAR | APR | MAY | JUN | JUL | AUG | SEP | OCT | NOV | DEC | YEAR |
|---|---|---|---|---|---|---|---|---|---|---|---|---|---|
| Maximum Temp °F | 72.6 | 75.2 | 79.5 | 84.8 | 88.9 | 91.0 | 92.1 | 91.8 | 90.2 | 85.4 | 79.5 | 74.7 | 83.8 |
| Minimum Temp °F | 48.0 | 50.2 | 54.6 | 59.2 | 64.9 | 70.0 | 71.4 | 71.7 | 70.7 | 64.5 | 57.0 | 51.0 | 61.1 |
| Mean Temp °F | 60.3 | 62.7 | 67.1 | 72.0 | 76.9 | 80.5 | 81.8 | 81.8 | 80.5 | 75.0 | 68.2 | 62.9 | 72.5 |
| Days Max Temp ≥ 90 °F | 0 | 0 | 1 | 6 | 14 | 21 | 26 | 26 | 20 | 6 | 0 | 0 | 120 |
| Days Max Temp ≤ 32 °F | 0 | 0 | 0 | 0 | 0 | 0 | 0 | 0 | 0 | 0 | 0 | 0 | 0 |
| Days Min Temp ≤ 32 °F | 3 | 1 | 0 | 0 | 0 | 0 | 0 | 0 | 0 | 0 | 0 | 1 | 5 |
| Days Min Temp ≤ 0 °F | 0 | 0 | 0 | 0 | 0 | 0 | 0 | 0 | 0 | 0 | 0 | 0 | 0 |
| Heating Degree Days | 193 | 123 | 61 | 11 | 0 | 0 | 0 | 0 | 0 | 4 | 46 | 135 | 573 |
| Cooling Degree Days | 49 | 74 | 125 | 199 | 360 | 474 | 525 | 525 | 462 | 317 | 158 | 74 | 3342 |
| Total Precipitation (") | 2.31 | 2.59 | 3.03 | 1.90 | 3.78 | 8.95 | 7.57 | 7.28 | 6.03 | 3.00 | 1.77 | 1.71 | 49.92 |
| Days ≥ 0.1" Precip | 4 | 4 | 4 | 3 | 6 | 11 | 11 | 12 | 9 | 5 | 4 | 3 | 76 |
| Total Snowfall (") | 0.0 | 0.0 | 0.0 | 0.0 | 0.0 | 0.0 | 0.0 | 0.0 | 0.0 | 0.0 | 0.0 | 0.0 | 0.0 |
| Days ≥ 1" Snow Depth | 0 | 0 | 0 | 0 | 0 | 0 | 0 | 0 | 0 | 0 | 0 | 0 | 0 |

## BARTOW *Polk County* ELEVATION 121 ft LAT/LONG 27° 54 ' N / 81° 50 ' W

| | JAN | FEB | MAR | APR | MAY | JUN | JUL | AUG | SEP | OCT | NOV | DEC | YEAR |
|---|---|---|---|---|---|---|---|---|---|---|---|---|---|
| Maximum Temp °F | 73.3 | 74.8 | 79.8 | 84.6 | 88.9 | 91.3 | 92.3 | 92.2 | 90.3 | 85.3 | 79.5 | 74.5 | 83.9 |
| Minimum Temp °F | 49.8 | 50.9 | 55.4 | 60.1 | 65.7 | 70.8 | 72.1 | 72.3 | 71.3 | 64.5 | 57.9 | 51.7 | 61.9 |
| Mean Temp °F | 61.6 | 62.9 | 67.6 | 72.4 | 77.3 | 81.1 | 82.2 | 82.3 | 80.8 | 75.0 | 68.7 | 63.1 | 72.9 |
| Days Max Temp ≥ 90 °F | 0 | 0 | 1 | 5 | 14 | 23 | 27 | 27 | 21 | 4 | 0 | 0 | 122 |
| Days Max Temp ≤ 32 °F | 0 | 0 | 0 | 0 | 0 | 0 | 0 | 0 | 0 | 0 | 0 | 0 | 0 |
| Days Min Temp ≤ 32 °F | 2 | 1 | 0 | 0 | 0 | 0 | 0 | 0 | 0 | 0 | 0 | 1 | 4 |
| Days Min Temp ≤ 0 °F | 0 | 0 | 0 | 0 | 0 | 0 | 0 | 0 | 0 | 0 | 0 | 0 | 0 |
| Heating Degree Days | 160 | 121 | 51 | 6 | 0 | 0 | 0 | 0 | 0 | 3 | 41 | 126 | 508 |
| Cooling Degree Days | 64 | 84 | 144 | 225 | 391 | 510 | 557 | 563 | 495 | 332 | 185 | 83 | 3633 |
| Total Precipitation (") | 2.49 | 2.90 | 3.21 | 2.21 | 4.18 | 6.98 | 8.26 | 7.01 | 6.25 | 2.60 | 2.02 | 2.17 | 50.28 |
| Days ≥ 0.1" Precip | 4 | 4 | 5 | 3 | 6 | 9 | 12 | 11 | 9 | 4 | 4 | 4 | 75 |
| Total Snowfall (") | 0.0 | 0.0 | 0.0 | 0.0 | 0.0 | 0.0 | 0.0 | 0.0 | 0.0 | 0.0 | 0.0 | 0.0 | 0.0 |
| Days ≥ 1" Snow Depth | 0 | 0 | 0 | 0 | 0 | 0 | 0 | 0 | 0 | 0 | 0 | 0 | 0 |

## BELLE GLADE EXP STN *Palm Beach County* ELEVATION 16 ft LAT/LONG 26° 40 ' N / 80° 38 ' W

| | JAN | FEB | MAR | APR | MAY | JUN | JUL | AUG | SEP | OCT | NOV | DEC | YEAR |
|---|---|---|---|---|---|---|---|---|---|---|---|---|---|
| Maximum Temp °F | 74.7 | 75.9 | 79.6 | 83.4 | 86.8 | 89.4 | 90.8 | 90.9 | 89.4 | 85.6 | 80.4 | 76.0 | 83.6 |
| Minimum Temp °F | 51.6 | 52.2 | 56.2 | 59.2 | 64.6 | 69.7 | 70.9 | 70.9 | 70.3 | 65.4 | 59.4 | 53.8 | 62.0 |
| Mean Temp °F | 63.2 | 64.1 | 68.0 | 71.4 | 75.7 | 79.6 | 80.9 | 80.9 | 79.9 | 75.5 | 69.9 | 64.9 | 72.8 |
| Days Max Temp ≥ 90 °F | 0 | 0 | 1 | 2 | 6 | 16 | 23 | 24 | 17 | 4 | 0 | 0 | 93 |
| Days Max Temp ≤ 32 °F | 0 | 0 | 0 | 0 | 0 | 0 | 0 | 0 | 0 | 0 | 0 | 0 | 0 |
| Days Min Temp ≤ 32 °F | 1 | 0 | 0 | 0 | 0 | 0 | 0 | 0 | 0 | 0 | 0 | 1 | 2 |
| Days Min Temp ≤ 0 °F | 0 | 0 | 0 | 0 | 0 | 0 | 0 | 0 | 0 | 0 | 0 | 0 | 0 |
| Heating Degree Days | 126 | 97 | 44 | 11 | 0 | 0 | 0 | 0 | 0 | 2 | 27 | 92 | 399 |
| Cooling Degree Days | 77 | 92 | 144 | 194 | 330 | 457 | 505 | 508 | 451 | 343 | 201 | 101 | 3403 |
| Total Precipitation (") | 2.62 | 2.02 | 2.90 | 2.41 | 5.15 | 7.97 | 7.83 | 7.27 | 7.25 | 3.78 | 2.34 | 1.69 | 53.23 |
| Days ≥ 0.1" Precip | 4 | 4 | 4 | 3 | 7 | 11 | 11 | 11 | 11 | 6 | 4 | 3 | 79 |
| Total Snowfall (") | 0.0 | 0.0 | 0.0 | 0.0 | 0.0 | 0.0 | 0.0 | 0.0 | 0.0 | 0.0 | 0.0 | 0.0 | 0.0 |
| Days ≥ 1" Snow Depth | 0 | 0 | 0 | 0 | 0 | 0 | 0 | 0 | 0 | 0 | 0 | 0 | 0 |

## BRADENTON 5 ESE *Manatee County* ELEVATION 20 ft LAT/LONG 27° 27 ' N / 82° 28 ' W

| | JAN | FEB | MAR | APR | MAY | JUN | JUL | AUG | SEP | OCT | NOV | DEC | YEAR |
|---|---|---|---|---|---|---|---|---|---|---|---|---|---|
| Maximum Temp °F | 72.3 | 73.5 | 77.6 | 82.3 | 87.1 | 90.1 | 91.2 | 91.2 | 89.7 | 85.1 | 79.2 | 74.3 | 82.8 |
| Minimum Temp °F | 50.0 | 51.1 | 55.6 | 59.3 | 64.8 | 70.5 | 72.2 | 72.6 | 71.6 | 64.9 | 57.7 | 52.0 | 61.9 |
| Mean Temp °F | 61.2 | 62.3 | 66.6 | 70.9 | 76.0 | 80.3 | 81.7 | 81.9 | 80.7 | 75.0 | 68.5 | 63.2 | 72.4 |
| Days Max Temp ≥ 90 °F | 0 | 0 | 0 | 1 | 8 | 19 | 24 | 25 | 18 | 4 | 0 | 0 | 99 |
| Days Max Temp ≤ 32 °F | 0 | 0 | 0 | 0 | 0 | 0 | 0 | 0 | 0 | 0 | 0 | 0 | 0 |
| Days Min Temp ≤ 32 °F | 2 | 1 | 0 | 0 | 0 | 0 | 0 | 0 | 0 | 0 | 0 | 1 | 4 |
| Days Min Temp ≤ 0 °F | 0 | 0 | 0 | 0 | 0 | 0 | 0 | 0 | 0 | 0 | 0 | 0 | 0 |
| Heating Degree Days | 174 | 130 | 62 | 12 | 0 | 0 | 0 | 0 | 0 | 3 | 46 | 128 | 555 |
| Cooling Degree Days | 63 | 78 | 128 | 198 | 359 | 487 | 541 | 552 | 488 | 336 | 186 | 86 | 3502 |
| Total Precipitation (") | 2.72 | 2.75 | 3.29 | 1.46 | 2.78 | 8.07 | 8.92 | 9.59 | 7.27 | 2.92 | 1.94 | 2.21 | 53.92 |
| Days ≥ 0.1" Precip | 4 | 4 | 4 | 3 | 4 | 9 | 12 | 13 | 10 | 4 | 3 | 3 | 73 |
| Total Snowfall (") | 0.0 | 0.0 | 0.0 | 0.0 | 0.0 | 0.0 | 0.0 | 0.0 | 0.0 | 0.0 | 0.0 | 0.0 | 0.0 |
| Days ≥ 1" Snow Depth | 0 | 0 | 0 | 0 | 0 | 0 | 0 | 0 | 0 | 0 | 0 | 0 | 0 |

## BROOKSVILLE CHIN HIL *Hernando County* ELEVATION 279 ft LAT/LONG 28° 37 ' N / 82° 22 ' W

| | JAN | FEB | MAR | APR | MAY | JUN | JUL | AUG | SEP | OCT | NOV | DEC | YEAR |
|---|---|---|---|---|---|---|---|---|---|---|---|---|---|
| Maximum Temp °F | 70.6 | 72.0 | 77.9 | 82.6 | *87.5* | 89.8 | 90.5 | 90.0 | 88.9 | 83.7 | 77.6 | 72.5 | 82.0 |
| Minimum Temp °F | 48.6 | 49.3 | 54.8 | 59.4 | *64.7* | 70.0 | 71.6 | 71.6 | 70.4 | 63.8 | 56.0 | 50.7 | 60.9 |
| Mean Temp °F | 59.6 | 60.7 | 66.4 | 71.0 | *76.2* | 79.9 | 81.0 | 80.9 | 79.7 | 73.8 | 66.8 | 61.6 | 71.5 |
| Days Max Temp ≥ 90 °F | 0 | 0 | 0 | 2 | 9 | *17* | *19* | 21 | 14 | 1 | 0 | 0 | 83 |
| Days Max Temp ≤ 32 °F | 0 | 0 | 0 | 0 | 0 | 0 | 0 | 0 | 0 | 0 | 0 | 0 | 0 |
| Days Min Temp ≤ 32 °F | 2 | 1 | 0 | 0 | 0 | 0 | 0 | 0 | 0 | 0 | 0 | 1 | 4 |
| Days Min Temp ≤ 0 °F | 0 | 0 | 0 | 0 | 0 | 0 | 0 | 0 | 0 | 0 | 0 | 0 | 0 |
| Heating Degree Days | 209 | 160 | 68 | 13 | *0* | 0 | 0 | 0 | 0 | 4 | 63 | 157 | 674 |
| Cooling Degree Days | 45 | 53 | 118 | 185 | na | 461 | 509 | *509* | *450* | *284* | 142 | 62 | na |
| Total Precipitation (") | 3.09 | 3.55 | 4.27 | 2.44 | 3.69 | 7.49 | 7.18 | 8.84 | 6.00 | 2.15 | 2.19 | 2.46 | 53.35 |
| Days ≥ 0.1" Precip | 5 | 4 | 5 | 3 | 5 | 9 | 10 | 12 | 8 | 4 | 3 | 4 | 72 |
| Total Snowfall (") | 0.0 | 0.0 | 0.0 | 0.0 | 0.0 | 0.0 | 0.0 | 0.0 | 0.0 | 0.0 | 0.0 | 0.0 | 0.0 |
| Days ≥ 1" Snow Depth | 0 | 0 | 0 | 0 | 0 | 0 | 0 | 0 | 0 | 0 | 0 | 0 | 0 |

## BUSHNELL 2 E *Sumter County* ELEVATION 75 ft LAT/LONG 28° 40 ' N / 82° 5 ' W

| | JAN | FEB | MAR | APR | MAY | JUN | JUL | AUG | SEP | OCT | NOV | DEC | YEAR |
|---|---|---|---|---|---|---|---|---|---|---|---|---|---|
| Maximum Temp °F | 70.8 | 73.0 | 78.6 | 83.8 | 88.1 | 90.6 | 91.7 | 91.1 | 89.7 | 84.6 | 78.6 | 72.9 | 82.8 |
| Minimum Temp °F | 45.4 | 46.5 | 51.8 | 56.3 | 62.2 | 68.8 | 70.2 | *70.6* | 69.0 | 60.9 | 53.6 | 48.5 | 58.7 |
| Mean Temp °F | 58.1 | 59.8 | 65.2 | 70.1 | 75.2 | 79.7 | 81.0 | *80.9* | 79.3 | 72.7 | 66.1 | 60.7 | 70.7 |
| Days Max Temp ≥ 90 °F | 0 | 0 | 0 | 3 | 11 | 20 | 25 | 23 | 18 | 4 | 1 | 0 | 105 |
| Days Max Temp ≤ 32 °F | 0 | 0 | 0 | 0 | 0 | 0 | 0 | 0 | 0 | 0 | 0 | 0 | 0 |
| Days Min Temp ≤ 32 °F | 5 | 4 | 1 | 0 | 0 | 0 | 0 | 0 | 0 | 0 | 1 | 3 | 14 |
| Days Min Temp ≤ 0 °F | 0 | 0 | 0 | 0 | 0 | 0 | 0 | 0 | 0 | 0 | 0 | 0 | 0 |
| Heating Degree Days | 245 | 182 | 88 | 19 | 1 | 0 | 0 | *0* | 0 | 10 | 74 | 181 | 800 |
| Cooling Degree Days | 28 | 50 | 98 | 169 | *323* | 463 | 510 | na | 441 | 262 | 141 | 62 | na |
| Total Precipitation (") | 3.11 | 3.44 | 3.91 | 2.20 | 4.11 | 6.67 | 6.87 | 8.03 | 5.50 | 2.15 | 2.14 | 2.33 | 50.46 |
| Days ≥ 0.1" Precip | 5 | 4 | 5 | 3 | 5 | 9 | 10 | 11 | 8 | 4 | 3 | 4 | 71 |
| Total Snowfall (") | 0.0 | 0.0 | 0.0 | 0.0 | 0.0 | 0.0 | 0.0 | 0.0 | 0.0 | 0.0 | 0.0 | 0.0 | 0.0 |
| Days ≥ 1" Snow Depth | 0 | 0 | 0 | 0 | 0 | 0 | 0 | 0 | 0 | 0 | 0 | 0 | 0 |

## CANAL POINT USDA *Palm Beach County* ELEVATION 30 ft LAT/LONG 26° 52 ' N / 80° 38 ' W

| | JAN | FEB | MAR | APR | MAY | JUN | JUL | AUG | SEP | OCT | NOV | DEC | YEAR |
|---|---|---|---|---|---|---|---|---|---|---|---|---|---|
| Maximum Temp °F | 74.3 | 75.5 | 79.6 | 84.1 | 87.8 | 90.2 | 91.7 | 91.4 | 90.1 | 86.2 | 80.9 | 76.0 | 84.0 |
| Minimum Temp °F | 52.6 | 53.6 | 57.3 | 60.6 | 65.1 | 69.6 | 70.5 | 70.8 | 70.3 | 66.8 | 60.7 | 55.0 | 62.7 |
| Mean Temp °F | 63.5 | 64.6 | 68.4 | 72.4 | 76.5 | 79.9 | 81.1 | 81.1 | 80.3 | 76.5 | 70.8 | 65.5 | 73.4 |
| Days Max Temp ≥ 90 °F | 0 | 0 | 1 | 3 | 10 | 19 | 26 | 26 | 20 | 7 | 0 | 0 | 112 |
| Days Max Temp ≤ 32 °F | 0 | 0 | 0 | 0 | 0 | 0 | 0 | 0 | 0 | 0 | 0 | 0 | 0 |
| Days Min Temp ≤ 32 °F | 0 | 0 | 0 | 0 | 0 | 0 | 0 | 0 | 0 | 0 | 0 | 0 | 0 |
| Days Min Temp ≤ 0 °F | 0 | 0 | 0 | 0 | 0 | 0 | 0 | 0 | 0 | 0 | 0 | 0 | 0 |
| Heating Degree Days | 109 | 81 | 32 | 4 | 0 | 0 | 0 | 0 | 0 | 1 | 17 | 76 | 320 |
| Cooling Degree Days | 73 | 98 | 154 | 230 | 362 | 470 | 516 | 518 | 473 | 381 | 228 | 111 | 3614 |
| Total Precipitation (") | 2.48 | 2.45 | 3.36 | 2.12 | 4.87 | 7.92 | 6.39 | 6.80 | 7.21 | 4.28 | 2.52 | 1.87 | 52.27 |
| Days ≥ 0.1" Precip | 4 | 4 | 5 | 3 | 6 | 10 | 11 | 11 | 10 | 6 | 4 | 4 | 78 |
| Total Snowfall (") | 0.0 | 0.0 | 0.0 | 0.0 | 0.0 | 0.0 | 0.0 | 0.0 | 0.0 | 0.0 | 0.0 | 0.0 | 0.0 |
| Days ≥ 1" Snow Depth | 0 | 0 | 0 | 0 | 0 | 0 | 0 | 0 | 0 | 0 | 0 | 0 | 0 |

## CHIPLEY 3 E *Washington County* ELEVATION 130 ft LAT/LONG 30° 47 ' N / 85° 29 ' W

| | JAN | FEB | MAR | APR | MAY | JUN | JUL | AUG | SEP | OCT | NOV | DEC | YEAR |
|---|---|---|---|---|---|---|---|---|---|---|---|---|---|
| Maximum Temp °F | 60.6 | 64.3 | 71.8 | 79.2 | 85.0 | 89.8 | 91.2 | 90.5 | 87.6 | 79.7 | 70.8 | 63.9 | 77.9 |
| Minimum Temp °F | 37.9 | 40.2 | 46.8 | 53.4 | 60.6 | 67.7 | 70.5 | 69.9 | 65.6 | 53.8 | 45.4 | 39.8 | 54.3 |
| Mean Temp °F | 49.3 | 52.3 | 59.3 | 66.3 | 72.8 | 78.8 | 80.9 | 80.2 | 76.7 | 66.8 | 58.1 | 51.9 | 66.1 |
| Days Max Temp ≥ 90 °F | 0 | 0 | 0 | 1 | 6 | 17 | 22 | 20 | 13 | 1 | 0 | 0 | 80 |
| Days Max Temp ≤ 32 °F | 0 | 0 | 0 | 0 | 0 | 0 | 0 | 0 | 0 | 0 | 0 | 0 | 0 |
| Days Min Temp ≤ 32 °F | 11 | 8 | 2 | 0 | 0 | 0 | 0 | 0 | 0 | 0 | 4 | 10 | 35 |
| Days Min Temp ≤ 0 °F | 0 | 0 | 0 | 0 | 0 | 0 | 0 | 0 | 0 | 0 | 0 | 0 | 0 |
| Heating Degree Days | 488 | 362 | 209 | 65 | 5 | 0 | 0 | 0 | 3 | 66 | 234 | 416 | 1848 |
| Cooling Degree Days | 4 | 11 | 35 | 90 | 247 | 422 | 502 | 480 | 338 | 125 | 36 | 13 | 2303 |
| Total Precipitation (") | 5.76 | 5.13 | 5.98 | 3.47 | 3.96 | 5.60 | 6.85 | 5.53 | 4.08 | 3.02 | 3.49 | 3.94 | 56.81 |
| Days ≥ 0.1" Precip | 7 | 6 | 7 | 5 | 6 | 8 | 10 | 9 | 6 | 4 | 5 | 6 | 79 |
| Total Snowfall (") | 0.1 | 0.0 | 0.0 | 0.0 | 0.0 | 0.0 | 0.0 | 0.0 | 0.0 | 0.0 | 0.0 | 0.0 | 0.1 |
| Days ≥ 1" Snow Depth | 0 | 0 | 0 | 0 | 0 | 0 | 0 | 0 | 0 | 0 | 0 | 0 | 0 |

## CLERMONT 7 S *Lake County* ELEVATION 131 ft LAT/LONG 28° 29 ' N / 81° 47 ' W

| | JAN | FEB | MAR | APR | MAY | JUN | JUL | AUG | SEP | OCT | NOV | DEC | YEAR |
|---|---|---|---|---|---|---|---|---|---|---|---|---|---|
| Maximum Temp °F | 70.1 | 72.3 | 78.1 | 83.6 | 87.8 | 90.1 | 91.4 | 91.0 | 89.0 | 83.4 | 76.8 | 71.6 | 82.1 |
| Minimum Temp °F | 49.4 | 50.2 | 55.2 | 59.5 | 65.1 | 70.4 | 71.9 | 72.3 | 71.4 | 65.1 | 57.5 | 51.5 | 61.6 |
| Mean Temp °F | 59.8 | 61.3 | 66.7 | 71.6 | 76.5 | 80.3 | 81.7 | 81.7 | 80.3 | 74.3 | 67.2 | 61.5 | 71.9 |
| Days Max Temp ≥ 90 °F | 0 | 0 | 0 | 3 | 11 | 18 | 23 | 22 | 16 | 2 | 0 | 0 | 95 |
| Days Max Temp ≤ 32 °F | 0 | 0 | 0 | 0 | 0 | 0 | 0 | 0 | 0 | 0 | 0 | 0 | 0 |
| Days Min Temp ≤ 32 °F | 2 | 1 | 0 | 0 | 0 | 0 | 0 | 0 | 0 | 0 | 0 | 1 | 4 |
| Days Min Temp ≤ 0 °F | 0 | 0 | 0 | 0 | 0 | 0 | 0 | 0 | 0 | 0 | 0 | 0 | 0 |
| Heating Degree Days | 200 | 148 | 63 | 11 | 0 | 0 | 0 | 0 | 0 | 5 | 57 | 156 | 640 |
| Cooling Degree Days | 45 | 60 | 118 | 188 | 347 | 468 | 533 | 526 | 459 | 301 | 151 | 62 | 3258 |
| Total Precipitation (") | 2.99 | 2.92 | 3.75 | 1.94 | 3.79 | 7.96 | 7.34 | 7.63 | 6.11 | 2.41 | 2.38 | 2.11 | 51.33 |
| Days ≥ 0.1" Precip | 5 | 5 | 5 | 3 | 6 | 10 | 12 | 11 | 8 | 4 | 3 | 4 | 76 |
| Total Snowfall (") | 0.0 | 0.0 | 0.0 | 0.0 | 0.0 | 0.0 | 0.0 | 0.0 | 0.0 | 0.0 | 0.0 | 0.0 | 0.0 |
| Days ≥ 1" Snow Depth | 0 | 0 | 0 | 0 | 0 | 0 | 0 | 0 | 0 | 0 | 0 | 0 | 0 |

## CLEWISTON US ENG *Hendry County* ELEVATION 20 ft LAT/LONG 26° 45 ' N / 80° 55 ' W

| | JAN | FEB | MAR | APR | MAY | JUN | JUL | AUG | SEP | OCT | NOV | DEC | YEAR |
|---|---|---|---|---|---|---|---|---|---|---|---|---|---|
| Maximum Temp °F | 73.4 | 74.9 | 79.3 | 83.8 | 87.5 | 89.9 | 91.6 | 91.3 | 89.6 | 85.1 | 79.6 | 74.7 | 83.4 |
| Minimum Temp °F | 54.4 | 55.0 | 59.0 | 62.9 | 67.5 | 71.7 | 72.6 | 72.9 | 73.1 | 69.3 | 63.1 | 56.9 | 64.9 |
| Mean Temp °F | 63.9 | 65.0 | 69.2 | 73.4 | 77.5 | 80.8 | 82.1 | 82.1 | 81.4 | 77.2 | 71.4 | 65.9 | 74.2 |
| Days Max Temp ≥ 90 °F | 0 | 0 | 2 | 5 | 9 | 17 | 25 | 25 | 17 | 4 | 1 | 0 | 105 |
| Days Max Temp ≤ 32 °F | 0 | 0 | 0 | 0 | 0 | 0 | 0 | 0 | 0 | 0 | 0 | 0 | 0 |
| Days Min Temp ≤ 32 °F | 0 | 0 | 0 | 0 | 0 | 0 | 0 | 0 | 0 | 0 | 0 | 0 | 0 |
| Days Min Temp ≤ 0 °F | 0 | 0 | 0 | 0 | 0 | 0 | 0 | 0 | 0 | 0 | 0 | 0 | 0 |
| Heating Degree Days | 115 | 86 | 35 | 5 | 0 | 0 | 0 | 0 | 0 | 1 | 20 | 81 | 343 |
| Cooling Degree Days | 87 | 112 | 175 | 256 | 403 | 500 | 549 | 550 | 502 | 397 | 245 | 119 | 3895 |
| Total Precipitation (") | 2.09 | 2.21 | 3.02 | 1.85 | 4.71 | 7.33 | 6.41 | 6.61 | 5.10 | 3.04 | 2.26 | 1.44 | 46.07 |
| Days ≥ 0.1" Precip | 4 | 4 | 4 | 3 | 6 | 10 | 10 | 10 | 9 | 5 | 3 | 3 | 71 |
| Total Snowfall (") | 0.0 | 0.0 | 0.0 | 0.0 | 0.0 | 0.0 | 0.0 | 0.0 | 0.0 | 0.0 | 0.0 | 0.0 | 0.0 |
| Days ≥ 1" Snow Depth | 0 | 0 | 0 | 0 | 0 | 0 | 0 | 0 | 0 | 0 | 0 | 0 | 0 |

## CROSS CITY 2 WNW *Dixie County* ELEVATION 49 ft LAT/LONG 29° 38 ' N / 83° 7 ' W

| | JAN | FEB | MAR | APR | MAY | JUN | JUL | AUG | SEP | OCT | NOV | DEC | YEAR |
|---|---|---|---|---|---|---|---|---|---|---|---|---|---|
| Maximum Temp °F | 64.9 | 67.4 | 73.8 | 79.8 | 85.5 | 89.6 | 90.2 | ***90.1*** | 88.0 | 81.8 | 74.5 | 68.0 | 79.5 |
| Minimum Temp °F | 39.6 | 42.1 | 48.1 | 53.7 | 60.0 | 67.1 | 69.9 | 70.0 | 67.4 | 56.9 | 48.5 | 41.6 | 55.4 |
| Mean Temp °F | 52.4 | 54.8 | 61.0 | 66.8 | 72.7 | 78.5 | 80.1 | ***80.0*** | 77.7 | ***69.6*** | 61.7 | 55.0 | 67.5 |
| Days Max Temp ≥ 90 °F | 0 | 0 | 0 | 0 | 5 | 17 | ***19*** | ***19*** | 13 | 2 | 0 | 0 | 75 |
| Days Max Temp ≤ 32 °F | 0 | 0 | 0 | 0 | 0 | 0 | 0 | 0 | 0 | 0 | 0 | 0 | 0 |
| Days Min Temp ≤ 32 °F | 10 | 5 | 2 | 0 | 0 | 0 | 0 | 0 | 0 | 0 | 3 | 8 | 28 |
| Days Min Temp ≤ 0 °F | 0 | 0 | 0 | 0 | 0 | 0 | 0 | 0 | 0 | 0 | 0 | 0 | 0 |
| Heating Degree Days | 395 | 294 | 165 | 50 | 4 | 0 | 0 | ***0*** | 1 | ***36*** | 156 | 328 | 1429 |
| Cooling Degree Days | 8 | 18 | 47 | 93 | 251 | 418 | 478 | ***481*** | 379 | 191 | 74 | 22 | 2460 |
| Total Precipitation (") | 4.34 | 3.75 | 4.64 | 2.96 | 3.69 | 6.13 | 9.45 | 9.50 | 5.84 | 2.74 | 2.30 | 3.54 | 58.88 |
| Days ≥ 0.1" Precip | 6 | 6 | 5 | 4 | 5 | 9 | 12 | 12 | 8 | 4 | 4 | 5 | 80 |
| Total Snowfall (") | 0.0 | 0.0 | 0.0 | 0.0 | 0.0 | 0.0 | 0.0 | 0.0 | 0.0 | 0.0 | 0.0 | 0.0 | 0.0 |
| Days ≥ 1" Snow Depth | 0 | 0 | 0 | 0 | 0 | 0 | 0 | 0 | 0 | 0 | 0 | 0 | 0 |

## DAYTONA BEACH REG AP *Volusia County* ELEVATION 33 ft LAT/LONG 29° 11 ' N / 81° 3 ' W

| | JAN | FEB | MAR | APR | MAY | JUN | JUL | AUG | SEP | OCT | NOV | DEC | YEAR |
|---|---|---|---|---|---|---|---|---|---|---|---|---|---|
| Maximum Temp °F | 68.5 | 69.7 | 74.8 | 79.9 | 84.5 | 88.0 | 90.2 | 89.3 | 87.0 | 81.8 | 75.7 | 70.7 | 80.0 |
| Minimum Temp °F | 47.5 | 48.7 | 53.8 | 58.9 | 64.8 | 70.9 | 72.6 | 72.8 | 71.9 | 65.5 | 56.7 | 50.1 | 61.2 |
| Mean Temp °F | 58.0 | 59.3 | 64.3 | 69.4 | 74.7 | 79.5 | 81.4 | 81.1 | 79.5 | 73.7 | 66.2 | 60.4 | 70.6 |
| Days Max Temp ≥ 90 °F | 0 | 0 | 0 | 2 | 4 | 10 | 18 | 14 | 6 | 1 | 0 | 0 | 55 |
| Days Max Temp ≤ 32 °F | 0 | 0 | 0 | 0 | 0 | 0 | 0 | 0 | 0 | 0 | 0 | 0 | 0 |
| Days Min Temp ≤ 32 °F | 2 | 1 | 0 | 0 | 0 | 0 | 0 | 0 | 0 | 0 | 0 | 1 | 4 |
| Days Min Temp ≤ 0 °F | 0 | 0 | 0 | 0 | 0 | 0 | 0 | 0 | 0 | 0 | 0 | 0 | 0 |
| Heating Degree Days | 244 | 193 | 100 | 23 | 1 | 0 | 0 | 0 | 0 | 6 | 73 | 185 | 825 |
| Cooling Degree Days | 31 | 45 | 83 | 143 | 295 | 449 | 523 | 516 | 437 | 280 | 133 | 53 | 2988 |
| Total Precipitation (") | 2.87 | 2.98 | 3.24 | 2.42 | 3.51 | 6.16 | 5.21 | 6.06 | 6.07 | 4.34 | 2.88 | 2.60 | 48.34 |
| Days ≥ 0.1" Precip | 5 | 5 | 5 | 3 | 6 | 8 | 8 | 9 | 8 | 6 | 4 | 5 | 72 |
| Total Snowfall (") | 0.0 | 0.0 | 0.0 | 0.0 | 0.0 | 0.0 | 0.0 | 0.0 | 0.0 | 0.0 | 0.0 | 0.0 | 0.0 |
| Days ≥ 1" Snow Depth | 0 | 0 | 0 | 0 | 0 | 0 | 0 | 0 | 0 | 0 | 0 | 0 | 0 |

## DE FUNIAK SPRINGS *Walton County* ELEVATION 259 ft LAT/LONG 30° 43 ' N / 86° 7 ' W

| | JAN | FEB | MAR | APR | MAY | JUN | JUL | AUG | SEP | OCT | NOV | DEC | YEAR |
|---|---|---|---|---|---|---|---|---|---|---|---|---|---|
| Maximum Temp °F | 62.5 | 65.6 | 72.7 | 79.9 | 85.5 | 90.5 | 91.7 | 90.9 | 87.9 | 80.5 | 71.4 | 64.9 | 78.7 |
| Minimum Temp °F | 37.6 | 39.9 | 46.2 | 52.3 | 59.5 | 66.3 | 69.4 | 68.6 | 64.5 | 52.9 | 44.1 | 39.8 | 53.4 |
| Mean Temp °F | 50.1 | 52.8 | 59.5 | 66.3 | 72.5 | 78.4 | 80.5 | 79.8 | 76.2 | 66.7 | 57.7 | 52.4 | 66.1 |
| Days Max Temp ≥ 90 °F | 0 | 0 | 0 | 1 | 6 | 19 | 23 | 21 | 14 | 1 | 0 | 0 | 85 |
| Days Max Temp ≤ 32 °F | 0 | 0 | 0 | 0 | 0 | 0 | 0 | 0 | 0 | 0 | 0 | 0 | 0 |
| Days Min Temp ≤ 32 °F | 12 | 8 | 3 | 0 | 0 | 0 | 0 | 0 | 0 | 1 | 6 | 10 | 40 |
| Days Min Temp ≤ 0 °F | 0 | 0 | 0 | 0 | 0 | 0 | 0 | 0 | 0 | 0 | 0 | 0 | 0 |
| Heating Degree Days | 463 | 348 | 203 | 62 | 5 | 0 | 0 | 0 | 3 | 61 | 241 | 401 | 1787 |
| Cooling Degree Days | 3 | 11 | 34 | 88 | ***242*** | ***422*** | ***496*** | ***474*** | ***338*** | ***131*** | ***30*** | ***16*** | 2285 |
| Total Precipitation (") | 5.07 | 5.74 | 6.10 | 3.66 | 4.68 | 6.82 | 8.09 | 6.78 | 5.28 | 3.77 | 4.10 | 4.42 | 64.51 |
| Days ≥ 0.1" Precip | 7 | 7 | 7 | 5 | 6 | 9 | 11 | 10 | 7 | 3 | 5 | 6 | 83 |
| Total Snowfall (") | 0.0 | 0.1 | 0.0 | 0.0 | 0.0 | 0.0 | 0.0 | 0.0 | 0.0 | 0.0 | 0.0 | 0.0 | 0.1 |
| Days ≥ 1" Snow Depth | 0 | 0 | 0 | 0 | 0 | 0 | 0 | 0 | 0 | 0 | 0 | 0 | 0 |

## DELAND 1 SSE *Volusia County* ELEVATION 30 ft LAT/LONG 29° 1 ' N / 81° 18 ' W

| | JAN | FEB | MAR | APR | MAY | JUN | JUL | AUG | SEP | OCT | NOV | DEC | YEAR |
|---|---|---|---|---|---|---|---|---|---|---|---|---|---|
| Maximum Temp °F | 70.0 | 71.5 | 77.0 | 82.2 | 86.3 | 89.3 | 91.3 | 91.0 | 88.8 | ***83.0*** | 76.9 | 71.9 | 81.6 |
| Minimum Temp °F | 45.0 | 46.2 | 50.8 | 55.7 | 62.1 | 69.0 | 70.6 | 70.8 | 69.2 | 61.3 | 53.1 | 47.0 | 58.4 |
| Mean Temp °F | 57.5 | 58.8 | 63.9 | 69.0 | 74.2 | 79.1 | 81.0 | 81.0 | 79.0 | ***72.1*** | 65.0 | 59.5 | 70.0 |
| Days Max Temp ≥ 90 °F | 0 | 0 | 0 | 3 | 8 | 15 | 23 | 22 | 14 | 3 | 0 | 0 | 88 |
| Days Max Temp ≤ 32 °F | 0 | 0 | 0 | 0 | 0 | 0 | 0 | 0 | 0 | 0 | 0 | 0 | 0 |
| Days Min Temp ≤ 32 °F | 4 | 4 | 1 | 0 | 0 | 0 | 0 | 0 | 0 | 0 | 1 | 3 | 13 |
| Days Min Temp ≤ 0 °F | 0 | 0 | 0 | 0 | 0 | 0 | 0 | 0 | 0 | 0 | 0 | 0 | 0 |
| Heating Degree Days | 256 | 200 | 106 | 27 | 1 | 0 | 0 | 0 | 0 | ***11*** | 83 | 199 | 883 |
| Cooling Degree Days | 22 | 37 | 79 | 137 | 280 | 440 | 517 | 511 | 425 | ***251*** | ***105*** | 40 | 2844 |
| Total Precipitation (") | 3.17 | 3.41 | 3.63 | 2.71 | 4.51 | 8.06 | 7.81 | 7.85 | 7.19 | 4.28 | 2.61 | 2.44 | 57.67 |
| Days ≥ 0.1" Precip | 5 | 5 | 6 | 4 | 7 | 11 | 12 | 11 | 10 | 6 | 4 | 4 | 85 |
| Total Snowfall (") | 0.0 | 0.0 | 0.0 | 0.0 | 0.0 | 0.0 | 0.0 | 0.0 | 0.0 | 0.0 | 0.0 | 0.0 | 0.0 |
| Days ≥ 1" Snow Depth | 0 | 0 | 0 | 0 | 0 | 0 | 0 | 0 | 0 | 0 | 0 | 0 | 0 |

## DEVILS GARDEN *Hendry County* ELEVATION 20 ft LAT/LONG 26° 36' N / 81° 8' W

| | JAN | FEB | MAR | APR | MAY | JUN | JUL | AUG | SEP | OCT | NOV | DEC | YEAR |
|---|---|---|---|---|---|---|---|---|---|---|---|---|---|
| Maximum Temp °F | ***76.0*** | ***77.7*** | 81.2 | ***85.9*** | 89.3 | 91.3 | ***92.2*** | ***92.3*** | 90.4 | 86.7 | ***81.9*** | ***77.7*** | 85.2 |
| Minimum Temp °F | ***50.1*** | ***51.2*** | 54.3 | ***58.0*** | ***62.6*** | 68.5 | ***70.6*** | ***70.9*** | 70.3 | 65.9 | ***59.0*** | ***52.5*** | 61.2 |
| Mean Temp °F | ***63.1*** | ***64.5*** | 67.8 | ***72.0*** | ***75.9*** | 79.9 | ***81.4*** | ***81.6*** | 80.4 | 76.1 | ***70.5*** | ***65.1*** | 73.2 |
| Days Max Temp ≥ 90 °F | 0 | 0 | 2 | ***7*** | na | na | ***27*** | ***26*** | na | ***8*** | 1 | 0 | na |
| Days Max Temp ≤ 32 °F | 0 | 0 | 0 | 0 | 0 | 0 | 0 | 0 | 0 | 0 | 0 | 0 | 0 |
| Days Min Temp ≤ 32 °F | ***1*** | 1 | 0 | 0 | 0 | 0 | 0 | 0 | 0 | 0 | 0 | 1 | 3 |
| Days Min Temp ≤ 0 °F | 0 | 0 | 0 | 0 | 0 | 0 | 0 | 0 | 0 | 0 | 0 | 0 | 0 |
| Heating Degree Days | ***127*** | ***92*** | 44 | ***8*** | 0 | 0 | ***0*** | 0 | 0 | 1 | ***23*** | na | na |
| Cooling Degree Days | 82 | 93 | 139 | 211 | 340 | 456 | 507 | 525 | 473 | 354 | 219 | 100 | 3499 |
| Total Precipitation (") | 2.38 | 2.24 | 3.23 | 2.26 | 4.70 | 9.25 | 7.59 | 7.60 | 6.31 | 3.63 | 2.40 | 1.61 | 53.20 |
| Days ≥ 0.1" Precip | 4 | 4 | 4 | 3 | 6 | 12 | 13 | ***12*** | ***10*** | 6 | 3 | 3 | 80 |
| Total Snowfall (") | 0.0 | 0.0 | 0.0 | 0.0 | 0.0 | 0.0 | 0.0 | 0.0 | 0.0 | 0.0 | 0.0 | 0.0 | 0.0 |
| Days ≥ 1" Snow Depth | 0 | 0 | 0 | 0 | 0 | 0 | 0 | 0 | 0 | 0 | 0 | 0 | 0 |

## EVERGLADES *Collier County* ELEVATION 6 ft LAT/LONG 25° 51' N / 81° 23' W

| | JAN | FEB | MAR | APR | MAY | JUN | JUL | AUG | SEP | OCT | NOV | DEC | YEAR |
|---|---|---|---|---|---|---|---|---|---|---|---|---|---|
| Maximum Temp °F | 75.1 | 75.4 | 78.7 | 82.4 | 85.7 | 88.1 | 89.7 | 89.8 | 89.4 | 86.1 | 81.2 | 76.6 | 83.2 |
| Minimum Temp °F | 53.7 | 54.0 | 58.4 | 62.7 | 67.3 | 72.0 | 73.2 | 73.6 | 73.1 | 68.2 | 61.6 | 55.6 | 64.5 |
| Mean Temp °F | 64.4 | 64.8 | 68.5 | 72.6 | 76.5 | 80.1 | 81.5 | 81.7 | 81.3 | 77.2 | 71.4 | 66.1 | 73.8 |
| Days Max Temp ≥ 90 °F | 0 | 0 | 0 | 0 | ***4*** | ***9*** | ***17*** | 17 | 15 | 6 | 0 | 0 | 68 |
| Days Max Temp ≤ 32 °F | 0 | 0 | 0 | 0 | 0 | 0 | 0 | 0 | 0 | 0 | 0 | 0 | 0 |
| Days Min Temp ≤ 32 °F | 1 | 0 | 0 | 0 | 0 | 0 | 0 | 0 | 0 | 0 | 0 | 0 | 1 |
| Days Min Temp ≤ 0 °F | 0 | 0 | 0 | 0 | 0 | 0 | 0 | 0 | 0 | 0 | 0 | 0 | 0 |
| Heating Degree Days | 98 | 80 | 33 | 4 | 0 | 0 | 0 | 0 | 0 | 0 | 15 | 70 | 300 |
| Cooling Degree Days | 85 | 99 | 160 | 236 | 364 | 474 | 520 | 528 | 497 | 385 | 234 | ***125*** | 3707 |
| Total Precipitation (") | 1.70 | 1.67 | 1.86 | 1.86 | 3.58 | 10.35 | 7.96 | 8.20 | 8.06 | 3.92 | 1.35 | 1.44 | 51.95 |
| Days ≥ 0.1" Precip | 3 | 3 | 3 | 3 | 5 | 11 | ***13*** | 13 | 12 | 5 | 3 | 2 | 76 |
| Total Snowfall (") | 0.0 | 0.0 | 0.0 | 0.0 | 0.0 | 0.0 | 0.0 | 0.0 | 0.0 | 0.0 | 0.0 | 0.0 | 0.0 |
| Days ≥ 1" Snow Depth | 0 | 0 | 0 | 0 | 0 | 0 | 0 | 0 | 0 | 0 | 0 | 0 | 0 |

## FEDERAL POINT *Putnam County* ELEVATION 10 ft LAT/LONG 29° 44' N / 81° 32' W

| | JAN | FEB | MAR | APR | MAY | JUN | JUL | AUG | SEP | OCT | NOV | DEC | YEAR |
|---|---|---|---|---|---|---|---|---|---|---|---|---|---|
| Maximum Temp °F | 65.6 | 68.9 | 75.6 | 81.6 | 86.6 | 90.0 | 91.9 | na | 88.1 | 81.6 | 74.3 | 68.3 | na |
| Minimum Temp °F | 45.8 | 47.3 | 52.7 | 57.4 | 63.5 | 69.4 | ***71.5*** | na | 69.6 | 62.7 | 55.0 | 48.7 | na |
| Mean Temp °F | 55.7 | 58.1 | 64.2 | 69.5 | 75.1 | 79.7 | ***81.8*** | na | 78.8 | 72.2 | 64.8 | 58.4 | na |
| Days Max Temp ≥ 90 °F | 0 | 0 | 0 | ***2*** | ***9*** | ***16*** | na | na | na | ***1*** | 0 | 0 | na |
| Days Max Temp ≤ 32 °F | 0 | 0 | 0 | 0 | 0 | 0 | 0 | 0 | 0 | 0 | 0 | 0 | 0 |
| Days Min Temp ≤ 32 °F | ***2*** | 1 | 0 | 0 | 0 | 0 | 0 | 0 | 0 | 0 | ***0*** | 1 | 4 |
| Days Min Temp ≤ 0 °F | 0 | 0 | 0 | 0 | 0 | 0 | 0 | 0 | 0 | 0 | 0 | 0 | 0 |
| Heating Degree Days | 294 | 212 | 93 | 20 | 0 | 0 | 0 | 0 | 0 | 9 | 88 | 226 | 942 |
| Cooling Degree Days | 12 | 26 | 70 | 152 | 324 | 469 | 539 | 518 | 410 | 227 | 87 | 24 | 2858 |
| Total Precipitation (") | 3.28 | 3.57 | 3.83 | 2.23 | 4.03 | 6.84 | 6.42 | 6.89 | 6.85 | 3.44 | 2.62 | 2.86 | 52.86 |
| Days ≥ 0.1" Precip | 5 | 5 | 5 | 3 | 5 | 9 | 9 | na | 8 | 5 | 4 | 4 | na |
| Total Snowfall (") | 0.0 | 0.0 | 0.0 | 0.0 | 0.0 | 0.0 | 0.0 | 0.0 | 0.0 | 0.0 | 0.0 | 0.0 | 0.0 |
| Days ≥ 1" Snow Depth | 0 | 0 | 0 | 0 | 0 | 0 | 0 | 0 | 0 | 0 | 0 | 0 | 0 |

## FERNANDINA BEACH *Nassau County* ELEVATION 20 ft LAT/LONG 30° 40' N / 81° 27' W

| | JAN | FEB | MAR | APR | MAY | JUN | JUL | AUG | SEP | OCT | NOV | DEC | YEAR |
|---|---|---|---|---|---|---|---|---|---|---|---|---|---|
| Maximum Temp °F | 62.4 | 65.1 | 71.4 | 77.5 | 82.9 | 87.4 | 90.2 | 88.9 | 85.6 | 78.8 | ***71.8*** | 65.6 | 77.3 |
| Minimum Temp °F | 42.7 | 45.3 | 50.6 | 57.5 | 64.3 | 70.7 | 73.0 | 73.0 | 71.1 | 63.0 | 54.1 | 46.6 | 59.3 |
| Mean Temp °F | 52.6 | 55.2 | 61.0 | 67.5 | 73.5 | 79.1 | 81.6 | 81.0 | 78.4 | 70.9 | ***62.9*** | 56.1 | 68.3 |
| Days Max Temp ≥ 90 °F | 0 | 0 | 0 | 1 | 3 | 10 | 16 | 12 | 5 | 1 | 0 | 0 | 48 |
| Days Max Temp ≤ 32 °F | 0 | 0 | 0 | 0 | 0 | 0 | 0 | 0 | 0 | 0 | 0 | 0 | 0 |
| Days Min Temp ≤ 32 °F | 5 | 3 | 1 | 0 | 0 | 0 | 0 | 0 | 0 | 0 | 1 | 2 | 12 |
| Days Min Temp ≤ 0 °F | 0 | 0 | 0 | 0 | 0 | 0 | 0 | 0 | 0 | 0 | 0 | 0 | 0 |
| Heating Degree Days | 389 | 284 | 159 | 42 | 2 | 0 | 0 | 0 | 0 | 19 | ***124*** | 289 | 1308 |
| Cooling Degree Days | 8 | 17 | 43 | 105 | 269 | 449 | 538 | 514 | 412 | 211 | 78 | 21 | 2665 |
| Total Precipitation (") | 3.74 | 3.39 | 4.09 | 2.43 | 3.41 | 5.27 | 6.01 | 5.42 | 7.02 | 4.28 | 3.21 | 2.64 | 50.91 |
| Days ≥ 0.1" Precip | 7 | 6 | 6 | 4 | 5 | 8 | 9 | 8 | 8 | 5 | ***4*** | 4 | 74 |
| Total Snowfall (") | 0.0 | 0.0 | 0.0 | 0.0 | 0.0 | 0.0 | 0.0 | 0.0 | 0.0 | 0.0 | 0.0 | 0.0 | 0.0 |
| Days ≥ 1" Snow Depth | 0 | 0 | 0 | 0 | 0 | 0 | 0 | 0 | 0 | 0 | 0 | 0 | 0 |

## FLAMINGO RANGER STN *Monroe County* ELEVATION 3 ft LAT/LONG 25° 9 ' N / 80° 56 ' W

| | JAN | FEB | MAR | APR | MAY | JUN | JUL | AUG | SEP | OCT | NOV | DEC | YEAR |
|---|---|---|---|---|---|---|---|---|---|---|---|---|---|
| Maximum Temp °F | 76.1 | 76.6 | 79.1 | 82.6 | 85.4 | 87.9 | 88.9 | 89.4 | 88.6 | 85.7 | 81.7 | 77.7 | 83.3 |
| Minimum Temp °F | 56.2 | 57.0 | 60.4 | 64.4 | 68.7 | 73.4 | 74.1 | 74.0 | 73.2 | 69.1 | 63.8 | 58.5 | 66.1 |
| Mean Temp °F | 66.2 | 66.8 | 69.8 | 73.5 | 77.1 | 80.7 | 81.5 | 81.7 | 80.9 | 77.4 | 72.8 | 68.1 | 74.7 |
| Days Max Temp ≥ 90 °F | 0 | 0 | 0 | 0 | 2 | *8* | 14 | 16 | 10 | 2 | 0 | 0 | 52 |
| Days Max Temp ≤ 32 °F | 0 | 0 | 0 | 0 | 0 | 0 | 0 | 0 | 0 | 0 | 0 | 0 | 0 |
| Days Min Temp ≤ 32 °F | 0 | 0 | 0 | 0 | 0 | 0 | 0 | 0 | 0 | 0 | 0 | 0 | 0 |
| Days Min Temp ≤ 0 °F | 0 | 0 | 0 | 0 | 0 | 0 | 0 | 0 | 0 | 0 | 0 | 0 | 0 |
| Heating Degree Days | 70 | 54 | 23 | 2 | 0 | 0 | 0 | 0 | 0 | 1 | 8 | 40 | 198 |
| Cooling Degree Days | 116 | 133 | 184 | 262 | 385 | 492 | 528 | 529 | 482 | 402 | 269 | 162 | 3944 |
| Total Precipitation (") | 1.93 | 1.79 | 1.66 | 1.89 | 5.55 | 8.03 | 4.73 | 7.17 | 7.74 | 4.52 | 2.16 | 1.51 | 48.68 |
| Days ≥ 0.1" Precip | 4 | 3 | 3 | 3 | 7 | 10 | 9 | 10 | 11 | 7 | 4 | 3 | 74 |
| Total Snowfall (") | 0.0 | 0.0 | 0.0 | 0.0 | 0.0 | 0.0 | 0.0 | 0.0 | 0.0 | 0.0 | 0.0 | 0.0 | 0.0 |
| Days ≥ 1" Snow Depth | 0 | 0 | 0 | 0 | 0 | 0 | 0 | 0 | 0 | 0 | 0 | 0 | 0 |

## FORT DRUM 5 NW *Okeechobee County* ELEVATION 59 ft LAT/LONG 27° 32 ' N / 80° 48 ' W

| | JAN | FEB | MAR | APR | MAY | JUN | JUL | AUG | SEP | OCT | NOV | DEC | YEAR |
|---|---|---|---|---|---|---|---|---|---|---|---|---|---|
| Maximum Temp °F | 74.6 | 75.5 | 79.4 | 84.0 | 87.9 | 90.0 | 91.3 | 91.2 | 89.6 | 85.4 | 80.0 | 75.6 | 83.7 |
| Minimum Temp °F | 50.1 | 50.6 | 54.4 | 57.6 | 63.0 | 69.0 | 70.5 | 71.1 | 70.1 | 64.8 | 57.6 | 51.9 | 60.9 |
| Mean Temp °F | 62.3 | 63.0 | 66.9 | 70.8 | 75.5 | 79.5 | 80.9 | 81.2 | 79.9 | 75.1 | 68.8 | 63.8 | 72.3 |
| Days Max Temp ≥ 90 °F | 0 | 0 | 0 | 3 | 10 | 17 | 22 | 23 | 18 | 4 | 0 | 0 | 97 |
| Days Max Temp ≤ 32 °F | 0 | 0 | 0 | 0 | 0 | 0 | 0 | 0 | 0 | 0 | 0 | 0 | 0 |
| Days Min Temp ≤ 32 °F | 2 | 1 | 0 | 0 | 0 | 0 | 0 | 0 | 0 | 0 | 0 | 1 | 4 |
| Days Min Temp ≤ 0 °F | 0 | 0 | 0 | 0 | 0 | 0 | 0 | 0 | 0 | 0 | 0 | 0 | 0 |
| Heating Degree Days | 140 | 113 | 54 | 9 | 0 | 0 | 0 | 0 | 0 | 2 | 36 | 111 | 465 |
| Cooling Degree Days | 71 | 80 | 129 | *188* | 336 | 464 | *522* | *528* | 461 | 342 | 186 | 88 | 3395 |
| Total Precipitation (") | 2.18 | 2.65 | 3.26 | 2.04 | 4.40 | 8.13 | 7.77 | 6.59 | 6.36 | 3.59 | 2.09 | 1.75 | 50.81 |
| Days ≥ 0.1" Precip | 3 | 4 | 4 | 3 | 5 | 10 | 10 | 9 | 8 | 5 | 3 | 3 | 67 |
| Total Snowfall (") | 0.0 | 0.0 | 0.0 | 0.0 | 0.0 | 0.0 | 0.0 | 0.0 | 0.0 | 0.0 | 0.0 | 0.0 | 0.0 |
| Days ≥ 1" Snow Depth | 0 | 0 | 0 | 0 | 0 | 0 | 0 | 0 | 0 | 0 | 0 | 0 | 0 |

## FORT LAUDERDALE *Broward County* ELEVATION 10 ft LAT/LONG 26° 6 ' N / 80° 9 ' W

| | JAN | FEB | MAR | APR | MAY | JUN | JUL | AUG | SEP | OCT | NOV | DEC | YEAR |
|---|---|---|---|---|---|---|---|---|---|---|---|---|---|
| Maximum Temp °F | 75.9 | 76.4 | 79.1 | 82.4 | 85.4 | 88.1 | 89.4 | 89.7 | 88.5 | 85.4 | 80.8 | 77.3 | 83.2 |
| Minimum Temp °F | 58.0 | 58.2 | 62.1 | 66.0 | 70.3 | 73.8 | 74.9 | 75.3 | 74.5 | 71.0 | 65.5 | 60.3 | 67.5 |
| Mean Temp °F | 67.0 | 67.3 | 70.6 | 74.2 | 77.9 | 81.0 | 82.2 | 82.5 | 81.5 | 78.2 | 73.2 | 68.8 | 75.4 |
| Days Max Temp ≥ 90 °F | 0 | 0 | 0 | 1 | 4 | 10 | 15 | 17 | 9 | 2 | 0 | 0 | 58 |
| Days Max Temp ≤ 32 °F | 0 | 0 | 0 | 0 | 0 | 0 | 0 | 0 | 0 | 0 | 0 | 0 | 0 |
| Days Min Temp ≤ 32 °F | 0 | 0 | 0 | 0 | 0 | 0 | 0 | 0 | 0 | 0 | 0 | 0 | 0 |
| Days Min Temp ≤ 0 °F | 0 | 0 | 0 | 0 | 0 | 0 | 0 | 0 | 0 | 0 | 0 | 0 | 0 |
| Heating Degree Days | 65 | 53 | 21 | 2 | 0 | 0 | 0 | 0 | 0 | 0 | 10 | 43 | 194 |
| Cooling Degree Days | 146 | 155 | 212 | 286 | 411 | 498 | 551 | 564 | 510 | 440 | 296 | 188 | 4257 |
| Total Precipitation (") | 2.71 | 2.88 | 2.91 | 3.49 | 6.37 | 9.96 | 6.94 | 6.85 | 7.76 | 6.69 | 4.41 | 2.22 | 63.19 |
| Days ≥ 0.1" Precip | 5 | 4 | 4 | 4 | 8 | 11 | 11 | 11 | 12 | 9 | 6 | 4 | 89 |
| Total Snowfall (") | 0.0 | 0.0 | 0.0 | 0.0 | 0.0 | 0.0 | 0.0 | 0.0 | 0.0 | 0.0 | 0.0 | 0.0 | 0.0 |
| Days ≥ 1" Snow Depth | 0 | 0 | 0 | 0 | 0 | 0 | 0 | 0 | 0 | 0 | 0 | 0 | 0 |

## FORT MYERS PAGE FLD *Lee County* ELEVATION 15 ft LAT/LONG 26° 35 ' N / 81° 52 ' W

| | JAN | FEB | MAR | APR | MAY | JUN | JUL | AUG | SEP | OCT | NOV | DEC | YEAR |
|---|---|---|---|---|---|---|---|---|---|---|---|---|---|
| Maximum Temp °F | 74.8 | 75.7 | 79.7 | 84.6 | 88.7 | 90.7 | 91.4 | 91.5 | 90.1 | 86.1 | 80.9 | 76.4 | 84.2 |
| Minimum Temp °F | 53.9 | 54.6 | 58.6 | 62.5 | 67.9 | 73.1 | 74.5 | 74.7 | 74.2 | 68.9 | 61.5 | 55.6 | 65.0 |
| Mean Temp °F | 64.4 | 65.2 | 69.2 | 73.5 | 78.3 | 81.9 | 83.0 | 83.1 | 82.1 | 77.5 | 71.2 | 66.0 | 74.6 |
| Days Max Temp ≥ 90 °F | 0 | 0 | 1 | 3 | 13 | 20 | 25 | 25 | 19 | 6 | 1 | 0 | 113 |
| Days Max Temp ≤ 32 °F | 0 | 0 | 0 | 0 | 0 | 0 | 0 | 0 | 0 | 0 | 0 | 0 | 0 |
| Days Min Temp ≤ 32 °F | 0 | 0 | 0 | 0 | 0 | 0 | 0 | 0 | 0 | 0 | 0 | 0 | 0 |
| Days Min Temp ≤ 0 °F | 0 | 0 | 0 | 0 | 0 | 0 | 0 | 0 | 0 | 0 | 0 | 0 | 0 |
| Heating Degree Days | 107 | 80 | 30 | 3 | 0 | 0 | 0 | 0 | 0 | 1 | 21 | 78 | 320 |
| Cooling Degree Days | 95 | 115 | 180 | 271 | 428 | 543 | 579 | 582 | 535 | 410 | 243 | 126 | 4107 |
| Total Precipitation (") | 2.20 | 2.22 | 3.13 | 1.47 | 3.84 | 9.76 | 8.85 | 9.55 | 7.60 | 3.11 | 1.47 | 1.62 | 54.82 |
| Days ≥ 0.1" Precip | 3 | 4 | 4 | 3 | 5 | 11 | 13 | 14 | 10 | 5 | 3 | 3 | 78 |
| Total Snowfall (") | 0.0 | 0.0 | 0.0 | 0.0 | 0.0 | 0.0 | 0.0 | 0.0 | 0.0 | 0.0 | 0.0 | 0.0 | 0.0 |
| Days ≥ 1" Snow Depth | 0 | 0 | 0 | 0 | 0 | 0 | 0 | 0 | 0 | 0 | 0 | 0 | 0 |

## FORT PIERCE *St. Lucie County* ELEVATION 10 ft LAT/LONG 27° 26' N / 80° 20' W

| | JAN | FEB | MAR | APR | MAY | JUN | JUL | AUG | SEP | OCT | NOV | DEC | YEAR |
|---|---|---|---|---|---|---|---|---|---|---|---|---|---|
| Maximum Temp °F | 73.9 | 74.6 | 78.3 | 82.0 | 85.9 | 88.8 | 90.7 | 90.5 | 88.9 | 84.9 | 79.8 | 75.7 | 82.8 |
| Minimum Temp °F | 51.8 | 52.7 | 57.1 | 61.7 | 67.1 | 71.1 | 72.2 | 72.4 | 72.0 | 67.5 | 60.5 | 54.1 | 63.4 |
| Mean Temp °F | 62.9 | 63.7 | 67.7 | 71.9 | 76.5 | 80.0 | 81.5 | 81.5 | 80.5 | 76.3 | 70.2 | 64.9 | 73.1 |
| Days Max Temp ≥ 90 °F | 0 | 0 | 1 | 2 | 5 | 12 | 21 | 21 | 14 | 3 | 0 | 0 | 79 |
| Days Max Temp ≤ 32 °F | 0 | 0 | 0 | 0 | 0 | 0 | 0 | 0 | 0 | 0 | 0 | 0 | 0 |
| Days Min Temp ≤ 32 °F | 1 | 0 | 0 | 0 | 0 | 0 | 0 | 0 | 0 | 0 | 0 | 1 | 2 |
| Days Min Temp ≤ 0 °F | 0 | 0 | 0 | 0 | 0 | 0 | 0 | 0 | 0 | 0 | 0 | 0 | 0 |
| Heating Degree Days | 134 | 106 | 52 | 8 | 0 | 0 | 0 | 0 | 0 | 2 | 29 | 98 | 429 |
| Cooling Degree Days | 72 | 87 | 139 | 210 | 367 | 473 | 532 | 526 | 475 | 364 | 209 | 105 | 3559 |
| Total Precipitation (") | 2.49 | 3.17 | 3.22 | 2.21 | 4.77 | 6.15 | 5.81 | 5.49 | 7.46 | 5.82 | 3.36 | 2.31 | 52.26 |
| Days ≥ 0.1" Precip | 4 | 5 | 5 | 4 | 7 | 9 | 9 | 9 | 10 | 8 | 5 | 4 | 79 |
| Total Snowfall (") | 0.0 | 0.0 | 0.0 | 0.0 | 0.0 | 0.0 | 0.0 | 0.0 | 0.0 | 0.0 | 0.0 | 0.0 | 0.0 |
| Days ≥ 1" Snow Depth | 0 | 0 | 0 | 0 | 0 | 0 | 0 | 0 | 0 | 0 | 0 | 0 | 0 |

## GLEN ST MARY 1 W *Baker County* ELEVATION 141 ft LAT/LONG 30° 15' N / 82° 10' W

| | JAN | FEB | MAR | APR | MAY | JUN | JUL | AUG | SEP | OCT | NOV | DEC | YEAR |
|---|---|---|---|---|---|---|---|---|---|---|---|---|---|
| Maximum Temp °F | 64.9 | 68.2 | ***74.4*** | ***80.7*** | ***85.6*** | 89.7 | 91.9 | 90.9 | 88.4 | 81.3 | 74.3 | 67.5 | 79.8 |
| Minimum Temp °F | 38.8 | 41.4 | ***46.3*** | ***52.5*** | ***59.2*** | 66.2 | 69.4 | 69.2 | 66.8 | 56.8 | 48.2 | 41.2 | 54.7 |
| Mean Temp °F | 51.9 | 54.8 | ***60.3*** | ***66.6*** | ***72.4*** | 77.9 | 80.7 | 80.1 | 77.6 | 69.1 | 61.3 | 54.3 | 67.2 |
| Days Max Temp ≥ 90 °F | 0 | 0 | 0 | 1 | 6 | ***17*** | ***24*** | na | ***13*** | 2 | 0 | 0 | na |
| Days Max Temp ≤ 32 °F | 0 | 0 | 0 | 0 | 0 | 0 | 0 | 0 | 0 | 0 | 0 | 0 | 0 |
| Days Min Temp ≤ 32 °F | ***9*** | 6 | ***2*** | 0 | 0 | 0 | 0 | 0 | 0 | 0 | 3 | 8 | 28 |
| Days Min Temp ≤ 0 °F | 0 | 0 | 0 | 0 | 0 | 0 | 0 | 0 | 0 | 0 | 0 | 0 | 0 |
| Heating Degree Days | 409 | 290 | ***181*** | ***56*** | ***6*** | 0 | 0 | 0 | 0 | 36 | 158 | 341 | 1477 |
| Cooling Degree Days | 7 | 14 | 35 | ***90*** | ***236*** | 404 | 501 | 476 | 366 | 160 | 59 | 14 | 2362 |
| Total Precipitation (") | 4.28 | 3.67 | 4.70 | 3.00 | 4.17 | 6.81 | 6.88 | 7.96 | 5.41 | 2.68 | 2.32 | 3.00 | 54.88 |
| Days ≥ 0.1" Precip | 7 | 5 | ***6*** | ***4*** | 6 | 9 | 10 | ***11*** | 8 | ***4*** | 4 | 5 | 79 |
| Total Snowfall (") | 0.0 | 0.0 | 0.0 | 0.0 | 0.0 | 0.0 | 0.0 | 0.0 | 0.0 | 0.0 | 0.0 | 0.0 | 0.0 |
| Days ≥ 1" Snow Depth | 0 | 0 | 0 | 0 | 0 | 0 | 0 | 0 | 0 | 0 | 0 | 0 | 0 |

## HIALEAH *Dade County* ELEVATION 12 ft LAT/LONG 25° 50' N / 80° 17' W

| | JAN | FEB | MAR | APR | MAY | JUN | JUL | AUG | SEP | OCT | NOV | DEC | YEAR |
|---|---|---|---|---|---|---|---|---|---|---|---|---|---|
| Maximum Temp °F | 76.3 | 76.6 | 79.7 | 82.7 | 85.8 | 88.3 | 89.9 | 90.0 | 88.7 | 85.4 | 81.1 | 77.4 | 83.5 |
| Minimum Temp °F | 57.6 | 58.4 | 62.8 | 66.4 | 70.5 | 73.7 | 75.1 | 75.1 | 74.3 | 70.3 | 65.3 | 59.7 | 67.4 |
| Mean Temp °F | 67.2 | 67.5 | 71.3 | 74.5 | 78.2 | 81.1 | 82.5 | 82.6 | 81.5 | 77.9 | 73.2 | 68.6 | 75.5 |
| Days Max Temp ≥ 90 °F | 0 | 0 | 0 | 1 | 5 | 12 | 19 | 19 | 13 | 4 | 0 | 0 | 73 |
| Days Max Temp ≤ 32 °F | 0 | 0 | 0 | 0 | 0 | 0 | 0 | 0 | 0 | 0 | 0 | 0 | 0 |
| Days Min Temp ≤ 32 °F | 0 | 0 | 0 | 0 | 0 | 0 | 0 | 0 | 0 | 0 | 0 | 0 | 0 |
| Days Min Temp ≤ 0 °F | 0 | 0 | 0 | 0 | 0 | 0 | 0 | 0 | 0 | 0 | 0 | 0 | 0 |
| Heating Degree Days | 62 | 52 | 20 | 2 | 0 | 0 | 0 | 0 | 0 | 0 | 10 | 45 | 191 |
| Cooling Degree Days | 156 | 163 | 250 | 302 | 434 | 520 | 575 | 580 | 525 | 429 | 298 | 189 | 4421 |
| Total Precipitation (") | 2.73 | 2.44 | 2.95 | 3.73 | 6.42 | 10.27 | 7.54 | 8.59 | 8.83 | 6.92 | 3.59 | 2.07 | 66.08 |
| Days ≥ 0.1" Precip | 5 | 4 | 4 | 4 | 8 | 13 | 12 | 12 | 12 | 9 | 6 | 4 | 93 |
| Total Snowfall (") | 0.0 | 0.0 | 0.0 | 0.0 | 0.0 | 0.0 | 0.0 | 0.0 | 0.0 | 0.0 | 0.0 | 0.0 | 0.0 |
| Days ≥ 1" Snow Depth | 0 | 0 | 0 | 0 | 0 | 0 | 0 | 0 | 0 | 0 | 0 | 0 | 0 |

## HIGH SPRINGS *Alachua County* ELEVATION 69 ft LAT/LONG 29° 50' N / 82° 36' W

| | JAN | FEB | MAR | APR | MAY | JUN | JUL | AUG | SEP | OCT | NOV | DEC | YEAR |
|---|---|---|---|---|---|---|---|---|---|---|---|---|---|
| Maximum Temp °F | 67.4 | 70.8 | 77.4 | 83.3 | 88.4 | 91.2 | 92.1 | 91.5 | 89.5 | 83.3 | 76.4 | 70.4 | 81.8 |
| Minimum Temp °F | ***40.2*** | 42.7 | ***48.8*** | ***54.2*** | ***61.6*** | 68.6 | 70.6 | 70.5 | 67.7 | ***57.5*** | 49.1 | 42.8 | 56.2 |
| Mean Temp °F | ***53.8*** | 56.8 | ***63.2*** | ***68.8*** | ***75.0*** | 79.9 | 81.3 | 81.1 | 78.6 | ***70.3*** | 62.8 | 56.6 | 69.0 |
| Days Max Temp ≥ 90 °F | 0 | 0 | 0 | 3 | ***14*** | ***22*** | ***25*** | ***23*** | ***17*** | 3 | 0 | 0 | 107 |
| Days Max Temp ≤ 32 °F | 0 | 0 | 0 | 0 | 0 | 0 | 0 | 0 | 0 | 0 | 0 | 0 | 0 |
| Days Min Temp ≤ 32 °F | ***9*** | ***5*** | 1 | 0 | 0 | 0 | 0 | 0 | 0 | 0 | 2 | 7 | 24 |
| Days Min Temp ≤ 0 °F | 0 | 0 | 0 | 0 | 0 | 0 | 0 | 0 | 0 | 0 | 0 | 0 | 0 |
| Heating Degree Days | ***357*** | 245 | ***120*** | ***33*** | ***1*** | 0 | 0 | 0 | 0 | ***24*** | 131 | 276 | 1187 |
| Cooling Degree Days | 14 | 26 | 67 | 134 | 326 | 470 | 528 | 524 | 415 | 191 | 85 | 25 | 2805 |
| Total Precipitation (") | 4.14 | 4.00 | 4.22 | 2.86 | 4.05 | 6.58 | 7.45 | 8.20 | 4.38 | 2.76 | 2.15 | 2.91 | 53.70 |
| Days ≥ 0.1" Precip | 6 | 5 | 5 | 4 | 6 | 8 | 11 | 11 | 7 | 4 | 4 | 4 | 75 |
| Total Snowfall (") | 0.1 | 0.0 | 0.0 | 0.0 | 0.0 | 0.0 | 0.0 | 0.0 | 0.0 | 0.0 | 0.0 | 0.0 | 0.1 |
| Days ≥ 1" Snow Depth | 0 | 0 | 0 | 0 | 0 | 0 | 0 | 0 | 0 | 0 | 0 | 0 | 0 |

## IMMOKALEE 3 NNW *Collier County* ELEVATION 39 ft LAT/LONG 26° 28 ' N / 81° 26 ' W

| | JAN | FEB | MAR | APR | MAY | JUN | JUL | AUG | SEP | OCT | NOV | DEC | YEAR |
|---|---|---|---|---|---|---|---|---|---|---|---|---|---|
| Maximum Temp °F | 76.5 | 78.1 | 81.7 | 85.1 | 88.9 | 90.8 | 91.5 | 91.4 | 89.9 | 86.4 | 81.6 | 77.6 | 85.0 |
| Minimum Temp °F | 52.0 | 52.8 | 56.4 | 58.9 | 64.4 | 69.8 | 71.5 | 72.3 | 71.8 | 66.2 | 60.1 | 54.1 | 62.5 |
| Mean Temp °F | 64.3 | 65.5 | 69.0 | 72.0 | 76.7 | 80.3 | 81.5 | 81.9 | 80.9 | 76.3 | 70.9 | 65.9 | 73.8 |
| Days Max Temp ≥ 90 °F | 0 | 0 | 1 | 5 | 14 | 21 | 26 | 25 | 19 | 6 | 0 | 0 | 117 |
| Days Max Temp ≤ 32 °F | 0 | 0 | 0 | 0 | 0 | 0 | 0 | 0 | 0 | 0 | 0 | 0 | 0 |
| Days Min Temp ≤ 32 °F | 1 | 0 | 0 | 0 | 0 | 0 | 0 | 0 | 0 | 0 | 0 | 1 | 2 |
| Days Min Temp ≤ 0 °F | 0 | 0 | 0 | 0 | 0 | 0 | 0 | 0 | 0 | 0 | 0 | 0 | 0 |
| Heating Degree Days | 106 | 78 | 32 | 6 | 0 | 0 | 0 | 0 | 0 | 1 | 22 | 75 | 320 |
| Cooling Degree Days | 87 | 109 | 163 | 226 | 363 | 479 | 523 | 537 | 483 | 365 | 216 | 109 | 3660 |
| Total Precipitation (") | 2.45 | 2.37 | 3.10 | 2.33 | 4.31 | 8.24 | 7.02 | 7.41 | 6.54 | 2.29 | 2.22 | 1.66 | 49.94 |
| Days ≥ 0.1" Precip | 4 | 4 | 4 | 3 | 6 | 11 | 12 | 13 | 10 | 5 | 3 | 3 | 78 |
| Total Snowfall (") | 0.0 | 0.0 | 0.0 | 0.0 | 0.0 | 0.0 | 0.0 | 0.0 | 0.0 | 0.0 | 0.0 | 0.0 | 0.0 |
| Days ≥ 1" Snow Depth | 0 | 0 | 0 | 0 | 0 | 0 | 0 | 0 | 0 | 0 | 0 | 0 | 0 |

## INVERNESS 3 SE *Citrus County* ELEVATION 69 ft LAT/LONG 28° 50 ' N / 82° 20 ' W

| | JAN | FEB | MAR | APR | MAY | JUN | JUL | AUG | SEP | OCT | NOV | DEC | YEAR |
|---|---|---|---|---|---|---|---|---|---|---|---|---|---|
| Maximum Temp °F | 69.7 | 71.4 | 77.0 | 82.7 | 87.6 | 90.5 | 91.6 | 91.0 | 89.5 | 83.6 | 77.1 | 72.0 | 82.0 |
| Minimum Temp °F | 45.0 | 46.0 | 51.8 | 57.2 | 63.7 | 69.9 | 71.6 | 71.8 | 70.1 | 62.0 | 53.4 | 47.2 | 59.1 |
| Mean Temp °F | 57.4 | 58.7 | 64.4 | 70.0 | 75.7 | 80.2 | 81.6 | 81.4 | 79.8 | 72.8 | 65.3 | 59.6 | 70.6 |
| Days Max Temp ≥ 90 °F | 0 | 0 | 0 | 3 | 10 | 20 | 24 | 23 | 17 | 3 | 0 | 0 | 100 |
| Days Max Temp ≤ 32 °F | 0 | 0 | 0 | 0 | 0 | 0 | 0 | 0 | 0 | 0 | 0 | 0 | 0 |
| Days Min Temp ≤ 32 °F | 5 | 3 | 1 | 0 | 0 | 0 | 0 | 0 | 0 | 0 | 0 | 3 | 12 |
| Days Min Temp ≤ 0 °F | 0 | 0 | 0 | 0 | 0 | 0 | 0 | 0 | 0 | 0 | 0 | 0 | 0 |
| Heating Degree Days | 264 | 205 | 100 | 24 | 1 | 0 | 0 | 0 | 0 | 9 | 88 | 203 | 894 |
| Cooling Degree Days | 26 | 35 | 79 | 143 | 317 | 458 | 512 | 513 | 435 | 250 | 112 | 41 | 2921 |
| Total Precipitation (") | 3.36 | 3.28 | 3.96 | 2.33 | 3.70 | 7.71 | 7.51 | 8.57 | 5.93 | 2.58 | 2.13 | 2.44 | 53.50 |
| Days ≥ 0.1" Precip | 5 | 5 | 5 | 3 | 5 | 10 | 12 | 13 | 9 | 4 | 4 | 4 | 79 |
| Total Snowfall (") | 0.0 | 0.0 | 0.0 | 0.0 | 0.0 | 0.0 | 0.0 | 0.0 | 0.0 | 0.0 | 0.0 | 0.0 | 0.0 |
| Days ≥ 1" Snow Depth | 0 | 0 | 0 | 0 | 0 | 0 | 0 | 0 | 0 | 0 | 0 | 0 | 0 |

## JACKSONVILLE BEACH *Duval County* ELEVATION 10 ft LAT/LONG 30° 17 ' N / 81° 24 ' W

| | JAN | FEB | MAR | APR | MAY | JUN | JUL | AUG | SEP | OCT | NOV | DEC | YEAR |
|---|---|---|---|---|---|---|---|---|---|---|---|---|---|
| Maximum Temp °F | 63.7 | 65.2 | 71.4 | 77.3 | 82.8 | 87.0 | 89.7 | 88.4 | 85.9 | 79.8 | 72.2 | 66.5 | 77.5 |
| Minimum Temp °F | 45.7 | 47.2 | 53.0 | 59.4 | 65.7 | 71.5 | 73.2 | 73.6 | 72.6 | 65.0 | 55.1 | 49.1 | 60.9 |
| Mean Temp °F | 54.7 | 56.3 | 62.2 | 68.4 | 74.3 | 79.3 | 81.4 | 81.0 | 79.2 | 72.4 | 63.6 | 57.9 | 69.2 |
| Days Max Temp ≥ 90 °F | 0 | 0 | 0 | 1 | 2 | 8 | 14 | 9 | 3 | 1 | 0 | 0 | 38 |
| Days Max Temp ≤ 32 °F | 0 | 0 | 0 | 0 | 0 | 0 | 0 | 0 | 0 | 0 | 0 | 0 | 0 |
| Days Min Temp ≤ 32 °F | 3 | 2 | 0 | 0 | 0 | 0 | 0 | 0 | 0 | 0 | 0 | 1 | 6 |
| Days Min Temp ≤ 0 °F | 0 | 0 | 0 | 0 | 0 | 0 | 0 | 0 | 0 | 0 | 0 | 0 | 0 |
| Heating Degree Days | 325 | 254 | 132 | 27 | 0 | 0 | 0 | 0 | 0 | 11 | 108 | 242 | 1099 |
| Cooling Degree Days | 9 | 16 | 52 | 127 | 300 | 448 | 533 | 525 | 440 | 252 | 79 | 30 | 2811 |
| Total Precipitation (") | 3.62 | 3.51 | 3.93 | 2.39 | 3.42 | 5.57 | 5.69 | 5.44 | 7.22 | 4.89 | 2.42 | 2.83 | 50.93 |
| Days ≥ 0.1" Precip | 6 | 5 | 5 | 4 | 5 | 8 | 8 | 8 | 8 | 6 | 4 | 5 | 72 |
| Total Snowfall (") | 0.0 | 0.0 | 0.0 | 0.0 | 0.0 | 0.0 | 0.0 | 0.0 | 0.0 | 0.0 | 0.0 | 0.1 | 0.1 |
| Days ≥ 1" Snow Depth | 0 | 0 | 0 | 0 | 0 | 0 | 0 | 0 | 0 | 0 | 0 | 0 | 0 |

## JACKSONVILLE INTL AP *Duval County* ELEVATION 26 ft LAT/LONG 30° 30 ' N / 81° 42 ' W

| | JAN | FEB | MAR | APR | MAY | JUN | JUL | AUG | SEP | OCT | NOV | DEC | YEAR |
|---|---|---|---|---|---|---|---|---|---|---|---|---|---|
| Maximum Temp °F | 64.3 | 67.0 | 73.8 | 79.9 | 85.1 | 89.5 | 92.0 | 90.8 | 87.4 | 80.2 | 73.3 | 67.0 | 79.2 |
| Minimum Temp °F | 41.9 | 43.9 | 50.0 | 55.7 | 62.9 | 69.7 | 72.6 | 72.4 | 69.6 | 60.1 | 50.8 | 44.3 | 57.8 |
| Mean Temp °F | 53.2 | 55.5 | 61.9 | 67.8 | 74.0 | 79.6 | 82.3 | 81.6 | 78.5 | 70.2 | 62.1 | 55.7 | 68.5 |
| Days Max Temp ≥ 90 °F | 0 | 0 | 0 | 2 | 6 | 16 | 23 | 21 | 10 | 1 | 0 | 0 | 79 |
| Days Max Temp ≤ 32 °F | 0 | 0 | 0 | 0 | 0 | 0 | 0 | 0 | 0 | 0 | 0 | 0 | 0 |
| Days Min Temp ≤ 32 °F | 7 | 4 | 1 | 0 | 0 | 0 | 0 | 0 | 0 | 0 | 1 | 4 | 17 |
| Days Min Temp ≤ 0 °F | 0 | 0 | 0 | 0 | 0 | 0 | 0 | 0 | 0 | 0 | 0 | 0 | 0 |
| Heating Degree Days | 374 | 282 | 153 | 44 | 3 | 0 | 0 | 0 | 0 | 27 | 146 | 304 | 1333 |
| Cooling Degree Days | 11 | 24 | 62 | 121 | 283 | 466 | 559 | 536 | 409 | 192 | 77 | 24 | 2764 |
| Total Precipitation (") | 3.60 | 3.57 | 4.06 | 2.80 | 3.86 | 5.84 | 6.02 | 7.57 | 7.39 | 3.75 | 2.22 | 2.58 | 53.26 |
| Days ≥ 0.1" Precip | 6 | 5 | 6 | 4 | 5 | 9 | 10 | 10 | 9 | 5 | 4 | 5 | 78 |
| Total Snowfall (") | 0.0 | 0.0 | 0.0 | 0.0 | 0.0 | 0.0 | 0.0 | 0.0 | 0.0 | 0.0 | 0.0 | 0.0 | 0.0 |
| Days ≥ 1" Snow Depth | 0 | 0 | 0 | 0 | 0 | 0 | 0 | 0 | 0 | 0 | 0 | 0 | 0 |

## JASPER *Hamilton County* ELEVATION 147 ft LAT/LONG 30° 31 ' N / 82° 57 ' W

| | JAN | FEB | MAR | APR | MAY | JUN | JUL | AUG | SEP | OCT | NOV | DEC | YEAR |
|---|---|---|---|---|---|---|---|---|---|---|---|---|---|
| Maximum Temp °F | 63.9 | 66.6 | 74.2 | 80.4 | 85.7 | 90.1 | 91.6 | 91.3 | 88.8 | 81.2 | 73.6 | 66.5 | 79.5 |
| Minimum Temp °F | 38.6 | 40.8 | 47.2 | 53.1 | 59.9 | 67.0 | 69.9 | 69.4 | 65.9 | 55.3 | 46.7 | 40.5 | 54.5 |
| Mean Temp °F | 51.2 | 53.7 | 60.7 | 66.7 | 72.9 | 78.6 | 80.8 | 80.4 | 77.4 | 68.3 | 60.2 | 53.5 | 67.0 |
| Days Max Temp ≥ 90 °F | 0 | 0 | 0 | 1 | 7 | 18 | 24 | 23 | 15 | 3 | 0 | 0 | 91 |
| Days Max Temp ≤ 32 °F | 0 | 0 | 0 | 0 | 0 | 0 | 0 | 0 | 0 | 0 | 0 | 0 | 0 |
| Days Min Temp ≤ 32 °F | 11 | 7 | 2 | 0 | 0 | 0 | 0 | 0 | 0 | 0 | 4 | 9 | 33 |
| Days Min Temp ≤ 0 °F | 0 | 0 | 0 | 0 | 0 | 0 | 0 | 0 | 0 | 0 | 0 | 0 | 0 |
| Heating Degree Days | 429 | 326 | 173 | 57 | 5 | 0 | 0 | 0 | 1 | 47 | 186 | 366 | 1590 |
| Cooling Degree Days | 5 | 15 | 46 | 99 | 251 | 425 | 502 | 488 | 368 | 150 | 57 | 18 | 2424 |
| Total Precipitation (") | 4.84 | 4.26 | 5.08 | 3.27 | 3.95 | 6.77 | 6.03 | 6.64 | 3.80 | 2.45 | 2.70 | 3.78 | 53.57 |
| Days ≥ 0.1" Precip | 7 | 6 | 6 | 4 | 6 | 9 | 10 | 10 | 6 | 3 | 4 | 5 | 76 |
| Total Snowfall (") | 0.0 | 0.0 | 0.0 | 0.0 | 0.0 | 0.0 | 0.0 | 0.0 | 0.0 | 0.0 | 0.0 | 0.1 | 0.1 |
| Days ≥ 1" Snow Depth | 0 | 0 | 0 | 0 | 0 | 0 | 0 | 0 | 0 | 0 | 0 | 0 | 0 |

## KEY WEST INTL ARPT *Monroe County* ELEVATION 4 ft LAT/LONG 24° 33 ' N / 81° 45 ' W

| | JAN | FEB | MAR | APR | MAY | JUN | JUL | AUG | SEP | OCT | NOV | DEC | YEAR |
|---|---|---|---|---|---|---|---|---|---|---|---|---|---|
| Maximum Temp °F | 75.2 | 75.6 | 78.5 | 81.9 | 85.1 | 87.8 | 89.2 | 89.3 | 88.0 | 84.6 | 80.2 | 76.6 | 82.7 |
| Minimum Temp °F | 65.3 | 65.7 | 68.8 | 72.5 | 75.9 | 78.7 | 79.7 | 79.4 | 78.5 | 75.7 | 71.5 | 67.2 | 73.2 |
| Mean Temp °F | 70.3 | 70.7 | 73.7 | 77.2 | 80.5 | 83.2 | 84.5 | 84.4 | 83.3 | 80.2 | 75.9 | 71.9 | 78.0 |
| Days Max Temp ≥ 90 °F | 0 | 0 | 0 | 0 | 1 | 7 | 14 | 14 | 6 | 0 | 0 | 0 | 42 |
| Days Max Temp ≤ 32 °F | 0 | 0 | 0 | 0 | 0 | 0 | 0 | 0 | 0 | 0 | 0 | 0 | 0 |
| Days Min Temp ≤ 32 °F | 0 | 0 | 0 | 0 | 0 | 0 | 0 | 0 | 0 | 0 | 0 | 0 | 0 |
| Days Min Temp ≤ 0 °F | 0 | 0 | 0 | 0 | 0 | 0 | 0 | 0 | 0 | 0 | 0 | 0 | 0 |
| Heating Degree Days | 23 | 16 | 5 | 0 | 0 | 0 | 0 | 0 | 0 | 0 | 1 | 12 | 57 |
| Cooling Degree Days | 185 | 202 | 277 | 358 | 484 | 568 | 616 | 612 | 557 | 484 | 354 | 235 | 4932 |
| Total Precipitation (") | 2.33 | 1.77 | 1.83 | 1.87 | 3.76 | 5.22 | 3.63 | 4.94 | 5.91 | 4.69 | 2.74 | 1.96 | 40.65 |
| Days ≥ 0.1" Precip | 4 | 3 | 3 | 2 | 5 | 7 | 7 | 8 | 10 | 7 | 4 | 3 | 63 |
| Total Snowfall (") | 0.0 | 0.0 | 0.0 | 0.0 | 0.0 | 0.0 | 0.0 | 0.0 | 0.0 | 0.0 | 0.0 | 0.0 | 0.0 |
| Days ≥ 1" Snow Depth | 0 | 0 | 0 | 0 | 0 | 0 | 0 | 0 | 0 | 0 | 0 | 0 | 0 |

## KISSIMMEE 2 *Osceola County* ELEVATION 69 ft LAT/LONG 28° 18 ' N / 81° 24 ' W

| | JAN | FEB | MAR | APR | MAY | JUN | JUL | AUG | SEP | OCT | NOV | DEC | YEAR |
|---|---|---|---|---|---|---|---|---|---|---|---|---|---|
| Maximum Temp °F | 73.1 | 74.3 | 78.7 | 83.3 | 87.3 | 90.2 | 91.4 | 91.1 | 89.4 | 84.6 | 79.1 | 74.4 | 83.1 |
| Minimum Temp °F | 49.5 | 50.3 | 55.0 | 59.5 | 64.9 | 70.6 | 72.3 | 72.5 | 71.4 | 65.0 | 57.4 | 50.9 | 61.6 |
| Mean Temp °F | 61.3 | 62.3 | 66.9 | 71.4 | 76.1 | 80.4 | 81.8 | 81.8 | 80.4 | 74.8 | 68.3 | 62.7 | 72.4 |
| Days Max Temp ≥ 90 °F | 0 | 0 | 0 | 2 | 9 | 18 | 24 | 24 | 16 | 3 | 0 | 0 | 96 |
| Days Max Temp ≤ 32 °F | 0 | 0 | 0 | 0 | 0 | 0 | 0 | 0 | 0 | 0 | 0 | 0 | 0 |
| Days Min Temp ≤ 32 °F | 2 | 1 | 0 | 0 | 0 | 0 | 0 | 0 | 0 | 0 | 0 | 1 | 4 |
| Days Min Temp ≤ 0 °F | 0 | 0 | 0 | 0 | 0 | 0 | 0 | 0 | 0 | 0 | 0 | 0 | 0 |
| Heating Degree Days | 165 | 126 | 59 | 10 | 0 | 0 | 0 | 0 | 0 | 4 | 44 | 133 | 541 |
| Cooling Degree Days | 56 | 72 | 126 | 196 | 349 | 482 | 541 | 540 | 469 | 323 | 174 | ***74*** | 3402 |
| Total Precipitation (") | 2.24 | 2.91 | 3.15 | 1.86 | 3.82 | 6.55 | 6.87 | 6.73 | 6.07 | 3.02 | 2.18 | 2.11 | 47.51 |
| Days ≥ 0.1" Precip | 4 | 5 | 5 | 3 | 6 | 9 | 11 | 11 | 9 | 5 | 3 | 4 | 75 |
| Total Snowfall (") | 0.0 | 0.0 | 0.0 | 0.0 | 0.0 | 0.0 | 0.0 | 0.0 | 0.0 | 0.0 | 0.0 | 0.0 | 0.0 |
| Days ≥ 1" Snow Depth | 0 | 0 | 0 | 0 | 0 | 0 | 0 | 0 | 0 | 0 | 0 | 0 | 0 |

## LA BELLE *Hendry County* ELEVATION 12 ft LAT/LONG 26° 46 ' N / 81° 26 ' W

| | JAN | FEB | MAR | APR | MAY | JUN | JUL | AUG | SEP | OCT | NOV | DEC | YEAR |
|---|---|---|---|---|---|---|---|---|---|---|---|---|---|
| Maximum Temp °F | ***75.5*** | 77.4 | 81.5 | 86.3 | ***90.2*** | 91.8 | ***92.8*** | ***92.5*** | ***90.6*** | ***86.6*** | na | na | na |
| Minimum Temp °F | ***49.5*** | 51.0 | 54.8 | 58.5 | 63.8 | 69.3 | ***70.6*** | na | ***70.7*** | ***65.1*** | na | na | na |
| Mean Temp °F | ***62.3*** | 64.2 | 68.2 | 72.4 | ***77.0*** | 80.6 | ***81.7*** | na | ***80.7*** | ***75.9*** | na | na | na |
| Days Max Temp ≥ 90 °F | 0 | 1 | 3 | 8 | 15 | 23 | 26 | ***26*** | 20 | 7 | 0 | 0 | 129 |
| Days Max Temp ≤ 32 °F | 0 | 0 | 0 | 0 | 0 | 0 | 0 | 0 | 0 | 0 | 0 | 0 | 0 |
| Days Min Temp ≤ 32 °F | 2 | 1 | 0 | 0 | 0 | 0 | 0 | 0 | 0 | 0 | 0 | 1 | 4 |
| Days Min Temp ≤ 0 °F | 0 | 0 | 0 | 0 | 0 | 0 | 0 | 0 | 0 | 0 | 0 | 0 | 0 |
| Heating Degree Days | ***146*** | 100 | 49 | 9 | ***0*** | 0 | ***0*** | na | ***0*** | ***1*** | na | na | na |
| Cooling Degree Days | ***73*** | 100 | ***156*** | 223 | ***377*** | 492 | na | ***533*** | ***487*** | 347 | na | na | na |
| Total Precipitation (") | 2.17 | 2.21 | 3.35 | 1.97 | 4.25 | 9.19 | 7.47 | 7.76 | 6.30 | 3.39 | 2.00 | 1.65 | 51.71 |
| Days ≥ 0.1" Precip | 3 | 4 | 4 | 3 | 5 | 10 | 11 | 10 | 9 | 5 | 2 | 2 | 68 |
| Total Snowfall (") | 0.0 | 0.0 | 0.0 | 0.0 | 0.0 | 0.0 | 0.0 | 0.0 | 0.0 | 0.0 | 0.0 | 0.0 | 0.0 |
| Days ≥ 1" Snow Depth | 0 | 0 | 0 | 0 | 0 | 0 | 0 | 0 | 0 | 0 | 0 | 0 | 0 |

## LAKE ALFRED EXP STN *Polk County* ELEVATION 151 ft LAT/LONG 28° 5' N / 81° 44' W

| | JAN | FEB | MAR | APR | MAY | JUN | JUL | AUG | SEP | OCT | NOV | DEC | YEAR |
|---|---|---|---|---|---|---|---|---|---|---|---|---|---|
| Maximum Temp °F | 71.8 | 73.5 | 78.3 | 83.6 | 88.2 | 90.9 | 92.4 | 92.2 | 90.5 | 85.4 | 79.1 | 73.5 | 83.3 |
| Minimum Temp °F | 47.1 | 48.5 | 53.9 | 58.5 | 64.2 | 70.1 | 71.8 | 71.8 | 70.1 | 63.2 | 55.9 | 49.4 | 60.4 |
| Mean Temp °F | 59.5 | 61.0 | 66.1 | 71.1 | 76.2 | 80.6 | 82.1 | 82.0 | 80.3 | 74.3 | 67.5 | 61.5 | 71.9 |
| Days Max Temp ≥ 90 °F | 0 | 0 | 1 | 4 | 12 | 21 | 26 | 26 | 21 | 5 | 0 | 0 | 116 |
| Days Max Temp ≤ 32 °F | 0 | 0 | 0 | 0 | 0 | 0 | 0 | 0 | 0 | 0 | 0 | 0 | 0 |
| Days Min Temp ≤ 32 °F | 3 | 1 | 0 | 0 | 0 | 0 | 0 | 0 | 0 | 0 | 0 | 2 | 6 |
| Days Min Temp ≤ 0 °F | 0 | 0 | 0 | 0 | 0 | 0 | 0 | 0 | 0 | 0 | 0 | 0 | 0 |
| Heating Degree Days | 213 | 158 | 77 | 16 | 0 | 0 | 0 | 0 | 0 | 5 | 56 | 166 | 691 |
| Cooling Degree Days | 46 | 63 | 118 | 188 | 355 | 487 | 548 | 547 | 470 | 311 | 162 | 68 | 3363 |
| Total Precipitation (") | 2.42 | 2.87 | 3.42 | 1.78 | 4.22 | 7.50 | 7.36 | 7.46 | 6.23 | 2.81 | 2.05 | 1.94 | 50.06 |
| Days ≥ 0.1" Precip | 4 | 5 | 5 | 3 | 6 | 10 | 12 | 11 | 9 | 4 | 3 | 3 | 75 |
| Total Snowfall (") | 0.0 | 0.0 | 0.0 | 0.0 | 0.0 | 0.0 | 0.0 | 0.0 | 0.0 | 0.0 | 0.0 | 0.0 | 0.0 |
| Days ≥ 1" Snow Depth | 0 | 0 | 0 | 0 | 0 | 0 | 0 | 0 | 0 | 0 | 0 | 0 | 0 |

## LAKE CITY 2 E *Columbia County* ELEVATION 210 ft LAT/LONG 30° 11' N / 82° 36' W

| | JAN | FEB | MAR | APR | MAY | JUN | JUL | AUG | SEP | OCT | NOV | DEC | YEAR |
|---|---|---|---|---|---|---|---|---|---|---|---|---|---|
| Maximum Temp °F | 64.1 | 66.9 | 73.8 | 80.1 | 85.6 | 89.6 | 91.1 | 90.6 | 87.7 | 80.6 | 73.5 | 67.0 | 79.2 |
| Minimum Temp °F | 41.4 | 43.5 | 49.2 | 54.9 | 61.4 | 68.1 | 70.6 | 70.4 | 67.7 | 58.3 | 50.3 | 44.0 | 56.7 |
| Mean Temp °F | 52.8 | 55.2 | 61.6 | 67.4 | 73.5 | 78.9 | 80.9 | 80.5 | 77.7 | 69.5 | 62.0 | 55.5 | 68.0 |
| Days Max Temp ≥ 90 °F | 0 | 0 | 0 | 1 | 6 | 17 | 22 | 22 | 12 | 1 | 0 | 0 | 81 |
| Days Max Temp ≤ 32 °F | 0 | 0 | 0 | 0 | 0 | 0 | 0 | 0 | 0 | 0 | 0 | 0 | 0 |
| Days Min Temp ≤ 32 °F | 8 | 5 | 1 | 0 | 0 | 0 | 0 | 0 | 0 | 0 | 2 | 5 | 21 |
| Days Min Temp ≤ 0 °F | 0 | 0 | 0 | 0 | 0 | 0 | 0 | 0 | 0 | 0 | 0 | 0 | 0 |
| Heating Degree Days | 384 | 288 | 158 | 48 | 4 | 0 | 0 | 0 | 0 | 33 | 151 | 311 | 1377 |
| Cooling Degree Days | 10 | 22 | 59 | 116 | 279 | 449 | 517 | 504 | 384 | 179 | 80 | 27 | 2626 |
| Total Precipitation (") | 4.38 | 4.04 | 4.99 | 2.95 | 4.33 | 6.83 | 7.01 | 7.38 | 4.69 | 2.60 | 2.34 | 3.34 | 54.88 |
| Days ≥ 0.1" Precip | 7 | 6 | 6 | 4 | 6 | 9 | 11 | 11 | 7 | 4 | 4 | 5 | 80 |
| Total Snowfall (") | 0.0 | 0.0 | 0.0 | 0.0 | 0.0 | 0.0 | 0.0 | 0.0 | 0.0 | 0.0 | 0.0 | 0.0 | 0.0 |
| Days ≥ 1" Snow Depth | 0 | 0 | 0 | 0 | 0 | 0 | 0 | 0 | 0 | 0 | 0 | 0 | 0 |

## LAKELAND *Polk County* ELEVATION 151 ft LAT/LONG 28° 1' N / 81° 55' W

| | JAN | FEB | MAR | APR | MAY | JUN | JUL | AUG | SEP | OCT | NOV | DEC | YEAR |
|---|---|---|---|---|---|---|---|---|---|---|---|---|---|
| Maximum Temp °F | 71.7 | 73.7 | 78.7 | 83.7 | 88.3 | 91.0 | 92.3 | 91.9 | 90.2 | 84.7 | 78.2 | 73.3 | 83.1 |
| Minimum Temp °F | 50.5 | 51.6 | 56.3 | 60.9 | 66.4 | 71.2 | 72.6 | 72.9 | 71.8 | 65.6 | 58.3 | 52.6 | 62.6 |
| Mean Temp °F | 61.1 | 62.7 | 67.5 | 72.3 | 77.4 | 81.2 | 82.5 | 82.5 | 81.0 | 75.2 | 68.3 | 62.9 | 72.9 |
| Days Max Temp ≥ 90 °F | 0 | 0 | 1 | 3 | 13 | 20 | 25 | 25 | 18 | 5 | 0 | 0 | 110 |
| Days Max Temp ≤ 32 °F | 0 | 0 | 0 | 0 | 0 | 0 | 0 | 0 | 0 | 0 | 0 | 0 | 0 |
| Days Min Temp ≤ 32 °F | 1 | 0 | 0 | 0 | 0 | 0 | 0 | 0 | 0 | 0 | 0 | 1 | 2 |
| Days Min Temp ≤ 0 °F | 0 | 0 | 0 | 0 | 0 | 0 | 0 | 0 | 0 | 0 | 0 | 0 | 0 |
| Heating Degree Days | 173 | 126 | 54 | 6 | 0 | 0 | 0 | 0 | 0 | 3 | 49 | 132 | 543 |
| Cooling Degree Days | 64 | 86 | 153 | 231 | 410 | 523 | 576 | 576 | 508 | 351 | 184 | 81 | 3743 |
| Total Precipitation (") | 2.50 | 2.81 | 3.38 | 1.82 | 4.26 | 7.10 | 7.58 | 7.52 | 6.12 | 2.31 | 2.00 | 2.14 | 49.54 |
| Days ≥ 0.1" Precip | 4 | 4 | 5 | 3 | 5 | 10 | 12 | 11 | 9 | 4 | 3 | 4 | 74 |
| Total Snowfall (") | 0.0 | 0.0 | 0.0 | 0.0 | 0.0 | 0.0 | 0.0 | 0.0 | 0.0 | 0.0 | 0.0 | 0.0 | 0.0 |
| Days ≥ 1" Snow Depth | 0 | 0 | 0 | 0 | 0 | 0 | 0 | 0 | 0 | 0 | 0 | 0 | 0 |

## LISBON *Lake County* ELEVATION 69 ft LAT/LONG 28° 53' N / 81° 48' W

| | JAN | FEB | MAR | APR | MAY | JUN | JUL | AUG | SEP | OCT | NOV | DEC | YEAR |
|---|---|---|---|---|---|---|---|---|---|---|---|---|---|
| Maximum Temp °F | 69.1 | 71.2 | 76.8 | 82.0 | 86.8 | 90.1 | 91.4 | 91.3 | 89.3 | 83.4 | 76.7 | 70.8 | 81.6 |
| Minimum Temp °F | 46.6 | 48.0 | 53.2 | 58.3 | 64.5 | 70.2 | 71.9 | 71.9 | 70.3 | 63.1 | 54.8 | 48.7 | 60.1 |
| Mean Temp °F | 57.9 | 59.8 | 65.0 | 70.1 | 75.7 | 80.2 | 81.7 | 81.7 | 79.8 | 73.3 | 65.8 | 59.8 | 70.9 |
| Days Max Temp ≥ 90 °F | 0 | 0 | 0 | 2 | 8 | 18 | 23 | 24 | 16 | 3 | 0 | 0 | 94 |
| Days Max Temp ≤ 32 °F | 0 | 0 | 0 | 0 | 0 | 0 | 0 | 0 | 0 | 0 | 0 | 0 | 0 |
| Days Min Temp ≤ 32 °F | 3 | 2 | 0 | 0 | 0 | 0 | 0 | 0 | 0 | 0 | 0 | 2 | 7 |
| Days Min Temp ≤ 0 °F | 0 | 0 | 0 | 0 | 0 | 0 | 0 | 0 | 0 | 0 | 0 | 0 | 0 |
| Heating Degree Days | 249 | 181 | 90 | 19 | 1 | 0 | 0 | 0 | 0 | 7 | 77 | 199 | 823 |
| Cooling Degree Days | 28 | 42 | 86 | 156 | 331 | 476 | 543 | 541 | 453 | 264 | 115 | 37 | 3072 |
| Total Precipitation (") | 3.11 | 3.22 | 3.85 | 2.57 | 4.19 | 6.28 | 6.25 | 6.39 | 5.12 | 2.55 | 2.25 | 2.50 | 48.28 |
| Days ≥ 0.1" Precip | 5 | 5 | 5 | 3 | 6 | 9 | 11 | 11 | 8 | 5 | 4 | 4 | 76 |
| Total Snowfall (") | 0.0 | 0.0 | 0.0 | 0.0 | 0.0 | 0.0 | 0.0 | 0.0 | 0.0 | 0.0 | 0.0 | 0.0 | 0.0 |
| Days ≥ 1" Snow Depth | 0 | 0 | 0 | 0 | 0 | 0 | 0 | 0 | 0 | 0 | 0 | 0 | 0 |

## LIVE OAK *Suwannee County* ELEVATION 102 ft LAT/LONG 30° 17' N / 82° 58' W

| | JAN | FEB | MAR | APR | MAY | JUN | JUL | AUG | SEP | OCT | NOV | DEC | YEAR |
|---|---|---|---|---|---|---|---|---|---|---|---|---|---|
| Maximum Temp °F | 66.5 | 69.6 | 76.2 | 82.3 | 87.8 | 91.6 | 92.6 | 92.0 | 89.5 | 82.8 | 75.5 | 69.0 | 81.3 |
| Minimum Temp °F | 41.8 | 43.2 | 49.3 | 54.8 | 61.6 | 68.3 | 70.8 | 70.6 | 67.7 | 58.0 | 50.0 | 43.7 | 56.7 |
| Mean Temp °F | 54.2 | 56.4 | 62.8 | 68.6 | 74.8 | 80.0 | 81.7 | 81.3 | 78.6 | 70.4 | 62.8 | 56.4 | 69.0 |
| Days Max Temp ≥ 90 °F | 0 | 0 | 0 | 2 | 11 | 22 | 25 | 25 | 18 | 4 | 0 | 0 | 107 |
| Days Max Temp ≤ 32 °F | 0 | 0 | 0 | 0 | 0 | 0 | 0 | 0 | 0 | 0 | 0 | 0 | 0 |
| Days Min Temp ≤ 32 °F | 8 | 6 | 2 | 0 | 0 | 0 | 0 | 0 | 0 | 0 | 3 | 6 | 25 |
| Days Min Temp ≤ 0 °F | 0 | 0 | 0 | 0 | 0 | 0 | 0 | 0 | 0 | 0 | 0 | 0 | 0 |
| Heating Degree Days | 345 | 260 | 132 | 35 | 2 | 0 | 0 | 0 | 0 | 28 | 137 | 287 | 1226 |
| Cooling Degree Days | 13 | 30 | 72 | 143 | 318 | 471 | 540 | 528 | 414 | 209 | 89 | 30 | 2857 |
| Total Precipitation (") | 4.72 | 3.96 | 4.93 | 3.40 | 3.86 | 6.25 | 6.88 | 6.88 | 3.87 | 2.71 | 2.39 | 3.54 | 53.39 |
| Days ≥ 0.1" Precip | 7 | 5 | 6 | 4 | 6 | 9 | 10 | 10 | 6 | 3 | 4 | 5 | 75 |
| Total Snowfall (") | 0.0 | 0.0 | 0.0 | 0.0 | 0.0 | 0.0 | 0.0 | 0.0 | 0.0 | 0.0 | 0.0 | 0.1 | 0.1 |
| Days ≥ 1" Snow Depth | 0 | 0 | 0 | 0 | 0 | 0 | 0 | 0 | 0 | 0 | 0 | 0 | 0 |

## MADISON 4 N *Madison County* ELEVATION 180 ft LAT/LONG 30° 32' N / 83° 26' W

| | JAN | FEB | MAR | APR | MAY | JUN | JUL | AUG | SEP | OCT | NOV | DEC | YEAR |
|---|---|---|---|---|---|---|---|---|---|---|---|---|---|
| Maximum Temp °F | 63.9 | ***67.5*** | 75.0 | ***81.4*** | 86.9 | 91.2 | 91.9 | 91.5 | 88.7 | 81.5 | 73.4 | 66.2 | 79.9 |
| Minimum Temp °F | 41.7 | ***43.9*** | 50.3 | ***56.1*** | 62.7 | 68.3 | 70.8 | 70.6 | 67.3 | 57.8 | 50.3 | 43.6 | 56.9 |
| Mean Temp °F | 52.8 | ***55.8*** | 62.6 | ***68.8*** | 74.8 | 79.8 | 81.4 | 81.0 | 78.0 | 69.7 | 61.9 | 54.9 | 68.5 |
| Days Max Temp ≥ 90 °F | 0 | 0 | 0 | ***2*** | 9 | 20 | 24 | 24 | 16 | 3 | 0 | 0 | 98 |
| Days Max Temp ≤ 32 °F | 0 | 0 | 0 | 0 | 0 | 0 | 0 | 0 | 0 | 0 | 0 | 0 | 0 |
| Days Min Temp ≤ 32 °F | 7 | ***4*** | 1 | ***0*** | 0 | 0 | 0 | 0 | 0 | 0 | 1 | 5 | 18 |
| Days Min Temp ≤ 0 °F | 0 | 0 | 0 | 0 | 0 | 0 | 0 | 0 | 0 | 0 | 0 | 0 | 0 |
| Heating Degree Days | 380 | ***274*** | 133 | ***30*** | 1 | 0 | 0 | 0 | 1 | 30 | 147 | 323 | 1319 |
| Cooling Degree Days | 8 | 23 | 64 | 133 | 310 | 460 | 521 | 511 | 403 | 197 | 74 | 22 | 2726 |
| Total Precipitation (") | 5.27 | 4.21 | 5.57 | 2.98 | 3.43 | 5.62 | 6.52 | 5.41 | 3.74 | 2.39 | 3.19 | 4.30 | 52.63 |
| Days ≥ 0.1" Precip | 7 | ***6*** | 6 | ***4*** | 5 | 8 | 10 | ***9*** | 6 | 3 | 4 | 6 | 74 |
| Total Snowfall (") | 0.0 | 0.0 | 0.0 | 0.0 | 0.0 | 0.0 | 0.0 | 0.0 | 0.0 | 0.0 | 0.0 | 0.0 | 0.0 |
| Days ≥ 1" Snow Depth | 0 | 0 | 0 | 0 | 0 | 0 | 0 | 0 | 0 | 0 | 0 | 0 | 0 |

## MAYO *Lafayette County* ELEVATION 65 ft LAT/LONG 30° 3' N / 83° 10' W

| | JAN | FEB | MAR | APR | MAY | JUN | JUL | AUG | SEP | OCT | NOV | DEC | YEAR |
|---|---|---|---|---|---|---|---|---|---|---|---|---|---|
| Maximum Temp °F | 64.6 | 67.6 | 74.7 | 81.1 | 86.6 | 90.5 | 91.6 | 91.3 | 88.9 | 81.9 | 74.3 | 67.2 | 80.0 |
| Minimum Temp °F | 40.2 | 42.3 | 48.8 | 54.3 | 61.5 | 68.4 | 71.0 | 70.7 | 67.2 | 56.2 | 48.9 | 42.3 | 56.0 |
| Mean Temp °F | 52.4 | 55.0 | 61.8 | 67.7 | 74.1 | 79.5 | 81.3 | 81.0 | 78.1 | 69.1 | 61.5 | 54.8 | 68.0 |
| Days Max Temp ≥ 90 °F | 0 | 0 | 0 | 1 | 8 | 20 | 23 | 23 | 17 | 3 | 0 | 0 | 95 |
| Days Max Temp ≤ 32 °F | 0 | 0 | 0 | 0 | 0 | 0 | 0 | 0 | 0 | 0 | 0 | 0 | 0 |
| Days Min Temp ≤ 32 °F | 9 | 6 | 2 | 0 | 0 | 0 | 0 | 0 | 0 | 0 | 3 | 7 | 27 |
| Days Min Temp ≤ 0 °F | 0 | 0 | 0 | 0 | 0 | 0 | 0 | 0 | 0 | 0 | 0 | 0 | 0 |
| Heating Degree Days | 397 | 295 | 155 | 46 | 3 | 0 | 0 | 0 | 1 | 42 | 162 | 332 | 1433 |
| Cooling Degree Days | 10 | 21 | 60 | 122 | 292 | 457 | 527 | 514 | 399 | 187 | 77 | 27 | 2693 |
| Total Precipitation (") | 4.68 | 3.77 | 4.88 | 2.87 | 3.59 | 5.86 | 7.97 | 8.32 | 4.47 | 2.59 | 2.47 | 3.54 | 55.01 |
| Days ≥ 0.1" Precip | 6 | 6 | 5 | 4 | 6 | 9 | 11 | 11 | 7 | 4 | 4 | 5 | 78 |
| Total Snowfall (") | 0.0 | 0.0 | 0.0 | 0.0 | 0.0 | 0.0 | 0.0 | 0.0 | 0.0 | 0.0 | 0.0 | 0.0 | 0.0 |
| Days ≥ 1" Snow Depth | 0 | 0 | 0 | 0 | 0 | 0 | 0 | 0 | 0 | 0 | 0 | 0 | 0 |

## MELBOURNE REGIONL AP *Brevard County* ELEVATION 35 ft LAT/LONG 28° 7' N / 80° 39' W

| | JAN | FEB | MAR | APR | MAY | JUN | JUL | AUG | SEP | OCT | NOV | DEC | YEAR |
|---|---|---|---|---|---|---|---|---|---|---|---|---|---|
| Maximum Temp °F | 71.3 | 72.3 | 76.5 | 80.5 | 84.6 | 88.1 | 90.0 | 89.4 | 87.6 | 83.0 | 77.6 | 73.1 | 81.2 |
| Minimum Temp °F | 51.2 | 51.5 | 56.1 | 60.9 | 66.6 | 71.0 | 72.2 | 72.7 | 72.1 | 67.4 | 60.0 | 53.2 | 62.9 |
| Mean Temp °F | 61.3 | 61.9 | 66.2 | 70.7 | 75.6 | 79.5 | 81.1 | 81.1 | 79.8 | 75.2 | 68.8 | 63.2 | 72.0 |
| Days Max Temp ≥ 90 °F | 0 | 0 | 1 | 2 | 4 | 11 | 17 | 15 | 7 | 2 | 0 | 0 | 59 |
| Days Max Temp ≤ 32 °F | 0 | 0 | 0 | 0 | 0 | 0 | 0 | 0 | 0 | 0 | 0 | 0 | 0 |
| Days Min Temp ≤ 32 °F | 1 | 1 | 0 | 0 | 0 | 0 | 0 | 0 | 0 | 0 | 0 | 1 | 3 |
| Days Min Temp ≤ 0 °F | 0 | 0 | 0 | 0 | 0 | 0 | 0 | 0 | 0 | 0 | 0 | 0 | 0 |
| Heating Degree Days | 167 | 136 | 69 | 15 | 0 | 0 | 0 | 0 | 0 | 3 | 42 | 129 | 561 |
| Cooling Degree Days | 56 | 70 | 111 | 182 | 336 | 460 | 515 | 513 | 454 | 342 | 185 | 83 | 3307 |
| Total Precipitation (") | 2.30 | 2.65 | 2.96 | 1.82 | 3.92 | 6.64 | 5.10 | 5.03 | 6.50 | 4.49 | 2.90 | 2.08 | 46.39 |
| Days ≥ 0.1" Precip | 4 | 5 | 4 | 3 | 6 | 9 | 9 | 8 | 8 | 7 | 4 | 4 | 71 |
| Total Snowfall (") | 0.0 | 0.0 | 0.0 | 0.0 | 0.0 | 0.0 | 0.0 | 0.0 | 0.0 | 0.0 | 0.0 | 0.0 | 0.0 |
| Days ≥ 1" Snow Depth | 0 | 0 | 0 | 0 | 0 | 0 | 0 | 0 | 0 | 0 | 0 | 0 | 0 |

## MIAMI BEACH *Dade County* ELEVATION 5 ft LAT/LONG 25° 47' N / 80° 8' W

| | JAN | FEB | MAR | APR | MAY | JUN | JUL | AUG | SEP | OCT | NOV | DEC | YEAR |
|---|---|---|---|---|---|---|---|---|---|---|---|---|---|
| Maximum Temp °F | 74.0 | 74.3 | 76.3 | 79.2 | 82.1 | 85.4 | 87.0 | 87.2 | 86.1 | 83.0 | 78.8 | 75.4 | 80.7 |
| Minimum Temp °F | 62.8 | 63.0 | 66.6 | 70.2 | 73.8 | 76.5 | 78.2 | 78.3 | 77.4 | 74.4 | 69.7 | 65.1 | 71.3 |
| Mean Temp °F | 68.4 | 68.7 | 71.4 | 74.7 | 78.0 | 81.0 | 82.6 | 82.8 | 81.8 | 78.7 | 74.3 | 70.2 | 76.1 |
| Days Max Temp ≥ 90 °F | 0 | 0 | 0 | 0 | 1 | 2 | 3 | 3 | 2 | 1 | 0 | 0 | 12 |
| Days Max Temp ≤ 32 °F | 0 | 0 | 0 | 0 | 0 | 0 | 0 | 0 | 0 | 0 | 0 | 0 | 0 |
| Days Min Temp ≤ 32 °F | 0 | 0 | 0 | 0 | 0 | 0 | 0 | 0 | 0 | 0 | 0 | 0 | 0 |
| Days Min Temp ≤ 0 °F | 0 | 0 | 0 | 0 | 0 | 0 | 0 | 0 | 0 | 0 | 0 | 0 | 0 |
| Heating Degree Days | 48 | 36 | 16 | 2 | 0 | 0 | 0 | 0 | 0 | 0 | 6 | 30 | 138 |
| Cooling Degree Days | 165 | 170 | 225 | 294 | 411 | 499 | 566 | 573 | 518 | 446 | 317 | 213 | 4397 |
| Total Precipitation (") | 2.59 | 2.26 | 2.06 | 2.32 | 5.44 | 7.20 | 3.76 | 4.67 | 6.52 | 4.92 | 3.21 | 1.65 | 46.60 |
| Days ≥ 0.1" Precip | 4 | 4 | 4 | 4 | 6 | 9 | 8 | 9 | 9 | 7 | 5 | 3 | 72 |
| Total Snowfall (") | 0.0 | 0.0 | 0.0 | 0.0 | 0.0 | 0.0 | 0.0 | 0.0 | 0.0 | 0.0 | 0.0 | 0.0 | 0.0 |
| Days ≥ 1" Snow Depth | 0 | 0 | 0 | 0 | 0 | 0 | 0 | 0 | 0 | 0 | 0 | 0 | 0 |

## MIAMI INTL ARPT *Dade County* ELEVATION 23 ft LAT/LONG 25° 49' N / 80° 17' W

| | JAN | FEB | MAR | APR | MAY | JUN | JUL | AUG | SEP | OCT | NOV | DEC | YEAR |
|---|---|---|---|---|---|---|---|---|---|---|---|---|---|
| Maximum Temp °F | 75.7 | 76.5 | 79.3 | 82.6 | 85.5 | 87.9 | 89.3 | 89.3 | 88.0 | 84.9 | 80.6 | 77.2 | 83.1 |
| Minimum Temp °F | 59.9 | 60.7 | 64.3 | 68.2 | 72.3 | 75.3 | 76.5 | 76.8 | 76.0 | 72.5 | 67.2 | 62.1 | 69.3 |
| Mean Temp °F | 67.8 | 68.6 | 71.8 | 75.4 | 78.9 | 81.6 | 83.0 | 83.1 | 82.0 | 78.7 | 73.9 | 69.7 | 76.2 |
| Days Max Temp ≥ 90 °F | 0 | 0 | 0 | 1 | 4 | 10 | 15 | 15 | 10 | 2 | 0 | 0 | 57 |
| Days Max Temp ≤ 32 °F | 0 | 0 | 0 | 0 | 0 | 0 | 0 | 0 | 0 | 0 | 0 | 0 | 0 |
| Days Min Temp ≤ 32 °F | 0 | 0 | 0 | 0 | 0 | 0 | 0 | 0 | 0 | 0 | 0 | 0 | 0 |
| Days Min Temp ≤ 0 °F | 0 | 0 | 0 | 0 | 0 | 0 | 0 | 0 | 0 | 0 | 0 | 0 | 0 |
| Heating Degree Days | 58 | 40 | 17 | 1 | 0 | 0 | 0 | 0 | 0 | 0 | 7 | 37 | 160 |
| Cooling Degree Days | 154 | 171 | 237 | 310 | 441 | 529 | 584 | 587 | 531 | 453 | 310 | 201 | 4508 |
| Total Precipitation (") | 2.16 | 2.25 | 2.62 | 3.10 | 5.98 | 9.04 | 6.00 | 7.94 | 8.20 | 6.33 | 3.31 | 1.72 | 58.65 |
| Days ≥ 0.1" Precip | 4 | 4 | 4 | 4 | 8 | 11 | 11 | 12 | 12 | 8 | 5 | 3 | 86 |
| Total Snowfall (") | 0.0 | 0.0 | 0.0 | 0.0 | 0.0 | 0.0 | 0.0 | 0.0 | 0.0 | 0.0 | 0.0 | 0.0 | 0.0 |
| Days ≥ 1" Snow Depth | 0 | 0 | 0 | 0 | 0 | 0 | 0 | 0 | 0 | 0 | 0 | 0 | 0 |

## MILTON EXPERIMENT ST *Santa Rosa County* ELEVATION 217 ft LAT/LONG 30° 47' N / 87° 9' W

| | JAN | FEB | MAR | APR | MAY | JUN | JUL | AUG | SEP | OCT | NOV | DEC | YEAR |
|---|---|---|---|---|---|---|---|---|---|---|---|---|---|
| Maximum Temp °F | 61.0 | 64.6 | 71.5 | 78.8 | 84.8 | 90.1 | 91.2 | 90.8 | 87.8 | 80.3 | 71.2 | 64.0 | 78.0 |
| Minimum Temp °F | 39.1 | 41.2 | 48.2 | 54.7 | 61.8 | 68.4 | 70.8 | 70.2 | 66.1 | 54.9 | 47.1 | 41.6 | 55.3 |
| Mean Temp °F | 50.0 | 52.9 | 59.9 | 66.8 | 73.3 | 79.3 | 81.0 | 80.5 | 77.0 | 67.6 | 59.1 | 52.9 | 66.7 |
| Days Max Temp ≥ 90 °F | 0 | 0 | 0 | 0 | 5 | 18 | 22 | 22 | 13 | 1 | 0 | 0 | 81 |
| Days Max Temp ≤ 32 °F | 0 | 0 | 0 | 0 | 0 | 0 | 0 | 0 | 0 | 0 | 0 | 0 | 0 |
| Days Min Temp ≤ 32 °F | 10 | 7 | 2 | 0 | 0 | 0 | 0 | 0 | 0 | 0 | 3 | 8 | 30 |
| Days Min Temp ≤ 0 °F | 0 | 0 | 0 | 0 | 0 | 0 | 0 | 0 | 0 | 0 | 0 | 0 | 0 |
| Heating Degree Days | 465 | 345 | 192 | 55 | 3 | 0 | 0 | 0 | 3 | 55 | 210 | 386 | 1714 |
| Cooling Degree Days | 4 | 10 | 38 | 97 | 263 | 439 | 511 | 499 | 366 | 142 | 45 | 14 | 2428 |
| Total Precipitation (") | 5.88 | 5.38 | 6.83 | 3.85 | 4.96 | 7.45 | 8.09 | 6.95 | 5.57 | 3.79 | 4.66 | 4.68 | 68.09 |
| Days ≥ 0.1" Precip | 7 | 7 | 7 | 5 | 6 | 8 | 11 | 11 | 7 | 4 | 6 | 6 | 85 |
| Total Snowfall (") | 0.1 | 0.1 | 0.1 | 0.0 | 0.0 | 0.0 | 0.0 | 0.0 | 0.0 | 0.0 | 0.0 | 0.0 | 0.3 |
| Days ≥ 1" Snow Depth | 0 | 0 | 0 | 0 | 0 | 0 | 0 | 0 | 0 | 0 | 0 | 0 | 0 |

## MONTICELLO 3 W *Jefferson County* ELEVATION 210 ft LAT/LONG 30° 32' N / 83° 52' W

| | JAN | FEB | MAR | APR | MAY | JUN | JUL | AUG | SEP | OCT | NOV | DEC | YEAR |
|---|---|---|---|---|---|---|---|---|---|---|---|---|---|
| Maximum Temp °F | 61.7 | 64.8 | 71.9 | 78.6 | 84.5 | 89.3 | 90.4 | 89.8 | 87.0 | 79.7 | 71.8 | 64.8 | 77.9 |
| Minimum Temp °F | 37.5 | 39.6 | 46.8 | 52.9 | 59.9 | 66.8 | 69.5 | 69.0 | 65.2 | 53.6 | 45.7 | 39.4 | 53.8 |
| Mean Temp °F | 49.6 | 52.2 | 59.4 | 65.8 | 72.2 | 78.1 | 79.9 | 79.4 | 76.1 | 66.7 | 58.8 | 52.1 | 65.9 |
| Days Max Temp ≥ 90 °F | 0 | 0 | 0 | 0 | 4 | 15 | 19 | 18 | 10 | 1 | 0 | 0 | 67 |
| Days Max Temp ≤ 32 °F | 0 | 0 | 0 | 0 | 0 | 0 | 0 | 0 | 0 | 0 | 0 | 0 | 0 |
| Days Min Temp ≤ 32 °F | 12 | 8 | 2 | 0 | 0 | 0 | 0 | 0 | 0 | 0 | 4 | 11 | 37 |
| Days Min Temp ≤ 0 °F | 0 | 0 | 0 | 0 | 0 | 0 | 0 | 0 | 0 | 0 | 0 | 0 | 0 |
| Heating Degree Days | 475 | 364 | 205 | 69 | 7 | 0 | 0 | 0 | 2 | 67 | 220 | 405 | 1814 |
| Cooling Degree Days | 4 | 12 | 39 | 88 | 246 | 418 | 488 | 470 | 343 | 133 | 49 | 15 | 2305 |
| Total Precipitation (") | 5.61 | 4.82 | 5.84 | 3.67 | 4.25 | 5.65 | 6.61 | 6.72 | 3.86 | 3.11 | 3.42 | 4.14 | 57.70 |
| Days ≥ 0.1" Precip | 7 | 6 | 6 | 4 | 6 | 8 | 10 | 10 | 6 | 4 | 5 | 6 | 78 |
| Total Snowfall (") | 0.0 | 0.0 | 0.0 | 0.0 | 0.0 | 0.0 | 0.0 | 0.0 | 0.0 | 0.0 | 0.0 | 0.0 | 0.0 |
| Days ≥ 1" Snow Depth | 0 | 0 | 0 | 0 | 0 | 0 | 0 | 0 | 0 | 0 | 0 | 0 | 0 |

## MOORE HAVEN LOCK 1 *Glades County* ELEVATION 20 ft LAT/LONG 26° 50 ' N / 81° 5 ' W

| | JAN | FEB | MAR | APR | MAY | JUN | JUL | AUG | SEP | OCT | NOV | DEC | YEAR |
|---|---|---|---|---|---|---|---|---|---|---|---|---|---|
| Maximum Temp °F | 73.9 | 75.1 | 79.2 | 83.5 | 87.4 | 90.0 | 91.4 | 91.1 | 89.3 | 85.0 | 79.6 | 75.1 | 83.4 |
| Minimum Temp °F | 51.5 | 51.9 | 56.4 | 60.7 | 65.7 | 70.7 | 72.1 | 72.5 | 72.2 | 67.0 | 60.1 | 53.8 | 62.9 |
| Mean Temp °F | 62.7 | 63.5 | 67.8 | 72.2 | 76.6 | 80.4 | 81.8 | 81.8 | 80.8 | 76.0 | 69.8 | 64.5 | 73.2 |
| Days Max Temp ≥ 90 °F | 0 | 0 | 1 | 4 | 10 | 17 | 24 | 23 | 16 | 5 | 0 | 0 | 100 |
| Days Max Temp ≤ 32 °F | 0 | 0 | 0 | 0 | 0 | 0 | 0 | 0 | 0 | 0 | 0 | 0 | 0 |
| Days Min Temp ≤ 32 °F | 1 | 0 | 0 | 0 | 0 | 0 | 0 | 0 | 0 | 0 | 0 | 1 | 2 |
| Days Min Temp ≤ 0 °F | 0 | 0 | 0 | 0 | 0 | 0 | 0 | 0 | 0 | 0 | 0 | 0 | 0 |
| Heating Degree Days | 139 | 109 | 49 | 8 | 0 | 0 | 0 | 0 | 0 | 2 | 30 | 102 | 439 |
| Cooling Degree Days | 77 | 92 | 149 | 217 | 370 | 485 | 537 | 542 | 484 | 359 | 204 | 98 | 3614 |
| Total Precipitation (") | 2.08 | 2.13 | 3.13 | 1.89 | 3.98 | 7.76 | 6.88 | 6.45 | 6.29 | 3.39 | 1.76 | 1.63 | 47.37 |
| Days ≥ 0.1" Precip | 4 | 4 | 4 | 3 | 6 | 11 | 11 | 10 | 9 | 5 | 3 | 3 | 73 |
| Total Snowfall (") | 0.0 | 0.0 | 0.0 | 0.0 | 0.0 | 0.0 | 0.0 | 0.0 | 0.0 | 0.0 | 0.0 | 0.0 | 0.0 |
| Days ≥ 1" Snow Depth | 0 | 0 | 0 | 0 | 0 | 0 | 0 | 0 | 0 | 0 | 0 | 0 | 0 |

## MOUNTAIN LAKE *Polk County* ELEVATION 161 ft LAT/LONG 27° 56 ' N / 81° 36 ' W

| | JAN | FEB | MAR | APR | MAY | JUN | JUL | AUG | SEP | OCT | NOV | DEC | YEAR |
|---|---|---|---|---|---|---|---|---|---|---|---|---|---|
| Maximum Temp °F | 74.0 | 75.8 | 80.5 | 85.4 | 89.7 | 91.9 | 92.8 | 92.5 | 90.7 | 85.9 | 79.9 | 75.3 | 84.5 |
| Minimum Temp °F | 48.6 | 49.5 | 54.3 | 58.6 | 63.8 | 69.3 | 70.6 | 71.1 | 70.1 | 63.4 | 56.2 | 50.6 | 60.5 |
| Mean Temp °F | 61.3 | 62.7 | 67.4 | 72.1 | 76.8 | 80.6 | 81.7 | 81.8 | 80.4 | 74.7 | 68.0 | 63.0 | 72.5 |
| Days Max Temp ≥ 90 °F | 0 | 0 | 3 | 7 | 17 | 22 | 27 | 27 | 21 | 6 | 0 | 0 | 130 |
| Days Max Temp ≤ 32 °F | 0 | 0 | 0 | 0 | 0 | 0 | 0 | 0 | 0 | 0 | 0 | 0 | 0 |
| Days Min Temp ≤ 32 °F | 3 | 2 | 0 | 0 | 0 | 0 | 0 | 0 | 0 | 0 | 0 | 2 | 7 |
| Days Min Temp ≤ 0 °F | 0 | 0 | 0 | 0 | 0 | 0 | 0 | 0 | 0 | 0 | 0 | 0 | 0 |
| Heating Degree Days | 172 | 128 | 57 | 9 | 0 | 0 | 0 | 0 | 0 | 4 | 49 | 130 | 549 |
| Cooling Degree Days | 59 | 76 | 131 | 210 | 366 | 482 | 521 | 534 | 466 | 312 | 160 | 74 | 3391 |
| Total Precipitation (") | 2.44 | 2.72 | 3.19 | 1.79 | 4.25 | 7.74 | 7.57 | 6.91 | 5.77 | 2.48 | 2.09 | 2.09 | 49.04 |
| Days ≥ 0.1" Precip | 4 | 4 | 5 | 3 | 6 | 10 | 12 | 11 | 8 | 4 | 4 | 3 | 74 |
| Total Snowfall (") | 0.0 | 0.0 | 0.0 | 0.0 | 0.0 | 0.0 | 0.0 | 0.0 | 0.0 | 0.0 | 0.0 | 0.0 | 0.0 |
| Days ≥ 1" Snow Depth | 0 | 0 | 0 | 0 | 0 | 0 | 0 | 0 | 0 | 0 | 0 | 0 | 0 |

## MYAKKA RIVER STATE P *Sarasota County* ELEVATION 20 ft LAT/LONG 27° 15 ' N / 82° 19 ' W

| | JAN | FEB | MAR | APR | MAY | JUN | JUL | AUG | SEP | OCT | NOV | DEC | YEAR |
|---|---|---|---|---|---|---|---|---|---|---|---|---|---|
| Maximum Temp °F | 74.2 | 76.0 | 80.4 | 85.8 | 90.6 | 91.8 | 92.5 | 92.4 | 91.2 | 86.8 | 80.7 | 75.8 | 84.9 |
| Minimum Temp °F | 48.8 | 49.8 | 54.2 | 57.6 | 62.7 | 68.8 | 70.8 | 72.0 | 71.0 | 64.4 | 57.2 | 51.2 | 60.7 |
| Mean Temp °F | 61.5 | 62.9 | 67.3 | 71.7 | 76.6 | 80.3 | 81.7 | 82.3 | 81.1 | 75.6 | 69.0 | 63.5 | 72.8 |
| Days Max Temp ≥ 90 °F | 0 | 0 | 2 | 9 | 20 | 24 | 27 | 26 | 23 | 10 | 1 | 0 | 142 |
| Days Max Temp ≤ 32 °F | 0 | 0 | 0 | 0 | 0 | 0 | 0 | 0 | 0 | 0 | 0 | 0 | 0 |
| Days Min Temp ≤ 32 °F | 2 | 1 | 0 | 0 | 0 | 0 | 0 | 0 | 0 | 0 | 0 | 1 | 4 |
| Days Min Temp ≤ 0 °F | 0 | 0 | 0 | 0 | 0 | 0 | 0 | 0 | 0 | 0 | 0 | 0 | 0 |
| Heating Degree Days | 160 | 120 | 56 | 10 | 0 | 0 | 0 | 0 | 0 | 2 | 41 | 121 | 510 |
| Cooling Degree Days | 67 | 87 | 142 | 211 | 369 | 483 | 527 | 550 | 496 | 354 | 191 | 94 | 3571 |
| Total Precipitation (") | 2.93 | 2.81 | 3.47 | 1.79 | 3.41 | 9.45 | 9.51 | 9.58 | 7.52 | 3.04 | 1.97 | 2.04 | 57.52 |
| Days ≥ 0.1" Precip | 4 | 4 | 4 | 3 | 5 | 11 | 13 | 13 | 10 | 4 | 4 | 3 | 78 |
| Total Snowfall (") | 0.0 | 0.0 | 0.0 | 0.0 | 0.0 | 0.0 | 0.0 | 0.0 | 0.0 | 0.0 | 0.0 | 0.0 | 0.0 |
| Days ≥ 1" Snow Depth | 0 | 0 | 0 | 0 | 0 | 0 | 0 | 0 | 0 | 0 | 0 | 0 | 0 |

## NAPLES *Collier County* ELEVATION 10 ft LAT/LONG 26° 9 ' N / 81° 49 ' W

| | JAN | FEB | MAR | APR | MAY | JUN | JUL | AUG | SEP | OCT | NOV | DEC | YEAR |
|---|---|---|---|---|---|---|---|---|---|---|---|---|---|
| Maximum Temp °F | 76.6 | 77.3 | 80.6 | 84.5 | 87.9 | 90.1 | 91.3 | 91.7 | 90.7 | 87.4 | 82.7 | 78.2 | 84.9 |
| Minimum Temp °F | 53.7 | 54.2 | 58.0 | 61.6 | 66.5 | 71.4 | 72.3 | 72.9 | 72.4 | 67.7 | 61.1 | 55.5 | 63.9 |
| Mean Temp °F | 65.2 | 65.7 | 69.3 | 73.0 | 77.2 | 80.8 | 81.8 | 82.3 | 81.6 | 77.6 | 71.9 | 66.8 | 74.4 |
| Days Max Temp ≥ 90 °F | 0 | 0 | 0 | 2 | 8 | 19 | 26 | 27 | 23 | 8 | 1 | 0 | 114 |
| Days Max Temp ≤ 32 °F | 0 | 0 | 0 | 0 | 0 | 0 | 0 | 0 | 0 | 0 | 0 | 0 | 0 |
| Days Min Temp ≤ 32 °F | 1 | 0 | 0 | 0 | 0 | 0 | 0 | 0 | 0 | 0 | 0 | 0 | 1 |
| Days Min Temp ≤ 0 °F | 0 | 0 | 0 | 0 | 0 | 0 | 0 | 0 | 0 | 0 | 0 | 0 | 0 |
| Heating Degree Days | 88 | 68 | 27 | 4 | 0 | 0 | 0 | 0 | 0 | 0 | 15 | 63 | 265 |
| Cooling Degree Days | 99 | 114 | 174 | 250 | 388 | 496 | 539 | 552 | 508 | 411 | 253 | 134 | 3918 |
| Total Precipitation (") | 2.08 | 2.35 | 2.38 | 1.72 | 4.05 | 8.56 | 8.45 | 8.43 | 8.00 | 3.45 | 1.74 | 1.44 | 52.65 |
| Days ≥ 0.1" Precip | 4 | 4 | 3 | 2 | 5 | 10 | 12 | 13 | 11 | 5 | 3 | 3 | 75 |
| Total Snowfall (") | 0.0 | 0.0 | 0.0 | 0.0 | 0.0 | 0.0 | 0.0 | 0.0 | 0.0 | 0.0 | 0.0 | 0.0 | 0.0 |
| Days ≥ 1" Snow Depth | 0 | 0 | 0 | 0 | 0 | 0 | 0 | 0 | 0 | 0 | 0 | 0 | 0 |

## NICEVILLE *Okaloosa County* ELEVATION 60 ft LAT/LONG 30° 31' N / 86° 30' W

| | JAN | FEB | MAR | APR | MAY | JUN | JUL | AUG | SEP | OCT | NOV | DEC | YEAR |
|---|---|---|---|---|---|---|---|---|---|---|---|---|---|
| Maximum Temp °F | 60.2 | 63.5 | 70.2 | 78.2 | 84.2 | 89.9 | 91.1 | 90.5 | 87.9 | 80.3 | 71.0 | 64.3 | 77.6 |
| Minimum Temp °F | 36.9 | 39.3 | 46.2 | 53.1 | 60.9 | 67.8 | 70.8 | 70.5 | 66.2 | 54.2 | 46.0 | 39.6 | 54.3 |
| Mean Temp °F | 48.6 | 51.5 | 58.2 | 65.6 | 72.6 | 78.7 | 81.0 | 80.5 | 77.1 | 67.3 | 58.6 | 51.9 | 66.0 |
| Days Max Temp ≥ 90 °F | 0 | 0 | 0 | 0 | 3 | 17 | 22 | 20 | 12 | 1 | 0 | 0 | 75 |
| Days Max Temp ≤ 32 °F | 0 | 0 | 0 | 0 | 0 | 0 | 0 | 0 | 0 | 0 | 0 | 0 | 0 |
| Days Min Temp ≤ 32 °F | 12 | 8 | 3 | 0 | 0 | 0 | 0 | 0 | 0 | 0 | 3 | 10 | 36 |
| Days Min Temp ≤ 0 °F | 0 | 0 | 0 | 0 | 0 | 0 | 0 | 0 | 0 | 0 | 0 | 0 | 0 |
| Heating Degree Days | 504 | 380 | 228 | 64 | 3 | 0 | 0 | 0 | 2 | 57 | 218 | 407 | 1863 |
| Cooling Degree Days | na | 7 | na | 78 | na | na | 515 | 505 | 372 | 141 | na | na | na |
| Total Precipitation (") | 5.19 | 5.72 | 6.41 | 3.59 | 4.47 | 6.45 | 9.39 | 7.14 | 5.63 | 4.71 | 4.05 | 4.55 | 67.30 |
| Days ≥ 0.1" Precip | 6 | 6 | 6 | 4 | 5 | 7 | 10 | 9 | 6 | 3 | 5 | 5 | 72 |
| Total Snowfall (") | 0.0 | 0.0 | 0.0 | 0.0 | 0.0 | 0.0 | 0.0 | 0.0 | 0.0 | 0.0 | 0.0 | 0.0 | 0.0 |
| Days ≥ 1" Snow Depth | 0 | 0 | 0 | 0 | 0 | 0 | 0 | 0 | 0 | 0 | 0 | 0 | 0 |

## OCALA *Marion County* ELEVATION 89 ft LAT/LONG 29° 11' N / 82° 8' W

| | JAN | FEB | MAR | APR | MAY | JUN | JUL | AUG | SEP | OCT | NOV | DEC | YEAR |
|---|---|---|---|---|---|---|---|---|---|---|---|---|---|
| Maximum Temp °F | 70.0 | 72.3 | 78.2 | 83.6 | 88.4 | 91.1 | 92.2 | 91.8 | 89.9 | 84.2 | 77.4 | 72.1 | 82.6 |
| Minimum Temp °F | 45.6 | 46.6 | 52.3 | 56.5 | 62.9 | 69.0 | 70.9 | 70.7 | 68.6 | 61.3 | 53.5 | 47.3 | 58.8 |
| Mean Temp °F | 57.8 | 59.5 | 65.2 | 70.1 | 75.6 | 80.1 | 81.6 | 81.2 | 79.3 | 72.7 | 65.5 | 59.7 | 70.7 |
| Days Max Temp ≥ 90 °F | 0 | 0 | 1 | 4 | 13 | 21 | 26 | 25 | 19 | 4 | 0 | 0 | 113 |
| Days Max Temp ≤ 32 °F | 0 | 0 | 0 | 0 | 0 | 0 | 0 | 0 | 0 | 0 | 0 | 0 | 0 |
| Days Min Temp ≤ 32 °F | 5 | 3 | 1 | 0 | 0 | 0 | 0 | 0 | 0 | 0 | 1 | 3 | 13 |
| Days Min Temp ≤ 0 °F | 0 | 0 | 0 | 0 | 0 | 0 | 0 | 0 | 0 | 0 | 0 | 0 | 0 |
| Heating Degree Days | 251 | 189 | 90 | 22 | 0 | 0 | 0 | 0 | 0 | 11 | 83 | 202 | 848 |
| Cooling Degree Days | 30 | 45 | 96 | 158 | 323 | 462 | 520 | 511 | 421 | 248 | 117 | 44 | 2975 |
| Total Precipitation (") | 3.33 | 3.53 | 3.76 | 2.74 | 4.38 | 7.32 | 7.10 | 6.57 | 5.00 | 2.64 | 2.23 | 2.50 | 51.10 |
| Days ≥ 0.1" Precip | 6 | 5 | 5 | 3 | 6 | 10 | 12 | 12 | 8 | 4 | 4 | 4 | 79 |
| Total Snowfall (") | 0.0 | 0.0 | 0.0 | 0.0 | 0.0 | 0.0 | 0.0 | 0.0 | 0.0 | 0.0 | 0.0 | 0.0 | 0.0 |
| Days ≥ 1" Snow Depth | 0 | 0 | 0 | 0 | 0 | 0 | 0 | 0 | 0 | 0 | 0 | 0 | 0 |

## OKEECHOBEE HRCN GATE *Okeechobee County* ELEVATION 30 ft LAT/LONG 27° 13' N / 80° 48' W

| | JAN | FEB | MAR | APR | MAY | JUN | JUL | AUG | SEP | OCT | NOV | DEC | YEAR |
|---|---|---|---|---|---|---|---|---|---|---|---|---|---|
| Maximum Temp °F | 73.6 | 75.1 | 78.1 | 82.6 | 86.6 | 89.0 | 90.4 | 90.3 | 89.2 | 84.8 | 80.1 | 74.7 | 82.9 |
| Minimum Temp °F | 51.1 | 53.2 | 57.4 | 62.0 | 67.2 | 71.7 | 73.6 | 73.6 | 72.5 | 66.8 | 60.1 | 53.5 | 63.6 |
| Mean Temp °F | 62.4 | 64.2 | 67.8 | 72.4 | 76.9 | 80.4 | 82.0 | 82.0 | 80.9 | 75.8 | 70.1 | 64.2 | 73.3 |
| Days Max Temp ≥ 90 °F | 0 | 0 | 0 | 1 | 7 | 15 | 19 | 20 | 16 | 4 | 0 | 0 | 82 |
| Days Max Temp ≤ 32 °F | 0 | 0 | 0 | 0 | 0 | 0 | 0 | 0 | 0 | 0 | 0 | 0 | 0 |
| Days Min Temp ≤ 32 °F | 1 | 0 | 0 | 0 | 0 | 0 | 0 | 0 | 0 | 0 | 0 | 1 | 2 |
| Days Min Temp ≤ 0 °F | 0 | 0 | 0 | 0 | 0 | 0 | 0 | 0 | 0 | 0 | 0 | 0 | 0 |
| Heating Degree Days | 139 | 91 | 46 | 7 | 0 | 0 | 0 | 0 | 0 | 1 | 23 | 106 | 413 |
| Cooling Degree Days | 79 | 97 | 156 | 224 | 382 | 481 | 541 | 548 | 482 | 341 | 200 | 102 | 3633 |
| Total Precipitation (") | 2.22 | 2.33 | 2.94 | 2.07 | 3.73 | 6.48 | 6.29 | 6.54 | 5.76 | 3.77 | 2.31 | 1.59 | 46.03 |
| Days ≥ 0.1" Precip | 4 | 4 | 5 | 3 | 6 | 10 | 10 | 11 | 10 | 6 | 4 | 3 | 76 |
| Total Snowfall (") | 0.0 | 0.0 | 0.0 | 0.0 | 0.0 | 0.0 | 0.0 | 0.0 | 0.0 | 0.0 | 0.0 | 0.0 | 0.0 |
| Days ≥ 1" Snow Depth | 0 | 0 | 0 | 0 | 0 | 0 | 0 | 0 | 0 | 0 | 0 | 0 | 0 |

## PARRISH *Manatee County* ELEVATION 39 ft LAT/LONG 27° 35' N / 82° 25' W

| | JAN | FEB | MAR | APR | MAY | JUN | JUL | AUG | SEP | OCT | NOV | DEC | YEAR |
|---|---|---|---|---|---|---|---|---|---|---|---|---|---|
| Maximum Temp °F | 73.4 | 74.5 | 78.6 | 83.5 | 88.0 | 90.6 | 91.4 | 91.3 | 90.1 | 85.7 | 79.8 | 74.8 | 83.5 |
| Minimum Temp °F | 50.1 | 50.6 | 55.0 | 58.7 | 64.2 | 69.9 | 71.6 | 72.0 | 71.1 | 64.4 | 57.4 | 51.5 | 61.4 |
| Mean Temp °F | 61.8 | 62.6 | 66.8 | 71.1 | 76.1 | 80.3 | 81.5 | 81.7 | 80.6 | 75.1 | 68.6 | 63.2 | 72.5 |
| Days Max Temp ≥ 90 °F | 0 | 0 | 0 | 3 | 11 | 21 | 26 | 25 | 21 | 6 | 0 | 0 | 113 |
| Days Max Temp ≤ 32 °F | 0 | 0 | 0 | 0 | 0 | 0 | 0 | 0 | 0 | 0 | 0 | 0 | 0 |
| Days Min Temp ≤ 32 °F | 2 | 1 | 0 | 0 | 0 | 0 | 0 | 0 | 0 | 0 | 0 | 1 | 4 |
| Days Min Temp ≤ 0 °F | 0 | 0 | 0 | 0 | 0 | 0 | 0 | 0 | 0 | 0 | 0 | 0 | 0 |
| Heating Degree Days | 160 | 123 | 61 | 14 | 1 | 0 | 0 | 0 | 0 | 4 | 42 | 128 | 533 |
| Cooling Degree Days | 56 | 66 | 111 | 175 | 331 | 465 | 515 | 518 | 465 | 318 | 166 | 77 | 3263 |
| Total Precipitation (") | 2.56 | 3.13 | 3.17 | 1.77 | 3.17 | 7.62 | 7.91 | 9.50 | 7.27 | 2.79 | 1.95 | 2.02 | 52.86 |
| Days ≥ 0.1" Precip | 4 | 4 | 4 | 3 | 4 | 9 | 11 | 13 | 9 | 4 | 3 | 3 | 71 |
| Total Snowfall (") | 0.0 | 0.0 | 0.0 | 0.0 | 0.0 | 0.0 | 0.0 | 0.0 | 0.0 | 0.0 | 0.0 | 0.0 | 0.0 |
| Days ≥ 1" Snow Depth | 0 | 0 | 0 | 0 | 0 | 0 | 0 | 0 | 0 | 0 | 0 | 0 | 0 |

## PENSACOLA REGIONL AP *Escambia County* ELEVATION 112 ft LAT/LONG 30° 28 ' N / 87° 12 ' W

| | JAN | FEB | MAR | APR | MAY | JUN | JUL | AUG | SEP | OCT | NOV | DEC | YEAR |
|---|---|---|---|---|---|---|---|---|---|---|---|---|---|
| Maximum Temp °F | 60.3 | 63.3 | 69.5 | 76.7 | 83.0 | 88.8 | 90.2 | 89.4 | 86.7 | 79.2 | 70.2 | 63.4 | 76.7 |
| Minimum Temp °F | 42.2 | 44.3 | 51.2 | 58.4 | 65.6 | 71.9 | 74.3 | 73.8 | 70.3 | 59.6 | 50.8 | 44.9 | 58.9 |
| Mean Temp °F | 51.2 | 53.8 | 60.4 | 67.6 | 74.3 | 80.4 | 82.3 | 81.6 | 78.5 | 69.4 | 60.5 | 54.2 | 67.9 |
| Days Max Temp ≥ 90 °F | 0 | 0 | 0 | 0 | 2 | 13 | 18 | 16 | 9 | 0 | 0 | 0 | 58 |
| Days Max Temp ≤ 32 °F | 0 | 0 | 0 | 0 | 0 | 0 | 0 | 0 | 0 | 0 | 0 | 0 | 0 |
| Days Min Temp ≤ 32 °F | 6 | 4 | 1 | 0 | 0 | 0 | 0 | 0 | 0 | 0 | 1 | 4 | 16 |
| Days Min Temp ≤ 0 °F | 0 | 0 | 0 | 0 | 0 | 0 | 0 | 0 | 0 | 0 | 0 | 0 | 0 |
| Heating Degree Days | 428 | 320 | 177 | 39 | 1 | 0 | 0 | 0 | 2 | 33 | 177 | 346 | 1523 |
| Cooling Degree Days | 4 | 10 | 40 | 106 | 292 | 473 | 552 | 534 | 408 | 175 | 52 | 16 | 2662 |
| Total Precipitation (") | 4.85 | 5.17 | 5.99 | 3.42 | 4.54 | 6.78 | 7.35 | 7.49 | 5.43 | 4.43 | 3.47 | 4.03 | 62.95 |
| Days ≥ 0.1" Precip | 7 | 6 | 6 | 4 | 5 | 7 | 10 | 9 | 6 | 4 | 5 | 6 | 75 |
| Total Snowfall (") | 0.1 | 0.1 | 0.0 | 0.0 | 0.0 | 0.0 | 0.0 | 0.0 | 0.0 | 0.0 | 0.0 | 0.0 | 0.2 |
| Days ≥ 1" Snow Depth | 0 | 0 | 0 | 0 | 0 | 0 | 0 | 0 | 0 | 0 | 0 | 0 | 0 |

## PERRY *Taylor County* ELEVATION 46 ft LAT/LONG 30° 5 ' N / 83° 34 ' W

| | JAN | FEB | MAR | APR | MAY | JUN | JUL | AUG | SEP | OCT | NOV | DEC | YEAR |
|---|---|---|---|---|---|---|---|---|---|---|---|---|---|
| Maximum Temp °F | 66.5 | 69.3 | 75.7 | 82.1 | 87.6 | 91.5 | 92.4 | 92.1 | 90.2 | 83.4 | 76.1 | 69.6 | 81.4 |
| Minimum Temp °F | 41.4 | 42.7 | 48.6 | 54.0 | 61.4 | 67.9 | 70.6 | 70.3 | 67.2 | 57.0 | 48.7 | 43.0 | 56.1 |
| Mean Temp °F | 54.0 | 56.0 | 62.2 | 68.1 | 74.5 | 79.7 | 81.5 | 81.2 | 78.7 | 70.3 | 62.4 | 56.3 | 68.7 |
| Days Max Temp ≥ 90 °F | 0 | 0 | 0 | 2 | 10 | 22 | 25 | 25 | 20 | 4 | 0 | 0 | 108 |
| Days Max Temp ≤ 32 °F | 0 | 0 | 0 | 0 | 0 | 0 | 0 | 0 | 0 | 0 | 0 | 0 | 0 |
| Days Min Temp ≤ 32 °F | 8 | 6 | 2 | 0 | 0 | 0 | 0 | 0 | 0 | 0 | 3 | 7 | 26 |
| Days Min Temp ≤ 0 °F | 0 | 0 | 0 | 0 | 0 | 0 | 0 | 0 | 0 | 0 | 0 | 0 | 0 |
| Heating Degree Days | 348 | 265 | 139 | 39 | 2 | 0 | 0 | 0 | 0 | 29 | 143 | 287 | 1252 |
| Cooling Degree Days | 11 | 24 | 62 | 127 | 306 | 453 | 526 | 519 | 415 | 205 | 88 | 29 | 2765 |
| Total Precipitation (") | 4.73 | 4.10 | 5.29 | 3.34 | 3.73 | 6.25 | 8.45 | 8.94 | 4.92 | 2.73 | 2.64 | 3.60 | 58.72 |
| Days ≥ 0.1" Precip | 7 | 6 | 6 | 4 | 6 | 8 | 12 | 12 | 7 | 4 | 4 | 5 | 81 |
| Total Snowfall (") | 0.0 | 0.0 | 0.0 | 0.0 | 0.0 | 0.0 | 0.0 | 0.0 | 0.0 | 0.0 | 0.0 | 0.0 | 0.0 |
| Days ≥ 1" Snow Depth | 0 | 0 | 0 | 0 | 0 | 0 | 0 | 0 | 0 | 0 | 0 | 0 | 0 |

## PLANT CITY *Hillsborough County* ELEVATION 120 ft LAT/LONG 28° 1 ' N / 82° 8 ' W

| | JAN | FEB | MAR | APR | MAY | JUN | JUL | AUG | SEP | OCT | NOV | DEC | YEAR |
|---|---|---|---|---|---|---|---|---|---|---|---|---|---|
| Maximum Temp °F | 73.2 | 74.6 | 79.5 | 84.3 | 88.8 | 90.9 | 91.9 | 91.6 | 90.4 | 85.8 | 79.7 | 75.0 | 83.8 |
| Minimum Temp °F | 49.0 | 49.9 | 54.6 | 58.7 | 64.3 | 70.0 | 71.6 | 71.9 | 70.5 | 64.0 | 56.4 | 50.8 | 61.0 |
| Mean Temp °F | 61.1 | 62.2 | 67.1 | 71.5 | 76.6 | 80.4 | 81.7 | 81.8 | 80.5 | 74.9 | 68.1 | 62.9 | 72.4 |
| Days Max Temp ≥ 90 °F | 0 | 0 | 1 | 5 | 13 | 22 | 26 | 25 | 21 | 6 | 0 | 0 | 119 |
| Days Max Temp ≤ 32 °F | 0 | 0 | 0 | 0 | 0 | 0 | 0 | 0 | 0 | 0 | 0 | 0 | 0 |
| Days Min Temp ≤ 32 °F | 3 | 2 | 0 | 0 | 0 | 0 | 0 | 0 | 0 | 0 | 0 | 2 | 7 |
| Days Min Temp ≤ 0 °F | 0 | 0 | 0 | 0 | 0 | 0 | 0 | 0 | 0 | 0 | 0 | 0 | 0 |
| Heating Degree Days | 175 | 133 | 62 | 13 | 0 | 0 | 0 | 0 | 0 | 4 | 51 | 135 | 573 |
| Cooling Degree Days | 57 | 76 | 128 | 198 | 363 | 486 | 534 | 538 | 472 | 320 | 166 | 78 | 3416 |
| Total Precipitation (") | 2.57 | 3.22 | 3.49 | 1.83 | 3.85 | 7.60 | 7.56 | 8.63 | 6.37 | 2.48 | 1.87 | 2.20 | 51.67 |
| Days ≥ 0.1" Precip | 4 | 4 | 5 | 3 | 5 | 10 | 12 | 12 | 9 | 4 | 3 | 4 | 75 |
| Total Snowfall (") | 0.0 | 0.0 | 0.0 | 0.0 | 0.0 | 0.0 | 0.0 | 0.0 | 0.0 | 0.0 | 0.0 | 0.0 | 0.0 |
| Days ≥ 1" Snow Depth | 0 | 0 | 0 | 0 | 0 | 0 | 0 | 0 | 0 | 0 | 0 | 0 | 0 |

## POMPANO BEACH *Broward County* ELEVATION 15 ft LAT/LONG 26° 14 ' N / 80° 9 ' W

| | JAN | FEB | MAR | APR | MAY | JUN | JUL | AUG | SEP | OCT | NOV | DEC | YEAR |
|---|---|---|---|---|---|---|---|---|---|---|---|---|---|
| Maximum Temp °F | 77.2 | 78.1 | 80.9 | 84.4 | 87.1 | 89.9 | 91.9 | 92.0 | 90.8 | 87.0 | *82.3* | 78.4 | 85.0 |
| Minimum Temp °F | 58.5 | 58.6 | 62.4 | 65.8 | 70.1 | 73.0 | 74.1 | 74.8 | 74.1 | 70.6 | *65.8* | 60.8 | 67.4 |
| Mean Temp °F | 67.9 | 68.4 | 71.7 | 75.1 | 78.6 | 81.5 | 82.9 | 83.4 | 82.4 | 78.8 | *74.1* | 69.6 | 76.2 |
| Days Max Temp ≥ 90 °F | 0 | 0 | 1 | 3 | 7 | 17 | 25 | 27 | 21 | 7 | 1 | 0 | 109 |
| Days Max Temp ≤ 32 °F | 0 | 0 | 0 | 0 | 0 | 0 | 0 | 0 | 0 | 0 | 0 | 0 | 0 |
| Days Min Temp ≤ 32 °F | 0 | 0 | 0 | 0 | 0 | 0 | 0 | 0 | 0 | 0 | 0 | 0 | 0 |
| Days Min Temp ≤ 0 °F | 0 | 0 | 0 | 0 | 0 | 0 | 0 | 0 | 0 | 0 | 0 | 0 | 0 |
| Heating Degree Days | 57 | 42 | 17 | 1 | 0 | 0 | 0 | 0 | 0 | 0 | *7* | 38 | 162 |
| Cooling Degree Days | 156 | 167 | 238 | 307 | 437 | 524 | 581 | 600 | 549 | 457 | 312 | 200 | 4528 |
| Total Precipitation (") | 2.81 | 2.91 | 3.10 | 3.17 | 6.20 | 8.51 | 6.32 | 6.28 | 6.90 | 7.59 | 4.17 | 2.16 | 60.12 |
| Days ≥ 0.1" Precip | 5 | 4 | 4 | 4 | 7 | 10 | 10 | 10 | 10 | 8 | 5 | 3 | 80 |
| Total Snowfall (") | 0.0 | 0.0 | 0.0 | 0.0 | 0.0 | 0.0 | 0.0 | 0.0 | 0.0 | 0.0 | 0.0 | 0.0 | 0.0 |
| Days ≥ 1" Snow Depth | 0 | 0 | 0 | 0 | 0 | 0 | 0 | 0 | 0 | 0 | 0 | 0 | 0 |

## PUNTA GORDA 4 ESE *Charlotte County* ELEVATION 10 ft LAT/LONG 26° 58' N / 81° 58' W

| | JAN | FEB | MAR | APR | MAY | JUN | JUL | AUG | SEP | OCT | NOV | DEC | YEAR |
|---|---|---|---|---|---|---|---|---|---|---|---|---|---|
| Maximum Temp °F | 74.0 | 75.3 | 79.8 | 84.6 | 88.9 | 91.4 | 92.0 | 91.9 | 90.5 | 86.2 | 80.3 | 76.0 | 84.2 |
| Minimum Temp °F | 51.4 | 52.2 | 56.5 | 60.1 | 65.6 | 71.0 | 72.6 | 72.9 | 72.0 | 66.2 | 58.9 | 53.4 | 62.7 |
| Mean Temp °F | 62.7 | 63.8 | 68.2 | 72.4 | 77.3 | 81.2 | 82.4 | 82.4 | 81.3 | 76.2 | 69.6 | 64.7 | 73.5 |
| Days Max Temp ≥ 90 °F | 0 | 0 | 0 | 3 | 13 | 23 | 27 | 27 | 22 | 6 | 0 | 0 | 121 |
| Days Max Temp ≤ 32 °F | 0 | 0 | 0 | 0 | 0 | 0 | 0 | 0 | 0 | 0 | 0 | 0 | 0 |
| Days Min Temp ≤ 32 °F | 1 | 0 | 0 | 0 | 0 | 0 | 0 | 0 | 0 | 0 | 0 | 1 | 2 |
| Days Min Temp ≤ 0 °F | 0 | 0 | 0 | 0 | 0 | 0 | 0 | 0 | 0 | 0 | 0 | 0 | 0 |
| Heating Degree Days | 135 | 99 | 38 | 4 | 0 | 0 | 0 | 0 | 0 | 2 | 31 | 95 | 404 |
| Cooling Degree Days | 79 | 93 | 156 | 235 | 392 | 514 | 561 | 567 | 510 | 373 | 207 | 105 | 3792 |
| Total Precipitation (") | 2.20 | 2.39 | 2.92 | 1.42 | 3.45 | 8.30 | 7.51 | 7.85 | 6.27 | 2.91 | 1.73 | 1.79 | 48.74 |
| Days ≥ 0.1" Precip | 4 | 4 | 4 | 2 | 5 | 10 | 13 | 12 | 10 | 5 | 3 | 3 | 75 |
| Total Snowfall (") | 0.0 | 0.0 | 0.0 | 0.0 | 0.0 | 0.0 | 0.0 | 0.0 | 0.0 | 0.0 | 0.0 | 0.0 | 0.0 |
| Days ≥ 1" Snow Depth | 0 | 0 | 0 | 0 | 0 | 0 | 0 | 0 | 0 | 0 | 0 | 0 | 0 |

## QUINCY 3 SSW *Gadsden County* ELEVATION 249 ft LAT/LONG 30° 36' N / 84° 33' W

| | JAN | FEB | MAR | APR | MAY | JUN | JUL | AUG | SEP | OCT | NOV | DEC | YEAR |
|---|---|---|---|---|---|---|---|---|---|---|---|---|---|
| Maximum Temp °F | 61.3 | 65.1 | 71.9 | 78.6 | 84.6 | 89.7 | 90.6 | 89.8 | 87.3 | 79.8 | 71.6 | 64.8 | 77.9 |
| Minimum Temp °F | 39.4 | 41.5 | 48.2 | 53.9 | 61.6 | 68.3 | 70.6 | 70.3 | 67.0 | 56.4 | 48.5 | 42.1 | 55.7 |
| Mean Temp °F | 50.4 | 53.3 | 60.0 | 66.3 | 73.1 | 79.0 | 80.6 | 80.1 | 77.2 | 68.2 | 60.1 | 53.5 | 66.8 |
| Days Max Temp ≥ 90 °F | 0 | 0 | 0 | 0 | 4 | 17 | *21* | 18 | *11* | 1 | 0 | 0 | 72 |
| Days Max Temp ≤ 32 °F | 0 | 0 | 0 | 0 | 0 | 0 | 0 | 0 | 0 | 0 | 0 | 0 | 0 |
| Days Min Temp ≤ 32 °F | 9 | 6 | 2 | 0 | 0 | 0 | 0 | 0 | 0 | 0 | 2 | 6 | 25 |
| Days Min Temp ≤ 0 °F | 0 | 0 | 0 | 0 | 0 | 0 | 0 | 0 | 0 | 0 | 0 | 0 | 0 |
| Heating Degree Days | 453 | 336 | 191 | 62 | 4 | 0 | 0 | 0 | 1 | 46 | 189 | 366 | 1648 |
| Cooling Degree Days | 5 | 13 | 42 | 98 | 265 | 437 | 503 | 487 | 371 | 156 | 54 | 16 | 2447 |
| Total Precipitation (") | 5.68 | 4.56 | 6.33 | 3.78 | 5.05 | 5.53 | 7.49 | 5.91 | 4.22 | 3.04 | 3.27 | 3.74 | 58.60 |
| Days ≥ 0.1" Precip | 7 | *6* | 7 | 4 | 6 | 8 | 11 | 10 | 6 | 4 | 4 | 5 | 78 |
| Total Snowfall (") | 0.0 | 0.1 | 0.0 | 0.0 | 0.0 | 0.0 | 0.0 | 0.0 | 0.0 | 0.0 | 0.0 | 0.0 | 0.1 |
| Days ≥ 1" Snow Depth | 0 | 0 | 0 | 0 | 0 | 0 | 0 | 0 | 0 | 0 | 0 | 0 | 0 |

## ST LEO *Pasco County* ELEVATION 200 ft LAT/LONG 28° 20' N / 82° 15' W

| | JAN | FEB | MAR | APR | MAY | JUN | JUL | AUG | SEP | OCT | NOV | DEC | YEAR |
|---|---|---|---|---|---|---|---|---|---|---|---|---|---|
| Maximum Temp °F | 71.5 | 73.6 | 78.8 | 84.0 | 88.9 | 91.1 | 92.0 | 91.9 | 90.4 | 85.2 | 78.8 | 73.9 | 83.3 |
| Minimum Temp °F | 48.5 | 50.0 | 54.8 | 59.3 | 64.7 | 70.0 | 71.5 | 71.7 | 70.3 | 63.7 | 56.6 | 50.9 | 61.0 |
| Mean Temp °F | 60.0 | 61.9 | 66.8 | 71.7 | 76.8 | 80.6 | 81.8 | 81.8 | 80.4 | 74.5 | 67.7 | 62.5 | 72.2 |
| Days Max Temp ≥ 90 °F | 0 | 0 | 1 | 4 | 14 | 22 | 26 | 26 | 21 | 6 | 0 | 0 | 120 |
| Days Max Temp ≤ 32 °F | 0 | 0 | 0 | 0 | 0 | 0 | 0 | 0 | 0 | 0 | 0 | 0 | 0 |
| Days Min Temp ≤ 32 °F | 2 | 1 | 0 | 0 | 0 | 0 | 0 | 0 | 0 | 0 | 0 | 1 | 4 |
| Days Min Temp ≤ 0 °F | 0 | 0 | 0 | 0 | 0 | 0 | 0 | 0 | 0 | 0 | 0 | 0 | 0 |
| Heating Degree Days | 197 | 141 | 64 | 9 | 0 | 0 | 0 | 0 | 0 | 4 | 54 | 142 | 611 |
| Cooling Degree Days | 52 | 74 | 131 | 210 | 384 | 494 | 544 | 544 | 474 | 311 | 166 | 75 | 3459 |
| Total Precipitation (") | 3.20 | 3.58 | 4.04 | 2.25 | 4.51 | 6.78 | 7.87 | 8.24 | 6.28 | 2.55 | 2.44 | 2.47 | 54.21 |
| Days ≥ 0.1" Precip | 5 | 5 | 5 | 3 | 5 | 9 | 12 | 12 | 9 | 4 | 4 | 4 | 77 |
| Total Snowfall (") | 0.0 | 0.0 | 0.0 | 0.0 | 0.0 | 0.0 | 0.0 | 0.0 | 0.0 | 0.0 | 0.0 | 0.0 | 0.0 |
| Days ≥ 1" Snow Depth | 0 | 0 | 0 | 0 | 0 | 0 | 0 | 0 | 0 | 0 | 0 | 0 | 0 |

## ST PETERSBURG WHITTD *Pinellas County* ELEVATION 8 ft LAT/LONG 27° 46' N / 82° 38' W

| | JAN | FEB | MAR | APR | MAY | JUN | JUL | AUG | SEP | OCT | NOV | DEC | YEAR |
|---|---|---|---|---|---|---|---|---|---|---|---|---|---|
| Maximum Temp °F | 70.0 | 71.2 | 75.7 | 81.3 | 86.3 | 89.5 | 90.4 | 90.1 | 88.7 | 83.8 | 77.3 | 72.1 | 81.4 |
| Minimum Temp °F | 54.3 | 55.4 | 60.1 | 65.4 | 70.8 | 75.3 | 76.6 | 76.5 | 75.5 | 69.9 | 62.6 | 56.6 | 66.6 |
| Mean Temp °F | 62.2 | 63.3 | 68.0 | 73.4 | 78.6 | 82.4 | 83.5 | 83.3 | 82.1 | 76.9 | 70.0 | 64.3 | 74.0 |
| Days Max Temp ≥ 90 °F | 0 | 0 | 0 | 0 | 5 | 16 | 22 | 20 | 13 | 2 | 0 | 0 | 78 |
| Days Max Temp ≤ 32 °F | 0 | 0 | 0 | 0 | 0 | 0 | 0 | 0 | 0 | 0 | 0 | 0 | 0 |
| Days Min Temp ≤ 32 °F | 0 | 0 | 0 | 0 | 0 | 0 | 0 | 0 | 0 | 0 | 0 | 0 | 0 |
| Days Min Temp ≤ 0 °F | 0 | 0 | 0 | 0 | 0 | 0 | 0 | 0 | 0 | 0 | 0 | 0 | 0 |
| Heating Degree Days | 138 | 100 | 40 | 3 | 0 | 0 | 0 | 0 | 0 | 1 | 27 | 96 | 405 |
| Cooling Degree Days | 52 | 68 | 133 | 238 | 422 | 535 | 588 | 585 | 518 | 377 | 196 | 86 | 3798 |
| Total Precipitation (") | 2.44 | 2.79 | 3.41 | 1.51 | 2.97 | 5.93 | 6.60 | 8.11 | 6.99 | 2.59 | 1.96 | 2.26 | 47.56 |
| Days ≥ 0.1" Precip | 5 | 4 | 5 | 2 | 4 | 8 | 10 | 12 | 9 | 4 | 3 | 4 | 70 |
| Total Snowfall (") | 0.0 | 0.0 | 0.0 | 0.0 | 0.0 | 0.0 | 0.0 | 0.0 | 0.0 | 0.0 | 0.0 | 0.0 | 0.0 |
| Days ≥ 1" Snow Depth | 0 | 0 | 0 | 0 | 0 | 0 | 0 | 0 | 0 | 0 | 0 | 0 | 0 |

## STEINHATCHEE 6 ENE *Dixie County* ELEVATION 30 ft LAT/LONG 29° 42 ' N / 83° 21 ' W

| | JAN | FEB | MAR | APR | MAY | JUN | JUL | AUG | SEP | OCT | NOV | DEC | YEAR |
|---|---|---|---|---|---|---|---|---|---|---|---|---|---|
| Maximum Temp °F | 65.5 | 67.8 | 74.8 | 80.7 | 86.6 | 90.9 | 91.4 | 91.1 | 89.2 | 82.3 | 74.7 | 68.1 | 80.3 |
| Minimum Temp °F | 40.4 | 42.3 | 49.5 | 54.8 | 61.2 | 68.3 | 71.1 | 71.1 | 68.0 | 58.3 | 49.2 | 42.5 | 56.4 |
| Mean Temp °F | 53.0 | 55.1 | 62.3 | 67.8 | 73.9 | 79.6 | 81.3 | 81.1 | 78.7 | 70.3 | 62.0 | 55.4 | 68.4 |
| Days Max Temp ≥ 90 °F | 0 | 0 | 0 | 1 | 8 | 22 | 24 | ***23*** | ***17*** | 2 | 0 | 0 | 97 |
| Days Max Temp ≤ 32 °F | 0 | 0 | 0 | 0 | 0 | 0 | 0 | 0 | 0 | 0 | 0 | 0 | 0 |
| Days Min Temp ≤ 32 °F | ***8*** | 5 | 1 | 0 | 0 | 0 | 0 | 0 | 0 | 0 | 3 | 7 | 24 |
| Days Min Temp ≤ 0 °F | 0 | 0 | 0 | 0 | 0 | 0 | 0 | 0 | 0 | 0 | 0 | 0 | 0 |
| Heating Degree Days | ***380*** | 288 | 138 | 44 | 4 | 0 | 0 | 0 | 1 | 27 | 150 | 316 | 1348 |
| Cooling Degree Days | 8 | 17 | 54 | 116 | 266 | 428 | 511 | 512 | 404 | 195 | 78 | 22 | 2611 |
| Total Precipitation (") | 4.55 | 4.03 | 4.63 | 2.99 | 3.55 | 6.70 | 9.31 | 9.87 | 5.61 | 3.20 | 2.41 | 3.47 | 60.32 |
| Days ≥ 0.1" Precip | ***6*** | 6 | 5 | 3 | 5 | ***9*** | 12 | 12 | 8 | 4 | 4 | 5 | 79 |
| Total Snowfall (") | 0.0 | 0.0 | 0.0 | 0.0 | 0.0 | 0.0 | 0.0 | 0.0 | 0.0 | 0.0 | 0.0 | 0.0 | 0.0 |
| Days ≥ 1" Snow Depth | 0 | 0 | 0 | 0 | 0 | 0 | 0 | 0 | 0 | 0 | 0 | 0 | 0 |

## STUART 1 N *Martin County* ELEVATION 10 ft LAT/LONG 27° 12 ' N / 80° 15 ' W

| | JAN | FEB | MAR | APR | MAY | JUN | JUL | AUG | SEP | OCT | NOV | DEC | YEAR |
|---|---|---|---|---|---|---|---|---|---|---|---|---|---|
| Maximum Temp °F | 75.0 | 75.8 | 78.9 | 82.1 | 85.3 | 88.2 | 89.7 | 90.0 | 88.7 | 85.1 | 80.1 | 76.3 | 82.9 |
| Minimum Temp °F | 54.9 | 55.2 | 59.8 | 63.7 | 68.6 | 72.3 | 73.6 | 74.1 | 73.7 | 69.1 | 63.0 | 57.5 | 65.5 |
| Mean Temp °F | 65.0 | 65.5 | 69.3 | 73.0 | 77.0 | 80.2 | 81.7 | 82.1 | 81.2 | 77.1 | 71.6 | 66.9 | 74.2 |
| Days Max Temp ≥ 90 °F | 0 | 0 | 0 | 2 | 5 | 10 | 16 | 17 | 12 | 3 | 0 | 0 | 65 |
| Days Max Temp ≤ 32 °F | 0 | 0 | 0 | 0 | 0 | 0 | 0 | 0 | 0 | 0 | 0 | 0 | 0 |
| Days Min Temp ≤ 32 °F | 1 | 0 | 0 | 0 | 0 | 0 | 0 | 0 | 0 | 0 | 0 | 0 | 1 |
| Days Min Temp ≤ 0 °F | 0 | 0 | 0 | 0 | 0 | 0 | 0 | 0 | 0 | 0 | 0 | 0 | 0 |
| Heating Degree Days | 89 | 72 | 30 | 5 | 0 | 0 | 0 | 0 | 0 | 1 | 16 | 58 | 271 |
| Cooling Degree Days | 101 | 111 | 179 | 246 | 390 | 485 | 542 | 553 | 507 | 405 | 254 | 142 | 3915 |
| Total Precipitation (") | 2.96 | 3.19 | 4.16 | 2.53 | 5.30 | 7.19 | 6.46 | 6.11 | 7.97 | 5.99 | 3.98 | 2.57 | 58.41 |
| Days ≥ 0.1" Precip | 5 | 5 | 6 | 5 | 8 | 11 | 11 | 10 | 11 | 10 | 6 | 4 | 92 |
| Total Snowfall (") | 0.0 | 0.0 | 0.0 | 0.0 | 0.0 | 0.0 | 0.0 | 0.0 | 0.0 | 0.0 | 0.0 | 0.0 | 0.0 |
| Days ≥ 1" Snow Depth | 0 | 0 | 0 | 0 | 0 | 0 | 0 | 0 | 0 | 0 | 0 | 0 | 0 |

## TALLAHASSEE MUNI AP *Leon County* ELEVATION 55 ft LAT/LONG 30° 23 ' N / 84° 22 ' W

| | JAN | FEB | MAR | APR | MAY | JUN | JUL | AUG | SEP | OCT | NOV | DEC | YEAR |
|---|---|---|---|---|---|---|---|---|---|---|---|---|---|
| Maximum Temp °F | 63.2 | 66.5 | 73.6 | 80.6 | 86.1 | 90.8 | 91.6 | 91.0 | 88.5 | 81.2 | 73.0 | 66.2 | 79.4 |
| Minimum Temp °F | 38.9 | 40.4 | 46.7 | 52.3 | 60.8 | 68.4 | 71.1 | 71.3 | 67.8 | 56.0 | 46.6 | 40.8 | 55.1 |
| Mean Temp °F | 51.1 | 53.5 | 60.2 | 66.5 | 73.5 | 79.6 | 81.4 | 81.2 | 78.2 | 68.7 | 59.8 | 53.6 | 67.3 |
| Days Max Temp ≥ 90 °F | 0 | 0 | 0 | 2 | 7 | 20 | 23 | 22 | 15 | 2 | 0 | 0 | 91 |
| Days Max Temp ≤ 32 °F | 0 | 0 | 0 | 0 | 0 | 0 | 0 | 0 | 0 | 0 | 0 | 0 | 0 |
| Days Min Temp ≤ 32 °F | 11 | 8 | 3 | 0 | 0 | 0 | 0 | 0 | 0 | 0 | 4 | 9 | 35 |
| Days Min Temp ≤ 0 °F | 0 | 0 | 0 | 0 | 0 | 0 | 0 | 0 | 0 | 0 | 0 | 0 | 0 |
| Heating Degree Days | 432 | 330 | 188 | 60 | 4 | 0 | 0 | 0 | 1 | 45 | 197 | 364 | 1621 |
| Cooling Degree Days | 5 | 14 | 46 | 98 | 275 | 455 | 524 | 518 | 401 | 175 | 59 | 16 | 2586 |
| Total Precipitation (") | 5.58 | 5.20 | 6.65 | 3.68 | 5.10 | 7.05 | 8.50 | 7.44 | 5.40 | 3.25 | 3.78 | 4.45 | 66.08 |
| Days ≥ 0.1" Precip | 7 | 6 | 6 | 4 | 6 | 9 | 12 | 10 | 7 | 4 | 4 | 6 | 81 |
| Total Snowfall (") | 0.0 | 0.0 | 0.0 | 0.0 | 0.0 | 0.0 | 0.0 | 0.0 | 0.0 | 0.0 | 0.0 | 0.0 | 0.0 |
| Days ≥ 1" Snow Depth | 0 | 0 | 0 | 0 | 0 | 0 | 0 | 0 | 0 | 0 | 0 | 0 | 0 |

## TAMIAMI TRAIL 40 MI *Dade County* ELEVATION 10 ft LAT/LONG 25° 45 ' N / 80° 50 ' W

| | JAN | FEB | MAR | APR | MAY | JUN | JUL | AUG | SEP | OCT | NOV | DEC | YEAR |
|---|---|---|---|---|---|---|---|---|---|---|---|---|---|
| Maximum Temp °F | 77.5 | 78.5 | 81.8 | 85.6 | 88.8 | 90.6 | 91.9 | 92.1 | 90.8 | ***86.9*** | 82.4 | 78.6 | 85.5 |
| Minimum Temp °F | 56.6 | 56.5 | 59.3 | 62.0 | 66.5 | 71.6 | 73.8 | 74.5 | 74.3 | ***70.6*** | 64.6 | 58.6 | 65.7 |
| Mean Temp °F | 67.1 | 67.5 | 70.6 | 73.8 | 77.7 | 81.1 | 82.8 | 83.4 | 82.6 | ***78.8*** | 73.5 | 68.6 | 75.6 |
| Days Max Temp ≥ 90 °F | 0 | 0 | 1 | 5 | 12 | 21 | 27 | 27 | 22 | 9 | 1 | 0 | 125 |
| Days Max Temp ≤ 32 °F | 0 | 0 | 0 | 0 | 0 | 0 | 0 | 0 | 0 | 0 | 0 | 0 | 0 |
| Days Min Temp ≤ 32 °F | 0 | 0 | 0 | 0 | 0 | 0 | 0 | 0 | 0 | 0 | 0 | 0 | 0 |
| Days Min Temp ≤ 0 °F | 0 | 0 | 0 | 0 | 0 | 0 | 0 | 0 | 0 | 0 | 0 | 0 | 0 |
| Heating Degree Days | 61 | 45 | 18 | 3 | 0 | 0 | 0 | 0 | 0 | 0 | 7 | 40 | 174 |
| Cooling Degree Days | 129 | 143 | 210 | 279 | 404 | 506 | 574 | 593 | 539 | ***440*** | 291 | 167 | 4275 |
| Total Precipitation (") | 1.95 | 1.78 | 2.06 | 2.22 | 4.99 | 9.64 | 7.39 | 6.82 | 6.65 | 4.29 | 2.14 | 1.44 | 51.37 |
| Days ≥ 0.1" Precip | 3 | 3 | 3 | 3 | 7 | 12 | 11 | 11 | 10 | 6 | 4 | 2 | 75 |
| Total Snowfall (") | 0.0 | 0.0 | 0.0 | 0.0 | 0.0 | 0.0 | 0.0 | 0.0 | 0.0 | 0.0 | 0.0 | 0.0 | 0.0 |
| Days ≥ 1" Snow Depth | 0 | 0 | 0 | 0 | 0 | 0 | 0 | 0 | 0 | 0 | 0 | 0 | 0 |

## TAMPA INTL ARPT *Hillsborough County* ELEVATION 36 ft LAT/LONG 27° 58 ' N / 82° 32 ' W

| | JAN | FEB | MAR | APR | MAY | JUN | JUL | AUG | SEP | OCT | NOV | DEC | YEAR |
|---|---|---|---|---|---|---|---|---|---|---|---|---|---|
| Maximum Temp °F | 70.3 | 71.6 | 76.4 | 81.8 | 87.1 | 89.7 | 90.4 | 90.5 | 89.2 | 84.2 | 77.9 | 72.5 | 81.8 |
| Minimum Temp °F | 50.0 | 51.1 | 56.2 | 60.9 | 67.3 | 73.0 | 74.4 | 74.3 | 72.9 | 65.7 | 57.8 | 52.0 | 63.0 |
| Mean Temp °F | 60.2 | 61.4 | 66.3 | 71.4 | 77.2 | 81.4 | 82.4 | 82.5 | 81.1 | 75.0 | 67.9 | 62.3 | 72.4 |
| Days Max Temp ≥ 90 °F | 0 | 0 | 0 | 1 | 8 | 17 | 21 | 22 | 16 | 3 | 0 | 0 | 88 |
| Days Max Temp ≤ 32 °F | 0 | 0 | 0 | 0 | 0 | 0 | 0 | 0 | 0 | 0 | 0 | 0 | 0 |
| Days Min Temp ≤ 32 °F | 2 | 1 | 0 | 0 | 0 | 0 | 0 | 0 | 0 | 0 | 0 | 1 | 4 |
| Days Min Temp ≤ 0 °F | 0 | 0 | 0 | 0 | 0 | 0 | 0 | 0 | 0 | 0 | 0 | 0 | 0 |
| Heating Degree Days | 195 | 148 | 69 | 11 | 0 | 0 | 0 | 0 | 0 | 4 | 56 | 149 | 632 |
| Cooling Degree Days | 54 | 68 | 122 | 200 | 391 | 513 | 557 | 557 | 491 | 327 | 175 | 80 | 3535 |
| Total Precipitation (") | 2.02 | 2.69 | 2.93 | 1.35 | 3.06 | 5.48 | 6.55 | 7.54 | 6.02 | 2.22 | 1.59 | 1.97 | 43.42 |
| Days ≥ 0.1" Precip | 4 | 4 | 4 | 3 | 4 | 8 | 10 | 12 | 8 | 4 | 3 | 4 | 68 |
| Total Snowfall (") | 0.0 | 0.0 | 0.0 | 0.0 | 0.0 | 0.0 | 0.0 | 0.0 | 0.0 | 0.0 | 0.0 | 0.0 | 0.0 |
| Days ≥ 1" Snow Depth | 0 | 0 | 0 | 0 | 0 | 0 | 0 | 0 | 0 | 0 | 0 | 0 | 0 |

## TARPON SPNGS SWG PLT *Pinellas County* ELEVATION 20 ft LAT/LONG 28° 8 ' N / 82° 47 ' W

| | JAN | FEB | MAR | APR | MAY | JUN | JUL | AUG | SEP | OCT | NOV | DEC | YEAR |
|---|---|---|---|---|---|---|---|---|---|---|---|---|---|
| Maximum Temp °F | 70.5 | 71.7 | 76.4 | 81.2 | 86.1 | 89.4 | 90.6 | 90.7 | 89.6 | 84.6 | 78.1 | 72.9 | 81.8 |
| Minimum Temp °F | 50.0 | 51.2 | 56.1 | 61.0 | 66.6 | 71.8 | 73.3 | 73.2 | 71.7 | 65.0 | 57.8 | 51.7 | 62.5 |
| Mean Temp °F | 60.3 | 61.5 | 66.2 | 71.1 | 76.4 | 80.7 | 82.0 | 82.0 | 80.7 | 74.8 | 68.0 | 62.3 | 72.2 |
| Days Max Temp ≥ 90 °F | 0 | 0 | 0 | 1 | 5 | 15 | 21 | 22 | 17 | 3 | 0 | 0 | 84 |
| Days Max Temp ≤ 32 °F | 0 | 0 | 0 | 0 | 0 | 0 | 0 | 0 | 0 | 0 | 0 | 0 | 0 |
| Days Min Temp ≤ 32 °F | 2 | 1 | 0 | 0 | 0 | 0 | 0 | 0 | 0 | 0 | 0 | 1 | 4 |
| Days Min Temp ≤ 0 °F | 0 | 0 | 0 | 0 | 0 | 0 | 0 | 0 | 0 | 0 | 0 | 0 | 0 |
| Heating Degree Days | 190 | 146 | 66 | 9 | 0 | 0 | 0 | 0 | 0 | 4 | 51 | 146 | 612 |
| Cooling Degree Days | 54 | 67 | 118 | 201 | 375 | 499 | 553 | 555 | 490 | 331 | 179 | 79 | 3501 |
| Total Precipitation (") | 3.06 | 3.24 | 3.81 | 1.67 | 3.34 | 5.47 | 7.14 | 8.93 | 6.90 | 2.95 | 2.15 | 2.78 | 51.44 |
| Days ≥ 0.1" Precip | 5 | 4 | 5 | 3 | 4 | 8 | 10 | 11 | 8 | 5 | 4 | 4 | 71 |
| Total Snowfall (") | 0.0 | 0.0 | 0.0 | 0.0 | 0.0 | 0.0 | 0.0 | 0.0 | 0.0 | 0.0 | 0.0 | 0.0 | 0.0 |
| Days ≥ 1" Snow Depth | 0 | 0 | 0 | 0 | 0 | 0 | 0 | 0 | 0 | 0 | 0 | 0 | 0 |

## TAVERNIER *Monroe County* ELEVATION 7 ft LAT/LONG 25° 1 ' N / 80° 31 ' W

| | JAN | FEB | MAR | APR | MAY | JUN | JUL | AUG | SEP | OCT | NOV | DEC | YEAR |
|---|---|---|---|---|---|---|---|---|---|---|---|---|---|
| Maximum Temp °F | 76.6 | 77.2 | 80.2 | 83.4 | 86.4 | 88.5 | 90.5 | 90.4 | 89.0 | 85.5 | 81.2 | 77.7 | 83.9 |
| Minimum Temp °F | 63.5 | 63.7 | 67.0 | 70.3 | 73.8 | 76.4 | 78.0 | 78.0 | 76.9 | 73.8 | 69.8 | 65.4 | 71.4 |
| Mean Temp °F | 70.1 | 70.5 | 73.6 | 76.9 | 80.1 | 82.4 | 84.3 | 84.2 | 83.0 | 79.7 | 75.5 | 71.6 | 77.7 |
| Days Max Temp ≥ 90 °F | 0 | 0 | 0 | 1 | 4 | 11 | 18 | 20 | 13 | 3 | 0 | 0 | 70 |
| Days Max Temp ≤ 32 °F | 0 | 0 | 0 | 0 | 0 | 0 | 0 | 0 | 0 | 0 | 0 | 0 | 0 |
| Days Min Temp ≤ 32 °F | 0 | 0 | 0 | 0 | 0 | 0 | 0 | 0 | 0 | 0 | 0 | 0 | 0 |
| Days Min Temp ≤ 0 °F | 0 | 0 | 0 | 0 | 0 | 0 | 0 | 0 | 0 | 0 | 0 | 0 | 0 |
| Heating Degree Days | 29 | 20 | 8 | 0 | 0 | 0 | 0 | 0 | 0 | 0 | 2 | 19 | 78 |
| Cooling Degree Days | 203 | 212 | 292 | 362 | 486 | 556 | *624* | 615 | 563 | 476 | 351 | 241 | 4981 |
| Total Precipitation (") | 2.53 | 2.01 | 1.87 | 1.98 | 4.33 | 7.96 | 3.23 | 4.67 | 6.45 | 5.74 | 2.67 | 1.79 | 45.23 |
| Days ≥ 0.1" Precip | 3 | 3 | 3 | 3 | 6 | 9 | 6 | 8 | 10 | 7 | 4 | 3 | 65 |
| Total Snowfall (") | 0.0 | 0.0 | 0.0 | 0.0 | 0.0 | 0.0 | 0.0 | 0.0 | 0.0 | 0.0 | 0.0 | 0.0 | 0.0 |
| Days ≥ 1" Snow Depth | 0 | 0 | 0 | 0 | 0 | 0 | 0 | 0 | 0 | 0 | 0 | 0 | 0 |

## TITUSVILLE *Brevard County* ELEVATION 13 ft LAT/LONG 28° 37 ' N / 80° 49 ' W

| | JAN | FEB | MAR | APR | MAY | JUN | JUL | AUG | SEP | OCT | NOV | DEC | YEAR |
|---|---|---|---|---|---|---|---|---|---|---|---|---|---|
| Maximum Temp °F | 70.6 | 72.2 | 77.5 | 82.3 | 86.8 | 89.7 | 91.7 | 91.4 | 88.9 | 83.4 | 77.5 | 72.7 | 82.1 |
| Minimum Temp °F | 48.0 | 49.1 | 54.4 | 58.9 | 64.6 | 69.7 | 71.1 | 71.3 | 70.6 | 64.6 | 57.0 | 50.6 | 60.8 |
| Mean Temp °F | 59.3 | 60.7 | 66.1 | 70.6 | 75.7 | 79.7 | 81.5 | 81.4 | 79.8 | 74.1 | 67.2 | 61.7 | 71.5 |
| Days Max Temp ≥ 90 °F | 0 | 0 | 1 | 3 | 9 | 17 | 24 | 24 | 14 | 3 | 0 | 0 | 95 |
| Days Max Temp ≤ 32 °F | 0 | 0 | 0 | 0 | 0 | 0 | 0 | 0 | 0 | 0 | 0 | 0 | 0 |
| Days Min Temp ≤ 32 °F | 2 | 1 | 0 | 0 | 0 | 0 | 0 | 0 | 0 | 0 | 0 | 1 | 4 |
| Days Min Temp ≤ 0 °F | 0 | 0 | 0 | 0 | 0 | 0 | 0 | 0 | 0 | 0 | 0 | 0 | 0 |
| Heating Degree Days | 213 | 160 | 74 | 16 | 0 | 0 | 0 | 0 | 0 | 5 | 58 | 159 | 685 |
| Cooling Degree Days | 31 | 46 | 109 | 166 | 329 | 451 | 522 | 514 | 437 | 289 | 145 | 62 | 3101 |
| Total Precipitation (") | 2.35 | 3.18 | 3.27 | 2.69 | 3.80 | 7.29 | 7.62 | 7.35 | 6.64 | 4.90 | 3.31 | 2.22 | 54.62 |
| Days ≥ 0.1" Precip | 4 | 5 | 5 | 3 | 6 | 10 | 10 | 11 | 10 | 8 | 5 | 4 | 81 |
| Total Snowfall (") | 0.0 | 0.0 | 0.0 | 0.0 | 0.0 | 0.0 | 0.0 | 0.0 | 0.0 | 0.0 | 0.0 | 0.0 | 0.0 |
| Days ≥ 1" Snow Depth | 0 | 0 | 0 | 0 | 0 | 0 | 0 | 0 | 0 | 0 | 0 | 0 | 0 |

## USHER TOWER *Levy County* ELEVATION 30 ft LAT/LONG 29° 25 ' N / 82° 49 ' W

| | JAN | FEB | MAR | APR | MAY | JUN | JUL | AUG | SEP | OCT | NOV | DEC | YEAR |
|---|---|---|---|---|---|---|---|---|---|---|---|---|---|
| Maximum Temp °F | 68.1 | 70.4 | 76.7 | 82.7 | 87.9 | 91.2 | 91.7 | 91.3 | 89.7 | 83.8 | 76.1 | 70.3 | 81.7 |
| Minimum Temp °F | 42.9 | 44.5 | 49.9 | 54.9 | 60.8 | 67.4 | 69.6 | 70.1 | 67.9 | 58.6 | 50.3 | 44.4 | 56.8 |
| Mean Temp °F | 55.5 | 57.5 | 63.3 | 68.8 | 74.4 | 79.3 | 80.7 | 80.7 | 78.8 | 71.2 | 63.2 | 57.4 | 69.2 |
| Days Max Temp ≥ 90 °F | 0 | 0 | 0 | 2 | 11 | 22 | 24 | 24 | 18 | 3 | 0 | 0 | 104 |
| Days Max Temp ≤ 32 °F | 0 | 0 | 0 | 0 | 0 | 0 | 0 | 0 | 0 | 0 | 0 | 0 | 0 |
| Days Min Temp ≤ 32 °F | 7 | 5 | 1 | 0 | 0 | 0 | 0 | 0 | 0 | 0 | 2 | 6 | 21 |
| Days Min Temp ≤ 0 °F | 0 | 0 | 0 | 0 | 0 | 0 | 0 | 0 | 0 | 0 | 0 | 0 | 0 |
| Heating Degree Days | 305 | 230 | 117 | 26 | 1 | 0 | 0 | 0 | 0 | 18 | 125 | 258 | 1080 |
| Cooling Degree Days | 15 | 30 | 76 | 146 | 306 | 450 | 511 | 510 | 425 | 227 | 97 | 32 | 2825 |
| Total Precipitation (") | 4.32 | 3.83 | 4.51 | 3.18 | 3.40 | 6.64 | 8.70 | 10.29 | 6.34 | 2.52 | 2.42 | 3.26 | 59.41 |
| Days ≥ 0.1" Precip | 7 | 5 | 6 | 4 | 5 | 9 | 13 | 13 | 9 | 4 | 4 | 5 | 84 |
| Total Snowfall (") | 0.0 | 0.0 | 0.0 | 0.0 | 0.0 | 0.0 | 0.0 | 0.0 | 0.0 | 0.0 | 0.0 | 0.0 | 0.0 |
| Days ≥ 1" Snow Depth | 0 | 0 | 0 | 0 | 0 | 0 | 0 | 0 | 0 | 0 | 0 | 0 | 0 |

## VENICE *Sarasota County* ELEVATION 10 ft LAT/LONG 27° 6 ' N / 82° 27 ' W

| | JAN | FEB | MAR | APR | MAY | JUN | JUL | AUG | SEP | OCT | NOV | DEC | YEAR |
|---|---|---|---|---|---|---|---|---|---|---|---|---|---|
| Maximum Temp °F | 71.6 | 72.8 | 76.7 | 81.4 | 85.9 | 88.9 | 90.5 | 90.6 | 89.5 | ***85.2*** | ***79.3*** | 74.0 | 82.2 |
| Minimum Temp °F | 51.0 | 51.9 | 56.9 | 61.1 | 66.4 | 71.7 | 72.9 | 73.2 | 72.4 | ***65.8*** | ***58.6*** | ***53.7*** | 63.0 |
| Mean Temp °F | 61.3 | 62.4 | 66.8 | 71.3 | 76.2 | 80.4 | 81.7 | 82.0 | 81.0 | ***75.5*** | ***69.0*** | ***63.9*** | 72.6 |
| Days Max Temp ≥ 90 °F | 0 | 0 | 0 | 1 | ***4*** | ***12*** | ***21*** | ***22*** | ***16*** | ***3*** | ***0*** | ***0*** | 79 |
| Days Max Temp ≤ 32 °F | 0 | 0 | 0 | 0 | 0 | 0 | ***0*** | 0 | 0 | ***0*** | ***0*** | 0 | 0 |
| Days Min Temp ≤ 32 °F | 1 | 0 | 0 | 0 | 0 | 0 | ***0*** | 0 | 0 | ***0*** | ***0*** | ***1*** | 2 |
| Days Min Temp ≤ 0 °F | 0 | 0 | 0 | 0 | 0 | 0 | ***0*** | 0 | 0 | ***0*** | ***0*** | 0 | 0 |
| Heating Degree Days | 166 | 122 | 53 | 9 | 0 | 0 | 0 | 0 | 0 | ***2*** | ***35*** | ***119*** | 506 |
| Cooling Degree Days | ***53*** | ***70*** | ***123*** | ***186*** | ***345*** | ***486*** | ***543*** | ***552*** | ***507*** | na | na | ***109*** | na |
| Total Precipitation (") | 2.44 | 2.33 | 3.34 | 1.57 | 2.53 | 7.11 | 6.58 | 8.29 | 7.26 | 2.98 | 1.92 | 2.09 | 48.44 |
| Days ≥ 0.1" Precip | 4 | 4 | 4 | 2 | 4 | 8 | 10 | 12 | 10 | 5 | 3 | 3 | 69 |
| Total Snowfall (") | 0.0 | 0.0 | 0.0 | 0.0 | 0.0 | 0.0 | 0.0 | 0.0 | 0.0 | 0.0 | 0.0 | 0.0 | 0.0 |
| Days ≥ 1" Snow Depth | 0 | 0 | 0 | 0 | 0 | 0 | 0 | 0 | 0 | 0 | 0 | 0 | 0 |

## VERO BEACH 4 W *Indian River County* ELEVATION 20 ft LAT/LONG 27° 38 ' N / 80° 27 ' W

| | JAN | FEB | MAR | APR | MAY | JUN | JUL | AUG | SEP | OCT | NOV | DEC | YEAR |
|---|---|---|---|---|---|---|---|---|---|---|---|---|---|
| Maximum Temp °F | 73.0 | 73.8 | 77.6 | 81.8 | 85.6 | 88.4 | 90.3 | 90.3 | 88.6 | 84.4 | 79.2 | 74.5 | 82.3 |
| Minimum Temp °F | 50.6 | 51.3 | 55.9 | 60.2 | 65.5 | 70.3 | 71.6 | 71.9 | 71.4 | 66.3 | 59.7 | 53.0 | 62.3 |
| Mean Temp °F | 61.8 | 62.6 | 66.8 | 71.0 | 75.5 | 79.4 | 81.0 | 81.1 | 80.0 | 75.4 | 69.4 | 63.8 | 72.3 |
| Days Max Temp ≥ 90 °F | 0 | 0 | 0 | 2 | 5 | 12 | 20 | 21 | 12 | 2 | 0 | 0 | 74 |
| Days Max Temp ≤ 32 °F | 0 | 0 | 0 | 0 | 0 | 0 | 0 | 0 | 0 | 0 | 0 | 0 | 0 |
| Days Min Temp ≤ 32 °F | 1 | 1 | 0 | 0 | 0 | 0 | 0 | 0 | 0 | 0 | 0 | 0 | 2 |
| Days Min Temp ≤ 0 °F | 0 | 0 | 0 | 0 | 0 | 0 | 0 | 0 | 0 | 0 | 0 | 0 | 0 |
| Heating Degree Days | 153 | 124 | 59 | 12 | 0 | 0 | 0 | 0 | 0 | 3 | 35 | 113 | 499 |
| Cooling Degree Days | 63 | 77 | 128 | 191 | 332 | 454 | 520 | 521 | 464 | 344 | 198 | 92 | 3384 |
| Total Precipitation (") | 2.66 | 3.01 | 3.62 | 2.29 | 4.65 | 7.32 | 6.50 | 6.62 | 7.32 | 5.84 | 3.84 | 2.19 | 55.86 |
| Days ≥ 0.1" Precip | 5 | 5 | 5 | 4 | 7 | 10 | 10 | 10 | 11 | 8 | 5 | 4 | 84 |
| Total Snowfall (") | 0.0 | 0.0 | 0.0 | 0.0 | 0.0 | 0.0 | 0.0 | 0.0 | 0.0 | 0.0 | 0.0 | 0.0 | 0.0 |
| Days ≥ 1" Snow Depth | 0 | 0 | 0 | 0 | 0 | 0 | 0 | 0 | 0 | 0 | 0 | 0 | 0 |

## W PALM BEACH INTL AP *Palm Beach County* ELEVATION 20 ft LAT/LONG 26° 41 ' N / 80° 6 ' W

| | JAN | FEB | MAR | APR | MAY | JUN | JUL | AUG | SEP | OCT | NOV | DEC | YEAR |
|---|---|---|---|---|---|---|---|---|---|---|---|---|---|
| Maximum Temp °F | 74.9 | 75.8 | 78.8 | 82.2 | 85.6 | 88.2 | 89.9 | 90.0 | 88.4 | 84.9 | 80.1 | 76.4 | 82.9 |
| Minimum Temp °F | 56.8 | 57.5 | 61.3 | 65.2 | 70.2 | 73.3 | 74.6 | 75.1 | 74.4 | 70.8 | 64.9 | 59.3 | 66.9 |
| Mean Temp °F | 65.9 | 66.7 | 70.0 | 73.8 | 77.9 | 80.8 | 82.3 | 82.6 | 81.5 | 77.8 | 72.5 | 67.9 | 75.0 |
| Days Max Temp ≥ 90 °F | 0 | 0 | 0 | 2 | 4 | 10 | 18 | 19 | 10 | 2 | 0 | 0 | 65 |
| Days Max Temp ≤ 32 °F | 0 | 0 | 0 | 0 | 0 | 0 | 0 | 0 | 0 | 0 | 0 | 0 | 0 |
| Days Min Temp ≤ 32 °F | 0 | 0 | 0 | 0 | 0 | 0 | 0 | 0 | 0 | 0 | 0 | 0 | 0 |
| Days Min Temp ≤ 0 °F | 0 | 0 | 0 | 0 | 0 | 0 | 0 | 0 | 0 | 0 | 0 | 0 | 0 |
| Heating Degree Days | 84 | 63 | 28 | 3 | 0 | 0 | 0 | 0 | 0 | 1 | 15 | 58 | 252 |
| Cooling Degree Days | 121 | 141 | 196 | 264 | 407 | 500 | 558 | 564 | 510 | 420 | 275 | 162 | 4118 |
| Total Precipitation (") | 3.46 | 2.83 | 3.93 | 3.26 | 6.00 | 8.57 | 6.29 | 6.13 | 8.42 | 6.23 | 5.22 | 2.79 | 63.13 |
| Days ≥ 0.1" Precip | 5 | 5 | 5 | 4 | 7 | 11 | 10 | 10 | 12 | 8 | 6 | 4 | 87 |
| Total Snowfall (") | 0.0 | 0.0 | 0.0 | 0.0 | 0.0 | 0.0 | 0.0 | 0.0 | 0.0 | 0.0 | 0.0 | 0.0 | 0.0 |
| Days ≥ 1" Snow Depth | 0 | 0 | 0 | 0 | 0 | 0 | 0 | 0 | 0 | 0 | 0 | 0 | 0 |

## WAUCHULA 2 N *Hardee County* ELEVATION 121 ft LAT/LONG 27° 34' N / 81° 49' W

| | JAN | FEB | MAR | APR | MAY | JUN | JUL | AUG | SEP | OCT | NOV | DEC | YEAR |
|---|---|---|---|---|---|---|---|---|---|---|---|---|---|
| Maximum Temp °F | 74.2 | 75.7 | 80.2 | 85.0 | 89.4 | 91.3 | 92.3 | 92.4 | 90.8 | 86.0 | 80.1 | 75.5 | 84.4 |
| Minimum Temp °F | 49.9 | 50.4 | 54.6 | 58.2 | 64.1 | 69.8 | 71.4 | 71.8 | 70.7 | 64.5 | 57.0 | 51.4 | 61.2 |
| Mean Temp °F | 62.0 | 63.0 | 67.4 | 71.6 | 76.8 | 80.6 | 81.9 | 82.1 | 80.8 | 75.3 | 68.6 | 63.5 | 72.8 |
| Days Max Temp ≥ 90 °F | 0 | 0 | 1 | 6 | 16 | 22 | 27 | 27 | 23 | 6 | 0 | 0 | 128 |
| Days Max Temp ≤ 32 °F | 0 | 0 | 0 | 0 | 0 | 0 | 0 | 0 | 0 | 0 | 0 | 0 | 0 |
| Days Min Temp ≤ 32 °F | 2 | 1 | 0 | 0 | 0 | 0 | 0 | 0 | 0 | 0 | 0 | 1 | 4 |
| Days Min Temp ≤ 0 °F | 0 | 0 | 0 | 0 | 0 | 0 | 0 | 0 | 0 | 0 | 0 | 0 | 0 |
| Heating Degree Days | 154 | 119 | 54 | 10 | 0 | 0 | 0 | 0 | 0 | 3 | 44 | 121 | 505 |
| Cooling Degree Days | 66 | 84 | 134 | 198 | 364 | 489 | 542 | 550 | 484 | 338 | 177 | 85 | 3511 |
| Total Precipitation (") | 2.30 | 2.77 | 3.22 | 1.92 | 3.94 | 8.81 | 8.45 | 7.00 | 5.92 | 2.45 | 1.66 | 1.85 | 50.29 |
| Days ≥ 0.1" Precip | 4 | 4 | 5 | 3 | 5 | 10 | 12 | 12 | 9 | 4 | 3 | 3 | 74 |
| Total Snowfall (") | 0.0 | 0.0 | 0.0 | 0.0 | 0.0 | 0.0 | 0.0 | 0.0 | 0.0 | 0.0 | 0.0 | 0.0 | 0.0 |
| Days ≥ 1" Snow Depth | 0 | 0 | 0 | 0 | 0 | 0 | 0 | 0 | 0 | 0 | 0 | 0 | 0 |

## WEWAHITCHKA *Gulf County* ELEVATION 49 ft LAT/LONG 30° 7' N / 85° 12' W

| | JAN | FEB | MAR | APR | MAY | JUN | JUL | AUG | SEP | OCT | NOV | DEC | YEAR |
|---|---|---|---|---|---|---|---|---|---|---|---|---|---|
| Maximum Temp °F | 63.9 | 66.6 | 73.3 | 80.3 | 86.3 | 90.7 | 91.3 | 91.2 | 88.7 | 81.5 | 73.5 | 66.6 | 79.5 |
| Minimum Temp °F | 41.0 | 42.0 | 48.7 | 54.5 | 61.1 | 68.1 | 70.9 | 70.5 | 67.3 | 55.8 | 48.1 | 42.0 | 55.8 |
| Mean Temp °F | 52.5 | 54.3 | 61.1 | 67.4 | 73.6 | 79.4 | 81.1 | 80.9 | 78.0 | 68.7 | 60.8 | 54.4 | 67.7 |
| Days Max Temp ≥ 90 °F | 0 | 0 | 0 | 1 | 8 | 21 | 22 | 23 | 16 | 3 | 0 | 0 | 94 |
| Days Max Temp ≤ 32 °F | 0 | 0 | 0 | 0 | 0 | 0 | 0 | 0 | 0 | 0 | 0 | 0 | 0 |
| Days Min Temp ≤ 32 °F | 8 | 6 | 2 | 0 | 0 | 0 | 0 | 0 | 0 | 0 | 2 | 7 | 25 |
| Days Min Temp ≤ 0 °F | 0 | 0 | 0 | 0 | 0 | 0 | 0 | 0 | 0 | 0 | 0 | 0 | 0 |
| Heating Degree Days | 392 | 307 | 165 | 48 | 2 | 0 | 0 | 0 | 1 | 42 | 168 | 344 | 1469 |
| Cooling Degree Days | 7 | 16 | 52 | 112 | 274 | 444 | *509* | *506* | 396 | 157 | 68 | 19 | 2560 |
| Total Precipitation (") | 5.91 | 5.11 | 6.11 | 3.16 | 4.10 | 7.37 | 9.66 | 9.96 | 6.44 | 3.71 | 3.39 | 4.34 | 69.26 |
| Days ≥ 0.1" Precip | 8 | 6 | 6 | 4 | 6 | 9 | 14 | 13 | 8 | 4 | 5 | 6 | 89 |
| Total Snowfall (") | 0.0 | 0.0 | 0.0 | 0.0 | 0.0 | 0.0 | 0.0 | 0.0 | 0.0 | 0.0 | 0.0 | 0.0 | 0.0 |
| Days ≥ 1" Snow Depth | 0 | 0 | 0 | 0 | 0 | 0 | 0 | 0 | 0 | 0 | 0 | 0 | 0 |

## WINTER HAVEN *Polk County* ELEVATION 180 ft LAT/LONG 28° 1' N / 81° 44' W

| | JAN | FEB | MAR | APR | MAY | JUN | JUL | AUG | SEP | OCT | NOV | DEC | YEAR |
|---|---|---|---|---|---|---|---|---|---|---|---|---|---|
| Maximum Temp °F | 73.7 | 75.2 | 79.9 | 84.8 | 88.9 | 91.2 | 92.4 | 92.2 | 90.5 | 85.3 | 79.5 | 74.8 | 84.0 |
| Minimum Temp °F | 50.2 | 50.1 | 55.4 | 59.5 | 65.2 | 69.8 | 71.4 | 72.1 | 71.4 | 65.2 | 58.0 | 52.2 | 61.7 |
| Mean Temp °F | 62.0 | 62.7 | 67.7 | 72.2 | 77.1 | 80.6 | 81.9 | 82.1 | 80.9 | 75.3 | 68.8 | 63.5 | 72.9 |
| Days Max Temp ≥ 90 °F | 0 | 0 | 1 | 5 | 13 | 22 | 28 | 27 | 20 | 4 | 0 | 0 | 120 |
| Days Max Temp ≤ 32 °F | 0 | 0 | 0 | 0 | 0 | 0 | 0 | 0 | 0 | 0 | 0 | 0 | 0 |
| Days Min Temp ≤ 32 °F | 2 | 1 | 0 | 0 | 0 | 0 | 0 | 0 | 0 | 0 | 0 | 1 | 4 |
| Days Min Temp ≤ 0 °F | 0 | 0 | 0 | 0 | 0 | 0 | 0 | 0 | 0 | 0 | 0 | 0 | 0 |
| Heating Degree Days | 156 | 122 | 51 | 8 | 0 | 0 | 0 | 0 | 0 | 3 | 39 | 118 | 497 |
| Cooling Degree Days | 77 | 77 | 147 | 225 | 396 | 501 | 553 | 558 | 497 | 336 | 185 | 86 | 3638 |
| Total Precipitation (") | 2.35 | 2.58 | 3.34 | 1.83 | 4.04 | 7.23 | 8.05 | 7.53 | 5.72 | 2.42 | 2.28 | 1.95 | 49.32 |
| Days ≥ 0.1" Precip | 4 | 4 | 5 | 3 | 5 | 10 | 13 | 12 | 9 | 4 | 3 | 3 | 75 |
| Total Snowfall (") | 0.0 | 0.0 | 0.0 | 0.0 | 0.0 | 0.0 | 0.0 | 0.0 | 0.0 | 0.0 | 0.0 | 0.0 | 0.0 |
| Days ≥ 1" Snow Depth | 0 | 0 | 0 | 0 | 0 | 0 | 0 | 0 | 0 | 0 | 0 | 0 | 0 |

## JANUARY MINIMUM TEMPERATURE °F

| LOWEST | | | HIGHEST | | |
|---|---|---|---|---|---|
| 1 | Niceville | 36.9 | 1 | Key West | 65.3 |
| 2 | Monticello | 37.5 | 2 | Tavernier | 63.5 |
| 3 | De Funiak Springs | 37.6 | 3 | Miami Beach | 62.8 |
| 4 | Chipley | 37.9 | 4 | Miami | 59.9 |
| 5 | Jasper | 38.6 | 5 | Pompano Beach | 58.5 |
| 6 | Glen st. Mary | 38.8 | 6 | Fort Lauderdale | 58.0 |
| 7 | Tallahassee | 38.9 | 7 | Hialeah | 57.6 |
| 8 | Milton | 39.1 | 8 | West Palm Beach | 56.8 |
| 9 | Quincy | 39.4 | 9 | Tamiami Trail | 56.6 |
| 10 | Cross City | 39.6 | 10 | Flamingo | 56.2 |
| 11 | High Springs | 40.2 | 11 | Stuart | 54.9 |
| | Mayo | 40.2 | 12 | Clewiston | 54.4 |
| 13 | Steinhatchee | 40.4 | 13 | St. Petersburg | 54.3 |
| 14 | Wewahitchka | 41.0 | 14 | Fort Myers | 53.9 |
| 15 | Lake City | 41.4 | 15 | Everglades | 53.7 |
| | Perry | 41.4 | | Naples | 53.7 |
| 17 | Madison | 41.7 | 17 | Canal Point | 52.6 |
| 18 | Live Oak | 41.8 | 18 | Immokalee | 52.0 |
| 19 | Jacksonville | 41.9 | 19 | Fort Pierce | 51.8 |
| 20 | Pensacola | 42.2 | 20 | Belle Glade | 51.6 |
| 21 | Fernandina Beach | 42.7 | 21 | Moore Haven Lck | 51.5 |
| 22 | Usher | 42.9 | 22 | Punta Gorda | 51.4 |
| 23 | Apalachicola | 44.3 | 23 | Melbourne | 51.2 |
| 24 | Deland | 45.0 | 24 | Okeechobee | 51.1 |
| | Inverness | 45.0 | 25 | Venice | 51.0 |

## JULY MAXIMUM TEMPERATURE °F

| HIGHEST | | | LOWEST | | |
|---|---|---|---|---|---|
| 1 | La Belle | 92.8 | 1 | Miami Beach | 87.0 |
| | Mountain Lake | 92.8 | 2 | Apalachicola | 88.9 |
| 3 | Live Oak | 92.6 | | Flamingo | 88.9 |
| 4 | Myakka River | 92.5 | 4 | Key West | 89.2 |
| 5 | Lake Alfred | 92.4 | 5 | Miami | 89.3 |
| | Perry | 92.4 | 6 | Fort Lauderdale | 89.4 |
| | Winter Haven | 92.4 | 7 | Everglades | 89.7 |
| 8 | Bartow | 92.3 | | Jacksonville Bch | 89.7 |
| | Lakeland | 92.3 | | Stuart | 89.7 |
| | Wauchula | 92.3 | 10 | Hialeah | 89.9 |
| 11 | Devils Garden | 92.2 | | West Palm Beach | 89.9 |
| | Ocala | 92.2 | 12 | Melbourne | 90.0 |
| 13 | Avon Park | 92.1 | 13 | Cross City | 90.2 |
| | High Springs | 92.1 | | Daytona Beach | 90.2 |
| 15 | Jacksonville | 92.0 | | Fernandina Beach | 90.2 |
| | Punta Gorda | 92.0 | | Pensacola | 90.2 |
| | St. Leo | 92.0 | 17 | Vero Beach | 90.3 |
| 18 | Federal Point | 91.9 | 18 | Monticello | 90.4 |
| | Glen st. Mary | 91.9 | | Okeechobee | 90.4 |
| | Madison | 91.9 | | St. Petersburg | 90.4 |
| | Plant City | 91.9 | | Tampa | 90.4 |
| | Pompano Beach | 91.9 | 22 | Brooksville | 90.5 |
| | Tamiami Trail | 91.9 | | Tavernier | 90.5 |
| 24 | Arcadia | 91.8 | | Venice | 90.5 |
| 25 | Bushnell | 91.7 | 25 | Quincy | 90.6 |

## ANNUAL PRECIPITATION (")

| HIGHEST | | | LOWEST | | |
|---|---|---|---|---|---|
| 1 | Wewahitchka | 69.26 | 1 | Key West | 40.65 |
| 2 | Milton | 68.09 | 2 | Tampa | 43.42 |
| 3 | Niceville | 67.30 | 3 | Tavernier | 45.23 |
| 4 | Hialeah | 66.08 | 4 | Okeechobee | 46.03 |
| | Tallahassee | 66.08 | 5 | Clewiston | 46.07 |
| 6 | De Funiak Springs | 64.51 | 6 | Melbourne | 46.39 |
| 7 | Fort Lauderdale | 63.19 | 7 | Miami Beach | 46.60 |
| 8 | West Palm Beach | 63.13 | 8 | Moore Haven Lck | 47.37 |
| 9 | Pensacola | 62.95 | 9 | Kissimmee | 47.51 |
| 10 | Steinhatchee | 60.32 | 10 | St. Petersburg | 47.56 |
| 11 | Pompano Beach | 60.12 | 11 | Lisbon | 48.28 |
| 12 | Usher | 59.41 | 12 | Daytona Beach | 48.34 |
| 13 | Cross City | 58.88 | 13 | Venice | 48.44 |
| 14 | Perry | 58.72 | 14 | Flamingo | 48.68 |
| 15 | Miami | 58.65 | 15 | Punta Gorda | 48.74 |
| 16 | Quincy | 58.60 | 16 | Mountain Lake | 49.04 |
| 17 | Stuart | 58.41 | 17 | Winter Haven | 49.32 |
| 18 | Monticello | 57.70 | 18 | Lakeland | 49.54 |
| 19 | Deland | 57.67 | 19 | Avon Park | 49.92 |
| 20 | Apalachicola | 57.62 | 20 | Immokalee | 49.94 |
| 21 | Myakka River | 57.52 | 21 | Lake Alfred | 50.06 |
| 22 | Chipley | 56.81 | 22 | Bartow | 50.28 |
| 23 | Vero Beach | 55.86 | 23 | Wauchula | 50.29 |
| 24 | Mayo | 55.01 | 24 | Bushnell | 50.46 |
| 25 | Glen st. Mary | 54.88 | 25 | Fort Drum | 50.81 |

## ANNUAL SNOWFALL (")

| HIGHEST* | | | LOWEST* | | |
|---|---|---|---|---|---|
| 1 | Milton | 0.3 | 1 | Chipley | 0.1 |
| 2 | Pensacola | 0.2 | | De Funiak Springs | 0.1 |
| 3 | Chipley | 0.1 | | High Springs | 0.1 |
| | De Funiak Springs | 0.1 | | Jacksonville Bch | 0.1 |
| | High Springs | 0.1 | | Jasper | 0.1 |
| | Jacksonville Bch | 0.1 | | Live Oak | 0.1 |
| | Jasper | 0.1 | | Quincy | 0.1 |
| | Live Oak | 0.1 | 8 | Pensacola | 0.2 |
| | Quincy | 0.1 | 9 | Milton | 0.3 |

* Does not include stations which receive no snowfall.

# GEORGIA

PHYSICAL FEATURES. Georgia is located roughly between latitudes 30° and 35° N. and longitudes 81° and 86° W. From north to south its length is 320 miles, and its maximum width is about 250 miles. With an area of almost 59,000 square miles, it is the largest State east of the Mississippi River. Its elevation ranges from near sea level along the southeast coast to almost 5,000 feet at its highest point in the northeast.

Georgia's land area is made up of four principal physiographic provinces: the Blue Ridge or Mountain Province, the Valley and Ridge Province, the Piedmont Province, and the Coastal Plain Province.

The Blue Ridge or Mountain Province is located in the northeastern part of the State. The terrain in this area is characterized by forest-covered mountains and narrow valleys with rapidly flowing streams. The average elevation of the area is less than 2,000 feet, but the higher mountains reach altitudes between 4,000 and 5,000 feet above sea level. The Valley and Ridge Province, located in northwest Georgia, is composed of wide, flat valleys separated by narrow, steep, wooded ridges than run more or less northeast-southwest. The elevation of the valleys ranges mostly between 500 and 800 feet above sea level, with the ridges rising to heights of 600 to 2,000 feet.

The Piedmont Plateau Province is a wide area extending from the foothills of the Appalachian Mountains to the Coastal Plain and comprising nearly one-third of the area of the State. The terrain is mostly hilly in the north to rolling in the south, where it merges with the Coastal Plain. Elevations range from near 1,200 feet in the north to less than 500 feet in the south. The boundary between the Piedmont Province and the Coastal Plain is called the Fall Line, because of the steep fall of rivers as they cross this boundary. The Fall Line extends across the State from west-southwest to east-northeast. The Coastal Plain Province includes all of Georgia south of the Fall Line and comprises about three-fifths of the total area of the State. The terrain is slightly rolling to level and ranges in altitude from near sea level along the coast to a maximum of 600 feet. The low-lying coastal sections are rather marshy and the large slow-moving streams are bordered by wide, swampy, densely wooded areas.

Georgia streams are divided into two main groups -- those flowing southeastward into the Atlantic and those flowing southward directly into the Gulf of Mexico, or indirectly into the Gulf through the Alabama-Mobile and Tennessee River systems. The Chattahoochee Ridge marks the dividing line between the parts of the State that are drained into the Atlantic and into the Gulf. The main streams in the Atlantic drainage system are the Savannah and Altamaha Rivers. The Savannah and its headwater streams form the boundary between Georgia and South Carolina throughout its entire length. The Altamaha drains a large area of central Georgia. The Chattahoochee and Flint River systems constitute the major streams of west Georgia, which drain directly into the Gulf of Mexico.

GENERAL CLIMATE. Georgia's climate is determined primarily by its latitude, the proximity of the Gulf of Mexico and the Atlantic Ocean, and by the altitude.

Average annual rainfall in Georgia ranges from more than 75 inches in the extreme northeast corner to about 40 inches in a small area of the East Central Division. Total rainfall varies greatly from year to year in all parts of the State, and most stations with several years of record show more than twice as much rain in their wettest year as in their driest. The distribution of rainfall throughout the year is also highly variable in all parts of the State, but the extremes occur at different seasons in different areas.

Most of the State shows two maxima and two minima in the annual rainfall curve. One maximum occurs in winter and early spring and the other in midsummer. The driest season for all the State is autumn, with most areas showing a secondary minimum about May. In the northern third of the State, the cool season rainfall maximum predominates, with either January or March normally the wettest month. This is due to the greater influence in that area of the cyclonic storms that move across the country with regularity during winter and early spring. The mountains of north Georgia add enough lift to the moist air that is drawn into the forward side of these storms from the Gulf to add materially to the total annual rainfall of the area. Most sections of central and south Georgia have their greatest rainfall in midsummer, with a secondary maximum about March. The lower east coastal area has its highest normal rainfall in

September, due to the occasional extremely heavy rains that occur with late summer and autumn tropical storms. October is normally the driest month in most of the State. Snowfall is light in Georgia and of no significance at all in most of the State.

Due to its latitude and proximity to the warm waters of the Gulf of Mexico and Atlantic Ocean, most of Georgia has warm, humid summers and short, mild winters. However, in the northern part of the State, altitude becomes the more predominant influence with resulting cool summers and colder, but not severe, winters. All four seasons are apparent, but spring is usually short and blustery with rather frequent periods of storminess of varying intensity. In autumn long periods of mild, sunny weather are the rule for all of Georgia.

TEMPERATURE. Average summer temperatures range from about 73° F. in the extreme north to nearly 82° F. in parts of south Georgia. There is little difference in summer averages over the southern two-thirds of the State, where they range between 80° and 82° F. Summer days are characteristically warm and humid in this area, with high temperatures exceeding 90° F. on most days and reaching 100° F. during most years. Temperatures usually drop to the middle or low 70's, or even below 70° F. by early morning, giving some relief from the daytime warmth. The flow of moist air from the Gulf over the warm land surface results in frequent afternoon thundershowers in south and central Georgia during summer. These showers not only provide most of the summer rainfall, but oftentimes bring welcome relief from the afternoon heat. All parts of the State have experienced 100° F. weather at one time or another during the period of official records, but such occurrences are highly unusual in the mountain section of the north.

Winter temperatures show more variation from north to south than do those of summer. There is also a much greater variation in winter from day to day in all sections of the State. The average temperature for the three winter months ranges from 41° F. in the north to about 56° F. on the lower east coast, with the increase being almost uniform from north to south. All of Georgia experiences freezing temperatures almost every year, but the frequency of such occurrences varies greatly from the mountains to the coast. The average annual number of days with a temperature of 32° F. or less ranges from 110 in the north to about 10 in the lower coastal region.

Georgia winters are characterized by frequent and sometimes large fluctuations in temperature. The cold snaps, which usually occur with regularity from mid-November to mid-March, alternate with longer periods of mild weather. Daytime temperatures almost always rise to above freezing in the southern three-fourths of the State, even during the coldest weather. There is approximately 4 months difference in the average length of the freeze-free growing season from north to south, ranging from about 170 days in the northernmost areas to near 300 days on the lower coast.

Relative humidity averages are moderately high in most of Georgia, as would be expected from its location in relation to the Gulf of Mexico and the Atlantic Ocean and from the high frequency of wind flow from the direction of these warm waters. Year-round averages at about 7:00 a.m. are approximately 85%, or slightly higher in the south. By 1:00 p.m. the average drops to about 55%. Monthly averages for both morning and afternoon are higher in summer than in other seasons in all sections of the State.

STORMS. Several tornadoes may be expected in Georgia each year. These storms have occurred during every month of the year, but have their highest frequency in spring. Approximately 50% of Georgia's tornadoes have occurred in March and April. Local windstorms, other than tornadoes, occur frequently in spring and early summer. These storms usually occur in connection with thunderstorms, the more severe of which may also produce hail. The southeast Georgia coast has been battered by hurricane winds on a few occasions; but, since most of these storms do not reach the State or move into the State after having traveled over land areas, they usually produce only moderate winds and heavy to copious rains. Tropical storm rainfall contributes materially to the precipitation normals for the late summer and fall months in southeast Georgia and to a lesser extent in other areas of the State.

## COUNTY INDEX

*Appling County*
SURRENCY 2 WNW

*Bacon County*
ALMA BACON CO AP

*Baldwin County*
MILLEDGEVILLE

*Barrow County*
WINDER 1 SSE

*Bartow County*
CARTERSVILLE

*Ben Hill County*
FITZGERALD

*Bibb County*
MACON LEWIS B WILSON

*Brantley County*
NAHUNTA 3 E

*Brooks County*
QUITMAN 2 NW

*Bulloch County*
BROOKLET 1 W

*Burke County*
MIDVILLE EXPERIMENT
WAYNESBORO 2 NE

*Candler County*
METTER

*Carroll County*
CARROLLTON

*Charlton County*
FOLKSTON 9 SW

*Chatham County*
SAVANNAH INTL AP

*Clarke County*
ATHENS MUNI AP

*Clinch County*
HOMERVILLE 3 WSW

*Coffee County*
DOUGLAS

*Colquitt County*
MOULTRIE 2 ESE

*Columbia County*
APPLING 2 NW

*Coweta County*
NEWNAN 4 NE

*Crisp County*
CORDELE

*Dodge County*
EASTMAN 1 W

*Dougherty County*
ALBANY 3 SE

*Elbert County*
ELBERTON 2 N

*Emanuel County*
SWAINSBORO

*Floyd County*
ROME

*Fulton County*
ALPHARETTA 4 SSW
ATLANTA HARTSFIELD

*Glynn County*
BRUNSWICK
BRUNSWICK MCKINNON

*Gordon County*
CALHOUN EXPERIMENT S

*Greene County*
SILOAM 3 N

*Habersham County*
CORNELIA

*Hall County*
GAINESVILLE

*Hart County*
HARTWELL

*Jackson County*
COMMERCE 4 NNW

*Jasper County*
MONTICELLO

*Jefferson County*
LOUISVILLE 1 E

*Jenkins County*
MILLEN 4 N

*Laurens County*
DUBLIN

*Liberty County*
FORT STEWART

*Lumpkin County*
DAHLONEGA 2 NW

*McIntosh County*
SAPELO ISLAND

*Miller County*
COLQUITT 2 W

*Mitchell County*
CAMILLA 3 SE

*Monroe County*
FORSYTH 6 NNW

*Muscogee County*
COLUMBUS METRO AP

*Newton County*
COVINGTON

*Paulding County*
DALLAS 7 NE

***Pickens County***
JASPER 1 NNW

***Pierce County***
WAYCROSS 4 NE

***Polk County***
CEDARTOWN 3 NE

***Pulaski County***
HAWKINSVILLE

***Rabun County***
CLAYTON 1 SSW

***Randolph County***
CUTHBERT

***Richmond County***
AUGUSTA BUSH FIELD

***Spalding County***
EXPERIMENT

***Stephens County***
TOCCOA

***Stewart County***
LUMPKIN 2 SE

***Sumter County***
AMERICUS 3 SW
PLAINS SW GA EXP STN

***Talbot County***
TALBOTTON

***Tattnall County***
GLENNVILLE

***Telfair County***
LUMBER CITY

***Thomas County***
THOMASVILLE 3 NE

***Tift County***
TIFTON EXP STN

***Troup County***
LA GRANGE
WEST POINT

***Turner County***
ASHBURN 3 ENE

***Union County***
BLAIRSVILLE EXP STA

***Upson County***
THOMASTON 2 S

***Walker County***
LAFAYETTE 5 SW

***Warren County***
WARRENTON

***Washington County***
SANDERSVILLE

***Wayne County***
JESUP 8 S

***White County***
HELEN

***Whitfield County***
DALTON

***Wilkes County***
WASHINGTON 2 ESE

***Wilkinson County***
IRWINTON 4 WNW

## ELEVATION INDEX

| FEET | STATION NAME |
|---|---|
| 10 | BRUNSWICK |
| 10 | SAPELO ISLAND |
| 23 | BRUNSWICK MCKINNON |
| 46 | SAVANNAH INTL AP |
| 78 | NAHUNTA 3 E |
| 89 | FORT STEWART |
| 100 | JESUP 8 S |
| 120 | FOLKSTON 9 SW |
| 120 | LUMBER CITY |
| 120 | METTER |
| 148 | AUGUSTA BUSH FIELD |
| 151 | WAYCROSS 4 NE |
| 161 | COLQUITT 2 W |
| 180 | CAMILLA 3 SE |
| 180 | GLENNVILLE |
| 180 | HOMERVILLE 3 WSW |
| 180 | QUITMAN 2 NW |
| 190 | BROOKLET 1 W |
| 200 | ALBANY 3 SE |
| 200 | MILLEN 4 N |
| 200 | SURRENCY 2 WNW |
| 203 | ALMA BACON CO AP |
| 215 | DUBLIN |
| 249 | DOUGLAS |
| 249 | HAWKINSVILLE |
| 259 | WAYNESBORO 2 NE |
| 260 | THOMASVILLE 3 NE |
| 279 | MIDVILLE EXPERIMENT |
| 312 | CORDELE |
| 322 | LOUISVILLE 1 E |
| 322 | MILLEDGEVILLE |
| 322 | SWAINSBORO |
| 331 | MOULTRIE 2 ESE |
| 354 | MACON LEWIS B WILSON |
| 361 | APPLING 2 NW |
| 361 | EASTMAN 1 W |
| 371 | FITZGERALD |
| 371 | TIFTON EXP STN |
| 387 | COLUMBUS METRO AP |
| 410 | SANDERSVILLE |
| 430 | ASHBURN 3 ENE |
| 449 | IRWINTON 4 WNW |
| 461 | CUTHBERT |
| 479 | AMERICUS 3 SW |
| 500 | PLAINS SW GA EXP STN |
| 502 | LUMPKIN 2 SE |
| 502 | WARRENTON |
| 581 | WEST POINT |
| 600 | FORSYTH 6 NNW |
| 620 | ROME |
| 630 | WASHINGTON 2 ESE |
| 650 | CALHOUN EXPERIMENT S |
| 665 | THOMASTON 2 S |
| 679 | MONTICELLO |
| 689 | ELBERTON 2 N |
| 690 | SILOAM 3 N |
| 702 | TALBOTTON |
| 715 | LA GRANGE |

| FEET | STATION NAME |
|---|---|
| 741 | CARTERSVILLE |
| 741 | COMMERCE 4 NNW |
| 751 | COVINGTON |
| 761 | DALTON |
| 791 | HARTWELL |
| 801 | ATHENS MUNI AP |
| 801 | CEDARTOWN 3 NE |
| 810 | LAFAYETTE 5 SW |
| 951 | EXPERIMENT |
| 981 | NEWNAN 4 NE |
| 981 | WINDER 1 SSE |
| 1004 | ATLANTA HARTSFIELD |
| 1020 | TOCCOA |
| 1040 | ALPHARETTA 4 SSW |
| 1079 | DALLAS 7 NE |
| 1122 | CARROLLTON |
| 1250 | GAINESVILLE |
| 1480 | JASPER 1 NNW |
| 1489 | HELEN |
| 1503 | CORNELIA |
| 1522 | DAHLONEGA 2 NW |
| 1942 | BLAIRSVILLE EXP STA |
| 2001 | CLAYTON 1 SSW |

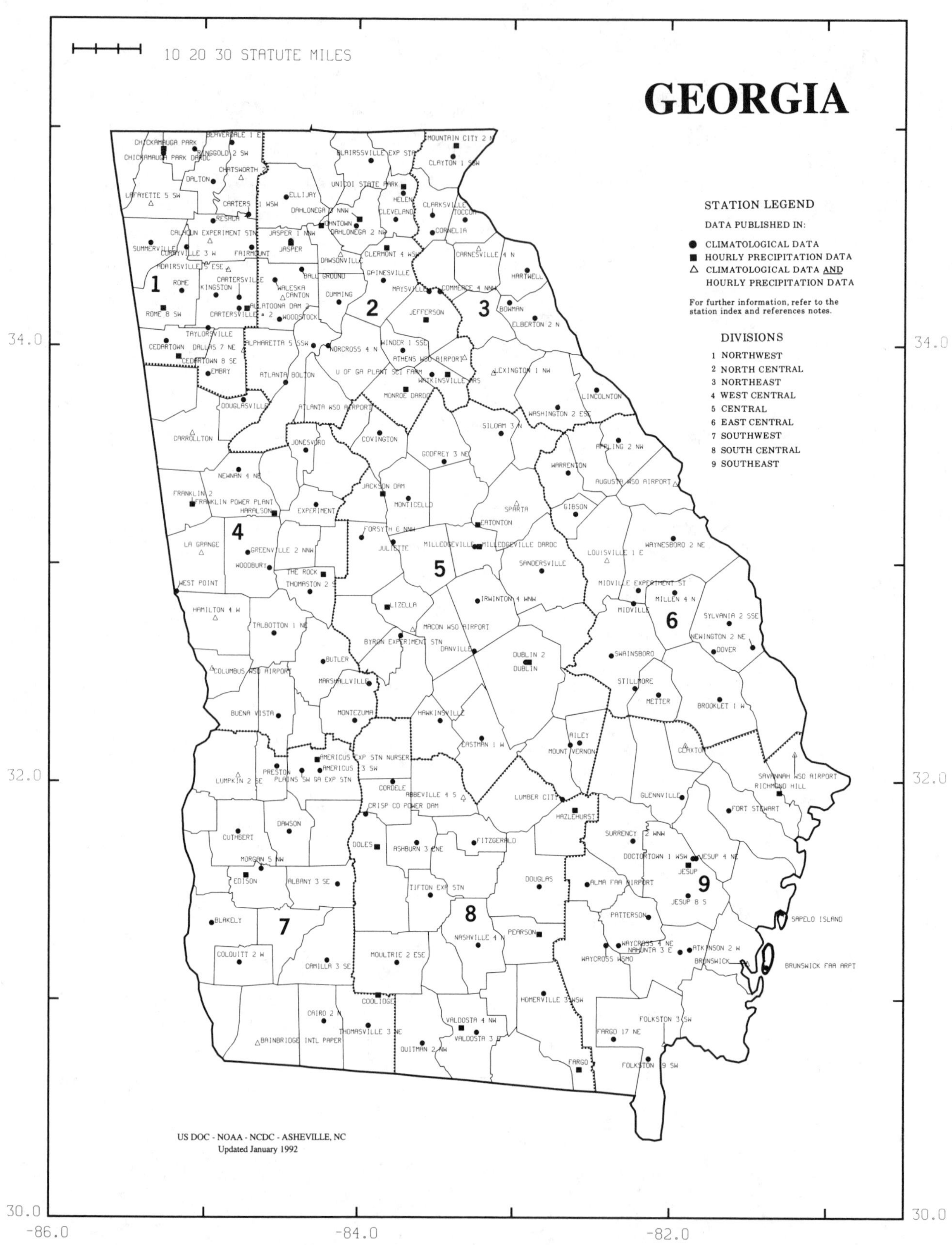
10 20 30 STATUTE MILES
GEORGIA
STATION LEGEND
DATA PUBLISHED IN:
● CLIMATOLOGICAL DATA
■ HOURLY PRECIPITATION DATA
△ CLIMATOLOGICAL DATA AND HOURLY PRECIPITATION DATA
For further information, refer to the station index and references notes.
DIVISIONS
1 NORTHWEST
2 NORTH CENTRAL
3 NORTHEAST
4 WEST CENTRAL
5 CENTRAL
6 EAST CENTRAL
7 SOUTHWEST
8 SOUTH CENTRAL
9 SOUTHEAST
34.0
32.0
30.0
-86.0
-84.0
-82.0
US DOC - NOAA - NCDC - ASHEVILLE, NC
Updated January 1992

## ALBANY 3 SE *Dougherty County* ELEVATION 200 ft LAT/LONG 31° 34 ' N / 84° 9 ' W

| | JAN | FEB | MAR | APR | MAY | JUN | JUL | AUG | SEP | OCT | NOV | DEC | YEAR |
|---|---|---|---|---|---|---|---|---|---|---|---|---|---|
| Maximum Temp °F | 59.4 | 63.1 | 71.0 | 78.9 | 85.1 | 90.1 | 92.1 | 91.5 | 88.2 | 80.1 | 71.1 | 63.2 | 77.8 |
| Minimum Temp °F | 34.2 | 36.3 | 43.3 | 50.1 | 58.5 | 66.2 | 69.6 | 69.3 | 64.5 | 52.1 | 42.8 | 36.8 | 52.0 |
| Mean Temp °F | 46.8 | 49.7 | 57.2 | 64.5 | 71.8 | 78.2 | 80.9 | 80.4 | 76.3 | 66.1 | 56.9 | 50.0 | 64.9 |
| Days Max Temp ≥ 90 °F | 0 | 0 | 0 | 1 | 7 | 18 | 23 | 23 | 15 | 2 | 0 | 0 | 89 |
| Days Max Temp ≤ 32 °F | 0 | 0 | 0 | 0 | 0 | 0 | 0 | 0 | 0 | 0 | 0 | 0 | 0 |
| Days Min Temp ≤ 32 °F | 15 | 11 | 5 | 1 | 0 | 0 | 0 | 0 | 0 | 1 | 6 | 12 | 51 |
| Days Min Temp ≤ 0 °F | 0 | 0 | 0 | 0 | 0 | 0 | 0 | 0 | 0 | 0 | 0 | 0 | 0 |
| Heating Degree Days | 559 | 432 | 260 | 91 | 9 | 0 | 0 | 0 | 3 | 77 | 262 | 468 | 2161 |
| Cooling Degree Days | 1 | 10 | 25 | 68 | 225 | 418 | 516 | 500 | 352 | 125 | 31 | 10 | 2281 |
| Total Precipitation (") | 5.94 | 4.86 | 5.82 | 3.40 | 4.03 | 5.31 | 6.12 | 4.39 | 3.43 | 2.21 | 3.43 | 3.77 | 52.71 |
| Days ≥ 0.1" Precip | 8 | 7 | 8 | 5 | 6 | 8 | 9 | 7 | 5 | 3 | 5 | 6 | 77 |
| Total Snowfall (") | 0.0 | 0.1 | 0.0 | 0.0 | 0.0 | 0.0 | 0.0 | 0.0 | 0.0 | 0.0 | 0.0 | 0.0 | 0.1 |
| Days ≥ 1" Snow Depth | 0 | 0 | 0 | 0 | 0 | 0 | 0 | 0 | 0 | 0 | 0 | 0 | 0 |

## ALMA BACON CO AP *Bacon County* ELEVATION 203 ft LAT/LONG 31° 32 ' N / 82° 31 ' W

| | JAN | FEB | MAR | APR | MAY | JUN | JUL | AUG | SEP | OCT | NOV | DEC | YEAR |
|---|---|---|---|---|---|---|---|---|---|---|---|---|---|
| Maximum Temp °F | 61.0 | 64.4 | 72.0 | 79.2 | 84.9 | 89.6 | 91.8 | 90.7 | 86.9 | 79.1 | 71.2 | 64.0 | 77.9 |
| Minimum Temp °F | 39.8 | 41.8 | 48.0 | 54.0 | 61.2 | 67.9 | 71.0 | 70.6 | 67.0 | 56.7 | 48.1 | 42.2 | 55.7 |
| Mean Temp °F | 50.4 | 53.2 | 60.0 | 66.6 | 73.1 | 78.8 | 81.4 | 80.7 | 77.0 | 68.0 | 59.7 | 53.1 | 66.8 |
| Days Max Temp ≥ 90 °F | 0 | 0 | 0 | 1 | 5 | 16 | 23 | 20 | 11 | 1 | 0 | 0 | 77 |
| Days Max Temp ≤ 32 °F | 0 | 0 | 0 | 0 | 0 | 0 | 0 | 0 | 0 | 0 | 0 | 0 | 0 |
| Days Min Temp ≤ 32 °F | 9 | 6 | 2 | 0 | 0 | 0 | 0 | 0 | 0 | 0 | 2 | 7 | 26 |
| Days Min Temp ≤ 0 °F | 0 | 0 | 0 | 0 | 0 | 0 | 0 | 0 | 0 | 0 | 0 | 0 | 0 |
| Heating Degree Days | 452 | 340 | 189 | 57 | 4 | 0 | 0 | 0 | 1 | 50 | 195 | 375 | 1663 |
| Cooling Degree Days | 5 | 16 | 41 | 105 | 264 | 440 | 532 | 503 | 365 | 152 | 49 | 15 | 2487 |
| Total Precipitation (") | 4.66 | 3.86 | 4.99 | 2.98 | 3.58 | 5.36 | 5.89 | 6.03 | 3.15 | 2.58 | 2.51 | 3.82 | 49.41 |
| Days ≥ 0.1" Precip | 7 | 6 | 6 | 4 | 6 | 8 | 10 | 9 | 6 | 4 | 4 | 6 | 76 |
| Total Snowfall (") | 0.0 | 0.2 | 0.0 | 0.0 | 0.0 | 0.0 | 0.0 | 0.0 | 0.0 | 0.0 | 0.0 | 0.1 | 0.3 |
| Days ≥ 1" Snow Depth | 0 | 0 | 0 | 0 | 0 | 0 | 0 | 0 | 0 | 0 | 0 | 0 | 0 |

## ALPHARETTA 4 SSW *Fulton County* ELEVATION 1040 ft LAT/LONG 34° 0 ' N / 84° 18 ' W

| | JAN | FEB | MAR | APR | MAY | JUN | JUL | AUG | SEP | OCT | NOV | DEC | YEAR |
|---|---|---|---|---|---|---|---|---|---|---|---|---|---|
| Maximum Temp °F | 49.2 | 54.0 | 62.8 | 71.6 | 77.7 | 84.3 | 87.3 | 86.2 | 80.8 | 71.2 | 62.1 | 53.6 | 70.1 |
| Minimum Temp °F | 27.9 | 30.0 | 36.7 | 44.5 | 53.0 | 61.1 | 65.7 | 64.9 | 58.9 | 46.2 | 37.2 | 31.1 | 46.4 |
| Mean Temp °F | 38.6 | 42.0 | 49.8 | 58.0 | 65.4 | 72.7 | 76.5 | 75.6 | 69.9 | 58.8 | 49.7 | 42.4 | 58.3 |
| Days Max Temp ≥ 90 °F | 0 | 0 | 0 | 0 | 0 | 5 | 11 | 7 | 2 | 0 | 0 | 0 | 25 |
| Days Max Temp ≤ 32 °F | 2 | 1 | 0 | 0 | 0 | 0 | 0 | 0 | 0 | 0 | 0 | 1 | 4 |
| Days Min Temp ≤ 32 °F | 21 | 18 | 11 | 4 | 0 | 0 | 0 | 0 | 0 | 2 | 12 | 19 | 87 |
| Days Min Temp ≤ 0 °F | 0 | 0 | 0 | 0 | 0 | 0 | 0 | 0 | 0 | 0 | 0 | 0 | 0 |
| Heating Degree Days | 812 | 644 | 470 | 226 | 66 | 6 | 0 | 0 | 25 | 210 | 456 | 696 | 3611 |
| Cooling Degree Days | 0 | 0 | 5 | 24 | 90 | 271 | 392 | 358 | 190 | 29 | 3 | 1 | 1363 |
| Total Precipitation (") | 5.15 | 4.59 | 5.64 | 4.14 | 4.96 | 3.54 | 4.62 | 4.41 | 3.64 | 3.61 | 3.87 | 4.34 | 52.51 |
| Days ≥ 0.1" Precip | 9 | 7 | 7 | 6 | 7 | 6 | 7 | 6 | 5 | 5 | 6 | 7 | 78 |
| Total Snowfall (") | 0.6 | 0.3 | 0.0 | 0.0 | 0.0 | 0.0 | 0.0 | 0.0 | 0.0 | 0.0 | 0.0 | 0.0 | 0.9 |
| Days ≥ 1" Snow Depth | 0 | 0 | 0 | 0 | 0 | 0 | 0 | 0 | 0 | 0 | 0 | 0 | 0 |

## AMERICUS 3 SW *Sumter County* ELEVATION 479 ft LAT/LONG 32° 4 ' N / 84° 14 ' W

| | JAN | FEB | MAR | APR | MAY | JUN | JUL | AUG | SEP | OCT | NOV | DEC | YEAR |
|---|---|---|---|---|---|---|---|---|---|---|---|---|---|
| Maximum Temp °F | ***57.2*** | 61.7 | 69.7 | 78.1 | 83.6 | 89.0 | ***90.7*** | 90.3 | 86.4 | 78.4 | 68.9 | ***61.1*** | 76.3 |
| Minimum Temp °F | ***36.1*** | 37.5 | 44.4 | 51.2 | 59.0 | 66.0 | ***69.3*** | 68.6 | 63.9 | 53.0 | 44.0 | ***38.5*** | 52.6 |
| Mean Temp °F | ***46.7*** | 49.6 | 57.1 | 64.7 | 71.3 | 77.5 | ***80.0*** | 79.5 | 75.2 | 65.7 | 56.5 | ***49.9*** | 64.5 |
| Days Max Temp ≥ 90 °F | 0 | 0 | 0 | 1 | 4 | 15 | ***18*** | 18 | 10 | 1 | 0 | 0 | 67 |
| Days Max Temp ≤ 32 °F | 0 | 0 | 0 | 0 | 0 | 0 | 0 | 0 | 0 | 0 | 0 | 0 | 0 |
| Days Min Temp ≤ 32 °F | 13 | 10 | 4 | 0 | 0 | 0 | 0 | 0 | 0 | 0 | 5 | ***10*** | 42 |
| Days Min Temp ≤ 0 °F | 0 | 0 | 0 | 0 | 0 | 0 | 0 | 0 | 0 | 0 | 0 | 0 | 0 |
| Heating Degree Days | ***561*** | 432 | 264 | 88 | 10 | 0 | ***0*** | 0 | 5 | 80 | 268 | ***470*** | 2178 |
| Cooling Degree Days | ***0*** | 5 | 20 | 64 | 198 | 391 | na | 459 | 311 | 101 | 22 | ***7*** | na |
| Total Precipitation (") | 5.34 | 4.68 | 5.29 | 2.94 | 3.64 | 4.18 | 5.81 | 4.26 | 2.88 | 2.10 | 3.25 | 4.10 | 48.47 |
| Days ≥ 0.1" Precip | 8 | 6 | 8 | 5 | 6 | 7 | 9 | 7 | 5 | 4 | 5 | 6 | 76 |
| Total Snowfall (") | 0.0 | 0.3 | 0.1 | 0.0 | 0.0 | 0.0 | 0.0 | 0.0 | 0.0 | 0.0 | 0.0 | 0.1 | 0.5 |
| Days ≥ 1" Snow Depth | 0 | 0 | 0 | 0 | 0 | 0 | 0 | 0 | 0 | 0 | 0 | 0 | 0 |

## APPLING 2 NW *Columbia County* ELEVATION 361 ft LAT/LONG 33° 33' N / 82° 20' W

| | JAN | FEB | MAR | APR | MAY | JUN | JUL | AUG | SEP | OCT | NOV | DEC | YEAR |
|---|---|---|---|---|---|---|---|---|---|---|---|---|---|
| Maximum Temp °F | 53.7 | 58.5 | 66.8 | 75.4 | 82.1 | 88.3 | 91.4 | 89.9 | 85.2 | 75.8 | 66.6 | 57.6 | 74.3 |
| Minimum Temp °F | 30.2 | 32.2 | 39.2 | 46.6 | 54.8 | 63.3 | 67.3 | 66.4 | 60.2 | 47.4 | 39.0 | 32.7 | 48.3 |
| Mean Temp °F | 42.0 | 45.4 | 53.0 | 61.0 | 68.5 | 75.8 | 79.4 | 78.2 | 72.7 | 61.6 | 52.8 | 45.2 | 61.3 |
| Days Max Temp ≥ 90 °F | 0 | 0 | 0 | 1 | 3 | 13 | 21 | 18 | 8 | 1 | 0 | 0 | 65 |
| Days Max Temp ≤ 32 °F | 1 | 0 | 0 | 0 | 0 | 0 | 0 | 0 | 0 | 0 | 0 | 0 | 1 |
| Days Min Temp ≤ 32 °F | 19 | 16 | 9 | 2 | 0 | 0 | 0 | 0 | 0 | 2 | 9 | 17 | 74 |
| Days Min Temp ≤ 0 °F | 0 | 0 | 0 | 0 | 0 | 0 | 0 | 0 | 0 | 0 | 0 | 0 | 0 |
| Heating Degree Days | 707 | 548 | 373 | 160 | 35 | 2 | 0 | 0 | 12 | 147 | 366 | 609 | 2959 |
| Cooling Degree Days | 0 | 1 | 9 | 40 | 143 | 346 | 469 | 417 | 249 | 47 | 8 | 1 | 1730 |
| Total Precipitation (") | 4.62 | 4.07 | 5.00 | 3.41 | 3.92 | 4.12 | 4.19 | 4.25 | 3.04 | 3.52 | 3.17 | 3.90 | 47.21 |
| Days ≥ 0.1" Precip | 8 | 7 | 7 | 5 | 6 | 6 | 8 | 7 | 5 | 4 | 5 | 7 | 75 |
| Total Snowfall (") | 0.2 | 0.6 | 0.2 | 0.0 | 0.0 | 0.0 | 0.0 | 0.0 | 0.0 | 0.0 | 0.0 | 0.0 | 1.0 |
| Days ≥ 1" Snow Depth | *0* | 0 | 0 | 0 | 0 | 0 | 0 | 0 | 0 | 0 | 0 | 0 | 0 |

## ASHBURN 3 ENE *Turner County* ELEVATION 430 ft LAT/LONG 31° 42' N / 83° 39' W

| | JAN | FEB | MAR | APR | MAY | JUN | JUL | AUG | SEP | OCT | NOV | DEC | YEAR |
|---|---|---|---|---|---|---|---|---|---|---|---|---|---|
| Maximum Temp °F | 57.4 | *60.7* | 69.0 | 77.1 | 83.3 | 88.7 | 90.3 | 90.0 | 86.2 | 78.1 | 68.9 | 61.3 | 75.9 |
| Minimum Temp °F | 36.3 | *38.9* | 46.3 | 53.9 | 61.1 | 67.3 | 70.4 | 69.6 | 65.2 | 54.1 | 45.4 | 39.3 | 54.0 |
| Mean Temp °F | 46.9 | *49.9* | 57.7 | 65.5 | 72.3 | 78.0 | 80.4 | 79.8 | 75.7 | 66.1 | 57.2 | 50.4 | 65.0 |
| Days Max Temp ≥ 90 °F | 0 | 0 | 0 | 1 | 5 | 14 | 19 | 18 | 9 | 1 | 0 | 0 | 67 |
| Days Max Temp ≤ 32 °F | 1 | 0 | 0 | 0 | 0 | 0 | 0 | 0 | 0 | 0 | 0 | 0 | 1 |
| Days Min Temp ≤ 32 °F | 12 | 7 | 2 | 0 | 0 | 0 | 0 | 0 | 0 | 0 | 3 | 8 | 32 |
| Days Min Temp ≤ 0 °F | 0 | 0 | 0 | 0 | 0 | 0 | 0 | 0 | 0 | 0 | 0 | 0 | 0 |
| Heating Degree Days | 555 | *426* | 247 | 77 | 8 | 0 | 0 | 0 | 3 | 71 | 250 | 454 | 2091 |
| Cooling Degree Days | 1 | *5* | 28 | 92 | 243 | *423* | 506 | 483 | 333 | 120 | 24 | 7 | 2265 |
| Total Precipitation (") | 5.23 | 4.62 | 5.17 | 2.92 | 3.58 | 4.85 | 4.92 | 4.62 | 3.17 | 2.36 | 3.16 | 3.56 | 48.16 |
| Days ≥ 0.1" Precip | 7 | 6 | 7 | 5 | 5 | 8 | 9 | 8 | 5 | 4 | 4 | 6 | 74 |
| Total Snowfall (") | 0.1 | 0.1 | 0.0 | 0.0 | 0.0 | 0.0 | 0.0 | 0.0 | 0.0 | 0.0 | 0.0 | 0.0 | 0.2 |
| Days ≥ 1" Snow Depth | 0 | 0 | 0 | 0 | 0 | 0 | 0 | 0 | 0 | 0 | 0 | 0 | 0 |

## ATHENS MUNI AP *Clarke County* ELEVATION 801 ft LAT/LONG 33° 57' N / 83° 19' W

| | JAN | FEB | MAR | APR | MAY | JUN | JUL | AUG | SEP | OCT | NOV | DEC | YEAR |
|---|---|---|---|---|---|---|---|---|---|---|---|---|---|
| Maximum Temp °F | 51.8 | 56.5 | 65.2 | 73.8 | 80.4 | 87.1 | 90.0 | 88.1 | 82.9 | 73.4 | 64.2 | 55.4 | 72.4 |
| Minimum Temp °F | 32.5 | 35.0 | 42.2 | 49.9 | 58.1 | 65.9 | 69.8 | 68.9 | 63.1 | 51.0 | 42.3 | 35.7 | 51.2 |
| Mean Temp °F | 42.2 | 45.8 | 53.7 | 61.9 | 69.3 | 76.5 | 79.9 | 78.6 | 73.0 | 62.2 | 53.3 | 45.6 | 61.8 |
| Days Max Temp ≥ 90 °F | 0 | 0 | 0 | 0 | 2 | 10 | 17 | 13 | 5 | 0 | 0 | 0 | 47 |
| Days Max Temp ≤ 32 °F | 1 | 0 | 0 | 0 | 0 | 0 | 0 | 0 | 0 | 0 | 0 | 0 | 1 |
| Days Min Temp ≤ 32 °F | 16 | 12 | 5 | 0 | 0 | 0 | 0 | 0 | 0 | 0 | 5 | 13 | 51 |
| Days Min Temp ≤ 0 °F | 0 | 0 | 0 | 0 | 0 | 0 | 0 | 0 | 0 | 0 | 0 | 0 | 0 |
| Heating Degree Days | 701 | 537 | 351 | 140 | 27 | 1 | 0 | 0 | 9 | 133 | 352 | 598 | 2849 |
| Cooling Degree Days | 0 | 1 | 9 | 53 | 179 | 389 | 503 | 450 | 270 | 61 | 8 | 2 | 1925 |
| Total Precipitation (") | 4.50 | 4.18 | 5.23 | 3.36 | 4.27 | 4.10 | 4.70 | 3.87 | 3.43 | 3.42 | 3.70 | 3.89 | 48.65 |
| Days ≥ 0.1" Precip | 8 | 7 | 7 | 6 | 6 | 6 | 7 | 6 | 5 | 5 | 6 | 7 | 76 |
| Total Snowfall (") | 1.3 | 0.9 | 0.5 | 0.0 | 0.0 | 0.0 | 0.0 | 0.0 | 0.0 | 0.0 | 0.1 | 0.2 | 3.0 |
| Days ≥ 1" Snow Depth | 1 | 1 | 0 | 0 | 0 | 0 | 0 | 0 | 0 | 0 | 0 | 0 | 2 |

## ATLANTA HARTSFIELD *Fulton County* ELEVATION 1004 ft LAT/LONG 33° 39' N / 84° 25' W

| | JAN | FEB | MAR | APR | MAY | JUN | JUL | AUG | SEP | OCT | NOV | DEC | YEAR |
|---|---|---|---|---|---|---|---|---|---|---|---|---|---|
| Maximum Temp °F | 50.9 | 55.6 | 64.5 | 73.3 | 79.6 | 86.1 | 88.6 | 87.3 | 82.1 | 72.6 | 63.4 | 54.7 | 71.6 |
| Minimum Temp °F | 32.4 | 35.2 | 42.7 | 50.7 | 58.9 | 66.6 | 70.0 | 69.4 | 64.0 | 52.3 | 43.0 | 35.9 | 51.8 |
| Mean Temp °F | 41.6 | 45.4 | 53.6 | 62.0 | 69.3 | 76.3 | 79.3 | 78.4 | 73.1 | 62.5 | 53.2 | 45.3 | 61.7 |
| Days Max Temp ≥ 90 °F | 0 | 0 | 0 | 0 | 1 | 9 | 13 | 10 | 3 | 0 | 0 | 0 | 36 |
| Days Max Temp ≤ 32 °F | 1 | 0 | 0 | 0 | 0 | 0 | 0 | 0 | 0 | 0 | 0 | 0 | 1 |
| Days Min Temp ≤ 32 °F | 15 | 12 | 5 | 1 | 0 | 0 | 0 | 0 | 0 | 0 | 5 | 12 | 50 |
| Days Min Temp ≤ 0 °F | 0 | 0 | 0 | 0 | 0 | 0 | 0 | 0 | 0 | 0 | 0 | 0 | 0 |
| Heating Degree Days | 717 | 548 | 355 | 139 | 26 | 1 | 0 | 0 | 11 | 129 | 355 | 604 | 2885 |
| Cooling Degree Days | 0 | 1 | 14 | 65 | 188 | 395 | 503 | 456 | 281 | 68 | 10 | 2 | 1983 |
| Total Precipitation (") | 4.73 | 4.48 | 5.53 | 3.83 | 4.46 | 3.61 | 5.44 | 3.84 | 3.71 | 3.20 | 4.03 | 4.00 | 50.86 |
| Days ≥ 0.1" Precip | 8 | 7 | 7 | 6 | 7 | 6 | 9 | 7 | 5 | 4 | 6 | 6 | 78 |
| Total Snowfall (") | 1.0 | 0.6 | 0.6 | 0.0 | 0.0 | 0.0 | 0.0 | 0.0 | 0.0 | 0.0 | 0.1 | 0.2 | 2.5 |
| Days ≥ 1" Snow Depth | 0 | 0 | 0 | 0 | 0 | 0 | 0 | 0 | 0 | 0 | 0 | 0 | 0 |

## AUGUSTA BUSH FIELD *Richmond County* ELEVATION 148 ft LAT/LONG 33° 22 ' N / 81° 58 ' W

| | JAN | FEB | MAR | APR | MAY | JUN | JUL | AUG | SEP | OCT | NOV | DEC | YEAR |
|---|---|---|---|---|---|---|---|---|---|---|---|---|---|
| Maximum Temp °F | 56.0 | 60.4 | 68.9 | 76.9 | 83.4 | 89.5 | 92.2 | 90.5 | 85.9 | 77.0 | 68.1 | 59.7 | 75.7 |
| Minimum Temp °F | 32.7 | 34.9 | 42.2 | 48.7 | 57.5 | 65.7 | 70.1 | 69.1 | 63.2 | 50.7 | 41.6 | 35.3 | 51.0 |
| Mean Temp °F | 44.4 | 47.7 | 55.5 | 62.8 | 70.5 | 77.6 | 81.2 | 79.8 | 74.6 | 63.9 | 54.9 | 47.5 | 63.4 |
| Days Max Temp ≥ 90 °F | 0 | 0 | 0 | 1 | 5 | 16 | 22 | 19 | 9 | 1 | 0 | 0 | 73 |
| Days Max Temp ≤ 32 °F | 1 | 0 | 0 | 0 | 0 | 0 | 0 | 0 | 0 | 0 | 0 | 0 | 1 |
| Days Min Temp ≤ 32 °F | 16 | 12 | 5 | 1 | 0 | 0 | 0 | 0 | 0 | 1 | 7 | 14 | 56 |
| Days Min Temp ≤ 0 °F | 0 | 0 | 0 | 0 | 0 | 0 | 0 | 0 | 0 | 0 | 0 | 0 | 0 |
| Heating Degree Days | 632 | 484 | 302 | 117 | 19 | 0 | 0 | 0 | 5 | 107 | 312 | 539 | 2517 |
| Cooling Degree Days | 1 | 3 | 17 | 58 | 204 | 423 | 545 | 489 | 308 | 91 | 19 | 4 | 2162 |
| Total Precipitation (") | 4.06 | 4.00 | 4.87 | 2.83 | 3.64 | 4.12 | 4.47 | 4.56 | 2.93 | 3.17 | 2.60 | 3.23 | 44.48 |
| Days ≥ 0.1" Precip | 7 | 6 | 7 | 5 | 6 | 6 | 7 | 7 | 4 | 4 | 4 | 6 | 69 |
| Total Snowfall (") | 0.4 | 1.0 | 0.0 | 0.0 | 0.0 | 0.0 | 0.0 | 0.0 | 0.0 | 0.0 | 0.0 | 0.1 | 1.5 |
| Days ≥ 1" Snow Depth | 0 | 0 | 0 | 0 | 0 | 0 | 0 | 0 | 0 | 0 | 0 | 0 | 0 |

## BLAIRSVILLE EXP STA *Union County* ELEVATION 1942 ft LAT/LONG 34° 51 ' N / 83° 56 ' W

| | JAN | FEB | MAR | APR | MAY | JUN | JUL | AUG | SEP | OCT | NOV | DEC | YEAR |
|---|---|---|---|---|---|---|---|---|---|---|---|---|---|
| Maximum Temp °F | 47.4 | 51.1 | 59.1 | 68.3 | 74.7 | 81.3 | 84.3 | 83.2 | 77.9 | 69.1 | 59.9 | 51.5 | 67.3 |
| Minimum Temp °F | 23.9 | 25.7 | 33.4 | 40.8 | 48.9 | 57.0 | 61.5 | 60.5 | 54.6 | 41.5 | 33.8 | 27.3 | 42.4 |
| Mean Temp °F | 35.7 | 38.4 | 46.3 | 54.6 | 61.8 | 69.1 | 72.9 | 71.8 | 66.3 | 55.3 | 46.9 | 39.4 | 54.9 |
| Days Max Temp ≥ 90 °F | 0 | 0 | 0 | 0 | 0 | 1 | 5 | 2 | 0 | 0 | 0 | 0 | 8 |
| Days Max Temp ≤ 32 °F | 3 | 2 | 0 | 0 | 0 | 0 | 0 | 0 | 0 | 0 | 0 | 1 | 6 |
| Days Min Temp ≤ 32 °F | 23 | 21 | 16 | 7 | 1 | 0 | 0 | 0 | 0 | 7 | 15 | 22 | 112 |
| Days Min Temp ≤ 0 °F | 1 | 0 | 0 | 0 | 0 | 0 | 0 | 0 | 0 | 0 | 0 | 0 | 1 |
| Heating Degree Days | 902 | 743 | 574 | 312 | 127 | 19 | 1 | 2 | 53 | 301 | 537 | 786 | 4357 |
| Cooling Degree Days | 0 | 0 | 1 | 5 | 40 | 170 | 278 | 230 | 102 | 10 | 1 | 0 | 837 |
| Total Precipitation (") | 5.15 | 5.05 | 6.34 | 4.55 | 4.88 | 4.26 | 4.70 | 4.84 | 4.29 | 3.72 | 4.78 | 4.82 | 57.38 |
| Days ≥ 0.1" Precip | 8 | 8 | 9 | 8 | 8 | 8 | 9 | 8 | 7 | 5 | 7 | 8 | 93 |
| Total Snowfall (") | *2.1* | *1.5* | 1.7 | 0.4 | 0.0 | 0.0 | 0.0 | 0.0 | 0.0 | 0.0 | 0.2 | 0.4 | 6.3 |
| Days ≥ 1" Snow Depth | *0* | *0* | *0* | 0 | 0 | 0 | 0 | 0 | 0 | 0 | 0 | 0 | 0 |

## BROOKLET 1 W *Bulloch County* ELEVATION 190 ft LAT/LONG 32° 23 ' N / 81° 41 ' W

| | JAN | FEB | MAR | APR | MAY | JUN | JUL | AUG | SEP | OCT | NOV | DEC | YEAR |
|---|---|---|---|---|---|---|---|---|---|---|---|---|---|
| Maximum Temp °F | 59.1 | 63.1 | 70.6 | 78.5 | 84.3 | 89.2 | 91.9 | 90.1 | 85.9 | 78.0 | 70.0 | 62.6 | 76.9 |
| Minimum Temp °F | 37.7 | 39.9 | 46.5 | 53.3 | 60.6 | 67.0 | 70.2 | 69.4 | 65.1 | 54.9 | 46.3 | 40.3 | 54.3 |
| Mean Temp °F | 48.4 | 51.5 | 58.6 | 65.9 | 72.5 | 78.1 | 81.1 | 79.8 | 75.5 | 66.5 | 58.2 | 51.5 | 65.6 |
| Days Max Temp ≥ 90 °F | 0 | 0 | 0 | 1 | 6 | 15 | 23 | 19 | 9 | 1 | 0 | 0 | 74 |
| Days Max Temp ≤ 32 °F | 0 | 0 | 0 | 0 | 0 | 0 | 0 | 0 | 0 | 0 | 0 | 0 | 0 |
| Days Min Temp ≤ 32 °F | 11 | 7 | 2 | 0 | 0 | 0 | 0 | 0 | 0 | 0 | 3 | 8 | 31 |
| Days Min Temp ≤ 0 °F | 0 | 0 | 0 | 0 | 0 | 0 | 0 | 0 | 0 | 0 | 0 | 0 | 0 |
| Heating Degree Days | 510 | 383 | 222 | 67 | 6 | 0 | 0 | 0 | 2 | 66 | 228 | 421 | 1905 |
| Cooling Degree Days | 3 | 11 | 29 | 94 | 250 | 437 | 538 | 485 | 332 | 127 | 36 | 11 | 2353 |
| Total Precipitation (") | 4.40 | 3.65 | 4.27 | 2.81 | 4.26 | 4.60 | 5.41 | 5.92 | 3.58 | 2.65 | 2.74 | 3.26 | 47.55 |
| Days ≥ 0.1" Precip | 7 | 6 | 7 | 5 | 7 | 7 | 9 | 8 | 6 | 4 | 4 | 5 | 75 |
| Total Snowfall (") | 0.2 | 0.4 | 0.0 | 0.0 | 0.0 | 0.0 | 0.0 | 0.0 | 0.0 | 0.0 | 0.0 | 0.1 | 0.7 |
| Days ≥ 1" Snow Depth | 0 | 0 | 0 | 0 | 0 | 0 | 0 | 0 | 0 | 0 | 0 | 0 | 0 |

## BRUNSWICK *Glynn County* ELEVATION 10 ft LAT/LONG 31° 9 ' N / 81° 30 ' W

| | JAN | FEB | MAR | APR | MAY | JUN | JUL | AUG | SEP | OCT | NOV | DEC | YEAR |
|---|---|---|---|---|---|---|---|---|---|---|---|---|---|
| Maximum Temp °F | 62.2 | 65.5 | 72.1 | 78.7 | 84.0 | 89.3 | 92.2 | 90.7 | 86.7 | *79.7* | *72.2* | *65.4* | 78.2 |
| Minimum Temp °F | 40.2 | 42.9 | 49.7 | 56.4 | 63.5 | 70.3 | 72.6 | 72.7 | 69.5 | *59.8* | *51.3* | 44.1 | 57.7 |
| Mean Temp °F | 51.2 | 54.2 | 60.9 | 67.6 | 73.8 | 79.8 | 82.4 | 81.7 | 78.1 | *69.8* | *61.8* | *54.7* | 68.0 |
| Days Max Temp ≥ 90 °F | 0 | 0 | 0 | 1 | 4 | 15 | 23 | 18 | 8 | 1 | 0 | 0 | 70 |
| Days Max Temp ≤ 32 °F | 0 | 0 | 0 | 0 | 0 | 0 | 0 | 0 | 0 | 0 | 0 | 0 | 0 |
| Days Min Temp ≤ 32 °F | 7 | 4 | 1 | 0 | 0 | 0 | 0 | 0 | 0 | 0 | *0* | *4* | 16 |
| Days Min Temp ≤ 0 °F | 0 | 0 | 0 | 0 | 0 | 0 | 0 | 0 | 0 | 0 | 0 | 0 | 0 |
| Heating Degree Days | 426 | 310 | 165 | 47 | 4 | 0 | 0 | 0 | 0 | *30* | *145* | *331* | 1458 |
| Cooling Degree Days | 2 | 17 | 43 | 120 | 289 | 472 | 567 | 533 | 392 | 190 | 60 | 18 | 2703 |
| Total Precipitation (") | 4.05 | 3.52 | 4.01 | 2.60 | 3.43 | 5.09 | 5.42 | 7.03 | 5.77 | 3.44 | 2.61 | 2.94 | 49.91 |
| Days ≥ 0.1" Precip | *6* | 6 | 5 | 4 | 5 | 7 | 8 | 9 | 7 | 5 | 4 | 5 | 71 |
| Total Snowfall (") | 0.0 | 0.0 | 0.0 | 0.0 | 0.0 | 0.0 | 0.0 | 0.0 | 0.0 | 0.0 | 0.0 | 0.0 | 0.0 |
| Days ≥ 1" Snow Depth | 0 | 0 | 0 | 0 | 0 | 0 | 0 | 0 | 0 | 0 | 0 | 0 | 0 |

## BRUNSWICK MCKINNON *Glynn County* ELEVATION 23 ft LAT/LONG 31° 9 ' N / 81° 23 ' W

| | JAN | FEB | MAR | APR | MAY | JUN | JUL | AUG | SEP | OCT | NOV | DEC | YEAR |
|---|---|---|---|---|---|---|---|---|---|---|---|---|---|
| Maximum Temp °F | 59.8 | 61.9 | 68.6 | 75.4 | 81.4 | 86.7 | 89.9 | 88.2 | 84.6 | 77.1 | 69.7 | 62.9 | 75.5 |
| Minimum Temp °F | 42.2 | 44.3 | 51.1 | 58.2 | 66.0 | 72.3 | 74.7 | 74.3 | 71.4 | 61.8 | 52.5 | 45.4 | 59.5 |
| Mean Temp °F | 51.0 | 53.1 | 59.9 | 66.8 | 73.7 | 79.5 | 82.3 | 81.3 | 78.0 | 69.5 | 61.1 | 54.2 | 67.5 |
| Days Max Temp ≥ 90 °F | 0 | 0 | 0 | 1 | 2 | 8 | 16 | 10 | 4 | 0 | 0 | 0 | 41 |
| Days Max Temp ≤ 32 °F | 0 | 0 | 0 | 0 | 0 | 0 | 0 | 0 | 0 | 0 | 0 | 0 | 0 |
| Days Min Temp ≤ 32 °F | 6 | 3 | 0 | 0 | 0 | 0 | 0 | 0 | 0 | 0 | 0 | 3 | 12 |
| Days Min Temp ≤ 0 °F | 0 | 0 | 0 | 0 | 0 | 0 | 0 | 0 | 0 | 0 | 0 | 0 | 0 |
| Heating Degree Days | 432 | 336 | 186 | 53 | 3 | 0 | 0 | 0 | 0 | 33 | 161 | 343 | 1547 |
| Cooling Degree Days | 3 | 7 | 30 | 98 | 275 | 460 | 560 | 519 | 391 | 180 | 60 | 15 | 2598 |
| Total Precipitation (") | 3.75 | 3.56 | 4.16 | 2.52 | 3.28 | 4.51 | 5.27 | 6.92 | 5.81 | 3.74 | 2.56 | 2.95 | 49.03 |
| Days ≥ 0.1" Precip | 6 | 5 | 5 | 4 | 5 | 7 | 8 | 9 | 7 | 5 | 4 | 5 | 70 |
| Total Snowfall (") | 0.0 | 0.0 | 0.0 | 0.0 | 0.0 | 0.0 | 0.0 | 0.0 | 0.0 | 0.0 | 0.0 | 0.1 | 0.1 |
| Days ≥ 1" Snow Depth | 0 | 0 | 0 | 0 | 0 | 0 | 0 | 0 | 0 | 0 | 0 | 0 | 0 |

## CALHOUN EXPERIMENT S *Gordon County* ELEVATION 650 ft LAT/LONG 34° 29 ' N / 84° 58 ' W

| | JAN | FEB | MAR | APR | MAY | JUN | JUL | AUG | SEP | OCT | NOV | DEC | YEAR |
|---|---|---|---|---|---|---|---|---|---|---|---|---|---|
| Maximum Temp °F | 50.0 | 55.0 | 63.9 | 73.2 | 79.7 | 87.0 | 90.0 | 88.7 | 83.1 | 73.2 | 63.2 | 53.8 | 71.7 |
| Minimum Temp °F | 27.9 | 30.5 | 38.4 | 45.7 | 53.9 | 61.9 | 66.2 | 65.2 | 58.9 | 45.7 | 37.7 | 31.3 | 46.9 |
| Mean Temp °F | 39.0 | 42.8 | 51.2 | 59.5 | 66.8 | 74.5 | 78.1 | 77.0 | 71.0 | 59.5 | 50.5 | 42.6 | 59.4 |
| Days Max Temp ≥ 90 °F | 0 | 0 | 0 | 0 | 1 | 9 | 17 | 14 | 5 | 0 | 0 | 0 | 46 |
| Days Max Temp ≤ 32 °F | 2 | 0 | 0 | 0 | 0 | 0 | 0 | 0 | 0 | 0 | 0 | 1 | 3 |
| Days Min Temp ≤ 32 °F | 20 | 17 | 9 | 2 | 0 | 0 | 0 | 0 | 0 | 3 | 11 | 18 | 80 |
| Days Min Temp ≤ 0 °F | 0 | 0 | 0 | 0 | 0 | 0 | 0 | 0 | 0 | 0 | 0 | 0 | 0 |
| Heating Degree Days | 800 | 621 | 426 | 189 | 50 | 3 | 0 | 0 | 20 | 195 | 433 | 689 | 3426 |
| Cooling Degree Days | 0 | 0 | 5 | 24 | 109 | 306 | 434 | 384 | 203 | 30 | 3 | 1 | 1499 |
| Total Precipitation (") | 5.05 | 4.89 | 6.21 | 4.46 | 4.76 | 3.95 | 4.48 | 3.79 | 4.26 | 3.30 | 4.33 | 4.82 | 54.30 |
| Days ≥ 0.1" Precip | 8 | 7 | 9 | 6 | 7 | 7 | 7 | 6 | 6 | 5 | 7 | 7 | 82 |
| Total Snowfall (") | 0.8 | 0.3 | 0.6 | 0.2 | 0.0 | 0.0 | 0.0 | 0.0 | 0.0 | 0.0 | 0.0 | 0.0 | 1.9 |
| Days ≥ 1" Snow Depth | 1 | 0 | 0 | 0 | 0 | 0 | 0 | 0 | 0 | 0 | 0 | 0 | 1 |

## CAMILLA 3 SE *Mitchell County* ELEVATION 180 ft LAT/LONG 31° 14 ' N / 84° 13 ' W

| | JAN | FEB | MAR | APR | MAY | JUN | JUL | AUG | SEP | OCT | NOV | DEC | YEAR |
|---|---|---|---|---|---|---|---|---|---|---|---|---|---|
| Maximum Temp °F | 61.5 | 65.4 | 73.2 | 80.6 | 86.5 | 91.1 | 92.7 | 92.0 | 88.7 | 80.6 | 71.6 | 64.3 | 79.0 |
| Minimum Temp °F | 39.0 | 41.2 | 47.6 | 53.9 | 61.2 | 68.0 | 70.7 | 70.3 | 66.5 | 55.3 | 46.9 | 41.4 | 55.2 |
| Mean Temp °F | 50.3 | 53.4 | 60.4 | 67.3 | 73.9 | 79.6 | 81.7 | 81.2 | 77.6 | 68.0 | 59.3 | 52.8 | 67.1 |
| Days Max Temp ≥ 90 °F | 0 | 0 | 0 | 2 | 9 | 20 | 24 | 23 | 16 | 3 | 0 | 0 | 97 |
| Days Max Temp ≤ 32 °F | 0 | 0 | 0 | 0 | 0 | 0 | 0 | 0 | 0 | 0 | 0 | 0 | 0 |
| Days Min Temp ≤ 32 °F | 10 | 6 | 2 | 0 | 0 | 0 | 0 | 0 | 0 | 0 | 4 | 8 | 30 |
| Days Min Temp ≤ 0 °F | 0 | 0 | 0 | 0 | 0 | 0 | 0 | 0 | 0 | 0 | 0 | 0 | 0 |
| Heating Degree Days | 454 | 333 | 180 | 50 | 3 | 0 | 0 | 0 | 1 | 52 | 204 | 384 | 1661 |
| Cooling Degree Days | 4 | 16 | 45 | 112 | 292 | 464 | 542 | 521 | 385 | 158 | 51 | 14 | 2604 |
| Total Precipitation (") | 5.70 | 4.78 | 6.29 | 3.68 | 4.11 | 5.46 | 6.04 | 4.69 | 2.90 | 2.31 | 2.96 | 4.09 | 53.01 |
| Days ≥ 0.1" Precip | 8 | 6 | 7 | 5 | 6 | 7 | 9 | 7 | 4 | 3 | 5 | 6 | 73 |
| Total Snowfall (") | 0.0 | 0.1 | 0.0 | 0.0 | 0.0 | 0.0 | 0.0 | 0.0 | 0.0 | 0.0 | 0.0 | 0.0 | 0.1 |
| Days ≥ 1" Snow Depth | 0 | 0 | 0 | 0 | 0 | 0 | 0 | 0 | 0 | 0 | 0 | 0 | 0 |

## CARROLLTON *Carroll County* ELEVATION 1122 ft LAT/LONG 33° 35 ' N / 85° 4 ' W

| | JAN | FEB | MAR | APR | MAY | JUN | JUL | AUG | SEP | OCT | NOV | DEC | YEAR |
|---|---|---|---|---|---|---|---|---|---|---|---|---|---|
| Maximum Temp °F | 52.5 | 57.2 | 66.3 | 74.9 | 80.6 | 86.3 | 88.6 | 87.7 | 82.6 | 73.7 | 65.0 | 56.1 | 72.6 |
| Minimum Temp °F | 30.0 | 32.5 | 39.4 | 46.3 | 54.4 | 61.6 | 66.1 | 65.2 | 59.5 | 47.0 | 39.3 | 33.0 | 47.9 |
| Mean Temp °F | 41.3 | 44.9 | 52.9 | 60.6 | 67.5 | 74.0 | 77.4 | 76.5 | 71.1 | 60.3 | 52.2 | 44.6 | 60.3 |
| Days Max Temp ≥ 90 °F | 0 | 0 | 0 | 0 | 1 | 7 | 14 | 11 | 3 | 0 | 0 | 0 | 36 |
| Days Max Temp ≤ 32 °F | 1 | 0 | 0 | 0 | 0 | 0 | 0 | 0 | 0 | 0 | 0 | 0 | 1 |
| Days Min Temp ≤ 32 °F | 18 | 15 | 8 | 2 | 0 | 0 | 0 | 0 | 0 | 2 | 8 | 16 | 69 |
| Days Min Temp ≤ 0 °F | 0 | 0 | 0 | 0 | 0 | 0 | 0 | 0 | 0 | 0 | 0 | 0 | 0 |
| Heating Degree Days | 719 | 561 | 375 | 156 | 33 | 2 | 0 | 0 | 16 | 168 | 381 | 628 | 3039 |
| Cooling Degree Days | 0 | 0 | 4 | 28 | 117 | 292 | 400 | 367 | 202 | 33 | 3 | 2 | 1448 |
| Total Precipitation (") | 5.18 | 4.75 | 5.82 | 4.40 | 4.79 | 3.92 | 4.80 | 3.67 | 3.30 | 3.37 | 4.39 | 4.61 | 53.00 |
| Days ≥ 0.1" Precip | 8 | 7 | 8 | 6 | 6 | 7 | 7 | 7 | 5 | 4 | 6 | 7 | 78 |
| Total Snowfall (") | 0.2 | 0.0 | 0.0 | 0.0 | 0.0 | 0.0 | 0.0 | 0.0 | 0.0 | 0.0 | 0.0 | 0.0 | 0.2 |
| Days ≥ 1" Snow Depth | 0 | 0 | 0 | 0 | 0 | 0 | 0 | 0 | 0 | 0 | 0 | 0 | 0 |

## CARTERSVILLE *Bartow County* ELEVATION 741 ft LAT/LONG 34° 9' N / 84° 48' W

| | JAN | FEB | MAR | APR | MAY | JUN | JUL | AUG | SEP | OCT | NOV | DEC | YEAR |
|---|---|---|---|---|---|---|---|---|---|---|---|---|---|
| Maximum Temp °F | 50.7 | 56.2 | 65.6 | 74.3 | 80.0 | 86.5 | 89.2 | 87.8 | 82.5 | 73.3 | 63.3 | 54.2 | 72.0 |
| Minimum Temp °F | 31.0 | 33.1 | 40.3 | 47.3 | 55.4 | 62.9 | 66.8 | 66.2 | 60.8 | 48.5 | 40.3 | 34.3 | 48.9 |
| Mean Temp °F | 40.9 | 44.7 | 53.0 | 60.8 | 67.7 | 74.7 | 78.0 | 77.0 | 71.7 | 60.9 | 51.8 | 44.3 | 60.5 |
| Days Max Temp ≥ 90 °F | 0 | 0 | 0 | 0 | 1 | 8 | 14 | 11 | 3 | 0 | 0 | 0 | 37 |
| Days Max Temp ≤ 32 °F | 2 | 0 | 0 | 0 | 0 | 0 | 0 | 0 | 0 | 0 | 0 | 0 | 2 |
| Days Min Temp ≤ 32 °F | 17 | 14 | 8 | 3 | 0 | 0 | 0 | 0 | 0 | 2 | 9 | 15 | 68 |
| Days Min Temp ≤ 0 °F | 0 | 0 | 0 | 0 | 0 | 0 | 0 | 0 | 0 | 0 | 0 | 0 | 0 |
| Heating Degree Days | 741 | 568 | 373 | 161 | 41 | 2 | 0 | 0 | 17 | 163 | 396 | 638 | 3100 |
| Cooling Degree Days | 0 | 1 | 11 | 42 | 137 | 323 | 432 | 391 | 229 | 51 | 8 | 2 | 1627 |
| Total Precipitation (") | 4.00 | 4.41 | 5.43 | 4.34 | 3.86 | 3.15 | 4.08 | 3.38 | 3.28 | 2.62 | 3.49 | 4.07 | 46.11 |
| Days ≥ 0.1" Precip | 6 | 6 | 6 | 5 | 6 | 5 | 6 | 5 | 5 | 4 | 5 | 6 | 65 |
| Total Snowfall (") | 0.7 | 0.7 | 0.1 | 0.0 | 0.0 | 0.0 | 0.0 | 0.0 | 0.0 | 0.0 | 0.0 | 0.2 | 1.7 |
| Days ≥ 1" Snow Depth | 0 | 0 | 0 | 0 | 0 | 0 | 0 | 0 | 0 | 0 | 0 | 0 | 0 |

## CEDARTOWN 3 NE *Polk County* ELEVATION 801 ft LAT/LONG 34° 1' N / 85° 15' W

| | JAN | FEB | MAR | APR | MAY | JUN | JUL | AUG | SEP | OCT | NOV | DEC | YEAR |
|---|---|---|---|---|---|---|---|---|---|---|---|---|---|
| Maximum Temp °F | 51.9 | 56.8 | 66.1 | 74.9 | 80.4 | 86.8 | 89.4 | 88.6 | 83.1 | 73.9 | 64.3 | 55.7 | 72.7 |
| Minimum Temp °F | 30.8 | 32.5 | 39.7 | 47.1 | 54.9 | 63.0 | 67.6 | 66.5 | 60.5 | 47.7 | 39.4 | 33.5 | 48.6 |
| Mean Temp °F | 41.4 | 44.7 | 52.9 | 61.1 | 67.7 | 74.9 | 78.5 | 77.6 | 71.8 | 60.8 | 51.8 | 44.6 | 60.7 |
| Days Max Temp ≥ 90 °F | 0 | 0 | 0 | 0 | 1 | 9 | 16 | 13 | 5 | 0 | 0 | 0 | 44 |
| Days Max Temp ≤ 32 °F | 1 | 0 | 0 | 0 | 0 | 0 | 0 | 0 | 0 | 0 | 0 | 0 | 1 |
| Days Min Temp ≤ 32 °F | 18 | 15 | 9 | 2 | 0 | 0 | 0 | 0 | 0 | 2 | 10 | 16 | 72 |
| Days Min Temp ≤ 0 °F | 0 | 0 | 0 | 0 | 0 | 0 | 0 | 0 | 0 | 0 | 0 | 0 | 0 |
| Heating Degree Days | 726 | 568 | 376 | 154 | 37 | 2 | 0 | 0 | 15 | 163 | 394 | 626 | 3061 |
| Cooling Degree Days | 0 | 1 | 11 | 40 | 129 | 327 | 451 | 410 | 227 | 47 | 6 | 2 | 1651 |
| Total Precipitation (") | 5.05 | 4.57 | 6.13 | 4.96 | 4.44 | 4.31 | 4.50 | 4.05 | 3.85 | 3.12 | 4.00 | 4.28 | 53.26 |
| Days ≥ 0.1" Precip | 8 | 6 | 8 | 7 | 7 | 7 | 7 | 6 | 6 | 5 | 7 | 7 | 81 |
| Total Snowfall (") | 1.3 | 0.8 | 0.7 | 0.2 | 0.0 | 0.0 | 0.0 | 0.0 | 0.0 | 0.0 | 0.1 | 0.1 | 3.2 |
| Days ≥ 1" Snow Depth | 1 | 0 | 0 | 0 | 0 | 0 | 0 | 0 | 0 | 0 | 0 | 0 | 1 |

## CLAYTON 1 SSW *Rabun County* ELEVATION 2001 ft LAT/LONG 34° 52' N / 83° 23' W

| | JAN | FEB | MAR | APR | MAY | JUN | JUL | AUG | SEP | OCT | NOV | DEC | YEAR |
|---|---|---|---|---|---|---|---|---|---|---|---|---|---|
| Maximum Temp °F | 50.6 | 54.5 | 62.8 | 71.4 | 76.7 | 82.7 | 85.2 | 84.0 | 79.2 | 71.2 | 62.3 | 53.7 | 69.5 |
| Minimum Temp °F | 27.3 | 28.9 | 35.3 | 42.0 | 50.1 | 57.6 | 62.1 | 61.3 | 55.6 | 43.8 | 36.0 | 30.0 | 44.2 |
| Mean Temp °F | 39.0 | 41.7 | 49.1 | 56.7 | 63.4 | 70.1 | 73.7 | 72.7 | 67.4 | 57.6 | 49.2 | 41.9 | 56.9 |
| Days Max Temp ≥ 90 °F | 0 | 0 | 0 | 0 | 0 | 2 | 5 | 3 | 1 | 0 | 0 | 0 | 11 |
| Days Max Temp ≤ 32 °F | 1 | 0 | 0 | 0 | 0 | 0 | 0 | 0 | 0 | 0 | 0 | 0 | 1 |
| Days Min Temp ≤ 32 °F | 21 | 19 | 13 | 5 | 1 | 0 | 0 | 0 | 0 | 4 | 13 | 20 | 96 |
| Days Min Temp ≤ 0 °F | 0 | 0 | 0 | 0 | 0 | 0 | 0 | 0 | 0 | 0 | 0 | 0 | 0 |
| Heating Degree Days | 801 | 651 | 486 | 251 | 89 | 9 | 1 | 2 | 38 | 235 | 469 | 710 | 3742 |
| Cooling Degree Days | 0 | 0 | 0 | 8 | 45 | 178 | 294 | 255 | 125 | 13 | 1 | 0 | 919 |
| Total Precipitation (") | 6.52 | 6.02 | 7.31 | 5.04 | 6.99 | 5.44 | 5.69 | 6.36 | 5.77 | 5.09 | 6.21 | 6.51 | 72.95 |
| Days ≥ 0.1" Precip | 9 | 7 | 9 | 7 | 9 | 8 | 9 | 8 | 7 | 6 | 8 | 8 | 95 |
| Total Snowfall (") | 1.9 | 1.8 | 1.0 | 0.0 | 0.0 | 0.0 | 0.0 | 0.0 | 0.0 | 0.0 | 0.1 | 0.6 | 5.4 |
| Days ≥ 1" Snow Depth | 1 | 0 | 0 | 0 | 0 | 0 | 0 | 0 | 0 | 0 | 0 | 0 | 1 |

## COLQUITT 2 W *Miller County* ELEVATION 161 ft LAT/LONG 31° 10' N / 84° 44' W

| | JAN | FEB | MAR | APR | MAY | JUN | JUL | AUG | SEP | OCT | NOV | DEC | YEAR |
|---|---|---|---|---|---|---|---|---|---|---|---|---|---|
| Maximum Temp °F | 60.9 | 64.5 | 72.5 | 80.1 | 85.7 | 90.8 | 92.2 | 91.4 | 88.5 | 80.7 | 71.7 | 64.3 | 78.6 |
| Minimum Temp °F | 38.1 | 40.1 | 47.0 | 53.5 | 60.5 | 67.5 | 70.3 | 69.9 | 65.9 | 54.3 | 45.9 | 40.1 | 54.4 |
| Mean Temp °F | 49.7 | 52.3 | 59.8 | 66.9 | 73.1 | 79.2 | 81.3 | 80.7 | 77.2 | 67.5 | 58.8 | 52.2 | 66.6 |
| Days Max Temp ≥ 90 °F | 0 | 0 | 0 | 1 | 7 | 19 | 24 | 22 | 14 | 3 | 0 | 0 | 90 |
| Days Max Temp ≤ 32 °F | 0 | 0 | 0 | 0 | 0 | 0 | 0 | 0 | 0 | 0 | 0 | 0 | 0 |
| Days Min Temp ≤ 32 °F | 10 | 7 | 2 | 0 | 0 | 0 | 0 | 0 | 0 | 0 | 4 | 8 | 31 |
| Days Min Temp ≤ 0 °F | 0 | 0 | 0 | 0 | 0 | 0 | 0 | 0 | 0 | 0 | 0 | 0 | 0 |
| Heating Degree Days | 471 | 361 | 193 | 54 | 4 | 0 | 0 | 0 | 2 | 52 | 210 | 401 | 1748 |
| Cooling Degree Days | 3 | 11 | 43 | 102 | 269 | 458 | 535 | 508 | 367 | 151 | 44 | 13 | 2504 |
| Total Precipitation (") | 5.75 | 4.81 | 6.18 | 3.57 | 3.85 | 4.78 | 5.20 | 4.64 | 3.53 | 2.31 | 3.29 | 4.47 | 52.38 |
| Days ≥ 0.1" Precip | 7 | 6 | 7 | 5 | 6 | 7 | 8 | 7 | 5 | 3 | 4 | 6 | 71 |
| Total Snowfall (") | 0.1 | 0.1 | 0.0 | 0.0 | 0.0 | 0.0 | 0.0 | 0.0 | 0.0 | 0.0 | 0.0 | 0.1 | 0.3 |
| Days ≥ 1" Snow Depth | 0 | 0 | 0 | 0 | 0 | 0 | 0 | 0 | 0 | 0 | 0 | 0 | 0 |

## COLUMBUS METRO AP *Muscogee County* ELEVATION 387 ft LAT/LONG 32° 31 ' N / 84° 56 ' W

| | JAN | FEB | MAR | APR | MAY | JUN | JUL | AUG | SEP | OCT | NOV | DEC | YEAR |
|---|---|---|---|---|---|---|---|---|---|---|---|---|---|
| Maximum Temp °F | 56.5 | 60.8 | 69.1 | 77.5 | 83.3 | 89.7 | 91.4 | 90.6 | 86.0 | 77.1 | 67.9 | 60.0 | 75.8 |
| Minimum Temp °F | 35.9 | 38.1 | 45.1 | 52.3 | 60.8 | 68.5 | 72.1 | 71.4 | 66.3 | 54.3 | 45.0 | 38.8 | 54.0 |
| Mean Temp °F | 46.2 | 49.5 | 57.1 | 64.9 | 72.1 | 79.1 | 81.8 | 81.0 | 76.2 | 65.8 | 56.5 | 49.4 | 65.0 |
| Days Max Temp ≥ 90 °F | 0 | 0 | 0 | 1 | 4 | 17 | 21 | 20 | 10 | 1 | 0 | 0 | 74 |
| Days Max Temp ≤ 32 °F | 0 | 0 | 0 | 0 | 0 | 0 | 0 | 0 | 0 | 0 | 0 | 0 | 0 |
| Days Min Temp ≤ 32 °F | 13 | 9 | 3 | 0 | 0 | 0 | 0 | 0 | 0 | 0 | 4 | 10 | 39 |
| Days Min Temp ≤ 0 °F | 0 | 0 | 0 | 0 | 0 | 0 | 0 | 0 | 0 | 0 | 0 | 0 | 0 |
| Heating Degree Days | 576 | 436 | 261 | 84 | 8 | 0 | 0 | 0 | 3 | 76 | 269 | 483 | 2196 |
| Cooling Degree Days | 1 | 5 | 26 | 80 | 237 | 458 | 551 | 525 | 357 | 114 | 26 | 6 | 2386 |
| Total Precipitation (") | 4.62 | 4.60 | 5.89 | 3.83 | 4.29 | 3.98 | 5.49 | 4.10 | 3.14 | 2.31 | 3.93 | 4.70 | 50.88 |
| Days ≥ 0.1" Precip | 7 | 7 | 7 | 5 | 6 | 7 | 9 | 7 | 5 | 4 | 5 | 6 | 75 |
| Total Snowfall (") | 0.2 | 0.5 | 0.1 | 0.0 | 0.0 | 0.0 | 0.0 | 0.0 | 0.0 | 0.0 | 0.0 | 0.1 | 0.9 |
| Days ≥ 1" Snow Depth | 0 | 0 | 0 | 0 | 0 | 0 | 0 | 0 | 0 | 0 | 0 | 0 | 0 |

## COMMERCE 4 NNW *Jackson County* ELEVATION 741 ft LAT/LONG 34° 13 ' N / 83° 26 ' W

| | JAN | FEB | MAR | APR | MAY | JUN | JUL | AUG | SEP | OCT | NOV | DEC | YEAR |
|---|---|---|---|---|---|---|---|---|---|---|---|---|---|
| Maximum Temp °F | 51.2 | 55.7 | 64.3 | 73.6 | 79.7 | 86.5 | 89.5 | 88.0 | 82.5 | 73.1 | 63.7 | 55.3 | 71.9 |
| Minimum Temp °F | 29.1 | 31.0 | 38.2 | 46.2 | 53.9 | 62.7 | 66.7 | 65.7 | 60.0 | 46.9 | 38.3 | 32.0 | 47.6 |
| Mean Temp °F | 40.2 | 43.2 | 51.3 | 59.9 | 66.8 | 74.7 | 78.2 | 76.9 | 71.3 | 60.0 | 51.0 | 43.7 | 59.8 |
| Days Max Temp ≥ 90 °F | 0 | 0 | 0 | 0 | 1 | 9 | 16 | 12 | 4 | 0 | 0 | 0 | 42 |
| Days Max Temp ≤ 32 °F | 1 | 0 | 0 | 0 | 0 | 0 | 0 | 0 | 0 | 0 | 0 | 0 | 1 |
| Days Min Temp ≤ 32 °F | 19 | 16 | 10 | 2 | 0 | 0 | 0 | 0 | 0 | 2 | 11 | 17 | 77 |
| Days Min Temp ≤ 0 °F | 0 | 0 | 0 | 0 | 0 | 0 | 0 | 0 | 0 | 0 | 0 | 0 | 0 |
| Heating Degree Days | 762 | 609 | 422 | 179 | 49 | 3 | 0 | 0 | 16 | 178 | 417 | 655 | 3290 |
| Cooling Degree Days | 0 | 0 | 5 | 33 | 118 | 326 | 445 | 393 | 215 | 38 | 4 | 1 | 1578 |
| Total Precipitation (") | 5.29 | 4.50 | 5.77 | 3.82 | 4.60 | 4.28 | 4.31 | 4.01 | 3.68 | 4.04 | 4.17 | 4.31 | 52.78 |
| Days ≥ 0.1" Precip | 8 | 7 | 8 | 6 | 6 | 6 | 7 | 6 | 6 | 5 | 6 | 7 | 78 |
| Total Snowfall (") | 1.0 | 0.6 | 0.2 | 0.0 | 0.0 | 0.0 | 0.0 | 0.0 | 0.0 | 0.0 | 0.0 | 0.1 | 1.9 |
| Days ≥ 1" Snow Depth | 0 | 0 | 0 | 0 | 0 | 0 | 0 | 0 | 0 | 0 | 0 | 0 | 0 |

## CORDELE *Crisp County* ELEVATION 312 ft LAT/LONG 31° 58 ' N / 83° 47 ' W

| | JAN | FEB | MAR | APR | MAY | JUN | JUL | AUG | SEP | OCT | NOV | DEC | YEAR |
|---|---|---|---|---|---|---|---|---|---|---|---|---|---|
| Maximum Temp °F | 59.4 | 63.6 | 72.1 | 80.3 | 86.4 | 91.5 | 93.4 | 92.3 | 88.0 | 79.6 | 70.5 | 63.0 | 78.3 |
| Minimum Temp °F | 37.2 | 39.6 | 46.5 | 53.3 | 61.0 | 67.9 | 71.0 | 70.2 | 65.6 | 54.0 | 46.1 | 40.2 | 54.4 |
| Mean Temp °F | 48.3 | 51.6 | 59.4 | 66.8 | 73.7 | 79.7 | 82.2 | 81.3 | 76.8 | 66.9 | 58.3 | 51.6 | 66.4 |
| Days Max Temp ≥ 90 °F | 0 | 0 | 0 | 2 | 9 | 21 | 25 | 24 | 14 | 2 | 0 | 0 | 97 |
| Days Max Temp ≤ 32 °F | 0 | 0 | 0 | 0 | 0 | 0 | 0 | 0 | 0 | 0 | 0 | 0 | 0 |
| Days Min Temp ≤ 32 °F | 12 | 8 | 3 | 0 | 0 | 0 | 0 | 0 | 0 | 0 | 4 | 8 | 35 |
| Days Min Temp ≤ 0 °F | 0 | 0 | 0 | 0 | 0 | 0 | 0 | 0 | 0 | 0 | 0 | 0 | 0 |
| Heating Degree Days | 512 | 379 | 207 | 59 | 3 | 0 | 0 | 0 | 2 | 63 | 226 | 417 | 1868 |
| Cooling Degree Days | 3 | 10 | 40 | 116 | 291 | 486 | 575 | 531 | 374 | 141 | 39 | 10 | 2616 |
| Total Precipitation (") | 4.72 | 4.38 | 5.03 | 2.83 | 3.41 | 4.08 | 5.13 | 3.92 | 2.96 | 1.99 | 3.17 | 3.60 | 45.22 |
| Days ≥ 0.1" Precip | 7 | 6 | 7 | 5 | 6 | 7 | 8 | 7 | 5 | 3 | 5 | 5 | 71 |
| Total Snowfall (") | 0.0 | 0.1 | 0.0 | 0.0 | 0.0 | 0.0 | 0.0 | 0.0 | 0.0 | 0.0 | 0.0 | 0.0 | 0.1 |
| Days ≥ 1" Snow Depth | 0 | 0 | 0 | 0 | 0 | 0 | 0 | 0 | 0 | 0 | 0 | 0 | 0 |

## CORNELIA *Habersham County* ELEVATION 1503 ft LAT/LONG 34° 31 ' N / 83° 32 ' W

| | JAN | FEB | MAR | APR | MAY | JUN | JUL | AUG | SEP | OCT | NOV | DEC | YEAR |
|---|---|---|---|---|---|---|---|---|---|---|---|---|---|
| Maximum Temp °F | 48.8 | 53.5 | 62.3 | 70.9 | 76.5 | 82.8 | 85.6 | 84.0 | 78.6 | 69.6 | 60.3 | 51.9 | 68.7 |
| Minimum Temp °F | 28.6 | 31.0 | 37.9 | 45.0 | 52.6 | 60.4 | 64.6 | 63.6 | 57.9 | 46.1 | 38.1 | 31.8 | 46.5 |
| Mean Temp °F | 38.7 | 42.3 | 50.1 | 58.0 | 64.6 | 71.6 | 75.2 | 73.8 | 68.3 | 57.9 | 49.2 | 41.9 | 57.6 |
| Days Max Temp ≥ 90 °F | 0 | 0 | 0 | 0 | 0 | 3 | 8 | 4 | 1 | 0 | 0 | 0 | 16 |
| Days Max Temp ≤ 32 °F | 1 | 1 | 0 | 0 | 0 | 0 | 0 | 0 | 0 | 0 | 0 | 1 | 3 |
| Days Min Temp ≤ 32 °F | 20 | 16 | 10 | 3 | 0 | 0 | 0 | 0 | 0 | 2 | 10 | 17 | 78 |
| Days Min Temp ≤ 0 °F | 0 | 0 | 0 | 0 | 0 | 0 | 0 | 0 | 0 | 0 | 0 | 0 | 0 |
| Heating Degree Days | 809 | 635 | 455 | 222 | 75 | 7 | 1 | 1 | 34 | 229 | 468 | 710 | 3646 |
| Cooling Degree Days | 0 | 0 | 2 | 18 | 74 | 242 | 350 | 295 | 151 | 18 | 1 | 0 | 1151 |
| Total Precipitation (") | 5.92 | 5.31 | 6.47 | 4.47 | 5.10 | 4.53 | 5.15 | 5.31 | 4.22 | 4.31 | 4.75 | 4.98 | 60.52 |
| Days ≥ 0.1" Precip | 8 | 7 | 8 | 7 | 8 | 7 | 7 | 8 | 6 | 5 | 6 | 8 | 85 |
| Total Snowfall (") | 0.4 | 0.0 | 0.3 | 0.0 | 0.0 | 0.0 | 0.0 | 0.0 | 0.0 | 0.0 | 0.0 | 0.0 | 0.7 |
| Days ≥ 1" Snow Depth | 0 | 0 | 0 | 0 | 0 | 0 | 0 | 0 | 0 | 0 | 0 | 0 | 0 |

## COVINGTON *Newton County* ELEVATION 751 ft LAT/LONG 33° 36' N / 83° 52' W

| | JAN | FEB | MAR | APR | MAY | JUN | JUL | AUG | SEP | OCT | NOV | DEC | YEAR |
|---|---|---|---|---|---|---|---|---|---|---|---|---|---|
| Maximum Temp °F | 52.3 | 57.5 | 66.5 | 74.7 | 80.8 | 86.8 | 89.5 | 88.0 | 83.0 | 73.4 | 64.3 | 55.6 | 72.7 |
| Minimum Temp °F | 32.2 | 34.5 | 41.8 | 48.9 | 57.1 | 64.6 | 68.4 | 67.7 | 62.3 | 50.4 | 41.7 | 35.3 | 50.4 |
| Mean Temp °F | 42.3 | 46.0 | 54.2 | 61.8 | 68.9 | 75.7 | 79.0 | 77.9 | 72.7 | 61.9 | 53.0 | 45.5 | 61.6 |
| Days Max Temp ≥ 90 °F | 0 | 0 | 0 | 0 | 2 | 9 | 16 | 12 | 5 | 0 | 0 | 0 | 44 |
| Days Max Temp ≤ 32 °F | 1 | 0 | 0 | 0 | 0 | 0 | 0 | 0 | 0 | 0 | 0 | 0 | 1 |
| Days Min Temp ≤ 32 °F | 16 | 13 | 6 | 1 | 0 | 0 | 0 | 0 | 0 | 1 | 7 | 13 | 57 |
| Days Min Temp ≤ 0 °F | 0 | 0 | 0 | 0 | 0 | 0 | 0 | 0 | 0 | 0 | 0 | 0 | 0 |
| Heating Degree Days | 697 | 531 | 338 | 139 | 26 | 1 | 0 | 0 | 11 | 139 | 360 | 599 | 2841 |
| Cooling Degree Days | 0 | 0 | 10 | 48 | 162 | 361 | 472 | 424 | 256 | 55 | 9 | 1 | 1798 |
| Total Precipitation (") | 4.90 | 4.37 | 5.35 | 3.47 | 4.10 | 3.78 | 4.92 | 3.96 | 3.23 | 3.33 | 3.73 | 4.06 | 49.20 |
| Days ≥ 0.1" Precip | 8 | 7 | 7 | 6 | 7 | 6 | 7 | 7 | 5 | 5 | 6 | 7 | 78 |
| Total Snowfall (") | 0.7 | 0.4 | 0.3 | 0.0 | 0.0 | 0.0 | 0.0 | 0.0 | 0.0 | 0.0 | 0.0 | 0.0 | 1.4 |
| Days ≥ 1" Snow Depth | 1 | 0 | 0 | 0 | 0 | 0 | 0 | 0 | 0 | 0 | 0 | 0 | 1 |

## CUTHBERT *Randolph County* ELEVATION 461 ft LAT/LONG 31° 46' N / 84° 47' W

| | JAN | FEB | MAR | APR | MAY | JUN | JUL | AUG | SEP | OCT | NOV | DEC | YEAR |
|---|---|---|---|---|---|---|---|---|---|---|---|---|---|
| Maximum Temp °F | 59.3 | 63.4 | 71.9 | 79.9 | 85.7 | 90.9 | 92.3 | 91.5 | 87.7 | 79.2 | 69.9 | 62.5 | 77.9 |
| Minimum Temp °F | 38.1 | 40.0 | 46.7 | 53.7 | 61.4 | 68.1 | 70.9 | 70.4 | 66.2 | 55.5 | 46.9 | 41.0 | 54.9 |
| Mean Temp °F | 48.7 | 51.8 | 59.3 | 66.8 | 73.5 | 79.5 | 81.7 | 81.0 | 77.0 | 67.3 | 58.6 | 51.8 | 66.4 |
| Days Max Temp ≥ 90 °F | 0 | 0 | 0 | 1 | 7 | 20 | 24 | 23 | 13 | 1 | 0 | 0 | 89 |
| Days Max Temp ≤ 32 °F | 0 | 0 | 0 | 0 | 0 | 0 | 0 | 0 | 0 | 0 | 0 | 0 | 0 |
| Days Min Temp ≤ 32 °F | 10 | 7 | 3 | 0 | 0 | 0 | 0 | 0 | 0 | 0 | 3 | 8 | 31 |
| Days Min Temp ≤ 0 °F | 0 | 0 | 0 | 0 | 0 | 0 | 0 | 0 | 0 | 0 | 0 | 0 | 0 |
| Heating Degree Days | 501 | 373 | 206 | 57 | 3 | 0 | 0 | 0 | 2 | 56 | 215 | 411 | 1824 |
| Cooling Degree Days | 2 | 9 | 36 | 104 | 274 | 462 | 539 | 512 | 369 | 138 | 32 | 9 | 2486 |
| Total Precipitation (") | 5.27 | 4.86 | 5.79 | 3.54 | 3.93 | 4.75 | 6.42 | 3.60 | 3.22 | 2.59 | 3.52 | 4.39 | 51.88 |
| Days ≥ 0.1" Precip | 7 | 7 | 8 | 5 | 6 | 7 | 9 | 7 | 5 | 4 | 5 | 6 | 76 |
| Total Snowfall (") | 0.0 | 0.3 | 0.1 | 0.0 | 0.0 | 0.0 | 0.0 | 0.0 | 0.0 | 0.0 | 0.0 | 0.0 | 0.4 |
| Days ≥ 1" Snow Depth | 0 | 0 | 0 | 0 | 0 | 0 | 0 | 0 | 0 | 0 | 0 | 0 | 0 |

## DAHLONEGA 2 NW *Lumpkin County* ELEVATION 1522 ft LAT/LONG 34° 32' N / 83° 59' W

| | JAN | FEB | MAR | APR | MAY | JUN | JUL | AUG | SEP | OCT | NOV | DEC | YEAR |
|---|---|---|---|---|---|---|---|---|---|---|---|---|---|
| Maximum Temp °F | 48.4 | 52.7 | 60.9 | 70.6 | 76.7 | 83.1 | 86.1 | 84.5 | 78.7 | 69.6 | 60.8 | 52.3 | 68.7 |
| Minimum Temp °F | 27.9 | 29.9 | 36.6 | 44.0 | 51.6 | 60.2 | 64.4 | 64.0 | 58.0 | 45.7 | 37.4 | 31.1 | 45.9 |
| Mean Temp °F | 38.2 | 41.3 | 48.8 | 57.3 | 64.2 | 71.7 | 75.3 | 74.3 | 68.3 | 57.7 | 49.1 | 41.7 | 57.3 |
| Days Max Temp ≥ 90 °F | 0 | 0 | 0 | 0 | 0 | 3 | 8 | 5 | 1 | 0 | 0 | 0 | 17 |
| Days Max Temp ≤ 32 °F | 2 | 1 | 0 | 0 | 0 | 0 | 0 | 0 | 0 | 0 | 0 | 1 | 4 |
| Days Min Temp ≤ 32 °F | 21 | 18 | 11 | 3 | 0 | 0 | 0 | 0 | 0 | 2 | 10 | 19 | 84 |
| Days Min Temp ≤ 0 °F | 0 | 0 | 0 | 0 | 0 | 0 | 0 | 0 | 0 | 0 | 0 | 0 | 0 |
| Heating Degree Days | 824 | 663 | 497 | 241 | 80 | 7 | 0 | 0 | 32 | 235 | 471 | 715 | 3765 |
| Cooling Degree Days | 0 | 0 | 2 | 14 | 54 | 225 | 340 | 298 | 136 | 20 | 1 | 0 | 1090 |
| Total Precipitation (") | 6.44 | 5.88 | 7.08 | 4.97 | 5.72 | 4.24 | 5.47 | 5.40 | 4.64 | 4.23 | 5.26 | 5.66 | 64.99 |
| Days ≥ 0.1" Precip | 8 | *8* | 8 | 7 | 8 | 7 | 9 | 8 | 7 | 5 | 7 | 8 | 90 |
| Total Snowfall (") | *0.7* | *0.2* | 0.1 | 0.0 | 0.0 | 0.0 | 0.0 | 0.0 | 0.0 | 0.0 | 0.0 | 0.1 | 1.1 |
| Days ≥ 1" Snow Depth | *0* | *0* | 0 | 0 | 0 | 0 | 0 | 0 | 0 | 0 | 0 | 0 | 0 |

## DALLAS 7 NE *Paulding County* ELEVATION 1079 ft LAT/LONG 33° 58' N / 84° 47' W

| | JAN | FEB | MAR | APR | MAY | JUN | JUL | AUG | SEP | OCT | NOV | DEC | YEAR |
|---|---|---|---|---|---|---|---|---|---|---|---|---|---|
| Maximum Temp °F | 49.4 | 54.2 | 63.4 | 72.7 | 78.8 | 85.6 | 88.7 | 87.7 | 82.3 | 72.6 | 63.1 | 53.7 | 71.0 |
| Minimum Temp °F | 28.0 | 30.2 | 37.9 | 45.7 | 53.8 | 61.7 | 66.0 | 64.9 | 58.6 | 46.2 | 38.2 | 31.3 | 46.9 |
| Mean Temp °F | 38.7 | 42.3 | 50.5 | 59.3 | 66.4 | 73.7 | 77.4 | 76.3 | 70.4 | 59.4 | 50.7 | 42.5 | 59.0 |
| Days Max Temp ≥ 90 °F | 0 | 0 | 0 | 0 | 1 | 8 | 14 | 12 | 5 | 0 | 0 | 0 | 40 |
| Days Max Temp ≤ 32 °F | 2 | 1 | 0 | 0 | 0 | 0 | 0 | 0 | 0 | 0 | 0 | 1 | 4 |
| Days Min Temp ≤ 32 °F | 21 | 18 | 10 | 3 | 0 | 0 | 0 | 0 | 0 | 2 | 10 | 19 | 83 |
| Days Min Temp ≤ 0 °F | 0 | 0 | 0 | 0 | 0 | 0 | 0 | 0 | 0 | 0 | 0 | 0 | 0 |
| Heating Degree Days | 808 | 635 | 447 | 200 | 59 | 4 | 0 | 0 | 24 | 198 | 426 | 691 | 3492 |
| Cooling Degree Days | 0 | 0 | 7 | 35 | 110 | 297 | 416 | 369 | 199 | 35 | 5 | 1 | 1474 |
| Total Precipitation (") | 5.44 | 4.76 | 5.89 | 4.72 | 4.77 | 4.21 | 4.57 | 4.41 | 3.29 | 3.46 | 4.03 | 4.56 | 54.11 |
| Days ≥ 0.1" Precip | 9 | 7 | 8 | 7 | 7 | 7 | 7 | 7 | 5 | 5 | 6 | 7 | 82 |
| Total Snowfall (") | 1.5 | 0.7 | 1.0 | 0.2 | 0.0 | 0.0 | 0.0 | 0.0 | 0.0 | 0.0 | 0.1 | 0.1 | 3.6 |
| Days ≥ 1" Snow Depth | 1 | 0 | 0 | 0 | 0 | 0 | 0 | 0 | 0 | 0 | 0 | 0 | 1 |

## DALTON *Whitfield County* ELEVATION 761 ft LAT/LONG 34° 46 ' N / 84° 58 ' W

| | JAN | FEB | MAR | APR | MAY | JUN | JUL | AUG | SEP | OCT | NOV | DEC | YEAR |
|---|---|---|---|---|---|---|---|---|---|---|---|---|---|
| Maximum Temp °F | 48.7 | 52.8 | 62.3 | 72.4 | 79.2 | 86.1 | 89.0 | 88.1 | 82.7 | 72.4 | 62.2 | 52.6 | 70.7 |
| Minimum Temp °F | 28.5 | 30.8 | 38.3 | 46.7 | 54.9 | 63.2 | 67.5 | 66.8 | 60.6 | 47.1 | 39.1 | 32.5 | 48.0 |
| Mean Temp °F | 38.5 | 41.8 | 50.2 | 59.7 | 67.1 | 74.6 | 78.3 | 77.4 | 71.7 | 59.8 | 50.7 | 42.6 | 59.4 |
| Days Max Temp ≥ 90 °F | 0 | 0 | 0 | 0 | 1 | 8 | 15 | 12 | 4 | 0 | 0 | 0 | 40 |
| Days Max Temp ≤ 32 °F | 3 | 1 | 0 | 0 | 0 | 0 | 0 | 0 | 0 | 0 | 0 | 1 | 5 |
| Days Min Temp ≤ 32 °F | 20 | 17 | 9 | 1 | 0 | 0 | 0 | 0 | 0 | 1 | 9 | 18 | 75 |
| Days Min Temp ≤ 0 °F | 0 | 0 | 0 | 0 | 0 | 0 | 0 | 0 | 0 | 0 | 0 | 0 | 0 |
| Heating Degree Days | 816 | 648 | 456 | 190 | 51 | 2 | 0 | 0 | 17 | 189 | 426 | 688 | 3483 |
| Cooling Degree Days | 0 | 1 | 6 | 31 | 124 | 324 | 449 | 413 | 235 | 39 | 4 | 0 | 1626 |
| Total Precipitation (") | 5.29 | 4.84 | 6.15 | 4.28 | 4.55 | 4.06 | 5.02 | 4.26 | 5.12 | 3.55 | 4.42 | 5.08 | 56.62 |
| Days ≥ 0.1" Precip | 9 | 7 | 9 | 7 | 7 | 7 | 8 | 7 | 7 | 5 | 7 | 8 | 88 |
| Total Snowfall (") | 1.1 | 1.0 | 0.8 | 0.1 | 0.0 | 0.0 | 0.0 | 0.0 | 0.0 | 0.0 | 0.0 | 0.1 | 3.1 |
| Days ≥ 1" Snow Depth | *1* | 0 | 0 | 0 | 0 | 0 | 0 | 0 | 0 | 0 | 0 | 0 | 1 |

## DOUGLAS *Coffee County* ELEVATION 249 ft LAT/LONG 31° 30 ' N / 82° 51 ' W

| | JAN | FEB | MAR | APR | MAY | JUN | JUL | AUG | SEP | OCT | NOV | DEC | YEAR |
|---|---|---|---|---|---|---|---|---|---|---|---|---|---|
| Maximum Temp °F | 59.8 | 63.4 | 71.9 | 79.6 | 85.5 | 90.6 | 92.6 | 91.6 | 88.0 | 79.3 | 70.7 | 62.7 | 78.0 |
| Minimum Temp °F | 35.4 | 37.4 | 44.5 | 51.4 | 58.9 | 66.0 | 69.2 | 68.5 | 64.1 | 52.9 | 44.1 | 37.0 | 52.5 |
| Mean Temp °F | 47.6 | 50.4 | 58.2 | 65.5 | 72.2 | 78.3 | 80.9 | 80.1 | 76.1 | 66.1 | 57.4 | 49.9 | 65.2 |
| Days Max Temp ≥ 90 °F | 0 | 0 | 0 | 2 | 8 | 20 | 25 | 24 | 14 | 2 | 0 | 0 | 95 |
| Days Max Temp ≤ 32 °F | 0 | 0 | 0 | 0 | 0 | 0 | 0 | 0 | 0 | 0 | 0 | 0 | 0 |
| Days Min Temp ≤ 32 °F | 14 | 10 | 3 | 0 | 0 | 0 | 0 | 0 | 0 | 0 | 5 | 12 | 44 |
| Days Min Temp ≤ 0 °F | 0 | 0 | 0 | 0 | 0 | 0 | 0 | 0 | 0 | 0 | 0 | 0 | 0 |
| Heating Degree Days | 537 | 413 | 235 | 75 | 7 | 0 | 0 | 0 | 2 | 75 | 249 | 469 | 2062 |
| Cooling Degree Days | 2 | 9 | 31 | 90 | 254 | 430 | 520 | 483 | 338 | 121 | 31 | 7 | 2316 |
| Total Precipitation (") | 4.94 | 4.11 | 4.97 | 3.13 | 3.87 | 4.89 | 6.52 | 6.39 | 3.70 | 2.70 | 2.52 | 3.97 | 51.71 |
| Days ≥ 0.1" Precip | 7 | 6 | 7 | 5 | 6 | 8 | 10 | 9 | 6 | 4 | 4 | 6 | 78 |
| Total Snowfall (") | 0.0 | 0.1 | 0.0 | 0.0 | 0.0 | 0.0 | 0.0 | 0.0 | 0.0 | 0.0 | 0.0 | 0.1 | 0.2 |
| Days ≥ 1" Snow Depth | 0 | 0 | 0 | 0 | 0 | 0 | 0 | 0 | 0 | 0 | 0 | 0 | 0 |

## DUBLIN *Laurens County* ELEVATION 215 ft LAT/LONG 32° 30 ' N / 82° 54 ' W

| | JAN | FEB | MAR | APR | MAY | JUN | JUL | AUG | SEP | OCT | NOV | DEC | YEAR |
|---|---|---|---|---|---|---|---|---|---|---|---|---|---|
| Maximum Temp °F | 57.1 | 61.3 | 70.2 | 78.6 | 85.3 | 91.1 | 93.6 | 92.2 | 87.6 | 78.8 | 69.4 | 60.9 | 77.2 |
| Minimum Temp °F | 33.3 | 35.2 | 42.2 | 49.2 | 57.3 | 65.0 | 68.6 | 67.6 | 62.4 | 50.4 | 41.4 | 35.4 | 50.7 |
| Mean Temp °F | 45.2 | 48.3 | 56.2 | 63.9 | 71.3 | 78.1 | 81.1 | 79.9 | 75.0 | 64.6 | 55.4 | 48.2 | 63.9 |
| Days Max Temp ≥ 90 °F | 0 | 0 | 0 | 2 | 8 | 20 | 26 | 23 | 14 | 2 | 0 | 0 | 95 |
| Days Max Temp ≤ 32 °F | 0 | 0 | 0 | 0 | 0 | 0 | 0 | 0 | 0 | 0 | 0 | 0 | 0 |
| Days Min Temp ≤ 32 °F | 16 | 12 | 5 | 0 | 0 | 0 | 0 | 0 | 0 | 1 | 7 | 14 | 55 |
| Days Min Temp ≤ 0 °F | 0 | 0 | 0 | 0 | 0 | 0 | 0 | 0 | 0 | 0 | 0 | 0 | 0 |
| Heating Degree Days | 608 | 470 | 287 | 105 | 13 | 0 | 0 | 0 | 5 | 97 | 298 | 520 | 2403 |
| Cooling Degree Days | 1 | 5 | 23 | 68 | 219 | 430 | 540 | 486 | 319 | 99 | 18 | 6 | 2214 |
| Total Precipitation (") | 4.94 | 4.43 | 5.10 | 3.15 | 3.25 | 4.07 | 4.71 | 5.05 | 3.07 | 2.63 | 3.30 | 3.69 | 47.39 |
| Days ≥ 0.1" Precip | 8 | 7 | 7 | 5 | 6 | 7 | 8 | 7 | 5 | 4 | 5 | 6 | 75 |
| Total Snowfall (") | 0.1 | 0.5 | 0.0 | 0.0 | 0.0 | 0.0 | 0.0 | 0.0 | 0.0 | 0.0 | 0.0 | 0.0 | 0.6 |
| Days ≥ 1" Snow Depth | *0* | 0 | 0 | 0 | 0 | 0 | 0 | 0 | 0 | 0 | 0 | 0 | 0 |

## EASTMAN 1 W *Dodge County* ELEVATION 361 ft LAT/LONG 32° 12 ' N / 83° 11 ' W

| | JAN | FEB | MAR | APR | MAY | JUN | JUL | AUG | SEP | OCT | NOV | DEC | YEAR |
|---|---|---|---|---|---|---|---|---|---|---|---|---|---|
| Maximum Temp °F | 57.1 | 61.3 | 69.7 | 77.8 | 84.2 | 89.9 | 91.6 | 91.0 | 87.1 | 78.7 | 69.4 | 61.1 | 76.6 |
| Minimum Temp °F | 35.6 | 37.8 | 45.0 | 52.0 | 59.6 | 66.8 | 70.0 | 69.2 | 64.3 | 53.1 | 44.5 | 38.3 | 53.0 |
| Mean Temp °F | 46.4 | 49.6 | 57.4 | 64.9 | 72.0 | 78.4 | 80.8 | 80.1 | 75.7 | 65.9 | 57.0 | 49.8 | 64.8 |
| Days Max Temp ≥ 90 °F | 0 | 0 | 0 | 1 | 5 | 17 | 21 | 21 | 12 | 2 | 0 | 0 | 79 |
| Days Max Temp ≤ 32 °F | 0 | 0 | 0 | 0 | 0 | 0 | 0 | 0 | 0 | 0 | 0 | 0 | 0 |
| Days Min Temp ≤ 32 °F | 13 | 9 | 3 | 0 | 0 | 0 | 0 | 0 | 0 | 0 | 4 | 10 | 39 |
| Days Min Temp ≤ 0 °F | 0 | 0 | 0 | 0 | 0 | 0 | 0 | 0 | 0 | 0 | 0 | 0 | 0 |
| Heating Degree Days | 572 | 435 | 258 | 87 | 8 | 0 | 0 | 0 | 4 | 75 | 256 | 472 | 2167 |
| Cooling Degree Days | 1 | 6 | 27 | 83 | 235 | 438 | 524 | 490 | 332 | 114 | 21 | 7 | 2278 |
| Total Precipitation (") | 4.85 | 4.36 | 4.98 | 3.20 | 3.48 | 4.39 | 5.44 | 4.27 | 2.75 | 2.63 | 3.11 | 3.62 | 47.08 |
| Days ≥ 0.1" Precip | 7 | 6 | 7 | 5 | 6 | 7 | 9 | 7 | 5 | 4 | 5 | 6 | 74 |
| Total Snowfall (") | 0.1 | 0.0 | 0.0 | 0.0 | 0.0 | 0.0 | 0.0 | 0.0 | 0.0 | 0.0 | 0.0 | 0.0 | 0.1 |
| Days ≥ 1" Snow Depth | 0 | 0 | 0 | 0 | 0 | 0 | 0 | 0 | 0 | 0 | 0 | 0 | 0 |

## ELBERTON 2 N *Elbert County* ELEVATION 689 ft LAT/LONG 34° 7 ' N / 82° 52 ' W

| | JAN | FEB | MAR | APR | MAY | JUN | JUL | AUG | SEP | OCT | NOV | DEC | YEAR |
|---|---|---|---|---|---|---|---|---|---|---|---|---|---|
| Maximum Temp °F | 52.6 | 57.8 | 67.2 | 75.7 | 81.3 | 87.3 | 90.0 | 88.5 | 83.2 | 73.8 | 63.9 | 55.3 | 73.1 |
| Minimum Temp °F | 29.5 | 31.2 | 38.0 | 45.0 | 53.5 | 61.6 | 65.9 | 65.0 | 58.9 | 45.9 | 37.5 | 31.6 | 47.0 |
| Mean Temp °F | 41.1 | 44.5 | 52.6 | 60.4 | 67.4 | 74.5 | 78.0 | 76.8 | 71.1 | 59.9 | 50.7 | 43.5 | 60.0 |
| Days Max Temp ≥ 90 °F | 0 | 0 | 0 | 0 | 2 | 10 | 17 | 13 | 4 | 0 | 0 | 0 | 46 |
| Days Max Temp ≤ 32 °F | 1 | 0 | 0 | 0 | 0 | 0 | 0 | 0 | 0 | 0 | 0 | 0 | 1 |
| Days Min Temp ≤ 32 °F | 19 | 17 | 11 | 3 | 0 | 0 | 0 | 0 | 0 | 4 | 12 | 18 | 84 |
| Days Min Temp ≤ 0 °F | 0 | 0 | 0 | 0 | 0 | 0 | 0 | 0 | 0 | 0 | 0 | 0 | 0 |
| Heating Degree Days | 735 | 575 | 382 | 166 | 40 | 2 | 0 | 0 | 17 | 185 | 426 | 660 | 3188 |
| Cooling Degree Days | 0 | 0 | 6 | 32 | 122 | 316 | 433 | 383 | 212 | 37 | 3 | 1 | 1545 |
| Total Precipitation (") | 5.03 | 4.37 | 5.24 | 3.38 | 4.29 | 4.09 | 4.60 | 3.98 | 3.23 | 3.47 | 3.50 | 4.01 | 49.19 |
| Days ≥ 0.1" Precip | 8 | 7 | 8 | 6 | 7 | 6 | 7 | 6 | 6 | 5 | 6 | 7 | 79 |
| Total Snowfall (") | ***0.0*** | ***0.0*** | 0.1 | 0.0 | 0.0 | 0.0 | 0.0 | 0.0 | 0.0 | 0.0 | 0.0 | 0.0 | 0.1 |
| Days ≥ 1" Snow Depth | ***0*** | 0 | 0 | 0 | 0 | 0 | 0 | 0 | 0 | 0 | 0 | 0 | 0 |

## EXPERIMENT *Spalding County* ELEVATION 951 ft LAT/LONG 33° 16 ' N / 84° 17 ' W

| | JAN | FEB | MAR | APR | MAY | JUN | JUL | AUG | SEP | OCT | NOV | DEC | YEAR |
|---|---|---|---|---|---|---|---|---|---|---|---|---|---|
| Maximum Temp °F | 51.7 | 56.2 | 64.6 | 73.1 | 79.3 | 86.1 | 88.6 | 87.3 | 82.6 | 73.3 | 64.5 | 55.8 | 71.9 |
| Minimum Temp °F | 32.0 | 34.4 | 42.0 | 49.6 | 57.2 | 64.7 | 68.1 | 67.1 | 61.5 | 49.7 | 42.1 | 35.3 | 50.3 |
| Mean Temp °F | 41.9 | 45.3 | 53.3 | 61.4 | 68.3 | 75.5 | 78.3 | 77.2 | 72.0 | 61.5 | 53.3 | 45.6 | 61.1 |
| Days Max Temp ≥ 90 °F | 0 | 0 | 0 | 0 | 1 | 7 | 14 | 10 | 3 | 0 | 0 | 0 | 35 |
| Days Max Temp ≤ 32 °F | 1 | 0 | 0 | 0 | 0 | 0 | 0 | 0 | 0 | 0 | 0 | 0 | 1 |
| Days Min Temp ≤ 32 °F | 17 | 13 | 6 | 1 | 0 | 0 | 0 | 0 | 0 | 1 | 6 | 14 | 58 |
| Days Min Temp ≤ 0 °F | 0 | 0 | 0 | 0 | 0 | 0 | 0 | 0 | 0 | 0 | 0 | 0 | 0 |
| Heating Degree Days | 710 | 550 | 364 | 151 | 34 | 2 | 0 | 1 | 14 | 149 | 352 | 597 | 2924 |
| Cooling Degree Days | 0 | 1 | 9 | 40 | 134 | 341 | 432 | 385 | 222 | 48 | 7 | 1 | 1620 |
| Total Precipitation (") | 4.78 | 4.42 | 5.53 | 4.06 | 4.17 | 4.05 | 5.28 | 4.31 | 3.17 | 3.17 | 3.72 | 4.46 | 51.12 |
| Days ≥ 0.1" Precip | 8 | 7 | 7 | 6 | 7 | 6 | 8 | 7 | 5 | 5 | 6 | 7 | 79 |
| Total Snowfall (") | 0.3 | 0.2 | 0.3 | 0.0 | 0.0 | 0.0 | 0.0 | 0.0 | 0.0 | 0.0 | 0.0 | 0.1 | 0.9 |
| Days ≥ 1" Snow Depth | 0 | 0 | 0 | 0 | 0 | 0 | 0 | 0 | 0 | 0 | 0 | 0 | 0 |

## FITZGERALD *Ben Hill County* ELEVATION 371 ft LAT/LONG 31° 43 ' N / 83° 16 ' W

| | JAN | FEB | MAR | APR | MAY | JUN | JUL | AUG | SEP | OCT | NOV | DEC | YEAR |
|---|---|---|---|---|---|---|---|---|---|---|---|---|---|
| Maximum Temp °F | 59.1 | 62.8 | 70.8 | 78.5 | 84.6 | 89.7 | 92.0 | 90.9 | 86.9 | 78.6 | 69.9 | 62.3 | 77.2 |
| Minimum Temp °F | 38.8 | 40.8 | 47.5 | 54.6 | 62.0 | 68.7 | 71.7 | 71.1 | 66.5 | 56.1 | 47.5 | 41.4 | 55.6 |
| Mean Temp °F | 49.0 | 51.8 | 59.2 | 66.6 | 73.3 | 79.3 | 81.9 | 81.0 | 76.7 | 67.4 | 58.7 | 51.9 | 66.4 |
| Days Max Temp ≥ 90 °F | 0 | 0 | 0 | 1 | 5 | 17 | 24 | 21 | 11 | 1 | 0 | 0 | 80 |
| Days Max Temp ≤ 32 °F | 0 | 0 | 0 | 0 | 0 | 0 | 0 | 0 | 0 | 0 | 0 | 0 | 0 |
| Days Min Temp ≤ 32 °F | 9 | 6 | 2 | 0 | 0 | 0 | 0 | 0 | 0 | 0 | 2 | 6 | 25 |
| Days Min Temp ≤ 0 °F | 0 | 0 | 0 | 0 | 0 | 0 | 0 | 0 | 0 | 0 | 0 | 0 | 0 |
| Heating Degree Days | 492 | 374 | 212 | 63 | 5 | 0 | 0 | 0 | 2 | 57 | 215 | 410 | 1830 |
| Cooling Degree Days | 1 | 10 | 30 | 91 | 248 | 437 | 533 | 498 | 342 | 129 | 33 | 9 | 2361 |
| Total Precipitation (") | 4.94 | 3.92 | 5.05 | 2.69 | 3.51 | 4.53 | 4.78 | 5.07 | 3.02 | 2.38 | 3.00 | 3.52 | 46.41 |
| Days ≥ 0.1" Precip | 7 | 6 | 7 | 4 | 5 | 7 | 8 | 8 | 5 | 4 | 4 | 5 | 70 |
| Total Snowfall (") | 0.1 | 0.0 | 0.0 | 0.0 | 0.0 | 0.0 | 0.0 | 0.0 | 0.0 | 0.0 | 0.0 | 0.0 | 0.1 |
| Days ≥ 1" Snow Depth | 0 | 0 | 0 | 0 | 0 | 0 | 0 | 0 | 0 | 0 | 0 | 0 | 0 |

## FOLKSTON 9 SW *Charlton County* ELEVATION 120 ft LAT/LONG 30° 44 ' N / 82° 8 ' W

| | JAN | FEB | MAR | APR | MAY | JUN | JUL | AUG | SEP | OCT | NOV | DEC | YEAR |
|---|---|---|---|---|---|---|---|---|---|---|---|---|---|
| Maximum Temp °F | 64.9 | 68.3 | 75.7 | 82.5 | 87.5 | 91.7 | 93.7 | 92.6 | 89.0 | 81.7 | 74.2 | 67.5 | 80.8 |
| Minimum Temp °F | 40.2 | 42.4 | 48.3 | 53.4 | 60.2 | 66.7 | 69.6 | 69.6 | 66.6 | 57.3 | 49.2 | 42.7 | 55.5 |
| Mean Temp °F | 52.6 | 55.5 | 62.1 | 68.0 | 73.9 | 79.2 | 81.7 | 81.2 | 77.8 | 69.5 | 61.7 | 55.1 | 68.2 |
| Days Max Temp ≥ 90 °F | 0 | 0 | 0 | 3 | 12 | 21 | 27 | 25 | 15 | 2 | 0 | 0 | 105 |
| Days Max Temp ≤ 32 °F | 0 | 0 | 0 | 0 | 0 | 0 | 0 | 0 | 0 | 0 | 0 | 0 | 0 |
| Days Min Temp ≤ 32 °F | 8 | 6 | 2 | 0 | 0 | 0 | 0 | 0 | 0 | 0 | 2 | 6 | 24 |
| Days Min Temp ≤ 0 °F | 0 | 0 | 0 | 0 | 0 | 0 | 0 | 0 | 0 | 0 | 0 | 0 | 0 |
| Heating Degree Days | 388 | 281 | 148 | 40 | 2 | 0 | 0 | 0 | 0 | 31 | 151 | 320 | 1361 |
| Cooling Degree Days | 7 | 23 | ***68*** | 138 | 292 | 460 | 548 | 523 | 395 | 188 | 75 | 22 | 2739 |
| Total Precipitation (") | 3.99 | 3.59 | 4.59 | 2.84 | 3.98 | 6.13 | 7.51 | 7.27 | 4.47 | 2.99 | 2.61 | 2.95 | 52.92 |
| Days ≥ 0.1" Precip | 7 | 5 | 6 | 4 | 6 | 8 | 11 | 10 | 7 | 4 | 4 | 5 | 77 |
| Total Snowfall (") | 0.0 | 0.0 | 0.0 | 0.0 | 0.0 | 0.0 | 0.0 | 0.0 | 0.0 | 0.0 | 0.0 | 0.0 | 0.0 |
| Days ≥ 1" Snow Depth | 0 | 0 | 0 | 0 | 0 | 0 | 0 | 0 | 0 | 0 | 0 | 0 | 0 |

## FORSYTH 6 NNW *Monroe County* ELEVATION 600 ft LAT/LONG 33° 7 ' N / 83° 59 ' W

| | JAN | FEB | MAR | APR | MAY | JUN | JUL | AUG | SEP | OCT | NOV | DEC | YEAR |
|---|---|---|---|---|---|---|---|---|---|---|---|---|---|
| Maximum Temp °F | 56.0 | 60.4 | 69.0 | 77.2 | 83.2 | 88.8 | 90.9 | 89.8 | 84.9 | 76.2 | 67.1 | 59.6 | 75.3 |
| Minimum Temp °F | 32.6 | 33.5 | 40.9 | 47.9 | 55.6 | 63.6 | 67.8 | 66.9 | 61.7 | 49.4 | 41.2 | 34.8 | 49.7 |
| Mean Temp °F | 44.3 | 47.0 | 55.0 | 62.6 | 69.3 | 76.2 | 79.4 | 78.4 | 73.3 | 62.8 | 54.2 | 47.3 | 62.5 |
| Days Max Temp ≥ 90 °F | 0 | 0 | 0 | 1 | 3 | 14 | 19 | 17 | 7 | 0 | 0 | 0 | 61 |
| Days Max Temp ≤ 32 °F | 0 | 0 | 0 | 0 | 0 | 0 | 0 | 0 | 0 | 0 | 0 | 0 | 0 |
| Days Min Temp ≤ 32 °F | 16 | 14 | 8 | 2 | 0 | 0 | 0 | 0 | 0 | 2 | 8 | 14 | 64 |
| Days Min Temp ≤ 0 °F | 0 | 0 | 0 | 0 | 0 | 0 | 0 | 0 | 0 | 0 | 0 | 0 | 0 |
| Heating Degree Days | 635 | 502 | 318 | 123 | 21 | 1 | 0 | 0 | 9 | 125 | 327 | 546 | 2607 |
| Cooling Degree Days | 0 | 0 | *12* | 44 | *147* | 355 | 471 | 427 | 255 | 58 | *10* | 2 | 1781 |
| Total Precipitation (") | 4.39 | 4.63 | 5.37 | 3.82 | 3.80 | 3.48 | 4.96 | 4.13 | 3.10 | 2.71 | 3.69 | 4.01 | 48.09 |
| Days ≥ 0.1" Precip | 7 | 6 | 8 | 5 | 6 | 6 | 8 | 6 | 5 | 4 | 6 | 6 | 73 |
| Total Snowfall (") | 0.4 | 0.5 | 0.1 | 0.0 | 0.0 | 0.0 | 0.0 | 0.0 | 0.0 | 0.0 | 0.0 | 0.0 | 1.0 |
| Days ≥ 1" Snow Depth | 0 | 0 | 0 | 0 | 0 | 0 | 0 | 0 | 0 | 0 | 0 | 0 | 0 |

## FORT STEWART *Liberty County* ELEVATION 89 ft LAT/LONG 31° 52 ' N / 81° 37 ' W

| | JAN | FEB | MAR | APR | MAY | JUN | JUL | AUG | SEP | OCT | NOV | DEC | YEAR |
|---|---|---|---|---|---|---|---|---|---|---|---|---|---|
| Maximum Temp °F | 62.0 | 65.6 | 72.8 | 79.8 | 85.5 | 90.1 | 93.0 | 91.1 | 87.1 | 79.4 | 71.7 | 64.7 | 78.6 |
| Minimum Temp °F | 39.6 | 41.7 | 48.0 | 54.3 | 61.6 | 68.3 | 71.3 | 71.1 | 67.4 | 57.5 | 48.8 | 42.3 | 56.0 |
| Mean Temp °F | 50.8 | 53.7 | 60.5 | 67.1 | 73.5 | 79.2 | 82.2 | 81.1 | 77.3 | 68.5 | 60.3 | 53.5 | 67.3 |
| Days Max Temp ≥ 90 °F | 0 | 0 | 0 | 2 | 7 | 17 | 25 | 21 | 10 | 1 | 0 | 0 | 83 |
| Days Max Temp ≤ 32 °F | 0 | 0 | 0 | 0 | 0 | 0 | 0 | 0 | 0 | 0 | 0 | 0 | 0 |
| Days Min Temp ≤ 32 °F | 9 | 6 | 2 | 0 | 0 | 0 | 0 | 0 | 0 | 0 | 2 | 6 | 25 |
| Days Min Temp ≤ 0 °F | 0 | 0 | 0 | 0 | 0 | 0 | 0 | 0 | 0 | 0 | 0 | 0 | 0 |
| Heating Degree Days | 438 | 326 | 178 | 51 | 3 | 0 | 0 | 0 | 1 | 43 | 179 | 364 | 1583 |
| Cooling Degree Days | 4 | 15 | 42 | 111 | 281 | 465 | 564 | 519 | 378 | 162 | 55 | 14 | 2610 |
| Total Precipitation (") | 4.16 | 3.07 | 4.21 | 2.73 | 4.28 | 4.77 | 6.29 | 6.10 | 4.42 | 2.72 | 2.55 | 3.12 | 48.42 |
| Days ≥ 0.1" Precip | 7 | 5 | 6 | 5 | 6 | 8 | 10 | 9 | 7 | 4 | 4 | 5 | 76 |
| Total Snowfall (") | 0.0 | 0.0 | 0.0 | 0.0 | 0.0 | 0.0 | 0.0 | 0.0 | 0.0 | 0.0 | 0.0 | 0.1 | 0.1 |
| Days ≥ 1" Snow Depth | 0 | 0 | 0 | 0 | 0 | 0 | 0 | 0 | 0 | 0 | 0 | 0 | 0 |

## GAINESVILLE *Hall County* ELEVATION 1250 ft LAT/LONG 34° 19 ' N / 83° 50 ' W

| | JAN | FEB | MAR | APR | MAY | JUN | JUL | AUG | SEP | OCT | NOV | DEC | YEAR |
|---|---|---|---|---|---|---|---|---|---|---|---|---|---|
| Maximum Temp °F | 49.1 | 53.9 | 62.8 | 71.9 | 76.7 | 83.3 | 86.6 | 85.2 | 79.5 | 70.4 | 61.8 | 53.1 | 69.5 |
| Minimum Temp °F | 30.1 | 32.2 | 39.1 | 47.2 | 55.0 | 63.0 | 67.5 | 66.5 | 60.6 | 48.9 | 40.7 | 33.6 | 48.7 |
| Mean Temp °F | 39.6 | 43.1 | 51.0 | 59.6 | 65.9 | 73.2 | 77.0 | 75.9 | 70.1 | 59.7 | 51.3 | 43.4 | 59.2 |
| Days Max Temp ≥ 90 °F | 0 | 0 | 0 | 0 | 0 | 4 | 10 | 6 | 1 | 0 | 0 | 0 | 21 |
| Days Max Temp ≤ 32 °F | 2 | 1 | 0 | 0 | 0 | 0 | 0 | 0 | 0 | 0 | 0 | 0 | 3 |
| Days Min Temp ≤ 32 °F | 18 | 15 | 8 | 1 | 0 | 0 | 0 | 0 | 0 | 1 | 6 | 15 | 64 |
| Days Min Temp ≤ 0 °F | 0 | 0 | 0 | 0 | 0 | 0 | 0 | 0 | 0 | 0 | 0 | 0 | 0 |
| Heating Degree Days | 780 | 614 | 431 | 189 | 58 | 5 | 0 | 0 | 21 | 185 | 407 | 664 | 3354 |
| Cooling Degree Days | 0 | 0 | 5 | 35 | 112 | 299 | 429 | 378 | 203 | 39 | 3 | 1 | 1504 |
| Total Precipitation (") | 5.74 | 4.78 | 6.17 | 4.22 | 4.58 | 3.86 | 4.57 | 4.28 | 4.19 | 4.13 | 4.40 | 4.70 | 55.62 |
| Days ≥ 0.1" Precip | 9 | 7 | 9 | 6 | 7 | 6 | 7 | 6 | 6 | 5 | 6 | 8 | 82 |
| Total Snowfall (") | 1.7 | 0.6 | 0.3 | 0.0 | 0.0 | 0.0 | 0.0 | 0.0 | 0.0 | 0.0 | 0.1 | 0.1 | 2.8 |
| Days ≥ 1" Snow Depth | 1 | 0 | 0 | 0 | 0 | 0 | 0 | 0 | 0 | 0 | 0 | 0 | 1 |

## GLENNVILLE *Tattnall County* ELEVATION 180 ft LAT/LONG 31° 56 ' N / 81° 55 ' W

| | JAN | FEB | MAR | APR | MAY | JUN | JUL | AUG | SEP | OCT | NOV | DEC | YEAR |
|---|---|---|---|---|---|---|---|---|---|---|---|---|---|
| Maximum Temp °F | 60.7 | 64.5 | 72.2 | 79.3 | 85.3 | 90.0 | 92.4 | 90.8 | 86.9 | 79.0 | 70.7 | 63.8 | 78.0 |
| Minimum Temp °F | 38.9 | 41.0 | 47.5 | 53.9 | 61.5 | 68.2 | 71.3 | 70.9 | 67.0 | 56.8 | 48.0 | 41.6 | 55.6 |
| Mean Temp °F | 49.8 | 52.8 | 59.9 | 66.6 | 73.4 | 79.1 | 81.9 | 80.9 | 77.0 | 67.9 | 59.4 | 52.7 | 66.8 |
| Days Max Temp ≥ 90 °F | 0 | 0 | 0 | 2 | 7 | 18 | 24 | 21 | 10 | 1 | 0 | 0 | 83 |
| Days Max Temp ≤ 32 °F | 0 | 0 | 0 | 0 | 0 | 0 | 0 | 0 | 0 | 0 | 0 | 0 | 0 |
| Days Min Temp ≤ 32 °F | 9 | 7 | 2 | 0 | 0 | 0 | 0 | 0 | 0 | 0 | 2 | 7 | 27 |
| Days Min Temp ≤ 0 °F | 0 | 0 | 0 | 0 | 0 | 0 | 0 | 0 | 0 | 0 | 0 | 0 | 0 |
| Heating Degree Days | 468 | 350 | 193 | 58 | 4 | 0 | 0 | 0 | 1 | 49 | 199 | 385 | 1707 |
| Cooling Degree Days | 3 | 14 | 37 | 104 | 268 | 458 | 551 | 506 | 363 | 143 | 43 | 11 | 2501 |
| Total Precipitation (") | 4.34 | 3.47 | 4.28 | 2.92 | 4.04 | 4.61 | 5.69 | 5.59 | 4.06 | 2.99 | 2.63 | 3.56 | 48.18 |
| Days ≥ 0.1" Precip | 7 | 6 | 7 | 5 | 6 | 7 | 10 | 9 | 6 | 4 | 4 | 5 | 76 |
| Total Snowfall (") | 0.1 | 0.2 | 0.0 | 0.0 | 0.0 | 0.0 | 0.0 | 0.0 | 0.0 | 0.0 | 0.0 | 0.1 | 0.4 |
| Days ≥ 1" Snow Depth | 0 | 0 | 0 | 0 | 0 | 0 | 0 | 0 | 0 | 0 | 0 | 0 | 0 |

## HARTWELL *Hart County* ELEVATION 791 ft LAT/LONG 34° 24 ' N / 82° 55 ' W

| | JAN | FEB | MAR | APR | MAY | JUN | JUL | AUG | SEP | OCT | NOV | DEC | YEAR |
|---|---|---|---|---|---|---|---|---|---|---|---|---|---|
| Maximum Temp °F | 52.6 | 57.6 | 66.9 | 75.7 | 81.6 | 87.5 | 90.7 | 88.8 | 83.5 | 74.2 | 65.5 | 56.9 | 73.5 |
| Minimum Temp °F | 32.6 | 34.7 | 42.1 | 49.4 | 57.7 | 65.3 | 69.4 | 68.3 | 62.5 | 51.0 | 41.9 | 35.7 | 50.9 |
| Mean Temp °F | 42.6 | 46.3 | 54.5 | 62.6 | 69.6 | 76.4 | 80.0 | 78.6 | 73.0 | 62.6 | 53.7 | 46.2 | 62.2 |
| Days Max Temp ≥ 90 °F | 0 | 0 | 0 | 1 | 3 | 12 | 19 | 15 | 6 | 0 | 0 | 0 | 56 |
| Days Max Temp ≤ 32 °F | 1 | 0 | 0 | 0 | 0 | 0 | 0 | 0 | 0 | 0 | 0 | 0 | 1 |
| Days Min Temp ≤ 32 °F | 15 | 12 | 5 | 0 | 0 | 0 | 0 | 0 | 0 | 0 | 5 | 12 | 49 |
| Days Min Temp ≤ 0 °F | 0 | 0 | 0 | 0 | 0 | 0 | 0 | 0 | 0 | 0 | 0 | 0 | 0 |
| Heating Degree Days | 688 | 523 | 330 | 124 | 19 | 1 | 0 | 0 | 8 | 121 | 340 | 573 | 2727 |
| Cooling Degree Days | 0 | 0 | 9 | 51 | *165* | 360 | 491 | 430 | *242* | 55 | 8 | 2 | 1813 |
| Total Precipitation (") | 5.23 | 4.38 | 5.94 | 3.43 | 4.41 | 4.29 | 4.32 | 3.88 | 3.89 | 3.84 | 3.68 | 4.64 | 51.93 |
| Days ≥ 0.1" Precip | *7* | 7 | 8 | 5 | 6 | 6 | 7 | 6 | 5 | 5 | 5 | 7 | 74 |
| Total Snowfall (") | 0.9 | *0.4* | 0.2 | 0.0 | 0.0 | 0.0 | 0.0 | 0.0 | 0.0 | 0.0 | 0.0 | 0.0 | 1.5 |
| Days ≥ 1" Snow Depth | 1 | *0* | 0 | 0 | 0 | 0 | 0 | 0 | 0 | 0 | 0 | 0 | 1 |

## HAWKINSVILLE *Pulaski County* ELEVATION 249 ft LAT/LONG 32° 17 ' N / 83° 28 ' W

| | JAN | FEB | MAR | APR | MAY | JUN | JUL | AUG | SEP | OCT | NOV | DEC | YEAR |
|---|---|---|---|---|---|---|---|---|---|---|---|---|---|
| Maximum Temp °F | 58.3 | 62.1 | 70.6 | 78.9 | 85.0 | 91.0 | 92.8 | *91.9* | 88.1 | 80.0 | 70.5 | 62.1 | 77.6 |
| Minimum Temp °F | 33.4 | 35.3 | 42.6 | 50.4 | 58.2 | 66.1 | 69.0 | *68.3* | 62.8 | 51.0 | 42.1 | 35.9 | 51.3 |
| Mean Temp °F | 45.9 | 48.7 | 56.6 | 64.7 | 71.6 | 78.6 | 80.9 | *80.1* | 75.5 | 65.5 | 56.3 | 49.0 | 64.5 |
| Days Max Temp ≥ 90 °F | 0 | 0 | 0 | 2 | 8 | 19 | 24 | *23* | 14 | 3 | 0 | 0 | 93 |
| Days Max Temp ≤ 32 °F | 0 | 0 | 0 | 0 | 0 | 0 | 0 | 0 | 0 | 0 | 0 | 0 | 0 |
| Days Min Temp ≤ 32 °F | 16 | 13 | 5 | 0 | 0 | 0 | 0 | 0 | 0 | 1 | 6 | 13 | 54 |
| Days Min Temp ≤ 0 °F | 0 | 0 | 0 | 0 | 0 | 0 | 0 | 0 | 0 | 0 | 0 | 0 | 0 |
| Heating Degree Days | 587 | 456 | 274 | 90 | 11 | 0 | 0 | *0* | 6 | 83 | 276 | 494 | 2277 |
| Cooling Degree Days | 1 | 5 | 23 | 81 | 231 | 436 | 529 | 495 | 322 | 109 | 25 | 6 | 2263 |
| Total Precipitation (") | 5.03 | 4.51 | 4.69 | 3.20 | 3.34 | 4.11 | 4.69 | 3.87 | 3.18 | 2.50 | 3.19 | 3.89 | 46.20 |
| Days ≥ 0.1" Precip | 7 | 7 | 7 | 5 | 6 | 7 | 8 | 6 | 5 | 4 | 5 | 7 | 74 |
| Total Snowfall (") | *0.1* | 0.5 | 0.1 | 0.0 | 0.0 | 0.0 | 0.0 | 0.0 | 0.0 | 0.0 | 0.0 | 0.0 | 0.7 |
| Days ≥ 1" Snow Depth | *0* | 0 | 0 | 0 | 0 | 0 | 0 | 0 | 0 | 0 | 0 | 0 | 0 |

## HELEN *White County* ELEVATION 1489 ft LAT/LONG 34° 42 ' N / 83° 44 ' W

| | JAN | FEB | MAR | APR | MAY | JUN | JUL | AUG | SEP | OCT | NOV | DEC | YEAR |
|---|---|---|---|---|---|---|---|---|---|---|---|---|---|
| Maximum Temp °F | *50.5* | *55.0* | *63.5* | *72.6* | *77.8* | *83.4* | *85.9* | *84.3* | *79.5* | *71.2* | *61.7* | *53.5* | 69.9 |
| Minimum Temp °F | *29.1* | *30.6* | *37.0* | *43.7* | *51.6* | *59.3* | *63.6* | *63.1* | *57.6* | *45.9* | *37.7* | *32.1* | 45.9 |
| Mean Temp °F | *39.8* | *42.8* | *50.3* | *58.2* | *64.7* | *71.4* | *74.8* | *73.7* | *68.6* | *58.6* | *49.7* | *42.8* | 58.0 |
| Days Max Temp ≥ 90 °F | *0* | *0* | *0* | *0* | *0* | *3* | *8* | *4* | *1* | *0* | *0* | *0* | 16 |
| Days Max Temp ≤ 32 °F | *1* | *0* | *0* | *0* | *0* | *0* | *0* | *0* | *0* | *0* | *0* | *0* | 1 |
| Days Min Temp ≤ 32 °F | *20* | *17* | *11* | *4* | *0* | *0* | *0* | *0* | *0* | *3* | *11* | *17* | 83 |
| Days Min Temp ≤ 0 °F | *0* | *0* | *0* | *0* | *0* | *0* | *0* | *0* | *0* | *0* | *0* | *0* | 0 |
| Heating Degree Days | *773* | *621* | *450* | *214* | *67* | *8* | *0* | *1* | *27* | *213* | *454* | *680* | 3508 |
| Cooling Degree Days | na | na | *2* | *19* | *80* | *248* | *348* | *308* | *159* | *29* | *2* | *0* | na |
| Total Precipitation (") | *6.45* | *5.96* | *7.36* | *4.83* | *6.76* | *5.14* | *6.46* | *6.24* | *5.45* | *4.89* | *5.95* | *6.25* | 71.74 |
| Days ≥ 0.1" Precip | *9* | *8* | *9* | *7* | *8* | *8* | *10* | *8* | *7* | *6* | *7* | *9* | 96 |
| Total Snowfall (") | na | *1.5* | *0.4* | *0.0* | *0.0* | *0.0* | *0.0* | *0.0* | *0.0* | *0.0* | *0.0* | *0.5* | na |
| Days ≥ 1" Snow Depth | na | na | *0* | *0* | *0* | *0* | *0* | *0* | *0* | *0* | *0* | *0* | na |

## HOMERVILLE 3 WSW *Clinch County* ELEVATION 180 ft LAT/LONG 31° 2 ' N / 82° 45 ' W

| | JAN | FEB | MAR | APR | MAY | JUN | JUL | AUG | SEP | OCT | NOV | DEC | YEAR |
|---|---|---|---|---|---|---|---|---|---|---|---|---|---|
| Maximum Temp °F | 62.1 | 65.6 | 73.1 | 80.2 | 85.8 | 90.4 | 92.2 | 91.1 | 87.5 | 79.7 | 71.6 | 64.8 | 78.7 |
| Minimum Temp °F | 38.7 | 40.5 | 46.5 | 52.1 | 59.0 | 66.0 | 68.8 | 68.8 | 65.6 | 55.2 | 46.5 | 40.6 | 54.0 |
| Mean Temp °F | 50.5 | 53.1 | 59.8 | 66.2 | 72.4 | 78.2 | 80.5 | 80.0 | 76.6 | 67.5 | 59.0 | 52.7 | 66.4 |
| Days Max Temp ≥ 90 °F | 0 | 0 | 0 | 1 | 6 | 19 | 25 | 23 | 12 | 1 | 0 | 0 | 87 |
| Days Max Temp ≤ 32 °F | 0 | 0 | 0 | 0 | 0 | 0 | 0 | 0 | 0 | 0 | 0 | 0 | 0 |
| Days Min Temp ≤ 32 °F | 10 | 7 | 3 | 0 | 0 | 0 | 0 | 0 | 0 | 0 | 4 | 9 | 33 |
| Days Min Temp ≤ 0 °F | 0 | 0 | 0 | 0 | 0 | 0 | 0 | 0 | 0 | 0 | 0 | 0 | 0 |
| Heating Degree Days | 450 | 341 | 192 | 60 | 6 | 0 | 0 | 0 | 1 | 57 | 209 | 387 | 1703 |
| Cooling Degree Days | 4 | 11 | 32 | 91 | 240 | 413 | 498 | 470 | 339 | 130 | 42 | 11 | 2281 |
| Total Precipitation (") | 5.01 | 4.16 | 4.98 | 3.41 | 3.54 | 5.63 | 6.32 | 6.32 | 4.26 | 2.60 | 2.74 | 4.01 | 52.98 |
| Days ≥ 0.1" Precip | 7 | 6 | 6 | 4 | 6 | 8 | 10 | 9 | 6 | 4 | 4 | 6 | 76 |
| Total Snowfall (") | 0.0 | 0.0 | 0.0 | 0.0 | 0.0 | 0.0 | 0.0 | 0.0 | 0.0 | 0.0 | 0.0 | 0.1 | 0.1 |
| Days ≥ 1" Snow Depth | 0 | 0 | 0 | 0 | 0 | 0 | 0 | 0 | 0 | 0 | 0 | 0 | 0 |

## IRWINTON 4 WNW *Wilkinson County* ELEVATION 449 ft LAT/LONG 32° 49 ' N / 83° 10 ' W

| | JAN | FEB | MAR | APR | MAY | JUN | JUL | AUG | SEP | OCT | NOV | DEC | YEAR |
|---|---|---|---|---|---|---|---|---|---|---|---|---|---|
| Maximum Temp °F | ***57.5*** | 62.3 | 71.3 | 79.5 | 84.0 | 90.2 | 92.5 | 91.1 | 86.9 | 78.0 | 69.9 | 61.5 | 77.1 |
| Minimum Temp °F | ***35.0*** | 37.9 | 44.3 | 51.2 | 58.5 | 65.6 | 69.4 | 68.7 | 63.7 | 52.8 | 43.5 | 38.1 | 52.4 |
| Mean Temp °F | ***46.3*** | 50.1 | 57.8 | 65.4 | 71.3 | 77.9 | 81.0 | 79.9 | 75.3 | 65.4 | 56.8 | 49.8 | 64.7 |
| Days Max Temp ≥ 90 °F | 0 | 0 | 0 | 1 | 5 | 18 | 23 | 20 | 11 | 1 | 0 | 0 | 79 |
| Days Max Temp ≤ 32 °F | 0 | 0 | 0 | 0 | 0 | 0 | 0 | 0 | 0 | 0 | 0 | 0 | 0 |
| Days Min Temp ≤ 32 °F | 13 | 9 | 3 | 0 | 0 | 0 | 0 | 0 | 0 | 0 | 4 | 10 | 39 |
| Days Min Temp ≤ 0 °F | 0 | 0 | 0 | 0 | 0 | 0 | 0 | 0 | 0 | 0 | 0 | 0 | 0 |
| Heating Degree Days | ***574*** | 417 | 240 | 74 | 10 | 1 | 0 | 0 | 4 | 80 | 254 | 466 | 2120 |
| Cooling Degree Days | ***1*** | 5 | 27 | 89 | 224 | 422 | 527 | 485 | 329 | 107 | 17 | 4 | 2237 |
| Total Precipitation (") | 4.64 | 4.36 | 5.01 | 2.90 | 3.32 | 3.62 | 5.03 | 4.13 | 3.22 | 2.73 | 3.19 | 4.25 | 46.40 |
| Days ≥ 0.1" Precip | ***7*** | 6 | 6 | 5 | 6 | 6 | 8 | 7 | 5 | 4 | 5 | 5 | 70 |
| Total Snowfall (") | 0.2 | 0.7 | 0.1 | 0.0 | 0.0 | 0.0 | 0.0 | 0.0 | 0.0 | 0.0 | 0.0 | 0.0 | 1.0 |
| Days ≥ 1" Snow Depth | ***0*** | 0 | 0 | 0 | 0 | 0 | 0 | 0 | 0 | 0 | 0 | 0 | 0 |

## JASPER 1 NNW *Pickens County* ELEVATION 1480 ft LAT/LONG 34° 28 ' N / 84° 26 ' W

| | JAN | FEB | MAR | APR | MAY | JUN | JUL | AUG | SEP | OCT | NOV | DEC | YEAR |
|---|---|---|---|---|---|---|---|---|---|---|---|---|---|
| Maximum Temp °F | 47.6 | 52.6 | 61.5 | 70.4 | 76.7 | 83.4 | 86.4 | 85.0 | 79.6 | 69.7 | 60.2 | 51.6 | 68.7 |
| Minimum Temp °F | 29.4 | 31.7 | 39.7 | 46.8 | 54.3 | 61.6 | 65.3 | 64.5 | 58.9 | 47.6 | 40.1 | 33.5 | 47.8 |
| Mean Temp °F | 38.5 | 42.2 | 50.6 | 58.6 | 65.5 | 72.5 | 75.9 | 74.8 | 69.3 | 58.7 | 50.2 | 42.6 | 58.3 |
| Days Max Temp ≥ 90 °F | 0 | 0 | 0 | 0 | 0 | 3 | 9 | 5 | 1 | 0 | 0 | 0 | 18 |
| Days Max Temp ≤ 32 °F | 3 | 1 | 0 | 0 | 0 | 0 | 0 | 0 | 0 | 0 | 0 | 1 | 5 |
| Days Min Temp ≤ 32 °F | 19 | 15 | 8 | 2 | 0 | 0 | 0 | 0 | 0 | 1 | 8 | 15 | 68 |
| Days Min Temp ≤ 0 °F | 0 | 0 | 0 | 0 | 0 | 0 | 0 | 0 | 0 | 0 | 0 | 0 | 0 |
| Heating Degree Days | 814 | 638 | 443 | 208 | 62 | 5 | 0 | 0 | 27 | 210 | 439 | 688 | 3534 |
| Cooling Degree Days | 0 | 0 | 3 | 20 | 83 | 254 | 370 | 321 | 165 | 23 | 2 | 0 | 1241 |
| Total Precipitation (") | 5.71 | 5.21 | 6.64 | 5.13 | 5.11 | 4.27 | 5.48 | 4.81 | 3.89 | 3.82 | 4.78 | 5.11 | 59.96 |
| Days ≥ 0.1" Precip | 9 | 7 | 8 | 7 | 8 | 7 | 9 | 7 | 6 | 5 | 7 | 8 | 88 |
| Total Snowfall (") | 0.8 | 1.0 | 0.4 | 0.1 | 0.0 | 0.0 | 0.0 | 0.0 | 0.0 | 0.0 | 0.1 | 0.1 | 2.5 |
| Days ≥ 1" Snow Depth | ***0*** | 0 | 0 | 0 | 0 | 0 | 0 | 0 | 0 | 0 | 0 | 0 | 0 |

## JESUP 8 S *Wayne County* ELEVATION 100 ft LAT/LONG 31° 29 ' N / 81° 53 ' W

| | JAN | FEB | MAR | APR | MAY | JUN | JUL | AUG | SEP | OCT | NOV | DEC | YEAR |
|---|---|---|---|---|---|---|---|---|---|---|---|---|---|
| Maximum Temp °F | 62.6 | 66.4 | 73.6 | 80.7 | 86.2 | 90.4 | 93.2 | 91.5 | 87.5 | 80.3 | 72.1 | 65.3 | 79.1 |
| Minimum Temp °F | 38.0 | 39.7 | 45.7 | 51.5 | 59.0 | 66.4 | 69.8 | 69.5 | 65.5 | 54.8 | 46.5 | 40.4 | 53.9 |
| Mean Temp °F | 50.3 | 53.1 | 59.7 | 66.1 | 72.6 | 78.4 | 81.5 | 80.5 | 76.6 | 67.6 | 59.3 | 52.9 | 66.6 |
| Days Max Temp ≥ 90 °F | 0 | 0 | 0 | 2 | 8 | 18 | 25 | 23 | 11 | 2 | 0 | 0 | 89 |
| Days Max Temp ≤ 32 °F | 0 | 0 | 0 | 0 | 0 | 0 | 0 | 0 | 0 | 0 | 0 | 0 | 0 |
| Days Min Temp ≤ 32 °F | 11 | 8 | 3 | 0 | 0 | 0 | 0 | 0 | 0 | 0 | 4 | 9 | 35 |
| Days Min Temp ≤ 0 °F | 0 | 0 | 0 | 0 | 0 | 0 | 0 | 0 | 0 | 0 | 0 | 0 | 0 |
| Heating Degree Days | 453 | 342 | 198 | 63 | 5 | 0 | 0 | 0 | 1 | 58 | 203 | 383 | 1706 |
| Cooling Degree Days | 3 | 15 | 35 | 98 | 251 | 436 | 541 | 500 | 353 | 148 | 45 | 14 | 2439 |
| Total Precipitation (") | 4.32 | 3.50 | 4.63 | 2.67 | 4.05 | 5.60 | 6.71 | 7.00 | 3.68 | 2.93 | 2.50 | 3.31 | 50.90 |
| Days ≥ 0.1" Precip | 7 | 6 | 6 | 4 | 7 | 8 | 10 | 9 | 7 | 4 | 4 | 5 | 77 |
| Total Snowfall (") | 0.0 | 0.2 | 0.0 | 0.0 | 0.0 | 0.0 | 0.0 | 0.0 | 0.0 | 0.0 | 0.0 | 0.0 | 0.2 |
| Days ≥ 1" Snow Depth | 0 | 0 | 0 | 0 | 0 | 0 | 0 | 0 | 0 | 0 | 0 | 0 | 0 |

## LA GRANGE *Troup County* ELEVATION 715 ft LAT/LONG 33° 2 ' N / 85° 2 ' W

| | JAN | FEB | MAR | APR | MAY | JUN | JUL | AUG | SEP | OCT | NOV | DEC | YEAR |
|---|---|---|---|---|---|---|---|---|---|---|---|---|---|
| Maximum Temp °F | 54.8 | 59.8 | 69.0 | 77.1 | 82.5 | 88.4 | 90.1 | 88.7 | 83.9 | 75.1 | 66.3 | 58.1 | 74.5 |
| Minimum Temp °F | 32.7 | 34.8 | 41.6 | 48.5 | 56.4 | 64.0 | 68.2 | 67.3 | 61.7 | 49.3 | 41.2 | 35.5 | 50.1 |
| Mean Temp °F | 43.8 | 47.3 | 55.3 | 62.8 | 69.5 | 76.2 | 79.2 | 78.0 | 72.8 | 62.3 | 53.8 | 46.8 | 62.3 |
| Days Max Temp ≥ 90 °F | 0 | 0 | 0 | 0 | 3 | 12 | 17 | 14 | 5 | 0 | 0 | 0 | 51 |
| Days Max Temp ≤ 32 °F | 1 | 0 | 0 | 0 | 0 | 0 | 0 | 0 | 0 | 0 | 0 | 0 | 1 |
| Days Min Temp ≤ 32 °F | 16 | 13 | 7 | 1 | 0 | 0 | 0 | 0 | 0 | 1 | 8 | 14 | 60 |
| Days Min Temp ≤ 0 °F | 0 | 0 | 0 | 0 | 0 | 0 | 0 | 0 | 0 | 0 | 0 | 0 | 0 |
| Heating Degree Days | 652 | 494 | 310 | 119 | 21 | 1 | 0 | 0 | 9 | 132 | 338 | 558 | 2634 |
| Cooling Degree Days | 0 | 1 | 18 | 53 | 163 | 368 | 466 | 427 | 253 | 61 | 10 | 3 | 1823 |
| Total Precipitation (") | 5.02 | 4.92 | 6.04 | 4.55 | 4.18 | 4.00 | 5.54 | 4.00 | 3.38 | 3.19 | 4.23 | 5.02 | 54.07 |
| Days ≥ 0.1" Precip | 8 | 7 | 7 | 6 | 7 | 7 | 9 | 7 | 5 | 4 | 6 | 7 | 80 |
| Total Snowfall (") | 0.4 | 0.2 | 0.2 | 0.0 | 0.0 | 0.0 | 0.0 | 0.0 | 0.0 | 0.0 | 0.0 | 0.0 | 0.8 |
| Days ≥ 1" Snow Depth | 0 | 0 | 0 | 0 | 0 | 0 | 0 | 0 | 0 | 0 | 0 | 0 | 0 |

## LAFAYETTE 5 SW *Walker County* ELEVATION 810 ft LAT/LONG 34° 42 ' N / 85° 17 ' W

| | JAN | FEB | MAR | APR | MAY | JUN | JUL | AUG | SEP | OCT | NOV | DEC | YEAR |
|---|---|---|---|---|---|---|---|---|---|---|---|---|---|
| Maximum Temp °F | 48.1 | 53.3 | 62.9 | 72.1 | 78.7 | 85.4 | 88.9 | 88.0 | 81.9 | 72.0 | 61.9 | 52.2 | 70.5 |
| Minimum Temp °F | 26.8 | 29.2 | 36.7 | 44.4 | 52.4 | 61.0 | 65.6 | 64.5 | 58.2 | 45.0 | 36.9 | 30.0 | 45.9 |
| Mean Temp °F | 37.5 | 41.3 | 49.8 | 58.3 | 65.6 | 73.3 | 77.3 | 76.3 | 70.1 | 58.5 | 49.4 | 41.2 | 58.2 |
| Days Max Temp ≥ 90 °F | 0 | 0 | 0 | 0 | 1 | 7 | 15 | 11 | 4 | 0 | 0 | 0 | 38 |
| Days Max Temp ≤ 32 °F | 3 | 1 | 0 | 0 | 0 | 0 | 0 | 0 | 0 | 0 | 0 | 1 | 5 |
| Days Min Temp ≤ 32 °F | 21 | 18 | 11 | 3 | 0 | 0 | 0 | 0 | 0 | 3 | 12 | 19 | 87 |
| Days Min Temp ≤ 0 °F | 0 | 0 | 0 | 0 | 0 | 0 | 0 | 0 | 0 | 0 | 0 | 0 | 0 |
| Heating Degree Days | 847 | 664 | 467 | 220 | 69 | 5 | 0 | 0 | 27 | 216 | 463 | 735 | 3713 |
| Cooling Degree Days | 0 | 0 | 4 | 19 | 93 | 267 | 405 | 359 | 187 | 26 | 2 | 1 | 1363 |
| Total Precipitation (") | 5.21 | 5.20 | 6.40 | 4.28 | 4.76 | 4.32 | 5.24 | 3.56 | 5.10 | 3.54 | 4.74 | 5.67 | 58.02 |
| Days ≥ 0.1" Precip | 8 | 7 | 8 | 6 | 7 | 7 | 8 | 6 | 6 | 5 | 7 | 8 | 83 |
| Total Snowfall (") | ***1.1*** | 0.2 | 0.7 | 0.1 | 0.0 | 0.0 | 0.0 | 0.0 | 0.0 | 0.0 | 0.0 | 0.1 | 2.2 |
| Days ≥ 1" Snow Depth | ***0*** | 0 | 0 | 0 | 0 | 0 | 0 | 0 | 0 | 0 | 0 | 0 | 0 |

## LOUISVILLE 1 E *Jefferson County* ELEVATION 322 ft LAT/LONG 33° 1 ' N / 82° 24 ' W

| | JAN | FEB | MAR | APR | MAY | JUN | JUL | AUG | SEP | OCT | NOV | DEC | YEAR |
|---|---|---|---|---|---|---|---|---|---|---|---|---|---|
| Maximum Temp °F | 57.9 | 62.4 | 70.8 | 78.5 | 84.5 | 89.5 | 92.0 | 90.4 | 86.3 | 78.0 | 69.3 | 61.0 | 76.7 |
| Minimum Temp °F | 35.4 | 37.5 | 44.1 | 51.0 | 58.9 | 66.6 | 69.9 | 69.1 | 64.1 | 52.7 | 43.8 | 37.5 | 52.6 |
| Mean Temp °F | 46.7 | 50.0 | 57.5 | 64.8 | 71.7 | 78.1 | 81.0 | 79.7 | 75.3 | 65.3 | 56.6 | 49.3 | 64.7 |
| Days Max Temp ≥ 90 °F | 0 | 0 | 0 | 1 | 6 | 15 | 21 | 19 | 9 | 1 | 0 | 0 | 72 |
| Days Max Temp ≤ 32 °F | 0 | 0 | 0 | 0 | 0 | 0 | 0 | 0 | 0 | 0 | 0 | 0 | 0 |
| Days Min Temp ≤ 32 °F | 13 | 10 | 4 | 0 | 0 | 0 | 0 | 0 | 0 | 0 | 5 | 11 | 43 |
| Days Min Temp ≤ 0 °F | 0 | 0 | 0 | 0 | 0 | 0 | 0 | 0 | 0 | 0 | 0 | 0 | 0 |
| Heating Degree Days | 563 | 423 | 251 | 84 | 11 | 0 | 0 | 0 | 3 | 82 | 265 | 485 | 2167 |
| Cooling Degree Days | 2 | 6 | 23 | 77 | 228 | 424 | 536 | 483 | 328 | 107 | 25 | 5 | 2244 |
| Total Precipitation (") | 4.44 | 4.04 | 4.95 | 2.86 | 3.55 | 4.35 | 4.41 | 4.92 | 2.91 | 2.98 | 2.71 | 3.46 | 45.58 |
| Days ≥ 0.1" Precip | 7 | 6 | 7 | 5 | 5 | 6 | 7 | 7 | 5 | 4 | 5 | 6 | 70 |
| Total Snowfall (") | 0.1 | 0.6 | 0.0 | 0.0 | 0.0 | 0.0 | 0.0 | 0.0 | 0.0 | 0.0 | 0.0 | 0.1 | 0.8 |
| Days ≥ 1" Snow Depth | 0 | 0 | 0 | 0 | 0 | 0 | 0 | 0 | 0 | 0 | 0 | 0 | 0 |

## LUMBER CITY *Telfair County* ELEVATION 120 ft LAT/LONG 31° 55 ' N / 82° 41 ' W

| | JAN | FEB | MAR | APR | MAY | JUN | JUL | AUG | SEP | OCT | NOV | DEC | YEAR |
|---|---|---|---|---|---|---|---|---|---|---|---|---|---|
| Maximum Temp °F | ***59.7*** | 63.0 | ***70.6*** | 78.8 | 84.8 | ***90.0*** | 92.4 | 91.1 | 87.3 | ***79.6*** | ***70.8*** | na | na |
| Minimum Temp °F | ***34.2*** | 36.7 | ***43.4*** | 50.6 | 58.0 | ***65.7*** | 69.8 | 69.1 | 63.8 | ***52.3*** | ***43.1*** | na | na |
| Mean Temp °F | ***46.7*** | 49.9 | ***57.1*** | 64.7 | 71.4 | ***77.9*** | 81.1 | 80.2 | 75.6 | ***65.9*** | ***57.0*** | na | na |
| Days Max Temp ≥ 90 °F | 0 | 0 | 0 | 1 | 5 | 16 | 22 | 21 | 11 | ***2*** | ***0*** | 0 | 78 |
| Days Max Temp ≤ 32 °F | 0 | 0 | 0 | 0 | 0 | 0 | 0 | 0 | 0 | 0 | 0 | ***0*** | 0 |
| Days Min Temp ≤ 32 °F | ***15*** | 11 | ***4*** | 0 | 0 | 0 | 0 | 0 | 0 | ***0*** | ***5*** | ***13*** | 48 |
| Days Min Temp ≤ 0 °F | 0 | 0 | 0 | 0 | 0 | 0 | 0 | 0 | 0 | 0 | 0 | ***0*** | 0 |
| Heating Degree Days | ***563*** | 427 | ***261*** | 86 | 9 | ***1*** | 0 | 0 | 3 | ***81*** | ***258*** | na | na |
| Cooling Degree Days | ***1*** | ***8*** | ***22*** | 75 | ***209*** | ***420*** | 535 | ***497*** | ***330*** | ***130*** | ***26*** | ***3*** | 2256 |
| Total Precipitation (") | 4.43 | 3.89 | 4.63 | 2.60 | 3.24 | 3.96 | 5.58 | 5.45 | 3.29 | 2.41 | 3.04 | 3.62 | 46.14 |
| Days ≥ 0.1" Precip | ***6*** | ***5*** | ***7*** | 4 | 5 | 6 | 8 | 7 | 5 | ***3*** | 4 | ***5*** | 65 |
| Total Snowfall (") | ***0.0*** | 0.2 | 0.0 | 0.0 | 0.0 | 0.0 | 0.0 | 0.0 | 0.0 | 0.0 | 0.0 | 0.0 | 0.2 |
| Days ≥ 1" Snow Depth | ***0*** | 0 | 0 | 0 | 0 | 0 | 0 | 0 | 0 | 0 | 0 | 0 | 0 |

## LUMPKIN 2 SE *Stewart County* ELEVATION 502 ft LAT/LONG 32° 4 ' N / 84° 47 ' W

| | JAN | FEB | MAR | APR | MAY | JUN | JUL | AUG | SEP | OCT | NOV | DEC | YEAR |
|---|---|---|---|---|---|---|---|---|---|---|---|---|---|
| Maximum Temp °F | 59.1 | 63.5 | 71.2 | 78.8 | 84.5 | 89.8 | ***91.7*** | ***91.1*** | 87.2 | 78.9 | 70.4 | 62.6 | 77.4 |
| Minimum Temp °F | 35.4 | 37.1 | ***42.4*** | ***49.5*** | ***57.5*** | 64.2 | ***67.8*** | ***67.0*** | 62.7 | ***50.9*** | 42.3 | ***37.4*** | 51.2 |
| Mean Temp °F | 47.3 | 50.4 | ***56.7*** | ***64.1*** | ***70.9*** | 77.0 | ***79.8*** | ***79.1*** | 74.8 | ***64.8*** | 56.4 | ***49.9*** | 64.3 |
| Days Max Temp ≥ 90 °F | 0 | 0 | 0 | 1 | 5 | 16 | 21 | 21 | 11 | 1 | 0 | 0 | 76 |
| Days Max Temp ≤ 32 °F | 0 | 0 | 0 | 0 | 0 | 0 | 0 | 0 | 0 | 0 | 0 | 0 | 0 |
| Days Min Temp ≤ 32 °F | 14 | 10 | 6 | 1 | ***0*** | 0 | 0 | 0 | 0 | 1 | ***7*** | ***11*** | 50 |
| Days Min Temp ≤ 0 °F | 0 | 0 | 0 | 0 | 0 | 0 | 0 | 0 | 0 | 0 | 0 | 0 | 0 |
| Heating Degree Days | 545 | ***418*** | ***271*** | ***95*** | ***12*** | 0 | ***0*** | ***0*** | 5 | ***89*** | 271 | ***466*** | 2172 |
| Cooling Degree Days | 1 | 5 | ***22*** | 63 | 205 | ***393*** | ***496*** | ***470*** | 311 | ***100*** | 20 | 4 | 2090 |
| Total Precipitation (") | 4.88 | 4.70 | 5.62 | 3.21 | 3.52 | 4.20 | 5.44 | 3.90 | 2.73 | 2.23 | 3.36 | 3.90 | 47.69 |
| Days ≥ 0.1" Precip | 6 | 6 | 7 | 5 | ***5*** | ***6*** | 8 | 6 | 4 | 4 | 4 | 5 | 66 |
| Total Snowfall (") | 0.0 | 0.1 | 0.0 | 0.0 | 0.0 | 0.0 | 0.0 | 0.0 | 0.0 | 0.0 | 0.0 | 0.0 | 0.1 |
| Days ≥ 1" Snow Depth | 0 | 0 | 0 | 0 | 0 | 0 | 0 | 0 | 0 | 0 | 0 | 0 | 0 |

## MACON LEWIS B WILSON *Bibb County* ELEVATION 354 ft LAT/LONG 32° 42 ' N / 83° 39 ' W

| | JAN | FEB | MAR | APR | MAY | JUN | JUL | AUG | SEP | OCT | NOV | DEC | YEAR |
|---|---|---|---|---|---|---|---|---|---|---|---|---|---|
| Maximum Temp °F | 56.9 | 61.0 | 69.5 | 77.8 | 84.2 | 90.1 | 92.3 | 91.1 | 86.4 | 77.9 | 68.7 | 60.5 | 76.4 |
| Minimum Temp °F | 35.0 | 37.3 | 44.4 | 51.4 | 59.5 | 67.3 | 71.1 | 70.2 | 64.7 | 52.4 | 43.5 | 37.8 | 52.9 |
| Mean Temp °F | 45.9 | 49.2 | 57.0 | 64.6 | 71.9 | 78.8 | 81.7 | 80.7 | 75.6 | 65.1 | 56.1 | 49.2 | 64.7 |
| Days Max Temp ≥ 90 °F | 0 | 0 | 0 | 1 | 6 | 18 | 22 | 21 | 11 | 1 | 0 | 0 | 80 |
| Days Max Temp ≤ 32 °F | 0 | 0 | 0 | 0 | 0 | 0 | 0 | 0 | 0 | 0 | 0 | 0 | 0 |
| Days Min Temp ≤ 32 °F | 14 | 10 | 4 | 0 | 0 | 0 | 0 | 0 | 0 | 0 | 5 | 11 | 44 |
| Days Min Temp ≤ 0 °F | 0 | 0 | 0 | 0 | 0 | 0 | 0 | 0 | 0 | 0 | 0 | 0 | 0 |
| Heating Degree Days | 585 | 443 | 265 | 90 | 11 | 0 | 0 | 0 | 5 | 87 | 278 | 489 | 2253 |
| Cooling Degree Days | 1 | 5 | 25 | 79 | 236 | 448 | 551 | 511 | 332 | 104 | 23 | 6 | 2321 |
| Total Precipitation (") | 4.58 | 4.58 | 5.07 | 3.02 | 3.35 | 3.73 | 4.60 | 3.99 | 2.69 | 2.52 | 2.98 | 4.06 | 45.17 |
| Days ≥ 0.1" Precip | 7 | 7 | 7 | 5 | 5 | 6 | 8 | 7 | 5 | 4 | 5 | 6 | 72 |
| Total Snowfall (") | 0.3 | 0.8 | 0.1 | 0.0 | 0.0 | 0.0 | 0.0 | 0.0 | 0.0 | 0.0 | 0.0 | 0.1 | 1.3 |
| Days ≥ 1" Snow Depth | 0 | 0 | 0 | 0 | 0 | 0 | 0 | 0 | 0 | 0 | 0 | 0 | 0 |

## METTER *Candler County* ELEVATION 120 ft LAT/LONG 32° 24 ' N / 82° 4 ' W

| | JAN | FEB | MAR | APR | MAY | JUN | JUL | AUG | SEP | OCT | NOV | DEC | YEAR |
|---|---|---|---|---|---|---|---|---|---|---|---|---|---|
| Maximum Temp °F | ***59.7*** | 64.5 | 72.7 | 80.8 | ***86.3*** | ***91.2*** | ***93.0*** | ***91.7*** | ***87.8*** | ***79.8*** | 71.5 | ***63.8*** | 78.6 |
| Minimum Temp °F | ***37.3*** | 40.3 | ***46.7*** | 53.6 | 61.2 | ***67.9*** | ***70.7*** | ***70.7*** | ***65.9*** | ***55.2*** | ***46.9*** | ***40.6*** | 54.8 |
| Mean Temp °F | ***48.5*** | 52.4 | ***59.7*** | 67.3 | ***73.7*** | ***79.6*** | ***81.9*** | ***81.2*** | ***76.9*** | ***67.5*** | ***59.1*** | ***52.3*** | 66.7 |
| Days Max Temp ≥ 90 °F | ***0*** | 0 | 0 | 2 | 8 | ***21*** | ***25*** | ***23*** | ***13*** | ***2*** | 0 | 0 | 94 |
| Days Max Temp ≤ 32 °F | 0 | 0 | 0 | 0 | 0 | 0 | 0 | 0 | ***0*** | 0 | 0 | 0 | 0 |
| Days Min Temp ≤ 32 °F | ***10*** | 6 | ***2*** | 0 | 0 | 0 | 0 | 0 | ***0*** | 0 | 3 | ***7*** | 28 |
| Days Min Temp ≤ 0 °F | 0 | 0 | 0 | 0 | 0 | 0 | 0 | 0 | ***0*** | 0 | 0 | 0 | 0 |
| Heating Degree Days | ***506*** | 358 | ***197*** | 51 | ***3*** | ***0*** | ***0*** | ***0*** | ***1*** | ***51*** | ***205*** | ***396*** | 1768 |
| Cooling Degree Days | ***3*** | 14 | 43 | ***118*** | ***282*** | 464 | 547 | 506 | 350 | 137 | 43 | 10 | 2517 |
| Total Precipitation (") | ***4.96*** | 3.59 | 4.54 | 2.92 | 4.43 | 4.30 | ***6.52*** | ***5.70*** | ***3.33*** | ***3.19*** | 2.97 | ***3.53*** | 49.98 |
| Days ≥ 0.1" Precip | ***7*** | ***6*** | ***6*** | 4 | 6 | ***6*** | ***9*** | ***8*** | ***5*** | ***4*** | 4 | ***5*** | 70 |
| Total Snowfall (") | 0.1 | 0.4 | 0.0 | 0.0 | 0.0 | 0.0 | 0.0 | 0.0 | 0.0 | 0.0 | 0.0 | 0.1 | 0.6 |
| Days ≥ 1" Snow Depth | ***0*** | ***0*** | 0 | 0 | 0 | 0 | 0 | 0 | 0 | 0 | 0 | 0 | 0 |

## MIDVILLE EXPERIMENT *Burke County* ELEVATION 279 ft LAT/LONG 32° 53 ' N / 82° 12 ' W

| | JAN | FEB | MAR | APR | MAY | JUN | JUL | AUG | SEP | OCT | NOV | DEC | YEAR |
|---|---|---|---|---|---|---|---|---|---|---|---|---|---|
| Maximum Temp °F | ***58.2*** | 61.7 | 69.5 | 77.4 | 83.8 | 89.6 | 92.0 | 90.3 | 86.4 | 77.9 | ***69.1*** | ***60.6*** | 76.4 |
| Minimum Temp °F | ***34.4*** | 36.8 | 44.2 | 51.9 | 59.8 | 66.2 | 69.2 | 68.3 | ***63.6*** | ***52.7*** | ***43.9*** | ***37.2*** | 52.4 |
| Mean Temp °F | na | 49.4 | ***57.0*** | 64.8 | 71.8 | 78.0 | 80.6 | 79.4 | ***75.0*** | ***65.3*** | ***56.5*** | ***48.8*** | na |
| Days Max Temp ≥ 90 °F | 0 | 0 | 0 | 1 | 5 | 16 | 22 | 19 | 10 | 1 | 0 | 0 | 74 |
| Days Max Temp ≤ 32 °F | 0 | 0 | 0 | 0 | 0 | 0 | 0 | 0 | 0 | 0 | 0 | 0 | 0 |
| Days Min Temp ≤ 32 °F | 14 | 10 | 3 | 0 | 0 | 0 | 0 | 0 | 0 | 1 | 5 | 10 | 43 |
| Days Min Temp ≤ 0 °F | 0 | 0 | 0 | 0 | 0 | 0 | 0 | 0 | 0 | 0 | 0 | 0 | 0 |
| Heating Degree Days | na | ***439*** | ***260*** | ***84*** | ***12*** | 1 | 0 | 0 | ***4*** | ***82*** | ***266*** | na | na |
| Cooling Degree Days | ***0*** | 5 | 19 | ***70*** | 220 | 417 | 506 | 462 | 318 | 98 | 22 | ***4*** | 2141 |
| Total Precipitation (") | 4.62 | 4.00 | 4.61 | 2.79 | 3.67 | 3.83 | 4.43 | 5.22 | 2.96 | 2.76 | 2.81 | 3.28 | 44.98 |
| Days ≥ 0.1" Precip | 6 | 6 | 6 | 5 | 6 | 6 | 7 | 7 | 4 | 3 | 4 | ***5*** | 65 |
| Total Snowfall (") | 0.1 | 0.5 | 0.0 | 0.0 | 0.0 | 0.0 | 0.0 | 0.0 | 0.0 | 0.0 | 0.0 | 0.0 | 0.6 |
| Days ≥ 1" Snow Depth | 0 | 0 | 0 | 0 | 0 | 0 | 0 | 0 | 0 | 0 | 0 | 0 | 0 |

## MILLEDGEVILLE *Baldwin County* ELEVATION 322 ft LAT/LONG 33° 5 ' N / 83° 13 ' W

| | JAN | FEB | MAR | APR | MAY | JUN | JUL | AUG | SEP | OCT | NOV | DEC | YEAR |
|---|---|---|---|---|---|---|---|---|---|---|---|---|---|
| Maximum Temp °F | 55.6 | 60.2 | 68.8 | 77.0 | 82.5 | 89.0 | 91.8 | 90.1 | 85.7 | 76.9 | 67.9 | 59.3 | 75.4 |
| Minimum Temp °F | 31.3 | 32.9 | 39.6 | 47.1 | 55.6 | 64.1 | 68.2 | 67.6 | 62.2 | 49.1 | 40.1 | 33.7 | 49.3 |
| Mean Temp °F | 43.5 | 46.6 | 54.3 | 62.2 | 69.1 | 76.5 | 80.0 | 78.9 | 74.0 | 63.0 | 54.0 | 46.5 | 62.4 |
| Days Max Temp ≥ 90 °F | 0 | 0 | 0 | 1 | 4 | 14 | 21 | 18 | 10 | 1 | 0 | 0 | 69 |
| Days Max Temp ≤ 32 °F | 1 | 0 | 0 | 0 | 0 | 0 | 0 | 0 | 0 | 0 | 0 | 0 | 1 |
| Days Min Temp ≤ 32 °F | 18 | 15 | 8 | 1 | 0 | 0 | 0 | 0 | 0 | 1 | 8 | 15 | 66 |
| Days Min Temp ≤ 0 °F | 0 | 0 | 0 | 0 | 0 | 0 | 0 | 0 | 0 | 0 | 0 | 0 | 0 |
| Heating Degree Days | 661 | 516 | 339 | 132 | 27 | 1 | 0 | 0 | 7 | 119 | 334 | 568 | 2704 |
| Cooling Degree Days | 0 | 2 | 14 | 55 | 169 | 391 | 502 | 452 | 291 | 72 | 13 | 3 | 1964 |
| Total Precipitation (") | 4.52 | 4.31 | 5.08 | 3.08 | 3.54 | 3.80 | 4.04 | 4.84 | 2.94 | 2.89 | 3.33 | 3.79 | 46.16 |
| Days ≥ 0.1" Precip | 7 | 6 | 7 | 5 | 6 | 6 | 8 | 7 | 5 | 4 | 5 | 6 | 72 |
| Total Snowfall (") | 0.1 | 0.8 | 0.1 | 0.0 | 0.0 | 0.0 | 0.0 | 0.0 | 0.0 | 0.0 | 0.0 | 0.1 | 1.1 |
| Days ≥ 1" Snow Depth | 0 | 0 | 0 | 0 | 0 | 0 | 0 | 0 | 0 | 0 | 0 | 0 | 0 |

## MILLEN 4 N *Jenkins County* ELEVATION 200 ft LAT/LONG 32° 48 ' N / 81° 56 ' W

| | JAN | FEB | MAR | APR | MAY | JUN | JUL | AUG | SEP | OCT | NOV | DEC | YEAR |
|---|---|---|---|---|---|---|---|---|---|---|---|---|---|
| Maximum Temp °F | 58.5 | 63.1 | 71.4 | 78.7 | 84.8 | 90.2 | 92.4 | 91.3 | 87.3 | 78.5 | 70.0 | 62.1 | 77.4 |
| Minimum Temp °F | 34.7 | 36.7 | 43.6 | 49.3 | 57.5 | 65.0 | 68.5 | 68.3 | 63.0 | 51.6 | 42.3 | 36.9 | 51.5 |
| Mean Temp °F | 46.6 | 50.0 | 57.5 | 64.1 | 71.2 | 77.6 | 80.5 | 79.8 | 75.2 | 65.1 | 56.2 | 49.5 | 64.4 |
| Days Max Temp ≥ 90 °F | 0 | 0 | 0 | 2 | 6 | 18 | 24 | 22 | 12 | 1 | 0 | 0 | 85 |
| Days Max Temp ≤ 32 °F | 0 | 0 | 0 | 0 | 0 | 0 | 0 | 0 | 0 | 0 | 0 | 0 | 0 |
| Days Min Temp ≤ 32 °F | 14 | *11* | 5 | 1 | 0 | 0 | 0 | 0 | 0 | 1 | 7 | 12 | 51 |
| Days Min Temp ≤ 0 °F | 0 | 0 | 0 | 0 | 0 | 0 | 0 | 0 | 0 | 0 | 0 | 0 | 0 |
| Heating Degree Days | 564 | 422 | 249 | 91 | 13 | 0 | 0 | 0 | 4 | 89 | 281 | 480 | 2193 |
| Cooling Degree Days | 2 | 8 | 29 | 69 | 217 | 411 | 520 | 484 | 316 | 105 | 27 | 9 | 2197 |
| Total Precipitation (") | 4.12 | 3.99 | 4.35 | 2.58 | 3.79 | 4.25 | 4.61 | 4.56 | 2.76 | 2.87 | 2.55 | 3.51 | 43.94 |
| Days ≥ 0.1" Precip | 7 | 6 | *7* | 5 | 6 | 7 | 8 | 7 | 5 | 4 | 4 | 6 | 72 |
| Total Snowfall (") | 0.0 | 0.6 | 0.0 | 0.0 | 0.0 | 0.0 | 0.0 | 0.0 | 0.0 | 0.0 | 0.0 | 0.1 | 0.7 |
| Days ≥ 1" Snow Depth | 0 | 0 | 0 | 0 | 0 | 0 | 0 | 0 | 0 | 0 | 0 | 0 | 0 |

## MONTICELLO *Jasper County* ELEVATION 679 ft LAT/LONG 33° 18 ' N / 83° 41 ' W

| | JAN | FEB | MAR | APR | MAY | JUN | JUL | AUG | SEP | OCT | NOV | DEC | YEAR |
|---|---|---|---|---|---|---|---|---|---|---|---|---|---|
| Maximum Temp °F | 53.8 | 58.5 | 67.4 | 75.5 | 81.4 | 87.6 | 90.4 | 89.1 | 84.5 | 75.3 | 66.3 | 57.6 | 74.0 |
| Minimum Temp °F | 33.5 | 35.5 | 42.6 | 50.6 | 58.2 | 66.0 | 69.5 | 68.8 | 63.5 | 51.9 | 42.9 | 36.1 | 51.6 |
| Mean Temp °F | 43.6 | 47.0 | 55.0 | 63.1 | 69.8 | 76.8 | 80.0 | 79.0 | 74.0 | 63.6 | 54.6 | 46.9 | 62.8 |
| Days Max Temp ≥ 90 °F | 0 | 0 | 0 | 0 | 2 | 12 | 19 | 16 | 7 | 0 | 0 | 0 | 56 |
| Days Max Temp ≤ 32 °F | 1 | 0 | 0 | 0 | 0 | 0 | 0 | 0 | 0 | 0 | 0 | 0 | 1 |
| Days Min Temp ≤ 32 °F | 15 | 12 | 5 | 0 | 0 | 0 | 0 | 0 | 0 | 0 | 5 | 12 | 49 |
| Days Min Temp ≤ 0 °F | 0 | 0 | 0 | 0 | 0 | 0 | 0 | 0 | 0 | 0 | 0 | 0 | 0 |
| Heating Degree Days | 656 | 504 | 317 | 117 | 21 | 1 | 0 | 0 | 6 | 105 | 314 | 556 | 2597 |
| Cooling Degree Days | 0 | 2 | 16 | 63 | 181 | 379 | 489 | 449 | 287 | 80 | 11 | 1 | 1958 |
| Total Precipitation (") | 4.35 | 4.31 | 5.37 | 3.45 | 3.88 | 3.46 | 4.56 | 4.42 | 2.86 | 2.87 | 3.35 | 3.95 | 46.83 |
| Days ≥ 0.1" Precip | 7 | 6 | 7 | 5 | 6 | 6 | 6 | 6 | 5 | 4 | 5 | 6 | 69 |
| Total Snowfall (") | 0.3 | 0.6 | 0.1 | 0.0 | 0.0 | 0.0 | 0.0 | 0.0 | 0.0 | 0.0 | 0.0 | 0.0 | 1.0 |
| Days ≥ 1" Snow Depth | 0 | 0 | 0 | 0 | 0 | 0 | 0 | 0 | 0 | 0 | 0 | 0 | 0 |

## MOULTRIE 2 ESE *Colquitt County* ELEVATION 331 ft LAT/LONG 31° 11 ' N / 83° 47 ' W

| | JAN | FEB | MAR | APR | MAY | JUN | JUL | AUG | SEP | OCT | NOV | DEC | YEAR |
|---|---|---|---|---|---|---|---|---|---|---|---|---|---|
| Maximum Temp °F | 61.0 | 64.7 | 72.5 | 79.7 | 85.4 | 90.3 | 92.2 | 91.2 | 87.7 | 79.9 | 71.6 | 64.5 | 78.4 |
| Minimum Temp °F | 39.1 | 41.0 | 47.8 | 53.8 | 61.0 | 67.3 | 70.3 | 69.8 | 65.9 | 55.7 | 47.3 | 41.4 | 55.0 |
| Mean Temp °F | 50.1 | 52.9 | 60.2 | 66.8 | 73.3 | 78.8 | 81.3 | 80.5 | 76.8 | 67.8 | 59.5 | 52.9 | 66.7 |
| Days Max Temp ≥ 90 °F | 0 | 0 | 0 | 1 | 6 | 18 | 24 | 23 | 13 | 2 | 0 | 0 | 87 |
| Days Max Temp ≤ 32 °F | 0 | 0 | 0 | 0 | 0 | 0 | 0 | 0 | 0 | 0 | 0 | 0 | 0 |
| Days Min Temp ≤ 32 °F | 9 | 6 | 2 | 0 | 0 | 0 | 0 | 0 | 0 | 0 | 3 | 7 | 27 |
| Days Min Temp ≤ 0 °F | 0 | 0 | 0 | 0 | 0 | 0 | 0 | 0 | 0 | 0 | 0 | 0 | 0 |
| Heating Degree Days | 459 | 344 | 185 | 53 | 3 | 0 | 0 | 0 | 1 | 49 | 198 | 380 | 1672 |
| Cooling Degree Days | 3 | 12 | 42 | 103 | 266 | 447 | 540 | 509 | 366 | 155 | 52 | 14 | 2509 |
| Total Precipitation (") | 5.51 | 4.68 | 5.48 | 3.11 | 3.92 | 5.55 | 5.91 | 5.11 | 3.19 | 2.38 | 3.00 | 4.01 | 51.85 |
| Days ≥ 0.1" Precip | 7 | 6 | 7 | 5 | 6 | 8 | 10 | 9 | 5 | 4 | 5 | 6 | 78 |
| Total Snowfall (") | 0.0 | 0.0 | 0.0 | 0.0 | 0.0 | 0.0 | 0.0 | 0.0 | 0.0 | 0.0 | 0.0 | 0.0 | 0.0 |
| Days ≥ 1" Snow Depth | 0 | 0 | 0 | 0 | 0 | 0 | 0 | 0 | 0 | 0 | 0 | 0 | 0 |

## NAHUNTA 3 E *Brantley County* ELEVATION 78 ft LAT/LONG 31° 13 ' N / 81° 56 ' W

| | JAN | FEB | MAR | APR | MAY | JUN | JUL | AUG | SEP | OCT | NOV | DEC | YEAR |
|---|---|---|---|---|---|---|---|---|---|---|---|---|---|
| Maximum Temp °F | 63.6 | 66.5 | 73.5 | 80.2 | 85.1 | 89.2 | *91.3* | 89.9 | 86.6 | 80.0 | 72.7 | 66.3 | 78.7 |
| Minimum Temp °F | 38.6 | 40.6 | 46.8 | 52.2 | 59.3 | 66.9 | *70.0* | 70.0 | 66.6 | *56.5* | 47.3 | 41.4 | 54.7 |
| Mean Temp °F | 51.3 | 53.5 | 60.2 | 66.2 | 72.2 | 78.1 | *80.7* | 80.0 | 76.6 | *68.3* | 60.0 | 54.0 | 66.8 |
| Days Max Temp ≥ 90 °F | 0 | 0 | 0 | 1 | 5 | 14 | 21 | 18 | 8 | 1 | 0 | 0 | 68 |
| Days Max Temp ≤ 32 °F | 0 | 0 | 0 | 0 | 0 | 0 | 0 | 0 | 0 | 0 | 0 | 0 | 0 |
| Days Min Temp ≤ 32 °F | 10 | 7 | 3 | 0 | 0 | 0 | 0 | 0 | 0 | 0 | 3 | 7 | 30 |
| Days Min Temp ≤ 0 °F | 0 | 0 | 0 | 0 | 0 | 0 | 0 | 0 | 0 | 0 | 0 | 0 | 0 |
| Heating Degree Days | 424 | 329 | 183 | 59 | 5 | 0 | *0* | 0 | 1 | *45* | 186 | 348 | 1580 |
| Cooling Degree Days | *3* | 12 | 37 | 92 | 233 | 422 | *518* | 481 | *353* | na | *51* | *15* | na |
| Total Precipitation (") | 4.13 | 3.69 | 4.89 | 2.61 | 3.92 | 5.88 | 6.45 | 8.14 | 3.91 | 2.86 | 2.58 | 3.21 | 52.27 |
| Days ≥ 0.1" Precip | 6 | 5 | 6 | 4 | 5 | 8 | 8 | 9 | 6 | 4 | 4 | 5 | 70 |
| Total Snowfall (") | 0.0 | 0.0 | 0.0 | 0.0 | 0.0 | 0.0 | 0.0 | 0.0 | 0.0 | 0.0 | 0.0 | 0.0 | 0.0 |
| Days ≥ 1" Snow Depth | 0 | 0 | 0 | 0 | 0 | 0 | 0 | 0 | 0 | 0 | 0 | 0 | 0 |

## NEWNAN 4 NE *Coweta County* ELEVATION 981 ft LAT/LONG 33° 23 ' N / 84° 48 ' W

| | JAN | FEB | MAR | APR | MAY | JUN | JUL | AUG | SEP | OCT | NOV | DEC | YEAR |
|---|---|---|---|---|---|---|---|---|---|---|---|---|---|
| Maximum Temp °F | 53.2 | 58.3 | 67.4 | 75.8 | 81.4 | 87.2 | 89.6 | 88.4 | 83.6 | 74.4 | 65.1 | 56.7 | 73.4 |
| Minimum Temp °F | 31.9 | 34.2 | 41.5 | 48.5 | 56.2 | 63.5 | 67.1 | 66.2 | 60.9 | 49.5 | 41.3 | 35.2 | 49.7 |
| Mean Temp °F | 42.6 | 46.3 | 54.5 | 62.2 | 68.8 | 75.4 | 78.4 | 77.3 | 72.3 | 61.9 | 53.3 | 46.0 | 61.6 |
| Days Max Temp ≥ 90 °F | 0 | 0 | 0 | 0 | 2 | 9 | 16 | 13 | 5 | 0 | 0 | 0 | 45 |
| Days Max Temp ≤ 32 °F | 1 | 0 | 0 | 0 | 0 | 0 | 0 | 0 | 0 | 0 | 0 | 0 | 1 |
| Days Min Temp ≤ 32 °F | 17 | 13 | 7 | 2 | 0 | 0 | 0 | 0 | 0 | 1 | 7 | 14 | 61 |
| Days Min Temp ≤ 0 °F | 0 | 0 | 0 | 0 | 0 | 0 | 0 | 0 | 0 | 0 | 0 | 0 | 0 |
| Heating Degree Days | 688 | 523 | 330 | 130 | 25 | 1 | 0 | 0 | 12 | 138 | 351 | 584 | 2782 |
| Cooling Degree Days | 0 | 1 | 10 | 43 | 137 | 328 | 434 | 394 | 236 | 54 | 6 | 2 | 1645 |
| Total Precipitation (") | 5.17 | 4.87 | 5.74 | 4.28 | 4.72 | 3.99 | 5.12 | 4.08 | 2.98 | 3.28 | 4.10 | 4.40 | 52.73 |
| Days ≥ 0.1" Precip | 9 | 7 | 8 | 6 | 7 | 7 | 8 | 6 | 5 | 4 | 6 | 7 | 80 |
| Total Snowfall (") | 0.7 | 0.3 | 0.5 | 0.0 | 0.0 | 0.0 | 0.0 | 0.0 | 0.0 | 0.0 | 0.0 | 0.1 | 1.6 |
| Days ≥ 1" Snow Depth | 0 | 0 | 0 | 0 | 0 | 0 | 0 | 0 | 0 | 0 | 0 | 0 | 0 |

## PLAINS SW GA EXP STN *Sumter County* ELEVATION 500 ft LAT/LONG 32° 3 ' N / 84° 22 ' W

| | JAN | FEB | MAR | APR | MAY | JUN | JUL | AUG | SEP | OCT | NOV | DEC | YEAR |
|---|---|---|---|---|---|---|---|---|---|---|---|---|---|
| Maximum Temp °F | 56.4 | 60.5 | 68.4 | 76.8 | 83.1 | 88.8 | 90.6 | 89.8 | 86.0 | 77.6 | 68.5 | 60.3 | 75.6 |
| Minimum Temp °F | 34.5 | 36.9 | 44.5 | 51.9 | 59.7 | 66.6 | 69.6 | 68.7 | 64.0 | 52.4 | 44.3 | 37.6 | 52.6 |
| Mean Temp °F | 45.5 | 48.7 | 56.5 | 64.4 | 71.4 | 77.8 | 80.1 | 79.3 | 75.0 | 65.0 | 56.4 | 49.0 | 64.1 |
| Days Max Temp ≥ 90 °F | 0 | 0 | 0 | 0 | 3 | 15 | 19 | 17 | 9 | 1 | 0 | 0 | 64 |
| Days Max Temp ≤ 32 °F | 1 | 0 | 0 | 0 | 0 | 0 | 0 | 0 | 0 | 0 | 0 | 0 | 1 |
| Days Min Temp ≤ 32 °F | 15 | 11 | 3 | 0 | 0 | 0 | 0 | 0 | 0 | 0 | 4 | 11 | 44 |
| Days Min Temp ≤ 0 °F | 0 | 0 | 0 | 0 | 0 | 0 | 0 | 0 | 0 | 0 | 0 | 0 | 0 |
| Heating Degree Days | 600 | 457 | 280 | 94 | 11 | 0 | 0 | 0 | 5 | 88 | 272 | 496 | 2303 |
| Cooling Degree Days | 1 | 5 | 24 | 73 | 219 | 411 | 495 | 464 | 311 | 102 | 24 | 6 | 2135 |
| Total Precipitation (") | 5.19 | 4.66 | 5.39 | 2.96 | 3.64 | 4.85 | 5.68 | 4.13 | 2.77 | 2.26 | 3.48 | 3.93 | 48.94 |
| Days ≥ 0.1" Precip | 8 | 7 | 7 | 5 | 6 | 7 | 9 | 7 | 5 | 4 | 5 | 6 | 76 |
| Total Snowfall (") | 0.1 | 0.3 | 0.0 | 0.0 | 0.0 | 0.0 | 0.0 | 0.0 | 0.0 | 0.0 | 0.0 | 0.0 | 0.4 |
| Days ≥ 1" Snow Depth | 0 | 0 | 0 | 0 | 0 | 0 | 0 | 0 | 0 | 0 | 0 | 0 | 0 |

## QUITMAN 2 NW *Brooks County* ELEVATION 180 ft LAT/LONG 30° 47 ' N / 83° 33 ' W

| | JAN | FEB | MAR | APR | MAY | JUN | JUL | AUG | SEP | OCT | NOV | DEC | YEAR |
|---|---|---|---|---|---|---|---|---|---|---|---|---|---|
| Maximum Temp °F | 61.6 | 65.6 | 72.8 | 80.0 | 85.9 | 90.7 | 91.9 | 91.5 | 88.7 | 80.6 | 72.6 | 64.7 | 78.9 |
| Minimum Temp °F | 37.5 | 39.9 | 46.6 | 52.8 | 60.1 | 66.9 | 69.5 | 68.9 | 64.7 | 53.6 | 45.6 | 39.6 | 53.8 |
| Mean Temp °F | 49.5 | 52.8 | 59.7 | 66.5 | 73.0 | 78.8 | 80.7 | 80.2 | 76.7 | 67.1 | 59.1 | 52.2 | 66.4 |
| Days Max Temp ≥ 90 °F | 0 | 0 | 0 | 1 | 7 | 20 | 25 | 24 | 16 | 3 | 0 | 0 | 96 |
| Days Max Temp ≤ 32 °F | 0 | 0 | 0 | 0 | 0 | 0 | 0 | 0 | 0 | 0 | 0 | 0 | 0 |
| Days Min Temp ≤ 32 °F | 11 | 7 | 2 | 0 | 0 | 0 | 0 | 0 | 0 | 0 | 3 | 9 | 32 |
| Days Min Temp ≤ 0 °F | 0 | 0 | 0 | 0 | 0 | 0 | 0 | 0 | 0 | 0 | 0 | 0 | 0 |
| Heating Degree Days | 480 | 346 | 195 | 57 | 4 | 0 | 0 | 0 | 1 | 58 | 207 | 404 | 1752 |
| Cooling Degree Days | 3 | 12 | 36 | 94 | 258 | 436 | 499 | 478 | 340 | 131 | 43 | 12 | 2342 |
| Total Precipitation (") | 5.55 | 4.60 | 5.42 | 3.34 | 3.65 | 5.38 | 6.60 | 5.58 | 3.17 | 2.62 | 3.08 | 4.11 | 53.10 |
| Days ≥ 0.1" Precip | *6* | *6* | *5* | 4 | 5 | 7 | 9 | 6 | 4 | 3 | *4* | *5* | 64 |
| Total Snowfall (") | 0.0 | 0.0 | 0.0 | 0.0 | 0.0 | 0.0 | 0.0 | 0.0 | 0.0 | 0.0 | 0.0 | 0.1 | 0.1 |
| Days ≥ 1" Snow Depth | 0 | 0 | 0 | 0 | 0 | 0 | 0 | 0 | 0 | 0 | 0 | 0 | 0 |

## ROME *Floyd County* ELEVATION 620 ft LAT/LONG 34° 15 ' N / 85° 10 ' W

| | JAN | FEB | MAR | APR | MAY | JUN | JUL | AUG | SEP | OCT | NOV | DEC | YEAR |
|---|---|---|---|---|---|---|---|---|---|---|---|---|---|
| Maximum Temp °F | 50.3 | 55.8 | 65.0 | 73.5 | 78.8 | 85.0 | 87.8 | 86.8 | 81.8 | 72.3 | 63.2 | 54.0 | 71.2 |
| Minimum Temp °F | 28.8 | 31.0 | 38.6 | 46.1 | 54.4 | 62.5 | 67.0 | 66.1 | 60.0 | 46.8 | 38.5 | 31.8 | 47.6 |
| Mean Temp °F | 39.6 | 43.4 | 51.8 | 59.9 | 66.7 | 73.7 | 77.4 | 76.5 | 70.9 | 59.6 | 50.9 | 42.9 | 59.4 |
| Days Max Temp ≥ 90 °F | 0 | 0 | 0 | 0 | 0 | 5 | 11 | 9 | 3 | 0 | 0 | 0 | 28 |
| Days Max Temp ≤ 32 °F | 1 | 0 | 0 | 0 | 0 | 0 | 0 | 0 | 0 | 0 | 0 | 0 | 1 |
| Days Min Temp ≤ 32 °F | 20 | 17 | 9 | 2 | 0 | 0 | 0 | 0 | 0 | 2 | 10 | 18 | 78 |
| Days Min Temp ≤ 0 °F | 0 | 0 | 0 | 0 | 0 | 0 | 0 | 0 | 0 | 0 | 0 | 0 | 0 |
| Heating Degree Days | 781 | 604 | 406 | 180 | 48 | 3 | 0 | 0 | 20 | 190 | 420 | 679 | 3331 |
| Cooling Degree Days | 0 | 0 | 4 | 25 | 104 | 287 | 412 | 366 | 194 | 27 | 3 | 1 | 1423 |
| Total Precipitation (") | 5.09 | 4.89 | 6.56 | 4.77 | 4.55 | 3.96 | 4.73 | 4.58 | 3.90 | 3.15 | 4.23 | 4.61 | 55.02 |
| Days ≥ 0.1" Precip | 9 | 7 | 8 | 7 | 7 | 6 | 7 | 6 | 6 | 5 | 7 | 7 | 82 |
| Total Snowfall (") | 0.7 | 0.3 | 0.5 | 0.2 | 0.0 | 0.0 | 0.0 | 0.0 | 0.0 | 0.0 | 0.0 | 0.1 | 1.8 |
| Days ≥ 1" Snow Depth | 0 | 0 | 0 | 0 | 0 | 0 | 0 | 0 | 0 | 0 | 0 | 0 | 0 |

## SANDERSVILLE *Washington County* ELEVATION 410 ft LAT/LONG 32° 57 ' N / 82° 46 ' W

| | JAN | FEB | MAR | APR | MAY | JUN | JUL | AUG | SEP | OCT | NOV | DEC | YEAR |
|---|---|---|---|---|---|---|---|---|---|---|---|---|---|
| Maximum Temp °F | 55.2 | 60.0 | 68.2 | 76.1 | 82.3 | 88.2 | 90.4 | 88.8 | 84.3 | 75.6 | 66.5 | 58.7 | 74.5 |
| Minimum Temp °F | 33.9 | 35.9 | 43.0 | 50.1 | 57.8 | 65.4 | 69.3 | 68.3 | 63.1 | 51.5 | 42.4 | 36.5 | 51.4 |
| Mean Temp °F | 44.6 | 48.0 | 55.6 | 63.1 | 70.1 | 76.8 | 79.9 | 78.6 | 73.7 | 63.6 | 54.5 | 47.6 | 63.0 |
| Days Max Temp ≥ 90 °F | 0 | 0 | 0 | 0 | 3 | 12 | 18 | 14 | 6 | 0 | 0 | 0 | 53 |
| Days Max Temp ≤ 32 °F | 1 | 0 | 0 | 0 | 0 | 0 | 0 | 0 | 0 | 0 | 0 | 0 | 1 |
| Days Min Temp ≤ 32 °F | 15 | 11 | 5 | 1 | 0 | 0 | 0 | 0 | 0 | 1 | 6 | 12 | 51 |
| Days Min Temp ≤ 0 °F | 0 | 0 | 0 | 0 | 0 | 0 | 0 | 0 | 0 | 0 | 0 | 0 | 0 |
| Heating Degree Days | 628 | 476 | 301 | 115 | 21 | 1 | 0 | 0 | 8 | 113 | 320 | 536 | 2519 |
| Cooling Degree Days | 0 | 3 | 14 | 54 | 171 | 376 | 491 | 438 | 273 | 77 | 12 | 3 | 1912 |
| Total Precipitation (") | 4.57 | 4.34 | 5.01 | 3.24 | 3.13 | 4.14 | 4.77 | 4.89 | 3.02 | 2.61 | 3.00 | 3.62 | 46.34 |
| Days ≥ 0.1" Precip | 8 | 7 | 7 | 5 | 6 | 7 | 8 | 7 | 5 | 4 | 5 | 6 | 75 |
| Total Snowfall (") | 0.3 | 0.8 | 0.1 | 0.0 | 0.0 | 0.0 | 0.0 | 0.0 | 0.0 | 0.0 | 0.0 | 0.1 | 1.3 |
| Days ≥ 1" Snow Depth | 0 | 0 | 0 | 0 | 0 | 0 | 0 | 0 | 0 | 0 | 0 | 0 | 0 |

## SAPELO ISLAND *McIntosh County* ELEVATION 10 ft LAT/LONG 31° 24 ' N / 81° 17 ' W

| | JAN | FEB | MAR | APR | MAY | JUN | JUL | AUG | SEP | OCT | NOV | DEC | YEAR |
|---|---|---|---|---|---|---|---|---|---|---|---|---|---|
| Maximum Temp °F | 60.5 | 62.8 | 69.1 | 76.1 | 81.4 | 86.6 | 89.8 | 88.5 | 84.9 | 77.8 | 70.2 | 63.2 | 75.9 |
| Minimum Temp °F | 40.7 | 42.2 | 48.9 | 55.7 | 63.4 | 69.9 | 72.6 | 72.4 | 69.6 | 60.3 | 50.9 | 43.7 | 57.5 |
| Mean Temp °F | 50.6 | 52.5 | 59.0 | 65.9 | 72.5 | 78.3 | 81.2 | 80.5 | 77.3 | 69.1 | 60.5 | 53.5 | 66.7 |
| Days Max Temp ≥ 90 °F | 0 | 0 | 0 | 0 | 2 | 8 | 16 | 12 | 4 | 0 | 0 | 0 | 42 |
| Days Max Temp ≤ 32 °F | 0 | 0 | 0 | 0 | 0 | 0 | 0 | 0 | 0 | 0 | 0 | 0 | 0 |
| Days Min Temp ≤ 32 °F | 7 | 5 | 1 | 0 | 0 | 0 | 0 | 0 | 0 | 0 | 1 | 4 | 18 |
| Days Min Temp ≤ 0 °F | 0 | 0 | 0 | 0 | 0 | 0 | 0 | 0 | 0 | 0 | 0 | 0 | 0 |
| Heating Degree Days | 443 | 350 | 202 | 57 | 4 | 0 | 0 | 0 | 0 | 32 | 166 | 357 | 1611 |
| Cooling Degree Days | 1 | 5 | 22 | 76 | 232 | 423 | 531 | 497 | 372 | 163 | 47 | 10 | 2379 |
| Total Precipitation (") | 4.31 | 3.47 | 4.36 | 2.81 | 3.41 | 4.61 | 5.74 | 7.32 | 6.91 | 3.81 | 2.83 | 3.14 | 52.72 |
| Days ≥ 0.1" Precip | 6 | 6 | 6 | 4 | 5 | 7 | 8 | 9 | 8 | 5 | 4 | 5 | 73 |
| Total Snowfall (") | 0.0 | 0.1 | 0.0 | 0.0 | 0.0 | 0.0 | 0.0 | 0.0 | 0.0 | 0.0 | 0.0 | 0.2 | 0.3 |
| Days ≥ 1" Snow Depth | 0 | 0 | 0 | 0 | 0 | 0 | 0 | 0 | 0 | 0 | 0 | 0 | 0 |

## SAVANNAH INTL AP *Chatham County* ELEVATION 46 ft LAT/LONG 32° 8 ' N / 81° 12 ' W

| | JAN | FEB | MAR | APR | MAY | JUN | JUL | AUG | SEP | OCT | NOV | DEC | YEAR |
|---|---|---|---|---|---|---|---|---|---|---|---|---|---|
| Maximum Temp °F | 59.7 | 63.2 | 70.7 | 78.1 | 84.1 | 89.2 | 91.9 | 90.1 | 86.1 | 78.2 | 70.4 | 63.0 | 77.1 |
| Minimum Temp °F | 38.4 | 40.9 | 47.8 | 54.3 | 62.4 | 69.3 | 72.7 | 72.3 | 68.1 | 57.1 | 47.7 | 41.2 | 56.0 |
| Mean Temp °F | 49.1 | 52.1 | 59.3 | 66.2 | 73.3 | 79.3 | 82.3 | 81.2 | 77.1 | 67.7 | 59.1 | 52.1 | 66.6 |
| Days Max Temp ≥ 90 °F | 0 | 0 | 0 | 1 | 5 | 14 | 22 | 18 | 8 | 1 | 0 | 0 | 69 |
| Days Max Temp ≤ 32 °F | 0 | 0 | 0 | 0 | 0 | 0 | 0 | 0 | 0 | 0 | 0 | 0 | 0 |
| Days Min Temp ≤ 32 °F | 10 | 7 | 2 | 0 | 0 | 0 | 0 | 0 | 0 | 0 | 2 | 7 | 28 |
| Days Min Temp ≤ 0 °F | 0 | 0 | 0 | 0 | 0 | 0 | 0 | 0 | 0 | 0 | 0 | 0 | 0 |
| Heating Degree Days | 491 | 369 | 209 | 65 | 5 | 0 | 0 | 0 | 1 | 55 | 210 | 404 | 1809 |
| Cooling Degree Days | 4 | 14 | 39 | 105 | 276 | 469 | 582 | 530 | 378 | 150 | 47 | 13 | 2607 |
| Total Precipitation (") | 3.89 | 2.83 | 3.99 | 3.03 | 4.20 | 5.58 | 6.00 | 7.22 | 4.65 | 3.05 | 2.43 | 2.94 | 49.81 |
| Days ≥ 0.1" Precip | 7 | 5 | 6 | 4 | 6 | 8 | 9 | 9 | 6 | 4 | 4 | 5 | 73 |
| Total Snowfall (") | 0.1 | 0.3 | 0.0 | 0.0 | 0.0 | 0.0 | 0.0 | 0.0 | 0.0 | 0.0 | 0.0 | 0.1 | 0.5 |
| Days ≥ 1" Snow Depth | 0 | 0 | 0 | 0 | 0 | 0 | 0 | 0 | 0 | 0 | 0 | 0 | 0 |

## SILOAM 3 N *Greene County* ELEVATION 690 ft LAT/LONG 33° 32 ' N / 83° 6 ' W

| | JAN | FEB | MAR | APR | MAY | JUN | JUL | AUG | SEP | OCT | NOV | DEC | YEAR |
|---|---|---|---|---|---|---|---|---|---|---|---|---|---|
| Maximum Temp °F | 53.9 | 58.6 | 67.2 | 75.3 | 81.7 | 88.4 | 90.9 | 89.4 | 84.6 | 75.3 | 66.1 | 57.4 | 74.1 |
| Minimum Temp °F | 33.1 | 35.3 | 42.5 | 49.5 | 57.1 | 64.6 | 68.6 | 67.9 | 62.4 | 50.9 | 42.7 | 36.1 | 50.9 |
| Mean Temp °F | 43.5 | 47.0 | 54.9 | 62.4 | 69.4 | 76.6 | 79.8 | 78.7 | 73.6 | 63.1 | 54.4 | 46.8 | 62.5 |
| Days Max Temp ≥ 90 °F | 0 | 0 | 0 | 0 | 3 | 12 | 18 | 15 | 8 | 0 | 0 | 0 | 56 |
| Days Max Temp ≤ 32 °F | 1 | 0 | 0 | 0 | 0 | 0 | 0 | 0 | 0 | 0 | 0 | 0 | 1 |
| Days Min Temp ≤ 32 °F | 15 | 12 | 6 | 1 | 0 | 0 | 0 | 0 | 0 | 0 | 6 | 13 | 53 |
| Days Min Temp ≤ 0 °F | 0 | 0 | 0 | 0 | 0 | 0 | 0 | 0 | 0 | 0 | 0 | 0 | 0 |
| Heating Degree Days | 660 | 504 | 321 | 130 | 26 | 1 | 0 | 0 | 10 | 120 | 322 | 560 | 2654 |
| Cooling Degree Days | 0 | 1 | 12 | 52 | 163 | 372 | 486 | 433 | 266 | 68 | 12 | 2 | 1867 |
| Total Precipitation (") | 4.92 | 4.47 | 5.12 | 3.54 | 3.94 | 3.28 | 5.31 | 4.06 | 3.11 | 2.98 | 3.35 | 3.74 | 47.82 |
| Days ≥ 0.1" Precip | 8 | 7 | 7 | 5 | 6 | 5 | 7 | 6 | 6 | 4 | 5 | 6 | 72 |
| Total Snowfall (") | 0.8 | 1.0 | 0.2 | 0.0 | 0.0 | 0.0 | 0.0 | 0.0 | 0.0 | 0.0 | 0.0 | 0.0 | 2.0 |
| Days ≥ 1" Snow Depth | 0 | 0 | 0 | 0 | 0 | 0 | 0 | 0 | 0 | 0 | 0 | 0 | 0 |

## SURRENCY 2 WNW *Appling County* ELEVATION 200 ft LAT/LONG 31° 43 ' N / 82° 12 ' W

| | JAN | FEB | MAR | APR | MAY | JUN | JUL | AUG | SEP | OCT | NOV | DEC | YEAR |
|---|---|---|---|---|---|---|---|---|---|---|---|---|---|
| Maximum Temp °F | 61.6 | 65.6 | 73.1 | 80.2 | 85.6 | 90.1 | 92.3 | 90.9 | 87.4 | 79.9 | 71.6 | 64.3 | 78.6 |
| Minimum Temp °F | 37.5 | 39.4 | 45.6 | 51.2 | 58.4 | 65.7 | 69.0 | 68.5 | 64.6 | 54.0 | 45.3 | 39.5 | 53.2 |
| Mean Temp °F | 49.6 | 52.5 | 59.4 | 65.7 | 72.1 | 77.9 | 80.7 | 79.7 | 76.0 | 67.0 | 58.5 | 51.9 | 65.9 |
| Days Max Temp ≥ 90 °F | 0 | 0 | 0 | 2 | 6 | 17 | 24 | 22 | 12 | 1 | 0 | 0 | 84 |
| Days Max Temp ≤ 32 °F | 0 | 0 | 0 | 0 | 0 | 0 | 0 | 0 | 0 | 0 | 0 | 0 | 0 |
| Days Min Temp ≤ 32 °F | 11 | 8 | 3 | 0 | 0 | 0 | 0 | 0 | 0 | 0 | 5 | 9 | 36 |
| Days Min Temp ≤ 0 °F | 0 | 0 | 0 | 0 | 0 | 0 | 0 | 0 | 0 | 0 | 0 | 0 | 0 |
| Heating Degree Days | 476 | 356 | 204 | 67 | 6 | 0 | 0 | 0 | 1 | 60 | 222 | 408 | 1800 |
| Cooling Degree Days | 3 | 10 | 34 | 86 | 230 | 410 | 509 | 463 | 329 | 129 | 41 | 11 | 2255 |
| Total Precipitation (") | 4.43 | 3.56 | 4.49 | 2.59 | 3.68 | 4.92 | 5.90 | 6.00 | 3.02 | 2.66 | 2.32 | 3.64 | 47.21 |
| Days ≥ 0.1" Precip | 7 | 6 | 6 | 5 | 6 | 8 | 9 | 9 | 6 | 4 | 4 | 5 | 75 |
| Total Snowfall (") | 0.1 | 0.2 | 0.0 | 0.0 | 0.0 | 0.0 | 0.0 | 0.0 | 0.0 | 0.0 | 0.0 | 0.1 | 0.4 |
| Days ≥ 1" Snow Depth | 0 | 0 | 0 | 0 | 0 | 0 | 0 | 0 | 0 | 0 | 0 | 0 | 0 |

## SWAINSBORO *Emanuel County* ELEVATION 322 ft LAT/LONG 32° 36 ' N / 82° 20 ' W

| | JAN | FEB | MAR | APR | MAY | JUN | JUL | AUG | SEP | OCT | NOV | DEC | YEAR |
|---|---|---|---|---|---|---|---|---|---|---|---|---|---|
| Maximum Temp °F | 58.6 | 62.9 | 71.3 | 79.0 | 84.8 | 90.3 | 92.5 | 90.8 | 86.5 | 78.3 | 69.8 | 62.1 | 77.2 |
| Minimum Temp °F | 36.6 | 38.4 | 45.0 | 51.4 | 59.3 | 66.2 | 69.5 | 69.0 | 64.7 | 53.9 | 45.0 | 39.1 | 53.2 |
| Mean Temp °F | 47.6 | 50.7 | 58.2 | 65.3 | 72.1 | 78.3 | 81.0 | 79.9 | 75.6 | 66.2 | 57.4 | 50.6 | 65.2 |
| Days Max Temp ≥ 90 °F | 0 | 0 | 0 | 1 | 5 | 17 | 23 | 20 | 10 | 1 | 0 | 0 | 77 |
| Days Max Temp ≤ 32 °F | 0 | 0 | 0 | 0 | 0 | 0 | 0 | 0 | 0 | 0 | 0 | 0 | 0 |
| Days Min Temp ≤ 32 °F | 12 | 9 | 4 | 0 | 0 | 0 | 0 | 0 | 0 | 0 | 5 | 10 | 40 |
| Days Min Temp ≤ 0 °F | 0 | 0 | 0 | 0 | 0 | 0 | 0 | 0 | 0 | 0 | 0 | 0 | 0 |
| Heating Degree Days | 533 | 404 | 234 | 80 | 8 | 0 | 0 | 0 | 2 | 70 | 246 | 447 | 2024 |
| Cooling Degree Days | 2 | 10 | 30 | 86 | ***242*** | 437 | 531 | 479 | 330 | 121 | 33 | 8 | 2309 |
| Total Precipitation (") | 4.21 | 4.03 | 4.64 | 2.88 | 3.22 | 4.00 | 4.62 | 5.19 | 2.99 | 2.36 | 2.76 | 3.30 | 44.20 |
| Days ≥ 0.1" Precip | 7 | 5 | 6 | 4 | 5 | 6 | 7 | 7 | 5 | 3 | 4 | 5 | 64 |
| Total Snowfall (") | 0.0 | 0.3 | 0.0 | 0.0 | 0.0 | 0.0 | 0.0 | 0.0 | 0.0 | 0.0 | 0.0 | 0.1 | 0.4 |
| Days ≥ 1" Snow Depth | 0 | 0 | 0 | 0 | 0 | 0 | 0 | 0 | 0 | 0 | 0 | 0 | 0 |

## TALBOTTON *Talbot County* ELEVATION 702 ft LAT/LONG 32° 40 ' N / 84° 32 ' W

| | JAN | FEB | MAR | APR | MAY | JUN | JUL | AUG | SEP | OCT | NOV | DEC | YEAR |
|---|---|---|---|---|---|---|---|---|---|---|---|---|---|
| Maximum Temp °F | 56.8 | 61.7 | 70.1 | 77.9 | 83.3 | 88.6 | 90.4 | 90.0 | 86.1 | 77.3 | 68.8 | 60.5 | 76.0 |
| Minimum Temp °F | 32.7 | 34.9 | 42.1 | 49.0 | 56.2 | 63.8 | 67.7 | 66.9 | 61.7 | ***49.6*** | 41.3 | 35.6 | 50.1 |
| Mean Temp °F | 44.8 | 48.3 | 56.2 | 63.4 | 69.8 | 76.2 | 79.1 | 78.5 | 73.9 | ***63.4*** | 55.1 | 48.1 | 63.1 |
| Days Max Temp ≥ 90 °F | 0 | 0 | 0 | 1 | 4 | 14 | 18 | 18 | 9 | 0 | 0 | 0 | 64 |
| Days Max Temp ≤ 32 °F | 0 | 0 | 0 | 0 | 0 | 0 | 0 | 0 | 0 | 0 | 0 | 0 | 0 |
| Days Min Temp ≤ 32 °F | 16 | 13 | 7 | 1 | 0 | 0 | 0 | 0 | 0 | 1 | 7 | 13 | 58 |
| Days Min Temp ≤ 0 °F | 0 | 0 | 0 | 0 | 0 | 0 | 0 | 0 | 0 | 0 | 0 | 0 | 0 |
| Heating Degree Days | 621 | 466 | 286 | 106 | 19 | 2 | 1 | 0 | 7 | ***112*** | 302 | 521 | 2443 |
| Cooling Degree Days | 0 | 2 | 17 | 55 | 161 | 351 | 448 | 431 | ***277*** | 73 | 13 | 2 | 1830 |
| Total Precipitation (") | 4.76 | 4.81 | 6.08 | 3.94 | 3.92 | 4.20 | 5.17 | 4.47 | 3.20 | 2.74 | 3.75 | 4.67 | 51.71 |
| Days ≥ 0.1" Precip | 7 | 6 | 7 | 5 | 6 | 6 | 8 | 6 | 5 | 4 | 5 | 6 | 71 |
| Total Snowfall (") | 0.3 | 0.5 | 0.1 | 0.0 | 0.0 | 0.0 | 0.0 | 0.0 | 0.0 | 0.0 | 0.0 | 0.0 | 0.9 |
| Days ≥ 1" Snow Depth | 0 | 0 | 0 | 0 | 0 | 0 | 0 | 0 | 0 | 0 | 0 | 0 | 0 |

## THOMASTON 2 S *Upson County* ELEVATION 665 ft LAT/LONG 32° 52 ' N / 84° 19 ' W

| | JAN | FEB | MAR | APR | MAY | JUN | JUL | AUG | SEP | OCT | NOV | DEC | YEAR |
|---|---|---|---|---|---|---|---|---|---|---|---|---|---|
| Maximum Temp °F | 57.8 | 62.7 | 71.1 | 78.8 | 83.9 | 89.9 | 91.4 | 90.3 | 86.3 | 78.4 | 69.6 | 61.4 | 76.8 |
| Minimum Temp °F | 33.3 | 35.0 | 41.6 | 48.7 | 56.8 | 64.1 | 68.3 | 67.8 | 62.7 | 51.3 | 42.5 | 36.1 | 50.7 |
| Mean Temp °F | 45.5 | 48.9 | 56.4 | 63.7 | 70.4 | 77.0 | 79.9 | 79.1 | 74.6 | 64.9 | 56.1 | 48.8 | 63.8 |
| Days Max Temp ≥ 90 °F | 0 | 0 | 0 | 1 | 5 | 16 | 20 | 18 | 10 | 1 | 0 | 0 | 71 |
| Days Max Temp ≤ 32 °F | 0 | 0 | 0 | 0 | 0 | 0 | 0 | 0 | 0 | 0 | 0 | 0 | 0 |
| Days Min Temp ≤ 32 °F | 16 | 13 | 6 | 1 | 0 | 0 | 0 | 0 | 0 | 0 | 6 | 12 | 54 |
| Days Min Temp ≤ 0 °F | 0 | 0 | 0 | 0 | 0 | 0 | 0 | 0 | 0 | 0 | 0 | 0 | 0 |
| Heating Degree Days | 597 | 451 | 276 | 96 | 12 | 0 | 0 | 0 | 4 | 82 | 275 | 500 | 2293 |
| Cooling Degree Days | 0 | 2 | 17 | 58 | 178 | 380 | 482 | 447 | 295 | 90 | 16 | 3 | 1968 |
| Total Precipitation (") | 4.70 | 4.68 | 6.05 | 3.91 | 3.88 | 3.79 | 5.52 | 4.20 | 3.18 | 2.59 | 3.63 | 4.43 | 50.56 |
| Days ≥ 0.1" Precip | 7 | 6 | 7 | 5 | 6 | 6 | 8 | 7 | 5 | 4 | 5 | 6 | 72 |
| Total Snowfall (") | 0.5 | 0.8 | 0.1 | 0.0 | 0.0 | 0.0 | 0.0 | 0.0 | 0.0 | 0.0 | 0.0 | 0.0 | 1.4 |
| Days ≥ 1" Snow Depth | 0 | 0 | 0 | 0 | 0 | 0 | 0 | 0 | 0 | 0 | 0 | 0 | 0 |

## THOMASVILLE 3 NE *Thomas County* ELEVATION 260 ft LAT/LONG 30° 53 ' N / 83° 56 ' W

| | JAN | FEB | MAR | APR | MAY | JUN | JUL | AUG | SEP | OCT | NOV | DEC | YEAR |
|---|---|---|---|---|---|---|---|---|---|---|---|---|---|
| Maximum Temp °F | *63.1* | 66.4 | *74.0* | *80.8* | *86.4* | *91.2* | na | *91.5* | *88.7* | na | na | na | na |
| Minimum Temp °F | *39.0* | 41.2 | 47.8 | *53.7* | *61.7* | *67.6* | na | *69.7* | *65.8* | na | na | na | na |
| Mean Temp °F | *51.1* | 53.8 | *60.9* | *67.3* | *74.1* | *79.4* | na | *80.7* | *77.3* | na | na | na | na |
| Days Max Temp ≥ 90 °F | 0 | 0 | 0 | 1 | 7 | *18* | *21* | *20* | 13 | 3 | *0* | 0 | 83 |
| Days Max Temp ≤ 32 °F | 0 | 0 | 0 | 0 | 0 | 0 | 0 | 0 | 0 | 0 | 0 | 0 | 0 |
| Days Min Temp ≤ 32 °F | 9 | 6 | 2 | 0 | 0 | 0 | 0 | 0 | 0 | 0 | *2* | *6* | 25 |
| Days Min Temp ≤ 0 °F | 0 | 0 | 0 | 0 | 0 | 0 | 0 | 0 | 0 | 0 | 0 | 0 | 0 |
| Heating Degree Days | *433* | 316 | *172* | *47* | *2* | *0* | na | *0* | *1* | na | na | na | na |
| Cooling Degree Days | na | *12* | *47* | *108* | *294* | *455* | na | na | na | na | na | na | na |
| Total Precipitation (") | 5.55 | 4.73 | 5.78 | 3.29 | 4.09 | 5.53 | 6.11 | 5.25 | 3.48 | 2.51 | 3.04 | 3.80 | 53.16 |
| Days ≥ 0.1" Precip | 6 | 6 | 6 | 4 | 6 | 7 | 9 | 7 | 5 | 3 | *4* | 5 | 68 |
| Total Snowfall (") | 0.0 | 0.0 | 0.0 | 0.0 | 0.0 | 0.0 | 0.0 | 0.0 | 0.0 | 0.0 | 0.0 | 0.0 | 0.0 |
| Days ≥ 1" Snow Depth | 0 | 0 | 0 | 0 | 0 | 0 | 0 | 0 | 0 | 0 | *0* | 0 | 0 |

## TIFTON EXP STN *Tift County* ELEVATION 371 ft LAT/LONG 31° 28 ' N / 83° 32 ' W

| | JAN | FEB | MAR | APR | MAY | JUN | JUL | AUG | SEP | OCT | NOV | DEC | YEAR |
|---|---|---|---|---|---|---|---|---|---|---|---|---|---|
| Maximum Temp °F | 58.5 | 62.0 | 69.7 | 77.5 | 83.6 | 88.8 | 90.8 | 90.2 | 86.8 | 78.6 | 70.0 | 62.2 | 76.6 |
| Minimum Temp °F | 37.4 | 39.6 | 47.1 | 54.0 | 61.5 | 68.0 | 70.8 | 70.3 | 66.0 | 55.0 | 47.0 | 40.1 | 54.7 |
| Mean Temp °F | 48.0 | 50.8 | 58.4 | 65.8 | 72.5 | 78.5 | 80.8 | 80.3 | 76.4 | 66.8 | 58.5 | 51.2 | 65.7 |
| Days Max Temp ≥ 90 °F | 0 | 0 | 0 | 1 | 4 | 14 | 21 | 20 | 11 | 1 | 0 | 0 | 72 |
| Days Max Temp ≤ 32 °F | 0 | 0 | 0 | 0 | 0 | 0 | 0 | 0 | 0 | 0 | 0 | 0 | 0 |
| Days Min Temp ≤ 32 °F | 11 | 7 | 2 | 0 | 0 | 0 | 0 | 0 | 0 | 0 | 2 | 8 | 30 |
| Days Min Temp ≤ 0 °F | 0 | 0 | 0 | 0 | 0 | 0 | 0 | 0 | 0 | 0 | 0 | 0 | 0 |
| Heating Degree Days | 524 | 400 | 229 | 72 | 7 | 0 | 0 | 0 | 2 | 65 | 221 | 430 | 1950 |
| Cooling Degree Days | 2 | 10 | 32 | 89 | 250 | 429 | 515 | 488 | 345 | 130 | 38 | 10 | 2338 |
| Total Precipitation (") | 5.29 | 4.43 | 5.16 | 3.31 | 3.73 | 4.26 | 4.95 | 4.60 | 2.89 | 2.59 | 3.02 | 3.69 | 47.92 |
| Days ≥ 0.1" Precip | 8 | 6 | 7 | 5 | 5 | 7 | 8 | 7 | 5 | 4 | 5 | 6 | 73 |
| Total Snowfall (") | 0.0 | 0.0 | 0.0 | 0.0 | 0.0 | 0.0 | 0.0 | 0.0 | 0.0 | 0.0 | 0.0 | 0.0 | 0.0 |
| Days ≥ 1" Snow Depth | 0 | 0 | 0 | 0 | 0 | 0 | 0 | 0 | 0 | 0 | 0 | 0 | 0 |

## TOCCOA *Stephens County* ELEVATION 1020 ft LAT/LONG 34° 35 ' N / 83° 19 ' W

| | JAN | FEB | MAR | APR | MAY | JUN | JUL | AUG | SEP | OCT | NOV | DEC | YEAR |
|---|---|---|---|---|---|---|---|---|---|---|---|---|---|
| Maximum Temp °F | 51.8 | 56.1 | 65.0 | 73.9 | 79.7 | 85.8 | 88.8 | 87.1 | 81.8 | 72.8 | 63.3 | 54.6 | 71.7 |
| Minimum Temp °F | 31.5 | 33.6 | 40.4 | 47.6 | 55.4 | 63.0 | 67.3 | 66.2 | 60.3 | 49.3 | 41.1 | 34.7 | 49.2 |
| Mean Temp °F | 41.6 | 44.9 | 52.7 | 60.8 | 67.5 | 74.4 | 78.1 | 76.7 | 71.1 | 61.1 | 52.2 | 44.6 | 60.5 |
| Days Max Temp ≥ 90 °F | 0 | 0 | 0 | 0 | 1 | 8 | 14 | 10 | 3 | 0 | 0 | 0 | 36 |
| Days Max Temp ≤ 32 °F | 1 | 0 | 0 | 0 | 0 | 0 | 0 | 0 | 0 | 0 | 0 | 0 | 1 |
| Days Min Temp ≤ 32 °F | 17 | 13 | 7 | 2 | 0 | 0 | 0 | 0 | 0 | 1 | 6 | 14 | 60 |
| Days Min Temp ≤ 0 °F | 0 | 0 | 0 | 0 | 0 | 0 | 0 | 0 | 0 | 0 | 0 | 0 | 0 |
| Heating Degree Days | 717 | 562 | 379 | 162 | 39 | 2 | 0 | 0 | 16 | 154 | 380 | 624 | 3035 |
| Cooling Degree Days | 0 | 0 | 7 | 41 | 126 | 315 | 438 | 383 | 214 | 45 | 3 | 1 | 1573 |
| Total Precipitation (") | 5.83 | 5.17 | 6.28 | 4.50 | 5.32 | 4.58 | 5.31 | 5.43 | 4.17 | 4.27 | 4.69 | 5.19 | 60.74 |
| Days ≥ 0.1" Precip | 8 | 7 | 8 | 7 | 8 | 7 | 8 | 8 | 6 | 5 | 7 | 8 | 87 |
| Total Snowfall (") | 1.9 | 1.0 | 0.7 | 0.0 | 0.0 | 0.0 | 0.0 | 0.0 | 0.0 | 0.0 | 0.1 | 0.1 | 3.8 |
| Days ≥ 1" Snow Depth | *0* | 0 | 0 | 0 | 0 | 0 | 0 | 0 | 0 | 0 | 0 | 0 | 0 |

## WARRENTON *Warren County* ELEVATION 502 ft LAT/LONG 33° 24 ' N / 82° 40 ' W

| | JAN | FEB | MAR | APR | MAY | JUN | JUL | AUG | SEP | OCT | NOV | DEC | YEAR |
|---|---|---|---|---|---|---|---|---|---|---|---|---|---|
| Maximum Temp °F | 53.7 | 58.6 | 67.2 | 75.7 | 82.4 | 88.6 | 91.2 | 89.7 | 84.4 | 74.8 | 65.9 | 57.3 | 74.1 |
| Minimum Temp °F | 32.1 | 34.0 | 40.7 | 48.0 | 56.9 | 64.4 | 68.4 | 67.9 | 62.5 | 50.4 | 41.4 | 34.6 | 50.1 |
| Mean Temp °F | 43.0 | 46.3 | 54.0 | 61.9 | 69.7 | 76.5 | 79.9 | 78.8 | 73.5 | 62.6 | 53.7 | 46.0 | 62.2 |
| Days Max Temp ≥ 90 °F | 0 | 0 | 0 | 1 | 4 | 13 | 20 | 17 | 7 | 0 | 0 | 0 | 62 |
| Days Max Temp ≤ 32 °F | 1 | 0 | 0 | 0 | 0 | 0 | 0 | 0 | 0 | 0 | 0 | 0 | 1 |
| Days Min Temp ≤ 32 °F | 17 | 13 | 7 | 1 | 0 | 0 | 0 | 0 | 0 | 0 | 6 | 14 | 58 |
| Days Min Temp ≤ 0 °F | 0 | 0 | 0 | 0 | 0 | 0 | 0 | 0 | 0 | 0 | 0 | 0 | 0 |
| Heating Degree Days | 675 | 522 | 345 | 140 | 22 | 1 | 0 | 0 | 9 | 130 | 342 | 585 | 2771 |
| Cooling Degree Days | 0 | 1 | 8 | 41 | 159 | 359 | 474 | 426 | 260 | 64 | 12 | 2 | 1806 |
| Total Precipitation (") | 4.91 | 4.40 | 5.56 | 3.50 | 4.03 | 3.90 | 4.23 | 4.76 | 3.21 | 3.37 | 3.28 | 4.09 | 49.24 |
| Days ≥ 0.1" Precip | 8 | 6 | 7 | 5 | 6 | 6 | 6 | 7 | 5 | 4 | 5 | 6 | 71 |
| Total Snowfall (") | 0.1 | 0.8 | 0.1 | 0.0 | 0.0 | 0.0 | 0.0 | 0.0 | 0.0 | 0.0 | 0.0 | 0.0 | 1.0 |
| Days ≥ 1" Snow Depth | 0 | 0 | 0 | 0 | 0 | 0 | 0 | 0 | 0 | 0 | 0 | 0 | 0 |

## WASHINGTON 2 ESE *Wilkes County* ELEVATION 630 ft LAT/LONG 33° 44 ' N / 82° 44 ' W

| | JAN | FEB | MAR | APR | MAY | JUN | JUL | AUG | SEP | OCT | NOV | DEC | YEAR |
|---|---|---|---|---|---|---|---|---|---|---|---|---|---|
| Maximum Temp °F | 52.5 | 57.3 | 65.6 | 74.2 | 80.6 | 86.8 | 90.1 | 88.6 | 83.9 | 74.4 | 65.3 | 56.2 | 73.0 |
| Minimum Temp °F | 30.2 | 32.2 | 39.1 | 47.2 | 55.4 | 63.5 | 67.6 | 66.7 | 60.6 | 48.5 | 39.9 | 33.1 | 48.7 |
| Mean Temp °F | 41.4 | 44.8 | 52.4 | 60.7 | 68.0 | 75.2 | 78.9 | 77.7 | 72.3 | 61.5 | 52.6 | 44.7 | 60.9 |
| Days Max Temp ≥ 90 °F | 0 | 0 | 0 | 0 | 2 | 10 | 17 | 15 | 6 | 0 | 0 | 0 | 50 |
| Days Max Temp ≤ 32 °F | 1 | 0 | 0 | 0 | 0 | 0 | 0 | 0 | 0 | 0 | 0 | 0 | 1 |
| Days Min Temp ≤ 32 °F | 18 | 15 | 8 | 1 | 0 | 0 | 0 | 0 | 0 | 1 | 8 | 16 | 67 |
| Days Min Temp ≤ 0 °F | 0 | 0 | 0 | 0 | 0 | 0 | 0 | 0 | 0 | 0 | 0 | 0 | 0 |
| Heating Degree Days | 725 | 565 | 390 | 164 | 39 | 2 | 0 | 0 | 13 | 152 | 371 | 625 | 3046 |
| Cooling Degree Days | 0 | 1 | 8 | 46 | 154 | 348 | 472 | 425 | 257 | 61 | 8 | 2 | 1782 |
| Total Precipitation (") | 4.89 | 4.17 | 4.96 | 3.59 | 4.20 | 3.83 | 4.89 | 3.82 | 3.07 | 3.35 | 3.32 | 3.80 | 47.89 |
| Days ≥ 0.1" Precip | 8 | 7 | 7 | 6 | 6 | 6 | 7 | 6 | 5 | 5 | 5 | 7 | 75 |
| Total Snowfall (") | 0.1 | ***0.2*** | 0.0 | 0.0 | 0.0 | 0.0 | 0.0 | 0.0 | 0.0 | 0.0 | 0.0 | 0.0 | 0.3 |
| Days ≥ 1" Snow Depth | 0 | 0 | 0 | 0 | 0 | 0 | 0 | 0 | 0 | 0 | 0 | 0 | 0 |

## WAYCROSS 4 NE *Pierce County* ELEVATION 151 ft LAT/LONG 31° 15 ' N / 82° 19 ' W

| | JAN | FEB | MAR | APR | MAY | JUN | JUL | AUG | SEP | OCT | NOV | DEC | YEAR |
|---|---|---|---|---|---|---|---|---|---|---|---|---|---|
| Maximum Temp °F | 62.0 | 65.4 | 73.2 | 80.8 | 86.7 | 91.3 | 93.7 | 92.4 | 88.8 | 81.0 | 73.1 | 65.1 | 79.5 |
| Minimum Temp °F | 35.1 | 36.9 | 43.7 | 49.8 | 57.6 | 65.7 | 69.1 | 69.1 | 64.4 | 52.3 | 43.6 | 36.9 | 52.0 |
| Mean Temp °F | 48.4 | 51.1 | 58.4 | 65.3 | 72.2 | 78.5 | 81.4 | 80.7 | 76.6 | 66.7 | 58.4 | 51.1 | 65.7 |
| Days Max Temp ≥ 90 °F | 0 | 0 | 0 | 3 | 11 | 21 | 26 | 25 | 16 | 3 | 0 | 0 | 105 |
| Days Max Temp ≤ 32 °F | 0 | 0 | 0 | 0 | 0 | 0 | 0 | 0 | 0 | 0 | 0 | 0 | 0 |
| Days Min Temp ≤ 32 °F | 15 | 11 | 5 | 1 | 0 | 0 | 0 | 0 | 0 | 1 | 6 | 13 | 52 |
| Days Min Temp ≤ 0 °F | 0 | 0 | 0 | 0 | 0 | 0 | 0 | 0 | 0 | 0 | 0 | 0 | 0 |
| Heating Degree Days | 515 | 396 | 229 | 82 | 9 | 0 | 0 | 0 | 1 | 71 | 230 | 435 | 1968 |
| Cooling Degree Days | 3 | 11 | 28 | 86 | 241 | 431 | 534 | ***501*** | 343 | 131 | 46 | 11 | 2366 |
| Total Precipitation (") | 4.79 | 3.53 | 4.91 | 2.88 | 3.98 | 5.60 | 6.17 | 6.15 | 3.47 | 2.72 | 2.76 | 3.40 | 50.36 |
| Days ≥ 0.1" Precip | 7 | 6 | 6 | 4 | 7 | 9 | 10 | 9 | 6 | 4 | 4 | 6 | 78 |
| Total Snowfall (") | 0.0 | 0.0 | 0.0 | 0.0 | 0.0 | 0.0 | 0.0 | 0.0 | 0.0 | 0.0 | 0.0 | 0.0 | 0.0 |
| Days ≥ 1" Snow Depth | 0 | 0 | 0 | 0 | 0 | 0 | 0 | 0 | 0 | 0 | 0 | 0 | 0 |

## WAYNESBORO 2 NE *Burke County* ELEVATION 259 ft LAT/LONG 33° 5 ' N / 82° 1 ' W

| | JAN | FEB | MAR | APR | MAY | JUN | JUL | AUG | SEP | OCT | NOV | DEC | YEAR |
|---|---|---|---|---|---|---|---|---|---|---|---|---|---|
| Maximum Temp °F | 55.8 | 60.1 | 68.4 | 76.6 | 83.0 | 88.8 | 91.5 | 89.8 | 85.4 | 76.9 | 68.5 | ***59.3*** | 75.3 |
| Minimum Temp °F | 31.6 | 33.6 | 40.7 | 47.7 | 55.9 | 63.9 | 68.2 | 67.0 | 61.3 | 48.7 | 39.7 | ***33.7*** | 49.3 |
| Mean Temp °F | 43.7 | 46.9 | 54.6 | 62.2 | ***69.5*** | 76.4 | 79.8 | 78.4 | 73.4 | 62.8 | 54.2 | ***46.5*** | 62.4 |
| Days Max Temp ≥ 90 °F | 0 | 0 | 0 | 1 | 4 | 14 | 21 | 17 | 8 | 1 | 0 | 0 | 66 |
| Days Max Temp ≤ 32 °F | 1 | 0 | 0 | 0 | 0 | 0 | 0 | 0 | 0 | 0 | 0 | 0 | 1 |
| Days Min Temp ≤ 32 °F | 17 | 13 | 6 | 1 | 0 | 0 | 0 | 0 | 0 | 1 | 9 | ***15*** | 62 |
| Days Min Temp ≤ 0 °F | 0 | 0 | 0 | 0 | 0 | 0 | 0 | 0 | 0 | 0 | 0 | 0 | 0 |
| Heating Degree Days | 653 | 507 | 330 | 134 | ***23*** | 1 | 0 | 0 | 7 | 127 | 330 | ***570*** | 2682 |
| Cooling Degree Days | 1 | 2 | 10 | 45 | ***163*** | 372 | 481 | 428 | 266 | 63 | 13 | 3 | 1847 |
| Total Precipitation (") | 4.19 | 4.08 | 4.78 | 2.98 | 3.45 | 4.28 | 4.40 | 4.95 | 3.13 | 3.04 | 2.70 | 3.62 | 45.60 |
| Days ≥ 0.1" Precip | 7 | 6 | 6 | 5 | 5 | 6 | 7 | 7 | 5 | 4 | 4 | ***6*** | 68 |
| Total Snowfall (") | 0.0 | 1.0 | 0.0 | 0.0 | 0.0 | 0.0 | 0.0 | 0.0 | 0.0 | 0.0 | 0.0 | 0.0 | 1.0 |
| Days ≥ 1" Snow Depth | 0 | ***0*** | 0 | 0 | 0 | 0 | 0 | 0 | 0 | 0 | 0 | 0 | 0 |

## WEST POINT *Troup County* ELEVATION 581 ft LAT/LONG 32° 52 ' N / 85° 11 ' W

| | JAN | FEB | MAR | APR | MAY | JUN | JUL | AUG | SEP | OCT | NOV | DEC | YEAR |
|---|---|---|---|---|---|---|---|---|---|---|---|---|---|
| Maximum Temp °F | 54.7 | 58.9 | 67.4 | 75.8 | 81.7 | 88.4 | 90.7 | 89.7 | 85.1 | 75.9 | 66.5 | 58.2 | 74.4 |
| Minimum Temp °F | 31.3 | 33.5 | 40.7 | 48.4 | 56.3 | 64.3 | 68.7 | 67.9 | 61.9 | 49.4 | 40.4 | 34.2 | 49.8 |
| Mean Temp °F | 43.0 | 46.2 | 54.0 | 62.1 | 69.0 | 76.4 | 79.7 | 78.8 | 73.5 | 62.7 | 53.5 | 46.2 | 62.1 |
| Days Max Temp ≥ 90 °F | 0 | 0 | 0 | 0 | 3 | 14 | 20 | 17 | 8 | 1 | 0 | 0 | 63 |
| Days Max Temp ≤ 32 °F | 1 | 0 | 0 | 0 | 0 | 0 | 0 | 0 | 0 | 0 | 0 | 0 | 1 |
| Days Min Temp ≤ 32 °F | 18 | 15 | 7 | 1 | 0 | 0 | 0 | 0 | 0 | 1 | 8 | 15 | 65 |
| Days Min Temp ≤ 0 °F | 0 | 0 | 0 | 0 | 0 | 0 | 0 | 0 | 0 | 0 | 0 | 0 | 0 |
| Heating Degree Days | 674 | 525 | 343 | 133 | 26 | 1 | 0 | 0 | 9 | 126 | 348 | 577 | 2762 |
| Cooling Degree Days | 0 | 1 | 9 | 42 | 153 | 365 | 477 | 446 | 276 | 63 | 10 | 2 | 1844 |
| Total Precipitation (") | 4.66 | 4.94 | 5.55 | 4.40 | 3.99 | 3.55 | 5.97 | 3.50 | 3.39 | 2.85 | 3.98 | 4.85 | 51.63 |
| Days ≥ 0.1" Precip | 8 | 6 | 7 | 5 | 6 | 6 | 8 | 6 | 5 | 4 | 5 | 7 | 73 |
| Total Snowfall (") | 0.1 | 0.0 | 0.0 | 0.0 | 0.0 | 0.0 | 0.0 | 0.0 | 0.0 | 0.0 | 0.0 | 0.0 | 0.1 |
| Days ≥ 1" Snow Depth | 0 | 0 | 0 | 0 | 0 | 0 | 0 | 0 | 0 | 0 | 0 | 0 | 0 |

## WINDER 1 SSE *Barrow County* ELEVATION 981 ft LAT/LONG 33° 58 ' N / 83° 43 ' W

| | JAN | FEB | MAR | APR | MAY | JUN | JUL | AUG | SEP | OCT | NOV | DEC | YEAR |
|---|---|---|---|---|---|---|---|---|---|---|---|---|---|
| Maximum Temp °F | 52.3 | 57.2 | 65.9 | 74.6 | 80.7 | 86.9 | 89.4 | 88.0 | 82.5 | 73.2 | 64.3 | 55.6 | 72.6 |
| Minimum Temp °F | 31.3 | 33.2 | 39.8 | 47.0 | 54.9 | 62.3 | 66.5 | 65.9 | 60.7 | 49.1 | 41.0 | 34.2 | 48.8 |
| Mean Temp °F | 41.8 | 45.3 | 52.9 | 60.8 | 67.8 | 74.6 | 78.0 | 77.0 | 71.6 | 61.2 | 52.7 | 44.9 | 60.7 |
| Days Max Temp ≥ 90 °F | 0 | 0 | 0 | 0 | 1 | 9 | 15 | 11 | 3 | 0 | 0 | 0 | 39 |
| Days Max Temp ≤ 32 °F | 1 | 0 | 0 | 0 | 0 | 0 | 0 | 0 | 0 | 0 | 0 | 0 | 1 |
| Days Min Temp ≤ 32 °F | 17 | 14 | 7 | 2 | 0 | 0 | 0 | 0 | 0 | 1 | 7 | 14 | 62 |
| Days Min Temp ≤ 0 °F | 0 | 0 | 0 | 0 | 0 | 0 | 0 | 0 | 0 | 0 | 0 | 0 | 0 |
| Heating Degree Days | 712 | 552 | 375 | 157 | 33 | 2 | 0 | 0 | 14 | 152 | 367 | 616 | 2980 |
| Cooling Degree Days | 0 | 0 | 6 | 40 | 127 | 317 | 424 | 383 | 217 | 46 | 4 | 1 | 1565 |
| Total Precipitation (") | 5.11 | 4.10 | 5.53 | 3.82 | 4.37 | 3.74 | 4.39 | 3.83 | 3.48 | 3.79 | 3.55 | 4.12 | 49.83 |
| Days ≥ 0.1" Precip | 8 | 6 | 8 | 5 | 6 | 6 | 8 | 6 | 6 | 5 | 5 | 7 | 76 |
| Total Snowfall (") | 0.3 | 0.3 | 0.1 | 0.0 | 0.0 | 0.0 | 0.0 | 0.0 | 0.0 | 0.0 | 0.0 | 0.0 | 0.7 |
| Days ≥ 1" Snow Depth | 0 | 0 | 0 | 0 | 0 | 0 | 0 | 0 | 0 | 0 | 0 | 0 | 0 |

## JANUARY MINIMUM TEMPERATURE °F

### LOWEST

| | | |
|---|---|---|
| 1 | Blairsville | 23.9 |
| 2 | Lafayette | 26.8 |
| 3 | Clayton | 27.3 |
| 4 | Alpharetta | 27.9 |
| | Calhoun | 27.9 |
| | Dahlonega | 27.9 |
| 7 | Dallas | 28.0 |
| 8 | Dalton | 28.5 |
| 9 | Cornelia | 28.6 |
| 10 | Rome | 28.8 |
| 11 | Commerce | 29.1 |
| | Helen | 29.1 |
| 13 | Jasper | 29.4 |
| 14 | Elberton | 29.5 |
| 15 | Carrollton | 30.0 |
| 16 | Gainesville | 30.1 |
| 17 | Appling | 30.2 |
| | Washington | 30.2 |
| 19 | Cedartown | 30.8 |
| 20 | Cartersville | 31.0 |
| 21 | Milledgeville | 31.3 |
| | West Point | 31.3 |
| | Winder | 31.3 |
| 24 | Toccoa | 31.5 |
| 25 | Waynesboro | 31.6 |

### HIGHEST

| | | |
|---|---|---|
| 1 | Brunswick-McK | 42.2 |
| 2 | Sapelo Island | 40.7 |
| 3 | Brunswick | 40.2 |
| | Folkston | 40.2 |
| 5 | Alma | 39.8 |
| 6 | Fort Stewart | 39.6 |
| 7 | Moultrie | 39.1 |
| 8 | Camilla | 39.0 |
| | Thomasville | 39.0 |
| 10 | Glennville | 38.9 |
| 11 | Fitzgerald | 38.8 |
| 12 | Homerville | 38.7 |
| 13 | Nahunta | 38.6 |
| 14 | Savannah | 38.4 |
| 15 | Colquitt | 38.1 |
| | Cuthbert | 38.1 |
| 17 | Jesup | 38.0 |
| 18 | Brooklet | 37.7 |
| 19 | Quitman | 37.5 |
| | Surrency | 37.5 |
| 21 | Tifton | 37.4 |
| 22 | Metter | 37.3 |
| 23 | Cordele | 37.2 |
| 24 | Swainsboro | 36.6 |
| 25 | Ashburn | 36.3 |

## JULY MAXIMUM TEMPERATURE °F

### HIGHEST

| | | |
|---|---|---|
| 1 | Folkston | 93.7 |
| | Waycross | 93.7 |
| 3 | Dublin | 93.6 |
| 4 | Cordele | 93.4 |
| 5 | Jesup | 93.2 |
| 6 | Fort Stewart | 93.0 |
| | Metter | 93.0 |
| 8 | Hawkinsville | 92.8 |
| 9 | Camilla | 92.7 |
| 10 | Douglas | 92.6 |
| 11 | Irwinton | 92.5 |
| | Swainsboro | 92.5 |
| 13 | Glennville | 92.4 |
| | Lumber City | 92.4 |
| | Millen | 92.4 |
| 16 | Cuthbert | 92.3 |
| | Macon | 92.3 |
| | Surrency | 92.3 |
| 19 | Augusta | 92.2 |
| | Brunswick | 92.2 |
| | Colquitt | 92.2 |
| | Homerville | 92.2 |
| | Moultrie | 92.2 |
| 24 | Albany | 92.1 |
| 25 | Fitzgerald | 92.0 |

### LOWEST

| | | |
|---|---|---|
| 1 | Blairsville | 84.3 |
| 2 | Clayton | 85.2 |
| 3 | Cornelia | 85.6 |
| 4 | Helen | 85.9 |
| 5 | Dahlonega | 86.1 |
| 6 | Jasper | 86.4 |
| 7 | Gainesville | 86.6 |
| 8 | Alpharetta | 87.3 |
| 9 | Rome | 87.8 |
| 10 | Atlanta | 88.6 |
| | Carrollton | 88.6 |
| | Experiment | 88.6 |
| 13 | Dallas | 88.7 |
| 14 | Toccoa | 88.8 |
| 15 | Lafayette | 88.9 |
| 16 | Dalton | 89.0 |
| 17 | Cartersville | 89.2 |
| 18 | Cedartown | 89.4 |
| | Winder | 89.4 |
| 20 | Commerce | 89.5 |
| | Covington | 89.5 |
| 22 | Newnan | 89.6 |
| 23 | Sapelo Island | 89.8 |
| 24 | Brunswick-McK | 89.9 |
| 25 | Athens | 90.0 |

## ANNUAL PRECIPITATION (")

### HIGHEST

| | | |
|---|---|---|
| 1 | Clayton | 72.95 |
| 2 | Helen | 71.74 |
| 3 | Dahlonega | 64.99 |
| 4 | Toccoa | 60.74 |
| 5 | Cornelia | 60.52 |
| 6 | Jasper | 59.96 |
| 7 | Lafayette | 58.02 |
| 8 | Blairsville | 57.38 |
| 9 | Dalton | 56.62 |
| 10 | Gainesville | 55.62 |
| 11 | Rome | 55.02 |
| 12 | Calhoun | 54.30 |
| 13 | Dallas | 54.11 |
| 14 | La Grange | 54.07 |
| 15 | Cedartown | 53.26 |
| 16 | Thomasville | 53.16 |
| 17 | Quitman | 53.10 |
| 18 | Camilla | 53.01 |
| 19 | Carrollton | 53.00 |
| 20 | Homerville | 52.98 |
| 21 | Folkston | 52.92 |
| 22 | Commerce | 52.78 |
| 23 | Newnan | 52.73 |
| 24 | Sapelo Island | 52.72 |
| 25 | Albany | 52.71 |

### LOWEST

| | | |
|---|---|---|
| 1 | Millen | 43.94 |
| 2 | Swainsboro | 44.20 |
| 3 | Augusta | 44.48 |
| 4 | Midville | 44.98 |
| 5 | Macon | 45.17 |
| 6 | Cordele | 45.22 |
| 7 | Louisville | 45.58 |
| 8 | Waynesboro | 45.60 |
| 9 | Cartersville | 46.11 |
| 10 | Lumber City | 46.14 |
| 11 | Milledgeville | 46.16 |
| 12 | Hawkinsville | 46.20 |
| 13 | Sandersville | 46.34 |
| 14 | Irwinton | 46.40 |
| 15 | Fitzgerald | 46.41 |
| 16 | Monticello | 46.83 |
| 17 | Eastman | 47.08 |
| 18 | Appling | 47.21 |
| | Surrency | 47.21 |
| 20 | Dublin | 47.39 |
| 21 | Brooklet | 47.55 |
| 22 | Lumpkin | 47.69 |
| 23 | Siloam | 47.82 |
| 24 | Washington | 47.89 |
| 25 | Tifton | 47.92 |

## ANNUAL SNOWFALL (")

### HIGHEST

| | | |
|---|---|---|
| 1 | Blairsville | 6.3 |
| 2 | Clayton | 5.4 |
| 3 | Toccoa | 3.8 |
| 4 | Dallas | 3.6 |
| 5 | Cedartown | 3.2 |
| 6 | Dalton | 3.1 |
| 7 | Athens | 3.0 |
| 8 | Gainesville | 2.8 |
| 9 | Atlanta | 2.5 |
| | Jasper | 2.5 |
| 11 | Lafayette | 2.2 |
| 12 | Siloam | 2.0 |
| 13 | Calhoun | 1.9 |
| | Commerce | 1.9 |
| 15 | Rome | 1.8 |
| 16 | Cartersville | 1.7 |
| 17 | Newnan | 1.6 |
| 18 | Augusta | 1.5 |
| | Hartwell | 1.5 |
| 20 | Covington | 1.4 |
| | Thomaston | 1.4 |
| 22 | Macon | 1.3 |
| | Sandersville | 1.3 |
| 24 | Dahlonega | 1.1 |
| | Milledgeville | 1.1 |

### LOWEST

| | | |
|---|---|---|
| 1 | Brunswick | 0.0 |
| | Folkston | 0.0 |
| | Moultrie | 0.0 |
| | Nahunta | 0.0 |
| | Thomasville | 0.0 |
| | Tifton | 0.0 |
| | Waycross | 0.0 |
| 8 | Albany | 0.1 |
| | Brunswick-McK | 0.1 |
| | Camilla | 0.1 |
| | Cordele | 0.1 |
| | Eastman | 0.1 |
| | Elberton | 0.1 |
| | Fitzgerald | 0.1 |
| | Fort Stewart | 0.1 |
| | Homerville | 0.1 |
| | Lumpkin | 0.1 |
| | Quitman | 0.1 |
| | West Point | 0.1 |
| 20 | Ashburn | 0.2 |
| | Carrollton | 0.2 |
| | Douglas | 0.2 |
| | Jesup | 0.2 |
| | Lumber City | 0.2 |
| 25 | Alma | 0.3 |

# HAWAII

PHYSICAL FEATURES. West and south of California, 2,100 miles away, lies Hawaii. Among the 50 states it is the only one surrounded by the ocean. It is the only state within the tropics. Both of these facts contribute significantly to its climate, as do also its division into separate, widely-spaced islands and its topographic diversity.

The islands of the State are the easternmost members of the Hawaiian Island Chain. This Chain extends for a distance of 2,000 miles from the Kure and Midway Islands at the northwest to the Island of Hawaii at the extreme southeast end. In longitude, the Hawaiian Chain reaches from 178° to 154° W.; in latitude, from 28° to 19° N. The islands of the State of Hawaii cover a far lesser range: from 160° to 154° W. and from 22° to 19° N. They occupy a narrow zone 430 miles long. There are six major islands in the State. From west to east these are Kauai, Oahu, Molokai, Lanai, Maui, and Hawaii. Taken together with the much smaller islands of Niihau and Kahoolawe, their total area is 6,424 square miles. The islands are terrestrial, summit portions of the long range of volcanic mountains that comprise the Hawaiian Chain. The mountainous nature of Hawaii is indicated by the fact that 50 percent of the State lies above an elevation of 2,000 feet and 10 percent lies above 7,000 feet. Almost half of the area of Hawaii lies within 5 miles of the coast. Because of this extreme insularity the marine influence upon the climate is very great, yet the mountains, especially the massive ones on Hawaii and Maui, strongly modify the marine effect and result in conditions that are semi-continental in some localities.

GENERAL CLIMATE. The most prominent feature of the circulation of air across the tropical Pacific is the trade-wind flow in a general east-to-west direction. In the central North Pacific the trade winds blow from the northeast quadrant, and represent the outflow of air from the great region of high pressure, the Pacific Anticyclone, whose typical location is well north and east of the Hawaiian Island Chain. The Pacific High, and with it the trade-wind zone, moves north and south with the sun, so that it reaches its northernmost position in the summer half-year. This brings the heart of the trade winds across Hawaii during the period May through September, when the trades are prevalent 80 to 95 percent of the time. From October through April, Hawaii is located to the north of the heart of the trade winds. Nevertheless, the trades still blow across the islands much of the time.

The dominance of the trades and the influence of terrain give special character to the climate of the Islands. Completely cloudless skies are extremely rare, even though much of the time the dense cloud cover is confined to the mountain areas and windward slopes, while the leeward lowlands have only a few scattered clouds. Showers are very common; yet while some of these are very heavy, the vast majority are light and brief -- a sudden sprinkle of rain and that is all. Even the heavy showers are of a special character, in that they are seldom accompanied by thunder and lightning. Finally, the trade winds provide a system of natural ventilation much of the time throughout most of the State and bring to the land, at least in the lower lying regions, the mildly warm temperatures that are characteristic of air that has moved great distances across the tropical seas.

The relatively slight variations in the length of the daylight period in Hawaii, together with the smaller annual variations in the altitude of the sun above the horizon, result in relatively small variations in the amount of incoming solar energy from one time of the year to another. This small variation partly explains why seasonal changes in temperature are so slight throughout much of Hawaii. The other principal reason for the slightness of the variation is the virtually constant flow of fresh ocean air across the islands. Just as the temperature of the ocean surface varies comparatively little from season to season, so also does the temperature of air that has moved great distances across the ocean.

The rugged configuration of the islands produces marked variations in conditions from one locality to another. Air swept inland on the trade winds or as part of storm circulations is shunted one way and another by the mountains and valleys and great open slopes. This complex three-dimensional flow of air results in striking differences from place to place in windspeed, cloudiness, and rainfall. Together with variations in the elevation of the land, it results in differences in air temperature.

The native Hawaiians recognized only two seasons. KAU was the fruitful season, the season when the sun was directly or almost directly overhead, when the weather was warmer, and when the trade winds were most reliable. HOO-ILO was the season when the sun was in the south, when the weather was cooler, and when the trade winds were most often interrupted by other winds. Modern analysis of the climatic records shows the soundness of this Hawaiian system of seasons.

In terms of variations in climatic conditions from one part of the State to another, the most striking contrasts are those in rainfall. At one extreme the annual rainfall averages 20 inches and less in leeward coastal areas and near the summits of the very high mountains, Mauna Loa and Mauna Kea. At the other extreme the annual average exceeds 300 inches along the lower windward slopes of these high mountains and of Haleakala and at or near the summit of the lower mountains.

In general the Hawaiian climate is characterized by a 2-season year, by mild and fairly uniform temperature conditions everywhere but at high elevations, by strikingly marked geographic differences in rainfall, by generally humid conditions and high cloudiness except on the driest coasts and at high elevations, and by a general dominance of trade-wind flow especially at elevations below a few thousand feet.

The surface waters of the open ocean around Hawaii have an average temperature that ranges from a minimum of 73° or 74° F. between late February and early April to a maximum of 79° or 80° F. in late September or early October. With temperatures as mild as these -- and with temperatures almost as mild for hundreds of miles around, even to the north -- the air that reaches Hawaii is neither very hot nor cold. The mild, equable temperatures of the ocean give rise to mild, equable temperatures in the air that moves across the oceans and onto the islands of Hawaii.

PRECIPITATION. If the islands of the State of Hawaii did not exist, the average annual rainfall upon the water where the islands actually lie would be about 25 inches. Instead, the actual average is about 70 inches. Thus the islands extract from the air that passes across them about 45 inches of rainfall that otherwise would not fall. The mountains are dominantly responsible for this added water bonus. The driest areas are on the upper slopes of the high mountains, on leeward coasts, or in leeward locations in the interior of the islands. In the driest of these areas the average annual rainfall is less than 10 inches. The contrast in rainfall between the rainier winter season and the drier summer season is generally most pronounced at low elevations in the areas with low annual rainfall. In the lowlands at all times of the year, rainfall is most likely to occur during the nighttime or in the morning hours, and least likely to occur during midafternoon.

In most parts of the tropics the rainfall is highly variable from one year to another. Hawaii is no exception. Even in areas where the rainfall is very high and the monthly averages are all above 10 inches, the rainfall of particular months may vary by 200 to 300 percent from one year to another and there may be very occasional months with only 1 or 2 inches of rain. With such wide swings in rainfall it is inevitable that there are occasional droughts.

STORMS. Intense local rainstorms other than those that occur under trade-wind conditions are small features that seldom cover more than a few square miles and sometimes less than a single square mile. They occur most typically in the late afternoon or early evening. In some areas in which there are well developed sea breezes, they are common occurrences, especially in summer. In most areas, however, they are apt to occur on only a few days per year when the overall winds are light and variable or when there is a gentle flow of air from a southerly direction.

Intense local storms are sometimes accompanied by lightning and thunder. Lightning and thunder also occasionally accompany very intense rainfall along a cold front moving across the islands. Thunderstorms are reported from somewhere in the State on 20 to 30 days a year, and more often in winter than in summer. Waterspouts and other funnel clouds are not uncommon in the Hawaiian area, about 20 of them being sighted in the average year. Often they are accompanied by towering cumulus clouds and rain, although they have also been observed under trade wind conditions. Hail falls somewhere in Hawaii between 5 and 10 times in the average year. Almost always it is quite small -- ¼ inch or less in diameter -- but on several occasions hail the size of marbles, and discs about five-eighths of an inch in diameter, have been reported.

Kona storms, like cold front storms, are features of the winter season. They are so-called because they often bring winds from "kona" or leeward directions. Kona rains last from several hours to several days.

## COUNTY INDEX

### *Hawaii County*
HAWAII VLC NP HQ 54
HILO GEN LYMAN FIELD
KAINALIU 73.2
KULANI CAMP 79
MAUNA LOA SLOPE OBS
NAALEHU 14
OPIHIHALE 2 24.1

### *Honolulu County*
HONOLULU INTL AP
KANEOHE MAUKA 781
WAIALUA 847
WAIKIKI 717.2

### *Kauai County*
KANALOHULUHULU 1075
LIHUE AP

### *Maui County*
HALEAKALA RS 338
HANA AP 355
KAHULUI AP
KAILUA 446
LAHAINA 361
LANAI CITY 672
MOLOKAI AP 524

## ELEVATION INDEX

| FEET | STATION NAME |
|---|---|
| 10 | WAIKIKI 717.2 |
| 30 | WAIALUA 847 |
| 39 | HILO GEN LYMAN FIELD |
| 39 | HONOLULU INTL AP |
| 40 | KAHULUI AP |
| 49 | LAHAINA 361 |
| 66 | HANA AP 355 |
| 103 | LIHUE AP |
| 200 | KANEOHE MAUKA 781 |
| 453 | MOLOKAI AP 524 |
| 669 | NAALEHU 14 |
| 702 | KAILUA 446 |
| 1302 | OPIHIHALE 2 24.1 |
| 1503 | KAINALIU 73.2 |
| 1621 | LANAI CITY 672 |
| 3602 | KANALOHULUHULU 1075 |
| 3971 | HAWAII VLC NP HQ 54 |
| 5194 | KULANI CAMP 79 |
| 7034 | HALEAKALA RS 338 |
| 10968 | MAUNA LOA SLOPE OBS |

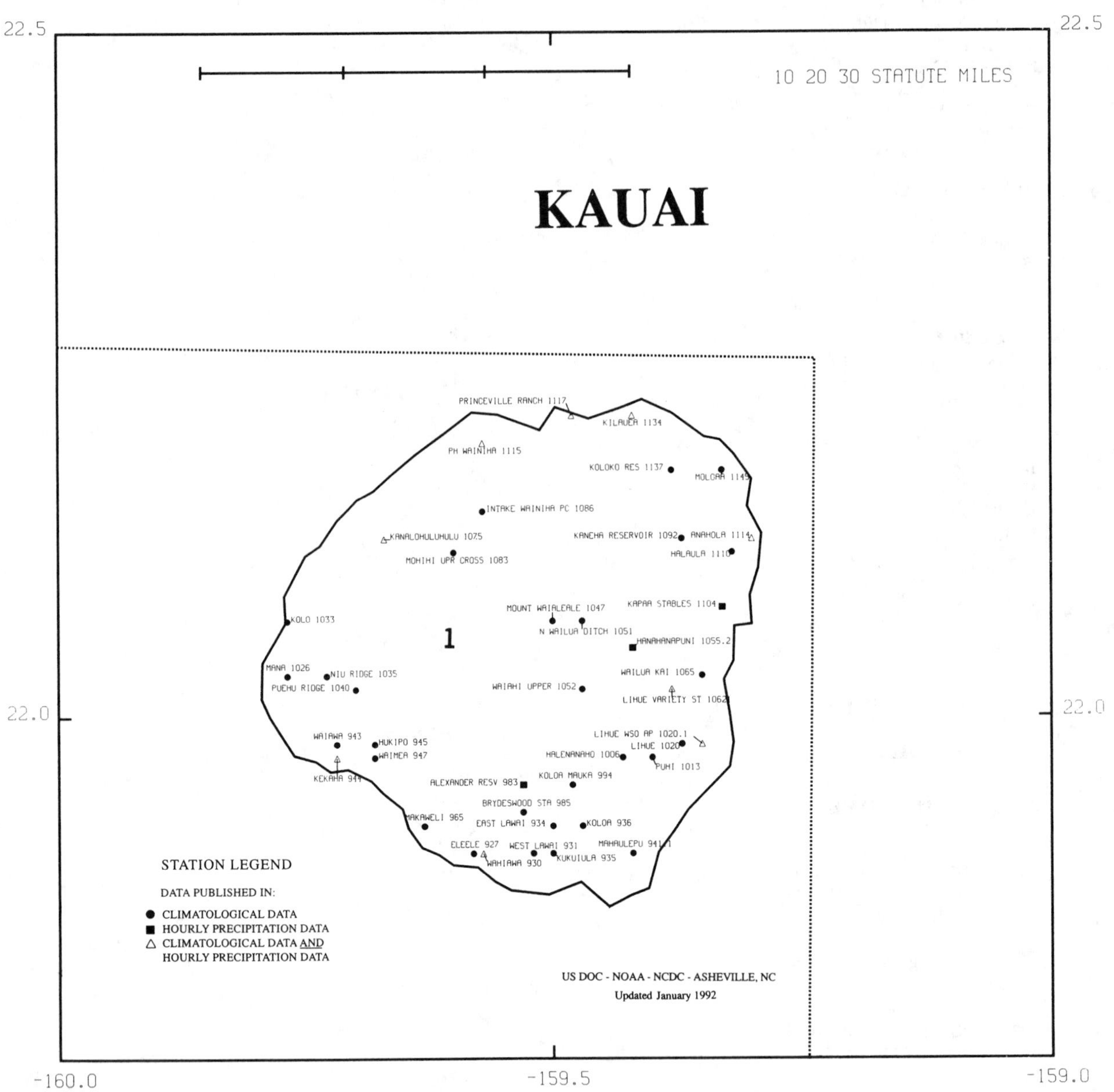
22.5
22.5
10 20 30 STATUTE MILES
KAUAI
PRINCEVILLE RANCH 1117
KILAUEA 1134
PH WAINIHA 1115
KOLOKO RES 1137
MOLOAA 114
INTAKE WAINIHA PC 1086
KANALOHULUHULU 1075
KANEHA RESERVOIR 1092
ANAHOLA 1114
HALAULA 1110
MOHIHI UPR CROSS 1083
MOUNT WAIALEALE 1047
KAPAA STABLES 1104
KOLO 1033
N WAILUA DITCH 1051
1
HANAHANAPUNI 1055.2
MANA 1026
NIU RIDGE 1035
WAILUA KAI 1065
PUEHU RIDGE 1040
WAIAHI UPPER 1052
LIHUE VARIETY ST 1062
22.0
22.0
WAIAWA 943
HUKIPO 945
LIHUE WSO AP 1020.1
LIHUE 1020
WAIMEA 947
HALENANAHO 1006
PUHI 1013
KEKAHA 944
ALEXANDER RESV 983
KOLOA MAUKA 994
BRYDESWOOD STA 985
MAKAWELI 965
EAST LAWAI 934
KOLOA 936
ELEELE 927
WEST LAWAI 931
MAHAULEPU 941
WAHIAWA 930
KUKUIULA 935
STATION LEGEND
DATA PUBLISHED IN:
CLIMATOLOGICAL DATA
HOURLY PRECIPITATION DATA
CLIMATOLOGICAL DATA AND
HOURLY PRECIPITATION DATA
US DOC - NOAA - NCDC - ASHEVILLE, NC
Updated January 1992
-160.0
-159.5
-159.0

10 2
21.8
21.8
OAHU
KAHUKU 912
KII-KAHUKU 911
PUPUKEA HEIGHTS 896.4
B Y U LAIE 903.1
WAIMEA ARBORETUM
WAIMEA 892
STATION LEGEND
DATA PUBLISHED IN:
CLIMATOLOGICAL DATA
HOURLY PRECIPITATION DATA
CLIMATOLOGICAL DATA AND HOURLY PRECIPITATION DATA
For further information, refer to the station index and references notes.
KAPAKA FARM 904.1
CAMP ERDMAN 841.16
KAWAIHAPAI 841
WAIALUA 847
OPAEULA 870
PUNALUU PUMP 905.2
KAENA POINT 841.3
KEMOO CAMP 8 855
MOUNT KAALA 844
UPPER WAHIAWA 874.3
KAHANA 883
21.5
21.5
KOOLAU DAM 833
WAHIAWA DAM 863
MAKAHA COUNTRY CLUB 800
WAIAWA 836
WAIAHOLE 837
WAIHEE 837.5
WAIANAE 798.2
2
COCONUT ISLAND 840.1
AHUIMANU LOOP 839.12
CAMP 84 807
KANEOHE MAUKA 781
KANEOHE 838.1
PEARL COUNTRY CL 760.2
HOKULOA 725.2
AIEA HEIGHTS 764.6
LULUKU 781.7
KAILUA FIRE STA 791.3
PALI GOLF COURSE 788.1
WAIPAHU 750
HALAWA SHAFT 771.2
PUU MANAWAHUA 725.6
KALIHI RES SITE 777
NUUANU RES 4 783
KAPALAMA 773
MAUNAWILI 787.1
EWA PLANTATION 741
MOANALUA 770
NUUANU RESERVR 5 775
PAUOA FLATS 784
MANOA TUNNEL 2 716
HONOLULU WSFO AP 703
DOWSETT 775.4
PALOLO VALLEY 718
TANTALUS 2 780.5
MANOA LYON ARBORETUM
WAIMANALO NONOKI 795.2
HONOLULU OBSERV 702.2
WAIMANALO EXP FM 795.1
CAMPBELL IND PK 702.5
BERETANIA PUMPING STN
PUNCHBOWL CRATER 709
MANOA 712.1
UNIV OF HAWAII 713
HAWAII KAI G C 724.19
WILHELMINA RISE 721
WAILUPE VLY SCH 723.6
WAIKIKI 717.2
US DOC - NOAA - NCDC - ASHEVILLE, NC Updated January 1992
WAIALAE KAHALA 715
21.3
21.3
-158.3
-158.0
-157.8

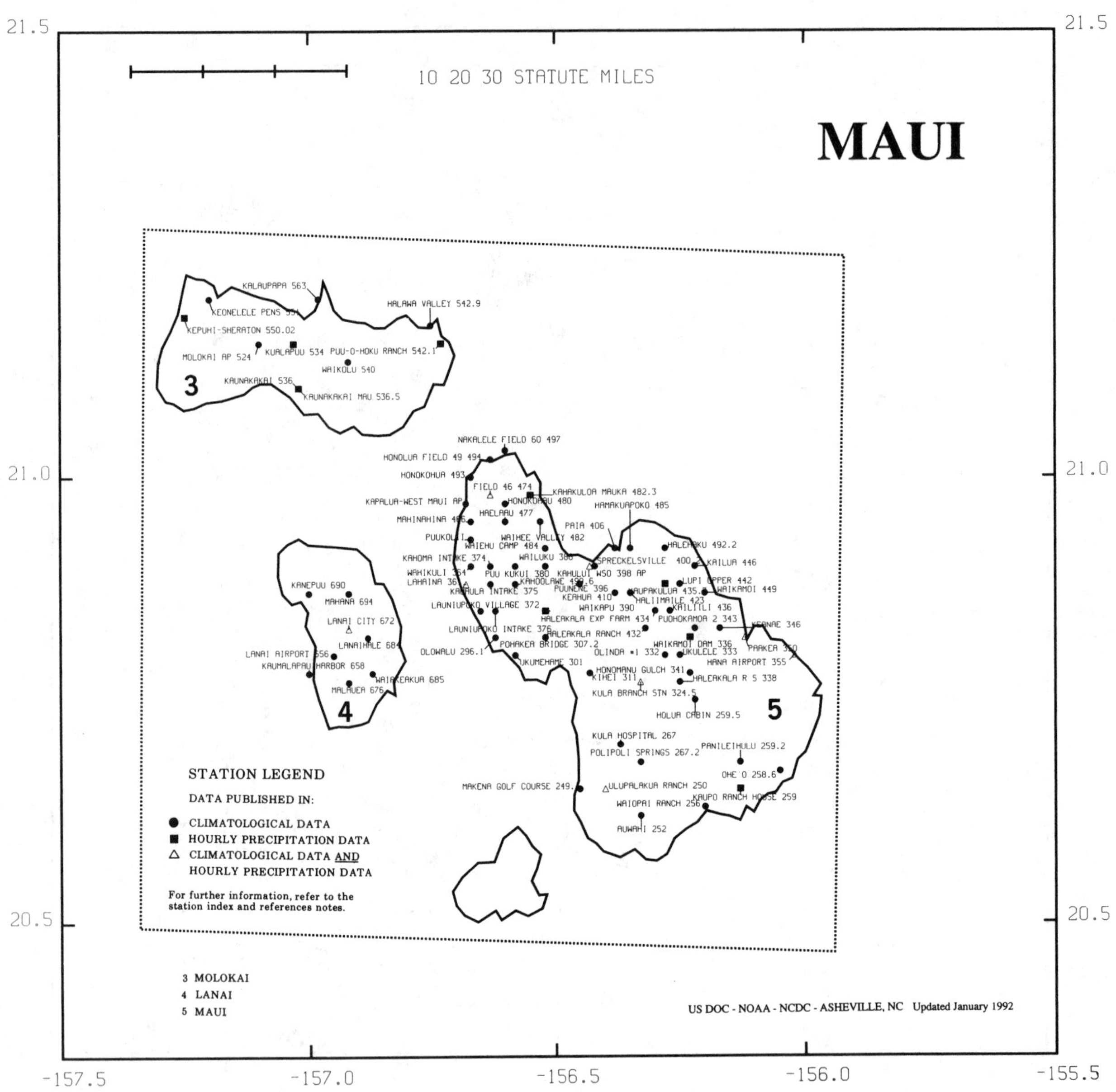
21.5
21.0
20.5
-157.5
-157.0
-156.5
-156.0
-155.5
10 20 30 STATUTE MILES
MAUI
KALAUPAPA 563
HALAWA VALLEY 542.9
KEPUHI-SHERATON 550.02
MOLOKAI AP 524
KUALAPUU 534
PUU-O-HOKU RANCH 542.1
WAIKOLU 540
KAUNAKAKAI 536
KAUNAKAKAI MAU 536.5
3
NAKALELE FIELD 60 497
HONOLUA FIELD 49 494
HONOKOHUA 493
FIELD 46 474
KAHAKULOA MAUKA 482.3
KAPALUA-WEST MAUI AP
HAELAAU 477
HAMAKUAPOKO 485
PAIA 406
WAIEHU CAMP 484
HALEHAKU 492.2
SPRECKELSVILLE 400.2
KAILUA 446
PUU KUKUI 380
KAHULUI WSO 398 AP
LUPI UPPER 442
PUUNENE 396
WAIKAMOI 449
KEAHUA 410
HALIIMAILE 423
LAUNIUPOKO VILLAGE 372
WAIKAPU 390
KAILIILI 436
HALEAKALA EXP FARM 434
PUOHOKAMOA 2 343
LAUNIUPOKO INTAKE 376
HALEAKALA RANCH 432
WAIKAMOI DAM 336
OLOWALU 296.1
POHAKEA BRIDGE 307.2
OLINDA #1 332
UKULELE 333
UKUMEHAME 301
HANA AIRPORT 355
HONOMANU GULCH 341
KIHEI 311
HALEAKALA R S 338
KULA BRANCH STN 324.5
HOLUA CABIN 259.5
5
KULA HOSPITAL 267
POLIPOLI SPRINGS 267.2
PANILEIHULU 259.2
OHE'O 258.6
MAKENA GOLF COURSE 249.
ULUPALAKUA RANCH 250
KAUPO RANCH HOUSE 259
WAIOPAI RANCH 256
AUWAHI 252
KANEPUU 690
MAHANA 694
LANAI CITY 672
LANAIHALE 684
LANAI AIRPORT 656
KAUMALAPAU HARBOR 658
WAIAKEAKUA 685
MALAUEA 676
4
STATION LEGEND
DATA PUBLISHED IN:
CLIMATOLOGICAL DATA
HOURLY PRECIPITATION DATA
CLIMATOLOGICAL DATA AND HOURLY PRECIPITATION DATA
For further information, refer to the station index and references notes.
3 MOLOKAI
4 LANAI
5 MAUI
US DOC - NOAA - NCDC - ASHEVILLE, NC Updated January 1992

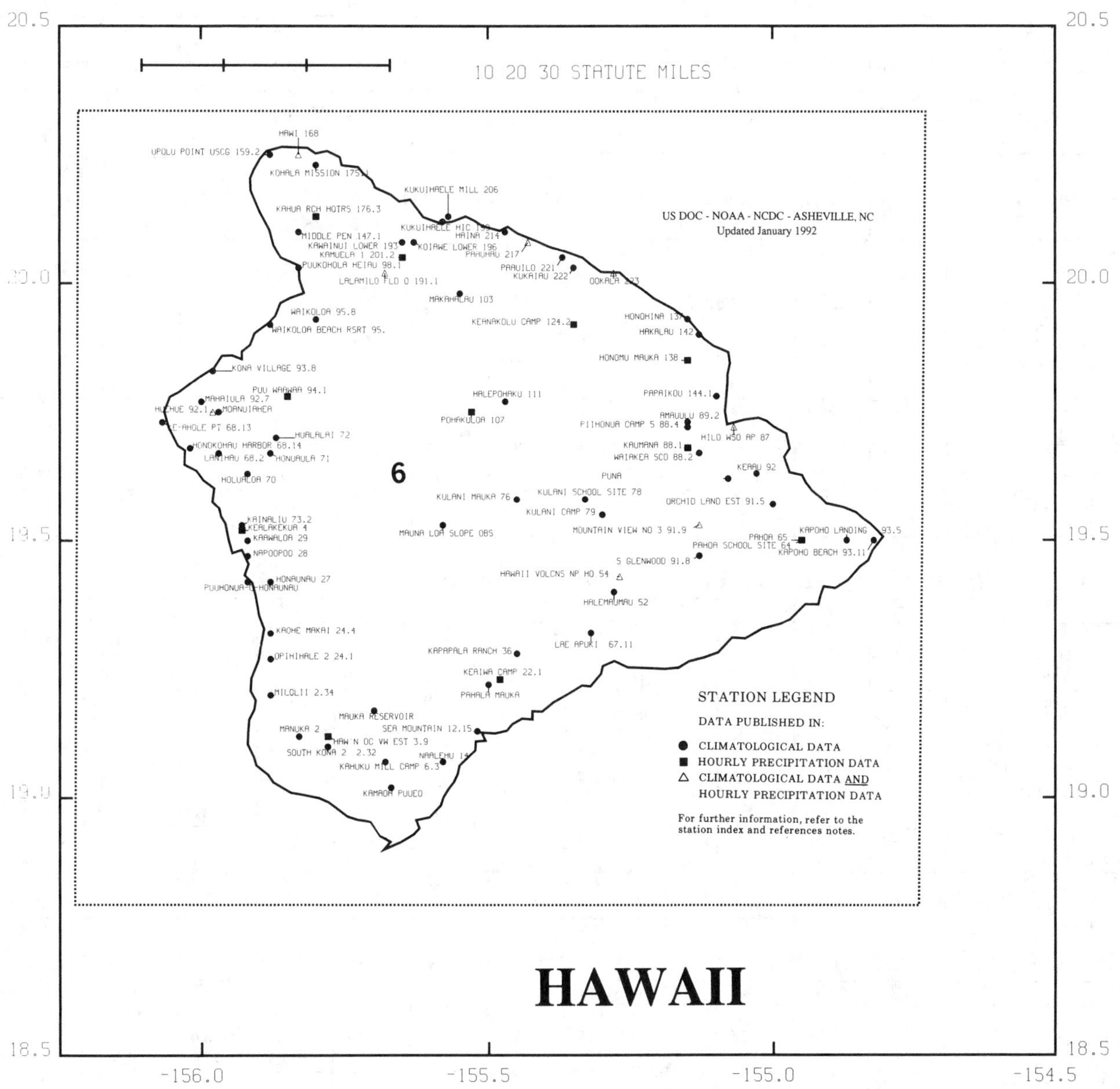
10 20 30 STATUTE MILES
US DOC - NOAA - NCDC - ASHEVILLE, NC
Updated January 1992
STATION LEGEND
DATA PUBLISHED IN:
● CLIMATOLOGICAL DATA
■ HOURLY PRECIPITATION DATA
△ CLIMATOLOGICAL DATA AND HOURLY PRECIPITATION DATA
For further information, refer to the station index and references notes.
HAWAII

## HALEAKALA RS 338 *Maui County* ELEVATION 7034 ft LAT/LONG 20° 46 ' N / 156° 15 ' W

| | JAN | FEB | MAR | APR | MAY | JUN | JUL | AUG | SEP | OCT | NOV | DEC | YEAR |
|---|---|---|---|---|---|---|---|---|---|---|---|---|---|
| Maximum Temp °F | 59.4 | 58.8 | 59.6 | 60.3 | 62.3 | 65.6 | 65.5 | 66.0 | 64.9 | 64.2 | 63.0 | 60.7 | 62.5 |
| Minimum Temp °F | 41.4 | 41.0 | 41.6 | 42.3 | 43.7 | 46.4 | 47.4 | 47.5 | 46.4 | 45.9 | 45.3 | 43.3 | 44.3 |
| Mean Temp °F | 50.4 | 49.9 | 50.7 | 51.3 | 53.0 | 56.0 | 56.5 | 56.8 | 55.7 | 55.1 | 54.2 | 52.0 | 53.5 |
| Days Max Temp ≥ 90 °F | 0 | 0 | 0 | 0 | 0 | 0 | 0 | 0 | 0 | 0 | 0 | 0 | 0 |
| Days Max Temp ≤ 32 °F | 0 | 0 | 0 | 0 | 0 | 0 | 0 | 0 | 0 | 0 | 0 | 0 | 0 |
| Days Min Temp ≤ 32 °F | 1 | 1 | 0 | 0 | 0 | 0 | 0 | 0 | 0 | 0 | 0 | 0 | 2 |
| Days Min Temp ≤ 0 °F | 0 | 0 | 0 | 0 | 0 | 0 | 0 | 0 | 0 | 0 | 0 | 0 | 0 |
| Heating Degree Days | na | na | na | na | na | na | na | na | na | na | na | na | na |
| Cooling Degree Days | 0 | 0 | 0 | 0 | 0 | 1 | 0 | 0 | 0 | 0 | 0 | 0 | 1 |
| Total Precipitation (") | 10.02 | 6.61 | 8.23 | 5.66 | 2.20 | 1.18 | 2.37 | 2.41 | 2.12 | 2.96 | 5.94 | 7.27 | 56.97 |
| Days ≥ 0.1" Precip | 8 | 7 | 9 | 8 | 5 | 3 | 6 | 5 | 6 | 6 | 7 | 7 | 77 |
| Total Snowfall (") | 0.0 | 0.0 | 0.0 | 0.0 | 0.0 | 0.0 | 0.0 | 0.0 | 0.0 | 0.0 | 0.0 | 0.0 | 0.0 |
| Days ≥ 1" Snow Depth | 0 | 0 | 0 | 0 | 0 | 0 | 0 | 0 | 0 | 0 | 0 | 0 | 0 |

## HANA AP 355 *Maui County* ELEVATION 66 ft LAT/LONG 20° 48 ' N / 156° 1 ' W

| | JAN | FEB | MAR | APR | MAY | JUN | JUL | AUG | SEP | OCT | NOV | DEC | YEAR |
|---|---|---|---|---|---|---|---|---|---|---|---|---|---|
| Maximum Temp °F | 78.2 | 78.2 | 78.5 | 79.3 | 81.0 | 82.6 | 82.9 | 83.8 | 84.1 | 82.9 | 80.8 | 78.6 | 80.9 |
| Minimum Temp °F | 63.9 | 64.1 | 65.3 | 66.7 | 67.8 | 69.1 | 70.2 | 70.8 | 70.2 | 69.6 | 68.7 | 66.3 | 67.7 |
| Mean Temp °F | 71.0 | 71.2 | 71.9 | 73.1 | 74.5 | 75.8 | 76.6 | 77.3 | 77.1 | 76.3 | 74.7 | 72.5 | 74.3 |
| Days Max Temp ≥ 90 °F | 0 | 0 | 0 | 0 | 0 | 0 | 0 | 0 | 0 | 0 | 0 | 0 | 0 |
| Days Max Temp ≤ 32 °F | 0 | 0 | 0 | 0 | 0 | 0 | 0 | 0 | 0 | 0 | 0 | 0 | 0 |
| Days Min Temp ≤ 32 °F | 0 | 0 | 0 | 0 | 0 | 0 | 0 | 0 | 0 | 0 | 0 | 0 | 0 |
| Days Min Temp ≤ 0 °F | 0 | 0 | 0 | 0 | 0 | 0 | 0 | 0 | 0 | 0 | 0 | 0 | 0 |
| Heating Degree Days | na | na | na | na | na | na | na | na | na | na | na | na | na |
| Cooling Degree Days | 198 | 179 | 216 | 245 | 285 | 331 | 367 | 391 | 370 | 355 | 304 | 240 | 3481 |
| Total Precipitation (") | 8.97 | 6.82 | 10.03 | 8.68 | 6.05 | 3.83 | 5.92 | 6.17 | 5.57 | 7.13 | 8.81 | 6.38 | 84.36 |
| Days ≥ 0.1" Precip | *12* | 10 | 13 | 14 | 13 | 12 | 16 | 15 | 13 | 14 | 15 | *13* | 160 |
| Total Snowfall (") | 0.0 | 0.0 | 0.0 | 0.0 | 0.0 | 0.0 | 0.0 | 0.0 | 0.0 | 0.0 | 0.0 | 0.0 | 0.0 |
| Days ≥ 1" Snow Depth | 0 | 0 | 0 | 0 | 0 | 0 | 0 | 0 | 0 | 0 | 0 | 0 | 0 |

## HAWAII VLC NP HQ 54 *Hawaii County* ELEVATION 3971 ft LAT/LONG 19° 26 ' N / 155° 16 ' W

| | JAN | FEB | MAR | APR | MAY | JUN | JUL | AUG | SEP | OCT | NOV | DEC | YEAR |
|---|---|---|---|---|---|---|---|---|---|---|---|---|---|
| Maximum Temp °F | 67.0 | 66.8 | 66.6 | 66.9 | 68.9 | 70.1 | 71.2 | 72.6 | 72.7 | 71.7 | 69.4 | 67.1 | 69.3 |
| Minimum Temp °F | 49.5 | 49.4 | 50.2 | 51.5 | 52.5 | 53.8 | 55.1 | 55.5 | 55.2 | 54.7 | 53.5 | 51.1 | 52.7 |
| Mean Temp °F | 58.3 | 58.1 | 58.4 | 59.2 | 60.7 | 62.0 | 63.2 | 64.1 | 64.0 | 63.3 | 61.4 | 59.1 | 61.0 |
| Days Max Temp ≥ 90 °F | 0 | 0 | 0 | 0 | 0 | 0 | 0 | 0 | 0 | 0 | 0 | 0 | 0 |
| Days Max Temp ≤ 32 °F | 0 | 0 | 0 | 0 | 0 | 0 | 0 | 0 | 0 | 0 | 0 | 0 | 0 |
| Days Min Temp ≤ 32 °F | 0 | 0 | 0 | 0 | 0 | 0 | 0 | 0 | 0 | 0 | 0 | 0 | 0 |
| Days Min Temp ≤ 0 °F | 0 | 0 | 0 | 0 | 0 | 0 | 0 | 0 | 0 | 0 | 0 | 0 | 0 |
| Heating Degree Days | na | na | na | na | na | na | na | na | na | na | na | na | na |
| Cooling Degree Days | 0 | 0 | 0 | 0 | 2 | 5 | 14 | 25 | 22 | 17 | 6 | 0 | 91 |
| Total Precipitation (") | 10.77 | 9.34 | 13.61 | 11.39 | 7.08 | 4.91 | 7.00 | 6.91 | 6.15 | 6.37 | 14.04 | 12.63 | 110.20 |
| Days ≥ 0.1" Precip | *11* | 10 | 17 | 18 | 16 | 13 | 14 | 12 | *12* | 14 | 14 | 15 | 166 |
| Total Snowfall (") | 0.0 | 0.0 | 0.0 | 0.0 | 0.0 | 0.0 | 0.0 | 0.0 | 0.0 | 0.0 | 0.0 | 0.0 | 0.0 |
| Days ≥ 1" Snow Depth | 0 | 0 | 0 | 0 | 0 | 0 | 0 | 0 | 0 | 0 | 0 | 0 | 0 |

## HILO GEN LYMAN FIELD *Hawaii County* ELEVATION 39 ft LAT/LONG 19° 43 ' N / 155° 4 ' W

| | JAN | FEB | MAR | APR | MAY | JUN | JUL | AUG | SEP | OCT | NOV | DEC | YEAR |
|---|---|---|---|---|---|---|---|---|---|---|---|---|---|
| Maximum Temp °F | 79.6 | 79.8 | 79.5 | 79.8 | 81.4 | 82.8 | 83.1 | 83.7 | 84.0 | 83.5 | 81.5 | 80.0 | 81.6 |
| Minimum Temp °F | 63.6 | 63.6 | 64.5 | 65.6 | 66.7 | 67.9 | 68.9 | 69.3 | 68.9 | 68.4 | 67.2 | 65.0 | 66.6 |
| Mean Temp °F | 71.6 | 71.7 | 72.1 | 72.7 | 74.1 | 75.3 | 76.1 | 76.5 | 76.5 | 76.0 | 74.4 | 72.5 | 74.1 |
| Days Max Temp ≥ 90 °F | 0 | 0 | 0 | 0 | 0 | 0 | 0 | 0 | 0 | 0 | 0 | 0 | 0 |
| Days Max Temp ≤ 32 °F | 0 | 0 | 0 | 0 | 0 | 0 | 0 | 0 | 0 | 0 | 0 | 0 | 0 |
| Days Min Temp ≤ 32 °F | 0 | 0 | 0 | 0 | 0 | 0 | 0 | 0 | 0 | 0 | 0 | 0 | 0 |
| Days Min Temp ≤ 0 °F | 0 | 0 | 0 | 0 | 0 | 0 | 0 | 0 | 0 | 0 | 0 | 0 | 0 |
| Heating Degree Days | na | na | na | na | na | na | na | na | na | na | na | na | na |
| Cooling Degree Days | 211 | 194 | 221 | 234 | 283 | 322 | 356 | 365 | 353 | 346 | 291 | 244 | 3420 |
| Total Precipitation (") | 9.78 | 10.06 | 14.59 | 14.72 | 8.80 | 6.60 | 10.56 | 10.67 | 9.02 | 8.99 | 15.06 | 12.38 | 131.23 |
| Days ≥ 0.1" Precip | 12 | 11 | 17 | 19 | 16 | 16 | 19 | 18 | 15 | 16 | 17 | 15 | 191 |
| Total Snowfall (") | 0.0 | 0.0 | 0.0 | 0.0 | 0.0 | 0.0 | 0.0 | 0.0 | 0.0 | 0.0 | 0.0 | 0.0 | 0.0 |
| Days ≥ 1" Snow Depth | 0 | 0 | 0 | 0 | 0 | 0 | 0 | 0 | 0 | 0 | 0 | 0 | 0 |

## HONOLULU INTL AP *Honolulu County* ELEVATION 39 ft LAT/LONG 21° 21 ' N / 157° 56 ' W

| | JAN | FEB | MAR | APR | MAY | JUN | JUL | AUG | SEP | OCT | NOV | DEC | YEAR |
|---|---|---|---|---|---|---|---|---|---|---|---|---|---|
| Maximum Temp °F | 80.1 | 80.5 | 81.6 | 83.0 | 84.9 | 86.9 | 87.8 | 89.0 | 88.9 | 87.2 | 84.4 | 81.5 | 84.6 |
| Minimum Temp °F | 65.2 | 65.3 | 67.2 | 68.7 | 70.3 | 72.4 | 73.7 | 74.4 | 73.7 | 72.4 | 70.5 | 67.3 | 70.1 |
| Mean Temp °F | 72.7 | 72.9 | 74.4 | 75.9 | 77.7 | 79.7 | 80.8 | 81.7 | 81.3 | 79.9 | 77.5 | 74.4 | 77.4 |
| Days Max Temp ≥ 90 °F | 0 | 0 | 0 | 0 | 0 | 2 | 5 | 13 | 11 | 5 | 0 | 0 | 36 |
| Days Max Temp ≤ 32 °F | 0 | 0 | 0 | 0 | 0 | 0 | 0 | 0 | 0 | 0 | 0 | 0 | 0 |
| Days Min Temp ≤ 32 °F | 0 | 0 | 0 | 0 | 0 | 0 | 0 | 0 | 0 | 0 | 0 | 0 | 0 |
| Days Min Temp ≤ 0 °F | 0 | 0 | 0 | 0 | 0 | 0 | 0 | 0 | 0 | 0 | 0 | 0 | 0 |
| Heating Degree Days | na | na | na | na | na | na | na | na | na | na | na | na | na |
| Cooling Degree Days | 251 | 228 | 301 | 338 | 402 | 459 | 508 | 535 | 512 | 477 | 397 | 308 | 4716 |
| Total Precipitation (") | 3.01 | 2.41 | 2.02 | 1.21 | 1.06 | 0.48 | 0.61 | 0.48 | 0.76 | 2.43 | 2.95 | 3.70 | 21.12 |
| Days ≥ 0.1" Precip | 4 | 4 | 3 | 3 | 2 | 1 | 1 | 1 | 2 | 3 | 4 | 5 | 33 |
| Total Snowfall (") | 0.0 | 0.0 | 0.0 | 0.0 | 0.0 | 0.0 | 0.0 | 0.0 | 0.0 | 0.0 | 0.0 | 0.0 | 0.0 |
| Days ≥ 1" Snow Depth | 0 | 0 | 0 | 0 | 0 | 0 | 0 | 0 | 0 | 0 | 0 | 0 | 0 |

## KAHULUI AP *Maui County* ELEVATION 40 ft LAT/LONG 20° 54 ' N / 156° 26 ' W

| | JAN | FEB | MAR | APR | MAY | JUN | JUL | AUG | SEP | OCT | NOV | DEC | YEAR |
|---|---|---|---|---|---|---|---|---|---|---|---|---|---|
| Maximum Temp °F | 79.9 | 80.1 | 81.1 | 82.2 | 84.4 | 86.0 | 86.9 | 87.8 | 88.0 | 86.8 | 84.1 | 81.3 | 84.1 |
| Minimum Temp °F | 63.3 | 63.2 | 64.6 | 66.0 | 67.0 | 69.2 | 70.8 | 71.2 | 70.1 | 69.4 | 67.9 | 65.4 | 67.3 |
| Mean Temp °F | 71.6 | 71.7 | 72.9 | 74.1 | 75.7 | 77.6 | 78.9 | 79.5 | 79.1 | 78.1 | 76.0 | 73.4 | 75.7 |
| Days Max Temp ≥ 90 °F | 0 | 0 | 0 | 0 | 1 | 2 | 3 | 6 | 7 | 5 | 1 | 0 | 25 |
| Days Max Temp ≤ 32 °F | 0 | 0 | 0 | 0 | 0 | 0 | 0 | 0 | 0 | 0 | 0 | 0 | 0 |
| Days Min Temp ≤ 32 °F | 0 | 0 | 0 | 0 | 0 | 0 | 0 | 0 | 0 | 0 | 0 | 0 | 0 |
| Days Min Temp ≤ 0 °F | 0 | 0 | 0 | 0 | 0 | 0 | 0 | 0 | 0 | 0 | 0 | 0 | 0 |
| Heating Degree Days | na | na | na | na | na | na | na | na | na | na | na | na | na |
| Cooling Degree Days | 228 | 208 | 258 | 282 | 347 | 399 | 451 | 467 | 442 | 419 | 351 | 275 | 4127 |
| Total Precipitation (") | 3.91 | 2.92 | 2.74 | 1.80 | 0.74 | 0.29 | 0.49 | 0.55 | 0.44 | 1.14 | 2.54 | 3.24 | 20.80 |
| Days ≥ 0.1" Precip | 6 | 5 | 5 | 4 | 2 | 1 | 1 | 1 | 1 | 2 | 4 | 5 | 37 |
| Total Snowfall (") | 0.0 | 0.0 | 0.0 | 0.0 | 0.0 | 0.0 | 0.0 | 0.0 | 0.0 | 0.0 | 0.0 | 0.0 | 0.0 |
| Days ≥ 1" Snow Depth | 0 | 0 | 0 | 0 | 0 | 0 | 0 | 0 | 0 | 0 | 0 | 0 | 0 |

## KAILUA 446 *Maui County* ELEVATION 702 ft LAT/LONG 20° 54 ' N / 156° 13 ' W

| | JAN | FEB | MAR | APR | MAY | JUN | JUL | AUG | SEP | OCT | NOV | DEC | YEAR |
|---|---|---|---|---|---|---|---|---|---|---|---|---|---|
| Maximum Temp °F | na | 75.1 | *74.9* | *75.2* | *76.8* | 78.0 | na | 79.3 | *80.2* | *79.3* | na | na | na |
| Minimum Temp °F | na | 61.6 | *62.5* | *63.6* | *65.1* | 66.3 | *67.5* | 67.7 | *67.6* | *66.7* | na | na | na |
| Mean Temp °F | na | 68.4 | *68.7* | *69.5* | *71.0* | 72.2 | *73.0* | 73.6 | *73.9* | *73.0* | na | na | na |
| Days Max Temp ≥ 90 °F | 0 | 0 | 0 | 0 | 0 | 0 | 0 | 0 | 0 | 0 | 0 | 0 | 0 |
| Days Max Temp ≤ 32 °F | 0 | 0 | 0 | 0 | 0 | 0 | 0 | 0 | 0 | 0 | 0 | 0 | 0 |
| Days Min Temp ≤ 32 °F | 0 | 0 | 0 | 0 | 0 | 0 | 0 | 0 | 0 | 0 | 0 | 0 | 0 |
| Days Min Temp ≤ 0 °F | 0 | 0 | 0 | 0 | 0 | 0 | 0 | 0 | 0 | 0 | 0 | 0 | 0 |
| Heating Degree Days | na | na | na | na | na | na | na | na | na | na | na | na | na |
| Cooling Degree Days | na | *112* | *128* | *152* | *187* | 231 | na | *277* | *276* | *258* | na | na | na |
| Total Precipitation (") | 11.07 | 10.44 | 16.26 | 13.89 | 8.84 | 6.98 | 10.59 | 9.73 | 7.54 | 9.03 | 13.61 | 10.04 | 128.02 |
| Days ≥ 0.1" Precip | *10* | *9* | *13* | *14* | *11* | *12* | *15* | *14* | na | na | na | na | na |
| Total Snowfall (") | 0.0 | 0.0 | 0.0 | 0.0 | 0.0 | 0.0 | 0.0 | 0.0 | 0.0 | 0.0 | 0.0 | 0.0 | 0.0 |
| Days ≥ 1" Snow Depth | 0 | 0 | 0 | 0 | 0 | 0 | 0 | 0 | 0 | 0 | 0 | 0 | 0 |

## KAINALIU 73.2 *Hawaii County* ELEVATION 1503 ft LAT/LONG 19° 32 ' N / 155° 56 ' W

| | JAN | FEB | MAR | APR | MAY | JUN | JUL | AUG | SEP | OCT | NOV | DEC | YEAR |
|---|---|---|---|---|---|---|---|---|---|---|---|---|---|
| Maximum Temp °F | *76.2* | 76.4 | na | *76.5* | *76.7* | *77.1* | na | *79.3* | *79.9* | *80.0* | na | na | na |
| Minimum Temp °F | *58.9* | 58.9 | na | *60.7* | *61.9* | *63.1* | *63.8* | *64.1* | *63.9* | *63.6* | na | na | na |
| Mean Temp °F | *67.6* | 67.7 | na | *68.6* | *69.3* | *70.1* | na | *71.7* | *71.9* | *71.8* | na | na | na |
| Days Max Temp ≥ 90 °F | 0 | 0 | 0 | 0 | 0 | 0 | 0 | 0 | 0 | 0 | 0 | 0 | 0 |
| Days Max Temp ≤ 32 °F | 0 | 0 | 0 | 0 | 0 | 0 | 0 | 0 | 0 | 0 | 0 | 0 | 0 |
| Days Min Temp ≤ 32 °F | 0 | 0 | 0 | 0 | 0 | 0 | 0 | 0 | 0 | 0 | 0 | 0 | 0 |
| Days Min Temp ≤ 0 °F | 0 | 0 | 0 | 0 | 0 | 0 | 0 | 0 | 0 | 0 | 0 | 0 | 0 |
| Heating Degree Days | na | na | na | na | na | na | na | na | na | na | na | na | na |
| Cooling Degree Days | na | *96* | na | na | na | na | na | na | na | na | na | na | na |
| Total Precipitation (") | 4.32 | 2.99 | 3.56 | 4.65 | 4.97 | 6.08 | 7.20 | 6.27 | 6.64 | 4.80 | 3.13 | 2.90 | 57.51 |
| Days ≥ 0.1" Precip | 4 | 3 | 5 | 6 | 8 | *10* | *10* | *10* | *9* | *6* | *4* | *4* | 79 |
| Total Snowfall (") | 0.0 | 0.0 | 0.0 | 0.0 | 0.0 | 0.0 | 0.0 | 0.0 | 0.0 | 0.0 | 0.0 | 0.0 | 0.0 |
| Days ≥ 1" Snow Depth | 0 | 0 | 0 | 0 | 0 | 0 | 0 | 0 | 0 | 0 | 0 | 0 | 0 |

## KANALOHULUHULU 1075 *Kauai County* ELEVATION 3602 ft LAT/LONG 22° 8' N / 159° 40' W

| | JAN | FEB | MAR | APR | MAY | JUN | JUL | AUG | SEP | OCT | NOV | DEC | YEAR |
|---|---|---|---|---|---|---|---|---|---|---|---|---|---|
| Maximum Temp °F | 62.8 | 63.7 | 64.8 | 65.9 | 68.1 | 70.3 | 71.0 | 72.2 | 72.1 | 70.1 | 66.9 | 63.6 | 67.6 |
| Minimum Temp °F | 46.7 | 46.9 | 48.5 | 50.0 | 50.9 | 53.4 | 55.2 | 55.5 | 53.5 | ***52.8*** | 51.3 | 48.7 | 51.1 |
| Mean Temp °F | 54.8 | 55.4 | 56.7 | 58.0 | 59.6 | 61.9 | 63.1 | 63.8 | 62.8 | 61.5 | 59.1 | 56.2 | 59.4 |
| Days Max Temp ≥ 90 °F | 0 | 0 | 0 | 0 | 0 | 0 | 0 | 0 | 0 | 0 | 0 | 0 | 0 |
| Days Max Temp ≤ 32 °F | 0 | 0 | 0 | 0 | 0 | 0 | 0 | 0 | 0 | 0 | 0 | 0 | 0 |
| Days Min Temp ≤ 32 °F | 0 | 0 | 0 | 0 | 0 | 0 | 0 | 0 | 0 | 0 | 0 | 0 | 0 |
| Days Min Temp ≤ 0 °F | 0 | 0 | 0 | 0 | 0 | 0 | 0 | 0 | 0 | 0 | 0 | 0 | 0 |
| Heating Degree Days | na | na | na | na | na | na | na | na | na | na | na | na | na |
| Cooling Degree Days | ***0*** | 0 | 0 | 1 | 0 | 3 | 8 | 10 | 7 | ***2*** | 0 | 0 | 31 |
| Total Precipitation (") | 11.91 | 8.40 | 6.67 | 4.52 | 3.45 | 1.77 | 2.21 | 2.18 | 2.40 | 4.94 | 7.58 | 10.93 | 66.96 |
| Days ≥ 0.1" Precip | 11 | 10 | 10 | 9 | 7 | 4 | 6 | 5 | 5 | 6 | 9 | 11 | 93 |
| Total Snowfall (") | 0.0 | 0.0 | 0.0 | 0.0 | 0.0 | 0.0 | 0.0 | 0.0 | 0.0 | 0.0 | 0.0 | 0.0 | 0.0 |
| Days ≥ 1" Snow Depth | 0 | 0 | 0 | 0 | 0 | 0 | 0 | 0 | 0 | 0 | 0 | 0 | 0 |

## KANEOHE MAUKA 781 *Honolulu County* ELEVATION 200 ft LAT/LONG 21° 25' N / 157° 49' W

| | JAN | FEB | MAR | APR | MAY | JUN | JUL | AUG | SEP | OCT | NOV | DEC | YEAR |
|---|---|---|---|---|---|---|---|---|---|---|---|---|---|
| Maximum Temp °F | 76.8 | 77.2 | 77.3 | 78.1 | 79.9 | 81.6 | 82.2 | 83.1 | 83.5 | 82.3 | 79.7 | 78.1 | 80.0 |
| Minimum Temp °F | 65.5 | 65.6 | 66.8 | 67.7 | 69.7 | 71.3 | 72.1 | 72.6 | 72.2 | 71.2 | 69.1 | 67.5 | 69.3 |
| Mean Temp °F | 71.2 | 71.4 | 72.1 | 73.0 | 74.8 | 76.5 | 77.1 | 77.9 | 77.9 | 76.8 | ***74.5*** | 72.9 | 74.7 |
| Days Max Temp ≥ 90 °F | 0 | 0 | 0 | 0 | 0 | 0 | 0 | 0 | 0 | 0 | 0 | 0 | 0 |
| Days Max Temp ≤ 32 °F | 0 | 0 | 0 | 0 | 0 | 0 | 0 | 0 | 0 | 0 | 0 | 0 | 0 |
| Days Min Temp ≤ 32 °F | 0 | 0 | 0 | 0 | 0 | 0 | 0 | 0 | 0 | 0 | 0 | 0 | 0 |
| Days Min Temp ≤ 0 °F | 0 | 0 | 0 | 0 | 0 | 0 | 0 | 0 | 0 | 0 | 0 | 0 | 0 |
| Heating Degree Days | na | na | na | na | na | na | na | na | na | na | na | na | na |
| Cooling Degree Days | 198 | 197 | 238 | 252 | 314 | 359 | 391 | ***414*** | 398 | 372 | ***302*** | 258 | 3693 |
| Total Precipitation (") | 8.52 | 7.02 | 6.83 | 6.59 | 5.25 | 4.00 | 5.11 | 4.44 | 4.90 | 7.08 | 11.20 | 8.00 | 78.94 |
| Days ≥ 0.1" Precip | 11 | 10 | 11 | 12 | 11 | 9 | 13 | 11 | 10 | 11 | 14 | 12 | 135 |
| Total Snowfall (") | 0.0 | 0.0 | 0.0 | 0.0 | 0.0 | 0.0 | 0.0 | 0.0 | 0.0 | 0.0 | 0.0 | 0.0 | 0.0 |
| Days ≥ 1" Snow Depth | 0 | 0 | 0 | 0 | 0 | 0 | 0 | 0 | 0 | 0 | 0 | 0 | 0 |

## KULANI CAMP 79 *Hawaii County* ELEVATION 5194 ft LAT/LONG 19° 33' N / 155° 18' W

| | JAN | FEB | MAR | APR | MAY | JUN | JUL | AUG | SEP | OCT | NOV | DEC | YEAR |
|---|---|---|---|---|---|---|---|---|---|---|---|---|---|
| Maximum Temp °F | 62.7 | 62.1 | 61.1 | 61.4 | 62.9 | 64.3 | 65.4 | 66.1 | 66.3 | 66.1 | 64.1 | 62.7 | 63.8 |
| Minimum Temp °F | 43.0 | 43.6 | 44.5 | 46.2 | 47.3 | 48.8 | 50.2 | 50.4 | 50.3 | 49.6 | 47.9 | 45.2 | 47.3 |
| Mean Temp °F | 52.9 | 52.9 | 52.8 | 53.8 | 55.1 | 56.6 | 57.8 | 58.3 | 58.3 | 57.9 | 56.1 | 53.9 | 55.5 |
| Days Max Temp ≥ 90 °F | 0 | 0 | 0 | 0 | 0 | 0 | 0 | 0 | 0 | 0 | 0 | 0 | 0 |
| Days Max Temp ≤ 32 °F | 0 | 0 | 0 | 0 | 0 | 0 | 0 | 0 | 0 | 0 | 0 | 0 | 0 |
| Days Min Temp ≤ 32 °F | 0 | 0 | 0 | 0 | 0 | 0 | 0 | 0 | 0 | 0 | 0 | 0 | 0 |
| Days Min Temp ≤ 0 °F | 0 | 0 | 0 | 0 | 0 | 0 | 0 | 0 | 0 | 0 | 0 | 0 | 0 |
| Heating Degree Days | na | na | na | na | na | na | na | na | na | na | na | na | na |
| Cooling Degree Days | 0 | 0 | 0 | 0 | 0 | 1 | 0 | 3 | 0 | 1 | 0 | 0 | 5 |
| Total Precipitation (") | 9.91 | 8.03 | 14.50 | 12.30 | 7.08 | 4.31 | 8.49 | 8.09 | 7.34 | 6.24 | 14.31 | 11.79 | 112.39 |
| Days ≥ 0.1" Precip | 10 | 9 | ***15*** | 20 | ***16*** | 11 | 14 | ***11*** | 12 | 14 | 14 | ***13*** | 159 |
| Total Snowfall (") | 0.0 | 0.0 | 0.0 | 0.0 | 0.0 | 0.0 | 0.0 | 0.0 | 0.0 | 0.0 | 0.0 | 0.0 | 0.0 |
| Days ≥ 1" Snow Depth | 0 | 0 | 0 | 0 | 0 | 0 | 0 | 0 | 0 | 0 | 0 | 0 | 0 |

## LAHAINA 361 *Maui County* ELEVATION 49 ft LAT/LONG 20° 53' N / 156° 40' W

| | JAN | FEB | MAR | APR | MAY | JUN | JUL | AUG | SEP | OCT | NOV | DEC | YEAR |
|---|---|---|---|---|---|---|---|---|---|---|---|---|---|
| Maximum Temp °F | 81.6 | 81.7 | 82.8 | 83.8 | 85.1 | 86.7 | 87.6 | 88.4 | 88.7 | 87.7 | 85.6 | 83.0 | 85.2 |
| Minimum Temp °F | 63.9 | 63.4 | 64.4 | 65.6 | 67.2 | 68.3 | 69.5 | 70.3 | 70.2 | 69.2 | 67.6 | 65.1 | 67.1 |
| Mean Temp °F | 72.7 | 72.6 | 73.6 | 74.7 | 76.2 | 77.5 | 78.6 | 79.4 | 79.4 | 78.5 | 76.6 | 74.1 | 76.2 |
| Days Max Temp ≥ 90 °F | 0 | 0 | 0 | 0 | 0 | 1 | 3 | 7 | 9 | 5 | 1 | 0 | 26 |
| Days Max Temp ≤ 32 °F | 0 | 0 | 0 | 0 | 0 | 0 | 0 | 0 | 0 | 0 | 0 | 0 | 0 |
| Days Min Temp ≤ 32 °F | 0 | 0 | 0 | 0 | 0 | 0 | 0 | 0 | 0 | 0 | 0 | 0 | 0 |
| Days Min Temp ≤ 0 °F | 0 | 0 | 0 | 0 | 0 | 0 | 0 | 0 | 0 | 0 | 0 | 0 | 0 |
| Heating Degree Days | na | na | na | na | na | na | na | na | na | na | na | na | na |
| Cooling Degree Days | 259 | 230 | 284 | ***311*** | ***363*** | 403 | 450 | 474 | 459 | 436 | 377 | ***303*** | 4349 |
| Total Precipitation (") | 3.19 | 2.44 | 1.55 | 0.99 | 0.60 | 0.08 | 0.12 | 0.20 | 0.38 | 1.11 | 1.80 | 2.89 | 15.35 |
| Days ≥ 0.1" Precip | 4 | 3 | 2 | 1 | 1 | 0 | 0 | 1 | 1 | 1 | 2 | 2 | 18 |
| Total Snowfall (") | 0.0 | 0.0 | 0.0 | 0.0 | 0.0 | 0.0 | 0.0 | 0.0 | 0.0 | 0.0 | 0.0 | 0.0 | 0.0 |
| Days ≥ 1" Snow Depth | 0 | 0 | 0 | 0 | 0 | 0 | 0 | 0 | 0 | 0 | 0 | 0 | 0 |

## LANAI CITY 672 *Maui County* ELEVATION 1621 ft LAT/LONG 20° 50 ' N / 156° 55 ' W

| | JAN | FEB | MAR | APR | MAY | JUN | JUL | AUG | SEP | OCT | NOV | DEC | YEAR |
|---|---|---|---|---|---|---|---|---|---|---|---|---|---|
| Maximum Temp °F | na | *73.3* | *73.4* | na | na | *76.7* | *77.3* | 78.2 | *78.8* | na | na | na | na |
| Minimum Temp °F | na | *59.3* | *60.6* | na | na | *64.1* | *65.2* | 66.0 | *65.5* | na | na | na | na |
| Mean Temp °F | na | *66.3* | *67.0* | na | na | *70.4* | *71.3* | 72.1 | *72.2* | na | na | na | na |
| Days Max Temp ≥ 90 °F | 0 | 0 | 0 | 0 | 0 | 0 | 0 | 0 | 0 | 0 | 0 | 0 | 0 |
| Days Max Temp ≤ 32 °F | 0 | 0 | 0 | 0 | 0 | 0 | 0 | 0 | 0 | 0 | 0 | 0 | 0 |
| Days Min Temp ≤ 32 °F | 0 | 0 | 0 | 0 | 0 | 0 | 0 | 0 | 0 | 0 | 0 | 0 | 0 |
| Days Min Temp ≤ 0 °F | 0 | 0 | 0 | 0 | 0 | 0 | 0 | 0 | 0 | 0 | 0 | 0 | 0 |
| Heating Degree Days | na | na | na | na | na | na | na | na | na | na | na | na | na |
| Cooling Degree Days | na | na | *77* | na | na | 170 | *203* | 233 | na | na | na | na | na |
| Total Precipitation (") | 5.55 | 4.28 | 3.09 | 3.08 | 2.69 | 1.45 | 1.95 | 1.52 | 2.48 | 2.80 | 3.64 | 4.41 | 36.94 |
| Days ≥ 0.1" Precip | na | na | na | na | *4* | *2* | *3* | *3* | na | na | na | na | na |
| Total Snowfall (") | 0.0 | 0.0 | 0.0 | 0.0 | 0.0 | 0.0 | 0.0 | 0.0 | 0.0 | 0.0 | 0.0 | 0.0 | 0.0 |
| Days ≥ 1" Snow Depth | 0 | 0 | 0 | 0 | 0 | 0 | 0 | 0 | 0 | 0 | 0 | 0 | 0 |

## LIHUE AP *Kauai County* ELEVATION 103 ft LAT/LONG 21° 59 ' N / 159° 21 ' W

| | JAN | FEB | MAR | APR | MAY | JUN | JUL | AUG | SEP | OCT | NOV | DEC | YEAR |
|---|---|---|---|---|---|---|---|---|---|---|---|---|---|
| Maximum Temp °F | 77.6 | 77.8 | 78.2 | 79.2 | 81.0 | 82.9 | 83.9 | 84.8 | 84.8 | 83.2 | 80.7 | 78.7 | 81.1 |
| Minimum Temp °F | 65.1 | 65.2 | 67.0 | 68.5 | 70.3 | 72.6 | 73.9 | 74.5 | 73.8 | 72.3 | 70.6 | 67.5 | 70.1 |
| Mean Temp °F | 71.4 | 71.5 | 72.6 | 73.9 | 75.7 | 77.8 | 78.9 | 79.6 | 79.3 | 77.8 | 75.7 | 73.1 | 75.6 |
| Days Max Temp ≥ 90 °F | 0 | 0 | 0 | 0 | 0 | 0 | 0 | 0 | 0 | 0 | 0 | 0 | 0 |
| Days Max Temp ≤ 32 °F | 0 | 0 | 0 | 0 | 0 | 0 | 0 | 0 | 0 | 0 | 0 | 0 | 0 |
| Days Min Temp ≤ 32 °F | 0 | 0 | 0 | 0 | 0 | 0 | 0 | 0 | 0 | 0 | 0 | 0 | 0 |
| Days Min Temp ≤ 0 °F | 0 | 0 | 0 | 0 | 0 | 0 | 0 | 0 | 0 | 0 | 0 | 0 | 0 |
| Heating Degree Days | na | na | na | na | na | na | na | na | na | na | na | na | na |
| Cooling Degree Days | 210 | 194 | 251 | 282 | 334 | 395 | 446 | 469 | 443 | 406 | 333 | 261 | 4024 |
| Total Precipitation (") | 4.94 | 3.51 | 3.66 | 2.99 | 3.00 | 1.69 | 2.09 | 1.76 | 2.44 | 4.74 | 5.21 | 5.73 | 41.76 |
| Days ≥ 0.1" Precip | 7 | 6 | 6 | 7 | 6 | 4 | 6 | 5 | 6 | 8 | 8 | 8 | 77 |
| Total Snowfall (") | 0.0 | 0.0 | 0.0 | 0.0 | 0.0 | 0.0 | 0.0 | 0.0 | 0.0 | 0.0 | 0.0 | 0.0 | 0.0 |
| Days ≥ 1" Snow Depth | 0 | 0 | 0 | 0 | 0 | 0 | 0 | 0 | 0 | 0 | 0 | 0 | 0 |

## MAUNA LOA SLOPE OBS *Hawaii County* ELEVATION 10968 ft LAT/LONG 19° 33 ' N / 155° 35 ' W

| | JAN | FEB | MAR | APR | MAY | JUN | JUL | AUG | SEP | OCT | NOV | DEC | YEAR |
|---|---|---|---|---|---|---|---|---|---|---|---|---|---|
| Maximum Temp °F | 49.7 | 49.6 | 50.5 | 51.6 | 53.8 | 57.1 | 56.2 | 56.2 | 55.3 | 54.6 | 52.6 | 50.4 | 53.1 |
| Minimum Temp °F | 33.8 | 33.5 | 33.9 | 34.9 | 37.1 | 40.0 | 39.2 | 39.4 | 38.9 | 38.3 | *37.0* | 34.8 | 36.7 |
| Mean Temp °F | 41.8 | 41.6 | 42.2 | 43.3 | 45.4 | 48.6 | 47.8 | 47.8 | 47.1 | 46.5 | 44.8 | 42.6 | 45.0 |
| Days Max Temp ≥ 90 °F | 0 | 0 | 0 | 0 | 0 | 0 | 0 | 0 | 0 | 0 | 0 | 0 | 0 |
| Days Max Temp ≤ 32 °F | 0 | 0 | 0 | 0 | 0 | 0 | 0 | 0 | 0 | 0 | 0 | 0 | 0 |
| Days Min Temp ≤ 32 °F | 11 | 11 | 11 | 8 | 3 | 1 | 1 | 0 | 0 | 1 | 3 | 8 | 58 |
| Days Min Temp ≤ 0 °F | 0 | 0 | 0 | 0 | 0 | 0 | 0 | 0 | 0 | 0 | 0 | 0 | 0 |
| Heating Degree Days | na | na | na | na | na | na | na | na | na | na | na | na | na |
| Cooling Degree Days | *0* | 0 | 0 | 0 | 0 | 0 | 0 | 0 | 0 | 0 | *0* | 0 | 0 |
| Total Precipitation (") | 3.04 | 1.70 | 2.34 | 1.58 | 1.14 | 0.54 | 1.40 | 1.54 | 1.59 | 1.14 | 2.40 | 2.27 | 20.68 |
| Days ≥ 0.1" Precip | 2 | 2 | 3 | 3 | 2 | 1 | 2 | 2 | *2* | 2 | 2 | 2 | 25 |
| Total Snowfall (") | na | na | na | na | na | na | na | na | na | na | na | na | na |
| Days ≥ 1" Snow Depth | na | na | na | na | na | na | na | na | na | na | na | na | na |

## MOLOKAI AP 524 *Maui County* ELEVATION 453 ft LAT/LONG 21° 19 ' N / 157° 6 ' W

| | JAN | FEB | MAR | APR | MAY | JUN | JUL | AUG | SEP | OCT | NOV | DEC | YEAR |
|---|---|---|---|---|---|---|---|---|---|---|---|---|---|
| Maximum Temp °F | 77.6 | 77.6 | 78.7 | 79.8 | 81.9 | 83.5 | 84.5 | 85.5 | 85.9 | 84.6 | 82.1 | 79.1 | 81.7 |
| Minimum Temp °F | 63.8 | 63.1 | 64.3 | 65.9 | 67.4 | 69.6 | 70.7 | 71.3 | 71.0 | 70.2 | 68.2 | 65.4 | 67.6 |
| Mean Temp °F | 70.7 | 70.4 | 71.6 | 72.9 | 74.7 | 76.6 | 77.6 | 78.4 | 78.5 | 77.4 | 75.2 | 72.3 | 74.7 |
| Days Max Temp ≥ 90 °F | 0 | 0 | 0 | 0 | 0 | 0 | 0 | 1 | 2 | 1 | 0 | 0 | 4 |
| Days Max Temp ≤ 32 °F | 0 | 0 | 0 | 0 | 0 | 0 | 0 | 0 | 0 | 0 | 0 | 0 | 0 |
| Days Min Temp ≤ 32 °F | 0 | 0 | 0 | 0 | 0 | 0 | 0 | 0 | 0 | 0 | 0 | 0 | 0 |
| Days Min Temp ≤ 0 °F | 0 | 0 | *0* | 0 | 0 | *0* | 0 | 0 | 0 | 0 | 0 | 0 | 0 |
| Heating Degree Days | na | na | na | na | na | na | na | na | na | na | na | na | na |
| Cooling Degree Days | 185 | 162 | 205 | 240 | 296 | 356 | 404 | 429 | 418 | 395 | 319 | 235 | 3644 |
| Total Precipitation (") | 4.14 | 3.92 | 2.55 | 2.15 | 1.73 | 0.66 | 0.68 | 0.65 | 0.82 | 2.21 | 3.26 | 4.45 | 27.22 |
| Days ≥ 0.1" Precip | 7 | *6* | 5 | 4 | 3 | 2 | 2 | 2 | 3 | 3 | 5 | 7 | 49 |
| Total Snowfall (") | 0.0 | 0.0 | 0.0 | 0.0 | 0.0 | 0.0 | 0.0 | 0.0 | 0.0 | 0.0 | 0.0 | 0.0 | 0.0 |
| Days ≥ 1" Snow Depth | 0 | 0 | 0 | 0 | 0 | 0 | 0 | 0 | 0 | 0 | 0 | 0 | 0 |

## NAALEHU 14 *Hawaii County* ELEVATION 669 ft LAT/LONG 19° 4 ' N / 155° 35 ' W

| | JAN | FEB | MAR | APR | MAY | JUN | JUL | AUG | SEP | OCT | NOV | DEC | YEAR |
|---|---|---|---|---|---|---|---|---|---|---|---|---|---|
| Maximum Temp °F | ***77.4*** | 77.9 | 77.8 | 77.8 | 78.7 | 79.7 | ***80.8*** | 81.8 | 82.2 | 81.4 | ***79.5*** | 77.8 | 79.4 |
| Minimum Temp °F | ***62.4*** | 62.5 | 63.2 | 64.0 | 65.3 | 66.3 | ***67.6*** | 67.8 | 67.8 | 67.5 | ***66.6*** | ***64.3*** | 65.4 |
| Mean Temp °F | ***69.9*** | 70.2 | 70.5 | 70.9 | 72.0 | 73.0 | ***74.3*** | 74.8 | 75.0 | 74.4 | ***73.1*** | ***71.0*** | 72.4 |
| Days Max Temp ≥ 90 °F | 0 | 0 | 0 | 0 | 0 | 0 | 0 | 0 | 0 | 0 | 0 | 0 | 0 |
| Days Max Temp ≤ 32 °F | 0 | 0 | 0 | 0 | 0 | 0 | 0 | 0 | 0 | 0 | 0 | 0 | 0 |
| Days Min Temp ≤ 32 °F | 0 | 0 | 0 | 0 | 0 | 0 | 0 | 0 | 0 | 0 | 0 | 0 | 0 |
| Days Min Temp ≤ 0 °F | 0 | 0 | 0 | 0 | 0 | 0 | 0 | 0 | 0 | 0 | 0 | 0 | 0 |
| Heating Degree Days | na | na | na | na | na | na | na | na | na | na | na | na | na |
| Cooling Degree Days | 156 | 163 | 180 | 184 | 219 | 249 | 296 | 310 | 301 | 295 | 249 | 196 | 2798 |
| Total Precipitation (") | 6.50 | 3.96 | 5.03 | 3.80 | 2.63 | 1.53 | 3.12 | 3.64 | 3.78 | 4.12 | 6.75 | 6.64 | 51.50 |
| Days ≥ 0.1" Precip | ***6*** | ***4*** | ***5*** | na | na | na | na | ***5*** | na | ***6*** | na | na | na |
| Total Snowfall (") | 0.0 | 0.0 | 0.0 | 0.0 | 0.0 | 0.0 | 0.0 | 0.0 | 0.0 | 0.0 | 0.0 | 0.0 | 0.0 |
| Days ≥ 1" Snow Depth | 0 | 0 | 0 | 0 | 0 | 0 | 0 | 0 | 0 | 0 | 0 | 0 | 0 |

## OPIHIHALE 2 24.1 *Hawaii County* ELEVATION 1302 ft LAT/LONG 19° 16 ' N / 155° 53 ' W

| | JAN | FEB | MAR | APR | MAY | JUN | JUL | AUG | SEP | OCT | NOV | DEC | YEAR |
|---|---|---|---|---|---|---|---|---|---|---|---|---|---|
| Maximum Temp °F | 76.2 | 76.5 | 76.8 | 77.4 | 78.1 | 79.0 | 80.1 | 80.9 | 81.0 | 80.6 | 79.1 | 77.4 | 78.6 |
| Minimum Temp °F | 57.3 | 57.4 | 58.5 | 59.7 | 61.2 | 62.5 | 63.3 | 63.8 | 63.2 | 62.3 | 60.9 | 58.4 | 60.7 |
| Mean Temp °F | 66.8 | 67.0 | 67.7 | 68.6 | 69.7 | 70.8 | 71.8 | 72.4 | 72.1 | 71.5 | 70.0 | 67.9 | 69.7 |
| Days Max Temp ≥ 90 °F | 0 | 0 | 0 | 0 | 0 | 0 | 0 | 0 | 0 | 0 | 0 | 0 | 0 |
| Days Max Temp ≤ 32 °F | 0 | 0 | 0 | 0 | 0 | 0 | 0 | 0 | 0 | 0 | 0 | 0 | 0 |
| Days Min Temp ≤ 32 °F | 0 | 0 | 0 | 0 | 0 | 0 | 0 | 0 | 0 | 0 | 0 | 0 | 0 |
| Days Min Temp ≤ 0 °F | 0 | 0 | 0 | 0 | 0 | 0 | 0 | 0 | 0 | 0 | 0 | 0 | 0 |
| Heating Degree Days | na | na | na | na | na | na | na | na | na | na | na | na | na |
| Cooling Degree Days | 88 | 86 | 110 | 130 | 169 | 202 | 237 | 257 | 248 | 232 | 183 | 124 | 2066 |
| Total Precipitation (") | 3.47 | 2.90 | 3.47 | 3.27 | 3.49 | 3.22 | 3.86 | 3.96 | 4.37 | 3.60 | 3.08 | 2.87 | 41.56 |
| Days ≥ 0.1" Precip | 5 | 4 | 6 | 7 | 9 | 8 | 9 | 9 | 9 | 7 | 6 | 5 | 84 |
| Total Snowfall (") | 0.0 | 0.0 | 0.0 | 0.0 | 0.0 | 0.0 | 0.0 | 0.0 | 0.0 | 0.0 | 0.0 | 0.0 | 0.0 |
| Days ≥ 1" Snow Depth | 0 | 0 | 0 | 0 | 0 | 0 | 0 | 0 | 0 | 0 | 0 | 0 | 0 |

## WAIALUA 847 *Honolulu County* ELEVATION 30 ft LAT/LONG 21° 35 ' N / 158° 7 ' W

| | JAN | FEB | MAR | APR | MAY | JUN | JUL | AUG | SEP | OCT | NOV | DEC | YEAR |
|---|---|---|---|---|---|---|---|---|---|---|---|---|---|
| Maximum Temp °F | na | 78.9 | 79.8 | ***80.3*** | ***82.8*** | 84.6 | ***85.7*** | ***86.8*** | ***87.1*** | ***85.3*** | na | na | na |
| Minimum Temp °F | na | 60.0 | 61.0 | ***62.7*** | ***63.5*** | 65.5 | ***66.5*** | 66.8 | ***66.4*** | ***65.7*** | na | na | na |
| Mean Temp °F | na | 69.5 | 70.4 | ***71.5*** | ***73.1*** | 75.1 | ***76.2*** | 76.8 | ***76.8*** | ***75.6*** | na | na | na |
| Days Max Temp ≥ 90 °F | 0 | 0 | 0 | 0 | 0 | 0 | 1 | 3 | 4 | 2 | 0 | ***0*** | 10 |
| Days Max Temp ≤ 32 °F | 0 | 0 | 0 | 0 | 0 | 0 | 0 | 0 | 0 | 0 | 0 | ***0*** | 0 |
| Days Min Temp ≤ 32 °F | 0 | 0 | 0 | 0 | 0 | 0 | 0 | 0 | 0 | 0 | 0 | ***0*** | 0 |
| Days Min Temp ≤ 0 °F | 0 | 0 | 0 | 0 | 0 | 0 | 0 | 0 | 0 | 0 | 0 | ***0*** | 0 |
| Heating Degree Days | na | na | na | na | na | na | na | na | na | na | na | na | na |
| Cooling Degree Days | na | ***145*** | ***188*** | na | na | 310 | na | ***380*** | na | na | na | na | na |
| Total Precipitation (") | 5.44 | 3.99 | 3.12 | 2.29 | 1.39 | 0.95 | 1.29 | 0.84 | 1.22 | 2.65 | 3.71 | 3.96 | 30.85 |
| Days ≥ 0.1" Precip | ***5*** | 5 | 4 | 4 | 2 | 1 | 2 | 2 | 2 | 3 | ***4*** | ***4*** | 38 |
| Total Snowfall (") | 0.0 | 0.0 | 0.0 | 0.0 | 0.0 | 0.0 | 0.0 | 0.0 | 0.0 | 0.0 | 0.0 | 0.0 | 0.0 |
| Days ≥ 1" Snow Depth | 0 | 0 | 0 | 0 | 0 | 0 | 0 | 0 | 0 | 0 | 0 | 0 | 0 |

## WAIKIKI 717.2 *Honolulu County* ELEVATION 10 ft LAT/LONG 21° 16 ' N / 157° 49 ' W

| | JAN | FEB | MAR | APR | MAY | JUN | JUL | AUG | SEP | OCT | NOV | DEC | YEAR |
|---|---|---|---|---|---|---|---|---|---|---|---|---|---|
| Maximum Temp °F | 81.1 | 81.2 | 82.1 | 83.2 | 84.9 | 86.5 | 87.4 | 88.3 | 88.4 | 87.0 | 84.5 | 82.0 | 84.7 |
| Minimum Temp °F | 64.2 | 64.2 | 66.0 | 67.3 | 68.6 | 70.7 | 72.1 | 72.5 | 71.7 | 70.7 | 69.2 | 66.3 | 68.6 |
| Mean Temp °F | 72.6 | 72.7 | 74.1 | 75.3 | 76.8 | 78.6 | 79.7 | 80.4 | 80.1 | 78.9 | 76.9 | 74.2 | 76.7 |
| Days Max Temp ≥ 90 °F | 0 | 0 | 0 | 0 | 0 | 2 | 5 | 8 | 9 | 3 | 0 | 0 | 27 |
| Days Max Temp ≤ 32 °F | 0 | 0 | 0 | 0 | 0 | 0 | 0 | 0 | 0 | 0 | 0 | 0 | 0 |
| Days Min Temp ≤ 32 °F | 0 | 0 | 0 | 0 | 0 | 0 | 0 | 0 | 0 | 0 | 0 | 0 | 0 |
| Days Min Temp ≤ 0 °F | 0 | 0 | 0 | 0 | 0 | 0 | 0 | 0 | 0 | 0 | 0 | 0 | 0 |
| Heating Degree Days | na | na | na | na | na | na | na | na | na | na | na | na | na |
| Cooling Degree Days | 253 | 231 | 298 | 328 | 381 | 430 | 478 | 492 | 471 | 443 | 379 | 300 | 4484 |
| Total Precipitation (") | 3.64 | 2.71 | 2.17 | 1.41 | 1.50 | 0.72 | 0.90 | 0.70 | 1.09 | 2.39 | 3.39 | 4.27 | 24.89 |
| Days ≥ 0.1" Precip | 5 | 4 | 4 | 4 | 3 | 1 | 2 | 1 | 2 | 3 | 5 | 6 | 40 |
| Total Snowfall (") | 0.0 | 0.0 | 0.0 | 0.0 | 0.0 | 0.0 | 0.0 | 0.0 | 0.0 | 0.0 | 0.0 | 0.0 | 0.0 |
| Days ≥ 1" Snow Depth | 0 | 0 | 0 | 0 | 0 | 0 | 0 | 0 | 0 | 0 | 0 | 0 | 0 |

## JANUARY MINIMUM TEMPERATURE °F

| LOWEST | | | HIGHEST | | |
|---|---|---|---|---|---|
| 1 | Mauna Loa | 33.8 | 1 | Kaneohe Mauka | 65.5 |
| 2 | Haleakala | 41.4 | 2 | Honolulu | 65.2 |
| 3 | Kulani Camp | 43.0 | 3 | Lihue | 65.1 |
| 4 | Kanalohuluhulu | 46.7 | 4 | Waikiki | 64.2 |
| 5 | Hawaii Volcano | 49.5 | 5 | Hana | 63.9 |
| 6 | Opihihale | 57.3 | | Lahaina | 63.9 |
| 7 | Kainaliu | 58.9 | 7 | Molokai | 63.8 |
| 8 | Naalehu | 62.4 | 8 | Hilo | 63.6 |
| 9 | Kahului | 63.3 | 9 | Kahului | 63.3 |
| 10 | Hilo | 63.6 | 10 | Naalehu | 62.4 |
| 11 | Molokai | 63.8 | 11 | Kainaliu | 58.9 |
| 12 | Hana | 63.9 | 12 | Opihihale | 57.3 |
| | Lahaina | 63.9 | 13 | Hawaii Volcano | 49.5 |
| 14 | Waikiki | 64.2 | 14 | Kanalohuluhulu | 46.7 |
| 15 | Lihue | 65.1 | 15 | Kulani Camp | 43.0 |
| 16 | Honolulu | 65.2 | 16 | Haleakala | 41.4 |
| 17 | Kaneohe Mauka | 65.5 | 17 | Mauna Loa | 33.8 |

## JULY MAXIMUM TEMPERATURE °F

| HIGHEST | | | LOWEST | | |
|---|---|---|---|---|---|
| 1 | Honolulu | 87.8 | 1 | Mauna Loa | 56.2 |
| 2 | Lahaina | 87.6 | 2 | Kulani Camp | 65.4 |
| 3 | Waikiki | 87.4 | 3 | Haleakala | 65.5 |
| 4 | Kahului | 86.9 | 4 | Kanalohuluhulu | 71.0 |
| 5 | Waialua | 85.7 | 5 | Hawaii Volcano | 71.2 |
| 6 | Molokai | 84.5 | 6 | Lanai City | 77.3 |
| 7 | Lihue | 83.9 | 7 | Opihihale | 80.1 |
| 8 | Hilo | 83.1 | 8 | Naalehu | 80.8 |
| 9 | Hana | 82.9 | 9 | Kaneohe Mauka | 82.2 |
| 10 | Kaneohe Mauka | 82.2 | 10 | Hana | 82.9 |
| 11 | Naalehu | 80.8 | 11 | Hilo | 83.1 |
| 12 | Opihihale | 80.1 | 12 | Lihue | 83.9 |
| 13 | Lanai City | 77.3 | 13 | Molokai | 84.5 |
| 14 | Hawaii Volcano | 71.2 | 14 | Waialua | 85.7 |
| 15 | Kanalohuluhulu | 71.0 | 15 | Kahului | 86.9 |
| 16 | Haleakala | 65.5 | 16 | Waikiki | 87.4 |
| 17 | Kulani Camp | 65.4 | 17 | Lahaina | 87.6 |
| 18 | Mauna Loa | 56.2 | 18 | Honolulu | 87.8 |

## ANNUAL PRECIPITATION (")

| HIGHEST | | | LOWEST | | |
|---|---|---|---|---|---|
| 1 | Hilo | 131.23 | 1 | Lahaina | 15.35 |
| 2 | Kailua | 128.02 | 2 | Mauna Loa | 20.68 |
| 3 | Kulani Camp | 112.39 | 3 | Kahului | 20.80 |
| 4 | Hawaii Volcano | 110.20 | 4 | Honolulu | 21.12 |
| 5 | Hana | 84.36 | 5 | Waikiki | 24.89 |
| 6 | Kaneohe Mauka | 78.94 | 6 | Molokai | 27.22 |
| 7 | Kanalohuluhulu | 66.96 | 7 | Waialua | 30.85 |
| 8 | Kainaliu | 57.51 | 8 | Lanai City | 36.94 |
| 9 | Haleakala | 56.97 | 9 | Opihihale | 41.56 |
| 10 | Naalehu | 51.50 | 10 | Lihue | 41.76 |
| 11 | Lihue | 41.76 | 11 | Naalehu | 51.50 |
| 12 | Opihihale | 41.56 | 12 | Haleakala | 56.97 |
| 13 | Lanai City | 36.94 | 13 | Kainaliu | 57.51 |
| 14 | Waialua | 30.85 | 14 | Kanalohuluhulu | 66.96 |
| 15 | Molokai | 27.22 | 15 | Kaneohe Mauka | 78.94 |
| 16 | Waikiki | 24.89 | 16 | Hana | 84.36 |
| 17 | Honolulu | 21.12 | 17 | Hawaii Volcano | 110.20 |
| 18 | Kahului | 20.80 | 18 | Kulani Camp | 112.39 |
| 19 | Mauna Loa | 20.68 | 19 | Kailua | 128.02 |
| 20 | Lahaina | 15.35 | 20 | Hilo | 131.23 |

## ANNUAL SNOWFALL (")

| HIGHEST* | | | LOWEST* | | |
|---|---|---|---|---|---|

* All stations in Hawaii receive no snowfall.

# IDAHO

PHYSICAL FEATURES. Idaho lies entirely west of the Continental Divide, which forms its boundary for some distance westward from Yellowstone National Park. With a maximum north-south extent of 7° of latitude, its east-west extent is 6° of longitude at latitude 42° N., but only 1° of longitude at 49° N. The northern part of the State averages lower in elevation than the much larger central and southern portions, where numerous mountain ranges form barriers to the free flow of air from all points of the compass. In the north the main barrier is the rugged chain of Bitterroot Mountains forming much of the boundary between Idaho and Montana. The extreme range of elevation in the State is from 738 feet at the confluence of the Clearwater and Snake Rivers to 12,655 feet at Mt. Borah in Custer County.

GENERAL CLIMATE. Comprising rugged mountain ranges, canyons, high grassy valleys, arid plains, and fertile lowlands, the State reflects in its topography and vegetation a wide range of climates. Located some 300 miles from the Pacific Ocean, Idaho is, nevertheless, influenced by maritime air borne eastward on the prevailing westerly winds. Particularly in winter, the maritime influence is noticeable in the greater average cloudiness, greater frequency of precipitation, and mean temperatures, which are above those at the same latitude and altitude in midcontinent. This maritime influence is most marked in the northern part of the State, where the air arrives via the Columbia River Gorge with a greater burden of moisture than at lower latitudes. Eastern Idaho's climate has a more continental character than the west and north, a fact quite evident not only in the somewhat greater range between winter and summer temperatures, but also in the reversal of the wet winter-dry summer patterns.

The pattern of average annual temperatures for the State indicates the effect both of latitude and altitude. The highest annual averages are found in the lower elevations of the Clearwater and Little Salmon River Basins, and in the stretch of the Snake River Valley from the vicinity of Bliss downstream to Lewiston, including the open valleys of the Boise, Payette, and Weiser Rivers. The diurnal range of temperature is, of course, most extreme in high valleys and in the semiarid plains of the Snake River Valley. The magnitude of diurnal range varies with the season, being lowest in winter when cloudiness is much more prevalent, and greatest in the warmer part of the year. In summer, periods of extreme heat extending beyond a week are quite rare, and the same can be said of periods of extremely low temperatures in winter. In both cases the normal progress of weather systems across the State usually results in a change at rather frequent intervals.

PRECIPITATION. To a large extent the source of moisture for precipitation in Idaho is the Pacific Ocean. In summer there are some exceptions to this when moisture-laden air is brought in from the south at high levels to produce thundershower activity, particularly in the eastern part of Idaho. The source of this moisture from the south is apparently the Gulf of Mexico and Caribbean region. The average precipitation map for Idaho is as complex as the physiography of the State. Partly because of the greater moisture supply in the west winds over the northern part of the State, (less formidable barriers to the west) and partly because of greater frequency of cyclonic activity in the north, the average valley precipitation is considerably greater than in southern sections. Peaks on the average annual precipitation map are found, however, in nearly all parts of the State at higher elevations. Sizable areas in the Clearwater, Payette, and Boise River Basins receive an average of 40 to 50 inches per year, with a few points or small areas receiving in excess of 60 inches. Large areas including the northeastern valleys, much of the Upper Snake River Plains, Central Plains, and the lower elevations of the Southwestern Valleys receive less than 10 inches annually.

Seasonal distribution of precipitation shows a very marked pattern of winter maximum and midsummer minimum in the northern and western portions of the State. In the eastern part of the State, however, many reporting stations show maximum monthly amounts in summer and minimum amounts in winter. In the divisions called Northeastern Valleys and Eastern Highlands, more than 50 percent of the annual rainfall occurs during the period April through September. Over nearly all of the northern part of the State, however, less than 40 percent of the annual rainfall occurs in this same period, and in portions of the Boise, Payette, and Weiser River drainages less than 30 percent of the annual amount comes in that 6-month period.

SNOW. Snowfall distribution is affected both by availability of moisture and by elevation. The major mountain ranges of the State accumulate a deep snow cover during the winter months, and the release of water from a melting snowpack in late spring furnishes irrigation water for more than 2 million acres, mainly within the Snake River Basin above Weiser.

FLOODS. Floods in Idaho occur most often during the period of seasonal snowmelt in spring, particularly in April and May. A few areas in the State are actually flooded or threatened by flood waters nearly every year. So-called "out-of-season" floods do occur occasionally at a number of points in the State. Flash floods on small streams, or occasionally in ravines or dry gulches, occur a few times each year as the result of heavy rains associated with thunderstorms.

HUMIDITY. The diurnal range of relative humidity generally follows a pattern which is the reverse of the diurnal temperature curve. Precipitation or fog interferes with such a pattern, but the averages show maximum humidity at the time of minimum temperature and vice versa. In winter, average relative humidities are considerably higher than during hot weather. Human comfort during the summer months is greatly affected by the moisture content of the air. In Idaho, where maximum temperatures above 90° are not uncommon in July and August, humidity at the time of maximum temperature is usually below 25 percent, and often down to 15 percent or lower. With any kind of air movement the higher temperatures are quite within the range of adjustment of the human system.

FOG. Fogs in Idaho are extremely variable. At Boise, heavy fog (visibility ¼ mile or less) is experienced on an average of 17 days per year, with a maximum of 6 occurrences in December. The year-to-year variation is considerable, however.

STORMS. Windstorms are not uncommon in Idaho, but the State has no such destructive storms as hurricanes, and an extremely small incidence of tornadoes. Windstorms associated with cyclonic systems, and their cold fronts, do some damage each year. Storms of this type may occur at any time from October into June, while during the remaining 3 months of the year strong winds almost invariably come with thunderstorms. Hail damage in Idaho is very small in comparison with such damage in other areas of the central part of the United States. Often the hail that occurs does not grow to a size larger than ½ inch in diameter, and the areas affected are usually limited in size. Quite often hail comes in springtime storms, when it is mostly of the small, soft variety with a limited damaging effect. The incidence of summer thunderstorms is greatest in mountainous areas, with an important influence on the economy of the State, resulting from the lightning-caused forest and range fires.

SUNSHINE. The annual average percentage of possible sunshine ranges from about 50 in the north to about 70 in the south. Winter, with its frequent periods of cloudy weather, has about 40 percent of possible sunshine in the large open valleys of the south and less than 30 percent in the north, but in July and August the average percentage rises to the upper 80s in the southwest and to near 80 in the east and north.

## COUNTY INDEX

### *Teton County*

DRIGGS

### *Twin Falls County*

CASTLEFORD 2 N
TWIN FALLS WSO

### *Valley County*

CASCADE 1 NW
MCCALL

### *Washington County*

CAMBRIDGE

## ELEVATION INDEX

| FEET | STATION NAME |
|---|---|
| 1421 | LEWISTON NEZ PERCE |
| 1601 | FENN RANGER STN |
| 1801 | PORTHILL |
| 1801 | RIGGINS |
| 1847 | BONNERS FERRY |
| 2070 | BAYVIEW MODEL BASIN |
| 2103 | SANDPOINT EXP STN |
| 2142 | SAINT MARIES 1 W |
| 2162 | PAYETTE |
| 2182 | CABINET GORGE |
| 2224 | PARMA EXPERIMENT STN |
| 2313 | KELLOGG |
| 2323 | SWAN FALLS POWER HOU |
| 2372 | CALDWELL |
| 2382 | GRAND VIEW 2 W |
| 2382 | PRIEST RIVER EXP STN |
| 2392 | EMMETT 2 E |
| 2503 | POTLATCH 3 NNE |
| 2513 | DEER FLAT DAM |
| 2572 | GLENNS FERRY |
| 2631 | MOSCOW U OF I |
| 2651 | CAMBRIDGE |
| 2690 | KUNA |
| 2858 | BOISE AIR TERMINAL |
| 2913 | ELK RIVER 1 S |
| 2953 | WALLACE WOODLAND PAR |
| 3002 | BRUNEAU |
| 3143 | HEADQUARTERS |
| 3153 | COUNCIL |
| 3153 | GARDEN VALLEY R S |
| 3153 | MOUNTAIN HOME |
| 3192 | PIERCE |
| 3241 | ARROWROCK DAM |
| 3251 | NEZPERCE |
| 3275 | BLISS 4 NW |
| 3353 | GRANGEVILLE |
| 3402 | SHOUP |
| 3632 | POWELL |
| 3773 | JEROME |
| 3832 | CASTLEFORD 2 N |
| 3862 | NEW MEADOWS RANGER S |
| 3881 | ANDERSON DAM |
| 3914 | REYNOLDS |
| 3944 | IDAHO CITY |
| 3963 | TWIN FALLS WSO |
| 3973 | SHOSHONE 1 WNW |
| 3983 | ELK CITY RANGER STN |
| 4062 | HAZELTON |
| 4157 | BURLEY MUNICIPAL AP |
| 4180 | WINCHESTER |
| 4203 | PAUL 1 ENE |
| 4213 | MINIDOKA DAM |
| 4314 | RICHFIELD |
| 4324 | AMERICAN FALLS 1 SW |
| 4403 | ABERDEEN EXPRMNT STN |
| 4468 | POCATELLO MUNICIPAL |
| 4475 | MALAD CITY |
| 4505 | BLACKFOOT 4 NNE |
| 4505 | FORT HALL 1 NNE |
| 4514 | GIBBONSVILLE |
| 4544 | MALTA 4 ESE |
| 4603 | OAKLEY |
| 4744 | IDAHO FALLS FANNING |
| 4774 | IDAHO FALLS 2 ESE |
| 4803 | HAMER 4 NW |
| 4823 | HOWE |
| 4872 | CASCADE 1 NW |
| 4882 | PICABO |
| 4934 | IDAHO FALLS 46 W |
| 4974 | SAINT ANTHONY 1 WNW |
| 5003 | HILL CITY 1 W |
| 5033 | MCCALL |
| 5072 | FAIRFIELD R S |
| 5105 | ASHTON |
| 5174 | ARBON 2 NW |
| 5174 | CHALLIS |
| 5243 | SWAN VALLEY 2 E |
| 5463 | DUBOIS EXPERIMENT ST |
| 5549 | GRACE |
| 5613 | DIXIE |
| 5715 | IDAHO FALLS 16 SE |
| 5905 | CRATERS OF MOON NM |
| 5905 | MACKAY RANGER STN |
| 5905 | WARREN |
| 5935 | LIFTON PUMPING STN |
| 6106 | DRIGGS |
| 6106 | GROUSE |
| 6184 | CHILLY BARTON FLAT |
| 6243 | STANLEY |
| 6306 | ISLAND PARK |

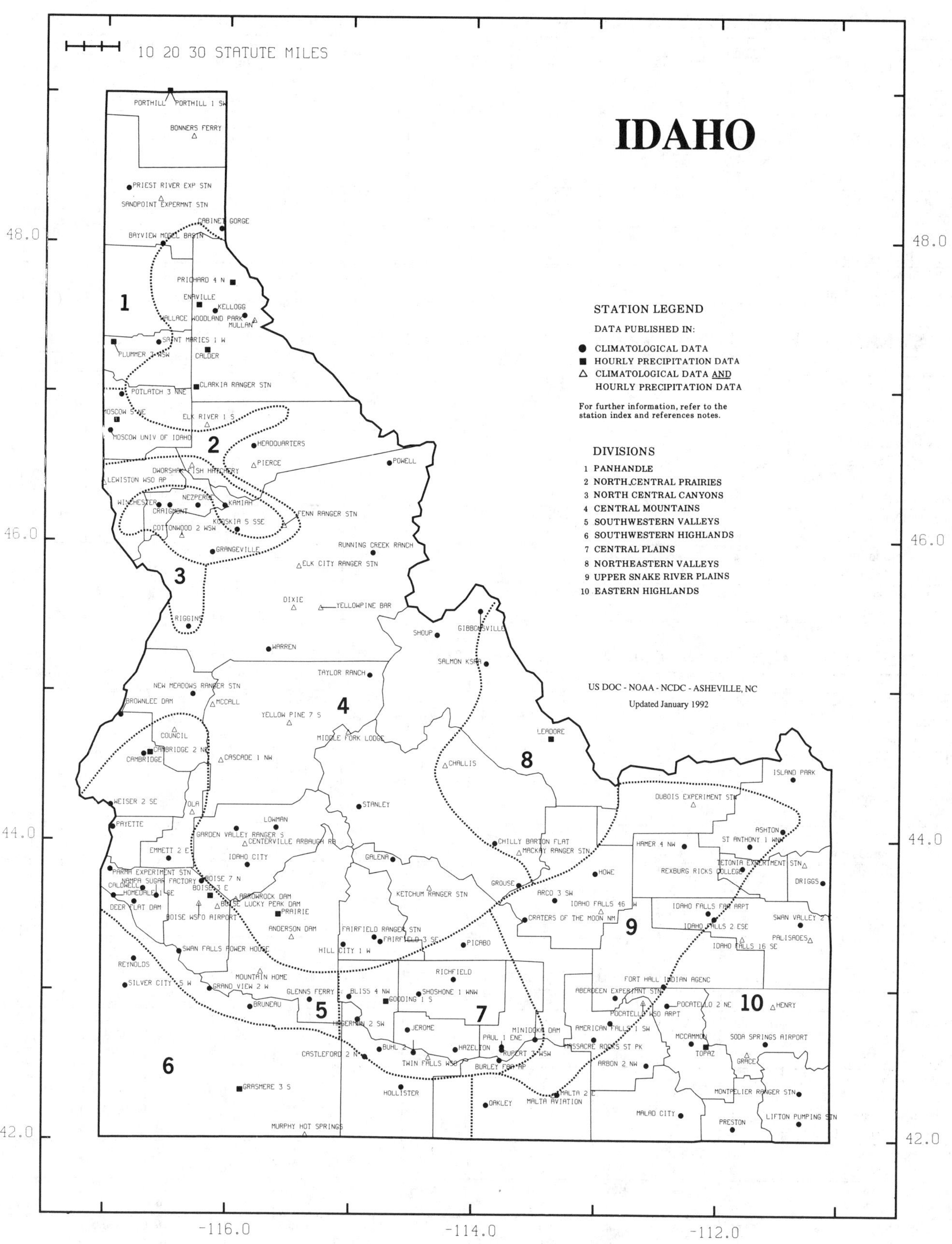
IDAHO
10 20 30 STATUTE MILES
STATION LEGEND
DATA PUBLISHED IN:
CLIMATOLOGICAL DATA
HOURLY PRECIPITATION DATA
CLIMATOLOGICAL DATA AND HOURLY PRECIPITATION DATA
For further information, refer to the station index and references notes.
DIVISIONS
1 PANHANDLE
2 NORTH CENTRAL PRAIRIES
3 NORTH CENTRAL CANYONS
4 CENTRAL MOUNTAINS
5 SOUTHWESTERN VALLEYS
6 SOUTHWESTERN HIGHLANDS
7 CENTRAL PLAINS
8 NORTHEASTERN VALLEYS
9 UPPER SNAKE RIVER PLAINS
10 EASTERN HIGHLANDS
US DOC - NOAA - NCDC - ASHEVILLE, NC
Updated January 1992
48.0
46.0
44.0
42.0
-116.0
-114.0
-112.0

## ABERDEEN EXPRMNT STN *Bingham County* ELEVATION 4403 ft LAT/LONG 42° 57 ' N / 112° 50 ' W

| | JAN | FEB | MAR | APR | MAY | JUN | JUL | AUG | SEP | OCT | NOV | DEC | YEAR |
|---|---|---|---|---|---|---|---|---|---|---|---|---|---|
| Maximum Temp °F | 31.1 | 37.3 | 46.8 | 57.6 | 67.3 | 76.4 | 85.3 | 84.8 | 74.3 | 61.6 | 43.9 | 32.7 | 58.3 |
| Minimum Temp °F | 11.6 | 16.3 | 23.5 | 29.8 | 37.8 | 44.9 | 49.3 | 46.8 | 37.7 | 28.2 | 21.1 | 12.0 | 29.9 |
| Mean Temp °F | 21.3 | 26.8 | 35.2 | 43.7 | 52.6 | 60.7 | 67.3 | 65.8 | 56.0 | 44.9 | 32.5 | 22.3 | 44.1 |
| Days Max Temp ≥ 90 °F | 0 | 0 | 0 | 0 | 0 | 2 | 10 | 9 | 1 | 0 | 0 | 0 | 22 |
| Days Max Temp ≤ 32 °F | 15 | 8 | 2 | 0 | 0 | 0 | 0 | 0 | 0 | 0 | 4 | 14 | 43 |
| Days Min Temp ≤ 32 °F | 30 | 27 | 29 | 20 | 7 | 0 | 0 | 0 | 7 | 23 | 28 | 30 | 201 |
| Days Min Temp ≤ 0 °F | 7 | 3 | 1 | 0 | 0 | 0 | 0 | 0 | 0 | 0 | 1 | 5 | 17 |
| Heating Degree Days | 1347 | 1071 | 919 | 633 | 381 | 161 | 37 | 57 | 271 | 616 | 968 | 1318 | 7779 |
| Cooling Degree Days | 0 | 0 | 0 | 0 | 4 | 39 | 108 | 96 | 9 | 0 | 0 | 0 | 256 |
| Total Precipitation (") | 0.70 | 0.67 | 0.77 | 0.77 | 1.12 | 1.00 | 0.54 | 0.63 | 0.67 | 0.81 | 0.82 | 0.69 | 9.19 |
| Days ≥ 0.1" Precip | 2 | 2 | 3 | 3 | 4 | 3 | 2 | 2 | 2 | 3 | 3 | 2 | 31 |
| Total Snowfall (") | na | na | *1.8* | 1.5 | 0.3 | 0.0 | 0.0 | 0.0 | 0.0 | 0.3 | *2.5* | na | na |
| Days ≥ 1" Snow Depth | *16* | 9 | 4 | 1 | 0 | 0 | 0 | 0 | 0 | 0 | 4 | *13* | 47 |

## AMERICAN FALLS 1 SW *Power County* ELEVATION 4324 ft LAT/LONG 42° 47 ' N / 112° 52 ' W

| | JAN | FEB | MAR | APR | MAY | JUN | JUL | AUG | SEP | OCT | NOV | DEC | YEAR |
|---|---|---|---|---|---|---|---|---|---|---|---|---|---|
| Maximum Temp °F | 33.1 | 39.2 | 49.0 | 59.3 | 68.9 | 78.7 | 87.5 | 86.4 | 76.0 | 62.5 | 45.2 | 34.4 | 60.0 |
| Minimum Temp °F | 17.0 | 20.9 | 27.8 | 34.5 | 41.9 | 49.1 | 54.9 | 53.6 | 44.9 | 35.7 | 27.3 | 18.6 | 35.5 |
| Mean Temp °F | 25.1 | 30.1 | 38.4 | 46.9 | 55.5 | 64.0 | 71.2 | 70.0 | 60.5 | 49.2 | 36.2 | 26.5 | 47.8 |
| Days Max Temp ≥ 90 °F | 0 | 0 | 0 | 0 | 0 | 4 | 14 | 12 | 2 | 0 | 0 | 0 | 32 |
| Days Max Temp ≤ 32 °F | 14 | 6 | 1 | 0 | 0 | 0 | 0 | 0 | 0 | 0 | 3 | 12 | 36 |
| Days Min Temp ≤ 32 °F | 28 | 24 | 22 | 12 | 2 | 0 | 0 | 0 | 2 | 11 | 21 | 29 | 151 |
| Days Min Temp ≤ 0 °F | 4 | 2 | 0 | 0 | 0 | 0 | 0 | 0 | 0 | 0 | 0 | 2 | 8 |
| Heating Degree Days | 1230 | 980 | 817 | 537 | 298 | 99 | 13 | 21 | 165 | 485 | 856 | 1186 | 6687 |
| Cooling Degree Days | 0 | 0 | 0 | 1 | 14 | 85 | 220 | 213 | 43 | 1 | 0 | 0 | 577 |
| Total Precipitation (") | 1.02 | 0.89 | 1.24 | 1.25 | 1.54 | 1.00 | 0.67 | 0.75 | 0.83 | 0.92 | 1.23 | 0.97 | 12.31 |
| Days ≥ 0.1" Precip | 4 | 3 | 4 | 4 | 4 | 3 | 2 | 2 | 3 | 3 | 4 | 3 | 39 |
| Total Snowfall (") | *7.0* | *4.6* | *2.2* | 0.6 | 0.5 | 0.0 | 0.0 | 0.0 | 0.0 | 0.8 | *3.4* | *7.0* | 26.1 |
| Days ≥ 1" Snow Depth | *18* | *10* | 3 | 0 | 0 | 0 | 0 | 0 | 0 | 0 | 3 | *12* | 46 |

## ANDERSON DAM *Elmore County* ELEVATION 3881 ft LAT/LONG 43° 21 ' N / 115° 28 ' W

| | JAN | FEB | MAR | APR | MAY | JUN | JUL | AUG | SEP | OCT | NOV | DEC | YEAR |
|---|---|---|---|---|---|---|---|---|---|---|---|---|---|
| Maximum Temp °F | *35.9* | *41.1* | 48.9 | 59.5 | *70.1* | *80.4* | *90.7* | 88.5 | *77.9* | *64.2* | na | na | na |
| Minimum Temp °F | *19.8* | *21.2* | 27.0 | 33.7 | *41.1* | *49.0* | *55.5* | 54.6 | *45.7* | *37.0* | na | *21.2* | na |
| Mean Temp °F | na | *31.1* | 38.0 | 46.6 | *55.6* | *64.7* | *73.1* | *71.5* | *61.9* | *50.7* | na | na | na |
| Days Max Temp ≥ 90 °F | 0 | 0 | 0 | 0 | 1 | 5 | 17 | 15 | 3 | 0 | 0 | 0 | 41 |
| Days Max Temp ≤ 32 °F | 8 | 3 | 0 | 0 | 0 | 0 | 0 | 0 | 0 | 0 | 1 | 7 | 19 |
| Days Min Temp ≤ 32 °F | 25 | 23 | 24 | 13 | 2 | 0 | 0 | 0 | 1 | 7 | 18 | 25 | 138 |
| Days Min Temp ≤ 0 °F | 2 | 1 | 0 | 0 | 0 | 0 | 0 | 0 | 0 | 0 | 0 | 1 | 4 |
| Heating Degree Days | na | *948* | 831 | 545 | *297* | *96* | *10* | *18* | *138* | *437* | na | na | na |
| Cooling Degree Days | na | na | *0* | *1* | na | na | na | na | na | *1* | na | na | na |
| Total Precipitation (") | 3.42 | 2.09 | 1.98 | 1.37 | 1.08 | 1.07 | 0.55 | 0.56 | 0.89 | 1.14 | 3.01 | 3.07 | 20.23 |
| Days ≥ 0.1" Precip | 6 | 5 | 4 | 3 | 3 | 3 | 2 | 1 | 2 | 2 | 5 | 6 | 42 |
| Total Snowfall (") | *17.7* | 11.0 | 3.3 | 0.1 | 0.0 | 0.0 | 0.0 | 0.0 | 0.0 | 0.3 | 7.1 | 17.4 | 56.9 |
| Days ≥ 1" Snow Depth | 24 | 19 | 9 | 1 | 0 | 0 | 0 | 0 | 0 | 0 | 5 | 19 | 77 |

## ARBON 2 NW *Power County* ELEVATION 5174 ft LAT/LONG 42° 30 ' N / 112° 34 ' W

| | JAN | FEB | MAR | APR | MAY | JUN | JUL | AUG | SEP | OCT | NOV | DEC | YEAR |
|---|---|---|---|---|---|---|---|---|---|---|---|---|---|
| Maximum Temp °F | 30.5 | 35.8 | 45.5 | 56.7 | 67.2 | 76.9 | 86.0 | 85.1 | 75.0 | 61.5 | 43.1 | 32.1 | 58.0 |
| Minimum Temp °F | 14.5 | 17.8 | 24.2 | 30.2 | 36.6 | 43.6 | 49.3 | 48.2 | 39.8 | 31.1 | 23.1 | 14.3 | 31.1 |
| Mean Temp °F | 22.7 | 26.9 | 34.8 | 43.5 | 51.9 | 60.3 | 67.7 | 66.7 | 57.5 | 46.3 | 33.1 | 23.2 | 44.6 |
| Days Max Temp ≥ 90 °F | 0 | 0 | 0 | 0 | 0 | 3 | 11 | 9 | 1 | 0 | 0 | 0 | 24 |
| Days Max Temp ≤ 32 °F | 18 | 9 | 2 | 0 | 0 | 0 | 0 | 0 | 0 | 0 | 5 | 15 | 49 |
| Days Min Temp ≤ 32 °F | 30 | 26 | 26 | 19 | 8 | 1 | 0 | 0 | 6 | 17 | 25 | 30 | 188 |
| Days Min Temp ≤ 0 °F | 4 | 2 | 1 | 0 | 0 | 0 | 0 | 0 | 0 | 0 | 1 | 4 | 12 |
| Heating Degree Days | 1305 | 1071 | 931 | 639 | 401 | 166 | 30 | 44 | 233 | 572 | 950 | 1289 | 7631 |
| Cooling Degree Days | 0 | 0 | 0 | 0 | 1 | 31 | 112 | 109 | 16 | 0 | 0 | 0 | 269 |
| Total Precipitation (") | 1.70 | 1.40 | 1.55 | 1.34 | 1.67 | 1.34 | 1.00 | 0.94 | 0.95 | 1.13 | 1.52 | 1.37 | 15.91 |
| Days ≥ 0.1" Precip | 6 | 5 | 5 | 5 | 5 | 4 | 3 | 3 | 3 | 4 | 6 | 5 | 54 |
| Total Snowfall (") | na | na | na | na | *0.5* | 0.0 | 0.0 | 0.0 | 0.1 | *1.0* | *8.5* | na | na |
| Days ≥ 1" Snow Depth | na | na | na | *0* | 0 | 0 | 0 | 0 | 0 | 0 | na | na | na |

## ARROWROCK DAM *Elmore County* ELEVATION 3241 ft LAT/LONG 43° 36 ' N / 115° 55 ' W

| | JAN | FEB | MAR | APR | MAY | JUN | JUL | AUG | SEP | OCT | NOV | DEC | YEAR |
|---|---|---|---|---|---|---|---|---|---|---|---|---|---|
| Maximum Temp °F | 34.0 | 41.1 | 50.3 | 59.0 | 69.3 | 79.0 | 88.9 | 88.9 | 77.0 | 62.9 | 44.9 | 35.0 | 60.9 |
| Minimum Temp °F | 20.1 | 23.2 | 28.9 | 34.9 | 42.0 | 49.3 | 55.3 | 54.7 | 45.0 | 36.3 | 28.7 | 21.3 | 36.6 |
| Mean Temp °F | 27.1 | 32.1 | 39.6 | 47.0 | 55.7 | 64.2 | 72.1 | 71.8 | 61.1 | 49.7 | 36.8 | 28.2 | 48.8 |
| Days Max Temp ≥ 90 °F | 0 | 0 | 0 | 0 | 1 | 5 | 17 | 16 | 3 | 0 | 0 | 0 | 42 |
| Days Max Temp ≤ 32 °F | 12 | 4 | 0 | 0 | 0 | 0 | 0 | 0 | 0 | 0 | 2 | 9 | 27 |
| Days Min Temp ≤ 32 °F | 28 | 25 | 23 | 11 | 2 | 0 | 0 | 0 | 1 | 8 | 20 | 27 | 145 |
| Days Min Temp ≤ 0 °F | 2 | 1 | 0 | 0 | 0 | 0 | 0 | 0 | 0 | 0 | 0 | 1 | 4 |
| Heating Degree Days | 1169 | 922 | 781 | 534 | 298 | 107 | 16 | 21 | 158 | 470 | 839 | 1136 | 6451 |
| Cooling Degree Days | 0 | 0 | 0 | 2 | 21 | 95 | 224 | 249 | 51 | 2 | 0 | 0 | 644 |
| Total Precipitation (") | 2.89 | 2.14 | 1.86 | 1.53 | 1.11 | 1.01 | 0.37 | 0.46 | 0.86 | 1.06 | 2.73 | 2.65 | 18.67 |
| Days ≥ 0.1" Precip | 7 | 6 | 6 | 5 | 4 | 3 | 1 | 1 | 3 | 3 | 7 | 7 | 53 |
| Total Snowfall (") | 12.0 | 7.1 | 2.1 | 0.1 | 0.0 | 0.0 | 0.0 | 0.0 | 0.0 | 0.2 | 6.0 | 12.7 | 40.2 |
| Days ≥ 1" Snow Depth | 22 | 17 | 5 | 0 | 0 | 0 | 0 | 0 | 0 | 0 | 4 | 17 | 65 |

## ASHTON *Fremont County* ELEVATION 5105 ft LAT/LONG 44° 5 ' N / 111° 27 ' W

| | JAN | FEB | MAR | APR | MAY | JUN | JUL | AUG | SEP | OCT | NOV | DEC | YEAR |
|---|---|---|---|---|---|---|---|---|---|---|---|---|---|
| Maximum Temp °F | 28.1 | 33.6 | 41.4 | 53.1 | 64.5 | 73.0 | 81.2 | 80.4 | 70.5 | 57.6 | 39.1 | 28.9 | 54.3 |
| Minimum Temp °F | 10.2 | 13.4 | 20.3 | 28.6 | 36.3 | 42.9 | 47.8 | 46.0 | 38.0 | 29.7 | 20.1 | 10.8 | 28.7 |
| Mean Temp °F | 19.2 | 23.6 | 30.9 | 40.9 | 50.4 | 58.0 | 64.5 | 63.2 | 54.3 | 43.7 | 29.7 | 19.9 | 41.5 |
| Days Max Temp ≥ 90 °F | 0 | 0 | 0 | 0 | 0 | 1 | 2 | 1 | 0 | 0 | 0 | 0 | 4 |
| Days Max Temp ≤ 32 °F | 21 | 11 | 3 | 0 | 0 | 0 | 0 | 0 | 0 | 0 | 8 | 20 | 63 |
| Days Min Temp ≤ 32 °F | 31 | 28 | 29 | 22 | 10 | 1 | 0 | 0 | 6 | 20 | 28 | 31 | 206 |
| Days Min Temp ≤ 0 °F | 8 | 5 | 1 | 0 | 0 | 0 | 0 | 0 | 0 | 0 | 2 | 6 | 22 |
| Heating Degree Days | 1413 | 1163 | 1051 | 717 | 446 | 218 | 61 | 86 | 318 | 654 | 1054 | 1392 | 8573 |
| Cooling Degree Days | 0 | 0 | 0 | 0 | 1 | 18 | 64 | 52 | 5 | 0 | 0 | 0 | 140 |
| Total Precipitation (") | 2.30 | 1.72 | 1.51 | 1.58 | 2.18 | 1.85 | 1.06 | 1.15 | 1.30 | 1.47 | 2.15 | 2.28 | 20.55 |
| Days ≥ 0.1" Precip | 8 | 6 | 5 | 5 | 6 | 5 | 3 | 3 | 4 | 4 | 6 | 7 | 62 |
| Total Snowfall (") | 23.4 | 15.9 | 10.2 | 5.5 | 2.0 | 0.0 | 0.0 | 0.0 | 0.3 | 2.7 | 16.8 | 24.9 | 101.7 |
| Days ≥ 1" Snow Depth | 31 | 28 | 27 | 7 | 0 | 0 | 0 | 0 | 0 | 2 | 15 | 30 | 140 |

## BAYVIEW MODEL BASIN *Kootenai County* ELEVATION 2070 ft LAT/LONG 47° 59 ' N / 116° 33 ' W

| | JAN | FEB | MAR | APR | MAY | JUN | JUL | AUG | SEP | OCT | NOV | DEC | YEAR |
|---|---|---|---|---|---|---|---|---|---|---|---|---|---|
| Maximum Temp °F | 34.8 | 38.2 | 45.8 | 54.4 | 64.0 | 71.3 | 78.4 | 78.5 | 67.7 | 55.5 | 42.4 | 35.6 | 55.6 |
| Minimum Temp °F | 21.4 | 23.9 | 27.2 | 32.5 | 38.6 | 45.3 | 48.8 | 48.1 | 40.5 | 32.6 | 28.6 | 22.8 | 34.2 |
| Mean Temp °F | 28.1 | 31.1 | 36.5 | 43.5 | 51.3 | 58.3 | 63.4 | 63.3 | 54.2 | 44.1 | 35.5 | 29.2 | 44.9 |
| Days Max Temp ≥ 90 °F | 0 | 0 | 0 | 0 | 0 | 0 | 3 | 2 | 0 | 0 | 0 | 0 | 5 |
| Days Max Temp ≤ 32 °F | 10 | 4 | 1 | 0 | 0 | 0 | 0 | 0 | 0 | 0 | 2 | 9 | 26 |
| Days Min Temp ≤ 32 °F | 28 | 24 | 26 | 17 | 5 | 0 | 0 | 0 | 4 | 15 | 21 | 27 | 167 |
| Days Min Temp ≤ 0 °F | 1 | 1 | 0 | 0 | 0 | 0 | 0 | 0 | 0 | 0 | 0 | 1 | 3 |
| Heating Degree Days | 1136 | 950 | 876 | 639 | 418 | 214 | 94 | 94 | 321 | 642 | 878 | 1102 | 7364 |
| Cooling Degree Days | 0 | 0 | 0 | 0 | 4 | 16 | 44 | 46 | 2 | 0 | 0 | 0 | 112 |
| Total Precipitation (") | 2.96 | 2.25 | 1.98 | 1.91 | 2.09 | 1.94 | 1.20 | 1.25 | 1.37 | 1.73 | 3.00 | 3.24 | 24.92 |
| Days ≥ 0.1" Precip | *8* | 7 | 7 | 6 | 6 | 5 | 3 | 3 | 4 | 5 | 9 | 9 | 72 |
| Total Snowfall (") | *13.9* | na | *1.9* | *0.2* | 0.0 | 0.0 | 0.0 | 0.0 | 0.0 | 0.1 | *3.1* | *11.9* | na |
| Days ≥ 1" Snow Depth | *19* | *13* | 4 | 0 | 0 | 0 | 0 | 0 | 0 | 0 | *3* | na | na |

## BLACKFOOT 4 NNE *Bingham County* ELEVATION 4505 ft LAT/LONG 43° 11 ' N / 112° 21 ' W

| | JAN | FEB | MAR | APR | MAY | JUN | JUL | AUG | SEP | OCT | NOV | DEC | YEAR |
|---|---|---|---|---|---|---|---|---|---|---|---|---|---|
| Maximum Temp °F | 31.4 | 37.9 | 48.5 | 59.0 | 69.2 | 78.2 | 86.0 | 84.9 | 75.1 | 62.0 | *44.6* | *32.3* | 59.1 |
| Minimum Temp °F | 14.5 | 18.3 | 25.0 | 31.4 | 39.1 | 45.9 | 51.3 | 49.3 | 40.9 | 32.0 | *23.9* | *14.8* | 32.2 |
| Mean Temp °F | 23.0 | 28.1 | 36.8 | 45.2 | 54.2 | 62.1 | 68.9 | 67.1 | 58.0 | 47.0 | *34.2* | *23.6* | 45.7 |
| Days Max Temp ≥ 90 °F | 0 | 0 | 0 | 0 | 0 | 3 | 10 | 8 | 1 | 0 | *0* | *0* | 22 |
| Days Max Temp ≤ 32 °F | 14 | 7 | 1 | 0 | 0 | 0 | 0 | 0 | 0 | 0 | *4* | *15* | 41 |
| Days Min Temp ≤ 32 °F | 28 | 27 | 26 | 16 | 6 | 0 | 0 | 0 | 4 | 16 | *25* | *29* | 177 |
| Days Min Temp ≤ 0 °F | 5 | 2 | 0 | 0 | 0 | 0 | 0 | 0 | 0 | 0 | *1* | *4* | 12 |
| Heating Degree Days | 1295 | 1035 | 868 | 585 | 333 | 126 | 18 | 33 | 217 | 550 | *918* | *1276* | 7254 |
| Cooling Degree Days | 0 | 0 | 0 | *0* | 4 | *50* | 136 | 116 | *13* | 0 | na | *0* | na |
| Total Precipitation (") | 0.93 | 0.78 | 0.92 | 0.93 | 1.22 | 0.97 | 0.54 | 0.51 | 0.65 | 0.77 | 1.00 | 0.99 | 10.21 |
| Days ≥ 0.1" Precip | *3* | *2* | *3* | 3 | 4 | *3* | 1 | 1 | 2 | *2* | na | *3* | na |
| Total Snowfall (") | na | na | na | *0.9* | 0.0 | 0.0 | 0.0 | 0.0 | 0.1 | *0.1* | na | na | na |
| Days ≥ 1" Snow Depth | na | na | na | *0* | 0 | 0 | 0 | 0 | 0 | 0 | na | na | na |

## BLISS 4 NW *Gooding County* ELEVATION 3275 ft LAT/LONG 42° 57 ' N / 115° 0 ' W

| | JAN | FEB | MAR | APR | MAY | JUN | JUL | AUG | SEP | OCT | NOV | DEC | YEAR |
|---|---|---|---|---|---|---|---|---|---|---|---|---|---|
| Maximum Temp °F | 36.8 | 44.5 | 54.4 | 64.0 | 73.5 | 82.7 | 91.1 | 89.7 | 79.2 | 66.8 | 48.9 | 37.8 | 64.1 |
| Minimum Temp °F | 20.3 | 24.5 | 29.3 | 34.8 | 41.9 | 49.4 | 54.4 | 52.8 | 44.3 | 35.3 | 27.8 | 20.5 | 36.3 |
| Mean Temp °F | 28.5 | 34.3 | 41.9 | 49.4 | 57.7 | 66.1 | 72.8 | 71.2 | 61.8 | 51.1 | 38.4 | 29.1 | 50.2 |
| Days Max Temp ≥ 90 °F | 0 | 0 | 0 | 0 | 1 | 8 | 21 | 18 | 4 | 0 | 0 | 0 | 52 |
| Days Max Temp ≤ 32 °F | 9 | 3 | 0 | 0 | 0 | 0 | 0 | 0 | 0 | 0 | 2 | 8 | 22 |
| Days Min Temp ≤ 32 °F | 27 | 23 | 21 | 11 | 3 | 0 | 0 | 0 | 2 | *11* | 21 | 28 | 147 |
| Days Min Temp ≤ 0 °F | 2 | 0 | 0 | 0 | 0 | 0 | 0 | 0 | 0 | 0 | 0 | 2 | 4 |
| Heating Degree Days | 1124 | 860 | 710 | 463 | 243 | 71 | 9 | 17 | 139 | 424 | 792 | 1105 | 5957 |
| Cooling Degree Days | 0 | 0 | 0 | 4 | 30 | 118 | 239 | 217 | 46 | *2* | 0 | 0 | 656 |
| Total Precipitation (") | 1.40 | 1.03 | 1.02 | 0.75 | 0.73 | 0.78 | 0.26 | 0.36 | 0.53 | 0.65 | 1.49 | 1.28 | 10.28 |
| Days ≥ 0.1" Precip | 5 | 4 | 3 | 3 | 2 | 2 | 1 | 1 | 2 | 2 | 5 | 4 | 34 |
| Total Snowfall (") | 5.6 | *3.9* | *0.8* | *0.2* | 0.0 | 0.0 | 0.0 | 0.0 | 0.0 | 0.1 | 3.6 | *5.1* | 19.3 |
| Days ≥ 1" Snow Depth | 12 | 6 | 1 | 0 | 0 | 0 | 0 | 0 | 0 | 0 | 3 | 10 | 32 |

## BOISE AIR TERMINAL *Ada County* ELEVATION 2858 ft LAT/LONG 43° 34 ' N / 116° 13 ' W

| | JAN | FEB | MAR | APR | MAY | JUN | JUL | AUG | SEP | OCT | NOV | DEC | YEAR |
|---|---|---|---|---|---|---|---|---|---|---|---|---|---|
| Maximum Temp °F | 36.7 | 44.6 | 53.7 | 61.6 | 71.6 | 80.9 | 90.0 | 88.4 | 77.4 | 64.7 | 48.0 | 37.6 | 62.9 |
| Minimum Temp °F | 22.1 | 27.4 | 32.1 | 37.2 | 44.2 | 51.9 | 57.5 | 56.8 | 48.2 | 38.7 | 30.4 | 22.2 | 39.1 |
| Mean Temp °F | 29.4 | 36.1 | 42.9 | 49.4 | 57.9 | 66.5 | 73.8 | 72.6 | 62.8 | 51.7 | 39.2 | 29.9 | 51.0 |
| Days Max Temp ≥ 90 °F | 0 | 0 | 0 | 0 | 2 | 7 | 18 | 16 | 3 | 0 | 0 | 0 | 46 |
| Days Max Temp ≤ 32 °F | 10 | 3 | 0 | 0 | 0 | 0 | 0 | 0 | 0 | 0 | 1 | 8 | 22 |
| Days Min Temp ≤ 32 °F | 26 | 20 | 16 | 8 | 2 | 0 | 0 | 0 | 1 | 6 | 17 | 26 | 122 |
| Days Min Temp ≤ 0 °F | 2 | 1 | 0 | 0 | 0 | 0 | 0 | 0 | 0 | 0 | 0 | 2 | 5 |
| Heating Degree Days | 1096 | 811 | 678 | 463 | 244 | 77 | 10 | 17 | 129 | 409 | 767 | 1081 | 5782 |
| Cooling Degree Days | 0 | 0 | 0 | 4 | 37 | 127 | 265 | 261 | 71 | 5 | 0 | 0 | 770 |
| Total Precipitation (") | 1.43 | 1.05 | 1.28 | 1.29 | 1.00 | 0.82 | 0.35 | 0.39 | 0.73 | 0.72 | 1.41 | 1.32 | 11.79 |
| Days ≥ 0.1" Precip | 5 | 4 | 5 | 4 | 3 | 3 | 1 | 1 | 2 | 3 | 6 | 4 | 41 |
| Total Snowfall (") | 5.6 | 3.1 | 1.2 | 0.9 | 0.1 | 0.0 | 0.0 | 0.0 | 0.0 | 0.2 | 2.8 | 6.7 | 20.6 |
| Days ≥ 1" Snow Depth | 12 | 5 | 1 | 0 | 0 | 0 | 0 | 0 | 0 | 0 | 2 | 8 | 28 |

## BONNERS FERRY *Boundary County* ELEVATION 1847 ft LAT/LONG 48° 42 ' N / 116° 18 ' W

| | JAN | FEB | MAR | APR | MAY | JUN | JUL | AUG | SEP | OCT | NOV | DEC | YEAR |
|---|---|---|---|---|---|---|---|---|---|---|---|---|---|
| Maximum Temp °F | 33.1 | 39.5 | 49.3 | 59.7 | 69.4 | 76.2 | 83.2 | 83.4 | 72.2 | 57.3 | 41.3 | 33.6 | 58.2 |
| Minimum Temp °F | 20.0 | 24.0 | 28.0 | 34.0 | 40.9 | 47.3 | 50.0 | 49.3 | 41.5 | 33.4 | 27.8 | 21.7 | 34.8 |
| Mean Temp °F | 26.6 | 31.8 | 38.7 | 46.9 | 55.2 | 61.8 | 66.6 | 66.3 | 56.8 | 45.4 | 34.6 | 27.7 | 46.5 |
| Days Max Temp ≥ 90 °F | 0 | 0 | 0 | 0 | 0 | 2 | 8 | 8 | 0 | 0 | 0 | 0 | 18 |
| Days Max Temp ≤ 32 °F | 12 | 4 | 1 | 0 | 0 | 0 | 0 | 0 | 0 | 0 | 4 | 13 | 34 |
| Days Min Temp ≤ 32 °F | 27 | 24 | 22 | 13 | 3 | 0 | 0 | 0 | 3 | *14* | 21 | 28 | 155 |
| Days Min Temp ≤ 0 °F | 2 | 1 | 0 | 0 | 0 | 0 | 0 | 0 | 0 | 0 | 0 | 1 | 4 |
| Heating Degree Days | 1185 | 931 | 811 | 537 | 303 | 132 | 45 | 47 | 245 | 601 | 906 | 1150 | 6893 |
| Cooling Degree Days | 0 | 0 | 0 | 0 | 9 | 44 | 98 | 96 | 7 | 0 | 0 | 0 | 254 |
| Total Precipitation (") | 2.98 | 1.84 | 1.56 | 1.53 | 1.60 | 1.73 | 0.97 | 1.17 | 1.28 | 1.65 | 3.17 | 3.26 | 22.74 |
| Days ≥ 0.1" Precip | 8 | 6 | 5 | 5 | 5 | 5 | 3 | 3 | 4 | 5 | 8 | 9 | 66 |
| Total Snowfall (") | 22.7 | 10.4 | 3.6 | 0.3 | 0.0 | 0.0 | 0.0 | 0.0 | 0.0 | 0.5 | 7.8 | 20.9 | 66.2 |
| Days ≥ 1" Snow Depth | 22 | *15* | na | 0 | 0 | 0 | 0 | 0 | 0 | 0 | 5 | 18 | na |

## BRUNEAU *Owyhee County* ELEVATION 3002 ft LAT/LONG 42° 53 ' N / 115° 48 ' W

| | JAN | FEB | MAR | APR | MAY | JUN | JUL | AUG | SEP | OCT | NOV | DEC | YEAR |
|---|---|---|---|---|---|---|---|---|---|---|---|---|---|
| Maximum Temp °F | 40.0 | 48.6 | 58.2 | 66.6 | 76.3 | 84.9 | 93.5 | 91.9 | 81.4 | 68.6 | 50.9 | 40.0 | 66.7 |
| Minimum Temp °F | 22.7 | 26.6 | 31.1 | 36.5 | 44.2 | 51.6 | 56.9 | 55.2 | 45.6 | 36.5 | 29.3 | 21.9 | 38.2 |
| Mean Temp °F | 31.5 | 37.5 | 44.7 | 51.6 | 60.3 | 68.3 | 75.2 | 73.6 | 63.5 | 52.6 | 40.2 | 31.0 | 52.5 |
| Days Max Temp ≥ 90 °F | 0 | 0 | 0 | 0 | 3 | 10 | 23 | 21 | 7 | 0 | 0 | 0 | 64 |
| Days Max Temp ≤ 32 °F | 7 | 2 | 0 | 0 | 0 | 0 | 0 | 0 | 0 | 0 | 1 | 6 | 16 |
| Days Min Temp ≤ 32 °F | 25 | 21 | 17 | *8* | 2 | 0 | 0 | 0 | 1 | 9 | 19 | 26 | 128 |
| Days Min Temp ≤ 0 °F | 1 | 0 | 0 | 0 | 0 | 0 | 0 | 0 | 0 | 0 | 0 | 1 | 2 |
| Heating Degree Days | 1032 | 771 | 622 | 401 | 182 | 47 | 3 | 9 | 107 | 380 | 739 | 1047 | 5340 |
| Cooling Degree Days | 0 | 0 | 0 | 6 | 48 | 150 | 306 | 291 | 71 | 2 | 0 | 0 | 874 |
| Total Precipitation (") | 0.76 | 0.57 | 0.75 | 0.69 | 0.60 | 0.82 | 0.19 | 0.32 | 0.53 | 0.52 | 0.98 | 0.71 | 7.44 |
| Days ≥ 0.1" Precip | 2 | 2 | 3 | 2 | 2 | 2 | 1 | 1 | 1 | 2 | 3 | *2* | 23 |
| Total Snowfall (") | na | *0.9* | na | 0.0 | 0.0 | 0.0 | 0.0 | 0.0 | 0.0 | 0.0 | 0.7 | na | na |
| Days ≥ 1" Snow Depth | na | na | *0* | 0 | 0 | 0 | 0 | 0 | 0 | 0 | 0 | na | na |

## BURLEY MUNICIPAL AP *Cassia County* ELEVATION 4157 ft LAT/LONG 42° 32 ' N / 113° 46 ' W

| | JAN | FEB | MAR | APR | MAY | JUN | JUL | AUG | SEP | OCT | NOV | DEC | YEAR |
|---|---|---|---|---|---|---|---|---|---|---|---|---|---|
| Maximum Temp °F | 35.6 | 42.6 | 51.1 | 59.9 | 69.0 | 78.2 | 86.3 | 85.2 | 75.2 | 63.4 | 46.8 | 36.9 | 60.9 |
| Minimum Temp °F | 18.6 | 22.9 | 27.9 | 33.7 | 41.3 | 48.6 | 53.9 | 51.8 | 43.2 | 34.0 | 26.1 | 18.7 | 35.1 |
| Mean Temp °F | 27.1 | 32.8 | 39.6 | 46.9 | 55.2 | 63.4 | 70.1 | 68.5 | 59.2 | 48.7 | 36.5 | 27.9 | 48.0 |
| Days Max Temp ≥ 90 °F | 0 | 0 | 0 | 0 | 0 | 4 | 12 | 10 | 2 | 0 | 0 | 0 | 28 |
| Days Max Temp ≤ 32 °F | 11 | 4 | 1 | 0 | 0 | 0 | 0 | 0 | 0 | 0 | 3 | 9 | 28 |
| Days Min Temp ≤ 32 °F | 27 | 25 | 23 | 13 | 3 | 0 | 0 | 0 | 2 | 13 | 23 | 29 | 158 |
| Days Min Temp ≤ 0 °F | 3 | 1 | 0 | 0 | 0 | 0 | 0 | 0 | 0 | 0 | 0 | 2 | 6 |
| Heating Degree Days | 1166 | 903 | 782 | 539 | 306 | 111 | 18 | 31 | 192 | 499 | 849 | 1145 | 6541 |
| Cooling Degree Days | 0 | 0 | 0 | 2 | 14 | 79 | 187 | 166 | 31 | 1 | 0 | 0 | 480 |
| Total Precipitation (") | 1.03 | 0.75 | 0.97 | 0.96 | 1.06 | 0.86 | 0.38 | 0.55 | 0.56 | 0.61 | 1.02 | 0.89 | 9.64 |
| Days ≥ 0.1" Precip | 3 | 3 | 3 | 3 | 3 | 3 | 1 | 1 | 2 | 2 | 4 | 3 | 31 |
| Total Snowfall (") | 5.4 | 3.6 | 2.3 | 0.9 | 0.3 | 0.0 | 0.0 | 0.0 | 0.0 | 0.1 | 3.2 | 6.4 | 22.2 |
| Days ≥ 1" Snow Depth | 14 | 8 | 3 | 0 | 0 | 0 | 0 | 0 | 0 | 0 | 3 | 11 | 39 |

## CABINET GORGE *Bonner County* ELEVATION 2182 ft LAT/LONG 48° 5 ' N / 116° 4 ' W

| | JAN | FEB | MAR | APR | MAY | JUN | JUL | AUG | SEP | OCT | NOV | DEC | YEAR |
|---|---|---|---|---|---|---|---|---|---|---|---|---|---|
| Maximum Temp °F | 32.7 | 39.1 | 47.8 | 57.6 | 67.7 | 74.5 | 82.7 | 82.6 | 71.5 | 57.6 | 40.6 | 33.2 | 57.3 |
| Minimum Temp °F | 21.4 | 23.9 | 27.7 | 33.5 | 40.0 | 46.1 | 49.5 | 49.3 | 42.7 | 35.4 | 29.3 | 23.1 | 35.2 |
| Mean Temp °F | 27.1 | 31.5 | 37.7 | 45.6 | 53.9 | 60.3 | 66.1 | 66.0 | 57.2 | 46.5 | 35.0 | 28.2 | 46.3 |
| Days Max Temp ≥ 90 °F | 0 | 0 | 0 | 0 | 0 | 2 | 8 | 7 | 1 | 0 | 0 | 0 | 18 |
| Days Max Temp ≤ 32 °F | 12 | 4 | 1 | 0 | 0 | 0 | 0 | 0 | 0 | 0 | 4 | 12 | 33 |
| Days Min Temp ≤ 32 °F | 29 | 26 | 24 | 14 | 3 | 0 | 0 | 0 | 1 | 10 | 20 | 27 | 154 |
| Days Min Temp ≤ 0 °F | 2 | 1 | 0 | 0 | 0 | 0 | 0 | 0 | 0 | 0 | 0 | 1 | 4 |
| Heating Degree Days | 1169 | 939 | 838 | 576 | 343 | 161 | 51 | 53 | 237 | 566 | 894 | 1135 | 6962 |
| Cooling Degree Days | 0 | 0 | 0 | 0 | 5 | 30 | 88 | 91 | 7 | 0 | 0 | 0 | 221 |
| Total Precipitation (") | 4.24 | 2.98 | 2.38 | 2.15 | 2.10 | 2.46 | 1.16 | 1.49 | 1.60 | 2.15 | 4.16 | 4.22 | 31.09 |
| Days ≥ 0.1" Precip | 11 | 8 | 8 | 6 | 6 | 7 | 3 | 4 | 5 | 6 | 10 | 12 | 86 |
| Total Snowfall (") | 28.4 | 14.3 | 4.2 | 0.5 | 0.0 | 0.0 | 0.0 | 0.0 | 0.0 | 0.2 | 6.5 | 22.8 | 76.9 |
| Days ≥ 1" Snow Depth | 28 | 23 | 11 | 1 | 0 | 0 | 0 | 0 | 0 | 0 | 4 | 20 | 87 |

## CALDWELL *Canyon County* ELEVATION 2372 ft LAT/LONG 43° 39 ' N / 116° 41 ' W

| | JAN | FEB | MAR | APR | MAY | JUN | JUL | AUG | SEP | OCT | NOV | DEC | YEAR |
|---|---|---|---|---|---|---|---|---|---|---|---|---|---|
| Maximum Temp °F | 37.0 | 46.1 | 57.2 | 66.0 | 75.7 | 84.2 | 92.9 | 91.4 | 80.4 | 66.5 | 49.3 | 38.1 | 65.4 |
| Minimum Temp °F | 21.7 | 26.6 | 32.0 | 37.8 | 45.5 | 52.7 | 58.0 | 55.9 | 46.1 | 36.7 | 29.3 | 22.1 | 38.7 |
| Mean Temp °F | 29.4 | 36.4 | 44.6 | 51.9 | 60.6 | 68.5 | 75.5 | 73.7 | 63.2 | 51.6 | 39.3 | 30.1 | 52.1 |
| Days Max Temp ≥ 90 °F | 0 | 0 | 0 | 0 | 3 | 10 | 22 | 20 | 6 | 0 | 0 | 0 | 61 |
| Days Max Temp ≤ 32 °F | 9 | 2 | 0 | 0 | 0 | 0 | 0 | 0 | 0 | 0 | 1 | 7 | 19 |
| Days Min Temp ≤ 32 °F | 27 | 22 | 17 | 8 | 1 | 0 | 0 | 0 | 1 | 9 | 20 | 27 | 132 |
| Days Min Temp ≤ 0 °F | 1 | 1 | 0 | 0 | 0 | 0 | 0 | 0 | 0 | 0 | 0 | 2 | 4 |
| Heating Degree Days | 1098 | 802 | 625 | 390 | 178 | 49 | 5 | 10 | 113 | 410 | 765 | 1074 | 5519 |
| Cooling Degree Days | 0 | 0 | 0 | 7 | 54 | 158 | 314 | 292 | 67 | 3 | 0 | 0 | 895 |
| Total Precipitation (") | 1.40 | 1.03 | 1.15 | 1.02 | 0.82 | 0.80 | 0.24 | 0.46 | 0.63 | 0.70 | 1.29 | 1.31 | 10.85 |
| Days ≥ 0.1" Precip | 5 | 3 | 4 | 3 | 3 | 3 | 1 | 1 | 2 | 2 | 5 | 5 | 37 |
| Total Snowfall (") | ***4.2*** | 1.8 | 0.2 | 0.0 | 0.0 | 0.0 | 0.0 | 0.0 | 0.0 | 0.1 | ***2.2*** | ***4.5*** | 13.0 |
| Days ≥ 1" Snow Depth | na | ***1*** | 0 | 0 | 0 | 0 | 0 | 0 | 0 | 0 | 0 | ***4*** | na |

## CAMBRIDGE *Washington County* ELEVATION 2651 ft LAT/LONG 44° 34 ' N / 116° 41 ' W

| | JAN | FEB | MAR | APR | MAY | JUN | JUL | AUG | SEP | OCT | NOV | DEC | YEAR |
|---|---|---|---|---|---|---|---|---|---|---|---|---|---|
| Maximum Temp °F | 30.8 | 38.4 | 51.5 | 62.8 | 72.6 | 81.2 | 90.7 | 89.6 | 79.0 | 64.9 | 45.5 | 33.0 | 61.7 |
| Minimum Temp °F | 14.2 | 18.9 | 27.7 | 34.2 | 40.9 | 48.2 | 53.4 | 51.3 | 41.8 | 32.4 | 25.8 | 16.2 | 33.8 |
| Mean Temp °F | 22.5 | 28.7 | 39.6 | 48.5 | 56.8 | 64.7 | 72.0 | 70.5 | 60.5 | 48.7 | 35.7 | 24.6 | 47.7 |
| Days Max Temp ≥ 90 °F | 0 | 0 | 0 | 0 | 1 | 6 | 20 | 18 | 5 | 0 | 0 | 0 | 50 |
| Days Max Temp ≤ 32 °F | 16 | 6 | 1 | 0 | 0 | 0 | 0 | 0 | 0 | 0 | 2 | 12 | 37 |
| Days Min Temp ≤ 32 °F | 30 | 26 | 23 | 13 | 4 | 0 | 0 | 0 | 3 | 17 | 23 | 30 | 169 |
| Days Min Temp ≤ 0 °F | 5 | 2 | 0 | 0 | 0 | 0 | 0 | 0 | 0 | 0 | 1 | 3 | 11 |
| Heating Degree Days | 1309 | 1020 | 780 | 489 | 263 | 90 | 13 | 22 | 162 | 500 | 875 | 1245 | 6768 |
| Cooling Degree Days | 0 | 0 | 0 | 1 | 21 | 95 | 229 | 219 | 40 | 1 | 0 | 0 | 606 |
| Total Precipitation (") | 3.02 | 2.28 | 2.05 | 1.30 | 1.17 | 1.12 | 0.32 | 0.53 | 0.84 | 1.16 | 2.74 | 3.02 | 19.55 |
| Days ≥ 0.1" Precip | 8 | 6 | 6 | 4 | 4 | 4 | 1 | 2 | 2 | 3 | 8 | 8 | 56 |
| Total Snowfall (") | na | na | 2.3 | 0.1 | 0.0 | 0.0 | 0.0 | 0.0 | 0.0 | 0.1 | ***8.4*** | na | na |
| Days ≥ 1" Snow Depth | na | na | na | 0 | 0 | 0 | 0 | 0 | 0 | 0 | na | na | na |

## CASCADE 1 NW *Valley County* ELEVATION 4872 ft LAT/LONG 44° 32' N / 116° 3' W

| | JAN | FEB | MAR | APR | MAY | JUN | JUL | AUG | SEP | OCT | NOV | DEC | YEAR |
|---|---|---|---|---|---|---|---|---|---|---|---|---|---|
| Maximum Temp °F | 29.4 | 35.5 | 43.0 | 52.3 | 63.1 | 71.9 | 81.3 | 80.9 | 70.4 | 57.3 | 39.4 | 30.0 | 54.5 |
| Minimum Temp °F | 10.7 | 13.3 | 19.3 | 26.2 | 33.1 | 39.7 | 44.0 | 42.5 | 34.4 | 27.1 | 20.9 | 12.3 | 27.0 |
| Mean Temp °F | 20.1 | 24.5 | 31.2 | 39.3 | 48.1 | 55.8 | 62.7 | 61.7 | 52.4 | 42.2 | 30.2 | 21.2 | 40.8 |
| Days Max Temp ≥ 90 °F | 0 | 0 | 0 | 0 | 0 | 1 | 4 | 4 | 0 | 0 | 0 | 0 | 9 |
| Days Max Temp ≤ 32 °F | 18 | 8 | 2 | 0 | 0 | 0 | 0 | 0 | 0 | 0 | 6 | 18 | 52 |
| Days Min Temp ≤ 32 °F | 30 | 28 | 30 | 26 | 15 | 5 | 1 | 2 | 12 | 25 | 27 | 30 | 231 |
| Days Min Temp ≤ 0 °F | 8 | 5 | 1 | 0 | 0 | 0 | 0 | 0 | 0 | 0 | 1 | 6 | 21 |
| Heating Degree Days | 1387 | 1138 | 1042 | 766 | 517 | 279 | 109 | 126 | 373 | 699 | 1037 | 1353 | 8826 |
| Cooling Degree Days | 0 | 0 | 0 | 0 | 1 | 7 | 30 | 23 | 0 | 0 | 0 | 0 | 61 |
| Total Precipitation (") | 2.86 | 2.08 | 2.04 | 1.78 | 1.60 | 1.70 | 0.61 | 0.86 | 1.13 | 1.46 | 2.66 | 2.83 | 21.61 |
| Days ≥ 0.1" Precip | 8 | 6 | 7 | 6 | 5 | 5 | 2 | 2 | 3 | 4 | 8 | 8 | 64 |
| Total Snowfall (") | 24.5 | 16.3 | 10.0 | 4.2 | 0.5 | 0.0 | 0.0 | 0.0 | 0.1 | 1.3 | 12.9 | 22.4 | 92.2 |
| Days ≥ 1" Snow Depth | 30 | 27 | 24 | 5 | 0 | 0 | 0 | 0 | 0 | 1 | 11 | 27 | 125 |

## CASTLEFORD 2 N *Twin Falls County* ELEVATION 3832 ft LAT/LONG 42° 33' N / 114° 52' W

| | JAN | FEB | MAR | APR | MAY | JUN | JUL | AUG | SEP | OCT | NOV | DEC | YEAR |
|---|---|---|---|---|---|---|---|---|---|---|---|---|---|
| Maximum Temp °F | 36.5 | 44.0 | 53.7 | 62.6 | 71.8 | 80.5 | 87.8 | 86.3 | 76.7 | 64.3 | 47.6 | 37.0 | 62.4 |
| Minimum Temp °F | 18.9 | 23.4 | 28.3 | 33.7 | 40.6 | 47.7 | 53.0 | 51.8 | 43.5 | 34.8 | 26.7 | 18.8 | 35.1 |
| Mean Temp °F | 27.7 | 33.6 | 41.0 | 48.2 | 56.2 | 64.2 | 70.5 | 69.0 | 60.2 | 49.5 | 37.2 | 28.0 | 48.8 |
| Days Max Temp ≥ 90 °F | 0 | 0 | 0 | 0 | 1 | 5 | 15 | 11 | 1 | 0 | 0 | 0 | 33 |
| Days Max Temp ≤ 32 °F | 10 | 3 | 0 | 0 | 0 | 0 | 0 | 0 | 0 | 0 | 2 | *9* | 24 |
| Days Min Temp ≤ 32 °F | 29 | 25 | 23 | 13 | 4 | 0 | 0 | 0 | 2 | 11 | 22 | 29 | 158 |
| Days Min Temp ≤ 0 °F | 2 | 1 | 0 | 0 | 0 | 0 | 0 | 0 | 0 | 0 | 0 | 2 | 5 |
| Heating Degree Days | 1149 | 880 | 736 | 500 | 277 | 96 | 17 | 28 | 167 | 473 | 829 | 1141 | 6293 |
| Cooling Degree Days | 0 | 0 | 0 | 1 | 16 | 79 | 183 | 156 | 29 | 1 | 0 | 0 | 465 |
| Total Precipitation (") | 1.19 | 0.82 | 0.98 | 0.92 | 1.08 | 1.07 | 0.24 | 0.49 | 0.57 | 0.63 | 1.15 | 1.06 | 10.20 |
| Days ≥ 0.1" Precip | 4 | 3 | 4 | 3 | 4 | 3 | 1 | 1 | 2 | 2 | 4 | *3* | 34 |
| Total Snowfall (") | *4.6* | 3.1 | 1.0 | 0.4 | 0.3 | 0.0 | 0.0 | 0.0 | 0.0 | 0.1 | 2.1 | *4.2* | 15.8 |
| Days ≥ 1" Snow Depth | na | 5 | 1 | 0 | 0 | 0 | 0 | 0 | 0 | 0 | 1 | na | na |

## CHALLIS *Custer County* ELEVATION 5174 ft LAT/LONG 44° 30' N / 114° 14' W

| | JAN | FEB | MAR | APR | MAY | JUN | JUL | AUG | SEP | OCT | NOV | DEC | YEAR |
|---|---|---|---|---|---|---|---|---|---|---|---|---|---|
| Maximum Temp °F | 30.8 | 38.1 | 47.8 | 57.4 | 67.3 | 76.5 | 85.3 | 83.6 | 73.6 | 60.4 | 42.3 | 30.9 | 57.8 |
| Minimum Temp °F | 11.3 | 16.1 | 24.1 | 31.3 | 39.3 | 46.3 | 51.5 | 49.6 | 40.9 | 31.6 | 21.6 | 11.4 | 31.3 |
| Mean Temp °F | 21.1 | 27.1 | 36.0 | 44.4 | 53.3 | 61.4 | 68.5 | 66.6 | 57.2 | 46.0 | 32.0 | 21.2 | 44.6 |
| Days Max Temp ≥ 90 °F | 0 | 0 | 0 | 0 | 0 | 3 | 10 | 8 | 1 | 0 | 0 | 0 | 22 |
| Days Max Temp ≤ 32 °F | 16 | 8 | 1 | 0 | 0 | 0 | 0 | 0 | 0 | 0 | 4 | 16 | 45 |
| Days Min Temp ≤ 32 °F | 30 | 27 | 28 | 17 | 5 | 1 | 0 | 0 | 4 | 17 | 26 | 30 | 185 |
| Days Min Temp ≤ 0 °F | 6 | 2 | 0 | 0 | 0 | 0 | 0 | 0 | 0 | 0 | 1 | 5 | 14 |
| Heating Degree Days | 1356 | 1064 | 892 | 612 | 360 | 149 | 33 | 51 | 239 | 581 | 984 | 1356 | 7677 |
| Cooling Degree Days | 0 | 0 | 0 | 0 | 6 | 49 | 140 | 110 | 12 | 0 | 0 | 0 | 317 |
| Total Precipitation (") | 0.48 | 0.34 | 0.52 | 0.56 | 1.00 | 1.08 | 0.75 | 0.68 | 0.71 | 0.45 | 0.58 | 0.47 | 7.62 |
| Days ≥ 0.1" Precip | 2 | 1 | 2 | 2 | 4 | 4 | 2 | 2 | 2 | 1 | 2 | 2 | 26 |
| Total Snowfall (") | *4.2* | *2.2* | *2.1* | 0.7 | 0.1 | 0.0 | 0.0 | 0.0 | 0.1 | 0.1 | *3.0* | *4.3* | 16.8 |
| Days ≥ 1" Snow Depth | na | na | na | 0 | 0 | 0 | 0 | 0 | 0 | 0 | na | na | na |

## CHILLY BARTON FLAT *Custer County* ELEVATION 6184 ft LAT/LONG 44° 0' N / 113° 48' W

| | JAN | FEB | MAR | APR | MAY | JUN | JUL | AUG | SEP | OCT | NOV | DEC | YEAR |
|---|---|---|---|---|---|---|---|---|---|---|---|---|---|
| Maximum Temp °F | 29.9 | 35.1 | 42.4 | 53.1 | 63.2 | 72.5 | 81.4 | 80.1 | 70.7 | 58.3 | 39.8 | 29.8 | 54.7 |
| Minimum Temp °F | 4.1 | 7.2 | 17.2 | 25.1 | 32.4 | 39.3 | 44.3 | 42.5 | 33.8 | 25.5 | 15.3 | 4.7 | 24.3 |
| Mean Temp °F | 17.0 | 21.2 | 29.8 | 39.1 | 47.8 | 55.9 | 62.9 | 61.2 | 52.3 | 42.0 | 27.4 | 17.3 | 39.5 |
| Days Max Temp ≥ 90 °F | 0 | 0 | 0 | 0 | 0 | 1 | 2 | 2 | 0 | 0 | 0 | 0 | 5 |
| Days Max Temp ≤ 32 °F | 18 | 10 | 3 | 0 | 0 | 0 | 0 | 0 | 0 | 0 | 6 | 19 | 56 |
| Days Min Temp ≤ 32 °F | 31 | 28 | 31 | 26 | 16 | 5 | 1 | 1 | 11 | 26 | 29 | 31 | 236 |
| Days Min Temp ≤ 0 °F | 12 | 8 | 2 | 0 | 0 | 0 | 0 | 0 | 0 | 0 | 3 | 11 | 36 |
| Heating Degree Days | 1481 | 1231 | 1084 | 769 | 527 | 274 | 92 | 130 | 375 | 707 | 1121 | 1472 | 9263 |
| Cooling Degree Days | 0 | 0 | 0 | 0 | 0 | 10 | 39 | 19 | 1 | 0 | 0 | 0 | 69 |
| Total Precipitation (") | 0.37 | 0.27 | 0.46 | 0.58 | 1.06 | 1.36 | 0.97 | 0.85 | 0.80 | 0.51 | 0.49 | 0.34 | 8.06 |
| Days ≥ 0.1" Precip | 1 | 1 | 1 | 2 | 4 | 4 | 3 | 3 | 2 | 2 | 2 | 1 | 26 |
| Total Snowfall (") | na | na | na | *0.7* | 0.1 | 0.0 | 0.0 | 0.0 | 0.2 | 0.4 | *1.2* | na | na |
| Days ≥ 1" Snow Depth | na | na | na | 0 | 0 | 0 | 0 | 0 | 0 | 0 | *1* | na | na |

## COUNCIL *Adams County* ELEVATION 3153 ft LAT/LONG 44° 45' N / 116° 25' W

| | JAN | FEB | MAR | APR | MAY | JUN | JUL | AUG | SEP | OCT | NOV | DEC | YEAR |
|---|---|---|---|---|---|---|---|---|---|---|---|---|---|
| Maximum Temp °F | 33.6 | 40.2 | 50.6 | 61.9 | 71.6 | 80.8 | 90.7 | 90.0 | 78.4 | 65.2 | na | 35.6 | na |
| Minimum Temp °F | 16.6 | 20.3 | 28.2 | 34.6 | 41.3 | 48.6 | 55.2 | 54.0 | 44.1 | 34.2 | na | 18.8 | na |
| Mean Temp °F | 25.1 | 30.5 | 39.4 | 48.2 | 56.6 | 64.7 | 73.0 | 72.1 | 61.3 | 49.7 | na | 27.3 | na |
| Days Max Temp ≥ 90 °F | 0 | 0 | 0 | 0 | 1 | 6 | 18 | 18 | 5 | 0 | 0 | 0 | 48 |
| Days Max Temp ≤ 32 °F | 11 | 3 | 0 | 0 | 0 | 0 | 0 | 0 | 0 | 0 | 1 | 8 | 23 |
| Days Min Temp ≤ 32 °F | 26 | 23 | 21 | 11 | 4 | 0 | 0 | 0 | 2 | 13 | 19 | 26 | 145 |
| Days Min Temp ≤ 0 °F | 4 | 2 | 0 | 0 | 0 | 0 | 0 | 0 | 0 | 0 | 0 | 3 | 9 |
| Heating Degree Days | 1230 | 969 | 787 | 498 | 272 | 97 | 14 | 19 | 155 | 468 | na | 1163 | na |
| Cooling Degree Days | na | 0 | na | na | na | 90 | na | 248 | na | na | na | na | na |
| Total Precipitation (") | 3.62 | 2.63 | 2.27 | 1.87 | 1.80 | 1.62 | 0.59 | 0.66 | 1.10 | 1.55 | 3.38 | 3.34 | 24.43 |
| Days ≥ 0.1" Precip | 7 | 6 | 5 | 4 | 4 | 4 | 1 | 2 | 3 | 3 | 6 | 7 | 52 |
| Total Snowfall (") | na | na | na | 0.0 | 0.0 | 0.0 | 0.0 | 0.0 | 0.0 | 0.3 | na | na | na |
| Days ≥ 1" Snow Depth | 19 | 16 | 3 | 0 | 0 | 0 | 0 | 0 | 0 | 0 | 3 | 13 | 54 |

## CRATERS OF MOON NM *Butte County* ELEVATION 5905 ft LAT/LONG 43° 28' N / 113° 34' W

| | JAN | FEB | MAR | APR | MAY | JUN | JUL | AUG | SEP | OCT | NOV | DEC | YEAR |
|---|---|---|---|---|---|---|---|---|---|---|---|---|---|
| Maximum Temp °F | 29.0 | 33.9 | 41.9 | 53.8 | 65.1 | 74.4 | 84.0 | 82.5 | 71.4 | 59.0 | 39.7 | 29.5 | 55.4 |
| Minimum Temp °F | 10.3 | 13.7 | 20.6 | 28.3 | 36.8 | 44.4 | 51.6 | 50.2 | 40.5 | 31.0 | 19.9 | 10.7 | 29.8 |
| Mean Temp °F | 19.7 | 23.8 | 31.3 | 40.8 | 51.0 | 59.5 | 67.8 | 66.4 | 56.0 | 45.0 | 29.8 | 20.1 | 42.6 |
| Days Max Temp ≥ 90 °F | 0 | 0 | 0 | 0 | 0 | 1 | 5 | 4 | 0 | 0 | 0 | 0 | 10 |
| Days Max Temp ≤ 32 °F | 19 | 11 | 4 | 0 | 0 | 0 | 0 | 0 | 0 | 0 | 7 | 19 | 60 |
| Days Min Temp ≤ 32 °F | 30 | 28 | 29 | 22 | 9 | 2 | 0 | 0 | 5 | 16 | 27 | 30 | 198 |
| Days Min Temp ≤ 0 °F | 6 | 3 | 1 | 0 | 0 | 0 | 0 | 0 | 0 | 0 | 1 | 5 | 16 |
| Heating Degree Days | 1399 | 1158 | 1039 | 719 | 430 | 195 | 35 | 57 | 277 | 612 | 1049 | 1385 | 8355 |
| Cooling Degree Days | 0 | 0 | 0 | 0 | 5 | 45 | 143 | 127 | 17 | na | 0 | 0 | na |
| Total Precipitation (") | 2.22 | 1.45 | 1.31 | 1.13 | 1.57 | 1.28 | 0.79 | 0.89 | 0.81 | 0.89 | 1.51 | 1.83 | 15.68 |
| Days ≥ 0.1" Precip | 4 | 4 | 4 | 3 | 5 | 4 | 2 | 3 | 2 | 2 | 4 | 5 | 42 |
| Total Snowfall (") | 23.0 | 18.0 | 10.5 | 4.1 | 2.5 | 0.0 | 0.0 | 0.0 | 0.2 | 1.5 | 11.8 | 21.8 | 93.4 |
| Days ≥ 1" Snow Depth | 28 | 23 | 21 | 6 | 1 | 0 | 0 | 0 | 0 | 1 | 10 | 25 | 115 |

## DEER FLAT DAM *Canyon County* ELEVATION 2513 ft LAT/LONG 43° 35' N / 116° 45' W

| | JAN | FEB | MAR | APR | MAY | JUN | JUL | AUG | SEP | OCT | NOV | DEC | YEAR |
|---|---|---|---|---|---|---|---|---|---|---|---|---|---|
| Maximum Temp °F | 37.0 | 45.5 | 56.1 | 64.0 | 73.0 | 80.6 | 87.9 | 87.0 | 78.2 | 65.7 | 49.1 | 38.1 | 63.5 |
| Minimum Temp °F | 21.6 | 26.5 | 32.3 | 38.2 | 45.7 | 52.3 | 57.1 | 55.8 | 47.6 | 37.6 | 30.0 | 22.2 | 38.9 |
| Mean Temp °F | 29.3 | 36.0 | 44.2 | 51.1 | 59.3 | 66.5 | 72.5 | 71.4 | 62.9 | 51.7 | 39.6 | 30.2 | 51.2 |
| Days Max Temp ≥ 90 °F | 0 | 0 | 0 | 0 | 1 | 5 | 14 | 13 | 1 | 0 | 0 | 0 | 34 |
| Days Max Temp ≤ 32 °F | 9 | 2 | 0 | 0 | 0 | 0 | 0 | 0 | 0 | 0 | 1 | 7 | 19 |
| Days Min Temp ≤ 32 °F | 26 | 22 | 16 | 7 | 1 | 0 | 0 | 0 | 1 | 8 | 18 | 26 | 125 |
| Days Min Temp ≤ 0 °F | 2 | 1 | 0 | 0 | 0 | 0 | 0 | 0 | 0 | 0 | 0 | 2 | 5 |
| Heating Degree Days | 1099 | 812 | 637 | 412 | 200 | 63 | 9 | 14 | 113 | 407 | 756 | 1073 | 5595 |
| Cooling Degree Days | 0 | 0 | 0 | 3 | 39 | 121 | 241 | 230 | 64 | 2 | 0 | 0 | 700 |
| Total Precipitation (") | 1.08 | 0.81 | 1.09 | 1.02 | 0.86 | 0.89 | 0.27 | 0.46 | 0.54 | 0.62 | 1.15 | 1.05 | 9.84 |
| Days ≥ 0.1" Precip | 4 | 3 | 4 | 3 | 3 | 3 | 1 | 1 | 2 | 2 | 4 | 4 | 34 |
| Total Snowfall (") | na | na | 0.3 | 0.1 | 0.0 | 0.0 | 0.0 | 0.0 | 0.0 | 0.0 | 0.6 | na | na |
| Days ≥ 1" Snow Depth | 10 | 3 | 0 | 0 | 0 | 0 | 0 | 0 | 0 | 0 | 0 | na | na |

## DIXIE *Idaho County* ELEVATION 5613 ft LAT/LONG 45° 33' N / 115° 28' W

| | JAN | FEB | MAR | APR | MAY | JUN | JUL | AUG | SEP | OCT | NOV | DEC | YEAR |
|---|---|---|---|---|---|---|---|---|---|---|---|---|---|
| Maximum Temp °F | 30.2 | 35.1 | 40.1 | 46.7 | 56.4 | 65.5 | 74.7 | 75.5 | 65.8 | 53.1 | 37.2 | 30.2 | 50.9 |
| Minimum Temp °F | 4.3 | 6.5 | 13.1 | 21.1 | 28.5 | 35.1 | 37.4 | 35.5 | 28.7 | 22.3 | 13.6 | 4.8 | 20.9 |
| Mean Temp °F | 17.3 | 20.8 | 26.6 | 33.9 | 42.5 | 50.3 | 56.1 | 55.6 | 47.3 | 37.7 | 25.4 | 17.5 | 35.9 |
| Days Max Temp ≥ 90 °F | 0 | 0 | 0 | 0 | 0 | 0 | 0 | 1 | 0 | 0 | 0 | 0 | 1 |
| Days Max Temp ≤ 32 °F | 18 | 10 | 5 | 1 | 0 | 0 | 0 | 0 | 0 | 1 | 9 | 18 | 62 |
| Days Min Temp ≤ 32 °F | 31 | 28 | 31 | 29 | 25 | 11 | 5 | 9 | 23 | 30 | 29 | 31 | 282 |
| Days Min Temp ≤ 0 °F | 12 | 9 | 4 | 0 | 0 | 0 | 0 | 0 | 0 | 0 | 4 | 12 | 41 |
| Heating Degree Days | 1474 | 1241 | 1183 | 925 | 691 | 435 | 275 | 289 | 524 | 841 | 1181 | 1465 | 10524 |
| Cooling Degree Days | 0 | 0 | 0 | 0 | 0 | 1 | 3 | 3 | 0 | 0 | 0 | 0 | 7 |
| Total Precipitation (") | 3.78 | 2.55 | 2.46 | 2.05 | 2.07 | 2.43 | 1.19 | 1.26 | 1.50 | 1.67 | 2.82 | 3.29 | 27.07 |
| Days ≥ 0.1" Precip | 11 | 8 | 8 | 7 | 7 | 7 | 4 | 4 | 4 | 5 | 9 | 11 | 85 |
| Total Snowfall (") | 42.9 | 28.7 | 27.0 | 16.8 | 6.6 | 0.6 | 0.0 | 0.0 | 0.8 | 5.6 | 31.0 | 39.0 | 199.0 |
| Days ≥ 1" Snow Depth | 31 | 28 | 31 | 28 | 13 | 0 | 0 | 0 | 0 | 5 | 23 | 31 | 190 |

## DRIGGS *Teton County* ELEVATION 6106 ft LAT/LONG 43° 44 ' N / 111° 7 ' W

| | JAN | FEB | MAR | APR | MAY | JUN | JUL | AUG | SEP | OCT | NOV | DEC | YEAR |
|---|---|---|---|---|---|---|---|---|---|---|---|---|---|
| Maximum Temp °F | 29.3 | 34.8 | 41.7 | 52.5 | 63.4 | 72.1 | 80.3 | 79.4 | 70.0 | 57.9 | 40.2 | 30.7 | 54.4 |
| Minimum Temp °F | 7.3 | 10.1 | 17.7 | 25.9 | 33.4 | 40.7 | 46.2 | 44.3 | 36.3 | 27.0 | 17.8 | 8.0 | 26.2 |
| Mean Temp °F | 18.3 | 22.5 | 29.7 | 39.2 | 48.4 | 56.4 | 63.3 | 61.8 | 53.2 | 42.5 | 29.1 | 19.4 | 40.3 |
| Days Max Temp ≥ 90 °F | 0 | 0 | 0 | 0 | 0 | 0 | 1 | 1 | 0 | 0 | 0 | 0 | 2 |
| Days Max Temp ≤ 32 °F | 19 | 9 | 3 | 0 | 0 | 0 | 0 | 0 | 0 | 1 | 7 | 18 | 57 |
| Days Min Temp ≤ 32 °F | 30 | 28 | 30 | 24 | 13 | 3 | 0 | 1 | 9 | 24 | 28 | 31 | 221 |
| Days Min Temp ≤ 0 °F | 10 | 7 | 2 | 0 | 0 | 0 | 0 | 0 | 0 | 0 | 3 | 9 | 31 |
| Heating Degree Days | 1442 | 1195 | 1087 | 767 | 506 | 259 | 83 | 113 | 349 | 691 | 1072 | 1409 | 8973 |
| Cooling Degree Days | 0 | 0 | 0 | 0 | 0 | 10 | 30 | 20 | 2 | 0 | 0 | 0 | 62 |
| Total Precipitation (") | 1.36 | 0.94 | 1.08 | 1.31 | 2.00 | 1.65 | 1.24 | 1.12 | 1.25 | 1.18 | 1.19 | 1.32 | 15.64 |
| Days ≥ 0.1" Precip | 5 | 4 | 4 | 5 | 6 | 5 | 4 | 4 | 4 | 4 | 5 | 5 | 55 |
| Total Snowfall (") | 16.3 | ***10.8*** | ***9.7*** | ***5.4*** | ***2.9*** | 0.4 | 0.0 | 0.0 | 0.4 | ***2.7*** | 11.1 | ***15.6*** | 75.3 |
| Days ≥ 1" Snow Depth | na | na | na | na | ***0*** | 0 | 0 | 0 | 0 | ***1*** | na | na | na |

## DUBOIS EXPERIMENT ST *Clark County* ELEVATION 5463 ft LAT/LONG 44° 15 ' N / 112° 12 ' W

| | JAN | FEB | MAR | APR | MAY | JUN | JUL | AUG | SEP | OCT | NOV | DEC | YEAR |
|---|---|---|---|---|---|---|---|---|---|---|---|---|---|
| Maximum Temp °F | 27.4 | 32.5 | 41.2 | 54.0 | 65.3 | 74.6 | 84.2 | 83.3 | 72.3 | 57.8 | 38.5 | 28.4 | 55.0 |
| Minimum Temp °F | 11.1 | 14.7 | 21.9 | 30.0 | 38.3 | 45.5 | 52.0 | 50.9 | 42.1 | 32.5 | 21.4 | 12.0 | 31.0 |
| Mean Temp °F | 19.2 | 23.6 | 31.5 | 42.0 | 51.8 | 60.1 | 68.1 | 67.1 | 57.2 | 45.2 | 30.0 | 20.3 | 43.0 |
| Days Max Temp ≥ 90 °F | 0 | 0 | 0 | 0 | 0 | 2 | 7 | 6 | 0 | 0 | 0 | 0 | 15 |
| Days Max Temp ≤ 32 °F | 22 | 14 | 4 | 0 | 0 | 0 | 0 | 0 | 0 | 1 | 8 | 20 | 69 |
| Days Min Temp ≤ 32 °F | 31 | 28 | 29 | 19 | 7 | 1 | 0 | 0 | 3 | 15 | 27 | 31 | 191 |
| Days Min Temp ≤ 0 °F | 6 | 3 | 1 | 0 | 0 | 0 | 0 | 0 | 0 | 0 | 1 | 4 | 15 |
| Heating Degree Days | 1412 | 1162 | 1030 | 684 | 405 | 176 | 33 | 47 | 244 | 608 | 1044 | 1381 | 8226 |
| Cooling Degree Days | 0 | 0 | 0 | 0 | 3 | 38 | 128 | 125 | 19 | 0 | 0 | 0 | 313 |
| Total Precipitation (") | 0.76 | 0.64 | 0.84 | 1.03 | 1.63 | 1.77 | 1.11 | 1.09 | 0.99 | 0.82 | 1.15 | 0.92 | 12.75 |
| Days ≥ 0.1" Precip | 3 | 2 | 3 | 4 | 5 | 5 | 3 | 3 | 3 | 2 | 4 | 3 | 40 |
| Total Snowfall (") | 9.6 | 7.2 | 5.2 | 2.5 | 1.1 | 0.1 | 0.0 | 0.0 | 0.2 | 1.7 | 8.3 | 11.9 | 47.8 |
| Days ≥ 1" Snow Depth | 30 | 28 | 19 | 3 | 0 | 0 | 0 | 0 | 0 | 1 | 10 | 27 | 118 |

## ELK CITY RANGER STN *Idaho County* ELEVATION 3983 ft LAT/LONG 45° 49 ' N / 115° 26 ' W

| | JAN | FEB | MAR | APR | MAY | JUN | JUL | AUG | SEP | OCT | NOV | DEC | YEAR |
|---|---|---|---|---|---|---|---|---|---|---|---|---|---|
| Maximum Temp °F | 34.0 | 41.0 | 46.5 | 53.4 | 62.5 | 70.8 | 80.1 | 80.8 | 70.9 | 59.2 | 42.0 | 33.5 | 56.2 |
| Minimum Temp °F | 11.2 | 13.8 | 19.1 | 26.1 | 32.6 | 39.2 | 40.8 | 38.7 | 32.2 | 25.8 | 20.5 | 11.3 | 25.9 |
| Mean Temp °F | 22.6 | 27.4 | 32.9 | 39.8 | 47.6 | 55.0 | 60.5 | 59.8 | 51.6 | 42.5 | 31.3 | 22.4 | 41.1 |
| Days Max Temp ≥ 90 °F | 0 | 0 | 0 | 0 | 0 | 1 | 4 | 6 | 1 | 0 | 0 | 0 | 12 |
| Days Max Temp ≤ 32 °F | 11 | 4 | 2 | 0 | 0 | 0 | 0 | 0 | 0 | 0 | 4 | 13 | 34 |
| Days Min Temp ≤ 32 °F | 31 | 28 | 31 | 27 | 16 | 5 | 2 | 5 | 16 | 26 | 29 | 31 | 247 |
| Days Min Temp ≤ 0 °F | 7 | 4 | 2 | 0 | 0 | 0 | 0 | 0 | 0 | 0 | 2 | 6 | 21 |
| Heating Degree Days | 1307 | 1054 | 989 | 750 | 534 | 302 | 155 | 173 | 397 | 690 | 1005 | 1314 | 8670 |
| Cooling Degree Days | 0 | 0 | 0 | 0 | 1 | 8 | 20 | 18 | 1 | 0 | 0 | 0 | 48 |
| Total Precipitation (") | 3.53 | 2.45 | 2.54 | 2.65 | 3.14 | 3.04 | 1.60 | 1.42 | 1.84 | 1.98 | 2.91 | 3.08 | 30.18 |
| Days ≥ 0.1" Precip | 10 | 7 | 9 | 9 | 9 | 8 | 5 | 4 | 5 | 6 | 9 | 9 | 90 |
| Total Snowfall (") | 31.7 | 19.2 | 18.5 | 10.9 | 1.9 | 0.0 | 0.0 | 0.0 | 0.0 | 2.6 | 18.3 | 27.1 | 130.2 |
| Days ≥ 1" Snow Depth | 30 | 27 | 26 | 12 | 1 | 0 | 0 | 0 | 0 | 2 | 15 | 29 | 142 |

## ELK RIVER 1 S *Clearwater County* ELEVATION 2913 ft LAT/LONG 46° 47 ' N / 116° 10 ' W

| | JAN | FEB | MAR | APR | MAY | JUN | JUL | AUG | SEP | OCT | NOV | DEC | YEAR |
|---|---|---|---|---|---|---|---|---|---|---|---|---|---|
| Maximum Temp °F | 34.2 | 39.9 | 46.6 | 54.4 | 63.7 | 71.7 | 80.3 | 81.2 | 71.0 | 58.3 | 41.8 | 34.0 | 56.4 |
| Minimum Temp °F | 17.4 | 19.8 | 24.4 | 30.8 | 36.6 | 43.2 | 45.4 | 44.2 | 37.0 | 29.7 | 25.4 | 18.2 | 31.0 |
| Mean Temp °F | 25.8 | 29.9 | 35.5 | 42.6 | 50.2 | 57.4 | 62.8 | 62.7 | 54.0 | 44.0 | 33.6 | 26.1 | 43.7 |
| Days Max Temp ≥ 90 °F | 0 | 0 | 0 | 0 | 0 | 1 | 5 | 6 | 1 | 0 | 0 | 0 | 13 |
| Days Max Temp ≤ 32 °F | 10 | 4 | 1 | 0 | 0 | 0 | 0 | 0 | 0 | 0 | 3 | 11 | 29 |
| Days Min Temp ≤ 32 °F | 30 | 27 | 29 | 19 | 8 | 1 | 0 | 1 | 7 | 22 | 26 | 30 | 200 |
| Days Min Temp ≤ 0 °F | 4 | 2 | 1 | 0 | 0 | 0 | 0 | 0 | 0 | 0 | 0 | 3 | 10 |
| Heating Degree Days | 1208 | 984 | 907 | 665 | 455 | 240 | 112 | 111 | 327 | 644 | 936 | 1198 | 7787 |
| Cooling Degree Days | 0 | 0 | 0 | 0 | 5 | 21 | 46 | 51 | 5 | 0 | 0 | 0 | 128 |
| Total Precipitation (") | 5.19 | 3.91 | 3.22 | 2.42 | 2.78 | 2.36 | 1.27 | 1.14 | 1.78 | 2.35 | 4.34 | 4.61 | 35.37 |
| Days ≥ 0.1" Precip | 12 | 10 | 9 | 7 | 7 | 7 | 4 | 3 | 5 | 6 | 11 | 11 | 92 |
| Total Snowfall (") | 33.3 | 18.5 | 10.3 | 2.7 | 0.3 | 0.0 | 0.0 | 0.0 | 0.0 | 0.2 | 13.4 | 27.4 | 106.1 |
| Days ≥ 1" Snow Depth | 30 | 28 | 25 | 9 | 0 | 0 | 0 | 0 | 0 | 0 | ***12*** | 29 | 133 |

## EMMETT 2 E *Gem County* ELEVATION 2392 ft LAT/LONG 43° 52 ' N / 116° 28 ' W

| | JAN | FEB | MAR | APR | MAY | JUN | JUL | AUG | SEP | OCT | NOV | DEC | YEAR |
|---|---|---|---|---|---|---|---|---|---|---|---|---|---|
| Maximum Temp °F | 37.5 | 45.8 | 56.0 | 64.5 | 74.1 | 83.0 | 91.2 | 89.8 | 79.4 | 66.5 | 49.3 | 38.6 | 64.6 |
| Minimum Temp °F | 21.5 | 26.4 | 31.3 | 36.3 | 42.9 | 50.2 | 55.0 | 54.0 | 45.4 | 36.7 | 29.2 | 22.1 | 37.6 |
| Mean Temp °F | 29.6 | 36.1 | 43.7 | 50.4 | 58.6 | 66.6 | 73.2 | 71.9 | 62.4 | 51.7 | 39.3 | 30.4 | 51.2 |
| Days Max Temp ≥ 90 °F | 0 | 0 | 0 | 0 | 2 | 8 | 20 | 18 | 4 | 0 | 0 | 0 | 52 |
| Days Max Temp ≤ 32 °F | 9 | 3 | 0 | 0 | 0 | 0 | 0 | 0 | 0 | 0 | 1 | 6 | 19 |
| Days Min Temp ≤ 32 °F | 27 | 22 | 18 | 10 | 3 | 0 | 0 | 0 | 1 | 9 | 20 | 27 | 137 |
| Days Min Temp ≤ 0 °F | 2 | 1 | 0 | 0 | 0 | 0 | 0 | 0 | 0 | 0 | 0 | 2 | 5 |
| Heating Degree Days | 1090 | 809 | 655 | 432 | 224 | 69 | 11 | 15 | 127 | 409 | 765 | 1066 | 5672 |
| Cooling Degree Days | 0 | 0 | 0 | 3 | 33 | 118 | 249 | 238 | 55 | 2 | 0 | 0 | 698 |
| Total Precipitation (") | 1.69 | 1.41 | 1.35 | 1.18 | 1.03 | 0.94 | 0.23 | 0.44 | 0.76 | 0.85 | 1.80 | 1.53 | 13.21 |
| Days ≥ 0.1" Precip | 5 | 5 | 5 | 4 | 3 | 3 | 1 | 1 | 2 | 3 | 6 | 5 | 43 |
| Total Snowfall (") | na | na | *0.2* | 0.0 | 0.0 | 0.0 | 0.0 | 0.0 | 0.0 | 0.0 | *1.1* | na | na |
| Days ≥ 1" Snow Depth | na | na | *0* | 0 | 0 | 0 | 0 | 0 | 0 | 0 | *1* | na | na |

## FAIRFIELD R S *Camas County* ELEVATION 5072 ft LAT/LONG 43° 21 ' N / 114° 48 ' W

| | JAN | FEB | MAR | APR | MAY | JUN | JUL | AUG | SEP | OCT | NOV | DEC | YEAR |
|---|---|---|---|---|---|---|---|---|---|---|---|---|---|
| Maximum Temp °F | 29.9 | 35.1 | 43.6 | 55.4 | 67.4 | 75.9 | 85.3 | 84.3 | 74.8 | 63.0 | 43.1 | 31.3 | 57.4 |
| Minimum Temp °F | 6.1 | 8.0 | 17.3 | 27.1 | 34.4 | 40.2 | 45.6 | 43.9 | 34.8 | 26.3 | 17.8 | 7.2 | 25.7 |
| Mean Temp °F | 18.0 | 21.6 | 30.5 | 41.2 | 50.9 | 58.1 | 65.5 | 64.1 | 54.7 | 44.7 | 30.4 | 19.3 | 41.6 |
| Days Max Temp ≥ 90 °F | 0 | 0 | 0 | 0 | 0 | 2 | 8 | 9 | 1 | 0 | 0 | 0 | 20 |
| Days Max Temp ≤ 32 °F | 18 | 9 | 2 | 0 | 0 | 0 | 0 | 0 | 0 | 0 | 5 | 16 | 50 |
| Days Min Temp ≤ 32 °F | 31 | 28 | 30 | 24 | 12 | 4 | 0 | 2 | 11 | 24 | 28 | 31 | 225 |
| Days Min Temp ≤ 0 °F | 11 | 9 | 2 | 0 | 0 | 0 | 0 | 0 | 0 | 0 | 2 | 9 | 33 |
| Heating Degree Days | 1449 | 1219 | 1063 | 706 | 431 | 220 | 56 | 82 | 307 | 622 | 1029 | 1410 | 8594 |
| Cooling Degree Days | 0 | 0 | 0 | 0 | 3 | 21 | 78 | 63 | 4 | 0 | 0 | 0 | 169 |
| Total Precipitation (") | 2.42 | 1.66 | 1.34 | 1.09 | 1.05 | 1.03 | 0.69 | 0.62 | 0.71 | 0.85 | 1.89 | 2.18 | 15.53 |
| Days ≥ 0.1" Precip | 6 | 5 | 4 | 3 | 3 | 3 | 2 | 2 | 2 | 2 | 6 | 6 | 44 |
| Total Snowfall (") | 19.7 | 12.6 | 5.6 | 2.6 | 0.5 | 0.0 | 0.0 | 0.0 | 0.3 | *0.6* | 9.7 | *20.5* | 72.1 |
| Days ≥ 1" Snow Depth | 28 | 25 | 20 | 3 | 0 | 0 | 0 | 0 | 0 | 0 | 8 | 24 | 108 |

## FENN RANGER STN *Idaho County* ELEVATION 1601 ft LAT/LONG 46° 6 ' N / 115° 33 ' W

| | JAN | FEB | MAR | APR | MAY | JUN | JUL | AUG | SEP | OCT | NOV | DEC | YEAR |
|---|---|---|---|---|---|---|---|---|---|---|---|---|---|
| Maximum Temp °F | 36.1 | 42.8 | 52.0 | 61.3 | 70.7 | 78.2 | 87.7 | 88.3 | 75.9 | 60.7 | 45.2 | 36.4 | 61.3 |
| Minimum Temp °F | 23.8 | 26.4 | 30.5 | 35.3 | 41.3 | 47.8 | 51.2 | 50.1 | 43.7 | 36.2 | 30.8 | 25.2 | 36.9 |
| Mean Temp °F | 30.0 | 34.6 | 41.3 | 48.4 | 56.1 | 63.1 | 69.4 | 69.3 | 59.8 | 48.5 | 38.0 | 30.8 | 49.1 |
| Days Max Temp ≥ 90 °F | 0 | 0 | 0 | 0 | 2 | 5 | 14 | 16 | 3 | 0 | 0 | 0 | 40 |
| Days Max Temp ≤ 32 °F | 7 | 2 | 0 | 0 | 0 | 0 | 0 | 0 | 0 | 0 | 1 | 5 | 15 |
| Days Min Temp ≤ 32 °F | 29 | 25 | 20 | 9 | 1 | 0 | 0 | 0 | 1 | 7 | 17 | 28 | 137 |
| Days Min Temp ≤ 0 °F | 1 | 0 | 0 | 0 | 0 | 0 | 0 | 0 | 0 | 0 | 0 | 1 | 2 |
| Heating Degree Days | 1079 | 852 | 728 | 493 | 283 | 111 | 22 | 25 | 173 | 505 | 803 | 1052 | 6126 |
| Cooling Degree Days | 0 | 0 | 0 | 1 | 19 | *70* | 160 | 170 | 27 | 0 | 0 | 0 | 447 |
| Total Precipitation (") | 4.92 | 3.35 | 3.59 | 3.60 | 3.31 | 3.13 | 1.22 | 1.40 | 2.25 | 2.79 | 4.30 | 4.01 | 37.87 |
| Days ≥ 0.1" Precip | 11 | 9 | 10 | 10 | 9 | 7 | 4 | 3 | 5 | 7 | 11 | 10 | 96 |
| Total Snowfall (") | 19.3 | 9.3 | 2.3 | 0.2 | 0.0 | 0.0 | 0.0 | 0.0 | 0.0 | 0.3 | 6.7 | 16.6 | 54.7 |
| Days ≥ 1" Snow Depth | 25 | 19 | 5 | 0 | 0 | 0 | 0 | 0 | 0 | 0 | 4 | 16 | 69 |

## FORT HALL 1 NNE *Bingham County* ELEVATION 4505 ft LAT/LONG 43° 2 ' N / 112° 26 ' W

| | JAN | FEB | MAR | APR | MAY | JUN | JUL | AUG | SEP | OCT | NOV | DEC | YEAR |
|---|---|---|---|---|---|---|---|---|---|---|---|---|---|
| Maximum Temp °F | 32.9 | 39.6 | 49.1 | 59.1 | 68.8 | 78.0 | 86.1 | 85.3 | 75.3 | 62.7 | 44.7 | 33.8 | 59.6 |
| Minimum Temp °F | 13.8 | 18.3 | 25.4 | 31.9 | 39.7 | 46.9 | 51.8 | 50.0 | 41.2 | 31.7 | 23.4 | 14.3 | 32.4 |
| Mean Temp °F | 23.3 | 28.9 | 37.3 | 45.5 | 54.3 | 62.5 | 69.0 | 67.7 | 58.3 | 47.2 | 34.1 | 24.1 | 46.0 |
| Days Max Temp ≥ 90 °F | 0 | 0 | 0 | 0 | 0 | 3 | 11 | 9 | 1 | 0 | 0 | 0 | 24 |
| Days Max Temp ≤ 32 °F | 14 | 6 | 1 | 0 | 0 | 0 | 0 | 0 | 0 | 0 | 4 | 13 | 38 |
| Days Min Temp ≤ 32 °F | 29 | 26 | 25 | *17* | 5 | 0 | 0 | 0 | 4 | 17 | 24 | 29 | 176 |
| Days Min Temp ≤ 0 °F | 6 | 2 | 0 | 0 | 0 | 0 | 0 | 0 | 0 | 0 | 1 | 4 | 13 |
| Heating Degree Days | 1285 | 1011 | 852 | 578 | 330 | 121 | 19 | 34 | 212 | 546 | 920 | 1260 | 7168 |
| Cooling Degree Days | 0 | 0 | 0 | 1 | 5 | 48 | 134 | 127 | 15 | 0 | 0 | 0 | 330 |
| Total Precipitation (") | 0.85 | 0.85 | 1.13 | 1.08 | 1.51 | 1.12 | 0.69 | 0.90 | 0.81 | 1.03 | 1.06 | 0.92 | 11.95 |
| Days ≥ 0.1" Precip | *3* | 3 | 4 | *3* | *3* | 3 | 2 | 2 | 2 | 3 | 3 | 3 | 34 |
| Total Snowfall (") | na | na | *2.6* | *1.0* | 0.3 | 0.0 | 0.0 | 0.0 | 0.0 | 0.8 | 3.2 | na | na |
| Days ≥ 1" Snow Depth | na | na | *3* | 0 | 0 | 0 | 0 | 0 | 0 | 0 | 4 | *12* | na |

## GARDEN VALLEY R S *Boise County* ELEVATION 3153 ft LAT/LONG 44° 4' N / 115° 55' W

| | JAN | FEB | MAR | APR | MAY | JUN | JUL | AUG | SEP | OCT | NOV | DEC | YEAR |
|---|---|---|---|---|---|---|---|---|---|---|---|---|---|
| Maximum Temp °F | 34.7 | 42.2 | 51.4 | 61.0 | 72.5 | 80.7 | 90.2 | 89.6 | 78.9 | 65.5 | 44.6 | 34.1 | 62.1 |
| Minimum Temp °F | 17.7 | 20.5 | 25.4 | 30.7 | 37.4 | 43.8 | 47.2 | 45.9 | 38.3 | 30.6 | 25.4 | 18.4 | 31.8 |
| Mean Temp °F | 26.2 | 31.4 | 38.4 | 45.9 | 55.0 | 62.3 | 68.7 | 67.8 | 58.6 | 48.1 | 35.0 | 26.3 | 47.0 |
| Days Max Temp ≥ 90 °F | 0 | 0 | 0 | 0 | 1 | 6 | 18 | 18 | 5 | 0 | 0 | 0 | 48 |
| Days Max Temp ≤ 32 °F | 10 | 2 | 0 | 0 | 0 | 0 | 0 | 0 | 0 | 0 | 2 | 10 | 24 |
| Days Min Temp ≤ 32 °F | 27 | 26 | 25 | 18 | 8 | 1 | 0 | 1 | 6 | 19 | 23 | 28 | 182 |
| Days Min Temp ≤ 0 °F | 3 | 1 | 0 | 0 | 0 | 0 | 0 | 0 | 0 | 0 | 0 | 2 | 6 |
| Heating Degree Days | 1191 | 948 | 822 | 567 | 313 | 126 | 29 | 39 | 204 | 518 | 893 | 1193 | 6843 |
| Cooling Degree Days | 0 | 0 | 0 | 0 | 9 | 45 | 117 | 104 | 12 | 0 | 0 | 0 | 287 |
| Total Precipitation (") | 3.80 | 2.39 | 2.13 | 1.59 | 1.32 | 1.47 | 0.57 | 0.61 | 1.17 | 1.39 | 3.18 | 3.43 | 23.05 |
| Days ≥ 0.1" Precip | 8 | 6 | 6 | 5 | 4 | 4 | 1 | 2 | 3 | 4 | 8 | 8 | 59 |
| Total Snowfall (") | 17.0 | na | na | 0.1 | 0.0 | 0.0 | 0.0 | 0.0 | 0.0 | 0.3 | 6.6 | na | na |
| Days ≥ 1" Snow Depth | 24 | na | na | 0 | 0 | 0 | 0 | 0 | 0 | 0 | na | 16 | na |

## GIBBONSVILLE *Lemhi County* ELEVATION 4514 ft LAT/LONG 45° 34' N / 113° 55' W

| | JAN | FEB | MAR | APR | MAY | JUN | JUL | AUG | SEP | OCT | NOV | DEC | YEAR |
|---|---|---|---|---|---|---|---|---|---|---|---|---|---|
| Maximum Temp °F | 28.7 | 36.4 | 46.7 | 57.3 | 66.8 | 75.4 | 85.3 | 83.8 | 73.3 | 59.5 | 40.0 | 28.2 | 56.8 |
| Minimum Temp °F | 9.1 | 13.1 | 21.1 | 28.1 | 34.3 | 40.8 | 45.0 | 43.6 | 36.1 | 28.0 | 20.4 | 8.7 | 27.4 |
| Mean Temp °F | 18.9 | 24.8 | 33.9 | 42.7 | 50.6 | 58.1 | 65.1 | 63.7 | 54.8 | 43.8 | 30.2 | 18.5 | 42.1 |
| Days Max Temp ≥ 90 °F | 0 | 0 | 0 | 0 | 0 | 2 | 11 | 9 | 1 | 0 | 0 | 0 | 23 |
| Days Max Temp ≤ 32 °F | 19 | 8 | 1 | 0 | 0 | 0 | 0 | 0 | 0 | 0 | 5 | 20 | 53 |
| Days Min Temp ≤ 32 °F | 31 | 28 | 30 | 23 | 12 | 3 | 0 | 1 | 8 | 24 | 28 | 31 | 219 |
| Days Min Temp ≤ 0 °F | 8 | 4 | 1 | 0 | 0 | 0 | 0 | 0 | 0 | 0 | 1 | 7 | 21 |
| Heating Degree Days | 1423 | 1130 | 957 | 662 | 442 | 216 | 59 | 82 | 304 | 651 | 1036 | 1436 | 8398 |
| Cooling Degree Days | 0 | 0 | 0 | 0 | 2 | 22 | 72 | 46 | 5 | 0 | 0 | 0 | 147 |
| Total Precipitation (") | 1.99 | 1.15 | 0.99 | 1.10 | 1.53 | 1.63 | 0.88 | 1.11 | 1.03 | 0.76 | 1.38 | 1.60 | 15.15 |
| Days ≥ 0.1" Precip | 6 | 4 | 4 | 4 | 5 | 4 | 3 | 3 | 3 | 2 | 5 | 6 | 49 |
| Total Snowfall (") | 24.8 | 12.0 | 7.0 | 2.5 | 0.3 | 0.0 | 0.0 | 0.0 | 0.0 | 1.0 | 13.3 | 20.9 | 81.8 |
| Days ≥ 1" Snow Depth | 31 | 28 | 25 | 3 | 0 | 0 | 0 | 0 | 0 | 1 | 11 | 29 | 128 |

## GLENNS FERRY *Elmore County* ELEVATION 2572 ft LAT/LONG 42° 57' N / 115° 18' W

| | JAN | FEB | MAR | APR | MAY | JUN | JUL | AUG | SEP | OCT | NOV | DEC | YEAR |
|---|---|---|---|---|---|---|---|---|---|---|---|---|---|
| Maximum Temp °F | 39.6 | 48.3 | 57.7 | 66.9 | 76.3 | 86.0 | 95.9 | 93.2 | 82.1 | 68.4 | 51.2 | 40.1 | 67.1 |
| Minimum Temp °F | 21.4 | 25.0 | 29.3 | 35.0 | 42.2 | 49.9 | 55.7 | 52.7 | 42.5 | 32.9 | 26.6 | 20.7 | 36.2 |
| Mean Temp °F | 30.4 | 36.7 | 43.5 | 51.0 | 59.3 | 68.0 | 75.8 | 73.0 | 62.3 | 50.7 | 38.9 | 30.4 | 51.7 |
| Days Max Temp ≥ 90 °F | 0 | 0 | 0 | 0 | 3 | 12 | 25 | 23 | 8 | 0 | 0 | 0 | 71 |
| Days Max Temp ≤ 32 °F | 6 | 1 | 0 | 0 | 0 | 0 | 0 | 0 | 0 | 0 | 1 | 5 | 13 |
| Days Min Temp ≤ 32 °F | 26 | 23 | 20 | 10 | 3 | 0 | 0 | 0 | 2 | 14 | 19 | 26 | 143 |
| Days Min Temp ≤ 0 °F | 1 | 0 | 0 | 0 | 0 | 0 | 0 | 0 | 0 | 0 | 0 | 2 | 3 |
| Heating Degree Days | 1064 | 793 | 659 | 416 | 204 | 51 | 2 | 12 | 126 | 437 | 775 | 1065 | 5604 |
| Cooling Degree Days | 0 | 0 | 0 | 5 | 38 | 150 | 340 | 272 | 41 | na | 0 | 0 | na |
| Total Precipitation (") | 1.44 | 0.98 | 0.96 | 0.67 | 0.66 | 0.77 | 0.25 | 0.36 | 0.48 | 0.62 | 1.40 | 1.27 | 9.86 |
| Days ≥ 0.1" Precip | 4 | 3 | 3 | 2 | 2 | 2 | 1 | 1 | 2 | 2 | 4 | 4 | 30 |
| Total Snowfall (") | na | 1.4 | 0.5 | 0.1 | 0.0 | 0.0 | 0.0 | 0.0 | 0.0 | 0.0 | 1.5 | na | na |
| Days ≥ 1" Snow Depth | na | 1 | 0 | 0 | 0 | 0 | 0 | 0 | 0 | 0 | 1 | na | na |

## GRACE *Caribou County* ELEVATION 5549 ft LAT/LONG 42° 35' N / 111° 44' W

| | JAN | FEB | MAR | APR | MAY | JUN | JUL | AUG | SEP | OCT | NOV | DEC | YEAR |
|---|---|---|---|---|---|---|---|---|---|---|---|---|---|
| Maximum Temp °F | 30.4 | 35.6 | 44.6 | 55.7 | 66.4 | 75.3 | 83.4 | 82.8 | 73.5 | 61.1 | 42.5 | 31.9 | 56.9 |
| Minimum Temp °F | 10.8 | 13.1 | 21.3 | 28.6 | 36.1 | 42.4 | 47.4 | 45.9 | 37.8 | 29.6 | 21.0 | 12.1 | 28.8 |
| Mean Temp °F | 20.7 | 24.4 | 32.9 | 42.2 | 51.2 | 58.9 | 65.4 | 64.4 | 55.7 | 45.3 | 31.9 | 22.1 | 42.9 |
| Days Max Temp ≥ 90 °F | 0 | 0 | 0 | 0 | 0 | 1 | 4 | 4 | 0 | 0 | 0 | 0 | 9 |
| Days Max Temp ≤ 32 °F | 18 | 9 | 2 | 0 | 0 | 0 | 0 | 0 | 0 | 0 | 5 | 16 | 50 |
| Days Min Temp ≤ 32 °F | 30 | 27 | 28 | 21 | 9 | 1 | 0 | 1 | 7 | 20 | 27 | 30 | 201 |
| Days Min Temp ≤ 0 °F | 8 | 6 | 1 | 0 | 0 | 0 | 0 | 0 | 0 | 0 | 1 | 6 | 22 |
| Heating Degree Days | 1369 | 1141 | 987 | 678 | 420 | 193 | 45 | 67 | 277 | 603 | 988 | 1324 | 8092 |
| Cooling Degree Days | 0 | 0 | 0 | 0 | 0 | 18 | 64 | 61 | 5 | 0 | 0 | 0 | 148 |
| Total Precipitation (") | 1.22 | 1.09 | 1.33 | 1.43 | 1.74 | 1.58 | 1.01 | 1.21 | 1.32 | 1.33 | 1.25 | 1.18 | 15.69 |
| Days ≥ 0.1" Precip | 4 | 4 | 5 | 5 | 5 | 4 | 3 | 4 | 4 | 4 | 5 | 4 | 51 |
| Total Snowfall (") | na | na | na | 2.4 | 0.0 | 0.0 | 0.0 | 0.0 | 0.0 | 0.5 | na | na | na |
| Days ≥ 1" Snow Depth | na | na | na | 1 | 0 | 0 | 0 | 0 | 0 | 0 | na | na | na |

## GRAND VIEW 2 W *Owyhee County* ELEVATION 2382 ft LAT/LONG 43° 0 ' N / 116° 10 ' W

| | JAN | FEB | MAR | APR | MAY | JUN | JUL | AUG | SEP | OCT | NOV | DEC | YEAR |
|---|---|---|---|---|---|---|---|---|---|---|---|---|---|
| Maximum Temp °F | 39.5 | 48.5 | 58.9 | 67.2 | 76.4 | 84.6 | 92.5 | 91.2 | 80.8 | 67.9 | 50.9 | 39.6 | 66.5 |
| Minimum Temp °F | 20.9 | 25.1 | 29.9 | 36.5 | 44.2 | 51.6 | 56.1 | 53.7 | 44.1 | 34.7 | 27.5 | 20.0 | 37.0 |
| Mean Temp °F | 30.2 | 36.8 | 44.4 | 51.9 | 60.3 | 68.1 | 74.3 | 72.5 | 62.7 | 51.3 | 39.2 | 29.8 | 51.8 |
| Days Max Temp ≥ 90 °F | 0 | 0 | 0 | 0 | 3 | 10 | 22 | 20 | 5 | 0 | 0 | 0 | 60 |
| Days Max Temp ≤ 32 °F | 7 | 2 | 0 | 0 | 0 | 0 | 0 | 0 | 0 | 0 | 1 | 6 | 16 |
| Days Min Temp ≤ 32 °F | 27 | 23 | 19 | 10 | 2 | 0 | 0 | 0 | 2 | 13 | 21 | 28 | 145 |
| Days Min Temp ≤ 0 °F | 2 | 0 | 0 | 0 | 0 | 0 | 0 | 0 | 0 | 0 | 0 | 2 | 4 |
| Heating Degree Days | 1071 | 788 | 630 | 391 | 177 | 47 | 4 | 12 | 116 | 420 | 767 | 1083 | 5506 |
| Cooling Degree Days | 0 | 0 | 0 | 6 | 49 | 150 | 290 | 274 | 55 | 2 | 0 | 0 | 826 |
| Total Precipitation (") | 0.65 | 0.53 | 0.71 | 0.71 | 0.68 | 0.78 | 0.22 | 0.29 | 0.57 | 0.49 | 0.84 | 0.55 | 7.02 |
| Days ≥ 0.1" Precip | 3 | 2 | 3 | 2 | 2 | 3 | 1 | 1 | 2 | 2 | 3 | 2 | 26 |
| Total Snowfall (") | na | *1.9* | *0.1* | 0.0 | 0.0 | 0.0 | 0.0 | 0.0 | 0.0 | 0.0 | 0.9 | na | na |
| Days ≥ 1" Snow Depth | na | *0* | *0* | 0 | 0 | 0 | 0 | 0 | 0 | 0 | *0* | na | na |

## GRANGEVILLE *Idaho County* ELEVATION 3353 ft LAT/LONG 45° 55 ' N / 116° 8 ' W

| | JAN | FEB | MAR | APR | MAY | JUN | JUL | AUG | SEP | OCT | NOV | DEC | YEAR |
|---|---|---|---|---|---|---|---|---|---|---|---|---|---|
| Maximum Temp °F | 37.2 | 43.1 | 49.3 | 56.2 | 64.4 | 71.9 | 81.2 | 81.6 | 71.2 | 59.3 | 44.2 | 37.5 | 58.1 |
| Minimum Temp °F | 21.9 | 24.8 | 28.1 | 32.8 | 39.1 | 45.8 | 49.8 | 49.3 | 41.5 | 34.1 | 27.9 | 22.3 | 34.8 |
| Mean Temp °F | 29.6 | 34.0 | 38.7 | 44.5 | 51.8 | 58.9 | 65.5 | 65.5 | 56.4 | 46.7 | 36.1 | 29.9 | 46.5 |
| Days Max Temp ≥ 90 °F | 0 | 0 | 0 | 0 | 0 | 1 | 6 | 6 | 1 | 0 | 0 | 0 | 14 |
| Days Max Temp ≤ 32 °F | 9 | 3 | 1 | 0 | 0 | 0 | 0 | 0 | 0 | 0 | 3 | 8 | 24 |
| Days Min Temp ≤ 32 °F | 26 | 23 | 23 | 16 | 5 | 0 | 0 | 0 | 3 | 13 | 21 | 27 | 157 |
| Days Min Temp ≤ 0 °F | 2 | 1 | 0 | 0 | 0 | 0 | 0 | 0 | 0 | 0 | 0 | 1 | 4 |
| Heating Degree Days | 1091 | 869 | 807 | 609 | 407 | 205 | 68 | 71 | 264 | 561 | 860 | 1081 | 6893 |
| Cooling Degree Days | 0 | 0 | 0 | 0 | 7 | 29 | 83 | 95 | 13 | 1 | 0 | 0 | 228 |
| Total Precipitation (") | 1.59 | 1.19 | 2.33 | 2.70 | 3.31 | 2.80 | 1.47 | 1.18 | 1.66 | 1.80 | 1.71 | 1.46 | 23.20 |
| Days ≥ 0.1" Precip | 5 | 4 | 7 | 8 | 8 | 7 | 3 | 3 | 4 | 5 | 6 | 5 | 65 |
| Total Snowfall (") | 11.0 | 7.2 | 9.0 | 3.4 | 0.3 | 0.0 | 0.0 | 0.0 | 0.0 | 1.8 | 6.1 | 11.0 | 49.8 |
| Days ≥ 1" Snow Depth | 16 | 9 | 5 | 1 | 0 | 0 | 0 | 0 | 0 | 1 | 4 | 13 | 49 |

## GROUSE *Custer County* ELEVATION 6106 ft LAT/LONG 43° 42 ' N / 113° 37 ' W

| | JAN | FEB | MAR | APR | MAY | JUN | JUL | AUG | SEP | OCT | NOV | DEC | YEAR |
|---|---|---|---|---|---|---|---|---|---|---|---|---|---|
| Maximum Temp °F | 28.0 | 33.7 | 41.5 | 51.6 | 62.4 | 70.5 | 79.6 | 78.8 | 69.7 | 57.1 | 39.3 | 28.6 | 53.4 |
| Minimum Temp °F | -0.4 | 2.3 | 12.8 | 23.2 | 31.1 | 37.0 | 40.4 | 39.3 | 31.2 | 22.5 | 12.4 | 0.7 | 21.0 |
| Mean Temp °F | 13.9 | 18.0 | 27.2 | 37.4 | 46.8 | 53.8 | 60.0 | 59.1 | 50.5 | 39.8 | 25.8 | 14.7 | 37.3 |
| Days Max Temp ≥ 90 °F | 0 | 0 | 0 | 0 | 0 | 0 | 1 | 1 | 0 | 0 | 0 | 0 | 2 |
| Days Max Temp ≤ 32 °F | 20 | 12 | 3 | 0 | 0 | 0 | 0 | 0 | 0 | 0 | 7 | 19 | 61 |
| Days Min Temp ≤ 32 °F | 31 | 28 | 31 | 27 | 18 | 6 | 3 | 4 | 17 | 28 | 29 | 31 | 253 |
| Days Min Temp ≤ 0 °F | 17 | 12 | 5 | 0 | 0 | 0 | 0 | 0 | 0 | 0 | 5 | 15 | 54 |
| Heating Degree Days | 1581 | 1321 | 1165 | 821 | 558 | 330 | 157 | 184 | 429 | 773 | 1169 | 1553 | 10041 |
| Cooling Degree Days | 0 | 0 | 0 | 0 | 0 | 1 | 13 | 10 | 0 | 0 | 0 | 0 | 24 |
| Total Precipitation (") | 1.34 | 1.03 | 1.19 | 1.01 | 1.37 | 1.69 | 0.97 | 1.08 | 0.81 | 0.71 | 1.15 | 1.23 | 13.58 |
| Days ≥ 0.1" Precip | 4 | 4 | 4 | 3 | 4 | 5 | 3 | 3 | 3 | 3 | 4 | 4 | 44 |
| Total Snowfall (") | *15.5* | *15.0* | *11.3* | 4.4 | *1.5* | 0.0 | 0.0 | 0.0 | 0.7 | *1.4* | *10.5* | *17.3* | 77.6 |
| Days ≥ 1" Snow Depth | na | na | na | *10* | 1 | 0 | 0 | 0 | 0 | 0 | na | na | na |

## HAMER 4 NW *Jefferson County* ELEVATION 4803 ft LAT/LONG 43° 59 ' N / 112° 15 ' W

| | JAN | FEB | MAR | APR | MAY | JUN | JUL | AUG | SEP | OCT | NOV | DEC | YEAR |
|---|---|---|---|---|---|---|---|---|---|---|---|---|---|
| Maximum Temp °F | 28.2 | 35.0 | 46.6 | 59.6 | 70.2 | 79.0 | 87.5 | 86.2 | 75.5 | 61.9 | 42.3 | 30.1 | 58.5 |
| Minimum Temp °F | 2.6 | 8.4 | 18.5 | 26.5 | 35.5 | 42.5 | 46.8 | 44.7 | *35.3* | 25.3 | 15.6 | 4.7 | 25.5 |
| Mean Temp °F | 15.4 | 21.7 | 32.6 | 43.2 | 52.9 | 60.8 | 67.2 | 65.5 | *55.6* | 43.7 | 28.9 | 17.4 | 42.1 |
| Days Max Temp ≥ 90 °F | 0 | 0 | 0 | 0 | 0 | 3 | *13* | 11 | 1 | 0 | 0 | 0 | 28 |
| Days Max Temp ≤ 32 °F | 21 | 10 | 2 | 0 | 0 | 0 | 0 | 0 | 0 | 0 | 5 | 18 | 56 |
| Days Min Temp ≤ 32 °F | 30 | 28 | 30 | 23 | 10 | 2 | 0 | 1 | 10 | *25* | 28 | 30 | 217 |
| Days Min Temp ≤ 0 °F | 14 | 8 | 2 | 0 | 0 | 0 | 0 | 0 | 0 | 0 | 3 | 12 | 39 |
| Heating Degree Days | 1532 | 1216 | 998 | 650 | 371 | 152 | 32 | 56 | *283* | 654 | 1070 | 1469 | 8483 |
| Cooling Degree Days | 0 | 0 | 0 | 0 | 1 | 31 | 92 | 76 | na | 0 | 0 | 0 | na |
| Total Precipitation (") | 0.63 | 0.44 | 0.63 | 0.82 | 1.31 | 1.18 | 0.88 | 0.80 | 0.57 | 0.65 | 0.81 | 0.67 | 9.39 |
| Days ≥ 0.1" Precip | 2 | 2 | 2 | 3 | 4 | 4 | 2 | 2 | 2 | 2 | 3 | 2 | 30 |
| Total Snowfall (") | 6.6 | 5.4 | 3.0 | 1.1 | 0.2 | 0.1 | 0.0 | 0.0 | 0.1 | 0.6 | 4.7 | *9.1* | 30.9 |
| Days ≥ 1" Snow Depth | 27 | 20 | 8 | 0 | 0 | 0 | 0 | 0 | 0 | 0 | 5 | 21 | 81 |

## HAZELTON *Jerome County* ELEVATION 4062 ft LAT/LONG 42° 36 ' N / 114° 8 ' W

| | JAN | FEB | MAR | APR | MAY | JUN | JUL | AUG | SEP | OCT | NOV | DEC | YEAR |
|---|---|---|---|---|---|---|---|---|---|---|---|---|---|
| Maximum Temp °F | 35.3 | 42.3 | 51.6 | 61.2 | 70.4 | 79.9 | 88.6 | 87.5 | 77.5 | 64.7 | 46.8 | 36.6 | 61.9 |
| Minimum Temp °F | 18.0 | 22.1 | 27.6 | 33.0 | 40.7 | 48.0 | 53.8 | 52.0 | 42.4 | 33.4 | 25.7 | 18.2 | 34.6 |
| Mean Temp °F | 26.7 | 32.2 | 39.6 | 47.2 | 55.6 | 64.0 | 71.2 | 69.7 | 60.0 | 49.1 | 36.3 | 27.4 | 48.2 |
| Days Max Temp ≥ 90 °F | 0 | 0 | 0 | 0 | 1 | 5 | 16 | 14 | 3 | 0 | 0 | 0 | 39 |
| Days Max Temp ≤ 32 °F | 11 | 4 | 1 | 0 | 0 | 0 | 0 | 0 | 0 | 0 | 3 | 10 | 29 |
| Days Min Temp ≤ 32 °F | 29 | 26 | 24 | 14 | 4 | 0 | 0 | 0 | 2 | 14 | 24 | 30 | 167 |
| Days Min Temp ≤ 0 °F | 2 | 1 | 0 | 0 | 0 | 0 | 0 | 0 | 0 | 0 | 0 | 2 | 5 |
| Heating Degree Days | 1181 | 919 | 780 | 529 | 298 | 105 | 14 | 25 | 175 | 489 | 855 | 1159 | 6529 |
| Cooling Degree Days | 0 | 0 | 0 | 2 | 19 | 91 | 210 | 185 | 31 | 1 | 0 | 0 | 539 |
| Total Precipitation (") | 1.16 | 0.91 | 0.98 | 0.78 | 0.92 | 0.74 | 0.26 | 0.40 | 0.60 | 0.66 | 1.32 | 1.10 | 9.83 |
| Days ≥ 0.1" Precip | 4 | 3 | 3 | 2 | 3 | 3 | 1 | 1 | 2 | 2 | 4 | 4 | 32 |
| Total Snowfall (") | *4.4* | *2.4* | *1.4* | 0.6 | 0.4 | 0.0 | 0.0 | 0.0 | 0.0 | 0.1 | *3.1* | *4.9* | 17.3 |
| Days ≥ 1" Snow Depth | na | *4* | *1* | 0 | 0 | 0 | 0 | 0 | 0 | 0 | *2* | *7* | na |

## HEADQUARTERS *Clearwater County* ELEVATION 3143 ft LAT/LONG 46° 38 ' N / 115° 48 ' W

| | JAN | FEB | MAR | APR | MAY | JUN | JUL | AUG | SEP | OCT | NOV | DEC | YEAR |
|---|---|---|---|---|---|---|---|---|---|---|---|---|---|
| Maximum Temp °F | 34.8 | 40.1 | *45.9* | *54.4* | *63.7* | 72.0 | *80.2* | *81.6* | *70.9* | *58.1* | na | na | na |
| Minimum Temp °F | 17.2 | 19.2 | *23.0* | *28.5* | *34.4* | *41.9* | *44.6* | *44.0* | *36.1* | *29.4* | na | na | na |
| Mean Temp °F | 26.0 | 29.7 | *34.5* | *41.5* | *49.1* | *57.0* | *62.5* | *62.8* | *53.5* | *43.5* | na | na | na |
| Days Max Temp ≥ 90 °F | 0 | 0 | 0 | 0 | 0 | 1 | *5* | 6 | 1 | 0 | 0 | 0 | 13 |
| Days Max Temp ≤ 32 °F | 8 | 4 | 1 | 0 | 0 | 0 | 0 | 0 | 0 | 0 | *2* | 8 | 23 |
| Days Min Temp ≤ 32 °F | 28 | 26 | 27 | *23* | *12* | 2 | 0 | 1 | *8* | *21* | *24* | 28 | 200 |
| Days Min Temp ≤ 0 °F | 3 | 1 | 1 | 0 | 0 | 0 | 0 | 0 | 0 | 0 | 0 | 2 | 7 |
| Heating Degree Days | *1196* | *994* | *942* | *702* | *497* | *251* | *118* | *111* | *343* | *661* | na | na | na |
| Cooling Degree Days | na | na | na | na | na | na | na | na | na | na | na | na | na |
| Total Precipitation (") | 5.28 | 3.64 | 3.30 | 2.91 | 3.11 | 2.61 | 1.31 | 1.43 | 1.82 | 2.66 | 4.56 | 4.63 | 37.26 |
| Days ≥ 0.1" Precip | *11* | 9 | 9 | *7* | *7* | *6* | 3 | 3 | 4 | 6 | 10 | *11* | 86 |
| Total Snowfall (") | *30.3* | *17.2* | 10.5 | *3.9* | 0.3 | 0.0 | 0.0 | 0.0 | 0.0 | *0.6* | *15.7* | na | na |
| Days ≥ 1" Snow Depth | 27 | 25 | 23 | 10 | 0 | 0 | 0 | 0 | 0 | 0 | *10* | 26 | 121 |

## HILL CITY 1 W *Camas County* ELEVATION 5003 ft LAT/LONG 43° 18 ' N / 115° 3 ' W

| | JAN | FEB | MAR | APR | MAY | JUN | JUL | AUG | SEP | OCT | NOV | DEC | YEAR |
|---|---|---|---|---|---|---|---|---|---|---|---|---|---|
| Maximum Temp °F | 29.6 | 34.2 | 42.2 | 54.4 | 66.2 | 75.1 | 85.1 | 84.6 | 74.9 | 62.0 | 42.2 | 30.7 | 56.8 |
| Minimum Temp °F | 6.9 | 9.4 | 18.5 | 27.7 | 34.7 | 39.7 | 44.7 | 43.6 | 35.0 | 26.7 | 18.9 | 7.7 | 26.1 |
| Mean Temp °F | 18.3 | 21.8 | 30.4 | 41.1 | 50.5 | 57.4 | 64.9 | 64.1 | 55.0 | 44.4 | 30.5 | 19.2 | 41.5 |
| Days Max Temp ≥ 90 °F | 0 | 0 | 0 | 0 | 0 | 2 | 9 | 9 | 1 | 0 | 0 | 0 | 21 |
| Days Max Temp ≤ 32 °F | 18 | 10 | 2 | 0 | 0 | 0 | 0 | 0 | 0 | 0 | 5 | 17 | 52 |
| Days Min Temp ≤ 32 °F | 31 | 28 | 30 | 22 | 12 | 4 | 1 | 2 | 11 | 23 | 27 | 31 | 222 |
| Days Min Temp ≤ 0 °F | 10 | 8 | 2 | 0 | 0 | 0 | 0 | 0 | 0 | 0 | 2 | 9 | 31 |
| Heating Degree Days | 1443 | 1213 | 1066 | 712 | 445 | 234 | 66 | 84 | 298 | 633 | 1027 | 1412 | 8633 |
| Cooling Degree Days | 0 | 0 | 0 | 0 | 1 | 14 | 67 | 60 | 5 | 0 | 0 | 0 | 147 |
| Total Precipitation (") | 2.23 | 1.36 | 1.09 | 1.01 | 0.88 | 0.96 | 0.49 | 0.48 | 0.68 | 0.85 | 1.69 | 1.93 | 13.65 |
| Days ≥ 0.1" Precip | 6 | 4 | 3 | 3 | 3 | 3 | 1 | 2 | 2 | 3 | 5 | 6 | 41 |
| Total Snowfall (") | na | na | na | 0.8 | 0.2 | 0.0 | 0.0 | 0.0 | 0.1 | *0.4* | na | na | na |
| Days ≥ 1" Snow Depth | *25* | *24* | na | 5 | 0 | 0 | 0 | 0 | 0 | 0 | na | *19* | na |

## HOWE *Butte County* ELEVATION 4823 ft LAT/LONG 43° 47 ' N / 113° 0 ' W

| | JAN | FEB | MAR | APR | MAY | JUN | JUL | AUG | SEP | OCT | NOV | DEC | YEAR |
|---|---|---|---|---|---|---|---|---|---|---|---|---|---|
| Maximum Temp °F | 29.7 | 36.0 | 46.9 | 59.7 | 68.7 | 77.2 | 86.1 | 84.5 | 73.8 | 60.5 | 42.4 | 30.6 | 58.0 |
| Minimum Temp °F | 6.3 | 11.4 | 22.0 | 30.0 | 38.4 | 45.4 | 49.7 | 48.0 | 38.5 | 28.7 | 17.7 | 6.6 | 28.6 |
| Mean Temp °F | 18.0 | 23.7 | 34.5 | 44.9 | 53.5 | 61.3 | 68.0 | 66.3 | 56.2 | 44.6 | 29.9 | 18.6 | 43.3 |
| Days Max Temp ≥ 90 °F | 0 | 0 | 0 | 0 | 0 | 3 | 10 | 8 | 1 | 0 | 0 | 0 | 22 |
| Days Max Temp ≤ 32 °F | 17 | 9 | 2 | 1 | 0 | 0 | 0 | 0 | 0 | 0 | 4 | 16 | 49 |
| Days Min Temp ≤ 32 °F | 30 | 28 | 28 | 19 | 6 | 1 | 0 | 0 | 6 | 22 | 28 | 30 | 198 |
| Days Min Temp ≤ 0 °F | 11 | 6 | 1 | 0 | 0 | 0 | 0 | 0 | 0 | 0 | 1 | *9* | 28 |
| Heating Degree Days | 1450 | 1160 | 940 | 597 | 352 | 146 | 34 | 48 | 266 | 625 | 1045 | 1432 | 8095 |
| Cooling Degree Days | 0 | 0 | 0 | 0 | 3 | 47 | 134 | 100 | 9 | 0 | 0 | 0 | 293 |
| Total Precipitation (") | 0.64 | 0.60 | 0.55 | 0.58 | 1.04 | 1.26 | 0.76 | 0.87 | 0.56 | 0.56 | 0.74 | 0.74 | 8.90 |
| Days ≥ 0.1" Precip | 2 | 2 | 2 | 2 | 3 | 4 | 2 | 2 | 1 | 2 | 2 | 2 | 26 |
| Total Snowfall (") | 4.0 | 3.2 | 1.6 | 0.5 | 0.1 | 0.0 | 0.0 | 0.0 | 0.0 | 0.4 | *1.9* | *6.0* | 17.7 |
| Days ≥ 1" Snow Depth | *15* | na | *4* | 0 | 0 | 0 | 0 | 0 | 0 | 0 | *2* | *12* | na |

## IDAHO CITY *Boise County* ELEVATION 3944 ft LAT/LONG 43° 50' N / 115° 50' W

| | JAN | FEB | MAR | APR | MAY | JUN | JUL | AUG | SEP | OCT | NOV | DEC | YEAR |
|---|---|---|---|---|---|---|---|---|---|---|---|---|---|
| Maximum Temp °F | 35.2 | 41.7 | 48.8 | 58.0 | 68.2 | 77.0 | 86.5 | 86.3 | 75.7 | 63.2 | 44.4 | 35.2 | 60.0 |
| Minimum Temp °F | 14.1 | 16.5 | 22.4 | 28.0 | 34.7 | 40.8 | 45.1 | 44.3 | 36.2 | 28.6 | 22.6 | 14.1 | 29.0 |
| Mean Temp °F | 24.7 | 29.1 | 35.7 | 43.1 | 51.5 | 59.0 | 65.9 | 65.3 | 56.0 | 45.9 | 33.5 | 24.7 | 44.5 |
| Days Max Temp ≥ 90 °F | 0 | 0 | 0 | 0 | 0 | 4 | 13 | 12 | 2 | 0 | 0 | 0 | 31 |
| Days Max Temp ≤ 32 °F | 9 | 3 | 0 | 0 | 0 | 0 | 0 | 0 | 0 | 0 | 2 | 10 | 24 |
| Days Min Temp ≤ 32 °F | 31 | 28 | 29 | 23 | 12 | 3 | 0 | 1 | 9 | 23 | 27 | 31 | 217 |
| Days Min Temp ≤ 0 °F | 6 | 3 | 1 | 0 | 0 | 0 | 0 | 0 | 0 | 0 | 1 | 5 | 16 |
| Heating Degree Days | 1243 | 1007 | 903 | 652 | 415 | 199 | 58 | 68 | 273 | 584 | 937 | 1243 | 7582 |
| Cooling Degree Days | 0 | 0 | 0 | 0 | 4 | 29 | 86 | 91 | 9 | 0 | 0 | 0 | 219 |
| Total Precipitation (") | 3.87 | 2.67 | 2.32 | 1.87 | 1.45 | 1.36 | 0.56 | 0.66 | 1.22 | 1.36 | 3.12 | 3.33 | 23.79 |
| Days ≥ 0.1" Precip | 8 | 7 | 7 | 6 | 4 | 4 | 1 | 2 | 3 | 4 | 8 | 9 | 63 |
| Total Snowfall (") | 23.4 | 12.3 | 4.7 | 0.7 | 0.0 | 0.0 | 0.0 | 0.0 | 0.0 | 1.2 | 11.6 | *22.3* | 76.2 |
| Days ≥ 1" Snow Depth | 31 | 27 | 21 | 3 | 0 | 0 | 0 | 0 | 0 | 0 | 10 | 29 | 121 |

## IDAHO FALLS 16 SE *Bonneville County* ELEVATION 5715 ft LAT/LONG 43° 21' N / 111° 47' W

| | JAN | FEB | MAR | APR | MAY | JUN | JUL | AUG | SEP | OCT | NOV | DEC | YEAR |
|---|---|---|---|---|---|---|---|---|---|---|---|---|---|
| Maximum Temp °F | 30.9 | 35.6 | 42.1 | 52.1 | 62.4 | 71.5 | 80.0 | 78.7 | 69.3 | 57.2 | 40.7 | 31.7 | 54.4 |
| Minimum Temp °F | 10.8 | 13.6 | 20.6 | 27.7 | 34.4 | 40.8 | 46.0 | 44.3 | 36.4 | 27.9 | 19.7 | 11.2 | 27.8 |
| Mean Temp °F | 20.9 | 24.6 | 31.4 | 39.9 | 48.4 | 56.1 | 63.0 | 61.6 | 52.9 | 42.6 | 30.2 | 21.5 | 41.1 |
| Days Max Temp ≥ 90 °F | 0 | 0 | 0 | 0 | 0 | 0 | 1 | 1 | 0 | 0 | 0 | 0 | 2 |
| Days Max Temp ≤ 32 °F | 17 | 9 | 3 | 0 | 0 | 0 | 0 | 0 | 0 | 0 | 6 | 17 | 52 |
| Days Min Temp ≤ 32 °F | 30 | 27 | 28 | 23 | 12 | 3 | 0 | 1 | 9 | 22 | 27 | 30 | 212 |
| Days Min Temp ≤ 0 °F | 7 | 4 | 2 | 0 | 0 | 0 | 0 | 0 | 0 | 0 | 2 | 7 | 22 |
| Heating Degree Days | 1361 | 1135 | 1036 | 745 | 506 | 267 | 89 | 123 | 359 | 689 | 1034 | 1343 | 8687 |
| Cooling Degree Days | 0 | 0 | 0 | 0 | 0 | 7 | 30 | 23 | 1 | 0 | 0 | 0 | 61 |
| Total Precipitation (") | 1.47 | 1.14 | 1.35 | 1.48 | 1.86 | 1.45 | 0.96 | 0.95 | 1.14 | 1.13 | 1.57 | 1.34 | 15.84 |
| Days ≥ 0.1" Precip | 5 | 4 | 5 | 5 | 6 | 5 | 3 | 3 | 4 | 4 | 5 | 5 | 54 |
| Total Snowfall (") | 17.4 | 12.4 | 10.7 | 7.3 | 1.8 | 0.2 | 0.0 | 0.0 | 0.7 | 2.2 | 10.2 | 17.6 | 80.5 |
| Days ≥ 1" Snow Depth | 31 | 27 | 18 | 4 | 0 | 0 | 0 | 0 | 0 | 1 | 10 | 24 | 115 |

## IDAHO FALLS 2 ESE *Bonneville County* ELEVATION 4774 ft LAT/LONG 43° 29' N / 112° 1' W

| | JAN | FEB | MAR | APR | MAY | JUN | JUL | AUG | SEP | OCT | NOV | DEC | YEAR |
|---|---|---|---|---|---|---|---|---|---|---|---|---|---|
| Maximum Temp °F | 29.5 | 36.9 | 47.2 | 58.6 | 68.8 | 77.8 | 86.2 | 85.5 | 74.8 | 61.3 | 43.0 | 31.3 | 58.4 |
| Minimum Temp °F | 12.1 | 17.0 | 24.6 | 31.8 | 39.4 | 46.6 | 51.9 | 50.0 | 41.1 | 31.7 | 23.3 | 12.9 | 31.9 |
| Mean Temp °F | 20.9 | 27.0 | 35.9 | 45.2 | 54.2 | 62.3 | 69.1 | 67.8 | 58.0 | 46.5 | 33.2 | 22.2 | 45.2 |
| Days Max Temp ≥ 90 °F | 0 | 0 | 0 | 0 | 0 | 3 | 10 | 10 | 1 | 0 | 0 | 0 | 24 |
| Days Max Temp ≤ 32 °F | 18 | 8 | 2 | 0 | 0 | 0 | 0 | 0 | 0 | 0 | 5 | 16 | 49 |
| Days Min Temp ≤ 32 °F | 30 | 27 | 27 | 17 | 5 | 0 | 0 | 0 | 4 | 17 | 25 | 30 | 182 |
| Days Min Temp ≤ 0 °F | 7 | 3 | 0 | 0 | 0 | 0 | 0 | 0 | 0 | 0 | 1 | 5 | 16 |
| Heating Degree Days | 1361 | 1066 | 894 | 587 | 334 | 130 | 21 | 34 | 219 | 566 | 948 | 1320 | 7480 |
| Cooling Degree Days | 0 | 0 | 0 | 1 | 6 | 63 | 146 | 140 | 17 | 0 | 0 | 0 | 373 |
| Total Precipitation (") | 1.18 | 0.92 | 1.06 | 1.17 | 1.66 | 1.30 | 0.68 | 0.89 | 0.89 | 1.03 | 1.12 | 1.15 | 13.05 |
| Days ≥ 0.1" Precip | na | *3* | *4* | *4* | 4 | 3 | 2 | 2 | 3 | 3 | 4 | *3* | na |
| Total Snowfall (") | na | na | na | na | 0.1 | 0.0 | 0.0 | 0.0 | 0.1 | 0.2 | na | na | na |
| Days ≥ 1" Snow Depth | na | na | na | *0* | 0 | 0 | 0 | 0 | 0 | 0 | na | na | na |

## IDAHO FALLS 46 W *Butte County* ELEVATION 4934 ft LAT/LONG 43° 32' N / 112° 57' W

| | JAN | FEB | MAR | APR | MAY | JUN | JUL | AUG | SEP | OCT | NOV | DEC | YEAR |
|---|---|---|---|---|---|---|---|---|---|---|---|---|---|
| Maximum Temp °F | 27.7 | 34.0 | 44.5 | 56.6 | 66.9 | 76.7 | 86.7 | 85.2 | 74.1 | 60.7 | 41.4 | 29.3 | 57.0 |
| Minimum Temp °F | 4.8 | 9.4 | 20.1 | 27.8 | 36.2 | 43.6 | 48.9 | 47.2 | 37.3 | 26.6 | 17.2 | 5.4 | 27.0 |
| Mean Temp °F | 16.3 | 21.7 | 32.3 | 42.2 | 51.5 | 60.1 | 67.8 | 66.2 | 55.7 | 43.7 | 29.3 | 17.4 | 42.0 |
| Days Max Temp ≥ 90 °F | 0 | 0 | 0 | 0 | 0 | 3 | 13 | 11 | 1 | 0 | 0 | 0 | 28 |
| Days Max Temp ≤ 32 °F | 21 | 12 | 3 | 0 | 0 | 0 | 0 | 0 | 0 | 0 | 6 | 19 | 61 |
| Days Min Temp ≤ 32 °F | 31 | 28 | 29 | 22 | 10 | 2 | 0 | 1 | 8 | 23 | 28 | 31 | 213 |
| Days Min Temp ≤ 0 °F | 12 | 7 | 2 | 0 | 0 | 0 | 0 | 0 | 0 | 0 | 2 | 10 | 33 |
| Heating Degree Days | 1506 | 1217 | 1007 | 677 | 412 | 174 | 33 | 52 | 279 | 654 | 1064 | 1471 | 8546 |
| Cooling Degree Days | 0 | 0 | 0 | 0 | 3 | 39 | 131 | 108 | 8 | 0 | 0 | 0 | 289 |
| Total Precipitation (") | 0.68 | 0.55 | 0.60 | 0.75 | 1.11 | 1.18 | 0.61 | 0.53 | 0.68 | 0.54 | 0.75 | 0.69 | 8.67 |
| Days ≥ 0.1" Precip | 2 | 2 | 2 | 3 | 4 | 3 | 2 | 2 | 2 | 2 | 3 | 2 | 29 |
| Total Snowfall (") | 5.7 | 4.7 | 2.4 | 1.1 | 0.9 | 0.0 | 0.0 | 0.0 | 0.1 | 0.6 | 4.0 | 6.8 | 26.3 |
| Days ≥ 1" Snow Depth | 26 | 20 | 9 | 1 | 0 | 0 | 0 | 0 | 0 | 0 | 5 | 19 | 80 |

## IDAHO FALLS FANNING *Bonneville County* ELEVATION 4744 ft LAT/LONG 43° 31 ' N / 112° 4 ' W

| | JAN | FEB | MAR | APR | MAY | JUN | JUL | AUG | SEP | OCT | NOV | DEC | YEAR |
|---|---|---|---|---|---|---|---|---|---|---|---|---|---|
| Maximum Temp °F | 27.1 | 33.6 | 44.7 | 56.5 | 66.6 | 76.4 | 85.4 | 84.2 | 73.1 | 59.6 | 41.7 | 29.1 | 56.5 |
| Minimum Temp °F | 10.9 | 15.4 | 24.5 | 31.3 | 38.7 | 45.9 | 50.9 | 49.0 | 40.5 | 30.8 | 22.5 | 11.8 | 31.0 |
| Mean Temp °F | 19.0 | 24.5 | 34.6 | 43.9 | 52.7 | 61.2 | 68.2 | 66.6 | 56.8 | 45.2 | 32.1 | 20.5 | 43.8 |
| Days Max Temp ≥ 90 °F | 0 | 0 | 0 | 0 | 0 | 3 | 10 | 9 | 1 | 0 | 0 | 0 | 23 |
| Days Max Temp ≤ 32 °F | 21 | 12 | 3 | 0 | 0 | 0 | 0 | 0 | 0 | 0 | 6 | 18 | 60 |
| Days Min Temp ≤ 32 °F | 30 | 27 | 27 | 18 | 6 | 0 | 0 | 0 | 4 | 18 | 26 | 30 | 186 |
| Days Min Temp ≤ 0 °F | 7 | 4 | 0 | 0 | 0 | 0 | 0 | 0 | 0 | 0 | 1 | 5 | 17 |
| Heating Degree Days | 1419 | 1137 | 935 | 626 | 378 | 147 | 24 | 44 | 250 | 607 | 979 | 1374 | 7920 |
| Cooling Degree Days | 0 | 0 | 0 | 0 | 3 | 39 | 120 | 107 | 11 | 0 | 0 | 0 | 280 |
| Total Precipitation (") | 0.82 | 0.71 | 0.86 | 1.00 | 1.47 | 1.25 | 0.66 | 0.74 | 0.80 | 0.84 | 0.95 | 0.86 | 10.96 |
| Days ≥ 0.1" Precip | 3 | 2 | 3 | 3 | 4 | 4 | 2 | 2 | 2 | 3 | 3 | 3 | 34 |
| Total Snowfall (") | 8.6 | 6.0 | 3.6 | 2.4 | 0.3 | 0.0 | 0.0 | 0.0 | 0.0 | 0.8 | 5.2 | 9.4 | 36.3 |
| Days ≥ 1" Snow Depth | 25 | 19 | 8 | 2 | 0 | 0 | 0 | 0 | 0 | 0 | 6 | 18 | 78 |

## ISLAND PARK *Fremont County* ELEVATION 6306 ft LAT/LONG 44° 25 ' N / 111° 24 ' W

| | JAN | FEB | MAR | APR | MAY | JUN | JUL | AUG | SEP | OCT | NOV | DEC | YEAR |
|---|---|---|---|---|---|---|---|---|---|---|---|---|---|
| Maximum Temp °F | 26.5 | 31.9 | 38.5 | 47.4 | 59.1 | 69.4 | 78.0 | ***77.4*** | 67.3 | 54.1 | 35.9 | 26.1 | 51.0 |
| Minimum Temp °F | 2.8 | 3.9 | 10.9 | 20.9 | 30.8 | 37.9 | 43.0 | ***40.9*** | 32.5 | 24.5 | 14.4 | 3.3 | 22.2 |
| Mean Temp °F | 14.6 | 18.0 | 24.7 | 34.2 | 45.0 | 53.7 | 60.5 | ***59.2*** | 49.9 | 39.4 | 25.2 | 14.6 | 36.6 |
| Days Max Temp ≥ 90 °F | 0 | 0 | 0 | 0 | 0 | 0 | 0 | ***0*** | 0 | 0 | 0 | 0 | 0 |
| Days Max Temp ≤ 32 °F | 23 | 15 | 7 | 1 | 0 | 0 | 0 | ***0*** | 0 | 1 | 11 | 23 | 81 |
| Days Min Temp ≤ 32 °F | 31 | 28 | 31 | 28 | 19 | 5 | 1 | ***3*** | 15 | 27 | 29 | 30 | 247 |
| Days Min Temp ≤ 0 °F | 14 | 11 | 7 | 1 | 0 | 0 | 0 | ***0*** | 0 | 0 | 4 | 13 | 50 |
| Heating Degree Days | 1555 | 1321 | 1242 | 918 | 616 | 336 | 143 | ***181*** | 447 | 788 | 1189 | 1557 | 10293 |
| Cooling Degree Days | 0 | 0 | 0 | 0 | ***1*** | ***1*** | ***11*** | na | ***1*** | ***0*** | ***0*** | ***0*** | na |
| Total Precipitation (") | 3.95 | 2.82 | 2.48 | 2.10 | 2.21 | 2.60 | 1.56 | ***1.65*** | 1.85 | 1.73 | 2.52 | 3.42 | 28.89 |
| Days ≥ 0.1" Precip | 10 | 6 | 6 | 6 | 5 | 7 | 5 | ***4*** | 5 | 4 | 7 | 8 | 73 |
| Total Snowfall (") | 49.3 | 33.9 | 27.2 | 15.1 | 4.3 | ***0.4*** | 0.0 | ***0.0*** | 1.3 | 6.2 | ***25.9*** | 46.7 | 210.3 |
| Days ≥ 1" Snow Depth | 29 | 26 | 29 | 25 | 8 | 0 | 0 | ***0*** | 0 | ***3*** | ***20*** | 30 | 170 |

## JEROME *Jerome County* ELEVATION 3773 ft LAT/LONG 42° 43 ' N / 114° 32 ' W

| | JAN | FEB | MAR | APR | MAY | JUN | JUL | AUG | SEP | OCT | NOV | DEC | YEAR |
|---|---|---|---|---|---|---|---|---|---|---|---|---|---|
| Maximum Temp °F | 35.8 | 42.8 | 52.4 | 61.5 | 71.8 | 81.5 | 90.5 | 89.3 | 78.3 | 65.3 | 47.8 | 36.9 | 62.8 |
| Minimum Temp °F | 18.8 | 22.6 | 28.0 | 33.6 | 41.4 | 49.1 | 55.3 | 54.0 | 44.9 | 35.7 | 26.8 | 19.1 | 35.8 |
| Mean Temp °F | 27.3 | 32.7 | 40.2 | 47.6 | 56.6 | 65.3 | 73.0 | 71.7 | 61.6 | 50.5 | 37.3 | 28.1 | 49.3 |
| Days Max Temp ≥ 90 °F | 0 | 0 | 0 | 0 | 1 | 7 | 19 | 18 | 4 | 0 | 0 | 0 | 49 |
| Days Max Temp ≤ 32 °F | 10 | 4 | 1 | 0 | 0 | 0 | 0 | 0 | 0 | 0 | 2 | 9 | 26 |
| Days Min Temp ≤ 32 °F | 29 | 26 | 23 | 14 | 3 | 0 | 0 | 0 | 1 | 10 | 23 | 29 | 158 |
| Days Min Temp ≤ 0 °F | 2 | 1 | 0 | 0 | 0 | 0 | 0 | 0 | 0 | 0 | 0 | 2 | 5 |
| Heating Degree Days | 1161 | 905 | 761 | 516 | 274 | 93 | 12 | 20 | 149 | 444 | 825 | 1139 | 6299 |
| Cooling Degree Days | 0 | 0 | 0 | 2 | 24 | 104 | 243 | 232 | 51 | 2 | 0 | 0 | 658 |
| Total Precipitation (") | 1.22 | 0.99 | 1.15 | 0.88 | 0.85 | 0.77 | 0.25 | 0.42 | 0.50 | 0.70 | 1.32 | 1.21 | 10.26 |
| Days ≥ 0.1" Precip | 4 | 3 | 4 | 3 | 3 | 3 | 1 | 1 | 2 | 2 | 4 | 4 | 34 |
| Total Snowfall (") | na | na | ***1.4*** | 0.5 | 0.2 | 0.0 | 0.0 | 0.0 | 0.1 | 0.1 | na | na | na |
| Days ≥ 1" Snow Depth | na | na | ***1*** | 0 | 0 | 0 | 0 | 0 | 0 | 0 | ***1*** | na | na |

## KELLOGG *Shoshone County* ELEVATION 2313 ft LAT/LONG 47° 32 ' N / 116° 8 ' W

| | JAN | FEB | MAR | APR | MAY | JUN | JUL | AUG | SEP | OCT | NOV | DEC | YEAR |
|---|---|---|---|---|---|---|---|---|---|---|---|---|---|
| Maximum Temp °F | 35.4 | 41.8 | 49.7 | 58.9 | 68.9 | 76.2 | 84.7 | 85.0 | 74.2 | 60.0 | 43.3 | 34.9 | 59.4 |
| Minimum Temp °F | 21.4 | 24.3 | 28.6 | 34.0 | 40.6 | 46.7 | 50.1 | 49.0 | 41.7 | 33.6 | 28.3 | 21.7 | 35.0 |
| Mean Temp °F | 28.4 | 33.1 | 39.2 | 46.4 | 54.8 | 61.5 | 67.4 | 67.0 | 58.0 | 46.8 | 35.8 | 28.3 | 47.2 |
| Days Max Temp ≥ 90 °F | 0 | 0 | 0 | 0 | 1 | 4 | 11 | 11 | 2 | 0 | 0 | 0 | 29 |
| Days Max Temp ≤ 32 °F | 9 | 3 | 0 | 0 | 0 | 0 | 0 | 0 | 0 | 0 | 2 | 11 | 25 |
| Days Min Temp ≤ 32 °F | 28 | 24 | 22 | 13 | 3 | 0 | 0 | 0 | 2 | 13 | 21 | 27 | 153 |
| Days Min Temp ≤ 0 °F | 2 | 1 | 0 | 0 | 0 | 0 | 0 | 0 | 0 | 0 | 0 | 1 | 4 |
| Heating Degree Days | 1127 | 895 | 794 | 550 | 323 | 150 | 50 | 55 | 220 | 557 | 869 | 1130 | 6720 |
| Cooling Degree Days | 0 | 0 | 0 | 1 | 18 | 57 | 128 | 123 | 18 | 0 | 0 | 0 | 345 |
| Total Precipitation (") | 3.85 | 2.58 | 2.59 | 2.38 | 2.48 | 2.28 | 1.20 | 1.45 | 1.68 | 2.09 | 3.63 | 3.91 | 30.12 |
| Days ≥ 0.1" Precip | 10 | 8 | 9 | 7 | 7 | 6 | 3 | 4 | 4 | 7 | ***10*** | ***11*** | 86 |
| Total Snowfall (") | 15.6 | ***8.2*** | ***3.6*** | 1.0 | 0.0 | 0.0 | 0.0 | 0.0 | 0.0 | 0.1 | ***5.7*** | ***14.4*** | 48.6 |
| Days ≥ 1" Snow Depth | na | na | na | 0 | 0 | 0 | 0 | 0 | 0 | 0 | ***4*** | ***14*** | na |

## KUNA *Ada County* ELEVATION 2690 ft LAT/LONG 43° 31 ' N / 116° 24 ' W

| | JAN | FEB | MAR | APR | MAY | JUN | JUL | AUG | SEP | OCT | NOV | DEC | YEAR |
|---|---|---|---|---|---|---|---|---|---|---|---|---|---|
| Maximum Temp °F | 37.0 | 45.4 | 55.8 | *64.0* | 73.2 | 81.2 | 88.2 | 87.2 | 77.0 | 65.8 | 48.9 | 37.8 | 63.5 |
| Minimum Temp °F | 21.1 | 25.0 | 29.5 | *34.1* | 41.6 | 48.6 | *52.4* | *50.8* | *43.1* | *34.9* | 28.5 | 21.2 | 35.9 |
| Mean Temp °F | 29.1 | 35.2 | 42.7 | *49.1* | 57.4 | 64.9 | *70.2* | *68.9* | *60.0* | *50.4* | 38.7 | 29.5 | 49.7 |
| Days Max Temp ≥ 90 °F | 0 | 0 | 0 | *0* | 1 | *7* | 14 | *12* | *2* | 0 | 0 | 0 | 36 |
| Days Max Temp ≤ 32 °F | 10 | 2 | 0 | *0* | 0 | 0 | 0 | 0 | 0 | 0 | 1 | 8 | 21 |
| Days Min Temp ≤ 32 °F | 27 | *23* | 19 | *13* | 4 | 0 | *0* | *0* | *2* | *11* | *20* | *27* | 146 |
| Days Min Temp ≤ 0 °F | 1 | 1 | 0 | *0* | 0 | 0 | *0* | *0* | *0* | *0* | 0 | 2 | 4 |
| Heating Degree Days | 1107 | 834 | 685 | *472* | 249 | 87 | *16* | *30* | *175* | *446* | 781 | 1093 | 5975 |
| Cooling Degree Days | *0* | *0* | *0* | *3* | *33* | *113* | *177* | *165* | *33* | *3* | *0* | *0* | 527 |
| Total Precipitation (") | 1.10 | *0.76* | *0.94* | *1.01* | 0.89 | 0.85 | 0.28 | 0.40 | 0.62 | 0.58 | *1.34* | 1.15 | 9.92 |
| Days ≥ 0.1" Precip | 4 | 3 | *3* | *3* | 3 | 3 | 1 | 1 | 2 | 2 | *5* | 4 | 34 |
| Total Snowfall (") | 3.3 | *1.6* | *0.4* | *0.2* | 0.0 | *0.0* | *0.0* | *0.0* | *0.0* | *0.0* | *1.6* | *4.2* | 11.3 |
| Days ≥ 1" Snow Depth | 10 | *3* | *0* | *0* | 0 | 0 | 0 | 0 | 0 | 0 | *1* | *7* | 21 |

## LEWISTON NEZ PERCE *Nez Perce County* ELEVATION 1421 ft LAT/LONG 46° 23 ' N / 117° 1 ' W

| | JAN | FEB | MAR | APR | MAY | JUN | JUL | AUG | SEP | OCT | NOV | DEC | YEAR |
|---|---|---|---|---|---|---|---|---|---|---|---|---|---|
| Maximum Temp °F | 39.8 | 46.6 | 54.7 | 62.3 | 71.4 | 79.5 | 88.7 | 88.4 | 77.5 | 63.3 | 47.8 | 40.3 | 63.4 |
| Minimum Temp °F | 27.6 | 31.0 | 35.1 | 40.1 | 46.9 | 53.8 | 59.2 | 59.0 | 50.7 | 41.2 | 33.9 | 28.4 | 42.2 |
| Mean Temp °F | 33.7 | 38.8 | 44.9 | 51.2 | 59.2 | 66.7 | 74.0 | 73.8 | 64.1 | 52.3 | 40.9 | 34.4 | 52.8 |
| Days Max Temp ≥ 90 °F | 0 | 0 | 0 | 0 | 1 | 6 | 16 | 15 | 3 | 0 | 0 | 0 | 41 |
| Days Max Temp ≤ 32 °F | 7 | 2 | 0 | 0 | 0 | 0 | 0 | 0 | 0 | 0 | 1 | 5 | 15 |
| Days Min Temp ≤ 32 °F | 20 | 15 | 10 | 3 | 0 | 0 | 0 | 0 | 0 | 3 | 11 | 20 | 82 |
| Days Min Temp ≤ 0 °F | 1 | 0 | 0 | 0 | 0 | 0 | 0 | 0 | 0 | 0 | 0 | 1 | 2 |
| Heating Degree Days | 963 | 732 | 616 | 410 | 205 | 62 | 9 | 10 | 103 | 391 | 717 | 943 | 5161 |
| Cooling Degree Days | 0 | 0 | 0 | 4 | 38 | 115 | 269 | 287 | 86 | 4 | 0 | 0 | 803 |
| Total Precipitation (") | 1.28 | 0.86 | 1.02 | 1.24 | 1.38 | 1.22 | 0.69 | 0.76 | 0.74 | 0.86 | 1.11 | 1.04 | 12.20 |
| Days ≥ 0.1" Precip | 4 | 3 | 4 | 4 | 5 | 4 | 2 | 2 | 3 | 3 | 4 | 4 | 42 |
| Total Snowfall (") | 5.4 | 2.4 | 0.9 | 0.1 | 0.0 | 0.0 | 0.0 | 0.0 | 0.0 | 0.1 | 1.6 | 4.0 | 14.5 |
| Days ≥ 1" Snow Depth | 7 | 3 | 0 | 0 | 0 | 0 | 0 | 0 | 0 | 0 | 1 | 4 | 15 |

## LIFTON PUMPING STN *Bear Lake County* ELEVATION 5935 ft LAT/LONG 42° 7 ' N / 111° 18 ' W

| | JAN | FEB | MAR | APR | MAY | JUN | JUL | AUG | SEP | OCT | NOV | DEC | YEAR |
|---|---|---|---|---|---|---|---|---|---|---|---|---|---|
| Maximum Temp °F | 29.5 | 32.6 | 40.3 | 51.1 | 62.2 | 72.1 | 80.9 | 79.6 | 69.4 | 56.8 | 41.0 | 31.1 | 53.9 |
| Minimum Temp °F | 6.0 | 5.9 | 16.9 | 29.2 | 39.2 | 46.5 | 51.3 | 47.8 | 38.5 | 29.1 | 19.8 | 9.4 | 28.3 |
| Mean Temp °F | 17.8 | 19.3 | 28.6 | 40.2 | 50.7 | 59.3 | 66.1 | 63.8 | 53.9 | 43.0 | 30.4 | 20.2 | 41.1 |
| Days Max Temp ≥ 90 °F | 0 | 0 | 0 | 0 | 0 | 0 | 1 | 1 | 0 | 0 | 0 | 0 | 2 |
| Days Max Temp ≤ 32 °F | 18 | 12 | 4 | 0 | 0 | 0 | 0 | 0 | 0 | 0 | 6 | 17 | 57 |
| Days Min Temp ≤ 32 °F | 31 | 28 | 30 | 21 | 5 | 0 | 0 | 0 | 6 | 22 | 29 | 31 | 203 |
| Days Min Temp ≤ 0 °F | 10 | 10 | 3 | 0 | 0 | 0 | 0 | 0 | 0 | 0 | 1 | 6 | 30 |
| Heating Degree Days | 1458 | 1285 | 1122 | 738 | 436 | 183 | 35 | 75 | 327 | 675 | 1031 | 1382 | 8747 |
| Cooling Degree Days | 0 | 0 | 0 | 0 | 0 | 22 | 77 | 50 | 2 | 0 | 0 | 0 | 151 |
| Total Precipitation (") | 0.68 | 0.67 | 0.76 | 0.99 | 1.37 | 1.02 | 0.83 | 0.88 | 1.15 | 1.10 | 0.84 | 0.60 | 10.89 |
| Days ≥ 0.1" Precip | 2 | 2 | 3 | 3 | 4 | 3 | 2 | 3 | 3 | 3 | 3 | 2 | 33 |
| Total Snowfall (") | 7.6 | 6.8 | 4.7 | 2.3 | 0.4 | 0.0 | 0.0 | 0.0 | 0.0 | 1.5 | 6.1 | 7.3 | 36.7 |
| Days ≥ 1" Snow Depth | 29 | 24 | 15 | 2 | 0 | 0 | 0 | 0 | 0 | 1 | 9 | 21 | 101 |

## MACKAY RANGER STN *Custer County* ELEVATION 5905 ft LAT/LONG 43° 55 ' N / 113° 37 ' W

| | JAN | FEB | MAR | APR | MAY | JUN | JUL | AUG | SEP | OCT | NOV | DEC | YEAR |
|---|---|---|---|---|---|---|---|---|---|---|---|---|---|
| Maximum Temp °F | 29.6 | 35.3 | 43.3 | 54.8 | 65.4 | 74.8 | 83.6 | 82.7 | 73.5 | 59.6 | 40.9 | 30.0 | 56.1 |
| Minimum Temp °F | 6.1 | 10.1 | 19.2 | 27.6 | 35.5 | 42.3 | 47.4 | 45.7 | 37.4 | 28.8 | 18.0 | 6.9 | 27.1 |
| Mean Temp °F | 17.8 | 22.7 | 31.3 | 41.2 | 50.5 | 58.6 | 65.5 | 64.2 | 55.5 | 44.0 | 29.4 | 18.5 | 41.6 |
| Days Max Temp ≥ 90 °F | 0 | 0 | 0 | 0 | 0 | 1 | 6 | 4 | 0 | 0 | 0 | 0 | 11 |
| Days Max Temp ≤ 32 °F | 19 | 10 | 3 | 0 | 0 | 0 | 0 | 0 | 0 | 0 | 6 | 19 | 57 |
| Days Min Temp ≤ 32 °F | 31 | 28 | 30 | 23 | 9 | 2 | 0 | 1 | *7* | *21* | 29 | 31 | 212 |
| Days Min Temp ≤ 0 °F | 10 | 5 | 1 | 0 | 0 | 0 | 0 | 0 | 0 | 0 | 1 | 8 | 25 |
| Heating Degree Days | 1455 | 1188 | 1037 | 706 | 444 | 203 | 54 | 77 | 281 | 643 | 1059 | 1433 | 8580 |
| Cooling Degree Days | 0 | 0 | 0 | 0 | 1 | 17 | 67 | 51 | 3 | 0 | 0 | 0 | 139 |
| Total Precipitation (") | 0.79 | 0.46 | 0.71 | 0.66 | 1.09 | 1.36 | 1.07 | 0.93 | 0.73 | 0.49 | 0.71 | 0.72 | 9.72 |
| Days ≥ 0.1" Precip | 3 | 2 | 2 | 2 | 3 | 4 | 3 | 3 | 2 | 2 | 2 | 2 | 30 |
| Total Snowfall (") | na | na | na | na | 0.0 | 0.0 | 0.0 | 0.0 | 0.0 | 0.2 | na | na | na |
| Days ≥ 1" Snow Depth | na | na | na | 0 | 0 | 0 | 0 | 0 | 0 | *0* | na | na | na |

## MALAD CITY *Oneida County* ELEVATION 4475 ft LAT/LONG 42° 10 ' N / 112° 17 ' W

| | JAN | FEB | MAR | APR | MAY | JUN | JUL | AUG | SEP | OCT | NOV | DEC | YEAR |
|---|---|---|---|---|---|---|---|---|---|---|---|---|---|
| Maximum Temp °F | 32.0 | 38.8 | 49.7 | 60.0 | 70.0 | 80.2 | 89.1 | 88.0 | 77.7 | 64.2 | 46.0 | 33.6 | 60.8 |
| Minimum Temp °F | 11.2 | 15.1 | 23.8 | 30.2 | 37.3 | 43.7 | 49.1 | 47.8 | 38.8 | 29.5 | 22.1 | 12.4 | 30.1 |
| Mean Temp °F | 21.6 | 27.0 | 36.8 | 45.2 | 53.7 | 62.0 | 69.1 | 67.9 | 58.3 | 46.9 | 34.0 | 23.0 | 45.5 |
| Days Max Temp ≥ 90 °F | 0 | 0 | 0 | 0 | 0 | 5 | 16 | 15 | 2 | 0 | 0 | 0 | 38 |
| Days Max Temp ≤ 32 °F | 15 | 6 | 1 | 0 | 0 | 0 | 0 | 0 | 0 | 0 | 3 | 13 | 38 |
| Days Min Temp ≤ 32 °F | 30 | 27 | 27 | 19 | 7 | 1 | 0 | 0 | 6 | 21 | 27 | 30 | 195 |
| Days Min Temp ≤ 0 °F | 8 | 4 | 0 | 0 | 0 | 0 | 0 | 0 | 0 | 0 | 1 | 5 | 18 |
| Heating Degree Days | 1338 | 1069 | 869 | 589 | 346 | 126 | 16 | 28 | 210 | 555 | 922 | 1295 | 7363 |
| Cooling Degree Days | 0 | 0 | 0 | 0 | 2 | 46 | 140 | 129 | 15 | 0 | 0 | 0 | 332 |
| Total Precipitation (") | 1.18 | 1.01 | 1.14 | 1.21 | 1.76 | 1.31 | 1.10 | 0.93 | 1.05 | 1.20 | 1.18 | 1.04 | 14.11 |
| Days ≥ 0.1" Precip | 4 | 3 | 4 | 4 | 5 | 4 | 2 | 3 | 3 | 3 | 4 | 4 | 43 |
| Total Snowfall (") | 10.1 | 7.8 | 4.3 | 1.8 | 0.2 | 0.0 | 0.0 | 0.0 | 0.0 | 0.5 | 6.3 | 10.2 | 41.2 |
| Days ≥ 1" Snow Depth | 26 | 19 | 8 | 0 | 0 | 0 | 0 | 0 | 0 | 0 | 5 | 22 | 80 |

## MALTA 4 ESE *Cassia County* ELEVATION 4544 ft LAT/LONG 42° 19 ' N / 113° 21 ' W

| | JAN | FEB | MAR | APR | MAY | JUN | JUL | AUG | SEP | OCT | NOV | DEC | YEAR |
|---|---|---|---|---|---|---|---|---|---|---|---|---|---|
| Maximum Temp °F | 36.7 | 42.9 | 51.8 | 61.2 | 69.9 | 80.0 | 88.9 | 87.8 | 77.4 | 65.0 | 47.4 | 37.8 | 62.2 |
| Minimum Temp °F | 15.9 | 20.7 | 26.0 | 31.5 | 38.0 | 44.0 | 49.6 | 48.2 | 39.5 | 31.5 | 23.7 | 16.2 | 32.1 |
| Mean Temp °F | 26.3 | 31.8 | 39.0 | 46.3 | 54.0 | 62.0 | 69.3 | 68.0 | 58.5 | 48.3 | 35.6 | 27.0 | 47.2 |
| Days Max Temp ≥ 90 °F | 0 | 0 | 0 | 0 | 0 | *4* | 17 | 15 | 3 | 0 | 0 | 0 | 39 |
| Days Max Temp ≤ 32 °F | 9 | 4 | 1 | 0 | 0 | 0 | 0 | 0 | 0 | 0 | 2 | 8 | 24 |
| Days Min Temp ≤ 32 °F | 29 | 25 | 24 | *16* | 7 | 1 | 0 | 0 | 5 | 16 | 25 | 30 | 178 |
| Days Min Temp ≤ 0 °F | 3 | 2 | 0 | 0 | 0 | 0 | 0 | 0 | 0 | 0 | 1 | 3 | 9 |
| Heating Degree Days | 1192 | 932 | 801 | 553 | 340 | 133 | 21 | 37 | 209 | 512 | 874 | 1170 | 6774 |
| Cooling Degree Days | 0 | 0 | 0 | 0 | 5 | 52 | 150 | 149 | 22 | 0 | 0 | 0 | 378 |
| Total Precipitation (") | 0.70 | 0.62 | 0.87 | 1.11 | 1.56 | 1.20 | 0.94 | 1.04 | 0.84 | 0.77 | 0.79 | 0.61 | 11.05 |
| Days ≥ 0.1" Precip | 2 | 3 | 3 | 4 | 5 | 4 | 3 | 3 | 3 | 3 | 4 | 3 | 40 |
| Total Snowfall (") | na | na | *2.1* | 0.9 | 0.4 | 0.0 | 0.0 | 0.0 | 0.0 | 0.0 | *2.3* | na | na |
| Days ≥ 1" Snow Depth | na | na | *2* | 0 | 0 | 0 | 0 | 0 | 0 | 0 | *4* | *11* | na |

## MCCALL *Valley County* ELEVATION 5033 ft LAT/LONG 44° 54 ' N / 116° 7 ' W

| | JAN | FEB | MAR | APR | MAY | JUN | JUL | AUG | SEP | OCT | NOV | DEC | YEAR |
|---|---|---|---|---|---|---|---|---|---|---|---|---|---|
| Maximum Temp °F | 31.7 | 37.3 | 43.6 | 52.0 | 62.3 | 70.9 | 80.6 | 80.6 | 70.4 | 58.2 | 40.6 | 31.9 | 55.0 |
| Minimum Temp °F | 12.7 | 14.5 | 19.9 | 26.7 | 33.6 | 39.6 | 42.8 | 40.9 | 33.6 | 27.0 | 22.1 | 14.8 | 27.4 |
| Mean Temp °F | 22.2 | 25.9 | 31.8 | 39.4 | 48.0 | 55.3 | 61.7 | 60.8 | 52.0 | 42.6 | 31.4 | 23.4 | 41.2 |
| Days Max Temp ≥ 90 °F | 0 | 0 | 0 | 0 | 0 | 0 | 3 | 3 | 0 | 0 | 0 | 0 | 6 |
| Days Max Temp ≤ 32 °F | 15 | 6 | 2 | 0 | 0 | 0 | 0 | 0 | 0 | 0 | 5 | 15 | 43 |
| Days Min Temp ≤ 32 °F | 30 | 28 | 30 | 25 | 14 | 5 | 1 | 3 | 13 | 25 | 27 | 30 | 231 |
| Days Min Temp ≤ 0 °F | 6 | 4 | 1 | 0 | 0 | 0 | 0 | 0 | 0 | 0 | 1 | 3 | 15 |
| Heating Degree Days | 1320 | 1097 | 1022 | 763 | 523 | 292 | 126 | 149 | 383 | 688 | 1003 | 1283 | 8649 |
| Cooling Degree Days | 0 | 0 | 0 | 0 | 1 | 6 | 24 | 24 | 1 | 0 | 0 | 0 | 56 |
| Total Precipitation (") | 3.67 | 2.71 | 2.43 | 2.07 | 2.04 | 2.00 | 0.88 | 1.10 | 1.64 | 1.73 | 3.14 | 3.38 | 26.79 |
| Days ≥ 0.1" Precip | 10 | 8 | 8 | 6 | 6 | 6 | 2 | 3 | 4 | 5 | 9 | 10 | 77 |
| Total Snowfall (") | 39.3 | 25.3 | 17.9 | 7.2 | 0.9 | 0.0 | 0.0 | 0.0 | 0.0 | *2.2* | 21.7 | 34.3 | 148.8 |
| Days ≥ 1" Snow Depth | 31 | 28 | 30 | 11 | 0 | 0 | 0 | 0 | 0 | 1 | 15 | 30 | 146 |

## MINIDOKA DAM *Minidoka County* ELEVATION 4213 ft LAT/LONG 42° 40 ' N / 113° 30 ' W

| | JAN | FEB | MAR | APR | MAY | JUN | JUL | AUG | SEP | OCT | NOV | DEC | YEAR |
|---|---|---|---|---|---|---|---|---|---|---|---|---|---|
| Maximum Temp °F | 34.4 | 40.8 | 50.6 | 59.6 | 69.2 | 79.2 | 87.6 | 86.8 | 76.7 | 63.8 | 46.3 | 35.3 | 60.9 |
| Minimum Temp °F | 16.7 | 20.6 | 27.2 | 33.1 | 41.0 | 48.8 | 54.7 | 53.3 | 44.4 | 35.1 | 26.4 | 18.0 | 34.9 |
| Mean Temp °F | 25.6 | 30.7 | 38.9 | 46.4 | 55.1 | 64.0 | 71.2 | 70.1 | 60.6 | 49.5 | 36.4 | 26.7 | 47.9 |
| Days Max Temp ≥ 90 °F | 0 | 0 | 0 | 0 | 0 | 4 | 14 | 13 | 2 | 0 | 0 | 0 | 33 |
| Days Max Temp ≤ 32 °F | 12 | 5 | 1 | 0 | 0 | 0 | 0 | 0 | 0 | 0 | 2 | 10 | 30 |
| Days Min Temp ≤ 32 °F | 29 | 26 | 24 | 14 | 3 | 0 | 0 | 0 | 2 | 10 | 23 | 29 | 160 |
| Days Min Temp ≤ 0 °F | 3 | 1 | 0 | 0 | 0 | 0 | 0 | 0 | 0 | 0 | 0 | 3 | 7 |
| Heating Degree Days | 1216 | 962 | 802 | 552 | 308 | 102 | 14 | 23 | 165 | 475 | 853 | 1180 | 6652 |
| Cooling Degree Days | 0 | 0 | 0 | 1 | 13 | 85 | 199 | 200 | 44 | 1 | 0 | 0 | 543 |
| Total Precipitation (") | 0.92 | 0.76 | 0.98 | 0.94 | 1.04 | 0.88 | 0.35 | 0.50 | 0.57 | 0.69 | 1.08 | 0.88 | 9.59 |
| Days ≥ 0.1" Precip | 3 | 3 | 3 | 3 | 3 | 3 | 1 | 1 | 2 | 2 | 4 | 3 | 31 |
| Total Snowfall (") | 6.7 | 4.5 | 2.9 | 1.3 | 0.3 | 0.0 | 0.0 | 0.0 | 0.0 | 0.2 | 3.5 | 7.2 | 26.6 |
| Days ≥ 1" Snow Depth | 17 | 10 | 3 | 0 | 0 | 0 | 0 | 0 | 0 | 0 | 3 | 14 | 47 |

## MOSCOW U OF I *Latah County* ELEVATION 2631 ft LAT/LONG 46° 44 ' N / 117° 0 ' W

| | JAN | FEB | MAR | APR | MAY | JUN | JUL | AUG | SEP | OCT | NOV | DEC | YEAR |
|---|---|---|---|---|---|---|---|---|---|---|---|---|---|
| Maximum Temp °F | 35.2 | 41.2 | 48.5 | 57.0 | 66.2 | 73.6 | 82.8 | 83.8 | 74.1 | 60.5 | 43.1 | 35.4 | 58.5 |
| Minimum Temp °F | 22.7 | 26.8 | 30.8 | 35.1 | 40.4 | 45.3 | 48.2 | 48.6 | 42.9 | 36.2 | 29.8 | 23.5 | 35.9 |
| Mean Temp °F | 29.0 | 34.0 | 39.7 | 46.1 | 53.3 | 59.5 | 65.5 | 66.2 | 58.5 | 48.4 | 36.5 | 29.5 | 47.2 |
| Days Max Temp ≥ 90 °F | 0 | 0 | 0 | 0 | 0 | 1 | 7 | 9 | 1 | 0 | 0 | 0 | 18 |
| Days Max Temp ≤ 32 °F | 10 | 3 | 1 | 0 | 0 | 0 | 0 | 0 | 0 | 0 | 3 | 10 | 27 |
| Days Min Temp ≤ 32 °F | 26 | 22 | 19 | 11 | 4 | 0 | 0 | 0 | 2 | 9 | 19 | 26 | 138 |
| Days Min Temp ≤ 0 °F | 2 | 1 | 0 | 0 | 0 | 0 | 0 | 0 | 0 | 0 | 0 | 1 | 4 |
| Heating Degree Days | 1111 | 868 | 778 | 562 | 363 | 186 | 67 | 61 | 212 | 510 | 848 | 1094 | 6660 |
| Cooling Degree Days | 0 | 0 | 0 | 1 | 11 | 24 | 82 | 104 | 25 | 2 | 0 | 0 | 249 |
| Total Precipitation (") | 3.05 | 2.15 | 2.36 | 2.33 | 2.33 | 1.84 | 1.03 | 1.16 | 1.20 | 1.89 | 3.22 | 2.86 | 25.42 |
| Days ≥ 0.1" Precip | 8 | 7 | 7 | 7 | 6 | 5 | 3 | 3 | 4 | 5 | 9 | 8 | 72 |
| Total Snowfall (") | 17.2 | 7.9 | 4.0 | 1.1 | 0.0 | 0.0 | 0.0 | 0.0 | 0.0 | 0.3 | 6.2 | 13.7 | 50.4 |
| Days ≥ 1" Snow Depth | 18 | 9 | 4 | 0 | 0 | 0 | 0 | 0 | 0 | 0 | 5 | 15 | 51 |

## MOUNTAIN HOME *Elmore County* ELEVATION 3153 ft LAT/LONG 43° 8 ' N / 115° 43 ' W

| | JAN | FEB | MAR | APR | MAY | JUN | JUL | AUG | SEP | OCT | NOV | DEC | YEAR |
|---|---|---|---|---|---|---|---|---|---|---|---|---|---|
| Maximum Temp °F | 37.6 | 44.9 | 53.8 | 62.4 | 72.3 | 82.8 | 91.9 | 90.8 | 79.5 | 66.4 | 48.7 | 38.4 | 64.1 |
| Minimum Temp °F | 20.1 | 24.4 | 29.1 | 34.7 | 42.2 | 50.4 | 56.2 | 55.0 | 45.1 | 34.9 | 27.2 | 20.2 | 36.6 |
| Mean Temp °F | 28.9 | 34.7 | 41.5 | 48.6 | 57.3 | 66.6 | 74.1 | 72.9 | 62.3 | 50.7 | 38.0 | 29.3 | 50.4 |
| Days Max Temp ≥ 90 °F | 0 | 0 | 0 | 0 | 2 | 9 | 21 | 20 | 5 | 0 | 0 | 0 | 57 |
| Days Max Temp ≤ 32 °F | 8 | 3 | 0 | 0 | 0 | 0 | 0 | 0 | 0 | 0 | 2 | 7 | 20 |
| Days Min Temp ≤ 32 °F | 28 | 24 | 21 | 13 | 3 | 0 | 0 | 0 | 2 | 12 | 22 | 28 | 153 |
| Days Min Temp ≤ 0 °F | 1 | 1 | 0 | 0 | 0 | 0 | 0 | 0 | 0 | 0 | 0 | 1 | 3 |
| Heating Degree Days | 1113 | 849 | 722 | 490 | 262 | 81 | 10 | 17 | 137 | 438 | 803 | 1100 | 6022 |
| Cooling Degree Days | 0 | 0 | 0 | 6 | 46 | 149 | 301 | 293 | 75 | 4 | 0 | 0 | 874 |
| Total Precipitation (") | 1.37 | 0.93 | 1.08 | 0.99 | 0.67 | 0.81 | 0.42 | 0.36 | 0.66 | 0.72 | 1.44 | 1.35 | 10.80 |
| Days ≥ 0.1" Precip | 4 | 3 | 4 | 3 | 2 | 2 | 1 | 1 | 2 | 2 | 5 | 5 | 34 |
| Total Snowfall (") | na | ***1.8*** | na | 0.1 | 0.0 | 0.0 | 0.0 | 0.0 | 0.0 | 0.0 | ***1.7*** | na | na |
| Days ≥ 1" Snow Depth | na | na | ***0*** | 0 | 0 | 0 | 0 | 0 | 0 | 0 | ***1*** | na | na |

## NEW MEADOWS RANGER S *Adams County* ELEVATION 3862 ft LAT/LONG 44° 58 ' N / 116° 17 ' W

| | JAN | FEB | MAR | APR | MAY | JUN | JUL | AUG | SEP | OCT | NOV | DEC | YEAR |
|---|---|---|---|---|---|---|---|---|---|---|---|---|---|
| Maximum Temp °F | 30.7 | 37.4 | 45.7 | 55.1 | 65.3 | 73.6 | 83.6 | 83.7 | 73.1 | 60.2 | 41.9 | ***31.2*** | 56.8 |
| Minimum Temp °F | 8.3 | 10.6 | 18.8 | 26.6 | 32.7 | 39.2 | 42.3 | 40.8 | 32.6 | 24.5 | 19.8 | ***9.8*** | 25.5 |
| Mean Temp °F | 19.5 | 24.0 | 32.3 | 40.9 | 49.0 | 56.5 | 63.0 | 62.3 | 52.9 | 42.4 | 30.9 | ***20.5*** | 41.2 |
| Days Max Temp ≥ 90 °F | 0 | 0 | 0 | 0 | 0 | 2 | 8 | 8 | 1 | 0 | 0 | 0 | 19 |
| Days Max Temp ≤ 32 °F | 16 | 5 | 1 | 0 | 0 | 0 | 0 | 0 | 0 | 0 | 4 | 14 | 40 |
| Days Min Temp ≤ 32 °F | 29 | 27 | 29 | 24 | 14 | 5 | 1 | 3 | 16 | 26 | 27 | 28 | 229 |
| Days Min Temp ≤ 0 °F | 10 | 7 | 2 | 0 | 0 | 0 | 0 | 0 | 0 | 0 | 2 | 7 | 28 |
| Heating Degree Days | ***1403*** | 1152 | 1007 | 717 | 489 | 265 | 104 | 117 | 358 | 694 | 1017 | ***1374*** | 8697 |
| Cooling Degree Days | na | 0 | 0 | ***0*** | 1 | 10 | 32 | 36 | 2 | ***0*** | ***0*** | na | na |
| Total Precipitation (") | 3.25 | 2.38 | 2.33 | 2.01 | 1.90 | 1.82 | 0.75 | 0.79 | 1.34 | 1.53 | 2.74 | 3.21 | 24.05 |
| Days ≥ 0.1" Precip | 7 | 6 | 6 | 5 | 5 | 5 | 2 | 2 | 3 | 4 | 7 | 7 | 59 |
| Total Snowfall (") | 23.8 | 14.2 | 6.2 | 1.9 | 0.3 | 0.0 | 0.0 | 0.0 | 0.1 | 0.5 | 10.2 | 22.8 | 80.0 |
| Days ≥ 1" Snow Depth | 27 | 24 | 16 | 2 | 0 | 0 | 0 | 0 | 0 | 0 | 8 | 23 | 100 |

## NEZPERCE *Lewis County* ELEVATION 3251 ft LAT/LONG 46° 15 ' N / 116° 12 ' W

| | JAN | FEB | MAR | APR | MAY | JUN | JUL | AUG | SEP | OCT | NOV | DEC | YEAR |
|---|---|---|---|---|---|---|---|---|---|---|---|---|---|
| Maximum Temp °F | 34.7 | 41.0 | 47.6 | 55.2 | 63.8 | 71.2 | 80.1 | 81.1 | 70.9 | 58.2 | 42.4 | 35.0 | 56.8 |
| Minimum Temp °F | 21.5 | 25.0 | 28.7 | 33.3 | 39.1 | 45.0 | 48.5 | 48.6 | 41.8 | 34.7 | 27.9 | 22.1 | 34.7 |
| Mean Temp °F | 28.1 | 33.0 | 38.1 | 44.3 | 51.5 | 58.1 | 64.4 | 64.9 | 56.3 | 46.5 | 35.2 | 28.6 | 45.8 |
| Days Max Temp ≥ 90 °F | 0 | 0 | 0 | 0 | 0 | 1 | 4 | 6 | 1 | 0 | 0 | 0 | 12 |
| Days Max Temp ≤ 32 °F | 11 | 4 | 1 | 0 | 0 | 0 | 0 | 0 | 0 | 0 | 3 | 11 | 30 |
| Days Min Temp ≤ 32 °F | 27 | 23 | 23 | 15 | 5 | 0 | 0 | 0 | 3 | 11 | 21 | 27 | 155 |
| Days Min Temp ≤ 0 °F | 2 | 1 | 0 | 0 | 0 | 0 | 0 | 0 | 0 | 0 | 0 | 1 | 4 |
| Heating Degree Days | 1137 | 895 | 825 | 615 | 415 | 219 | 84 | 79 | 263 | 568 | 887 | 1123 | 7110 |
| Cooling Degree Days | 0 | 0 | 0 | 0 | 6 | 19 | 60 | 79 | 11 | 0 | 0 | 0 | 175 |
| Total Precipitation (") | 1.67 | 1.27 | 1.81 | 2.19 | 2.76 | 2.03 | 1.23 | 1.25 | 1.34 | 1.53 | 1.80 | 1.49 | 20.37 |
| Days ≥ 0.1" Precip | 5 | 5 | 7 | 7 | 8 | 6 | 4 | 3 | 4 | 5 | 7 | 5 | 66 |
| Total Snowfall (") | 12.7 | 7.6 | 6.3 | 3.6 | 0.3 | 0.0 | 0.0 | 0.0 | 0.0 | 1.4 | 7.1 | 10.6 | 49.6 |
| Days ≥ 1" Snow Depth | 20 | 11 | 5 | 1 | 0 | 0 | 0 | 0 | 0 | 1 | 6 | 17 | 61 |

## OAKLEY *Cassia County* ELEVATION 4603 ft LAT/LONG 42° 15 ' N / 113° 53 ' W

| | JAN | FEB | MAR | APR | MAY | JUN | JUL | AUG | SEP | OCT | NOV | DEC | YEAR |
|---|---|---|---|---|---|---|---|---|---|---|---|---|---|
| Maximum Temp °F | 38.1 | 44.0 | 51.2 | 59.1 | 67.9 | 76.7 | 84.3 | 83.9 | 74.9 | 63.6 | 47.6 | 38.9 | 60.9 |
| Minimum Temp °F | 19.7 | 23.7 | 28.4 | 33.3 | 40.4 | 47.9 | 54.6 | 53.7 | 44.8 | 36.1 | 27.5 | 20.2 | 35.9 |
| Mean Temp °F | 28.9 | 33.9 | 39.8 | 46.2 | 54.2 | 62.3 | 69.5 | 68.8 | 59.9 | 49.9 | 37.6 | 29.6 | 48.4 |
| Days Max Temp ≥ 90 °F | 0 | 0 | 0 | 0 | 0 | 2 | 6 | 6 | 1 | 0 | 0 | 0 | 15 |
| Days Max Temp ≤ 32 °F | 8 | 3 | 1 | 0 | 0 | 0 | 0 | 0 | 0 | 0 | 2 | 7 | 21 |
| Days Min Temp ≤ 32 °F | 27 | 24 | 22 | 14 | 4 | 0 | 0 | 0 | 2 | 9 | 20 | 28 | 150 |
| Days Min Temp ≤ 0 °F | 2 | 1 | 0 | 0 | 0 | 0 | 0 | 0 | 0 | 0 | 0 | 1 | 4 |
| Heating Degree Days | 1112 | 872 | 774 | 557 | 336 | 131 | 22 | 32 | 180 | 463 | 816 | 1092 | 6387 |
| Cooling Degree Days | 0 | 0 | 0 | 0 | 9 | 57 | 153 | 159 | 33 | 1 | 0 | 0 | 412 |
| Total Precipitation (") | 0.76 | 0.60 | 1.01 | 1.12 | 1.58 | 1.29 | 0.77 | 0.94 | 0.89 | 0.73 | 0.82 | 0.69 | 11.20 |
| Days ≥ 0.1" Precip | 3 | 2 | 3 | 4 | 5 | 4 | 3 | 3 | 3 | 2 | 3 | 3 | 38 |
| Total Snowfall (") | 7.3 | 4.5 | 3.9 | 1.7 | 0.8 | 0.0 | 0.0 | 0.0 | 0.1 | 0.0 | 4.0 | 6.2 | 28.5 |
| Days ≥ 1" Snow Depth | 8 | *3* | *1* | 0 | 0 | 0 | 0 | 0 | 0 | 0 | 2 | *8* | 22 |

## PARMA EXPERIMENT STN *Canyon County* ELEVATION 2224 ft LAT/LONG 43° 47 ' N / 116° 57 ' W

| | JAN | FEB | MAR | APR | MAY | JUN | JUL | AUG | SEP | OCT | NOV | DEC | YEAR |
|---|---|---|---|---|---|---|---|---|---|---|---|---|---|
| Maximum Temp °F | 35.3 | 44.1 | 55.6 | 64.2 | 73.4 | 81.8 | 90.8 | 90.2 | 79.3 | 66.2 | 48.7 | 37.3 | 63.9 |
| Minimum Temp °F | 18.4 | 23.8 | 29.1 | 35.2 | 43.1 | 49.7 | 53.7 | 51.6 | 42.4 | 32.8 | 26.5 | 19.7 | 35.5 |
| Mean Temp °F | 26.9 | 34.0 | 42.4 | 49.7 | 58.3 | 65.8 | 72.3 | 70.9 | 60.9 | 49.5 | 37.6 | 28.5 | 49.7 |
| Days Max Temp ≥ 90 °F | 0 | 0 | 0 | 0 | 2 | 7 | 19 | 19 | 5 | 0 | 0 | 0 | 52 |
| Days Max Temp ≤ 32 °F | 11 | 3 | 0 | 0 | 0 | 0 | 0 | 0 | 0 | 0 | 1 | 8 | 23 |
| Days Min Temp ≤ 32 °F | 29 | 25 | 21 | 11 | 2 | 0 | 0 | 0 | 2 | 15 | 23 | 28 | 156 |
| Days Min Temp ≤ 0 °F | 3 | 1 | 0 | 0 | 0 | 0 | 0 | 0 | 0 | 0 | 0 | 2 | 6 |
| Heating Degree Days | 1175 | 869 | 696 | 454 | 234 | 81 | 15 | 23 | 156 | 473 | 814 | 1124 | 6114 |
| Cooling Degree Days | 0 | 0 | 0 | 3 | 35 | 107 | 225 | 208 | 40 | 0 | 0 | 0 | 618 |
| Total Precipitation (") | 1.44 | 0.94 | 1.16 | 0.97 | 0.92 | 1.00 | 0.31 | 0.54 | 0.66 | 0.71 | 1.34 | 1.34 | 11.33 |
| Days ≥ 0.1" Precip | 5 | 3 | 4 | 4 | 3 | 3 | 1 | 1 | 2 | 2 | 4 | 4 | 36 |
| Total Snowfall (") | 6.9 | 2.1 | 0.8 | 0.1 | 0.1 | 0.0 | 0.0 | 0.0 | 0.0 | 0.0 | 2.5 | 5.3 | 17.8 |
| Days ≥ 1" Snow Depth | 15 | 6 | 1 | 0 | 0 | 0 | 0 | 0 | 0 | 0 | 2 | 8 | 32 |

## PAUL 1 ENE *Minidoka County* ELEVATION 4203 ft LAT/LONG 42° 37 ' N / 113° 45 ' W

| | JAN | FEB | MAR | APR | MAY | JUN | JUL | AUG | SEP | OCT | NOV | DEC | YEAR |
|---|---|---|---|---|---|---|---|---|---|---|---|---|---|
| Maximum Temp °F | 34.8 | 41.6 | 50.2 | 59.2 | 68.6 | 78.1 | 87.1 | 86.3 | 75.7 | 63.4 | 46.6 | 36.5 | 60.7 |
| Minimum Temp °F | 17.0 | 21.1 | 26.2 | 32.1 | 39.9 | 47.3 | 52.7 | 50.2 | 41.2 | 32.3 | 24.9 | 17.4 | 33.5 |
| Mean Temp °F | 25.9 | 31.4 | 38.2 | 45.7 | 54.3 | 62.7 | 69.9 | 68.3 | 58.5 | 47.9 | 35.8 | 26.9 | 47.1 |
| Days Max Temp ≥ 90 °F | 0 | 0 | 0 | 0 | 0 | 4 | 14 | 12 | 2 | 0 | 0 | 0 | 32 |
| Days Max Temp ≤ 32 °F | 12 | 5 | 1 | 0 | 0 | 0 | 0 | 0 | 0 | 0 | 3 | 10 | 31 |
| Days Min Temp ≤ 32 °F | 29 | 26 | 26 | 16 | 4 | 0 | 0 | 0 | 3 | 16 | 25 | 30 | 175 |
| Days Min Temp ≤ 0 °F | 3 | 1 | 0 | 0 | 0 | 0 | 0 | 0 | 0 | 0 | 0 | 2 | 6 |
| Heating Degree Days | 1205 | 944 | 823 | 574 | 336 | 129 | 23 | 39 | 210 | 524 | 870 | 1173 | 6850 |
| Cooling Degree Days | 0 | 0 | 0 | 1 | 14 | 77 | 185 | 166 | 26 | 1 | 0 | 0 | 470 |
| Total Precipitation (") | 0.90 | 0.65 | 0.91 | 0.86 | 1.08 | 0.91 | 0.41 | 0.49 | 0.62 | 0.68 | 0.98 | 0.83 | 9.32 |
| Days ≥ 0.1" Precip | 3 | 2 | 3 | 3 | 3 | 3 | 1 | 1 | 2 | 2 | 3 | 3 | 29 |
| Total Snowfall (") | 4.8 | *2.6* | 2.0 | 0.8 | 0.5 | 0.0 | 0.0 | 0.0 | 0.0 | 0.1 | 2.5 | 5.4 | 18.7 |
| Days ≥ 1" Snow Depth | na | *2* | *1* | 0 | 0 | 0 | 0 | 0 | 0 | 0 | *2* | *7* | na |

## PAYETTE *Payette County* ELEVATION 2162 ft LAT/LONG 44° 4 ' N / 116° 56 ' W

| | JAN | FEB | MAR | APR | MAY | JUN | JUL | AUG | SEP | OCT | NOV | DEC | YEAR |
|---|---|---|---|---|---|---|---|---|---|---|---|---|---|
| Maximum Temp °F | 36.5 | 46.1 | 58.2 | 67.0 | 75.9 | 83.9 | 92.2 | 91.2 | 81.3 | 68.0 | 50.5 | 38.7 | 65.8 |
| Minimum Temp °F | 19.7 | 24.7 | 30.8 | 36.3 | 44.1 | 51.4 | 56.3 | 54.6 | 45.2 | 35.2 | 28.0 | 20.9 | 37.3 |
| Mean Temp °F | 28.1 | 35.4 | 44.5 | 51.7 | 60.1 | 67.7 | 74.2 | 72.9 | 63.3 | 51.6 | 39.3 | 29.8 | 51.5 |
| Days Max Temp ≥ 90 °F | 0 | 0 | 0 | 0 | 2 | 9 | 22 | 20 | 6 | 0 | 0 | 0 | 59 |
| Days Max Temp ≤ 32 °F | 9 | 3 | 0 | 0 | 0 | 0 | 0 | 0 | 0 | 0 | 1 | 6 | 19 |
| Days Min Temp ≤ 32 °F | 27 | 23 | 18 | 10 | 2 | 0 | 0 | 0 | 2 | 11 | 21 | 28 | 142 |
| Days Min Temp ≤ 0 °F | 3 | 1 | 0 | 0 | 0 | 0 | 0 | 0 | 0 | 0 | 0 | 1 | 5 |
| Heating Degree Days | 1137 | 829 | 628 | 395 | 183 | 51 | 6 | 11 | 111 | 410 | 765 | 1084 | 5610 |
| Cooling Degree Days | 0 | 0 | 0 | 4 | 45 | 145 | 292 | 276 | 76 | 3 | 0 | 0 | 841 |
| Total Precipitation (") | 1.48 | 1.08 | 1.05 | 0.82 | 0.76 | 0.85 | 0.23 | 0.45 | 0.47 | 0.67 | 1.55 | 1.59 | 11.00 |
| Days ≥ 0.1" Precip | 5 | 4 | 4 | 3 | 3 | 3 | 1 | 1 | 1 | 2 | 5 | 5 | 37 |
| Total Snowfall (") | na | na | *0.2* | 0.0 | 0.0 | 0.0 | 0.0 | 0.0 | 0.0 | 0.0 | *1.4* | na | na |
| Days ≥ 1" Snow Depth | *14* | na | *0* | 0 | 0 | 0 | 0 | 0 | 0 | 0 | *0* | na | na |

## PICABO *Blaine County* ELEVATION 4882 ft LAT/LONG 43° 18 ' N / 114° 4 ' W

| | JAN | FEB | MAR | APR | MAY | JUN | JUL | AUG | SEP | OCT | NOV | DEC | YEAR |
|---|---|---|---|---|---|---|---|---|---|---|---|---|---|
| Maximum Temp °F | 30.3 | 36.0 | 44.8 | 56.5 | 67.0 | 76.4 | 86.1 | 85.1 | 73.8 | 61.3 | 42.8 | 31.9 | 57.7 |
| Minimum Temp °F | 6.8 | 10.5 | 19.6 | 27.2 | 34.7 | 41.5 | 46.8 | 45.6 | 37.1 | 28.1 | 18.9 | 8.8 | 27.1 |
| Mean Temp °F | 18.6 | 23.3 | 32.2 | 41.9 | 50.9 | 59.0 | 66.5 | 65.3 | 55.5 | 44.7 | 30.9 | 20.4 | 42.4 |
| Days Max Temp ≥ 90 °F | 0 | 0 | 0 | 0 | 0 | 3 | 11 | 10 | 1 | 0 | 0 | 0 | 25 |
| Days Max Temp ≤ 32 °F | 17 | 9 | 2 | 0 | 0 | 0 | 0 | 0 | 0 | 0 | 5 | 15 | 48 |
| Days Min Temp ≤ 32 °F | 31 | 28 | 30 | 24 | 12 | 3 | 0 | 0 | 7 | 22 | 28 | 31 | 216 |
| Days Min Temp ≤ 0 °F | 10 | 6 | 1 | 0 | 0 | 0 | 0 | 0 | 0 | 0 | 1 | 7 | 25 |
| Heating Degree Days | 1432 | 1172 | 1010 | 687 | 432 | 200 | 49 | 66 | 286 | 622 | 1018 | 1376 | 8350 |
| Cooling Degree Days | 0 | 0 | 0 | 0 | 3 | 29 | 92 | 87 | 5 | 0 | 0 | 0 | 216 |
| Total Precipitation (") | 1.73 | 1.24 | 1.18 | 0.89 | 1.13 | 1.08 | 0.43 | 0.53 | 0.73 | 0.86 | 1.56 | 1.60 | 12.96 |
| Days ≥ 0.1" Precip | 5 | 4 | 4 | 4 | 4 | 3 | 1 | 2 | 2 | 3 | 5 | 5 | 42 |
| Total Snowfall (") | 11.2 | 9.2 | 3.3 | 1.4 | 0.6 | 0.0 | 0.0 | 0.0 | 0.0 | 0.4 | 6.3 | 11.6 | 44.0 |
| Days ≥ 1" Snow Depth | 29 | 24 | 13 | 2 | 0 | 0 | 0 | 0 | 0 | 0 | 8 | 21 | 97 |

## PIERCE *Clearwater County* ELEVATION 3192 ft LAT/LONG 46° 30 ' N / 115° 48 ' W

| | JAN | FEB | MAR | APR | MAY | JUN | JUL | AUG | SEP | OCT | NOV | DEC | YEAR |
|---|---|---|---|---|---|---|---|---|---|---|---|---|---|
| Maximum Temp °F | na | *37.3* | *45.6* | *52.7* | *63.6* | 71.5 | 80.9 | 81.0 | 70.2 | *56.9* | *40.0* | na | na |
| Minimum Temp °F | na | *18.3* | *22.9* | *28.5* | *34.4* | 41.0 | 43.3 | 41.4 | 34.0 | 27.6 | *24.4* | na | na |
| Mean Temp °F | na | *28.0* | *34.4* | *40.6* | *49.0* | 56.2 | 62.1 | 61.2 | 52.2 | *42.3* | *32.3* | na | na |
| Days Max Temp ≥ 90 °F | 0 | 0 | 0 | 0 | 0 | 1 | 5 | 5 | 1 | 0 | 0 | 0 | 12 |
| Days Max Temp ≤ 32 °F | *11* | 4 | *1* | 0 | 0 | 0 | 0 | 0 | 0 | 0 | *3* | *13* | 32 |
| Days Min Temp ≤ 32 °F | *29* | 27 | 29 | *25* | 12 | 2 | 1 | 2 | 13 | 24 | 25 | *28* | 217 |
| Days Min Temp ≤ 0 °F | *4* | 2 | 0 | 0 | 0 | 0 | 0 | 0 | 0 | 0 | 0 | 2 | 8 |
| Heating Degree Days | na | *1037* | *943* | *726* | *492* | 269 | 121 | 138 | 381 | *696* | *977* | na | na |
| Cooling Degree Days | na | *0* | 0 | 0 | *3* | 13 | 34 | 32 | 2 | *0* | *0* | na | na |
| Total Precipitation (") | 5.23 | 4.06 | 3.78 | 3.33 | 3.68 | 2.95 | 1.55 | 1.48 | 2.14 | 2.94 | 4.74 | 4.96 | 40.84 |
| Days ≥ 0.1" Precip | *11* | *10* | *10* | *9* | *8* | 7 | 4 | 4 | 5 | 7 | *11* | *11* | 97 |
| Total Snowfall (") | *31.6* | *21.0* | *13.2* | *5.6* | 0.6 | 0.0 | 0.0 | 0.0 | 0.0 | 0.7 | na | na | na |
| Days ≥ 1" Snow Depth | 28 | 27 | 27 | 16 | 1 | 0 | 0 | 0 | 0 | 1 | *13* | *28* | 141 |

## POCATELLO MUNICIPAL *Power County* ELEVATION 4468 ft LAT/LONG 42° 55 ' N / 112° 32 ' W

| | JAN | FEB | MAR | APR | MAY | JUN | JUL | AUG | SEP | OCT | NOV | DEC | YEAR |
|---|---|---|---|---|---|---|---|---|---|---|---|---|---|
| Maximum Temp °F | 32.3 | 38.8 | 47.9 | 57.9 | 68.0 | 78.0 | 87.6 | 86.5 | 75.4 | 62.0 | 44.3 | 33.6 | 59.4 |
| Minimum Temp °F | 15.1 | 19.8 | 26.5 | 32.7 | 39.9 | 46.9 | 52.4 | 50.8 | 42.1 | 33.0 | 24.6 | 15.6 | 33.3 |
| Mean Temp °F | 23.7 | 29.3 | 37.2 | 45.3 | 54.0 | 62.5 | 70.0 | 68.7 | 58.8 | 47.5 | 34.5 | 24.6 | 46.3 |
| Days Max Temp ≥ 90 °F | 0 | 0 | 0 | 0 | 0 | 4 | 14 | 13 | 2 | 0 | 0 | 0 | 33 |
| Days Max Temp ≤ 32 °F | 14 | 7 | 2 | 0 | 0 | 0 | 0 | 0 | 0 | 0 | 4 | 13 | 40 |
| Days Min Temp ≤ 32 °F | 28 | 25 | 24 | 15 | 4 | 0 | 0 | 0 | 3 | 15 | 23 | 29 | 166 |
| Days Min Temp ≤ 0 °F | 5 | 2 | 0 | 0 | 0 | 0 | 0 | 0 | 0 | 0 | 1 | 4 | 12 |
| Heating Degree Days | 1274 | 1001 | 855 | 584 | 339 | 125 | 18 | 29 | 201 | 536 | 909 | 1245 | 7116 |
| Cooling Degree Days | 0 | 0 | 0 | 0 | 6 | 58 | 163 | 158 | 21 | 0 | 0 | 0 | 406 |
| Total Precipitation (") | 1.04 | 0.94 | 1.27 | 1.16 | 1.38 | 1.00 | 0.68 | 0.72 | 0.78 | 0.94 | 1.17 | 1.06 | 12.14 |
| Days ≥ 0.1" Precip | 4 | 3 | 4 | 4 | 4 | 3 | 2 | 2 | 2 | 3 | 4 | 4 | 39 |
| Total Snowfall (") | 8.9 | 7.1 | 6.0 | 4.3 | 0.8 | 0.0 | 0.0 | 0.0 | 0.1 | 1.9 | 6.2 | 10.2 | 45.5 |
| Days ≥ 1" Snow Depth | 17 | 11 | 5 | 1 | 0 | 0 | 0 | 0 | 0 | 1 | 5 | 14 | 54 |

## PORTHILL *Boundary County* ELEVATION 1801 ft LAT/LONG 49° 0 ' N / 116° 30 ' W

| | JAN | FEB | MAR | APR | MAY | JUN | JUL | AUG | SEP | OCT | NOV | DEC | YEAR |
|---|---|---|---|---|---|---|---|---|---|---|---|---|---|
| Maximum Temp °F | 32.8 | 38.5 | 47.7 | 58.5 | 68.1 | 74.5 | 81.4 | 81.6 | 70.7 | 56.2 | 41.2 | 33.3 | 57.0 |
| Minimum Temp °F | 18.0 | 21.6 | 26.8 | 33.8 | 40.9 | 47.5 | 50.7 | 49.0 | 40.7 | 32.4 | 26.6 | 19.9 | 34.0 |
| Mean Temp °F | 25.4 | 30.1 | 37.3 | 46.2 | 54.6 | 61.0 | 66.1 | 65.3 | 55.7 | 44.3 | 33.9 | 26.6 | 45.5 |
| Days Max Temp ≥ 90 °F | 0 | 0 | 0 | 0 | 0 | 2 | 6 | 6 | 0 | 0 | 0 | 0 | 14 |
| Days Max Temp ≤ 32 °F | 13 | 5 | 1 | 0 | 0 | 0 | 0 | 0 | 0 | 0 | 4 | 13 | 36 |
| Days Min Temp ≤ 32 °F | 29 | 26 | 25 | 13 | 2 | 0 | 0 | 0 | 3 | 17 | 23 | 29 | 167 |
| Days Min Temp ≤ 0 °F | 3 | 1 | 0 | 0 | 0 | 0 | 0 | 0 | 0 | 0 | 0 | 2 | 6 |
| Heating Degree Days | 1221 | 980 | 853 | 557 | 322 | 149 | 58 | 67 | 277 | 635 | 926 | 1185 | 7230 |
| Cooling Degree Days | 0 | 0 | 0 | 0 | 10 | 40 | 96 | 82 | 4 | 0 | 0 | 0 | 232 |
| Total Precipitation (") | 2.27 | 1.60 | 1.40 | 1.44 | 1.64 | 1.87 | 1.14 | 1.25 | 1.37 | 1.33 | 2.58 | 2.45 | 20.34 |
| Days ≥ 0.1" Precip | 8 | 5 | 5 | 5 | 6 | 6 | 4 | 4 | 5 | 5 | 8 | 8 | 69 |
| Total Snowfall (") | na | 6.4 | 2.1 | 0.1 | 0.0 | 0.0 | 0.0 | 0.0 | 0.0 | 0.1 | *5.6* | na | na |
| Days ≥ 1" Snow Depth | 21 | 15 | 5 | 0 | 0 | 0 | 0 | 0 | 0 | 0 | 5 | 17 | 63 |

## POTLATCH 3 NNE *Latah County* ELEVATION 2503 ft LAT/LONG 46° 58 ' N / 116° 53 ' W

| | JAN | FEB | MAR | APR | MAY | JUN | JUL | AUG | SEP | OCT | NOV | DEC | YEAR |
|---|---|---|---|---|---|---|---|---|---|---|---|---|---|
| Maximum Temp °F | 35.5 | 41.7 | 48.3 | 56.2 | 65.2 | 72.2 | 80.6 | 81.6 | 72.1 | 59.6 | 43.1 | 35.9 | 57.7 |
| Minimum Temp °F | 21.4 | 25.2 | 28.5 | 32.6 | 37.4 | 42.5 | 44.2 | 43.2 | 36.9 | 31.0 | 28.1 | 22.3 | 32.8 |
| Mean Temp °F | 28.5 | 33.5 | 38.4 | 44.5 | 51.3 | 57.4 | 62.4 | 62.4 | 54.5 | 45.3 | 35.6 | 29.1 | 45.2 |
| Days Max Temp ≥ 90 °F | 0 | 0 | 0 | 0 | 0 | 1 | 5 | 6 | 1 | 0 | 0 | 0 | 13 |
| Days Max Temp ≤ 32 °F | 10 | 3 | 1 | 0 | 0 | 0 | 0 | 0 | 0 | 0 | 2 | 10 | 26 |
| Days Min Temp ≤ 32 °F | 26 | 23 | 23 | 16 | 8 | 2 | 1 | 1 | 8 | 18 | 22 | 26 | 174 |
| Days Min Temp ≤ 0 °F | 3 | 1 | 0 | 0 | 0 | 0 | 0 | 0 | 0 | 0 | 0 | 2 | 6 |
| Heating Degree Days | 1126 | 884 | 817 | 610 | 419 | 233 | 111 | 116 | 311 | 603 | 874 | 1106 | 7210 |
| Cooling Degree Days | 0 | 0 | 0 | 0 | 3 | 9 | 35 | 45 | 4 | 1 | 0 | 0 | 97 |
| Total Precipitation (") | 3.02 | 2.37 | 2.34 | 2.19 | 2.31 | 1.86 | 1.15 | 1.10 | 1.22 | 1.64 | 2.99 | 3.08 | 25.27 |
| Days ≥ 0.1" Precip | 9 | 7 | 8 | 6 | 6 | 5 | 3 | 3 | 4 | 5 | 9 | 8 | 73 |
| Total Snowfall (") | 15.4 | 7.3 | 4.5 | 1.2 | 0.2 | 0.0 | 0.0 | 0.0 | 0.0 | 0.3 | 4.8 | *11.3* | 45.0 |
| Days ≥ 1" Snow Depth | na | na | na | 0 | 0 | 0 | 0 | 0 | 0 | 0 | *1* | na | na |

## POWELL *Idaho County* ELEVATION 3632 ft LAT/LONG 46° 31 ' N / 114° 42 ' W

| | JAN | FEB | MAR | APR | MAY | JUN | JUL | AUG | SEP | OCT | NOV | DEC | YEAR |
|---|---|---|---|---|---|---|---|---|---|---|---|---|---|
| Maximum Temp °F | 32.4 | 38.6 | 46.0 | 54.1 | 64.6 | 73.2 | 82.6 | 82.7 | 71.5 | 58.2 | 39.8 | 31.4 | 56.3 |
| Minimum Temp °F | 15.8 | 17.8 | 22.4 | 28.0 | 33.9 | 41.2 | 44.1 | 42.6 | 35.5 | 28.8 | 23.1 | 16.0 | 29.1 |
| Mean Temp °F | 24.1 | 28.2 | 34.2 | 41.1 | 49.3 | 57.3 | 63.4 | 62.7 | 53.5 | 43.5 | 31.5 | 23.7 | 42.7 |
| Days Max Temp ≥ 90 °F | 0 | 0 | 0 | 0 | 0 | 2 | 8 | 8 | 1 | 0 | 0 | 0 | 19 |
| Days Max Temp ≤ 32 °F | 13 | 5 | 1 | 0 | 0 | 0 | 0 | 0 | 0 | 0 | 4 | 14 | 37 |
| Days Min Temp ≤ 32 °F | 31 | 28 | 31 | 26 | 14 | 2 | 0 | 1 | 9 | 24 | 28 | 30 | 224 |
| Days Min Temp ≤ 0 °F | 4 | 2 | 1 | 0 | 0 | 0 | 0 | 0 | 0 | 0 | 1 | 3 | 11 |
| Heating Degree Days | 1261 | 1032 | 947 | 710 | 483 | 245 | 98 | 110 | 342 | 659 | 998 | 1273 | 8158 |
| Cooling Degree Days | 0 | 0 | 0 | 0 | 2 | 19 | 50 | 44 | 5 | 0 | 0 | 0 | 120 |
| Total Precipitation (") | 5.66 | 3.77 | 3.19 | 2.65 | 2.84 | 2.92 | 1.42 | 1.56 | 2.23 | 2.71 | 4.34 | 4.91 | 38.20 |
| Days ≥ 0.1" Precip | 14 | 11 | 10 | 8 | 8 | 8 | 5 | 4 | 6 | 7 | 12 | 12 | 105 |
| Total Snowfall (") | *51.7* | *31.2* | 18.3 | 8.3 | 1.1 | 0.0 | 0.0 | 0.0 | 0.0 | 2.6 | *21.0* | *41.7* | 175.9 |
| Days ≥ 1" Snow Depth | *31* | *27* | 25 | 14 | 1 | 0 | 0 | 0 | 0 | 2 | *15* | *29* | 144 |

## PRIEST RIVER EXP STN *Bonner County* ELEVATION 2382 ft LAT/LONG 48° 21 ' N / 116° 50 ' W

| | JAN | FEB | MAR | APR | MAY | JUN | JUL | AUG | SEP | OCT | NOV | DEC | YEAR |
|---|---|---|---|---|---|---|---|---|---|---|---|---|---|
| Maximum Temp °F | 29.9 | 36.4 | 45.6 | 56.2 | 66.8 | 73.8 | 81.4 | 81.5 | 71.0 | 55.6 | 37.4 | 30.2 | 55.5 |
| Minimum Temp °F | 19.9 | 22.4 | 25.9 | 31.4 | 38.7 | 44.9 | 47.5 | 46.6 | 39.4 | 32.5 | 27.4 | 21.7 | 33.2 |
| Mean Temp °F | 24.9 | 29.4 | 35.8 | 43.8 | 52.8 | 59.4 | 64.5 | 64.1 | 55.2 | 44.0 | 32.4 | 26.0 | 44.4 |
| Days Max Temp ≥ 90 °F | 0 | 0 | 0 | 0 | 0 | 1 | 6 | 6 | 1 | 0 | 0 | 0 | 14 |
| Days Max Temp ≤ 32 °F | 17 | 6 | 1 | 0 | 0 | 0 | 0 | 0 | 0 | 0 | 7 | 18 | 49 |
| Days Min Temp ≤ 32 °F | 29 | 26 | 26 | 18 | 5 | 0 | 0 | 0 | 5 | 16 | 23 | 29 | 177 |
| Days Min Temp ≤ 0 °F | 3 | 1 | 0 | 0 | 0 | 0 | 0 | 0 | 0 | 0 | 0 | 1 | 5 |
| Heating Degree Days | 1236 | 1000 | 899 | 629 | 376 | 185 | 76 | 82 | 291 | 643 | 971 | 1204 | 7592 |
| Cooling Degree Days | 0 | 0 | 0 | 0 | 5 | 20 | 56 | 54 | 3 | 0 | 0 | 0 | 138 |
| Total Precipitation (") | 4.01 | 2.94 | 2.57 | 2.16 | 2.25 | 2.16 | 1.31 | 1.37 | 1.42 | 1.88 | 4.09 | 4.35 | 30.51 |
| Days ≥ 0.1" Precip | 10 | 8 | 7 | 6 | 7 | 6 | 3 | 4 | 4 | 5 | 10 | 11 | 81 |
| Total Snowfall (") | 25.5 | 13.8 | 4.7 | 0.2 | 0.0 | 0.0 | 0.0 | 0.0 | 0.0 | 0.3 | 10.1 | 23.3 | 77.9 |
| Days ≥ 1" Snow Depth | 30 | 27 | 21 | 2 | 0 | 0 | 0 | 0 | 0 | 0 | 10 | 27 | 117 |

## REYNOLDS *Owyhee County* ELEVATION 3914 ft LAT/LONG 43° 12 ' N / 116° 45 ' W

| | JAN | FEB | MAR | APR | MAY | JUN | JUL | AUG | SEP | OCT | NOV | DEC | YEAR |
|---|---|---|---|---|---|---|---|---|---|---|---|---|---|
| Maximum Temp °F | 38.2 | 43.4 | 50.5 | 58.2 | 67.7 | 76.7 | 85.7 | 85.1 | 74.6 | 63.2 | 47.8 | 39.3 | 60.9 |
| Minimum Temp °F | 19.4 | 23.0 | 27.5 | 32.5 | 39.5 | 46.4 | 52.4 | 51.4 | 41.9 | 32.8 | 25.5 | 18.8 | 34.3 |
| Mean Temp °F | 28.8 | 33.2 | 39.0 | 45.4 | 53.6 | 61.5 | 69.1 | 68.3 | 58.3 | 48.0 | 36.7 | 29.1 | 47.6 |
| Days Max Temp ≥ 90 °F | 0 | 0 | 0 | 0 | 0 | 3 | 11 | 11 | 1 | 0 | 0 | 0 | 26 |
| Days Max Temp ≤ 32 °F | 8 | 2 | 1 | 0 | 0 | 0 | 0 | 0 | 0 | 0 | 1 | 6 | 18 |
| Days Min Temp ≤ 32 °F | 28 | 25 | 24 | 15 | 5 | 1 | 0 | 0 | 3 | 15 | 24 | 28 | 168 |
| Days Min Temp ≤ 0 °F | 2 | 1 | 0 | 0 | 0 | 0 | 0 | 0 | 0 | 0 | 0 | 2 | 5 |
| Heating Degree Days | 1114 | 892 | 798 | 583 | 358 | 157 | 37 | 47 | 221 | 520 | 843 | 1106 | 6676 |
| Cooling Degree Days | 0 | 0 | 0 | 1 | 14 | 66 | 162 | 163 | 30 | 1 | 0 | 0 | 437 |
| Total Precipitation (") | 1.12 | 0.76 | 0.98 | 1.03 | 1.03 | 1.19 | 0.33 | 0.60 | 0.58 | 0.76 | 1.16 | 1.05 | 10.59 |
| Days ≥ 0.1" Precip | 3 | 3 | 3 | 3 | 3 | 4 | 1 | 2 | 2 | 2 | 4 | 3 | 33 |
| Total Snowfall (") | na | na | *1.2* | 0.8 | 0.0 | 0.0 | 0.0 | 0.0 | 0.0 | 0.4 | *1.1* | na | na |
| Days ≥ 1" Snow Depth | na | na | 1 | 0 | 0 | 0 | 0 | 0 | 0 | 0 | 1 | *5* | na |

## RICHFIELD *Lincoln County* ELEVATION 4314 ft LAT/LONG 43° 3 ' N / 114° 9 ' W

| | JAN | FEB | MAR | APR | MAY | JUN | JUL | AUG | SEP | OCT | NOV | DEC | YEAR |
|---|---|---|---|---|---|---|---|---|---|---|---|---|---|
| Maximum Temp °F | 31.1 | 37.8 | 48.3 | 59.6 | 69.0 | 77.8 | 86.5 | 85.6 | 75.6 | 62.7 | 43.8 | 32.4 | 59.2 |
| Minimum Temp °F | 13.7 | 17.5 | 24.4 | 30.3 | 38.3 | 45.7 | 51.8 | 50.6 | 41.5 | 32.0 | 23.5 | 14.6 | 32.0 |
| Mean Temp °F | 22.4 | 27.6 | 36.4 | 45.0 | 53.6 | 61.8 | 69.1 | 68.1 | 58.6 | 47.4 | 33.7 | 23.5 | 45.6 |
| Days Max Temp ≥ 90 °F | 0 | 0 | 0 | 0 | 0 | 4 | 11 | 10 | 2 | 0 | 0 | 0 | 27 |
| Days Max Temp ≤ 32 °F | 16 | 8 | 1 | 0 | 0 | 0 | 0 | 0 | 0 | 0 | 4 | 14 | 43 |
| Days Min Temp ≤ 32 °F | 30 | 27 | 27 | 19 | 7 | 1 | 0 | 0 | 3 | 16 | 26 | 30 | 186 |
| Days Min Temp ≤ 0 °F | 5 | 2 | 0 | 0 | 0 | 0 | 0 | 0 | 0 | 0 | 1 | 4 | 12 |
| Heating Degree Days | 1314 | 1048 | 881 | 593 | 353 | 142 | 26 | 40 | 210 | 540 | 933 | 1280 | 7360 |
| Cooling Degree Days | 0 | 0 | 0 | 1 | 12 | 60 | 162 | 167 | 30 | 1 | 0 | 0 | 433 |
| Total Precipitation (") | 1.56 | 1.17 | 1.07 | 0.75 | 0.90 | 0.70 | 0.38 | 0.50 | 0.56 | 0.67 | 1.43 | 1.44 | 11.13 |
| Days ≥ 0.1" Precip | 5 | 4 | 4 | 3 | 3 | 2 | 1 | 1 | 2 | 2 | 5 | 5 | 37 |
| Total Snowfall (") | na | *5.4* | *2.1* | *0.0* | 0.5 | 0.0 | 0.0 | 0.0 | 0.0 | 0.2 | 5.8 | 9.5 | na |
| Days ≥ 1" Snow Depth | 25 | 19 | 8 | 0 | 0 | 0 | 0 | 0 | 0 | 0 | 4 | 17 | 73 |

## RIGGINS *Idaho County* ELEVATION 1801 ft LAT/LONG 45° 25 ' N / 116° 19 ' W

| | JAN | FEB | MAR | APR | MAY | JUN | JUL | AUG | SEP | OCT | NOV | DEC | YEAR |
|---|---|---|---|---|---|---|---|---|---|---|---|---|---|
| Maximum Temp °F | 41.9 | 49.8 | 57.8 | 65.9 | 74.7 | 82.5 | 92.0 | 92.4 | 80.4 | 67.4 | 50.4 | 41.7 | 66.4 |
| Minimum Temp °F | 28.3 | 31.3 | 35.1 | 39.8 | 46.3 | 53.0 | 58.0 | 58.4 | 49.1 | 41.1 | 34.3 | 28.5 | 41.9 |
| Mean Temp °F | 35.1 | 40.6 | 46.5 | 52.9 | 60.5 | 67.8 | 75.0 | 75.4 | 64.8 | 54.3 | 42.4 | 35.1 | 54.2 |
| Days Max Temp ≥ 90 °F | 0 | 0 | 0 | 0 | 3 | 8 | 20 | 21 | 6 | 0 | 0 | 0 | 58 |
| Days Max Temp ≤ 32 °F | 3 | 1 | 0 | 0 | 0 | 0 | 0 | 0 | 0 | 0 | 1 | 3 | 8 |
| Days Min Temp ≤ 32 °F | 20 | 15 | 10 | 3 | 0 | 0 | 0 | 0 | 0 | 3 | 12 | 21 | 84 |
| Days Min Temp ≤ 0 °F | 0 | 0 | 0 | 0 | 0 | 0 | 0 | 0 | 0 | 0 | 0 | 1 | 1 |
| Heating Degree Days | 920 | 683 | 567 | 362 | 176 | 55 | 7 | 9 | 97 | 333 | 673 | 920 | 4802 |
| Cooling Degree Days | 0 | 0 | 0 | 8 | *49* | 142 | *295* | 327 | *96* | *10* | 0 | 0 | 927 |
| Total Precipitation (") | 1.35 | 1.04 | 1.62 | 1.72 | 1.96 | 1.81 | 0.88 | 0.92 | 1.04 | 1.11 | 1.48 | 1.29 | 16.22 |
| Days ≥ 0.1" Precip | *4* | 4 | 5 | 5 | 5 | 6 | 3 | 3 | 3 | 3 | *5* | *4* | 50 |
| Total Snowfall (") | *2.8* | na | *0.2* | *0.0* | 0.0 | 0.0 | 0.0 | 0.0 | 0.0 | *0.0* | *0.5* | na | na |
| Days ≥ 1" Snow Depth | na | 1 | *0* | 0 | 0 | 0 | 0 | 0 | 0 | 0 | *0* | na | na |

## SAINT ANTHONY 1 WNW *Fremont County* ELEVATION 4974 ft LAT/LONG 43° 58 ' N / 111° 40 ' W

| | JAN | FEB | MAR | APR | MAY | JUN | JUL | AUG | SEP | OCT | NOV | DEC | YEAR |
|---|---|---|---|---|---|---|---|---|---|---|---|---|---|
| Maximum Temp °F | 28.8 | 33.7 | 43.4 | 55.4 | 66.2 | 74.1 | 82.2 | 82.0 | 72.1 | 59.7 | 41.1 | 30.3 | 55.8 |
| Minimum Temp °F | 8.3 | 10.7 | 19.2 | 27.4 | 35.1 | 42.2 | 46.7 | 44.8 | 36.5 | 28.1 | 19.0 | 8.5 | 27.2 |
| Mean Temp °F | 18.8 | 22.3 | 31.3 | 41.4 | 50.7 | 58.1 | 64.5 | 63.4 | 54.4 | 44.0 | 30.1 | 19.5 | 41.5 |
| Days Max Temp ≥ 90 °F | 0 | 0 | 0 | 0 | 0 | 1 | 3 | 3 | 0 | 0 | 0 | 0 | 7 |
| Days Max Temp ≤ 32 °F | 19 | 10 | 2 | 0 | 0 | 0 | 0 | 0 | 0 | 0 | 6 | 18 | 55 |
| Days Min Temp ≤ 32 °F | 30 | 28 | 30 | 23 | 11 | 1 | 0 | 1 | 9 | 22 | 28 | 30 | 213 |
| Days Min Temp ≤ 0 °F | 9 | 6 | 2 | 0 | 0 | 0 | 0 | 0 | 0 | 0 | 2 | 8 | 27 |
| Heating Degree Days | 1427 | 1201 | 1039 | 702 | 440 | 214 | 66 | 86 | 315 | 642 | 1042 | 1423 | 8597 |
| Cooling Degree Days | 0 | 0 | 0 | 0 | 1 | 15 | 47 | 39 | 3 | 0 | 0 | 0 | 105 |
| Total Precipitation (") | 1.18 | 0.86 | 0.95 | 1.15 | 1.63 | 1.57 | 0.93 | 0.81 | 0.98 | 1.00 | 1.35 | 1.34 | 13.75 |
| Days ≥ 0.1" Precip | 4 | 3 | *3* | 4 | 5 | 4 | 3 | 2 | 3 | 3 | 5 | 5 | 44 |
| Total Snowfall (") | na | na | na | *1.0* | 0.1 | 0.0 | 0.0 | 0.0 | 0.0 | *0.6* | na | na | na |
| Days ≥ 1" Snow Depth | na | na | na | 1 | 0 | 0 | 0 | 0 | 0 | 0 | *2* | na | na |

## SAINT MARIES 1 W *Benewah County* ELEVATION 2142 ft LAT/LONG 47° 19 ' N / 116° 34 ' W

| | JAN | FEB | MAR | APR | MAY | JUN | JUL | AUG | SEP | OCT | NOV | DEC | YEAR |
|---|---|---|---|---|---|---|---|---|---|---|---|---|---|
| Maximum Temp °F | 34.5 | 41.7 | 50.2 | 58.9 | 68.0 | 75.6 | 84.3 | 84.7 | 74.2 | 59.4 | 41.8 | 34.2 | 59.0 |
| Minimum Temp °F | 23.0 | 26.2 | 29.7 | 34.4 | 40.7 | 46.9 | 50.1 | 49.4 | 42.1 | 34.4 | 29.6 | 23.8 | 35.9 |
| Mean Temp °F | 28.8 | 33.9 | 40.0 | 46.7 | 54.4 | 61.3 | 67.2 | 67.1 | 58.2 | 46.9 | 35.7 | 29.1 | 47.4 |
| Days Max Temp ≥ 90 °F | 0 | 0 | 0 | 0 | 1 | 3 | 11 | 11 | 2 | 0 | 0 | 0 | 28 |
| Days Max Temp ≤ 32 °F | 10 | 3 | 0 | 0 | 0 | 0 | 0 | 0 | 0 | 0 | 3 | 10 | 26 |
| Days Min Temp ≤ 32 °F | 26 | 22 | 21 | 13 | 3 | 0 | 0 | 0 | 3 | 12 | 19 | 26 | 145 |
| Days Min Temp ≤ 0 °F | 2 | 1 | 0 | 0 | 0 | 0 | 0 | 0 | 0 | 0 | 0 | 1 | 4 |
| Heating Degree Days | 1115 | 870 | 769 | 545 | 330 | 149 | 47 | 49 | 218 | 554 | 872 | 1108 | 6626 |
| Cooling Degree Days | 0 | 0 | 0 | 2 | 11 | 43 | 116 | 125 | 20 | 0 | 0 | 0 | 317 |
| Total Precipitation (") | 4.23 | 2.75 | 2.51 | 2.22 | 2.19 | 1.98 | 1.18 | 1.28 | 1.35 | 1.84 | 3.69 | 3.81 | 29.03 |
| Days ≥ 0.1" Precip | 8 | 7 | 7 | 6 | 5 | 5 | 3 | 3 | 3 | 5 | 8 | 9 | 69 |
| Total Snowfall (") | *17.4* | *7.7* | *2.5* | 0.1 | 0.0 | 0.0 | 0.0 | 0.0 | 0.0 | 0.3 | *6.2* | *14.2* | 48.4 |
| Days ≥ 1" Snow Depth | 19 | 10 | 4 | 0 | 0 | 0 | 0 | 0 | 0 | 0 | *5* | *13* | 51 |

## SANDPOINT EXP STN *Bonner County* ELEVATION 2103 ft LAT/LONG 48° 17 ' N / 116° 34 ' W

| | JAN | FEB | MAR | APR | MAY | JUN | JUL | AUG | SEP | OCT | NOV | DEC | YEAR |
|---|---|---|---|---|---|---|---|---|---|---|---|---|---|
| Maximum Temp °F | 32.7 | 38.5 | 47.1 | 57.0 | 66.5 | 73.2 | 80.6 | 80.8 | 70.1 | 56.6 | 40.7 | 33.4 | 56.4 |
| Minimum Temp °F | 21.3 | 24.0 | 28.1 | 34.1 | 40.3 | 46.3 | 48.8 | 47.8 | 41.1 | 33.2 | 28.4 | 22.4 | 34.7 |
| Mean Temp °F | 27.1 | 31.3 | 37.6 | 45.6 | 53.4 | 59.8 | 64.7 | 64.4 | 55.7 | 45.0 | 34.5 | 27.9 | 45.6 |
| Days Max Temp ≥ 90 °F | 0 | 0 | 0 | 0 | 0 | 1 | 4 | 5 | 0 | 0 | 0 | 0 | 10 |
| Days Max Temp ≤ 32 °F | 13 | 4 | 1 | 0 | 0 | 0 | 0 | 0 | 0 | 0 | 3 | 12 | 33 |
| Days Min Temp ≤ 32 °F | 27 | 24 | 23 | 13 | 4 | 0 | 0 | 0 | 3 | 15 | 21 | 27 | 157 |
| Days Min Temp ≤ 0 °F | 2 | 1 | 0 | 0 | 0 | 0 | 0 | 0 | 0 | 0 | 0 | 1 | 4 |
| Heating Degree Days | 1169 | 946 | 842 | 577 | 357 | 177 | 69 | 77 | 278 | 614 | 907 | 1143 | 7156 |
| Cooling Degree Days | 0 | 0 | 0 | 0 | 7 | 29 | 69 | 66 | 4 | 0 | 0 | 0 | 175 |
| Total Precipitation (") | 4.14 | 3.09 | 2.65 | 2.16 | 2.45 | 2.48 | 1.54 | 1.53 | 1.64 | 2.23 | 4.63 | 4.57 | 33.11 |
| Days ≥ 0.1" Precip | 10 | 8 | 8 | 6 | 7 | 7 | 4 | 4 | 4 | 6 | 10 | 11 | 85 |
| Total Snowfall (") | 23.7 | 12.6 | 3.2 | 0.3 | 0.0 | 0.0 | 0.0 | 0.0 | 0.0 | 0.1 | 6.7 | 19.6 | 66.2 |
| Days ≥ 1" Snow Depth | 25 | 19 | 9 | 0 | 0 | 0 | 0 | 0 | 0 | 0 | 5 | 19 | 77 |

## SHOSHONE 1 WNW *Lincoln County* ELEVATION 3973 ft LAT/LONG 42° 56 ' N / 114° 24 ' W

| | JAN | FEB | MAR | APR | MAY | JUN | JUL | AUG | SEP | OCT | NOV | DEC | YEAR |
|---|---|---|---|---|---|---|---|---|---|---|---|---|---|
| Maximum Temp °F | 33.7 | 40.6 | 51.1 | 62.3 | 73.3 | 83.3 | 92.0 | 90.5 | 78.7 | 64.7 | 45.5 | 34.7 | 62.5 |
| Minimum Temp °F | 16.6 | 20.6 | 26.8 | 33.2 | 41.5 | 49.2 | 56.1 | 55.0 | 45.1 | 35.1 | 25.9 | 17.4 | 35.2 |
| Mean Temp °F | 25.2 | 30.7 | 39.0 | 47.8 | 57.5 | 66.2 | 74.1 | 72.8 | 62.0 | 50.0 | 35.7 | 26.1 | 48.9 |
| Days Max Temp ≥ 90 °F | 0 | 0 | 0 | 0 | 2 | 8 | 22 | 20 | 4 | 0 | 0 | 0 | 56 |
| Days Max Temp ≤ 32 °F | 12 | 5 | 1 | 0 | 0 | 0 | 0 | 0 | 0 | 0 | 3 | 11 | 32 |
| Days Min Temp ≤ 32 °F | 29 | 26 | 24 | 14 | *4* | 0 | 0 | 0 | 2 | 11 | 23 | 30 | 163 |
| Days Min Temp ≤ 0 °F | 3 | 1 | 0 | 0 | 0 | 0 | 0 | 0 | 0 | 0 | 0 | 2 | 6 |
| Heating Degree Days | 1227 | 962 | 799 | 512 | 254 | 80 | 10 | 17 | 144 | 463 | 870 | 1201 | 6539 |
| Cooling Degree Days | 0 | 0 | 0 | 3 | 35 | 133 | 272 | 268 | 58 | 2 | 0 | 0 | 771 |
| Total Precipitation (") | 1.24 | 1.01 | 1.18 | 0.72 | 0.73 | 0.69 | 0.27 | 0.43 | 0.53 | 0.63 | 1.34 | 1.12 | 9.89 |
| Days ≥ 0.1" Precip | 4 | 3 | 4 | 2 | 2 | 2 | 1 | 1 | 2 | 2 | 4 | 4 | 31 |
| Total Snowfall (") | na | na | na | 0.1 | 0.1 | 0.0 | 0.0 | 0.0 | 0.2 | 0.0 | *3.4* | na | na |
| Days ≥ 1" Snow Depth | na | na | na | 0 | 0 | 0 | 0 | 0 | 0 | 0 | *1* | na | na |

## SHOUP *Lemhi County* ELEVATION 3402 ft LAT/LONG 45° 22 ' N / 114° 18 ' W

| | JAN | FEB | MAR | APR | MAY | JUN | JUL | AUG | SEP | OCT | NOV | DEC | YEAR |
|---|---|---|---|---|---|---|---|---|---|---|---|---|---|
| Maximum Temp °F | 30.9 | 39.4 | 51.5 | 61.8 | 71.6 | 79.9 | 88.8 | 87.6 | 76.5 | 60.6 | 41.9 | 30.4 | 60.1 |
| Minimum Temp °F | 15.2 | 19.7 | 27.5 | 33.2 | 39.5 | 45.8 | 50.8 | 49.6 | 41.9 | 32.8 | 25.2 | 16.2 | 33.1 |
| Mean Temp °F | 23.1 | 29.6 | 39.5 | 47.5 | 55.5 | 62.9 | 69.8 | 68.6 | 59.2 | 46.7 | 33.6 | 23.3 | 46.6 |
| Days Max Temp ≥ 90 °F | 0 | 0 | 0 | 0 | 1 | 5 | 17 | 14 | 3 | 0 | 0 | 0 | 40 |
| Days Max Temp ≤ 32 °F | 16 | 5 | 0 | 0 | 0 | 0 | 0 | 0 | 0 | 0 | 3 | 16 | 40 |
| Days Min Temp ≤ 32 °F | 30 | 27 | 25 | 15 | 3 | 0 | 0 | 0 | 2 | 15 | 26 | 30 | 173 |
| Days Min Temp ≤ 0 °F | 4 | 1 | 0 | 0 | 0 | 0 | 0 | 0 | 0 | 0 | 0 | 2 | 7 |
| Heating Degree Days | 1294 | 994 | 784 | 518 | 294 | 110 | 19 | 28 | 187 | 561 | 936 | 1286 | 7011 |
| Cooling Degree Days | 0 | 0 | 0 | 0 | 11 | 57 | 170 | 158 | 22 | 0 | 0 | 0 | 418 |
| Total Precipitation (") | *1.36* | 1.28 | 0.93 | 1.07 | 1.56 | 1.81 | 0.96 | 0.94 | 1.08 | 0.89 | 1.41 | *1.54* | 14.83 |
| Days ≥ 0.1" Precip | *5* | 4 | 4 | 4 | 5 | 5 | 3 | 3 | 3 | 3 | 5 | *5* | 49 |
| Total Snowfall (") | *11.4* | 5.1 | *1.2* | *0.1* | 0.0 | 0.0 | 0.0 | 0.0 | 0.0 | 0.0 | *3.6* | na | na |
| Days ≥ 1" Snow Depth | *28* | *19* | *5* | 0 | 0 | 0 | 0 | 0 | 0 | 0 | 3 | na | na |

## STANLEY *Custer County* ELEVATION 6243 ft LAT/LONG 44° 13 ' N / 114° 56 ' W

| | JAN | FEB | MAR | APR | MAY | JUN | JUL | AUG | SEP | OCT | NOV | DEC | YEAR |
|---|---|---|---|---|---|---|---|---|---|---|---|---|---|
| Maximum Temp °F | 27.1 | 34.0 | *42.7* | *50.3* | *60.4* | 68.7 | *77.8* | *77.8* | *68.2* | *56.8* | *38.2* | *26.0* | 52.3 |
| Minimum Temp °F | -1.6 | 1.2 | *10.0* | *19.9* | *28.2* | 33.9 | *35.9* | *34.1* | *27.2* | *20.8* | *11.4* | *-2.5* | 18.2 |
| Mean Temp °F | 12.7 | 17.6 | *26.4* | *35.1* | *44.3* | 51.3 | *56.8* | *56.0* | *47.8* | *38.9* | *24.9* | *11.8* | 35.3 |
| Days Max Temp ≥ 90 °F | 0 | 0 | 0 | 0 | 0 | 0 | *1* | *1* | *0* | *0* | 0 | 0 | 2 |
| Days Max Temp ≤ 32 °F | *22* | *11* | *3* | 0 | 0 | 0 | 0 | 0 | *0* | *0* | *8* | na | na |
| Days Min Temp ≤ 32 °F | 31 | 28 | 31 | 28 | na | *13* | *9* | na | *24* | *29* | *30* | *31* | na |
| Days Min Temp ≤ 0 °F | *17* | *13* | *7* | 1 | 0 | 0 | 0 | 0 | *0* | *0* | *5* | na | na |
| Heating Degree Days | 1615 | 1334 | *1190* | *889* | *634* | 405 | *250* | *276* | *511* | *804* | *1199* | *1644* | 10751 |
| Cooling Degree Days | 0 | 0 | 0 | 0 | 0 | 1 | 5 | 3 | 0 | 0 | 0 | 0 | 9 |
| Total Precipitation (") | *1.99* | *1.54* | *1.20* | 1.15 | *1.33* | *1.33* | 0.78 | 0.85 | 1.04 | *1.12* | 1.68 | *1.94* | 15.95 |
| Days ≥ 0.1" Precip | na | na | na | *4* | *4* | *4* | 2 | 3 | *3* | na | na | na | na |
| Total Snowfall (") | na | na | *11.7* | *4.2* | *1.1* | *0.1* | 0.0 | 0.0 | *0.6* | na | na | na | na |
| Days ≥ 1" Snow Depth | na | na | *26* | *15* | *1* | 0 | 0 | 0 | 0 | na | na | na | na |

## SWAN FALLS POWER HOU *Ada County* ELEVATION 2323 ft LAT/LONG 43° 15' N / 116° 23' W

| | JAN | FEB | MAR | APR | MAY | JUN | JUL | AUG | SEP | OCT | NOV | DEC | YEAR |
|---|---|---|---|---|---|---|---|---|---|---|---|---|---|
| Maximum Temp °F | 40.5 | 49.2 | 59.4 | 67.8 | 78.0 | 87.2 | 95.9 | 94.7 | 83.9 | 70.1 | 51.9 | 40.8 | 68.3 |
| Minimum Temp °F | 24.7 | 29.2 | 35.0 | 40.6 | 48.5 | 56.1 | 62.7 | 60.9 | 51.2 | 41.4 | 32.5 | 24.7 | 42.3 |
| Mean Temp °F | 32.6 | 39.3 | 47.2 | 54.2 | 63.3 | 71.7 | 79.3 | 77.8 | 67.6 | 55.8 | 42.2 | 32.7 | 55.3 |
| Days Max Temp ≥ 90 °F | 0 | 0 | 0 | 0 | 4 | 13 | 25 | 24 | 10 | 0 | 0 | 0 | 76 |
| Days Max Temp ≤ 32 °F | 6 | 2 | 0 | 0 | 0 | 0 | 0 | 0 | 0 | 0 | 1 | 5 | 14 |
| Days Min Temp ≤ 32 °F | 25 | 19 | 11 | 3 | 0 | 0 | 0 | 0 | 0 | 3 | 15 | 24 | 100 |
| Days Min Temp ≤ 0 °F | 1 | 0 | 0 | 0 | 0 | 0 | 0 | 0 | 0 | 0 | 0 | 1 | 2 |
| Heating Degree Days | 996 | 721 | 545 | 326 | 126 | 27 | 1 | 4 | 56 | 290 | 676 | 996 | 4764 |
| Cooling Degree Days | 0 | 0 | 0 | 13 | 85 | 237 | 427 | 410 | 140 | 14 | 0 | 0 | 1326 |
| Total Precipitation (") | 0.83 | 0.54 | 0.83 | 1.07 | 0.95 | 0.77 | 0.27 | 0.32 | 0.53 | 0.52 | 0.90 | 0.74 | 8.27 |
| Days ≥ 0.1" Precip | 3 | 2 | 3 | 3 | 3 | 2 | 1 | 1 | 2 | 2 | 3 | 3 | 28 |
| Total Snowfall (") | *1.9* | *0.6* | 0.1 | 0.0 | 0.0 | 0.0 | 0.0 | 0.0 | 0.0 | 0.0 | 0.4 | 1.3 | 4.3 |
| Days ≥ 1" Snow Depth | 2 | 0 | 0 | 0 | 0 | 0 | 0 | 0 | 0 | 0 | 1 | 2 | 5 |

## SWAN VALLEY 2 E *Bonneville County* ELEVATION 5243 ft LAT/LONG 43° 27' N / 111° 22' W

| | JAN | FEB | MAR | APR | MAY | JUN | JUL | AUG | SEP | OCT | NOV | DEC | YEAR |
|---|---|---|---|---|---|---|---|---|---|---|---|---|---|
| Maximum Temp °F | 30.0 | 35.8 | 44.2 | 54.7 | 65.3 | 74.9 | 84.2 | 83.1 | 73.1 | 59.7 | 41.2 | 30.9 | 56.4 |
| Minimum Temp °F | 10.4 | 12.5 | 20.8 | 27.6 | 34.5 | 40.4 | 44.7 | 43.4 | 35.7 | 27.3 | 20.5 | 11.4 | 27.4 |
| Mean Temp °F | 20.2 | 24.2 | 32.6 | 41.2 | 49.9 | 57.7 | 64.5 | 63.3 | 54.4 | 43.5 | 30.8 | 21.1 | 42.0 |
| Days Max Temp ≥ 90 °F | 0 | 0 | 0 | 0 | 0 | 1 | 7 | 5 | 1 | 0 | 0 | 0 | 14 |
| Days Max Temp ≤ 32 °F | 17 | 9 | 2 | 0 | 0 | 0 | 0 | 0 | 0 | 0 | 6 | 17 | 51 |
| Days Min Temp ≤ 32 °F | 30 | 27 | 28 | 22 | 12 | 3 | 1 | 2 | 11 | 23 | 27 | 30 | 216 |
| Days Min Temp ≤ 0 °F | 8 | 6 | 2 | 0 | 0 | 0 | 0 | 0 | 0 | 0 | 2 | 6 | 24 |
| Heating Degree Days | 1383 | 1145 | 999 | 708 | 461 | 225 | 66 | 88 | 315 | 659 | 1020 | 1356 | 8425 |
| Cooling Degree Days | 0 | 0 | 0 | 0 | 1 | 18 | 61 | 49 | 7 | 0 | 0 | 0 | 136 |
| Total Precipitation (") | 1.45 | 0.94 | 1.06 | 1.52 | 2.41 | 1.60 | 1.31 | 1.37 | 1.44 | 1.31 | 1.53 | 1.17 | 17.11 |
| Days ≥ 0.1" Precip | 5 | 3 | 4 | 5 | 7 | 5 | 3 | 4 | 4 | 4 | 5 | 5 | 54 |
| Total Snowfall (") | na | *10.1* | *7.7* | *4.9* | 1.4 | 0.1 | 0.0 | 0.0 | 0.2 | 1.2 | na | *14.1* | na |
| Days ≥ 1" Snow Depth | na | *23* | *16* | 3 | 0 | 0 | 0 | 0 | 0 | 0 | *8* | na | na |

## TWIN FALLS WSO *Twin Falls County* ELEVATION 3963 ft LAT/LONG 42° 33' N / 114° 21' W

| | JAN | FEB | MAR | APR | MAY | JUN | JUL | AUG | SEP | OCT | NOV | DEC | YEAR |
|---|---|---|---|---|---|---|---|---|---|---|---|---|---|
| Maximum Temp °F | 35.2 | 41.7 | 50.3 | 58.5 | 67.7 | 76.5 | 84.6 | 83.6 | 74.1 | 62.6 | 46.7 | 36.7 | 59.9 |
| Minimum Temp °F | 18.8 | 23.0 | 28.3 | 33.5 | 41.2 | 48.2 | 53.0 | 51.3 | 42.6 | 33.9 | 26.6 | 18.9 | 34.9 |
| Mean Temp °F | 27.1 | 32.4 | 39.3 | 46.0 | 54.5 | 62.4 | 68.8 | 67.5 | 58.3 | 48.3 | 36.7 | 27.8 | 47.4 |
| Days Max Temp ≥ 90 °F | 0 | 0 | 0 | 0 | 0 | 3 | 8 | 7 | 1 | 0 | 0 | 0 | 19 |
| Days Max Temp ≤ 32 °F | 11 | 4 | 1 | 0 | 0 | 0 | 0 | 0 | 0 | 0 | 2 | 9 | 27 |
| Days Min Temp ≤ 32 °F | 28 | 25 | 23 | 14 | 3 | 0 | 0 | 0 | 2 | 12 | 23 | 29 | 159 |
| Days Min Temp ≤ 0 °F | 2 | 1 | 0 | 0 | 0 | 0 | 0 | 0 | 0 | 0 | 0 | 2 | 5 |
| Heating Degree Days | 1170 | 914 | 789 | 563 | 330 | 134 | 30 | 43 | 214 | 513 | 843 | 1146 | 6689 |
| Cooling Degree Days | 0 | 0 | 0 | 1 | 12 | 65 | 145 | 129 | 20 | 1 | 0 | 0 | 373 |
| Total Precipitation (") | 1.06 | 0.84 | 1.11 | 0.98 | 1.07 | 0.81 | 0.29 | 0.45 | 0.66 | 0.73 | 1.19 | 1.03 | 10.22 |
| Days ≥ 0.1" Precip | 3 | 3 | 4 | 3 | 3 | 3 | 1 | 1 | 2 | 2 | 4 | 4 | 33 |
| Total Snowfall (") | 6.3 | 5.1 | 3.2 | 1.6 | 0.8 | 0.0 | 0.0 | 0.0 | 0.2 | 0.3 | 4.1 | 7.0 | 28.6 |
| Days ≥ 1" Snow Depth | 13 | 7 | 2 | 1 | 0 | 0 | 0 | 0 | 0 | 0 | 3 | 11 | 37 |

## WALLACE WOODLAND PAR *Shoshone County* ELEVATION 2953 ft LAT/LONG 47° 30' N / 115° 53' W

| | JAN | FEB | MAR | APR | MAY | JUN | JUL | AUG | SEP | OCT | NOV | DEC | YEAR |
|---|---|---|---|---|---|---|---|---|---|---|---|---|---|
| Maximum Temp °F | 33.5 | 39.3 | 46.3 | 54.6 | 64.0 | 71.3 | 79.9 | 80.8 | 70.4 | 58.0 | 41.0 | 33.4 | 56.0 |
| Minimum Temp °F | 19.3 | 22.3 | 26.1 | 32.2 | 38.4 | 44.7 | 48.1 | 47.7 | 40.4 | 33.0 | 27.3 | 20.6 | 33.3 |
| Mean Temp °F | 26.4 | 30.8 | 36.2 | 43.4 | 51.2 | 58.0 | 64.0 | 64.3 | 55.4 | 45.5 | 34.2 | 27.0 | 44.7 |
| Days Max Temp ≥ 90 °F | 0 | 0 | 0 | 0 | 0 | 1 | 6 | 6 | 1 | 0 | 0 | 0 | 14 |
| Days Max Temp ≤ 32 °F | 12 | 4 | 1 | 0 | 0 | 0 | 0 | 0 | 0 | 0 | 3 | 12 | 32 |
| Days Min Temp ≤ 32 °F | 29 | 25 | 25 | 16 | 5 | 0 | 0 | 0 | 3 | 15 | 23 | 28 | 169 |
| Days Min Temp ≤ 0 °F | 3 | 1 | 0 | 0 | 0 | 0 | 0 | 0 | 0 | 0 | 0 | 2 | 6 |
| Heating Degree Days | 1189 | 958 | 884 | 641 | 424 | 227 | 95 | 88 | 290 | 597 | 918 | 1170 | 7481 |
| Cooling Degree Days | 0 | 0 | 0 | 0 | 7 | 25 | 67 | 71 | 9 | 0 | 0 | 0 | 179 |
| Total Precipitation (") | 5.51 | 3.80 | 3.49 | 2.85 | 2.81 | 2.72 | 1.43 | 1.43 | 1.87 | 2.65 | 4.92 | 5.12 | 38.60 |
| Days ≥ 0.1" Precip | 12 | 10 | 10 | 8 | 8 | 7 | 4 | 4 | 5 | 7 | 12 | 12 | 99 |
| Total Snowfall (") | 23.6 | 15.4 | 8.8 | 2.7 | 0.2 | 0.0 | 0.0 | 0.0 | 0.0 | 0.6 | 9.8 | 20.9 | 82.0 |
| Days ≥ 1" Snow Depth | 28 | 23 | 18 | 2 | 0 | 0 | 0 | 0 | 0 | 0 | 9 | 23 | 103 |

## WARREN *Idaho County* ELEVATION 5905 ft LAT/LONG 45° 16 ' N / 115° 40 ' W

| | JAN | FEB | MAR | APR | MAY | JUN | JUL | AUG | SEP | OCT | NOV | DEC | YEAR |
|---|---|---|---|---|---|---|---|---|---|---|---|---|---|
| Maximum Temp °F | 33.1 | 38.6 | 42.8 | 49.2 | 58.5 | 67.7 | 76.4 | 76.2 | 66.8 | 55.7 | 39.4 | 32.0 | 53.0 |
| Minimum Temp °F | 7.3 | 9.3 | 13.6 | 20.6 | 27.6 | 33.4 | 36.2 | 34.6 | 28.7 | 23.1 | 15.8 | 8.1 | 21.5 |
| Mean Temp °F | 20.3 | 24.0 | 28.2 | 34.9 | 43.1 | 50.6 | 56.3 | 55.4 | 47.8 | 39.4 | 27.6 | 20.1 | 37.3 |
| Days Max Temp ≥ 90 °F | 0 | 0 | 0 | 0 | 0 | 0 | 0 | 0 | 0 | 0 | 0 | 0 | 0 |
| Days Max Temp ≤ 32 °F | 13 | 6 | 3 | 0 | 0 | 0 | 0 | 0 | 0 | 0 | 6 | 15 | 43 |
| Days Min Temp ≤ 32 °F | 31 | 28 | 31 | 29 | 25 | 14 | 8 | 11 | 23 | 29 | 29 | 31 | 289 |
| Days Min Temp ≤ 0 °F | 10 | 7 | 4 | 0 | 0 | 0 | 0 | 0 | 0 | 0 | 3 | 8 | 32 |
| Heating Degree Days | 1379 | 1153 | 1133 | 895 | 673 | 426 | 264 | 291 | 510 | 785 | 1114 | 1385 | 10008 |
| Cooling Degree Days | 0 | 0 | 0 | 0 | 0 | 1 | 4 | 3 | 0 | 0 | 0 | 0 | 8 |
| Total Precipitation (") | 3.14 | 1.95 | 2.50 | 2.26 | 2.24 | 2.51 | 1.25 | 1.30 | 1.64 | 1.81 | 2.62 | 2.68 | 25.90 |
| Days ≥ 0.1" Precip | 8 | 6 | 7 | 6 | 7 | 7 | 3 | 4 | 4 | 5 | 8 | 8 | 73 |
| Total Snowfall (") | 37.1 | 23.0 | 26.2 | 15.0 | 5.0 | 0.7 | 0.0 | 0.0 | 0.7 | 6.2 | 23.0 | 32.3 | 169.2 |
| Days ≥ 1" Snow Depth | 30 | 28 | 29 | 24 | 8 | 0 | 0 | 0 | 0 | 3 | 19 | 30 | 171 |

## WINCHESTER *Lewis County* ELEVATION 4180 ft LAT/LONG 46° 13 ' N / 116° 38 ' W

| | JAN | FEB | MAR | APR | MAY | JUN | JUL | AUG | SEP | OCT | NOV | DEC | YEAR |
|---|---|---|---|---|---|---|---|---|---|---|---|---|---|
| Maximum Temp °F | 34.6 | 39.6 | 44.4 | 51.4 | 60.0 | 67.6 | 76.4 | 77.5 | 68.2 | 56.9 | 41.5 | 34.8 | 54.4 |
| Minimum Temp °F | 18.4 | 21.5 | 25.1 | 30.3 | 36.2 | 41.9 | 45.1 | 44.8 | 38.4 | 32.0 | 25.1 | 19.1 | 31.5 |
| Mean Temp °F | 26.5 | 30.6 | 34.8 | 40.9 | 48.1 | 54.8 | 60.8 | 61.2 | 53.4 | 44.5 | 33.3 | 26.9 | 43.0 |
| Days Max Temp ≥ 90 °F | 0 | 0 | 0 | 0 | 0 | 0 | 1 | 3 | 0 | 0 | 0 | 0 | 4 |
| Days Max Temp ≤ 32 °F | 12 | 5 | 2 | 0 | 0 | 0 | 0 | 0 | 0 | 0 | 4 | 11 | 34 |
| Days Min Temp ≤ 32 °F | 28 | 25 | 26 | 20 | 9 | 2 | 1 | 1 | 6 | 17 | 24 | 28 | 187 |
| Days Min Temp ≤ 0 °F | 3 | 1 | 1 | 0 | 0 | 0 | 0 | 0 | 0 | 0 | 1 | 2 | 8 |
| Heating Degree Days | 1186 | 965 | 930 | 716 | 518 | 307 | 155 | 150 | 348 | 630 | 944 | 1174 | 8023 |
| Cooling Degree Days | 0 | 0 | 0 | 0 | 3 | 6 | 24 | 38 | 5 | 1 | 0 | 0 | 77 |
| Total Precipitation (") | 2.16 | 1.66 | 2.50 | 2.68 | 2.95 | 2.18 | 1.37 | 1.32 | 1.49 | 1.85 | 2.37 | 2.00 | 24.53 |
| Days ≥ 0.1" Precip | 7 | 6 | 8 | 8 | 8 | 7 | 4 | 3 | 4 | 5 | 8 | 7 | 75 |
| Total Snowfall (") | 20.1 | 13.3 | 16.3 | 10.3 | 2.3 | 0.0 | 0.0 | 0.0 | 0.2 | 2.4 | 14.2 | 18.7 | 97.8 |
| Days ≥ 1" Snow Depth | 25 | 17 | 15 | 4 | 0 | 0 | 0 | 0 | 0 | 1 | 10 | 22 | 94 |

## JANUARY MINIMUM TEMPERATURE °F

### LOWEST

| | | |
|---|---|---|
| 1 | Stanley | -1.6 |
| 2 | Grouse | -0.4 |
| 3 | Hamer | 2.6 |
| 4 | Island Park | 2.8 |
| 5 | Chilly Barton Flat | 4.1 |
| 6 | Dixie | 4.3 |
| 7 | Idaho Falls-46 W | 4.8 |
| 8 | Lifton | 6.0 |
| 9 | Fairfield | 6.1 |
| | Mackay | 6.1 |
| 11 | Howe | 6.3 |
| 12 | Picabo | 6.8 |
| 13 | Hill City | 6.9 |
| 14 | Driggs | 7.3 |
| | Warren | 7.3 |
| 16 | New Meadows | 8.3 |
| | St. Anthony | 8.3 |
| 18 | Gibbonsville | 9.1 |
| 19 | Ashton | 10.2 |
| 20 | Craters of the Mn | 10.3 |
| 21 | Swan Valley | 10.4 |
| 22 | Cascade | 10.7 |
| 23 | Grace | 10.8 |
| | Idaho Falls-16 SE | 10.8 |
| 25 | Idaho Falls-Fann | 10.9 |

### HIGHEST

| | | |
|---|---|---|
| 1 | Riggins | 28.3 |
| 2 | Lewiston | 27.6 |
| 3 | Swan Falls | 24.7 |
| 4 | Fenn | 23.8 |
| 5 | St. Maries | 23.0 |
| 6 | Bruneau | 22.7 |
| | Moscow | 22.7 |
| 8 | Boise | 22.1 |
| 9 | Grangeville | 21.9 |
| 10 | Caldwell | 21.7 |
| 11 | Deer Flat Dam | 21.6 |
| 12 | Emmett | 21.5 |
| | Nezperce | 21.5 |
| 14 | Bayview | 21.4 |
| | Cabinet Gorge | 21.4 |
| | Glenns Ferry | 21.4 |
| | Kellogg | 21.4 |
| | Potlatch | 21.4 |
| 19 | Sandpoint | 21.3 |
| 20 | Kuna | 21.1 |
| 21 | Grand View | 20.9 |
| 22 | Bliss | 20.3 |
| 23 | Arrowrock Dam | 20.1 |
| | Mountain Home | 20.1 |
| 25 | Bonners Ferry | 20.0 |

## JULY MAXIMUM TEMPERATURE °F

### HIGHEST

| | | |
|---|---|---|
| 1 | Glenns Ferry | 95.9 |
| | Swan Falls | 95.9 |
| 3 | Bruneau | 93.5 |
| 4 | Caldwell | 92.9 |
| 5 | Grand View | 92.5 |
| 6 | Payette | 92.2 |
| 7 | Riggins | 92.0 |
| | Shoshone | 92.0 |
| 9 | Mountain Home | 91.9 |
| 10 | Emmett | 91.2 |
| 11 | Bliss | 91.1 |
| 12 | Parma | 90.8 |
| 13 | Anderson Dam | 90.7 |
| | Cambridge | 90.7 |
| | Council | 90.7 |
| 16 | Jerome | 90.5 |
| 17 | Garden Valley | 90.2 |
| 18 | Boise | 90.0 |
| 19 | Malad City | 89.1 |
| 20 | Arrowrock Dam | 88.9 |
| | Malta | 88.9 |
| 22 | Shoup | 88.8 |
| 23 | Lewiston | 88.7 |
| 24 | Hazelton | 88.6 |
| 25 | Kuna | 88.2 |

### LOWEST

| | | |
|---|---|---|
| 1 | Dixie | 74.7 |
| 2 | Warren | 76.4 |
| | Winchester | 76.4 |
| 4 | Stanley | 77.8 |
| 5 | Island Park | 78.0 |
| 6 | Bayview | 78.4 |
| 7 | Grouse | 79.6 |
| 8 | Wallace Wdlnd Pk | 79.9 |
| 9 | Idaho Falls-16 SE | 80.0 |
| 10 | Elk City | 80.1 |
| | Nezperce | 80.1 |
| 12 | Headquarters | 80.2 |
| 13 | Driggs | 80.3 |
| | Elk River | 80.3 |
| 15 | McCall | 80.6 |
| | Potlatch | 80.6 |
| | Sandpoint | 80.6 |
| 18 | Lifton | 80.9 |
| | Pierce | 80.9 |
| 20 | Ashton | 81.2 |
| | Grangeville | 81.2 |
| 22 | Cascade | 81.3 |
| 23 | Chilly Barton Flat | 81.4 |
| | Porthill | 81.4 |
| | Priest River | 81.4 |

## ANNUAL PRECIPITATION (")

### HIGHEST

| | | |
|---|---|---|
| 1 | Pierce | 40.84 |
| 2 | Wallace Wdlnd Pk | 38.60 |
| 3 | Powell | 38.20 |
| 4 | Fenn | 37.87 |
| 5 | Headquarters | 37.26 |
| 6 | Elk River | 35.37 |
| 7 | Sandpoint | 33.11 |
| 8 | Cabinet Gorge | 31.09 |
| 9 | Priest River | 30.51 |
| 10 | Elk City | 30.18 |
| 11 | Kellogg | 30.12 |
| 12 | St. Maries | 29.03 |
| 13 | Island Park | 28.89 |
| 14 | Dixie | 27.07 |
| 15 | McCall | 26.79 |
| 16 | Warren | 25.90 |
| 17 | Moscow | 25.42 |
| 18 | Potlatch | 25.27 |
| 19 | Bayview | 24.92 |
| 20 | Winchester | 24.53 |
| 21 | Council | 24.43 |
| 22 | New Meadows | 24.05 |
| 23 | Idaho City | 23.79 |
| 24 | Grangeville | 23.20 |
| 25 | Garden Valley | 23.05 |

### LOWEST

| | | |
|---|---|---|
| 1 | Grand View | 7.02 |
| 2 | Bruneau | 7.44 |
| 3 | Challis | 7.62 |
| 4 | Chilly Barton Flat | 8.06 |
| 5 | Swan Falls | 8.27 |
| 6 | Idaho Falls-46 W | 8.67 |
| 7 | Howe | 8.90 |
| 8 | Aberdeen | 9.19 |
| 9 | Paul | 9.32 |
| 10 | Hamer | 9.39 |
| 11 | Minidoka Dam | 9.59 |
| 12 | Burley | 9.64 |
| 13 | Mackay | 9.72 |
| 14 | Hazelton | 9.83 |
| 15 | Deer Flat Dam | 9.84 |
| 16 | Glenns Ferry | 9.86 |
| 17 | Shoshone | 9.89 |
| 18 | Kuna | 9.92 |
| 19 | Castleford | 10.20 |
| 20 | Blackfoot | 10.21 |
| 21 | Twin Falls | 10.22 |
| 22 | Jerome | 10.26 |
| 23 | Bliss | 10.28 |
| 24 | Reynolds | 10.59 |
| 25 | Mountain Home | 10.80 |

## ANNUAL SNOWFALL (")

### HIGHEST

| | | |
|---|---|---|
| 1 | Island Park | 210.3 |
| 2 | Dixie | 199.0 |
| 3 | Powell | 175.9 |
| 4 | Warren | 169.2 |
| 5 | McCall | 148.8 |
| 6 | Elk City | 130.2 |
| 7 | Elk River | 106.1 |
| 8 | Ashton | 101.7 |
| 9 | Winchester | 97.8 |
| 10 | Craters of the Mn | 93.4 |
| 11 | Cascade | 92.2 |
| 12 | Wallace Wdlnd Pk | 82.0 |
| 13 | Gibbonsville | 81.8 |
| 14 | Idaho Falls-16 SE | 80.5 |
| 15 | New Meadows | 80.0 |
| 16 | Priest River | 77.9 |
| 17 | Grouse | 77.6 |
| 18 | Cabinet Gorge | 76.9 |
| 19 | Idaho City | 76.2 |
| 20 | Driggs | 75.3 |
| 21 | Fairfield | 72.1 |
| 22 | Bonners Ferry | 66.2 |
| | Sandpoint | 66.2 |
| 24 | Anderson Dam | 56.9 |
| 25 | Fenn | 54.7 |

### LOWEST

| | | |
|---|---|---|
| 1 | Swan Falls | 4.3 |
| 2 | Kuna | 11.3 |
| 3 | Caldwell | 13.0 |
| 4 | Lewiston | 14.5 |
| 5 | Castleford | 15.8 |
| 6 | Challis | 16.8 |
| 7 | Hazelton | 17.3 |
| 8 | Howe | 17.7 |
| 9 | Parma | 17.8 |
| 10 | Paul | 18.7 |
| 11 | Bliss | 19.3 |
| 12 | Boise | 20.6 |
| 13 | Burley | 22.2 |
| 14 | American Falls | 26.1 |
| 15 | Idaho Falls-46 W | 26.3 |
| 16 | Minidoka Dam | 26.6 |
| 17 | Oakley | 28.5 |
| 18 | Twin Falls | 28.6 |
| 19 | Hamer | 30.9 |
| 20 | Idaho Falls-Fann | 36.3 |
| 21 | Lifton | 36.7 |
| 22 | Arrowrock Dam | 40.2 |
| 23 | Malad City | 41.2 |
| 24 | Picabo | 44.0 |
| 25 | Potlatch | 45.0 |

# ILLINOIS

PHYSICAL FEATURES. Illinois lies midway between the Continental Divide and the Atlantic Ocean and some 500 miles north of the Gulf of Mexico. Its climate is typically continental with cold winters, warm summers, and frequent short period fluctuations in temperature, humidity, cloudiness, and wind direction.

The irregularly-shaped area of the State has a width of less than 200 miles at most points, but extends for 385 miles in the north-south direction. Except for a few low hills in the extreme south and a small unglaciated area in the extreme northwest, the terrain is flat. Differences in elevation have no significant influence on the climate. River drainage is mainly toward the Mississippi River, which forms the entire western boundary of the State. From north to south the principal rivers entering the Mississippi are the Rock, Illinois, Kaskaskia, and the Big Muddy. Approximately one-seventh of the State area drains southeastward into the Wabash and Ohio Rivers. Only a small area drains into Lake Michigan.

GENERAL CLIMATE. Without the protection of natural barriers, such as mountain ranges, Illinois experiences the full sweep of the winds which are constantly bringing in the climates of other areas. Southeast and easterly winds bring mild and wet weather; southerly winds are warm and showery; westerly winds are dry with moderate temperatures, and winds from the northwest and north are cool and dry. Winds are controlled by the storm systems and weather fronts which move eastward and northeastward through this area.

Storm systems move through the State most frequently during the winter and spring months and cause a maximum of cloudiness during those seasons. Summer-season storm systems tend to be weaker and to stay farther north, leaving Illinois with much sunshine interspersed with thunderstorm situations of comparatively short duration. The retreat of the sun in autumn is associated with variable periods of pleasant dry weather of the Indian summer variety. This season ends rather abruptly with the returning storminess which usually begins in November.

TEMPERATURE. Because Illinois extends so far in a north-south direction, the contrasts in winter temperature conditions are rather strong. The extreme north has frequent snows and temperatures drop to below zero several times each winter. The soil freezes to a depth of about 3 feet and occasionally remains snow-covered for weeks at a time. In the extreme south snow falls only occasionally and leaves after a few days, while temperatures drop to zero on an average of only about 1 day each winter. The soil freezes, but only to a depth of 8 to 12 inches, with great variation in the duration of soil-frost periods. The north-south range in winter mean temperatures is approximately 14° F.

During the summer season the sun heats the entire State quite strongly and uniformly. The north-south range of mean temperatures in July is only about 6° F. The annual average of days with temperatures of 90° F. or higher is near 20 in the north and near 50 in the south and west-central. Summer also brings periods of uncomfortably hot and humid weather, which are most persistent in the south. In the north the heat is usually broken after a few days by the arrival of cool air from Canada, but this cooling does not always penetrate to the southern portions of the State.

PRECIPITATION. Latitude is the principal control for both temperature and precipitation, with the northern counties averaging cooler and drier than the south. Distance from the Gulf of Mexico and lower airmass temperatures both tend to reduce the amounts of precipitation in the northern portion. Annual precipitation is approximately one and one-half times as great in the extreme south as in the extreme north, but most of the excess in the southern portion falls during winter and early spring. Mean total precipitation for the 4-month period of December through March ranges from near 7 inches in the extreme northwest to more than 14 inches in the extreme southeast. Precipitation during the warm season is more uniform. Totals for the 6-month period of April through September range from 21 to 24 inches throughout the State. The driest month is February. The wettest months are May and June.

Precipitation during fall, winter, and spring tends to fall uniformly over large areas. In contrast, summer rainfall occurs principally as brief showers affecting relatively small areas. The erratic occurrence of summer showers results in uneven distribution. The high rates of summer rainfall also cause runoff and soil erosion. Summer showers are usually thundery and may be accompanied by hail or destructive windstorms.

Floods occur nearly every year in at least some part of the State. The spring and early summer flood season results from a tendency for heavy general rainfall at that time of the year. The extreme north frequently has late winter or early spring flooding with the breakup of river ice, especially if there is an appreciable snow cover which is taken off by rain. River stages tend to decline during late summer, but local flash floods in minor streams, due to heavy thunderstorm rains, are common throughout the warm season. The interior rivers in the central and south have flat beds and sluggish currents so that they rise slowly and remain in flood for relatively long periods.

SNOWFALL. The annual average of snowfall ranges from near 30 inches in the extreme north to only 10 inches in the extreme south. In the extreme north the most likely form of winter precipitation is snow. In contrast, more than 90 percent of Cairo's winter precipitation falls as rain. In a large number of winter storm situations, only a slight change in the temperature pattern would suffice to change rain to snow or vice versa. For this reason, Illinois snowfall records show great variability. Snowfalls of 1 inch or more occur on an average of 10 to 12 days per year in the extreme north and decrease to 3 or 4 in the far south. The two northern divisions average about 50 days annually when the ground is covered with 1 inch or more of snow, and this average decreases to about 15 days in the two southern divisions.

STORMS. Heavy snows of 4-6 inches or more average one or two per year in the north and less frequently in the south. Strong winds will drift snow and make driving hazardous. Moderate to heavy ice storms average about once every four or five years and can be quite damaging. Thunderstorms average about 35 to 50 annually, but most are quite harmless. On occasion they provide the source for hail, damaging winds, and tornadoes. Hail falls on an average of two or three days annually in the same locality, but usually causes little damage.

More than 65 percent of Illinois tornadoes occur during the months of March, April, May, and June. This "tornado season" is marked by a rapid increase in activity during March, a peak in April and May, and a decline during June. Tornadoes have occurred during each of the twelve months of the year.

INFLUENCE OF LAKE MICHIGAN. Because prevailing winds are westerly and storm systems move from the same direction, the influence of the lake on Illinois weather is not large. When the wind blows from the lake toward the shore, which it does for approximately one-fourth of the time during spring and summer and for about one-eighth of the time during fall and winter, the result is a moderation of temperature. In addition to the general occurrence of onshore winds, there is the local "sea breeze" effect on summer afternoons which is usually observable in a narrow strip near the lake shore.

## COUNTY INDEX

***Montgomery County***
HILLSBORO

***Morgan County***
JACKSONVILLE 2 E

***Peoria County***
PEORIA GTR PEORIA AP
PRINCEVILLE

***Perry County***
DU QUOIN 4 SE

***Pike County***
GRIGGSVILLE

***Putnam County***
HENNEPIN POWER PLANT

***Randolph County***
SPARTA 3 N

***Richland County***
OLNEY 2 S

***Rock Island County***
MOLINE QUAD CITY AP

***St. Clair County***
BELLEVILLE SIU RSRCH

***Saline County***
HARRISBURG

***Sangamon County***
SPRINGFIELD CAPTL AP

***Schuyler County***
RUSHVILLE

***Shelby County***
MOWEAQUA
WINDSOR

***Union County***
ANNA 1 E

***Vermilion County***
DANVILLE
HOOPESTON 1 NE

***Warren County***
MONMOUTH

***Washington County***
NASHVILLE 4 NE

***Wayne County***
FAIRFIELD RADIO WFIW

***Whiteside County***
FULTON LOCK&DAM #13
MORRISON

***Williamson County***
MARION 4 NNE

***Winnebago County***
ROCKFORD GREATER AP

***Woodford County***
MINONK

## ELEVATION INDEX

| FEET | STATION NAME |
|---|---|
| 310 | CAIRO 3 N |
| 341 | BROOKPORT DAM 52 |
| 365 | HARRISBURG |
| 420 | CARBONDALE SEWAGE PL |
| 430 | ALTON MELVIN PRICE |
| 449 | ALBION |
| 449 | BELLEVILLE SIU RSRCH |
| 449 | FAIRFIELD RADIO WFIW |
| 449 | MARION 4 NNE |
| 459 | DU QUOIN 4 SE |
| 459 | HAVANA 4 NNE |
| 459 | HENNEPIN POWER PLANT |
| 459 | PERU |
| 479 | LACON 1 N |
| 479 | OLNEY 2 S |
| 490 | MT VERNON 3 NE |
| 502 | CARLYLE RESERVOIR |
| 502 | FLORA 5 NW |
| 502 | MCLEANSBORO 2 ENE |
| 502 | NEWTON 6 SSE |
| 512 | TISKILWA 2 SE |
| 514 | PALESTINE |
| 520 | SPARTA 3 N |
| 522 | NASHVILLE 4 NE |
| 525 | OTTAWA 4 SW |
| 550 | SALEM |
| 581 | MASON CITY 1 W |
| 581 | WHITE HALL 1 E |
| 582 | MOLINE QUAD CITY AP |
| 588 | SPRINGFIELD CAPTL AP |
| 591 | FULTON LOCK&DAM #13 |
| 591 | LINCOLN |
| 594 | CHICAGO UNIVERSITY |
| 595 | EFFINGHAM |
| 600 | DANVILLE |
| 610 | JACKSONVILLE 2 E |
| 620 | CHICAGO MIDWAY AP |
| 620 | DECATUR |
| 620 | MOWEAQUA |
| 630 | CARLINVILLE |
| 630 | HILLSBORO |
| 630 | WATSEKA 2 NW |
| 640 | ANNA 1 E |
| 640 | GENESEO |
| 640 | JERSEYVILLE 2 SW |
| 650 | PONTIAC |
| 650 | TUSCOLA |
| 658 | CHICAGO OHARE INT AP |
| 659 | PEORIA GTR PEORIA AP |
| 659 | RUSHVILLE |
| 669 | MORRISON |
| 669 | VIRDEN |
| 670 | PIPER CITY |
| 679 | WAUKEGAN |
| 680 | PARIS WATERWORKS |
| 689 | CHARLESTON |
| 689 | GRIGGSVILLE |
| 689 | LA HARPE |
| 689 | WINDSOR |
| 702 | DIXON 1 NW |
| 702 | PANA |
| 712 | CHENOA |
| 712 | PARK FOREST |
| 712 | WALNUT |
| 722 | HOOPESTON 1 NE |
| 722 | MATTOON |
| 724 | ROCKFORD GREATER AP |
| 732 | FARMER CITY |
| 732 | GOLDEN |
| 740 | RANTOUL |
| 741 | ALEDO |
| 741 | AURORA |
| 741 | URBANA |
| 751 | MINONK |

751 PRINCEVILLE

751 WHEATON 3 SE
768 QUINCY MUNI BALDWIN
770 MONMOUTH
771 ANTIOCH
781 GALESBURG

800 BARRINGTON
817 MOUNT CARROLL
820 MARENGO
850 GALVA
850 KEWANEE 1 E

932 PAW PAW 1 E
1001 STOCKTON 1 N

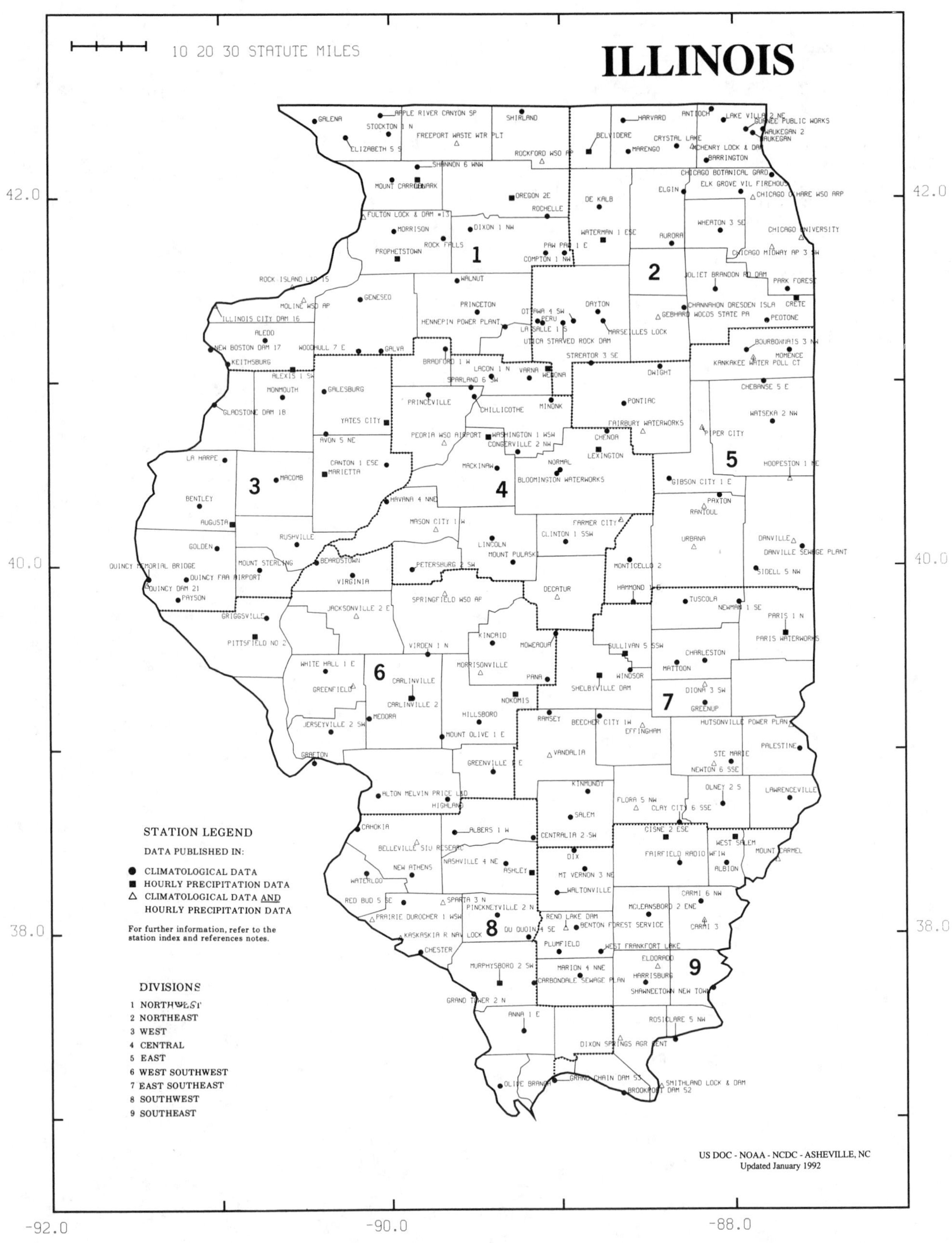
10 20 30 STATUTE MILES
ILLINOIS
STATION LEGEND
DATA PUBLISHED IN:
● CLIMATOLOGICAL DATA
■ HOURLY PRECIPITATION DATA
△ CLIMATOLOGICAL DATA AND HOURLY PRECIPITATION DATA
For further information, refer to the station index and references notes.
DIVISIONS
1 NORTHWEST
2 NORTHEAST
3 WEST
4 CENTRAL
5 EAST
6 WEST SOUTHWEST
7 EAST SOUTHEAST
8 SOUTHWEST
9 SOUTHEAST
US DOC - NOAA - NCDC - ASHEVILLE, NC
Updated January 1992
42.0
40.0
38.0
-92.0
-90.0
-88.0

## ALBION *Edwards County* ELEVATION 449 ft LAT/LONG 38° 22 ' N / 88° 3 ' W

| | JAN | FEB | MAR | APR | MAY | JUN | JUL | AUG | SEP | OCT | NOV | DEC | YEAR |
|---|---|---|---|---|---|---|---|---|---|---|---|---|---|
| Maximum Temp °F | 39.8 | 45.0 | 56.5 | 68.1 | 77.3 | 86.4 | 90.2 | 88.2 | 81.5 | 70.3 | 55.7 | 44.0 | 66.9 |
| Minimum Temp °F | 22.9 | 26.2 | 35.8 | 45.8 | 55.1 | 64.2 | 67.8 | 65.6 | 58.6 | 46.9 | 37.2 | 27.9 | 46.2 |
| Mean Temp °F | 31.4 | 35.6 | 46.2 | 56.9 | 66.2 | 75.3 | 79.0 | 76.9 | 70.1 | 58.7 | 46.5 | 35.9 | 56.6 |
| Days Max Temp ≥ 90 °F | 0 | 0 | 0 | 0 | 2 | 11 | 18 | 13 | 5 | 0 | 0 | 0 | 49 |
| Days Max Temp ≤ 32 °F | 8 | 5 | 0 | 0 | 0 | 0 | 0 | 0 | 0 | 0 | 0 | 4 | 17 |
| Days Min Temp ≤ 32 °F | 25 | 20 | 13 | 2 | 0 | 0 | 0 | 0 | 0 | 2 | 10 | 21 | 93 |
| Days Min Temp ≤ 0 °F | 2 | 1 | 0 | 0 | 0 | 0 | 0 | 0 | 0 | 0 | 0 | 1 | 4 |
| Heating Degree Days | 1037 | 824 | 582 | 267 | 77 | 4 | 0 | 1 | 38 | 227 | 551 | 894 | 4502 |
| Cooling Degree Days | 0 | 0 | 5 | 34 | 125 | 333 | 460 | 398 | 208 | 37 | 2 | 0 | 1602 |
| Total Precipitation (") | 2.37 | 2.83 | 3.93 | 4.90 | 4.64 | 3.92 | 3.89 | 3.16 | 3.17 | 3.45 | 4.45 | 3.59 | 44.30 |
| Days ≥ 0.1" Precip | 4 | *5* | 6 | 8 | 7 | 6 | 6 | 5 | 5 | 5 | 7 | 6 | 70 |
| Total Snowfall (") | *3.5* | 3.7 | 1.9 | 0.1 | 0.0 | 0.0 | 0.0 | 0.0 | 0.0 | 0.1 | 0.2 | 2.2 | 11.7 |
| Days ≥ 1" Snow Depth | *5* | 4 | 1 | 0 | 0 | 0 | 0 | 0 | 0 | 0 | 0 | *1* | 11 |

## ALEDO *Mercer County* ELEVATION 741 ft LAT/LONG 41° 12 ' N / 90° 45 ' W

| | JAN | FEB | MAR | APR | MAY | JUN | JUL | AUG | SEP | OCT | NOV | DEC | YEAR |
|---|---|---|---|---|---|---|---|---|---|---|---|---|---|
| Maximum Temp °F | 30.3 | 35.2 | 48.4 | 63.0 | 74.1 | 82.8 | 86.2 | 83.8 | 76.4 | 64.6 | 48.3 | 34.9 | 60.7 |
| Minimum Temp °F | 13.4 | 17.7 | 29.2 | 40.4 | 50.7 | 60.0 | 63.6 | 61.3 | 53.2 | 42.4 | 30.9 | 19.6 | 40.2 |
| Mean Temp °F | 21.9 | 26.5 | 39.0 | 51.8 | 62.4 | 71.5 | 74.8 | 72.6 | 64.7 | 53.5 | 39.6 | 27.3 | 50.5 |
| Days Max Temp ≥ 90 °F | 0 | 0 | 0 | 0 | 1 | 4 | 9 | 6 | 2 | 0 | 0 | 0 | 22 |
| Days Max Temp ≤ 32 °F | 16 | 11 | 3 | 0 | 0 | 0 | 0 | 0 | 0 | 0 | 2 | 12 | 44 |
| Days Min Temp ≤ 32 °F | 29 | 25 | 19 | 6 | 0 | 0 | 0 | 0 | 0 | 5 | 18 | 28 | 130 |
| Days Min Temp ≤ 0 °F | 6 | 3 | 0 | 0 | 0 | 0 | 0 | 0 | 0 | 0 | 0 | 3 | 12 |
| Heating Degree Days | 1328 | 1082 | 802 | 404 | 147 | 16 | 3 | 9 | 102 | 363 | 754 | 1163 | 6173 |
| Cooling Degree Days | 0 | 0 | 2 | 17 | 71 | 220 | 314 | 275 | 107 | 9 | 0 | 0 | 1015 |
| Total Precipitation (") | 1.24 | 1.03 | 2.41 | 3.61 | 3.61 | 4.46 | 4.33 | 4.36 | 3.58 | 2.87 | 2.46 | 2.07 | 36.03 |
| Days ≥ 0.1" Precip | 3 | 3 | 6 | 7 | 7 | 6 | 7 | 6 | 6 | 6 | 6 | 4 | 67 |
| Total Snowfall (") | 7.5 | 6.5 | 4.4 | 0.9 | 0.0 | 0.0 | 0.0 | 0.0 | 0.0 | 0.2 | 2.2 | 6.0 | 27.7 |
| Days ≥ 1" Snow Depth | na | na | *2* | 0 | 0 | 0 | 0 | 0 | 0 | 0 | 1 | na | na |

## ALTON MELVIN PRICE *Madison County* ELEVATION 430 ft LAT/LONG 38° 53 ' N / 90° 11 ' W

| | JAN | FEB | MAR | APR | MAY | JUN | JUL | AUG | SEP | OCT | NOV | DEC | YEAR |
|---|---|---|---|---|---|---|---|---|---|---|---|---|---|
| Maximum Temp °F | 35.6 | 40.4 | 52.3 | 64.7 | 74.6 | 84.1 | 88.2 | 86.0 | 79.1 | 67.0 | 53.4 | 41.4 | 63.9 |
| Minimum Temp °F | 18.5 | 22.4 | 33.2 | 44.9 | 54.6 | 64.3 | 68.5 | 65.8 | 58.0 | 45.8 | 35.5 | 25.3 | 44.7 |
| Mean Temp °F | 27.1 | 31.4 | 42.8 | 54.8 | 64.6 | 74.2 | 78.4 | 75.9 | 68.6 | 56.5 | 44.5 | 33.4 | 54.4 |
| Days Max Temp ≥ 90 °F | 0 | 0 | 0 | 0 | 1 | 6 | 14 | 9 | 3 | 0 | 0 | 0 | 33 |
| Days Max Temp ≤ 32 °F | 12 | 8 | 1 | 0 | 0 | 0 | 0 | 0 | 0 | 0 | 1 | 6 | 28 |
| Days Min Temp ≤ 32 °F | 28 | 24 | 16 | 2 | 0 | 0 | 0 | 0 | 0 | 1 | 12 | 23 | 106 |
| Days Min Temp ≤ 0 °F | 3 | 1 | 0 | 0 | 0 | 0 | 0 | 0 | 0 | 0 | 0 | 1 | 5 |
| Heating Degree Days | 1169 | 942 | 685 | 318 | 100 | 7 | 1 | 1 | 50 | 279 | 609 | 974 | 5135 |
| Cooling Degree Days | 0 | 0 | 3 | 19 | 93 | 298 | 439 | 361 | 174 | 19 | 1 | 0 | 1407 |
| Total Precipitation (") | 1.94 | 2.07 | 3.32 | 4.01 | 4.00 | 3.25 | 3.65 | 3.18 | 3.36 | 2.77 | 3.76 | 3.28 | 38.59 |
| Days ≥ 0.1" Precip | 4 | 4 | 7 | 7 | 7 | 5 | 6 | 5 | 5 | 6 | 7 | 6 | 69 |
| Total Snowfall (") | na | na | *2.8* | 0.4 | 0.0 | 0.0 | 0.0 | 0.0 | 0.0 | 0.0 | 0.3 | *1.0* | na |
| Days ≥ 1" Snow Depth | *8* | *5* | 2 | 0 | 0 | 0 | 0 | 0 | 0 | 0 | 0 | 2 | 17 |

## ANNA 1 E *Union County* ELEVATION 640 ft LAT/LONG 37° 28 ' N / 89° 15 ' W

| | JAN | FEB | MAR | APR | MAY | JUN | JUL | AUG | SEP | OCT | NOV | DEC | YEAR |
|---|---|---|---|---|---|---|---|---|---|---|---|---|---|
| Maximum Temp °F | 40.7 | 46.2 | 57.2 | 68.6 | 77.1 | 85.7 | 89.1 | 87.3 | 80.3 | 69.8 | 56.8 | 45.4 | 67.0 |
| Minimum Temp °F | 23.1 | 26.6 | 36.3 | 46.6 | 54.9 | 63.5 | 67.6 | 65.3 | 58.6 | 46.6 | 37.8 | 28.4 | 46.3 |
| Mean Temp °F | 32.0 | 36.4 | 46.8 | 57.6 | 66.0 | 74.6 | 78.4 | 76.3 | 69.5 | 58.2 | 47.3 | 36.9 | 56.7 |
| Days Max Temp ≥ 90 °F | 0 | 0 | 0 | 0 | 1 | 8 | 16 | 11 | 3 | 0 | 0 | 0 | 39 |
| Days Max Temp ≤ 32 °F | 8 | 4 | 1 | 0 | 0 | 0 | 0 | 0 | 0 | 0 | 0 | 4 | 17 |
| Days Min Temp ≤ 32 °F | 24 | 19 | 12 | 2 | 0 | 0 | 0 | 0 | 0 | 2 | 10 | 20 | 89 |
| Days Min Temp ≤ 0 °F | 1 | 0 | 0 | 0 | 0 | 0 | 0 | 0 | 0 | 0 | 0 | 1 | 2 |
| Heating Degree Days | 1017 | 800 | 563 | 245 | 71 | 3 | 0 | 1 | 38 | 232 | 526 | 863 | 4359 |
| Cooling Degree Days | 0 | 0 | 4 | 30 | 113 | 308 | 440 | 382 | 185 | 29 | 2 | 0 | 1493 |
| Total Precipitation (") | 3.23 | 3.37 | 4.64 | 4.83 | 5.04 | 4.02 | 3.56 | 3.75 | 3.55 | 3.53 | 4.70 | 4.34 | 48.56 |
| Days ≥ 0.1" Precip | 6 | 6 | 8 | 8 | 7 | 6 | 6 | 5 | 5 | 6 | 7 | 7 | 77 |
| Total Snowfall (") | 5.3 | 5.3 | 2.6 | 0.1 | 0.0 | 0.0 | 0.0 | 0.0 | 0.0 | 0.1 | 0.6 | 2.5 | 16.5 |
| Days ≥ 1" Snow Depth | 6 | 5 | 1 | 0 | 0 | 0 | 0 | 0 | 0 | 0 | 0 | 3 | 15 |

## ANTIOCH *Lake County* ELEVATION 771 ft LAT/LONG 42° 29 ' N / 88° 6 ' W

| | JAN | FEB | MAR | APR | MAY | JUN | JUL | AUG | SEP | OCT | NOV | DEC | YEAR |
|---|---|---|---|---|---|---|---|---|---|---|---|---|---|
| Maximum Temp °F | 27.6 | 31.8 | 43.1 | 56.6 | 68.5 | *77.8* | 82.1 | 80.5 | 72.7 | 60.9 | 46.3 | 33.4 | 56.8 |
| Minimum Temp °F | 9.6 | 12.7 | 24.2 | 35.4 | 44.7 | *54.4* | 59.7 | 58.3 | 51.1 | 39.7 | 29.0 | 17.4 | 36.4 |
| Mean Temp °F | 18.6 | 22.3 | 33.7 | 46.0 | 56.6 | *66.1* | 70.9 | 69.4 | 61.9 | 50.3 | 37.7 | 25.4 | 46.6 |
| Days Max Temp ≥ 90 °F | 0 | 0 | 0 | 0 | 0 | 2 | 4 | 3 | 1 | 0 | 0 | 0 | 10 |
| Days Max Temp ≤ 32 °F | 19 | 14 | 5 | 0 | 0 | 0 | 0 | 0 | 0 | 0 | 3 | 13 | 54 |
| Days Min Temp ≤ 32 °F | 30 | 27 | 25 | 12 | 2 | 0 | 0 | 0 | 0 | 7 | 20 | 29 | 152 |
| Days Min Temp ≤ 0 °F | 9 | 7 | 1 | 0 | 0 | 0 | 0 | 0 | 0 | 0 | 0 | 3 | 20 |
| Heating Degree Days | 1433 | 1200 | 965 | 568 | 284 | *71* | 16 | 27 | 147 | 454 | 814 | 1220 | 7199 |
| Cooling Degree Days | 0 | *0* | *1* | *9* | *30* | na | *204* | 184 | 66 | 2 | 0 | 0 | na |
| Total Precipitation (") | 1.69 | *1.32* | 2.37 | 3.65 | 3.35 | *4.05* | 4.11 | 4.33 | 3.70 | 2.61 | 2.79 | 2.23 | 36.20 |
| Days ≥ 0.1" Precip | 4 | 4 | 5 | 7 | 7 | 7 | 7 | 6 | 6 | 5 | 6 | 5 | 69 |
| Total Snowfall (") | 13.2 | 9.0 | 7.8 | 2.8 | 0.0 | 0.0 | 0.0 | 0.0 | 0.0 | 0.2 | 2.6 | 11.4 | 47.0 |
| Days ≥ 1" Snow Depth | 22 | 21 | 9 | 1 | 0 | 0 | 0 | 0 | 0 | 0 | 2 | 14 | 69 |

## AURORA *Kane County* ELEVATION 741 ft LAT/LONG 41° 45 ' N / 88° 20 ' W

| | JAN | FEB | MAR | APR | MAY | JUN | JUL | AUG | SEP | OCT | NOV | DEC | YEAR |
|---|---|---|---|---|---|---|---|---|---|---|---|---|---|
| Maximum Temp °F | 29.2 | 34.1 | 46.3 | 59.9 | 71.8 | 81.0 | 84.5 | 82.2 | 75.0 | 62.9 | 47.7 | 34.3 | 59.1 |
| Minimum Temp °F | 11.3 | 15.6 | 27.2 | 38.0 | 47.9 | 57.3 | 62.2 | 60.3 | 52.3 | 40.4 | 30.1 | 18.5 | 38.4 |
| Mean Temp °F | 20.1 | 25.0 | 36.8 | 49.0 | 59.8 | 69.2 | 73.4 | 71.2 | 63.7 | 51.6 | 39.0 | 26.3 | 48.8 |
| Days Max Temp ≥ 90 °F | 0 | 0 | 0 | 0 | 1 | 4 | 6 | 4 | 1 | 0 | 0 | 0 | 16 |
| Days Max Temp ≤ 32 °F | 17 | 12 | 3 | 0 | 0 | 0 | 0 | 0 | 0 | 0 | 2 | 11 | 45 |
| Days Min Temp ≤ 32 °F | 30 | 26 | 23 | 8 | 1 | 0 | 0 | 0 | 0 | 6 | 19 | 28 | 141 |
| Days Min Temp ≤ 0 °F | 8 | 4 | 0 | 0 | 0 | 0 | 0 | 0 | 0 | 0 | 0 | 3 | 15 |
| Heating Degree Days | 1386 | 1121 | 867 | 481 | 207 | 36 | 5 | 13 | 117 | 418 | 774 | 1194 | 6619 |
| Cooling Degree Days | 0 | 0 | 0 | 9 | 53 | 167 | 279 | 222 | 83 | 4 | 0 | 0 | 817 |
| Total Precipitation (") | 1.60 | 1.38 | 2.58 | 3.89 | 3.78 | 4.35 | 3.76 | 4.35 | 3.67 | 2.66 | 3.25 | 2.60 | 37.87 |
| Days ≥ 0.1" Precip | 4 | 4 | 5 | 7 | 7 | 7 | 6 | 7 | 6 | 5 | 6 | 6 | 70 |
| Total Snowfall (") | 9.0 | 7.7 | 3.7 | 1.0 | 0.0 | 0.0 | 0.0 | 0.0 | 0.0 | 0.2 | 1.2 | *7.3* | 30.1 |
| Days ≥ 1" Snow Depth | 20 | 15 | 5 | 0 | 0 | 0 | 0 | 0 | 0 | 0 | 1 | 11 | 52 |

## BARRINGTON *Lake County* ELEVATION 800 ft LAT/LONG 42° 12 ' N / 88° 9 ' W

| | JAN | FEB | MAR | APR | MAY | JUN | JUL | AUG | SEP | OCT | NOV | DEC | YEAR |
|---|---|---|---|---|---|---|---|---|---|---|---|---|---|
| Maximum Temp °F | *27.9* | 32.5 | *43.4* | *57.8* | 69.4 | *79.3* | 83.2 | *81.2* | 73.5 | 61.5 | *46.5* | *33.5* | 57.5 |
| Minimum Temp °F | *10.1* | *14.1* | na | *37.3* | *46.4* | *56.4* | *61.4* | *59.6* | 51.5 | 40.0 | *29.1* | *17.6* | na |
| Mean Temp °F | *19.0* | *23.3* | na | *47.8* | *57.8* | *67.9* | *72.3* | *70.4* | 62.5 | 50.8 | *37.8* | *25.6* | na |
| Days Max Temp ≥ 90 °F | 0 | 0 | 0 | 0 | 0 | *3* | 6 | *3* | 1 | 0 | *0* | 0 | 13 |
| Days Max Temp ≤ 32 °F | 19 | 14 | *5* | 0 | 0 | 0 | 0 | *0* | 0 | 0 | *2* | 13 | 53 |
| Days Min Temp ≤ 32 °F | 30 | *27* | *22* | *9* | 2 | 0 | 0 | *0* | 0 | 8 | *19* | 28 | 145 |
| Days Min Temp ≤ 0 °F | 9 | 6 | 0 | 0 | 0 | 0 | 0 | 0 | 0 | 0 | *0* | 4 | 19 |
| Heating Degree Days | *1421* | *1171* | na | *520* | *254* | *55* | *12* | *20* | 140 | 442 | *809* | *1216* | na |
| Cooling Degree Days | 0 | 0 | na | *12* | 37 | *152* | *270* | *210* | 70 | 3 | *0* | *0* | na |
| Total Precipitation (") | 1.53 | 1.21 | *2.44* | *3.25* | 3.53 | *4.23* | 3.63 | *4.46* | 7.12 | 2.53 | *2.82* | 2.19 | 38.94 |
| Days ≥ 0.1" Precip | *4* | 4 | *5* | *6* | *7* | 6 | 6 | *6* | 6 | 5 | *6* | *5* | 66 |
| Total Snowfall (") | na | *7.4* | *4.9* | *1.1* | 0.0 | 0.0 | *0.0* | *0.0* | *0.0* | *0.3* | na | *8.1* | na |
| Days ≥ 1" Snow Depth | *18* | 15 | *5* | *1* | 0 | 0 | 0 | *0* | 0 | 0 | *2* | *13* | 54 |

## BELLEVILLE SIU RSRCH *St. Clair County* ELEVATION 449 ft LAT/LONG 38° 30 ' N / 89° 51 ' W

| | JAN | FEB | MAR | APR | MAY | JUN | JUL | AUG | SEP | OCT | NOV | DEC | YEAR |
|---|---|---|---|---|---|---|---|---|---|---|---|---|---|
| Maximum Temp °F | 39.4 | 44.7 | 56.6 | 68.4 | 77.1 | 85.7 | 89.5 | 87.2 | 81.0 | 70.1 | 56.0 | 44.1 | 66.7 |
| Minimum Temp °F | 21.1 | 24.8 | 34.6 | 44.5 | 52.8 | 61.7 | 65.6 | 62.4 | 55.6 | 44.0 | 35.9 | 26.8 | 44.2 |
| Mean Temp °F | 30.3 | 34.8 | 45.7 | 56.4 | 65.0 | 73.7 | 77.5 | 74.9 | 68.3 | 57.1 | 46.0 | 35.5 | 55.4 |
| Days Max Temp ≥ 90 °F | 0 | 0 | 0 | 0 | 1 | 9 | 16 | 11 | 4 | 0 | 0 | 0 | 41 |
| Days Max Temp ≤ 32 °F | 10 | 6 | 1 | 0 | 0 | 0 | 0 | 0 | 0 | 0 | 0 | 4 | 21 |
| Days Min Temp ≤ 32 °F | 26 | 21 | 14 | 4 | 0 | 0 | 0 | 0 | 0 | 4 | 13 | 22 | 104 |
| Days Min Temp ≤ 0 °F | 2 | 1 | 0 | 0 | 0 | 0 | 0 | 0 | 0 | 0 | 0 | 1 | 4 |
| Heating Degree Days | 1070 | 847 | 599 | 279 | 92 | 6 | 1 | 3 | 54 | 266 | 565 | 908 | 4690 |
| Cooling Degree Days | 0 | 0 | 6 | 31 | 100 | 282 | 416 | 337 | 167 | 28 | 2 | 0 | 1369 |
| Total Precipitation (") | 1.89 | 2.08 | 3.32 | 3.70 | 3.78 | 3.75 | 3.51 | 3.39 | 3.61 | 2.95 | 3.91 | 3.13 | 39.02 |
| Days ≥ 0.1" Precip | *4* | 4 | 7 | 7 | 7 | 6 | 5 | 5 | 5 | 5 | 6 | *5* | 66 |
| Total Snowfall (") | *5.6* | *4.0* | 2.5 | 0.8 | 0.0 | 0.0 | 0.0 | 0.0 | 0.0 | 0.0 | *0.9* | *2.5* | 16.3 |
| Days ≥ 1" Snow Depth | *9* | *5* | 1 | 0 | 0 | 0 | 0 | 0 | 0 | 0 | *1* | *4* | 20 |

## BROOKPORT DAM 52 *Massac County* ELEVATION 341 ft LAT/LONG 37° 8 ' N / 88° 39 ' W

| | JAN | FEB | MAR | APR | MAY | JUN | JUL | AUG | SEP | OCT | NOV | DEC | YEAR |
|---|---|---|---|---|---|---|---|---|---|---|---|---|---|
| Maximum Temp °F | 42.2 | 47.3 | 58.0 | 69.3 | 77.7 | 86.1 | 89.7 | 88.2 | 81.3 | 70.5 | 57.9 | 47.0 | 67.9 |
| Minimum Temp °F | 24.2 | 27.8 | 37.0 | 47.0 | 55.5 | 63.7 | 68.1 | 65.5 | 58.8 | 46.5 | 38.1 | 29.3 | 46.8 |
| Mean Temp °F | 33.2 | 37.5 | 47.5 | 58.2 | 66.6 | 75.0 | 78.9 | 76.9 | 70.0 | 58.7 | 48.0 | 38.1 | 57.4 |
| Days Max Temp ≥ 90 °F | 0 | 0 | 0 | 0 | 1 | 9 | 17 | 13 | 4 | 0 | 0 | 0 | 44 |
| Days Max Temp ≤ 32 °F | 6 | 3 | 0 | 0 | 0 | 0 | 0 | 0 | 0 | 0 | 0 | 3 | 12 |
| Days Min Temp ≤ 32 °F | 23 | 18 | 12 | 2 | 0 | 0 | 0 | 0 | 0 | 2 | 10 | 19 | 86 |
| Days Min Temp ≤ 0 °F | 1 | 1 | 0 | 0 | 0 | 0 | 0 | 0 | 0 | 0 | 0 | 0 | 2 |
| Heating Degree Days | 977 | 768 | 537 | 226 | 61 | 3 | 0 | 0 | 37 | 226 | 506 | 826 | 4167 |
| Cooling Degree Days | 0 | 0 | 4 | 25 | 115 | 304 | 449 | 385 | 191 | 31 | 2 | 0 | 1506 |
| Total Precipitation (") | 3.50 | 3.83 | 4.44 | 4.84 | 4.54 | 3.78 | 4.12 | 3.13 | 3.60 | 3.18 | 4.36 | 4.66 | 47.98 |
| Days ≥ 0.1" Precip | 6 | 6 | 8 | 7 | 8 | 6 | 6 | 5 | 6 | 6 | 7 | 7 | 78 |
| Total Snowfall (") | 3.4 | 3.6 | 1.9 | 0.0 | 0.0 | 0.0 | 0.0 | 0.0 | 0.0 | 0.0 | 0.1 | *1.0* | 10.0 |
| Days ≥ 1" Snow Depth | 6 | 5 | 1 | 0 | 0 | 0 | 0 | 0 | 0 | 0 | 0 | 1 | 13 |

## CAIRO 3 N *Alexander County* ELEVATION 310 ft LAT/LONG 37° 3 ' N / 89° 11 ' W

| | JAN | FEB | MAR | APR | MAY | JUN | JUL | AUG | SEP | OCT | NOV | DEC | YEAR |
|---|---|---|---|---|---|---|---|---|---|---|---|---|---|
| Maximum Temp °F | 40.6 | 46.1 | 57.0 | 69.0 | 78.0 | 86.8 | 89.7 | 87.0 | 80.4 | 69.8 | 56.9 | 46.0 | 67.3 |
| Minimum Temp °F | 26.0 | 30.2 | 40.1 | 50.6 | 59.2 | 67.5 | 71.4 | 68.8 | 61.7 | 49.9 | 40.5 | 31.2 | 49.8 |
| Mean Temp °F | 33.3 | 38.2 | 48.6 | 59.8 | 68.6 | 77.2 | 80.6 | 77.9 | 71.0 | 59.9 | 48.7 | 38.6 | 58.5 |
| Days Max Temp ≥ 90 °F | 0 | 0 | 0 | 0 | 2 | 10 | 17 | 11 | 3 | 0 | 0 | 0 | 43 |
| Days Max Temp ≤ 32 °F | 8 | 4 | 0 | 0 | 0 | 0 | 0 | 0 | 0 | 0 | 0 | 4 | 16 |
| Days Min Temp ≤ 32 °F | 23 | 17 | 8 | 1 | 0 | 0 | 0 | 0 | 0 | 1 | 6 | 17 | 73 |
| Days Min Temp ≤ 0 °F | 1 | 0 | 0 | 0 | 0 | 0 | 0 | 0 | 0 | 0 | 0 | 0 | 1 |
| Heating Degree Days | 975 | 750 | 510 | 198 | 43 | 2 | 0 | 1 | 28 | 197 | 485 | 811 | 4000 |
| Cooling Degree Days | *0* | 0 | *9* | *51* | 153 | 371 | 497 | 412 | 207 | 37 | 2 | 1 | 1740 |
| Total Precipitation (") | 3.05 | 3.58 | 4.24 | 4.88 | 5.07 | 3.64 | 3.98 | 3.89 | 3.72 | 3.30 | 4.33 | 4.33 | 48.01 |
| Days ≥ 0.1" Precip | 5 | 5 | 8 | 7 | 8 | 6 | 6 | 5 | 5 | 5 | 6 | 6 | 72 |
| Total Snowfall (") | 4.4 | 3.0 | 2.2 | 0.0 | 0.0 | 0.0 | 0.0 | 0.0 | 0.0 | 0.1 | 0.3 | 1.2 | 11.2 |
| Days ≥ 1" Snow Depth | 7 | 5 | 1 | 0 | 0 | 0 | 0 | 0 | 0 | 0 | 0 | 1 | 14 |

## CARBONDALE SEWAGE PL *Jackson County* ELEVATION 420 ft LAT/LONG 37° 43 ' N / 89° 13 ' W

| | JAN | FEB | MAR | APR | MAY | JUN | JUL | AUG | SEP | OCT | NOV | DEC | YEAR |
|---|---|---|---|---|---|---|---|---|---|---|---|---|---|
| Maximum Temp °F | 40.0 | 45.1 | 55.8 | 68.0 | 76.9 | 85.4 | 89.1 | 87.5 | 80.6 | 69.5 | 56.4 | 45.0 | 66.6 |
| Minimum Temp °F | 20.8 | 23.7 | 34.1 | 44.3 | 52.7 | 62.1 | 66.3 | 63.0 | 55.6 | 42.4 | 35.3 | 26.2 | 43.9 |
| Mean Temp °F | 30.4 | 34.4 | 45.0 | 56.2 | 64.8 | 73.8 | 77.7 | 75.3 | 68.1 | 56.0 | 45.9 | 35.7 | 55.3 |
| Days Max Temp ≥ 90 °F | 0 | 0 | 0 | 0 | 1 | 8 | 15 | 11 | 4 | 0 | 0 | 0 | 39 |
| Days Max Temp ≤ 32 °F | 9 | 5 | 1 | 0 | 0 | 0 | 0 | 0 | 0 | 0 | 0 | 4 | 19 |
| Days Min Temp ≤ 32 °F | 26 | 22 | 15 | 4 | 0 | 0 | 0 | 0 | 0 | 6 | 13 | 23 | 109 |
| Days Min Temp ≤ 0 °F | 2 | 1 | 0 | 0 | 0 | 0 | 0 | 0 | 0 | 0 | 0 | 0 | 3 |
| Heating Degree Days | 1067 | 857 | 617 | 286 | 94 | 6 | 1 | 2 | 56 | 296 | 569 | 904 | 4755 |
| Cooling Degree Days | 0 | 0 | 4 | 23 | *80* | 269 | 402 | 327 | 146 | *16* | 1 | *0* | 1268 |
| Total Precipitation (") | 2.84 | 2.83 | 4.13 | 4.24 | 4.25 | 4.28 | 3.54 | 4.09 | 3.44 | 2.90 | 4.49 | 3.77 | 44.80 |
| Days ≥ 0.1" Precip | 6 | 5 | 7 | 8 | 7 | 6 | 5 | 6 | 6 | 5 | 7 | 7 | 75 |
| Total Snowfall (") | *4.6* | *4.5* | 2.2 | 0.3 | 0.0 | 0.0 | 0.0 | 0.0 | 0.0 | 0.1 | 0.6 | *2.2* | 14.5 |
| Days ≥ 1" Snow Depth | 6 | 5 | 2 | 0 | 0 | 0 | 0 | 0 | 0 | 0 | 0 | *2* | 15 |

## CARLINVILLE *Macoupin County* ELEVATION 630 ft LAT/LONG 39° 17 ' N / 89° 49 ' W

| | JAN | FEB | MAR | APR | MAY | JUN | JUL | AUG | SEP | OCT | NOV | DEC | YEAR |
|---|---|---|---|---|---|---|---|---|---|---|---|---|---|
| Maximum Temp °F | 35.5 | 41.3 | 53.8 | 66.7 | 76.0 | 84.6 | 88.2 | 85.9 | 79.5 | 67.7 | 53.1 | 40.8 | 64.4 |
| Minimum Temp °F | 18.2 | 22.6 | 32.8 | 43.7 | 52.4 | 61.6 | 65.6 | 63.1 | 56.0 | 44.7 | 34.7 | 24.7 | 43.3 |
| Mean Temp °F | 26.9 | 31.9 | 43.5 | 55.2 | 64.2 | 73.1 | 76.9 | 74.5 | 67.8 | 56.2 | 43.9 | 32.8 | 53.9 |
| Days Max Temp ≥ 90 °F | 0 | 0 | 0 | 0 | 1 | 7 | 13 | 9 | 3 | 0 | 0 | 0 | 33 |
| Days Max Temp ≤ 32 °F | 12 | 8 | 1 | 0 | 0 | 0 | 0 | 0 | 0 | 0 | 1 | 7 | 29 |
| Days Min Temp ≤ 32 °F | 27 | 22 | 16 | 4 | 0 | 0 | 0 | 0 | 0 | 4 | 13 | 24 | 110 |
| Days Min Temp ≤ 0 °F | 3 | 2 | 0 | 0 | 0 | 0 | 0 | 0 | 0 | 0 | 0 | 1 | 6 |
| Heating Degree Days | 1174 | 927 | 663 | 311 | 109 | 10 | 1 | 3 | 62 | 289 | 626 | 992 | 5167 |
| Cooling Degree Days | 0 | 0 | 4 | 26 | 96 | 271 | 389 | 324 | 157 | 21 | 2 | 0 | 1290 |
| Total Precipitation (") | 1.84 | 1.93 | 3.30 | 4.19 | 4.22 | 3.68 | 3.98 | 3.44 | 3.62 | 2.81 | 3.47 | 3.13 | 39.61 |
| Days ≥ 0.1" Precip | 4 | 4 | 7 | 8 | 7 | 6 | 6 | 5 | 6 | 6 | 7 | 6 | 72 |
| Total Snowfall (") | 5.9 | 4.8 | 3.7 | 0.7 | 0.0 | 0.0 | 0.0 | 0.0 | 0.0 | 0.0 | 1.3 | 3.7 | 20.1 |
| Days ≥ 1" Snow Depth | 10 | 6 | 2 | 0 | 0 | 0 | 0 | 0 | 0 | 0 | 1 | 4 | 23 |

## CARLYLE RESERVOIR *Clinton County* ELEVATION 502 ft LAT/LONG 38° 38 ' N / 89° 20 ' W

| | JAN | FEB | MAR | APR | MAY | JUN | JUL | AUG | SEP | OCT | NOV | DEC | YEAR |
|---|---|---|---|---|---|---|---|---|---|---|---|---|---|
| Maximum Temp °F | na | na | 52.5 | 65.0 | 74.7 | 83.8 | 88.0 | 85.9 | 78.7 | 67.0 | 54.0 | ***42.1*** | na |
| Minimum Temp °F | na | ***21.1*** | ***32.9*** | 44.4 | 53.8 | 63.0 | 67.2 | 64.7 | 57.0 | 44.6 | 35.4 | ***25.2*** | na |
| Mean Temp °F | na | na | ***42.6*** | 54.7 | 64.3 | 73.4 | 77.6 | 75.3 | 67.9 | 55.8 | 44.7 | ***33.6*** | na |
| Days Max Temp ≥ 90 °F | 0 | 0 | 0 | 0 | 1 | 6 | 13 | ***9*** | 3 | 0 | 0 | 0 | 32 |
| Days Max Temp ≤ 32 °F | 12 | ***8*** | 2 | 0 | 0 | 0 | 0 | 0 | 0 | 0 | 1 | 6 | 29 |
| Days Min Temp ≤ 32 °F | ***26*** | 20 | 15 | 3 | 0 | 0 | 0 | 0 | 0 | 3 | 13 | ***23*** | 103 |
| Days Min Temp ≤ 0 °F | 3 | 1 | 0 | 0 | 0 | 0 | 0 | 0 | 0 | 0 | 0 | ***1*** | 5 |
| Heating Degree Days | na | na | ***689*** | 324 | 108 | 9 | 1 | 2 | 58 | 302 | 603 | ***965*** | na |
| Cooling Degree Days | na | na | na | ***20*** | ***93*** | ***282*** | ***419*** | ***357*** | ***156*** | ***23*** | ***2*** | na | na |
| Total Precipitation (") | 1.89 | 2.30 | 3.76 | 4.08 | 3.58 | 4.02 | 3.72 | 3.01 | 3.62 | 3.05 | 3.99 | 3.39 | 40.41 |
| Days ≥ 0.1" Precip | ***4*** | 4 | 6 | 8 | 6 | 6 | 6 | 5 | 5 | 6 | 6 | ***6*** | 68 |
| Total Snowfall (") | na | ***2.7*** | ***2.3*** | 0.4 | 0.0 | 0.0 | 0.0 | 0.0 | 0.0 | 0.0 | ***1.0*** | ***2.8*** | na |
| Days ≥ 1" Snow Depth | ***8*** | ***6*** | ***2*** | 0 | 0 | 0 | 0 | 0 | 0 | 0 | ***0*** | ***3*** | 19 |

## CHARLESTON *Coles County* ELEVATION 689 ft LAT/LONG 39° 29 ' N / 88° 10 ' W

| | JAN | FEB | MAR | APR | MAY | JUN | JUL | AUG | SEP | OCT | NOV | DEC | YEAR |
|---|---|---|---|---|---|---|---|---|---|---|---|---|---|
| Maximum Temp °F | 34.7 | 40.0 | 52.2 | 65.0 | 74.6 | 83.2 | 86.4 | 84.1 | 78.0 | 66.3 | 52.2 | 40.1 | 63.1 |
| Minimum Temp °F | 18.6 | 22.5 | 33.1 | 43.7 | 53.3 | 62.2 | 66.3 | 63.9 | 56.8 | 45.4 | 35.4 | 25.0 | 43.8 |
| Mean Temp °F | 26.7 | 31.3 | 42.6 | 54.4 | 64.0 | 72.8 | 76.3 | 74.0 | 67.4 | 55.9 | 43.8 | 32.6 | 53.5 |
| Days Max Temp ≥ 90 °F | 0 | 0 | 0 | 0 | 0 | 5 | 9 | 6 | 1 | 0 | 0 | 0 | 21 |
| Days Max Temp ≤ 32 °F | 13 | 8 | 2 | 0 | 0 | 0 | 0 | 0 | 0 | 0 | 1 | 7 | 31 |
| Days Min Temp ≤ 32 °F | 27 | 22 | 16 | 4 | 0 | 0 | 0 | 0 | 0 | 3 | 13 | 24 | 109 |
| Days Min Temp ≤ 0 °F | 3 | 2 | 0 | 0 | 0 | 0 | 0 | 0 | 0 | 0 | 0 | 1 | 6 |
| Heating Degree Days | 1182 | 947 | 688 | 331 | 115 | 11 | 1 | 3 | 62 | 297 | 629 | 999 | 5265 |
| Cooling Degree Days | 0 | 0 | 3 | 20 | 88 | 255 | 367 | 297 | 146 | 21 | 1 | 0 | 1198 |
| Total Precipitation (") | 2.04 | 2.13 | 3.03 | 4.00 | 3.96 | 3.63 | 4.76 | 3.19 | 3.58 | 3.22 | 3.88 | 3.48 | 40.90 |
| Days ≥ 0.1" Precip | 5 | 5 | 7 | 8 | 7 | 6 | 7 | 5 | 5 | 5 | 7 | 6 | 73 |
| Total Snowfall (") | 7.7 | 4.7 | 3.2 | 0.2 | 0.0 | 0.0 | 0.0 | 0.0 | 0.0 | 0.0 | 1.6 | 3.5 | 20.9 |
| Days ≥ 1" Snow Depth | ***12*** | 9 | 3 | 0 | 0 | 0 | 0 | 0 | 0 | 0 | 1 | ***6*** | 31 |

## CHENOA *McLean County* ELEVATION 712 ft LAT/LONG 40° 44 ' N / 88° 44 ' W

| | JAN | FEB | MAR | APR | MAY | JUN | JUL | AUG | SEP | OCT | NOV | DEC | YEAR |
|---|---|---|---|---|---|---|---|---|---|---|---|---|---|
| Maximum Temp °F | 31.4 | 36.0 | 48.9 | 63.1 | 74.4 | 83.6 | 85.8 | 83.4 | 77.4 | 65.1 | 49.7 | 36.4 | 61.3 |
| Minimum Temp °F | 14.6 | 18.4 | 29.5 | 40.0 | 49.9 | 59.5 | 63.2 | 60.6 | 53.6 | 42.4 | 32.0 | 21.4 | 40.4 |
| Mean Temp °F | 23.1 | 27.2 | 39.2 | 51.6 | 62.2 | 71.6 | 74.5 | 72.0 | 65.5 | 53.8 | 40.9 | 29.0 | 50.9 |
| Days Max Temp ≥ 90 °F | 0 | 0 | 0 | 0 | 2 | 6 | 8 | 5 | 2 | 0 | 0 | 0 | 23 |
| Days Max Temp ≤ 32 °F | 16 | 11 | 3 | 0 | 0 | 0 | 0 | 0 | 0 | 0 | 2 | 10 | 42 |
| Days Min Temp ≤ 32 °F | 29 | 25 | 20 | 7 | 1 | 0 | 0 | 0 | 0 | 5 | 17 | 26 | 130 |
| Days Min Temp ≤ 0 °F | 6 | 3 | 0 | 0 | 0 | 0 | 0 | 0 | 0 | 0 | 0 | 2 | 11 |
| Heating Degree Days | 1293 | 1061 | 792 | 408 | 155 | 17 | 3 | 9 | 89 | 356 | 718 | 1108 | 6009 |
| Cooling Degree Days | 0 | 0 | 2 | 15 | 77 | 228 | 317 | 250 | 116 | 11 | 0 | 0 | 1016 |
| Total Precipitation (") | 1.61 | 1.35 | 2.82 | 3.63 | 3.60 | 3.95 | 3.55 | 3.36 | 3.62 | 2.76 | 2.83 | 2.79 | 35.87 |
| Days ≥ 0.1" Precip | 4 | 4 | 6 | 7 | 7 | 7 | 6 | 6 | 6 | 6 | 6 | 5 | 70 |
| Total Snowfall (") | 7.3 | 5.9 | 3.1 | 0.9 | 0.0 | 0.0 | 0.0 | 0.0 | 0.0 | 0.1 | 1.7 | 5.3 | 24.3 |
| Days ≥ 1" Snow Depth | 14 | 11 | 3 | 0 | 0 | 0 | 0 | 0 | 0 | 0 | 1 | 7 | 36 |

## CHICAGO MIDWAY AP *Cook County* ELEVATION 620 ft LAT/LONG 41° 44 ' N / 87° 46 ' W

| | JAN | FEB | MAR | APR | MAY | JUN | JUL | AUG | SEP | OCT | NOV | DEC | YEAR |
|---|---|---|---|---|---|---|---|---|---|---|---|---|---|
| Maximum Temp °F | 30.2 | 34.7 | 46.5 | 59.2 | 70.8 | 80.5 | 84.5 | 82.3 | 75.1 | 63.2 | 48.4 | 36.0 | 59.3 |
| Minimum Temp °F | 15.5 | 19.8 | 30.1 | 40.5 | 50.3 | 60.3 | 65.8 | 64.4 | 56.5 | 44.5 | 33.7 | 22.5 | 42.0 |
| Mean Temp °F | 22.9 | 27.3 | 38.4 | 49.9 | 60.6 | 70.5 | 75.2 | 73.3 | 65.8 | 53.8 | 41.1 | 29.3 | 50.7 |
| Days Max Temp ≥ 90 °F | 0 | 0 | 0 | 0 | 1 | 4 | 8 | 5 | 2 | 0 | 0 | 0 | 20 |
| Days Max Temp ≤ 32 °F | 16 | 12 | 4 | 0 | 0 | 0 | 0 | 0 | 0 | 0 | 2 | 10 | 44 |
| Days Min Temp ≤ 32 °F | 28 | 24 | 19 | 5 | 0 | 0 | 0 | 0 | 0 | 2 | 14 | 25 | 117 |
| Days Min Temp ≤ 0 °F | 5 | 2 | 0 | 0 | 0 | 0 | 0 | 0 | 0 | 0 | 0 | 2 | 9 |
| Heating Degree Days | 1299 | 1060 | 819 | 460 | 198 | 31 | 3 | 6 | 84 | 355 | 712 | 1101 | 6128 |
| Cooling Degree Days | 0 | 0 | 2 | 15 | 73 | 204 | 340 | 280 | 120 | 11 | 0 | 0 | 1045 |
| Total Precipitation (") | 1.83 | 1.59 | 2.87 | 3.95 | 3.57 | 4.18 | 3.87 | 3.90 | 3.54 | 2.86 | 3.31 | 2.94 | 38.41 |
| Days ≥ 0.1" Precip | 5 | 5 | 7 | 7 | 7 | 6 | 6 | 6 | 6 | 6 | 7 | 6 | 74 |
| Total Snowfall (") | 12.6 | 11.3 | 6.6 | 1.8 | 0.0 | 0.0 | 0.0 | 0.0 | 0.0 | 0.3 | 2.2 | 9.4 | 44.2 |
| Days ≥ 1" Snow Depth | 19 | 14 | 5 | 1 | 0 | 0 | 0 | 0 | 0 | 0 | 1 | 10 | 50 |

## CHICAGO OHARE INT AP *Cook County* ELEVATION 658 ft LAT/LONG 41° 59 ' N / 87° 54 ' W

| | JAN | FEB | MAR | APR | MAY | JUN | JUL | AUG | SEP | OCT | NOV | DEC | YEAR |
|---|---|---|---|---|---|---|---|---|---|---|---|---|---|
| Maximum Temp °F | 29.4 | 33.8 | 45.8 | 58.6 | 70.1 | 79.4 | 83.9 | 81.8 | 74.5 | 62.7 | 48.0 | 35.3 | 58.6 |
| Minimum Temp °F | 13.7 | 18.0 | 28.5 | 38.5 | 47.6 | 57.3 | 63.0 | 61.8 | 54.0 | 42.1 | 31.9 | 20.8 | 39.8 |
| Mean Temp °F | 21.5 | 25.9 | 37.2 | 48.6 | 58.9 | 68.4 | 73.5 | 71.8 | 64.3 | 52.4 | 40.0 | 28.1 | 49.2 |
| Days Max Temp ≥ 90 °F | 0 | 0 | 0 | 0 | 1 | 4 | 7 | 4 | 1 | 0 | 0 | 0 | 17 |
| Days Max Temp ≤ 32 °F | 17 | 13 | 4 | 0 | 0 | 0 | 0 | 0 | 0 | 0 | 2 | 10 | 46 |
| Days Min Temp ≤ 32 °F | 29 | 25 | 21 | 7 | 1 | 0 | 0 | 0 | 0 | 5 | 16 | 26 | 130 |
| Days Min Temp ≤ 0 °F | 6 | 3 | 0 | 0 | 0 | 0 | 0 | 0 | 0 | 0 | 0 | 2 | 11 |
| Heating Degree Days | 1341 | 1098 | 856 | 495 | 232 | 48 | 6 | 12 | 109 | 396 | 745 | 1139 | 6477 |
| Cooling Degree Days | 0 | 0 | 1 | 12 | 50 | 152 | 286 | 232 | 94 | 6 | 0 | 0 | 833 |
| Total Precipitation (") | 1.65 | 1.44 | 2.71 | 3.61 | 3.26 | 4.01 | 3.46 | 4.50 | 3.57 | 2.74 | 3.15 | 2.57 | 36.67 |
| Days ≥ 0.1" Precip | 5 | 4 | 6 | 7 | 7 | 6 | 6 | 7 | 6 | 5 | 7 | 6 | 72 |
| Total Snowfall (") | 10.9 | 8.7 | 6.4 | 1.7 | 0.1 | 0.0 | 0.0 | 0.0 | 0.0 | 0.5 | 1.8 | 8.4 | 38.5 |
| Days ≥ 1" Snow Depth | 18 | 14 | 5 | 1 | 0 | 0 | 0 | 0 | 0 | 0 | 1 | 8 | 47 |

## CHICAGO UNIVERSITY *Cook County* ELEVATION 594 ft LAT/LONG 41° 47 ' N / 87° 36 ' W

| | JAN | FEB | MAR | APR | MAY | JUN | JUL | AUG | SEP | OCT | NOV | DEC | YEAR |
|---|---|---|---|---|---|---|---|---|---|---|---|---|---|
| Maximum Temp °F | 31.4 | 35.9 | 46.6 | 58.6 | 69.8 | 80.0 | 84.2 | 82.6 | 75.8 | 63.9 | 49.1 | 36.9 | 59.6 |
| Minimum Temp °F | 17.8 | 22.0 | 31.5 | 41.4 | 50.8 | 60.6 | 66.6 | 65.3 | 58.2 | 46.9 | 35.9 | 24.2 | 43.4 |
| Mean Temp °F | 24.7 | 29.0 | 39.0 | 50.0 | 60.3 | 70.3 | 75.4 | 74.0 | 67.0 | 55.4 | 42.5 | 30.6 | 51.5 |
| Days Max Temp ≥ 90 °F | 0 | 0 | 0 | 0 | 1 | 4 | 7 | 5 | 1 | 0 | 0 | 0 | 18 |
| Days Max Temp ≤ 32 °F | 16 | 11 | 3 | 0 | 0 | 0 | 0 | 0 | 0 | 0 | 1 | 9 | 40 |
| Days Min Temp ≤ 32 °F | 27 | 23 | 17 | 4 | 0 | 0 | 0 | 0 | 0 | 1 | 11 | 24 | 107 |
| Days Min Temp ≤ 0 °F | 4 | 1 | 0 | 0 | 0 | 0 | 0 | 0 | 0 | 0 | 0 | 1 | 6 |
| Heating Degree Days | 1245 | 1009 | 798 | 457 | 201 | 36 | 3 | 3 | 64 | 310 | 669 | 1060 | 5855 |
| Cooling Degree Days | 0 | 0 | 2 | 16 | 72 | 211 | 350 | 309 | 140 | 15 | 0 | 0 | 1115 |
| Total Precipitation (") | 2.02 | 1.64 | 2.94 | 3.76 | 3.45 | 4.21 | 3.57 | 4.00 | 3.38 | 2.73 | 3.36 | 2.89 | 37.95 |
| Days ≥ 0.1" Precip | 5 | 5 | 7 | 7 | 7 | 7 | 6 | 6 | 6 | 5 | 7 | 6 | 74 |
| Total Snowfall (") | na | na | na | na | na | na | na | na | na | na | na | na | na |
| Days ≥ 1" Snow Depth | na | na | na | na | na | na | na | na | na | na | na | na | na |

## DANVILLE *Vermilion County* ELEVATION 600 ft LAT/LONG 40° 8 ' N / 87° 38 ' W

| | JAN | FEB | MAR | APR | MAY | JUN | JUL | AUG | SEP | OCT | NOV | DEC | YEAR |
|---|---|---|---|---|---|---|---|---|---|---|---|---|---|
| Maximum Temp °F | 34.1 | 39.1 | 51.6 | 64.9 | 75.3 | 84.0 | 86.7 | 84.2 | 78.3 | 66.3 | 52.0 | 39.4 | 63.0 |
| Minimum Temp °F | 16.8 | 20.5 | 31.0 | 41.3 | 50.3 | 59.6 | 64.0 | 61.8 | 54.7 | 42.9 | 33.9 | 23.7 | 41.7 |
| Mean Temp °F | 25.5 | 29.8 | 41.3 | 53.1 | 62.8 | 71.8 | 75.3 | 73.0 | 66.5 | 54.6 | 43.0 | 31.6 | 52.4 |
| Days Max Temp ≥ 90 °F | 0 | 0 | 0 | 0 | 1 | 6 | 9 | 5 | 2 | 0 | 0 | 0 | 23 |
| Days Max Temp ≤ 32 °F | 14 | 9 | 2 | 0 | 0 | 0 | 0 | 0 | 0 | 0 | 1 | 7 | 33 |
| Days Min Temp ≤ 32 °F | 28 | 24 | 18 | 6 | 1 | 0 | 0 | 0 | 0 | 5 | 15 | 24 | 121 |
| Days Min Temp ≤ 0 °F | 4 | 3 | 0 | 0 | 0 | 0 | 0 | 0 | 0 | 0 | 0 | 2 | 9 |
| Heating Degree Days | 1219 | 987 | 729 | 366 | 138 | 15 | 2 | 5 | 74 | 332 | 654 | 1030 | 5551 |
| Cooling Degree Days | 0 | 0 | 2 | 17 | 80 | 233 | 342 | 274 | 129 | 17 | 1 | 0 | 1095 |
| Total Precipitation (") | 1.98 | 1.87 | 2.90 | 4.23 | 4.27 | 3.95 | 4.41 | 4.15 | 3.36 | 3.20 | 3.51 | 3.14 | 40.97 |
| Days ≥ 0.1" Precip | 5 | 4 | 7 | 8 | 8 | 7 | 7 | 6 | 5 | 6 | 7 | 6 | 76 |
| Total Snowfall (") | *5.9* | *5.0* | 3.2 | 0.3 | 0.0 | 0.0 | 0.0 | 0.0 | 0.0 | 0.1 | *1.1* | *5.2* | 20.8 |
| Days ≥ 1" Snow Depth | *9* | *7* | 2 | 0 | 0 | 0 | 0 | 0 | 0 | 0 | 0 | *5* | 23 |

## DECATUR *Macon County* ELEVATION 620 ft LAT/LONG 39° 50 ' N / 89° 1 ' W

| | JAN | FEB | MAR | APR | MAY | JUN | JUL | AUG | SEP | OCT | NOV | DEC | YEAR |
|---|---|---|---|---|---|---|---|---|---|---|---|---|---|
| Maximum Temp °F | 34.1 | 39.6 | 51.9 | 65.3 | 76.0 | 84.7 | 87.8 | 85.6 | 79.6 | 67.2 | 52.1 | 39.9 | 63.7 |
| Minimum Temp °F | 16.6 | 20.7 | 31.3 | 42.1 | 51.3 | 60.2 | 64.3 | 62.2 | 54.8 | 43.4 | 33.4 | 23.4 | 42.0 |
| Mean Temp °F | 25.4 | 30.2 | 41.6 | 53.8 | 63.7 | 72.5 | 76.1 | 73.9 | 67.2 | 55.3 | 42.8 | 31.7 | 52.9 |
| Days Max Temp ≥ 90 °F | 0 | 0 | 0 | 0 | 2 | 8 | 12 | 8 | 3 | 0 | 0 | 0 | 33 |
| Days Max Temp ≤ 32 °F | 13 | 8 | 2 | 0 | 0 | 0 | 0 | 0 | 0 | 0 | 1 | 7 | 31 |
| Days Min Temp ≤ 32 °F | 28 | 23 | 18 | 5 | 0 | 0 | 0 | 0 | 0 | 5 | 15 | 25 | 119 |
| Days Min Temp ≤ 0 °F | 4 | 3 | 0 | 0 | 0 | 0 | 0 | 0 | 0 | 0 | 0 | 1 | 8 |
| Heating Degree Days | 1220 | 977 | 720 | 348 | 124 | 13 | 1 | 5 | 67 | 314 | 660 | 1025 | 5474 |
| Cooling Degree Days | 0 | 0 | 2 | 20 | 84 | 243 | 355 | 296 | 142 | 19 | 1 | 0 | 1162 |
| Total Precipitation (") | 2.11 | 1.91 | 3.23 | 3.92 | 4.23 | 3.89 | 4.63 | 4.11 | 3.56 | 2.87 | 3.13 | 3.30 | 40.89 |
| Days ≥ 0.1" Precip | 5 | 4 | 7 | 7 | 6 | 6 | 6 | 6 | 5 | 5 | 6 | 6 | 69 |
| Total Snowfall (") | 7.4 | 5.9 | na | *0.4* | 0.0 | 0.0 | 0.0 | 0.0 | 0.0 | 0.0 | 1.3 | *5.7* | na |
| Days ≥ 1" Snow Depth | 14 | 10 | *4* | 0 | 0 | 0 | 0 | 0 | 0 | 0 | 1 | 6 | 35 |

## DIXON 1 NW *Lee County* ELEVATION 702 ft LAT/LONG 41° 51 ' N / 89° 29 ' W

| | JAN | FEB | MAR | APR | MAY | JUN | JUL | AUG | SEP | OCT | NOV | DEC | YEAR |
|---|---|---|---|---|---|---|---|---|---|---|---|---|---|
| Maximum Temp °F | 27.9 | 32.8 | 45.7 | 60.4 | 72.2 | 81.4 | 84.7 | 82.5 | 74.9 | 63.0 | 47.0 | 33.4 | 58.8 |
| Minimum Temp °F | 9.9 | 14.6 | 26.9 | 38.2 | 48.6 | 58.1 | 62.6 | 59.8 | 51.7 | 39.9 | 29.3 | 17.6 | 38.1 |
| Mean Temp °F | 18.9 | 23.7 | 36.3 | 49.3 | 60.5 | 69.8 | 73.7 | 71.2 | 63.3 | 51.5 | 38.2 | 25.5 | 48.5 |
| Days Max Temp ≥ 90 °F | 0 | 0 | 0 | 0 | 1 | 4 | 7 | 5 | 2 | 0 | 0 | 0 | 19 |
| Days Max Temp ≤ 32 °F | 19 | 13 | 4 | 0 | 0 | 0 | 0 | 0 | 0 | 0 | 3 | 13 | 52 |
| Days Min Temp ≤ 32 °F | 30 | 26 | 22 | 8 | 1 | 0 | 0 | 0 | 0 | 7 | 20 | 28 | 142 |
| Days Min Temp ≤ 0 °F | 9 | 5 | 0 | 0 | 0 | 0 | 0 | 0 | 0 | 0 | 0 | 4 | 18 |
| Heating Degree Days | 1423 | 1159 | 882 | 474 | 193 | 29 | 5 | 16 | 125 | 424 | 799 | 1217 | 6746 |
| Cooling Degree Days | 0 | 0 | 1 | 10 | 50 | 160 | 266 | 206 | 73 | 6 | 0 | 0 | 772 |
| Total Precipitation (") | 1.46 | 1.22 | 2.54 | 3.50 | 3.88 | 4.72 | 3.35 | 4.14 | 3.84 | 2.71 | 2.85 | 2.21 | 36.42 |
| Days ≥ 0.1" Precip | 4 | 3 | 6 | 7 | 7 | 7 | 6 | 6 | 6 | 6 | 6 | 5 | 69 |
| Total Snowfall (") | 9.6 | 7.4 | 4.9 | 1.2 | 0.0 | 0.0 | 0.0 | 0.0 | 0.0 | 0.2 | 1.6 | 8.7 | 33.6 |
| Days ≥ 1" Snow Depth | 19 | 15 | 6 | 0 | 0 | 0 | 0 | 0 | 0 | 0 | 1 | 12 | 53 |

## DU QUOIN 4 SE *Perry County* ELEVATION 459 ft LAT/LONG 38° 1 ' N / 89° 14 ' W

| | JAN | FEB | MAR | APR | MAY | JUN | JUL | AUG | SEP | OCT | NOV | DEC | YEAR |
|---|---|---|---|---|---|---|---|---|---|---|---|---|---|
| Maximum Temp °F | 39.1 | 44.8 | 56.4 | 68.1 | 77.0 | 85.9 | 89.8 | 87.7 | 81.0 | 69.8 | 55.9 | 43.8 | 66.6 |
| Minimum Temp °F | 21.3 | 25.2 | 35.3 | 45.4 | 54.2 | 62.8 | 67.0 | 63.8 | 57.0 | 44.9 | 36.0 | 26.7 | 45.0 |
| Mean Temp °F | 30.2 | 35.0 | 45.9 | 56.8 | 65.6 | 74.4 | 78.4 | 75.8 | 69.0 | 57.4 | 46.0 | 35.1 | 55.8 |
| Days Max Temp ≥ 90 °F | 0 | 0 | 0 | 0 | 2 | 9 | 17 | 12 | 4 | 0 | 0 | 0 | 44 |
| Days Max Temp ≤ 32 °F | 9 | 5 | 1 | 0 | 0 | 0 | 0 | 0 | 0 | 0 | 0 | 4 | 19 |
| Days Min Temp ≤ 32 °F | 26 | 20 | 13 | 3 | 0 | 0 | 0 | 0 | 0 | 4 | 12 | 22 | 100 |
| Days Min Temp ≤ 0 °F | 2 | 1 | 0 | 0 | 0 | 0 | 0 | 0 | 0 | 0 | 0 | 1 | 4 |
| Heating Degree Days | 1072 | 841 | 590 | 270 | 82 | 5 | 0 | 1 | 45 | 260 | 566 | 920 | 4652 |
| Cooling Degree Days | 0 | 0 | 4 | 29 | 107 | 301 | 435 | 358 | 169 | 29 | 2 | 0 | 1434 |
| Total Precipitation (") | 2.45 | 2.49 | 4.13 | 3.97 | 4.25 | 3.51 | 3.74 | 3.17 | 3.61 | 3.22 | 4.50 | 3.66 | 42.70 |
| Days ≥ 0.1" Precip | 5 | 5 | 7 | 8 | 7 | 6 | 5 | 5 | 5 | 5 | 6 | 6 | 70 |
| Total Snowfall (") | *5.2* | *4.2* | *1.2* | 0.4 | 0.0 | 0.0 | 0.0 | 0.0 | 0.0 | 0.2 | *0.8* | na | na |
| Days ≥ 1" Snow Depth | *7* | na | *1* | 0 | 0 | 0 | 0 | 0 | 0 | 0 | *0* | *3* | na |

## EFFINGHAM *Effingham County* ELEVATION 595 ft LAT/LONG 39° 8 ' N / 88° 32 ' W

| | JAN | FEB | MAR | APR | MAY | JUN | JUL | AUG | SEP | OCT | NOV | DEC | YEAR |
|---|---|---|---|---|---|---|---|---|---|---|---|---|---|
| Maximum Temp °F | 34.8 | 39.9 | 51.7 | 64.6 | 74.6 | 83.9 | 87.6 | 85.3 | 78.6 | 66.5 | 52.7 | 40.5 | 63.4 |
| Minimum Temp °F | 17.6 | 21.0 | 31.9 | 42.8 | 51.9 | 61.2 | 65.6 | 62.9 | 55.4 | 43.3 | 34.2 | 23.9 | 42.6 |
| Mean Temp °F | 26.2 | 30.5 | 41.8 | 53.7 | 63.3 | 72.6 | 76.6 | 74.1 | 67.0 | 54.9 | 43.5 | 32.3 | 53.0 |
| Days Max Temp ≥ 90 °F | 0 | 0 | 0 | 0 | 1 | 7 | 12 | 9 | 3 | 0 | 0 | 0 | 32 |
| Days Max Temp ≤ 32 °F | 13 | 8 | 2 | 0 | 0 | 0 | 0 | 0 | 0 | 0 | 1 | 7 | 31 |
| Days Min Temp ≤ 32 °F | 27 | 23 | 17 | 4 | 0 | 0 | 0 | 0 | 0 | 5 | 14 | 24 | 114 |
| Days Min Temp ≤ 0 °F | 4 | 2 | 0 | 0 | 0 | 0 | 0 | 0 | 0 | 0 | 0 | 1 | 7 |
| Heating Degree Days | 1195 | 968 | 714 | 350 | 128 | 13 | 1 | 4 | 71 | 326 | 640 | 1009 | 5419 |
| Cooling Degree Days | 0 | 0 | 3 | 17 | 77 | 253 | 377 | 298 | 138 | 18 | 1 | 0 | 1182 |
| Total Precipitation (") | 2.15 | 2.31 | 3.49 | 4.06 | 4.05 | 4.26 | 4.32 | 2.83 | 3.55 | 2.95 | 3.94 | 3.54 | 41.45 |
| Days ≥ 0.1" Precip | 5 | 5 | 7 | 8 | 7 | 7 | 7 | 5 | 5 | 5 | 7 | 6 | 74 |
| Total Snowfall (") | 6.8 | 5.4 | 3.2 | 0.2 | 0.0 | 0.0 | 0.0 | 0.0 | 0.0 | 0.0 | 1.4 | 4.0 | 21.0 |
| Days ≥ 1" Snow Depth | 10 | *8* | 3 | 0 | 0 | 0 | 0 | 0 | 0 | 0 | 1 | 4 | 26 |

## FAIRFIELD RADIO WFIW *Wayne County* ELEVATION 449 ft LAT/LONG 38° 23 ' N / 88° 22 ' W

| | JAN | FEB | MAR | APR | MAY | JUN | JUL | AUG | SEP | OCT | NOV | DEC | YEAR |
|---|---|---|---|---|---|---|---|---|---|---|---|---|---|
| Maximum Temp °F | 37.7 | 42.9 | 54.9 | 66.8 | 76.0 | 84.9 | 88.2 | 86.4 | 79.9 | 68.3 | 54.6 | 43.0 | 65.3 |
| Minimum Temp °F | 20.8 | 24.3 | 34.3 | 44.2 | 52.9 | 61.5 | 65.6 | 63.0 | 56.3 | 44.4 | 36.0 | 26.5 | 44.2 |
| Mean Temp °F | 29.3 | 33.6 | 44.6 | 55.5 | 64.5 | 73.2 | 76.9 | 74.7 | 68.1 | 56.4 | 45.3 | 34.8 | 54.7 |
| Days Max Temp ≥ 90 °F | 0 | 0 | 0 | 0 | 1 | 7 | 13 | 9 | 3 | 0 | 0 | 0 | 33 |
| Days Max Temp ≤ 32 °F | 11 | 6 | 1 | 0 | 0 | 0 | 0 | 0 | 0 | 0 | 1 | 5 | 24 |
| Days Min Temp ≤ 32 °F | 26 | 21 | 15 | 4 | 0 | 0 | 0 | 0 | 0 | 4 | 12 | 22 | 104 |
| Days Min Temp ≤ 0 °F | 2 | 1 | 0 | 0 | 0 | 0 | 0 | 0 | 0 | 0 | 0 | 1 | 4 |
| Heating Degree Days | 1100 | 880 | 629 | 301 | 102 | 9 | 1 | 2 | 57 | 285 | 585 | 930 | 4881 |
| Cooling Degree Days | 0 | 0 | 3 | 22 | 91 | 263 | 385 | 321 | 154 | 24 | 1 | 0 | 1264 |
| Total Precipitation (") | 2.47 | 2.62 | 4.40 | 4.73 | 4.60 | 3.71 | 3.94 | 3.26 | 3.36 | 3.48 | 4.39 | 3.71 | 44.67 |
| Days ≥ 0.1" Precip | 5 | 5 | 8 | 8 | 7 | 6 | 6 | 5 | 6 | 6 | 7 | 7 | 76 |
| Total Snowfall (") | 4.9 | 4.2 | 2.3 | 0.3 | 0.0 | 0.0 | 0.0 | 0.0 | 0.0 | 0.1 | 0.7 | 2.5 | 15.0 |
| Days ≥ 1" Snow Depth | 7 | 6 | 1 | 0 | 0 | 0 | 0 | 0 | 0 | 0 | 0 | 3 | 17 |

## FARMER CITY *De Witt County* ELEVATION 732 ft LAT/LONG 40° 14' N / 88° 38' W

| | JAN | FEB | MAR | APR | MAY | JUN | JUL | AUG | SEP | OCT | NOV | DEC | YEAR |
|---|---|---|---|---|---|---|---|---|---|---|---|---|---|
| Maximum Temp °F | 32.9 | ***37.3*** | 49.7 | 63.7 | 74.6 | 83.5 | 86.4 | 84.2 | 78.8 | ***65.9*** | 50.4 | 38.0 | 62.1 |
| Minimum Temp °F | 15.1 | ***19.2*** | 29.6 | 40.4 | 50.8 | 59.6 | 63.6 | 61.2 | 53.9 | ***42.4*** | 32.0 | 22.1 | 40.8 |
| Mean Temp °F | 24.0 | ***28.3*** | 39.7 | 52.1 | 62.8 | 71.6 | 75.0 | 72.7 | 66.4 | ***54.2*** | 41.2 | 30.0 | 51.5 |
| Days Max Temp ≥ 90 °F | 0 | 0 | 0 | 0 | 1 | 6 | ***9*** | ***6*** | 2 | 0 | 0 | 0 | 24 |
| Days Max Temp ≤ 32 °F | 14 | ***10*** | 2 | 0 | 0 | 0 | 0 | 0 | 0 | 0 | 1 | 8 | 35 |
| Days Min Temp ≤ 32 °F | 29 | ***24*** | 20 | 6 | 1 | 0 | 0 | 0 | 0 | ***5*** | 17 | 26 | 128 |
| Days Min Temp ≤ 0 °F | 5 | ***3*** | 0 | 0 | 0 | 0 | 0 | 0 | 0 | 0 | 0 | 2 | 10 |
| Heating Degree Days | 1264 | ***1028*** | 779 | 391 | 141 | 16 | 1 | 8 | 75 | ***345*** | 708 | 1077 | 5833 |
| Cooling Degree Days | 0 | ***0*** | 1 | 10 | 76 | ***209*** | 319 | ***266*** | 125 | ***13*** | ***0*** | ***0*** | 1019 |
| Total Precipitation (") | 1.73 | 1.50 | 2.89 | 3.70 | 3.89 | 3.42 | 4.05 | 3.74 | 3.21 | 2.83 | 2.91 | ***3.03*** | 36.90 |
| Days ≥ 0.1" Precip | 4 | 4 | 6 | ***7*** | 7 | 6 | 6 | 5 | 5 | ***6*** | ***6*** | ***6*** | 68 |
| Total Snowfall (") | 5.5 | ***5.4*** | 2.9 | 0.7 | 0.0 | 0.0 | 0.0 | 0.0 | 0.0 | 0.1 | 1.2 | ***5.9*** | 21.7 |
| Days ≥ 1" Snow Depth | 12 | ***12*** | 4 | 1 | 0 | 0 | 0 | 0 | 0 | 0 | 1 | 7 | 37 |

## FLORA 5 NW *Clay County* ELEVATION 502 ft LAT/LONG 38° 41' N / 88° 34' W

| | JAN | FEB | MAR | APR | MAY | JUN | JUL | AUG | SEP | OCT | NOV | DEC | YEAR |
|---|---|---|---|---|---|---|---|---|---|---|---|---|---|
| Maximum Temp °F | 38.0 | 43.0 | 55.1 | 67.4 | 76.7 | 85.7 | 89.3 | 87.3 | 80.6 | 69.1 | 55.2 | 43.0 | 65.9 |
| Minimum Temp °F | 20.3 | 23.7 | 34.0 | 43.8 | 52.4 | 61.3 | 65.3 | 62.6 | 55.7 | 44.3 | 35.5 | 25.9 | 43.7 |
| Mean Temp °F | 29.1 | 33.4 | 44.6 | 55.6 | 64.6 | 73.5 | 77.3 | 75.0 | 68.2 | 56.8 | 45.4 | 34.5 | 54.8 |
| Days Max Temp ≥ 90 °F | 0 | 0 | 0 | 0 | 1 | 8 | 15 | 11 | 4 | 0 | 0 | 0 | 39 |
| Days Max Temp ≤ 32 °F | 11 | 6 | 1 | 0 | 0 | 0 | 0 | 0 | 0 | 0 | 0 | 5 | 23 |
| Days Min Temp ≤ 32 °F | 26 | 21 | 15 | 4 | 1 | 0 | 0 | 0 | 0 | 4 | 12 | 23 | 106 |
| Days Min Temp ≤ 0 °F | 3 | 2 | 0 | 0 | 0 | 0 | 0 | 0 | 0 | 0 | 0 | 1 | 6 |
| Heating Degree Days | 1104 | 886 | 631 | 300 | 102 | 8 | 1 | 3 | 56 | 276 | 585 | 939 | 4891 |
| Cooling Degree Days | 0 | 0 | 5 | 27 | 99 | 285 | 409 | 338 | 164 | 30 | 2 | 0 | 1359 |
| Total Precipitation (") | 2.19 | 2.33 | 3.89 | 4.22 | 3.80 | 3.83 | 3.83 | 3.25 | 3.53 | 3.14 | 4.14 | 3.63 | 41.78 |
| Days ≥ 0.1" Precip | 5 | 5 | 8 | 8 | 7 | 6 | 6 | 5 | 6 | 5 | 7 | 7 | 75 |
| Total Snowfall (") | ***2.7*** | ***2.8*** | 1.6 | 0.2 | 0.0 | 0.0 | 0.0 | 0.0 | 0.0 | 0.0 | 0.8 | ***1.8*** | 9.9 |
| Days ≥ 1" Snow Depth | 8 | 6 | 1 | 0 | 0 | 0 | 0 | 0 | 0 | 0 | 1 | 3 | 19 |

## FULTON LOCK&DAM #13 *Whiteside County* ELEVATION 591 ft LAT/LONG 41° 54' N / 90° 9' W

| | JAN | FEB | MAR | APR | MAY | JUN | JUL | AUG | SEP | OCT | NOV | DEC | YEAR |
|---|---|---|---|---|---|---|---|---|---|---|---|---|---|
| Maximum Temp °F | 28.4 | 32.3 | 44.9 | 59.3 | 71.2 | 80.4 | 84.2 | 82.1 | 74.7 | 63.1 | 47.2 | 33.6 | 58.5 |
| Minimum Temp °F | 11.0 | 14.8 | 27.0 | 39.5 | 50.6 | 59.8 | 64.6 | 62.3 | 54.0 | 42.6 | 31.0 | 18.4 | 39.6 |
| Mean Temp °F | 19.7 | 23.6 | 36.0 | 49.5 | 60.9 | 70.1 | 74.4 | 72.2 | 64.4 | 52.9 | 39.1 | 26.0 | 49.1 |
| Days Max Temp ≥ 90 °F | 0 | 0 | 0 | 0 | 0 | 3 | 6 | 4 | 1 | 0 | 0 | 0 | 14 |
| Days Max Temp ≤ 32 °F | 18 | 13 | 4 | 0 | 0 | 0 | 0 | 0 | 0 | 0 | 2 | 13 | 50 |
| Days Min Temp ≤ 32 °F | 30 | 27 | 23 | 5 | 0 | 0 | 0 | 0 | 0 | 4 | 17 | 28 | 134 |
| Days Min Temp ≤ 0 °F | 8 | 5 | 0 | 0 | 0 | 0 | 0 | 0 | 0 | 0 | 0 | 3 | 16 |
| Heating Degree Days | 1400 | 1164 | 892 | 470 | 177 | 21 | 2 | 6 | 103 | 381 | 770 | 1201 | 6587 |
| Cooling Degree Days | 0 | 0 | 0 | 13 | 52 | ***178*** | 301 | 249 | 96 | 8 | 0 | 0 | 897 |
| Total Precipitation (") | 1.27 | 1.04 | 2.36 | 3.35 | 3.56 | 4.39 | 3.43 | 4.65 | 3.63 | 2.75 | 2.64 | 1.97 | 35.04 |
| Days ≥ 0.1" Precip | 3 | 3 | 5 | 7 | 7 | 7 | 6 | 6 | 6 | 5 | 5 | 4 | 64 |
| Total Snowfall (") | ***6.0*** | ***4.7*** | 2.9 | 0.6 | 0.0 | 0.0 | 0.0 | 0.0 | 0.0 | 0.0 | 1.0 | ***5.0*** | 20.2 |
| Days ≥ 1" Snow Depth | 17 | 15 | 6 | 0 | 0 | 0 | 0 | 0 | 0 | 0 | 1 | 11 | 50 |

## GALESBURG *Knox County* ELEVATION 781 ft LAT/LONG 40° 57' N / 90° 22' W

| | JAN | FEB | MAR | APR | MAY | JUN | JUL | AUG | SEP | OCT | NOV | DEC | YEAR |
|---|---|---|---|---|---|---|---|---|---|---|---|---|---|
| Maximum Temp °F | 29.6 | 34.7 | 47.4 | 61.5 | 72.5 | 81.6 | 84.9 | 82.4 | 75.3 | 63.4 | 48.0 | 34.8 | 59.7 |
| Minimum Temp °F | 12.9 | 17.6 | 28.7 | 40.3 | 51.0 | 60.5 | 65.0 | 62.4 | 54.5 | 42.5 | 30.9 | 19.6 | 40.5 |
| Mean Temp °F | 21.3 | 26.2 | 38.1 | 50.9 | 61.8 | 71.1 | 75.0 | 72.4 | 64.9 | 53.0 | 39.5 | 27.3 | 50.1 |
| Days Max Temp ≥ 90 °F | 0 | 0 | 0 | 0 | 0 | 4 | 8 | 4 | 1 | 0 | 0 | 0 | 17 |
| Days Max Temp ≤ 32 °F | 17 | 12 | 4 | 0 | 0 | 0 | 0 | 0 | 0 | 0 | 2 | 12 | 47 |
| Days Min Temp ≤ 32 °F | 29 | 26 | 20 | 6 | 0 | 0 | 0 | 0 | 0 | 5 | 18 | 27 | 131 |
| Days Min Temp ≤ 0 °F | 7 | 4 | 0 | 0 | 0 | 0 | 0 | 0 | 0 | 0 | 0 | 3 | 14 |
| Heating Degree Days | 1350 | 1090 | 829 | 427 | 162 | 18 | 3 | 9 | 99 | 378 | 759 | 1163 | 6287 |
| Cooling Degree Days | 0 | 0 | 2 | 15 | 68 | 216 | 332 | 259 | 107 | 9 | 0 | 0 | 1008 |
| Total Precipitation (") | 1.44 | 1.32 | 2.86 | 4.03 | 3.64 | 4.03 | 4.63 | 4.05 | 3.77 | 2.54 | 2.71 | 2.41 | 37.43 |
| Days ≥ 0.1" Precip | 4 | 4 | 6 | 7 | 7 | 6 | 7 | 6 | 6 | 5 | 6 | 5 | 69 |
| Total Snowfall (") | 7.2 | 5.5 | 3.1 | 1.3 | 0.0 | 0.0 | 0.0 | 0.0 | 0.0 | 0.1 | 1.8 | 5.1 | 24.1 |
| Days ≥ 1" Snow Depth | 16 | 11 | 4 | 1 | 0 | 0 | 0 | 0 | 0 | 0 | 1 | 8 | 41 |

## GALVA *Henry County* ELEVATION 850 ft LAT/LONG 41° 10 ' N / 90° 2 ' W

| | JAN | FEB | MAR | APR | MAY | JUN | JUL | AUG | SEP | OCT | NOV | DEC | YEAR |
|---|---|---|---|---|---|---|---|---|---|---|---|---|---|
| Maximum Temp °F | 28.4 | 33.2 | 46.3 | 60.7 | 72.3 | 81.7 | 84.9 | 82.6 | 75.5 | 63.1 | 47.3 | 33.9 | 59.2 |
| Minimum Temp °F | 11.7 | 16.1 | 27.2 | 38.9 | 49.8 | 59.3 | 63.8 | 61.1 | 53.2 | 41.2 | 30.3 | 18.3 | 39.2 |
| Mean Temp °F | 20.1 | 24.7 | 36.8 | 49.8 | 61.1 | 70.5 | 74.4 | 71.9 | 64.4 | 52.2 | 38.8 | 26.1 | 49.2 |
| Days Max Temp ≥ 90 °F | 0 | 0 | 0 | 0 | 0 | 3 | 8 | 5 | 1 | 0 | 0 | 0 | 17 |
| Days Max Temp ≤ 32 °F | 19 | 13 | 4 | 0 | 0 | 0 | 0 | 0 | 0 | 0 | 3 | 13 | 52 |
| Days Min Temp ≤ 32 °F | 30 | 26 | 22 | 8 | 1 | 0 | 0 | 0 | 0 | 5 | 18 | 28 | 138 |
| Days Min Temp ≤ 0 °F | 8 | 5 | 0 | 0 | 0 | 0 | 0 | 0 | 0 | 0 | 0 | 3 | 16 |
| Heating Degree Days | 1388 | 1133 | 870 | 459 | 180 | 23 | 4 | 11 | 108 | 402 | 779 | 1199 | 6556 |
| Cooling Degree Days | 0 | 0 | 1 | 12 | 64 | 200 | 314 | 249 | 101 | 8 | 0 | 0 | 949 |
| Total Precipitation (") | 1.50 | 1.29 | 2.82 | 4.11 | 3.97 | 4.45 | 4.28 | 4.10 | 3.94 | 2.70 | 2.95 | 2.49 | 38.60 |
| Days ≥ 0.1" Precip | 4 | 4 | 6 | 7 | 8 | 6 | 7 | 6 | 6 | 6 | 6 | 5 | 71 |
| Total Snowfall (") | 8.2 | 6.7 | 4.7 | 1.6 | 0.0 | 0.0 | 0.0 | 0.0 | 0.0 | 0.3 | 2.2 | 6.3 | 30.0 |
| Days ≥ 1" Snow Depth | 21 | 15 | 6 | 1 | 0 | 0 | 0 | 0 | 0 | 0 | 2 | 12 | 57 |

## GENESEO *Henry County* ELEVATION 640 ft LAT/LONG 41° 27 ' N / 90° 10 ' W

| | JAN | FEB | MAR | APR | MAY | JUN | JUL | AUG | SEP | OCT | NOV | DEC | YEAR |
|---|---|---|---|---|---|---|---|---|---|---|---|---|---|
| Maximum Temp °F | 28.7 | 33.6 | 46.9 | 61.4 | 73.2 | 82.5 | 85.8 | 83.1 | 75.5 | 63.1 | 47.3 | 34.0 | 59.6 |
| Minimum Temp °F | 12.1 | 16.9 | 28.3 | 39.8 | 50.6 | 60.3 | 64.5 | 61.7 | 53.6 | 42.0 | 30.5 | 18.8 | 39.9 |
| Mean Temp °F | 20.4 | 25.3 | 37.6 | 50.6 | 61.9 | 71.4 | 75.2 | 72.5 | 64.5 | 52.6 | 38.9 | 26.4 | 49.8 |
| Days Max Temp ≥ 90 °F | 0 | 0 | 0 | 0 | 1 | 5 | 9 | 6 | 2 | 0 | 0 | 0 | 23 |
| Days Max Temp ≤ 32 °F | 18 | 12 | 3 | 0 | 0 | 0 | 0 | 0 | 0 | 0 | 3 | 12 | 48 |
| Days Min Temp ≤ 32 °F | 30 | 26 | 22 | 6 | 0 | 0 | 0 | 0 | 0 | 5 | 18 | 28 | 135 |
| Days Min Temp ≤ 0 °F | 7 | 4 | 0 | 0 | 0 | 0 | 0 | 0 | 0 | 0 | 0 | 3 | 14 |
| Heating Degree Days | 1376 | 1115 | 842 | 438 | 161 | 17 | 3 | 9 | 105 | 392 | 776 | 1189 | 6423 |
| Cooling Degree Days | 0 | 0 | 1 | 16 | 72 | 221 | 337 | 257 | 102 | 8 | 0 | 0 | 1014 |
| Total Precipitation (") | 1.44 | 1.27 | 2.61 | 3.72 | 3.92 | 4.31 | 4.42 | 4.27 | 3.37 | 2.96 | 2.81 | 2.20 | 37.30 |
| Days ≥ 0.1" Precip | 4 | 3 | 6 | 7 | 7 | 7 | 7 | 6 | 6 | 5 | 6 | 5 | 69 |
| Total Snowfall (") | 6.7 | 5.5 | 3.6 | 1.5 | 0.0 | 0.0 | 0.0 | 0.0 | 0.0 | 0.1 | 2.1 | *6.4* | 25.9 |
| Days ≥ 1" Snow Depth | *18* | *14* | 5 | 1 | 0 | 0 | 0 | 0 | 0 | 0 | 1 | *9* | 48 |

## GOLDEN *Adams County* ELEVATION 732 ft LAT/LONG 40° 6 ' N / 91° 1 ' W

| | JAN | FEB | MAR | APR | MAY | JUN | JUL | AUG | SEP | OCT | NOV | DEC | YEAR |
|---|---|---|---|---|---|---|---|---|---|---|---|---|---|
| Maximum Temp °F | 32.6 | 37.8 | 50.8 | 64.1 | 74.2 | 83.4 | 87.2 | 84.8 | 78.0 | 66.6 | 51.0 | 37.8 | 62.4 |
| Minimum Temp °F | 15.5 | 19.4 | 30.8 | 42.0 | 51.9 | 61.3 | 65.2 | 62.3 | 54.6 | 43.5 | 32.6 | 21.6 | 41.7 |
| Mean Temp °F | 24.1 | 28.6 | 40.8 | 53.1 | 63.1 | 72.4 | 76.3 | 73.6 | 66.3 | 55.1 | 41.8 | 29.7 | 52.1 |
| Days Max Temp ≥ 90 °F | 0 | 0 | 0 | 0 | 1 | 6 | 12 | 8 | 3 | 0 | 0 | 0 | 30 |
| Days Max Temp ≤ 32 °F | 14 | 10 | 2 | 0 | 0 | 0 | 0 | 0 | 0 | 0 | 2 | 9 | 37 |
| Days Min Temp ≤ 32 °F | 29 | 25 | 19 | 5 | 0 | 0 | 0 | 0 | 0 | 5 | 16 | 26 | 125 |
| Days Min Temp ≤ 0 °F | 5 | 3 | 0 | 0 | 0 | 0 | 0 | 0 | 0 | 0 | 0 | 2 | 10 |
| Heating Degree Days | 1262 | 1020 | 744 | 369 | 131 | 14 | 2 | 6 | 79 | 322 | 690 | 1086 | 5725 |
| Cooling Degree Days | 0 | 0 | 1 | 18 | 72 | 250 | 363 | 294 | 127 | 15 | 0 | 0 | 1140 |
| Total Precipitation (") | 1.25 | 1.40 | 2.66 | 3.79 | 4.38 | 3.89 | 4.68 | 3.37 | 4.14 | 3.22 | 2.81 | 2.30 | 37.89 |
| Days ≥ 0.1" Precip | 4 | 3 | 6 | 7 | 8 | 7 | 6 | 5 | 6 | 6 | 6 | 5 | 69 |
| Total Snowfall (") | 5.6 | 5.4 | 2.8 | 0.7 | 0.0 | 0.0 | 0.0 | 0.0 | 0.0 | 0.1 | 1.8 | *4.6* | 21.0 |
| Days ≥ 1" Snow Depth | *10* | 7 | 2 | 0 | 0 | 0 | 0 | 0 | 0 | 0 | *1* | 6 | 26 |

## GRIGGSVILLE *Pike County* ELEVATION 689 ft LAT/LONG 39° 43 ' N / 90° 44 ' W

| | JAN | FEB | MAR | APR | MAY | JUN | JUL | AUG | SEP | OCT | NOV | DEC | YEAR |
|---|---|---|---|---|---|---|---|---|---|---|---|---|---|
| Maximum Temp °F | *33.5* | *38.6* | *51.1* | *64.8* | *74.1* | *82.6* | *87.2* | *84.7* | *77.9* | *66.4* | *51.8* | *39.1* | 62.7 |
| Minimum Temp °F | *15.9* | *20.1* | *31.2* | *43.3* | *52.6* | *61.8* | *65.8* | *62.9* | *55.3* | *44.1* | *33.3* | *22.7* | 42.4 |
| Mean Temp °F | *24.7* | *29.4* | *41.2* | *54.0* | *63.4* | *72.2* | *76.6* | *73.8* | *66.6* | *55.3* | *42.6* | *30.9* | 52.6 |
| Days Max Temp ≥ 90 °F | *0* | *0* | *0* | *0* | *0* | *4* | *11* | *7* | *2* | *0* | *0* | *0* | 24 |
| Days Max Temp ≤ 32 °F | *14* | *9* | *2* | *0* | *0* | *0* | *0* | *0* | *0* | *0* | *1* | *9* | 35 |
| Days Min Temp ≤ 32 °F | *28* | *24* | *18* | *4* | *0* | *0* | *0* | *0* | *0* | *4* | *15* | *25* | 118 |
| Days Min Temp ≤ 0 °F | *4* | *3* | *0* | *0* | *0* | *0* | *0* | *0* | *0* | *0* | *0* | *2* | 9 |
| Heating Degree Days | *1241* | *1001* | *734* | *343* | *124* | *13* | *2* | *6* | *74* | *316* | *667* | *1050* | 5571 |
| Cooling Degree Days | na | na | na | na | na | na | na | na | na | na | na | na | na |
| Total Precipitation (") | *1.64* | *1.53* | *3.12* | *3.90* | *4.21* | *3.72* | *4.24* | *3.40* | *3.85* | *3.41* | *3.06* | *2.92* | 39.00 |
| Days ≥ 0.1" Precip | *4* | *4* | *6* | *7* | *7* | *6* | *6* | *6* | *6* | *6* | *6* | *6* | 70 |
| Total Snowfall (") | *6.6* | *6.9* | *4.2* | *0.9* | *0.0* | *0.0* | *0.0* | *0.0* | *0.0* | *0.0* | *2.0* | *5.0* | 25.6 |
| Days ≥ 1" Snow Depth | *13* | *13* | *5* | *0* | *0* | *0* | *0* | *0* | *0* | *0* | *1* | *7* | 39 |

## HARRISBURG *Saline County* ELEVATION 365 ft LAT/LONG 37° 44 ' N / 88° 31 ' W

| | JAN | FEB | MAR | APR | MAY | JUN | JUL | AUG | SEP | OCT | NOV | DEC | YEAR |
|---|---|---|---|---|---|---|---|---|---|---|---|---|---|
| Maximum Temp °F | 40.3 | 46.3 | 57.9 | 70.0 | 79.0 | 87.2 | 90.8 | 88.9 | 82.1 | 70.8 | 57.4 | 46.4 | 68.1 |
| Minimum Temp °F | 22.4 | 26.2 | 36.0 | 46.2 | 54.7 | 63.3 | 67.6 | 64.8 | 57.9 | 45.3 | 37.7 | 28.8 | 45.9 |
| Mean Temp °F | 31.4 | 36.3 | 47.0 | 58.1 | 66.9 | 75.3 | 79.3 | 76.9 | 70.0 | 58.1 | 47.6 | 37.7 | 57.1 |
| Days Max Temp ≥ 90 °F | 0 | 0 | 0 | 0 | 3 | 11 | 19 | 14 | 5 | 0 | 0 | 0 | 52 |
| Days Max Temp ≤ 32 °F | 9 | 4 | 1 | 0 | 0 | 0 | 0 | 0 | 0 | 0 | 0 | 3 | 17 |
| Days Min Temp ≤ 32 °F | 25 | 20 | 13 | 2 | 0 | 0 | 0 | 0 | 0 | 4 | 10 | 20 | 94 |
| Days Min Temp ≤ 0 °F | 2 | 1 | 0 | 0 | 0 | 0 | 0 | 0 | 0 | 0 | 0 | 0 | 3 |
| Heating Degree Days | 1036 | 805 | 558 | 238 | 65 | 4 | 0 | 1 | 39 | 243 | 519 | 841 | 4349 |
| Cooling Degree Days | 0 | 0 | 4 | 32 | 124 | 307 | *456* | 376 | 194 | 31 | 3 | 1 | 1528 |
| Total Precipitation (") | 2.88 | 2.82 | 4.28 | 4.51 | 4.58 | 3.98 | 3.86 | 3.30 | 3.32 | 3.19 | 4.18 | 4.18 | 45.08 |
| Days ≥ 0.1" Precip | 5 | 6 | 8 | 8 | 7 | 6 | 6 | 5 | 6 | 5 | 7 | 7 | 76 |
| Total Snowfall (") | *5.8* | *5.1* | 2.9 | 0.4 | 0.0 | 0.0 | 0.0 | 0.0 | 0.0 | 0.0 | 0.7 | 2.6 | 17.5 |
| Days ≥ 1" Snow Depth | *8* | 5 | 1 | 0 | 0 | 0 | 0 | 0 | 0 | 0 | 0 | 2 | 16 |

## HAVANA 4 NNE *Mason County* ELEVATION 459 ft LAT/LONG 40° 21 ' N / 90° 1 ' W

| | JAN | FEB | MAR | APR | MAY | JUN | JUL | AUG | SEP | OCT | NOV | DEC | YEAR |
|---|---|---|---|---|---|---|---|---|---|---|---|---|---|
| Maximum Temp °F | 32.7 | 37.8 | 51.4 | 64.6 | 74.5 | 84.4 | 88.2 | 85.7 | 79.3 | 66.9 | 50.9 | 37.8 | 62.8 |
| Minimum Temp °F | 15.0 | 18.9 | 30.7 | 41.6 | 51.2 | 61.2 | 65.1 | 62.2 | 54.2 | 42.6 | 32.3 | 21.1 | 41.3 |
| Mean Temp °F | 23.8 | 28.4 | 41.1 | 53.1 | 62.8 | 72.8 | 76.6 | 74.0 | 66.8 | 54.8 | 41.6 | 29.6 | 52.1 |
| Days Max Temp ≥ 90 °F | 0 | 0 | 0 | 0 | 2 | 8 | 13 | 9 | 4 | 0 | 0 | 0 | 36 |
| Days Max Temp ≤ 32 °F | 15 | 9 | 2 | 0 | 0 | 0 | 0 | 0 | 0 | 0 | 2 | 9 | 37 |
| Days Min Temp ≤ 32 °F | 29 | 25 | 18 | 5 | 0 | 0 | 0 | 0 | 0 | 5 | 16 | 26 | 124 |
| Days Min Temp ≤ 0 °F | 6 | 4 | 0 | 0 | 0 | 0 | 0 | 0 | 0 | 0 | 0 | 2 | 12 |
| Heating Degree Days | 1270 | 1029 | 738 | 369 | 137 | 14 | 1 | 6 | 75 | 332 | 696 | 1091 | 5758 |
| Cooling Degree Days | 0 | 0 | 2 | 16 | 75 | 250 | 373 | 297 | 130 | 14 | 1 | 0 | 1158 |
| Total Precipitation (") | 1.76 | 1.70 | 2.85 | 3.56 | 3.95 | 3.73 | 4.02 | 3.55 | 3.90 | 3.13 | 3.06 | 2.85 | 38.06 |
| Days ≥ 0.1" Precip | 5 | 4 | 6 | 7 | 8 | 6 | 7 | 5 | 6 | 5 | 6 | 6 | 71 |
| Total Snowfall (") | *8.5* | *8.3* | 3.6 | 0.8 | 0.0 | 0.0 | 0.0 | 0.0 | 0.0 | 0.0 | 1.4 | 5.4 | 28.0 |
| Days ≥ 1" Snow Depth | 14 | 10 | *3* | 0 | 0 | 0 | 0 | 0 | 0 | 0 | 1 | *7* | 35 |

## HENNEPIN POWER PLANT *Putnam County* ELEVATION 459 ft LAT/LONG 41° 18 ' N / 89° 19 ' W

| | JAN | FEB | MAR | APR | MAY | JUN | JUL | AUG | SEP | OCT | NOV | DEC | YEAR |
|---|---|---|---|---|---|---|---|---|---|---|---|---|---|
| Maximum Temp °F | 30.2 | 35.3 | 48.0 | 62.4 | 74.0 | 82.9 | 86.1 | 83.8 | 77.2 | 64.8 | 49.4 | 36.0 | 60.8 |
| Minimum Temp °F | 10.6 | 15.4 | 27.6 | 38.4 | 48.2 | 57.6 | 61.6 | 59.3 | 51.5 | 39.6 | 29.4 | 18.1 | 38.1 |
| Mean Temp °F | 20.4 | 25.3 | 37.9 | 50.4 | 61.1 | 70.3 | 73.9 | 71.6 | 64.4 | 52.2 | 39.4 | 27.1 | 49.5 |
| Days Max Temp ≥ 90 °F | 0 | 0 | 0 | 0 | 2 | 5 | 9 | 6 | 2 | 0 | 0 | 0 | 24 |
| Days Max Temp ≤ 32 °F | 16 | 11 | 3 | 0 | 0 | 0 | 0 | 0 | 0 | 0 | 2 | 10 | 42 |
| Days Min Temp ≤ 32 °F | 30 | 27 | 22 | 8 | 1 | 0 | 0 | 0 | 1 | 8 | 20 | 28 | 145 |
| Days Min Temp ≤ 0 °F | 8 | 4 | 0 | 0 | 0 | 0 | 0 | 0 | 0 | 0 | 0 | 4 | 16 |
| Heating Degree Days | 1379 | 1112 | 835 | 442 | 174 | 23 | 5 | 13 | 106 | 403 | 760 | 1168 | 6420 |
| Cooling Degree Days | 0 | 0 | 2 | 13 | 59 | 179 | 285 | 227 | 91 | 6 | 0 | 0 | 862 |
| Total Precipitation (") | *1.30* | 1.07 | 2.27 | 3.59 | 3.47 | 4.11 | 3.84 | 4.26 | 3.95 | 2.42 | 2.68 | *2.17* | 35.13 |
| Days ≥ 0.1" Precip | na | 3 | *5* | 7 | 6 | 7 | 6 | 6 | 6 | 5 | 5 | *4* | na |
| Total Snowfall (") | na | na | na | *0.5* | 0.0 | 0.0 | 0.0 | 0.0 | 0.0 | 0.0 | *0.7* | na | na |
| Days ≥ 1" Snow Depth | na | na | na | 0 | 0 | 0 | 0 | 0 | 0 | 0 | *1* | *6* | na |

## HILLSBORO *Montgomery County* ELEVATION 630 ft LAT/LONG 39° 9 ' N / 89° 29 ' W

| | JAN | FEB | MAR | APR | MAY | JUN | JUL | AUG | SEP | OCT | NOV | DEC | YEAR |
|---|---|---|---|---|---|---|---|---|---|---|---|---|---|
| Maximum Temp °F | 36.8 | 42.4 | 54.6 | 67.5 | 76.9 | 85.7 | 89.2 | 87.0 | 80.4 | 68.8 | 54.4 | 41.9 | 65.5 |
| Minimum Temp °F | 18.6 | 22.6 | 33.0 | 43.9 | 53.0 | 62.0 | 65.8 | 62.8 | 55.6 | 44.4 | 35.0 | 24.7 | 43.4 |
| Mean Temp °F | 27.7 | 32.5 | 43.8 | 55.7 | 65.0 | 73.9 | 77.6 | 74.9 | 68.1 | 56.6 | 44.7 | 33.3 | 54.5 |
| Days Max Temp ≥ 90 °F | 0 | 0 | 0 | 0 | 1 | 9 | 15 | 10 | 4 | 0 | 0 | 0 | 39 |
| Days Max Temp ≤ 32 °F | 11 | 7 | 1 | 0 | 0 | 0 | 0 | 0 | 0 | 0 | 1 | 6 | 26 |
| Days Min Temp ≤ 32 °F | 27 | 22 | 16 | 4 | 0 | 0 | 0 | 0 | 0 | 4 | 13 | 24 | 110 |
| Days Min Temp ≤ 0 °F | 3 | 1 | 0 | 0 | 0 | 0 | 0 | 0 | 0 | 0 | 0 | 1 | 5 |
| Heating Degree Days | 1150 | 910 | 652 | 297 | 97 | 8 | 1 | 3 | 56 | 276 | 603 | 977 | 5030 |
| Cooling Degree Days | 0 | 0 | 5 | 29 | 120 | 320 | 447 | 363 | 168 | 26 | 2 | 0 | 1480 |
| Total Precipitation (") | 2.02 | 1.94 | 3.34 | 4.29 | 4.18 | 4.06 | 3.83 | 3.42 | 3.74 | 2.99 | 3.70 | 3.46 | 40.97 |
| Days ≥ 0.1" Precip | 4 | 4 | 7 | 7 | 7 | 6 | 6 | 5 | 6 | 5 | 7 | 6 | 70 |
| Total Snowfall (") | 4.8 | 4.8 | *3.5* | 0.5 | 0.0 | 0.0 | 0.0 | 0.0 | 0.0 | 0.0 | 1.1 | *4.1* | 18.8 |
| Days ≥ 1" Snow Depth | na | na | na | 0 | 0 | 0 | 0 | 0 | 0 | 0 | *0* | na | na |

## HOOPESTON 1 NE *Vermilion County* ELEVATION 722 ft LAT/LONG 40° 28' N / 87° 40' W

| | JAN | FEB | MAR | APR | MAY | JUN | JUL | AUG | SEP | OCT | NOV | DEC | YEAR |
|---|---|---|---|---|---|---|---|---|---|---|---|---|---|
| Maximum Temp °F | 31.6 | 36.4 | 49.1 | 62.5 | 73.6 | 82.7 | 85.3 | 83.0 | 76.9 | 64.2 | 49.8 | 37.1 | 61.0 |
| Minimum Temp °F | 16.3 | 20.0 | 30.7 | 40.9 | 50.9 | 60.4 | 64.2 | 61.7 | 54.7 | 43.5 | 33.8 | 23.0 | 41.7 |
| Mean Temp °F | 24.0 | 28.2 | 39.9 | 51.7 | 62.3 | 71.6 | 74.8 | 72.3 | 65.8 | 53.9 | 41.8 | 30.1 | 51.4 |
| Days Max Temp ≥ 90 °F | 0 | 0 | 0 | 0 | 1 | 5 | 8 | 4 | 1 | 0 | 0 | 0 | 19 |
| Days Max Temp ≤ 32 °F | 16 | 10 | 3 | 0 | 0 | 0 | 0 | 0 | 0 | 0 | 1 | 9 | 39 |
| Days Min Temp ≤ 32 °F | 28 | 24 | 19 | 6 | 1 | 0 | 0 | 0 | 0 | 4 | 14 | 26 | 122 |
| Days Min Temp ≤ 0 °F | 5 | 3 | 0 | 0 | 0 | 0 | 0 | 0 | 0 | 0 | 0 | 2 | 10 |
| Heating Degree Days | 1266 | 1031 | 772 | 404 | 153 | 18 | 3 | 8 | 85 | 351 | 689 | 1075 | 5855 |
| Cooling Degree Days | 0 | 0 | 2 | 13 | 81 | 235 | 329 | 259 | 122 | 13 | 0 | 0 | 1054 |
| Total Precipitation (") | 1.60 | 1.55 | 2.76 | 3.89 | 3.94 | 3.63 | 3.87 | 3.85 | 3.48 | 3.02 | 3.14 | 2.72 | 37.45 |
| Days ≥ 0.1" Precip | 4 | 4 | 7 | 8 | 8 | 7 | 6 | 6 | 6 | 5 | 7 | 6 | 74 |
| Total Snowfall (") | 4.8 | 4.5 | 2.6 | 0.6 | 0.0 | 0.0 | 0.0 | 0.0 | 0.0 | 0.1 | 1.2 | 4.3 | 18.1 |
| Days ≥ 1" Snow Depth | 15 | 12 | 4 | 0 | 0 | 0 | 0 | 0 | 0 | 0 | 1 | 7 | 39 |

## JACKSONVILLE 2 E *Morgan County* ELEVATION 610 ft LAT/LONG 39° 44' N / 90° 14' W

| | JAN | FEB | MAR | APR | MAY | JUN | JUL | AUG | SEP | OCT | NOV | DEC | YEAR |
|---|---|---|---|---|---|---|---|---|---|---|---|---|---|
| Maximum Temp °F | 33.5 | 38.3 | 50.8 | 63.9 | 73.6 | 82.8 | 86.3 | 83.9 | 78.0 | 66.3 | 51.9 | 39.0 | 62.4 |
| Minimum Temp °F | 15.7 | 19.0 | 29.7 | 40.9 | 50.4 | 59.9 | 63.8 | 60.8 | 53.3 | 42.0 | 32.1 | 21.9 | 40.8 |
| Mean Temp °F | 24.6 | 28.7 | 40.3 | 52.4 | 62.0 | 71.4 | 75.1 | 72.4 | 65.7 | 54.2 | 42.0 | 30.5 | 51.6 |
| Days Max Temp ≥ 90 °F | 0 | 0 | 0 | 0 | 0 | 5 | 9 | 6 | 2 | 0 | 0 | 0 | 22 |
| Days Max Temp ≤ 32 °F | 14 | 9 | 3 | 0 | 0 | 0 | 0 | 0 | 0 | 0 | 1 | 9 | 36 |
| Days Min Temp ≤ 32 °F | 29 | 25 | 20 | 5 | 0 | 0 | 0 | 0 | 0 | 6 | 17 | 26 | 128 |
| Days Min Temp ≤ 0 °F | 4 | 3 | 0 | 0 | 0 | 0 | 0 | 0 | 0 | 0 | 0 | 2 | 9 |
| Heating Degree Days | 1245 | 1020 | 762 | 385 | 153 | 18 | 3 | 10 | 88 | 345 | 683 | 1065 | 5777 |
| Cooling Degree Days | 0 | 0 | 3 | 14 | 63 | 216 | 328 | 258 | 121 | 15 | 0 | 0 | 1018 |
| Total Precipitation (") | 1.36 | 1.59 | 2.98 | 4.09 | 4.89 | 4.24 | 4.38 | 3.47 | 4.25 | 3.00 | 3.36 | 2.79 | 40.40 |
| Days ≥ 0.1" Precip | 3 | 3 | 6 | 8 | 7 | 6 | 6 | 6 | 6 | 5 | 6 | 5 | 67 |
| Total Snowfall (") | 6.0 | 6.3 | 3.4 | 0.5 | 0.0 | 0.0 | 0.0 | 0.0 | 0.0 | 0.0 | 1.4 | 5.0 | 22.6 |
| Days ≥ 1" Snow Depth | 12 | 10 | 3 | 0 | 0 | 0 | 0 | 0 | 0 | 0 | 1 | 6 | 32 |

## JERSEYVILLE 2 SW *Jersey County* ELEVATION 640 ft LAT/LONG 39° 7' N / 90° 20' W

| | JAN | FEB | MAR | APR | MAY | JUN | JUL | AUG | SEP | OCT | NOV | DEC | YEAR |
|---|---|---|---|---|---|---|---|---|---|---|---|---|---|
| Maximum Temp °F | 36.0 | 40.9 | 52.9 | 65.6 | 74.8 | 83.6 | 88.0 | 85.7 | 78.9 | 67.5 | 53.3 | 41.0 | 64.0 |
| Minimum Temp °F | 17.3 | 21.1 | 32.1 | 42.8 | 51.8 | 61.1 | 65.1 | 62.1 | 54.5 | 42.8 | 33.6 | 23.4 | 42.3 |
| Mean Temp °F | 26.7 | 31.0 | 42.5 | 54.2 | 63.3 | 72.4 | 76.6 | 73.9 | 66.7 | 55.2 | 43.5 | 32.2 | 53.2 |
| Days Max Temp ≥ 90 °F | 0 | 0 | 0 | 0 | 0 | 6 | 13 | 9 | 3 | 0 | 0 | 0 | 31 |
| Days Max Temp ≤ 32 °F | 12 | 7 | 2 | 0 | 0 | 0 | 0 | 0 | 0 | 0 | 1 | 7 | 29 |
| Days Min Temp ≤ 32 °F | 28 | 23 | 17 | 5 | 0 | 0 | 0 | 0 | 0 | 5 | 15 | 25 | 118 |
| Days Min Temp ≤ 0 °F | 4 | 2 | 0 | 0 | 0 | 0 | 0 | 0 | 0 | 0 | 0 | 1 | 7 |
| Heating Degree Days | 1182 | 955 | 693 | 338 | 126 | 13 | 2 | 5 | 76 | 317 | 640 | 1011 | 5358 |
| Cooling Degree Days | 0 | 0 | 4 | 22 | 82 | 251 | 381 | 303 | 139 | 16 | 1 | 0 | 1199 |
| Total Precipitation (") | 1.69 | 1.82 | 3.24 | 4.21 | 3.77 | 3.52 | 3.67 | 3.22 | 3.72 | 2.87 | 3.56 | 2.92 | 38.21 |
| Days ≥ 0.1" Precip | 4 | 4 | 6 | 7 | 7 | 6 | 6 | 5 | 6 | 6 | 6 | 5 | 68 |
| Total Snowfall (") | 4.6 | 4.4 | 3.6 | 0.5 | 0.0 | 0.0 | 0.0 | 0.0 | 0.0 | 0.0 | 1.2 | 3.7 | 18.0 |
| Days ≥ 1" Snow Depth | 10 | 7 | 3 | 0 | 0 | 0 | 0 | 0 | 0 | 0 | 1 | 5 | 26 |

## KEWANEE 1 E *Henry County* ELEVATION 850 ft LAT/LONG 41° 14' N / 89° 55' W

| | JAN | FEB | MAR | APR | MAY | JUN | JUL | AUG | SEP | OCT | NOV | DEC | YEAR |
|---|---|---|---|---|---|---|---|---|---|---|---|---|---|
| Maximum Temp °F | ***28.5*** | 32.9 | 46.4 | 61.3 | 72.6 | 81.8 | 85.1 | 82.7 | 75.6 | 63.2 | 47.8 | ***34.9*** | 59.4 |
| Minimum Temp °F | ***11.0*** | 15.3 | 27.6 | 38.5 | 49.3 | 58.9 | 63.3 | 60.7 | 52.4 | 40.8 | 30.5 | ***19.1*** | 38.9 |
| Mean Temp °F | ***19.9*** | 24.1 | 37.0 | 49.9 | 60.9 | 70.4 | 74.2 | 71.7 | 64.0 | 52.0 | 39.2 | ***27.0*** | 49.2 |
| Days Max Temp ≥ 90 °F | 0 | 0 | 0 | 0 | 1 | 4 | 7 | 5 | 1 | 0 | 0 | 0 | 18 |
| Days Max Temp ≤ 32 °F | ***18*** | 14 | 4 | 0 | 0 | 0 | 0 | 0 | 0 | 0 | 2 | ***12*** | 50 |
| Days Min Temp ≤ 32 °F | ***30*** | 26 | 22 | ***8*** | 1 | 0 | 0 | 0 | 0 | 6 | 19 | ***27*** | 139 |
| Days Min Temp ≤ 0 °F | ***8*** | ***5*** | 0 | 0 | 0 | 0 | 0 | 0 | 0 | 0 | 0 | ***3*** | 16 |
| Heating Degree Days | ***1393*** | 1148 | 862 | 454 | 178 | 26 | 5 | 13 | 110 | 406 | 769 | ***1169*** | 6533 |
| Cooling Degree Days | ***0*** | ***0*** | 1 | ***11*** | ***48*** | ***185*** | ***289*** | ***229*** | ***82*** | 6 | ***0*** | ***0*** | 851 |
| Total Precipitation (") | ***1.70*** | 1.14 | 2.40 | ***3.61*** | 3.61 | 4.60 | ***4.27*** | 4.07 | 3.63 | 2.59 | 2.86 | ***2.48*** | 36.96 |
| Days ≥ 0.1" Precip | ***4*** | ***3*** | 5 | ***7*** | 7 | ***7*** | 7 | 6 | 6 | 5 | 6 | ***5*** | 68 |
| Total Snowfall (") | na | ***6.8*** | ***4.1*** | ***1.4*** | ***0.0*** | ***0.0*** | ***0.0*** | ***0.0*** | 0.0 | 0.3 | ***2.3*** | na | na |
| Days ≥ 1" Snow Depth | na | na | ***4*** | ***0*** | ***0*** | ***0*** | ***0*** | ***0*** | 0 | 0 | ***1*** | na | na |

## LA HARPE *Hancock County* ELEVATION 689 ft LAT/LONG 40° 35' N / 90° 58' W

| | JAN | FEB | MAR | APR | MAY | JUN | JUL | AUG | SEP | OCT | NOV | DEC | YEAR |
|---|---|---|---|---|---|---|---|---|---|---|---|---|---|
| Maximum Temp °F | 32.2 | 37.6 | 50.2 | 63.7 | 74.3 | 83.4 | 87.4 | 84.8 | 77.5 | 66.0 | 50.2 | 37.4 | 62.1 |
| Minimum Temp °F | 13.0 | 17.4 | 28.6 | 39.8 | 50.0 | 59.3 | 63.2 | 60.5 | 52.7 | 41.2 | 30.3 | 19.4 | 39.6 |
| Mean Temp °F | 22.6 | 27.5 | 39.4 | 51.7 | 62.2 | 71.4 | 75.3 | 72.7 | 65.1 | 53.6 | 40.3 | 28.4 | 50.8 |
| Days Max Temp ≥ 90 °F | 0 | 0 | 0 | 0 | 1 | 5 | 11 | 7 | 2 | 0 | 0 | 0 | 26 |
| Days Max Temp ≤ 32 °F | 15 | 10 | 2 | 0 | 0 | 0 | 0 | 0 | 0 | 0 | 2 | 10 | 39 |
| Days Min Temp ≤ 32 °F | 30 | 25 | 20 | 6 | 1 | 0 | 0 | 0 | 0 | 7 | 19 | 28 | 136 |
| Days Min Temp ≤ 0 °F | 7 | 4 | 0 | 0 | 0 | 0 | 0 | 0 | 0 | 0 | 0 | 3 | 14 |
| Heating Degree Days | 1308 | 1053 | 787 | 404 | 148 | 16 | 3 | 7 | 94 | 360 | 735 | 1129 | 6044 |
| Cooling Degree Days | 0 | 0 | 2 | 13 | 63 | 220 | 335 | 263 | 106 | 9 | 0 | 0 | 1011 |
| Total Precipitation (") | 1.51 | 1.40 | 2.89 | 3.93 | 3.91 | 4.39 | 4.35 | 3.88 | 4.44 | 3.02 | 3.03 | 2.40 | 39.15 |
| Days ≥ 0.1" Precip | 4 | 4 | 6 | 7 | 8 | 7 | 6 | 6 | 6 | 5 | 6 | 5 | 70 |
| Total Snowfall (") | 6.9 | 5.4 | 3.9 | 1.0 | 0.0 | 0.0 | 0.0 | 0.0 | 0.0 | 0.1 | 1.9 | 5.6 | 24.8 |
| Days ≥ 1" Snow Depth | 16 | 11 | 4 | 0 | 0 | 0 | 0 | 0 | 0 | 0 | 1 | 9 | 41 |

## LACON 1 N *Marshall County* ELEVATION 479 ft LAT/LONG 41° 2' N / 89° 25' W

| | JAN | FEB | MAR | APR | MAY | JUN | JUL | AUG | SEP | OCT | NOV | DEC | YEAR |
|---|---|---|---|---|---|---|---|---|---|---|---|---|---|
| Maximum Temp °F | 31.5 | 36.7 | 49.6 | 63.8 | 74.6 | 83.7 | 86.8 | 84.8 | 78.1 | 66.2 | 50.6 | 37.4 | 62.0 |
| Minimum Temp °F | 14.5 | 18.6 | 29.7 | 40.8 | 50.4 | 59.4 | 63.9 | 61.7 | 54.2 | 43.1 | 32.6 | 21.5 | 40.9 |
| Mean Temp °F | 23.2 | 27.7 | 39.7 | 52.3 | 62.4 | 71.5 | 75.4 | 73.3 | 66.1 | 54.7 | 41.6 | 29.5 | 51.5 |
| Days Max Temp ≥ 90 °F | 0 | 0 | 0 | 0 | 1 | 5 | 10 | 7 | 2 | 0 | 0 | 0 | 25 |
| Days Max Temp ≤ 32 °F | 15 | 11 | 3 | 0 | 0 | 0 | 0 | 0 | 0 | 0 | 1 | 9 | 39 |
| Days Min Temp ≤ 32 °F | 28 | 24 | 20 | 7 | 1 | 0 | 0 | 0 | 0 | 4 | 16 | 26 | 126 |
| Days Min Temp ≤ 0 °F | 6 | 3 | 0 | 0 | 0 | 0 | 0 | 0 | 0 | 0 | 0 | 2 | 11 |
| Heating Degree Days | 1289 | 1048 | 780 | 390 | 147 | 16 | 2 | 6 | 81 | 332 | 696 | 1094 | 5881 |
| Cooling Degree Days | *0* | 0 | 2 | 18 | 76 | 230 | 349 | 289 | 127 | 15 | 1 | 0 | 1107 |
| Total Precipitation (") | 1.54 | 1.33 | 3.06 | 4.10 | 3.95 | 4.25 | 4.18 | 3.63 | 3.94 | 3.06 | 3.09 | 2.53 | 38.66 |
| Days ≥ 0.1" Precip | 4 | 3 | 7 | 7 | 7 | 6 | 7 | 6 | 6 | 6 | 6 | 5 | 70 |
| Total Snowfall (") | 6.7 | 5.0 | 2.7 | 0.5 | 0.0 | 0.0 | 0.0 | 0.0 | 0.0 | 0.1 | 1.5 | 5.2 | 21.7 |
| Days ≥ 1" Snow Depth | 14 | 12 | 3 | 0 | 0 | 0 | 0 | 0 | 0 | 0 | 1 | 8 | 38 |

## LINCOLN *Logan County* ELEVATION 591 ft LAT/LONG 40° 9' N / 89° 22' W

| | JAN | FEB | MAR | APR | MAY | JUN | JUL | AUG | SEP | OCT | NOV | DEC | YEAR |
|---|---|---|---|---|---|---|---|---|---|---|---|---|---|
| Maximum Temp °F | 32.5 | 37.9 | 50.6 | 64.3 | 74.8 | 83.6 | 86.4 | 83.8 | 77.9 | 66.0 | 51.2 | 38.7 | 62.3 |
| Minimum Temp °F | 15.6 | 19.8 | 30.6 | 41.5 | 51.4 | 61.0 | 65.0 | 62.3 | 54.8 | 42.8 | 33.0 | 22.6 | 41.7 |
| Mean Temp °F | 24.1 | 28.9 | 40.7 | 52.9 | 63.1 | 72.3 | 75.7 | 73.1 | 66.4 | 54.4 | 42.1 | 30.6 | 52.0 |
| Days Max Temp ≥ 90 °F | 0 | 0 | 0 | 0 | 1 | 6 | 9 | 5 | 2 | 0 | 0 | 0 | 23 |
| Days Max Temp ≤ 32 °F | 15 | 9 | 2 | 0 | 0 | 0 | 0 | 0 | 0 | 0 | 1 | 8 | 35 |
| Days Min Temp ≤ 32 °F | 29 | 24 | 19 | 5 | 0 | 0 | 0 | 0 | 0 | 5 | 16 | 26 | 124 |
| Days Min Temp ≤ 0 °F | 4 | 3 | 0 | 0 | 0 | 0 | 0 | 0 | 0 | 0 | 0 | 2 | 9 |
| Heating Degree Days | 1263 | 1013 | 750 | 371 | 133 | 14 | 2 | 6 | 75 | 336 | 679 | 1061 | 5703 |
| Cooling Degree Days | 0 | 0 | 2 | 16 | 81 | 248 | 350 | 279 | 129 | 13 | 0 | 0 | 1118 |
| Total Precipitation (") | 1.75 | 1.36 | 2.95 | 3.72 | 4.09 | 3.92 | 4.41 | 3.80 | 3.59 | 3.02 | 2.97 | 2.96 | 38.54 |
| Days ≥ 0.1" Precip | 4 | 4 | 7 | 7 | 7 | 6 | 6 | 6 | 6 | 5 | 6 | 6 | 70 |
| Total Snowfall (") | *5.8* | 5.4 | 2.8 | 0.6 | 0.0 | 0.0 | 0.0 | 0.0 | 0.0 | 0.0 | 1.2 | 4.2 | 20.0 |
| Days ≥ 1" Snow Depth | *12* | *8* | 3 | 0 | 0 | 0 | 0 | 0 | 0 | 0 | 1 | *5* | 29 |

## MARENGO *McHenry County* ELEVATION 820 ft LAT/LONG 42° 15' N / 88° 36' W

| | JAN | FEB | MAR | APR | MAY | JUN | JUL | AUG | SEP | OCT | NOV | DEC | YEAR |
|---|---|---|---|---|---|---|---|---|---|---|---|---|---|
| Maximum Temp °F | 27.7 | 32.6 | 45.0 | 59.3 | 72.1 | 81.4 | 84.9 | 82.3 | 75.0 | 62.7 | 46.8 | 33.2 | 58.6 |
| Minimum Temp °F | 9.5 | 13.8 | 25.2 | 36.0 | 46.0 | 55.8 | 60.7 | 58.4 | 50.2 | 38.7 | 28.6 | 16.6 | 36.6 |
| Mean Temp °F | 18.6 | 23.2 | 35.1 | 47.7 | 59.1 | 68.7 | 72.8 | 70.4 | 62.6 | 50.8 | 37.7 | 24.9 | 47.6 |
| Days Max Temp ≥ 90 °F | 0 | 0 | 0 | 0 | 1 | 5 | 8 | 5 | 1 | 0 | 0 | 0 | 20 |
| Days Max Temp ≤ 32 °F | 19 | 13 | 4 | 0 | 0 | 0 | 0 | 0 | 0 | 0 | 2 | 13 | 51 |
| Days Min Temp ≤ 32 °F | 30 | 27 | 24 | 11 | 2 | 0 | 0 | 0 | 1 | 9 | 20 | 29 | 153 |
| Days Min Temp ≤ 0 °F | 9 | 5 | 1 | 0 | 0 | 0 | 0 | 0 | 0 | 0 | 0 | 4 | 19 |
| Heating Degree Days | 1433 | 1174 | 921 | 520 | 228 | 43 | 9 | 21 | 137 | 441 | 812 | 1236 | 6975 |
| Cooling Degree Days | 0 | 0 | 0 | 11 | 50 | 155 | 258 | 192 | 68 | 4 | 0 | 0 | 738 |
| Total Precipitation (") | 1.53 | 1.17 | 2.27 | 3.75 | 3.41 | 4.66 | 4.01 | 4.50 | 3.80 | 2.73 | 2.80 | 2.14 | 36.77 |
| Days ≥ 0.1" Precip | 4 | 4 | 6 | 7 | 7 | 7 | 7 | 7 | 7 | 6 | 6 | 5 | 73 |
| Total Snowfall (") | 11.2 | 8.1 | 6.0 | 1.7 | 0.2 | 0.0 | 0.0 | 0.0 | 0.0 | 0.2 | 2.1 | *8.8* | 38.3 |
| Days ≥ 1" Snow Depth | *20* | *16* | *7* | *1* | 0 | 0 | 0 | 0 | 0 | 0 | *1* | *12* | 57 |

## MARION 4 NNE *Williamson County* ELEVATION 449 ft LAT/LONG 37° 44 ' N / 88° 57 ' W

| | JAN | FEB | MAR | APR | MAY | JUN | JUL | AUG | SEP | OCT | NOV | DEC | YEAR |
|---|---|---|---|---|---|---|---|---|---|---|---|---|---|
| Maximum Temp °F | 38.0 | 43.6 | 54.9 | 66.4 | 75.2 | 83.5 | 87.6 | 86.0 | 79.5 | 68.2 | 55.4 | 43.3 | 65.1 |
| Minimum Temp °F | 19.7 | 23.3 | 33.5 | 43.6 | 52.6 | 61.5 | 66.2 | 63.4 | 56.3 | 43.8 | 35.4 | 25.4 | 43.7 |
| Mean Temp °F | 28.9 | 33.5 | 44.2 | 55.0 | 63.9 | 72.6 | 76.9 | 74.8 | 67.9 | 56.1 | 45.4 | 34.3 | 54.5 |
| Days Max Temp ≥ 90 °F | 0 | 0 | 0 | 0 | 0 | 5 | 12 | 9 | 3 | 0 | 0 | 0 | 29 |
| Days Max Temp ≤ 32 °F | 10 | 6 | 1 | 0 | 0 | 0 | 0 | 0 | 0 | 0 | 1 | 5 | 23 |
| Days Min Temp ≤ 32 °F | 27 | 22 | 15 | 4 | 0 | 0 | 0 | 0 | 0 | 5 | 13 | 23 | 109 |
| Days Min Temp ≤ 0 °F | 2 | 2 | 0 | 0 | 0 | 0 | 0 | 0 | 0 | 0 | 0 | 1 | 5 |
| Heating Degree Days | 1112 | 883 | 640 | 315 | 111 | 11 | 1 | 3 | 61 | 294 | 583 | 944 | 4958 |
| Cooling Degree Days | 0 | 0 | 3 | 21 | 85 | 254 | 391 | 321 | 150 | 23 | 3 | 0 | 1251 |
| Total Precipitation (") | 2.99 | 3.06 | 4.29 | 4.47 | 4.35 | 3.89 | 3.77 | 3.77 | 3.41 | 3.13 | 4.70 | 3.86 | 45.69 |
| Days ≥ 0.1" Precip | 5 | 5 | 8 | 8 | 8 | 6 | 6 | 6 | 5 | 6 | 7 | 7 | 77 |
| Total Snowfall (") | 5.8 | 6.3 | 2.8 | 0.4 | 0.0 | 0.0 | 0.0 | 0.0 | 0.0 | 0.2 | 0.8 | 3.1 | 19.4 |
| Days ≥ 1" Snow Depth | 7 | 6 | 2 | 0 | 0 | 0 | 0 | 0 | 0 | 0 | 0 | 3 | 18 |

## MASON CITY 1 W *Mason County* ELEVATION 581 ft LAT/LONG 40° 12 ' N / 89° 42 ' W

| | JAN | FEB | MAR | APR | MAY | JUN | JUL | AUG | SEP | OCT | NOV | DEC | YEAR |
|---|---|---|---|---|---|---|---|---|---|---|---|---|---|
| Maximum Temp °F | 32.9 | 37.6 | 51.0 | 64.8 | 75.1 | 83.9 | 86.7 | 84.4 | 78.4 | 66.3 | 51.0 | 37.7 | 62.5 |
| Minimum Temp °F | 16.1 | 20.0 | 31.4 | 41.9 | 52.0 | 61.0 | 64.9 | 62.3 | 55.0 | 43.4 | 33.0 | 22.3 | 41.9 |
| Mean Temp °F | 24.5 | 28.8 | 41.2 | 53.4 | 63.6 | 72.5 | 75.8 | 73.4 | 66.7 | 54.9 | 42.0 | 30.1 | 52.2 |
| Days Max Temp ≥ 90 °F | 0 | 0 | 0 | 0 | 1 | 6 | 10 | 6 | 2 | 0 | 0 | 0 | 25 |
| Days Max Temp ≤ 32 °F | 14 | 10 | 2 | 0 | 0 | 0 | 0 | 0 | 0 | 0 | 1 | 9 | 36 |
| Days Min Temp ≤ 32 °F | 28 | 24 | 17 | 5 | 0 | 0 | 0 | 0 | 0 | 5 | 16 | 25 | 120 |
| Days Min Temp ≤ 0 °F | 5 | 3 | 0 | 0 | 0 | 0 | 0 | 0 | 0 | 0 | 0 | 2 | 10 |
| Heating Degree Days | 1249 | 1016 | 735 | 359 | 125 | 12 | 2 | 6 | 72 | 326 | 683 | 1077 | 5662 |
| Cooling Degree Days | 0 | 0 | 4 | 20 | 89 | 254 | 360 | 289 | 137 | 17 | 1 | 0 | 1171 |
| Total Precipitation (") | 1.47 | 1.41 | 2.62 | 3.54 | 3.87 | 3.44 | 4.04 | 3.57 | 3.65 | 3.00 | 2.87 | 2.69 | 36.17 |
| Days ≥ 0.1" Precip | 3 | 4 | 6 | 7 | 7 | 6 | 6 | 6 | 6 | 5 | 6 | 5 | 67 |
| Total Snowfall (") | 6.0 | 5.3 | 2.2 | 0.6 | 0.0 | 0.0 | 0.0 | 0.0 | 0.0 | 0.0 | 1.5 | 4.4 | 20.0 |
| Days ≥ 1" Snow Depth | 13 | 11 | 2 | 0 | 0 | 0 | 0 | 0 | 0 | 0 | 1 | 7 | 34 |

## MATTOON *Coles County* ELEVATION 722 ft LAT/LONG 39° 28 ' N / 88° 21 ' W

| | JAN | FEB | MAR | APR | MAY | JUN | JUL | AUG | SEP | OCT | NOV | DEC | YEAR |
|---|---|---|---|---|---|---|---|---|---|---|---|---|---|
| Maximum Temp °F | 33.4 | 38.3 | 50.5 | 63.7 | 74.2 | 83.6 | 86.5 | 84.4 | 78.1 | 65.6 | 51.1 | 38.7 | 62.3 |
| Minimum Temp °F | 17.0 | 21.0 | 31.4 | 42.1 | 52.0 | 61.4 | 65.4 | 62.8 | 55.5 | 43.8 | 33.8 | 23.1 | 42.4 |
| Mean Temp °F | 25.2 | 29.7 | 41.0 | 52.9 | 63.1 | 72.5 | 76.0 | 73.6 | 66.9 | 54.7 | 42.4 | 30.9 | 52.4 |
| Days Max Temp ≥ 90 °F | 0 | 0 | 0 | 0 | 1 | 6 | 10 | 7 | 2 | 0 | 0 | 0 | 26 |
| Days Max Temp ≤ 32 °F | 14 | 9 | 2 | 0 | 0 | 0 | 0 | 0 | 0 | 0 | 1 | 8 | 34 |
| Days Min Temp ≤ 32 °F | 28 | 24 | 18 | 5 | 0 | 0 | 0 | 0 | 0 | 4 | 15 | 25 | 119 |
| Days Min Temp ≤ 0 °F | 4 | 2 | 0 | 0 | 0 | 0 | 0 | 0 | 0 | 0 | 0 | 2 | 8 |
| Heating Degree Days | 1227 | 990 | 740 | 371 | 133 | 12 | 1 | 4 | 72 | 329 | 670 | 1049 | 5598 |
| Cooling Degree Days | 0 | 0 | 2 | 15 | 85 | 260 | 364 | 293 | 140 | 16 | 0 | 0 | 1175 |
| Total Precipitation (") | 1.91 | 1.92 | 2.90 | 4.09 | 3.85 | 3.60 | 4.42 | 3.20 | 3.44 | 3.00 | 3.69 | 3.19 | 39.21 |
| Days ≥ 0.1" Precip | 5 | 4 | 7 | 8 | 7 | 6 | 7 | 5 | 6 | 6 | 6 | 6 | 73 |
| Total Snowfall (") | 5.8 | 3.9 | 2.4 | 0.1 | 0.0 | 0.0 | 0.0 | 0.0 | 0.0 | 0.0 | 1.2 | 4.1 | 17.5 |
| Days ≥ 1" Snow Depth | 11 | 8 | 2 | 0 | 0 | 0 | 0 | 0 | 0 | 0 | 1 | 5 | 27 |

## MCLEANSBORO 2 ENE *Hamilton County* ELEVATION 502 ft LAT/LONG 38° 5 ' N / 88° 32 ' W

| | JAN | FEB | MAR | APR | MAY | JUN | JUL | AUG | SEP | OCT | NOV | DEC | YEAR |
|---|---|---|---|---|---|---|---|---|---|---|---|---|---|
| Maximum Temp °F | 38.2 | 43.3 | 54.6 | 66.5 | 75.9 | 85.0 | 88.8 | 87.2 | 80.6 | 68.7 | 55.3 | 43.6 | 65.6 |
| Minimum Temp °F | 20.5 | 23.8 | 34.0 | 44.5 | 53.2 | 62.0 | 66.0 | 62.9 | 55.8 | 43.6 | 35.6 | 25.8 | 44.0 |
| Mean Temp °F | 29.4 | 33.6 | 44.3 | 55.5 | 64.6 | 73.6 | 77.5 | 75.1 | 68.3 | 56.2 | 45.5 | 34.7 | 54.9 |
| Days Max Temp ≥ 90 °F | 0 | 0 | 0 | 0 | 1 | 8 | 15 | 11 | 5 | 0 | 0 | 0 | 40 |
| Days Max Temp ≤ 32 °F | 10 | 6 | 1 | 0 | 0 | 0 | 0 | 0 | 0 | 0 | 0 | 5 | 22 |
| Days Min Temp ≤ 32 °F | 26 | 22 | 15 | 3 | 0 | 0 | 0 | 0 | 0 | 4 | 12 | 23 | 105 |
| Days Min Temp ≤ 0 °F | 2 | 1 | 0 | 0 | 0 | 0 | 0 | 0 | 0 | 0 | 0 | 1 | 4 |
| Heating Degree Days | 1098 | 882 | 639 | 302 | 101 | 8 | 1 | 3 | 55 | 291 | 580 | 931 | 4891 |
| Cooling Degree Days | 0 | 0 | 3 | 24 | 90 | 273 | 402 | 332 | 160 | 23 | 2 | 0 | 1309 |
| Total Precipitation (") | 2.70 | 2.69 | 4.20 | 4.44 | 4.25 | 3.51 | 3.58 | 2.98 | 3.22 | 3.12 | 4.61 | 3.71 | 43.01 |
| Days ≥ 0.1" Precip | 5 | 5 | 8 | 8 | 8 | 6 | 6 | 6 | 5 | 6 | 7 | 7 | 77 |
| Total Snowfall (") | 4.6 | 5.0 | 2.5 | 0.4 | 0.0 | 0.0 | 0.0 | 0.0 | 0.0 | 0.2 | 0.7 | 2.9 | 16.3 |
| Days ≥ 1" Snow Depth | 7 | 6 | 2 | 0 | 0 | 0 | 0 | 0 | 0 | 0 | 0 | 3 | 18 |

## MINONK *Woodford County* ELEVATION 751 ft LAT/LONG 40° 54 ' N / 89° 2 ' W

| | JAN | FEB | MAR | APR | MAY | JUN | JUL | AUG | SEP | OCT | NOV | DEC | YEAR |
|---|---|---|---|---|---|---|---|---|---|---|---|---|---|
| Maximum Temp °F | 30.0 | 35.2 | 47.8 | 62.5 | 73.7 | 83.8 | 86.2 | 83.9 | 77.7 | 65.2 | 48.9 | 35.7 | 60.9 |
| Minimum Temp °F | 12.8 | 17.5 | 28.6 | 39.1 | 48.8 | 58.6 | 62.3 | 59.5 | 52.3 | 40.8 | 30.7 | 19.5 | 39.2 |
| Mean Temp °F | 21.4 | 26.3 | 38.2 | 50.8 | 61.3 | 71.2 | 74.3 | 71.7 | 65.0 | 53.0 | 39.8 | 27.6 | 50.1 |
| Days Max Temp ≥ 90 °F | 0 | 0 | 0 | 0 | 1 | 7 | 9 | 6 | 2 | 0 | 0 | 0 | 25 |
| Days Max Temp ≤ 32 °F | 17 | 11 | 3 | 0 | 0 | 0 | 0 | 0 | 0 | 0 | 2 | 11 | 44 |
| Days Min Temp ≤ 32 °F | 29 | 26 | 21 | 7 | 1 | 0 | 0 | 0 | 0 | 6 | 18 | 27 | 135 |
| Days Min Temp ≤ 0 °F | 7 | 4 | 0 | 0 | 0 | 0 | 0 | 0 | 0 | 0 | 0 | 3 | 14 |
| Heating Degree Days | 1345 | 1086 | 825 | 431 | 176 | 22 | 4 | 13 | 98 | 379 | 749 | 1152 | 6280 |
| Cooling Degree Days | 0 | 0 | 2 | 15 | 67 | 210 | 300 | 228 | 101 | 10 | 0 | 0 | 933 |
| Total Precipitation (") | 1.68 | 1.72 | 3.12 | 3.77 | 3.96 | 3.61 | 4.10 | 3.35 | 3.69 | 3.01 | 3.42 | 2.63 | 38.06 |
| Days ≥ 0.1" Precip | 5 | *4* | 7 | 8 | 7 | 7 | 7 | 6 | 5 | 6 | 7 | 6 | 75 |
| Total Snowfall (") | 8.3 | 7.2 | 3.7 | 0.5 | 0.0 | 0.0 | 0.0 | 0.0 | 0.0 | 0.1 | 1.4 | 6.4 | 27.6 |
| Days ≥ 1" Snow Depth | *14* | *10* | *4* | 0 | 0 | 0 | 0 | 0 | 0 | 0 | 1 | *9* | 38 |

## MOLINE QUAD CITY AP *Rock Island County* ELEVATION 582 ft LAT/LONG 41° 27 ' N / 90° 30 ' W

| | JAN | FEB | MAR | APR | MAY | JUN | JUL | AUG | SEP | OCT | NOV | DEC | YEAR |
|---|---|---|---|---|---|---|---|---|---|---|---|---|---|
| Maximum Temp °F | 28.7 | 34.0 | 47.2 | 61.6 | 73.0 | 82.4 | 85.8 | 83.4 | 75.8 | 63.7 | 47.7 | 34.4 | 59.8 |
| Minimum Temp °F | 11.7 | 16.5 | 28.3 | 39.6 | 49.7 | 59.5 | 64.4 | 61.8 | 53.3 | 41.4 | 30.4 | 18.8 | 39.6 |
| Mean Temp °F | 20.3 | 25.2 | 37.8 | 50.6 | 61.4 | 71.0 | 75.1 | 72.6 | 64.5 | 52.6 | 39.1 | 26.7 | 49.7 |
| Days Max Temp ≥ 90 °F | 0 | 0 | 0 | 0 | 1 | 5 | 9 | 6 | 2 | 0 | 0 | 0 | 23 |
| Days Max Temp ≤ 32 °F | 18 | 13 | 3 | 0 | 0 | 0 | 0 | 0 | 0 | 0 | 2 | 12 | 48 |
| Days Min Temp ≤ 32 °F | 29 | 25 | 21 | 7 | 1 | 0 | 0 | 0 | 0 | 6 | 18 | 27 | 134 |
| Days Min Temp ≤ 0 °F | 8 | 4 | 0 | 0 | 0 | 0 | 0 | 0 | 0 | 0 | 0 | 3 | 15 |
| Heating Degree Days | 1382 | 1116 | 839 | 438 | 173 | 18 | 3 | 9 | 106 | 393 | 771 | 1182 | 6430 |
| Cooling Degree Days | 0 | 0 | 2 | 17 | 66 | 212 | 334 | 263 | 106 | 9 | 0 | 0 | 1009 |
| Total Precipitation (") | 1.55 | 1.29 | 2.94 | 3.88 | 4.18 | 4.64 | 4.79 | 4.36 | 3.77 | 2.93 | 2.73 | 2.37 | 39.43 |
| Days ≥ 0.1" Precip | 4 | 4 | 6 | 7 | 7 | 7 | 7 | 7 | 6 | 6 | 6 | 5 | 72 |
| Total Snowfall (") | 8.9 | 7.1 | 5.2 | 1.5 | 0.0 | 0.0 | 0.0 | 0.0 | 0.0 | 0.4 | 2.8 | 7.3 | 33.2 |
| Days ≥ 1" Snow Depth | 17 | 13 | 5 | 0 | 0 | 0 | 0 | 0 | 0 | 0 | 1 | 10 | 46 |

## MONMOUTH *Warren County* ELEVATION 770 ft LAT/LONG 40° 55 ' N / 90° 38 ' W

| | JAN | FEB | MAR | APR | MAY | JUN | JUL | AUG | SEP | OCT | NOV | DEC | YEAR |
|---|---|---|---|---|---|---|---|---|---|---|---|---|---|
| Maximum Temp °F | 31.3 | 36.8 | 49.8 | 64.0 | 74.7 | 83.2 | 86.7 | 84.2 | 77.1 | 65.2 | 49.5 | 36.2 | 61.6 |
| Minimum Temp °F | 14.2 | 18.9 | 29.7 | 40.8 | 50.6 | 59.8 | 63.9 | 61.1 | 53.6 | 42.8 | 31.5 | 20.7 | 40.6 |
| Mean Temp °F | 22.8 | 27.9 | 39.8 | 52.4 | 62.6 | 71.5 | 75.3 | 72.7 | 65.4 | 54.0 | 40.5 | 28.5 | 51.1 |
| Days Max Temp ≥ 90 °F | 0 | 0 | 0 | 0 | 1 | 5 | 10 | 6 | 2 | 0 | 0 | 0 | 24 |
| Days Max Temp ≤ 32 °F | 15 | 10 | 2 | 0 | 0 | 0 | 0 | 0 | 0 | 0 | 2 | 10 | 39 |
| Days Min Temp ≤ 32 °F | 29 | 25 | 19 | 6 | 0 | 0 | 0 | 0 | 0 | 5 | 17 | 27 | 128 |
| Days Min Temp ≤ 0 °F | 6 | 3 | 0 | 0 | 0 | 0 | 0 | 0 | 0 | 0 | 0 | 2 | 11 |
| Heating Degree Days | 1302 | 1041 | 777 | 385 | 142 | 16 | 3 | 7 | 90 | 348 | 728 | 1126 | 5965 |
| Cooling Degree Days | 0 | 0 | 2 | 19 | 77 | 235 | 344 | 267 | 114 | 10 | 0 | 0 | 1068 |
| Total Precipitation (") | 1.56 | 1.50 | 2.84 | 3.79 | 3.91 | 4.04 | 4.54 | 3.98 | 3.78 | 2.96 | 2.66 | 2.44 | 38.00 |
| Days ≥ 0.1" Precip | 4 | 4 | 7 | 7 | 8 | 6 | 7 | 6 | 6 | 6 | 6 | 5 | 72 |
| Total Snowfall (") | 7.5 | 6.2 | 4.1 | 1.1 | 0.0 | 0.0 | 0.0 | 0.0 | 0.0 | 0.2 | 2.3 | 6.4 | 27.8 |
| Days ≥ 1" Snow Depth | 19 | 13 | 4 | 0 | 0 | 0 | 0 | 0 | 0 | 0 | 2 | 9 | 47 |

## MORRISON *Whiteside County* ELEVATION 669 ft LAT/LONG 41° 49 ' N / 89° 58 ' W

| | JAN | FEB | MAR | APR | MAY | JUN | JUL | AUG | SEP | OCT | NOV | DEC | YEAR |
|---|---|---|---|---|---|---|---|---|---|---|---|---|---|
| Maximum Temp °F | 28.8 | 33.9 | 46.7 | 61.0 | 72.8 | 81.9 | 85.2 | 82.8 | 75.8 | 64.1 | 48.0 | 34.4 | 59.6 |
| Minimum Temp °F | 10.1 | 14.6 | 27.0 | 38.0 | 48.7 | 57.9 | 62.3 | 59.3 | 51.2 | 39.6 | 29.1 | 17.4 | 37.9 |
| Mean Temp °F | 19.5 | 24.3 | 36.9 | 49.5 | 60.8 | 69.9 | 73.8 | 71.1 | 63.5 | 51.9 | 38.6 | 25.9 | 48.8 |
| Days Max Temp ≥ 90 °F | 0 | 0 | 0 | 0 | 1 | 4 | 7 | 5 | 1 | 0 | 0 | 0 | 18 |
| Days Max Temp ≤ 32 °F | 18 | 12 | 3 | 0 | 0 | 0 | 0 | 0 | 0 | 0 | 2 | 12 | 47 |
| Days Min Temp ≤ 32 °F | 30 | 27 | 23 | 8 | 1 | 0 | 0 | 0 | 1 | 8 | 20 | 28 | 146 |
| Days Min Temp ≤ 0 °F | 9 | 5 | 0 | 0 | 0 | 0 | 0 | 0 | 0 | 0 | 0 | 4 | 18 |
| Heating Degree Days | 1406 | 1144 | 865 | 468 | 183 | 26 | 5 | 14 | 120 | 412 | 786 | 1205 | 6634 |
| Cooling Degree Days | 0 | 0 | 1 | 11 | 56 | 181 | 285 | 216 | 79 | 4 | 0 | 0 | 833 |
| Total Precipitation (") | 1.43 | 1.29 | 2.73 | 3.58 | 4.34 | 4.33 | 3.68 | 4.71 | 3.41 | 2.76 | 2.91 | 2.27 | 37.44 |
| Days ≥ 0.1" Precip | 4 | 4 | 6 | 7 | 8 | 7 | 6 | 7 | 6 | 5 | 6 | 5 | 71 |
| Total Snowfall (") | 9.2 | 7.4 | 5.1 | 1.2 | 0.0 | 0.0 | 0.0 | 0.0 | 0.0 | 0.2 | 1.7 | 8.2 | 33.0 |
| Days ≥ 1" Snow Depth | 21 | 16 | 6 | 1 | 0 | 0 | 0 | 0 | 0 | 0 | 1 | 14 | 59 |

## MOUNT CARROLL *Carroll County* ELEVATION 817 ft LAT/LONG 42° 5 ' N / 89° 58 ' W

| | JAN | FEB | MAR | APR | MAY | JUN | JUL | AUG | SEP | OCT | NOV | DEC | YEAR |
|---|---|---|---|---|---|---|---|---|---|---|---|---|---|
| Maximum Temp °F | 27.7 | 32.9 | 45.9 | 60.1 | 71.9 | 80.9 | 84.6 | 82.2 | 74.6 | 62.7 | 46.7 | 33.3 | 58.6 |
| Minimum Temp °F | 7.4 | 11.7 | 24.6 | 35.0 | 45.7 | 54.3 | 58.8 | 56.2 | 47.9 | 36.4 | 27.0 | 14.8 | 35.0 |
| Mean Temp °F | 17.6 | 22.3 | 35.3 | 47.5 | 58.9 | 67.6 | 71.6 | 69.3 | 61.3 | 49.6 | 36.9 | 24.1 | 46.8 |
| Days Max Temp ≥ 90 °F | 0 | 0 | 0 | 0 | 1 | 4 | 7 | 4 | 1 | 0 | 0 | 0 | 17 |
| Days Max Temp ≤ 32 °F | 19 | 13 | 4 | 0 | 0 | 0 | 0 | 0 | 0 | 0 | 2 | 13 | 51 |
| Days Min Temp ≤ 32 °F | 30 | 27 | 24 | 12 | 3 | 0 | 0 | 0 | 2 | 13 | 21 | 30 | 162 |
| Days Min Temp ≤ 0 °F | 11 | 7 | 1 | 0 | 0 | 0 | 0 | 0 | 0 | 0 | 0 | 5 | 24 |
| Heating Degree Days | 1464 | 1199 | 914 | 524 | 227 | 48 | 12 | 28 | 165 | 479 | 837 | 1263 | 7160 |
| Cooling Degree Days | 0 | 0 | 0 | 9 | 38 | 119 | 206 | 154 | 60 | 2 | 0 | 0 | 588 |
| Total Precipitation (") | 1.38 | 1.28 | 2.62 | 3.62 | 4.17 | 4.72 | 3.66 | 4.37 | 3.83 | 2.70 | 2.83 | 2.13 | 37.31 |
| Days ≥ 0.1" Precip | 4 | 4 | 6 | 7 | 7 | 7 | 6 | 7 | 6 | 5 | 6 | 5 | 70 |
| Total Snowfall (") | 9.2 | 7.2 | 4.3 | 1.5 | 0.0 | 0.0 | 0.0 | 0.0 | 0.0 | 0.2 | 2.3 | 7.5 | 32.2 |
| Days ≥ 1" Snow Depth | 22 | 17 | 6 | 1 | 0 | 0 | 0 | 0 | 0 | 0 | 1 | 14 | 61 |

## MOWEAQUA *Shelby County* ELEVATION 620 ft LAT/LONG 39° 38 ' N / 89° 2 ' W

| | JAN | FEB | MAR | APR | MAY | JUN | JUL | AUG | SEP | OCT | NOV | DEC | YEAR |
|---|---|---|---|---|---|---|---|---|---|---|---|---|---|
| Maximum Temp °F | 34.3 | 39.5 | 52.1 | 65.5 | 75.5 | 84.3 | 87.4 | 85.2 | 79.4 | 67.5 | 52.5 | 39.7 | 63.6 |
| Minimum Temp °F | 17.0 | 20.5 | 31.2 | 41.9 | 51.4 | 60.8 | 64.6 | 61.7 | 54.7 | 43.4 | 33.7 | 23.2 | 42.0 |
| Mean Temp °F | 25.7 | 30.0 | 41.7 | 53.7 | 63.5 | 72.5 | 76.0 | 73.5 | 67.1 | 55.5 | 43.2 | 31.5 | 52.8 |
| Days Max Temp ≥ 90 °F | 0 | 0 | 0 | 0 | 1 | 7 | 11 | 7 | 3 | 0 | 0 | 0 | 29 |
| Days Max Temp ≤ 32 °F | 13 | 9 | 2 | 0 | 0 | 0 | 0 | 0 | 0 | 0 | 1 | 7 | 32 |
| Days Min Temp ≤ 32 °F | 28 | 24 | 18 | 5 | 0 | 0 | 0 | 0 | 0 | 5 | 15 | 25 | 120 |
| Days Min Temp ≤ 0 °F | 4 | 2 | 0 | 0 | 0 | 0 | 0 | 0 | 0 | 0 | 0 | 2 | 8 |
| Heating Degree Days | 1211 | 982 | 719 | 349 | 121 | 12 | 1 | 5 | 69 | 310 | 650 | 1027 | 5456 |
| Cooling Degree Days | 0 | 0 | 3 | 17 | 76 | 252 | 355 | 285 | 140 | 18 | 0 | 0 | 1146 |
| Total Precipitation (") | 1.81 | 1.90 | 2.82 | 3.99 | 4.03 | 3.94 | 4.41 | 3.47 | 3.35 | 2.91 | 3.46 | 2.63 | 38.72 |
| Days ≥ 0.1" Precip | *4* | *4* | 6 | 8 | 7 | 7 | 7 | 6 | 5 | 6 | 6 | 5 | 71 |
| Total Snowfall (") | na | na | *1.4* | *0.2* | 0.0 | 0.0 | 0.0 | 0.0 | 0.0 | 0.0 | 0.6 | na | na |
| Days ≥ 1" Snow Depth | na | na | na | 0 | 0 | 0 | 0 | 0 | 0 | 0 | 0 | *3* | na |

## MT VERNON 3 NE *Jefferson County* ELEVATION 490 ft LAT/LONG 38° 21 ' N / 88° 52 ' W

| | JAN | FEB | MAR | APR | MAY | JUN | JUL | AUG | SEP | OCT | NOV | DEC | YEAR |
|---|---|---|---|---|---|---|---|---|---|---|---|---|---|
| Maximum Temp °F | 37.8 | 42.8 | 54.5 | 66.5 | 75.7 | 85.0 | 88.8 | 86.7 | 79.8 | 68.4 | 54.8 | 43.0 | 65.3 |
| Minimum Temp °F | 19.7 | 23.1 | 33.7 | 44.3 | 52.7 | 61.9 | 66.4 | 63.6 | 56.4 | 43.8 | 35.6 | 25.3 | 43.9 |
| Mean Temp °F | 28.8 | 33.0 | 44.1 | 55.4 | 64.2 | 73.5 | 77.6 | 75.2 | 68.1 | 56.1 | 45.2 | 34.2 | 54.6 |
| Days Max Temp ≥ 90 °F | 0 | 0 | 0 | 0 | 1 | 8 | 15 | 11 | 3 | 0 | 0 | 0 | 38 |
| Days Max Temp ≤ 32 °F | 11 | 6 | 1 | 0 | 0 | 0 | 0 | 0 | 0 | 0 | 1 | 5 | 24 |
| Days Min Temp ≤ 32 °F | 27 | 22 | 15 | 4 | 0 | 0 | 0 | 0 | 0 | 5 | 13 | 24 | 110 |
| Days Min Temp ≤ 0 °F | 2 | 1 | 0 | 0 | 0 | 0 | 0 | 0 | 0 | 0 | 0 | 1 | 4 |
| Heating Degree Days | 1117 | 899 | 644 | 305 | 107 | 8 | 0 | 2 | 58 | 293 | 589 | 950 | 4972 |
| Cooling Degree Days | 0 | 0 | 3 | 22 | 76 | 259 | 390 | 317 | 147 | 21 | 2 | 0 | 1237 |
| Total Precipitation (") | 2.25 | 2.54 | 3.85 | 4.39 | 4.14 | 3.33 | 3.77 | 3.14 | 3.61 | 3.08 | 4.39 | 3.45 | 41.94 |
| Days ≥ 0.1" Precip | 5 | 5 | 7 | 8 | 7 | 6 | 6 | 5 | 5 | 5 | 7 | 6 | 72 |
| Total Snowfall (") | 5.7 | 5.7 | 2.4 | 0.4 | 0.0 | 0.0 | 0.0 | 0.0 | 0.0 | 0.2 | 0.5 | *2.8* | 17.7 |
| Days ≥ 1" Snow Depth | *7* | *5* | 2 | 0 | 0 | 0 | 0 | 0 | 0 | 0 | 0 | *3* | 17 |

## NASHVILLE 4 NE *Washington County* ELEVATION 522 ft LAT/LONG 38° 23 ' N / 89° 20 ' W

| | JAN | FEB | MAR | APR | MAY | JUN | JUL | AUG | SEP | OCT | NOV | DEC | YEAR |
|---|---|---|---|---|---|---|---|---|---|---|---|---|---|
| Maximum Temp °F | 37.7 | 42.8 | 55.1 | 66.7 | 75.9 | 85.0 | 88.4 | 86.4 | 79.7 | 68.4 | 54.5 | 42.8 | 65.3 |
| Minimum Temp °F | 21.2 | 25.0 | 35.0 | 45.5 | 54.6 | 63.7 | 67.5 | 64.8 | 58.0 | 46.7 | 36.8 | 27.0 | 45.5 |
| Mean Temp °F | 29.5 | 33.9 | 45.1 | 56.1 | 65.3 | 74.4 | 78.0 | 75.6 | 68.9 | 57.5 | 45.7 | 34.9 | 55.4 |
| Days Max Temp ≥ 90 °F | 0 | 0 | 0 | 0 | 1 | 7 | 13 | 9 | 3 | 0 | 0 | 0 | 33 |
| Days Max Temp ≤ 32 °F | 11 | 6 | 1 | 0 | 0 | 0 | 0 | 0 | 0 | 0 | 1 | 5 | 24 |
| Days Min Temp ≤ 32 °F | 26 | 21 | 14 | 3 | 0 | 0 | 0 | 0 | 0 | 2 | 11 | 22 | 99 |
| Days Min Temp ≤ 0 °F | 2 | 1 | 0 | 0 | 0 | 0 | 0 | 0 | 0 | 0 | 0 | 1 | 4 |
| Heating Degree Days | 1094 | 870 | 615 | 284 | 89 | 5 | 1 | 2 | 48 | 254 | 575 | 926 | 4763 |
| Cooling Degree Days | 0 | 0 | 3 | 26 | 107 | 302 | 422 | 357 | 172 | 29 | 1 | 0 | 1419 |
| Total Precipitation (") | 1.98 | 2.32 | 3.44 | 3.88 | 3.70 | 3.34 | 3.58 | 2.80 | 3.61 | 3.18 | 3.87 | 3.30 | 39.00 |
| Days ≥ 0.1" Precip | 5 | 5 | 7 | 7 | 7 | 6 | 6 | 5 | 5 | 6 | 7 | 6 | 72 |
| Total Snowfall (") | 5.2 | 4.9 | 2.3 | 0.6 | 0.0 | 0.0 | 0.0 | 0.0 | 0.0 | 0.1 | 0.9 | 3.4 | 17.4 |
| Days ≥ 1" Snow Depth | 9 | 7 | 2 | 0 | 0 | 0 | 0 | 0 | 0 | 0 | 1 | 3 | 22 |

## NEWTON 6 SSE *Jasper County* ELEVATION 502 ft LAT/LONG 38° 57' N / 88° 7' W

| | JAN | FEB | MAR | APR | MAY | JUN | JUL | AUG | SEP | OCT | NOV | DEC | YEAR |
|---|---|---|---|---|---|---|---|---|---|---|---|---|---|
| Maximum Temp °F | 35.3 | 40.4 | 52.2 | 64.4 | 74.5 | 83.9 | 87.6 | 85.6 | 79.5 | 67.4 | 53.3 | 40.9 | 63.8 |
| Minimum Temp °F | 18.4 | 21.7 | 32.5 | 42.7 | 51.7 | 61.3 | 65.2 | 61.9 | 54.7 | 42.8 | 34.6 | 24.6 | 42.7 |
| Mean Temp °F | 26.9 | 31.1 | 42.4 | 53.6 | 63.1 | 72.6 | 76.4 | 73.8 | 67.1 | 55.2 | 44.0 | 32.8 | 53.3 |
| Days Max Temp ≥ 90 °F | 0 | 0 | 0 | 0 | 1 | 6 | 11 | 8 | 3 | 0 | 0 | 0 | 29 |
| Days Max Temp ≤ 32 °F | 13 | 8 | 2 | 0 | 0 | 0 | 0 | 0 | 0 | 0 | 1 | 6 | 30 |
| Days Min Temp ≤ 32 °F | 27 | 23 | 17 | 4 | 0 | 0 | 0 | 0 | 0 | 5 | 14 | 24 | 114 |
| Days Min Temp ≤ 0 °F | 3 | 2 | 0 | 0 | 0 | 0 | 0 | 0 | 0 | 0 | 0 | 1 | 6 |
| Heating Degree Days | 1176 | 951 | 696 | 352 | 127 | 13 | 1 | 4 | 68 | 319 | 625 | 991 | 5323 |
| Cooling Degree Days | 0 | 0 | 2 | 14 | 74 | 250 | 366 | 281 | 134 | 20 | 1 | 0 | 1142 |
| Total Precipitation (") | 2.10 | 2.18 | 3.64 | 3.90 | 4.01 | 3.29 | 4.22 | 3.50 | 3.57 | 2.90 | 4.03 | 3.22 | 40.56 |
| Days ≥ 0.1" Precip | 5 | 5 | 7 | 8 | 7 | 6 | 7 | 5 | 6 | 5 | 7 | 6 | 74 |
| Total Snowfall (") | 5.0 | *4.2* | 2.3 | 0.1 | 0.0 | 0.0 | 0.0 | 0.0 | 0.0 | 0.1 | 0.7 | 3.6 | 16.0 |
| Days ≥ 1" Snow Depth | 9 | 8 | 2 | 0 | 0 | 0 | 0 | 0 | 0 | 0 | 0 | 5 | 24 |

## OLNEY 2 S *Richland County* ELEVATION 479 ft LAT/LONG 38° 44' N / 88° 5' W

| | JAN | FEB | MAR | APR | MAY | JUN | JUL | AUG | SEP | OCT | NOV | DEC | YEAR |
|---|---|---|---|---|---|---|---|---|---|---|---|---|---|
| Maximum Temp °F | 37.3 | 42.5 | 54.6 | 66.8 | 76.3 | 85.2 | 88.5 | 86.6 | 80.5 | 68.8 | 54.9 | 42.6 | 65.4 |
| Minimum Temp °F | 20.5 | 24.0 | 34.3 | 44.5 | 53.0 | 62.0 | 66.0 | 63.4 | 56.7 | 44.8 | 36.1 | 26.4 | 44.3 |
| Mean Temp °F | 28.9 | 33.3 | 44.5 | 55.7 | 64.7 | 73.6 | 77.3 | 75.0 | 68.7 | 56.8 | 45.5 | 34.6 | 54.9 |
| Days Max Temp ≥ 90 °F | 0 | 0 | 0 | 0 | 1 | 8 | 13 | 9 | 4 | 0 | 0 | 0 | 35 |
| Days Max Temp ≤ 32 °F | 11 | 6 | 1 | 0 | 0 | 0 | 0 | 0 | 0 | 0 | 1 | 5 | 24 |
| Days Min Temp ≤ 32 °F | 26 | 21 | 15 | 4 | 0 | 0 | 0 | 0 | 0 | 4 | 12 | 22 | 104 |
| Days Min Temp ≤ 0 °F | 3 | 2 | 0 | 0 | 0 | 0 | 0 | 0 | 0 | 0 | 0 | 1 | 6 |
| Heating Degree Days | 1112 | 889 | 632 | 297 | 100 | 7 | 1 | 1 | 51 | 275 | 580 | 937 | 4882 |
| Cooling Degree Days | 0 | 0 | 3 | 23 | 102 | 283 | 402 | 335 | 171 | 26 | 1 | 0 | 1346 |
| Total Precipitation (") | 2.35 | 2.65 | 4.09 | 4.44 | 4.01 | 3.49 | 3.62 | 3.33 | 3.24 | 3.39 | 4.42 | 3.74 | 42.77 |
| Days ≥ 0.1" Precip | 5 | 5 | 8 | 8 | 7 | 6 | 6 | 5 | 6 | 5 | 7 | 7 | 75 |
| Total Snowfall (") | 3.6 | 4.3 | 2.1 | 0.2 | 0.0 | 0.0 | 0.0 | 0.0 | 0.0 | 0.0 | 0.7 | 2.7 | 13.6 |
| Days ≥ 1" Snow Depth | na | na | 1 | 0 | 0 | 0 | 0 | 0 | 0 | 0 | 0 | *1* | na |

## OTTAWA 4 SW *La Salle County* ELEVATION 525 ft LAT/LONG 41° 20' N / 88° 55' W

| | JAN | FEB | MAR | APR | MAY | JUN | JUL | AUG | SEP | OCT | NOV | DEC | YEAR |
|---|---|---|---|---|---|---|---|---|---|---|---|---|---|
| Maximum Temp °F | 30.7 | 36.2 | 48.6 | 62.8 | 74.1 | 82.8 | 85.6 | 83.6 | 77.2 | 65.3 | 49.7 | 36.5 | 61.1 |
| Minimum Temp °F | 14.2 | 18.5 | 29.8 | 40.5 | 50.7 | 60.5 | 64.6 | 62.3 | 54.7 | 42.8 | 32.6 | 21.1 | 41.0 |
| Mean Temp °F | 22.5 | 27.4 | 39.3 | 51.7 | 62.4 | 71.7 | 75.1 | 73.0 | 66.0 | 54.1 | 41.1 | 28.8 | 51.1 |
| Days Max Temp ≥ 90 °F | 0 | 0 | 0 | 0 | 2 | 5 | 8 | 6 | 2 | 0 | 0 | 0 | 23 |
| Days Max Temp ≤ 32 °F | 16 | 10 | 2 | 0 | 0 | 0 | 0 | 0 | 0 | 0 | 1 | 10 | 39 |
| Days Min Temp ≤ 32 °F | 29 | 25 | 19 | 6 | 1 | 0 | 0 | 0 | 0 | 4 | 16 | 26 | 126 |
| Days Min Temp ≤ 0 °F | 6 | 3 | 0 | 0 | 0 | 0 | 0 | 0 | 0 | 0 | 0 | 2 | 11 |
| Heating Degree Days | 1311 | 1056 | 792 | 407 | 150 | 17 | 3 | 8 | 82 | 348 | 710 | 1116 | 6000 |
| Cooling Degree Days | 0 | 0 | 2 | 16 | 75 | 207 | 310 | 253 | 109 | 10 | 0 | 0 | 982 |
| Total Precipitation (") | 1.54 | 1.25 | 2.72 | 3.75 | 3.87 | 4.13 | 3.71 | 3.92 | 3.83 | 2.54 | 3.05 | 2.45 | 36.76 |
| Days ≥ 0.1" Precip | 4 | 3 | 6 | 7 | 7 | 6 | 6 | 6 | 6 | 5 | 6 | 5 | 67 |
| Total Snowfall (") | 8.1 | 5.5 | 3.4 | 0.7 | 0.0 | 0.0 | 0.0 | 0.0 | 0.0 | 0.1 | 1.0 | 5.5 | 24.3 |
| Days ≥ 1" Snow Depth | 19 | 13 | 4 | 0 | 0 | 0 | 0 | 0 | 0 | 0 | 1 | 8 | 45 |

## PALESTINE *Crawford County* ELEVATION 514 ft LAT/LONG 39° 0' N / 87° 37' W

| | JAN | FEB | MAR | APR | MAY | JUN | JUL | AUG | SEP | OCT | NOV | DEC | YEAR |
|---|---|---|---|---|---|---|---|---|---|---|---|---|---|
| Maximum Temp °F | 36.5 | 41.6 | 53.8 | 66.2 | 76.0 | 85.0 | 88.1 | 85.9 | 79.4 | 67.8 | 53.8 | 41.9 | 64.7 |
| Minimum Temp °F | 20.1 | 24.2 | 34.8 | 44.9 | 53.7 | 62.0 | 66.4 | 64.8 | 55.9 | 45.0 | 36.6 | 26.7 | 44.6 |
| Mean Temp °F | 28.3 | 32.9 | 44.3 | 55.6 | 64.9 | 73.5 | 77.2 | 75.4 | 67.7 | 56.3 | 45.2 | 34.3 | 54.6 |
| Days Max Temp ≥ 90 °F | 0 | 0 | 0 | 0 | 1 | 8 | 12 | 9 | 3 | 0 | 0 | 0 | 33 |
| Days Max Temp ≤ 32 °F | 11 | 6 | 1 | 0 | 0 | 0 | 0 | 0 | 0 | 0 | 0 | 5 | 23 |
| Days Min Temp ≤ 32 °F | 26 | 21 | 13 | 3 | 0 | 0 | 0 | 0 | 0 | 3 | 12 | 21 | 99 |
| Days Min Temp ≤ 0 °F | 3 | 2 | 0 | 0 | 0 | 0 | 0 | 0 | 0 | 0 | 0 | 1 | 6 |
| Heating Degree Days | 1130 | 900 | 639 | 302 | 101 | 10 | 1 | 1 | 60 | 285 | 588 | 945 | 4962 |
| Cooling Degree Days | 0 | 0 | 4 | 25 | 109 | 278 | 400 | 360 | 140 | 20 | 1 | 0 | 1337 |
| Total Precipitation (") | 2.25 | 2.47 | 3.74 | 4.16 | 4.26 | 3.65 | 4.46 | 3.51 | 3.60 | 3.05 | 3.97 | 3.34 | 42.46 |
| Days ≥ 0.1" Precip | 5 | 5 | 7 | 8 | 7 | 6 | 6 | 6 | 6 | 5 | 7 | 6 | 74 |
| Total Snowfall (") | 5.9 | 5.4 | 3.0 | 0.2 | 0.0 | 0.0 | 0.0 | 0.0 | 0.0 | 0.1 | 1.0 | 3.9 | 19.5 |
| Days ≥ 1" Snow Depth | 10 | 7 | 2 | 0 | 0 | 0 | 0 | 0 | 0 | 0 | 1 | 3 | 23 |

## PANA *Christian County* ELEVATION 702 ft LAT/LONG 39° 23 ' N / 89° 5 ' W

| | JAN | FEB | MAR | APR | MAY | JUN | JUL | AUG | SEP | OCT | NOV | DEC | YEAR |
|---|---|---|---|---|---|---|---|---|---|---|---|---|---|
| Maximum Temp °F | 34.7 | 39.9 | 52.3 | 65.6 | 74.8 | 83.5 | 86.8 | 84.3 | 78.1 | 66.3 | 52.3 | 40.1 | 63.2 |
| Minimum Temp °F | 18.3 | 22.4 | 32.7 | 43.2 | 52.7 | 61.8 | 65.8 | 63.2 | 56.1 | 44.8 | 34.8 | 24.6 | 43.4 |
| Mean Temp °F | 26.5 | 31.2 | 42.5 | 54.5 | 63.8 | 72.7 | 76.3 | 73.8 | 67.2 | 55.6 | 43.5 | 32.4 | 53.3 |
| Days Max Temp ≥ 90 °F | 0 | 0 | 0 | 0 | 1 | 6 | 10 | 6 | 2 | 0 | 0 | 0 | 25 |
| Days Max Temp ≤ 32 °F | 13 | 8 | 2 | 0 | 0 | 0 | 0 | 0 | 0 | 0 | 1 | 7 | 31 |
| Days Min Temp ≤ 32 °F | 27 | 23 | 16 | 4 | 0 | 0 | 0 | 0 | 0 | 3 | 13 | 24 | 110 |
| Days Min Temp ≤ 0 °F | 3 | 1 | 0 | 0 | 0 | 0 | 0 | 0 | 0 | 0 | 0 | 1 | 5 |
| Heating Degree Days | 1187 | 949 | 692 | 328 | 116 | 10 | 1 | 4 | 65 | 304 | 637 | 1005 | 5298 |
| Cooling Degree Days | 0 | 0 | 3 | 20 | 90 | 270 | 384 | 308 | 146 | 19 | 1 | 0 | 1241 |
| Total Precipitation (") | 2.06 | 2.07 | 3.39 | 4.14 | 3.92 | 4.10 | 4.28 | 3.23 | 3.63 | 3.01 | 3.71 | 3.53 | 41.07 |
| Days ≥ 0.1" Precip | 5 | 5 | 8 | 8 | 7 | 6 | 7 | 6 | 6 | 6 | 7 | 6 | 77 |
| Total Snowfall (") | 7.5 | 5.3 | 4.5 | 0.8 | 0.0 | 0.0 | 0.0 | 0.0 | 0.0 | 0.1 | 1.9 | 5.5 | 25.6 |
| Days ≥ 1" Snow Depth | 14 | 10 | 4 | 0 | 0 | 0 | 0 | 0 | 0 | 0 | 1 | 6 | 35 |

## PARIS WATERWORKS *Edgar County* ELEVATION 680 ft LAT/LONG 39° 38 ' N / 87° 42 ' W

| | JAN | FEB | MAR | APR | MAY | JUN | JUL | AUG | SEP | OCT | NOV | DEC | YEAR |
|---|---|---|---|---|---|---|---|---|---|---|---|---|---|
| Maximum Temp °F | 34.3 | 39.5 | 51.7 | 64.6 | 74.9 | 83.7 | 86.8 | 84.3 | 78.3 | 66.1 | 52.0 | 39.7 | 63.0 |
| Minimum Temp °F | 17.9 | 21.8 | 32.5 | 43.1 | 52.8 | 62.0 | 65.8 | 63.3 | 56.5 | 44.9 | 34.9 | 24.5 | 43.3 |
| Mean Temp °F | 26.1 | 30.7 | 42.1 | 53.9 | 63.9 | 72.9 | 76.3 | 73.8 | 67.4 | 55.5 | 43.5 | 32.1 | 53.2 |
| Days Max Temp ≥ 90 °F | 0 | 0 | 0 | 0 | 1 | 6 | 10 | 6 | 2 | 0 | 0 | 0 | 25 |
| Days Max Temp ≤ 32 °F | 13 | 8 | 2 | 0 | 0 | 0 | 0 | 0 | 0 | 0 | 1 | 7 | 31 |
| Days Min Temp ≤ 32 °F | 27 | 22 | 17 | 4 | 0 | 0 | 0 | 0 | 0 | 3 | 13 | 24 | 110 |
| Days Min Temp ≤ 0 °F | 4 | 2 | 0 | 0 | 0 | 0 | 0 | 0 | 0 | 0 | 0 | 1 | 7 |
| Heating Degree Days | 1199 | 963 | 705 | 345 | 119 | 12 | 1 | 4 | 65 | 308 | 640 | 1012 | 5373 |
| Cooling Degree Days | 0 | 0 | 3 | 18 | 96 | 263 | 367 | 292 | 146 | 20 | 1 | 0 | 1206 |
| Total Precipitation (") | 2.14 | 2.02 | 3.05 | 4.09 | 4.15 | 3.63 | 4.82 | 4.01 | 3.38 | 2.87 | 3.71 | 3.56 | 41.43 |
| Days ≥ 0.1" Precip | 5 | 5 | 7 | 8 | 7 | 6 | 7 | 6 | 6 | 5 | 7 | 7 | 76 |
| Total Snowfall (") | 9.6 | 6.9 | 3.8 | 0.3 | 0.0 | 0.0 | 0.0 | 0.0 | 0.0 | 0.1 | 1.4 | 6.0 | 28.1 |
| Days ≥ 1" Snow Depth | 13 | 9 | 3 | 0 | 0 | 0 | 0 | 0 | 0 | 0 | 1 | 6 | 32 |

## PARK FOREST *Cook County* ELEVATION 712 ft LAT/LONG 41° 30 ' N / 87° 41 ' W

| | JAN | FEB | MAR | APR | MAY | JUN | JUL | AUG | SEP | OCT | NOV | DEC | YEAR |
|---|---|---|---|---|---|---|---|---|---|---|---|---|---|
| Maximum Temp °F | 29.1 | 33.8 | 45.4 | 58.8 | 70.1 | 80.2 | 83.8 | 81.5 | 74.5 | 62.8 | 47.8 | 35.2 | 58.6 |
| Minimum Temp °F | 12.9 | 16.9 | 27.9 | 38.4 | 48.5 | 58.4 | 63.4 | 61.4 | 53.8 | 41.8 | 31.5 | 20.0 | 39.6 |
| Mean Temp °F | 21.0 | 25.4 | 36.7 | 48.7 | 59.3 | 69.3 | 73.6 | 71.5 | 64.2 | 52.3 | 39.7 | 27.6 | 49.1 |
| Days Max Temp ≥ 90 °F | 0 | 0 | 0 | 0 | 1 | 4 | 6 | 3 | 1 | 0 | 0 | 0 | 15 |
| Days Max Temp ≤ 32 °F | 18 | 13 | 4 | 0 | 0 | 0 | 0 | 0 | 0 | 0 | 2 | 11 | 48 |
| Days Min Temp ≤ 32 °F | 30 | 26 | 22 | 8 | 1 | 0 | 0 | 0 | 0 | 5 | 17 | 27 | 136 |
| Days Min Temp ≤ 0 °F | 7 | 4 | 0 | 0 | 0 | 0 | 0 | 0 | 0 | 0 | 0 | 2 | 13 |
| Heating Degree Days | 1358 | 1113 | 872 | 493 | 226 | 41 | 7 | 14 | 112 | 398 | 753 | 1152 | 6539 |
| Cooling Degree Days | 0 | 0 | 1 | 12 | 58 | 179 | 289 | 232 | 93 | 8 | 0 | 0 | 872 |
| Total Precipitation (") | 1.60 | 1.46 | 2.71 | 4.00 | 3.89 | 4.43 | 3.85 | 3.91 | 3.57 | 2.76 | 3.47 | 2.68 | 38.33 |
| Days ≥ 0.1" Precip | 4 | 4 | 6 | 8 | 7 | 7 | 6 | 6 | 7 | 6 | 6 | 5 | 72 |
| Total Snowfall (") | ***10.3*** | ***8.4*** | ***4.1*** | 0.9 | 0.0 | 0.0 | 0.0 | 0.0 | 0.0 | 0.2 | ***1.0*** | ***5.7*** | 30.6 |
| Days ≥ 1" Snow Depth | 19 | ***15*** | ***5*** | 1 | 0 | 0 | 0 | 0 | 0 | 0 | ***1*** | ***9*** | 50 |

## PAW PAW 1 E *Lee County* ELEVATION 932 ft LAT/LONG 41° 41 ' N / 88° 59 ' W

| | JAN | FEB | MAR | APR | MAY | JUN | JUL | AUG | SEP | OCT | NOV | DEC | YEAR |
|---|---|---|---|---|---|---|---|---|---|---|---|---|---|
| Maximum Temp °F | 26.2 | 30.8 | 43.1 | 58.0 | 70.5 | 80.1 | 83.3 | 80.3 | 73.6 | 61.5 | 45.7 | 32.2 | 57.1 |
| Minimum Temp °F | 10.0 | 14.0 | 25.3 | 36.4 | 47.3 | 57.2 | 61.6 | 59.0 | 51.5 | 39.2 | 28.7 | 16.9 | 37.3 |
| Mean Temp °F | 18.1 | 22.4 | 34.3 | 47.2 | 58.9 | 68.7 | 72.5 | 69.7 | 62.6 | 50.4 | 37.2 | 24.6 | 47.2 |
| Days Max Temp ≥ 90 °F | 0 | 0 | 0 | 0 | 1 | 3 | 5 | 2 | 1 | 0 | 0 | 0 | 12 |
| Days Max Temp ≤ 32 °F | 21 | 15 | 6 | 0 | 0 | 0 | 0 | 0 | 0 | 0 | 3 | 15 | 60 |
| Days Min Temp ≤ 32 °F | 30 | 27 | 24 | 10 | 1 | 0 | 0 | 0 | 0 | 7 | 20 | 29 | 148 |
| Days Min Temp ≤ 0 °F | 9 | 6 | 0 | 0 | 0 | 0 | 0 | 0 | 0 | 0 | 0 | 4 | 19 |
| Heating Degree Days | 1448 | 1196 | 946 | 532 | 230 | 40 | 9 | 22 | 135 | 453 | 828 | 1245 | 7084 |
| Cooling Degree Days | 0 | 0 | 0 | 6 | 45 | 160 | 251 | 175 | 71 | 4 | 0 | 0 | 712 |
| Total Precipitation (") | 1.49 | 1.02 | 2.32 | 3.79 | 4.05 | 4.61 | 3.67 | 4.23 | 4.15 | 2.68 | 3.14 | 2.32 | 37.47 |
| Days ≥ 0.1" Precip | 4 | 3 | 6 | 8 | 7 | 7 | 6 | 6 | 6 | 5 | 6 | 5 | 69 |
| Total Snowfall (") | 9.0 | 6.5 | ***4.8*** | 1.0 | 0.0 | 0.0 | 0.0 | 0.0 | 0.0 | 0.3 | 2.3 | 7.6 | 31.5 |
| Days ≥ 1" Snow Depth | ***21*** | ***14*** | 7 | 0 | 0 | 0 | 0 | 0 | 0 | 0 | 2 | 12 | 56 |

## PEORIA GTR PEORIA AP *Peoria County* ELEVATION 659 ft LAT/LONG 40° 40 ' N / 89° 41 ' W

| | JAN | FEB | MAR | APR | MAY | JUN | JUL | AUG | SEP | OCT | NOV | DEC | YEAR |
|---|---|---|---|---|---|---|---|---|---|---|---|---|---|
| Maximum Temp °F | 30.4 | 35.3 | 48.6 | 62.3 | 73.0 | 82.3 | 85.6 | 83.1 | 76.3 | 63.9 | 48.9 | 35.8 | 60.5 |
| Minimum Temp °F | 14.0 | 18.3 | 29.7 | 40.8 | 50.6 | 60.1 | 64.8 | 62.5 | 54.5 | 42.4 | 31.9 | 20.7 | 40.9 |
| Mean Temp °F | 22.2 | 26.8 | 39.2 | 51.5 | 61.8 | 71.2 | 75.2 | 72.8 | 65.4 | 53.2 | 40.4 | 28.3 | 50.7 |
| Days Max Temp ≥ 90 °F | 0 | 0 | 0 | 0 | 0 | 4 | 9 | 5 | 2 | 0 | 0 | 0 | 20 |
| Days Max Temp ≤ 32 °F | 17 | 12 | 3 | 0 | 0 | 0 | 0 | 0 | 0 | 0 | 2 | 11 | 45 |
| Days Min Temp ≤ 32 °F | 29 | 25 | 19 | 5 | 0 | 0 | 0 | 0 | 0 | 5 | 17 | 26 | 126 |
| Days Min Temp ≤ 0 °F | 6 | 3 | 0 | 0 | 0 | 0 | 0 | 0 | 0 | 0 | 0 | 2 | 11 |
| Heating Degree Days | 1322 | 1072 | 796 | 410 | 160 | 18 | 2 | 8 | 92 | 372 | 732 | 1131 | 6115 |
| Cooling Degree Days | 0 | 0 | 2 | 15 | 73 | 223 | 345 | 276 | 121 | 9 | 0 | 0 | 1064 |
| Total Precipitation (") | 1.60 | 1.50 | 2.79 | 3.74 | 3.75 | 3.95 | 4.42 | 3.27 | 3.82 | 2.90 | 2.89 | 2.60 | 37.23 |
| Days ≥ 0.1" Precip | 4 | 4 | 6 | 7 | 7 | 6 | 6 | 6 | 6 | 6 | 6 | 5 | 69 |
| Total Snowfall (") | 7.6 | 5.9 | 3.5 | 1.1 | 0.0 | 0.0 | 0.0 | 0.0 | 0.0 | 0.1 | 2.1 | 6.1 | 26.4 |
| Days ≥ 1" Snow Depth | 15 | 12 | 4 | 0 | 0 | 0 | 0 | 0 | 0 | 0 | 1 | 8 | 40 |

## PERU *La Salle County* ELEVATION 459 ft LAT/LONG 41° 20 ' N / 89° 8 ' W

| | JAN | FEB | MAR | APR | MAY | JUN | JUL | AUG | SEP | OCT | NOV | DEC | YEAR |
|---|---|---|---|---|---|---|---|---|---|---|---|---|---|
| Maximum Temp °F | 29.4 | 34.5 | 47.0 | 61.6 | 73.2 | 82.6 | 85.8 | 83.2 | 76.6 | 64.7 | 48.6 | 35.6 | 60.2 |
| Minimum Temp °F | 12.2 | 16.6 | 28.0 | 38.7 | 49.0 | 58.4 | 63.0 | 60.5 | 52.3 | 40.9 | 31.0 | 19.8 | 39.2 |
| Mean Temp °F | 20.8 | 25.6 | 37.5 | 50.2 | 61.1 | 70.5 | 74.4 | 71.9 | 64.4 | 52.8 | 39.8 | 27.7 | 49.7 |
| Days Max Temp ≥ 90 °F | 0 | 0 | 0 | 0 | 1 | 5 | 9 | 5 | 2 | 0 | 0 | 0 | 22 |
| Days Max Temp ≤ 32 °F | 17 | 12 | 4 | 0 | 0 | 0 | 0 | 0 | 0 | 0 | 2 | 11 | 46 |
| Days Min Temp ≤ 32 °F | 30 | 26 | 21 | 7 | 1 | 0 | 0 | 0 | 0 | 6 | 18 | 28 | 137 |
| Days Min Temp ≤ 0 °F | 8 | 4 | 0 | 0 | 0 | 0 | 0 | 0 | 0 | 0 | 0 | 3 | 15 |
| Heating Degree Days | 1363 | 1106 | 845 | 449 | 180 | 25 | 5 | 14 | 106 | 382 | 748 | 1150 | 6373 |
| Cooling Degree Days | 0 | 0 | 1 | 11 | 69 | 202 | 311 | 244 | 99 | 8 | 0 | 0 | 945 |
| Total Precipitation (") | 1.41 | 1.24 | 2.87 | 3.94 | 4.51 | 4.01 | 4.11 | 4.22 | 3.84 | 3.05 | 2.90 | 2.48 | 38.58 |
| Days ≥ 0.1" Precip | 3 | 3 | 6 | 7 | 7 | 7 | 6 | 6 | 6 | 6 | 6 | 5 | 68 |
| Total Snowfall (") | *7.1* | *4.8* | *3.4* | 0.5 | 0.0 | 0.0 | 0.0 | 0.0 | 0.0 | 0.1 | 1.2 | *5.5* | 22.6 |
| Days ≥ 1" Snow Depth | *18* | *11* | 3 | 0 | 0 | 0 | 0 | 0 | 0 | 0 | 1 | *8* | 41 |

## PIPER CITY *Ford County* ELEVATION 670 ft LAT/LONG 40° 45 ' N / 88° 11 ' W

| | JAN | FEB | MAR | APR | MAY | JUN | JUL | AUG | SEP | OCT | NOV | DEC | YEAR |
|---|---|---|---|---|---|---|---|---|---|---|---|---|---|
| Maximum Temp °F | 30.5 | 35.8 | 48.8 | 63.2 | 75.5 | 84.1 | 86.4 | 84.1 | 78.1 | 65.3 | 49.3 | 36.3 | 61.4 |
| Minimum Temp °F | 13.7 | 18.4 | 29.1 | 38.8 | 49.5 | 58.9 | 62.9 | 59.8 | 52.6 | 41.3 | 31.7 | 20.3 | 39.8 |
| Mean Temp °F | 22.1 | 27.1 | 38.9 | 51.0 | 62.5 | 71.5 | 74.7 | 72.0 | 65.4 | 53.3 | 40.5 | 28.3 | 50.6 |
| Days Max Temp ≥ 90 °F | 0 | 0 | 0 | 0 | 3 | 8 | 10 | 6 | 2 | 0 | 0 | 0 | 29 |
| Days Max Temp ≤ 32 °F | 16 | 11 | 3 | 0 | 0 | 0 | 0 | 0 | 0 | 0 | 2 | 10 | 42 |
| Days Min Temp ≤ 32 °F | 30 | 25 | 21 | 8 | 1 | 0 | 0 | 0 | 0 | 6 | 17 | 27 | 135 |
| Days Min Temp ≤ 0 °F | 6 | 4 | 0 | 0 | 0 | 0 | 0 | 0 | 0 | 0 | 0 | 2 | 12 |
| Heating Degree Days | 1324 | 1062 | 802 | 425 | 150 | 20 | 3 | 11 | 92 | 369 | 729 | 1129 | 6116 |
| Cooling Degree Days | 0 | 0 | 2 | 12 | 74 | 225 | 324 | 239 | 105 | 11 | 0 | 0 | 992 |
| Total Precipitation (") | *1.48* | *1.43* | 2.72 | 3.27 | 3.82 | 3.54 | 4.11 | 3.67 | 3.29 | 2.77 | 2.99 | 2.44 | 35.53 |
| Days ≥ 0.1" Precip | 4 | 3 | 6 | 7 | 7 | 7 | 6 | 5 | 6 | 5 | 6 | 5 | 67 |
| Total Snowfall (") | 8.3 | 6.6 | 3.1 | 1.0 | 0.0 | 0.0 | 0.0 | 0.0 | 0.0 | 0.2 | 1.6 | 5.7 | 26.5 |
| Days ≥ 1" Snow Depth | *14* | *11* | *5* | 0 | 0 | 0 | 0 | 0 | 0 | 0 | 1 | *5* | 36 |

## PONTIAC *Livingston County* ELEVATION 650 ft LAT/LONG 40° 53 ' N / 88° 37 ' W

| | JAN | FEB | MAR | APR | MAY | JUN | JUL | AUG | SEP | OCT | NOV | DEC | YEAR |
|---|---|---|---|---|---|---|---|---|---|---|---|---|---|
| Maximum Temp °F | 30.4 | 35.8 | 48.4 | 62.6 | 73.9 | 82.9 | 85.5 | 83.2 | 77.2 | 65.0 | 49.3 | 36.1 | 60.9 |
| Minimum Temp °F | 14.4 | 18.5 | 29.5 | 40.1 | 50.5 | 60.3 | 64.3 | 61.8 | 54.4 | 42.4 | 32.3 | 21.2 | 40.8 |
| Mean Temp °F | 22.4 | 27.2 | 39.0 | 51.4 | 62.2 | 71.6 | 74.9 | 72.5 | 65.8 | 53.7 | 40.8 | 28.7 | 50.8 |
| Days Max Temp ≥ 90 °F | 0 | 0 | 0 | 0 | 1 | 6 | 8 | 5 | 2 | 0 | 0 | 0 | 22 |
| Days Max Temp ≤ 32 °F | 17 | 11 | 3 | 0 | 0 | 0 | 0 | 0 | 0 | 0 | 2 | 10 | 43 |
| Days Min Temp ≤ 32 °F | 29 | 25 | 20 | 6 | 0 | 0 | 0 | 0 | 0 | 5 | 17 | 26 | 128 |
| Days Min Temp ≤ 0 °F | 6 | 3 | 0 | 0 | 0 | 0 | 0 | 0 | 0 | 0 | 0 | 2 | 11 |
| Heating Degree Days | 1314 | 1061 | 802 | 415 | 154 | 19 | 2 | 8 | 84 | 358 | 720 | 1118 | 6055 |
| Cooling Degree Days | 0 | 0 | 2 | 14 | 68 | 211 | 313 | 247 | 111 | 10 | 0 | 0 | 976 |
| Total Precipitation (") | 1.63 | 1.31 | 2.71 | 3.55 | 3.85 | 3.93 | 4.03 | 3.41 | 3.53 | 2.74 | 3.00 | 2.67 | 36.36 |
| Days ≥ 0.1" Precip | 4 | 4 | 6 | 7 | 7 | 6 | 6 | 6 | 6 | 5 | 6 | 5 | 68 |
| Total Snowfall (") | 7.8 | 6.2 | 3.7 | 0.7 | 0.0 | 0.0 | 0.0 | 0.0 | 0.0 | 0.1 | 1.7 | 5.4 | 25.6 |
| Days ≥ 1" Snow Depth | *11* | 8 | 4 | 0 | 0 | 0 | 0 | 0 | 0 | 0 | 1 | 5 | 29 |

## PRINCEVILLE *Peoria County* ELEVATION 751 ft LAT/LONG 40° 57 ' N / 89° 47 ' W

| | JAN | FEB | MAR | APR | MAY | JUN | JUL | AUG | SEP | OCT | NOV | DEC | YEAR |
|---|---|---|---|---|---|---|---|---|---|---|---|---|---|
| Maximum Temp °F | 30.0 | 35.6 | 48.4 | 62.7 | 73.1 | 81.7 | 84.9 | 82.8 | 76.3 | 64.4 | 49.2 | 35.9 | 60.4 |
| Minimum Temp °F | 11.5 | 16.2 | 27.4 | 38.3 | 48.5 | 57.7 | 61.4 | 59.0 | 51.1 | 40.2 | 29.6 | 18.5 | 38.3 |
| Mean Temp °F | 20.8 | 25.9 | 37.9 | 50.6 | 60.8 | 69.7 | 73.2 | 70.9 | 63.7 | 52.4 | 39.4 | 27.3 | 49.4 |
| Days Max Temp ≥ 90 °F | 0 | 0 | 0 | 0 | 0 | 4 | 8 | 5 | 2 | 0 | 0 | 0 | 19 |
| Days Max Temp ≤ 32 °F | 17 | 11 | 3 | 0 | 0 | 0 | 0 | 0 | 0 | 0 | 2 | 10 | 43 |
| Days Min Temp ≤ 32 °F | 30 | 26 | 22 | 9 | 1 | 0 | 0 | 0 | 1 | 8 | 19 | 28 | 144 |
| Days Min Temp ≤ 0 °F | 8 | 4 | 0 | 0 | 0 | 0 | 0 | 0 | 0 | 0 | 0 | 3 | 15 |
| Heating Degree Days | 1365 | 1096 | 832 | 438 | 179 | 29 | 6 | 18 | 120 | 398 | 761 | 1162 | 6404 |
| Cooling Degree Days | 0 | 0 | 1 | 10 | 44 | 155 | 246 | 188 | 76 | 6 | 0 | 0 | 726 |
| Total Precipitation (") | 1.59 | 1.40 | 3.03 | 3.86 | 3.98 | 3.96 | 4.05 | 3.70 | 3.86 | 2.86 | 3.05 | 2.51 | 37.85 |
| Days ≥ 0.1" Precip | 4 | 3 | 6 | 7 | 7 | 7 | 7 | 6 | 6 | 6 | 6 | 5 | 70 |
| Total Snowfall (") | 6.3 | 5.7 | 2.9 | 0.7 | 0.0 | 0.0 | 0.0 | 0.0 | 0.0 | 0.2 | 1.9 | 5.4 | 23.1 |
| Days ≥ 1" Snow Depth | *11* | *6* | *1* | 0 | 0 | 0 | 0 | 0 | 0 | 0 | 1 | na | na |

## QUINCY MUNI BALDWIN *Adams County* ELEVATION 768 ft LAT/LONG 39° 56 ' N / 91° 12 ' W

| | JAN | FEB | MAR | APR | MAY | JUN | JUL | AUG | SEP | OCT | NOV | DEC | YEAR |
|---|---|---|---|---|---|---|---|---|---|---|---|---|---|
| Maximum Temp °F | 32.1 | 37.1 | 49.8 | 63.2 | 72.9 | 82.2 | 86.5 | 83.7 | 76.9 | 65.1 | 50.3 | 37.4 | 61.4 |
| Minimum Temp °F | 16.3 | 20.6 | 31.8 | 43.2 | 52.8 | 62.1 | 66.6 | 63.7 | 55.9 | 44.3 | 33.4 | 22.6 | 42.8 |
| Mean Temp °F | 24.2 | 28.9 | 40.9 | 53.3 | 62.9 | 72.2 | 76.5 | 73.8 | 66.4 | 54.7 | 41.9 | 30.0 | 52.1 |
| Days Max Temp ≥ 90 °F | 0 | 0 | 0 | 0 | 0 | 5 | 10 | 7 | 2 | 0 | 0 | 0 | 24 |
| Days Max Temp ≤ 32 °F | 15 | 10 | 3 | 0 | 0 | 0 | 0 | 0 | 0 | 0 | 2 | 10 | 40 |
| Days Min Temp ≤ 32 °F | 28 | 23 | 17 | 4 | 0 | 0 | 0 | 0 | 0 | 3 | 14 | 26 | 115 |
| Days Min Temp ≤ 0 °F | 5 | 2 | 0 | 0 | 0 | 0 | 0 | 0 | 0 | 0 | 0 | 2 | 9 |
| Heating Degree Days | 1256 | 1014 | 744 | 362 | 134 | 13 | 1 | 6 | 80 | 330 | 688 | 1077 | 5705 |
| Cooling Degree Days | 0 | 0 | 4 | 20 | 72 | 241 | 374 | 301 | 137 | 14 | 1 | 0 | 1164 |
| Total Precipitation (") | 1.41 | 1.52 | 3.06 | 3.88 | 4.76 | 3.92 | 4.51 | 3.57 | 4.59 | 3.62 | 3.11 | 2.50 | 40.45 |
| Days ≥ 0.1" Precip | 3 | 3 | 6 | 7 | 7 | 6 | 6 | 5 | 6 | 6 | 6 | 5 | 66 |
| Total Snowfall (") | 6.4 | 6.1 | 3.1 | 0.6 | 0.0 | 0.0 | 0.0 | 0.0 | 0.0 | 0.0 | 2.1 | 4.5 | 22.8 |
| Days ≥ 1" Snow Depth | 13 | 11 | 4 | 0 | 0 | 0 | 0 | 0 | 0 | 0 | 1 | 6 | 35 |

## RANTOUL *Champaign County* ELEVATION 740 ft LAT/LONG 40° 19 ' N / 88° 10 ' W

| | JAN | FEB | MAR | APR | MAY | JUN | JUL | AUG | SEP | OCT | NOV | DEC | YEAR |
|---|---|---|---|---|---|---|---|---|---|---|---|---|---|
| Maximum Temp °F | 32.3 | 36.4 | 48.6 | 62.8 | 74.1 | 83.7 | 86.6 | 84.1 | 78.3 | 65.7 | 50.5 | 37.9 | 61.8 |
| Minimum Temp °F | 14.6 | 18.0 | 29.2 | 39.9 | 50.3 | 60.0 | 63.8 | 61.3 | 53.6 | 41.9 | 31.9 | 21.7 | 40.5 |
| Mean Temp °F | 23.5 | 27.2 | 38.8 | 51.4 | 62.2 | 71.8 | 75.3 | 72.7 | 66.0 | 53.8 | 41.2 | 29.8 | 51.1 |
| Days Max Temp ≥ 90 °F | 0 | 0 | 0 | 0 | 2 | 7 | 10 | 6 | 3 | 0 | 0 | 0 | 28 |
| Days Max Temp ≤ 32 °F | *15* | 11 | 2 | 0 | 0 | 0 | 0 | 0 | 0 | 0 | 1 | 8 | 37 |
| Days Min Temp ≤ 32 °F | 29 | 25 | 20 | 6 | 0 | 0 | 0 | 0 | 0 | 5 | 17 | 26 | 128 |
| Days Min Temp ≤ 0 °F | 5 | 3 | 0 | 0 | 0 | 0 | 0 | 0 | 0 | 0 | 0 | 2 | 10 |
| Heating Degree Days | 1281 | 1061 | 806 | 413 | 154 | 19 | 3 | 8 | 85 | 355 | 707 | 1084 | 5976 |
| Cooling Degree Days | 0 | 0 | 2 | 13 | 67 | 228 | 330 | 253 | 112 | 11 | 0 | 0 | 1016 |
| Total Precipitation (") | 1.73 | 1.67 | 2.96 | 4.02 | 3.97 | 3.66 | 4.50 | 4.07 | 3.65 | 2.95 | 3.21 | 2.97 | 39.36 |
| Days ≥ 0.1" Precip | 4 | 4 | 6 | 8 | 7 | 6 | 6 | 6 | 6 | 6 | 6 | 6 | 71 |
| Total Snowfall (") | na | na | na | *0.5* | *0.0* | *0.0* | 0.0 | 0.0 | 0.0 | *0.0* | *0.5* | na | na |
| Days ≥ 1" Snow Depth | na | na | na | *0* | *0* | *0* | 0 | 0 | 0 | *0* | *0* | na | na |

## ROCKFORD GREATER AP *Winnebago County* ELEVATION 724 ft LAT/LONG 42° 12 ' N / 89° 6 ' W

| | JAN | FEB | MAR | APR | MAY | JUN | JUL | AUG | SEP | OCT | NOV | DEC | YEAR |
|---|---|---|---|---|---|---|---|---|---|---|---|---|---|
| Maximum Temp °F | 26.9 | 31.7 | 44.7 | 58.9 | 71.0 | 80.3 | 83.8 | 81.4 | 74.1 | 61.7 | 45.8 | 32.3 | 57.7 |
| Minimum Temp °F | 10.2 | 14.8 | 26.5 | 37.2 | 47.4 | 57.5 | 62.6 | 60.4 | 51.9 | 40.1 | 29.2 | 17.1 | 37.9 |
| Mean Temp °F | 18.5 | 23.2 | 35.6 | 48.1 | 59.2 | 68.9 | 73.3 | 70.9 | 63.0 | 50.9 | 37.5 | 24.7 | 47.8 |
| Days Max Temp ≥ 90 °F | 0 | 0 | 0 | 0 | 1 | 3 | 6 | 4 | 1 | 0 | 0 | 0 | 15 |
| Days Max Temp ≤ 32 °F | 19 | 14 | 4 | 0 | 0 | 0 | 0 | 0 | 0 | 0 | 3 | 14 | 54 |
| Days Min Temp ≤ 32 °F | 30 | 26 | 23 | 9 | 1 | 0 | 0 | 0 | 1 | 7 | 20 | 28 | 145 |
| Days Min Temp ≤ 0 °F | 9 | 5 | 0 | 0 | 0 | 0 | 0 | 0 | 0 | 0 | 0 | 4 | 18 |
| Heating Degree Days | 1435 | 1173 | 903 | 510 | 222 | 36 | 6 | 15 | 130 | 438 | 818 | 1242 | 6928 |
| Cooling Degree Days | 0 | 0 | 0 | 10 | 53 | 168 | 281 | 216 | 80 | 4 | 0 | 0 | 812 |
| Total Precipitation (") | 1.28 | 1.23 | 2.47 | 3.68 | 3.64 | 4.98 | 3.89 | 4.15 | 3.67 | 2.87 | 2.76 | 2.18 | 36.80 |
| Days ≥ 0.1" Precip | 4 | 3 | 5 | 7 | 7 | 7 | 6 | 7 | 6 | 5 | 6 | 5 | 68 |
| Total Snowfall (") | 8.8 | 8.0 | 5.4 | 1.7 | 0.0 | 0.0 | 0.0 | 0.0 | 0.0 | 0.2 | 2.5 | 10.3 | 36.9 |
| Days ≥ 1" Snow Depth | 22 | 17 | 6 | 1 | 0 | 0 | 0 | 0 | 0 | 0 | 1 | 14 | 61 |

## RUSHVILLE *Schuyler County* ELEVATION 659 ft LAT/LONG 40° 7 ' N / 90° 34 ' W

| | JAN | FEB | MAR | APR | MAY | JUN | JUL | AUG | SEP | OCT | NOV | DEC | YEAR |
|---|---|---|---|---|---|---|---|---|---|---|---|---|---|
| Maximum Temp °F | 32.6 | 38.4 | 50.8 | 64.4 | 74.4 | 83.3 | 87.6 | 85.1 | 78.1 | 66.1 | 51.0 | 38.1 | 62.5 |
| Minimum Temp °F | 15.2 | 19.6 | 30.5 | 42.1 | 51.1 | 60.4 | 65.0 | 62.1 | 54.5 | 43.0 | 32.1 | 21.4 | 41.4 |
| Mean Temp °F | 23.9 | 29.0 | 40.7 | 53.3 | 62.8 | 71.9 | 76.3 | 73.7 | 66.3 | 54.6 | 41.6 | 29.8 | 52.0 |
| Days Max Temp ≥ 90 °F | 0 | 0 | 0 | 0 | 1 | 5 | 12 | 7 | 2 | 0 | 0 | 0 | 27 |
| Days Max Temp ≤ 32 °F | 15 | 9 | 2 | 0 | 0 | 0 | 0 | 0 | 0 | 0 | 2 | 9 | 37 |
| Days Min Temp ≤ 32 °F | 29 | 24 | 19 | 4 | 0 | 0 | 0 | 0 | 0 | 4 | 16 | 26 | 122 |
| Days Min Temp ≤ 0 °F | 5 | 3 | 0 | 0 | 0 | 0 | 0 | 0 | 0 | 0 | 0 | 2 | 10 |
| Heating Degree Days | 1268 | 1009 | 752 | 363 | 137 | 16 | 2 | 5 | 77 | 331 | 697 | 1083 | 5740 |
| Cooling Degree Days | 0 | 0 | 3 | 20 | 71 | 232 | 369 | 293 | 130 | 14 | 0 | 0 | 1132 |
| Total Precipitation (") | 1.54 | 1.68 | 3.09 | 4.19 | 4.58 | 4.01 | 4.17 | 3.63 | 4.51 | 3.52 | 3.04 | 2.65 | 40.61 |
| Days ≥ 0.1" Precip | 4 | 4 | 6 | 7 | 7 | 6 | 6 | 6 | 6 | 5 | 6 | 5 | 68 |
| Total Snowfall (") | 5.7 | 5.3 | 2.6 | 0.8 | 0.0 | 0.0 | 0.0 | 0.0 | 0.0 | 0.0 | 1.2 | 4.6 | 20.2 |
| Days ≥ 1" Snow Depth | na | na | na | na | na | na | na | na | na | na | na | na | na |

## SALEM *Marion County* ELEVATION 550 ft LAT/LONG 38° 38 ' N / 88° 57 ' W

| | JAN | FEB | MAR | APR | MAY | JUN | JUL | AUG | SEP | OCT | NOV | DEC | YEAR |
|---|---|---|---|---|---|---|---|---|---|---|---|---|---|
| Maximum Temp °F | 37.8 | 43.2 | 55.2 | 67.6 | 76.8 | 85.8 | 89.3 | 87.2 | 80.4 | 68.8 | 54.9 | 43.0 | 65.8 |
| Minimum Temp °F | 20.1 | 23.9 | 34.0 | 44.4 | 53.2 | 62.2 | 66.4 | 63.8 | 56.7 | 44.7 | 35.5 | 25.9 | 44.2 |
| Mean Temp °F | 29.0 | 33.5 | 44.7 | 56.0 | 65.0 | 74.0 | 77.9 | 75.5 | 68.6 | 56.8 | 45.2 | 34.5 | 55.1 |
| Days Max Temp ≥ 90 °F | 0 | 0 | 0 | 0 | 1 | 8 | 15 | 11 | 4 | 0 | 0 | 0 | 39 |
| Days Max Temp ≤ 32 °F | 10 | 6 | 1 | 0 | 0 | 0 | 0 | 0 | 0 | 0 | 1 | 5 | 23 |
| Days Min Temp ≤ 32 °F | 27 | 21 | 15 | 4 | 0 | 0 | 0 | 0 | 0 | 4 | 13 | 23 | 107 |
| Days Min Temp ≤ 0 °F | 2 | 1 | 0 | 0 | 0 | 0 | 0 | 0 | 0 | 0 | 0 | 1 | 4 |
| Heating Degree Days | 1109 | 882 | 629 | 290 | 93 | 6 | 1 | 1 | 50 | 275 | 588 | 939 | 4863 |
| Cooling Degree Days | 0 | 0 | 4 | 27 | 104 | 298 | 427 | 353 | 168 | 26 | 1 | 0 | 1408 |
| Total Precipitation (") | 2.03 | 2.32 | 3.76 | 4.13 | 3.78 | 3.88 | 3.99 | 3.19 | 3.52 | 3.08 | 4.04 | 3.59 | 41.31 |
| Days ≥ 0.1" Precip | 5 | 5 | 7 | 8 | 7 | 6 | 6 | 6 | 5 | 5 | 7 | 6 | 73 |
| Total Snowfall (") | 4.8 | 4.2 | 1.6 | 0.3 | 0.0 | 0.0 | 0.0 | 0.0 | 0.0 | 0.1 | 0.5 | 3.1 | 14.6 |
| Days ≥ 1" Snow Depth | 8 | 6 | 1 | 0 | 0 | 0 | 0 | 0 | 0 | 0 | 0 | 4 | 19 |

## SPARTA 3 N *Randolph County* ELEVATION 520 ft LAT/LONG 38° 10 ' N / 89° 42 ' W

| | JAN | FEB | MAR | APR | MAY | JUN | JUL | AUG | SEP | OCT | NOV | DEC | YEAR |
|---|---|---|---|---|---|---|---|---|---|---|---|---|---|
| Maximum Temp °F | 39.3 | 45.3 | 56.9 | 68.8 | 77.5 | 86.4 | 90.0 | 87.9 | 81.0 | 69.9 | 55.9 | 44.2 | 66.9 |
| Minimum Temp °F | 22.0 | 26.0 | 35.8 | 45.8 | 54.2 | 63.6 | 67.4 | 64.9 | 58.0 | 46.4 | 37.2 | 27.5 | 45.7 |
| Mean Temp °F | 30.7 | 35.7 | 46.4 | 57.3 | 65.9 | 75.0 | 78.7 | 76.4 | 69.5 | 58.2 | 46.6 | 35.9 | 56.4 |
| Days Max Temp ≥ 90 °F | 0 | 0 | 0 | 0 | 2 | 9 | 17 | 12 | 4 | 0 | 0 | 0 | 44 |
| Days Max Temp ≤ 32 °F | 10 | 5 | 1 | 0 | 0 | 0 | 0 | 0 | 0 | 0 | 0 | 4 | 20 |
| Days Min Temp ≤ 32 °F | 26 | 20 | 13 | 3 | 0 | 0 | 0 | 0 | 0 | 2 | 11 | 21 | 96 |
| Days Min Temp ≤ 0 °F | 2 | 0 | 0 | 0 | 0 | 0 | 0 | 0 | 0 | 0 | 0 | 1 | 3 |
| Heating Degree Days | 1057 | 821 | 575 | 259 | 80 | 4 | 0 | 2 | 43 | 239 | 549 | 897 | 4526 |
| Cooling Degree Days | 0 | 0 | 6 | 35 | 116 | 330 | 447 | 383 | 192 | 34 | 3 | 0 | 1546 |
| Total Precipitation (") | 2.00 | 2.45 | 3.96 | 4.12 | 4.19 | 3.23 | 4.16 | 3.26 | 3.43 | 3.28 | 4.20 | 3.49 | 41.77 |
| Days ≥ 0.1" Precip | 4 | 5 | 7 | 7 | 7 | 6 | 6 | 5 | 5 | 5 | 6 | 6 | 69 |
| Total Snowfall (") | 4.9 | 4.5 | 2.3 | 0.5 | 0.0 | 0.0 | 0.0 | 0.0 | 0.0 | 0.0 | 0.9 | 2.9 | 16.0 |
| Days ≥ 1" Snow Depth | 8 | 6 | 1 | 0 | 0 | 0 | 0 | 0 | 0 | 0 | 0 | 3 | 18 |

## SPRINGFIELD CAPTL AP *Sangamon County* ELEVATION 588 ft LAT/LONG 39° 50 ' N / 89° 40 ' W

| | JAN | FEB | MAR | APR | MAY | JUN | JUL | AUG | SEP | OCT | NOV | DEC | YEAR |
|---|---|---|---|---|---|---|---|---|---|---|---|---|---|
| Maximum Temp °F | 32.8 | 37.6 | 50.5 | 64.0 | 74.7 | 83.9 | 87.1 | 84.5 | 78.3 | 66.3 | 51.3 | 38.4 | 62.4 |
| Minimum Temp °F | 16.6 | 20.8 | 31.9 | 42.9 | 52.4 | 61.8 | 66.1 | 63.4 | 55.7 | 44.2 | 34.0 | 23.3 | 42.8 |
| Mean Temp °F | 24.7 | 29.2 | 41.2 | 53.5 | 63.6 | 72.9 | 76.6 | 73.9 | 67.0 | 55.2 | 42.6 | 30.8 | 52.6 |
| Days Max Temp ≥ 90 °F | 0 | 0 | 0 | 0 | 1 | 7 | 11 | 7 | 3 | 0 | 0 | 0 | 29 |
| Days Max Temp ≤ 32 °F | 15 | 10 | 3 | 0 | 0 | 0 | 0 | 0 | 0 | 0 | 1 | 8 | 37 |
| Days Min Temp ≤ 32 °F | 28 | 23 | 17 | 4 | 0 | 0 | 0 | 0 | 0 | 4 | 14 | 24 | 114 |
| Days Min Temp ≤ 0 °F | 4 | 3 | 0 | 0 | 0 | 0 | 0 | 0 | 0 | 0 | 0 | 2 | 9 |
| Heating Degree Days | 1241 | 1004 | 732 | 359 | 128 | 12 | 1 | 5 | 73 | 320 | 664 | 1052 | 5591 |
| Cooling Degree Days | 0 | 0 | 3 | 21 | 89 | 258 | 377 | 309 | 147 | 20 | 1 | 0 | 1225 |
| Total Precipitation (") | 1.58 | 1.75 | 3.00 | 3.73 | 3.64 | 3.68 | 3.62 | 3.25 | 3.49 | 2.75 | 2.71 | 2.84 | 36.04 |
| Days ≥ 0.1" Precip | 4 | 4 | 6 | 7 | 7 | 6 | 6 | 5 | 6 | 5 | 6 | 6 | 68 |
| Total Snowfall (") | 6.7 | 6.8 | 3.8 | 0.8 | 0.0 | 0.0 | 0.0 | 0.0 | 0.0 | 0.0 | 1.6 | 5.2 | 24.9 |
| Days ≥ 1" Snow Depth | 12 | 10 | 4 | 0 | 0 | 0 | 0 | 0 | 0 | 0 | 1 | 6 | 33 |

## STOCKTON 1 N *Jo Daviess County* ELEVATION 1001 ft LAT/LONG 42° 21 ' N / 90° 0 ' W

| | JAN | FEB | MAR | APR | MAY | JUN | JUL | AUG | SEP | OCT | NOV | DEC | YEAR |
|---|---|---|---|---|---|---|---|---|---|---|---|---|---|
| Maximum Temp °F | 25.7 | 30.8 | 43.1 | 58.1 | 70.1 | 78.9 | 82.3 | 79.9 | 72.1 | 60.5 | 44.4 | 31.1 | 56.4 |
| Minimum Temp °F | 8.7 | 13.7 | 25.5 | 36.9 | 47.6 | 57.1 | 61.5 | 59.3 | 51.0 | 39.6 | 28.3 | 15.9 | 37.1 |
| Mean Temp °F | 17.4 | 22.2 | 34.3 | 47.5 | 58.9 | 68.1 | 71.9 | 69.6 | 61.6 | 50.1 | 36.4 | 23.5 | 46.8 |
| Days Max Temp ≥ 90 °F | 0 | 0 | 0 | 0 | 0 | 2 | 4 | 2 | 0 | 0 | 0 | 0 | 8 |
| Days Max Temp ≤ 32 °F | 21 | 15 | 5 | 0 | 0 | 0 | 0 | 0 | 0 | 0 | 4 | 16 | 61 |
| Days Min Temp ≤ 32 °F | 31 | 27 | 24 | 10 | 1 | 0 | 0 | 0 | 1 | 8 | 20 | 29 | 151 |
| Days Min Temp ≤ 0 °F | 10 | 6 | 0 | 0 | 0 | 0 | 0 | 0 | 0 | 0 | 0 | 4 | 20 |
| Heating Degree Days | 1471 | 1201 | 945 | 525 | 227 | 42 | 10 | 24 | 154 | 463 | 853 | 1279 | 7194 |
| Cooling Degree Days | 0 | 0 | 0 | 8 | 44 | 137 | 222 | 171 | 57 | 4 | 0 | 0 | 643 |
| Total Precipitation (") | 1.23 | 1.19 | 2.35 | 3.38 | 3.52 | 4.63 | 3.28 | 4.37 | 4.15 | 2.81 | 2.77 | 1.82 | 35.50 |
| Days ≥ 0.1" Precip | 3 | 3 | 5 | 7 | 7 | 7 | 6 | 7 | 7 | 6 | 6 | 4 | 68 |
| Total Snowfall (") | *7.8* | 6.7 | *5.1* | 1.9 | 0.1 | 0.0 | 0.0 | 0.0 | 0.0 | 0.2 | 2.2 | *7.4* | 31.4 |
| Days ≥ 1" Snow Depth | na | *14* | *5* | 1 | 0 | 0 | 0 | 0 | 0 | 0 | 2 | *11* | na |

## TISKILWA 2 SE *Bureau County* ELEVATION 512 ft LAT/LONG 41° 17 ' N / 89° 30 ' W

| | JAN | FEB | MAR | APR | MAY | JUN | JUL | AUG | SEP | OCT | NOV | DEC | YEAR |
|---|---|---|---|---|---|---|---|---|---|---|---|---|---|
| Maximum Temp °F | 30.3 | 35.4 | 48.3 | 62.9 | *74.0* | 83.0 | 86.5 | *84.3* | *77.6* | *64.9* | *49.4* | *35.6* | 61.0 |
| Minimum Temp °F | 12.6 | 16.8 | *28.2* | 39.1 | 48.6 | 58.1 | *62.4* | *60.1* | *52.3* | *41.1* | *31.0* | *19.1* | 39.1 |
| Mean Temp °F | 21.5 | 26.1 | *38.2* | 51.0 | *61.3* | 70.6 | *74.5* | *72.2* | *65.0* | *53.0* | *40.1* | *27.4* | 50.1 |
| Days Max Temp ≥ 90 °F | 0 | 0 | 0 | 0 | 1 | 5 | *9* | 6 | 2 | *0* | *0* | *0* | 23 |
| Days Max Temp ≤ 32 °F | 16 | 11 | 3 | 0 | 0 | 0 | 0 | 0 | 0 | *0* | *2* | *11* | 43 |
| Days Min Temp ≤ 32 °F | 30 | 25 | 21 | 8 | 1 | 0 | 0 | 0 | 0 | *6* | *18* | *27* | 136 |
| Days Min Temp ≤ 0 °F | 7 | 4 | *0* | 0 | 0 | 0 | 0 | 0 | 0 | *0* | *0* | *3* | 14 |
| Heating Degree Days | 1342 | 1091 | *825* | 425 | *169* | 21 | *4* | *10* | *93* | *379* | *739* | *1160* | 6258 |
| Cooling Degree Days | *0* | *0* | na | *15* | na | *189* | na | na | na | na | na | na | na |
| Total Precipitation (") | 1.42 | 1.27 | 2.73 | 3.84 | 3.97 | 4.32 | 3.80 | 4.01 | 3.99 | *2.77* | *2.53* | *2.28* | 36.93 |
| Days ≥ 0.1" Precip | 4 | 4 | 6 | *7* | 7 | 7 | *6* | *6* | *6* | *6* | *5* | *5* | 69 |
| Total Snowfall (") | na | na | na | *0.9* | 0.0 | 0.0 | 0.0 | 0.0 | 0.0 | *0.1* | *1.6* | na | na |
| Days ≥ 1" Snow Depth | *17* | *15* | na | *0* | 0 | 0 | 0 | *0* | *0* | *0* | *1* | *10* | na |

## TUSCOLA *Douglas County* ELEVATION 650 ft LAT/LONG 39° 48 ' N / 88° 17 ' W

| | JAN | FEB | MAR | APR | MAY | JUN | JUL | AUG | SEP | OCT | NOV | DEC | YEAR |
|---|---|---|---|---|---|---|---|---|---|---|---|---|---|
| Maximum Temp °F | 34.1 | 39.4 | 52.2 | 65.8 | 76.5 | 85.6 | 88.1 | 85.8 | 80.3 | 67.8 | 52.8 | 40.4 | 64.1 |
| Minimum Temp °F | 17.0 | 20.9 | 31.6 | 41.8 | 51.8 | 61.0 | 64.5 | 61.8 | 54.9 | 43.5 | 33.8 | 24.3 | 42.2 |
| Mean Temp °F | 25.6 | 30.2 | 41.9 | 53.9 | 64.2 | 73.3 | 76.3 | 73.8 | 67.6 | 55.7 | 43.3 | 32.4 | 53.2 |
| Days Max Temp ≥ 90 °F | 0 | 0 | 0 | 0 | 2 | 9 | 13 | 8 | 4 | 0 | 0 | 0 | 36 |
| Days Max Temp ≤ 32 °F | 13 | 8 | 2 | 0 | 0 | 0 | 0 | 0 | 0 | 0 | 1 | 6 | 30 |
| Days Min Temp ≤ 32 °F | 28 | 24 | 18 | 5 | 0 | 0 | 0 | 0 | 0 | 4 | 14 | 24 | 117 |
| Days Min Temp ≤ 0 °F | 4 | 2 | 0 | 0 | 0 | 0 | 0 | 0 | 0 | 0 | 0 | 1 | 7 |
| Heating Degree Days | 1216 | 977 | 710 | 344 | 114 | 10 | 1 | 3 | 60 | 303 | 647 | 1005 | 5390 |
| Cooling Degree Days | 0 | 0 | 2 | 18 | 100 | 279 | 372 | 298 | 151 | 19 | 0 | 0 | 1239 |
| Total Precipitation (") | 2.15 | 1.97 | 2.99 | 4.09 | 3.71 | 3.75 | 4.78 | 3.68 | 3.45 | 2.95 | 3.70 | 3.40 | 40.62 |
| Days ≥ 0.1" Precip | 5 | 5 | 7 | 8 | 7 | 6 | 6 | 5 | 5 | 6 | 7 | 6 | 73 |
| Total Snowfall (") | 7.5 | 5.4 | 2.9 | 0.3 | 0.0 | 0.0 | 0.0 | 0.0 | 0.0 | 0.0 | 1.8 | *4.9* | 22.8 |
| Days ≥ 1" Snow Depth | 11 | *8* | *2* | 0 | 0 | 0 | 0 | 0 | 0 | 0 | *0* | *4* | 25 |

## URBANA *Champaign County* ELEVATION 741 ft LAT/LONG 40° 8 ' N / 88° 13 ' W

| | JAN | FEB | MAR | APR | MAY | JUN | JUL | AUG | SEP | OCT | NOV | DEC | YEAR |
|---|---|---|---|---|---|---|---|---|---|---|---|---|---|
| Maximum Temp °F | 31.8 | 36.5 | 49.0 | 62.5 | 73.6 | 82.9 | 85.3 | 82.9 | 77.3 | 64.6 | 49.7 | 37.3 | 61.1 |
| Minimum Temp °F | 16.4 | 20.6 | 31.0 | 41.4 | 51.7 | 60.8 | 64.8 | 62.3 | 54.9 | 43.4 | 33.5 | 23.0 | 42.0 |
| Mean Temp °F | 24.1 | 28.6 | 40.0 | 52.2 | 62.7 | 71.9 | 75.1 | 72.7 | 66.1 | 54.0 | 41.7 | 30.2 | 51.6 |
| Days Max Temp ≥ 90 °F | 0 | 0 | 0 | 0 | 1 | 6 | 7 | 4 | 2 | 0 | 0 | 0 | 20 |
| Days Max Temp ≤ 32 °F | 16 | 11 | 3 | 0 | 0 | 0 | 0 | 0 | 0 | 0 | 1 | 9 | 40 |
| Days Min Temp ≤ 32 °F | 29 | 24 | 18 | 5 | 0 | 0 | 0 | 0 | 0 | 3 | 15 | 25 | 119 |
| Days Min Temp ≤ 0 °F | 4 | 2 | 0 | 0 | 0 | 0 | 0 | 0 | 0 | 0 | 0 | 2 | 8 |
| Heating Degree Days | 1261 | 1022 | 768 | 395 | 144 | 15 | 1 | 6 | 79 | 348 | 694 | 1072 | 5805 |
| Cooling Degree Days | 0 | 0 | 2 | 12 | 77 | 234 | 330 | 259 | 120 | 12 | 0 | 0 | 1046 |
| Total Precipitation (") | 1.88 | 1.88 | 3.04 | 3.96 | 4.17 | 3.88 | 4.70 | 4.30 | 3.64 | 2.86 | 3.43 | 3.10 | 40.84 |
| Days ≥ 0.1" Precip | 5 | 4 | 7 | 8 | 7 | 6 | 7 | 6 | 6 | 6 | 6 | 6 | 74 |
| Total Snowfall (") | 7.8 | 6.1 | 3.8 | 0.6 | 0.0 | 0.0 | 0.0 | 0.0 | 0.0 | 0.1 | 2.2 | 5.4 | 26.0 |
| Days ≥ 1" Snow Depth | 13 | 10 | 4 | 0 | 0 | 0 | 0 | 0 | 0 | 0 | 1 | 6 | 34 |

## VIRDEN *Macoupin County* ELEVATION 669 ft LAT/LONG 39° 30 ' N / 89° 46 ' W

| | JAN | FEB | MAR | APR | MAY | JUN | JUL | AUG | SEP | OCT | NOV | DEC | YEAR |
|---|---|---|---|---|---|---|---|---|---|---|---|---|---|
| Maximum Temp °F | 34.8 | 39.5 | 52.7 | 66.0 | 76.0 | 84.7 | 88.1 | 86.0 | 79.9 | 67.7 | 52.7 | 40.0 | 64.0 |
| Minimum Temp °F | 17.7 | 21.7 | 32.8 | 43.8 | 52.9 | 62.0 | 65.1 | 62.4 | 55.5 | 44.4 | 34.4 | 24.0 | 43.1 |
| Mean Temp °F | 26.3 | 30.6 | 42.8 | 54.9 | 64.5 | 73.3 | 76.6 | 74.2 | 67.7 | 56.1 | 43.6 | 32.0 | 53.6 |
| Days Max Temp ≥ 90 °F | 0 | 0 | 0 | 0 | 1 | 7 | 13 | 9 | 4 | 0 | 0 | 0 | 34 |
| Days Max Temp ≤ 32 °F | 12 | 9 | 2 | 0 | 0 | 0 | 0 | 0 | 0 | 0 | 1 | ***7*** | 31 |
| Days Min Temp ≤ 32 °F | 28 | 22 | 16 | 4 | 0 | 0 | 0 | 0 | 0 | 4 | 14 | 25 | 113 |
| Days Min Temp ≤ 0 °F | 4 | 2 | 0 | 0 | 0 | 0 | 0 | 0 | 0 | 0 | 0 | 1 | 7 |
| Heating Degree Days | 1194 | 966 | 685 | 320 | 107 | 10 | 2 | 4 | 63 | 293 | 637 | 1017 | 5298 |
| Cooling Degree Days | 0 | 0 | 3 | 26 | 104 | 284 | 395 | 327 | 165 | 23 | 1 | 0 | 1328 |
| Total Precipitation (") | 1.84 | 1.97 | 3.14 | 4.02 | 4.25 | 3.77 | 3.92 | 3.27 | 3.58 | 2.62 | 3.25 | 2.79 | 38.42 |
| Days ≥ 0.1" Precip | 4 | 4 | 7 | 7 | 7 | 6 | 5 | 5 | 5 | 5 | 6 | 5 | 66 |
| Total Snowfall (") | 5.6 | ***7.3*** | 4.2 | 0.7 | 0.0 | 0.0 | 0.0 | 0.0 | 0.0 | 0.0 | 1.5 | 4.6 | 23.9 |
| Days ≥ 1" Snow Depth | 11 | ***9*** | 3 | 0 | 0 | 0 | 0 | 0 | 0 | 0 | 1 | ***6*** | 30 |

## WALNUT *Bureau County* ELEVATION 712 ft LAT/LONG 41° 33 ' N / 89° 35 ' W

| | JAN | FEB | MAR | APR | MAY | JUN | JUL | AUG | SEP | OCT | NOV | DEC | YEAR |
|---|---|---|---|---|---|---|---|---|---|---|---|---|---|
| Maximum Temp °F | 28.6 | 33.5 | 46.5 | 61.2 | 73.3 | 82.2 | 85.3 | 82.9 | 75.9 | 64.0 | 47.4 | 34.2 | 59.6 |
| Minimum Temp °F | 10.8 | 15.8 | 27.5 | 38.6 | 49.6 | 58.8 | 62.9 | 60.3 | 52.6 | 40.9 | 30.0 | 18.0 | 38.8 |
| Mean Temp °F | 19.8 | 24.7 | 37.1 | 49.9 | 61.5 | 70.5 | 74.1 | 71.7 | 64.2 | 52.5 | 38.7 | 26.1 | 49.2 |
| Days Max Temp ≥ 90 °F | 0 | 0 | 0 | 0 | 1 | 4 | 8 | 4 | 1 | 0 | 0 | 0 | 18 |
| Days Max Temp ≤ 32 °F | 18 | 13 | 4 | 0 | 0 | 0 | 0 | 0 | 0 | 0 | 2 | 12 | 49 |
| Days Min Temp ≤ 32 °F | 30 | 26 | 23 | 8 | 1 | 0 | 0 | 0 | 1 | 6 | 19 | 28 | 142 |
| Days Min Temp ≤ 0 °F | 8 | 5 | 0 | 0 | 0 | 0 | 0 | 0 | 0 | 0 | 0 | 4 | 17 |
| Heating Degree Days | 1396 | 1132 | 860 | 457 | 171 | 22 | 5 | 13 | 108 | 394 | 781 | 1199 | 6538 |
| Cooling Degree Days | 0 | 0 | 1 | 12 | 69 | 190 | 297 | 229 | 87 | 6 | 0 | 0 | 891 |
| Total Precipitation (") | 1.33 | 1.18 | 2.80 | 3.80 | 3.99 | 4.44 | 3.80 | 4.32 | 3.91 | 2.70 | 2.69 | 2.14 | 37.10 |
| Days ≥ 0.1" Precip | 4 | 4 | 7 | 8 | 8 | 7 | 7 | 7 | 7 | 6 | 6 | 5 | 76 |
| Total Snowfall (") | 9.0 | 7.4 | 4.4 | 1.2 | 0.0 | 0.0 | 0.0 | 0.0 | 0.0 | 0.3 | 1.7 | 7.5 | 31.5 |
| Days ≥ 1" Snow Depth | 19 | 13 | 3 | 0 | 0 | 0 | 0 | 0 | 0 | 0 | 1 | 10 | 46 |

## WATSEKA 2 NW *Iroquois County* ELEVATION 630 ft LAT/LONG 40° 46 ' N / 87° 44 ' W

| | JAN | FEB | MAR | APR | MAY | JUN | JUL | AUG | SEP | OCT | NOV | DEC | YEAR |
|---|---|---|---|---|---|---|---|---|---|---|---|---|---|
| Maximum Temp °F | 30.6 | 35.3 | 47.7 | 60.8 | 72.3 | 82.4 | 84.9 | 82.6 | 76.8 | 64.1 | 49.4 | 36.3 | 60.3 |
| Minimum Temp °F | 14.3 | 17.4 | 28.9 | 38.9 | 48.8 | 58.3 | 62.3 | 59.4 | 52.3 | 40.6 | 31.9 | 21.1 | 39.5 |
| Mean Temp °F | 22.5 | 26.4 | 38.3 | 49.9 | 60.6 | 70.4 | 73.6 | 71.0 | 64.6 | 52.4 | 40.6 | 28.8 | 49.9 |
| Days Max Temp ≥ 90 °F | 0 | 0 | 0 | 0 | 1 | 5 | 8 | 3 | 1 | 0 | 0 | 0 | 18 |
| Days Max Temp ≤ 32 °F | 16 | 11 | 3 | 0 | 0 | 0 | 0 | 0 | 0 | 0 | 2 | 10 | 42 |
| Days Min Temp ≤ 32 °F | 29 | 25 | 21 | 8 | 1 | 0 | 0 | 0 | 0 | 7 | 17 | 27 | 135 |
| Days Min Temp ≤ 0 °F | 6 | 4 | 0 | 0 | 0 | 0 | 0 | 0 | 0 | 0 | 0 | 2 | 12 |
| Heating Degree Days | 1311 | 1086 | 821 | 455 | 191 | 27 | 5 | 14 | 103 | 397 | 724 | 1117 | 6251 |
| Cooling Degree Days | ***0*** | 0 | 1 | 7 | 62 | 198 | 289 | 216 | 96 | ***10*** | 0 | ***0*** | 879 |
| Total Precipitation (") | 1.76 | 1.65 | 3.48 | 4.36 | 4.11 | 4.24 | 4.26 | 3.58 | 3.90 | 3.09 | 3.66 | 3.19 | 41.28 |
| Days ≥ 0.1" Precip | 4 | 4 | 7 | 8 | 7 | 7 | 7 | 6 | 6 | 5 | 7 | 6 | 74 |
| Total Snowfall (") | ***6.4*** | 4.8 | 2.3 | 0.9 | 0.0 | 0.0 | 0.0 | 0.0 | 0.0 | 0.0 | 1.2 | 5.0 | 20.6 |
| Days ≥ 1" Snow Depth | 13 | 11 | 3 | 0 | 0 | 0 | 0 | 0 | 0 | 0 | 1 | 7 | 35 |

## WAUKEGAN *Lake County* ELEVATION 679 ft LAT/LONG 42° 22 ' N / 87° 52 ' W

| | JAN | FEB | MAR | APR | MAY | JUN | JUL | AUG | SEP | OCT | NOV | DEC | YEAR |
|---|---|---|---|---|---|---|---|---|---|---|---|---|---|
| Maximum Temp °F | 28.5 | 32.0 | 42.6 | 54.7 | 66.7 | 76.6 | 81.1 | 79.4 | 72.1 | 60.8 | 46.9 | 34.0 | 56.3 |
| Minimum Temp °F | 12.1 | 15.4 | 25.6 | 35.9 | 45.6 | 55.2 | 61.2 | 60.0 | 52.6 | 41.0 | 30.6 | 19.0 | 37.9 |
| Mean Temp °F | 20.3 | 23.7 | 34.1 | 45.3 | 56.2 | 65.9 | 71.2 | 69.7 | 62.4 | 50.9 | 38.7 | 26.5 | 47.1 |
| Days Max Temp ≥ 90 °F | 0 | 0 | 0 | 0 | 0 | 2 | 4 | 3 | 1 | 0 | 0 | 0 | 10 |
| Days Max Temp ≤ 32 °F | 19 | 14 | 5 | 0 | 0 | 0 | 0 | 0 | 0 | 0 | 2 | 13 | 53 |
| Days Min Temp ≤ 32 °F | 30 | 27 | 24 | 10 | 1 | 0 | 0 | 0 | 0 | 5 | 18 | 27 | 142 |
| Days Min Temp ≤ 0 °F | 7 | 4 | 0 | 0 | 0 | 0 | 0 | 0 | 0 | 0 | 0 | 3 | 14 |
| Heating Degree Days | 1379 | 1159 | 951 | 589 | 300 | 80 | 17 | 24 | 136 | 438 | 781 | 1187 | 7041 |
| Cooling Degree Days | 0 | 0 | 1 | 7 | 35 | 108 | 217 | 185 | 65 | 4 | ***0*** | 0 | 622 |
| Total Precipitation (") | 1.63 | 1.22 | 2.33 | 3.67 | 3.16 | 3.89 | 3.55 | 4.15 | 3.55 | 2.48 | 2.75 | 2.28 | 34.66 |
| Days ≥ 0.1" Precip | 4 | 4 | 5 | 7 | 6 | 6 | 6 | 6 | 6 | 5 | 6 | 5 | 66 |
| Total Snowfall (") | 12.5 | 10.7 | 6.3 | 1.7 | 0.0 | 0.0 | 0.0 | 0.0 | 0.0 | 0.2 | 1.5 | 8.5 | 41.4 |
| Days ≥ 1" Snow Depth | na | na | 5 | 1 | 0 | 0 | 0 | 0 | 0 | 0 | ***1*** | na | na |

## WHEATON 3 SE *DuPage County* ELEVATION 751 ft LAT/LONG 41° 52 ' N / 88° 6 ' W

| | JAN | FEB | MAR | APR | MAY | JUN | JUL | AUG | SEP | OCT | NOV | DEC | YEAR |
|---|---|---|---|---|---|---|---|---|---|---|---|---|---|
| Maximum Temp °F | 30.4 | 35.6 | 48.3 | 61.9 | 73.8 | 83.0 | 86.2 | 84.2 | 77.2 | 64.9 | 49.0 | 36.4 | 60.9 |
| Minimum Temp °F | 13.0 | 16.7 | 27.7 | 37.6 | 47.0 | 56.9 | 62.0 | 60.4 | 52.8 | 41.4 | 31.5 | 20.1 | 38.9 |
| Mean Temp °F | 21.7 | 26.2 | 38.0 | 49.8 | 60.4 | 69.9 | 74.1 | 72.3 | 65.0 | 53.2 | 40.3 | 28.2 | 49.9 |
| Days Max Temp ≥ 90 °F | 0 | 0 | 0 | 0 | 1 | 6 | 9 | 6 | 2 | 0 | 0 | 0 | 24 |
| Days Max Temp ≤ 32 °F | 16 | 11 | 2 | 0 | 0 | 0 | 0 | 0 | 0 | 0 | 1 | 10 | 40 |
| Days Min Temp ≤ 32 °F | 29 | 25 | 22 | 9 | 2 | 0 | 0 | 0 | 0 | 7 | 17 | 27 | 138 |
| Days Min Temp ≤ 0 °F | 7 | 4 | 0 | 0 | 0 | 0 | 0 | 0 | 0 | 0 | 0 | 2 | 13 |
| Heating Degree Days | 1336 | 1089 | 830 | 461 | 194 | 31 | 5 | 10 | 97 | 374 | 735 | 1135 | 6297 |
| Cooling Degree Days | 0 | 0 | 1 | 14 | 64 | 197 | 318 | 269 | 114 | 10 | 0 | 0 | 987 |
| Total Precipitation (") | 1.72 | 1.44 | 2.63 | 3.80 | 3.71 | 3.92 | 3.82 | 4.53 | 3.62 | 2.61 | 3.27 | 2.57 | 37.64 |
| Days ≥ 0.1" Precip | 4 | 4 | 6 | 7 | 7 | 7 | 7 | 6 | 6 | 5 | 6 | 5 | 70 |
| Total Snowfall (") | 11.0 | 8.6 | 4.5 | 1.0 | 0.1 | 0.0 | 0.0 | 0.0 | 0.0 | 0.2 | 1.9 | 6.2 | 33.5 |
| Days ≥ 1" Snow Depth | *17* | *13* | 4 | 0 | 0 | 0 | 0 | 0 | 0 | 0 | 1 | 8 | 43 |

## WHITE HALL 1 E *Greene County* ELEVATION 581 ft LAT/LONG 39° 26 ' N / 90° 23 ' W

| | JAN | FEB | MAR | APR | MAY | JUN | JUL | AUG | SEP | OCT | NOV | DEC | YEAR |
|---|---|---|---|---|---|---|---|---|---|---|---|---|---|
| Maximum Temp °F | 35.3 | 40.5 | 52.8 | 65.6 | 75.0 | 83.8 | 87.5 | 85.0 | 78.6 | 67.5 | 53.1 | 40.5 | 63.8 |
| Minimum Temp °F | 16.8 | 20.4 | 31.3 | 42.8 | 51.9 | 61.3 | 65.4 | 62.8 | 55.3 | 43.7 | 33.4 | 23.0 | 42.3 |
| Mean Temp °F | 26.1 | 30.5 | 42.1 | 54.2 | 63.5 | 72.6 | 76.5 | 73.9 | 67.0 | 55.7 | 43.3 | 31.8 | 53.1 |
| Days Max Temp ≥ 90 °F | 0 | 0 | 0 | 0 | 1 | 6 | 12 | 7 | 3 | 0 | 0 | 0 | 29 |
| Days Max Temp ≤ 32 °F | 12 | 8 | 2 | 0 | 0 | 0 | 0 | 0 | 0 | 0 | 1 | 7 | 30 |
| Days Min Temp ≤ 32 °F | 28 | 24 | 18 | 4 | 0 | 0 | 0 | 0 | 0 | 4 | 15 | 25 | 118 |
| Days Min Temp ≤ 0 °F | 4 | 2 | 0 | 0 | 0 | 0 | 0 | 0 | 0 | 0 | 0 | 2 | 8 |
| Heating Degree Days | 1201 | 968 | 707 | 337 | 120 | 11 | 1 | 5 | 69 | 305 | 645 | 1023 | 5392 |
| Cooling Degree Days | 0 | 0 | 3 | 20 | 73 | 246 | 366 | 294 | 133 | 18 | 1 | 0 | 1154 |
| Total Precipitation (") | 1.52 | 1.68 | 3.06 | 3.86 | 4.16 | 3.49 | 3.65 | 2.92 | 3.61 | 2.75 | 3.14 | 2.84 | 36.68 |
| Days ≥ 0.1" Precip | 4 | 4 | 6 | 7 | 6 | 6 | 6 | 6 | 6 | 6 | 6 | 5 | 68 |
| Total Snowfall (") | na | na | ***2.4*** | 0.4 | 0.0 | 0.0 | 0.0 | 0.0 | 0.0 | 0.0 | ***0.6*** | ***2.9*** | na |
| Days ≥ 1" Snow Depth | na | na | *1* | 0 | 0 | 0 | 0 | 0 | 0 | 0 | 0 | na | na |

## WINDSOR *Shelby County* ELEVATION 689 ft LAT/LONG 39° 26 ' N / 88° 36 ' W

| | JAN | FEB | MAR | APR | MAY | JUN | JUL | AUG | SEP | OCT | NOV | DEC | YEAR |
|---|---|---|---|---|---|---|---|---|---|---|---|---|---|
| Maximum Temp °F | 34.6 | 39.9 | 52.2 | 65.4 | 75.2 | 83.9 | 87.2 | 85.1 | 79.6 | 67.2 | 52.5 | 40.1 | 63.6 |
| Minimum Temp °F | 18.1 | 22.1 | 32.5 | 43.1 | 52.7 | 61.6 | 65.5 | 63.0 | 56.3 | 44.8 | 34.8 | 24.5 | 43.3 |
| Mean Temp °F | 26.4 | 31.0 | 42.4 | 54.3 | 64.0 | 72.8 | 76.4 | 74.1 | 68.0 | 56.0 | 43.7 | 32.3 | 53.5 |
| Days Max Temp ≥ 90 °F | 0 | 0 | 0 | 0 | 1 | 6 | 11 | 7 | 2 | 0 | 0 | 0 | 27 |
| Days Max Temp ≤ 32 °F | 13 | 8 | 2 | 0 | 0 | 0 | 0 | 0 | 0 | 0 | 1 | 7 | 31 |
| Days Min Temp ≤ 32 °F | 27 | 22 | 17 | 4 | 0 | 0 | 0 | 0 | 0 | 3 | 13 | 24 | 110 |
| Days Min Temp ≤ 0 °F | 4 | 2 | 0 | 0 | 0 | 0 | 0 | 0 | 0 | 0 | 0 | 1 | 7 |
| Heating Degree Days | 1191 | 953 | 696 | 333 | 114 | 10 | 1 | 4 | 57 | 293 | 633 | 1005 | 5290 |
| Cooling Degree Days | 0 | 0 | 2 | 19 | 93 | 264 | 375 | 305 | 152 | 23 | 0 | 0 | 1233 |
| Total Precipitation (") | 1.86 | 1.84 | 3.05 | 3.95 | 3.85 | 3.63 | 4.24 | 3.09 | 3.47 | 3.02 | 3.73 | 3.38 | 39.11 |
| Days ≥ 0.1" Precip | 4 | 4 | 7 | 8 | 7 | 6 | 7 | 5 | 5 | 5 | 6 | 6 | 70 |
| Total Snowfall (") | 6.7 | 5.0 | 3.6 | 0.3 | 0.0 | 0.0 | 0.0 | 0.0 | 0.0 | 0.0 | 1.3 | 4.9 | 21.8 |
| Days ≥ 1" Snow Depth | 12 | 9 | 3 | 0 | 0 | 0 | 0 | 0 | 0 | 0 | 1 | 5 | 30 |

## JANUARY MINIMUM TEMPERATURE °F

### LOWEST

| | | |
|---|---|---|
| 1 | Mount Carroll | 7.4 |
| 2 | Stockton | 8.7 |
| 3 | Marengo | 9.5 |
| 4 | Antioch | 9.6 |
| 5 | Dixon | 9.9 |
| 6 | Paw Paw | 10.0 |
| 7 | Barrington | 10.1 |
| | Morrison | 10.1 |
| 9 | Rockford | 10.2 |
| 10 | Hennepin | 10.6 |
| 11 | Walnut | 10.8 |
| 12 | Fulton Lk & Dam | 11.0 |
| | Kewanee | 11.0 |
| 14 | Aurora | 11.3 |
| 15 | Princeville | 11.5 |
| 16 | Galva | 11.7 |
| | Moline | 11.7 |
| 18 | Geneseo | 12.1 |
| | Waukegan | 12.1 |
| 20 | Peru | 12.2 |
| 21 | Tiskilwa | 12.6 |
| 22 | Minonk | 12.8 |
| 23 | Galesburg | 12.9 |
| | Park Forest | 12.9 |
| 25 | La Harpe | 13.0 |

### HIGHEST

| | | |
|---|---|---|
| 1 | Cairo | 26.0 |
| 2 | Brookport Dam | 24.2 |
| 3 | Anna | 23.1 |
| 4 | Albion | 22.9 |
| 5 | Harrisburg | 22.4 |
| 6 | Sparta | 22.0 |
| 7 | Du Quoin | 21.3 |
| 8 | Nashville | 21.2 |
| 9 | Belleville | 21.1 |
| 10 | Carbondale | 20.8 |
| | Fairfield | 20.8 |
| 12 | McLeansboro | 20.5 |
| | Olney | 20.5 |
| 14 | Flora | 20.3 |
| 15 | Palestine | 20.1 |
| | Salem | 20.1 |
| 17 | Marion | 19.7 |
| | Mt. Vernon | 19.7 |
| 19 | Charleston | 18.6 |
| | Hillsboro | 18.6 |
| 21 | Alton | 18.5 |
| 22 | Newton | 18.4 |
| 23 | Pana | 18.3 |
| 24 | Carlinville | 18.2 |
| 25 | Windsor | 18.1 |

## JULY MAXIMUM TEMPERATURE °F

### HIGHEST

| | | |
|---|---|---|
| 1 | Harrisburg | 90.8 |
| 2 | Albion | 90.2 |
| 3 | Sparta | 90.0 |
| 4 | Du Quoin | 89.8 |
| 5 | Brookport Dam | 89.7 |
| | Cairo | 89.7 |
| 7 | Belleville | 89.5 |
| 8 | Flora | 89.3 |
| | Salem | 89.3 |
| 10 | Hillsboro | 89.2 |
| 11 | Anna | 89.1 |
| | Carbondale | 89.1 |
| 13 | McLeansboro | 88.8 |
| | Mt. Vernon | 88.8 |
| 15 | Olney | 88.5 |
| 16 | Nashville | 88.4 |
| 17 | Alton | 88.2 |
| | Carlinville | 88.2 |
| | Fairfield | 88.2 |
| | Havana | 88.2 |
| 21 | Palestine | 88.1 |
| | Tuscola | 88.1 |
| | Virden | 88.1 |
| 24 | Carlyle Reservoir | 88.0 |
| | Jerseyville | 88.0 |

### LOWEST

| | | |
|---|---|---|
| 1 | Waukegan | 81.1 |
| 2 | Antioch | 82.1 |
| 3 | Stockton | 82.3 |
| 4 | Barrington | 83.2 |
| 5 | Paw Paw | 83.3 |
| 6 | Park Forest | 83.8 |
| | Rockford | 83.8 |
| 8 | Chicago-O'Hare | 83.9 |
| 9 | Chicago-Univ | 84.2 |
| | Fulton Lk & Dam | 84.2 |
| 11 | Aurora | 84.5 |
| | Chicago-Midway | 84.5 |
| 13 | Mount Carroll | 84.6 |
| 14 | Dixon | 84.7 |
| 15 | Galesburg | 84.9 |
| | Galva | 84.9 |
| | Marengo | 84.9 |
| | Princeville | 84.9 |
| | Watseka | 84.9 |
| 20 | Kewanee | 85.1 |
| 21 | Morrison | 85.2 |
| 22 | Hoopeston | 85.3 |
| | Urbana | 85.3 |
| | Walnut | 85.3 |
| 25 | Pontiac | 85.5 |

## ANNUAL PRECIPITATION (")

### HIGHEST

| | | |
|---|---|---|
| 1 | Anna | 48.56 |
| 2 | Cairo | 48.01 |
| 3 | Brookport Dam | 47.98 |
| 4 | Marion | 45.69 |
| 5 | Harrisburg | 45.08 |
| 6 | Carbondale | 44.80 |
| 7 | Fairfield | 44.67 |
| 8 | Albion | 44.30 |
| 9 | McLeansboro | 43.01 |
| 10 | Olney | 42.77 |
| 11 | Du Quoin | 42.70 |
| 12 | Palestine | 42.46 |
| 13 | Mt. Vernon | 41.94 |
| 14 | Flora | 41.78 |
| 15 | Sparta | 41.77 |
| 16 | Effingham | 41.45 |
| 17 | Paris | 41.43 |
| 18 | Salem | 41.31 |
| 19 | Watseka | 41.28 |
| 20 | Pana | 41.07 |
| 21 | Danville | 40.97 |
| | Hillsboro | 40.97 |
| 23 | Charleston | 40.90 |
| 24 | Decatur | 40.89 |
| 25 | Urbana | 40.84 |

### LOWEST

| | | |
|---|---|---|
| 1 | Waukegan | 34.66 |
| 2 | Fulton Lk & Dam | 35.04 |
| 3 | Hennepin | 35.13 |
| 4 | Stockton | 35.50 |
| 5 | Piper City | 35.53 |
| 6 | Chenoa | 35.87 |
| 7 | Aledo | 36.03 |
| 8 | Springfield | 36.04 |
| 9 | Mason City | 36.17 |
| 10 | Antioch | 36.20 |
| 11 | Pontiac | 36.36 |
| 12 | Dixon | 36.42 |
| 13 | Chicago-O'Hare | 36.67 |
| 14 | White Hall | 36.68 |
| 15 | Ottawa | 36.76 |
| 16 | Marengo | 36.77 |
| 17 | Rockford | 36.80 |
| 18 | Farmer City | 36.90 |
| 19 | Tiskilwa | 36.93 |
| 20 | Kewanee | 36.96 |
| 21 | Walnut | 37.10 |
| 22 | Peoria | 37.23 |
| 23 | Geneseo | 37.30 |
| 24 | Mount Carroll | 37.31 |
| 25 | Galesburg | 37.43 |

## ANNUAL SNOWFALL (")

### HIGHEST

| | | |
|---|---|---|
| 1 | Antioch | 47.0 |
| 2 | Chicago-Midway | 44.2 |
| 3 | Waukegan | 41.4 |
| 4 | Chicago-O'Hare | 38.5 |
| 5 | Marengo | 38.3 |
| 6 | Rockford | 36.9 |
| 7 | Dixon | 33.6 |
| 8 | Wheaton | 33.5 |
| 9 | Moline | 33.2 |
| 10 | Morrison | 33.0 |
| 11 | Mount Carroll | 32.2 |
| 12 | Paw Paw | 31.5 |
| | Walnut | 31.5 |
| 14 | Stockton | 31.4 |
| 15 | Park Forest | 30.6 |
| 16 | Aurora | 30.1 |
| 17 | Galva | 30.0 |
| 18 | Paris | 28.1 |
| 19 | Havana | 28.0 |
| 20 | Monmouth | 27.8 |
| 21 | Aledo | 27.7 |
| 22 | Minonk | 27.6 |
| 23 | Piper City | 26.5 |
| 24 | Peoria | 26.4 |
| 25 | Urbana | 26.0 |

### LOWEST

| | | |
|---|---|---|
| 1 | Flora | 9.9 |
| 2 | Brookport Dam | 10.0 |
| 3 | Cairo | 11.2 |
| 4 | Albion | 11.7 |
| 5 | Olney | 13.6 |
| 6 | Carbondale | 14.5 |
| 7 | Salem | 14.6 |
| 8 | Fairfield | 15.0 |
| 9 | Newton | 16.0 |
| | Sparta | 16.0 |
| 11 | Belleville | 16.3 |
| | McLeansboro | 16.3 |
| 13 | Anna | 16.5 |
| 14 | Nashville | 17.4 |
| 15 | Harrisburg | 17.5 |
| | Mattoon | 17.5 |
| 17 | Mt. Vernon | 17.7 |
| 18 | Jerseyville | 18.0 |
| 19 | Hoopeston | 18.1 |
| 20 | Hillsboro | 18.8 |
| 21 | Marion | 19.4 |
| 22 | Palestine | 19.5 |
| 23 | Lincoln | 20.0 |
| | Mason City | 20.0 |
| 25 | Carlinville | 20.1 |

# INDIANA

PHYSICAL FEATURES AND GENERAL CLIMATE. Indiana has an invigorating climate of warm summers and cool winters, because of its location in the middle latitudes in the interior of a large continent. Imposed on the well-known daily and seasonal changes of temperature are changes occurring every few days as surges of polar air move southeastward or air of tropical origin moves northeastward. These outbreaks are more frequent and pronounced in the winter than in the summer. A winter may be unusually cold or a summer cool if the influence of polar air is rather continuous. Likewise, a summer may be unusually warm or a winter mild if air of tropical origin predominates. The action between these two air masses with a contrast in temperature and density fosters the development of low pressure centers which in moving generally eastward frequently pass through or near Indiana, resulting in normally abundant rain. The cyclones are least active and frequently pass north of Indiana in midsummer. Thunderstorms, often local in areal coverage, are important at such times when evaporation and loss of moisture from the soil and vegetation exceeds rainfall. Major climatological variations within the State are caused by differences of latitude, elevation, terrain, soil, and lakes.

The effect of the Great Lakes and more specifically, Lake Michigan, on the climate of northern Indiana is most pronounced just inland from the Lake Michigan shore and diminishes to insignificance in central Indiana. The result of cold air passing over the warmer lake water of Lake Michigan induces precipitation in the lee of Lake Michigan in fall and winter. Average daily minimum temperatures in the fall are higher and daily maximum temperatures in the spring are lower in northwestern Indiana than farther south. Winter precipitation, especially snowfall, is several times greater in the counties of Lake, Porter, and LaPorte as the result of this phenomena. Lake related snowfall and cloudiness often extends to central Indiana in the winter. Very local severe snowstorms have occurred just inland from Lake Michigan.

Another important variable in the composition of Indiana weather is the topography of the State. Elevations range from a little more than 300 feet at the mouth of the Wabash in the southwest corner of the State, to a little over 1,200 feet in the east-central portion (Randolph County) and northeastern section (Steuben County). Differences of terrain affect the climate considerably. South-central Indiana is unglaciated and has the most rugged relief. The Kankakee Valley in the northwest has but little slope to the west and drains what was formerly marshlands. Many small lakes abound in northeastern Indiana among numerous glacial moraines and hills. Most of the north, central, and southwest is rolling country.

TEMPERATURE. Variations of temperature and precipitation occur in short distances where terrain is hilly. On calm, clear nights the valley bottoms have lower temperatures than the slopes and tops of the surrounding hills. Mean maximum as well as mean minimum temperatures decrease from south to north with latitude and decrease from west to east with elevation. Near Lake Michigan temperatures average higher than expected for the latitude in the fall and winter, and lower than expected for the latitude in the spring and summer.

The average date of the last freezing temperature in the spring ranges from the first week of April in the Ohio River Valley of the southwest to the second week of May in the extreme northeast. The usual trend of a later date toward the north is reversed in extreme northwestern Indiana, where the average date is about April 30 near Lake Michigan. In the fall the average date of the first temperature of 32° F. or colder is from October 7 in the extreme northeast to October 26 along the Ohio River in the southwest.

Spring freezes are later in valleys and hollows and fall freezes are earlier. Longer freeze-free periods occur on ridges and hills. Southern Indiana has much of this type of terrain. The gradual slope upward from southwestern Indiana to northeastern Indiana results in lower minimum temperatures and shorter growing seasons in the east compared to the west at the same latitude. In the Kankakee Valley, peat or muck lands experience late spring and early fall frosts because of the radiative characteristic of the soil.

PRECIPITATION. Average annual rainfall ranges from 36 inches in northern Indiana to 43 inches in southern Indiana. July rainfall averages about the same in all areas. The greater precipitation in the south compared with the north comes in the winter months. Southern Indiana has the greatest rainfall in March and the least in October. The wettest month in northern and central Indiana is June and the driest is February. A drought occasionally occurs in the summer when evaporation is highest and dependence on rainfall is greatest.

Most of the state is drained by the Wabash River system. Other river basins are the Maumee in the extreme northeast, the St. Joseph (Lake Michigan) and Kankakee (Illinois River) in the north-central and northwest, and some Ohio River drainage in the extreme south and southeast. Floods occur in some part of the State nearly every year and have occurred in every month of the year. The season of greatest flood frequency is during the winter and spring months. The primary cause of floods is prolonged periods of heavy rains, although occasionally the rains falling on a snow cover and the formation of ice jams are an added factor. The most common type of flood-producing storm in the area is that having a quasi-stationary front oriented from west-southwest to east-northeast with a series of waves or perturbations moving to the east along the front.

Average annual snowfall increases from about 10 inches in southern Indiana to 40 inches in the northern portion of the State and higher in the three county areas along Lake Michigan. From year to year snowfall varies greatly, depending both on temperatures and the frequency of winter storms. At a given latitude in central and southern Indiana snowfall is greatest toward the east because of higher elevation.

OTHER CLIMATIC ELEMENTS. Cloudiness is least in the fall and greatest in the winter. The north is cloudier than the south, particularly in the winter when the Great Lakes have the greatest effect upon the weather.

Average relative humidity differs very little at night over Indiana. During the day relative humidity is usually lower in the south than in the north. This is true for all seasons. However, the simultaneous occurrence of high temperatures and high relative humidity is most frequent in the south.

Prevailing winds are from the southwest quadrant throughout most of the year. Winds from the northern quadrant occur in the winter and persist for a longer time in the north. Along the shore of Lake Michigan the sea-breeze effect is observed in the summer when winds in central United States are light or calm. Vertical currents from the heating of land during the day cause wind near the ground to flow from over water to land reducing the maximum temperature of the day. At night the breezes are in the opposite direction or from the land to water because of land cooling. These breezes are important in limiting extremely high temperatures of a summer day and account for rapid changes in short distances within a mile or so of the lake shore. Winds meet less friction passing over water so off-lake winds have a considerably higher speed than those off or over land.

Severe storms are most frequent in the spring. About one-half of the tornadoes occur between 2 p.m. and 6 p.m. and nearly three-fourths between 10 a.m. and 10 p.m. Hail falls occasionally in very local areas.

## COUNTY INDEX

### *Shelby County*
SHELBYVILLE SEWAGE P

### *Spencer County*
SAINT MEINRAD

### *Steuben County*
ANGOLA

### *Switzerland County*
VEVAY

### *Tippecanoe County*
LAFAYETTE 5 S
WEST LAFAYETTE 6 NW
WEST LAFAYETTE AP

### *Vanderburgh County*
EVANSVILLE MUSEUM
EVANSVILLE REG AP

### *Vigo County*
TERRE HAUTE

### *Wabash County*
WABASH

### *Washington County*
SALEM

### *Wayne County*
CAMBRIDGE CITY 3 N

### *Whitley County*
COLUMBIA CITY

## ELEVATION INDEX

| FEET | STATION NAME |
|---|---|
| 387 | EVANSVILLE REG AP |
| 390 | EVANSVILLE MUSEUM |
| 390 | TELL CITY |
| 410 | MOUNT VERNON |
| 440 | SPURGEON 2 N |
| 459 | MADISON SEWAGE PLANT |
| 470 | VEVAY |
| 479 | PRINCETON 1 W |
| 479 | WASHINGTON |
| 522 | SHOALS HIWAY 50 BRDG |
| 525 | CHARLESTOWN ORD PLNT |
| 541 | SAINT MEINRAD |
| 561 | SPENCER |
| 561 | TERRE HAUTE |
| 571 | SCOTTSBURG |
| 571 | SEYMOUR 2 N |
| 600 | LAFAYETTE 5 S |
| 600 | MARTINSVILLE 2 SW |
| 610 | PAOLI |
| 620 | HOBART 2 WNW |
| 630 | BROOKVILLE |
| 630 | COLUMBUS |
| 636 | WEST LAFAYETTE AP |
| 650 | OOLITIC PURDUE EXP F |
| 650 | RENSSELAER |
| 669 | DELPHI 3 NNE |
| 669 | WHEATFIELD 4 NNW |
| 689 | DUBOIS S IND FORAGE |
| 689 | KENTLAND 4 NNW |
| 689 | ROCKVILLE |
| 690 | WINAMAC 2 SSE |
| 712 | LOWELL |
| 712 | WEST LAFAYETTE 6 NW |
| 722 | SALEM |
| 741 | WANATAH 2 WNW |
| 751 | GREENCASTLE 1 SE |
| 751 | INDIANAPOLIS SE SIDE |
| 761 | CRANE NAVAL DEPOT |
| 761 | WABASH |
| 770 | ROCHESTER |
| 771 | SHELBYVILLE SEWAGE P |
| 794 | OAKLANDON GEIST RSVR |
| 801 | GOSHEN COLLEGE |
| 801 | SOUTH BEND MICHIANA |
| 804 | FORT WAYNE BAER FLD |
| 807 | LA PORTE |
| 810 | INDIANAPOLIS INTL AP |
| 810 | VALPARAISO WATERWORK |
| 810 | WARSAW |
| 830 | BLOOMINGTON IND UN |
| 830 | FRANKFORT DISPOSAL P |
| 830 | WHITESTOWN |
| 850 | MARION 2 N |
| 853 | ELWOOD WASTEWATER PL |
| 860 | BERNE |
| 860 | KOKOMO 3 WSW |
| 879 | COLUMBIA CITY |
| 879 | HARTFORD CITY 4 ESE |
| 889 | ANDERSON SEWAGE PLT |
| 902 | LAGRANGE SEWAGE PLAN |
| 935 | GREENSBURG |
| 961 | MUNCIE BALL STATE UN |
| 971 | FARMLAND 5 NNW |
| 971 | NEW CASTLE 4 N |
| 991 | GREENFIELD |
| 991 | RUSHVILLE SEWGE PLT |
| 1000 | CAMBRIDGE CITY 3 N |
| 1060 | ANGOLA |
| 1089 | WINCHESTER AAP 3 |

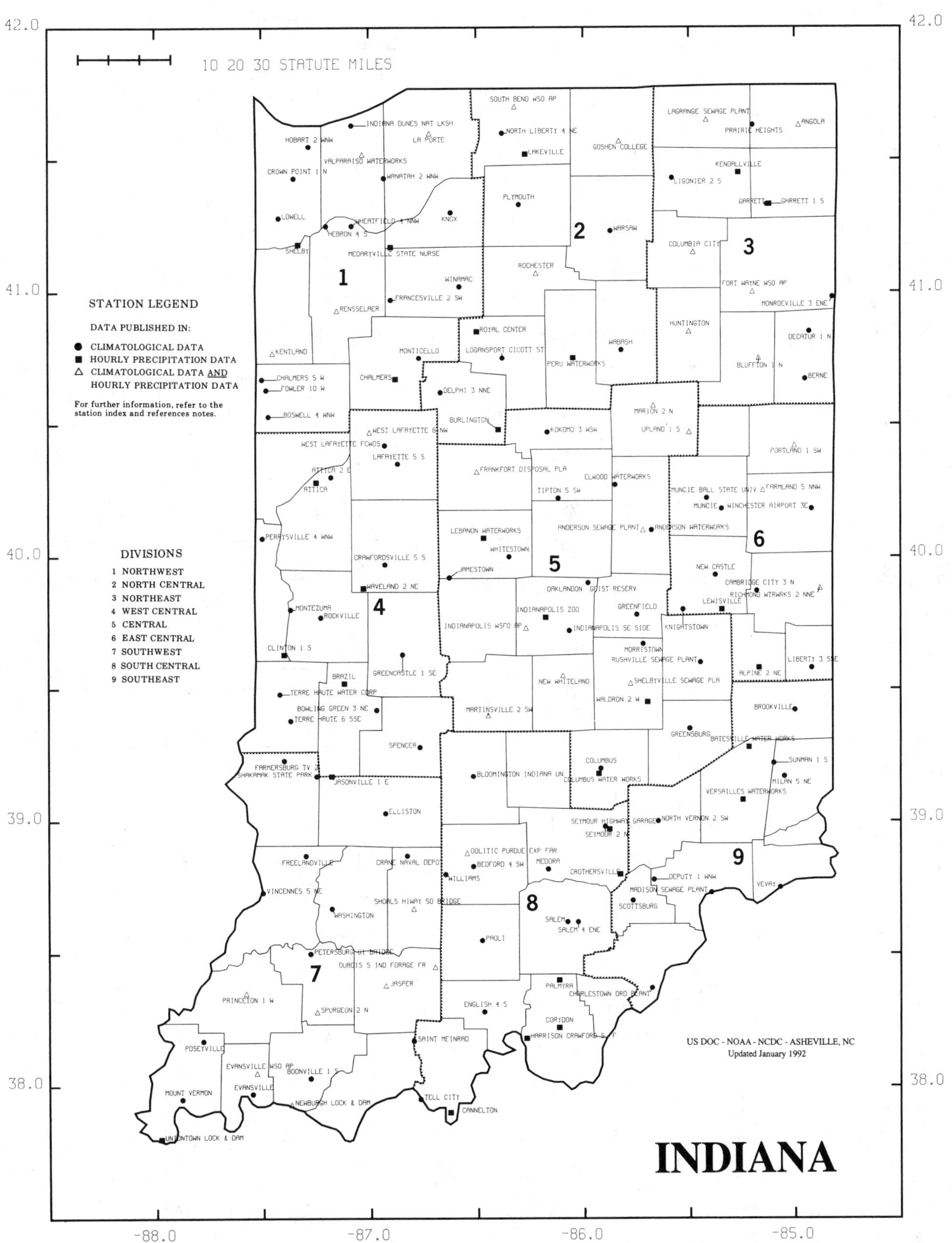
10 20 30 STATUTE MILES
STATION LEGEND
DATA PUBLISHED IN:
● CLIMATOLOGICAL DATA
■ HOURLY PRECIPITATION DATA
△ CLIMATOLOGICAL DATA AND HOURLY PRECIPITATION DATA
For further information, refer to the station index and references notes.
DIVISIONS
1 NORTHWEST
2 NORTH CENTRAL
3 NORTHEAST
4 WEST CENTRAL
5 CENTRAL
6 EAST CENTRAL
7 SOUTHWEST
8 SOUTH CENTRAL
9 SOUTHEAST
42.0
41.0
40.0
39.0
38.0
-88.0
-87.0
-86.0
-85.0
US DOC - NOAA - NCDC - ASHEVILLE, NC
Updated January 1992
INDIANA

## ANDERSON SEWAGE PLT *Madison County* ELEVATION 889 ft LAT/LONG 40° 6 ' N / 85° 41 ' W

| | JAN | FEB | MAR | APR | MAY | JUN | JUL | AUG | SEP | OCT | NOV | DEC | YEAR |
|---|---|---|---|---|---|---|---|---|---|---|---|---|---|
| Maximum Temp °F | 33.1 | 37.2 | 48.2 | 60.5 | 71.2 | 80.4 | 83.6 | 81.6 | 75.1 | 63.6 | 50.1 | 37.9 | 60.2 |
| Minimum Temp °F | 18.1 | 21.4 | 31.1 | 40.8 | 50.4 | 59.9 | 64.0 | 61.6 | 54.6 | 43.3 | 34.8 | 24.5 | 42.0 |
| Mean Temp °F | 25.6 | 29.3 | 39.6 | 50.6 | 60.8 | 70.2 | 73.8 | 71.6 | 64.8 | 53.5 | 42.5 | 31.2 | 51.1 |
| Days Max Temp ≥ 90 °F | 0 | 0 | 0 | 0 | 0 | 3 | 5 | 3 | 1 | 0 | 0 | 0 | 12 |
| Days Max Temp ≤ 32 °F | 14 | 10 | 4 | 0 | 0 | 0 | 0 | 0 | 0 | 0 | 1 | 9 | 38 |
| Days Min Temp ≤ 32 °F | 28 | 23 | 19 | 7 | 1 | 0 | 0 | 0 | 0 | 4 | 14 | 24 | 120 |
| Days Min Temp ≤ 0 °F | 4 | 2 | 0 | 0 | 0 | 0 | 0 | 0 | 0 | 0 | 0 | 1 | 7 |
| Heating Degree Days | 1213 | 1001 | 780 | 433 | 183 | 26 | 4 | 11 | 99 | 365 | 669 | 1041 | 5825 |
| Cooling Degree Days | 0 | 0 | 2 | 12 | 72 | 201 | 312 | 238 | 108 | 14 | 1 | 0 | 960 |
| Total Precipitation (") | 2.03 | 2.27 | 3.12 | 3.84 | 3.91 | 3.67 | 4.17 | 3.28 | 3.22 | 2.76 | 3.77 | 3.12 | 39.16 |
| Days ≥ 0.1" Precip | 5 | 5 | 7 | 8 | 8 | 6 | 6 | 6 | 6 | 6 | 7 | 7 | 77 |
| Total Snowfall (") | 5.7 | *5.5* | 2.4 | 0.2 | 0.0 | 0.0 | 0.0 | 0.0 | 0.0 | 0.1 | 0.9 | 3.5 | 18.3 |
| Days ≥ 1" Snow Depth | *9* | *6* | 2 | 0 | 0 | 0 | 0 | 0 | 0 | 0 | 0 | *3* | 20 |

## ANGOLA *Steuben County* ELEVATION 1060 ft LAT/LONG 41° 38 ' N / 85° 0 ' W

| | JAN | FEB | MAR | APR | MAY | JUN | JUL | AUG | SEP | OCT | NOV | DEC | YEAR |
|---|---|---|---|---|---|---|---|---|---|---|---|---|---|
| Maximum Temp °F | 29.2 | 32.0 | *42.9* | *56.8* | *69.0* | 78.3 | 82.1 | 80.1 | 73.4 | 60.7 | 46.9 | 34.7 | 57.2 |
| Minimum Temp °F | 13.3 | 13.9 | *23.9* | *35.7* | *46.6* | 56.0 | 60.7 | 58.4 | 51.2 | 39.3 | 30.3 | 20.4 | 37.5 |
| Mean Temp °F | 21.3 | 22.9 | *33.4* | *46.3* | *57.9* | 67.2 | 71.4 | 69.3 | 62.3 | 50.0 | 38.6 | 27.5 | 47.3 |
| Days Max Temp ≥ 90 °F | 0 | 0 | *0* | *0* | 0 | 2 | 4 | 2 | 0 | 0 | 0 | 0 | 8 |
| Days Max Temp ≤ 32 °F | 18 | 15 | *6* | *0* | 0 | 0 | 0 | 0 | 0 | 0 | 3 | 13 | 55 |
| Days Min Temp ≤ 32 °F | 30 | 27 | *26* | *12* | 2 | 0 | 0 | 0 | 0 | 8 | 19 | 28 | 152 |
| Days Min Temp ≤ 0 °F | 5 | 5 | *1* | *0* | 0 | 0 | 0 | 0 | 0 | 0 | 0 | 2 | 13 |
| Heating Degree Days | 1349 | 1180 | *972* | *559* | *252* | 59 | 14 | 27 | 141 | 463 | 785 | 1154 | 6955 |
| Cooling Degree Days | 0 | 0 | 0 | 5 | 40 | 129 | 228 | 165 | 59 | 2 | 0 | 0 | 628 |
| Total Precipitation (") | 1.82 | 1.60 | *2.51* | *3.54* | *3.54* | 3.58 | 3.97 | 3.66 | 3.40 | 2.80 | 3.30 | 3.03 | 36.75 |
| Days ≥ 0.1" Precip | 5 | 4 | *6* | *8* | *7* | *6* | 7 | 7 | 7 | 7 | *7* | 7 | 78 |
| Total Snowfall (") | *8.4* | *8.5* | *4.8* | *1.0* | 0.1 | 0.0 | 0.0 | 0.0 | 0.1 | 0.3 | 2.3 | *8.0* | 33.5 |
| Days ≥ 1" Snow Depth | 19 | 16 | *8* | *1* | 0 | 0 | 0 | 0 | 0 | 0 | 2 | 9 | 55 |

## BERNE *Adams County* ELEVATION 860 ft LAT/LONG 40° 40 ' N / 84° 57 ' W

| | JAN | FEB | MAR | APR | MAY | JUN | JUL | AUG | SEP | OCT | NOV | DEC | YEAR |
|---|---|---|---|---|---|---|---|---|---|---|---|---|---|
| Maximum Temp °F | 32.1 | 36.2 | 48.2 | 60.9 | 72.2 | 81.3 | 84.9 | 82.7 | 76.1 | 63.6 | 49.7 | 37.7 | 60.5 |
| Minimum Temp °F | 16.9 | 19.5 | 29.5 | 39.8 | 50.0 | 59.5 | 63.7 | 61.3 | 54.5 | 43.0 | 34.0 | 23.7 | 41.3 |
| Mean Temp °F | 24.5 | 27.9 | 38.9 | 50.4 | 61.1 | 70.4 | 74.3 | 72.1 | 65.3 | 53.3 | 41.8 | 30.8 | 50.9 |
| Days Max Temp ≥ 90 °F | 0 | 0 | 0 | 0 | 1 | 4 | 7 | 4 | 1 | 0 | 0 | 0 | 17 |
| Days Max Temp ≤ 32 °F | 16 | 11 | 3 | 0 | 0 | 0 | 0 | 0 | 0 | 0 | 1 | 9 | 40 |
| Days Min Temp ≤ 32 °F | 28 | 24 | 20 | 7 | 1 | 0 | 0 | 0 | 0 | 4 | 15 | 25 | 124 |
| Days Min Temp ≤ 0 °F | 4 | 3 | 0 | 0 | 0 | 0 | 0 | 0 | 0 | 0 | 0 | 1 | 8 |
| Heating Degree Days | 1250 | 1044 | 806 | 444 | 180 | 26 | 3 | 9 | 92 | 369 | 691 | 1057 | 5971 |
| Cooling Degree Days | 0 | 0 | 2 | 11 | 68 | 197 | 315 | 245 | 110 | 12 | 0 | 0 | 960 |
| Total Precipitation (") | 2.05 | 2.01 | 2.86 | 3.67 | 3.68 | 4.27 | 3.74 | 3.41 | 3.35 | 2.63 | 3.34 | 2.93 | 37.94 |
| Days ≥ 0.1" Precip | 5 | 5 | 7 | 8 | 8 | 8 | 6 | 6 | 6 | 6 | 7 | 6 | 78 |
| Total Snowfall (") | 8.1 | 8.1 | 5.3 | 1.1 | 0.0 | 0.0 | 0.0 | 0.0 | 0.0 | 0.4 | 2.5 | 6.3 | 31.8 |
| Days ≥ 1" Snow Depth | 15 | 13 | 5 | 0 | 0 | 0 | 0 | 0 | 0 | 0 | 1 | 7 | 41 |

## BLOOMINGTON IND UN *Monroe County* ELEVATION 830 ft LAT/LONG 39° 10 ' N / 86° 31 ' W

| | JAN | FEB | MAR | APR | MAY | JUN | JUL | AUG | SEP | OCT | NOV | DEC | YEAR |
|---|---|---|---|---|---|---|---|---|---|---|---|---|---|
| Maximum Temp °F | 36.3 | 41.1 | 52.2 | 64.2 | 73.9 | 82.0 | 86.1 | 84.0 | 77.7 | 66.5 | 53.7 | 41.7 | 63.3 |
| Minimum Temp °F | 19.0 | 22.1 | 32.1 | 42.7 | 52.4 | 61.7 | 66.1 | 63.4 | 56.4 | 44.6 | 35.8 | 25.3 | 43.5 |
| Mean Temp °F | 27.7 | 31.6 | 42.1 | 53.5 | 63.2 | 71.9 | 76.2 | 73.7 | 67.1 | 55.6 | 44.8 | 33.4 | 53.4 |
| Days Max Temp ≥ 90 °F | 0 | 0 | 0 | 0 | 0 | 3 | 8 | 5 | 2 | 0 | 0 | 0 | 18 |
| Days Max Temp ≤ 32 °F | 12 | 7 | 2 | 0 | 0 | 0 | 0 | 0 | 0 | 0 | 1 | 6 | 28 |
| Days Min Temp ≤ 32 °F | 27 | 23 | 17 | 4 | 0 | 0 | 0 | 0 | 0 | 3 | 13 | 23 | 110 |
| Days Min Temp ≤ 0 °F | 3 | 1 | 0 | 0 | 0 | 0 | 0 | 0 | 0 | 0 | 0 | 1 | 5 |
| Heating Degree Days | 1149 | 934 | 705 | 358 | 130 | 15 | 1 | 4 | 66 | 306 | 602 | 973 | 5243 |
| Cooling Degree Days | 0 | 0 | 3 | 19 | 79 | 238 | 367 | 296 | 135 | 20 | 1 | 0 | 1158 |
| Total Precipitation (") | 2.48 | 2.60 | 3.56 | 4.05 | 4.66 | 3.60 | 4.72 | 4.03 | 3.82 | 3.17 | 4.02 | 3.56 | 44.27 |
| Days ≥ 0.1" Precip | 6 | 5 | 7 | 8 | 8 | 7 | 7 | 6 | 6 | 5 | 7 | 7 | 79 |
| Total Snowfall (") | na | na | na | 0.0 | 0.0 | 0.0 | 0.0 | 0.0 | 0.0 | 0.2 | *0.2* | *0.8* | na |
| Days ≥ 1" Snow Depth | na | na | na | 0 | 0 | 0 | 0 | 0 | 0 | 0 | 0 | na | na |

## BROOKVILLE *Franklin County* ELEVATION 630 ft LAT/LONG 39° 25 ' N / 85° 1 ' W

| | JAN | FEB | MAR | APR | MAY | JUN | JUL | AUG | SEP | OCT | NOV | DEC | YEAR |
|---|---|---|---|---|---|---|---|---|---|---|---|---|---|
| Maximum Temp °F | 36.2 | 40.6 | 52.0 | 64.1 | 74.2 | 83.0 | 86.5 | 84.6 | 78.4 | 66.3 | 53.1 | 41.5 | 63.4 |
| Minimum Temp °F | 16.7 | 18.9 | 28.1 | 37.9 | 47.7 | 57.1 | 61.9 | 59.7 | 52.2 | 39.6 | 31.8 | 23.2 | 39.6 |
| Mean Temp °F | 26.5 | 29.8 | 40.1 | 51.0 | 61.0 | 70.1 | 74.2 | 72.2 | 65.3 | 53.0 | 42.5 | 32.4 | 51.5 |
| Days Max Temp ≥ 90 °F | 0 | 0 | 0 | 0 | 1 | 5 | 9 | 7 | 2 | 0 | 0 | 0 | 24 |
| Days Max Temp ≤ 32 °F | 12 | 8 | 2 | 0 | 0 | 0 | 0 | 0 | 0 | 0 | 1 | 6 | 29 |
| Days Min Temp ≤ 32 °F | 28 | 24 | 21 | 10 | 1 | 0 | 0 | 0 | 0 | 8 | 17 | 25 | 134 |
| Days Min Temp ≤ 0 °F | 4 | 3 | 0 | 0 | 0 | 0 | 0 | 0 | 0 | 0 | 0 | 1 | 8 |
| Heating Degree Days | 1187 | 986 | 765 | 420 | 175 | 25 | 3 | 9 | 90 | 378 | 668 | 1005 | 5711 |
| Cooling Degree Days | 0 | 0 | 1 | 9 | 59 | 189 | 317 | 252 | 107 | 13 | 1 | 0 | 948 |
| Total Precipitation (") | 2.70 | 2.52 | 3.56 | 3.85 | 4.61 | 3.71 | 4.72 | 3.93 | 2.85 | 3.10 | 3.74 | 3.31 | 42.60 |
| Days ≥ 0.1" Precip | 6 | 5 | 8 | 8 | 8 | 7 | 7 | 6 | 6 | 6 | 7 | 7 | 81 |
| Total Snowfall (") | na | ***4.0*** | 2.4 | 0.3 | 0.0 | 0.0 | 0.0 | 0.0 | 0.0 | 0.1 | 1.5 | 2.7 | na |
| Days ≥ 1" Snow Depth | ***10*** | 7 | 2 | 0 | 0 | 0 | 0 | 0 | 0 | 0 | 1 | 3 | 23 |

## CAMBRIDGE CITY 3 N *Wayne County* ELEVATION 1000 ft LAT/LONG 39° 52 ' N / 85° 11 ' W

| | JAN | FEB | MAR | APR | MAY | JUN | JUL | AUG | SEP | OCT | NOV | DEC | YEAR |
|---|---|---|---|---|---|---|---|---|---|---|---|---|---|
| Maximum Temp °F | 33.1 | 37.1 | 48.7 | 61.2 | 71.6 | 80.4 | 83.9 | 82.1 | 76.0 | 64.2 | 50.8 | 38.9 | 60.7 |
| Minimum Temp °F | 14.9 | 17.2 | 27.6 | 37.4 | 48.0 | 57.2 | 61.2 | 58.5 | 51.2 | 38.7 | 31.4 | 22.0 | 38.8 |
| Mean Temp °F | 24.0 | 27.2 | 38.2 | 49.3 | 59.8 | 68.8 | 72.6 | 70.3 | 63.7 | 51.5 | 41.1 | 30.4 | 49.7 |
| Days Max Temp ≥ 90 °F | 0 | 0 | 0 | 0 | 0 | 2 | 6 | 3 | 1 | 0 | 0 | 0 | 12 |
| Days Max Temp ≤ 32 °F | 15 | 10 | 3 | 0 | 0 | 0 | 0 | 0 | 0 | 0 | 1 | 8 | 37 |
| Days Min Temp ≤ 32 °F | 29 | 25 | 22 | 10 | 1 | 0 | 0 | 0 | 0 | 9 | 18 | 26 | 140 |
| Days Min Temp ≤ 0 °F | 6 | 4 | 0 | 0 | 0 | 0 | 0 | 0 | 0 | 0 | 0 | 2 | 12 |
| Heating Degree Days | 1265 | 1062 | 826 | 469 | 202 | 37 | 6 | 18 | 116 | 418 | 711 | 1065 | 6195 |
| Cooling Degree Days | 0 | 0 | 1 | 5 | 42 | 151 | 245 | 185 | 76 | 8 | 0 | 0 | 713 |
| Total Precipitation (") | 2.21 | 2.26 | 3.29 | 3.99 | 4.73 | 3.87 | 4.34 | 3.65 | 2.92 | 2.85 | 3.73 | 3.14 | 40.98 |
| Days ≥ 0.1" Precip | 6 | 5 | 7 | 8 | 8 | 7 | 7 | 6 | 6 | 6 | 7 | 7 | 80 |
| Total Snowfall (") | 5.9 | 5.9 | 3.6 | 0.4 | 0.0 | 0.0 | 0.0 | 0.0 | 0.0 | 0.1 | 1.4 | 4.2 | 21.5 |
| Days ≥ 1" Snow Depth | 12 | 10 | 4 | 0 | 0 | 0 | 0 | 0 | 0 | 0 | 1 | 5 | 32 |

## CHARLESTOWN ORD PLNT *Clark County* ELEVATION 525 ft LAT/LONG 38° 22 ' N / 85° 41 ' W

| | JAN | FEB | MAR | APR | MAY | JUN | JUL | AUG | SEP | OCT | NOV | DEC | YEAR |
|---|---|---|---|---|---|---|---|---|---|---|---|---|---|
| Maximum Temp °F | 41.5 | 46.4 | ***57.3*** | ***68.3*** | ***77.4*** | ***85.4*** | 88.8 | 87.1 | 80.1 | 69.9 | ***56.5*** | na | na |
| Minimum Temp °F | 22.7 | 26.0 | ***34.6*** | ***44.1*** | ***53.4*** | ***62.1*** | 66.8 | 64.6 | 57.0 | 45.3 | ***37.0*** | ***28.1*** | 45.1 |
| Mean Temp °F | 32.3 | 36.2 | ***46.2*** | ***56.1*** | ***65.5*** | ***73.7*** | 77.8 | 75.9 | 68.6 | 57.7 | ***46.7*** | na | na |
| Days Max Temp ≥ 90 °F | 0 | 0 | 0 | 0 | ***1*** | ***7*** | 14 | 11 | 3 | 0 | 0 | 0 | 36 |
| Days Max Temp ≤ 32 °F | 7 | 4 | 0 | 0 | ***0*** | 0 | 0 | 0 | 0 | 0 | 0 | ***3*** | 14 |
| Days Min Temp ≤ 32 °F | 24 | 21 | 14 | ***3*** | ***0*** | 0 | 0 | 0 | 0 | 3 | ***10*** | ***20*** | 95 |
| Days Min Temp ≤ 0 °F | 2 | 1 | 0 | 0 | ***0*** | 0 | 0 | 0 | 0 | 0 | 0 | ***0*** | 3 |
| Heating Degree Days | 1005 | 807 | ***580*** | ***286*** | ***87*** | ***5*** | 0 | 2 | 47 | 249 | ***544*** | na | na |
| Cooling Degree Days | 0 | 0 | 4 | 25 | 122 | ***284*** | 434 | 365 | 172 | 32 | 2 | 0 | 1440 |
| Total Precipitation (") | 2.86 | 2.91 | 4.18 | 4.42 | 4.63 | 3.78 | 4.39 | 4.26 | 3.34 | 3.08 | 3.90 | 3.81 | 45.56 |
| Days ≥ 0.1" Precip | ***6*** | 6 | ***8*** | 8 | ***8*** | 6 | 7 | 6 | 6 | 6 | ***8*** | ***8*** | 83 |
| Total Snowfall (") | na | na | na | 0.0 | 0.0 | 0.0 | 0.0 | 0.0 | 0.0 | 0.0 | na | na | na |
| Days ≥ 1" Snow Depth | na | na | na | 0 | 0 | 0 | 0 | 0 | 0 | 0 | na | na | na |

## COLUMBIA CITY *Whitley County* ELEVATION 879 ft LAT/LONG 41° 9 ' N / 85° 30 ' W

| | JAN | FEB | MAR | APR | MAY | JUN | JUL | AUG | SEP | OCT | NOV | DEC | YEAR |
|---|---|---|---|---|---|---|---|---|---|---|---|---|---|
| Maximum Temp °F | 30.3 | 34.0 | 45.7 | 58.8 | 70.4 | 79.6 | 83.2 | 81.2 | 74.7 | 62.2 | 48.4 | 36.1 | 58.7 |
| Minimum Temp °F | 13.9 | 16.0 | 26.2 | 36.7 | 47.1 | 56.5 | 60.7 | 58.1 | 51.1 | 39.4 | 31.5 | 21.3 | 38.2 |
| Mean Temp °F | 22.2 | 25.0 | 36.0 | 47.8 | 58.8 | 68.1 | 72.0 | 69.7 | 62.9 | 50.8 | 40.0 | 28.7 | 48.5 |
| Days Max Temp ≥ 90 °F | 0 | 0 | 0 | 0 | 0 | 2 | 5 | 2 | 1 | 0 | 0 | 0 | 10 |
| Days Max Temp ≤ 32 °F | 17 | 13 | 4 | 0 | 0 | 0 | 0 | 0 | 0 | 0 | 2 | 11 | 47 |
| Days Min Temp ≤ 32 °F | 29 | 26 | 23 | 10 | 1 | 0 | 0 | 0 | 0 | 7 | 18 | 27 | 141 |
| Days Min Temp ≤ 0 °F | 6 | 4 | 0 | 0 | 0 | 0 | 0 | 0 | 0 | 0 | 0 | 2 | 12 |
| Heating Degree Days | 1321 | 1121 | 893 | 516 | 227 | 46 | 9 | 22 | 131 | 439 | 744 | 1119 | 6588 |
| Cooling Degree Days | 0 | 0 | 0 | 6 | 41 | 147 | 247 | 180 | 72 | 4 | 0 | 0 | 697 |
| Total Precipitation (") | 2.12 | 1.76 | 2.78 | 3.71 | 3.55 | 4.25 | 3.69 | 3.46 | 3.67 | 2.85 | 3.63 | 3.18 | 38.65 |
| Days ≥ 0.1" Precip | 5 | 5 | 7 | 8 | 7 | 7 | 7 | 7 | 7 | 7 | 7 | 7 | 81 |
| Total Snowfall (") | 8.8 | 8.3 | 4.8 | 1.1 | 0.0 | 0.0 | 0.0 | 0.0 | 0.0 | 0.2 | 2.1 | 7.2 | 32.5 |
| Days ≥ 1" Snow Depth | 17 | 14 | 6 | 1 | 0 | 0 | 0 | 0 | 0 | 0 | 2 | 10 | 50 |

## COLUMBUS *Bartholomew County* ELEVATION 630 ft LAT/LONG 39° 13' N / 85° 54' W

| | JAN | FEB | MAR | APR | MAY | JUN | JUL | AUG | SEP | OCT | NOV | DEC | YEAR |
|---|---|---|---|---|---|---|---|---|---|---|---|---|---|
| Maximum Temp °F | 36.2 | 40.6 | 52.0 | 64.1 | 73.8 | 82.4 | 86.0 | 84.3 | 78.3 | 66.3 | 53.4 | 41.7 | 63.3 |
| Minimum Temp °F | 18.0 | 20.6 | 30.5 | 40.8 | 50.8 | 60.2 | 64.4 | 61.6 | 54.3 | 41.5 | 33.7 | 24.6 | 41.8 |
| Mean Temp °F | 27.1 | 30.6 | 41.3 | 52.5 | 62.3 | 71.3 | 75.2 | 73.0 | 66.3 | 53.9 | 43.6 | 33.2 | 52.5 |
| Days Max Temp ≥ 90 °F | 0 | 0 | 0 | 0 | 0 | 4 | 8 | 6 | 2 | 0 | 0 | 0 | 20 |
| Days Max Temp ≤ 32 °F | 12 | 7 | 1 | 0 | 0 | 0 | 0 | 0 | 0 | 0 | 1 | 6 | 27 |
| Days Min Temp ≤ 32 °F | 27 | 24 | 19 | 6 | 0 | 0 | 0 | 0 | 0 | 6 | 15 | 24 | 121 |
| Days Min Temp ≤ 0 °F | 4 | 2 | 0 | 0 | 0 | 0 | 0 | 0 | 0 | 0 | 0 | 1 | 7 |
| Heating Degree Days | 1168 | 964 | 730 | 381 | 144 | 18 | 2 | 5 | 75 | 351 | 636 | 979 | 5453 |
| Cooling Degree Days | 0 | 0 | 1 | 14 | 71 | 229 | 350 | 277 | 123 | 15 | 0 | 0 | 1080 |
| Total Precipitation (") | 2.48 | 2.47 | 3.58 | 4.26 | 4.45 | 2.81 | 4.31 | 3.63 | 2.98 | 2.86 | 3.86 | 3.28 | 40.97 |
| Days ≥ 0.1" Precip | 5 | 5 | 8 | 8 | 8 | 6 | 7 | 6 | 6 | 5 | 7 | 7 | 78 |
| Total Snowfall (") | na | *4.6* | *1.6* | 0.0 | 0.0 | 0.0 | 0.0 | 0.0 | 0.0 | 0.1 | 0.9 | *1.7* | na |
| Days ≥ 1" Snow Depth | na | na | *1* | 0 | 0 | 0 | 0 | 0 | 0 | 0 | *0* | na | na |

## CRANE NAVAL DEPOT *Martin County* ELEVATION 761 ft LAT/LONG 38° 52' N / 86° 50' W

| | JAN | FEB | MAR | APR | MAY | JUN | JUL | AUG | SEP | OCT | NOV | DEC | YEAR |
|---|---|---|---|---|---|---|---|---|---|---|---|---|---|
| Maximum Temp °F | 39.0 | 44.1 | 55.8 | 67.8 | 76.6 | 84.5 | 88.0 | 86.2 | 80.2 | 69.1 | 56.0 | *44.6* | 66.0 |
| Minimum Temp °F | 21.4 | 24.7 | 34.2 | 44.5 | 53.4 | 62.0 | 66.5 | 64.4 | 57.9 | 46.0 | 37.1 | *27.3* | 44.9 |
| Mean Temp °F | 30.2 | 34.4 | 45.0 | 56.2 | 65.0 | 73.3 | 77.3 | 75.3 | 69.1 | 57.6 | 46.6 | *36.0* | 55.5 |
| Days Max Temp ≥ 90 °F | 0 | 0 | 0 | 0 | 1 | 5 | 12 | 7 | 2 | 0 | 0 | 0 | 27 |
| Days Max Temp ≤ 32 °F | 9 | 6 | 1 | 0 | 0 | 0 | 0 | 0 | 0 | 0 | 0 | 4 | 20 |
| Days Min Temp ≤ 32 °F | 25 | 21 | 14 | 4 | 0 | 0 | 0 | 0 | 0 | 2 | 11 | 20 | 97 |
| Days Min Temp ≤ 0 °F | 2 | 1 | 0 | 0 | 0 | 0 | 0 | 0 | 0 | 0 | 0 | 1 | 4 |
| Heating Degree Days | 1073 | 856 | 607 | 286 | 97 | 8 | 1 | 1 | 42 | 255 | 548 | *891* | 4665 |
| Cooling Degree Days | 0 | 0 | 6 | 28 | 101 | 263 | 406 | 332 | 168 | 31 | 2 | *0* | 1337 |
| Total Precipitation (") | 2.81 | 2.66 | 4.01 | 4.41 | 4.81 | 3.47 | 5.29 | 4.01 | 3.51 | 3.49 | 4.17 | 3.78 | 46.42 |
| Days ≥ 0.1" Precip | 5 | 5 | 7 | 8 | 8 | 6 | 7 | 5 | 5 | 5 | 6 | 6 | 73 |
| Total Snowfall (") | na | na | *1.4* | 0.0 | 0.0 | 0.0 | 0.0 | 0.0 | 0.0 | 0.1 | *0.2* | na | na |
| Days ≥ 1" Snow Depth | na | na | *0* | 0 | 0 | 0 | 0 | 0 | 0 | 0 | 0 | na | na |

## DELPHI 3 NNE *Carroll County* ELEVATION 669 ft LAT/LONG 40° 35' N / 86° 40' W

| | JAN | FEB | MAR | APR | MAY | JUN | JUL | AUG | SEP | OCT | NOV | DEC | YEAR |
|---|---|---|---|---|---|---|---|---|---|---|---|---|---|
| Maximum Temp °F | 33.0 | 37.5 | 50.4 | 63.7 | 74.3 | 82.9 | 85.8 | 83.4 | 77.4 | 65.2 | 51.2 | 38.7 | 62.0 |
| Minimum Temp °F | 16.6 | 19.7 | 30.6 | 39.9 | 49.7 | 59.1 | 62.8 | 60.4 | 53.7 | 42.4 | 33.9 | 23.6 | 41.0 |
| Mean Temp °F | 24.8 | 28.6 | 40.5 | 51.8 | 62.0 | 71.0 | 74.3 | 71.9 | 65.6 | 53.8 | 42.6 | 31.2 | 51.5 |
| Days Max Temp ≥ 90 °F | 0 | 0 | 0 | 0 | 1 | 5 | 8 | 4 | 1 | 0 | 0 | 0 | 19 |
| Days Max Temp ≤ 32 °F | 14 | 10 | 2 | 0 | 0 | 0 | 0 | 0 | 0 | 0 | 1 | 8 | 35 |
| Days Min Temp ≤ 32 °F | 28 | 24 | 19 | 8 | 1 | 0 | 0 | 0 | 0 | 5 | 15 | 25 | 125 |
| Days Min Temp ≤ 0 °F | 5 | 3 | 0 | 0 | 0 | 0 | 0 | 0 | 0 | 0 | 0 | 1 | 9 |
| Heating Degree Days | 1240 | 1020 | 754 | 401 | 154 | 19 | 3 | 8 | 84 | 352 | 667 | 1042 | 5744 |
| Cooling Degree Days | 0 | 0 | 2 | 13 | 73 | 218 | 319 | 248 | 114 | 13 | 1 | 0 | 1001 |
| Total Precipitation (") | 1.89 | 1.82 | 2.70 | 3.76 | 3.94 | 3.91 | 4.02 | 4.18 | 3.13 | 2.71 | 3.27 | 3.04 | 38.37 |
| Days ≥ 0.1" Precip | 5 | 4 | 6 | 8 | 7 | 6 | 6 | 6 | 5 | 6 | 6 | 6 | 71 |
| Total Snowfall (") | 5.7 | 5.3 | 2.9 | 0.8 | 0.0 | 0.0 | 0.0 | 0.0 | 0.0 | 0.2 | 1.1 | 5.5 | 21.5 |
| Days ≥ 1" Snow Depth | 12 | 10 | 3 | 0 | 0 | 0 | 0 | 0 | 0 | 0 | 1 | 6 | 32 |

## DUBOIS S IND FORAGE *Dubois County* ELEVATION 689 ft LAT/LONG 38° 27' N / 86° 42' W

| | JAN | FEB | MAR | APR | MAY | JUN | JUL | AUG | SEP | OCT | NOV | DEC | YEAR |
|---|---|---|---|---|---|---|---|---|---|---|---|---|---|
| Maximum Temp °F | 37.0 | 41.8 | 53.1 | 64.8 | 73.7 | 82.1 | 85.9 | 84.4 | 78.0 | 66.6 | 54.1 | 42.8 | 63.7 |
| Minimum Temp °F | 19.3 | 22.1 | 32.7 | 43.5 | 52.1 | 61.1 | 65.4 | 63.2 | 56.5 | 43.6 | 35.4 | 25.3 | 43.4 |
| Mean Temp °F | 28.2 | 32.0 | 42.9 | 54.2 | 62.9 | 71.6 | 75.6 | 73.8 | 67.3 | 55.1 | 44.8 | 34.1 | 53.5 |
| Days Max Temp ≥ 90 °F | 0 | 0 | 0 | 0 | 0 | 3 | 8 | 6 | 2 | 0 | 0 | 0 | 19 |
| Days Max Temp ≤ 32 °F | 11 | 7 | 2 | 0 | 0 | 0 | 0 | 0 | 0 | 0 | 1 | 5 | 26 |
| Days Min Temp ≤ 32 °F | 27 | 22 | 17 | 4 | 0 | 0 | 0 | 0 | 0 | 4 | 14 | 22 | 110 |
| Days Min Temp ≤ 0 °F | 3 | 2 | 0 | 0 | 0 | 0 | 0 | 0 | 0 | 0 | 0 | 1 | 6 |
| Heating Degree Days | 1135 | 926 | 681 | 341 | 134 | 18 | 2 | 5 | 68 | 321 | 601 | 951 | 5183 |
| Cooling Degree Days | 0 | 0 | 4 | 24 | 82 | 238 | 370 | 309 | 151 | 25 | 1 | 0 | 1204 |
| Total Precipitation (") | 2.76 | 2.61 | 3.99 | 4.37 | 4.67 | 4.13 | 4.49 | 4.24 | 3.72 | 3.32 | 4.34 | 3.46 | 46.10 |
| Days ≥ 0.1" Precip | 5 | 5 | 7 | 8 | 8 | 7 | 7 | 6 | 6 | 6 | 7 | 6 | 78 |
| Total Snowfall (") | na | na | na | 0.0 | 0.0 | 0.0 | 0.0 | 0.0 | 0.0 | 0.2 | 0.3 | na | na |
| Days ≥ 1" Snow Depth | na | na | *1* | 0 | 0 | 0 | 0 | 0 | 0 | 0 | 0 | na | na |

## ELWOOD WASTEWATER PL *Madison County* ELEVATION 853 ft LAT/LONG 40° 16 ' N / 85° 50 ' W

| | JAN | FEB | MAR | APR | MAY | JUN | JUL | AUG | SEP | OCT | NOV | DEC | YEAR |
|---|---|---|---|---|---|---|---|---|---|---|---|---|---|
| Maximum Temp °F | 32.0 | 36.2 | 47.8 | 61.0 | 71.9 | 81.2 | 84.8 | 82.6 | 76.6 | 64.5 | 50.4 | 37.9 | 60.6 |
| Minimum Temp °F | 14.5 | 17.1 | 27.3 | 37.6 | 47.9 | 57.3 | 61.3 | 58.5 | 51.5 | 39.5 | 31.7 | 21.8 | 38.8 |
| Mean Temp °F | 23.3 | 26.7 | 37.6 | 49.3 | 60.0 | 69.3 | 73.1 | 70.6 | 64.1 | 52.0 | 41.1 | 29.9 | 49.8 |
| Days Max Temp ≥ 90 °F | 0 | 0 | 0 | 0 | 0 | 4 | 7 | 4 | 1 | 0 | 0 | 0 | 16 |
| Days Max Temp ≤ 32 °F | 15 | 11 | 3 | 0 | 0 | 0 | 0 | 0 | 0 | 0 | 1 | 9 | 39 |
| Days Min Temp ≤ 32 °F | 29 | 26 | 22 | 10 | 1 | 0 | 0 | 0 | 0 | 9 | 17 | 26 | 140 |
| Days Min Temp ≤ 0 °F | 5 | 4 | 0 | 0 | 0 | 0 | 0 | 0 | 0 | 0 | 0 | 1 | 10 |
| Heating Degree Days | 1288 | 1076 | 844 | 472 | 201 | 33 | 5 | 17 | 110 | 406 | 712 | 1082 | 6246 |
| Cooling Degree Days | 0 | 0 | 1 | 8 | 52 | 177 | 278 | 206 | 93 | 10 | 0 | 0 | 825 |
| Total Precipitation (") | 2.13 | 1.92 | 2.98 | 3.78 | 3.94 | 3.87 | 4.22 | 3.89 | 3.67 | 2.79 | 3.82 | 3.22 | 40.23 |
| Days ≥ 0.1" Precip | 5 | 4 | 7 | 8 | 8 | 7 | 7 | 6 | 6 | 6 | 7 | 7 | 78 |
| Total Snowfall (") | na | na | na | *0.0* | 0.0 | 0.0 | 0.0 | 0.0 | 0.0 | 0.0 | *0.6* | na | na |
| Days ≥ 1" Snow Depth | na | na | na | 0 | 0 | 0 | 0 | 0 | 0 | 0 | *0* | na | na |

## EVANSVILLE MUSEUM *Vanderburgh County* ELEVATION 390 ft LAT/LONG 37° 58 ' N / 87° 33 ' W

| | JAN | FEB | MAR | APR | MAY | JUN | JUL | AUG | SEP | OCT | NOV | DEC | YEAR |
|---|---|---|---|---|---|---|---|---|---|---|---|---|---|
| Maximum Temp °F | 40.5 | 46.2 | 57.9 | 69.6 | 78.2 | 86.7 | 89.6 | 88.1 | 81.5 | 70.2 | 57.1 | 45.8 | 67.6 |
| Minimum Temp °F | 24.4 | 27.9 | 37.4 | 47.1 | 55.7 | 64.5 | 68.7 | 66.7 | 59.8 | 47.4 | 39.2 | 30.0 | 47.4 |
| Mean Temp °F | 32.5 | 37.1 | 47.6 | 58.3 | 66.9 | 75.6 | 79.2 | 77.4 | 70.7 | 58.8 | 48.2 | 37.9 | 57.5 |
| Days Max Temp ≥ 90 °F | 0 | 0 | 0 | 0 | 2 | 10 | 16 | 13 | 5 | 0 | 0 | 0 | 46 |
| Days Max Temp ≤ 32 °F | 8 | 4 | 0 | 0 | 0 | 0 | 0 | 0 | 0 | 0 | 0 | 4 | 16 |
| Days Min Temp ≤ 32 °F | 23 | 18 | 11 | 2 | 0 | 0 | 0 | 0 | 0 | 1 | 8 | 18 | 81 |
| Days Min Temp ≤ 0 °F | 1 | 0 | 0 | 0 | 0 | 0 | 0 | 0 | 0 | 0 | 0 | 0 | 1 |
| Heating Degree Days | 1002 | 782 | 537 | 232 | 65 | 3 | 0 | 1 | 31 | 222 | 502 | 833 | 4210 |
| Cooling Degree Days | 0 | 0 | 7 | 39 | 139 | 344 | 470 | 419 | 217 | 39 | 3 | 1 | 1678 |
| Total Precipitation (") | 3.00 | 3.36 | 4.44 | 4.28 | 4.30 | 3.68 | 4.64 | 3.25 | 3.28 | 3.15 | 4.43 | 3.77 | 45.58 |
| Days ≥ 0.1" Precip | 6 | 6 | 8 | 8 | 8 | 6 | 6 | 5 | 5 | 5 | 7 | 7 | 77 |
| Total Snowfall (") | 4.7 | 4.2 | 2.7 | 0.4 | 0.0 | 0.0 | 0.0 | 0.0 | 0.0 | 0.0 | 0.5 | 1.8 | 14.3 |
| Days ≥ 1" Snow Depth | 7 | 6 | 1 | 0 | 0 | 0 | 0 | 0 | 0 | 0 | 0 | 2 | 16 |

## EVANSVILLE REG AP *Vanderburgh County* ELEVATION 387 ft LAT/LONG 38° 2 ' N / 87° 32 ' W

| | JAN | FEB | MAR | APR | MAY | JUN | JUL | AUG | SEP | OCT | NOV | DEC | YEAR |
|---|---|---|---|---|---|---|---|---|---|---|---|---|---|
| Maximum Temp °F | 38.9 | 43.9 | 55.9 | 67.5 | 76.8 | 86.0 | 89.0 | 86.9 | 80.5 | 69.1 | 55.5 | 44.3 | 66.2 |
| Minimum Temp °F | 22.1 | 25.5 | 35.6 | 45.2 | 54.3 | 63.6 | 68.1 | 65.0 | 57.9 | 45.0 | 36.9 | 27.9 | 45.6 |
| Mean Temp °F | 30.6 | 34.7 | 45.7 | 56.4 | 65.6 | 74.8 | 78.5 | 76.0 | 69.2 | 57.1 | 46.3 | 36.2 | 55.9 |
| Days Max Temp ≥ 90 °F | 0 | 0 | 0 | 0 | 2 | 10 | 15 | 11 | 4 | 0 | 0 | 0 | 42 |
| Days Max Temp ≤ 32 °F | 10 | 6 | 1 | 0 | 0 | 0 | 0 | 0 | 0 | 0 | 0 | 5 | 22 |
| Days Min Temp ≤ 32 °F | 25 | 20 | 14 | 3 | 0 | 0 | 0 | 0 | 0 | 3 | 11 | 21 | 97 |
| Days Min Temp ≤ 0 °F | 2 | 1 | 0 | 0 | 0 | 0 | 0 | 0 | 0 | 0 | 0 | 1 | 4 |
| Heating Degree Days | 1061 | 848 | 594 | 279 | 85 | 5 | 0 | 1 | 43 | 265 | 558 | 888 | 4627 |
| Cooling Degree Days | 0 | 0 | 4 | 26 | 115 | 318 | 451 | 373 | 183 | 29 | 2 | 1 | 1502 |
| Total Precipitation (") | 2.74 | 3.02 | 4.15 | 4.10 | 4.37 | 3.54 | 4.15 | 3.06 | 3.08 | 3.10 | 4.09 | 3.62 | 43.02 |
| Days ≥ 0.1" Precip | 5 | 5 | 8 | 8 | 8 | 6 | 6 | 5 | 5 | 6 | 7 | 7 | 76 |
| Total Snowfall (") | 4.7 | 4.5 | 2.9 | 0.4 | 0.0 | 0.0 | 0.0 | 0.0 | 0.0 | 0.2 | 0.6 | 2.3 | 15.6 |
| Days ≥ 1" Snow Depth | 6 | 6 | 1 | 0 | 0 | 0 | 0 | 0 | 0 | 0 | 0 | 2 | 15 |

## FARMLAND 5 NNW *Randolph County* ELEVATION 971 ft LAT/LONG 40° 15 ' N / 85° 9 ' W

| | JAN | FEB | MAR | APR | MAY | JUN | JUL | AUG | SEP | OCT | NOV | DEC | YEAR |
|---|---|---|---|---|---|---|---|---|---|---|---|---|---|
| Maximum Temp °F | 31.7 | 35.5 | 47.2 | 60.0 | 71.1 | 80.3 | 84.0 | 81.8 | 75.8 | 63.6 | 49.9 | 37.8 | 59.9 |
| Minimum Temp °F | 14.9 | 16.9 | 27.3 | 38.0 | 48.5 | 58.0 | 61.9 | 58.8 | 51.8 | 39.9 | 32.1 | 22.0 | 39.2 |
| Mean Temp °F | 23.3 | 26.2 | 37.3 | 49.0 | 59.9 | 69.2 | 73.0 | 70.3 | 63.8 | 51.8 | 41.0 | 29.9 | 49.6 |
| Days Max Temp ≥ 90 °F | 0 | 0 | 0 | 0 | 0 | 3 | 6 | 3 | 1 | 0 | 0 | 0 | 13 |
| Days Max Temp ≤ 32 °F | 15 | 12 | 4 | 0 | 0 | 0 | 0 | 0 | 0 | 0 | 2 | 9 | 42 |
| Days Min Temp ≤ 32 °F | 29 | 26 | 22 | 10 | 1 | 0 | 0 | 0 | 0 | 8 | 17 | 26 | 139 |
| Days Min Temp ≤ 0 °F | 6 | 4 | 0 | 0 | 0 | 0 | 0 | 0 | 0 | 0 | 0 | 2 | 12 |
| Heating Degree Days | 1286 | 1089 | 854 | 479 | 205 | 38 | 7 | 21 | 119 | 414 | 714 | 1080 | 6306 |
| Cooling Degree Days | 0 | 0 | 1 | 9 | 56 | 181 | 283 | 207 | 96 | 12 | 0 | 0 | 845 |
| Total Precipitation (") | 1.77 | 1.70 | 2.61 | 3.41 | 3.87 | 4.21 | 4.00 | 3.64 | 3.22 | 2.71 | 3.32 | 2.69 | 37.15 |
| Days ≥ 0.1" Precip | 4 | 4 | 6 | 8 | 8 | 7 | 7 | 6 | 6 | 6 | 7 | 6 | 75 |
| Total Snowfall (") | 7.5 | 6.8 | 4.4 | 0.6 | 0.0 | 0.0 | 0.0 | 0.0 | 0.0 | 0.2 | 1.6 | 5.6 | 26.7 |
| Days ≥ 1" Snow Depth | 14 | 11 | 5 | 0 | 0 | 0 | 0 | 0 | 0 | 0 | 1 | 6 | 37 |

## FORT WAYNE BAER FLD *Allen County* ELEVATION 804 ft LAT/LONG 41° 0 ' N / 85° 13 ' W

| | JAN | FEB | MAR | APR | MAY | JUN | JUL | AUG | SEP | OCT | NOV | DEC | YEAR |
|---|---|---|---|---|---|---|---|---|---|---|---|---|---|
| Maximum Temp °F | 30.7 | 34.3 | 46.5 | 60.0 | 71.4 | 80.7 | 84.4 | 81.9 | 75.3 | 62.8 | 48.7 | 36.5 | 59.4 |
| Minimum Temp °F | 15.8 | 18.3 | 28.7 | 38.9 | 49.2 | 59.0 | 63.0 | 60.6 | 53.5 | 41.9 | 33.1 | 23.0 | 40.4 |
| Mean Temp °F | 23.3 | 26.3 | 37.6 | 49.4 | 60.3 | 69.9 | 73.7 | 71.3 | 64.4 | 52.3 | 40.9 | 29.8 | 49.9 |
| Days Max Temp ≥ 90 °F | 0 | 0 | 0 | 0 | 1 | 4 | 6 | 4 | 1 | 0 | 0 | 0 | 16 |
| Days Max Temp ≤ 32 °F | 16 | 12 | 4 | 0 | 0 | 0 | 0 | 0 | 0 | 0 | 2 | 11 | 45 |
| Days Min Temp ≤ 32 °F | 28 | 25 | 21 | 8 | 1 | 0 | 0 | 0 | 0 | 5 | 15 | 25 | 128 |
| Days Min Temp ≤ 0 °F | 5 | 3 | 0 | 0 | 0 | 0 | 0 | 0 | 0 | 0 | 0 | 1 | 9 |
| Heating Degree Days | 1287 | 1085 | 842 | 469 | 193 | 28 | 4 | 12 | 104 | 394 | 715 | 1085 | 6218 |
| Cooling Degree Days | 0 | 0 | 1 | 9 | 59 | 189 | 303 | 227 | 93 | 8 | 0 | 0 | 889 |
| Total Precipitation (") | 2.03 | 1.85 | 2.78 | 3.51 | 3.60 | 3.71 | 3.49 | 3.43 | 2.85 | 2.80 | 3.19 | 3.02 | 36.26 |
| Days ≥ 0.1" Precip | 5 | 5 | 7 | 8 | 7 | 7 | 6 | 6 | 6 | 6 | 7 | 7 | 77 |
| Total Snowfall (") | 8.9 | 8.0 | 5.1 | 1.3 | 0.0 | 0.0 | 0.0 | 0.0 | 0.0 | 0.4 | 3.2 | 7.5 | 34.4 |
| Days ≥ 1" Snow Depth | 16 | 13 | 5 | 0 | 0 | 0 | 0 | 0 | 0 | 0 | 2 | 8 | 44 |

## FRANKFORT DISPOSAL P *Clinton County* ELEVATION 830 ft LAT/LONG 40° 19 ' N / 86° 30 ' W

| | JAN | FEB | MAR | APR | MAY | JUN | JUL | AUG | SEP | OCT | NOV | DEC | YEAR |
|---|---|---|---|---|---|---|---|---|---|---|---|---|---|
| Maximum Temp °F | 32.4 | 36.5 | 48.5 | 61.5 | 72.4 | 81.5 | 84.6 | 82.1 | 76.1 | 63.9 | 49.9 | 37.7 | 60.6 |
| Minimum Temp °F | 16.0 | 18.7 | 29.3 | 39.4 | 49.4 | 58.8 | 62.7 | 60.2 | 53.5 | 42.0 | 32.9 | 22.6 | 40.5 |
| Mean Temp °F | 24.2 | 27.6 | 38.9 | 50.5 | 61.0 | 70.1 | 73.7 | 71.2 | 64.8 | 53.0 | 41.4 | 30.2 | 50.6 |
| Days Max Temp ≥ 90 °F | 0 | 0 | 0 | 0 | 0 | 4 | 7 | 3 | 1 | 0 | 0 | 0 | 15 |
| Days Max Temp ≤ 32 °F | 15 | 10 | 3 | 0 | 0 | 0 | 0 | 0 | 0 | 0 | 1 | 9 | 38 |
| Days Min Temp ≤ 32 °F | 28 | 25 | 20 | 8 | 1 | 0 | 0 | 0 | 0 | 6 | 16 | 26 | 130 |
| Days Min Temp ≤ 0 °F | 5 | 3 | 0 | 0 | 0 | 0 | 0 | 0 | 0 | 0 | 0 | 2 | 10 |
| Heating Degree Days | 1257 | 1049 | 802 | 436 | 179 | 26 | 5 | 13 | 99 | 377 | 701 | 1072 | 6016 |
| Cooling Degree Days | 0 | 0 | 1 | 9 | 61 | 191 | 294 | 223 | 99 | 10 | 0 | 0 | 888 |
| Total Precipitation (") | 1.89 | 1.92 | 2.97 | 3.80 | 3.90 | 3.95 | 4.08 | 4.17 | 3.39 | 2.97 | 3.55 | 3.23 | 39.82 |
| Days ≥ 0.1" Precip | 5 | 5 | 7 | 8 | 7 | 6 | 7 | 6 | 6 | 6 | 7 | 7 | 77 |
| Total Snowfall (") | 7.3 | 6.8 | 3.6 | 0.7 | 0.0 | 0.0 | 0.0 | 0.0 | 0.0 | 0.4 | 1.3 | 6.2 | 26.3 |
| Days ≥ 1" Snow Depth | 13 | 10 | 3 | 0 | 0 | 0 | 0 | 0 | 0 | 0 | 1 | 7 | 34 |

## GOSHEN COLLEGE *Elkhart County* ELEVATION 801 ft LAT/LONG 41° 35 ' N / 85° 50 ' W

| | JAN | FEB | MAR | APR | MAY | JUN | JUL | AUG | SEP | OCT | NOV | DEC | YEAR |
|---|---|---|---|---|---|---|---|---|---|---|---|---|---|
| Maximum Temp °F | 30.6 | 34.6 | 46.6 | 60.1 | 71.2 | 80.6 | 83.9 | 81.4 | 74.4 | 62.0 | 48.3 | 36.1 | 59.2 |
| Minimum Temp °F | 16.2 | 18.5 | 28.6 | 38.9 | 48.6 | 58.2 | 62.5 | 60.3 | 53.5 | 42.3 | 33.1 | 23.2 | 40.3 |
| Mean Temp °F | 23.4 | 26.6 | 37.6 | 49.5 | 59.9 | 69.4 | 73.2 | 70.8 | 64.0 | 52.2 | 40.8 | 29.7 | 49.8 |
| Days Max Temp ≥ 90 °F | 0 | 0 | 0 | 0 | 0 | 3 | 6 | 3 | 1 | 0 | 0 | 0 | 13 |
| Days Max Temp ≤ 32 °F | 17 | 12 | 3 | 0 | 0 | 0 | 0 | 0 | 0 | 0 | 2 | 11 | 45 |
| Days Min Temp ≤ 32 °F | 29 | 25 | 21 | 8 | 1 | 0 | 0 | 0 | 0 | 5 | 15 | 26 | 130 |
| Days Min Temp ≤ 0 °F | 4 | 3 | 0 | 0 | 0 | 0 | 0 | 0 | 0 | 0 | 0 | 1 | 8 |
| Heating Degree Days | 1282 | 1078 | 842 | 465 | 204 | 32 | 5 | 15 | 110 | 398 | 720 | 1089 | 6240 |
| Cooling Degree Days | 0 | 0 | 1 | 9 | 60 | 177 | 284 | 213 | 87 | 6 | 0 | 0 | 837 |
| Total Precipitation (") | 1.69 | 1.60 | 2.67 | 3.44 | 3.11 | 3.83 | 3.52 | 3.87 | 3.89 | 3.02 | 2.94 | 2.85 | 36.43 |
| Days ≥ 0.1" Precip | 4 | 5 | 6 | 8 | 6 | 7 | 6 | 6 | 7 | 6 | 7 | 7 | 75 |
| Total Snowfall (") | 9.8 | 8.5 | 5.4 | 1.4 | 0.0 | 0.0 | 0.0 | 0.0 | 0.0 | 0.4 | 3.8 | 8.8 | 38.1 |
| Days ≥ 1" Snow Depth | 19 | 15 | 5 | 0 | 0 | 0 | 0 | 0 | 0 | 0 | 2 | 11 | 52 |

## GREENCASTLE 1 SE *Putnam County* ELEVATION 751 ft LAT/LONG 39° 39 ' N / 86° 51 ' W

| | JAN | FEB | MAR | APR | MAY | JUN | JUL | AUG | SEP | OCT | NOV | DEC | YEAR |
|---|---|---|---|---|---|---|---|---|---|---|---|---|---|
| Maximum Temp °F | 33.7 | 38.9 | 50.7 | 63.6 | 74.1 | 82.9 | 86.6 | 84.5 | 78.2 | 65.9 | 51.8 | 39.4 | 62.5 |
| Minimum Temp °F | 16.9 | 20.4 | 31.0 | 41.8 | 51.6 | 60.9 | 64.5 | 62.3 | 55.6 | 43.8 | 34.4 | 23.7 | 42.2 |
| Mean Temp °F | 25.3 | 29.7 | 40.8 | 52.8 | 62.9 | 71.9 | 75.6 | 73.4 | 66.9 | 54.9 | 43.1 | 31.6 | 52.4 |
| Days Max Temp ≥ 90 °F | 0 | 0 | 0 | 0 | 1 | 5 | 11 | 7 | 3 | 0 | 0 | 0 | 27 |
| Days Max Temp ≤ 32 °F | 14 | 9 | 2 | 0 | 0 | 0 | 0 | 0 | 0 | 0 | 1 | 8 | 34 |
| Days Min Temp ≤ 32 °F | 28 | 24 | 18 | 5 | 0 | 0 | 0 | 0 | 0 | 4 | 14 | 25 | 118 |
| Days Min Temp ≤ 0 °F | 4 | 2 | 0 | 0 | 0 | 0 | 0 | 0 | 0 | 0 | 0 | 1 | 7 |
| Heating Degree Days | 1223 | 990 | 744 | 379 | 141 | 17 | 2 | 6 | 73 | 328 | 650 | 1030 | 5583 |
| Cooling Degree Days | 0 | 0 | 2 | 19 | 80 | 235 | 349 | 271 | 137 | 20 | 1 | 0 | 1114 |
| Total Precipitation (") | 2.37 | 2.37 | 3.37 | 3.72 | 4.57 | 3.87 | 5.24 | 4.31 | 3.60 | 3.22 | 4.11 | 3.32 | 44.07 |
| Days ≥ 0.1" Precip | 5 | 5 | 7 | 8 | 8 | 7 | 7 | 6 | 6 | 6 | 7 | 7 | 79 |
| Total Snowfall (") | 8.4 | 7.3 | 3.8 | 0.5 | 0.0 | 0.0 | 0.0 | 0.0 | 0.0 | 0.2 | 1.4 | 5.6 | 27.2 |
| Days ≥ 1" Snow Depth | 11 | 9 | 3 | 0 | 0 | 0 | 0 | 0 | 0 | 0 | 1 | 5 | 29 |

## GREENFIELD *Hancock County* ELEVATION 991 ft LAT/LONG 39° 47 ' N / 85° 46 ' W

| | JAN | FEB | MAR | APR | MAY | JUN | JUL | AUG | SEP | OCT | NOV | DEC | YEAR |
|---|---|---|---|---|---|---|---|---|---|---|---|---|---|
| Maximum Temp °F | 33.4 | 37.6 | 49.6 | 62.2 | 73.2 | 82.4 | 85.7 | 83.7 | 77.6 | 65.3 | 51.1 | 39.1 | 61.7 |
| Minimum Temp °F | 16.1 | 18.7 | 29.7 | 40.4 | 50.6 | 59.8 | 63.9 | 61.3 | 54.5 | 42.2 | 33.2 | 23.0 | 41.1 |
| Mean Temp °F | 24.8 | 28.1 | 39.7 | 51.3 | 61.9 | 71.1 | 74.8 | 72.6 | 66.1 | 53.8 | 42.2 | 31.1 | 51.5 |
| Days Max Temp ≥ 90 °F | 0 | 0 | 0 | 0 | 1 | 5 | 8 | 5 | 2 | 0 | 0 | 0 | 21 |
| Days Max Temp ≤ 32 °F | 14 | 10 | 3 | 0 | 0 | 0 | 0 | 0 | 0 | 0 | 1 | 8 | 36 |
| Days Min Temp ≤ 32 °F | 28 | 25 | 20 | 6 | 1 | 0 | 0 | 0 | 0 | 5 | 16 | 25 | 126 |
| Days Min Temp ≤ 0 °F | 4 | 3 | 0 | 0 | 0 | 0 | 0 | 0 | 0 | 0 | 0 | 1 | 8 |
| Heating Degree Days | 1242 | 1033 | 780 | 415 | 161 | 23 | 3 | 8 | 80 | 357 | 681 | 1045 | 5828 |
| Cooling Degree Days | 0 | 0 | 2 | 13 | 74 | 226 | 333 | 265 | 123 | 16 | 0 | 0 | 1052 |
| Total Precipitation (") | 2.51 | 2.45 | 3.15 | 3.97 | 4.34 | 4.03 | 5.25 | 4.11 | 3.30 | 3.22 | 3.99 | 3.30 | 43.62 |
| Days ≥ 0.1" Precip | 6 | 5 | 7 | 8 | 9 | 7 | 8 | 6 | 6 | 6 | 7 | 7 | 82 |
| Total Snowfall (") | na | na | *1.8* | 0.2 | 0.0 | 0.0 | 0.0 | 0.0 | 0.0 | 0.1 | 0.8 | *2.9* | na |
| Days ≥ 1" Snow Depth | *8* | *7* | 2 | 0 | 0 | 0 | 0 | 0 | 0 | 0 | *0* | 4 | 21 |

## GREENSBURG *Decatur County* ELEVATION 935 ft LAT/LONG 39° 21 ' N / 85° 30 ' W

| | JAN | FEB | MAR | APR | MAY | JUN | JUL | AUG | SEP | OCT | NOV | DEC | YEAR |
|---|---|---|---|---|---|---|---|---|---|---|---|---|---|
| Maximum Temp °F | 35.2 | 40.1 | 51.5 | 63.4 | 73.2 | 81.9 | 85.2 | 83.3 | 77.1 | 65.3 | 52.1 | 40.5 | 62.4 |
| Minimum Temp °F | 18.6 | 21.6 | 31.8 | 42.0 | 51.3 | 60.0 | 63.6 | 61.2 | 54.9 | 42.9 | 34.5 | 24.9 | 42.3 |
| Mean Temp °F | 26.9 | 30.9 | 41.7 | 52.7 | 62.3 | 71.0 | 74.4 | 72.3 | 66.1 | 54.2 | 43.3 | 32.8 | 52.4 |
| Days Max Temp ≥ 90 °F | 0 | 0 | 0 | 0 | 0 | 3 | 7 | 4 | 1 | 0 | 0 | 0 | 15 |
| Days Max Temp ≤ 32 °F | 13 | 8 | 2 | 0 | 0 | 0 | 0 | 0 | 0 | 0 | 1 | 7 | 31 |
| Days Min Temp ≤ 32 °F | 27 | 22 | 17 | 6 | 1 | 0 | 0 | 0 | 0 | 5 | 14 | 23 | 115 |
| Days Min Temp ≤ 0 °F | 4 | 2 | 0 | 0 | 0 | 0 | 0 | 0 | 0 | 0 | 0 | 1 | 7 |
| Heating Degree Days | 1174 | 957 | 719 | 376 | 148 | 19 | 3 | 8 | 81 | 345 | 644 | 994 | 5468 |
| Cooling Degree Days | 0 | 0 | 2 | 18 | 78 | 224 | 338 | 267 | 128 | 18 | 1 | 0 | 1074 |
| Total Precipitation (") | 2.28 | 2.25 | 3.59 | 4.20 | 4.68 | 3.70 | 4.37 | 3.94 | 3.14 | 3.06 | 3.92 | 3.31 | 42.44 |
| Days ≥ 0.1" Precip | 5 | 5 | 7 | 9 | 8 | 7 | 7 | 6 | 6 | 6 | 7 | 7 | 80 |
| Total Snowfall (") | 5.5 | 4.5 | 2.6 | 0.5 | 0.0 | 0.0 | 0.0 | 0.0 | 0.0 | 0.2 | 1.3 | 3.3 | 17.9 |
| Days ≥ 1" Snow Depth | 9 | 7 | 2 | 0 | 0 | 0 | 0 | 0 | 0 | 0 | 1 | 3 | 22 |

## HARTFORD CITY 4 ESE *Blackford County* ELEVATION 879 ft LAT/LONG 40° 26 ' N / 85° 22 ' W

| | JAN | FEB | MAR | APR | MAY | JUN | JUL | AUG | SEP | OCT | NOV | DEC | YEAR |
|---|---|---|---|---|---|---|---|---|---|---|---|---|---|
| Maximum Temp °F | 30.1 | 34.3 | 46.2 | *59.5* | 70.6 | 79.6 | 83.1 | 81.1 | 74.9 | 62.5 | 49.0 | 36.6 | 59.0 |
| Minimum Temp °F | 13.8 | 16.9 | 28.1 | *38.7* | 49.0 | 58.4 | 62.3 | 59.7 | 52.8 | 40.8 | 32.2 | 21.8 | 39.5 |
| Mean Temp °F | 22.0 | 25.6 | 37.2 | *49.1* | 59.8 | 69.1 | 72.8 | 70.4 | 63.9 | 51.7 | 40.7 | 29.2 | 49.3 |
| Days Max Temp ≥ 90 °F | 0 | 0 | 0 | *0* | 0 | 2 | 4 | 2 | 0 | 0 | 0 | 0 | 8 |
| Days Max Temp ≤ 32 °F | 17 | 12 | 4 | *0* | 0 | 0 | 0 | 0 | 0 | 0 | 2 | 11 | 46 |
| Days Min Temp ≤ 32 °F | 29 | 25 | 21 | *9* | 1 | 0 | 0 | 0 | 0 | 6 | 17 | 26 | 134 |
| Days Min Temp ≤ 0 °F | 6 | 4 | 0 | *0* | 0 | 0 | 0 | 0 | 0 | 0 | 0 | 2 | 12 |
| Heating Degree Days | 1327 | 1107 | 855 | *478* | 203 | 36 | 7 | 18 | 111 | 414 | 724 | 1103 | 6383 |
| Cooling Degree Days | *0* | *0* | *1* | na | *52* | *179* | *280* | *220* | *89* | *9* | *0* | *0* | na |
| Total Precipitation (") | 1.84 | 1.83 | 2.63 | *3.27* | 3.54 | 4.31 | 3.81 | 3.82 | 3.15 | 2.53 | 3.34 | 2.80 | 36.87 |
| Days ≥ 0.1" Precip | 5 | 4 | 6 | *7* | 7 | 7 | 6 | *6* | 6 | 6 | 7 | 7 | 74 |
| Total Snowfall (") | 6.0 | 6.4 | 3.6 | *1.0* | 0.0 | 0.0 | 0.0 | 0.0 | 0.0 | 0.4 | 1.6 | 5.6 | 24.6 |
| Days ≥ 1" Snow Depth | 18 | 14 | 6 | *1* | 0 | 0 | 0 | 0 | 0 | 0 | 2 | 10 | 51 |

## HOBART 2 WNW *Lake County* ELEVATION 620 ft LAT/LONG 41° 32 ' N / 87° 15 ' W

| | JAN | FEB | MAR | APR | MAY | JUN | JUL | AUG | SEP | OCT | NOV | DEC | YEAR |
|---|---|---|---|---|---|---|---|---|---|---|---|---|---|
| Maximum Temp °F | 32.0 | 36.3 | 48.1 | 60.7 | 72.0 | 81.9 | 85.3 | 83.1 | 76.9 | 65.3 | 50.1 | 37.2 | 60.7 |
| Minimum Temp °F | 15.7 | 18.9 | 28.8 | 38.6 | 47.5 | 57.5 | 62.8 | 60.9 | 54.1 | 43.0 | 33.1 | 22.4 | 40.3 |
| Mean Temp °F | 23.9 | 27.6 | 38.4 | 49.7 | 59.8 | 69.7 | 74.0 | 72.0 | 65.6 | 54.2 | 41.7 | 29.8 | 50.5 |
| Days Max Temp ≥ 90 °F | 0 | 0 | 0 | 0 | 1 | 5 | 9 | 5 | 2 | 0 | 0 | 0 | 22 |
| Days Max Temp ≤ 32 °F | 15 | 11 | 2 | 0 | 0 | 0 | 0 | 0 | 0 | 0 | 1 | 9 | 38 |
| Days Min Temp ≤ 32 °F | 29 | 25 | 21 | 8 | 1 | 0 | 0 | 0 | 0 | 4 | 16 | 25 | 129 |
| Days Min Temp ≤ 0 °F | 5 | 3 | 0 | 0 | 0 | 0 | 0 | 0 | 0 | 0 | 0 | 2 | 10 |
| Heating Degree Days | 1268 | 1049 | 817 | 465 | 214 | 39 | 6 | 11 | 91 | 342 | 693 | 1083 | 6078 |
| Cooling Degree Days | 0 | 0 | 2 | 14 | 57 | 184 | 297 | 238 | 110 | 8 | 0 | 0 | 910 |
| Total Precipitation (") | 1.72 | 1.45 | 2.52 | 3.68 | 3.65 | 4.27 | 3.44 | 3.84 | 3.98 | 3.10 | 3.62 | 2.66 | 37.93 |
| Days ≥ 0.1" Precip | 5 | 4 | 6 | 8 | 7 | 6 | 6 | 7 | 7 | 6 | 7 | 6 | 75 |
| Total Snowfall (") | *9.8* | 8.5 | 4.4 | 0.9 | 0.0 | 0.0 | 0.0 | 0.0 | 0.0 | 0.1 | 1.2 | 6.3 | 31.2 |
| Days ≥ 1" Snow Depth | 18 | 13 | 4 | 0 | 0 | 0 | 0 | 0 | 0 | 0 | 1 | 9 | 45 |

## INDIANAPOLIS INTL AP *Marion County* ELEVATION 810 ft LAT/LONG 39° 44' N / 86° 16' W

| | JAN | FEB | MAR | APR | MAY | JUN | JUL | AUG | SEP | OCT | NOV | DEC | YEAR |
|---|---|---|---|---|---|---|---|---|---|---|---|---|---|
| Maximum Temp °F | 34.0 | 38.7 | 51.0 | 63.4 | 73.6 | 82.5 | 85.6 | 83.6 | 77.3 | 65.3 | 51.6 | 39.6 | 62.2 |
| Minimum Temp °F | 17.9 | 21.4 | 31.9 | 41.8 | 51.7 | 61.2 | 65.5 | 63.0 | 55.6 | 43.5 | 34.4 | 24.6 | 42.7 |
| Mean Temp °F | 26.0 | 30.1 | 41.5 | 52.7 | 62.7 | 71.9 | 75.6 | 73.3 | 66.4 | 54.4 | 43.0 | 32.1 | 52.5 |
| Days Max Temp ≥ 90 °F | 0 | 0 | 0 | 0 | 1 | 4 | 8 | 5 | 1 | 0 | 0 | 0 | 19 |
| Days Max Temp ≤ 32 °F | 13 | 9 | 2 | 0 | 0 | 0 | 0 | 0 | 0 | 0 | 1 | 7 | 32 |
| Days Min Temp ≤ 32 °F | 27 | 23 | 17 | 5 | 0 | 0 | 0 | 0 | 0 | 4 | 14 | 24 | 114 |
| Days Min Temp ≤ 0 °F | 4 | 2 | 0 | 0 | 0 | 0 | 0 | 0 | 0 | 0 | 0 | 1 | 7 |
| Heating Degree Days | 1203 | 980 | 725 | 376 | 139 | 13 | 1 | 4 | 75 | 338 | 653 | 1013 | 5520 |
| Cooling Degree Days | 0 | 0 | 2 | 12 | 72 | 230 | 350 | 277 | 127 | 15 | 0 | 0 | 1085 |
| Total Precipitation (") | 2.32 | 2.41 | 3.25 | 3.66 | 3.87 | 3.63 | 4.54 | 3.73 | 2.95 | 2.85 | 3.67 | 3.31 | 40.19 |
| Days ≥ 0.1" Precip | 6 | 5 | 7 | 8 | 8 | 6 | 7 | 6 | 5 | 5 | 7 | 7 | 77 |
| Total Snowfall (") | 8.3 | 7.0 | 3.1 | 0.4 | 0.0 | 0.0 | 0.0 | 0.0 | 0.0 | 0.4 | 1.7 | 5.5 | 26.4 |
| Days ≥ 1" Snow Depth | 11 | 9 | 3 | 0 | 0 | 0 | 0 | 0 | 0 | 0 | 1 | 5 | 29 |

## INDIANAPOLIS SE SIDE *Marion County* ELEVATION 751 ft LAT/LONG 39° 45' N / 86° 7' W

| | JAN | FEB | MAR | APR | MAY | JUN | JUL | AUG | SEP | OCT | NOV | DEC | YEAR |
|---|---|---|---|---|---|---|---|---|---|---|---|---|---|
| Maximum Temp °F | 33.9 | 38.3 | 49.8 | 62.3 | 73.0 | 81.9 | 85.4 | 83.4 | 77.2 | 65.2 | 51.5 | 39.7 | 61.8 |
| Minimum Temp °F | 17.7 | 20.5 | 30.7 | 41.3 | 51.5 | 61.1 | 65.5 | 62.9 | 55.5 | 43.2 | 34.3 | 24.1 | 42.4 |
| Mean Temp °F | 25.8 | 29.4 | 40.3 | 51.8 | 62.3 | 71.6 | 75.5 | 73.2 | 66.4 | 54.2 | 43.0 | 31.9 | 52.1 |
| Days Max Temp ≥ 90 °F | 0 | 0 | 0 | 0 | 0 | 4 | 8 | 5 | 2 | 0 | 0 | 0 | 19 |
| Days Max Temp ≤ 32 °F | 14 | 9 | 3 | 0 | 0 | 0 | 0 | 0 | 0 | 0 | 1 | 7 | 34 |
| Days Min Temp ≤ 32 °F | 28 | 24 | 19 | 5 | 0 | 0 | 0 | 0 | 0 | 4 | 15 | 24 | 119 |
| Days Min Temp ≤ 0 °F | 4 | 2 | 0 | 0 | 0 | 0 | 0 | 0 | 0 | 0 | 0 | 1 | 7 |
| Heating Degree Days | 1208 | 997 | 761 | 400 | 151 | 19 | 1 | 6 | 76 | 343 | 656 | 1018 | 5636 |
| Cooling Degree Days | 0 | 0 | 2 | 14 | 73 | 227 | 342 | 272 | 124 | 17 | 0 | 0 | 1071 |
| Total Precipitation (") | 2.03 | 2.05 | 2.91 | 3.78 | 4.27 | 3.50 | 5.02 | 3.75 | 2.79 | 2.94 | 3.80 | 2.96 | 39.80 |
| Days ≥ 0.1" Precip | 5 | 4 | 7 | 8 | 8 | 6 | 7 | 6 | 5 | 6 | 6 | 7 | 75 |
| Total Snowfall (") | na | na | na | 0.1 | 0.0 | 0.0 | 0.0 | 0.0 | 0.0 | 0.2 | *0.2* | na | na |
| Days ≥ 1" Snow Depth | 10 | 8 | 2 | 0 | 0 | 0 | 0 | 0 | 0 | 0 | 1 | 4 | 25 |

## KENTLAND 4 NNW *Newton County* ELEVATION 689 ft LAT/LONG 40° 46' N / 87° 27' W

| | JAN | FEB | MAR | APR | MAY | JUN | JUL | AUG | SEP | OCT | NOV | DEC | YEAR |
|---|---|---|---|---|---|---|---|---|---|---|---|---|---|
| Maximum Temp °F | 31.4 | 36.1 | 48.8 | 62.7 | 74.0 | 83.5 | 85.9 | 83.8 | 77.9 | 65.7 | 50.0 | 37.0 | 61.4 |
| Minimum Temp °F | 15.1 | 18.9 | 30.1 | 40.0 | 50.2 | 59.7 | 63.5 | 60.7 | 53.8 | 42.4 | 32.9 | 21.9 | 40.8 |
| Mean Temp °F | 23.3 | 27.5 | 39.5 | 51.3 | 62.1 | 71.6 | 74.7 | 72.3 | 65.9 | 54.1 | 41.5 | 29.5 | 51.1 |
| Days Max Temp ≥ 90 °F | 0 | 0 | 0 | 0 | 2 | 6 | 9 | 5 | 2 | 0 | 0 | 0 | 24 |
| Days Max Temp ≤ 32 °F | 16 | 10 | 3 | 0 | 0 | 0 | 0 | 0 | 0 | 0 | 1 | 9 | 39 |
| Days Min Temp ≤ 32 °F | 29 | 24 | 20 | 7 | 1 | 0 | 0 | 0 | 0 | 5 | 16 | 26 | 128 |
| Days Min Temp ≤ 0 °F | 6 | 4 | 0 | 0 | 0 | 0 | 0 | 0 | 0 | 0 | 0 | 2 | 12 |
| Heating Degree Days | 1287 | 1053 | 785 | 417 | 160 | 19 | 3 | 9 | 86 | 349 | 699 | 1094 | 5961 |
| Cooling Degree Days | 0 | 0 | 1 | 13 | 78 | 227 | 319 | 248 | 115 | 14 | 0 | 0 | 1015 |
| Total Precipitation (") | 1.65 | 1.58 | 2.80 | 3.59 | 3.96 | 4.09 | 3.99 | 3.85 | 3.52 | 2.95 | 3.34 | 2.80 | 38.12 |
| Days ≥ 0.1" Precip | 4 | 4 | 7 | 8 | 8 | 7 | 7 | 6 | 6 | 6 | 7 | 6 | 76 |
| Total Snowfall (") | 7.3 | 7.2 | 3.9 | 0.9 | 0.0 | 0.0 | 0.0 | 0.0 | 0.0 | 0.1 | 1.9 | 5.9 | 27.2 |
| Days ≥ 1" Snow Depth | 14 | 12 | 4 | 0 | 0 | 0 | 0 | 0 | 0 | 0 | 1 | 8 | 39 |

## KOKOMO 3 WSW *Howard County* ELEVATION 860 ft LAT/LONG 40° 25' N / 86° 3' W

| | JAN | FEB | MAR | APR | MAY | JUN | JUL | AUG | SEP | OCT | NOV | DEC | YEAR |
|---|---|---|---|---|---|---|---|---|---|---|---|---|---|
| Maximum Temp °F | 29.9 | *34.1* | *46.4* | *59.8* | *71.2* | *80.5* | 84.1 | 81.7 | 75.7 | 63.4 | 48.9 | 35.9 | 59.3 |
| Minimum Temp °F | 13.8 | *16.6* | *27.3* | *37.6* | *48.2* | *57.6* | 61.8 | 58.8 | 51.6 | 40.0 | 31.7 | 20.9 | 38.8 |
| Mean Temp °F | 21.9 | *25.4* | *36.9* | *48.7* | *59.7* | *69.1* | 73.0 | 70.3 | 63.7 | 51.7 | 40.3 | 28.4 | 49.1 |
| Days Max Temp ≥ 90 °F | 0 | *0* | *0* | *0* | *1* | *4* | 7 | 3 | 1 | 0 | 0 | 0 | 16 |
| Days Max Temp ≤ 32 °F | 17 | *13* | *4* | *0* | *0* | *0* | 0 | 0 | 0 | 0 | 2 | 11 | 47 |
| Days Min Temp ≤ 32 °F | 29 | *26* | *22* | *10* | *1* | *0* | 0 | 0 | 1 | 7 | 17 | 27 | 140 |
| Days Min Temp ≤ 0 °F | 7 | *4* | *0* | *0* | *0* | *0* | 0 | 0 | 0 | 0 | 0 | 2 | 13 |
| Heating Degree Days | 1332 | *1113* | *866* | *491* | *206* | *41* | 8 | 21 | 120 | 415 | 734 | 1129 | 6476 |
| Cooling Degree Days | 0 | 0 | 1 | 8 | 43 | 173 | 270 | 202 | 87 | 10 | 0 | 0 | 794 |
| Total Precipitation (") | 2.44 | *2.27* | *3.14* | *3.88* | *3.84* | *3.64* | 4.44 | 4.50 | 3.38 | 3.27 | 3.92 | 3.21 | 41.93 |
| Days ≥ 0.1" Precip | 7 | *6* | *8* | *8* | *7* | *7* | 7 | 7 | 5 | 6 | 7 | 8 | 83 |
| Total Snowfall (") | 12.7 | *11.7* | *6.0* | *1.6* | *0.0* | *0.0* | 0.0 | 0.0 | 0.0 | 0.4 | 2.1 | 10.1 | 44.6 |
| Days ≥ 1" Snow Depth | 17 | *14* | *6* | *1* | *0* | *0* | 0 | 0 | 0 | 0 | 1 | 9 | 48 |

## LA PORTE *La Porte County* ELEVATION 807 ft LAT/LONG 41° 35' N / 86° 44' W

| | JAN | FEB | MAR | APR | MAY | JUN | JUL | AUG | SEP | OCT | NOV | DEC | YEAR |
|---|---|---|---|---|---|---|---|---|---|---|---|---|---|
| Maximum Temp °F | 30.0 | 34.1 | 46.0 | 58.8 | 70.6 | 79.8 | 83.3 | 81.1 | 74.0 | 62.0 | 47.7 | 35.5 | 58.6 |
| Minimum Temp °F | 15.5 | 19.0 | 28.9 | 39.0 | 49.2 | 58.7 | 63.4 | 61.6 | 54.5 | 43.1 | 33.2 | 22.4 | 40.7 |
| Mean Temp °F | 22.8 | 26.5 | 37.4 | 48.9 | 60.0 | 69.3 | 73.4 | 71.3 | 64.3 | 52.6 | 40.5 | 29.0 | 49.7 |
| Days Max Temp ≥ 90 °F | 0 | 0 | 0 | 0 | 0 | 3 | 5 | 3 | 1 | 0 | 0 | 0 | 12 |
| Days Max Temp ≤ 32 °F | 17 | 12 | 4 | 0 | 0 | 0 | 0 | 0 | 0 | 0 | 2 | 11 | 46 |
| Days Min Temp ≤ 32 °F | 29 | 25 | 21 | 7 | 1 | 0 | 0 | 0 | 0 | 3 | 15 | 26 | 127 |
| Days Min Temp ≤ 0 °F | 5 | 2 | 0 | 0 | 0 | 0 | 0 | 0 | 0 | 0 | 0 | 2 | 9 |
| Heating Degree Days | 1302 | 1079 | 848 | 484 | 207 | 36 | 4 | 11 | 102 | 388 | 730 | 1109 | 6300 |
| Cooling Degree Days | 0 | 0 | 1 | 11 | 59 | 169 | 276 | 221 | 87 | 6 | 0 | 0 | 830 |
| Total Precipitation (") | 2.24 | 1.83 | 3.09 | 3.66 | 3.31 | 4.20 | 3.74 | 3.95 | 4.34 | 3.45 | 3.86 | 3.38 | 41.05 |
| Days ≥ 0.1" Precip | 6 | 5 | 7 | 8 | 7 | 7 | 7 | 7 | 7 | 7 | 8 | 8 | 84 |
| Total Snowfall (") | *21.4* | 13.9 | *7.9* | 1.4 | 0.1 | 0.0 | 0.0 | 0.0 | 0.0 | 0.3 | 4.9 | 14.0 | 63.9 |
| Days ≥ 1" Snow Depth | *21* | *16* | *8* | 0 | 0 | 0 | 0 | 0 | 0 | 0 | *2* | *12* | 59 |

## LAFAYETTE 5 S *Tippecanoe County* ELEVATION 600 ft LAT/LONG 40° 21' N / 86° 52' W

| | JAN | FEB | MAR | APR | MAY | JUN | JUL | AUG | SEP | OCT | NOV | DEC | YEAR |
|---|---|---|---|---|---|---|---|---|---|---|---|---|---|
| Maximum Temp °F | 31.1 | 35.6 | 47.7 | 60.7 | 72.1 | 81.3 | 84.5 | 82.2 | 76.3 | 64.1 | 49.9 | 37.3 | 60.2 |
| Minimum Temp °F | 14.5 | 17.4 | 28.9 | 39.4 | 49.6 | 58.9 | 62.9 | 60.3 | 53.2 | 41.3 | 32.4 | 21.8 | 40.1 |
| Mean Temp °F | 22.8 | 26.5 | 38.4 | 50.1 | 60.9 | 70.1 | 73.7 | 71.3 | 64.8 | 52.8 | 41.2 | 29.5 | 50.2 |
| Days Max Temp ≥ 90 °F | 0 | 0 | 0 | 0 | 1 | 4 | 6 | 3 | 1 | 0 | 0 | 0 | 15 |
| Days Max Temp ≤ 32 °F | 16 | 11 | 4 | 0 | 0 | 0 | 0 | 0 | 0 | 0 | 2 | 10 | 43 |
| Days Min Temp ≤ 32 °F | 29 | 25 | 21 | 8 | 1 | 0 | 0 | 0 | 0 | 7 | 16 | 26 | 133 |
| Days Min Temp ≤ 0 °F | 6 | 4 | 0 | 0 | 0 | 0 | 0 | 0 | 0 | 0 | 0 | 2 | 12 |
| Heating Degree Days | 1301 | 1080 | 820 | 451 | 185 | 31 | 7 | 16 | 104 | 386 | 708 | 1092 | 6181 |
| Cooling Degree Days | 0 | 0 | 1 | 13 | 66 | 205 | 305 | 232 | 110 | 13 | 0 | 0 | 945 |
| Total Precipitation (") | 1.80 | 1.62 | 2.79 | 3.55 | 3.88 | 3.76 | 3.97 | 3.62 | 3.06 | 2.51 | 3.23 | 2.81 | 36.60 |
| Days ≥ 0.1" Precip | 4 | 4 | 7 | 8 | 7 | 7 | 6 | 6 | 6 | 5 | 6 | 6 | 72 |
| Total Snowfall (") | 6.3 | 5.5 | 2.6 | 0.7 | 0.0 | 0.0 | 0.0 | 0.0 | 0.0 | 0.5 | 0.9 | 5.0 | 21.5 |
| Days ≥ 1" Snow Depth | 13 | 10 | 4 | 0 | 0 | 0 | 0 | 0 | 0 | 0 | 1 | 6 | 34 |

## LAGRANGE SEWAGE PLAN *Lagrange County* ELEVATION 902 ft LAT/LONG 41° 39' N / 85° 25' W

| | JAN | FEB | MAR | APR | MAY | JUN | JUL | AUG | SEP | OCT | NOV | DEC | YEAR |
|---|---|---|---|---|---|---|---|---|---|---|---|---|---|
| Maximum Temp °F | 29.1 | 32.5 | 44.2 | 57.7 | 69.4 | 79.1 | 82.7 | 80.6 | 73.5 | 61.0 | 46.9 | 34.8 | 57.6 |
| Minimum Temp °F | 13.8 | 15.5 | 25.7 | 36.7 | 46.9 | 56.4 | 60.6 | 58.4 | 51.5 | 40.1 | 31.2 | 20.8 | 38.1 |
| Mean Temp °F | 21.6 | 24.0 | 35.0 | 47.2 | 58.2 | 67.8 | 71.7 | 69.5 | 62.5 | 50.5 | 39.0 | 27.8 | 47.9 |
| Days Max Temp ≥ 90 °F | 0 | 0 | 0 | 0 | 0 | 3 | 5 | 2 | 1 | 0 | 0 | 0 | 11 |
| Days Max Temp ≤ 32 °F | 18 | 14 | 5 | 0 | 0 | 0 | 0 | 0 | 0 | 0 | 3 | 12 | 52 |
| Days Min Temp ≤ 32 °F | 30 | 26 | 24 | 11 | 2 | 0 | 0 | 0 | 0 | 7 | 18 | 27 | 145 |
| Days Min Temp ≤ 0 °F | 5 | 4 | 0 | 0 | 0 | 0 | 0 | 0 | 0 | 0 | 0 | 2 | 11 |
| Heating Degree Days | 1341 | 1150 | 923 | 534 | 250 | 55 | 14 | 28 | 142 | 449 | 772 | 1146 | 6804 |
| Cooling Degree Days | 0 | 0 | 1 | 9 | 48 | 148 | 248 | 186 | 73 | 5 | 0 | 0 | 718 |
| Total Precipitation (") | 1.84 | 1.70 | 2.64 | 3.55 | 3.41 | 3.93 | 3.79 | 3.86 | 3.61 | 2.88 | 3.09 | 2.89 | 37.19 |
| Days ≥ 0.1" Precip | 5 | 4 | 6 | 8 | 7 | 7 | 7 | 7 | 7 | 7 | 7 | 7 | 79 |
| Total Snowfall (") | 11.0 | 10.2 | 6.6 | 1.6 | 0.0 | 0.0 | 0.0 | 0.0 | 0.0 | 0.5 | 4.6 | 10.9 | 45.4 |
| Days ≥ 1" Snow Depth | 20 | 17 | 7 | 1 | 0 | 0 | 0 | 0 | 0 | 0 | 3 | 14 | 62 |

## LOWELL *Lake County* ELEVATION 712 ft LAT/LONG 41° 17' N / 87° 26' W

| | JAN | FEB | MAR | APR | MAY | JUN | JUL | AUG | SEP | OCT | NOV | DEC | YEAR |
|---|---|---|---|---|---|---|---|---|---|---|---|---|---|
| Maximum Temp °F | 29.6 | 34.0 | 46.4 | 59.8 | 71.7 | 81.3 | 84.4 | 81.9 | 75.6 | 63.6 | 48.4 | 35.4 | 59.3 |
| Minimum Temp °F | 12.8 | 16.1 | 27.5 | 37.7 | 47.6 | 57.3 | 61.5 | 59.1 | 52.0 | 40.0 | 31.0 | 19.8 | 38.5 |
| Mean Temp °F | 21.2 | 25.1 | 37.0 | 48.8 | 59.7 | 69.3 | 73.0 | 70.5 | 63.8 | 51.8 | 39.7 | 27.4 | 48.9 |
| Days Max Temp ≥ 90 °F | 0 | 0 | 0 | 0 | 1 | 5 | 7 | 3 | 1 | 0 | 0 | 0 | 17 |
| Days Max Temp ≤ 32 °F | 17 | 12 | 3 | 0 | 0 | 0 | 0 | 0 | 0 | 0 | 2 | 11 | 45 |
| Days Min Temp ≤ 32 °F | 29 | 26 | 23 | 9 | 1 | 0 | 0 | 0 | 0 | 7 | 18 | 27 | 140 |
| Days Min Temp ≤ 0 °F | 7 | 5 | 0 | 0 | 0 | 0 | 0 | 0 | 0 | 0 | 0 | 3 | 15 |
| Heating Degree Days | 1352 | 1121 | 861 | 489 | 214 | 38 | 7 | 18 | 116 | 412 | 753 | 1157 | 6538 |
| Cooling Degree Days | 0 | 0 | 1 | 10 | 58 | 178 | 271 | 206 | 87 | 7 | 0 | 0 | 818 |
| Total Precipitation (") | 1.81 | 1.62 | 2.96 | 4.37 | 4.25 | 4.20 | 3.81 | 3.98 | 3.82 | 3.29 | 3.81 | 3.05 | 40.97 |
| Days ≥ 0.1" Precip | 4 | 4 | 7 | 8 | 8 | 7 | 7 | 6 | 6 | 6 | 7 | 6 | 76 |
| Total Snowfall (") | 9.8 | 10.2 | 4.6 | 0.5 | 0.0 | 0.0 | 0.0 | 0.0 | 0.0 | 0.2 | 1.8 | 6.6 | 33.7 |
| Days ≥ 1" Snow Depth | na | na | na | 0 | 0 | 0 | 0 | 0 | 0 | 0 | *1* | na | na |

## MADISON SEWAGE PLANT *Jefferson County* ELEVATION 459 ft LAT/LONG 38° 44 ' N / 85° 23 ' W

| | JAN | FEB | MAR | APR | MAY | JUN | JUL | AUG | SEP | OCT | NOV | DEC | YEAR |
|---|---|---|---|---|---|---|---|---|---|---|---|---|---|
| Maximum Temp °F | 40.1 | 45.2 | 56.5 | 67.6 | 75.9 | 84.2 | 87.3 | 85.8 | 79.8 | 68.9 | 56.3 | 44.9 | 66.0 |
| Minimum Temp °F | 23.1 | 25.6 | 34.5 | 43.7 | 53.0 | 62.0 | 66.6 | 64.8 | 58.5 | 46.3 | 37.6 | 28.0 | 45.3 |
| Mean Temp °F | 31.6 | 35.4 | 45.5 | 55.7 | 64.5 | 73.1 | 77.0 | 75.3 | 69.2 | 57.6 | 47.0 | 36.5 | 55.7 |
| Days Max Temp ≥ 90 °F | 0 | 0 | 0 | 0 | 0 | 5 | 11 | 8 | 3 | 0 | 0 | 0 | 27 |
| Days Max Temp ≤ 32 °F | 8 | 4 | 0 | 0 | 0 | 0 | 0 | 0 | 0 | 0 | 0 | 4 | 16 |
| Days Min Temp ≤ 32 °F | 25 | 21 | 14 | 3 | 0 | 0 | 0 | 0 | 0 | 2 | 10 | 21 | 96 |
| Days Min Temp ≤ 0 °F | 1 | 0 | 0 | 0 | 0 | 0 | 0 | 0 | 0 | 0 | 0 | 0 | 1 |
| Heating Degree Days | 1027 | 830 | 600 | 291 | 98 | 8 | 0 | 1 | 41 | 248 | 536 | 878 | 4558 |
| Cooling Degree Days | 0 | 0 | 2 | 15 | 88 | 251 | 392 | 331 | 163 | 23 | 1 | 1 | 1267 |
| Total Precipitation (") | 2.81 | 2.85 | 4.07 | 4.46 | 4.38 | 3.83 | 4.36 | 3.90 | 3.04 | 3.19 | 3.90 | 3.50 | 44.29 |
| Days ≥ 0.1" Precip | 6 | 6 | 8 | 9 | 9 | 7 | 7 | 6 | 5 | 6 | 7 | 7 | 83 |
| Total Snowfall (") | *5.1* | *3.8* | na | 0.1 | 0.0 | 0.0 | 0.0 | 0.0 | 0.0 | 0.1 | 0.1 | *2.7* | na |
| Days ≥ 1" Snow Depth | na | na | na | 0 | 0 | 0 | 0 | 0 | 0 | 0 | 0 | *1* | na |

## MARION 2 N *Grant County* ELEVATION 850 ft LAT/LONG 40° 31 ' N / 85° 41 ' W

| | JAN | FEB | MAR | APR | MAY | JUN | JUL | AUG | SEP | OCT | NOV | DEC | YEAR |
|---|---|---|---|---|---|---|---|---|---|---|---|---|---|
| Maximum Temp °F | 31.4 | 35.3 | 47.2 | 60.3 | 71.5 | 80.9 | 84.4 | 82.0 | 76.1 | 63.8 | 49.7 | 37.3 | 60.0 |
| Minimum Temp °F | 15.4 | 17.5 | 27.6 | 37.6 | 48.1 | 57.9 | 62.3 | 59.7 | 52.7 | 40.8 | 32.8 | 22.5 | 39.6 |
| Mean Temp °F | 23.5 | 26.4 | 37.4 | 49.0 | 59.8 | 69.4 | 73.4 | 70.9 | 64.4 | 52.3 | 41.3 | 30.0 | 49.8 |
| Days Max Temp ≥ 90 °F | 0 | 0 | 0 | 0 | 0 | 4 | 6 | 3 | 1 | 0 | 0 | 0 | 14 |
| Days Max Temp ≤ 32 °F | 16 | 12 | 4 | 0 | 0 | 0 | 0 | 0 | 0 | 0 | 2 | 9 | 43 |
| Days Min Temp ≤ 32 °F | 29 | 25 | 22 | 10 | 1 | 0 | 0 | 0 | 0 | 6 | 17 | 25 | 135 |
| Days Min Temp ≤ 0 °F | 5 | 4 | 0 | 0 | 0 | 0 | 0 | 0 | 0 | 0 | 0 | 2 | 11 |
| Heating Degree Days | 1281 | 1083 | 849 | 483 | 208 | 35 | 5 | 15 | 106 | 396 | 706 | 1080 | 6247 |
| Cooling Degree Days | 0 | 0 | 1 | 9 | 54 | 182 | 289 | 216 | 98 | 11 | 0 | 0 | 860 |
| Total Precipitation (") | 2.11 | 2.00 | 2.94 | 3.59 | 3.99 | 3.61 | 4.28 | 3.49 | 3.04 | 2.67 | 3.53 | 3.06 | 38.31 |
| Days ≥ 0.1" Precip | 5 | 5 | 7 | 8 | 7 | 7 | 7 | 6 | 5 | 6 | 7 | 7 | 77 |
| Total Snowfall (") | 6.8 | 7.3 | 3.6 | 1.0 | 0.0 | 0.0 | 0.0 | 0.0 | 0.0 | 0.4 | 1.1 | 5.9 | 26.1 |
| Days ≥ 1" Snow Depth | 15 | 11 | 4 | 0 | 0 | 0 | 0 | 0 | 0 | 0 | 1 | 7 | 38 |

## MARTINSVILLE 2 SW *Morgan County* ELEVATION 600 ft LAT/LONG 39° 26 ' N / 86° 25 ' W

| | JAN | FEB | MAR | APR | MAY | JUN | JUL | AUG | SEP | OCT | NOV | DEC | YEAR |
|---|---|---|---|---|---|---|---|---|---|---|---|---|---|
| Maximum Temp °F | 35.5 | 39.5 | 50.8 | 63.2 | 73.0 | 81.6 | 85.5 | 83.8 | 77.3 | 66.1 | 53.0 | 40.9 | 62.5 |
| Minimum Temp °F | 17.5 | 19.6 | 30.0 | 40.2 | 49.4 | 58.9 | 63.1 | 60.2 | 52.5 | 40.5 | 33.0 | 23.8 | 40.7 |
| Mean Temp °F | 26.5 | 29.5 | 40.3 | 51.7 | 61.2 | 70.2 | 74.4 | 72.0 | 64.9 | 53.3 | 43.0 | 32.4 | 51.6 |
| Days Max Temp ≥ 90 °F | 0 | 0 | 0 | 0 | 0 | 3 | 7 | 5 | 1 | 0 | 0 | 0 | 16 |
| Days Max Temp ≤ 32 °F | 12 | 8 | 2 | 0 | 0 | 0 | 0 | 0 | 0 | 0 | 1 | 6 | 29 |
| Days Min Temp ≤ 32 °F | 28 | 24 | 20 | 7 | 1 | 0 | 0 | 0 | 0 | 8 | 16 | 24 | 128 |
| Days Min Temp ≤ 0 °F | 4 | 3 | 0 | 0 | 0 | 0 | 0 | 0 | 0 | 0 | 0 | 1 | 8 |
| Heating Degree Days | 1186 | 996 | 759 | 402 | 166 | 26 | 4 | 11 | 99 | 369 | 653 | 1004 | 5675 |
| Cooling Degree Days | 0 | 0 | 1 | 11 | 53 | 185 | 315 | 237 | 99 | 12 | 0 | 0 | 913 |
| Total Precipitation (") | 2.33 | 2.40 | 3.22 | 4.00 | 4.54 | 3.47 | 4.38 | 3.90 | 3.38 | 3.09 | 4.06 | 3.22 | 41.99 |
| Days ≥ 0.1" Precip | 5 | 5 | 7 | 8 | 8 | 7 | 7 | 6 | 5 | 5 | 7 | *7* | 77 |
| Total Snowfall (") | na | na | *2.3* | *0.1* | *0.0* | *0.0* | 0.0 | *0.0* | *0.0* | *0.2* | na | na | na |
| Days ≥ 1" Snow Depth | na | na | na | *0* | 0 | *0* | 0 | *0* | *0* | *0* | na | na | na |

## MOUNT VERNON *Posey County* ELEVATION 410 ft LAT/LONG 37° 56 ' N / 87° 54 ' W

| | JAN | FEB | MAR | APR | MAY | JUN | JUL | AUG | SEP | OCT | NOV | DEC | YEAR |
|---|---|---|---|---|---|---|---|---|---|---|---|---|---|
| Maximum Temp °F | 38.5 | 43.2 | 54.6 | 66.7 | 76.2 | 85.1 | 88.8 | 86.9 | 80.8 | 69.3 | 55.9 | 44.2 | 65.9 |
| Minimum Temp °F | 21.8 | 24.9 | 35.0 | 45.5 | 54.3 | 63.3 | 67.4 | 64.6 | 57.6 | 45.1 | 37.0 | 27.4 | 45.3 |
| Mean Temp °F | 30.1 | 34.0 | 44.8 | 56.1 | 65.3 | 74.2 | 78.1 | 75.8 | 69.3 | 57.2 | 46.5 | 35.8 | 55.6 |
| Days Max Temp ≥ 90 °F | 0 | 0 | 0 | 0 | 1 | 8 | 15 | 11 | 5 | 0 | 0 | 0 | 40 |
| Days Max Temp ≤ 32 °F | 10 | 6 | 1 | 0 | 0 | 0 | 0 | 0 | 0 | 0 | 1 | 5 | 23 |
| Days Min Temp ≤ 32 °F | 25 | 21 | 14 | 2 | 0 | 0 | 0 | 0 | 0 | 2 | 11 | 21 | 96 |
| Days Min Temp ≤ 0 °F | 2 | 1 | 0 | 0 | 0 | 0 | 0 | 0 | 0 | 0 | 0 | 1 | 4 |
| Heating Degree Days | 1074 | 867 | 622 | 288 | 90 | 7 | 1 | 2 | 46 | 264 | 551 | 897 | 4709 |
| Cooling Degree Days | 0 | 0 | 3 | 27 | 104 | 297 | 428 | 356 | 182 | 29 | 2 | 0 | 1428 |
| Total Precipitation (") | 3.12 | 3.10 | 4.45 | 4.22 | 4.75 | 3.62 | 4.48 | 2.92 | 2.97 | 3.06 | 4.44 | 3.75 | 44.88 |
| Days ≥ 0.1" Precip | 6 | 6 | 8 | 8 | 8 | 6 | 6 | 5 | 5 | 5 | 7 | 7 | 77 |
| Total Snowfall (") | 4.5 | 4.3 | 2.3 | 0.4 | 0.0 | 0.0 | 0.0 | 0.0 | 0.0 | 0.1 | 0.6 | 2.1 | 14.3 |
| Days ≥ 1" Snow Depth | 7 | 6 | 2 | 0 | 0 | 0 | 0 | 0 | 0 | 0 | 0 | 2 | 17 |

## MUNCIE BALL STATE UN *Delaware County* ELEVATION 961 ft LAT/LONG 40° 12 ' N / 85° 25 ' W

| | JAN | FEB | MAR | APR | MAY | JUN | JUL | AUG | SEP | OCT | NOV | DEC | YEAR |
|---|---|---|---|---|---|---|---|---|---|---|---|---|---|
| Maximum Temp °F | 32.3 | 36.6 | 48.2 | 61.4 | 72.3 | 81.3 | 85.1 | 83.0 | 76.8 | 64.5 | 50.3 | 38.3 | 60.8 |
| Minimum Temp °F | 16.8 | 19.9 | 29.9 | 40.6 | 51.0 | 60.3 | 64.3 | 61.8 | 54.7 | 42.8 | 33.9 | 23.6 | 41.6 |
| Mean Temp °F | 24.6 | 28.3 | 39.1 | 51.0 | 61.6 | 70.8 | 74.8 | 72.4 | 65.7 | 53.6 | 42.1 | 31.0 | 51.3 |
| Days Max Temp ≥ 90 °F | 0 | 0 | 0 | 0 | 0 | 4 | 7 | 4 | 1 | 0 | 0 | 0 | 16 |
| Days Max Temp ≤ 32 °F | 15 | 11 | 3 | 0 | 0 | 0 | 0 | 0 | 0 | 0 | 1 | 9 | 39 |
| Days Min Temp ≤ 32 °F | 28 | 24 | 20 | 7 | 1 | 0 | 0 | 0 | 0 | 5 | 15 | 25 | 125 |
| Days Min Temp ≤ 0 °F | 4 | 3 | 0 | 0 | 0 | 0 | 0 | 0 | 0 | 0 | 0 | 1 | 8 |
| Heating Degree Days | 1247 | 1029 | 797 | 426 | 169 | 25 | 3 | 9 | 88 | 365 | 681 | 1049 | 5888 |
| Cooling Degree Days | 0 | 0 | 1 | 11 | 65 | 193 | 302 | 232 | 103 | 13 | 0 | 0 | 920 |
| Total Precipitation (") | 2.16 | 2.12 | 3.08 | 3.62 | 3.90 | 3.89 | 3.56 | 3.62 | 3.26 | 2.70 | 3.46 | 3.33 | 38.70 |
| Days ≥ 0.1" Precip | 6 | 6 | 7 | 7 | 8 | 7 | 6 | 6 | 6 | 6 | 7 | 8 | 80 |
| Total Snowfall (") | 7.0 | *6.9* | 3.6 | 0.5 | 0.0 | 0.0 | 0.0 | 0.0 | 0.0 | 0.2 | 1.9 | *6.4* | 26.5 |
| Days ≥ 1" Snow Depth | 13 | *10* | 3 | 0 | 0 | 0 | 0 | 0 | 0 | 0 | 1 | *7* | 34 |

## NEW CASTLE 4 N *Henry County* ELEVATION 971 ft LAT/LONG 39° 56 ' N / 85° 23 ' W

| | JAN | FEB | MAR | APR | MAY | JUN | JUL | AUG | SEP | OCT | NOV | DEC | YEAR |
|---|---|---|---|---|---|---|---|---|---|---|---|---|---|
| Maximum Temp °F | 32.6 | 36.5 | 48.3 | 60.8 | 71.9 | 81.1 | 84.5 | 82.6 | 76.0 | 63.7 | 50.1 | 38.4 | 60.5 |
| Minimum Temp °F | 15.7 | 18.0 | 28.5 | 38.1 | 48.2 | 57.7 | 61.9 | 59.6 | 52.6 | 40.7 | 32.5 | 22.5 | 39.7 |
| Mean Temp °F | 24.2 | 27.3 | 38.4 | 49.5 | 60.1 | 69.4 | 73.2 | 71.1 | 64.3 | 52.2 | 41.4 | 30.5 | 50.1 |
| Days Max Temp ≥ 90 °F | 0 | 0 | 0 | 0 | 0 | 4 | 6 | 4 | 1 | 0 | 0 | 0 | 15 |
| Days Max Temp ≤ 32 °F | 15 | 11 | 3 | 0 | 0 | 0 | 0 | 0 | 0 | 0 | 1 | 9 | 39 |
| Days Min Temp ≤ 32 °F | 29 | 25 | 21 | 10 | 1 | 0 | 0 | 0 | 0 | 7 | 17 | 26 | 136 |
| Days Min Temp ≤ 0 °F | 5 | 3 | 0 | 0 | 0 | 0 | 0 | 0 | 0 | 0 | 0 | 2 | 10 |
| Heating Degree Days | 1259 | 1059 | 817 | 466 | 198 | 33 | 4 | 13 | 108 | 400 | 703 | 1063 | 6123 |
| Cooling Degree Days | 0 | 0 | 1 | 7 | 50 | 172 | 275 | 219 | 94 | 11 | 0 | 0 | 829 |
| Total Precipitation (") | 2.05 | 2.15 | 2.94 | 3.88 | 4.43 | 4.22 | 4.59 | 3.76 | 2.93 | 3.11 | 3.83 | 2.96 | 40.85 |
| Days ≥ 0.1" Precip | 5 | 5 | 7 | 8 | 8 | 7 | 7 | 6 | 6 | 6 | 7 | 7 | 79 |
| Total Snowfall (") | na | na | *1.7* | 0.3 | 0.0 | 0.0 | 0.0 | 0.0 | 0.0 | 0.0 | *0.6* | na | na |
| Days ≥ 1" Snow Depth | na | na | *1* | 0 | 0 | 0 | 0 | 0 | 0 | 0 | *0* | na | na |

## OAKLANDON GEIST RSVR *Marion County* ELEVATION 794 ft LAT/LONG 39° 54 ' N / 85° 59 ' W

| | JAN | FEB | MAR | APR | MAY | JUN | JUL | AUG | SEP | OCT | NOV | DEC | YEAR |
|---|---|---|---|---|---|---|---|---|---|---|---|---|---|
| Maximum Temp °F | 33.5 | 37.8 | 49.2 | 62.1 | 72.5 | 81.2 | 84.9 | 82.8 | 76.7 | 64.7 | 51.1 | 38.9 | 61.3 |
| Minimum Temp °F | 16.2 | 18.5 | 28.9 | 39.3 | 49.7 | 59.1 | 63.4 | 60.6 | 53.6 | 41.3 | 32.9 | 22.9 | 40.5 |
| Mean Temp °F | 24.9 | 28.1 | 39.1 | 50.8 | 61.1 | 70.2 | 74.2 | 71.7 | 65.2 | 53.0 | 42.0 | 30.9 | 50.9 |
| Days Max Temp ≥ 90 °F | 0 | 0 | 0 | 0 | 0 | 3 | 7 | 4 | 1 | 0 | 0 | 0 | 15 |
| Days Max Temp ≤ 32 °F | 15 | 10 | 3 | 0 | 0 | 0 | 0 | 0 | 0 | 0 | 1 | 8 | 37 |
| Days Min Temp ≤ 32 °F | 28 | 25 | 21 | 8 | 1 | 0 | 0 | 0 | 0 | 7 | 16 | 25 | 131 |
| Days Min Temp ≤ 0 °F | 5 | 3 | 0 | 0 | 0 | 0 | 0 | 0 | 0 | 0 | 0 | 2 | 10 |
| Heating Degree Days | 1236 | 1034 | 795 | 431 | 173 | 27 | 4 | 10 | 92 | 377 | 683 | 1050 | 5912 |
| Cooling Degree Days | 0 | 0 | 1 | 10 | 56 | 193 | 305 | 232 | 102 | 13 | 0 | 0 | 912 |
| Total Precipitation (") | 2.36 | 2.38 | 3.17 | 4.00 | 4.45 | 3.52 | 4.67 | 4.04 | 3.48 | 3.16 | 3.87 | 3.33 | 42.43 |
| Days ≥ 0.1" Precip | 5 | 5 | 7 | 9 | 8 | 6 | 7 | 6 | 6 | 6 | 7 | 7 | 79 |
| Total Snowfall (") | *7.0* | 6.9 | 2.9 | 0.4 | 0.0 | 0.0 | 0.0 | 0.0 | 0.0 | 0.3 | 1.1 | 5.2 | 23.8 |
| Days ≥ 1" Snow Depth | 10 | 8 | 3 | 0 | 0 | 0 | 0 | 0 | 0 | 0 | 0 | 5 | 26 |

## OOLITIC PURDUE EXP F *Lawrence County* ELEVATION 650 ft LAT/LONG 38° 53 ' N / 86° 32 ' W

| | JAN | FEB | MAR | APR | MAY | JUN | JUL | AUG | SEP | OCT | NOV | DEC | YEAR |
|---|---|---|---|---|---|---|---|---|---|---|---|---|---|
| Maximum Temp °F | 36.4 | 41.2 | 52.4 | 64.2 | 73.6 | 82.0 | 85.9 | 84.1 | 78.0 | 66.3 | 53.6 | 42.1 | 63.3 |
| Minimum Temp °F | 17.3 | 19.7 | 30.0 | 40.3 | 49.6 | 58.9 | 63.3 | 60.9 | 53.3 | 40.3 | 32.9 | 23.4 | 40.8 |
| Mean Temp °F | 26.9 | 30.5 | 41.2 | 52.3 | 61.6 | 70.5 | 74.6 | 72.5 | 65.7 | 53.3 | 43.3 | 32.8 | 52.1 |
| Days Max Temp ≥ 90 °F | 0 | 0 | 0 | 0 | 0 | 4 | 8 | 6 | 2 | 0 | 0 | 0 | 20 |
| Days Max Temp ≤ 32 °F | 12 | 7 | 2 | 0 | 0 | 0 | 0 | 0 | 0 | 0 | 1 | 6 | 28 |
| Days Min Temp ≤ 32 °F | 28 | 24 | 20 | 7 | 1 | 0 | 0 | 0 | 0 | 8 | 16 | 24 | 128 |
| Days Min Temp ≤ 0 °F | 4 | 3 | 0 | 0 | 0 | 0 | 0 | 0 | 0 | 0 | 0 | 1 | 8 |
| Heating Degree Days | 1176 | 969 | 732 | 386 | 158 | 22 | 3 | 7 | 86 | 368 | 646 | 992 | 5545 |
| Cooling Degree Days | 0 | 0 | 2 | 12 | 61 | 201 | 329 | 261 | 117 | 14 | 0 | 0 | 997 |
| Total Precipitation (") | 2.61 | 2.53 | 3.72 | 4.19 | 4.67 | 3.63 | 4.62 | 4.12 | 3.11 | 3.41 | 4.01 | 3.41 | 44.03 |
| Days ≥ 0.1" Precip | 5 | 5 | 7 | 8 | 8 | 6 | 7 | 6 | 5 | 6 | 7 | 6 | 76 |
| Total Snowfall (") | 5.7 | 4.6 | 2.7 | 0.1 | 0.0 | 0.0 | 0.0 | 0.0 | 0.0 | 0.1 | 0.8 | 2.7 | 16.7 |
| Days ≥ 1" Snow Depth | 9 | 7 | 2 | 0 | 0 | 0 | 0 | 0 | 0 | 0 | 1 | 3 | 22 |

## PAOLI *Orange County* ELEVATION 610 ft LAT/LONG 38° 34 ' N / 86° 28 ' W

| | JAN | FEB | MAR | APR | MAY | JUN | JUL | AUG | SEP | OCT | NOV | DEC | YEAR |
|---|---|---|---|---|---|---|---|---|---|---|---|---|---|
| Maximum Temp °F | 39.2 | 44.2 | 55.4 | 66.6 | 76.1 | 83.8 | 87.7 | 85.8 | 79.3 | 68.6 | 55.4 | 44.1 | 65.5 |
| Minimum Temp °F | 18.7 | 21.0 | 31.4 | 41.3 | 50.2 | 59.3 | 63.8 | 61.3 | 54.0 | 41.3 | 33.5 | 24.2 | 41.7 |
| Mean Temp °F | 28.9 | 32.6 | 43.4 | 53.9 | 63.2 | 71.6 | 75.8 | 73.6 | 66.7 | 55.0 | 44.5 | 34.2 | 53.6 |
| Days Max Temp ≥ 90 °F | 0 | 0 | 0 | 0 | 1 | 6 | 12 | 8 | 3 | 0 | 0 | 0 | 30 |
| Days Max Temp ≤ 32 °F | 9 | 5 | 1 | 0 | 0 | 0 | 0 | 0 | 0 | 0 | 1 | 5 | 21 |
| Days Min Temp ≤ 32 °F | 27 | 23 | 17 | 7 | 1 | 0 | 0 | 0 | 0 | 7 | 15 | 24 | 121 |
| Days Min Temp ≤ 0 °F | 3 | 2 | 0 | 0 | 0 | 0 | 0 | 0 | 0 | 0 | 0 | 1 | 6 |
| Heating Degree Days | 1110 | 908 | 665 | 344 | 128 | 17 | 2 | 5 | 74 | 324 | 609 | 950 | 5136 |
| Cooling Degree Days | 0 | ***0*** | 4 | 20 | 85 | 239 | 384 | 302 | 136 | 24 | 1 | 1 | 1196 |
| Total Precipitation (") | 2.99 | 2.93 | 4.33 | 4.90 | 4.70 | 3.77 | 4.55 | 4.09 | 3.29 | 3.22 | 4.39 | 3.59 | 46.75 |
| Days ≥ 0.1" Precip | 6 | ***6*** | 8 | 9 | 8 | 6 | 7 | 6 | 5 | 6 | 7 | 6 | 80 |
| Total Snowfall (") | na | na | na | 0.1 | 0.0 | 0.0 | 0.0 | 0.0 | 0.0 | 0.0 | 0.7 | na | na |
| Days ≥ 1" Snow Depth | na | na | na | 0 | 0 | 0 | 0 | 0 | 0 | 0 | 0 | ***1*** | na |

## PRINCETON 1 W *Gibson County* ELEVATION 479 ft LAT/LONG 38° 22 ' N / 87° 34 ' W

| | JAN | FEB | MAR | APR | MAY | JUN | JUL | AUG | SEP | OCT | NOV | DEC | YEAR |
|---|---|---|---|---|---|---|---|---|---|---|---|---|---|
| Maximum Temp °F | 37.6 | 43.0 | 54.5 | 66.4 | 75.9 | 85.1 | 88.4 | 86.4 | 80.1 | 68.5 | 54.6 | 43.0 | 65.3 |
| Minimum Temp °F | 21.5 | 24.8 | 34.7 | 45.1 | 54.2 | 63.2 | 67.4 | 64.7 | 58.1 | 46.1 | 37.2 | 27.3 | 45.4 |
| Mean Temp °F | 29.5 | 33.9 | 44.7 | 55.8 | 65.1 | 74.2 | 78.0 | 75.6 | 69.1 | 57.3 | 45.9 | 35.2 | 55.4 |
| Days Max Temp ≥ 90 °F | 0 | 0 | 0 | 0 | 1 | 8 | 13 | 10 | 4 | 0 | 0 | 0 | 36 |
| Days Max Temp ≤ 32 °F | 11 | 6 | 1 | 0 | 0 | 0 | 0 | 0 | 0 | 0 | 0 | 5 | 23 |
| Days Min Temp ≤ 32 °F | 26 | 20 | 14 | 3 | 0 | 0 | 0 | 0 | 0 | 2 | 11 | 21 | 97 |
| Days Min Temp ≤ 0 °F | 2 | 1 | 0 | 0 | 0 | 0 | 0 | 0 | 0 | 0 | 0 | 1 | 4 |
| Heating Degree Days | 1092 | 872 | 627 | 296 | 98 | 7 | 0 | 1 | 48 | 263 | 568 | 918 | 4790 |
| Cooling Degree Days | 0 | 0 | 3 | 25 | 108 | 298 | 428 | 354 | 180 | 30 | 1 | 0 | 1427 |
| Total Precipitation (") | 2.66 | 2.67 | 4.24 | 4.49 | 4.93 | 3.74 | 4.15 | 4.02 | 3.29 | 3.41 | 4.54 | 3.85 | 45.99 |
| Days ≥ 0.1" Precip | 6 | 5 | 8 | 8 | 7 | 6 | 6 | 6 | 5 | 5 | 7 | 7 | 76 |
| Total Snowfall (") | na | na | ***0.8*** | 0.3 | 0.0 | 0.0 | 0.0 | 0.0 | 0.0 | 0.1 | 0.4 | ***1.4*** | na |
| Days ≥ 1" Snow Depth | na | na | 1 | 0 | 0 | 0 | 0 | 0 | 0 | 0 | 0 | ***1*** | na |

## RENSSELAER *Jasper County* ELEVATION 650 ft LAT/LONG 40° 56 ' N / 87° 9 ' W

| | JAN | FEB | MAR | APR | MAY | JUN | JUL | AUG | SEP | OCT | NOV | DEC | YEAR |
|---|---|---|---|---|---|---|---|---|---|---|---|---|---|
| Maximum Temp °F | ***31.0*** | ***35.0*** | ***46.8*** | ***59.9*** | ***72.7*** | ***81.5*** | ***84.6*** | ***82.2*** | ***75.4*** | ***63.2*** | ***48.9*** | ***35.5*** | 59.7 |
| Minimum Temp °F | ***15.0*** | ***17.6*** | ***28.9*** | ***38.9*** | ***50.1*** | ***59.1*** | ***63.3*** | ***60.6*** | ***52.7*** | ***40.7*** | ***32.2*** | ***20.9*** | 40.0 |
| Mean Temp °F | ***23.0*** | ***26.3*** | ***37.9*** | ***49.3*** | ***61.4*** | ***70.4*** | ***74.0*** | ***71.5*** | ***64.1*** | ***52.0*** | ***40.6*** | ***28.1*** | 49.9 |
| Days Max Temp ≥ 90 °F | ***0*** | ***0*** | ***0*** | ***0*** | ***2*** | ***5*** | ***8*** | ***4*** | ***1*** | ***0*** | ***0*** | ***0*** | 20 |
| Days Max Temp ≤ 32 °F | ***16*** | ***12*** | ***4*** | ***0*** | ***0*** | ***0*** | ***0*** | ***0*** | ***0*** | ***0*** | ***2*** | ***10*** | 44 |
| Days Min Temp ≤ 32 °F | ***29*** | ***25*** | ***20*** | ***7*** | ***0*** | ***0*** | ***0*** | ***0*** | ***0*** | ***6*** | ***16*** | ***26*** | 129 |
| Days Min Temp ≤ 0 °F | ***6*** | ***4*** | ***0*** | ***0*** | ***0*** | ***0*** | ***0*** | ***0*** | ***0*** | ***0*** | ***0*** | ***2*** | 12 |
| Heating Degree Days | ***1297*** | ***1086*** | ***836*** | ***476*** | ***175*** | ***30*** | ***6*** | ***13*** | ***114*** | ***405*** | ***725*** | ***1136*** | 6299 |
| Cooling Degree Days | 0 | 0 | 1 | 12 | 67 | 204 | 302 | 229 | 97 | 7 | 0 | 0 | 919 |
| Total Precipitation (") | ***1.74*** | ***1.73*** | ***3.18*** | ***3.36*** | ***4.04*** | ***4.03*** | ***3.75*** | ***3.71*** | ***3.76*** | ***3.22*** | ***3.49*** | ***2.87*** | 38.88 |
| Days ≥ 0.1" Precip | na | na | na | ***7*** | ***7*** | ***7*** | ***6*** | ***6*** | ***7*** | ***6*** | ***7*** | ***7*** | na |
| Total Snowfall (") | na | na | na | na | na | ***0.0*** | ***0.0*** | ***0.0*** | ***0.0*** | ***0.1*** | na | na | na |
| Days ≥ 1" Snow Depth | na | na | na | ***0*** | ***0*** | ***0*** | ***0*** | ***0*** | ***0*** | ***0*** | na | na | na |

## ROCHESTER *Fulton County* ELEVATION 770 ft LAT/LONG 41° 4 ' N / 86° 13 ' W

| | JAN | FEB | MAR | APR | MAY | JUN | JUL | AUG | SEP | OCT | NOV | DEC | YEAR |
|---|---|---|---|---|---|---|---|---|---|---|---|---|---|
| Maximum Temp °F | 30.1 | 34.3 | 46.1 | 59.5 | 71.1 | 80.4 | 84.1 | 81.8 | 75.2 | 62.8 | 48.6 | 36.2 | 59.2 |
| Minimum Temp °F | 13.4 | 16.0 | 26.5 | 37.2 | 47.8 | 57.4 | 61.6 | 58.8 | 51.6 | 39.8 | 31.3 | 20.8 | 38.5 |
| Mean Temp °F | 21.8 | 25.2 | 36.3 | 48.4 | 59.5 | 68.9 | 72.8 | 70.3 | 63.4 | 51.3 | 40.0 | 28.5 | 48.9 |
| Days Max Temp ≥ 90 °F | 0 | 0 | 0 | 0 | 1 | 4 | 7 | 4 | 1 | 0 | 0 | 0 | 17 |
| Days Max Temp ≤ 32 °F | 17 | 13 | 4 | 0 | 0 | 0 | 0 | 0 | 0 | 0 | 2 | 10 | 46 |
| Days Min Temp ≤ 32 °F | 29 | 26 | 23 | 10 | 1 | 0 | 0 | 0 | 0 | 7 | 18 | 27 | 141 |
| Days Min Temp ≤ 0 °F | 6 | 4 | 0 | 0 | 0 | 0 | 0 | 0 | 0 | 0 | 0 | 2 | 12 |
| Heating Degree Days | 1334 | 1118 | 883 | 500 | 217 | 41 | 8 | 21 | 128 | 426 | 745 | 1124 | 6545 |
| Cooling Degree Days | 0 | 0 | 1 | 10 | 55 | 168 | 270 | 202 | 84 | 6 | 0 | 0 | 796 |
| Total Precipitation (") | 1.91 | 1.59 | 2.59 | 3.86 | 3.86 | 4.00 | 3.59 | 3.55 | 3.76 | 2.97 | 3.64 | 3.08 | 38.40 |
| Days ≥ 0.1" Precip | 5 | 5 | 6 | 8 | 7 | 7 | 7 | 6 | 7 | 6 | 7 | 7 | 78 |
| Total Snowfall (") | ***7.9*** | 8.7 | 4.0 | 1.2 | 0.0 | 0.0 | 0.0 | 0.0 | 0.0 | 0.3 | 2.8 | 6.6 | 31.5 |
| Days ≥ 1" Snow Depth | na | na | na | 0 | 0 | 0 | 0 | 0 | 0 | 0 | ***1*** | na | na |

## ROCKVILLE *Parke County* ELEVATION 689 ft LAT/LONG 39° 46 ' N / 87° 14 ' W

| | JAN | FEB | MAR | APR | MAY | JUN | JUL | AUG | SEP | OCT | NOV | DEC | YEAR |
|---|---|---|---|---|---|---|---|---|---|---|---|---|---|
| Maximum Temp °F | 35.4 | 40.8 | 53.1 | 66.0 | 75.7 | 84.4 | 87.4 | 84.9 | 78.9 | 67.3 | 53.3 | 40.8 | 64.0 |
| Minimum Temp °F | 17.5 | 21.3 | 31.7 | 42.0 | 51.2 | 60.3 | 64.6 | 61.9 | 55.6 | 43.7 | 34.8 | 24.3 | 42.4 |
| Mean Temp °F | 26.5 | 31.1 | 42.4 | 54.0 | 63.5 | 72.4 | 76.0 | 73.5 | 67.2 | 55.5 | 44.0 | 32.6 | 53.2 |
| Days Max Temp ≥ 90 °F | 0 | 0 | 0 | 0 | 1 | 6 | 11 | 6 | 2 | 0 | 0 | 0 | 26 |
| Days Max Temp ≤ 32 °F | 13 | 8 | 1 | 0 | 0 | 0 | 0 | 0 | 0 | 0 | 1 | 6 | 29 |
| Days Min Temp ≤ 32 °F | 27 | 23 | 17 | 6 | 1 | 0 | 0 | 0 | 0 | 4 | 14 | 24 | 116 |
| Days Min Temp ≤ 0 °F | 4 | 3 | 0 | 0 | 0 | 0 | 0 | 0 | 0 | 0 | 0 | 1 | 8 |
| Heating Degree Days | 1189 | 952 | 696 | 341 | 123 | 12 | 1 | 6 | 63 | 305 | 625 | 997 | 5310 |
| Cooling Degree Days | 0 | 0 | 3 | 20 | 86 | 254 | 367 | 291 | 138 | 19 | 1 | 0 | 1179 |
| Total Precipitation (") | 2.39 | 2.10 | 3.40 | 4.21 | 4.51 | 3.71 | 4.94 | 3.99 | 3.55 | 3.11 | 4.35 | 3.96 | 44.22 |
| Days ≥ 0.1" Precip | 5 | 5 | 7 | 8 | 8 | 7 | 6 | 6 | 6 | 6 | 7 | 7 | 78 |
| Total Snowfall (") | *4.5* | *4.3* | 2.2 | 0.2 | 0.0 | 0.0 | 0.0 | 0.0 | 0.0 | 0.0 | 0.6 | 5.5 | 17.3 |
| Days ≥ 1" Snow Depth | *10* | *8* | 2 | 0 | 0 | 0 | 0 | 0 | 0 | 0 | 0 | 5 | 25 |

## RUSHVILLE SEWGE PLT *Rush County* ELEVATION 991 ft LAT/LONG 39° 37 ' N / 85° 26 ' W

| | JAN | FEB | MAR | APR | MAY | JUN | JUL | AUG | SEP | OCT | NOV | DEC | YEAR |
|---|---|---|---|---|---|---|---|---|---|---|---|---|---|
| Maximum Temp °F | 33.4 | 37.8 | 49.0 | 62.0 | 72.5 | 81.2 | 84.3 | 82.5 | 76.5 | 64.5 | 50.9 | 39.2 | 61.2 |
| Minimum Temp °F | 15.8 | 18.3 | 29.0 | 39.5 | 50.0 | 59.1 | 62.9 | 60.0 | 53.0 | 40.6 | 32.6 | 23.1 | 40.3 |
| Mean Temp °F | 24.6 | 28.1 | 38.9 | 50.8 | 61.3 | 70.2 | 73.6 | 71.3 | 64.8 | 52.6 | 41.8 | 31.2 | 50.8 |
| Days Max Temp ≥ 90 °F | 0 | 0 | 0 | 0 | 0 | 3 | 6 | 3 | 1 | 0 | 0 | 0 | 13 |
| Days Max Temp ≤ 32 °F | 14 | 10 | 3 | 0 | 0 | 0 | 0 | 0 | 0 | 0 | 1 | 8 | 36 |
| Days Min Temp ≤ 32 °F | 28 | 25 | 21 | 8 | 1 | 0 | 0 | 0 | 0 | 7 | 17 | 25 | 132 |
| Days Min Temp ≤ 0 °F | 5 | 3 | 0 | 0 | 0 | 0 | 0 | 0 | 0 | 0 | 0 | 2 | 10 |
| Heating Degree Days | 1248 | 1036 | 802 | 428 | 170 | 26 | 4 | 11 | 97 | 387 | 690 | 1040 | 5939 |
| Cooling Degree Days | 0 | 0 | 1 | 10 | 66 | 199 | 298 | 222 | 100 | 11 | 0 | 0 | 907 |
| Total Precipitation (") | 2.39 | 2.49 | 3.07 | 3.94 | 4.82 | 4.01 | 4.72 | 3.65 | 3.06 | 2.91 | 3.77 | 3.19 | 42.02 |
| Days ≥ 0.1" Precip | 6 | 5 | 7 | 8 | 8 | 7 | 7 | 6 | 6 | 6 | 7 | 7 | 80 |
| Total Snowfall (") | 5.1 | 4.8 | 1.9 | 0.3 | 0.0 | 0.0 | 0.0 | 0.0 | 0.0 | 0.3 | 0.9 | 3.1 | 16.4 |
| Days ≥ 1" Snow Depth | 9 | 7 | 2 | 0 | 0 | 0 | 0 | 0 | 0 | 0 | 1 | 4 | 23 |

## SAINT MEINRAD *Spencer County* ELEVATION 541 ft LAT/LONG 38° 10 ' N / 86° 48 ' W

| | JAN | FEB | MAR | APR | MAY | JUN | JUL | AUG | SEP | OCT | NOV | DEC | YEAR |
|---|---|---|---|---|---|---|---|---|---|---|---|---|---|
| Maximum Temp °F | 40.3 | 45.8 | 57.2 | 68.5 | 76.7 | 84.6 | 87.6 | 86.2 | 80.2 | 69.4 | 56.9 | 45.4 | 66.6 |
| Minimum Temp °F | 22.6 | 25.7 | 35.4 | 44.9 | 52.9 | 61.9 | 66.0 | 64.1 | 57.7 | 45.2 | 37.3 | 28.0 | 45.1 |
| Mean Temp °F | 31.5 | 35.8 | 46.3 | 56.7 | 64.9 | 73.3 | 76.8 | 75.2 | 69.0 | 57.3 | 47.1 | 36.8 | 55.9 |
| Days Max Temp ≥ 90 °F | 0 | 0 | 0 | 0 | 0 | 5 | 11 | 9 | 3 | 0 | 0 | 0 | 28 |
| Days Max Temp ≤ 32 °F | 8 | 5 | 0 | 0 | 0 | 0 | 0 | 0 | 0 | 0 | 0 | 4 | 17 |
| Days Min Temp ≤ 32 °F | 24 | 20 | 14 | 4 | 0 | 0 | 0 | 0 | 0 | 4 | 11 | 20 | 97 |
| Days Min Temp ≤ 0 °F | 2 | 1 | 0 | 0 | 0 | 0 | 0 | 0 | 0 | 0 | 0 | 1 | 4 |
| Heating Degree Days | 1033 | 818 | 576 | 267 | 92 | 7 | 0 | 2 | 45 | 260 | 534 | 869 | 4503 |
| Cooling Degree Days | 0 | 0 | 5 | 25 | 94 | 266 | 385 | 334 | 170 | 30 | 3 | 1 | 1313 |
| Total Precipitation (") | 3.10 | 3.00 | 4.02 | 4.48 | 4.30 | 3.63 | 4.60 | 3.99 | 3.47 | 3.04 | 4.04 | 3.60 | 45.27 |
| Days ≥ 0.1" Precip | 6 | 6 | 8 | 8 | 8 | 7 | 7 | 6 | 6 | 6 | 7 | 7 | 82 |
| Total Snowfall (") | *3.6* | *3.3* | *1.5* | 0.0 | 0.0 | 0.0 | 0.0 | 0.0 | 0.0 | 0.0 | 0.4 | 1.6 | 10.4 |
| Days ≥ 1" Snow Depth | *5* | 6 | *1* | 0 | 0 | 0 | 0 | 0 | 0 | 0 | 0 | 2 | 14 |

## SALEM *Washington County* ELEVATION 722 ft LAT/LONG 38° 36 ' N / 86° 6 ' W

| | JAN | FEB | MAR | APR | MAY | JUN | JUL | AUG | SEP | OCT | NOV | DEC | YEAR |
|---|---|---|---|---|---|---|---|---|---|---|---|---|---|
| Maximum Temp °F | 39.0 | 44.0 | 55.4 | 67.0 | 75.5 | 84.3 | 87.1 | 85.5 | 79.6 | 68.2 | 55.2 | 44.3 | 65.4 |
| Minimum Temp °F | 21.3 | 23.3 | 33.2 | 42.9 | 51.4 | 60.4 | 64.7 | 62.2 | 55.9 | 43.9 | 35.9 | 27.2 | 43.5 |
| Mean Temp °F | 30.2 | 33.7 | 44.3 | 55.0 | 63.5 | 72.4 | 75.9 | 73.9 | 67.8 | 56.1 | 45.5 | 35.8 | 54.5 |
| Days Max Temp ≥ 90 °F | 0 | 0 | 0 | 0 | 0 | 5 | 10 | 7 | 2 | 0 | 0 | 0 | 24 |
| Days Max Temp ≤ 32 °F | 10 | 6 | 1 | 0 | 0 | 0 | 0 | 0 | 0 | 0 | 0 | 4 | 21 |
| Days Min Temp ≤ 32 °F | 25 | 22 | 16 | 5 | 1 | 0 | 0 | 0 | 0 | 5 | 12 | 21 | 107 |
| Days Min Temp ≤ 0 °F | 2 | 2 | 0 | 0 | 0 | 0 | 0 | 0 | 0 | 0 | 0 | 1 | 5 |
| Heating Degree Days | 1073 | 877 | 636 | 313 | 119 | 11 | 1 | 4 | 59 | 294 | 578 | 899 | 4864 |
| Cooling Degree Days | 0 | 0 | 4 | 19 | 82 | 235 | 361 | 298 | 142 | 23 | 1 | 0 | 1165 |
| Total Precipitation (") | 3.11 | 2.87 | 4.18 | 4.51 | 4.57 | 3.41 | 4.89 | 3.59 | 3.14 | 3.06 | 4.17 | *3.80* | 45.30 |
| Days ≥ 0.1" Precip | 6 | 6 | 8 | 9 | 8 | 6 | 7 | 6 | 5 | 6 | 7 | 7 | 81 |
| Total Snowfall (") | 5.4 | 5.8 | 3.6 | 0.2 | 0.0 | 0.0 | 0.0 | 0.0 | 0.0 | 0.2 | 1.1 | 2.4 | 18.7 |
| Days ≥ 1" Snow Depth | 8 | 7 | 1 | 0 | 0 | 0 | 0 | 0 | 0 | 0 | *1* | 3 | 20 |

## SCOTTSBURG *Scott County* ELEVATION 571 ft LAT/LONG 38° 42 ' N / 85° 46 ' W

| | JAN | FEB | MAR | APR | MAY | JUN | JUL | AUG | SEP | OCT | NOV | DEC | YEAR |
|---|---|---|---|---|---|---|---|---|---|---|---|---|---|
| Maximum Temp °F | 38.1 | 43.0 | 54.2 | 65.9 | 75.4 | 83.9 | 87.5 | 85.9 | 79.6 | 68.0 | 54.9 | 43.6 | 65.0 |
| Minimum Temp °F | 19.1 | 21.5 | 31.5 | 41.8 | 51.4 | 60.6 | 64.8 | 62.1 | 54.9 | 41.9 | 34.0 | 25.3 | 42.4 |
| Mean Temp °F | 28.6 | 32.3 | 42.9 | 53.9 | 63.4 | 72.3 | 76.2 | 74.0 | 67.3 | 55.0 | 44.5 | 34.5 | 53.7 |
| Days Max Temp ≥ 90 °F | 0 | 0 | 0 | 0 | 1 | 6 | 12 | 9 | 3 | 0 | 0 | 0 | 31 |
| Days Max Temp ≤ 32 °F | 10 | 6 | 1 | 0 | 0 | 0 | 0 | 0 | 0 | 0 | 1 | 5 | 23 |
| Days Min Temp ≤ 32 °F | 27 | 23 | 18 | 5 | 0 | 0 | 0 | 0 | 0 | 6 | 14 | 23 | 116 |
| Days Min Temp ≤ 0 °F | 3 | 2 | 0 | 0 | 0 | 0 | 0 | 0 | 0 | 0 | 0 | 1 | 6 |
| Heating Degree Days | 1122 | 918 | 682 | 346 | 126 | 15 | 1 | 4 | 66 | 325 | 610 | 941 | 5156 |
| Cooling Degree Days | 0 | 0 | 2 | 19 | 85 | 241 | 373 | 299 | 141 | 21 | 0 | 0 | 1181 |
| Total Precipitation (") | 2.90 | 2.63 | 4.04 | 4.36 | 4.26 | 3.67 | 4.67 | 4.30 | 3.25 | 3.02 | 3.78 | 3.23 | 44.11 |
| Days ≥ 0.1" Precip | 6 | 6 | 8 | 8 | 8 | 7 | 7 | 7 | 6 | 6 | 7 | 7 | 83 |
| Total Snowfall (") | 5.1 | 4.7 | 3.7 | 0.1 | 0.0 | 0.0 | 0.0 | 0.0 | 0.0 | 0.1 | 1.0 | 2.3 | 17.0 |
| Days ≥ 1" Snow Depth | 7 | 6 | 2 | 0 | 0 | 0 | 0 | 0 | 0 | 0 | 0 | 3 | 18 |

## SEYMOUR 2 N *Jackson County* ELEVATION 571 ft LAT/LONG 38° 59 ' N / 85° 54 ' W

| | JAN | FEB | MAR | APR | MAY | JUN | JUL | AUG | SEP | OCT | NOV | DEC | YEAR |
|---|---|---|---|---|---|---|---|---|---|---|---|---|---|
| Maximum Temp °F | 37.1 | 41.6 | 52.9 | 64.9 | 74.2 | 82.3 | 85.6 | 84.2 | 78.4 | 67.1 | 53.9 | 42.2 | 63.7 |
| Minimum Temp °F | 17.9 | 20.3 | 30.0 | 40.2 | 50.1 | 59.4 | 63.1 | 60.1 | 52.5 | 39.9 | 32.7 | 24.1 | 40.9 |
| Mean Temp °F | 27.5 | 31.0 | 41.5 | 52.6 | 62.2 | 70.8 | 74.4 | 72.1 | 65.5 | 53.5 | 43.3 | 33.2 | 52.3 |
| Days Max Temp ≥ 90 °F | 0 | 0 | 0 | 0 | 1 | 4 | 8 | 6 | 2 | 0 | 0 | 0 | 21 |
| Days Max Temp ≤ 32 °F | 11 | 7 | 1 | 0 | 0 | 0 | 0 | 0 | 0 | 0 | 1 | 6 | 26 |
| Days Min Temp ≤ 32 °F | 27 | 24 | 20 | 7 | 0 | 0 | 0 | 0 | 0 | 8 | 16 | 25 | 127 |
| Days Min Temp ≤ 0 °F | 4 | 2 | 0 | 0 | 0 | 0 | 0 | 0 | 0 | 0 | 0 | 1 | 7 |
| Heating Degree Days | 1155 | 954 | 723 | 379 | 150 | 20 | 2 | 8 | 88 | 362 | 644 | 980 | 5465 |
| Cooling Degree Days | 0 | 0 | 1 | 13 | 66 | 205 | 312 | 243 | 108 | 15 | 0 | 0 | 963 |
| Total Precipitation (") | 2.76 | 2.54 | 3.57 | 4.19 | 4.50 | 3.50 | 4.64 | 3.99 | 2.99 | 3.30 | 4.04 | 3.28 | 43.30 |
| Days ≥ 0.1" Precip | 6 | 6 | 8 | 8 | 8 | 7 | 7 | 7 | 6 | 6 | 8 | 7 | 84 |
| Total Snowfall (") | na | na | *1.8* | 0.1 | 0.0 | 0.0 | 0.0 | 0.0 | 0.0 | 0.2 | 0.9 | *1.7* | na |
| Days ≥ 1" Snow Depth | 8 | 6 | 2 | 0 | 0 | 0 | 0 | 0 | 0 | 0 | 0 | 2 | 18 |

## SHELBYVILLE SEWAGE P *Shelby County* ELEVATION 771 ft LAT/LONG 39° 31 ' N / 85° 46 ' W

| | JAN | FEB | MAR | APR | MAY | JUN | JUL | AUG | SEP | OCT | NOV | DEC | YEAR |
|---|---|---|---|---|---|---|---|---|---|---|---|---|---|
| Maximum Temp °F | 34.1 | 38.8 | 51.3 | 63.0 | 73.3 | 82.4 | 85.8 | 83.3 | 77.6 | 65.6 | ***52.1*** | 39.9 | 62.3 |
| Minimum Temp °F | 15.9 | 18.7 | 30.3 | 40.6 | 50.7 | 59.9 | 63.7 | 60.5 | 54.0 | 41.1 | ***32.8*** | 23.0 | 40.9 |
| Mean Temp °F | 25.0 | 28.7 | 40.8 | 51.8 | 62.0 | 71.2 | 74.8 | 71.9 | 65.8 | 53.4 | ***42.5*** | 31.4 | 51.6 |
| Days Max Temp ≥ 90 °F | 0 | 0 | 0 | 0 | 0 | 4 | 8 | 4 | 1 | 0 | ***0*** | 0 | 17 |
| Days Max Temp ≤ 32 °F | 13 | 9 | 2 | 0 | 0 | 0 | 0 | 0 | 0 | 0 | ***1*** | 7 | 32 |
| Days Min Temp ≤ 32 °F | 28 | 25 | 20 | 7 | 1 | 0 | 0 | 0 | 0 | 7 | ***16*** | 25 | 129 |
| Days Min Temp ≤ 0 °F | 5 | 3 | 0 | 0 | 0 | 0 | 0 | 0 | 0 | 0 | ***0*** | 1 | 9 |
| Heating Degree Days | 1232 | 1017 | 745 | 401 | 153 | 21 | 3 | 9 | 84 | 366 | ***669*** | 1034 | 5734 |
| Cooling Degree Days | 0 | 0 | 1 | 14 | 66 | 229 | 335 | 230 | 118 | 12 | na | ***0*** | na |
| Total Precipitation (") | 2.36 | 2.35 | 3.26 | 3.73 | 4.41 | 3.39 | 4.32 | 3.47 | 2.97 | 2.82 | 3.65 | 2.82 | 39.55 |
| Days ≥ 0.1" Precip | 5 | 5 | 7 | 8 | 7 | 7 | 7 | 6 | 6 | 5 | *7* | 6 | 76 |
| Total Snowfall (") | na | na | na | 0.0 | 0.0 | 0.0 | 0.0 | 0.0 | 0.0 | 0.0 | na | na | na |
| Days ≥ 1" Snow Depth | na | na | na | 0 | 0 | 0 | 0 | 0 | 0 | 0 | na | na | na |

## SHOALS HIWAY 50 BRDG *Martin County* ELEVATION 522 ft LAT/LONG 38° 40 ' N / 86° 48 ' W

| | JAN | FEB | MAR | APR | MAY | JUN | JUL | AUG | SEP | OCT | NOV | DEC | YEAR |
|---|---|---|---|---|---|---|---|---|---|---|---|---|---|
| Maximum Temp °F | 37.7 | 42.7 | 54.1 | 66.2 | 75.7 | 83.9 | 87.2 | 85.6 | 79.4 | 68.0 | 54.8 | 43.1 | 64.9 |
| Minimum Temp °F | 18.3 | 20.7 | 30.7 | 40.6 | 49.4 | 58.8 | 63.4 | 61.3 | 54.2 | 41.4 | 34.0 | 24.5 | 41.4 |
| Mean Temp °F | 28.0 | 31.8 | 42.4 | 53.5 | 62.6 | 71.4 | 75.4 | 73.5 | 66.8 | 54.7 | 44.4 | 33.8 | 53.2 |
| Days Max Temp ≥ 90 °F | 0 | 0 | 0 | 0 | 1 | 7 | 11 | 9 | 3 | 0 | 0 | 0 | 31 |
| Days Max Temp ≤ 32 °F | 10 | 6 | 1 | 0 | 0 | 0 | 0 | 0 | 0 | 0 | 1 | 5 | 23 |
| Days Min Temp ≤ 32 °F | 27 | 23 | 19 | 8 | 1 | 0 | 0 | 0 | 0 | 7 | 15 | 23 | 123 |
| Days Min Temp ≤ 0 °F | 4 | 2 | 0 | 0 | 0 | 0 | 0 | 0 | 0 | 0 | 0 | 1 | 7 |
| Heating Degree Days | 1139 | 932 | 695 | 359 | 143 | 18 | 2 | 4 | 72 | 332 | 612 | 959 | 5267 |
| Cooling Degree Days | 0 | 0 | 3 | 19 | 73 | 220 | 348 | 286 | 133 | 19 | 1 | 0 | 1102 |
| Total Precipitation (") | 2.97 | 2.88 | 4.28 | 4.26 | 5.00 | 3.75 | 4.82 | 3.72 | 3.45 | 3.37 | 4.52 | 3.62 | 46.64 |
| Days ≥ 0.1" Precip | 6 | 6 | 8 | 8 | 8 | 6 | 7 | 6 | 6 | 6 | 7 | 7 | 81 |
| Total Snowfall (") | 5.6 | 4.4 | 2.7 | 0.1 | 0.0 | 0.0 | 0.0 | 0.0 | 0.0 | 0.1 | 0.6 | 2.7 | 16.2 |
| Days ≥ 1" Snow Depth | 9 | 7 | 2 | 0 | 0 | 0 | 0 | 0 | 0 | 0 | 0 | 3 | 21 |

## SOUTH BEND MICHIANA *St. Joseph County* ELEVATION 801 ft LAT/LONG 41° 45 ' N / 86° 10 ' W

| | JAN | FEB | MAR | APR | MAY | JUN | JUL | AUG | SEP | OCT | NOV | DEC | YEAR |
|---|---|---|---|---|---|---|---|---|---|---|---|---|---|
| Maximum Temp °F | 30.7 | 34.3 | 45.9 | 58.8 | 70.2 | 79.5 | 83.0 | 80.7 | 73.7 | 61.8 | 48.1 | 36.2 | 58.6 |
| Minimum Temp °F | 16.5 | 19.1 | 29.0 | 39.0 | 48.8 | 58.7 | 63.2 | 61.4 | 54.3 | 43.3 | 33.9 | 23.4 | 40.9 |
| Mean Temp °F | 23.6 | 26.7 | 37.5 | 48.9 | 59.6 | 69.1 | 73.1 | 71.1 | 64.0 | 52.6 | 41.0 | 29.8 | 49.8 |
| Days Max Temp ≥ 90 °F | 0 | 0 | 0 | 0 | 1 | 3 | 5 | 3 | 1 | 0 | 0 | 0 | 13 |
| Days Max Temp ≤ 32 °F | 17 | 13 | 4 | 0 | 0 | 0 | 0 | 0 | 0 | 0 | 2 | 10 | 46 |
| Days Min Temp ≤ 32 °F | 28 | 24 | 20 | 8 | 1 | 0 | 0 | 0 | 0 | 3 | 14 | 25 | 123 |
| Days Min Temp ≤ 0 °F | 4 | 3 | 0 | 0 | 0 | 0 | 0 | 0 | 0 | 0 | 0 | 1 | 8 |
| Heating Degree Days | 1276 | 1074 | 848 | 486 | 216 | 40 | 6 | 16 | 110 | 388 | 713 | 1083 | 6256 |
| Cooling Degree Days | 0 | 0 | 1 | 13 | 59 | 175 | 289 | 228 | 91 | 7 | 0 | 0 | 863 |
| Total Precipitation (") | 2.30 | 1.93 | 2.95 | 3.76 | 3.26 | 4.22 | 3.64 | 3.78 | 4.04 | 3.40 | 3.52 | 3.32 | 40.12 |
| Days ≥ 0.1" Precip | 6 | 5 | 7 | 8 | 6 | 7 | 7 | 7 | 7 | 7 | 8 | 8 | 83 |
| Total Snowfall (") | 22.8 | 16.8 | 9.5 | 2.2 | 0.0 | 0.0 | 0.0 | 0.0 | 0.0 | 0.7 | 8.4 | 19.1 | 79.5 |
| Days ≥ 1" Snow Depth | 22 | 18 | 8 | 1 | 0 | 0 | 0 | 0 | 0 | 0 | 4 | 15 | 68 |

## SPENCER *Owen County* ELEVATION 561 ft LAT/LONG 39° 17 ' N / 86° 46 ' W

| | JAN | FEB | MAR | APR | MAY | JUN | JUL | AUG | SEP | OCT | NOV | DEC | YEAR |
|---|---|---|---|---|---|---|---|---|---|---|---|---|---|
| Maximum Temp °F | 35.2 | 40.0 | 51.5 | 63.7 | 73.6 | 82.1 | 85.8 | 83.8 | 77.5 | 66.0 | 52.8 | 41.0 | 62.8 |
| Minimum Temp °F | 16.2 | 18.7 | 28.9 | 38.8 | 48.5 | 58.5 | 62.8 | 60.3 | 52.6 | 39.6 | 32.2 | 23.0 | 40.0 |
| Mean Temp °F | 25.7 | 29.4 | 40.2 | 51.3 | 61.1 | 70.4 | 74.3 | 72.1 | 65.1 | 52.8 | 42.5 | 32.0 | 51.4 |
| Days Max Temp ≥ 90 °F | 0 | 0 | 0 | 0 | 0 | 4 | 8 | 5 | 1 | 0 | 0 | 0 | 18 |
| Days Max Temp ≤ 32 °F | 13 | 8 | 2 | 0 | 0 | 0 | 0 | 0 | 0 | 0 | 1 | 7 | 31 |
| Days Min Temp ≤ 32 °F | 28 | 24 | 21 | 9 | 1 | 0 | 0 | 0 | 0 | 9 | 17 | 25 | 134 |
| Days Min Temp ≤ 0 °F | 5 | 3 | 0 | 0 | 0 | 0 | 0 | 0 | 0 | 0 | 0 | 2 | 10 |
| Heating Degree Days | 1210 | 999 | 762 | 415 | 170 | 23 | 3 | 9 | 94 | 383 | 668 | 1016 | 5752 |
| Cooling Degree Days | 0 | 0 | 1 | 8 | 46 | 182 | 295 | 227 | 95 | 9 | 0 | 0 | 863 |
| Total Precipitation (") | 2.48 | 2.52 | 3.70 | 4.17 | 4.77 | 3.95 | 4.85 | 4.30 | 3.45 | 3.18 | 4.22 | 3.45 | 45.04 |
| Days ≥ 0.1" Precip | 6 | 5 | 7 | 8 | 8 | 7 | 7 | 6 | 5 | 5 | 7 | 7 | 78 |
| Total Snowfall (") | *5.1* | 5.6 | 2.4 | 0.3 | 0.0 | 0.0 | 0.0 | 0.0 | 0.0 | 0.1 | 0.6 | 2.7 | 16.8 |
| Days ≥ 1" Snow Depth | 10 | 7 | 2 | 0 | 0 | 0 | 0 | 0 | 0 | 0 | 1 | 3 | 23 |

## SPURGEON 2 N *Pike County* ELEVATION 440 ft LAT/LONG 38° 17 ' N / 87° 15 ' W

| | JAN | FEB | MAR | APR | MAY | JUN | JUL | AUG | SEP | OCT | NOV | DEC | YEAR |
|---|---|---|---|---|---|---|---|---|---|---|---|---|---|
| Maximum Temp °F | 38.1 | *42.9* | 54.6 | 66.6 | 75.9 | 85.1 | 88.8 | 86.8 | *79.9* | 69.3 | 55.4 | 43.8 | 65.6 |
| Minimum Temp °F | 20.0 | 22.1 | 32.8 | 42.9 | 51.7 | 61.6 | 66.0 | 63.1 | *55.6* | 43.1 | 34.8 | 25.3 | 43.3 |
| Mean Temp °F | 29.1 | *32.5* | 43.7 | 54.7 | 63.9 | 73.4 | 77.5 | 75.0 | *67.8* | 56.2 | 45.1 | 34.6 | 54.5 |
| Days Max Temp ≥ 90 °F | 0 | 0 | 0 | 0 | 2 | 9 | 14 | 10 | 4 | 0 | 0 | 0 | 39 |
| Days Max Temp ≤ 32 °F | 9 | 6 | 1 | 0 | 0 | 0 | 0 | 0 | 0 | 0 | 0 | 5 | 21 |
| Days Min Temp ≤ 32 °F | 25 | 21 | 16 | 5 | 0 | 0 | 0 | 0 | 0 | 6 | 13 | 22 | 108 |
| Days Min Temp ≤ 0 °F | 2 | 2 | 0 | 0 | 0 | 0 | 0 | 0 | 0 | 0 | 0 | 1 | 5 |
| Heating Degree Days | 1107 | *912* | 657 | 328 | 119 | 11 | 1 | 3 | *67* | 293 | 591 | 935 | 5024 |
| Cooling Degree Days | *0* | *0* | *4* | 23 | *84* | 268 | *407* | 331 | *156* | 27 | *2* | *1* | 1303 |
| Total Precipitation (") | 2.85 | 2.83 | 4.19 | 4.55 | 4.86 | 4.06 | 4.21 | 3.69 | 3.47 | 3.07 | 4.47 | 3.72 | 45.97 |
| Days ≥ 0.1" Precip | 5 | 5 | 8 | 7 | 7 | 6 | 6 | 5 | 6 | 5 | 7 | 6 | 73 |
| Total Snowfall (") | na | na | *1.8* | 0.0 | 0.0 | 0.0 | 0.0 | 0.0 | 0.0 | 0.1 | *0.3* | na | na |
| Days ≥ 1" Snow Depth | na | na | *1* | 0 | 0 | 0 | 0 | 0 | 0 | 0 | *0* | na | na |

## TELL CITY *Perry County* ELEVATION 390 ft LAT/LONG 37° 57 ' N / 86° 46 ' W

| | JAN | FEB | MAR | APR | MAY | JUN | JUL | AUG | SEP | OCT | NOV | DEC | YEAR |
|---|---|---|---|---|---|---|---|---|---|---|---|---|---|
| Maximum Temp °F | 39.4 | 44.2 | 55.3 | 67.1 | 75.9 | 84.3 | 88.2 | 86.5 | 80.4 | 68.9 | 56.5 | 45.1 | 66.0 |
| Minimum Temp °F | 22.4 | 25.1 | 34.4 | 44.3 | 53.2 | 62.6 | 67.3 | 65.0 | 58.3 | 45.2 | 37.5 | 28.2 | 45.3 |
| Mean Temp °F | 30.9 | 34.6 | 44.9 | 55.7 | 64.6 | 73.5 | 77.8 | 75.8 | 69.4 | 57.0 | 47.0 | 36.7 | 55.7 |
| Days Max Temp ≥ 90 °F | 0 | 0 | 0 | 0 | 1 | 7 | 14 | 10 | 4 | 0 | 0 | 0 | 36 |
| Days Max Temp ≤ 32 °F | 9 | 5 | 1 | 0 | 0 | 0 | 0 | 0 | 0 | 0 | 0 | 4 | 19 |
| Days Min Temp ≤ 32 °F | 25 | 21 | 15 | 3 | 0 | 0 | 0 | 0 | 0 | 3 | 10 | 20 | 97 |
| Days Min Temp ≤ 0 °F | 2 | 0 | 0 | 0 | 0 | 0 | 0 | 0 | 0 | 0 | 0 | 0 | 2 |
| Heating Degree Days | 1049 | 851 | 620 | 297 | 101 | 8 | 0 | 1 | 43 | 269 | 535 | 873 | 4647 |
| Cooling Degree Days | 0 | 0 | 4 | 25 | 96 | 279 | 425 | 358 | 183 | 29 | 2 | 1 | 1402 |
| Total Precipitation (") | 3.16 | 3.15 | 4.25 | 4.87 | 4.65 | 4.05 | 4.49 | 3.79 | 3.55 | 3.18 | 4.22 | 4.00 | 47.36 |
| Days ≥ 0.1" Precip | 6 | 6 | 8 | 7 | 8 | 7 | 7 | 6 | 6 | 6 | 7 | 7 | 81 |
| Total Snowfall (") | *5.5* | *3.5* | *1.4* | 0.1 | 0.0 | 0.0 | 0.0 | 0.0 | 0.0 | 0.0 | 0.8 | na | na |
| Days ≥ 1" Snow Depth | *5* | 5 | 1 | 0 | 0 | 0 | 0 | 0 | 0 | 0 | 0 | *1* | 12 |

## TERRE HAUTE *Vigo County* ELEVATION 561 ft LAT/LONG 39° 21 ' N / 87° 25 ' W

| | JAN | FEB | MAR | APR | MAY | JUN | JUL | AUG | SEP | OCT | NOV | DEC | YEAR |
|---|---|---|---|---|---|---|---|---|---|---|---|---|---|
| Maximum Temp °F | 34.8 | 39.2 | 51.9 | 63.9 | 74.6 | 83.2 | 86.8 | 84.2 | ***78.5*** | ***67.0*** | 52.7 | 40.9 | 63.1 |
| Minimum Temp °F | 16.7 | 19.6 | 31.5 | 41.9 | 51.3 | 60.3 | 64.5 | 61.8 | ***55.0*** | ***42.7*** | 33.6 | 23.5 | 41.9 |
| Mean Temp °F | 25.7 | 29.4 | 41.7 | 52.9 | 63.0 | 71.8 | 75.7 | 73.0 | ***66.8*** | ***54.9*** | 43.2 | 32.2 | 52.5 |
| Days Max Temp ≥ 90 °F | 0 | 0 | 0 | 0 | 1 | 6 | 9 | 6 | ***2*** | ***0*** | 0 | 0 | 24 |
| Days Max Temp ≤ 32 °F | 13 | 8 | 2 | 0 | 0 | 0 | 0 | 0 | 0 | ***0*** | 1 | 6 | 30 |
| Days Min Temp ≤ 32 °F | 28 | 24 | 18 | 5 | 0 | 0 | 0 | 0 | 0 | ***5*** | 15 | 24 | 119 |
| Days Min Temp ≤ 0 °F | ***5*** | 3 | 0 | 0 | 0 | 0 | 0 | 0 | 0 | ***0*** | 0 | 1 | 9 |
| Heating Degree Days | 1211 | 998 | 715 | 371 | 134 | 15 | 1 | 5 | ***73*** | ***326*** | 647 | 1007 | 5503 |
| Cooling Degree Days | ***0*** | ***0*** | ***2*** | ***17*** | ***78*** | ***223*** | ***349*** | ***259*** | ***135*** | na | ***0*** | ***0*** | na |
| Total Precipitation (") | 2.16 | 2.33 | 3.47 | 4.08 | 4.16 | 3.70 | 4.98 | 3.72 | 3.53 | ***2.82*** | 3.93 | 3.11 | 41.99 |
| Days ≥ 0.1" Precip | 6 | 5 | 7 | 8 | 8 | 6 | 8 | 6 | ***5*** | ***6*** | 7 | 6 | 78 |
| Total Snowfall (") | na | na | ***1.9*** | ***0.1*** | 0.0 | 0.0 | 0.0 | 0.0 | 0.0 | ***0.0*** | ***0.4*** | na | na |
| Days ≥ 1" Snow Depth | na | na | ***1*** | 0 | 0 | 0 | 0 | 0 | 0 | ***0*** | ***0*** | ***3*** | na |

## VALPARAISO WATERWORK *Porter County* ELEVATION 810 ft LAT/LONG 41° 31 ' N / 87° 2 ' W

| | JAN | FEB | MAR | APR | MAY | JUN | JUL | AUG | SEP | OCT | NOV | DEC | YEAR |
|---|---|---|---|---|---|---|---|---|---|---|---|---|---|
| Maximum Temp °F | 30.5 | 34.9 | 46.9 | 60.1 | 71.4 | 80.2 | 83.3 | 80.9 | 74.4 | 63.0 | 48.7 | 36.2 | 59.2 |
| Minimum Temp °F | 14.9 | 18.3 | 28.7 | 38.6 | 48.3 | 57.8 | 62.4 | 60.3 | 53.7 | 42.7 | 33.1 | 22.5 | 40.1 |
| Mean Temp °F | 22.8 | 26.6 | 37.8 | 49.4 | 59.9 | 69.0 | 72.9 | 70.5 | 64.1 | 52.9 | 40.9 | 29.3 | 49.7 |
| Days Max Temp ≥ 90 °F | 0 | 0 | 0 | 0 | 0 | 3 | 4 | 2 | 1 | 0 | 0 | 0 | 10 |
| Days Max Temp ≤ 32 °F | 17 | 12 | 3 | 0 | 0 | 0 | 0 | 0 | 0 | 0 | 1 | 10 | 43 |
| Days Min Temp ≤ 32 °F | 29 | 25 | 21 | 8 | 1 | 0 | 0 | 0 | 0 | 4 | 16 | 26 | 130 |
| Days Min Temp ≤ 0 °F | 6 | 3 | 0 | 0 | 0 | 0 | 0 | 0 | 0 | 0 | 0 | 2 | 11 |
| Heating Degree Days | 1304 | 1078 | 837 | 471 | 206 | 37 | 7 | 15 | 108 | 378 | 716 | 1098 | 6255 |
| Cooling Degree Days | 0 | 0 | 2 | 10 | 57 | 165 | 270 | 196 | 88 | 8 | 0 | 0 | 796 |
| Total Precipitation (") | 2.00 | 1.70 | 2.97 | 3.86 | 3.90 | 4.52 | 3.86 | 3.97 | 4.17 | 3.52 | 3.74 | 3.07 | 41.28 |
| Days ≥ 0.1" Precip | 5 | 5 | 7 | 8 | 7 | 7 | 7 | 7 | 7 | 7 | 8 | 6 | 81 |
| Total Snowfall (") | 12.0 | 10.5 | 6.5 | 1.2 | 0.0 | 0.0 | 0.0 | 0.0 | 0.0 | 0.3 | 3.4 | 8.4 | 42.3 |
| Days ≥ 1" Snow Depth | 20 | 14 | 6 | 0 | 0 | 0 | 0 | 0 | 0 | 0 | 2 | 11 | 53 |

## VEVAY *Switzerland County* ELEVATION 470 ft LAT/LONG 38° 45 ' N / 85° 5 ' W

| | JAN | FEB | MAR | APR | MAY | JUN | JUL | AUG | SEP | OCT | NOV | DEC | YEAR |
|---|---|---|---|---|---|---|---|---|---|---|---|---|---|
| Maximum Temp °F | 39.9 | 44.9 | 56.7 | 68.3 | 77.4 | 85.4 | 88.4 | 86.6 | 79.9 | 68.3 | 55.7 | 45.0 | 66.4 |
| Minimum Temp °F | 22.1 | 24.4 | 33.6 | 42.6 | 51.6 | 60.4 | 65.1 | 63.3 | 56.9 | 44.5 | 36.1 | 27.7 | 44.0 |
| Mean Temp °F | 31.0 | 34.7 | 45.2 | 55.5 | 64.5 | 72.9 | 76.8 | 75.0 | 68.4 | 56.4 | 46.0 | 36.4 | 55.2 |
| Days Max Temp ≥ 90 °F | 0 | 0 | 0 | 0 | 1 | 7 | 13 | 9 | 3 | 0 | 0 | 0 | 33 |
| Days Max Temp ≤ 32 °F | 9 | 4 | 0 | 0 | 0 | 0 | 0 | 0 | 0 | 0 | 0 | 4 | 17 |
| Days Min Temp ≤ 32 °F | 25 | 21 | 15 | 5 | 0 | 0 | 0 | 0 | 0 | 3 | 12 | 21 | 102 |
| Days Min Temp ≤ 0 °F | 2 | 1 | 0 | 0 | 0 | 0 | 0 | 0 | 0 | 0 | 0 | 1 | 4 |
| Heating Degree Days | 1047 | 849 | 610 | 295 | 98 | 9 | 0 | 2 | 46 | 279 | 565 | 881 | 4681 |
| Cooling Degree Days | 0 | 0 | 2 | 18 | 92 | 259 | 398 | 334 | 161 | 24 | 1 | 0 | 1289 |
| Total Precipitation (") | 2.81 | 2.93 | 3.96 | 4.24 | 4.18 | 4.00 | 3.85 | 4.04 | 3.19 | 2.96 | 3.78 | 3.57 | 43.51 |
| Days ≥ 0.1" Precip | 6 | 6 | 8 | 8 | 8 | 7 | 7 | 6 | 6 | 5 | 7 | 7 | 81 |
| Total Snowfall (") | 5.7 | 5.1 | 3.2 | 0.1 | 0.0 | 0.0 | 0.0 | 0.0 | 0.0 | 0.3 | 0.8 | 2.5 | 17.7 |
| Days ≥ 1" Snow Depth | 8 | 6 | 1 | 0 | 0 | 0 | 0 | 0 | 0 | 0 | 0 | 2 | 17 |

## WABASH *Wabash County* ELEVATION 761 ft LAT/LONG 40° 48 ' N / 85° 50 ' W

| | JAN | FEB | MAR | APR | MAY | JUN | JUL | AUG | SEP | OCT | NOV | DEC | YEAR |
|---|---|---|---|---|---|---|---|---|---|---|---|---|---|
| Maximum Temp °F | 31.2 | 34.7 | 46.8 | 60.5 | 71.7 | 80.8 | 84.3 | 81.9 | 76.2 | 63.8 | 49.5 | 36.7 | 59.8 |
| Minimum Temp °F | 13.8 | 15.1 | 26.3 | 36.8 | 46.8 | 56.5 | 60.6 | 57.8 | 50.7 | 38.6 | 31.1 | 20.4 | 37.9 |
| Mean Temp °F | 22.5 | 24.9 | 36.6 | 48.7 | 59.3 | 68.7 | 72.5 | 69.9 | 63.5 | 51.2 | 40.3 | 28.5 | 48.9 |
| Days Max Temp ≥ 90 °F | 0 | 0 | 0 | 0 | 1 | 4 | 6 | 3 | 1 | 0 | 0 | 0 | 15 |
| Days Max Temp ≤ 32 °F | 16 | 12 | 4 | 0 | 0 | 0 | 0 | 0 | 0 | 0 | 2 | 9 | 43 |
| Days Min Temp ≤ 32 °F | 29 | 26 | 23 | 11 | 2 | 0 | 0 | 0 | 0 | 9 | 18 | 27 | 145 |
| Days Min Temp ≤ 0 °F | 6 | 5 | 0 | 0 | 0 | 0 | 0 | 0 | 0 | 0 | 0 | 2 | 13 |
| Heating Degree Days | 1310 | 1124 | 873 | 490 | 216 | 41 | 7 | 20 | 120 | 427 | 735 | 1125 | 6488 |
| Cooling Degree Days | ***0*** | ***0*** | 1 | 7 | 40 | 160 | 255 | 191 | 79 | 6 | ***0*** | ***0*** | 739 |
| Total Precipitation (") | 2.31 | 1.88 | 2.77 | 3.63 | 4.08 | 3.77 | 4.17 | 4.11 | 3.26 | 2.91 | 3.28 | 3.07 | 39.24 |
| Days ≥ 0.1" Precip | ***6*** | 5 | 7 | 8 | 9 | 7 | 7 | 6 | 5 | 6 | 7 | 7 | 80 |
| Total Snowfall (") | na | ***6.9*** | 4.6 | 1.1 | 0.0 | 0.0 | 0.0 | 0.0 | 0.0 | 0.2 | ***1.3*** | ***6.1*** | na |
| Days ≥ 1" Snow Depth | na | na | 4 | 1 | 0 | 0 | 0 | 0 | 0 | 0 | ***1*** | ***5*** | na |

## WANATAH 2 WNW *Porter County* ELEVATION 741 ft LAT/LONG 41° 26 ' N / 86° 56 ' W

| | JAN | FEB | MAR | APR | MAY | JUN | JUL | AUG | SEP | OCT | NOV | DEC | YEAR |
|---|---|---|---|---|---|---|---|---|---|---|---|---|---|
| Maximum Temp °F | 29.5 | 33.6 | 45.1 | 58.5 | 69.9 | 80.0 | 83.2 | 81.1 | 74.6 | 62.6 | 48.1 | 35.4 | 58.5 |
| Minimum Temp °F | 13.5 | 16.8 | 27.8 | 38.0 | 47.5 | 57.5 | 61.3 | 58.8 | 51.9 | 40.4 | 31.8 | 20.8 | 38.8 |
| Mean Temp °F | 21.5 | 25.3 | 36.5 | 48.3 | 58.7 | 68.7 | 72.3 | 70.0 | 63.3 | 51.5 | 40.0 | 28.1 | 48.7 |
| Days Max Temp ≥ 90 °F | 0 | 0 | 0 | 0 | 1 | 4 | 6 | 3 | 1 | 0 | 0 | 0 | 15 |
| Days Max Temp ≤ 32 °F | 18 | 13 | 4 | 0 | 0 | 0 | 0 | 0 | 0 | 0 | 2 | 11 | 48 |
| Days Min Temp ≤ 32 °F | 29 | 26 | 22 | 9 | 1 | 0 | 0 | 0 | 0 | 6 | 17 | 27 | 137 |
| Days Min Temp ≤ 0 °F | 6 | 4 | 0 | 0 | 0 | 0 | 0 | 0 | 0 | 0 | 0 | 2 | 12 |
| Heating Degree Days | 1342 | 1116 | 878 | 504 | 240 | 47 | 13 | 25 | 128 | 419 | 744 | 1136 | 6592 |
| Cooling Degree Days | 0 | 0 | 1 | 12 | 54 | 171 | 258 | 195 | 83 | 6 | 0 | 0 | 780 |
| Total Precipitation (") | 1.61 | 1.46 | 2.73 | 3.83 | 3.63 | 4.11 | 3.95 | 3.79 | 4.02 | 3.10 | 3.48 | 2.71 | 38.42 |
| Days ≥ 0.1" Precip | 4 | 4 | 7 | 8 | 7 | 7 | 7 | 7 | 7 | 7 | 7 | 6 | 78 |
| Total Snowfall (") | 13.5 | 11.9 | 7.2 | 1.2 | 0.0 | 0.0 | 0.0 | 0.0 | 0.0 | 0.3 | 3.6 | 9.2 | 46.9 |
| Days ≥ 1" Snow Depth | 20 | 16 | 7 | 1 | 0 | 0 | 0 | 0 | 0 | 0 | 2 | 11 | 57 |

## WARSAW *Kosciusko County* ELEVATION 810 ft LAT/LONG 41° 14 ' N / 85° 51 ' W

| | JAN | FEB | MAR | APR | MAY | JUN | JUL | AUG | SEP | OCT | NOV | DEC | YEAR |
|---|---|---|---|---|---|---|---|---|---|---|---|---|---|
| Maximum Temp °F | ***30.6*** | 35.1 | 46.3 | 59.5 | 71.3 | 80.0 | 83.3 | 81.0 | ***74.9*** | 62.4 | ***48.9*** | ***36.7*** | 59.2 |
| Minimum Temp °F | 14.1 | 16.7 | 27.3 | 38.2 | 48.6 | 57.2 | 61.8 | 59.2 | 52.5 | 41.1 | ***31.9*** | ***21.6*** | 39.2 |
| Mean Temp °F | ***22.5*** | 26.2 | 36.8 | 48.9 | 60.0 | 68.6 | 72.5 | ***70.2*** | ***63.7*** | 51.8 | ***40.4*** | ***29.3*** | 49.2 |
| Days Max Temp ≥ 90 °F | 0 | 0 | 0 | 0 | 0 | 3 | 5 | 2 | 1 | ***0*** | 0 | 0 | 11 |
| Days Max Temp ≤ 32 °F | ***17*** | 12 | 4 | 0 | 0 | 0 | 0 | 0 | 0 | 0 | ***2*** | ***10*** | 45 |
| Days Min Temp ≤ 32 °F | ***29*** | 25 | 22 | 9 | 1 | 0 | 0 | 0 | 0 | ***6*** | ***17*** | ***26*** | 135 |
| Days Min Temp ≤ 0 °F | 6 | 4 | 0 | 0 | 0 | 0 | 0 | 0 | 0 | 0 | 0 | 2 | 12 |
| Heating Degree Days | ***1312*** | 1093 | 869 | 487 | 205 | 42 | 9 | ***23*** | ***118*** | 410 | ***731*** | ***1101*** | 6400 |
| Cooling Degree Days | 0 | 0 | 1 | 10 | 56 | 167 | 274 | 218 | 87 | 7 | 0 | 0 | 820 |
| Total Precipitation (") | 1.65 | 1.38 | 1.93 | 3.23 | 3.45 | 4.12 | 3.50 | 3.58 | 3.28 | 3.23 | 3.07 | 2.94 | 35.36 |
| Days ≥ 0.1" Precip | ***4*** | 4 | 5 | 7 | 6 | 7 | 7 | 6 | 6 | ***7*** | ***6*** | 6 | 71 |
| Total Snowfall (") | na | na | na | 0.2 | 0.0 | 0.0 | 0.0 | 0.0 | 0.0 | 0.0 | na | na | na |
| Days ≥ 1" Snow Depth | na | na | na | 0 | 0 | 0 | 0 | 0 | 0 | 0 | na | na | na |

## WASHINGTON *Daviess County* ELEVATION 479 ft LAT/LONG 38° 40 ' N / 87° 11 ' W

| | JAN | FEB | MAR | APR | MAY | JUN | JUL | AUG | SEP | OCT | NOV | DEC | YEAR |
|---|---|---|---|---|---|---|---|---|---|---|---|---|---|
| Maximum Temp °F | 38.7 | 44.3 | 56.1 | 67.7 | 76.6 | 85.3 | 88.3 | 86.1 | 79.7 | 68.1 | 55.1 | 43.6 | 65.8 |
| Minimum Temp °F | 22.6 | 26.1 | 35.8 | 45.4 | 54.5 | 63.3 | 67.2 | 65.0 | 58.4 | 46.5 | 37.9 | 28.4 | 45.9 |
| Mean Temp °F | 30.7 | 35.2 | 45.9 | 56.6 | 65.6 | 74.3 | 77.8 | 75.6 | 69.1 | 57.3 | 46.5 | 36.0 | 55.9 |
| Days Max Temp ≥ 90 °F | 0 | 0 | 0 | 0 | 1 | 7 | 13 | 9 | 3 | 0 | 0 | 0 | 33 |
| Days Max Temp ≤ 32 °F | 9 | 5 | 1 | 0 | 0 | 0 | 0 | 0 | 0 | 0 | 0 | 5 | 20 |
| Days Min Temp ≤ 32 °F | 24 | 19 | 13 | 3 | 0 | 0 | 0 | 0 | 0 | 3 | 10 | 20 | 92 |
| Days Min Temp ≤ 0 °F | 2 | 1 | 0 | 0 | 0 | 0 | 0 | 0 | 0 | 0 | 0 | 1 | 4 |
| Heating Degree Days | 1057 | 835 | 588 | 275 | 88 | 6 | 0 | 1 | 46 | 261 | 550 | 891 | 4598 |
| Cooling Degree Days | 0 | 0 | 5 | 28 | 112 | 297 | 421 | 350 | 177 | 30 | 2 | 0 | 1422 |
| Total Precipitation (") | 2.58 | 2.71 | 4.13 | 3.93 | 4.90 | 3.59 | 5.04 | 3.85 | 3.05 | 3.23 | 4.33 | 3.62 | 44.96 |
| Days ≥ 0.1" Precip | 6 | 5 | 8 | 8 | 8 | 6 | 7 | 5 | 5 | 5 | 7 | 7 | 77 |
| Total Snowfall (") | 4.2 | 3.7 | 1.7 | 0.1 | 0.0 | 0.0 | 0.0 | 0.0 | 0.0 | 0.1 | 0.5 | 2.0 | 12.3 |
| Days ≥ 1" Snow Depth | 6 | 5 | 1 | 0 | 0 | 0 | 0 | 0 | 0 | 0 | 0 | 2 | 14 |

## WEST LAFAYETTE 6 NW *Tippecanoe County* ELEVATION 712 ft LAT/LONG 40° 28 ' N / 87° 0 ' W

| | JAN | FEB | MAR | APR | MAY | JUN | JUL | AUG | SEP | OCT | NOV | DEC | YEAR |
|---|---|---|---|---|---|---|---|---|---|---|---|---|---|
| Maximum Temp °F | 30.8 | 35.2 | 47.3 | 60.5 | 71.8 | 80.9 | 84.2 | 81.8 | 76.1 | 63.8 | 49.7 | 37.1 | 59.9 |
| Minimum Temp °F | 14.5 | 17.4 | 28.8 | 39.5 | 49.8 | 59.1 | 62.8 | 60.1 | 53.1 | 41.3 | 32.5 | 21.7 | 40.1 |
| Mean Temp °F | 22.6 | 26.3 | 38.1 | 50.0 | 60.8 | 70.0 | 73.5 | 71.0 | 64.6 | 52.6 | 41.1 | 29.5 | 50.0 |
| Days Max Temp ≥ 90 °F | 0 | 0 | 0 | 0 | 0 | 4 | 6 | 3 | 1 | 0 | 0 | 0 | 14 |
| Days Max Temp ≤ 32 °F | 16 | 12 | 4 | 0 | 0 | 0 | 0 | 0 | 0 | 0 | 1 | 10 | 43 |
| Days Min Temp ≤ 32 °F | 29 | 25 | 21 | 8 | 1 | 0 | 0 | 0 | 0 | 6 | 16 | 26 | 132 |
| Days Min Temp ≤ 0 °F | 6 | 4 | 0 | 0 | 0 | 0 | 0 | 0 | 0 | 0 | 0 | 2 | 12 |
| Heating Degree Days | 1307 | 1086 | 829 | 453 | 184 | 30 | 7 | 17 | 107 | 391 | 710 | 1096 | 6217 |
| Cooling Degree Days | 0 | 0 | 2 | 12 | 66 | 200 | 288 | 218 | 104 | 11 | 0 | 0 | 901 |
| Total Precipitation (") | 1.76 | 1.40 | 2.67 | 3.86 | 3.94 | 3.74 | 3.92 | 3.89 | 3.33 | 2.82 | 3.24 | 2.68 | 37.25 |
| Days ≥ 0.1" Precip | 4 | 3 | 6 | 8 | 7 | 7 | 6 | 6 | 6 | 5 | 6 | 6 | 70 |
| Total Snowfall (") | 6.4 | 5.5 | 2.5 | 0.7 | 0.0 | 0.0 | 0.0 | 0.0 | 0.0 | 0.2 | 1.1 | 5.0 | 21.4 |
| Days ≥ 1" Snow Depth | 14 | 10 | 4 | 0 | 0 | 0 | 0 | 0 | 0 | 0 | 1 | 7 | 36 |

## WEST LAFAYETTE AP *Tippecanoe County* ELEVATION 636 ft LAT/LONG 40° 25 ' N / 86° 56 ' W

| | JAN | FEB | MAR | APR | MAY | JUN | JUL | AUG | SEP | OCT | NOV | DEC | YEAR |
|---|---|---|---|---|---|---|---|---|---|---|---|---|---|
| Maximum Temp °F | 31.6 | 36.2 | 48.7 | 62.0 | 73.3 | 82.6 | 86.1 | ***83.4*** | ***77.1*** | 64.6 | 50.1 | 38.2 | 61.2 |
| Minimum Temp °F | 16.0 | 19.7 | 30.4 | 40.1 | 49.6 | 59.2 | 64.1 | ***61.8*** | ***54.7*** | 42.6 | 33.8 | 23.9 | 41.3 |
| Mean Temp °F | 23.8 | 28.0 | 39.6 | 51.0 | 61.5 | 71.0 | 75.2 | ***72.6*** | ***65.9*** | 53.6 | 41.9 | 31.1 | 51.3 |
| Days Max Temp ≥ 90 °F | 0 | 0 | 0 | 0 | 1 | 6 | 9 | 5 | 1 | 0 | 0 | 0 | 22 |
| Days Max Temp ≤ 32 °F | 16 | 11 | 3 | 0 | 0 | 0 | 0 | 0 | 0 | 0 | 1 | 9 | 40 |
| Days Min Temp ≤ 32 °F | 28 | 24 | 19 | 7 | 1 | 0 | 0 | 0 | 0 | 5 | 15 | 24 | 123 |
| Days Min Temp ≤ 0 °F | 5 | 3 | 0 | 0 | 0 | 0 | 0 | 0 | 0 | 0 | 0 | 1 | 9 |
| Heating Degree Days | 1268 | 1039 | 782 | 424 | 167 | 22 | 3 | ***8*** | ***82*** | 361 | 685 | 1045 | 5886 |
| Cooling Degree Days | 0 | 0 | 2 | 12 | 64 | 211 | ***338*** | ***267*** | ***119*** | 13 | 0 | 0 | 1026 |
| Total Precipitation (") | 1.82 | 1.49 | 2.61 | 3.71 | 3.67 | 4.03 | 3.81 | 4.08 | 3.24 | 2.45 | 3.07 | 2.78 | 36.76 |
| Days ≥ 0.1" Precip | 4 | 4 | 7 | 8 | 7 | 6 | 6 | 6 | 5 | 6 | 6 | 6 | 71 |
| Total Snowfall (") | 6.9 | 5.6 | 2.3 | 0.6 | 0.0 | 0.0 | 0.0 | 0.0 | 0.0 | 0.2 | 1.1 | 5.4 | 22.1 |
| Days ≥ 1" Snow Depth | 13 | 9 | 3 | 0 | 0 | 0 | 0 | 0 | 0 | 0 | 1 | 6 | 32 |

## WHEATFIELD 4 NNW *Jasper County* ELEVATION 669 ft LAT/LONG 41° 11 ' N / 87° 4 ' W

| | JAN | FEB | MAR | APR | MAY | JUN | JUL | AUG | SEP | OCT | NOV | DEC | YEAR |
|---|---|---|---|---|---|---|---|---|---|---|---|---|---|
| Maximum Temp °F | 30.2 | 34.9 | ***45.8*** | 60.3 | 71.6 | 81.2 | 84.4 | ***83.0*** | 76.8 | 64.2 | 49.8 | 36.2 | 59.9 |
| Minimum Temp °F | ***12.0*** | ***16.0*** | ***27.0*** | 37.5 | 47.7 | ***57.0*** | 61.0 | ***58.7*** | 50.9 | 39.7 | 31.2 | 19.9 | 38.2 |
| Mean Temp °F | ***21.0*** | ***25.5*** | ***36.4*** | 48.9 | 59.7 | ***69.1*** | 72.7 | ***70.9*** | 63.8 | 52.0 | 40.5 | 28.1 | 49.1 |
| Days Max Temp ≥ 90 °F | 0 | 0 | 0 | 0 | 1 | 4 | 6 | 4 | 1 | 0 | 0 | 0 | 16 |
| Days Max Temp ≤ 32 °F | 17 | 11 | 3 | 0 | 0 | 0 | 0 | 0 | 0 | 0 | 1 | 10 | 42 |
| Days Min Temp ≤ 32 °F | 28 | 25 | 21 | 10 | 1 | 0 | 0 | 0 | 1 | 8 | 17 | 26 | 137 |
| Days Min Temp ≤ 0 °F | 7 | 4 | 0 | 0 | 0 | 0 | 0 | 0 | 0 | 0 | 0 | 2 | 13 |
| Heating Degree Days | ***1360*** | ***1111*** | ***879*** | 486 | 214 | ***40*** | 9 | ***18*** | 117 | 408 | 727 | ***1139*** | 6508 |
| Cooling Degree Days | ***0*** | ***0*** | na | 11 | 51 | ***154*** | 254 | na | ***84*** | ***8*** | 0 | ***0*** | na |
| Total Precipitation (") | 1.65 | 1.31 | 2.74 | 3.82 | 3.70 | 4.26 | 3.83 | 3.78 | 3.67 | 2.79 | 2.98 | 2.77 | 37.30 |
| Days ≥ 0.1" Precip | ***4*** | ***3*** | ***6*** | 7 | 7 | 6 | 6 | 5 | 6 | 5 | 6 | 5 | 66 |
| Total Snowfall (") | na | na | na | 0.6 | 0.0 | 0.0 | 0.0 | 0.0 | 0.0 | 0.2 | 1.3 | ***4.6*** | na |
| Days ≥ 1" Snow Depth | ***15*** | ***10*** | 4 | 0 | 0 | 0 | 0 | 0 | 0 | 0 | 1 | 7 | 37 |

## WHITESTOWN *Boone County* ELEVATION 830 ft LAT/LONG 40° 0 ' N / 86° 20 ' W

| | JAN | FEB | MAR | APR | MAY | JUN | JUL | AUG | SEP | OCT | NOV | DEC | YEAR |
|---|---|---|---|---|---|---|---|---|---|---|---|---|---|
| Maximum Temp °F | 32.6 | 37.6 | 49.7 | 62.8 | 73.5 | 82.6 | 85.8 | 83.6 | 77.6 | 65.2 | 51.0 | 38.5 | 61.7 |
| Minimum Temp °F | 15.1 | 18.1 | 29.0 | 39.5 | 49.6 | 58.8 | 62.5 | 59.8 | 52.6 | 41.2 | 32.5 | 22.6 | 40.1 |
| Mean Temp °F | 23.9 | 27.9 | 39.4 | 51.2 | 61.6 | 70.7 | 74.2 | 71.7 | 65.1 | 53.2 | 41.8 | 30.6 | 50.9 |
| Days Max Temp ≥ 90 °F | 0 | 0 | 0 | 0 | 1 | 5 | 8 | 5 | 2 | 0 | 0 | 0 | 21 |
| Days Max Temp ≤ 32 °F | 15 | 9 | 3 | 0 | 0 | 0 | 0 | 0 | 0 | 0 | 1 | 8 | 36 |
| Days Min Temp ≤ 32 °F | 28 | 25 | 20 | 8 | 1 | 0 | 0 | 0 | 0 | 6 | 16 | 25 | 129 |
| Days Min Temp ≤ 0 °F | 6 | 4 | 0 | 0 | 0 | 0 | 0 | 0 | 0 | 0 | 0 | 2 | 12 |
| Heating Degree Days | 1269 | 1042 | 789 | 419 | 167 | 24 | 4 | 11 | 95 | 370 | 691 | 1061 | 5942 |
| Cooling Degree Days | 0 | 0 | 2 | 14 | 70 | 218 | 317 | 246 | 115 | 13 | 0 | 0 | 995 |
| Total Precipitation (") | 2.29 | 2.25 | 3.16 | 3.99 | 4.03 | 3.85 | 4.67 | 3.73 | 3.24 | 3.05 | 3.87 | 3.35 | 41.48 |
| Days ≥ 0.1" Precip | 5 | 5 | 7 | 9 | 9 | 6 | 7 | 6 | 6 | 6 | 7 | 7 | 80 |
| Total Snowfall (") | 8.3 | 6.8 | 3.0 | 0.3 | 0.0 | 0.0 | 0.0 | 0.0 | 0.0 | 0.3 | 1.1 | 5.5 | 25.3 |
| Days ≥ 1" Snow Depth | 12 | 10 | 3 | 0 | 0 | 0 | 0 | 0 | 0 | 0 | 1 | 5 | 31 |

## WINAMAC 2 SSE *Pulaski County* ELEVATION 690 ft LAT/LONG 41° 1 ' N / 86° 35 ' W

| | JAN | FEB | MAR | APR | MAY | JUN | JUL | AUG | SEP | OCT | NOV | DEC | YEAR |
|---|---|---|---|---|---|---|---|---|---|---|---|---|---|
| Maximum Temp °F | 31.3 | 35.7 | 48.1 | 61.9 | 72.6 | 80.8 | 83.7 | 81.2 | 74.5 | 63.1 | 49.3 | 36.7 | 59.9 |
| Minimum Temp °F | 14.3 | 17.3 | 28.1 | 38.7 | 49.3 | 58.5 | 62.4 | 60.2 | 53.4 | 41.7 | 32.0 | 21.4 | 39.8 |
| Mean Temp °F | 22.8 | 26.5 | 38.1 | 50.3 | 60.9 | 69.6 | 73.1 | 70.7 | 64.0 | 52.4 | 40.7 | 29.1 | 49.9 |
| Days Max Temp ≥ 90 °F | 0 | 0 | 0 | 0 | 1 | 3 | 5 | 2 | 0 | 0 | 0 | 0 | 11 |
| Days Max Temp ≤ 32 °F | 16 | 11 | 3 | 0 | 0 | 0 | 0 | 0 | 0 | 0 | 2 | 9 | 41 |
| Days Min Temp ≤ 32 °F | 29 | 25 | 22 | 8 | 1 | 0 | 0 | 0 | 0 | 5 | 17 | 27 | 134 |
| Days Min Temp ≤ 0 °F | 6 | 4 | 0 | 0 | 0 | 0 | 0 | 0 | 0 | 0 | 0 | 2 | 12 |
| Heating Degree Days | 1301 | 1079 | 827 | 444 | 181 | 31 | 6 | 15 | 108 | 390 | 723 | 1108 | 6213 |
| Cooling Degree Days | 0 | 0 | 1 | 13 | 63 | 183 | 280 | 214 | 91 | 7 | 0 | 0 | 852 |
| Total Precipitation (") | 1.91 | 1.60 | 2.63 | 3.71 | 3.61 | 4.03 | 3.81 | 3.87 | 3.52 | 3.02 | 3.27 | 2.99 | 37.97 |
| Days ≥ 0.1" Precip | 5 | 5 | 7 | 8 | 7 | 7 | 6 | 6 | 6 | 6 | 7 | 6 | 76 |
| Total Snowfall (") | 7.7 | 6.1 | 3.4 | 1.1 | 0.0 | 0.0 | 0.0 | 0.0 | 0.0 | 0.1 | 2.3 | 5.4 | 26.1 |
| Days ≥ 1" Snow Depth | 16 | 13 | 4 | 0 | 0 | 0 | 0 | 0 | 0 | 0 | 1 | 8 | 42 |

**WINCHESTER AAP 3** ***Randolph County*** ELEVATION **1089 ft** LAT/LONG **40° 10 ' N / 84° 58 ' W**

| | JAN | FEB | MAR | APR | MAY | JUN | JUL | AUG | SEP | OCT | NOV | DEC | YEAR |
|---|---|---|---|---|---|---|---|---|---|---|---|---|---|
| Maximum Temp °F | 31.8 | 35.4 | 47.1 | 59.9 | 70.8 | 80.0 | 83.3 | 81.3 | 75.3 | 62.9 | 49.3 | 37.5 | 59.5 |
| Minimum Temp °F | 15.8 | 18.1 | 28.7 | 39.3 | 50.0 | 59.1 | 62.7 | 60.3 | 53.7 | 42.1 | 32.9 | 22.9 | 40.5 |
| Mean Temp °F | 23.8 | 26.7 | 37.9 | 49.6 | 60.4 | 69.6 | 73.0 | 70.9 | 64.5 | 52.6 | 41.1 | 30.2 | 50.0 |
| Days Max Temp ≥ 90 °F | 0 | 0 | 0 | 0 | 0 | 2 | 4 | 2 | 1 | 0 | 0 | 0 | 9 |
| Days Max Temp ≤ 32 °F | 16 | 11 | 4 | 0 | 0 | 0 | 0 | 0 | 0 | 0 | 2 | 10 | 43 |
| Days Min Temp ≤ 32 °F | 29 | 25 | 21 | 8 | 1 | 0 | 0 | 0 | 0 | 5 | 16 | 26 | 131 |
| Days Min Temp ≤ 0 °F | 5 | 3 | 0 | 0 | 0 | 0 | 0 | 0 | 0 | 0 | 0 | 2 | 10 |
| Heating Degree Days | 1270 | 1073 | 834 | 463 | 189 | 33 | 5 | 16 | 105 | 390 | 710 | 1072 | 6160 |
| Cooling Degree Days | 0 | 0 | 1 | 11 | 61 | 182 | 278 | 212 | 100 | 13 | 0 | 0 | 858 |
| Total Precipitation (") | 1.77 | 1.63 | 2.80 | 3.66 | 3.95 | 4.19 | 4.13 | 3.61 | 2.94 | 2.65 | 3.46 | 2.94 | 37.73 |
| Days ≥ 0.1" Precip | *4* | 4 | 6 | 7 | 7 | 7 | 7 | 6 | 6 | 6 | 7 | *6* | 73 |
| Total Snowfall (") | *4.3* | 5.4 | 2.7 | 0.4 | 0.0 | 0.0 | 0.0 | 0.0 | 0.0 | 0.1 | 0.8 | 2.8 | 16.5 |
| Days ≥ 1" Snow Depth | *11* | 10 | 3 | 0 | 0 | 0 | 0 | 0 | 0 | 0 | *1* | 5 | 30 |

## JANUARY MINIMUM TEMPERATURE °F

### LOWEST

| Rank | Place | Value |
|---|---|---|
| 1 | Wheatfield | 12.0 |
| 2 | Lowell | 12.8 |
| 3 | Angola | 13.3 |
| 4 | Rochester | 13.4 |
| 5 | Wanatah | 13.5 |
| 6 | Hartford City | 13.8 |
| | Kokomo | 13.8 |
| | LaGrange | 13.8 |
| | Wabash | 13.8 |
| 10 | Columbia City | 13.9 |
| 11 | Warsaw | 14.1 |
| 12 | Winamac | 14.3 |
| 13 | Elwood | 14.5 |
| | Lafayette | 14.5 |
| | W Lafayette-6NW | 14.5 |
| 16 | Cambridge City | 14.9 |
| | Farmland | 14.9 |
| | Valparaiso | 14.9 |
| 19 | Rensselaer | 15.0 |
| 20 | Kentland | 15.1 |
| | Whitestown | 15.1 |
| 22 | Marion | 15.4 |
| 23 | La Porte | 15.5 |
| 24 | Hobart | 15.7 |
| | New Castle | 15.7 |

### HIGHEST

| Rank | Place | Value |
|---|---|---|
| 1 | Evansville-Musm | 24.4 |
| 2 | Madison | 23.1 |
| 3 | Charlestown | 22.7 |
| 4 | St. Meinrad | 22.6 |
| | Washington | 22.6 |
| 6 | Tell City | 22.4 |
| 7 | Evansville-Reg | 22.1 |
| | Vevay | 22.1 |
| 9 | Mount Vernon | 21.8 |
| 10 | Princeton | 21.5 |
| 11 | Crane | 21.4 |
| 12 | Salem | 21.3 |
| 13 | Spurgeon | 20.0 |
| 14 | Dubois | 19.3 |
| 15 | Scottsburg | 19.1 |
| 16 | Bloomington | 19.0 |
| 17 | Paoli | 18.7 |
| 18 | Greensburg | 18.6 |
| 19 | Shoals | 18.3 |
| 20 | Anderson | 18.1 |
| 21 | Columbus | 18.0 |
| 22 | Indianapolis-Intl | 17.9 |
| | Seymour | 17.9 |
| 24 | Indianapolis-SE | 17.7 |
| 25 | Martinsville | 17.5 |

## JULY MAXIMUM TEMPERATURE °F

### HIGHEST

| Rank | Place | Value |
|---|---|---|
| 1 | Evansville-Musm | 89.6 |
| 2 | Evansville-Reg | 89.0 |
| 3 | Charlestown | 88.8 |
| | Mount Vernon | 88.8 |
| | Spurgeon | 88.8 |
| 6 | Princeton | 88.4 |
| | Vevay | 88.4 |
| 8 | Washington | 88.3 |
| 9 | Tell City | 88.2 |
| 10 | Crane | 88.0 |
| 11 | Paoli | 87.7 |
| 12 | St. Meinrad | 87.6 |
| 13 | Scottsburg | 87.5 |
| 14 | Rockville | 87.4 |
| 15 | Madison | 87.3 |
| 16 | Shoals | 87.2 |
| 17 | Salem | 87.1 |
| 18 | Terre Haute | 86.8 |
| 19 | Greencastle | 86.6 |
| 20 | Brookville | 86.5 |
| 21 | Bloomington | 86.1 |
| | Wst Lafayette-Ap | 86.1 |
| 23 | Columbus | 86.0 |
| 24 | Dubois | 85.9 |
| | Kentland | 85.9 |

### LOWEST

| Rank | Place | Value |
|---|---|---|
| 1 | Angola | 82.1 |
| 2 | LaGrange | 82.7 |
| 3 | South Bend | 83.0 |
| 4 | Hartford City | 83.1 |
| 5 | Columbia City | 83.2 |
| | Wanatah | 83.2 |
| 7 | La Porte | 83.3 |
| | Valparaiso | 83.3 |
| | Warsaw | 83.3 |
| | Winchester | 83.3 |
| 11 | Anderson | 83.6 |
| 12 | Winamac | 83.7 |
| 13 | Cambridge City | 83.9 |
| | Goshen | 83.9 |
| 15 | Farmland | 84.0 |
| 16 | Kokomo | 84.1 |
| | Rochester | 84.1 |
| 18 | W Lafayette-6NW | 84.2 |
| 19 | Rushville | 84.3 |
| | Wabash | 84.3 |
| 21 | Fort Wayne | 84.4 |
| | Lowell | 84.4 |
| | Marion | 84.4 |
| | Wheatfield | 84.4 |
| 25 | Lafayette | 84.5 |

## ANNUAL PRECIPITATION (")

### HIGHEST

| Rank | Place | Value |
|---|---|---|
| 1 | Tell City | 47.36 |
| 2 | Paoli | 46.75 |
| 3 | Shoals | 46.64 |
| 4 | Crane | 46.42 |
| 5 | Dubois | 46.10 |
| 6 | Princeton | 45.99 |
| 7 | Spurgeon | 45.97 |
| 8 | Evansville-Musm | 45.58 |
| 9 | Charlestown | 45.56 |
| 10 | Salem | 45.30 |
| 11 | St. Meinrad | 45.27 |
| 12 | Spencer | 45.04 |
| 13 | Washington | 44.96 |
| 14 | Mount Vernon | 44.88 |
| 15 | Madison | 44.29 |
| 16 | Bloomington | 44.27 |
| 17 | Rockville | 44.22 |
| 18 | Scottsburg | 44.11 |
| 19 | Greencastle | 44.07 |
| 20 | Oolitic | 44.03 |
| 21 | Greenfield | 43.62 |
| 22 | Vevay | 43.51 |
| 23 | Seymour | 43.30 |
| 24 | Evansville-Reg | 43.02 |
| 25 | Brookville | 42.60 |

### LOWEST

| Rank | Place | Value |
|---|---|---|
| 1 | Warsaw | 35.36 |
| 2 | Fort Wayne | 36.26 |
| 3 | Goshen | 36.43 |
| 4 | Lafayette | 36.60 |
| 5 | Angola | 36.75 |
| 6 | Wst Lafayette-Ap | 36.76 |
| 7 | Hartford City | 36.87 |
| 8 | Farmland | 37.15 |
| 9 | LaGrange | 37.19 |
| 10 | W Lafayette-6NW | 37.25 |
| 11 | Wheatfield | 37.30 |
| 12 | Winchester | 37.73 |
| 13 | Hobart | 37.93 |
| 14 | Berne | 37.94 |
| 15 | Winamac | 37.97 |
| 16 | Kentland | 38.12 |
| 17 | Marion | 38.31 |
| 18 | Delphi | 38.37 |
| 19 | Rochester | 38.40 |
| 20 | Wanatah | 38.42 |
| 21 | Columbia City | 38.65 |
| 22 | Muncie | 38.70 |
| 23 | Rensselaer | 38.88 |
| 24 | Anderson | 39.16 |
| 25 | Wabash | 39.24 |

## ANNUAL SNOWFALL (")

### HIGHEST

| Rank | Place | Value |
|---|---|---|
| 1 | South Bend | 79.5 |
| 2 | La Porte | 63.9 |
| 3 | Wanatah | 46.9 |
| 4 | LaGrange | 45.4 |
| 5 | Kokomo | 44.6 |
| 6 | Valparaiso | 42.3 |
| 7 | Goshen | 38.1 |
| 8 | Fort Wayne | 34.4 |
| 9 | Lowell | 33.7 |
| 10 | Angola | 33.5 |
| 11 | Columbia City | 32.5 |
| 12 | Berne | 31.8 |
| 13 | Rochester | 31.5 |
| 14 | Hobart | 31.2 |
| 15 | Greencastle | 27.2 |
| | Kentland | 27.2 |
| 17 | Farmland | 26.7 |
| 18 | Muncie | 26.5 |
| 19 | Indianapolis-Intl | 26.4 |
| 20 | Frankfort | 26.3 |
| 21 | Marion | 26.1 |
| | Winamac | 26.1 |
| 23 | Whitestown | 25.3 |
| 24 | Hartford City | 24.6 |
| 25 | Oaklandon Geist r | 23.8 |

### LOWEST

| Rank | Place | Value |
|---|---|---|
| 1 | St. Meinrad | 10.4 |
| 2 | Washington | 12.3 |
| 3 | Evansville-Musm | 14.3 |
| | Mount Vernon | 14.3 |
| 5 | Evansville-Reg | 15.6 |
| 6 | Shoals | 16.2 |
| 7 | Rushville | 16.4 |
| 8 | Winchester | 16.5 |
| 9 | Oolitic | 16.7 |
| 10 | Spencer | 16.8 |
| 11 | Scottsburg | 17.0 |
| 12 | Rockville | 17.3 |
| 13 | Vevay | 17.7 |
| 14 | Greensburg | 17.9 |
| 15 | Anderson | 18.3 |
| 16 | Salem | 18.7 |
| 17 | W Lafayette-6NW | 21.4 |
| 18 | Cambridge City | 21.5 |
| | Delphi | 21.5 |
| | Lafayette | 21.5 |
| 21 | Wst Lafayette-Ap | 22.1 |
| 22 | Oaklandon Geist | 23.8 |
| 23 | Hartford City | 24.6 |
| 24 | Whitestown | 25.3 |
| 25 | Marion | 26.1 |

# IOWA

PHYSICAL FEATURES. The State of Iowa comprises 56,290 square miles, primarily of rolling prairie, located in the middle latitudes between the Upper Mississippi and the Missouri Rivers. The interior continental location is 800 to 1,000 miles distant from the Gulf of Mexico, North Atlantic, and Hudson Bay. The North Pacific Ocean is approximately 1,300 miles west and the Rocky Mountains shield is some 400 to 700 miles west of Iowa.

The extreme north-south distance across Iowa is 205 miles; the extreme east-west distance, 310 miles. Elevational changes are small across the State, varying from 1,675 feet on Ocheyedan Mound in the northwest to 477 feet at the mouth of the Des Moines River in the southeast. There is some rugged terrain, mostly of forest soils, in the northeast. Most of the State's lakes are located in the northwest.

GENERAL CLIMATE. Iowa's climate, because of latitude and interior continental location, is characterized by marked seasonal variations. During the 6 warm months of the year the prevailing moist, southerly flow from the Gulf of Mexico produces a summer rainfall maximum. The prevailing northwesterly flow of dry Canadian air in the winter causes this season to be cold and relatively dry. At intervals throughout the year, airmasses from the Pacific Ocean moving across the western United States reach Iowa, producing comparatively mild and dry weather. The autumnal "Indian Summers" are a result of the dominance of these modified Pacific airmasses. Hot, dry winds, originating in the desert southwest United States, occasionally sweep into Iowa during the summer, producing unusually high temperatures.

TEMPERATURE. The average annual temperatures range from 46° in the northern counties to 52° in the southeastern counties. In July, the hottest month, the average daily maximum is around 85° and the daily minima are mostly in the lower sixties. In January daily maxima range from 24° to 34°, north to south, and the minima from 4° to 14°. In almost every year at some location in the State, a maximum exceeds 100° and a minimum of less than -20° occurs. In half the years the maximum exceeds 104° and the minimum -31°. The average number of days with temperatures 90° or higher range from 47 to 6. The number of days with zero or lower temperatures range from about 10 per year in the south to 30 in the north.

PRECIPITATION. Precipitation averages around 31 inches per year for the State, ranging from 25 inches in the extreme northwest to about 34 inches in the East Central and Southeast Divisions. However, annual totals vary widely from year to year and locality to locality. Nearly two-thirds of the annual precipitation is measured during the six months of April through September. Measurable rain occurs on about 100 days per year; the frequency of a tenth of an inch or more increases southeastward across the State from 44 days per year to 69 days. Half an inch or more of rain per day varies from 15 days in the extreme northwest to near 25 in the southeast.

SNOWFALL. The average seasonal snowfall varies from near 20 inches at Keokuk to 35-45 inches over northern counties. The season normally extends from October or November to April but measurable snow has fallen as late in the season as May and as early as September. The average number of days with snow cover one inch or deeper per season varies from about 40 days along the southern border to around 90 in the northernmost counties. The average date of the first 1-inch snowfall varies from November 25 in the north to December 10 in the southeast. The first trace of snow occurs about one month earlier. In about half the years a daily snowfall of 5 inches or more occurs over southern Iowa, 6 or more over central counties, and 7 or 8 over northern counties. Late winter snowstorms have produced as much as 31 inches of snow in a single storm and 24-hour amounts have exceeded 20 inches.

STORMS. Around 80 percent of the 40 to 50 thunderstorms per year occur in the warm half. Occasionally hail, high winds, heavy rains, and even tornadoes, are associated with the thunderstorms. The probability of occurrence is highest in late spring and early summer. Tornado frequency is highest in May and June in the afternoon and early evening. Tornado occurrences average about 15 per year on 8 days. Damaging hailstorms, reaching a maximum in early summer, average 58 per year. Severe hailstorms are slightly more frequent over northwestern counties. In any locality hail usually occurs from two to six times a year.

Floods are most frequent in June at the normal maximum rainfall period, but also occur near the end of March, usually as a consequence of rain on frozen ground, or rain and rapid snowmelt. Ice jams often contribute to the spring flooding.

High winds at 15 feet above the ground (house-top level) reach 50 m.p.h. in about half the years. Winds to 75 m.p.h. at the 15-foot-level, excluding gusts, may be expected once in 50 years.

Drought occurs periodically in Iowa.

Sunshine increases from northeast to southwest. The percent of the possible sunshine varies from 40-52 in December, the cloudiest month, to 72-76 in July, the sunniest month. Available solar energy is four times as abundant in July as in December. The growing season for warm weather crops extends from mid-May to early October. The spring growing season, suitable for hardy crops, lasts approximately six weeks and the autumn season about seven weeks.

## COUNTY INDEX

***Jackson County***
BELLEVUE L&D 12

***Jasper County***
NEWTON

***Jefferson County***
FAIRFIELD

***Johnson County***
IOWA CITY

***Jones County***
ANAMOSA 1 WNW

***Keokuk County***
SIGOURNEY

***Kossuth County***
ALGONA 3 W
SWEA CITY

***Lee County***
FORT MADISON

***Linn County***
CEDAR RAPIDS MUNI AP
CEDAR RAPIDS NO 1

***Louisa County***
COLUMBUS JUNCT 2 SSW

***Lucas County***
CHARITON 1 E

***Lyon County***
ROCK RAPIDS

***Madison County***
WINTERSET 2 NNW

***Mahaska County***
OSKALOOSA

***Marion County***
KNOXVILLE

***Marshall County***
MARSHALLTOWN

***Mills County***
GLENWOOD 3 SW

***Mitchell County***
OSAGE

***Monona County***
CASTANA EXPRMNT FARM
MAPLETON NO 2
ONAWA

***Monroe County***
ALBIA 3 NNE

***Montgomery County***
RED OAK

***Muscatine County***
MUSCATINE

***O'Brien County***
PRIMGHAR
SANBORN
SHELDON

***Osceola County***
SIBLEY 5 NNE

***Page County***
CLARINDA
SHENANDOAH

***Palo Alto County***
EMMETSBURG

***Plymouth County***
LE MARS

***Pocahontas County***
POCAHONTAS

***Polk County***
ANKENY
DES MOINES INTL AP

***Pottawattamie County***
OAKLAND 4 WSW

***Poweshiek County***
GRINNELL 3 SW

***Ringgold County***
BEACONSFIELD 2 N
MOUNT AYR 4 SW

***Sac County***
SAC CITY

***Scott County***
LE CLAIRE L & D 14

***Shelby County***
HARLAN

***Sioux County***
HAWARDEN
SIOUX CENTER 2 SE

***Story County***
COLO

***Tama County***
TOLEDO

***Taylor County***
BEDFORD

***Union County***
CRESTON 2 SW

***Van Buren County***
KEOSAUQUA

***Wapello County***
OTTUMWA INDUSTRL AP

***Warren County***
INDIANOLA

***Washington County***
WASHINGTON

***Webster County***
FORT DODGE

***Winnebago County***
FOREST CITY 2 NNE

***Winneshiek County***
DECORAH

### *Woodbury County*
SIOUX CITY MUNI AP

### *Worth County*
NORTHWOOD

### *Wright County*
CLARION

## ELEVATION INDEX

| FEET | STATION NAME |
|---|---|
| 530 | FORT MADISON |
| 549 | MUSCATINE |
| 581 | LE CLAIRE L & D 14 |
| 585 | CLINTON 1 |
| 603 | BELLEVUE L&D 12 |
| 610 | COLUMBUS JUNCT 2 SSW |
| 620 | DUBUQUE L&D 11 |
| 620 | GUTTENBERG L&D 10 |
| 680 | MAQUOKETA 3 S |
| 702 | BURLINGTN RADIO KBUR |
| 712 | KEOSAUQUA |
| 732 | ELKADER 5 SSW |
| 732 | MOUNT PLEASANT 1 SSW |
| 751 | WASHINGTON |
| 771 | IOWA CITY |
| 781 | FAIRFIELD |
| 781 | SIGOURNEY |
| 801 | WILLIAMSBURG |
| 810 | OSKALOOSA |
| 810 | TIPTON 4 NE |
| 810 | VINTON |
| 820 | CEDAR RAPIDS NO 1 |
| 840 | BLOOMFIELD 1 WNW |
| 846 | OTTUMWA INDUSTRL AP |
| 855 | BELLE PLAINE |
| 860 | DECORAH |
| 869 | ANAMOSA 1 WNW |
| 869 | WATERLOO MUNI AP |
| 870 | MARSHALLTOWN |
| 879 | CASCADE |
| 880 | ALBIA 3 NNE |
| 902 | CEDAR RAPIDS MUNI AP |
| 912 | KNOXVILLE |
| 932 | TOLEDO |
| 932 | TRIPOLI |
| 938 | DES MOINES INTL AP |
| 940 | ANKENY |
| 940 | INDIANOLA |
| 942 | CHARITON 1 E |
| 951 | NEWTON |
| 965 | PERRY |
| 971 | SHENANDOAH |
| 980 | CENTERVILLE |
| 980 | GLENWOOD 3 SW |
| 991 | LEON 6 ESE |
| 1001 | CLARINDA |
| 1001 | COLO |
| 1001 | FAYETTE |
| 1001 | GRINNELL 3 SW |
| 1020 | CHARLES CITY |
| 1020 | GRUNDY CENTER |
| 1030 | OELWEIN 2 S |
| 1040 | RED OAK |
| 1040 | WEBSTER CITY |
| 1050 | ONAWA |
| 1050 | SIDNEY |
| 1060 | ALLISON |
| 1060 | JEFFERSON |
| 1070 | DUBUQUE MUNICPAL AP |
| 1075 | GUTHRIE CENTER |
| 1094 | ELDORA |
| 1095 | SIOUX CITY MUNI AP |
| 1102 | AMES 8 WSW |
| 1110 | HUMBOLDT 3 W |
| 1110 | OSCEOLA |
| 1112 | ATLANTIC 1 NE |
| 1115 | FORT DODGE |
| 1122 | LOGAN |
| 1132 | BOONE |
| 1132 | MASON CITY |
| 1132 | WINTERSET 2 NNW |
| 1142 | IOWA FALLS |
| 1142 | LAMONI |
| 1150 | OAKLAND 4 WSW |
| 1161 | NEW HAMPTON |
| 1171 | CLARION |
| 1171 | OSAGE |
| 1190 | MAPLETON NO 2 |
| 1191 | HAWARDEN |
| 1200 | SWEA CITY |
| 1201 | ALGONA 3 W |
| 1211 | BEDFORD |
| 1211 | HARLAN |
| 1211 | MASON CITY AP |
| 1212 | CHEROKEE |
| 1220 | BEACONSFIELD 2 N |
| 1220 | HAMPTON |
| 1220 | NORTHWOOD |
| 1220 | POCAHONTAS |
| 1220 | ROCKWELL CITY |
| 1230 | BRITT |
| 1230 | EMMETSBURG |
| 1230 | LE MARS |
| 1240 | CARROLL |
| 1240 | WAUKON |
| 1270 | MOUNT AYR 4 SW |
| 1270 | SAC CITY |
| 1280 | CORNING |
| 1280 | CRESCO 1 NE |
| 1289 | AUDUBON 1 SSE |
| 1298 | ESTHERVILLE 2 N |
| 1300 | FOREST CITY 2 NNE |
| 1302 | DENISON |
| 1302 | GREENFIELD |
| 1312 | CRESTON 2 SW |
| 1320 | IDA GROVE 5 NW |
| 1326 | SPENCER |
| 1342 | ROCK RAPIDS |
| 1401 | MILFORD 4 NW |
| 1420 | SIOUX RAPIDS 4 E |
| 1421 | SHELDON |
| 1430 | STORM LAKE 2 E |
| 1440 | CASTANA EXPRMNT FARM |
| 1450 | SIOUX CENTER 2 SE |
| 1480 | LAKE PARK |
| 1489 | SIBLEY 5 NNE |
| 1522 | PRIMGHAR |
| 1552 | SANBORN |

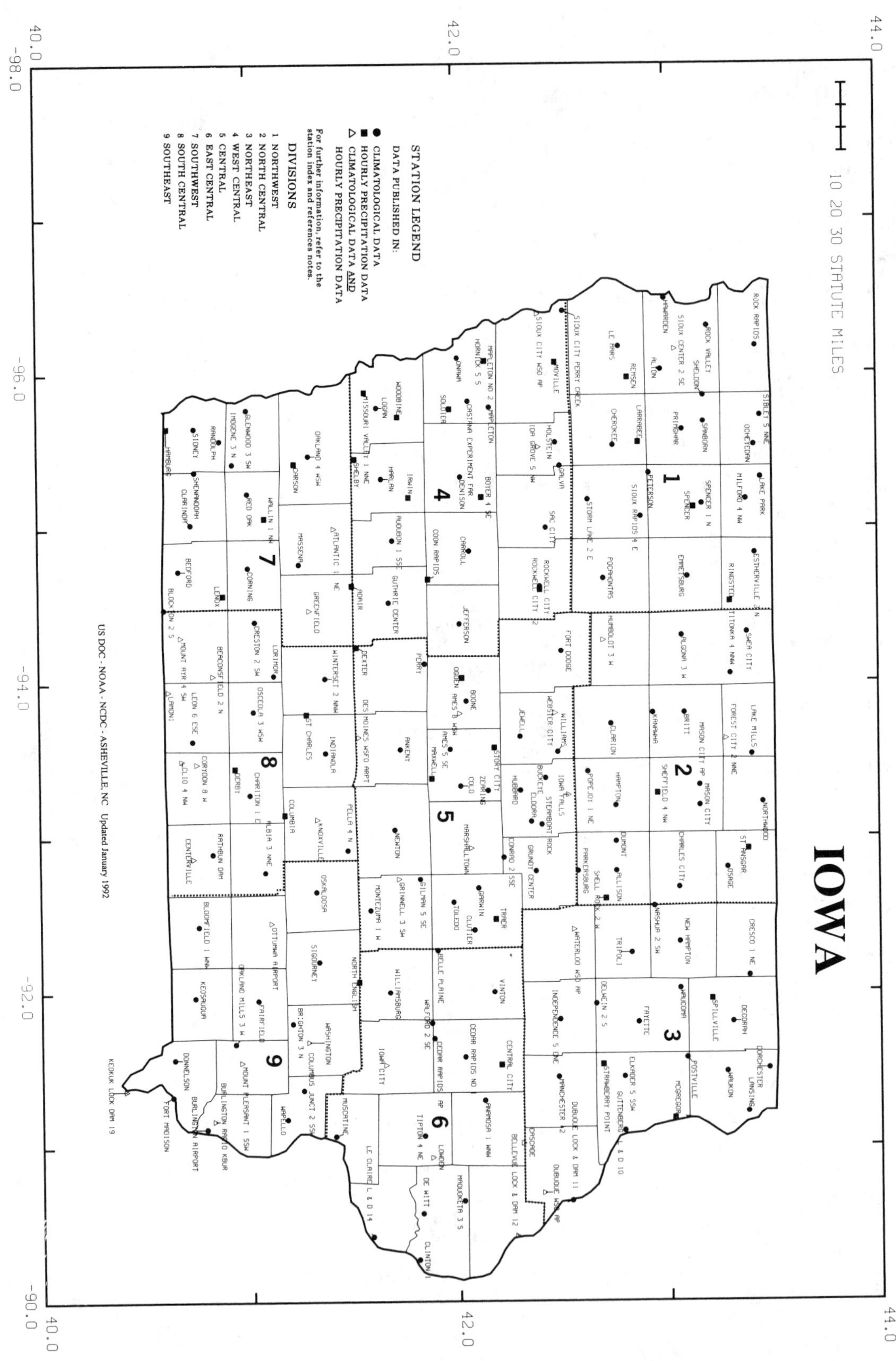
IOWA
STATION LEGEND
DATA PUBLISHED IN:
● CLIMATOLOGICAL DATA
■ HOURLY PRECIPITATION DATA
△ CLIMATOLOGICAL DATA AND HOURLY PRECIPITATION DATA
For further information, refer to the station index and reference notes.
DIVISIONS
1 NORTHWEST
2 NORTH CENTRAL
3 NORTHEAST
4 WEST CENTRAL
5 CENTRAL
6 EAST CENTRAL
7 SOUTHWEST
8 SOUTH CENTRAL
9 SOUTHEAST
10 20 30 STATUTE MILES
US DOC - NOAA - NCDC - ASHEVILLE, NC Updated January 1992

## ALBIA 3 NNE *Monroe County* ELEVATION 880 ft LAT/LONG 41° 4 ' N / 92° 47 ' W

| | JAN | FEB | MAR | APR | MAY | JUN | JUL | AUG | SEP | OCT | NOV | DEC | YEAR |
|---|---|---|---|---|---|---|---|---|---|---|---|---|---|
| Maximum Temp °F | 31.8 | 37.2 | 50.0 | 63.6 | 73.7 | 82.5 | 86.9 | 84.6 | 76.9 | 65.2 | 49.1 | 35.8 | 61.4 |
| Minimum Temp °F | 13.9 | 18.1 | 29.6 | 41.1 | 51.1 | 60.4 | 65.3 | 62.8 | 54.7 | 43.3 | 31.1 | 19.4 | 40.9 |
| Mean Temp °F | 22.9 | 27.7 | 39.8 | 52.4 | 62.4 | 71.4 | 76.1 | 73.7 | 65.8 | 54.3 | 40.1 | 27.6 | 51.2 |
| Days Max Temp ≥ 90 °F | 0 | 0 | 0 | 0 | 0 | 4 | 11 | 7 | 2 | 0 | 0 | 0 | 24 |
| Days Max Temp ≤ 32 °F | 14 | 11 | 3 | 0 | 0 | 0 | 0 | 0 | 0 | 0 | 2 | 11 | 41 |
| Days Min Temp ≤ 32 °F | 28 | 25 | 19 | 6 | 0 | 0 | 0 | 0 | 0 | 4 | 17 | 27 | 126 |
| Days Min Temp ≤ 0 °F | 6 | 3 | 0 | 0 | 0 | 0 | 0 | 0 | 0 | 0 | 0 | 3 | 12 |
| Heating Degree Days | 1301 | 1047 | 777 | 388 | 140 | 14 | 2 | 7 | 87 | 342 | 741 | 1152 | 5998 |
| Cooling Degree Days | ***0*** | 0 | 3 | 20 | 60 | 212 | 348 | 288 | 124 | 12 | 1 | 0 | 1068 |
| Total Precipitation (") | 1.13 | 1.11 | 2.33 | 3.89 | 4.36 | 4.44 | 5.00 | 3.73 | 4.87 | 2.69 | 2.60 | 1.56 | 37.71 |
| Days ≥ 0.1" Precip | 3 | 3 | 6 | 8 | 8 | 8 | 7 | 6 | 7 | 5 | 5 | 4 | 70 |
| Total Snowfall (") | 6.6 | 6.4 | 3.7 | 2.1 | 0.0 | 0.0 | 0.0 | 0.0 | 0.0 | 0.4 | 2.9 | 5.0 | 27.1 |
| Days ≥ 1" Snow Depth | 15 | 12 | 5 | 1 | 0 | 0 | 0 | 0 | 0 | 0 | 2 | 10 | 45 |

## ALGONA 3 W *Kossuth County* ELEVATION 1201 ft LAT/LONG 43° 4 ' N / 94° 14 ' W

| | JAN | FEB | MAR | APR | MAY | JUN | JUL | AUG | SEP | OCT | NOV | DEC | YEAR |
|---|---|---|---|---|---|---|---|---|---|---|---|---|---|
| Maximum Temp °F | 23.8 | 29.2 | 42.8 | 59.2 | 72.3 | 81.4 | 84.0 | 81.5 | 73.6 | 61.3 | 42.5 | 28.1 | 56.6 |
| Minimum Temp °F | 5.4 | 11.1 | 24.2 | 36.2 | 47.8 | 57.5 | 61.6 | 58.8 | 50.2 | 38.7 | 25.1 | 11.8 | 35.7 |
| Mean Temp °F | 14.7 | 20.2 | 33.5 | 47.7 | 60.1 | 69.5 | 72.8 | 70.2 | 61.9 | 50.0 | 33.8 | 20.0 | 46.2 |
| Days Max Temp ≥ 90 °F | 0 | 0 | 0 | 0 | 1 | 5 | 7 | 4 | 1 | 0 | 0 | 0 | 18 |
| Days Max Temp ≤ 32 °F | 22 | 16 | 6 | 0 | 0 | 0 | 0 | 0 | 0 | 0 | 6 | 19 | 69 |
| Days Min Temp ≤ 32 °F | 31 | 27 | 25 | 11 | 2 | 0 | 0 | 0 | 1 | 8 | 24 | 30 | 159 |
| Days Min Temp ≤ 0 °F | 12 | 8 | 1 | 0 | 0 | 0 | 0 | 0 | 0 | 0 | 1 | 6 | 28 |
| Heating Degree Days | 1556 | 1260 | 969 | 519 | 197 | 27 | 7 | 20 | 148 | 462 | 928 | 1391 | 7484 |
| Cooling Degree Days | 0 | 0 | 0 | 8 | 51 | 169 | 247 | 186 | 68 | 3 | 0 | 0 | 732 |
| Total Precipitation (") | 0.76 | 0.77 | 1.86 | 2.82 | 3.73 | 4.79 | 4.06 | 3.61 | 3.27 | 2.28 | 1.75 | 0.97 | 30.67 |
| Days ≥ 0.1" Precip | 2 | 2 | 4 | 6 | 7 | 7 | 7 | 6 | 6 | 4 | 4 | 3 | 58 |
| Total Snowfall (") | 8.5 | 7.2 | 7.0 | 1.9 | 0.0 | 0.0 | 0.0 | 0.0 | 0.0 | 0.2 | 4.4 | 8.0 | 37.2 |
| Days ≥ 1" Snow Depth | 24 | 21 | 11 | 2 | 0 | 0 | 0 | 0 | 0 | 0 | 5 | 20 | 83 |

## ALLISON *Butler County* ELEVATION 1060 ft LAT/LONG 42° 46 ' N / 92° 47 ' W

| | JAN | FEB | MAR | APR | MAY | JUN | JUL | AUG | SEP | OCT | NOV | DEC | YEAR |
|---|---|---|---|---|---|---|---|---|---|---|---|---|---|
| Maximum Temp °F | 24.9 | 30.4 | 43.7 | 60.1 | 72.7 | 81.7 | 84.6 | 82.5 | 74.4 | 62.4 | 44.2 | 29.8 | 57.6 |
| Minimum Temp °F | 6.9 | 12.3 | 24.8 | 37.3 | 48.5 | 58.1 | 61.9 | 59.6 | 51.2 | 39.9 | 27.0 | 14.0 | 36.8 |
| Mean Temp °F | 15.9 | 21.4 | 34.3 | 48.7 | 60.6 | 70.0 | 73.3 | 71.1 | 62.8 | 51.2 | 35.7 | 22.0 | 47.3 |
| Days Max Temp ≥ 90 °F | 0 | 0 | 0 | 0 | 1 | 4 | 7 | 5 | 1 | 0 | 0 | 0 | 18 |
| Days Max Temp ≤ 32 °F | 21 | 14 | 6 | 0 | 0 | 0 | 0 | 0 | 0 | 0 | 5 | 17 | 63 |
| Days Min Temp ≤ 32 °F | 31 | 27 | 24 | 9 | 1 | 0 | 0 | 0 | 0 | 8 | 21 | 30 | 151 |
| Days Min Temp ≤ 0 °F | 11 | 7 | 1 | 0 | 0 | 0 | 0 | 0 | 0 | 0 | 0 | 5 | 24 |
| Heating Degree Days | 1517 | 1226 | 945 | 490 | 183 | 23 | 5 | 15 | 130 | 430 | 874 | 1329 | 7167 |
| Cooling Degree Days | 0 | 0 | 0 | 11 | 50 | 179 | 265 | 208 | 76 | 4 | 0 | 0 | 793 |
| Total Precipitation (") | 0.81 | 0.91 | 1.82 | 3.12 | 4.13 | 5.04 | 4.78 | 4.21 | 3.79 | 2.62 | 2.01 | 1.29 | 34.53 |
| Days ≥ 0.1" Precip | *2* | 3 | 4 | 6 | 8 | 7 | 7 | 6 | 6 | 5 | 4 | 4 | 62 |
| Total Snowfall (") | *7.7* | 7.1 | 4.6 | 1.7 | 0.0 | 0.0 | 0.0 | 0.0 | 0.0 | 0.0 | 4.0 | 7.4 | 32.5 |
| Days ≥ 1" Snow Depth | na | na | na | 1 | 0 | 0 | 0 | 0 | 0 | 0 | *4* | na | na |

## AMES 8 WSW *Boone County* ELEVATION 1102 ft LAT/LONG 42° 2 ' N / 93° 48 ' W

| | JAN | FEB | MAR | APR | MAY | JUN | JUL | AUG | SEP | OCT | NOV | DEC | YEAR |
|---|---|---|---|---|---|---|---|---|---|---|---|---|---|
| Maximum Temp °F | 27.3 | 33.0 | 46.5 | 62.0 | 73.4 | 82.0 | 84.8 | 82.5 | 75.7 | 63.7 | 45.7 | 32.0 | 59.1 |
| Minimum Temp °F | 8.9 | 14.2 | 26.5 | 38.2 | 49.4 | 58.8 | 62.9 | 60.1 | 51.8 | 40.1 | 27.4 | 15.2 | 37.8 |
| Mean Temp °F | 18.1 | 23.6 | 36.5 | 50.1 | 61.4 | 70.4 | 73.9 | 71.3 | 63.8 | 52.0 | 36.6 | 23.6 | 48.4 |
| Days Max Temp ≥ 90 °F | 0 | 0 | 0 | 0 | 1 | 4 | 7 | 5 | 2 | 0 | 0 | 0 | 19 |
| Days Max Temp ≤ 32 °F | 19 | 13 | 5 | 0 | 0 | 0 | 0 | 0 | 0 | 0 | 4 | 15 | 56 |
| Days Min Temp ≤ 32 °F | 31 | 27 | 23 | 8 | 1 | 0 | 0 | 0 | 0 | 7 | 21 | 30 | 148 |
| Days Min Temp ≤ 0 °F | 10 | 5 | 1 | 0 | 0 | 0 | 0 | 0 | 0 | 0 | 0 | 4 | 20 |
| Heating Degree Days | 1447 | 1163 | 877 | 452 | 164 | 18 | 4 | 13 | 114 | 407 | 846 | 1277 | 6782 |
| Cooling Degree Days | 0 | 0 | 1 | 16 | 58 | 195 | 287 | 222 | 98 | 6 | 0 | 0 | 883 |
| Total Precipitation (") | 0.76 | 0.77 | 2.12 | 3.48 | 4.23 | 5.25 | 4.16 | 4.22 | 3.45 | 2.69 | 1.78 | 1.13 | 34.04 |
| Days ≥ 0.1" Precip | 2 | 2 | 5 | 7 | 8 | 7 | 7 | 6 | 6 | 5 | 4 | 3 | 62 |
| Total Snowfall (") | 6.8 | 6.6 | 5.2 | 1.9 | 0.0 | 0.0 | 0.0 | 0.0 | 0.0 | 0.2 | 2.8 | 6.3 | 29.8 |
| Days ≥ 1" Snow Depth | 19 | 14 | 7 | 1 | 0 | 0 | 0 | 0 | 0 | 0 | 3 | 14 | 58 |

## ANAMOSA 1 WNW *Jones County* ELEVATION 869 ft LAT/LONG 42° 7' N / 91° 18' W

| | JAN | FEB | MAR | APR | MAY | JUN | JUL | AUG | SEP | OCT | NOV | DEC | YEAR |
|---|---|---|---|---|---|---|---|---|---|---|---|---|---|
| Maximum Temp °F | 27.3 | 32.7 | 46.0 | 61.3 | 73.2 | 81.8 | 85.4 | 82.8 | 75.2 | 63.3 | 46.3 | 32.8 | 59.0 |
| Minimum Temp °F | 8.2 | 13.4 | 25.7 | 37.0 | 47.0 | 56.3 | 61.1 | 58.4 | 50.3 | 39.2 | 27.8 | 15.8 | 36.7 |
| Mean Temp °F | 17.8 | 23.0 | 35.9 | 49.2 | 60.1 | 69.0 | 73.3 | 70.7 | 62.8 | 51.3 | 37.1 | 24.3 | 47.9 |
| Days Max Temp ≥ 90 °F | 0 | 0 | 0 | 0 | 1 | 3 | 7 | 5 | 1 | 0 | 0 | 0 | 17 |
| Days Max Temp ≤ 32 °F | 19 | 13 | 4 | 0 | 0 | 0 | 0 | 0 | 0 | 0 | 3 | 13 | 52 |
| Days Min Temp ≤ 32 °F | 30 | 27 | 23 | 11 | 3 | 0 | 0 | 0 | 1 | 9 | 21 | 29 | 154 |
| Days Min Temp ≤ 0 °F | 10 | 6 | 1 | 0 | 0 | 0 | 0 | 0 | 0 | 0 | 0 | 4 | 21 |
| Heating Degree Days | 1459 | 1179 | 896 | 476 | 197 | 34 | 6 | 20 | 138 | 429 | 832 | 1255 | 6921 |
| Cooling Degree Days | 0 | 0 | 1 | 14 | 50 | 172 | 278 | 211 | 88 | 6 | 0 | 0 | 820 |
| Total Precipitation (") | 1.12 | 1.14 | 2.38 | 3.34 | 3.86 | 4.62 | 4.15 | 4.53 | 4.02 | 2.35 | 2.41 | 1.61 | 35.53 |
| Days ≥ 0.1" Precip | 3 | 3 | 5 | 7 | 7 | 7 | 7 | 7 | 7 | 5 | 5 | 4 | 67 |
| Total Snowfall (") | *6.3* | 5.3 | 3.6 | 1.4 | 0.0 | 0.0 | 0.0 | 0.0 | 0.0 | 0.1 | *1.5* | *5.6* | 23.8 |
| Days ≥ 1" Snow Depth | *21* | *15* | *5* | 0 | 0 | 0 | 0 | 0 | 0 | 0 | 2 | na | na |

## ANKENY *Polk County* ELEVATION 940 ft LAT/LONG 41° 44' N / 93° 34' W

| | JAN | FEB | MAR | APR | MAY | JUN | JUL | AUG | SEP | OCT | NOV | DEC | YEAR |
|---|---|---|---|---|---|---|---|---|---|---|---|---|---|
| Maximum Temp °F | 27.9 | 33.6 | 46.7 | 61.5 | 72.7 | 81.3 | 85.5 | 82.8 | 75.3 | 63.2 | 46.6 | 32.9 | 59.2 |
| Minimum Temp °F | 9.2 | 14.2 | 26.6 | 39.0 | 50.1 | 59.6 | 63.8 | 60.9 | 52.4 | 40.0 | 27.8 | 15.4 | 38.2 |
| Mean Temp °F | 18.5 | 23.9 | 36.7 | 50.3 | 61.4 | 70.5 | 74.7 | 71.8 | 63.8 | 51.6 | 37.2 | 24.2 | 48.7 |
| Days Max Temp ≥ 90 °F | 0 | 0 | 0 | 0 | 1 | 3 | 9 | 6 | 2 | 0 | 0 | 0 | 21 |
| Days Max Temp ≤ 32 °F | 18 | 13 | 5 | 0 | 0 | 0 | 0 | 0 | 0 | 0 | 4 | 14 | 54 |
| Days Min Temp ≤ 32 °F | 30 | 26 | 22 | 8 | 1 | 0 | 0 | 0 | 0 | 7 | 21 | 30 | 145 |
| Days Min Temp ≤ 0 °F | 9 | 5 | 0 | 0 | 0 | 0 | 0 | 0 | 0 | 0 | 0 | 4 | 18 |
| Heating Degree Days | 1434 | 1154 | 873 | 449 | 168 | 21 | 5 | 13 | 118 | 419 | 827 | 1260 | 6741 |
| Cooling Degree Days | 0 | 0 | 2 | 17 | 57 | 197 | 316 | 245 | 104 | 5 | 0 | 0 | 943 |
| Total Precipitation (") | 0.72 | 0.91 | 2.11 | 3.26 | 4.04 | 5.22 | 4.18 | 4.19 | 3.27 | 2.45 | 1.81 | 1.14 | 33.30 |
| Days ≥ 0.1" Precip | 2 | 3 | 5 | 6 | 8 | 7 | 6 | 6 | 6 | 5 | 4 | 3 | 61 |
| Total Snowfall (") | 6.2 | 6.1 | 3.8 | 1.1 | 0.0 | 0.0 | 0.0 | 0.0 | 0.0 | 0.4 | 2.1 | 5.4 | 25.1 |
| Days ≥ 1" Snow Depth | 19 | 14 | 5 | 0 | 0 | 0 | 0 | 0 | 0 | 0 | 2 | 12 | 52 |

## ATLANTIC 1 NE *Cass County* ELEVATION 1112 ft LAT/LONG 41° 24' N / 94° 59' W

| | JAN | FEB | MAR | APR | MAY | JUN | JUL | AUG | SEP | OCT | NOV | DEC | YEAR |
|---|---|---|---|---|---|---|---|---|---|---|---|---|---|
| Maximum Temp °F | 29.2 | 34.4 | 47.9 | 62.8 | 73.6 | 82.7 | 85.8 | 83.2 | 75.7 | 63.8 | 46.5 | 33.3 | 59.9 |
| Minimum Temp °F | 8.9 | 13.4 | 25.8 | 37.5 | 48.7 | 58.4 | 62.6 | 59.7 | 50.8 | 38.8 | 26.6 | 14.9 | 37.2 |
| Mean Temp °F | 19.1 | 23.9 | 36.9 | 50.2 | 61.2 | 70.6 | 74.2 | 71.5 | 63.3 | 51.3 | 36.6 | 24.1 | 48.6 |
| Days Max Temp ≥ 90 °F | 0 | 0 | 0 | 0 | 1 | 5 | 9 | 5 | 2 | 0 | 0 | 0 | 22 |
| Days Max Temp ≤ 32 °F | 17 | 12 | 4 | 0 | 0 | 0 | 0 | 0 | 0 | 0 | 4 | 14 | 51 |
| Days Min Temp ≤ 32 °F | 30 | 27 | 23 | 10 | 1 | 0 | 0 | 0 | 1 | 10 | 22 | 30 | 154 |
| Days Min Temp ≤ 0 °F | 9 | 6 | 1 | 0 | 0 | 0 | 0 | 0 | 0 | 0 | 0 | 4 | 20 |
| Heating Degree Days | 1417 | 1153 | 865 | 450 | 172 | 22 | 5 | 18 | 131 | 427 | 847 | 1261 | 6768 |
| Cooling Degree Days | 0 | 0 | 1 | 17 | 57 | 202 | 302 | 238 | 100 | 7 | 0 | 0 | 924 |
| Total Precipitation (") | 0.78 | 0.81 | 2.28 | 3.12 | 3.85 | 4.60 | 4.29 | 3.75 | 4.09 | 2.65 | 1.60 | 1.15 | 32.97 |
| Days ≥ 0.1" Precip | 2 | 2 | 5 | 7 | 8 | 6 | 7 | 6 | 6 | 5 | 4 | 3 | 61 |
| Total Snowfall (") | 5.6 | *5.5* | 3.3 | 0.9 | 0.0 | 0.0 | 0.0 | 0.0 | 0.0 | 0.3 | *2.4* | 4.7 | 22.7 |
| Days ≥ 1" Snow Depth | na | na | *1* | 0 | 0 | 0 | 0 | 0 | 0 | 0 | *1* | na | na |

## AUDUBON 1 SSE *Audubon County* ELEVATION 1289 ft LAT/LONG 41° 42' N / 94° 57' W

| | JAN | FEB | MAR | APR | MAY | JUN | JUL | AUG | SEP | OCT | NOV | DEC | YEAR |
|---|---|---|---|---|---|---|---|---|---|---|---|---|---|
| Maximum Temp °F | 29.3 | 34.9 | 48.4 | 63.2 | 74.2 | 83.2 | 86.5 | 84.2 | 76.2 | 64.2 | 46.8 | 33.5 | 60.4 |
| Minimum Temp °F | 9.6 | 14.6 | 26.2 | 38.1 | 49.2 | 58.9 | 63.5 | 60.9 | 52.1 | 40.3 | 27.4 | 15.6 | 38.0 |
| Mean Temp °F | 19.5 | 24.8 | 37.3 | 50.7 | 61.7 | 71.0 | 75.0 | 72.6 | 64.2 | 52.3 | 37.1 | 24.6 | 49.2 |
| Days Max Temp ≥ 90 °F | 0 | 0 | 0 | 0 | 1 | 5 | 10 | 7 | 2 | 0 | 0 | 0 | 25 |
| Days Max Temp ≤ 32 °F | 17 | 12 | 4 | 0 | 0 | 0 | 0 | 0 | 0 | 0 | 3 | 13 | 49 |
| Days Min Temp ≤ 32 °F | 30 | 27 | 23 | 9 | 1 | 0 | 0 | 0 | 1 | 7 | 21 | 30 | 149 |
| Days Min Temp ≤ 0 °F | 9 | 5 | 1 | 0 | 0 | 0 | 0 | 0 | 0 | 0 | 0 | 4 | 19 |
| Heating Degree Days | 1405 | 1129 | 853 | 436 | 157 | 16 | 3 | 10 | 111 | 397 | 831 | 1246 | 6594 |
| Cooling Degree Days | 0 | 0 | 1 | 17 | 60 | 207 | 309 | 251 | 102 | 6 | 0 | 0 | 953 |
| Total Precipitation (") | 0.92 | 0.91 | 2.27 | 3.20 | 4.07 | 4.25 | 4.14 | 3.75 | 3.83 | 2.71 | 1.77 | 1.20 | 33.02 |
| Days ≥ 0.1" Precip | 3 | 3 | 5 | 6 | 8 | 7 | 6 | 6 | 6 | 5 | 4 | 4 | 63 |
| Total Snowfall (") | 6.7 | 6.3 | 4.9 | 1.3 | 0.0 | 0.0 | 0.0 | 0.0 | 0.0 | 0.3 | 2.9 | 6.6 | 29.0 |
| Days ≥ 1" Snow Depth | na | na | na | 0 | 0 | 0 | 0 | 0 | 0 | 0 | na | na | na |

## BEACONSFIELD 2 N *Ringgold County* ELEVATION 1220 ft LAT/LONG 40° 49 ' N / 94° 3 ' W

| | JAN | FEB | MAR | APR | MAY | JUN | JUL | AUG | SEP | OCT | NOV | DEC | YEAR |
|---|---|---|---|---|---|---|---|---|---|---|---|---|---|
| Maximum Temp °F | 29.9 | 36.5 | 49.2 | 62.4 | 72.3 | 81.5 | 85.8 | 83.7 | 75.8 | 64.5 | 48.1 | 35.2 | 60.4 |
| Minimum Temp °F | 11.8 | 17.0 | 28.0 | 39.7 | 50.2 | 59.1 | 64.1 | 61.3 | 52.8 | 41.3 | 29.1 | 17.9 | 39.4 |
| Mean Temp °F | 20.9 | 26.8 | 38.6 | 51.0 | 61.2 | 70.3 | 74.9 | 72.5 | 64.3 | 52.9 | 38.6 | 26.6 | 49.9 |
| Days Max Temp ≥ 90 °F | 0 | 0 | 0 | 0 | 0 | 3 | 8 | 6 | 1 | 0 | 0 | 0 | 18 |
| Days Max Temp ≤ 32 °F | 17 | 11 | 3 | 0 | 0 | 0 | 0 | 0 | 0 | 0 | 3 | 12 | 46 |
| Days Min Temp ≤ 32 °F | 30 | 26 | 20 | 7 | 1 | 0 | 0 | 0 | 1 | 6 | 19 | 29 | 139 |
| Days Min Temp ≤ 0 °F | 7 | 4 | 0 | 0 | 0 | 0 | 0 | 0 | 0 | 0 | 0 | 3 | 14 |
| Heating Degree Days | 1362 | 1073 | 811 | 422 | 161 | 20 | 3 | 10 | 109 | 379 | 784 | 1184 | 6318 |
| Cooling Degree Days | 0 | 0 | 1 | 13 | 47 | 183 | 316 | 257 | 109 | 7 | 0 | 0 | 933 |
| Total Precipitation (") | 0.78 | 0.81 | 2.31 | 3.45 | 4.06 | 4.43 | 4.74 | 3.89 | 4.46 | 2.88 | 2.10 | 1.23 | 35.14 |
| Days ≥ 0.1" Precip | 2 | 2 | 5 | 7 | 8 | 7 | 6 | 6 | 6 | 5 | 4 | 3 | 61 |
| Total Snowfall (") | 5.5 | 4.4 | 3.0 | 0.9 | 0.1 | 0.0 | 0.0 | 0.0 | 0.0 | 0.3 | 2.4 | 4.5 | 21.1 |
| Days ≥ 1" Snow Depth | 17 | 11 | 3 | 0 | 0 | 0 | 0 | 0 | 0 | 0 | 2 | 10 | 43 |

## BEDFORD *Taylor County* ELEVATION 1211 ft LAT/LONG 40° 41 ' N / 94° 44 ' W

| | JAN | FEB | MAR | APR | MAY | JUN | JUL | AUG | SEP | OCT | NOV | DEC | YEAR |
|---|---|---|---|---|---|---|---|---|---|---|---|---|---|
| Maximum Temp °F | 31.9 | 38.3 | 51.4 | 64.3 | 74.2 | 83.1 | 87.2 | 85.0 | 77.4 | 65.7 | 49.8 | 36.7 | 62.1 |
| Minimum Temp °F | 12.2 | 17.0 | 28.8 | 40.2 | 50.8 | 60.0 | 64.6 | 61.9 | 53.5 | 41.5 | 29.8 | 18.4 | 39.9 |
| Mean Temp °F | 22.1 | 27.7 | 40.1 | 52.3 | 62.5 | 71.6 | 75.9 | 73.5 | 65.4 | 53.6 | 39.8 | 27.6 | 51.0 |
| Days Max Temp ≥ 90 °F | 0 | 0 | 0 | 0 | 0 | 5 | 11 | 8 | 2 | 0 | 0 | 0 | 26 |
| Days Max Temp ≤ 32 °F | 15 | 9 | 2 | 0 | 0 | 0 | 0 | 0 | 0 | 0 | 2 | 10 | 38 |
| Days Min Temp ≤ 32 °F | 30 | 25 | 19 | 7 | 1 | 0 | 0 | 0 | 1 | 6 | 18 | 28 | 135 |
| Days Min Temp ≤ 0 °F | 7 | 4 | 0 | 0 | 0 | 0 | 0 | 0 | 0 | 0 | 0 | 3 | 14 |
| Heating Degree Days | 1326 | 1049 | 764 | 389 | 137 | 15 | 2 | 6 | 91 | 359 | 749 | 1153 | 6040 |
| Cooling Degree Days | 0 | 0 | 2 | 20 | 63 | 230 | 356 | 289 | 129 | 8 | 0 | 0 | 1097 |
| Total Precipitation (") | 0.92 | 0.87 | 2.15 | 3.25 | 4.48 | 4.70 | 5.09 | 4.00 | 4.03 | 3.03 | 2.09 | 1.19 | 35.80 |
| Days ≥ 0.1" Precip | 3 | 3 | 4 | 6 | 8 | 7 | 6 | 6 | 6 | 5 | 4 | 3 | 61 |
| Total Snowfall (") | 5.8 | 5.5 | 3.6 | 1.3 | 0.0 | 0.0 | 0.0 | 0.0 | 0.0 | 0.2 | 2.2 | *4.3* | 22.9 |
| Days ≥ 1" Snow Depth | na | na | na | 0 | 0 | 0 | 0 | 0 | 0 | 0 | *1* | na | na |

## BELLE PLAINE *Benton County* ELEVATION 855 ft LAT/LONG 41° 54 ' N / 92° 16 ' W

| | JAN | FEB | MAR | APR | MAY | JUN | JUL | AUG | SEP | OCT | NOV | DEC | YEAR |
|---|---|---|---|---|---|---|---|---|---|---|---|---|---|
| Maximum Temp °F | 28.0 | 33.4 | 46.3 | 61.1 | 72.6 | 81.4 | 84.9 | 82.6 | 74.8 | 63.1 | 46.5 | 32.8 | 59.0 |
| Minimum Temp °F | 9.4 | 14.3 | 26.7 | 38.5 | 49.3 | 58.9 | 63.2 | 60.5 | 52.2 | 40.3 | 28.5 | 16.3 | 38.2 |
| Mean Temp °F | 18.7 | 23.9 | 36.5 | 49.8 | 61.0 | 70.2 | 74.1 | 71.6 | 63.5 | 51.7 | 37.5 | 24.6 | 48.6 |
| Days Max Temp ≥ 90 °F | 0 | 0 | 0 | 0 | 1 | 3 | 8 | 5 | 1 | 0 | 0 | 0 | 18 |
| Days Max Temp ≤ 32 °F | 18 | 13 | 4 | 0 | 0 | 0 | 0 | 0 | 0 | 0 | 3 | 14 | 52 |
| Days Min Temp ≤ 32 °F | 30 | 26 | 22 | 8 | 1 | 0 | 0 | 0 | 0 | 8 | 20 | 29 | 144 |
| Days Min Temp ≤ 0 °F | 9 | 5 | 0 | 0 | 0 | 0 | 0 | 0 | 0 | 0 | 0 | 4 | 18 |
| Heating Degree Days | 1429 | 1154 | 876 | 459 | 174 | 22 | 4 | 13 | 122 | 415 | 818 | 1247 | 6733 |
| Cooling Degree Days | 0 | 0 | 1 | 14 | 56 | 192 | 301 | 235 | 93 | 6 | 0 | 0 | 898 |
| Total Precipitation (") | 1.06 | 1.05 | 2.44 | 3.79 | 3.75 | 4.70 | 4.47 | 4.98 | 3.89 | 2.68 | 2.16 | 1.54 | 36.51 |
| Days ≥ 0.1" Precip | 3 | 3 | 5 | 7 | 8 | 7 | 7 | 6 | 6 | 5 | 4 | 4 | 65 |
| Total Snowfall (") | 7.3 | 6.7 | 4.2 | 2.1 | 0.0 | 0.0 | 0.0 | 0.0 | 0.0 | 0.2 | 2.6 | 6.8 | 29.9 |
| Days ≥ 1" Snow Depth | 20 | 16 | 7 | 1 | 0 | 0 | 0 | 0 | 0 | 0 | 3 | 13 | 60 |

## BELLEVUE L&D 12 *Jackson County* ELEVATION 603 ft LAT/LONG 42° 16 ' N / 90° 25 ' W

| | JAN | FEB | MAR | APR | MAY | JUN | JUL | AUG | SEP | OCT | NOV | DEC | YEAR |
|---|---|---|---|---|---|---|---|---|---|---|---|---|---|
| Maximum Temp °F | 27.6 | 32.6 | 44.8 | 59.4 | 71.5 | 80.6 | 84.3 | 82.0 | 74.1 | 62.3 | 46.2 | 33.1 | 58.2 |
| Minimum Temp °F | 9.5 | 13.8 | 26.2 | 37.5 | 48.0 | 57.5 | 62.4 | 60.0 | 51.8 | 40.3 | 29.0 | 17.2 | 37.8 |
| Mean Temp °F | 18.6 | 23.2 | 35.5 | 48.5 | 59.8 | 69.1 | 73.4 | 71.0 | 63.0 | 51.4 | 37.6 | 25.2 | 48.0 |
| Days Max Temp ≥ 90 °F | 0 | 0 | 0 | 0 | 0 | 3 | 7 | 4 | 1 | 0 | 0 | 0 | 15 |
| Days Max Temp ≤ 32 °F | 19 | 13 | 4 | 0 | 0 | 0 | 0 | 0 | 0 | 0 | 3 | 13 | 52 |
| Days Min Temp ≤ 32 °F | 30 | 26 | 23 | 9 | 2 | 0 | 0 | 0 | 1 | 7 | 20 | 29 | 147 |
| Days Min Temp ≤ 0 °F | 9 | 5 | 0 | 0 | 0 | 0 | 0 | 0 | 0 | 0 | 0 | 4 | 18 |
| Heating Degree Days | 1435 | 1173 | 908 | 497 | 205 | 32 | 6 | 17 | 132 | 426 | 815 | 1229 | 6875 |
| Cooling Degree Days | 0 | 0 | 0 | 11 | 45 | 158 | 274 | 211 | 80 | 4 | 0 | 0 | 783 |
| Total Precipitation (") | 1.10 | 1.06 | 2.34 | 3.20 | 3.42 | 4.55 | 3.34 | 4.35 | 4.16 | 2.55 | 2.55 | 1.76 | 34.38 |
| Days ≥ 0.1" Precip | 3 | 3 | 5 | 7 | 7 | 7 | 6 | 7 | 6 | 5 | 5 | 4 | 65 |
| Total Snowfall (") | 8.6 | 6.1 | 4.7 | 1.3 | 0.1 | 0.0 | 0.0 | 0.0 | 0.0 | 0.0 | 1.6 | 7.6 | 30.0 |
| Days ≥ 1" Snow Depth | 22 | 19 | 7 | 1 | 0 | 0 | 0 | 0 | 0 | 0 | 2 | 16 | 67 |

## BLOOMFIELD 1 WNW *Davis County* ELEVATION 840 ft LAT/LONG 40° 48 ' N / 92° 25 ' W

| | JAN | FEB | MAR | APR | MAY | JUN | JUL | AUG | SEP | OCT | NOV | DEC | YEAR |
|---|---|---|---|---|---|---|---|---|---|---|---|---|---|
| Maximum Temp °F | 32.4 | 37.9 | 50.8 | 64.1 | 73.9 | 82.8 | 87.3 | 85.0 | 77.2 | 66.0 | 50.5 | 37.2 | 62.1 |
| Minimum Temp °F | 13.4 | 17.9 | 29.5 | 41.2 | 51.0 | 60.3 | 64.5 | 61.9 | 54.0 | 42.9 | 31.1 | 19.8 | 40.6 |
| Mean Temp °F | 22.9 | 27.9 | 40.1 | 52.7 | 62.4 | 71.6 | 76.0 | 73.5 | 65.7 | 54.5 | 40.8 | 28.5 | 51.4 |
| Days Max Temp ≥ 90 °F | 0 | 0 | 0 | 0 | 0 | 4 | 11 | 8 | 1 | 0 | 0 | 0 | 24 |
| Days Max Temp ≤ 32 °F | 14 | 10 | 2 | 0 | 0 | 0 | 0 | 0 | 0 | 0 | 2 | 10 | 38 |
| Days Min Temp ≤ 32 °F | 29 | 25 | 19 | 5 | 0 | 0 | 0 | 0 | 0 | 5 | 17 | 28 | 128 |
| Days Min Temp ≤ 0 °F | 7 | 4 | 0 | 0 | 0 | 0 | 0 | 0 | 0 | 0 | 0 | 3 | 14 |
| Heating Degree Days | 1298 | 1041 | 766 | 379 | 141 | 15 | 3 | 7 | 89 | 333 | 719 | 1124 | 5915 |
| Cooling Degree Days | 0 | 0 | 2 | 22 | 70 | 239 | 374 | 315 | 136 | 12 | 0 | 0 | 1170 |
| Total Precipitation (") | 1.15 | 1.02 | 2.51 | 3.73 | 4.37 | 4.33 | 4.89 | 4.53 | 4.45 | 2.99 | 2.48 | 1.64 | 38.09 |
| Days ≥ 0.1" Precip | 3 | 3 | 5 | 7 | 8 | 7 | 7 | 7 | 7 | 6 | 5 | 3 | 68 |
| Total Snowfall (") | 7.5 | 6.7 | 3.7 | 2.0 | 0.0 | 0.0 | 0.0 | 0.0 | 0.0 | 0.3 | 2.3 | 5.4 | 27.9 |
| Days ≥ 1" Snow Depth | 17 | 13 | 5 | 1 | 0 | 0 | 0 | 0 | 0 | 0 | 2 | 9 | 47 |

## BOONE *Boone County* ELEVATION 1132 ft LAT/LONG 42° 4 ' N / 93° 53 ' W

| | JAN | FEB | MAR | APR | MAY | JUN | JUL | AUG | SEP | OCT | NOV | DEC | YEAR |
|---|---|---|---|---|---|---|---|---|---|---|---|---|---|
| Maximum Temp °F | 27.9 | 33.5 | 46.4 | 60.5 | 72.7 | 81.5 | 85.6 | 83.2 | 75.9 | 64.0 | 46.3 | 32.5 | 59.2 |
| Minimum Temp °F | 5.8 | 10.9 | 24.5 | 36.6 | 47.5 | 56.7 | 61.4 | 58.5 | 49.5 | 37.3 | 25.4 | 12.9 | 35.6 |
| Mean Temp °F | 16.9 | 22.2 | 35.5 | 48.6 | 60.2 | 69.2 | 73.5 | 70.9 | 62.7 | 50.7 | 35.9 | 22.7 | 47.4 |
| Days Max Temp ≥ 90 °F | 0 | 0 | 0 | 0 | 1 | 4 | 9 | 6 | 2 | 0 | 0 | 0 | 22 |
| Days Max Temp ≤ 32 °F | 19 | 13 | 4 | 0 | 0 | 0 | 0 | 0 | 0 | 0 | 4 | 14 | 54 |
| Days Min Temp ≤ 32 °F | 31 | 28 | 25 | 10 | 1 | 0 | 0 | 0 | 1 | 11 | 23 | 30 | 160 |
| Days Min Temp ≤ 0 °F | 11 | 7 | 1 | 0 | 0 | 0 | 0 | 0 | 0 | 0 | 1 | 5 | 25 |
| Heating Degree Days | 1487 | 1202 | 909 | 495 | 193 | 31 | 6 | 17 | 134 | 447 | 865 | 1305 | 7091 |
| Cooling Degree Days | 0 | 0 | 0 | 14 | 43 | 157 | 263 | 200 | 77 | 5 | 0 | 0 | 759 |
| Total Precipitation (") | 1.04 | 1.03 | 2.37 | 3.42 | 4.32 | 5.37 | 4.14 | 4.19 | 3.60 | 2.70 | 2.02 | 1.44 | 35.64 |
| Days ≥ 0.1" Precip | 3 | 3 | 5 | 7 | 8 | 8 | 7 | 6 | 6 | 5 | 4 | 3 | 65 |
| Total Snowfall (") | 7.1 | 6.7 | 4.5 | 1.4 | 0.0 | 0.0 | 0.0 | 0.0 | 0.0 | 0.1 | 2.8 | 6.3 | 28.9 |
| Days ≥ 1" Snow Depth | 21 | 16 | 7 | 1 | 0 | 0 | 0 | 0 | 0 | 0 | 4 | 14 | 63 |

## BRITT *Hancock County* ELEVATION 1230 ft LAT/LONG 43° 4 ' N / 93° 49 ' W

| | JAN | FEB | MAR | APR | MAY | JUN | JUL | AUG | SEP | OCT | NOV | DEC | YEAR |
|---|---|---|---|---|---|---|---|---|---|---|---|---|---|
| Maximum Temp °F | 23.9 | 29.2 | ***42.3*** | 58.3 | 72.1 | 81.2 | 83.4 | 81.3 | 73.5 | 61.7 | 43.0 | 28.9 | 56.6 |
| Minimum Temp °F | 5.3 | 10.5 | ***23.3*** | 36.4 | 48.3 | 58.4 | 61.5 | 59.2 | 50.5 | 39.1 | 25.7 | 12.0 | 35.9 |
| Mean Temp °F | 14.6 | 19.9 | ***32.6*** | 47.3 | 60.2 | 69.8 | 72.4 | 70.2 | 62.0 | 50.4 | 34.3 | 20.3 | 46.2 |
| Days Max Temp ≥ 90 °F | 0 | 0 | 0 | 0 | 1 | 4 | 6 | 3 | 1 | 0 | 0 | 0 | 15 |
| Days Max Temp ≤ 32 °F | 22 | 16 | ***7*** | 0 | 0 | 0 | 0 | 0 | 0 | 0 | 6 | 19 | 70 |
| Days Min Temp ≤ 32 °F | 31 | 27 | ***25*** | 11 | 2 | 0 | 0 | 0 | 1 | 9 | 22 | 30 | 158 |
| Days Min Temp ≤ 0 °F | 12 | 8 | ***2*** | 0 | 0 | 0 | 0 | 0 | 0 | 0 | 1 | 6 | 29 |
| Heating Degree Days | 1558 | 1274 | ***998*** | 536 | 198 | 28 | 8 | 20 | 149 | 454 | 915 | 1380 | 7518 |
| Cooling Degree Days | 0 | 0 | ***0*** | ***10*** | ***48*** | 174 | 227 | 189 | 80 | 4 | 0 | 0 | 732 |
| Total Precipitation (") | 0.99 | 0.80 | 1.87 | 3.00 | 3.74 | 4.97 | 4.18 | 3.61 | 3.41 | 2.13 | 1.68 | 1.42 | 31.80 |
| Days ≥ 0.1" Precip | 3 | 3 | ***4*** | 7 | 7 | 7 | 7 | 6 | 6 | 4 | 4 | 3 | 61 |
| Total Snowfall (") | ***9.3*** | 7.4 | ***6.5*** | ***1.6*** | 0.0 | 0.0 | 0.0 | 0.0 | 0.0 | 0.3 | 3.7 | ***10.0*** | 38.8 |
| Days ≥ 1" Snow Depth | na | na | na | ***1*** | 0 | 0 | 0 | 0 | 0 | 0 | na | na | na |

## BURLINGTN RADIO KBUR *Des Moines County* ELEVATION 702 ft LAT/LONG 40° 49 ' N / 91° 10 ' W

| | JAN | FEB | MAR | APR | MAY | JUN | JUL | AUG | SEP | OCT | NOV | DEC | YEAR |
|---|---|---|---|---|---|---|---|---|---|---|---|---|---|
| Maximum Temp °F | 30.5 | 35.7 | 48.4 | 62.0 | 72.6 | 81.8 | 85.5 | 82.8 | 75.5 | 63.9 | 48.7 | 35.5 | 60.2 |
| Minimum Temp °F | 13.7 | 18.4 | 29.7 | 41.5 | 51.9 | 61.3 | 65.8 | 63.2 | 55.2 | 43.5 | 31.6 | 20.1 | 41.3 |
| Mean Temp °F | 22.1 | 27.0 | 39.1 | 51.8 | 62.3 | 71.6 | 75.6 | 73.0 | 65.4 | 53.7 | 40.2 | 27.8 | 50.8 |
| Days Max Temp ≥ 90 °F | 0 | 0 | 0 | 0 | 0 | 4 | 8 | 5 | 1 | 0 | 0 | 0 | 18 |
| Days Max Temp ≤ 32 °F | 17 | 11 | 3 | 0 | 0 | 0 | 0 | 0 | 0 | 0 | 2 | 11 | 44 |
| Days Min Temp ≤ 32 °F | 29 | 25 | 19 | 5 | 0 | 0 | 0 | 0 | 0 | 3 | 16 | 27 | 124 |
| Days Min Temp ≤ 0 °F | 6 | 3 | 0 | 0 | 0 | 0 | 0 | 0 | 0 | 0 | 0 | 3 | 12 |
| Heating Degree Days | 1323 | 1065 | 798 | 403 | 148 | 16 | 2 | 8 | 91 | 356 | 739 | 1146 | 6095 |
| Cooling Degree Days | 0 | 0 | 3 | 18 | 74 | 243 | 359 | 288 | 122 | 11 | 0 | 0 | 1118 |
| Total Precipitation (") | 1.22 | 1.20 | 2.85 | 3.63 | 3.88 | 4.18 | 4.46 | 4.05 | 3.91 | 2.94 | 2.62 | 2.16 | 37.10 |
| Days ≥ 0.1" Precip | 3 | 3 | 6 | 7 | 8 | 7 | 7 | 6 | 6 | 5 | 5 | 4 | 67 |
| Total Snowfall (") | 6.7 | 5.6 | 3.4 | 1.1 | 0.0 | 0.0 | 0.0 | 0.0 | 0.0 | 0.1 | 2.2 | 5.4 | 24.5 |
| Days ≥ 1" Snow Depth | 15 | 10 | 3 | 0 | 0 | 0 | 0 | 0 | 0 | 0 | 1 | 7 | 36 |

## CARROLL *Carroll County* ELEVATION 1240 ft LAT/LONG 42° 4 ' N / 94° 51 ' W

| | JAN | FEB | MAR | APR | MAY | JUN | JUL | AUG | SEP | OCT | NOV | DEC | YEAR |
|---|---|---|---|---|---|---|---|---|---|---|---|---|---|
| Maximum Temp °F | 27.0 | 32.7 | 46.1 | 61.4 | 73.1 | 82.5 | 85.7 | 83.3 | 75.3 | 62.7 | 45.0 | 31.6 | 58.9 |
| Minimum Temp °F | 8.1 | 13.2 | 24.9 | 36.4 | 47.8 | 57.6 | 62.0 | 59.2 | 50.4 | 38.7 | 25.9 | 14.0 | 36.5 |
| Mean Temp °F | 17.6 | 23.0 | 35.5 | 48.9 | 60.5 | 70.1 | 73.9 | 71.3 | 62.9 | 50.7 | 35.4 | 22.9 | 47.7 |
| Days Max Temp ≥ 90 °F | 0 | 0 | 0 | 0 | 1 | 6 | 9 | 6 | 2 | 0 | 0 | 0 | 24 |
| Days Max Temp ≤ 32 °F | 19 | 13 | 5 | 0 | 0 | 0 | 0 | 0 | 0 | 0 | 4 | 15 | 56 |
| Days Min Temp ≤ 32 °F | 30 | 27 | 24 | 11 | 2 | 0 | 0 | 0 | 1 | 9 | 23 | 30 | 157 |
| Days Min Temp ≤ 0 °F | 10 | 6 | 1 | 0 | 0 | 0 | 0 | 0 | 0 | 0 | 1 | 5 | 23 |
| Heating Degree Days | 1464 | 1180 | 908 | 487 | 189 | 25 | 5 | 17 | 135 | 444 | 880 | 1300 | 7034 |
| Cooling Degree Days | 0 | 0 | 1 | 14 | 51 | 181 | 282 | 211 | 80 | 5 | 0 | 0 | 825 |
| Total Precipitation (") | 0.81 | 0.78 | 2.22 | 3.24 | 4.11 | 4.80 | 4.62 | 3.82 | 3.73 | 2.50 | 1.62 | 1.01 | 33.26 |
| Days ≥ 0.1" Precip | 2 | 2 | 5 | 6 | 8 | 7 | 6 | 6 | 6 | 4 | 4 | 3 | 59 |
| Total Snowfall (") | 6.5 | 5.7 | 5.4 | 1.7 | 0.0 | 0.0 | 0.0 | 0.0 | 0.0 | 0.4 | 3.2 | 6.4 | 29.3 |
| Days ≥ 1" Snow Depth | 16 | 13 | 6 | 1 | 0 | 0 | 0 | 0 | 0 | 0 | *4* | *13* | 53 |

## CASCADE *Dubuque County* ELEVATION 879 ft LAT/LONG 42° 18 ' N / 91° 0 ' W

| | JAN | FEB | MAR | APR | MAY | JUN | JUL | AUG | SEP | OCT | NOV | DEC | YEAR |
|---|---|---|---|---|---|---|---|---|---|---|---|---|---|
| Maximum Temp °F | 26.0 | 31.6 | 44.5 | 59.2 | 71.6 | 80.7 | 84.8 | 82.0 | 73.7 | 61.7 | 45.2 | 31.9 | 57.7 |
| Minimum Temp °F | 7.0 | 11.9 | 25.0 | 36.4 | 47.6 | 56.8 | 61.2 | 58.4 | 49.7 | 38.2 | 27.0 | 14.6 | 36.2 |
| Mean Temp °F | 16.5 | 21.8 | 34.8 | 47.8 | 59.6 | 68.8 | 73.0 | 70.3 | 61.7 | 50.0 | 36.1 | 23.3 | 47.0 |
| Days Max Temp ≥ 90 °F | 0 | 0 | 0 | 0 | 0 | 3 | 7 | 4 | 1 | 0 | 0 | 0 | 15 |
| Days Max Temp ≤ 32 °F | 20 | 14 | 4 | 0 | 0 | 0 | 0 | 0 | 0 | 0 | 4 | 15 | 57 |
| Days Min Temp ≤ 32 °F | 30 | 27 | 24 | 11 | 2 | 0 | 0 | 0 | 1 | 10 | 22 | 29 | 156 |
| Days Min Temp ≤ 0 °F | 11 | 7 | 1 | 0 | 0 | 0 | 0 | 0 | 0 | 0 | 0 | 5 | 24 |
| Heating Degree Days | 1499 | 1216 | 931 | 517 | 208 | 34 | 7 | 20 | 154 | 466 | 860 | 1288 | 7200 |
| Cooling Degree Days | 0 | 0 | 0 | 10 | 43 | 154 | 273 | 195 | 66 | 4 | 0 | 0 | 745 |
| Total Precipitation (") | 1.18 | 1.12 | 2.29 | 3.07 | 3.38 | 4.57 | 3.38 | 4.51 | 4.03 | 2.29 | 2.39 | 1.68 | 33.89 |
| Days ≥ 0.1" Precip | 4 | 3 | 5 | 7 | 7 | 7 | 6 | 6 | 7 | 5 | 5 | 4 | 66 |
| Total Snowfall (") | 8.9 | 7.5 | 5.0 | 1.9 | 0.0 | 0.0 | 0.0 | 0.0 | 0.0 | 0.0 | 2.5 | 7.9 | 33.7 |
| Days ≥ 1" Snow Depth | 21 | 19 | 7 | 1 | 0 | 0 | 0 | 0 | 0 | 0 | 2 | 14 | 64 |

## CASTANA EXPRMNT FARM *Monona County* ELEVATION 1440 ft LAT/LONG 42° 4 ' N / 95° 49 ' W

| | JAN | FEB | MAR | APR | MAY | JUN | JUL | AUG | SEP | OCT | NOV | DEC | YEAR |
|---|---|---|---|---|---|---|---|---|---|---|---|---|---|
| Maximum Temp °F | 28.7 | 34.5 | 47.5 | 62.5 | 73.1 | 82.1 | 85.4 | 83.3 | 75.4 | 63.8 | 45.8 | 32.7 | 59.6 |
| Minimum Temp °F | 9.6 | 14.9 | 26.2 | 38.1 | 49.1 | 58.6 | 63.2 | 60.8 | 51.9 | 40.3 | 27.1 | 15.2 | 37.9 |
| Mean Temp °F | 19.2 | 24.7 | 36.9 | 50.3 | 61.1 | 70.4 | 74.3 | 72.1 | 63.6 | 52.0 | 36.5 | 24.0 | 48.8 |
| Days Max Temp ≥ 90 °F | 0 | 0 | 0 | 0 | 1 | 4 | 8 | 6 | 2 | 0 | 0 | 0 | 21 |
| Days Max Temp ≤ 32 °F | 17 | 12 | 4 | 0 | 0 | 0 | 0 | 0 | 0 | 0 | 4 | 14 | 51 |
| Days Min Temp ≤ 32 °F | 30 | 27 | 23 | 9 | 1 | 0 | 0 | 0 | 1 | 7 | 21 | 30 | 149 |
| Days Min Temp ≤ 0 °F | 9 | 5 | 1 | 0 | 0 | 0 | 0 | 0 | 0 | 0 | 0 | 4 | 19 |
| Heating Degree Days | 1415 | 1130 | 866 | 447 | 170 | 23 | 4 | 11 | 119 | 403 | 848 | 1266 | 6702 |
| Cooling Degree Days | 0 | 0 | 1 | 19 | 57 | 202 | 300 | 243 | 98 | 7 | 0 | 0 | 927 |
| Total Precipitation (") | 0.59 | 0.60 | 2.12 | 2.92 | 4.01 | 4.49 | 3.52 | 3.45 | 3.53 | 2.52 | 1.42 | 0.88 | 30.05 |
| Days ≥ 0.1" Precip | 2 | 2 | 4 | 6 | 7 | 7 | 6 | 5 | 6 | 4 | 3 | 2 | 54 |
| Total Snowfall (") | 6.8 | 6.0 | 6.2 | 1.6 | 0.0 | 0.0 | 0.0 | 0.0 | 0.0 | 1.0 | 4.5 | 7.5 | 33.6 |
| Days ≥ 1" Snow Depth | 19 | 14 | 8 | 1 | 0 | 0 | 0 | 0 | 0 | 0 | 5 | 15 | 62 |

## CEDAR RAPIDS MUNI AP *Linn County* ELEVATION 902 ft LAT/LONG 41° 53 ' N / 91° 42 ' W

| | JAN | FEB | MAR | APR | MAY | JUN | JUL | AUG | SEP | OCT | NOV | DEC | YEAR |
|---|---|---|---|---|---|---|---|---|---|---|---|---|---|
| Maximum Temp °F | 26.1 | 31.6 | 44.7 | 59.7 | 71.7 | 80.8 | 84.3 | 81.5 | 73.9 | 62.0 | 45.3 | 31.7 | 57.8 |
| Minimum Temp °F | 9.3 | 14.6 | 26.8 | 38.7 | 49.8 | 59.2 | 63.6 | 60.8 | 52.1 | 40.4 | 28.1 | 15.8 | 38.3 |
| Mean Temp °F | 17.7 | 23.2 | 35.8 | 49.2 | 60.8 | 70.0 | 74.0 | 71.1 | 63.1 | 51.2 | 36.8 | 23.8 | 48.1 |
| Days Max Temp ≥ 90 °F | 0 | 0 | 0 | 0 | 1 | 3 | 7 | 4 | 1 | 0 | 0 | 0 | 16 |
| Days Max Temp ≤ 32 °F | 20 | 14 | 5 | 0 | 0 | 0 | 0 | 0 | 0 | 0 | 4 | 15 | 58 |
| Days Min Temp ≤ 32 °F | 30 | 27 | 22 | 7 | 0 | 0 | 0 | 0 | 0 | 7 | 21 | 29 | 143 |
| Days Min Temp ≤ 0 °F | 9 | 5 | 0 | 0 | 0 | 0 | 0 | 0 | 0 | 0 | 0 | 4 | 18 |
| Heating Degree Days | 1459 | 1176 | 899 | 477 | 183 | 24 | 4 | 15 | 130 | 431 | 841 | 1272 | 6911 |
| Cooling Degree Days | 0 | 0 | 1 | 13 | 55 | 188 | 297 | 226 | 85 | 7 | 0 | 0 | 872 |
| Total Precipitation (") | 1.06 | 0.98 | 2.26 | 3.29 | 3.71 | 4.60 | 4.32 | 4.45 | 3.83 | 2.32 | 2.09 | 1.61 | 34.52 |
| Days ≥ 0.1" Precip | 3 | 3 | 5 | 7 | 7 | 7 | 7 | 7 | 6 | 5 | 5 | 4 | 66 |
| Total Snowfall (") | 7.0 | 6.3 | 3.6 | 1.6 | 0.0 | 0.0 | 0.0 | 0.0 | 0.0 | 0.1 | 2.5 | 7.2 | 28.3 |
| Days ≥ 1" Snow Depth | 19 | 15 | 5 | 1 | 0 | 0 | 0 | 0 | 0 | 0 | 2 | 11 | 53 |

## CEDAR RAPIDS NO 1 *Linn County* ELEVATION 820 ft LAT/LONG 42° 2 ' N / 91° 35 ' W

| | JAN | FEB | MAR | APR | MAY | JUN | JUL | AUG | SEP | OCT | NOV | DEC | YEAR |
|---|---|---|---|---|---|---|---|---|---|---|---|---|---|
| Maximum Temp °F | 28.0 | 33.8 | 46.7 | 61.8 | 73.1 | 81.8 | 85.2 | 82.8 | 75.1 | 63.6 | 46.7 | 33.0 | 59.3 |
| Minimum Temp °F | 10.2 | 15.3 | 27.2 | 38.9 | 49.8 | 59.4 | 63.8 | 61.2 | 53.1 | 41.8 | 29.3 | 17.1 | 38.9 |
| Mean Temp °F | 19.1 | 24.6 | 37.0 | 50.3 | 61.5 | 70.6 | 74.5 | 72.0 | 64.1 | 52.7 | 38.0 | 25.1 | 49.1 |
| Days Max Temp ≥ 90 °F | 0 | 0 | 0 | 0 | 1 | 3 | 7 | 5 | 1 | 0 | 0 | 0 | 17 |
| Days Max Temp ≤ 32 °F | 19 | 13 | 4 | 0 | 0 | 0 | 0 | 0 | 0 | 0 | 3 | 14 | 53 |
| Days Min Temp ≤ 32 °F | 30 | 26 | 22 | 9 | 1 | 0 | 0 | 0 | 0 | 6 | 19 | 29 | 142 |
| Days Min Temp ≤ 0 °F | 9 | 5 | 0 | 0 | 0 | 0 | 0 | 0 | 0 | 0 | 0 | 4 | 18 |
| Heating Degree Days | 1417 | 1136 | 863 | 445 | 164 | 19 | 3 | 10 | 110 | 387 | 803 | 1231 | 6588 |
| Cooling Degree Days | 0 | 0 | 1 | 16 | 61 | 202 | 314 | 248 | 99 | 9 | 0 | 0 | 950 |
| Total Precipitation (") | 1.16 | 0.98 | 2.19 | 3.50 | 4.28 | 5.13 | 4.80 | 4.79 | 4.12 | 2.54 | 2.38 | 1.65 | 37.52 |
| Days ≥ 0.1" Precip | 3 | 3 | 5 | 7 | 8 | 7 | 7 | 7 | 7 | 5 | 5 | 4 | 68 |
| Total Snowfall (") | 7.3 | 6.3 | 4.3 | 2.1 | 0.0 | 0.0 | 0.0 | 0.0 | 0.0 | 0.1 | 3.1 | 7.2 | 30.4 |
| Days ≥ 1" Snow Depth | 20 | 17 | 6 | 1 | 0 | 0 | 0 | 0 | 0 | 0 | 2 | 13 | 59 |

## CENTERVILLE *Appanoose County* ELEVATION 980 ft LAT/LONG 40° 44 ' N / 92° 52 ' W

| | JAN | FEB | MAR | APR | MAY | JUN | JUL | AUG | SEP | OCT | NOV | DEC | YEAR |
|---|---|---|---|---|---|---|---|---|---|---|---|---|---|
| Maximum Temp °F | 31.5 | 37.1 | 49.9 | 63.1 | 73.3 | 82.0 | 86.9 | 84.3 | 76.1 | 64.6 | 49.1 | 36.1 | 61.2 |
| Minimum Temp °F | 13.9 | 18.2 | 29.6 | 40.9 | 51.4 | 60.6 | 65.6 | 62.8 | 54.6 | 43.3 | 30.9 | 19.6 | 41.0 |
| Mean Temp °F | 22.8 | 27.7 | 39.7 | 52.0 | 62.4 | 71.4 | 76.2 | 73.6 | 65.4 | 54.0 | 40.0 | 27.9 | 51.1 |
| Days Max Temp ≥ 90 °F | 0 | 0 | 0 | 0 | 0 | 3 | 10 | 6 | 2 | 0 | 0 | 0 | 21 |
| Days Max Temp ≤ 32 °F | 15 | 10 | 3 | 0 | 0 | 0 | 0 | 0 | 0 | 0 | 2 | 11 | 41 |
| Days Min Temp ≤ 32 °F | 29 | 25 | 19 | 5 | 0 | 0 | 0 | 0 | 0 | 4 | 18 | 28 | 128 |
| Days Min Temp ≤ 0 °F | 6 | 3 | 0 | 0 | 0 | 0 | 0 | 0 | 0 | 0 | 0 | 3 | 12 |
| Heating Degree Days | 1304 | 1046 | 778 | 398 | 139 | 15 | 1 | 6 | 93 | 349 | 744 | 1144 | 6017 |
| Cooling Degree Days | 0 | 0 | 2 | 19 | 58 | 214 | ***361*** | 293 | 119 | 9 | 0 | 0 | 1075 |
| Total Precipitation (") | 0.94 | 0.86 | 2.30 | 3.62 | 4.17 | 4.48 | 5.06 | 4.25 | 4.15 | 2.89 | 2.22 | 1.33 | 36.27 |
| Days ≥ 0.1" Precip | 2 | *3* | 5 | 6 | 7 | 7 | 6 | 6 | 6 | 5 | 4 | 3 | 60 |
| Total Snowfall (") | 5.1 | ***5.8*** | ***2.6*** | 1.4 | 0.0 | 0.0 | 0.0 | 0.0 | 0.0 | 0.2 | 1.8 | 3.8 | 20.7 |
| Days ≥ 1" Snow Depth | na | na | *2* | 0 | 0 | 0 | 0 | 0 | 0 | 0 | *1* | *6* | na |

## CHARITON 1 E *Lucas County* ELEVATION 942 ft LAT/LONG 41° 0 ' N / 93° 15 ' W

| | JAN | FEB | MAR | APR | MAY | JUN | JUL | AUG | SEP | OCT | NOV | DEC | YEAR |
|---|---|---|---|---|---|---|---|---|---|---|---|---|---|
| Maximum Temp °F | 31.3 | 36.8 | 49.8 | 63.6 | 73.6 | 82.4 | 86.9 | 84.8 | 77.0 | 65.6 | 49.2 | 36.0 | 61.4 |
| Minimum Temp °F | 11.3 | 15.5 | 27.6 | 39.0 | 48.9 | 58.3 | 63.4 | 60.7 | 52.3 | 40.4 | 28.8 | 17.5 | 38.6 |
| Mean Temp °F | 21.3 | 26.2 | 38.7 | 51.4 | 61.3 | 70.4 | 75.2 | 72.8 | 64.7 | 53.0 | 39.0 | 26.8 | 50.1 |
| Days Max Temp ≥ 90 °F | 0 | 0 | 0 | 0 | 0 | 4 | 11 | 8 | 2 | 0 | 0 | 0 | 25 |
| Days Max Temp ≤ 32 °F | 15 | 11 | 3 | 0 | 0 | 0 | 0 | 0 | 0 | 0 | 2 | 11 | 42 |
| Days Min Temp ≤ 32 °F | 30 | 26 | 21 | 8 | 1 | 0 | 0 | 0 | 1 | 7 | 20 | 28 | 142 |
| Days Min Temp ≤ 0 °F | 8 | 5 | 0 | 0 | 0 | 0 | 0 | 0 | 0 | 0 | 0 | 3 | 16 |
| Heating Degree Days | 1349 | 1089 | 809 | 416 | 165 | 22 | 4 | 10 | 107 | 378 | 774 | 1179 | 6302 |
| Cooling Degree Days | 0 | 0 | 1 | 18 | 49 | 192 | 326 | 271 | 114 | 8 | 0 | 0 | 979 |
| Total Precipitation (") | 0.94 | 0.97 | 2.14 | 3.74 | 4.00 | 4.63 | 4.96 | 4.05 | 4.78 | 2.88 | 2.21 | 1.36 | 36.66 |
| Days ≥ 0.1" Precip | 3 | 3 | 5 | 7 | 8 | 7 | 7 | 6 | 7 | 5 | 5 | 3 | 66 |
| Total Snowfall (") | 7.1 | 6.1 | 3.8 | 2.0 | 0.0 | 0.0 | 0.0 | 0.0 | 0.0 | 0.2 | 2.2 | 5.2 | 26.6 |
| Days ≥ 1" Snow Depth | 17 | 11 | 4 | 1 | 0 | 0 | 0 | 0 | 0 | 0 | 2 | 10 | 45 |

## CHARLES CITY *Floyd County* ELEVATION 1020 ft LAT/LONG 43° 4 ' N / 92° 40 ' W

| | JAN | FEB | MAR | APR | MAY | JUN | JUL | AUG | SEP | OCT | NOV | DEC | YEAR |
|---|---|---|---|---|---|---|---|---|---|---|---|---|---|
| Maximum Temp °F | 24.0 | 29.6 | 42.8 | 59.0 | 71.8 | 80.9 | 84.1 | 81.9 | 74.0 | 61.6 | 43.1 | 28.8 | 56.8 |
| Minimum Temp °F | 5.5 | 11.2 | 24.3 | 36.7 | 47.9 | 57.3 | 61.3 | 58.8 | 50.2 | 39.0 | 26.1 | 12.8 | 35.9 |
| Mean Temp °F | 14.7 | 20.5 | 33.6 | 47.9 | 59.9 | 69.1 | 72.7 | 70.4 | 62.1 | 50.3 | 34.6 | 20.8 | 46.4 |
| Days Max Temp ≥ 90 °F | 0 | 0 | 0 | 0 | 1 | 3 | 6 | 4 | 1 | 0 | 0 | 0 | 15 |
| Days Max Temp ≤ 32 °F | 22 | 15 | 6 | 0 | 0 | 0 | 0 | 0 | 0 | 0 | 5 | 19 | 67 |
| Days Min Temp ≤ 32 °F | 31 | 27 | 25 | 10 | 1 | 0 | 0 | 0 | 1 | 9 | 22 | 30 | 156 |
| Days Min Temp ≤ 0 °F | 12 | 7 | 1 | 0 | 0 | 0 | 0 | 0 | 0 | 0 | 1 | 6 | 27 |
| Heating Degree Days | 1553 | 1252 | 968 | 515 | 200 | 30 | 7 | 19 | 145 | 455 | 904 | 1363 | 7411 |
| Cooling Degree Days | 0 | 0 | 0 | 10 | 46 | 162 | 249 | 190 | 70 | 3 | 0 | 0 | 730 |
| Total Precipitation (") | 0.91 | 0.85 | 2.09 | 3.49 | 3.95 | 5.06 | 4.27 | 4.06 | 3.82 | 2.60 | 2.02 | 1.24 | 34.36 |
| Days ≥ 0.1" Precip | 3 | 2 | 5 | 8 | 7 | 8 | 6 | 6 | 6 | 5 | 4 | 3 | 63 |
| Total Snowfall (") | 8.9 | 6.5 | 6.0 | 2.6 | 0.0 | 0.0 | 0.0 | 0.0 | 0.0 | 0.1 | 3.9 | 8.4 | 36.4 |
| Days ≥ 1" Snow Depth | 25 | 23 | 13 | 2 | 0 | 0 | 0 | 0 | 0 | 0 | 4 | 20 | 87 |

## CHEROKEE *Cherokee County* ELEVATION 1212 ft LAT/LONG 42° 44 ' N / 95° 33 ' W

| | JAN | FEB | MAR | APR | MAY | JUN | JUL | AUG | SEP | OCT | NOV | DEC | YEAR |
|---|---|---|---|---|---|---|---|---|---|---|---|---|---|
| Maximum Temp °F | 25.6 | 31.1 | 43.9 | 59.6 | 72.3 | 81.3 | 85.2 | 82.8 | 74.4 | 62.1 | 44.2 | 30.1 | 57.7 |
| Minimum Temp °F | 4.0 | 9.3 | 22.6 | 35.1 | 46.5 | 56.6 | 61.4 | 58.4 | 48.1 | 35.3 | 23.2 | 10.3 | 34.2 |
| Mean Temp °F | 14.8 | 20.2 | 33.3 | 47.4 | 59.4 | 69.0 | 73.3 | 70.6 | 61.3 | 48.7 | 33.7 | 20.3 | 46.0 |
| Days Max Temp ≥ 90 °F | 0 | 0 | 0 | 0 | 1 | 5 | 9 | 6 | 2 | 0 | 0 | 0 | 23 |
| Days Max Temp ≤ 32 °F | 20 | 15 | 6 | 0 | 0 | 0 | 0 | 0 | 0 | 0 | 5 | 16 | 62 |
| Days Min Temp ≤ 32 °F | 31 | 28 | 25 | 12 | 2 | 0 | 0 | 0 | 2 | 13 | 25 | 31 | 169 |
| Days Min Temp ≤ 0 °F | 13 | 9 | 2 | 0 | 0 | 0 | 0 | 0 | 0 | 0 | 1 | 7 | 32 |
| Heating Degree Days | 1553 | 1258 | 977 | 531 | 218 | 39 | 8 | 22 | 168 | 503 | 932 | 1381 | 7590 |
| Cooling Degree Days | 0 | 0 | 0 | 12 | 50 | 172 | 268 | 205 | 71 | 3 | 0 | 0 | 781 |
| Total Precipitation (") | 0.57 | 0.64 | 1.91 | 2.64 | 3.69 | 4.50 | 3.84 | 3.61 | 3.44 | 2.24 | 1.54 | 0.86 | 29.48 |
| Days ≥ 0.1" Precip | 2 | 2 | 4 | 5 | 7 | 7 | 6 | 6 | 6 | 4 | 3 | 2 | 54 |
| Total Snowfall (") | *5.7* | 6.3 | *5.6* | 1.1 | 0.0 | 0.0 | 0.0 | 0.0 | 0.0 | 0.2 | *4.6* | 7.7 | 31.2 |
| Days ≥ 1" Snow Depth | na | na | *4* | 0 | 0 | 0 | 0 | 0 | 0 | 0 | na | na | na |

## CLARINDA *Page County* ELEVATION 1001 ft LAT/LONG 40° 44 ' N / 95° 1 ' W

| | JAN | FEB | MAR | APR | MAY | JUN | JUL | AUG | SEP | OCT | NOV | DEC | YEAR |
|---|---|---|---|---|---|---|---|---|---|---|---|---|---|
| Maximum Temp °F | 31.9 | 37.6 | 50.7 | 64.1 | 74.4 | 83.6 | 87.4 | 84.9 | 76.9 | 65.2 | 48.8 | 36.1 | 61.8 |
| Minimum Temp °F | 12.2 | 16.8 | 28.2 | 39.8 | 50.5 | 60.0 | 64.5 | 61.8 | 53.2 | 41.3 | 29.0 | 17.9 | 39.6 |
| Mean Temp °F | 22.1 | 27.2 | 39.5 | 52.0 | 62.4 | 71.8 | 76.0 | 73.4 | 65.1 | 53.3 | 38.9 | 27.0 | 50.7 |
| Days Max Temp ≥ 90 °F | 0 | 0 | 0 | 0 | 1 | 6 | 12 | 8 | 3 | 0 | 0 | 0 | 30 |
| Days Max Temp ≤ 32 °F | 15 | 10 | 3 | 0 | 0 | 0 | 0 | 0 | 0 | 0 | 3 | 11 | 42 |
| Days Min Temp ≤ 32 °F | 30 | 26 | 20 | 7 | 1 | 0 | 0 | 0 | 0 | 6 | 20 | 29 | 139 |
| Days Min Temp ≤ 0 °F | 7 | 4 | 0 | 0 | 0 | 0 | 0 | 0 | 0 | 0 | 0 | 3 | 14 |
| Heating Degree Days | 1325 | 1060 | 785 | 400 | 139 | 13 | 2 | 7 | 99 | 369 | 777 | 1170 | 6146 |
| Cooling Degree Days | 0 | 0 | 1 | 20 | 60 | 223 | 345 | 269 | 116 | 8 | 0 | 0 | 1042 |
| Total Precipitation (") | 0.95 | 0.88 | 2.40 | 3.20 | 4.48 | 4.80 | 5.01 | 4.32 | 4.16 | 2.75 | 2.07 | 1.27 | 36.29 |
| Days ≥ 0.1" Precip | 3 | 2 | 5 | 6 | 8 | 7 | 7 | 6 | 6 | 5 | 4 | 3 | 62 |
| Total Snowfall (") | 7.0 | 5.7 | 3.2 | 0.8 | 0.0 | 0.0 | 0.0 | 0.0 | 0.0 | 0.0 | 2.5 | 5.0 | 24.2 |
| Days ≥ 1" Snow Depth | 14 | 11 | 4 | 0 | 0 | 0 | 0 | 0 | 0 | 0 | 2 | 9 | 40 |

## CLARION *Wright County* ELEVATION 1171 ft LAT/LONG 42° 44 ' N / 93° 44 ' W

| | JAN | FEB | MAR | APR | MAY | JUN | JUL | AUG | SEP | OCT | NOV | DEC | YEAR |
|---|---|---|---|---|---|---|---|---|---|---|---|---|---|
| Maximum Temp °F | 23.3 | 28.9 | 41.9 | 58.1 | 71.6 | 80.6 | 83.5 | 80.9 | 73.5 | 61.2 | 43.2 | 28.4 | 56.3 |
| Minimum Temp °F | 4.2 | 9.9 | 23.1 | 35.7 | 47.4 | 57.3 | 61.3 | 58.1 | 49.0 | 37.1 | 24.6 | 10.8 | 34.9 |
| Mean Temp °F | 13.7 | 19.4 | 32.5 | 46.9 | 59.5 | 69.0 | 72.4 | 69.5 | 61.3 | 49.2 | 33.9 | 19.6 | 45.6 |
| Days Max Temp ≥ 90 °F | 0 | 0 | 0 | 0 | 1 | 4 | 6 | 3 | 1 | 0 | 0 | 0 | 15 |
| Days Max Temp ≤ 32 °F | 23 | 16 | 7 | 0 | 0 | 0 | 0 | 0 | 0 | 0 | 5 | 19 | 70 |
| Days Min Temp ≤ 32 °F | 31 | 28 | 26 | 11 | 2 | 0 | 0 | 0 | 1 | 11 | 23 | 31 | 164 |
| Days Min Temp ≤ 0 °F | 12 | 8 | 1 | 0 | 0 | 0 | 0 | 0 | 0 | 0 | 1 | 7 | 29 |
| Heating Degree Days | 1585 | 1281 | 1000 | 544 | 214 | 36 | 9 | 27 | 166 | 490 | 925 | 1400 | 7677 |
| Cooling Degree Days | 0 | 0 | 0 | 10 | 47 | 167 | 239 | 178 | 63 | 4 | 0 | 0 | 708 |
| Total Precipitation (") | 0.72 | 0.76 | 1.94 | 3.12 | 4.02 | 5.11 | 4.29 | 3.88 | 3.51 | 2.47 | 1.84 | 1.19 | 32.85 |
| Days ≥ 0.1" Precip | 2 | 2 | 5 | 6 | 8 | 8 | 6 | 6 | 6 | 5 | 4 | 3 | 61 |
| Total Snowfall (") | 7.8 | 5.9 | 6.2 | 1.7 | 0.0 | 0.0 | 0.0 | 0.0 | 0.0 | 0.3 | 3.2 | 8.0 | 33.1 |
| Days ≥ 1" Snow Depth | 22 | 18 | 9 | 1 | 0 | 0 | 0 | 0 | 0 | 0 | 4 | 17 | 71 |

## CLINTON 1 *Clinton County* ELEVATION 585 ft LAT/LONG 41° 48 ' N / 90° 16 ' W

| | JAN | FEB | MAR | APR | MAY | JUN | JUL | AUG | SEP | OCT | NOV | DEC | YEAR |
|---|---|---|---|---|---|---|---|---|---|---|---|---|---|
| Maximum Temp °F | 28.7 | 34.1 | 47.3 | 62.2 | 73.6 | 82.2 | 85.5 | 83.1 | 75.8 | 63.9 | 47.7 | 34.2 | 59.9 |
| Minimum Temp °F | 11.6 | 16.3 | 28.2 | 39.6 | 50.3 | 59.8 | 64.2 | 61.6 | 53.5 | 42.1 | 30.6 | 18.8 | 39.7 |
| Mean Temp °F | 20.2 | 25.2 | 37.8 | 50.9 | 62.0 | 71.0 | 74.9 | 72.4 | 64.7 | 53.0 | 39.2 | 26.5 | 49.8 |
| Days Max Temp ≥ 90 °F | 0 | 0 | 0 | 0 | 1 | 4 | 8 | 5 | 1 | 0 | 0 | 0 | 19 |
| Days Max Temp ≤ 32 °F | 18 | 12 | 3 | 0 | 0 | 0 | 0 | 0 | 0 | 0 | 2 | 12 | 47 |
| Days Min Temp ≤ 32 °F | 30 | 26 | 21 | 7 | 1 | 0 | 0 | 0 | 0 | 6 | 18 | 28 | 137 |
| Days Min Temp ≤ 0 °F | 8 | 4 | 0 | 0 | 0 | 0 | 0 | 0 | 0 | 0 | 0 | 3 | 15 |
| Heating Degree Days | 1384 | 1118 | 836 | 430 | 159 | 17 | 3 | 9 | 102 | 378 | 768 | 1186 | 6390 |
| Cooling Degree Days | 0 | 0 | 1 | 17 | 72 | 208 | 325 | 255 | 104 | 8 | 0 | 0 | 990 |
| Total Precipitation (") | 1.41 | 1.18 | 2.47 | 3.39 | 3.67 | 4.55 | 3.57 | 4.56 | 3.50 | 2.74 | 2.38 | 2.16 | 35.58 |
| Days ≥ 0.1" Precip | 4 | 4 | 6 | 7 | 7 | 7 | 6 | 7 | 6 | 5 | 6 | 5 | 70 |
| Total Snowfall (") | 8.8 | 6.4 | 3.8 | 1.2 | 0.0 | 0.0 | 0.0 | 0.0 | 0.0 | 0.2 | 1.6 | 6.8 | 28.8 |
| Days ≥ 1" Snow Depth | 19 | 14 | 5 | 1 | 0 | 0 | 0 | 0 | 0 | 0 | 1 | 11 | 51 |

## COLO *Story County* ELEVATION 1001 ft LAT/LONG 42° 2 ' N / 93° 18 ' W

| | JAN | FEB | MAR | APR | MAY | JUN | JUL | AUG | SEP | OCT | NOV | DEC | YEAR |
|---|---|---|---|---|---|---|---|---|---|---|---|---|---|
| Maximum Temp °F | 25.6 | 31.0 | 44.0 | 59.1 | 71.2 | 80.2 | 83.7 | 81.2 | 73.7 | 61.8 | 44.8 | 30.6 | 57.2 |
| Minimum Temp °F | 7.2 | 12.3 | 24.7 | 36.9 | 48.3 | 58.1 | 62.4 | 59.4 | 50.7 | 38.5 | 26.1 | 13.9 | 36.5 |
| Mean Temp °F | 16.4 | 21.6 | 34.4 | 48.0 | 59.7 | 69.2 | 73.1 | 70.3 | 62.2 | 50.2 | 35.5 | 22.3 | 46.9 |
| Days Max Temp ≥ 90 °F | 0 | 0 | 0 | 0 | 0 | 3 | 6 | 4 | 1 | 0 | 0 | 0 | 14 |
| Days Max Temp ≤ 32 °F | 21 | 15 | 6 | 0 | 0 | 0 | 1 | 0 | 0 | 0 | 5 | 16 | 64 |
| Days Min Temp ≤ 32 °F | 31 | 27 | 24 | 10 | 1 | 0 | 0 | 0 | 1 | 9 | 23 | 30 | 156 |
| Days Min Temp ≤ 0 °F | 10 | 7 | 1 | 0 | 0 | 0 | 0 | 0 | 0 | 0 | 1 | 5 | 24 |
| Heating Degree Days | 1501 | 1219 | 943 | 511 | 205 | 29 | 6 | 20 | 145 | 460 | 879 | 1317 | 7235 |
| Cooling Degree Days | 0 | 0 | 1 | 10 | 44 | 164 | 261 | 200 | 76 | 5 | 0 | 0 | 761 |
| Total Precipitation (") | 0.83 | 0.86 | 2.14 | 3.38 | 4.30 | 5.38 | 4.78 | 4.67 | 3.41 | 2.54 | 1.87 | 1.18 | 35.34 |
| Days ≥ 0.1" Precip | 3 | 3 | 5 | 7 | 8 | 8 | 7 | 6 | 6 | 5 | 4 | 3 | 65 |
| Total Snowfall (") | 5.9 | 5.0 | 4.6 | 1.4 | 0.0 | 0.0 | 0.0 | 0.0 | 0.0 | 0.2 | 2.8 | 6.2 | 26.1 |
| Days ≥ 1" Snow Depth | 20 | 15 | 7 | 1 | 0 | 0 | 0 | 0 | 0 | 0 | 4 | 14 | 61 |

## COLUMBUS JUNCT 2 SSW *Louisa County* ELEVATION 610 ft LAT/LONG 41° 17 ' N / 91° 22 ' W

| | JAN | FEB | MAR | APR | MAY | JUN | JUL | AUG | SEP | OCT | NOV | DEC | YEAR |
|---|---|---|---|---|---|---|---|---|---|---|---|---|---|
| Maximum Temp °F | 30.1 | 35.6 | 48.9 | 63.5 | 74.4 | 83.1 | 86.8 | 84.2 | 76.9 | 65.3 | 48.8 | 35.3 | 61.1 |
| Minimum Temp °F | 12.3 | 17.5 | 28.9 | 40.5 | 50.5 | 59.8 | 64.1 | 61.5 | 53.6 | 42.2 | 30.7 | 19.1 | 40.1 |
| Mean Temp °F | 21.2 | 26.6 | 39.0 | 52.0 | 62.5 | 71.5 | 75.5 | 72.9 | 65.3 | 53.8 | 39.8 | 27.2 | 50.6 |
| Days Max Temp ≥ 90 °F | 0 | 0 | 0 | 0 | 1 | 5 | 11 | 7 | 2 | 0 | 0 | 0 | 26 |
| Days Max Temp ≤ 32 °F | 16 | 11 | 3 | 0 | 0 | 0 | 0 | 0 | 0 | 0 | 2 | 11 | 43 |
| Days Min Temp ≤ 32 °F | 29 | 25 | 19 | 6 | 0 | 0 | 0 | 0 | 0 | 6 | 18 | 28 | 131 |
| Days Min Temp ≤ 0 °F | 8 | 4 | 0 | 0 | 0 | 0 | 0 | 0 | 0 | 0 | 0 | 3 | 15 |
| Heating Degree Days | 1351 | 1077 | 801 | 399 | 145 | 16 | 2 | 8 | 94 | 358 | 749 | 1164 | 6164 |
| Cooling Degree Days | 0 | 0 | 2 | 19 | 72 | 227 | 346 | 277 | 118 | 9 | 0 | 0 | 1070 |
| Total Precipitation (") | 1.25 | 1.10 | 2.70 | 3.75 | 4.12 | 4.41 | 4.65 | 4.71 | 4.43 | 3.21 | 2.70 | 2.08 | 39.11 |
| Days ≥ 0.1" Precip | 3 | 3 | 6 | 7 | 8 | 7 | 7 | 7 | 7 | 6 | 6 | 4 | 71 |
| Total Snowfall (") | 9.1 | 7.9 | 5.9 | 2.1 | 0.0 | 0.0 | 0.0 | 0.0 | 0.0 | 0.5 | 3.5 | 8.5 | 37.5 |
| Days ≥ 1" Snow Depth | 19 | 15 | 5 | 1 | 0 | 0 | 0 | 0 | 0 | 0 | 2 | 12 | 54 |

## CORNING *Adams County* ELEVATION 1280 ft LAT/LONG 41° 0 ' N / 94° 43 ' W

| | JAN | FEB | MAR | APR | MAY | JUN | JUL | AUG | SEP | OCT | NOV | DEC | YEAR |
|---|---|---|---|---|---|---|---|---|---|---|---|---|---|
| Maximum Temp °F | 30.3 | 35.8 | 48.5 | 61.9 | 72.4 | 81.6 | 86.3 | 84.0 | 76.1 | 64.5 | 48.2 | 35.3 | 60.4 |
| Minimum Temp °F | 8.7 | 13.4 | 25.3 | 37.1 | 47.7 | 57.6 | 62.2 | 59.4 | 50.4 | 38.4 | 26.3 | 15.2 | 36.8 |
| Mean Temp °F | 19.5 | 24.6 | 36.8 | 49.5 | 60.1 | 69.6 | 74.3 | 71.6 | 63.2 | 51.4 | 37.3 | 25.3 | 48.6 |
| Days Max Temp ≥ 90 °F | 0 | 0 | 0 | 0 | 0 | 4 | 10 | 7 | 2 | 0 | 0 | 0 | 23 |
| Days Max Temp ≤ 32 °F | 16 | 12 | 3 | 0 | 0 | 0 | 0 | 0 | 0 | 0 | 3 | 12 | 46 |
| Days Min Temp ≤ 32 °F | 30 | 27 | 24 | 9 | 1 | 0 | 0 | 0 | 1 | 9 | 22 | 30 | 153 |
| Days Min Temp ≤ 0 °F | 9 | 6 | 0 | 0 | 0 | 0 | 0 | 0 | 0 | 0 | 1 | 4 | 20 |
| Heating Degree Days | 1405 | 1135 | 867 | 466 | 192 | 29 | 4 | 15 | 130 | 425 | 825 | 1226 | 6719 |
| Cooling Degree Days | 0 | 0 | 1 | 14 | 43 | 185 | 303 | 234 | 92 | 6 | 0 | 0 | 878 |
| Total Precipitation (") | 0.96 | 0.86 | 2.35 | 3.38 | 4.30 | 4.86 | 4.90 | 4.47 | 4.62 | 2.63 | 2.06 | 1.30 | 36.69 |
| Days ≥ 0.1" Precip | 2 | 3 | 5 | 6 | 8 | 7 | 7 | 6 | 7 | 5 | 4 | 3 | 63 |
| Total Snowfall (") | *5.3* | 5.5 | 2.9 | 1.1 | 0.1 | 0.0 | 0.0 | 0.0 | 0.0 | 0.1 | 2.4 | 4.9 | 22.3 |
| Days ≥ 1" Snow Depth | na | na | *2* | 0 | 0 | 0 | 0 | 0 | 0 | 0 | 1 | na | na |

## CRESCO 1 NE *Howard County* ELEVATION 1280 ft LAT/LONG 43° 22 ' N / 92° 7 ' W

| | JAN | FEB | MAR | APR | MAY | JUN | JUL | AUG | SEP | OCT | NOV | DEC | YEAR |
|---|---|---|---|---|---|---|---|---|---|---|---|---|---|
| Maximum Temp °F | 21.0 | 26.5 | 39.6 | 55.6 | 68.6 | 77.7 | 81.5 | 79.2 | 70.6 | 58.4 | 41.2 | 26.8 | 53.9 |
| Minimum Temp °F | 2.7 | 8.1 | 21.7 | 34.6 | 45.8 | 55.6 | 59.7 | 57.2 | 48.2 | 36.7 | 24.1 | 10.3 | 33.7 |
| Mean Temp °F | 11.9 | 17.3 | 30.7 | 45.1 | 57.2 | 66.7 | 70.6 | 68.2 | 59.4 | 47.6 | 32.7 | 18.6 | 43.8 |
| Days Max Temp ≥ 90 °F | 0 | 0 | 0 | 0 | 0 | 2 | 3 | 2 | 0 | 0 | 0 | 0 | 7 |
| Days Max Temp ≤ 32 °F | 24 | 18 | 8 | 1 | 0 | 0 | 0 | 0 | 0 | 0 | 7 | 21 | 79 |
| Days Min Temp ≤ 32 °F | 31 | 28 | 26 | 13 | 2 | 0 | 0 | 0 | 1 | 12 | 24 | 30 | 167 |
| Days Min Temp ≤ 0 °F | 14 | 9 | 2 | 0 | 0 | 0 | 0 | 0 | 0 | 0 | 1 | 8 | 34 |
| Heating Degree Days | 1642 | 1341 | 1057 | 595 | 265 | 58 | 16 | 37 | 202 | 537 | 963 | 1434 | 8147 |
| Cooling Degree Days | 0 | 0 | 0 | 7 | 29 | 113 | 189 | 142 | 44 | 1 | 0 | 0 | 525 |
| Total Precipitation (") | 0.99 | 0.86 | 2.22 | 3.63 | 3.85 | 4.52 | 4.36 | 4.70 | 4.46 | 2.63 | 2.19 | 1.38 | 35.79 |
| Days ≥ 0.1" Precip | 3 | 3 | 5 | 7 | 8 | 7 | 6 | 6 | 7 | 5 | 4 | 4 | 65 |
| Total Snowfall (") | 9.8 | 7.4 | 6.4 | 2.5 | 0.0 | 0.0 | 0.0 | 0.0 | 0.0 | 0.4 | 4.5 | 9.7 | 40.7 |
| Days ≥ 1" Snow Depth | na | na | na | *0* | 0 | 0 | 0 | 0 | 0 | 0 | *1* | na | na |

## CRESTON 2 SW *Union County* ELEVATION 1312 ft LAT/LONG 41° 2 ' N / 94° 24 ' W

| | JAN | FEB | MAR | APR | MAY | JUN | JUL | AUG | SEP | OCT | NOV | DEC | YEAR |
|---|---|---|---|---|---|---|---|---|---|---|---|---|---|
| Maximum Temp °F | 29.8 | 35.8 | 49.2 | 62.8 | 73.2 | 82.3 | 85.8 | 83.9 | 76.1 | 64.2 | 47.4 | 34.7 | 60.4 |
| Minimum Temp °F | 11.4 | 16.5 | 28.0 | 39.3 | 50.1 | 59.6 | 64.0 | 61.5 | 53.2 | 41.8 | 29.3 | 17.6 | 39.4 |
| Mean Temp °F | 20.5 | 26.2 | 38.6 | 51.0 | 61.7 | 71.0 | 75.0 | 72.7 | 64.7 | 53.0 | 38.4 | 26.2 | 49.9 |
| Days Max Temp ≥ 90 °F | 0 | 0 | 0 | 0 | 0 | 4 | 9 | 6 | 2 | 0 | 0 | 0 | 21 |
| Days Max Temp ≤ 32 °F | 17 | 11 | 3 | 0 | 0 | 0 | 0 | 0 | 0 | 0 | 3 | 12 | 46 |
| Days Min Temp ≤ 32 °F | 30 | 26 | 20 | 7 | 1 | 0 | 0 | 0 | 0 | 5 | 19 | 29 | 137 |
| Days Min Temp ≤ 0 °F | 7 | 5 | 0 | 0 | 0 | 0 | 0 | 0 | 0 | 0 | 0 | 3 | 15 |
| Heating Degree Days | 1375 | 1089 | 813 | 425 | 154 | 16 | 3 | 10 | 104 | 377 | 792 | 1198 | 6356 |
| Cooling Degree Days | 0 | 0 | 2 | 17 | 56 | 208 | 328 | 268 | 113 | 9 | 0 | 0 | 1001 |
| Total Precipitation (") | 0.86 | 1.00 | 2.11 | 3.38 | 4.38 | 4.48 | 4.83 | 3.82 | 4.40 | 2.58 | 2.12 | 1.18 | 35.14 |
| Days ≥ 0.1" Precip | 2 | 2 | 5 | 6 | 7 | 7 | 6 | 6 | 6 | 5 | 4 | *3* | 59 |
| Total Snowfall (") | *5.8* | na | 2.8 | 0.7 | 0.0 | 0.0 | 0.0 | 0.0 | 0.0 | 0.2 | *2.7* | na | na |
| Days ≥ 1" Snow Depth | na | na | *2* | *0* | 0 | 0 | 0 | 0 | 0 | 0 | *1* | na | na |

## DECORAH *Winneshiek County* ELEVATION 860 ft LAT/LONG 43° 18 ' N / 91° 48 ' W

| | JAN | FEB | MAR | APR | MAY | JUN | JUL | AUG | SEP | OCT | NOV | DEC | YEAR |
|---|---|---|---|---|---|---|---|---|---|---|---|---|---|
| Maximum Temp °F | 23.7 | 29.3 | 42.4 | 58.3 | 70.8 | 79.6 | 83.3 | 81.1 | 72.6 | 60.5 | 43.0 | 29.0 | 56.1 |
| Minimum Temp °F | 5.4 | 10.8 | 23.8 | 36.4 | 47.5 | 56.7 | 61.3 | 59.0 | 50.6 | 39.5 | 26.6 | 13.2 | 35.9 |
| Mean Temp °F | 14.6 | 20.1 | 33.1 | 47.4 | 59.2 | 68.2 | 72.3 | 70.1 | 61.6 | 50.0 | 34.8 | 21.1 | 46.0 |
| Days Max Temp ≥ 90 °F | 0 | 0 | 0 | 0 | 0 | 3 | 5 | 3 | 1 | 0 | 0 | 0 | 12 |
| Days Max Temp ≤ 32 °F | 22 | 16 | 6 | 0 | 0 | 0 | 0 | 0 | 0 | 0 | 5 | 18 | 67 |
| Days Min Temp ≤ 32 °F | 31 | 27 | 24 | 11 | 2 | 0 | 0 | 0 | 1 | 8 | 21 | 30 | 155 |
| Days Min Temp ≤ 0 °F | 12 | 8 | 1 | 0 | 0 | 0 | 0 | 0 | 0 | 0 | 1 | 6 | 28 |
| Heating Degree Days | 1559 | 1263 | 981 | 530 | 218 | 40 | 8 | 22 | 157 | 464 | 898 | 1354 | 7494 |
| Cooling Degree Days | 0 | 0 | 0 | 11 | 42 | 149 | 255 | 197 | 74 | 3 | 0 | 0 | 731 |
| Total Precipitation (") | 0.87 | 0.81 | 1.84 | 3.71 | 3.83 | 4.53 | 4.10 | 4.05 | 3.92 | 2.29 | 1.94 | 1.24 | 33.13 |
| Days ≥ 0.1" Precip | 3 | 3 | 4 | 7 | 8 | 7 | 7 | 6 | 7 | 5 | 4 | 3 | 64 |
| Total Snowfall (") | 9.0 | 7.6 | 6.2 | 1.6 | 0.0 | 0.0 | 0.0 | 0.0 | 0.0 | 0.2 | 4.0 | 9.5 | 38.1 |
| Days ≥ 1" Snow Depth | na | na | 7 | 0 | 0 | 0 | 0 | 0 | 0 | 0 | na | na | na |

## DENISON *Crawford County* ELEVATION 1302 ft LAT/LONG 42° 0 ' N / 95° 22 ' W

| | JAN | FEB | MAR | APR | MAY | JUN | JUL | AUG | SEP | OCT | NOV | DEC | YEAR |
|---|---|---|---|---|---|---|---|---|---|---|---|---|---|
| Maximum Temp °F | 27.7 | 33.1 | 46.0 | 60.8 | 71.9 | 81.0 | 84.3 | 81.9 | 74.3 | 62.3 | 45.0 | 31.9 | 58.4 |
| Minimum Temp °F | 9.3 | 14.6 | 26.1 | 38.3 | 49.8 | 59.3 | 63.8 | 61.2 | 52.5 | 40.9 | 27.2 | 15.1 | 38.2 |
| Mean Temp °F | 18.5 | 23.9 | 36.1 | 49.6 | 60.8 | 70.2 | 74.1 | 71.6 | 63.4 | 51.6 | 36.1 | 23.6 | 48.3 |
| Days Max Temp ≥ 90 °F | 0 | 0 | 0 | 0 | 1 | 3 | 6 | 4 | 1 | 0 | 0 | 0 | 15 |
| Days Max Temp ≤ 32 °F | 18 | 13 | 5 | 0 | 0 | 0 | 0 | 0 | 0 | 0 | 5 | 16 | 57 |
| Days Min Temp ≤ 32 °F | 30 | 27 | 23 | 8 | 1 | 0 | 0 | 0 | 0 | 6 | 21 | 30 | 146 |
| Days Min Temp ≤ 0 °F | 9 | 5 | 1 | 0 | 0 | 0 | 0 | 0 | 0 | 0 | 0 | 4 | 19 |
| Heating Degree Days | 1437 | 1155 | 890 | 469 | 178 | 24 | 4 | 12 | 122 | 416 | 859 | 1279 | 6845 |
| Cooling Degree Days | 0 | 0 | 1 | 17 | 57 | 191 | 290 | 223 | 93 | 6 | 0 | 0 | 878 |
| Total Precipitation (") | 0.81 | 0.81 | 2.07 | 2.73 | 3.98 | 4.25 | 3.64 | 3.35 | 3.78 | 2.31 | 1.48 | 1.06 | 30.27 |
| Days ≥ 0.1" Precip | 3 | 3 | 5 | 6 | 7 | 7 | 6 | 6 | 6 | 4 | 4 | 3 | 60 |
| Total Snowfall (") | 7.5 | 6.7 | 6.1 | 1.7 | 0.0 | 0.0 | 0.0 | 0.0 | 0.0 | 0.7 | 3.9 | 7.8 | 34.4 |
| Days ≥ 1" Snow Depth | 21 | 17 | 9 | 1 | 0 | 0 | 0 | 0 | 0 | 0 | 5 | 14 | 67 |

## DES MOINES INTL AP *Polk County* ELEVATION 938 ft LAT/LONG 41° 32 ' N / 93° 39 ' W

| | JAN | FEB | MAR | APR | MAY | JUN | JUL | AUG | SEP | OCT | NOV | DEC | YEAR |
|---|---|---|---|---|---|---|---|---|---|---|---|---|---|
| Maximum Temp °F | 28.4 | 34.3 | 47.6 | 61.7 | 72.8 | 82.0 | 86.2 | 83.8 | 75.5 | 63.4 | 47.1 | 33.7 | 59.7 |
| Minimum Temp °F | 11.2 | 16.2 | 28.2 | 40.3 | 51.5 | 61.3 | 66.3 | 63.5 | 54.3 | 42.3 | 29.5 | 17.4 | 40.2 |
| Mean Temp °F | 19.8 | 25.2 | 37.9 | 51.0 | 62.2 | 71.7 | 76.2 | 73.7 | 64.9 | 52.9 | 38.3 | 25.6 | 49.9 |
| Days Max Temp ≥ 90 °F | 0 | 0 | 0 | 0 | 1 | 4 | 10 | 7 | 2 | 0 | 0 | 0 | 24 |
| Days Max Temp ≤ 32 °F | 17 | 13 | 4 | 0 | 0 | 0 | 0 | 0 | 0 | 0 | 3 | 13 | 50 |
| Days Min Temp ≤ 32 °F | 30 | 25 | 20 | 6 | 0 | 0 | 0 | 0 | 0 | 5 | 19 | 29 | 134 |
| Days Min Temp ≤ 0 °F | 8 | 4 | 0 | 0 | 0 | 0 | 0 | 0 | 0 | 0 | 0 | 3 | 15 |
| Heating Degree Days | 1394 | 1117 | 834 | 427 | 150 | 14 | 1 | 7 | 101 | 383 | 794 | 1216 | 6438 |
| Cooling Degree Days | 0 | 0 | 2 | 18 | 65 | 224 | 359 | 292 | 115 | 7 | 0 | 0 | 1082 |
| Total Precipitation (") | 1.04 | 1.11 | 2.29 | 3.52 | 3.80 | 4.53 | 3.97 | 4.44 | 3.41 | 2.59 | 1.98 | 1.45 | 34.13 |
| Days ≥ 0.1" Precip | 3 | 3 | 5 | 6 | 8 | 7 | 6 | 6 | 6 | 5 | 4 | 4 | 63 |
| Total Snowfall (") | 7.6 | 7.6 | 4.4 | 2.4 | 0.0 | 0.0 | 0.0 | 0.0 | 0.0 | 0.4 | 3.9 | 7.1 | 33.4 |
| Days ≥ 1" Snow Depth | 19 | 14 | 5 | 1 | 0 | 0 | 0 | 0 | 0 | 0 | 3 | 11 | 53 |

## DUBUQUE L&D 11 *Dubuque County* ELEVATION 620 ft LAT/LONG 42° 32 ' N / 90° 39 ' W

| | JAN | FEB | MAR | APR | MAY | JUN | JUL | AUG | SEP | OCT | NOV | DEC | YEAR |
|---|---|---|---|---|---|---|---|---|---|---|---|---|---|
| Maximum Temp °F | 27.3 | 32.3 | 44.8 | 59.4 | 71.9 | 80.7 | 84.4 | 82.1 | 73.9 | 62.1 | 45.8 | 32.6 | 58.1 |
| Minimum Temp °F | 10.4 | 15.0 | 27.6 | 40.1 | 51.5 | 61.0 | 65.6 | 63.2 | 55.2 | 44.0 | 31.2 | 18.3 | 40.3 |
| Mean Temp °F | 18.9 | 23.7 | 36.2 | 49.8 | 61.8 | 70.9 | 75.0 | 72.7 | 64.6 | 53.1 | 38.6 | 25.5 | 49.2 |
| Days Max Temp ≥ 90 °F | 0 | 0 | 0 | 0 | 0 | 3 | 6 | 4 | 1 | 0 | 0 | 0 | 14 |
| Days Max Temp ≤ 32 °F | 19 | 13 | 4 | 0 | 0 | 0 | 0 | 0 | 0 | 0 | 2 | 13 | 51 |
| Days Min Temp ≤ 32 °F | 30 | 26 | 21 | 5 | 0 | 0 | 0 | 0 | 0 | 2 | 17 | 28 | 129 |
| Days Min Temp ≤ 0 °F | 9 | 5 | 0 | 0 | 0 | 0 | 0 | 0 | 0 | 0 | 0 | 3 | 17 |
| Heating Degree Days | 1424 | 1161 | 886 | 460 | 160 | 17 | 2 | 7 | 100 | 373 | 787 | 1219 | 6596 |
| Cooling Degree Days | 0 | 0 | 0 | 14 | 72 | 217 | 339 | 270 | 105 | 7 | 0 | 0 | 1024 |
| Total Precipitation (") | 1.11 | 0.97 | 2.16 | 3.13 | 3.40 | 4.23 | 3.95 | 4.35 | 4.27 | 2.31 | 2.25 | 1.61 | 33.74 |
| Days ≥ 0.1" Precip | 3 | 3 | 5 | 7 | 7 | 7 | 6 | 6 | 7 | 5 | 5 | 4 | 65 |
| Total Snowfall (") | 9.0 | 6.7 | 4.4 | 1.4 | 0.0 | 0.0 | 0.0 | 0.0 | 0.0 | 0.0 | 2.1 | 8.2 | 31.8 |
| Days ≥ 1" Snow Depth | 23 | 20 | 8 | 1 | 0 | 0 | 0 | 0 | 0 | 0 | 3 | 16 | 71 |

## DUBUQUE MUNICPAL AP *Dubuque County* ELEVATION 1070 ft LAT/LONG 42° 24 ' N / 90° 42 ' W

| | JAN | FEB | MAR | APR | MAY | JUN | JUL | AUG | SEP | OCT | NOV | DEC | YEAR |
|---|---|---|---|---|---|---|---|---|---|---|---|---|---|
| Maximum Temp °F | 24.5 | 29.7 | 42.9 | 57.9 | 69.3 | 78.7 | 82.3 | 79.7 | 71.7 | 60.1 | 43.8 | 30.2 | 55.9 |
| Minimum Temp °F | 8.5 | 13.3 | 25.5 | 37.7 | 48.0 | 57.5 | 62.0 | 59.5 | 51.5 | 40.2 | 27.7 | 15.4 | 37.2 |
| Mean Temp °F | 16.6 | 21.5 | 34.3 | 47.8 | 58.6 | 68.2 | 72.2 | 69.7 | 61.6 | 50.1 | 35.7 | 22.8 | 46.6 |
| Days Max Temp ≥ 90 °F | 0 | 0 | 0 | 0 | 0 | 1 | 4 | 2 | 1 | 0 | 0 | 0 | 8 |
| Days Max Temp ≤ 32 °F | 22 | 16 | 6 | 0 | 0 | 0 | 0 | 0 | 0 | 0 | 5 | 17 | 66 |
| Days Min Temp ≤ 32 °F | 31 | 27 | 23 | 8 | 1 | 0 | 0 | 0 | 0 | 7 | 21 | 29 | 147 |
| Days Min Temp ≤ 0 °F | 10 | 6 | 0 | 0 | 0 | 0 | 0 | 0 | 0 | 0 | 0 | 5 | 21 |
| Heating Degree Days | 1497 | 1222 | 946 | 516 | 230 | 38 | 8 | 22 | 156 | 461 | 871 | 1302 | 7269 |
| Cooling Degree Days | 0 | 0 | 0 | 10 | 40 | 156 | 249 | 184 | 68 | 4 | 0 | 0 | 711 |
| Total Precipitation (") | 1.31 | 1.40 | 2.73 | 3.62 | 3.91 | 4.39 | 3.75 | 4.82 | 4.53 | 2.71 | 2.61 | 2.05 | 37.83 |
| Days ≥ 0.1" Precip | 4 | 4 | 6 | 8 | 8 | 7 | 7 | 7 | 7 | 6 | 5 | 4 | 73 |
| Total Snowfall (") | 9.4 | 8.8 | 7.5 | 3.1 | 0.1 | 0.0 | 0.0 | 0.0 | 0.0 | 0.2 | 4.2 | 10.3 | 43.6 |
| Days ≥ 1" Snow Depth | 22 | 19 | 8 | 1 | 0 | 0 | 0 | 0 | 0 | 0 | 3 | 15 | 68 |

## ELDORA *Hardin County* ELEVATION 1094 ft LAT/LONG 42° 22 ' N / 93° 9 ' W

| | JAN | FEB | MAR | APR | MAY | JUN | JUL | AUG | SEP | OCT | NOV | DEC | YEAR |
|---|---|---|---|---|---|---|---|---|---|---|---|---|---|
| Maximum Temp °F | 24.9 | 30.6 | 43.9 | 58.8 | 71.6 | 80.8 | 84.6 | 82.0 | 74.3 | 62.0 | 44.5 | 30.2 | 57.4 |
| Minimum Temp °F | 6.6 | 11.9 | 24.7 | 36.8 | 48.2 | 57.8 | 62.1 | 59.2 | 50.5 | 38.7 | 26.2 | 13.5 | 36.4 |
| Mean Temp °F | 15.8 | 21.2 | 34.3 | 47.8 | 59.9 | 69.3 | 73.4 | 70.6 | 62.4 | 50.4 | 35.4 | 21.9 | 46.9 |
| Days Max Temp ≥ 90 °F | 0 | 0 | 0 | 0 | 1 | 4 | 8 | 5 | 1 | 0 | 0 | 0 | 19 |
| Days Max Temp ≤ 32 °F | 21 | 15 | 6 | 0 | 0 | 0 | 0 | 0 | 0 | 0 | 5 | 17 | 64 |
| Days Min Temp ≤ 32 °F | 31 | 27 | 24 | 10 | 1 | 0 | 0 | 0 | 1 | 9 | 22 | 30 | 155 |
| Days Min Temp ≤ 0 °F | 11 | 7 | 1 | 0 | 0 | 0 | 0 | 0 | 0 | 0 | 1 | 5 | 25 |
| Heating Degree Days | 1521 | 1230 | 945 | 518 | 201 | 29 | 6 | 17 | 140 | 455 | 883 | 1332 | 7277 |
| Cooling Degree Days | 0 | 0 | 0 | 11 | 51 | 179 | 281 | 216 | 82 | 6 | 0 | 0 | 826 |
| Total Precipitation (") | 0.93 | 0.90 | 2.09 | 3.21 | 4.27 | 5.32 | 4.38 | 4.58 | 3.42 | 2.59 | 1.99 | 1.23 | 34.91 |
| Days ≥ 0.1" Precip | 2 | 2 | 5 | 7 | 8 | 7 | 6 | 6 | 6 | 5 | 4 | 3 | 61 |
| Total Snowfall (") | 7.5 | 7.0 | 5.3 | 1.6 | 0.0 | 0.0 | 0.0 | 0.0 | 0.0 | 0.0 | 3.2 | 6.9 | 31.5 |
| Days ≥ 1" Snow Depth | 24 | 20 | 10 | 1 | 0 | 0 | 0 | 0 | 0 | 0 | 4 | 17 | 76 |

## ELKADER 5 SSW *Clayton County* ELEVATION 732 ft LAT/LONG 42° 51 ' N / 91° 24 ' W

| | JAN | FEB | MAR | APR | MAY | JUN | JUL | AUG | SEP | OCT | NOV | DEC | YEAR |
|---|---|---|---|---|---|---|---|---|---|---|---|---|---|
| Maximum Temp °F | 26.8 | 32.6 | 44.7 | 60.5 | 72.7 | 81.2 | 85.0 | 82.6 | 74.4 | 62.4 | 45.3 | 31.1 | 58.3 |
| Minimum Temp °F | 5.7 | 11.0 | 23.1 | 35.1 | 45.6 | 54.5 | 59.5 | 57.2 | 48.8 | 37.4 | 25.7 | 12.8 | 34.7 |
| Mean Temp °F | 16.2 | 21.7 | 33.8 | 47.9 | 59.2 | 67.9 | 72.2 | 70.0 | 61.7 | 49.9 | 35.5 | 22.0 | 46.5 |
| Days Max Temp ≥ 90 °F | 0 | 0 | 0 | 0 | 1 | 4 | 8 | 5 | 1 | 0 | 0 | 0 | 19 |
| Days Max Temp ≤ 32 °F | 20 | 13 | 5 | 0 | 0 | 0 | 0 | 0 | 0 | 0 | 4 | 15 | 57 |
| Days Min Temp ≤ 32 °F | 31 | 28 | 25 | *13* | 3 | 0 | 0 | 0 | 2 | 11 | 22 | 30 | 165 |
| Days Min Temp ≤ 0 °F | 11 | 7 | 1 | 0 | 0 | 0 | 0 | 0 | 0 | 0 | 1 | 6 | 26 |
| Heating Degree Days | 1506 | 1213 | 958 | 515 | 215 | 46 | 9 | 25 | 161 | 468 | 880 | 1327 | 7323 |
| Cooling Degree Days | 0 | 0 | 0 | 10 | 40 | 136 | 242 | 190 | 67 | 5 | 0 | 0 | 690 |
| Total Precipitation (") | 1.06 | 1.10 | 2.04 | 3.37 | 3.76 | 4.41 | 4.31 | 4.36 | 3.52 | 2.38 | 2.24 | 1.42 | 33.97 |
| Days ≥ 0.1" Precip | 3 | 3 | 5 | 7 | 8 | 8 | 7 | 6 | 7 | 5 | 5 | 4 | 68 |
| Total Snowfall (") | 8.6 | 7.1 | 4.9 | 1.8 | 0.0 | 0.0 | 0.0 | 0.0 | 0.0 | 0.0 | 3.8 | 9.5 | 35.7 |
| Days ≥ 1" Snow Depth | na | na | na | 0 | 0 | 0 | 0 | 0 | 0 | 0 | *1* | na | na |

## EMMETSBURG *Palo Alto County* ELEVATION 1230 ft LAT/LONG 43° 6 ' N / 94° 40 ' W

| | JAN | FEB | MAR | APR | MAY | JUN | JUL | AUG | SEP | OCT | NOV | DEC | YEAR |
|---|---|---|---|---|---|---|---|---|---|---|---|---|---|
| Maximum Temp °F | 24.1 | 29.9 | 42.8 | 59.3 | 72.8 | 81.5 | 84.4 | 81.9 | 74.1 | 61.8 | 42.6 | 29.0 | 57.0 |
| Minimum Temp °F | 5.4 | 11.4 | 24.0 | 36.5 | 48.1 | 58.0 | 62.2 | 59.2 | 50.5 | 38.3 | 24.7 | 12.1 | 35.9 |
| Mean Temp °F | 14.7 | 20.7 | 33.5 | 47.9 | 60.5 | 69.8 | 73.3 | 70.6 | 62.4 | 50.1 | 33.7 | 20.6 | 46.5 |
| Days Max Temp ≥ 90 °F | 0 | 0 | 0 | 0 | 1 | 4 | 7 | 4 | 1 | 0 | 0 | 0 | 17 |
| Days Max Temp ≤ 32 °F | 22 | 16 | 7 | 0 | 0 | 0 | 0 | 0 | 0 | 0 | 6 | 18 | 69 |
| Days Min Temp ≤ 32 °F | 31 | 28 | 25 | 10 | 1 | 0 | 0 | 0 | 1 | 8 | 24 | 31 | 159 |
| Days Min Temp ≤ 0 °F | 12 | 7 | 1 | 0 | 0 | 0 | 0 | 0 | 0 | 0 | 1 | 6 | 27 |
| Heating Degree Days | 1554 | 1245 | 972 | 515 | 186 | 26 | 5 | 17 | 140 | 461 | 934 | 1371 | 7426 |
| Cooling Degree Days | 0 | 0 | 0 | 10 | 56 | 181 | 269 | 200 | 75 | 3 | 0 | 0 | 794 |
| Total Precipitation (") | 0.89 | 0.74 | 2.14 | 3.03 | 3.68 | 4.62 | 3.98 | 3.86 | 3.03 | 2.26 | 1.69 | 1.16 | 31.08 |
| Days ≥ 0.1" Precip | 3 | 2 | 4 | 6 | 8 | 7 | 6 | 6 | 5 | 4 | 4 | 3 | 58 |
| Total Snowfall (") | 7.7 | 5.8 | 6.7 | 2.0 | 0.0 | 0.0 | 0.0 | 0.0 | 0.0 | 0.2 | 4.6 | ***8.0*** | 35.0 |
| Days ≥ 1" Snow Depth | *23* | *17* | *10* | *1* | 0 | 0 | 0 | 0 | 0 | 0 | na | na | na |

## ESTHERVILLE 2 N *Emmet County* ELEVATION 1298 ft LAT/LONG 43° 24 ' N / 94° 50 ' W

| | JAN | FEB | MAR | APR | MAY | JUN | JUL | AUG | SEP | OCT | NOV | DEC | YEAR |
|---|---|---|---|---|---|---|---|---|---|---|---|---|---|
| Maximum Temp °F | 22.2 | 27.5 | 40.2 | 56.4 | 70.4 | 79.1 | 82.8 | 80.1 | 71.9 | 59.8 | 41.4 | 27.0 | 54.9 |
| Minimum Temp °F | 3.1 | 8.7 | 21.4 | 34.3 | 46.3 | 56.3 | 60.4 | 57.5 | 48.0 | 36.0 | 23.2 | 9.8 | 33.8 |
| Mean Temp °F | 12.7 | 18.1 | 30.8 | 45.4 | 58.4 | 67.7 | 71.6 | 68.8 | 60.0 | 47.9 | 32.3 | 18.4 | 44.3 |
| Days Max Temp ≥ 90 °F | 0 | 0 | 0 | 0 | 1 | 3 | 5 | 3 | 1 | 0 | 0 | 0 | 13 |
| Days Max Temp ≤ 32 °F | 23 | 18 | 9 | 1 | 0 | 0 | 0 | 0 | 0 | 0 | 8 | 20 | 79 |
| Days Min Temp ≤ 32 °F | 31 | 28 | 27 | 13 | 2 | 0 | 0 | 0 | 1 | 12 | 25 | 31 | 170 |
| Days Min Temp ≤ 0 °F | 14 | 9 | 2 | 0 | 0 | 0 | 0 | 0 | 0 | 0 | 1 | 8 | 34 |
| Heating Degree Days | 1619 | 1318 | 1053 | 586 | 239 | 49 | 12 | 33 | 192 | 526 | 974 | 1439 | 8040 |
| Cooling Degree Days | 0 | 0 | 0 | 7 | 40 | 135 | 216 | 156 | 51 | 2 | 0 | 0 | 607 |
| Total Precipitation (") | 0.73 | 0.65 | 1.76 | 3.02 | 3.37 | 4.77 | 3.34 | 3.54 | 3.05 | 2.07 | 1.30 | 0.73 | 28.33 |
| Days ≥ 0.1" Precip | 2 | 2 | 4 | 6 | 7 | 8 | 6 | 6 | 6 | 4 | 3 | 2 | 56 |
| Total Snowfall (") | 7.3 | 6.3 | 6.3 | 2.3 | 0.0 | 0.0 | 0.0 | 0.0 | 0.0 | 0.6 | 5.2 | 7.1 | 35.1 |
| Days ≥ 1" Snow Depth | na | na | na | *1* | 0 | 0 | 0 | 0 | 0 | 0 | 2 | na | na |

## FAIRFIELD *Jefferson County* ELEVATION 781 ft LAT/LONG 41° 1 ' N / 91° 57 ' W

| | JAN | FEB | MAR | APR | MAY | JUN | JUL | AUG | SEP | OCT | NOV | DEC | YEAR |
|---|---|---|---|---|---|---|---|---|---|---|---|---|---|
| Maximum Temp °F | 30.4 | 35.8 | 49.0 | 63.2 | 74.1 | 83.3 | 87.3 | 84.5 | 76.4 | 64.4 | 48.4 | 35.4 | 61.0 |
| Minimum Temp °F | 12.5 | 17.2 | 28.8 | 40.6 | 50.9 | 60.5 | 65.0 | 62.2 | 54.3 | 42.9 | 31.0 | 19.1 | 40.4 |
| Mean Temp °F | 21.5 | 26.5 | 38.9 | 51.9 | 62.5 | 71.9 | 76.2 | 73.4 | 65.3 | 53.7 | 39.7 | 27.3 | 50.7 |
| Days Max Temp ≥ 90 °F | 0 | 0 | 0 | 0 | 1 | 5 | 12 | 8 | 2 | 0 | 0 | 0 | 28 |
| Days Max Temp ≤ 32 °F | 16 | 11 | 3 | 0 | 0 | 0 | 0 | 0 | 0 | 0 | 2 | 11 | 43 |
| Days Min Temp ≤ 32 °F | 29 | 25 | 20 | 6 | 0 | 0 | 0 | 0 | 0 | 5 | 17 | 28 | 130 |
| Days Min Temp ≤ 0 °F | 7 | 4 | 0 | 0 | 0 | 0 | 0 | 0 | 0 | 0 | 0 | 3 | 14 |
| Heating Degree Days | 1343 | 1081 | 805 | 403 | 144 | 14 | 2 | 7 | 94 | 359 | 752 | 1162 | 6166 |
| Cooling Degree Days | 0 | 0 | 3 | 21 | 75 | 241 | 368 | 290 | 126 | 12 | 0 | 0 | 1136 |
| Total Precipitation (") | 1.17 | 0.97 | 2.40 | 3.41 | 4.03 | 3.76 | 5.06 | 4.32 | 4.18 | 3.01 | 2.37 | 1.82 | 36.50 |
| Days ≥ 0.1" Precip | 3 | 3 | 5 | 7 | 8 | 7 | 7 | 7 | 6 | 6 | 5 | 4 | 68 |
| Total Snowfall (") | 7.2 | 7.1 | 3.8 | 1.4 | 0.0 | 0.0 | 0.0 | 0.0 | 0.0 | 0.1 | 2.1 | 5.9 | 27.6 |
| Days ≥ 1" Snow Depth | 16 | 11 | 4 | 0 | 0 | 0 | 0 | 0 | 0 | 0 | 1 | 7 | 39 |

## FAYETTE *Fayette County* ELEVATION 1001 ft LAT/LONG 42° 50 ' N / 91° 48 ' W

| | JAN | FEB | MAR | APR | MAY | JUN | JUL | AUG | SEP | OCT | NOV | DEC | YEAR |
|---|---|---|---|---|---|---|---|---|---|---|---|---|---|
| Maximum Temp °F | 24.2 | 29.9 | 43.2 | 58.9 | 71.1 | 79.8 | 83.6 | 81.4 | 73.3 | 61.1 | 43.4 | 29.7 | 56.6 |
| Minimum Temp °F | 5.3 | 10.2 | 23.4 | 35.3 | 46.4 | 55.8 | 60.4 | 57.7 | 49.3 | 38.0 | 25.7 | 12.7 | 35.0 |
| Mean Temp °F | 14.8 | 20.1 | 33.3 | 47.1 | 58.8 | 67.9 | 72.0 | 69.6 | 61.3 | 49.6 | 34.6 | 21.2 | 45.9 |
| Days Max Temp ≥ 90 °F | 0 | 0 | 0 | 0 | 0 | 2 | 6 | 3 | 1 | 0 | 0 | 0 | 12 |
| Days Max Temp ≤ 32 °F | 22 | 15 | 5 | 0 | 0 | 0 | 0 | 0 | 0 | 0 | 5 | 17 | 64 |
| Days Min Temp ≤ 32 °F | 31 | 27 | 25 | 13 | 3 | 0 | 0 | 0 | 1 | 10 | 22 | 30 | 162 |
| Days Min Temp ≤ 0 °F | 12 | 8 | 2 | 0 | 0 | 0 | 0 | 0 | 0 | 0 | 1 | 6 | 29 |
| Heating Degree Days | 1552 | 1263 | 975 | 537 | 227 | 44 | 10 | 28 | 164 | 478 | 906 | 1351 | 7535 |
| Cooling Degree Days | 0 | 0 | 0 | 9 | 36 | 135 | 233 | 183 | 68 | 3 | 0 | 0 | 667 |
| Total Precipitation (") | 1.08 | 1.07 | 2.07 | 3.53 | 4.05 | 4.64 | 4.38 | 4.47 | 3.76 | 2.46 | 2.16 | 1.48 | 35.15 |
| Days ≥ 0.1" Precip | 4 | 3 | 5 | 7 | 8 | 7 | 7 | 6 | 7 | 5 | 5 | 4 | 68 |
| Total Snowfall (") | 9.4 | 8.0 | 6.2 | 2.1 | 0.0 | 0.0 | 0.0 | 0.0 | 0.0 | 0.1 | 4.0 | 9.8 | 39.6 |
| Days ≥ 1" Snow Depth | *23* | 19 | *9* | 1 | 0 | 0 | 0 | 0 | 0 | 0 | 4 | *17* | 73 |

## FOREST CITY 2 NNE *Winnebago County* ELEVATION 1300 ft LAT/LONG 43° 17 ' N / 93° 38 ' W

| | JAN | FEB | MAR | APR | MAY | JUN | JUL | AUG | SEP | OCT | NOV | DEC | YEAR |
|---|---|---|---|---|---|---|---|---|---|---|---|---|---|
| Maximum Temp °F | 22.4 | 27.9 | 42.0 | 57.8 | 71.3 | 79.9 | 83.0 | 80.3 | 72.4 | 60.4 | 41.4 | 28.1 | 55.6 |
| Minimum Temp °F | 3.9 | 9.8 | 23.5 | 35.8 | 47.8 | 57.0 | 61.7 | 58.7 | 49.9 | 38.2 | 24.4 | 11.7 | 35.2 |
| Mean Temp °F | 13.2 | 18.9 | 32.8 | 46.9 | 59.6 | 68.5 | 72.4 | 69.5 | 61.2 | 49.3 | 32.9 | 19.9 | 45.4 |
| Days Max Temp ≥ 90 °F | 0 | 0 | 0 | 0 | 1 | 3 | 5 | 3 | 1 | 0 | 0 | 0 | 13 |
| Days Max Temp ≤ 32 °F | 23 | 17 | 7 | 0 | 0 | 0 | 0 | 0 | 0 | 0 | 7 | *20* | 74 |
| Days Min Temp ≤ 32 °F | 31 | 28 | 25 | 11 | 1 | 0 | 0 | 0 | 1 | 9 | 24 | *31* | 161 |
| Days Min Temp ≤ 0 °F | 13 | 8 | 2 | 0 | 0 | 0 | 0 | 0 | 0 | 0 | 1 | *7* | 31 |
| Heating Degree Days | 1602 | 1295 | 993 | 544 | 208 | 35 | 8 | 25 | 163 | 485 | 955 | 1391 | 7704 |
| Cooling Degree Days | 0 | 0 | 0 | 9 | 48 | 152 | 229 | 163 | 57 | 3 | 0 | 0 | 661 |
| Total Precipitation (") | 0.89 | 0.66 | 1.83 | 3.04 | 3.76 | 4.81 | 4.24 | 4.39 | 3.13 | 2.28 | 1.54 | 1.15 | 31.72 |
| Days ≥ 0.1" Precip | 2 | 2 | 4 | 7 | 7 | 7 | 6 | 6 | 5 | 5 | 4 | 3 | 58 |
| Total Snowfall (") | 10.3 | 6.3 | 6.1 | 2.3 | 0.0 | 0.0 | 0.0 | 0.0 | 0.0 | 0.2 | 4.2 | 9.7 | 39.1 |
| Days ≥ 1" Snow Depth | *22* | *18* | 9 | 1 | 0 | 0 | 0 | 0 | 0 | 0 | 4 | *17* | 71 |

## FORT DODGE *Webster County* ELEVATION 1115 ft LAT/LONG 42° 30 ' N / 94° 12 ' W

| | JAN | FEB | MAR | APR | MAY | JUN | JUL | AUG | SEP | OCT | NOV | DEC | YEAR |
|---|---|---|---|---|---|---|---|---|---|---|---|---|---|
| Maximum Temp °F | 25.8 | 31.8 | 45.0 | 60.5 | 72.9 | 81.9 | 85.0 | 82.4 | 74.9 | 62.4 | 44.8 | 30.6 | 58.2 |
| Minimum Temp °F | 6.8 | 12.2 | 24.8 | 37.0 | 48.4 | 58.2 | 62.6 | 59.8 | 50.7 | 38.9 | 26.0 | 13.3 | 36.6 |
| Mean Temp °F | 16.3 | 22.0 | 34.9 | 48.7 | 60.7 | 70.1 | 73.8 | 71.1 | 62.8 | 50.7 | 35.4 | 22.0 | 47.4 |
| Days Max Temp ≥ 90 °F | 0 | 0 | 0 | 0 | 1 | 5 | 8 | 5 | 2 | 0 | 0 | 0 | 21 |
| Days Max Temp ≤ 32 °F | 20 | 14 | 5 | 0 | 0 | 0 | 0 | 0 | 0 | 0 | 5 | 16 | 60 |
| Days Min Temp ≤ 32 °F | 31 | 27 | 24 | 10 | 1 | 0 | 0 | 0 | 1 | 8 | 22 | 30 | 154 |
| Days Min Temp ≤ 0 °F | 11 | 7 | 1 | 0 | 0 | 0 | 0 | 0 | 0 | 0 | 1 | 6 | 26 |
| Heating Degree Days | 1504 | 1208 | 925 | 492 | 185 | 26 | 5 | 16 | 136 | 443 | 881 | 1327 | 7148 |
| Cooling Degree Days | 0 | 0 | 0 | 14 | 60 | 197 | 292 | 225 | 89 | 5 | 0 | 0 | 882 |
| Total Precipitation (") | 0.90 | 0.80 | 2.17 | 3.22 | 4.13 | 5.24 | 4.56 | 4.51 | 3.88 | 2.56 | 1.82 | 1.18 | 34.97 |
| Days ≥ 0.1" Precip | 3 | 3 | 4 | 6 | 8 | 8 | 6 | 6 | 6 | 4 | 4 | 3 | 61 |
| Total Snowfall (") | 8.1 | 7.4 | 6.2 | 1.7 | 0.0 | 0.0 | 0.0 | 0.0 | 0.0 | 0.2 | 4.2 | 8.2 | 36.0 |
| Days ≥ 1" Snow Depth | na | na | *6* | 1 | 0 | 0 | 0 | 0 | 0 | 0 | *4* | na | na |

## FORT MADISON *Lee County* ELEVATION 530 ft LAT/LONG 40° 37 ' N / 91° 20 ' W

| | JAN | FEB | MAR | APR | MAY | JUN | JUL | AUG | SEP | OCT | NOV | DEC | YEAR |
|---|---|---|---|---|---|---|---|---|---|---|---|---|---|
| Maximum Temp °F | 32.2 | 36.6 | 49.2 | 63.0 | 73.1 | 82.2 | 86.6 | 84.1 | 77.0 | 65.0 | 49.9 | 36.9 | 61.3 |
| Minimum Temp °F | 13.6 | 18.2 | 30.1 | 41.9 | 52.2 | 61.5 | 66.3 | 63.3 | 55.3 | 43.3 | 31.9 | 20.1 | 41.5 |
| Mean Temp °F | 22.9 | 27.4 | 39.7 | 52.5 | 62.6 | 71.9 | 76.4 | 73.6 | 66.2 | 54.2 | 40.9 | 28.6 | 51.4 |
| Days Max Temp ≥ 90 °F | 0 | 0 | 0 | 0 | 0 | 4 | 10 | 6 | 2 | 0 | 0 | 0 | 22 |
| Days Max Temp ≤ 32 °F | 14 | 10 | 2 | 0 | 0 | 0 | 0 | 0 | 0 | 0 | 2 | 10 | 38 |
| Days Min Temp ≤ 32 °F | 29 | 25 | 18 | 5 | 0 | 0 | 0 | 0 | 0 | 3 | 16 | 27 | 123 |
| Days Min Temp ≤ 0 °F | 6 | 3 | 0 | 0 | 0 | 0 | 0 | 0 | 0 | 0 | 0 | 3 | 12 |
| Heating Degree Days | 1300 | 1054 | 779 | 381 | 141 | 17 | 2 | 6 | 79 | 343 | 717 | 1123 | 5942 |
| Cooling Degree Days | 0 | 0 | 0 | 14 | 71 | 252 | 372 | 308 | 131 | 9 | 0 | 0 | 1157 |
| Total Precipitation (") | 1.26 | 1.30 | 2.91 | 3.64 | 4.39 | 4.16 | 4.54 | 3.83 | 4.30 | 3.12 | 2.89 | 2.29 | 38.63 |
| Days ≥ 0.1" Precip | 3 | 4 | 6 | 8 | 8 | 7 | 7 | 6 | 7 | 5 | 5 | 5 | 71 |
| Total Snowfall (") | *5.1* | *4.8* | 1.9 | 0.2 | 0.0 | 0.0 | 0.0 | 0.0 | 0.0 | 0.0 | 0.7 | na | na |
| Days ≥ 1" Snow Depth | na | na | *0* | 0 | 0 | 0 | 0 | 0 | 0 | 0 | *0* | na | na |

## GLENWOOD 3 SW *Mills County* ELEVATION 980 ft LAT/LONG 41° 0 ' N / 95° 46 ' W

| | JAN | FEB | MAR | APR | MAY | JUN | JUL | AUG | SEP | OCT | NOV | DEC | YEAR |
|---|---|---|---|---|---|---|---|---|---|---|---|---|---|
| Maximum Temp °F | 32.2 | 38.1 | 51.4 | 65.1 | 75.3 | 84.2 | 87.6 | 85.4 | 77.9 | 66.7 | 49.5 | 36.3 | 62.5 |
| Minimum Temp °F | 11.2 | 16.2 | 28.2 | 39.8 | 50.6 | 59.9 | 64.6 | 62.0 | 52.9 | 40.6 | 28.5 | 17.0 | 39.3 |
| Mean Temp °F | 21.7 | 27.2 | 39.8 | 52.5 | 63.0 | 72.0 | 76.1 | 73.7 | 65.4 | 53.7 | 39.0 | 26.6 | 50.9 |
| Days Max Temp ≥ 90 °F | 0 | 0 | 0 | 1 | 1 | 7 | 12 | 9 | 3 | 0 | 0 | 0 | 33 |
| Days Max Temp ≤ 32 °F | 15 | 10 | 3 | 0 | 0 | 0 | 0 | 0 | 0 | 0 | 3 | 11 | 42 |
| Days Min Temp ≤ 32 °F | 30 | 26 | 20 | 8 | 1 | 0 | 0 | 0 | 0 | 7 | 20 | 29 | 141 |
| Days Min Temp ≤ 0 °F | 7 | 4 | 0 | 0 | 0 | 0 | 0 | 0 | 0 | 0 | 0 | 3 | 14 |
| Heating Degree Days | 1336 | 1062 | 777 | 389 | 133 | 15 | 2 | 8 | 97 | 361 | 772 | 1183 | 6135 |
| Cooling Degree Days | 0 | 0 | 2 | 25 | 75 | 235 | 345 | 281 | 127 | 9 | 0 | 0 | 1099 |
| Total Precipitation (") | 0.79 | 0.74 | 2.03 | 3.04 | 4.59 | 4.78 | 4.38 | 3.77 | 3.80 | 2.51 | 1.68 | 1.20 | 33.31 |
| Days ≥ 0.1" Precip | 2 | 2 | 4 | 6 | 8 | 7 | 6 | 6 | 6 | 4 | 4 | 3 | 58 |
| Total Snowfall (") | 6.1 | 6.1 | 3.5 | 0.4 | 0.3 | 0.0 | 0.0 | 0.0 | 0.0 | 0.1 | 2.3 | 5.5 | 24.3 |
| Days ≥ 1" Snow Depth | 17 | 11 | 3 | 0 | 0 | 0 | 0 | 0 | 0 | 0 | 2 | 9 | 42 |

## GREENFIELD *Adair County* ELEVATION 1302 ft LAT/LONG 41° 19 ' N / 94° 28 ' W

| | JAN | FEB | MAR | APR | MAY | JUN | JUL | AUG | SEP | OCT | NOV | DEC | YEAR |
|---|---|---|---|---|---|---|---|---|---|---|---|---|---|
| Maximum Temp °F | 30.2 | 36.0 | 49.5 | 63.7 | 74.1 | 82.9 | 86.5 | 84.2 | 77.0 | 65.3 | 48.0 | 35.0 | 61.0 |
| Minimum Temp °F | 11.3 | 16.4 | 27.6 | 39.6 | 50.2 | 59.6 | 64.5 | 61.7 | 53.4 | 41.8 | 29.1 | 17.5 | 39.4 |
| Mean Temp °F | 20.8 | 26.2 | 38.6 | 51.6 | 62.2 | 71.3 | 75.5 | 72.9 | 65.3 | 53.6 | 38.5 | 26.3 | 50.2 |
| Days Max Temp ≥ 90 °F | 0 | 0 | 0 | 0 | 0 | 5 | 10 | 6 | 2 | 0 | 0 | 0 | 23 |
| Days Max Temp ≤ 32 °F | 16 | 11 | 3 | 0 | 0 | 0 | 0 | 0 | 0 | 0 | 3 | 12 | 45 |
| Days Min Temp ≤ 32 °F | 30 | 26 | 21 | 7 | 0 | 0 | 0 | 0 | 0 | 6 | 19 | 29 | 138 |
| Days Min Temp ≤ 0 °F | 8 | 4 | 0 | 0 | 0 | 0 | 0 | 0 | 0 | 0 | 0 | 3 | 15 |
| Heating Degree Days | 1365 | 1087 | 814 | 409 | 145 | 15 | 2 | 8 | 94 | 361 | 787 | 1194 | 6281 |
| Cooling Degree Days | 0 | 0 | 2 | 21 | 62 | 215 | 326 | 253 | 120 | 9 | 0 | 0 | 1008 |
| Total Precipitation (") | 0.94 | 0.96 | 2.33 | 3.58 | 3.96 | 4.48 | 4.75 | 3.85 | 4.13 | 2.58 | 2.00 | 1.37 | 34.93 |
| Days ≥ 0.1" Precip | 3 | 3 | 5 | 7 | 8 | 7 | 6 | 6 | 6 | 5 | 4 | 3 | 63 |
| Total Snowfall (") | 7.0 | 6.6 | 2.9 | 0.9 | 0.0 | 0.0 | 0.0 | 0.0 | 0.0 | 0.1 | 2.7 | 5.2 | 25.4 |
| Days ≥ 1" Snow Depth | *20* | 15 | 5 | 0 | 0 | 0 | 0 | 0 | 0 | 0 | 3 | *12* | 55 |

## GRINNELL 3 SW *Poweshiek County* ELEVATION 1001 ft LAT/LONG 41° 44 ' N / 92° 44 ' W

| | JAN | FEB | MAR | APR | MAY | JUN | JUL | AUG | SEP | OCT | NOV | DEC | YEAR |
|---|---|---|---|---|---|---|---|---|---|---|---|---|---|
| Maximum Temp °F | 26.8 | 32.1 | 45.2 | 59.4 | 70.8 | 80.0 | 84.2 | 81.9 | 74.2 | 62.4 | 46.2 | 32.4 | 58.0 |
| Minimum Temp °F | 6.9 | 11.2 | 23.6 | 35.2 | 45.7 | 55.8 | 60.8 | 57.4 | 48.7 | 36.6 | 25.5 | 14.0 | 35.1 |
| Mean Temp °F | 16.9 | 21.7 | 34.4 | 47.3 | 58.2 | 67.9 | 72.5 | 69.7 | 61.5 | 49.5 | 35.9 | 23.2 | 46.6 |
| Days Max Temp ≥ 90 °F | 0 | 0 | 0 | 0 | 0 | 2 | 7 | 4 | 1 | 0 | 0 | 0 | 14 |
| Days Max Temp ≤ 32 °F | 19 | 15 | 5 | 0 | 0 | 0 | 0 | 0 | 0 | 0 | 4 | 14 | 57 |
| Days Min Temp ≤ 32 °F | 31 | 27 | 24 | 12 | 2 | 0 | 0 | 0 | 1 | 12 | 23 | 30 | 162 |
| Days Min Temp ≤ 0 °F | 11 | 7 | 1 | 0 | 0 | 0 | 0 | 0 | 0 | 0 | 0 | 5 | 24 |
| Heating Degree Days | 1486 | 1218 | 941 | 531 | 241 | 44 | 10 | 28 | 163 | 480 | 866 | 1288 | 7296 |
| Cooling Degree Days | 0 | 0 | 0 | 10 | 38 | 145 | 256 | 197 | 73 | 5 | 0 | 0 | 724 |
| Total Precipitation (") | 1.21 | 1.25 | 2.50 | 3.85 | 3.93 | 4.85 | 4.08 | 4.66 | 4.02 | 2.70 | 2.33 | 1.64 | 37.02 |
| Days ≥ 0.1" Precip | 3 | 3 | 6 | 7 | 8 | 7 | 7 | 7 | 6 | 5 | 5 | 4 | 68 |
| Total Snowfall (") | 7.5 | 7.3 | 3.9 | 1.5 | 0.0 | 0.0 | 0.0 | 0.0 | 0.0 | 0.3 | 2.6 | 6.8 | 29.9 |
| Days ≥ 1" Snow Depth | na | *14* | 5 | 1 | 0 | 0 | 0 | 0 | 0 | 0 | *2* | *11* | na |

## GRUNDY CENTER *Grundy County* ELEVATION 1020 ft LAT/LONG 42° 22 ' N / 92° 47 ' W

| | JAN | FEB | MAR | APR | MAY | JUN | JUL | AUG | SEP | OCT | NOV | DEC | YEAR |
|---|---|---|---|---|---|---|---|---|---|---|---|---|---|
| Maximum Temp °F | 24.5 | 29.9 | 43.1 | 58.2 | 70.5 | 79.6 | 83.0 | 80.6 | 73.3 | 61.4 | 44.5 | 29.9 | 56.5 |
| Minimum Temp °F | 5.9 | 11.1 | 24.1 | 35.9 | 47.2 | 57.3 | 61.3 | 58.3 | 49.4 | 37.6 | 25.5 | 12.7 | 35.5 |
| Mean Temp °F | 15.2 | 20.5 | 33.6 | 47.1 | 58.9 | 68.5 | 72.2 | 69.5 | 61.4 | 49.5 | 35.0 | 21.3 | 46.1 |
| Days Max Temp ≥ 90 °F | 0 | 0 | 0 | 0 | 0 | 3 | 5 | 3 | 1 | 0 | 0 | 0 | 12 |
| Days Max Temp ≤ 32 °F | 21 | 16 | 6 | 0 | 0 | 0 | 0 | 0 | 0 | 0 | 5 | 17 | 65 |
| Days Min Temp ≤ 32 °F | 31 | 27 | 24 | 11 | 1 | 0 | 0 | 0 | 1 | 11 | 22 | 30 | 158 |
| Days Min Temp ≤ 0 °F | 12 | 7 | 1 | 0 | 0 | 0 | 0 | 0 | 0 | 0 | 0 | 6 | 26 |
| Heating Degree Days | 1538 | 1250 | 965 | 538 | 224 | 37 | 8 | 27 | 162 | 480 | 892 | 1348 | 7469 |
| Cooling Degree Days | 0 | 0 | 0 | 9 | 40 | 156 | 245 | 188 | 69 | 4 | 0 | 0 | 711 |
| Total Precipitation (") | 0.94 | 0.98 | 2.33 | 3.30 | 4.18 | 4.98 | 4.55 | 4.20 | 3.36 | 2.61 | 2.07 | 1.35 | 34.85 |
| Days ≥ 0.1" Precip | 3 | 3 | 5 | 7 | 7 | 7 | 7 | 6 | 6 | 5 | 4 | 3 | 63 |
| Total Snowfall (") | 8.3 | 7.5 | 6.1 | 2.3 | 0.0 | 0.0 | 0.0 | 0.0 | 0.0 | 0.1 | 3.6 | 8.3 | 36.2 |
| Days ≥ 1" Snow Depth | 24 | 19 | 10 | 1 | 0 | 0 | 0 | 0 | 0 | 0 | 4 | 19 | 77 |

## GUTHRIE CENTER *Guthrie County* ELEVATION 1075 ft LAT/LONG 41° 41 ' N / 94° 31 ' W

| | JAN | FEB | MAR | APR | MAY | JUN | JUL | AUG | SEP | OCT | NOV | DEC | YEAR |
|---|---|---|---|---|---|---|---|---|---|---|---|---|---|
| Maximum Temp °F | 28.1 | 33.8 | 46.2 | 60.9 | 72.9 | 81.9 | 86.0 | 83.5 | 75.9 | 63.3 | 46.5 | 32.4 | 59.3 |
| Minimum Temp °F | 7.6 | 12.9 | 24.9 | 37.3 | 48.2 | 57.6 | 62.5 | 59.3 | 49.8 | 37.7 | 25.9 | 13.6 | 36.4 |
| Mean Temp °F | 17.8 | 23.3 | 35.6 | 49.1 | 60.6 | 69.8 | 74.3 | 71.5 | 62.9 | 50.5 | 36.2 | 23.1 | 47.9 |
| Days Max Temp ≥ 90 °F | 0 | 0 | 0 | 0 | 1 | 4 | 10 | 7 | 2 | 0 | 0 | 0 | 24 |
| Days Max Temp ≤ 32 °F | 18 | 13 | 5 | 0 | 0 | 0 | 0 | 0 | 0 | 0 | 4 | 14 | 54 |
| Days Min Temp ≤ 32 °F | 31 | 27 | 24 | 9 | 1 | 0 | 0 | 0 | 1 | 9 | 22 | 30 | 154 |
| Days Min Temp ≤ 0 °F | 10 | 6 | 0 | 0 | 0 | 0 | 0 | 0 | 0 | 0 | 1 | 5 | 22 |
| Heating Degree Days | 1457 | 1169 | 906 | 481 | 188 | 28 | 5 | 18 | 137 | 451 | 856 | 1294 | 6990 |
| Cooling Degree Days | 0 | 0 | 2 | 15 | 50 | 178 | 291 | 219 | 88 | 5 | 0 | 0 | 848 |
| Total Precipitation (") | 0.89 | 1.06 | 2.43 | 3.24 | 4.29 | 4.93 | 3.95 | 4.56 | 3.75 | 2.68 | 1.86 | 1.35 | 34.99 |
| Days ≥ 0.1" Precip | 3 | 3 | 5 | 6 | 8 | 7 | 7 | 6 | 6 | 5 | 4 | 3 | 63 |
| Total Snowfall (") | 7.2 | *6.8* | 4.2 | 1.4 | 0.0 | 0.0 | 0.0 | 0.0 | 0.0 | 0.1 | 2.7 | 6.3 | 28.7 |
| Days ≥ 1" Snow Depth | na | na | na | 0 | 0 | 0 | 0 | 0 | 0 | 0 | na | na | na |

## GUTTENBERG L&D 10 *Clayton County* ELEVATION 620 ft LAT/LONG 42° 47 ' N / 91° 6 ' W

| | JAN | FEB | MAR | APR | MAY | JUN | JUL | AUG | SEP | OCT | NOV | DEC | YEAR |
|---|---|---|---|---|---|---|---|---|---|---|---|---|---|
| Maximum Temp °F | 26.6 | 32.1 | 44.7 | 59.4 | 71.7 | 80.5 | 84.2 | 82.0 | 73.4 | 61.7 | 44.8 | 31.9 | 57.8 |
| Minimum Temp °F | 8.4 | 13.1 | 26.3 | 39.2 | 50.5 | 59.7 | 64.1 | 61.7 | 53.3 | 41.9 | 29.1 | 16.4 | 38.6 |
| Mean Temp °F | 17.5 | 22.6 | 35.5 | 49.3 | 61.1 | 70.1 | 74.2 | 71.9 | 63.4 | 51.8 | 37.0 | 24.2 | 48.2 |
| Days Max Temp ≥ 90 °F | 0 | 0 | 0 | 0 | 0 | 3 | 6 | 4 | 1 | 0 | 0 | 0 | 14 |
| Days Max Temp ≤ 32 °F | 20 | 14 | 4 | 0 | 0 | 0 | 0 | 0 | 0 | 0 | 3 | 14 | 55 |
| Days Min Temp ≤ 32 °F | 30 | 27 | 23 | 7 | 0 | 0 | 0 | 0 | 0 | 5 | 19 | 29 | 140 |
| Days Min Temp ≤ 0 °F | 10 | 6 | 1 | 0 | 0 | 0 | 0 | 0 | 0 | 0 | 0 | 4 | 21 |
| Heating Degree Days | 1467 | 1191 | 907 | 472 | 171 | 22 | 3 | 10 | 122 | 411 | 833 | 1260 | 6869 |
| Cooling Degree Days | 0 | 0 | 0 | 13 | 59 | 189 | 306 | 241 | 91 | 5 | 0 | 0 | 904 |
| Total Precipitation (") | 0.99 | 1.04 | 1.92 | 3.07 | 3.51 | 4.34 | 4.33 | 4.20 | 3.32 | 2.27 | 2.13 | 1.45 | 32.57 |
| Days ≥ 0.1" Precip | 3 | 3 | 5 | 7 | 7 | 7 | 7 | 7 | 6 | 5 | 5 | 4 | 66 |
| Total Snowfall (") | ***9.3*** | ***5.8*** | ***3.8*** | 1.2 | 0.0 | 0.0 | 0.0 | 0.0 | 0.0 | 0.0 | ***1.9*** | ***7.8*** | 29.8 |
| Days ≥ 1" Snow Depth | ***24*** | ***22*** | ***8*** | 1 | 0 | 0 | 0 | 0 | 0 | 0 | ***2*** | ***15*** | 72 |

## HAMPTON *Franklin County* ELEVATION 1220 ft LAT/LONG 42° 45 ' N / 93° 12 ' W

| | JAN | FEB | MAR | APR | MAY | JUN | JUL | AUG | SEP | OCT | NOV | DEC | YEAR |
|---|---|---|---|---|---|---|---|---|---|---|---|---|---|
| Maximum Temp °F | 23.8 | 29.4 | 42.6 | 58.1 | 71.3 | 80.4 | 83.2 | 80.9 | 73.5 | 61.4 | 43.3 | 29.2 | 56.4 |
| Minimum Temp °F | 6.5 | 11.4 | 24.3 | 36.3 | 48.1 | 58.0 | 61.9 | 58.9 | 50.3 | 39.0 | 25.8 | 13.1 | 36.1 |
| Mean Temp °F | 15.3 | 20.4 | 33.5 | 47.3 | 59.7 | 69.2 | 72.6 | 69.9 | 61.9 | 50.2 | 34.6 | 21.1 | 46.3 |
| Days Max Temp ≥ 90 °F | 0 | 0 | 0 | 0 | 1 | 3 | 5 | 3 | 1 | 0 | 0 | 0 | 13 |
| Days Max Temp ≤ 32 °F | 22 | 16 | 7 | 0 | 0 | 0 | 0 | 0 | 0 | 0 | 5 | 18 | 68 |
| Days Min Temp ≤ 32 °F | 31 | 27 | 24 | 11 | 1 | 0 | 0 | 0 | 1 | 8 | 23 | 30 | 156 |
| Days Min Temp ≤ 0 °F | 11 | 7 | 1 | 0 | 0 | 0 | 0 | 0 | 0 | 0 | 1 | 6 | 26 |
| Heating Degree Days | 1534 | 1253 | 970 | 533 | 206 | 32 | 8 | 23 | 149 | 459 | 905 | 1354 | 7426 |
| Cooling Degree Days | 0 | 0 | 0 | 10 | 48 | 168 | 247 | 181 | 67 | 3 | 0 | 0 | 724 |
| Total Precipitation (") | 0.97 | 0.87 | 2.29 | 3.21 | 4.38 | 5.40 | 5.07 | 4.38 | 3.63 | 2.59 | 1.83 | 1.35 | 35.97 |
| Days ≥ 0.1" Precip | 2 | 2 | 5 | 7 | 8 | 7 | 7 | 6 | 6 | 5 | 4 | 3 | 62 |
| Total Snowfall (") | 7.8 | 6.8 | 6.2 | 2.5 | 0.0 | 0.0 | 0.0 | 0.0 | 0.0 | 0.4 | 4.0 | 8.0 | 35.7 |
| Days ≥ 1" Snow Depth | ***23*** | 20 | 10 | 1 | 0 | 0 | 0 | 0 | 0 | 0 | 4 | 18 | 76 |

## HARLAN *Shelby County* ELEVATION 1211 ft LAT/LONG 41° 39 ' N / 95° 19 ' W

| | JAN | FEB | MAR | APR | MAY | JUN | JUL | AUG | SEP | OCT | NOV | DEC | YEAR |
|---|---|---|---|---|---|---|---|---|---|---|---|---|---|
| Maximum Temp °F | 29.3 | 35.3 | 48.2 | 63.0 | 73.4 | 82.5 | 85.4 | 83.0 | 75.0 | 63.2 | 46.5 | 33.5 | 59.9 |
| Minimum Temp °F | 9.4 | 15.1 | 26.4 | 38.2 | 49.3 | 59.2 | 63.6 | 61.1 | 52.3 | 40.1 | 27.2 | 15.5 | 38.1 |
| Mean Temp °F | 19.3 | 25.2 | 37.3 | 50.6 | 61.4 | 70.9 | 74.5 | 72.0 | 63.7 | 51.6 | 36.9 | 24.5 | 49.0 |
| Days Max Temp ≥ 90 °F | 0 | 0 | 0 | 0 | 1 | 5 | 8 | 5 | 1 | 0 | 0 | 0 | 20 |
| Days Max Temp ≤ 32 °F | 17 | 11 | 4 | 0 | 0 | 0 | 0 | 0 | 0 | 0 | 3 | 13 | 48 |
| Days Min Temp ≤ 32 °F | 31 | 27 | 22 | 8 | 1 | 0 | 0 | 0 | 0 | 7 | 22 | 30 | 148 |
| Days Min Temp ≤ 0 °F | 9 | 5 | 1 | 0 | 0 | 0 | 0 | 0 | 0 | 0 | 0 | 4 | 19 |
| Heating Degree Days | 1409 | 1117 | 852 | 437 | 163 | 19 | 3 | 11 | 117 | 414 | 837 | 1248 | 6627 |
| Cooling Degree Days | 0 | 0 | 1 | 17 | 60 | 212 | 305 | 245 | 98 | 5 | 0 | 0 | 943 |
| Total Precipitation (") | 0.73 | 0.75 | 2.08 | 3.07 | 3.97 | 4.50 | 3.83 | 3.57 | 4.80 | 2.76 | 1.60 | 1.05 | 32.71 |
| Days ≥ 0.1" Precip | 2 | 2 | 5 | 6 | 8 | 7 | 6 | 6 | 6 | 5 | 4 | 3 | 60 |
| Total Snowfall (") | 7.4 | 6.5 | 4.8 | 1.3 | 0.0 | 0.0 | 0.0 | 0.0 | 0.0 | 0.6 | 3.0 | 6.5 | 30.1 |
| Days ≥ 1" Snow Depth | 19 | 15 | 6 | 0 | 0 | 0 | 0 | 0 | 0 | 0 | 3 | 11 | 54 |

## HAWARDEN *Sioux County* ELEVATION 1191 ft LAT/LONG 43° 0 ' N / 96° 29 ' W

| | JAN | FEB | MAR | APR | MAY | JUN | JUL | AUG | SEP | OCT | NOV | DEC | YEAR |
|---|---|---|---|---|---|---|---|---|---|---|---|---|---|
| Maximum Temp °F | 26.8 | 32.8 | 45.9 | 61.6 | 73.3 | 82.4 | 85.8 | 83.4 | 75.0 | 62.7 | 44.2 | 30.9 | 58.7 |
| Minimum Temp °F | 5.9 | 12.2 | 24.3 | 36.6 | 48.4 | 58.6 | 63.1 | 60.5 | 50.3 | 37.7 | 24.1 | 11.6 | 36.1 |
| Mean Temp °F | 16.4 | 22.5 | 35.1 | 49.1 | 60.9 | 70.5 | 74.5 | 72.0 | 62.7 | 50.2 | 34.2 | 21.3 | 47.4 |
| Days Max Temp ≥ 90 °F | 0 | 0 | 0 | 0 | 1 | 5 | 9 | 6 | 2 | 0 | 0 | 0 | 23 |
| Days Max Temp ≤ 32 °F | 19 | 13 | 4 | 0 | 0 | 0 | 0 | 0 | 0 | 0 | 5 | 16 | 57 |
| Days Min Temp ≤ 32 °F | 31 | 28 | 24 | 11 | 2 | 0 | 0 | 0 | 1 | 10 | 24 | 31 | 162 |
| Days Min Temp ≤ 0 °F | 11 | 6 | 1 | 0 | 0 | 0 | 0 | 0 | 0 | 0 | 1 | 6 | 25 |
| Heating Degree Days | 1502 | 1196 | 919 | 479 | 179 | 24 | 4 | 14 | 140 | 457 | 918 | 1350 | 7182 |
| Cooling Degree Days | 0 | 0 | 0 | 13 | 61 | 197 | 293 | 237 | 82 | 4 | 0 | 0 | 887 |
| Total Precipitation (") | 0.52 | 0.64 | 1.76 | 2.64 | 3.32 | 4.00 | 3.54 | 3.26 | 3.17 | 2.07 | 1.32 | 0.79 | 27.03 |
| Days ≥ 0.1" Precip | 2 | 2 | 4 | 5 | 7 | 6 | 6 | 6 | 6 | 4 | 3 | 2 | 53 |
| Total Snowfall (") | 6.5 | 5.2 | 5.8 | 1.6 | 0.0 | 0.0 | 0.0 | 0.0 | 0.0 | 1.1 | 5.8 | 7.5 | 33.5 |
| Days ≥ 1" Snow Depth | 21 | 16 | 7 | 1 | 0 | 0 | 0 | 0 | 0 | 0 | 5 | 14 | 64 |

## HUMBOLDT 3 W *Humboldt County* ELEVATION 1110 ft LAT/LONG 42° 42 ' N / 94° 16 ' W

| | JAN | FEB | MAR | APR | MAY | JUN | JUL | AUG | SEP | OCT | NOV | DEC | YEAR |
|---|---|---|---|---|---|---|---|---|---|---|---|---|---|
| Maximum Temp °F | 24.7 | 30.5 | 43.7 | 59.6 | 72.6 | 81.3 | 84.2 | 81.4 | 73.8 | 61.5 | 43.5 | 29.4 | 57.2 |
| Minimum Temp °F | 5.8 | 11.7 | 24.5 | 36.9 | 48.7 | 58.1 | 62.0 | 58.8 | 49.8 | 38.2 | 25.3 | 12.3 | 36.0 |
| Mean Temp °F | 15.2 | 20.9 | 34.1 | 48.4 | 60.8 | 69.7 | 73.1 | 70.1 | 61.8 | 49.9 | 34.4 | 20.8 | 46.6 |
| Days Max Temp ≥ 90 °F | 0 | 0 | 0 | 0 | 1 | 4 | 7 | 4 | 1 | 0 | 0 | 0 | 17 |
| Days Max Temp ≤ 32 °F | 21 | 15 | 6 | 0 | 0 | 0 | 0 | 0 | 0 | 0 | 5 | 17 | 64 |
| Days Min Temp ≤ 32 °F | 31 | 28 | 24 | 10 | 1 | 0 | 0 | 0 | 1 | 9 | 23 | 31 | 158 |
| Days Min Temp ≤ 0 °F | 12 | 7 | 1 | 0 | 0 | 0 | 0 | 0 | 0 | 0 | 1 | 6 | 27 |
| Heating Degree Days | 1540 | 1238 | 950 | 502 | 184 | 26 | 6 | 22 | 152 | 469 | 911 | 1363 | 7363 |
| Cooling Degree Days | 0 | 0 | 0 | 13 | 60 | 182 | 261 | 192 | 74 | 4 | 0 | 0 | 786 |
| Total Precipitation (") | 0.88 | 0.74 | 2.07 | 3.11 | 3.83 | 4.70 | 3.97 | 3.93 | 3.52 | 2.33 | 1.61 | 1.09 | 31.78 |
| Days ≥ 0.1" Precip | 3 | 2 | 5 | 6 | 7 | 8 | 6 | 6 | 6 | 4 | 3 | 3 | 59 |
| Total Snowfall (") | 6.6 | 5.7 | 4.9 | 1.2 | 0.0 | 0.0 | 0.0 | 0.0 | 0.0 | 0.2 | 3.0 | 6.9 | 28.5 |
| Days ≥ 1" Snow Depth | 22 | 16 | 7 | 1 | 0 | 0 | 0 | 0 | 0 | 0 | 4 | 15 | 65 |

## IDA GROVE 5 NW *Ida County* ELEVATION 1320 ft LAT/LONG 42° 24 ' N / 95° 31 ' W

| | JAN | FEB | MAR | APR | MAY | JUN | JUL | AUG | SEP | OCT | NOV | DEC | YEAR |
|---|---|---|---|---|---|---|---|---|---|---|---|---|---|
| Maximum Temp °F | 27.0 | 32.5 | 46.1 | 61.6 | 72.8 | 82.3 | 85.4 | 82.9 | 75.1 | 63.1 | 45.1 | 31.2 | 58.8 |
| Minimum Temp °F | 7.0 | 12.0 | 24.1 | 36.1 | 47.5 | 57.8 | 62.4 | 59.6 | 50.6 | 38.3 | 25.1 | 13.1 | 36.1 |
| Mean Temp °F | 16.9 | 22.1 | 35.1 | 48.9 | 60.3 | 70.0 | 73.9 | 71.1 | 62.8 | 50.7 | 35.1 | 22.1 | 47.4 |
| Days Max Temp ≥ 90 °F | 0 | 0 | 0 | 0 | 1 | 5 | 9 | 5 | 2 | 0 | 0 | 0 | 22 |
| Days Max Temp ≤ 32 °F | 19 | 14 | 5 | 0 | 0 | 0 | 0 | 0 | 0 | 0 | 4 | 15 | 57 |
| Days Min Temp ≤ 32 °F | 31 | 27 | 24 | 11 | 1 | 0 | 0 | 0 | 1 | 9 | 23 | 30 | 157 |
| Days Min Temp ≤ 0 °F | 11 | 7 | 1 | 0 | 0 | 0 | 0 | 0 | 0 | 0 | 1 | 5 | 25 |
| Heating Degree Days | 1488 | 1204 | 920 | 485 | 190 | 27 | 5 | 15 | 137 | 443 | 889 | 1325 | 7128 |
| Cooling Degree Days | 0 | 0 | 0 | ***16*** | ***47*** | ***182*** | 278 | ***198*** | 88 | 4 | 0 | 0 | 813 |
| Total Precipitation (") | 0.79 | 0.71 | 1.94 | 2.89 | 3.94 | 4.68 | 3.69 | 3.94 | 3.26 | 2.42 | 1.32 | 0.92 | 30.50 |
| Days ≥ 0.1" Precip | 2 | 2 | 4 | 6 | 7 | 7 | 6 | 6 | 5 | 4 | 3 | 2 | 54 |
| Total Snowfall (") | 6.9 | 7.2 | 6.2 | 1.3 | 0.0 | 0.0 | 0.0 | 0.0 | 0.0 | 0.4 | 4.1 | 7.9 | 34.0 |
| Days ≥ 1" Snow Depth | 20 | 16 | 8 | 1 | 0 | 0 | 0 | 0 | 0 | 0 | 4 | 14 | 63 |

## INDIANOLA *Warren County* ELEVATION 940 ft LAT/LONG 41° 22 ' N / 93° 33 ' W

| | JAN | FEB | MAR | APR | MAY | JUN | JUL | AUG | SEP | OCT | NOV | DEC | YEAR |
|---|---|---|---|---|---|---|---|---|---|---|---|---|---|
| Maximum Temp °F | 31.0 | 37.0 | 49.6 | 63.9 | 74.0 | 83.0 | 87.0 | 85.1 | 77.2 | 65.9 | 48.8 | 35.7 | 61.5 |
| Minimum Temp °F | 10.9 | 16.1 | 27.8 | 39.8 | 49.7 | 59.0 | 63.4 | 60.4 | 51.9 | 40.5 | 28.0 | 17.1 | 38.7 |
| Mean Temp °F | 21.0 | 26.6 | 38.7 | 51.9 | 61.9 | 71.1 | 75.2 | 72.8 | 64.6 | 53.3 | 38.5 | 26.4 | 50.2 |
| Days Max Temp ≥ 90 °F | 0 | 0 | 0 | 0 | 0 | 4 | 11 | 8 | 2 | 0 | 0 | 0 | 25 |
| Days Max Temp ≤ 32 °F | 15 | 11 | 3 | 0 | 0 | 0 | 0 | 0 | 0 | 0 | 2 | 11 | 42 |
| Days Min Temp ≤ 32 °F | 30 | 26 | 20 | 8 | 1 | 0 | 0 | 0 | 1 | 8 | 20 | 28 | 142 |
| Days Min Temp ≤ 0 °F | 8 | 5 | 0 | 0 | 0 | 0 | 0 | 0 | 0 | 0 | 0 | 4 | 17 |
| Heating Degree Days | 1359 | 1078 | 809 | 403 | 151 | 18 | 3 | 9 | 108 | 373 | 789 | 1189 | 6289 |
| Cooling Degree Days | 0 | 0 | 2 | 22 | 55 | 205 | 318 | 265 | 112 | 8 | 0 | 0 | 987 |
| Total Precipitation (") | 1.03 | 1.07 | 2.20 | 3.80 | 4.24 | 4.54 | 4.13 | 3.62 | 3.92 | 2.83 | 1.99 | 1.46 | 34.83 |
| Days ≥ 0.1" Precip | 3 | 3 | 5 | 7 | 8 | 7 | 7 | 6 | 6 | 5 | 5 | 4 | 66 |
| Total Snowfall (") | 5.8 | 6.5 | 2.3 | 1.0 | 0.0 | 0.0 | 0.0 | 0.0 | 0.0 | 0.3 | 2.6 | 4.6 | 23.1 |
| Days ≥ 1" Snow Depth | ***16*** | ***12*** | 3 | 0 | 0 | 0 | 0 | 0 | 0 | 0 | 2 | 8 | 41 |

## IOWA CITY *Johnson County* ELEVATION 771 ft LAT/LONG 41° 40 ' N / 91° 32 ' W

| | JAN | FEB | MAR | APR | MAY | JUN | JUL | AUG | SEP | OCT | NOV | DEC | YEAR |
|---|---|---|---|---|---|---|---|---|---|---|---|---|---|
| Maximum Temp °F | 29.8 | 35.5 | 48.7 | 63.5 | 74.8 | 83.8 | 87.4 | 85.0 | 77.4 | 65.7 | 48.7 | 34.9 | 61.3 |
| Minimum Temp °F | 12.4 | 17.3 | 29.0 | 40.7 | 51.3 | 61.0 | 65.6 | 63.0 | 54.8 | 43.1 | 31.3 | 19.2 | 40.7 |
| Mean Temp °F | 21.1 | 26.4 | 38.9 | 52.1 | 63.1 | 72.4 | 76.5 | 74.0 | 66.1 | 54.5 | 40.0 | 27.1 | 51.0 |
| Days Max Temp ≥ 90 °F | 0 | 0 | 0 | 0 | 1 | 6 | 11 | 7 | 2 | 0 | 0 | 0 | 27 |
| Days Max Temp ≤ 32 °F | 17 | 11 | 3 | 0 | 0 | 0 | 0 | 0 | 0 | 0 | 2 | 11 | 44 |
| Days Min Temp ≤ 32 °F | 30 | 25 | 19 | 5 | 0 | 0 | 0 | 0 | 0 | 4 | 17 | 28 | 128 |
| Days Min Temp ≤ 0 °F | 7 | 3 | 0 | 0 | 0 | 0 | 0 | 0 | 0 | 0 | 0 | 3 | 13 |
| Heating Degree Days | 1354 | 1083 | 804 | 394 | 132 | 10 | 1 | 4 | 79 | 336 | 743 | 1168 | 6108 |
| Cooling Degree Days | 0 | 0 | 1 | 20 | 82 | 255 | 383 | 311 | 134 | 12 | 0 | 0 | 1198 |
| Total Precipitation (") | 1.08 | 0.98 | 2.42 | 3.77 | 4.09 | 4.92 | 4.80 | 4.78 | 4.02 | 2.74 | 2.40 | 1.66 | 37.66 |
| Days ≥ 0.1" Precip | 3 | 3 | 5 | 7 | 7 | 7 | 7 | 6 | 6 | 6 | 5 | 4 | 66 |
| Total Snowfall (") | 6.9 | 5.6 | 3.5 | 1.7 | 0.0 | 0.0 | 0.0 | 0.0 | 0.0 | 0.3 | 1.6 | 6.2 | 25.8 |
| Days ≥ 1" Snow Depth | 18 | 15 | 4 | 1 | 0 | 0 | 0 | 0 | 0 | 0 | 2 | 11 | 51 |

## IOWA FALLS *Hardin County* ELEVATION 1142 ft LAT/LONG 42° 32 ' N / 93° 16 ' W

| | JAN | FEB | MAR | APR | MAY | JUN | JUL | AUG | SEP | OCT | NOV | DEC | YEAR |
|---|---|---|---|---|---|---|---|---|---|---|---|---|---|
| Maximum Temp °F | 25.6 | 30.9 | 44.5 | 60.0 | 72.8 | 81.5 | 84.8 | 81.9 | 74.6 | 62.5 | 44.3 | 30.3 | 57.8 |
| Minimum Temp °F | 7.1 | 12.0 | 25.0 | 37.0 | 48.5 | 58.0 | 62.0 | 58.9 | 50.6 | 39.1 | 26.3 | 13.4 | 36.5 |
| Mean Temp °F | 16.3 | 21.5 | 34.8 | 48.5 | 60.7 | 69.8 | 73.4 | 70.4 | 62.7 | 50.8 | 35.4 | 21.9 | 47.2 |
| Days Max Temp ≥ 90 °F | 0 | 0 | 0 | 0 | 1 | 4 | 7 | 4 | 1 | 0 | 0 | 0 | 17 |
| Days Max Temp ≤ 32 °F | 20 | 15 | 5 | 0 | 0 | 0 | 0 | 0 | 0 | 0 | 5 | 16 | 61 |
| Days Min Temp ≤ 32 °F | 31 | 28 | 24 | 10 | 1 | 0 | 0 | 0 | 1 | 9 | 22 | 30 | 156 |
| Days Min Temp ≤ 0 °F | 11 | 7 | 1 | 0 | 0 | 0 | 0 | 0 | 0 | 0 | 0 | 5 | 24 |
| Heating Degree Days | 1504 | 1223 | 931 | 498 | 182 | 24 | 6 | 18 | 136 | 440 | 883 | 1329 | 7174 |
| Cooling Degree Days | 0 | *0* | 0 | 10 | 56 | 181 | 290 | 209 | 84 | 5 | 0 | 0 | 835 |
| Total Precipitation (") | 1.02 | 1.00 | 2.12 | 3.36 | 4.03 | 5.31 | 4.28 | 4.40 | 3.70 | 2.72 | 1.98 | 1.33 | 35.25 |
| Days ≥ 0.1" Precip | 3 | 3 | 5 | 7 | 7 | 8 | 6 | 6 | 6 | 5 | 4 | 3 | 63 |
| Total Snowfall (") | 8.2 | 7.2 | 5.7 | 2.0 | 0.0 | 0.0 | 0.0 | 0.0 | 0.0 | 0.1 | 3.9 | 7.8 | 34.9 |
| Days ≥ 1" Snow Depth | na | na | na | 1 | 0 | 0 | 0 | 0 | 0 | 0 | *2* | na | na |

## JEFFERSON *Greene County* ELEVATION 1060 ft LAT/LONG 42° 1 ' N / 94° 23 ' W

| | JAN | FEB | MAR | APR | MAY | JUN | JUL | AUG | SEP | OCT | NOV | DEC | YEAR |
|---|---|---|---|---|---|---|---|---|---|---|---|---|---|
| Maximum Temp °F | 29.1 | 34.8 | 47.9 | 63.9 | 74.8 | 83.4 | 86.9 | 84.4 | 77.1 | 64.8 | 46.9 | 33.4 | 60.6 |
| Minimum Temp °F | 9.9 | 14.8 | 26.7 | 38.7 | 49.9 | 59.6 | 64.0 | 61.3 | 52.6 | 40.7 | 28.0 | 15.9 | 38.5 |
| Mean Temp °F | 19.6 | 24.8 | 37.3 | 51.3 | 62.4 | 71.5 | 75.4 | 72.9 | 64.9 | 52.8 | 37.5 | 24.7 | 49.6 |
| Days Max Temp ≥ 90 °F | 0 | 0 | 0 | 1 | 1 | 5 | 11 | 7 | 2 | 0 | 0 | 0 | 27 |
| Days Max Temp ≤ 32 °F | 17 | 12 | 4 | 0 | 0 | 0 | 0 | 0 | 0 | 0 | 3 | 13 | 49 |
| Days Min Temp ≤ 32 °F | 30 | 26 | 23 | 8 | 1 | 0 | 0 | 0 | 1 | 7 | 20 | 30 | 146 |
| Days Min Temp ≤ 0 °F | 9 | 5 | 1 | 0 | 0 | 0 | 0 | 0 | 0 | 0 | 0 | 4 | 19 |
| Heating Degree Days | 1403 | 1128 | 852 | 420 | 145 | 14 | 2 | 9 | 101 | 384 | 819 | 1243 | 6520 |
| Cooling Degree Days | 0 | 0 | 1 | 22 | 73 | 229 | 343 | 270 | 118 | 8 | 0 | 0 | 1064 |
| Total Precipitation (") | 0.90 | 0.89 | 2.15 | 3.15 | 3.94 | 4.61 | 3.98 | 3.85 | 3.40 | 2.56 | 1.81 | 1.22 | 32.46 |
| Days ≥ 0.1" Precip | 3 | 3 | 5 | 6 | 8 | 7 | 6 | 6 | 5 | 5 | 4 | 3 | 61 |
| Total Snowfall (") | 6.7 | 6.6 | 5.1 | 1.7 | 0.0 | 0.0 | 0.0 | 0.0 | 0.0 | 0.1 | 2.8 | 6.2 | 29.2 |
| Days ≥ 1" Snow Depth | 21 | 18 | 9 | 1 | 0 | 0 | 0 | 0 | 0 | 0 | 3 | 15 | 67 |

## KEOSAUQUA *Van Buren County* ELEVATION 712 ft LAT/LONG 40° 43 ' N / 91° 58 ' W

| | JAN | FEB | MAR | APR | MAY | JUN | JUL | AUG | SEP | OCT | NOV | DEC | YEAR |
|---|---|---|---|---|---|---|---|---|---|---|---|---|---|
| Maximum Temp °F | 32.9 | 38.5 | 51.6 | 65.0 | 75.4 | 84.1 | 88.2 | 86.0 | 78.2 | 66.6 | 50.7 | 37.4 | 62.9 |
| Minimum Temp °F | 13.5 | 18.0 | 29.4 | 40.6 | 50.2 | 59.5 | 64.4 | 61.7 | 53.7 | 42.2 | 31.1 | 19.9 | 40.4 |
| Mean Temp °F | 23.2 | 28.3 | 40.5 | 52.8 | 62.8 | 71.8 | 76.3 | 73.9 | 66.0 | 54.4 | 40.9 | 28.7 | 51.6 |
| Days Max Temp ≥ 90 °F | 0 | 0 | 0 | 0 | 1 | 6 | 13 | 10 | 3 | 0 | 0 | 0 | 33 |
| Days Max Temp ≤ 32 °F | 14 | 9 | 2 | 0 | 0 | 0 | 0 | 0 | 0 | 0 | 2 | 9 | 36 |
| Days Min Temp ≤ 32 °F | 29 | 25 | 19 | 6 | 0 | 0 | 0 | 0 | 1 | 6 | 17 | 27 | 130 |
| Days Min Temp ≤ 0 °F | 6 | 3 | 0 | 0 | 0 | 0 | 0 | 0 | 0 | 0 | 0 | 3 | 12 |
| Heating Degree Days | 1288 | 1030 | 755 | 375 | 135 | 14 | 2 | 6 | 85 | 338 | 715 | 1119 | 5862 |
| Cooling Degree Days | 0 | 0 | 3 | 23 | 72 | 233 | 368 | 300 | 130 | 13 | 1 | 0 | 1143 |
| Total Precipitation (") | 1.36 | 1.17 | 2.67 | 3.69 | 4.17 | 4.29 | 5.15 | 3.94 | 4.34 | 3.08 | 2.78 | 2.09 | 38.73 |
| Days ≥ 0.1" Precip | 4 | 3 | 6 | 7 | 8 | 7 | 7 | 7 | 7 | 6 | 6 | 5 | 73 |
| Total Snowfall (") | 6.9 | 6.1 | 3.0 | 1.2 | 0.0 | 0.0 | 0.0 | 0.0 | 0.0 | 0.2 | 2.0 | 5.5 | 24.9 |
| Days ≥ 1" Snow Depth | na | *7* | 2 | 0 | 0 | 0 | 0 | 0 | 0 | 0 | 1 | *5* | na |

## KNOXVILLE *Marion County* ELEVATION 912 ft LAT/LONG 41° 19 ' N / 93° 6 ' W

| | JAN | FEB | MAR | APR | MAY | JUN | JUL | AUG | SEP | OCT | NOV | DEC | YEAR |
|---|---|---|---|---|---|---|---|---|---|---|---|---|---|
| Maximum Temp °F | 30.5 | 36.3 | 49.1 | 63.0 | 74.1 | 82.9 | 87.4 | 84.6 | 76.7 | 64.8 | 48.5 | 35.3 | 61.1 |
| Minimum Temp °F | 12.2 | 17.2 | 28.7 | 40.7 | 51.7 | 60.8 | 65.6 | 62.8 | 54.4 | 42.8 | 30.3 | 18.3 | 40.5 |
| Mean Temp °F | 21.4 | 26.7 | 38.9 | 51.9 | 62.9 | 71.9 | 76.5 | 73.7 | 65.6 | 53.9 | 39.4 | 26.9 | 50.8 |
| Days Max Temp ≥ 90 °F | 0 | 0 | 0 | 0 | 0 | 5 | 12 | 8 | 2 | 0 | 0 | 0 | 27 |
| Days Max Temp ≤ 32 °F | 16 | 11 | 3 | 0 | 0 | 0 | 0 | 0 | 0 | 0 | 3 | 12 | 45 |
| Days Min Temp ≤ 32 °F | 29 | 25 | 20 | 6 | 0 | 0 | 0 | 0 | 0 | 5 | 18 | 28 | 131 |
| Days Min Temp ≤ 0 °F | 8 | 4 | 0 | 0 | 0 | 0 | 0 | 0 | 0 | 0 | 0 | 3 | 15 |
| Heating Degree Days | 1346 | 1074 | 803 | 401 | 134 | 13 | 2 | 7 | 92 | 353 | 762 | 1177 | 6164 |
| Cooling Degree Days | 0 | 0 | 2 | 22 | 76 | 238 | 381 | 303 | 129 | 10 | 0 | 0 | 1161 |
| Total Precipitation (") | 0.92 | 1.03 | 2.04 | 4.16 | 3.89 | 4.26 | 4.20 | 4.17 | 3.96 | 2.73 | 2.03 | 1.35 | 34.74 |
| Days ≥ 0.1" Precip | 3 | 3 | 5 | 7 | 8 | 7 | 6 | 6 | 6 | 5 | 4 | 3 | 63 |
| Total Snowfall (") | *5.5* | 6.0 | 1.9 | 1.1 | 0.0 | 0.0 | 0.0 | 0.0 | 0.0 | 0.1 | 1.8 | na | na |
| Days ≥ 1" Snow Depth | na | *10* | 4 | 0 | 0 | 0 | 0 | 0 | 0 | 0 | 1 | *7* | na |

## LAKE PARK *Dickinson County* ELEVATION 1480 ft LAT/LONG 43° 27 ' N / 95° 19 ' W

| | JAN | FEB | MAR | APR | MAY | JUN | JUL | AUG | SEP | OCT | NOV | DEC | YEAR |
|---|---|---|---|---|---|---|---|---|---|---|---|---|---|
| Maximum Temp °F | 22.2 | 27.4 | 40.0 | 56.3 | 69.8 | 79.0 | 83.1 | 80.3 | 71.8 | 59.1 | 40.8 | 26.9 | 54.7 |
| Minimum Temp °F | 2.7 | 8.2 | 21.2 | 34.8 | 46.6 | 56.7 | 61.2 | 58.2 | 48.7 | 36.5 | 23.2 | 9.7 | 34.0 |
| Mean Temp °F | 12.4 | 17.8 | 30.6 | 45.6 | 58.2 | 67.9 | 72.2 | 69.3 | 60.3 | 47.8 | 32.0 | 18.3 | 44.4 |
| Days Max Temp ≥ 90 °F | 0 | 0 | 0 | 0 | 0 | 3 | 6 | 3 | 1 | 0 | 0 | 0 | 13 |
| Days Max Temp ≤ 32 °F | 23 | 17 | 9 | 1 | 0 | 0 | 0 | 0 | 0 | 0 | 8 | 20 | 78 |
| Days Min Temp ≤ 32 °F | 31 | 28 | 27 | 12 | 2 | 0 | 0 | 0 | 1 | 10 | 25 | 31 | 167 |
| Days Min Temp ≤ 0 °F | 14 | 9 | 2 | 0 | 0 | 0 | 0 | 0 | 0 | 0 | 1 | 8 | 34 |
| Heating Degree Days | 1625 | 1326 | 1059 | 581 | 241 | 45 | 9 | 28 | 184 | 528 | 982 | 1441 | 8049 |
| Cooling Degree Days | 0 | 0 | 0 | 6 | 36 | 141 | 229 | 167 | 50 | 2 | 0 | 0 | 631 |
| Total Precipitation (") | 0.67 | 0.69 | 1.80 | 2.76 | 3.36 | 4.72 | 3.74 | 3.73 | 3.07 | 1.93 | 1.43 | 0.93 | 28.83 |
| Days ≥ 0.1" Precip | 2 | 2 | 4 | 6 | 7 | 7 | 5 | 5 | 6 | 4 | 3 | 2 | 53 |
| Total Snowfall (") | 7.4 | 6.0 | 7.8 | 2.5 | 0.0 | 0.0 | 0.0 | 0.0 | 0.0 | 0.6 | 5.4 | 8.8 | 38.5 |
| Days ≥ 1" Snow Depth | na | na | na | 1 | 0 | 0 | 0 | 0 | 0 | 0 | ***2*** | na | na |

## LAMONI *Decatur County* ELEVATION 1142 ft LAT/LONG 40° 37 ' N / 93° 57 ' W

| | JAN | FEB | MAR | APR | MAY | JUN | JUL | AUG | SEP | OCT | NOV | DEC | YEAR |
|---|---|---|---|---|---|---|---|---|---|---|---|---|---|
| Maximum Temp °F | 31.0 | 36.8 | 49.8 | 62.4 | 72.7 | 82.1 | 86.6 | 84.4 | 75.9 | 64.0 | 48.4 | 36.1 | 60.9 |
| Minimum Temp °F | 13.1 | 17.9 | 29.4 | 40.9 | 51.5 | 61.1 | 65.7 | 63.1 | 54.8 | 42.6 | 30.5 | 19.4 | 40.8 |
| Mean Temp °F | 22.1 | 27.4 | 39.6 | 51.7 | 62.1 | 71.6 | 76.2 | 73.8 | 65.3 | 53.3 | 39.5 | 27.7 | 50.9 |
| Days Max Temp ≥ 90 °F | 0 | 0 | 0 | 0 | 0 | 5 | 11 | 8 | 2 | 0 | 0 | 0 | 26 |
| Days Max Temp ≤ 32 °F | 16 | 10 | 2 | 0 | 0 | 0 | 0 | 0 | 0 | 0 | 2 | 11 | 41 |
| Days Min Temp ≤ 32 °F | 30 | 25 | 19 | 5 | 0 | 0 | 0 | 0 | 0 | 4 | 18 | 28 | 129 |
| Days Min Temp ≤ 0 °F | 7 | 3 | 0 | 0 | 0 | 0 | 0 | 0 | 0 | 0 | 0 | 3 | 13 |
| Heating Degree Days | 1324 | 1055 | 781 | 407 | 144 | 15 | 1 | 6 | 95 | 368 | 760 | ***1147*** | 6103 |
| Cooling Degree Days | 0 | 0 | 1 | 18 | 60 | 232 | 371 | 300 | 121 | 8 | 0 | 0 | 1111 |
| Total Precipitation (") | 1.00 | 1.12 | 2.57 | 3.91 | 4.24 | 4.17 | 5.26 | 4.46 | 4.61 | 3.14 | 2.35 | 1.58 | 38.41 |
| Days ≥ 0.1" Precip | 3 | 3 | 5 | 7 | 7 | 7 | 6 | 6 | 6 | 5 | 5 | 3 | 63 |
| Total Snowfall (") | 5.8 | 5.2 | 3.3 | 1.5 | 0.1 | 0.0 | 0.0 | 0.0 | 0.0 | 0.3 | 2.0 | 4.7 | 22.9 |
| Days ≥ 1" Snow Depth | 17 | 12 | 4 | 1 | 0 | 0 | 0 | 0 | 0 | 0 | 2 | ***8*** | 44 |

## LE CLAIRE L & D 14 *Scott County* ELEVATION 581 ft LAT/LONG 41° 35 ' N / 90° 25 ' W

| | JAN | FEB | MAR | APR | MAY | JUN | JUL | AUG | SEP | OCT | NOV | DEC | YEAR |
|---|---|---|---|---|---|---|---|---|---|---|---|---|---|
| Maximum Temp °F | 29.0 | 33.8 | 46.1 | 60.4 | 72.1 | 81.1 | 84.6 | 82.4 | 75.0 | 63.2 | 47.4 | 34.4 | 59.1 |
| Minimum Temp °F | 12.1 | 16.8 | 28.5 | 40.7 | 52.1 | 61.8 | 66.4 | 63.7 | 55.7 | 44.0 | 31.4 | 19.3 | 41.0 |
| Mean Temp °F | 20.6 | 25.3 | 37.3 | 50.6 | 62.1 | 71.5 | 75.5 | 73.1 | 65.4 | 53.6 | 39.5 | 26.9 | 50.1 |
| Days Max Temp ≥ 90 °F | 0 | 0 | 0 | 0 | 0 | 3 | 7 | 4 | 1 | 0 | 0 | 0 | 15 |
| Days Max Temp ≤ 32 °F | 18 | 13 | 4 | 0 | 0 | 0 | 0 | 0 | 0 | 0 | 2 | 12 | 49 |
| Days Min Temp ≤ 32 °F | 30 | 26 | 21 | 5 | 0 | 0 | 0 | 0 | 0 | 3 | 16 | 28 | 129 |
| Days Min Temp ≤ 0 °F | 8 | 4 | 0 | 0 | 0 | 0 | 0 | 0 | 0 | 0 | 0 | 3 | 15 |
| Heating Degree Days | 1373 | 1115 | 853 | 438 | 153 | 15 | 2 | 7 | 88 | 361 | 761 | 1178 | 6344 |
| Cooling Degree Days | 0 | 0 | 0 | 11 | 70 | 218 | 340 | 271 | 106 | 7 | 0 | 0 | 1023 |
| Total Precipitation (") | 1.11 | 1.02 | 2.32 | 3.14 | 3.63 | 4.73 | 3.93 | 4.25 | 3.32 | 2.56 | 2.40 | 2.07 | 34.48 |
| Days ≥ 0.1" Precip | 3 | 3 | 5 | 7 | 7 | 7 | 6 | 6 | 5 | 5 | 5 | 4 | 63 |
| Total Snowfall (") | na | ***4.0*** | 2.7 | 0.8 | 0.0 | 0.0 | 0.0 | 0.0 | 0.0 | 0.2 | ***0.4*** | 3.5 | na |
| Days ≥ 1" Snow Depth | ***18*** | 13 | 5 | 0 | 0 | 0 | 0 | 0 | 0 | 0 | ***1*** | 8 | 45 |

## LE MARS *Plymouth County* ELEVATION 1230 ft LAT/LONG 42° 48 ' N / 96° 10 ' W

| | JAN | FEB | MAR | APR | MAY | JUN | JUL | AUG | SEP | OCT | NOV | DEC | YEAR |
|---|---|---|---|---|---|---|---|---|---|---|---|---|---|
| Maximum Temp °F | 26.8 | 32.7 | 46.5 | 62.4 | 74.4 | 83.6 | 86.9 | 84.5 | 76.3 | 63.9 | 44.6 | 31.0 | 59.5 |
| Minimum Temp °F | 6.6 | 12.2 | 24.4 | 36.3 | 48.4 | 58.6 | 62.9 | 60.3 | 50.6 | 38.1 | 24.7 | 12.5 | 36.3 |
| Mean Temp °F | 16.7 | 22.5 | 35.5 | 49.4 | 61.4 | 71.1 | 74.9 | 72.4 | 63.5 | 51.0 | 34.7 | 21.8 | 47.9 |
| Days Max Temp ≥ 90 °F | 0 | 0 | 0 | 0 | 2 | 7 | 11 | 8 | 2 | 0 | 0 | 0 | 30 |
| Days Max Temp ≤ 32 °F | 19 | 13 | 4 | 0 | 0 | 0 | 0 | 0 | 0 | 0 | 5 | 16 | 57 |
| Days Min Temp ≤ 32 °F | 31 | 27 | 24 | 11 | 1 | 0 | 0 | 0 | 1 | 10 | 24 | 30 | 159 |
| Days Min Temp ≤ 0 °F | 11 | 7 | 1 | 0 | 0 | 0 | 0 | 0 | 0 | 0 | 1 | 6 | 26 |
| Heating Degree Days | 1491 | 1194 | 908 | 472 | 170 | 21 | 4 | 11 | 128 | 435 | 903 | 1334 | 7071 |
| Cooling Degree Days | 0 | 0 | 0 | 15 | 65 | 210 | 310 | 247 | 95 | 4 | 0 | 0 | 946 |
| Total Precipitation (") | 0.62 | 0.60 | 1.88 | 2.56 | 3.36 | 3.98 | 3.20 | 3.43 | 2.93 | 2.09 | 1.28 | 0.84 | 26.77 |
| Days ≥ 0.1" Precip | 2 | 2 | 4 | 6 | 7 | 6 | 5 | 5 | 5 | 4 | 3 | 2 | 51 |
| Total Snowfall (") | 7.1 | 5.1 | 6.6 | 1.4 | 0.0 | 0.0 | 0.0 | 0.0 | 0.0 | 0.7 | 3.9 | 7.1 | 31.9 |
| Days ≥ 1" Snow Depth | ***19*** | ***15*** | 10 | 1 | 0 | 0 | 0 | 0 | 0 | 0 | 4 | 12 | 61 |

## LEON 6 ESE *Decatur County* ELEVATION 991 ft LAT/LONG 40° 44 ' N / 93° 39 ' W

| | JAN | FEB | MAR | APR | MAY | JUN | JUL | AUG | SEP | OCT | NOV | DEC | YEAR |
|---|---|---|---|---|---|---|---|---|---|---|---|---|---|
| Maximum Temp °F | 31.5 | 37.4 | 50.3 | 63.8 | 73.8 | 82.6 | 87.4 | 85.1 | 77.0 | 65.3 | 48.9 | 36.1 | 61.6 |
| Minimum Temp °F | 12.3 | 16.6 | 28.0 | 39.3 | 49.4 | 58.7 | 64.0 | 61.1 | 52.7 | 41.1 | 29.4 | 17.8 | 39.2 |
| Mean Temp °F | 21.9 | 27.0 | 39.2 | 51.6 | 61.7 | 70.7 | 75.7 | 73.1 | 64.9 | 53.3 | 39.1 | 27.0 | 50.4 |
| Days Max Temp ≥ 90 °F | 0 | 0 | 0 | 0 | 0 | 4 | 11 | 8 | 2 | 0 | 0 | 0 | 25 |
| Days Max Temp ≤ 32 °F | 15 | 10 | 3 | 0 | 0 | 0 | 0 | 0 | 0 | 0 | 3 | 11 | 42 |
| Days Min Temp ≤ 32 °F | 30 | 26 | 21 | 8 | 1 | 0 | 0 | 0 | 1 | 7 | 19 | 29 | 142 |
| Days Min Temp ≤ 0 °F | 7 | 4 | 0 | 0 | 0 | 0 | 0 | 0 | 0 | 0 | 0 | 3 | 14 |
| Heating Degree Days | 1330 | 1065 | 794 | 411 | 155 | 20 | 2 | 10 | 103 | 371 | 769 | 1172 | 6202 |
| Cooling Degree Days | 0 | 0 | 1 | 18 | 46 | 185 | 326 | 266 | 107 | 8 | 0 | 0 | 957 |
| Total Precipitation (") | 1.05 | 1.02 | 2.26 | 3.85 | 4.39 | 4.29 | 4.73 | 4.36 | 4.28 | 2.97 | 2.22 | 1.47 | 36.89 |
| Days ≥ 0.1" Precip | 3 | 3 | 5 | 7 | 8 | 7 | 7 | 6 | 6 | 5 | 5 | 4 | 66 |
| Total Snowfall (") | 6.0 | 5.3 | 3.5 | 1.2 | 0.0 | 0.0 | 0.0 | 0.0 | 0.0 | 0.0 | 2.5 | 5.3 | 23.8 |
| Days ≥ 1" Snow Depth | 17 | 11 | 4 | 1 | 0 | 0 | 0 | 0 | 0 | 0 | 2 | 10 | 45 |

## LOGAN *Harrison County* ELEVATION 1122 ft LAT/LONG 41° 38 ' N / 95° 48 ' W

| | JAN | FEB | MAR | APR | MAY | JUN | JUL | AUG | SEP | OCT | NOV | DEC | YEAR |
|---|---|---|---|---|---|---|---|---|---|---|---|---|---|
| Maximum Temp °F | 30.5 | 36.2 | 49.3 | 64.3 | 75.0 | 84.2 | 87.6 | 85.2 | 77.0 | 65.5 | 47.8 | 34.5 | 61.4 |
| Minimum Temp °F | 9.6 | 15.1 | 26.9 | 38.9 | 49.8 | 59.5 | 64.0 | 61.3 | 52.1 | 39.7 | 27.1 | 15.2 | 38.3 |
| Mean Temp °F | 20.1 | 25.7 | 38.1 | 51.6 | 62.4 | 71.9 | 75.8 | 73.3 | 64.6 | 52.6 | 37.5 | 24.9 | 49.9 |
| Days Max Temp ≥ 90 °F | 0 | 0 | 0 | 1 | 2 | 8 | 12 | 9 | 3 | 0 | 0 | 0 | 35 |
| Days Max Temp ≤ 32 °F | 16 | 11 | 3 | 0 | 0 | 0 | 0 | 0 | 0 | 0 | 3 | 13 | 46 |
| Days Min Temp ≤ 32 °F | 30 | 27 | 22 | 8 | 1 | 0 | 0 | 0 | 1 | 8 | 22 | 30 | 149 |
| Days Min Temp ≤ 0 °F | 9 | 5 | 1 | 0 | 0 | 0 | 0 | 0 | 0 | 0 | 1 | 4 | 20 |
| Heating Degree Days | 1387 | 1103 | 827 | 411 | 147 | 17 | 3 | 10 | 109 | 388 | 819 | 1237 | 6458 |
| Cooling Degree Days | 0 | 0 | 1 | 23 | 69 | 228 | 335 | 266 | 112 | 5 | 0 | 0 | 1039 |
| Total Precipitation (") | 0.85 | 0.79 | 2.14 | 2.93 | 4.33 | 4.47 | 3.81 | 3.32 | 3.89 | 2.59 | 1.54 | 1.11 | 31.77 |
| Days ≥ 0.1" Precip | 2 | 2 | 5 | 7 | 9 | 7 | 6 | 6 | 6 | 5 | 4 | 3 | 62 |
| Total Snowfall (") | 7.2 | 7.1 | 5.6 | 1.4 | 0.0 | 0.0 | 0.0 | 0.0 | 0.0 | 0.7 | 3.7 | 7.1 | 32.8 |
| Days ≥ 1" Snow Depth | 19 | 15 | 7 | 1 | 0 | 0 | 0 | 0 | 0 | 0 | 4 | 12 | 58 |

## MAPLETON NO 2 *Monona County* ELEVATION 1190 ft LAT/LONG 42° 10 ' N / 95° 47 ' W

| | JAN | FEB | MAR | APR | MAY | JUN | JUL | AUG | SEP | OCT | NOV | DEC | YEAR |
|---|---|---|---|---|---|---|---|---|---|---|---|---|---|
| Maximum Temp °F | 28.5 | 34.4 | 47.2 | 63.2 | 73.8 | 82.1 | 85.0 | 83.0 | 75.4 | 64.1 | 45.9 | 32.5 | 59.6 |
| Minimum Temp °F | 8.9 | 14.3 | 25.8 | 38.1 | 49.3 | 59.1 | 63.4 | 61.1 | 51.9 | 39.7 | 26.6 | 14.3 | 37.7 |
| Mean Temp °F | 18.7 | 24.4 | 36.5 | 50.6 | 61.6 | 70.6 | 74.2 | 72.1 | 63.7 | 51.9 | 36.3 | 23.4 | 48.7 |
| Days Max Temp ≥ 90 °F | 0 | 0 | 0 | 0 | 1 | 4 | 8 | 6 | 1 | 0 | 0 | 0 | 20 |
| Days Max Temp ≤ 32 °F | 18 | 12 | 4 | 0 | 0 | 0 | 0 | 0 | 0 | 0 | 4 | 14 | 52 |
| Days Min Temp ≤ 32 °F | 30 | 27 | 22 | 9 | 1 | 0 | 0 | 0 | 1 | 8 | 22 | 30 | 150 |
| Days Min Temp ≤ 0 °F | 10 | 6 | 1 | 0 | 0 | 0 | 0 | 0 | 0 | 0 | 1 | 5 | 23 |
| Heating Degree Days | 1429 | 1141 | 877 | 437 | 163 | 21 | 4 | 12 | 122 | 409 | 856 | 1283 | 6754 |
| Cooling Degree Days | 0 | 0 | 0 | 20 | 64 | 208 | 304 | 250 | 100 | 7 | 0 | 0 | 953 |
| Total Precipitation (") | 0.65 | 0.69 | 2.03 | 2.80 | 3.96 | 4.36 | 3.45 | 3.48 | 3.16 | 2.40 | 1.39 | 0.92 | 29.29 |
| Days ≥ 0.1" Precip | 2 | 2 | 4 | 6 | 7 | 7 | 6 | 6 | 6 | 4 | 3 | 3 | 56 |
| Total Snowfall (") | 7.0 | 6.7 | 5.8 | 1.0 | 0.0 | 0.0 | 0.0 | 0.0 | 0.0 | 0.7 | 3.6 | 7.2 | 32.0 |
| Days ≥ 1" Snow Depth | 19 | 14 | 8 | 1 | 0 | 0 | 0 | 0 | 0 | 0 | 4 | 15 | 61 |

## MAQUOKETA 3 S *Clinton County* ELEVATION 680 ft LAT/LONG 42° 4 ' N / 90° 40 ' W

| | JAN | FEB | MAR | APR | MAY | JUN | JUL | AUG | SEP | OCT | NOV | DEC | YEAR |
|---|---|---|---|---|---|---|---|---|---|---|---|---|---|
| Maximum Temp °F | 27.4 | 32.7 | 45.6 | 60.3 | 72.2 | 81.1 | 84.7 | 82.2 | 74.4 | 62.5 | 46.2 | 33.0 | 58.5 |
| Minimum Temp °F | 9.4 | 14.0 | 26.3 | 37.5 | 48.0 | 57.5 | 61.8 | 59.2 | 50.9 | 39.4 | 28.3 | 16.8 | 37.4 |
| Mean Temp °F | 18.4 | 23.4 | 36.0 | 48.9 | 60.1 | 69.3 | 73.4 | 70.7 | 62.7 | 51.0 | 37.2 | 24.9 | 48.0 |
| Days Max Temp ≥ 90 °F | 0 | 0 | 0 | 0 | 1 | 4 | 8 | 4 | 1 | 0 | 0 | 0 | 18 |
| Days Max Temp ≤ 32 °F | 19 | 13 | 4 | 0 | 0 | 0 | 0 | 0 | 0 | 0 | 3 | 13 | 52 |
| Days Min Temp ≤ 32 °F | 30 | 27 | 23 | 10 | 1 | 0 | 0 | 0 | 1 | 8 | 20 | 29 | 149 |
| Days Min Temp ≤ 0 °F | 9 | 6 | 1 | 0 | 0 | 0 | 0 | 0 | 0 | 0 | 0 | 4 | 20 |
| Heating Degree Days | 1437 | 1167 | 892 | 485 | 195 | 28 | 6 | 17 | 134 | 438 | 827 | 1236 | 6862 |
| Cooling Degree Days | 0 | 0 | 1 | 10 | 43 | 148 | 256 | 188 | 68 | 3 | 0 | 0 | 717 |
| Total Precipitation (") | 1.18 | 1.11 | 2.33 | 3.27 | 3.76 | 4.65 | 3.42 | 4.58 | 4.22 | 2.46 | 2.44 | 1.88 | 35.30 |
| Days ≥ 0.1" Precip | 3 | 3 | 5 | 7 | 8 | 7 | 7 | 7 | 7 | 6 | 5 | 4 | 69 |
| Total Snowfall (") | 7.9 | 5.6 | 3.7 | 1.6 | 0.1 | 0.0 | 0.0 | 0.0 | 0.0 | 0.1 | 1.8 | 6.0 | 26.8 |
| Days ≥ 1" Snow Depth | *17* | *13* | *4* | 0 | 0 | 0 | 0 | 0 | 0 | 0 | 1 | *11* | 46 |

## MARSHALLTOWN *Marshall County* ELEVATION 870 ft LAT/LONG 42° 4 ' N / 92° 56 ' W

| | JAN | FEB | MAR | APR | MAY | JUN | JUL | AUG | SEP | OCT | NOV | DEC | YEAR |
|---|---|---|---|---|---|---|---|---|---|---|---|---|---|
| Maximum Temp °F | 26.8 | 32.3 | 45.2 | 60.0 | 71.7 | 81.1 | 84.6 | 82.1 | 74.4 | 62.4 | 45.8 | 31.8 | 58.2 |
| Minimum Temp °F | 6.9 | 12.4 | 25.5 | 37.7 | 48.6 | 58.4 | 62.3 | 58.9 | 50.0 | 38.4 | 26.5 | 14.1 | 36.6 |
| Mean Temp °F | 16.9 | 22.4 | 35.4 | 48.9 | 60.2 | 69.8 | 73.5 | 70.5 | 62.1 | 50.5 | 36.2 | 23.0 | 47.5 |
| Days Max Temp ≥ 90 °F | 0 | 0 | 0 | 0 | 1 | 3 | 7 | 5 | 1 | 0 | 0 | 0 | 17 |
| Days Max Temp ≤ 32 °F | 19 | 14 | 4 | 0 | 0 | 0 | 0 | 0 | 0 | 0 | 4 | 15 | 56 |
| Days Min Temp ≤ 32 °F | 31 | 27 | 24 | 9 | 1 | 0 | 0 | 0 | 1 | 10 | 22 | 30 | 155 |
| Days Min Temp ≤ 0 °F | 11 | 6 | 1 | 0 | 0 | 0 | 0 | 0 | 0 | 0 | 0 | 5 | 23 |
| Heating Degree Days | 1486 | 1198 | 913 | 488 | 195 | 26 | 5 | 21 | 146 | 453 | 858 | 1296 | 7085 |
| Cooling Degree Days | 0 | 0 | 1 | 13 | 48 | 178 | 276 | 212 | 78 | 6 | 0 | 0 | 812 |
| Total Precipitation (") | 0.94 | 1.02 | 2.48 | 3.28 | 4.01 | 5.08 | 4.36 | 4.97 | 3.73 | 2.58 | 2.00 | 1.34 | 35.79 |
| Days ≥ 0.1" Precip | 3 | 3 | 5 | 7 | 8 | 8 | 7 | 7 | 6 | 5 | 4 | 4 | 67 |
| Total Snowfall (") | 6.7 | 6.6 | 4.6 | 1.2 | 0.0 | 0.0 | 0.0 | 0.0 | 0.0 | 0.2 | 2.1 | 6.6 | 28.0 |
| Days ≥ 1" Snow Depth | 17 | 15 | 7 | 1 | 0 | 0 | 0 | 0 | 0 | 0 | 2 | 12 | 54 |

## MASON CITY *Cerro Gordo County* ELEVATION 1132 ft LAT/LONG 43° 9 ' N / 93° 12 ' W

| | JAN | FEB | MAR | APR | MAY | JUN | JUL | AUG | SEP | OCT | NOV | DEC | YEAR |
|---|---|---|---|---|---|---|---|---|---|---|---|---|---|
| Maximum Temp °F | 23.2 | 29.2 | 41.8 | 58.2 | 71.4 | 80.5 | 83.5 | 81.2 | 73.4 | 60.7 | 42.5 | 28.3 | 56.2 |
| Minimum Temp °F | 4.1 | 10.4 | 22.8 | 35.6 | 47.0 | 56.5 | 61.0 | 57.9 | 49.5 | 37.9 | 24.5 | 11.3 | 34.9 |
| Mean Temp °F | 13.7 | 19.8 | 32.4 | 46.9 | 59.2 | 68.5 | 72.3 | 69.6 | 61.5 | 49.3 | 33.6 | 19.8 | 45.5 |
| Days Max Temp ≥ 90 °F | 0 | 0 | 0 | 0 | 1 | 4 | 6 | 4 | 1 | 0 | 0 | 0 | 16 |
| Days Max Temp ≤ 32 °F | 22 | 16 | 7 | 0 | 0 | 0 | 0 | 0 | 0 | 0 | 6 | 19 | 70 |
| Days Min Temp ≤ 32 °F | 31 | 28 | 26 | 12 | 2 | 0 | 0 | 0 | 1 | 10 | 24 | 30 | 164 |
| Days Min Temp ≤ 0 °F | 13 | 8 | 1 | 0 | 0 | 0 | 0 | 0 | 0 | 0 | 1 | 7 | 30 |
| Heating Degree Days | 1587 | 1270 | 1006 | 544 | 220 | 39 | 10 | 28 | 161 | 486 | 937 | 1395 | 7683 |
| Cooling Degree Days | 0 | 0 | 0 | 10 | 40 | 146 | 238 | 177 | 66 | 2 | 0 | 0 | 679 |
| Total Precipitation (") | 0.85 | 0.77 | 1.97 | 3.39 | 4.24 | 5.18 | 4.48 | 4.48 | 3.97 | 2.58 | 1.91 | 1.19 | 35.01 |
| Days ≥ 0.1" Precip | 3 | 2 | 5 | 7 | 8 | 8 | 7 | 6 | 7 | 5 | 4 | 3 | 65 |
| Total Snowfall (") | ***8.2*** | ***5.6*** | 5.7 | 0.8 | 0.0 | 0.0 | 0.0 | 0.0 | 0.0 | 0.2 | ***2.6*** | na | na |
| Days ≥ 1" Snow Depth | ***22*** | ***18*** | ***9*** | 1 | 0 | 0 | 0 | 0 | 0 | 0 | ***2*** | na | na |

## MASON CITY AP *Cerro Gordo County* ELEVATION 1211 ft LAT/LONG 43° 9 ' N / 93° 20 ' W

| | JAN | FEB | MAR | APR | MAY | JUN | JUL | AUG | SEP | OCT | NOV | DEC | YEAR |
|---|---|---|---|---|---|---|---|---|---|---|---|---|---|
| Maximum Temp °F | 22.0 | 27.5 | 40.7 | 57.1 | 70.4 | 79.8 | 83.2 | 80.6 | 72.3 | 59.7 | 41.9 | 27.1 | 55.2 |
| Minimum Temp °F | 4.1 | 10.2 | 23.3 | 35.8 | 47.0 | 57.1 | 61.3 | 58.4 | 48.9 | 37.4 | 24.3 | 11.0 | 34.9 |
| Mean Temp °F | 13.1 | 18.9 | 32.1 | 46.5 | 58.7 | 68.5 | 72.3 | 69.6 | 60.6 | 48.6 | 33.2 | 19.1 | 45.1 |
| Days Max Temp ≥ 90 °F | 0 | 0 | 0 | 0 | 1 | 4 | 6 | 4 | 1 | 0 | 0 | 0 | 16 |
| Days Max Temp ≤ 32 °F | 23 | 17 | 8 | 1 | 0 | 0 | 0 | 0 | 0 | 0 | 6 | 20 | 75 |
| Days Min Temp ≤ 32 °F | 30 | 27 | 25 | 11 | 1 | 0 | 0 | 0 | 1 | 10 | 24 | 30 | 159 |
| Days Min Temp ≤ 0 °F | 12 | 8 | 1 | 0 | 0 | 0 | 0 | 0 | 0 | 0 | 1 | 7 | 29 |
| Heating Degree Days | 1606 | 1297 | 1015 | 554 | 229 | 39 | 9 | 30 | 177 | 508 | 949 | 1417 | 7830 |
| Cooling Degree Days | 0 | 0 | 0 | 9 | 40 | 153 | 242 | 181 | 56 | 3 | 0 | 0 | 684 |
| Total Precipitation (") | 0.98 | 0.87 | 2.08 | 3.24 | 4.30 | 4.94 | 4.18 | 4.12 | 3.63 | 2.55 | 1.81 | 1.29 | 33.99 |
| Days ≥ 0.1" Precip | 3 | 3 | 5 | 7 | 8 | 8 | 6 | 6 | 6 | 5 | 4 | 4 | 65 |
| Total Snowfall (") | 9.4 | 6.9 | 6.2 | 2.1 | 0.0 | 0.0 | 0.0 | 0.0 | 0.0 | 0.5 | 4.3 | 9.1 | 38.5 |
| Days ≥ 1" Snow Depth | 24 | 21 | 11 | 2 | 0 | 0 | 0 | 0 | 0 | 0 | 4 | 17 | 79 |

## MILFORD 4 NW *Dickinson County* ELEVATION 1401 ft LAT/LONG 43° 23 ' N / 95° 11 ' W

| | JAN | FEB | MAR | APR | MAY | JUN | JUL | AUG | SEP | OCT | NOV | DEC | YEAR |
|---|---|---|---|---|---|---|---|---|---|---|---|---|---|
| Maximum Temp °F | 22.5 | 28.5 | 41.5 | 58.0 | 71.4 | 80.3 | 83.8 | 81.0 | 72.3 | 60.1 | 41.3 | 27.2 | 55.7 |
| Minimum Temp °F | 3.7 | 9.8 | 22.5 | 35.2 | 47.1 | 56.8 | 61.2 | 58.4 | 49.6 | 37.9 | 23.8 | 10.4 | 34.7 |
| Mean Temp °F | 13.1 | 19.2 | 32.0 | 46.6 | 59.3 | 68.6 | 72.5 | 69.8 | 60.9 | 49.0 | 32.6 | 18.8 | 45.2 |
| Days Max Temp ≥ 90 °F | 0 | 0 | 0 | 0 | 1 | 3 | 6 | 3 | 1 | 0 | 0 | 0 | 14 |
| Days Max Temp ≤ 32 °F | 23 | 17 | 7 | 0 | 0 | 0 | 0 | 0 | 0 | 0 | 7 | 20 | 74 |
| Days Min Temp ≤ 32 °F | 31 | 28 | 25 | 12 | 2 | 0 | 0 | 0 | 1 | 9 | 25 | 31 | 164 |
| Days Min Temp ≤ 0 °F | 14 | 8 | 2 | 0 | 0 | 0 | 0 | 0 | 0 | 0 | 1 | 7 | 32 |
| Heating Degree Days | 1606 | 1288 | 1015 | 551 | 214 | 37 | 8 | 23 | 168 | 493 | 966 | 1426 | 7795 |
| Cooling Degree Days | 0 | 0 | 0 | 8 | 45 | 149 | 229 | 166 | 51 | 2 | 0 | 0 | 650 |
| Total Precipitation (") | 0.57 | 0.69 | 1.84 | 3.09 | 3.64 | 4.74 | 3.74 | 3.82 | 3.24 | 2.19 | 1.61 | 0.83 | 30.00 |
| Days ≥ 0.1" Precip | 2 | 2 | 4 | 6 | 7 | 7 | 6 | 5 | 6 | 4 | 4 | 2 | 55 |
| Total Snowfall (") | na | ***6.5*** | 6.2 | 0.6 | 0.0 | 0.0 | 0.0 | 0.0 | 0.0 | 0.5 | ***2.8*** | ***8.2*** | na |
| Days ≥ 1" Snow Depth | na | na | na | ***0*** | 0 | 0 | 0 | 0 | 0 | 0 | ***3*** | na | na |

## MOUNT AYR 4 SW *Ringgold County* ELEVATION 1270 ft LAT/LONG 40° 39 ' N / 94° 18 ' W

| | JAN | FEB | MAR | APR | MAY | JUN | JUL | AUG | SEP | OCT | NOV | DEC | YEAR |
|---|---|---|---|---|---|---|---|---|---|---|---|---|---|
| Maximum Temp °F | ***30.8*** | 36.7 | 49.7 | 62.9 | 72.3 | 81.1 | 85.5 | ***83.2*** | 75.9 | 65.0 | 48.7 | ***36.2*** | 60.7 |
| Minimum Temp °F | ***11.8*** | 16.3 | 28.0 | 39.6 | 50.2 | 59.9 | 64.4 | 61.6 | 53.4 | 41.6 | 29.3 | ***18.7*** | 39.6 |
| Mean Temp °F | ***21.3*** | 26.6 | 38.8 | 51.3 | 61.2 | 70.5 | 74.9 | ***72.4*** | 64.7 | 53.3 | 39.0 | ***27.4*** | 50.1 |
| Days Max Temp ≥ 90 °F | 0 | 0 | 0 | 0 | 0 | 3 | 8 | 5 | 1 | 0 | 0 | 0 | 17 |
| Days Max Temp ≤ 32 °F | ***16*** | 11 | 3 | 0 | 0 | 0 | 0 | 0 | 0 | 0 | 3 | ***10*** | 43 |
| Days Min Temp ≤ 32 °F | ***30*** | 26 | 21 | 7 | 1 | 0 | 0 | 0 | 0 | 6 | 19 | ***28*** | 138 |
| Days Min Temp ≤ 0 °F | ***7*** | 4 | 0 | 0 | 0 | 0 | 0 | 0 | 0 | 0 | 0 | 3 | 14 |
| Heating Degree Days | ***1349*** | 1078 | 806 | 418 | 163 | 20 | 3 | ***11*** | 105 | 366 | 772 | ***1158*** | 6249 |
| Cooling Degree Days | 0 | 0 | 1 | 16 | 46 | 202 | 311 | ***238*** | 110 | 8 | 0 | 0 | 932 |
| Total Precipitation (") | 0.82 | 0.84 | 2.26 | 3.19 | 4.13 | 4.48 | 5.14 | 4.34 | 4.11 | 2.98 | 2.20 | 1.39 | 35.88 |
| Days ≥ 0.1" Precip | ***2*** | 3 | 5 | 6 | 7 | 6 | 6 | 6 | 6 | 5 | 4 | ***3*** | 59 |
| Total Snowfall (") | ***4.1*** | ***4.5*** | ***2.4*** | 0.7 | 0.0 | 0.0 | 0.0 | 0.0 | 0.0 | 0.2 | ***1.4*** | na | na |
| Days ≥ 1" Snow Depth | na | na | ***1*** | 0 | 0 | 0 | 0 | 0 | 0 | 0 | ***1*** | na | na |

## MOUNT PLEASANT 1 SSW *Henry County* ELEVATION 732 ft LAT/LONG 40° 57 ' N / 91° 32 ' W

| | JAN | FEB | MAR | APR | MAY | JUN | JUL | AUG | SEP | OCT | NOV | DEC | YEAR |
|---|---|---|---|---|---|---|---|---|---|---|---|---|---|
| Maximum Temp °F | 30.6 | 36.0 | 49.1 | 63.0 | 73.1 | 82.0 | 86.0 | 83.5 | 76.3 | 64.9 | 49.1 | 35.6 | 60.8 |
| Minimum Temp °F | 12.6 | 17.3 | 28.9 | 40.6 | 50.7 | 60.1 | 64.5 | 61.8 | 53.9 | 42.8 | 30.9 | 19.2 | 40.3 |
| Mean Temp °F | 21.6 | 26.7 | 39.0 | 51.8 | 61.9 | 71.1 | 75.2 | 72.7 | 65.2 | 53.9 | 40.0 | 27.4 | 50.5 |
| Days Max Temp ≥ 90 °F | 0 | 0 | 0 | 0 | 0 | 3 | 9 | 6 | 2 | 0 | 0 | 0 | 20 |
| Days Max Temp ≤ 32 °F | 16 | 11 | 3 | 0 | 0 | 0 | 0 | 0 | 0 | 0 | 2 | 11 | 43 |
| Days Min Temp ≤ 32 °F | 29 | 25 | 19 | 6 | 0 | 0 | 0 | 0 | 0 | 5 | 17 | 28 | 129 |
| Days Min Temp ≤ 0 °F | 7 | 4 | 0 | 0 | 0 | 0 | 0 | 0 | 0 | 0 | 0 | 3 | 14 |
| Heating Degree Days | 1338 | 1076 | 799 | 402 | 152 | 17 | 3 | 9 | 95 | 353 | 743 | 1159 | 6146 |
| Cooling Degree Days | 0 | 0 | 2 | 17 | 58 | 207 | 330 | 266 | 113 | 11 | 0 | 0 | 1004 |
| Total Precipitation (") | 1.35 | 1.11 | 2.56 | 3.33 | 3.95 | 4.17 | 4.96 | 4.49 | 4.78 | 2.84 | 2.65 | 2.00 | 38.19 |
| Days ≥ 0.1" Precip | 4 | 3 | 6 | 7 | 8 | 6 | 7 | 6 | 7 | 6 | 5 | 5 | 70 |
| Total Snowfall (") | 7.2 | 6.0 | 2.6 | 0.9 | 0.0 | 0.0 | 0.0 | 0.0 | 0.0 | 0.1 | 1.7 | 6.1 | 24.6 |
| Days ≥ 1" Snow Depth | ***12*** | ***9*** | 4 | 1 | 0 | 0 | 0 | 0 | 0 | 0 | 1 | ***6*** | 33 |

## MUSCATINE *Muscatine County* ELEVATION 549 ft LAT/LONG 41° 24 ' N / 91° 4 ' W

| | JAN | FEB | MAR | APR | MAY | JUN | JUL | AUG | SEP | OCT | NOV | DEC | YEAR |
|---|---|---|---|---|---|---|---|---|---|---|---|---|---|
| Maximum Temp °F | 30.2 | 36.0 | 48.7 | 62.9 | 73.9 | 82.7 | 86.2 | 83.9 | 76.6 | 65.0 | 48.7 | 35.1 | 60.8 |
| Minimum Temp °F | 12.4 | 17.2 | 28.8 | 40.4 | 51.2 | 60.7 | 65.3 | 62.8 | 54.4 | 42.8 | 31.1 | 19.2 | 40.5 |
| Mean Temp °F | 21.3 | 26.6 | 38.7 | 51.7 | 62.5 | 71.7 | 75.8 | 73.4 | 65.5 | 53.9 | 39.9 | 27.1 | 50.7 |
| Days Max Temp ≥ 90 °F | 0 | 0 | 0 | 0 | 1 | 4 | 9 | 5 | 2 | 0 | 0 | 0 | 21 |
| Days Max Temp ≤ 32 °F | 16 | 11 | 2 | 0 | 0 | 0 | 0 | 0 | 0 | 0 | 2 | 11 | 42 |
| Days Min Temp ≤ 32 °F | 30 | 25 | 20 | 6 | 0 | 0 | 0 | 0 | 0 | 5 | 17 | 28 | 131 |
| Days Min Temp ≤ 0 °F | 7 | 4 | 0 | 0 | 0 | 0 | 0 | 0 | 0 | 0 | 0 | 3 | 14 |
| Heating Degree Days | 1348 | 1077 | 808 | 407 | 145 | 14 | 2 | 5 | 88 | 352 | 745 | 1169 | 6160 |
| Cooling Degree Days | 0 | 0 | 2 | 18 | 77 | 235 | 362 | 289 | 114 | 8 | 1 | 0 | 1106 |
| Total Precipitation (") | 1.30 | 1.11 | 2.67 | 3.45 | 3.88 | 4.60 | 4.64 | 4.21 | 3.82 | 2.70 | 2.54 | 2.09 | 37.01 |
| Days ≥ 0.1" Precip | 3 | 3 | 6 | 7 | 7 | 7 | 7 | 6 | 6 | 5 | 5 | 4 | 66 |
| Total Snowfall (") | 7.7 | 6.1 | 3.1 | 0.8 | 0.0 | 0.0 | 0.0 | 0.0 | 0.0 | 0.3 | 2.0 | 6.6 | 26.6 |
| Days ≥ 1" Snow Depth | 19 | 14 | 4 | 0 | 0 | 0 | 0 | 0 | 0 | 0 | 1 | 11 | 49 |

## NEW HAMPTON *Chickasaw County* ELEVATION 1161 ft LAT/LONG 43° 4 ' N / 92° 18 ' W

| | JAN | FEB | MAR | APR | MAY | JUN | JUL | AUG | SEP | OCT | NOV | DEC | YEAR |
|---|---|---|---|---|---|---|---|---|---|---|---|---|---|
| Maximum Temp °F | 23.5 | 29.4 | 42.2 | 58.0 | 70.5 | 79.3 | 82.5 | 80.4 | 72.4 | 60.6 | 42.8 | 28.7 | 55.9 |
| Minimum Temp °F | 5.9 | 11.6 | 24.3 | 36.8 | 48.2 | 57.5 | 62.0 | 59.4 | 51.0 | 39.7 | 26.4 | 13.0 | 36.3 |
| Mean Temp °F | 14.6 | 20.5 | 33.3 | 47.4 | 59.4 | 68.4 | 72.2 | 69.9 | 61.7 | 50.1 | 34.6 | 20.8 | 46.1 |
| Days Max Temp ≥ 90 °F | 0 | 0 | 0 | 0 | 0 | 2 | 4 | 2 | 1 | 0 | 0 | 0 | 9 |
| Days Max Temp ≤ 32 °F | 22 | 16 | 6 | 0 | 0 | 0 | 0 | 0 | 0 | 0 | 5 | 19 | 68 |
| Days Min Temp ≤ 32 °F | 31 | 27 | 24 | 10 | 1 | 0 | 0 | 0 | 0 | 7 | 22 | 30 | 152 |
| Days Min Temp ≤ 0 °F | 11 | 7 | 1 | 0 | 0 | 0 | 0 | 0 | 0 | 0 | 1 | 6 | 26 |
| Heating Degree Days | 1558 | 1250 | 978 | 528 | 211 | 35 | 9 | 22 | 153 | 460 | 905 | 1365 | 7474 |
| Cooling Degree Days | 0 | 0 | 0 | 9 | 46 | 161 | 261 | 202 | 75 | 3 | 0 | 0 | 757 |
| Total Precipitation (") | 1.01 | 0.93 | 2.18 | 3.66 | 4.09 | 4.82 | 4.53 | 4.43 | 3.83 | 2.61 | 2.23 | 1.44 | 35.76 |
| Days ≥ 0.1" Precip | 3 | 3 | 5 | 7 | 7 | 8 | 7 | 6 | 6 | 5 | 4 | 4 | 65 |
| Total Snowfall (") | 9.3 | 6.9 | 6.9 | 2.4 | 0.0 | 0.0 | 0.0 | 0.0 | 0.0 | 0.2 | 4.7 | 9.4 | 39.8 |
| Days ≥ 1" Snow Depth | 26 | 23 | 12 | 2 | 0 | 0 | 0 | 0 | 0 | 0 | 5 | 20 | 88 |

## NEWTON *Jasper County* ELEVATION 951 ft LAT/LONG 41° 42 ' N / 93° 4 ' W

| | JAN | FEB | MAR | APR | MAY | JUN | JUL | AUG | SEP | OCT | NOV | DEC | YEAR |
|---|---|---|---|---|---|---|---|---|---|---|---|---|---|
| Maximum Temp °F | 28.3 | 34.2 | 47.6 | 62.5 | 73.6 | 82.5 | 86.0 | 83.6 | 76.0 | 63.9 | 46.7 | 33.3 | 59.9 |
| Minimum Temp °F | 10.4 | 15.3 | 27.4 | 39.5 | 50.1 | 59.8 | 64.1 | 61.6 | 53.3 | 41.6 | 28.7 | 16.9 | 39.1 |
| Mean Temp °F | 19.4 | 24.8 | 37.5 | 51.0 | 61.9 | 71.2 | 75.1 | 72.6 | 64.7 | 52.8 | 37.7 | 25.1 | 49.5 |
| Days Max Temp ≥ 90 °F | 0 | 0 | 0 | 0 | 1 | 4 | 9 | 6 | 1 | 0 | 0 | 0 | 21 |
| Days Max Temp ≤ 32 °F | 18 | 13 | 4 | 0 | 0 | 0 | 0 | 0 | 0 | 0 | 3 | 14 | 52 |
| Days Min Temp ≤ 32 °F | 30 | 26 | 21 | 7 | 0 | 0 | 0 | 0 | 0 | 6 | 20 | 29 | 139 |
| Days Min Temp ≤ 0 °F | 9 | 5 | 0 | 0 | 0 | 0 | 0 | 0 | 0 | 0 | 0 | 3 | 17 |
| Heating Degree Days | 1409 | 1129 | 846 | 424 | 153 | 16 | 3 | 8 | 102 | 385 | 811 | 1229 | 6515 |
| Cooling Degree Days | 0 | 0 | 2 | 17 | 68 | 221 | 336 | 278 | 115 | 8 | 0 | 0 | 1045 |
| Total Precipitation (") | 0.94 | 0.99 | 2.23 | 3.43 | 4.33 | 4.63 | 3.88 | 4.06 | 3.97 | 2.65 | 2.10 | 1.22 | 34.43 |
| Days ≥ 0.1" Precip | 2 | 3 | 5 | 6 | 8 | 7 | 6 | 6 | 6 | 5 | 4 | 3 | 61 |
| Total Snowfall (") | 6.1 | 6.4 | *3.8* | 1.3 | 0.0 | 0.0 | 0.0 | 0.0 | 0.0 | 0.2 | *2.2* | 6.1 | 26.1 |
| Days ≥ 1" Snow Depth | na | na | na | 0 | 0 | 0 | 0 | 0 | 0 | 0 | *1* | na | na |

## NORTHWOOD *Worth County* ELEVATION 1220 ft LAT/LONG 43° 27 ' N / 93° 13 ' W

| | JAN | FEB | MAR | APR | MAY | JUN | JUL | AUG | SEP | OCT | NOV | DEC | YEAR |
|---|---|---|---|---|---|---|---|---|---|---|---|---|---|
| Maximum Temp °F | 22.1 | 27.9 | 41.1 | 57.8 | 71.2 | 80.5 | 83.7 | 81.3 | 72.9 | 60.2 | 41.5 | 27.2 | 55.6 |
| Minimum Temp °F | 3.8 | 9.6 | 23.0 | 35.7 | 47.6 | 57.5 | 61.5 | 58.7 | 49.6 | 38.1 | 24.8 | 11.1 | 35.1 |
| Mean Temp °F | 13.0 | 18.8 | 32.1 | 46.8 | 59.4 | 69.0 | 72.7 | 70.0 | 61.2 | 49.2 | 33.2 | 19.2 | 45.4 |
| Days Max Temp ≥ 90 °F | 0 | 0 | 0 | 0 | 1 | 4 | 6 | 4 | 1 | 0 | 0 | 0 | 16 |
| Days Max Temp ≤ 32 °F | 24 | 17 | 7 | 1 | 0 | 0 | 0 | 0 | 0 | 0 | 7 | 21 | 77 |
| Days Min Temp ≤ 32 °F | 31 | 27 | 25 | 11 | 1 | 0 | 0 | 0 | 1 | 9 | 24 | 30 | 159 |
| Days Min Temp ≤ 0 °F | 13 | 8 | 2 | 0 | 0 | 0 | 0 | 0 | 0 | 0 | 1 | 7 | 31 |
| Heating Degree Days | 1609 | 1299 | 1014 | 546 | 213 | 34 | 8 | 22 | 163 | 489 | 948 | 1414 | 7759 |
| Cooling Degree Days | 0 | 0 | 0 | 9 | 45 | 166 | 245 | 181 | 60 | 2 | 0 | 0 | 708 |
| Total Precipitation (") | 0.92 | 0.72 | 2.01 | 3.17 | 3.81 | 4.39 | 4.29 | 4.46 | 3.93 | 2.27 | 1.80 | 1.21 | 32.98 |
| Days ≥ 0.1" Precip | 3 | 2 | 4 | 7 | 7 | 7 | 6 | 6 | 7 | 5 | 4 | 3 | 61 |
| Total Snowfall (") | 9.1 | 6.2 | 5.4 | 1.8 | 0.0 | 0.0 | 0.0 | 0.0 | 0.0 | 0.4 | 4.7 | 10.6 | 38.2 |
| Days ≥ 1" Snow Depth | na | na | na | *0* | 0 | 0 | 0 | 0 | 0 | 0 | *3* | na | na |

## OAKLAND 4 WSW *Pottawattamie County* ELEVATION 1150 ft LAT/LONG 41° 18 ' N / 95° 28 ' W

| | JAN | FEB | MAR | APR | MAY | JUN | JUL | AUG | SEP | OCT | NOV | DEC | YEAR |
|---|---|---|---|---|---|---|---|---|---|---|---|---|---|
| Maximum Temp °F | 30.5 | 35.9 | 49.0 | 63.4 | 73.8 | 83.0 | 85.9 | 83.3 | 76.6 | 64.9 | 47.5 | 34.3 | 60.7 |
| Minimum Temp °F | 9.8 | 15.0 | 26.4 | 38.0 | 49.3 | 59.1 | 63.1 | 60.2 | 51.7 | 39.4 | 26.9 | 15.5 | 37.9 |
| Mean Temp °F | 20.2 | 25.5 | 37.7 | 50.7 | 61.6 | 71.1 | 74.5 | 71.8 | 64.2 | 52.1 | 37.2 | 24.9 | 49.3 |
| Days Max Temp ≥ 90 °F | 0 | 0 | 0 | 0 | 1 | 5 | 8 | 5 | 2 | 0 | 0 | 0 | 21 |
| Days Max Temp ≤ 32 °F | 16 | 11 | 3 | 0 | 0 | 0 | 0 | 0 | 0 | 0 | 3 | 12 | 45 |
| Days Min Temp ≤ 32 °F | 30 | 26 | 22 | 9 | 1 | 0 | 0 | 0 | 1 | 9 | 21 | 30 | 149 |
| Days Min Temp ≤ 0 °F | 9 | 5 | 1 | 0 | 0 | 0 | 0 | 0 | 0 | 0 | 0 | 4 | 19 |
| Heating Degree Days | 1384 | 1110 | 838 | 435 | 161 | 18 | 5 | 13 | 109 | 403 | 826 | 1236 | 6538 |
| Cooling Degree Days | 0 | 0 | 1 | 16 | 55 | *215* | *318* | *235* | *107* | *6* | *0* | 0 | 953 |
| Total Precipitation (") | 0.77 | 0.77 | 2.26 | 3.01 | 4.18 | 4.86 | 3.95 | 3.86 | 3.86 | 2.67 | 1.53 | 1.04 | 32.76 |
| Days ≥ 0.1" Precip | 2 | 2 | 5 | 6 | 8 | 7 | 6 | 6 | 6 | 4 | 3 | 3 | 58 |
| Total Snowfall (") | 6.0 | 5.7 | 4.0 | 1.1 | 0.0 | 0.0 | 0.0 | 0.0 | 0.0 | 0.3 | 2.5 | 4.6 | 24.2 |
| Days ≥ 1" Snow Depth | *16* | 12 | 4 | 0 | 0 | 0 | 0 | 0 | 0 | 0 | 2 | 10 | 44 |

## OELWEIN 2 S *Fayette County* ELEVATION 1030 ft LAT/LONG 42° 40 ' N / 91° 55 ' W

| | JAN | FEB | MAR | APR | MAY | JUN | JUL | AUG | SEP | OCT | NOV | DEC | YEAR |
|---|---|---|---|---|---|---|---|---|---|---|---|---|---|
| Maximum Temp °F | 24.2 | 29.8 | 42.9 | 58.7 | 70.9 | 79.8 | 82.8 | 80.6 | 72.8 | 61.0 | 43.7 | 29.9 | 56.4 |
| Minimum Temp °F | 5.9 | 11.5 | 24.7 | 36.9 | 48.1 | 57.4 | 61.5 | 59.0 | 50.5 | 39.1 | 26.3 | 13.6 | 36.2 |
| Mean Temp °F | 15.0 | 20.7 | 33.8 | 47.8 | 59.5 | 68.6 | 72.2 | 69.8 | 61.7 | 50.1 | 35.0 | 21.7 | 46.3 |
| Days Max Temp ≥ 90 °F | 0 | 0 | 0 | 0 | 0 | 2 | 4 | 3 | 1 | 0 | 0 | 0 | 10 |
| Days Max Temp ≤ 32 °F | 22 | 15 | 6 | 0 | 0 | 0 | 0 | 0 | 0 | 0 | 5 | 17 | 65 |
| Days Min Temp ≤ 32 °F | 31 | 27 | 24 | 10 | 1 | 0 | 0 | 0 | 1 | 9 | 22 | 30 | 155 |
| Days Min Temp ≤ 0 °F | 12 | 7 | 1 | 0 | 0 | 0 | 0 | 0 | 0 | 0 | 1 | 6 | 27 |
| Heating Degree Days | 1546 | 1245 | 960 | 518 | 210 | 34 | 9 | 24 | 156 | 463 | 894 | 1336 | 7395 |
| Cooling Degree Days | 0 | 0 | 0 | 11 | 43 | 157 | 255 | 193 | 74 | 4 | 0 | 0 | 737 |
| Total Precipitation (") | 1.13 | 1.06 | 1.94 | 3.34 | 3.60 | 4.53 | 4.44 | 4.73 | 3.82 | 2.48 | 1.97 | 1.55 | 34.59 |
| Days ≥ 0.1" Precip | 4 | 3 | 5 | 7 | 7 | 7 | 7 | 6 | 7 | 5 | 5 | 4 | 67 |
| Total Snowfall (") | 8.1 | 5.9 | 4.8 | 1.1 | 0.0 | 0.0 | 0.0 | 0.0 | 0.0 | 0.1 | 3.6 | 7.7 | 31.3 |
| Days ≥ 1" Snow Depth | na | *17* | 7 | 0 | 0 | 0 | 0 | 0 | 0 | 0 | 3 | *14* | na |

## ONAWA *Monona County* ELEVATION 1050 ft LAT/LONG 42° 2 ' N / 96° 6 ' W

| | JAN | FEB | MAR | APR | MAY | JUN | JUL | AUG | SEP | OCT | NOV | DEC | YEAR |
|---|---|---|---|---|---|---|---|---|---|---|---|---|---|
| Maximum Temp °F | 29.8 | 35.9 | 49.1 | 64.2 | 74.7 | 83.7 | 87.0 | 84.6 | 76.5 | 64.8 | 46.9 | 33.7 | 60.9 |
| Minimum Temp °F | 9.5 | 15.2 | 26.6 | 38.7 | 50.0 | 59.8 | 64.1 | 61.7 | 52.4 | 40.3 | 27.1 | 15.2 | 38.4 |
| Mean Temp °F | 19.7 | 25.6 | 37.9 | 51.5 | 62.4 | 71.7 | 75.6 | 73.2 | 64.5 | 52.6 | 37.0 | 24.4 | 49.7 |
| Days Max Temp ≥ 90 °F | 0 | 0 | 0 | 1 | 1 | 7 | 11 | 7 | 2 | 0 | 0 | 0 | 29 |
| Days Max Temp ≤ 32 °F | 16 | 11 | 3 | 0 | 0 | 0 | 0 | 0 | 0 | 0 | 3 | 13 | 46 |
| Days Min Temp ≤ 32 °F | 30 | 26 | 22 | 9 | 1 | 0 | 0 | 0 | 1 | 7 | 21 | 30 | 147 |
| Days Min Temp ≤ 0 °F | 9 | 5 | 1 | 0 | 0 | 0 | 0 | 0 | 0 | 0 | 0 | 4 | 19 |
| Heating Degree Days | 1400 | 1107 | 834 | 415 | 148 | 17 | 3 | 9 | 109 | 389 | 832 | 1250 | 6513 |
| Cooling Degree Days | 0 | 0 | 1 | 21 | 68 | 222 | 322 | 266 | 110 | 7 | 0 | 0 | 1017 |
| Total Precipitation (") | 0.65 | 0.72 | 2.17 | 2.75 | 4.02 | 4.48 | 3.85 | 3.09 | 3.26 | 2.51 | 1.45 | 0.95 | 29.90 |
| Days ≥ 0.1" Precip | 2 | 2 | 4 | 6 | 7 | 7 | 6 | 5 | 6 | 4 | 3 | 2 | 54 |
| Total Snowfall (") | 7.2 | 6.7 | 6.0 | 1.3 | 0.0 | 0.0 | 0.0 | 0.0 | 0.0 | 0.7 | 3.7 | 7.2 | 32.8 |
| Days ≥ 1" Snow Depth | 19 | 13 | 7 | 0 | 0 | 0 | 0 | 0 | 0 | 0 | 4 | 12 | 55 |

## OSAGE *Mitchell County* ELEVATION 1171 ft LAT/LONG 43° 17 ' N / 92° 48 ' W

| | JAN | FEB | MAR | APR | MAY | JUN | JUL | AUG | SEP | OCT | NOV | DEC | YEAR |
|---|---|---|---|---|---|---|---|---|---|---|---|---|---|
| Maximum Temp °F | 23.3 | 28.9 | 42.0 | 58.0 | 70.8 | 79.8 | 82.9 | 80.6 | 72.3 | 60.5 | 42.7 | 28.4 | 55.9 |
| Minimum Temp °F | 5.5 | 11.3 | 24.4 | 37.0 | 48.5 | 58.1 | 62.4 | 59.7 | 51.1 | 39.7 | 26.3 | 12.6 | 36.4 |
| Mean Temp °F | 14.4 | 20.1 | 33.2 | 47.5 | 59.7 | 69.0 | 72.7 | 70.2 | 61.7 | 50.1 | 34.5 | 20.5 | 46.1 |
| Days Max Temp ≥ 90 °F | 0 | 0 | 0 | 0 | 0 | 3 | 5 | 2 | 1 | 0 | 0 | 0 | 11 |
| Days Max Temp ≤ 32 °F | 23 | 16 | 7 | 0 | 0 | 0 | 0 | 0 | 0 | 0 | 5 | 19 | 70 |
| Days Min Temp ≤ 32 °F | 31 | 27 | 24 | 9 | 1 | 0 | 0 | 0 | 0 | 7 | 22 | 30 | 151 |
| Days Min Temp ≤ 0 °F | 12 | 7 | 1 | 0 | 0 | 0 | 0 | 0 | 0 | 0 | 1 | 6 | 27 |
| Heating Degree Days | 1564 | 1261 | 978 | 526 | 206 | 32 | 7 | 19 | 151 | 462 | 907 | 1373 | 7486 |
| Cooling Degree Days | 0 | 0 | 0 | 9 | 48 | 170 | 259 | 199 | 69 | 4 | 0 | 0 | 758 |
| Total Precipitation (") | 0.98 | 0.86 | 2.05 | 3.51 | 3.97 | 4.73 | 4.23 | 4.63 | 4.13 | 2.46 | 1.93 | 1.34 | 34.82 |
| Days ≥ 0.1" Precip | 3 | 2 | 5 | 7 | 8 | 7 | 6 | 7 | 7 | 4 | 4 | 3 | 63 |
| Total Snowfall (") | 9.2 | 6.4 | 5.7 | 2.2 | 0.0 | 0.0 | 0.0 | 0.0 | 0.0 | 0.2 | 3.8 | 8.8 | 36.3 |
| Days ≥ 1" Snow Depth | 25 | 23 | 10 | 1 | 0 | 0 | 0 | 0 | 0 | 0 | 4 | 18 | 81 |

## OSCEOLA *Clarke County* ELEVATION 1110 ft LAT/LONG 41° 1 ' N / 93° 49 ' W

| | JAN | FEB | MAR | APR | MAY | JUN | JUL | AUG | SEP | OCT | NOV | DEC | YEAR |
|---|---|---|---|---|---|---|---|---|---|---|---|---|---|
| Maximum Temp °F | 30.3 | ***35.3*** | 49.1 | ***62.9*** | ***73.1*** | 81.8 | 86.2 | 83.8 | 76.0 | 64.3 | 47.7 | 35.3 | 60.5 |
| Minimum Temp °F | 10.5 | 14.9 | 27.1 | 39.3 | 49.9 | 59.1 | 63.6 | 60.8 | 52.4 | 40.4 | 28.0 | 17.0 | 38.6 |
| Mean Temp °F | 20.6 | ***25.2*** | 38.2 | ***51.3*** | ***61.5*** | 70.6 | 75.0 | 72.3 | 64.2 | 52.4 | 37.8 | 26.2 | 49.6 |
| Days Max Temp ≥ 90 °F | 0 | 0 | 0 | ***0*** | ***0*** | 4 | ***10*** | ***7*** | ***1*** | 0 | 0 | 0 | 22 |
| Days Max Temp ≤ 32 °F | 16 | 11 | 4 | ***0*** | ***0*** | 0 | 0 | ***0*** | 0 | 0 | ***3*** | ***11*** | 45 |
| Days Min Temp ≤ 32 °F | 30 | 27 | 21 | 8 | ***1*** | 0 | 0 | ***0*** | 0 | ***7*** | 20 | 29 | 143 |
| Days Min Temp ≤ 0 °F | 8 | 5 | 0 | 0 | 0 | 0 | 0 | ***0*** | 0 | 0 | 0 | 3 | 16 |
| Heating Degree Days | ***1369*** | 1112 | 826 | ***418*** | ***159*** | 19 | 3 | 9 | 107 | 395 | 808 | 1197 | 6422 |
| Cooling Degree Days | ***0*** | ***0*** | ***2*** | na | na | ***190*** | ***316*** | ***230*** | ***89*** | ***4*** | ***0*** | ***0*** | na |
| Total Precipitation (") | 0.86 | 0.93 | 2.19 | 3.41 | 4.36 | 4.57 | 4.89 | 4.07 | 4.50 | 2.69 | 2.04 | 1.28 | 35.79 |
| Days ≥ 0.1" Precip | 3 | 3 | 5 | 6 | 8 | 7 | 7 | 6 | 6 | 5 | 4 | 3 | 63 |
| Total Snowfall (") | 5.9 | 6.1 | 3.6 | 1.5 | 0.0 | 0.0 | 0.0 | 0.0 | 0.0 | 0.3 | 2.3 | ***4.0*** | 23.7 |
| Days ≥ 1" Snow Depth | ***15*** | ***10*** | 4 | 1 | 0 | 0 | 0 | 0 | 0 | 0 | 1 | ***8*** | 39 |

## OSKALOOSA *Mahaska County* ELEVATION 810 ft LAT/LONG 41° 17 ' N / 92° 40 ' W

| | JAN | FEB | MAR | APR | MAY | JUN | JUL | AUG | SEP | OCT | NOV | DEC | YEAR |
|---|---|---|---|---|---|---|---|---|---|---|---|---|---|
| Maximum Temp °F | 30.2 | 35.7 | 48.8 | 62.8 | 73.5 | 82.4 | 86.5 | 84.0 | 76.4 | 64.4 | 48.2 | 35.0 | 60.7 |
| Minimum Temp °F | 11.5 | 16.7 | 28.1 | 40.0 | 50.5 | 60.2 | 64.4 | 61.5 | 53.3 | 41.8 | 29.3 | 18.4 | 39.6 |
| Mean Temp °F | 20.8 | 26.2 | 38.5 | 51.4 | 62.0 | 71.3 | 75.5 | 72.8 | 64.9 | 53.1 | 38.7 | 26.7 | 50.2 |
| Days Max Temp ≥ 90 °F | 0 | 0 | 0 | 0 | 0 | 4 | 11 | 7 | 2 | 0 | 0 | 0 | 24 |
| Days Max Temp ≤ 32 °F | 17 | 12 | 3 | 0 | 0 | 0 | 0 | 0 | 0 | 0 | 3 | 12 | 47 |
| Days Min Temp ≤ 32 °F | 30 | 26 | 20 | 7 | 0 | 0 | 0 | 0 | 0 | 6 | 19 | 28 | 136 |
| Days Min Temp ≤ 0 °F | 8 | 4 | 0 | 0 | 0 | 0 | 0 | 0 | 0 | 0 | 0 | 3 | 15 |
| Heating Degree Days | 1365 | 1088 | 818 | 413 | 148 | 15 | 2 | 9 | 101 | 375 | 781 | 1180 | 6295 |
| Cooling Degree Days | 0 | 0 | 2 | 16 | 54 | 204 | 321 | 258 | 107 | 8 | 0 | 0 | 970 |
| Total Precipitation (") | 1.10 | 1.05 | 2.24 | 3.89 | 4.07 | 4.32 | 4.47 | 3.97 | 4.46 | 2.93 | 2.69 | 1.58 | 36.77 |
| Days ≥ 0.1" Precip | 3 | 3 | 5 | 7 | 8 | 7 | 7 | 6 | 7 | 6 | 5 | 4 | 68 |
| Total Snowfall (") | 6.9 | 6.6 | 3.3 | 1.6 | 0.0 | 0.0 | 0.0 | 0.0 | 0.0 | 0.2 | 2.5 | 6.3 | 27.4 |
| Days ≥ 1" Snow Depth | ***15*** | na | 3 | 1 | 0 | 0 | 0 | 0 | 0 | 0 | ***1*** | ***7*** | na |

## OTTUMWA INDUSTRL AP *Wapello County* ELEVATION 846 ft LAT/LONG 41° 6 ' N / 92° 27 ' W

| | JAN | FEB | MAR | APR | MAY | JUN | JUL | AUG | SEP | OCT | NOV | DEC | YEAR |
|---|---|---|---|---|---|---|---|---|---|---|---|---|---|
| Maximum Temp °F | 29.7 | 35.1 | 48.0 | 61.8 | 72.9 | 82.3 | 86.3 | 83.8 | 75.5 | 63.7 | 47.9 | 34.7 | 60.1 |
| Minimum Temp °F | 12.5 | 17.4 | 29.1 | 41.1 | 52.2 | 61.7 | 66.3 | 63.5 | 54.8 | 43.2 | 31.0 | 19.0 | 41.0 |
| Mean Temp °F | 21.1 | 26.3 | 38.6 | 51.5 | 62.6 | 72.0 | 76.3 | 73.6 | 65.2 | 53.5 | 39.5 | 26.9 | 50.6 |
| Days Max Temp ≥ 90 °F | 0 | 0 | 0 | 0 | 1 | 4 | 10 | 7 | 2 | 0 | 0 | 0 | 24 |
| Days Max Temp ≤ 32 °F | 17 | 12 | 4 | 0 | 0 | 0 | 0 | 0 | 0 | 0 | 3 | 12 | 48 |
| Days Min Temp ≤ 32 °F | 29 | 25 | 19 | 5 | 0 | 0 | 0 | 0 | 0 | 4 | 17 | 28 | 127 |
| Days Min Temp ≤ 0 °F | 7 | 4 | 0 | 0 | 0 | 0 | 0 | 0 | 0 | 0 | 0 | 3 | 14 |
| Heating Degree Days | 1355 | 1087 | 813 | 414 | 143 | 13 | 2 | 6 | 96 | 364 | 759 | 1175 | 6227 |
| Cooling Degree Days | 0 | 0 | 2 | 19 | 74 | 244 | 371 | 299 | 119 | 10 | 0 | 0 | 1138 |
| Total Precipitation (") | 1.00 | 0.92 | 2.22 | 3.37 | 3.99 | 4.15 | 4.53 | 3.96 | 4.30 | 2.67 | 2.30 | 1.46 | 34.87 |
| Days ≥ 0.1" Precip | 3 | 3 | 5 | 6 | 7 | 7 | 7 | 6 | 6 | 5 | 5 | 4 | 64 |
| Total Snowfall (") | 6.6 | 5.8 | 3.5 | 1.6 | 0.0 | 0.0 | 0.0 | 0.0 | 0.0 | 0.3 | 2.2 | 5.8 | 25.8 |
| Days ≥ 1" Snow Depth | 17 | 12 | 5 | 1 | 0 | 0 | 0 | 0 | 0 | 0 | 2 | 9 | 46 |

## PERRY *Dallas County* ELEVATION 965 ft LAT/LONG 41° 51 ' N / 94° 7 ' W

| | JAN | FEB | MAR | APR | MAY | JUN | JUL | AUG | SEP | OCT | NOV | DEC | YEAR |
|---|---|---|---|---|---|---|---|---|---|---|---|---|---|
| Maximum Temp °F | 27.1 | 32.5 | 45.7 | 60.2 | 72.2 | 81.1 | 85.1 | 82.5 | 75.1 | 63.1 | 46.0 | 32.1 | 58.6 |
| Minimum Temp °F | 7.6 | 12.5 | 25.5 | 37.6 | 48.6 | 58.5 | 62.7 | 59.4 | 50.3 | 38.1 | 26.2 | 14.4 | 36.8 |
| Mean Temp °F | 17.4 | 22.5 | 35.7 | 48.9 | 60.5 | 69.9 | 73.9 | 71.0 | 62.7 | 50.6 | 36.1 | 23.3 | 47.7 |
| Days Max Temp ≥ 90 °F | 0 | 0 | 0 | 0 | 1 | 4 | 8 | 5 | 2 | 0 | 0 | 0 | 20 |
| Days Max Temp ≤ 32 °F | 19 | 14 | 5 | 0 | 0 | 0 | 0 | 0 | 0 | 0 | 4 | 14 | 56 |
| Days Min Temp ≤ 32 °F | 31 | 27 | 23 | 9 | 1 | 0 | 0 | 0 | 1 | 10 | 22 | 30 | 154 |
| Days Min Temp ≤ 0 °F | 10 | 7 | 1 | 0 | 0 | 0 | 0 | 0 | 0 | 0 | 0 | 5 | 23 |
| Heating Degree Days | 1470 | 1192 | 904 | 486 | 190 | 26 | 5 | 19 | 137 | 448 | 859 | 1287 | 7023 |
| Cooling Degree Days | 0 | 0 | 0 | 15 | 52 | 184 | 288 | 220 | 87 | 6 | 0 | 0 | 852 |
| Total Precipitation (") | 0.77 | 0.69 | 2.02 | 3.12 | 4.13 | 4.80 | 3.84 | 3.85 | 3.26 | 2.46 | 1.60 | 1.10 | 31.64 |
| Days ≥ 0.1" Precip | 2 | 2 | 4 | 6 | 7 | 7 | 6 | 6 | 6 | 5 | 4 | 2 | 57 |
| Total Snowfall (") | ***4.6*** | ***5.3*** | ***4.2*** | 0.3 | 0.0 | 0.0 | 0.0 | 0.0 | 0.0 | 0.1 | ***2.6*** | ***4.4*** | 21.5 |
| Days ≥ 1" Snow Depth | na | na | na | 0 | 0 | 0 | 0 | 0 | 0 | 0 | ***1*** | na | na |

## POCAHONTAS *Pocahontas County* ELEVATION 1220 ft LAT/LONG 42° 43 ' N / 94° 39 ' W

| | JAN | FEB | MAR | APR | MAY | JUN | JUL | AUG | SEP | OCT | NOV | DEC | YEAR |
|---|---|---|---|---|---|---|---|---|---|---|---|---|---|
| Maximum Temp °F | 24.6 | 29.7 | 43.0 | 59.3 | 72.7 | 81.7 | 84.3 | 81.6 | 74.1 | 61.9 | 43.6 | 29.0 | 57.1 |
| Minimum Temp °F | 5.1 | 10.2 | 23.2 | 35.9 | 47.5 | 57.8 | 61.3 | 58.2 | 49.2 | 37.0 | 24.2 | 11.1 | 35.1 |
| Mean Temp °F | 14.9 | 20.0 | 33.1 | 47.6 | 60.1 | 69.8 | 72.8 | 69.9 | 61.7 | 49.5 | 33.9 | 20.1 | 46.1 |
| Days Max Temp ≥ 90 °F | 0 | 0 | 0 | 0 | 2 | 5 | 7 | 4 | 1 | 0 | 0 | 0 | 19 |
| Days Max Temp ≤ 32 °F | 21 | 16 | 7 | 0 | 0 | 0 | 0 | 0 | 0 | 0 | 5 | 18 | 67 |
| Days Min Temp ≤ 32 °F | 31 | 28 | 25 | 11 | 1 | 0 | 0 | 0 | 1 | 11 | 24 | 31 | 163 |
| Days Min Temp ≤ 0 °F | 12 | 8 | 1 | 0 | 0 | 0 | 0 | 0 | 0 | 0 | 1 | 6 | 28 |
| Heating Degree Days | 1548 | 1266 | 982 | 524 | 201 | 29 | 6 | 22 | 157 | 481 | 926 | 1388 | 7530 |
| Cooling Degree Days | 0 | 0 | 0 | ***13*** | ***59*** | ***187*** | ***234*** | 174 | 71 | 4 | 0 | 0 | 742 |
| Total Precipitation (") | 0.76 | 0.69 | 2.06 | 2.91 | 3.58 | 4.55 | 4.19 | 4.36 | 3.62 | 2.20 | 1.69 | 1.07 | 31.68 |
| Days ≥ 0.1" Precip | 2 | 2 | 5 | 6 | ***7*** | 7 | ***6*** | 6 | 6 | 4 | 3 | 3 | 57 |
| Total Snowfall (") | ***7.0*** | ***6.3*** | 6.5 | ***1.5*** | 0.0 | 0.0 | 0.0 | 0.0 | 0.0 | 0.3 | ***4.7*** | 8.0 | 34.3 |
| Days ≥ 1" Snow Depth | na | na | na | 1 | 0 | 0 | 0 | 0 | 0 | 0 | ***2*** | na | na |

## PRIMGHAR *O'Brien County* ELEVATION 1522 ft LAT/LONG 43° 5 ' N / 95° 38 ' W

| | JAN | FEB | MAR | APR | MAY | JUN | JUL | AUG | SEP | OCT | NOV | DEC | YEAR |
|---|---|---|---|---|---|---|---|---|---|---|---|---|---|
| Maximum Temp °F | 24.8 | 31.0 | 44.5 | 60.8 | 73.5 | 82.4 | 85.6 | 83.1 | 75.2 | 63.0 | 43.4 | 29.2 | 58.0 |
| Minimum Temp °F | 6.1 | 12.2 | 24.3 | 36.8 | 48.6 | 58.4 | 62.7 | 60.1 | 50.8 | 39.0 | 25.3 | 12.1 | 36.4 |
| Mean Temp °F | 15.4 | 21.6 | 34.4 | 48.8 | 61.1 | 70.4 | 74.2 | 71.7 | 63.1 | 51.0 | 34.4 | 20.7 | 47.2 |
| Days Max Temp ≥ 90 °F | 0 | 0 | 0 | 0 | 1 | 5 | 9 | 6 | 1 | 0 | 0 | 0 | 22 |
| Days Max Temp ≤ 32 °F | 21 | 15 | 6 | 0 | 0 | 0 | 0 | 0 | 0 | 0 | 6 | 18 | 66 |
| Days Min Temp ≤ 32 °F | 31 | 27 | 24 | 10 | 1 | 0 | 0 | 0 | 1 | 8 | 23 | 30 | 155 |
| Days Min Temp ≤ 0 °F | 11 | 7 | 1 | 0 | 0 | 0 | 0 | 0 | 0 | 0 | 1 | 6 | 26 |
| Heating Degree Days | 1532 | 1219 | 942 | 489 | 177 | 23 | 5 | 13 | 129 | 437 | 910 | 1369 | 7245 |
| Cooling Degree Days | 0 | 0 | 1 | 12 | 62 | 196 | 284 | 220 | 82 | 5 | 0 | 0 | 862 |
| Total Precipitation (") | 0.63 | 0.60 | 1.80 | 2.83 | 3.72 | 4.90 | 4.10 | 4.06 | 3.02 | 2.21 | 1.37 | 0.80 | 30.04 |
| Days ≥ 0.1" Precip | 2 | 2 | 4 | 6 | 7 | 8 | 6 | 5 | 6 | 4 | 3 | 2 | 55 |
| Total Snowfall (") | 6.4 | 4.3 | ***5.9*** | 0.7 | 0.0 | 0.0 | 0.0 | 0.0 | 0.0 | 0.3 | 3.7 | 6.7 | 28.0 |
| Days ≥ 1" Snow Depth | na | na | ***8*** | 1 | 0 | 0 | 0 | 0 | 0 | 0 | ***2*** | na | na |

## RED OAK *Montgomery County* ELEVATION 1040 ft LAT/LONG 41° 1 ' N / 95° 13 ' W

| | JAN | FEB | MAR | APR | MAY | JUN | JUL | AUG | SEP | OCT | NOV | DEC | YEAR |
|---|---|---|---|---|---|---|---|---|---|---|---|---|---|
| Maximum Temp °F | 31.9 | 37.8 | 51.0 | 65.1 | 75.8 | 85.0 | 88.5 | 86.4 | 78.4 | 66.3 | 48.9 | 36.6 | 62.6 |
| Minimum Temp °F | 11.6 | 16.3 | 28.2 | 39.7 | 50.4 | 59.6 | 64.2 | 61.5 | 52.7 | 40.7 | 28.5 | 17.7 | 39.3 |
| Mean Temp °F | 21.8 | 27.1 | 39.6 | 52.5 | 63.1 | 72.3 | 76.4 | 73.9 | 65.6 | 53.5 | 38.7 | 27.2 | 51.0 |
| Days Max Temp ≥ 90 °F | 0 | 0 | 0 | 1 | 1 | 8 | 14 | 10 | 3 | 0 | 0 | 0 | 37 |
| Days Max Temp ≤ 32 °F | 15 | 10 | 2 | 0 | 0 | 0 | 0 | 0 | 0 | 0 | 3 | 10 | 40 |
| Days Min Temp ≤ 32 °F | 30 | 26 | 21 | 8 | 1 | 0 | 0 | 0 | 1 | 8 | 20 | 29 | 144 |
| Days Min Temp ≤ 0 °F | 7 | 4 | 0 | 0 | 0 | 0 | 0 | 0 | 0 | 0 | 0 | 3 | 14 |
| Heating Degree Days | 1333 | 1063 | 782 | 387 | 130 | 12 | 2 | 8 | 94 | 365 | 782 | 1165 | 6123 |
| Cooling Degree Days | 0 | 0 | 2 | 24 | 78 | 243 | 364 | 290 | 128 | 11 | 0 | 0 | 1140 |
| Total Precipitation (") | 0.88 | 1.02 | 2.21 | 3.43 | 4.26 | 5.33 | 4.68 | 4.11 | 4.24 | 2.63 | 1.87 | 1.33 | 35.99 |
| Days ≥ 0.1" Precip | 2 | 3 | 5 | 7 | 7 | 7 | 6 | 6 | 6 | 5 | 4 | 4 | 62 |
| Total Snowfall (") | 7.2 | 7.7 | 4.1 | 1.4 | 0.1 | 0.0 | 0.0 | 0.0 | 0.0 | 0.2 | 3.0 | 5.8 | 29.5 |
| Days ≥ 1" Snow Depth | 16 | 14 | 4 | 0 | 0 | 0 | 0 | 0 | 0 | 0 | 2 | ***8*** | 44 |

## ROCK RAPIDS *Lyon County* ELEVATION 1342 ft LAT/LONG 43° 26 ' N / 96° 10 ' W

| | JAN | FEB | MAR | APR | MAY | JUN | JUL | AUG | SEP | OCT | NOV | DEC | YEAR |
|---|---|---|---|---|---|---|---|---|---|---|---|---|---|
| Maximum Temp °F | 24.3 | 30.2 | 43.1 | 59.3 | 72.2 | 81.4 | 86.0 | 83.5 | 74.0 | 61.7 | 42.9 | 28.9 | 57.3 |
| Minimum Temp °F | 2.4 | 8.4 | 21.6 | 34.5 | 46.2 | 56.8 | 61.1 | 58.1 | 47.7 | 34.9 | 22.1 | 8.9 | 33.6 |
| Mean Temp °F | 13.4 | 19.3 | 32.3 | 47.0 | 59.2 | 69.1 | 73.6 | 70.8 | 60.8 | 48.3 | 32.5 | 18.7 | 45.4 |
| Days Max Temp ≥ 90 °F | 0 | 0 | 0 | 0 | 1 | 5 | 11 | 7 | 2 | 0 | 0 | 0 | 26 |
| Days Max Temp ≤ 32 °F | 21 | 15 | 6 | 0 | 0 | 0 | 0 | 0 | 0 | 0 | 6 | 18 | 66 |
| Days Min Temp ≤ 32 °F | 31 | 28 | 26 | 13 | 2 | 0 | 0 | 0 | 2 | 13 | 26 | 31 | 172 |
| Days Min Temp ≤ 0 °F | 14 | 9 | 2 | 0 | 0 | 0 | 0 | 0 | 0 | 0 | 1 | 8 | 34 |
| Heating Degree Days | 1597 | 1285 | 1009 | 541 | 219 | 37 | 7 | 21 | 174 | 515 | 969 | 1428 | 7802 |
| Cooling Degree Days | 0 | 0 | 0 | 9 | 44 | 168 | 266 | 197 | 56 | 2 | 0 | 0 | 742 |
| Total Precipitation (") | 0.54 | 0.60 | 1.77 | 2.50 | 3.06 | 4.39 | 3.47 | 3.76 | 2.85 | 1.95 | 1.45 | 0.81 | 27.15 |
| Days ≥ 0.1" Precip | 2 | 2 | 4 | 5 | 7 | 6 | 5 | 5 | 6 | 4 | 3 | 2 | 51 |
| Total Snowfall (") | 5.8 | ***5.6*** | 6.3 | 1.3 | 0.0 | 0.0 | 0.0 | 0.0 | 0.0 | 0.6 | 4.2 | 6.1 | 29.9 |
| Days ≥ 1" Snow Depth | 24 | 19 | 10 | 1 | 0 | 0 | 0 | 0 | 0 | 0 | 5 | 17 | 76 |

## ROCKWELL CITY *Calhoun County* ELEVATION 1220 ft LAT/LONG 42° 24 ' N / 94° 37 ' W

| | JAN | FEB | MAR | APR | MAY | JUN | JUL | AUG | SEP | OCT | NOV | DEC | YEAR |
|---|---|---|---|---|---|---|---|---|---|---|---|---|---|
| Maximum Temp °F | 26.7 | 32.4 | 45.8 | 61.3 | 73.4 | 82.5 | 85.3 | 82.8 | 75.4 | 63.1 | 44.9 | 31.2 | 58.7 |
| Minimum Temp °F | 7.8 | 13.3 | 25.6 | 37.6 | 49.2 | 59.1 | 62.9 | 60.2 | 51.6 | 39.9 | 26.5 | 14.0 | 37.3 |
| Mean Temp °F | 17.3 | 22.9 | 35.7 | 49.5 | 61.3 | 70.8 | 74.1 | 71.5 | 63.5 | 51.5 | 35.7 | 22.6 | 48.0 |
| Days Max Temp ≥ 90 °F | 0 | 0 | 0 | 0 | 1 | 5 | 9 | 5 | 1 | 0 | 0 | 0 | 21 |
| Days Max Temp ≤ 32 °F | 20 | 13 | 5 | 0 | 0 | 0 | 0 | 0 | 0 | 0 | 4 | 16 | 58 |
| Days Min Temp ≤ 32 °F | 31 | 27 | 24 | 9 | 1 | 0 | 0 | 0 | 1 | 8 | 22 | 31 | 154 |
| Days Min Temp ≤ 0 °F | 10 | 6 | 1 | 0 | 0 | 0 | 0 | 0 | 0 | 0 | 1 | 5 | 23 |
| Heating Degree Days | 1474 | 1182 | 902 | 469 | 168 | 17 | 3 | 12 | 118 | 420 | 872 | 1307 | 6944 |
| Cooling Degree Days | 0 | 0 | 1 | 15 | 58 | 202 | 285 | 223 | 91 | 5 | 0 | 0 | 880 |
| Total Precipitation (") | 0.71 | 0.64 | 1.88 | 2.85 | 4.19 | 4.72 | 4.13 | 3.93 | 3.80 | 2.54 | 1.47 | 0.94 | 31.80 |
| Days ≥ 0.1" Precip | 2 | 2 | 4 | 6 | 8 | 8 | 7 | 6 | 6 | 5 | 4 | 3 | 61 |
| Total Snowfall (") | 6.9 | 6.2 | 6.0 | 2.5 | 0.0 | 0.0 | 0.0 | 0.0 | 0.0 | 0.7 | 4.2 | 7.3 | 33.8 |
| Days ≥ 1" Snow Depth | 23 | 18 | 9 | 1 | 0 | 0 | 0 | 0 | 0 | 0 | 5 | 16 | 72 |

## SAC CITY *Sac County* ELEVATION 1270 ft LAT/LONG 42° 26 ' N / 94° 59 ' W

| | JAN | FEB | MAR | APR | MAY | JUN | JUL | AUG | SEP | OCT | NOV | DEC | YEAR |
|---|---|---|---|---|---|---|---|---|---|---|---|---|---|
| Maximum Temp °F | 26.4 | 31.9 | 44.9 | 60.6 | 72.7 | 81.6 | 85.1 | 82.6 | 74.4 | 62.3 | 44.3 | 31.1 | 58.2 |
| Minimum Temp °F | 7.4 | 11.9 | 24.5 | 36.4 | 47.4 | 57.4 | 62.2 | 59.4 | 50.2 | 38.3 | 25.4 | 13.5 | 36.2 |
| Mean Temp °F | 16.9 | 21.9 | 34.7 | 48.5 | 60.0 | 69.5 | 73.7 | 71.0 | 62.3 | 50.3 | 34.9 | 22.3 | 47.2 |
| Days Max Temp ≥ 90 °F | 0 | 0 | 0 | 0 | 1 | 4 | 8 | 5 | 1 | 0 | 0 | 0 | 19 |
| Days Max Temp ≤ 32 °F | 20 | 14 | 5 | 0 | 0 | 0 | 0 | 0 | 0 | 0 | 5 | 16 | 60 |
| Days Min Temp ≤ 32 °F | 31 | 28 | 25 | 10 | 1 | 0 | 0 | 0 | 1 | 9 | 24 | 30 | 159 |
| Days Min Temp ≤ 0 °F | 10 | 7 | 1 | 0 | 0 | 0 | 0 | 0 | 0 | 0 | 1 | 5 | 24 |
| Heating Degree Days | 1485 | 1210 | 932 | 497 | 196 | 29 | 5 | 18 | 142 | 455 | 897 | 1317 | 7183 |
| Cooling Degree Days | 0 | 0 | 0 | 12 | 42 | 159 | 258 | 193 | 65 | 2 | 0 | 0 | 731 |
| Total Precipitation (") | 0.75 | 0.83 | 2.37 | 3.07 | 4.17 | 4.82 | 3.88 | 3.81 | 3.83 | 2.50 | 1.66 | 1.22 | 32.91 |
| Days ≥ 0.1" Precip | 2 | 3 | 5 | 7 | 7 | 7 | 6 | 6 | 6 | 4 | 4 | 3 | 60 |
| Total Snowfall (") | 6.1 | 6.4 | 5.9 | 1.4 | 0.0 | 0.0 | 0.0 | 0.0 | 0.0 | 0.4 | 2.6 | 7.2 | 30.0 |
| Days ≥ 1" Snow Depth | ***20*** | ***16*** | ***7*** | 1 | 0 | 0 | 0 | 0 | 0 | 0 | 4 | ***13*** | 61 |

## SANBORN *O'Brien County* ELEVATION 1552 ft LAT/LONG 43° 11' N / 95° 39' W

| | JAN | FEB | MAR | APR | MAY | JUN | JUL | AUG | SEP | OCT | NOV | DEC | YEAR |
|---|---|---|---|---|---|---|---|---|---|---|---|---|---|
| Maximum Temp °F | 23.2 | 29.0 | 42.2 | 58.5 | 71.7 | 80.6 | 83.8 | 81.3 | 73.2 | 60.7 | 41.9 | 27.7 | 56.2 |
| Minimum Temp °F | 3.9 | 9.9 | 22.3 | 34.7 | 46.9 | 57.1 | 61.6 | 58.8 | 49.4 | 37.4 | 23.5 | 10.3 | 34.7 |
| Mean Temp °F | 13.6 | 19.5 | 32.2 | 46.6 | 59.3 | 68.9 | 72.7 | 70.1 | 61.3 | 49.0 | 32.7 | 19.1 | 45.4 |
| Days Max Temp ≥ 90 °F | 0 | 0 | 0 | 0 | 1 | 4 | 7 | 4 | 1 | 0 | 0 | 0 | 17 |
| Days Max Temp ≤ 32 °F | 22 | 16 | 7 | 0 | 0 | 0 | 0 | 0 | 0 | 0 | 7 | 19 | 71 |
| Days Min Temp ≤ 32 °F | 31 | 28 | 26 | 13 | 2 | 0 | 0 | 0 | 1 | 10 | 24 | 31 | 166 |
| Days Min Temp ≤ 0 °F | 13 | 8 | 2 | 0 | 0 | 0 | 0 | 0 | 0 | 0 | 1 | 7 | 31 |
| Heating Degree Days | 1590 | 1280 | 1008 | 550 | 215 | 34 | 8 | 20 | 162 | 493 | 962 | 1418 | 7740 |
| Cooling Degree Days | 0 | 0 | 0 | 7 | 47 | 161 | 246 | 182 | 63 | 2 | 0 | 0 | 708 |
| Total Precipitation (") | 0.61 | 0.67 | 1.68 | 2.58 | 3.49 | 4.27 | 3.48 | 3.89 | 3.11 | 2.07 | 1.40 | 0.77 | 28.02 |
| Days ≥ 0.1" Precip | 2 | 2 | 4 | 6 | 7 | 7 | 6 | 6 | 6 | 4 | 3 | 2 | 55 |
| Total Snowfall (") | 8.6 | 7.4 | 8.2 | 2.2 | 0.0 | 0.0 | 0.0 | 0.0 | 0.0 | 0.8 | 5.7 | 8.0 | 40.9 |
| Days ≥ 1" Snow Depth | *23* | *16* | *9* | 1 | 0 | 0 | 0 | 0 | 0 | 0 | 6 | *18* | 73 |

## SHELDON *O'Brien County* ELEVATION 1421 ft LAT/LONG 43° 11' N / 95° 51' W

| | JAN | FEB | MAR | APR | MAY | JUN | JUL | AUG | SEP | OCT | NOV | DEC | YEAR |
|---|---|---|---|---|---|---|---|---|---|---|---|---|---|
| Maximum Temp °F | 24.0 | 29.6 | 43.1 | 59.2 | 72.0 | 80.6 | 83.9 | 81.3 | 73.0 | 60.6 | 42.1 | 28.4 | 56.5 |
| Minimum Temp °F | 4.3 | 10.4 | 23.1 | 35.0 | 46.9 | 56.7 | 60.9 | 58.3 | 49.1 | 37.0 | 23.4 | 10.7 | 34.7 |
| Mean Temp °F | 14.2 | 20.0 | 33.1 | 47.1 | 59.5 | 68.7 | 72.4 | 69.8 | 61.1 | 48.8 | 32.8 | 19.6 | 45.6 |
| Days Max Temp ≥ 90 °F | 0 | 0 | 0 | 0 | 1 | 4 | 7 | 4 | 1 | 0 | 0 | 0 | 17 |
| Days Max Temp ≤ 32 °F | 21 | 16 | 6 | 0 | 0 | 0 | 0 | 0 | 0 | 0 | 7 | 19 | 69 |
| Days Min Temp ≤ 32 °F | 31 | 28 | 26 | 13 | 2 | 0 | 0 | 0 | 1 | 11 | 25 | 31 | 168 |
| Days Min Temp ≤ 0 °F | 13 | 8 | 2 | 0 | 0 | 0 | 0 | 0 | 0 | 0 | 1 | 7 | 31 |
| Heating Degree Days | 1570 | 1263 | 981 | 537 | 207 | 35 | 9 | 24 | 165 | 499 | 961 | 1403 | 7654 |
| Cooling Degree Days | 0 | 0 | 0 | 9 | 44 | 152 | 239 | 179 | 57 | 2 | 0 | 0 | 682 |
| Total Precipitation (") | 0.72 | 0.70 | 2.01 | 2.75 | 3.56 | 4.34 | 3.60 | 3.86 | 2.85 | 2.08 | 1.40 | 0.91 | 28.78 |
| Days ≥ 0.1" Precip | 2 | 2 | 4 | 6 | 7 | 7 | 6 | 6 | 5 | 4 | 3 | 2 | 54 |
| Total Snowfall (") | 7.4 | 6.1 | 8.2 | 2.1 | 0.0 | 0.0 | 0.0 | 0.0 | 0.0 | 0.9 | 5.0 | 7.7 | 37.4 |
| Days ≥ 1" Snow Depth | 24 | 19 | 10 | 1 | 0 | 0 | 0 | 0 | 0 | 0 | 6 | 17 | 77 |

## SHENANDOAH *Page County* ELEVATION 971 ft LAT/LONG 40° 46' N / 95° 23' W

| | JAN | FEB | MAR | APR | MAY | JUN | JUL | AUG | SEP | OCT | NOV | DEC | YEAR |
|---|---|---|---|---|---|---|---|---|---|---|---|---|---|
| Maximum Temp °F | 32.1 | 38.2 | 51.7 | 65.3 | 75.5 | 84.5 | 87.9 | 86.0 | 78.5 | 67.2 | 49.8 | 36.7 | 62.8 |
| Minimum Temp °F | 12.8 | 17.7 | 29.4 | 40.9 | 51.3 | 60.8 | 65.2 | 62.6 | 54.0 | 42.2 | 29.7 | 18.7 | 40.4 |
| Mean Temp °F | 22.5 | 27.9 | 40.6 | 53.1 | 63.4 | 72.7 | 76.6 | 74.3 | 66.2 | 54.7 | 39.7 | 27.8 | 51.6 |
| Days Max Temp ≥ 90 °F | 0 | 0 | 0 | 0 | 1 | 8 | 13 | 10 | 3 | 0 | 0 | 0 | 35 |
| Days Max Temp ≤ 32 °F | 14 | 10 | 3 | 0 | 0 | 0 | 0 | 0 | 0 | 0 | 2 | 10 | 39 |
| Days Min Temp ≤ 32 °F | 30 | 25 | 19 | 7 | 0 | 0 | 0 | 0 | 0 | 6 | 19 | 29 | 135 |
| Days Min Temp ≤ 0 °F | 7 | 4 | 0 | 0 | 0 | 0 | 0 | 0 | 0 | 0 | 0 | 3 | 14 |
| Heating Degree Days | 1313 | 1039 | 753 | 371 | 124 | 11 | 2 | 5 | 88 | 331 | 751 | 1148 | 5936 |
| Cooling Degree Days | 0 | 0 | 3 | 25 | 72 | 242 | 368 | 302 | 147 | 13 | 0 | 0 | 1172 |
| Total Precipitation (") | 0.72 | 0.80 | 2.20 | 2.99 | 3.96 | 4.53 | 4.75 | 3.82 | 3.76 | 2.68 | 1.80 | 1.20 | 33.21 |
| Days ≥ 0.1" Precip | 2 | 2 | 5 | 6 | 7 | 6 | 6 | 6 | 5 | 5 | 4 | 3 | 57 |
| Total Snowfall (") | 5.9 | 5.3 | 3.2 | 1.0 | 0.0 | 0.0 | 0.0 | 0.0 | 0.0 | 0.1 | 1.6 | 4.7 | 21.8 |
| Days ≥ 1" Snow Depth | *14* | *11* | *3* | 0 | 0 | 0 | 0 | 0 | 0 | 0 | *1* | 7 | 36 |

## SIBLEY 5 NNE *Osceola County* ELEVATION 1489 ft LAT/LONG 43° 24' N / 95° 45' W

| | JAN | FEB | MAR | APR | MAY | JUN | JUL | AUG | SEP | OCT | NOV | DEC | YEAR |
|---|---|---|---|---|---|---|---|---|---|---|---|---|---|
| Maximum Temp °F | 23.1 | 28.9 | 41.9 | 58.1 | 71.2 | 79.8 | 83.4 | 80.8 | 72.8 | 60.9 | 41.8 | 27.7 | 55.9 |
| Minimum Temp °F | 3.4 | 9.2 | 21.6 | 33.9 | 45.4 | 55.5 | 59.9 | 57.0 | 47.9 | 36.1 | 22.6 | 9.7 | 33.5 |
| Mean Temp °F | 13.2 | 19.1 | 31.7 | 46.0 | 58.3 | 67.7 | 71.7 | 69.0 | 60.4 | 48.5 | 32.2 | 18.7 | 44.7 |
| Days Max Temp ≥ 90 °F | 0 | 0 | 0 | 0 | 1 | 4 | 6 | 4 | 1 | 0 | 0 | 0 | 16 |
| Days Max Temp ≤ 32 °F | 22 | 16 | 7 | 1 | 0 | 0 | 0 | 0 | 0 | 0 | 7 | 19 | 72 |
| Days Min Temp ≤ 32 °F | 31 | 28 | 27 | 14 | 2 | 0 | 0 | 0 | 2 | 12 | 25 | 30 | 171 |
| Days Min Temp ≤ 0 °F | 13 | 8 | 2 | 0 | 0 | 0 | 0 | 0 | 0 | 0 | 1 | 8 | 32 |
| Heating Degree Days | 1601 | 1291 | 1024 | 568 | 238 | 49 | 12 | 30 | 182 | 508 | 977 | 1429 | 7909 |
| Cooling Degree Days | 0 | 0 | 0 | 7 | 33 | 125 | 195 | 138 | 46 | 2 | 0 | 0 | 546 |
| Total Precipitation (") | 0.59 | 0.64 | 1.89 | 2.79 | 3.43 | 4.40 | 3.37 | 4.33 | 3.21 | 2.04 | 1.28 | 0.81 | 28.78 |
| Days ≥ 0.1" Precip | 2 | 2 | 4 | 6 | 7 | 8 | 5 | 6 | 6 | 4 | 3 | 2 | 55 |
| Total Snowfall (") | 7.3 | 5.1 | 8.0 | 2.8 | 0.0 | 0.0 | 0.0 | 0.0 | 0.0 | 1.0 | 5.3 | 7.3 | 36.8 |
| Days ≥ 1" Snow Depth | na | na | na | *2* | 0 | 0 | 0 | 0 | 0 | 0 | *5* | na | na |

## SIDNEY *Fremont County* ELEVATION 1050 ft LAT/LONG 40° 45' N / 95° 38' W

| | JAN | FEB | MAR | APR | MAY | JUN | JUL | AUG | SEP | OCT | NOV | DEC | YEAR |
|---|---|---|---|---|---|---|---|---|---|---|---|---|---|
| Maximum Temp °F | 33.0 | 39.2 | 52.4 | 65.7 | 75.6 | 84.7 | 88.3 | 86.3 | 78.6 | 67.6 | 49.9 | 37.1 | 63.2 |
| Minimum Temp °F | 13.5 | 18.6 | 29.9 | 41.5 | 52.4 | 61.6 | 66.0 | 63.4 | 54.8 | 43.4 | 30.3 | 19.1 | 41.2 |
| Mean Temp °F | 23.3 | 28.9 | 41.2 | 53.7 | 64.0 | 73.2 | 77.2 | 74.9 | 66.8 | 55.5 | 40.1 | 28.1 | 52.2 |
| Days Max Temp ≥ 90 °F | 0 | 0 | 0 | 1 | 1 | 8 | 13 | 11 | 4 | 0 | 0 | 0 | 38 |
| Days Max Temp ≤ 32 °F | 14 | 10 | 2 | 0 | 0 | 0 | 0 | 0 | 0 | 0 | 2 | 10 | 38 |
| Days Min Temp ≤ 32 °F | 29 | 25 | 19 | 5 | 0 | 0 | 0 | 0 | 0 | 4 | 18 | 29 | 129 |
| Days Min Temp ≤ 0 °F | 6 | 3 | 0 | 0 | 0 | 0 | 0 | 0 | 0 | 0 | 0 | 3 | 12 |
| Heating Degree Days | 1287 | 1013 | 735 | 356 | 112 | 9 | 1 | 4 | 78 | 308 | 740 | 1136 | 5779 |
| Cooling Degree Days | 0 | 0 | 3 | 29 | 84 | 262 | 376 | 307 | 147 | 15 | 0 | 0 | 1223 |
| Total Precipitation (") | 0.77 | 0.85 | 2.36 | 3.11 | 4.01 | 4.42 | 5.31 | 4.12 | 3.89 | 2.89 | 1.77 | 1.23 | 34.73 |
| Days ≥ 0.1" Precip | 2 | 2 | 5 | 6 | 7 | 6 | 6 | 6 | 6 | 5 | 4 | 3 | 58 |
| Total Snowfall (") | 6.6 | 6.2 | 5.5 | 2.1 | 0.0 | 0.0 | 0.0 | 0.0 | 0.0 | 0.5 | 3.4 | 6.6 | 30.9 |
| Days ≥ 1" Snow Depth | 14 | 11 | 4 | 1 | 0 | 0 | 0 | 0 | 0 | 0 | 2 | 9 | 41 |

## SIGOURNEY *Keokuk County* ELEVATION 781 ft LAT/LONG 41° 20' N / 92° 12' W

| | JAN | FEB | MAR | APR | MAY | JUN | JUL | AUG | SEP | OCT | NOV | DEC | YEAR |
|---|---|---|---|---|---|---|---|---|---|---|---|---|---|
| Maximum Temp °F | 29.9 | 35.7 | 48.4 | 62.4 | 73.4 | 82.3 | 86.6 | 84.2 | 76.1 | 64.3 | 48.4 | 35.1 | 60.6 |
| Minimum Temp °F | 11.7 | 16.4 | 28.3 | 40.0 | 51.1 | 60.6 | 65.3 | 62.4 | 54.2 | 42.4 | 30.3 | 18.4 | 40.1 |
| Mean Temp °F | 20.8 | 26.1 | 38.4 | 51.2 | 62.3 | 71.5 | 76.0 | 73.3 | 65.1 | 53.4 | 39.4 | 26.8 | 50.4 |
| Days Max Temp ≥ 90 °F | 0 | 0 | 0 | 0 | 1 | 4 | 10 | 7 | 2 | 0 | 0 | 0 | 24 |
| Days Max Temp ≤ 32 °F | 16 | 11 | 3 | 0 | 0 | 0 | 0 | 0 | 0 | 0 | 2 | 12 | 44 |
| Days Min Temp ≤ 32 °F | 30 | 26 | 20 | 6 | 0 | 0 | 0 | 0 | 0 | 5 | 18 | 28 | 133 |
| Days Min Temp ≤ 0 °F | 8 | 4 | 0 | 0 | 0 | 0 | 0 | 0 | 0 | 0 | 0 | 3 | 15 |
| Heating Degree Days | 1363 | 1093 | 820 | 421 | 149 | 16 | 3 | 8 | 99 | 366 | 762 | 1177 | 6277 |
| Cooling Degree Days | 0 | 0 | 1 | 17 | 67 | 219 | 345 | 281 | 111 | 7 | 0 | 0 | 1048 |
| Total Precipitation (") | 1.03 | 0.87 | 2.24 | 3.82 | 3.84 | 4.05 | 4.57 | 4.08 | 4.16 | 2.84 | 2.52 | 1.45 | 35.47 |
| Days ≥ 0.1" Precip | *2* | 3 | 5 | 7 | 8 | 7 | 6 | 6 | 7 | 6 | 5 | 4 | 66 |
| Total Snowfall (") | *6.2* | *6.2* | *2.0* | 1.8 | 0.0 | 0.0 | 0.0 | 0.0 | 0.0 | 0.3 | *2.1* | *5.2* | 23.8 |
| Days ≥ 1" Snow Depth | na | na | na | *1* | 0 | 0 | 0 | 0 | 0 | 0 | *1* | na | na |

## SIOUX CENTER 2 SE *Sioux County* ELEVATION 1450 ft LAT/LONG 43° 5' N / 96° 10' W

| | JAN | FEB | MAR | APR | MAY | JUN | JUL | AUG | SEP | OCT | NOV | DEC | YEAR |
|---|---|---|---|---|---|---|---|---|---|---|---|---|---|
| Maximum Temp °F | 25.3 | 31.4 | 45.2 | 61.8 | 74.0 | 82.9 | 85.3 | 82.9 | 75.4 | 63.2 | 43.4 | 29.6 | 58.4 |
| Minimum Temp °F | 5.7 | 11.6 | 23.8 | 35.8 | 47.2 | 57.2 | 61.4 | 58.8 | 49.5 | 37.6 | 24.2 | 11.9 | 35.4 |
| Mean Temp °F | 15.5 | 21.5 | 34.5 | 48.8 | 60.6 | 70.0 | 73.4 | 70.9 | 62.5 | 50.4 | 33.9 | 20.8 | 46.9 |
| Days Max Temp ≥ 90 °F | 0 | 0 | 0 | 0 | 2 | 6 | 9 | 5 | 2 | 0 | 0 | 0 | 24 |
| Days Max Temp ≤ 32 °F | 20 | 14 | 5 | 0 | 0 | 0 | 0 | 0 | 0 | 0 | 6 | 17 | 62 |
| Days Min Temp ≤ 32 °F | 31 | 27 | 24 | 11 | 2 | 0 | 0 | 0 | 1 | 10 | 24 | 31 | 161 |
| Days Min Temp ≤ 0 °F | 12 | 7 | 1 | 0 | 0 | 0 | 0 | 0 | 0 | 0 | 1 | 6 | 27 |
| Heating Degree Days | 1530 | 1221 | 938 | 489 | 188 | 27 | 6 | 18 | 141 | 451 | 927 | 1364 | 7300 |
| Cooling Degree Days | 0 | 0 | 0 | 13 | 66 | 194 | 265 | 210 | 81 | 4 | 0 | 0 | 833 |
| Total Precipitation (") | 0.69 | 0.68 | 1.91 | 2.63 | 3.52 | 4.39 | 3.71 | 3.25 | 2.96 | 2.12 | 1.36 | 0.91 | 28.13 |
| Days ≥ 0.1" Precip | 2 | 2 | 4 | 6 | 7 | 7 | 6 | 5 | 6 | 4 | 4 | 3 | 56 |
| Total Snowfall (") | 7.3 | 6.1 | 7.7 | 2.5 | 0.0 | 0.0 | 0.0 | 0.0 | 0.0 | 1.3 | 5.5 | 7.6 | 38.0 |
| Days ≥ 1" Snow Depth | 25 | 19 | 11 | 1 | 0 | 0 | 0 | 0 | 0 | 0 | 6 | 19 | 81 |

## SIOUX CITY MUNI AP *Woodbury County* ELEVATION 1095 ft LAT/LONG 42° 23' N / 96° 22' W

| | JAN | FEB | MAR | APR | MAY | JUN | JUL | AUG | SEP | OCT | NOV | DEC | YEAR |
|---|---|---|---|---|---|---|---|---|---|---|---|---|---|
| Maximum Temp °F | 27.7 | 33.4 | 46.6 | 62.1 | 73.0 | 82.0 | 85.9 | 83.5 | 75.1 | 63.2 | 45.4 | 31.6 | 59.1 |
| Minimum Temp °F | 8.1 | 14.3 | 26.0 | 38.2 | 49.8 | 59.5 | 64.3 | 61.8 | 51.7 | 39.1 | 25.9 | 13.7 | 37.7 |
| Mean Temp °F | 17.9 | 23.8 | 36.4 | 50.1 | 61.4 | 70.7 | 75.1 | 72.6 | 63.4 | 51.2 | 35.6 | 22.7 | 48.4 |
| Days Max Temp ≥ 90 °F | 0 | 0 | 0 | 1 | 1 | 5 | 10 | 7 | 2 | 0 | 0 | 0 | 26 |
| Days Max Temp ≤ 32 °F | 18 | 13 | 5 | 0 | 0 | 0 | 0 | 0 | 0 | 0 | 4 | 15 | 55 |
| Days Min Temp ≤ 32 °F | 31 | 27 | 23 | 8 | 1 | 0 | 0 | 0 | 1 | 8 | 23 | 30 | 152 |
| Days Min Temp ≤ 0 °F | 9 | 5 | 0 | 0 | 0 | 0 | 0 | 0 | 0 | 0 | 1 | 5 | 20 |
| Heating Degree Days | 1454 | 1156 | 882 | 450 | 165 | 21 | 2 | 10 | 125 | 429 | 874 | 1306 | 6874 |
| Cooling Degree Days | 0 | 0 | 0 | 18 | 61 | 204 | 314 | 251 | 92 | 5 | 0 | 0 | 945 |
| Total Precipitation (") | 0.58 | 0.68 | 1.91 | 2.45 | 3.49 | 3.58 | 3.20 | 2.88 | 2.90 | 2.14 | 1.27 | 0.80 | 25.88 |
| Days ≥ 0.1" Precip | 2 | 2 | 4 | 5 | 7 | 6 | 6 | 5 | 5 | 4 | 3 | 2 | 51 |
| Total Snowfall (") | 6.3 | 5.6 | 6.2 | 1.5 | 0.0 | 0.0 | 0.0 | 0.0 | 0.0 | 1.2 | 4.2 | 6.7 | 31.7 |
| Days ≥ 1" Snow Depth | 19 | 13 | 8 | 1 | 0 | 0 | 0 | 0 | 0 | 0 | 4 | 15 | 60 |

## SIOUX RAPIDS 4 E *Buena Vista County* ELEVATION 1420 ft LAT/LONG 42° 53 ' N / 95° 3 ' W

| | JAN | FEB | MAR | APR | MAY | JUN | JUL | AUG | SEP | OCT | NOV | DEC | YEAR |
|---|---|---|---|---|---|---|---|---|---|---|---|---|---|
| Maximum Temp °F | 24.6 | 30.2 | 42.8 | 59.1 | 72.2 | 81.0 | 84.2 | 81.8 | 73.5 | 61.3 | 42.9 | 28.4 | 56.8 |
| Minimum Temp °F | 5.3 | 10.5 | 23.1 | 35.7 | 47.3 | 57.0 | 61.3 | 58.4 | 49.2 | 37.1 | 24.1 | 11.0 | 35.0 |
| Mean Temp °F | 15.0 | 20.4 | 33.0 | 47.4 | 59.8 | 69.0 | 72.8 | 70.1 | 61.4 | 49.2 | 33.6 | 19.7 | 45.9 |
| Days Max Temp ≥ 90 °F | 0 | 0 | 0 | 0 | 1 | 4 | 7 | 5 | 1 | 0 | 0 | 0 | 18 |
| Days Max Temp ≤ 32 °F | 21 | 16 | 6 | 0 | 0 | 0 | 0 | 0 | 0 | 0 | 6 | 18 | 67 |
| Days Min Temp ≤ 32 °F | 31 | 27 | 26 | 11 | 1 | 0 | 0 | 0 | 1 | 10 | 24 | 30 | 161 |
| Days Min Temp ≤ 0 °F | 12 | 8 | 2 | 0 | 0 | 0 | 0 | 0 | 0 | 0 | 1 | 7 | 30 |
| Heating Degree Days | 1545 | 1251 | 986 | 528 | 206 | 34 | 7 | 21 | 160 | 489 | 937 | 1399 | 7563 |
| Cooling Degree Days | 0 | 0 | 0 | 10 | 53 | 158 | 231 | *180* | 57 | 3 | 0 | 0 | 692 |
| Total Precipitation (") | 0.60 | 0.62 | 1.97 | 2.98 | 3.52 | 4.56 | 3.72 | 4.22 | 3.56 | 2.41 | 1.56 | 0.90 | 30.62 |
| Days ≥ 0.1" Precip | 2 | 2 | 4 | 6 | 7 | 7 | 6 | 6 | 6 | 4 | 4 | 2 | 56 |
| Total Snowfall (") | *7.2* | *5.8* | *5.8* | 1.6 | 0.0 | 0.0 | 0.0 | 0.0 | 0.0 | 0.3 | na | *6.1* | na |
| Days ≥ 1" Snow Depth | na | na | *6* | 1 | 0 | 0 | 0 | 0 | 0 | 0 | *6* | na | na |

## SPENCER *Clay County* ELEVATION 1326 ft LAT/LONG 43° 10 ' N / 95° 9 ' W

| | JAN | FEB | MAR | APR | MAY | JUN | JUL | AUG | SEP | OCT | NOV | DEC | YEAR |
|---|---|---|---|---|---|---|---|---|---|---|---|---|---|
| Maximum Temp °F | 24.0 | 29.4 | 42.1 | 58.3 | 71.3 | 80.7 | 84.0 | 81.5 | 72.7 | 60.7 | 41.7 | *28.1* | 56.2 |
| Minimum Temp °F | 4.6 | 10.5 | 22.8 | 35.3 | 47.1 | 57.3 | 61.5 | 58.5 | 48.8 | 36.7 | 23.1 | *10.6* | 34.7 |
| Mean Temp °F | 14.3 | 19.9 | 32.5 | 46.8 | 59.2 | 69.0 | 72.8 | 70.0 | 60.8 | 48.7 | 32.4 | *19.4* | 45.5 |
| Days Max Temp ≥ 90 °F | 0 | 0 | 0 | 0 | 1 | 5 | 7 | 5 | 1 | 0 | 0 | 0 | 19 |
| Days Max Temp ≤ 32 °F | 22 | 16 | 7 | 0 | 0 | 0 | 0 | 0 | 0 | 0 | 7 | *19* | 71 |
| Days Min Temp ≤ 32 °F | 31 | 28 | 26 | 12 | 2 | 0 | 0 | 0 | 1 | 11 | 25 | *31* | 167 |
| Days Min Temp ≤ 0 °F | 12 | 8 | 2 | 0 | 0 | 0 | 0 | 0 | 0 | 0 | 1 | *7* | 30 |
| Heating Degree Days | 1566 | 1266 | 1000 | 546 | 222 | 35 | 7 | 25 | 180 | 504 | 971 | *1409* | 7731 |
| Cooling Degree Days | 0 | 0 | 0 | 10 | 51 | 163 | 253 | 190 | 66 | 4 | 0 | 0 | 737 |
| Total Precipitation (") | 0.54 | 0.55 | 1.89 | 3.06 | 3.72 | 4.14 | 4.02 | 4.02 | 3.10 | 2.12 | 1.27 | 0.70 | 29.13 |
| Days ≥ 0.1" Precip | 2 | 1 | 4 | 6 | 8 | 7 | 6 | 6 | 6 | 4 | 3 | 2 | 55 |
| Total Snowfall (") | na | na | na | *0.8* | 0.0 | 0.0 | 0.0 | 0.0 | 0.0 | 0.3 | na | na | na |
| Days ≥ 1" Snow Depth | na | na | na | 1 | 0 | 0 | 0 | 0 | 0 | 0 | *4* | na | na |

## STORM LAKE 2 E *Buena Vista County* ELEVATION 1430 ft LAT/LONG 42° 38 ' N / 95° 11 ' W

| | JAN | FEB | MAR | APR | MAY | JUN | JUL | AUG | SEP | OCT | NOV | DEC | YEAR |
|---|---|---|---|---|---|---|---|---|---|---|---|---|---|
| Maximum Temp °F | 24.4 | 29.6 | 42.5 | 58.1 | 70.5 | 79.5 | 83.2 | 80.7 | 73.0 | 60.9 | 43.2 | 29.1 | 56.2 |
| Minimum Temp °F | 4.6 | 9.7 | 21.9 | 34.4 | 46.2 | 56.2 | 60.7 | 58.1 | 49.0 | 37.2 | 23.9 | 11.1 | 34.4 |
| Mean Temp °F | 14.5 | 19.7 | 32.3 | 46.3 | 58.4 | 67.9 | 72.0 | 69.5 | 61.0 | 49.1 | 33.6 | 20.1 | 45.4 |
| Days Max Temp ≥ 90 °F | 0 | 0 | 0 | 0 | 0 | 3 | 6 | 3 | 1 | 0 | 0 | 0 | 13 |
| Days Max Temp ≤ 32 °F | 21 | 16 | 7 | 1 | 0 | 0 | 0 | 0 | 0 | 0 | 6 | 18 | 69 |
| Days Min Temp ≤ 32 °F | 31 | 28 | 27 | 13 | 2 | 0 | 0 | 0 | 1 | 10 | 24 | 31 | 167 |
| Days Min Temp ≤ 0 °F | 12 | 9 | 2 | 0 | 0 | 0 | 0 | 0 | 0 | 0 | 1 | 7 | 31 |
| Heating Degree Days | 1561 | 1274 | 1009 | 561 | 237 | 44 | 9 | 28 | 166 | 490 | 935 | 1386 | 7700 |
| Cooling Degree Days | 0 | 0 | 0 | 9 | 40 | 144 | 226 | 173 | 60 | 3 | 0 | 0 | 655 |
| Total Precipitation (") | 0.68 | 0.59 | 2.04 | 3.22 | 4.02 | 4.86 | 4.46 | 4.38 | 3.84 | 2.40 | 1.52 | 0.85 | 32.86 |
| Days ≥ 0.1" Precip | 2 | 2 | 4 | 6 | 8 | 7 | 7 | 6 | 6 | 4 | 3 | 2 | 57 |
| Total Snowfall (") | 7.4 | 7.3 | 7.2 | 2.2 | 0.0 | 0.0 | 0.0 | 0.0 | 0.0 | 0.4 | 4.3 | 7.9 | 36.7 |
| Days ≥ 1" Snow Depth | 24 | 18 | 11 | 2 | 0 | 0 | 0 | 0 | 0 | 0 | 5 | 18 | 78 |

## SWEA CITY *Kossuth County* ELEVATION 1200 ft LAT/LONG 43° 23 ' N / 94° 15 ' W

| | JAN | FEB | MAR | APR | MAY | JUN | JUL | AUG | SEP | OCT | NOV | DEC | YEAR |
|---|---|---|---|---|---|---|---|---|---|---|---|---|---|
| Maximum Temp °F | *23.0* | 27.9 | *41.1* | 57.6 | 71.6 | 80.6 | 83.2 | 80.9 | 73.3 | 61.2 | 42.6 | 27.6 | 55.9 |
| Minimum Temp °F | *4.4* | 9.7 | *22.8* | 35.2 | 47.1 | 57.1 | 60.8 | 57.9 | 49.2 | 37.5 | 24.8 | 10.5 | 34.7 |
| Mean Temp °F | *13.7* | 18.9 | *32.0* | 46.4 | 59.4 | 68.9 | 72.0 | 69.4 | 61.3 | 49.4 | 33.8 | 19.1 | 45.4 |
| Days Max Temp ≥ 90 °F | 0 | 0 | 0 | 0 | 1 | 4 | 5 | 3 | 1 | 0 | 0 | 0 | 14 |
| Days Max Temp ≤ 32 °F | *22* | 18 | *7* | 1 | 0 | 0 | 0 | 0 | 0 | 0 | 6 | 20 | 74 |
| Days Min Temp ≤ 32 °F | *31* | 28 | *26* | 12 | 2 | 0 | 0 | 0 | 1 | 10 | 24 | 30 | 164 |
| Days Min Temp ≤ 0 °F | *13* | 8 | *2* | 0 | 0 | 0 | 0 | 0 | 0 | 0 | 1 | 8 | 32 |
| Heating Degree Days | *1587* | 1299 | *1017* | 555 | 213 | 36 | 10 | 27 | 161 | 480 | 931 | 1419 | 7735 |
| Cooling Degree Days | 0 | 0 | 0 | 7 | 53 | 179 | 253 | 185 | 70 | 3 | 0 | 0 | 750 |
| Total Precipitation (") | 0.76 | 0.76 | 1.90 | 3.07 | 3.81 | 4.62 | 3.90 | 3.89 | 3.23 | 2.17 | 1.76 | 0.94 | 30.81 |
| Days ≥ 0.1" Precip | *2* | 3 | *4* | *7* | 7 | 7 | 7 | 6 | 6 | 5 | 4 | 3 | 61 |
| Total Snowfall (") | *9.6* | *7.1* | *6.5* | 1.5 | 0.0 | 0.0 | 0.0 | 0.0 | 0.0 | 0.5 | 4.9 | 8.8 | 38.9 |
| Days ≥ 1" Snow Depth | na | *21* | *12* | *2* | 0 | 0 | 0 | 0 | 0 | 0 | *4* | *18* | na |

## TIPTON 4 NE *Cedar County* ELEVATION 810 ft LAT/LONG 41° 46 ' N / 91° 8 ' W

| | JAN | FEB | MAR | APR | MAY | JUN | JUL | AUG | SEP | OCT | NOV | DEC | YEAR |
|---|---|---|---|---|---|---|---|---|---|---|---|---|---|
| Maximum Temp °F | 26.8 | 32.5 | 46.1 | 60.6 | 72.5 | 81.2 | 84.7 | 82.3 | 74.9 | 63.3 | 46.9 | 33.0 | 58.7 |
| Minimum Temp °F | 8.1 | 13.5 | 26.1 | 37.5 | 48.6 | 58.3 | 62.2 | 59.3 | 50.8 | 39.3 | 28.2 | 15.7 | 37.3 |
| Mean Temp °F | 17.5 | 23.1 | 36.1 | 49.1 | 60.6 | 69.8 | 73.5 | 70.8 | 62.9 | 51.3 | 37.6 | 24.4 | 48.1 |
| Days Max Temp ≥ 90 °F | 0 | 0 | 0 | 0 | 1 | 4 | 8 | 5 | 1 | 0 | 0 | 0 | 19 |
| Days Max Temp ≤ 32 °F | 20 | 13 | 4 | 0 | 0 | 0 | 0 | 0 | 0 | 0 | 3 | 14 | 54 |
| Days Min Temp ≤ 32 °F | 30 | 27 | 23 | 9 | 1 | 0 | 0 | 0 | 1 | 8 | 21 | 29 | 149 |
| Days Min Temp ≤ 0 °F | 10 | 6 | 0 | 0 | 0 | 0 | 0 | 0 | 0 | 0 | 0 | 4 | 20 |
| Heating Degree Days | 1468 | 1178 | 889 | 480 | 186 | 27 | 5 | 16 | 133 | 427 | 816 | 1252 | 6877 |
| Cooling Degree Days | 0 | 0 | 1 | 13 | 51 | 177 | 278 | 211 | 80 | 4 | 0 | 0 | 815 |
| Total Precipitation (") | 1.25 | 1.26 | 2.48 | 3.72 | 4.24 | 4.68 | 4.42 | 4.81 | 4.19 | 2.72 | 2.65 | 2.12 | 38.54 |
| Days ≥ 0.1" Precip | 3 | 3 | 6 | 7 | 8 | 7 | 7 | 6 | 7 | 6 | 6 | 5 | 71 |
| Total Snowfall (") | 5.7 | 6.0 | 2.8 | 1.1 | 0.0 | 0.0 | 0.0 | 0.0 | 0.0 | 0.2 | *1.5* | *6.7* | 24.0 |
| Days ≥ 1" Snow Depth | *17* | na | *3* | 0 | 0 | 0 | 0 | 0 | 0 | 0 | *1* | na | na |

## TOLEDO *Tama County* ELEVATION 932 ft LAT/LONG 42° 0 ' N / 92° 34 ' W

| | JAN | FEB | MAR | APR | MAY | JUN | JUL | AUG | SEP | OCT | NOV | DEC | YEAR |
|---|---|---|---|---|---|---|---|---|---|---|---|---|---|
| Maximum Temp °F | 26.3 | 32.0 | 44.6 | 59.1 | 71.0 | 80.6 | 84.9 | 82.3 | 74.8 | 62.7 | 46.3 | 32.0 | 58.1 |
| Minimum Temp °F | 6.9 | 11.9 | 24.8 | 36.6 | 47.5 | 57.6 | 61.8 | 59.3 | 49.9 | 38.0 | 26.7 | 14.1 | 36.3 |
| Mean Temp °F | 16.6 | 22.0 | 34.7 | 47.9 | 59.3 | 69.1 | 73.4 | 70.8 | 62.4 | 50.4 | 36.5 | 23.0 | 47.2 |
| Days Max Temp ≥ 90 °F | 0 | 0 | 0 | 0 | 0 | 3 | 8 | 5 | 1 | 0 | 0 | 0 | 17 |
| Days Max Temp ≤ 32 °F | 20 | 14 | 5 | 0 | 0 | 0 | 0 | 0 | 0 | 0 | 4 | 14 | 57 |
| Days Min Temp ≤ 32 °F | 31 | 27 | 24 | 10 | 1 | 0 | 0 | 0 | 1 | 10 | 22 | 30 | 156 |
| Days Min Temp ≤ 0 °F | 11 | 7 | 1 | 0 | 0 | 0 | 0 | 0 | 0 | 0 | 0 | 5 | 24 |
| Heating Degree Days | 1495 | 1208 | 933 | 514 | 217 | 33 | 7 | 19 | 144 | 456 | 848 | 1296 | 7170 |
| Cooling Degree Days | 0 | 0 | 1 | 11 | 46 | 173 | 286 | 228 | 83 | 8 | 0 | 0 | 836 |
| Total Precipitation (") | 1.00 | 0.91 | 2.32 | 3.54 | 4.40 | 4.89 | 4.59 | 4.85 | 3.67 | 2.44 | 2.08 | 1.33 | 36.02 |
| Days ≥ 0.1" Precip | 3 | 3 | 5 | 7 | 8 | 7 | 7 | 6 | 7 | 5 | 4 | 3 | 65 |
| Total Snowfall (") | 7.1 | 5.5 | 3.6 | 1.1 | 0.0 | 0.0 | 0.0 | 0.0 | 0.0 | 0.0 | 2.0 | 6.4 | 25.7 |
| Days ≥ 1" Snow Depth | 17 | 14 | 6 | 1 | 0 | 0 | 0 | 0 | 0 | 0 | 2 | 12 | 52 |

## TRIPOLI *Bremer County* ELEVATION 932 ft LAT/LONG 42° 49 ' N / 92° 15 ' W

| | JAN | FEB | MAR | APR | MAY | JUN | JUL | AUG | SEP | OCT | NOV | DEC | YEAR |
|---|---|---|---|---|---|---|---|---|---|---|---|---|---|
| Maximum Temp °F | 23.5 | 29.0 | 43.0 | 58.6 | 71.4 | 80.1 | 83.3 | 80.9 | 73.1 | 60.9 | 43.3 | 29.4 | 56.4 |
| Minimum Temp °F | 4.9 | 10.4 | 24.3 | 36.5 | 47.6 | 57.1 | 61.3 | 58.4 | 50.1 | 38.5 | 26.2 | 12.9 | 35.7 |
| Mean Temp °F | 14.2 | 19.7 | 33.7 | 47.6 | 59.5 | 68.6 | 72.3 | 69.6 | 61.6 | 49.7 | 34.8 | 21.2 | 46.0 |
| Days Max Temp ≥ 90 °F | 0 | 0 | 0 | 0 | 1 | 3 | 5 | 3 | 1 | 0 | 0 | 0 | 13 |
| Days Max Temp ≤ 32 °F | 22 | 16 | 6 | 0 | 0 | 0 | 0 | 0 | 0 | 0 | 5 | 18 | 67 |
| Days Min Temp ≤ 32 °F | 31 | 27 | 24 | 10 | 1 | 0 | 0 | 0 | 1 | 9 | 22 | 30 | 155 |
| Days Min Temp ≤ 0 °F | 12 | 8 | 1 | 0 | 0 | 0 | 0 | 0 | 0 | 0 | 1 | 6 | 28 |
| Heating Degree Days | 1569 | 1272 | 964 | 523 | 209 | 35 | 8 | 25 | 155 | 474 | 900 | 1351 | 7485 |
| Cooling Degree Days | 0 | 0 | *0* | 9 | 42 | 145 | *232* | *163* | 62 | 2 | *0* | 0 | 655 |
| Total Precipitation (") | 1.02 | 0.97 | 2.14 | 3.65 | 4.13 | 4.58 | 4.82 | 4.99 | 3.53 | 2.54 | 2.17 | 1.30 | 35.84 |
| Days ≥ 0.1" Precip | 4 | 3 | 5 | 8 | 8 | 8 | 7 | 7 | 6 | 5 | 5 | 4 | 70 |
| Total Snowfall (") | 8.2 | 7.2 | 5.4 | 2.2 | 0.0 | 0.0 | 0.0 | 0.0 | 0.0 | 0.0 | 4.4 | 8.5 | 35.9 |
| Days ≥ 1" Snow Depth | 17 | *16* | *7* | 1 | 0 | 0 | 0 | 0 | 0 | 0 | 3 | 12 | 56 |

## VINTON *Benton County* ELEVATION 810 ft LAT/LONG 42° 10 ' N / 92° 1 ' W

| | JAN | FEB | MAR | APR | MAY | JUN | JUL | AUG | SEP | OCT | NOV | DEC | YEAR |
|---|---|---|---|---|---|---|---|---|---|---|---|---|---|
| Maximum Temp °F | 27.0 | 32.1 | 46.1 | 61.6 | 72.9 | 82.0 | 85.6 | 83.0 | 75.3 | 63.0 | 45.8 | 31.9 | 58.9 |
| Minimum Temp °F | 8.3 | 13.8 | 26.6 | 38.1 | 48.6 | 57.9 | 61.8 | 59.4 | 51.0 | 39.8 | 27.7 | 15.6 | 37.4 |
| Mean Temp °F | 17.7 | 22.8 | 36.4 | 50.1 | 60.8 | 70.1 | 73.8 | 71.2 | 63.3 | 51.5 | 36.8 | 23.6 | 48.2 |
| Days Max Temp ≥ 90 °F | 0 | 0 | 0 | 0 | 1 | 4 | 8 | 6 | 1 | 0 | 0 | 0 | 20 |
| Days Max Temp ≤ 32 °F | 19 | 13 | 4 | 0 | 0 | 0 | 0 | 0 | 0 | 0 | 3 | 14 | 53 |
| Days Min Temp ≤ 32 °F | 30 | 27 | 22 | 9 | 1 | 0 | 0 | 0 | 1 | 8 | 21 | 29 | 148 |
| Days Min Temp ≤ 0 °F | 10 | 6 | 1 | 0 | 0 | 0 | 0 | 0 | 0 | 0 | 0 | 4 | 21 |
| Heating Degree Days | 1463 | 1185 | 879 | 451 | 179 | 23 | 4 | 14 | 125 | 421 | 839 | 1276 | 6859 |
| Cooling Degree Days | *0* | *0* | *1* | *15* | 53 | 189 | 293 | 222 | 88 | 5 | 0 | 0 | 866 |
| Total Precipitation (") | 1.00 | 0.94 | 2.18 | 3.33 | 3.79 | 4.36 | 4.07 | 4.44 | 3.84 | 2.36 | 2.17 | 1.44 | 33.92 |
| Days ≥ 0.1" Precip | 3 | 3 | 5 | 7 | 8 | 7 | 6 | 6 | 6 | 5 | 5 | 3 | 64 |
| Total Snowfall (") | 7.4 | 6.8 | 4.5 | 1.6 | 0.0 | 0.0 | 0.0 | 0.0 | 0.0 | 0.0 | 3.0 | 6.9 | 30.2 |
| Days ≥ 1" Snow Depth | 22 | *18* | 6 | 1 | 0 | 0 | 0 | 0 | 0 | 0 | 3 | 13 | 63 |

## WASHINGTON *Washington County* ELEVATION 751 ft LAT/LONG 41° 17' N / 91° 41' W

| | JAN | FEB | MAR | APR | MAY | JUN | JUL | AUG | SEP | OCT | NOV | DEC | YEAR |
|---|---|---|---|---|---|---|---|---|---|---|---|---|---|
| Maximum Temp °F | 30.8 | 36.2 | 49.5 | 64.1 | 75.0 | 83.8 | 87.7 | 85.1 | 77.7 | 66.0 | 49.2 | 35.7 | 61.7 |
| Minimum Temp °F | 13.0 | 18.1 | 29.6 | 41.2 | 51.7 | 61.1 | 65.4 | 62.5 | 54.7 | 43.3 | 31.4 | 19.6 | 41.0 |
| Mean Temp °F | 21.9 | 27.2 | 39.6 | 52.7 | 63.4 | 72.5 | 76.5 | 73.8 | 66.2 | 54.7 | 40.3 | 27.7 | 51.4 |
| Days Max Temp ≥ 90 °F | 0 | 0 | 0 | 0 | 1 | 5 | 12 | 8 | 2 | 0 | 0 | 0 | 28 |
| Days Max Temp ≤ 32 °F | 16 | 11 | 3 | 0 | 0 | 0 | 0 | 0 | 0 | 0 | 2 | 11 | 43 |
| Days Min Temp ≤ 32 °F | 29 | 25 | 18 | 5 | 0 | 0 | 0 | 0 | 0 | 4 | 17 | 27 | 125 |
| Days Min Temp ≤ 0 °F | 7 | 3 | 0 | 0 | 0 | 0 | 0 | 0 | 0 | 0 | 0 | 3 | 13 |
| Heating Degree Days | 1330 | 1061 | 782 | 380 | 126 | 10 | 1 | 5 | 80 | 331 | 733 | 1151 | 5990 |
| Cooling Degree Days | 0 | 0 | 2 | 21 | 82 | 250 | 374 | 301 | 131 | 12 | 0 | 0 | 1173 |
| Total Precipitation (") | 1.13 | 0.88 | 2.25 | 3.13 | 3.85 | 4.31 | 4.25 | 4.00 | 4.22 | 2.64 | 2.24 | 1.77 | 34.67 |
| Days ≥ 0.1" Precip | 3 | 3 | 5 | 6 | 7 | 7 | 7 | 6 | 7 | 5 | 5 | 4 | 65 |
| Total Snowfall (") | 6.0 | 4.9 | 2.7 | 1.0 | 0.0 | 0.0 | 0.0 | 0.0 | 0.0 | 0.2 | 1.3 | 4.9 | 21.0 |
| Days ≥ 1" Snow Depth | 18 | 13 | 4 | 0 | 0 | 0 | 0 | 0 | 0 | 0 | 2 | 10 | 47 |

## WATERLOO MUNI AP *Black Hawk County* ELEVATION 869 ft LAT/LONG 42° 33' N / 92° 24' W

| | JAN | FEB | MAR | APR | MAY | JUN | JUL | AUG | SEP | OCT | NOV | DEC | YEAR |
|---|---|---|---|---|---|---|---|---|---|---|---|---|---|
| Maximum Temp °F | 23.8 | 29.4 | 43.3 | 58.9 | 71.1 | 80.5 | 83.7 | 81.3 | 73.3 | 60.9 | 44.1 | 30.1 | 56.7 |
| Minimum Temp °F | 5.6 | 11.1 | 24.6 | 36.4 | 47.9 | 57.8 | 62.1 | 59.0 | 50.0 | 38.1 | 25.8 | 13.2 | 36.0 |
| Mean Temp °F | 14.7 | 20.3 | 34.0 | 47.7 | 59.5 | 69.2 | 72.9 | 70.2 | 61.7 | 49.5 | 35.0 | 21.7 | 46.4 |
| Days Max Temp ≥ 90 °F | 0 | 0 | 0 | 0 | 1 | 4 | 6 | 4 | 1 | 0 | 0 | 0 | 16 |
| Days Max Temp ≤ 32 °F | 22 | 15 | 6 | 0 | 0 | 0 | 0 | 0 | 0 | 0 | 5 | 17 | 65 |
| Days Min Temp ≤ 32 °F | 31 | 27 | 24 | 11 | 1 | 0 | 0 | 0 | 1 | 10 | 22 | 30 | 157 |
| Days Min Temp ≤ 0 °F | 12 | 8 | 1 | 0 | 0 | 0 | 0 | 0 | 0 | 0 | 0 | 6 | 27 |
| Heating Degree Days | 1555 | 1257 | 956 | 521 | 212 | 30 | 8 | 25 | 160 | 480 | 894 | 1337 | 7435 |
| Cooling Degree Days | 0 | 0 | 0 | 12 | 53 | 173 | 280 | 209 | 76 | 4 | 0 | 0 | 807 |
| Total Precipitation (") | 0.88 | 1.04 | 2.24 | 3.35 | 4.07 | 4.66 | 4.72 | 4.03 | 3.31 | 2.54 | 1.98 | 1.38 | 34.20 |
| Days ≥ 0.1" Precip | 2 | 3 | 5 | 7 | 7 | 7 | 7 | 6 | 6 | 5 | 4 | 4 | 63 |
| Total Snowfall (") | 7.1 | 6.9 | 5.0 | 1.8 | 0.0 | 0.0 | 0.0 | 0.0 | 0.0 | 0.1 | 4.0 | 7.8 | 32.7 |
| Days ≥ 1" Snow Depth | 22 | 18 | 8 | 1 | 0 | 0 | 0 | 0 | 0 | 0 | 4 | 17 | 70 |

## WAUKON *Allamakee County* ELEVATION 1240 ft LAT/LONG 43° 16' N / 91° 29' W

| | JAN | FEB | MAR | APR | MAY | JUN | JUL | AUG | SEP | OCT | NOV | DEC | YEAR |
|---|---|---|---|---|---|---|---|---|---|---|---|---|---|
| Maximum Temp °F | 22.9 | 28.6 | 41.7 | 57.2 | 69.1 | 77.2 | 81.2 | 78.8 | 71.2 | 59.7 | 42.3 | 28.3 | 54.9 |
| Minimum Temp °F | 6.1 | 11.6 | 23.9 | 36.6 | 47.9 | 57.2 | 61.8 | 59.5 | 51.3 | 39.7 | 26.2 | 13.2 | 36.3 |
| Mean Temp °F | 14.5 | 20.1 | 32.8 | 47.0 | 58.5 | 67.2 | 71.5 | 69.3 | 61.3 | 49.7 | 34.3 | 20.8 | 45.6 |
| Days Max Temp ≥ 90 °F | 0 | 0 | 0 | 0 | 0 | 1 | 2 | 1 | 0 | 0 | 0 | 0 | 4 |
| Days Max Temp ≤ 32 °F | 23 | 16 | 6 | 0 | 0 | 0 | 0 | 0 | 0 | 0 | 5 | 19 | 69 |
| Days Min Temp ≤ 32 °F | 31 | 27 | 25 | 10 | 1 | 0 | 0 | 0 | 0 | 8 | 23 | 30 | 155 |
| Days Min Temp ≤ 0 °F | 11 | 7 | 1 | 0 | 0 | 0 | 0 | 0 | 0 | 0 | 1 | 6 | 26 |
| Heating Degree Days | 1561 | 1261 | 991 | 539 | 228 | 45 | 9 | 23 | 158 | 473 | 915 | 1364 | 7567 |
| Cooling Degree Days | 0 | 0 | 0 | 8 | 32 | 124 | 221 | 170 | 63 | 2 | 0 | 0 | 620 |
| Total Precipitation (") | 0.59 | 0.48 | 1.54 | 3.43 | 3.45 | 4.58 | 4.02 | 4.15 | 3.73 | 2.24 | 1.87 | 0.92 | 31.00 |
| Days ≥ 0.1" Precip | 2 | 1 | 4 | 7 | 7 | 7 | 6 | 6 | 6 | 5 | 4 | 3 | 58 |
| Total Snowfall (") | 8.0 | 6.3 | 5.8 | 1.8 | 0.0 | 0.0 | 0.0 | 0.0 | 0.0 | 0.0 | 3.3 | ***8.1*** | 33.3 |
| Days ≥ 1" Snow Depth | na | na | ***9*** | 1 | 0 | 0 | 0 | 0 | 0 | 0 | na | na | na |

## WEBSTER CITY *Hamilton County* ELEVATION 1040 ft LAT/LONG 42° 28' N / 93° 49' W

| | JAN | FEB | MAR | APR | MAY | JUN | JUL | AUG | SEP | OCT | NOV | DEC | YEAR |
|---|---|---|---|---|---|---|---|---|---|---|---|---|---|
| Maximum Temp °F | 26.0 | 31.7 | 44.9 | 60.6 | 72.8 | 81.3 | 84.4 | 81.8 | 74.4 | 62.5 | 44.7 | 30.7 | 58.0 |
| Minimum Temp °F | 7.3 | 12.7 | 25.2 | 37.0 | 48.1 | 57.8 | 62.1 | 59.5 | 50.8 | 39.1 | 26.3 | 13.5 | 36.6 |
| Mean Temp °F | 16.7 | 22.3 | 35.1 | 48.8 | 60.5 | 69.6 | 73.3 | 70.7 | 62.6 | 50.8 | 35.5 | 22.1 | 47.3 |
| Days Max Temp ≥ 90 °F | 0 | 0 | 0 | 0 | 0 | 3 | 6 | 4 | 1 | 0 | 0 | 0 | 14 |
| Days Max Temp ≤ 32 °F | 20 | 14 | 5 | 0 | 0 | 0 | 0 | 0 | 0 | 0 | 5 | 16 | 60 |
| Days Min Temp ≤ 32 °F | 31 | 27 | 24 | 10 | 1 | 0 | 0 | 0 | 1 | 9 | 22 | 30 | 155 |
| Days Min Temp ≤ 0 °F | 11 | 7 | 1 | 0 | 0 | 0 | 0 | 0 | 0 | 0 | 0 | 5 | 24 |
| Heating Degree Days | 1493 | 1201 | 921 | 488 | 186 | 26 | 6 | 18 | 136 | 440 | 878 | 1323 | 7116 |
| Cooling Degree Days | 0 | 0 | 0 | 11 | 49 | 176 | 269 | 206 | 78 | 4 | 0 | 0 | 793 |
| Total Precipitation (") | 0.75 | 0.80 | 1.85 | 2.88 | 3.86 | 4.94 | 4.44 | 4.50 | 3.45 | 2.53 | 1.64 | 1.25 | 32.89 |
| Days ≥ 0.1" Precip | 2 | 2 | 4 | 6 | 8 | 8 | 7 | 6 | 6 | 5 | 4 | 3 | 61 |
| Total Snowfall (") | 7.3 | 6.6 | 5.0 | 1.7 | 0.0 | 0.0 | 0.0 | 0.0 | 0.0 | 0.1 | 3.5 | 7.3 | 31.5 |
| Days ≥ 1" Snow Depth | 22 | 18 | 7 | 1 | 0 | 0 | 0 | 0 | 0 | 0 | 4 | 17 | 69 |

## WILLIAMSBURG *Iowa County* ELEVATION 801 ft LAT/LONG 41° 39 ' N / 92° 1 ' W

| | JAN | FEB | MAR | APR | MAY | JUN | JUL | AUG | SEP | OCT | NOV | DEC | YEAR |
|---|---|---|---|---|---|---|---|---|---|---|---|---|---|
| Maximum Temp °F | 28.6 | 34.2 | 47.3 | 62.1 | 73.2 | 81.9 | 85.4 | 83.2 | 75.8 | 64.2 | 47.5 | 33.9 | 59.8 |
| Minimum Temp °F | 10.2 | 15.1 | 27.3 | 38.5 | 49.6 | 59.4 | 63.8 | 61.0 | 52.7 | 41.0 | 29.0 | 17.0 | 38.7 |
| Mean Temp °F | 19.4 | 24.7 | 37.4 | 50.3 | 61.4 | 70.7 | 74.6 | 72.1 | 64.3 | 52.6 | 38.3 | 25.5 | 49.3 |
| Days Max Temp ≥ 90 °F | 0 | 0 | 0 | 0 | 1 | 4 | 8 | 6 | 2 | 0 | 0 | 0 | 21 |
| Days Max Temp ≤ 32 °F | 18 | 12 | 4 | 0 | 0 | 0 | 0 | 0 | 0 | 0 | 3 | 13 | 50 |
| Days Min Temp ≤ 32 °F | 30 | 26 | 22 | 9 | 1 | 0 | 0 | 0 | 0 | 7 | 19 | 29 | 143 |
| Days Min Temp ≤ 0 °F | 9 | 5 | 0 | 0 | 0 | 0 | 0 | 0 | 0 | 0 | 0 | 3 | 17 |
| Heating Degree Days | 1407 | 1133 | 851 | 445 | 165 | 19 | 4 | 12 | 110 | 389 | 795 | 1219 | 6549 |
| Cooling Degree Days | 0 | 0 | 1 | 15 | 56 | 197 | 308 | 245 | 99 | 6 | 0 | 0 | 927 |
| Total Precipitation (") | 1.08 | 0.87 | 2.18 | 3.70 | 4.19 | 4.84 | 4.76 | 4.89 | 4.41 | 2.54 | 2.45 | 1.55 | 37.46 |
| Days ≥ 0.1" Precip | 3 | 3 | 5 | 7 | 8 | 7 | 7 | 6 | 7 | 5 | 5 | 4 | 67 |
| Total Snowfall (") | 7.9 | 6.4 | 3.6 | 1.9 | 0.1 | 0.0 | 0.0 | 0.0 | 0.0 | 0.3 | 2.3 | 6.2 | 28.7 |
| Days ≥ 1" Snow Depth | 20 | 14 | 6 | 1 | 0 | 0 | 0 | 0 | 0 | 0 | 2 | 11 | 54 |

## WINTERSET 2 NNW *Madison County* ELEVATION 1132 ft LAT/LONG 41° 22 ' N / 94° 1 ' W

| | JAN | FEB | MAR | APR | MAY | JUN | JUL | AUG | SEP | OCT | NOV | DEC | YEAR |
|---|---|---|---|---|---|---|---|---|---|---|---|---|---|
| Maximum Temp °F | 29.9 | 35.3 | 48.6 | 62.9 | 73.2 | 81.9 | 85.9 | 83.6 | 75.6 | 64.3 | 47.6 | 35.0 | 60.3 |
| Minimum Temp °F | 10.2 | 14.7 | 26.8 | 38.4 | 49.1 | 58.2 | 63.0 | 60.5 | 51.8 | 40.6 | 27.9 | 16.9 | 38.2 |
| Mean Temp °F | 20.1 | 25.0 | 37.7 | 50.7 | 61.2 | 70.1 | 74.5 | 72.1 | 63.7 | 52.5 | 37.8 | 26.0 | 49.3 |
| Days Max Temp ≥ 90 °F | 0 | 0 | 0 | 0 | 0 | 4 | 9 | 5 | 1 | 0 | 0 | 0 | 19 |
| Days Max Temp ≤ 32 °F | 16 | 12 | 4 | 0 | 0 | 0 | 0 | 0 | 0 | 0 | 3 | 12 | 47 |
| Days Min Temp ≤ 32 °F | 30 | 26 | 22 | 9 | 1 | 0 | 0 | 0 | 1 | 7 | 20 | 29 | 145 |
| Days Min Temp ≤ 0 °F | 9 | 5 | 1 | 0 | 0 | 0 | 0 | 0 | 0 | 0 | 0 | 4 | 19 |
| Heating Degree Days | 1387 | 1124 | 841 | 435 | 168 | 24 | 5 | 12 | 119 | 392 | 809 | 1203 | 6519 |
| Cooling Degree Days | 0 | *0* | *3* | *20* | 60 | 199 | 323 | 257 | 109 | 8 | 0 | 0 | 979 |
| Total Precipitation (") | 0.97 | 0.93 | 2.20 | 3.49 | 4.00 | 4.67 | 4.00 | 4.19 | 3.96 | 2.57 | 2.02 | 1.15 | 34.15 |
| Days ≥ 0.1" Precip | 3 | 3 | 5 | 6 | 8 | 7 | 7 | 6 | 6 | 5 | 4 | 3 | 63 |
| Total Snowfall (") | 6.4 | 6.6 | 2.6 | 1.1 | 0.1 | 0.0 | 0.0 | 0.0 | 0.0 | 0.2 | 3.0 | 5.1 | 25.1 |
| Days ≥ 1" Snow Depth | 18 | 15 | 6 | 1 | 0 | 0 | 0 | 0 | 0 | 0 | 3 | 11 | 54 |

## JANUARY MINIMUM TEMPERATURE °F

| LOWEST | | | HIGHEST | | |
|---|---|---|---|---|---|
| 1 | Rock Rapids | 2.4 | 1 | Albia | 13.9 |
| 2 | Cresco | 2.7 | | Centerville | 13.9 |
| | Lake Park | 2.7 | 3 | Burlington | 13.7 |
| 4 | Estherville | 3.1 | 4 | Fort Madison | 13.6 |
| 5 | Sibley | 3.4 | 5 | Keosauqua | 13.5 |
| 6 | Milford | 3.7 | | Sidney | 13.5 |
| 7 | Northwood | 3.8 | 7 | Bloomfield | 13.4 |
| 8 | Forest City | 3.9 | 8 | Lamoni | 13.1 |
| | Sanborn | 3.9 | 9 | Washington | 13.0 |
| 10 | Cherokee | 4.0 | 10 | Shenandoah | 12.8 |
| 11 | Mason City | 4.1 | 11 | Mount Pleasant | 12.6 |
| | Mason City-Ap | 4.1 | 12 | Fairfield | 12.5 |
| 13 | Clarion | 4.2 | | Ottumwa | 12.5 |
| 14 | Sheldon | 4.3 | 14 | Iowa City | 12.4 |
| 15 | Swea City | 4.4 | | Muscatine | 12.4 |
| 16 | Spencer | 4.6 | 16 | Columbus Jnction | 12.3 |
| | Storm Lake | 4.6 | | Leon | 12.3 |
| 18 | Tripoli | 4.9 | 18 | Bedford | 12.2 |
| 19 | Pocahontas | 5.1 | | Clarinda | 12.2 |
| 20 | Britt | 5.3 | | Knoxville | 12.2 |
| | Fayette | 5.3 | 21 | Le Claire L & D | 12.1 |
| | Sioux Rapids | 5.3 | 22 | Beaconsfield | 11.8 |
| 23 | Algona | 5.4 | | Mount Ayr | 11.8 |
| | Decorah | 5.4 | 24 | Sigourney | 11.7 |
| | Emmetsburg | 5.4 | 25 | Clinton | 11.6 |

## JULY MAXIMUM TEMPERATURE °F

| HIGHEST | | | LOWEST | | |
|---|---|---|---|---|---|
| 1 | Red Oak | 88.5 | 1 | Waukon | 81.2 |
| 2 | Sidney | 88.3 | 2 | Cresco | 81.5 |
| 3 | Keosauqua | 88.2 | 3 | Dubuque | 82.3 |
| 4 | Shenandoah | 87.9 | 4 | New Hampton | 82.5 |
| 5 | Washington | 87.7 | 5 | Estherville | 82.8 |
| 6 | Glenwood | 87.6 | | Oelwein | 82.8 |
| | Logan | 87.6 | 7 | Osage | 82.9 |
| 8 | Clarinda | 87.4 | 8 | Forest City | 83.0 |
| | Iowa City | 87.4 | | Grundy Center | 83.0 |
| | Knoxville | 87.4 | 10 | Lake Park | 83.1 |
| | Leon | 87.4 | 11 | Hampton | 83.2 |
| 12 | Bloomfield | 87.3 | | Mason City-Ap | 83.2 |
| | Fairfield | 87.3 | | Storm Lake | 83.2 |
| 14 | Bedford | 87.2 | | Swea City | 83.2 |
| 15 | Indianola | 87.0 | 15 | Decorah | 83.3 |
| | Onawa | 87.0 | | Tripoli | 83.3 |
| 17 | Albia | 86.9 | 17 | Britt | 83.4 |
| | Centerville | 86.9 | | Sibley | 83.4 |
| | Chariton | 86.9 | 19 | Clarion | 83.5 |
| | Jefferson | 86.9 | | Mason City | 83.5 |
| | Le Mars | 86.9 | 21 | Fayette | 83.6 |
| 22 | Columbus Jnction | 86.8 | 22 | Colo | 83.7 |
| 23 | Fort Madison | 86.6 | | Northwood | 83.7 |
| | Lamoni | 86.6 | | Waterloo | 83.7 |
| | Sigourney | 86.6 | 25 | Milford | 83.8 |

## ANNUAL PRECIPITATION (")

| HIGHEST | | | LOWEST | | |
|---|---|---|---|---|---|
| 1 | Columbus Jnction | 39.11 | 1 | Sioux City | 25.88 |
| 2 | Keosauqua | 38.73 | 2 | Le Mars | 26.77 |
| 3 | Fort Madison | 38.63 | 3 | Hawarden | 27.03 |
| 4 | Tipton | 38.54 | 4 | Rock Rapids | 27.15 |
| 5 | Lamoni | 38.41 | 5 | Sanborn | 28.02 |
| 6 | Mount Pleasant | 38.19 | 6 | Sioux Center | 28.13 |
| 7 | Bloomfield | 38.09 | 7 | Estherville | 28.33 |
| 8 | Dubuque | 37.83 | 8 | Sheldon | 28.78 |
| 9 | Albia | 37.71 | | Sibley | 28.78 |
| 10 | Iowa City | 37.66 | 10 | Lake Park | 28.83 |
| 11 | Cedar Rpids-No 1 | 37.52 | 11 | Spencer | 29.13 |
| 12 | Williamsburg | 37.46 | 12 | Mapleton | 29.29 |
| 13 | Burlington | 37.10 | 13 | Cherokee | 29.48 |
| 14 | Grinnell | 37.02 | 14 | Onawa | 29.90 |
| 15 | Muscatine | 37.01 | 15 | Milford | 30.00 |
| 16 | Leon | 36.89 | 16 | Primghar | 30.04 |
| 17 | Oskaloosa | 36.77 | 17 | Castana | 30.05 |
| 18 | Corning | 36.69 | 18 | Denison | 30.27 |
| 19 | Chariton | 36.66 | 19 | Ida Grove | 30.50 |
| 20 | Belle Plaine | 36.51 | 20 | Sioux Rapids | 30.62 |
| 21 | Fairfield | 36.50 | 21 | Algona | 30.67 |
| 22 | Clarinda | 36.29 | 22 | Swea City | 30.81 |
| 23 | Centerville | 36.27 | 23 | Waukon | 31.00 |
| 24 | Toledo | 36.02 | 24 | Emmetsburg | 31.08 |
| 25 | Red Oak | 35.99 | 25 | Perry | 31.64 |

## ANNUAL SNOWFALL (")

| HIGHEST | | | LOWEST | | |
|---|---|---|---|---|---|
| 1 | Dubuque | 43.6 | 1 | Centerville | 20.7 |
| 2 | Sanborn | 40.9 | 2 | Washington | 21.0 |
| 3 | Cresco | 40.7 | 3 | Beaconsfield | 21.1 |
| 4 | New Hampton | 39.8 | 4 | Perry | 21.5 |
| 5 | Fayette | 39.6 | 5 | Shenandoah | 21.8 |
| 6 | Forest City | 39.1 | 6 | Corning | 22.3 |
| 7 | Swea City | 38.9 | 7 | Atlantic | 22.7 |
| 8 | Britt | 38.8 | 8 | Bedford | 22.9 |
| 9 | Lake Park | 38.5 | | Lamoni | 22.9 |
| | Mason City-Ap | 38.5 | 10 | Indianola | 23.1 |
| 11 | Northwood | 38.2 | 11 | Osceola | 23.7 |
| 12 | Decorah | 38.1 | 12 | Anamosa | 23.8 |
| 13 | Sioux Center | 38.0 | | Leon | 23.8 |
| 14 | Columbus Jnction | 37.5 | | Sigourney | 23.8 |
| 15 | Sheldon | 37.4 | 15 | Tipton | 24.0 |
| 16 | Algona | 37.2 | 16 | Clarinda | 24.2 |
| 17 | Sibley | 36.8 | | Oakland | 24.2 |
| 18 | Storm Lake | 36.7 | 18 | Glenwood | 24.3 |
| 19 | Charles City | 36.4 | 19 | Burlington | 24.5 |
| 20 | Osage | 36.3 | 20 | Mount Pleasant | 24.6 |
| 21 | Grundy Center | 36.2 | 21 | Keosauqua | 24.9 |
| 22 | Fort Dodge | 36.0 | 22 | Ankeny | 25.1 |
| 23 | Tripoli | 35.9 | | Winterset | 25.1 |
| 24 | Elkader | 35.7 | 24 | Greenfield | 25.4 |
| | Hampton | 35.7 | 25 | Toledo | 25.7 |

# KANSAS

PHYSICAL FEATURES AND GENERAL CLIMATE. Located at the geographical center of the contiguous 48 states, Kansas has a distinctly continental climate with characteristically changeable temperature and precipitation.

Kansas weather is affected largely by two physical features, both some distance from the State: the Rocky Mountains to the west and the Gulf of Mexico to the south. The mountains on the west prevent the importation of moisture from the Pacific Ocean, while the Gulf is the feeding source for much of the State's precipitation.

A third factor, differences in elevation, also influences the climate. Elevation changes are quite gradual rising from 800 or 1,000 feet above sea level in a number of extreme eastern and southeastern counties to approximately 1,500 feet about the center of the State, north to south, and to 3,500 feet at the Colorado line. Quite coincident with these gradations is a change in climate.

PRECIPITATION. Average annual precipitation totals range from slightly more than 40 inches in the southeastern counties to 30 to 35 inches in the northeast, decreasing gradually westward to the Colorado line where the average is from 16 to 18 inches. Distribution of rainfall through the year favors crop production, with an average of about 75 percent of the year's total falling in the crop growing season, April to September. January, the month of least precipitation, has an average of 1 to 2 inches at the more eastern stations, decreasing to less than an inch over the western three-fourths of the State and to near a quarter inch in the extreme west. May and June, in contrast, are the months of greatest rain with between 4 and 5 inches on the average in the eastern three-fifths of the State to between 2 and 3 inches in the western counties. In addition to the seasonal changes of precipitation amounts over the State, there is a secondary fluctuation in the average rainfall which is quite pronounced in the east. In this area a noticeable decrease in the average rainfall occurs during the latter part of July and the forepart of August, with an increase again in September.

Precipitation is most frequent in the extreme east where on the average measurable amounts are recorded on 90 to 100 days of the year. The average annual number of days with an inch or more of precipitation is 60 to 80 over two-thirds of the central portion and about 70 in the northwestern portion, but decreases to near 50 in the southwestern section.

All parts of Kansas may receive 24-hour rainfalls of 5 to 10 inches, with the more frequent occurrence of heavy rains in the eastern area during the month of September. Almost one-half of the total rain falls in daily amounts of 0.75 inch or less. Monthly precipitation totals of 20 inches or more have been recorded at some eastern stations. Protracted periods of successive days with rain are occasionally recorded. On the other hand, all parts of the State have experienced from 50 to 75 successive days without more than 0.25 inch of rain on any day during the period April to September.

Snowfall averages near 10 inches a year in the south-central counties and increases gradually in other parts of the State to the largest average of 24 inches in the northwest. Snow has been recorded in all months except July and August. The greatest average fall is in February with March snows only slightly less. Falls of 12 to 24 inches in 24 hours have been recorded in most sections. Ordinarily snow remains on the ground only a short time, but during the winter the ground may be snow-covered from 10 to 15 days in the south and from 30 to 35 days in the north. In rare instances snow has covered the ground continuously in western and northern sections from 40 to 60 days.

Wet and dry trends or periods are noted in the longer records. Dry periods may persist for several years with an occasional interim of a month or two of above average rainfall. There appears to be some indication of recurring patterns of years with similar trends.

The river drainage in Kansas is about equally divided between the Missouri and Arkansas Rivers, the northern half of the State draining into the Missouri, the southern half into the Arkansas. The Kansas River basin, occupying the north half, has a total drainage area of more than 60,000 square miles above its confluence with the Missouri River at Kansas City. The Arkansas River above the Oklahoma line drains an area of over 45,000 square miles.

Floods in Kansas are generally due to torrential and often prolonged rains. They are seldom caused by melting snow, except in the case of the Arkansas River, where melting snow and heavy rains near its source in the mountains of Colorado have caused flooding. Overflows in the Kansas River and tributaries are practically unknown during the winter season, November to February, but they do occur in the Marias des Cygnes River and in the Arkansas River basin in southeastern Kansas during that period.

For the State as a whole the period of greatest flood frequency is during the spring and summer, from general heavy and prolonged rains. Intense local convective storms also cause damaging flash floods in the smaller streams during the warmer season.

TEMPERATURE. The annual mean temperature ranges from about 58° F. along the south-central and southeastern border to 52° F. in the extreme northwest. Monthly mean temperatures in the northwest range from about 28° F. in July, and in the southeast and south-central from 34° F. in January to 80° F. or 81° F. in July. Daily temperature ranges, on the average, increase from 20° in the east to 30° in the higher and drier elevations of the northwest.

Temperatures of 100° F. or higher occur on an average of 10 days per year in the east and west and about 15 days in the central portion. In some of the hotter summers the number of days with 100° F. or higher has totalled 50 to 60 in the central and south-central counties. The number of days with zero or lower averages 2 to 4 days per year at the southeastern stations and 8 to 10 days in the northwest. Freezing temperatures have been recorded somewhere in the State all months of the year.

During much of the year there is a progressive increase in mean temperature from the higher northwestern counties to the southeastern area. The exception is during the warm summer months when the higher mean temperatures are found in the central and south-central counties.

WINDS AND STORMS. The prevailing winds are from a southerly direction with the exception of the cold months of December through March, which have considerable wind from the north or northwest. Generally the extreme winds are from a northerly direction. In the western part of the State wind speeds are higher and average about 15 m.p.h., approximately 5 m.p.h. faster than in the eastern sections.

Although storms occasionally result in considerable damage, they are for the most part of short duration. In dry periods duststorms may recur frequently in the west, and at intervals a blizzard or severe snowstorm lasts for 36 to 48 hours. The damaging winds, hail, and tornadoes, however, are generally of short duration, although very severe. Seldom are they of any great areal extent.

## COUNTY INDEX

*Labette County*
MOUND VALLEY 3 WSW
PARSONS 2 NW

*Lane County*
HEALY

*Leavenworth County*
LEAVENWORTH

*Lincoln County*
LINCOLN 1 ESE

*Linn County*
MOUND CITY

*Logan County*
OAKLEY 4 W
RUSSELL SPRINGS
WINONA

*McPherson County*
MCPHERSON

*Marion County*
FLORENCE

*Marshall County*
MARYSVILLE

*Meade County*
MEADE

*Miami County*
PAOLA

*Mitchell County*
BELOIT
GLEN ELDER LAKE

*Montgomery County*
INDEPENDENCE

*Morris County*
COUNCIL GROVE LAKE

*Morton County*
ELKHART

*Nemaha County*
CENTRALIA

*Neosho County*
CHANUTE M JOHNSON AP

*Ness County*
NESS CITY

*Norton County*
NORTON 9 SSE
NORTON DAM

*Osage County*
POMONA LAKE

*Osborne County*
ALTON 6 ESE

*Ottawa County*
MINNEAPOLIS

*Pawnee County*
LARNED

*Phillips County*
KIRWIN DAM
PHILLIPSBURG 1 SSE

*Pottawatomie County*
WAMEGO

*Pratt County*
PRATT 4 W

*Rawlins County*
ATWOOD 2 SW
MC DONALD

*Reno County*
HUTCHINSON 10 SW

*Republic County*
BELLEVILLE

*Rice County*
STERLING

*Riley County*
MANHATTAN
TUTTLE CREEK LAKE

*Rooks County*
PLAINVILLE 4 WNW
WEBSTER DAM

*Rush County*
BISON

*Russell County*
RUSSELL MUNI ARPT
WILSON LAKE

*Saline County*
SALINA FCWOS

*Scott County*
SCOTT CITY

*Sedgwick County*
WICHITA MID-CNTNT AP

*Seward County*
LIBERAL

*Shawnee County*
TOPEKA MUNI ARPT

*Sheridan County*
HOXIE

*Sherman County*
BREWSTER 4 W
GOODLAND RENNER FLD

*Smith County*
SMITH CENTER

*Stafford County*
HUDSON

*Stevens County*
HUGOTON

*Sumner County*
WELLINGTON 2 S

*Thomas County*
COLBY 1 SW

*Trego County*
WAKEENEY

*Wabaunsee County*
ESKRIDGE

### *Wallace County*
SHARON SPRINGS

### *Washington County*
WASHINGTON

### *Wichita County*
LEOTI 1 W

### *Wilson County*
FREDONIA

### *Woodson County*
TORONTO LAKE
YATES CENTER

## ELEVATION INDEX

| FEET | STATION NAME |
|---|---|
| 801 | INDEPENDENCE |
| 801 | MOUND VALLEY 3 WSW |
| 830 | SEDAN |
| 850 | FORT SCOTT |
| 860 | LAWRENCE |
| 860 | MOUND CITY |
| 869 | PAOLA |
| 877 | TOPEKA MUNI ARPT |
| 889 | OTTAWA |
| 910 | LEAVENWORTH |
| 912 | COLUMBUS 1 SW |
| 912 | PARSONS 2 NW |
| 951 | IOLA 1 W |
| 961 | TORONTO LAKE |
| 971 | ATCHISON |
| 981 | CHANUTE M JOHNSON AP |
| 991 | FREDONIA |
| 991 | GIRARD |
| 1010 | WAMEGO |
| 1020 | FALL RIVER LAKE |
| 1020 | HORTON |
| 1030 | OLATHE 3 E |
| 1040 | GARNETT 1 E |
| 1060 | POMONA LAKE |
| 1060 | TROY 2 E |
| 1060 | TUTTLE CREEK LAKE |
| 1070 | MANHATTAN |
| 1089 | EUREKA |
| 1089 | JOHN REDMOND LAKE |
| 1089 | YATES CENTER |
| 1102 | HOWARD 5 NE |
| 1112 | OSKALOOSA |
| 1122 | HOLTON 1 S |
| 1170 | ABILENE 2 W |
| 1180 | MARYSVILLE |
| 1191 | COTTONWOOD FALLS |
| 1201 | CLAY CENTER |
| 1201 | WINFIELD NO. 1 |
| 1211 | MILFORD LAKE |
| 1220 | WELLINGTON 2 S |
| 1250 | MINNEAPOLIS |
| 1260 | FLORENCE |
| 1261 | SALINA FCWOS |
| 1270 | CENTRALIA |
| 1289 | EL DORADO |
| 1302 | WASHINGTON |
| 1321 | WICHITA MID-CNTNT AP |
| 1322 | COUNCIL GROVE LAKE |
| 1332 | ANTHONY |
| 1332 | HERINGTON |
| 1381 | BELOIT |
| 1401 | LINCOLN 1 ESE |
| 1411 | ESKRIDGE |
| 1450 | NEWTON 2 SW |
| 1460 | CASSODAY |
| 1470 | CONCORDIA BLOSSER AP |
| 1489 | KANOPOLIS LAKE |
| 1489 | NORWICH |
| 1495 | MCPHERSON |
| 1503 | GLEN ELDER LAKE |
| 1503 | KINGMAN |
| 1512 | BELLEVILLE |
| 1512 | WILSON LAKE |
| 1532 | ELLSWORTH |
| 1540 | MEDICINE LODGE |
| 1572 | HUTCHINSON 10 SW |
| 1601 | LOVEWELL LAKE |
| 1620 | ALTON 6 ESE |
| 1640 | STERLING |
| 1703 | KIRWIN DAM |
| 1781 | MANKATO |
| 1801 | SMITH CENTER |
| 1850 | GREAT BEND |
| 1860 | HUDSON |
| 1860 | WEBSTER DAM |
| 1870 | RUSSELL MUNI ARPT |
| 1880 | PHILLIPSBURG 1 SSE |
| 1941 | PRATT 4 W |
| 1982 | ASHLAND |
| 2001 | HAYS 1 S |
| 2011 | BISON |
| 2080 | COLDWATER |
| 2090 | LARNED |
| 2152 | KINSLEY |
| 2152 | PLAINVILLE 4 WNW |
| 2241 | GREENSBURG |
| 2260 | NESS CITY |
| 2340 | NORTON DAM |
| 2342 | NORTON 9 SSE |
| 2461 | WAKEENEY |
| 2503 | MEADE |
| 2533 | OBERLIN 1 E |
| 2594 | DODGE CITY MUNI AP |
| 2631 | CIMARRON |
| 2661 | QUINTER |
| 2703 | HOXIE |
| 2841 | LIBERAL |
| 2844 | GARDEN CITY MUNI AP |
| 2851 | HEALY |
| 2868 | GARDEN CITY EXP STN |
| 2890 | ATWOOD 2 SW |
| 2913 | SUBLETTE |
| 2960 | RUSSELL SPRINGS |
| 2972 | SCOTT CITY |
| 2992 | LAKIN |
| 3041 | OAKLEY 4 W |
| 3051 | ULYSSES 1 SE |
| 3104 | HUGOTON |
| 3173 | COLBY 1 SW |
| 3261 | SYRACUSE |
| 3304 | LEOTI 1 W |
| 3304 | SAINT FRANCIS |
| 3323 | WINONA |
| 3373 | MC DONALD |
| 3437 | BREWSTER 4 W |
| 3442 | SHARON SPRINGS |
| 3612 | TRIBUNE 1 W |
| 3620 | ELKHART |
| 3658 | GOODLAND RENNER FLD |

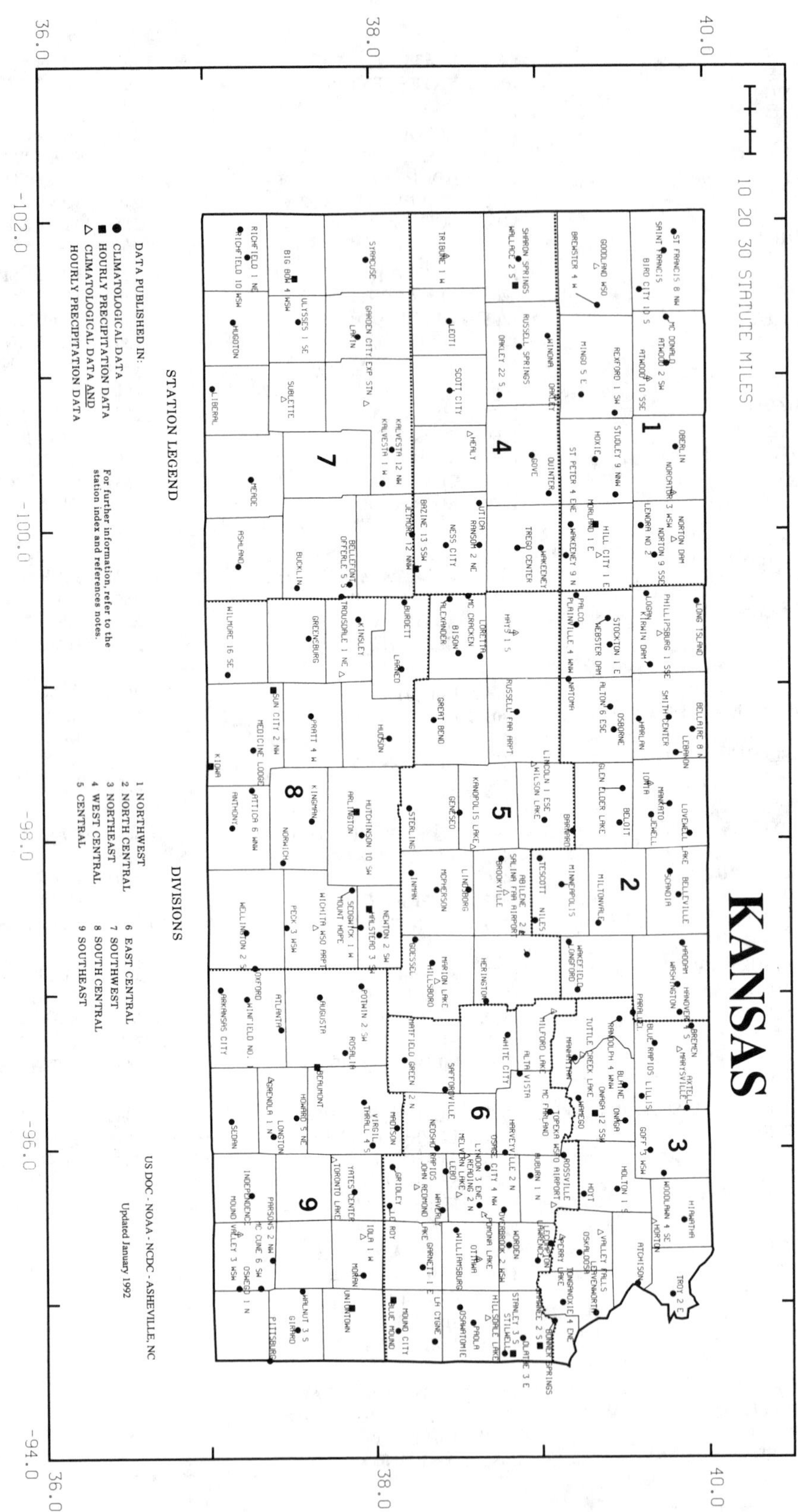
KANSAS
10 20 30 STATUTE MILES
STATION LEGEND
DATA PUBLISHED IN:
● CLIMATOLOGICAL DATA
■ HOURLY PRECIPITATION DATA
△ CLIMATOLOGICAL DATA AND HOURLY PRECIPITATION DATA
For further information, refer to the station index and references notes.
DIVISIONS
1 NORTHWEST
2 NORTH CENTRAL
3 NORTHEAST
4 WEST CENTRAL
5 CENTRAL
6 EAST CENTRAL
7 SOUTHWEST
8 SOUTH CENTRAL
9 SOUTHEAST
US DOC - NOAA - NCDC - ASHEVILLE, NC
Updated January 1992
36.0
38.0
40.0
-102.0
-100.0
-98.0
-96.0
-94.0

## ABILENE 2 W *Dickinson County* ELEVATION 1170 ft LAT/LONG 38° 55 ' N / 97° 15 ' W

| | JAN | FEB | MAR | APR | MAY | JUN | JUL | AUG | SEP | OCT | NOV | DEC | YEAR |
|---|---|---|---|---|---|---|---|---|---|---|---|---|---|
| Maximum Temp °F | 39.2 | 45.1 | 56.7 | 68.1 | 76.9 | 87.1 | 93.0 | 90.7 | 82.0 | 70.2 | 54.4 | 43.3 | 67.2 |
| Minimum Temp °F | 16.8 | 20.8 | 31.3 | 42.0 | 52.5 | 62.3 | 67.7 | 65.2 | 56.2 | 43.7 | 31.6 | 21.6 | 42.6 |
| Mean Temp °F | 28.0 | 33.1 | 44.0 | 55.1 | 64.7 | 74.7 | 80.4 | 78.0 | 69.1 | 57.0 | 43.0 | 32.5 | 55.0 |
| Days Max Temp ≥ 90 °F | 0 | 0 | 0 | 1 | 2 | 12 | 21 | 18 | 7 | 1 | 0 | 0 | 62 |
| Days Max Temp ≤ 32 °F | 9 | 6 | 1 | 0 | 0 | 0 | 0 | 0 | 0 | 0 | 1 | 5 | 22 |
| Days Min Temp ≤ 32 °F | 29 | 24 | 17 | 5 | 0 | 0 | 0 | 0 | 0 | 4 | 17 | 27 | 123 |
| Days Min Temp ≤ 0 °F | 3 | 2 | 0 | 0 | 0 | 0 | 0 | 0 | 0 | 0 | 0 | 1 | 6 |
| Heating Degree Days | 1140 | 891 | 647 | 314 | 99 | 8 | 0 | 1 | 58 | 273 | 653 | 1001 | 5085 |
| Cooling Degree Days | 0 | 0 | 3 | 32 | 103 | 323 | 508 | 435 | 216 | 32 | 1 | 0 | 1653 |
| Total Precipitation (") | 0.74 | 0.89 | 2.17 | 2.67 | 3.85 | 4.82 | 4.00 | 3.51 | 2.91 | 2.70 | 1.57 | 1.07 | 30.90 |
| Days ≥ 0.1" Precip | 2 | 2 | 4 | 5 | 7 | 7 | 6 | 5 | 4 | 4 | 3 | 2 | 51 |
| Total Snowfall (") | na | na | *1.3* | 0.0 | 0.0 | 0.0 | 0.0 | 0.0 | 0.0 | 0.1 | 0.8 | na | na |
| Days ≥ 1" Snow Depth | na | na | na | 0 | 0 | 0 | 0 | 0 | 0 | 0 | 0 | na | na |

## ALTON 6 ESE *Osborne County* ELEVATION 1620 ft LAT/LONG 39° 26 ' N / 98° 51 ' W

| | JAN | FEB | MAR | APR | MAY | JUN | JUL | AUG | SEP | OCT | NOV | DEC | YEAR |
|---|---|---|---|---|---|---|---|---|---|---|---|---|---|
| Maximum Temp °F | 39.6 | 45.8 | 57.5 | 69.2 | 77.0 | 87.7 | 93.5 | 90.6 | 82.5 | 71.3 | 54.4 | 44.0 | 67.8 |
| Minimum Temp °F | 13.8 | 17.2 | 27.8 | 38.7 | 49.3 | 59.7 | 65.3 | 62.3 | 52.6 | 39.6 | 26.9 | 17.9 | 39.3 |
| Mean Temp °F | 26.7 | 31.5 | 42.7 | 54.0 | 63.2 | 73.7 | 79.4 | 76.5 | 67.6 | 55.5 | 40.6 | 31.0 | 53.5 |
| Days Max Temp ≥ 90 °F | 0 | 0 | 0 | 1 | 3 | 13 | *22* | *18* | 8 | 1 | 0 | 0 | 66 |
| Days Max Temp ≤ 32 °F | 9 | 6 | 1 | 0 | 0 | 0 | 0 | 0 | 0 | 0 | 2 | 6 | 24 |
| Days Min Temp ≤ 32 °F | 30 | 26 | 21 | 8 | 1 | 0 | 0 | 0 | 1 | 7 | 22 | *30* | 146 |
| Days Min Temp ≤ 0 °F | 4 | 3 | 0 | 0 | 0 | 0 | 0 | 0 | 0 | 0 | 0 | 2 | 9 |
| Heating Degree Days | 1180 | 937 | 688 | 342 | 128 | 13 | 1 | 4 | 77 | 309 | 725 | 1048 | 5452 |
| Cooling Degree Days | 0 | 0 | 1 | 22 | 70 | 278 | 441 | 352 | 170 | 14 | 0 | 0 | 1348 |
| Total Precipitation (") | 0.56 | 0.79 | 2.01 | 2.43 | 3.67 | 3.46 | 3.94 | 3.05 | 2.56 | 1.68 | 1.19 | 0.81 | 26.15 |
| Days ≥ 0.1" Precip | 2 | 2 | 3 | 5 | 7 | 6 | 6 | 5 | 4 | 4 | 3 | 2 | 49 |
| Total Snowfall (") | 5.1 | 5.2 | 3.0 | 0.5 | 0.0 | 0.0 | 0.0 | 0.0 | 0.1 | 0.1 | 2.0 | *4.2* | 20.2 |
| Days ≥ 1" Snow Depth | na | *8* | *2* | 0 | 0 | 0 | 0 | 0 | 0 | 0 | 2 | na | na |

## ANTHONY *Harper County* ELEVATION 1332 ft LAT/LONG 37° 9 ' N / 98° 1 ' W

| | JAN | FEB | MAR | APR | MAY | JUN | JUL | AUG | SEP | OCT | NOV | DEC | YEAR |
|---|---|---|---|---|---|---|---|---|---|---|---|---|---|
| Maximum Temp °F | 43.3 | 50.2 | 60.8 | 71.0 | 79.6 | 90.1 | 95.7 | 93.9 | 84.6 | 73.6 | 57.0 | 46.3 | 70.5 |
| Minimum Temp °F | 22.5 | 26.3 | 35.2 | 45.5 | 54.9 | 64.5 | 69.4 | 67.8 | 59.7 | 47.5 | 34.9 | 24.8 | 46.1 |
| Mean Temp °F | 33.3 | 38.3 | 48.0 | 58.3 | 67.2 | 77.3 | 82.6 | 80.8 | 72.1 | 60.6 | 46.0 | 35.6 | 58.3 |
| Days Max Temp ≥ 90 °F | 0 | 0 | 0 | 0 | 3 | 17 | 26 | 23 | 9 | 1 | 0 | 0 | 79 |
| Days Max Temp ≤ 32 °F | 7 | 3 | 1 | 0 | 0 | 0 | 0 | 0 | 0 | 0 | 1 | 4 | 16 |
| Days Min Temp ≤ 32 °F | 26 | 21 | 13 | 2 | 0 | 0 | 0 | 0 | 0 | 1 | 12 | *25* | 100 |
| Days Min Temp ≤ 0 °F | 1 | 0 | 0 | 0 | 0 | 0 | 0 | 0 | 0 | 0 | 0 | 1 | 2 |
| Heating Degree Days | 976 | 748 | 523 | 223 | 56 | 3 | 0 | 0 | 29 | 182 | 565 | *911* | 4216 |
| Cooling Degree Days | 0 | 0 | 4 | 26 | 137 | 374 | 558 | 521 | 262 | *45* | *2* | *0* | 1929 |
| Total Precipitation (") | 0.90 | 0.95 | 2.52 | 3.19 | 4.16 | 4.59 | 3.09 | 2.70 | 2.89 | 1.88 | 1.77 | 1.02 | 29.66 |
| Days ≥ 0.1" Precip | 2 | 2 | 4 | 5 | 6 | 6 | 5 | 4 | 4 | 3 | 3 | 2 | 46 |
| Total Snowfall (") | *3.0* | *4.5* | *1.7* | 0.0 | 0.0 | 0.0 | 0.0 | 0.0 | 0.0 | 0.1 | 1.1 | *1.9* | 12.3 |
| Days ≥ 1" Snow Depth | na | na | *1* | 0 | 0 | 0 | 0 | 0 | 0 | 0 | 0 | na | na |

## ASHLAND *Clark County* ELEVATION 1982 ft LAT/LONG 37° 12 ' N / 99° 47 ' W

| | JAN | FEB | MAR | APR | MAY | JUN | JUL | AUG | SEP | OCT | NOV | DEC | YEAR |
|---|---|---|---|---|---|---|---|---|---|---|---|---|---|
| Maximum Temp °F | 44.5 | 50.2 | 59.9 | 70.6 | 78.3 | 88.6 | 94.7 | 92.8 | 83.9 | 73.1 | 57.9 | 47.8 | 70.2 |
| Minimum Temp °F | 15.7 | 19.7 | 28.9 | 39.4 | 50.0 | 60.4 | 65.2 | 63.5 | 54.3 | 40.1 | 27.6 | 19.0 | 40.3 |
| Mean Temp °F | 30.1 | 35.0 | 44.4 | 55.0 | 64.2 | 74.5 | 80.0 | 78.2 | 69.1 | 56.6 | 42.8 | 33.4 | 55.3 |
| Days Max Temp ≥ 90 °F | 0 | 0 | 0 | 1 | 4 | 15 | 25 | 22 | 10 | 2 | 0 | 0 | 79 |
| Days Max Temp ≤ 32 °F | 7 | 4 | 1 | 0 | 0 | 0 | 0 | 0 | 0 | 0 | 1 | 4 | 17 |
| Days Min Temp ≤ 32 °F | 30 | 26 | 21 | 7 | 0 | 0 | 0 | 0 | 0 | 6 | 21 | 29 | 140 |
| Days Min Temp ≤ 0 °F | 3 | 1 | 0 | 0 | 0 | 0 | 0 | 0 | 0 | 0 | 0 | 1 | 5 |
| Heating Degree Days | 1075 | 841 | 632 | 311 | 107 | 10 | 1 | 1 | 56 | 277 | 660 | 972 | 4943 |
| Cooling Degree Days | 0 | 0 | 1 | 21 | 90 | 308 | 468 | 433 | 200 | 19 | 1 | 0 | 1541 |
| Total Precipitation (") | 0.44 | 0.62 | 1.51 | 1.76 | 3.45 | 3.33 | 2.52 | 2.71 | 2.07 | 1.53 | 0.90 | 0.72 | 21.56 |
| Days ≥ 0.1" Precip | 1 | 2 | 3 | 3 | 6 | 5 | 4 | 5 | 4 | 3 | 2 | 2 | 40 |
| Total Snowfall (") | *3.3* | *2.4* | 1.1 | 0.5 | 0.0 | 0.0 | 0.0 | 0.0 | 0.0 | 0.2 | 0.2 | *2.1* | 9.8 |
| Days ≥ 1" Snow Depth | na | na | *0* | 0 | 0 | 0 | 0 | 0 | 0 | 0 | 0 | na | na |

## ATCHISON *Atchison County* ELEVATION 971 ft LAT/LONG 39° 34 ' N / 95° 7 ' W

| | JAN | FEB | MAR | APR | MAY | JUN | JUL | AUG | SEP | OCT | NOV | DEC | YEAR |
|---|---|---|---|---|---|---|---|---|---|---|---|---|---|
| Maximum Temp °F | 36.3 | 42.6 | 55.1 | 66.9 | 75.7 | 84.0 | 88.9 | 84.1 | 79.1 | 68.2 | 52.7 | 40.6 | 64.5 |
| Minimum Temp °F | 17.9 | 22.4 | 33.5 | 44.6 | 54.2 | 63.0 | 67.7 | 63.0 | 56.9 | 46.1 | 33.8 | 23.2 | 43.9 |
| Mean Temp °F | 27.1 | 32.5 | 44.3 | 55.8 | 65.0 | 73.5 | 78.3 | 76.2 | 68.1 | 57.1 | 43.3 | 31.9 | 54.4 |
| Days Max Temp ≥ 90 °F | 0 | 0 | 0 | 0 | 0 | 6 | 14 | 11 | 3 | 0 | 0 | 0 | 34 |
| Days Max Temp ≤ 32 °F | 12 | 7 | 1 | 0 | 0 | 0 | 0 | 0 | 0 | 0 | 1 | 7 | 28 |
| Days Min Temp ≤ 32 °F | 27 | 22 | 15 | 3 | 0 | 0 | 0 | 0 | 0 | 2 | 13 | 26 | 108 |
| Days Min Temp ≤ 0 °F | 3 | 1 | 0 | 0 | 0 | 0 | 0 | 0 | 0 | 0 | 0 | 1 | 5 |
| Heating Degree Days | 1168 | 912 | 636 | 298 | 86 | 7 | 1 | 1 | 59 | 263 | 647 | 1020 | 5098 |
| Cooling Degree Days | 0 | 0 | 4 | 33 | 91 | 283 | 435 | 367 | 176 | 25 | 2 | 0 | 1416 |
| Total Precipitation (") | 1.12 | 0.87 | 2.32 | 3.29 | 4.56 | 4.86 | 4.52 | 4.00 | 4.77 | 3.15 | 2.03 | 1.54 | 37.03 |
| Days ≥ 0.1" Precip | 3 | 3 | 5 | 6 | 7 | 7 | 6 | 6 | 7 | 5 | 4 | 3 | 62 |
| Total Snowfall (") | 6.6 | 5.4 | 3.1 | 0.7 | 0.0 | 0.0 | 0.0 | 0.0 | 0.0 | 0.0 | 1.2 | 4.0 | 21.0 |
| Days ≥ 1" Snow Depth | 11 | 9 | 3 | 0 | 0 | 0 | 0 | 0 | 0 | 0 | 1 | 6 | 30 |

## ATWOOD 2 SW *Rawlins County* ELEVATION 2890 ft LAT/LONG 39° 48 ' N / 101° 3 ' W

| | JAN | FEB | MAR | APR | MAY | JUN | JUL | AUG | SEP | OCT | NOV | DEC | YEAR |
|---|---|---|---|---|---|---|---|---|---|---|---|---|---|
| Maximum Temp °F | 39.1 | 45.3 | 54.8 | 65.5 | 74.4 | 85.6 | 90.9 | 89.1 | 80.4 | 67.9 | 51.6 | 42.1 | 65.6 |
| Minimum Temp °F | 12.9 | 17.2 | 25.4 | 35.6 | 46.0 | 55.8 | 61.3 | 59.1 | 48.8 | 35.1 | 23.8 | 15.8 | 36.4 |
| Mean Temp °F | 26.1 | 31.3 | 40.1 | 50.6 | 60.2 | 70.7 | 76.1 | 74.1 | 64.9 | 51.5 | 37.7 | 28.9 | 51.0 |
| Days Max Temp ≥ 90 °F | 0 | 0 | 0 | 0 | 2 | 11 | 19 | 17 | 7 | 1 | 0 | 0 | 57 |
| Days Max Temp ≤ 32 °F | 10 | 6 | 2 | 0 | 0 | 0 | 0 | 0 | 0 | 0 | 2 | 7 | 27 |
| Days Min Temp ≤ 32 °F | 31 | 27 | 24 | *12* | 2 | 0 | 0 | 0 | 1 | 12 | 26 | 30 | 165 |
| Days Min Temp ≤ 0 °F | 5 | 3 | 1 | 0 | 0 | 0 | 0 | 0 | 0 | 0 | 0 | 2 | 11 |
| Heating Degree Days | 1201 | 946 | 766 | 432 | 189 | 26 | 2 | 8 | 101 | 416 | 811 | 1111 | 6009 |
| Cooling Degree Days | 0 | 0 | 0 | 6 | 43 | 198 | 337 | 291 | 98 | 3 | 0 | 0 | 976 |
| Total Precipitation (") | 0.69 | 0.58 | 1.49 | 2.01 | 3.70 | 3.24 | 3.34 | 2.57 | 1.68 | 1.36 | 0.98 | 0.62 | 22.26 |
| Days ≥ 0.1" Precip | 2 | 2 | 4 | 4 | 7 | 6 | 6 | 5 | 3 | 3 | 2 | 1 | 45 |
| Total Snowfall (") | *8.5* | *4.6* | *5.3* | na | 0.0 | 0.0 | 0.0 | 0.0 | 0.0 | 1.1 | *3.8* | *4.4* | na |
| Days ≥ 1" Snow Depth | na | na | na | *0* | 0 | 0 | 0 | 0 | 0 | 0 | *2* | na | na |

## BELLEVILLE *Republic County* ELEVATION 1512 ft LAT/LONG 39° 50 ' N / 97° 37 ' W

| | JAN | FEB | MAR | APR | MAY | JUN | JUL | AUG | SEP | OCT | NOV | DEC | YEAR |
|---|---|---|---|---|---|---|---|---|---|---|---|---|---|
| Maximum Temp °F | 36.2 | 41.6 | 53.6 | 65.3 | 74.5 | 84.7 | 90.1 | 87.6 | 78.9 | 67.2 | 51.4 | 40.0 | 64.3 |
| Minimum Temp °F | 14.7 | 19.0 | 29.5 | 40.7 | 51.3 | 61.0 | 66.3 | 63.9 | 54.6 | 42.3 | 29.1 | 19.2 | 41.0 |
| Mean Temp °F | 25.5 | 30.4 | 41.6 | 53.0 | 62.9 | 72.9 | 78.2 | 75.8 | 66.8 | 54.8 | 40.3 | 29.6 | 52.7 |
| Days Max Temp ≥ 90 °F | 0 | 0 | 0 | 0 | 1 | 9 | 16 | 13 | 5 | 0 | 0 | 0 | 44 |
| Days Max Temp ≤ 32 °F | 12 | 8 | 2 | 0 | 0 | 0 | 0 | 0 | 0 | 0 | 2 | 8 | 32 |
| Days Min Temp ≤ 32 °F | 30 | 25 | 19 | 6 | 0 | 0 | 0 | 0 | 0 | 4 | 20 | 29 | 133 |
| Days Min Temp ≤ 0 °F | 5 | 3 | 0 | 0 | 0 | 0 | 0 | 0 | 0 | 0 | 0 | 2 | 10 |
| Heating Degree Days | 1220 | 972 | 720 | 369 | 126 | 13 | 1 | 3 | 80 | 327 | 735 | 1090 | 5656 |
| Cooling Degree Days | 0 | 0 | 1 | 21 | 67 | 258 | 417 | 342 | 153 | 13 | 1 | 0 | 1273 |
| Total Precipitation (") | 0.67 | 0.82 | 2.30 | 2.66 | 4.19 | 4.78 | 4.42 | 3.98 | 3.17 | 2.18 | 1.37 | 1.07 | 31.61 |
| Days ≥ 0.1" Precip | 2 | 2 | 5 | 5 | 8 | 6 | 6 | 6 | 5 | 4 | 3 | 3 | 55 |
| Total Snowfall (") | *4.9* | *4.6* | *1.3* | 0.3 | 0.0 | 0.0 | 0.0 | 0.0 | 0.0 | 0.3 | 2.2 | *3.2* | 16.8 |
| Days ≥ 1" Snow Depth | na | na | *1* | 0 | 0 | 0 | 0 | 0 | 0 | 0 | 1 | na | na |

## BELOIT *Mitchell County* ELEVATION 1381 ft LAT/LONG 39° 28 ' N / 98° 6 ' W

| | JAN | FEB | MAR | APR | MAY | JUN | JUL | AUG | SEP | OCT | NOV | DEC | YEAR |
|---|---|---|---|---|---|---|---|---|---|---|---|---|---|
| Maximum Temp °F | 37.7 | 43.8 | 55.7 | 67.0 | 76.2 | 87.0 | 93.0 | 90.2 | 81.3 | 69.2 | 52.8 | 41.7 | 66.3 |
| Minimum Temp °F | 15.4 | 19.6 | 29.9 | 40.8 | 51.2 | 61.5 | 67.0 | 64.8 | 55.2 | 42.4 | 29.3 | 19.7 | 41.4 |
| Mean Temp °F | 26.6 | 31.7 | 42.9 | 53.9 | 63.7 | 74.3 | 80.0 | 77.6 | 68.3 | 55.9 | 41.1 | 30.8 | 53.9 |
| Days Max Temp ≥ 90 °F | 0 | 0 | 0 | 1 | 2 | 12 | 21 | 18 | 7 | 1 | 0 | 0 | 62 |
| Days Max Temp ≤ 32 °F | 11 | 7 | 1 | 0 | 0 | 0 | 0 | 0 | 0 | 0 | 2 | 7 | 28 |
| Days Min Temp ≤ 32 °F | 30 | 25 | 19 | 6 | 0 | 0 | 0 | 0 | 0 | 4 | 19 | 29 | 132 |
| Days Min Temp ≤ 0 °F | 4 | 2 | 0 | 0 | 0 | 0 | 0 | 0 | 0 | 0 | 0 | 2 | 8 |
| Heating Degree Days | 1184 | 934 | 682 | 345 | 115 | 10 | 0 | 1 | 67 | 299 | 712 | 1055 | 5404 |
| Cooling Degree Days | 0 | 0 | 2 | 25 | 79 | 303 | 476 | 403 | 189 | 19 | 1 | 0 | 1497 |
| Total Precipitation (") | 0.69 | 0.76 | 2.06 | 2.29 | 3.89 | 4.26 | 3.69 | 3.02 | 2.57 | 2.11 | 1.11 | 0.88 | 27.33 |
| Days ≥ 0.1" Precip | 2 | 2 | 4 | 5 | 7 | 6 | 6 | 5 | 5 | 4 | 3 | 2 | 51 |
| Total Snowfall (") | 5.6 | 6.2 | 2.8 | 0.5 | 0.0 | 0.0 | 0.0 | 0.0 | 0.0 | 0.2 | 1.4 | 3.7 | 20.4 |
| Days ≥ 1" Snow Depth | na | na | *0* | 0 | 0 | 0 | 0 | 0 | 0 | 0 | *1* | na | na |

## BISON *Rush County* ELEVATION 2011 ft LAT/LONG 38° 31 ' N / 99° 12 ' W

| | JAN | FEB | MAR | APR | MAY | JUN | JUL | AUG | SEP | OCT | NOV | DEC | YEAR |
|---|---|---|---|---|---|---|---|---|---|---|---|---|---|
| Maximum Temp °F | 41.2 | 46.7 | 57.5 | 68.4 | 77.0 | 87.8 | 94.4 | 91.9 | 83.2 | 71.5 | 55.1 | 44.7 | 68.3 |
| Minimum Temp °F | 16.0 | 19.8 | 29.2 | 39.8 | 50.3 | 60.5 | 65.5 | 63.4 | 54.4 | 41.6 | 28.6 | 19.4 | 40.7 |
| Mean Temp °F | 28.6 | 33.3 | 43.4 | 54.1 | 63.7 | 74.2 | 80.0 | 77.7 | 68.8 | 56.6 | 41.9 | 32.1 | 54.5 |
| Days Max Temp ≥ 90 °F | 0 | 0 | 0 | 1 | 3 | 14 | 23 | 20 | 9 | 2 | 0 | 0 | 72 |
| Days Max Temp ≤ 32 °F | 8 | 5 | 1 | 0 | 0 | 0 | 0 | 0 | 0 | 0 | 2 | 6 | 22 |
| Days Min Temp ≤ 32 °F | 30 | 26 | 20 | 7 | 1 | 0 | 0 | 0 | 0 | 5 | 20 | 29 | 138 |
| Days Min Temp ≤ 0 °F | 3 | 2 | 0 | 0 | 0 | 0 | 0 | 0 | 0 | 0 | 0 | 2 | 7 |
| Heating Degree Days | 1122 | 889 | 666 | 339 | 118 | 14 | 0 | 3 | 63 | 285 | 688 | 1014 | 5201 |
| Cooling Degree Days | 0 | 0 | 1 | 22 | 78 | 294 | 458 | 406 | 196 | 23 | 0 | 0 | 1478 |
| Total Precipitation (") | 0.57 | 0.77 | 1.85 | 2.18 | 3.39 | 3.58 | 3.24 | 2.61 | 1.98 | 1.56 | 1.01 | 0.78 | 23.52 |
| Days ≥ 0.1" Precip | 2 | 2 | 4 | 4 | 6 | 6 | 5 | 5 | 4 | 3 | 2 | 2 | 45 |
| Total Snowfall (") | 5.4 | 5.0 | 4.3 | 1.4 | 0.0 | 0.0 | 0.0 | 0.0 | 0.0 | 0.4 | 2.0 | 4.6 | 23.1 |
| Days ≥ 1" Snow Depth | 8 | *6* | 2 | 0 | 0 | 0 | 0 | 0 | 0 | 0 | 1 | 5 | 22 |

## BREWSTER 4 W *Sherman County* ELEVATION 3437 ft LAT/LONG 39° 22 ' N / 101° 27 ' W

| | JAN | FEB | MAR | APR | MAY | JUN | JUL | AUG | SEP | OCT | NOV | DEC | YEAR |
|---|---|---|---|---|---|---|---|---|---|---|---|---|---|
| Maximum Temp °F | *40.9* | *46.2* | *55.5* | *66.5* | *75.0* | *86.4* | *91.3* | *88.6* | *80.4* | *68.8* | *51.9* | *43.2* | 66.2 |
| Minimum Temp °F | *13.5* | *17.0* | *24.6* | *35.3* | *45.1* | *55.3* | *60.6* | *58.2* | *48.6* | *36.1* | *24.3* | *16.3* | 36.2 |
| Mean Temp °F | *27.3* | *31.6* | *40.1* | *50.9* | *60.0* | *70.8* | *76.0* | *73.4* | *64.4* | *52.5* | *38.1* | *29.8* | 51.2 |
| Days Max Temp ≥ 90 °F | *0* | *0* | *0* | *0* | *2* | *12* | *19* | *16* | *7* | *1* | *0* | *0* | 57 |
| Days Max Temp ≤ 32 °F | *8* | *5* | *2* | *0* | *0* | *0* | *0* | *0* | *0* | *0* | *2* | *6* | 23 |
| Days Min Temp ≤ 32 °F | *31* | *27* | *25* | *11* | *2* | *0* | *0* | *0* | *1* | *10* | *25* | *30* | 162 |
| Days Min Temp ≤ 0 °F | *4* | *3* | na | *0* | *0* | *0* | *0* | *0* | *0* | *0* | *0* | *2* | na |
| Heating Degree Days | *1150* | *923* | *764* | *420* | *190* | *26* | *4* | *9* | *109* | *387* | *799* | *1087* | 5868 |
| Cooling Degree Days | na | na | na | na | na | na | na | na | na | na | na | na | na |
| Total Precipitation (") | *0.35* | *0.38* | *1.24* | *1.43* | *3.34* | *3.60* | *3.42* | *2.47* | *1.47* | *1.19* | *0.73* | *0.33* | 19.95 |
| Days ≥ 0.1" Precip | na | *1* | na | *3* | *6* | *6* | *6* | *5* | *3* | *2* | na | *1* | na |
| Total Snowfall (") | na | *3.4* | na | na | *0.1* | *0.0* | *0.0* | *0.0* | *0.0* | *1.2* | na | *3.2* | na |
| Days ≥ 1" Snow Depth | na | na | na | *0* | *0* | *0* | *0* | *0* | *0* | na | na | na | na |

## CASSODAY *Butler County* ELEVATION 1460 ft LAT/LONG 38° 2 ' N / 96° 37 ' W

| | JAN | FEB | MAR | APR | MAY | JUN | JUL | AUG | SEP | OCT | NOV | DEC | YEAR |
|---|---|---|---|---|---|---|---|---|---|---|---|---|---|
| Maximum Temp °F | 39.9 | 45.7 | 57.8 | 68.7 | 76.3 | 85.2 | 91.5 | 90.0 | 81.6 | 70.3 | 55.3 | 43.8 | 67.2 |
| Minimum Temp °F | 17.3 | 21.6 | 31.9 | 43.2 | 52.8 | 62.2 | 67.1 | 65.0 | 56.6 | 44.9 | 32.0 | 22.1 | 43.1 |
| Mean Temp °F | 28.6 | 33.7 | 45.0 | 56.0 | 64.6 | 73.7 | 79.3 | 77.5 | 69.1 | 57.6 | 43.7 | 33.0 | 55.2 |
| Days Max Temp ≥ 90 °F | 0 | 0 | 0 | 0 | 1 | 8 | 19 | 17 | 6 | 0 | 0 | 0 | 51 |
| Days Max Temp ≤ 32 °F | 9 | 5 | 1 | 0 | 0 | 0 | 0 | 0 | 0 | 0 | 1 | 5 | 21 |
| Days Min Temp ≤ 32 °F | 29 | 23 | 16 | 4 | 0 | 0 | 0 | 0 | 0 | 4 | 16 | 27 | 119 |
| Days Min Temp ≤ 0 °F | 3 | 1 | 0 | 0 | 0 | 0 | 0 | 0 | 0 | 0 | 0 | 1 | 5 |
| Heating Degree Days | 1123 | 880 | 617 | 289 | 96 | 8 | 1 | 2 | 55 | 255 | 635 | 986 | 4947 |
| Cooling Degree Days | 0 | 0 | 2 | 26 | 84 | 286 | 456 | 406 | 208 | 27 | 1 | 0 | 1496 |
| Total Precipitation (") | 0.78 | 0.84 | 2.21 | 3.26 | 4.22 | 5.41 | 3.65 | 3.44 | 3.33 | 2.60 | 1.82 | 1.16 | 32.72 |
| Days ≥ 0.1" Precip | 2 | 2 | 4 | 5 | 7 | 6 | 5 | 4 | 5 | 4 | 3 | 3 | 50 |
| Total Snowfall (") | na | na | *1.0* | 0.4 | 0.0 | 0.0 | 0.0 | 0.0 | 0.0 | 0.0 | 0.9 | na | na |
| Days ≥ 1" Snow Depth | na | na | na | 0 | 0 | 0 | 0 | 0 | 0 | 0 | 0 | na | na |

## CENTRALIA *Nemaha County* ELEVATION 1270 ft LAT/LONG 39° 43 ' N / 96° 8 ' W

| | JAN | FEB | MAR | APR | MAY | JUN | JUL | AUG | SEP | OCT | NOV | DEC | YEAR |
|---|---|---|---|---|---|---|---|---|---|---|---|---|---|
| Maximum Temp °F | 35.8 | 42.0 | 54.3 | 66.1 | 75.4 | 84.6 | 89.8 | 87.9 | 79.6 | 68.3 | 51.8 | 40.0 | 64.6 |
| Minimum Temp °F | 15.7 | 19.9 | 30.8 | 41.8 | 52.1 | 61.8 | 66.6 | 64.4 | 55.7 | 43.8 | 31.1 | 20.9 | 42.0 |
| Mean Temp °F | 25.8 | 31.0 | 42.6 | 54.0 | 63.8 | 73.2 | 78.2 | 76.2 | 67.7 | 56.1 | 41.5 | 30.5 | 53.4 |
| Days Max Temp ≥ 90 °F | 0 | 0 | 0 | 0 | 1 | 7 | 16 | 13 | 5 | 0 | 0 | 0 | 42 |
| Days Max Temp ≤ 32 °F | 12 | 8 | 2 | 0 | 0 | 0 | 0 | 0 | 0 | 0 | 2 | 8 | 32 |
| Days Min Temp ≤ 32 °F | 29 | 24 | 18 | 6 | 0 | 0 | 0 | 0 | 0 | 4 | 17 | 27 | 125 |
| Days Min Temp ≤ 0 °F | 5 | 2 | 0 | 0 | 0 | 0 | 0 | 0 | 0 | 0 | 0 | 2 | 9 |
| Heating Degree Days | 1211 | 954 | 692 | 347 | 113 | 10 | 1 | 3 | 71 | 294 | 699 | 1065 | 5460 |
| Cooling Degree Days | 0 | 0 | 2 | 29 | 75 | 264 | 415 | 356 | 171 | 20 | 1 | 0 | 1333 |
| Total Precipitation (") | 0.90 | 0.87 | 2.45 | 3.08 | 4.56 | 4.96 | 4.56 | 3.78 | 4.14 | 2.79 | 1.81 | 1.17 | 35.07 |
| Days ≥ 0.1" Precip | 2 | 2 | 5 | 6 | 7 | 7 | 7 | 6 | 6 | 5 | 4 | 3 | 60 |
| Total Snowfall (") | 8.3 | 7.3 | 5.0 | 1.7 | 0.0 | 0.0 | 0.0 | 0.0 | 0.0 | 0.3 | 2.9 | 5.8 | 31.3 |
| Days ≥ 1" Snow Depth | 11 | 10 | 3 | 1 | 0 | 0 | 0 | 0 | 0 | 0 | 2 | 6 | 33 |

## CHANUTE M JOHNSON AP *Neosho County* ELEVATION 981 ft LAT/LONG 37° 40 ' N / 95° 29 ' W

| | JAN | FEB | MAR | APR | MAY | JUN | JUL | AUG | SEP | OCT | NOV | DEC | YEAR |
|---|---|---|---|---|---|---|---|---|---|---|---|---|---|
| Maximum Temp °F | 40.3 | 46.1 | 57.5 | 68.4 | 75.9 | 84.8 | 90.6 | 88.9 | 80.5 | 69.7 | 55.5 | 44.5 | 66.9 |
| Minimum Temp °F | 20.7 | 24.9 | 34.8 | 45.2 | 54.6 | 63.8 | 68.2 | 66.1 | 58.3 | 46.4 | 34.9 | 25.4 | 45.3 |
| Mean Temp °F | 30.5 | 35.5 | 46.2 | 56.9 | 65.3 | 74.3 | 79.4 | 77.5 | 69.4 | 58.1 | 45.2 | 35.0 | 56.1 |
| Days Max Temp ≥ 90 °F | 0 | 0 | 0 | 0 | 0 | 7 | 18 | 15 | 5 | 0 | 0 | 0 | 45 |
| Days Max Temp ≤ 32 °F | 9 | 5 | 1 | 0 | 0 | 0 | 0 | 0 | 0 | 0 | 1 | 5 | 21 |
| Days Min Temp ≤ 32 °F | 27 | 21 | 13 | 3 | 0 | 0 | 0 | 0 | 0 | 2 | 13 | 24 | 103 |
| Days Min Temp ≤ 0 °F | 2 | 1 | 0 | 0 | 0 | 0 | 0 | 0 | 0 | 0 | 0 | 1 | 4 |
| Heating Degree Days | 1063 | 825 | 580 | 263 | 79 | 5 | 1 | 1 | 50 | 240 | 587 | 924 | 4618 |
| Cooling Degree Days | 0 | 0 | 3 | 27 | 95 | 304 | 466 | 416 | 204 | 28 | 2 | 0 | 1545 |
| Total Precipitation (") | 1.35 | 1.65 | 3.24 | 4.23 | 5.43 | 5.10 | 4.36 | 4.07 | 4.33 | 4.08 | 2.88 | 1.91 | 42.63 |
| Days ≥ 0.1" Precip | 3 | 3 | 5 | 7 | 8 | 7 | 5 | 5 | 6 | 5 | 5 | 4 | 63 |
| Total Snowfall (") | 3.8 | 3.8 | 1.9 | 0.1 | 0.0 | 0.0 | 0.0 | 0.0 | 0.0 | 0.0 | 1.2 | 2.1 | 12.9 |
| Days ≥ 1" Snow Depth | 7 | 5 | 1 | 0 | 0 | 0 | 0 | 0 | 0 | 0 | 1 | 3 | 17 |

## CIMARRON *Gray County* ELEVATION 2631 ft LAT/LONG 37° 48 ' N / 100° 21 ' W

| | JAN | FEB | MAR | APR | MAY | JUN | JUL | AUG | SEP | OCT | NOV | DEC | YEAR |
|---|---|---|---|---|---|---|---|---|---|---|---|---|---|
| Maximum Temp °F | 43.3 | 49.6 | 58.8 | 69.1 | 76.8 | 87.5 | 92.8 | 90.5 | 82.8 | 71.7 | 56.0 | 46.4 | 68.8 |
| Minimum Temp °F | 16.4 | 20.6 | 28.9 | 39.1 | 49.3 | 59.3 | 64.1 | 61.9 | 53.4 | 40.8 | 27.6 | 19.3 | 40.1 |
| Mean Temp °F | 29.8 | 35.2 | 43.9 | 54.1 | 63.1 | 73.4 | 78.5 | 76.2 | 68.1 | 56.3 | 41.8 | 32.9 | 54.4 |
| Days Max Temp ≥ 90 °F | 0 | 0 | 0 | 1 | 2 | 13 | 22 | 19 | 9 | 1 | 0 | 0 | 67 |
| Days Max Temp ≤ 32 °F | 8 | 4 | 2 | 0 | 0 | 0 | 0 | 0 | 0 | 0 | 1 | 5 | 20 |
| Days Min Temp ≤ 32 °F | 30 | 26 | 20 | 7 | 1 | 0 | 0 | 0 | 0 | 5 | 22 | 30 | 141 |
| Days Min Temp ≤ 0 °F | 3 | 1 | 0 | 0 | 0 | 0 | 0 | 0 | 0 | 0 | 0 | 1 | 5 |
| Heating Degree Days | 1083 | 836 | 649 | 336 | 123 | 16 | 1 | 3 | 66 | 286 | 689 | 988 | 5076 |
| Cooling Degree Days | 0 | 0 | 1 | 19 | 63 | 266 | 419 | 371 | 175 | 17 | 0 | 0 | 1331 |
| Total Precipitation (") | 0.60 | 0.68 | 1.66 | 1.97 | 3.43 | 3.71 | 3.29 | 2.82 | 1.61 | 1.39 | 0.99 | 0.61 | 22.76 |
| Days ≥ 0.1" Precip | 2 | 2 | 3 | 4 | 6 | 6 | 5 | 5 | 4 | 2 | 2 | 2 | 43 |
| Total Snowfall (") | ***4.9*** | ***3.4*** | ***4.5*** | 1.0 | 0.0 | 0.0 | 0.0 | 0.0 | 0.0 | 0.3 | 1.9 | ***4.1*** | 20.1 |
| Days ≥ 1" Snow Depth | na | ***3*** | na | 1 | 0 | 0 | 0 | 0 | 0 | 0 | 1 | ***5*** | na |

## CLAY CENTER *Clay County* ELEVATION 1201 ft LAT/LONG 39° 23 ' N / 97° 8 ' W

| | JAN | FEB | MAR | APR | MAY | JUN | JUL | AUG | SEP | OCT | NOV | DEC | YEAR |
|---|---|---|---|---|---|---|---|---|---|---|---|---|---|
| Maximum Temp °F | 38.4 | 45.0 | 57.3 | 69.2 | 77.9 | 87.5 | 92.9 | 90.8 | 82.0 | 70.5 | 53.8 | 42.2 | 67.3 |
| Minimum Temp °F | 17.1 | 21.3 | 32.1 | 43.1 | 53.1 | 62.9 | 68.1 | 66.0 | 56.8 | 44.6 | 31.8 | 21.8 | 43.2 |
| Mean Temp °F | 27.7 | 33.2 | 44.7 | 56.2 | 65.5 | 75.2 | 80.5 | 78.4 | 69.5 | 57.6 | 42.8 | 32.0 | 55.3 |
| Days Max Temp ≥ 90 °F | 0 | 0 | 0 | 1 | 2 | 12 | 21 | 18 | 7 | 0 | 0 | 0 | 61 |
| Days Max Temp ≤ 32 °F | 10 | 6 | 1 | 0 | 0 | 0 | 0 | 0 | 0 | 0 | 1 | 6 | 24 |
| Days Min Temp ≤ 32 °F | 29 | 24 | 16 | 4 | 0 | 0 | 0 | 0 | 0 | 3 | 16 | 27 | 119 |
| Days Min Temp ≤ 0 °F | 3 | 2 | 0 | 0 | 0 | 0 | 0 | 0 | 0 | 0 | 0 | 1 | 6 |
| Heating Degree Days | 1147 | 892 | 625 | 288 | 88 | 7 | 0 | 1 | 55 | 255 | 659 | 1015 | 5032 |
| Cooling Degree Days | 0 | 0 | 4 | 38 | 110 | 329 | 497 | 431 | 214 | 28 | 1 | 0 | 1652 |
| Total Precipitation (") | 0.74 | 0.80 | 2.24 | 2.62 | 4.35 | 4.03 | 4.01 | 3.71 | 3.40 | 2.52 | 1.44 | 1.06 | 30.92 |
| Days ≥ 0.1" Precip | 2 | 2 | 4 | 6 | 7 | 7 | 6 | 6 | 5 | 5 | 3 | 3 | 56 |
| Total Snowfall (") | 6.6 | 6.2 | 2.9 | 0.6 | 0.0 | 0.0 | 0.0 | 0.0 | 0.0 | 0.1 | 1.7 | 4.0 | 22.1 |
| Days ≥ 1" Snow Depth | 12 | 8 | 2 | 0 | 0 | 0 | 0 | 0 | 0 | 0 | 2 | 5 | 29 |

## COLBY 1 SW *Thomas County* ELEVATION 3173 ft LAT/LONG 39° 23 ' N / 101° 4 ' W

| | JAN | FEB | MAR | APR | MAY | JUN | JUL | AUG | SEP | OCT | NOV | DEC | YEAR |
|---|---|---|---|---|---|---|---|---|---|---|---|---|---|
| Maximum Temp °F | 38.8 | 43.6 | 52.3 | 63.3 | 72.3 | 83.8 | 89.8 | 87.5 | 78.6 | 66.8 | 50.7 | 41.4 | 64.1 |
| Minimum Temp °F | 12.7 | 16.4 | 24.3 | 35.0 | 45.6 | 55.7 | 61.4 | 58.8 | 48.7 | 35.7 | 23.9 | 15.3 | 36.1 |
| Mean Temp °F | 25.8 | 30.0 | 38.3 | 49.2 | 59.0 | 69.8 | 75.6 | 73.2 | 63.7 | 51.3 | 37.3 | 28.3 | 50.1 |
| Days Max Temp ≥ 90 °F | 0 | 0 | 0 | 0 | 1 | 9 | 17 | 14 | 6 | 1 | 0 | 0 | 48 |
| Days Max Temp ≤ 32 °F | 10 | 7 | 4 | 0 | 0 | 0 | 0 | 0 | 0 | 0 | 3 | 8 | 32 |
| Days Min Temp ≤ 32 °F | 31 | 28 | 26 | 12 | 2 | 0 | 0 | 0 | 1 | 10 | 26 | 31 | 167 |
| Days Min Temp ≤ 0 °F | 5 | 3 | 1 | 0 | 0 | 0 | 0 | 0 | 0 | 0 | 0 | 2 | 11 |
| Heating Degree Days | 1210 | 979 | 820 | 473 | 217 | 40 | 4 | 11 | 127 | 423 | 826 | 1131 | 6261 |
| Cooling Degree Days | 0 | 0 | 0 | 6 | 32 | 196 | 332 | 283 | 105 | 4 | 0 | 0 | 958 |
| Total Precipitation (") | 0.39 | 0.42 | 1.11 | 1.78 | 3.53 | 3.15 | 3.58 | 2.34 | 1.52 | 1.18 | 0.72 | 0.40 | 20.12 |
| Days ≥ 0.1" Precip | 1 | 2 | 3 | 4 | 7 | 6 | 6 | 4 | 3 | 3 | 2 | 1 | 42 |
| Total Snowfall (") | 5.3 | 4.3 | 7.1 | 3.1 | 0.3 | 0.0 | 0.0 | 0.0 | 0.1 | 1.1 | 3.6 | 4.7 | 29.6 |
| Days ≥ 1" Snow Depth | 11 | 7 | 4 | 1 | 0 | 0 | 0 | 0 | 0 | 0 | 4 | 9 | 36 |

## COLDWATER *Comanche County* ELEVATION 2080 ft LAT/LONG 37° 16 ' N / 99° 20 ' W

| | JAN | FEB | MAR | APR | MAY | JUN | JUL | AUG | SEP | OCT | NOV | DEC | YEAR |
|---|---|---|---|---|---|---|---|---|---|---|---|---|---|
| Maximum Temp °F | 44.4 | 51.3 | 60.9 | 71.3 | 78.9 | 88.7 | 94.3 | 92.4 | 83.8 | 73.0 | 57.2 | 47.5 | 70.3 |
| Minimum Temp °F | 20.4 | 24.8 | 33.6 | 43.9 | 53.0 | 62.8 | 67.6 | 65.7 | 57.3 | 45.7 | 32.8 | 23.8 | 44.3 |
| Mean Temp °F | 32.5 | 38.1 | 47.3 | 57.6 | 66.0 | 75.9 | 81.0 | 79.1 | 70.6 | 59.4 | 45.0 | 35.7 | 57.4 |
| Days Max Temp ≥ 90 °F | 0 | 0 | 0 | 1 | 3 | 14 | 24 | 21 | 9 | 1 | 0 | 0 | 73 |
| Days Max Temp ≤ 32 °F | 7 | 3 | 1 | 0 | 0 | 0 | 0 | 0 | 0 | 0 | 1 | 4 | 16 |
| Days Min Temp ≤ 32 °F | 28 | 21 | 14 | 3 | 0 | 0 | 0 | 0 | 0 | 2 | 15 | 26 | 109 |
| Days Min Temp ≤ 0 °F | 2 | 1 | 0 | 0 | 0 | 0 | 0 | 0 | 0 | 0 | 0 | 1 | 4 |
| Heating Degree Days | 1000 | 752 | 548 | 247 | 79 | 6 | 0 | 0 | 41 | 213 | 595 | 902 | 4383 |
| Cooling Degree Days | 0 | 0 | 5 | 38 | 110 | 340 | 507 | 469 | 240 | 42 | 2 | 0 | 1753 |
| Total Precipitation (") | 0.61 | 0.92 | 1.64 | 2.03 | 3.50 | 3.96 | 2.80 | 3.30 | 2.23 | 1.99 | 1.20 | 0.90 | 25.08 |
| Days ≥ 0.1" Precip | 2 | 2 | 4 | 4 | 6 | 5 | 4 | 4 | 4 | 3 | 2 | 2 | 42 |
| Total Snowfall (") | 4.0 | 4.3 | 3.4 | 0.9 | 0.0 | 0.0 | 0.0 | 0.0 | 0.0 | 0.3 | 2.0 | 3.3 | 18.2 |
| Days ≥ 1" Snow Depth | na | 2 | 1 | 0 | 0 | 0 | 0 | 0 | 0 | 0 | 1 | 2 | na |

## COLUMBUS 1 SW *Cherokee County* ELEVATION 912 ft LAT/LONG 37° 10 ' N / 94° 50 ' W

| | JAN | FEB | MAR | APR | MAY | JUN | JUL | AUG | SEP | OCT | NOV | DEC | YEAR |
|---|---|---|---|---|---|---|---|---|---|---|---|---|---|
| Maximum Temp °F | 42.4 | 48.4 | 58.8 | 69.1 | 77.0 | 85.4 | 91.1 | 89.7 | 81.5 | 71.3 | 57.4 | 46.7 | 68.2 |
| Minimum Temp °F | 22.6 | 26.8 | 36.1 | 45.3 | 54.8 | 63.8 | 68.2 | 65.8 | 58.4 | 47.0 | 36.5 | 27.1 | 46.0 |
| Mean Temp °F | 32.5 | 37.6 | 47.5 | 57.2 | 65.9 | 74.6 | 79.7 | 77.8 | 70.0 | 59.2 | 47.0 | 36.9 | 57.2 |
| Days Max Temp ≥ 90 °F | 0 | 0 | 0 | 0 | 0 | 8 | 20 | 17 | 5 | 0 | 0 | 0 | 50 |
| Days Max Temp ≤ 32 °F | 7 | 4 | 1 | 0 | 0 | 0 | 0 | 0 | 0 | 0 | 1 | 4 | 17 |
| Days Min Temp ≤ 32 °F | 26 | 20 | 12 | 3 | 0 | 0 | 0 | 0 | 0 | 2 | 11 | 22 | 96 |
| Days Min Temp ≤ 0 °F | 1 | 1 | 0 | 0 | 0 | 0 | 0 | 0 | 0 | 0 | 0 | 1 | 3 |
| Heating Degree Days | 1002 | 766 | 541 | 255 | 69 | 4 | 0 | 1 | 43 | 214 | 538 | 864 | 4297 |
| Cooling Degree Days | 0 | 0 | 5 | 25 | 102 | 307 | 473 | 422 | 209 | 35 | 4 | 0 | 1582 |
| Total Precipitation (") | 1.67 | 1.92 | 3.31 | 3.97 | 5.14 | 4.66 | 3.51 | 3.80 | 5.10 | 3.97 | 3.84 | 2.35 | 43.24 |
| Days ≥ 0.1" Precip | 4 | 4 | 6 | 7 | 8 | 7 | 5 | 5 | 6 | 6 | 5 | 5 | 68 |
| Total Snowfall (") | 3.3 | 2.8 | 1.9 | 0.0 | 0.0 | 0.0 | 0.0 | 0.0 | 0.0 | 0.0 | 0.6 | 2.3 | 10.9 |
| Days ≥ 1" Snow Depth | 7 | 3 | 1 | 0 | 0 | 0 | 0 | 0 | 0 | 0 | 0 | 3 | 14 |

## CONCORDIA BLOSSER AP *Cloud County* ELEVATION 1470 ft LAT/LONG 39° 33 ' N / 97° 39 ' W

| | JAN | FEB | MAR | APR | MAY | JUN | JUL | AUG | SEP | OCT | NOV | DEC | YEAR |
|---|---|---|---|---|---|---|---|---|---|---|---|---|---|
| Maximum Temp °F | 35.9 | 41.6 | 53.8 | 65.0 | 74.2 | 84.9 | 90.8 | 88.1 | 79.3 | 67.5 | 51.2 | 40.1 | 64.4 |
| Minimum Temp °F | 16.3 | 20.5 | 30.6 | 41.4 | 51.7 | 61.6 | 67.2 | 65.1 | 55.7 | 43.6 | 30.5 | 20.9 | 42.1 |
| Mean Temp °F | 26.1 | 31.1 | 42.2 | 53.3 | 63.0 | 73.3 | 79.0 | 76.6 | 67.5 | 55.6 | 40.9 | 30.5 | 53.3 |
| Days Max Temp ≥ 90 °F | 0 | 0 | 0 | 0 | 1 | 9 | 17 | 14 | 5 | 0 | 0 | 0 | 46 |
| Days Max Temp ≤ 32 °F | 12 | 8 | 2 | 0 | 0 | 0 | 0 | 0 | 0 | 0 | 2 | 8 | 32 |
| Days Min Temp ≤ 32 °F | 29 | 24 | 18 | 5 | 0 | 0 | 0 | 0 | 0 | 3 | 18 | 28 | 125 |
| Days Min Temp ≤ 0 °F | 4 | 2 | 0 | 0 | 0 | 0 | 0 | 0 | 0 | 0 | 0 | 2 | 8 |
| Heating Degree Days | 1198 | 951 | 701 | 362 | 126 | 13 | 1 | 2 | 75 | 308 | 717 | 1062 | 5516 |
| Cooling Degree Days | 0 | 0 | 2 | 24 | 72 | 277 | 456 | 381 | 176 | 19 | 1 | 0 | 1408 |
| Total Precipitation (") | 0.63 | 0.78 | 2.14 | 2.31 | 4.12 | 4.45 | 4.34 | 3.51 | 2.67 | 1.92 | 1.18 | 0.92 | 28.97 |
| Days ≥ 0.1" Precip | 2 | 2 | 4 | 5 | 7 | 6 | 6 | 6 | 5 | 4 | 3 | 2 | 52 |
| Total Snowfall (") | 5.5 | 5.8 | 3.4 | 0.7 | 0.0 | 0.0 | 0.0 | 0.0 | 0.0 | 0.3 | 2.2 | 4.7 | 22.6 |
| Days ≥ 1" Snow Depth | 12 | 9 | 3 | 0 | 0 | 0 | 0 | 0 | 0 | 0 | 2 | 6 | 32 |

## COTTONWOOD FALLS *Chase County* ELEVATION 1191 ft LAT/LONG 38° 22 ' N / 96° 33 ' W

| | JAN | FEB | MAR | APR | MAY | JUN | JUL | AUG | SEP | OCT | NOV | DEC | YEAR |
|---|---|---|---|---|---|---|---|---|---|---|---|---|---|
| Maximum Temp °F | 40.6 | 46.5 | 58.1 | 68.8 | 76.8 | 85.4 | 91.5 | 90.5 | 82.0 | 71.1 | 55.8 | 44.5 | 67.6 |
| Minimum Temp °F | 18.5 | 22.9 | 33.2 | 44.2 | 53.8 | 62.9 | 67.7 | 65.7 | 57.1 | 45.3 | 33.0 | 23.2 | 44.0 |
| Mean Temp °F | 29.6 | 34.7 | 45.7 | 56.6 | 65.4 | 74.2 | 79.6 | 78.1 | 69.6 | 58.2 | 44.4 | 33.9 | 55.8 |
| Days Max Temp ≥ 90 °F | 0 | 0 | 0 | 1 | 1 | 8 | 19 | 17 | 7 | 1 | 0 | 0 | 54 |
| Days Max Temp ≤ 32 °F | 9 | 5 | 1 | 0 | 0 | 0 | 0 | 0 | 0 | 0 | 1 | 5 | 21 |
| Days Min Temp ≤ 32 °F | 28 | 23 | 16 | 4 | 0 | 0 | 0 | 0 | 0 | 3 | 15 | 26 | 115 |
| Days Min Temp ≤ 0 °F | 2 | 1 | 0 | 0 | 0 | 0 | 0 | 0 | 0 | 0 | 0 | 1 | 4 |
| Heating Degree Days | 1091 | 848 | 596 | 278 | 85 | 7 | 0 | 1 | 52 | 243 | 612 | 957 | 4770 |
| Cooling Degree Days | 0 | 0 | 4 | 33 | 94 | 296 | 466 | 418 | 213 | 33 | 2 | 0 | 1559 |
| Total Precipitation (") | 0.91 | 0.95 | 2.53 | 3.04 | 4.48 | 5.41 | 4.16 | 3.33 | 3.55 | 2.81 | 2.09 | 1.35 | 34.61 |
| Days ≥ 0.1" Precip | 2 | 3 | 5 | 5 | 7 | 7 | 5 | 5 | 5 | 4 | 4 | 3 | 55 |
| Total Snowfall (") | 4.7 | 4.2 | 1.4 | 0.6 | 0.0 | 0.0 | 0.0 | 0.0 | 0.0 | 0.0 | 1.0 | 3.6 | 15.5 |
| Days ≥ 1" Snow Depth | 7 | 6 | 1 | 0 | 0 | 0 | 0 | 0 | 0 | 0 | 0 | 3 | 17 |

## COUNCIL GROVE LAKE *Morris County* ELEVATION 1322 ft LAT/LONG 38° 41 ' N / 96° 31 ' W

| | JAN | FEB | MAR | APR | MAY | JUN | JUL | AUG | SEP | OCT | NOV | DEC | YEAR |
|---|---|---|---|---|---|---|---|---|---|---|---|---|---|
| Maximum Temp °F | 36.3 | 42.3 | 54.5 | 65.8 | 74.3 | 83.5 | 89.8 | 88.4 | 80.0 | 68.6 | 53.3 | 41.5 | 64.9 |
| Minimum Temp °F | 15.6 | 20.0 | 31.2 | 42.6 | 52.4 | 61.9 | 66.9 | 64.5 | 55.4 | 43.1 | 31.6 | 21.0 | 42.2 |
| Mean Temp °F | 26.0 | 31.2 | 42.9 | 54.2 | 63.4 | 72.8 | 78.4 | 76.5 | 67.7 | 55.9 | 42.4 | 31.3 | 53.6 |
| Days Max Temp ≥ 90 °F | 0 | 0 | 0 | 1 | 1 | 6 | 16 | 14 | 6 | 1 | 0 | 0 | 45 |
| Days Max Temp ≤ 32 °F | 12 | 8 | 2 | 0 | 0 | 0 | 0 | 0 | 0 | 0 | 2 | 7 | 31 |
| Days Min Temp ≤ 32 °F | 30 | 24 | 17 | 4 | 0 | 0 | 0 | 0 | 0 | 5 | 17 | 28 | 125 |
| Days Min Temp ≤ 0 °F | 4 | 2 | 0 | 0 | 0 | 0 | 0 | 0 | 0 | 0 | 0 | 1 | 7 |
| Heating Degree Days | 1203 | 944 | 683 | 342 | 120 | 14 | 2 | 2 | 71 | 302 | 671 | 1040 | 5394 |
| Cooling Degree Days | 0 | 0 | 3 | 29 | 72 | 268 | 440 | 382 | 179 | 23 | 1 | 0 | 1397 |
| Total Precipitation (") | 0.80 | 0.86 | 2.41 | 3.31 | 4.38 | 4.54 | 3.85 | 3.27 | 3.35 | 2.57 | 1.84 | 1.24 | 32.42 |
| Days ≥ 0.1" Precip | 2 | 2 | 4 | 6 | 7 | 6 | 5 | 5 | 5 | 5 | 3 | 3 | 53 |
| Total Snowfall (") | 4.7 | 3.8 | 1.3 | 0.4 | 0.0 | 0.0 | 0.0 | 0.0 | 0.0 | 0.0 | 1.1 | 3.4 | 14.7 |
| Days ≥ 1" Snow Depth | 10 | 6 | 1 | 0 | 0 | 0 | 0 | 0 | 0 | 0 | 1 | 4 | 22 |

## DODGE CITY MUNI AP *Ford County* ELEVATION 2594 ft LAT/LONG 37° 46 ' N / 99° 58 ' W

| | JAN | FEB | MAR | APR | MAY | JUN | JUL | AUG | SEP | OCT | NOV | DEC | YEAR |
|---|---|---|---|---|---|---|---|---|---|---|---|---|---|
| Maximum Temp °F | 41.3 | 47.2 | 57.0 | 67.7 | 75.9 | 87.0 | 92.9 | 90.5 | 81.7 | 70.1 | 54.4 | 44.6 | 67.5 |
| Minimum Temp °F | 18.6 | 22.7 | 30.8 | 41.4 | 51.7 | 61.8 | 67.2 | 65.4 | 56.4 | 43.8 | 30.5 | 22.0 | 42.7 |
| Mean Temp °F | 30.0 | 35.0 | 43.9 | 54.5 | 63.8 | 74.4 | 80.1 | 78.0 | 69.0 | 57.0 | 42.5 | 33.3 | 55.1 |
| Days Max Temp ≥ 90 °F | 0 | 0 | 0 | 1 | 2 | 13 | 22 | 19 | 8 | 1 | 0 | 0 | 66 |
| Days Max Temp ≤ 32 °F | 9 | 5 | 2 | 0 | 0 | 0 | 0 | 0 | 0 | 0 | 2 | 6 | 24 |
| Days Min Temp ≤ 32 °F | 29 | 23 | 18 | 5 | 0 | 0 | 0 | 0 | 0 | 3 | 18 | 28 | 124 |
| Days Min Temp ≤ 0 °F | 2 | 1 | 0 | 0 | 0 | 0 | 0 | 0 | 0 | 0 | 0 | 1 | 4 |
| Heating Degree Days | 1079 | 841 | 649 | 327 | 115 | 11 | 1 | 2 | 62 | 275 | 669 | 976 | 5007 |
| Cooling Degree Days | 0 | 0 | 2 | 23 | 86 | 300 | 475 | 426 | 210 | 27 | 1 | 0 | 1550 |
| Total Precipitation (") | 0.50 | 0.70 | 1.58 | 2.13 | 2.93 | 3.06 | 3.05 | 2.75 | 1.71 | 1.33 | 0.88 | 0.75 | 21.37 |
| Days ≥ 0.1" Precip | 2 | 2 | 4 | 4 | 6 | 5 | 5 | 5 | 3 | 2 | 2 | 2 | 42 |
| Total Snowfall (") | 4.6 | 4.5 | 4.5 | 0.9 | 0.0 | 0.0 | 0.0 | 0.0 | 0.0 | 0.3 | 1.9 | 3.4 | 20.1 |
| Days ≥ 1" Snow Depth | 8 | 6 | 3 | 0 | 0 | 0 | 0 | 0 | 0 | 0 | 1 | 5 | 23 |

## EL DORADO *Butler County* ELEVATION 1289 ft LAT/LONG 37° 49 ' N / 96° 51 ' W

| | JAN | FEB | MAR | APR | MAY | JUN | JUL | AUG | SEP | OCT | NOV | DEC | YEAR |
|---|---|---|---|---|---|---|---|---|---|---|---|---|---|
| Maximum Temp °F | 41.5 | 47.7 | 59.1 | 69.6 | 77.3 | 85.8 | 91.5 | 89.7 | 81.7 | 71.1 | 56.2 | 45.3 | 68.0 |
| Minimum Temp °F | 20.0 | 24.1 | 34.3 | 45.0 | 54.3 | 63.6 | 68.4 | 66.1 | 58.3 | 46.5 | 34.2 | 24.3 | 44.9 |
| Mean Temp °F | 30.8 | 35.9 | 46.7 | 57.3 | 65.8 | 74.7 | 80.0 | 78.0 | 70.0 | 58.8 | 45.2 | 34.8 | 56.5 |
| Days Max Temp ≥ 90 °F | 0 | 0 | 0 | 0 | 1 | 9 | 20 | 17 | 6 | 0 | 0 | 0 | 53 |
| Days Max Temp ≤ 32 °F | 8 | 4 | 1 | 0 | 0 | 0 | 0 | 0 | 0 | 0 | 1 | 4 | 18 |
| Days Min Temp ≤ 32 °F | 27 | 22 | 14 | 3 | 0 | 0 | 0 | 0 | 0 | 2 | 14 | 25 | 107 |
| Days Min Temp ≤ 0 °F | 2 | 1 | 0 | 0 | 0 | 0 | 0 | 0 | 0 | 0 | 0 | 1 | 4 |
| Heating Degree Days | 1054 | 813 | 564 | 255 | 75 | 5 | 0 | 1 | 44 | 224 | 588 | 927 | 4550 |
| Cooling Degree Days | 0 | 0 | 4 | 33 | 109 | 320 | 483 | 419 | 221 | 34 | 2 | 0 | 1625 |
| Total Precipitation (") | 0.87 | 1.17 | 2.53 | 3.15 | 4.32 | 5.60 | 3.42 | 3.59 | 3.49 | 2.80 | 2.07 | 1.46 | 34.47 |
| Days ≥ 0.1" Precip | 2 | 3 | 5 | 5 | 7 | 7 | 5 | 5 | 6 | 5 | 4 | 3 | 57 |
| Total Snowfall (") | *2.9* | *3.9* | 1.4 | 0.0 | 0.0 | 0.0 | 0.0 | 0.0 | 0.0 | 0.0 | 0.9 | *2.6* | 11.7 |
| Days ≥ 1" Snow Depth | na | na | 1 | 0 | 0 | 0 | 0 | 0 | 0 | 0 | 0 | *1* | na |

## ELKHART *Morton County* ELEVATION 3620 ft LAT/LONG 37° 0 ' N / 101° 54 ' W

| | JAN | FEB | MAR | APR | MAY | JUN | JUL | AUG | SEP | OCT | NOV | DEC | YEAR |
|---|---|---|---|---|---|---|---|---|---|---|---|---|---|
| Maximum Temp °F | 47.0 | 52.4 | 60.2 | 70.6 | 78.8 | 89.4 | 93.6 | 90.7 | 82.4 | 72.2 | 57.1 | 48.3 | 70.2 |
| Minimum Temp °F | 19.4 | 23.1 | 30.1 | 39.7 | 49.2 | 59.1 | 64.4 | 62.5 | 53.8 | 41.5 | 29.3 | 21.5 | 41.1 |
| Mean Temp °F | 33.3 | 37.8 | 45.2 | 55.1 | 64.0 | 74.3 | 79.0 | 76.6 | 68.1 | 56.8 | 43.2 | 34.9 | 55.7 |
| Days Max Temp ≥ 90 °F | 0 | 0 | 0 | 1 | 4 | 16 | 24 | 19 | 8 | 1 | 0 | 0 | 73 |
| Days Max Temp ≤ 32 °F | 5 | 3 | 1 | 0 | 0 | 0 | 0 | 0 | 0 | 0 | 1 | 4 | 14 |
| Days Min Temp ≤ 32 °F | 29 | 24 | 19 | 6 | 1 | 0 | 0 | 0 | 0 | 5 | 20 | 29 | 133 |
| Days Min Temp ≤ 0 °F | 1 | 1 | 0 | 0 | 0 | 0 | 0 | 0 | 0 | 0 | 0 | 1 | 3 |
| Heating Degree Days | 978 | 764 | 607 | 303 | 105 | 10 | 1 | 3 | 57 | 269 | 646 | 925 | 4668 |
| Cooling Degree Days | 0 | 0 | 1 | 13 | 73 | 295 | 448 | 386 | 161 | 18 | 0 | 0 | 1395 |
| Total Precipitation (") | 0.49 | 0.50 | 1.17 | 1.68 | 2.71 | 2.40 | 2.80 | 2.71 | 1.85 | 0.87 | 0.78 | 0.46 | 18.42 |
| Days ≥ 0.1" Precip | 2 | 2 | 3 | 3 | 5 | 5 | 5 | 5 | 3 | 2 | 2 | *1* | 38 |
| Total Snowfall (") | *4.9* | *4.2* | *4.3* | 1.0 | 0.1 | 0.0 | 0.0 | 0.0 | 0.1 | 0.6 | *1.5* | 3.5 | 20.2 |
| Days ≥ 1" Snow Depth | na | na | na | 0 | 0 | 0 | 0 | 0 | 0 | 0 | *0* | na | na |

## ELLSWORTH *Ellsworth County* ELEVATION 1532 ft LAT/LONG 38° 44 ' N / 98° 14 ' W

| | JAN | FEB | MAR | APR | MAY | JUN | JUL | AUG | SEP | OCT | NOV | DEC | YEAR |
|---|---|---|---|---|---|---|---|---|---|---|---|---|---|
| Maximum Temp °F | 40.8 | 46.9 | 58.1 | 69.2 | 77.4 | 87.6 | 93.3 | 91.2 | 82.6 | 71.4 | 55.0 | 44.5 | 68.2 |
| Minimum Temp °F | 15.9 | 20.1 | 30.7 | 41.4 | 51.6 | 61.8 | 67.0 | 64.8 | 55.6 | 42.8 | 29.6 | 19.8 | 41.8 |
| Mean Temp °F | 28.3 | 33.5 | 44.4 | 55.3 | 64.5 | 74.7 | 80.1 | 78.0 | 69.1 | 57.1 | 42.3 | 32.2 | 55.0 |
| Days Max Temp ≥ 90 °F | 0 | 0 | 0 | 1 | 2 | 13 | 22 | 19 | 8 | 1 | 0 | 0 | 66 |
| Days Max Temp ≤ 32 °F | 9 | 5 | 1 | 0 | 0 | 0 | 0 | 0 | 0 | 0 | 1 | 5 | 21 |
| Days Min Temp ≤ 32 °F | 29 | 25 | 18 | 6 | 1 | 0 | 0 | 0 | 0 | 5 | 19 | 28 | 131 |
| Days Min Temp ≤ 0 °F | 3 | 2 | 0 | 0 | 0 | 0 | 0 | 0 | 0 | 0 | 0 | 1 | 6 |
| Heating Degree Days | 1129 | 882 | 634 | 308 | 101 | 9 | 0 | 2 | 59 | 269 | 675 | 1011 | 5079 |
| Cooling Degree Days | 0 | 0 | 3 | 26 | 85 | 302 | 466 | 406 | 199 | 25 | 1 | 0 | 1513 |
| Total Precipitation (") | 0.67 | 0.85 | 2.15 | 2.48 | 4.32 | 4.07 | 3.56 | 3.26 | 3.04 | 2.36 | 1.08 | 0.85 | 28.69 |
| Days ≥ 0.1" Precip | 2 | 2 | 4 | 5 | 7 | 6 | 5 | 5 | 4 | 4 | 3 | 2 | 49 |
| Total Snowfall (") | *5.9* | *3.9* | 2.1 | 0.4 | 0.0 | 0.0 | 0.0 | 0.0 | 0.0 | 0.3 | 1.0 | 4.3 | 17.9 |
| Days ≥ 1" Snow Depth | *11* | 8 | 2 | 0 | 0 | 0 | 0 | 0 | 0 | 0 | 1 | 5 | 27 |

## ESKRIDGE *Wabaunsee County* ELEVATION 1411 ft LAT/LONG 38° 52 ' N / 96° 7 ' W

| | JAN | FEB | MAR | APR | MAY | JUN | JUL | AUG | SEP | OCT | NOV | DEC | YEAR |
|---|---|---|---|---|---|---|---|---|---|---|---|---|---|
| Maximum Temp °F | 37.2 | 43.1 | 55.5 | 66.4 | 75.0 | 83.7 | 89.6 | 88.0 | 79.9 | 68.7 | 53.3 | 41.5 | 65.2 |
| Minimum Temp °F | 17.0 | 21.3 | 31.5 | 42.7 | 52.7 | 61.8 | 66.4 | 64.1 | 55.9 | 44.6 | 31.9 | 21.9 | 42.7 |
| Mean Temp °F | 27.1 | 32.2 | 43.5 | 54.6 | 63.9 | 72.8 | 78.0 | 76.1 | 67.9 | 56.7 | 42.7 | 31.7 | 53.9 |
| Days Max Temp ≥ 90 °F | 0 | 0 | 0 | 0 | 0 | 6 | 16 | 14 | 5 | 0 | 0 | 0 | 41 |
| Days Max Temp ≤ 32 °F | 11 | 7 | 1 | 0 | 0 | 0 | 0 | 0 | 0 | 0 | 1 | 6 | 26 |
| Days Min Temp ≤ 32 °F | 29 | 24 | 17 | 4 | 0 | 0 | 0 | 0 | 0 | 3 | 15 | 26 | 118 |
| Days Min Temp ≤ 0 °F | 3 | 1 | 0 | 0 | 0 | 0 | 0 | 0 | 0 | 0 | 0 | 2 | 6 |
| Heating Degree Days | 1169 | 918 | 661 | 325 | 105 | 11 | 1 | 3 | 63 | 276 | 664 | 1026 | 5222 |
| Cooling Degree Days | 0 | 0 | 1 | 23 | 61 | 242 | 402 | 342 | 161 | 17 | 1 | 0 | 1250 |
| Total Precipitation (") | 0.92 | 0.96 | 2.49 | 3.54 | 4.76 | 5.01 | 3.83 | 3.85 | 4.19 | 2.79 | 2.12 | 1.43 | 35.89 |
| Days ≥ 0.1" Precip | 3 | 3 | 5 | 6 | 7 | 7 | 5 | 6 | 6 | 5 | 4 | 3 | 60 |
| Total Snowfall (") | 6.3 | 5.2 | 2.8 | 0.8 | 0.0 | 0.0 | 0.0 | 0.0 | 0.0 | 0.0 | 2.0 | 4.1 | 21.2 |
| Days ≥ 1" Snow Depth | 11 | 7 | 2 | 0 | 0 | 0 | 0 | 0 | 0 | 0 | 1 | 6 | 27 |

## EUREKA *Greenwood County* ELEVATION 1089 ft LAT/LONG 37° 49 ' N / 96° 17 ' W

| | JAN | FEB | MAR | APR | MAY | JUN | JUL | AUG | SEP | OCT | NOV | DEC | YEAR |
|---|---|---|---|---|---|---|---|---|---|---|---|---|---|
| Maximum Temp °F | 42.0 | 48.3 | 60.8 | 71.1 | 78.4 | 86.7 | 92.9 | 90.8 | 82.6 | 71.9 | 57.0 | 46.1 | 69.1 |
| Minimum Temp °F | 20.0 | 24.0 | 34.2 | 44.8 | 54.2 | 63.3 | 68.4 | 65.8 | 57.7 | 45.6 | 33.9 | 24.5 | 44.7 |
| Mean Temp °F | 31.0 | 36.2 | 47.5 | 58.0 | 66.3 | 75.0 | 80.7 | 78.3 | 70.2 | 58.8 | 45.5 | 35.3 | 56.9 |
| Days Max Temp ≥ 90 °F | 0 | 0 | 0 | 1 | 1 | 11 | 21 | 18 | 6 | 1 | 0 | 0 | 59 |
| Days Max Temp ≤ 32 °F | 8 | 4 | 1 | 0 | 0 | 0 | 0 | 0 | 0 | 0 | 1 | 4 | 18 |
| Days Min Temp ≤ 32 °F | 27 | 22 | 14 | 3 | 0 | 0 | 0 | 0 | 0 | 3 | 14 | 25 | 108 |
| Days Min Temp ≤ 0 °F | 2 | 1 | 0 | 0 | 0 | 0 | 0 | 0 | 0 | 0 | 0 | 1 | 4 |
| Heating Degree Days | 1046 | 807 | 540 | 241 | 66 | 4 | 0 | 0 | 42 | 225 | 582 | 914 | 4467 |
| Cooling Degree Days | 0 | 0 | 5 | 39 | 113 | 328 | 509 | 432 | 220 | 35 | 3 | 0 | 1684 |
| Total Precipitation (") | 1.18 | 1.38 | 2.58 | 3.49 | 4.53 | 5.19 | 3.72 | 4.26 | 3.69 | 3.20 | 2.49 | 1.73 | 37.44 |
| Days ≥ 0.1" Precip | 3 | 4 | 5 | 6 | 7 | 7 | 6 | 6 | 6 | 5 | 4 | 3 | 62 |
| Total Snowfall (") | 5.5 | 5.8 | 2.9 | 0.3 | 0.0 | 0.0 | 0.0 | 0.0 | 0.0 | 0.0 | 1.6 | 4.0 | 20.1 |
| Days ≥ 1" Snow Depth | 8 | 6 | 1 | 0 | 0 | 0 | 0 | 0 | 0 | 0 | 1 | 3 | 19 |

## FALL RIVER LAKE *Greenwood County* ELEVATION 1020 ft LAT/LONG 37° 39 ' N / 96° 5 ' W

| | JAN | FEB | MAR | APR | MAY | JUN | JUL | AUG | SEP | OCT | NOV | DEC | YEAR |
|---|---|---|---|---|---|---|---|---|---|---|---|---|---|
| Maximum Temp °F | 39.9 | 45.6 | 57.2 | 68.7 | 75.7 | 84.1 | 90.4 | 89.6 | 81.3 | 70.7 | 56.7 | 45.2 | 67.1 |
| Minimum Temp °F | 17.9 | 22.3 | 32.9 | 44.6 | 53.7 | 62.7 | 67.6 | 65.2 | 57.0 | 44.5 | 33.1 | 22.8 | 43.7 |
| Mean Temp °F | 28.9 | 34.0 | 45.1 | 56.7 | 64.8 | 73.4 | 79.0 | 77.5 | 69.2 | 57.7 | 44.9 | 34.0 | 55.4 |
| Days Max Temp ≥ 90 °F | 0 | 0 | 0 | 1 | 0 | 6 | 17 | 16 | 6 | 1 | 0 | 0 | 47 |
| Days Max Temp ≤ 32 °F | 9 | 6 | 1 | 0 | 0 | 0 | 0 | 0 | 0 | 0 | 1 | 4 | 21 |
| Days Min Temp ≤ 32 °F | 29 | 24 | 15 | 3 | 0 | 0 | 0 | 0 | 0 | 2 | 14 | 25 | 112 |
| Days Min Temp ≤ 0 °F | 2 | 2 | 0 | 0 | 0 | 0 | 0 | 0 | 0 | 0 | 0 | 1 | 5 |
| Heating Degree Days | 1112 | 869 | 614 | 272 | 90 | 8 | 1 | 1 | 52 | 254 | 599 | 960 | 4832 |
| Cooling Degree Days | 0 | 0 | 3 | 31 | 84 | 273 | 447 | 406 | 190 | 25 | 3 | *0* | 1462 |
| Total Precipitation (") | 0.92 | 1.09 | 2.56 | 3.53 | 4.67 | 4.97 | 3.92 | 3.66 | 3.71 | 3.39 | 2.51 | 1.54 | 36.47 |
| Days ≥ 0.1" Precip | 3 | 3 | 5 | 6 | 7 | 7 | 6 | 5 | 6 | 5 | 4 | 3 | 60 |
| Total Snowfall (") | na | na | *0.7* | 0.0 | 0.0 | 0.0 | 0.0 | 0.0 | 0.0 | 0.0 | 0.5 | na | na |
| Days ≥ 1" Snow Depth | na | na | *0* | 0 | 0 | 0 | 0 | 0 | 0 | 0 | 0 | na | na |

## FLORENCE *Marion County* ELEVATION 1260 ft LAT/LONG 38° 15 ' N / 96° 56 ' W

| | JAN | FEB | MAR | APR | MAY | JUN | JUL | AUG | SEP | OCT | NOV | DEC | YEAR |
|---|---|---|---|---|---|---|---|---|---|---|---|---|---|
| Maximum Temp °F | 40.7 | 46.8 | 58.5 | 69.3 | 77.3 | 86.6 | 92.6 | 91.0 | 82.3 | 71.4 | 55.8 | 44.8 | 68.1 |
| Minimum Temp °F | 18.7 | 22.8 | 33.1 | 44.0 | 53.2 | 62.7 | 67.8 | 65.7 | 57.1 | 45.2 | 32.8 | 23.1 | 43.9 |
| Mean Temp °F | 29.7 | 34.9 | 45.8 | 56.7 | 65.3 | 74.7 | 80.2 | 78.4 | 69.7 | 58.3 | 44.3 | 34.0 | 56.0 |
| Days Max Temp ≥ 90 °F | 0 | 0 | 0 | 0 | 1 | 10 | 21 | 19 | 7 | 1 | 0 | 0 | 59 |
| Days Max Temp ≤ 32 °F | 9 | 5 | 1 | 0 | 0 | 0 | 0 | 0 | 0 | 0 | 1 | 4 | 20 |
| Days Min Temp ≤ 32 °F | 28 | 22 | 16 | 4 | 0 | 0 | 0 | 0 | 0 | 4 | 16 | 26 | 116 |
| Days Min Temp ≤ 0 °F | 2 | 2 | 0 | 0 | 0 | 0 | 0 | 0 | 0 | 0 | 0 | 1 | 5 |
| Heating Degree Days | 1088 | 844 | 592 | 273 | 88 | 7 | 0 | 1 | 51 | 239 | 613 | 955 | 4751 |
| Cooling Degree Days | 0 | 0 | 3 | 32 | 98 | 309 | 484 | 427 | 216 | 34 | 2 | 0 | 1605 |
| Total Precipitation (") | 0.78 | 0.89 | 2.37 | 2.93 | 4.53 | 5.03 | 3.70 | 2.98 | 3.42 | 2.70 | 1.90 | 1.17 | 32.40 |
| Days ≥ 0.1" Precip | 2 | 2 | 4 | 5 | 7 | 6 | 5 | 5 | 5 | 4 | 3 | 3 | 51 |
| Total Snowfall (") | na | na | ***0.7*** | 0.2 | 0.0 | 0.0 | 0.0 | 0.0 | 0.0 | 0.0 | 0.7 | na | na |
| Days ≥ 1" Snow Depth | na | na | ***0*** | 0 | 0 | 0 | 0 | 0 | 0 | 0 | 0 | na | na |

## FORT SCOTT *Bourbon County* ELEVATION 850 ft LAT/LONG 37° 51 ' N / 94° 42 ' W

| | JAN | FEB | MAR | APR | MAY | JUN | JUL | AUG | SEP | OCT | NOV | DEC | YEAR |
|---|---|---|---|---|---|---|---|---|---|---|---|---|---|
| Maximum Temp °F | 40.6 | 47.0 | 58.7 | 70.0 | 78.1 | 86.7 | 92.4 | 90.9 | 82.6 | 71.6 | 56.4 | 44.8 | 68.3 |
| Minimum Temp °F | 21.2 | 25.9 | 36.0 | 46.3 | 55.9 | 65.3 | 70.3 | 67.7 | 59.3 | 47.5 | 36.3 | 26.3 | 46.5 |
| Mean Temp °F | 30.9 | 36.5 | 47.4 | 58.2 | 67.1 | 76.0 | 81.4 | 79.3 | 71.0 | 59.6 | 46.4 | 35.6 | 57.5 |
| Days Max Temp ≥ 90 °F | 0 | 0 | 0 | 0 | 1 | 11 | 22 | 19 | 7 | 1 | 0 | 0 | 61 |
| Days Max Temp ≤ 32 °F | 9 | 5 | 1 | 0 | 0 | 0 | 0 | 0 | 0 | 0 | 1 | 5 | 21 |
| Days Min Temp ≤ 32 °F | 26 | 20 | 13 | 3 | 0 | 0 | 0 | 0 | 0 | 2 | 12 | 23 | 99 |
| Days Min Temp ≤ 0 °F | 2 | 1 | 0 | 0 | 0 | 0 | 0 | 0 | 0 | 0 | 0 | 1 | 4 |
| Heating Degree Days | 1051 | 800 | 547 | 240 | 61 | 4 | 0 | 1 | 41 | 211 | 555 | 906 | 4417 |
| Cooling Degree Days | 0 | 0 | 6 | 39 | 126 | 347 | 513 | 452 | 227 | 38 | 5 | 0 | 1753 |
| Total Precipitation (") | 1.57 | 1.66 | 3.12 | 4.11 | 4.67 | 5.65 | 4.06 | 3.64 | 4.48 | 4.29 | 3.25 | 2.10 | 42.60 |
| Days ≥ 0.1" Precip | 4 | 4 | 6 | 6 | 7 | 7 | 6 | 5 | 6 | 6 | 5 | 4 | 66 |
| Total Snowfall (") | 4.8 | 4.8 | 1.9 | 0.1 | 0.0 | 0.0 | 0.0 | 0.0 | 0.0 | 0.0 | 1.3 | 3.1 | 16.0 |
| Days ≥ 1" Snow Depth | 7 | 5 | 1 | 0 | 0 | 0 | 0 | 0 | 0 | 0 | 0 | 4 | 17 |

## FREDONIA *Wilson County* ELEVATION 991 ft LAT/LONG 37° 32 ' N / 95° 50 ' W

| | JAN | FEB | MAR | APR | MAY | JUN | JUL | AUG | SEP | OCT | NOV | DEC | YEAR |
|---|---|---|---|---|---|---|---|---|---|---|---|---|---|
| Maximum Temp °F | 42.4 | 48.8 | 60.6 | 70.8 | ***78.2*** | 86.9 | 92.6 | 91.5 | 83.5 | 72.2 | 57.9 | 46.8 | 69.4 |
| Minimum Temp °F | 20.5 | 25.1 | 34.9 | 45.4 | ***54.3*** | 63.3 | 68.1 | 65.9 | 58.5 | 46.6 | 35.1 | 25.7 | 45.3 |
| Mean Temp °F | 31.5 | 37.0 | 47.8 | 58.1 | ***66.3*** | 75.1 | 80.4 | 78.7 | 71.0 | 59.4 | 46.6 | 36.3 | 57.3 |
| Days Max Temp ≥ 90 °F | 0 | 0 | 0 | 0 | 1 | 10 | 21 | 19 | 7 | 1 | 0 | 0 | 59 |
| Days Max Temp ≤ 32 °F | 7 | 4 | 0 | 0 | 0 | 0 | 0 | 0 | 0 | 0 | 0 | ***3*** | 14 |
| Days Min Temp ≤ 32 °F | 27 | 20 | 14 | 3 | 0 | 0 | 0 | 0 | 0 | 2 | ***13*** | ***24*** | 103 |
| Days Min Temp ≤ 0 °F | 2 | 1 | 0 | 0 | 0 | 0 | 0 | 0 | 0 | 0 | 0 | 1 | 4 |
| Heating Degree Days | 1034 | 784 | 532 | 232 | ***66*** | 3 | 0 | 0 | 38 | 210 | 548 | 883 | 4330 |
| Cooling Degree Days | ***0*** | 0 | 6 | 34 | ***101*** | 323 | 496 | 447 | ***255*** | ***44*** | ***2*** | ***0*** | 1708 |
| Total Precipitation (") | 1.23 | 1.37 | 2.82 | 3.55 | 5.30 | 5.07 | 4.33 | 3.85 | 4.06 | 3.96 | 2.77 | 1.72 | 40.03 |
| Days ≥ 0.1" Precip | 3 | 3 | 5 | 6 | 7 | 6 | 5 | 5 | 5 | 5 | 4 | ***3*** | 57 |
| Total Snowfall (") | ***3.1*** | 3.4 | 1.3 | 0.0 | 0.0 | 0.0 | 0.0 | 0.0 | 0.0 | 0.0 | ***0.7*** | ***1.7*** | 10.2 |
| Days ≥ 1" Snow Depth | ***7*** | 4 | 1 | 0 | 0 | 0 | 0 | 0 | 0 | 0 | 0 | ***2*** | 14 |

## GARDEN CITY EXP STN *Finney County* ELEVATION 2868 ft LAT/LONG 37° 59 ' N / 100° 49 ' W

| | JAN | FEB | MAR | APR | MAY | JUN | JUL | AUG | SEP | OCT | NOV | DEC | YEAR |
|---|---|---|---|---|---|---|---|---|---|---|---|---|---|
| Maximum Temp °F | 41.2 | 46.4 | 56.1 | 66.8 | 75.1 | 86.1 | 91.2 | 88.3 | 80.5 | 69.6 | 54.1 | 44.4 | 66.7 |
| Minimum Temp °F | 14.6 | 18.3 | 27.3 | 38.1 | 48.6 | 58.8 | 63.6 | 61.6 | 52.3 | 38.7 | 26.4 | 17.8 | 38.8 |
| Mean Temp °F | 27.9 | 32.4 | 41.7 | 52.4 | 61.9 | 72.4 | 77.4 | 74.9 | 66.4 | 54.2 | 40.3 | 31.1 | 52.8 |
| Days Max Temp ≥ 90 °F | 0 | 0 | 0 | 1 | 2 | 12 | 20 | 16 | 7 | 1 | 0 | 0 | 59 |
| Days Max Temp ≤ 32 °F | 9 | 5 | 3 | 0 | 0 | 0 | 0 | 0 | 0 | 0 | 2 | 7 | 26 |
| Days Min Temp ≤ 32 °F | 31 | 27 | 22 | 8 | 1 | 0 | 0 | 0 | 0 | 7 | 24 | 30 | 150 |
| Days Min Temp ≤ 0 °F | 3 | 2 | 0 | 0 | 0 | 0 | 0 | 0 | 0 | 0 | 0 | 2 | 7 |
| Heating Degree Days | 1142 | 912 | 715 | 382 | 154 | 22 | 2 | 5 | 86 | 341 | 734 | 1043 | 5538 |
| Cooling Degree Days | 0 | 0 | 0 | 15 | 66 | 256 | 393 | 336 | 155 | 9 | 0 | 0 | 1230 |
| Total Precipitation (") | 0.39 | 0.51 | 1.20 | 1.62 | 3.20 | 2.95 | 2.56 | 2.49 | 1.48 | 1.01 | 0.76 | 0.43 | 18.60 |
| Days ≥ 0.1" Precip | 1 | 1 | 3 | 3 | 6 | 6 | 5 | 5 | 3 | 2 | 2 | 2 | 39 |
| Total Snowfall (") | 4.4 | 3.7 | 4.4 | 1.1 | 0.0 | 0.0 | 0.0 | 0.0 | 0.0 | 0.4 | 2.2 | ***3.4*** | 19.6 |
| Days ≥ 1" Snow Depth | na | ***3*** | na | 0 | 0 | 0 | 0 | 0 | 0 | 0 | 1 | na | na |

## GARDEN CITY MUNI AP *Finney County* ELEVATION 2844 ft LAT/LONG 37° 58 ' N / 100° 49 ' W

| | JAN | FEB | MAR | APR | MAY | JUN | JUL | AUG | SEP | OCT | NOV | DEC | YEAR |
|---|---|---|---|---|---|---|---|---|---|---|---|---|---|
| Maximum Temp °F | 42.1 | 48.0 | 57.7 | 68.6 | 76.6 | 87.5 | 93.3 | 90.7 | 82.3 | 70.7 | 54.9 | 44.8 | 68.1 |
| Minimum Temp °F | 16.0 | 20.1 | 28.8 | 39.2 | 49.8 | 60.1 | 65.3 | 63.4 | 54.0 | 40.6 | 27.3 | 18.6 | 40.3 |
| Mean Temp °F | 29.1 | 34.1 | 43.3 | 53.9 | 63.2 | 73.9 | 79.3 | 77.1 | 68.2 | 55.7 | 41.1 | 31.7 | 54.2 |
| Days Max Temp ≥ 90 °F | 0 | 0 | 0 | 1 | 3 | 13 | 22 | 19 | 9 | 1 | 0 | 0 | 68 |
| Days Max Temp ≤ 32 °F | 8 | 5 | 2 | 0 | 0 | 0 | 0 | 0 | 0 | 0 | 2 | 6 | 23 |
| Days Min Temp ≤ 32 °F | 30 | 26 | 20 | 7 | 1 | 0 | 0 | 0 | 0 | 5 | 22 | 30 | 141 |
| Days Min Temp ≤ 0 °F | 3 | 1 | 0 | 0 | 0 | 0 | 0 | 0 | 0 | 0 | 0 | 2 | 6 |
| Heating Degree Days | 1106 | 866 | 668 | 341 | 126 | 14 | 1 | 2 | 68 | 304 | 710 | 1024 | 5230 |
| Cooling Degree Days | 0 | 0 | 0 | 17 | 74 | 277 | 432 | 387 | 182 | 18 | 0 | 0 | 1387 |
| Total Precipitation (") | 0.40 | 0.55 | 1.48 | 1.79 | 3.03 | 3.19 | 3.09 | 2.53 | 1.44 | 1.20 | 0.79 | 0.48 | 19.97 |
| Days ≥ 0.1" Precip | 2 | 2 | 3 | 4 | 6 | 5 | 5 | 5 | 3 | 2 | 2 | 2 | 41 |
| Total Snowfall (") | 4.9 | 3.6 | 4.4 | 1.3 | 0.0 | 0.0 | 0.0 | 0.0 | 0.0 | 0.5 | 2.2 | 3.5 | 20.4 |
| Days ≥ 1" Snow Depth | 10 | 5 | 3 | 1 | 0 | 0 | 0 | 0 | 0 | 0 | 2 | 6 | 27 |

## GARNETT 1 E *Anderson County* ELEVATION 1040 ft LAT/LONG 38° 17 ' N / 95° 14 ' W

| | JAN | FEB | MAR | APR | MAY | JUN | JUL | AUG | SEP | OCT | NOV | DEC | YEAR |
|---|---|---|---|---|---|---|---|---|---|---|---|---|---|
| Maximum Temp °F | 40.3 | 46.2 | 58.5 | 69.2 | 77.1 | 85.3 | 91.0 | 89.3 | 81.5 | 70.5 | 55.5 | 44.4 | 67.4 |
| Minimum Temp °F | 19.8 | 23.8 | 34.3 | 44.8 | 54.1 | 63.1 | 68.0 | 65.9 | 57.9 | 46.4 | 34.8 | 24.9 | 44.8 |
| Mean Temp °F | 30.1 | 35.0 | 46.4 | 57.0 | 65.6 | 74.2 | 79.6 | 77.6 | 69.7 | 58.5 | 45.2 | 34.7 | 56.1 |
| Days Max Temp ≥ 90 °F | 0 | 0 | 0 | 0 | 0 | 8 | 19 | 16 | 5 | 0 | 0 | 0 | 48 |
| Days Max Temp ≤ 32 °F | 9 | 5 | 1 | 0 | 0 | 0 | 0 | 0 | 0 | 0 | 1 | 5 | 21 |
| Days Min Temp ≤ 32 °F | 27 | 22 | 14 | 3 | 0 | 0 | 0 | 0 | 0 | 2 | 13 | 24 | 105 |
| Days Min Temp ≤ 0 °F | 2 | 1 | 0 | 0 | 0 | 0 | 0 | 0 | 0 | 0 | 0 | 1 | 4 |
| Heating Degree Days | 1077 | 841 | 574 | 263 | 77 | 7 | 0 | 1 | 45 | 230 | 590 | 935 | 4640 |
| Cooling Degree Days | 0 | 0 | 3 | 33 | 96 | 295 | 462 | 408 | 206 | 29 | 2 | 0 | 1534 |
| Total Precipitation (") | 1.39 | 1.26 | 2.84 | 3.88 | 4.39 | 5.68 | 3.91 | 4.49 | 4.08 | 4.01 | 2.52 | 1.86 | 40.31 |
| Days ≥ 0.1" Precip | 3 | 3 | 5 | 6 | 7 | 7 | 5 | 5 | 5 | 6 | 4 | 3 | 59 |
| Total Snowfall (") | 5.2 | 4.8 | 2.3 | 0.1 | 0.0 | 0.0 | 0.0 | 0.0 | 0.0 | 0.0 | 1.3 | 4.1 | 17.8 |
| Days ≥ 1" Snow Depth | 7 | 5 | *1* | 0 | 0 | 0 | 0 | 0 | 0 | 0 | 0 | *3* | 16 |

## GIRARD *Crawford County* ELEVATION 991 ft LAT/LONG 37° 31 ' N / 94° 50 ' W

| | JAN | FEB | MAR | APR | MAY | JUN | JUL | AUG | SEP | OCT | NOV | DEC | YEAR |
|---|---|---|---|---|---|---|---|---|---|---|---|---|---|
| Maximum Temp °F | 41.5 | 47.4 | 58.3 | 69.5 | 77.2 | 86.3 | 91.5 | 90.2 | 81.6 | 71.1 | 56.7 | 45.9 | 68.1 |
| Minimum Temp °F | 21.6 | 26.0 | 35.7 | 46.1 | 55.3 | 64.5 | 69.0 | 66.7 | 59.1 | 47.4 | 36.1 | 26.4 | 46.2 |
| Mean Temp °F | 31.6 | 36.7 | 47.0 | 57.8 | 66.3 | 75.4 | 80.3 | 78.5 | 70.4 | 59.2 | 46.4 | 36.2 | 57.2 |
| Days Max Temp ≥ 90 °F | 0 | 0 | 0 | 0 | 1 | 9 | 20 | 18 | 5 | 0 | 0 | 0 | 53 |
| Days Max Temp ≤ 32 °F | 7 | 4 | 1 | 0 | 0 | 0 | 0 | 0 | 0 | 0 | 1 | 4 | 17 |
| Days Min Temp ≤ 32 °F | 27 | 21 | 12 | 2 | 0 | 0 | 0 | 0 | 0 | 1 | 11 | 23 | 97 |
| Days Min Temp ≤ 0 °F | 1 | 1 | 0 | 0 | 0 | 0 | 0 | 0 | 0 | 0 | 0 | 1 | 3 |
| Heating Degree Days | 1033 | 793 | 555 | 239 | 65 | 2 | 0 | 1 | 40 | 215 | 553 | 887 | 4383 |
| Cooling Degree Days | 0 | 0 | 4 | 30 | 104 | 331 | 482 | 436 | 208 | 35 | 3 | 0 | 1633 |
| Total Precipitation (") | 1.63 | 1.85 | 3.57 | 4.23 | 5.35 | 5.36 | 4.08 | 3.76 | 5.09 | 4.43 | 3.50 | 2.28 | 45.13 |
| Days ≥ 0.1" Precip | 3 | 3 | 6 | 6 | 8 | 7 | 5 | 5 | 6 | 5 | 5 | 4 | 63 |
| Total Snowfall (") | ***2.0*** | ***2.6*** | 1.0 | 0.0 | 0.0 | 0.0 | 0.0 | 0.0 | 0.0 | 0.0 | 0.2 | ***1.4*** | 7.2 |
| Days ≥ 1" Snow Depth | na | na | *0* | 0 | 0 | 0 | 0 | 0 | 0 | 0 | 0 | na | na |

## GLEN ELDER LAKE *Mitchell County* ELEVATION 1503 ft LAT/LONG 39° 30 ' N / 98° 19 ' W

| | JAN | FEB | MAR | APR | MAY | JUN | JUL | AUG | SEP | OCT | NOV | DEC | YEAR |
|---|---|---|---|---|---|---|---|---|---|---|---|---|---|
| Maximum Temp °F | 36.2 | 41.7 | 53.6 | 65.1 | 74.2 | 85.0 | 91.3 | 89.0 | 80.0 | 68.3 | 52.0 | 41.1 | 64.8 |
| Minimum Temp °F | 13.3 | 16.9 | 27.8 | 39.1 | 49.5 | 59.9 | 65.6 | 63.2 | 53.4 | 40.6 | 28.0 | 18.7 | 39.7 |
| Mean Temp °F | 24.8 | 29.4 | 40.8 | 52.1 | 61.9 | 72.5 | 78.5 | 76.1 | 66.7 | 54.4 | 40.1 | 29.9 | 52.3 |
| Days Max Temp ≥ 90 °F | 0 | 0 | 0 | 0 | 1 | 9 | 19 | 16 | 6 | 1 | 0 | 0 | 52 |
| Days Max Temp ≤ 32 °F | 12 | 8 | 2 | 0 | 0 | 0 | 0 | 0 | 0 | 0 | 2 | 7 | 31 |
| Days Min Temp ≤ 32 °F | 31 | 27 | 22 | 7 | 1 | 0 | 0 | 0 | 0 | 6 | 22 | 30 | 146 |
| Days Min Temp ≤ 0 °F | 5 | 3 | 0 | 0 | 0 | 0 | 0 | 0 | 0 | 0 | 0 | 2 | 10 |
| Heating Degree Days | 1240 | 999 | 745 | 392 | 152 | 18 | 1 | 4 | 86 | 337 | 741 | 1080 | 5795 |
| Cooling Degree Days | 0 | 0 | 1 | 17 | 61 | 252 | 425 | 355 | 160 | 13 | 0 | 0 | 1284 |
| Total Precipitation (") | 0.56 | 0.63 | 1.86 | 2.26 | 3.56 | 3.93 | 3.94 | 2.88 | 2.71 | 1.98 | 1.08 | 0.74 | 26.13 |
| Days ≥ 0.1" Precip | 2 | 1 | 3 | 5 | 7 | 6 | 5 | 5 | 5 | 4 | 3 | 2 | 48 |
| Total Snowfall (") | na | na | *2.1* | 0.2 | 0.0 | 0.0 | 0.0 | 0.0 | 0.0 | 0.0 | 1.6 | na | na |
| Days ≥ 1" Snow Depth | na | na | *1* | 0 | 0 | 0 | 0 | 0 | 0 | 0 | 2 | *5* | na |

## GOODLAND RENNER FLD *Sherman County* ELEVATION 3658 ft LAT/LONG 39° 22 ' N / 101° 42 ' W

| | JAN | FEB | MAR | APR | MAY | JUN | JUL | AUG | SEP | OCT | NOV | DEC | YEAR |
|---|---|---|---|---|---|---|---|---|---|---|---|---|---|
| Maximum Temp °F | 40.5 | 45.0 | 53.2 | 63.6 | 72.2 | 83.6 | 89.5 | 87.1 | 78.4 | 66.4 | 50.9 | 42.4 | 64.4 |
| Minimum Temp °F | 15.5 | 19.0 | 25.9 | 35.4 | 45.7 | 55.6 | 61.2 | 59.4 | 49.9 | 37.5 | 25.4 | 17.8 | 37.4 |
| Mean Temp °F | 28.1 | 32.0 | 39.6 | 49.5 | 59.0 | 69.6 | 75.4 | 73.3 | 64.2 | 52.0 | 38.2 | 30.1 | 50.9 |
| Days Max Temp ≥ 90 °F | 0 | 0 | 0 | 0 | 1 | 9 | 17 | 14 | 6 | 0 | 0 | 0 | 47 |
| Days Max Temp ≤ 32 °F | 9 | 6 | 3 | 0 | 0 | 0 | 0 | 0 | 0 | 0 | 3 | 7 | 28 |
| Days Min Temp ≤ 32 °F | 30 | 27 | 24 | 11 | 2 | 0 | 0 | 0 | 1 | 8 | 24 | 30 | 157 |
| Days Min Temp ≤ 0 °F | 4 | 2 | 0 | 0 | 0 | 0 | 0 | 0 | 0 | 0 | 0 | 2 | 8 |
| Heating Degree Days | 1139 | 924 | 781 | 463 | 211 | 34 | 4 | 10 | 115 | 402 | 798 | 1075 | 5956 |
| Cooling Degree Days | 0 | 0 | 0 | 7 | 32 | 188 | 330 | 288 | 107 | 5 | 0 | 0 | 957 |
| Total Precipitation (") | 0.42 | 0.40 | 1.17 | 1.35 | 3.41 | 3.28 | 3.10 | 2.29 | 1.33 | 1.07 | 0.76 | 0.45 | 19.03 |
| Days ≥ 0.1" Precip | 1 | 1 | 3 | 3 | 6 | 5 | 6 | 4 | 3 | 3 | 2 | 1 | 38 |
| Total Snowfall (") | 7.2 | 5.2 | 9.1 | 4.8 | 0.6 | 0.1 | 0.0 | 0.0 | 0.2 | 3.1 | 5.7 | 6.2 | 42.2 |
| Days ≥ 1" Snow Depth | 11 | 7 | 5 | 2 | 0 | 0 | 0 | 0 | 0 | 1 | 4 | 8 | 38 |

## GREAT BEND *Barton County* ELEVATION 1850 ft LAT/LONG 38° 21 ' N / 98° 46 ' W

| | JAN | FEB | MAR | APR | MAY | JUN | JUL | AUG | SEP | OCT | NOV | DEC | YEAR |
|---|---|---|---|---|---|---|---|---|---|---|---|---|---|
| Maximum Temp °F | 41.7 | 47.9 | 58.5 | 69.6 | 77.8 | 88.5 | 94.1 | 91.8 | 83.1 | 72.0 | 55.5 | 44.8 | 68.8 |
| Minimum Temp °F | 19.7 | 23.6 | 33.0 | 43.5 | 53.4 | 63.4 | 68.5 | 66.3 | 57.4 | 45.5 | 32.6 | 23.2 | 44.2 |
| Mean Temp °F | 30.7 | 35.8 | 45.8 | 56.6 | 65.6 | 76.0 | 81.3 | 79.1 | 70.3 | 58.8 | 44.1 | 34.0 | 56.5 |
| Days Max Temp ≥ 90 °F | 0 | 0 | 0 | 1 | 3 | 15 | 24 | 20 | 9 | 1 | 0 | 0 | 73 |
| Days Max Temp ≤ 32 °F | 7 | 4 | 1 | 0 | 0 | 0 | 0 | 0 | 0 | 0 | 1 | 5 | 18 |
| Days Min Temp ≤ 32 °F | 28 | 23 | 15 | 3 | 0 | 0 | 0 | 0 | 0 | 2 | 15 | 27 | 113 |
| Days Min Temp ≤ 0 °F | 2 | 1 | 0 | 0 | 0 | 0 | 0 | 0 | 0 | 0 | 0 | 1 | 4 |
| Heating Degree Days | 1056 | 818 | 593 | 274 | 85 | 7 | 0 | 1 | 47 | 227 | 623 | 953 | 4684 |
| Cooling Degree Days | 0 | 0 | 3 | 35 | 111 | 346 | 507 | 449 | 229 | 34 | 1 | 0 | 1715 |
| Total Precipitation (") | 0.64 | 0.87 | 1.83 | 2.29 | 3.48 | 4.13 | 3.48 | 3.07 | 2.24 | 2.16 | 1.08 | 0.95 | 26.22 |
| Days ≥ 0.1" Precip | 2 | 2 | 4 | 4 | 6 | 6 | 5 | 5 | 4 | 3 | 2 | 2 | 45 |
| Total Snowfall (") | 5.5 | 5.3 | 2.6 | 0.7 | 0.0 | 0.0 | 0.0 | 0.0 | 0.0 | 0.4 | 1.0 | *3.1* | 18.6 |
| Days ≥ 1" Snow Depth | *10* | *6* | 2 | 0 | 0 | 0 | 0 | 0 | 0 | 0 | 1 | 5 | 24 |

## GREENSBURG *Kiowa County* ELEVATION 2241 ft LAT/LONG 37° 36 ' N / 99° 18 ' W

| | JAN | FEB | MAR | APR | MAY | JUN | JUL | AUG | SEP | OCT | NOV | DEC | YEAR |
|---|---|---|---|---|---|---|---|---|---|---|---|---|---|
| Maximum Temp °F | 41.4 | 47.2 | 57.5 | 68.1 | 76.6 | 87.1 | 92.9 | 90.6 | 81.1 | 70.0 | 54.4 | 44.4 | 67.6 |
| Minimum Temp °F | 18.7 | 22.7 | 31.5 | 42.0 | 52.2 | 62.1 | 67.5 | 64.9 | 56.1 | 43.9 | 30.6 | 21.6 | 42.8 |
| Mean Temp °F | 30.1 | 34.9 | 44.5 | 55.1 | 64.4 | 74.6 | 80.2 | 77.8 | 68.6 | 57.0 | 42.5 | 33.0 | 55.2 |
| Days Max Temp ≥ 90 °F | 0 | 0 | 0 | 1 | 3 | 13 | 21 | 18 | 7 | 1 | 0 | 0 | 64 |
| Days Max Temp ≤ 32 °F | 8 | 5 | 1 | 0 | 0 | 0 | 0 | 0 | 0 | 0 | 2 | 6 | 22 |
| Days Min Temp ≤ 32 °F | 29 | 24 | 17 | 4 | 0 | 0 | 0 | 0 | 0 | 3 | 18 | 28 | 123 |
| Days Min Temp ≤ 0 °F | 2 | 1 | 0 | 0 | 0 | 0 | 0 | 0 | 0 | 0 | 0 | 1 | 4 |
| Heating Degree Days | 1075 | 842 | 630 | 316 | 104 | 12 | 0 | 2 | 64 | 273 | 669 | 985 | 4972 |
| Cooling Degree Days | 0 | 0 | 3 | 30 | 94 | 304 | 484 | 414 | 190 | 24 | 1 | 0 | 1544 |
| Total Precipitation (") | 0.53 | 0.75 | 1.64 | 2.04 | 3.29 | 4.12 | 3.01 | 2.91 | 2.29 | 1.76 | 0.99 | 0.80 | 24.13 |
| Days ≥ 0.1" Precip | 2 | 2 | 4 | 4 | 6 | 6 | 5 | 5 | 4 | 3 | 3 | 2 | 46 |
| Total Snowfall (") | 4.3 | 4.0 | 2.6 | 0.5 | 0.0 | 0.0 | 0.0 | 0.0 | 0.0 | 0.4 | 1.3 | 3.2 | 16.3 |
| Days ≥ 1" Snow Depth | 7 | 4 | 2 | 0 | 0 | 0 | 0 | 0 | 0 | 0 | 1 | *3* | 17 |

## HAYS 1 S *Ellis County* ELEVATION 2001 ft LAT/LONG 38° 52 ' N / 99° 20 ' W

| | JAN | FEB | MAR | APR | MAY | JUN | JUL | AUG | SEP | OCT | NOV | DEC | YEAR |
|---|---|---|---|---|---|---|---|---|---|---|---|---|---|
| Maximum Temp °F | 39.2 | 45.0 | 55.3 | 66.7 | 74.9 | 86.0 | 92.3 | 89.9 | 81.1 | 69.7 | 53.6 | 43.1 | 66.4 |
| Minimum Temp °F | 14.3 | 18.5 | 28.0 | 39.8 | 50.1 | 60.6 | 65.7 | 63.3 | 53.6 | 40.3 | 27.5 | 18.3 | 40.0 |
| Mean Temp °F | 26.8 | 31.8 | 41.7 | 53.3 | 62.5 | 73.4 | 79.0 | 76.6 | 67.4 | 55.0 | 40.5 | 30.7 | 53.2 |
| Days Max Temp ≥ 90 °F | 0 | 0 | 0 | 1 | 2 | 11 | 20 | 17 | 8 | 1 | 0 | 0 | 60 |
| Days Max Temp ≤ 32 °F | 10 | 7 | 2 | 0 | 0 | 0 | 0 | 0 | 0 | 0 | 2 | 7 | 28 |
| Days Min Temp ≤ 32 °F | 31 | 27 | 21 | 6 | 1 | 0 | 0 | 0 | 1 | 6 | 22 | 30 | 145 |
| Days Min Temp ≤ 0 °F | 4 | 2 | 0 | 0 | 0 | 0 | 0 | 0 | 0 | 0 | 0 | 2 | 8 |
| Heating Degree Days | 1179 | 931 | 716 | 363 | 142 | 19 | 1 | 4 | 83 | 323 | 727 | 1056 | 5544 |
| Cooling Degree Days | 0 | 0 | 1 | 23 | 76 | 294 | 452 | 389 | 183 | 19 | 0 | 0 | 1437 |
| Total Precipitation (") | 0.44 | 0.63 | 1.76 | 2.04 | 2.81 | 3.10 | 3.64 | 2.68 | 1.78 | 1.49 | 0.97 | 0.65 | 21.99 |
| Days ≥ 0.1" Precip | 1 | 2 | 3 | 4 | 6 | 5 | 5 | 5 | 4 | 3 | 2 | 2 | 42 |
| Total Snowfall (") | 4.7 | 4.5 | 4.3 | 0.9 | 0.0 | 0.0 | 0.0 | 0.0 | 0.1 | 0.2 | 1.7 | 4.1 | 20.5 |
| Days ≥ 1" Snow Depth | 10 | 8 | 3 | 1 | 0 | 0 | 0 | 0 | 0 | 0 | 2 | 6 | 30 |

## HEALY *Lane County* ELEVATION 2851 ft LAT/LONG 38° 36 ' N / 100° 37 ' W

| | JAN | FEB | MAR | APR | MAY | JUN | JUL | AUG | SEP | OCT | NOV | DEC | YEAR |
|---|---|---|---|---|---|---|---|---|---|---|---|---|---|
| Maximum Temp °F | 42.0 | 47.6 | 57.6 | 68.1 | 76.5 | 87.8 | 93.4 | 90.6 | 81.8 | 70.3 | 54.0 | 44.7 | 67.9 |
| Minimum Temp °F | 15.3 | 19.2 | 27.4 | 37.7 | 48.2 | 59.0 | 63.8 | 62.0 | 52.2 | 39.3 | 26.7 | 18.3 | 39.1 |
| Mean Temp °F | 28.7 | 33.4 | 42.5 | 52.9 | 62.4 | 73.4 | 78.7 | 76.3 | 67.0 | 54.9 | 40.4 | 31.6 | 53.5 |
| Days Max Temp ≥ 90 °F | 0 | 0 | 0 | 1 | 3 | 14 | 22 | 19 | 8 | 1 | 0 | 0 | 68 |
| Days Max Temp ≤ 32 °F | 8 | 5 | 2 | 0 | 0 | 0 | 0 | 0 | 0 | 0 | 2 | 6 | 23 |
| Days Min Temp ≤ 32 °F | 30 | 26 | 22 | 9 | 1 | 0 | 0 | 0 | 1 | 6 | 23 | 30 | 148 |
| Days Min Temp ≤ 0 °F | 3 | 2 | 0 | 0 | 0 | 0 | 0 | 0 | 0 | 0 | 0 | 2 | 7 |
| Heating Degree Days | 1120 | 885 | 690 | 368 | 140 | 17 | 1 | 3 | 78 | 323 | 732 | 1030 | 5387 |
| Cooling Degree Days | 0 | 0 | 1 | 13 | 59 | 274 | 408 | 357 | 152 | 13 | 0 | 0 | 1277 |
| Total Precipitation (") | 0.57 | 0.67 | 1.43 | 1.89 | 3.12 | 3.10 | 2.74 | 2.65 | 1.76 | 1.32 | 1.03 | 0.57 | 20.85 |
| Days ≥ 0.1" Precip | 2 | 2 | 3 | 4 | 6 | 5 | 5 | 4 | 3 | 3 | 2 | 2 | 41 |
| Total Snowfall (") | 5.2 | 4.9 | 5.2 | 2.5 | 0.0 | 0.0 | 0.0 | 0.0 | 0.0 | 0.9 | 3.6 | 4.4 | 26.7 |
| Days ≥ 1" Snow Depth | 12 | 8 | 4 | 1 | 0 | 0 | 0 | 0 | 0 | 0 | 3 | 8 | 36 |

## HERINGTON *Dickinson County* ELEVATION 1332 ft LAT/LONG 38° 40 ' N / 96° 59 ' W

| | JAN | FEB | MAR | APR | MAY | JUN | JUL | AUG | SEP | OCT | NOV | DEC | YEAR |
|---|---|---|---|---|---|---|---|---|---|---|---|---|---|
| Maximum Temp °F | 38.8 | 45.1 | 57.1 | 68.3 | 76.6 | 86.1 | 91.8 | 90.0 | 81.5 | 69.9 | 54.0 | 42.7 | 66.8 |
| Minimum Temp °F | 17.3 | 21.6 | 32.1 | 42.7 | 52.2 | 62.1 | 66.9 | 65.0 | 56.1 | 44.3 | 31.8 | 21.8 | 42.8 |
| Mean Temp °F | 28.1 | 33.4 | 44.6 | 55.5 | 64.5 | 74.1 | 79.4 | 77.5 | 68.8 | 57.1 | 42.9 | 32.3 | 54.9 |
| Days Max Temp ≥ 90 °F | 0 | 0 | 0 | 0 | 1 | 9 | 20 | 17 | 6 | 0 | 0 | 0 | 53 |
| Days Max Temp ≤ 32 °F | 10 | 6 | 1 | 0 | 0 | 0 | 0 | 0 | 0 | 0 | 1 | 5 | 23 |
| Days Min Temp ≤ 32 °F | 29 | 23 | 16 | 5 | 0 | 0 | 0 | 0 | 0 | 4 | 16 | 27 | 120 |
| Days Min Temp ≤ 0 °F | 3 | 2 | 0 | 0 | 0 | 0 | 0 | 0 | 0 | 0 | 0 | 1 | 6 |
| Heating Degree Days | 1138 | 886 | 628 | 301 | 100 | 9 | 1 | 1 | 60 | 265 | 656 | 1007 | 5052 |
| Cooling Degree Days | 0 | 0 | 3 | 30 | 86 | 302 | 464 | 406 | 197 | 26 | 2 | 0 | 1516 |
| Total Precipitation (") | 0.94 | 1.11 | 2.65 | 3.23 | 4.38 | 5.62 | 3.75 | 3.41 | 3.59 | 3.01 | 2.00 | 1.37 | 35.06 |
| Days ≥ 0.1" Precip | 3 | 3 | 5 | 6 | 8 | 7 | 6 | 5 | 5 | 4 | 4 | 3 | 59 |
| Total Snowfall (") | 6.9 | 5.9 | 3.7 | 1.6 | 0.0 | 0.0 | 0.0 | 0.0 | 0.0 | 0.2 | 2.6 | 5.2 | 26.1 |
| Days ≥ 1" Snow Depth | 12 | 7 | 2 | 1 | 0 | 0 | 0 | 0 | 0 | 0 | 1 | 5 | 28 |

## HOLTON 1 S *Jackson County* ELEVATION 1122 ft LAT/LONG 39° 28 ' N / 95° 46 ' W

| | JAN | FEB | MAR | APR | MAY | JUN | JUL | AUG | SEP | OCT | NOV | DEC | YEAR |
|---|---|---|---|---|---|---|---|---|---|---|---|---|---|
| Maximum Temp °F | 37.3 | 43.3 | 55.7 | 67.4 | 76.3 | 84.8 | 90.0 | 88.2 | 80.1 | 69.0 | 53.1 | 41.3 | 65.5 |
| Minimum Temp °F | 16.2 | 20.6 | 31.2 | 42.5 | 52.3 | 61.8 | 66.4 | 63.8 | 55.3 | 43.7 | 31.5 | 21.2 | 42.2 |
| Mean Temp °F | 26.8 | 32.0 | 43.4 | 55.0 | 64.3 | 73.3 | 78.2 | 76.1 | 67.7 | 56.4 | 42.3 | 31.3 | 53.9 |
| Days Max Temp ≥ 90 °F | 0 | 0 | 0 | 0 | 1 | 7 | 17 | 14 | 5 | 0 | 0 | 0 | 44 |
| Days Max Temp ≤ 32 °F | 11 | 7 | 1 | 0 | 0 | 0 | 0 | 0 | 0 | 0 | 1 | 7 | 27 |
| Days Min Temp ≤ 32 °F | 28 | 23 | 17 | 5 | 0 | 0 | 0 | 0 | 0 | 4 | 16 | 27 | 120 |
| Days Min Temp ≤ 0 °F | 4 | 2 | 0 | 0 | 0 | 0 | 0 | 0 | 0 | 0 | 0 | 2 | 8 |
| Heating Degree Days | 1178 | 926 | 664 | 321 | 101 | 10 | 1 | 2 | 69 | 286 | 674 | 1038 | 5270 |
| Cooling Degree Days | 0 | 0 | 5 | 33 | 87 | 282 | 436 | 378 | 184 | 24 | 1 | 0 | 1430 |
| Total Precipitation (") | 0.97 | 0.89 | 2.35 | 3.44 | 4.39 | 5.50 | 4.21 | 4.11 | 4.67 | 3.13 | 2.00 | 1.49 | 37.15 |
| Days ≥ 0.1" Precip | 3 | 3 | 5 | 6 | 7 | 7 | 6 | 6 | 5 | 5 | 3 | 3 | 59 |
| Total Snowfall (") | 6.1 | 4.8 | 2.2 | 0.3 | 0.0 | 0.0 | 0.0 | 0.0 | 0.0 | 0.0 | 1.1 | 3.2 | 17.7 |
| Days ≥ 1" Snow Depth | na | na | *1* | 0 | 0 | 0 | 0 | 0 | 0 | 0 | 1 | na | na |

## HORTON *Brown County* ELEVATION 1020 ft LAT/LONG 39° 40 ' N / 95° 31 ' W

| | JAN | FEB | MAR | APR | MAY | JUN | JUL | AUG | SEP | OCT | NOV | DEC | YEAR |
|---|---|---|---|---|---|---|---|---|---|---|---|---|---|
| Maximum Temp °F | 36.7 | 42.9 | 55.8 | 68.0 | 77.5 | 86.3 | 91.2 | 89.2 | 81.0 | 69.3 | 52.9 | 40.9 | 66.0 |
| Minimum Temp °F | 15.4 | 19.9 | 31.3 | 42.6 | 52.4 | 62.2 | 66.8 | 64.4 | 56.0 | 43.9 | 31.6 | 21.0 | 42.3 |
| Mean Temp °F | 26.1 | 31.4 | 43.6 | 55.3 | 65.0 | 74.3 | 79.0 | 76.8 | 68.5 | 56.7 | 42.3 | 31.0 | 54.2 |
| Days Max Temp ≥ 90 °F | 0 | 0 | 0 | 1 | 2 | 10 | 19 | 15 | 6 | 0 | 0 | 0 | 53 |
| Days Max Temp ≤ 32 °F | 11 | 7 | 1 | 0 | 0 | 0 | 0 | 0 | 0 | 0 | 1 | 7 | 27 |
| Days Min Temp ≤ 32 °F | 29 | 24 | 18 | 5 | 0 | 0 | 0 | 0 | 0 | 4 | 16 | 27 | 123 |
| Days Min Temp ≤ 0 °F | 5 | 2 | 0 | 0 | 0 | 0 | 0 | 0 | 0 | 0 | 0 | 2 | 9 |
| Heating Degree Days | 1200 | 942 | 661 | 311 | 93 | 7 | 1 | 1 | 58 | 278 | 676 | 1048 | 5276 |
| Cooling Degree Days | 0 | 0 | 4 | 34 | 102 | 311 | 462 | 394 | 191 | 25 | 1 | 0 | 1524 |
| Total Precipitation (") | 0.99 | 0.98 | 2.35 | 3.47 | 4.52 | 5.39 | 4.24 | 3.82 | 4.53 | 3.05 | 1.97 | 1.41 | 36.72 |
| Days ≥ 0.1" Precip | 3 | 3 | 5 | 6 | 7 | 7 | 7 | 6 | 6 | 5 | 4 | 3 | 62 |
| Total Snowfall (") | *5.1* | *5.1* | 2.3 | 0.4 | 0.0 | 0.0 | 0.0 | 0.0 | 0.0 | 0.0 | 0.9 | na | na |
| Days ≥ 1" Snow Depth | *10* | 8 | 2 | 0 | 0 | 0 | 0 | 0 | 0 | 0 | 1 | *5* | 26 |

## HOWARD 5 NE *Elk County* ELEVATION 1102 ft LAT/LONG 37° 31 ' N / 96° 12 ' W

| | JAN | FEB | MAR | APR | MAY | JUN | JUL | AUG | SEP | OCT | NOV | DEC | YEAR |
|---|---|---|---|---|---|---|---|---|---|---|---|---|---|
| Maximum Temp °F | 42.7 | 48.8 | 60.1 | 70.7 | 77.7 | 85.9 | 92.3 | 91.3 | 83.1 | 72.5 | 57.5 | 46.5 | 69.1 |
| Minimum Temp °F | 20.1 | 24.2 | 34.2 | 45.0 | 54.0 | 63.1 | 67.4 | 65.2 | 57.4 | 45.7 | 34.1 | 24.6 | 44.6 |
| Mean Temp °F | 31.4 | 36.5 | 47.2 | 57.9 | 65.9 | 74.6 | 79.9 | 78.3 | 70.3 | 59.1 | 45.8 | 35.6 | 56.9 |
| Days Max Temp ≥ 90 °F | 0 | 0 | 0 | 1 | 1 | 8 | 21 | 19 | 7 | 1 | 0 | 0 | 58 |
| Days Max Temp ≤ 32 °F | 7 | 4 | 1 | 0 | 0 | 0 | 0 | 0 | 0 | 0 | 1 | 3 | 16 |
| Days Min Temp ≤ 32 °F | 27 | 21 | 14 | 3 | 0 | 0 | 0 | 0 | 0 | 3 | 14 | 25 | 107 |
| Days Min Temp ≤ 0 °F | 2 | 1 | 0 | 0 | 0 | 0 | 0 | 0 | 0 | 0 | 0 | 1 | 4 |
| Heating Degree Days | 1035 | 798 | 550 | 241 | 70 | 4 | 0 | 0 | 41 | 215 | 572 | 907 | 4433 |
| Cooling Degree Days | 0 | 0 | 4 | 37 | 108 | 313 | 481 | 439 | 226 | 36 | 3 | 0 | 1647 |
| Total Precipitation (") | 1.08 | 1.25 | 2.62 | 3.60 | 4.98 | 4.84 | 3.81 | 3.41 | 3.91 | 3.37 | 2.55 | 1.63 | 37.05 |
| Days ≥ 0.1" Precip | 3 | 3 | 5 | 6 | 7 | 7 | 6 | 5 | 6 | 5 | 4 | 3 | 60 |
| Total Snowfall (") | 5.0 | 3.8 | 1.8 | 0.1 | 0.0 | 0.0 | 0.0 | 0.0 | 0.0 | 0.0 | 0.9 | 1.8 | 13.4 |
| Days ≥ 1" Snow Depth | 8 | 5 | 1 | 0 | 0 | 0 | 0 | 0 | 0 | 0 | 1 | 3 | 18 |

## HOXIE *Sheridan County* ELEVATION 2703 ft LAT/LONG 39° 21 ' N / 100° 27 ' W

| | JAN | FEB | MAR | APR | MAY | JUN | JUL | AUG | SEP | OCT | NOV | DEC | YEAR |
|---|---|---|---|---|---|---|---|---|---|---|---|---|---|
| Maximum Temp °F | 41.5 | 47.4 | 57.3 | 68.4 | 76.8 | 87.5 | 92.9 | 90.6 | 82.2 | 70.7 | 53.3 | 43.6 | 67.7 |
| Minimum Temp °F | 16.0 | 19.8 | 27.8 | 38.4 | 48.8 | 58.5 | 63.9 | 61.9 | 52.2 | 39.5 | 26.9 | 19.0 | 39.4 |
| Mean Temp °F | 28.8 | 33.6 | 42.6 | 53.4 | 62.8 | 73.0 | 78.4 | 76.3 | 67.2 | 55.1 | 40.1 | 31.3 | 53.6 |
| Days Max Temp ≥ 90 °F | 0 | 0 | 0 | 1 | 2 | 13 | 21 | 19 | 8 | 1 | 0 | 0 | 65 |
| Days Max Temp ≤ 32 °F | 8 | 5 | 2 | 0 | 0 | 0 | 0 | 0 | 0 | 0 | 2 | 6 | 23 |
| Days Min Temp ≤ 32 °F | 30 | 26 | 22 | 8 | 1 | 0 | 0 | 0 | 1 | 7 | 22 | 29 | 146 |
| Days Min Temp ≤ 0 °F | 3 | 2 | 0 | 0 | 0 | 0 | 0 | 0 | 0 | 0 | 0 | 2 | 7 |
| Heating Degree Days | 1115 | 880 | 690 | 355 | 132 | 16 | 1 | 4 | 77 | 315 | 741 | 1037 | 5363 |
| Cooling Degree Days | 0 | 0 | 1 | 18 | 65 | 264 | 407 | 360 | 157 | 12 | 0 | 0 | 1284 |
| Total Precipitation (") | 0.51 | 0.54 | 1.42 | 2.00 | 3.45 | 2.70 | 3.24 | 2.87 | 1.60 | 1.24 | 0.89 | 0.53 | 20.99 |
| Days ≥ 0.1" Precip | 1 | 2 | 4 | 4 | 7 | 6 | 6 | 5 | 4 | 3 | 2 | 2 | 46 |
| Total Snowfall (") | *5.3* | 4.4 | 5.5 | 1.9 | 0.0 | 0.0 | 0.0 | 0.0 | 0.1 | 0.8 | 2.6 | *4.1* | 24.7 |
| Days ≥ 1" Snow Depth | *9* | 6 | *2* | 1 | 0 | 0 | 0 | 0 | 0 | 0 | 2 | *6* | 26 |

## HUDSON *Stafford County* ELEVATION 1860 ft LAT/LONG 38° 6 ' N / 98° 39 ' W

| | JAN | FEB | MAR | APR | MAY | JUN | JUL | AUG | SEP | OCT | NOV | DEC | YEAR |
|---|---|---|---|---|---|---|---|---|---|---|---|---|---|
| Maximum Temp °F | 41.4 | 47.5 | 58.4 | 69.2 | 77.5 | 88.6 | 94.1 | 91.6 | 82.4 | 71.0 | 55.1 | 44.7 | 68.5 |
| Minimum Temp °F | 19.8 | 24.0 | 33.4 | 43.8 | 53.4 | 63.1 | 68.3 | 66.1 | 57.8 | 45.8 | 33.0 | 23.8 | 44.4 |
| Mean Temp °F | 30.6 | 35.8 | 45.9 | 56.5 | 65.5 | 75.9 | 81.2 | 78.9 | 70.1 | 58.5 | 44.1 | 34.3 | 56.4 |
| Days Max Temp ≥ 90 °F | 0 | 0 | 0 | 1 | 3 | 15 | 24 | 20 | 8 | 1 | 0 | 0 | 72 |
| Days Max Temp ≤ 32 °F | 8 | 5 | 1 | 0 | 0 | 0 | 0 | 0 | 0 | 0 | 1 | 5 | 20 |
| Days Min Temp ≤ 32 °F | 28 | 22 | 14 | 3 | 0 | 0 | 0 | 0 | 0 | 2 | 15 | 26 | 110 |
| Days Min Temp ≤ 0 °F | 2 | 1 | 0 | 0 | 0 | 0 | 0 | 0 | 0 | 0 | 0 | 1 | 4 |
| Heating Degree Days | 1060 | 818 | 589 | 275 | 86 | 6 | 0 | 1 | 47 | 234 | 622 | 945 | 4683 |
| Cooling Degree Days | 0 | 0 | 3 | 37 | 112 | 361 | 529 | 472 | 241 | 35 | 1 | 0 | 1791 |
| Total Precipitation (") | 0.55 | 0.83 | 1.83 | 2.47 | 3.65 | 3.77 | 2.80 | 2.49 | 2.32 | 2.12 | 1.07 | 0.95 | 24.85 |
| Days ≥ 0.1" Precip | 2 | 2 | 3 | 4 | 6 | 6 | 5 | 5 | 4 | 3 | 3 | 2 | 45 |
| Total Snowfall (") | 5.1 | 4.9 | 2.7 | 0.9 | 0.0 | 0.0 | 0.0 | 0.0 | 0.0 | 0.4 | 1.2 | 3.9 | 19.1 |
| Days ≥ 1" Snow Depth | *7* | 5 | 1 | 0 | 0 | 0 | 0 | 0 | 0 | 0 | 1 | *4* | 18 |

## HUGOTON *Stevens County* ELEVATION 3104 ft LAT/LONG 37° 11 ' N / 101° 21 ' W

| | JAN | FEB | MAR | APR | MAY | JUN | JUL | AUG | SEP | OCT | NOV | DEC | YEAR |
|---|---|---|---|---|---|---|---|---|---|---|---|---|---|
| Maximum Temp °F | 45.4 | 50.7 | 59.2 | 69.2 | 77.3 | 88.1 | 92.7 | 89.9 | 81.5 | 71.5 | 56.8 | 47.7 | 69.2 |
| Minimum Temp °F | 18.2 | 22.3 | 30.0 | 40.4 | 50.4 | 60.6 | 65.7 | 63.5 | 54.8 | 42.3 | 29.0 | 20.8 | 41.5 |
| Mean Temp °F | 31.9 | 36.6 | 44.6 | 54.8 | 63.9 | 74.4 | 79.3 | 76.7 | 68.1 | 56.9 | 43.0 | 34.2 | 55.4 |
| Days Max Temp ≥ 90 °F | 0 | 0 | 0 | 1 | 4 | 14 | 22 | 18 | 8 | 1 | 0 | 0 | 68 |
| Days Max Temp ≤ 32 °F | 6 | 3 | 1 | 0 | 0 | 0 | 0 | 0 | 0 | 0 | 1 | 4 | 15 |
| Days Min Temp ≤ 32 °F | 30 | 25 | 18 | 5 | 0 | 0 | 0 | 0 | 0 | 4 | 20 | 29 | 131 |
| Days Min Temp ≤ 0 °F | 2 | 1 | 0 | 0 | 0 | 0 | 0 | 0 | 0 | 0 | 0 | 1 | 4 |
| Heating Degree Days | 1020 | 795 | 626 | 316 | 115 | 12 | 0 | 2 | 61 | 270 | 653 | 947 | 4817 |
| Cooling Degree Days | 0 | 0 | 1 | 21 | 82 | 291 | 441 | 383 | 166 | 18 | 0 | 0 | 1403 |
| Total Precipitation (") | 0.39 | 0.42 | 1.08 | 1.60 | 3.16 | 3.09 | 2.65 | 2.61 | 1.76 | 0.92 | 0.88 | 0.41 | 18.97 |
| Days ≥ 0.1" Precip | 1 | 1 | 2 | 3 | 5 | 5 | 5 | 4 | 4 | 2 | 2 | 1 | 35 |
| Total Snowfall (") | na | na | *2.2* | 0.3 | 0.1 | 0.0 | 0.0 | 0.0 | 0.0 | 0.2 | 0.6 | na | na |
| Days ≥ 1" Snow Depth | na | na | *1* | 0 | 0 | 0 | 0 | 0 | 0 | 0 | 0 | na | na |

## HUTCHINSON 10 SW *Reno County* ELEVATION 1572 ft LAT/LONG 37° 56 ' N / 98° 2 ' W

| | JAN | FEB | MAR | APR | MAY | JUN | JUL | AUG | SEP | OCT | NOV | DEC | YEAR |
|---|---|---|---|---|---|---|---|---|---|---|---|---|---|
| Maximum Temp °F | 40.7 | 46.7 | 57.6 | 68.0 | 76.2 | 87.7 | 93.7 | 91.6 | 82.5 | 71.0 | 54.8 | 44.3 | 67.9 |
| Minimum Temp °F | 18.8 | 22.8 | 32.8 | 42.8 | 52.7 | 62.7 | 67.6 | 65.5 | 56.8 | 44.6 | 32.2 | 22.9 | 43.5 |
| Mean Temp °F | 29.7 | 34.8 | 45.2 | 55.4 | 64.5 | 75.2 | 80.7 | 78.6 | 69.7 | 57.8 | 43.5 | 33.6 | 55.7 |
| Days Max Temp ≥ 90 °F | 0 | 0 | 0 | 0 | 1 | 13 | 23 | 20 | 8 | 1 | 0 | 0 | 66 |
| Days Max Temp ≤ 32 °F | 9 | 5 | 1 | 0 | 0 | 0 | 0 | 0 | 0 | 0 | 1 | 5 | 21 |
| Days Min Temp ≤ 32 °F | 29 | 23 | 15 | 4 | 0 | 0 | 0 | 0 | 0 | 3 | 16 | 27 | 117 |
| Days Min Temp ≤ 0 °F | 2 | 1 | 0 | 0 | 0 | 0 | 0 | 0 | 0 | 0 | 0 | 1 | 4 |
| Heating Degree Days | 1086 | 846 | 608 | 301 | 98 | 8 | 0 | 1 | 52 | 249 | 637 | 965 | 4851 |
| Cooling Degree Days | 0 | 0 | 1 | 24 | 83 | 318 | 484 | 435 | 211 | 26 | 1 | 0 | 1583 |
| Total Precipitation (") | 0.62 | 0.97 | 2.22 | 2.91 | 4.10 | 4.34 | 3.34 | 3.00 | 3.03 | 2.46 | 1.35 | 1.01 | 29.35 |
| Days ≥ 0.1" Precip | 2 | 2 | 4 | 5 | 7 | 6 | 5 | 4 | 4 | 3 | 3 | 3 | 48 |
| Total Snowfall (") | 4.2 | 4.1 | 2.2 | 0.8 | 0.0 | 0.0 | 0.0 | 0.0 | 0.0 | 0.2 | 0.9 | 2.7 | 15.1 |
| Days ≥ 1" Snow Depth | 8 | 5 | 1 | 0 | 0 | 0 | 0 | 0 | 0 | 0 | 1 | 4 | 19 |

## INDEPENDENCE *Montgomery County* ELEVATION 801 ft LAT/LONG 37° 13 ' N / 95° 43 ' W

| | JAN | FEB | MAR | APR | MAY | JUN | JUL | AUG | SEP | OCT | NOV | DEC | YEAR |
|---|---|---|---|---|---|---|---|---|---|---|---|---|---|
| Maximum Temp °F | 42.9 | 48.9 | 60.0 | 70.8 | 77.7 | 86.4 | 92.1 | 90.9 | 82.3 | 71.8 | 58.0 | 47.0 | 69.1 |
| Minimum Temp °F | 21.7 | 26.0 | 36.0 | 46.4 | 55.5 | 64.6 | 69.2 | 67.0 | 59.3 | 47.3 | 35.8 | 26.1 | 46.2 |
| Mean Temp °F | 32.3 | 37.5 | 48.0 | 58.6 | 66.7 | 75.5 | 80.7 | 79.0 | 70.8 | 59.6 | 46.9 | 36.6 | 57.7 |
| Days Max Temp ≥ 90 °F | 0 | 0 | 0 | 1 | 1 | 10 | 21 | 19 | 7 | 1 | 0 | 0 | 60 |
| Days Max Temp ≤ 32 °F | 7 | 4 | 1 | 0 | 0 | 0 | 0 | 0 | 0 | 0 | 0 | 4 | 16 |
| Days Min Temp ≤ 32 °F | 26 | 21 | 12 | 2 | 0 | 0 | 0 | 0 | 0 | 1 | 12 | 23 | 97 |
| Days Min Temp ≤ 0 °F | 1 | 1 | 0 | 0 | 0 | 0 | 0 | 0 | 0 | 0 | 0 | 1 | 3 |
| Heating Degree Days | 1006 | 770 | 527 | 222 | 61 | 3 | 0 | 0 | 38 | 206 | 539 | 874 | 4246 |
| Cooling Degree Days | 0 | 0 | 6 | 39 | 119 | 347 | 510 | 467 | 235 | 39 | 4 | 0 | 1766 |
| Total Precipitation (") | 1.43 | 1.67 | 3.50 | 3.89 | 5.54 | 5.28 | 3.53 | 3.74 | 4.72 | 3.78 | 3.03 | 1.98 | 42.09 |
| Days ≥ 0.1" Precip | 3 | 3 | 6 | 7 | 8 | 6 | 5 | 6 | 6 | 5 | 4 | 4 | 63 |
| Total Snowfall (") | 3.3 | 3.7 | 1.8 | 0.0 | 0.0 | 0.0 | 0.0 | 0.0 | 0.0 | 0.0 | 0.5 | ***2.5*** | 11.8 |
| Days ≥ 1" Snow Depth | 6 | 4 | 1 | 0 | 0 | 0 | 0 | 0 | 0 | 0 | 0 | 3 | 14 |

## IOLA 1 W *Allen County* ELEVATION 951 ft LAT/LONG 37° 55 ' N / 95° 26 ' W

| | JAN | FEB | MAR | APR | MAY | JUN | JUL | AUG | SEP | OCT | NOV | DEC | YEAR |
|---|---|---|---|---|---|---|---|---|---|---|---|---|---|
| Maximum Temp °F | 41.0 | 47.0 | 58.9 | 70.0 | 77.4 | 85.9 | 91.4 | 89.5 | 81.6 | 71.0 | 56.2 | 45.1 | 67.9 |
| Minimum Temp °F | 21.2 | 25.5 | 35.6 | 46.2 | 55.6 | 64.9 | 69.4 | 67.0 | 59.1 | 47.7 | 36.1 | 26.2 | 46.2 |
| Mean Temp °F | 31.1 | 36.3 | 47.2 | 58.1 | 66.5 | 75.4 | 80.4 | 78.3 | 70.4 | 59.4 | 46.2 | 35.7 | 57.1 |
| Days Max Temp ≥ 90 °F | 0 | 0 | 0 | 0 | 0 | 8 | 20 | 16 | 5 | 0 | 0 | 0 | 49 |
| Days Max Temp ≤ 32 °F | 8 | 4 | 1 | 0 | 0 | 0 | 0 | 0 | 0 | 0 | 1 | 4 | 18 |
| Days Min Temp ≤ 32 °F | 26 | 21 | 13 | 2 | 0 | 0 | 0 | 0 | 0 | 2 | 12 | 23 | 99 |
| Days Min Temp ≤ 0 °F | 2 | 1 | 0 | 0 | 0 | 0 | 0 | 0 | 0 | 0 | 0 | 1 | 4 |
| Heating Degree Days | 1043 | 803 | 547 | 234 | 63 | 2 | 0 | 0 | 39 | 210 | 559 | 903 | 4403 |
| Cooling Degree Days | 0 | 0 | 4 | 37 | 118 | 336 | 495 | 435 | 225 | 39 | 3 | 0 | 1692 |
| Total Precipitation (") | 1.26 | 1.27 | 3.12 | 3.84 | 4.80 | 5.43 | 4.54 | 4.26 | 4.41 | 3.94 | 2.93 | 1.74 | 41.54 |
| Days ≥ 0.1" Precip | 3 | 3 | 6 | 6 | 7 | 7 | 6 | 6 | 6 | 5 | 4 | 4 | 63 |
| Total Snowfall (") | na | na | ***0.9*** | 0.0 | 0.0 | 0.0 | 0.0 | 0.0 | 0.0 | 0.0 | 0.6 | ***1.0*** | na |
| Days ≥ 1" Snow Depth | na | na | ***0*** | 0 | 0 | 0 | 0 | 0 | 0 | 0 | 0 | ***2*** | na |

## JOHN REDMOND LAKE *Coffey County* ELEVATION 1089 ft LAT/LONG 38° 15 ' N / 95° 45 ' W

| | JAN | FEB | MAR | APR | MAY | JUN | JUL | AUG | SEP | OCT | NOV | DEC | YEAR |
|---|---|---|---|---|---|---|---|---|---|---|---|---|---|
| Maximum Temp °F | ***38.9*** | 43.4 | 56.0 | 67.2 | 75.4 | 84.4 | 90.3 | 88.8 | 80.6 | 69.4 | ***54.9*** | na | na |
| Minimum Temp °F | ***16.8*** | 20.7 | 31.6 | 42.9 | 52.8 | 62.2 | 66.9 | 64.5 | 55.9 | 43.4 | ***32.3*** | ***22.8*** | 42.7 |
| Mean Temp °F | ***27.9*** | 31.9 | 43.8 | 55.1 | 64.1 | 73.4 | 78.7 | 76.7 | 68.3 | 56.5 | ***43.6*** | na | na |
| Days Max Temp ≥ 90 °F | 0 | 0 | 0 | 0 | 1 | 7 | 18 | 15 | 5 | 0 | 0 | 0 | 46 |
| Days Max Temp ≤ 32 °F | 9 | 6 | 1 | 0 | 0 | 0 | 0 | 0 | 0 | 0 | 1 | ***5*** | 22 |
| Days Min Temp ≤ 32 °F | 28 | 24 | 17 | 4 | 0 | 0 | 0 | 0 | 0 | 4 | ***15*** | 25 | 117 |
| Days Min Temp ≤ 0 °F | 2 | 2 | 0 | 0 | 0 | 0 | 0 | 0 | 0 | 0 | 0 | 1 | 5 |
| Heating Degree Days | ***1151*** | 930 | 656 | 314 | 104 | 10 | 2 | 2 | 61 | 284 | ***631*** | na | na |
| Cooling Degree Days | na | 0 | ***2*** | 22 | 77 | 274 | 437 | 376 | 181 | 21 | ***2*** | ***0*** | na |
| Total Precipitation (") | 0.96 | 0.84 | 2.43 | 3.32 | 4.06 | 5.39 | 4.48 | 3.49 | 3.98 | 3.02 | 2.09 | 1.45 | 35.51 |
| Days ≥ 0.1" Precip | 2 | 3 | 5 | 6 | 7 | 7 | 6 | 5 | 5 | 5 | 4 | ***3*** | 58 |
| Total Snowfall (") | ***3.4*** | ***2.6*** | ***1.6*** | 0.0 | 0.0 | 0.0 | 0.0 | 0.0 | 0.0 | 0.0 | 0.9 | na | na |
| Days ≥ 1" Snow Depth | na | ***3*** | 1 | 0 | 0 | 0 | 0 | 0 | 0 | 0 | 0 | na | na |

## KANOPOLIS LAKE *Ellsworth County* ELEVATION 1489 ft LAT/LONG 38° 37 ' N / 97° 57 ' W

| | JAN | FEB | MAR | APR | MAY | JUN | JUL | AUG | SEP | OCT | NOV | DEC | YEAR |
|---|---|---|---|---|---|---|---|---|---|---|---|---|---|
| Maximum Temp °F | 38.0 | 43.4 | 54.8 | 66.2 | 74.9 | 85.6 | 92.0 | 90.0 | 80.9 | 69.6 | 53.7 | 42.4 | 66.0 |
| Minimum Temp °F | 15.1 | 19.2 | 29.6 | 41.0 | 51.0 | 61.0 | 65.9 | 63.4 | 54.0 | 42.1 | 29.8 | 20.1 | 41.0 |
| Mean Temp °F | 26.6 | 31.3 | 42.3 | 53.6 | 63.0 | 73.3 | 79.0 | 76.7 | 67.5 | 55.9 | 41.8 | 31.2 | 53.5 |
| Days Max Temp ≥ 90 °F | 0 | 0 | 0 | 0 | 1 | 10 | 19 | 17 | 7 | 1 | 0 | 0 | 55 |
| Days Max Temp ≤ 32 °F | 11 | 7 | 2 | 0 | 0 | 0 | 0 | 0 | 0 | 0 | 1 | 7 | 28 |
| Days Min Temp ≤ 32 °F | 30 | 25 | 20 | 5 | 0 | 0 | 0 | 0 | 0 | 5 | 19 | 29 | 133 |
| Days Min Temp ≤ 0 °F | 4 | 2 | 0 | 0 | 0 | 0 | 0 | 0 | 0 | 0 | 0 | 2 | 8 |
| Heating Degree Days | 1185 | 944 | 699 | 352 | 129 | 15 | 1 | 3 | 78 | 303 | 691 | 1039 | 5439 |
| Cooling Degree Days | 0 | 0 | 2 | 21 | 71 | 280 | 446 | 384 | 181 | 23 | 0 | 0 | 1408 |
| Total Precipitation (") | 0.47 | 0.73 | 2.06 | 2.35 | 3.71 | 3.91 | 3.32 | 3.19 | 2.75 | 2.18 | 1.22 | 0.81 | 26.70 |
| Days ≥ 0.1" Precip | 1 | 2 | 4 | 5 | 6 | 6 | 5 | 4 | 4 | 3 | 3 | 2 | 45 |
| Total Snowfall (") | 3.5 | 3.3 | 1.8 | 0.2 | 0.0 | 0.0 | 0.0 | 0.0 | 0.0 | 0.2 | 0.3 | *2.8* | 12.1 |
| Days ≥ 1" Snow Depth | 11 | 7 | *2* | 0 | 0 | 0 | 0 | 0 | 0 | 0 | 1 | 4 | 25 |

## KINGMAN *Kingman County* ELEVATION 1503 ft LAT/LONG 37° 39 ' N / 98° 7 ' W

| | JAN | FEB | MAR | APR | MAY | JUN | JUL | AUG | SEP | OCT | NOV | DEC | YEAR |
|---|---|---|---|---|---|---|---|---|---|---|---|---|---|
| Maximum Temp °F | 42.7 | 48.9 | 59.7 | 70.4 | 78.5 | 89.1 | 94.7 | 92.9 | 83.8 | 72.5 | 56.4 | 45.9 | 69.6 |
| Minimum Temp °F | 20.4 | 24.4 | 33.7 | 44.7 | 54.0 | 63.7 | 68.6 | 66.8 | 58.1 | 45.9 | 33.3 | 24.0 | 44.8 |
| Mean Temp °F | 31.6 | 36.7 | 46.8 | 57.6 | 66.3 | 76.4 | 81.7 | 79.9 | 71.0 | 59.2 | 44.9 | 35.0 | 57.3 |
| Days Max Temp ≥ 90 °F | 0 | 0 | 0 | 1 | 2 | 16 | 24 | 21 | 9 | 1 | 0 | 0 | 74 |
| Days Max Temp ≤ 32 °F | 7 | 4 | 1 | 0 | 0 | 0 | 0 | 0 | 0 | 0 | 1 | 4 | 17 |
| Days Min Temp ≤ 32 °F | 28 | 22 | 15 | 3 | 0 | 0 | 0 | 0 | 0 | 2 | 14 | 26 | 110 |
| Days Min Temp ≤ 0 °F | 2 | 1 | 0 | 0 | 0 | 0 | 0 | 0 | 0 | 0 | 0 | 1 | 4 |
| Heating Degree Days | 1030 | 793 | 563 | 248 | 72 | 4 | 0 | 0 | 40 | 215 | 598 | 924 | 4487 |
| Cooling Degree Days | 0 | 0 | 4 | 36 | 121 | 363 | 529 | 487 | 244 | 37 | 2 | 0 | 1823 |
| Total Precipitation (") | 0.63 | 0.97 | 2.31 | 2.76 | 4.03 | 4.19 | 3.05 | 3.06 | 2.95 | 2.48 | 1.68 | 0.99 | 29.10 |
| Days ≥ 0.1" Precip | 2 | 3 | 4 | 5 | 7 | 6 | 5 | 4 | 4 | 4 | 3 | 3 | 50 |
| Total Snowfall (") | 4.2 | 4.0 | 2.0 | 0.3 | 0.0 | 0.0 | 0.0 | 0.0 | 0.0 | 0.1 | 1.2 | 2.8 | 14.6 |
| Days ≥ 1" Snow Depth | 7 | 5 | 1 | 0 | 0 | 0 | 0 | 0 | 0 | 0 | 1 | 3 | 17 |

## KINSLEY *Edwards County* ELEVATION 2152 ft LAT/LONG 37° 56 ' N / 99° 24 ' W

| | JAN | FEB | MAR | APR | MAY | JUN | JUL | AUG | SEP | OCT | NOV | DEC | YEAR |
|---|---|---|---|---|---|---|---|---|---|---|---|---|---|
| Maximum Temp °F | 41.6 | 46.9 | 57.2 | 68.2 | 76.6 | 87.5 | 93.5 | 91.2 | 82.4 | 71.3 | 55.3 | 45.2 | 68.1 |
| Minimum Temp °F | 17.2 | 20.6 | 30.2 | 40.8 | 50.9 | 61.3 | 66.5 | 64.3 | 55.2 | 42.4 | 29.8 | 20.7 | 41.7 |
| Mean Temp °F | 29.4 | 33.8 | 43.7 | 54.5 | 63.8 | 74.4 | 80.0 | 77.8 | 68.8 | 56.9 | 42.5 | 33.0 | 54.9 |
| Days Max Temp ≥ 90 °F | 0 | 0 | 0 | 1 | 3 | 14 | 23 | 20 | 9 | 2 | 0 | 0 | 72 |
| Days Max Temp ≤ 32 °F | 9 | 5 | 2 | 0 | 0 | 0 | 0 | 0 | 0 | 0 | 1 | 6 | 23 |
| Days Min Temp ≤ 32 °F | 30 | 25 | 18 | 5 | 0 | 0 | 0 | 0 | 0 | 4 | 19 | 29 | 130 |
| Days Min Temp ≤ 0 °F | 3 | 2 | 0 | 0 | 0 | 0 | 0 | 0 | 0 | 0 | 0 | 1 | 6 |
| Heating Degree Days | 1097 | 874 | 654 | 328 | 118 | 13 | 1 | 2 | 63 | 274 | 667 | 986 | 5077 |
| Cooling Degree Days | 0 | 0 | 2 | 23 | 74 | 293 | 460 | 407 | 196 | 23 | 0 | 0 | 1478 |
| Total Precipitation (") | 0.60 | 0.87 | 1.98 | 2.25 | 3.43 | 3.48 | 3.24 | 3.25 | 1.91 | 1.77 | 1.05 | 1.00 | 24.83 |
| Days ≥ 0.1" Precip | 2 | 2 | 4 | 4 | 6 | 6 | 5 | 5 | 4 | 3 | 3 | 2 | 46 |
| Total Snowfall (") | na | na | *2.1* | 0.7 | 0.0 | 0.0 | 0.0 | 0.0 | 0.0 | 0.2 | 0.7 | na | na |
| Days ≥ 1" Snow Depth | na | na | na | 0 | 0 | 0 | 0 | 0 | 0 | 0 | 0 | na | na |

## KIRWIN DAM *Phillips County* ELEVATION 1703 ft LAT/LONG 39° 40 ' N / 99° 7 ' W

| | JAN | FEB | MAR | APR | MAY | JUN | JUL | AUG | SEP | OCT | NOV | DEC | YEAR |
|---|---|---|---|---|---|---|---|---|---|---|---|---|---|
| Maximum Temp °F | 37.6 | 43.8 | 54.3 | 66.0 | 75.1 | 85.9 | 91.8 | 89.6 | 79.9 | 69.2 | 53.0 | *43.3* | 65.8 |
| Minimum Temp °F | 11.6 | 15.7 | 25.7 | 37.3 | 47.5 | 57.8 | 63.1 | 60.3 | 49.6 | 37.1 | 25.3 | *17.3* | 37.4 |
| Mean Temp °F | 24.6 | 29.8 | 40.0 | 51.7 | 61.3 | 71.9 | 77.5 | 75.0 | 64.8 | 53.2 | 39.2 | *30.3* | 51.6 |
| Days Max Temp ≥ 90 °F | 0 | 0 | 0 | 1 | 2 | 11 | 19 | 16 | 6 | 1 | 0 | 0 | 56 |
| Days Max Temp ≤ 32 °F | 10 | 7 | 2 | 0 | 0 | 0 | 0 | 0 | 0 | 0 | 2 | 6 | 27 |
| Days Min Temp ≤ 32 °F | 29 | 26 | 24 | 9 | 1 | 0 | 0 | 0 | 1 | 10 | 23 | 27 | 150 |
| Days Min Temp ≤ 0 °F | 5 | 3 | 0 | 0 | 0 | 0 | 0 | 0 | 0 | 0 | 0 | 2 | 10 |
| Heating Degree Days | 1245 | 981 | 761 | 403 | 164 | 21 | 2 | 6 | 109 | 371 | 768 | *1068* | 5899 |
| Cooling Degree Days | *0* | 0 | 0 | 11 | *51* | 237 | 390 | 330 | *116* | 12 | *0* | na | na |
| Total Precipitation (") | 0.45 | 0.63 | 1.92 | 2.24 | 3.91 | 3.18 | 3.10 | 2.82 | 2.36 | 1.71 | 1.00 | 0.58 | 23.90 |
| Days ≥ 0.1" Precip | 1 | 1 | 4 | 5 | 7 | 5 | 5 | 5 | 4 | 3 | 2 | 1 | 43 |
| Total Snowfall (") | *3.6* | *4.1* | *1.8* | 0.2 | 0.0 | 0.0 | 0.0 | 0.0 | 0.0 | 0.0 | 1.1 | na | na |
| Days ≥ 1" Snow Depth | na | *5* | *1* | 0 | 0 | 0 | 0 | 0 | 0 | 0 | 0 | *3* | na |

## LAKIN *Kearny County* ELEVATION 2992 ft LAT/LONG 37° 54' N / 101° 16' W

| | JAN | FEB | MAR | APR | MAY | JUN | JUL | AUG | SEP | OCT | NOV | DEC | YEAR |
|---|---|---|---|---|---|---|---|---|---|---|---|---|---|
| Maximum Temp °F | 42.5 | 48.1 | 57.2 | 67.1 | 76.3 | 87.5 | 93.0 | 90.2 | 81.6 | 70.7 | 55.3 | 45.4 | 67.9 |
| Minimum Temp °F | 15.6 | 19.8 | 28.2 | 38.0 | 48.8 | 58.6 | 63.9 | 61.6 | 52.6 | 39.7 | 27.1 | 18.6 | 39.4 |
| Mean Temp °F | 29.1 | 33.9 | 42.7 | 52.5 | 62.6 | 73.1 | 78.5 | 75.9 | 67.1 | 55.2 | 41.2 | 32.1 | 53.7 |
| Days Max Temp ≥ 90 °F | 0 | 0 | 0 | 1 | 3 | 13 | 22 | 18 | 8 | 1 | 0 | 0 | 66 |
| Days Max Temp ≤ 32 °F | 8 | 5 | 2 | 0 | 0 | 0 | 0 | 0 | 0 | 0 | 2 | 6 | 23 |
| Days Min Temp ≤ 32 °F | 31 | 26 | 21 | 8 | 0 | 0 | 0 | 0 | 0 | 6 | 22 | 30 | 144 |
| Days Min Temp ≤ 0 °F | 3 | 2 | 0 | 0 | 0 | 0 | 0 | 0 | 0 | 0 | 0 | 2 | 7 |
| Heating Degree Days | 1106 | 870 | 685 | 381 | 140 | 19 | 1 | 4 | 78 | 313 | 706 | 1016 | 5319 |
| Cooling Degree Days | 0 | 0 | 0 | 12 | 70 | 274 | 427 | 369 | 163 | 14 | 0 | 0 | 1329 |
| Total Precipitation (") | 0.30 | 0.43 | 1.00 | 1.49 | 2.72 | 2.98 | 2.75 | 2.67 | 1.77 | 0.86 | 0.73 | 0.38 | 18.08 |
| Days ≥ 0.1" Precip | 1 | 1 | 3 | 3 | 5 | 5 | 5 | 4 | 3 | 2 | 2 | 1 | 35 |
| Total Snowfall (") | ***4.0*** | ***2.9*** | 3.3 | 0.6 | 0.0 | 0.0 | 0.0 | 0.0 | 0.0 | 0.1 | 0.8 | 2.2 | 13.9 |
| Days ≥ 1" Snow Depth | na | na | ***0*** | 0 | 0 | 0 | 0 | 0 | 0 | 0 | 0 | ***1*** | na |

## LARNED *Pawnee County* ELEVATION 2090 ft LAT/LONG 38° 11' N / 99° 6' W

| | JAN | FEB | MAR | APR | MAY | JUN | JUL | AUG | SEP | OCT | NOV | DEC | YEAR |
|---|---|---|---|---|---|---|---|---|---|---|---|---|---|
| Maximum Temp °F | 41.5 | 47.8 | 58.5 | 69.6 | 77.6 | 88.1 | 93.5 | 91.2 | 82.3 | 71.2 | 55.2 | 44.9 | 68.5 |
| Minimum Temp °F | 18.8 | 22.7 | 31.8 | 42.1 | 52.2 | 62.2 | 67.2 | 65.0 | 56.2 | 44.2 | 31.2 | 22.1 | 43.0 |
| Mean Temp °F | 30.2 | 35.3 | 45.2 | 55.9 | 65.0 | 75.2 | 80.4 | 78.2 | 69.3 | 57.7 | 43.2 | 33.5 | 55.8 |
| Days Max Temp ≥ 90 °F | 0 | 0 | 0 | 1 | 3 | 14 | 22 | 20 | 8 | 1 | 0 | 0 | 69 |
| Days Max Temp ≤ 32 °F | 8 | 5 | 1 | 0 | 0 | 0 | 0 | 0 | 0 | 0 | 1 | 5 | 20 |
| Days Min Temp ≤ 32 °F | 29 | 24 | 17 | 4 | 0 | 0 | 0 | 0 | 0 | 3 | 17 | 28 | 122 |
| Days Min Temp ≤ 0 °F | 2 | 1 | 0 | 0 | 0 | 0 | 0 | 0 | 0 | 0 | 0 | 1 | 4 |
| Heating Degree Days | 1073 | 833 | 611 | 294 | 97 | 9 | 0 | 2 | 57 | 255 | 649 | 969 | 4849 |
| Cooling Degree Days | 0 | 0 | 2 | 30 | 92 | 310 | 472 | 420 | 203 | 27 | 1 | 0 | 1557 |
| Total Precipitation (") | 0.52 | 0.75 | 1.65 | 2.06 | 2.93 | 3.88 | 3.40 | 3.07 | 2.27 | 1.64 | 1.05 | 0.84 | 24.06 |
| Days ≥ 0.1" Precip | 2 | 2 | 3 | 4 | 6 | 6 | 5 | 5 | 4 | 3 | 2 | 2 | 44 |
| Total Snowfall (") | 4.4 | 4.5 | 3.6 | 1.2 | 0.0 | 0.0 | 0.0 | 0.0 | 0.0 | 0.5 | 0.7 | 3.4 | 18.3 |
| Days ≥ 1" Snow Depth | 8 | 5 | 2 | 1 | 0 | 0 | 0 | 0 | 0 | 0 | 1 | 4 | 21 |

## LAWRENCE *Douglas County* ELEVATION 860 ft LAT/LONG 38° 58' N / 95° 14' W

| | JAN | FEB | MAR | APR | MAY | JUN | JUL | AUG | SEP | OCT | NOV | DEC | YEAR |
|---|---|---|---|---|---|---|---|---|---|---|---|---|---|
| Maximum Temp °F | 39.2 | 44.9 | 57.2 | 68.4 | 77.0 | 85.5 | 91.0 | 89.1 | 81.1 | 70.1 | 54.7 | 43.1 | 66.8 |
| Minimum Temp °F | 19.9 | 24.4 | 34.6 | 45.7 | 55.3 | 64.4 | 69.2 | 66.9 | 58.8 | 47.8 | 35.6 | 25.3 | 45.7 |
| Mean Temp °F | 29.5 | 34.7 | 45.9 | 57.2 | 66.2 | 75.0 | 80.1 | 78.0 | 70.0 | 59.0 | 45.2 | 34.2 | 56.3 |
| Days Max Temp ≥ 90 °F | 0 | 0 | 0 | 0 | 1 | 8 | 19 | 16 | 5 | 0 | 0 | 0 | 49 |
| Days Max Temp ≤ 32 °F | 10 | 6 | 1 | 0 | 0 | 0 | 0 | 0 | 0 | 0 | 1 | 6 | 24 |
| Days Min Temp ≤ 32 °F | 26 | 21 | 14 | 3 | 0 | 0 | 0 | 0 | 0 | 1 | 12 | 23 | 100 |
| Days Min Temp ≤ 0 °F | 2 | 1 | 0 | 0 | 0 | 0 | 0 | 0 | 0 | 0 | 0 | 1 | 4 |
| Heating Degree Days | 1092 | 849 | 588 | 263 | 72 | 5 | 0 | 1 | 44 | 221 | 591 | 948 | 4674 |
| Cooling Degree Days | 0 | 0 | 5 | 42 | 117 | 326 | 490 | 429 | 220 | 38 | 3 | 0 | 1670 |
| Total Precipitation (") | 1.28 | 1.08 | 2.59 | 3.84 | 4.87 | 5.89 | 4.23 | 3.83 | 4.52 | 3.37 | 2.21 | 1.94 | 39.65 |
| Days ≥ 0.1" Precip | 3 | 3 | 5 | 7 | 7 | 7 | 6 | 5 | 6 | 6 | 4 | 4 | 63 |
| Total Snowfall (") | 5.9 | 5.0 | 1.9 | 0.5 | 0.0 | 0.0 | 0.0 | 0.0 | 0.0 | 0.0 | 0.9 | 4.0 | 18.2 |
| Days ≥ 1" Snow Depth | 10 | 6 | 1 | 0 | 0 | 0 | 0 | 0 | 0 | 0 | 1 | 4 | 22 |

## LEAVENWORTH *Leavenworth County* ELEVATION 910 ft LAT/LONG 39° 19' N / 94° 56' W

| | JAN | FEB | MAR | APR | MAY | JUN | JUL | AUG | SEP | OCT | NOV | DEC | YEAR |
|---|---|---|---|---|---|---|---|---|---|---|---|---|---|
| Maximum Temp °F | 37.6 | 43.8 | 56.1 | ***67.8*** | 76.5 | 85.0 | 89.9 | 88.0 | 79.9 | 68.9 | 53.6 | 42.0 | 65.8 |
| Minimum Temp °F | 17.3 | 21.7 | 31.8 | 43.2 | 53.1 | 62.0 | 67.1 | 64.2 | 56.1 | 44.4 | 32.6 | 22.6 | 43.0 |
| Mean Temp °F | 27.6 | 32.8 | 44.0 | ***55.6*** | 64.8 | 73.5 | 78.5 | 76.1 | 68.0 | 56.7 | 43.1 | 32.3 | 54.4 |
| Days Max Temp ≥ 90 °F | 0 | 0 | 0 | 0 | 1 | 8 | 17 | 13 | 4 | 0 | 0 | 0 | 43 |
| Days Max Temp ≤ 32 °F | 10 | 7 | 1 | 0 | 0 | 0 | 0 | 0 | 0 | 0 | 1 | ***6*** | 25 |
| Days Min Temp ≤ 32 °F | 28 | 23 | 17 | ***4*** | 0 | 0 | 0 | 0 | 0 | 3 | 16 | 26 | 117 |
| Days Min Temp ≤ 0 °F | 3 | 2 | 0 | 0 | 0 | 0 | 0 | 0 | 0 | 0 | 0 | 2 | 7 |
| Heating Degree Days | 1152 | 903 | 646 | ***302*** | 92 | 8 | 1 | 2 | 61 | 277 | 650 | 1007 | 5101 |
| Cooling Degree Days | 0 | 0 | 4 | 32 | 87 | 277 | 437 | 366 | 172 | 23 | 1 | 0 | 1399 |
| Total Precipitation (") | 1.12 | 1.05 | 2.57 | 3.68 | 5.04 | 4.99 | 4.85 | 4.18 | 5.07 | 3.74 | 2.33 | 1.67 | 40.29 |
| Days ≥ 0.1" Precip | 3 | 3 | 5 | 7 | 7 | 7 | 6 | 6 | 5 | 5 | 4 | 3 | 61 |
| Total Snowfall (") | na | na | na | 0.0 | 0.0 | 0.0 | 0.0 | 0.0 | 0.0 | 0.0 | ***0.4*** | na | na |
| Days ≥ 1" Snow Depth | na | na | na | 0 | 0 | 0 | 0 | 0 | 0 | 0 | 0 | na | na |

## LEOTI 1 W *Wichita County* ELEVATION 3304 ft LAT/LONG 38° 29 ' N / 101° 22 ' W

| | JAN | FEB | MAR | APR | MAY | JUN | JUL | AUG | SEP | OCT | NOV | DEC | YEAR |
|---|---|---|---|---|---|---|---|---|---|---|---|---|---|
| Maximum Temp °F | 41.5 | 47.3 | 56.6 | 67.0 | 76.0 | 86.9 | 92.2 | 88.9 | 81.0 | 69.5 | 53.6 | 44.1 | 67.1 |
| Minimum Temp °F | 15.1 | 19.1 | 26.6 | 36.3 | 46.9 | 57.0 | 62.2 | 60.0 | 50.5 | 37.7 | 25.7 | 17.5 | 37.9 |
| Mean Temp °F | 28.3 | 33.2 | 41.6 | 51.7 | 61.5 | 71.9 | 77.2 | 74.5 | 65.8 | 53.7 | 39.7 | 30.8 | 52.5 |
| Days Max Temp ≥ 90 °F | 0 | 0 | 0 | 0 | 3 | 13 | 21 | 17 | 7 | 1 | 0 | 0 | 62 |
| Days Max Temp ≤ 32 °F | 9 | 5 | 2 | 0 | 0 | 0 | 0 | 0 | 0 | 0 | 2 | 6 | 24 |
| Days Min Temp ≤ 32 °F | 31 | 27 | 23 | 10 | 1 | 0 | 0 | 0 | 1 | 8 | 25 | 30 | 156 |
| Days Min Temp ≤ 0 °F | 3 | 2 | 0 | 0 | 0 | 0 | 0 | 0 | 0 | 0 | 0 | 2 | 7 |
| Heating Degree Days | 1131 | 892 | 717 | 401 | 158 | 21 | 1 | 6 | 85 | 353 | 754 | 1053 | 5572 |
| Cooling Degree Days | 0 | 0 | 0 | 9 | 48 | 221 | 368 | 319 | 122 | 5 | 0 | 0 | 1092 |
| Total Precipitation (") | 0.41 | 0.44 | 1.12 | 1.43 | 2.71 | 2.68 | 2.78 | 2.52 | 1.64 | 1.08 | 0.69 | 0.36 | 17.86 |
| Days ≥ 0.1" Precip | 2 | 1 | 3 | 3 | 5 | 5 | 5 | 5 | 3 | 2 | 2 | 1 | 37 |
| Total Snowfall (") | *6.2* | *3.5* | *5.0* | 2.4 | 0.1 | 0.0 | 0.0 | 0.0 | 0.1 | 1.1 | 2.5 | *4.1* | 25.0 |
| Days ≥ 1" Snow Depth | na | *6* | *4* | 1 | 0 | 0 | 0 | 0 | 0 | 0 | *2* | 7 | na |

## LIBERAL *Seward County* ELEVATION 2841 ft LAT/LONG 37° 2 ' N / 100° 55 ' W

| | JAN | FEB | MAR | APR | MAY | JUN | JUL | AUG | SEP | OCT | NOV | DEC | YEAR |
|---|---|---|---|---|---|---|---|---|---|---|---|---|---|
| Maximum Temp °F | 47.0 | 52.3 | 61.4 | 72.0 | 79.3 | 90.1 | 95.1 | 92.9 | 84.6 | 73.8 | 58.4 | 48.9 | 71.3 |
| Minimum Temp °F | 20.3 | 24.2 | 31.9 | 42.0 | 51.7 | 61.7 | 66.8 | 64.8 | 56.1 | 43.5 | 31.1 | 22.9 | 43.1 |
| Mean Temp °F | 33.7 | 38.2 | 46.7 | 57.0 | 65.5 | 75.9 | 81.0 | 78.8 | 70.4 | 58.7 | 44.8 | 35.9 | 57.2 |
| Days Max Temp ≥ 90 °F | 0 | 0 | 0 | 1 | 5 | 17 | 25 | 22 | 11 | 3 | 0 | 0 | 84 |
| Days Max Temp ≤ 32 °F | 6 | 3 | 1 | 0 | 0 | 0 | 0 | 0 | 0 | 0 | 1 | 4 | 15 |
| Days Min Temp ≤ 32 °F | 28 | 23 | 16 | 4 | 0 | 0 | 0 | 0 | 0 | 2 | 17 | 28 | 118 |
| Days Min Temp ≤ 0 °F | 1 | 1 | 0 | 0 | 0 | 0 | 0 | 0 | 0 | 0 | 0 | 1 | 3 |
| Heating Degree Days | 964 | 748 | 563 | 258 | 89 | 9 | 0 | 2 | 44 | 227 | 601 | 894 | 4399 |
| Cooling Degree Days | 0 | 0 | 2 | 30 | 104 | 331 | 499 | 456 | 227 | 32 | 1 | 0 | 1682 |
| Total Precipitation (") | 0.54 | 0.63 | 1.30 | 1.48 | 3.09 | 2.77 | 2.68 | 2.35 | 1.73 | 1.06 | 0.90 | 0.54 | 19.07 |
| Days ≥ 0.1" Precip | 2 | 2 | 3 | 3 | 5 | 5 | 5 | 4 | 3 | 2 | 2 | 2 | 38 |
| Total Snowfall (") | *4.5* | 3.9 | *5.1* | 1.5 | 0.0 | 0.0 | 0.0 | 0.0 | 0.1 | 0.6 | 1.8 | 3.9 | 21.4 |
| Days ≥ 1" Snow Depth | na | na | *1* | 0 | 0 | 0 | 0 | 0 | 0 | 0 | 0 | *2* | na |

## LINCOLN 1 ESE *Lincoln County* ELEVATION 1401 ft LAT/LONG 39° 2 ' N / 98° 7 ' W

| | JAN | FEB | MAR | APR | MAY | JUN | JUL | AUG | SEP | OCT | NOV | DEC | YEAR |
|---|---|---|---|---|---|---|---|---|---|---|---|---|---|
| Maximum Temp °F | 40.3 | 46.4 | 58.2 | 69.3 | 77.5 | 88.8 | 94.9 | 92.4 | 83.6 | 71.8 | 54.6 | 44.1 | 68.5 |
| Minimum Temp °F | 15.4 | 19.7 | 30.2 | 40.7 | 51.0 | 61.0 | 66.6 | 64.4 | 55.0 | 42.4 | 29.1 | 19.6 | 41.3 |
| Mean Temp °F | 27.9 | 33.1 | 44.2 | 55.0 | 64.2 | 74.9 | 80.8 | 78.4 | 69.3 | 57.1 | 41.8 | 31.9 | 54.9 |
| Days Max Temp ≥ 90 °F | 0 | 0 | 0 | 1 | 2 | 15 | 24 | 21 | 9 | 1 | 0 | 0 | 73 |
| Days Max Temp ≤ 32 °F | 9 | 5 | 1 | 0 | 0 | 0 | 0 | 0 | 0 | 0 | 1 | 5 | 21 |
| Days Min Temp ≤ 32 °F | 30 | 25 | 19 | 6 | 1 | 0 | 0 | 0 | 0 | 5 | 19 | 29 | 134 |
| Days Min Temp ≤ 0 °F | 4 | 2 | 0 | 0 | 0 | 0 | 0 | 0 | 0 | 0 | 0 | 2 | 8 |
| Heating Degree Days | 1144 | 894 | 640 | 314 | 109 | 9 | 0 | 2 | 56 | 271 | 690 | 1020 | 5149 |
| Cooling Degree Days | 0 | 0 | 3 | 26 | 89 | 315 | 487 | 420 | 201 | 27 | 1 | 0 | 1569 |
| Total Precipitation (") | 0.73 | 0.85 | 2.19 | 2.34 | 4.19 | 3.36 | 3.67 | 3.53 | 2.67 | 2.14 | 1.35 | 0.88 | 27.90 |
| Days ≥ 0.1" Precip | 2 | 2 | 4 | 5 | 7 | 6 | 6 | 5 | 5 | 4 | 3 | 2 | 51 |
| Total Snowfall (") | 5.7 | 5.5 | 2.4 | 0.5 | 0.0 | 0.0 | 0.0 | 0.0 | 0.0 | 0.3 | 1.2 | 3.7 | 19.3 |
| Days ≥ 1" Snow Depth | 11 | 8 | 2 | 0 | 0 | 0 | 0 | 0 | 0 | 0 | 1 | 5 | 27 |

## LOVEWELL LAKE *Jewell County* ELEVATION 1601 ft LAT/LONG 39° 54 ' N / 98° 2 ' W

| | JAN | FEB | MAR | APR | MAY | JUN | JUL | AUG | SEP | OCT | NOV | DEC | YEAR |
|---|---|---|---|---|---|---|---|---|---|---|---|---|---|
| Maximum Temp °F | 35.1 | 40.6 | 52.5 | 64.3 | 73.5 | 83.7 | 89.8 | 87.3 | 78.8 | 67.2 | 50.8 | 39.7 | 63.6 |
| Minimum Temp °F | 12.0 | 15.8 | 27.0 | 38.6 | 49.4 | 59.4 | 64.4 | 61.9 | 52.1 | 39.6 | 26.9 | 17.1 | 38.7 |
| Mean Temp °F | 23.6 | 28.2 | 39.8 | 51.5 | 61.5 | 71.6 | 77.1 | 74.6 | 65.5 | 53.4 | 38.9 | 28.4 | 51.2 |
| Days Max Temp ≥ 90 °F | 0 | 0 | 0 | 0 | 1 | 8 | 16 | 13 | 5 | 0 | 0 | 0 | 43 |
| Days Max Temp ≤ 32 °F | 13 | 9 | 3 | 0 | 0 | 0 | 0 | 0 | 0 | 0 | 2 | 9 | 36 |
| Days Min Temp ≤ 32 °F | 31 | 27 | 22 | 7 | 1 | 0 | 0 | 0 | 0 | 7 | *22* | 30 | 147 |
| Days Min Temp ≤ 0 °F | 6 | 4 | 0 | 0 | 0 | 0 | 0 | 0 | 0 | 0 | 0 | 2 | 12 |
| Heating Degree Days | 1276 | 1033 | 775 | 408 | 156 | 20 | 2 | 5 | 101 | 364 | 776 | 1128 | 6044 |
| Cooling Degree Days | 0 | 0 | 1 | 15 | 54 | 229 | 381 | 308 | 138 | 8 | 0 | 0 | 1134 |
| Total Precipitation (") | 0.60 | 0.74 | 2.01 | 2.47 | 3.87 | 3.80 | 4.05 | 3.37 | 3.23 | 2.04 | 1.18 | 0.82 | 28.18 |
| Days ≥ 0.1" Precip | 2 | 2 | 4 | 5 | 7 | 6 | 6 | 6 | 5 | 4 | 3 | 2 | 52 |
| Total Snowfall (") | 5.8 | 5.4 | 3.0 | 0.4 | 0.1 | 0.0 | 0.0 | 0.0 | 0.1 | 0.1 | 1.4 | *5.1* | 21.4 |
| Days ≥ 1" Snow Depth | 14 | 9 | 3 | 0 | 0 | 0 | 0 | 0 | 0 | 0 | 3 | 8 | 37 |

## MANHATTAN *Riley County* ELEVATION 1070 ft LAT/LONG 39° 11' N / 96° 34' W

| | JAN | FEB | MAR | APR | MAY | JUN | JUL | AUG | SEP | OCT | NOV | DEC | YEAR |
|---|---|---|---|---|---|---|---|---|---|---|---|---|---|
| Maximum Temp °F | 38.7 | 45.0 | 57.0 | 68.1 | 76.7 | 86.1 | 91.3 | 89.5 | 81.0 | 69.8 | 54.3 | 42.8 | 66.7 |
| Minimum Temp °F | 17.7 | 22.0 | 32.5 | 43.6 | 53.6 | 63.2 | 68.2 | 66.2 | 57.3 | 45.0 | 32.6 | 22.5 | 43.7 |
| Mean Temp °F | 28.2 | 33.5 | 44.8 | 55.9 | 65.2 | 74.7 | 79.7 | 77.9 | 69.2 | 57.4 | 43.5 | 32.7 | 55.2 |
| Days Max Temp ≥ 90 °F | 0 | 0 | 0 | 1 | 1 | 10 | 19 | 16 | 6 | 1 | 0 | 0 | 54 |
| Days Max Temp ≤ 32 °F | 10 | 6 | 1 | 0 | 0 | 0 | 0 | 0 | 0 | 0 | 1 | 5 | 23 |
| Days Min Temp ≤ 32 °F | 28 | 23 | 16 | 5 | 0 | 0 | 0 | 0 | 0 | 3 | 16 | 27 | 118 |
| Days Min Temp ≤ 0 °F | 3 | 2 | 0 | 0 | 0 | 0 | 0 | 0 | 0 | 0 | 0 | 1 | 6 |
| Heating Degree Days | 1135 | 882 | 624 | 300 | 92 | 7 | 1 | 1 | 58 | 262 | 641 | 996 | 4999 |
| Cooling Degree Days | 0 | 0 | 5 | 38 | 100 | 312 | 470 | 414 | 210 | 30 | 1 | 0 | 1580 |
| Total Precipitation (") | 0.90 | 0.91 | 2.36 | 2.97 | 4.60 | 5.49 | 4.30 | 3.36 | 3.95 | 2.90 | 1.83 | 1.20 | 34.77 |
| Days ≥ 0.1" Precip | 3 | 2 | 5 | 6 | 8 | 7 | 6 | 5 | 5 | 5 | 4 | 3 | 59 |
| Total Snowfall (") | 5.9 | 5.4 | 2.0 | 0.5 | 0.0 | 0.0 | 0.0 | 0.0 | 0.0 | 0.0 | 1.1 | 3.7 | 18.6 |
| Days ≥ 1" Snow Depth | 11 | 7 | 1 | 0 | 0 | 0 | 0 | 0 | 0 | 0 | 1 | 5 | 25 |

## MANKATO *Jewell County* ELEVATION 1781 ft LAT/LONG 39° 47' N / 98° 13' W

| | JAN | FEB | MAR | APR | MAY | JUN | JUL | AUG | SEP | OCT | NOV | DEC | YEAR |
|---|---|---|---|---|---|---|---|---|---|---|---|---|---|
| Maximum Temp °F | 35.0 | 40.4 | 52.3 | 64.3 | 73.5 | 84.2 | 90.3 | 87.9 | 79.0 | 66.9 | 50.8 | 39.2 | 63.7 |
| Minimum Temp °F | 12.9 | 17.3 | 27.6 | 39.1 | 49.9 | 59.8 | 64.9 | 62.6 | 52.7 | 40.3 | 27.8 | 17.2 | 39.3 |
| Mean Temp °F | 24.0 | 28.9 | 40.0 | 51.7 | 61.7 | 72.0 | 77.7 | 75.3 | 65.8 | 53.5 | 39.5 | 28.1 | 51.5 |
| Days Max Temp ≥ 90 °F | 0 | 0 | 0 | 0 | 1 | 9 | 16 | 13 | 5 | 0 | 0 | 0 | 44 |
| Days Max Temp ≤ 32 °F | 13 | 9 | 2 | 0 | 0 | 0 | 0 | 0 | 0 | 0 | 2 | 9 | 35 |
| Days Min Temp ≤ 32 °F | 30 | 27 | 21 | 7 | 1 | 0 | 0 | 0 | 0 | 6 | 20 | 30 | 142 |
| Days Min Temp ≤ 0 °F | 6 | 3 | 0 | 0 | 0 | 0 | 0 | 0 | 0 | 0 | 0 | 2 | 11 |
| Heating Degree Days | 1261 | 1014 | 770 | 404 | 156 | 21 | 2 | 5 | 93 | 363 | 761 | 1139 | 5989 |
| Cooling Degree Days | 0 | 0 | 1 | 17 | 56 | 232 | 386 | 320 | 133 | 11 | 0 | 0 | 1156 |
| Total Precipitation (") | 0.70 | 0.76 | 1.93 | 2.51 | 3.97 | 3.84 | 3.88 | 3.43 | 2.98 | 2.02 | 1.25 | 0.96 | 28.23 |
| Days ≥ 0.1" Precip | 2 | 2 | 4 | 5 | 7 | 6 | 6 | 6 | 5 | 4 | 3 | 2 | 52 |
| Total Snowfall (") | 6.4 | 6.9 | *3.4* | 0.5 | 0.0 | 0.0 | 0.0 | 0.0 | 0.0 | 0.0 | 2.6 | *5.7* | 25.5 |
| Days ≥ 1" Snow Depth | na | na | *2* | 0 | 0 | 0 | 0 | 0 | 0 | 0 | 1 | *5* | na |

## MARYSVILLE *Marshall County* ELEVATION 1180 ft LAT/LONG 39° 51' N / 96° 38' W

| | JAN | FEB | MAR | APR | MAY | JUN | JUL | AUG | SEP | OCT | NOV | DEC | YEAR |
|---|---|---|---|---|---|---|---|---|---|---|---|---|---|
| Maximum Temp °F | 34.9 | 41.0 | 53.4 | 65.7 | 75.5 | 85.1 | 90.1 | 87.9 | 79.4 | 67.5 | 51.3 | 39.6 | 64.3 |
| Minimum Temp °F | 13.4 | 17.6 | 29.2 | 40.4 | 51.4 | 61.5 | 66.3 | 63.8 | 54.1 | 41.3 | 28.9 | 18.5 | 40.5 |
| Mean Temp °F | 24.2 | 29.3 | 41.3 | 53.1 | 63.5 | 73.3 | 78.2 | 75.9 | 66.7 | 54.4 | 40.1 | 29.1 | 52.4 |
| Days Max Temp ≥ 90 °F | 0 | 0 | 0 | 1 | 1 | 9 | 17 | 13 | 5 | 0 | 0 | 0 | 46 |
| Days Max Temp ≤ 32 °F | 13 | 8 | 2 | 0 | 0 | 0 | 0 | 0 | 0 | 0 | 2 | 8 | 33 |
| Days Min Temp ≤ 32 °F | 30 | 26 | 20 | 7 | 0 | 0 | 0 | 0 | 0 | 6 | 20 | 29 | 138 |
| Days Min Temp ≤ 0 °F | 5 | 3 | 0 | 0 | 0 | 0 | 0 | 0 | 0 | 0 | 0 | 2 | 10 |
| Heating Degree Days | 1258 | 1001 | 730 | 372 | 124 | 11 | 1 | 4 | 85 | 341 | 741 | 1106 | 5774 |
| Cooling Degree Days | 0 | 0 | 2 | 27 | 81 | 273 | 419 | 348 | 158 | 14 | 1 | 0 | 1323 |
| Total Precipitation (") | 0.67 | 0.74 | 2.23 | 2.67 | 4.25 | 4.72 | 4.70 | 3.70 | 3.40 | 2.49 | 1.41 | 1.01 | 31.99 |
| Days ≥ 0.1" Precip | 2 | 2 | 5 | 6 | 8 | 7 | 6 | 6 | 6 | 4 | 3 | 3 | 58 |
| Total Snowfall (") | na | 4.8 | 2.5 | 0.3 | 0.0 | 0.0 | 0.0 | 0.0 | 0.0 | 0.0 | 0.6 | *1.9* | na |
| Days ≥ 1" Snow Depth | na | na | *0* | 0 | 0 | 0 | 0 | 0 | 0 | 0 | 0 | na | na |

## MC DONALD *Rawlins County* ELEVATION 3373 ft LAT/LONG 39° 47' N / 101° 23' W

| | JAN | FEB | MAR | APR | MAY | JUN | JUL | AUG | SEP | OCT | NOV | DEC | YEAR |
|---|---|---|---|---|---|---|---|---|---|---|---|---|---|
| Maximum Temp °F | 40.2 | 45.1 | 54.3 | 64.4 | 73.5 | 84.4 | 90.3 | 88.1 | 79.4 | 68.0 | 51.1 | 42.4 | 65.1 |
| Minimum Temp °F | 15.1 | 19.2 | 26.9 | 36.1 | 46.6 | 56.5 | 62.2 | 59.9 | 50.3 | 38.2 | 25.9 | 17.9 | 37.9 |
| Mean Temp °F | 27.6 | 32.2 | 40.6 | 50.3 | 60.1 | 70.5 | 76.3 | 74.0 | 64.9 | 53.2 | 38.5 | 30.2 | 51.5 |
| Days Max Temp ≥ 90 °F | 0 | 0 | 0 | 0 | 1 | 9 | 18 | 15 | 6 | 1 | 0 | 0 | 50 |
| Days Max Temp ≤ 32 °F | 9 | 6 | 2 | 0 | 0 | 0 | 0 | 0 | 0 | 0 | 3 | 7 | 27 |
| Days Min Temp ≤ 32 °F | 30 | 26 | 23 | 10 | 1 | 0 | 0 | 0 | 1 | 7 | 24 | 30 | 152 |
| Days Min Temp ≤ 0 °F | 4 | 2 | 0 | 0 | 0 | 0 | 0 | 0 | 0 | 0 | 0 | 2 | 8 |
| Heating Degree Days | 1152 | 921 | 748 | 440 | 189 | 30 | 3 | 8 | 106 | 369 | 788 | 1073 | 5827 |
| Cooling Degree Days | 0 | 0 | 0 | 8 | 38 | 205 | 350 | 299 | 117 | 7 | 0 | 0 | 1024 |
| Total Precipitation (") | 0.57 | 0.55 | 1.45 | 1.79 | 3.84 | 3.63 | 3.26 | 2.29 | 1.43 | 1.32 | 0.90 | 0.53 | 21.56 |
| Days ≥ 0.1" Precip | 2 | 2 | 4 | 4 | 6 | 6 | 6 | 5 | 3 | 3 | 2 | 1 | 44 |
| Total Snowfall (") | 8.3 | 6.1 | 9.9 | 4.7 | 0.3 | 0.0 | 0.0 | 0.0 | 0.3 | 2.6 | 5.9 | 6.5 | 44.6 |
| Days ≥ 1" Snow Depth | 15 | 9 | 5 | 2 | 0 | 0 | 0 | 0 | 0 | 1 | 5 | 10 | 47 |

## MCPHERSON *McPherson County* ELEVATION 1495 ft LAT/LONG 38° 22 ' N / 97° 40 ' W

| | JAN | FEB | MAR | APR | MAY | JUN | JUL | AUG | SEP | OCT | NOV | DEC | YEAR |
|---|---|---|---|---|---|---|---|---|---|---|---|---|---|
| Maximum Temp °F | 39.9 | 46.3 | 58.1 | 68.7 | 76.9 | 87.5 | 93.5 | 91.4 | 82.7 | 71.4 | 54.9 | 43.8 | 67.9 |
| Minimum Temp °F | 19.4 | 23.5 | 33.2 | 43.6 | 52.9 | 63.0 | 68.1 | 66.2 | 57.7 | 46.0 | 33.3 | 23.6 | 44.2 |
| Mean Temp °F | 29.7 | 34.9 | 45.7 | 56.2 | 65.0 | 75.2 | 80.8 | 78.9 | 70.2 | 58.7 | 44.1 | 33.7 | 56.1 |
| Days Max Temp ≥ 90 °F | 0 | 0 | 0 | 0 | 1 | 13 | 22 | 19 | 8 | 1 | 0 | 0 | 64 |
| Days Max Temp ≤ 32 °F | 9 | 5 | 1 | 0 | 0 | 0 | 0 | 0 | 0 | 0 | 1 | 5 | 21 |
| Days Min Temp ≤ 32 °F | 28 | 22 | 15 | 4 | 0 | 0 | 0 | 0 | 0 | 2 | 15 | 26 | 112 |
| Days Min Temp ≤ 0 °F | 2 | 1 | 0 | 0 | 0 | 0 | 0 | 0 | 0 | 0 | 0 | 1 | 4 |
| Heating Degree Days | 1088 | 842 | 596 | 283 | 90 | 7 | 0 | 1 | 46 | 229 | 621 | 962 | 4765 |
| Cooling Degree Days | 0 | 0 | 3 | 32 | 99 | 334 | 507 | 454 | 236 | 39 | 2 | 0 | 1706 |
| Total Precipitation (") | 0.62 | 0.96 | 2.34 | 2.84 | 4.10 | 4.86 | 3.74 | 3.38 | 3.11 | 2.34 | 1.42 | 1.00 | 30.71 |
| Days ≥ 0.1" Precip | 2 | 2 | 4 | 5 | 6 | 7 | 5 | 5 | 4 | 4 | 3 | 2 | 49 |
| Total Snowfall (") | 4.5 | 4.4 | 2.2 | 1.0 | 0.0 | 0.0 | 0.0 | 0.0 | 0.0 | 0.0 | 0.9 | 3.1 | 16.1 |
| Days ≥ 1" Snow Depth | 10 | 7 | 2 | 1 | 0 | 0 | 0 | 0 | 0 | 0 | 1 | 4 | 25 |

## MEADE *Meade County* ELEVATION 2503 ft LAT/LONG 37° 17 ' N / 100° 20 ' W

| | JAN | FEB | MAR | APR | MAY | JUN | JUL | AUG | SEP | OCT | NOV | DEC | YEAR |
|---|---|---|---|---|---|---|---|---|---|---|---|---|---|
| Maximum Temp °F | 46.0 | 52.1 | 61.3 | 71.8 | 79.3 | 89.4 | 94.9 | 92.9 | 84.6 | 74.0 | 57.9 | 48.5 | 71.1 |
| Minimum Temp °F | 18.7 | 22.4 | 30.9 | 41.2 | 50.9 | 61.1 | 65.7 | 64.0 | 54.9 | 42.5 | 29.7 | 21.5 | 42.0 |
| Mean Temp °F | 32.4 | 37.3 | 46.1 | 56.6 | 65.1 | 75.2 | 80.3 | 78.5 | 69.9 | 58.2 | 43.8 | 35.0 | 56.5 |
| Days Max Temp ≥ 90 °F | 0 | 0 | 0 | 1 | 4 | 16 | 25 | 22 | 11 | 2 | 0 | 0 | 81 |
| Days Max Temp ≤ 32 °F | 6 | 4 | 1 | 0 | 0 | 0 | 0 | 0 | 0 | 0 | 1 | 4 | 16 |
| Days Min Temp ≤ 32 °F | 29 | 25 | 18 | 4 | 0 | 0 | 0 | 0 | 0 | 4 | 19 | 29 | 128 |
| Days Min Temp ≤ 0 °F | 2 | 1 | 0 | 0 | 0 | 0 | 0 | 0 | 0 | 0 | 0 | 1 | 4 |
| Heating Degree Days | 1004 | 775 | 583 | 271 | 90 | 8 | 1 | 1 | 50 | 238 | 628 | 924 | 4573 |
| Cooling Degree Days | 0 | 0 | 1 | 27 | 95 | 316 | 476 | 439 | 211 | 26 | 1 | 0 | 1592 |
| Total Precipitation (") | 0.60 | 0.67 | 1.57 | 1.76 | 3.33 | 2.77 | 2.79 | 2.41 | 2.12 | 1.37 | 0.87 | 0.71 | 20.97 |
| Days ≥ 0.1" Precip | 2 | 2 | 3 | 3 | 6 | 5 | 4 | 4 | 4 | 3 | 2 | 2 | 40 |
| Total Snowfall (") | 4.3 | 4.6 | 4.5 | 0.9 | 0.0 | 0.0 | 0.0 | 0.0 | 0.0 | 0.3 | 1.9 | 3.7 | 20.2 |
| Days ≥ 1" Snow Depth | na | na | 2 | 0 | 0 | 0 | 0 | 0 | 0 | 0 | 1 | na | na |

## MEDICINE LODGE *Barber County* ELEVATION 1540 ft LAT/LONG 37° 18 ' N / 98° 35 ' W

| | JAN | FEB | MAR | APR | MAY | JUN | JUL | AUG | SEP | OCT | NOV | DEC | YEAR |
|---|---|---|---|---|---|---|---|---|---|---|---|---|---|
| Maximum Temp °F | 43.8 | 50.3 | 60.7 | 71.3 | 79.2 | 89.5 | 95.0 | 93.1 | 84.5 | 73.3 | 57.8 | 47.7 | 70.5 |
| Minimum Temp °F | 19.7 | 23.4 | 33.1 | 44.0 | 53.3 | 63.1 | 67.4 | 65.9 | 57.5 | 44.8 | 32.4 | 23.2 | 44.0 |
| Mean Temp °F | 31.7 | 36.9 | 46.9 | 57.7 | 66.3 | 76.3 | 81.3 | 79.5 | 71.0 | 59.1 | 45.1 | 35.5 | 57.3 |
| Days Max Temp ≥ 90 °F | 0 | 0 | 0 | 1 | 3 | 16 | 25 | 22 | 10 | 2 | 0 | 0 | 79 |
| Days Max Temp ≤ 32 °F | 7 | 3 | 1 | 0 | 0 | 0 | 0 | 0 | 0 | 0 | 0 | 3 | 14 |
| Days Min Temp ≤ 32 °F | 28 | 23 | 15 | 4 | 0 | 0 | 0 | 0 | 0 | 3 | 16 | 27 | 116 |
| Days Min Temp ≤ 0 °F | 2 | 1 | 0 | 0 | 0 | 0 | 0 | 0 | 0 | 0 | 0 | 1 | 4 |
| Heating Degree Days | 1023 | 788 | 557 | 243 | 70 | 4 | 0 | 0 | 37 | 218 | 591 | 909 | 4440 |
| Cooling Degree Days | 0 | 0 | 2 | 31 | 117 | 359 | 514 | 473 | 245 | 38 | 2 | 0 | 1781 |
| Total Precipitation (") | 0.61 | 0.97 | 2.11 | 2.60 | 3.58 | 3.97 | 2.58 | 3.21 | 2.36 | 2.19 | 1.45 | 0.87 | 26.50 |
| Days ≥ 0.1" Precip | 2 | 2 | 4 | 5 | 6 | 6 | 4 | 5 | 4 | 3 | 3 | 2 | 46 |
| Total Snowfall (") | 3.4 | 4.2 | 2.6 | 0.5 | 0.0 | 0.0 | 0.0 | 0.0 | 0.0 | 0.1 | 0.9 | *2.9* | 14.6 |
| Days ≥ 1" Snow Depth | na | na | 1 | 0 | 0 | 0 | 0 | 0 | 0 | 0 | 0 | na | na |

## MILFORD LAKE *Geary County* ELEVATION 1211 ft LAT/LONG 39° 5 ' N / 96° 53 ' W

| | JAN | FEB | MAR | APR | MAY | JUN | JUL | AUG | SEP | OCT | NOV | DEC | YEAR |
|---|---|---|---|---|---|---|---|---|---|---|---|---|---|
| Maximum Temp °F | 36.4 | 42.3 | 55.1 | 65.9 | 74.8 | 84.4 | 90.6 | 88.8 | 80.2 | 68.7 | 53.2 | 41.0 | 65.1 |
| Minimum Temp °F | 15.4 | 19.5 | 30.8 | 41.7 | 51.5 | 61.3 | 66.6 | 64.5 | 55.1 | 43.0 | 31.3 | 20.7 | 41.8 |
| Mean Temp °F | 25.9 | 30.9 | 43.0 | 53.8 | 63.2 | 72.9 | 78.6 | 76.7 | 67.8 | 55.9 | 42.3 | 30.9 | 53.5 |
| Days Max Temp ≥ 90 °F | 0 | 0 | 0 | 0 | 1 | 8 | 17 | 14 | 5 | 0 | 0 | 0 | 45 |
| Days Max Temp ≤ 32 °F | 12 | 8 | 1 | 0 | 0 | 0 | 0 | 0 | 0 | 0 | 1 | 7 | 29 |
| Days Min Temp ≤ 32 °F | 30 | 25 | 18 | 5 | 0 | 0 | 0 | 0 | 0 | 4 | 17 | 28 | 127 |
| Days Min Temp ≤ 0 °F | 4 | 3 | 0 | 0 | 0 | 0 | 0 | 0 | 0 | 0 | 0 | 2 | 9 |
| Heating Degree Days | 1204 | 957 | 681 | 348 | 126 | 14 | 1 | 2 | 71 | 300 | 675 | 1052 | 5431 |
| Cooling Degree Days | 0 | 0 | 3 | 26 | 78 | 262 | 440 | 384 | 176 | 20 | 1 | 0 | 1390 |
| Total Precipitation (") | 0.73 | 0.77 | 2.29 | 2.73 | 4.37 | 4.57 | 4.25 | 3.40 | 3.76 | 2.57 | 1.57 | 1.14 | 32.15 |
| Days ≥ 0.1" Precip | 2 | 2 | 4 | 5 | 7 | 6 | 6 | 5 | 5 | 4 | 3 | 2 | 51 |
| Total Snowfall (") | *4.8* | *3.6* | 1.3 | 0.5 | 0.0 | 0.0 | 0.0 | 0.0 | 0.0 | 0.0 | 0.6 | *2.3* | 13.1 |
| Days ≥ 1" Snow Depth | 10 | *8* | 1 | 0 | 0 | 0 | 0 | 0 | 0 | 0 | 1 | *5* | 25 |

## MINNEAPOLIS *Ottawa County* ELEVATION 1250 ft LAT/LONG 39° 8' N / 97° 42' W

| | JAN | FEB | MAR | APR | MAY | JUN | JUL | AUG | SEP | OCT | NOV | DEC | YEAR |
|---|---|---|---|---|---|---|---|---|---|---|---|---|---|
| Maximum Temp °F | 39.6 | 45.7 | 57.8 | 69.3 | 77.9 | 88.4 | 94.1 | 91.6 | 82.8 | 71.1 | 54.6 | 44.0 | 68.1 |
| Minimum Temp °F | 18.3 | 21.8 | 32.3 | 43.0 | 52.6 | 62.6 | 67.8 | 65.6 | 56.5 | 44.5 | 32.0 | 23.3 | 43.4 |
| Mean Temp °F | 29.0 | 33.7 | 45.1 | 56.2 | 65.3 | 75.5 | 80.9 | 78.7 | 69.7 | 57.8 | 43.3 | 33.7 | 55.7 |
| Days Max Temp ≥ 90 °F | 0 | 0 | 0 | 1 | 2 | 14 | 23 | 20 | 8 | 1 | 0 | 0 | 69 |
| Days Max Temp ≤ 32 °F | 9 | 5 | 1 | 0 | 0 | 0 | 0 | 0 | 0 | 0 | 1 | 5 | 21 |
| Days Min Temp ≤ 32 °F | 28 | 24 | 16 | 4 | 0 | 0 | 0 | 0 | 0 | 3 | 16 | 27 | 118 |
| Days Min Temp ≤ 0 °F | 2 | 2 | 0 | 0 | 0 | 0 | 0 | 0 | 0 | 0 | 0 | 1 | 5 |
| Heating Degree Days | 1110 | 875 | 614 | 285 | 91 | 7 | 0 | 1 | 47 | 251 | 644 | 964 | 4889 |
| Cooling Degree Days | ***0*** | ***0*** | ***2*** | ***36*** | ***106*** | ***336*** | ***505*** | ***435*** | ***216*** | ***33*** | ***2*** | ***0*** | 1671 |
| Total Precipitation (") | 0.73 | 0.80 | 2.03 | 2.01 | 4.37 | 3.88 | 4.48 | 3.29 | 2.95 | 2.20 | 1.31 | 0.92 | 28.97 |
| Days ≥ 0.1" Precip | 2 | 2 | 4 | 5 | 7 | 6 | 6 | 5 | 5 | 3 | 3 | 2 | 50 |
| Total Snowfall (") | 5.9 | 5.2 | 2.3 | 0.4 | 0.0 | 0.0 | 0.0 | 0.0 | 0.0 | 0.2 | ***1.5*** | 3.0 | 18.5 |
| Days ≥ 1" Snow Depth | ***9*** | 8 | 2 | 0 | 0 | 0 | 0 | 0 | 0 | 0 | ***1*** | 4 | 24 |

## MOUND CITY *Linn County* ELEVATION 860 ft LAT/LONG 38° 9' N / 94° 49' W

| | JAN | FEB | MAR | APR | MAY | JUN | JUL | AUG | SEP | OCT | NOV | DEC | YEAR |
|---|---|---|---|---|---|---|---|---|---|---|---|---|---|
| Maximum Temp °F | 40.1 | 46.2 | 57.9 | 68.8 | 76.8 | 85.3 | 91.2 | 89.5 | 81.3 | 70.5 | 55.8 | 44.6 | 67.3 |
| Minimum Temp °F | 19.4 | 23.5 | 34.0 | 44.1 | 53.6 | 62.9 | 67.7 | 65.3 | 57.0 | 45.3 | 34.6 | 24.5 | 44.3 |
| Mean Temp °F | 29.8 | 34.9 | 46.0 | 56.5 | 65.2 | 74.2 | 79.5 | 77.4 | 69.2 | 58.0 | 45.2 | 34.6 | 55.9 |
| Days Max Temp ≥ 90 °F | 0 | 0 | 0 | 0 | 0 | 8 | 19 | 16 | 6 | 0 | 0 | 0 | 49 |
| Days Max Temp ≤ 32 °F | 9 | 5 | 1 | 0 | 0 | 0 | 0 | 0 | 0 | 0 | 1 | 5 | 21 |
| Days Min Temp ≤ 32 °F | 28 | 22 | 15 | 4 | 0 | 0 | 0 | 0 | 0 | 4 | 14 | 24 | 111 |
| Days Min Temp ≤ 0 °F | 2 | 1 | 0 | 0 | 0 | 0 | 0 | 0 | 0 | 0 | 0 | 1 | 4 |
| Heating Degree Days | 1085 | 843 | 587 | 280 | 86 | 7 | 1 | 3 | 56 | 248 | 590 | 937 | 4723 |
| Cooling Degree Days | 0 | 0 | 3 | 31 | 86 | 282 | 450 | 391 | 190 | 28 | 3 | 0 | 1464 |
| Total Precipitation (") | 1.50 | 1.51 | 3.28 | 4.13 | 4.85 | 5.05 | 3.69 | 4.36 | 4.65 | 3.86 | 3.03 | 1.85 | 41.76 |
| Days ≥ 0.1" Precip | 3 | 3 | 5 | 6 | 8 | 7 | 5 | 5 | 6 | 6 | 5 | 4 | 63 |
| Total Snowfall (") | 4.8 | 4.4 | 1.6 | 0.0 | 0.0 | 0.0 | 0.0 | 0.0 | 0.0 | 0.1 | 0.8 | 2.7 | 14.4 |
| Days ≥ 1" Snow Depth | na | ***2*** | 1 | 0 | 0 | 0 | 0 | 0 | 0 | 0 | 0 | na | na |

## MOUND VALLEY 3 WSW *Labette County* ELEVATION 801 ft LAT/LONG 37° 11' N / 95° 27' W

| | JAN | FEB | MAR | APR | MAY | JUN | JUL | AUG | SEP | OCT | NOV | DEC | YEAR |
|---|---|---|---|---|---|---|---|---|---|---|---|---|---|
| Maximum Temp °F | 43.2 | 48.3 | 59.7 | 70.3 | 77.7 | 86.2 | 92.1 | 91.2 | 82.3 | 72.1 | 57.8 | 46.9 | 69.0 |
| Minimum Temp °F | 20.5 | 24.6 | 34.9 | 45.2 | 54.5 | 63.8 | 68.1 | 65.6 | 57.9 | 45.4 | 34.7 | 25.4 | 45.1 |
| Mean Temp °F | 32.0 | 36.6 | 47.3 | 57.8 | 66.1 | 75.0 | 80.1 | 78.4 | 70.1 | 58.8 | 46.3 | 36.3 | 57.1 |
| Days Max Temp ≥ 90 °F | 0 | 0 | 0 | 0 | 1 | 9 | 22 | 19 | 7 | 1 | 0 | 0 | 59 |
| Days Max Temp ≤ 32 °F | 7 | 4 | 1 | 0 | 0 | 0 | 0 | 0 | 0 | 0 | 1 | 4 | 17 |
| Days Min Temp ≤ 32 °F | 28 | 22 | 14 | 4 | 0 | 0 | 0 | 0 | 0 | 3 | 14 | 25 | 110 |
| Days Min Temp ≤ 0 °F | 2 | 1 | 0 | 0 | 0 | 0 | 0 | 0 | 0 | 0 | 0 | 1 | 4 |
| Heating Degree Days | 1015 | 796 | 546 | 244 | 69 | 5 | 0 | 1 | 45 | 227 | 559 | 885 | 4392 |
| Cooling Degree Days | 0 | 0 | 3 | 28 | 94 | 304 | 447 | 410 | 197 | 26 | 3 | 0 | 1512 |
| Total Precipitation (") | 1.61 | 1.67 | 3.56 | 4.15 | 5.71 | 5.27 | 3.72 | 4.06 | 5.13 | 4.21 | 3.39 | 2.17 | 44.65 |
| Days ≥ 0.1" Precip | 3 | 3 | 5 | 6 | 8 | 7 | 4 | 5 | 6 | 5 | 5 | 4 | 61 |
| Total Snowfall (") | na | ***3.1*** | 1.8 | 0.0 | 0.0 | 0.0 | 0.0 | 0.0 | 0.0 | 0.0 | 0.5 | ***2.3*** | na |
| Days ≥ 1" Snow Depth | na | ***2*** | 0 | 0 | 0 | 0 | 0 | 0 | 0 | 0 | 0 | na | na |

## NESS CITY *Ness County* ELEVATION 2260 ft LAT/LONG 38° 27' N / 99° 54' W

| | JAN | FEB | MAR | APR | MAY | JUN | JUL | AUG | SEP | OCT | NOV | DEC | YEAR |
|---|---|---|---|---|---|---|---|---|---|---|---|---|---|
| Maximum Temp °F | 41.6 | 48.0 | 57.9 | 68.9 | 77.6 | 88.5 | 93.9 | 91.5 | 82.7 | 71.1 | 54.9 | 44.7 | 68.4 |
| Minimum Temp °F | 15.9 | 20.3 | 29.1 | 39.6 | 50.0 | 60.2 | 65.3 | 62.9 | 53.3 | 40.1 | 27.7 | 19.2 | 40.3 |
| Mean Temp °F | 28.8 | 34.2 | 43.5 | 54.3 | 63.8 | 74.4 | 79.6 | 77.2 | 68.0 | 55.6 | 41.3 | 32.0 | 54.4 |
| Days Max Temp ≥ 90 °F | 0 | 0 | 0 | 1 | 3 | 15 | 23 | 20 | 9 | 1 | 0 | 0 | 72 |
| Days Max Temp ≤ 32 °F | 8 | 5 | 1 | 0 | 0 | 0 | 0 | 0 | 0 | 0 | 1 | 6 | 21 |
| Days Min Temp ≤ 32 °F | 30 | 26 | 19 | 6 | 1 | 0 | 0 | 0 | 0 | 6 | 21 | 29 | 138 |
| Days Min Temp ≤ 0 °F | 3 | 2 | 0 | 0 | 0 | 0 | 0 | 0 | 0 | 0 | 0 | 1 | 6 |
| Heating Degree Days | 1116 | 864 | 661 | 333 | 114 | 13 | 1 | 2 | 69 | 304 | 704 | 1017 | 5198 |
| Cooling Degree Days | 0 | 0 | 1 | 23 | 84 | 302 | 457 | 394 | 180 | 16 | 0 | 0 | 1457 |
| Total Precipitation (") | 0.47 | 0.69 | 1.68 | 1.99 | 2.88 | 3.32 | 3.09 | 2.60 | 1.76 | 1.36 | 1.02 | 0.61 | 21.47 |
| Days ≥ 0.1" Precip | 1 | 2 | 3 | 4 | 6 | 6 | 5 | 5 | 3 | 2 | 2 | 2 | 41 |
| Total Snowfall (") | na | na | ***3.0*** | 0.4 | 0.0 | 0.0 | 0.0 | 0.0 | 0.0 | 0.1 | 1.5 | na | na |
| Days ≥ 1" Snow Depth | na | na | ***1*** | 0 | 0 | 0 | 0 | 0 | 0 | 0 | ***1*** | na | na |

## NEWTON 2 SW *Harvey County* ELEVATION 1450 ft LAT/LONG 38° 3 ' N / 97° 20 ' W

| | JAN | FEB | MAR | APR | MAY | JUN | JUL | AUG | SEP | OCT | NOV | DEC | YEAR |
|---|---|---|---|---|---|---|---|---|---|---|---|---|---|
| Maximum Temp °F | 40.0 | 46.6 | 57.8 | 68.5 | 77.1 | 87.6 | 93.7 | 91.7 | 82.6 | 70.8 | 54.7 | 43.7 | 67.9 |
| Minimum Temp °F | 19.6 | 24.0 | 33.7 | 44.2 | 53.8 | 63.6 | 68.7 | 66.8 | 58.1 | 46.5 | 33.6 | 23.8 | 44.7 |
| Mean Temp °F | 29.8 | 35.3 | 45.8 | 56.4 | 65.5 | 75.6 | 81.2 | 79.3 | 70.4 | 58.7 | 44.2 | 33.8 | 56.3 |
| Days Max Temp ≥ 90 °F | 0 | 0 | 0 | 0 | 2 | 13 | 23 | 20 | 8 | 1 | 0 | 0 | 67 |
| Days Max Temp ≤ 32 °F | 9 | 5 | 1 | 0 | 0 | 0 | 0 | 0 | 0 | 0 | 1 | 5 | 21 |
| Days Min Temp ≤ 32 °F | 28 | 22 | 14 | 3 | 0 | 0 | 0 | 0 | 0 | 2 | 14 | 26 | 109 |
| Days Min Temp ≤ 0 °F | 2 | 1 | 0 | 0 | 0 | 0 | 0 | 0 | 0 | 0 | 0 | 1 | 4 |
| Heating Degree Days | 1083 | 831 | 591 | 279 | 84 | 6 | 0 | 1 | 48 | 231 | 619 | 962 | 4735 |
| Cooling Degree Days | 0 | 0 | 3 | 32 | 106 | 345 | 518 | 465 | 233 | 37 | 1 | 0 | 1740 |
| Total Precipitation (") | 0.69 | 0.89 | 2.38 | 2.85 | 4.35 | 4.83 | 3.39 | 3.03 | 3.23 | 2.44 | 1.78 | 1.23 | 31.09 |
| Days ≥ 0.1" Precip | 2 | 2 | 4 | 5 | 6 | 6 | 5 | 5 | 5 | 4 | 3 | 3 | 50 |
| Total Snowfall (") | *2.9* | 4.1 | 1.3 | 0.3 | 0.0 | 0.0 | 0.0 | 0.0 | 0.0 | 0.1 | 0.9 | 2.7 | 12.3 |
| Days ≥ 1" Snow Depth | na | na | 1 | 0 | 0 | 0 | 0 | 0 | 0 | 0 | 0 | *2* | na |

## NORTON 9 SSE *Norton County* ELEVATION 2342 ft LAT/LONG 39° 41 ' N / 99° 51 ' W

| | JAN | FEB | MAR | APR | MAY | JUN | JUL | AUG | SEP | OCT | NOV | DEC | YEAR |
|---|---|---|---|---|---|---|---|---|---|---|---|---|---|
| Maximum Temp °F | 38.7 | 44.2 | 55.1 | 66.5 | 74.8 | 86.1 | 91.8 | 89.6 | 80.8 | 69.0 | 52.2 | 41.9 | 65.9 |
| Minimum Temp °F | 14.6 | 18.5 | 27.1 | 37.7 | 48.1 | 57.7 | 63.0 | 60.9 | 51.3 | 39.3 | 26.8 | 17.9 | 38.6 |
| Mean Temp °F | 26.7 | 31.4 | 41.1 | 52.1 | 61.5 | 71.9 | 77.4 | 75.3 | 66.1 | 54.2 | 39.5 | 29.9 | 52.3 |
| Days Max Temp ≥ 90 °F | 0 | 0 | 0 | 1 | 2 | 11 | 19 | 17 | 7 | 1 | 0 | 0 | 58 |
| Days Max Temp ≤ 32 °F | 10 | 7 | 2 | 0 | 0 | 0 | 0 | 0 | 0 | 0 | 2 | 7 | 28 |
| Days Min Temp ≤ 32 °F | 30 | 26 | 22 | 8 | 1 | 0 | 0 | 0 | 1 | 7 | 22 | 30 | 147 |
| Days Min Temp ≤ 0 °F | 4 | 2 | 0 | 0 | 0 | 0 | 0 | 0 | 0 | 0 | 0 | 2 | 8 |
| Heating Degree Days | 1181 | 942 | 735 | 392 | 157 | 20 | 1 | 5 | 90 | 343 | 759 | 1081 | 5706 |
| Cooling Degree Days | 0 | 0 | 0 | 15 | 49 | 244 | 392 | 337 | 142 | 13 | 0 | 0 | 1192 |
| Total Precipitation (") | 0.59 | 0.58 | 1.58 | 2.03 | 3.55 | 2.84 | 3.48 | 2.92 | 1.92 | 1.43 | 0.97 | 0.54 | 22.43 |
| Days ≥ 0.1" Precip | 1 | 2 | 3 | 4 | 7 | 6 | 6 | 5 | 4 | 3 | 2 | 2 | 45 |
| Total Snowfall (") | 5.4 | 4.7 | 5.4 | 1.7 | 0.0 | 0.0 | 0.0 | 0.0 | 0.3 | 0.5 | 3.5 | 5.1 | 26.6 |
| Days ≥ 1" Snow Depth | 12 | 8 | 4 | 1 | 0 | 0 | 0 | 0 | 0 | 0 | 3 | 8 | 36 |

## NORTON DAM *Norton County* ELEVATION 2340 ft LAT/LONG 39° 49 ' N / 99° 56 ' W

| | JAN | FEB | MAR | APR | MAY | JUN | JUL | AUG | SEP | OCT | NOV | DEC | YEAR |
|---|---|---|---|---|---|---|---|---|---|---|---|---|---|
| Maximum Temp °F | 36.9 | 42.5 | 52.8 | 64.5 | 73.0 | 83.8 | *90.3* | *87.9* | *78.9* | *67.9* | 51.0 | *41.3* | 64.2 |
| Minimum Temp °F | 11.7 | 16.0 | 25.1 | 36.7 | 46.9 | 57.2 | *63.0* | *60.6* | 50.0 | 37.5 | 24.6 | 15.5 | 37.1 |
| Mean Temp °F | 24.3 | 29.2 | 39.0 | 50.6 | 60.0 | 70.5 | *76.6* | *74.2* | *64.5* | *52.7* | 37.8 | *28.7* | 50.7 |
| Days Max Temp ≥ 90 °F | 0 | 0 | 0 | 0 | 1 | 8 | *17* | 13 | 6 | 1 | 0 | 0 | 46 |
| Days Max Temp ≤ 32 °F | 12 | 7 | *3* | 0 | 0 | 0 | *0* | 0 | 0 | 0 | 2 | 8 | 32 |
| Days Min Temp ≤ 32 °F | 30 | 26 | *25* | 9 | 1 | 0 | *0* | 0 | 1 | 8 | 24 | 29 | 153 |
| Days Min Temp ≤ 0 °F | 6 | 3 | 0 | 0 | 0 | 0 | *0* | 0 | 0 | 0 | 0 | 2 | 11 |
| Heating Degree Days | 1255 | 994 | 803 | 433 | 193 | 31 | *3* | *7* | *114* | *383* | 808 | *1117* | 6141 |
| Cooling Degree Days | *0* | *0* | *0* | *12* | *47* | *222* | na | na | na | na | *0* | na | na |
| Total Precipitation (") | 0.39 | 0.45 | 1.51 | 2.48 | 3.99 | 3.63 | *3.78* | 3.32 | 2.12 | 1.73 | 0.98 | 0.45 | 24.83 |
| Days ≥ 0.1" Precip | 1 | 1 | *3* | 5 | 7 | 6 | *6* | 5 | 4 | 3 | 2 | 1 | 44 |
| Total Snowfall (") | *4.5* | *3.5* | *3.5* | 1.2 | 0.0 | 0.0 | *0.0* | 0.0 | 0.0 | 0.4 | 1.1 | *3.4* | 17.6 |
| Days ≥ 1" Snow Depth | 10 | 7 | *4* | 1 | 0 | 0 | *0* | 0 | 0 | 0 | 3 | 8 | 33 |

## NORWICH *Kingman County* ELEVATION 1489 ft LAT/LONG 37° 28 ' N / 97° 52 ' W

| | JAN | FEB | MAR | APR | MAY | JUN | JUL | AUG | SEP | OCT | NOV | DEC | YEAR |
|---|---|---|---|---|---|---|---|---|---|---|---|---|---|
| Maximum Temp °F | 42.0 | 48.1 | 59.4 | 69.6 | 77.9 | 88.5 | 94.9 | 93.3 | 83.8 | 72.1 | 56.2 | 45.2 | 69.3 |
| Minimum Temp °F | 21.8 | 25.8 | 34.9 | 45.3 | 54.7 | 64.2 | 69.2 | 67.6 | 59.5 | 47.9 | 35.2 | 25.7 | 46.0 |
| Mean Temp °F | 31.9 | 37.0 | 47.2 | 57.5 | 66.3 | 76.4 | 82.1 | 80.5 | 71.7 | 60.0 | 45.7 | 35.4 | 57.6 |
| Days Max Temp ≥ 90 °F | 0 | 0 | 0 | 0 | 2 | 15 | 24 | 22 | 9 | 1 | 0 | 0 | 73 |
| Days Max Temp ≤ 32 °F | 8 | 4 | 1 | 0 | 0 | 0 | 0 | 0 | 0 | 0 | 1 | 4 | 18 |
| Days Min Temp ≤ 32 °F | 27 | 21 | 12 | 2 | 0 | 0 | 0 | 0 | 0 | 1 | 12 | 24 | 99 |
| Days Min Temp ≤ 0 °F | 1 | 1 | 0 | 0 | 0 | 0 | 0 | 0 | 0 | 0 | 0 | 1 | 3 |
| Heating Degree Days | 1018 | 784 | 548 | 246 | 67 | 4 | 0 | 0 | 33 | 196 | 573 | 910 | 4379 |
| Cooling Degree Days | 0 | 0 | 4 | 31 | 118 | 359 | 553 | 505 | 264 | 45 | 2 | 0 | 1881 |
| Total Precipitation (") | 0.65 | 1.00 | 2.24 | 2.65 | 4.08 | 4.14 | 2.69 | 2.93 | 3.18 | 2.27 | 1.64 | 0.99 | 28.46 |
| Days ≥ 0.1" Precip | 2 | 3 | 4 | 5 | 6 | 6 | 5 | 4 | 5 | 4 | 3 | 2 | 49 |
| Total Snowfall (") | 2.9 | 4.1 | 1.7 | 0.3 | 0.0 | 0.0 | 0.0 | 0.0 | 0.0 | 0.0 | 0.4 | 2.1 | 11.5 |
| Days ≥ 1" Snow Depth | na | *2* | 1 | 0 | 0 | 0 | 0 | 0 | 0 | 0 | 1 | *2* | na |

## OAKLEY 4 W *Logan County* ELEVATION 3041 ft LAT/LONG 39° 8 ' N / 100° 51 ' W

| | JAN | FEB | MAR | APR | MAY | JUN | JUL | AUG | SEP | OCT | NOV | DEC | YEAR |
|---|---|---|---|---|---|---|---|---|---|---|---|---|---|
| Maximum Temp °F | 41.4 | 46.9 | 56.6 | 67.2 | 74.9 | 85.8 | 90.5 | 88.2 | 80.1 | 69.1 | 53.1 | 44.0 | 66.5 |
| Minimum Temp °F | 16.4 | 19.8 | 27.4 | 37.7 | 47.8 | 57.6 | 63.2 | 61.1 | 51.7 | 39.7 | 27.5 | 19.4 | 39.1 |
| Mean Temp °F | 28.8 | 33.3 | 42.1 | 52.5 | 61.4 | 71.7 | 76.9 | 74.7 | 65.9 | 54.4 | 40.4 | 31.7 | 52.8 |
| Days Max Temp ≥ 90 °F | 0 | 0 | 0 | 0 | 2 | 11 | 18 | 15 | 6 | 1 | 0 | 0 | 53 |
| Days Max Temp ≤ 32 °F | 8 | 5 | 2 | 0 | 0 | 0 | 0 | 0 | 0 | 0 | 2 | 6 | 23 |
| Days Min Temp ≤ 32 °F | 30 | 26 | 22 | 8 | 1 | 0 | 0 | 0 | 0 | 6 | 21 | 29 | 143 |
| Days Min Temp ≤ 0 °F | 3 | 2 | 1 | 0 | 0 | 0 | 0 | 0 | 0 | 0 | 0 | 1 | 7 |
| Heating Degree Days | 1115 | 887 | 705 | 379 | 156 | 22 | 2 | 6 | 89 | 333 | 733 | 1024 | 5451 |
| Cooling Degree Days | *0* | *0* | *0* | 12 | 43 | 222 | 353 | 307 | 122 | 7 | 0 | *0* | 1066 |
| Total Precipitation (") | 0.46 | 0.47 | 1.10 | 1.68 | 3.26 | 2.68 | 3.54 | 2.52 | 1.49 | 1.07 | 0.80 | 0.48 | 19.55 |
| Days ≥ 0.1" Precip | 2 | 2 | 3 | 3 | 6 | 6 | 5 | 4 | 3 | 2 | 2 | 2 | 40 |
| Total Snowfall (") | 5.1 | 4.5 | 5.4 | 2.2 | 0.0 | 0.0 | 0.0 | 0.0 | 0.2 | 1.2 | 2.2 | 5.2 | 26.0 |
| Days ≥ 1" Snow Depth | 9 | 6 | 3 | 1 | 0 | 0 | 0 | 0 | 0 | 0 | 2 | 6 | 27 |

## OBERLIN 1 E *Decatur County* ELEVATION 2533 ft LAT/LONG 39° 49 ' N / 100° 31 ' W

| | JAN | FEB | MAR | APR | MAY | JUN | JUL | AUG | SEP | OCT | NOV | DEC | YEAR |
|---|---|---|---|---|---|---|---|---|---|---|---|---|---|
| Maximum Temp °F | 40.6 | 46.2 | 56.0 | 67.1 | 75.8 | 86.8 | 92.0 | 89.9 | 81.6 | 69.8 | 52.4 | 43.1 | 66.8 |
| Minimum Temp °F | 14.0 | 17.9 | 26.1 | 36.7 | 47.4 | 57.4 | 63.0 | 60.5 | 50.1 | 37.0 | 24.9 | 16.4 | 37.6 |
| Mean Temp °F | 27.3 | 32.1 | 41.1 | 51.9 | 61.6 | 72.1 | 77.5 | 75.2 | 65.9 | 53.4 | 38.7 | 29.8 | 52.2 |
| Days Max Temp ≥ 90 °F | 0 | 0 | 0 | 1 | 2 | 12 | 20 | 18 | 8 | 1 | 0 | 0 | 62 |
| Days Max Temp ≤ 32 °F | 9 | 6 | 2 | 0 | 0 | 0 | 0 | 0 | 0 | 0 | 2 | 6 | 25 |
| Days Min Temp ≤ 32 °F | 31 | 27 | 24 | 10 | 2 | 0 | 0 | 0 | 1 | 9 | 25 | 30 | 159 |
| Days Min Temp ≤ 0 °F | 4 | 2 | 1 | 0 | 0 | 0 | 0 | 0 | 0 | 0 | 0 | 2 | 9 |
| Heating Degree Days | 1162 | 924 | 736 | 396 | 156 | 20 | 2 | 5 | 94 | 361 | 781 | 1086 | 5723 |
| Cooling Degree Days | 0 | 0 | 0 | 12 | 48 | 233 | 378 | 325 | 128 | 6 | 0 | 0 | 1130 |
| Total Precipitation (") | 0.53 | 0.59 | 1.54 | 2.07 | 3.62 | 3.59 | 3.87 | 2.65 | 1.82 | 1.30 | 0.96 | 0.54 | 23.08 |
| Days ≥ 0.1" Precip | 1 | 2 | 3 | 4 | 6 | 6 | 6 | 5 | 4 | 3 | 2 | 2 | 44 |
| Total Snowfall (") | 7.5 | 5.1 | 7.0 | 2.5 | 0.0 | 0.0 | 0.0 | 0.0 | 0.3 | 1.0 | 4.3 | 6.5 | 34.2 |
| Days ≥ 1" Snow Depth | na | na | na | 1 | 0 | 0 | 0 | 0 | 0 | 0 | *1* | na | na |

## OLATHE 3 E *Johnson County* ELEVATION 1030 ft LAT/LONG 38° 54 ' N / 94° 46 ' W

| | JAN | FEB | MAR | APR | MAY | JUN | JUL | AUG | SEP | OCT | NOV | DEC | YEAR |
|---|---|---|---|---|---|---|---|---|---|---|---|---|---|
| Maximum Temp °F | 37.7 | 43.3 | 55.6 | 66.5 | 75.3 | 83.5 | 88.7 | 86.9 | 79.3 | 68.4 | 53.4 | 42.1 | 65.1 |
| Minimum Temp °F | 19.1 | 23.6 | 34.0 | 44.8 | 54.4 | 63.4 | 68.0 | 65.6 | 57.7 | 46.6 | 34.5 | 24.4 | 44.7 |
| Mean Temp °F | 28.4 | 33.5 | 44.8 | 55.7 | 64.8 | 73.5 | 78.4 | 76.3 | 68.5 | 57.5 | 44.0 | 33.3 | 54.9 |
| Days Max Temp ≥ 90 °F | 0 | 0 | 0 | 0 | 0 | 5 | 14 | 11 | 4 | 0 | 0 | 0 | 34 |
| Days Max Temp ≤ 32 °F | 10 | 6 | 1 | 0 | 0 | 0 | 0 | 0 | 0 | 0 | 1 | 6 | 24 |
| Days Min Temp ≤ 32 °F | 27 | 21 | 15 | 3 | 0 | 0 | 0 | 0 | 0 | 2 | 13 | 24 | 105 |
| Days Min Temp ≤ 0 °F | 3 | 1 | 0 | 0 | 0 | 0 | 0 | 0 | 0 | 0 | 0 | 1 | 5 |
| Heating Degree Days | 1129 | 884 | 622 | 298 | 89 | 8 | 1 | 2 | 55 | 256 | 624 | 977 | 4945 |
| Cooling Degree Days | 0 | 0 | 3 | 28 | 85 | 281 | 436 | 373 | 183 | 28 | 1 | 0 | 1418 |
| Total Precipitation (") | 1.24 | 1.06 | 2.63 | 3.93 | 4.91 | 5.47 | 3.94 | 3.63 | 4.86 | 3.60 | 2.64 | 1.75 | 39.66 |
| Days ≥ 0.1" Precip | 3 | 3 | 5 | 7 | 8 | 7 | 6 | 6 | 6 | 5 | 5 | 4 | 65 |
| Total Snowfall (") | 5.5 | 4.6 | 2.8 | 0.5 | 0.0 | 0.0 | 0.0 | 0.0 | 0.0 | 0.0 | 1.4 | 2.9 | 17.7 |
| Days ≥ 1" Snow Depth | 11 | 8 | 3 | 0 | 0 | 0 | 0 | 0 | 0 | 0 | 1 | 5 | 28 |

## OSKALOOSA *Jefferson County* ELEVATION 1112 ft LAT/LONG 39° 13 ' N / 95° 19 ' W

| | JAN | FEB | MAR | APR | MAY | JUN | JUL | AUG | SEP | OCT | NOV | DEC | YEAR |
|---|---|---|---|---|---|---|---|---|---|---|---|---|---|
| Maximum Temp °F | 37.1 | 43.1 | 55.8 | 67.3 | 76.3 | 84.7 | 90.1 | 88.5 | 80.0 | 68.6 | 52.7 | 41.4 | 65.5 |
| Minimum Temp °F | 17.7 | 22.3 | 32.8 | 44.0 | 53.4 | 62.5 | 67.3 | 64.9 | 56.7 | 45.5 | 33.1 | 22.9 | 43.6 |
| Mean Temp °F | 27.5 | 32.7 | 44.3 | 55.7 | 64.9 | 73.6 | 78.8 | 76.7 | 68.3 | 57.1 | 42.9 | 32.1 | 54.6 |
| Days Max Temp ≥ 90 °F | 0 | 0 | 0 | 0 | 1 | 7 | 17 | 15 | 4 | 0 | 0 | 0 | 44 |
| Days Max Temp ≤ 32 °F | 11 | 7 | 1 | 0 | 0 | 0 | 0 | 0 | 0 | 0 | 1 | 7 | 27 |
| Days Min Temp ≤ 32 °F | 28 | 23 | 16 | 4 | 0 | 0 | 0 | 0 | 0 | 2 | 14 | 26 | 113 |
| Days Min Temp ≤ 0 °F | 3 | 1 | 0 | 0 | 0 | 0 | 0 | 0 | 0 | 0 | 0 | 1 | 5 |
| Heating Degree Days | 1157 | 904 | 637 | 301 | 90 | 8 | 1 | 1 | 58 | 266 | 656 | 1010 | 5089 |
| Cooling Degree Days | 0 | 0 | 3 | 30 | 81 | 281 | 440 | 378 | 173 | 22 | 1 | 0 | 1409 |
| Total Precipitation (") | 1.07 | 0.95 | 2.49 | 3.56 | 5.10 | 5.49 | 4.50 | 3.73 | 5.00 | 3.25 | 2.11 | 1.57 | 38.82 |
| Days ≥ 0.1" Precip | 3 | 3 | 5 | 6 | 8 | 8 | 6 | 6 | 6 | 5 | 4 | 3 | 63 |
| Total Snowfall (") | 5.4 | 4.6 | 2.6 | 0.6 | 0.0 | 0.0 | 0.0 | 0.0 | 0.0 | 0.0 | 1.1 | 3.7 | 18.0 |
| Days ≥ 1" Snow Depth | 13 | 8 | 2 | 0 | 0 | 0 | 0 | 0 | 0 | 0 | 1 | 6 | 30 |

## OTTAWA *Franklin County* ELEVATION 889 ft LAT/LONG 38° 38 ' N / 95° 15 ' W

| | JAN | FEB | MAR | APR | MAY | JUN | JUL | AUG | SEP | OCT | NOV | DEC | YEAR |
|---|---|---|---|---|---|---|---|---|---|---|---|---|---|
| Maximum Temp °F | 39.7 | 45.4 | 57.6 | 68.7 | 77.1 | 85.4 | 91.0 | 89.4 | 81.4 | 70.3 | 55.3 | 43.7 | 67.1 |
| Minimum Temp °F | 19.2 | 23.5 | 34.2 | 45.0 | 54.2 | 63.7 | 68.2 | 65.6 | 57.5 | 46.0 | 34.3 | 24.5 | 44.7 |
| Mean Temp °F | 29.5 | 34.5 | 46.0 | 56.9 | 65.6 | 74.6 | 79.6 | 77.5 | 69.5 | 58.2 | 44.8 | 34.1 | 55.9 |
| Days Max Temp ≥ 90 °F | 0 | 0 | 0 | 0 | 0 | 8 | 19 | 16 | 5 | 0 | 0 | 0 | 48 |
| Days Max Temp ≤ 32 °F | 9 | 5 | 1 | 0 | 0 | 0 | 0 | 0 | 0 | 0 | 1 | 5 | 21 |
| Days Min Temp ≤ 32 °F | 27 | 22 | 14 | 3 | 0 | 0 | 0 | 0 | 0 | 3 | 14 | 24 | 107 |
| Days Min Temp ≤ 0 °F | 3 | 1 | 0 | 0 | 0 | 0 | 0 | 0 | 0 | 0 | 0 | 1 | 5 |
| Heating Degree Days | 1094 | 855 | 588 | 269 | 78 | 6 | 0 | 1 | 47 | 239 | 602 | 950 | 4729 |
| Cooling Degree Days | 0 | 0 | 4 | 36 | 103 | 311 | 474 | 411 | 207 | 32 | 2 | 0 | 1580 |
| Total Precipitation (") | 1.33 | 1.17 | 2.87 | 3.73 | 4.81 | 5.70 | 3.78 | 3.77 | 4.21 | 3.41 | 2.49 | 1.74 | 39.01 |
| Days ≥ 0.1" Precip | 3 | 3 | 5 | 7 | 8 | 7 | 6 | 6 | 6 | 6 | 5 | 4 | 66 |
| Total Snowfall (") | 5.7 | 5.4 | 1.8 | 0.2 | 0.0 | 0.0 | 0.0 | 0.0 | 0.0 | 0.0 | 1.0 | 3.5 | 17.6 |
| Days ≥ 1" Snow Depth | na | *2* | *1* | 0 | 0 | 0 | 0 | 0 | 0 | 0 | 0 | *2* | na |

## PAOLA *Miami County* ELEVATION 869 ft LAT/LONG 38° 34 ' N / 94° 52 ' W

| | JAN | FEB | MAR | APR | MAY | JUN | JUL | AUG | SEP | OCT | NOV | DEC | YEAR |
|---|---|---|---|---|---|---|---|---|---|---|---|---|---|
| Maximum Temp °F | 39.4 | 45.3 | 57.3 | 68.8 | 77.0 | 85.2 | 90.6 | 88.8 | 80.6 | 69.6 | 54.7 | 43.0 | 66.7 |
| Minimum Temp °F | 19.3 | 23.5 | 33.9 | 44.2 | 53.5 | 62.9 | 67.5 | 64.9 | 56.9 | 45.3 | 34.2 | 24.2 | 44.2 |
| Mean Temp °F | 29.4 | 34.4 | 45.6 | 56.5 | 65.3 | 74.1 | 79.1 | 76.9 | 68.8 | 57.5 | 44.5 | 33.7 | 55.5 |
| Days Max Temp ≥ 90 °F | 0 | 0 | 0 | 0 | 0 | 7 | 18 | 15 | 4 | 0 | 0 | 0 | 44 |
| Days Max Temp ≤ 32 °F | 9 | 5 | 1 | 0 | 0 | 0 | 0 | 0 | 0 | 0 | 1 | 5 | 21 |
| Days Min Temp ≤ 32 °F | 27 | 22 | 15 | 3 | 0 | 0 | 0 | 0 | 0 | 3 | 13 | 25 | 108 |
| Days Min Temp ≤ 0 °F | 3 | 1 | 0 | 0 | 0 | 0 | 0 | 0 | 0 | 0 | 0 | 1 | 5 |
| Heating Degree Days | 1098 | 857 | 599 | 278 | 83 | 6 | 1 | 1 | 54 | 256 | 612 | 965 | 4810 |
| Cooling Degree Days | 0 | 0 | 4 | 32 | 97 | 301 | 457 | 393 | 192 | 27 | 2 | 0 | 1505 |
| Total Precipitation (") | 1.47 | 1.21 | 2.68 | 4.01 | 4.99 | 6.52 | 3.74 | 3.89 | 4.44 | 3.79 | 2.53 | 1.71 | 40.98 |
| Days ≥ 0.1" Precip | 4 | 3 | 6 | 6 | 8 | 7 | 6 | 6 | 6 | 6 | 5 | 3 | 66 |
| Total Snowfall (") | 5.2 | 4.9 | 2.3 | 0.3 | 0.0 | 0.0 | 0.0 | 0.0 | 0.0 | 0.0 | 0.8 | *2.8* | 16.3 |
| Days ≥ 1" Snow Depth | 7 | 5 | 2 | 0 | 0 | 0 | 0 | 0 | 0 | 0 | 1 | 2 | 17 |

## PARSONS 2 NW *Labette County* ELEVATION 912 ft LAT/LONG 37° 20 ' N / 95° 16 ' W

| | JAN | FEB | MAR | APR | MAY | JUN | JUL | AUG | SEP | OCT | NOV | DEC | YEAR |
|---|---|---|---|---|---|---|---|---|---|---|---|---|---|
| Maximum Temp °F | 41.3 | 47.3 | 58.4 | 69.1 | 76.8 | 85.7 | 91.8 | 90.4 | 81.6 | 70.9 | 56.5 | 45.6 | 68.0 |
| Minimum Temp °F | 21.4 | 25.9 | 35.8 | 45.9 | 55.1 | 64.1 | 68.8 | 66.3 | 58.7 | 46.8 | 36.1 | 26.3 | 45.9 |
| Mean Temp °F | 31.3 | 36.7 | 47.1 | 57.5 | 66.0 | 74.9 | 80.3 | 78.4 | 70.2 | 58.9 | 46.3 | 35.9 | 57.0 |
| Days Max Temp ≥ 90 °F | 0 | 0 | 0 | 0 | 0 | 9 | 20 | 18 | 5 | 0 | 0 | 0 | 52 |
| Days Max Temp ≤ 32 °F | 8 | 5 | 1 | 0 | 0 | 0 | 0 | 0 | 0 | 0 | 1 | 4 | 19 |
| Days Min Temp ≤ 32 °F | 27 | 21 | 12 | 2 | 0 | 0 | 0 | 0 | 0 | 2 | 12 | 23 | 99 |
| Days Min Temp ≤ 0 °F | 1 | 1 | 0 | 0 | 0 | 0 | 0 | 0 | 0 | 0 | 0 | 1 | 3 |
| Heating Degree Days | 1037 | 794 | 552 | 249 | 70 | 4 | 0 | 1 | 45 | 223 | 556 | 894 | 4425 |
| Cooling Degree Days | 0 | 0 | 3 | 25 | 89 | 303 | 466 | 413 | 196 | 26 | 2 | 0 | 1523 |
| Total Precipitation (") | 1.40 | 1.59 | 3.28 | 4.11 | 5.28 | 4.36 | 3.72 | 3.44 | 4.90 | 4.04 | 3.13 | 2.01 | 41.26 |
| Days ≥ 0.1" Precip | 3 | 3 | 5 | 6 | 8 | 6 | 5 | 5 | 6 | 5 | 5 | 4 | 61 |
| Total Snowfall (") | *3.0* | *3.0* | 1.4 | 0.0 | 0.0 | 0.0 | 0.0 | 0.0 | 0.0 | 0.0 | 0.4 | *1.7* | 9.5 |
| Days ≥ 1" Snow Depth | na | 2 | 1 | 0 | 0 | 0 | 0 | 0 | 0 | 0 | 0 | *2* | na |

## PHILLIPSBURG 1 SSE *Phillips County* ELEVATION 1880 ft LAT/LONG 39° 44 ' N / 99° 19 ' W

| | JAN | FEB | MAR | APR | MAY | JUN | JUL | AUG | SEP | OCT | NOV | DEC | YEAR |
|---|---|---|---|---|---|---|---|---|---|---|---|---|---|
| Maximum Temp °F | 38.7 | 44.4 | 55.4 | 67.2 | 75.7 | 87.2 | 93.3 | 91.0 | 81.8 | 69.9 | 52.8 | 42.0 | 66.6 |
| Minimum Temp °F | 13.6 | 17.7 | 27.8 | 39.0 | 49.3 | 59.7 | 65.1 | 62.6 | 52.1 | 39.3 | 26.6 | 17.2 | 39.2 |
| Mean Temp °F | 26.2 | 31.1 | 41.6 | 53.1 | 62.6 | 73.5 | 79.3 | 76.8 | 66.9 | 54.6 | 39.7 | 29.6 | 52.9 |
| Days Max Temp ≥ 90 °F | 0 | 0 | 0 | 1 | 2 | 13 | 21 | 18 | 8 | 1 | 0 | 0 | 64 |
| Days Max Temp ≤ 32 °F | 11 | 7 | 2 | 0 | 0 | 0 | 0 | 0 | 0 | 0 | 2 | 7 | 29 |
| Days Min Temp ≤ 32 °F | 30 | 26 | 21 | 7 | 1 | 0 | 0 | 0 | 1 | 7 | 22 | 30 | 145 |
| Days Min Temp ≤ 0 °F | 5 | 3 | 0 | 0 | 0 | 0 | 0 | 0 | 0 | 0 | 0 | 2 | 10 |
| Heating Degree Days | 1197 | 950 | 721 | 366 | 138 | 15 | 1 | 3 | 81 | 330 | 753 | 1089 | 5644 |
| Cooling Degree Days | 0 | 0 | 1 | 18 | 60 | 272 | 441 | 383 | 155 | 11 | 0 | 0 | 1341 |
| Total Precipitation (") | 0.41 | 0.51 | 1.78 | 2.26 | 3.76 | 3.41 | 3.47 | 3.02 | 2.25 | 1.56 | 0.92 | 0.48 | 23.83 |
| Days ≥ 0.1" Precip | 1 | 1 | 3 | 5 | 7 | 6 | 5 | 5 | 4 | 3 | 2 | 2 | 44 |
| Total Snowfall (") | 4.9 | 4.7 | 4.1 | 0.8 | 0.1 | 0.0 | 0.0 | 0.0 | 0.2 | 0.2 | 2.0 | 3.6 | 20.6 |
| Days ≥ 1" Snow Depth | na | *4* | *2* | 0 | 0 | 0 | 0 | 0 | 0 | 0 | 1 | *2* | na |

## PLAINVILLE 4 WNW *Rooks County* ELEVATION 2152 ft LAT/LONG 39° 14 ' N / 99° 18 ' W

| | JAN | FEB | MAR | APR | MAY | JUN | JUL | AUG | SEP | OCT | NOV | DEC | YEAR |
|---|---|---|---|---|---|---|---|---|---|---|---|---|---|
| Maximum Temp °F | 38.4 | 43.9 | 55.5 | 66.7 | 74.6 | 86.1 | 91.5 | *89.0* | 80.6 | 69.2 | 53.1 | 43.2 | 66.0 |
| Minimum Temp °F | 16.0 | 19.1 | 28.9 | 39.7 | 50.0 | 60.3 | 65.5 | *62.9* | 53.8 | 41.1 | 28.4 | 20.2 | 40.5 |
| Mean Temp °F | 27.2 | 31.6 | 42.2 | 53.2 | 62.4 | 73.2 | 78.5 | *76.0* | 67.2 | 55.2 | 40.7 | 31.7 | 53.3 |
| Days Max Temp ≥ 90 °F | 0 | 0 | 0 | 1 | 1 | 11 | 19 | 15 | 6 | 1 | 0 | 0 | 54 |
| Days Max Temp ≤ 32 °F | 10 | 7 | 2 | 0 | 0 | 0 | 0 | 0 | 0 | 0 | 2 | 6 | 27 |
| Days Min Temp ≤ 32 °F | 30 | 26 | 19 | 6 | 1 | 0 | 0 | 0 | 0 | 5 | 21 | 29 | 137 |
| Days Min Temp ≤ 0 °F | 3 | 2 | 0 | 0 | 0 | 0 | 0 | 0 | 0 | 0 | 0 | 1 | 6 |
| Heating Degree Days | 1164 | 937 | 695 | 361 | 140 | 16 | 2 | *4* | 71 | 316 | 721 | 1025 | 5452 |
| Cooling Degree Days | *0* | *0* | *1* | *21* | *60* | *268* | *404* | na | *151* | *12* | *0* | *0* | na |
| Total Precipitation (") | 0.49 | 0.66 | 1.94 | 2.05 | 3.57 | 2.92 | 4.19 | *2.88* | 2.20 | 1.49 | 1.06 | 0.60 | 24.05 |
| Days ≥ 0.1" Precip | 1 | 2 | 4 | 4 | 7 | 6 | 6 | 5 | 4 | 3 | 2 | 2 | 46 |
| Total Snowfall (") | 5.1 | 5.7 | 4.9 | 1.0 | 0.0 | 0.0 | 0.0 | 0.0 | 0.0 | 0.4 | 2.2 | 4.3 | 23.6 |
| Days ≥ 1" Snow Depth | *11* | *8* | *4* | 1 | 0 | 0 | 0 | 0 | 0 | 0 | 2 | *6* | 32 |

## POMONA LAKE *Osage County* ELEVATION 1060 ft LAT/LONG 38° 39 ' N / 95° 34 ' W

| | JAN | FEB | MAR | APR | MAY | JUN | JUL | AUG | SEP | OCT | NOV | DEC | YEAR |
|---|---|---|---|---|---|---|---|---|---|---|---|---|---|
| Maximum Temp °F | 36.7 | 42.5 | 55.1 | 66.1 | 74.7 | 83.5 | 89.3 | 88.0 | 79.8 | 68.7 | 53.7 | 41.8 | 65.0 |
| Minimum Temp °F | 16.6 | 21.0 | 32.2 | 43.6 | 53.7 | 63.2 | 67.9 | 65.4 | 56.8 | 44.3 | 32.8 | 22.3 | 43.3 |
| Mean Temp °F | 26.6 | 31.8 | 43.7 | 54.9 | 64.2 | 73.4 | 78.6 | 76.6 | 68.3 | 56.5 | 43.3 | 32.1 | 54.2 |
| Days Max Temp ≥ 90 °F | 0 | 0 | 0 | 0 | 0 | 6 | 15 | 13 | 5 | 0 | 0 | 0 | 39 |
| Days Max Temp ≤ 32 °F | 11 | 7 | 1 | 0 | 0 | 0 | 0 | 0 | 0 | 0 | 1 | 7 | 27 |
| Days Min Temp ≤ 32 °F | 29 | 24 | 16 | 3 | 0 | 0 | 0 | 0 | 0 | 3 | 15 | 26 | 116 |
| Days Min Temp ≤ 0 °F | 3 | 2 | 0 | 0 | 0 | 0 | 0 | 0 | 0 | 0 | 0 | 1 | 6 |
| Heating Degree Days | 1183 | 931 | 657 | 324 | 104 | 10 | 1 | 2 | 60 | 285 | 647 | 1013 | 5217 |
| Cooling Degree Days | 0 | 0 | 3 | 30 | 87 | 286 | 446 | 393 | 186 | 26 | 2 | 0 | 1459 |
| Total Precipitation (") | 1.10 | 0.97 | 2.58 | 3.61 | 4.59 | 5.14 | 3.35 | 3.49 | 4.18 | 2.81 | 2.31 | 1.49 | 35.62 |
| Days ≥ 0.1" Precip | 3 | 3 | 5 | 6 | 7 | 7 | 5 | 5 | 6 | 5 | 4 | 3 | 59 |
| Total Snowfall (") | na | *2.4* | *0.9* | 0.1 | 0.0 | 0.0 | 0.0 | 0.0 | 0.0 | 0.0 | 0.5 | na | na |
| Days ≥ 1" Snow Depth | na | *5* | *1* | 0 | 0 | 0 | 0 | 0 | 0 | 0 | 0 | na | na |

## PRATT 4 W *Pratt County* ELEVATION 1941 ft LAT/LONG 37° 38 ' N / 98° 48 ' W

| | JAN | FEB | MAR | APR | MAY | JUN | JUL | AUG | SEP | OCT | NOV | DEC | YEAR |
|---|---|---|---|---|---|---|---|---|---|---|---|---|---|
| Maximum Temp °F | 43.0 | 49.4 | 59.8 | 70.2 | 77.8 | 88.1 | 93.7 | 91.7 | 83.3 | 72.4 | 56.1 | 45.9 | 69.3 |
| Minimum Temp °F | 19.4 | 23.2 | 32.2 | 42.6 | 52.2 | 61.9 | 66.6 | 64.8 | 56.6 | 44.6 | 31.9 | 22.5 | 43.2 |
| Mean Temp °F | 31.2 | 36.3 | 46.0 | 56.4 | 65.0 | 75.0 | 80.2 | 78.3 | 70.0 | 58.5 | 44.0 | 34.3 | 56.3 |
| Days Max Temp ≥ 90 °F | 0 | 0 | 0 | 1 | 2 | 14 | 23 | 20 | 8 | 1 | 0 | 0 | 69 |
| Days Max Temp ≤ 32 °F | 7 | 4 | 1 | 0 | 0 | 0 | 0 | 0 | 0 | 0 | 1 | 4 | 17 |
| Days Min Temp ≤ 32 °F | 29 | 23 | 16 | 4 | 0 | 0 | 0 | 0 | 0 | 3 | 15 | 26 | 116 |
| Days Min Temp ≤ 0 °F | 2 | 1 | 0 | 0 | 0 | 0 | 0 | 0 | 0 | 0 | 0 | 1 | 4 |
| Heating Degree Days | 1040 | 803 | 585 | 275 | 88 | 7 | 0 | 1 | 45 | 233 | 623 | 947 | 4647 |
| Cooling Degree Days | 0 | 0 | 4 | 25 | 81 | 305 | 465 | 425 | 209 | 30 | *0* | *0* | 1544 |
| Total Precipitation (") | 0.56 | 0.92 | 1.98 | 2.64 | 3.63 | 3.67 | 3.16 | 2.95 | 2.50 | 2.23 | 1.19 | 0.99 | 26.42 |
| Days ≥ 0.1" Precip | 2 | 2 | 4 | 4 | 6 | 6 | 4 | 5 | 4 | 3 | 2 | 2 | 44 |
| Total Snowfall (") | 4.6 | 3.8 | 3.3 | 0.8 | 0.0 | 0.0 | 0.0 | 0.0 | 0.0 | 0.0 | 1.2 | 2.8 | 16.5 |
| Days ≥ 1" Snow Depth | *6* | *4* | 1 | 0 | 0 | 0 | 0 | 0 | 0 | 0 | 1 | *3* | 15 |

## QUINTER *Gove County* ELEVATION 2661 ft LAT/LONG 39° 4 ' N / 100° 14 ' W

| | JAN | FEB | MAR | APR | MAY | JUN | JUL | AUG | SEP | OCT | NOV | DEC | YEAR |
|---|---|---|---|---|---|---|---|---|---|---|---|---|---|
| Maximum Temp °F | 39.3 | 44.3 | 54.6 | 65.5 | 74.1 | 85.1 | 91.1 | 88.7 | 79.8 | 68.3 | 52.1 | 42.2 | 65.4 |
| Minimum Temp °F | 14.8 | 18.8 | 26.3 | 37.5 | 47.8 | 57.9 | 63.5 | 61.6 | 51.9 | 39.4 | 26.8 | 18.1 | 38.7 |
| Mean Temp °F | 27.1 | 31.6 | 40.5 | 51.5 | 61.0 | 71.5 | 77.3 | 75.2 | 65.9 | 53.9 | 39.5 | 30.1 | 52.1 |
| Days Max Temp ≥ 90 °F | 0 | 0 | 0 | 1 | 2 | 10 | 19 | 17 | 6 | 1 | 0 | 0 | 56 |
| Days Max Temp ≤ 32 °F | 10 | 7 | 2 | 0 | 0 | 0 | 0 | 0 | 0 | 0 | 3 | 8 | 30 |
| Days Min Temp ≤ 32 °F | 30 | 27 | 23 | 8 | 1 | 0 | 0 | 0 | 1 | 7 | 23 | 30 | 150 |
| Days Min Temp ≤ 0 °F | 4 | 2 | 0 | 0 | 0 | 0 | 0 | 0 | 0 | 0 | 0 | 2 | 8 |
| Heating Degree Days | 1168 | 937 | 754 | 407 | 168 | 26 | 2 | 6 | 96 | 350 | 759 | 1074 | 5747 |
| Cooling Degree Days | 0 | 0 | 0 | 14 | 48 | 239 | 389 | 344 | 142 | 11 | 0 | 0 | 1187 |
| Total Precipitation (") | 0.55 | 0.72 | 1.67 | 2.08 | 3.80 | 2.99 | 3.44 | 3.11 | 1.86 | 1.43 | 1.00 | 0.65 | 23.30 |
| Days ≥ 0.1" Precip | 2 | 2 | 4 | 4 | 6 | 6 | 6 | 5 | 4 | 3 | 2 | 2 | 46 |
| Total Snowfall (") | 5.3 | *5.0* | *5.8* | 1.5 | 0.1 | 0.0 | 0.0 | 0.0 | 0.1 | 1.0 | 2.2 | *5.8* | 26.8 |
| Days ≥ 1" Snow Depth | *5* | *3* | na | 0 | 0 | 0 | 0 | 0 | 0 | 0 | 1 | na | na |

## RUSSELL MUNI ARPT *Russell County* ELEVATION 1870 ft LAT/LONG 38° 52 ' N / 98° 49 ' W

| | JAN | FEB | MAR | APR | MAY | JUN | JUL | AUG | SEP | OCT | NOV | DEC | YEAR |
|---|---|---|---|---|---|---|---|---|---|---|---|---|---|
| Maximum Temp °F | 38.7 | 44.4 | 55.2 | 66.1 | 74.9 | 86.3 | 92.1 | 89.4 | 80.4 | 68.6 | 52.9 | 42.4 | 66.0 |
| Minimum Temp °F | 16.7 | 20.7 | 30.4 | 41.3 | 51.7 | 62.0 | 67.3 | 65.3 | 55.8 | 43.2 | 30.0 | 20.6 | 42.1 |
| Mean Temp °F | 27.8 | 32.6 | 42.8 | 53.7 | 63.3 | 74.2 | 79.7 | 77.3 | 68.1 | 56.0 | 41.4 | 31.5 | 54.0 |
| Days Max Temp ≥ 90 °F | 0 | 0 | 0 | 1 | 2 | 12 | 20 | 16 | 7 | 1 | 0 | 0 | 59 |
| Days Max Temp ≤ 32 °F | 10 | 6 | 2 | 0 | 0 | 0 | 0 | 0 | 0 | 0 | 2 | 7 | 27 |
| Days Min Temp ≤ 32 °F | 30 | 25 | 18 | 5 | 0 | 0 | 0 | 0 | 0 | 4 | 19 | 29 | 130 |
| Days Min Temp ≤ 0 °F | 3 | 2 | 0 | 0 | 0 | 0 | 0 | 0 | 0 | 0 | 0 | 1 | 6 |
| Heating Degree Days | 1146 | 909 | 682 | 350 | 127 | 14 | 1 | 2 | 72 | 300 | 700 | 1032 | 5335 |
| Cooling Degree Days | 0 | 0 | 2 | 24 | 80 | 301 | 466 | 397 | 192 | 21 | 1 | 0 | 1484 |
| Total Precipitation (") | 0.60 | 0.77 | 2.13 | 2.78 | 3.69 | 3.49 | 3.71 | 3.67 | 2.65 | 1.64 | 1.10 | 0.91 | 27.14 |
| Days ≥ 0.1" Precip | 2 | 2 | 4 | 5 | 7 | 6 | 5 | 6 | 4 | 3 | 2 | 2 | 48 |
| Total Snowfall (") | 5.5 | 5.2 | 5.1 | 1.2 | 0.0 | 0.0 | 0.0 | 0.0 | 0.0 | 0.4 | 2.0 | 4.5 | 23.9 |
| Days ≥ 1" Snow Depth | 10 | 7 | 3 | 1 | 0 | 0 | 0 | 0 | 0 | 0 | 2 | 6 | 29 |

## RUSSELL SPRINGS *Logan County* ELEVATION 2960 ft LAT/LONG 38° 54 ' N / 101° 11 ' W

| | JAN | FEB | MAR | APR | MAY | JUN | JUL | AUG | SEP | OCT | NOV | DEC | YEAR |
|---|---|---|---|---|---|---|---|---|---|---|---|---|---|
| Maximum Temp °F | 43.1 | 48.8 | 57.9 | 68.6 | 76.9 | 87.6 | 93.3 | 91.0 | 82.6 | 71.1 | 54.8 | 45.3 | 68.4 |
| Minimum Temp °F | 14.4 | 19.0 | 27.0 | 37.6 | 47.5 | 57.1 | 62.9 | 60.5 | 50.6 | 37.2 | 24.6 | 16.3 | 37.9 |
| Mean Temp °F | 28.8 | 34.0 | 42.4 | 53.1 | 62.3 | 72.4 | 78.1 | 75.8 | 66.6 | 54.2 | 39.7 | 30.8 | 53.2 |
| Days Max Temp ≥ 90 °F | 0 | 0 | 0 | 1 | 3 | 13 | 22 | 19 | 9 | 1 | 0 | 0 | 68 |
| Days Max Temp ≤ 32 °F | 7 | 4 | 1 | 0 | 0 | 0 | 0 | 0 | 0 | 0 | 1 | 5 | 18 |
| Days Min Temp ≤ 32 °F | 31 | 27 | *22* | *8* | 1 | 0 | 0 | 0 | 1 | 9 | 24 | 30 | 153 |
| Days Min Temp ≤ 0 °F | 4 | 2 | 0 | 0 | 0 | 0 | 0 | 0 | 0 | 0 | 0 | 2 | 8 |
| Heating Degree Days | 1118 | 872 | 696 | 364 | 141 | 17 | 1 | 4 | 81 | 340 | 753 | 1055 | 5442 |
| Cooling Degree Days | 0 | 0 | 1 | 17 | 65 | 267 | 418 | 363 | 153 | 8 | 0 | 0 | 1292 |
| Total Precipitation (") | 0.43 | 0.43 | 1.23 | 1.48 | 3.14 | 2.77 | 2.89 | 2.53 | 1.50 | 1.07 | 0.78 | 0.40 | 18.65 |
| Days ≥ 0.1" Precip | 1 | 1 | 3 | 3 | 6 | 5 | 6 | 5 | 3 | 2 | 2 | 1 | 38 |
| Total Snowfall (") | 5.6 | 3.9 | 5.0 | 1.7 | 0.0 | 0.0 | 0.0 | 0.0 | 0.2 | 0.8 | 2.8 | 5.3 | 25.3 |
| Days ≥ 1" Snow Depth | 10 | 5 | 3 | 0 | 0 | 0 | 0 | 0 | 0 | 0 | 2 | 6 | 26 |

## SAINT FRANCIS *Cheyenne County* ELEVATION 3304 ft LAT/LONG 39° 46 ' N / 101° 48 ' W

| | JAN | FEB | MAR | APR | MAY | JUN | JUL | AUG | SEP | OCT | NOV | DEC | YEAR |
|---|---|---|---|---|---|---|---|---|---|---|---|---|---|
| Maximum Temp °F | 41.9 | 47.2 | 55.7 | 66.2 | 75.1 | 85.9 | 91.3 | 89.3 | 81.2 | 69.6 | 53.1 | 43.9 | 66.7 |
| Minimum Temp °F | 14.8 | 18.8 | 26.4 | 36.1 | 46.6 | 56.3 | 62.1 | 59.7 | 49.7 | 36.9 | 25.0 | 17.0 | 37.4 |
| Mean Temp °F | 28.4 | 33.0 | 41.1 | 51.2 | 60.8 | 71.1 | 76.8 | 74.5 | 65.5 | 53.3 | 39.0 | 30.5 | 52.1 |
| Days Max Temp ≥ 90 °F | 0 | 0 | 0 | 1 | 2 | 12 | 19 | 17 | 8 | 1 | 0 | 0 | 60 |
| Days Max Temp ≤ 32 °F | 7 | 5 | 2 | 0 | 0 | 0 | 0 | 0 | 0 | 0 | 2 | 6 | 22 |
| Days Min Temp ≤ 32 °F | 30 | 27 | 23 | 10 | 1 | 0 | 0 | 0 | 1 | 9 | 25 | 30 | 156 |
| Days Min Temp ≤ 0 °F | 4 | 2 | 0 | 0 | 0 | 0 | 0 | 0 | 0 | 0 | 0 | 2 | 8 |
| Heating Degree Days | 1129 | 896 | 735 | 415 | 172 | 26 | 3 | 6 | 95 | 363 | 772 | 1064 | 5676 |
| Cooling Degree Days | 0 | 0 | 0 | 8 | 41 | 206 | 350 | 296 | 107 | 4 | 0 | 0 | 1012 |
| Total Precipitation (") | 0.51 | 0.47 | 1.17 | 1.59 | 3.08 | 2.72 | 2.93 | 2.04 | 1.17 | 1.15 | 0.73 | 0.46 | 18.02 |
| Days ≥ 0.1" Precip | 2 | 2 | 3 | 4 | 7 | 5 | 5 | 4 | 3 | 3 | 2 | 1 | 41 |
| Total Snowfall (") | 6.9 | 5.0 | 7.2 | 3.0 | 0.3 | 0.0 | 0.0 | 0.0 | 0.1 | 2.6 | 4.7 | 5.6 | 35.4 |
| Days ≥ 1" Snow Depth | 11 | 6 | 4 | 1 | 0 | 0 | 0 | 0 | 0 | 1 | 3 | 7 | 33 |

## SALINA FCWOS *Saline County* ELEVATION 1261 ft LAT/LONG 38° 48 ' N / 97° 39 ' W

| | JAN | FEB | MAR | APR | MAY | JUN | JUL | AUG | SEP | OCT | NOV | DEC | YEAR |
|---|---|---|---|---|---|---|---|---|---|---|---|---|---|
| Maximum Temp °F | 38.5 | 44.3 | 55.8 | 66.6 | 75.6 | 86.6 | 92.5 | 90.2 | 81.0 | 69.1 | 53.3 | 42.5 | 66.3 |
| Minimum Temp °F | 18.4 | 22.3 | 32.8 | 43.2 | 53.3 | 63.5 | 69.1 | 67.1 | 57.8 | 45.3 | 32.4 | 22.7 | 44.0 |
| Mean Temp °F | 28.4 | 33.4 | 44.3 | 54.9 | 64.5 | 75.1 | 80.8 | 78.7 | 69.4 | 57.2 | 42.9 | 32.6 | 55.2 |
| Days Max Temp ≥ 90 °F | 0 | 0 | 0 | 0 | 2 | 11 | 21 | 17 | 7 | 1 | 0 | 0 | 59 |
| Days Max Temp ≤ 32 °F | 10 | 6 | 1 | 0 | 0 | 0 | 0 | 0 | 0 | 0 | 1 | 6 | 24 |
| Days Min Temp ≤ 32 °F | 28 | 23 | 15 | 4 | 0 | 0 | 0 | 0 | 0 | 2 | 16 | 27 | 115 |
| Days Min Temp ≤ 0 °F | 2 | 2 | 0 | 0 | 0 | 0 | 0 | 0 | 0 | 0 | 0 | 1 | 5 |
| Heating Degree Days | 1124 | 886 | 637 | 318 | 101 | 8 | 0 | 1 | 57 | 268 | 658 | 997 | 5055 |
| Cooling Degree Days | 0 | 0 | 3 | 27 | 96 | 331 | 512 | 448 | 220 | 31 | 1 | 0 | 1669 |
| Total Precipitation (") | 0.74 | 0.98 | 2.35 | 2.97 | 4.39 | 4.34 | 4.02 | 3.39 | 2.89 | 2.57 | 1.39 | 1.00 | 31.03 |
| Days ≥ 0.1" Precip | 2 | 2 | 4 | 5 | 7 | 6 | 5 | 5 | 4 | 4 | 3 | 3 | 50 |
| Total Snowfall (") | 6.0 | 4.9 | 2.0 | 0.5 | 0.0 | 0.0 | 0.0 | 0.0 | 0.0 | 0.3 | 1.0 | 3.6 | 18.3 |
| Days ≥ 1" Snow Depth | 11 | 8 | 2 | 0 | 0 | 0 | 0 | 0 | 0 | 0 | 1 | 4 | 26 |

## SCOTT CITY *Scott County* ELEVATION 2972 ft LAT/LONG 38° 29 ' N / 100° 54 ' W

| | JAN | FEB | MAR | APR | MAY | JUN | JUL | AUG | SEP | OCT | NOV | DEC | YEAR |
|---|---|---|---|---|---|---|---|---|---|---|---|---|---|
| Maximum Temp °F | 42.9 | 48.5 | 57.9 | 68.5 | 76.8 | 87.7 | 92.4 | 90.0 | 82.2 | 71.1 | 54.5 | 45.3 | 68.2 |
| Minimum Temp °F | 16.2 | 20.2 | 27.9 | 37.9 | 48.1 | 58.3 | 63.4 | 61.4 | 52.2 | 39.6 | 27.1 | 19.0 | 39.3 |
| Mean Temp °F | 29.6 | 34.4 | 42.9 | 53.2 | 62.5 | 73.0 | 77.9 | 75.7 | 67.2 | 55.4 | 40.9 | 32.2 | 53.7 |
| Days Max Temp ≥ 90 °F | 0 | 0 | 0 | 1 | 3 | 14 | 22 | 18 | 8 | 1 | 0 | 0 | 67 |
| Days Max Temp ≤ 32 °F | 7 | 4 | 2 | 0 | 0 | 0 | 0 | 0 | 0 | 0 | 2 | 5 | 20 |
| Days Min Temp ≤ 32 °F | 30 | 26 | 22 | 8 | 1 | 0 | 0 | 0 | 0 | 6 | 22 | 30 | 145 |
| Days Min Temp ≤ 0 °F | 3 | 1 | 0 | 0 | 0 | 0 | 0 | 0 | 0 | 0 | 0 | 1 | 5 |
| Heating Degree Days | 1091 | 857 | 679 | 357 | 135 | 16 | 1 | 3 | 72 | 306 | 717 | 1011 | 5245 |
| Cooling Degree Days | 0 | 0 | 0 | 12 | 58 | 262 | 404 | 355 | 155 | 11 | 0 | 0 | 1257 |
| Total Precipitation (") | 0.71 | 0.68 | 1.40 | 1.74 | 2.97 | 3.25 | 2.99 | 2.47 | 1.75 | 1.18 | 1.00 | 0.67 | 20.81 |
| Days ≥ 0.1" Precip | 2 | 2 | 3 | 4 | 6 | 5 | 5 | 5 | 3 | 3 | 2 | 2 | 42 |
| Total Snowfall (") | 6.7 | 5.4 | 5.3 | 2.1 | 0.0 | 0.0 | 0.0 | 0.0 | 0.0 | 0.9 | 2.7 | 5.2 | 28.3 |
| Days ≥ 1" Snow Depth | 12 | 7 | 4 | 1 | 0 | 0 | 0 | 0 | 0 | 0 | 3 | 7 | 34 |

## SEDAN *Chautauqua County* ELEVATION 830 ft LAT/LONG 37° 8 ' N / 96° 11 ' W

| | JAN | FEB | MAR | APR | MAY | JUN | JUL | AUG | SEP | OCT | NOV | DEC | YEAR |
|---|---|---|---|---|---|---|---|---|---|---|---|---|---|
| Maximum Temp °F | 43.5 | 49.5 | 60.3 | 70.9 | 78.0 | 86.4 | 92.6 | 91.9 | 82.9 | 72.6 | 58.8 | 47.7 | 69.6 |
| Minimum Temp °F | 19.4 | 23.6 | 33.7 | 44.5 | 53.8 | 63.4 | 67.7 | 65.5 | 57.8 | 44.6 | 33.6 | 23.9 | 44.3 |
| Mean Temp °F | 31.5 | 36.6 | 47.0 | 57.7 | 65.9 | 74.9 | 80.2 | 78.7 | 70.3 | 58.6 | 46.2 | 35.8 | 57.0 |
| Days Max Temp ≥ 90 °F | 0 | 0 | 0 | 1 | 1 | 10 | 22 | 20 | 8 | 1 | 0 | 0 | 63 |
| Days Max Temp ≤ 32 °F | 7 | 4 | 0 | 0 | 0 | 0 | 0 | 0 | 0 | 0 | 0 | 4 | 15 |
| Days Min Temp ≤ 32 °F | 28 | 22 | 15 | 3 | 0 | 0 | 0 | 0 | 0 | 4 | 15 | 25 | 112 |
| Days Min Temp ≤ 0 °F | 2 | 1 | 0 | 0 | 0 | 0 | 0 | 0 | 0 | 0 | 0 | 1 | 4 |
| Heating Degree Days | 1032 | 796 | 556 | 245 | 72 | 4 | 0 | 1 | 45 | 230 | 560 | 899 | 4440 |
| Cooling Degree Days | 0 | 0 | 4 | 34 | 106 | 319 | 495 | 458 | 231 | 34 | 4 | 0 | 1685 |
| Total Precipitation (") | 1.32 | 1.54 | 3.02 | 3.78 | 5.70 | 4.78 | 3.19 | 3.05 | 4.50 | 3.74 | 3.00 | 1.81 | 39.43 |
| Days ≥ 0.1" Precip | 3 | 3 | 5 | 6 | 7 | 6 | 5 | 5 | 6 | 5 | 4 | 3 | 58 |
| Total Snowfall (") | 3.4 | 3.6 | 1.9 | 0.0 | 0.0 | 0.0 | 0.0 | 0.0 | 0.0 | 0.0 | 0.5 | 2.1 | 11.5 |
| Days ≥ 1" Snow Depth | 5 | 4 | 1 | 0 | 0 | 0 | 0 | 0 | 0 | 0 | 0 | 2 | 12 |

## SHARON SPRINGS *Wallace County* ELEVATION 3442 ft LAT/LONG 38° 54 ' N / 101° 45 ' W

| | JAN | FEB | MAR | APR | MAY | JUN | JUL | AUG | SEP | OCT | NOV | DEC | YEAR |
|---|---|---|---|---|---|---|---|---|---|---|---|---|---|
| Maximum Temp °F | 44.6 | 49.7 | 58.2 | 68.9 | 77.6 | 88.6 | 93.7 | 91.1 | 83.0 | 71.5 | 55.0 | 45.9 | 69.0 |
| Minimum Temp °F | 16.9 | 20.6 | 28.0 | 38.0 | 48.2 | 58.0 | 63.7 | 61.5 | 52.1 | 39.3 | 26.8 | 19.4 | 39.4 |
| Mean Temp °F | 30.8 | 35.2 | 43.1 | 53.5 | 62.9 | 73.3 | 78.7 | 76.3 | 67.6 | 55.4 | 40.9 | 32.7 | 54.2 |
| Days Max Temp ≥ 90 °F | 0 | 0 | 0 | 1 | 3 | 15 | 23 | 20 | 8 | 1 | 0 | 0 | 71 |
| Days Max Temp ≤ 32 °F | 6 | 4 | 1 | 0 | 0 | 0 | 0 | 0 | 0 | 0 | 1 | 5 | 17 |
| Days Min Temp ≤ 32 °F | 30 | 26 | 21 | 8 | 1 | 0 | 0 | 0 | 0 | 6 | 23 | 29 | 144 |
| Days Min Temp ≤ 0 °F | 3 | 1 | 0 | 0 | 0 | 0 | 0 | 0 | 0 | 0 | 0 | 1 | 5 |
| Heating Degree Days | 1054 | 836 | 673 | 350 | 130 | 13 | 1 | 3 | 65 | 302 | 716 | 996 | 5139 |
| Cooling Degree Days | 0 | 0 | 1 | 13 | 68 | 268 | 417 | 368 | 160 | 11 | 0 | 0 | 1306 |
| Total Precipitation (") | 0.45 | 0.53 | 1.41 | 1.41 | 3.29 | 3.05 | 3.01 | 2.20 | 1.56 | 1.18 | 0.84 | 0.52 | 19.45 |
| Days ≥ 0.1" Precip | 1 | 2 | 3 | 3 | 6 | 6 | 5 | 5 | 3 | 3 | 2 | 2 | 41 |
| Total Snowfall (") | 5.0 | 3.6 | 7.2 | 1.9 | 0.2 | 0.0 | 0.0 | 0.0 | 0.1 | 1.1 | 2.7 | *4.3* | 26.1 |
| Days ≥ 1" Snow Depth | na | na | na | 0 | 0 | 0 | 0 | 0 | 0 | 0 | *1* | na | na |

## SMITH CENTER *Smith County* ELEVATION 1801 ft LAT/LONG 39° 47 ' N / 98° 47 ' W

| | JAN | FEB | MAR | APR | MAY | JUN | JUL | AUG | SEP | OCT | NOV | DEC | YEAR |
|---|---|---|---|---|---|---|---|---|---|---|---|---|---|
| Maximum Temp °F | 38.0 | 44.2 | 55.6 | 67.8 | 76.6 | 87.6 | 93.2 | 90.7 | 82.0 | 69.7 | 52.1 | 41.3 | 66.6 |
| Minimum Temp °F | 15.6 | 19.9 | 29.8 | 41.2 | 51.3 | 61.4 | 66.7 | 64.5 | 54.8 | 42.5 | 29.0 | 19.6 | 41.4 |
| Mean Temp °F | 26.8 | 32.1 | 42.7 | 54.5 | 63.9 | 74.5 | 80.0 | 77.6 | 68.4 | 56.1 | 40.6 | 30.5 | 54.0 |
| Days Max Temp ≥ 90 °F | 0 | 0 | 0 | 1 | 2 | 13 | 21 | 18 | 8 | 1 | 0 | 0 | 64 |
| Days Max Temp ≤ 32 °F | 10 | 6 | 2 | 0 | 0 | 0 | 0 | 0 | 0 | 0 | 2 | 7 | 27 |
| Days Min Temp ≤ 32 °F | 30 | 25 | 19 | 5 | 1 | 0 | 0 | 0 | 0 | 4 | 20 | 29 | 133 |
| Days Min Temp ≤ 0 °F | 4 | 2 | 0 | 0 | 0 | 0 | 0 | 0 | 0 | 0 | 0 | 2 | 8 |
| Heating Degree Days | 1178 | 923 | 685 | 328 | 111 | 11 | 0 | 2 | 65 | 292 | 725 | 1063 | 5383 |
| Cooling Degree Days | 0 | 0 | 1 | 26 | 82 | 309 | 466 | 401 | 191 | 19 | 0 | 0 | 1495 |
| Total Precipitation (") | 0.45 | 0.53 | 1.86 | 2.37 | 3.63 | 3.47 | 3.17 | 3.12 | 2.51 | 1.76 | 1.02 | 0.63 | 24.52 |
| Days ≥ 0.1" Precip | 1 | 2 | 4 | 5 | 7 | 6 | 6 | 5 | 4 | 3 | 3 | 2 | 48 |
| Total Snowfall (") | 4.2 | 4.3 | 3.7 | 0.5 | 0.0 | 0.0 | 0.0 | 0.0 | 0.2 | 0.1 | 1.9 | 3.7 | 18.6 |
| Days ≥ 1" Snow Depth | 12 | 8 | 3 | 0 | 0 | 0 | 0 | 0 | 0 | 0 | 2 | 6 | 31 |

## STERLING *Rice County* ELEVATION 1640 ft LAT/LONG 38° 13' N / 98° 12' W

| | JAN | FEB | MAR | APR | MAY | JUN | JUL | AUG | SEP | OCT | NOV | DEC | YEAR |
|---|---|---|---|---|---|---|---|---|---|---|---|---|---|
| Maximum Temp °F | 40.4 | 46.4 | 57.1 | 67.9 | 76.5 | 87.6 | 92.9 | 90.7 | 81.8 | 70.1 | 54.6 | 43.9 | 67.5 |
| Minimum Temp °F | 18.8 | 22.6 | 32.1 | 42.7 | 53.2 | 63.1 | 68.3 | 66.2 | 57.1 | 44.8 | 31.9 | 22.6 | 43.6 |
| Mean Temp °F | 29.6 | 34.5 | 44.7 | 55.3 | 64.9 | 75.3 | 80.7 | 78.5 | 69.5 | 57.5 | 43.2 | 33.3 | 55.6 |
| Days Max Temp ≥ 90 °F | 0 | 0 | 0 | 1 | 2 | 13 | 22 | 19 | 8 | 1 | 0 | 0 | 66 |
| Days Max Temp ≤ 32 °F | 9 | 5 | 1 | 0 | 0 | 0 | 0 | 0 | 0 | 0 | 1 | 6 | 22 |
| Days Min Temp ≤ 32 °F | 29 | 23 | 16 | 3 | 0 | 0 | 0 | 0 | 0 | 2 | 17 | 27 | 117 |
| Days Min Temp ≤ 0 °F | 2 | 1 | 0 | 0 | 0 | 0 | 0 | 0 | 0 | 0 | 0 | 1 | 4 |
| Heating Degree Days | 1089 | 854 | 626 | 306 | 95 | 8 | 0 | 1 | 54 | 259 | 647 | 976 | 4915 |
| Cooling Degree Days | 0 | 0 | 2 | 29 | 99 | 336 | 501 | 440 | 217 | 30 | 2 | 0 | 1656 |
| Total Precipitation (") | 0.63 | 1.02 | 2.34 | 2.43 | 3.82 | 4.14 | 3.53 | 3.04 | 2.51 | 2.32 | 1.24 | 1.06 | 28.08 |
| Days ≥ 0.1" Precip | 2 | 2 | 5 | 5 | 7 | 7 | 5 | 5 | 4 | 4 | 3 | 3 | 52 |
| Total Snowfall (") | 4.5 | *4.7* | 1.5 | 1.3 | 0.0 | 0.0 | 0.0 | 0.0 | 0.0 | 0.2 | 1.0 | 3.3 | 16.5 |
| Days ≥ 1" Snow Depth | 8 | 6 | 1 | 0 | 0 | 0 | 0 | 0 | 0 | 0 | 1 | 4 | 20 |

## SUBLETTE *Haskell County* ELEVATION 2913 ft LAT/LONG 37° 29' N / 100° 50' W

| | JAN | FEB | MAR | APR | MAY | JUN | JUL | AUG | SEP | OCT | NOV | DEC | YEAR |
|---|---|---|---|---|---|---|---|---|---|---|---|---|---|
| Maximum Temp °F | 44.9 | 50.9 | 60.3 | 70.2 | 77.9 | 88.4 | 93.1 | 90.6 | 83.0 | 72.4 | 56.4 | 47.1 | 69.6 |
| Minimum Temp °F | 17.8 | 21.9 | 29.9 | 39.5 | 49.6 | 59.5 | 64.2 | 62.5 | 53.9 | 41.4 | 28.8 | 20.6 | 40.8 |
| Mean Temp °F | 31.4 | 36.4 | 45.1 | 54.9 | 63.8 | 74.0 | 78.7 | 76.5 | 68.4 | 56.9 | 42.6 | 33.9 | 55.2 |
| Days Max Temp ≥ 90 °F | 0 | 0 | 0 | 1 | 3 | 15 | 23 | 19 | 9 | 1 | 0 | 0 | 71 |
| Days Max Temp ≤ 32 °F | 6 | 3 | 1 | 0 | 0 | 0 | 0 | 0 | 0 | 0 | 1 | 4 | 15 |
| Days Min Temp ≤ 32 °F | 30 | 25 | 19 | 6 | 1 | 0 | 0 | 0 | 0 | 5 | 20 | 29 | 135 |
| Days Min Temp ≤ 0 °F | 2 | 1 | 0 | 0 | 0 | 0 | 0 | 0 | 0 | 0 | 0 | 1 | 4 |
| Heating Degree Days | 1036 | 801 | 610 | 312 | 109 | 11 | 0 | 2 | 56 | 267 | 665 | 958 | 4827 |
| Cooling Degree Days | 0 | 0 | 1 | 16 | 72 | 278 | 419 | 379 | 171 | 17 | 0 | 0 | 1353 |
| Total Precipitation (") | 0.41 | 0.50 | 1.21 | 1.44 | 3.14 | 3.16 | 2.48 | 2.49 | 1.87 | 1.11 | 0.85 | 0.43 | 19.09 |
| Days ≥ 0.1" Precip | 1 | 1 | 3 | 3 | 6 | 6 | 4 | 4 | 3 | 2 | 2 | 2 | 37 |
| Total Snowfall (") | 4.4 | 3.7 | 4.0 | 0.9 | 0.1 | 0.0 | 0.0 | 0.0 | 0.0 | 0.5 | 2.2 | 3.2 | 19.0 |
| Days ≥ 1" Snow Depth | 8 | 5 | 3 | 1 | 0 | 0 | 0 | 0 | 0 | 0 | 2 | 6 | 25 |

## SYRACUSE *Hamilton County* ELEVATION 3261 ft LAT/LONG 37° 59' N / 101° 46' W

| | JAN | FEB | MAR | APR | MAY | JUN | JUL | AUG | SEP | OCT | NOV | DEC | YEAR |
|---|---|---|---|---|---|---|---|---|---|---|---|---|---|
| Maximum Temp °F | 44.5 | 50.9 | 59.8 | 69.9 | 78.0 | 89.3 | 94.0 | 91.6 | 83.8 | 72.8 | 56.7 | 47.0 | 69.9 |
| Minimum Temp °F | 13.8 | 18.9 | 27.2 | 37.4 | 47.8 | 57.8 | 63.4 | 61.2 | 51.6 | 37.8 | 24.8 | 16.3 | 38.2 |
| Mean Temp °F | 29.2 | 34.9 | 43.5 | 53.7 | 62.9 | 73.6 | 78.8 | 76.4 | 67.7 | 55.3 | 40.8 | 31.7 | 54.0 |
| Days Max Temp ≥ 90 °F | 0 | 0 | 0 | 1 | 4 | 15 | 23 | 20 | 10 | 2 | 0 | 0 | 75 |
| Days Max Temp ≤ 32 °F | 7 | 3 | 1 | 0 | 0 | 0 | 0 | 0 | 0 | 0 | 1 | 4 | 16 |
| Days Min Temp ≤ 32 °F | 31 | 27 | 22 | 9 | 1 | 0 | 0 | 0 | 1 | 9 | 24 | 30 | 154 |
| Days Min Temp ≤ 0 °F | 4 | 1 | 0 | 0 | 0 | 0 | 0 | 0 | 0 | 0 | 0 | 2 | 7 |
| Heating Degree Days | 1104 | 842 | 660 | 345 | 131 | 14 | 1 | 3 | 66 | 306 | 720 | 1027 | 5219 |
| Cooling Degree Days | 0 | 0 | 1 | 13 | 76 | 284 | 440 | 388 | 171 | 13 | 0 | 0 | 1386 |
| Total Precipitation (") | 0.40 | 0.45 | 1.00 | 1.28 | 2.24 | 2.74 | 2.42 | 2.20 | 1.52 | 1.00 | 0.66 | 0.41 | 16.32 |
| Days ≥ 0.1" Precip | 1 | 1 | 3 | 3 | 5 | 5 | 4 | 4 | 3 | 2 | 2 | 2 | 35 |
| Total Snowfall (") | 4.2 | 2.6 | 4.6 | 1.4 | 0.0 | 0.0 | 0.0 | 0.0 | 0.1 | 0.5 | 2.1 | 3.8 | 19.3 |
| Days ≥ 1" Snow Depth | na | *4* | *2* | 1 | 0 | 0 | 0 | 0 | 0 | 0 | 2 | *6* | na |

## TOPEKA MUNI ARPT *Shawnee County* ELEVATION 877 ft LAT/LONG 39° 4' N / 95° 38' W

| | JAN | FEB | MAR | APR | MAY | JUN | JUL | AUG | SEP | OCT | NOV | DEC | YEAR |
|---|---|---|---|---|---|---|---|---|---|---|---|---|---|
| Maximum Temp °F | 37.0 | 42.5 | 55.6 | 66.8 | 75.6 | 84.3 | 89.1 | 87.5 | 79.9 | 68.6 | 53.3 | 41.4 | 65.1 |
| Minimum Temp °F | 16.9 | 21.6 | 32.5 | 43.0 | 53.1 | 63.0 | 67.4 | 64.8 | 55.8 | 43.7 | 32.0 | 22.0 | 43.0 |
| Mean Temp °F | 27.0 | 32.1 | 44.0 | 54.9 | 64.4 | 73.7 | 78.3 | 76.2 | 67.9 | 56.2 | 42.7 | 31.7 | 54.1 |
| Days Max Temp ≥ 90 °F | 0 | 0 | 0 | 0 | 1 | 7 | 15 | 13 | 5 | 0 | 0 | 0 | 41 |
| Days Max Temp ≤ 32 °F | 11 | 7 | 1 | 0 | 0 | 0 | 0 | 0 | 0 | 0 | 1 | 7 | 27 |
| Days Min Temp ≤ 32 °F | 29 | 23 | 16 | 4 | 0 | 0 | 0 | 0 | 0 | 4 | 16 | 27 | 119 |
| Days Min Temp ≤ 0 °F | 3 | 2 | 0 | 0 | 0 | 0 | 0 | 0 | 0 | 0 | 0 | 2 | 7 |
| Heating Degree Days | 1171 | 923 | 646 | 321 | 103 | 8 | 1 | 1 | 69 | 293 | 663 | 1025 | 5224 |
| Cooling Degree Days | 0 | 0 | 3 | 29 | 88 | 289 | 438 | 358 | 176 | 24 | 1 | 0 | 1406 |
| Total Precipitation (") | 0.95 | 1.08 | 2.43 | 3.33 | 4.45 | 5.33 | 3.79 | 3.83 | 3.88 | 3.06 | 1.95 | 1.56 | 35.64 |
| Days ≥ 0.1" Precip | 2 | 3 | 5 | 6 | 8 | 7 | 6 | 6 | 5 | 5 | 4 | 4 | 61 |
| Total Snowfall (") | 6.1 | 5.3 | 2.5 | 0.7 | 0.0 | 0.0 | 0.0 | 0.0 | 0.0 | 0.0 | 1.5 | 5.2 | 21.3 |
| Days ≥ 1" Snow Depth | 11 | 7 | 2 | 0 | 0 | 0 | 0 | 0 | 0 | 0 | 1 | 5 | 26 |

## TORONTO LAKE *Woodson County* ELEVATION 961 ft LAT/LONG 37° 45' N / 95° 56' W

| | JAN | FEB | MAR | APR | MAY | JUN | JUL | AUG | SEP | OCT | NOV | DEC | YEAR |
|---|---|---|---|---|---|---|---|---|---|---|---|---|---|
| Maximum Temp °F | 39.6 | 45.2 | 56.7 | 67.9 | 75.9 | 84.7 | 90.8 | 89.7 | 81.4 | 70.4 | 55.9 | 44.7 | 66.9 |
| Minimum Temp °F | 18.2 | 22.4 | 33.5 | 44.8 | 53.9 | 63.3 | 68.0 | 65.6 | 57.1 | 44.5 | 33.8 | 23.6 | 44.1 |
| Mean Temp °F | 28.9 | 33.9 | 45.1 | 56.4 | 64.9 | 74.0 | 79.4 | 77.7 | 69.3 | 57.5 | 44.8 | 34.1 | 55.5 |
| Days Max Temp ≥ 90 °F | 0 | 0 | 0 | 0 | 0 | 7 | 18 | 16 | 6 | 0 | 0 | 0 | 47 |
| Days Max Temp ≤ 32 °F | 9 | 6 | 1 | 0 | 0 | 0 | 0 | 0 | 0 | 0 | 1 | 5 | 22 |
| Days Min Temp ≤ 32 °F | 29 | 24 | 14 | 3 | 0 | 0 | 0 | 0 | 0 | 3 | 13 | 26 | 112 |
| Days Min Temp ≤ 0 °F | 2 | 2 | 0 | 0 | 0 | 0 | 0 | 0 | 0 | 0 | 0 | 1 | 5 |
| Heating Degree Days | 1110 | 872 | 612 | 279 | 86 | 7 | 1 | 1 | 48 | 256 | 600 | 950 | 4822 |
| Cooling Degree Days | 0 | 0 | 3 | 28 | 91 | 301 | 468 | 417 | 199 | 25 | 2 | 0 | 1534 |
| Total Precipitation (") | 1.06 | 1.20 | 2.77 | 3.70 | 4.66 | 5.00 | 3.90 | 4.03 | 3.88 | 3.48 | 2.58 | 1.66 | 37.92 |
| Days ≥ 0.1" Precip | 3 | 3 | 5 | 6 | 7 | 6 | 6 | 5 | 5 | 4 | 4 | 3 | 57 |
| Total Snowfall (") | na | na | *0.3* | 0.0 | 0.0 | 0.0 | 0.0 | 0.0 | 0.0 | 0.0 | 0.2 | na | na |
| Days ≥ 1" Snow Depth | na | na | *0* | 0 | 0 | 0 | 0 | 0 | 0 | 0 | 0 | na | na |

## TRIBUNE 1 W *Greeley County* ELEVATION 3612 ft LAT/LONG 38° 28' N / 101° 46' W

| | JAN | FEB | MAR | APR | MAY | JUN | JUL | AUG | SEP | OCT | NOV | DEC | YEAR |
|---|---|---|---|---|---|---|---|---|---|---|---|---|---|
| Maximum Temp °F | 42.7 | 48.4 | 56.7 | 67.2 | 75.5 | 86.5 | 92.0 | 89.2 | 81.3 | 69.7 | 53.7 | 44.8 | 67.3 |
| Minimum Temp °F | 14.5 | 18.4 | 25.7 | 35.1 | 45.2 | 55.4 | 60.8 | 58.8 | 49.3 | 36.8 | 24.8 | 16.7 | 36.8 |
| Mean Temp °F | 28.6 | 33.4 | 41.2 | 51.1 | 60.3 | 71.0 | 76.4 | 74.0 | 65.3 | 53.3 | 39.3 | 30.8 | 52.1 |
| Days Max Temp ≥ 90 °F | 0 | 0 | 0 | 0 | 2 | 12 | 21 | 17 | 7 | 1 | 0 | 0 | 60 |
| Days Max Temp ≤ 32 °F | 8 | 4 | 2 | 0 | 0 | 0 | 0 | 0 | 0 | 0 | 2 | 6 | 22 |
| Days Min Temp ≤ 32 °F | 31 | 27 | 25 | 11 | 2 | 0 | 0 | 0 | 1 | 9 | 25 | 30 | 161 |
| Days Min Temp ≤ 0 °F | 4 | 2 | 0 | 0 | 0 | 0 | 0 | 0 | 0 | 0 | 0 | 2 | 8 |
| Heating Degree Days | 1121 | 885 | 731 | 414 | 177 | 24 | 2 | 6 | 91 | 363 | 763 | 1054 | 5631 |
| Cooling Degree Days | 0 | 0 | 0 | 5 | 40 | 211 | 350 | 304 | 114 | 5 | 0 | 0 | 1029 |
| Total Precipitation (") | 0.42 | 0.45 | 1.06 | 1.24 | 2.46 | 2.98 | 2.79 | 2.19 | 1.40 | 0.89 | 0.58 | 0.35 | 16.81 |
| Days ≥ 0.1" Precip | 1 | 1 | 2 | 3 | 5 | 5 | 5 | 4 | 3 | 2 | 2 | 1 | 34 |
| Total Snowfall (") | 5.1 | 3.8 | 5.7 | 2.6 | 0.1 | 0.0 | 0.0 | 0.0 | 0.1 | 0.6 | 2.5 | 3.8 | 24.3 |
| Days ≥ 1" Snow Depth | 9 | 5 | 3 | 1 | 0 | 0 | 0 | 0 | 0 | 0 | 2 | 6 | 26 |

## TROY 2 E *Doniphan County* ELEVATION 1060 ft LAT/LONG 39° 47' N / 95° 3' W

| | JAN | FEB | MAR | APR | MAY | JUN | JUL | AUG | SEP | OCT | NOV | DEC | YEAR |
|---|---|---|---|---|---|---|---|---|---|---|---|---|---|
| Maximum Temp °F | 35.0 | 41.0 | 54.1 | 66.4 | 75.6 | 84.2 | 88.2 | 86.4 | 79.3 | 68.3 | 51.8 | 39.5 | 64.1 |
| Minimum Temp °F | 16.4 | 21.0 | 31.8 | 43.2 | 53.6 | 62.5 | 66.6 | 63.9 | 55.8 | 44.8 | 32.5 | 22.0 | 42.8 |
| Mean Temp °F | 25.7 | 31.0 | 43.0 | 54.9 | 64.6 | 73.4 | 77.4 | 75.2 | 67.6 | 56.5 | 42.2 | 30.7 | 53.5 |
| Days Max Temp ≥ 90 °F | 0 | 0 | 0 | 0 | 1 | 6 | 13 | 11 | 3 | 0 | 0 | 0 | 34 |
| Days Max Temp ≤ 32 °F | 12 | 8 | 1 | 0 | 0 | 0 | 0 | 0 | 0 | 0 | 2 | 8 | 31 |
| Days Min Temp ≤ 32 °F | 28 | 23 | 17 | 4 | 0 | 0 | 0 | 0 | 0 | 3 | 15 | 26 | 116 |
| Days Min Temp ≤ 0 °F | 4 | 2 | 0 | 0 | 0 | 0 | 0 | 0 | 0 | 0 | 0 | 2 | 8 |
| Heating Degree Days | 1210 | 952 | 678 | 323 | 97 | 7 | 1 | 3 | 64 | 280 | 678 | 1055 | 5348 |
| Cooling Degree Days | 0 | 0 | 3 | 31 | 87 | 272 | 406 | 340 | 165 | 23 | 1 | 0 | 1328 |
| Total Precipitation (") | 0.95 | 0.93 | 2.37 | 3.25 | 4.74 | 4.92 | 4.65 | 4.08 | 4.93 | 3.05 | 1.92 | 1.39 | 37.18 |
| Days ≥ 0.1" Precip | 3 | 3 | 5 | 6 | 7 | 6 | 6 | 6 | 6 | 5 | 4 | 3 | 60 |
| Total Snowfall (") | 5.4 | 4.2 | 3.2 | 0.8 | 0.0 | 0.0 | 0.0 | 0.0 | 0.0 | 0.0 | 1.0 | 3.5 | 18.1 |
| Days ≥ 1" Snow Depth | 12 | 8 | 2 | 0 | 0 | 0 | 0 | 0 | 0 | 0 | 1 | 6 | 29 |

## TUTTLE CREEK LAKE *Riley County* ELEVATION 1060 ft LAT/LONG 39° 15' N / 96° 36' W

| | JAN | FEB | MAR | APR | MAY | JUN | JUL | AUG | SEP | OCT | NOV | DEC | YEAR |
|---|---|---|---|---|---|---|---|---|---|---|---|---|---|
| Maximum Temp °F | 36.3 | 41.9 | 54.4 | 65.8 | 74.8 | 84.4 | 90.2 | 88.2 | 79.9 | 68.6 | 52.9 | 41.1 | 64.9 |
| Minimum Temp °F | 13.6 | 17.8 | 29.5 | 41.0 | 50.9 | 61.0 | 66.2 | 63.3 | 53.9 | 41.0 | 29.4 | 19.0 | 40.6 |
| Mean Temp °F | 25.0 | 29.8 | 42.0 | 53.4 | 62.9 | 72.7 | 78.2 | 75.8 | 66.9 | 54.8 | 41.2 | 30.1 | 52.7 |
| Days Max Temp ≥ 90 °F | 0 | 0 | 0 | 0 | 1 | 8 | 17 | 14 | 6 | 0 | 0 | 0 | 46 |
| Days Max Temp ≤ 32 °F | 12 | 8 | 2 | 0 | 0 | 0 | 0 | 0 | 0 | 0 | 2 | 7 | 31 |
| Days Min Temp ≤ 32 °F | 30 | 26 | 19 | 6 | 0 | 0 | 0 | 0 | 0 | 6 | 20 | 29 | 136 |
| Days Min Temp ≤ 0 °F | 4 | 3 | 0 | 0 | 0 | 0 | 0 | 0 | 0 | 0 | 0 | 2 | 9 |
| Heating Degree Days | 1233 | 986 | 709 | 362 | 134 | 16 | 2 | 3 | 84 | 331 | 709 | 1075 | 5644 |
| Cooling Degree Days | 0 | 0 | 4 | 28 | 76 | 261 | 425 | 365 | 167 | 18 | 1 | 0 | 1345 |
| Total Precipitation (") | 0.61 | 0.72 | 2.16 | 2.77 | 4.32 | 5.10 | 3.89 | 3.29 | 4.02 | 2.77 | 1.60 | 0.98 | 32.23 |
| Days ≥ 0.1" Precip | 2 | 2 | 5 | 5 | 7 | 7 | 6 | 5 | 5 | 4 | 3 | 2 | 53 |
| Total Snowfall (") | 3.6 | *3.6* | 1.7 | 0.0 | 0.0 | 0.0 | 0.0 | 0.0 | 0.0 | 0.0 | 0.5 | *2.3* | 11.7 |
| Days ≥ 1" Snow Depth | *9* | 6 | *1* | 0 | 0 | 0 | 0 | 0 | 0 | 0 | 1 | *4* | 21 |

## ULYSSES 1 SE *Grant County* ELEVATION 3051 ft LAT/LONG 37° 35' N / 101° 21' W

| | JAN | FEB | MAR | APR | MAY | JUN | JUL | AUG | SEP | OCT | NOV | DEC | YEAR |
|---|---|---|---|---|---|---|---|---|---|---|---|---|---|
| Maximum Temp °F | ***45.9*** | 50.3 | ***60.6*** | 70.2 | 78.6 | 88.8 | 93.9 | 91.0 | 83.8 | 72.7 | 57.0 | ***47.5*** | 70.0 |
| Minimum Temp °F | ***15.7*** | 19.6 | ***28.0*** | 37.4 | 48.4 | 59.1 | 63.9 | 61.9 | 52.5 | 39.0 | 26.3 | ***17.8*** | 39.1 |
| Mean Temp °F | ***30.8*** | 35.0 | ***44.3*** | 53.8 | 63.5 | 74.0 | 78.9 | 76.5 | 68.2 | 55.9 | 41.7 | ***32.6*** | 54.6 |
| Days Max Temp ≥ 90 °F | 0 | 0 | ***0*** | 1 | ***3*** | ***15*** | 23 | ***19*** | ***10*** | 2 | 0 | ***0*** | 73 |
| Days Max Temp ≤ 32 °F | ***6*** | 4 | ***1*** | 0 | 0 | 0 | 0 | 0 | 0 | 0 | 1 | ***5*** | 17 |
| Days Min Temp ≤ 32 °F | ***30*** | ***27*** | na | ***8*** | 1 | 0 | 0 | 0 | 0 | 6 | 23 | ***30*** | na |
| Days Min Temp ≤ 0 °F | ***2*** | ***1*** | ***0*** | 0 | 0 | 0 | 0 | 0 | 0 | 0 | 0 | ***2*** | 5 |
| Heating Degree Days | ***1054*** | 841 | ***638*** | ***341*** | 116 | 13 | 1 | 3 | 62 | 293 | 690 | ***997*** | 5049 |
| Cooling Degree Days | 0 | 0 | 0 | 14 | 70 | 271 | 425 | 373 | 165 | 12 | 0 | 0 | 1330 |
| Total Precipitation (") | 0.42 | 0.47 | 0.94 | 1.43 | 2.66 | 2.94 | 2.12 | 2.74 | 1.66 | 0.86 | 0.75 | 0.37 | 17.36 |
| Days ≥ 0.1" Precip | 1 | ***1*** | ***2*** | 3 | ***5*** | 5 | 4 | ***5*** | 3 | 2 | 2 | ***1*** | 34 |
| Total Snowfall (") | 5.7 | ***3.5*** | ***3.0*** | 0.8 | 0.0 | 0.0 | 0.0 | 0.0 | 0.1 | 0.3 | 2.0 | ***4.0*** | 19.4 |
| Days ≥ 1" Snow Depth | ***7*** | ***4*** | ***2*** | 0 | 0 | 0 | 0 | 0 | 0 | 0 | 1 | ***5*** | 19 |

## WAKEENEY *Trego County* ELEVATION 2461 ft LAT/LONG 39° 1' N / 99° 53' W

| | JAN | FEB | MAR | APR | MAY | JUN | JUL | AUG | SEP | OCT | NOV | DEC | YEAR |
|---|---|---|---|---|---|---|---|---|---|---|---|---|---|
| Maximum Temp °F | 39.4 | 44.6 | 54.7 | 65.6 | 74.2 | 85.2 | 91.2 | 88.8 | 79.9 | 68.7 | 52.5 | 42.5 | 65.6 |
| Minimum Temp °F | 15.4 | 19.4 | 28.3 | 39.1 | 49.4 | 59.5 | 64.8 | 62.5 | 52.7 | 40.3 | 27.8 | 19.1 | 39.9 |
| Mean Temp °F | 27.4 | 32.0 | 41.5 | 52.4 | 61.8 | 72.4 | 78.1 | 75.7 | 66.3 | 54.5 | 40.1 | 30.8 | 52.8 |
| Days Max Temp ≥ 90 °F | 0 | 0 | 0 | 0 | 2 | 10 | 19 | 16 | 6 | 1 | 0 | 0 | 54 |
| Days Max Temp ≤ 32 °F | 10 | 7 | 3 | 0 | 0 | 0 | 0 | 0 | 0 | 0 | 2 | 7 | 29 |
| Days Min Temp ≤ 32 °F | 30 | 26 | 21 | 6 | 1 | 0 | 0 | 0 | 0 | 6 | 21 | 30 | 141 |
| Days Min Temp ≤ 0 °F | 4 | 2 | 0 | 0 | 0 | 0 | 0 | 0 | 0 | 0 | 0 | 2 | 8 |
| Heating Degree Days | 1157 | 925 | 723 | 384 | 155 | 24 | 1 | 5 | 92 | 336 | 739 | 1054 | 5595 |
| Cooling Degree Days | 0 | 0 | 1 | 18 | 66 | 272 | 427 | 363 | 158 | 16 | 0 | 0 | 1321 |
| Total Precipitation (") | 0.64 | 0.71 | 1.70 | 2.09 | 3.41 | 2.71 | 3.42 | 2.87 | 1.92 | 1.38 | 1.08 | 0.71 | 22.64 |
| Days ≥ 0.1" Precip | 2 | 2 | 4 | 4 | 6 | 6 | 5 | 5 | 4 | 3 | 2 | 2 | 45 |
| Total Snowfall (") | 5.5 | ***5.1*** | 5.3 | 1.6 | 0.0 | 0.0 | 0.0 | 0.0 | 0.1 | 0.3 | 2.7 | 5.4 | 26.0 |
| Days ≥ 1" Snow Depth | 11 | 8 | 3 | 0 | 0 | 0 | 0 | 0 | 0 | 0 | 2 | 6 | 30 |

## WAMEGO *Pottawatomie County* ELEVATION 1010 ft LAT/LONG 39° 13' N / 96° 19' W

| | JAN | FEB | MAR | APR | MAY | JUN | JUL | AUG | SEP | OCT | NOV | DEC | YEAR |
|---|---|---|---|---|---|---|---|---|---|---|---|---|---|
| Maximum Temp °F | 38.8 | 45.1 | 57.5 | 69.2 | 77.4 | 85.8 | 91.0 | 89.5 | 81.3 | 70.1 | 54.5 | 42.9 | 66.9 |
| Minimum Temp °F | 17.4 | 21.8 | 32.6 | 43.7 | 53.4 | 63.0 | 67.5 | 65.3 | 56.8 | 44.8 | 32.5 | 22.3 | 43.4 |
| Mean Temp °F | 28.1 | 33.5 | 45.0 | 56.5 | 65.4 | 74.4 | 79.3 | 77.4 | 69.1 | 57.6 | 43.5 | 32.6 | 55.2 |
| Days Max Temp ≥ 90 °F | 0 | 0 | 0 | 1 | 1 | 9 | 19 | 16 | 6 | 0 | 0 | 0 | 52 |
| Days Max Temp ≤ 32 °F | 10 | 6 | 1 | 0 | 0 | 0 | 0 | 0 | 0 | 0 | 1 | 5 | 23 |
| Days Min Temp ≤ 32 °F | 28 | 23 | 16 | 4 | 0 | 0 | 0 | 0 | 0 | 3 | 15 | 27 | 116 |
| Days Min Temp ≤ 0 °F | 3 | 2 | 0 | 0 | 0 | 0 | 0 | 0 | 0 | 0 | 0 | 2 | 7 |
| Heating Degree Days | 1137 | 884 | 616 | 281 | 84 | 6 | 0 | 1 | 53 | 256 | 639 | 997 | 4954 |
| Cooling Degree Days | 0 | 0 | 5 | 38 | 98 | 298 | 457 | 401 | 200 | 31 | 1 | 0 | 1529 |
| Total Precipitation (") | 0.91 | 0.88 | 2.34 | 2.83 | 4.43 | 5.26 | 4.40 | 3.53 | 3.96 | 2.73 | 1.89 | 1.37 | 34.53 |
| Days ≥ 0.1" Precip | 3 | 2 | 4 | 5 | 7 | 7 | 6 | 6 | 5 | 5 | 4 | 3 | 57 |
| Total Snowfall (") | 5.7 | 4.5 | 1.4 | 0.4 | 0.0 | 0.0 | 0.0 | 0.0 | 0.0 | 0.0 | 1.2 | 2.9 | 16.1 |
| Days ≥ 1" Snow Depth | ***12*** | 7 | 1 | 0 | 0 | 0 | 0 | 0 | 0 | 0 | 1 | 4 | 25 |

## WASHINGTON *Washington County* ELEVATION 1302 ft LAT/LONG 39° 49' N / 97° 3' W

| | JAN | FEB | MAR | APR | MAY | JUN | JUL | AUG | SEP | OCT | NOV | DEC | YEAR |
|---|---|---|---|---|---|---|---|---|---|---|---|---|---|
| Maximum Temp °F | 37.9 | 44.1 | 56.4 | 68.5 | 77.5 | 87.0 | 91.9 | 89.5 | 81.3 | 70.2 | 53.3 | 41.7 | 66.6 |
| Minimum Temp °F | 15.3 | 19.2 | 30.3 | 41.4 | 51.9 | 61.8 | 66.7 | 64.5 | 55.4 | 43.2 | 30.2 | 20.3 | 41.7 |
| Mean Temp °F | 26.6 | 31.7 | 43.4 | 55.0 | 64.7 | 74.4 | 79.3 | 77.0 | 68.4 | 56.7 | 41.7 | 31.0 | 54.2 |
| Days Max Temp ≥ 90 °F | 0 | 0 | 0 | 1 | 1 | 11 | 19 | 17 | 6 | 0 | 0 | 0 | 55 |
| Days Max Temp ≤ 32 °F | 10 | 6 | 1 | 0 | 0 | 0 | 0 | 0 | 0 | 0 | 1 | 6 | 24 |
| Days Min Temp ≤ 32 °F | 29 | 25 | 19 | 6 | 0 | 0 | 0 | 0 | 0 | 5 | 18 | 28 | 130 |
| Days Min Temp ≤ 0 °F | 4 | 3 | 0 | 0 | 0 | 0 | 0 | 0 | 0 | 0 | 0 | 2 | 9 |
| Heating Degree Days | 1183 | 934 | 665 | 319 | 98 | 7 | 0 | 1 | 62 | 276 | 692 | 1048 | 5285 |
| Cooling Degree Days | 0 | 0 | 3 | 32 | 92 | 296 | 451 | 389 | 186 | 22 | 1 | 0 | 1472 |
| Total Precipitation (") | 0.74 | 0.85 | 2.24 | 2.92 | 4.36 | 4.70 | 4.36 | 3.72 | 3.74 | 2.26 | 1.53 | 1.04 | 32.46 |
| Days ≥ 0.1" Precip | 3 | 2 | 4 | 6 | 8 | 7 | 6 | 6 | 5 | 4 | 3 | 2 | 56 |
| Total Snowfall (") | 6.9 | 6.3 | 2.4 | 0.2 | 0.0 | 0.0 | 0.0 | 0.0 | 0.0 | 0.1 | 1.3 | ***4.3*** | 21.5 |
| Days ≥ 1" Snow Depth | ***12*** | 8 | 2 | 0 | 0 | 0 | 0 | 0 | 0 | 0 | 1 | 5 | 28 |

## WEBSTER DAM *Rooks County* ELEVATION 1860 ft LAT/LONG 39° 25' N / 99° 25' W

| | JAN | FEB | MAR | APR | MAY | JUN | JUL | AUG | SEP | OCT | NOV | DEC | YEAR |
|---|---|---|---|---|---|---|---|---|---|---|---|---|---|
| Maximum Temp °F | 39.0 | 44.0 | 55.2 | 66.8 | 75.6 | 86.7 | 93.0 | 90.2 | 81.3 | 69.5 | 53.1 | 42.6 | 66.4 |
| Minimum Temp °F | 13.0 | 16.7 | 27.0 | 38.3 | 48.6 | 58.8 | 64.3 | 61.6 | 51.6 | 38.6 | 26.4 | 17.3 | 38.5 |
| Mean Temp °F | 26.1 | 30.4 | 41.3 | 52.6 | 62.2 | 72.8 | 78.7 | 76.0 | 66.5 | 54.1 | 39.8 | 30.0 | 52.5 |
| Days Max Temp ≥ 90 °F | 0 | 0 | 0 | 1 | 3 | 12 | 21 | 18 | 8 | 2 | 0 | 0 | 65 |
| Days Max Temp ≤ 32 °F | 10 | 8 | 2 | 0 | 0 | 0 | 0 | 0 | 0 | 0 | 2 | 7 | 29 |
| Days Min Temp ≤ 32 °F | 31 | 27 | 22 | 8 | 1 | 0 | 0 | 0 | 1 | 7 | 23 | 30 | 150 |
| Days Min Temp ≤ 0 °F | 5 | 3 | 0 | 0 | 0 | 0 | 0 | 0 | 0 | 0 | 0 | 2 | 10 |
| Heating Degree Days | 1201 | 970 | 729 | 383 | 150 | 20 | 1 | 5 | 92 | 346 | 751 | 1078 | 5726 |
| Cooling Degree Days | 0 | 0 | 1 | 22 | 66 | 266 | 424 | 349 | 157 | 14 | 0 | 0 | 1299 |
| Total Precipitation (") | 0.51 | 0.65 | 1.69 | 2.21 | 3.71 | 2.69 | 3.59 | 3.25 | 2.03 | 1.57 | 1.00 | 0.66 | 23.56 |
| Days ≥ 0.1" Precip | 1 | 2 | 3 | 4 | 7 | 5 | 6 | 5 | 4 | 3 | 2 | 2 | 44 |
| Total Snowfall (") | 5.1 | 5.6 | 4.9 | 0.6 | 0.0 | 0.0 | 0.0 | 0.0 | 0.1 | 0.1 | 2.0 | *4.1* | 22.5 |
| Days ≥ 1" Snow Depth | na | *5* | 3 | 0 | 0 | 0 | 0 | 0 | 0 | 0 | 1 | na | na |

## WELLINGTON 2 S *Sumner County* ELEVATION 1220 ft LAT/LONG 37° 16' N / 97° 25' W

| | JAN | FEB | MAR | APR | MAY | JUN | JUL | AUG | SEP | OCT | NOV | DEC | YEAR |
|---|---|---|---|---|---|---|---|---|---|---|---|---|---|
| Maximum Temp °F | 41.1 | 47.0 | 58.3 | 68.7 | 76.9 | 87.4 | 93.6 | 92.3 | 83.2 | 71.5 | 56.2 | 45.2 | 68.5 |
| Minimum Temp °F | 19.8 | 23.8 | 33.9 | 44.4 | 53.8 | 63.8 | 68.7 | 67.0 | 58.3 | 45.2 | 33.8 | 23.8 | 44.7 |
| Mean Temp °F | 30.4 | 35.4 | 46.1 | 56.6 | 65.4 | 75.6 | 81.2 | 79.7 | 70.8 | 58.3 | 45.0 | 34.5 | 56.6 |
| Days Max Temp ≥ 90 °F | 0 | 0 | 0 | 0 | 2 | 13 | 23 | 21 | 9 | 1 | 0 | 0 | 69 |
| Days Max Temp ≤ 32 °F | 8 | 5 | 1 | 0 | 0 | 0 | 0 | 0 | 0 | 0 | 1 | 4 | 19 |
| Days Min Temp ≤ 32 °F | 29 | 22 | 14 | 3 | 0 | 0 | 0 | 0 | 0 | 3 | 14 | 26 | 111 |
| Days Min Temp ≤ 0 °F | 2 | 1 | 0 | 0 | 0 | 0 | 0 | 0 | 0 | 0 | 0 | 1 | 4 |
| Heating Degree Days | 1064 | 828 | 582 | 272 | 85 | 6 | 0 | 1 | 46 | 237 | 594 | 938 | 4653 |
| Cooling Degree Days | 0 | 0 | 3 | 24 | 96 | 336 | 514 | 475 | 239 | 30 | 2 | 0 | 1719 |
| Total Precipitation (") | 0.88 | 1.10 | 2.51 | 3.21 | 4.62 | 4.76 | 3.07 | 3.15 | 3.11 | 2.09 | 1.87 | 1.24 | 31.61 |
| Days ≥ 0.1" Precip | 2 | 3 | 5 | 5 | 7 | 6 | 5 | 4 | 5 | 3 | 4 | 3 | 52 |
| Total Snowfall (") | 3.7 | 4.0 | 1.7 | 0.1 | 0.0 | 0.0 | 0.0 | 0.0 | 0.0 | 0.0 | 1.0 | 2.6 | 13.1 |
| Days ≥ 1" Snow Depth | 7 | 4 | 1 | 0 | 0 | 0 | 0 | 0 | 0 | 0 | 0 | 3 | 15 |

## WICHITA MID-CNTNT AP *Sedgwick County* ELEVATION 1321 ft LAT/LONG 37° 39' N / 97° 26' W

| | JAN | FEB | MAR | APR | MAY | JUN | JUL | AUG | SEP | OCT | NOV | DEC | YEAR |
|---|---|---|---|---|---|---|---|---|---|---|---|---|---|
| Maximum Temp °F | 39.9 | 46.3 | 57.6 | 68.0 | 76.3 | 87.2 | 93.2 | 91.1 | 81.7 | 70.1 | 54.9 | 43.9 | 67.5 |
| Minimum Temp °F | 19.9 | 24.0 | 34.0 | 44.4 | 54.1 | 64.4 | 69.5 | 67.5 | 58.9 | 46.2 | 33.6 | 23.9 | 45.0 |
| Mean Temp °F | 29.9 | 35.2 | 45.8 | 56.2 | 65.2 | 75.8 | 81.4 | 79.4 | 70.3 | 58.2 | 44.3 | 33.9 | 56.3 |
| Days Max Temp ≥ 90 °F | 0 | 0 | 0 | 0 | 1 | 12 | 22 | 19 | 7 | 1 | 0 | 0 | 62 |
| Days Max Temp ≤ 32 °F | 9 | 5 | 1 | 0 | 0 | 0 | 0 | 0 | 0 | 0 | 1 | 5 | 21 |
| Days Min Temp ≤ 32 °F | 28 | 22 | 14 | 3 | 0 | 0 | 0 | 0 | 0 | 1 | 15 | 26 | 109 |
| Days Min Temp ≤ 0 °F | 2 | 1 | 0 | 0 | 0 | 0 | 0 | 0 | 0 | 0 | 0 | 1 | 4 |
| Heating Degree Days | 1081 | 836 | 591 | 280 | 87 | 5 | 0 | 0 | 46 | 241 | 617 | 957 | 4741 |
| Cooling Degree Days | 0 | 0 | 3 | 25 | 100 | 348 | 528 | 473 | 236 | 33 | 2 | 0 | 1748 |
| Total Precipitation (") | 0.77 | 0.97 | 2.39 | 2.50 | 3.85 | 4.32 | 3.09 | 2.93 | 3.06 | 2.17 | 1.58 | 1.26 | 28.89 |
| Days ≥ 0.1" Precip | 2 | 3 | 5 | 5 | 7 | 7 | 5 | 5 | 5 | 4 | 3 | 3 | 54 |
| Total Snowfall (") | 4.3 | 4.3 | 1.9 | 0.3 | 0.0 | 0.0 | 0.0 | 0.0 | 0.0 | 0.1 | 1.5 | 3.3 | 15.7 |
| Days ≥ 1" Snow Depth | 7 | 5 | 1 | 0 | 0 | 0 | 0 | 0 | 0 | 0 | 1 | 3 | 17 |

## WILSON LAKE *Russell County* ELEVATION 1512 ft LAT/LONG 38° 58' N / 98° 29' W

| | JAN | FEB | MAR | APR | MAY | JUN | JUL | AUG | SEP | OCT | NOV | DEC | YEAR |
|---|---|---|---|---|---|---|---|---|---|---|---|---|---|
| Maximum Temp °F | 38.4 | 43.8 | 55.0 | 66.8 | 75.5 | 86.4 | 92.8 | 90.3 | 81.3 | 69.7 | 53.6 | 42.7 | 66.4 |
| Minimum Temp °F | 16.1 | 19.8 | 29.9 | 41.1 | 51.5 | 61.4 | 66.8 | 64.3 | 54.6 | 42.6 | 30.2 | 20.8 | 41.6 |
| Mean Temp °F | 27.3 | 31.8 | 42.5 | 54.0 | 63.5 | 73.9 | 79.8 | 77.3 | 68.0 | 56.2 | 41.9 | 31.8 | 54.0 |
| Days Max Temp ≥ 90 °F | 0 | 0 | 0 | 0 | 2 | 12 | 21 | 18 | 8 | 1 | 0 | 0 | 62 |
| Days Max Temp ≤ 32 °F | 10 | 7 | 2 | 0 | 0 | 0 | 0 | 0 | 0 | 0 | 2 | 6 | 27 |
| Days Min Temp ≤ 32 °F | 30 | 25 | 18 | 5 | 0 | 0 | 0 | 0 | 0 | 4 | 18 | 28 | 128 |
| Days Min Temp ≤ 0 °F | 3 | 2 | 0 | 0 | 0 | 0 | 0 | 0 | 0 | 0 | 0 | 2 | 7 |
| Heating Degree Days | 1163 | 929 | 692 | 343 | 122 | 15 | 1 | 3 | 73 | 293 | 686 | 1024 | 5344 |
| Cooling Degree Days | 0 | 0 | 2 | 24 | 83 | 295 | 469 | 395 | 187 | 21 | 1 | 0 | 1477 |
| Total Precipitation (") | 0.46 | 0.61 | 1.83 | 2.22 | 3.64 | 3.47 | 3.21 | 3.44 | 2.45 | 1.95 | 1.05 | 0.66 | 24.99 |
| Days ≥ 0.1" Precip | 1 | 2 | 3 | 5 | 7 | 6 | 5 | 5 | 4 | 3 | 2 | 2 | 45 |
| Total Snowfall (") | 5.1 | 4.9 | 2.1 | 0.5 | 0.0 | 0.0 | 0.0 | 0.0 | 0.0 | 0.0 | 0.9 | *2.8* | 16.3 |
| Days ≥ 1" Snow Depth | 9 | *7* | 1 | 0 | 0 | 0 | 0 | 0 | 0 | 0 | 1 | *4* | 22 |

## WINFIELD NO. 1 *Cowley County* ELEVATION 1201 ft LAT/LONG 37° 15 ' N / 97° 0 ' W

| | JAN | FEB | MAR | APR | MAY | JUN | JUL | AUG | SEP | OCT | NOV | DEC | YEAR |
|---|---|---|---|---|---|---|---|---|---|---|---|---|---|
| Maximum Temp °F | 43.3 | 49.6 | 60.6 | 70.9 | 78.5 | 87.8 | 93.6 | 92.3 | 83.6 | 72.8 | 57.7 | 47.1 | 69.8 |
| Minimum Temp °F | 21.5 | 25.7 | 35.6 | 46.0 | 54.7 | 64.3 | 69.0 | 67.1 | 59.2 | 46.8 | 35.2 | 25.6 | 45.9 |
| Mean Temp °F | 32.4 | 37.7 | 48.1 | 58.5 | 66.6 | 76.1 | 81.3 | 79.7 | 71.4 | 59.8 | 46.4 | 36.4 | 57.9 |
| Days Max Temp ≥ 90 °F | 0 | 0 | 0 | 1 | 1 | 13 | 23 | 21 | 8 | 1 | 0 | 0 | 68 |
| Days Max Temp ≤ 32 °F | 7 | 4 | 1 | 0 | 0 | 0 | 0 | 0 | 0 | 0 | 1 | 3 | 16 |
| Days Min Temp ≤ 32 °F | 27 | 20 | 13 | 3 | 0 | 0 | 0 | 0 | 0 | 2 | 13 | 24 | 102 |
| Days Min Temp ≤ 0 °F | 1 | 1 | 0 | 0 | 0 | 0 | 0 | 0 | 0 | 0 | 0 | 1 | 3 |
| Heating Degree Days | 1004 | 765 | 524 | 227 | 65 | 3 | 0 | 0 | 34 | 202 | 553 | 879 | 4256 |
| Cooling Degree Days | 0 | 0 | 6 | 38 | 123 | 347 | 520 | 481 | 244 | 40 | 3 | 0 | 1802 |
| Total Precipitation (") | 1.08 | 1.42 | 2.50 | 3.22 | 4.90 | 4.66 | 3.30 | 3.48 | 3.56 | 2.73 | 2.42 | 1.62 | 34.89 |
| Days ≥ 0.1" Precip | 3 | 3 | 4 | 5 | 7 | 6 | 5 | 5 | 5 | 4 | 4 | 3 | 54 |
| Total Snowfall (") | *2.8* | 4.0 | 1.2 | 0.0 | 0.0 | 0.0 | 0.0 | 0.0 | 0.0 | 0.0 | 0.9 | 2.0 | 10.9 |
| Days ≥ 1" Snow Depth | *6* | 4 | 1 | 0 | 0 | 0 | 0 | 0 | 0 | 0 | 0 | 2 | 13 |

## WINONA *Logan County* ELEVATION 3323 ft LAT/LONG 39° 4 ' N / 101° 15 ' W

| | JAN | FEB | MAR | APR | MAY | JUN | JUL | AUG | SEP | OCT | NOV | DEC | YEAR |
|---|---|---|---|---|---|---|---|---|---|---|---|---|---|
| Maximum Temp °F | 40.2 | 45.5 | 54.6 | 65.3 | 73.8 | 85.1 | 91.0 | 89.1 | 80.3 | 68.3 | 52.2 | 43.0 | 65.7 |
| Minimum Temp °F | 15.1 | 18.8 | 25.8 | 35.8 | 46.3 | 56.3 | 62.2 | 60.5 | 50.9 | 38.6 | 26.1 | 18.1 | 37.9 |
| Mean Temp °F | 27.7 | 32.2 | 40.3 | 50.6 | 60.1 | 70.8 | 76.6 | 74.8 | 65.6 | 53.5 | 39.2 | 30.6 | 51.8 |
| Days Max Temp ≥ 90 °F | 0 | 0 | 0 | 0 | 1 | 11 | 18 | *17* | 7 | 1 | 0 | 0 | 55 |
| Days Max Temp ≤ 32 °F | 9 | 6 | 3 | 0 | 0 | 0 | 0 | 0 | 0 | 0 | 2 | 7 | 27 |
| Days Min Temp ≤ 32 °F | 31 | 27 | 24 | 10 | 1 | 0 | 0 | 0 | 1 | 7 | 24 | 30 | 155 |
| Days Min Temp ≤ 0 °F | 3 | 2 | 0 | 0 | 0 | 0 | 0 | 0 | 0 | 0 | 0 | 1 | 6 |
| Heating Degree Days | 1150 | 918 | 762 | 434 | 189 | 33 | 3 | 7 | 95 | 362 | 767 | 1061 | 5781 |
| Cooling Degree Days | 0 | 0 | 1 | 11 | 42 | 215 | 366 | 340 | 137 | 9 | 0 | 0 | 1121 |
| Total Precipitation (") | 0.37 | 0.34 | 1.15 | 1.43 | 3.27 | 3.03 | 3.09 | 2.65 | 1.41 | 1.33 | 0.71 | 0.46 | 19.24 |
| Days ≥ 0.1" Precip | *1* | 1 | 3 | 3 | 6 | 6 | 5 | *5* | 3 | 2 | 2 | 1 | 38 |
| Total Snowfall (") | *5.5* | 3.6 | *6.1* | *2.9* | 0.1 | 0.0 | 0.0 | 0.0 | 0.2 | 1.0 | 3.3 | *5.4* | 28.1 |
| Days ≥ 1" Snow Depth | na | na | na | 1 | 0 | 0 | 0 | 0 | 0 | 1 | *1* | na | na |

## YATES CENTER *Woodson County* ELEVATION 1089 ft LAT/LONG 37° 53 ' N / 95° 44 ' W

| | JAN | FEB | MAR | APR | MAY | JUN | JUL | AUG | SEP | OCT | NOV | DEC | YEAR |
|---|---|---|---|---|---|---|---|---|---|---|---|---|---|
| Maximum Temp °F | 41.4 | 47.4 | 58.9 | 69.7 | 77.0 | 85.8 | 91.4 | 90.0 | 81.8 | 71.2 | 56.6 | 45.4 | 68.1 |
| Minimum Temp °F | 20.8 | 25.1 | 34.9 | 45.6 | 54.4 | 63.6 | 67.9 | 65.6 | 58.1 | 47.1 | 35.3 | 25.3 | 45.3 |
| Mean Temp °F | 31.1 | 36.3 | 46.9 | 57.6 | 65.7 | 74.7 | 79.7 | 77.8 | 70.0 | 59.1 | 45.9 | 35.4 | 56.7 |
| Days Max Temp ≥ 90 °F | 0 | 0 | 0 | 0 | 1 | 9 | 19 | 17 | 6 | 1 | 0 | 0 | 53 |
| Days Max Temp ≤ 32 °F | 8 | 4 | 1 | 0 | 0 | 0 | 0 | 0 | 0 | 0 | 1 | 4 | 18 |
| Days Min Temp ≤ 32 °F | 27 | 21 | 13 | 3 | 0 | 0 | 0 | 0 | 0 | 1 | 12 | 24 | 101 |
| Days Min Temp ≤ 0 °F | 1 | 1 | 0 | 0 | 0 | 0 | 0 | 0 | 0 | 0 | 0 | 1 | 3 |
| Heating Degree Days | 1044 | 804 | 558 | 245 | 72 | 4 | 0 | 0 | 42 | 213 | 568 | 911 | 4461 |
| Cooling Degree Days | 0 | 0 | 4 | 31 | 96 | 319 | 466 | 416 | 210 | 31 | 2 | 0 | 1575 |
| Total Precipitation (") | 1.21 | 1.37 | 3.08 | 4.20 | 4.57 | 5.62 | 4.54 | 4.27 | 4.53 | 3.90 | 2.86 | 1.83 | 41.98 |
| Days ≥ 0.1" Precip | 3 | 3 | 6 | 6 | 8 | 7 | 6 | 5 | 6 | 6 | 5 | 4 | 65 |
| Total Snowfall (") | *4.7* | 4.9 | 2.4 | 0.1 | 0.0 | 0.0 | 0.0 | 0.0 | 0.0 | 0.0 | 0.9 | *3.4* | 16.4 |
| Days ≥ 1" Snow Depth | na | na | *0* | 0 | 0 | 0 | 0 | 0 | 0 | 0 | 0 | na | na |

## JANUARY MINIMUM TEMPERATURE °F

### LOWEST

| | | |
|---|---|---|
| 1 | Kirwin Dam | 11.6 |
| 2 | Norton Dam | 11.7 |
| 3 | Lovewell Lake | 12.0 |
| 4 | Colby | 12.7 |
| 5 | Atwood | 12.9 |
| | Mankato | 12.9 |
| 7 | Webster Dam | 13.0 |
| 8 | Glen Elder Lake | 13.3 |
| 9 | Marysville | 13.4 |
| 10 | Brewster | 13.5 |
| 11 | Phillipsburg | 13.6 |
| | Tuttle Creek Lake | 13.6 |
| 13 | Alton | 13.8 |
| | Syracuse | 13.8 |
| 15 | Oberlin | 14.0 |
| 16 | Hays | 14.3 |
| 17 | Russell Springs | 14.4 |
| 18 | Tribune | 14.5 |
| 19 | Garden City-Exp | 14.6 |
| | Norton | 14.6 |
| 21 | Belleville | 14.7 |
| 22 | Quinter | 14.8 |
| | St. Francis | 14.8 |
| 24 | Kanopolis Lake | 15.1 |
| | Leoti | 15.1 |

### HIGHEST

| | | |
|---|---|---|
| 1 | Columbus | 22.6 |
| 2 | Anthony | 22.5 |
| 3 | Norwich | 21.8 |
| 4 | Independence | 21.7 |
| 5 | Girard | 21.6 |
| 6 | Winfield | 21.5 |
| 7 | Parsons | 21.4 |
| 8 | Fort Scott | 21.2 |
| | Iola | 21.2 |
| 10 | Yates Center | 20.8 |
| 11 | Chanute | 20.7 |
| 12 | Fredonia | 20.5 |
| | Mound Valley | 20.5 |
| 14 | Coldwater | 20.4 |
| | Kingman | 20.4 |
| 16 | Liberal | 20.3 |
| 17 | Howard | 20.1 |
| 18 | El Dorado | 20.0 |
| | Eureka | 20.0 |
| 20 | Lawrence | 19.9 |
| | Wichita | 19.9 |
| 22 | Garnett | 19.8 |
| | Hudson | 19.8 |
| | Wellington | 19.8 |
| 25 | Great Bend | 19.7 |

## JULY MAXIMUM TEMPERATURE °F

### HIGHEST

| | | |
|---|---|---|
| 1 | Anthony | 95.7 |
| 2 | Liberal | 95.1 |
| 3 | Medicine Lodge | 95.0 |
| 4 | Lincoln | 94.9 |
| | Meade | 94.9 |
| | Norwich | 94.9 |
| 7 | Ashland | 94.7 |
| | Kingman | 94.7 |
| 9 | Bison | 94.4 |
| 10 | Coldwater | 94.3 |
| 11 | Great Bend | 94.1 |
| | Hudson | 94.1 |
| | Minneapolis | 94.1 |
| 14 | Syracuse | 94.0 |
| 15 | Ness City | 93.9 |
| | Ulysses | 93.9 |
| 17 | Hutchinson | 93.7 |
| | Newton | 93.7 |
| | Pratt | 93.7 |
| | Sharon Springs | 93.7 |
| 21 | Elkhart | 93.6 |
| | Wellington | 93.6 |
| | Winfield | 93.6 |
| 24 | Alton | 93.5 |
| | Kinsley | 93.5 |

### LOWEST

| | | |
|---|---|---|
| 1 | Troy | 88.2 |
| 2 | Olathe | 88.7 |
| 3 | Atchison | 88.9 |
| 4 | Topeka | 89.1 |
| 5 | Pomona Lake | 89.3 |
| 6 | Goodland | 89.5 |
| 7 | Eskridge | 89.6 |
| 8 | Centralia | 89.8 |
| | Colby | 89.8 |
| | Council Grove Lk | 89.8 |
| | Lovewell Lake | 89.8 |
| 12 | Leavenworth | 89.9 |
| 13 | Holton | 90.0 |
| 14 | Belleville | 90.1 |
| | Marysville | 90.1 |
| | Oskaloosa | 90.1 |
| 17 | Tuttle Creek Lake | 90.2 |
| 18 | John Redmond Lk | 90.3 |
| | Mankato | 90.3 |
| | McDonald | 90.3 |
| | Norton Dam | 90.3 |
| 22 | Fall River Lake | 90.4 |
| 23 | Oakley | 90.5 |
| 24 | Chanute | 90.6 |
| | Milford Lake | 90.6 |

## ANNUAL PRECIPITATION (")

### HIGHEST

| | | |
|---|---|---|
| 1 | Girard | 45.13 |
| 2 | Mound Valley | 44.65 |
| 3 | Columbus | 43.24 |
| 4 | Chanute | 42.63 |
| 5 | Fort Scott | 42.60 |
| 6 | Independence | 42.09 |
| 7 | Yates Center | 41.98 |
| 8 | Mound City | 41.76 |
| 9 | Iola | 41.54 |
| 10 | Parsons | 41.26 |
| 11 | Paola | 40.98 |
| 12 | Garnett | 40.31 |
| 13 | Leavenworth | 40.29 |
| 14 | Fredonia | 40.03 |
| 15 | Olathe | 39.66 |
| 16 | Lawrence | 39.65 |
| 17 | Sedan | 39.43 |
| 18 | Ottawa | 39.01 |
| 19 | Oskaloosa | 38.82 |
| 20 | Toronto Lake | 37.92 |
| 21 | Eureka | 37.44 |
| 22 | Troy | 37.18 |
| 23 | Holton | 37.15 |
| 24 | Howard | 37.05 |
| 25 | Atchison | 37.03 |

### LOWEST

| | | |
|---|---|---|
| 1 | Syracuse | 16.32 |
| 2 | Tribune | 16.81 |
| 3 | Ulysses | 17.36 |
| 4 | Leoti | 17.86 |
| 5 | St. Francis | 18.02 |
| 6 | Lakin | 18.08 |
| 7 | Elkhart | 18.42 |
| 8 | Garden City-Exp | 18.60 |
| 9 | Russell Springs | 18.65 |
| 10 | Hugoton | 18.97 |
| 11 | Goodland | 19.03 |
| 12 | Liberal | 19.07 |
| 13 | Sublette | 19.09 |
| 14 | Winona | 19.24 |
| 15 | Sharon Springs | 19.45 |
| 16 | Oakley | 19.55 |
| 17 | Brewster | 19.95 |
| 18 | Garden City-Muni | 19.97 |
| 19 | Colby | 20.12 |
| 20 | Scott City | 20.81 |
| 21 | Healy | 20.85 |
| 22 | Meade | 20.97 |
| 23 | Hoxie | 20.99 |
| 24 | Dodge City | 21.37 |
| 25 | Ness City | 21.47 |

## ANNUAL SNOWFALL (")

### HIGHEST

| | | |
|---|---|---|
| 1 | McDonald | 44.6 |
| 2 | Goodland | 42.2 |
| 3 | St. Francis | 35.4 |
| 4 | Oberlin | 34.2 |
| 5 | Centralia | 31.3 |
| 6 | Colby | 29.6 |
| 7 | Scott City | 28.3 |
| 8 | Winona | 28.1 |
| 9 | Quinter | 26.8 |
| 10 | Healy | 26.7 |
| 11 | Norton | 26.6 |
| 12 | Herington | 26.1 |
| | Sharon Springs | 26.1 |
| 14 | Oakley | 26.0 |
| | Wakeeney | 26.0 |
| 16 | Mankato | 25.5 |
| 17 | Russell Springs | 25.3 |
| 18 | Leoti | 25.0 |
| 19 | Hoxie | 24.7 |
| 20 | Tribune | 24.3 |
| 21 | Russell | 23.9 |
| 22 | Plainville | 23.6 |
| 23 | Bison | 23.1 |
| 24 | Concordia | 22.6 |
| 25 | Webster Dam | 22.5 |

### LOWEST

| | | |
|---|---|---|
| 1 | Girard | 7.2 |
| 2 | Parsons | 9.5 |
| 3 | Ashland | 9.8 |
| 4 | Fredonia | 10.2 |
| 5 | Columbus | 10.9 |
| | Winfield | 10.9 |
| 7 | Norwich | 11.5 |
| | Sedan | 11.5 |
| 9 | El Dorado | 11.7 |
| | Tuttle Creek Lake | 11.7 |
| 11 | Independence | 11.8 |
| 12 | Kanopolis Lake | 12.1 |
| 13 | Anthony | 12.3 |
| | Newton | 12.3 |
| 15 | Chanute | 12.9 |
| 16 | Milford Lake | 13.1 |
| | Wellington | 13.1 |
| 18 | Howard | 13.4 |
| 19 | Lakin | 13.9 |
| 20 | Mound City | 14.4 |
| 21 | Kingman | 14.6 |
| | Medicine Lodge | 14.6 |
| 23 | Council Grove Lk | 14.7 |
| 24 | Hutchinson | 15.1 |
| 25 | Cottonwood Falls | 15.5 |

# KENTUCKY

PHYSICAL FEATURES. Kentucky has a land surface of 40,109 square miles. It is essentially an eroded plateau that slopes downward gradually to the southwest, with elevations ranging from about 400 feet above sea level at the western edge to 1,000 feet in the central districts, to above 4,000 feet near the southeastern border. There are seven major physiographic or natural regions.

The Bluegrass Region comprises about one-fifth of the State. The central area of this region is undulating to gently rolling. The outer area is more rolling and less uniform. Separating the two areas is a terrain that is hilly, with winding ridges and valleys and steep slopes.

The Knobs Region, named for its conical and flat-topped hills, comprises about one-tenth of the State. It forms a narrow crescent encircling the Bluegrass on the east, south, and west. Towards the Bluegrass the terrain is flat to rolling with scattered knobs and wide valleys, while the outer margin is rough.

The Eastern Mountains, also called the Cumberland Plateau, extends over the entire eastern fourth of the State. Ridges are high and sharp-crested; there is little level land, and valleys are narrow. In the southeast the Pine and Cumberland Mountains comprise the highest and most rugged part of Kentucky.

The Pennyroyal Region or the Mississippean Plateaus Region is one of the three largest regions. Much of the surface is quite uniform, but as a whole is rather diverse. Much of the terrain is undulating to rolling. In some places it is hilly or cavernous. Subsurface drainage has created limestone sinks and karst terrain in much of the area.

The Western Coal Field is a small region. This area has extensive bottom lands in the valleys of the Ohio, Green, and Tradewater Rivers and many of their tributaries. There is also some undulating to gently rolling uplands.

The Cumberland-Tennessee Rivers Area is the smallest region. The topography is hilly and rough, except for the wide bottoms along the two major streams.

The Jackson Purchase is the extreme western area of Kentucky. In both elevation and relief it is lower than the other regions of the State, but it also has a varied surface. It is largely an upland plain which is mostly undulating to gently rolling, but is also level in places and hilly in others.

GENERAL CLIMATE. The climate of Kentucky is essentially continental in character, with rather wide extremes of temperature and precipitation. The State lies within the path of storms, in the belt of the westerly winds. The temperature generally varies as the storms move across the State. Thus in winter and summer, there are occasional cold and hot spells of short duration. In the spring and fall, the systems have a smaller frequency, temperatures are more consistent, and fewer extremes are experienced. Precipitation occurs with the systems which generally move from the west to east, or from summer thunderstorms. However, the greater portion of precipitation is due to the moisture-bearing low-pressure formations which move from southwest to northeast from the western Gulf of Mexico and frequently cross Kentucky. With warm moist tropical air predominating during the summer months, relative humidity remains consistently high during that season.

TEMPERATURE. The mean annual temperature ranges from 54° F. in the extreme north to 59° F. in the southwestern counties. July is usually the warmest month and January, the coldest. Extreme summer temperatures nearly always reach 100° or higher at most locations, but the frequencies of these high temperatures are low. Minimum temperatures of 0° or below can be expected during the months of December, January, and February at most locations, but the number of days with such temperatures is relatively small. Because of the State's geographic locations with reference to the center of the continent, the mid-winter cold waves from the Canadian Northwest usually have their intensity considerably modified by the time they reach Kentucky. In summer when the high pressure off the Florida coast is displaced westward from its normal position, extended periods of hot, sultry weather will occur. The spring and fall months are usually pleasant.

PRECIPITATION. Precipitation is generally plentiful. The fall season is generally the driest and the spring season the wettest. Approximately half of the average annual total occurs during the warm months of April to September. The average annual total in the State ranges from 36 inches in northern counties to 50 inches in the southern. Thunderstorms with high intensity rainfall are common during the spring and summer months, and rainfall during these storms in a 24-hour period frequently exceeds 2 to 3 inches, occasionally reaching 5 to 6 inches. Flash floods frequently result from the high intensity showers. Snowfall occurrence also varies from year to year but is common from November through March. Some snow has also been reported in the months of October and April. In some sections, the ground seldom remains covered with snow for more than a few days. The average annual snowfall for the State ranges from 6 to 10 inches in the southwest to 15 to 20 inches in the southeast.

WINDS. Winds in the State have an average velocity of 7 to 12 m.p.h., and the prevailing direction is from south to southwest for the year. During the fall season some areas show a prevailing direction having a northerly component. The highest wind speeds usually range from 50 to 70 m.p.h., but in some storms (generally squalls attending thunderstorms), winds in gusts may occasionally exceed these speeds. A number of years may pass without a tornado, or several may visit the State in a single year. On the average, about 1 per year occurs somewhere in the State.

Thunderstorms may occur in any month, but they occur most frequently during the months March through September. The mean number of days with thunderstorms ranges approximately between 45 and 60. They are occasionally attended by damaging hail, but the area thus affected is nearly always small.

OTHER CLIMATIC ELEMENTS. Heavy fogs are rather rare in the State. The average number of days with heavy fogs varies between 8 and 17 during the year with the majority occurring during the months of September through March inclusive.

The average date of the last spring freeze ranges from April 4 in the extreme west to May 5 in the mountain region in the extreme southeast; that of the first fall freeze, from October 11 in the Pennyroyal Region to October 30 near the lower Ohio River. The average length of the freeze-free period varies from 166 days on the southeastern plateau to 210 near the lower Ohio River.

The average number of days with clear and with partly cloudy skies is about the same and ranges between 115 and 120 days over the State. The number of days with cloudy skies averages about 130. The extreme northern section shows the greatest number of days with cloudy skies. The percentage of possible sunshine averages 35 to 50 for the winter months, 50 to 65 in the spring, 65 to 75 in the summer, and 55 to 65 in the fall. The largest percentage of possible sunshine is recorded in the extreme western section of the State.

The Ohio River forms the northern boundary and the Mississippi the western. All of the State is in the Ohio River Basin, except for a small section in the Jackson Purchase area that drains directly into the Mississippi. Kentucky lies in the path of rain producing lows moving from the west Gulf area northeastward. The flood season is in the winter and spring. Numerous flash floods occur from excessive rains and thunderstorms, particularly in the mountains of the eastern portion.

## COUNTY INDEX

***Allen County***
BARREN RIVER LAKE
SCOTTSVILLE 3 SSW

***Ballard County***
LOVELACEVILLE

***Barren County***
GLASGOW

***Bell County***
MIDDLESBORO 2 N

***Boone County***
CINCI-NORTHERN KY AP

***Boyd County***
ASHLAND

***Boyle County***
DANVILLE

***Breckinridge County***
ROUGH RIVER DAM

***Caldwell County***
PRINCETON 1 SE

***Calloway County***
MURRAY

***Carlisle County***
BARDWELL 2 E

***Carroll County***
CARROLLTON LOCK 1

***Carter County***
GRAYSON 3 SW

***Christian County***
HOPKINSVILLE

***Clay County***
MANCHESTER 4 W

***Daviess County***
OWENSBORO 3 W

***Edmonson County***
MAMMOTH CAVE
NOLIN RIVER LAKE

***Fayette County***
LEXINGTON BLUEGRASS

***Franklin County***
FRANKFORT LOCK 4

***Gallatin County***
WARSAW MARKLAND DAM

***Grant County***
WILLIAMSTOWN 3 NW

***Grayson County***
LEITCHFIELD 2 N

***Green County***
GREENSBURG

***Harlan County***
BAXTER

***Henderson County***
HENDERSON 7 SSW

***Hopkins County***
MADISONVILLE 1 SE

***Jefferson County***
LOUISVILLE STANDIFRD

***Knox County***
BARBOURVILLE

***Larue County***
HODGENVILLE-LINCOLN

***Laurel County***
LONDON-CORBIN AP

***Lee County***
HEIDELBERG

***McCracken County***
PADUCAH BARKLEY FLD
PADUCAH WALKER

***Madison County***
BEREA COLLEGE

***Marion County***
BRADFORDSVILLE

***Mason County***
MAYSVILLE SEWAGE PLA

***Mercer County***
DIX DAM

***Metcalfe County***
SUMMER SHADE

***Montgomery County***
MOUNT STERLING 1 NW

***Morgan County***
WEST LIBERTY

***Nelson County***
BARDSTOWN 5 E

***Ohio County***
BEAVER DAM
ROCHESTER FERRY

***Pendleton County***
FALMOUTH

***Pulaski County***
SOMERSET 2 N

***Rockcastle County***
MOUNT VERNON

***Rowan County***
FARMERS 2 S

***Shelby County***
SHELBYVILLE 1 E

***Trigg County***
GOLDEN POND 8 N

### *Warren County*
BOWLING GREEN WARREN

### *Wayne County*
MONTICELLO 3 NE

## ELEVATION INDEX

| FEET | STATION NAME |
|---|---|
| 331 | BARDWELL 2 E |
| 340 | PADUCAH WALKER |
| 381 | HENDERSON 7 SSW |
| 390 | GOLDEN POND 8 N |
| 397 | PADUCAH BARKLEY FLD |
| 400 | ROCHESTER FERRY |
| 410 | MURRAY |
| 420 | OWENSBORO 3 W |
| 440 | BEAVER DAM |
| 440 | MADISONVILLE 1 SE |
| 466 | WARSAW MARKLAND DAM |
| 479 | CARROLLTON LOCK 1 |
| 479 | LOVELACEVILLE |
| 485 | LOUISVILLE STANDIFRD |
| 502 | FRANKFORT LOCK 4 |
| 502 | PRINCETON 1 SE |
| 522 | HOPKINSVILLE |
| 522 | MAYSVILLE SEWAGE PLA |
| 538 | BOWLING GREEN WARREN |
| 551 | ASHLAND |
| 551 | ROUGH RIVER DAM |
| 581 | GREENSBURG |
| 591 | BARREN RIVER LAKE |
| 600 | FALMOUTH |
| 610 | NOLIN RIVER LAKE |
| 640 | LEITCHFIELD 2 N |
| 659 | BRADFORDSVILLE |
| 659 | HEIDELBERG |
| 669 | FARMERS 2 S |
| 700 | GRAYSON 3 SW |
| 728 | SHELBYVILLE 1 E |
| 771 | SCOTTSVILLE 3 SSW |
| 780 | BARDSTOWN 5 E |
| 788 | HODGENVILLE-LINCOLN |
| 801 | MAMMOTH CAVE |
| 801 | WEST LIBERTY |
| 810 | GLASGOW |
| 860 | SUMMER SHADE |
| 870 | MANCHESTER 4 W |

| FEET | STATION NAME |
|---|---|
| 889 | CINCI-NORTHERN KY AP |
| 914 | WILLIAMSTOWN 3 NW |
| 922 | DIX DAM |
| 932 | MOUNT STERLING 1 NW |
| 961 | DANVILLE |
| 979 | MONTICELLO 3 NE |
| 981 | LEXINGTON BLUEGRASS |
| 1001 | BARBOURVILLE |
| 1020 | SOMERSET 2 N |
| 1070 | BEREA COLLEGE |
| 1128 | MIDDLESBORO 2 N |
| 1142 | MOUNT VERNON |
| 1161 | BAXTER |
| 1191 | LONDON-CORBIN AP |

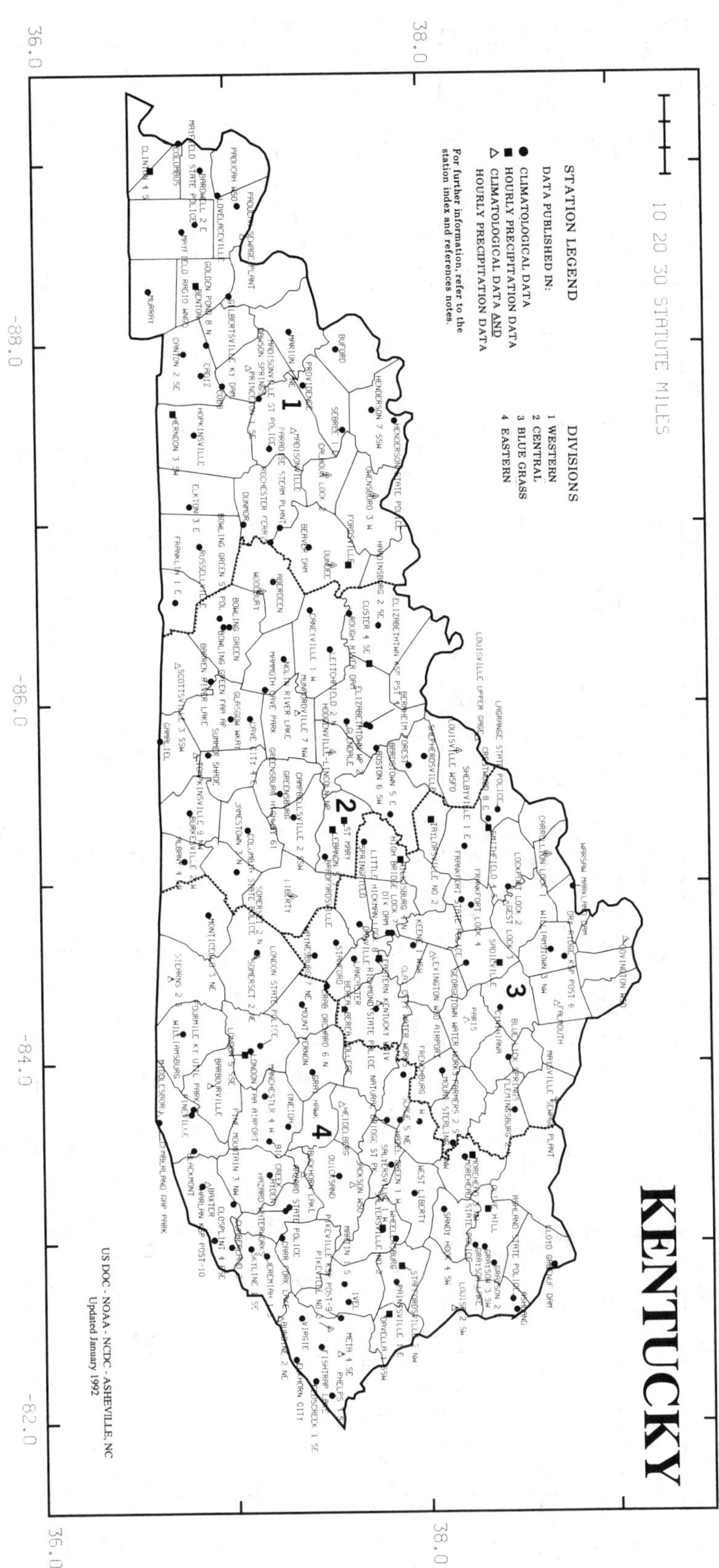
KENTUCKY
10 20 30 STATUTE MILES
STATION LEGEND
DATA PUBLISHED IN:
● CLIMATOLOGICAL DATA
■ HOURLY PRECIPITATION DATA
△ CLIMATOLOGICAL DATA AND HOURLY PRECIPITATION DATA
For further information, refer to the station index and references notes.
DIVISIONS
1 WESTERN
2 CENTRAL
3 BLUE GRASS
4 EASTERN
36.0
38.0
-88.0
-86.0
-84.0
-82.0
US DOC - NOAA - NCDC - ASHEVILLE, NC
Updated January 1992

## ASHLAND *Boyd County* ELEVATION 551 ft LAT/LONG 38° 27 ' N / 82° 36 ' W

| | JAN | FEB | MAR | APR | MAY | JUN | JUL | AUG | SEP | OCT | NOV | DEC | YEAR |
|---|---|---|---|---|---|---|---|---|---|---|---|---|---|
| Maximum Temp °F | 41.2 | 45.3 | 56.6 | 67.9 | 76.6 | 84.2 | 87.6 | 86.0 | 79.8 | 68.9 | 57.1 | 46.2 | 66.5 |
| Minimum Temp °F | 18.2 | 19.9 | 28.7 | 37.3 | 46.5 | 55.9 | 61.0 | 59.1 | 52.6 | 40.3 | 31.0 | 23.5 | 39.5 |
| Mean Temp °F | 29.5 | 32.7 | 42.6 | 52.6 | 61.6 | 70.1 | 74.3 | 72.6 | 66.2 | 54.7 | 44.1 | 34.9 | 53.0 |
| Days Max Temp ≥ 90 °F | 0 | 0 | 0 | 0 | 1 | 7 | 12 | 9 | 3 | 0 | 0 | 0 | 32 |
| Days Max Temp ≤ 32 °F | 8 | 5 | 1 | 0 | 0 | 0 | 0 | 0 | 0 | 0 | 0 | 4 | 18 |
| Days Min Temp ≤ 32 °F | 27 | 25 | 21 | 10 | 2 | 0 | 0 | 0 | 0 | 7 | 18 | 25 | 135 |
| Days Min Temp ≤ 0 °F | 3 | 2 | 0 | 0 | 0 | 0 | 0 | 0 | 0 | 0 | 0 | 1 | 6 |
| Heating Degree Days | 1093 | 907 | 687 | 376 | 155 | 26 | 2 | 6 | 72 | 325 | 621 | 928 | 5198 |
| Cooling Degree Days | 0 | 0 | 1 | 10 | 57 | 182 | 312 | 244 | 110 | 12 | 0 | 0 | 928 |
| Total Precipitation (") | 2.83 | 2.88 | 3.71 | 3.57 | 4.28 | 3.60 | 4.66 | 3.89 | 2.81 | 3.06 | 3.41 | 3.64 | 42.34 |
| Days ≥ 0.1" Precip | *7* | 7 | 9 | 8 | 9 | 7 | 8 | 7 | 5 | 6 | 8 | 7 | 88 |
| Total Snowfall (") | na | na | na | 0.0 | 0.0 | 0.0 | 0.0 | 0.0 | 0.0 | 0.0 | *0.2* | *1.4* | na |
| Days ≥ 1" Snow Depth | *7* | *6* | *1* | 0 | 0 | 0 | 0 | 0 | 0 | 0 | *0* | *2* | 16 |

## BARBOURVILLE *Knox County* ELEVATION 1001 ft LAT/LONG 36° 52 ' N / 83° 53 ' W

| | JAN | FEB | MAR | APR | MAY | JUN | JUL | AUG | SEP | OCT | NOV | DEC | YEAR |
|---|---|---|---|---|---|---|---|---|---|---|---|---|---|
| Maximum Temp °F | 45.3 | 49.4 | 59.7 | 70.0 | 77.1 | 84.2 | 87.4 | 85.8 | 80.3 | 70.3 | 59.8 | 49.4 | 68.2 |
| Minimum Temp °F | 21.9 | 24.8 | 32.7 | 40.9 | 49.9 | 59.0 | 63.8 | 62.8 | 56.1 | 42.4 | 34.0 | 27.0 | 42.9 |
| Mean Temp °F | 33.7 | 37.1 | 46.2 | 55.5 | 63.5 | 71.6 | 75.6 | 74.4 | 68.1 | 56.3 | 46.9 | 38.3 | 55.6 |
| Days Max Temp ≥ 90 °F | 0 | 0 | 0 | 0 | 0 | 5 | 11 | 7 | 2 | 0 | 0 | 0 | 25 |
| Days Max Temp ≤ 32 °F | 4 | 2 | 0 | 0 | 0 | 0 | 0 | 0 | 0 | 0 | 0 | 2 | 8 |
| Days Min Temp ≤ 32 °F | 24 | 21 | 16 | 7 | 0 | 0 | 0 | 0 | 0 | 5 | 15 | 22 | 110 |
| Days Min Temp ≤ 0 °F | 2 | 1 | 0 | 0 | 0 | 0 | 0 | 0 | 0 | 0 | 0 | 0 | 3 |
| Heating Degree Days | 964 | 781 | 578 | 296 | 109 | 12 | 0 | 3 | 43 | 286 | 538 | 821 | 4431 |
| Cooling Degree Days | 0 | 0 | 2 | 13 | 76 | 229 | 353 | 298 | 135 | 16 | 2 | 0 | 1124 |
| Total Precipitation (") | 3.84 | 3.84 | 4.84 | 4.12 | 5.02 | 4.23 | 4.94 | 4.05 | 3.81 | 3.13 | 4.15 | 4.42 | 50.39 |
| Days ≥ 0.1" Precip | 8 | 7 | 9 | 8 | 8 | 8 | 8 | 7 | 6 | 6 | 7 | 7 | 89 |
| Total Snowfall (") | na | na | na | 0.4 | 0.0 | 0.0 | 0.0 | 0.0 | 0.0 | 0.0 | na | na | na |
| Days ≥ 1" Snow Depth | na | na | *0* | 0 | 0 | 0 | 0 | 0 | 0 | 0 | na | na | na |

## BARDSTOWN 5 E *Nelson County* ELEVATION 780 ft LAT/LONG 37° 49 ' N / 85° 23 ' W

| | JAN | FEB | MAR | APR | MAY | JUN | JUL | AUG | SEP | OCT | NOV | DEC | YEAR |
|---|---|---|---|---|---|---|---|---|---|---|---|---|---|
| Maximum Temp °F | *42.1* | 47.6 | 58.0 | 68.9 | 76.7 | 84.6 | 88.1 | 86.8 | 80.7 | 70.3 | 57.3 | 47.6 | 67.4 |
| Minimum Temp °F | *23.3* | 26.2 | 35.5 | 44.5 | 52.5 | 61.0 | 65.2 | 63.2 | 56.7 | 44.9 | 36.4 | 29.0 | 44.9 |
| Mean Temp °F | *32.7* | 36.9 | 46.8 | 56.7 | 64.6 | 72.8 | 76.7 | 75.0 | 68.7 | 57.6 | 46.9 | 38.3 | 56.1 |
| Days Max Temp ≥ 90 °F | 0 | 0 | 0 | 0 | 1 | 6 | 13 | 10 | 4 | 0 | 0 | 0 | 34 |
| Days Max Temp ≤ 32 °F | 7 | 4 | 0 | 0 | 0 | 0 | 0 | 0 | 0 | 0 | 0 | 3 | 14 |
| Days Min Temp ≤ 32 °F | 23 | 20 | 14 | 4 | 0 | 0 | 0 | 0 | 0 | 4 | 12 | 20 | 97 |
| Days Min Temp ≤ 0 °F | 2 | 1 | 0 | 0 | 0 | 0 | 0 | 0 | 0 | 0 | 0 | 0 | 3 |
| Heating Degree Days | *992* | 786 | 564 | 268 | 98 | 8 | 0 | 2 | 48 | 251 | 540 | 819 | 4376 |
| Cooling Degree Days | na | *0* | *4* | *26* | *85* | *249* | *393* | *330* | *162* | *30* | *2* | *1* | na |
| Total Precipitation (") | *3.48* | 3.81 | 4.63 | 5.02 | 5.22 | 4.27 | 5.21 | 3.65 | 3.73 | 2.97 | 3.76 | 4.71 | 50.46 |
| Days ≥ 0.1" Precip | *6* | 6 | *9* | 8 | 8 | 7 | 7 | 6 | 6 | 6 | *7* | 7 | 83 |
| Total Snowfall (") | na | na | na | *0.0* | 0.0 | 0.0 | 0.0 | 0.0 | 0.0 | 0.0 | *0.7* | na | na |
| Days ≥ 1" Snow Depth | na | na | na | 0 | 0 | 0 | 0 | 0 | 0 | 0 | *0* | na | na |

## BARDWELL 2 E *Carlisle County* ELEVATION 331 ft LAT/LONG 36° 52 ' N / 88° 57 ' W

| | JAN | FEB | MAR | APR | MAY | JUN | JUL | AUG | SEP | OCT | NOV | DEC | YEAR |
|---|---|---|---|---|---|---|---|---|---|---|---|---|---|
| Maximum Temp °F | 42.9 | 48.1 | 59.0 | 70.3 | 78.4 | 86.9 | 90.3 | 88.3 | 81.7 | 71.4 | 58.6 | 47.8 | 68.6 |
| Minimum Temp °F | 24.3 | 28.0 | 37.4 | 47.0 | 55.0 | 63.3 | 67.3 | 64.4 | 57.7 | 45.7 | 37.8 | 29.1 | 46.4 |
| Mean Temp °F | 33.6 | 38.0 | 48.2 | 58.7 | 66.8 | 75.1 | 78.8 | 76.4 | 69.7 | 58.5 | 48.2 | 38.5 | 57.5 |
| Days Max Temp ≥ 90 °F | 0 | 0 | 0 | 0 | 2 | 11 | 19 | 13 | 4 | 0 | 0 | 0 | 49 |
| Days Max Temp ≤ 32 °F | 7 | 3 | 0 | 0 | 0 | 0 | 0 | 0 | 0 | 0 | 0 | 2 | 12 |
| Days Min Temp ≤ 32 °F | 24 | 18 | 11 | 2 | 0 | 0 | 0 | 0 | 0 | 3 | 10 | 20 | 88 |
| Days Min Temp ≤ 0 °F | 1 | 0 | 0 | 0 | 0 | 0 | 0 | 0 | 0 | 0 | 0 | 0 | 1 |
| Heating Degree Days | 967 | 754 | 518 | 217 | 61 | 2 | 0 | 1 | 37 | 228 | 500 | 815 | 4100 |
| Cooling Degree Days | 0 | 0 | 7 | 33 | 122 | 321 | 460 | 386 | 193 | 32 | 3 | 1 | 1558 |
| Total Precipitation (") | 3.48 | 4.03 | 4.71 | 5.36 | 4.88 | 3.82 | 4.41 | 3.78 | 3.71 | 3.44 | 4.77 | 5.02 | 51.41 |
| Days ≥ 0.1" Precip | 6 | 6 | 8 | 8 | 8 | 6 | 6 | 5 | 5 | 5 | 7 | 7 | 77 |
| Total Snowfall (") | 4.8 | 4.2 | 2.3 | 0.0 | 0.0 | 0.0 | 0.0 | 0.0 | 0.0 | 0.1 | 0.3 | 1.9 | 13.6 |
| Days ≥ 1" Snow Depth | 7 | 4 | 1 | 0 | 0 | 0 | 0 | 0 | 0 | 0 | 0 | 2 | 14 |

## BARREN RIVER LAKE *Allen County* ELEVATION 591 ft LAT/LONG 36° 54' N / 86° 8' W

| | JAN | FEB | MAR | APR | MAY | JUN | JUL | AUG | SEP | OCT | NOV | DEC | YEAR |
|---|---|---|---|---|---|---|---|---|---|---|---|---|---|
| Maximum Temp °F | 42.7 | 47.8 | 58.6 | 69.7 | 77.6 | 85.8 | 89.6 | 88.0 | 81.7 | 70.7 | 58.9 | 48.5 | 68.3 |
| Minimum Temp °F | 22.9 | 25.4 | 34.7 | 44.3 | 52.7 | 61.5 | 66.1 | 64.2 | 58.0 | 45.0 | 36.8 | 28.4 | 45.0 |
| Mean Temp °F | 32.8 | 36.6 | 46.7 | 57.0 | 65.2 | 73.7 | 77.9 | 76.1 | 69.8 | 57.9 | 47.8 | 38.5 | 56.7 |
| Days Max Temp ≥ 90 °F | 0 | 0 | 0 | 0 | 2 | 9 | 17 | 13 | 5 | 0 | 0 | 0 | 46 |
| Days Max Temp ≤ 32 °F | 7 | 3 | 0 | 0 | 0 | 0 | 0 | 0 | 0 | 0 | 0 | 3 | 13 |
| Days Min Temp ≤ 32 °F | 24 | 21 | 14 | 4 | 0 | 0 | 0 | 0 | 0 | 4 | 12 | 20 | 99 |
| Days Min Temp ≤ 0 °F | 2 | 1 | 0 | 0 | 0 | 0 | 0 | 0 | 0 | 0 | 0 | 0 | 3 |
| Heating Degree Days | 990 | 796 | 567 | 266 | 93 | 9 | 0 | 1 | 40 | 248 | 512 | 816 | 4338 |
| Cooling Degree Days | 0 | 0 | 6 | 35 | 112 | 288 | 433 | 367 | 192 | 36 | 4 | 1 | 1474 |
| Total Precipitation (") | 3.56 | 3.84 | 4.50 | 4.23 | 5.14 | 4.34 | 4.89 | 4.33 | 3.87 | 3.24 | 4.26 | 4.80 | 51.00 |
| Days ≥ 0.1" Precip | 7 | 7 | 8 | 8 | 8 | 7 | 7 | 6 | 6 | 6 | 7 | 7 | 84 |
| Total Snowfall (") | 3.9 | 2.8 | 1.4 | 0.0 | 0.0 | 0.0 | 0.0 | 0.0 | 0.0 | 0.0 | 0.6 | 1.0 | 9.7 |
| Days ≥ 1" Snow Depth | 6 | 4 | 1 | 0 | 0 | 0 | 0 | 0 | 0 | 0 | 0 | 1 | 12 |

## BAXTER *Harlan County* ELEVATION 1161 ft LAT/LONG 36° 51' N / 83° 20' W

| | JAN | FEB | MAR | APR | MAY | JUN | JUL | AUG | SEP | OCT | NOV | DEC | YEAR |
|---|---|---|---|---|---|---|---|---|---|---|---|---|---|
| Maximum Temp °F | 43.6 | 47.9 | 57.9 | 68.6 | 75.9 | 82.8 | 86.2 | 84.8 | 78.8 | 68.3 | 58.0 | 48.0 | 66.7 |
| Minimum Temp °F | 23.0 | 24.8 | 32.3 | 40.3 | 49.8 | 58.4 | 63.3 | 62.3 | 56.0 | 42.9 | 33.7 | 27.3 | 42.8 |
| Mean Temp °F | 33.3 | 36.4 | 45.1 | 54.5 | 62.9 | 70.6 | 74.7 | 73.6 | 67.4 | 55.6 | 45.9 | 37.7 | 54.8 |
| Days Max Temp ≥ 90 °F | 0 | 0 | 0 | 0 | 0 | 3 | 8 | 6 | 1 | 0 | 0 | 0 | 18 |
| Days Max Temp ≤ 32 °F | 5 | 3 | 0 | 0 | 0 | 0 | 0 | 0 | 0 | 0 | 0 | 2 | 10 |
| Days Min Temp ≤ 32 °F | 25 | 21 | 17 | 7 | 1 | 0 | 0 | 0 | 0 | 5 | 15 | 22 | 113 |
| Days Min Temp ≤ 0 °F | 2 | 1 | 0 | 0 | 0 | 0 | 0 | 0 | 0 | 0 | 0 | 1 | 4 |
| Heating Degree Days | 975 | 801 | 611 | 320 | 120 | 15 | 1 | 3 | 48 | 296 | 567 | 840 | 4597 |
| Cooling Degree Days | 0 | 0 | 1 | 9 | 62 | 197 | 324 | 274 | 120 | 12 | 0 | 0 | 999 |
| Total Precipitation (") | 4.08 | 3.98 | 4.77 | 4.08 | 4.76 | 4.50 | 5.04 | 4.34 | 3.30 | 3.25 | 3.93 | 4.41 | 50.44 |
| Days ≥ 0.1" Precip | 8 | 8 | 9 | 8 | 9 | 8 | 9 | 8 | 6 | 6 | 8 | 8 | 95 |
| Total Snowfall (") | na | na | 1.6 | 0.7 | 0.0 | 0.0 | 0.0 | 0.0 | 0.0 | 0.0 | *0.7* | na | na |
| Days ≥ 1" Snow Depth | *4* | 4 | 1 | 0 | 0 | 0 | 0 | 0 | 0 | 0 | 0 | 1 | 10 |

## BEAVER DAM *Ohio County* ELEVATION 440 ft LAT/LONG 37° 25' N / 86° 52' W

| | JAN | FEB | MAR | APR | MAY | JUN | JUL | AUG | SEP | OCT | NOV | DEC | YEAR |
|---|---|---|---|---|---|---|---|---|---|---|---|---|---|
| Maximum Temp °F | 42.9 | 48.6 | 59.7 | 70.7 | 78.0 | 85.6 | 88.7 | 87.2 | 81.4 | 71.0 | 58.7 | 48.1 | 68.4 |
| Minimum Temp °F | 24.0 | 27.3 | 36.5 | 45.7 | 53.8 | 62.5 | 66.5 | 64.5 | 58.0 | 45.5 | 37.7 | 29.3 | 45.9 |
| Mean Temp °F | 33.5 | 38.0 | 48.1 | 58.2 | 65.9 | 74.1 | 77.6 | 75.9 | 69.7 | 58.3 | 48.2 | 38.7 | 57.2 |
| Days Max Temp ≥ 90 °F | 0 | 0 | 0 | 0 | 1 | 7 | 15 | 11 | 4 | 0 | 0 | 0 | 38 |
| Days Max Temp ≤ 32 °F | 7 | 3 | 0 | 0 | 0 | 0 | 0 | 0 | 0 | 0 | 0 | 3 | 13 |
| Days Min Temp ≤ 32 °F | 23 | 19 | 13 | 3 | 0 | 0 | 0 | 0 | 0 | 4 | 10 | 19 | 91 |
| Days Min Temp ≤ 0 °F | 2 | 1 | 0 | 0 | 0 | 0 | 0 | 0 | 0 | 0 | 0 | 0 | 3 |
| Heating Degree Days | 971 | 757 | 524 | 232 | 74 | 5 | 0 | 1 | 37 | 235 | 500 | 809 | 4145 |
| Cooling Degree Days | 0 | 0 | 8 | 35 | 110 | 293 | 422 | 363 | 191 | 38 | 4 | 1 | 1465 |
| Total Precipitation (") | 3.37 | 4.17 | 4.22 | 4.44 | 4.76 | 3.49 | 4.47 | 3.29 | 3.93 | 3.10 | 4.31 | 4.48 | 48.03 |
| Days ≥ 0.1" Precip | 6 | 6 | 8 | 7 | 8 | 6 | 7 | 5 | 5 | 5 | 7 | 7 | 77 |
| Total Snowfall (") | na | na | na | 0.0 | 0.0 | 0.0 | 0.0 | 0.0 | 0.0 | 0.1 | 0.3 | *0.9* | na |
| Days ≥ 1" Snow Depth | na | na | 0 | 0 | 0 | 0 | 0 | 0 | 0 | 0 | 0 | *1* | na |

## BEREA COLLEGE *Madison County* ELEVATION 1070 ft LAT/LONG 37° 34' N / 84° 18' W

| | JAN | FEB | MAR | APR | MAY | JUN | JUL | AUG | SEP | OCT | NOV | DEC | YEAR |
|---|---|---|---|---|---|---|---|---|---|---|---|---|---|
| Maximum Temp °F | 43.0 | 48.1 | 58.7 | 69.2 | 76.6 | 83.6 | 87.0 | 85.6 | 79.6 | 68.8 | 57.6 | 48.0 | 67.2 |
| Minimum Temp °F | 25.3 | 28.2 | 37.0 | 46.0 | 54.4 | 62.2 | 66.0 | 64.6 | 58.8 | 47.5 | 39.2 | 30.3 | 46.6 |
| Mean Temp °F | 34.2 | 38.2 | 47.9 | 57.6 | 65.5 | 72.9 | 76.5 | 75.1 | 69.2 | 58.1 | 48.4 | 39.2 | 56.9 |
| Days Max Temp ≥ 90 °F | 0 | 0 | 0 | 0 | 0 | 3 | 9 | 7 | 2 | 0 | 0 | 0 | 21 |
| Days Max Temp ≤ 32 °F | 6 | 4 | 0 | 0 | 0 | 0 | 0 | 0 | 0 | 0 | 0 | 2 | 12 |
| Days Min Temp ≤ 32 °F | 22 | 18 | 12 | 3 | 0 | 0 | 0 | 0 | 0 | 2 | 9 | 18 | 84 |
| Days Min Temp ≤ 0 °F | 1 | 0 | 0 | 0 | 0 | 0 | 0 | 0 | 0 | 0 | 0 | 0 | 1 |
| Heating Degree Days | 949 | 752 | 531 | 249 | 84 | 8 | 0 | 1 | 38 | 237 | 494 | 792 | 4135 |
| Cooling Degree Days | 0 | 0 | 6 | 37 | 116 | *265* | 390 | 327 | 171 | 32 | 3 | 1 | 1348 |
| Total Precipitation (") | 2.87 | 2.99 | 4.28 | 4.17 | 4.65 | 3.91 | 4.30 | 4.31 | 4.05 | 3.00 | 3.81 | 4.21 | 46.55 |
| Days ≥ 0.1" Precip | 6 | 6 | 8 | 7 | 8 | 7 | 7 | 7 | 6 | 6 | 7 | *7* | 82 |
| Total Snowfall (") | *6.6* | *3.9* | 1.2 | 0.1 | 0.0 | 0.0 | 0.0 | 0.0 | 0.0 | 0.0 | 0.5 | na | na |
| Days ≥ 1" Snow Depth | *7* | 4 | 1 | 0 | 0 | 0 | 0 | 0 | 0 | 0 | 0 | *1* | 13 |

## BOWLING GREEN WARREN *Warren County* ELEVATION 538 ft LAT/LONG 36° 58 ' N / 86° 26 ' W

| | JAN | FEB | MAR | APR | MAY | JUN | JUL | AUG | SEP | OCT | NOV | DEC | YEAR |
|---|---|---|---|---|---|---|---|---|---|---|---|---|---|
| Maximum Temp °F | 42.3 | 47.4 | 58.2 | 69.0 | 77.0 | 85.5 | 89.0 | 87.3 | 80.8 | 69.7 | 57.5 | 47.6 | 67.6 |
| Minimum Temp °F | 24.4 | 27.4 | 36.4 | 45.6 | 54.3 | 63.0 | 67.5 | 65.3 | 58.4 | 45.4 | 37.3 | 29.2 | 46.2 |
| Mean Temp °F | 33.4 | 37.4 | 47.3 | 57.3 | 65.7 | 74.2 | 78.3 | 76.3 | 69.6 | 57.6 | 47.4 | 38.4 | 56.9 |
| Days Max Temp ≥ 90 °F | 0 | 0 | 0 | 0 | 1 | 9 | 15 | 12 | 4 | 0 | 0 | 0 | 41 |
| Days Max Temp ≤ 32 °F | 7 | 4 | 0 | 0 | 0 | 0 | 0 | 0 | 0 | 0 | 0 | 3 | 14 |
| Days Min Temp ≤ 32 °F | 23 | 20 | 12 | 3 | 0 | 0 | 0 | 0 | 0 | 2 | 11 | 20 | 91 |
| Days Min Temp ≤ 0 °F | 1 | 0 | 0 | 0 | 0 | 0 | 0 | 0 | 0 | 0 | 0 | 0 | 1 |
| Heating Degree Days | 972 | 772 | 547 | 250 | 81 | 5 | 0 | 1 | 38 | 251 | 524 | 817 | 4258 |
| Cooling Degree Days | 0 | 0 | 5 | 24 | 105 | 288 | 431 | 361 | 173 | 25 | 3 | 1 | 1416 |
| Total Precipitation (") | 3.91 | 4.17 | 4.95 | 4.31 | 5.08 | 4.20 | 4.73 | 3.42 | 3.89 | 3.08 | 4.47 | 5.14 | 51.35 |
| Days ≥ 0.1" Precip | 7 | 7 | 8 | 8 | 8 | 7 | 6 | 6 | 6 | 6 | 7 | 8 | 84 |
| Total Snowfall (") | 4.4 | 4.0 | 1.4 | 0.1 | 0.0 | 0.0 | 0.0 | 0.0 | 0.0 | 0.0 | 0.4 | 1.4 | 11.7 |
| Days ≥ 1" Snow Depth | 5 | 4 | 1 | 0 | 0 | 0 | 0 | 0 | 0 | 0 | 0 | 1 | 11 |

## BRADFORDSVILLE *Marion County* ELEVATION 659 ft LAT/LONG 37° 29 ' N / 85° 9 ' W

| | JAN | FEB | MAR | APR | MAY | JUN | JUL | AUG | SEP | OCT | NOV | DEC | YEAR |
|---|---|---|---|---|---|---|---|---|---|---|---|---|---|
| Maximum Temp °F | 42.3 | 46.9 | 57.5 | 68.6 | 76.4 | 84.4 | 88.1 | 86.7 | 81.3 | 70.3 | 58.5 | 47.8 | 67.4 |
| Minimum Temp °F | 21.7 | 24.0 | 32.7 | 41.8 | 50.2 | 59.1 | 64.2 | 61.9 | 55.4 | 42.0 | 34.8 | 27.1 | 42.9 |
| Mean Temp °F | 32.1 | 35.5 | 45.1 | 55.2 | 63.3 | 71.8 | 76.2 | 74.3 | 68.4 | 56.2 | 46.7 | 37.5 | 55.2 |
| Days Max Temp ≥ 90 °F | 0 | 0 | 0 | 0 | 1 | 6 | 13 | 10 | 4 | 0 | 0 | 0 | 34 |
| Days Max Temp ≤ 32 °F | 7 | 4 | 0 | 0 | 0 | 0 | 0 | 0 | 0 | 0 | 0 | 3 | 14 |
| Days Min Temp ≤ 32 °F | 24 | 22 | 17 | 6 | 1 | 0 | 0 | 0 | 0 | 7 | 13 | 22 | 112 |
| Days Min Temp ≤ 0 °F | 3 | 1 | 0 | 0 | 0 | 0 | 0 | 0 | 0 | 0 | 0 | 1 | 5 |
| Heating Degree Days | 1015 | 827 | 613 | 306 | 123 | 16 | 1 | 3 | 53 | 291 | 546 | 847 | 4641 |
| Cooling Degree Days | 0 | 0 | 3 | 17 | 80 | 232 | 377 | 310 | 148 | 23 | 2 | 1 | 1193 |
| Total Precipitation (") | 3.81 | 3.96 | 4.95 | 4.82 | 5.20 | 4.04 | 5.08 | 4.03 | 3.85 | 3.01 | 4.32 | 4.90 | 51.97 |
| Days ≥ 0.1" Precip | *8* | 7 | 8 | 8 | 8 | 7 | 7 | 7 | 5 | 6 | 8 | 8 | 87 |
| Total Snowfall (") | na | na | na | 0.1 | 0.0 | 0.0 | 0.0 | 0.0 | 0.0 | 0.0 | 0.4 | *1.6* | na |
| Days ≥ 1" Snow Depth | *8* | *7* | *1* | 0 | 0 | 0 | 0 | 0 | 0 | 0 | 0 | *2* | 18 |

## CARROLLTON LOCK 1 *Carroll County* ELEVATION 479 ft LAT/LONG 38° 41 ' N / 85° 11 ' W

| | JAN | FEB | MAR | APR | MAY | JUN | JUL | AUG | SEP | OCT | NOV | DEC | YEAR |
|---|---|---|---|---|---|---|---|---|---|---|---|---|---|
| Maximum Temp °F | 41.4 | 46.5 | 57.4 | 68.3 | 76.3 | 83.9 | 87.0 | 85.4 | 79.8 | 69.5 | 57.5 | 46.7 | 66.6 |
| Minimum Temp °F | 22.6 | 24.9 | 34.1 | 43.4 | 52.3 | 60.6 | 65.3 | 63.5 | 57.0 | 44.4 | 36.5 | 28.0 | 44.4 |
| Mean Temp °F | 32.0 | 35.7 | 45.8 | 55.9 | 64.5 | 72.3 | 76.2 | 74.5 | 68.4 | 57.0 | 47.0 | 37.4 | 55.6 |
| Days Max Temp ≥ 90 °F | 0 | 0 | 0 | 0 | 0 | 5 | 9 | 8 | 2 | 0 | 0 | 0 | 24 |
| Days Max Temp ≤ 32 °F | 7 | 3 | 0 | 0 | 0 | 0 | 0 | 0 | 0 | 0 | 0 | 3 | 13 |
| Days Min Temp ≤ 32 °F | 24 | 21 | 15 | 4 | 0 | 0 | 0 | 0 | 0 | 4 | 12 | 20 | 100 |
| Days Min Temp ≤ 0 °F | 2 | 1 | 0 | 0 | 0 | 0 | 0 | 0 | 0 | 0 | 0 | 1 | 4 |
| Heating Degree Days | 1016 | 821 | 592 | 288 | 97 | 10 | 0 | 3 | 46 | 267 | 535 | 851 | 4526 |
| Cooling Degree Days | 0 | 0 | 2 | 21 | 94 | 240 | 379 | 319 | 164 | 29 | 2 | 1 | 1251 |
| Total Precipitation (") | 2.78 | 2.93 | 4.03 | 4.26 | 4.35 | 3.82 | 3.98 | 3.89 | 3.16 | 3.00 | 3.59 | 3.49 | 43.28 |
| Days ≥ 0.1" Precip | 6 | 5 | 8 | 8 | 8 | 7 | 7 | 6 | 5 | 6 | 7 | 7 | 80 |
| Total Snowfall (") | *4.6* | *3.9* | *2.0* | 0.2 | 0.0 | 0.0 | 0.0 | 0.0 | 0.0 | 0.0 | *0.2* | *1.4* | 12.3 |
| Days ≥ 1" Snow Depth | 7 | 6 | 2 | 0 | 0 | 0 | 0 | 0 | 0 | 0 | 0 | 2 | 17 |

## CINCI-NORTHERN KY AP *Boone County* ELEVATION 889 ft LAT/LONG 39° 3 ' N / 84° 40 ' W

| | JAN | FEB | MAR | APR | MAY | JUN | JUL | AUG | SEP | OCT | NOV | DEC | YEAR |
|---|---|---|---|---|---|---|---|---|---|---|---|---|---|
| Maximum Temp °F | 37.1 | 41.3 | 52.7 | 64.4 | 74.0 | 82.2 | 85.8 | 84.0 | 77.4 | 65.5 | 53.0 | 42.2 | 63.3 |
| Minimum Temp °F | 20.2 | 23.2 | 32.9 | 42.6 | 52.1 | 60.9 | 65.7 | 63.6 | 56.8 | 44.4 | 35.6 | 26.5 | 43.7 |
| Mean Temp °F | 28.7 | 32.3 | 42.9 | 53.5 | 63.0 | 71.6 | 75.8 | 73.8 | 67.1 | 55.0 | 44.4 | 34.4 | 53.5 |
| Days Max Temp ≥ 90 °F | 0 | 0 | 0 | 0 | 0 | 4 | 8 | 5 | 1 | 0 | 0 | 0 | 18 |
| Days Max Temp ≤ 32 °F | 11 | 8 | 1 | 0 | 0 | 0 | 0 | 0 | 0 | 0 | 1 | 6 | 27 |
| Days Min Temp ≤ 32 °F | 26 | 22 | 16 | 4 | 0 | 0 | 0 | 0 | 0 | 3 | 12 | 21 | 104 |
| Days Min Temp ≤ 0 °F | 3 | 1 | 0 | 0 | 0 | 0 | 0 | 0 | 0 | 0 | 0 | 1 | 5 |
| Heating Degree Days | 1120 | 918 | 682 | 353 | 132 | 16 | 1 | 3 | 65 | 322 | 613 | 943 | 5168 |
| Cooling Degree Days | 0 | 0 | 3 | 16 | 82 | 226 | 359 | 290 | 133 | 20 | 2 | 1 | 1132 |
| Total Precipitation (") | 2.66 | 2.61 | 3.74 | 3.86 | 4.14 | 3.89 | 4.11 | 3.69 | 3.00 | 2.90 | 3.59 | 3.19 | 41.38 |
| Days ≥ 0.1" Precip | 6 | 5 | 8 | 8 | 8 | 7 | 7 | 7 | 6 | 5 | 7 | 7 | 81 |
| Total Snowfall (") | 6.6 | 5.8 | 4.3 | 0.5 | 0.0 | 0.0 | 0.0 | 0.0 | 0.0 | 0.4 | 1.7 | 3.3 | 22.6 |
| Days ≥ 1" Snow Depth | 8 | 7 | 2 | 0 | 0 | 0 | 0 | 0 | 0 | 0 | 1 | 3 | 21 |

## DANVILLE *Boyle County* ELEVATION 961 ft LAT/LONG 37° 39 ' N / 84° 46 ' W

| | JAN | FEB | MAR | APR | MAY | JUN | JUL | AUG | SEP | OCT | NOV | DEC | YEAR |
|---|---|---|---|---|---|---|---|---|---|---|---|---|---|
| Maximum Temp °F | 40.3 | 44.5 | 55.5 | 66.6 | 74.6 | 82.4 | 86.4 | 85.3 | 79.4 | 67.9 | 56.1 | 45.6 | 65.4 |
| Minimum Temp °F | 22.0 | 24.6 | 33.5 | 43.3 | 52.4 | 60.9 | 65.3 | 63.6 | 57.1 | 44.7 | 36.2 | 27.4 | 44.3 |
| Mean Temp °F | 31.2 | 34.5 | 44.5 | 54.9 | 63.5 | 71.7 | 75.9 | 74.5 | 68.2 | 56.3 | 46.2 | 36.5 | 54.8 |
| Days Max Temp ≥ 90 °F | 0 | 0 | 0 | 0 | 0 | 3 | 9 | 7 | 3 | 0 | 0 | 0 | 22 |
| Days Max Temp ≤ 32 °F | 8 | 5 | 1 | 0 | 0 | 0 | 0 | 0 | 0 | 0 | 0 | 4 | 18 |
| Days Min Temp ≤ 32 °F | 25 | 22 | 16 | 4 | 0 | 0 | 0 | 0 | 0 | 3 | 12 | 21 | 103 |
| Days Min Temp ≤ 0 °F | 2 | 1 | 0 | 0 | 0 | 0 | 0 | 0 | 0 | 0 | 0 | 0 | 3 |
| Heating Degree Days | 1041 | 853 | 631 | 316 | 118 | 15 | 1 | 2 | 50 | 284 | 560 | 877 | 4748 |
| Cooling Degree Days | 0 | 0 | 2 | 21 | 80 | 233 | 373 | 316 | 152 | 24 | 2 | 0 | 1203 |
| Total Precipitation (") | 3.37 | 3.68 | 4.48 | 4.20 | 4.63 | 4.09 | 4.70 | 3.58 | 3.55 | 3.04 | 3.82 | 4.35 | 47.49 |
| Days ≥ 0.1" Precip | 7 | 6 | 9 | 8 | 8 | 7 | 7 | 6 | 5 | 6 | 7 | 7 | 83 |
| Total Snowfall (") | 4.8 | 4.4 | 2.2 | 0.1 | 0.0 | 0.0 | 0.0 | 0.0 | 0.0 | 0.0 | 0.8 | 1.5 | 13.8 |
| Days ≥ 1" Snow Depth | 7 | 6 | 1 | 0 | 0 | 0 | 0 | 0 | 0 | 0 | 1 | 2 | 17 |

## DIX DAM *Mercer County* ELEVATION 922 ft LAT/LONG 37° 48 ' N / 84° 43 ' W

| | JAN | FEB | MAR | APR | MAY | JUN | JUL | AUG | SEP | OCT | NOV | DEC | YEAR |
|---|---|---|---|---|---|---|---|---|---|---|---|---|---|
| Maximum Temp °F | 42.4 | 47.3 | 58.2 | 68.5 | 76.2 | 83.9 | 87.3 | 86.0 | 80.2 | 69.8 | 58.1 | 47.6 | 67.1 |
| Minimum Temp °F | 24.7 | 27.1 | 35.8 | 44.8 | 53.5 | 62.1 | 66.3 | 64.6 | 58.5 | 46.9 | 38.5 | 30.2 | 46.1 |
| Mean Temp °F | 33.6 | 37.2 | 47.0 | 56.7 | 64.9 | 73.0 | 76.8 | 75.3 | 69.4 | 58.4 | 48.3 | 38.9 | 56.6 |
| Days Max Temp ≥ 90 °F | 0 | 0 | 0 | 0 | 0 | 4 | 10 | 8 | 2 | 0 | 0 | 0 | 24 |
| Days Max Temp ≤ 32 °F | 6 | 4 | 0 | 0 | 0 | 0 | 0 | 0 | 0 | 0 | 0 | 3 | 13 |
| Days Min Temp ≤ 32 °F | 22 | 19 | 13 | 3 | 0 | 0 | 0 | 0 | 0 | 2 | 9 | 19 | 87 |
| Days Min Temp ≤ 0 °F | 1 | 0 | 0 | 0 | 0 | 0 | 0 | 0 | 0 | 0 | 0 | 0 | 1 |
| Heating Degree Days | 965 | 778 | 556 | 267 | 92 | 8 | 0 | 2 | 37 | 231 | 498 | 803 | 4237 |
| Cooling Degree Days | 0 | 0 | 2 | 25 | 97 | 261 | 393 | 338 | 173 | 36 | 2 | 1 | 1328 |
| Total Precipitation (") | 3.13 | 3.44 | 4.10 | 4.13 | 4.58 | 3.63 | 4.65 | 4.26 | 3.40 | 3.01 | 3.64 | 3.96 | 45.93 |
| Days ≥ 0.1" Precip | 6 | *6* | 8 | 7 | 8 | 7 | 7 | 6 | 5 | 6 | 7 | 7 | 80 |
| Total Snowfall (") | *4.7* | na | na | 0.0 | 0.0 | 0.0 | 0.0 | 0.0 | 0.0 | 0.0 | 0.4 | *1.0* | na |
| Days ≥ 1" Snow Depth | *7* | na | *1* | 0 | 0 | 0 | 0 | 0 | 0 | 0 | 0 | *1* | na |

## FALMOUTH *Pendleton County* ELEVATION 600 ft LAT/LONG 38° 40 ' N / 84° 20 ' W

| | JAN | FEB | MAR | APR | MAY | JUN | JUL | AUG | SEP | OCT | NOV | DEC | YEAR |
|---|---|---|---|---|---|---|---|---|---|---|---|---|---|
| Maximum Temp °F | 39.2 | 43.3 | 54.6 | 66.0 | 75.1 | 83.3 | 86.8 | 85.2 | 79.3 | 67.7 | 55.0 | 44.7 | 65.0 |
| Minimum Temp °F | 19.0 | 21.0 | 30.1 | 39.5 | 48.9 | 57.8 | 62.8 | 60.6 | 53.8 | 40.5 | 32.9 | 24.5 | 41.0 |
| Mean Temp °F | 29.1 | 32.2 | 42.3 | 52.8 | 62.1 | 70.6 | 74.8 | 72.9 | 66.6 | 54.1 | 44.0 | 34.6 | 53.0 |
| Days Max Temp ≥ 90 °F | 0 | 0 | 0 | 0 | 1 | 5 | 10 | 7 | 2 | 0 | 0 | 0 | 25 |
| Days Max Temp ≤ 32 °F | 9 | 6 | 1 | 0 | 0 | 0 | 0 | 0 | 0 | 0 | 0 | 4 | 20 |
| Days Min Temp ≤ 32 °F | 27 | 23 | 19 | 7 | 1 | 0 | 0 | 0 | 0 | 7 | 15 | 24 | 123 |
| Days Min Temp ≤ 0 °F | 3 | 2 | 0 | 0 | 0 | 0 | 0 | 0 | 0 | 0 | 0 | 1 | 6 |
| Heating Degree Days | 1106 | 920 | 698 | 368 | 144 | 21 | 2 | 6 | 69 | 342 | 624 | 935 | 5235 |
| Cooling Degree Days | 0 | 0 | 1 | 9 | 64 | 197 | 344 | 279 | 122 | 14 | 1 | 0 | 1031 |
| Total Precipitation (") | 2.69 | 2.58 | 3.89 | 4.10 | 4.27 | 3.80 | 4.57 | 3.93 | 3.32 | 2.83 | 3.61 | 3.28 | 42.87 |
| Days ≥ 0.1" Precip | 6 | 6 | 7 | 8 | 8 | 7 | 7 | 7 | 5 | 6 | 7 | 7 | 81 |
| Total Snowfall (") | *3.2* | *3.5* | *2.1* | 0.0 | 0.0 | 0.0 | 0.0 | 0.0 | 0.0 | 0.0 | 0.4 | 1.4 | 10.6 |
| Days ≥ 1" Snow Depth | *6* | *6* | 1 | 0 | 0 | 0 | 0 | 0 | 0 | 0 | 0 | *2* | 15 |

## FARMERS 2 S *Rowan County* ELEVATION 669 ft LAT/LONG 38° 9 ' N / 83° 33 ' W

| | JAN | FEB | MAR | APR | MAY | JUN | JUL | AUG | SEP | OCT | NOV | DEC | YEAR |
|---|---|---|---|---|---|---|---|---|---|---|---|---|---|
| Maximum Temp °F | 41.2 | 45.8 | 56.8 | 67.8 | 75.9 | 83.6 | 87.0 | 85.7 | 79.6 | 68.6 | 57.2 | 46.7 | 66.3 |
| Minimum Temp °F | 21.4 | 23.4 | 32.2 | 41.2 | 49.8 | 58.1 | 63.0 | 61.4 | 54.8 | 42.4 | 34.8 | 26.9 | 42.4 |
| Mean Temp °F | 31.3 | 34.6 | 44.5 | 54.6 | 62.9 | 70.9 | 75.0 | 73.6 | 67.2 | 55.5 | 46.0 | 36.8 | 54.4 |
| Days Max Temp ≥ 90 °F | 0 | 0 | 0 | 0 | 1 | 5 | 10 | 8 | 3 | 0 | 0 | 0 | 27 |
| Days Max Temp ≤ 32 °F | 7 | 5 | 1 | 0 | 0 | 0 | 0 | 0 | 0 | 0 | 0 | 3 | 16 |
| Days Min Temp ≤ 32 °F | 25 | 22 | 17 | 7 | 1 | 0 | 0 | 0 | 0 | 6 | 13 | 22 | 113 |
| Days Min Temp ≤ 0 °F | 3 | 1 | 0 | 0 | 0 | 0 | 0 | 0 | 0 | 0 | 0 | 1 | 5 |
| Heating Degree Days | 1037 | 852 | 631 | 325 | 130 | 19 | 1 | 4 | 61 | 306 | 566 | 867 | 4799 |
| Cooling Degree Days | 0 | 0 | 2 | 19 | 78 | 208 | 345 | 282 | 137 | 23 | 2 | 0 | 1096 |
| Total Precipitation (") | 2.94 | 3.12 | 3.95 | 4.10 | 4.52 | 3.97 | 5.66 | 4.17 | 3.47 | 3.04 | 3.54 | 4.05 | 46.53 |
| Days ≥ 0.1" Precip | 7 | 6 | 8 | 8 | 9 | 8 | 8 | 7 | 6 | 6 | 7 | 8 | 88 |
| Total Snowfall (") | *4.5* | na | *1.4* | 0.0 | 0.0 | 0.0 | 0.0 | 0.0 | 0.0 | 0.0 | *0.3* | *1.5* | na |
| Days ≥ 1" Snow Depth | *7* | *5* | 1 | 0 | 0 | 0 | 0 | 0 | 0 | 0 | 0 | 2 | 15 |

## FRANKFORT LOCK 4 *Franklin County* ELEVATION 502 ft LAT/LONG 38° 14 ' N / 84° 52 ' W

| | JAN | FEB | MAR | APR | MAY | JUN | JUL | AUG | SEP | OCT | NOV | DEC | YEAR |
|---|---|---|---|---|---|---|---|---|---|---|---|---|---|
| Maximum Temp °F | 40.6 | 44.9 | 55.6 | 66.8 | 75.2 | 83.2 | 87.5 | 86.0 | 79.9 | 68.7 | 56.6 | 45.8 | 65.9 |
| Minimum Temp °F | 20.5 | 22.7 | 31.1 | 40.5 | 49.9 | 58.9 | 64.1 | 62.2 | 55.3 | 42.2 | 34.2 | 26.0 | 42.3 |
| Mean Temp °F | 30.5 | 33.8 | 43.4 | 53.6 | 62.6 | 71.1 | 75.8 | 74.1 | 67.6 | 55.5 | 45.5 | 35.9 | 54.1 |
| Days Max Temp ≥ 90 °F | 0 | 0 | 0 | 0 | 0 | 4 | 11 | 9 | 3 | 0 | 0 | 0 | 27 |
| Days Max Temp ≤ 32 °F | 7 | 5 | 1 | 0 | 0 | 0 | 0 | 0 | 0 | 0 | 0 | 4 | 17 |
| Days Min Temp ≤ 32 °F | 26 | 23 | 19 | 6 | 0 | 0 | 0 | 0 | 0 | 5 | 14 | 23 | 116 |
| Days Min Temp ≤ 0 °F | 2 | 1 | 0 | 0 | 0 | 0 | 0 | 0 | 0 | 0 | 0 | 1 | 4 |
| Heating Degree Days | 1063 | 876 | 665 | 345 | 134 | 17 | 1 | 2 | 54 | 307 | 582 | 896 | 4942 |
| Cooling Degree Days | 0 | 0 | 1 | 10 | 63 | 203 | 358 | 297 | 134 | 18 | 1 | 0 | 1085 |
| Total Precipitation (") | 2.78 | 2.91 | 3.82 | 3.79 | 4.42 | 3.73 | 4.30 | 3.64 | 3.28 | 2.65 | 3.39 | 3.54 | 42.25 |
| Days ≥ 0.1" Precip | 5 | 5 | 8 | 7 | 8 | 7 | 7 | 6 | 5 | 6 | 7 | 7 | 78 |
| Total Snowfall (") | na | 3.0 | 1.0 | 0.0 | 0.0 | 0.0 | 0.0 | 0.0 | 0.0 | 0.0 | 0.3 | 0.9 | na |
| Days ≥ 1" Snow Depth | 5 | 6 | 1 | 0 | 0 | 0 | 0 | 0 | 0 | 0 | 0 | 1 | 13 |

## GLASGOW *Barren County* ELEVATION 810 ft LAT/LONG 37° 0 ' N / 85° 55 ' W

| | JAN | FEB | MAR | APR | MAY | JUN | JUL | AUG | SEP | OCT | NOV | DEC | YEAR |
|---|---|---|---|---|---|---|---|---|---|---|---|---|---|
| Maximum Temp °F | 42.3 | 48.0 | 59.0 | 70.0 | 77.6 | 85.4 | 88.3 | 86.7 | 80.4 | 68.8 | 57.0 | 47.7 | 67.6 |
| Minimum Temp °F | 24.8 | 28.2 | 36.7 | 45.6 | 53.7 | 62.1 | 66.1 | 64.4 | 58.1 | 45.3 | 37.5 | 29.9 | 46.0 |
| Mean Temp °F | 33.6 | 38.1 | 47.9 | 57.8 | 65.7 | 73.8 | 77.2 | 75.6 | 69.3 | 57.1 | 47.3 | 38.9 | 56.9 |
| Days Max Temp ≥ 90 °F | 0 | 0 | 0 | 0 | 1 | 7 | 14 | 9 | 3 | 0 | 0 | 0 | 34 |
| Days Max Temp ≤ 32 °F | 7 | 4 | 0 | 0 | 0 | 0 | 0 | 0 | 0 | 0 | 0 | 3 | 14 |
| Days Min Temp ≤ 32 °F | 22 | 19 | 12 | 3 | 0 | 0 | 0 | 0 | 0 | 3 | 10 | 19 | 88 |
| Days Min Temp ≤ 0 °F | 2 | 0 | 0 | 0 | 0 | 0 | 0 | 0 | 0 | 0 | 0 | 0 | 2 |
| Heating Degree Days | 967 | 753 | 529 | 240 | 80 | 5 | 0 | 1 | 40 | 264 | 528 | 804 | 4211 |
| Cooling Degree Days | 0 | 0 | 6 | 33 | 121 | 297 | 419 | 354 | 183 | 28 | 2 | 1 | 1444 |
| Total Precipitation (") | 4.00 | 4.31 | 4.86 | 4.65 | 5.21 | 4.53 | 5.22 | 4.25 | 4.04 | 3.19 | 4.51 | 5.18 | 53.95 |
| Days ≥ 0.1" Precip | 7 | 7 | 9 | 8 | 8 | 7 | 7 | 6 | 6 | 6 | 8 | 8 | 87 |
| Total Snowfall (") | 5.4 | 4.2 | 1.5 | 0.1 | 0.0 | 0.0 | 0.0 | 0.0 | 0.0 | 0.0 | 0.9 | 1.7 | 13.8 |
| Days ≥ 1" Snow Depth | 6 | 3 | 0 | 0 | 0 | 0 | 0 | 0 | 0 | 0 | 0 | 1 | 10 |

## GOLDEN POND 8 N *Trigg County* ELEVATION 390 ft LAT/LONG 36° 54 ' N / 88° 0 ' W

| | JAN | FEB | MAR | APR | MAY | JUN | JUL | AUG | SEP | OCT | NOV | DEC | YEAR |
|---|---|---|---|---|---|---|---|---|---|---|---|---|---|
| Maximum Temp °F | 43.3 | 47.5 | 58.5 | 69.7 | 77.4 | 85.4 | 89.3 | 87.6 | 81.4 | 70.3 | 58.5 | 48.2 | 68.1 |
| Minimum Temp °F | 23.6 | 27.4 | 36.6 | 46.3 | 55.3 | 63.3 | 67.9 | 65.5 | 59.1 | 46.4 | 38.0 | 29.2 | 46.6 |
| Mean Temp °F | 33.7 | 37.5 | 47.4 | 58.0 | 66.4 | 74.4 | 78.6 | 76.6 | 70.2 | 58.4 | 48.3 | 38.8 | 57.4 |
| Days Max Temp ≥ 90 °F | 0 | 0 | 0 | 0 | 1 | 8 | 16 | 12 | 4 | 0 | 0 | 0 | 41 |
| Days Max Temp ≤ 32 °F | 6 | 3 | 0 | 0 | 0 | 0 | 0 | 0 | 0 | 0 | 0 | 3 | 12 |
| Days Min Temp ≤ 32 °F | 25 | 19 | 11 | 2 | 0 | 0 | 0 | 0 | 0 | 2 | 10 | 20 | 89 |
| Days Min Temp ≤ 0 °F | 1 | 0 | 0 | 0 | 0 | 0 | 0 | 0 | 0 | 0 | 0 | 0 | 1 |
| Heating Degree Days | 962 | 768 | 543 | 238 | 66 | 4 | 0 | 1 | 34 | 231 | 499 | 806 | 4152 |
| Cooling Degree Days | 0 | 0 | 6 | 32 | 113 | 298 | 447 | 374 | 197 | 31 | 3 | 1 | 1502 |
| Total Precipitation (") | 3.69 | 4.46 | 4.43 | 4.74 | 5.04 | 3.80 | 4.11 | 3.46 | 3.64 | 3.40 | 4.91 | 5.00 | 50.68 |
| Days ≥ 0.1" Precip | 6 | 6 | 8 | 7 | 8 | 6 | 6 | 5 | 6 | 5 | 7 | 7 | 77 |
| Total Snowfall (") | na | na | 0.7 | 0.0 | 0.0 | 0.0 | 0.0 | 0.0 | 0.0 | 0.0 | 0.1 | na | na |
| Days ≥ 1" Snow Depth | na | na | 0 | 0 | 0 | 0 | 0 | 0 | 0 | 0 | 0 | na | na |

## GRAYSON 3 SW *Carter County* ELEVATION 700 ft LAT/LONG 38° 17 ' N / 82° 58 ' W

| | JAN | FEB | MAR | APR | MAY | JUN | JUL | AUG | SEP | OCT | NOV | DEC | YEAR |
|---|---|---|---|---|---|---|---|---|---|---|---|---|---|
| Maximum Temp °F | 41.1 | 45.4 | 55.6 | 66.8 | 75.6 | 82.7 | 86.5 | 85.3 | 79.3 | 68.5 | 57.2 | 46.2 | 65.9 |
| Minimum Temp °F | 18.9 | 20.7 | 28.2 | 37.0 | 46.8 | 56.2 | 61.8 | 59.9 | 52.5 | 38.9 | 30.9 | 23.8 | 39.6 |
| Mean Temp °F | 30.0 | 33.1 | 41.9 | 51.9 | 61.2 | 69.5 | 74.2 | 72.6 | 65.9 | 53.7 | 44.1 | 35.0 | 52.8 |
| Days Max Temp ≥ 90 °F | 0 | 0 | 0 | 0 | 1 | 4 | 9 | 7 | 2 | 0 | 0 | 0 | 23 |
| Days Max Temp ≤ 32 °F | 8 | 4 | 1 | 0 | 0 | 0 | 0 | 0 | 0 | 0 | 0 | 3 | 16 |
| Days Min Temp ≤ 32 °F | 27 | 24 | 21 | 10 | 2 | 0 | 0 | 0 | 0 | 10 | 18 | 24 | 136 |
| Days Min Temp ≤ 0 °F | 3 | 2 | 0 | 0 | 0 | 0 | 0 | 0 | 0 | 0 | 0 | 1 | 6 |
| Heating Degree Days | 1077 | 895 | 712 | 393 | 165 | 28 | 2 | 6 | 78 | 355 | 621 | 922 | 5254 |
| Cooling Degree Days | 0 | 0 | 1 | 8 | 52 | 168 | 306 | 245 | 103 | 12 | 1 | 0 | 896 |
| Total Precipitation (") | 2.79 | 3.10 | 3.63 | 3.49 | 4.21 | 3.88 | 4.89 | 3.67 | 2.67 | 3.14 | 3.19 | 3.77 | 42.43 |
| Days ≥ 0.1" Precip | 7 | 6 | 8 | 7 | 9 | 7 | 8 | 6 | 5 | 6 | 7 | 7 | 83 |
| Total Snowfall (") | 5.7 | 4.3 | 2.8 | 0.0 | 0.0 | 0.0 | 0.0 | 0.0 | 0.0 | 0.0 | 0.6 | 1.5 | 14.9 |
| Days ≥ 1" Snow Depth | 7 | 5 | 1 | 0 | 0 | 0 | 0 | 0 | 0 | 0 | 0 | 2 | 15 |

## GREENSBURG *Green County* ELEVATION 581 ft LAT/LONG 37° 16 ' N / 85° 30 ' W

| | JAN | FEB | MAR | APR | MAY | JUN | JUL | AUG | SEP | OCT | NOV | DEC | YEAR |
|---|---|---|---|---|---|---|---|---|---|---|---|---|---|
| Maximum Temp °F | 43.0 | 48.0 | 58.4 | 69.5 | 77.6 | 85.5 | 89.4 | 87.9 | 82.1 | 70.8 | 58.8 | 48.4 | 68.3 |
| Minimum Temp °F | 22.2 | 24.4 | 33.5 | 42.8 | 51.6 | 61.3 | 65.8 | 63.5 | 56.6 | 42.9 | 35.3 | 27.3 | 43.9 |
| Mean Temp °F | 32.6 | 36.2 | 46.0 | 56.2 | 64.6 | 73.4 | 77.6 | 75.7 | 69.3 | 56.9 | 47.1 | 37.9 | 56.1 |
| Days Max Temp ≥ 90 °F | 0 | 0 | 0 | 0 | 1 | 9 | 16 | 13 | 6 | 0 | 0 | 0 | 45 |
| Days Max Temp ≤ 32 °F | 6 | 3 | 0 | 0 | 0 | 0 | 0 | 0 | 0 | 0 | 0 | 2 | 11 |
| Days Min Temp ≤ 32 °F | 25 | 22 | 16 | 5 | 0 | 0 | 0 | 0 | 0 | 5 | 13 | 22 | 108 |
| Days Min Temp ≤ 0 °F | 2 | 1 | 0 | 0 | 0 | 0 | 0 | 0 | 0 | 0 | 0 | 0 | 3 |
| Heating Degree Days | 997 | 806 | 586 | 283 | 101 | 9 | 0 | 1 | 40 | 271 | 534 | 835 | 4463 |
| Cooling Degree Days | 0 | 0 | 4 | 24 | 97 | 276 | 422 | 345 | 170 | 25 | 2 | 1 | 1366 |
| Total Precipitation (") | 3.81 | 4.17 | 4.82 | 4.34 | 5.49 | 4.32 | 4.98 | 4.53 | 4.07 | 3.13 | 4.24 | 4.95 | 52.85 |
| Days ≥ 0.1" Precip | 7 | 7 | 9 | 7 | 8 | 7 | 8 | 6 | 6 | 6 | 7 | 8 | 86 |
| Total Snowfall (") | *3.7* | na | 0.8 | 0.0 | 0.0 | 0.0 | 0.0 | 0.0 | 0.0 | 0.0 | 0.7 | *1.2* | na |
| Days ≥ 1" Snow Depth | na | na | 1 | 0 | 0 | 0 | 0 | 0 | 0 | 0 | 0 | *1* | na |

## HEIDELBERG *Lee County* ELEVATION 659 ft LAT/LONG 37° 33 ' N / 83° 46 ' W

| | JAN | FEB | MAR | APR | MAY | JUN | JUL | AUG | SEP | OCT | NOV | DEC | YEAR |
|---|---|---|---|---|---|---|---|---|---|---|---|---|---|
| Maximum Temp °F | *45.4* | *50.1* | *60.0* | *70.3* | *77.7* | *84.5* | 87.7 | *85.8* | 79.9 | 70.7 | *59.5* | *49.2* | 68.4 |
| Minimum Temp °F | *23.3* | *24.8* | *31.9* | *39.7* | *49.4* | *58.4* | 63.4 | *62.0* | 55.6 | 42.4 | *33.4* | *26.8* | 42.6 |
| Mean Temp °F | *34.5* | *37.5* | *46.0* | *55.0* | *63.6* | *71.5* | 75.6 | *73.9* | 67.8 | 56.5 | *46.5* | *38.0* | 55.5 |
| Days Max Temp ≥ 90 °F | 0 | 0 | 0 | 0 | 1 | *6* | *12* | 8 | 2 | 0 | 0 | 0 | 29 |
| Days Max Temp ≤ 32 °F | *4* | *2* | 0 | 0 | 0 | 0 | 0 | 0 | 0 | 0 | 0 | 2 | 8 |
| Days Min Temp ≤ 32 °F | *24* | 21 | *17* | 8 | 1 | 0 | 0 | 0 | 0 | 6 | 14 | *21* | 112 |
| Days Min Temp ≤ 0 °F | *1* | 1 | 0 | 0 | 0 | 0 | 0 | 0 | 0 | 0 | 0 | 0 | 2 |
| Heating Degree Days | *938* | *770* | *586* | *308* | *111* | *14* | *1* | *2* | 51 | *271* | *551* | *831* | 4434 |
| Cooling Degree Days | *0* | *0* | *1* | 11 | 75 | 212 | 348 | 288 | 132 | 19 | 0 | 0 | 1086 |
| Total Precipitation (") | 3.27 | 3.54 | 4.50 | 3.98 | 4.43 | 3.57 | 5.06 | 4.32 | 3.67 | 3.03 | 3.93 | 4.35 | 47.65 |
| Days ≥ 0.1" Precip | 7 | 7 | *9* | *8* | 8 | 7 | 8 | 7 | 6 | 6 | *7* | *8* | 88 |
| Total Snowfall (") | na | na | na | 0.5 | 0.0 | 0.0 | 0.0 | 0.0 | 0.0 | 0.0 | na | na | na |
| Days ≥ 1" Snow Depth | *8* | *5* | *1* | 0 | 0 | 0 | 0 | 0 | 0 | 0 | *0* | *3* | 17 |

## HENDERSON 7 SSW *Henderson County* ELEVATION 381 ft LAT/LONG 37° 45 ' N / 87° 38 ' W

| | JAN | FEB | MAR | APR | MAY | JUN | JUL | AUG | SEP | OCT | NOV | DEC | YEAR |
|---|---|---|---|---|---|---|---|---|---|---|---|---|---|
| Maximum Temp °F | 41.0 | 46.2 | 57.6 | 69.0 | 77.3 | 85.5 | 88.5 | 87.2 | 81.1 | 70.3 | 57.3 | 46.0 | 67.3 |
| Minimum Temp °F | 24.1 | 27.3 | 36.9 | 46.7 | 55.1 | 63.5 | 67.5 | 64.9 | 58.6 | 46.9 | 38.7 | 29.5 | 46.6 |
| Mean Temp °F | 32.6 | 36.8 | 47.3 | 57.9 | 66.2 | 74.5 | 78.0 | 76.1 | 69.9 | 58.6 | 48.0 | 37.8 | 57.0 |
| Days Max Temp ≥ 90 °F | 0 | 0 | 0 | 0 | 1 | 7 | 14 | 11 | 4 | 0 | 0 | 0 | 37 |
| Days Max Temp ≤ 32 °F | 8 | 4 | 0 | 0 | 0 | 0 | 0 | 0 | 0 | 0 | 0 | 4 | 16 |
| Days Min Temp ≤ 32 °F | 23 | 18 | 12 | 2 | 0 | 0 | 0 | 0 | 0 | 2 | 9 | 19 | 85 |
| Days Min Temp ≤ 0 °F | 1 | 1 | 0 | 0 | 0 | 0 | 0 | 0 | 0 | 0 | 0 | 0 | 2 |
| Heating Degree Days | 1000 | 790 | 549 | 242 | 70 | 4 | 0 | 1 | 36 | 230 | 507 | 838 | 4267 |
| Cooling Degree Days | 0 | 0 | 6 | 33 | 118 | 304 | 425 | 367 | 192 | 38 | 4 | 1 | 1488 |
| Total Precipitation (") | 2.90 | 3.28 | 4.23 | 4.39 | 4.37 | 3.71 | 4.27 | 3.06 | 3.61 | 3.02 | 4.23 | 3.76 | 44.83 |
| Days ≥ 0.1" Precip | 6 | 6 | 8 | 8 | 8 | 6 | 6 | 5 | 5 | 5 | 6 | 7 | 76 |
| Total Snowfall (") | 5.4 | 5.0 | 2.9 | 0.4 | 0.0 | 0.0 | 0.0 | 0.0 | 0.0 | 0.1 | 0.7 | 2.0 | 16.5 |
| Days ≥ 1" Snow Depth | 7 | 6 | 1 | 0 | 0 | 0 | 0 | 0 | 0 | 0 | 0 | 2 | 16 |

## HODGENVILLE-LINCOLN *Larue County* ELEVATION 788 ft LAT/LONG 37° 32 ' N / 85° 44 ' W

| | JAN | FEB | MAR | APR | MAY | JUN | JUL | AUG | SEP | OCT | NOV | DEC | YEAR |
|---|---|---|---|---|---|---|---|---|---|---|---|---|---|
| Maximum Temp °F | 42.7 | 48.4 | 59.5 | 70.1 | 76.9 | 84.3 | 87.5 | 86.3 | 80.9 | 70.2 | 58.0 | 47.5 | 67.7 |
| Minimum Temp °F | 23.6 | 26.5 | 35.8 | 44.6 | 52.8 | 61.2 | 65.4 | 63.7 | 57.4 | 45.5 | 37.5 | 28.9 | 45.2 |
| Mean Temp °F | 33.2 | 37.5 | 47.7 | 57.3 | 64.9 | 72.8 | 76.5 | 75.0 | 69.2 | 57.9 | 47.8 | 38.2 | 56.5 |
| Days Max Temp ≥ 90 °F | 0 | 0 | 0 | 0 | 0 | 5 | 11 | 8 | 3 | 0 | 0 | 0 | 27 |
| Days Max Temp ≤ 32 °F | 6 | 3 | 0 | 0 | 0 | 0 | 0 | 0 | 0 | 0 | 0 | 3 | 12 |
| Days Min Temp ≤ 32 °F | 23 | 20 | 13 | 3 | 0 | 0 | 0 | 0 | 0 | 3 | 10 | 20 | 92 |
| Days Min Temp ≤ 0 °F | 2 | 1 | 0 | 0 | 0 | 0 | 0 | 0 | 0 | 0 | 0 | 0 | 3 |
| Heating Degree Days | 980 | 771 | 527 | 252 | 91 | 10 | 1 | 2 | 41 | 242 | 512 | 822 | 4251 |
| Cooling Degree Days | 0 | 0 | 4 | 27 | 97 | 255 | 381 | 326 | 173 | 31 | 2 | 1 | 1297 |
| Total Precipitation (") | 3.39 | 4.21 | 4.50 | 4.66 | 5.13 | 3.92 | 4.93 | 4.09 | 3.89 | 3.30 | 4.66 | 4.72 | 51.40 |
| Days ≥ 0.1" Precip | 6 | 6 | 8 | *7* | 8 | 7 | 7 | 6 | 6 | 6 | 7 | *7* | 81 |
| Total Snowfall (") | na | *3.7* | 1.4 | 0.0 | 0.0 | 0.0 | 0.0 | 0.0 | 0.0 | 0.0 | 0.9 | *1.2* | na |
| Days ≥ 1" Snow Depth | *6* | *5* | 0 | 0 | 0 | 0 | 0 | 0 | 0 | 0 | 0 | *2* | 13 |

## HOPKINSVILLE *Christian County* ELEVATION 522 ft LAT/LONG 36° 52' N / 87° 27' W

| | JAN | FEB | MAR | APR | MAY | JUN | JUL | AUG | SEP | OCT | NOV | DEC | YEAR |
|---|---|---|---|---|---|---|---|---|---|---|---|---|---|
| Maximum Temp °F | 41.6 | 46.5 | 58.0 | 69.3 | 77.4 | 85.5 | 89.3 | 88.0 | 81.8 | 70.6 | 58.2 | 47.1 | 67.8 |
| Minimum Temp °F | 22.7 | 25.1 | 34.9 | 44.8 | 53.3 | 62.1 | 66.2 | 64.3 | 57.5 | 44.7 | 36.5 | 28.0 | 45.0 |
| Mean Temp °F | 32.2 | 35.8 | 46.5 | 57.1 | 65.4 | 73.8 | 77.8 | 76.1 | 69.7 | 57.7 | 47.4 | 37.6 | 56.4 |
| Days Max Temp ≥ 90 °F | 0 | 0 | 0 | 0 | 1 | 8 | 16 | 13 | 5 | 0 | 0 | 0 | 43 |
| Days Max Temp ≤ 32 °F | 8 | 4 | 0 | 0 | 0 | 0 | 0 | 0 | 0 | 0 | 0 | 3 | 15 |
| Days Min Temp ≤ 32 °F | 25 | 21 | 14 | 3 | 0 | 0 | 0 | 0 | 0 | 3 | 11 | 21 | 98 |
| Days Min Temp ≤ 0 °F | 1 | 0 | 0 | 0 | 0 | 0 | 0 | 0 | 0 | 0 | 0 | 0 | 1 |
| Heating Degree Days | 1009 | 817 | 571 | 263 | 84 | 6 | 0 | 1 | 40 | 249 | 524 | 843 | 4407 |
| Cooling Degree Days | 0 | 0 | 4 | 27 | 93 | 270 | 408 | 350 | 176 | 26 | 2 | 0 | 1356 |
| Total Precipitation (") | 4.00 | 4.24 | 4.94 | 4.67 | 4.97 | 3.60 | 4.30 | 3.45 | 3.43 | 3.32 | 5.00 | 5.23 | 51.15 |
| Days ≥ 0.1" Precip | 7 | 7 | 9 | 8 | 8 | 6 | 7 | 5 | 5 | 6 | 7 | 8 | 83 |
| Total Snowfall (") | 4.5 | 4.5 | 1.7 | 0.1 | 0.0 | 0.0 | 0.0 | 0.0 | 0.0 | 0.0 | 0.4 | 1.1 | 12.3 |
| Days ≥ 1" Snow Depth | *5* | *4* | 1 | 0 | 0 | 0 | 0 | 0 | 0 | 0 | 0 | 1 | 11 |

## LEITCHFIELD 2 N *Grayson County* ELEVATION 640 ft LAT/LONG 37° 29' N / 86° 18' W

| | JAN | FEB | MAR | APR | MAY | JUN | JUL | AUG | SEP | OCT | NOV | DEC | YEAR |
|---|---|---|---|---|---|---|---|---|---|---|---|---|---|
| Maximum Temp °F | 41.6 | 46.6 | 57.6 | 68.7 | 76.4 | 84.2 | 87.6 | 86.0 | 79.7 | 69.1 | 57.0 | 46.6 | 66.8 |
| Minimum Temp °F | 22.7 | 25.4 | 34.7 | 43.9 | 52.4 | 61.1 | 65.5 | 63.5 | 57.0 | 44.4 | 36.1 | 28.0 | 44.6 |
| Mean Temp °F | 32.1 | 36.0 | 46.2 | 56.3 | 64.5 | 72.7 | 76.5 | 74.8 | 68.4 | 56.8 | 46.6 | 37.3 | 55.7 |
| Days Max Temp ≥ 90 °F | 0 | 0 | 0 | 0 | 1 | 5 | 11 | 8 | 3 | 0 | 0 | 0 | 28 |
| Days Max Temp ≤ 32 °F | 7 | 4 | 0 | 0 | 0 | 0 | 0 | 0 | 0 | 0 | 0 | 3 | 14 |
| Days Min Temp ≤ 32 °F | 24 | 21 | 15 | 5 | 0 | 0 | 0 | 0 | 0 | 4 | 12 | 20 | 101 |
| Days Min Temp ≤ 0 °F | 2 | 1 | 0 | 0 | 0 | 0 | 0 | 0 | 0 | 0 | 0 | 0 | 3 |
| Heating Degree Days | 1012 | 808 | 580 | 274 | 97 | 9 | 1 | 2 | 47 | 270 | 547 | 851 | 4498 |
| Cooling Degree Days | 0 | 0 | 4 | 20 | 87 | 249 | 384 | 317 | 156 | 23 | 1 | 1 | 1242 |
| Total Precipitation (") | 3.30 | 4.13 | 4.28 | 4.32 | 4.81 | 3.75 | 4.82 | 3.68 | 3.64 | 3.13 | 4.28 | 4.58 | 48.72 |
| Days ≥ 0.1" Precip | 6 | 6 | 8 | 8 | 8 | 7 | 7 | 6 | 6 | 6 | 7 | 7 | 82 |
| Total Snowfall (") | 4.8 | 4.0 | *1.8* | 0.1 | 0.0 | 0.0 | 0.0 | 0.0 | 0.0 | 0.0 | 0.8 | 1.6 | 13.1 |
| Days ≥ 1" Snow Depth | na | na | 1 | 0 | 0 | 0 | 0 | 0 | 0 | 0 | *0* | *1* | na |

## LEXINGTON BLUEGRASS *Fayette County* ELEVATION 981 ft LAT/LONG 38° 2' N / 84° 36' W

| | JAN | FEB | MAR | APR | MAY | JUN | JUL | AUG | SEP | OCT | NOV | DEC | YEAR |
|---|---|---|---|---|---|---|---|---|---|---|---|---|---|
| Maximum Temp °F | 39.3 | 43.9 | 55.0 | 65.6 | 74.1 | 82.4 | 85.8 | 84.4 | 78.0 | 66.8 | 54.8 | 44.9 | 64.6 |
| Minimum Temp °F | 22.9 | 25.9 | 35.1 | 44.3 | 53.5 | 61.9 | 66.3 | 64.6 | 58.0 | 46.0 | 37.1 | 28.5 | 45.3 |
| Mean Temp °F | 31.1 | 34.9 | 45.1 | 55.0 | 63.8 | 72.2 | 76.1 | 74.5 | 68.0 | 56.4 | 46.0 | 36.7 | 55.0 |
| Days Max Temp ≥ 90 °F | 0 | 0 | 0 | 0 | 0 | 3 | 8 | 6 | 1 | 0 | 0 | 0 | 18 |
| Days Max Temp ≤ 32 °F | 9 | 6 | 1 | 0 | 0 | 0 | 0 | 0 | 0 | 0 | 0 | 5 | 21 |
| Days Min Temp ≤ 32 °F | 24 | 20 | 14 | 3 | 0 | 0 | 0 | 0 | 0 | 2 | 11 | 20 | 94 |
| Days Min Temp ≤ 0 °F | 2 | 1 | 0 | 0 | 0 | 0 | 0 | 0 | 0 | 0 | 0 | 0 | 3 |
| Heating Degree Days | 1043 | 843 | 614 | 312 | 114 | 12 | 1 | 2 | 51 | 283 | 566 | 871 | 4712 |
| Cooling Degree Days | 0 | 0 | 3 | 20 | 90 | 248 | 381 | 321 | 149 | 25 | 2 | 0 | 1239 |
| Total Precipitation (") | 2.95 | 3.17 | 4.16 | 3.90 | 4.43 | 3.77 | 4.93 | 4.09 | 3.21 | 2.68 | 3.46 | 4.08 | 44.83 |
| Days ≥ 0.1" Precip | 6 | 6 | 9 | 7 | 8 | 7 | 8 | 7 | 6 | 5 | 7 | 7 | 83 |
| Total Snowfall (") | 6.0 | 5.0 | 2.6 | 0.3 | 0.0 | 0.0 | 0.0 | 0.0 | 0.0 | 0.0 | 0.8 | 2.3 | 17.0 |
| Days ≥ 1" Snow Depth | 7 | 6 | 1 | 0 | 0 | 0 | 0 | 0 | 0 | 0 | 0 | 2 | 16 |

## LONDON-CORBIN AP *Laurel County* ELEVATION 1191 ft LAT/LONG 37° 5' N / 84° 5' W

| | JAN | FEB | MAR | APR | MAY | JUN | JUL | AUG | SEP | OCT | NOV | DEC | YEAR |
|---|---|---|---|---|---|---|---|---|---|---|---|---|---|
| Maximum Temp °F | 42.8 | 47.9 | 58.2 | 68.5 | 75.5 | 82.7 | 85.9 | 84.6 | 78.5 | 68.2 | 57.4 | 47.9 | 66.5 |
| Minimum Temp °F | 24.1 | 26.7 | 35.1 | 43.7 | 52.0 | 60.4 | 64.9 | 63.3 | 56.8 | 43.8 | 36.0 | 28.9 | 44.6 |
| Mean Temp °F | 33.5 | 37.3 | 46.7 | 56.1 | 63.7 | 71.5 | 75.5 | 74.0 | 67.7 | 56.0 | 46.7 | 38.5 | 55.6 |
| Days Max Temp ≥ 90 °F | 0 | 0 | 0 | 0 | 0 | 3 | 8 | 5 | 1 | 0 | 0 | 0 | 17 |
| Days Max Temp ≤ 32 °F | 7 | 4 | 1 | 0 | 0 | 0 | 0 | 0 | 0 | 0 | 0 | 3 | 15 |
| Days Min Temp ≤ 32 °F | 23 | 20 | 13 | 4 | 0 | 0 | 0 | 0 | 0 | 4 | 13 | 20 | 97 |
| Days Min Temp ≤ 0 °F | 2 | 1 | 0 | 0 | 0 | 0 | 0 | 0 | 0 | 0 | 0 | 0 | 3 |
| Heating Degree Days | 969 | 776 | 564 | 280 | 109 | 12 | 1 | 2 | 50 | 285 | 544 | 816 | 4408 |
| Cooling Degree Days | 0 | 0 | 3 | 21 | 88 | 239 | 365 | 307 | 144 | 18 | 2 | 1 | 1188 |
| Total Precipitation (") | 3.56 | 3.58 | 4.51 | 3.98 | 4.37 | 3.91 | 4.59 | 3.50 | 3.54 | 2.81 | 3.87 | 4.38 | 46.60 |
| Days ≥ 0.1" Precip | 8 | 8 | 9 | 8 | 8 | 7 | 7 | 7 | 6 | 5 | 7 | 7 | 87 |
| Total Snowfall (") | 6.1 | 5.0 | 1.5 | 0.4 | 0.0 | 0.0 | 0.0 | 0.0 | 0.0 | 0.0 | 0.6 | 2.3 | 15.9 |
| Days ≥ 1" Snow Depth | 7 | 6 | 1 | 0 | 0 | 0 | 0 | 0 | 0 | 0 | 0 | 2 | 16 |

## LOUISVILLE STANDIFRD *Jefferson County* ELEVATION 485 ft LAT/LONG 38° 11 ' N / 85° 44 ' W

| | JAN | FEB | MAR | APR | MAY | JUN | JUL | AUG | SEP | OCT | NOV | DEC | YEAR |
|---|---|---|---|---|---|---|---|---|---|---|---|---|---|
| Maximum Temp °F | 40.5 | 45.2 | 56.4 | 67.7 | 76.0 | 84.3 | 87.9 | 86.3 | 79.7 | 68.4 | 56.2 | 45.8 | 66.2 |
| Minimum Temp °F | 24.0 | 27.1 | 36.3 | 45.9 | 55.1 | 63.9 | 68.5 | 66.7 | 59.7 | 46.9 | 38.4 | 29.7 | 46.9 |
| Mean Temp °F | 32.2 | 36.2 | 46.4 | 56.8 | 65.6 | 74.1 | 78.3 | 76.5 | 69.7 | 57.7 | 47.3 | 37.8 | 56.6 |
| Days Max Temp ≥ 90 °F | 0 | 0 | 0 | 0 | 0 | 6 | 12 | 9 | 3 | 0 | 0 | 0 | 30 |
| Days Max Temp ≤ 32 °F | 9 | 5 | 1 | 0 | 0 | 0 | 0 | 0 | 0 | 0 | 0 | 4 | 19 |
| Days Min Temp ≤ 32 °F | 24 | 20 | 12 | 2 | 0 | 0 | 0 | 0 | 0 | 1 | 9 | 19 | 87 |
| Days Min Temp ≤ 0 °F | 1 | 0 | 0 | 0 | 0 | 0 | 0 | 0 | 0 | 0 | 0 | 0 | 1 |
| Heating Degree Days | 1009 | 807 | 575 | 266 | 85 | 6 | 0 | 1 | 37 | 250 | 526 | 837 | 4399 |
| Cooling Degree Days | 0 | 0 | 5 | 29 | 119 | 303 | 449 | 384 | 193 | 32 | 3 | 1 | 1518 |
| Total Precipitation (") | 2.98 | 3.21 | 4.15 | 4.35 | 4.41 | 3.31 | 4.49 | 3.76 | 3.18 | 2.74 | 3.79 | 3.62 | 43.99 |
| Days ≥ 0.1" Precip | 6 | 6 | 8 | 8 | 8 | 6 | 7 | 6 | 6 | 5 | 7 | 7 | 80 |
| Total Snowfall (") | 5.6 | 4.7 | 3.3 | 0.2 | 0.0 | 0.0 | 0.0 | 0.0 | 0.0 | 0.1 | 1.0 | 1.7 | 16.6 |
| Days ≥ 1" Snow Depth | 6 | 5 | 1 | 0 | 0 | 0 | 0 | 0 | 0 | 0 | 0 | 2 | 14 |

## LOVELACEVILLE *Ballard County* ELEVATION 479 ft LAT/LONG 36° 58 ' N / 88° 50 ' W

| | JAN | FEB | MAR | APR | MAY | JUN | JUL | AUG | SEP | OCT | NOV | DEC | YEAR |
|---|---|---|---|---|---|---|---|---|---|---|---|---|---|
| Maximum Temp °F | 43.4 | 49.3 | 60.0 | 71.0 | 79.0 | 87.4 | 90.8 | 89.0 | 82.4 | 72.3 | 59.2 | 48.4 | 69.4 |
| Minimum Temp °F | 24.6 | 27.4 | 36.5 | 46.1 | 54.4 | 62.7 | 66.9 | 64.5 | 57.7 | 45.3 | 37.0 | 29.1 | 46.0 |
| Mean Temp °F | 34.4 | 38.5 | 48.4 | 58.5 | 66.7 | 75.1 | 78.9 | 76.8 | 70.1 | 59.0 | 48.1 | 38.8 | 57.8 |
| Days Max Temp ≥ 90 °F | 0 | 0 | 0 | 0 | 2 | 12 | 20 | 15 | 6 | 0 | 0 | 0 | 55 |
| Days Max Temp ≤ 32 °F | 6 | 2 | 0 | 0 | 0 | 0 | 0 | 0 | 0 | 0 | 0 | 2 | 10 |
| Days Min Temp ≤ 32 °F | 23 | 19 | 12 | 3 | 0 | 0 | 0 | 0 | 0 | 3 | 11 | 20 | 91 |
| Days Min Temp ≤ 0 °F | 1 | 0 | 0 | 0 | 0 | 0 | 0 | 0 | 0 | 0 | 0 | 0 | 1 |
| Heating Degree Days | 943 | 741 | 515 | 223 | 58 | 3 | 0 | 0 | 35 | 217 | 504 | 805 | 4044 |
| Cooling Degree Days | 0 | 0 | 7 | 34 | 129 | 331 | 472 | 403 | 209 | 36 | 5 | 1 | 1627 |
| Total Precipitation (") | 3.57 | 4.04 | 4.45 | 5.29 | 4.51 | 3.82 | 4.46 | 2.93 | 3.48 | 3.28 | 4.46 | 4.80 | 49.09 |
| Days ≥ 0.1" Precip | 6 | 6 | 8 | 8 | 7 | 6 | 6 | 5 | 5 | 5 | 7 | 8 | 77 |
| Total Snowfall (") | *3.5* | na | *1.5* | 0.0 | 0.0 | 0.0 | 0.0 | 0.0 | 0.0 | 0.1 | 0.1 | *1.6* | na |
| Days ≥ 1" Snow Depth | na | *5* | 1 | 0 | 0 | 0 | 0 | 0 | 0 | 0 | 0 | *1* | na |

## MADISONVILLE 1 SE *Hopkins County* ELEVATION 440 ft LAT/LONG 37° 21 ' N / 87° 31 ' W

| | JAN | FEB | MAR | APR | MAY | JUN | JUL | AUG | SEP | OCT | NOV | DEC | YEAR |
|---|---|---|---|---|---|---|---|---|---|---|---|---|---|
| Maximum Temp °F | 43.2 | 48.3 | 59.5 | 71.0 | 78.6 | 86.4 | 89.8 | 88.4 | 82.4 | 71.3 | 58.6 | 48.0 | 68.8 |
| Minimum Temp °F | 24.1 | 27.2 | 37.0 | 46.6 | 54.7 | 62.9 | 67.0 | 64.7 | 58.2 | 46.1 | 37.9 | 29.2 | 46.3 |
| Mean Temp °F | 33.7 | 37.7 | 48.3 | 58.8 | 66.7 | 74.7 | 78.4 | 76.6 | 70.4 | 58.7 | 48.3 | 38.6 | 57.6 |
| Days Max Temp ≥ 90 °F | 0 | 0 | 0 | 0 | 1 | 9 | 18 | 13 | 5 | 0 | 0 | 0 | 46 |
| Days Max Temp ≤ 32 °F | 6 | 3 | 0 | 0 | 0 | 0 | 0 | 0 | 0 | 0 | 0 | 2 | 11 |
| Days Min Temp ≤ 32 °F | 23 | 18 | 12 | 2 | 0 | 0 | 0 | 0 | 0 | 3 | 10 | 20 | 88 |
| Days Min Temp ≤ 0 °F | 1 | 0 | 0 | 0 | 0 | 0 | 0 | 0 | 0 | 0 | 0 | 0 | 1 |
| Heating Degree Days | 962 | 764 | 520 | 220 | 65 | 3 | 0 | 1 | 32 | 224 | 498 | 812 | 4101 |
| Cooling Degree Days | 0 | 0 | 8 | 42 | 133 | 318 | 445 | 396 | 214 | 39 | 4 | 1 | 1600 |
| Total Precipitation (") | 3.41 | 3.80 | 4.20 | 5.12 | 4.69 | 3.57 | 4.25 | 3.47 | 3.51 | 3.28 | 4.25 | 4.28 | 47.83 |
| Days ≥ 0.1" Precip | 6 | 6 | 8 | 8 | 8 | 6 | 6 | 6 | 6 | 6 | 7 | 7 | 80 |
| Total Snowfall (") | *3.6* | *2.9* | 1.7 | 0.0 | 0.0 | 0.0 | 0.0 | 0.0 | 0.0 | 0.0 | 0.5 | *0.4* | 9.1 |
| Days ≥ 1" Snow Depth | *5* | *5* | 1 | 0 | 0 | 0 | 0 | 0 | 0 | 0 | 0 | *1* | 12 |

## MAMMOTH CAVE *Edmonson County* ELEVATION 801 ft LAT/LONG 37° 11 ' N / 86° 6 ' W

| | JAN | FEB | MAR | APR | MAY | JUN | JUL | AUG | SEP | OCT | NOV | DEC | YEAR |
|---|---|---|---|---|---|---|---|---|---|---|---|---|---|
| Maximum Temp °F | 43.1 | 48.9 | 59.6 | 70.6 | 77.7 | 84.6 | 88.0 | 86.5 | 80.7 | 70.4 | 58.3 | 48.4 | 68.1 |
| Minimum Temp °F | 24.0 | 27.2 | 35.9 | 44.8 | 52.3 | 60.3 | 64.6 | 63.0 | 56.8 | 44.9 | 37.4 | 29.3 | 45.0 |
| Mean Temp °F | 33.5 | 38.0 | 47.7 | 57.7 | 65.0 | 72.5 | 76.3 | 74.8 | 68.8 | 57.7 | 47.8 | 38.9 | 56.6 |
| Days Max Temp ≥ 90 °F | 0 | 0 | 0 | 0 | 0 | 5 | 12 | 9 | 3 | 0 | 0 | 0 | 29 |
| Days Max Temp ≤ 32 °F | 7 | 3 | 0 | 0 | 0 | 0 | 0 | 0 | 0 | 0 | 0 | 2 | 12 |
| Days Min Temp ≤ 32 °F | 23 | 19 | 13 | 4 | 0 | 0 | 0 | 0 | 0 | 4 | 11 | 19 | 93 |
| Days Min Temp ≤ 0 °F | 2 | 1 | 0 | 0 | 0 | 0 | 0 | 0 | 0 | 0 | 0 | 0 | 3 |
| Heating Degree Days | 967 | 753 | 535 | 246 | 88 | 9 | 1 | 2 | 43 | 248 | 512 | 803 | 4207 |
| Cooling Degree Days | 0 | 0 | 6 | *36* | 102 | 251 | 388 | 331 | 160 | 31 | 3 | 1 | 1309 |
| Total Precipitation (") | 3.85 | 3.93 | 4.92 | 4.44 | 5.20 | 4.42 | 4.81 | 3.70 | 4.30 | 3.40 | 4.54 | 5.09 | 52.60 |
| Days ≥ 0.1" Precip | 7 | 6 | 8 | 8 | 8 | 7 | 7 | 6 | 6 | 5 | 7 | 7 | 82 |
| Total Snowfall (") | 5.1 | 4.5 | 1.7 | 0.1 | 0.0 | 0.0 | 0.0 | 0.0 | 0.0 | 0.0 | 0.9 | 2.0 | 14.3 |
| Days ≥ 1" Snow Depth | 7 | 5 | 1 | 0 | 0 | 0 | 0 | 0 | 0 | 0 | 0 | 1 | 14 |

## MANCHESTER 4 W *Clay County* ELEVATION 870 ft LAT/LONG 37° 9 ' N / 83° 49 ' W

| | JAN | FEB | MAR | APR | MAY | JUN | JUL | AUG | SEP | OCT | NOV | DEC | YEAR |
|---|---|---|---|---|---|---|---|---|---|---|---|---|---|
| Maximum Temp °F | 45.5 | 50.0 | 60.5 | 70.6 | 76.8 | 82.9 | 86.2 | 84.7 | 79.5 | 70.3 | 60.0 | 50.4 | 68.1 |
| Minimum Temp °F | 22.7 | 24.7 | 32.7 | 40.8 | 49.8 | 58.0 | 63.0 | 62.0 | 55.2 | 41.7 | 33.7 | 27.1 | 42.6 |
| Mean Temp °F | 34.0 | 37.4 | 46.6 | 55.9 | 63.4 | 70.6 | 74.7 | 73.3 | 67.4 | 56.1 | 46.9 | 38.8 | 55.4 |
| Days Max Temp ≥ 90 °F | 0 | 0 | 0 | 0 | 0 | 2 | 8 | 5 | 1 | 0 | 0 | 0 | 16 |
| Days Max Temp ≤ 32 °F | 4 | 2 | 0 | 0 | 0 | 0 | 0 | 0 | 0 | 0 | 0 | 2 | 8 |
| Days Min Temp ≤ 32 °F | 23 | 21 | 16 | 7 | 1 | 0 | 0 | 0 | 0 | 6 | 15 | 22 | 111 |
| Days Min Temp ≤ 0 °F | 2 | 1 | 0 | 0 | 0 | 0 | 0 | 0 | 0 | 0 | 0 | 0 | 3 |
| Heating Degree Days | 953 | 772 | 566 | 284 | 109 | 15 | 1 | 3 | 48 | 283 | 540 | 805 | 4379 |
| Cooling Degree Days | 0 | 0 | 2 | *13* | 67 | 189 | *323* | *257* | 117 | 14 | 1 | 0 | 983 |
| Total Precipitation (") | 3.88 | 3.65 | 4.56 | 4.08 | 4.58 | 4.27 | 5.24 | 3.93 | 4.04 | 3.30 | 4.07 | 4.42 | 50.02 |
| Days ≥ 0.1" Precip | 8 | 7 | 9 | 7 | 8 | 8 | 8 | 7 | 6 | 6 | 7 | 7 | 88 |
| Total Snowfall (") | na | *3.2* | 1.9 | 0.3 | 0.0 | 0.0 | 0.0 | 0.0 | 0.0 | 0.0 | 0.7 | 1.7 | na |
| Days ≥ 1" Snow Depth | *4* | *4* | 1 | 0 | 0 | 0 | 0 | 0 | 0 | 0 | 0 | *1* | 10 |

## MAYSVILLE SEWAGE PLA *Mason County* ELEVATION 522 ft LAT/LONG 38° 38 ' N / 83° 42 ' W

| | JAN | FEB | MAR | APR | MAY | JUN | JUL | AUG | SEP | OCT | NOV | DEC | YEAR |
|---|---|---|---|---|---|---|---|---|---|---|---|---|---|
| Maximum Temp °F | 39.5 | 43.7 | 54.7 | 65.9 | 74.9 | 83.6 | 87.1 | 85.8 | 79.5 | 68.0 | 55.8 | 44.9 | 65.3 |
| Minimum Temp °F | 20.3 | 21.8 | 30.5 | 39.9 | 49.1 | 58.4 | 63.4 | 61.8 | 55.2 | 42.7 | 34.0 | 25.7 | 41.9 |
| Mean Temp °F | 30.2 | 32.8 | 42.5 | 52.9 | 62.0 | 71.0 | 75.3 | 73.8 | 67.4 | 55.4 | 45.0 | 35.3 | 53.6 |
| Days Max Temp ≥ 90 °F | 0 | 0 | 0 | 0 | 1 | 5 | 10 | 8 | 3 | 0 | 0 | 0 | 27 |
| Days Max Temp ≤ 32 °F | 9 | 6 | 1 | 0 | 0 | 0 | 0 | 0 | 0 | 0 | 0 | 4 | 20 |
| Days Min Temp ≤ 32 °F | 26 | 23 | 19 | 6 | 1 | 0 | 0 | 0 | 0 | 4 | 14 | 23 | 116 |
| Days Min Temp ≤ 0 °F | 2 | 1 | 0 | 0 | 0 | 0 | 0 | 0 | 0 | 0 | 0 | 1 | 4 |
| Heating Degree Days | 1073 | 904 | 691 | 367 | 146 | 18 | 1 | 4 | 58 | 308 | 595 | 914 | 5079 |
| Cooling Degree Days | 0 | 0 | 1 | 11 | 60 | 213 | 351 | 297 | 139 | 19 | 1 | 0 | 1092 |
| Total Precipitation (") | 3.17 | 2.96 | 3.93 | 4.32 | 4.58 | 3.54 | 4.70 | 4.01 | 3.25 | 2.85 | 3.60 | 3.89 | 44.80 |
| Days ≥ 0.1" Precip | 7 | 6 | 8 | 9 | 9 | 7 | 8 | 6 | 6 | 6 | 7 | 8 | 87 |
| Total Snowfall (") | na | na | na | 0.0 | 0.0 | 0.0 | 0.0 | 0.0 | 0.0 | 0.0 | *0.4* | na | na |
| Days ≥ 1" Snow Depth | na | na | *1* | 0 | 0 | 0 | 0 | 0 | 0 | 0 | *0* | na | na |

## MIDDLESBORO 2 N *Bell County* ELEVATION 1128 ft LAT/LONG 36° 37 ' N / 83° 43 ' W

| | JAN | FEB | MAR | APR | MAY | JUN | JUL | AUG | SEP | OCT | NOV | DEC | YEAR |
|---|---|---|---|---|---|---|---|---|---|---|---|---|---|
| Maximum Temp °F | 43.8 | 47.9 | 58.6 | 68.8 | 75.9 | 83.3 | 86.7 | 85.6 | 79.5 | 69.0 | 58.5 | 48.3 | 67.2 |
| Minimum Temp °F | 22.9 | 24.5 | 32.7 | 40.3 | 49.2 | 57.9 | 62.8 | 62.2 | 55.9 | 42.6 | 34.0 | 27.0 | 42.7 |
| Mean Temp °F | 33.4 | 36.2 | 45.7 | 54.6 | 62.5 | 70.6 | 74.8 | 73.9 | 67.7 | 55.8 | 46.3 | 37.7 | 54.9 |
| Days Max Temp ≥ 90 °F | 0 | 0 | 0 | 0 | 0 | 3 | 9 | 6 | 1 | 0 | 0 | 0 | 19 |
| Days Max Temp ≤ 32 °F | 5 | 3 | 0 | 0 | 0 | 0 | 0 | 0 | 0 | 0 | 0 | 2 | 10 |
| Days Min Temp ≤ 32 °F | 25 | 21 | 17 | 7 | 1 | 0 | 0 | 0 | 0 | 6 | 14 | *22* | 113 |
| Days Min Temp ≤ 0 °F | 2 | 1 | 0 | 0 | 0 | 0 | 0 | 0 | 0 | 0 | 0 | 1 | 4 |
| Heating Degree Days | 973 | 807 | 595 | 317 | 125 | 15 | 1 | 2 | 45 | 292 | 556 | 840 | 4568 |
| Cooling Degree Days | *0* | *0* | *3* | *9* | *60* | *207* | *338* | *293* | *132* | *20* | *1* | *0* | 1063 |
| Total Precipitation (") | 4.29 | 3.88 | 4.90 | 4.01 | 5.03 | 4.25 | 5.29 | 4.50 | 3.07 | 3.40 | 4.28 | 4.48 | 51.38 |
| Days ≥ 0.1" Precip | 8 | 8 | 8 | 8 | 9 | 7 | 8 | 8 | 6 | 6 | 7 | 8 | 91 |
| Total Snowfall (") | na | na | na | *0.0* | 0.0 | 0.0 | 0.0 | 0.0 | 0.0 | 0.0 | na | na | na |
| Days ≥ 1" Snow Depth | na | na | *1* | *0* | 0 | 0 | 0 | 0 | 0 | 0 | *0* | na | na |

## MONTICELLO 3 NE *Wayne County* ELEVATION 979 ft LAT/LONG 36° 52 ' N / 84° 50 ' W

| | JAN | FEB | MAR | APR | MAY | JUN | JUL | AUG | SEP | OCT | NOV | DEC | YEAR |
|---|---|---|---|---|---|---|---|---|---|---|---|---|---|
| Maximum Temp °F | 44.7 | 49.4 | 59.4 | 69.4 | 76.4 | 83.7 | 86.9 | 85.9 | 80.2 | 70.1 | 59.1 | 49.1 | 67.9 |
| Minimum Temp °F | 24.9 | 27.2 | 35.4 | 43.8 | 52.2 | 60.7 | 64.8 | 63.2 | 57.1 | 44.8 | 36.9 | 29.3 | 45.0 |
| Mean Temp °F | 34.8 | 38.3 | 47.4 | 56.6 | 64.3 | 72.2 | 75.9 | 74.6 | 68.7 | 57.4 | 48.0 | 39.2 | 56.5 |
| Days Max Temp ≥ 90 °F | 0 | 0 | 0 | 0 | 0 | 4 | 9 | 7 | 2 | 0 | 0 | 0 | 22 |
| Days Max Temp ≤ 32 °F | 5 | 2 | 0 | 0 | 0 | 0 | 0 | 0 | 0 | 0 | 0 | 2 | 9 |
| Days Min Temp ≤ 32 °F | 23 | 20 | 14 | 5 | 0 | 0 | 0 | 0 | 0 | 4 | 11 | 19 | 96 |
| Days Min Temp ≤ 0 °F | 2 | 1 | 0 | 0 | 0 | 0 | 0 | 0 | 0 | 0 | 0 | 0 | 3 |
| Heating Degree Days | 929 | 747 | 543 | 270 | 100 | 10 | 1 | 2 | 42 | 255 | 506 | 793 | 4198 |
| Cooling Degree Days | 0 | 0 | 5 | 20 | 84 | 233 | 364 | 308 | 154 | 29 | 3 | 1 | 1201 |
| Total Precipitation (") | 3.92 | 3.97 | 4.80 | 4.17 | 4.82 | 4.00 | 4.54 | 4.10 | 4.08 | 3.03 | 4.06 | 4.79 | 50.28 |
| Days ≥ 0.1" Precip | 8 | 7 | 9 | 8 | 9 | 7 | 8 | 7 | 6 | 6 | 7 | 8 | 90 |
| Total Snowfall (") | 6.5 | 4.9 | *2.2* | 0.1 | 0.0 | 0.0 | 0.0 | 0.0 | 0.0 | 0.0 | 0.9 | 2.1 | 16.7 |
| Days ≥ 1" Snow Depth | na | *3* | *1* | 0 | 0 | 0 | 0 | 0 | 0 | 0 | 0 | *2* | na |

## MOUNT STERLING 1 NW *Montgomery County* ELEVATION 932 ft LAT/LONG 38° 3 ' N / 83° 57 ' W

| | JAN | FEB | MAR | APR | MAY | JUN | JUL | AUG | SEP | OCT | NOV | DEC | YEAR |
|---|---|---|---|---|---|---|---|---|---|---|---|---|---|
| Maximum Temp °F | *42.0* | *45.8* | 57.2 | *68.0* | *75.6* | *82.7* | *85.7* | *84.2* | *78.3* | *68.0* | *56.8* | *46.8* | 65.9 |
| Minimum Temp °F | *23.9* | 25.2 | 34.9 | *43.7* | *52.9* | *61.3* | *65.7* | *64.0* | *57.4* | *44.8* | *36.8* | *28.6* | 44.9 |
| Mean Temp °F | *32.9* | *35.5* | 46.0 | *55.9* | *64.3* | *72.1* | *75.7* | *74.1* | *67.9* | *56.5* | *46.8* | *37.8* | 55.5 |
| Days Max Temp ≥ 90 °F | *0* | 0 | 0 | *0* | *0* | *3* | *7* | *4* | *1* | *0* | *0* | *0* | 15 |
| Days Max Temp ≤ 32 °F | *7* | 4 | 0 | *0* | *0* | *0* | *0* | *0* | *0* | *0* | *0* | *3* | 14 |
| Days Min Temp ≤ 32 °F | *24* | 21 | 15 | *4* | *0* | *0* | *0* | *0* | *0* | *3* | *12* | *20* | 99 |
| Days Min Temp ≤ 0 °F | *1* | 1 | 0 | *0* | *0* | *0* | *0* | *0* | *0* | *0* | *0* | *0* | 2 |
| Heating Degree Days | *986* | *827* | 586 | *290* | *104* | *12* | *1* | *2* | *48* | *280* | *540* | *837* | 4513 |
| Cooling Degree Days | na | na | *3* | na | na | na | na | na | na | na | na | na | na |
| Total Precipitation (") | *3.41* | 3.26 | 4.13 | *4.02* | *4.50* | *3.90* | *5.58* | *4.72* | *4.01* | *2.58* | *3.18* | *3.68* | 46.97 |
| Days ≥ 0.1" Precip | *7* | 7 | *8* | *8* | *8* | *7* | *7* | *6* | *6* | *5* | *6* | *7* | 82 |
| Total Snowfall (") | na | na | *3.0* | *0.1* | *0.0* | *0.0* | *0.0* | *0.0* | *0.0* | *0.0* | na | na | na |
| Days ≥ 1" Snow Depth | na | na | *2* | *0* | *0* | *0* | *0* | *0* | *0* | *0* | *1* | na | na |

## MOUNT VERNON *Rockcastle County* ELEVATION 1142 ft LAT/LONG 37° 21 ' N / 84° 20 ' W

| | JAN | FEB | MAR | APR | MAY | JUN | JUL | AUG | SEP | OCT | NOV | DEC | YEAR |
|---|---|---|---|---|---|---|---|---|---|---|---|---|---|
| Maximum Temp °F | 42.6 | 47.3 | 58.0 | 68.5 | 76.1 | 83.3 | 86.6 | 85.3 | 79.6 | 69.2 | 57.6 | 47.6 | 66.8 |
| Minimum Temp °F | 23.2 | 25.6 | 34.5 | 43.4 | 51.8 | 60.1 | 64.4 | 62.5 | 56.3 | 43.8 | 36.3 | 28.3 | 44.2 |
| Mean Temp °F | 32.9 | 36.5 | 46.2 | 56.0 | 64.0 | 71.7 | 75.5 | 74.0 | 68.0 | 56.5 | 47.0 | 37.9 | 55.5 |
| Days Max Temp ≥ 90 °F | 0 | 0 | 0 | 0 | 0 | 3 | 8 | 6 | 2 | 0 | 0 | 0 | 19 |
| Days Max Temp ≤ 32 °F | 7 | 4 | 1 | 0 | 0 | 0 | 0 | 0 | 0 | 0 | 0 | 3 | 15 |
| Days Min Temp ≤ 32 °F | 24 | 20 | 15 | 5 | 0 | 0 | 0 | 0 | 0 | 5 | 12 | 20 | 101 |
| Days Min Temp ≤ 0 °F | 2 | 1 | 0 | 0 | 0 | 0 | 0 | 0 | 0 | 0 | 0 | 0 | 3 |
| Heating Degree Days | 986 | 799 | 580 | 290 | 110 | 13 | 1 | 3 | 53 | 278 | 537 | 833 | 4483 |
| Cooling Degree Days | 0 | 0 | 3 | 23 | 82 | 220 | 348 | 287 | 139 | 20 | 2 | 1 | 1125 |
| Total Precipitation (") | 3.95 | 3.86 | 5.09 | 4.41 | 5.15 | 4.25 | 4.85 | 4.12 | 3.95 | 3.37 | 4.29 | 4.96 | 52.25 |
| Days ≥ 0.1" Precip | 8 | 7 | 9 | 8 | 9 | 8 | 8 | 7 | 6 | 6 | 8 | 8 | 92 |
| Total Snowfall (") | 8.0 | *6.1* | 1.7 | 0.2 | 0.0 | 0.0 | 0.0 | 0.0 | 0.0 | 0.0 | 1.0 | *2.0* | 19.0 |
| Days ≥ 1" Snow Depth | na | na | *1* | 0 | 0 | 0 | 0 | 0 | 0 | 0 | *0* | *2* | na |

## MURRAY *Calloway County* ELEVATION 410 ft LAT/LONG 36° 37 ' N / 88° 20 ' W

| | JAN | FEB | MAR | APR | MAY | JUN | JUL | AUG | SEP | OCT | NOV | DEC | YEAR |
|---|---|---|---|---|---|---|---|---|---|---|---|---|---|
| Maximum Temp °F | 42.8 | 48.3 | 59.0 | 70.2 | 77.5 | 86.3 | 89.5 | 88.0 | 81.4 | 70.7 | 58.6 | 48.0 | 68.4 |
| Minimum Temp °F | 25.7 | 29.0 | 38.3 | 48.2 | 56.1 | 64.7 | 68.7 | 66.6 | 60.0 | 47.7 | 39.1 | 30.4 | 47.9 |
| Mean Temp °F | 34.3 | 38.7 | 48.6 | 59.2 | 66.9 | 75.6 | 79.1 | 77.4 | 70.7 | 59.2 | 48.9 | 39.2 | 58.2 |
| Days Max Temp ≥ 90 °F | 0 | 0 | 0 | 0 | 1 | 9 | 16 | 12 | 4 | 0 | 0 | 0 | 42 |
| Days Max Temp ≤ 32 °F | 7 | 3 | 0 | 0 | 0 | 0 | 0 | 0 | 0 | 0 | 0 | 3 | 13 |
| Days Min Temp ≤ 32 °F | 23 | 18 | 11 | 1 | 0 | 0 | 0 | 0 | 0 | 1 | 9 | 18 | 81 |
| Days Min Temp ≤ 0 °F | 1 | 0 | 0 | 0 | 0 | 0 | 0 | 0 | 0 | 0 | 0 | 0 | 1 |
| Heating Degree Days | 945 | 737 | 507 | 208 | 58 | 2 | 0 | 0 | 32 | 209 | 481 | 794 | 3973 |
| Cooling Degree Days | 0 | 0 | *7* | *39* | *118* | *344* | 474 | *425* | *215* | 37 | 3 | 1 | 1663 |
| Total Precipitation (") | 4.21 | 4.36 | 5.06 | 5.41 | 5.05 | 4.35 | 4.63 | 3.48 | 3.76 | 3.64 | 5.31 | 5.43 | 54.69 |
| Days ≥ 0.1" Precip | 7 | 7 | 8 | 8 | *7* | 7 | 6 | 6 | 6 | 5 | 7 | 7 | 81 |
| Total Snowfall (") | 5.4 | 4.3 | 1.9 | 0.0 | 0.0 | 0.0 | 0.0 | 0.0 | 0.0 | 0.1 | 0.2 | *0.9* | 12.8 |
| Days ≥ 1" Snow Depth | 5 | 4 | 1 | 0 | 0 | 0 | 0 | 0 | 0 | 0 | 0 | *1* | 11 |

## NOLIN RIVER LAKE *Edmonson County* ELEVATION 610 ft LAT/LONG 37° 17 ' N / 86° 15 ' W

| | JAN | FEB | MAR | APR | MAY | JUN | JUL | AUG | SEP | OCT | NOV | DEC | YEAR |
|---|---|---|---|---|---|---|---|---|---|---|---|---|---|
| Maximum Temp °F | 42.1 | 47.0 | 57.7 | 69.0 | 77.0 | 85.2 | 89.0 | 87.7 | 81.6 | 70.3 | 58.2 | 47.9 | 67.7 |
| Minimum Temp °F | 20.6 | 22.8 | 32.0 | 41.3 | 49.4 | 58.2 | 62.7 | 61.2 | 54.9 | 42.2 | 34.4 | 26.5 | 42.2 |
| Mean Temp °F | 31.4 | 34.9 | 44.9 | 55.1 | 63.2 | 71.7 | 75.9 | 74.5 | 68.3 | 56.3 | 46.4 | 37.2 | 55.0 |
| Days Max Temp ≥ 90 °F | 0 | 0 | 0 | 0 | 1 | 8 | 15 | 12 | 5 | 0 | 0 | 0 | 41 |
| Days Max Temp ≤ 32 °F | 7 | 4 | 1 | 0 | 0 | 0 | 0 | 0 | 0 | 0 | 0 | 3 | 15 |
| Days Min Temp ≤ 32 °F | 26 | 23 | 17 | 6 | 1 | 0 | 0 | 0 | 0 | 6 | 14 | 22 | 115 |
| Days Min Temp ≤ 0 °F | 2 | 1 | 0 | 0 | 0 | 0 | 0 | 0 | 0 | 0 | 0 | 1 | 4 |
| Heating Degree Days | 1036 | 843 | 622 | 309 | 122 | 16 | 1 | 1 | 51 | 284 | 555 | 854 | 4694 |
| Cooling Degree Days | 0 | 0 | 4 | 18 | 73 | 228 | 360 | 308 | 146 | 21 | 2 | 1 | 1161 |
| Total Precipitation (") | 3.78 | 4.08 | 4.44 | 4.41 | 5.46 | 4.08 | 4.92 | 3.66 | 4.12 | 3.20 | 4.37 | 4.97 | 51.49 |
| Days ≥ 0.1" Precip | 7 | 7 | 8 | 8 | 8 | 7 | 7 | 6 | 6 | 6 | 8 | 8 | 86 |
| Total Snowfall (") | na | na | *0.6* | 0.0 | 0.0 | 0.0 | 0.0 | 0.0 | 0.0 | 0.0 | *0.1* | 0.7 | na |
| Days ≥ 1" Snow Depth | 6 | 5 | 1 | 0 | 0 | 0 | 0 | 0 | 0 | 0 | 0 | 1 | 13 |

## OWENSBORO 3 W *Daviess County* ELEVATION 420 ft LAT/LONG 37° 46 ' N / 87° 9 ' W

| | JAN | FEB | MAR | APR | MAY | JUN | JUL | AUG | SEP | OCT | NOV | DEC | YEAR |
|---|---|---|---|---|---|---|---|---|---|---|---|---|---|
| Maximum Temp °F | 41.5 | 46.9 | 57.9 | 69.6 | 77.9 | 86.3 | 89.4 | 88.1 | 82.0 | 71.2 | 57.6 | 46.5 | 67.9 |
| Minimum Temp °F | 24.0 | 27.3 | 36.3 | 46.0 | 54.6 | 63.2 | 67.1 | 64.8 | 58.3 | 45.9 | 37.4 | 29.1 | 46.2 |
| Mean Temp °F | 32.8 | 37.1 | 47.1 | 57.8 | 66.2 | 74.7 | 78.3 | 76.5 | 70.2 | 58.5 | 47.5 | 37.8 | 57.0 |
| Days Max Temp ≥ 90 °F | 0 | 0 | 0 | 0 | 2 | 9 | 16 | 13 | 5 | 0 | 0 | 0 | 45 |
| Days Max Temp ≤ 32 °F | 7 | 4 | 0 | 0 | 0 | 0 | 0 | 0 | 0 | 0 | 0 | 3 | 14 |
| Days Min Temp ≤ 32 °F | 24 | 19 | 12 | 3 | 0 | 0 | 0 | 0 | 0 | 3 | 10 | 20 | 91 |
| Days Min Temp ≤ 0 °F | 2 | 1 | 0 | 0 | 0 | 0 | 0 | 0 | 0 | 0 | 0 | 0 | 3 |
| Heating Degree Days | 991 | 781 | 551 | 240 | 71 | 4 | 0 | 1 | 35 | 228 | 521 | 836 | 4259 |
| Cooling Degree Days | 0 | 0 | 5 | 29 | 115 | 312 | 435 | 387 | 200 | 33 | 3 | 1 | 1520 |
| Total Precipitation (") | 3.28 | 3.74 | 4.37 | 4.79 | 4.40 | 3.69 | 4.01 | 3.57 | 3.78 | 3.21 | 4.29 | 3.98 | 47.11 |
| Days ≥ 0.1" Precip | 6 | 6 | 8 | 7 | 8 | 6 | 6 | 6 | 6 | 6 | 7 | 7 | 79 |
| Total Snowfall (") | *3.4* | 3.7 | 2.2 | 0.1 | 0.0 | 0.0 | 0.0 | 0.0 | 0.0 | 0.1 | 0.5 | 1.2 | 11.2 |
| Days ≥ 1" Snow Depth | na | 5 | *1* | 0 | 0 | 0 | 0 | 0 | 0 | 0 | 0 | 1 | na |

## PADUCAH BARKLEY FLD *McCracken County* ELEVATION 397 ft LAT/LONG 37° 4 ' N / 88° 46 ' W

| | JAN | FEB | MAR | APR | MAY | JUN | JUL | AUG | SEP | OCT | NOV | DEC | YEAR |
|---|---|---|---|---|---|---|---|---|---|---|---|---|---|
| Maximum Temp °F | 41.7 | 46.8 | 57.8 | 69.1 | 77.3 | 86.0 | 89.5 | 87.6 | 81.0 | 70.3 | 57.5 | 46.9 | 67.6 |
| Minimum Temp °F | 24.3 | 27.9 | 37.5 | 47.3 | 56.0 | 64.9 | 68.9 | 66.1 | 59.1 | 46.7 | 38.4 | 29.3 | 47.2 |
| Mean Temp °F | 33.0 | 37.4 | 47.7 | 58.2 | 66.7 | 75.5 | 79.2 | 76.9 | 70.1 | 58.5 | 48.0 | 38.1 | 57.4 |
| Days Max Temp ≥ 90 °F | 0 | 0 | 0 | 0 | 1 | 9 | 17 | 12 | 4 | 0 | 0 | 0 | 43 |
| Days Max Temp ≤ 32 °F | 7 | 4 | 1 | 0 | 0 | 0 | 0 | 0 | 0 | 0 | 0 | 3 | 15 |
| Days Min Temp ≤ 32 °F | 24 | 18 | 11 | 2 | 0 | 0 | 0 | 0 | 0 | 2 | 10 | 20 | 87 |
| Days Min Temp ≤ 0 °F | 1 | 0 | 0 | 0 | 0 | 0 | 0 | 0 | 0 | 0 | 0 | 0 | 1 |
| Heating Degree Days | 983 | 774 | 535 | 233 | 65 | 2 | 0 | 1 | 36 | 231 | 507 | 826 | 4193 |
| Cooling Degree Days | 0 | 0 | 7 | 38 | 128 | 337 | 472 | 398 | 201 | 38 | 4 | 1 | 1624 |
| Total Precipitation (") | 3.41 | 3.82 | 4.22 | 5.07 | 4.77 | 3.91 | 4.19 | 3.21 | 3.86 | 3.30 | 4.40 | 4.64 | 48.80 |
| Days ≥ 0.1" Precip | 6 | 6 | 8 | 8 | 7 | 6 | 6 | 5 | 5 | 5 | 7 | 7 | 76 |
| Total Snowfall (") | 4.3 | 3.7 | 1.9 | 0.1 | 0.0 | 0.0 | 0.0 | 0.0 | 0.0 | 0.1 | 0.1 | 1.5 | 11.7 |
| Days ≥ 1" Snow Depth | 6 | 5 | 1 | 0 | 0 | 0 | 0 | 0 | 0 | 0 | 0 | 1 | 13 |

## PADUCAH WALKER *McCracken County* ELEVATION 340 ft LAT/LONG 37° 4 ' N / 88° 35 ' W

| | JAN | FEB | MAR | APR | MAY | JUN | JUL | AUG | SEP | OCT | NOV | DEC | YEAR |
|---|---|---|---|---|---|---|---|---|---|---|---|---|---|
| Maximum Temp °F | *40.9* | *46.8* | *57.5* | *68.9* | *77.3* | *86.4* | *90.2* | *88.4* | *81.9* | *70.2* | *57.8* | *46.6* | 67.7 |
| Minimum Temp °F | *23.8* | *27.3* | *37.3* | *46.8* | *55.3* | *64.3* | *68.8* | *66.2* | *59.4* | *46.8* | *38.1* | *28.9* | 46.9 |
| Mean Temp °F | *32.4* | *37.1* | *47.4* | *57.9* | *66.3* | *75.3* | *79.5* | *77.3* | *70.7* | *58.5* | *48.0* | *37.8* | 57.4 |
| Days Max Temp ≥ 90 °F | *0* | *0* | *0* | *0* | *2* | *10* | *19* | *14* | *5* | *0* | *0* | *0* | 50 |
| Days Max Temp ≤ 32 °F | *8* | *4* | *1* | *0* | *0* | *0* | *0* | *0* | *0* | *0* | *0* | *3* | 16 |
| Days Min Temp ≤ 32 °F | *24* | *19* | *11* | *2* | *0* | *0* | *0* | *0* | *0* | *1* | *9* | *20* | 86 |
| Days Min Temp ≤ 0 °F | *1* | *0* | *0* | *0* | *0* | *0* | *0* | *0* | *0* | *0* | *0* | *0* | 1 |
| Heating Degree Days | *1004* | *782* | *543* | *245* | *70* | *3* | *0* | *1* | *34* | *230* | *508* | *837* | 4257 |
| Cooling Degree Days | 0 | 0 | 6 | 36 | 118 | 325 | 467 | 399 | 202 | 33 | 2 | 0 | 1588 |
| Total Precipitation (") | *3.06* | *3.84* | *4.40* | *4.66* | *4.45* | *3.57* | *3.96* | *3.12* | *3.46* | *3.33* | *4.53* | *4.40* | 46.78 |
| Days ≥ 0.1" Precip | *6* | *6* | *8* | *7* | *7* | *6* | *6* | *5* | *5* | *6* | *7* | *7* | 76 |
| Total Snowfall (") | na | na | *1.2* | *0.0* | *0.0* | *0.0* | *0.0* | *0.0* | *0.0* | *0.0* | *0.1* | *1.0* | na |
| Days ≥ 1" Snow Depth | na | na | *0* | *0* | *0* | *0* | *0* | *0* | *0* | *0* | *0* | *1* | na |

## PRINCETON 1 SE *Caldwell County* ELEVATION 502 ft LAT/LONG 37° 7 ' N / 87° 52 ' W

| | JAN | FEB | MAR | APR | MAY | JUN | JUL | AUG | SEP | OCT | NOV | DEC | YEAR |
|---|---|---|---|---|---|---|---|---|---|---|---|---|---|
| Maximum Temp °F | 43.3 | 48.9 | 59.9 | 70.7 | 78.0 | 86.1 | 89.6 | 88.2 | 82.1 | 71.5 | 59.1 | 48.5 | 68.8 |
| Minimum Temp °F | 24.7 | 28.3 | 37.5 | 47.2 | 55.0 | 63.6 | 67.6 | 65.5 | 59.0 | 47.1 | 39.0 | 30.0 | 47.0 |
| Mean Temp °F | 34.0 | 38.6 | 48.7 | 58.9 | 66.5 | 74.8 | 78.6 | 76.9 | 70.6 | 59.3 | 49.1 | 39.3 | 57.9 |
| Days Max Temp ≥ 90 °F | 0 | 0 | 0 | 0 | 1 | 8 | 17 | 13 | 5 | 0 | 0 | 0 | 44 |
| Days Max Temp ≤ 32 °F | 6 | 3 | 0 | 0 | 0 | 0 | 0 | 0 | 0 | 0 | 0 | 2 | 11 |
| Days Min Temp ≤ 32 °F | 23 | 17 | 12 | 2 | 0 | 0 | 0 | 0 | 0 | 3 | 9 | 18 | 84 |
| Days Min Temp ≤ 0 °F | 1 | 1 | 0 | 0 | 0 | 0 | 0 | 0 | 0 | 0 | 0 | 0 | 2 |
| Heating Degree Days | 955 | 739 | 507 | 217 | 65 | 3 | 0 | 1 | 31 | 213 | 477 | 792 | 4000 |
| Cooling Degree Days | 0 | 0 | 8 | 39 | 117 | 317 | 456 | 400 | 212 | 42 | 4 | 1 | 1596 |
| Total Precipitation (") | 3.81 | 4.40 | 4.54 | 4.69 | 4.88 | 3.81 | 4.21 | 3.85 | 3.33 | 3.27 | 4.77 | 5.11 | 50.67 |
| Days ≥ 0.1" Precip | 6 | 7 | 8 | 8 | 8 | 6 | 6 | 6 | 6 | 6 | 7 | 8 | 82 |
| Total Snowfall (") | 6.3 | 4.4 | 2.4 | 0.2 | 0.0 | 0.0 | 0.0 | 0.0 | 0.0 | 0.1 | 0.5 | 1.6 | 15.5 |
| Days ≥ 1" Snow Depth | na | *2* | 1 | 0 | 0 | 0 | 0 | 0 | 0 | 0 | 0 | 1 | na |

## ROCHESTER FERRY *Ohio County* ELEVATION 400 ft LAT/LONG 37° 13 ' N / 86° 54 ' W

| | JAN | FEB | MAR | APR | MAY | JUN | JUL | AUG | SEP | OCT | NOV | DEC | YEAR |
|---|---|---|---|---|---|---|---|---|---|---|---|---|---|
| Maximum Temp °F | 41.8 | 46.6 | 57.5 | 68.9 | 76.7 | 85.0 | 88.5 | 86.6 | 80.8 | 69.3 | *58.0* | *47.5* | 67.3 |
| Minimum Temp °F | 22.1 | 24.8 | 34.3 | 43.8 | 52.1 | 60.9 | 65.0 | 62.5 | 56.1 | 43.0 | *35.9* | 27.1 | 44.0 |
| Mean Temp °F | 32.0 | *35.8* | 46.0 | 56.4 | 64.4 | 73.0 | 76.8 | 74.6 | 68.5 | 56.0 | *47.0* | *37.6* | 55.7 |
| Days Max Temp ≥ 90 °F | 0 | 0 | 0 | 0 | 0 | *7* | *14* | 10 | 3 | 0 | 0 | 0 | 34 |
| Days Max Temp ≤ 32 °F | 7 | 4 | 0 | 0 | 0 | 0 | 0 | 0 | 0 | 0 | 0 | 3 | 14 |
| Days Min Temp ≤ 32 °F | 26 | *21* | *15* | 4 | 0 | 0 | 0 | 0 | 0 | 5 | *13* | 22 | 106 |
| Days Min Temp ≤ 0 °F | 2 | 1 | 0 | 0 | 0 | 0 | 0 | 0 | 0 | 0 | 0 | 0 | 3 |
| Heating Degree Days | 1021 | *814* | 590 | *279* | *94* | *9* | 1 | *2* | *50* | *291* | *541* | *841* | 4533 |
| Cooling Degree Days | *0* | na | *4* | *24* | *88* | *274* | 403 | *324* | *163* | *21* | na | na | na |
| Total Precipitation (") | 3.44 | 3.63 | 4.51 | 4.24 | 4.58 | 3.69 | 4.18 | 3.41 | 3.74 | 3.21 | 4.61 | 4.48 | 47.72 |
| Days ≥ 0.1" Precip | 6 | 6 | 7 | 8 | 8 | *6* | 6 | *6* | 5 | 6 | *7* | 7 | 78 |
| Total Snowfall (") | na | na | na | 0.0 | 0.0 | 0.0 | 0.0 | 0.0 | 0.0 | 0.0 | *0.3* | na | na |
| Days ≥ 1" Snow Depth | *7* | *5* | *1* | 0 | 0 | 0 | 0 | 0 | 0 | 0 | 0 | *1* | 14 |

## ROUGH RIVER DAM *Breckinridge County* ELEVATION 551 ft LAT/LONG 37° 37 ' N / 86° 30 ' W

| | JAN | FEB | MAR | APR | MAY | JUN | JUL | AUG | SEP | OCT | NOV | DEC | YEAR |
|---|---|---|---|---|---|---|---|---|---|---|---|---|---|
| Maximum Temp °F | 40.4 | 45.4 | 56.3 | 67.9 | 76.0 | 84.6 | 88.6 | 87.1 | 80.6 | 69.0 | 57.0 | 46.1 | 66.6 |
| Minimum Temp °F | 19.7 | 22.4 | 32.1 | 42.1 | 50.3 | 59.3 | 63.6 | 61.2 | 54.3 | 41.5 | 34.3 | 25.3 | 42.2 |
| Mean Temp °F | 30.1 | 33.9 | 44.2 | 55.0 | 63.1 | 72.0 | 76.1 | 74.2 | 67.5 | 55.3 | 45.7 | 35.7 | 54.4 |
| Days Max Temp ≥ 90 °F | 0 | 0 | 0 | 0 | 1 | 7 | 14 | 11 | 4 | 0 | 0 | 0 | 37 |
| Days Max Temp ≤ 32 °F | 8 | 5 | 1 | 0 | 0 | 0 | 0 | 0 | 0 | 0 | 0 | 4 | 18 |
| Days Min Temp ≤ 32 °F | 26 | 22 | 18 | 6 | 1 | 0 | 0 | 0 | 0 | 7 | 14 | 23 | 117 |
| Days Min Temp ≤ 0 °F | 3 | 1 | 0 | 0 | 0 | 0 | 0 | 0 | 0 | 0 | 0 | 1 | 5 |
| Heating Degree Days | 1076 | 871 | 641 | 314 | 127 | 14 | 1 | 3 | 63 | 315 | 573 | 903 | 4901 |
| Cooling Degree Days | 0 | 0 | 4 | 19 | 65 | 217 | 359 | 291 | 132 | 18 | 1 | 1 | 1107 |
| Total Precipitation (") | 3.28 | 3.86 | 4.20 | 4.43 | 4.93 | 3.75 | 4.74 | 3.55 | 3.70 | 3.21 | 3.99 | 4.21 | 47.85 |
| Days ≥ 0.1" Precip | 6 | 6 | 8 | 7 | 8 | 6 | 7 | 6 | 6 | 6 | 7 | 7 | 80 |
| Total Snowfall (") | *2.7* | *3.0* | 1.0 | 0.1 | 0.0 | 0.0 | 0.0 | 0.0 | 0.0 | 0.0 | 0.3 | *0.5* | 7.6 |
| Days ≥ 1" Snow Depth | *5* | 4 | 1 | 0 | 0 | 0 | 0 | 0 | 0 | 0 | 0 | *1* | 11 |

## SCOTTSVILLE 3 SSW *Allen County* ELEVATION 771 ft LAT/LONG 36° 45 ' N / 86° 12 ' W

| | JAN | FEB | MAR | APR | MAY | JUN | JUL | AUG | SEP | OCT | NOV | DEC | YEAR |
|---|---|---|---|---|---|---|---|---|---|---|---|---|---|
| Maximum Temp °F | 43.4 | 49.1 | 59.8 | 70.1 | 76.1 | 83.5 | 87.0 | 85.9 | 80.0 | 70.1 | 58.2 | 48.5 | 67.6 |
| Minimum Temp °F | 26.3 | 29.7 | 38.7 | 47.9 | 55.8 | 63.6 | 67.6 | 66.1 | 60.5 | 49.0 | 40.1 | 31.5 | 48.1 |
| Mean Temp °F | 34.9 | 39.4 | 49.3 | 59.0 | 66.0 | 73.6 | 77.3 | 76.0 | 70.2 | 59.6 | 49.2 | 40.1 | 57.9 |
| Days Max Temp ≥ 90 °F | 0 | 0 | 0 | 0 | 0 | 3 | 10 | 8 | 3 | 0 | 0 | 0 | 24 |
| Days Max Temp ≤ 32 °F | 6 | 3 | 0 | 0 | 0 | 0 | 0 | 0 | 0 | 0 | 0 | 2 | 11 |
| Days Min Temp ≤ 32 °F | 21 | 17 | 10 | 2 | 0 | 0 | 0 | 0 | 0 | 1 | 7 | 17 | 75 |
| Days Min Temp ≤ 0 °F | 1 | 0 | 0 | 0 | 0 | 0 | 0 | 0 | 0 | 0 | 0 | 0 | 1 |
| Heating Degree Days | 927 | 715 | 489 | 214 | 70 | 4 | 0 | 1 | 31 | 201 | 472 | 767 | 3891 |
| Cooling Degree Days | 0 | 0 | 8 | 40 | 108 | 273 | 407 | 360 | 196 | 41 | 4 | 1 | 1438 |
| Total Precipitation (") | 3.93 | 4.47 | 5.03 | 4.32 | 5.63 | 4.50 | 4.38 | 3.93 | 4.00 | 3.36 | 4.59 | 5.10 | 53.24 |
| Days ≥ 0.1" Precip | 8 | 8 | 9 | 8 | 9 | 7 | 8 | 6 | 6 | 6 | 7 | 8 | 90 |
| Total Snowfall (") | 4.9 | 4.5 | 1.5 | 0.1 | 0.0 | 0.0 | 0.0 | 0.0 | 0.0 | 0.0 | 0.8 | 1.5 | 13.3 |
| Days ≥ 1" Snow Depth | 7 | 4 | 1 | 0 | 0 | 0 | 0 | 0 | 0 | 0 | 0 | 1 | 13 |

## SHELBYVILLE 1 E *Shelby County* ELEVATION 728 ft LAT/LONG 38° 12 ' N / 85° 12 ' W

| | JAN | FEB | MAR | APR | MAY | JUN | JUL | AUG | SEP | OCT | NOV | DEC | YEAR |
|---|---|---|---|---|---|---|---|---|---|---|---|---|---|
| Maximum Temp °F | 38.9 | 44.1 | 54.8 | 65.8 | 74.9 | 82.9 | 86.9 | 85.3 | 79.2 | 67.7 | 55.3 | 44.4 | 65.0 |
| Minimum Temp °F | 18.8 | 21.3 | 30.3 | 39.5 | 49.2 | 57.9 | 62.6 | 60.5 | 53.4 | 40.3 | 33.0 | 24.4 | 40.9 |
| Mean Temp °F | 28.9 | 32.7 | 42.6 | 52.7 | 62.0 | 70.4 | 74.8 | 73.0 | 66.3 | 54.0 | 44.1 | 34.3 | 53.0 |
| Days Max Temp ≥ 90 °F | 0 | 0 | 0 | 0 | 0 | 4 | 10 | 7 | 2 | 0 | 0 | 0 | 23 |
| Days Max Temp ≤ 32 °F | 9 | 6 | 1 | 0 | 0 | 0 | 0 | 0 | 0 | 0 | 0 | 4 | 20 |
| Days Min Temp ≤ 32 °F | 27 | 23 | 19 | 8 | 1 | 0 | 0 | 0 | 0 | 8 | 15 | 23 | 124 |
| Days Min Temp ≤ 0 °F | 3 | 2 | 0 | 0 | 0 | 0 | 0 | 0 | 0 | 0 | 0 | 1 | 6 |
| Heating Degree Days | 1113 | 906 | 689 | 375 | 150 | 23 | 2 | 7 | 78 | 349 | 621 | 943 | 5256 |
| Cooling Degree Days | 0 | 0 | 1 | 11 | 64 | 188 | 323 | 254 | 113 | 13 | 1 | 0 | 968 |
| Total Precipitation (") | 3.18 | 3.49 | 4.23 | 4.26 | 4.77 | 3.90 | 4.87 | 3.83 | 3.17 | 2.99 | 3.89 | 3.92 | 46.50 |
| Days ≥ 0.1" Precip | 6 | 6 | 8 | 8 | 7 | 7 | 7 | 6 | 5 | 6 | 7 | 7 | 80 |
| Total Snowfall (") | 5.4 | 3.7 | 2.0 | 0.1 | 0.0 | 0.0 | 0.0 | 0.0 | 0.0 | 0.0 | 0.6 | 1.4 | 13.2 |
| Days ≥ 1" Snow Depth | 7 | 5 | 1 | 0 | 0 | 0 | 0 | 0 | 0 | 0 | 0 | 2 | 15 |

## SOMERSET 2 N *Pulaski County* ELEVATION 1020 ft LAT/LONG 37° 7 ' N / 84° 37 ' W

| | JAN | FEB | MAR | APR | MAY | JUN | JUL | AUG | SEP | OCT | NOV | DEC | YEAR |
|---|---|---|---|---|---|---|---|---|---|---|---|---|---|
| Maximum Temp °F | 43.1 | 48.0 | 58.5 | 68.9 | 75.9 | 83.3 | 86.4 | 85.1 | 79.0 | 69.0 | 57.4 | 48.0 | 66.9 |
| Minimum Temp °F | 24.3 | 26.5 | 35.1 | 43.6 | 51.9 | 60.1 | 64.4 | 62.6 | 56.3 | 43.9 | 36.3 | 29.0 | 44.5 |
| Mean Temp °F | 33.7 | 37.3 | 46.8 | 56.3 | 63.9 | 71.7 | 75.4 | 73.9 | 67.7 | 56.5 | 46.8 | 38.5 | 55.7 |
| Days Max Temp ≥ 90 °F | 0 | 0 | 0 | 0 | 0 | 3 | 8 | 6 | 2 | 0 | 0 | 0 | 19 |
| Days Max Temp ≤ 32 °F | 6 | 3 | 0 | 0 | 0 | 0 | 0 | 0 | 0 | 0 | 0 | 2 | 11 |
| Days Min Temp ≤ 32 °F | 22 | 20 | 14 | 5 | 0 | 0 | 0 | 0 | 0 | 5 | 12 | 20 | 98 |
| Days Min Temp ≤ 0 °F | 2 | 1 | 0 | 0 | 0 | 0 | 0 | 0 | 0 | 0 | 0 | 0 | 3 |
| Heating Degree Days | 962 | 776 | 561 | 275 | 102 | 11 | 1 | 3 | 51 | 274 | 540 | 815 | 4371 |
| Cooling Degree Days | 0 | 0 | 4 | 20 | 82 | 235 | 359 | 300 | 134 | 19 | 2 | 0 | 1155 |
| Total Precipitation (") | 4.03 | 3.65 | 4.63 | 4.28 | 5.01 | 4.22 | 4.73 | 4.08 | 4.02 | 3.54 | 4.22 | 4.78 | 51.19 |
| Days ≥ 0.1" Precip | 8 | 7 | 9 | 8 | 9 | 8 | 8 | 7 | 6 | 6 | 8 | 7 | 91 |
| Total Snowfall (") | *5.3* | *3.8* | *0.8* | 0.0 | 0.0 | 0.0 | 0.0 | 0.0 | 0.0 | 0.0 | 0.5 | na | na |
| Days ≥ 1" Snow Depth | na | na | 1 | 0 | 0 | 0 | 0 | 0 | 0 | 0 | 0 | na | na |

## SUMMER SHADE *Metcalfe County* ELEVATION 860 ft LAT/LONG 36° 53 ' N / 85° 43 ' W

| | JAN | FEB | MAR | APR | MAY | JUN | JUL | AUG | SEP | OCT | NOV | DEC | YEAR |
|---|---|---|---|---|---|---|---|---|---|---|---|---|---|
| Maximum Temp °F | 43.3 | 48.7 | 59.2 | 69.4 | 76.4 | 84.0 | 87.2 | 85.7 | 79.6 | 69.1 | 58.2 | 48.5 | 67.4 |
| Minimum Temp °F | 24.5 | 27.5 | 36.3 | 45.2 | 53.3 | 61.4 | 65.5 | 63.8 | 57.8 | 45.4 | 37.0 | 29.4 | 45.6 |
| Mean Temp °F | 33.9 | 38.1 | 47.7 | 57.3 | 64.9 | 72.7 | 76.4 | 74.8 | 68.7 | 57.3 | 47.6 | 39.0 | 56.5 |
| Days Max Temp ≥ 90 °F | 0 | 0 | 0 | 0 | 0 | 4 | 11 | 7 | 2 | 0 | 0 | 0 | 24 |
| Days Max Temp ≤ 32 °F | 6 | 3 | 0 | 0 | 0 | 0 | 0 | 0 | 0 | 0 | 0 | 2 | 11 |
| Days Min Temp ≤ 32 °F | 23 | 19 | 13 | 4 | 0 | 0 | 0 | 0 | 0 | 4 | 11 | 20 | 94 |
| Days Min Temp ≤ 0 °F | 2 | 1 | 0 | 0 | 0 | 0 | 0 | 0 | 0 | 0 | 0 | 0 | 3 |
| Heating Degree Days | 956 | 753 | 533 | 252 | 89 | 7 | 0 | 1 | 43 | 258 | 518 | 801 | 4211 |
| Cooling Degree Days | 0 | 0 | 5 | 28 | 101 | 264 | 390 | 330 | 165 | 29 | 2 | 1 | 1315 |
| Total Precipitation (") | 3.75 | 4.13 | 4.86 | 4.12 | 4.93 | 4.40 | 4.98 | 3.75 | 4.21 | 3.22 | 4.34 | 5.15 | 51.84 |
| Days ≥ 0.1" Precip | 7 | 7 | 9 | 8 | 8 | 7 | 7 | 6 | 6 | 6 | 8 | 8 | 87 |
| Total Snowfall (") | 5.1 | 4.0 | 1.1 | 0.1 | 0.0 | 0.0 | 0.0 | 0.0 | 0.0 | 0.0 | 0.7 | 1.6 | 12.6 |
| Days ≥ 1" Snow Depth | 7 | 4 | 1 | 0 | 0 | 0 | 0 | 0 | 0 | 0 | 0 | 1 | 13 |

## WARSAW MARKLAND DAM *Gallatin County* ELEVATION 466 ft LAT/LONG 38° 46 ' N / 84° 58 ' W

| | JAN | FEB | MAR | APR | MAY | JUN | JUL | AUG | SEP | OCT | NOV | DEC | YEAR |
|---|---|---|---|---|---|---|---|---|---|---|---|---|---|
| Maximum Temp °F | 39.3 | 43.4 | 54.2 | 65.8 | 75.0 | 82.9 | 86.7 | 85.4 | 79.2 | 67.9 | 55.6 | 44.3 | 65.0 |
| Minimum Temp °F | 19.2 | 21.8 | 30.7 | 39.9 | 49.1 | 58.3 | 63.1 | 61.3 | 54.9 | 42.2 | 33.9 | 25.3 | 41.6 |
| Mean Temp °F | 29.5 | 32.6 | 42.5 | 52.7 | 62.1 | 70.6 | 74.9 | 73.3 | 67.1 | 55.1 | 44.8 | 34.8 | 53.3 |
| Days Max Temp ≥ 90 °F | 0 | 0 | 0 | 0 | 0 | 5 | 10 | 7 | 2 | 0 | 0 | 0 | 24 |
| Days Max Temp ≤ 32 °F | 9 | 6 | 1 | 0 | 0 | 0 | 0 | 0 | 0 | 0 | 0 | 4 | 20 |
| Days Min Temp ≤ 32 °F | 27 | 23 | 18 | 7 | 1 | 0 | 0 | 0 | 0 | 5 | 14 | 23 | 118 |
| Days Min Temp ≤ 0 °F | 3 | 1 | 0 | 0 | 0 | 0 | 0 | 0 | 0 | 0 | 0 | 1 | 5 |
| Heating Degree Days | 1095 | 909 | 692 | 371 | 145 | 20 | 2 | 4 | 62 | 318 | 599 | 929 | 5146 |
| Cooling Degree Days | 0 | 0 | 1 | 10 | 69 | 206 | 339 | 283 | 126 | 17 | 1 | 0 | 1052 |
| Total Precipitation (") | 2.82 | 2.82 | 4.00 | 4.16 | 4.30 | 4.16 | 4.00 | 3.85 | 3.30 | 2.97 | 3.59 | 3.42 | 43.39 |
| Days ≥ 0.1" Precip | 6 | 6 | 8 | 8 | 8 | 7 | 7 | 6 | 5 | 6 | 7 | 7 | 81 |
| Total Snowfall (") | 4.2 | *3.9* | 2.2 | 0.0 | 0.0 | 0.0 | 0.0 | 0.0 | 0.0 | 0.0 | 0.6 | 1.3 | 12.2 |
| Days ≥ 1" Snow Depth | 6 | 6 | 1 | 0 | 0 | 0 | 0 | 0 | 0 | 0 | 0 | 2 | 15 |

## WEST LIBERTY *Morgan County* ELEVATION 801 ft LAT/LONG 37° 55 ' N / 83° 15 ' W

| | JAN | FEB | MAR | APR | MAY | JUN | JUL | AUG | SEP | OCT | NOV | DEC | YEAR |
|---|---|---|---|---|---|---|---|---|---|---|---|---|---|
| Maximum Temp °F | 41.9 | 46.7 | 57.0 | 68.4 | 76.2 | 83.2 | 86.8 | 85.4 | 79.5 | 69.4 | 57.7 | 47.0 | 66.6 |
| Minimum Temp °F | 18.6 | 20.6 | 28.8 | 37.7 | 47.2 | 56.2 | 61.4 | 59.6 | 52.2 | 38.9 | 30.9 | 23.8 | 39.7 |
| Mean Temp °F | 30.1 | 33.7 | 42.9 | 53.0 | 61.7 | 69.7 | 74.1 | 72.5 | 66.0 | 54.2 | 44.1 | 35.3 | 53.1 |
| Days Max Temp ≥ 90 °F | 0 | 0 | 0 | 0 | 1 | 5 | 10 | 7 | 2 | 0 | 0 | 0 | 25 |
| Days Max Temp ≤ 32 °F | 7 | 4 | 1 | 0 | 0 | 0 | 0 | 0 | 0 | 0 | 0 | 3 | 15 |
| Days Min Temp ≤ 32 °F | 27 | 24 | 20 | 10 | 2 | 0 | 0 | 0 | 0 | 9 | 18 | 24 | 134 |
| Days Min Temp ≤ 0 °F | 3 | 2 | 0 | 0 | 0 | 0 | 0 | 0 | 0 | 0 | 0 | 1 | 6 |
| Heating Degree Days | 1074 | 878 | 679 | 363 | 151 | 25 | 2 | 5 | 76 | 340 | 620 | 915 | 5128 |
| Cooling Degree Days | 0 | 0 | 1 | 9 | 56 | 175 | 307 | 246 | 107 | 11 | 1 | 0 | 913 |
| Total Precipitation (") | 3.00 | 2.91 | 3.98 | 3.85 | 4.42 | 3.47 | 5.10 | 3.90 | 3.14 | 2.99 | 3.32 | 4.03 | 44.11 |
| Days ≥ 0.1" Precip | 7 | 6 | 8 | 8 | 9 | 7 | 8 | 6 | 6 | 5 | 7 | 8 | 85 |
| Total Snowfall (") | na | na | *1.1* | 0.0 | 0.0 | 0.0 | 0.0 | 0.0 | 0.0 | 0.0 | 0.2 | *0.6* | na |
| Days ≥ 1" Snow Depth | 7 | 6 | 1 | 0 | 0 | 0 | 0 | 0 | 0 | 0 | 0 | 2 | 16 |

## WILLIAMSTOWN 3 NW *Grant County* ELEVATION 914 ft LAT/LONG 38° 38' N / 84° 38' W

| | JAN | FEB | MAR | APR | MAY | JUN | JUL | AUG | SEP | OCT | NOV | DEC | YEAR |
|---|---|---|---|---|---|---|---|---|---|---|---|---|---|
| Maximum Temp °F | 39.4 | 44.2 | 55.1 | 66.1 | 74.6 | 82.4 | 85.8 | 84.5 | 78.4 | 67.6 | 55.3 | 44.7 | 64.8 |
| Minimum Temp °F | 21.7 | 24.6 | 33.9 | 43.8 | 52.9 | 61.3 | 65.4 | 63.8 | 57.6 | 46.0 | 37.0 | 27.5 | 44.6 |
| Mean Temp °F | 30.6 | 34.4 | 44.6 | 55.0 | 63.8 | 71.8 | 75.6 | 74.1 | 68.1 | 56.8 | 46.2 | 36.1 | 54.8 |
| Days Max Temp ≥ 90 °F | 0 | 0 | 0 | 0 | 0 | 3 | 8 | 5 | 1 | 0 | 0 | 0 | 17 |
| Days Max Temp ≤ 32 °F | 9 | 5 | 1 | 0 | 0 | 0 | 0 | 0 | 0 | 0 | 0 | 4 | 19 |
| Days Min Temp ≤ 32 °F | 26 | 21 | 15 | 4 | 0 | 0 | 0 | 0 | 0 | 3 | 12 | 21 | 102 |
| Days Min Temp ≤ 0 °F | 2 | 1 | 0 | 0 | 0 | 0 | 0 | 0 | 0 | 0 | 0 | 1 | 4 |
| Heating Degree Days | 1060 | 856 | 630 | 313 | 112 | 12 | 1 | 3 | 50 | 270 | 561 | 887 | 4755 |
| Cooling Degree Days | 0 | 0 | 2 | 22 | 86 | 235 | 363 | 312 | 155 | 25 | 2 | 0 | 1202 |
| Total Precipitation (") | 2.71 | 2.67 | 4.17 | 4.39 | 4.28 | 3.81 | 4.40 | 3.90 | 3.34 | 3.03 | 3.64 | 3.45 | 43.79 |
| Days ≥ 0.1" Precip | 6 | 6 | 9 | 9 | 8 | 7 | 7 | 6 | 5 | 5 | 8 | 8 | 84 |
| Total Snowfall (") | 6.3 | 4.7 | 3.9 | 0.3 | 0.0 | 0.0 | 0.0 | 0.0 | 0.0 | 0.2 | 1.0 | 2.6 | 19.0 |
| Days ≥ 1" Snow Depth | 9 | 7 | 2 | 0 | 0 | 0 | 0 | 0 | 0 | 0 | 1 | 3 | 22 |

## JANUARY MINIMUM TEMPERATURE °F

LOWEST

| | | |
|---|---|---|
| 1 | Ashland | 18.2 |
| 2 | West Liberty | 18.6 |
| 3 | Shelbyville | 18.8 |
| 4 | Grayson | 18.9 |
| 5 | Falmouth | 19.0 |
| 6 | Warsaw Markland | 19.2 |
| 7 | Rough River Dam | 19.7 |
| 8 | Cincinnati | 20.2 |
| 9 | Maysville | 20.3 |
| 10 | Frankfort Lock | 20.5 |
| 11 | Nolin River Lake | 20.6 |
| 12 | Farmers | 21.4 |
| 13 | Bradfordsville | 21.7 |
| | Williamstown | 21.7 |
| 15 | Barbourville | 21.9 |
| 16 | Danville | 22.0 |
| 17 | Rochester Ferry | 22.1 |
| 18 | Greensburg | 22.2 |
| 19 | Carrollton Lock | 22.6 |
| 20 | Hopkinsville | 22.7 |
| | Leitchfield | 22.7 |
| | Manchester | 22.7 |
| 23 | Barren River Lake | 22.9 |
| | Lexington | 22.9 |
| | Middlesboro | 22.9 |

HIGHEST

| | | |
|---|---|---|
| 1 | Scottsville | 26.3 |
| 2 | Murray | 25.7 |
| 3 | Berea | 25.3 |
| 4 | Monticello | 24.9 |
| 5 | Glasgow | 24.8 |
| 6 | Dix Dam | 24.7 |
| | Princeton | 24.7 |
| 8 | Lovelaceville | 24.6 |
| 9 | Summer Shade | 24.5 |
| 10 | Bowling Green | 24.4 |
| 11 | Bardwell | 24.3 |
| | Paducah-Barkley | 24.3 |
| | Somerset | 24.3 |
| 14 | Henderson | 24.1 |
| | London | 24.1 |
| | Madisonville | 24.1 |
| 17 | Beaver Dam | 24.0 |
| | Louisville | 24.0 |
| | Mammoth Cave | 24.0 |
| | Owensboro | 24.0 |
| 21 | Mount Sterling | 23.9 |
| 22 | Paducah-Walker | 23.8 |
| 23 | Golden Pond | 23.6 |
| | Hodgenville | 23.6 |
| 25 | Bardstown | 23.3 |

## JULY MAXIMUM TEMPERATURE °F

HIGHEST

| | | |
|---|---|---|
| 1 | Lovelaceville | 90.8 |
| 2 | Bardwell | 90.3 |
| 3 | Paducah-Walker | 90.2 |
| 4 | Madisonville | 89.8 |
| 5 | Barren River Lake | 89.6 |
| | Princeton | 89.6 |
| 7 | Murray | 89.5 |
| | Paducah-Barkley | 89.5 |
| 9 | Greensburg | 89.4 |
| | Owensboro | 89.4 |
| 11 | Golden Pond | 89.3 |
| | Hopkinsville | 89.3 |
| 13 | Bowling Green | 89.0 |
| | Nolin River Lake | 89.0 |
| 15 | Beaver Dam | 88.7 |
| 16 | Rough River Dam | 88.6 |
| 17 | Henderson | 88.5 |
| | Rochester Ferry | 88.5 |
| 19 | Glasgow | 88.3 |
| 20 | Bardstown | 88.1 |
| | Bradfordsville | 88.1 |
| 22 | Mammoth Cave | 88.0 |
| 23 | Louisville | 87.9 |
| 24 | Heidelberg | 87.7 |
| 25 | Ashland | 87.6 |

LOWEST

| | | |
|---|---|---|
| 1 | Mount Sterling | 85.7 |
| 2 | Cincinnati | 85.8 |
| | Lexington | 85.8 |
| | Williamstown | 85.8 |
| 5 | London | 85.9 |
| 6 | Baxter | 86.2 |
| | Manchester | 86.2 |
| 8 | Danville | 86.4 |
| | Somerset | 86.4 |
| 10 | Grayson | 86.5 |
| 11 | Mount Vernon | 86.6 |
| 12 | Middlesboro | 86.7 |
| | Warsaw Markland | 86.7 |
| 14 | Falmouth | 86.8 |
| | West Liberty | 86.8 |
| 16 | Monticello | 86.9 |
| | Shelbyville | 86.9 |
| 18 | Berea | 87.0 |
| | Carrollton Lock | 87.0 |
| | Farmers | 87.0 |
| | Scottsville | 87.0 |
| 22 | Maysville | 87.1 |
| 23 | Summer Shade | 87.2 |
| 24 | Dix Dam | 87.3 |
| 25 | Barbourville | 87.4 |

## ANNUAL PRECIPITATION (")

HIGHEST

| | | |
|---|---|---|
| 1 | Murray | 54.69 |
| 2 | Glasgow | 53.95 |
| 3 | Scottsville | 53.24 |
| 4 | Greensburg | 52.85 |
| 5 | Mammoth Cave | 52.60 |
| 6 | Mount Vernon | 52.25 |
| 7 | Bradfordsville | 51.97 |
| 8 | Summer Shade | 51.84 |
| 9 | Nolin River Lake | 51.49 |
| 10 | Bardwell | 51.41 |
| 11 | Hodgenville | 51.40 |
| 12 | Middlesboro | 51.38 |
| 13 | Bowling Green | 51.35 |
| 14 | Somerset | 51.19 |
| 15 | Hopkinsville | 51.15 |
| 16 | Barren River Lake | 51.00 |
| 17 | Golden Pond | 50.68 |
| 18 | Princeton | 50.67 |
| 19 | Bardstown | 50.46 |
| 20 | Baxter | 50.44 |
| 21 | Barbourville | 50.39 |
| 22 | Monticello | 50.28 |
| 23 | Manchester | 50.02 |
| 24 | Lovelaceville | 49.09 |
| 25 | Paducah-Barkley | 48.80 |

LOWEST

| | | |
|---|---|---|
| 1 | Cincinnati | 41.38 |
| 2 | Frankfort Lock | 42.25 |
| 3 | Ashland | 42.34 |
| 4 | Grayson | 42.43 |
| 5 | Falmouth | 42.87 |
| 6 | Carrollton Lock | 43.28 |
| 7 | Warsaw Markland | 43.39 |
| 8 | Williamstown | 43.79 |
| 9 | Louisville | 43.99 |
| 10 | West Liberty | 44.11 |
| 11 | Maysville | 44.80 |
| 12 | Henderson | 44.83 |
| | Lexington | 44.83 |
| 14 | Dix Dam | 45.93 |
| 15 | Shelbyville | 46.50 |
| 16 | Farmers | 46.53 |
| 17 | Berea | 46.55 |
| 18 | London | 46.60 |
| 19 | Paducah-Walker | 46.78 |
| 20 | Mount Sterling | 46.97 |
| 21 | Owensboro | 47.11 |
| 22 | Danville | 47.49 |
| 23 | Heidelberg | 47.65 |
| 24 | Rochester Ferry | 47.72 |
| 25 | Madisonville | 47.83 |

## ANNUAL SNOWFALL (")

HIGHEST

| | | |
|---|---|---|
| 1 | Cincinnati | 22.6 |
| 2 | Mount Vernon | 19.0 |
| | Williamstown | 19.0 |
| 4 | Lexington | 17.0 |
| 5 | Monticello | 16.7 |
| 6 | Louisville | 16.6 |
| 7 | Henderson | 16.5 |
| 8 | London | 15.9 |
| 9 | Princeton | 15.5 |
| 10 | Grayson | 14.9 |
| 11 | Mammoth Cave | 14.3 |
| 12 | Danville | 13.8 |
| | Glasgow | 13.8 |
| 14 | Bardwell | 13.6 |
| 15 | Scottsville | 13.3 |
| 16 | Shelbyville | 13.2 |
| 17 | Leitchfield | 13.1 |
| 18 | Murray | 12.8 |
| 19 | Summer Shade | 12.6 |
| 20 | Carrollton Lock | 12.3 |
| | Hopkinsville | 12.3 |
| 22 | Warsaw Markland | 12.2 |
| 23 | Bowling Green | 11.7 |
| | Paducah-Barkley | 11.7 |
| 25 | Owensboro | 11.2 |

LOWEST

| | | |
|---|---|---|
| 1 | Rough River Dam | 7.6 |
| 2 | Madisonville | 9.1 |
| 3 | Barren River Lake | 9.7 |
| 4 | Falmouth | 10.6 |
| 5 | Owensboro | 11.2 |
| 6 | Bowling Green | 11.7 |
| | Paducah-Barkley | 11.7 |
| 8 | Warsaw Markland | 12.2 |
| 9 | Carrollton Lock | 12.3 |
| | Hopkinsville | 12.3 |
| 11 | Summer Shade | 12.6 |
| 12 | Murray | 12.8 |
| 13 | Leitchfield | 13.1 |
| 14 | Shelbyville | 13.2 |
| 15 | Scottsville | 13.3 |
| 16 | Bardwell | 13.6 |
| 17 | Danville | 13.8 |
| | Glasgow | 13.8 |
| 19 | Mammoth Cave | 14.3 |
| 20 | Grayson | 14.9 |
| 21 | Princeton | 15.5 |
| 22 | London | 15.9 |
| 23 | Henderson | 16.5 |
| 24 | Louisville | 16.6 |
| 25 | Monticello | 16.7 |

# LOUISIANA

PHYSICAL FEATURES. Louisiana extends roughly between latitudes 29.5° N. and 33° N. and from the 94th meridian eastward to the Mississippi River, and in the south to the Pearl River. Elevations increase gradually from the coast northward, rising to over 100 feet above sea level on uplands and 400 to 500 feet on some of the hills in the northwest.

Drainage in Louisiana is into the Gulf of Mexico. The Red River basin comprises the largest drainage area in the State. The Red joins with the Atchafalaya and Old Rivers, the latter forming an outlet to the Mississippi River. Southern Louisiana is mostly low and level with elevations generally less than 60 feet above mean Gulf level. The runoff is through numerous sluggish streams or bayous which flow through lakes and considerable marshland. The larger marshlands are mainly in the coastal area, extending farthest inland in the southeast. A great part of the southwestern region is drained through the Calcasieu River. The extreme southwestern part of the State drains into the Sabine River which forms more than half of the western boundary. The Pearl River drains a relatively small area in the southeast and forms the southeastern boundary.

GENERAL CLIMATE. The principal influences that determine the climate of Louisiana are its subtropical latitude and its proximity to the Gulf of Mexico. The marine tropical influence is evident from the fact that the average water temperatures of the Gulf along the Louisiana shore range from 64° F. in February to 84° in August.

In summer the prevailing southerly winds provide moist, semi-tropical weather often favorable for afternoon thundershowers. When westerly to northerly winds occur, periods of hotter and drier weather interrupt the prevailing moist condition. In the colder season the State is subjected alternately to tropical air and cold continental air, in periods of varying length. Although warmed by its southward journey, the cold air occasionally brings large and rather sudden drops in temperature, but conditions are usually less severe than farther west.

Louisiana is south of the usual track of winter storm centers, but occasionally one moves this far south. In some winters a succession of such centers will develop in the Gulf of Mexico and move over or near the State. The State is occasionally in the path of tropical storms or hurricanes.

From December to May the water of the Mississippi River is usually colder than the air temperature, which favors river fogs during this season, particularly with weak southerly winds. In the more southern sections, lakes also serve to modify the extremes of temperature and to increase fogginess over narrow strips along the shores.

PRECIPITATION. Mean annual precipitation ranges from 46 inches in Caddo Parish to as much as 66 inches in parts of St. Mary, Assumption, Terrebonne, and Lafourche Parishes. A median line of 56 inches per year runs from Hackberry northward to Leesville, Montgomery, Winona, Luna, and southward to Harrisonburg and Deerpark on the Mississippi River. This line separates areas of lower precipitation averages to the north from areas of higher precipitation to the south.

During the summer months, seasonal rainfall usually increases from the northwest toward the southeast. In the winter this pattern is reversed with the heaviest seasonal precipitation in the area extending from the Carroll Parishes southwestward to Winn and southward to St. Landry, with the least in the lower Delta. During the summer months the rich source of moist tropical air results in almost daily showers in the coastal parishes; however, shower frequency diminishes with distance from the Gulf coast toward the northern parishes. In the winter months the northern portion of the State is invaded by cold air which tends to stall and become stationary. This sometimes produces prolonged rains over that area, while clear weather continues in the southern parishes. The pattern of spring rains is similar to that of winter, while fall rains are distributed in the same manner as summer rains. However, fall (September, October, and November) is the driest season of the year, with precipitation ranging from 9 inches in the north to 15 inches in the southeast. Spring precipitation ranges from 13 inches on the coast to 18 inches in the central interior.

The heaviest rains of short duration are associated with thunderstorms, although tropical storms sometimes cause prolonged heavy rains. Rains of as much as 20 inches in 1 month have occurred at most stations, and falls of as much as 10 inches in 24 hours are not rare. Although Louisiana is one of the wettest states, droughts are not unknown, especially during the summer and fall. Snow and sleet are of little importance in Louisiana.

FLOODS. Flood producing rains may occur during any month of the year in Louisiana, although they are less likely during September, October, and November, the drier months, and are most frequent during late winter and early spring. Floods on the lower Mississippi and Atchafalaya result from runoff upstream, and rainfall within the State has little influence on these stages. Major floods can occur on the lower Red River from heavy rains in Louisiana. Heavy rains cause several minor floods each year on the Sabine, Calcasieu, and Mermentau Rivers. A major flood on the Sabine occurs about once in 4 years. In the upper portions of the Calcasieu and Mermentau a major flood occurs about once in 10 years and in the lower portions not more than once in 25 years.

TEMPERATURE. The average annual temperature ranges from 66° F. in northern divisions to 69° F. in southern divisions. The lowest January average is 49° F. in the northwest and north-central ranging upward to 57° F. in the southeast. The highest July average is 83° F. in the northwest and north-central, ranging downward to 81° F. in the east-central. This reversal of temperature distribution with warmer summers in the northern portion than in the southern portion, is due to the almost daily showers in the parishes near or on the Gulf of Mexico. This is further shown by the number of days with temperatures 90° F. or above. While Shreveport and Alexandria average 102 such days, Lake Charles and New Orleans average 86 and 57, respectively. In other words, at New Orleans, where there is a 50 percent expectancy of showers on any day in July, the temperature reaches or exceeds 90° F. about half as often as at Shreveport where summer showers are much less frequent. Temperatures above 110° and below 0° are rare. The average number of days with freezing temperatures (32° F.) or lower ranges from 24 at Shreveport to 4 at New Orleans. Near the mouth of the Mississippi River a freeze can be expected only about once in 7 years.

STORMS. Showers and thunderstorms occur on an average of 50 to 60 days a year in the northwest and north-central, 70 to 80 days a year in the south, and 60 to 70 days in central and northeast Louisiana. During fall, winter, and spring, these are often attended by high winds, but this is not the case during the summer. Thundershowers which move off Lake Pontchartrain at any season are usually attended by high winds. During late fall, winter, and early spring, thunder may occur at any time of day, but from late spring to early fall about 80 percent of all thundershowers occur between noon and midnight in the northern and between 6 a.m. and 6 p.m. in the southern half of the State.

Tropical cyclones are one of the hazards to life and property in Louisiana, especially in the parishes near the coast. About a third of the cyclones have been of hurricane intensity with winds of 74 miles per hour or more at some points. Most others are attended by gale winds. Almost one-half of all tropical cyclones and one-half of those reaching hurricane intensity have occurred in September.

Tornadoes have been reported in most years and in all months of the year. The largest number of tornadoes has occurred in April, followed by May, November, and March. Hurricanes and tornadoes affect only a relatively small area for a brief time and their frequency is quite low. Contrasting with these occasional adverse features are the mild and short winters, the abundant precipitation, the long growing (freeze-free) season, freedom from extreme summer heat, and the delightful spring and fall weather.

## COUNTY INDEX

***Acadia Parish***
CROWLEY 2 NE

***Allen Parish***
ELIZABETH
OBERLIN FIRE TOWER

***Assumption Parish***
DONALDSONVILLE 4 SW

***Avoyelles Parish***
BUNKIE

***Bossier Parish***
PLAIN DEALING

***Caddo Parish***
SHREVEPORT REGIONAL

***Calcasieu Parish***
LAKE CHARLES MUNI AP

***Cameron Parish***
HACKBERRY 8 SSW
ROCKEFELLR WL REFUGE

***Claiborne Parish***
HOMER 3 SSW

***De Soto Parish***
LOGANSPORT 4 ENE

***East Baton Rouge Parish***
BATON ROUGE LSU BEN
BATON ROUGE RYAN AP

***East Carroll Parish***
LAKE PROVIDENCE

***Franklin Parish***
WINNSBORO 5 SSE

***Iberia Parish***
JEANERETTE 5 NW
NEW IBERIA

***Iberville Parish***
CARVILLE 2 SW

***Jefferson Parish***
NEW ORLEANS INTL AP

***Jefferson Davis Parish***
JENNINGS

***Lafayette Parish***
LAFAYETTE REGL AP

***Lincoln Parish***
RUSTON LA TECH UNIV

***Madison Parish***
TALLULAH

***Morehouse Parish***
BASTROP

***Natchitoches Parish***
ASHLAND
NATCHITOCHES

***Orleans Parish***
NEW ORLEANS AUDUBON

***Ouachita Parish***
CALHOUN RESEARCH STN
MONROE MUNI ARPT

***Pointe Coupee Parish***
NEW ROADS 5 ESE

***Rapides Parish***
ALEXANDRIA

***Sabine Parish***
MANY

***St. Charles Parish***
PARADIS 7 S

***St. John the Baptist Parish***
RESERVE 4 WNW

***St. Landry Parish***
GRAND COTEAU

***St. Mary Parish***
FRANKLIN 3 NW
MORGAN CITY

***St. Tammany Parish***
COVINGTON 4 NNW
SLIDELL WSFO

***Tangipahoa Parish***
AMITE
HAMMOND

***Tensas Parish***
ST JOSEPH 3 N

***Terrebonne Parish***
HOUMA

***Vernon Parish***
LEESVILLE

***Washington Parish***
BOGALUSA
FRANKLINTON 3 SW

***Webster Parish***
MINDEN

***Winn Parish***
WINNFIELD 2 W

## ELEVATION INDEX

| FEET | STATION NAME |
|---|---|
| 0 | ROCKEFELLR WL REFUGE |
| 7 | NEW ORLEANS INTL AP |
| 9 | LAKE CHARLES MUNI AP |
| 10 | FRANKLIN 3 NW |
| 10 | HACKBERRY 8 SSW |
| 10 | HOUMA |
| 10 | NEW ORLEANS AUDUBON |
| 10 | PARADIS 7 S |
| 14 | MORGAN CITY |
| 15 | RESERVE 4 WNW |
| 16 | SLIDELL WSFO |
| 20 | BATON ROUGE LSU BEN |
| 20 | CROWLEY 2 NE |
| 20 | JEANERETTE 5 NW |
| 20 | NEW IBERIA |
| 30 | CARVILLE 2 SW |
| 30 | DONALDSONVILLE 4 SW |
| 30 | JENNINGS |

| FEET | STATION NAME |
|---|---|
| 39 | COVINGTON 4 NNW |
| 39 | HAMMOND |
| 43 | LAFAYETTE REGL AP |
| 49 | NEW ROADS 5 ESE |
| 59 | BUNKIE |
| 59 | GRAND COTEAU |
| 69 | BATON ROUGE RYAN AP |
| 69 | OBERLIN FIRE TOWER |
| 75 | WINNSBORO 5 SSE |
| 79 | ST JOSEPH 3 N |
| 82 | MONROE MUNI ARPT |
| 85 | ALEXANDRIA |
| 89 | TALLULAH |
| 102 | BOGALUSA |
| 102 | LAKE PROVIDENCE |
| 121 | NATCHITOCHES |
| 131 | AMITE |
| 141 | BASTROP |
| 141 | WINNFIELD 2 W |
| 151 | ELIZABETH |
| 151 | FRANKLINTON 3 SW |
| 180 | CALHOUN RESEARCH STN |
| 194 | SHREVEPORT REGIONAL |
| 210 | LOGANSPORT 4 ENE |
| 230 | ASHLAND |
| 239 | LEESVILLE |
| 239 | MINDEN |
| 259 | MANY |
| 279 | RUSTON LA TECH UNIV |
| 289 | PLAIN DEALING |
| 400 | HOMER 3 SSW |

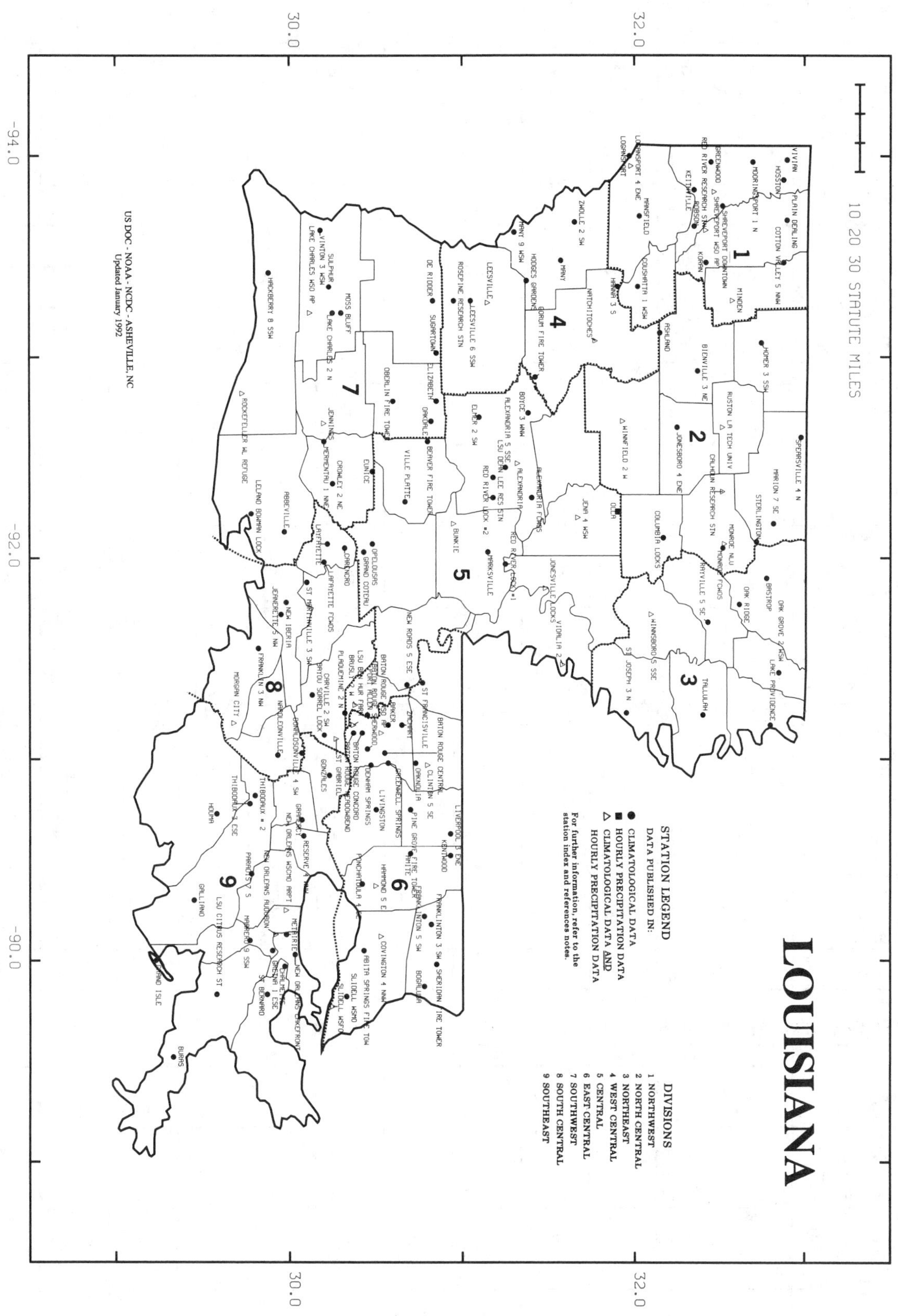
LOUISIANA
10 20 30 STATUTE MILES
STATION LEGEND
DATA PUBLISHED IN:
● CLIMATOLOGICAL DATA
■ HOURLY PRECIPITATION DATA
△ CLIMATOLOGICAL DATA AND HOURLY PRECIPITATION DATA
For further information, refer to the station index and references notes.
DIVISIONS
1 NORTHWEST
2 NORTH CENTRAL
3 NORTHEAST
4 WEST CENTRAL
5 CENTRAL
6 EAST CENTRAL
7 SOUTHWEST
8 SOUTH CENTRAL
9 SOUTHEAST
US DOC - NOAA - NCDC - ASHEVILLE, NC
Updated January 1992
30.0
32.0
-94.0
-92.0
-90.0

## ALEXANDRIA *Rapides Parish* ELEVATION 85 ft LAT/LONG 31° 19' N / 92° 32' W

| | JAN | FEB | MAR | APR | MAY | JUN | JUL | AUG | SEP | OCT | NOV | DEC | YEAR |
|---|---|---|---|---|---|---|---|---|---|---|---|---|---|
| Maximum Temp °F | 57.3 | 61.8 | 69.6 | 77.5 | 83.7 | 90.1 | 92.5 | 92.2 | 87.7 | 79.3 | 69.3 | 61.4 | 76.9 |
| Minimum Temp °F | 37.0 | 39.9 | 48.0 | 56.1 | 63.3 | 70.0 | 72.5 | 71.6 | 66.7 | 54.9 | 46.7 | 40.1 | 55.6 |
| Mean Temp °F | 47.1 | 50.9 | 58.8 | 66.8 | 73.5 | 80.1 | 82.5 | 81.9 | 77.2 | 67.2 | 58.1 | 50.8 | 66.2 |
| Days Max Temp ≥ 90 °F | 0 | 0 | 0 | 0 | 4 | 18 | 25 | 25 | 14 | 2 | 0 | 0 | 88 |
| Days Max Temp ≤ 32 °F | 0 | 0 | 0 | 0 | 0 | 0 | 0 | 0 | 0 | 0 | 0 | 0 | 0 |
| Days Min Temp ≤ 32 °F | 11 | 7 | 1 | 0 | 0 | 0 | 0 | 0 | 0 | 0 | 2 | 7 | 28 |
| Days Min Temp ≤ 0 °F | 0 | 0 | 0 | 0 | 0 | 0 | 0 | 0 | 0 | 0 | 0 | 0 | 0 |
| Heating Degree Days | 552 | 399 | 222 | 63 | 5 | 0 | 0 | 0 | 3 | 64 | 239 | 444 | 1991 |
| Cooling Degree Days | 3 | 9 | 38 | 116 | 282 | 472 | 564 | 551 | 391 | 144 | 40 | 11 | 2621 |
| Total Precipitation (") | 5.68 | 4.98 | 5.39 | 4.59 | 5.54 | 4.57 | 4.52 | 4.42 | 3.90 | 4.89 | 5.21 | 6.67 | 60.36 |
| Days ≥ 0.1" Precip | 8 | 6 | 6 | 5 | 7 | 6 | 6 | 7 | 5 | 5 | 6 | 8 | 75 |
| Total Snowfall (") | 0.3 | 0.1 | 0.0 | 0.0 | 0.0 | 0.0 | 0.0 | 0.0 | 0.0 | 0.0 | 0.0 | 0.0 | 0.4 |
| Days ≥ 1" Snow Depth | 0 | 0 | 0 | 0 | 0 | 0 | 0 | 0 | 0 | 0 | 0 | 0 | 0 |

## AMITE *Tangipahoa Parish* ELEVATION 131 ft LAT/LONG 30° 44' N / 90° 30' W

| | JAN | FEB | MAR | APR | MAY | JUN | JUL | AUG | SEP | OCT | NOV | DEC | YEAR |
|---|---|---|---|---|---|---|---|---|---|---|---|---|---|
| Maximum Temp °F | 59.8 | 63.9 | 71.4 | 79.0 | 84.9 | 91.0 | 92.2 | 91.9 | 88.4 | 80.5 | 71.1 | 63.7 | 78.2 |
| Minimum Temp °F | 38.0 | 40.4 | 47.6 | 54.9 | 61.9 | 68.2 | 70.9 | 70.3 | 66.1 | 54.5 | 46.7 | 40.7 | 55.0 |
| Mean Temp °F | 48.9 | 52.2 | 59.6 | 67.0 | 73.4 | 79.6 | 81.6 | 81.1 | 77.3 | 67.6 | 58.9 | 52.2 | 66.6 |
| Days Max Temp ≥ 90 °F | 0 | 0 | 0 | 0 | 6 | 21 | 25 | 24 | 15 | 2 | 0 | 0 | 93 |
| Days Max Temp ≤ 32 °F | 0 | 0 | 0 | 0 | 0 | 0 | 0 | 0 | 0 | 0 | 0 | 0 | 0 |
| Days Min Temp ≤ 32 °F | 11 | 7 | 2 | 0 | 0 | 0 | 0 | 0 | 0 | 0 | 3 | 8 | 31 |
| Days Min Temp ≤ 0 °F | 0 | 0 | 0 | 0 | 0 | 0 | 0 | 0 | 0 | 0 | 0 | 0 | 0 |
| Heating Degree Days | 498 | 366 | 208 | 59 | 5 | 0 | 0 | 0 | 3 | 59 | 221 | 405 | 1824 |
| Cooling Degree Days | 4 | 12 | 46 | 108 | 264 | 440 | 518 | 513 | 371 | 146 | 49 | 15 | 2486 |
| Total Precipitation (") | 6.07 | 6.01 | 5.91 | 6.24 | 5.67 | 4.59 | 6.31 | 5.62 | 5.02 | 3.69 | 4.33 | 5.92 | 65.38 |
| Days ≥ 0.1" Precip | 7 | 7 | 7 | 5 | 7 | 7 | 10 | 8 | 6 | 4 | 6 | 7 | 81 |
| Total Snowfall (") | 0.1 | 0.2 | 0.0 | 0.0 | 0.0 | 0.0 | 0.0 | 0.0 | 0.0 | 0.0 | 0.0 | 0.0 | 0.3 |
| Days ≥ 1" Snow Depth | 0 | 0 | 0 | 0 | 0 | 0 | 0 | 0 | 0 | 0 | 0 | 0 | 0 |

## ASHLAND *Natchitoches Parish* ELEVATION 230 ft LAT/LONG 32° 7' N / 93° 6' W

| | JAN | FEB | MAR | APR | MAY | JUN | JUL | AUG | SEP | OCT | NOV | DEC | YEAR |
|---|---|---|---|---|---|---|---|---|---|---|---|---|---|
| Maximum Temp °F | 54.1 | 59.8 | 68.2 | 76.7 | 82.3 | 89.2 | 92.7 | 92.4 | 86.7 | 77.4 | 67.0 | 58.2 | 75.4 |
| Minimum Temp °F | 32.2 | 35.0 | 42.5 | 50.9 | 58.6 | 66.0 | 69.0 | 67.7 | 62.5 | 50.2 | 41.7 | 34.9 | 50.9 |
| Mean Temp °F | 43.2 | 47.4 | 55.3 | 63.8 | 70.5 | 77.7 | 80.9 | 80.1 | 74.6 | 63.8 | 54.3 | 46.6 | 63.2 |
| Days Max Temp ≥ 90 °F | 0 | 0 | 0 | 0 | 3 | 16 | 25 | 24 | 12 | 1 | 0 | 0 | 81 |
| Days Max Temp ≤ 32 °F | 1 | 0 | 0 | 0 | 0 | 0 | 0 | 0 | 0 | 0 | 0 | 1 | 2 |
| Days Min Temp ≤ 32 °F | 17 | 13 | 5 | 1 | 0 | 0 | 0 | 0 | 0 | 1 | 7 | 15 | 59 |
| Days Min Temp ≤ 0 °F | 0 | 0 | 0 | 0 | 0 | 0 | 0 | 0 | 0 | 0 | 0 | 0 | 0 |
| Heating Degree Days | 673 | 494 | 314 | 112 | 19 | 0 | 0 | 0 | 11 | 121 | 330 | 568 | 2642 |
| Cooling Degree Days | 1 | 3 | 22 | 72 | 193 | 396 | 505 | 492 | 318 | 95 | 18 | 5 | 2120 |
| Total Precipitation (") | 4.66 | 5.32 | 4.59 | 4.19 | 5.90 | 4.30 | 4.36 | 3.09 | 3.52 | 4.08 | 4.10 | 5.31 | 53.42 |
| Days ≥ 0.1" Precip | 7 | 6 | 7 | 5 | 7 | 6 | 6 | 5 | 5 | 5 | 6 | 7 | 72 |
| Total Snowfall (") | 0.8 | 0.2 | 0.0 | 0.0 | 0.0 | 0.0 | 0.0 | 0.0 | 0.0 | 0.0 | 0.0 | 0.0 | 1.0 |
| Days ≥ 1" Snow Depth | 0 | 0 | 0 | 0 | 0 | 0 | 0 | 0 | 0 | 0 | 0 | 0 | 0 |

## BASTROP *Morehouse Parish* ELEVATION 141 ft LAT/LONG 32° 46' N / 91° 54' W

| | JAN | FEB | MAR | APR | MAY | JUN | JUL | AUG | SEP | OCT | NOV | DEC | YEAR |
|---|---|---|---|---|---|---|---|---|---|---|---|---|---|
| Maximum Temp °F | 55.5 | 60.8 | 69.7 | 77.9 | 84.3 | 90.9 | 93.6 | 93.0 | 88.1 | 78.9 | 67.7 | 59.0 | 76.6 |
| Minimum Temp °F | 34.7 | 37.5 | 45.5 | 53.7 | 61.3 | 68.8 | 71.8 | 70.3 | 64.7 | 53.4 | 44.4 | 37.8 | 53.7 |
| Mean Temp °F | 45.1 | 49.2 | 57.6 | 65.9 | 72.8 | 79.9 | 82.7 | 81.7 | 76.4 | 66.2 | 56.1 | 48.4 | 65.2 |
| Days Max Temp ≥ 90 °F | 0 | 0 | 0 | 1 | 6 | 20 | 26 | 25 | 15 | 2 | 0 | 0 | 95 |
| Days Max Temp ≤ 32 °F | 1 | 0 | 0 | 0 | 0 | 0 | 0 | 0 | 0 | 0 | 0 | 1 | 2 |
| Days Min Temp ≤ 32 °F | 14 | 10 | 3 | 0 | 0 | 0 | 0 | 0 | 0 | 0 | 4 | 10 | 41 |
| Days Min Temp ≤ 0 °F | 0 | 0 | 0 | 0 | 0 | 0 | 0 | 0 | 0 | 0 | 0 | 0 | 0 |
| Heating Degree Days | 614 | 445 | 255 | 80 | 9 | 0 | 0 | 0 | 7 | 81 | 287 | 512 | 2290 |
| Cooling Degree Days | 1 | 7 | 27 | 96 | 243 | 449 | 555 | 528 | 351 | *122* | 23 | 6 | 2408 |
| Total Precipitation (") | 5.06 | 5.35 | 5.48 | 5.39 | 5.90 | 4.35 | 3.87 | 2.96 | 3.18 | 4.31 | 4.70 | 5.52 | 56.07 |
| Days ≥ 0.1" Precip | 7 | 6 | 7 | 6 | 7 | 6 | 6 | 5 | 5 | 5 | 6 | 7 | 73 |
| Total Snowfall (") | *0.3* | 0.1 | 0.1 | 0.0 | 0.0 | 0.0 | 0.0 | 0.0 | 0.0 | 0.0 | 0.0 | 0.0 | 0.5 |
| Days ≥ 1" Snow Depth | 0 | 0 | 0 | 0 | 0 | 0 | 0 | 0 | 0 | 0 | 0 | 0 | 0 |

## BATON ROUGE LSU BEN *East Baton Rouge Parish* ELEVATION 20 ft LAT/LONG 30° 22 ' N / 91° 10 ' W

| | JAN | FEB | MAR | APR | MAY | JUN | JUL | AUG | SEP | OCT | NOV | DEC | YEAR |
|---|---|---|---|---|---|---|---|---|---|---|---|---|---|
| Maximum Temp °F | 59.7 | 63.1 | 70.6 | 78.4 | 84.2 | 89.6 | 90.9 | 90.7 | 87.5 | 80.1 | 71.0 | 63.8 | 77.5 |
| Minimum Temp °F | 39.4 | 41.6 | 48.8 | 56.2 | 63.5 | 69.0 | 71.6 | 70.8 | 66.9 | 55.3 | 47.7 | 42.2 | 56.1 |
| Mean Temp °F | 49.6 | 52.4 | 59.7 | 67.3 | 73.9 | 79.3 | 81.3 | 80.8 | 77.2 | 67.7 | 59.4 | 53.0 | 66.8 |
| Days Max Temp ≥ 90 °F | 0 | 0 | 0 | 0 | 3 | 18 | 23 | 22 | 13 | 2 | 0 | 0 | 81 |
| Days Max Temp ≤ 32 °F | 0 | 0 | 0 | 0 | 0 | 0 | 0 | 0 | 0 | 0 | 0 | 0 | 0 |
| Days Min Temp ≤ 32 °F | 9 | 6 | 1 | 0 | 0 | 0 | 0 | 0 | 0 | 0 | 3 | 7 | 26 |
| Days Min Temp ≤ 0 °F | 0 | 0 | 0 | 0 | 0 | 0 | 0 | 0 | 0 | 0 | 0 | 0 | 0 |
| Heating Degree Days | 481 | 360 | 200 | 57 | 5 | 0 | 0 | 0 | 3 | 58 | 214 | 382 | 1760 |
| Cooling Degree Days | 5 | 11 | 45 | 119 | 289 | 444 | 521 | 513 | 383 | 155 | 57 | 19 | 2561 |
| Total Precipitation (") | 5.49 | 5.51 | 4.75 | 5.05 | 5.36 | 5.37 | 5.68 | 5.74 | 4.28 | 3.25 | 4.18 | 5.24 | 59.90 |
| Days ≥ 0.1" Precip | 8 | 7 | 6 | 5 | 6 | 8 | 9 | 8 | 7 | 4 | 5 | 7 | 80 |
| Total Snowfall (") | 0.0 | 0.0 | 0.0 | 0.0 | 0.0 | 0.0 | 0.0 | 0.0 | 0.0 | 0.0 | 0.0 | 0.0 | 0.0 |
| Days ≥ 1" Snow Depth | 0 | 0 | 0 | 0 | 0 | 0 | 0 | 0 | 0 | 0 | 0 | 0 | 0 |

## BATON ROUGE RYAN AP *East Baton Rouge Parish* ELEVATION 69 ft LAT/LONG 30° 32 ' N / 91° 9 ' W

| | JAN | FEB | MAR | APR | MAY | JUN | JUL | AUG | SEP | OCT | NOV | DEC | YEAR |
|---|---|---|---|---|---|---|---|---|---|---|---|---|---|
| Maximum Temp °F | 60.1 | 63.8 | 71.5 | 78.9 | 84.5 | 90.2 | 91.2 | 90.7 | 87.2 | 79.7 | 70.6 | 63.7 | 77.7 |
| Minimum Temp °F | 40.2 | 42.6 | 49.8 | 57.4 | 64.4 | 70.6 | 73.1 | 72.3 | 68.2 | 56.9 | 48.3 | 42.9 | 57.2 |
| Mean Temp °F | 50.2 | 53.2 | 60.7 | 68.2 | 74.5 | 80.4 | 82.2 | 81.5 | 77.7 | 68.3 | 59.5 | 53.3 | 67.5 |
| Days Max Temp ≥ 90 °F | 0 | 0 | 0 | 0 | 5 | 19 | 24 | 22 | 12 | 2 | 0 | 0 | 84 |
| Days Max Temp ≤ 32 °F | 0 | 0 | 0 | 0 | 0 | 0 | 0 | 0 | 0 | 0 | 0 | 0 | 0 |
| Days Min Temp ≤ 32 °F | 8 | 5 | 1 | 0 | 0 | 0 | 0 | 0 | 0 | 0 | 2 | 6 | 22 |
| Days Min Temp ≤ 0 °F | 0 | 0 | 0 | 0 | 0 | 0 | 0 | 0 | 0 | 0 | 0 | 0 | 0 |
| Heating Degree Days | 463 | 339 | 183 | 44 | 2 | 0 | 0 | 0 | 2 | 48 | 210 | 377 | 1668 |
| Cooling Degree Days | 9 | 15 | 55 | 137 | 308 | 476 | 546 | 534 | 393 | 159 | 57 | 21 | 2710 |
| Total Precipitation (") | 5.50 | 5.52 | 4.77 | 5.77 | 5.29 | 4.79 | 6.52 | 6.34 | 4.65 | 3.50 | 4.19 | 5.32 | 62.16 |
| Days ≥ 0.1" Precip | 7 | 6 | 6 | 5 | 6 | 7 | 10 | 8 | 7 | 4 | 5 | 7 | 78 |
| Total Snowfall (") | 0.0 | 0.2 | 0.0 | 0.0 | 0.0 | 0.0 | 0.0 | 0.0 | 0.0 | 0.0 | 0.0 | 0.0 | 0.2 |
| Days ≥ 1" Snow Depth | 0 | 0 | 0 | 0 | 0 | 0 | 0 | 0 | 0 | 0 | 0 | 0 | 0 |

## BOGALUSA *Washington Parish* ELEVATION 102 ft LAT/LONG 30° 47 ' N / 89° 52 ' W

| | JAN | FEB | MAR | APR | MAY | JUN | JUL | AUG | SEP | OCT | NOV | DEC | YEAR |
|---|---|---|---|---|---|---|---|---|---|---|---|---|---|
| Maximum Temp °F | 59.6 | 63.4 | 70.8 | 78.4 | 84.3 | 90.5 | 91.8 | 91.2 | 87.7 | 79.6 | 70.3 | 63.3 | 77.6 |
| Minimum Temp °F | 37.9 | 39.9 | 47.4 | 54.9 | 62.2 | 69.0 | 71.4 | 70.8 | 66.2 | 54.1 | 46.1 | 40.2 | 55.0 |
| Mean Temp °F | 48.8 | 51.7 | 59.2 | 66.6 | 73.2 | 79.7 | 81.6 | 81.0 | 77.0 | 66.8 | 58.2 | 51.8 | 66.3 |
| Days Max Temp ≥ 90 °F | 0 | 0 | 0 | 0 | 5 | 18 | 24 | 23 | 13 | 2 | 0 | 0 | 85 |
| Days Max Temp ≤ 32 °F | 0 | 0 | 0 | 0 | 0 | 0 | 0 | 0 | 0 | 0 | 0 | 0 | 0 |
| Days Min Temp ≤ 32 °F | 11 | 8 | 2 | 0 | 0 | 0 | 0 | 0 | 0 | 0 | 3 | 9 | 33 |
| Days Min Temp ≤ 0 °F | 0 | 0 | 0 | 0 | 0 | 0 | 0 | 0 | 0 | 0 | 0 | 0 | 0 |
| Heating Degree Days | 504 | 380 | 216 | 67 | 6 | 0 | 0 | 0 | 3 | 69 | 238 | 419 | 1902 |
| Cooling Degree Days | 5 | 11 | 41 | 96 | 264 | 442 | 525 | 512 | 364 | 137 | 44 | 18 | 2459 |
| Total Precipitation (") | 5.78 | 6.03 | 5.98 | 5.33 | 5.65 | 5.25 | 5.88 | 5.53 | 4.46 | 3.21 | 4.49 | 5.30 | 62.89 |
| Days ≥ 0.1" Precip | 8 | 7 | 7 | 5 | 6 | 7 | 9 | 8 | 6 | 4 | 6 | 7 | 80 |
| Total Snowfall (") | 0.2 | 0.0 | 0.1 | 0.0 | 0.0 | 0.0 | 0.0 | 0.0 | 0.0 | 0.0 | 0.0 | 0.0 | 0.3 |
| Days ≥ 1" Snow Depth | 0 | 0 | 0 | 0 | 0 | 0 | 0 | 0 | 0 | 0 | 0 | 0 | 0 |

## BUNKIE *Avoyelles Parish* ELEVATION 59 ft LAT/LONG 30° 57 ' N / 92° 10 ' W

| | JAN | FEB | MAR | APR | MAY | JUN | JUL | AUG | SEP | OCT | NOV | DEC | YEAR |
|---|---|---|---|---|---|---|---|---|---|---|---|---|---|
| Maximum Temp °F | 57.2 | 61.5 | 69.6 | 77.7 | 84.0 | 90.1 | 91.9 | 91.6 | 87.3 | 79.5 | 69.5 | 61.5 | 76.8 |
| Minimum Temp °F | 38.1 | 40.6 | 48.3 | 56.0 | 63.4 | 69.9 | 72.1 | 70.8 | 65.3 | 54.3 | 47.2 | 40.9 | 55.6 |
| Mean Temp °F | 47.7 | 51.1 | 59.0 | 66.9 | 73.7 | 80.0 | 82.0 | 81.2 | 76.3 | 66.9 | 58.4 | 51.2 | 66.2 |
| Days Max Temp ≥ 90 °F | 0 | 0 | 0 | 0 | 5 | 18 | 24 | 23 | 12 | 2 | 0 | 0 | 84 |
| Days Max Temp ≤ 32 °F | 0 | 0 | 0 | 0 | 0 | 0 | 0 | 0 | 0 | 0 | 0 | 0 | 0 |
| Days Min Temp ≤ 32 °F | 11 | 6 | 2 | 0 | 0 | 0 | 0 | 0 | 0 | 0 | 2 | 7 | 28 |
| Days Min Temp ≤ 0 °F | 0 | 0 | 0 | 0 | 0 | 0 | 0 | 0 | 0 | 0 | 0 | 0 | 0 |
| Heating Degree Days | 536 | 394 | 220 | 62 | 5 | 0 | 0 | 0 | 4 | 70 | 234 | 432 | 1957 |
| Cooling Degree Days | 3 | 9 | 39 | 116 | 284 | 458 | 534 | 519 | 346 | 139 | 44 | 13 | 2504 |
| Total Precipitation (") | 5.87 | 4.92 | 5.68 | 5.21 | 6.63 | 4.77 | 4.23 | 4.36 | 4.19 | 4.82 | 5.25 | 6.70 | 62.63 |
| Days ≥ 0.1" Precip | 7 | 6 | 6 | 5 | 7 | 7 | 7 | 6 | 5 | 4 | 5 | 7 | 72 |
| Total Snowfall (") | 0.0 | 0.0 | 0.0 | 0.0 | 0.0 | 0.0 | 0.0 | 0.0 | 0.0 | 0.0 | 0.0 | 0.0 | 0.0 |
| Days ≥ 1" Snow Depth | 0 | 0 | 0 | 0 | 0 | 0 | 0 | 0 | 0 | 0 | 0 | 0 | 0 |

## CALHOUN RESEARCH STN *Ouachita Parish* ELEVATION 180 ft LAT/LONG 32° 30 ' N / 92° 20 ' W

| | JAN | FEB | MAR | APR | MAY | JUN | JUL | AUG | SEP | OCT | NOV | DEC | YEAR |
|---|---|---|---|---|---|---|---|---|---|---|---|---|---|
| Maximum Temp °F | 54.3 | 59.5 | 67.7 | 76.5 | 82.6 | 89.6 | 92.7 | 92.5 | 87.1 | 77.4 | 67.1 | 58.3 | 75.4 |
| Minimum Temp °F | 33.2 | 35.8 | 43.7 | 52.0 | 59.9 | 67.6 | 70.9 | 69.3 | 63.6 | 51.0 | 43.0 | 36.1 | 52.2 |
| Mean Temp °F | 43.8 | 47.7 | 55.8 | 64.3 | 71.3 | 78.6 | 81.8 | 81.0 | 75.4 | 64.2 | 55.1 | 47.2 | 63.9 |
| Days Max Temp ≥ 90 °F | 0 | 0 | 0 | 0 | 3 | 17 | 25 | 24 | 13 | 2 | 0 | 0 | 84 |
| Days Max Temp ≤ 32 °F | 1 | 0 | 0 | 0 | 0 | 0 | 0 | 0 | 0 | 0 | 0 | 1 | 2 |
| Days Min Temp ≤ 32 °F | 16 | 12 | 4 | 1 | 0 | 0 | 0 | 0 | 0 | 0 | 5 | 13 | 51 |
| Days Min Temp ≤ 0 °F | 0 | 0 | 0 | 0 | 0 | 0 | 0 | 0 | 0 | 0 | 0 | 0 | 0 |
| Heating Degree Days | 654 | 488 | 304 | 104 | 17 | 0 | 0 | 0 | 9 | 114 | 314 | 550 | 2554 |
| Cooling Degree Days | 1 | 6 | 27 | 83 | 220 | 421 | 538 | 519 | 335 | 99 | 24 | 7 | 2280 |
| Total Precipitation (") | 4.91 | 5.03 | 5.33 | 4.65 | 5.78 | 4.16 | 3.91 | 3.06 | 3.34 | 4.42 | 4.40 | 5.30 | 54.29 |
| Days ≥ 0.1" Precip | 7 | 6 | 7 | 6 | 7 | 6 | 6 | 5 | 5 | 5 | 6 | 7 | 73 |
| Total Snowfall (") | 0.5 | 0.1 | 0.0 | 0.0 | 0.0 | 0.0 | 0.0 | 0.0 | 0.0 | 0.0 | 0.0 | 0.0 | 0.6 |
| Days ≥ 1" Snow Depth | 0 | 0 | 0 | 0 | 0 | 0 | 0 | 0 | 0 | 0 | 0 | 0 | 0 |

## CARVILLE 2 SW *Iberville Parish* ELEVATION 30 ft LAT/LONG 30° 12 ' N / 91° 7 ' W

| | JAN | FEB | MAR | APR | MAY | JUN | JUL | AUG | SEP | OCT | NOV | DEC | YEAR |
|---|---|---|---|---|---|---|---|---|---|---|---|---|---|
| Maximum Temp °F | 60.4 | 64.2 | 71.3 | 78.6 | 84.5 | 90.0 | 91.3 | 91.0 | 87.6 | 80.1 | 71.0 | 64.0 | 77.8 |
| Minimum Temp °F | 40.3 | 42.7 | 49.6 | 57.2 | 64.5 | 70.6 | 72.8 | 72.4 | 68.5 | 57.8 | 49.6 | 43.7 | 57.5 |
| Mean Temp °F | 50.4 | 53.5 | 60.5 | 67.9 | 74.5 | 80.4 | 82.1 | 81.7 | 78.1 | 69.0 | 60.3 | 53.9 | 67.7 |
| Days Max Temp ≥ 90 °F | 0 | 0 | 0 | 0 | 3 | 18 | 22 | 22 | 12 | 1 | 0 | 0 | 78 |
| Days Max Temp ≤ 32 °F | 0 | 0 | 0 | 0 | 0 | 0 | 0 | 0 | 0 | 0 | 0 | 0 | 0 |
| Days Min Temp ≤ 32 °F | 8 | 4 | 1 | 0 | 0 | 0 | 0 | 0 | 0 | 0 | 1 | 4 | 18 |
| Days Min Temp ≤ 0 °F | 0 | 0 | 0 | 0 | 0 | 0 | 0 | 0 | 0 | 0 | 0 | 0 | 0 |
| Heating Degree Days | 452 | 327 | 170 | 39 | 1 | 0 | 0 | 0 | 1 | 37 | 186 | 352 | 1565 |
| Cooling Degree Days | 3 | 8 | 33 | 122 | 304 | 471 | 541 | 534 | 397 | 164 | 53 | 16 | 2646 |
| Total Precipitation (") | 5.62 | 5.28 | 4.81 | 5.04 | 5.04 | 5.55 | 6.14 | 6.02 | 4.46 | 3.27 | 4.07 | 5.44 | 60.74 |
| Days ≥ 0.1" Precip | 7 | 6 | 6 | 5 | 7 | 7 | 9 | 9 | 6 | 4 | 5 | 7 | 78 |
| Total Snowfall (") | 0.0 | 0.1 | 0.0 | 0.0 | 0.0 | 0.0 | 0.0 | 0.0 | 0.0 | 0.0 | 0.0 | 0.0 | 0.1 |
| Days ≥ 1" Snow Depth | 0 | 0 | 0 | 0 | 0 | 0 | 0 | 0 | 0 | 0 | 0 | 0 | 0 |

## COVINGTON 4 NNW *St. Tammany Parish* ELEVATION 39 ft LAT/LONG 30° 32 ' N / 90° 7 ' W

| | JAN | FEB | MAR | APR | MAY | JUN | JUL | AUG | SEP | OCT | NOV | DEC | YEAR |
|---|---|---|---|---|---|---|---|---|---|---|---|---|---|
| Maximum Temp °F | 61.2 | 65.4 | 72.2 | 79.1 | 84.8 | 90.2 | 91.5 | 91.0 | 87.4 | 80.0 | 70.8 | 64.8 | 78.2 |
| Minimum Temp °F | 39.4 | 41.4 | 48.0 | 54.9 | 61.7 | 67.8 | 70.4 | 70.1 | 66.3 | 55.1 | 47.4 | 42.4 | 55.4 |
| Mean Temp °F | 50.3 | 53.4 | 60.2 | 67.0 | 73.3 | 79.0 | 81.0 | 80.6 | 76.9 | 67.6 | 59.1 | 53.6 | 66.8 |
| Days Max Temp ≥ 90 °F | 0 | 0 | 0 | 0 | 3 | 18 | 24 | 23 | 11 | 1 | 0 | 0 | 80 |
| Days Max Temp ≤ 32 °F | 0 | 0 | 0 | 0 | 0 | 0 | 0 | 0 | 0 | 0 | 0 | 0 | 0 |
| Days Min Temp ≤ 32 °F | 10 | 7 | 3 | 0 | 0 | 0 | 0 | 0 | 0 | 0 | 4 | 8 | 32 |
| Days Min Temp ≤ 0 °F | 0 | 0 | 0 | 0 | 0 | 0 | 0 | 0 | 0 | 0 | 0 | 0 | 0 |
| Heating Degree Days | 455 | 333 | 188 | 53 | 4 | 0 | 0 | 0 | 2 | 54 | 214 | 365 | 1668 |
| Cooling Degree Days | 5 | 13 | 46 | 107 | 270 | 430 | 509 | 502 | 363 | 147 | 48 | 22 | 2462 |
| Total Precipitation (") | 5.48 | 5.94 | 5.89 | 4.92 | 5.34 | 5.09 | 6.62 | 5.95 | 4.76 | 3.32 | 4.23 | 5.45 | 62.99 |
| Days ≥ 0.1" Precip | 8 | 6 | 6 | 5 | 6 | 7 | 10 | 8 | 6 | 4 | 6 | 7 | 79 |
| Total Snowfall (") | 0.1 | 0.1 | 0.0 | 0.0 | 0.0 | 0.0 | 0.0 | 0.0 | 0.0 | 0.0 | 0.0 | 0.0 | 0.2 |
| Days ≥ 1" Snow Depth | 0 | 0 | 0 | 0 | 0 | 0 | 0 | 0 | 0 | 0 | 0 | 0 | 0 |

## CROWLEY 2 NE *Acadia Parish* ELEVATION 20 ft LAT/LONG 30° 15 ' N / 92° 22 ' W

| | JAN | FEB | MAR | APR | MAY | JUN | JUL | AUG | SEP | OCT | NOV | DEC | YEAR |
|---|---|---|---|---|---|---|---|---|---|---|---|---|---|
| Maximum Temp °F | 58.6 | 62.6 | 70.1 | 77.5 | 83.8 | 89.3 | 90.8 | 90.9 | 87.5 | 80.4 | 70.0 | 62.9 | 77.0 |
| Minimum Temp °F | 39.7 | 42.5 | 49.8 | 57.4 | 64.8 | 71.1 | 72.5 | 71.4 | 67.4 | 56.8 | 48.7 | 42.6 | 57.1 |
| Mean Temp °F | 49.2 | 52.6 | 60.0 | 67.5 | 74.3 | 80.2 | 81.7 | 81.2 | 77.5 | 68.6 | 59.4 | 52.8 | 67.1 |
| Days Max Temp ≥ 90 °F | 0 | 0 | 0 | 0 | 3 | 16 | 22 | 22 | 13 | 2 | 0 | 0 | 78 |
| Days Max Temp ≤ 32 °F | 0 | 0 | 0 | 0 | 0 | 0 | 0 | 0 | 0 | 0 | 0 | 0 | 0 |
| Days Min Temp ≤ 32 °F | 9 | 4 | 1 | 0 | 0 | 0 | 0 | 0 | 0 | 0 | 2 | 5 | 21 |
| Days Min Temp ≤ 0 °F | 0 | 0 | 0 | 0 | 0 | 0 | 0 | 0 | 0 | 0 | 0 | 0 | 0 |
| Heating Degree Days | 491 | 353 | 190 | 51 | 3 | 0 | 0 | 0 | 2 | 49 | 215 | 388 | 1742 |
| Cooling Degree Days | 5 | 11 | 44 | 132 | 307 | 471 | 529 | 523 | 393 | 174 | 58 | 19 | 2666 |
| Total Precipitation (") | 5.89 | 4.43 | 4.35 | 4.83 | 5.98 | 5.41 | 5.89 | 5.61 | 4.87 | 4.19 | 4.36 | 5.20 | 61.01 |
| Days ≥ 0.1" Precip | 7 | 6 | 6 | 5 | 7 | 7 | 9 | 8 | 7 | 5 | 6 | 7 | 80 |
| Total Snowfall (") | 0.1 | 0.0 | 0.0 | 0.0 | 0.0 | 0.0 | 0.0 | 0.0 | 0.0 | 0.0 | 0.0 | 0.0 | 0.1 |
| Days ≥ 1" Snow Depth | 0 | 0 | 0 | 0 | 0 | 0 | 0 | 0 | 0 | 0 | 0 | 0 | 0 |

## DONALDSONVILLE 4 SW *Assumption Parish* ELEVATION 30 ft LAT/LONG 30° 4 ' N / 91° 2 ' W

| | JAN | FEB | MAR | APR | MAY | JUN | JUL | AUG | SEP | OCT | NOV | DEC | YEAR |
|---|---|---|---|---|---|---|---|---|---|---|---|---|---|
| Maximum Temp °F | 61.2 | 64.4 | 71.6 | 79.0 | 84.5 | 89.8 | 91.2 | 91.0 | 87.9 | 80.3 | 71.9 | 64.9 | 78.1 |
| Minimum Temp °F | 41.2 | 43.6 | 50.2 | 58.1 | 64.9 | 71.4 | 73.4 | 72.7 | 69.1 | 58.0 | 50.3 | 44.0 | 58.1 |
| Mean Temp °F | 51.2 | 54.0 | 60.9 | 68.6 | 74.7 | 80.6 | 82.3 | 81.9 | 78.5 | 69.1 | 61.1 | 54.5 | 68.1 |
| Days Max Temp ≥ 90 °F | 0 | 0 | 0 | 1 | 4 | 18 | 24 | 23 | 13 | 2 | 0 | 0 | 85 |
| Days Max Temp ≤ 32 °F | 0 | 0 | 0 | 0 | 0 | 0 | 0 | 0 | 0 | 0 | 0 | 0 | 0 |
| Days Min Temp ≤ 32 °F | 7 | 3 | 1 | 0 | 0 | 0 | 0 | 0 | 0 | 0 | 1 | 4 | 16 |
| Days Min Temp ≤ 0 °F | 0 | 0 | 0 | 0 | 0 | 0 | 0 | 0 | 0 | 0 | 0 | 0 | 0 |
| Heating Degree Days | 432 | 316 | 171 | 39 | 2 | 0 | 0 | 0 | 1 | 40 | 174 | 341 | 1516 |
| Cooling Degree Days | 7 | 13 | 46 | 125 | 301 | 465 | 536 | 527 | 397 | 167 | 60 | 20 | 2664 |
| Total Precipitation (") | 5.19 | 5.25 | 4.94 | 5.09 | 4.76 | 4.88 | 6.94 | 5.93 | 5.08 | 3.28 | 3.80 | 5.40 | 60.54 |
| Days ≥ 0.1" Precip | 7 | 6 | 6 | 5 | 6 | 8 | 10 | 9 | 7 | 4 | 5 | 6 | 79 |
| Total Snowfall (") | 0.0 | 0.1 | 0.0 | 0.0 | 0.0 | 0.0 | 0.0 | 0.0 | 0.0 | 0.0 | 0.0 | 0.0 | 0.1 |
| Days ≥ 1" Snow Depth | 0 | 0 | 0 | 0 | 0 | 0 | 0 | 0 | 0 | 0 | 0 | 0 | 0 |

## ELIZABETH *Allen Parish* ELEVATION 151 ft LAT/LONG 30° 52 ' N / 92° 48 ' W

| | JAN | FEB | MAR | APR | MAY | JUN | JUL | AUG | SEP | OCT | NOV | DEC | YEAR |
|---|---|---|---|---|---|---|---|---|---|---|---|---|---|
| Maximum Temp °F | 58.7 | 63.1 | 71.0 | 78.5 | 84.1 | 90.0 | 92.6 | 92.3 | 88.1 | 80.1 | 70.7 | 63.1 | 77.7 |
| Minimum Temp °F | 36.6 | 39.2 | 46.9 | 54.7 | 61.9 | 68.2 | 70.6 | 69.9 | 65.1 | 53.5 | 45.7 | 39.5 | 54.3 |
| Mean Temp °F | 47.6 | 51.1 | 59.0 | 66.6 | 73.1 | 79.1 | 81.6 | 81.1 | 76.6 | 66.9 | 58.2 | 51.3 | 66.0 |
| Days Max Temp ≥ 90 °F | 0 | 0 | 0 | 0 | 4 | 18 | 25 | 24 | 15 | 2 | 0 | 0 | 88 |
| Days Max Temp ≤ 32 °F | 0 | 0 | 0 | 0 | 0 | 0 | 0 | 0 | 0 | 0 | 0 | 0 | 0 |
| Days Min Temp ≤ 32 °F | 12 | 8 | 2 | 0 | 0 | 0 | 0 | 0 | 0 | 0 | 3 | 9 | 34 |
| Days Min Temp ≤ 0 °F | 0 | 0 | 0 | 0 | 0 | 0 | 0 | 0 | 0 | 0 | 0 | 0 | 0 |
| Heating Degree Days | 536 | 394 | 217 | 63 | 6 | 0 | 0 | 0 | 4 | 70 | 236 | 428 | 1954 |
| Cooling Degree Days | 3 | 11 | 38 | 108 | 264 | 440 | 528 | 523 | 364 | 138 | 41 | 14 | 2472 |
| Total Precipitation (") | 5.86 | 4.68 | 5.18 | 4.44 | 6.33 | 5.57 | 4.72 | 4.22 | 4.44 | 5.06 | 5.25 | 6.86 | 62.61 |
| Days ≥ 0.1" Precip | 8 | 6 | 6 | 5 | 7 | 7 | 7 | 6 | 6 | 5 | 6 | 8 | 77 |
| Total Snowfall (") | 0.1 | 0.0 | 0.1 | 0.0 | 0.0 | 0.0 | 0.0 | 0.0 | 0.0 | 0.0 | 0.0 | 0.0 | 0.2 |
| Days ≥ 1" Snow Depth | 0 | 0 | 0 | 0 | 0 | 0 | 0 | 0 | 0 | 0 | 0 | 0 | 0 |

## FRANKLIN 3 NW *St. Mary Parish* ELEVATION 10 ft LAT/LONG 29° 48 ' N / 91° 30 ' W

| | JAN | FEB | MAR | APR | MAY | JUN | JUL | AUG | SEP | OCT | NOV | DEC | YEAR |
|---|---|---|---|---|---|---|---|---|---|---|---|---|---|
| Maximum Temp °F | 61.4 | 64.5 | 71.3 | 78.0 | 83.6 | 88.7 | 89.9 | 89.7 | 86.6 | 79.7 | 71.4 | 65.0 | 77.5 |
| Minimum Temp °F | 42.4 | 44.6 | 51.1 | 58.6 | 65.5 | 71.4 | 73.1 | 72.7 | 68.9 | 58.3 | 50.2 | 45.0 | 58.5 |
| Mean Temp °F | 51.9 | 54.6 | 61.2 | 68.3 | 74.6 | 80.1 | 81.6 | 81.2 | 77.8 | 69.0 | 60.9 | 55.0 | 68.0 |
| Days Max Temp ≥ 90 °F | 0 | 0 | 0 | 0 | 2 | 13 | 19 | 18 | 9 | 1 | 0 | 0 | 62 |
| Days Max Temp ≤ 32 °F | 0 | 0 | 0 | 0 | 0 | 0 | 0 | 0 | 0 | 0 | 0 | 0 | 0 |
| Days Min Temp ≤ 32 °F | 6 | 3 | 1 | 0 | 0 | 0 | 0 | 0 | 0 | 0 | 1 | 3 | 14 |
| Days Min Temp ≤ 0 °F | 0 | 0 | 0 | 0 | 0 | 0 | 0 | 0 | 0 | 0 | 0 | 0 | 0 |
| Heating Degree Days | 411 | 300 | 160 | 37 | 1 | 0 | 0 | 0 | 1 | 41 | 179 | 324 | 1454 |
| Cooling Degree Days | 8 | 12 | 52 | 129 | 304 | 457 | 521 | 516 | 386 | 169 | 62 | 22 | 2638 |
| Total Precipitation (") | 5.10 | 4.57 | 4.35 | 5.20 | 5.41 | 6.23 | 7.52 | 8.85 | 6.04 | 3.48 | 3.85 | 5.16 | 65.76 |
| Days ≥ 0.1" Precip | 7 | 6 | 6 | 5 | 6 | 8 | 11 | 11 | 8 | 4 | 5 | 7 | 84 |
| Total Snowfall (") | 0.0 | 0.0 | 0.0 | 0.0 | 0.0 | 0.0 | 0.0 | 0.0 | 0.0 | 0.0 | 0.0 | 0.1 | 0.1 |
| Days ≥ 1" Snow Depth | 0 | 0 | 0 | 0 | 0 | 0 | 0 | 0 | 0 | 0 | 0 | 0 | 0 |

## FRANKLINTON 3 SW *Washington Parish* ELEVATION 151 ft LAT/LONG 30° 48 ' N / 90° 8 ' W

| | JAN | FEB | MAR | APR | MAY | JUN | JUL | AUG | SEP | OCT | NOV | DEC | YEAR |
|---|---|---|---|---|---|---|---|---|---|---|---|---|---|
| Maximum Temp °F | 59.7 | 63.7 | 71.8 | 79.2 | 85.1 | 91.3 | 92.6 | 91.6 | 87.5 | 79.4 | 70.0 | 63.0 | 77.9 |
| Minimum Temp °F | 37.2 | 39.4 | 46.4 | 53.7 | 60.6 | 66.8 | 69.6 | 68.9 | 64.8 | 53.0 | 44.9 | 39.8 | 53.8 |
| Mean Temp °F | 48.5 | 51.6 | 59.1 | 66.5 | 72.9 | 79.1 | 81.1 | 80.3 | 76.1 | 66.3 | 57.5 | 51.4 | 65.9 |
| Days Max Temp ≥ 90 °F | 0 | 0 | 0 | 1 | 7 | 21 | 24 | 22 | 13 | 2 | 0 | 0 | 90 |
| Days Max Temp ≤ 32 °F | 0 | 0 | 0 | 0 | 0 | 0 | 0 | 0 | 0 | 0 | 0 | 0 | 0 |
| Days Min Temp ≤ 32 °F | 12 | 8 | 3 | 0 | 0 | 0 | 0 | 0 | 0 | 0 | 5 | 10 | 38 |
| Days Min Temp ≤ 0 °F | 0 | 0 | 0 | 0 | 0 | 0 | 0 | 0 | 0 | 0 | 0 | 0 | 0 |
| Heating Degree Days | 511 | 381 | 218 | 63 | 5 | 0 | 0 | 0 | 4 | 74 | 256 | 427 | 1939 |
| Cooling Degree Days | 6 | 11 | 43 | 107 | 267 | 442 | 524 | 502 | 350 | 134 | 44 | 15 | 2445 |
| Total Precipitation (") | 5.70 | 5.81 | 5.95 | 6.19 | 5.76 | 5.09 | 5.67 | 5.77 | 3.85 | 3.21 | 4.98 | 5.76 | 63.74 |
| Days ≥ 0.1" Precip | 7 | 7 | 7 | 6 | 7 | 7 | 10 | 8 | 6 | 4 | 6 | 7 | 82 |
| Total Snowfall (") | 0.2 | 0.2 | 0.1 | 0.0 | 0.0 | 0.0 | 0.0 | 0.0 | 0.0 | 0.0 | 0.0 | 0.0 | 0.5 |
| Days ≥ 1" Snow Depth | 0 | 0 | 0 | 0 | 0 | 0 | 0 | 0 | 0 | 0 | 0 | 0 | 0 |

## GRAND COTEAU *St. Landry Parish* ELEVATION 59 ft LAT/LONG 30° 26 ' N / 92° 2 ' W

| | JAN | FEB | MAR | APR | MAY | JUN | JUL | AUG | SEP | OCT | NOV | DEC | YEAR |
|---|---|---|---|---|---|---|---|---|---|---|---|---|---|
| Maximum Temp °F | 60.6 | 64.3 | 71.6 | 78.9 | 84.9 | 90.1 | 91.5 | 91.7 | 87.9 | 80.6 | 71.2 | 64.3 | 78.1 |
| Minimum Temp °F | 40.7 | 43.2 | 50.0 | 57.2 | 63.9 | 69.7 | 72.0 | 71.3 | 67.1 | 56.5 | 49.2 | 43.4 | 57.0 |
| Mean Temp °F | 50.7 | 53.8 | 60.8 | 68.1 | 74.4 | 79.9 | 81.8 | 81.6 | 77.5 | 68.6 | 60.2 | 53.9 | 67.6 |
| Days Max Temp ≥ 90 °F | 0 | 0 | 0 | 0 | 4 | 18 | 24 | 24 | 13 | 2 | 0 | 0 | 85 |
| Days Max Temp ≤ 32 °F | 0 | 0 | 0 | 0 | 0 | 0 | 0 | 0 | 0 | 0 | 0 | 0 | 0 |
| Days Min Temp ≤ 32 °F | 8 | 5 | 1 | 0 | 0 | 0 | 0 | 0 | 0 | 0 | 2 | 5 | 21 |
| Days Min Temp ≤ 0 °F | 0 | 0 | 0 | 0 | 0 | 0 | 0 | 0 | 0 | 0 | 0 | 0 | 0 |
| Heating Degree Days | 448 | 324 | 173 | 42 | 2 | 0 | 0 | 0 | 2 | 45 | 193 | 357 | 1586 |
| Cooling Degree Days | 8 | 14 | 50 | 137 | 306 | 461 | 537 | 548 | 388 | 167 | 59 | 20 | 2695 |
| Total Precipitation (") | 5.94 | 4.95 | 4.69 | 5.29 | 6.09 | 5.66 | 6.08 | 4.71 | 4.79 | 4.42 | 4.83 | 5.53 | 62.98 |
| Days ≥ 0.1" Precip | 8 | 7 | 6 | 5 | 7 | 7 | 8 | 7 | 7 | 5 | 6 | 7 | 80 |
| Total Snowfall (") | 0.0 | 0.1 | 0.0 | 0.0 | 0.0 | 0.0 | 0.0 | 0.0 | 0.0 | 0.0 | 0.0 | 0.0 | 0.1 |
| Days ≥ 1" Snow Depth | 0 | 0 | 0 | 0 | 0 | 0 | 0 | 0 | 0 | 0 | 0 | 0 | 0 |

## HACKBERRY 8 SSW *Cameron Parish* ELEVATION 10 ft LAT/LONG 29° 53 ' N / 93° 25 ' W

| | JAN | FEB | MAR | APR | MAY | JUN | JUL | AUG | SEP | OCT | NOV | DEC | YEAR |
|---|---|---|---|---|---|---|---|---|---|---|---|---|---|
| Maximum Temp °F | 58.4 | 61.9 | 68.6 | 75.8 | 81.9 | 87.4 | 89.7 | 89.7 | 86.4 | 79.0 | 69.9 | 62.4 | 75.9 |
| Minimum Temp °F | 42.5 | 45.4 | 52.8 | 61.3 | 68.1 | 74.3 | 75.4 | 74.8 | 71.2 | 61.6 | 52.5 | 45.8 | 60.5 |
| Mean Temp °F | 50.4 | 53.7 | 60.7 | 68.6 | 75.1 | 80.9 | 82.6 | 82.3 | 78.8 | 70.3 | 61.2 | 54.1 | 68.2 |
| Days Max Temp ≥ 90 °F | 0 | 0 | 0 | 0 | 1 | 8 | 19 | 19 | 7 | 0 | 0 | 0 | 54 |
| Days Max Temp ≤ 32 °F | 0 | 0 | 0 | 0 | 0 | 0 | 0 | 0 | 0 | 0 | 0 | 0 | 0 |
| Days Min Temp ≤ 32 °F | 5 | 2 | 0 | 0 | 0 | 0 | 0 | 0 | 0 | 0 | 1 | 3 | 11 |
| Days Min Temp ≤ 0 °F | 0 | 0 | 0 | 0 | 0 | 0 | 0 | 0 | 0 | 0 | 0 | 0 | 0 |
| Heating Degree Days | 450 | 319 | 163 | 32 | 1 | 0 | 0 | 0 | 1 | 32 | 168 | 342 | 1508 |
| Cooling Degree Days | 3 | 7 | 42 | 139 | 326 | 503 | 569 | 566 | 435 | 216 | 66 | 15 | 2887 |
| Total Precipitation (") | 5.48 | 3.57 | 3.87 | 4.17 | 5.27 | 5.99 | 6.64 | 5.87 | 5.40 | 4.24 | 4.07 | 4.63 | 59.20 |
| Days ≥ 0.1" Precip | 7 | 6 | 5 | 4 | 6 | 7 | 8 | 8 | 6 | 4 | 5 | 6 | 72 |
| Total Snowfall (") | 0.1 | 0.0 | 0.0 | 0.0 | 0.0 | 0.0 | 0.0 | 0.0 | 0.0 | 0.0 | 0.0 | 0.0 | 0.1 |
| Days ≥ 1" Snow Depth | 0 | 0 | 0 | 0 | 0 | 0 | 0 | 0 | 0 | 0 | 0 | 0 | 0 |

## HAMMOND *Tangipahoa Parish* ELEVATION 39 ft LAT/LONG 30° 32 ' N / 90° 29 ' W

| | JAN | FEB | MAR | APR | MAY | JUN | JUL | AUG | SEP | OCT | NOV | DEC | YEAR |
|---|---|---|---|---|---|---|---|---|---|---|---|---|---|
| Maximum Temp °F | *59.6* | *63.6* | *72.0* | *79.3* | *84.8* | *90.9* | *92.0* | *91.1* | *87.6* | *79.5* | *69.8* | *62.7* | 77.7 |
| Minimum Temp °F | *38.0* | *39.8* | *47.5* | *55.1* | *61.4* | *67.0* | *70.1* | *69.6* | *65.7* | *54.7* | *46.8* | *40.9* | 54.7 |
| Mean Temp °F | *48.8* | *51.7* | *59.8* | *67.3* | *73.1* | *79.0* | *81.1* | *80.4* | *76.7* | *67.1* | *58.3* | *51.8* | 66.3 |
| Days Max Temp ≥ 90 °F | *0* | *0* | *0* | *0* | na | *20* | *25* | *23* | *12* | *1* | *0* | *0* | na |
| Days Max Temp ≤ 32 °F | *0* | *0* | *0* | *0* | *0* | *0* | *0* | *0* | *0* | *0* | *0* | *0* | 0 |
| Days Min Temp ≤ 32 °F | *11* | na | *3* | *0* | *0* | *0* | *0* | *0* | *0* | *0* | *4* | *9* | na |
| Days Min Temp ≤ 0 °F | *0* | *0* | *0* | *0* | *0* | *0* | *0* | *0* | *0* | *0* | *0* | *0* | 0 |
| Heating Degree Days | *503* | *378* | *201* | *52* | *5* | *0* | *0* | *0* | *4* | *65* | *233* | *415* | 1856 |
| Cooling Degree Days | na | na | na | na | na | na | na | na | na | na | na | na | na |
| Total Precipitation (") | *5.33* | *5.76* | *5.50* | *6.53* | *5.83* | *4.51* | *7.56* | *5.55* | *5.15* | *3.98* | *3.84* | *6.12* | 65.66 |
| Days ≥ 0.1" Precip | *7* | *7* | *7* | *6* | *7* | *6* | *11* | *9* | *7* | *4* | *5* | *8* | 84 |
| Total Snowfall (") | *0.0* | *0.0* | *0.0* | *0.0* | *0.0* | *0.0* | *0.0* | *0.0* | *0.0* | *0.0* | *0.0* | *0.0* | 0.0 |
| Days ≥ 1" Snow Depth | *0* | *0* | *0* | *0* | *0* | *0* | *0* | *0* | *0* | *0* | *0* | *0* | 0 |

## HOMER 3 SSW *Claiborne Parish* ELEVATION 400 ft LAT/LONG 32° 45 ' N / 93° 4 ' W

| | JAN | FEB | MAR | APR | MAY | JUN | JUL | AUG | SEP | OCT | NOV | DEC | YEAR |
|---|---|---|---|---|---|---|---|---|---|---|---|---|---|
| Maximum Temp °F | 54.1 | 58.8 | 67.4 | 75.9 | 81.8 | 88.7 | 92.1 | 92.1 | 86.5 | 76.8 | 66.3 | 57.5 | 74.8 |
| Minimum Temp °F | 32.6 | 35.3 | 42.8 | 51.4 | 59.0 | 66.5 | 70.1 | 68.8 | 63.0 | 51.1 | 42.7 | 35.5 | 51.6 |
| Mean Temp °F | 43.3 | 47.0 | 55.1 | 63.7 | 70.5 | 77.6 | 81.1 | 80.4 | 74.7 | 64.0 | 54.5 | 46.5 | 63.2 |
| Days Max Temp ≥ 90 °F | 0 | 0 | 0 | 0 | 2 | 14 | 24 | 24 | 12 | 1 | 0 | 0 | 77 |
| Days Max Temp ≤ 32 °F | 1 | 0 | 0 | 0 | 0 | 0 | 0 | 0 | 0 | 0 | 0 | 1 | 2 |
| Days Min Temp ≤ 32 °F | 17 | 12 | 5 | 1 | 0 | 0 | 0 | 0 | 0 | 0 | 6 | 13 | 54 |
| Days Min Temp ≤ 0 °F | 0 | 0 | 0 | 0 | 0 | 0 | 0 | 0 | 0 | 0 | 0 | 0 | 0 |
| Heating Degree Days | 667 | 504 | 319 | 109 | 19 | 0 | 0 | 0 | 12 | 115 | 326 | 570 | 2641 |
| Cooling Degree Days | 1 | 3 | 18 | 67 | 185 | 383 | 507 | 504 | 319 | 89 | 19 | 5 | 2100 |
| Total Precipitation (") | 4.71 | 5.10 | 4.76 | 4.70 | 5.49 | 4.10 | 4.49 | 2.81 | 4.02 | 4.27 | 5.01 | 4.73 | 54.19 |
| Days ≥ 0.1" Precip | 7 | 6 | 7 | 6 | 7 | 6 | 6 | 5 | 5 | 5 | 6 | 7 | 73 |
| Total Snowfall (") | 0.6 | 0.2 | 0.1 | 0.0 | 0.0 | 0.0 | 0.0 | 0.0 | 0.0 | 0.0 | 0.0 | 0.2 | 1.1 |
| Days ≥ 1" Snow Depth | 0 | 0 | 0 | 0 | 0 | 0 | 0 | 0 | 0 | 0 | 0 | 0 | 0 |

## HOUMA *Terrebonne Parish* ELEVATION 10 ft LAT/LONG 29° 35' N / 90° 44' W

| | JAN | FEB | MAR | APR | MAY | JUN | JUL | AUG | SEP | OCT | NOV | DEC | YEAR |
|---|---|---|---|---|---|---|---|---|---|---|---|---|---|
| Maximum Temp °F | 62.4 | 65.2 | 71.6 | 78.2 | 83.9 | 88.9 | 90.3 | 89.9 | 86.8 | 79.9 | 72.3 | 66.0 | 77.9 |
| Minimum Temp °F | 42.8 | 45.1 | 52.3 | 59.2 | 66.0 | 71.0 | 72.7 | 72.3 | 69.1 | 58.4 | 51.1 | 45.3 | 58.8 |
| Mean Temp °F | 52.7 | 55.2 | 62.0 | 68.8 | 75.0 | 80.0 | 81.6 | 81.1 | 78.0 | 69.2 | 61.7 | 55.7 | 68.4 |
| Days Max Temp ≥ 90 °F | 0 | 0 | 0 | 0 | 2 | 14 | 21 | 19 | 10 | 1 | 0 | 0 | 67 |
| Days Max Temp ≤ 32 °F | 0 | 0 | 0 | 0 | 0 | 0 | 0 | 0 | 0 | 0 | 0 | 0 | 0 |
| Days Min Temp ≤ 32 °F | 6 | 3 | 1 | 0 | 0 | 0 | 0 | 0 | 0 | 0 | 1 | 4 | 15 |
| Days Min Temp ≤ 0 °F | 0 | 0 | 0 | 0 | 0 | 0 | 0 | 0 | 0 | 0 | 0 | 0 | 0 |
| Heating Degree Days | 391 | 288 | 146 | 36 | 1 | 0 | 0 | 0 | 1 | 39 | 161 | 309 | 1372 |
| Cooling Degree Days | 11 | 19 | 57 | 133 | 318 | 457 | 526 | 516 | 394 | 181 | 73 | 27 | 2712 |
| Total Precipitation (") | 5.32 | 4.88 | 4.51 | 4.90 | 5.96 | 5.45 | 7.81 | 7.27 | 6.57 | 3.07 | 3.87 | 5.13 | 64.74 |
| Days ≥ 0.1" Precip | 7 | 6 | 5 | 5 | 6 | 7 | 12 | 11 | 8 | 3 | 5 | 6 | 81 |
| Total Snowfall (") | 0.0 | 0.0 | 0.0 | 0.0 | 0.0 | 0.0 | 0.0 | 0.0 | 0.0 | 0.0 | 0.0 | 0.1 | 0.1 |
| Days ≥ 1" Snow Depth | 0 | 0 | 0 | 0 | 0 | 0 | 0 | 0 | 0 | 0 | 0 | 0 | 0 |

## JEANERETTE 5 NW *Iberia Parish* ELEVATION 20 ft LAT/LONG 29° 57' N / 91° 43' W

| | JAN | FEB | MAR | APR | MAY | JUN | JUL | AUG | SEP | OCT | NOV | DEC | YEAR |
|---|---|---|---|---|---|---|---|---|---|---|---|---|---|
| Maximum Temp °F | *59.4* | *62.2* | *69.6* | *77.4* | *83.2* | *88.5* | *90.0* | *89.7* | na | *79.1* | *71.2* | *63.5* | na |
| Minimum Temp °F | *40.3* | *42.8* | *49.5* | *57.9* | *64.9* | *70.9* | *72.6* | *71.8* | na | *56.5* | *49.8* | *43.4* | na |
| Mean Temp °F | *49.9* | *52.6* | *59.6* | *67.7* | *74.1* | *79.7* | *81.3* | *80.8* | na | *67.8* | *60.5* | *53.5* | na |
| Days Max Temp ≥ 90 °F | *0* | *0* | *0* | *0* | *1* | *12* | na | na | *8* | *1* | *0* | *0* | na |
| Days Max Temp ≤ 32 °F | *0* | *0* | *0* | *0* | *0* | *0* | *0* | *0* | *0* | *0* | *0* | *0* | 0 |
| Days Min Temp ≤ 32 °F | *8* | *4* | *1* | *0* | *0* | *0* | *0* | *0* | *0* | *0* | *1* | *5* | 19 |
| Days Min Temp ≤ 0 °F | *0* | *0* | *0* | *0* | *0* | *0* | *0* | *0* | *0* | *0* | *0* | *0* | 0 |
| Heating Degree Days | *468* | *353* | *200* | *44* | *2* | *0* | *0* | *0* | na | *56* | *183* | *369* | na |
| Cooling Degree Days | 4 | 10 | 43 | 119 | 290 | 449 | 515 | 505 | 370 | 157 | 57 | 21 | 2540 |
| Total Precipitation (") | 5.27 | 4.56 | 4.06 | 4.85 | 5.27 | 6.62 | 6.94 | 6.63 | 5.69 | 3.77 | 3.68 | 5.06 | 62.40 |
| Days ≥ 0.1" Precip | 7 | 6 | 6 | 5 | 6 | 8 | 10 | 10 | 8 | 5 | 5 | 7 | 83 |
| Total Snowfall (") | 0.0 | 0.1 | 0.0 | 0.0 | 0.0 | 0.0 | 0.0 | 0.0 | 0.0 | 0.0 | 0.0 | 0.0 | 0.1 |
| Days ≥ 1" Snow Depth | 0 | 0 | 0 | 0 | 0 | 0 | 0 | 0 | 0 | 0 | 0 | 0 | 0 |

## JENNINGS *Jefferson Davis Parish* ELEVATION 30 ft LAT/LONG 30° 13' N / 92° 39' W

| | JAN | FEB | MAR | APR | MAY | JUN | JUL | AUG | SEP | OCT | NOV | DEC | YEAR |
|---|---|---|---|---|---|---|---|---|---|---|---|---|---|
| Maximum Temp °F | 59.0 | 63.1 | 70.7 | 78.2 | 84.0 | 89.3 | 91.0 | 91.0 | 87.7 | 80.2 | 70.4 | 63.0 | 77.3 |
| Minimum Temp °F | 40.4 | 42.8 | 49.8 | 57.5 | 64.8 | 71.0 | 72.7 | 71.9 | 68.0 | 57.5 | 49.4 | 43.4 | 57.4 |
| Mean Temp °F | 49.7 | 53.0 | 60.3 | 67.9 | 74.4 | 80.2 | 81.9 | 81.5 | 77.8 | 68.9 | 59.9 | 53.2 | 67.4 |
| Days Max Temp ≥ 90 °F | 0 | 0 | 0 | 0 | 3 | 15 | 22 | 23 | 13 | 2 | 0 | 0 | 78 |
| Days Max Temp ≤ 32 °F | 0 | 0 | 0 | 0 | 0 | 0 | 0 | 0 | 0 | 0 | 0 | 0 | 0 |
| Days Min Temp ≤ 32 °F | 8 | 4 | 1 | 0 | 0 | 0 | 0 | 0 | 0 | 0 | 1 | 5 | 19 |
| Days Min Temp ≤ 0 °F | 0 | 0 | 0 | 0 | 0 | 0 | 0 | 0 | 0 | 0 | 0 | 0 | 0 |
| Heating Degree Days | 476 | 344 | 186 | 47 | 3 | 0 | 0 | 0 | 2 | 49 | 203 | 376 | 1686 |
| Cooling Degree Days | 4 | 10 | 39 | 114 | 291 | 453 | 520 | 515 | 385 | 167 | 56 | 18 | 2572 |
| Total Precipitation (") | 5.82 | 4.07 | 4.36 | 4.57 | 6.21 | 4.96 | 5.75 | 4.96 | 5.21 | 4.58 | 4.54 | 5.48 | 60.51 |
| Days ≥ 0.1" Precip | 7 | 6 | 6 | 4 | 7 | 6 | 9 | 7 | 7 | 5 | 5 | 7 | 76 |
| Total Snowfall (") | 0.1 | 0.1 | 0.0 | 0.0 | 0.0 | 0.0 | 0.0 | 0.0 | 0.0 | 0.0 | 0.0 | 0.0 | 0.2 |
| Days ≥ 1" Snow Depth | 0 | 0 | 0 | 0 | 0 | 0 | 0 | 0 | 0 | 0 | 0 | 0 | 0 |

## LAFAYETTE REGL AP *Lafayette Parish* ELEVATION 43 ft LAT/LONG 30° 12' N / 91° 59' W

| | JAN | FEB | MAR | APR | MAY | JUN | JUL | AUG | SEP | OCT | NOV | DEC | YEAR |
|---|---|---|---|---|---|---|---|---|---|---|---|---|---|
| Maximum Temp °F | 60.2 | 63.5 | 71.0 | 78.4 | 84.3 | 89.5 | 90.6 | 90.3 | 87.0 | 79.8 | 71.0 | 64.2 | 77.5 |
| Minimum Temp °F | 41.8 | 44.3 | 51.2 | 58.6 | 65.5 | 71.4 | 73.6 | 73.1 | 68.9 | 58.2 | 49.9 | 44.7 | 58.4 |
| Mean Temp °F | 51.0 | 53.9 | 61.1 | 68.5 | 74.9 | 80.5 | 82.2 | 81.7 | 78.0 | 69.0 | 60.5 | 54.4 | 68.0 |
| Days Max Temp ≥ 90 °F | 0 | 0 | 0 | 0 | 4 | 16 | 22 | 20 | 11 | 1 | 0 | 0 | 74 |
| Days Max Temp ≤ 32 °F | 0 | 0 | 0 | 0 | 0 | 0 | 0 | 0 | 0 | 0 | 0 | 0 | 0 |
| Days Min Temp ≤ 32 °F | 6 | 3 | 1 | 0 | 0 | 0 | 0 | 0 | 0 | 0 | 1 | 4 | 15 |
| Days Min Temp ≤ 0 °F | 0 | 0 | 0 | 0 | 0 | 0 | 0 | 0 | 0 | 0 | 0 | 0 | 0 |
| Heating Degree Days | 438 | 319 | 169 | 39 | 1 | 0 | 0 | 0 | 2 | 40 | 189 | 345 | 1542 |
| Cooling Degree Days | 7 | 14 | 53 | 139 | 316 | 478 | 550 | 544 | 398 | 172 | 62 | 23 | 2756 |
| Total Precipitation (") | 5.86 | 4.50 | 4.12 | 4.80 | 5.41 | 5.30 | 7.04 | 5.51 | 5.14 | 3.77 | 3.92 | 5.43 | 60.80 |
| Days ≥ 0.1" Precip | 7 | 6 | 6 | 5 | 6 | 7 | 10 | 8 | 7 | 4 | 6 | 7 | 79 |
| Total Snowfall (") | 0.1 | 0.1 | 0.0 | 0.0 | 0.0 | 0.0 | 0.0 | 0.0 | 0.0 | 0.0 | 0.0 | 0.0 | 0.2 |
| Days ≥ 1" Snow Depth | 0 | 0 | 0 | 0 | 0 | 0 | 0 | 0 | 0 | 0 | 0 | 0 | 0 |

## LAKE CHARLES MUNI AP *Calcasieu Parish* ELEVATION 9 ft LAT/LONG 30° 7 ' N / 93° 13 ' W

| | JAN | FEB | MAR | APR | MAY | JUN | JUL | AUG | SEP | OCT | NOV | DEC | YEAR |
|---|---|---|---|---|---|---|---|---|---|---|---|---|---|
| Maximum Temp °F | 59.9 | 63.5 | 70.7 | 77.9 | 83.7 | 88.9 | 90.8 | 90.7 | 87.2 | 80.4 | 70.8 | 63.7 | 77.4 |
| Minimum Temp °F | 41.8 | 44.3 | 50.9 | 58.8 | 65.8 | 72.0 | 73.8 | 73.0 | 68.6 | 58.2 | 49.9 | 44.4 | 58.5 |
| Mean Temp °F | 50.9 | 53.9 | 60.8 | 68.3 | 74.8 | 80.5 | 82.4 | 81.8 | 78.0 | 69.3 | 60.4 | 54.1 | 67.9 |
| Days Max Temp ≥ 90 °F | 0 | 0 | 0 | 0 | 2 | 14 | 23 | 22 | 11 | 1 | 0 | 0 | 73 |
| Days Max Temp ≤ 32 °F | 0 | 0 | 0 | 0 | 0 | 0 | 0 | 0 | 0 | 0 | 0 | 0 | 0 |
| Days Min Temp ≤ 32 °F | 6 | 3 | 1 | 0 | 0 | 0 | 0 | 0 | 0 | 0 | 1 | 4 | 15 |
| Days Min Temp ≤ 0 °F | 0 | 0 | 0 | 0 | 0 | 0 | 0 | 0 | 0 | 0 | 0 | 0 | 0 |
| Heating Degree Days | 440 | 316 | 170 | 37 | 1 | 0 | 0 | 0 | 1 | 37 | 189 | 351 | 1542 |
| Cooling Degree Days | 6 | 12 | 47 | 133 | 316 | 478 | 553 | 548 | 402 | 183 | 62 | 21 | 2761 |
| Total Precipitation (") | 5.09 | 3.47 | 3.45 | 3.86 | 6.28 | 5.71 | 5.13 | 4.95 | 5.44 | 4.21 | 4.06 | 4.99 | 56.64 |
| Days ≥ 0.1" Precip | 7 | 5 | 5 | 5 | 6 | 6 | 7 | 7 | 6 | 4 | 6 | 6 | 70 |
| Total Snowfall (") | 0.2 | 0.1 | 0.0 | 0.0 | 0.0 | 0.0 | 0.0 | 0.0 | 0.0 | 0.0 | 0.0 | 0.0 | 0.3 |
| Days ≥ 1" Snow Depth | 0 | 0 | 0 | 0 | 0 | 0 | 0 | 0 | 0 | 0 | 0 | 0 | 0 |

## LAKE PROVIDENCE *East Carroll Parish* ELEVATION 102 ft LAT/LONG 32° 48 ' N / 91° 11 ' W

| | JAN | FEB | MAR | APR | MAY | JUN | JUL | AUG | SEP | OCT | NOV | DEC | YEAR |
|---|---|---|---|---|---|---|---|---|---|---|---|---|---|
| Maximum Temp °F | 52.3 | 57.3 | 66.1 | 75.3 | 82.3 | 89.1 | 91.6 | 90.9 | 86.2 | 76.9 | 65.9 | 56.5 | 74.2 |
| Minimum Temp °F | 34.3 | 37.5 | 45.4 | 54.2 | 62.2 | 69.7 | 72.5 | 71.1 | 65.6 | 53.7 | 44.9 | 37.8 | 54.1 |
| Mean Temp °F | 43.3 | 47.4 | 55.8 | 64.8 | 72.3 | 79.4 | 82.1 | 81.0 | 75.9 | 65.3 | 55.4 | 47.2 | 64.2 |
| Days Max Temp ≥ 90 °F | 0 | 0 | 0 | 0 | 3 | 15 | 22 | 21 | 10 | 1 | 0 | 0 | 72 |
| Days Max Temp ≤ 32 °F | 1 | 1 | 0 | 0 | 0 | 0 | 0 | 0 | 0 | 0 | 0 | 1 | 3 |
| Days Min Temp ≤ 32 °F | 14 | 9 | 2 | 0 | 0 | 0 | 0 | 0 | 0 | 0 | 3 | 9 | 37 |
| Days Min Temp ≤ 0 °F | 0 | 0 | 0 | 0 | 0 | 0 | 0 | 0 | 0 | 0 | 0 | 0 | 0 |
| Heating Degree Days | 667 | 493 | 300 | 93 | 10 | 0 | 0 | 0 | 6 | 90 | 303 | 550 | 2512 |
| Cooling Degree Days | 0 | 4 | 19 | 84 | 238 | 440 | 547 | 518 | 342 | 105 | 21 | 4 | 2322 |
| Total Precipitation (") | 5.20 | 5.41 | 5.98 | 5.59 | 5.77 | 4.30 | 4.16 | 2.99 | 2.95 | 4.56 | 4.90 | 5.95 | 57.76 |
| Days ≥ 0.1" Precip | 7 | 6 | 7 | 6 | 8 | 6 | 6 | 4 | 5 | 5 | 6 | 7 | 73 |
| Total Snowfall (") | 0.6 | 0.2 | 0.2 | 0.0 | 0.0 | 0.0 | 0.0 | 0.0 | 0.0 | 0.0 | 0.0 | 0.1 | 1.1 |
| Days ≥ 1" Snow Depth | 1 | 0 | 0 | 0 | 0 | 0 | 0 | 0 | 0 | 0 | 0 | 0 | 1 |

## LEESVILLE *Vernon Parish* ELEVATION 239 ft LAT/LONG 31° 8 ' N / 93° 16 ' W

| | JAN | FEB | MAR | APR | MAY | JUN | JUL | AUG | SEP | OCT | NOV | DEC | YEAR |
|---|---|---|---|---|---|---|---|---|---|---|---|---|---|
| Maximum Temp °F | 58.1 | 62.7 | 70.8 | 78.2 | 84.0 | 90.0 | 92.7 | 92.4 | 88.1 | 79.6 | 69.5 | 61.6 | 77.3 |
| Minimum Temp °F | 36.5 | 38.8 | 46.4 | 54.2 | 61.2 | 67.7 | 70.5 | 69.4 | 64.5 | 52.9 | 45.3 | 39.0 | 53.9 |
| Mean Temp °F | 47.3 | 50.8 | 58.6 | 66.2 | 72.6 | 78.9 | 81.6 | 80.9 | 76.3 | 66.3 | 57.4 | 50.3 | 65.6 |
| Days Max Temp ≥ 90 °F | 0 | 0 | 0 | 0 | 4 | 19 | 25 | 25 | 14 | 2 | 0 | 0 | 89 |
| Days Max Temp ≤ 32 °F | 0 | 0 | 0 | 0 | 0 | 0 | 0 | 0 | 0 | 0 | 0 | 0 | 0 |
| Days Min Temp ≤ 32 °F | 13 | 9 | 3 | 0 | 0 | 0 | 0 | 0 | 0 | 0 | 4 | 11 | 40 |
| Days Min Temp ≤ 0 °F | 0 | 0 | 0 | 0 | 0 | 0 | 0 | 0 | 0 | 0 | 0 | 0 | 0 |
| Heating Degree Days | 547 | 404 | 230 | 73 | 8 | 0 | 0 | 0 | 5 | 79 | 257 | 459 | 2062 |
| Cooling Degree Days | 3 | 11 | 41 | 107 | 256 | 436 | 527 | 518 | 359 | 130 | 40 | 13 | 2441 |
| Total Precipitation (") | 5.27 | 5.13 | 4.65 | 4.06 | 5.62 | 4.69 | 4.55 | 3.55 | 3.71 | 4.03 | 4.72 | 6.48 | 56.46 |
| Days ≥ 0.1" Precip | 7 | 6 | 7 | 5 | 7 | 6 | 7 | 6 | 5 | 5 | 6 | 8 | 75 |
| Total Snowfall (") | 0.2 | 0.0 | 0.0 | 0.0 | 0.0 | 0.0 | 0.0 | 0.0 | 0.0 | 0.0 | 0.0 | 0.0 | 0.2 |
| Days ≥ 1" Snow Depth | 0 | 0 | 0 | 0 | 0 | 0 | 0 | 0 | 0 | 0 | 0 | 0 | 0 |

## LOGANSPORT 4 ENE *De Soto Parish* ELEVATION 210 ft LAT/LONG 31° 59 ' N / 93° 57 ' W

| | JAN | FEB | MAR | APR | MAY | JUN | JUL | AUG | SEP | OCT | NOV | DEC | YEAR |
|---|---|---|---|---|---|---|---|---|---|---|---|---|---|
| Maximum Temp °F | *57.4* | *62.9* | *70.9* | *78.1* | *84.1* | *90.3* | 93.7 | 93.9 | 88.4 | 79.3 | 67.9 | 60.4 | 77.3 |
| Minimum Temp °F | *34.6* | *37.7* | *44.6* | *52.1* | *60.2* | *67.1* | 69.9 | 69.0 | 63.6 | 52.0 | 42.9 | 36.8 | 52.5 |
| Mean Temp °F | *46.0* | *50.3* | *57.7* | *65.1* | *72.2* | *78.7* | 81.8 | 81.5 | 76.0 | 65.7 | 55.4 | 48.7 | 64.9 |
| Days Max Temp ≥ 90 °F | *0* | *0* | *0* | *0* | *5* | *19* | 27 | 27 | 15 | 2 | 0 | 0 | 95 |
| Days Max Temp ≤ 32 °F | *0* | *0* | *0* | *0* | *0* | *0* | 0 | 0 | 0 | 0 | 0 | 1 | 1 |
| Days Min Temp ≤ 32 °F | *15* | *10* | *4* | *1* | *0* | *0* | 0 | 0 | 0 | 1 | 7 | 13 | 51 |
| Days Min Temp ≤ 0 °F | *0* | *0* | *0* | *0* | *0* | *0* | 0 | 0 | 0 | 0 | 0 | 0 | 0 |
| Heating Degree Days | *589* | *414* | *248* | *84* | *9* | *0* | 0 | 0 | 7 | 90 | 307 | 506 | 2254 |
| Cooling Degree Days | 2 | 6 | 28 | 90 | 235 | 424 | 535 | 529 | 351 | 112 | 27 | 7 | 2346 |
| Total Precipitation (") | *4.50* | *4.35* | *4.15* | *4.22* | *5.64* | *4.36* | 3.76 | 2.68 | 3.65 | 4.52 | 4.47 | 4.72 | 51.02 |
| Days ≥ 0.1" Precip | *7* | *6* | *6* | *5* | *7* | *6* | *5* | 4 | 4 | 5 | 6 | 6 | 67 |
| Total Snowfall (") | *0.4* | *0.2* | *0.0* | *0.0* | *0.0* | *0.0* | 0.0 | 0.0 | 0.0 | 0.0 | 0.0 | 0.0 | 0.6 |
| Days ≥ 1" Snow Depth | *0* | *0* | *0* | *0* | *0* | *0* | 0 | 0 | 0 | 0 | 0 | 0 | 0 |

## MANY *Sabine Parish* ELEVATION 259 ft LAT/LONG 31° 35 ' N / 93° 28 ' W

| | JAN | FEB | MAR | APR | MAY | JUN | JUL | AUG | SEP | OCT | NOV | DEC | YEAR |
|---|---|---|---|---|---|---|---|---|---|---|---|---|---|
| Maximum Temp °F | 56.6 | 61.5 | 69.7 | 77.7 | 83.3 | 89.6 | 92.7 | 92.2 | 87.3 | 78.5 | 68.3 | 60.7 | 76.5 |
| Minimum Temp °F | 33.4 | 36.2 | 43.9 | 52.1 | 59.8 | 66.9 | 69.9 | 68.5 | 63.0 | 50.3 | 42.5 | 35.8 | 51.9 |
| Mean Temp °F | 45.0 | 48.9 | 56.8 | 65.0 | 71.6 | 78.2 | 81.3 | 80.4 | 75.2 | 64.4 | 55.4 | 48.3 | 64.2 |
| Days Max Temp ≥ 90 °F | 0 | 0 | 0 | 0 | 4 | 18 | 25 | 24 | 13 | 2 | 0 | 0 | 86 |
| Days Max Temp ≤ 32 °F | 1 | 0 | 0 | 0 | 0 | 0 | 0 | 0 | 0 | 0 | 0 | 0 | 1 |
| Days Min Temp ≤ 32 °F | 16 | 12 | 5 | 1 | 0 | 0 | 0 | 0 | 0 | 1 | 7 | 14 | 56 |
| Days Min Temp ≤ 0 °F | 0 | 0 | 0 | 0 | 0 | 0 | 0 | 0 | 0 | 0 | 0 | 0 | 0 |
| Heating Degree Days | 618 | 455 | 276 | 91 | 15 | 0 | 0 | 0 | 10 | 108 | 308 | 518 | 2399 |
| Cooling Degree Days | 2 | 7 | 31 | 91 | 229 | 416 | 517 | 504 | 332 | 106 | 26 | 9 | 2270 |
| Total Precipitation (") | 5.18 | 4.60 | 5.09 | 4.11 | 6.10 | 4.49 | 4.19 | 3.72 | 3.19 | 4.49 | 4.10 | 5.86 | 55.12 |
| Days ≥ 0.1" Precip | 7 | 6 | 6 | 5 | 7 | 6 | 6 | 6 | 5 | 5 | 5 | 7 | 71 |
| Total Snowfall (") | 0.3 | 0.0 | 0.0 | 0.0 | 0.0 | 0.0 | 0.0 | 0.0 | 0.0 | 0.0 | 0.0 | 0.0 | 0.3 |
| Days ≥ 1" Snow Depth | 0 | 0 | 0 | 0 | 0 | 0 | 0 | 0 | 0 | 0 | 0 | 0 | 0 |

## MINDEN *Webster Parish* ELEVATION 239 ft LAT/LONG 32° 38 ' N / 93° 17 ' W

| | JAN | FEB | MAR | APR | MAY | JUN | JUL | AUG | SEP | OCT | NOV | DEC | YEAR |
|---|---|---|---|---|---|---|---|---|---|---|---|---|---|
| Maximum Temp °F | 54.5 | 59.6 | 67.9 | 76.2 | 82.4 | 89.3 | 92.5 | 92.5 | 86.9 | 77.6 | 67.1 | 58.1 | 75.4 |
| Minimum Temp °F | 32.7 | 35.5 | 43.2 | 52.1 | 60.1 | 67.9 | 71.3 | 70.1 | 64.1 | 51.7 | 42.7 | 35.5 | 52.2 |
| Mean Temp °F | 43.6 | 47.6 | 55.6 | 64.2 | 71.3 | 78.6 | 81.9 | 81.3 | 75.5 | 64.7 | 54.9 | 46.8 | 63.8 |
| Days Max Temp ≥ 90 °F | 0 | 0 | 0 | 0 | 3 | 16 | 24 | 24 | 13 | 2 | 0 | 0 | 82 |
| Days Max Temp ≤ 32 °F | 1 | 0 | 0 | 0 | 0 | 0 | 0 | 0 | 0 | 0 | 0 | 1 | 2 |
| Days Min Temp ≤ 32 °F | 17 | 12 | 5 | 1 | 0 | 0 | 0 | 0 | 0 | 0 | 5 | 14 | 54 |
| Days Min Temp ≤ 0 °F | 0 | 0 | 0 | 0 | 0 | 0 | 0 | 0 | 0 | 0 | 0 | 0 | 0 |
| Heating Degree Days | 659 | 489 | 309 | 100 | 14 | 0 | 0 | 0 | 10 | 105 | 317 | 562 | 2565 |
| Cooling Degree Days | 1 | 3 | 21 | 72 | 216 | 417 | 536 | 524 | 337 | 99 | 21 | 6 | 2253 |
| Total Precipitation (") | 4.56 | 4.94 | 4.38 | 4.48 | 5.51 | 4.41 | 4.10 | 2.75 | 3.44 | 4.04 | 4.98 | 4.75 | 52.34 |
| Days ≥ 0.1" Precip | 7 | 6 | 7 | 6 | 7 | 6 | 6 | 5 | 5 | 5 | 6 | 7 | 73 |
| Total Snowfall (") | 0.6 | 0.1 | 0.2 | 0.0 | 0.0 | 0.0 | 0.0 | 0.0 | 0.0 | 0.0 | 0.0 | 0.1 | 1.0 |
| Days ≥ 1" Snow Depth | 0 | 0 | 0 | 0 | 0 | 0 | 0 | 0 | 0 | 0 | 0 | 0 | 0 |

## MONROE MUNI ARPT *Ouachita Parish* ELEVATION 82 ft LAT/LONG 32° 31 ' N / 92° 3 ' W

| | JAN | FEB | MAR | APR | MAY | JUN | JUL | AUG | SEP | OCT | NOV | DEC | YEAR |
|---|---|---|---|---|---|---|---|---|---|---|---|---|---|
| Maximum Temp °F | 53.9 | 59.3 | 67.6 | 76.6 | 83.3 | 90.6 | 92.6 | 91.5 | 86.7 | 77.9 | 66.9 | 57.9 | 75.4 |
| Minimum Temp °F | 35.4 | 38.9 | 46.3 | 54.7 | 62.9 | 69.9 | 72.3 | 70.5 | 64.8 | 52.4 | 44.4 | 38.1 | 54.2 |
| Mean Temp °F | 44.7 | 49.1 | 57.0 | 65.7 | 73.1 | 80.3 | 82.5 | 81.1 | 75.8 | 65.2 | 55.7 | 48.1 | 64.9 |
| Days Max Temp ≥ 90 °F | 0 | 0 | 0 | 1 | 6 | 19 | 24 | 22 | 12 | 2 | 0 | 0 | 86 |
| Days Max Temp ≤ 32 °F | 1 | 0 | 0 | 0 | 0 | 0 | 0 | 0 | 0 | 0 | 0 | 0 | 1 |
| Days Min Temp ≤ 32 °F | 14 | 8 | 2 | 0 | 0 | 0 | 0 | 0 | 0 | 0 | 4 | 10 | 38 |
| Days Min Temp ≤ 0 °F | 0 | 0 | 0 | 0 | 0 | 0 | 0 | 0 | 0 | 0 | 0 | 0 | 0 |
| Heating Degree Days | 626 | 447 | 272 | 80 | 7 | 0 | 0 | 0 | 7 | 96 | 298 | 525 | 2358 |
| Cooling Degree Days | 3 | 8 | 29 | 107 | 274 | 478 | 565 | 535 | 357 | 113 | 28 | 8 | 2505 |
| Total Precipitation (") | 4.81 | 4.64 | 5.14 | 4.64 | 5.80 | 3.91 | 3.54 | 2.75 | 3.59 | 4.03 | 4.36 | 5.19 | 52.40 |
| Days ≥ 0.1" Precip | 7 | 6 | 7 | 6 | 7 | 6 | 6 | 4 | 5 | 5 | 6 | 7 | 72 |
| Total Snowfall (") | 0.7 | 0.1 | 0.1 | 0.0 | 0.0 | 0.0 | 0.0 | 0.0 | 0.0 | 0.0 | 0.0 | 0.0 | 0.9 |
| Days ≥ 1" Snow Depth | 0 | 0 | 0 | 0 | 0 | 0 | 0 | 0 | 0 | 0 | 0 | 0 | 0 |

## MORGAN CITY *St. Mary Parish* ELEVATION 14 ft LAT/LONG 29° 42 ' N / 91° 12 ' W

| | JAN | FEB | MAR | APR | MAY | JUN | JUL | AUG | SEP | OCT | NOV | DEC | YEAR |
|---|---|---|---|---|---|---|---|---|---|---|---|---|---|
| Maximum Temp °F | 61.4 | 64.3 | 70.8 | 78.0 | 83.3 | 88.5 | 90.3 | 90.1 | 87.1 | 80.2 | 71.8 | 65.3 | 77.6 |
| Minimum Temp °F | 42.3 | 44.3 | 51.9 | 59.3 | 66.0 | 71.8 | 73.6 | 73.1 | 69.9 | 59.9 | 51.5 | 45.5 | 59.1 |
| Mean Temp °F | 51.7 | 54.3 | 61.4 | 68.7 | 74.7 | 80.2 | 82.0 | 81.7 | 78.5 | 70.1 | 61.7 | 55.4 | 68.4 |
| Days Max Temp ≥ 90 °F | 0 | 0 | 0 | 0 | 1 | 13 | 20 | 20 | 9 | 1 | 0 | 0 | 64 |
| Days Max Temp ≤ 32 °F | 0 | 0 | 0 | 0 | 0 | 0 | 0 | 0 | 0 | 0 | 0 | 0 | 0 |
| Days Min Temp ≤ 32 °F | 5 | 3 | 0 | 0 | 0 | 0 | 0 | 0 | 0 | 0 | 0 | 3 | 11 |
| Days Min Temp ≤ 0 °F | 0 | 0 | 0 | 0 | 0 | 0 | 0 | 0 | 0 | 0 | 0 | 0 | 0 |
| Heating Degree Days | 416 | 307 | 156 | 32 | 1 | 0 | 0 | 0 | 1 | 28 | 158 | 312 | 1411 |
| Cooling Degree Days | 6 | 12 | 50 | 125 | 302 | 457 | 531 | 530 | 414 | 190 | 68 | 22 | 2707 |
| Total Precipitation (") | 5.56 | 4.99 | 4.47 | 4.49 | 5.74 | 5.25 | 7.95 | 7.99 | 6.55 | 3.57 | 4.30 | 5.24 | 66.10 |
| Days ≥ 0.1" Precip | 7 | 6 | 6 | 4 | 6 | 7 | 11 | 11 | 8 | 4 | 5 | 6 | 81 |
| Total Snowfall (") | 0.0 | 0.0 | 0.0 | 0.0 | 0.0 | 0.0 | 0.0 | 0.0 | 0.0 | 0.0 | 0.0 | 0.0 | 0.0 |
| Days ≥ 1" Snow Depth | 0 | 0 | 0 | 0 | 0 | 0 | 0 | 0 | 0 | 0 | 0 | 0 | 0 |

## NATCHITOCHES *Natchitoches Parish* ELEVATION 121 ft LAT/LONG 31° 46' N / 93° 5' W

| | JAN | FEB | MAR | APR | MAY | JUN | JUL | AUG | SEP | OCT | NOV | DEC | YEAR |
|---|---|---|---|---|---|---|---|---|---|---|---|---|---|
| Maximum Temp °F | 57.1 | 62.2 | 70.4 | 78.5 | 84.3 | 90.8 | 93.6 | 93.2 | 88.2 | 79.3 | 68.9 | 60.6 | 77.3 |
| Minimum Temp °F | 36.1 | 38.9 | 46.4 | 54.6 | 62.1 | 69.3 | 72.3 | 71.2 | 65.8 | 54.0 | 45.7 | 38.9 | 54.6 |
| Mean Temp °F | 46.7 | 50.5 | 58.5 | 66.6 | 73.3 | 80.1 | 83.0 | 82.2 | 77.0 | 66.7 | 57.3 | 49.8 | 66.0 |
| Days Max Temp ≥ 90 °F | 0 | 0 | 0 | 1 | 6 | 20 | 26 | 26 | 15 | 3 | 0 | 0 | 97 |
| Days Max Temp ≤ 32 °F | 0 | 0 | 0 | 0 | 0 | 0 | 0 | 0 | 0 | 0 | 0 | 0 | 0 |
| Days Min Temp ≤ 32 °F | 12 | 8 | 2 | 0 | 0 | 0 | 0 | 0 | 0 | 0 | 3 | 9 | 34 |
| Days Min Temp ≤ 0 °F | 0 | 0 | 0 | 0 | 0 | 0 | 0 | 0 | 0 | 0 | 0 | 0 | 0 |
| Heating Degree Days | 569 | 409 | 232 | 66 | 6 | 0 | 0 | 0 | 4 | 73 | 257 | 474 | 2090 |
| Cooling Degree Days | 1 | 7 | 31 | 98 | 254 | 452 | 561 | 547 | 366 | 122 | 31 | 7 | 2477 |
| Total Precipitation (") | 5.05 | 4.74 | 4.99 | 4.48 | 6.48 | 4.38 | 3.55 | 3.40 | 3.17 | 4.42 | 4.11 | 5.93 | 54.70 |
| Days ≥ 0.1" Precip | 7 | 6 | 6 | 5 | 7 | 6 | 6 | 6 | 5 | 5 | 5 | 7 | 71 |
| Total Snowfall (") | 0.6 | 0.1 | 0.0 | 0.0 | 0.0 | 0.0 | 0.0 | 0.0 | 0.0 | 0.0 | 0.0 | 0.0 | 0.7 |
| Days ≥ 1" Snow Depth | 0 | 0 | 0 | 0 | 0 | 0 | 0 | 0 | 0 | 0 | 0 | 0 | 0 |

## NEW IBERIA *Iberia Parish* ELEVATION 20 ft LAT/LONG 29° 59' N / 91° 47' W

| | JAN | FEB | MAR | APR | MAY | JUN | JUL | AUG | SEP | OCT | NOV | DEC | YEAR |
|---|---|---|---|---|---|---|---|---|---|---|---|---|---|
| Maximum Temp °F | 60.7 | 63.9 | 71.0 | 78.4 | 84.0 | 89.1 | 90.8 | 90.4 | 87.1 | 79.8 | 71.4 | 64.7 | 77.6 |
| Minimum Temp °F | 41.2 | 43.7 | 50.7 | 58.7 | 65.5 | 71.1 | 73.0 | 72.5 | 68.7 | 58.1 | 49.9 | 44.4 | 58.1 |
| Mean Temp °F | 51.0 | 53.8 | 60.9 | 68.6 | 74.8 | 80.1 | 81.9 | 81.5 | 77.9 | 69.0 | 60.7 | 54.6 | 67.9 |
| Days Max Temp ≥ 90 °F | 0 | 0 | 0 | 0 | 3 | 16 | 22 | 21 | 11 | 1 | 0 | 0 | 74 |
| Days Max Temp ≤ 32 °F | 0 | 0 | 0 | 0 | 0 | 0 | 0 | 0 | 0 | 0 | 0 | 0 | 0 |
| Days Min Temp ≤ 32 °F | 7 | 4 | 1 | 0 | 0 | 0 | 0 | 0 | 0 | 0 | 1 | 4 | 17 |
| Days Min Temp ≤ 0 °F | 0 | 0 | 0 | 0 | 0 | 0 | 0 | 0 | 0 | 0 | 0 | 0 | 0 |
| Heating Degree Days | 439 | 321 | 169 | 37 | 2 | 0 | 0 | 0 | 1 | 41 | 182 | 338 | 1530 |
| Cooling Degree Days | 6 | 12 | 51 | 140 | 312 | 465 | 540 | 534 | 400 | 175 | 62 | 23 | 2720 |
| Total Precipitation (") | 4.93 | 4.62 | 4.22 | 4.74 | 5.04 | 6.11 | 7.56 | 6.71 | 5.58 | 4.00 | 3.85 | 5.06 | 62.42 |
| Days ≥ 0.1" Precip | 7 | 6 | 6 | 5 | 6 | 8 | 10 | 9 | 7 | 4 | 5 | 6 | 79 |
| Total Snowfall (") | 0.0 | 0.0 | 0.0 | 0.0 | 0.0 | 0.0 | 0.0 | 0.0 | 0.0 | 0.0 | 0.0 | 0.0 | 0.0 |
| Days ≥ 1" Snow Depth | 0 | 0 | 0 | 0 | 0 | 0 | 0 | 0 | 0 | 0 | 0 | 0 | 0 |

## NEW ORLEANS AUDUBON *Orleans Parish* ELEVATION 10 ft LAT/LONG 29° 56' N / 90° 8' W

| | JAN | FEB | MAR | APR | MAY | JUN | JUL | AUG | SEP | OCT | NOV | DEC | YEAR |
|---|---|---|---|---|---|---|---|---|---|---|---|---|---|
| Maximum Temp °F | 61.6 | 64.8 | 71.8 | 78.7 | 84.3 | 89.7 | 90.8 | 90.5 | 87.2 | 79.9 | 71.5 | 65.2 | 78.0 |
| Minimum Temp °F | 44.9 | 47.3 | 54.1 | 60.8 | 67.7 | 73.2 | 75.0 | 74.8 | 71.7 | 62.1 | 53.7 | 48.1 | 61.1 |
| Mean Temp °F | 53.3 | 56.1 | 62.9 | 69.8 | 76.0 | 81.5 | 83.0 | 82.7 | 79.5 | 71.0 | 62.6 | 56.7 | 69.6 |
| Days Max Temp ≥ 90 °F | 0 | 0 | 0 | 0 | 4 | 17 | 21 | 20 | 11 | 1 | 0 | 0 | 74 |
| Days Max Temp ≤ 32 °F | 0 | 0 | 0 | 0 | 0 | 0 | 0 | 0 | 0 | 0 | 0 | 0 | 0 |
| Days Min Temp ≤ 32 °F | 3 | 1 | 0 | 0 | 0 | 0 | 0 | 0 | 0 | 0 | 0 | 1 | 5 |
| Days Min Temp ≤ 0 °F | 0 | 0 | 0 | 0 | 0 | 0 | 0 | 0 | 0 | 0 | 0 | 0 | 0 |
| Heating Degree Days | 373 | 266 | 129 | 26 | 1 | 0 | 0 | 0 | 0 | 19 | 145 | 284 | 1243 |
| Cooling Degree Days | 14 | 23 | 76 | 168 | 372 | 514 | 580 | 578 | 454 | 229 | 91 | 38 | 3137 |
| Total Precipitation (") | 5.39 | 5.09 | 5.00 | 4.95 | 5.15 | 5.98 | 7.05 | 6.77 | 5.72 | 2.93 | 4.36 | 5.18 | 63.57 |
| Days ≥ 0.1" Precip | 7 | 6 | 6 | 5 | 6 | 8 | 11 | 10 | 7 | 4 | 5 | 6 | 81 |
| Total Snowfall (") | 0.0 | 0.0 | 0.0 | 0.0 | 0.0 | 0.0 | 0.0 | 0.0 | 0.0 | 0.0 | 0.0 | 0.0 | 0.0 |
| Days ≥ 1" Snow Depth | 0 | 0 | 0 | 0 | 0 | 0 | 0 | 0 | 0 | 0 | 0 | 0 | 0 |

## NEW ORLEANS INTL AP *Jefferson Parish* ELEVATION 7 ft LAT/LONG 29° 59' N / 90° 15' W

| | JAN | FEB | MAR | APR | MAY | JUN | JUL | AUG | SEP | OCT | NOV | DEC | YEAR |
|---|---|---|---|---|---|---|---|---|---|---|---|---|---|
| Maximum Temp °F | 61.1 | 64.3 | 71.4 | 78.4 | 84.3 | 89.4 | 90.8 | 90.3 | 86.7 | 79.5 | 71.0 | 64.7 | 77.7 |
| Minimum Temp °F | 42.6 | 44.8 | 51.6 | 58.4 | 65.3 | 71.1 | 73.3 | 72.8 | 69.6 | 59.2 | 50.8 | 45.3 | 58.7 |
| Mean Temp °F | 51.9 | 54.5 | 61.5 | 68.4 | 74.8 | 80.3 | 82.1 | 81.5 | 78.2 | 69.3 | 60.9 | 55.0 | 68.2 |
| Days Max Temp ≥ 90 °F | 0 | 0 | 0 | 0 | 3 | 16 | 21 | 20 | 9 | 1 | 0 | 0 | 70 |
| Days Max Temp ≤ 32 °F | 0 | 0 | 0 | 0 | 0 | 0 | 0 | 0 | 0 | 0 | 0 | 0 | 0 |
| Days Min Temp ≤ 32 °F | 5 | 3 | 1 | 0 | 0 | 0 | 0 | 0 | 0 | 0 | 1 | 4 | 14 |
| Days Min Temp ≤ 0 °F | 0 | 0 | 0 | 0 | 0 | 0 | 0 | 0 | 0 | 0 | 0 | 0 | 0 |
| Heating Degree Days | 414 | 306 | 161 | 38 | 1 | 0 | 0 | 0 | 1 | 32 | 178 | 330 | 1461 |
| Cooling Degree Days | 11 | 18 | 57 | 132 | 325 | 473 | 544 | 537 | 410 | 184 | 71 | 27 | 2789 |
| Total Precipitation (") | 5.48 | 5.87 | 5.13 | 4.99 | 5.01 | 6.11 | 6.32 | 6.46 | 5.33 | 3.09 | 4.40 | 5.65 | 63.84 |
| Days ≥ 0.1" Precip | 7 | 6 | 6 | 5 | 6 | 8 | 10 | 9 | 7 | 4 | 6 | 7 | 81 |
| Total Snowfall (") | 0.0 | 0.0 | 0.0 | 0.0 | 0.0 | 0.0 | 0.0 | 0.0 | 0.0 | 0.0 | 0.0 | 0.0 | 0.0 |
| Days ≥ 1" Snow Depth | 0 | 0 | 0 | 0 | 0 | 0 | 0 | 0 | 0 | 0 | 0 | 0 | 0 |

## NEW ROADS 5 ESE *Pointe Coupee Parish* ELEVATION 49 ft LAT/LONG 30° 41 ' N / 91° 22 ' W

| | JAN | FEB | MAR | APR | MAY | JUN | JUL | AUG | SEP | OCT | NOV | DEC | YEAR |
|---|---|---|---|---|---|---|---|---|---|---|---|---|---|
| Maximum Temp °F | 58.6 | na | 70.6 | 77.5 | 83.8 | 89.4 | 91.1 | 90.7 | 87.0 | 79.1 | 70.4 | 62.2 | na |
| Minimum Temp °F | 38.2 | na | 48.5 | 55.5 | 63.6 | 69.9 | 72.2 | 71.4 | 67.3 | 55.5 | 48.2 | 40.9 | na |
| Mean Temp °F | 48.4 | na | 59.6 | 66.5 | 73.8 | 79.7 | 81.6 | 81.1 | 77.1 | 67.4 | 59.3 | 51.6 | na |
| Days Max Temp ≥ 90 °F | 0 | 0 | 0 | 0 | 3 | 16 | 23 | 22 | 11 | 1 | 0 | 0 | 76 |
| Days Max Temp ≤ 32 °F | 0 | 0 | 0 | 0 | 0 | 0 | 0 | 0 | 0 | 0 | 0 | 0 | 0 |
| Days Min Temp ≤ 32 °F | 10 | 5 | 1 | 0 | 0 | 0 | 0 | 0 | 0 | 0 | 2 | 7 | 25 |
| Days Min Temp ≤ 0 °F | 0 | 0 | 0 | 0 | 0 | 0 | 0 | 0 | 0 | 0 | 0 | 0 | 0 |
| Heating Degree Days | 512 | na | 200 | 61 | 3 | 0 | 0 | 0 | 2 | 58 | 214 | 423 | na |
| Cooling Degree Days | 4 | 11 | 36 | 108 | 278 | 452 | 524 | 512 | 373 | 142 | 48 | 14 | 2502 |
| Total Precipitation (") | 6.57 | 6.02 | 5.01 | 5.34 | 5.68 | 4.76 | 5.11 | 5.65 | 5.21 | 3.81 | 5.25 | 5.48 | 63.89 |
| Days ≥ 0.1" Precip | 7 | 7 | 6 | 6 | 7 | 7 | 8 | 8 | 6 | 4 | 6 | 7 | 79 |
| Total Snowfall (") | 0.0 | 0.0 | 0.0 | 0.0 | 0.0 | 0.0 | 0.0 | 0.0 | 0.0 | 0.0 | 0.0 | 0.0 | 0.0 |
| Days ≥ 1" Snow Depth | 0 | 0 | 0 | 0 | 0 | 0 | 0 | 0 | 0 | 0 | 0 | 0 | 0 |

## OBERLIN FIRE TOWER *Allen Parish* ELEVATION 69 ft LAT/LONG 30° 36 ' N / 92° 47 ' W

| | JAN | FEB | MAR | APR | MAY | JUN | JUL | AUG | SEP | OCT | NOV | DEC | YEAR |
|---|---|---|---|---|---|---|---|---|---|---|---|---|---|
| Maximum Temp °F | 59.0 | 63.3 | 70.8 | 78.3 | 83.9 | 89.6 | 91.7 | 91.7 | 87.8 | 80.1 | 70.5 | 62.9 | 77.5 |
| Minimum Temp °F | 39.7 | 42.2 | 49.6 | 56.9 | 63.7 | 69.5 | 71.6 | 70.9 | 66.7 | 56.3 | 48.5 | 42.6 | 56.5 |
| Mean Temp °F | 49.4 | 52.8 | 60.2 | 67.6 | 73.8 | 79.6 | 81.7 | 81.3 | 77.2 | 68.3 | 59.5 | 52.8 | 67.0 |
| Days Max Temp ≥ 90 °F | 0 | 0 | 0 | 0 | 3 | 17 | 24 | 24 | 13 | 3 | 0 | 0 | 84 |
| Days Max Temp ≤ 32 °F | 0 | 0 | 0 | 0 | 0 | 0 | 0 | 0 | 0 | 0 | 0 | 0 | 0 |
| Days Min Temp ≤ 32 °F | 9 | 6 | 1 | 0 | 0 | 0 | 0 | 0 | 0 | 0 | 2 | 6 | 24 |
| Days Min Temp ≤ 0 °F | 0 | 0 | 0 | 0 | 0 | 0 | 0 | 0 | 0 | 0 | 0 | 0 | 0 |
| Heating Degree Days | 485 | 349 | 187 | 49 | 4 | 0 | 0 | 0 | 3 | 54 | 209 | 388 | 1728 |
| Cooling Degree Days | 4 | 12 | 43 | 121 | 279 | 445 | 520 | 520 | 373 | 159 | 50 | 16 | 2542 |
| Total Precipitation (") | 6.12 | 4.77 | 5.24 | 4.72 | 6.90 | 5.81 | 5.51 | 4.77 | 5.71 | 5.07 | 4.85 | 6.54 | 66.01 |
| Days ≥ 0.1" Precip | 8 | 7 | 6 | 5 | 7 | 7 | 9 | 7 | 7 | 5 | 6 | 8 | 82 |
| Total Snowfall (") | 0.0 | 0.0 | 0.0 | 0.0 | 0.0 | 0.0 | 0.0 | 0.0 | 0.0 | 0.0 | 0.0 | 0.0 | 0.0 |
| Days ≥ 1" Snow Depth | 0 | 0 | 0 | 0 | 0 | 0 | 0 | 0 | 0 | 0 | 0 | 0 | 0 |

## PARADIS 7 S *St. Charles Parish* ELEVATION 10 ft LAT/LONG 29° 47 ' N / 90° 26 ' W

| | JAN | FEB | MAR | APR | MAY | JUN | JUL | AUG | SEP | OCT | NOV | DEC | YEAR |
|---|---|---|---|---|---|---|---|---|---|---|---|---|---|
| Maximum Temp °F | 63.3 | 66.0 | 72.8 | 79.1 | 84.1 | 89.3 | 90.8 | 90.3 | 87.3 | 81.0 | 73.0 | 67.0 | 78.7 |
| Minimum Temp °F | 42.1 | 43.9 | 51.6 | 59.2 | 65.2 | 70.1 | 72.0 | 71.6 | 68.3 | 58.3 | 50.4 | 45.1 | 58.2 |
| Mean Temp °F | 52.7 | 55.0 | 62.2 | 69.2 | 74.7 | 79.7 | 81.4 | 81.0 | 77.8 | 69.6 | 61.7 | 56.0 | 68.4 |
| Days Max Temp ≥ 90 °F | 0 | 0 | 0 | 0 | 2 | 15 | 21 | 21 | 9 | 1 | 0 | 0 | 69 |
| Days Max Temp ≤ 32 °F | 0 | 0 | 0 | 0 | 0 | 0 | 0 | 0 | 0 | 0 | 0 | 0 | 0 |
| Days Min Temp ≤ 32 °F | 7 | 4 | 1 | 0 | 0 | 0 | 0 | 0 | 0 | 0 | 2 | 5 | 19 |
| Days Min Temp ≤ 0 °F | 0 | 0 | 0 | 0 | 0 | 0 | 0 | 0 | 0 | 0 | 0 | 0 | 0 |
| Heating Degree Days | 391 | 293 | 143 | 29 | 1 | 0 | 0 | 0 | 2 | 34 | 158 | 300 | 1351 |
| Cooling Degree Days | na | na | na | na | na | na | na | na | na | na | na | na | na |
| Total Precipitation (") | 5.62 | 5.57 | 4.92 | 4.88 | 6.11 | 4.92 | 7.06 | 6.58 | 5.93 | 3.75 | 4.39 | 5.58 | 65.31 |
| Days ≥ 0.1" Precip | 8 | 6 | 6 | 5 | 6 | 7 | 10 | 10 | 7 | 4 | 6 | 7 | 82 |
| Total Snowfall (") | 0.0 | 0.0 | 0.0 | 0.0 | 0.0 | 0.0 | 0.0 | 0.0 | 0.0 | 0.0 | 0.0 | 0.1 | 0.1 |
| Days ≥ 1" Snow Depth | 0 | 0 | 0 | 0 | 0 | 0 | 0 | 0 | 0 | 0 | 0 | 0 | 0 |

## PLAIN DEALING *Bossier Parish* ELEVATION 289 ft LAT/LONG 32° 54 ' N / 93° 41 ' W

| | JAN | FEB | MAR | APR | MAY | JUN | JUL | AUG | SEP | OCT | NOV | DEC | YEAR |
|---|---|---|---|---|---|---|---|---|---|---|---|---|---|
| Maximum Temp °F | 53.8 | 58.9 | 67.5 | 76.0 | 81.8 | 88.8 | 92.7 | 92.4 | 86.5 | 76.9 | 66.1 | 57.8 | 74.9 |
| Minimum Temp °F | 31.1 | 33.8 | 41.2 | 49.7 | 58.2 | 65.9 | 69.3 | 68.4 | 62.5 | 50.0 | 41.0 | 34.0 | 50.4 |
| Mean Temp °F | 42.5 | 46.4 | 54.4 | 62.9 | 70.0 | 77.3 | 81.0 | 80.4 | 74.5 | 63.5 | 53.6 | 45.9 | 62.7 |
| Days Max Temp ≥ 90 °F | 0 | 0 | 0 | 0 | 2 | 15 | 24 | 23 | 12 | 2 | 0 | 0 | 78 |
| Days Max Temp ≤ 32 °F | 1 | 0 | 0 | 0 | 0 | 0 | 0 | 0 | 0 | 0 | 0 | 1 | 2 |
| Days Min Temp ≤ 32 °F | 19 | 14 | 7 | 1 | 0 | 0 | 0 | 0 | 0 | 1 | 7 | 16 | 65 |
| Days Min Temp ≤ 0 °F | 0 | 0 | 0 | 0 | 0 | 0 | 0 | 0 | 0 | 0 | 0 | 0 | 0 |
| Heating Degree Days | 693 | 521 | 341 | 122 | 22 | 1 | 0 | 0 | 12 | 122 | 352 | 589 | 2775 |
| Cooling Degree Days | 1 | 3 | 18 | 65 | 192 | 393 | 511 | 509 | 327 | 93 | 20 | 4 | 2136 |
| Total Precipitation (") | 4.30 | 4.29 | 4.56 | 4.30 | 4.74 | 4.50 | 3.58 | 3.64 | 3.22 | 4.43 | 4.94 | 4.74 | 51.24 |
| Days ≥ 0.1" Precip | 7 | 6 | 6 | 6 | 7 | 6 | 6 | 4 | 5 | 5 | 6 | 7 | 71 |
| Total Snowfall (") | 0.7 | 0.4 | 0.2 | 0.0 | 0.0 | 0.0 | 0.0 | 0.0 | 0.0 | 0.0 | 0.0 | 0.3 | 1.6 |
| Days ≥ 1" Snow Depth | 0 | 0 | 0 | 0 | 0 | 0 | 0 | 0 | 0 | 0 | 0 | 0 | 0 |

## RESERVE 4 WNW *St. John the Baptist Parish* ELEVATION 15 ft LAT/LONG 30° 5 ' N / 90° 37 ' W

| | JAN | FEB | MAR | APR | MAY | JUN | JUL | AUG | SEP | OCT | NOV | DEC | YEAR |
|---|---|---|---|---|---|---|---|---|---|---|---|---|---|
| Maximum Temp °F | 60.8 | 64.1 | 71.0 | 78.3 | 83.9 | 89.3 | 90.9 | 90.4 | 87.0 | 79.5 | 70.9 | 64.5 | 77.6 |
| Minimum Temp °F | 40.6 | 43.0 | 50.2 | 57.3 | 64.1 | 70.3 | 72.3 | 71.9 | 68.3 | 57.5 | 49.6 | 43.4 | 57.4 |
| Mean Temp °F | 50.7 | 53.6 | 60.6 | 67.8 | 74.0 | 79.8 | 81.6 | 81.2 | 77.7 | 68.5 | 60.3 | 54.0 | 67.5 |
| Days Max Temp ≥ 90 °F | 0 | 0 | 0 | 0 | 2 | 16 | 23 | 20 | 9 | 1 | 0 | 0 | 71 |
| Days Max Temp ≤ 32 °F | 0 | 0 | 0 | 0 | 0 | 0 | 0 | 0 | 0 | 0 | 0 | 0 | 0 |
| Days Min Temp ≤ 32 °F | 7 | 4 | 1 | 0 | 0 | 0 | 0 | 0 | 0 | 0 | 1 | 5 | 18 |
| Days Min Temp ≤ 0 °F | 0 | 0 | 0 | 0 | 0 | 0 | 0 | 0 | 0 | 0 | 0 | 0 | 0 |
| Heating Degree Days | 443 | 334 | 179 | 44 | 2 | 0 | 0 | 0 | 1 | 44 | 186 | 354 | 1587 |
| Cooling Degree Days | 6 | 15 | 50 | 122 | 287 | 454 | 531 | 522 | 392 | 166 | 59 | 19 | 2623 |
| Total Precipitation (") | 5.97 | 6.00 | 5.49 | 4.90 | 5.24 | 5.96 | 6.26 | 5.80 | 5.63 | 3.62 | 4.04 | 5.03 | 63.94 |
| Days ≥ 0.1" Precip | 7 | 7 | 6 | 5 | 6 | 8 | 10 | 9 | 7 | 4 | 5 | 6 | 80 |
| Total Snowfall (") | 0.0 | 0.1 | 0.0 | 0.0 | 0.0 | 0.0 | 0.0 | 0.0 | 0.0 | 0.0 | 0.0 | 0.0 | 0.1 |
| Days ≥ 1" Snow Depth | 0 | 0 | 0 | 0 | 0 | 0 | 0 | 0 | 0 | 0 | 0 | 0 | 0 |

## ROCKEFELLR WL REFUGE *Cameron Parish* ELEVATION 0 ft LAT/LONG 29° 44 ' N / 92° 49 ' W

| | JAN | FEB | MAR | APR | MAY | JUN | JUL | AUG | SEP | OCT | NOV | DEC | YEAR |
|---|---|---|---|---|---|---|---|---|---|---|---|---|---|
| Maximum Temp °F | 59.4 | 62.7 | 69.6 | 76.6 | 82.8 | 88.3 | 90.1 | 90.3 | 87.1 | 80.1 | 70.9 | 63.8 | 76.8 |
| Minimum Temp °F | 42.2 | 44.9 | 51.9 | 60.1 | 66.8 | 72.6 | 74.4 | 73.6 | 69.8 | 59.6 | 51.9 | 45.6 | 59.5 |
| Mean Temp °F | 50.8 | 53.8 | 60.8 | 68.4 | 74.8 | 80.5 | 82.3 | 82.0 | 78.5 | 69.9 | 61.4 | 54.7 | 68.2 |
| Days Max Temp ≥ 90 °F | 0 | 0 | 0 | 0 | 1 | 12 | 20 | 22 | 10 | 1 | 0 | 0 | 66 |
| Days Max Temp ≤ 32 °F | 0 | 0 | 0 | 0 | 0 | 0 | 0 | 0 | 0 | 0 | 0 | 0 | 0 |
| Days Min Temp ≤ 32 °F | 5 | 3 | 0 | 0 | 0 | 0 | 0 | 0 | 0 | 0 | 0 | 3 | 11 |
| Days Min Temp ≤ 0 °F | 0 | 0 | 0 | 0 | 0 | 0 | 0 | 0 | 0 | 0 | 0 | 0 | 0 |
| Heating Degree Days | 440 | 318 | 165 | 35 | 1 | 0 | 0 | 0 | 1 | 33 | 166 | 332 | 1491 |
| Cooling Degree Days | 4 | 9 | 42 | 126 | 315 | 482 | 555 | 548 | 421 | 194 | 72 | 21 | 2789 |
| Total Precipitation (") | 5.55 | 3.73 | 3.50 | 3.80 | 5.11 | 5.03 | 7.34 | 6.64 | 6.05 | 4.82 | 4.31 | 5.09 | 60.97 |
| Days ≥ 0.1" Precip | 7 | 5 | 5 | 4 | 5 | 6 | 9 | 9 | 6 | 4 | 5 | 6 | 71 |
| Total Snowfall (") | 0.1 | 0.0 | 0.0 | 0.0 | 0.0 | 0.0 | 0.0 | 0.0 | 0.0 | 0.0 | 0.0 | 0.0 | 0.1 |
| Days ≥ 1" Snow Depth | 0 | 0 | 0 | 0 | 0 | 0 | 0 | 0 | 0 | 0 | 0 | 0 | 0 |

## RUSTON LA TECH UNIV *Lincoln Parish* ELEVATION 279 ft LAT/LONG 32° 31 ' N / 92° 41 ' W

| | JAN | FEB | MAR | APR | MAY | JUN | JUL | AUG | SEP | OCT | NOV | DEC | YEAR |
|---|---|---|---|---|---|---|---|---|---|---|---|---|---|
| Maximum Temp °F | 54.4 | 59.4 | 67.7 | 76.4 | 82.4 | 89.5 | 92.6 | 92.3 | 86.7 | 77.2 | 66.8 | 58.3 | 75.3 |
| Minimum Temp °F | 33.1 | 36.0 | 43.7 | 52.3 | 60.0 | 67.4 | 70.3 | 69.1 | 63.7 | 51.6 | 43.0 | 36.0 | 52.2 |
| Mean Temp °F | 43.8 | 47.7 | 55.8 | 64.4 | 71.2 | 78.4 | 81.5 | 80.7 | 75.2 | 64.4 | 54.9 | 47.2 | 63.8 |
| Days Max Temp ≥ 90 °F | 0 | 0 | 0 | 0 | 3 | 17 | 25 | 23 | 12 | 1 | 0 | 0 | 81 |
| Days Max Temp ≤ 32 °F | 1 | 0 | 0 | 0 | 0 | 0 | 0 | 0 | 0 | 0 | 0 | 1 | 2 |
| Days Min Temp ≤ 32 °F | 16 | 11 | 4 | 1 | 0 | 0 | 0 | 0 | 0 | 0 | 5 | 12 | 49 |
| Days Min Temp ≤ 0 °F | 0 | 0 | 0 | 0 | 0 | 0 | 0 | 0 | 0 | 0 | 0 | 0 | 0 |
| Heating Degree Days | 645 | 484 | 302 | 98 | 14 | 0 | 0 | 0 | 9 | 108 | 316 | 549 | 2525 |
| Cooling Degree Days | 1 | 4 | 22 | 80 | 215 | 417 | 525 | 514 | 334 | 98 | 20 | 5 | 2235 |
| Total Precipitation (") | 4.99 | 5.18 | 4.96 | 4.59 | 6.08 | 4.02 | 4.01 | 3.06 | 3.36 | 4.17 | 4.58 | 5.30 | 54.30 |
| Days ≥ 0.1" Precip | 7 | 6 | 7 | 6 | 7 | 6 | 6 | 5 | 5 | 5 | 6 | 7 | 73 |
| Total Snowfall (") | 0.6 | 0.1 | 0.2 | 0.0 | 0.0 | 0.0 | 0.0 | 0.0 | 0.0 | 0.0 | 0.0 | 0.0 | 0.9 |
| Days ≥ 1" Snow Depth | 0 | 0 | 0 | 0 | 0 | 0 | 0 | 0 | 0 | 0 | 0 | 0 | 0 |

## SHREVEPORT REGIONAL *Caddo Parish* ELEVATION 194 ft LAT/LONG 32° 33 ' N / 93° 46 ' W

| | JAN | FEB | MAR | APR | MAY | JUN | JUL | AUG | SEP | OCT | NOV | DEC | YEAR |
|---|---|---|---|---|---|---|---|---|---|---|---|---|---|
| Maximum Temp °F | 55.5 | 60.6 | 69.1 | 77.1 | 82.8 | 89.7 | 93.0 | 92.8 | 87.2 | 78.2 | 67.4 | 59.0 | 76.0 |
| Minimum Temp °F | 35.5 | 38.4 | 45.8 | 54.1 | 61.8 | 69.2 | 72.2 | 71.1 | 65.6 | 54.2 | 45.0 | 38.1 | 54.3 |
| Mean Temp °F | 45.5 | 49.5 | 57.5 | 65.6 | 72.3 | 79.5 | 82.6 | 82.0 | 76.4 | 66.2 | 56.2 | 48.5 | 65.2 |
| Days Max Temp ≥ 90 °F | 0 | 0 | 0 | 0 | 3 | 18 | 25 | 25 | 14 | 2 | 0 | 0 | 87 |
| Days Max Temp ≤ 32 °F | 1 | 0 | 0 | 0 | 0 | 0 | 0 | 0 | 0 | 0 | 0 | 0 | 1 |
| Days Min Temp ≤ 32 °F | 13 | 8 | 3 | 0 | 0 | 0 | 0 | 0 | 0 | 0 | 3 | 10 | 37 |
| Days Min Temp ≤ 0 °F | 0 | 0 | 0 | 0 | 0 | 0 | 0 | 0 | 0 | 0 | 0 | 0 | 0 |
| Heating Degree Days | 602 | 436 | 257 | 75 | 8 | 0 | 0 | 0 | 5 | 78 | 284 | 511 | 2256 |
| Cooling Degree Days | 2 | 7 | 30 | 96 | 246 | 451 | 563 | 556 | 367 | 123 | 30 | 9 | 2480 |
| Total Precipitation (") | 4.17 | 4.26 | 3.72 | 4.24 | 5.72 | 4.41 | 3.93 | 2.55 | 3.09 | 4.26 | 4.41 | 4.34 | 49.10 |
| Days ≥ 0.1" Precip | 7 | 5 | 6 | 5 | 7 | 5 | 5 | 4 | 4 | 5 | 6 | 6 | 65 |
| Total Snowfall (") | 0.8 | 0.5 | 0.2 | 0.0 | 0.0 | 0.0 | 0.0 | 0.0 | 0.0 | 0.0 | 0.0 | 0.2 | 1.7 |
| Days ≥ 1" Snow Depth | 1 | 0 | 0 | 0 | 0 | 0 | 0 | 0 | 0 | 0 | 0 | 0 | 1 |

## SLIDELL WSFO *St. Tammany Parish* ELEVATION 16 ft LAT/LONG 30° 17 ' N / 89° 46 ' W

| | JAN | FEB | MAR | APR | MAY | JUN | JUL | AUG | SEP | OCT | NOV | DEC | YEAR |
|---|---|---|---|---|---|---|---|---|---|---|---|---|---|
| Maximum Temp °F | 61.3 | 64.5 | 71.3 | 78.4 | 84.3 | 89.8 | 91.2 | 90.8 | 87.6 | 80.2 | 71.4 | 64.5 | 77.9 |
| Minimum Temp °F | 40.6 | 42.8 | 49.7 | 57.2 | 64.1 | 70.3 | 72.6 | 71.8 | 68.0 | 56.8 | 48.8 | 43.1 | 57.1 |
| Mean Temp °F | 51.0 | 53.7 | 60.5 | 67.8 | 74.2 | 80.1 | 81.9 | 81.3 | 77.8 | 68.6 | 60.1 | 53.8 | 67.6 |
| Days Max Temp ≥ 90 °F | 0 | 0 | 0 | 0 | 3 | 17 | 23 | 22 | 13 | 1 | 0 | 0 | 79 |
| Days Max Temp ≤ 32 °F | 0 | 0 | 0 | 0 | 0 | 0 | 0 | 0 | 0 | 0 | 0 | 0 | 0 |
| Days Min Temp ≤ 32 °F | 8 | 5 | 1 | 0 | 0 | 0 | 0 | 0 | 0 | 0 | 2 | 6 | 22 |
| Days Min Temp ≤ 0 °F | 0 | 0 | 0 | 0 | 0 | 0 | 0 | 0 | 0 | 0 | 0 | 0 | 0 |
| Heating Degree Days | 435 | 325 | 178 | 46 | 2 | 0 | 0 | 0 | 2 | 43 | 191 | 359 | 1581 |
| Cooling Degree Days | 5 | 11 | 47 | 123 | 303 | 467 | 542 | 528 | 396 | 166 | 55 | 20 | 2663 |
| Total Precipitation (") | 6.11 | 5.39 | 5.50 | 4.53 | 5.33 | 4.65 | 6.68 | 6.14 | 4.90 | 3.35 | 4.34 | 5.35 | 62.27 |
| Days ≥ 0.1" Precip | 8 | 7 | 6 | 5 | 6 | 7 | 8 | 9 | 7 | 4 | 6 | 7 | 80 |
| Total Snowfall (") | 0.0 | 0.1 | 0.1 | 0.0 | 0.0 | 0.0 | 0.0 | 0.0 | 0.0 | 0.0 | 0.0 | 0.0 | 0.2 |
| Days ≥ 1" Snow Depth | 0 | 0 | 0 | 0 | 0 | 0 | 0 | 0 | 0 | 0 | 0 | 0 | 0 |

## ST JOSEPH 3 N *Tensas Parish* ELEVATION 79 ft LAT/LONG 31° 55 ' N / 91° 14 ' W

| | JAN | FEB | MAR | APR | MAY | JUN | JUL | AUG | SEP | OCT | NOV | DEC | YEAR |
|---|---|---|---|---|---|---|---|---|---|---|---|---|---|
| Maximum Temp °F | 55.1 | 59.5 | 67.9 | 76.6 | 83.2 | 89.9 | 92.0 | 91.5 | 87.3 | 78.3 | 68.3 | 59.6 | 75.8 |
| Minimum Temp °F | 35.9 | 38.4 | 46.5 | 55.0 | 62.7 | 69.5 | 72.3 | 70.7 | 65.3 | 53.0 | 45.5 | 39.1 | 54.5 |
| Mean Temp °F | 45.5 | 49.0 | 57.2 | 65.8 | 72.9 | 79.7 | 82.1 | 81.1 | 76.3 | 65.7 | 56.9 | 49.4 | 65.1 |
| Days Max Temp ≥ 90 °F | 0 | 0 | 0 | 0 | 4 | 18 | 24 | 23 | 13 | 2 | 0 | 0 | 84 |
| Days Max Temp ≤ 32 °F | 1 | 0 | 0 | 0 | 0 | 0 | 0 | 0 | 0 | 0 | 0 | 0 | 1 |
| Days Min Temp ≤ 32 °F | 13 | 9 | 2 | 0 | 0 | 0 | 0 | 0 | 0 | 0 | 3 | 9 | 36 |
| Days Min Temp ≤ 0 °F | 0 | 0 | 0 | 0 | 0 | 0 | 0 | 0 | 0 | 0 | 0 | 0 | 0 |
| Heating Degree Days | 600 | 451 | 264 | 79 | 8 | 0 | 0 | 0 | 6 | 90 | 269 | 486 | 2253 |
| Cooling Degree Days | 2 | 7 | 32 | 109 | 279 | 465 | 554 | 528 | 364 | 134 | 39 | 10 | 2523 |
| Total Precipitation (") | 5.71 | 5.04 | 5.74 | 5.20 | 5.59 | 3.47 | 3.79 | 3.52 | 2.97 | 3.74 | 4.59 | 5.89 | 55.25 |
| Days ≥ 0.1" Precip | 8 | 6 | 7 | 5 | 7 | 5 | 7 | 5 | 5 | 5 | 6 | 7 | 73 |
| Total Snowfall (") | 0.4 | 0.1 | 0.0 | 0.0 | 0.0 | 0.0 | 0.0 | 0.0 | 0.0 | 0.0 | 0.0 | 0.0 | 0.5 |
| Days ≥ 1" Snow Depth | 0 | 3 | 0 | 0 | 0 | 0 | 0 | 0 | 0 | 0 | 0 | 0 | 3 |

## TALLULAH *Madison Parish* ELEVATION 89 ft LAT/LONG 32° 23 ' N / 91° 13 ' W

| | JAN | FEB | MAR | APR | MAY | JUN | JUL | AUG | SEP | OCT | NOV | DEC | YEAR |
|---|---|---|---|---|---|---|---|---|---|---|---|---|---|
| Maximum Temp °F | 53.6 | 58.5 | 67.1 | 75.9 | 82.4 | 89.4 | 91.6 | 91.0 | 86.3 | 77.3 | 66.9 | 57.9 | 74.8 |
| Minimum Temp °F | 34.0 | 37.1 | 45.1 | 53.4 | 61.3 | 68.5 | 71.2 | 69.8 | 64.1 | 51.2 | 43.7 | 36.9 | 53.0 |
| Mean Temp °F | 43.8 | 47.8 | 56.2 | 64.7 | 71.9 | 79.0 | 81.5 | 80.4 | 75.2 | 64.3 | 55.3 | 47.5 | 64.0 |
| Days Max Temp ≥ 90 °F | 0 | 0 | 0 | 0 | 4 | 17 | 23 | 22 | 11 | 1 | 0 | 0 | 78 |
| Days Max Temp ≤ 32 °F | 1 | 1 | 0 | 0 | 0 | 0 | 0 | 0 | 0 | 0 | 0 | 1 | 3 |
| Days Min Temp ≤ 32 °F | 15 | 10 | 3 | 0 | 0 | 0 | 0 | 0 | 0 | 0 | 5 | 12 | 45 |
| Days Min Temp ≤ 0 °F | 0 | 0 | 0 | 0 | 0 | 0 | 0 | 0 | 0 | 0 | 0 | 0 | 0 |
| Heating Degree Days | 652 | 483 | 295 | 99 | 14 | 0 | 0 | 0 | 9 | 109 | 307 | 542 | 2510 |
| Cooling Degree Days | 1 | 6 | 27 | 88 | 230 | 432 | 519 | 498 | 326 | 103 | 25 | 6 | 2261 |
| Total Precipitation (") | 5.49 | 5.34 | 6.08 | 5.13 | 6.04 | 4.64 | 3.97 | 3.30 | 3.21 | 3.83 | 4.74 | 6.35 | 58.12 |
| Days ≥ 0.1" Precip | 7 | 6 | 7 | 5 | 8 | 6 | 6 | 5 | 5 | 5 | 6 | 7 | 73 |
| Total Snowfall (") | 0.2 | 0.1 | 0.3 | 0.0 | 0.0 | 0.0 | 0.0 | 0.0 | 0.0 | 0.0 | 0.0 | 0.0 | 0.6 |
| Days ≥ 1" Snow Depth | 0 | 0 | 0 | 0 | 0 | 0 | 0 | 0 | 0 | 0 | 0 | 0 | 0 |

## WINNFIELD 2 W *Winn Parish* ELEVATION 141 ft LAT/LONG 31° 55 ' N / 92° 38 ' W

| | JAN | FEB | MAR | APR | MAY | JUN | JUL | AUG | SEP | OCT | NOV | DEC | YEAR |
|---|---|---|---|---|---|---|---|---|---|---|---|---|---|
| Maximum Temp °F | 56.9 | 62.4 | 70.6 | 78.3 | 83.6 | 89.9 | 92.5 | 92.4 | 87.5 | 78.8 | 68.4 | 60.3 | 76.8 |
| Minimum Temp °F | 35.2 | 37.8 | 44.8 | 53.0 | 60.0 | 66.9 | 69.6 | 68.2 | 63.1 | 51.3 | 43.2 | 37.5 | 52.6 |
| Mean Temp °F | 46.1 | 50.1 | 57.7 | 65.7 | 71.8 | 78.4 | 81.1 | 80.3 | 75.3 | 65.0 | 55.8 | 48.9 | 64.7 |
| Days Max Temp ≥ 90 °F | 0 | 0 | 0 | 0 | 4 | 18 | 26 | 24 | 13 | 1 | 0 | 0 | 86 |
| Days Max Temp ≤ 32 °F | 0 | 0 | 0 | 0 | 0 | 0 | 0 | 0 | 0 | 0 | 0 | 0 | 0 |
| Days Min Temp ≤ 32 °F | 14 | 10 | 4 | 1 | 0 | 0 | 0 | 0 | 0 | 1 | 6 | 12 | 48 |
| Days Min Temp ≤ 0 °F | 0 | 0 | 0 | 0 | 0 | 0 | 0 | 0 | 0 | 0 | 0 | 0 | 0 |
| Heating Degree Days | 584 | 420 | 250 | 75 | 9 | 0 | 0 | 0 | 7 | 94 | 293 | 498 | 2230 |
| Cooling Degree Days | 3 | 9 | 32 | 98 | 236 | 422 | 520 | 507 | 340 | 113 | 27 | 8 | 2315 |
| Total Precipitation (") | 5.27 | 4.83 | 5.53 | 5.31 | 6.41 | 5.09 | 4.34 | 3.49 | 3.61 | 4.44 | 4.68 | 6.28 | 59.28 |
| Days ≥ 0.1" Precip | 8 | 6 | 7 | 5 | 7 | 6 | 7 | 5 | 5 | 5 | 6 | 8 | 75 |
| Total Snowfall (") | 0.9 | 0.2 | 0.1 | 0.0 | 0.0 | 0.0 | 0.0 | 0.0 | 0.0 | 0.0 | 0.0 | 0.1 | 1.3 |
| Days ≥ 1" Snow Depth | 1 | 0 | 0 | 0 | 0 | 0 | 0 | 0 | 0 | 0 | 0 | 0 | 1 |

## WINNSBORO 5 SSE *Franklin Parish* ELEVATION 75 ft LAT/LONG 32° 6 ' N / 91° 43 ' W

| | JAN | FEB | MAR | APR | MAY | JUN | JUL | AUG | SEP | OCT | NOV | DEC | YEAR |
|---|---|---|---|---|---|---|---|---|---|---|---|---|---|
| Maximum Temp °F | 54.3 | 59.0 | 67.6 | 76.5 | 83.2 | 90.3 | 92.9 | 92.5 | 87.9 | 78.6 | 67.7 | 58.6 | 75.8 |
| Minimum Temp °F | 34.6 | 37.4 | 45.5 | 53.8 | 61.6 | 68.7 | 71.5 | 69.8 | 64.2 | 51.8 | 44.6 | 37.8 | 53.4 |
| Mean Temp °F | 44.5 | 48.3 | 56.6 | 65.2 | 72.4 | 79.5 | 82.2 | 81.2 | 76.1 | 65.2 | 56.1 | 48.2 | 64.6 |
| Days Max Temp ≥ 90 °F | 0 | 0 | 0 | 0 | 5 | 19 | 25 | 24 | 15 | 2 | 0 | 0 | 90 |
| Days Max Temp ≤ 32 °F | 1 | 0 | 0 | 0 | 0 | 0 | 0 | 0 | 0 | 0 | 0 | 1 | 2 |
| Days Min Temp ≤ 32 °F | 15 | 10 | 3 | 0 | 0 | 0 | 0 | 0 | 0 | 0 | 4 | 11 | 43 |
| Days Min Temp ≤ 0 °F | 0 | 0 | 0 | 0 | 0 | 0 | 0 | 0 | 0 | 0 | 0 | 0 | 0 |
| Heating Degree Days | 633 | 470 | 282 | 91 | 12 | 0 | 0 | 0 | 8 | 98 | 287 | 519 | 2400 |
| Cooling Degree Days | 1 | 6 | 28 | 99 | 257 | 455 | 548 | 526 | 356 | 120 | 31 | 7 | 2434 |
| Total Precipitation (") | 5.35 | 5.14 | 5.72 | 5.04 | 5.46 | 4.15 | 3.72 | 3.33 | 3.03 | 4.05 | 4.97 | 5.96 | 55.92 |
| Days ≥ 0.1" Precip | 7 | 6 | 7 | 5 | 7 | 6 | 6 | 5 | 5 | 5 | 6 | 7 | 72 |
| Total Snowfall (") | 0.1 | 0.0 | 0.0 | 0.0 | 0.0 | 0.0 | 0.0 | 0.0 | 0.0 | 0.0 | 0.0 | 0.0 | 0.1 |
| Days ≥ 1" Snow Depth | 0 | 0 | 0 | 0 | 0 | 0 | 0 | 0 | 0 | 0 | 0 | 0 | 0 |

## JANUARY MINIMUM TEMPERATURE °F

| LOWEST | | | HIGHEST | | |
|---|---|---|---|---|---|
| 1 | Plain Dealing | 31.1 | 1 | New Orleans-Aud | 44.9 |
| 2 | Ashland | 32.2 | 2 | Houma | 42.8 |
| 3 | Homer | 32.6 | 3 | New Orleans-Intl | 42.6 |
| 4 | Minden | 32.7 | 4 | Hackberry | 42.5 |
| 5 | Ruston | 33.1 | 5 | Franklin | 42.4 |
| 6 | Calhoun | 33.2 | 6 | Morgan City | 42.3 |
| 7 | Many | 33.4 | 7 | Rockefeller | 42.2 |
| 8 | Tallulah | 34.0 | 8 | Paradis | 42.1 |
| 9 | Lake Providence | 34.3 | 9 | Lafayette | 41.8 |
| 10 | Logansport | 34.6 | | Lake Charles | 41.8 |
| | Winnsboro | 34.6 | 11 | Donaldsonville | 41.2 |
| 12 | Bastrop | 34.7 | | New Iberia | 41.2 |
| 13 | Winnfield | 35.2 | 13 | Grand Coteau | 40.7 |
| 14 | Monroe | 35.4 | 14 | Reserve | 40.6 |
| 15 | Shreveport | 35.5 | | Slidell | 40.6 |
| 16 | St. Joseph | 35.9 | 16 | Jennings | 40.4 |
| 17 | Natchitoches | 36.1 | 17 | Carville | 40.3 |
| 18 | Leesville | 36.5 | | Jeanerette | 40.3 |
| 19 | Elizabeth | 36.6 | 19 | Baton Rouge-Ryn | 40.2 |
| 20 | Alexandria | 37.0 | 20 | Crowley | 39.7 |
| 21 | Franklinton | 37.2 | | Oberlin | 39.7 |
| 22 | Bogalusa | 37.9 | 22 | Baton Rouge-LSU | 39.4 |
| 23 | Amite | 38.0 | | Covington | 39.4 |
| | Hammond | 38.0 | 24 | New Roads | 38.2 |
| 25 | Bunkie | 38.1 | 25 | Bunkie | 38.1 |

## JULY MAXIMUM TEMPERATURE °F

| HIGHEST | | | LOWEST | | |
|---|---|---|---|---|---|
| 1 | Logansport | 93.7 | 1 | Hackberry | 89.7 |
| 2 | Bastrop | 93.6 | 2 | Franklin | 89.9 |
| | Natchitoches | 93.6 | 3 | Jeanerette | 90.0 |
| 4 | Shreveport | 93.0 | 4 | Rockefeller | 90.1 |
| 5 | Winnsboro | 92.9 | 5 | Houma | 90.3 |
| 6 | Ashland | 92.7 | | Morgan City | 90.3 |
| | Calhoun | 92.7 | 7 | Lafayette | 90.6 |
| | Leesville | 92.7 | 8 | Crowley | 90.8 |
| | Many | 92.7 | | Lake Charles | 90.8 |
| | Plain Dealing | 92.7 | | New Iberia | 90.8 |
| 11 | Elizabeth | 92.6 | | New Orleans-Aud | 90.8 |
| | Franklinton | 92.6 | | New Orleans-Intl | 90.8 |
| | Monroe | 92.6 | | Paradis | 90.8 |
| | Ruston | 92.6 | 14 | Baton Rouge-LSU | 90.9 |
| 15 | Alexandria | 92.5 | | Reserve | 90.9 |
| | Minden | 92.5 | 16 | Jennings | 91.0 |
| | Winnfield | 92.5 | 17 | New Roads | 91.1 |
| 18 | Amite | 92.2 | 18 | Baton Rouge-Ryn | 91.2 |
| 19 | Homer | 92.1 | | Donaldsonville | 91.2 |
| 20 | Hammond | 92.0 | | Slidell | 91.2 |
| | St. Joseph | 92.0 | 21 | Carville | 91.3 |
| 22 | Bunkie | 91.9 | 22 | Covington | 91.5 |
| 23 | Bogalusa | 91.8 | | Grand Coteau | 91.5 |
| 24 | Oberlin | 91.7 | 24 | Lake Providence | 91.6 |
| 25 | Lake Providence | 91.6 | | Tallulah | 91.6 |

## ANNUAL PRECIPITATION (")

| HIGHEST | | | LOWEST | | |
|---|---|---|---|---|---|
| 1 | Morgan City | 66.10 | 1 | Shreveport | 49.10 |
| 2 | Oberlin | 66.01 | 2 | Logansport | 51.02 |
| 3 | Franklin | 65.76 | 3 | Plain Dealing | 51.24 |
| 4 | Hammond | 65.66 | 4 | Minden | 52.34 |
| 5 | Amite | 65.38 | 5 | Monroe | 52.40 |
| 6 | Paradis | 65.31 | 6 | Ashland | 53.42 |
| 7 | Houma | 64.74 | 7 | Homer | 54.19 |
| 8 | Reserve | 63.94 | 8 | Calhoun | 54.29 |
| 9 | New Roads | 63.89 | 9 | Ruston | 54.30 |
| 10 | New Orleans-Intl | 63.84 | 10 | Natchitoches | 54.70 |
| 11 | Franklinton | 63.74 | 11 | Many | 55.12 |
| 12 | New Orleans-Aud | 63.57 | 12 | St. Joseph | 55.25 |
| 13 | Covington | 62.99 | 13 | Winnsboro | 55.92 |
| 14 | Grand Coteau | 62.98 | 14 | Bastrop | 56.07 |
| 15 | Bogalusa | 62.89 | 15 | Leesville | 56.46 |
| 16 | Bunkie | 62.63 | 16 | Lake Charles | 56.64 |
| 17 | Elizabeth | 62.61 | 17 | Lake Providence | 57.76 |
| 18 | New Iberia | 62.42 | 18 | Tallulah | 58.12 |
| 19 | Jeanerette | 62.40 | 19 | Hackberry | 59.20 |
| 20 | Slidell | 62.27 | 20 | Winnfield | 59.28 |
| 21 | Baton Rouge-Ryn | 62.16 | 21 | Baton Rouge-LSU | 59.90 |
| 22 | Crowley | 61.01 | 22 | Alexandria | 60.36 |
| 23 | Rockefeller | 60.97 | 23 | Jennings | 60.51 |
| 24 | Lafayette | 60.80 | 24 | Donaldsonville | 60.54 |
| 25 | Carville | 60.74 | 25 | Carville | 60.74 |

## ANNUAL SNOWFALL (")

| HIGHEST | | | LOWEST | | |
|---|---|---|---|---|---|
| 1 | Shreveport | 1.7 | 1 | Baton Rouge-LSU | 0.0 |
| 2 | Plain Dealing | 1.6 | | Bunkie | 0.0 |
| 3 | Winnfield | 1.3 | | Hammond | 0.0 |
| 4 | Homer | 1.1 | | Morgan City | 0.0 |
| | Lake Providence | 1.1 | | New Iberia | 0.0 |
| 6 | Ashland | 1.0 | | New Orleans-Aud | 0.0 |
| | Minden | 1.0 | | New Orleans-Intl | 0.0 |
| 8 | Monroe | 0.9 | | New Roads | 0.0 |
| | Ruston | 0.9 | | Oberlin | 0.0 |
| 10 | Natchitoches | 0.7 | 10 | Carville | 0.1 |
| 11 | Calhoun | 0.6 | | Crowley | 0.1 |
| | Logansport | 0.6 | | Donaldsonville | 0.1 |
| | Tallulah | 0.6 | | Franklin | 0.1 |
| 14 | Bastrop | 0.5 | | Grand Coteau | 0.1 |
| | Franklinton | 0.5 | | Hackberry | 0.1 |
| | St. Joseph | 0.5 | | Houma | 0.1 |
| 17 | Alexandria | 0.4 | | Jeanerette | 0.1 |
| 18 | Amite | 0.3 | | Paradis | 0.1 |
| | Bogalusa | 0.3 | | Reserve | 0.1 |
| | Lake Charles | 0.3 | | Rockefeller | 0.1 |
| | Many | 0.3 | | Winnsboro | 0.1 |
| 22 | Baton Rouge-Ryn | 0.2 | 22 | Baton Rouge-Ryn | 0.2 |
| | Covington | 0.2 | | Covington | 0.2 |
| | Elizabeth | 0.2 | | Elizabeth | 0.2 |
| | Jennings | 0.2 | | Jennings | 0.2 |

# MAINE

PHYSICAL FEATURES. Maine occupies 33,215 square miles. From near the 43d parallel, the State extends northward over 300 miles, spanning a full 4½° of latitude. Its width from the 67th meridian extends westward over 4° of longitude, a span of over 200 miles. The terrain is hilly. Elevations are generally less than 500 feet above sea level over the southeastern one-half of the State. The northwestern one-half is a plateau ranging in elevation from 1,000 to 1,500 feet, but sloping downward to 500 feet in the northeast from 1,000 feet in the north (Aroostook County). A number of mountain peaks, extensions of the Appalachian chain, rise to heights from 3,000 to 5,000 feet mostly in the western and central portions of the State. Mt. Katahdin is the highest point. Its summit, at 5,268 feet, rises nearly 4,500 feet from a relatively low base elevation.

The great glaciers of the ice age were all-important in the physical formation of the State. They left "horsebacks" or ridges of glacial deposits, some as long as 150 miles in length. These ridges furnish both natural highway routes and abundant material for roadbuilding. The glaciers formed or left over 1,600 lakes, spread abundantly over the entire State. The largest of these lakes is Moosehead. The total water area of the State exceeds 2,200 square miles. Some flatland is found near the coast, especially near the mouths of the Androscoggin and Kennebec Rivers. Other tracts of flatland, often marshy, lie near lakes.

The coastal portion of the State has many inlets, bays, channels, fine harbors, rocky islands, and promontories. The extreme irregularity of the coast stretches the total coastline to about 2,400 miles, more than 10 times the distance from Kittery to Eastport. The southwestern portion of the coast has many beaches. The mid-coastal portion has many rugged hills and small mountains, some of which rise abruptly from the water, such as Mount Desert Island.

GENERAL CLIMATE. Maine's chief climatic characteristics include: (1) changeableness of the weather, (2) large ranges of temperature, both diurnal and annual, (3) great differences between the same seasons in different years, (4) equable distribution of precipitation, and (5) considerable diversity from place to place. The regional climatic influences are modified in Maine by varying distances from the ocean, by elevations, and by types of terrain. These modifying factors divide the State into three natural climatological divisions. The Northern Division contains slightly more than one-half of the State's area, with its southern boundary nearly parallel to the coast. It represents that area of the State least affected by ocean influences and most affected by higher elevations. In contrast, the Coastal Division is a strip roughly 20 to 30 miles in width. It is most affected by maritime influences and has the lowest average elevation above sea level. The remainder, known conveniently as the Southern Interior Division, covers nearly one-third of the State's area.

Maine lies in the "prevailing westerlies"--the belt of generally eastward air movement which encircles the globe in the middle latitudes. Embedded in this circulation are extensive masses of air originating in higher or lower latitudes and interacting to produce storm systems. Relative to most other sections of the country, a large number of such storms pass over or near Maine. The majority of air masses affecting this State belong to three types: (1) cold, dry air pouring down from subarctic North America, (2) warm, moist air streaming up on a long overland journey from the Gulf of Mexico and from subtropical waters eastward, and (3) cool, damp air moving in from the North Atlantic. Because the atmospheric flow is usually offshore, Maine is influenced more by the first two types than it is by the third.

The procession of contrasting air masses and the relatively frequent passage of storms bring about a roughly twice-weekly alternation from fair to cloudy or stormy conditions, attended by often abrupt changes in temperature, moisture, sunshine, wind direction, and wind speed. There is no regular or persistent rhythm to this sequence. It is interrupted by periods of time during which the same weather patterns continue for several days, and infrequently for several weeks. Maine weather, however, is distinguished for variety rather than for monotony. Changeability is also one of its features on a longer time-scale; that is, the same month or season will exhibit varying characteristics over the years--sometimes in close alternation, sometimes arranged in similar groups for successive years.

TEMPERATURE. The average annual temperature ranges from near 40° in the Northern Division, to 44° in the Southern Interior Division, and to nearly 45° in the Coastal Division. Summer temperatures are delightfully cool and are reasonably uniform over the State. Average temperatures vary much more in winter than in summer.

PRECIPITATION. Maine has precipitation rather evenly distributed throughout the year. The distribution is most regular in the Southern Interior Division. Along the Atlantic coast, summer thunderstorm activity is somewhat suppressed by the effects of the cool ocean, while winter precipitation is increased by coastal storms or "northeasters." In the Northern Division, these effects are reversed with increased thunderstorm activity in summer and with very little effect of coastal storms in winter. Precipitation totals in this Division are greater in summer.

Storm systems are the principal year-round moisture producers. Such systems are less active in the summer, but bands or patches of thunderstorm or shower activity take over much of this function. Though brief and often of small areal extent, thunderstorms produce the heaviest local rainfall rates for short intervals. Many stations have received from 1 to 2 inches in an hour. Winter precipitation occurs mostly as snow, except in the Coastal Division where considerable rain or wet snow falls; stations in this Division, more than stations farther inland, are subject to occasional glazing, or "ice storm" conditions. Freezing rain coats streets, roads, and all exposed surfaces; on rare occasions, a heavy load of ice builds on trees and wires.

SNOWFALL. As a rule, average seasonal snowfall amounts increase northwestward from the coast. The Coastal Division snowfall totals range from 50 to 80 inches. The Southern Interior Division receives from 60 to 90 inches. The Northern Division totals range on the average from 90 to 110 inches. Local topography has a marked influence on snowfall, causing large variations within a short distance. The snowfall season usually begins in late October or in November and lasts into April and sometimes into May. Seasonal totals in the north do not vary as markedly as along the coast. Snow cover lasts throughout the season in the north. Along the coast, however, the snow cover may melt entirely in midwinter and then be replaced by a new cover. Melting is usually gradual enough to prevent serious flooding.

OTHER CLIMATIC FEATURES. The amount of possible sunshine averages from 50 to 60 percent in most of the southern half of the State. This percentage varies along the coast from near 50 to 60 percent. At higher elevations and over much of northern Maine, the average is near 45 percent. The average annual number of clear days ranges from 80 to 120 days in the southern half and from about 50 to 90 days in the northern half of the State.

Heavy fog is frequent and sometimes persistent along the coast, particularly in the eastern portion of the coast where it may occur on an average of 1 day out of 6. Fog frequency and duration diminish inland. But short-duration heavy ground fogs of early morning occur frequently at susceptible places inland.

Prolonged dry spells, quite frequent in late summer or fall, create serious forest-fire hazards. Low humidities and lack of precipitation during some late summers cause the forest litter to become extremely inflammable.

WINDS AND STORMS. On a yearly basis, the wind direction is mostly from the west. During winter, north to northwest winds tend to prevail. In the summer, they are more often from the southwest or south. Topography has a strong influence on the prevailing direction. Parts of a major river valley, for example, may have a prevailing wind paralleling the valley. Along the coast in spring and summer, the sea breeze is important. On-shore local winds, blowing from the cool ocean, may come as far inland as 10 miles. They tend to retard spring growth, but are pleasingly cooling in summer.

Coastal storms or "northeasters" sometimes seriously affect the Coastal Division. They generate very strong winds and heavy rain or snow. They can produce abnormally high wind-driven tides, affecting beaches and coastal installations. In winter, these storms produce some of the heavier snowfalls along the coast. Occasionally, in summer or fall, a storm of tropical origin affects Maine. Usually the storm will be similar to the northeasters. But a few such storms may retain near or full hurricane force. Tornadoes are a phenomena not common in Maine. It is likely that several occur on the average each year. Fortunately, most tornadoes are very small, affecting a very localized area. About 80 percent of Maine's tornadoes occur between May 15 and September 15. The peak month is July. Thunderstorms and hailstorms have a similar frequency maximum from midspring to early fall. Thunderstorms occur in a range from 10 to 20 days a year in the Coastal Division and from 15 to 30 days a year elsewhere. The most severe storms are attended by hail.

## COUNTY INDEX



## ELEVATION INDEX

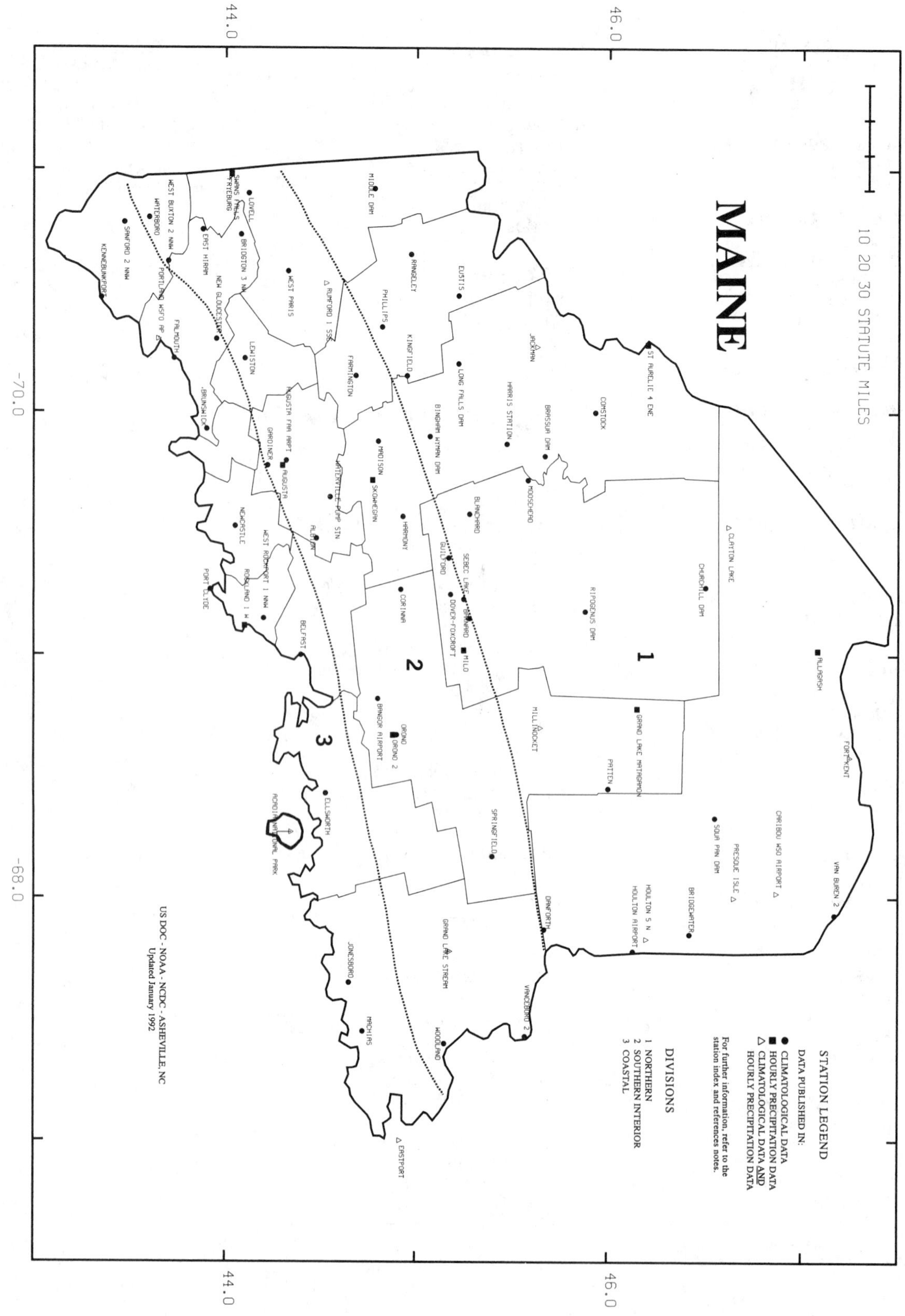
MAINE
10 20 30 STATUTE MILES
44.0
46.0
-70.0
-68.0
1
2
3
STATION LEGEND
DATA PUBLISHED IN:
● CLIMATOLOGICAL DATA
■ HOURLY PRECIPITATION DATA
△ CLIMATOLOGICAL DATA AND HOURLY PRECIPITATION DATA
For further information, refer to the station index and references notes.
DIVISIONS
1 NORTHERN
2 SOUTHERN INTERIOR
3 COASTAL
US DOC - NOAA - NCDC - ASHEVILLE, NC
Updated January 1992
ALLAGASH
FORT KENT
VAN BUREN 2
CARIBOU WSO AIRPORT
PRESQUE ISLE
SQUA PAN DAM
BRIDGEWATER
HOULTON 5 N
HOULTON AIRPORT
CLAYTON LAKE
CHURCHILL DAM
RIPOGENUS DAM
GRAND LAKE MATAGAMON
PATTEN
MILLINOCKET
DANFORTH
SPRINGFIELD
GRAND LAKE STREAM
VANCEBORO 2
WOODLAND
EASTPORT
MACHIAS
JONESBORO
ELLSWORTH
ACADIA NATIONAL PARK
BANGOR AIRPORT
ORONO
ORONO 2
CORINNA
MILO
DOVER-FOXCROFT
BROWNVILLE
SEBEC LAKE
GUILFORD
BLANCHARD
MOOSEHEAD
BRASSUA DAM
COMSTOCK
ST AURELIE 4 ENE
JACKMAN
HARRIS STATION
LONG FALLS DAM
BINGHAM WYMAN DAM
MADISON
SKOWHEGAN
HARMONY
WATERVILLE PUMP STN
ALBION
BELFAST
ROCKLAND 1 W
WEST ROCKPORT 1 NNW
NEWCASTLE
PORT CLYDE
AUGUSTA
AUGUSTA FAA ARPT
GARDINER
BRUNSWICK
LEWISTON
FARMINGTON
KINGFIELD
EUSTIS
RANGELEY
PHILLIPS
RUMFORD 1 SSE
WEST PARIS
MIDDLE DAM
LOVELL
SWANS FALLS
FRYEBURG
BRIDGTON 3 NW
EAST HIRAM
NEW GLOUCESTER
WEST BUXTON 2 NNW
WATERBORO
SANFORD 2 NNW
KENNEBUNKPORT
PORTLAND WSFO AP
FALMOUTH

## AUGUSTA AIRPORT *Kennebec County* ELEVATION 354 ft LAT/LONG 44° 19 ' N / 69° 48 ' W

| | JAN | FEB | MAR | APR | MAY | JUN | JUL | AUG | SEP | OCT | NOV | DEC | YEAR |
|---|---|---|---|---|---|---|---|---|---|---|---|---|---|
| Maximum Temp °F | 27.6 | 30.9 | 40.0 | 52.4 | 64.9 | 73.9 | 79.2 | 77.6 | 68.6 | 57.1 | 44.6 | 32.4 | 54.1 |
| Minimum Temp °F | 10.5 | 12.9 | 23.6 | 34.7 | 45.0 | 54.3 | 60.4 | 58.9 | 50.0 | 39.7 | 30.5 | 17.0 | 36.5 |
| Mean Temp °F | 19.1 | 21.9 | 31.8 | 43.6 | 54.9 | 64.1 | 69.9 | 68.3 | 59.3 | 48.4 | 37.6 | 24.7 | 45.3 |
| Days Max Temp ≥ 90 °F | 0 | 0 | 0 | 0 | 0 | 1 | 1 | 1 | 0 | 0 | 0 | 0 | 3 |
| Days Max Temp ≤ 32 °F | 20 | 15 | 6 | 0 | 0 | 0 | 0 | 0 | 0 | 0 | 3 | 15 | 59 |
| Days Min Temp ≤ 32 °F | 30 | 26 | 25 | 11 | 1 | 0 | 0 | 0 | 0 | 6 | 18 | 29 | 146 |
| Days Min Temp ≤ 0 °F | 7 | 4 | 1 | 0 | 0 | 0 | 0 | 0 | 0 | 0 | 0 | 2 | 14 |
| Heating Degree Days | 1417 | 1210 | 1021 | 635 | 313 | 85 | 14 | 30 | 189 | 508 | 817 | 1243 | 7482 |
| Cooling Degree Days | 0 | 0 | 0 | 0 | 10 | 62 | 174 | 138 | 28 | 1 | 0 | 0 | 413 |
| Total Precipitation (") | 2.79 | 2.79 | 3.42 | 3.75 | 3.79 | 3.45 | 3.15 | 3.46 | 3.20 | 3.67 | 4.23 | 3.71 | 41.41 |
| Days ≥ 0.1" Precip | 6 | 6 | 7 | 7 | 7 | 7 | 6 | 6 | 6 | 6 | 7 | 7 | 78 |
| Total Snowfall (") | 19.6 | 16.5 | 12.4 | 4.5 | 0.1 | 0.0 | 0.0 | 0.0 | 0.0 | 0.3 | 4.4 | 15.9 | 73.7 |
| Days ≥ 1" Snow Depth | 27 | 25 | 19 | 4 | 0 | 0 | 0 | 0 | 0 | 0 | 4 | 18 | 97 |

## BANGOR AIRPORT *Penobscot County* ELEVATION 194 ft LAT/LONG 44° 48 ' N / 68° 49 ' W

| | JAN | FEB | MAR | APR | MAY | JUN | JUL | AUG | SEP | OCT | NOV | DEC | YEAR |
|---|---|---|---|---|---|---|---|---|---|---|---|---|---|
| Maximum Temp °F | 26.4 | 29.4 | 38.7 | 51.4 | 64.0 | 73.3 | 78.6 | 77.1 | 68.1 | 56.7 | 44.3 | 31.9 | 53.3 |
| Minimum Temp °F | 7.7 | 10.3 | 21.7 | 32.8 | 43.1 | 52.7 | 58.4 | 57.2 | 48.4 | 38.4 | 29.3 | 15.1 | 34.6 |
| Mean Temp °F | 17.1 | 19.9 | 30.2 | 42.1 | 53.6 | 63.0 | 68.5 | 67.2 | 58.3 | 47.6 | 36.8 | 23.5 | 44.0 |
| Days Max Temp ≥ 90 °F | 0 | 0 | 0 | 0 | 0 | 1 | 2 | 1 | 0 | 0 | 0 | 0 | 4 |
| Days Max Temp ≤ 32 °F | 21 | 17 | 7 | 0 | 0 | 0 | 0 | 0 | 0 | 0 | 3 | 15 | 63 |
| Days Min Temp ≤ 32 °F | 30 | 27 | 26 | 15 | 1 | 0 | 0 | 0 | 0 | 8 | 20 | 29 | 156 |
| Days Min Temp ≤ 0 °F | 9 | 6 | 1 | 0 | 0 | 0 | 0 | 0 | 0 | 0 | 0 | 4 | 20 |
| Heating Degree Days | 1479 | 1269 | 1071 | 681 | 355 | 105 | 21 | 42 | 215 | 534 | 838 | 1279 | 7889 |
| Cooling Degree Days | 0 | 0 | 0 | 0 | 8 | 57 | 152 | 124 | 23 | 1 | 0 | 0 | 365 |
| Total Precipitation (") | 2.99 | 2.74 | 3.25 | 3.26 | 3.45 | 3.44 | 3.26 | 3.41 | 3.49 | 3.24 | 4.06 | 3.73 | 40.32 |
| Days ≥ 0.1" Precip | 7 | 6 | 7 | 7 | 7 | 7 | 7 | 6 | 6 | 7 | 8 | 7 | 82 |
| Total Snowfall (") | 18.7 | 16.8 | 12.1 | 4.1 | 0.1 | 0.0 | 0.0 | 0.0 | 0.0 | 0.4 | 4.2 | 15.3 | 71.7 |
| Days ≥ 1" Snow Depth | 26 | 24 | 19 | 3 | 0 | 0 | 0 | 0 | 0 | 0 | 3 | 17 | 92 |

## BELFAST *Waldo County* ELEVATION 20 ft LAT/LONG 44° 24 ' N / 69° 0 ' W

| | JAN | FEB | MAR | APR | MAY | JUN | JUL | AUG | SEP | OCT | NOV | DEC | YEAR |
|---|---|---|---|---|---|---|---|---|---|---|---|---|---|
| Maximum Temp °F | 31.6 | 34.4 | 42.8 | 53.7 | 65.4 | 74.5 | 80.0 | 78.5 | 70.3 | 59.2 | 47.3 | 35.9 | 56.1 |
| Minimum Temp °F | 10.3 | 11.9 | 22.4 | 32.3 | 42.3 | 51.3 | 57.1 | 56.0 | 48.1 | 38.5 | 30.4 | 17.1 | 34.8 |
| Mean Temp °F | 21.0 | 23.1 | 32.6 | 43.0 | 53.9 | 62.9 | 68.6 | 67.3 | 59.2 | 48.9 | 38.9 | 26.5 | 45.5 |
| Days Max Temp ≥ 90 °F | 0 | 0 | 0 | 0 | 0 | 1 | 2 | 1 | 0 | 0 | 0 | 0 | 4 |
| Days Max Temp ≤ 32 °F | 16 | 11 | 3 | 0 | 0 | 0 | 0 | 0 | 0 | 0 | 1 | 11 | 42 |
| Days Min Temp ≤ 32 °F | 30 | 27 | 27 | 16 | 2 | 0 | 0 | 0 | 0 | 9 | 19 | 29 | 159 |
| Days Min Temp ≤ 0 °F | 8 | 5 | 1 | 0 | 0 | 0 | 0 | 0 | 0 | 0 | 0 | 3 | 17 |
| Heating Degree Days | 1359 | 1175 | 996 | 652 | 343 | 100 | 17 | 32 | 186 | 493 | 777 | 1187 | 7317 |
| Cooling Degree Days | 0 | 0 | 0 | 0 | 5 | 39 | 124 | 91 | 15 | 0 | 0 | 0 | 274 |
| Total Precipitation (") | 3.57 | 3.31 | 4.16 | 4.31 | 3.98 | 3.67 | 3.04 | 3.26 | 3.85 | 4.25 | 5.09 | 4.76 | 47.25 |
| Days ≥ 0.1" Precip | 5 | *5* | 6 | 7 | 7 | 7 | 6 | 6 | 6 | 7 | 8 | 6 | 76 |
| Total Snowfall (") | na | na | na | *1.1* | 0.0 | 0.0 | 0.0 | 0.0 | 0.0 | 0.0 | *2.0* | na | na |
| Days ≥ 1" Snow Depth | *19* | *16* | *14* | *2* | 0 | 0 | 0 | 0 | 0 | 0 | 1 | 12 | 64 |

## BRASSUA DAM *Somerset County* ELEVATION 1060 ft LAT/LONG 45° 40 ' N / 69° 49 ' W

| | JAN | FEB | MAR | APR | MAY | JUN | JUL | AUG | SEP | OCT | NOV | DEC | YEAR |
|---|---|---|---|---|---|---|---|---|---|---|---|---|---|
| Maximum Temp °F | 21.3 | 24.5 | 34.6 | 46.2 | 60.6 | 70.3 | 75.3 | 73.6 | 64.7 | 52.4 | 39.1 | 26.2 | 49.1 |
| Minimum Temp °F | -1.5 | -0.4 | 10.8 | 25.3 | 36.8 | 48.2 | 53.4 | 51.9 | 43.0 | 33.5 | 23.9 | 7.6 | 27.7 |
| Mean Temp °F | 9.9 | 12.0 | 22.7 | 35.8 | 48.7 | 59.2 | 64.4 | 62.8 | 53.8 | 43.0 | 31.5 | 16.9 | 38.4 |
| Days Max Temp ≥ 90 °F | 0 | 0 | 0 | 0 | 0 | 0 | 0 | 0 | 0 | 0 | 0 | 0 | 0 |
| Days Max Temp ≤ 32 °F | 25 | 21 | 12 | 2 | 0 | 0 | 0 | 0 | 0 | 0 | 7 | 22 | 89 |
| Days Min Temp ≤ 32 °F | 31 | 28 | 30 | 25 | 10 | 0 | 0 | 0 | 3 | 15 | 25 | 31 | 198 |
| Days Min Temp ≤ 0 °F | 18 | 16 | 8 | 0 | 0 | 0 | 0 | 0 | 0 | 0 | 0 | 10 | 52 |
| Heating Degree Days | 1704 | 1491 | 1304 | 870 | 502 | 190 | 79 | 111 | 334 | 676 | 998 | 1485 | 9744 |
| Cooling Degree Days | 0 | 0 | 0 | 0 | 2 | 24 | 66 | 52 | 6 | 0 | 0 | 0 | 150 |
| Total Precipitation (") | 2.65 | 2.25 | 2.77 | 3.28 | 3.56 | 4.10 | 4.08 | 3.82 | 3.69 | 3.43 | 3.69 | 3.21 | 40.53 |
| Days ≥ 0.1" Precip | 6 | 6 | 7 | 7 | 8 | 9 | 8 | 8 | 7 | 7 | 8 | 7 | 88 |
| Total Snowfall (") | 24.2 | 20.6 | 19.3 | 9.6 | 0.4 | 0.0 | 0.0 | 0.0 | 0.0 | 0.9 | 10.3 | 23.2 | 108.5 |
| Days ≥ 1" Snow Depth | na | na | na | na | 0 | 0 | 0 | 0 | 0 | 0 | na | na | na |

## BRIDGEWATER *Aroostook County* ELEVATION 430 ft LAT/LONG 46° 26 ' N / 67° 51 ' W

| | JAN | FEB | MAR | APR | MAY | JUN | JUL | AUG | SEP | OCT | NOV | DEC | YEAR |
|---|---|---|---|---|---|---|---|---|---|---|---|---|---|
| Maximum Temp °F | 22.2 | 26.5 | 36.9 | 50.1 | 64.5 | 74.0 | 78.7 | 76.2 | 66.3 | 54.0 | 39.7 | 27.2 | 51.4 |
| Minimum Temp °F | -1.4 | 1.0 | 14.3 | 27.4 | 37.7 | 47.9 | 52.7 | 50.8 | 42.3 | 33.3 | 23.4 | 7.4 | 28.1 |
| Mean Temp °F | 10.4 | 13.8 | 25.6 | 38.8 | 51.1 | 61.0 | 65.7 | 63.5 | 54.3 | 43.6 | 31.6 | 17.3 | 39.7 |
| Days Max Temp ≥ 90 °F | 0 | 0 | 0 | 0 | 0 | 1 | 1 | 1 | 0 | 0 | 0 | 0 | 3 |
| Days Max Temp ≤ 32 °F | 24 | 20 | 9 | 1 | 0 | 0 | 0 | 0 | 0 | 0 | 6 | 21 | 81 |
| Days Min Temp ≤ 32 °F | 30 | 28 | 29 | *23* | 9 | 1 | 0 | 0 | 5 | 15 | 25 | 30 | 195 |
| Days Min Temp ≤ 0 °F | 17 | 15 | 6 | 0 | 0 | 0 | 0 | 0 | 0 | 0 | 1 | 10 | 49 |
| Heating Degree Days | 1688 | 1444 | 1213 | 780 | 428 | 151 | 57 | 101 | 321 | 656 | 996 | 1474 | 9309 |
| Cooling Degree Days | 0 | 0 | 0 | 0 | 5 | 29 | 81 | 68 | 9 | 0 | 0 | 0 | 192 |
| Total Precipitation (") | 3.12 | 2.28 | 2.80 | 3.07 | 3.76 | 3.58 | 3.87 | 3.99 | 3.84 | 3.47 | 4.11 | 3.53 | 41.42 |
| Days ≥ 0.1" Precip | 7 | 6 | 7 | 8 | 8 | 7 | 8 | 8 | 7 | 7 | 9 | 8 | 90 |
| Total Snowfall (") | *21.4* | 15.8 | *14.1* | na | 0.1 | 0.0 | 0.0 | 0.0 | 0.1 | 0.3 | 9.3 | 20.8 | na |
| Days ≥ 1" Snow Depth | na | na | na | na | *0* | 0 | 0 | *0* | 0 | 0 | na | na | na |

## BRIDGTON 3 NW *Cumberland County* ELEVATION 600 ft LAT/LONG 44° 4 ' N / 70° 43 ' W

| | JAN | FEB | MAR | APR | MAY | JUN | JUL | AUG | SEP | OCT | NOV | DEC | YEAR |
|---|---|---|---|---|---|---|---|---|---|---|---|---|---|
| Maximum Temp °F | 29.4 | 32.3 | 41.6 | 53.3 | 66.5 | 74.6 | 79.7 | 77.5 | 69.3 | 58.3 | 45.2 | 33.2 | 55.1 |
| Minimum Temp °F | 5.4 | 7.7 | 19.5 | 30.3 | 40.9 | 50.6 | 55.9 | 54.0 | 45.1 | 34.6 | 26.8 | 13.0 | 32.0 |
| Mean Temp °F | 17.4 | 20.0 | 30.5 | 41.8 | 53.7 | 62.6 | 67.8 | 65.8 | 57.2 | 46.5 | 36.0 | 23.1 | 43.5 |
| Days Max Temp ≥ 90 °F | 0 | 0 | 0 | 0 | 0 | 1 | 1 | 1 | 0 | 0 | 0 | 0 | 3 |
| Days Max Temp ≤ 32 °F | 19 | 14 | 4 | 0 | 0 | 0 | 0 | 0 | 0 | 0 | 2 | 13 | 52 |
| Days Min Temp ≤ 32 °F | 31 | 28 | 29 | *20* | 4 | 0 | 0 | 0 | 2 | 14 | 23 | 30 | 181 |
| Days Min Temp ≤ 0 °F | 11 | 9 | 2 | 0 | 0 | 0 | 0 | 0 | 0 | 0 | 0 | 5 | 27 |
| Heating Degree Days | 1469 | 1263 | 1061 | 690 | 353 | 116 | 31 | 61 | 244 | 568 | 865 | 1291 | 8012 |
| Cooling Degree Days | 0 | 0 | 0 | 0 | 7 | 39 | 112 | 73 | 12 | 1 | *0* | 0 | 244 |
| Total Precipitation (") | 3.90 | 3.30 | 3.96 | 3.89 | 3.56 | 4.08 | 3.72 | 4.03 | 3.51 | 3.97 | 4.34 | 4.09 | 46.35 |
| Days ≥ 0.1" Precip | *6* | *5* | *6* | *7* | 7 | 8 | 7 | 7 | 6 | 6 | 8 | *7* | 80 |
| Total Snowfall (") | *22.7* | *16.8* | *12.1* | *5.2* | 0.0 | 0.0 | 0.0 | 0.0 | 0.0 | 0.3 | *5.0* | na | na |
| Days ≥ 1" Snow Depth | na | na | na | na | 0 | 0 | 0 | 0 | 0 | 0 | na | na | na |

## BRUNSWICK *Cumberland County* ELEVATION 69 ft LAT/LONG 43° 54 ' N / 69° 56 ' W

| | JAN | FEB | MAR | APR | MAY | JUN | JUL | AUG | SEP | OCT | NOV | DEC | YEAR |
|---|---|---|---|---|---|---|---|---|---|---|---|---|---|
| Maximum Temp °F | 30.1 | 32.5 | 41.2 | 52.7 | 63.5 | 72.8 | 78.7 | 77.7 | 68.9 | 58.3 | 46.7 | 35.4 | 54.9 |
| Minimum Temp °F | 10.5 | 12.5 | 23.8 | 34.2 | 43.9 | 53.3 | 59.5 | 58.6 | 49.8 | 39.3 | 30.4 | 19.5 | 36.3 |
| Mean Temp °F | 20.3 | 22.5 | 32.5 | 43.5 | 53.7 | 63.1 | 69.1 | 68.2 | 59.4 | 48.8 | 38.6 | 27.5 | 45.6 |
| Days Max Temp ≥ 90 °F | 0 | 0 | 0 | 0 | 0 | 1 | 1 | 1 | 0 | 0 | 0 | 0 | 3 |
| Days Max Temp ≤ 32 °F | 17 | 13 | 5 | 0 | 0 | 0 | 0 | 0 | 0 | 0 | 1 | 12 | 48 |
| Days Min Temp ≤ 32 °F | 30 | 27 | 24 | 13 | 1 | 0 | 0 | 0 | 0 | 8 | 18 | 28 | 149 |
| Days Min Temp ≤ 0 °F | 7 | 5 | 1 | 0 | 0 | 0 | 0 | 0 | 0 | 0 | 0 | 2 | 15 |
| Heating Degree Days | 1378 | 1194 | 1000 | 640 | 347 | 103 | 15 | 29 | 188 | 495 | 787 | 1175 | 7351 |
| Cooling Degree Days | *0* | *0* | *0* | *0* | 7 | 56 | 169 | 145 | 29 | 1 | 0 | 0 | 407 |
| Total Precipitation (") | 3.54 | 3.47 | 4.43 | 4.22 | 3.86 | 3.56 | 2.97 | 3.47 | 3.35 | 3.66 | 5.14 | 4.76 | 46.43 |
| Days ≥ 0.1" Precip | 7 | 6 | 7 | 7 | 7 | 7 | 6 | 6 | 5 | 6 | 8 | 8 | 80 |
| Total Snowfall (") | 20.0 | 16.6 | 13.1 | 3.6 | 0.1 | 0.0 | 0.0 | 0.0 | 0.0 | 0.2 | 3.3 | *13.8* | 70.7 |
| Days ≥ 1" Snow Depth | 24 | 22 | *14* | 2 | 0 | 0 | 0 | 0 | 0 | 0 | *2* | *15* | 79 |

## CARIBOU MUNI ARPT *Aroostook County* ELEVATION 627 ft LAT/LONG 46° 52 ' N / 68° 1 ' W

| | JAN | FEB | MAR | APR | MAY | JUN | JUL | AUG | SEP | OCT | NOV | DEC | YEAR |
|---|---|---|---|---|---|---|---|---|---|---|---|---|---|
| Maximum Temp °F | 19.0 | 22.9 | 34.0 | 47.0 | 61.8 | 71.6 | 76.3 | 73.9 | 63.9 | 51.5 | 37.6 | 24.4 | 48.7 |
| Minimum Temp °F | -0.3 | 2.6 | 15.0 | 29.2 | 40.2 | 49.9 | 54.7 | 52.8 | 43.4 | 34.4 | 23.6 | 7.7 | 29.4 |
| Mean Temp °F | 9.4 | 12.7 | 24.5 | 38.1 | 51.0 | 60.8 | 65.6 | 63.4 | 53.7 | 43.0 | 30.7 | 16.1 | 39.1 |
| Days Max Temp ≥ 90 °F | 0 | 0 | 0 | 0 | 0 | 0 | 1 | 0 | 0 | 0 | 0 | 0 | 1 |
| Days Max Temp ≤ 32 °F | 26 | 22 | 12 | 2 | 0 | 0 | 0 | 0 | 0 | 0 | 10 | 23 | 95 |
| Days Min Temp ≤ 32 °F | 31 | 28 | 28 | 20 | 5 | 0 | 0 | 0 | 2 | 14 | 25 | 30 | 183 |
| Days Min Temp ≤ 0 °F | 17 | 13 | 5 | 0 | 0 | 0 | 0 | 0 | 0 | 0 | 0 | 10 | 45 |
| Heating Degree Days | 1720 | 1471 | 1249 | 799 | 433 | 158 | 58 | 102 | 341 | 676 | 1023 | 1512 | 9542 |
| Cooling Degree Days | 0 | 0 | 0 | 0 | 5 | 33 | 82 | 60 | 8 | 0 | 0 | 0 | 188 |
| Total Precipitation (") | 2.55 | 1.95 | 2.50 | 2.48 | 3.10 | 3.14 | 3.84 | 4.13 | 3.40 | 3.12 | 3.32 | 3.26 | 36.79 |
| Days ≥ 0.1" Precip | 6 | 5 | 6 | 7 | 8 | 8 | 8 | 8 | 7 | 7 | 8 | 8 | 86 |
| Total Snowfall (") | 25.0 | 19.3 | 18.9 | 9.0 | 0.9 | 0.0 | 0.0 | 0.0 | 0.1 | 0.8 | 12.8 | 25.8 | 112.6 |
| Days ≥ 1" Snow Depth | 31 | 28 | 30 | 13 | 0 | 0 | 0 | 0 | 0 | 0 | 11 | 28 | 141 |

## CORINNA *Penobscot County* ELEVATION 322 ft LAT/LONG 44° 55' N / 69° 16' W

| | JAN | FEB | MAR | APR | MAY | JUN | JUL | AUG | SEP | OCT | NOV | DEC | YEAR |
|---|---|---|---|---|---|---|---|---|---|---|---|---|---|
| Maximum Temp °F | 26.1 | 30.1 | 39.8 | 52.9 | 66.4 | 74.9 | 79.8 | 78.0 | 68.9 | 57.2 | 43.6 | 30.7 | 54.0 |
| Minimum Temp °F | 1.4 | 3.8 | 17.6 | 29.9 | 40.6 | 50.2 | 55.6 | 53.6 | 44.4 | 33.7 | 25.5 | 9.7 | 30.5 |
| Mean Temp °F | 13.7 | 17.0 | 28.7 | 41.4 | 53.5 | 62.6 | 67.7 | 65.8 | 56.6 | 45.5 | 34.6 | 20.2 | 42.3 |
| Days Max Temp ≥ 90 °F | 0 | 0 | 0 | 0 | 0 | 1 | 2 | 1 | 0 | 0 | 0 | 0 | 4 |
| Days Max Temp ≤ 32 °F | 22 | 16 | 6 | 0 | 0 | 0 | 0 | 0 | 0 | 0 | 3 | 17 | 64 |
| Days Min Temp ≤ 32 °F | 30 | 28 | 28 | 19 | 5 | 0 | 0 | 0 | 3 | 15 | 23 | 30 | 181 |
| Days Min Temp ≤ 0 °F | 16 | 12 | 3 | 0 | 0 | 0 | 0 | 0 | 0 | 0 | 0 | 8 | 39 |
| Heating Degree Days | 1583 | 1351 | 1119 | 702 | 357 | 117 | 31 | 61 | 258 | 598 | 907 | 1381 | 8465 |
| Cooling Degree Days | 0 | 0 | 0 | 0 | 9 | 48 | 125 | 93 | 12 | 0 | 0 | 0 | 287 |
| Total Precipitation (") | 2.94 | 2.66 | 3.32 | 3.56 | 3.61 | 3.72 | 3.26 | 3.79 | 3.75 | 3.61 | 4.10 | 3.66 | 41.98 |
| Days ≥ 0.1" Precip | 7 | 6 | 7 | 7 | 8 | 8 | 7 | 6 | 6 | 7 | 8 | 7 | 84 |
| Total Snowfall (") | 19.3 | 16.8 | 11.8 | 4.8 | 0.1 | 0.0 | 0.0 | 0.0 | 0.0 | 0.5 | 4.7 | 16.7 | 74.7 |
| Days ≥ 1" Snow Depth | 29 | 27 | 24 | 6 | 0 | 0 | 0 | 0 | 0 | 0 | 5 | 22 | 113 |

## EASTPORT *Washington County* ELEVATION 75 ft LAT/LONG 44° 54' N / 66° 59' W

| | JAN | FEB | MAR | APR | MAY | JUN | JUL | AUG | SEP | OCT | NOV | DEC | YEAR |
|---|---|---|---|---|---|---|---|---|---|---|---|---|---|
| Maximum Temp °F | 29.5 | 30.6 | 38.3 | 48.7 | 59.3 | 68.1 | 73.5 | 73.1 | 65.5 | 55.5 | 45.2 | 34.7 | 51.8 |
| Minimum Temp °F | 13.6 | 15.0 | 23.8 | 32.9 | 40.9 | 47.9 | 53.2 | 53.9 | 48.5 | 40.6 | 32.3 | 20.2 | 35.2 |
| Mean Temp °F | 21.6 | 22.8 | 31.1 | 40.8 | 50.1 | 58.0 | 63.4 | 63.5 | 57.0 | 48.1 | 38.8 | 27.4 | 43.6 |
| Days Max Temp ≥ 90 °F | 0 | 0 | 0 | 0 | 0 | 0 | 0 | 0 | 0 | 0 | 0 | 0 | 0 |
| Days Max Temp ≤ 32 °F | 17 | 16 | 6 | 0 | 0 | 0 | 0 | 0 | 0 | 0 | 2 | 12 | 53 |
| Days Min Temp ≤ 32 °F | 29 | 27 | 26 | 14 | 1 | 0 | 0 | 0 | 0 | 4 | 16 | 26 | 143 |
| Days Min Temp ≤ 0 °F | 5 | 3 | 0 | 0 | 0 | 0 | 0 | 0 | 0 | 0 | 0 | 1 | 9 |
| Heating Degree Days | 1340 | 1185 | 1043 | 718 | 455 | 212 | 78 | 75 | 237 | 518 | 779 | 1156 | 7796 |
| Cooling Degree Days | 0 | 0 | 0 | 0 | 2 | 10 | 37 | 38 | 5 | 0 | 0 | *0* | 92 |
| Total Precipitation (") | 3.76 | 3.11 | 3.71 | 3.39 | 3.84 | 3.17 | 2.97 | 3.15 | 3.58 | 3.79 | 4.57 | 4.62 | 43.66 |
| Days ≥ 0.1" Precip | 7 | 6 | 7 | 7 | 7 | 6 | 6 | 5 | 6 | 7 | 8 | 9 | 81 |
| Total Snowfall (") | 15.2 | 15.5 | 12.3 | 4.0 | 0.0 | 0.0 | 0.0 | 0.0 | 0.0 | 0.1 | 1.9 | 12.3 | 61.3 |
| Days ≥ 1" Snow Depth | *20* | 21 | 17 | 3 | 0 | 0 | 0 | 0 | 0 | 0 | 1 | *12* | 74 |

## ELLSWORTH *Hancock County* ELEVATION 20 ft LAT/LONG 44° 32' N / 68° 26' W

| | JAN | FEB | MAR | APR | MAY | JUN | JUL | AUG | SEP | OCT | NOV | DEC | YEAR |
|---|---|---|---|---|---|---|---|---|---|---|---|---|---|
| Maximum Temp °F | 29.9 | 32.3 | 40.6 | 51.9 | 63.7 | 72.8 | 78.3 | 77.5 | *68.4* | 57.4 | 45.8 | 34.1 | 54.4 |
| Minimum Temp °F | 10.4 | 11.8 | 22.6 | 32.4 | 42.1 | 51.1 | 57.1 | 56.6 | *48.6* | 39.3 | 30.5 | 16.9 | 35.0 |
| Mean Temp °F | 20.2 | 22.1 | 31.6 | 42.2 | 52.9 | 62.0 | 67.7 | 67.1 | *58.5* | 48.4 | 38.2 | 25.6 | 44.7 |
| Days Max Temp ≥ 90 °F | 0 | 0 | 0 | 0 | 0 | 1 | 1 | 1 | 0 | 0 | 0 | 0 | 3 |
| Days Max Temp ≤ 32 °F | 17 | 14 | 4 | 0 | 0 | 0 | 0 | 0 | 0 | 0 | 1 | 13 | 49 |
| Days Min Temp ≤ 32 °F | 30 | 27 | 26 | 15 | 2 | 0 | 0 | 0 | 0 | 7 | 19 | 28 | 154 |
| Days Min Temp ≤ 0 °F | 7 | 6 | 1 | 0 | 0 | 0 | 0 | 0 | 0 | 0 | 0 | 3 | 17 |
| Heating Degree Days | 1383 | 1204 | 1029 | 679 | 370 | 116 | 20 | 33 | *199* | 509 | 798 | 1216 | 7556 |
| Cooling Degree Days | 0 | 0 | 0 | 0 | 3 | 31 | 120 | 104 | *10* | 0 | 0 | 0 | 268 |
| Total Precipitation (") | 3.65 | 3.26 | 3.78 | 3.93 | 3.67 | 3.51 | 3.24 | 3.16 | 4.03 | 3.97 | 5.04 | 4.69 | 45.93 |
| Days ≥ 0.1" Precip | 7 | 6 | 7 | 7 | 7 | 7 | 6 | 5 | *6* | 7 | 9 | 8 | 82 |
| Total Snowfall (") | na | na | na | na | 0.1 | 0.0 | 0.0 | 0.0 | *0.0* | 0.2 | na | na | na |
| Days ≥ 1" Snow Depth | na | na | na | na | 0 | 0 | 0 | 0 | *0* | 0 | na | na | na |

## FARMINGTON *Franklin County* ELEVATION 390 ft LAT/LONG 44° 40' N / 70° 9' W

| | JAN | FEB | MAR | APR | MAY | JUN | JUL | AUG | SEP | OCT | NOV | DEC | YEAR |
|---|---|---|---|---|---|---|---|---|---|---|---|---|---|
| Maximum Temp °F | 25.5 | 29.0 | 38.2 | 51.0 | 64.9 | 73.6 | 78.9 | 77.1 | 68.4 | 56.9 | 43.2 | 30.5 | 53.1 |
| Minimum Temp °F | 0.3 | 2.8 | 15.5 | 28.5 | 38.4 | 48.0 | 53.3 | 51.3 | 41.7 | 32.0 | 23.8 | 9.4 | 28.8 |
| Mean Temp °F | 13.0 | 16.0 | 26.9 | 39.8 | 51.7 | 60.9 | 66.1 | 64.2 | 55.1 | 44.5 | 33.6 | 20.0 | 41.0 |
| Days Max Temp ≥ 90 °F | 0 | 0 | 0 | 0 | 0 | 1 | 1 | 1 | 0 | 0 | 0 | 0 | 3 |
| Days Max Temp ≤ 32 °F | 23 | 18 | 8 | 1 | 0 | 0 | 0 | 0 | 0 | 0 | 4 | 17 | 71 |
| Days Min Temp ≤ 32 °F | 31 | 28 | 29 | 22 | 8 | 1 | 0 | 0 | 5 | 17 | 25 | 30 | 196 |
| Days Min Temp ≤ 0 °F | 16 | 13 | 4 | 0 | 0 | 0 | 0 | 0 | 0 | 0 | 0 | 8 | 41 |
| Heating Degree Days | 1610 | 1379 | 1175 | 750 | 412 | 156 | 54 | 89 | 303 | 630 | 936 | 1390 | 8884 |
| Cooling Degree Days | 0 | 0 | 0 | 0 | 5 | 36 | 96 | 73 | 9 | 1 | 0 | 0 | 220 |
| Total Precipitation (") | 3.19 | 3.07 | 3.71 | 3.91 | 3.72 | 4.32 | 3.32 | 4.14 | 3.71 | 3.75 | 4.37 | 4.37 | 45.58 |
| Days ≥ 0.1" Precip | 6 | 5 | 7 | 7 | 8 | 9 | 8 | 8 | 6 | 7 | 8 | 8 | 87 |
| Total Snowfall (") | 21.8 | 18.5 | 14.1 | 6.9 | 0.1 | 0.0 | 0.0 | 0.0 | 0.0 | 0.7 | 7.0 | 20.2 | 89.3 |
| Days ≥ 1" Snow Depth | 29 | 26 | 25 | 7 | 0 | 0 | 0 | 0 | 0 | 0 | 5 | 22 | 114 |

## FORT KENT *Aroostook County* ELEVATION 522 ft LAT/LONG 47° 15 ' N / 68° 36 ' W

| | JAN | FEB | MAR | APR | MAY | JUN | JUL | AUG | SEP | OCT | NOV | DEC | YEAR |
|---|---|---|---|---|---|---|---|---|---|---|---|---|---|
| Maximum Temp °F | 19.1 | 23.3 | 34.4 | 46.4 | 61.9 | 72.1 | 76.2 | 74.2 | 64.3 | 52.0 | 38.0 | 24.5 | 48.9 |
| Minimum Temp °F | -5.4 | -4.0 | 9.8 | 25.9 | 37.7 | 48.0 | 52.8 | 51.0 | 41.6 | 32.6 | 22.2 | 4.5 | 26.4 |
| Mean Temp °F | 6.9 | 9.7 | 22.1 | 36.2 | 49.8 | 60.1 | 64.5 | 62.6 | 53.0 | 42.3 | 30.1 | 14.5 | 37.7 |
| Days Max Temp ≥ 90 °F | 0 | 0 | 0 | 0 | 0 | 0 | 1 | 0 | 0 | 0 | 0 | 0 | 1 |
| Days Max Temp ≤ 32 °F | 26 | 22 | 12 | 2 | 0 | 0 | 0 | 0 | 0 | 0 | 9 | 23 | 94 |
| Days Min Temp ≤ 32 °F | 31 | 28 | 29 | 24 | 8 | 1 | 0 | 0 | 4 | 16 | 26 | 31 | 198 |
| Days Min Temp ≤ 0 °F | 21 | 18 | 9 | 0 | 0 | 0 | 0 | 0 | 0 | 0 | 1 | 13 | 62 |
| Heating Degree Days | 1799 | 1559 | 1324 | 858 | 471 | 180 | 80 | 121 | 362 | 697 | 1040 | 1559 | 10050 |
| Cooling Degree Days | 0 | 0 | 0 | 0 | 4 | 29 | 69 | 51 | 7 | 0 | 0 | 0 | 160 |
| Total Precipitation (") | 2.33 | 1.79 | 2.22 | 2.57 | 3.18 | 3.43 | 3.62 | 4.15 | 3.53 | 3.23 | 3.11 | 2.91 | 36.07 |
| Days ≥ 0.1" Precip | 6 | 4 | 6 | 7 | 8 | 8 | 9 | 9 | 7 | 7 | 7 | 7 | 85 |
| Total Snowfall (") | 22.7 | 18.2 | 15.5 | 7.8 | 0.7 | 0.0 | 0.0 | 0.0 | 0.0 | 0.6 | 10.3 | 21.6 | 97.4 |
| Days ≥ 1" Snow Depth | 31 | 28 | ***30*** | ***15*** | 0 | 0 | 0 | 0 | 0 | 0 | 10 | 27 | 141 |

## GARDINER *Kennebec County* ELEVATION 141 ft LAT/LONG 44° 13 ' N / 69° 47 ' W

| | JAN | FEB | MAR | APR | MAY | JUN | JUL | AUG | SEP | OCT | NOV | DEC | YEAR |
|---|---|---|---|---|---|---|---|---|---|---|---|---|---|
| Maximum Temp °F | 29.2 | 32.4 | 41.1 | 53.4 | 66.0 | 74.8 | 80.1 | 78.8 | 70.0 | 58.6 | 46.0 | 33.8 | 55.4 |
| Minimum Temp °F | 5.8 | 8.2 | 20.5 | 31.6 | 42.3 | 51.4 | 57.0 | 55.8 | 47.0 | 36.6 | 28.2 | 13.9 | 33.2 |
| Mean Temp °F | 17.5 | 20.4 | 30.7 | 42.4 | 54.2 | 63.1 | 68.6 | 67.3 | 58.5 | 47.6 | 37.1 | 23.8 | 44.3 |
| Days Max Temp ≥ 90 °F | 0 | 0 | 0 | 0 | 0 | 1 | 2 | 1 | 0 | 0 | 0 | 0 | 4 |
| Days Max Temp ≤ 32 °F | 18 | 14 | 5 | 0 | 0 | 0 | 0 | 0 | 0 | 0 | 1 | 13 | 51 |
| Days Min Temp ≤ 32 °F | 31 | 27 | 28 | 17 | 3 | 0 | 0 | 0 | 1 | 11 | 21 | 30 | 169 |
| Days Min Temp ≤ 0 °F | 11 | 9 | 2 | 0 | 0 | 0 | 0 | 0 | 0 | 0 | 0 | 5 | 27 |
| Heating Degree Days | 1468 | 1250 | 1057 | 671 | 334 | 104 | 20 | 39 | 207 | 532 | 830 | 1268 | 7780 |
| Cooling Degree Days | 0 | 0 | 0 | 0 | 8 | 54 | 145 | 117 | 20 | 1 | 0 | 0 | 345 |
| Total Precipitation (") | 3.18 | 2.83 | 3.67 | 3.94 | 3.64 | 3.64 | 2.90 | 3.40 | 3.32 | 3.68 | 4.51 | 3.90 | 42.61 |
| Days ≥ 0.1" Precip | 6 | 6 | 6 | 7 | 7 | 7 | 6 | 6 | 6 | 6 | 7 | 6 | 76 |
| Total Snowfall (") | ***19.2*** | 16.4 | 10.7 | 3.2 | 0.1 | 0.0 | 0.0 | 0.0 | 0.0 | 0.3 | ***3.3*** | ***14.8*** | 68.0 |
| Days ≥ 1" Snow Depth | na | na | ***16*** | na | 0 | 0 | 0 | 0 | ***0*** | 0 | na | na | na |

## GRAND LAKE STREAM *Washington County* ELEVATION 89 ft LAT/LONG 45° 11 ' N / 67° 48 ' W

| | JAN | FEB | MAR | APR | MAY | JUN | JUL | AUG | SEP | OCT | NOV | DEC | YEAR |
|---|---|---|---|---|---|---|---|---|---|---|---|---|---|
| Maximum Temp °F | 25.9 | 29.0 | 38.5 | 50.2 | 63.3 | 73.0 | 78.5 | 77.0 | 68.0 | 56.2 | 43.9 | 31.1 | 52.9 |
| Minimum Temp °F | 3.7 | 5.8 | 17.4 | 29.3 | 39.6 | 49.6 | 55.4 | 54.2 | 45.0 | 34.6 | 26.0 | 11.9 | 31.0 |
| Mean Temp °F | 14.8 | 17.5 | 28.0 | 39.8 | 51.5 | 61.3 | 66.9 | 65.6 | 56.5 | 45.4 | 35.0 | 21.5 | 42.0 |
| Days Max Temp ≥ 90 °F | 0 | 0 | 0 | 0 | 0 | 1 | 1 | 1 | 0 | 0 | 0 | 0 | 3 |
| Days Max Temp ≤ 32 °F | 22 | 18 | 8 | 1 | 0 | 0 | 0 | 0 | 0 | 0 | 3 | 17 | 69 |
| Days Min Temp ≤ 32 °F | 31 | 28 | 29 | 21 | 4 | 0 | 0 | 0 | 1 | 14 | 24 | 30 | 182 |
| Days Min Temp ≤ 0 °F | 14 | 10 | 3 | 0 | 0 | 0 | 0 | 0 | 0 | 0 | 0 | 6 | 33 |
| Heating Degree Days | 1550 | 1336 | 1139 | 750 | 419 | 141 | 38 | 60 | 259 | 601 | 894 | 1342 | 8529 |
| Cooling Degree Days | 0 | 0 | 0 | 0 | 4 | 37 | 108 | 87 | 10 | 0 | 0 | 0 | 246 |
| Total Precipitation (") | 3.65 | 3.18 | 3.65 | 3.58 | 3.40 | 3.34 | 3.15 | 3.35 | 3.80 | 3.50 | 4.64 | 4.38 | 43.62 |
| Days ≥ 0.1" Precip | ***6*** | 6 | 7 | 7 | 7 | 7 | 6 | 6 | 6 | 7 | 9 | 8 | 82 |
| Total Snowfall (") | na | na | na | na | 0.0 | 0.0 | 0.0 | 0.0 | 0.0 | ***0.0*** | na | na | na |
| Days ≥ 1" Snow Depth | na | na | na | na | 0 | 0 | 0 | 0 | 0 | ***0*** | na | na | na |

## HOULTON INTL AP *Aroostook County* ELEVATION 466 ft LAT/LONG 46° 8 ' N / 67° 48 ' W

| | JAN | FEB | MAR | APR | MAY | JUN | JUL | AUG | SEP | OCT | NOV | DEC | YEAR |
|---|---|---|---|---|---|---|---|---|---|---|---|---|---|
| Maximum Temp °F | 22.1 | 25.8 | 36.1 | 48.8 | 63.7 | 73.7 | 78.7 | 76.3 | 65.8 | 53.8 | 40.3 | 27.0 | 51.0 |
| Minimum Temp °F | 0.0 | 1.8 | 14.6 | 27.3 | 38.4 | 48.4 | 53.8 | 52.0 | 42.0 | 32.9 | 23.5 | 7.9 | 28.6 |
| Mean Temp °F | 11.1 | 13.8 | 25.4 | 38.1 | 51.0 | 61.0 | 66.3 | 64.2 | 53.9 | 43.4 | 31.9 | 17.5 | 39.8 |
| Days Max Temp ≥ 90 °F | 0 | 0 | 0 | 0 | 0 | 1 | 2 | 1 | 0 | 0 | 0 | 0 | 4 |
| Days Max Temp ≤ 32 °F | 25 | 20 | 10 | 1 | 0 | 0 | 0 | 0 | 0 | 0 | 7 | 21 | 84 |
| Days Min Temp ≤ 32 °F | 30 | 28 | 28 | 23 | 9 | 1 | 0 | 0 | 5 | 16 | 24 | 30 | 194 |
| Days Min Temp ≤ 0 °F | 16 | 14 | 5 | 0 | 0 | 0 | 0 | 0 | 0 | 0 | 1 | 10 | 46 |
| Heating Degree Days | 1666 | 1441 | 1221 | 801 | 432 | 160 | 55 | 91 | 336 | 665 | 985 | 1467 | 9320 |
| Cooling Degree Days | ***0*** | 0 | 0 | 0 | ***4*** | ***41*** | ***90*** | ***75*** | ***10*** | ***0*** | ***0*** | 0 | 220 |
| Total Precipitation (") | 2.80 | 2.27 | 2.50 | 2.72 | 3.19 | 3.63 | 3.52 | 3.94 | 3.52 | 3.42 | 3.93 | 3.49 | 38.93 |
| Days ≥ 0.1" Precip | 7 | 6 | 7 | 6 | 8 | 8 | 7 | 7 | 6 | 7 | 8 | 8 | 85 |
| Total Snowfall (") | 22.2 | 17.5 | 16.2 | 8.0 | 0.7 | 0.0 | 0.0 | 0.0 | 0.0 | 0.6 | 9.4 | 22.3 | 96.9 |
| Days ≥ 1" Snow Depth | 30 | 27 | 29 | 11 | 0 | 0 | 0 | 0 | 0 | 0 | 8 | 26 | 131 |

## JACKMAN *Somerset County* ELEVATION 1181 ft LAT/LONG 45° 38 ' N / 70° 16 ' W

| | JAN | FEB | MAR | APR | MAY | JUN | JUL | AUG | SEP | OCT | NOV | DEC | YEAR |
|---|---|---|---|---|---|---|---|---|---|---|---|---|---|
| Maximum Temp °F | 21.1 | 24.4 | 34.7 | 47.1 | 62.0 | 71.4 | 76.3 | 74.5 | 65.5 | 53.1 | 39.3 | 26.1 | 49.6 |
| Minimum Temp °F | -2.4 | -1.0 | 10.3 | 24.9 | 36.2 | 46.8 | 51.7 | 50.0 | 40.6 | 31.3 | 21.9 | 6.2 | 26.4 |
| Mean Temp °F | 9.4 | 11.7 | 22.5 | 36.0 | 49.1 | 59.2 | 64.1 | 62.2 | 53.1 | 42.2 | 30.6 | 16.1 | 38.0 |
| Days Max Temp ≥ 90 °F | 0 | 0 | 0 | 0 | 0 | 0 | 1 | 0 | 0 | 0 | 0 | 0 | 1 |
| Days Max Temp ≤ 32 °F | 26 | 21 | 12 | 2 | 0 | 0 | 0 | 0 | 0 | 0 | 7 | 22 | 90 |
| Days Min Temp ≤ 32 °F | 31 | 28 | 30 | 25 | 11 | 1 | 0 | 0 | 5 | 18 | 26 | 31 | 206 |
| Days Min Temp ≤ 0 °F | 19 | 17 | 9 | 0 | 0 | 0 | 0 | 0 | 0 | 0 | 1 | 11 | 57 |
| Heating Degree Days | 1720 | 1499 | 1311 | 862 | 488 | 192 | 86 | 121 | 358 | 699 | 1026 | 1509 | 9871 |
| Cooling Degree Days | 0 | 0 | 0 | 0 | 2 | 21 | 60 | 40 | 5 | 0 | 0 | 0 | 128 |
| Total Precipitation (") | 2.55 | 2.10 | 2.44 | 2.95 | 3.19 | 3.86 | 3.75 | 3.87 | 3.62 | 3.17 | 3.52 | 3.26 | 38.28 |
| Days ≥ 0.1" Precip | 6 | 5 | 7 | 7 | 8 | 9 | 8 | 8 | 8 | 7 | 8 | 8 | 89 |
| Total Snowfall (") | 24.8 | 20.3 | *19.1* | 9.6 | 0.2 | 0.0 | 0.0 | 0.0 | 0.0 | 0.8 | *9.8* | 24.7 | 109.3 |
| Days ≥ 1" Snow Depth | na | na | na | *12* | 0 | 0 | 0 | 0 | 0 | 1 | *9* | *27* | na |

## JONESBORO *Washington County* ELEVATION 180 ft LAT/LONG 44° 39 ' N / 67° 39 ' W

| | JAN | FEB | MAR | APR | MAY | JUN | JUL | AUG | SEP | OCT | NOV | DEC | YEAR |
|---|---|---|---|---|---|---|---|---|---|---|---|---|---|
| Maximum Temp °F | 28.4 | 30.5 | 38.7 | 49.8 | 61.0 | 69.9 | 75.6 | 75.2 | 67.5 | 56.6 | 45.8 | 33.9 | 52.7 |
| Minimum Temp °F | 7.5 | 9.4 | 20.3 | 30.9 | 39.9 | 48.6 | 53.9 | 53.4 | 45.5 | 36.3 | 28.0 | 14.4 | 32.3 |
| Mean Temp °F | 18.0 | 20.0 | 29.5 | 40.4 | 50.5 | 59.3 | 64.8 | 64.4 | 56.5 | 46.5 | 36.9 | 24.2 | 42.6 |
| Days Max Temp ≥ 90 °F | 0 | 0 | 0 | 0 | 0 | 0 | 1 | 1 | 0 | 0 | 0 | 0 | 2 |
| Days Max Temp ≤ 32 °F | 19 | 16 | 6 | 0 | 0 | 0 | 0 | 0 | 0 | 0 | 2 | 13 | 56 |
| Days Min Temp ≤ 32 °F | 30 | 27 | 28 | 18 | 4 | 0 | 0 | 0 | 1 | 11 | 22 | 29 | 170 |
| Days Min Temp ≤ 0 °F | 10 | 7 | 1 | 0 | 0 | 0 | 0 | 0 | 0 | 0 | 0 | 4 | 22 |
| Heating Degree Days | 1452 | 1265 | 1093 | 732 | 444 | 182 | 57 | 66 | 255 | 568 | 836 | 1259 | 8209 |
| Cooling Degree Days | 0 | 0 | 0 | 0 | 2 | 18 | 59 | 54 | 6 | 0 | 0 | 0 | 139 |
| Total Precipitation (") | 4.49 | 3.59 | 4.30 | 4.43 | 4.47 | 3.64 | 3.37 | 3.21 | 3.99 | 4.32 | 5.18 | 5.33 | 50.32 |
| Days ≥ 0.1" Precip | 7 | 6 | 7 | 7 | 8 | 7 | 6 | 6 | 6 | 7 | 8 | 9 | 84 |
| Total Snowfall (") | 17.7 | 17.2 | 12.6 | 5.2 | 0.1 | 0.0 | 0.0 | 0.0 | 0.0 | 0.2 | 3.2 | 13.3 | 69.5 |
| Days ≥ 1" Snow Depth | 23 | *22* | 20 | *5* | 0 | 0 | 0 | 0 | 0 | 0 | 3 | 15 | 88 |

## LEWISTON *Androscoggin County* ELEVATION 180 ft LAT/LONG 44° 6 ' N / 70° 13 ' W

| | JAN | FEB | MAR | APR | MAY | JUN | JUL | AUG | SEP | OCT | NOV | DEC | YEAR |
|---|---|---|---|---|---|---|---|---|---|---|---|---|---|
| Maximum Temp °F | 29.1 | 32.3 | 41.2 | 53.5 | 66.1 | 75.5 | 81.0 | 79.1 | 69.9 | 58.6 | 45.6 | 33.6 | 55.5 |
| Minimum Temp °F | 11.0 | 13.3 | 24.1 | 34.5 | 45.2 | 54.9 | 60.9 | 59.6 | 51.1 | 40.5 | 31.4 | 18.7 | 37.1 |
| Mean Temp °F | 20.1 | 22.8 | 32.7 | 44.0 | 55.7 | 65.2 | 71.0 | 69.4 | 60.5 | 49.6 | 38.5 | 26.2 | 46.3 |
| Days Max Temp ≥ 90 °F | 0 | 0 | 0 | 0 | 0 | 2 | 3 | 2 | 0 | 0 | 0 | 0 | 7 |
| Days Max Temp ≤ 32 °F | 19 | 15 | 5 | 0 | 0 | 0 | 0 | 0 | 0 | 0 | 2 | 14 | 55 |
| Days Min Temp ≤ 32 °F | 30 | 27 | 26 | 12 | 1 | 0 | 0 | 0 | 0 | 5 | 18 | 29 | 148 |
| Days Min Temp ≤ 0 °F | 6 | 4 | 1 | 0 | 0 | 0 | 0 | 0 | 0 | 0 | 0 | 2 | 13 |
| Heating Degree Days | 1386 | 1184 | 995 | 624 | 297 | 70 | 8 | 21 | 162 | 472 | 789 | 1196 | 7204 |
| Cooling Degree Days | 0 | 0 | 0 | 0 | 15 | 86 | 212 | 168 | 38 | 1 | 0 | 0 | 520 |
| Total Precipitation (") | 3.50 | 3.33 | 4.38 | 4.03 | 3.62 | 3.81 | 3.37 | 3.26 | 3.27 | 3.71 | 4.62 | 4.47 | 45.37 |
| Days ≥ 0.1" Precip | 7 | 6 | 7 | 7 | 7 | 7 | 7 | 6 | 6 | 7 | 7 | 8 | 82 |
| Total Snowfall (") | 20.8 | 17.3 | 11.7 | 3.7 | 0.1 | 0.0 | 0.0 | 0.0 | 0.0 | 0.3 | 3.5 | 16.8 | 74.2 |
| Days ≥ 1" Snow Depth | na | na | na | na | 0 | 0 | 0 | 0 | 0 | 0 | na | na | na |

## LONG FALLS DAM *Somerset County* ELEVATION 1161 ft LAT/LONG 45° 13 ' N / 70° 12 ' W

| | JAN | FEB | MAR | APR | MAY | JUN | JUL | AUG | SEP | OCT | NOV | DEC | YEAR |
|---|---|---|---|---|---|---|---|---|---|---|---|---|---|
| Maximum Temp °F | 22.1 | 25.0 | 34.3 | 46.6 | 60.7 | 70.1 | 75.1 | 73.4 | 64.3 | 52.8 | 39.9 | 27.2 | 49.3 |
| Minimum Temp °F | 0.7 | 2.4 | 13.0 | 26.6 | 37.5 | 47.9 | 52.8 | 51.2 | 42.6 | 33.2 | 23.7 | 9.2 | 28.4 |
| Mean Temp °F | 11.4 | 13.7 | 23.7 | 36.6 | 49.1 | 59.0 | 64.0 | 62.3 | 53.5 | 43.0 | 31.8 | 18.2 | 38.9 |
| Days Max Temp ≥ 90 °F | 0 | 0 | 0 | 0 | 0 | 0 | 0 | 0 | 0 | 0 | 0 | 0 | 0 |
| Days Max Temp ≤ 32 °F | 25 | 21 | 13 | 2 | 0 | 0 | 0 | 0 | 0 | 0 | 7 | 21 | 89 |
| Days Min Temp ≤ 32 °F | 31 | 28 | 30 | 24 | 9 | 0 | 0 | 0 | 3 | 16 | 25 | 30 | 196 |
| Days Min Temp ≤ 0 °F | 16 | 14 | 6 | 0 | 0 | 0 | 0 | 0 | 0 | 0 | 0 | 8 | 44 |
| Heating Degree Days | 1657 | 1444 | 1274 | 846 | 490 | 195 | 84 | 120 | 347 | 675 | 989 | 1445 | 9566 |
| Cooling Degree Days | 0 | 0 | 0 | 0 | 4 | 25 | 65 | 50 | 6 | 0 | 0 | 0 | 150 |
| Total Precipitation (") | 2.61 | 2.32 | 3.04 | 3.18 | 3.33 | 3.55 | 3.39 | 3.54 | 3.23 | 3.14 | 3.81 | 3.34 | 38.48 |
| Days ≥ 0.1" Precip | 6 | 5 | 7 | 7 | 7 | 8 | 8 | 8 | 6 | 6 | 7 | 7 | 82 |
| Total Snowfall (") | 24.3 | 22.4 | 19.0 | 10.1 | 0.3 | 0.0 | 0.0 | 0.0 | 0.0 | 1.2 | 9.3 | 25.0 | 111.6 |
| Days ≥ 1" Snow Depth | na | na | na | na | 0 | 0 | 0 | 0 | 0 | 0 | na | na | na |

## MADISON *Somerset County* ELEVATION 259 ft LAT/LONG 44° 48' N / 69° 53' W

| | JAN | FEB | MAR | APR | MAY | JUN | JUL | AUG | SEP | OCT | NOV | DEC | YEAR |
|---|---|---|---|---|---|---|---|---|---|---|---|---|---|
| Maximum Temp °F | 26.6 | 30.1 | 39.1 | 51.4 | 64.6 | 73.7 | 79.2 | 77.5 | 68.4 | 56.8 | 43.8 | 31.2 | 53.5 |
| Minimum Temp °F | 2.1 | 4.2 | 17.0 | 30.3 | 40.4 | 50.2 | 55.6 | 53.5 | 44.6 | 34.6 | 26.1 | 11.3 | 30.8 |
| Mean Temp °F | 14.4 | 17.2 | 28.1 | 40.9 | 52.5 | 62.0 | 67.4 | 65.5 | 56.6 | 45.7 | 35.0 | 21.3 | 42.2 |
| Days Max Temp ≥ 90 °F | 0 | 0 | 0 | 0 | 0 | 1 | 2 | 1 | 0 | 0 | 0 | 0 | 4 |
| Days Max Temp ≤ 32 °F | 22 | 17 | 7 | 1 | 0 | 0 | 0 | 0 | 0 | 0 | 3 | 16 | 66 |
| Days Min Temp ≤ 32 °F | 31 | 28 | 28 | 19 | 4 | 0 | 0 | 0 | 1 | 14 | 23 | 30 | 178 |
| Days Min Temp ≤ 0 °F | 14 | 12 | 3 | 0 | 0 | 0 | 0 | 0 | 0 | 0 | 0 | 7 | 36 |
| Heating Degree Days | 1563 | 1345 | 1138 | 719 | 385 | 129 | 34 | 64 | 259 | 592 | 893 | 1349 | 8470 |
| Cooling Degree Days | 0 | 0 | 0 | 0 | 5 | 45 | 120 | 85 | 10 | 0 | 0 | 0 | 265 |
| Total Precipitation (") | 2.84 | 2.56 | 3.06 | 3.19 | 3.42 | 3.48 | 3.01 | 3.34 | 3.23 | 3.29 | 3.72 | 3.63 | 38.77 |
| Days ≥ 0.1" Precip | 6 | 6 | 6 | 6 | 7 | 8 | 7 | 7 | 6 | 7 | 7 | 8 | 81 |
| Total Snowfall (") | 21.3 | 17.6 | 12.4 | 6.2 | 0.1 | 0.0 | 0.0 | 0.0 | 0.0 | 0.2 | 5.2 | 19.4 | 82.4 |
| Days ≥ 1" Snow Depth | na | na | na | na | ***0*** | 0 | 0 | 0 | 0 | 0 | na | na | na |

## MIDDLE DAM *Oxford County* ELEVATION 1460 ft LAT/LONG 44° 47' N / 70° 55' W

| | JAN | FEB | MAR | APR | MAY | JUN | JUL | AUG | SEP | OCT | NOV | DEC | YEAR |
|---|---|---|---|---|---|---|---|---|---|---|---|---|---|
| Maximum Temp °F | 22.8 | 25.5 | 35.3 | 47.0 | 61.5 | 70.6 | 75.5 | 73.7 | 64.9 | 53.1 | 39.9 | 27.2 | 49.8 |
| Minimum Temp °F | -1.0 | -0.5 | 11.2 | 25.3 | 36.5 | 46.8 | 51.7 | 50.5 | 42.2 | 32.2 | 22.8 | 6.9 | 27.1 |
| Mean Temp °F | 10.8 | 12.6 | 23.3 | 36.2 | 49.0 | 58.7 | 63.6 | 62.1 | 53.6 | 42.6 | 31.4 | 17.0 | 38.4 |
| Days Max Temp ≥ 90 °F | 0 | 0 | 0 | 0 | 0 | 0 | 0 | 0 | 0 | 0 | 0 | 0 | 0 |
| Days Max Temp ≤ 32 °F | 24 | 21 | 12 | 2 | 0 | 0 | 0 | 0 | 0 | 0 | 7 | 21 | 87 |
| Days Min Temp ≤ 32 °F | 30 | 28 | 30 | 25 | 10 | 1 | 0 | 0 | 3 | 17 | 25 | 30 | 199 |
| Days Min Temp ≤ 0 °F | 17 | 17 | 8 | 0 | 0 | 0 | 0 | 0 | 0 | 0 | 0 | ***10*** | 52 |
| Heating Degree Days | 1677 | 1475 | 1287 | 859 | 494 | 202 | 91 | 125 | 343 | 687 | 1003 | 1483 | 9726 |
| Cooling Degree Days | 0 | 0 | 0 | 0 | 4 | 25 | 64 | 47 | 5 | 0 | 0 | 0 | 145 |
| Total Precipitation (") | 2.26 | 1.91 | 2.45 | 2.61 | 3.15 | 3.86 | 3.55 | 4.04 | 3.33 | 3.05 | 3.13 | 2.81 | 36.15 |
| Days ≥ 0.1" Precip | 5 | 5 | 6 | 7 | 7 | 9 | 7 | 8 | 7 | 7 | 8 | 7 | 83 |
| Total Snowfall (") | na | na | na | na | ***0.3*** | 0.0 | 0.0 | 0.0 | 0.0 | 0.5 | na | na | na |
| Days ≥ 1" Snow Depth | na | na | na | na | ***0*** | 0 | 0 | 0 | 0 | 0 | na | na | na |

## MILLINOCKET *Penobscot County* ELEVATION 390 ft LAT/LONG 45° 39' N / 68° 42' W

| | JAN | FEB | MAR | APR | MAY | JUN | JUL | AUG | SEP | OCT | NOV | DEC | YEAR |
|---|---|---|---|---|---|---|---|---|---|---|---|---|---|
| Maximum Temp °F | 23.2 | 26.7 | 36.9 | 49.4 | 63.6 | 73.2 | 78.4 | 76.5 | 66.9 | 54.5 | 41.5 | 28.4 | 51.6 |
| Minimum Temp °F | 1.9 | 3.9 | 16.3 | 29.7 | 40.8 | 51.5 | 57.0 | 55.2 | 45.5 | 35.0 | 26.0 | 10.5 | 31.1 |
| Mean Temp °F | 12.6 | 15.3 | 26.6 | 39.6 | 52.2 | 62.3 | 67.8 | 65.9 | 56.2 | 44.8 | 33.8 | 19.5 | 41.4 |
| Days Max Temp ≥ 90 °F | 0 | 0 | 0 | 0 | 0 | 1 | 2 | 1 | 0 | 0 | 0 | 0 | 4 |
| Days Max Temp ≤ 32 °F | 24 | 20 | 10 | 1 | 0 | 0 | 0 | 0 | 0 | 0 | 5 | 19 | 79 |
| Days Min Temp ≤ 32 °F | 31 | 28 | 29 | 20 | 4 | 0 | 0 | 0 | 1 | 13 | 23 | 30 | 179 |
| Days Min Temp ≤ 0 °F | 15 | 12 | 3 | 0 | 0 | 0 | 0 | 0 | 0 | 0 | 0 | 8 | 38 |
| Heating Degree Days | 1620 | 1397 | 1182 | 755 | 397 | 126 | 34 | 61 | 272 | 621 | 930 | 1405 | 8800 |
| Cooling Degree Days | 0 | 0 | 0 | 0 | 7 | 52 | 126 | 100 | 13 | 0 | 0 | 0 | 298 |
| Total Precipitation (") | 2.88 | 2.48 | 2.85 | 3.35 | 3.51 | 3.85 | 3.63 | 4.01 | 3.72 | 3.63 | 4.12 | 3.60 | 41.63 |
| Days ≥ 0.1" Precip | 6 | 5 | 6 | 7 | 8 | 8 | 8 | 8 | 7 | 7 | 8 | 7 | 85 |
| Total Snowfall (") | 22.7 | 19.1 | 15.9 | 6.9 | 0.2 | 0.0 | 0.0 | 0.0 | 0.0 | 0.3 | 7.3 | 21.1 | 93.5 |
| Days ≥ 1" Snow Depth | na | na | na | na | 0 | 0 | 0 | 0 | 0 | 0 | na | na | na |

## NEWCASTLE *Lincoln County* ELEVATION 190 ft LAT/LONG 44° 3' N / 69° 32' W

| | JAN | FEB | MAR | APR | MAY | JUN | JUL | AUG | SEP | OCT | NOV | DEC | YEAR |
|---|---|---|---|---|---|---|---|---|---|---|---|---|---|
| Maximum Temp °F | 29.5 | 32.5 | 41.1 | 52.9 | 64.8 | 73.1 | 78.5 | 76.9 | 67.7 | 57.0 | 45.4 | 33.9 | 54.4 |
| Minimum Temp °F | 12.4 | 14.4 | 24.0 | 33.7 | 43.4 | 52.5 | 58.4 | 57.6 | 49.8 | 40.0 | 31.2 | 18.4 | 36.3 |
| Mean Temp °F | 21.0 | 23.5 | 32.6 | 43.3 | 54.1 | 62.8 | 68.5 | 67.3 | 58.8 | 48.6 | 38.4 | 26.2 | 45.4 |
| Days Max Temp ≥ 90 °F | 0 | 0 | 0 | 0 | 0 | 0 | 1 | 0 | 0 | 0 | 0 | 0 | 1 |
| Days Max Temp ≤ 32 °F | 18 | 14 | 4 | 0 | 0 | 0 | 0 | 0 | 0 | 0 | 2 | 13 | 51 |
| Days Min Temp ≤ 32 °F | 30 | 26 | 25 | 14 | 1 | 0 | 0 | 0 | 0 | 6 | 18 | 28 | 148 |
| Days Min Temp ≤ 0 °F | 5 | 4 | 0 | 0 | 0 | 0 | 0 | 0 | 0 | 0 | 0 | 2 | 11 |
| Heating Degree Days | 1358 | 1166 | 998 | 645 | 336 | 104 | 17 | 33 | 200 | 504 | 793 | 1195 | 7349 |
| Cooling Degree Days | 0 | 0 | 0 | 0 | 7 | 41 | 129 | 103 | 17 | 0 | 0 | 0 | 297 |
| Total Precipitation (") | 3.88 | 3.27 | 4.33 | 4.22 | 3.78 | 3.50 | 2.75 | 3.03 | 3.60 | 3.89 | 4.98 | 5.05 | 46.28 |
| Days ≥ 0.1" Precip | 7 | 6 | 7 | 7 | 7 | 7 | 5 | 5 | 6 | 7 | 8 | 8 | 80 |
| Total Snowfall (") | 20.6 | 17.4 | 12.2 | 3.9 | 0.1 | 0.0 | 0.0 | 0.0 | 0.0 | 0.5 | 3.2 | 16.6 | 74.5 |
| Days ≥ 1" Snow Depth | 26 | 24 | 19 | 3 | 0 | 0 | 0 | 0 | 0 | 0 | 3 | 17 | 92 |

**WEATHER AMERICA:** The Latest Detailed Climatological Data for Over 4,000 Places — *With Rankings*

## ORONO *Penobscot County* ELEVATION 115 ft LAT/LONG 44° 54 ' N / 68° 40 ' W

| | JAN | FEB | MAR | APR | MAY | JUN | JUL | AUG | SEP | OCT | NOV | DEC | YEAR |
|---|---|---|---|---|---|---|---|---|---|---|---|---|---|
| Maximum Temp °F | 27.8 | 30.7 | 39.8 | 52.1 | 65.6 | 74.7 | 80.2 | 77.9 | 67.4 | 55.4 | 43.0 | 31.7 | 53.9 |
| Minimum Temp °F | 7.9 | 10.5 | 21.3 | 31.6 | 41.4 | 50.4 | 56.0 | 54.6 | 45.6 | 36.4 | 28.3 | 14.9 | 33.2 |
| Mean Temp °F | 17.9 | 20.6 | 30.6 | 41.9 | 53.5 | 62.6 | 68.1 | 66.3 | 56.5 | 45.9 | 35.7 | 23.3 | 43.6 |
| Days Max Temp ≥ 90 °F | 0 | 0 | 0 | 0 | 0 | 1 | 4 | 2 | 0 | 0 | 0 | 0 | 7 |
| Days Max Temp ≤ 32 °F | 21 | 17 | 6 | 0 | 0 | 0 | 0 | 0 | 0 | 0 | 3 | 17 | 64 |
| Days Min Temp ≤ 32 °F | 30 | 27 | 28 | 18 | 3 | 0 | 0 | 0 | 1 | 10 | 22 | 30 | 169 |
| Days Min Temp ≤ 0 °F | 9 | 5 | 1 | 0 | 0 | 0 | 0 | 0 | 0 | 0 | 0 | 4 | 19 |
| Heating Degree Days | 1453 | 1247 | 1061 | 688 | 358 | 115 | 28 | 52 | 262 | 584 | 873 | 1286 | 8007 |
| Cooling Degree Days | 0 | 0 | 0 | 0 | 7 | 45 | 129 | 93 | 11 | 0 | 0 | 0 | 285 |
| Total Precipitation (") | 3.06 | 2.58 | 2.87 | 3.02 | 3.24 | 3.47 | 3.40 | 3.52 | 3.70 | 3.20 | 3.79 | 3.65 | 39.50 |
| Days ≥ 0.1" Precip | 7 | 6 | 7 | 6 | 7 | 7 | 6 | 6 | 7 | 6 | 8 | 7 | 80 |
| Total Snowfall (") | 21.1 | 17.1 | 11.7 | 4.4 | 0.0 | 0.0 | 0.0 | 0.0 | 0.0 | 0.5 | 4.8 | 17.4 | 77.0 |
| Days ≥ 1" Snow Depth | na | na | na | na | 0 | 0 | 0 | 0 | 0 | 0 | na | na | na |

## PORTLAND INTL JETPRT *Cumberland County* ELEVATION 69 ft LAT/LONG 43° 39 ' N / 70° 19 ' W

| | JAN | FEB | MAR | APR | MAY | JUN | JUL | AUG | SEP | OCT | NOV | DEC | YEAR |
|---|---|---|---|---|---|---|---|---|---|---|---|---|---|
| Maximum Temp °F | 30.4 | 33.2 | 41.3 | 52.4 | 63.2 | 72.9 | 79.1 | 77.7 | 69.3 | 58.4 | 47.3 | 35.7 | 55.1 |
| Minimum Temp °F | 11.5 | 13.9 | 24.2 | 34.0 | 43.4 | 52.5 | 58.6 | 57.6 | 49.0 | 38.3 | 30.3 | 18.3 | 36.0 |
| Mean Temp °F | 21.0 | 23.5 | 32.8 | 43.2 | 53.3 | 62.7 | 68.9 | 67.7 | 59.2 | 48.4 | 38.8 | 27.0 | 45.5 |
| Days Max Temp ≥ 90 °F | 0 | 0 | 0 | 0 | 0 | 1 | 2 | 1 | 0 | 0 | 0 | 0 | 4 |
| Days Max Temp ≤ 32 °F | 17 | 13 | 4 | 0 | 0 | 0 | 0 | 0 | 0 | 0 | 1 | 11 | 46 |
| Days Min Temp ≤ 32 °F | 30 | 26 | 25 | 13 | 2 | 0 | 0 | 0 | 0 | 9 | 19 | 29 | 153 |
| Days Min Temp ≤ 0 °F | 6 | 4 | 0 | 0 | 0 | 0 | 0 | 0 | 0 | 0 | 0 | 2 | 12 |
| Heating Degree Days | 1358 | 1164 | 992 | 647 | 362 | 113 | 19 | 36 | 195 | 508 | 778 | 1171 | 7343 |
| Cooling Degree Days | 0 | 0 | 0 | 0 | 8 | 56 | 157 | 131 | 29 | 1 | 0 | 0 | 382 |
| Total Precipitation (") | 3.67 | 3.30 | 4.05 | 4.13 | 3.66 | 3.44 | 3.06 | 3.26 | 3.27 | 3.58 | 4.98 | 4.66 | 45.06 |
| Days ≥ 0.1" Precip | 7 | 5 | 7 | 7 | 7 | 7 | 6 | 5 | 6 | 6 | 8 | 8 | 79 |
| Total Snowfall (") | 19.6 | 15.8 | 11.8 | 3.4 | 0.1 | 0.0 | 0.0 | 0.0 | 0.0 | 0.2 | 3.0 | 15.7 | 69.6 |
| Days ≥ 1" Snow Depth | 25 | 22 | 15 | 2 | 0 | 0 | 0 | 0 | 0 | 0 | 2 | 15 | 81 |

## PRESQUE ISLE *Aroostook County* ELEVATION 610 ft LAT/LONG 46° 39 ' N / 68° 0 ' W

| | JAN | FEB | MAR | APR | MAY | JUN | JUL | AUG | SEP | OCT | NOV | DEC | YEAR |
|---|---|---|---|---|---|---|---|---|---|---|---|---|---|
| Maximum Temp °F | 20.9 | 24.6 | 35.2 | 48.2 | 63.9 | 73.6 | 77.8 | 75.6 | 65.9 | 53.1 | 38.7 | 25.7 | 50.3 |
| Minimum Temp °F | 1.4 | 4.1 | 15.8 | 28.8 | 39.9 | 49.9 | 54.7 | 53.1 | 44.3 | 35.2 | 24.7 | 8.8 | 30.1 |
| Mean Temp °F | 11.2 | 14.4 | 25.5 | 38.5 | 52.0 | 61.8 | 66.3 | 64.4 | 55.2 | 44.2 | 31.7 | 17.3 | 40.2 |
| Days Max Temp ≥ 90 °F | 0 | 0 | 0 | 0 | 0 | 0 | 1 | 0 | 0 | 0 | 0 | 0 | 1 |
| Days Max Temp ≤ 32 °F | 26 | 21 | 11 | 1 | 0 | 0 | 0 | 0 | 0 | 0 | 8 | 22 | 89 |
| Days Min Temp ≤ 32 °F | 31 | 28 | 29 | 21 | 6 | 0 | 0 | 0 | 3 | 13 | 24 | 30 | 185 |
| Days Min Temp ≤ 0 °F | 16 | 12 | 5 | 0 | 0 | 0 | 0 | 0 | 0 | 0 | 0 | 9 | 42 |
| Heating Degree Days | 1664 | 1425 | 1216 | 788 | 407 | 137 | 50 | 87 | 301 | 639 | 992 | 1473 | 9179 |
| Cooling Degree Days | 0 | 0 | 0 | 0 | 6 | 41 | 94 | 82 | 13 | 0 | 0 | 0 | 236 |
| Total Precipitation (") | 2.23 | 1.60 | 2.08 | 2.18 | 3.33 | 3.38 | 3.71 | 3.98 | 3.55 | 3.24 | 3.05 | 2.59 | 34.92 |
| Days ≥ 0.1" Precip | 6 | 4 | 6 | 6 | 8 | 8 | 8 | 8 | 7 | 7 | 7 | 7 | 82 |
| Total Snowfall (") | 21.2 | 16.9 | 16.4 | 6.4 | 0.7 | 0.0 | 0.0 | 0.0 | 0.1 | 0.7 | 9.2 | 18.6 | 90.2 |
| Days ≥ 1" Snow Depth | 30 | 28 | 30 | 10 | 0 | 0 | 0 | 0 | 0 | 0 | 10 | 26 | 134 |

## RIPOGENUS DAM *Piscataquis County* ELEVATION 965 ft LAT/LONG 45° 53 ' N / 69° 11 ' W

| | JAN | FEB | MAR | APR | MAY | JUN | JUL | AUG | SEP | OCT | NOV | DEC | YEAR |
|---|---|---|---|---|---|---|---|---|---|---|---|---|---|
| Maximum Temp °F | 20.8 | 24.4 | 34.7 | 46.5 | 61.1 | 71.3 | 76.4 | 74.6 | 65.5 | 53.0 | 39.3 | 26.0 | 49.5 |
| Minimum Temp °F | -1.7 | -0.1 | 10.8 | 26.0 | 37.9 | 48.6 | 53.9 | 52.7 | 43.9 | 33.9 | 23.9 | 7.5 | 28.1 |
| Mean Temp °F | 9.6 | 12.2 | 22.8 | 36.3 | 49.5 | 60.0 | 65.2 | 63.7 | 54.7 | 43.5 | 31.6 | 16.8 | 38.8 |
| Days Max Temp ≥ 90 °F | 0 | 0 | 0 | 0 | 0 | 0 | 0 | 0 | 0 | 0 | 0 | 0 | 0 |
| Days Max Temp ≤ 32 °F | 26 | 22 | 12 | 2 | 0 | 0 | 0 | 0 | 0 | 0 | 7 | 22 | 91 |
| Days Min Temp ≤ 32 °F | 31 | 28 | 30 | 25 | 8 | 0 | 0 | 0 | 2 | 15 | 25 | 30 | 194 |
| Days Min Temp ≤ 0 °F | 18 | 16 | 8 | 0 | 0 | 0 | 0 | 0 | 0 | 0 | 0 | 9 | 51 |
| Heating Degree Days | 1714 | 1486 | 1302 | 855 | 477 | 174 | 63 | 92 | 311 | 660 | 994 | 1491 | 9619 |
| Cooling Degree Days | 0 | 0 | 0 | 0 | 2 | 27 | 76 | 60 | 9 | 0 | 0 | 0 | 174 |
| Total Precipitation (") | 2.50 | 1.92 | 2.63 | 2.96 | 3.31 | 3.89 | 3.97 | 4.05 | 3.70 | 3.45 | 3.48 | 3.05 | 38.91 |
| Days ≥ 0.1" Precip | 7 | 5 | 7 | 8 | 8 | 9 | 9 | 9 | 8 | 7 | 8 | 7 | 92 |
| Total Snowfall (") | 26.6 | 21.2 | 20.2 | 8.0 | 0.2 | 0.0 | 0.0 | 0.0 | 0.0 | 0.7 | 9.2 | 25.2 | 111.3 |
| Days ≥ 1" Snow Depth | na | na | na | na | 0 | 0 | 0 | 0 | 0 | 0 | na | na | na |

## RUMFORD 1 SSE *Oxford County* ELEVATION 674 ft LAT/LONG 44° 31 ' N / 70° 34 ' W

| | JAN | FEB | MAR | APR | MAY | JUN | JUL | AUG | SEP | OCT | NOV | DEC | YEAR |
|---|---|---|---|---|---|---|---|---|---|---|---|---|---|
| Maximum Temp °F | 27.1 | 30.8 | 40.0 | 52.5 | 66.0 | 74.7 | 79.7 | 77.6 | 68.8 | 57.2 | 43.5 | 31.3 | 54.1 |
| Minimum Temp °F | 5.9 | 8.1 | 19.5 | 31.7 | 42.2 | 51.5 | 56.7 | 55.4 | 46.6 | 36.6 | 27.9 | 13.6 | 33.0 |
| Mean Temp °F | 16.5 | 19.5 | 29.8 | 42.1 | 54.1 | 63.1 | 68.2 | 66.5 | 57.7 | 46.9 | 35.7 | 22.4 | 43.5 |
| Days Max Temp ≥ 90 °F | 0 | 0 | 0 | 0 | 1 | 1 | 2 | 1 | 0 | 0 | 0 | 0 | 5 |
| Days Max Temp ≤ 32 °F | 21 | 16 | 6 | 0 | 0 | 0 | 0 | 0 | 0 | 0 | 3 | 16 | 62 |
| Days Min Temp ≤ 32 °F | 31 | 28 | 28 | 17 | 3 | 0 | 0 | 0 | 1 | 11 | 22 | 29 | 170 |
| Days Min Temp ≤ 0 °F | 11 | 8 | 2 | 0 | 0 | 0 | 0 | 0 | 0 | 0 | 0 | 5 | 26 |
| Heating Degree Days | 1497 | 1279 | 1085 | 680 | 341 | 106 | 28 | 53 | 231 | 555 | 872 | 1312 | 8039 |
| Cooling Degree Days | 0 | 0 | 0 | 1 | 12 | 55 | 143 | 110 | 20 | 0 | 0 | 0 | 341 |
| Total Precipitation (") | 2.90 | 2.55 | 3.43 | 3.71 | 3.63 | 4.17 | 3.52 | 4.18 | 3.52 | 3.80 | 4.41 | 3.75 | 43.57 |
| Days ≥ 0.1" Precip | 6 | 5 | 6 | 7 | 8 | 8 | 8 | 7 | 6 | 7 | 7 | 7 | 82 |
| Total Snowfall (") | 22.5 | 17.9 | 13.7 | 7.2 | 0.1 | 0.0 | 0.0 | 0.0 | 0.0 | 0.4 | 6.0 | 20.9 | 88.7 |
| Days ≥ 1" Snow Depth | 30 | 27 | 29 | 12 | 0 | 0 | 0 | 0 | 0 | 0 | 6 | 23 | 127 |

## SANFORD 2 NNW *York County* ELEVATION 279 ft LAT/LONG 43° 28 ' N / 70° 47 ' W

| | JAN | FEB | MAR | APR | MAY | JUN | JUL | AUG | SEP | OCT | NOV | DEC | YEAR |
|---|---|---|---|---|---|---|---|---|---|---|---|---|---|
| Maximum Temp °F | 32.2 | 35.7 | 45.0 | 57.4 | 70.1 | 78.5 | 83.3 | 81.5 | 72.9 | 61.9 | 48.2 | 36.5 | 58.6 |
| Minimum Temp °F | 10.6 | 12.6 | 22.9 | 32.3 | 42.6 | 52.0 | 57.6 | 56.4 | 47.9 | 37.2 | 29.3 | 17.7 | 34.9 |
| Mean Temp °F | 21.4 | 24.2 | 34.0 | 44.9 | 56.3 | 65.3 | 70.4 | 69.0 | 60.4 | 49.5 | 38.8 | 27.2 | 46.8 |
| Days Max Temp ≥ 90 °F | 0 | 0 | 0 | 0 | 1 | 2 | 4 | 3 | 0 | 0 | 0 | 0 | 10 |
| Days Max Temp ≤ 32 °F | 15 | 10 | 2 | 0 | 0 | 0 | 0 | 0 | 0 | 0 | 1 | 10 | 38 |
| Days Min Temp ≤ 32 °F | 30 | 27 | 27 | 16 | 3 | 0 | 0 | 0 | 1 | 11 | 20 | 29 | 164 |
| Days Min Temp ≤ 0 °F | 7 | 5 | 1 | 0 | 0 | 0 | 0 | 0 | 0 | 0 | 0 | 2 | 15 |
| Heating Degree Days | 1344 | 1146 | 954 | 597 | 280 | 71 | 12 | 26 | 171 | 474 | 781 | 1166 | 7022 |
| Cooling Degree Days | 0 | 0 | 0 | 1 | 18 | 83 | 195 | 157 | 42 | 2 | 0 | 0 | 498 |
| Total Precipitation (") | 3.55 | 3.67 | 4.20 | 4.11 | 3.86 | 3.55 | 3.33 | 3.94 | 3.63 | 3.95 | 4.96 | 4.62 | 47.37 |
| Days ≥ 0.1" Precip | 6 | 6 | 7 | 7 | 8 | 7 | 7 | 6 | 6 | 6 | 8 | 7 | 81 |
| Total Snowfall (") | na | na | na | *2.3* | 0.1 | 0.0 | 0.0 | 0.0 | 0.0 | 0.2 | *2.6* | na | na |
| Days ≥ 1" Snow Depth | *23* | *20* | *13* | 1 | 0 | 0 | 0 | 0 | 0 | 0 | *2* | na | na |

## SPRINGFIELD *Penobscot County* ELEVATION 440 ft LAT/LONG 45° 25 ' N / 68° 8 ' W

| | JAN | FEB | MAR | APR | MAY | JUN | JUL | AUG | SEP | OCT | NOV | DEC | YEAR |
|---|---|---|---|---|---|---|---|---|---|---|---|---|---|
| Maximum Temp °F | 25.2 | 28.7 | 38.2 | 50.8 | 64.6 | 73.5 | 78.3 | 76.8 | 67.7 | 55.5 | 43.1 | 30.1 | 52.7 |
| Minimum Temp °F | 2.4 | 4.9 | 16.6 | 28.7 | 39.6 | 49.5 | 55.0 | 53.5 | 44.1 | 34.3 | 25.5 | 10.3 | 30.4 |
| Mean Temp °F | 13.9 | 16.8 | 27.5 | 39.8 | 52.2 | 61.5 | 66.6 | 65.2 | 55.9 | 44.9 | 34.3 | 20.2 | 41.6 |
| Days Max Temp ≥ 90 °F | 0 | 0 | 0 | 0 | 0 | 1 | 1 | 1 | 0 | 0 | 0 | 0 | 3 |
| Days Max Temp ≤ 32 °F | 23 | 18 | 8 | 0 | 0 | 0 | 0 | 0 | 0 | 0 | 4 | 18 | 71 |
| Days Min Temp ≤ 32 °F | 31 | 28 | 29 | 21 | 6 | 0 | 0 | 0 | 3 | 15 | 23 | 30 | 186 |
| Days Min Temp ≤ 0 °F | 15 | 11 | 4 | 0 | 0 | 0 | 0 | 0 | 0 | 0 | 0 | 8 | 38 |
| Heating Degree Days | 1582 | 1353 | 1156 | 748 | 397 | 145 | 49 | 74 | 280 | 615 | 913 | 1382 | 8694 |
| Cooling Degree Days | 0 | 0 | 0 | 0 | 5 | 44 | 110 | 93 | 14 | 1 | 0 | 0 | 267 |
| Total Precipitation (") | 3.48 | 2.89 | 3.59 | 3.40 | 3.75 | 3.82 | 3.97 | 3.93 | 3.95 | 3.53 | 4.46 | 4.08 | 44.85 |
| Days ≥ 0.1" Precip | 7 | 6 | 8 | 8 | 9 | 8 | 8 | 7 | 7 | 7 | 9 | 8 | 92 |
| Total Snowfall (") | 25.2 | 21.3 | 18.3 | 7.8 | 0.2 | 0.0 | 0.0 | 0.0 | 0.0 | 0.9 | 6.1 | 21.9 | 101.7 |
| Days ≥ 1" Snow Depth | 29 | 28 | *29* | *11* | 0 | 0 | 0 | 0 | 0 | 0 | 5 | *21* | 123 |

## SQUA PAN DAM *Aroostook County* ELEVATION 610 ft LAT/LONG 46° 33 ' N / 68° 20 ' W

| | JAN | FEB | MAR | APR | MAY | JUN | JUL | AUG | SEP | OCT | NOV | DEC | YEAR |
|---|---|---|---|---|---|---|---|---|---|---|---|---|---|
| Maximum Temp °F | 20.5 | 24.3 | 35.1 | 46.9 | 62.0 | 72.6 | *77.6* | 74.9 | 65.0 | 52.4 | *38.5* | 25.2 | 49.6 |
| Minimum Temp °F | -6.7 | -6.2 | 7.4 | 24.1 | 35.1 | 45.8 | 50.1 | 48.0 | 38.6 | 30.5 | *20.6* | 2.8 | 24.2 |
| Mean Temp °F | 6.9 | 9.1 | 21.3 | 35.6 | 48.6 | 59.2 | *63.9* | 61.4 | 51.8 | 41.5 | *29.6* | 14.1 | 36.9 |
| Days Max Temp ≥ 90 °F | 0 | 0 | 0 | 0 | 0 | 1 | 1 | 1 | 0 | 0 | 0 | 0 | 3 |
| Days Max Temp ≤ 32 °F | 26 | 22 | 11 | 1 | 0 | 0 | 0 | 0 | 0 | 0 | 7 | 22 | 89 |
| Days Min Temp ≤ 32 °F | 31 | 28 | 30 | 25 | 13 | 2 | 0 | 1 | 8 | 18 | 24 | 31 | 211 |
| Days Min Temp ≤ 0 °F | 21 | 20 | 11 | 1 | 0 | 0 | 0 | 0 | 0 | 0 | 1 | 14 | 68 |
| Heating Degree Days | 1798 | 1575 | 1350 | 877 | 507 | 200 | *96* | 146 | 395 | 723 | *1056* | 1575 | 10298 |
| Cooling Degree Days | 0 | 0 | 0 | 0 | 3 | 31 | na | 48 | 6 | *0* | na | 0 | na |
| Total Precipitation (") | 2.49 | 1.95 | 2.35 | 2.56 | 3.25 | 3.40 | 3.55 | 3.88 | 3.45 | 3.34 | 3.25 | 3.12 | 36.59 |
| Days ≥ 0.1" Precip | 6 | 5 | 6 | 7 | 8 | 8 | 7 | 7 | 6 | 7 | 7 | 8 | 82 |
| Total Snowfall (") | 22.6 | 17.4 | 15.4 | 7.0 | 0.3 | 0.0 | 0.0 | 0.0 | 0.0 | 0.5 | 10.4 | 21.5 | 95.1 |
| Days ≥ 1" Snow Depth | 31 | 28 | 31 | 23 | 2 | 0 | 0 | 0 | 0 | 0 | 9 | 28 | 152 |

## VAN BUREN 2 *Aroostook County* ELEVATION 512 ft LAT/LONG 47° 10 ' N / 67° 57 ' W

| | JAN | FEB | MAR | APR | MAY | JUN | JUL | AUG | SEP | OCT | NOV | DEC | YEAR |
|---|---|---|---|---|---|---|---|---|---|---|---|---|---|
| Maximum Temp °F | 18.4 | 22.6 | 33.5 | 46.9 | 62.0 | 71.7 | 76.3 | na | 64.5 | 52.0 | 38.6 | 25.3 | na |
| Minimum Temp °F | -8.2 | -5.6 | 8.9 | 26.4 | 37.2 | 47.5 | na | na | 40.7 | 31.9 | 22.0 | 3.6 | na |
| Mean Temp °F | 5.1 | 8.5 | 21.2 | 36.7 | 49.6 | 59.6 | na | na | 52.7 | 42.0 | 30.3 | 14.5 | na |
| Days Max Temp ≥ 90 °F | 0 | 0 | 0 | 0 | 0 | 1 | 1 | 0 | 0 | 0 | 0 | 0 | 2 |
| Days Max Temp ≤ 32 °F | 26 | 23 | 13 | 2 | 0 | 0 | 0 | 0 | 0 | 0 | 8 | 22 | 94 |
| Days Min Temp ≤ 32 °F | 31 | 28 | 30 | 23 | 9 | 1 | 0 | 0 | 5 | 18 | 25 | 30 | 200 |
| Days Min Temp ≤ 0 °F | na | 19 | 9 | 0 | 0 | 0 | 0 | 0 | 0 | 0 | 1 | 13 | na |
| Heating Degree Days | 1855 | 1593 | 1351 | 843 | 475 | 189 | na | na | 371 | 707 | 1032 | 1563 | na |
| Cooling Degree Days | 0 | 0 | 0 | 0 | 3 | 32 | 70 | 52 | 6 | 0 | 0 | 0 | 163 |
| Total Precipitation (") | 2.50 | 1.74 | 2.23 | 2.55 | 3.20 | 3.46 | 3.96 | 4.17 | 3.48 | 3.42 | 3.26 | 3.11 | 37.08 |
| Days ≥ 0.1" Precip | 6 | 5 | 6 | 7 | 8 | 8 | 9 | na | 7 | 7 | 7 | 7 | na |
| Total Snowfall (") | 23.7 | 17.0 | 14.2 | 5.3 | 0.2 | 0.0 | 0.0 | 0.0 | 0.0 | 0.1 | 7.8 | 20.8 | 89.1 |
| Days ≥ 1" Snow Depth | 31 | 28 | 30 | 15 | 0 | 0 | 0 | 0 | 0 | 0 | 10 | 26 | 140 |

## VANCEBORO 2 *Washington County* ELEVATION 390 ft LAT/LONG 45° 34 ' N / 67° 26 ' W

| | JAN | FEB | MAR | APR | MAY | JUN | JUL | AUG | SEP | OCT | NOV | DEC | YEAR |
|---|---|---|---|---|---|---|---|---|---|---|---|---|---|
| Maximum Temp °F | 26.4 | 30.1 | 39.7 | 52.2 | 66.4 | 75.7 | 80.3 | 78.7 | 69.4 | 57.6 | 43.3 | 30.9 | 54.2 |
| Minimum Temp °F | 3.5 | 6.0 | 17.2 | 28.7 | 39.0 | 48.4 | 54.6 | 53.2 | 44.4 | 34.6 | 25.9 | 11.0 | 30.5 |
| Mean Temp °F | 15.0 | 18.1 | 28.5 | 40.5 | 52.7 | 62.1 | 67.5 | 65.9 | 56.9 | 46.1 | 34.6 | 21.0 | 42.4 |
| Days Max Temp ≥ 90 °F | 0 | 0 | 0 | 0 | 0 | 1 | 2 | 1 | 0 | 0 | 0 | 0 | 4 |
| Days Max Temp ≤ 32 °F | 22 | 17 | 6 | 0 | 0 | 0 | 0 | 0 | 0 | 0 | 3 | 17 | 65 |
| Days Min Temp ≤ 32 °F | 30 | 28 | 28 | 21 | 6 | 0 | 0 | 0 | 2 | 14 | 23 | 30 | 182 |
| Days Min Temp ≤ 0 °F | 14 | 10 | 3 | 0 | 0 | 0 | 0 | 0 | 0 | 0 | 0 | 7 | 34 |
| Heating Degree Days | 1547 | 1320 | 1125 | 730 | 379 | 122 | 31 | 55 | 249 | 580 | 905 | 1358 | 8401 |
| Cooling Degree Days | 0 | 0 | 0 | 0 | 4 | 39 | 113 | 96 | 12 | 1 | 0 | 0 | 265 |
| Total Precipitation (") | 3.33 | 2.55 | 2.76 | 3.29 | 3.99 | 3.75 | 3.90 | 3.99 | 4.16 | 3.81 | 4.38 | 3.57 | 43.48 |
| Days ≥ 0.1" Precip | 6 | 5 | 6 | 6 | 8 | 7 | 7 | 7 | 7 | 7 | 8 | 7 | 81 |
| Total Snowfall (") | 18.5 | 14.4 | 11.7 | 3.6 | 0.2 | 0.0 | 0.0 | 0.0 | 0.0 | 0.2 | 5.2 | 16.8 | 70.6 |
| Days ≥ 1" Snow Depth | 27 | 27 | 26 | 9 | 0 | 0 | 0 | 0 | 0 | 0 | 4 | 22 | 115 |

## WATERVILLE TRTMNT PL *Kennebec County* ELEVATION 89 ft LAT/LONG 44° 33 ' N / 69° 39 ' W

| | JAN | FEB | MAR | APR | MAY | JUN | JUL | AUG | SEP | OCT | NOV | DEC | YEAR |
|---|---|---|---|---|---|---|---|---|---|---|---|---|---|
| Maximum Temp °F | 30.5 | 34.5 | 43.5 | 56.0 | 69.0 | 77.5 | 82.3 | 80.6 | 71.7 | 60.3 | 46.7 | 34.6 | 57.3 |
| Minimum Temp °F | 7.1 | 9.7 | 21.0 | 31.4 | 41.8 | 51.4 | 57.0 | 55.7 | 47.2 | 37.0 | 28.2 | 14.7 | 33.5 |
| Mean Temp °F | 18.8 | 22.1 | 32.3 | 43.7 | 55.4 | 64.5 | 69.7 | 68.2 | 59.5 | 48.7 | 37.5 | 24.7 | 45.4 |
| Days Max Temp ≥ 90 °F | 0 | 0 | 0 | 0 | 1 | 2 | 3 | 2 | 0 | 0 | 0 | 0 | 8 |
| Days Max Temp ≤ 32 °F | 17 | 11 | 3 | 0 | 0 | 0 | 0 | 0 | 0 | 0 | 1 | 12 | 44 |
| Days Min Temp ≤ 32 °F | 30 | 27 | 27 | 17 | 3 | 0 | 0 | 0 | 1 | 11 | 21 | 29 | 166 |
| Days Min Temp ≤ 0 °F | 10 | 7 | 2 | 0 | 0 | 0 | 0 | 0 | 0 | 0 | 0 | 4 | 23 |
| Heating Degree Days | 1425 | 1205 | 1008 | 634 | 301 | 77 | 14 | 30 | 184 | 500 | 818 | 1243 | 7439 |
| Cooling Degree Days | 0 | 0 | 0 | 0 | 9 | 61 | 166 | 135 | 24 | 2 | 0 | 0 | 397 |
| Total Precipitation (") | 2.71 | 2.40 | 3.36 | 3.24 | 3.60 | 3.63 | 3.42 | 3.78 | 3.48 | 3.69 | 3.95 | 3.38 | 40.64 |
| Days ≥ 0.1" Precip | 6 | 5 | 6 | 7 | 7 | 7 | 7 | 7 | 6 | 6 | 7 | 7 | 78 |
| Total Snowfall (") | 16.7 | 14.3 | 8.2 | 2.8 | 0.1 | 0.0 | 0.0 | 0.0 | 0.0 | 0.1 | 2.6 | na | na |
| Days ≥ 1" Snow Depth | 28 | 25 | 16 | 2 | 0 | 0 | 0 | 0 | 0 | 0 | 3 | 19 | 93 |

## WEST BUXTON 2 NNW *York County* ELEVATION 151 ft LAT/LONG 43° 42 ' N / 70° 37 ' W

| | JAN | FEB | MAR | APR | MAY | JUN | JUL | AUG | SEP | OCT | NOV | DEC | YEAR |
|---|---|---|---|---|---|---|---|---|---|---|---|---|---|
| Maximum Temp °F | 30.4 | 33.8 | 42.3 | 54.5 | 66.1 | 75.0 | 80.0 | 78.2 | 69.6 | 58.8 | 46.6 | 35.0 | 55.9 |
| Minimum Temp °F | 5.2 | 7.0 | 19.4 | 29.7 | 39.9 | 49.9 | 55.6 | 53.8 | 44.8 | 33.8 | 25.6 | 13.2 | 31.5 |
| Mean Temp °F | 17.8 | 20.4 | 30.9 | 42.1 | 53.0 | 62.5 | 67.8 | 66.0 | 57.2 | 46.3 | 36.1 | 24.1 | 43.7 |
| Days Max Temp ≥ 90 °F | 0 | 0 | 0 | 0 | 0 | 1 | 2 | 1 | 0 | 0 | 0 | 0 | 4 |
| Days Max Temp ≤ 32 °F | 18 | 13 | 4 | 0 | 0 | 0 | 0 | 0 | 0 | 0 | 2 | 13 | 50 |
| Days Min Temp ≤ 32 °F | 31 | 28 | 29 | 20 | 6 | 0 | 0 | 0 | 2 | 15 | 23 | 30 | 184 |
| Days Min Temp ≤ 0 °F | 12 | 9 | 2 | 0 | 0 | 0 | 0 | 0 | 0 | 0 | 0 | 5 | 28 |
| Heating Degree Days | 1458 | 1253 | 1050 | 681 | 371 | 119 | 32 | 59 | 244 | 573 | 860 | 1261 | 7961 |
| Cooling Degree Days | 0 | 0 | 0 | 0 | 8 | 48 | 129 | 99 | 19 | 0 | 0 | 0 | 303 |
| Total Precipitation (") | 3.23 | 3.18 | 4.03 | 4.31 | 3.79 | 3.66 | 3.25 | 3.20 | 3.56 | 3.61 | 4.70 | 4.33 | 44.85 |
| Days ≥ 0.1" Precip | 6 | 5 | 7 | 7 | 7 | 7 | 6 | 6 | 6 | 6 | 7 | 7 | 77 |
| Total Snowfall (") | na | na | na | 1.8 | 0.1 | 0.0 | 0.0 | 0.0 | 0.0 | 0.2 | 1.9 | na | na |
| Days ≥ 1" Snow Depth | na | 26 | 21 | 2 | 0 | 0 | 0 | 0 | 0 | 0 | 3 | 16 | na |

## WOODLAND *Washington County* ELEVATION 141 ft LAT/LONG 45° 9 ' N / 67° 24 ' W

| | JAN | FEB | MAR | APR | MAY | JUN | JUL | AUG | SEP | OCT | NOV | DEC | YEAR |
|---|---|---|---|---|---|---|---|---|---|---|---|---|---|
| Maximum Temp °F | 27.9 | 30.3 | 39.9 | ***52.0*** | ***65.1*** | 74.5 | 80.3 | 78.8 | 69.9 | 57.7 | 45.7 | ***34.1*** | 54.7 |
| Minimum Temp °F | 4.6 | 5.8 | 17.8 | ***30.1*** | ***39.7*** | 49.5 | 55.2 | 53.8 | 44.6 | 35.3 | 26.7 | ***12.3*** | 31.3 |
| Mean Temp °F | 16.3 | 18.1 | 28.9 | ***41.1*** | ***52.4*** | 62.0 | 67.8 | 66.3 | 57.3 | 46.5 | 36.2 | ***23.5*** | 43.0 |
| Days Max Temp ≥ 90 °F | 0 | 0 | 0 | 0 | 0 | 1 | 3 | 2 | 0 | 0 | 0 | 0 | 6 |
| Days Max Temp ≤ 32 °F | 19 | 16 | 6 | 0 | ***0*** | 0 | 0 | 0 | 0 | 0 | 2 | na | na |
| Days Min Temp ≤ 32 °F | 30 | 28 | 28 | ***19*** | ***5*** | 0 | 0 | 0 | 1 | 13 | ***23*** | ***30*** | 177 |
| Days Min Temp ≤ 0 °F | 13 | 10 | 3 | 0 | ***0*** | 0 | 0 | 0 | 0 | 0 | 0 | ***6*** | 32 |
| Heating Degree Days | 1504 | 1318 | 1114 | ***712*** | ***389*** | 130 | 30 | 51 | 239 | 567 | 858 | ***1276*** | 8188 |
| Cooling Degree Days | 0 | 0 | 0 | 0 | 4 | 42 | 119 | 105 | 14 | 1 | 0 | ***0*** | 285 |
| Total Precipitation (") | 3.78 | 3.28 | 3.53 | 3.69 | 3.83 | 3.40 | 3.11 | 3.34 | 3.73 | 3.91 | 4.87 | 4.73 | 45.20 |
| Days ≥ 0.1" Precip | ***7*** | ***7*** | 7 | ***7*** | ***8*** | 8 | ***7*** | ***6*** | 6 | 7 | ***8*** | na | na |
| Total Snowfall (") | na | na | na | na | ***0.0*** | 0.0 | 0.0 | 0.0 | 0.0 | 0.0 | na | na | na |
| Days ≥ 1" Snow Depth | na | na | na | na | ***0*** | 0 | 0 | 0 | 0 | 0 | na | na | na |

## JANUARY MINIMUM TEMPERATURE °F

### LOWEST

| | | |
|---|---|---|
| 1 | Van Buren | -8.2 |
| 2 | Squa Pan Dam | -6.7 |
| 3 | Fort Kent | -5.4 |
| 4 | Jackman | -2.4 |
| 5 | Ripogenus Dam | -1.7 |
| 6 | Brassua Dam | -1.5 |
| 7 | Bridgewater | -1.4 |
| 8 | Middle Dam | -1.0 |
| 9 | Caribou | -0.3 |
| 10 | Houlton | 0.0 |
| 11 | Farmington | 0.3 |
| 12 | Long Falls Dam | 0.7 |
| 13 | Corinna | 1.4 |
| | Presque Isle | 1.4 |
| 15 | Millinocket | 1.9 |
| 16 | Madison | 2.1 |
| 17 | Springfield | 2.4 |
| 18 | Vanceboro | 3.5 |
| 19 | Grand Lake Strm | 3.7 |
| 20 | Woodland | 4.6 |
| 21 | West Buxton | 5.2 |
| 22 | Bridgton | 5.4 |
| 23 | Gardiner | 5.8 |
| 24 | Rumford | 5.9 |
| 25 | Waterville | 7.1 |

### HIGHEST

| | | |
|---|---|---|
| 1 | Eastport | 13.6 |
| 2 | Newcastle | 12.4 |
| 3 | Portland | 11.5 |
| 4 | Lewiston | 11.0 |
| 5 | Sanford | 10.6 |
| 6 | Augusta | 10.5 |
| | Brunswick | 10.5 |
| 8 | Ellsworth | 10.4 |
| 9 | Belfast | 10.3 |
| 10 | Orono | 7.9 |
| 11 | Bangor | 7.7 |
| 12 | Jonesboro | 7.5 |
| 13 | Waterville | 7.1 |
| 14 | Rumford | 5.9 |
| 15 | Gardiner | 5.8 |
| 16 | Bridgton | 5.4 |
| 17 | West Buxton | 5.2 |
| 18 | Woodland | 4.6 |
| 19 | Grand Lake Strm | 3.7 |
| 20 | Vanceboro | 3.5 |
| 21 | Springfield | 2.4 |
| 22 | Madison | 2.1 |
| 23 | Millinocket | 1.9 |
| 24 | Corinna | 1.4 |
| | Presque Isle | 1.4 |

## JULY MAXIMUM TEMPERATURE °F

### HIGHEST

| | | |
|---|---|---|
| 1 | Sanford | 83.3 |
| 2 | Waterville | 82.3 |
| 3 | Lewiston | 81.0 |
| 4 | Vanceboro | 80.3 |
| | Woodland | 80.3 |
| 6 | Orono | 80.2 |
| 7 | Gardiner | 80.1 |
| 8 | Belfast | 80.0 |
| | West Buxton | 80.0 |
| 10 | Corinna | 79.8 |
| 11 | Bridgton | 79.7 |
| | Rumford | 79.7 |
| 13 | Augusta | 79.2 |
| | Madison | 79.2 |
| 15 | Portland | 79.1 |
| 16 | Farmington | 78.9 |
| 17 | Bridgewater | 78.7 |
| | Brunswick | 78.7 |
| | Houlton | 78.7 |
| 20 | Bangor | 78.6 |
| 21 | Grand Lake Stram | 78.5 |
| | Newcastle | 78.5 |
| 23 | Millinocket | 78.4 |
| 24 | Ellsworth | 78.3 |
| | Springfield | 78.3 |

### LOWEST

| | | |
|---|---|---|
| 1 | Eastport | 73.5 |
| 2 | Long Falls Dam | 75.1 |
| 3 | Brassua Dam | 75.3 |
| 4 | Middle Dam | 75.5 |
| 5 | Jonesboro | 75.6 |
| 6 | Fort Kent | 76.2 |
| 7 | Caribou | 76.3 |
| | Jackman | 76.3 |
| | Van Buren | 76.3 |
| 10 | Ripogenus Dam | 76.4 |
| 11 | Squa Pan Dam | 77.6 |
| 12 | Presque Isle | 77.8 |
| 13 | Ellsworth | 78.3 |
| | Springfield | 78.3 |
| 15 | Millinocket | 78.4 |
| 16 | Grand Lake Strm | 78.5 |
| | Newcastle | 78.5 |
| 18 | Bangor | 78.6 |
| 19 | Bridgewater | 78.7 |
| | Brunswick | 78.7 |
| | Houlton | 78.7 |
| 22 | Farmington | 78.9 |
| 23 | Portland | 79.1 |
| 24 | Augusta | 79.2 |
| | Madison | 79.2 |

## ANNUAL PRECIPITATION (")

### HIGHEST

| | | |
|---|---|---|
| 1 | Jonesboro | 50.32 |
| 2 | Sanford | 47.37 |
| 3 | Belfast | 47.25 |
| 4 | Brunswick | 46.43 |
| 5 | Bridgton | 46.35 |
| 6 | Newcastle | 46.28 |
| 7 | Ellsworth | 45.93 |
| 8 | Farmington | 45.58 |
| 9 | Lewiston | 45.37 |
| 10 | Woodland | 45.20 |
| 11 | Portland | 45.06 |
| 12 | Springfield | 44.85 |
| | West Buxton | 44.85 |
| 14 | Eastport | 43.66 |
| 15 | Grand Lake Strm | 43.62 |
| 16 | Rumford | 43.57 |
| 17 | Vanceboro | 43.48 |
| 18 | Gardiner | 42.61 |
| 19 | Corinna | 41.98 |
| 20 | Millinocket | 41.63 |
| 21 | Bridgewater | 41.42 |
| 22 | Augusta | 41.41 |
| 23 | Waterville | 40.64 |
| 24 | Brassua Dam | 40.53 |
| 25 | Bangor | 40.32 |

### LOWEST

| | | |
|---|---|---|
| 1 | Presque Isle | 34.92 |
| 2 | Fort Kent | 36.07 |
| 3 | Middle Dam | 36.15 |
| 4 | Squa Pan Dam | 36.59 |
| 5 | Caribou | 36.79 |
| 6 | Van Buren | 37.08 |
| 7 | Jackman | 38.28 |
| 8 | Long Falls Dam | 38.48 |
| 9 | Madison | 38.77 |
| 10 | Ripogenus Dam | 38.91 |
| 11 | Houlton | 38.93 |
| 12 | Orono | 39.50 |
| 13 | Bangor | 40.32 |
| 14 | Brassua Dam | 40.53 |
| 15 | Waterville | 40.64 |
| 16 | Augusta | 41.41 |
| 17 | Bridgewater | 41.42 |
| 18 | Millinocket | 41.63 |
| 19 | Corinna | 41.98 |
| 20 | Gardiner | 42.61 |
| 21 | Vanceboro | 43.48 |
| 22 | Rumford | 43.57 |
| 23 | Grand Lake Strm | 43.62 |
| 24 | Eastport | 43.66 |
| 25 | Springfield | 44.85 |

## ANNUAL SNOWFALL (")

### HIGHEST

| | | |
|---|---|---|
| 1 | Caribou | 112.6 |
| 2 | Long Falls Dam | 111.6 |
| 3 | Ripogenus Dam | 111.3 |
| 4 | Jackman | 109.3 |
| 5 | Brassua Dam | 108.5 |
| 6 | Springfield | 101.7 |
| 7 | Fort Kent | 97.4 |
| 8 | Houlton | 96.9 |
| 9 | Squa Pan Dam | 95.1 |
| 10 | Millinocket | 93.5 |
| 11 | Presque Isle | 90.2 |
| 12 | Farmington | 89.3 |
| 13 | Van Buren | 89.1 |
| 14 | Rumford | 88.7 |
| 15 | Madison | 82.4 |
| 16 | Orono | 77.0 |
| 17 | Corinna | 74.7 |
| 18 | Newcastle | 74.5 |
| 19 | Lewiston | 74.2 |
| 20 | Augusta | 73.7 |
| 21 | Bangor | 71.7 |
| 22 | Brunswick | 70.7 |
| 23 | Vanceboro | 70.6 |
| 24 | Portland | 69.6 |
| 25 | Jonesboro | 69.5 |

### LOWEST

| | | |
|---|---|---|
| 1 | Eastport | 61.3 |
| 2 | Gardiner | 68.0 |
| 3 | Jonesboro | 69.5 |
| 4 | Portland | 69.6 |
| 5 | Vanceboro | 70.6 |
| 6 | Brunswick | 70.7 |
| 7 | Bangor | 71.7 |
| 8 | Augusta | 73.7 |
| 9 | Lewiston | 74.2 |
| 10 | Newcastle | 74.5 |
| 11 | Corinna | 74.7 |
| 12 | Orono | 77.0 |
| 13 | Madison | 82.4 |
| 14 | Rumford | 88.7 |
| 15 | Van Buren | 89.1 |
| 16 | Farmington | 89.3 |
| 17 | Presque Isle | 90.2 |
| 18 | Millinocket | 93.5 |
| 19 | Squa Pan Dam | 95.1 |
| 20 | Houlton | 96.9 |
| 21 | Fort Kent | 97.4 |
| 22 | Springfield | 101.7 |
| 23 | Brassua Dam | 108.5 |
| 24 | Jackman | 109.3 |
| 25 | Ripogenus Dam | 111.3 |

# MARYLAND & WASHINGTON, D.C.

PHYSICAL FEATURES. The State of Maryland is on the east coast of the United States and lies in an east-west position between longitudes 75° and 79° W., spanning a distance of 240 miles. The latitude varies from about 38° to nearly 40° N., with a latitudinal width of approximately 125 miles in eastern portions which gradually narrows to about 1½ miles in the Appalachian Mountain region near Hancock and increases again to 35 miles at the extreme western boundary. The total area of the State is 12,303 square miles, of which 9,887 square miles are land, 2,310 square miles are in the Chesapeake Bay and its tidal river waters, and 106 square miles are in Chincoteague Bay.

The Chesapeake Bay, elongated in a northerly direction, extends for about two-thirds of its length deep into Maryland. It virtually separates the State into two provinces except for a narrow neck of land about 10 miles wide in Cecil County, which bridges the gap between Chesapeake Bay and the State of Pennsylvania. That portion of the State east of Chesapeake Bay is commonly referred to as the Eastern Shore. The five southernmost counties between the Potomac River and Chesapeake Bay are commonly referred to as Southern Maryland. To the north and northwest of Southern Maryland, an area made up of six counties and located on the Piedmont, is an area commonly referred to as Northern-Central Maryland. The remainder of the State including roughly the Appalachian Mountain area or the three western counties is termed Western Maryland.

Although Maryland ranks as one of the smaller States with respect to size, it encompasses an extremely wide range of physiographic features which contribute to a comparatively wide range of climatic conditions. It extends across three well-defined physiographic belts which parallel the Atlantic coast in varying widths from New England to the southeastern United States. These physiographic provinces are the Coastal Plain, Piedmont province, and Appalachian province.

The land rises more or less gradually from the Atlantic Ocean across the Coastal Plain (which virtually includes the Eastern Shore and Southern Maryland) and then more rapidly across the Piedmont Plateau (northern-central Maryland) and the ridges of the Appalachian Mountains and finally reaches its highest point at 3,340 feet above mean sea level on Backbone Mountain in the Allegheny Plateau of Garrett County.

GENERAL CLIMATE. Since the general flow of the atmosphere in temperate latitudes is from west to east, the expansive North American Continent immediately to the west predisposes the Maryland area to a continental type of climate. This type of climate in middle latitudes is marked by well-defined seasons. Winter is the dormant season for plant growth based on low temperatures rather than drought. In spring and fall the changeableness of the weather is a striking feature. It is occasioned by a rapid succession of warm and cold fronts associated with cyclones and anticyclones which generally move from a westerly direction. Summers are warm to hot. The higher atmospheric humidity along the Atlantic coastal area causes the summer heat to be more oppressive and the winter cold more penetrating than for drier climates of the interior of the continent.

At times in winter the Appalachian Mountains afford a degree of protection from the icy blasts of cold Arctic air, particularly when a high pressure area attended by a coldwave approaches from the west. The modifying influences of the mountain barrier attending the passage of a storm area from the Ohio Valley is sometimes quite marked. The warming of the air as it descends the eastern slopes of the mountains may at times exceed 10° F.

The Allegheny Mountains contribute to the higher precipitation and heavy snowfall on the Allegheny Plateau. The formation of precipitation in the form of rain or snow is increased in storms or air masses which ascend the mountains from the Ohio Valley

TEMPERATURE. The winter climate on the Piedmont and Coastal plain sections of Maryland is intermediate between the cold of the Northeast and the mild weather of the South. Extremely cold air masses from the interior of the continent are moderated somewhat by passage over the Appalachian Mountains and in some instances by a short trajectory over the nearby ocean and bays. Weather on the Allegheny Plateau is frequently 10° to 15° F. colder than it is in eastern portions of the State and, at times, extremely low temperatures occur in winter. The average frost penetration ranges from about 5 inches or less in extreme southern portions of Maryland to more than 18 inches on the Allegheny Plateau.

Summer is characterized by considerable warm weather including at least several hot, humid periods. However, nights are usually quite comfortable. The average length of the freeze-free season based on a minimum temperature higher than 32° F. ranges from more than 225 days in extreme southern portions to fewer than 130 days on the Allegheny Plateau.

PRECIPITATION AND SNOWFALL. Although the heaviest precipitation occurs in the summer, this is the season when severe droughts are most frequent. Summer precipitation is less dependable and more variable than in winter. Average annual snowfall over Maryland ranges from a minimum of 8 to 10 inches along the coastal areas of the Southern Eastern Shore division to a maximum well over 70 in the Garrett County area. Snow flurries fall as early as September on the Allegheny Plateau, and in October in extreme eastern portions of the State. The last snowfall in eastern portions usually occurs in April and on the Allegheny Plateau in May. Even in the warmest winters snow falls in Maryland; however, averages for a climatological division may be less than 1 inch for the season.

FLOODS. All of the State lies in the Atlantic drainage except for a portion of Garrett County in the western end of Maryland which drains into the Ohio Basin. The largest river in the State, the Potomac, forms the southern boundary through most of its length. The far eastern area is drained by many small streams and tidal estuaries into Chesapeake Bay and the Atlantic Ocean.

Minor or local flood damage can be expected every year in streams above the tidewater areas. Floods do occur in all months of the year, but the greatest frequency is in late winter and spring. Snow-melt at times is a factor. Intense convectional storms in summer occasionally cause local flash floods. Storms of tropical origin passing through the area in late summer and fall produce high water and occasionally damaging floods, mostly in tidewater areas. These are due to the heavy rains or strong easterly winds accompanying the storm, or a combination of both. Flooding from wind-driven tides at times extends upstream in the Potomac to the District of Columbia area. High water also results from persistent northeast winds along the coast caused by extra-tropical storms.

STORMS. Thunderstorms occur at a given station on an average of 30 days per year in extreme eastern portions of Maryland, and 40 days per year in western portions. They occur in all months of the year, but during the 4-month cold season from November through February, an average of less than one storm per month is observed. May, June, July, and August make up the thunderstorm season and include from 75 to 80 percent of the thunderstorms which occur annually. July is the peak of the season with about 25 percent of the annual total number of thunderstorms.

Hail at a given station occurs on an average of 1 day per year in extreme eastern portions and about 2 days per year in extreme western portions. The total number of days on which hail is observed at one or more stations in Maryland averages about 18 to 20 per year. Hail has been observed in all months of the year; however, occurrences in the 7-month period from September through March are infrequent. The number of days with hail at one or more stations increases from an average of 1 in April to about 5 in July, the peak of the hail season, and then decreases to an average of 3 in August. Although spring thunderstorms are much fewer in number than summer thunderstorms, they have a much greater tendency to occur with hail. Virtually all hailstorms occur between 2 p.m. and 9 p.m. Severe, devastating hailstorms occur somewhere in the State about once every 5 years on the average.

Tornadoes occur infrequently in Maryland, and of the ones that do occur most are small. Most tornadoes in Maryland tend to travel in the usual southwest to northeast direction, but few have been reported to travel southeastward or in a southerly direction. Usually paths are not more than a few miles in length; however, 10 to 15 percent of these storms maintain paths 20 miles or more in length.

RELATIVE HUMIDITY. Average relative humidity is lowest in the winter and early spring from February through April, and highest in the late summer and early from August through October.

## COUNTY INDEX

### *Allegany County*
FROSTBURG 2

### *Anne Arundel County*
BALT-WASHGTN INTL AP

### *Baltimore County*
WOODSTOCK

### *Calvert County*
OWINGS FERRY LANDING

### *Charles County*
LA PLATA 1 W

### *Dorchester County*
VIENNA

### *Frederick County*
CATOCTIN MOUNTAIN PK
EMMITSBURG 2 SE
UNIONVILLE

### *Garrett County*
OAKLAND 1 SE
SAVAGE RIVER DAM

### *Harford County*
ABERDEEN PHILLIPS FI
BENSON POLICE BARRAC
CONOWINGO DAM

### *Kent County*
CHESTERTOWN
MILLINGTON 1 SE

### *Montgomery County*
ROCKVILLE 1 NE

### *Prince George's County*
BELTSVILLE
COLLEGE PARK
GLENN DALE BELL STN
LAUREL 3 W

### *Somerset County*
PRINCESS ANNE

### *Talbot County*
ROYAL OAK 2 SSW

### *Washington County*
HANCOCK

### *Wicomico County*
SALISBURY
SALISBURY WICOMICO

### *Worcester County*
SNOW HILL 4 N

### *Baltimore Independent City*
BALTIMORE CITY

### *District of Columbia*
DALECARLIA RESERVOIR
WASHINGTON NATL ARBO

## ELEVATION INDEX

| FEET | STATION NAME |
|---|---|
| 10 | ROYAL OAK 2 SSW |
| 10 | SALISBURY |
| 10 | VIENNA |
| 14 | BALTIMORE CITY |
| 20 | PRINCESS ANNE |
| 30 | MILLINGTON 1 SE |
| 30 | SNOW HILL 4 N |
| 39 | CHESTERTOWN |
| 39 | CONOWINGO DAM |
| 49 | SALISBURY WICOMICO |
| 62 | ABERDEEN PHILLIPS FI |
| 69 | WASHINGTON NATL ARBO |
| 70 | COLLEGE PARK |
| 120 | BELTSVILLE |
| 121 | OWINGS FERRY LANDING |
| 148 | BALT-WASHGTN INTL AP |
| 151 | DALECARLIA RESERVOIR |
| 151 | GLENN DALE BELL STN |
| 190 | LA PLATA 1 W |
| 365 | BENSON POLICE BARRAC |
| 400 | LAUREL 3 W |
| 420 | EMMITSBURG 2 SE |
| 420 | WOODSTOCK |
| 428 | HANCOCK |
| 430 | UNIONVILLE |
| 440 | ROCKVILLE 1 NE |
| 1460 | SAVAGE RIVER DAM |
| 1750 | CATOCTIN MOUNTAIN PK |
| 2152 | FROSTBURG 2 |
| 2421 | OAKLAND 1 SE |

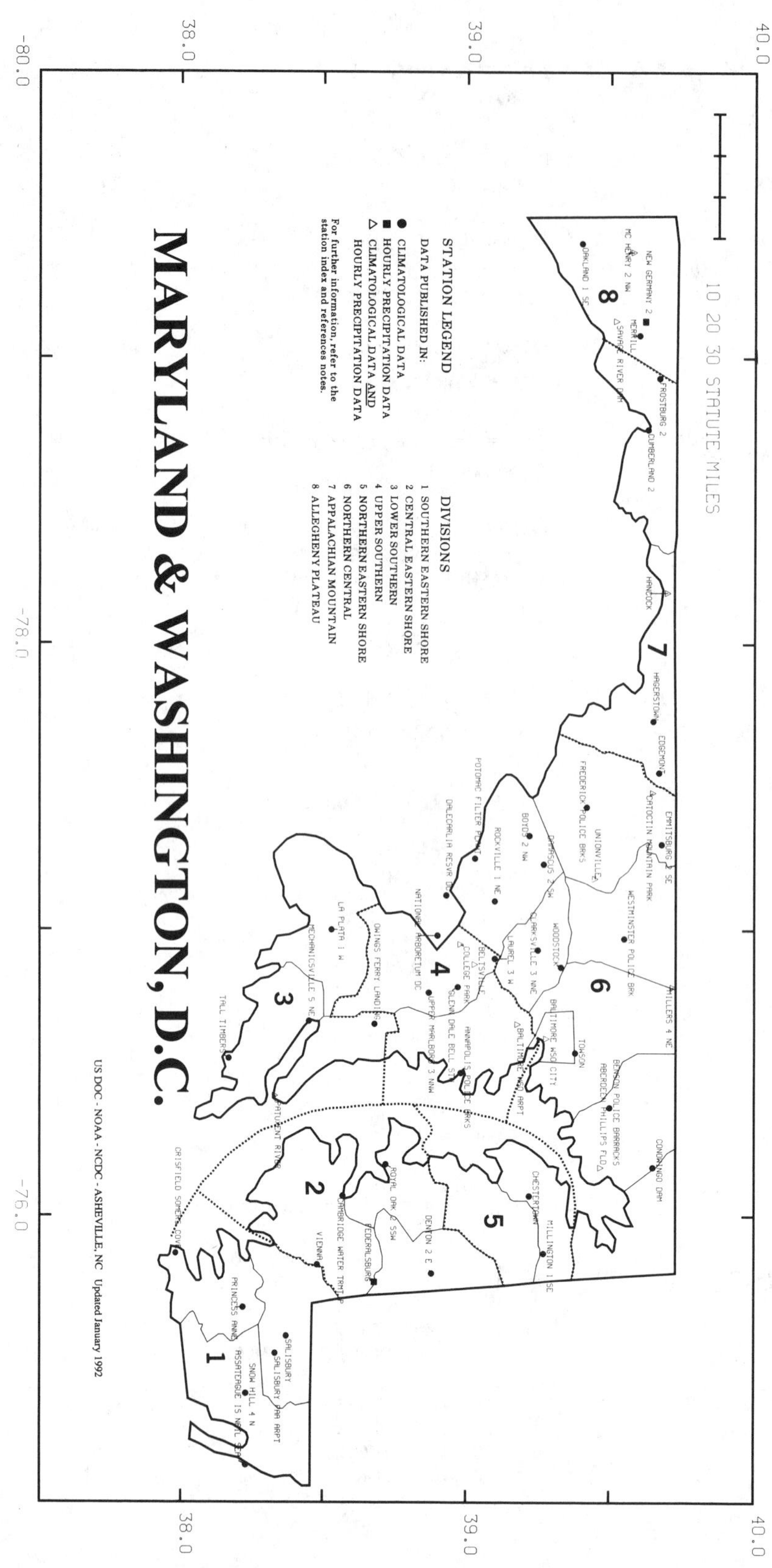
MARYLAND & WASHINGTON, D.C.
STATION LEGEND
DATA PUBLISHED IN:
● CLIMATOLOGICAL DATA
■ HOURLY PRECIPITATION DATA
△ CLIMATOLOGICAL DATA AND HOURLY PRECIPITATION DATA
For further information, refer to the station index and reference notes.
DIVISIONS
1 SOUTHERN EASTERN SHORE
2 CENTRAL EASTERN SHORE
3 LOWER SOUTHERN
4 UPPER SOUTHERN
5 NORTHERN EASTERN SHORE
6 NORTHERN CENTRAL
7 APPALACHIAN MOUNTAIN
8 ALLEGHENY PLATEAU
10 20 30 STATUTE MILES
US DOC - NOAA - NCDC - ASHEVILLE, NC Updated January 1992

## ABERDEEN PHILLIPS FI *Harford County* ELEVATION 62 ft LAT/LONG 39° 28 ' N / 76° 10 ' W

| | JAN | FEB | MAR | APR | MAY | JUN | JUL | AUG | SEP | OCT | NOV | DEC | YEAR |
|---|---|---|---|---|---|---|---|---|---|---|---|---|---|
| Maximum Temp °F | na | 44.5 | ***53.3*** | 64.7 | 74.1 | 82.6 | ***86.5*** | 84.9 | ***78.7*** | ***67.4*** | na | na | na |
| Minimum Temp °F | na | ***27.2*** | ***34.3*** | 43.1 | 52.6 | 61.7 | ***66.6*** | 65.4 | ***58.3*** | ***46.8*** | na | na | na |
| Mean Temp °F | na | ***35.9*** | ***43.8*** | 53.9 | 63.4 | 72.2 | ***76.6*** | ***75.2*** | ***68.6*** | na | na | na | na |
| Days Max Temp ≥ 90 °F | 0 | 0 | 0 | 0 | 1 | 4 | 8 | ***5*** | 2 | 0 | 0 | 0 | 20 |
| Days Max Temp ≤ 32 °F | 5 | 3 | 0 | 0 | 0 | 0 | 0 | 0 | 0 | 0 | 0 | 2 | 10 |
| Days Min Temp ≤ 32 °F | ***21*** | 18 | ***12*** | 3 | 0 | 0 | 0 | 0 | 0 | 2 | 8 | 16 | 80 |
| Days Min Temp ≤ 0 °F | 0 | 0 | 0 | 0 | 0 | 0 | 0 | 0 | 0 | 0 | 0 | 0 | 0 |
| Heating Degree Days | na | ***818*** | ***650*** | 337 | 113 | 10 | ***1*** | ***1*** | ***38*** | na | na | na | na |
| Cooling Degree Days | na | 0 | 2 | 13 | 86 | 243 | ***401*** | 335 | ***162*** | na | na | na | na |
| Total Precipitation (") | 2.88 | 2.62 | 3.37 | 3.41 | 4.39 | 4.14 | 4.32 | 4.53 | 4.21 | 3.13 | 3.09 | 3.62 | 43.71 |
| Days ≥ 0.1" Precip | ***5*** | ***4*** | ***5*** | ***5*** | ***6*** | 5 | ***5*** | 5 | 4 | 4 | ***4*** | ***5*** | 57 |
| Total Snowfall (") | ***3.9*** | 4.0 | ***0.6*** | 0.0 | 0.0 | 0.0 | 0.0 | 0.0 | 0.0 | 0.0 | 0.3 | 1.9 | 10.7 |
| Days ≥ 1" Snow Depth | ***4*** | 3 | ***0*** | 0 | 0 | 0 | 0 | 0 | 0 | 0 | 0 | 2 | 9 |

## BALT-WASHGTN INTL AP *Anne Arundel County* ELEVATION 148 ft LAT/LONG 39° 11 ' N / 76° 40 ' W

| | JAN | FEB | MAR | APR | MAY | JUN | JUL | AUG | SEP | OCT | NOV | DEC | YEAR |
|---|---|---|---|---|---|---|---|---|---|---|---|---|---|
| Maximum Temp °F | 40.6 | 44.2 | 53.6 | 64.6 | 74.2 | 83.3 | 87.5 | 85.6 | 78.8 | 67.2 | 56.7 | 46.0 | 65.2 |
| Minimum Temp °F | 24.0 | 26.2 | 34.0 | 42.9 | 52.6 | 61.9 | 67.4 | 65.9 | 58.8 | 46.2 | 37.4 | 29.1 | 45.5 |
| Mean Temp °F | 32.3 | 35.2 | 43.8 | 53.8 | 63.5 | 72.6 | 77.5 | 75.8 | 68.8 | 56.7 | 47.1 | 37.6 | 55.4 |
| Days Max Temp ≥ 90 °F | 0 | 0 | 0 | 0 | 1 | 6 | 12 | 8 | 3 | 0 | 0 | 0 | 30 |
| Days Max Temp ≤ 32 °F | 7 | 4 | 0 | 0 | 0 | 0 | 0 | 0 | 0 | 0 | 0 | 3 | 14 |
| Days Min Temp ≤ 32 °F | 25 | 21 | 14 | 3 | 0 | 0 | 0 | 0 | 0 | 2 | 10 | 20 | 95 |
| Days Min Temp ≤ 0 °F | 0 | 0 | 0 | 0 | 0 | 0 | 0 | 0 | 0 | 0 | 0 | 0 | 0 |
| Heating Degree Days | 1006 | 834 | 652 | 343 | 117 | 10 | 0 | 1 | 39 | 272 | 534 | 842 | 4650 |
| Cooling Degree Days | 0 | 0 | 3 | 14 | 85 | 254 | 417 | 331 | 155 | 22 | 1 | 0 | 1282 |
| Total Precipitation (") | 3.05 | 2.94 | 3.73 | 2.91 | 3.84 | 3.29 | 3.80 | 3.85 | 3.62 | 3.08 | 3.06 | 3.52 | 40.69 |
| Days ≥ 0.1" Precip | 6 | 6 | 7 | 6 | 7 | 6 | 6 | 6 | 5 | 5 | 6 | 6 | 72 |
| Total Snowfall (") | 6.0 | 6.6 | 2.6 | 0.0 | 0.0 | 0.0 | 0.0 | 0.0 | 0.0 | 0.0 | 1.0 | 2.8 | 19.0 |
| Days ≥ 1" Snow Depth | 6 | 5 | 1 | 0 | 0 | 0 | 0 | 0 | 0 | 0 | 0 | 2 | 14 |

## BALTIMORE CITY *Baltimore Independent City* ELEVATION 14 ft LAT/LONG 39° 17 ' N / 76° 37 ' W

| | JAN | FEB | MAR | APR | MAY | JUN | JUL | AUG | SEP | OCT | NOV | DEC | YEAR |
|---|---|---|---|---|---|---|---|---|---|---|---|---|---|
| Maximum Temp °F | 42.0 | 45.3 | 54.6 | 66.1 | 75.9 | 84.9 | 89.2 | 87.1 | 79.9 | 68.1 | 57.3 | 46.9 | 66.4 |
| Minimum Temp °F | 28.4 | 30.2 | 38.1 | 47.9 | 57.9 | 67.5 | 72.5 | 70.7 | 63.5 | 51.3 | 42.0 | 33.1 | 50.3 |
| Mean Temp °F | 35.2 | 37.8 | 46.4 | 57.1 | 66.9 | 76.2 | 80.9 | 78.9 | 71.7 | 59.7 | 49.7 | 40.0 | 58.4 |
| Days Max Temp ≥ 90 °F | 0 | 0 | 0 | 1 | 3 | 9 | 15 | 12 | 4 | 0 | 0 | 0 | 44 |
| Days Max Temp ≤ 32 °F | 6 | 3 | 0 | 0 | 0 | 0 | 0 | 0 | 0 | 0 | 0 | 2 | 11 |
| Days Min Temp ≤ 32 °F | 20 | 17 | 8 | 1 | 0 | 0 | 0 | 0 | 0 | 0 | 4 | 14 | 64 |
| Days Min Temp ≤ 0 °F | 0 | 0 | 0 | 0 | 0 | 0 | 0 | 0 | 0 | 0 | 0 | 0 | 0 |
| Heating Degree Days | 916 | 763 | 577 | 262 | 68 | 3 | 0 | 0 | 20 | 197 | 458 | 769 | 4033 |
| Cooling Degree Days | 0 | 0 | 8 | 34 | 149 | 363 | 526 | 435 | 220 | 40 | 4 | 0 | 1779 |
| Total Precipitation (") | 3.12 | 2.99 | 3.95 | 3.05 | 4.12 | 3.17 | 3.87 | 4.34 | 3.67 | 3.05 | 3.41 | 3.81 | 42.55 |
| Days ≥ 0.1" Precip | 6 | 6 | 6 | 6 | 7 | 6 | 6 | 6 | 5 | 5 | 6 | 6 | 71 |
| Total Snowfall (") | na | na | na | na | na | na | na | na | na | na | na | na | na |
| Days ≥ 1" Snow Depth | na | na | na | na | na | na | na | na | na | na | na | na | na |

## BELTSVILLE *Prince George's County* ELEVATION 120 ft LAT/LONG 39° 2 ' N / 76° 53 ' W

| | JAN | FEB | MAR | APR | MAY | JUN | JUL | AUG | SEP | OCT | NOV | DEC | YEAR |
|---|---|---|---|---|---|---|---|---|---|---|---|---|---|
| Maximum Temp °F | 40.7 | 43.8 | 53.4 | 64.1 | 73.9 | 82.5 | 87.0 | 85.4 | 78.8 | 67.2 | 56.7 | 45.6 | 64.9 |
| Minimum Temp °F | 21.5 | 24.1 | 32.1 | 40.6 | 50.7 | 59.6 | 64.9 | 63.4 | 56.0 | 42.9 | 35.0 | 26.8 | 43.1 |
| Mean Temp °F | 31.1 | 34.0 | 42.8 | 52.4 | 62.3 | 71.1 | 76.0 | 74.5 | 67.5 | 55.1 | 45.9 | 36.3 | 54.1 |
| Days Max Temp ≥ 90 °F | 0 | 0 | 0 | 0 | 1 | 5 | 11 | 8 | 3 | 0 | 0 | 0 | 28 |
| Days Max Temp ≤ 32 °F | 7 | 4 | 1 | 0 | 0 | 0 | 0 | 0 | 0 | 0 | 0 | 3 | 15 |
| Days Min Temp ≤ 32 °F | 27 | 23 | 17 | 6 | 0 | 0 | 0 | 0 | 0 | 4 | ***13*** | 24 | 114 |
| Days Min Temp ≤ 0 °F | 1 | 0 | 0 | 0 | 0 | 0 | 0 | 0 | 0 | 0 | 0 | 0 | 1 |
| Heating Degree Days | 1044 | 869 | 685 | 382 | 140 | 20 | 1 | 4 | 56 | 315 | 568 | 882 | 4966 |
| Cooling Degree Days | 0 | 0 | 3 | 10 | 75 | 225 | 383 | 306 | 141 | 19 | 2 | 0 | 1164 |
| Total Precipitation (") | 2.99 | 2.57 | 3.69 | 3.27 | 4.57 | 3.28 | 4.34 | 4.12 | 3.90 | 3.51 | 3.19 | 3.56 | 42.99 |
| Days ≥ 0.1" Precip | ***6*** | 5 | 7 | 7 | 8 | 6 | 7 | 7 | 5 | 5 | 5 | 6 | 74 |
| Total Snowfall (") | ***6.1*** | 4.2 | 1.4 | 0.0 | 0.0 | 0.0 | 0.0 | 0.0 | 0.0 | 0.0 | 0.5 | ***2.4*** | 14.6 |
| Days ≥ 1" Snow Depth | ***4*** | 4 | 1 | 0 | 0 | 0 | 0 | 0 | 0 | 0 | 0 | ***2*** | 11 |

## BENSON POLICE BARRAC *Harford County* ELEVATION 365 ft LAT/LONG 39° 30 ' N / 76° 23 ' W

| | JAN | FEB | MAR | APR | MAY | JUN | JUL | AUG | SEP | OCT | NOV | DEC | YEAR |
|---|---|---|---|---|---|---|---|---|---|---|---|---|---|
| Maximum Temp °F | 41.1 | 45.6 | 54.8 | 66.3 | 75.9 | 83.7 | 87.7 | 86.1 | 80.1 | 68.9 | 57.4 | 46.0 | 66.1 |
| Minimum Temp °F | 21.9 | 24.5 | 32.1 | 41.1 | 51.4 | 60.0 | 65.0 | 63.4 | 56.5 | 44.4 | 35.8 | 27.1 | 43.6 |
| Mean Temp °F | 31.5 | 35.1 | 43.4 | 53.6 | 63.7 | 71.9 | 76.4 | 74.8 | 68.3 | 56.5 | 46.5 | 36.7 | 54.9 |
| Days Max Temp ≥ 90 °F | 0 | 0 | 0 | 0 | 1 | 6 | 12 | 8 | 3 | 0 | 0 | 0 | 30 |
| Days Max Temp ≤ 32 °F | 6 | 3 | 0 | 0 | 0 | 0 | 0 | 0 | 0 | 0 | 0 | 2 | 11 |
| Days Min Temp ≤ 32 °F | 27 | 23 | 17 | 5 | 0 | 0 | 0 | 0 | 0 | 3 | 12 | 23 | 110 |
| Days Min Temp ≤ 0 °F | 1 | 0 | 0 | 0 | 0 | 0 | 0 | 0 | 0 | 0 | 0 | 0 | 1 |
| Heating Degree Days | 1033 | 839 | 662 | 346 | 109 | 12 | 0 | 2 | 42 | 273 | 550 | 871 | 4739 |
| Cooling Degree Days | 0 | 0 | 2 | 16 | 96 | 257 | 406 | 327 | 152 | 17 | 1 | 0 | 1274 |
| Total Precipitation (") | 3.35 | 2.86 | 4.18 | 3.73 | 4.86 | 4.14 | 4.74 | 4.87 | 4.17 | 3.48 | 3.88 | 3.93 | 48.19 |
| Days ≥ 0.1" Precip | 7 | 6 | 7 | 7 | 9 | 7 | 7 | 7 | 6 | 6 | 7 | 6 | 82 |
| Total Snowfall (") | 7.5 | 5.4 | 2.0 | 0.0 | 0.0 | 0.0 | 0.0 | 0.0 | 0.0 | 0.0 | 0.6 | *2.2* | 17.7 |
| Days ≥ 1" Snow Depth | 9 | 6 | 2 | 0 | 0 | 0 | 0 | 0 | 0 | 0 | 0 | 3 | 20 |

## CATOCTIN MOUNTAIN PK *Frederick County* ELEVATION 1750 ft LAT/LONG 39° 39 ' N / 77° 27 ' W

| | JAN | FEB | MAR | APR | MAY | JUN | JUL | AUG | SEP | OCT | NOV | DEC | YEAR |
|---|---|---|---|---|---|---|---|---|---|---|---|---|---|
| Maximum Temp °F | 37.0 | 39.0 | 49.1 | 61.9 | 70.9 | 77.1 | 80.2 | 78.4 | 72.0 | 61.9 | *51.3* | 40.2 | 59.9 |
| Minimum Temp °F | 21.7 | 22.6 | 30.5 | 40.3 | 50.5 | 59.0 | 63.6 | 62.4 | 56.0 | 44.8 | 35.8 | 25.7 | 42.7 |
| Mean Temp °F | 29.4 | 30.8 | 39.8 | 51.2 | 60.7 | 68.1 | 71.9 | 70.4 | 64.0 | 53.4 | *43.5* | 33.0 | 51.4 |
| Days Max Temp ≥ 90 °F | 0 | 0 | 0 | 0 | 0 | 0 | 1 | 0 | 0 | 0 | 0 | 0 | 1 |
| Days Max Temp ≤ 32 °F | 12 | 8 | 2 | 0 | 0 | 0 | 0 | 0 | 0 | 0 | 1 | 7 | 30 |
| Days Min Temp ≤ 32 °F | 27 | 23 | 18 | 6 | 0 | 0 | 0 | 0 | 0 | *2* | *12* | 24 | 112 |
| Days Min Temp ≤ 0 °F | 1 | 1 | 0 | 0 | 0 | 0 | 0 | 0 | 0 | 0 | 0 | 0 | 2 |
| Heating Degree Days | 1101 | 960 | 776 | 420 | 171 | 34 | 5 | 13 | 92 | 360 | *638* | 986 | 5556 |
| Cooling Degree Days | 0 | 0 | 3 | 12 | 52 | 142 | 248 | 186 | 72 | 8 | 1 | 0 | 724 |
| Total Precipitation (") | 3.41 | 3.24 | 4.24 | 4.05 | 5.06 | 4.49 | 3.62 | 3.78 | 4.56 | 4.04 | 4.15 | 3.64 | 48.28 |
| Days ≥ 0.1" Precip | 7 | 6 | 7 | 7 | 9 | 7 | 7 | 7 | 6 | *6* | 7 | 6 | 82 |
| Total Snowfall (") | 9.9 | 9.6 | 5.6 | 1.1 | 0.0 | 0.0 | 0.0 | 0.0 | 0.0 | 0.3 | 2.7 | 3.9 | 33.1 |
| Days ≥ 1" Snow Depth | 16 | 15 | *8* | 1 | 0 | 0 | 0 | 0 | 0 | 0 | *2* | 6 | 48 |

## CHESTERTOWN *Kent County* ELEVATION 39 ft LAT/LONG 39° 13 ' N / 76° 4 ' W

| | JAN | FEB | MAR | APR | MAY | JUN | JUL | AUG | SEP | OCT | NOV | DEC | YEAR |
|---|---|---|---|---|---|---|---|---|---|---|---|---|---|
| Maximum Temp °F | 40.3 | 43.6 | 53.1 | 64.3 | 74.3 | 83.0 | 87.1 | 85.7 | 79.0 | 67.6 | 56.7 | 45.7 | 65.0 |
| Minimum Temp °F | 24.2 | 26.4 | 34.2 | 43.1 | 53.0 | 62.0 | 67.1 | 65.5 | 58.4 | 46.5 | 37.8 | 29.3 | 45.6 |
| Mean Temp °F | 32.3 | 35.0 | 43.7 | 53.7 | 63.6 | 72.6 | 77.1 | 75.6 | 68.7 | 57.1 | 47.2 | 37.5 | 55.3 |
| Days Max Temp ≥ 90 °F | 0 | 0 | 0 | 0 | 1 | 5 | 11 | 8 | 2 | 0 | 0 | 0 | 27 |
| Days Max Temp ≤ 32 °F | 7 | 4 | 1 | 0 | 0 | 0 | 0 | 0 | 0 | 0 | 0 | 3 | 15 |
| Days Min Temp ≤ 32 °F | 25 | 22 | 14 | 2 | 0 | 0 | 0 | 0 | 0 | 1 | 9 | 21 | 94 |
| Days Min Temp ≤ 0 °F | 0 | 0 | 0 | 0 | 0 | 0 | 0 | 0 | 0 | 0 | 0 | 0 | 0 |
| Heating Degree Days | 1007 | 840 | 655 | 343 | 111 | 9 | 0 | 2 | 37 | 260 | 528 | 845 | 4637 |
| Cooling Degree Days | 0 | 0 | 2 | 12 | 88 | 262 | 413 | 342 | 156 | 25 | 2 | 0 | 1302 |
| Total Precipitation (") | 3.17 | 2.93 | 3.86 | 3.21 | 4.14 | 4.04 | 3.85 | 4.05 | 3.62 | 3.18 | 3.20 | 3.80 | 43.05 |
| Days ≥ 0.1" Precip | 6 | 6 | 7 | 7 | 7 | 6 | 6 | 6 | 5 | 5 | 6 | 6 | 73 |
| Total Snowfall (") | 5.8 | 6.3 | 2.0 | 0.1 | 0.0 | 0.0 | 0.0 | 0.0 | 0.0 | 0.0 | 0.7 | 2.5 | 17.4 |
| Days ≥ 1" Snow Depth | 7 | 5 | 1 | 0 | 0 | 0 | 0 | 0 | 0 | 0 | 0 | 2 | 15 |

## COLLEGE PARK *Prince George's County* ELEVATION 70 ft LAT/LONG 38° 59 ' N / 76° 56 ' W

| | JAN | FEB | MAR | APR | MAY | JUN | JUL | AUG | SEP | OCT | NOV | DEC | YEAR |
|---|---|---|---|---|---|---|---|---|---|---|---|---|---|
| Maximum Temp °F | 41.9 | 45.8 | 55.2 | 66.5 | 76.1 | 84.5 | 88.5 | 86.7 | 80.0 | 68.5 | 57.9 | 46.9 | 66.5 |
| Minimum Temp °F | 24.8 | 27.0 | 34.7 | 43.4 | 53.3 | 62.4 | 67.8 | 66.1 | 59.0 | 46.3 | 37.9 | 29.8 | 46.0 |
| Mean Temp °F | 33.4 | 36.4 | 45.0 | 55.0 | 64.8 | 73.5 | 78.2 | 76.4 | 69.5 | 57.4 | 47.9 | 38.4 | 56.3 |
| Days Max Temp ≥ 90 °F | 0 | 0 | 0 | 1 | 2 | 8 | 14 | 11 | 4 | 0 | 0 | 0 | 40 |
| Days Max Temp ≤ 32 °F | 6 | 3 | 0 | 0 | 0 | 0 | 0 | 0 | 0 | 0 | 0 | 2 | 11 |
| Days Min Temp ≤ 32 °F | 24 | 21 | 13 | 3 | 0 | 0 | 0 | 0 | 0 | 2 | 10 | 20 | 93 |
| Days Min Temp ≤ 0 °F | 0 | 0 | 0 | 0 | 0 | 0 | 0 | 0 | 0 | 0 | 0 | 1 | 1 |
| Heating Degree Days | 973 | 801 | 618 | 315 | 99 | 10 | 0 | 3 | 35 | 252 | 509 | 818 | 4433 |
| Cooling Degree Days | 0 | 0 | 6 | 22 | 109 | 287 | 451 | 362 | 181 | 28 | 2 | 0 | 1448 |
| Total Precipitation (") | 3.10 | 2.74 | 3.65 | 3.16 | 4.48 | 3.60 | 4.44 | 4.42 | 3.82 | 3.30 | 3.24 | 3.42 | 43.37 |
| Days ≥ 0.1" Precip | *6* | 6 | 7 | 6 | 8 | 6 | 6 | 7 | 5 | 5 | 5 | 6 | 73 |
| Total Snowfall (") | 5.7 | 5.4 | 1.9 | 0.0 | 0.0 | 0.0 | 0.0 | 0.0 | 0.0 | 0.0 | 0.9 | 1.9 | 15.8 |
| Days ≥ 1" Snow Depth | 7 | 4 | 1 | 0 | 0 | 0 | 0 | 0 | 0 | 0 | 0 | 1 | 13 |

## CONOWINGO DAM *Harford County* ELEVATION 39 ft LAT/LONG 39° 39 ' N / 76° 10 ' W

| | JAN | FEB | MAR | APR | MAY | JUN | JUL | AUG | SEP | OCT | NOV | DEC | YEAR |
|---|---|---|---|---|---|---|---|---|---|---|---|---|---|
| Maximum Temp °F | 38.6 | 42.2 | 52.1 | 63.4 | ***73.7*** | 82.6 | 87.1 | 85.5 | 78.4 | 66.7 | ***54.8*** | 44.1 | 64.1 |
| Minimum Temp °F | 20.5 | 22.7 | 31.0 | 40.3 | 50.7 | 59.6 | 64.3 | 63.4 | 56.1 | 43.7 | 34.4 | 26.5 | 42.8 |
| Mean Temp °F | 29.6 | ***32.6*** | 41.6 | 51.8 | ***62.2*** | 71.1 | 75.7 | 74.5 | 67.3 | 55.3 | ***44.8*** | 35.3 | 53.5 |
| Days Max Temp ≥ 90 °F | 0 | 0 | 0 | 0 | 1 | 5 | 11 | 8 | 2 | 0 | 0 | 0 | 27 |
| Days Max Temp ≤ 32 °F | 8 | 4 | 0 | 0 | 0 | 0 | 0 | 0 | 0 | 0 | 0 | 3 | 15 |
| Days Min Temp ≤ 32 °F | 28 | 24 | 18 | 5 | 0 | 0 | 0 | 0 | 0 | 4 | 13 | 24 | 116 |
| Days Min Temp ≤ 0 °F | 1 | 0 | 0 | 0 | 0 | 0 | 0 | 0 | 0 | 0 | 0 | 0 | 1 |
| Heating Degree Days | 1091 | ***908*** | 721 | 396 | ***134*** | 14 | 1 | 2 | 50 | 306 | ***599*** | 913 | 5135 |
| Cooling Degree Days | 0 | 0 | 0 | 9 | ***65*** | 219 | 370 | 308 | 124 | 15 | ***0*** | 0 | 1110 |
| Total Precipitation (") | 3.43 | 2.91 | 3.61 | 3.56 | 4.51 | 4.15 | 4.19 | 4.54 | 4.19 | 3.38 | 3.71 | 3.91 | 46.09 |
| Days ≥ 0.1" Precip | 7 | 6 | 7 | 7 | 8 | 7 | 6 | 7 | 6 | 5 | 6 | 6 | 78 |
| Total Snowfall (") | 5.1 | ***4.7*** | 1.0 | 0.0 | 0.0 | 0.0 | 0.0 | 0.0 | 0.0 | 0.0 | 0.6 | 1.9 | 13.3 |
| Days ≥ 1" Snow Depth | ***8*** | ***5*** | 1 | 0 | 0 | 0 | 0 | 0 | 0 | 0 | ***0*** | 2 | 16 |

## DALECARLIA RESERVOIR *District of Columbia* ELEVATION 151 ft LAT/LONG 38° 56 ' N / 77° 7 ' W

| | JAN | FEB | MAR | APR | MAY | JUN | JUL | AUG | SEP | OCT | NOV | DEC | YEAR |
|---|---|---|---|---|---|---|---|---|---|---|---|---|---|
| Maximum Temp °F | 43.3 | 47.5 | 57.4 | 68.6 | 77.5 | 85.3 | 88.8 | 87.1 | 80.8 | 69.7 | 59.0 | 48.2 | 67.8 |
| Minimum Temp °F | 22.8 | 25.5 | 33.1 | 42.2 | 52.1 | 61.0 | 66.1 | 64.8 | 57.5 | 44.7 | 35.8 | 27.6 | 44.4 |
| Mean Temp °F | 32.9 | 36.7 | 45.4 | 55.5 | 64.9 | 73.2 | 77.4 | 76.0 | 69.2 | 57.2 | 47.4 | 37.9 | 56.1 |
| Days Max Temp ≥ 90 °F | 0 | 0 | 0 | 0 | 2 | 8 | 14 | 10 | 3 | 0 | 0 | 0 | 37 |
| Days Max Temp ≤ 32 °F | 4 | 2 | 0 | 0 | 0 | 0 | 0 | 0 | 0 | 0 | 0 | 2 | 8 |
| Days Min Temp ≤ 32 °F | 26 | 22 | 15 | 4 | 0 | 0 | 0 | 0 | 0 | 3 | 12 | 22 | 104 |
| Days Min Temp ≤ 0 °F | 0 | 0 | 0 | 0 | 0 | 0 | 0 | 0 | 0 | 0 | 0 | 0 | 0 |
| Heating Degree Days | 988 | 793 | 606 | 299 | 93 | 9 | 0 | 2 | 37 | 258 | 524 | 834 | 4443 |
| Cooling Degree Days | 0 | 0 | 3 | ***20*** | 100 | 273 | 413 | 344 | 168 | 25 | 2 | 0 | 1348 |
| Total Precipitation (") | 3.09 | 2.87 | 3.86 | 3.46 | 4.28 | 3.47 | 4.43 | 4.22 | 3.78 | 3.61 | 3.45 | 3.62 | 44.14 |
| Days ≥ 0.1" Precip | 6 | 6 | 7 | 7 | 8 | 6 | 8 | 6 | 5 | 6 | 6 | 6 | 77 |
| Total Snowfall (") | ***3.6*** | ***3.4*** | 0.7 | 0.0 | 0.0 | 0.0 | 0.0 | 0.0 | 0.0 | 0.0 | 0.0 | ***0.7*** | 8.4 |
| Days ≥ 1" Snow Depth | na | na | ***0*** | 0 | 0 | 0 | 0 | 0 | 0 | 0 | 0 | ***0*** | na |

## EMMITSBURG 2 SE *Frederick County* ELEVATION 420 ft LAT/LONG 39° 41 ' N / 77° 18 ' W

| | JAN | FEB | MAR | APR | MAY | JUN | JUL | AUG | SEP | OCT | NOV | DEC | YEAR |
|---|---|---|---|---|---|---|---|---|---|---|---|---|---|
| Maximum Temp °F | 38.6 | 42.8 | 53.1 | 64.5 | 73.9 | 82.3 | 86.2 | 84.6 | 77.6 | 66.2 | 54.7 | 43.6 | 64.0 |
| Minimum Temp °F | 20.2 | 22.1 | 30.3 | 38.9 | 48.5 | 57.2 | 62.3 | 60.7 | 53.5 | 41.6 | 33.7 | 25.8 | 41.2 |
| Mean Temp °F | 29.4 | 32.5 | 41.7 | 51.7 | 61.3 | 69.8 | 74.3 | 72.6 | 65.5 | 53.9 | 44.2 | 34.8 | 52.6 |
| Days Max Temp ≥ 90 °F | 0 | 0 | 0 | 0 | 0 | 4 | 8 | 6 | 2 | 0 | 0 | 0 | 20 |
| Days Max Temp ≤ 32 °F | 9 | 4 | 1 | 0 | 0 | 0 | 0 | 0 | 0 | 0 | 0 | 3 | 17 |
| Days Min Temp ≤ 32 °F | 27 | 24 | 19 | 8 | 1 | 0 | 0 | 0 | 0 | 6 | 14 | 24 | 123 |
| Days Min Temp ≤ 0 °F | 2 | 1 | 0 | 0 | 0 | 0 | 0 | 0 | 0 | 0 | 0 | 0 | 3 |
| Heating Degree Days | 1095 | 912 | 715 | 398 | 155 | 23 | 2 | 8 | 80 | 346 | 616 | 931 | 5281 |
| Cooling Degree Days | 0 | 0 | 1 | 7 | 51 | 178 | 321 | 250 | 105 | 13 | 0 | 0 | 926 |
| Total Precipitation (") | 3.11 | 2.99 | 3.86 | 3.77 | 4.44 | 3.94 | 3.40 | 3.65 | 4.23 | 3.53 | 3.79 | 3.44 | 44.15 |
| Days ≥ 0.1" Precip | 7 | 6 | 7 | 7 | 9 | 7 | 6 | 6 | 6 | 5 | 7 | 6 | 79 |
| Total Snowfall (") | 10.8 | 9.7 | 5.5 | 0.8 | 0.0 | 0.0 | 0.0 | 0.0 | 0.0 | 0.1 | 1.7 | 4.8 | 33.4 |
| Days ≥ 1" Snow Depth | 13 | 9 | 3 | 0 | 0 | 0 | 0 | 0 | 0 | 0 | 1 | 4 | 30 |

## FROSTBURG 2 *Allegany County* ELEVATION 2152 ft LAT/LONG 39° 39 ' N / 78° 55 ' W

| | JAN | FEB | MAR | APR | MAY | JUN | JUL | AUG | SEP | OCT | NOV | DEC | YEAR |
|---|---|---|---|---|---|---|---|---|---|---|---|---|---|
| Maximum Temp °F | ***32.4*** | ***36.4*** | ***45.7*** | ***56.9*** | ***66.9*** | ***74.9*** | ***79.0*** | ***77.5*** | ***70.4*** | ***58.8*** | ***48.7*** | ***37.5*** | 57.1 |
| Minimum Temp °F | ***17.0*** | ***19.1*** | ***26.5*** | ***36.4*** | ***46.0*** | ***53.3*** | ***58.3*** | ***57.0*** | ***50.3*** | ***39.5*** | ***32.1*** | ***22.3*** | 38.2 |
| Mean Temp °F | ***24.7*** | ***27.8*** | ***36.1*** | ***46.7*** | ***56.5*** | ***64.1*** | ***68.6*** | ***67.3*** | ***60.4*** | ***49.2*** | ***40.4*** | ***29.9*** | 47.6 |
| Days Max Temp ≥ 90 °F | ***0*** | ***0*** | ***0*** | ***0*** | ***0*** | ***0*** | ***1*** | ***1*** | ***0*** | ***0*** | ***0*** | ***0*** | 2 |
| Days Max Temp ≤ 32 °F | ***16*** | ***11*** | ***5*** | ***0*** | ***0*** | ***0*** | ***0*** | ***0*** | ***0*** | ***0*** | ***2*** | ***11*** | 45 |
| Days Min Temp ≤ 32 °F | ***29*** | ***25*** | ***23*** | ***10*** | ***1*** | ***0*** | ***0*** | ***0*** | ***0*** | 7 | ***17*** | ***27*** | 139 |
| Days Min Temp ≤ 0 °F | ***3*** | ***1*** | ***0*** | ***0*** | ***0*** | ***0*** | ***0*** | ***0*** | ***0*** | ***0*** | ***0*** | ***1*** | 5 |
| Heating Degree Days | ***1242*** | ***1044*** | ***889*** | ***548*** | ***274*** | ***89*** | ***24*** | ***40*** | ***171*** | ***486*** | ***731*** | ***1081*** | 6619 |
| Cooling Degree Days | 0 | 0 | 1 | 4 | 18 | 72 | 157 | 117 | 41 | 2 | 0 | 0 | 412 |
| Total Precipitation (") | ***3.38*** | ***2.98*** | ***3.99*** | ***4.22*** | ***4.74*** | ***3.77*** | ***4.13*** | ***3.87*** | ***3.32*** | ***3.49*** | ***3.86*** | ***3.44*** | 45.19 |
| Days ≥ 0.1" Precip | ***8*** | ***7*** | ***8*** | ***8*** | ***9*** | ***8*** | ***8*** | ***7*** | ***7*** | ***7*** | ***8*** | ***8*** | 93 |
| Total Snowfall (") | ***28.9*** | ***20.3*** | ***17.6*** | ***3.7*** | ***0.0*** | ***0.0*** | ***0.0*** | ***0.0*** | ***0.0*** | ***0.5*** | ***6.7*** | ***18.2*** | 95.9 |
| Days ≥ 1" Snow Depth | ***23*** | ***20*** | ***14*** | ***2*** | ***0*** | ***0*** | ***0*** | ***0*** | ***0*** | ***0*** | ***5*** | ***16*** | 80 |

## GLENN DALE BELL STN *Prince George's County* ELEVATION 151 ft LAT/LONG 38° 58 ' N / 76° 48 ' W

| | JAN | FEB | MAR | APR | MAY | JUN | JUL | AUG | SEP | OCT | NOV | DEC | YEAR |
|---|---|---|---|---|---|---|---|---|---|---|---|---|---|
| Maximum Temp °F | 43.5 | 47.7 | 57.2 | 67.9 | 76.6 | 84.3 | 88.2 | 86.6 | 80.6 | 69.7 | 59.2 | 48.5 | 67.5 |
| Minimum Temp °F | 22.1 | 24.1 | 31.5 | 39.8 | 49.7 | 58.4 | 63.5 | 62.1 | 55.1 | 42.8 | 34.8 | 27.2 | 42.6 |
| Mean Temp °F | 32.8 | 36.0 | 44.4 | 53.9 | 63.2 | 71.4 | 75.9 | 74.4 | 67.9 | 56.2 | 47.1 | 37.8 | 55.1 |
| Days Max Temp ≥ 90 °F | 0 | 0 | 0 | 1 | 2 | 7 | 14 | 9 | 4 | 0 | 0 | 0 | 37 |
| Days Max Temp ≤ 32 °F | 4 | 2 | 0 | 0 | 0 | 0 | 0 | 0 | 0 | 0 | 0 | 2 | 8 |
| Days Min Temp ≤ 32 °F | 26 | 22 | 18 | 7 | 1 | 0 | 0 | 0 | 0 | 5 | 13 | 22 | 114 |
| Days Min Temp ≤ 0 °F | 1 | 0 | 0 | 0 | 0 | 0 | 0 | 0 | 0 | 0 | 0 | 0 | 1 |
| Heating Degree Days | 990 | 814 | 635 | 340 | 121 | 17 | 1 | 4 | 51 | 282 | 535 | 836 | 4626 |
| Cooling Degree Days | 0 | 0 | 2 | 14 | 74 | 221 | 368 | 296 | 144 | 20 | 2 | 0 | 1141 |
| Total Precipitation (") | 3.12 | 2.67 | 3.78 | 3.31 | 4.59 | 3.57 | 4.53 | 4.50 | 3.68 | 3.52 | 3.33 | 3.49 | 44.09 |
| Days ≥ 0.1" Precip | 6 | 5 | 7 | 7 | 8 | 6 | 7 | 7 | 5 | 6 | 6 | 7 | 77 |
| Total Snowfall (") | 6.1 | 5.7 | 2.3 | 0.0 | 0.0 | 0.0 | 0.0 | 0.0 | 0.0 | 0.1 | 1.3 | 2.5 | 18.0 |
| Days ≥ 1" Snow Depth | 7 | 4 | 1 | 0 | 0 | 0 | 0 | 0 | 0 | 0 | 0 | 2 | 14 |

## HANCOCK *Washington County* ELEVATION 428 ft LAT/LONG 39° 42 ' N / 78° 11 ' W

| | JAN | FEB | MAR | APR | MAY | JUN | JUL | AUG | SEP | OCT | NOV | DEC | YEAR |
|---|---|---|---|---|---|---|---|---|---|---|---|---|---|
| Maximum Temp °F | 38.3 | 41.8 | 52.2 | 64.3 | 73.6 | 81.6 | 85.4 | 83.7 | 76.6 | 65.2 | 54.5 | 42.7 | 63.3 |
| Minimum Temp °F | 20.1 | 22.1 | 29.8 | 38.4 | 47.8 | 56.3 | 61.7 | 60.2 | 53.1 | 40.5 | 33.2 | 25.6 | 40.7 |
| Mean Temp °F | 29.2 | 32.0 | 41.0 | 51.4 | 60.7 | 69.0 | 73.6 | 71.9 | 64.9 | 52.9 | 43.9 | 34.2 | 52.1 |
| Days Max Temp ≥ 90 °F | 0 | 0 | 0 | 0 | 1 | 4 | 8 | 5 | 2 | 0 | 0 | 0 | 20 |
| Days Max Temp ≤ 32 °F | 9 | 6 | 1 | 0 | 0 | 0 | 0 | 0 | 0 | 0 | 0 | 5 | 21 |
| Days Min Temp ≤ 32 °F | 27 | 24 | 19 | 9 | 1 | 0 | 0 | 0 | 0 | 8 | 15 | 23 | 126 |
| Days Min Temp ≤ 0 °F | 2 | 1 | 0 | 0 | 0 | 0 | 0 | 0 | 0 | 0 | 0 | 0 | 3 |
| Heating Degree Days | 1102 | 925 | 737 | 411 | 175 | 35 | 4 | 11 | 92 | 377 | 628 | 949 | 5446 |
| Cooling Degree Days | 0 | 0 | 1 | 10 | 54 | 168 | 301 | 236 | 95 | 10 | 0 | 0 | 875 |
| Total Precipitation (") | 2.39 | 2.11 | 3.14 | 3.21 | 3.75 | 3.24 | 3.87 | 3.39 | 3.16 | 3.15 | 3.07 | 2.80 | 37.28 |
| Days ≥ 0.1" Precip | 6 | 5 | 7 | 7 | 8 | 6 | 7 | 6 | 6 | 5 | 6 | 6 | 75 |
| Total Snowfall (") | 8.7 | 7.2 | 5.6 | 0.3 | 0.0 | 0.0 | 0.0 | 0.0 | 0.0 | 0.0 | 1.5 | 3.9 | 27.2 |
| Days ≥ 1" Snow Depth | 10 | 9 | 2 | 0 | 0 | 0 | 0 | 0 | 0 | 0 | 1 | 3 | 25 |

## LA PLATA 1 W *Charles County* ELEVATION 190 ft LAT/LONG 38° 32 ' N / 76° 58 ' W

| | JAN | FEB | MAR | APR | MAY | JUN | JUL | AUG | SEP | OCT | NOV | DEC | YEAR |
|---|---|---|---|---|---|---|---|---|---|---|---|---|---|
| Maximum Temp °F | 43.6 | 47.8 | 57.6 | 68.1 | 75.3 | 81.9 | 85.5 | 84.3 | 78.8 | 68.6 | 59.1 | 48.2 | 66.6 |
| Minimum Temp °F | 24.9 | 27.0 | 34.5 | 42.5 | 52.6 | 61.1 | 66.1 | 64.9 | 58.0 | 46.4 | 37.7 | 29.5 | 45.4 |
| Mean Temp °F | 34.2 | 37.4 | 46.1 | 55.3 | 64.0 | 71.5 | 75.8 | 74.6 | 68.4 | 57.5 | 48.4 | 38.9 | 56.0 |
| Days Max Temp ≥ 90 °F | 0 | 0 | 0 | 1 | 1 | 3 | 7 | 6 | 2 | 0 | 0 | 0 | 20 |
| Days Max Temp ≤ 32 °F | 4 | 2 | 0 | 0 | 0 | 0 | 0 | 0 | 0 | 0 | 0 | 2 | 8 |
| Days Min Temp ≤ 32 °F | 24 | 20 | 14 | 4 | 0 | 0 | 0 | 0 | 0 | 2 | 10 | 20 | 94 |
| Days Min Temp ≤ 0 °F | 0 | 0 | 0 | 0 | 0 | 0 | 0 | 0 | 0 | 0 | 0 | 0 | 0 |
| Heating Degree Days | 947 | 772 | 584 | 303 | 104 | 13 | 1 | 3 | 42 | 247 | 494 | 802 | 4312 |
| Cooling Degree Days | 0 | 0 | 7 | 19 | 84 | 219 | 358 | 298 | 150 | 26 | 3 | 0 | 1164 |
| Total Precipitation (") | 3.09 | 2.76 | 3.60 | 3.10 | 4.08 | 3.69 | 4.11 | 4.65 | 3.92 | 3.34 | 3.22 | 3.38 | 42.94 |
| Days ≥ 0.1" Precip | 6 | 6 | 7 | 6 | 7 | 6 | 6 | 6 | 5 | 5 | 6 | 6 | 72 |
| Total Snowfall (") | 5.9 | 5.7 | 2.1 | 0.1 | 0.0 | 0.0 | 0.0 | 0.0 | 0.0 | 0.0 | 0.8 | 2.7 | 17.3 |
| Days ≥ 1" Snow Depth | 7 | 5 | 1 | 0 | 0 | 0 | 0 | 0 | 0 | 0 | 0 | 2 | 15 |

## LAUREL 3 W *Prince George's County* ELEVATION 400 ft LAT/LONG 39° 6 ' N / 76° 54 ' W

| | JAN | FEB | MAR | APR | MAY | JUN | JUL | AUG | SEP | OCT | NOV | DEC | YEAR |
|---|---|---|---|---|---|---|---|---|---|---|---|---|---|
| Maximum Temp °F | 40.8 | 44.6 | 54.7 | 65.5 | 75.0 | 83.4 | 87.9 | 86.5 | 79.2 | 67.9 | 56.9 | 45.8 | 65.7 |
| Minimum Temp °F | 23.7 | 26.2 | 34.3 | 43.8 | 53.6 | 62.0 | 67.1 | 66.3 | 58.8 | 46.7 | 38.3 | 29.3 | 45.8 |
| Mean Temp °F | 32.2 | 35.4 | 44.6 | 54.7 | 64.3 | 72.8 | 77.6 | 76.5 | 69.0 | 57.3 | 47.6 | 37.6 | 55.8 |
| Days Max Temp ≥ 90 °F | 0 | 0 | 0 | 0 | 1 | 6 | 12 | 9 | 3 | 0 | 0 | 0 | 31 |
| Days Max Temp ≤ 32 °F | 6 | 4 | 0 | 0 | 0 | 0 | 0 | 0 | 0 | 0 | 0 | 3 | 13 |
| Days Min Temp ≤ 32 °F | 25 | 22 | 13 | 2 | 0 | 0 | 0 | 0 | 0 | 1 | 9 | 20 | 92 |
| Days Min Temp ≤ 0 °F | 0 | 0 | 0 | 0 | 0 | 0 | 0 | 0 | 0 | 0 | 0 | 0 | 0 |
| Heating Degree Days | 1011 | 831 | 630 | 321 | 102 | 10 | 0 | 1 | 36 | 252 | 518 | 844 | 4556 |
| Cooling Degree Days | 0 | 0 | 5 | 20 | 103 | 266 | 434 | *366* | 157 | 23 | 2 | 0 | 1376 |
| Total Precipitation (") | 2.98 | 2.84 | 3.94 | 3.48 | 4.58 | 3.52 | 4.24 | 3.95 | 3.90 | 3.49 | 3.75 | 3.67 | 44.34 |
| Days ≥ 0.1" Precip | 6 | 6 | 8 | 7 | 8 | 6 | 6 | 6 | 5 | 5 | 6 | 7 | 76 |
| Total Snowfall (") | na | na | 1.8 | 0.0 | 0.0 | 0.0 | 0.0 | 0.0 | 0.0 | 0.0 | 0.8 | *1.2* | na |
| Days ≥ 1" Snow Depth | na | na | *0* | 0 | 0 | 0 | 0 | 0 | 0 | 0 | 0 | na | na |

## MILLINGTON 1 SE *Kent County* ELEVATION 30 ft LAT/LONG 39° 15' N / 75° 50' W

| | JAN | FEB | MAR | APR | MAY | JUN | JUL | AUG | SEP | OCT | NOV | DEC | YEAR |
|---|---|---|---|---|---|---|---|---|---|---|---|---|---|
| Maximum Temp °F | 41.4 | 44.4 | 54.6 | 65.0 | 74.8 | 82.7 | 87.0 | 85.6 | 79.2 | 67.9 | 57.7 | 46.4 | 65.6 |
| Minimum Temp °F | 22.4 | 23.8 | 32.5 | 40.5 | 50.6 | 59.5 | 64.5 | 63.1 | 55.9 | 43.6 | 36.0 | ***27.2*** | 43.3 |
| Mean Temp °F | 31.9 | 34.1 | 43.6 | 52.7 | 62.7 | 71.1 | 75.7 | 74.3 | 67.5 | 55.7 | 46.9 | ***36.9*** | 54.4 |
| Days Max Temp ≥ 90 °F | 0 | 0 | 0 | 0 | 1 | 5 | 10 | 7 | 3 | 0 | 0 | 0 | 26 |
| Days Max Temp ≤ 32 °F | 6 | 4 | 0 | 0 | 0 | 0 | 0 | 0 | 0 | 0 | 0 | 2 | 12 |
| Days Min Temp ≤ 32 °F | 26 | 23 | 16 | 6 | 0 | 0 | 0 | 0 | 0 | 4 | 11 | 22 | 108 |
| Days Min Temp ≤ 0 °F | 1 | 1 | 0 | 0 | 0 | 0 | 0 | 0 | 0 | 0 | 0 | 0 | 2 |
| Heating Degree Days | 1019 | 865 | 659 | 372 | 125 | 17 | 1 | 3 | 50 | 295 | 536 | ***866*** | 4808 |
| Cooling Degree Days | ***0*** | ***0*** | ***3*** | ***11*** | ***74*** | ***222*** | ***377*** | ***292*** | ***133*** | ***17*** | ***3*** | ***0*** | 1132 |
| Total Precipitation (") | 3.21 | 2.75 | 3.66 | 3.32 | 4.21 | 3.82 | 3.93 | 4.36 | 3.86 | 3.20 | 3.06 | 3.78 | 43.16 |
| Days ≥ 0.1" Precip | 6 | 5 | 6 | 7 | 7 | 6 | 6 | 6 | 5 | 5 | 6 | 6 | 71 |
| Total Snowfall (") | ***5.0*** | 5.7 | 1.7 | 0.0 | 0.0 | 0.0 | 0.0 | 0.0 | 0.0 | 0.0 | 0.5 | 1.9 | 14.8 |
| Days ≥ 1" Snow Depth | ***6*** | ***4*** | 1 | 0 | 0 | 0 | 0 | 0 | 0 | 0 | 0 | ***1*** | 12 |

## OAKLAND 1 SE *Garrett County* ELEVATION 2421 ft LAT/LONG 39° 24' N / 79° 24' W

| | JAN | FEB | MAR | APR | MAY | JUN | JUL | AUG | SEP | OCT | NOV | DEC | YEAR |
|---|---|---|---|---|---|---|---|---|---|---|---|---|---|
| Maximum Temp °F | 35.6 | 38.7 | 48.9 | 60.1 | 69.0 | 76.3 | 79.3 | 78.1 | 72.0 | 61.8 | 50.6 | 40.7 | 59.3 |
| Minimum Temp °F | 16.3 | 17.8 | 25.8 | 34.6 | 43.9 | 51.8 | 57.1 | 55.5 | 49.3 | 37.4 | 29.9 | 21.8 | 36.8 |
| Mean Temp °F | 26.0 | 28.3 | 37.4 | 47.4 | 56.5 | 64.1 | 68.2 | 66.9 | 60.7 | 49.6 | 40.3 | 31.3 | 48.1 |
| Days Max Temp ≥ 90 °F | 0 | 0 | 0 | 0 | 0 | 0 | 0 | 1 | 0 | 0 | 0 | 0 | 1 |
| Days Max Temp ≤ 32 °F | 13 | 9 | 4 | 0 | 0 | 0 | 0 | 0 | 0 | 0 | 2 | 8 | 36 |
| Days Min Temp ≤ 32 °F | 28 | 25 | 23 | 14 | 4 | 0 | 0 | 0 | 1 | 10 | 19 | 26 | 150 |
| Days Min Temp ≤ 0 °F | 4 | 3 | 1 | 0 | 0 | 0 | 0 | 0 | 0 | 0 | 0 | 1 | 9 |
| Heating Degree Days | 1203 | 1031 | 849 | 522 | 269 | 83 | 24 | 37 | 157 | 472 | 735 | 1039 | 6421 |
| Cooling Degree Days | 0 | 0 | 1 | 1 | 15 | 72 | 155 | 111 | 37 | 2 | 0 | 0 | 394 |
| Total Precipitation (") | 3.26 | 2.93 | 4.04 | 4.20 | 4.72 | 4.17 | 5.25 | 4.08 | 3.38 | 3.12 | 3.53 | 3.85 | 46.53 |
| Days ≥ 0.1" Precip | 8 | 8 | 10 | 10 | 10 | 8 | 10 | 8 | 8 | 7 | 8 | 9 | 104 |
| Total Snowfall (") | 23.8 | 20.7 | 15.4 | 4.3 | 0.0 | 0.0 | 0.0 | 0.0 | 0.0 | 0.5 | 7.3 | 18.5 | 90.5 |
| Days ≥ 1" Snow Depth | 18 | 15 | 8 | 1 | 0 | 0 | 0 | 0 | 0 | 0 | 3 | 11 | 56 |

## OWINGS FERRY LANDING *Calvert County* ELEVATION 121 ft LAT/LONG 38° 42' N / 76° 41' W

| | JAN | FEB | MAR | APR | MAY | JUN | JUL | AUG | SEP | OCT | NOV | DEC | YEAR |
|---|---|---|---|---|---|---|---|---|---|---|---|---|---|
| Maximum Temp °F | 42.8 | 46.3 | 55.8 | 66.9 | 75.9 | 83.1 | 87.0 | 85.3 | 79.2 | 68.4 | ***58.4*** | ***47.7*** | 66.4 |
| Minimum Temp °F | 24.8 | 26.9 | 34.2 | 43.2 | 52.8 | 61.3 | 65.9 | 64.2 | 57.9 | 45.7 | ***37.8*** | ***29.6*** | 45.4 |
| Mean Temp °F | 33.8 | 36.6 | 44.9 | 55.1 | 64.4 | 72.2 | 76.5 | 74.8 | 68.6 | 57.1 | ***48.1*** | ***38.7*** | 55.9 |
| Days Max Temp ≥ 90 °F | 0 | 0 | 0 | 0 | 1 | 5 | 10 | 7 | 2 | 0 | 0 | 0 | 25 |
| Days Max Temp ≤ 32 °F | 5 | 3 | 0 | 0 | 0 | 0 | 0 | 0 | 0 | 0 | 0 | 2 | 10 |
| Days Min Temp ≤ 32 °F | 24 | 21 | ***14*** | 3 | 0 | 0 | 0 | 0 | 0 | 2 | ***9*** | ***20*** | 93 |
| Days Min Temp ≤ 0 °F | 0 | 0 | 0 | 0 | 0 | 0 | 0 | 0 | 0 | 0 | 0 | 0 | 0 |
| Heating Degree Days | 959 | 796 | 621 | 309 | 96 | 10 | 1 | 3 | 36 | 258 | ***503*** | ***810*** | 4402 |
| Cooling Degree Days | 0 | 0 | 5 | 18 | 88 | 236 | 396 | 307 | 156 | ***24*** | ***3*** | ***0*** | 1233 |
| Total Precipitation (") | 3.08 | 2.52 | 3.71 | 2.96 | 4.17 | 3.54 | 4.03 | 3.98 | 3.73 | 2.98 | 3.09 | 3.10 | 40.89 |
| Days ≥ 0.1" Precip | 6 | 5 | ***7*** | 6 | 7 | 6 | 6 | 6 | 5 | ***5*** | ***6*** | ***6*** | 71 |
| Total Snowfall (") | 5.3 | 5.6 | 2.4 | 0.0 | 0.0 | 0.0 | 0.0 | 0.0 | 0.0 | 0.0 | 0.8 | ***2.2*** | 16.3 |
| Days ≥ 1" Snow Depth | 6 | 3 | 1 | 0 | 0 | 0 | 0 | 0 | 0 | 0 | ***0*** | ***2*** | 12 |

## PRINCESS ANNE *Somerset County* ELEVATION 20 ft LAT/LONG 38° 12' N / 75° 40' W

| | JAN | FEB | MAR | APR | MAY | JUN | JUL | AUG | SEP | OCT | NOV | DEC | YEAR |
|---|---|---|---|---|---|---|---|---|---|---|---|---|---|
| Maximum Temp °F | 45.0 | 47.5 | 56.9 | 67.0 | 75.7 | 83.9 | 88.1 | 86.3 | 80.7 | 70.2 | 60.4 | 50.8 | 67.7 |
| Minimum Temp °F | 25.4 | 27.2 | 34.5 | 41.6 | 51.1 | 60.2 | 65.4 | 63.9 | 56.4 | 45.3 | 37.7 | 30.3 | 44.9 |
| Mean Temp °F | 35.2 | 37.4 | 45.7 | 54.4 | 63.4 | 72.1 | 76.7 | 75.1 | 68.6 | 57.7 | 49.1 | 40.6 | 56.3 |
| Days Max Temp ≥ 90 °F | 0 | 0 | 0 | 0 | 1 | 5 | 12 | 8 | 2 | 0 | 0 | 0 | 28 |
| Days Max Temp ≤ 32 °F | 4 | 2 | 0 | 0 | 0 | 0 | 0 | 0 | 0 | 0 | 0 | 1 | 7 |
| Days Min Temp ≤ 32 °F | 23 | 20 | 14 | 5 | 0 | 0 | 0 | 0 | 0 | 4 | 11 | 19 | 96 |
| Days Min Temp ≤ 0 °F | 0 | 0 | 0 | 0 | 0 | 0 | 0 | 0 | 0 | 0 | 0 | 0 | 0 |
| Heating Degree Days | 916 | 774 | 594 | 325 | 114 | 13 | 0 | 3 | 40 | 248 | 475 | 749 | 4251 |
| Cooling Degree Days | 0 | 0 | 3 | 12 | 80 | 245 | 394 | 309 | 150 | 30 | 3 | 0 | 1226 |
| Total Precipitation (") | 3.45 | 2.92 | 3.93 | 3.25 | 3.38 | 3.09 | 4.18 | 5.35 | 3.64 | 3.15 | 3.04 | 3.39 | 42.77 |
| Days ≥ 0.1" Precip | 6 | 6 | 7 | 6 | 6 | 6 | 7 | 6 | 5 | 5 | 6 | 6 | 72 |
| Total Snowfall (") | na | na | 0.7 | 0.0 | 0.0 | 0.0 | 0.0 | 0.0 | 0.0 | 0.0 | 0.1 | 0.9 | na |
| Days ≥ 1" Snow Depth | na | na | ***0*** | 0 | 0 | 0 | 0 | 0 | 0 | 0 | 0 | ***1*** | na |

## ROCKVILLE 1 NE *Montgomery County* ELEVATION 440 ft LAT/LONG 39° 5 ' N / 77° 9 ' W

| | JAN | FEB | MAR | APR | MAY | JUN | JUL | AUG | SEP | OCT | NOV | DEC | YEAR |
|---|---|---|---|---|---|---|---|---|---|---|---|---|---|
| Maximum Temp °F | 41.9 | 45.0 | 55.8 | 67.1 | 75.6 | 83.2 | 86.9 | 84.7 | 78.0 | 67.5 | 56.9 | 46.7 | 65.8 |
| Minimum Temp °F | 23.1 | 24.3 | 32.9 | 41.7 | 51.1 | 59.6 | 64.6 | 62.6 | 55.6 | 43.6 | 35.5 | 27.5 | 43.5 |
| Mean Temp °F | 32.5 | 34.7 | 44.4 | 54.4 | 63.4 | 71.5 | 75.8 | 73.7 | 66.8 | 55.6 | 46.2 | 37.1 | 54.7 |
| Days Max Temp ≥ 90 °F | 0 | 0 | 0 | 0 | 1 | 5 | 10 | 6 | 2 | 0 | 0 | 0 | 24 |
| Days Max Temp ≤ 32 °F | 5 | 3 | 0 | 0 | 0 | 0 | 0 | 0 | 0 | 0 | 0 | 2 | 10 |
| Days Min Temp ≤ 32 °F | 25 | 22 | *16* | 4 | 0 | 0 | 0 | 0 | 0 | 4 | *12* | *21* | 104 |
| Days Min Temp ≤ 0 °F | 1 | 0 | 0 | 0 | 0 | 0 | 0 | 0 | 0 | 0 | 0 | 0 | 1 |
| Heating Degree Days | 1001 | 849 | 637 | 328 | 117 | 17 | 1 | 4 | 61 | 299 | 563 | 859 | 4736 |
| Cooling Degree Days | 0 | 0 | 5 | 19 | 86 | 238 | 378 | 287 | 129 | 18 | 1 | 0 | 1161 |
| Total Precipitation (") | 2.90 | 2.73 | 3.70 | 2.98 | 4.27 | 3.53 | 4.11 | 4.02 | 3.53 | 3.24 | 3.26 | 3.08 | 41.35 |
| Days ≥ 0.1" Precip | 6 | *6* | 7 | 7 | 8 | 6 | 7 | 6 | 5 | *5* | *6* | 6 | 75 |
| Total Snowfall (") | *6.5* | *4.1* | 2.4 | 0.1 | 0.0 | 0.0 | 0.0 | 0.0 | 0.0 | 0.0 | 0.9 | *2.4* | 16.4 |
| Days ≥ 1" Snow Depth | *9* | *6* | 2 | 0 | 0 | 0 | 0 | 0 | 0 | 0 | 0 | 3 | 20 |

## ROYAL OAK 2 SSW *Talbot County* ELEVATION 10 ft LAT/LONG 38° 43 ' N / 76° 11 ' W

| | JAN | FEB | MAR | APR | MAY | JUN | JUL | AUG | SEP | OCT | NOV | DEC | YEAR |
|---|---|---|---|---|---|---|---|---|---|---|---|---|---|
| Maximum Temp °F | 42.3 | 45.5 | 54.8 | 65.3 | 74.8 | 83.1 | 87.1 | 85.6 | 79.6 | 68.7 | 58.3 | 47.7 | 66.1 |
| Minimum Temp °F | 26.7 | 28.5 | 36.2 | 45.2 | 55.2 | 64.1 | 68.7 | 66.9 | 60.2 | 49.3 | 40.7 | 31.9 | 47.8 |
| Mean Temp °F | 34.5 | 37.0 | 45.5 | 55.3 | 65.0 | 73.6 | 77.9 | 76.3 | 69.9 | 59.1 | 49.5 | 39.8 | 57.0 |
| Days Max Temp ≥ 90 °F | 0 | 0 | 0 | 0 | 1 | 4 | 10 | 7 | 2 | 0 | 0 | 0 | 24 |
| Days Max Temp ≤ 32 °F | 5 | 2 | 0 | 0 | 0 | 0 | 0 | 0 | 0 | 0 | 0 | 2 | 9 |
| Days Min Temp ≤ 32 °F | 23 | 19 | 11 | 1 | 0 | 0 | 0 | 0 | 0 | 1 | 6 | 16 | 77 |
| Days Min Temp ≤ 0 °F | 0 | 0 | 0 | 0 | 0 | 0 | 0 | 0 | 0 | 0 | 0 | 0 | 0 |
| Heating Degree Days | 938 | 783 | 598 | 298 | 82 | 6 | 0 | 1 | 27 | 209 | 461 | 773 | 4176 |
| Cooling Degree Days | 0 | 0 | 2 | 14 | 96 | 282 | 434 | 353 | 181 | 35 | 2 | 0 | 1399 |
| Total Precipitation (") | 3.60 | 3.21 | 4.21 | 3.37 | 4.21 | 3.34 | 4.12 | 4.57 | 3.71 | 3.17 | 3.22 | 3.76 | 44.49 |
| Days ≥ 0.1" Precip | 7 | 6 | 7 | 7 | 7 | 6 | 6 | 6 | 5 | 5 | 6 | 7 | 75 |
| Total Snowfall (") | 5.7 | 5.7 | 1.8 | 0.0 | 0.0 | 0.0 | 0.0 | 0.0 | 0.0 | 0.0 | 0.7 | 1.8 | 15.7 |
| Days ≥ 1" Snow Depth | 5 | 4 | 1 | 0 | 0 | 0 | 0 | 0 | 0 | 0 | 0 | 2 | 12 |

## SALISBURY *Wicomico County* ELEVATION 10 ft LAT/LONG 38° 22 ' N / 75° 35 ' W

| | JAN | FEB | MAR | APR | MAY | JUN | JUL | AUG | SEP | OCT | NOV | DEC | YEAR |
|---|---|---|---|---|---|---|---|---|---|---|---|---|---|
| Maximum Temp °F | 45.2 | 47.9 | 56.8 | 67.0 | 75.7 | 83.2 | 87.0 | 85.6 | 80.2 | 70.0 | 60.4 | 50.3 | 67.4 |
| Minimum Temp °F | 27.4 | 29.0 | 36.2 | 44.1 | 53.7 | 62.5 | 67.8 | 66.4 | 59.6 | 48.0 | 40.0 | 31.9 | 47.2 |
| Mean Temp °F | 36.3 | 38.5 | 46.5 | 55.6 | 64.7 | 72.9 | 77.5 | 76.1 | 69.9 | 59.0 | 50.3 | 41.1 | 57.4 |
| Days Max Temp ≥ 90 °F | 0 | 0 | 0 | 0 | 1 | 4 | 9 | 7 | 2 | 0 | 0 | 0 | 23 |
| Days Max Temp ≤ 32 °F | 4 | 2 | 0 | 0 | 0 | 0 | 0 | 0 | 0 | 0 | 0 | 2 | 8 |
| Days Min Temp ≤ 32 °F | 22 | 18 | 11 | 2 | 0 | 0 | 0 | 0 | 0 | 1 | 7 | 17 | 78 |
| Days Min Temp ≤ 0 °F | 0 | 0 | 0 | 0 | 0 | 0 | 0 | 0 | 0 | 0 | 0 | 0 | 0 |
| Heating Degree Days | 882 | 743 | 569 | 293 | 87 | 8 | 0 | 1 | 26 | 211 | 440 | 735 | 3995 |
| Cooling Degree Days | 0 | 0 | 4 | 20 | 93 | 266 | 425 | 350 | 180 | 34 | 4 | 0 | 1376 |
| Total Precipitation (") | 3.66 | 3.31 | 4.38 | 3.21 | 3.56 | 3.48 | 4.51 | 5.43 | 3.65 | 3.15 | 2.91 | 3.59 | 44.84 |
| Days ≥ 0.1" Precip | 7 | 6 | 8 | 6 | 7 | 6 | 7 | 6 | 5 | 5 | 6 | 7 | 76 |
| Total Snowfall (") | *3.5* | *2.6* | 0.9 | 0.0 | 0.0 | 0.0 | 0.0 | 0.0 | 0.0 | 0.0 | 0.2 | 1.2 | 8.4 |
| Days ≥ 1" Snow Depth | *4* | *2* | 0 | 0 | 0 | 0 | 0 | 0 | 0 | 0 | 0 | *1* | 7 |

## SALISBURY WICOMICO *Wicomico County* ELEVATION 49 ft LAT/LONG 38° 20 ' N / 75° 31 ' W

| | JAN | FEB | MAR | APR | MAY | JUN | JUL | AUG | SEP | OCT | NOV | DEC | YEAR |
|---|---|---|---|---|---|---|---|---|---|---|---|---|---|
| Maximum Temp °F | 43.6 | 46.1 | 54.9 | 65.2 | 74.0 | 82.4 | 86.6 | 84.9 | 78.9 | 68.3 | 59.1 | 49.1 | 66.1 |
| Minimum Temp °F | 25.8 | 27.4 | 34.8 | 42.8 | 52.4 | 61.7 | 67.6 | 66.0 | 58.4 | 46.3 | 38.3 | 30.4 | 46.0 |
| Mean Temp °F | 34.7 | 36.8 | 44.9 | 54.0 | 63.2 | 72.0 | 77.1 | 75.5 | 68.7 | 57.3 | 48.7 | 39.8 | 56.1 |
| Days Max Temp ≥ 90 °F | 0 | 0 | 0 | 0 | 1 | 4 | 10 | 6 | 2 | 0 | 0 | 0 | 23 |
| Days Max Temp ≤ 32 °F | 5 | 3 | 0 | 0 | 0 | 0 | 0 | 0 | 0 | 0 | 0 | 2 | 10 |
| Days Min Temp ≤ 32 °F | 23 | 20 | 14 | 4 | 0 | 0 | 0 | 0 | 0 | 3 | 10 | 19 | 93 |
| Days Min Temp ≤ 0 °F | 0 | 0 | 0 | 0 | 0 | 0 | 0 | 0 | 0 | 0 | 0 | 0 | 0 |
| Heating Degree Days | 931 | 790 | 620 | 337 | 121 | 13 | 0 | 2 | 39 | 258 | 487 | 775 | 4373 |
| Cooling Degree Days | 0 | 0 | 3 | 15 | 80 | 242 | 407 | 322 | 154 | 30 | 4 | 0 | 1257 |
| Total Precipitation (") | 3.64 | 3.34 | 4.30 | 3.19 | 3.72 | 3.52 | 4.59 | 5.74 | 3.69 | 3.40 | 3.05 | 3.82 | 46.00 |
| Days ≥ 0.1" Precip | 7 | 7 | 7 | 6 | 6 | 6 | 7 | 6 | 5 | 5 | 6 | 7 | 75 |
| Total Snowfall (") | 4.1 | 4.3 | 1.5 | 0.1 | 0.0 | 0.0 | 0.0 | 0.0 | 0.0 | 0.0 | 0.5 | 1.9 | 12.4 |
| Days ≥ 1" Snow Depth | 3 | 3 | 1 | 0 | 0 | 0 | 0 | 0 | 0 | 0 | 0 | 2 | 9 |

## SAVAGE RIVER DAM *Garrett County* ELEVATION 1460 ft LAT/LONG 39° 30' N / 79° 8' W

| | JAN | FEB | MAR | APR | MAY | JUN | JUL | AUG | SEP | OCT | NOV | DEC | YEAR |
|---|---|---|---|---|---|---|---|---|---|---|---|---|---|
| Maximum Temp °F | 34.5 | 37.9 | 47.9 | 59.7 | 69.5 | 77.1 | 80.8 | 79.7 | 73.3 | 61.5 | 50.3 | 39.1 | 59.3 |
| Minimum Temp °F | 17.0 | 18.7 | 26.8 | 36.5 | 45.8 | 53.8 | 58.9 | 56.9 | 50.6 | 38.6 | 31.3 | 22.9 | 38.2 |
| Mean Temp °F | 25.8 | 28.3 | 37.3 | 48.2 | 57.7 | 65.5 | 69.8 | 68.4 | 61.9 | 50.1 | 40.8 | 31.0 | 48.7 |
| Days Max Temp ≥ 90 °F | 0 | 0 | 0 | 0 | 0 | 0 | 2 | 2 | 0 | 0 | 0 | 0 | 4 |
| Days Max Temp ≤ 32 °F | 13 | 9 | 3 | 0 | 0 | 0 | 0 | 0 | 0 | 0 | 1 | 8 | 34 |
| Days Min Temp ≤ 32 °F | 29 | 25 | 23 | 10 | 2 | 0 | 0 | 0 | 0 | 8 | 18 | 27 | 142 |
| Days Min Temp ≤ 0 °F | 3 | 2 | 0 | 0 | 0 | 0 | 0 | 0 | 0 | 0 | 0 | 1 | 6 |
| Heating Degree Days | 1209 | 1028 | 851 | 502 | 243 | 68 | 14 | 28 | 137 | 457 | 719 | 1047 | 6303 |
| Cooling Degree Days | 0 | 0 | 0 | 2 | 24 | 91 | 190 | 145 | 52 | 3 | 0 | 0 | 507 |
| Total Precipitation (") | 2.49 | 2.23 | 3.32 | 3.33 | 3.98 | 3.24 | 4.22 | 3.33 | 3.09 | 2.92 | 2.91 | 2.87 | 37.93 |
| Days ≥ 0.1" Precip | 6 | 6 | 7 | 8 | 9 | 7 | 8 | 7 | 7 | 6 | 6 | 6 | 83 |
| Total Snowfall (") | na | *8.7* | 5.4 | 0.9 | 0.0 | 0.0 | 0.0 | 0.0 | 0.0 | 0.0 | *1.6* | na | na |
| Days ≥ 1" Snow Depth | 18 | 14 | 6 | 1 | 0 | 0 | 0 | 0 | 0 | 0 | 2 | 9 | 50 |

## SNOW HILL 4 N *Worcester County* ELEVATION 30 ft LAT/LONG 38° 14' N / 75° 23' W

| | JAN | FEB | MAR | APR | MAY | JUN | JUL | AUG | SEP | OCT | NOV | DEC | YEAR |
|---|---|---|---|---|---|---|---|---|---|---|---|---|---|
| Maximum Temp °F | 44.9 | 47.3 | 55.7 | 66.1 | 75.1 | 82.9 | 86.6 | 85.6 | 80.5 | 69.5 | 60.2 | 50.1 | 67.0 |
| Minimum Temp °F | 26.6 | 28.1 | 35.0 | 42.6 | 52.7 | 61.1 | 66.0 | 64.4 | 57.7 | 46.3 | 38.8 | 30.8 | 45.8 |
| Mean Temp °F | 35.8 | 37.8 | 45.4 | 54.4 | 63.9 | 72.0 | 76.3 | 75.0 | 69.1 | 58.0 | 49.5 | 40.5 | 56.5 |
| Days Max Temp ≥ 90 °F | 0 | 0 | 0 | 0 | 1 | 5 | 10 | 7 | 3 | 0 | 0 | 0 | 26 |
| Days Max Temp ≤ 32 °F | 4 | 2 | 0 | 0 | 0 | 0 | 0 | 0 | 0 | 0 | 0 | 1 | 7 |
| Days Min Temp ≤ 32 °F | 22 | 19 | 14 | 4 | 0 | 0 | 0 | 0 | 0 | 3 | 9 | 19 | 90 |
| Days Min Temp ≤ 0 °F | 0 | 0 | 0 | 0 | 0 | 0 | 0 | 0 | 0 | 0 | 0 | 0 | 0 |
| Heating Degree Days | 900 | 762 | 604 | 326 | 105 | 13 | 1 | 3 | 36 | 241 | 462 | 751 | 4204 |
| Cooling Degree Days | 0 | 0 | 4 | 17 | 92 | 257 | 395 | 324 | 173 | 34 | 4 | 0 | 1300 |
| Total Precipitation (") | 3.70 | 3.33 | 4.46 | 3.09 | 3.48 | 3.18 | 4.35 | 5.54 | 3.48 | 3.16 | 3.10 | 3.34 | 44.21 |
| Days ≥ 0.1" Precip | 7 | 6 | 7 | 6 | 6 | 6 | 8 | 6 | 5 | 5 | 6 | 7 | 75 |
| Total Snowfall (") | 4.8 | 5.4 | 1.6 | 0.1 | 0.0 | 0.0 | 0.0 | 0.0 | 0.0 | 0.0 | 0.3 | 1.5 | 13.7 |
| Days ≥ 1" Snow Depth | 3 | 3 | 1 | 0 | 0 | 0 | 0 | 0 | 0 | 0 | 0 | 1 | 8 |

## UNIONVILLE *Frederick County* ELEVATION 430 ft LAT/LONG 39° 27' N / 77° 11' W

| | JAN | FEB | MAR | APR | MAY | JUN | JUL | AUG | SEP | OCT | NOV | DEC | YEAR |
|---|---|---|---|---|---|---|---|---|---|---|---|---|---|
| Maximum Temp °F | 39.1 | 43.2 | 53.4 | 64.7 | 74.1 | 82.1 | 85.9 | 84.1 | 77.3 | 66.0 | 55.3 | 44.1 | 64.1 |
| Minimum Temp °F | 19.5 | 21.2 | 29.5 | 37.7 | 47.8 | 56.8 | 61.8 | 59.9 | 52.6 | 40.7 | 32.4 | 24.4 | 40.4 |
| Mean Temp °F | 29.5 | 32.2 | 41.5 | 51.2 | 61.0 | 69.5 | 73.9 | 72.0 | 65.0 | 53.4 | 43.8 | 34.3 | 52.3 |
| Days Max Temp ≥ 90 °F | 0 | 0 | 0 | 0 | 1 | 4 | 8 | 5 | 1 | 0 | 0 | 0 | 19 |
| Days Max Temp ≤ 32 °F | 8 | 4 | 1 | 0 | 0 | 0 | 0 | 0 | 0 | 0 | 0 | 3 | 16 |
| Days Min Temp ≤ 32 °F | 27 | 24 | 20 | 10 | 2 | 0 | 0 | 0 | 0 | 8 | 16 | 24 | 131 |
| Days Min Temp ≤ 0 °F | 2 | 1 | 0 | 0 | 0 | 0 | 0 | 0 | 0 | 0 | 0 | 1 | 4 |
| Heating Degree Days | 1095 | 919 | 724 | 413 | 162 | 25 | 2 | 10 | 87 | 361 | 628 | 945 | 5371 |
| Cooling Degree Days | 0 | 0 | 1 | 6 | 50 | 177 | 314 | 234 | 96 | 11 | 0 | 0 | 889 |
| Total Precipitation (") | 2.77 | 2.52 | 3.51 | 3.26 | 4.17 | 4.10 | 3.46 | 3.37 | 3.66 | 3.34 | 3.54 | 3.37 | 41.07 |
| Days ≥ 0.1" Precip | 6 | 5 | 7 | 6 | 8 | 7 | 6 | 6 | 5 | 5 | 6 | 6 | 73 |
| Total Snowfall (") | 6.6 | 6.5 | 3.4 | 0.1 | 0.0 | 0.0 | 0.0 | 0.0 | 0.0 | 0.1 | 1.0 | 4.3 | 22.0 |
| Days ≥ 1" Snow Depth | *9* | 4 | 1 | 0 | 0 | 0 | 0 | 0 | 0 | 0 | 0 | 2 | 16 |

## VIENNA *Dorchester County* ELEVATION 10 ft LAT/LONG 38° 29' N / 75° 50' W

| | JAN | FEB | MAR | APR | MAY | JUN | JUL | AUG | SEP | OCT | NOV | DEC | YEAR |
|---|---|---|---|---|---|---|---|---|---|---|---|---|---|
| Maximum Temp °F | 44.0 | 46.9 | 56.9 | 67.3 | 76.5 | 85.1 | 89.0 | 87.2 | 81.2 | 70.1 | 59.8 | 49.7 | 67.8 |
| Minimum Temp °F | 26.4 | 28.2 | 35.6 | 43.9 | 53.7 | 62.6 | 67.3 | 65.7 | 58.7 | 47.5 | 39.1 | 31.0 | 46.6 |
| Mean Temp °F | 35.2 | 37.6 | 46.3 | 55.6 | 65.1 | 74.0 | 78.2 | 76.5 | 69.9 | 58.8 | 49.5 | 40.4 | 57.3 |
| Days Max Temp ≥ 90 °F | 0 | 0 | 0 | 0 | 1 | 8 | 14 | 11 | 3 | 0 | 0 | 0 | 37 |
| Days Max Temp ≤ 32 °F | 5 | 2 | 0 | 0 | 0 | 0 | 0 | 0 | 0 | 0 | 0 | 2 | 9 |
| Days Min Temp ≤ 32 °F | 22 | 19 | 12 | 2 | 0 | 0 | 0 | 0 | 0 | 1 | 8 | 17 | 81 |
| Days Min Temp ≤ 0 °F | 0 | 0 | 0 | 0 | 0 | 0 | 0 | 0 | 0 | 0 | 0 | 0 | 0 |
| Heating Degree Days | 919 | 768 | 575 | 289 | 83 | 7 | 0 | 1 | 28 | 217 | 461 | 756 | 4104 |
| Cooling Degree Days | 0 | 0 | 3 | 17 | *107* | 305 | 443 | 352 | 177 | 33 | 3 | 0 | 1440 |
| Total Precipitation (") | 3.53 | 3.35 | 3.99 | 3.22 | 3.51 | 3.62 | 3.91 | 5.13 | 3.59 | 2.99 | 2.91 | 3.41 | 43.16 |
| Days ≥ 0.1" Precip | 7 | *6* | 7 | 6 | 6 | 6 | 7 | 6 | 5 | 5 | 5 | 6 | 72 |
| Total Snowfall (") | *4.2* | *3.4* | 0.9 | 0.0 | 0.0 | 0.0 | 0.0 | 0.0 | 0.0 | 0.0 | 0.2 | 1.0 | 9.7 |
| Days ≥ 1" Snow Depth | na | *2* | 0 | 0 | 0 | 0 | 0 | 0 | 0 | 0 | 0 | 1 | na |

## WASHINGTON NATL ARBO *District of Columbia* ELEVATION 69 ft LAT/LONG 38° 54 ' N / 76° 58 ' W

| | JAN | FEB | MAR | APR | MAY | JUN | JUL | AUG | SEP | OCT | NOV | DEC | YEAR |
|---|---|---|---|---|---|---|---|---|---|---|---|---|---|
| Maximum Temp °F | 42.6 | 46.1 | 55.8 | 66.5 | 76.0 | 84.2 | 88.1 | 86.6 | 80.2 | 69.0 | 58.4 | 47.7 | 66.8 |
| Minimum Temp °F | 24.2 | 26.3 | 34.2 | 43.2 | 52.6 | 61.8 | 66.9 | 65.0 | 57.4 | 45.0 | 37.0 | 29.1 | 45.2 |
| Mean Temp °F | 33.4 | 36.2 | 45.0 | 54.9 | 64.3 | 73.0 | 77.5 | 75.8 | 68.8 | 57.0 | 47.7 | 38.4 | 56.0 |
| Days Max Temp ≥ 90 °F | 0 | 0 | 0 | 0 | 1 | 7 | 13 | 10 | 4 | 0 | 0 | 0 | 35 |
| Days Max Temp ≤ 32 °F | 5 | 3 | 0 | 0 | 0 | 0 | 0 | 0 | 0 | 0 | 0 | 2 | 10 |
| Days Min Temp ≤ 32 °F | 25 | 21 | 14 | 3 | 0 | 0 | 0 | 0 | 0 | 3 | 10 | 21 | 97 |
| Days Min Temp ≤ 0 °F | 0 | 0 | 0 | 0 | 0 | 0 | 0 | 0 | 0 | 0 | 0 | 0 | 0 |
| Heating Degree Days | 973 | 805 | 614 | 314 | 100 | 9 | 0 | 2 | 40 | 262 | 516 | 817 | 4452 |
| Cooling Degree Days | 0 | 0 | 4 | 17 | 90 | 260 | 412 | 327 | 150 | 21 | 2 | 0 | 1283 |
| Total Precipitation (") | 3.16 | 2.82 | 3.65 | 3.19 | 4.19 | 3.51 | 4.60 | 4.28 | 3.76 | 3.35 | 3.25 | 3.45 | 43.21 |
| Days ≥ 0.1" Precip | 6 | 6 | 7 | 7 | 8 | 6 | 7 | 6 | 5 | 5 | 6 | 6 | 75 |
| Total Snowfall (") | *6.0* | 5.2 | 1.3 | 0.0 | 0.0 | 0.0 | 0.0 | 0.0 | 0.0 | 0.0 | 0.8 | 1.9 | 15.2 |
| Days ≥ 1" Snow Depth | na | *3* | 0 | 0 | 0 | 0 | 0 | 0 | 0 | 0 | 0 | 1 | na |

## WOODSTOCK *Baltimore County* ELEVATION 420 ft LAT/LONG 39° 20 ' N / 76° 52 ' W

| | JAN | FEB | MAR | APR | MAY | JUN | JUL | AUG | SEP | OCT | NOV | DEC | YEAR |
|---|---|---|---|---|---|---|---|---|---|---|---|---|---|
| Maximum Temp °F | 40.2 | 44.1 | 53.9 | 65.6 | 75.5 | 83.2 | 87.3 | 85.4 | 78.2 | 67.1 | 56.3 | 45.3 | 65.2 |
| Minimum Temp °F | 22.2 | 24.2 | 32.3 | 41.0 | 50.5 | 58.9 | 64.4 | 63.0 | 55.8 | 43.0 | 35.5 | 27.4 | 43.2 |
| Mean Temp °F | 31.2 | 34.1 | 43.3 | 53.3 | 63.0 | 71.0 | 75.7 | 74.2 | 66.7 | 55.0 | 45.7 | 36.1 | 54.1 |
| Days Max Temp ≥ 90 °F | 0 | 0 | 0 | 0 | 1 | *5* | 10 | 7 | 2 | 0 | 0 | 0 | 25 |
| Days Max Temp ≤ 32 °F | 7 | 4 | 1 | 0 | 0 | 0 | 0 | 0 | 0 | 0 | 0 | 3 | 15 |
| Days Min Temp ≤ 32 °F | 27 | 23 | *16* | *5* | 0 | 0 | 0 | 0 | 0 | 5 | 12 | *22* | 110 |
| Days Min Temp ≤ 0 °F | 1 | 0 | 0 | 0 | 0 | 0 | 0 | 0 | 0 | 0 | 0 | 0 | 1 |
| Heating Degree Days | 1041 | 863 | 671 | 359 | 122 | 17 | 1 | 4 | 61 | 315 | 568 | 881 | 4903 |
| Cooling Degree Days | 0 | 0 | 4 | 14 | 78 | 228 | 380 | 301 | 136 | 19 | 2 | 0 | 1162 |
| Total Precipitation (") | 3.27 | 2.90 | 4.11 | 3.43 | 4.66 | 3.74 | 3.67 | 3.76 | 3.95 | 3.25 | 3.53 | 3.63 | 43.90 |
| Days ≥ 0.1" Precip | 7 | 6 | 8 | 7 | 8 | 6 | 6 | 6 | 5 | 5 | 6 | 6 | 76 |
| Total Snowfall (") | 8.1 | 7.3 | 3.2 | 0.2 | 0.0 | 0.0 | 0.0 | 0.0 | 0.0 | 0.0 | 1.1 | 3.5 | 23.4 |
| Days ≥ 1" Snow Depth | 10 | 8 | 2 | 0 | 0 | 0 | 0 | 0 | 0 | 0 | 0 | 3 | 23 |

## JANUARY MINIMUM TEMPERATURE °F

### LOWEST

| | | |
|---|---|---|
| 1 | Oakland | 16.3 |
| 2 | Frostburg | 17.0 |
| | Savage River Dam | 17.0 |
| 4 | Unionville | 19.5 |
| 5 | Hancock | 20.1 |
| 6 | Emmitsburg | 20.2 |
| 7 | Conowingo Dam | 20.5 |
| 8 | Beltsville | 21.5 |
| 9 | Catoctin Mntn Pk | 21.7 |
| 10 | Benson | 21.9 |
| 11 | Glenn Dale Bell | 22.1 |
| 12 | Woodstock | 22.2 |
| 13 | Millington | 22.4 |
| 14 | Dalecarlia Rsrvr | 22.8 |
| 15 | Rockville | 23.1 |
| 16 | Laurel | 23.7 |
| 17 | Baltimore-Intl | 24.0 |
| 18 | Chestertown | 24.2 |
| | Washington | 24.2 |
| 20 | College Park | 24.8 |
| | Owings Ferry | 24.8 |
| 22 | La Plata | 24.9 |
| 23 | Princess Anne | 25.4 |
| 24 | Salisbury-Wicmic | 25.8 |
| 25 | Vienna | 26.4 |

### HIGHEST

| | | |
|---|---|---|
| 1 | Baltimore | 28.4 |
| 2 | Salisbury | 27.4 |
| 3 | Royal Oak | 26.7 |
| 4 | Snow Hill | 26.6 |
| 5 | Vienna | 26.4 |
| 6 | Salisbury-Wicmic | 25.8 |
| 7 | Princess Anne | 25.4 |
| 8 | La Plata | 24.9 |
| 9 | College Park | 24.8 |
| | Owings Ferry | 24.8 |
| 11 | Chestertown | 24.2 |
| | Washington | 24.2 |
| 13 | Baltimore-Intl | 24.0 |
| 14 | Laurel | 23.7 |
| 15 | Rockville | 23.1 |
| 16 | Dalecarlia Rsrvr | 22.8 |
| 17 | Millington | 22.4 |
| 18 | Woodstock | 22.2 |
| 19 | Glenn Dale Bell | 22.1 |
| 20 | Benson | 21.9 |
| 21 | Catoctin Mntn Pk | 21.7 |
| 22 | Beltsville | 21.5 |
| 23 | Conowingo Dam | 20.5 |
| 24 | Emmitsburg | 20.2 |
| 25 | Hancock | 20.1 |

## JULY MAXIMUM TEMPERATURE °F

### HIGHEST

| | | |
|---|---|---|
| 1 | Baltimore | 89.2 |
| 2 | Vienna | 89.0 |
| 3 | Dalecarlia Rsrvr | 88.8 |
| 4 | College Park | 88.5 |
| 5 | Glenn Dale Bell | 88.2 |
| 6 | Princess Anne | 88.1 |
| | Washington | 88.1 |
| 8 | Laurel | 87.9 |
| 9 | Benson | 87.7 |
| 10 | Baltimore-Intl | 87.5 |
| 11 | Woodstock | 87.3 |
| 12 | Chestertown | 87.1 |
| | Conowingo Dam | 87.1 |
| | Royal Oak | 87.1 |
| 15 | Beltsville | 87.0 |
| | Millington | 87.0 |
| | Owings Ferry | 87.0 |
| | Salisbury | 87.0 |
| 19 | Rockville | 86.9 |
| 20 | Salisbury-Wicmic | 86.6 |
| | Snow Hill | 86.6 |
| 22 | Aberdeen | 86.5 |
| 23 | Emmitsburg | 86.2 |
| 24 | Unionville | 85.9 |
| 25 | La Plata | 85.5 |

### LOWEST

| | | |
|---|---|---|
| 1 | Frostburg | 79.0 |
| 2 | Oakland | 79.3 |
| 3 | Catoctin Mntn Pk | 80.2 |
| 4 | Savage River Dam | 80.8 |
| 5 | Hancock | 85.4 |
| 6 | La Plata | 85.5 |
| 7 | Unionville | 85.9 |
| 8 | Emmitsburg | 86.2 |
| 9 | Aberdeen | 86.5 |
| 10 | Salisbury-Wicmic | 86.6 |
| | Snow Hill | 86.6 |
| 12 | Rockville | 86.9 |
| 13 | Beltsville | 87.0 |
| | Millington | 87.0 |
| | Owings Ferry | 87.0 |
| | Salisbury | 87.0 |
| 17 | Chestertown | 87.1 |
| | Conowingo Dam | 87.1 |
| | Royal Oak | 87.1 |
| 20 | Woodstock | 87.3 |
| 21 | Baltimore-Intl | 87.5 |
| 22 | Benson | 87.7 |
| 23 | Laurel | 87.9 |
| 24 | Princess Anne | 88.1 |
| | Washington | 88.1 |

## ANNUAL PRECIPITATION (")

### HIGHEST

| | | |
|---|---|---|
| 1 | Catoctin Mntn Pk | 48.28 |
| 2 | Benson | 48.19 |
| 3 | Oakland | 46.53 |
| 4 | Conowingo Dam | 46.09 |
| 5 | Salisbury-Wicmic | 46.00 |
| 6 | Frostburg | 45.19 |
| 7 | Salisbury | 44.84 |
| 8 | Royal Oak | 44.49 |
| 9 | Laurel | 44.34 |
| 10 | Snow Hill | 44.21 |
| 11 | Emmitsburg | 44.15 |
| 12 | Dalecarlia Rsrvr | 44.14 |
| 13 | Glenn Dale Bell | 44.09 |
| 14 | Woodstock | 43.90 |
| 15 | Aberdeen | 43.71 |
| 16 | College Park | 43.37 |
| 17 | Washington | 43.21 |
| 18 | Millington | 43.16 |
| | Vienna | 43.16 |
| 20 | Chestertown | 43.05 |
| 21 | Beltsville | 42.99 |
| 22 | La Plata | 42.94 |
| 23 | Princess Anne | 42.77 |
| 24 | Baltimore | 42.55 |
| 25 | Rockville | 41.35 |

### LOWEST

| | | |
|---|---|---|
| 1 | Hancock | 37.28 |
| 2 | Savage River Dam | 37.93 |
| 3 | Baltimore-Intl | 40.69 |
| 4 | Owings Ferry | 40.89 |
| 5 | Unionville | 41.07 |
| 6 | Rockville | 41.35 |
| 7 | Baltimore | 42.55 |
| 8 | Princess Anne | 42.77 |
| 9 | La Plata | 42.94 |
| 10 | Beltsville | 42.99 |
| 11 | Chestertown | 43.05 |
| 12 | Millington | 43.16 |
| | Vienna | 43.16 |
| 14 | Washington | 43.21 |
| 15 | College Park | 43.37 |
| 16 | Aberdeen | 43.71 |
| 17 | Woodstock | 43.90 |
| 18 | Glenn Dale Bell | 44.09 |
| 19 | Dalecarlia Rsrvr | 44.14 |
| 20 | Emmitsburg | 44.15 |
| 21 | Snow Hill | 44.21 |
| 22 | Laurel | 44.34 |
| 23 | Royal Oak | 44.49 |
| 24 | Salisbury | 44.84 |
| 25 | Frostburg | 45.19 |

## ANNUAL SNOWFALL (")

### HIGHEST

| | | |
|---|---|---|
| 1 | Frostburg | 95.9 |
| 2 | Oakland | 90.5 |
| 3 | Emmitsburg | 33.4 |
| 4 | Catoctin Mntn Pk | 33.1 |
| 5 | Hancock | 27.2 |
| 6 | Woodstock | 23.4 |
| 7 | Unionville | 22.0 |
| 8 | Baltimore-Intl | 19.0 |
| 9 | Glenn Dale Bell | 18.0 |
| 10 | Benson | 17.7 |
| 11 | Chestertown | 17.4 |
| 12 | La Plata | 17.3 |
| 13 | Rockville | 16.4 |
| 14 | Owings Ferry | 16.3 |
| 15 | College Park | 15.8 |
| 16 | Royal Oak | 15.7 |
| 17 | Washington | 15.2 |
| 18 | Millington | 14.8 |
| 19 | Beltsville | 14.6 |
| 20 | Snow Hill | 13.7 |
| 21 | Conowingo Dam | 13.3 |
| 22 | Salisbury-Wicmic | 12.4 |
| 23 | Aberdeen | 10.7 |
| 24 | Vienna | 9.7 |
| 25 | Dalecarlia Rsrvr | 8.4 |

### LOWEST

| | | |
|---|---|---|
| 1 | Dalecarlia Rsrvr | 8.4 |
| | Salisbury | 8.4 |
| 3 | Vienna | 9.7 |
| 4 | Aberdeen | 10.7 |
| 5 | Salisbury-Wicmic | 12.4 |
| 6 | Conowingo Dam | 13.3 |
| 7 | Snow Hill | 13.7 |
| 8 | Beltsville | 14.6 |
| 9 | Millington | 14.8 |
| 10 | Washington | 15.2 |
| 11 | Royal Oak | 15.7 |
| 12 | College Park | 15.8 |
| 13 | Owings Ferry | 16.3 |
| 14 | Rockville | 16.4 |
| 15 | La Plata | 17.3 |
| 16 | Chestertown | 17.4 |
| 17 | Benson | 17.7 |
| 18 | Glenn Dale Bell | 18.0 |
| 19 | Baltimore-Intl | 19.0 |
| 20 | Unionville | 22.0 |
| 21 | Woodstock | 23.4 |
| 22 | Hancock | 27.2 |
| 23 | Catoctin Mntn Pk | 33.1 |
| 24 | Emmitsburg | 33.4 |
| 25 | Oakland | 90.5 |

# MASSACHUSETTS

PHYSICAL FEATURES. Massachusetts occupies 8,266 square miles, nearly one-eighth of New England's total area. Most of the State lies just above the 42nd parallel of latitude. Its north-south width is, roughly, 50 miles, except 100 miles in the eastern, Atlantic coast, portion. The east-west extension is barely 150 miles, excepting "the Cape". This is the familiar name of the long arm of land which reaches around the southern and eastern shores of Cape Cod Bay. Including the Cape, the State is nearly 200 miles in length.

The land surface is mountainous along the western border and generally hilly elsewhere. However, the Cape and some other sections of the coastal area consist of flat land with numerous marshes and some small lakes and ponds. In the west Mt. Greylock rises 3,491 feet above sea level, the highest peak in Massachusetts. The elevation is mostly over 1,000 feet west of the Connecticut River Valley. A number of peaks reach above 2,000 feet. Most of central Massachusetts lies between 500 and 1,000 feet. Eastern Massachusetts and the Connecticut River Valley are mostly less than 500 feet.

GENERAL CLIMATE. Climatic characteristics of Massachusetts includes: (1) changeableness in the weather, (2) large ranges of temperature, both daily and annual, (3) great differences between the same seasons in different years, (4) equable distribution of precipitation, and (5) considerable diversity from place to place. The regional New England climatic influences are modified in Massachusetts by varying distances from the ocean, elevations, and types of terrain. These modifying factors divide the State into three climatological divisions (the Western Division, the Central Division, and the Coastal Division)

Massachusetts lies in the "prevailing westerlies", the belt of generally eastward air movement which encircles the globe in middle latitudes. Embedded in this circulation are extensive masses of air originating in higher or lower latitudes and interacting to produce storm systems. Relative to most other sections of the country, a large number of such storms pass over or near Massachusetts. The majority of air masses affecting this State belong to three types: (1) cold, dry air pouring down from subarctic North America, (2) warm, moist air streaming up on a long overland journey from the Gulf of Mexico and subtropical waters eastward, and (3) cool, damp air moving in from the North Atlantic. Because the atmospheric flow is usually offshore, Massachusetts is more influenced by the first two types than it is by the third. In other words, the adjacent ocean constitutes an important modifying factor, particularly on the immediate coast, but does not dominate the climate.

The procession of contrasting air masses and the relatively frequent passage of storm centers bring about a roughly twice-weekly alternation from fair to cloudy or stormy conditions, attended by often abrupt changes in temperature, moisture, sunshine, wind direction, and speed. There is no regular or persistent rhythm to this sequence, and it is interrupted by periods during which the weather patterns continue the same for several days, infrequently for several weeks. Massachusetts weather, however, is cited for variety rather than monotony. Changeability is also one of its features on a longer time-scale. That is, the same month or season will exhibit varying characteristics over the years, sometimes in close alternation, sometimes arranged in similar groups for successive years.

TEMPERATURE. Summer temperatures are delightfully comfortable for the most part, and summer averages are nearly uniform over the state. Hot days with maxima of 90° or higher generally average from 5 to 15 per year, varying not only from place to place but from year to year. They range, in frequency of occurrence, from only a few in cool summers to 25 or more in an occasional hot summer. The Cape and offshore islands are exceptions, averaging less than one day with a reading of 90° F. or higher per year.

Average temperatures vary from place to place more in winter than in summer. The diurnal temperature range in winter, though less than in summer, is still greater inland than along the coast. Days with subzero readings are rare on offshore islands. They average only a few per year near the coast, but increase in number of occurrences farther inland to from 5 to 15 annually.

PRECIPITATION. Massachusetts is fortunate in having its precipitation rather evenly distributed through the year. In this respect, the State is located in one of the relatively few areas of the world that does not have its "rainy" and "dry" seasons. Storm systems are the principal year-round moisture producers. But in the summer, when this activity ebbs, bands or patches of thunderstorms or showers tend to make up the difference. Though brief and often of small extent, the thunderstorms produce the heaviest local rainfall. Prolonged droughts are infrequent. Storms of a coastal nature make the Coastal Division the wettest in the winter season. Inland sections get the heavier rain in the warm season due, principally, to the higher frequency and greater intensity of convective showers and thunderstorms. The mountainous character of much of the Western Division is an additional cause for the heaviest annual totals being recorded in that part of the State.

SNOWFALL. Average annual amounts of snowfall increase rapidly from the coast westward. About 25 to 30 inches fall over Cape Cod, but up to 60 to 80 inches are recorded in the western part of the State. Topography has a marked influence on snowfall, causing much variation even in short distances. The average number of days with 1 inch or more of snowfall varies from about 8 to 15 in the Coastal Division to mostly 20 to 30 in the Western Division. Most winters will have at least one snowstorm of 5 inches or more. The average number of days with snow on the ground also increases from shore to interior and with rise in elevation. There is little lasting snow cover in the coastal lowlands. In the Western Division the cover usually extends well into spring. Maximum snow depths usually occur in the middle part of February. Water stored in the snow over the watersheds makes an important contribution to the water supply. Melting is usually too gradual to threaten serious flooding.

FLOODS. The Connecticut River, the largest river system in New England, drains most of the western half of the State. Second in size in Massachusetts is the Merrimack River which occupies the northeast portion. The rest of the rivers are relatively small, most of them with headwaters in the State and flowing southward through Connecticut and Rhode Island, or directly to the coast in the east and southeast. Flooding occurs most often in spring, caused by combination of rain and melting snow. The Connecticut River shows a regular annual rise as the result of the melting of high elevation snow in northern and central New England, but extensive flooding does not occur unless the rise is accompanied by heavy rains. High flows and major floods occur from rainfall alone but less frequently. Some of the severest floods caused by heavy rains have been those associated with hurricanes or storms of tropical origin in late summer or fall, normally the low water season.

OTHER CLIMATIC FEATURES. The percentage of possible sunshine averages from 50 to 60 in most sections. Higher elevations are cloudier, reducing the Berkshire average to between 45 and 50 percent. The average annual number of clear days is between 90 and 120 for most of the State, with less in the Berkshires. Heavy fog is frequent and sometimes persistent south of Cape Cod. Nantucket Island has heavy fog on nearly 1 day out of 4. Fog frequency diminishes along the Massachusetts coast north of the Cape. Duration of fog also diminishes inland. But the shorter duration heavy ground fogs of early morning occur frequently at susceptible places inland. These, plus the fewer occurrences of other heavy fog, produce a frequency that also approaches this 1 day out of 4 in many localities. The number of days with fog varies from as low as about 15 up to nearly 100 per year over the State.

WINDS AND STORMS. The prevailing wind, on a yearly basis, comes from a westerly direction. It is more northwesterly in winter and southwesterly in summer. Along the coast in spring and summer the sea breeze is important. These onshore winds, blowing from the cool ocean, may come inland for 10 miles or so. They tend to retard the spring growth, but they are pleasantly cooling in summer. Coastal storms or "northeasters" are one of the State's most serious weather hazards. They generate very strong winds and heavy rain or snow. They can produce abnormally high, wind-driven tides. In winter, these storms produce the heaviest snow. Occasionally in summer or fall a storm of tropical origin affects Massachusetts. Often these will be similar to the northeasters. The few which retain full hurricane force cause widespread damage. Storms of tropical origin seriously affect Massachusetts about once in 2 years, on the average. Two such storms in the same year may be expected once in 8 or 10 years.

Tornadoes are not common phenomena, yet, on a per unit area basis, Massachusetts ranks fairly high among the states. One or more may occur in Massachusetts each year. Four out of five tornadoes occur between May 15 and September 15. The peak month is July. Thunderstorms and hailstorms have a similar frequency maximum from midspring to early fall. Thunderstorms occur on about 20 to 30 days a year, and the most severe are attended by hail.

## COUNTY INDEX



## ELEVATION INDEX

| FEET | STATION NAME |
|---|---|
| 20 | EAST WAREHAM |
| 20 | EDGARTOWN |
| 20 | HYANNIS |
| 20 | TAUNTON |
| 30 | HINGHAM |
| 30 | PLYMOUTH-KINGSTON |
| 43 | BOSTON LOGAN INTL AP |
| 57 | LAWRENCE |
| 59 | ROCHESTER |
| 70 | NEW BEDFORD |
| 89 | MIDDLETON |
| 89 | READING |
| 112 | BROCKTON |
| 121 | HAVERHILL |
| 141 | BEDFORD |
| 161 | PEABODY |
| 210 | WEST MEDWAY |
| 220 | AMHERST |
| 502 | BUFFUMVILLE LAKE |
| 630 | KNIGHTVILLE DAM |
| 640 | MILTON BLUE HILL OBS |
| 679 | EAST BRIMFIELD LAKE |
| 689 | TULLY LAKE |
| 860 | BIRCH HILL DAM |
| 910 | BARRE FALLS DAM |
| 991 | WORCESTER MUNI AP |
| 1611 | CUMMINGTON HILL |

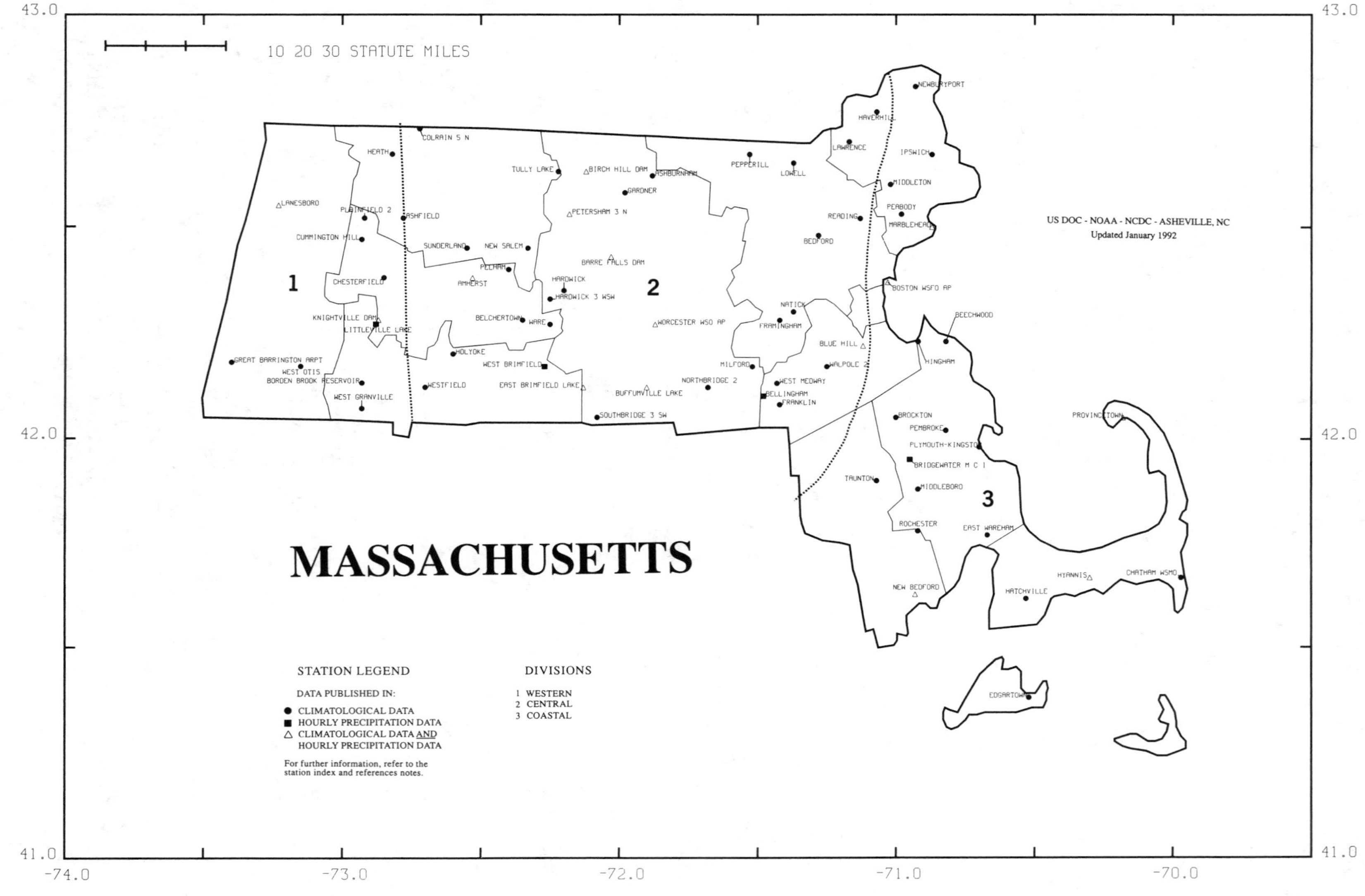

10 20 30 STATUTE MILES
MASSACHUSETTS
1
2
3
LANESBORO
PLAINFIELD 2
ASHFIELD
CUMMINGTON HILL
HEATH
COLRAIN 5 N
CHESTERFIELD
KNIGHTVILLE DAM
LITTLEVILLE LAKE
GREAT BARRINGTON ARPT
WEST OTIS
BORDEN BROOK RESERVOIR
WEST GRANVILLE
WESTFIELD
HOLYOKE
SUNDERLAND
NEW SALEM
PELHAM
AMHERST
BELCHERTOWN
WARE
HARDWICK
HARDWICK 3 WSW
WEST BRIMFIELD
EAST BRIMFIELD LAKE
SOUTHBRIDGE 3 SW
BUFFUMVILLE LAKE
TULLY LAKE
BIRCH HILL DAM
ASHBURNHAM
GARDNER
PETERSHAM 3 N
BARRE FALLS DAM
WORCESTER WSO AP
NORTHBRIDGE 2
MILFORD
PEPPERELL
LOWELL
LAWRENCE
HAVERHILL
NEWBURYPORT
IPSWICH
MIDDLETON
PEABODY
MARBLEHEAD
READING
BEDFORD
NATICK
FRAMINGHAM
BOSTON WSFO AP
BEECHWOOD
BLUE HILL
WALPOLE 2
WEST MEDWAY
BELLINGHAM
FRANKLIN
HINGHAM
BROCKTON
PEMBROKE
PLYMOUTH-KINGSTON
BRIDGEWATER M C I
TAUNTON
MIDDLEBORO
ROCHESTER
EAST WAREHAM
NEW BEDFORD
PROVINCETOWN
HYANNIS
CHATHAM WSMO
HATCHVILLE
EDGARTOWN
US DOC - NOAA - NCDC - ASHEVILLE, NC
Updated January 1992
STATION LEGEND
DATA PUBLISHED IN:
● CLIMATOLOGICAL DATA
■ HOURLY PRECIPITATION DATA
△ CLIMATOLOGICAL DATA AND HOURLY PRECIPITATION DATA
For further information, refer to the station index and references notes.
DIVISIONS
1 WESTERN
2 CENTRAL
3 COASTAL
43.0
42.0
41.0
-74.0
-73.0
-72.0
-71.0
-70.0

## AMHERST *Hampshire County* ELEVATION 220 ft LAT/LONG 42° 24 ' N / 72° 32 ' W

| | JAN | FEB | MAR | APR | MAY | JUN | JUL | AUG | SEP | OCT | NOV | DEC | YEAR |
|---|---|---|---|---|---|---|---|---|---|---|---|---|---|
| Maximum Temp °F | 34.0 | 37.2 | 46.6 | 59.7 | 71.4 | 79.4 | 84.2 | 82.3 | 74.4 | 63.4 | 51.0 | 38.5 | 60.2 |
| Minimum Temp °F | 12.0 | 14.8 | 24.6 | 34.1 | 44.4 | 53.6 | 58.5 | 56.9 | 48.6 | 37.7 | 30.1 | 19.3 | 36.2 |
| Mean Temp °F | 23.0 | 26.0 | 35.6 | 46.9 | 58.0 | 66.5 | 71.4 | 69.6 | 61.5 | 50.6 | 40.6 | 28.9 | 48.2 |
| Days Max Temp ≥ 90 °F | 0 | 0 | 0 | 0 | 1 | 2 | 5 | 3 | 1 | 0 | 0 | 0 | 12 |
| Days Max Temp ≤ 32 °F | 13 | 8 | 2 | 0 | 0 | 0 | 0 | 0 | 0 | 0 | 0 | 8 | 31 |
| Days Min Temp ≤ 32 °F | 30 | 26 | 25 | 14 | 3 | 0 | 0 | 0 | 1 | 11 | 19 | 28 | 157 |
| Days Min Temp ≤ 0 °F | 6 | 4 | 0 | 0 | 0 | 0 | 0 | 0 | 0 | 0 | 0 | 2 | 12 |
| Heating Degree Days | 1295 | 1094 | 906 | 539 | 238 | 59 | 10 | 24 | 152 | 444 | 727 | 1112 | 6600 |
| Cooling Degree Days | 0 | 0 | 0 | 2 | 28 | 103 | 213 | 161 | 51 | 4 | 0 | 0 | 562 |
| Total Precipitation (") | 3.15 | 2.80 | 3.49 | 3.72 | 4.12 | 3.73 | 3.94 | 4.08 | 3.67 | 3.61 | 4.03 | 3.82 | 44.16 |
| Days ≥ 0.1" Precip | 6 | 5 | 7 | 7 | 7 | 7 | 6 | 6 | 7 | 6 | 7 | 7 | 78 |
| Total Snowfall (") | 11.5 | 9.4 | 6.7 | 1.2 | 0.0 | 0.0 | 0.0 | 0.0 | 0.0 | 0.0 | 2.3 | 9.6 | 40.7 |
| Days ≥ 1" Snow Depth | 23 | 19 | 10 | 1 | 0 | 0 | 0 | 0 | 0 | 0 | 2 | 12 | 67 |

## BARRE FALLS DAM *Worcester County* ELEVATION 910 ft LAT/LONG 42° 26 ' N / 72° 2 ' W

| | JAN | FEB | MAR | APR | MAY | JUN | JUL | AUG | SEP | OCT | NOV | DEC | YEAR |
|---|---|---|---|---|---|---|---|---|---|---|---|---|---|
| Maximum Temp °F | 30.6 | 33.0 | 42.1 | 54.2 | 66.4 | 74.4 | 79.3 | 77.3 | 69.4 | 58.7 | ***47.0*** | ***34.8*** | 55.6 |
| Minimum Temp °F | 9.8 | 11.5 | 21.5 | 31.7 | 41.0 | 50.1 | 55.2 | 53.3 | 44.4 | 34.0 | ***27.6*** | ***15.7*** | 33.0 |
| Mean Temp °F | 20.2 | 22.2 | 31.8 | 43.0 | 53.7 | 62.3 | 67.3 | 65.3 | 56.9 | 46.3 | ***37.3*** | ***25.3*** | 44.3 |
| Days Max Temp ≥ 90 °F | 0 | 0 | 0 | 0 | 0 | 0 | 1 | 0 | 0 | 0 | 0 | 0 | 1 |
| Days Max Temp ≤ 32 °F | 17 | 13 | 5 | 0 | 0 | 0 | 0 | 0 | 0 | 0 | 2 | 11 | 48 |
| Days Min Temp ≤ 32 °F | 29 | 26 | 27 | 17 | 5 | 0 | 0 | 0 | 3 | 15 | 21 | 27 | 170 |
| Days Min Temp ≤ 0 °F | 7 | 5 | 1 | 0 | 0 | 0 | 0 | 0 | 0 | 0 | 0 | 3 | 16 |
| Heating Degree Days | 1383 | 1201 | 1022 | 655 | 352 | 126 | 39 | 70 | 255 | 575 | ***824*** | ***1225*** | 7727 |
| Cooling Degree Days | ***0*** | ***0*** | ***0*** | ***0*** | 10 | 46 | 118 | 89 | 20 | ***1*** | na | na | na |
| Total Precipitation (") | 3.14 | 2.66 | 3.37 | 3.62 | 4.01 | 3.81 | 3.77 | 4.47 | 3.59 | 3.63 | 3.89 | 3.63 | 43.59 |
| Days ≥ 0.1" Precip | 7 | 6 | 8 | 7 | 8 | 7 | 7 | 7 | 6 | 6 | 8 | 7 | 84 |
| Total Snowfall (") | 15.3 | 13.2 | 10.2 | 3.0 | 0.8 | 0.0 | 0.0 | 0.0 | 0.0 | 0.2 | 3.4 | 11.2 | 57.3 |
| Days ≥ 1" Snow Depth | 24 | 24 | 18 | 3 | 0 | 0 | 0 | 0 | 0 | 0 | 3 | 14 | 86 |

## BEDFORD *Middlesex County* ELEVATION 141 ft LAT/LONG 42° 28 ' N / 71° 17 ' W

| | JAN | FEB | MAR | APR | MAY | JUN | JUL | AUG | SEP | OCT | NOV | DEC | YEAR |
|---|---|---|---|---|---|---|---|---|---|---|---|---|---|
| Maximum Temp °F | 34.2 | 37.0 | 46.0 | 57.9 | 69.1 | 77.3 | 82.6 | 80.8 | 72.3 | 62.1 | 50.7 | 39.0 | 59.1 |
| Minimum Temp °F | 14.7 | 17.0 | 26.1 | 35.4 | 45.5 | 54.4 | 60.2 | 58.8 | 49.9 | 38.9 | 31.2 | 20.9 | 37.7 |
| Mean Temp °F | 24.5 | 27.0 | 36.1 | 46.7 | 57.4 | 65.9 | 71.4 | 69.8 | 61.1 | 50.5 | 40.9 | 29.9 | 48.4 |
| Days Max Temp ≥ 90 °F | 0 | 0 | 0 | 0 | 1 | 2 | 4 | 2 | 1 | 0 | 0 | 0 | 10 |
| Days Max Temp ≤ 32 °F | 13 | 9 | 2 | 0 | 0 | 0 | 0 | 0 | 0 | 0 | 1 | 8 | 33 |
| Days Min Temp ≤ 32 °F | 29 | 26 | 24 | 11 | 1 | 0 | 0 | 0 | 0 | 9 | 18 | 28 | 146 |
| Days Min Temp ≤ 0 °F | 4 | 2 | 0 | 0 | 0 | 0 | 0 | 0 | 0 | 0 | 0 | 1 | 7 |
| Heating Degree Days | 1250 | 1065 | 890 | 547 | 253 | 68 | 8 | 22 | 155 | 446 | 716 | 1080 | 6500 |
| Cooling Degree Days | 0 | 0 | 0 | 2 | 23 | 92 | 213 | 173 | 46 | 3 | 0 | 0 | 552 |
| Total Precipitation (") | 3.68 | 3.40 | 4.12 | 3.80 | 3.78 | 3.59 | 3.44 | 3.66 | 3.58 | 3.39 | 4.53 | 4.35 | 45.32 |
| Days ≥ 0.1" Precip | 7 | 6 | 8 | 7 | 7 | 7 | 6 | 6 | 6 | 6 | 8 | 7 | 81 |
| Total Snowfall (") | 14.9 | 14.0 | 9.9 | 2.0 | 0.3 | 0.0 | 0.0 | 0.0 | 0.0 | 0.1 | 2.9 | 11.4 | 55.5 |
| Days ≥ 1" Snow Depth | 20 | 18 | 11 | 1 | 0 | 0 | 0 | 0 | 0 | 0 | 2 | 12 | 64 |

## BIRCH HILL DAM *Worcester County* ELEVATION 860 ft LAT/LONG 42° 38 ' N / 72° 7 ' W

| | JAN | FEB | MAR | APR | MAY | JUN | JUL | AUG | SEP | OCT | NOV | DEC | YEAR |
|---|---|---|---|---|---|---|---|---|---|---|---|---|---|
| Maximum Temp °F | 30.8 | 33.9 | 42.8 | 55.3 | 67.6 | 76.1 | 81.0 | 79.0 | 70.8 | 59.8 | 47.4 | 35.0 | 56.6 |
| Minimum Temp °F | 7.5 | 10.1 | 20.7 | 31.2 | 41.2 | 49.9 | 55.2 | 53.2 | 44.3 | 33.6 | 26.9 | 14.7 | 32.4 |
| Mean Temp °F | 19.1 | 22.0 | 31.8 | 43.3 | 54.4 | 63.0 | 68.1 | 66.2 | 57.6 | 46.7 | 37.2 | ***24.7*** | 44.5 |
| Days Max Temp ≥ 90 °F | 0 | 0 | 0 | 0 | 0 | 1 | 3 | 1 | 0 | 0 | 0 | 0 | 5 |
| Days Max Temp ≤ 32 °F | 16 | 12 | 4 | 0 | 0 | 0 | 0 | 0 | 0 | 0 | 1 | 12 | 45 |
| Days Min Temp ≤ 32 °F | 29 | 26 | 27 | 18 | 6 | 0 | 0 | 0 | 4 | 16 | 22 | 27 | 175 |
| Days Min Temp ≤ 0 °F | 9 | 7 | 1 | 0 | 0 | 0 | 0 | 0 | 0 | 0 | 0 | 4 | 21 |
| Heating Degree Days | 1415 | 1207 | 1024 | 647 | 337 | 115 | 35 | 62 | 242 | 562 | 828 | ***1242*** | 7716 |
| Cooling Degree Days | ***0*** | ***0*** | 0 | 0 | ***13*** | 54 | 135 | 99 | 24 | ***2*** | ***0*** | na | na |
| Total Precipitation (") | 3.12 | 2.76 | 3.54 | 3.40 | 3.70 | 3.70 | 3.86 | 3.98 | 3.25 | 3.31 | 4.03 | 3.73 | 42.38 |
| Days ≥ 0.1" Precip | 7 | 6 | 7 | 7 | 8 | 7 | 7 | 6 | 6 | 6 | 8 | 8 | 83 |
| Total Snowfall (") | 14.9 | 13.7 | 9.5 | 2.3 | 0.2 | 0.0 | 0.0 | 0.0 | 0.0 | 0.1 | 4.1 | 11.9 | 56.7 |
| Days ≥ 1" Snow Depth | 23 | 23 | 16 | 2 | 0 | 0 | 0 | 0 | 0 | 0 | 4 | 16 | 84 |

## BOSTON LOGAN INTL AP *Suffolk County* ELEVATION 43 ft LAT/LONG 42° 22 ' N / 71° 2 ' W

| | JAN | FEB | MAR | APR | MAY | JUN | JUL | AUG | SEP | OCT | NOV | DEC | YEAR |
|---|---|---|---|---|---|---|---|---|---|---|---|---|---|
| Maximum Temp °F | 35.8 | 37.8 | 45.6 | 56.1 | 66.6 | 76.6 | 82.1 | 80.2 | 72.3 | 61.9 | 51.8 | 41.2 | 59.0 |
| Minimum Temp °F | 21.7 | 23.4 | 31.1 | 40.4 | 50.0 | 59.3 | 65.4 | 64.5 | 56.7 | 46.5 | 38.1 | 27.4 | 43.7 |
| Mean Temp °F | 28.7 | 30.6 | 38.4 | 48.3 | 58.4 | 68.0 | 73.8 | 72.4 | 64.5 | 54.2 | 45.0 | 34.3 | 51.4 |
| Days Max Temp ≥ 90 °F | 0 | 0 | 0 | 0 | 1 | 3 | 6 | 3 | 1 | 0 | 0 | 0 | 14 |
| Days Max Temp ≤ 32 °F | 11 | 8 | 2 | 0 | 0 | 0 | 0 | 0 | 0 | 0 | 0 | 6 | 27 |
| Days Min Temp ≤ 32 °F | 26 | 23 | 17 | 2 | 0 | 0 | 0 | 0 | 0 | 1 | 7 | 21 | 97 |
| Days Min Temp ≤ 0 °F | 1 | 0 | 0 | 0 | 0 | 0 | 0 | 0 | 0 | 0 | 0 | 0 | 1 |
| Heating Degree Days | 1117 | 964 | 818 | 498 | 230 | 47 | 3 | 8 | 84 | 335 | 595 | 944 | 5643 |
| Cooling Degree Days | 0 | 0 | 0 | 2 | 34 | 145 | 291 | 246 | 84 | 8 | 1 | 0 | 811 |
| Total Precipitation (") | 3.59 | 3.47 | 3.99 | 3.56 | 3.22 | 3.17 | 2.79 | 3.50 | 3.13 | 3.12 | 4.24 | 4.23 | 42.01 |
| Days ≥ 0.1" Precip | 7 | 6 | 7 | 6 | 7 | 6 | 6 | 6 | 6 | 6 | 7 | 7 | 77 |
| Total Snowfall (") | 12.5 | 12.1 | 8.2 | 1.2 | 0.0 | 0.0 | 0.0 | 0.0 | 0.0 | 0.0 | 1.2 | 7.5 | 42.7 |
| Days ≥ 1" Snow Depth | 14 | 11 | 6 | 0 | 0 | 0 | 0 | 0 | 0 | 0 | 1 | 5 | 37 |

## BROCKTON *Plymouth County* ELEVATION 112 ft LAT/LONG 42° 5 ' N / 71° 1 ' W

| | JAN | FEB | MAR | APR | MAY | JUN | JUL | AUG | SEP | OCT | NOV | DEC | YEAR |
|---|---|---|---|---|---|---|---|---|---|---|---|---|---|
| Maximum Temp °F | 37.0 | 39.1 | 47.2 | 57.7 | 68.5 | 77.1 | 82.6 | 81.0 | 73.2 | 63.2 | 52.7 | 41.7 | 60.1 |
| Minimum Temp °F | 17.1 | 19.0 | 27.3 | 35.6 | 45.1 | 54.4 | 60.6 | 59.5 | 50.3 | 39.7 | 32.4 | 23.0 | 38.7 |
| Mean Temp °F | 27.1 | 29.1 | 37.2 | 46.7 | 56.8 | 65.9 | 71.7 | 70.3 | 61.7 | 51.5 | 42.6 | 32.4 | 49.4 |
| Days Max Temp ≥ 90 °F | 0 | 0 | 0 | 0 | 1 | 2 | 4 | 3 | 1 | 0 | 0 | 0 | 11 |
| Days Max Temp ≤ 32 °F | 11 | 7 | 1 | 0 | 0 | 0 | 0 | 0 | 0 | 0 | 0 | 5 | 24 |
| Days Min Temp ≤ 32 °F | 29 | 26 | 22 | 11 | 1 | 0 | 0 | 0 | 1 | 8 | 17 | 26 | 141 |
| Days Min Temp ≤ 0 °F | 2 | 1 | 0 | 0 | 0 | 0 | 0 | 0 | 0 | 0 | 0 | 0 | 3 |
| Heating Degree Days | 1168 | 1008 | 854 | 544 | 267 | 66 | 9 | 19 | 144 | 418 | 667 | 1005 | 6169 |
| Cooling Degree Days | 0 | 0 | 0 | 1 | 26 | 112 | 250 | 208 | 60 | 7 | 0 | 0 | 664 |
| Total Precipitation (") | 3.49 | 3.59 | 4.32 | 3.86 | 3.37 | 3.39 | 3.16 | 4.10 | 3.65 | 3.83 | 4.57 | 4.69 | 46.02 |
| Days ≥ 0.1" Precip | 6 | 6 | 7 | 6 | 6 | 6 | 5 | 6 | 6 | 6 | 7 | 7 | 74 |
| Total Snowfall (") | na | na | na | na | 0.0 | 0.0 | 0.0 | 0.0 | 0.0 | 0.0 | ***0.6*** | na | na |
| Days ≥ 1" Snow Depth | na | na | na | na | 0 | 0 | 0 | 0 | 0 | 0 | ***0*** | na | na |

## BUFFUMVILLE LAKE *Worcester County* ELEVATION 502 ft LAT/LONG 42° 7 ' N / 71° 54 ' W

| | JAN | FEB | MAR | APR | MAY | JUN | JUL | AUG | SEP | OCT | NOV | DEC | YEAR |
|---|---|---|---|---|---|---|---|---|---|---|---|---|---|
| Maximum Temp °F | 32.8 | 35.1 | 44.3 | 56.0 | 68.1 | 76.3 | 81.4 | 79.5 | 71.7 | 61.3 | 49.7 | ***37.3*** | 57.8 |
| Minimum Temp °F | 10.6 | 12.9 | 23.2 | 33.5 | 43.6 | 52.9 | 58.5 | 56.8 | 47.7 | 36.5 | 29.3 | ***17.6*** | 35.3 |
| Mean Temp °F | 21.8 | 24.0 | 33.8 | 44.8 | 55.9 | 64.6 | 69.9 | 68.2 | 59.7 | 48.9 | 39.5 | ***27.5*** | 46.5 |
| Days Max Temp ≥ 90 °F | 0 | 0 | 0 | 0 | 0 | 1 | 3 | 1 | 0 | 0 | 0 | 0 | 5 |
| Days Max Temp ≤ 32 °F | 14 | 11 | 3 | 0 | 0 | 0 | 0 | 0 | 0 | 0 | 1 | 9 | 38 |
| Days Min Temp ≤ 32 °F | 28 | 25 | 25 | 14 | 2 | 0 | 0 | 0 | 1 | 11 | 19 | 27 | 152 |
| Days Min Temp ≤ 0 °F | 6 | 4 | 1 | 0 | 0 | 0 | 0 | 0 | 0 | 0 | 0 | 2 | 13 |
| Heating Degree Days | 1334 | 1150 | 966 | 601 | 292 | 85 | 16 | 34 | 185 | 494 | 759 | ***1156*** | 7072 |
| Cooling Degree Days | ***0*** | ***0*** | ***0*** | 1 | 18 | 79 | 185 | 141 | 35 | ***3*** | ***0*** | na | na |
| Total Precipitation (") | 3.81 | 3.24 | 4.18 | 3.95 | 3.83 | 3.94 | 3.89 | 4.14 | 3.99 | 3.91 | 4.62 | 4.29 | 47.79 |
| Days ≥ 0.1" Precip | 7 | 6 | 7 | 7 | 8 | 7 | 6 | 7 | 6 | 6 | 8 | 7 | 82 |
| Total Snowfall (") | ***13.2*** | ***12.0*** | 9.1 | 2.3 | 0.0 | 0.0 | 0.0 | 0.0 | 0.0 | 0.1 | 2.8 | 9.0 | 48.5 |
| Days ≥ 1" Snow Depth | 20 | 18 | 12 | ***1*** | 0 | 0 | 0 | 0 | 0 | 0 | 2 | ***11*** | 64 |

## CUMMINGTON HILL *Hampshire County* ELEVATION 1611 ft LAT/LONG 42° 28 ' N / 72° 56 ' W

| | JAN | FEB | MAR | APR | MAY | JUN | JUL | AUG | SEP | OCT | NOV | DEC | YEAR |
|---|---|---|---|---|---|---|---|---|---|---|---|---|---|
| Maximum Temp °F | 28.1 | 31.0 | 39.7 | 52.1 | 64.5 | 72.3 | 77.0 | 75.5 | 67.7 | 57.2 | 44.7 | 32.8 | 53.6 |
| Minimum Temp °F | 11.5 | 13.3 | 22.2 | 33.1 | 43.9 | 52.7 | 57.6 | 56.3 | 48.4 | 37.9 | 28.9 | 17.0 | 35.2 |
| Mean Temp °F | 19.8 | 22.2 | 30.9 | 42.7 | 54.2 | 62.5 | 67.3 | 65.9 | 58.1 | 47.6 | 36.8 | 24.9 | 44.4 |
| Days Max Temp ≥ 90 °F | 0 | 0 | 0 | 0 | 0 | 0 | 0 | 0 | 0 | 0 | 0 | 0 | 0 |
| Days Max Temp ≤ 32 °F | 20 | 15 | 8 | 0 | 0 | 0 | 0 | 0 | 0 | 0 | 3 | 16 | 62 |
| Days Min Temp ≤ 32 °F | 30 | 27 | 27 | 15 | 2 | 0 | 0 | 0 | 0 | 9 | 21 | 29 | 160 |
| Days Min Temp ≤ 0 °F | 6 | 4 | 1 | 0 | 0 | 0 | 0 | 0 | 0 | 0 | 0 | 2 | 13 |
| Heating Degree Days | 1395 | 1204 | 1052 | 666 | 339 | 119 | 35 | 56 | 226 | 536 | 839 | 1235 | 7702 |
| Cooling Degree Days | 0 | 0 | ***0*** | 1 | 16 | 45 | 116 | 88 | 25 | ***1*** | 0 | 0 | 292 |
| Total Precipitation (") | 3.06 | 3.08 | 3.46 | 3.98 | 4.71 | 4.00 | 4.34 | 4.44 | 3.85 | 3.89 | 4.46 | 3.81 | 47.08 |
| Days ≥ 0.1" Precip | 6 | 6 | 6 | 7 | 8 | 8 | 7 | 7 | 7 | 6 | 9 | 8 | 85 |
| Total Snowfall (") | 19.2 | 16.3 | 13.4 | 4.4 | 0.6 | 0.0 | 0.0 | 0.0 | 0.0 | 0.2 | 5.6 | 17.8 | 77.5 |
| Days ≥ 1" Snow Depth | 28 | 26 | 25 | 7 | 0 | 0 | 0 | 0 | 0 | 0 | 5 | 21 | 112 |

## EAST BRIMFIELD LAKE *Worcester County* ELEVATION 679 ft LAT/LONG 42° 7 ' N / 72° 8 ' W

| | JAN | FEB | MAR | APR | MAY | JUN | JUL | AUG | SEP | OCT | NOV | DEC | YEAR |
|---|---|---|---|---|---|---|---|---|---|---|---|---|---|
| Maximum Temp °F | 32.4 | 35.0 | 44.5 | 56.8 | 69.0 | 77.1 | 82.2 | 80.1 | 71.8 | 61.1 | 49.4 | 37.2 | 58.0 |
| Minimum Temp °F | 11.0 | 13.0 | 23.1 | 33.7 | 43.7 | 53.1 | 58.5 | 57.0 | 48.3 | 36.9 | 29.5 | 18.2 | 35.5 |
| Mean Temp °F | 21.7 | 24.0 | 33.8 | 45.3 | 56.4 | 65.1 | 70.4 | 68.6 | 60.1 | 49.0 | 39.4 | 27.7 | 46.8 |
| Days Max Temp ≥ 90 °F | 0 | 0 | 0 | 0 | 0 | 1 | 3 | 1 | 0 | 0 | 0 | 0 | 5 |
| Days Max Temp ≤ 32 °F | 15 | 11 | 3 | 0 | 0 | 0 | 0 | 0 | 0 | 0 | 1 | 9 | 39 |
| Days Min Temp ≤ 32 °F | 30 | 27 | 27 | 14 | 2 | 0 | 0 | 0 | 0 | 10 | 20 | 28 | 158 |
| Days Min Temp ≤ 0 °F | 7 | 4 | 1 | 0 | 0 | 0 | 0 | 0 | 0 | 0 | 0 | 2 | 14 |
| Heating Degree Days | 1337 | 1152 | 959 | 586 | 279 | 76 | 13 | 30 | 177 | 490 | 759 | 1150 | 7008 |
| Cooling Degree Days | 0 | 0 | 0 | 0 | 18 | 80 | 186 | 142 | 37 | 3 | 0 | 0 | 466 |
| Total Precipitation (") | 3.69 | 3.21 | 4.01 | 3.98 | 3.87 | 3.84 | 3.50 | 3.92 | 3.92 | 3.71 | 4.35 | 4.27 | 46.27 |
| Days ≥ 0.1" Precip | 7 | 6 | 8 | 7 | 8 | 7 | 7 | 7 | 6 | 6 | 8 | 8 | 85 |
| Total Snowfall (") | 14.6 | 13.4 | 9.9 | 2.5 | 0.2 | 0.0 | 0.0 | 0.0 | 0.0 | 0.1 | 2.6 | 11.5 | 54.8 |
| Days ≥ 1" Snow Depth | 21 | 19 | 12 | 1 | 0 | 0 | 0 | 0 | 0 | 0 | 2 | 12 | 67 |

## EAST WAREHAM *Plymouth County* ELEVATION 20 ft LAT/LONG 41° 46 ' N / 70° 40 ' W

| | JAN | FEB | MAR | APR | MAY | JUN | JUL | AUG | SEP | OCT | NOV | DEC | YEAR |
|---|---|---|---|---|---|---|---|---|---|---|---|---|---|
| Maximum Temp °F | 36.0 | 37.1 | 44.5 | 54.3 | 64.8 | 73.8 | 79.9 | 78.8 | 71.2 | 61.1 | 51.2 | 41.3 | 57.8 |
| Minimum Temp °F | 18.7 | 19.8 | 28.0 | 36.3 | 46.3 | 55.9 | 62.6 | 61.7 | 52.9 | 42.5 | 34.3 | 24.7 | 40.3 |
| Mean Temp °F | 27.4 | 28.5 | 36.2 | 45.3 | 55.6 | 64.9 | 71.3 | 70.3 | 62.1 | 51.8 | 42.7 | 33.0 | 49.1 |
| Days Max Temp ≥ 90 °F | 0 | 0 | 0 | 0 | 0 | 0 | 1 | 1 | 0 | 0 | 0 | 0 | 2 |
| Days Max Temp ≤ 32 °F | 11 | 8 | 2 | 0 | 0 | 0 | 0 | 0 | 0 | 0 | 0 | 6 | 27 |
| Days Min Temp ≤ 32 °F | 28 | 25 | 22 | 10 | 0 | 0 | 0 | 0 | 0 | 4 | 14 | 24 | 127 |
| Days Min Temp ≤ 0 °F | 2 | 1 | 0 | 0 | 0 | 0 | 0 | 0 | 0 | 0 | 0 | 0 | 3 |
| Heating Degree Days | 1159 | 1024 | 885 | 585 | 293 | 67 | 6 | 15 | 129 | 404 | 661 | 984 | 6212 |
| Cooling Degree Days | 0 | 0 | 0 | 1 | 9 | 70 | 206 | 186 | 53 | 5 | 0 | 0 | 530 |
| Total Precipitation (") | 4.15 | 3.76 | 4.65 | 4.19 | 3.78 | 3.54 | 3.00 | 4.00 | 3.65 | 3.51 | 4.79 | 4.81 | 47.83 |
| Days ≥ 0.1" Precip | 7 | 6 | 8 | 7 | 7 | 6 | 6 | 6 | 5 | 6 | 8 | 8 | 80 |
| Total Snowfall (") | 10.1 | 10.3 | 5.7 | 0.8 | 0.0 | 0.0 | 0.0 | 0.0 | 0.0 | 0.0 | 1.2 | 5.1 | 33.2 |
| Days ≥ 1" Snow Depth | 13 | 11 | 5 | 0 | 0 | 0 | 0 | 0 | 0 | 0 | 1 | 4 | 34 |

## EDGARTOWN *Dukes County* ELEVATION 20 ft LAT/LONG 41° 23 ' N / 70° 31 ' W

| | JAN | FEB | MAR | APR | MAY | JUN | JUL | AUG | SEP | OCT | NOV | DEC | YEAR |
|---|---|---|---|---|---|---|---|---|---|---|---|---|---|
| Maximum Temp °F | 37.7 | 38.5 | 45.2 | 53.9 | 63.6 | 72.4 | 78.3 | 78.0 | 71.5 | 62.5 | 53.2 | 43.3 | 58.2 |
| Minimum Temp °F | 21.5 | 22.1 | 29.2 | 37.2 | 46.6 | 55.8 | 62.0 | 61.9 | 55.0 | 45.1 | 37.3 | 27.3 | 41.8 |
| Mean Temp °F | 29.6 | 30.4 | 37.2 | 45.6 | 55.1 | 64.1 | 70.2 | 70.0 | 63.3 | 53.8 | 45.3 | 35.3 | 50.0 |
| Days Max Temp ≥ 90 °F | 0 | 0 | 0 | 0 | 0 | 0 | 1 | 0 | 0 | 0 | 0 | 0 | 1 |
| Days Max Temp ≤ 32 °F | 8 | 7 | 1 | 0 | 0 | 0 | 0 | 0 | 0 | 0 | 0 | 4 | 20 |
| Days Min Temp ≤ 32 °F | 27 | 24 | 21 | 7 | 0 | 0 | 0 | 0 | 0 | 2 | 10 | 22 | 113 |
| Days Min Temp ≤ 0 °F | 0 | 0 | 0 | 0 | 0 | 0 | 0 | 0 | 0 | 0 | 0 | 0 | 0 |
| Heating Degree Days | 1091 | 972 | 853 | 576 | 302 | 68 | 7 | 10 | 93 | 345 | 585 | 914 | 5816 |
| Cooling Degree Days | 0 | 0 | 0 | 0 | 5 | 59 | 180 | 168 | 51 | 4 | 0 | 0 | 467 |
| Total Precipitation (") | 3.76 | 3.41 | 4.31 | 3.98 | 3.75 | 3.41 | 2.96 | 3.66 | 3.33 | 3.54 | 4.49 | 4.60 | 45.20 |
| Days ≥ 0.1" Precip | 7 | 6 | 8 | 7 | 7 | 5 | 5 | 6 | 5 | 6 | 8 | 8 | 78 |
| Total Snowfall (") | na | na | na | *0.1* | 0.0 | 0.0 | 0.0 | 0.0 | 0.0 | 0.0 | *0.4* | na | na |
| Days ≥ 1" Snow Depth | na | na | na | na | 0 | 0 | 0 | 0 | 0 | 0 | *0* | na | na |

## HAVERHILL *Essex County* ELEVATION 121 ft LAT/LONG 42° 48 ' N / 71° 4 ' W

| | JAN | FEB | MAR | APR | MAY | JUN | JUL | AUG | SEP | OCT | NOV | DEC | YEAR |
|---|---|---|---|---|---|---|---|---|---|---|---|---|---|
| Maximum Temp °F | 34.5 | 37.4 | 46.5 | 58.2 | 69.5 | 78.4 | 84.0 | 82.0 | 73.4 | 62.6 | 50.8 | 39.2 | 59.7 |
| Minimum Temp °F | 15.5 | 17.6 | 26.5 | 35.6 | 45.7 | 55.0 | 61.2 | 59.9 | 51.1 | 40.4 | 32.3 | 21.4 | 38.5 |
| Mean Temp °F | 25.0 | 27.5 | 36.5 | 46.9 | 57.6 | 66.7 | 72.6 | 70.9 | 62.3 | 51.5 | 41.6 | 30.3 | 49.1 |
| Days Max Temp ≥ 90 °F | 0 | 0 | 0 | 0 | 1 | 3 | 6 | 4 | 1 | 0 | 0 | 0 | 15 |
| Days Max Temp ≤ 32 °F | 13 | 8 | 2 | 0 | 0 | 0 | 0 | 0 | 0 | 0 | 0 | 7 | 30 |
| Days Min Temp ≤ 32 °F | 29 | 26 | 23 | 11 | 1 | 0 | 0 | 0 | 0 | 7 | 16 | 27 | 140 |
| Days Min Temp ≤ 0 °F | 3 | 2 | 0 | 0 | 0 | 0 | 0 | 0 | 0 | 0 | 0 | 1 | 6 |
| Heating Degree Days | 1233 | 1052 | 876 | 538 | 249 | 59 | 6 | 18 | 135 | 416 | 697 | 1068 | 6347 |
| Cooling Degree Days | 0 | 0 | 0 | 1 | 21 | 88 | 210 | 178 | 47 | 3 | 0 | 0 | 548 |
| Total Precipitation (") | 3.43 | 3.37 | 4.04 | 4.01 | 3.61 | 3.64 | 3.29 | 3.55 | 3.65 | 3.63 | 4.59 | 4.46 | 45.27 |
| Days ≥ 0.1" Precip | 7 | 5 | 7 | 7 | 7 | 7 | 6 | 6 | 6 | 6 | 7 | 7 | 78 |
| Total Snowfall (") | 15.9 | 14.6 | 9.3 | 2.0 | 0.0 | 0.0 | 0.0 | 0.0 | 0.0 | 0.0 | 2.9 | 10.9 | 55.6 |
| Days ≥ 1" Snow Depth | 19 | 15 | 8 | 0 | 0 | 0 | 0 | 0 | 0 | 0 | 2 | 9 | 53 |

## HINGHAM *Plymouth County* ELEVATION 30 ft LAT/LONG 42° 14 ' N / 70° 55 ' W

| | JAN | FEB | MAR | APR | MAY | JUN | JUL | AUG | SEP | OCT | NOV | DEC | YEAR |
|---|---|---|---|---|---|---|---|---|---|---|---|---|---|
| Maximum Temp °F | 36.2 | 38.2 | 46.2 | 57.0 | 67.6 | 76.2 | 81.6 | 79.4 | 71.6 | 61.9 | 51.8 | 41.1 | 59.1 |
| Minimum Temp °F | 18.7 | 20.6 | 28.6 | 37.0 | 46.4 | 55.7 | 61.7 | 60.6 | 52.4 | 42.2 | 34.4 | 24.8 | 40.3 |
| Mean Temp °F | 27.5 | 29.4 | 37.4 | 47.0 | 57.0 | 66.0 | 71.7 | 70.0 | 62.0 | 52.1 | 43.1 | 33.0 | 49.7 |
| Days Max Temp ≥ 90 °F | 0 | 0 | 0 | 0 | 0 | 1 | 4 | 2 | 1 | 0 | 0 | 0 | 8 |
| Days Max Temp ≤ 32 °F | 11 | 7 | 2 | 0 | 0 | 0 | 0 | 0 | 0 | 0 | 0 | 6 | 26 |
| Days Min Temp ≤ 32 °F | 28 | 25 | 21 | 9 | 1 | 0 | 0 | 0 | 0 | 5 | 14 | 25 | 128 |
| Days Min Temp ≤ 0 °F | 2 | 1 | 0 | 0 | 0 | 0 | 0 | 0 | 0 | 0 | 0 | 0 | 3 |
| Heating Degree Days | 1156 | 998 | 848 | 536 | 261 | 67 | 8 | 21 | 136 | 399 | 649 | 986 | 6065 |
| Cooling Degree Days | 0 | 0 | 0 | 2 | 23 | 101 | 222 | 179 | 55 | 6 | 1 | 0 | 589 |
| Total Precipitation (") | 4.17 | 3.90 | 4.70 | 4.02 | 3.76 | 3.39 | 3.24 | 4.20 | 3.58 | 3.85 | 4.84 | 4.77 | 48.42 |
| Days ≥ 0.1" Precip | 7 | 6 | 8 | 6 | 7 | 6 | 6 | 6 | 6 | 6 | 8 | 8 | 80 |
| Total Snowfall (") | 13.8 | 13.5 | 8.6 | 1.1 | 0.0 | 0.0 | 0.0 | 0.0 | 0.0 | 0.0 | 1.7 | 8.8 | 47.5 |
| Days ≥ 1" Snow Depth | 17 | 16 | 9 | 0 | 0 | 0 | 0 | 0 | 0 | 0 | 1 | 8 | 51 |

## HYANNIS *Barnstable County* ELEVATION 20 ft LAT/LONG 41° 39 ' N / 70° 16 ' W

| | JAN | FEB | MAR | APR | MAY | JUN | JUL | AUG | SEP | OCT | NOV | DEC | YEAR |
|---|---|---|---|---|---|---|---|---|---|---|---|---|---|
| Maximum Temp °F | 37.4 | 38.3 | 44.7 | 53.3 | 63.5 | *72.9* | 78.6 | 77.9 | 71.1 | 61.7 | 52.5 | 43.1 | 57.9 |
| Minimum Temp °F | 20.4 | 21.5 | 29.0 | 37.1 | 46.5 | *56.4* | 62.4 | 62.1 | 54.1 | 43.8 | 35.8 | 25.8 | 41.2 |
| Mean Temp °F | 28.9 | 29.9 | 36.9 | 45.2 | 55.1 | *64.7* | 70.5 | 70.0 | 62.6 | 52.8 | 44.1 | 34.5 | 49.6 |
| Days Max Temp ≥ 90 °F | 0 | 0 | 0 | 0 | 0 | 0 | 1 | 0 | 0 | 0 | 0 | 0 | 1 |
| Days Max Temp ≤ 32 °F | 8 | 7 | 2 | 0 | 0 | 0 | 0 | 0 | 0 | 0 | 0 | *4* | 21 |
| Days Min Temp ≤ 32 °F | 27 | 24 | 20 | 10 | 1 | 0 | 0 | 0 | 0 | 3 | 11 | *23* | 119 |
| Days Min Temp ≤ 0 °F | 1 | 0 | 0 | 0 | 0 | 0 | 0 | 0 | 0 | 0 | 0 | 0 | 1 |
| Heating Degree Days | 1109 | 984 | 866 | 587 | 307 | *69* | 9 | 15 | 115 | 375 | 621 | 938 | 5995 |
| Cooling Degree Days | 0 | 0 | 0 | 0 | 5 | 66 | 182 | 172 | 48 | 5 | 0 | 0 | 478 |
| Total Precipitation (") | 3.67 | 3.43 | 3.87 | 3.62 | 3.55 | 3.33 | 2.92 | 3.54 | 3.23 | 3.45 | 4.37 | 4.52 | 43.50 |
| Days ≥ 0.1" Precip | 7 | *6* | 7 | 7 | 6 | *5* | *5* | *5* | 5 | *6* | *7* | *8* | 74 |
| Total Snowfall (") | *6.2* | na | na | *0.3* | 0.0 | 0.0 | *0.0* | 0.0 | 0.0 | *0.0* | *0.2* | na | na |
| Days ≥ 1" Snow Depth | *8* | na | na | *0* | 0 | 0 | 0 | 0 | 0 | 0 | *0* | na | na |

## KNIGHTVILLE DAM *Hampshire County* ELEVATION 630 ft LAT/LONG 42° 17 ' N / 72° 52 ' W

| | JAN | FEB | MAR | APR | MAY | JUN | JUL | AUG | SEP | OCT | NOV | DEC | YEAR |
|---|---|---|---|---|---|---|---|---|---|---|---|---|---|
| Maximum Temp °F | 31.6 | 34.4 | 43.3 | 56.4 | 68.9 | 76.8 | 81.6 | 79.7 | 71.7 | 60.8 | 48.5 | 35.8 | 57.5 |
| Minimum Temp °F | 9.0 | 11.3 | 21.2 | 32.0 | 41.7 | 50.9 | 55.6 | 54.0 | 45.3 | 34.3 | 27.7 | 16.4 | 33.3 |
| Mean Temp °F | 20.3 | 22.9 | 32.2 | 44.2 | 55.3 | 63.8 | 68.6 | 66.9 | 58.5 | 47.6 | 38.1 | 26.2 | 45.4 |
| Days Max Temp ≥ 90 °F | 0 | 0 | 0 | 0 | 0 | 2 | 3 | 2 | 0 | 0 | 0 | 0 | 7 |
| Days Max Temp ≤ 32 °F | 15 | 12 | 5 | 0 | 0 | 0 | 0 | 0 | 0 | 0 | 1 | 10 | 43 |
| Days Min Temp ≤ 32 °F | 30 | 27 | 28 | 17 | 4 | 0 | 0 | 0 | 2 | 14 | 21 | 28 | 171 |
| Days Min Temp ≤ 0 °F | 8 | 6 | 1 | 0 | 0 | 0 | 0 | 0 | 0 | 0 | 0 | 3 | 18 |
| Heating Degree Days | 1378 | 1182 | 1009 | 619 | 309 | 98 | 25 | 47 | 215 | 534 | 799 | 1197 | 7412 |
| Cooling Degree Days | *0* | *0* | 0 | 1 | 16 | 59 | 141 | 110 | 25 | *2* | *0* | *0* | 354 |
| Total Precipitation (") | 3.37 | 3.19 | 4.07 | 4.03 | 4.36 | 3.76 | 3.93 | 4.31 | 3.60 | 3.67 | 4.25 | 4.11 | 46.65 |
| Days ≥ 0.1" Precip | 6 | 6 | 7 | 7 | 8 | 7 | 7 | 6 | 6 | 6 | 7 | 7 | 80 |
| Total Snowfall (") | 13.7 | 10.8 | 8.7 | 1.4 | 0.1 | 0.0 | 0.0 | 0.0 | 0.0 | 0.1 | 2.9 | 11.5 | 49.2 |
| Days ≥ 1" Snow Depth | 25 | 22 | 18 | 3 | 0 | 0 | 0 | 0 | 0 | 0 | 3 | 16 | 87 |

## LAWRENCE *Essex County* ELEVATION 57 ft LAT/LONG 42° 42 ' N / 71° 10 ' W

| | JAN | FEB | MAR | APR | MAY | JUN | JUL | AUG | SEP | OCT | NOV | DEC | YEAR |
|---|---|---|---|---|---|---|---|---|---|---|---|---|---|
| Maximum Temp °F | 33.7 | 36.4 | 45.1 | 56.8 | 68.4 | 77.3 | 82.8 | 81.4 | 72.7 | 62.0 | 50.8 | 38.8 | 58.9 |
| Minimum Temp °F | 15.1 | 17.5 | 26.6 | 37.0 | 47.1 | 56.3 | 62.2 | 60.7 | 51.8 | 40.9 | 32.9 | 22.0 | 39.2 |
| Mean Temp °F | 24.4 | 27.0 | 35.9 | 46.9 | 57.8 | 66.8 | 72.5 | 71.1 | 62.3 | 51.5 | 41.9 | 30.5 | 49.1 |
| Days Max Temp ≥ 90 °F | 0 | 0 | 0 | 0 | 0 | 2 | 4 | 3 | 0 | 0 | 0 | 0 | 9 |
| Days Max Temp ≤ 32 °F | 14 | 9 | 2 | 0 | 0 | 0 | 0 | 0 | 0 | 0 | 1 | 8 | 34 |
| Days Min Temp ≤ 32 °F | 30 | 26 | 23 | 8 | 0 | 0 | 0 | 0 | 0 | 5 | 16 | 27 | 135 |
| Days Min Temp ≤ 0 °F | 3 | 1 | 0 | 0 | 0 | 0 | 0 | 0 | 0 | 0 | 0 | 1 | 5 |
| Heating Degree Days | 1249 | 1066 | 896 | 538 | 244 | 55 | 5 | 13 | 127 | 414 | 688 | 1065 | 6360 |
| Cooling Degree Days | 0 | 0 | 0 | 1 | 28 | 113 | 244 | 205 | 51 | 3 | 0 | 0 | 645 |
| Total Precipitation (") | 3.65 | 3.23 | 3.75 | 3.81 | 3.68 | 3.74 | 3.09 | 3.24 | 3.55 | 3.45 | 4.32 | 3.89 | 43.40 |
| Days ≥ 0.1" Precip | 7 | 6 | 7 | 7 | 7 | 7 | 6 | 5 | 6 | 5 | 7 | 6 | 76 |
| Total Snowfall (") | 15.0 | 12.7 | 6.9 | 1.2 | 0.1 | 0.0 | 0.0 | 0.0 | 0.0 | 0.0 | 2.1 | 9.8 | 47.8 |
| Days ≥ 1" Snow Depth | na | na | na | *0* | 0 | 0 | 0 | 0 | 0 | 0 | *1* | na | na |

## MIDDLETON *Essex County* ELEVATION 89 ft LAT/LONG 42° 36 ' N / 71° 1 ' W

| | JAN | FEB | MAR | APR | MAY | JUN | JUL | AUG | SEP | OCT | NOV | DEC | YEAR |
|---|---|---|---|---|---|---|---|---|---|---|---|---|---|
| Maximum Temp °F | 36.2 | 38.6 | 46.5 | 57.3 | 68.4 | 77.4 | 82.7 | 81.1 | 73.6 | 63.6 | 52.1 | 40.9 | 59.9 |
| Minimum Temp °F | 16.0 | 17.8 | 26.2 | 35.8 | 45.9 | 55.4 | 61.2 | 59.8 | 51.4 | 41.1 | 33.3 | 22.6 | 38.9 |
| Mean Temp °F | 26.2 | 28.2 | 36.4 | 46.6 | 57.2 | 66.4 | 72.0 | 70.5 | 62.6 | 52.4 | 42.7 | 31.7 | 49.4 |
| Days Max Temp ≥ 90 °F | 0 | 0 | 0 | 0 | 0 | 1 | 3 | 2 | 1 | 0 | 0 | 0 | 7 |
| Days Max Temp ≤ 32 °F | 11 | 7 | 1 | 0 | 0 | 0 | 0 | 0 | 0 | 0 | 0 | 6 | 25 |
| Days Min Temp ≤ 32 °F | 29 | 26 | 24 | 11 | 1 | 0 | 0 | 0 | 0 | 6 | 15 | 26 | 138 |
| Days Min Temp ≤ 0 °F | 3 | 2 | 0 | 0 | 0 | 0 | 0 | 0 | 0 | 0 | 0 | 1 | 6 |
| Heating Degree Days | 1197 | 1032 | 881 | 548 | 258 | 61 | 7 | 18 | 127 | 390 | 662 | 1024 | 6205 |
| Cooling Degree Days | 0 | 0 | 0 | 1 | 23 | 106 | 229 | 187 | 60 | 5 | 0 | 0 | 611 |
| Total Precipitation (") | 3.38 | 3.24 | 3.86 | 3.87 | 3.58 | 3.56 | 3.06 | 3.51 | 3.51 | 3.40 | 4.61 | 4.24 | 43.82 |
| Days ≥ 0.1" Precip | 6 | 6 | 7 | 7 | 8 | 7 | 6 | 6 | 6 | 6 | 7 | 7 | 79 |
| Total Snowfall (") | ***12.0*** | 13.9 | 6.9 | 0.8 | 0.0 | 0.0 | 0.0 | 0.0 | 0.0 | 0.1 | 1.9 | ***7.7*** | 43.3 |
| Days ≥ 1" Snow Depth | 14 | ***13*** | 7 | 0 | 0 | 0 | 0 | 0 | 0 | 0 | 1 | 8 | 43 |

## MILTON BLUE HILL OBS *Norfolk County* ELEVATION 640 ft LAT/LONG 42° 13 ' N / 71° 7 ' W

| | JAN | FEB | MAR | APR | MAY | JUN | JUL | AUG | SEP | OCT | NOV | DEC | YEAR |
|---|---|---|---|---|---|---|---|---|---|---|---|---|---|
| Maximum Temp °F | 33.1 | 35.3 | 44.0 | 55.4 | 66.7 | 75.2 | 80.9 | 78.9 | 70.6 | 60.2 | 49.3 | 38.1 | 57.3 |
| Minimum Temp °F | 17.5 | 19.2 | 27.4 | 36.9 | 46.6 | 55.6 | 61.9 | 60.9 | 53.0 | 43.0 | 34.3 | 23.3 | 40.0 |
| Mean Temp °F | 25.3 | 27.3 | 35.7 | 46.2 | 56.7 | 65.4 | 71.4 | 69.9 | 61.8 | 51.6 | 41.8 | 30.7 | 48.7 |
| Days Max Temp ≥ 90 °F | 0 | 0 | 0 | 0 | 0 | 1 | 3 | 1 | 0 | 0 | 0 | 0 | 5 |
| Days Max Temp ≤ 32 °F | 15 | 12 | 3 | 0 | 0 | 0 | 0 | 0 | 0 | 0 | 1 | 9 | 40 |
| Days Min Temp ≤ 32 °F | 29 | 26 | 22 | 8 | 0 | 0 | 0 | 0 | 0 | 3 | 13 | 26 | 127 |
| Days Min Temp ≤ 0 °F | 2 | 1 | 0 | 0 | 0 | 0 | 0 | 0 | 0 | 0 | 0 | 1 | 4 |
| Heating Degree Days | 1224 | 1059 | 900 | 560 | 272 | 75 | 9 | 22 | 137 | 412 | 689 | 1055 | 6414 |
| Cooling Degree Days | 0 | 0 | 0 | 1 | 23 | 95 | 220 | 175 | 51 | 4 | 0 | 0 | 569 |
| Total Precipitation (") | 4.17 | 4.24 | 4.82 | 4.08 | 3.77 | 3.58 | 3.47 | 4.12 | 3.72 | 3.79 | 4.87 | 5.01 | 49.64 |
| Days ≥ 0.1" Precip | 7 | 6 | 8 | 7 | 8 | 7 | 6 | 6 | 6 | 6 | 8 | 8 | 83 |
| Total Snowfall (") | 15.5 | 16.0 | 11.4 | 2.5 | 0.3 | 0.0 | 0.0 | 0.0 | 0.0 | 0.3 | 2.8 | 11.3 | 60.1 |
| Days ≥ 1" Snow Depth | 19 | 19 | 11 | 1 | 0 | 0 | 0 | 0 | 0 | 0 | 2 | 10 | 62 |

## NEW BEDFORD *Bristol County* ELEVATION 70 ft LAT/LONG 41° 38 ' N / 70° 56 ' W

| | JAN | FEB | MAR | APR | MAY | JUN | JUL | AUG | SEP | OCT | NOV | DEC | YEAR |
|---|---|---|---|---|---|---|---|---|---|---|---|---|---|
| Maximum Temp °F | 37.4 | 38.5 | 45.9 | 55.8 | 66.3 | 75.5 | 81.7 | 80.9 | 73.5 | 63.3 | 52.9 | 42.6 | 59.5 |
| Minimum Temp °F | 22.8 | 23.9 | 31.5 | 40.1 | 49.9 | 59.0 | 65.9 | 65.3 | 57.7 | 47.4 | 38.9 | 28.2 | 44.2 |
| Mean Temp °F | 30.1 | 31.2 | 38.7 | 48.0 | 58.1 | 67.3 | 73.8 | 73.1 | 65.6 | 55.3 | 45.9 | 35.4 | 51.9 |
| Days Max Temp ≥ 90 °F | 0 | 0 | 0 | 0 | 0 | 1 | 4 | 2 | 1 | 0 | 0 | 0 | 8 |
| Days Max Temp ≤ 32 °F | 10 | 7 | 1 | 0 | 0 | 0 | 0 | 0 | 0 | 0 | 0 | 4 | 22 |
| Days Min Temp ≤ 32 °F | 25 | 23 | 16 | 3 | 0 | 0 | 0 | 0 | 0 | 1 | 7 | 21 | 96 |
| Days Min Temp ≤ 0 °F | 1 | 0 | 0 | 0 | 0 | 0 | 0 | 0 | 0 | 0 | 0 | 0 | 1 |
| Heating Degree Days | 1074 | 947 | 809 | 505 | 225 | 40 | 2 | 5 | 67 | 303 | 566 | 913 | 5456 |
| Cooling Degree Days | 0 | 0 | 0 | 0 | 19 | 110 | 271 | 241 | 83 | 7 | 0 | 0 | 731 |
| Total Precipitation (") | 4.23 | 3.81 | 4.65 | 4.11 | 3.73 | 3.79 | 3.17 | 4.52 | 3.56 | 3.46 | 4.82 | 5.09 | 48.94 |
| Days ≥ 0.1" Precip | 7 | 6 | 8 | 7 | 7 | 6 | 5 | 6 | 5 | 6 | 8 | 9 | 80 |
| Total Snowfall (") | 10.0 | 10.5 | 5.3 | 0.6 | 0.1 | 0.0 | 0.0 | 0.0 | 0.0 | 0.0 | 1.0 | 5.3 | 32.8 |
| Days ≥ 1" Snow Depth | na | na | na | ***0*** | 0 | 0 | 0 | 0 | 0 | 0 | 0 | na | na |

## PEABODY *Essex County* ELEVATION 161 ft LAT/LONG 42° 32 ' N / 70° 59 ' W

| | JAN | FEB | MAR | APR | MAY | JUN | JUL | AUG | SEP | OCT | NOV | DEC | YEAR |
|---|---|---|---|---|---|---|---|---|---|---|---|---|---|
| Maximum Temp °F | 34.4 | 36.7 | 45.3 | 56.5 | 67.2 | 77.1 | 82.5 | 80.7 | 72.3 | 61.0 | 49.8 | 38.7 | 58.5 |
| Minimum Temp °F | 18.2 | 19.9 | 28.1 | 37.4 | 47.1 | 56.3 | 62.3 | 60.8 | 52.7 | 42.7 | 34.3 | 23.4 | 40.3 |
| Mean Temp °F | 26.3 | 28.4 | 36.7 | 47.0 | 57.2 | 66.7 | 72.5 | 70.8 | 62.5 | 51.9 | 42.1 | 31.1 | 49.4 |
| Days Max Temp ≥ 90 °F | 0 | 0 | 0 | 0 | 1 | 2 | 5 | 3 | 1 | 0 | 0 | 0 | 12 |
| Days Max Temp ≤ 32 °F | 13 | 9 | 2 | 0 | 0 | 0 | 0 | 0 | 0 | 0 | 1 | 8 | 33 |
| Days Min Temp ≤ 32 °F | 28 | 25 | 21 | 8 | 0 | 0 | 0 | 0 | 0 | 4 | 13 | 26 | 125 |
| Days Min Temp ≤ 0 °F | 2 | 1 | 0 | 0 | 0 | 0 | 0 | 0 | 0 | 0 | 0 | 1 | 4 |
| Heating Degree Days | 1193 | 1029 | 869 | 537 | 261 | 59 | 8 | 17 | 127 | 405 | 681 | 1045 | 6231 |
| Cooling Degree Days | 0 | 0 | 0 | 2 | 29 | 125 | 256 | 213 | 63 | 4 | 0 | 0 | 692 |
| Total Precipitation (") | 3.78 | 3.58 | 4.28 | 4.16 | 3.81 | 3.39 | 3.33 | 4.07 | 3.77 | 3.62 | 5.05 | 4.86 | 47.70 |
| Days ≥ 0.1" Precip | 6 | 6 | 7 | 7 | 8 | 7 | 6 | 6 | 6 | ***6*** | 7 | 8 | 80 |
| Total Snowfall (") | 13.6 | 13.7 | 8.4 | 1.5 | 0.0 | 0.0 | 0.0 | 0.0 | 0.0 | 0.0 | 2.1 | 9.4 | 48.7 |
| Days ≥ 1" Snow Depth | 18 | 15 | 10 | 1 | 0 | 0 | 0 | 0 | 0 | 0 | 2 | 9 | 55 |

## PLYMOUTH-KINGSTON *Plymouth County* ELEVATION 30 ft LAT/LONG 41° 59 ' N / 70° 42 ' W

| | JAN | FEB | MAR | APR | MAY | JUN | JUL | AUG | SEP | OCT | NOV | DEC | YEAR |
|---|---|---|---|---|---|---|---|---|---|---|---|---|---|
| Maximum Temp °F | 36.9 | 38.8 | 46.3 | 55.8 | 66.7 | 76.0 | 81.8 | 80.2 | 72.6 | 62.7 | 52.7 | 42.3 | 59.4 |
| Minimum Temp °F | 16.5 | 17.6 | 26.1 | 34.8 | 44.3 | 54.2 | 60.5 | 59.5 | 51.2 | 40.8 | 32.8 | 22.8 | 38.4 |
| Mean Temp °F | 26.7 | 28.2 | 36.2 | 45.3 | 55.5 | 65.2 | 71.2 | 69.9 | 61.9 | 51.8 | 42.8 | 32.5 | 48.9 |
| Days Max Temp ≥ 90 °F | 0 | 0 | 0 | 0 | 0 | 2 | 4 | 2 | 1 | 0 | 0 | 0 | 9 |
| Days Max Temp ≤ 32 °F | 10 | 7 | 1 | 0 | 0 | 0 | 0 | 0 | 0 | 0 | 0 | 5 | 23 |
| Days Min Temp ≤ 32 °F | 29 | 26 | 24 | 12 | 2 | 0 | 0 | 0 | 0 | 6 | 16 | 26 | 141 |
| Days Min Temp ≤ 0 °F | 2 | 2 | 0 | 0 | 0 | 0 | 0 | 0 | 0 | 0 | 0 | 1 | 5 |
| Heating Degree Days | 1180 | 1031 | 886 | 585 | 301 | 78 | 11 | 22 | 136 | 408 | 660 | 999 | 6297 |
| Cooling Degree Days | 0 | 0 | 0 | 1 | 15 | 89 | 215 | 181 | 53 | 6 | 0 | 0 | 560 |
| Total Precipitation (") | 4.18 | 3.93 | 4.61 | 4.26 | 3.87 | 3.57 | 3.01 | 4.09 | 4.14 | 3.70 | 4.98 | 4.86 | 49.20 |
| Days ≥ 0.1" Precip | 7 | 6 | 7 | 7 | 8 | 7 | 6 | 6 | 6 | 7 | 8 | 8 | 83 |
| Total Snowfall (") | 12.3 | 11.2 | 6.3 | 0.9 | 0.0 | 0.0 | 0.0 | 0.0 | 0.0 | 0.0 | 1.1 | 5.2 | 37.0 |
| Days ≥ 1" Snow Depth | 12 | 12 | 5 | 0 | 0 | 0 | 0 | 0 | 0 | 0 | 1 | 5 | 35 |

## READING *Middlesex County* ELEVATION 89 ft LAT/LONG 42° 31 ' N / 71° 8 ' W

| | JAN | FEB | MAR | APR | MAY | JUN | JUL | AUG | SEP | OCT | NOV | DEC | YEAR |
|---|---|---|---|---|---|---|---|---|---|---|---|---|---|
| Maximum Temp °F | 34.6 | 37.4 | 46.1 | 57.6 | 68.6 | 77.1 | 82.7 | 80.8 | 72.6 | 62.1 | 50.9 | 39.4 | 59.2 |
| Minimum Temp °F | 14.5 | 16.6 | 25.3 | 34.5 | 44.3 | 53.9 | 59.5 | 58.4 | 49.5 | 38.5 | 30.9 | 20.8 | 37.2 |
| Mean Temp °F | 24.6 | 27.0 | 35.7 | 46.1 | 56.5 | 65.5 | 71.1 | 69.7 | 61.1 | 50.3 | 40.9 | 30.1 | 48.2 |
| Days Max Temp ≥ 90 °F | 0 | 0 | 0 | 0 | 1 | 2 | 4 | 3 | 1 | 0 | 0 | 0 | 11 |
| Days Max Temp ≤ 32 °F | 13 | 9 | 2 | 0 | 0 | 0 | 0 | 0 | 0 | 0 | 0 | 7 | 31 |
| Days Min Temp ≤ 32 °F | 29 | 26 | 25 | 13 | 2 | 0 | 0 | 0 | 0 | 10 | 19 | 28 | 152 |
| Days Min Temp ≤ 0 °F | 4 | 2 | 0 | 0 | 0 | 0 | 0 | 0 | 0 | 0 | 0 | 1 | 7 |
| Heating Degree Days | 1246 | 1066 | 901 | 563 | 279 | 76 | 11 | 25 | 159 | 452 | 716 | 1075 | 6569 |
| Cooling Degree Days | 0 | 0 | 0 | 1 | 24 | 94 | 208 | 175 | 49 | 3 | 0 | 0 | 554 |
| Total Precipitation (") | 3.88 | 3.65 | 4.31 | 3.95 | 3.81 | 3.61 | 3.27 | 3.68 | 3.60 | 3.62 | 4.81 | 4.72 | 46.91 |
| Days ≥ 0.1" Precip | 7 | 6 | 8 | 7 | 8 | 7 | 6 | 6 | 6 | 6 | 8 | 8 | 83 |
| Total Snowfall (") | 16.7 | 15.7 | 11.3 | 2.5 | 0.3 | 0.0 | 0.0 | 0.0 | 0.0 | 0.1 | 2.9 | 12.7 | 62.2 |
| Days ≥ 1" Snow Depth | 21 | 19 | 12 | 1 | 0 | 0 | 0 | 0 | 0 | 0 | 2 | 12 | 67 |

## ROCHESTER *Plymouth County* ELEVATION 59 ft LAT/LONG 41° 47 ' N / 70° 55 ' W

| | JAN | FEB | MAR | APR | MAY | JUN | JUL | AUG | SEP | OCT | NOV | DEC | YEAR |
|---|---|---|---|---|---|---|---|---|---|---|---|---|---|
| Maximum Temp °F | 36.5 | 38.3 | 46.4 | 56.3 | 67.2 | 76.5 | 82.0 | 80.8 | 73.1 | 62.8 | 52.6 | 41.8 | 59.5 |
| Minimum Temp °F | 17.0 | 18.5 | 27.5 | 36.5 | 46.0 | 55.2 | 61.2 | 60.1 | 51.8 | 40.9 | 33.7 | 23.5 | 39.3 |
| Mean Temp °F | 26.8 | 28.4 | 37.0 | 46.4 | 56.7 | 65.9 | 71.6 | 70.5 | 62.5 | 51.9 | 43.2 | 32.6 | 49.5 |
| Days Max Temp ≥ 90 °F | 0 | 0 | 0 | 0 | 0 | 2 | 4 | 2 | 1 | 0 | 0 | 0 | 9 |
| Days Max Temp ≤ 32 °F | 11 | 8 | 2 | 0 | 0 | 0 | 0 | 0 | 0 | 0 | 0 | 5 | 26 |
| Days Min Temp ≤ 32 °F | 29 | 26 | 23 | 9 | 1 | 0 | 0 | 0 | 0 | 6 | 15 | 26 | 135 |
| Days Min Temp ≤ 0 °F | 3 | 2 | 0 | 0 | 0 | 0 | 0 | 0 | 0 | 0 | 0 | 1 | 6 |
| Heating Degree Days | 1178 | 1026 | 862 | 552 | 269 | 62 | 8 | 15 | 126 | 405 | 648 | 996 | 6147 |
| Cooling Degree Days | 0 | 0 | 0 | 1 | 22 | 109 | 241 | 204 | 66 | 7 | 0 | 0 | 650 |
| Total Precipitation (") | 3.96 | 3.31 | 4.40 | 4.24 | 3.73 | 3.78 | 3.54 | 4.43 | 3.86 | 3.41 | 4.84 | 5.04 | 48.54 |
| Days ≥ 0.1" Precip | 7 | 6 | 7 | 7 | 7 | 6 | 6 | 6 | 5 | 6 | 8 | 8 | 79 |
| Total Snowfall (") | *9.5* | 9.3 | 4.4 | 0.3 | 0.0 | 0.0 | 0.0 | 0.0 | 0.0 | 0.0 | 1.0 | *5.5* | 30.0 |
| Days ≥ 1" Snow Depth | *12* | *12* | 6 | 0 | 0 | 0 | 0 | 0 | 0 | 0 | 0 | *5* | 35 |

## TAUNTON *Bristol County* ELEVATION 20 ft LAT/LONG 41° 54 ' N / 71° 4 ' W

| | JAN | FEB | MAR | APR | MAY | JUN | JUL | AUG | SEP | OCT | NOV | DEC | YEAR |
|---|---|---|---|---|---|---|---|---|---|---|---|---|---|
| Maximum Temp °F | 36.2 | 38.4 | 47.0 | 58.1 | 69.0 | 77.6 | 82.8 | 81.1 | 73.2 | 62.4 | 51.9 | 40.9 | 59.9 |
| Minimum Temp °F | 16.1 | 18.1 | 26.6 | 34.9 | 44.5 | 54.1 | 59.9 | 58.9 | 49.9 | 39.1 | 31.8 | 21.9 | 38.0 |
| Mean Temp °F | 26.2 | 28.3 | 36.8 | 46.5 | 56.8 | 65.9 | 71.4 | 70.0 | 61.6 | 50.8 | 41.9 | 31.4 | 49.0 |
| Days Max Temp ≥ 90 °F | 0 | 0 | 0 | 0 | 1 | 2 | 4 | 3 | 0 | 0 | 0 | 0 | 10 |
| Days Max Temp ≤ 32 °F | 11 | 7 | 1 | 0 | 0 | 0 | 0 | 0 | 0 | 0 | 0 | 6 | 25 |
| Days Min Temp ≤ 32 °F | 29 | 26 | 23 | 12 | 2 | 0 | 0 | 0 | 1 | 9 | 18 | 27 | 147 |
| Days Min Temp ≤ 0 °F | 3 | 2 | 0 | 0 | 0 | 0 | 0 | 0 | 0 | 0 | 0 | 0 | 5 |
| Heating Degree Days | 1197 | 1031 | 867 | 551 | 266 | 65 | 10 | 22 | 149 | 438 | 687 | 1036 | 6319 |
| Cooling Degree Days | 0 | 0 | 0 | 1 | 21 | 108 | 226 | 195 | 60 | 5 | 0 | 0 | 616 |
| Total Precipitation (") | 3.76 | 3.55 | 4.36 | 3.72 | 3.96 | 3.33 | 3.63 | 4.15 | 3.86 | 3.42 | 4.66 | 4.62 | 47.02 |
| Days ≥ 0.1" Precip | 7 | 6 | 7 | 6 | 7 | 6 | 5 | 7 | 6 | 6 | 7 | 8 | 78 |
| Total Snowfall (") | 10.6 | 10.6 | 5.5 | 0.7 | 0.1 | 0.0 | 0.0 | 0.0 | 0.0 | 0.1 | 1.3 | 6.6 | 35.5 |
| Days ≥ 1" Snow Depth | 15 | 14 | 6 | 0 | 0 | 0 | 0 | 0 | 0 | 0 | 0 | 6 | 41 |

## TULLY LAKE *Worcester County* ELEVATION 689 ft LAT/LONG 42° 38 ' N / 72° 13 ' W

| | JAN | FEB | MAR | APR | MAY | JUN | JUL | AUG | SEP | OCT | NOV | DEC | YEAR |
|---|---|---|---|---|---|---|---|---|---|---|---|---|---|
| Maximum Temp °F | 30.0 | 33.4 | 43.0 | 56.1 | 69.0 | 77.2 | 82.4 | 80.1 | 71.0 | 59.4 | 46.6 | 34.3 | 56.9 |
| Minimum Temp °F | 8.0 | 10.3 | 21.4 | 32.3 | 42.4 | 51.5 | 56.8 | 55.2 | 46.7 | 35.6 | 28.3 | 16.2 | 33.7 |
| Mean Temp °F | 19.0 | 21.8 | 32.2 | 44.2 | 55.7 | 64.4 | 69.6 | 67.7 | 58.9 | 47.5 | 37.5 | 25.3 | 45.3 |
| Days Max Temp ≥ 90 °F | 0 | 0 | 0 | 0 | 1 | 2 | 4 | 3 | 0 | 0 | 0 | 0 | 10 |
| Days Max Temp ≤ 32 °F | 17 | 13 | 4 | 0 | 0 | 0 | 0 | 0 | 0 | 0 | 1 | 12 | 47 |
| Days Min Temp ≤ 32 °F | 30 | 27 | 27 | 17 | 3 | 0 | 0 | 0 | 1 | 13 | 21 | 28 | 167 |
| Days Min Temp ≤ 0 °F | 9 | 6 | 1 | 0 | 0 | 0 | 0 | 0 | 0 | 0 | 0 | 4 | 20 |
| Heating Degree Days | 1420 | 1212 | 1010 | 619 | 298 | 91 | 20 | 41 | 207 | 537 | 819 | 1226 | 7500 |
| Cooling Degree Days | 0 | 0 | 0 | 1 | 18 | 75 | 172 | 133 | 33 | 2 | 0 | 0 | 434 |
| Total Precipitation (") | 3.26 | 2.96 | 3.66 | 3.60 | 3.93 | 3.86 | 4.02 | 4.40 | 3.57 | 3.44 | 4.07 | 3.77 | 44.54 |
| Days ≥ 0.1" Precip | 7 | 6 | 7 | 7 | 8 | 7 | 7 | 7 | 6 | 6 | 8 | 7 | 83 |
| Total Snowfall (") | 15.5 | 13.1 | 9.5 | 1.8 | 0.1 | 0.0 | 0.0 | 0.0 | 0.0 | 0.0 | 3.2 | 11.8 | 55.0 |
| Days ≥ 1" Snow Depth | 25 | 24 | 19 | 2 | 0 | 0 | 0 | 0 | 0 | 0 | 3 | 15 | 88 |

## WEST MEDWAY *Norfolk County* ELEVATION 210 ft LAT/LONG 42° 8 ' N / 71° 26 ' W

| | JAN | FEB | MAR | APR | MAY | JUN | JUL | AUG | SEP | OCT | NOV | DEC | YEAR |
|---|---|---|---|---|---|---|---|---|---|---|---|---|---|
| Maximum Temp °F | 36.0 | 38.5 | 47.1 | 58.3 | 69.6 | 78.2 | 83.7 | 82.1 | 74.2 | 63.7 | 52.3 | 40.7 | 60.4 |
| Minimum Temp °F | 12.2 | 14.7 | 24.3 | 34.1 | 43.6 | 52.8 | 58.5 | 56.7 | 47.3 | 36.1 | 29.2 | 19.1 | 35.7 |
| Mean Temp °F | 24.2 | 26.6 | 35.7 | 46.2 | 56.6 | 65.6 | 71.1 | 69.4 | 60.8 | 49.9 | 40.8 | 29.9 | 48.1 |
| Days Max Temp ≥ 90 °F | 0 | 0 | 0 | 0 | 1 | 2 | 5 | 4 | 1 | 0 | 0 | 0 | 13 |
| Days Max Temp ≤ 32 °F | 11 | 8 | 2 | 0 | 0 | 0 | 0 | 0 | 0 | 0 | 0 | 7 | 28 |
| Days Min Temp ≤ 32 °F | 30 | 27 | 26 | 14 | 3 | 0 | 0 | 0 | 2 | 13 | 20 | 28 | 163 |
| Days Min Temp ≤ 0 °F | 5 | 3 | 0 | 0 | 0 | 0 | 0 | 0 | 0 | 0 | 0 | 2 | 10 |
| Heating Degree Days | 1260 | 1078 | 901 | 560 | 272 | 74 | 13 | 27 | 167 | 464 | 719 | 1082 | 6617 |
| Cooling Degree Days | 0 | 0 | 0 | 1 | 22 | 97 | 212 | 166 | 46 | 4 | 0 | 0 | 548 |
| Total Precipitation (") | 3.94 | 3.53 | 4.06 | 4.11 | 3.40 | 3.71 | 3.58 | 4.24 | 3.78 | 3.75 | 4.80 | 4.38 | 47.28 |
| Days ≥ 0.1" Precip | 6 | 6 | 7 | 6 | 7 | 7 | 6 | 6 | 6 | 6 | 8 | 7 | 78 |
| Total Snowfall (") | 12.2 | 11.7 | 8.1 | 1.4 | 0.2 | 0.0 | 0.0 | 0.0 | 0.0 | 0.0 | 1.6 | 8.1 | 43.3 |
| Days ≥ 1" Snow Depth | 17 | 15 | 8 | 1 | 0 | 0 | 0 | 0 | 0 | 0 | 1 | 9 | 51 |

## WORCESTER MUNI AP *Worcester County* ELEVATION 991 ft LAT/LONG 42° 16 ' N / 71° 52 ' W

| | JAN | FEB | MAR | APR | MAY | JUN | JUL | AUG | SEP | OCT | NOV | DEC | YEAR |
|---|---|---|---|---|---|---|---|---|---|---|---|---|---|
| Maximum Temp °F | 30.6 | 33.3 | 42.3 | 54.6 | 66.3 | 74.3 | 79.4 | 77.4 | 69.1 | 58.7 | 47.1 | 35.6 | 55.7 |
| Minimum Temp °F | 15.1 | 16.9 | 25.3 | 35.5 | 45.9 | 54.7 | 60.9 | 59.4 | 51.2 | 40.8 | 32.0 | 20.9 | 38.2 |
| Mean Temp °F | 22.9 | 25.1 | 33.8 | 45.1 | 56.1 | 64.5 | 70.2 | 68.4 | 60.2 | 49.7 | 39.6 | 28.3 | 47.0 |
| Days Max Temp ≥ 90 °F | 0 | 0 | 0 | 0 | 0 | 1 | 2 | 1 | 0 | 0 | 0 | 0 | 4 |
| Days Max Temp ≤ 32 °F | 18 | 14 | 5 | 0 | 0 | 0 | 0 | 0 | 0 | 0 | 2 | 12 | 51 |
| Days Min Temp ≤ 32 °F | 30 | 27 | 25 | 11 | 1 | 0 | 0 | 0 | 0 | 5 | 17 | 27 | 143 |
| Days Min Temp ≤ 0 °F | 3 | 2 | 0 | 0 | 0 | 0 | 0 | 0 | 0 | 0 | 0 | 1 | 6 |
| Heating Degree Days | 1299 | 1119 | 960 | 594 | 288 | 84 | 12 | 32 | 174 | 469 | 757 | 1132 | 6920 |
| Cooling Degree Days | 0 | 0 | 0 | 1 | 22 | 77 | 185 | 143 | 37 | 2 | 0 | 0 | 467 |
| Total Precipitation (") | 3.60 | 3.26 | 4.09 | 3.89 | 4.42 | 3.97 | 3.95 | 4.08 | 4.22 | 4.18 | 4.53 | 4.14 | 48.33 |
| Days ≥ 0.1" Precip | 7 | 6 | 7 | 7 | 8 | 7 | 6 | 7 | 7 | 6 | 8 | 7 | 83 |
| Total Snowfall (") | 16.7 | 14.8 | 12.0 | 3.4 | 0.4 | 0.0 | 0.0 | 0.0 | 0.0 | 0.3 | 4.2 | 12.9 | 64.7 |
| Days ≥ 1" Snow Depth | 21 | 19 | 13 | 2 | 0 | 0 | 0 | 0 | 0 | 0 | 3 | 12 | 70 |

## JANUARY MINIMUM TEMPERATURE °F

### LOWEST

| | | |
|---|---|---|
| 1 | Birch Hill Dam | 7.5 |
| 2 | Tully Lake | 8.0 |
| 3 | Knightville Dam | 9.0 |
| 4 | Barre Falls Dam | 9.8 |
| 5 | Buffumville Lake | 10.6 |
| 6 | East Brimfield Lk | 11.0 |
| 7 | Cummington Hill | 11.5 |
| 8 | Amherst | 12.0 |
| 9 | West Wedway | 12.2 |
| 10 | Reading | 14.5 |
| 11 | Bedford | 14.7 |
| 12 | Lawrence | 15.1 |
| | Worcester | 15.1 |
| 14 | Haverhill | 15.5 |
| 15 | Middleton | 16.0 |
| 16 | Taunton | 16.1 |
| 17 | Plymouth | 16.5 |
| 18 | Rochester | 17.0 |
| 19 | Brockton | 17.1 |
| 20 | Milton | 17.5 |
| 21 | Peabody | 18.2 |
| 22 | East Wareham | 18.7 |
| | Hingham | 18.7 |
| 24 | Hyannis | 20.4 |
| 25 | Edgartown | 21.5 |

### HIGHEST

| | | |
|---|---|---|
| 1 | New Bedford | 22.8 |
| 2 | Boston | 21.7 |
| 3 | Edgartown | 21.5 |
| 4 | Hyannis | 20.4 |
| 5 | East Wareham | 18.7 |
| | Hingham | 18.7 |
| 7 | Peabody | 18.2 |
| 8 | Milton | 17.5 |
| 9 | Brockton | 17.1 |
| 10 | Rochester | 17.0 |
| 11 | Plymouth | 16.5 |
| 12 | Taunton | 16.1 |
| 13 | Middleton | 16.0 |
| 14 | Haverhill | 15.5 |
| 15 | Lawrence | 15.1 |
| | Worcester | 15.1 |
| 17 | Bedford | 14.7 |
| 18 | Reading | 14.5 |
| 19 | West Wedway | 12.2 |
| 20 | Amherst | 12.0 |
| 21 | Cummington Hill | 11.5 |
| 22 | East Brimfield Lk | 11.0 |
| 23 | Buffumville Lake | 10.6 |
| 24 | Barre Falls Dam | 9.8 |
| 25 | Knightville Dam | 9.0 |

## JULY MAXIMUM TEMPERATURE °F

### HIGHEST

| | | |
|---|---|---|
| 1 | Amherst | 84.2 |
| 2 | Haverhill | 84.0 |
| 3 | West Wedway | 83.7 |
| 4 | Lawrence | 82.8 |
| | Taunton | 82.8 |
| 6 | Middleton | 82.7 |
| | Reading | 82.7 |
| 8 | Bedford | 82.6 |
| | Brockton | 82.6 |
| 10 | Peabody | 82.5 |
| 11 | Tully Lake | 82.4 |
| 12 | East Brimfield Lk | 82.2 |
| 13 | Boston | 82.1 |
| 14 | Rochester | 82.0 |
| 15 | Plymouth | 81.8 |
| 16 | New Bedford | 81.7 |
| 17 | Hingham | 81.6 |
| | Knightville Dam | 81.6 |
| 19 | Buffumville Lake | 81.4 |
| 20 | Birch Hill Dam | 81.0 |
| 21 | Milton | 80.9 |
| 22 | East Wareham | 79.9 |
| 23 | Worcester | 79.4 |
| 24 | Barre Falls Dam | 79.3 |
| 25 | Hyannis | 78.6 |

### LOWEST

| | | |
|---|---|---|
| 1 | Cummington Hill | 77.0 |
| 2 | Edgartown | 78.3 |
| 3 | Hyannis | 78.6 |
| 4 | Barre Falls Dam | 79.3 |
| 5 | Worcester | 79.4 |
| 6 | East Wareham | 79.9 |
| 7 | Milton | 80.9 |
| 8 | Birch Hill Dam | 81.0 |
| 9 | Buffumville Lake | 81.4 |
| 10 | Hingham | 81.6 |
| | Knightville Dam | 81.6 |
| 12 | New Bedford | 81.7 |
| 13 | Plymouth | 81.8 |
| 14 | Rochester | 82.0 |
| 15 | Boston | 82.1 |
| 16 | East Brimfield Lk | 82.2 |
| 17 | Tully Lake | 82.4 |
| 18 | Peabody | 82.5 |
| 19 | Bedford | 82.6 |
| | Brockton | 82.6 |
| 21 | Middleton | 82.7 |
| | Reading | 82.7 |
| 23 | Lawrence | 82.8 |
| | Taunton | 82.8 |
| 25 | West Wedway | 83.7 |

## ANNUAL PRECIPITATION (")

### HIGHEST

| | | |
|---|---|---|
| 1 | Milton | 49.64 |
| 2 | Plymouth | 49.20 |
| 3 | New Bedford | 48.94 |
| 4 | Rochester | 48.54 |
| 5 | Hingham | 48.42 |
| 6 | Worcester | 48.33 |
| 7 | East Wareham | 47.83 |
| 8 | Buffumville Lake | 47.79 |
| 9 | Peabody | 47.70 |
| 10 | West Wedway | 47.28 |
| 11 | Cummington Hill | 47.08 |
| 12 | Taunton | 47.02 |
| 13 | Reading | 46.91 |
| 14 | Knightville Dam | 46.65 |
| 15 | East Brimfield Lk | 46.27 |
| 16 | Brockton | 46.02 |
| 17 | Bedford | 45.32 |
| 18 | Haverhill | 45.27 |
| 19 | Edgartown | 45.20 |
| 20 | Tully Lake | 44.54 |
| 21 | Amherst | 44.16 |
| 22 | Middleton | 43.82 |
| 23 | Barre Falls Dam | 43.59 |
| 24 | Hyannis | 43.50 |
| 25 | Lawrence | 43.40 |

### LOWEST

| | | |
|---|---|---|
| 1 | Boston | 42.01 |
| 2 | Birch Hill Dam | 42.38 |
| 3 | Lawrence | 43.40 |
| 4 | Hyannis | 43.50 |
| 5 | Barre Falls Dam | 43.59 |
| 6 | Middleton | 43.82 |
| 7 | Amherst | 44.16 |
| 8 | Tully Lake | 44.54 |
| 9 | Edgartown | 45.20 |
| 10 | Haverhill | 45.27 |
| 11 | Bedford | 45.32 |
| 12 | Brockton | 46.02 |
| 13 | East Brimfield Lk | 46.27 |
| 14 | Knightville Dam | 46.65 |
| 15 | Reading | 46.91 |
| 16 | Taunton | 47.02 |
| 17 | Cummington Hill | 47.08 |
| 18 | West Wedway | 47.28 |
| 19 | Peabody | 47.70 |
| 20 | Buffumville Lake | 47.79 |
| 21 | East Wareham | 47.83 |
| 22 | Worcester | 48.33 |
| 23 | Hingham | 48.42 |
| 24 | Rochester | 48.54 |
| 25 | New Bedford | 48.94 |

## ANNUAL SNOWFALL (")

### HIGHEST

| | | |
|---|---|---|
| 1 | Cummington Hill | 77.5 |
| 2 | Worcester | 64.7 |
| 3 | Reading | 62.2 |
| 4 | Milton | 60.1 |
| 5 | Barre Falls Dam | 57.3 |
| 6 | Birch Hill Dam | 56.7 |
| 7 | Haverhill | 55.6 |
| 8 | Bedford | 55.5 |
| 9 | Tully Lake | 55.0 |
| 10 | East Brimfield Lk | 54.8 |
| 11 | Knightville Dam | 49.2 |
| 12 | Peabody | 48.7 |
| 13 | Buffumville Lake | 48.5 |
| 14 | Lawrence | 47.8 |
| 15 | Hingham | 47.5 |
| 16 | Middleton | 43.3 |
| | West Wedway | 43.3 |
| 18 | Boston | 42.7 |
| 19 | Amherst | 40.7 |
| 20 | Plymouth | 37.0 |
| 21 | Taunton | 35.5 |
| 22 | East Wareham | 33.2 |
| 23 | New Bedford | 32.8 |
| 24 | Rochester | 30.0 |
| | | |

### LOWEST

| | | |
|---|---|---|
| 1 | Rochester | 30.0 |
| 2 | New Bedford | 32.8 |
| 3 | East Wareham | 33.2 |
| 4 | Taunton | 35.5 |
| 5 | Plymouth | 37.0 |
| 6 | Amherst | 40.7 |
| 7 | Boston | 42.7 |
| 8 | Middleton | 43.3 |
| | West Wedway | 43.3 |
| 10 | Hingham | 47.5 |
| 11 | Lawrence | 47.8 |
| 12 | Buffumville Lake | 48.5 |
| 13 | Peabody | 48.7 |
| 14 | Knightville Dam | 49.2 |
| 15 | East Brimfield Lk | 54.8 |
| 16 | Tully Lake | 55.0 |
| 17 | Bedford | 55.5 |
| 18 | Haverhill | 55.6 |
| 19 | Birch Hill Dam | 56.7 |
| 20 | Barre Falls Dam | 57.3 |
| 21 | Milton | 60.1 |
| 22 | Reading | 62.2 |
| 23 | Worcester | 64.7 |
| 24 | Cummington Hill | 77.5 |
| | | |

# MICHIGAN

PHYSICAL FEATURES. Michigan is located in the heart of the Great Lakes region and is composed of two large peninsulas. Many smaller peninsulas jut from these two peninsulas into the world's largest bodies of fresh water to give most of Michigan a quasi-marine type climate in spite of its midcontinent location.

The Upper Peninsula is long and narrow, lying primarily between 45° and 47° N. latitude. It averages only 75 miles in width and extends from Northern Wisconsin eastward over 300 miles into Northern Lake Huron. Lake Superior lies to the north while the northern portion of Lake Michigan forms the boundary to the southeast. Isle Royale, separated from the mainland, is located in Lake Superior about 50 miles northwest of the tip of the Keweenaw Peninsula. The Lower Peninsula, shaped like a mitten and occupying about 70 percent of Michigan's total land area, extends northward nearly 300 miles from the Indiana and Ohio border or about 42° N. latitude to the eastern end of the Upper Peninsula. Lake Michigan extends the entire length of the Lower Peninsula on the west while Lakes Huron, St. Clair and Erie form the eastern boundary. The total coastline for the state exceeds 3,100 miles. In addition, Michigan has over 11,000 smaller lakes with a total surface area of over 1,000 square miles. These lakes are scattered throughout 81 of the 83 counties while more than 36,000 miles of streams wind their way across the state.

While latitude, by determining the amount of solar insolation, is the major climatic control, the Great Lakes and variations in elevation play an important role in the amelioration of Michigan's climate. Because of its mid-latitude location, prevailing winds are from a westerly direction. During the summer months winds are predominantly from the southwest when the semi-permanent Bermuda High Pressure Center is located over the southeastern United States. During the winter months the prevailing winds are west to northwest, but change quite frequently for short periods as migrating cyclones and anticyclones move through the area.

The eastern half of the Upper Peninsula varies from level to gently rolling hills with elevation generally between 600 and 1,000 feet above sea level. The western tablelands rise to elevations generally between 1,400 and 1,600 feet with Porcupine Mountain, the State's highest point, 2,023 feet, located in Ontonagan County overlooking Lake Superior. The rugged hills extend northeastward from Ontonagan County through the center of the Keweenaw Peninsula and play an important role in the larger precipitation amounts received in this area.

The Lower Peninsula features range from quite level terrain in the southeast to gently rolling hills in the southwest with elevations generally between 800 and 1,000 feet. A series of sand dunes along the Lake Michigan shoreline rise to heights of nearly 400 feet above the lake level. These are the result of the prevailing westerly winds which blow across the lake. Tablelands cover the northern part of the Lower Peninsula and reach a maximum elevation of 1,700 feet in Osecola County near Cadillac. In the northwestern section of the Lower Peninsula a number of finger-like peninsulas extend into Grand Traverse Bay and Lake Michigan.

GENERAL CLIMATE. The lake effect imparts many interesting departures to Michigan's climate which one would not ordinarily expect to find at a midcontinental location. Because of the lake waters' slow response to temperature changes and the dominating westerly winds, the arrival of both summer and winter are retarded. In the spring, the cooler temperatures slow the development of vegetation until the danger of frost is past. In the fall, the warmer lake waters temper the first outbreaks of cold air allowing additional time for crops to mature. With the first cold air outbreaks in the fall, Michigan experiences a considerable increase in cloudiness. When cold air passes over the warmer lake water, a shallow layer of unstable, moisture-laden air develops in the lower levels of the atmosphere. This air, when forced to rise, produces the increased cloudiness and frequent snow flurry activity observed in the fall and early winter months.

On warm, summer days when prevailing winds are generally light, the lake's shore area frequently develops a localized wind pattern which may extend inland for only a few miles. This is frequently referred to as the "lake breeze". It develops when the much warmer air over the land masses begins to rise, allowing the cooler air over the lakes to move inland. At night this pattern may be reversed creating what is known as a "land breeze". A wind of this type may also be observed, but on a much smaller scale, along the shores of the larger inland lakes.

The length of Michigan's growing season or freeze-free period does not decrease in the normal manner from south to north. Instead, isolines for the length of the growing season follow closely the contours of the lake shores. The shortest average growing season, about 60 days, occurs in the interior section of the Western Upper Peninsula. The growing season increases to between 140 and 160 days, as one goes towards the lake shores. A similar pattern exists in the Lower Peninsula where the growing season in the northern tablelands averages only 70 days, but increases rapidly to 140 days near the lakes. Michigan's maximum average growing season, 170 days, is found in the southwest and southeastern corners of the state.

PRECIPITATION. Michigan averages about 31 inches of precipitation per year. About 55-60 percent of the annual total is recorded during the normal growing season. Summer precipitation falls primarily in the form of showers or thunderstorms, while a more steady type of precipitation of lighter intensity dominates the winter months. The annual number of thunderstorms observed decreases from about 40 in the south to around 25 in the Upper Peninsula area with nearly 50 percent of these recorded during the summer months, June through August.

The frequency of floods is quite low in Michigan with the greatest likelihood occurring in late winter or early spring when sudden warming and rain may be combined with snowmelt. Mild meteorological drought conditions are not uncommon in Michigan, but meteorological droughts reaching severe conditions are infrequent and generally of short duration. The normally even distribution of precipitation and higher humidities observed in Michigan are helpful in reducing the high demands for moisture.

SNOWFALL. Michigan receives some of the heaviest snowfall totals east of the Rockies except for isolated points in the New England States. The maximum average annual snowfall amounts of over 170 inches are located along the escarpment which rises abruptly to an elevation of over 1,400 feet above Lake Superior, at the western end of the Upper Peninsula. Another area with amounts exceeding 120 inches is centered in the western section of the tableland region of the Lower Peninsula. The prevailing westerlies, passing over the Great Lakes, become moisture laden in the lower levels and when forced upward by the land masses, drop much of their excessive moisture in the form of snow squalls in these areas.

STORMS. Damaging or dangerous storms do not occur as frequently in Michigan as in the states to the south and west. Recorded tornado occurrences have averaged four per year. About 90 percent of these tornadoes occurred in the southern one-half of the Lower Peninsula. Damaging wind storms and blizzards are not as frequent but do cause considerable damage from time to time. Hail is most frequently observed in the spring months. A higher frequency of hail is noted in the fall months over the northwestern section of the Lower Peninsula. This is attributed mainly to the strong lake influence in this region.

## COUNTY INDEX

***Alger County***
GRAND MARAIS 2 E
MUNISING

***Allegan County***
ALLEGAN 5 NE

***Alpena County***
ALPENA PHELPS COL AP
ALPENA WASTEWATER PL

***Arenac County***
STANDISH 5 SW

***Baraga County***
ALBERTA FORD FOR CEN

***Barry County***
HASTINGS

***Benzie County***
FRANKFORT 2 NE

***Berrien County***
BENTON HRBR ROSS FLD
EAU CLAIRE 4 NE

***Branch County***
COLDWATER ST SCHOOL

***Calhoun County***
BATTLE CREEK 5 NW

***Cass County***
DOWAGIAC 1 W

***Charlevoix County***
BOYNE FALLS
EAST JORDAN
ST JAMES 2S BEAVR IS

***Cheboygan County***
CHEBOYGAN

***Chippewa County***
DETOUR VILLAGE
SAULT STE MARIE WSO
WHITEFISH POINT

***Clinton County***
LANSING CAPITAL CITY
ST JOHNS

***Crawford County***
GRAYLING

***Delta County***
ESCANABA
FAYETTE 4 SW

***Dickinson County***
IRON MTN-KINGSFORD W

***Eaton County***
CHARLOTTE

***Emmet County***
CROSS VILLAGE
PELLSTON EMMET CNTY
PETOSKEY

***Genesee County***
FLINT BISHOP ARPT

***Gladwin County***
GLADWIN

***Gogebic County***
IRONWOOD
WATERSMEET

***Grand Traverse County***
TRAVERSE CITY AP

***Gratiot County***
ALMA

***Hillsdale County***
HILLSDALE

***Houghton County***
HANCOCK HOUGHTON CO

***Huron County***
BAD AXE
HARBOR BEACH 1 SSE

***Ingham County***
EAST LANSING 4 S

***Ionia County***
IONIA 2 SSW

***Iosco County***
EAST TAWAS
HALE LOUD DAM

***Iron County***
STAMBAUGH 2 SSE

***Isabella County***
MT PLEASANT UNIV

***Jackson County***
JACKSON REYNOLDS FLD

***Kalamazoo County***
GULL LAKE BIOL STA
KALAMAZOO STATE HOSP

***Kent County***
GRAND RAPIDS KENT AP

***Lake County***
BALDWIN

***Lapeer County***
LAPEER WWTP

***Leelanau County***
MAPLE CITY

***Lenawee County***
ADRIAN 2 NNE

***Livingston County***
MILFORD GM PROVING G

***Luce County***
NEWBERRY 3 S

***Macomb County***
MT CLEMENS SELFRIDGE

***Manistee County***
MANISTEE 3 SE

***Marquette County***
BIG BAY 2 SE
CHAMPION VAN RIPER P
MARQUETTE
MARQUETTE CO AP

### *Mason County*
LUDINGTON 4 SE

### *Mecosta County*
BIG RAPIDS WATERWORK

### *Menominee County*
STEPHENSON 8 WNW

### *Missaukee County*
HOUGHTON LAKE 6 WSW
LAKE CITY EXP FARM

### *Monroe County*
MONROE

### *Montcalm County*
GREENVILLE 2 NNE

### *Muskegon County*
MONTAGUE 4 NW
MUSKEGON CO ARPT

### *Oakland County*
PONTIAC STATE HOSP

### *Oceana County*
HART
HESPERIA 4 WNW

### *Ogemaw County*
LUPTON 1 S
WEST BRANCH 3 SE

### *Ontonagon County*
BERGLAND DAM

### *Osceola County*
EVART

### *Oscoda County*
MIO HYDRO PLANT

### *Otsego County*
GAYLORD
VANDERBILT 11 ENE

### *Ottawa County*
GRAND HAVEN FIRE DEP
HOLLAND

### *Presque Isle County*
ONAWAY STATE PARK

### *Roscommon County*
HOUGHTON LAKE ROSCMN

### *Saginaw County*
SAGINAW TRI CITY AP
ST CHARLES

### *St. Clair County*
PORT HURON

### *St. Joseph County*
THREE RIVERS

### *Sanilac County*
SANDUSKY

### *Schoolcraft County*
MANISTIQUE
SENEY WILDLIFE REF

### *Shiawassee County*
OWOSSO 3 NNW

### *Tuscola County*
CARO REGIONAL CENTER

### *Van Buren County*
BLOOMINGDALE
SOUTH HAVEN

### *Washtenaw County*
ANN ARBOR UNIV OF MI
YPSILANTI E MICH UNI

### *Wayne County*
DEARBORN
DETROIT CITY AIRPORT
DETROIT METRO AP
GROSSE POINTE FARMS

### *Wexford County*
CADILLAC

## ELEVATION INDEX

| FEET | STATION NAME |
|---|---|
| 576 | MT CLEMENS SELFRIDGE |
| 591 | DETOUR VILLAGE |
| 591 | EAST TAWAS |
| 591 | SOUTH HAVEN |
| 600 | DEARBORN |
| 600 | MANISTEE 3 SE |
| 600 | MONROE |
| 600 | PORT HURON |
| 600 | ST CHARLES |
| 610 | CHEBOYGAN |
| 610 | EAST JORDAN |
| 610 | GROSSE POINTE FARMS |
| 610 | HOLLAND |
| 610 | LUDINGTON 4 SE |
| 610 | MANISTIQUE |
| 617 | ALPENA WASTEWATER PL |
| 617 | ESCANABA |
| 617 | WHITEFISH POINT |
| 618 | TRAVERSE CITY AP |
| 619 | DETROIT CITY AIRPORT |
| 620 | GRAND HAVEN FIRE DEP |
| 620 | HARBOR BEACH 1 SSE |
| 620 | MUNISING |
| 628 | BENTON HRBR ROSS FLD |
| 630 | GRAND MARAIS 2 E |
| 633 | DETROIT METRO AP |
| 633 | MUSKEGON CO ARPT |
| 645 | STANDISH 5 SW |
| 650 | ONAWAY STATE PARK |
| 659 | MONTAGUE 4 NW |
| 662 | SAGINAW TRI CITY AP |
| 669 | BIG BAY 2 SE |
| 669 | CROSS VILLAGE |
| 670 | ST JAMES 2S BEAVR IS |
| 677 | MARQUETTE |
| 679 | HART |
| 689 | ALPENA PHELPS COL AP |
| 689 | PETOSKEY |
| 690 | STEPHENSON 8 WNW |
| 709 | PELLSTON EMMET CNTY |
| 712 | CARO REGIONAL CENTER |
| 712 | SENEY WILDLIFE REF |
| 718 | SAULT STE MARIE WSO |
| 720 | FRANKFORT 2 NE |
| 732 | BOYNE FALLS |
| 732 | MAPLE CITY |

| FEET | STATION NAME |
|---|---|
| 741 | ALMA |
| 741 | DOWAGIAC 1 W |
| 741 | ST JOHNS |
| 750 | ALLEGAN 5 NE |
| 751 | BAD AXE |
| 751 | BLOOMINGDALE |
| 758 | IONIA 2 SSW |
| 765 | OWOSSO 3 NNW |
| 771 | ADRIAN 2 NNE |
| 771 | HALE LOUD DAM |
| 771 | SANDUSKY |
| 781 | HESPERIA 4 WNW |
| 784 | GRAND RAPIDS KENT AP |
| 787 | FLINT BISHOP ARPT |
| 791 | FAYETTE 4 SW |
| 791 | GLADWIN |
| 791 | HASTINGS |
| 791 | MT PLEASANT UNIV |
| 791 | YPSILANTI E MICH UNI |
| 820 | LAPEER WWTP |
| 820 | THREE RIVERS |
| 830 | BALDWIN |
| 840 | GREENVILLE 2 NNE |
| 869 | BATTLE CREEK 5 NW |
| 869 | EAST LANSING 4 S |
| 869 | EAU CLAIRE 4 NE |
| 873 | LANSING CAPITAL CITY |
| 885 | WEST BRANCH 3 SE |
| 889 | LUPTON 1 S |
| 889 | NEWBERRY 3 S |
| 932 | ANN ARBOR UNIV OF MI |
| 932 | BIG RAPIDS WATERWORK |
| 932 | KALAMAZOO STATE HOSP |
| 932 | VANDERBILT 11 ENE |
| 942 | CHARLOTTE |
| 942 | GULL LAKE BIOL STA |
| 961 | MIO HYDRO PLANT |
| 971 | PONTIAC STATE HOSP |
| 981 | COLDWATER ST SCHOOL |
| 1003 | JACKSON REYNOLDS FLD |
| 1030 | EVART |
| 1060 | IRON MTN-KINGSFORD W |
| 1060 | MILFORD GM PROVING G |
| 1086 | HANCOCK HOUGHTON CO |
| 1102 | HILLSDALE |
| 1142 | GRAYLING |
| 1142 | HOUGHTON LAKE 6 WSW |
| 1155 | HOUGHTON LAKE ROSCMN |
| 1230 | LAKE CITY EXP FARM |
| 1302 | BERGLAND DAM |

| FEET | STATION NAME |
|---|---|
| 1302 | CADILLAC |
| 1312 | ALBERTA FORD FOR CEN |
| 1381 | GAYLORD |
| 1414 | MARQUETTE CO AP |
| 1522 | IRONWOOD |
| 1565 | CHAMPION VAN RIPER P |
| 1611 | WATERSMEET |
| 1621 | STAMBAUGH 2 SSE |

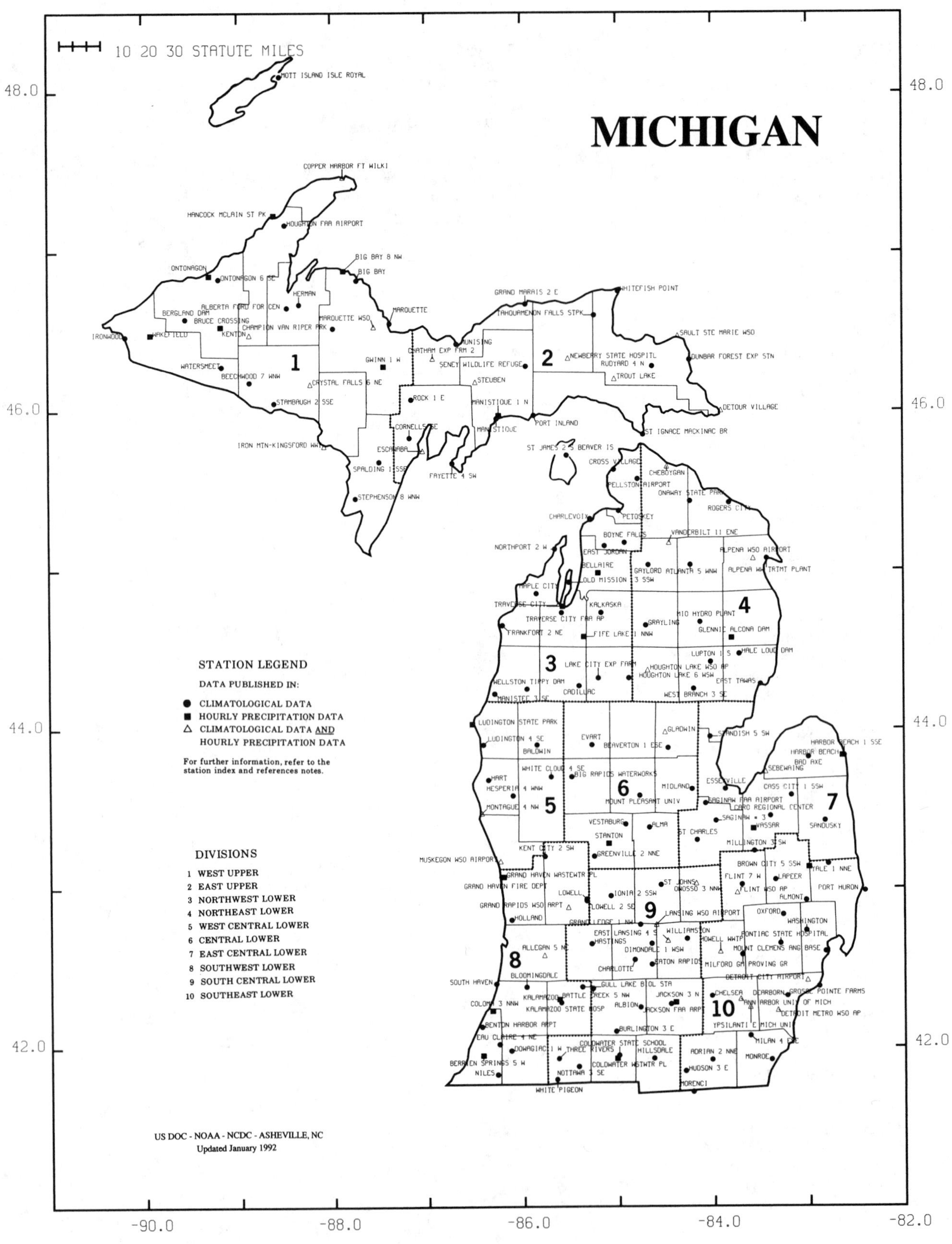

10 20 30 STATUTE MILES
MICHIGAN
STATION LEGEND
DATA PUBLISHED IN:
CLIMATOLOGICAL DATA
HOURLY PRECIPITATION DATA
CLIMATOLOGICAL DATA AND HOURLY PRECIPITATION DATA
For further information, refer to the station index and references notes.
DIVISIONS
1 WEST UPPER
2 EAST UPPER
3 NORTHWEST LOWER
4 NORTHEAST LOWER
5 WEST CENTRAL LOWER
6 CENTRAL LOWER
7 EAST CENTRAL LOWER
8 SOUTHWEST LOWER
9 SOUTH CENTRAL LOWER
10 SOUTHEAST LOWER
US DOC - NOAA - NCDC - ASHEVILLE, NC
Updated January 1992
48.0
46.0
44.0
42.0
-90.0
-88.0
-86.0
-84.0
-82.0

## ADRIAN 2 NNE *Lenawee County* ELEVATION 771 ft LAT/LONG 41° 54 ' N / 84° 2 ' W

| | JAN | FEB | MAR | APR | MAY | JUN | JUL | AUG | SEP | OCT | NOV | DEC | YEAR |
|---|---|---|---|---|---|---|---|---|---|---|---|---|---|
| Maximum Temp °F | 31.6 | 33.8 | 45.2 | 58.4 | 70.2 | 79.7 | 83.2 | 81.3 | 73.8 | 61.5 | 48.1 | 36.2 | 58.6 |
| Minimum Temp °F | 14.6 | 15.8 | 25.2 | 35.3 | 45.1 | 54.6 | 59.1 | 56.7 | 49.5 | 38.2 | 30.2 | 20.4 | 37.1 |
| Mean Temp °F | 23.1 | 24.7 | 35.3 | 46.9 | 57.7 | 67.2 | 71.2 | 69.0 | 61.7 | 49.9 | 39.2 | 28.3 | 47.8 |
| Days Max Temp ≥ 90 °F | 0 | 0 | 0 | 0 | 0 | 3 | 5 | 3 | 1 | 0 | 0 | 0 | 12 |
| Days Max Temp ≤ 32 °F | 16 | 13 | 4 | 0 | 0 | 0 | 0 | 0 | 0 | 0 | 2 | 10 | 45 |
| Days Min Temp ≤ 32 °F | 30 | 27 | 25 | 12 | 2 | 0 | 0 | 0 | 1 | 8 | 19 | 28 | 152 |
| Days Min Temp ≤ 0 °F | 4 | 4 | 0 | 0 | 0 | 0 | 0 | 0 | 0 | 0 | 0 | 2 | 10 |
| Heating Degree Days | 1291 | 1132 | 915 | 540 | 251 | 53 | 10 | 26 | 150 | 465 | 767 | 1132 | 6732 |
| Cooling Degree Days | 0 | 0 | 0 | 5 | 34 | 112 | 218 | 158 | 53 | 2 | 0 | 0 | 582 |
| Total Precipitation (") | 1.78 | 1.66 | 2.54 | 3.15 | 3.35 | 3.73 | 3.30 | 3.45 | 3.44 | 2.58 | 3.08 | 2.92 | 34.98 |
| Days ≥ 0.1" Precip | 5 | 4 | 7 | 7 | 7 | 7 | 6 | 6 | 6 | 6 | 7 | 7 | 75 |
| Total Snowfall (") | 7.3 | *7.6* | *4.7* | 0.9 | 0.0 | 0.0 | 0.0 | 0.0 | 0.0 | 0.1 | 2.3 | 7.2 | 30.1 |
| Days ≥ 1" Snow Depth | 19 | 17 | 6 | 1 | 0 | 0 | 0 | 0 | 0 | 0 | 2 | 11 | 56 |

## ALBERTA FORD FOR CEN *Baraga County* ELEVATION 1312 ft LAT/LONG 46° 39 ' N / 88° 29 ' W

| | JAN | FEB | MAR | APR | MAY | JUN | JUL | AUG | SEP | OCT | NOV | DEC | YEAR |
|---|---|---|---|---|---|---|---|---|---|---|---|---|---|
| Maximum Temp °F | 21.4 | 26.2 | 37.4 | 51.4 | 65.5 | 73.7 | 78.4 | 76.0 | 66.4 | 54.7 | 37.8 | 26.3 | 51.3 |
| Minimum Temp °F | 3.0 | 4.0 | 13.8 | 27.2 | 38.7 | 47.7 | 53.9 | 52.6 | 45.1 | 35.2 | 23.2 | 10.7 | 29.6 |
| Mean Temp °F | 12.3 | 15.1 | 25.6 | 39.3 | 52.1 | 60.7 | 66.2 | 64.3 | 55.8 | 45.0 | 30.5 | 18.5 | 40.5 |
| Days Max Temp ≥ 90 °F | 0 | 0 | 0 | 0 | 0 | 1 | 2 | 1 | 0 | 0 | 0 | 0 | 4 |
| Days Max Temp ≤ 32 °F | 26 | 20 | 10 | 2 | 0 | 0 | 0 | 0 | 0 | 0 | 10 | 22 | 90 |
| Days Min Temp ≤ 32 °F | 31 | 28 | 29 | 22 | 10 | 1 | 0 | 0 | 3 | 13 | 26 | 30 | 193 |
| Days Min Temp ≤ 0 °F | 13 | 12 | 7 | 1 | 0 | 0 | 0 | 0 | 0 | 0 | 1 | 7 | 41 |
| Heating Degree Days | 1631 | 1403 | 1215 | 766 | 412 | 169 | 65 | 99 | 293 | 616 | 1027 | 1434 | 9130 |
| Cooling Degree Days | 0 | 0 | 0 | 3 | 18 | 50 | 111 | 90 | 21 | 1 | 0 | 0 | 294 |
| Total Precipitation (") | 1.67 | 1.26 | 2.15 | 2.16 | 3.34 | 3.67 | 3.86 | 3.90 | 3.96 | 3.33 | 2.84 | 1.92 | 34.06 |
| Days ≥ 0.1" Precip | 6 | 4 | 5 | 6 | 7 | 8 | 7 | 7 | 8 | 8 | 7 | 6 | 79 |
| Total Snowfall (") | 34.0 | 23.9 | 22.1 | 8.2 | 0.9 | 0.0 | 0.0 | 0.0 | 0.0 | 3.9 | 22.5 | 30.2 | 145.7 |
| Days ≥ 1" Snow Depth | 31 | 27 | 29 | 13 | 0 | 0 | 0 | 0 | 0 | 2 | 16 | 30 | 148 |

## ALLEGAN 5 NE *Allegan County* ELEVATION 750 ft LAT/LONG 42° 35 ' N / 85° 47 ' W

| | JAN | FEB | MAR | APR | MAY | JUN | JUL | AUG | SEP | OCT | NOV | DEC | YEAR |
|---|---|---|---|---|---|---|---|---|---|---|---|---|---|
| Maximum Temp °F | 30.8 | 33.7 | 44.3 | 58.2 | 69.9 | 79.0 | 83.2 | 80.8 | 73.2 | 60.7 | 47.3 | 35.5 | 58.1 |
| Minimum Temp °F | 14.9 | 15.1 | 24.5 | 35.2 | 45.0 | 54.0 | 58.9 | 56.8 | 49.7 | 38.8 | 30.7 | 21.0 | 37.1 |
| Mean Temp °F | 22.9 | 24.5 | 34.4 | 46.7 | 57.5 | 66.4 | 71.1 | 68.8 | 61.4 | 49.7 | 39.0 | 28.3 | 47.6 |
| Days Max Temp ≥ 90 °F | 0 | 0 | 0 | 0 | 1 | 3 | 5 | 3 | 1 | 0 | 0 | 0 | 13 |
| Days Max Temp ≤ 32 °F | 17 | 13 | 5 | 0 | 0 | 0 | 0 | 0 | 0 | 0 | 2 | 10 | 47 |
| Days Min Temp ≤ 32 °F | 30 | 27 | 25 | 13 | 3 | 0 | 0 | 0 | 1 | 8 | 19 | 27 | 153 |
| Days Min Temp ≤ 0 °F | 4 | 3 | 0 | 0 | 0 | 0 | 0 | 0 | 0 | 0 | 0 | 1 | 8 |
| Heating Degree Days | 1299 | 1138 | 943 | 549 | 264 | 70 | 17 | 30 | 159 | 473 | 774 | 1131 | 6847 |
| Cooling Degree Days | 0 | 0 | 1 | 9 | 38 | 111 | 213 | 157 | 55 | 4 | 0 | 0 | 588 |
| Total Precipitation (") | 2.64 | 1.77 | 2.77 | 3.55 | 3.22 | 4.11 | 3.61 | 3.87 | 4.09 | 3.17 | 3.79 | 3.23 | 39.82 |
| Days ≥ 0.1" Precip | 8 | 6 | 7 | 8 | 7 | 6 | 6 | 7 | 7 | 8 | 9 | 8 | 87 |
| Total Snowfall (") | 24.0 | 15.9 | 9.0 | 1.8 | 0.0 | 0.0 | 0.0 | 0.0 | 0.0 | 0.3 | 7.8 | 21.0 | 79.8 |
| Days ≥ 1" Snow Depth | 26 | 21 | 11 | 1 | 0 | 0 | 0 | 0 | 0 | 0 | 4 | *18* | 81 |

## ALMA *Gratiot County* ELEVATION 741 ft LAT/LONG 43° 23 ' N / 84° 40 ' W

| | JAN | FEB | MAR | APR | MAY | JUN | JUL | AUG | SEP | OCT | NOV | DEC | YEAR |
|---|---|---|---|---|---|---|---|---|---|---|---|---|---|
| Maximum Temp °F | 28.6 | 31.6 | 42.2 | 56.5 | 69.8 | 78.5 | 83.3 | 80.7 | 72.4 | 59.9 | 45.6 | 33.7 | 56.9 |
| Minimum Temp °F | 13.2 | 14.0 | 23.1 | 34.4 | 44.8 | 54.2 | 58.9 | 56.7 | 49.2 | 38.7 | 29.8 | 20.1 | 36.4 |
| Mean Temp °F | 20.9 | 22.8 | 32.7 | 45.5 | 57.3 | 66.3 | 71.1 | 68.7 | 60.8 | 49.3 | 37.8 | 26.9 | 46.7 |
| Days Max Temp ≥ 90 °F | 0 | 0 | 0 | 0 | 1 | 2 | 5 | 3 | 0 | 0 | 0 | 0 | 11 |
| Days Max Temp ≤ 32 °F | 20 | 15 | 5 | 0 | 0 | 0 | 0 | 0 | 0 | 0 | 3 | 14 | 57 |
| Days Min Temp ≤ 32 °F | 30 | 27 | 26 | 14 | 2 | 0 | 0 | 0 | 1 | 8 | 20 | 29 | 157 |
| Days Min Temp ≤ 0 °F | 5 | 4 | 0 | 0 | 0 | 0 | 0 | 0 | 0 | 0 | 0 | 1 | 10 |
| Heating Degree Days | 1361 | 1183 | 995 | 584 | 266 | 67 | 13 | 31 | 169 | 485 | 810 | 1174 | 7138 |
| Cooling Degree Days | 0 | 0 | 0 | 7 | 33 | 94 | 205 | 143 | 41 | 1 | 0 | 0 | 524 |
| Total Precipitation (") | 1.70 | 1.30 | 2.32 | 3.08 | 2.81 | 3.17 | 2.54 | 3.67 | 3.83 | 2.89 | 2.85 | 2.25 | 32.41 |
| Days ≥ 0.1" Precip | 5 | 4 | 5 | 7 | 6 | 6 | 5 | 6 | 7 | 6 | 6 | 5 | 68 |
| Total Snowfall (") | 10.9 | 7.8 | 7.1 | 2.1 | 0.0 | 0.0 | 0.0 | 0.0 | 0.0 | 0.3 | 3.6 | 9.1 | 40.9 |
| Days ≥ 1" Snow Depth | 25 | 22 | 12 | 1 | 0 | 0 | 0 | 0 | 0 | 0 | 3 | 15 | 78 |

## ALPENA PHELPS COL AP *Alpena County* ELEVATION 689 ft LAT/LONG 45° 4 ' N / 83° 32 ' W

| | JAN | FEB | MAR | APR | MAY | JUN | JUL | AUG | SEP | OCT | NOV | DEC | YEAR |
|---|---|---|---|---|---|---|---|---|---|---|---|---|---|
| Maximum Temp °F | 26.5 | 28.4 | 37.9 | 51.7 | 64.9 | 74.3 | 79.8 | 77.0 | 68.1 | 56.1 | 42.6 | 31.6 | 53.2 |
| Minimum Temp °F | 9.3 | 9.0 | 18.4 | 30.4 | 39.6 | 48.4 | 54.3 | 53.0 | 45.7 | 36.4 | 27.6 | 17.4 | 32.5 |
| Mean Temp °F | 17.9 | 18.7 | 28.2 | 41.1 | 52.2 | 61.4 | 67.0 | 65.0 | 56.9 | 46.3 | 35.1 | 24.5 | 42.9 |
| Days Max Temp ≥ 90 °F | 0 | 0 | 0 | 0 | 0 | 2 | 3 | 1 | 0 | 0 | 0 | 0 | 6 |
| Days Max Temp ≤ 32 °F | 22 | 19 | 9 | 1 | 0 | 0 | 0 | 0 | 0 | 0 | 4 | 16 | 71 |
| Days Min Temp ≤ 32 °F | 30 | 27 | 28 | 18 | 7 | 0 | 0 | 0 | 2 | 11 | 22 | 29 | 174 |
| Days Min Temp ≤ 0 °F | 8 | 8 | 3 | 0 | 0 | 0 | 0 | 0 | 0 | 0 | 0 | 3 | 22 |
| Heating Degree Days | 1453 | 1301 | 1135 | 714 | 402 | 153 | 49 | 80 | 258 | 576 | 889 | 1248 | 8258 |
| Cooling Degree Days | 0 | 0 | 0 | 4 | 14 | 49 | 134 | 94 | 22 | 1 | 0 | 0 | 318 |
| Total Precipitation (") | 1.68 | 1.25 | 2.02 | 2.36 | 2.67 | 2.99 | 3.14 | 3.58 | 2.90 | 2.36 | 2.32 | 1.93 | 29.20 |
| Days ≥ 0.1" Precip | 5 | 4 | 5 | 6 | 6 | 6 | 6 | 7 | 6 | 6 | 7 | 5 | 69 |
| Total Snowfall (") | 23.0 | 16.0 | 13.2 | 5.2 | 0.2 | 0.0 | 0.0 | 0.0 | 0.0 | 0.6 | 9.0 | 20.2 | 87.4 |
| Days ≥ 1" Snow Depth | 30 | 27 | 22 | 4 | 0 | 0 | 0 | 0 | 0 | 0 | 6 | 22 | 111 |

## ALPENA WASTEWATER PL *Alpena County* ELEVATION 617 ft LAT/LONG 45° 4 ' N / 83° 26 ' W

| | JAN | FEB | MAR | APR | MAY | JUN | JUL | AUG | SEP | OCT | NOV | DEC | YEAR |
|---|---|---|---|---|---|---|---|---|---|---|---|---|---|
| Maximum Temp °F | 26.8 | 28.5 | 36.6 | 49.4 | 61.2 | 71.2 | 77.0 | 75.5 | 67.2 | 55.3 | 42.6 | 32.7 | 52.0 |
| Minimum Temp °F | 11.8 | 12.3 | 21.4 | 32.8 | 43.1 | 52.4 | 58.7 | 57.3 | 50.0 | 39.8 | 30.1 | 20.3 | 35.8 |
| Mean Temp °F | 19.3 | 20.4 | 29.0 | 41.1 | 52.2 | 61.8 | 67.9 | 66.4 | 58.6 | 47.6 | 36.4 | 26.5 | 43.9 |
| Days Max Temp ≥ 90 °F | 0 | 0 | 0 | 0 | 0 | 1 | 2 | 1 | 0 | 0 | 0 | 0 | 4 |
| Days Max Temp ≤ 32 °F | 22 | 19 | 9 | 1 | 0 | 0 | 0 | 0 | 0 | 0 | 3 | 14 | 68 |
| Days Min Temp ≤ 32 °F | 31 | 27 | 27 | 15 | 2 | 0 | 0 | 0 | 0 | 5 | 19 | 28 | 154 |
| Days Min Temp ≤ 0 °F | 5 | 5 | 1 | 0 | 0 | 0 | 0 | 0 | 0 | 0 | 0 | 1 | 12 |
| Heating Degree Days | 1409 | 1253 | 1110 | 711 | 398 | 137 | 35 | 54 | 210 | 533 | 851 | 1186 | 7887 |
| Cooling Degree Days | 0 | 0 | 0 | 2 | 10 | 49 | 143 | 112 | 27 | 0 | 0 | 0 | 343 |
| Total Precipitation (") | 1.39 | 1.10 | 1.71 | 2.20 | 2.82 | 2.90 | 3.03 | 3.61 | 3.18 | 2.51 | 2.27 | 1.78 | 28.50 |
| Days ≥ 0.1" Precip | 4 | 3 | 5 | 6 | 6 | 6 | 6 | 7 | 7 | 6 | 6 | 5 | 67 |
| Total Snowfall (") | 16.5 | 12.1 | 8.9 | 2.2 | 0.2 | 0.0 | 0.0 | 0.0 | 0.0 | 0.2 | 6.1 | 13.6 | 59.8 |
| Days ≥ 1" Snow Depth | 29 | 26 | 19 | 2 | 0 | 0 | 0 | 0 | 0 | 0 | 4 | 18 | 98 |

## ANN ARBOR UNIV OF MI *Washtenaw County* ELEVATION 932 ft LAT/LONG 42° 17 ' N / 83° 45 ' W

| | JAN | FEB | MAR | APR | MAY | JUN | JUL | AUG | SEP | OCT | NOV | DEC | YEAR |
|---|---|---|---|---|---|---|---|---|---|---|---|---|---|
| Maximum Temp °F | 30.2 | 33.6 | 45.3 | 58.9 | 71.1 | 79.6 | 83.4 | 81.2 | 73.9 | 61.3 | 47.7 | 35.4 | 58.5 |
| Minimum Temp °F | 16.4 | 18.1 | 27.1 | 37.8 | 48.3 | 57.7 | 62.3 | 60.7 | 53.7 | 42.5 | 33.1 | 22.8 | 40.0 |
| Mean Temp °F | 23.3 | 25.9 | 36.2 | 48.4 | 59.7 | 68.7 | 72.9 | 70.9 | 63.8 | 51.9 | 40.4 | 29.1 | 49.3 |
| Days Max Temp ≥ 90 °F | 0 | 0 | 0 | 0 | 0 | 2 | 5 | 2 | 1 | 0 | 0 | 0 | 10 |
| Days Max Temp ≤ 32 °F | 18 | 13 | 4 | 0 | 0 | 0 | 0 | 0 | 0 | 0 | 2 | 11 | 48 |
| Days Min Temp ≤ 32 °F | 29 | 26 | 23 | 9 | 1 | 0 | 0 | 0 | 0 | 4 | 15 | 27 | 134 |
| Days Min Temp ≤ 0 °F | 3 | 2 | 0 | 0 | 0 | 0 | 0 | 0 | 0 | 0 | 0 | 1 | 6 |
| Heating Degree Days | 1285 | 1099 | 886 | 500 | 205 | 40 | 6 | 14 | 112 | 407 | 731 | 1105 | 6390 |
| Cooling Degree Days | 0 | 0 | 0 | 10 | 50 | 142 | 259 | 200 | 78 | 5 | 0 | 0 | 744 |
| Total Precipitation (") | 2.00 | 1.79 | 2.63 | 3.20 | 2.83 | 3.53 | 3.12 | 3.46 | 3.18 | 2.40 | 3.07 | 2.95 | 34.16 |
| Days ≥ 0.1" Precip | 5 | 5 | 6 | 8 | 6 | 7 | 6 | 6 | 6 | 6 | 7 | 7 | 75 |
| Total Snowfall (") | 12.5 | 9.7 | 7.2 | 2.4 | 0.0 | 0.0 | 0.0 | 0.0 | 0.0 | 0.3 | 3.5 | 11.7 | 47.3 |
| Days ≥ 1" Snow Depth | 21 | 17 | 7 | 1 | 0 | 0 | 0 | 0 | 0 | 0 | 2 | 14 | 62 |

## BAD AXE *Huron County* ELEVATION 751 ft LAT/LONG 43° 48 ' N / 83° 1 ' W

| | JAN | FEB | MAR | APR | MAY | JUN | JUL | AUG | SEP | OCT | NOV | DEC | YEAR |
|---|---|---|---|---|---|---|---|---|---|---|---|---|---|
| Maximum Temp °F | 27.9 | 30.0 | 40.0 | 53.8 | 66.8 | 76.2 | 81.2 | 78.9 | 70.8 | 58.8 | 45.1 | 32.8 | 55.2 |
| Minimum Temp °F | 13.4 | 14.1 | 22.8 | 33.8 | 43.5 | 52.8 | 58.0 | 56.3 | 49.8 | 40.0 | 31.0 | 20.6 | 36.3 |
| Mean Temp °F | 20.7 | 22.0 | 31.4 | 43.8 | 55.2 | 64.6 | 69.7 | 67.6 | 60.3 | 49.4 | 38.1 | 26.8 | 45.8 |
| Days Max Temp ≥ 90 °F | 0 | 0 | 0 | 0 | 0 | 2 | 4 | 2 | 0 | 0 | 0 | 0 | 8 |
| Days Max Temp ≤ 32 °F | 21 | 17 | 8 | 0 | 0 | 0 | 0 | 0 | 0 | 0 | 2 | 14 | 62 |
| Days Min Temp ≤ 32 °F | 30 | 27 | 26 | 15 | 3 | 0 | 0 | 0 | 0 | 6 | 18 | 28 | 153 |
| Days Min Temp ≤ 0 °F | 4 | 4 | 1 | 0 | 0 | 0 | 0 | 0 | 0 | 0 | 0 | 1 | 10 |
| Heating Degree Days | 1367 | 1206 | 1034 | 634 | 323 | 98 | 23 | 45 | 178 | 481 | 800 | 1178 | 7367 |
| Cooling Degree Days | 0 | 0 | 0 | 6 | 22 | 77 | 172 | 118 | 35 | 1 | 0 | 0 | 431 |
| Total Precipitation (") | 1.78 | 1.45 | 2.44 | 2.95 | 2.60 | 3.07 | 2.90 | 3.49 | 3.55 | 2.72 | 2.97 | 2.23 | 32.15 |
| Days ≥ 0.1" Precip | 5 | 4 | 6 | 7 | 6 | 7 | 6 | 6 | 7 | 6 | 7 | 6 | 73 |
| Total Snowfall (") | 13.4 | 10.0 | 10.3 | 3.1 | 0.1 | 0.0 | 0.0 | 0.0 | 0.0 | 0.8 | 5.9 | 11.8 | 55.4 |
| Days ≥ 1" Snow Depth | 25 | 22 | 13 | 2 | 0 | 0 | 0 | 0 | 0 | 0 | 4 | 17 | 83 |

## BALDWIN *Lake County* ELEVATION 830 ft LAT/LONG 43° 54' N / 85° 51' W

| | JAN | FEB | MAR | APR | MAY | JUN | JUL | AUG | SEP | OCT | NOV | DEC | YEAR |
|---|---|---|---|---|---|---|---|---|---|---|---|---|---|
| Maximum Temp °F | 28.5 | 31.8 | 42.6 | 56.3 | 69.8 | 77.9 | 82.4 | 80.0 | 71.5 | 59.0 | 45.0 | 33.3 | 56.5 |
| Minimum Temp °F | 10.0 | 9.1 | 18.6 | 31.2 | 41.9 | 50.4 | 54.6 | 52.7 | 45.3 | 35.2 | 26.6 | 16.7 | 32.7 |
| Mean Temp °F | 19.3 | 20.5 | 30.6 | 43.8 | 56.0 | 64.2 | 68.5 | 66.4 | 58.4 | 47.1 | 35.9 | 25.0 | 44.6 |
| Days Max Temp ≥ 90 °F | 0 | 0 | 0 | 0 | 0 | 2 | 4 | 2 | 0 | 0 | 0 | 0 | 8 |
| Days Max Temp ≤ 32 °F | 20 | 15 | 5 | 0 | 0 | 0 | 0 | 0 | 0 | 0 | 3 | 14 | 57 |
| Days Min Temp ≤ 32 °F | 31 | 27 | 27 | 17 | 6 | 1 | 0 | 0 | 3 | 13 | 22 | 29 | 176 |
| Days Min Temp ≤ 0 °F | 7 | 8 | 3 | 0 | 0 | 0 | 0 | 0 | 0 | 0 | 0 | 3 | 21 |
| Heating Degree Days | 1411 | 1252 | 1059 | 633 | 301 | 99 | 32 | 59 | 224 | 549 | 868 | 1233 | 7720 |
| Cooling Degree Days | 0 | 0 | 0 | 6 | 25 | *67* | 145 | 105 | 27 | 0 | 0 | 0 | 375 |
| Total Precipitation (") | 2.32 | 1.59 | 2.21 | 3.19 | 2.84 | 3.54 | 2.76 | 3.88 | 3.85 | 3.34 | 3.33 | 2.49 | 35.34 |
| Days ≥ 0.1" Precip | 7 | 5 | 6 | 8 | 6 | 6 | 6 | 6 | 7 | 7 | 8 | 7 | 79 |
| Total Snowfall (") | 25.8 | 17.9 | 9.3 | 2.1 | 0.0 | 0.0 | 0.0 | 0.0 | 0.0 | 0.2 | 8.2 | 19.0 | 82.5 |
| Days ≥ 1" Snow Depth | 29 | 26 | 17 | 2 | 0 | 0 | 0 | 0 | 0 | 0 | 6 | 22 | 102 |

## BATTLE CREEK 5 NW *Calhoun County* ELEVATION 869 ft LAT/LONG 42° 20' N / 85° 11' W

| | JAN | FEB | MAR | APR | MAY | JUN | JUL | AUG | SEP | OCT | NOV | DEC | YEAR |
|---|---|---|---|---|---|---|---|---|---|---|---|---|---|
| Maximum Temp °F | 30.2 | 33.6 | 45.0 | 58.7 | 70.4 | 79.1 | 82.8 | 80.6 | 72.9 | 60.5 | 47.0 | 35.1 | 58.0 |
| Minimum Temp °F | 14.9 | 16.4 | 25.5 | 36.2 | 46.1 | 55.4 | 60.2 | 58.0 | 51.0 | 40.2 | 31.3 | 21.0 | 38.0 |
| Mean Temp °F | 22.6 | 25.0 | 35.3 | 47.5 | 58.2 | 67.3 | 71.5 | 69.3 | 61.9 | 50.4 | 39.2 | 28.1 | 48.0 |
| Days Max Temp ≥ 90 °F | 0 | 0 | 0 | 0 | 0 | 2 | 4 | 2 | 1 | 0 | 0 | 0 | 9 |
| Days Max Temp ≤ 32 °F | 18 | 13 | 4 | 0 | 0 | 0 | 0 | 0 | 0 | 0 | 2 | 12 | 49 |
| Days Min Temp ≤ 32 °F | 30 | 26 | 24 | 11 | 2 | 0 | 0 | 0 | 1 | 7 | 18 | 28 | 147 |
| Days Min Temp ≤ 0 °F | 4 | 3 | 0 | 0 | 0 | 0 | 0 | 0 | 0 | 0 | 0 | 1 | 8 |
| Heating Degree Days | 1308 | 1122 | 915 | 527 | 243 | 55 | 11 | 27 | 149 | 451 | 768 | 1137 | 6713 |
| Cooling Degree Days | 0 | 0 | 1 | 10 | 42 | 123 | 221 | 167 | 61 | 3 | 0 | 0 | 628 |
| Total Precipitation (") | 1.67 | 1.52 | 2.56 | 3.46 | 3.43 | 3.64 | 3.62 | 3.67 | 3.86 | 3.07 | 3.24 | 2.83 | 36.57 |
| Days ≥ 0.1" Precip | 4 | 4 | 6 | 8 | 7 | 7 | 7 | 7 | 7 | 7 | 7 | 7 | 78 |
| Total Snowfall (") | 13.9 | 10.7 | 6.9 | 2.6 | 0.0 | 0.0 | 0.0 | 0.0 | 0.0 | 0.3 | 5.9 | 13.1 | 53.4 |
| Days ≥ 1" Snow Depth | 23 | 18 | 9 | 1 | 0 | 0 | 0 | 0 | 0 | 0 | 4 | 17 | 72 |

## BENTON HRBR ROSS FLD *Berrien County* ELEVATION 628 ft LAT/LONG 42° 8' N / 86° 26' W

| | JAN | FEB | MAR | APR | MAY | JUN | JUL | AUG | SEP | OCT | NOV | DEC | YEAR |
|---|---|---|---|---|---|---|---|---|---|---|---|---|---|
| Maximum Temp °F | 31.7 | 34.9 | 45.8 | 58.0 | 69.1 | 78.8 | 82.4 | 80.6 | 73.9 | 62.3 | 48.9 | 37.0 | 58.6 |
| Minimum Temp °F | 17.6 | 18.8 | 27.6 | 37.3 | 46.5 | 56.4 | 60.9 | 58.4 | 51.7 | 41.6 | 32.7 | 23.1 | 39.4 |
| Mean Temp °F | 24.9 | 26.8 | 36.7 | 47.7 | 57.8 | 67.6 | 71.7 | 69.6 | 62.9 | 51.9 | 40.8 | 30.1 | 49.0 |
| Days Max Temp ≥ 90 °F | 0 | 0 | 0 | 0 | 0 | 3 | 5 | 3 | 1 | 0 | 0 | 0 | 12 |
| Days Max Temp ≤ 32 °F | 16 | 12 | 4 | 0 | 0 | 0 | 0 | 0 | 0 | 0 | 1 | 9 | 42 |
| Days Min Temp ≤ 32 °F | 29 | 26 | 22 | 9 | 2 | 0 | 0 | 0 | 1 | 5 | 16 | 26 | 136 |
| Days Min Temp ≤ 0 °F | 2 | 1 | 0 | 0 | 0 | 0 | 0 | 0 | 0 | 0 | 0 | 1 | 4 |
| Heating Degree Days | 1237 | 1073 | 872 | 524 | 263 | 59 | 14 | 27 | 131 | 406 | 721 | 1074 | 6401 |
| Cooling Degree Days | 0 | 0 | 1 | 11 | 47 | 133 | 226 | 166 | 68 | 4 | 0 | 0 | 656 |
| Total Precipitation (") | 2.42 | 1.69 | 2.53 | 3.64 | 3.12 | 3.63 | 3.24 | 3.56 | 4.51 | 3.18 | 3.59 | 2.92 | 38.03 |
| Days ≥ 0.1" Precip | *6* | 5 | 6 | 8 | 7 | 7 | 5 | 6 | 8 | 7 | 8 | 8 | 81 |
| Total Snowfall (") | *25.3* | *19.9* | *8.1* | 1.3 | 0.0 | 0.0 | 0.0 | 0.0 | 0.0 | 0.4 | *3.8* | 18.4 | 77.2 |
| Days ≥ 1" Snow Depth | 19 | *15* | 6 | 1 | 0 | 0 | 0 | 0 | 0 | 0 | *2* | 13 | 56 |

## BERGLAND DAM *Ontonagon County* ELEVATION 1302 ft LAT/LONG 46° 35' N / 89° 33' W

| | JAN | FEB | MAR | APR | MAY | JUN | JUL | AUG | SEP | OCT | NOV | DEC | YEAR |
|---|---|---|---|---|---|---|---|---|---|---|---|---|---|
| Maximum Temp °F | 20.0 | 24.5 | 36.2 | 50.0 | 64.2 | 72.7 | 77.9 | 75.6 | 65.6 | 53.3 | 37.0 | 25.0 | 50.2 |
| Minimum Temp °F | -1.1 | -0.9 | 9.9 | 25.2 | 36.8 | 46.5 | 51.9 | 49.7 | 42.1 | 32.6 | 20.9 | 7.4 | 26.8 |
| Mean Temp °F | 9.4 | 11.8 | 23.1 | 37.6 | 50.5 | 59.6 | 64.9 | 62.6 | 53.8 | 42.9 | 29.0 | 16.2 | 38.4 |
| Days Max Temp ≥ 90 °F | 0 | 0 | 0 | 0 | 0 | 0 | 2 | 1 | 0 | 0 | 0 | 0 | 3 |
| Days Max Temp ≤ 32 °F | 27 | 21 | 12 | 2 | 0 | 0 | 0 | 0 | 0 | 0 | 10 | 23 | 95 |
| Days Min Temp ≤ 32 °F | 31 | 28 | 30 | 24 | 11 | 1 | 0 | 0 | 4 | *17* | 27 | 31 | 204 |
| Days Min Temp ≤ 0 °F | 16 | 15 | 9 | 1 | 0 | 0 | 0 | 0 | 0 | 0 | 1 | 10 | 52 |
| Heating Degree Days | 1719 | 1498 | 1293 | 816 | 454 | 190 | 80 | 126 | 341 | 678 | 1073 | 1507 | 9775 |
| Cooling Degree Days | 0 | 0 | 0 | 2 | 13 | 37 | 76 | 58 | 10 | 0 | 0 | 0 | 196 |
| Total Precipitation (") | 2.61 | 1.54 | 2.24 | 2.39 | 3.31 | 4.09 | 3.72 | 4.06 | 3.81 | 3.54 | 3.42 | 2.76 | 37.49 |
| Days ≥ 0.1" Precip | 8 | 5 | 6 | 6 | 8 | 9 | 7 | 7 | 9 | 9 | 9 | 9 | 92 |
| Total Snowfall (") | 43.1 | 25.1 | 23.6 | 9.5 | 1.0 | 0.0 | 0.0 | 0.0 | 0.0 | 3.8 | 28.7 | 40.8 | 175.6 |
| Days ≥ 1" Snow Depth | 31 | 28 | 31 | 16 | 0 | 0 | 0 | 0 | 0 | 2 | 18 | 31 | 157 |

## BIG BAY 2 SE *Marquette County* ELEVATION 669 ft LAT/LONG 46° 49 ' N / 87° 44 ' W

| | JAN | FEB | MAR | APR | MAY | JUN | JUL | AUG | SEP | OCT | NOV | DEC | YEAR |
|---|---|---|---|---|---|---|---|---|---|---|---|---|---|
| Maximum Temp °F | 24.9 | ***28.0*** | 37.6 | ***50.2*** | ***63.6*** | 72.6 | ***77.9*** | ***76.1*** | 67.3 | 55.8 | 41.3 | ***29.3*** | 52.0 |
| Minimum Temp °F | 9.2 | 9.5 | 18.6 | ***29.3*** | ***38.6*** | 48.7 | ***55.0*** | ***54.4*** | 47.6 | 37.8 | 27.1 | ***15.6*** | 32.6 |
| Mean Temp °F | 17.1 | ***18.8*** | 28.1 | ***39.8*** | ***51.2*** | 60.7 | ***66.5*** | ***65.2*** | 57.6 | 46.8 | 34.3 | ***22.4*** | 42.4 |
| Days Max Temp ≥ 90 °F | 0 | 0 | 0 | 0 | 0 | 1 | 2 | 1 | 0 | 0 | 0 | 0 | 4 |
| Days Max Temp ≤ 32 °F | 23 | ***19*** | 10 | 1 | 0 | 0 | 0 | 0 | 0 | 0 | 5 | 17 | 75 |
| Days Min Temp ≤ 32 °F | 30 | 28 | 29 | 20 | ***8*** | 0 | 0 | 0 | 1 | ***9*** | 22 | 29 | 176 |
| Days Min Temp ≤ 0 °F | 7 | ***7*** | 2 | 0 | 0 | 0 | 0 | 0 | 0 | 0 | 0 | 3 | 19 |
| Heating Degree Days | 1480 | ***1302*** | 1136 | ***753*** | ***438*** | 172 | ***60*** | ***79*** | ***245*** | 567 | ***921*** | ***1311*** | 8464 |
| Cooling Degree Days | 0 | 0 | 0 | ***3*** | ***20*** | 53 | 116 | 89 | 19 | 1 | ***0*** | ***0*** | 301 |
| Total Precipitation (") | 1.65 | 1.24 | 2.15 | 2.00 | 2.60 | 3.38 | 2.84 | 3.06 | 3.51 | 3.06 | 2.32 | 2.00 | 29.81 |
| Days ≥ 0.1" Precip | ***5*** | ***4*** | ***5*** | ***5*** | ***6*** | ***7*** | ***6*** | ***6*** | ***8*** | 7 | 6 | ***6*** | 71 |
| Total Snowfall (") | ***31.5*** | 20.1 | ***19.5*** | 5.1 | 0.6 | 0.0 | 0.0 | 0.0 | 0.0 | 0.8 | 11.2 | ***27.2*** | 116.0 |
| Days ≥ 1" Snow Depth | ***30*** | ***28*** | ***28*** | ***10*** | ***1*** | 0 | 0 | 0 | 0 | 0 | ***10*** | 28 | 135 |

## BIG RAPIDS WATERWORK *Mecosta County* ELEVATION 932 ft LAT/LONG 43° 42 ' N / 85° 29 ' W

| | JAN | FEB | MAR | APR | MAY | JUN | JUL | AUG | SEP | OCT | NOV | DEC | YEAR |
|---|---|---|---|---|---|---|---|---|---|---|---|---|---|
| Maximum Temp °F | 28.0 | 31.4 | 41.4 | 55.3 | 68.6 | 77.3 | 81.9 | 79.2 | 70.3 | 57.9 | 44.1 | 32.9 | 55.7 |
| Minimum Temp °F | 10.5 | 11.4 | 20.7 | 32.3 | 42.7 | 51.5 | 56.6 | 54.5 | 46.7 | 36.3 | 28.0 | 17.9 | 34.1 |
| Mean Temp °F | 19.2 | 21.4 | 31.1 | 43.7 | 55.7 | 64.4 | 69.2 | 66.8 | 58.5 | 47.2 | 36.1 | 25.4 | 44.9 |
| Days Max Temp ≥ 90 °F | 0 | 0 | 0 | 0 | 0 | 1 | 4 | 2 | 0 | 0 | 0 | 0 | 7 |
| Days Max Temp ≤ 32 °F | 21 | 15 | 6 | 0 | 0 | 0 | 0 | 0 | 0 | 0 | 3 | 15 | 60 |
| Days Min Temp ≤ 32 °F | 30 | 28 | 27 | 17 | 5 | 0 | 0 | 0 | 2 | 11 | 22 | 29 | 171 |
| Days Min Temp ≤ 0 °F | 7 | 6 | 1 | 0 | 0 | 0 | 0 | 0 | 0 | 0 | 0 | 3 | 17 |
| Heating Degree Days | 1413 | 1223 | 1045 | 634 | 307 | 88 | 25 | 53 | 218 | 549 | 861 | 1221 | 7637 |
| Cooling Degree Days | 0 | 0 | 0 | 3 | 25 | 75 | 176 | 121 | 30 | 1 | 0 | 0 | 431 |
| Total Precipitation (") | 2.05 | 1.43 | 2.36 | 3.06 | 3.06 | 3.44 | 2.52 | 4.00 | 4.18 | 3.05 | 3.19 | 2.52 | 34.86 |
| Days ≥ 0.1" Precip | 6 | 4 | 6 | 7 | 7 | 7 | 5 | 6 | 8 | 7 | 8 | 7 | 78 |
| Total Snowfall (") | 20.4 | 13.4 | 9.5 | 2.3 | 0.0 | 0.0 | 0.0 | 0.0 | 0.0 | 0.4 | 6.6 | 16.9 | 69.5 |
| Days ≥ 1" Snow Depth | 29 | 24 | 15 | 2 | 0 | 0 | 0 | 0 | 0 | 0 | 5 | 21 | 96 |

## BLOOMINGDALE *Van Buren County* ELEVATION 751 ft LAT/LONG 42° 23 ' N / 85° 57 ' W

| | JAN | FEB | MAR | APR | MAY | JUN | JUL | AUG | SEP | OCT | NOV | DEC | YEAR |
|---|---|---|---|---|---|---|---|---|---|---|---|---|---|
| Maximum Temp °F | 30.1 | 33.3 | 43.7 | 57.2 | 69.7 | 78.2 | 82.5 | 80.7 | 73.0 | 60.9 | 47.3 | 35.4 | 57.7 |
| Minimum Temp °F | 14.7 | 15.2 | 23.8 | 35.0 | 44.6 | 53.4 | 57.8 | 56.0 | 49.1 | 38.8 | 30.5 | 20.9 | 36.7 |
| Mean Temp °F | 22.4 | 24.3 | 33.8 | 46.1 | 57.2 | 65.8 | 70.2 | 68.4 | 61.1 | 49.9 | 38.9 | 28.1 | 47.2 |
| Days Max Temp ≥ 90 °F | 0 | 0 | 0 | 0 | 0 | 3 | 4 | 3 | 1 | 0 | 0 | 0 | 11 |
| Days Max Temp ≤ 32 °F | 18 | 14 | 5 | 0 | 0 | 0 | 0 | 0 | 0 | 0 | 2 | 11 | 50 |
| Days Min Temp ≤ 32 °F | 30 | 26 | 25 | 13 | 4 | 0 | 0 | 0 | 1 | 8 | 20 | 28 | 155 |
| Days Min Temp ≤ 0 °F | 4 | 3 | 1 | 0 | 0 | 0 | 0 | 0 | 0 | 0 | 0 | 1 | 9 |
| Heating Degree Days | 1314 | 1143 | 962 | 568 | 276 | 80 | 21 | 38 | 169 | 468 | 776 | 1136 | 6951 |
| Cooling Degree Days | 0 | 0 | 1 | 10 | 38 | 109 | 194 | 156 | 56 | 4 | 0 | 0 | 568 |
| Total Precipitation (") | 2.56 | 1.70 | 2.79 | 3.54 | 3.35 | 3.91 | 4.10 | 3.66 | 4.63 | 3.21 | 3.88 | 3.51 | 40.84 |
| Days ≥ 0.1" Precip | 8 | 6 | 7 | 8 | 7 | 7 | 6 | 7 | 7 | 7 | 9 | 9 | 88 |
| Total Snowfall (") | 26.6 | 15.4 | 7.9 | 1.9 | 0.0 | 0.0 | 0.0 | 0.0 | 0.0 | 0.2 | 8.1 | 22.4 | 82.5 |
| Days ≥ 1" Snow Depth | 24 | 20 | 10 | 1 | 0 | 0 | 0 | 0 | 0 | 0 | 5 | 19 | 79 |

## BOYNE FALLS *Charlevoix County* ELEVATION 732 ft LAT/LONG 45° 10 ' N / 84° 55 ' W

| | JAN | FEB | MAR | APR | MAY | JUN | JUL | AUG | SEP | OCT | NOV | DEC | YEAR |
|---|---|---|---|---|---|---|---|---|---|---|---|---|---|
| Maximum Temp °F | 27.3 | 30.1 | 41.0 | 55.5 | 69.4 | 77.3 | 81.9 | 79.2 | 70.4 | 58.4 | 43.7 | 32.2 | 55.5 |
| Minimum Temp °F | 11.2 | 9.5 | 18.7 | 30.9 | 40.3 | 49.8 | 54.7 | 53.4 | 47.0 | 37.8 | 28.8 | 18.3 | 33.4 |
| Mean Temp °F | 19.3 | 19.8 | 29.9 | 43.2 | 54.9 | 63.6 | 68.3 | 66.3 | 58.7 | 48.2 | 36.3 | 25.2 | 44.5 |
| Days Max Temp ≥ 90 °F | 0 | 0 | 0 | 0 | 0 | 2 | 4 | 2 | 0 | 0 | 0 | 0 | 8 |
| Days Max Temp ≤ 32 °F | 22 | 17 | 7 | 0 | 0 | 0 | 0 | 0 | 0 | 0 | 3 | 16 | 65 |
| Days Min Temp ≤ 32 °F | 31 | 27 | 27 | 18 | 8 | 1 | 0 | 0 | 2 | 10 | 21 | 29 | 174 |
| Days Min Temp ≤ 0 °F | 6 | 8 | 3 | 0 | 0 | 0 | 0 | 0 | 0 | 0 | 0 | 2 | 19 |
| Heating Degree Days | 1411 | 1269 | 1083 | 652 | 332 | 112 | 37 | 63 | 217 | 519 | 856 | 1226 | 7777 |
| Cooling Degree Days | 0 | 0 | 0 | 7 | 26 | 70 | 150 | 118 | 37 | 1 | 0 | 0 | 409 |
| Total Precipitation (") | 2.20 | 1.37 | 1.91 | 2.43 | 2.56 | 2.77 | 3.03 | 3.46 | 4.30 | 3.58 | 3.15 | 2.47 | 33.23 |
| Days ≥ 0.1" Precip | 7 | 4 | 6 | 7 | 7 | 7 | 5 | 8 | 9 | 9 | 9 | 9 | 87 |
| Total Snowfall (") | 34.3 | 20.5 | 11.7 | 5.1 | 0.4 | 0.0 | 0.0 | 0.0 | 0.0 | 1.1 | 15.3 | 30.0 | 118.4 |
| Days ≥ 1" Snow Depth | 30 | 27 | 22 | 4 | 0 | 0 | 0 | 0 | 0 | 0 | 8 | 24 | 115 |

## CADILLAC *Wexford County* ELEVATION 1302 ft LAT/LONG 44° 16 ' N / 85° 24 ' W

| | JAN | FEB | MAR | APR | MAY | JUN | JUL | AUG | SEP | OCT | NOV | DEC | YEAR |
|---|---|---|---|---|---|---|---|---|---|---|---|---|---|
| Maximum Temp °F | 25.3 | 27.8 | 37.7 | 52.0 | 65.8 | 74.1 | 78.9 | 76.3 | 67.8 | 55.5 | 41.7 | 30.3 | 52.8 |
| Minimum Temp °F | 9.4 | 8.2 | 17.0 | 30.4 | 40.1 | 49.4 | 54.0 | 52.1 | 44.9 | 35.6 | 26.8 | 16.1 | 32.0 |
| Mean Temp °F | 17.4 | 17.9 | 27.2 | 41.2 | 53.0 | 61.8 | 66.6 | 64.3 | 56.4 | 45.6 | 34.3 | 23.2 | 42.4 |
| Days Max Temp ≥ 90 °F | 0 | 0 | 0 | 0 | 0 | 1 | 1 | 1 | 0 | 0 | 0 | 0 | 3 |
| Days Max Temp ≤ 32 °F | 24 | 19 | 10 | 1 | 0 | 0 | 0 | 0 | 0 | 0 | 6 | 19 | 79 |
| Days Min Temp ≤ 32 °F | 31 | 28 | 28 | 19 | 8 | 1 | 0 | 0 | 4 | 13 | 23 | 30 | 185 |
| Days Min Temp ≤ 0 °F | 8 | 8 | 4 | 0 | 0 | 0 | 0 | 0 | 0 | 0 | 0 | 4 | 24 |
| Heating Degree Days | 1470 | 1323 | 1165 | 710 | 383 | 146 | 59 | 97 | 275 | 597 | 915 | 1289 | 8429 |
| Cooling Degree Days | 0 | 0 | 0 | 5 | 16 | 54 | 121 | 81 | 20 | 0 | 0 | 0 | 297 |
| Total Precipitation (") | 1.69 | 1.26 | 1.96 | 2.91 | 2.63 | 3.22 | 3.09 | 3.59 | 4.25 | 3.32 | 2.94 | 2.09 | 32.95 |
| Days ≥ 0.1" Precip | 5 | 4 | 6 | 7 | 6 | 6 | 6 | 6 | 8 | 8 | 7 | 6 | 75 |
| Total Snowfall (") | 25.0 | *17.3* | 12.5 | 3.7 | 0.0 | 0.0 | 0.0 | 0.0 | 0.0 | 0.5 | 9.3 | 20.2 | 88.5 |
| Days ≥ 1" Snow Depth | *27* | *24* | *17* | 4 | 0 | 0 | 0 | 0 | 0 | 0 | 6 | *22* | 100 |

## CARO REGIONAL CENTER *Tuscola County* ELEVATION 712 ft LAT/LONG 43° 29 ' N / 83° 24 ' W

| | JAN | FEB | MAR | APR | MAY | JUN | JUL | AUG | SEP | OCT | NOV | DEC | YEAR |
|---|---|---|---|---|---|---|---|---|---|---|---|---|---|
| Maximum Temp °F | 29.1 | 32.0 | 43.4 | 57.8 | 70.9 | 79.5 | 84.0 | 81.2 | 73.3 | 60.8 | 46.6 | 34.3 | 57.7 |
| Minimum Temp °F | 13.1 | 14.2 | 23.8 | 34.2 | 43.7 | 52.7 | 57.7 | 55.9 | 49.2 | 39.2 | 30.9 | 20.5 | 36.3 |
| Mean Temp °F | 21.1 | 23.1 | 33.6 | 46.0 | 57.3 | 66.1 | 70.9 | 68.6 | 61.2 | 50.0 | 38.8 | 27.5 | 47.0 |
| Days Max Temp ≥ 90 °F | 0 | 0 | 0 | 0 | 1 | 3 | 6 | 3 | 1 | 0 | 0 | 0 | 14 |
| Days Max Temp ≤ 32 °F | 19 | 15 | 5 | 0 | 0 | 0 | 0 | 0 | 0 | 0 | 2 | 13 | 54 |
| Days Min Temp ≤ 32 °F | 30 | 27 | 25 | 15 | 4 | 0 | 0 | 0 | 1 | 8 | 18 | 27 | 155 |
| Days Min Temp ≤ 0 °F | 6 | 5 | 1 | 0 | 0 | 0 | 0 | 0 | 0 | 0 | 0 | 2 | 14 |
| Heating Degree Days | 1354 | 1176 | 967 | 571 | 270 | 72 | 17 | 36 | 163 | 465 | 781 | 1157 | 7029 |
| Cooling Degree Days | 0 | 0 | 1 | 8 | 41 | 106 | 216 | 156 | 53 | 3 | 0 | 0 | 584 |
| Total Precipitation (") | 1.58 | 1.07 | 2.27 | 2.86 | 2.74 | 3.35 | 2.93 | 3.11 | 4.08 | 2.47 | 2.71 | 2.04 | 31.21 |
| Days ≥ 0.1" Precip | 5 | 3 | 6 | 8 | 6 | 7 | 6 | 6 | 7 | 6 | 7 | 6 | 73 |
| Total Snowfall (") | 11.3 | 7.0 | 5.9 | 1.3 | 0.0 | 0.0 | 0.0 | 0.0 | 0.0 | 0.1 | 2.7 | 8.8 | 37.1 |
| Days ≥ 1" Snow Depth | 24 | 18 | 8 | 1 | 0 | 0 | 0 | 0 | 0 | 0 | 2 | 14 | 67 |

## CHAMPION VAN RIPER P *Marquette County* ELEVATION 1565 ft LAT/LONG 46° 31 ' N / 87° 59 ' W

| | JAN | FEB | MAR | APR | MAY | JUN | JUL | AUG | SEP | OCT | NOV | DEC | YEAR |
|---|---|---|---|---|---|---|---|---|---|---|---|---|---|
| Maximum Temp °F | 21.8 | 26.5 | 37.4 | 51.4 | 65.8 | 73.6 | 78.4 | 75.6 | 66.2 | 54.0 | 37.3 | 25.9 | 51.2 |
| Minimum Temp °F | 0.3 | 0.9 | 10.7 | 24.1 | 35.2 | 44.5 | 50.0 | 48.7 | 41.8 | 32.2 | 20.8 | 8.1 | 26.4 |
| Mean Temp °F | 11.1 | 13.7 | 24.1 | 37.8 | 50.5 | 59.0 | 64.2 | 62.2 | 54.0 | 43.1 | 29.1 | 17.0 | 38.8 |
| Days Max Temp ≥ 90 °F | 0 | 0 | 0 | 0 | 0 | 0 | 1 | 1 | 0 | 0 | 0 | 0 | 2 |
| Days Max Temp ≤ 32 °F | 26 | 20 | 10 | 1 | 0 | 0 | 0 | 0 | 0 | 0 | 10 | 23 | 90 |
| Days Min Temp ≤ 32 °F | 31 | 28 | 30 | 24 | 14 | 3 | 1 | 1 | 6 | 17 | 27 | 31 | 213 |
| Days Min Temp ≤ 0 °F | 15 | 13 | 8 | 1 | 0 | 0 | 0 | 0 | 0 | 0 | 1 | 9 | 47 |
| Heating Degree Days | 1667 | 1445 | 1263 | 810 | 454 | 201 | 93 | 134 | 336 | 672 | 1070 | 1482 | 9627 |
| Cooling Degree Days | 0 | 0 | 0 | 2 | 10 | 29 | 79 | 57 | 10 | 0 | 0 | 0 | 187 |
| Total Precipitation (") | 1.59 | 1.21 | 2.22 | 2.37 | 3.04 | 3.53 | 3.56 | 3.83 | 4.17 | 3.60 | 2.60 | 1.91 | 33.63 |
| Days ≥ 0.1" Precip | 5 | 4 | 6 | 5 | 7 | 8 | 7 | 7 | 9 | 8 | 7 | 6 | 79 |
| Total Snowfall (") | 28.5 | 19.6 | 22.1 | 9.0 | 1.3 | 0.0 | 0.0 | 0.0 | 0.1 | 5.2 | *22.6* | 29.2 | 137.6 |
| Days ≥ 1" Snow Depth | 31 | 28 | 30 | 17 | 1 | 0 | 0 | 0 | 0 | 2 | 18 | 31 | 158 |

## CHARLOTTE *Eaton County* ELEVATION 942 ft LAT/LONG 42° 31 ' N / 84° 50 ' W

| | JAN | FEB | MAR | APR | MAY | JUN | JUL | AUG | SEP | OCT | NOV | DEC | YEAR |
|---|---|---|---|---|---|---|---|---|---|---|---|---|---|
| Maximum Temp °F | 29.5 | 32.6 | 43.5 | 57.2 | 69.7 | 78.6 | 82.6 | 80.3 | 72.9 | 60.6 | 46.6 | 34.7 | 57.4 |
| Minimum Temp °F | 12.8 | 13.4 | 23.4 | 34.2 | 44.3 | 53.4 | 57.3 | 54.8 | 48.2 | 37.8 | 29.4 | 19.2 | 35.7 |
| Mean Temp °F | 21.2 | 23.0 | 33.5 | 45.7 | 57.0 | 66.0 | 70.0 | 67.6 | 60.6 | 49.2 | 38.0 | 26.9 | 46.6 |
| Days Max Temp ≥ 90 °F | 0 | 0 | 0 | 0 | 0 | 2 | 4 | 2 | 0 | 0 | 0 | 0 | 8 |
| Days Max Temp ≤ 32 °F | 18 | 14 | 5 | 0 | 0 | 0 | 0 | 0 | 0 | 0 | 2 | 12 | 51 |
| Days Min Temp ≤ 32 °F | 30 | 27 | 25 | 13 | 3 | 0 | 0 | 0 | 1 | 10 | 20 | 28 | 157 |
| Days Min Temp ≤ 0 °F | 6 | 5 | 1 | 0 | 0 | 0 | 0 | 0 | 0 | 0 | 0 | 2 | 14 |
| Heating Degree Days | 1352 | 1179 | 971 | 577 | 274 | 70 | 20 | 41 | 175 | 487 | 803 | 1172 | 7121 |
| Cooling Degree Days | 0 | 0 | 1 | 6 | 37 | 99 | 186 | 129 | 45 | 2 | 0 | 0 | 505 |
| Total Precipitation (") | 1.58 | 1.33 | 2.43 | 3.22 | 3.02 | 3.84 | 3.29 | 3.53 | 3.88 | 2.84 | 3.07 | 2.47 | 34.50 |
| Days ≥ 0.1" Precip | 4 | 4 | 6 | 7 | 6 | 7 | 7 | 7 | 7 | 7 | 7 | 6 | 75 |
| Total Snowfall (") | 13.2 | 10.3 | 7.7 | 2.4 | 0.0 | 0.0 | 0.0 | 0.0 | 0.0 | 0.3 | 3.8 | *12.3* | 50.0 |
| Days ≥ 1" Snow Depth | 22 | 19 | 9 | 1 | 0 | 0 | 0 | 0 | 0 | 0 | 3 | 16 | 70 |

## CHEBOYGAN *Cheboygan County* ELEVATION 610 ft LAT/LONG 45° 38 ' N / 84° 30 ' W

| | JAN | FEB | MAR | APR | MAY | JUN | JUL | AUG | SEP | OCT | NOV | DEC | YEAR |
|---|---|---|---|---|---|---|---|---|---|---|---|---|---|
| Maximum Temp °F | 26.4 | 28.4 | 37.1 | 49.6 | 62.7 | 72.0 | 78.0 | 76.2 | 67.7 | 56.2 | 42.6 | 31.6 | 52.4 |
| Minimum Temp °F | 9.3 | 8.6 | 17.7 | 30.3 | 40.7 | 50.3 | 57.0 | 56.0 | 48.7 | 38.5 | 29.0 | 17.8 | 33.7 |
| Mean Temp °F | 17.9 | 18.6 | 27.6 | 40.0 | 51.7 | 61.2 | 67.5 | 66.2 | 58.2 | 47.4 | 35.8 | 24.7 | 43.1 |
| Days Max Temp ≥ 90 °F | 0 | 0 | 0 | 0 | 0 | 1 | 1 | 1 | 0 | 0 | 0 | 0 | 3 |
| Days Max Temp ≤ 32 °F | 23 | 18 | 9 | 1 | 0 | 0 | 0 | 0 | 0 | 0 | 3 | 16 | 70 |
| Days Min Temp ≤ 32 °F | 30 | 28 | 29 | 18 | 4 | 0 | 0 | 0 | 0 | 7 | 21 | 29 | 166 |
| Days Min Temp ≤ 0 °F | 7 | 8 | 2 | 0 | 0 | 0 | 0 | 0 | 0 | 0 | 0 | 2 | 19 |
| Heating Degree Days | 1455 | 1306 | 1154 | 743 | 409 | 147 | 37 | 57 | 225 | 542 | 868 | 1242 | 8185 |
| Cooling Degree Days | 0 | 0 | *0* | 0 | 4 | 38 | 122 | 100 | 25 | 0 | 0 | 0 | 289 |
| Total Precipitation (") | 1.50 | 1.17 | 1.80 | 2.38 | 2.52 | 2.53 | 2.89 | 3.00 | 3.66 | 2.75 | 2.50 | 2.07 | 28.77 |
| Days ≥ 0.1" Precip | 5 | 3 | *4* | 6 | 6 | 5 | 5 | 6 | 7 | 7 | 6 | 5 | 65 |
| Total Snowfall (") | 23.4 | 15.9 | 10.4 | 3.5 | 0.1 | 0.0 | 0.0 | 0.0 | 0.0 | 0.1 | 8.5 | 21.0 | 82.9 |
| Days ≥ 1" Snow Depth | 29 | 27 | 21 | 4 | 0 | 0 | 0 | 0 | 0 | 0 | *7* | 22 | 110 |

## COLDWATER ST SCHOOL *Branch County* ELEVATION 981 ft LAT/LONG 41° 57 ' N / 85° 0 ' W

| | JAN | FEB | MAR | APR | MAY | JUN | JUL | AUG | SEP | OCT | NOV | DEC | YEAR |
|---|---|---|---|---|---|---|---|---|---|---|---|---|---|
| Maximum Temp °F | 30.1 | 33.5 | 44.4 | 57.5 | 69.5 | 78.6 | 82.3 | 80.3 | 72.8 | 60.5 | 46.9 | 35.1 | 57.6 |
| Minimum Temp °F | 14.0 | 15.8 | 25.3 | 36.0 | 46.4 | 55.6 | 59.7 | 57.4 | 50.3 | 39.5 | 30.7 | 20.3 | 37.6 |
| Mean Temp °F | 22.1 | 24.7 | 34.9 | 46.7 | 58.0 | 67.1 | 71.1 | 68.9 | 61.6 | 50.0 | 38.9 | 27.7 | 47.6 |
| Days Max Temp ≥ 90 °F | 0 | 0 | 0 | 0 | 0 | 2 | 4 | 2 | 0 | 0 | 0 | 0 | 8 |
| Days Max Temp ≤ 32 °F | 18 | 13 | 5 | 0 | 0 | 0 | 0 | 0 | 0 | 0 | 2 | 12 | 50 |
| Days Min Temp ≤ 32 °F | 30 | 26 | 24 | 12 | 2 | 0 | 0 | 0 | 0 | 7 | 19 | 28 | 148 |
| Days Min Temp ≤ 0 °F | 5 | 4 | 0 | 0 | 0 | 0 | 0 | 0 | 0 | 0 | 0 | 2 | 11 |
| Heating Degree Days | 1325 | 1133 | 928 | 547 | 250 | 56 | 12 | 30 | 156 | 463 | 778 | 1149 | 6827 |
| Cooling Degree Days | 0 | 0 | 1 | 7 | 43 | 127 | 223 | 167 | 60 | 4 | 0 | 0 | 632 |
| Total Precipitation (") | 1.60 | 1.55 | 2.35 | 3.26 | 3.68 | 3.66 | 3.82 | 3.61 | 3.59 | 2.81 | 2.79 | 2.68 | 35.40 |
| Days ≥ 0.1" Precip | 4 | 4 | 7 | 8 | 7 | 7 | 7 | 6 | 7 | 6 | 7 | 7 | 77 |
| Total Snowfall (") | 13.9 | 11.1 | 8.3 | 2.2 | 0.0 | 0.0 | 0.0 | 0.0 | 0.1 | 0.6 | 4.9 | 13.1 | 54.2 |
| Days ≥ 1" Snow Depth | 21 | 17 | 7 | 1 | 0 | 0 | 0 | 0 | 0 | 0 | 3 | 15 | 64 |

## CROSS VILLAGE *Emmet County* ELEVATION 669 ft LAT/LONG 45° 39 ' N / 85° 2 ' W

| | JAN | FEB | MAR | APR | MAY | JUN | JUL | AUG | SEP | OCT | NOV | DEC | YEAR |
|---|---|---|---|---|---|---|---|---|---|---|---|---|---|
| Maximum Temp °F | 26.9 | 29.3 | 38.7 | 52.1 | 65.2 | 71.9 | 76.8 | 75.9 | 68.3 | 56.8 | 43.3 | 31.9 | 53.1 |
| Minimum Temp °F | 12.7 | 11.0 | 19.5 | 31.2 | 41.1 | 50.2 | 57.2 | 57.3 | 50.6 | 41.0 | 31.2 | 20.7 | 35.3 |
| Mean Temp °F | 19.8 | 20.1 | 29.1 | 41.7 | 53.1 | 61.1 | 67.0 | 66.6 | 59.5 | 48.9 | 37.3 | 26.3 | 44.2 |
| Days Max Temp ≥ 90 °F | 0 | 0 | 0 | 0 | 0 | 0 | 0 | 0 | 0 | 0 | 0 | 0 | 0 |
| Days Max Temp ≤ 32 °F | 22 | 18 | 7 | 1 | 0 | 0 | 0 | 0 | 0 | 0 | 3 | 15 | 66 |
| Days Min Temp ≤ 32 °F | 30 | 27 | 27 | 17 | 6 | 0 | 0 | 0 | 0 | 4 | 17 | 28 | 156 |
| Days Min Temp ≤ 0 °F | 5 | 6 | 2 | 0 | 0 | 0 | 0 | 0 | 0 | 0 | 0 | 1 | 14 |
| Heating Degree Days | 1394 | 1260 | 1105 | 696 | 371 | 150 | 40 | 45 | 191 | 493 | 826 | 1192 | 7763 |
| Cooling Degree Days | 0 | 0 | 0 | 4 | 9 | 41 | 116 | 107 | 33 | 1 | 0 | 0 | 311 |
| Total Precipitation (") | 1.70 | 1.19 | 1.93 | 2.49 | 2.42 | 2.71 | 2.12 | 3.30 | 3.56 | 2.86 | 2.64 | 1.99 | 28.91 |
| Days ≥ 0.1" Precip | 6 | 4 | 5 | 6 | 6 | 6 | 5 | 7 | 7 | 7 | 7 | 6 | 72 |
| Total Snowfall (") | 23.3 | 15.2 | 10.6 | 4.6 | 0.3 | 0.0 | 0.0 | 0.0 | 0.0 | 0.4 | 7.5 | 17.3 | 79.2 |
| Days ≥ 1" Snow Depth | 29 | 28 | 23 | 5 | 0 | 0 | 0 | 0 | 0 | 0 | 6 | 21 | 112 |

## DEARBORN *Wayne County* ELEVATION 600 ft LAT/LONG 42° 18 ' N / 83° 14 ' W

| | JAN | FEB | MAR | APR | MAY | JUN | JUL | AUG | SEP | OCT | NOV | DEC | YEAR |
|---|---|---|---|---|---|---|---|---|---|---|---|---|---|
| Maximum Temp °F | 31.3 | 34.3 | 45.0 | 58.7 | 71.2 | 80.1 | 84.2 | 82.5 | 75.0 | 62.2 | 48.9 | 36.7 | 59.2 |
| Minimum Temp °F | 16.3 | 18.0 | 26.9 | 37.5 | 47.6 | 56.9 | 61.8 | 60.3 | 52.8 | 41.3 | 33.0 | 23.0 | 39.6 |
| Mean Temp °F | 23.8 | 26.2 | 36.0 | 48.2 | 59.4 | 68.5 | 73.0 | 71.4 | 63.9 | 51.8 | 41.0 | 29.9 | 49.4 |
| Days Max Temp ≥ 90 °F | 0 | 0 | 0 | 0 | 1 | 3 | 6 | 4 | 1 | 0 | 0 | 0 | 15 |
| Days Max Temp ≤ 32 °F | 16 | 12 | 4 | 0 | 0 | 0 | 0 | 0 | 0 | 0 | 1 | 10 | 43 |
| Days Min Temp ≤ 32 °F | 29 | 25 | 23 | 10 | 1 | 0 | 0 | 0 | 0 | 5 | 16 | 26 | 135 |
| Days Min Temp ≤ 0 °F | 3 | 2 | 0 | 0 | 0 | 0 | 0 | 0 | 0 | 0 | 0 | 1 | 6 |
| Heating Degree Days | 1270 | 1088 | 893 | 507 | 214 | 45 | 6 | 14 | 114 | 411 | 714 | 1082 | 6358 |
| Cooling Degree Days | 0 | 0 | 1 | 10 | 49 | 140 | 259 | 209 | 77 | 4 | 0 | 0 | 749 |
| Total Precipitation (") | 1.85 | 1.68 | 2.59 | 3.12 | 2.83 | 3.71 | 3.25 | 2.90 | 3.20 | 2.47 | 2.87 | 2.76 | 33.23 |
| Days ≥ 0.1" Precip | 5 | 4 | 6 | 7 | 6 | 7 | 6 | 6 | 6 | 6 | 7 | 7 | 73 |
| Total Snowfall (") | 10.0 | 7.7 | 4.8 | 0.8 | 0.0 | 0.0 | 0.0 | 0.0 | 0.0 | 0.1 | 1.9 | 7.8 | 33.1 |
| Days ≥ 1" Snow Depth | *20* | 17 | 6 | 0 | 0 | 0 | 0 | 0 | 0 | 0 | 1 | 11 | 55 |

## DETOUR VILLAGE *Chippewa County* ELEVATION 591 ft LAT/LONG 46° 1 ' N / 83° 55 ' W

| | JAN | FEB | MAR | APR | MAY | JUN | JUL | AUG | SEP | OCT | NOV | DEC | YEAR |
|---|---|---|---|---|---|---|---|---|---|---|---|---|---|
| Maximum Temp °F | 24.3 | 26.2 | 35.4 | 47.7 | 61.0 | 70.2 | 76.4 | 74.8 | 65.7 | 53.9 | 41.7 | 30.8 | 50.7 |
| Minimum Temp °F | 7.4 | 7.1 | 16.7 | 29.6 | 40.0 | 49.2 | 55.9 | 56.1 | 48.8 | 38.8 | 29.3 | 17.0 | 33.0 |
| Mean Temp °F | 15.9 | 16.7 | 26.0 | 38.7 | 50.5 | 59.7 | 66.2 | 65.5 | 57.2 | 46.4 | 35.5 | 23.9 | 41.8 |
| Days Max Temp ≥ 90 °F | 0 | 0 | 0 | 0 | 0 | 0 | 1 | 0 | 0 | 0 | 0 | 0 | 1 |
| Days Max Temp ≤ 32 °F | 23 | 20 | 10 | 1 | 0 | 0 | 0 | 0 | 0 | 0 | 4 | 16 | 74 |
| Days Min Temp ≤ 32 °F | 31 | 28 | 29 | 19 | 5 | 0 | 0 | 0 | 0 | 6 | 19 | 29 | 166 |
| Days Min Temp ≤ 0 °F | 10 | 10 | 3 | 0 | 0 | 0 | 0 | 0 | 0 | 0 | 0 | 3 | 26 |
| Heating Degree Days | 1517 | 1360 | 1202 | 783 | 446 | 174 | 47 | 59 | 241 | 572 | 878 | 1267 | 8546 |
| Cooling Degree Days | 0 | 0 | 0 | 0 | 3 | 26 | 102 | 95 | 16 | 0 | 0 | 0 | 242 |
| Total Precipitation (") | 1.72 | 1.20 | 2.11 | 2.36 | 2.48 | 2.64 | 3.02 | 2.89 | 3.88 | 2.68 | 2.68 | 1.97 | 29.63 |
| Days ≥ 0.1" Precip | 7 | 4 | 6 | 6 | 6 | 7 | 6 | 7 | 8 | 7 | 7 | 7 | 78 |
| Total Snowfall (") | 18.8 | 13.4 | 11.3 | 3.5 | 0.0 | 0.0 | 0.0 | 0.0 | 0.0 | 0.1 | 5.8 | 14.2 | 67.1 |
| Days ≥ 1" Snow Depth | 29 | 27 | 25 | 8 | 0 | 0 | 0 | 0 | 0 | 0 | 5 | 19 | 113 |

## DETROIT CITY AIRPORT *Wayne County* ELEVATION 619 ft LAT/LONG 42° 24 ' N / 83° 0 ' W

| | JAN | FEB | MAR | APR | MAY | JUN | JUL | AUG | SEP | OCT | NOV | DEC | YEAR |
|---|---|---|---|---|---|---|---|---|---|---|---|---|---|
| Maximum Temp °F | 30.7 | 33.4 | 44.0 | 57.2 | 69.7 | 78.8 | 83.1 | 81.0 | 73.4 | 60.7 | 48.0 | 36.3 | 58.0 |
| Minimum Temp °F | 18.9 | 20.6 | 28.9 | 39.4 | 50.0 | 59.7 | 65.2 | 63.7 | 56.3 | 45.1 | 35.5 | 25.4 | 42.4 |
| Mean Temp °F | 24.8 | 27.0 | 36.5 | 48.3 | 59.9 | 69.3 | 74.2 | 72.4 | 64.9 | 52.9 | 41.8 | 30.9 | 50.2 |
| Days Max Temp ≥ 90 °F | 0 | 0 | 0 | 0 | 1 | 3 | 5 | 3 | 1 | 0 | 0 | 0 | 13 |
| Days Max Temp ≤ 32 °F | 16 | 14 | 5 | 0 | 0 | 0 | 0 | 0 | 0 | 0 | 1 | 10 | 46 |
| Days Min Temp ≤ 32 °F | 27 | 24 | 20 | 6 | 0 | 0 | 0 | 0 | 0 | 2 | 12 | 23 | 114 |
| Days Min Temp ≤ 0 °F | 1 | 1 | 0 | 0 | 0 | 0 | 0 | 0 | 0 | 0 | 0 | 0 | 2 |
| Heating Degree Days | 1238 | 1066 | 878 | 502 | 208 | 38 | 4 | 8 | 95 | 378 | 690 | 1051 | 6156 |
| Cooling Degree Days | 0 | 0 | 0 | 10 | 63 | 165 | 312 | 249 | 96 | 7 | 0 | 0 | 902 |
| Total Precipitation (") | na | na | na | na | na | na | na | na | na | na | na | na | na |
| Days ≥ 0.1" Precip | na | na | na | na | na | na | na | na | na | na | na | na | na |
| Total Snowfall (") | na | na | na | na | na | na | na | na | na | na | na | na | na |
| Days ≥ 1" Snow Depth | na | na | na | na | na | na | na | na | na | na | na | na | na |

## DETROIT METRO AP *Wayne County* ELEVATION 633 ft LAT/LONG 42° 14 ' N / 83° 20 ' W

| | JAN | FEB | MAR | APR | MAY | JUN | JUL | AUG | SEP | OCT | NOV | DEC | YEAR |
|---|---|---|---|---|---|---|---|---|---|---|---|---|---|
| Maximum Temp °F | 30.6 | 33.6 | 44.6 | 58.0 | 69.9 | 78.9 | 83.3 | 81.4 | 73.7 | 61.1 | 47.9 | 36.0 | 58.3 |
| Minimum Temp °F | 16.1 | 18.1 | 27.1 | 37.4 | 47.4 | 56.8 | 61.9 | 60.3 | 52.8 | 41.2 | 32.6 | 22.6 | 39.5 |
| Mean Temp °F | 23.4 | 25.9 | 35.9 | 47.7 | 58.7 | 67.9 | 72.6 | 70.9 | 63.3 | 51.2 | 40.3 | 29.3 | 48.9 |
| Days Max Temp ≥ 90 °F | 0 | 0 | 0 | 0 | 0 | 3 | 5 | 3 | 1 | 0 | 0 | 0 | 12 |
| Days Max Temp ≤ 32 °F | 17 | 13 | 4 | 0 | 0 | 0 | 0 | 0 | 0 | 0 | 1 | 10 | 45 |
| Days Min Temp ≤ 32 °F | 29 | 25 | 22 | 9 | 1 | 0 | 0 | 0 | 0 | 5 | 16 | 26 | 133 |
| Days Min Temp ≤ 0 °F | 4 | 2 | 0 | 0 | 0 | 0 | 0 | 0 | 0 | 0 | 0 | 1 | 7 |
| Heating Degree Days | 1283 | 1098 | 896 | 518 | 229 | 46 | 6 | 13 | 120 | 429 | 735 | 1099 | 6472 |
| Cooling Degree Days | 0 | 0 | 0 | 7 | 46 | 141 | 271 | 209 | 76 | 4 | 0 | 0 | 754 |
| Total Precipitation (") | 1.91 | 1.72 | 2.55 | 2.94 | 2.93 | 3.67 | 3.20 | 3.17 | 2.93 | 2.27 | 2.84 | 2.89 | 33.02 |
| Days ≥ 0.1" Precip | 5 | 4 | 7 | 7 | 6 | 6 | 6 | 6 | 6 | 5 | 7 | 7 | 72 |
| Total Snowfall (") | 11.1 | 9.4 | 7.1 | 1.8 | 0.0 | 0.0 | 0.0 | 0.0 | 0.0 | 0.2 | 3.0 | 10.5 | 43.1 |
| Days ≥ 1" Snow Depth | 18 | 14 | 6 | 1 | 0 | 0 | 0 | 0 | 0 | 0 | 2 | 10 | 51 |

## DOWAGIAC 1 W *Cass County* ELEVATION 741 ft LAT/LONG 41° 59 ' N / 86° 8 ' W

| | JAN | FEB | MAR | APR | MAY | JUN | JUL | AUG | SEP | OCT | NOV | DEC | YEAR |
|---|---|---|---|---|---|---|---|---|---|---|---|---|---|
| Maximum Temp °F | 31.3 | 34.9 | 45.7 | 59.0 | 70.6 | 79.7 | 83.6 | 81.3 | 74.3 | 62.2 | 48.5 | 36.5 | 59.0 |
| Minimum Temp °F | 14.6 | 16.8 | 25.9 | 36.6 | 46.0 | 55.1 | 59.5 | 57.3 | 50.3 | 39.8 | 31.1 | 21.5 | 37.9 |
| Mean Temp °F | 23.0 | 26.1 | 35.8 | 47.8 | 58.3 | 67.5 | 71.6 | 69.4 | 62.3 | 51.0 | 39.8 | 29.0 | 48.5 |
| Days Max Temp ≥ 90 °F | 0 | 0 | 0 | 0 | 1 | 3 | 6 | 3 | 1 | 0 | 0 | 0 | 14 |
| Days Max Temp ≤ 32 °F | 17 | 12 | 4 | 0 | 0 | 0 | 0 | 0 | 0 | 0 | 1 | 10 | 44 |
| Days Min Temp ≤ 32 °F | 30 | 26 | 24 | 11 | 3 | 0 | 0 | 0 | 1 | 7 | 18 | 28 | 148 |
| Days Min Temp ≤ 0 °F | 4 | 3 | 0 | 0 | 0 | 0 | 0 | 0 | 0 | 0 | 0 | 1 | 8 |
| Heating Degree Days | 1295 | 1093 | 897 | 517 | 246 | 57 | 14 | 28 | 144 | 433 | 750 | 1109 | 6583 |
| Cooling Degree Days | 0 | 0 | 1 | 10 | 46 | 135 | 230 | 169 | 64 | 4 | 0 | 0 | 659 |
| Total Precipitation (") | 2.41 | 1.94 | 2.63 | 3.69 | 3.47 | 3.70 | 3.86 | 3.77 | 4.27 | 3.64 | 3.70 | 3.30 | 40.38 |
| Days ≥ 0.1" Precip | 7 | 6 | 7 | 8 | 7 | 7 | 7 | 7 | 7 | 7 | 8 | 9 | 87 |
| Total Snowfall (") | 20.0 | 14.2 | *7.5* | 1.5 | 0.0 | 0.0 | 0.0 | 0.0 | 0.0 | 0.0 | 6.6 | 17.0 | 66.8 |
| Days ≥ 1" Snow Depth | 23 | 18 | 7 | 1 | 0 | 0 | 0 | 0 | 0 | 0 | 4 | *15* | 68 |

## EAST JORDAN *Charlevoix County* ELEVATION 610 ft LAT/LONG 45° 10 ' N / 85° 7 ' W

| | JAN | FEB | MAR | APR | MAY | JUN | JUL | AUG | SEP | OCT | NOV | DEC | YEAR |
|---|---|---|---|---|---|---|---|---|---|---|---|---|---|
| Maximum Temp °F | 28.3 | 30.8 | 40.7 | 54.6 | 68.3 | 76.6 | 80.8 | 78.6 | 70.8 | 59.1 | 44.5 | 33.1 | 55.5 |
| Minimum Temp °F | 12.3 | 9.8 | 18.7 | 30.5 | 40.3 | 48.9 | 54.7 | 53.1 | 46.8 | 38.0 | 29.3 | 19.8 | 33.5 |
| Mean Temp °F | 20.3 | 20.3 | 29.7 | 42.6 | 54.4 | 62.8 | 67.8 | 65.9 | 58.8 | 48.6 | 36.9 | 26.4 | 44.5 |
| Days Max Temp ≥ 90 °F | 0 | 0 | 0 | 0 | 0 | 1 | 3 | 2 | 0 | 0 | 0 | 0 | 6 |
| Days Max Temp ≤ 32 °F | 21 | 16 | 7 | 1 | 0 | 0 | 0 | 0 | 0 | 0 | 3 | 14 | 62 |
| Days Min Temp ≤ 32 °F | 30 | 27 | 27 | 18 | 7 | 1 | 0 | 0 | 2 | 9 | 20 | 29 | 170 |
| Days Min Temp ≤ 0 °F | 5 | 7 | 3 | 0 | 0 | 0 | 0 | 0 | 0 | 0 | 0 | 1 | 16 |
| Heating Degree Days | 1380 | 1255 | 1087 | 670 | 344 | 121 | 40 | 66 | 210 | 507 | 835 | 1189 | 7704 |
| Cooling Degree Days | 0 | 0 | 0 | 4 | 23 | 56 | 143 | 109 | 33 | 2 | 0 | 0 | 370 |
| Total Precipitation (") | 1.94 | 1.19 | 1.55 | 2.53 | 2.43 | 2.80 | 2.86 | 3.22 | 4.27 | 3.70 | 3.04 | 2.32 | 31.85 |
| Days ≥ 0.1" Precip | 7 | 4 | 5 | 7 | 6 | 6 | 5 | 7 | 9 | 8 | 9 | 8 | 81 |
| Total Snowfall (") | 33.4 | 17.8 | 9.1 | 2.5 | 0.2 | 0.0 | 0.0 | 0.0 | 0.0 | 0.4 | 10.3 | 27.4 | 101.1 |
| Days ≥ 1" Snow Depth | 29 | 27 | 21 | 3 | 0 | 0 | 0 | 0 | 0 | 0 | 7 | 23 | 110 |

## EAST LANSING 4 S *Ingham County* ELEVATION 869 ft LAT/LONG 42° 43 ' N / 84° 28 ' W

| | JAN | FEB | MAR | APR | MAY | JUN | JUL | AUG | SEP | OCT | NOV | DEC | YEAR |
|---|---|---|---|---|---|---|---|---|---|---|---|---|---|
| Maximum Temp °F | 28.9 | 31.9 | 42.9 | 56.7 | 69.0 | 78.0 | 82.1 | 80.2 | 72.5 | 60.0 | 46.3 | 34.1 | 56.9 |
| Minimum Temp °F | 13.8 | 15.1 | 24.6 | 35.7 | 45.7 | 54.9 | 59.2 | 57.2 | 50.0 | 39.2 | 30.7 | 20.3 | 37.2 |
| Mean Temp °F | 21.4 | 23.5 | 33.8 | 46.2 | 57.4 | 66.5 | 70.7 | 68.8 | 61.3 | 49.7 | 38.5 | 27.3 | 47.1 |
| Days Max Temp ≥ 90 °F | 0 | 0 | 0 | 0 | 0 | 2 | 3 | 2 | 1 | 0 | 0 | 0 | 8 |
| Days Max Temp ≤ 32 °F | 19 | 15 | 6 | 0 | 0 | 0 | 0 | 0 | 0 | 0 | 2 | 14 | 56 |
| Days Min Temp ≤ 32 °F | 30 | 27 | 25 | 12 | 2 | 0 | 0 | 0 | 1 | 7 | 19 | 28 | 151 |
| Days Min Temp ≤ 0 °F | 4 | 4 | 0 | 0 | 0 | 0 | 0 | 0 | 0 | 0 | 0 | 1 | 9 |
| Heating Degree Days | 1345 | 1164 | 961 | 562 | 266 | 65 | 15 | 33 | 161 | 474 | 788 | 1164 | 6998 |
| Cooling Degree Days | 0 | 0 | 1 | 6 | 38 | 105 | 203 | 154 | 50 | 3 | 0 | 0 | 560 |
| Total Precipitation (") | 1.40 | 1.24 | 2.15 | 3.08 | 2.52 | 3.50 | 3.03 | 3.20 | 3.42 | 2.47 | 2.78 | 2.12 | 30.91 |
| Days ≥ 0.1" Precip | 4 | 4 | 6 | 8 | 5 | 6 | 6 | 6 | 7 | 6 | 6 | 5 | 69 |
| Total Snowfall (") | na | na | *5.1* | 1.6 | 0.0 | 0.0 | 0.0 | 0.0 | 0.0 | 0.1 | *2.0* | na | na |
| Days ≥ 1" Snow Depth | 22 | 18 | 9 | 2 | 0 | 0 | 0 | 0 | 0 | 0 | 2 | 15 | 68 |

## EAST TAWAS *Iosco County* ELEVATION 591 ft LAT/LONG 44° 17 ' N / 83° 29 ' W

| | JAN | FEB | MAR | APR | MAY | JUN | JUL | AUG | SEP | OCT | NOV | DEC | YEAR |
|---|---|---|---|---|---|---|---|---|---|---|---|---|---|
| Maximum Temp °F | 28.6 | 30.8 | 39.7 | 52.1 | 65.3 | 74.3 | 79.8 | 77.6 | 69.9 | 57.9 | 44.7 | 33.9 | 54.6 |
| Minimum Temp °F | 11.1 | 11.4 | 20.9 | 32.3 | 42.0 | 51.0 | 56.6 | 55.2 | 48.2 | 37.9 | 29.2 | 19.3 | 34.6 |
| Mean Temp °F | 19.9 | 21.1 | 30.3 | 42.3 | 53.7 | 62.7 | 68.2 | 66.4 | 59.2 | 48.0 | 37.0 | 26.6 | 44.6 |
| Days Max Temp ≥ 90 °F | 0 | 0 | 0 | 0 | 0 | 1 | 2 | 1 | 0 | 0 | 0 | 0 | 4 |
| Days Max Temp ≤ 32 °F | 19 | 15 | 6 | 0 | 0 | 0 | 0 | 0 | 0 | 0 | 2 | 13 | 55 |
| Days Min Temp ≤ 32 °F | 30 | 28 | 27 | *16* | 4 | 0 | 0 | 0 | 1 | 9 | 20 | 29 | 164 |
| Days Min Temp ≤ 0 °F | 7 | 6 | 2 | 0 | 0 | 0 | 0 | 0 | 0 | 0 | 0 | 2 | 17 |
| Heating Degree Days | 1392 | 1231 | 1068 | 676 | 359 | 121 | 33 | 55 | 198 | 520 | 833 | 1183 | 7669 |
| Cooling Degree Days | 0 | 0 | 0 | 1 | 16 | 60 | 149 | 109 | 30 | 0 | 0 | 0 | 365 |
| Total Precipitation (") | 1.91 | 1.23 | 2.14 | 2.76 | 2.82 | 3.18 | 2.64 | 3.36 | 3.43 | 2.67 | 2.67 | 2.20 | 31.01 |
| Days ≥ 0.1" Precip | 5 | *4* | 5 | *7* | 7 | 6 | 6 | 7 | 7 | 6 | 6 | 6 | 72 |
| Total Snowfall (") | na | *12.3* | *9.5* | *2.2* | 0.1 | 0.0 | 0.0 | 0.0 | 0.0 | 0.0 | *2.8* | *11.3* | na |
| Days ≥ 1" Snow Depth | *29* | *26* | *18* | *2* | 0 | 0 | 0 | 0 | 0 | 0 | *2* | 14 | 91 |

## EAU CLAIRE 4 NE *Berrien County* ELEVATION 869 ft LAT/LONG 42° 1 ' N / 86° 15 ' W

| | JAN | FEB | MAR | APR | MAY | JUN | JUL | AUG | SEP | OCT | NOV | DEC | YEAR |
|---|---|---|---|---|---|---|---|---|---|---|---|---|---|
| Maximum Temp °F | 30.4 | 34.0 | 45.6 | 59.0 | 70.6 | 79.7 | 83.4 | 81.2 | 73.7 | 61.5 | 47.9 | 35.7 | 58.6 |
| Minimum Temp °F | 16.7 | 19.0 | 28.2 | 38.4 | 48.4 | 58.0 | 62.6 | 61.0 | 54.4 | 43.5 | 33.5 | 23.1 | 40.6 |
| Mean Temp °F | 23.5 | 26.5 | 36.9 | 48.7 | 59.5 | 68.9 | 73.0 | 71.1 | 64.1 | 52.5 | 40.7 | 29.4 | 49.6 |
| Days Max Temp ≥ 90 °F | 0 | 0 | 0 | 0 | 0 | 3 | 5 | 3 | 1 | 0 | 0 | 0 | 12 |
| Days Max Temp ≤ 32 °F | 18 | 13 | 4 | 0 | 0 | 0 | 0 | 0 | 0 | 0 | 2 | 11 | 48 |
| Days Min Temp ≤ 32 °F | 29 | 25 | 22 | 8 | 1 | 0 | 0 | 0 | 0 | 3 | 15 | 27 | 130 |
| Days Min Temp ≤ 0 °F | 3 | 1 | 0 | 0 | 0 | 0 | 0 | 0 | 0 | 0 | 0 | 1 | 5 |
| Heating Degree Days | 1278 | 1081 | 865 | 491 | 218 | 42 | 5 | 13 | 106 | 388 | 722 | 1096 | 6305 |
| Cooling Degree Days | 0 | 0 | 1 | 12 | 61 | 170 | 277 | 222 | 89 | 6 | 0 | 0 | 838 |
| Total Precipitation (") | 1.95 | 1.49 | 2.53 | 3.53 | 3.29 | 3.50 | 3.42 | 3.51 | 3.95 | 3.19 | 3.45 | 2.86 | 36.67 |
| Days ≥ 0.1" Precip | 6 | 5 | 7 | 8 | 7 | 7 | 6 | 6 | 7 | 7 | 8 | 8 | 82 |
| Total Snowfall (") | 23.1 | 15.5 | 8.5 | 2.0 | 0.0 | 0.0 | 0.0 | 0.0 | 0.0 | 0.3 | 7.3 | 19.6 | 76.3 |
| Days ≥ 1" Snow Depth | 23 | 20 | 9 | 1 | 0 | 0 | 0 | 0 | 0 | 0 | 4 | 17 | 74 |

## ESCANABA *Delta County* ELEVATION 617 ft LAT/LONG 45° 45 ' N / 87° 3 ' W

| | JAN | FEB | MAR | APR | MAY | JUN | JUL | AUG | SEP | OCT | NOV | DEC | YEAR |
|---|---|---|---|---|---|---|---|---|---|---|---|---|---|
| Maximum Temp °F | 24.5 | 26.9 | 35.2 | 45.8 | 58.3 | 68.2 | 74.6 | 73.1 | 64.8 | 53.4 | ***40.5*** | ***29.4*** | 49.6 |
| Minimum Temp °F | 7.7 | 8.6 | 18.3 | 31.1 | 42.5 | 52.2 | 58.2 | 57.0 | 49.3 | 39.5 | ***27.8*** | ***16.0*** | 34.0 |
| Mean Temp °F | 16.2 | 17.7 | 26.8 | 38.5 | 50.4 | 60.2 | 66.4 | 65.1 | 57.1 | 46.5 | ***34.2*** | ***22.7*** | 41.8 |
| Days Max Temp ≥ 90 °F | 0 | 0 | 0 | 0 | 0 | 0 | 0 | 0 | 0 | 0 | ***0*** | ***0*** | 0 |
| Days Max Temp ≤ 32 °F | 24 | 19 | 10 | 1 | 0 | 0 | 0 | 0 | 0 | 0 | ***4*** | ***18*** | 76 |
| Days Min Temp ≤ 32 °F | 31 | 28 | 29 | 17 | 3 | 0 | 0 | 0 | 1 | 6 | ***21*** | ***30*** | 166 |
| Days Min Temp ≤ 0 °F | 9 | 8 | 2 | 0 | 0 | 0 | 0 | 0 | 0 | 0 | ***0*** | 4 | 23 |
| Heating Degree Days | 1506 | 1328 | 1177 | 788 | 445 | 160 | 44 | 64 | 243 | 564 | ***918*** | ***1304*** | 8541 |
| Cooling Degree Days | 0 | 0 | 0 | 0 | 2 | 24 | 96 | 80 | 10 | 0 | ***0*** | 0 | 212 |
| Total Precipitation (") | 1.37 | 0.97 | 1.80 | 2.20 | 2.84 | 3.12 | 3.01 | 3.21 | 3.39 | 2.54 | 2.48 | 1.69 | 28.62 |
| Days ≥ 0.1" Precip | 4 | 3 | 5 | 6 | 6 | 7 | 6 | 6 | 7 | 6 | ***6*** | ***5*** | 67 |
| Total Snowfall (") | ***14.1*** | ***8.6*** | ***9.0*** | 2.1 | 0.1 | 0.0 | 0.0 | 0.0 | 0.0 | ***0.0*** | na | ***13.8*** | na |
| Days ≥ 1" Snow Depth | na | na | na | na | 0 | 0 | 0 | 0 | 0 | ***0*** | na | na | na |

## EVART *Osceola County* ELEVATION 1030 ft LAT/LONG 43° 54 ' N / 85° 16 ' W

| | JAN | FEB | MAR | APR | MAY | JUN | JUL | AUG | SEP | OCT | NOV | DEC | YEAR |
|---|---|---|---|---|---|---|---|---|---|---|---|---|---|
| Maximum Temp °F | 26.9 | 30.1 | ***40.3*** | ***55.0*** | ***68.3*** | 76.7 | 81.7 | 79.1 | 70.2 | 57.9 | 43.6 | ***32.1*** | 55.2 |
| Minimum Temp °F | 8.7 | 8.8 | 19.1 | 31.3 | 41.0 | 49.8 | 54.7 | 53.0 | 45.4 | 35.0 | 27.0 | ***16.8*** | 32.6 |
| Mean Temp °F | 17.8 | 19.5 | ***29.7*** | ***43.1*** | ***54.7*** | 63.2 | 68.2 | 66.1 | 57.8 | 46.5 | 35.3 | ***24.5*** | 43.9 |
| Days Max Temp ≥ 90 °F | 0 | 0 | 0 | ***0*** | ***0*** | 1 | 4 | 1 | 0 | 0 | 0 | 0 | 6 |
| Days Max Temp ≤ 32 °F | 22 | 17 | 7 | ***1*** | ***0*** | 0 | 0 | 0 | 0 | 0 | 4 | 16 | 67 |
| Days Min Temp ≤ 32 °F | 31 | 28 | 27 | 17 | 7 | 0 | 0 | 0 | 3 | 14 | 22 | 29 | 178 |
| Days Min Temp ≤ 0 °F | 8 | 8 | 3 | 0 | 0 | 0 | 0 | 0 | 0 | 0 | 0 | 3 | 22 |
| Heating Degree Days | 1457 | 1281 | ***1088*** | ***650*** | ***332*** | 114 | 37 | 63 | 236 | 569 | 883 | ***1250*** | 7960 |
| Cooling Degree Days | ***0*** | ***0*** | na | na | na | ***61*** | ***169*** | ***118*** | ***23*** | ***1*** | ***0*** | na | na |
| Total Precipitation (") | 1.85 | 1.41 | 2.21 | 2.88 | 2.84 | 3.50 | 2.56 | 3.80 | 4.37 | 3.05 | 2.49 | 2.22 | 33.18 |
| Days ≥ 0.1" Precip | 6 | 5 | 5 | 7 | 6 | 6 | 5 | 6 | 7 | 7 | 7 | 5 | 72 |
| Total Snowfall (") | 16.9 | 10.3 | 7.3 | 1.7 | 0.0 | 0.0 | 0.0 | 0.0 | 0.0 | 0.2 | 4.8 | na | na |
| Days ≥ 1" Snow Depth | na | na | na | ***1*** | 0 | 0 | 0 | 0 | 0 | 0 | na | na | na |

## FAYETTE 4 SW *Delta County* ELEVATION 791 ft LAT/LONG 45° 41 ' N / 86° 42 ' W

| | JAN | FEB | MAR | APR | MAY | JUN | JUL | AUG | SEP | OCT | NOV | DEC | YEAR |
|---|---|---|---|---|---|---|---|---|---|---|---|---|---|
| Maximum Temp °F | 24.5 | 26.9 | 36.1 | 48.0 | 60.4 | 69.1 | 75.2 | 73.8 | 65.6 | 54.5 | 41.4 | 30.1 | 50.5 |
| Minimum Temp °F | 10.4 | 11.5 | 20.7 | 31.5 | 41.2 | 50.4 | 57.6 | 57.4 | 50.8 | 41.0 | 30.1 | 18.4 | 35.1 |
| Mean Temp °F | 17.5 | 19.3 | 28.4 | 39.8 | 50.8 | 59.8 | 66.4 | 65.6 | 58.3 | 47.8 | 35.8 | 24.3 | 42.8 |
| Days Max Temp ≥ 90 °F | 0 | 0 | 0 | 0 | 0 | 0 | 0 | 0 | 0 | 0 | 0 | 0 | 0 |
| Days Max Temp ≤ 32 °F | 24 | 19 | 9 | 1 | 0 | 0 | 0 | 0 | 0 | 0 | 4 | 18 | 75 |
| Days Min Temp ≤ 32 °F | 31 | 28 | 28 | 16 | 3 | 0 | 0 | 0 | 0 | 4 | 19 | 29 | 158 |
| Days Min Temp ≤ 0 °F | 6 | 5 | 1 | 0 | 0 | 0 | 0 | 0 | 0 | 0 | 0 | 2 | 14 |
| Heating Degree Days | 1467 | 1285 | 1128 | 751 | 433 | 169 | 39 | 50 | 210 | 528 | 871 | 1257 | 8188 |
| Cooling Degree Days | 0 | 0 | 0 | 0 | 1 | 24 | 102 | 90 | 15 | 0 | 0 | 0 | 232 |
| Total Precipitation (") | 1.51 | 1.03 | 1.92 | 2.38 | 2.82 | 2.85 | 2.65 | 3.43 | 3.34 | 2.70 | 2.31 | 1.85 | 28.79 |
| Days ≥ 0.1" Precip | 6 | 4 | 5 | 6 | 6 | 7 | 6 | 7 | 7 | 6 | 6 | ***5*** | 71 |
| Total Snowfall (") | 15.4 | 10.0 | 9.5 | 2.5 | 0.0 | 0.0 | 0.0 | 0.0 | 0.0 | 0.2 | ***4.0*** | 13.3 | 54.9 |
| Days ≥ 1" Snow Depth | 27 | 25 | 19 | 3 | 0 | 0 | 0 | 0 | 0 | 0 | ***4*** | ***16*** | 94 |

## FLINT BISHOP ARPT *Genesee County* ELEVATION 787 ft LAT/LONG 42° 58 ' N / 83° 44 ' W

| | JAN | FEB | MAR | APR | MAY | JUN | JUL | AUG | SEP | OCT | NOV | DEC | YEAR |
|---|---|---|---|---|---|---|---|---|---|---|---|---|---|
| Maximum Temp °F | 28.8 | 31.4 | 42.4 | 56.2 | 68.3 | 77.1 | 81.5 | 79.3 | 71.4 | 59.3 | 46.3 | 34.4 | 56.4 |
| Minimum Temp °F | 14.4 | 15.9 | 25.4 | 36.2 | 45.8 | 54.9 | 59.9 | 58.1 | 51.0 | 40.4 | 31.9 | 21.4 | 37.9 |
| Mean Temp °F | 21.6 | 23.7 | 33.9 | 46.2 | 57.1 | 66.0 | 70.7 | 68.7 | 61.2 | 49.8 | 39.1 | 27.9 | 47.2 |
| Days Max Temp ≥ 90 °F | 0 | 0 | 0 | 0 | 0 | 1 | 3 | 1 | 1 | 0 | 0 | 0 | 6 |
| Days Max Temp ≤ 32 °F | 19 | 15 | 5 | 0 | 0 | 0 | 0 | 0 | 0 | 0 | 2 | 12 | 53 |
| Days Min Temp ≤ 32 °F | 29 | 26 | 23 | 11 | 2 | 0 | 0 | 0 | 0 | 6 | 16 | 27 | 140 |
| Days Min Temp ≤ 0 °F | 5 | 4 | 1 | 0 | 0 | 0 | 0 | 0 | 0 | 0 | 0 | 2 | 12 |
| Heating Degree Days | 1338 | 1161 | 957 | 564 | 273 | 69 | 14 | 30 | 161 | 468 | 769 | 1143 | 6947 |
| Cooling Degree Days | 0 | 0 | 0 | 7 | 40 | 106 | 219 | 161 | 54 | 3 | 0 | 0 | 590 |
| Total Precipitation (") | 1.49 | 1.30 | 2.19 | 3.11 | 2.61 | 3.31 | 3.04 | 3.47 | 3.59 | 2.36 | 2.83 | 2.19 | 31.49 |
| Days ≥ 0.1" Precip | 4 | 4 | 6 | 7 | 6 | 7 | 6 | 6 | 7 | 6 | 7 | 6 | 72 |
| Total Snowfall (") | 13.1 | 10.4 | 8.2 | 2.8 | 0.0 | 0.0 | 0.0 | 0.0 | 0.0 | 0.3 | 4.0 | 11.1 | 49.9 |
| Days ≥ 1" Snow Depth | 22 | 17 | 9 | 1 | 0 | 0 | 0 | 0 | 0 | 0 | 3 | 15 | 67 |

## FRANKFORT 2 NE *Benzie County* ELEVATION 720 ft LAT/LONG 44° 39 ' N / 86° 13 ' W

| | JAN | FEB | MAR | APR | MAY | JUN | JUL | AUG | SEP | OCT | NOV | DEC | YEAR |
|---|---|---|---|---|---|---|---|---|---|---|---|---|---|
| Maximum Temp °F | 27.5 | 29.3 | 38.6 | 51.1 | 63.4 | 71.7 | 76.7 | 74.8 | 67.8 | 56.5 | 43.3 | 32.4 | 52.8 |
| Minimum Temp °F | 16.3 | 16.6 | 23.9 | 34.1 | 43.1 | 51.7 | 58.1 | 58.2 | 51.5 | 41.7 | 31.7 | 22.1 | 37.4 |
| Mean Temp °F | 21.9 | 23.0 | 31.3 | 42.6 | 53.3 | 61.8 | 67.4 | 66.5 | 59.6 | 49.1 | 37.5 | 27.3 | 45.1 |
| Days Max Temp ≥ 90 °F | 0 | 0 | 0 | 0 | 0 | 0 | 0 | 0 | 0 | 0 | 0 | 0 | 0 |
| Days Max Temp ≤ 32 °F | 21 | 17 | 9 | 1 | 0 | 0 | 0 | 0 | 0 | 0 | 3 | 15 | 66 |
| Days Min Temp ≤ 32 °F | 30 | 27 | 26 | 14 | 3 | 0 | 0 | 0 | 0 | 4 | 17 | 29 | 150 |
| Days Min Temp ≤ 0 °F | 2 | 1 | 0 | 0 | 0 | 0 | 0 | 0 | 0 | 0 | 0 | 0 | 3 |
| Heating Degree Days | 1329 | 1179 | 1039 | 667 | 368 | 137 | 31 | 42 | 186 | 487 | 819 | 1163 | 7447 |
| Cooling Degree Days | 0 | 0 | 0 | 3 | 13 | 48 | 127 | 106 | 32 | 0 | 0 | 0 | 329 |
| Total Precipitation (") | 2.68 | 2.03 | 2.21 | 2.85 | 2.57 | 3.45 | 2.80 | 3.10 | 4.29 | 3.19 | 3.03 | 2.79 | 34.99 |
| Days ≥ 0.1" Precip | 10 | 7 | 6 | 7 | 6 | 6 | 5 | 6 | 8 | 8 | 8 | 9 | 86 |
| Total Snowfall (") | 37.1 | 22.4 | 13.6 | 5.0 | 0.1 | 0.0 | 0.0 | 0.0 | 0.0 | 0.5 | 9.0 | 26.9 | 114.6 |
| Days ≥ 1" Snow Depth | 29 | 27 | 23 | 3 | 0 | 0 | 0 | 0 | 0 | 0 | 7 | 22 | 111 |

## GAYLORD *Otsego County* ELEVATION 1381 ft LAT/LONG 45° 2 ' N / 84° 41 ' W

| | JAN | FEB | MAR | APR | MAY | JUN | JUL | AUG | SEP | OCT | NOV | DEC | YEAR |
|---|---|---|---|---|---|---|---|---|---|---|---|---|---|
| Maximum Temp °F | 25.2 | 28.4 | 39.0 | 53.4 | 67.7 | 76.1 | 80.4 | 77.6 | 68.5 | 56.4 | 41.4 | 29.9 | 53.7 |
| Minimum Temp °F | 9.5 | 9.2 | 18.1 | 30.3 | 40.9 | 50.1 | 55.2 | 53.7 | 46.6 | 36.9 | 26.9 | 16.3 | 32.8 |
| Mean Temp °F | 17.4 | 18.8 | 28.6 | 41.9 | 54.3 | 63.1 | 67.9 | 65.7 | 57.6 | 46.7 | 34.2 | 23.1 | 43.3 |
| Days Max Temp ≥ 90 °F | 0 | 0 | 0 | 0 | 0 | 1 | 2 | 1 | 0 | 0 | 0 | 0 | 4 |
| Days Max Temp ≤ 32 °F | 24 | 19 | 8 | 1 | 0 | 0 | 0 | 0 | 0 | 0 | 6 | 19 | 77 |
| Days Min Temp ≤ 32 °F | 31 | 28 | 28 | 19 | 6 | 1 | 0 | 0 | 2 | 11 | 23 | 29 | 178 |
| Days Min Temp ≤ 0 °F | 7 | 7 | 3 | 0 | 0 | 0 | 0 | 0 | 0 | 0 | 0 | 3 | 20 |
| Heating Degree Days | 1470 | 1299 | 1123 | 689 | 345 | 119 | 40 | 69 | 242 | 564 | 917 | 1291 | 8168 |
| Cooling Degree Days | 0 | 0 | 0 | 4 | 20 | 60 | 137 | 98 | 24 | 0 | 0 | 0 | 343 |
| Total Precipitation (") | 2.85 | 1.92 | 2.37 | 2.57 | 2.74 | 2.87 | 3.15 | 3.58 | 4.11 | 3.63 | 3.55 | 3.12 | 36.46 |
| Days ≥ 0.1" Precip | 9 | 6 | 7 | 7 | 6 | 6 | 6 | 7 | 9 | 9 | 10 | 10 | 92 |
| Total Snowfall (") | 38.6 | 23.6 | 17.7 | 7.2 | 0.9 | 0.0 | 0.0 | 0.0 | 0.0 | 3.0 | 23.4 | 34.6 | 149.0 |
| Days ≥ 1" Snow Depth | 31 | 28 | 26 | 6 | 0 | 0 | 0 | 0 | 0 | 1 | 12 | 27 | 131 |

## GLADWIN *Gladwin County* ELEVATION 791 ft LAT/LONG 43° 59 ' N / 84° 29 ' W

| | JAN | FEB | MAR | APR | MAY | JUN | JUL | AUG | SEP | OCT | NOV | DEC | YEAR |
|---|---|---|---|---|---|---|---|---|---|---|---|---|---|
| Maximum Temp °F | 28.2 | 31.6 | 41.7 | 56.0 | 69.9 | 78.7 | 83.1 | 80.3 | 71.5 | 59.2 | 44.6 | 33.4 | 56.5 |
| Minimum Temp °F | 9.6 | 11.3 | 20.9 | 32.2 | 42.5 | 51.4 | 56.5 | 54.5 | 46.6 | 36.1 | 27.8 | 18.4 | 34.0 |
| Mean Temp °F | 18.9 | 21.5 | 31.3 | 44.1 | 56.2 | 65.1 | 69.8 | 67.4 | 59.1 | 47.8 | 36.2 | 25.9 | 45.3 |
| Days Max Temp ≥ 90 °F | 0 | 0 | 0 | 0 | 1 | 3 | 5 | 3 | 0 | 0 | 0 | 0 | 12 |
| Days Max Temp ≤ 32 °F | 21 | 16 | 6 | 0 | 0 | 0 | 0 | 0 | 0 | 0 | 3 | 14 | 60 |
| Days Min Temp ≤ 32 °F | 31 | 28 | 28 | 16 | 4 | 0 | 0 | 0 | 2 | 11 | 22 | 29 | 171 |
| Days Min Temp ≤ 0 °F | 8 | 6 | 2 | 0 | 0 | 0 | 0 | 0 | 0 | 0 | 0 | 2 | 18 |
| Heating Degree Days | 1422 | 1222 | 1037 | 623 | 295 | 81 | 22 | 44 | 205 | 529 | 858 | 1205 | 7543 |
| Cooling Degree Days | 0 | 0 | 0 | 5 | 32 | 86 | 190 | 131 | 32 | 1 | 0 | 0 | 477 |
| Total Precipitation (") | 1.71 | 1.19 | 2.22 | 2.66 | 2.89 | 3.35 | 2.65 | 3.53 | 3.76 | 2.84 | 2.68 | 2.26 | 31.74 |
| Days ≥ 0.1" Precip | *5* | *4* | 5 | 6 | 6 | 6 | 5 | 7 | 7 | 6 | 6 | *6* | 69 |
| Total Snowfall (") | 14.8 | *10.1* | 8.7 | 1.8 | 0.0 | 0.0 | 0.0 | 0.0 | 0.0 | 0.2 | 4.3 | *11.7* | 51.6 |
| Days ≥ 1" Snow Depth | *27* | na | na | *1* | 0 | 0 | 0 | 0 | 0 | 0 | *3* | na | na |

## GRAND HAVEN FIRE DEP *Ottawa County* ELEVATION 620 ft LAT/LONG 43° 4 ' N / 86° 13 ' W

| | JAN | FEB | MAR | APR | MAY | JUN | JUL | AUG | SEP | OCT | NOV | DEC | YEAR |
|---|---|---|---|---|---|---|---|---|---|---|---|---|---|
| Maximum Temp °F | 30.6 | 32.9 | 42.8 | 55.3 | 66.8 | 75.4 | 79.6 | 77.9 | 71.3 | 59.7 | 46.8 | 35.8 | 56.2 |
| Minimum Temp °F | 18.7 | 19.7 | 27.5 | 37.6 | 47.4 | 56.4 | 62.6 | 61.2 | 54.4 | 44.4 | 34.4 | 24.6 | 40.7 |
| Mean Temp °F | 24.7 | 26.3 | 35.2 | 46.5 | 57.1 | 65.9 | 71.1 | 69.6 | 62.9 | 52.1 | 40.6 | 30.2 | 48.5 |
| Days Max Temp ≥ 90 °F | 0 | 0 | 0 | 0 | 0 | 0 | 1 | 0 | 0 | 0 | 0 | 0 | 1 |
| Days Max Temp ≤ 32 °F | 17 | 13 | 4 | 0 | 0 | 0 | 0 | 0 | 0 | 0 | 1 | 9 | 44 |
| Days Min Temp ≤ 32 °F | 29 | 25 | 22 | 8 | 1 | 0 | 0 | 0 | 0 | 2 | 13 | 25 | 125 |
| Days Min Temp ≤ 0 °F | 1 | 1 | 0 | 0 | 0 | 0 | 0 | 0 | 0 | 0 | 0 | 0 | 2 |
| Heating Degree Days | 1244 | 1085 | 918 | 555 | 269 | 65 | 10 | 21 | 121 | 400 | 725 | 1072 | 6485 |
| Cooling Degree Days | 0 | 0 | 0 | 7 | 36 | 108 | 239 | 191 | 69 | 3 | 0 | 0 | 653 |
| Total Precipitation (") | 2.05 | 1.25 | 2.19 | 2.86 | 2.74 | 3.08 | 2.49 | 3.46 | 3.79 | 2.97 | 3.31 | 2.66 | 32.85 |
| Days ≥ 0.1" Precip | 6 | 4 | 5 | 7 | 6 | 6 | 5 | 6 | 6 | 7 | 7 | 7 | 72 |
| Total Snowfall (") | 23.9 | 13.7 | 6.4 | 1.6 | 0.0 | 0.0 | 0.0 | 0.0 | 0.0 | 0.3 | 4.8 | 19.0 | 69.7 |
| Days ≥ 1" Snow Depth | 24 | 20 | 9 | 1 | 0 | 0 | 0 | 0 | 0 | 0 | 3 | *16* | 73 |

## GRAND MARAIS 2 E *Alger County* ELEVATION 630 ft LAT/LONG 46° 40 ' N / 85° 59 ' W

| | JAN | FEB | MAR | APR | MAY | JUN | JUL | AUG | SEP | OCT | NOV | DEC | YEAR |
|---|---|---|---|---|---|---|---|---|---|---|---|---|---|
| Maximum Temp °F | 25.0 | 27.0 | 36.2 | 48.8 | 61.5 | 70.1 | 75.4 | 74.8 | 66.6 | 54.8 | 40.5 | 29.9 | 50.9 |
| Minimum Temp °F | 10.9 | 10.5 | 17.7 | 29.0 | 37.5 | 45.4 | 51.4 | 52.5 | 46.5 | 37.0 | 27.4 | 17.1 | 31.9 |
| Mean Temp °F | 18.0 | 18.8 | 27.0 | 38.9 | 49.6 | 57.8 | 63.4 | 63.6 | 56.6 | 46.0 | 33.9 | 23.6 | 41.4 |
| Days Max Temp ≥ 90 °F | 0 | 0 | 0 | 0 | 0 | 0 | 1 | 1 | 0 | 0 | 0 | 0 | 2 |
| Days Max Temp ≤ 32 °F | 25 | 20 | 11 | 1 | 0 | 0 | 0 | 0 | 0 | 0 | 5 | 18 | 80 |
| Days Min Temp ≤ 32 °F | 31 | 28 | 29 | 21 | 10 | 2 | 0 | 0 | 1 | 9 | 23 | 30 | 184 |
| Days Min Temp ≤ 0 °F | 5 | 5 | 2 | 0 | 0 | 0 | 0 | 0 | 0 | 0 | 0 | 2 | 14 |
| Heating Degree Days | 1452 | 1299 | 1172 | 778 | 481 | 238 | 114 | 104 | 267 | 585 | 926 | 1278 | 8694 |
| Cooling Degree Days | 0 | 0 | 0 | 2 | 10 | 29 | 73 | 69 | 18 | 0 | 0 | 0 | 201 |
| Total Precipitation (") | 2.24 | 1.28 | 1.52 | 1.57 | 2.53 | 3.02 | 2.74 | 3.05 | 3.55 | 3.24 | 2.50 | 2.27 | 29.51 |
| Days ≥ 0.1" Precip | 8 | 5 | 4 | 4 | 6 | 7 | 5 | 6 | 8 | 8 | 8 | 8 | 77 |
| Total Snowfall (") | 46.8 | 29.1 | 15.8 | 4.9 | 0.3 | 0.0 | 0.0 | 0.0 | 0.0 | 0.7 | 13.5 | 38.8 | 149.9 |
| Days ≥ 1" Snow Depth | 30 | 27 | 29 | 15 | 0 | 0 | 0 | 0 | 0 | 0 | 11 | 28 | 140 |

## GRAND RAPIDS KENT AP *Kent County* ELEVATION 784 ft LAT/LONG 42° 53 ' N / 85° 31 ' W

| | JAN | FEB | MAR | APR | MAY | JUN | JUL | AUG | SEP | OCT | NOV | DEC | YEAR |
|---|---|---|---|---|---|---|---|---|---|---|---|---|---|
| Maximum Temp °F | 29.1 | 31.7 | 42.6 | 56.7 | 69.5 | 78.5 | 82.7 | 80.2 | 72.1 | 59.8 | 45.9 | 34.1 | 56.9 |
| Minimum Temp °F | 15.0 | 16.1 | 25.1 | 35.7 | 45.8 | 55.3 | 60.4 | 58.4 | 51.0 | 40.1 | 31.3 | 21.4 | 38.0 |
| Mean Temp °F | 22.1 | 23.9 | 33.9 | 46.3 | 57.7 | 66.9 | 71.6 | 69.4 | 61.6 | 50.0 | 38.6 | 27.8 | 47.5 |
| Days Max Temp ≥ 90 °F | 0 | 0 | 0 | 0 | 1 | 2 | 5 | 2 | 0 | 0 | 0 | 0 | 10 |
| Days Max Temp ≤ 32 °F | 19 | 15 | 6 | 0 | 0 | 0 | 0 | 0 | 0 | 0 | 2 | 13 | 55 |
| Days Min Temp ≤ 32 °F | 29 | 26 | 24 | 12 | 2 | 0 | 0 | 0 | 0 | 6 | 18 | 28 | 145 |
| Days Min Temp ≤ 0 °F | 4 | 3 | 0 | 0 | 0 | 0 | 0 | 0 | 0 | 0 | 0 | 1 | 8 |
| Heating Degree Days | 1323 | 1153 | 958 | 563 | 259 | 59 | 10 | 24 | 155 | 466 | 785 | 1147 | 6902 |
| Cooling Degree Days | 0 | 0 | 0 | 9 | 43 | 120 | 233 | 173 | 57 | 3 | 0 | 0 | 638 |
| Total Precipitation (") | 1.93 | 1.44 | 2.66 | 3.50 | 3.09 | 3.87 | 3.67 | 3.97 | 4.35 | 3.11 | 3.75 | 2.93 | 38.27 |
| Days ≥ 0.1" Precip | 6 | 4 | 7 | 8 | 6 | 6 | 6 | 6 | 7 | 6 | 8 | 7 | 77 |
| Total Snowfall (") | 20.4 | 12.5 | 9.5 | 2.9 | 0.0 | 0.0 | 0.0 | 0.0 | 0.0 | 0.7 | 7.5 | 17.9 | 71.4 |
| Days ≥ 1" Snow Depth | 24 | 19 | 10 | 2 | 0 | 0 | 0 | 0 | 0 | 0 | 4 | 17 | 76 |

## GRAYLING *Crawford County* ELEVATION 1142 ft LAT/LONG 44° 40 ' N / 84° 42 ' W

| | JAN | FEB | MAR | APR | MAY | JUN | JUL | AUG | SEP | OCT | NOV | DEC | YEAR |
|---|---|---|---|---|---|---|---|---|---|---|---|---|---|
| Maximum Temp °F | 25.6 | 28.0 | 38.4 | 53.1 | 67.3 | 76.0 | 80.6 | 77.8 | 68.8 | 56.0 | 42.1 | 30.5 | 53.7 |
| Minimum Temp °F | 6.8 | 5.8 | 15.4 | 29.0 | 39.2 | 48.5 | 53.5 | 51.4 | 43.8 | 34.1 | 25.5 | 14.8 | 30.7 |
| Mean Temp °F | 16.2 | 16.9 | 26.9 | 41.1 | 53.3 | 62.3 | 67.1 | 64.6 | 56.3 | 45.1 | 33.8 | 22.6 | 42.2 |
| Days Max Temp ≥ 90 °F | 0 | 0 | 0 | 0 | 0 | 2 | 4 | 1 | 0 | 0 | 0 | 0 | 7 |
| Days Max Temp ≤ 32 °F | 23 | 19 | 9 | 1 | 0 | 0 | 0 | 0 | 0 | 0 | 5 | 18 | 75 |
| Days Min Temp ≤ 32 °F | 31 | 28 | 28 | 20 | 9 | 1 | 0 | 0 | 4 | 15 | 24 | 30 | 190 |
| Days Min Temp ≤ 0 °F | 10 | 11 | 5 | 0 | 0 | 0 | 0 | 0 | 0 | 0 | 0 | 4 | 30 |
| Heating Degree Days | 1507 | 1352 | 1174 | 716 | 377 | 139 | 54 | 91 | 276 | 613 | 926 | 1306 | 8531 |
| Cooling Degree Days | 0 | 0 | 0 | 4 | 19 | 55 | 124 | 83 | 19 | 0 | 0 | 0 | 304 |
| Total Precipitation (") | 1.70 | 1.24 | 1.89 | 2.77 | 3.08 | 3.53 | 3.63 | 3.90 | 4.21 | 3.45 | 2.61 | 1.91 | 33.92 |
| Days ≥ 0.1" Precip | 5 | 4 | 5 | 7 | 7 | 7 | 7 | 7 | 8 | 8 | 7 | 5 | 77 |
| Total Snowfall (") | *31.5* | 18.9 | 13.9 | 4.5 | 0.1 | 0.0 | 0.0 | 0.0 | 0.0 | 0.8 | 11.5 | 24.2 | 105.4 |
| Days ≥ 1" Snow Depth | 29 | 27 | 25 | 6 | 0 | 0 | 0 | 0 | 0 | 1 | 9 | 26 | 123 |

## GREENVILLE 2 NNE *Montcalm County* ELEVATION 840 ft LAT/LONG 43° 11 ' N / 85° 15 ' W

| | JAN | FEB | MAR | APR | MAY | JUN | JUL | AUG | SEP | OCT | NOV | DEC | YEAR |
|---|---|---|---|---|---|---|---|---|---|---|---|---|---|
| Maximum Temp °F | 28.5 | 31.9 | 43.1 | 57.9 | 70.4 | 79.2 | 83.4 | 81.0 | 72.8 | 60.4 | 45.9 | 33.7 | 57.4 |
| Minimum Temp °F | 12.9 | 14.0 | 23.2 | 34.4 | 44.7 | 53.7 | 58.0 | 56.2 | 49.2 | 38.8 | 29.5 | 19.3 | 36.2 |
| Mean Temp °F | 20.7 | 23.0 | 33.2 | 46.2 | 57.6 | 66.5 | 70.7 | 68.6 | 61.0 | 49.6 | 37.7 | 26.5 | 46.8 |
| Days Max Temp ≥ 90 °F | 0 | 0 | 0 | 0 | 0 | 2 | 5 | 3 | 0 | 0 | 0 | 0 | 10 |
| Days Max Temp ≤ 32 °F | 20 | 15 | 5 | 0 | 0 | 0 | 0 | 0 | 0 | 0 | 2 | 14 | 56 |
| Days Min Temp ≤ 32 °F | 31 | 27 | 26 | 14 | 3 | 0 | 0 | 0 | 1 | 9 | 20 | 29 | 160 |
| Days Min Temp ≤ 0 °F | 5 | 4 | 1 | 0 | 0 | 0 | 0 | 0 | 0 | 0 | 0 | 2 | 12 |
| Heating Degree Days | 1366 | 1180 | 979 | 564 | 258 | 62 | 13 | 31 | 165 | 475 | 811 | 1187 | 7091 |
| Cooling Degree Days | 0 | 0 | 0 | 6 | 34 | 100 | 203 | 147 | 48 | 2 | 0 | 0 | 540 |
| Total Precipitation (") | 1.72 | 1.29 | 2.38 | 3.16 | 3.10 | 3.53 | 2.74 | 4.18 | 3.87 | 3.12 | 3.28 | 2.64 | 35.01 |
| Days ≥ 0.1" Precip | 5 | 4 | 6 | 7 | 6 | 7 | 6 | 6 | 7 | 7 | 7 | 6 | 74 |
| Total Snowfall (") | 18.2 | 12.5 | 9.2 | 2.6 | 0.0 | 0.0 | 0.0 | 0.0 | 0.0 | 0.3 | 5.7 | 15.8 | 64.3 |
| Days ≥ 1" Snow Depth | 24 | 20 | 12 | 1 | 0 | 0 | 0 | 0 | 0 | 0 | 3 | 17 | 77 |

## GROSSE POINTE FARMS *Wayne County* ELEVATION 610 ft LAT/LONG 42° 23 ' N / 82° 54 ' W

| | JAN | FEB | MAR | APR | MAY | JUN | JUL | AUG | SEP | OCT | NOV | DEC | YEAR |
|---|---|---|---|---|---|---|---|---|---|---|---|---|---|
| Maximum Temp °F | 32.0 | 34.5 | 44.6 | 57.9 | 70.2 | 79.5 | 83.6 | 81.2 | 74.0 | 61.5 | 49.0 | 37.2 | 58.8 |
| Minimum Temp °F | 17.8 | 19.2 | 27.4 | 37.9 | 48.4 | 58.1 | 63.4 | 62.1 | 55.3 | 44.2 | 34.5 | 24.3 | 41.1 |
| Mean Temp °F | 25.0 | 26.9 | 36.1 | 47.9 | 59.3 | 68.8 | 73.5 | 71.7 | 64.7 | 52.8 | 41.7 | 30.8 | 49.9 |
| Days Max Temp ≥ 90 °F | 0 | 0 | 0 | 0 | 0 | 3 | 6 | 3 | 1 | 0 | 0 | 0 | 13 |
| Days Max Temp ≤ 32 °F | 15 | 12 | 3 | 0 | 0 | 0 | 0 | 0 | 0 | 0 | 1 | 8 | 39 |
| Days Min Temp ≤ 32 °F | 29 | 26 | 23 | 8 | 1 | 0 | 0 | 0 | 0 | 2 | 13 | 25 | 127 |
| Days Min Temp ≤ 0 °F | 2 | 1 | 0 | 0 | 0 | 0 | 0 | 0 | 0 | 0 | 0 | 0 | 3 |
| Heating Degree Days | 1233 | 1071 | 890 | 514 | 213 | 41 | 5 | 10 | 96 | 378 | 691 | 1054 | 6196 |
| Cooling Degree Days | 0 | 0 | 1 | 8 | 49 | 158 | 291 | 231 | 89 | 6 | 0 | 0 | 833 |
| Total Precipitation (") | 1.71 | 1.59 | 2.38 | 3.03 | 2.83 | 3.56 | 3.65 | 3.61 | 3.19 | 2.64 | 3.04 | 2.64 | 33.87 |
| Days ≥ 0.1" Precip | 5 | 4 | 6 | 7 | 6 | 7 | 6 | 6 | 6 | 6 | 7 | 7 | 73 |
| Total Snowfall (") | 8.1 | 7.4 | 3.6 | 0.6 | 0.0 | 0.0 | 0.0 | 0.0 | 0.0 | 0.0 | 1.3 | 6.4 | 27.4 |
| Days ≥ 1" Snow Depth | 19 | 17 | 7 | 1 | 0 | 0 | 0 | 0 | 0 | 0 | 1 | 10 | 55 |

## GULL LAKE BIOL STA *Kalamazoo County* ELEVATION 942 ft LAT/LONG 42° 24 ' N / 85° 23 ' W

| | JAN | FEB | MAR | APR | MAY | JUN | JUL | AUG | SEP | OCT | NOV | DEC | YEAR |
|---|---|---|---|---|---|---|---|---|---|---|---|---|---|
| Maximum Temp °F | 30.6 | 34.1 | 45.4 | 59.1 | 71.3 | 80.1 | 83.9 | 81.6 | 74.3 | 62.1 | 48.0 | 35.9 | 58.9 |
| Minimum Temp °F | 15.2 | 16.0 | 25.5 | 36.5 | 47.0 | 56.5 | 61.4 | 59.6 | 52.8 | 42.0 | 32.7 | 22.1 | 38.9 |
| Mean Temp °F | 22.9 | 25.1 | 35.5 | 47.9 | 59.2 | 68.3 | 72.6 | 70.6 | 63.5 | 52.1 | 40.3 | 29.0 | 48.9 |
| Days Max Temp ≥ 90 °F | 0 | 0 | 0 | 0 | 0 | 2 | 5 | 2 | 0 | 0 | 0 | 0 | 9 |
| Days Max Temp ≤ 32 °F | 17 | 13 | 3 | 0 | 0 | 0 | 0 | 0 | 0 | 0 | 1 | 11 | 45 |
| Days Min Temp ≤ 32 °F | 30 | 26 | 24 | 11 | 2 | 0 | 0 | 0 | 0 | 5 | 16 | 27 | 141 |
| Days Min Temp ≤ 0 °F | 4 | 3 | 0 | 0 | 0 | 0 | 0 | 0 | 0 | 0 | 0 | 1 | 8 |
| Heating Degree Days | 1298 | 1120 | 909 | 512 | 219 | 41 | 6 | 16 | 116 | 399 | 735 | 1110 | 6481 |
| Cooling Degree Days | 0 | 0 | 0 | 7 | 49 | 152 | 261 | 208 | 78 | 4 | 0 | 0 | 759 |
| Total Precipitation (") | 1.83 | 1.62 | 2.61 | 3.74 | 3.25 | 3.87 | 3.77 | 3.91 | 4.01 | 3.26 | 3.51 | 3.05 | 38.43 |
| Days ≥ 0.1" Precip | 5 | 4 | 6 | 9 | 7 | 7 | 7 | 7 | 7 | 7 | 8 | 8 | 82 |
| Total Snowfall (") | 16.9 | 12.3 | 6.3 | 1.7 | 0.0 | 0.0 | 0.0 | 0.0 | 0.0 | 0.3 | 4.5 | 15.1 | 57.1 |
| Days ≥ 1" Snow Depth | 22 | 18 | 8 | 1 | 0 | 0 | 0 | 0 | 0 | 0 | 3 | 16 | 68 |

## HALE LOUD DAM *Iosco County* ELEVATION 771 ft LAT/LONG 44° 27 ' N / 83° 41 ' W

| | JAN | FEB | MAR | APR | MAY | JUN | JUL | AUG | SEP | OCT | NOV | DEC | YEAR |
|---|---|---|---|---|---|---|---|---|---|---|---|---|---|
| Maximum Temp °F | 28.1 | 30.7 | 41.1 | 54.9 | 68.0 | 76.2 | 80.7 | 78.3 | 70.1 | 58.2 | 44.1 | 32.4 | 55.2 |
| Minimum Temp °F | 9.1 | 8.8 | 18.7 | 31.0 | 41.3 | 50.7 | 56.0 | 54.5 | 47.3 | 37.1 | 28.4 | 17.5 | 33.4 |
| Mean Temp °F | 18.6 | 19.8 | 29.9 | 43.0 | 54.7 | 63.5 | 68.4 | 66.4 | 58.7 | 47.7 | 36.3 | 25.0 | 44.3 |
| Days Max Temp ≥ 90 °F | 0 | 0 | 0 | 0 | 0 | 1 | 3 | 1 | 0 | 0 | 0 | 0 | 5 |
| Days Max Temp ≤ 32 °F | 21 | 16 | 7 | 0 | 0 | 0 | 0 | 0 | 0 | 0 | 3 | 15 | 62 |
| Days Min Temp ≤ 32 °F | 31 | 28 | 28 | 17 | 6 | 0 | 0 | 0 | 1 | 10 | 21 | 29 | 171 |
| Days Min Temp ≤ 0 °F | 9 | 9 | 3 | 0 | 0 | 0 | 0 | 0 | 0 | 0 | 0 | 3 | 24 |
| Heating Degree Days | 1432 | 1271 | 1080 | 656 | 331 | 106 | 31 | 53 | 211 | 531 | 855 | 1232 | 7789 |
| Cooling Degree Days | 0 | 0 | 0 | 4 | 22 | 65 | 157 | 112 | 31 | 1 | 0 | 0 | 392 |
| Total Precipitation (") | 1.63 | 1.06 | 1.68 | 2.28 | 2.62 | 2.99 | 2.96 | 3.58 | 3.40 | 2.56 | 2.40 | 1.72 | 28.88 |
| Days ≥ 0.1" Precip | 5 | 4 | 5 | 6 | 6 | 6 | 6 | 6 | 7 | 6 | 6 | 5 | 68 |
| Total Snowfall (") | 14.8 | 9.3 | 8.0 | 2.0 | 0.2 | 0.0 | 0.0 | 0.0 | 0.0 | 0.1 | 4.0 | 10.0 | 48.4 |
| Days ≥ 1" Snow Depth | 30 | 27 | 23 | 6 | 0 | 0 | 0 | 0 | 0 | 0 | 5 | 20 | 111 |

## HANCOCK HOUGHTON CO *Houghton County* ELEVATION 1086 ft LAT/LONG 47° 10 ' N / 88° 30 ' W

| | JAN | FEB | MAR | APR | MAY | JUN | JUL | AUG | SEP | OCT | NOV | DEC | YEAR |
|---|---|---|---|---|---|---|---|---|---|---|---|---|---|
| Maximum Temp °F | 20.6 | 22.5 | 32.2 | 46.2 | 60.9 | 69.8 | 75.4 | 72.9 | 62.8 | 51.1 | 36.3 | 25.5 | 48.0 |
| Minimum Temp °F | 8.0 | 8.0 | 17.3 | 29.9 | 40.5 | 49.0 | 55.5 | 54.8 | 46.9 | 37.0 | 25.6 | 14.5 | 32.3 |
| Mean Temp °F | 14.4 | 15.3 | 24.8 | 38.0 | 50.7 | 59.4 | 65.5 | 63.9 | 54.9 | 44.1 | 31.0 | 20.0 | 40.2 |
| Days Max Temp ≥ 90 °F | 0 | 0 | 0 | 0 | 0 | 0 | 1 | 0 | 0 | 0 | 0 | 0 | 1 |
| Days Max Temp ≤ 32 °F | 27 | 23 | 15 | 3 | 0 | 0 | 0 | 0 | 0 | 1 | 10 | 24 | 103 |
| Days Min Temp ≤ 32 °F | 30 | 28 | 29 | 19 | 4 | 0 | 0 | 0 | 1 | 9 | 25 | 30 | 175 |
| Days Min Temp ≤ 0 °F | 7 | 7 | 2 | 0 | 0 | 0 | 0 | 0 | 0 | 0 | 0 | 3 | 19 |
| Heating Degree Days | 1565 | 1397 | 1240 | 803 | 447 | 194 | 68 | 100 | 311 | 642 | 1014 | 1387 | 9168 |
| Cooling Degree Days | 0 | 0 | 0 | 2 | 15 | 36 | 98 | 77 | 13 | 0 | 0 | 0 | 241 |
| Total Precipitation (") | 4.05 | 2.28 | 2.31 | 1.83 | 2.46 | 2.89 | 3.09 | 2.83 | 3.30 | 2.83 | 2.81 | 3.49 | 34.17 |
| Days ≥ 0.1" Precip | 13 | 8 | 6 | 5 | 6 | 7 | 6 | 6 | 7 | 7 | 8 | 12 | 91 |
| Total Snowfall (") | 66.2 | 35.0 | 22.4 | 7.8 | 0.9 | 0.0 | 0.0 | 0.0 | 0.0 | 4.3 | 24.7 | 54.8 | 216.1 |
| Days ≥ 1" Snow Depth | 31 | 28 | 30 | 15 | 0 | 0 | 0 | 0 | 0 | 2 | 15 | 30 | 151 |

## HARBOR BEACH 1 SSE *Huron County* ELEVATION 620 ft LAT/LONG 43° 52 ' N / 82° 41 ' W

| | JAN | FEB | MAR | APR | MAY | JUN | JUL | AUG | SEP | OCT | NOV | DEC | YEAR |
|---|---|---|---|---|---|---|---|---|---|---|---|---|---|
| Maximum Temp °F | 28.4 | 30.2 | 38.8 | 51.1 | 62.9 | 72.3 | 78.0 | 76.8 | 69.9 | 57.9 | 45.3 | 33.9 | 53.8 |
| Minimum Temp °F | 13.8 | 14.7 | 23.3 | 33.8 | 43.2 | 52.4 | 58.4 | 57.7 | 51.1 | 40.7 | 31.4 | 21.3 | 36.8 |
| Mean Temp °F | 21.1 | 22.4 | 31.1 | 42.5 | 53.1 | 62.4 | 68.2 | 67.3 | 60.5 | 49.3 | 38.4 | 27.6 | 45.3 |
| Days Max Temp ≥ 90 °F | 0 | 0 | 0 | 0 | 0 | 1 | 2 | 1 | 0 | 0 | 0 | 0 | 4 |
| Days Max Temp ≤ 32 °F | 20 | 16 | 8 | 1 | 0 | 0 | 0 | 0 | 0 | 0 | 2 | 13 | 60 |
| Days Min Temp ≤ 32 °F | 30 | 27 | 27 | 14 | 2 | 0 | 0 | 0 | 0 | 5 | 18 | 28 | 151 |
| Days Min Temp ≤ 0 °F | 4 | 3 | 1 | 0 | 0 | 0 | 0 | 0 | 0 | 0 | 0 | 1 | 9 |
| Heating Degree Days | 1353 | 1195 | 1044 | 670 | 379 | 135 | 34 | 41 | 167 | 482 | 793 | 1152 | 7445 |
| Cooling Degree Days | 0 | 0 | 0 | 2 | 17 | 53 | 141 | 110 | 33 | 1 | 0 | 0 | 357 |
| Total Precipitation (") | 2.75 | 2.07 | 2.45 | 2.92 | 2.76 | 2.99 | 2.76 | 3.47 | 3.76 | 2.81 | 3.19 | 2.97 | 34.90 |
| Days ≥ 0.1" Precip | 9 | 7 | 7 | 7 | 7 | 7 | 6 | 6 | 7 | 7 | 8 | 9 | 87 |
| Total Snowfall (") | ***25.2*** | 16.3 | 12.2 | 3.8 | 0.3 | 0.0 | 0.0 | 0.0 | 0.0 | 0.4 | 5.5 | ***19.9*** | 83.6 |
| Days ≥ 1" Snow Depth | 26 | 23 | 16 | 2 | 0 | 0 | 0 | 0 | 0 | 0 | 3 | ***16*** | 86 |

## HART *Oceana County* ELEVATION 679 ft LAT/LONG 43° 42 ' N / 86° 22 ' W

| | JAN | FEB | MAR | APR | MAY | JUN | JUL | AUG | SEP | OCT | NOV | DEC | YEAR |
|---|---|---|---|---|---|---|---|---|---|---|---|---|---|
| Maximum Temp °F | 28.8 | 31.2 | 41.1 | 54.5 | 67.1 | 75.8 | 80.5 | 78.3 | 70.4 | 58.4 | 45.3 | 34.2 | 55.5 |
| Minimum Temp °F | 15.6 | 15.6 | 23.8 | 34.7 | 44.0 | 53.4 | 58.5 | 57.4 | 50.4 | 40.5 | 31.4 | 22.2 | 37.3 |
| Mean Temp °F | 22.2 | 23.4 | 32.5 | 44.6 | 55.5 | 64.6 | 69.5 | 67.8 | 60.4 | 49.5 | 38.4 | 28.2 | 46.4 |
| Days Max Temp ≥ 90 °F | 0 | 0 | 0 | 0 | 0 | 1 | 2 | 1 | 0 | 0 | 0 | 0 | 4 |
| Days Max Temp ≤ 32 °F | 19 | 15 | 6 | 0 | 0 | 0 | 0 | 0 | 0 | 0 | 2 | 12 | 54 |
| Days Min Temp ≤ 32 °F | 30 | 27 | 25 | 13 | 3 | 0 | 0 | 0 | 1 | 6 | 18 | 27 | 150 |
| Days Min Temp ≤ 0 °F | 3 | 3 | 1 | 0 | 0 | 0 | 0 | 0 | 0 | 0 | 0 | 1 | 8 |
| Heating Degree Days | 1319 | 1167 | 1000 | 608 | 312 | 90 | 23 | 39 | 177 | 479 | 792 | 1134 | 7140 |
| Cooling Degree Days | 0 | 0 | 0 | 5 | 24 | 74 | 166 | 127 | 41 | 2 | 0 | 0 | 439 |
| Total Precipitation (") | 2.51 | 1.78 | 2.33 | 3.19 | 2.71 | 3.28 | 2.70 | 3.58 | 4.06 | 3.87 | 3.58 | 2.80 | 36.39 |
| Days ≥ 0.1" Precip | 9 | 5 | 6 | 7 | 6 | 7 | 5 | 6 | 8 | 8 | 8 | 8 | 83 |
| Total Snowfall (") | 34.4 | 21.3 | 10.0 | 2.6 | 0.0 | 0.0 | 0.0 | 0.0 | 0.0 | 0.2 | 5.8 | 23.6 | 97.9 |
| Days ≥ 1" Snow Depth | 27 | 25 | 15 | 1 | 0 | 0 | 0 | 0 | 0 | 0 | 4 | 19 | 91 |

## HASTINGS *Barry County* ELEVATION 791 ft LAT/LONG 42° 39 ' N / 85° 17 ' W

| | JAN | FEB | MAR | APR | MAY | JUN | JUL | AUG | SEP | OCT | NOV | DEC | YEAR |
|---|---|---|---|---|---|---|---|---|---|---|---|---|---|
| Maximum Temp °F | 30.0 | 33.3 | 44.1 | 58.0 | 70.1 | 78.9 | 82.9 | 80.8 | 73.1 | 60.9 | 47.0 | 34.8 | 57.8 |
| Minimum Temp °F | 13.7 | 14.3 | 24.1 | 35.5 | 45.3 | 54.6 | 59.0 | 56.9 | 49.6 | 38.7 | 30.5 | 20.3 | 36.9 |
| Mean Temp °F | 21.9 | 23.8 | 34.1 | 46.8 | 57.8 | 66.8 | 71.0 | 68.9 | 61.4 | 49.8 | 38.8 | 27.5 | 47.4 |
| Days Max Temp ≥ 90 °F | 0 | 0 | 0 | 0 | 0 | 2 | 4 | 2 | 0 | 0 | 0 | 0 | 8 |
| Days Max Temp ≤ 32 °F | 19 | 14 | 5 | 0 | 0 | 0 | 0 | 0 | 0 | 0 | 2 | 12 | 52 |
| Days Min Temp ≤ 32 °F | 30 | 27 | 25 | 13 | 3 | 0 | 0 | 0 | 1 | 9 | 20 | 28 | 156 |
| Days Min Temp ≤ 0 °F | 5 | 4 | 1 | 0 | 0 | 0 | 0 | 0 | 0 | 0 | 0 | 2 | 12 |
| Heating Degree Days | 1329 | 1156 | 951 | 548 | 256 | 64 | 14 | 31 | 163 | 470 | 780 | 1155 | 6917 |
| Cooling Degree Days | 0 | 0 | 1 | 11 | 39 | 116 | 213 | 158 | 54 | 3 | 0 | 0 | 595 |
| Total Precipitation (") | 1.80 | 1.35 | 2.40 | 3.21 | 2.77 | 4.15 | 3.05 | 3.78 | 3.93 | 3.01 | 3.23 | 2.56 | 35.24 |
| Days ≥ 0.1" Precip | 5 | 4 | 6 | 8 | 6 | 7 | 6 | 7 | 7 | 7 | 7 | 6 | 76 |
| Total Snowfall (") | 15.0 | 10.9 | 7.4 | 2.4 | 0.0 | 0.0 | 0.0 | 0.0 | 0.0 | 0.5 | 4.6 | 12.3 | 53.1 |
| Days ≥ 1" Snow Depth | 24 | 20 | 9 | 1 | 0 | 0 | 0 | 0 | 0 | 0 | 3 | 17 | 74 |

## HESPERIA 4 WNW *Oceana County* ELEVATION 781 ft LAT/LONG 43° 35 ' N / 86° 6 ' W

| | JAN | FEB | MAR | APR | MAY | JUN | JUL | AUG | SEP | OCT | NOV | DEC | YEAR |
|---|---|---|---|---|---|---|---|---|---|---|---|---|---|
| Maximum Temp °F | 28.6 | 31.7 | 41.7 | 55.8 | 68.8 | 77.2 | 82.0 | 79.4 | 71.3 | 59.0 | 45.2 | 33.4 | 56.2 |
| Minimum Temp °F | 12.9 | 13.3 | 21.5 | 32.8 | 42.7 | 51.5 | 56.7 | 54.3 | 47.4 | 37.5 | 28.7 | 19.4 | 34.9 |
| Mean Temp °F | 20.8 | 22.4 | 31.6 | 44.3 | 55.8 | 64.3 | 69.4 | 66.8 | 59.4 | 48.3 | 37.0 | 26.4 | 45.5 |
| Days Max Temp ≥ 90 °F | 0 | 0 | 0 | 0 | 0 | 1 | 3 | 1 | 0 | 0 | 0 | 0 | 5 |
| Days Max Temp ≤ 32 °F | 19 | 15 | 6 | 0 | 0 | 0 | 0 | 0 | 0 | 0 | 2 | 14 | 56 |
| Days Min Temp ≤ 32 °F | 30 | 27 | 26 | 15 | 5 | 0 | 0 | 0 | 2 | 10 | 21 | 29 | 165 |
| Days Min Temp ≤ 0 °F | 5 | 5 | 2 | 0 | 0 | 0 | 0 | 0 | 0 | 0 | 0 | 2 | 14 |
| Heating Degree Days | 1365 | 1193 | 1028 | 618 | 305 | 94 | 25 | 53 | 196 | 516 | 835 | 1190 | 7418 |
| Cooling Degree Days | 0 | 0 | ***0*** | ***6*** | 26 | ***78*** | ***193*** | 124 | 33 | 0 | 0 | 0 | 460 |
| Total Precipitation (") | 2.24 | 1.56 | 2.43 | ***3.10*** | 2.81 | 3.29 | 2.41 | 3.94 | 3.78 | 3.37 | 3.08 | 2.90 | 34.91 |
| Days ≥ 0.1" Precip | 7 | 5 | 6 | 7 | 6 | 6 | 5 | 6 | 7 | 7 | 8 | 8 | 78 |
| Total Snowfall (") | 21.9 | 13.4 | 9.5 | 2.5 | 0.0 | 0.0 | 0.0 | 0.0 | 0.0 | 0.2 | 6.7 | 18.7 | 72.9 |
| Days ≥ 1" Snow Depth | 29 | 26 | ***18*** | 2 | 0 | 0 | 0 | 0 | 0 | 0 | 5 | 21 | 101 |

## HILLSDALE *Hillsdale County* ELEVATION 1102 ft LAT/LONG 41° 55' N / 84° 38' W

| | JAN | FEB | MAR | APR | MAY | JUN | JUL | AUG | SEP | OCT | NOV | DEC | YEAR |
|---|---|---|---|---|---|---|---|---|---|---|---|---|---|
| Maximum Temp °F | 29.7 | 33.1 | 44.1 | 57.6 | 69.4 | 78.5 | 82.1 | 80.2 | 72.8 | 60.5 | 46.9 | 34.9 | 57.5 |
| Minimum Temp °F | 13.1 | 14.5 | 24.4 | 35.2 | 45.0 | 54.4 | 58.5 | 56.1 | 49.5 | 38.6 | 29.8 | 19.8 | 36.6 |
| Mean Temp °F | 21.5 | 23.8 | 34.3 | 46.4 | 57.2 | 66.5 | 70.3 | 68.2 | 61.2 | 49.6 | 38.4 | 27.4 | 47.1 |
| Days Max Temp ≥ 90 °F | 0 | 0 | 0 | 0 | 0 | 2 | 4 | 2 | 0 | 0 | 0 | 0 | 8 |
| Days Max Temp ≤ 32 °F | 18 | 14 | 5 | 0 | 0 | 0 | 0 | 0 | 0 | 0 | 2 | 12 | 51 |
| Days Min Temp ≤ 32 °F | 30 | 26 | 25 | 12 | 3 | 0 | 0 | 0 | 1 | 9 | 19 | 28 | 153 |
| Days Min Temp ≤ 0 °F | 5 | 4 | 1 | 0 | 0 | 0 | 0 | 0 | 0 | 0 | 0 | 2 | 12 |
| Heating Degree Days | 1344 | 1156 | 946 | 556 | 267 | 63 | 19 | 34 | 163 | 474 | 791 | 1159 | 6972 |
| Cooling Degree Days | 0 | 0 | 0 | 7 | 35 | 111 | 204 | 138 | 53 | 2 | 0 | 0 | 550 |
| Total Precipitation (") | 2.05 | 1.79 | 2.82 | 3.49 | 3.73 | 4.08 | 3.89 | 3.50 | 3.53 | 2.89 | 3.24 | 3.05 | 38.06 |
| Days ≥ 0.1" Precip | 6 | 5 | 7 | 9 | 8 | 7 | 8 | 7 | 7 | 7 | 8 | 7 | 86 |
| Total Snowfall (") | 13.7 | 12.2 | 8.2 | 1.9 | 0.1 | 0.0 | 0.0 | 0.0 | 0.1 | 0.3 | 5.1 | 12.5 | 54.1 |
| Days ≥ 1" Snow Depth | 20 | *17* | *8* | 1 | 0 | 0 | 0 | 0 | 0 | 0 | *4* | *13* | 63 |

## HOLLAND *Ottawa County* ELEVATION 610 ft LAT/LONG 42° 48' N / 86° 7' W

| | JAN | FEB | MAR | APR | MAY | JUN | JUL | AUG | SEP | OCT | NOV | DEC | YEAR |
|---|---|---|---|---|---|---|---|---|---|---|---|---|---|
| Maximum Temp °F | 30.9 | 34.0 | 44.5 | 57.4 | 69.4 | 78.5 | 82.5 | 80.6 | 73.0 | 60.8 | 47.7 | 36.2 | 58.0 |
| Minimum Temp °F | 17.1 | 18.0 | 26.1 | 36.3 | 45.9 | 54.9 | 59.8 | 58.2 | 52.0 | 41.6 | 32.8 | 22.8 | 38.8 |
| Mean Temp °F | 24.1 | 26.0 | 35.3 | 46.9 | 57.7 | 66.7 | 71.1 | 69.5 | 62.5 | 51.2 | 40.3 | 29.5 | 48.4 |
| Days Max Temp ≥ 90 °F | 0 | 0 | 0 | 0 | 0 | 2 | 4 | 2 | 0 | 0 | 0 | 0 | 8 |
| Days Max Temp ≤ 32 °F | 17 | 12 | 4 | 0 | 0 | 0 | 0 | 0 | 0 | 0 | 1 | 10 | 44 |
| Days Min Temp ≤ 32 °F | 29 | 26 | 24 | 11 | 2 | 0 | 0 | 0 | 0 | 5 | 16 | 27 | 140 |
| Days Min Temp ≤ 0 °F | 2 | 2 | 0 | 0 | 0 | 0 | 0 | 0 | 0 | 0 | 0 | 0 | 4 |
| Heating Degree Days | 1263 | 1095 | 914 | 547 | 262 | 67 | 17 | 28 | 137 | 429 | 736 | 1094 | 6589 |
| Cooling Degree Days | 0 | 0 | 1 | 12 | 48 | 132 | 239 | 190 | 71 | 5 | 0 | 0 | 698 |
| Total Precipitation (") | 2.17 | 1.24 | 2.18 | 3.40 | 3.01 | 3.85 | 3.57 | 3.53 | 4.17 | 3.17 | 3.48 | 3.00 | 36.77 |
| Days ≥ 0.1" Precip | 6 | 4 | 6 | 7 | 6 | 6 | 6 | 6 | 8 | 7 | 8 | 8 | 78 |
| Total Snowfall (") | 29.6 | 17.6 | 8.3 | 2.1 | 0.0 | 0.0 | 0.0 | 0.0 | 0.0 | 0.3 | 5.7 | 21.9 | 85.5 |
| Days ≥ 1" Snow Depth | 25 | 21 | 10 | 1 | 0 | 0 | 0 | 0 | 0 | 0 | 4 | 18 | 79 |

## HOUGHTON LAKE 6 WSW *Missaukee County* ELEVATION 1142 ft LAT/LONG 44° 20' N / 84° 49' W

| | JAN | FEB | MAR | APR | MAY | JUN | JUL | AUG | SEP | OCT | NOV | DEC | YEAR |
|---|---|---|---|---|---|---|---|---|---|---|---|---|---|
| Maximum Temp °F | 26.0 | 29.0 | 39.4 | 53.4 | 67.2 | 75.7 | 80.7 | 77.8 | 68.9 | 56.4 | 42.5 | 30.7 | 54.0 |
| Minimum Temp °F | 8.1 | 7.4 | 17.2 | 30.5 | 40.0 | 48.8 | 53.2 | 51.0 | 43.9 | 34.4 | 26.1 | 15.5 | 31.3 |
| Mean Temp °F | 17.1 | 18.2 | 28.4 | 42.0 | 53.7 | 62.3 | 67.0 | 64.4 | 56.4 | 45.4 | 34.3 | 23.1 | 42.7 |
| Days Max Temp ≥ 90 °F | 0 | 0 | 0 | 0 | 0 | 1 | 3 | 1 | 0 | 0 | 0 | 0 | 5 |
| Days Max Temp ≤ 32 °F | 23 | 18 | 8 | 1 | 0 | 0 | 0 | 0 | 0 | 0 | 5 | 18 | 73 |
| Days Min Temp ≤ 32 °F | 31 | 28 | 28 | 18 | 8 | 1 | 0 | 0 | 4 | 15 | 24 | 30 | 187 |
| Days Min Temp ≤ 0 °F | 9 | 10 | 3 | 0 | 0 | 0 | 0 | 0 | 0 | 0 | 0 | 4 | 26 |
| Heating Degree Days | 1481 | 1315 | 1130 | 688 | 364 | 136 | 52 | 92 | 270 | 603 | 913 | 1293 | 8337 |
| Cooling Degree Days | 0 | 0 | 0 | 4 | 17 | 50 | 116 | 78 | 17 | 1 | 0 | 0 | 283 |
| Total Precipitation (") | 1.45 | 1.14 | 1.53 | 2.34 | 2.57 | 3.18 | 2.78 | 3.64 | 3.64 | 2.93 | 2.49 | 1.80 | 29.49 |
| Days ≥ 0.1" Precip | 4 | 4 | 5 | 6 | 6 | 6 | 5 | 7 | 7 | 7 | 7 | *5* | 69 |
| Total Snowfall (") | *14.3* | 10.3 | *8.1* | *2.2* | 0.0 | 0.0 | 0.0 | 0.0 | 0.0 | 0.4 | *7.7* | na | na |
| Days ≥ 1" Snow Depth | *30* | *27* | *20* | *3* | 0 | 0 | 0 | 0 | 0 | 0 | na | na | na |

## HOUGHTON LAKE ROSCMN *Roscommon County* ELEVATION 1155 ft LAT/LONG 44° 22' N / 84° 41' W

| | JAN | FEB | MAR | APR | MAY | JUN | JUL | AUG | SEP | OCT | NOV | DEC | YEAR |
|---|---|---|---|---|---|---|---|---|---|---|---|---|---|
| Maximum Temp °F | 25.4 | 28.3 | 38.6 | 52.7 | 66.5 | 74.8 | 79.5 | 76.6 | 67.7 | 55.4 | 41.7 | 30.3 | 53.1 |
| Minimum Temp °F | 8.8 | 8.7 | 18.4 | 31.8 | 42.2 | 50.8 | 55.7 | 54.0 | 46.9 | 37.3 | 28.2 | 16.9 | 33.3 |
| Mean Temp °F | 17.1 | 18.5 | 28.5 | 42.3 | 54.4 | 62.8 | 67.6 | 65.3 | 57.3 | 46.4 | 35.0 | 23.6 | 43.2 |
| Days Max Temp ≥ 90 °F | 0 | 0 | 0 | 0 | 0 | 1 | 2 | 1 | 0 | 0 | 0 | 0 | 4 |
| Days Max Temp ≤ 32 °F | 24 | 19 | 9 | 1 | 0 | 0 | 0 | 0 | 0 | 0 | 5 | 19 | 77 |
| Days Min Temp ≤ 32 °F | 31 | 28 | 28 | 17 | 5 | 0 | 0 | 0 | 2 | 9 | 22 | 30 | 172 |
| Days Min Temp ≤ 0 °F | 9 | 8 | 3 | 0 | 0 | 0 | 0 | 0 | 0 | 0 | 0 | 3 | 23 |
| Heating Degree Days | 1479 | 1306 | 1125 | 677 | 343 | 120 | 41 | 74 | 247 | 572 | 894 | 1275 | 8153 |
| Cooling Degree Days | 0 | 0 | 0 | 4 | 23 | 59 | 143 | 98 | 25 | 0 | 0 | 0 | 352 |
| Total Precipitation (") | 1.54 | 1.16 | 2.02 | 2.40 | 2.53 | 3.04 | 2.61 | 3.51 | 3.26 | 2.41 | 2.35 | 1.97 | 28.80 |
| Days ≥ 0.1" Precip | 5 | 4 | 6 | 6 | 6 | 6 | 5 | 7 | 7 | 6 | 6 | 5 | 69 |
| Total Snowfall (") | 19.3 | 12.9 | 11.4 | 4.1 | 0.3 | 0.0 | 0.0 | 0.0 | 0.0 | 0.7 | 9.4 | 16.4 | 74.5 |
| Days ≥ 1" Snow Depth | 30 | 27 | 19 | 3 | 0 | 0 | 0 | 0 | 0 | 0 | 7 | 23 | 109 |

## IONIA 2 SSW *Ionia County* ELEVATION 758 ft LAT/LONG 42° 59' N / 85° 3' W

| | JAN | FEB | MAR | APR | MAY | JUN | JUL | AUG | SEP | OCT | NOV | DEC | YEAR |
|---|---|---|---|---|---|---|---|---|---|---|---|---|---|
| Maximum Temp °F | 29.5 | 32.5 | 43.7 | 57.7 | 70.6 | 79.5 | 83.4 | 81.1 | 73.3 | 60.7 | 46.5 | 34.4 | 57.7 |
| Minimum Temp °F | 13.1 | 14.1 | 23.4 | 34.2 | 44.7 | 53.9 | 58.3 | 56.0 | 48.9 | 38.6 | 29.9 | 20.0 | 36.3 |
| Mean Temp °F | 21.3 | 23.3 | 33.5 | 46.0 | 57.7 | 66.7 | 70.9 | 68.6 | 61.1 | 49.7 | 38.2 | 27.2 | 47.0 |
| Days Max Temp ≥ 90 °F | 0 | 0 | 0 | 0 | 1 | 2 | 5 | 3 | 1 | 0 | 0 | 0 | 12 |
| Days Max Temp ≤ 32 °F | 18 | 14 | 5 | 0 | 0 | 0 | 0 | 0 | 0 | 0 | 2 | 13 | 52 |
| Days Min Temp ≤ 32 °F | 30 | 27 | 25 | 14 | 3 | 0 | 0 | 0 | 1 | 8 | 20 | 28 | 156 |
| Days Min Temp ≤ 0 °F | 5 | 4 | 0 | 0 | 0 | 0 | 0 | 0 | 0 | 0 | 0 | 2 | 11 |
| Heating Degree Days | 1347 | 1170 | 971 | 570 | 255 | 61 | 14 | 33 | 161 | 473 | 798 | 1165 | 7018 |
| Cooling Degree Days | *0* | *0* | *0* | *8* | *33* | 107 | 202 | 139 | 44 | 2 | 0 | 0 | 535 |
| Total Precipitation (") | 2.02 | 1.68 | 2.68 | 3.22 | 2.95 | 3.72 | 2.90 | 4.22 | 4.00 | 3.07 | 3.16 | 2.72 | 36.34 |
| Days ≥ 0.1" Precip | 6 | 4 | 6 | 8 | 6 | 6 | 6 | 7 | 6 | 7 | 7 | *7* | 76 |
| Total Snowfall (") | 13.3 | 10.2 | 7.9 | 1.9 | 0.0 | 0.0 | 0.0 | 0.0 | 0.0 | 0.4 | 4.8 | 12.8 | 51.3 |
| Days ≥ 1" Snow Depth | 25 | 20 | 11 | *2* | 0 | 0 | 0 | 0 | 0 | 0 | 3 | *19* | 80 |

## IRON MTN-KINGSFORD W *Dickinson County* ELEVATION 1060 ft LAT/LONG 45° 47' N / 88° 5' W

| | JAN | FEB | MAR | APR | MAY | JUN | JUL | AUG | SEP | OCT | NOV | DEC | YEAR |
|---|---|---|---|---|---|---|---|---|---|---|---|---|---|
| Maximum Temp °F | 23.4 | 28.3 | 38.9 | 53.7 | 67.4 | 75.5 | 80.0 | 77.5 | 67.8 | 55.8 | 39.9 | 27.8 | 53.0 |
| Minimum Temp °F | 1.4 | 4.9 | 16.3 | 29.3 | 40.3 | 49.9 | 55.2 | 53.5 | 45.0 | 34.8 | 23.3 | 10.4 | 30.4 |
| Mean Temp °F | 12.4 | 16.6 | 27.6 | 41.5 | 53.9 | 62.7 | 67.6 | 65.5 | 56.4 | 45.3 | 31.6 | 19.1 | 41.7 |
| Days Max Temp ≥ 90 °F | 0 | 0 | 0 | 0 | 0 | 1 | 2 | 1 | 0 | 0 | 0 | 0 | 4 |
| Days Max Temp ≤ 32 °F | 25 | 19 | 8 | 1 | 0 | 0 | 0 | 0 | 0 | 0 | 7 | 21 | 81 |
| Days Min Temp ≤ 32 °F | 31 | 28 | 29 | 20 | 7 | 0 | 0 | 0 | 2 | 13 | 26 | 31 | 187 |
| Days Min Temp ≤ 0 °F | 15 | 11 | 4 | 0 | 0 | 0 | 0 | 0 | 0 | 0 | 1 | 7 | 38 |
| Heating Degree Days | 1625 | 1361 | 1152 | 699 | 356 | 123 | 42 | 71 | 271 | 604 | 995 | 1415 | 8714 |
| Cooling Degree Days | 0 | 0 | 0 | 2 | 19 | 64 | 132 | 102 | 18 | 0 | 0 | 0 | 337 |
| Total Precipitation (") | 1.18 | 0.82 | 1.60 | 2.28 | 3.19 | 3.59 | 3.18 | 3.51 | 3.77 | 2.69 | 2.05 | 1.49 | 29.35 |
| Days ≥ 0.1" Precip | 4 | 3 | 4 | 6 | 6 | 8 | 7 | 7 | 7 | 6 | 5 | *4* | 67 |
| Total Snowfall (") | 15.9 | 9.4 | 10.7 | 4.4 | 0.8 | 0.0 | 0.0 | 0.0 | 0.1 | 0.4 | 7.0 | 15.1 | 63.8 |
| Days ≥ 1" Snow Depth | 31 | 28 | 22 | 4 | 0 | 0 | 0 | 0 | 0 | 0 | 8 | 27 | 120 |

## IRONWOOD *Gogebic County* ELEVATION 1522 ft LAT/LONG 46° 27' N / 90° 11' W

| | JAN | FEB | MAR | APR | MAY | JUN | JUL | AUG | SEP | OCT | NOV | DEC | YEAR |
|---|---|---|---|---|---|---|---|---|---|---|---|---|---|
| Maximum Temp °F | 19.6 | 24.9 | 36.0 | 50.8 | 64.7 | 72.9 | 77.5 | 75.0 | 65.2 | 53.3 | 37.0 | 24.6 | 50.1 |
| Minimum Temp °F | -0.6 | 2.1 | 13.8 | 28.6 | 40.2 | 49.0 | 54.3 | 52.2 | 44.3 | 34.3 | 21.4 | 7.6 | 28.9 |
| Mean Temp °F | 9.5 | 13.6 | 24.9 | 39.7 | 52.5 | 61.0 | 66.0 | 63.6 | 54.7 | 43.8 | 29.3 | 16.1 | 39.6 |
| Days Max Temp ≥ 90 °F | 0 | 0 | 0 | 0 | 0 | 0 | 1 | 1 | 0 | 0 | 0 | 0 | 2 |
| Days Max Temp ≤ 32 °F | 26 | 21 | 12 | 2 | 0 | 0 | 0 | 0 | 0 | 1 | 11 | 24 | 97 |
| Days Min Temp ≤ 32 °F | 31 | 28 | 29 | 21 | 8 | 1 | 0 | 0 | 3 | 14 | 26 | 31 | 192 |
| Days Min Temp ≤ 0 °F | 16 | 12 | 6 | 0 | 0 | 0 | 0 | 0 | 0 | 0 | 1 | 9 | 44 |
| Heating Degree Days | 1717 | 1448 | 1235 | 753 | 399 | 164 | 67 | 109 | 319 | 651 | 1063 | 1510 | 9435 |
| Cooling Degree Days | 0 | 0 | 0 | 3 | 17 | 46 | 92 | 67 | 12 | 0 | 0 | 0 | 237 |
| Total Precipitation (") | 1.91 | 1.18 | 1.98 | 2.18 | 2.97 | 4.12 | 3.88 | 3.87 | 3.95 | 3.53 | 2.99 | 2.11 | 34.67 |
| Days ≥ 0.1" Precip | 6 | 4 | 6 | 6 | 7 | 8 | 7 | 8 | 9 | 8 | 8 | 6 | 83 |
| Total Snowfall (") | 42.5 | 24.6 | 21.9 | 10.6 | 1.5 | 0.0 | 0.0 | 0.0 | 0.0 | 6.1 | 27.7 | 39.8 | 174.7 |
| Days ≥ 1" Snow Depth | 31 | 28 | 29 | 11 | 1 | 0 | 0 | 0 | 0 | 3 | 18 | 31 | 152 |

## JACKSON REYNOLDS FLD *Jackson County* ELEVATION 1003 ft LAT/LONG 42° 16' N / 84° 28' W

| | JAN | FEB | MAR | APR | MAY | JUN | JUL | AUG | SEP | OCT | NOV | DEC | YEAR |
|---|---|---|---|---|---|---|---|---|---|---|---|---|---|
| Maximum Temp °F | 29.0 | 32.2 | 43.8 | 57.6 | 69.5 | 79.0 | 82.9 | 80.6 | 72.7 | 60.0 | 46.8 | 34.4 | 57.4 |
| Minimum Temp °F | 14.9 | 16.3 | 26.0 | 36.9 | 46.7 | 56.3 | 61.3 | 59.1 | 51.8 | 40.5 | 31.7 | 21.3 | 38.6 |
| Mean Temp °F | 22.0 | 24.3 | 34.9 | 47.3 | 58.2 | 67.6 | 72.2 | 69.9 | 62.3 | 50.3 | 39.3 | 27.9 | 48.0 |
| Days Max Temp ≥ 90 °F | 0 | 0 | 0 | 0 | 0 | 3 | 5 | 3 | 1 | 0 | 0 | 0 | 12 |
| Days Max Temp ≤ 32 °F | 19 | 14 | 6 | 0 | 0 | 0 | 0 | 0 | 0 | 0 | 3 | 13 | 55 |
| Days Min Temp ≤ 32 °F | 29 | 26 | 23 | 11 | 2 | 0 | 0 | 0 | 0 | 6 | 17 | 27 | 141 |
| Days Min Temp ≤ 0 °F | 5 | 4 | 0 | 0 | 0 | 0 | 0 | 0 | 0 | 0 | 0 | 2 | 11 |
| Heating Degree Days | 1325 | 1143 | 926 | 533 | 248 | 54 | 10 | 26 | 147 | 456 | 765 | 1143 | 6776 |
| Cooling Degree Days | 0 | 0 | 1 | 9 | 39 | 127 | 235 | 179 | 64 | 4 | 0 | 0 | 658 |
| Total Precipitation (") | 1.35 | 1.23 | 2.04 | 2.74 | 2.94 | 3.36 | 3.41 | 3.12 | 3.40 | 2.27 | 2.63 | 2.16 | 30.65 |
| Days ≥ 0.1" Precip | 4 | 4 | 5 | 7 | 6 | 7 | 6 | 6 | 7 | 6 | 6 | 6 | 70 |
| Total Snowfall (") | 10.9 | 7.7 | 6.3 | 1.8 | 0.0 | 0.0 | 0.0 | 0.0 | 0.0 | 0.1 | 2.7 | 8.8 | 38.3 |
| Days ≥ 1" Snow Depth | 23 | 18 | 9 | 1 | 0 | 0 | 0 | 0 | 0 | 0 | 3 | 15 | 69 |

## KALAMAZOO STATE HOSP *Kalamazoo County* ELEVATION 932 ft LAT/LONG 42° 17' N / 85° 36' W

| | JAN | FEB | MAR | APR | MAY | JUN | JUL | AUG | SEP | OCT | NOV | DEC | YEAR |
|---|---|---|---|---|---|---|---|---|---|---|---|---|---|
| Maximum Temp °F | 30.1 | 35.1 | 46.6 | 60.6 | 72.0 | 80.9 | 84.6 | 82.4 | 75.0 | 62.9 | 48.8 | 36.5 | 59.6 |
| Minimum Temp °F | 15.7 | 18.7 | 27.5 | 38.4 | 48.5 | 57.8 | 62.1 | 60.5 | 53.6 | 43.1 | 33.1 | 23.0 | 40.2 |
| Mean Temp °F | 23.0 | 26.9 | 37.1 | 49.5 | 60.3 | 69.4 | 73.4 | 71.5 | 64.3 | 53.0 | 41.0 | 29.8 | 49.9 |
| Days Max Temp ≥ 90 °F | 0 | 0 | 0 | 0 | 1 | 3 | 6 | 3 | 1 | 0 | 0 | 0 | 14 |
| Days Max Temp ≤ 32 °F | 19 | 12 | 3 | 0 | 0 | 0 | 0 | 0 | 0 | 0 | 1 | 10 | 45 |
| Days Min Temp ≤ 32 °F | 29 | 25 | 22 | 9 | 1 | 0 | 0 | 0 | 0 | 4 | 15 | 27 | 132 |
| Days Min Temp ≤ 0 °F | 3 | 2 | 0 | 0 | 0 | 0 | 0 | 0 | 0 | 0 | 0 | 0 | 5 |
| Heating Degree Days | 1297 | 1070 | 861 | 469 | 197 | 35 | 5 | 13 | 106 | 375 | 716 | 1085 | 6229 |
| Cooling Degree Days | 0 | na | na | 17 | na | na | 278 | 227 | 95 | na | na | 0 | na |
| Total Precipitation (") | 2.06 | 1.58 | 2.41 | 3.53 | 3.33 | 3.98 | 3.82 | 3.76 | 3.94 | 3.05 | 3.03 | 2.99 | 37.48 |
| Days ≥ 0.1" Precip | 6 | 4 | 7 | 9 | 7 | 7 | 6 | 7 | 7 | 7 | 7 | 8 | 82 |
| Total Snowfall (") | 18.1 | 11.3 | 7.3 | 2.4 | 0.0 | 0.0 | 0.0 | 0.0 | 0.0 | 0.3 | na | na | na |
| Days ≥ 1" Snow Depth | 19 | 16 | 5 | 1 | 0 | 0 | 0 | 0 | 0 | 0 | na | na | na |

## LAKE CITY EXP FARM *Missaukee County* ELEVATION 1230 ft LAT/LONG 44° 18' N / 85° 12' W

| | JAN | FEB | MAR | APR | MAY | JUN | JUL | AUG | SEP | OCT | NOV | DEC | YEAR |
|---|---|---|---|---|---|---|---|---|---|---|---|---|---|
| Maximum Temp °F | 25.7 | 28.7 | 38.5 | 52.8 | 66.5 | 75.4 | 80.2 | 77.5 | 68.6 | 56.1 | 42.1 | 30.5 | 53.6 |
| Minimum Temp °F | 8.4 | 7.7 | 16.9 | 30.6 | 40.3 | 49.2 | 53.9 | 52.2 | 44.9 | 35.2 | 26.4 | 15.4 | 31.8 |
| Mean Temp °F | 17.1 | 18.2 | 27.7 | 41.8 | 53.5 | 62.3 | 67.1 | 64.9 | 56.8 | 45.7 | 34.2 | 23.0 | 42.7 |
| Days Max Temp ≥ 90 °F | 0 | 0 | 0 | 0 | 0 | 1 | 2 | 1 | 0 | 0 | 0 | 0 | 4 |
| Days Max Temp ≤ 32 °F | 24 | 18 | 9 | 1 | 0 | 0 | 0 | 0 | 0 | 0 | 5 | 18 | 75 |
| Days Min Temp ≤ 32 °F | 31 | 28 | 28 | 18 | 7 | 1 | 0 | 0 | 3 | 14 | 23 | 30 | 183 |
| Days Min Temp ≤ 0 °F | 8 | 9 | 4 | 0 | 0 | 0 | 0 | 0 | 0 | 0 | 0 | 4 | 25 |
| Heating Degree Days | 1479 | 1316 | 1148 | 692 | 366 | 130 | 47 | 80 | 261 | 594 | 916 | 1295 | 8324 |
| Cooling Degree Days | 0 | 0 | 0 | 3 | 14 | 50 | 118 | 78 | 17 | 0 | 0 | 0 | 280 |
| Total Precipitation (") | 1.41 | 1.15 | 1.89 | 2.90 | 2.65 | 3.09 | 2.71 | 3.56 | 3.93 | 3.06 | 2.74 | 1.88 | 30.97 |
| Days ≥ 0.1" Precip | 4 | 3 | 5 | 7 | 6 | 6 | 6 | 7 | 8 | 7 | 7 | 5 | 71 |
| Total Snowfall (") | 20.1 | 15.2 | 11.1 | 4.2 | 0.4 | 0.0 | 0.0 | 0.0 | 0.0 | 1.1 | 9.0 | 15.4 | 76.5 |
| Days ≥ 1" Snow Depth | 30 | 27 | 22 | 4 | 0 | 0 | 0 | 0 | 0 | 0 | 8 | 24 | 115 |

## LANSING CAPITAL CITY *Clinton County* ELEVATION 873 ft LAT/LONG 42° 47' N / 84° 36' W

| | JAN | FEB | MAR | APR | MAY | JUN | JUL | AUG | SEP | OCT | NOV | DEC | YEAR |
|---|---|---|---|---|---|---|---|---|---|---|---|---|---|
| Maximum Temp °F | 28.9 | 31.6 | 42.8 | 56.7 | 69.2 | 78.2 | 82.6 | 80.4 | 72.2 | 59.6 | 46.0 | 34.1 | 56.9 |
| Minimum Temp °F | 13.5 | 14.6 | 24.2 | 35.2 | 44.8 | 54.4 | 59.0 | 57.0 | 49.6 | 39.0 | 30.6 | 20.3 | 36.9 |
| Mean Temp °F | 21.2 | 23.1 | 33.5 | 46.0 | 57.0 | 66.4 | 70.8 | 68.7 | 60.9 | 49.4 | 38.3 | 27.2 | 46.9 |
| Days Max Temp ≥ 90 °F | 0 | 0 | 0 | 0 | 0 | 2 | 4 | 2 | 1 | 0 | 0 | 0 | 9 |
| Days Max Temp ≤ 32 °F | 19 | 15 | 6 | 0 | 0 | 0 | 0 | 0 | 0 | 0 | 3 | 13 | 56 |
| Days Min Temp ≤ 32 °F | 29 | 26 | 24 | 13 | 3 | 0 | 0 | 0 | 1 | 8 | 18 | 27 | 149 |
| Days Min Temp ≤ 0 °F | 6 | 5 | 1 | 0 | 0 | 0 | 0 | 0 | 0 | 0 | 0 | 2 | 14 |
| Heating Degree Days | 1350 | 1176 | 969 | 570 | 276 | 69 | 16 | 35 | 173 | 484 | 793 | 1165 | 7076 |
| Cooling Degree Days | 0 | 0 | 0 | 8 | 39 | 110 | 217 | 161 | 54 | 3 | 0 | 0 | 592 |
| Total Precipitation (") | 1.56 | 1.37 | 2.34 | 3.06 | 2.57 | 3.77 | 2.68 | 3.33 | 3.62 | 2.33 | 2.86 | 2.38 | 31.87 |
| Days ≥ 0.1" Precip | 4 | 4 | 6 | 8 | 5 | 7 | 5 | 6 | 7 | 6 | 7 | 6 | 71 |
| Total Snowfall (") | 13.7 | 10.8 | 8.9 | 3.0 | 0.0 | 0.0 | 0.0 | 0.0 | 0.0 | 0.6 | 5.0 | 12.6 | 54.6 |
| Days ≥ 1" Snow Depth | 23 | 18 | 10 | 2 | 0 | 0 | 0 | 0 | 0 | 0 | 3 | 16 | 72 |

## LAPEER WWTP *Lapeer County* ELEVATION 820 ft LAT/LONG 43° 2' N / 83° 20' W

| | JAN | FEB | MAR | APR | MAY | JUN | JUL | AUG | SEP | OCT | NOV | DEC | YEAR |
|---|---|---|---|---|---|---|---|---|---|---|---|---|---|
| Maximum Temp °F | 29.3 | 32.7 | 43.9 | 58.0 | 69.8 | 78.6 | 82.8 | 80.4 | 73.2 | 60.7 | 46.8 | 34.5 | 57.6 |
| Minimum Temp °F | 13.1 | 14.9 | 23.8 | 35.2 | 44.9 | 53.7 | 58.4 | 56.2 | 49.8 | 39.4 | 30.6 | 20.1 | 36.7 |
| Mean Temp °F | 21.2 | 23.9 | 33.9 | 46.6 | 57.3 | 66.2 | 70.7 | 68.3 | 61.5 | 50.2 | 38.7 | 27.3 | 47.2 |
| Days Max Temp ≥ 90 °F | 0 | 0 | 0 | 0 | 0 | 2 | 4 | 2 | 1 | 0 | 0 | 0 | 9 |
| Days Max Temp ≤ 32 °F | 19 | 14 | 5 | 0 | 0 | 0 | 0 | 0 | 0 | 0 | 2 | 13 | 53 |
| Days Min Temp ≤ 32 °F | 30 | 27 | 25 | 13 | 3 | 0 | 0 | 0 | 1 | 8 | 18 | 28 | 153 |
| Days Min Temp ≤ 0 °F | 5 | 4 | 1 | 0 | 0 | 0 | 0 | 0 | 0 | 0 | 0 | 2 | 12 |
| Heating Degree Days | 1350 | 1157 | 957 | 552 | 267 | 70 | 17 | 37 | 156 | 458 | 785 | 1161 | 6967 |
| Cooling Degree Days | 0 | 0 | 1 | 9 | 45 | 98 | 213 | 148 | 55 | 2 | 0 | 0 | 571 |
| Total Precipitation (") | 1.29 | 1.10 | 1.92 | 2.87 | 2.77 | 3.38 | 2.94 | 3.41 | 3.62 | 2.67 | 2.86 | 1.89 | 30.72 |
| Days ≥ 0.1" Precip | 4 | 4 | 6 | 7 | 6 | 7 | 6 | 7 | 7 | 6 | 7 | 5 | 72 |
| Total Snowfall (") | 8.8 | 9.5 | 7.7 | 1.5 | 0.0 | 0.0 | 0.0 | 0.0 | 0.0 | 0.1 | 3.1 | 9.8 | 40.5 |
| Days ≥ 1" Snow Depth | 20 | 18 | 7 | 1 | 0 | 0 | 0 | 0 | 0 | 0 | 2 | 16 | 64 |

## LUDINGTON 4 SE *Mason County* ELEVATION 610 ft LAT/LONG 43° 55' N / 86° 25' W

| | JAN | FEB | MAR | APR | MAY | JUN | JUL | AUG | SEP | OCT | NOV | DEC | YEAR |
|---|---|---|---|---|---|---|---|---|---|---|---|---|---|
| Maximum Temp °F | 29.2 | 32.2 | 42.1 | 55.4 | 67.0 | 75.5 | 80.6 | ***78.0*** | ***70.5*** | 58.6 | 45.2 | 34.1 | 55.7 |
| Minimum Temp °F | 15.7 | 15.7 | 23.5 | 34.0 | 42.7 | 52.0 | 57.7 | ***56.7*** | ***50.5*** | 40.6 | 31.3 | 21.5 | 36.8 |
| Mean Temp °F | 22.5 | 24.0 | 32.8 | 44.7 | 54.8 | 63.8 | 69.2 | ***67.4*** | ***60.5*** | 49.7 | 38.3 | 27.8 | 46.3 |
| Days Max Temp ≥ 90 °F | 0 | 0 | 0 | 0 | 0 | 1 | 1 | 1 | 0 | 0 | 0 | 0 | 3 |
| Days Max Temp ≤ 32 °F | 19 | 13 | 5 | 0 | 0 | 0 | 0 | 0 | 0 | 0 | 2 | 12 | 51 |
| Days Min Temp ≤ 32 °F | 30 | 27 | 24 | 14 | 4 | 0 | 0 | 0 | 0 | 5 | 17 | 28 | 149 |
| Days Min Temp ≤ 0 °F | 3 | 2 | 1 | 0 | 0 | 0 | 0 | 0 | 0 | 0 | 0 | 0 | 6 |
| Heating Degree Days | 1311 | 1151 | 991 | 607 | 327 | 102 | 26 | ***45*** | ***170*** | 473 | 796 | 1146 | 7145 |
| Cooling Degree Days | 0 | 0 | 0 | 6 | 18 | 80 | 181 | ***140*** | na | 1 | 0 | 0 | na |
| Total Precipitation (") | 1.76 | 1.26 | 1.95 | 2.88 | 2.61 | 3.29 | 2.39 | 4.06 | 3.95 | 3.70 | 3.39 | 2.38 | 33.62 |
| Days ≥ 0.1" Precip | 5 | 3 | 5 | 7 | 5 | 6 | ***4*** | 6 | 7 | 7 | ***8*** | 6 | 69 |
| Total Snowfall (") | na | ***19.8*** | ***9.4*** | ***3.0*** | 0.1 | 0.0 | 0.0 | 0.0 | 0.0 | 0.3 | na | na | na |
| Days ≥ 1" Snow Depth | 27 | 23 | 14 | 1 | 0 | 0 | 0 | 0 | 0 | 0 | 5 | 20 | 90 |

## LUPTON 1 S *Ogemaw County* ELEVATION 889 ft LAT/LONG 44° 25' N / 84° 2' W

| | JAN | FEB | MAR | APR | MAY | JUN | JUL | AUG | SEP | OCT | NOV | DEC | YEAR |
|---|---|---|---|---|---|---|---|---|---|---|---|---|---|
| Maximum Temp °F | 26.3 | 30.6 | 40.7 | 55.3 | 68.6 | 77.1 | 82.1 | 79.3 | 70.5 | 58.1 | 43.6 | 31.6 | 55.3 |
| Minimum Temp °F | 4.9 | 5.8 | 16.2 | 28.8 | 38.0 | 46.9 | 51.7 | 49.8 | 42.8 | 32.8 | 24.8 | 14.4 | 29.7 |
| Mean Temp °F | 15.7 | 18.2 | 28.5 | 42.0 | 53.3 | 62.0 | 66.9 | 64.6 | 56.7 | 45.5 | 34.2 | 23.0 | 42.6 |
| Days Max Temp ≥ 90 °F | 0 | 0 | 0 | 0 | 1 | 2 | 4 | 2 | 0 | 0 | 0 | 0 | 9 |
| Days Max Temp ≤ 32 °F | 22 | 16 | 6 | 0 | 0 | 0 | 0 | 0 | 0 | 0 | 3 | 16 | 63 |
| Days Min Temp ≤ 32 °F | 30 | 27 | 28 | 20 | 10 | 1 | 0 | 0 | 5 | 17 | 24 | 30 | 192 |
| Days Min Temp ≤ 0 °F | 12 | 10 | 4 | 0 | 0 | 0 | 0 | 0 | 0 | 0 | 0 | 5 | 31 |
| Heating Degree Days | 1523 | 1318 | 1126 | 685 | 370 | 137 | 49 | 83 | 262 | 599 | 916 | 1295 | 8363 |
| Cooling Degree Days | ***0*** | 0 | 0 | 3 | ***14*** | 48 | 129 | 82 | 16 | 1 | ***0*** | 0 | 293 |
| Total Precipitation (") | 1.62 | 1.26 | 2.04 | 2.52 | 2.58 | 3.22 | 2.88 | 3.67 | 3.54 | 2.56 | 2.40 | 2.02 | 30.31 |
| Days ≥ 0.1" Precip | 4 | 4 | 5 | 7 | 6 | 7 | 5 | 7 | 7 | 6 | 6 | 6 | 70 |
| Total Snowfall (") | 17.0 | 10.3 | 9.2 | 3.0 | 0.2 | 0.0 | 0.0 | 0.0 | 0.0 | 0.4 | ***4.7*** | 12.5 | 57.3 |
| Days ≥ 1" Snow Depth | 29 | 27 | 22 | 4 | 0 | 0 | 0 | 0 | 0 | 0 | 5 | 22 | 109 |

## MANISTEE 3 SE *Manistee County* ELEVATION 600 ft LAT/LONG 44° 13' N / 86° 18' W

| | JAN | FEB | MAR | APR | MAY | JUN | JUL | AUG | SEP | OCT | NOV | DEC | YEAR |
|---|---|---|---|---|---|---|---|---|---|---|---|---|---|
| Maximum Temp °F | 29.2 | 31.9 | 41.1 | 54.2 | 66.1 | 75.5 | 80.1 | 77.8 | 70.6 | 58.7 | 44.9 | ***33.9*** | 55.3 |
| Minimum Temp °F | 16.5 | 16.2 | 23.8 | 34.7 | 43.6 | 52.5 | 58.5 | 57.6 | 51.3 | 41.8 | 32.0 | 22.3 | 37.6 |
| Mean Temp °F | 22.9 | 23.9 | 32.4 | 44.4 | 54.9 | 64.0 | 69.3 | 67.7 | 61.0 | 50.3 | 38.4 | ***28.3*** | 46.5 |
| Days Max Temp ≥ 90 °F | 0 | 0 | 0 | 0 | 0 | 1 | 2 | 1 | 0 | 0 | 0 | 0 | 4 |
| Days Max Temp ≤ 32 °F | 19 | 14 | 6 | 0 | 0 | 0 | 0 | 0 | 0 | 0 | 2 | ***12*** | 53 |
| Days Min Temp ≤ 32 °F | 29 | 27 | 25 | 13 | 3 | 0 | 0 | 0 | 0 | 4 | 16 | 27 | 144 |
| Days Min Temp ≤ 0 °F | 2 | 2 | 1 | 0 | 0 | 0 | 0 | 0 | 0 | 0 | 0 | 0 | 5 |
| Heating Degree Days | 1299 | 1159 | 1004 | 615 | 328 | 102 | 30 | 43 | 160 | 453 | 791 | ***1132*** | 7116 |
| Cooling Degree Days | ***0*** | ***0*** | ***0*** | ***7*** | 22 | 72 | ***167*** | ***130*** | 41 | ***0*** | 0 | ***0*** | 439 |
| Total Precipitation (") | 2.10 | 1.43 | 1.98 | 2.99 | 2.46 | 3.45 | 2.85 | 3.52 | 3.96 | 3.33 | 3.02 | 2.35 | 33.44 |
| Days ≥ 0.1" Precip | 6 | 4 | 5 | 7 | 6 | 6 | 5 | 6 | 8 | 7 | 8 | 6 | 74 |
| Total Snowfall (") | ***31.4*** | 17.3 | ***8.0*** | ***1.3*** | 0.0 | 0.0 | 0.0 | 0.0 | 0.0 | 0.2 | ***3.7*** | na | na |
| Days ≥ 1" Snow Depth | 25 | ***23*** | ***12*** | 1 | 0 | 0 | 0 | 0 | 0 | 0 | 3 | ***17*** | 81 |

## MANISTIQUE *Schoolcraft County* ELEVATION 610 ft LAT/LONG 45° 58' N / 86° 15' W

| | JAN | FEB | MAR | APR | MAY | JUN | JUL | AUG | SEP | OCT | NOV | DEC | YEAR |
|---|---|---|---|---|---|---|---|---|---|---|---|---|---|
| Maximum Temp °F | 25.1 | 26.8 | 35.4 | 46.8 | 58.9 | 68.2 | 74.0 | 73.4 | 64.9 | 53.0 | 40.8 | 30.1 | 49.8 |
| Minimum Temp °F | 7.1 | 7.7 | 17.3 | 29.4 | 39.6 | 48.2 | 54.5 | 53.8 | 46.4 | 36.6 | 26.4 | 14.8 | 31.8 |
| Mean Temp °F | 16.1 | 17.3 | 26.4 | 38.1 | 49.3 | 58.2 | 64.3 | 63.6 | 55.7 | 44.9 | 33.6 | 22.5 | 40.8 |
| Days Max Temp ≥ 90 °F | 0 | 0 | 0 | 0 | 0 | 0 | 0 | 0 | 0 | 0 | 0 | 0 | 0 |
| Days Max Temp ≤ 32 °F | 23 | 20 | 9 | 1 | 0 | 0 | 0 | 0 | 0 | 0 | 5 | 17 | 75 |
| Days Min Temp ≤ 32 °F | 31 | 28 | 29 | 19 | 6 | 0 | 0 | 0 | 2 | 10 | 23 | 29 | 177 |
| Days Min Temp ≤ 0 °F | 10 | 8 | 3 | 0 | 0 | 0 | 0 | 0 | 0 | 0 | 0 | 5 | 26 |
| Heating Degree Days | 1511 | 1339 | 1192 | 802 | 480 | 209 | 73 | 85 | 279 | 617 | 938 | 1311 | 8836 |
| Cooling Degree Days | 0 | 0 | 0 | 1 | 1 | 14 | 56 | 58 | 4 | 0 | 0 | 0 | 134 |
| Total Precipitation (") | ***1.73*** | 1.17 | 1.95 | 2.21 | 2.56 | 3.18 | 3.06 | 3.28 | 3.59 | 3.01 | 2.79 | 1.90 | 30.43 |
| Days ≥ 0.1" Precip | ***4*** | 4 | ***4*** | 6 | ***6*** | 7 | 6 | 7 | 8 | ***7*** | ***6*** | ***5*** | 70 |
| Total Snowfall (") | na | ***12.2*** | ***8.5*** | 2.3 | 0.0 | 0.0 | 0.0 | 0.0 | 0.0 | 0.3 | ***5.8*** | ***16.9*** | na |
| Days ≥ 1" Snow Depth | 29 | 28 | 25 | 5 | 0 | 0 | 0 | 0 | 0 | 0 | ***6*** | ***23*** | 116 |

## MAPLE CITY *Leelanau County* ELEVATION 732 ft LAT/LONG 44° 51 ' N / 85° 51 ' W

| | JAN | FEB | MAR | APR | MAY | JUN | JUL | AUG | SEP | OCT | NOV | DEC | YEAR |
|---|---|---|---|---|---|---|---|---|---|---|---|---|---|
| Maximum Temp °F | 28.1 | 30.8 | 40.6 | 54.3 | 67.5 | 76.7 | 81.3 | 78.9 | 70.7 | 58.9 | 44.4 | 33.2 | 55.5 |
| Minimum Temp °F | 14.6 | 13.8 | 21.2 | 32.1 | 41.1 | 50.9 | 56.9 | 56.1 | 49.6 | 40.0 | 30.5 | 20.8 | 35.6 |
| Mean Temp °F | 21.4 | 22.3 | 30.9 | 43.2 | 54.3 | 63.8 | 69.1 | 67.5 | 60.2 | 49.4 | 37.5 | 27.0 | 45.5 |
| Days Max Temp ≥ 90 °F | 0 | 0 | 0 | 0 | 0 | 2 | 3 | 2 | 0 | 0 | 0 | 0 | 7 |
| Days Max Temp ≤ 32 °F | 21 | 16 | 7 | 1 | 0 | 0 | 0 | 0 | 0 | 0 | 2 | 14 | 61 |
| Days Min Temp ≤ 32 °F | 30 | 27 | 27 | 16 | 7 | 0 | 0 | 0 | 1 | 6 | 19 | 28 | 161 |
| Days Min Temp ≤ 0 °F | 3 | 4 | 1 | 0 | 0 | 0 | 0 | 0 | 0 | 0 | 0 | 1 | 9 |
| Heating Degree Days | 1344 | 1198 | 1049 | 652 | 350 | 110 | 29 | 47 | 182 | 481 | 818 | 1171 | 7431 |
| Cooling Degree Days | 0 | 0 | 0 | 8 | 28 | 83 | 170 | 141 | 45 | 2 | 0 | 0 | 477 |
| Total Precipitation (") | 2.81 | 1.86 | 2.09 | 2.77 | 2.59 | 3.29 | 2.83 | 3.32 | 4.35 | 3.48 | 3.37 | 2.82 | 35.58 |
| Days ≥ 0.1" Precip | 10 | 6 | 6 | 7 | 6 | 7 | 6 | 7 | 9 | 8 | 10 | 10 | 92 |
| Total Snowfall (") | 50.7 | 30.1 | 15.5 | 3.8 | 0.0 | 0.0 | 0.0 | 0.0 | 0.0 | 0.4 | 13.1 | 36.7 | 150.3 |
| Days ≥ 1" Snow Depth | 28 | 27 | 22 | 4 | 0 | 0 | 0 | 0 | 0 | 0 | 7 | 22 | 110 |

## MARQUETTE *Marquette County* ELEVATION 677 ft LAT/LONG 46° 34 ' N / 87° 24 ' W

| | JAN | FEB | MAR | APR | MAY | JUN | JUL | AUG | SEP | OCT | NOV | DEC | YEAR |
|---|---|---|---|---|---|---|---|---|---|---|---|---|---|
| Maximum Temp °F | 24.9 | 27.6 | 36.1 | 48.1 | 60.2 | 69.1 | 75.7 | 74.4 | 65.8 | 54.5 | 40.1 | 29.3 | 50.5 |
| Minimum Temp °F | 10.7 | 12.1 | 20.8 | 31.6 | 40.7 | 49.5 | 56.7 | 56.5 | 49.1 | 39.2 | 28.3 | 17.3 | 34.4 |
| Mean Temp °F | 17.8 | 19.9 | 28.5 | 39.9 | 50.5 | 59.3 | 66.2 | 65.5 | 57.4 | 46.9 | 34.2 | 23.3 | 42.4 |
| Days Max Temp ≥ 90 °F | 0 | 0 | 0 | 0 | 0 | 1 | 2 | 1 | 0 | 0 | 0 | 0 | 4 |
| Days Max Temp ≤ 32 °F | 24 | 19 | 11 | 2 | 0 | 0 | 0 | 0 | 0 | 0 | 6 | 19 | 81 |
| Days Min Temp ≤ 32 °F | 31 | 28 | 28 | 17 | 4 | 0 | 0 | 0 | 0 | 6 | 21 | 30 | 165 |
| Days Min Temp ≤ 0 °F | 5 | 5 | 1 | 0 | 0 | 0 | 0 | 0 | 0 | 0 | 0 | 2 | 13 |
| Heating Degree Days | 1456 | 1268 | 1124 | 749 | 454 | 203 | 65 | 75 | 246 | 558 | 917 | 1285 | 8400 |
| Cooling Degree Days | 0 | 0 | 0 | 2 | 10 | 34 | 102 | 94 | 22 | 1 | 0 | 0 | 265 |
| Total Precipitation (") | 1.83 | 1.35 | 2.15 | 2.42 | 2.69 | 2.88 | 2.57 | 3.06 | 3.84 | 3.25 | 2.66 | 1.99 | 30.69 |
| Days ≥ 0.1" Precip | 6 | 4 | 5 | 6 | 6 | 7 | 6 | 7 | 8 | 8 | 7 | 6 | 76 |
| Total Snowfall (") | 28.0 | 20.7 | 19.8 | 6.7 | 1.1 | 0.0 | 0.0 | 0.0 | 0.2 | 2.0 | 12.7 | 24.7 | 115.9 |
| Days ≥ 1" Snow Depth | 31 | 28 | 29 | 11 | 0 | 0 | 0 | 0 | 0 | 1 | 11 | 29 | 140 |

## MARQUETTE CO AP *Marquette County* ELEVATION 1414 ft LAT/LONG 46° 32 ' N / 87° 34 ' W

| | JAN | FEB | MAR | APR | MAY | JUN | JUL | AUG | SEP | OCT | NOV | DEC | YEAR |
|---|---|---|---|---|---|---|---|---|---|---|---|---|---|
| Maximum Temp °F | 20.6 | 24.2 | 33.9 | 47.5 | 62.1 | 70.8 | 76.3 | 73.5 | 63.9 | 51.6 | 36.5 | 25.3 | 48.8 |
| Minimum Temp °F | 3.4 | 4.4 | 13.9 | 27.3 | 38.7 | 47.7 | 53.5 | 51.9 | 44.0 | 34.4 | 22.9 | 11.0 | 29.4 |
| Mean Temp °F | 12.0 | 14.3 | 24.0 | 37.4 | 50.4 | 59.3 | 64.9 | 62.7 | 54.0 | 43.0 | 29.7 | 18.2 | 39.2 |
| Days Max Temp ≥ 90 °F | 0 | 0 | 0 | 0 | 0 | 0 | 1 | 1 | 0 | 0 | 0 | 0 | 2 |
| Days Max Temp ≤ 32 °F | 27 | 22 | 14 | 3 | 0 | 0 | 0 | 0 | 0 | 1 | 11 | 24 | 102 |
| Days Min Temp ≤ 32 °F | 31 | 28 | 29 | 22 | 10 | 1 | 0 | 0 | 3 | 15 | 27 | 31 | 197 |
| Days Min Temp ≤ 0 °F | 13 | 11 | 5 | 0 | 0 | 0 | 0 | 0 | 0 | 0 | 1 | 7 | 37 |
| Heating Degree Days | 1637 | 1427 | 1266 | 824 | 462 | 200 | 84 | 127 | 338 | 675 | 1053 | 1446 | 9539 |
| Cooling Degree Days | 0 | 0 | 0 | 2 | 16 | 38 | 94 | 68 | 12 | 0 | 0 | 0 | 230 |
| Total Precipitation (") | 2.22 | 1.71 | 2.87 | 2.76 | 3.08 | 3.42 | 2.99 | 3.40 | 4.04 | 3.79 | 3.17 | 2.55 | 36.00 |
| Days ≥ 0.1" Precip | 7 | 5 | 7 | 6 | 6 | 7 | 6 | 7 | 8 | 8 | 7 | 6 | 80 |
| Total Snowfall (") | 36.6 | 26.5 | 28.4 | 11.0 | 1.5 | 0.0 | 0.0 | 0.0 | 0.1 | 5.7 | 21.6 | 35.0 | 166.4 |
| Days ≥ 1" Snow Depth | 31 | 28 | 30 | 16 | 1 | 0 | 0 | 0 | 0 | 3 | 17 | 30 | 156 |

## MILFORD GM PROVING G *Livingston County* ELEVATION 1060 ft LAT/LONG 42° 34 ' N / 83° 41 ' W

| | JAN | FEB | MAR | APR | MAY | JUN | JUL | AUG | SEP | OCT | NOV | DEC | YEAR |
|---|---|---|---|---|---|---|---|---|---|---|---|---|---|
| Maximum Temp °F | 28.5 | 30.9 | 41.9 | 55.3 | 68.2 | 76.8 | 81.2 | 79.1 | 70.9 | 58.4 | 45.5 | 33.7 | 55.9 |
| Minimum Temp °F | 13.7 | 14.8 | 24.3 | 35.6 | 46.6 | 55.8 | 60.7 | 58.7 | 51.5 | 40.4 | 31.1 | 20.5 | 37.8 |
| Mean Temp °F | 21.1 | 23.0 | 33.1 | 45.5 | 57.3 | 66.4 | 71.0 | 69.0 | 61.2 | 49.4 | 38.3 | 27.1 | 46.9 |
| Days Max Temp ≥ 90 °F | 0 | 0 | 0 | 0 | 0 | 2 | 3 | 1 | 0 | 0 | 0 | 0 | 6 |
| Days Max Temp ≤ 32 °F | 20 | 16 | 7 | 1 | 0 | 0 | 0 | 0 | 0 | 0 | 3 | 14 | 61 |
| Days Min Temp ≤ 32 °F | 30 | 27 | 24 | 12 | 1 | 0 | 0 | 0 | 1 | 6 | 17 | 27 | 145 |
| Days Min Temp ≤ 0 °F | 5 | 4 | 1 | 0 | 0 | 0 | 0 | 0 | 0 | 0 | 0 | 2 | 12 |
| Heating Degree Days | 1354 | 1182 | 982 | 584 | 265 | 68 | 13 | 29 | 162 | 482 | 794 | 1169 | 7084 |
| Cooling Degree Days | 0 | *0* | 0 | 5 | *37* | 100 | 201 | 148 | 49 | 3 | 0 | 0 | 543 |
| Total Precipitation (") | 1.51 | 1.67 | 2.00 | 2.85 | 2.66 | 3.31 | 2.58 | 3.08 | 2.98 | 2.19 | 2.57 | 2.33 | 29.73 |
| Days ≥ 0.1" Precip | 4 | 4 | 5 | 7 | 6 | 6 | 5 | 6 | 6 | 5 | 6 | 6 | 66 |
| Total Snowfall (") | *11.9* | 8.7 | 4.8 | 1.4 | 0.0 | 0.0 | 0.0 | 0.0 | 0.0 | 0.1 | 3.5 | 9.8 | 40.2 |
| Days ≥ 1" Snow Depth | na | na | *6* | 1 | 0 | 0 | 0 | 0 | 0 | 0 | 2 | *12* | na |

## MIO HYDRO PLANT *Oscoda County* ELEVATION 961 ft LAT/LONG 44° 40 ' N / 84° 7 ' W

| | JAN | FEB | MAR | APR | MAY | JUN | JUL | AUG | SEP | OCT | NOV | DEC | YEAR |
|---|---|---|---|---|---|---|---|---|---|---|---|---|---|
| Maximum Temp °F | 27.4 | 30.4 | 40.6 | 54.6 | 68.2 | 76.5 | 81.8 | 78.9 | 70.0 | 57.9 | 43.8 | 32.2 | 55.2 |
| Minimum Temp °F | 8.3 | 8.1 | 17.8 | 30.5 | 40.5 | 49.6 | 55.0 | 53.1 | 45.9 | 36.2 | 27.7 | 16.4 | 32.4 |
| Mean Temp °F | 17.9 | 19.3 | 29.2 | 42.6 | 54.3 | 63.1 | 68.4 | 66.1 | 57.9 | 47.1 | 35.8 | 24.3 | 43.8 |
| Days Max Temp ≥ 90 °F | 0 | 0 | 0 | 0 | 1 | 2 | 4 | 2 | 0 | 0 | 0 | 0 | 9 |
| Days Max Temp ≤ 32 °F | 23 | 17 | 7 | 0 | 0 | 0 | 0 | 0 | 0 | 0 | 3 | 16 | 66 |
| Days Min Temp ≤ 32 °F | 30 | 28 | 28 | 18 | 7 | 1 | 0 | 0 | 2 | 11 | 22 | 30 | 177 |
| Days Min Temp ≤ 0 °F | 9 | 9 | 3 | 0 | 0 | 0 | 0 | 0 | 0 | 0 | 0 | 4 | 25 |
| Heating Degree Days | 1454 | 1285 | 1103 | 669 | 346 | 118 | 35 | 62 | 230 | 551 | 869 | 1255 | 7977 |
| Cooling Degree Days | 0 | 0 | 0 | 4 | 21 | 62 | 155 | 104 | 22 | 0 | 0 | 0 | 368 |
| Total Precipitation (") | 1.60 | 1.19 | 1.77 | 2.20 | 2.37 | 2.73 | 2.86 | 3.66 | 2.92 | 2.39 | 2.21 | 1.78 | 27.68 |
| Days ≥ 0.1" Precip | 5 | 4 | 5 | 6 | 6 | 6 | 6 | 7 | 7 | 6 | 6 | 6 | 70 |
| Total Snowfall (") | 17.0 | 9.2 | 10.4 | 3.2 | 0.3 | 0.0 | 0.0 | 0.0 | 0.0 | 0.5 | 5.8 | 13.4 | 59.8 |
| Days ≥ 1" Snow Depth | 30 | 26 | 23 | 6 | 0 | 0 | 0 | 0 | 0 | 0 | ***6*** | 22 | 113 |

## MONROE *Monroe County* ELEVATION 600 ft LAT/LONG 41° 55 ' N / 83° 23 ' W

| | JAN | FEB | MAR | APR | MAY | JUN | JUL | AUG | SEP | OCT | NOV | DEC | YEAR |
|---|---|---|---|---|---|---|---|---|---|---|---|---|---|
| Maximum Temp °F | 31.3 | 34.3 | 44.6 | 57.9 | 70.3 | 80.6 | 85.0 | 83.0 | 75.1 | 61.6 | 48.3 | 36.7 | 59.1 |
| Minimum Temp °F | 16.4 | 18.2 | 27.4 | 38.2 | 48.9 | 58.9 | 63.5 | 61.6 | 54.0 | 42.1 | 33.0 | 22.9 | 40.4 |
| Mean Temp °F | 23.9 | 26.3 | 36.0 | 48.1 | 59.6 | 69.8 | 74.3 | 72.3 | 64.6 | 51.9 | 40.7 | 29.8 | 49.8 |
| Days Max Temp ≥ 90 °F | 0 | 0 | 0 | 0 | 1 | 5 | 8 | 5 | 1 | 0 | 0 | 0 | 20 |
| Days Max Temp ≤ 32 °F | 16 | 12 | 4 | 0 | 0 | 0 | 0 | 0 | 0 | 0 | 1 | 10 | 43 |
| Days Min Temp ≤ 32 °F | 29 | 26 | 23 | 8 | 0 | 0 | 0 | 0 | 0 | 4 | 16 | 26 | 132 |
| Days Min Temp ≤ 0 °F | 3 | 1 | 0 | 0 | 0 | 0 | 0 | 0 | 0 | 0 | 0 | 1 | 5 |
| Heating Degree Days | 1268 | 1087 | 892 | 508 | 210 | 35 | 4 | 9 | 99 | 408 | 723 | 1084 | 6327 |
| Cooling Degree Days | 0 | 0 | 1 | 11 | 59 | 198 | 330 | 258 | 94 | 6 | 0 | 0 | 957 |
| Total Precipitation (") | 1.82 | 1.64 | 2.62 | 3.07 | 3.20 | 3.55 | 3.13 | 3.25 | 3.00 | 2.43 | 2.98 | 2.88 | 33.57 |
| Days ≥ 0.1" Precip | 5 | 4 | 6 | 8 | 7 | 6 | 6 | 6 | 6 | 5 | 7 | 7 | 73 |
| Total Snowfall (") | na | ***6.8*** | ***5.6*** | ***1.1*** | 0.0 | 0.0 | 0.0 | 0.0 | 0.0 | 0.0 | na | na | na |
| Days ≥ 1" Snow Depth | na | na | na | ***0*** | 0 | 0 | 0 | 0 | 0 | 0 | na | na | na |

## MONTAGUE 4 NW *Muskegon County* ELEVATION 659 ft LAT/LONG 43° 28 ' N / 86° 25 ' W

| | JAN | FEB | MAR | APR | MAY | JUN | JUL | AUG | SEP | OCT | NOV | DEC | YEAR |
|---|---|---|---|---|---|---|---|---|---|---|---|---|---|
| Maximum Temp °F | 29.7 | 32.4 | 42.6 | 55.4 | 67.2 | 75.4 | 79.7 | 77.8 | 70.4 | 58.9 | 45.8 | 34.9 | 55.9 |
| Minimum Temp °F | 16.1 | 16.5 | 23.9 | 33.4 | 42.4 | 51.4 | 57.1 | 55.9 | 49.9 | 40.2 | 31.6 | 22.3 | 36.7 |
| Mean Temp °F | 22.9 | 24.6 | 33.4 | 44.4 | 54.9 | 63.5 | 68.4 | 66.9 | 60.2 | 49.5 | 38.7 | 28.6 | 46.3 |
| Days Max Temp ≥ 90 °F | 0 | 0 | 0 | 0 | 0 | 0 | 1 | 0 | 0 | 0 | 0 | 0 | 1 |
| Days Max Temp ≤ 32 °F | 18 | 13 | 5 | 0 | 0 | 0 | 0 | 0 | 0 | 0 | 2 | 11 | 49 |
| Days Min Temp ≤ 32 °F | 30 | 27 | 25 | 14 | 5 | 0 | 0 | 0 | 1 | 7 | 17 | 27 | 153 |
| Days Min Temp ≤ 0 °F | 3 | 2 | 1 | 0 | 0 | 0 | 0 | 0 | 0 | 0 | 0 | 0 | 6 |
| Heating Degree Days | 1297 | 1134 | 971 | 613 | 325 | 104 | 29 | 49 | 178 | 476 | 781 | 1120 | 7077 |
| Cooling Degree Days | 0 | 0 | 0 | 3 | 16 | 59 | 144 | 114 | 39 | 1 | 0 | 0 | 376 |
| Total Precipitation (") | 2.05 | 1.12 | 2.39 | 3.33 | 2.57 | 2.85 | 2.72 | 4.02 | 3.73 | 3.55 | 3.35 | 2.13 | 33.81 |
| Days ≥ 0.1" Precip | ***6*** | ***3*** | 6 | 7 | 5 | 6 | 5 | 6 | 7 | 7 | 8 | 6 | 72 |
| Total Snowfall (") | 28.4 | 17.1 | 6.3 | 1.8 | 0.0 | 0.0 | 0.0 | 0.0 | 0.0 | 0.2 | 4.7 | 19.4 | 77.9 |
| Days ≥ 1" Snow Depth | 26 | 23 | 11 | 1 | 0 | 0 | 0 | 0 | 0 | 0 | 3 | 17 | 81 |

## MT CLEMENS SELFRIDGE *Macomb County* ELEVATION 576 ft LAT/LONG 42° 37 ' N / 82° 49 ' W

| | JAN | FEB | MAR | APR | MAY | JUN | JUL | AUG | SEP | OCT | NOV | DEC | YEAR |
|---|---|---|---|---|---|---|---|---|---|---|---|---|---|
| Maximum Temp °F | 29.8 | 31.9 | 42.4 | 55.5 | 66.7 | 76.8 | 81.2 | 79.0 | 71.6 | 59.1 | 46.5 | 35.1 | 56.3 |
| Minimum Temp °F | 17.0 | 17.8 | 26.5 | ***36.7*** | ***45.7*** | 55.9 | 61.4 | ***59.6*** | 52.9 | ***42.0*** | 33.2 | ***23.2*** | 39.3 |
| Mean Temp °F | 23.4 | 25.0 | 34.5 | ***46.2*** | ***56.3*** | 66.5 | 71.3 | ***69.3*** | 62.3 | ***50.5*** | 40.0 | ***29.2*** | 47.9 |
| Days Max Temp ≥ 90 °F | 0 | 0 | 0 | 0 | 0 | 2 | 3 | 2 | 0 | 0 | 0 | 0 | 7 |
| Days Max Temp ≤ 32 °F | 18 | 15 | 5 | 0 | 0 | 0 | 0 | 0 | 0 | 0 | 2 | 11 | 51 |
| Days Min Temp ≤ 32 °F | 29 | 25 | 23 | ***10*** | ***1*** | 0 | 0 | ***0*** | 0 | ***4*** | 14 | 25 | 131 |
| Days Min Temp ≤ 0 °F | 3 | 1 | 0 | ***0*** | ***0*** | 0 | 0 | ***0*** | 0 | 0 | 0 | ***1*** | 5 |
| Heating Degree Days | 1283 | 1122 | 940 | ***561*** | ***286*** | 66 | 11 | ***25*** | 139 | ***448*** | 745 | ***1104*** | 6730 |
| Cooling Degree Days | 0 | 0 | 0 | 5 | 28 | 115 | 230 | 178 | 60 | 3 | 0 | 0 | 619 |
| Total Precipitation (") | 1.80 | 1.68 | 2.28 | ***2.88*** | ***2.69*** | 3.40 | 3.63 | ***3.05*** | 3.10 | 2.27 | 2.76 | 2.66 | 32.20 |
| Days ≥ 0.1" Precip | 5 | 4 | 6 | ***7*** | ***6*** | 7 | 6 | ***5*** | 6 | ***6*** | 7 | ***7*** | 72 |
| Total Snowfall (") | 10.4 | 8.6 | 5.4 | ***1.1*** | 0.0 | 0.0 | 0.0 | 0.0 | 0.0 | 0.1 | 1.7 | ***8.9*** | 36.2 |
| Days ≥ 1" Snow Depth | 18 | 15 | ***6*** | 1 | 0 | 0 | 0 | 0 | 0 | 0 | 1 | ***11*** | 52 |

## MT PLEASANT UNIV *Isabella County* ELEVATION 791 ft LAT/LONG 43° 36' N / 84° 47' W

| | JAN | FEB | MAR | APR | MAY | JUN | JUL | AUG | SEP | OCT | NOV | DEC | YEAR |
|---|---|---|---|---|---|---|---|---|---|---|---|---|---|
| Maximum Temp °F | 28.1 | 30.8 | 41.4 | 55.8 | 69.0 | 78.0 | 82.6 | 80.3 | 71.7 | 59.3 | 45.6 | 33.4 | 56.3 |
| Minimum Temp °F | 13.1 | 13.8 | 22.5 | 34.1 | 44.4 | 54.0 | 58.8 | 56.8 | 49.1 | 38.9 | 29.9 | 20.0 | 36.3 |
| Mean Temp °F | 20.6 | 22.3 | 32.0 | 45.0 | 56.8 | 66.0 | 70.7 | 68.6 | 60.5 | 49.2 | 37.9 | 26.7 | 46.4 |
| Days Max Temp ≥ 90 °F | 0 | 0 | 0 | 0 | 0 | 2 | 4 | 2 | 0 | 0 | 0 | 0 | 8 |
| Days Max Temp ≤ 32 °F | 20 | 16 | 6 | 1 | 0 | 0 | 0 | 0 | 0 | 0 | 3 | 14 | 60 |
| Days Min Temp ≤ 32 °F | 30 | 27 | 27 | 14 | 2 | 0 | 0 | 0 | 1 | 7 | 20 | 29 | 157 |
| Days Min Temp ≤ 0 °F | 4 | 4 | 1 | 0 | 0 | 0 | 0 | 0 | 0 | 0 | 0 | 1 | 10 |
| Heating Degree Days | 1370 | 1198 | 1018 | 599 | 280 | 69 | 15 | 32 | 175 | 486 | 807 | 1180 | 7229 |
| Cooling Degree Days | 0 | 0 | 0 | 5 | 33 | 96 | 206 | 147 | 40 | 1 | 0 | 0 | 528 |
| Total Precipitation (") | 1.55 | 1.14 | 2.15 | 3.22 | 2.82 | 3.46 | 2.51 | 3.56 | 3.70 | 2.92 | 2.76 | 2.03 | 31.82 |
| Days ≥ 0.1" Precip | 4 | 3 | 5 | 7 | 6 | 6 | 5 | 6 | 6 | 6 | 6 | 5 | 65 |
| Total Snowfall (") | na | 5.2 | na | *1.5* | 0.0 | 0.0 | 0.0 | 0.0 | 0.0 | 0.3 | *2.2* | na | na |
| Days ≥ 1" Snow Depth | *17* | *19* | *9* | 1 | 0 | 0 | 0 | 0 | 0 | 0 | *2* | na | na |

## MUNISING *Alger County* ELEVATION 620 ft LAT/LONG 46° 24' N / 86° 39' W

| | JAN | FEB | MAR | APR | MAY | JUN | JUL | AUG | SEP | OCT | NOV | DEC | YEAR |
|---|---|---|---|---|---|---|---|---|---|---|---|---|---|
| Maximum Temp °F | *24.0* | *25.9* | *35.7* | 47.9 | *61.6* | 69.5 | 75.6 | 75.0 | 66.0 | 54.5 | 40.3 | *29.5* | 50.5 |
| Minimum Temp °F | *8.0* | *8.1* | *17.4* | 28.8 | *38.6* | 46.7 | 53.5 | 53.6 | 47.2 | 37.4 | 27.0 | *16.0* | 31.9 |
| Mean Temp °F | *16.0* | *17.0* | *26.4* | 38.4 | *50.1* | 58.2 | 64.6 | 64.3 | 56.6 | 46.0 | 33.7 | *22.8* | 41.2 |
| Days Max Temp ≥ 90 °F | *0* | 0 | 0 | 0 | 0 | 0 | 2 | 1 | 0 | 0 | 0 | 0 | 3 |
| Days Max Temp ≤ 32 °F | *25* | *21* | *11* | 2 | 0 | 0 | 0 | 0 | 0 | 0 | 5 | *18* | 82 |
| Days Min Temp ≤ 32 °F | *31* | 28 | *29* | 21 | *9* | 1 | 0 | 0 | 1 | 9 | *23* | *30* | 182 |
| Days Min Temp ≤ 0 °F | na | *7* | *3* | 0 | 0 | 0 | 0 | 0 | 0 | 0 | 0 | *3* | na |
| Heating Degree Days | *1511* | *1347* | *1190* | 792 | *463* | 228 | 90 | 95 | 267 | 585 | 933 | *1302* | 8803 |
| Cooling Degree Days | *0* | 0 | *0* | *2* | *8* | 29 | 88 | 88 | 22 | 0 | *0* | 0 | 237 |
| Total Precipitation (") | *3.17* | *1.94* | *2.10* | *2.07* | *2.95* | 3.25 | 3.11 | 3.23 | *3.88* | *4.24* | *3.05* | *3.26* | 36.25 |
| Days ≥ 0.1" Precip | na | *7* | *6* | *6* | *7* | *7* | *6* | *6* | *8* | *10* | *9* | *11* | na |
| Total Snowfall (") | *43.7* | *28.2* | *17.8* | *4.9* | *0.6* | 0.0 | 0.0 | 0.0 | 0.0 | *2.2* | *12.1* | *35.9* | 145.4 |
| Days ≥ 1" Snow Depth | na | na | na | na | *0* | 0 | 0 | 0 | 0 | *1* | na | na | na |

## MUSKEGON CO ARPT *Muskegon County* ELEVATION 633 ft LAT/LONG 43° 10' N / 86° 14' W

| | JAN | FEB | MAR | APR | MAY | JUN | JUL | AUG | SEP | OCT | NOV | DEC | YEAR |
|---|---|---|---|---|---|---|---|---|---|---|---|---|---|
| Maximum Temp °F | 29.1 | 31.1 | 41.7 | 54.8 | 67.0 | 75.8 | 80.4 | 78.4 | 70.6 | 58.5 | 45.5 | 34.3 | 55.6 |
| Minimum Temp °F | 17.1 | 17.6 | 25.6 | 35.8 | 45.4 | 54.6 | 60.5 | 59.0 | 51.6 | 41.5 | 32.6 | 23.2 | 38.7 |
| Mean Temp °F | 23.1 | 24.4 | 33.7 | 45.3 | 56.2 | 65.2 | 70.5 | 68.7 | 61.1 | 50.0 | 39.1 | 28.8 | 47.2 |
| Days Max Temp ≥ 90 °F | 0 | 0 | 0 | 0 | 0 | 1 | 1 | 1 | 0 | 0 | 0 | 0 | 3 |
| Days Max Temp ≤ 32 °F | 18 | 15 | 6 | 0 | 0 | 0 | 0 | 0 | 0 | 0 | 2 | 12 | 53 |
| Days Min Temp ≤ 32 °F | 29 | 26 | 24 | 11 | 2 | 0 | 0 | 0 | 0 | 6 | 16 | 26 | 140 |
| Days Min Temp ≤ 0 °F | 2 | 2 | 0 | 0 | 0 | 0 | 0 | 0 | 0 | 0 | 0 | 0 | 4 |
| Heating Degree Days | 1291 | 1141 | 965 | 587 | 292 | 75 | 13 | 27 | 158 | 461 | 771 | 1115 | 6896 |
| Cooling Degree Days | 0 | 0 | 0 | 6 | 27 | 85 | 202 | 157 | 47 | 2 | 0 | 0 | 526 |
| Total Precipitation (") | 2.29 | 1.53 | 2.51 | 2.97 | 2.64 | 2.62 | 2.32 | 3.70 | 3.74 | 2.99 | 3.34 | 2.90 | 33.55 |
| Days ≥ 0.1" Precip | 7 | 5 | 7 | 7 | 6 | 6 | 5 | 6 | 7 | 7 | 8 | 8 | 79 |
| Total Snowfall (") | 36.8 | 19.3 | 11.4 | 3.2 | 0.0 | 0.0 | 0.0 | 0.0 | 0.0 | 0.6 | 8.6 | 27.6 | 107.5 |
| Days ≥ 1" Snow Depth | 26 | 23 | 12 | 1 | 0 | 0 | 0 | 0 | 0 | 0 | 4 | 18 | 84 |

## NEWBERRY 3 S *Luce County* ELEVATION 889 ft LAT/LONG 46° 20' N / 85° 30' W

| | JAN | FEB | MAR | APR | MAY | JUN | JUL | AUG | SEP | OCT | NOV | DEC | YEAR |
|---|---|---|---|---|---|---|---|---|---|---|---|---|---|
| Maximum Temp °F | 22.9 | 25.5 | 34.9 | 48.7 | 62.7 | 70.8 | 76.3 | 73.9 | 64.8 | 52.8 | 39.0 | 27.9 | 50.0 |
| Minimum Temp °F | 7.1 | 7.6 | 16.6 | 28.9 | 38.9 | 47.1 | 52.9 | 52.6 | 45.1 | 35.9 | 25.8 | 14.3 | 31.1 |
| Mean Temp °F | 15.0 | 16.6 | 25.8 | 38.8 | 50.8 | 59.0 | 64.6 | 63.3 | 55.0 | 44.4 | 32.5 | 21.1 | 40.6 |
| Days Max Temp ≥ 90 °F | 0 | 0 | 0 | 0 | 0 | 0 | 1 | 0 | 0 | 0 | 0 | 0 | 1 |
| Days Max Temp ≤ 32 °F | 26 | 21 | 12 | 2 | 0 | 0 | 0 | 0 | 0 | 0 | 7 | 21 | 89 |
| Days Min Temp ≤ 32 °F | 31 | 28 | 29 | 21 | 8 | 1 | 0 | 0 | 2 | 12 | 24 | 30 | 186 |
| Days Min Temp ≤ 0 °F | 9 | 8 | 3 | 0 | 0 | 0 | 0 | 0 | 0 | 0 | 0 | 4 | 24 |
| Heating Degree Days | 1543 | 1362 | 1209 | 779 | 440 | 197 | 79 | 106 | 307 | 634 | 970 | 1354 | 8980 |
| Cooling Degree Days | 0 | 0 | 0 | 0 | 6 | 22 | 80 | 61 | 12 | 0 | 0 | 0 | 181 |
| Total Precipitation (") | 2.17 | 1.49 | 2.08 | 2.16 | 2.88 | 3.24 | 3.01 | 3.62 | 3.57 | 3.26 | 2.67 | 2.34 | 32.49 |
| Days ≥ 0.1" Precip | 7 | 5 | 5 | 6 | 6 | 7 | 6 | 7 | 8 | 8 | 8 | 7 | 80 |
| Total Snowfall (") | 31.4 | 21.6 | 14.4 | 5.9 | 0.3 | 0.0 | 0.0 | 0.0 | 0.1 | 0.9 | 13.9 | 26.1 | 114.6 |
| Days ≥ 1" Snow Depth | 30 | *27* | 26 | 8 | 0 | 0 | 0 | 0 | 0 | 1 | *8* | 25 | 125 |

## ONAWAY STATE PARK *Presque Isle County* ELEVATION 650 ft LAT/LONG 45° 25' N / 84° 14' W

| | JAN | FEB | MAR | APR | MAY | JUN | JUL | AUG | SEP | OCT | NOV | DEC | YEAR |
|---|---|---|---|---|---|---|---|---|---|---|---|---|---|
| Maximum Temp °F | 27.1 | 30.0 | 40.2 | 54.2 | 68.3 | 77.0 | 81.9 | 79.2 | 70.5 | 58.6 | 43.9 | 32.3 | 55.3 |
| Minimum Temp °F | 9.8 | 9.1 | 18.2 | 31.1 | 41.8 | 50.6 | 56.3 | 55.0 | 48.1 | 38.7 | 29.5 | 18.3 | 33.9 |
| Mean Temp °F | 18.5 | 19.6 | 29.2 | 42.7 | 55.1 | 63.8 | 69.1 | 67.1 | 59.4 | 48.7 | 36.7 | 25.2 | 44.6 |
| Days Max Temp ≥ 90 °F | 0 | 0 | 0 | 0 | 0 | 1 | 4 | 2 | 0 | 0 | 0 | 0 | 7 |
| Days Max Temp ≤ 32 °F | 22 | 17 | 6 | 0 | 0 | 0 | 0 | 0 | 0 | 0 | 3 | ***15*** | 63 |
| Days Min Temp ≤ 32 °F | 31 | 28 | 28 | 18 | 5 | 0 | 0 | 0 | 1 | 7 | 20 | 29 | 167 |
| Days Min Temp ≤ 0 °F | 8 | 8 | 3 | 0 | 0 | 0 | 0 | 0 | 0 | 0 | 0 | 2 | 21 |
| Heating Degree Days | 1435 | 1277 | 1103 | 668 | 322 | 98 | 23 | 46 | 198 | 502 | 842 | 1225 | 7739 |
| Cooling Degree Days | 0 | 0 | 0 | 6 | 22 | 72 | 172 | 134 | 37 | 1 | 0 | 0 | 444 |
| Total Precipitation (") | 1.83 | 1.32 | 2.12 | 2.50 | 2.74 | 2.74 | 3.16 | 3.57 | 3.61 | 2.86 | 2.65 | 2.11 | 31.21 |
| Days ≥ 0.1" Precip | 5 | 4 | 5 | 6 | 6 | 6 | 6 | 7 | 8 | 7 | 7 | 5 | 72 |
| Total Snowfall (") | 26.7 | 18.1 | 14.8 | 5.3 | 0.2 | 0.0 | 0.0 | 0.0 | 0.1 | 0.2 | 11.5 | ***20.5*** | 97.4 |
| Days ≥ 1" Snow Depth | 30 | 27 | 23 | 6 | 0 | 0 | 0 | 0 | 0 | 0 | 7 | 23 | 116 |

## OWOSSO 3 NNW *Shiawassee County* ELEVATION 765 ft LAT/LONG 42° 58' N / 84° 12' W

| | JAN | FEB | MAR | APR | MAY | JUN | JUL | AUG | SEP | OCT | NOV | DEC | YEAR |
|---|---|---|---|---|---|---|---|---|---|---|---|---|---|
| Maximum Temp °F | 28.2 | 31.2 | 42.7 | 56.8 | 69.3 | 78.1 | 82.4 | 79.9 | 72.1 | 60.0 | 46.4 | 34.5 | 56.8 |
| Minimum Temp °F | 12.8 | 14.4 | 24.2 | 35.1 | 45.4 | 54.4 | 59.1 | 57.1 | 50.3 | 40.0 | 30.9 | 20.6 | 37.0 |
| Mean Temp °F | 20.6 | 22.8 | 33.5 | 46.0 | 57.4 | 66.3 | 70.8 | 68.5 | 61.2 | 50.0 | 38.7 | 27.5 | 46.9 |
| Days Max Temp ≥ 90 °F | 0 | 0 | 0 | 0 | 0 | 2 | 4 | 2 | 0 | 0 | 0 | 0 | 8 |
| Days Max Temp ≤ 32 °F | 20 | 15 | 6 | 0 | 0 | 0 | 0 | 0 | 0 | 0 | 2 | 13 | 56 |
| Days Min Temp ≤ 32 °F | 30 | 27 | 25 | 13 | 2 | 0 | 0 | 0 | 0 | 7 | 19 | 28 | 151 |
| Days Min Temp ≤ 0 °F | 5 | 4 | 0 | 0 | 0 | 0 | 0 | 0 | 0 | 0 | 0 | 1 | 10 |
| Heating Degree Days | 1371 | 1184 | 970 | 569 | 265 | 68 | 15 | 34 | 162 | 463 | 783 | 1154 | 7038 |
| Cooling Degree Days | ***0*** | 0 | 0 | 5 | ***38*** | 100 | 205 | ***147*** | 53 | 3 | 0 | 0 | 551 |
| Total Precipitation (") | 1.47 | 1.18 | 2.07 | 2.85 | 2.54 | 3.31 | 2.60 | 3.50 | 3.67 | 2.52 | 2.64 | 2.20 | 30.55 |
| Days ≥ 0.1" Precip | 4 | 4 | 5 | 7 | 5 | 6 | 5 | 6 | 6 | 6 | 6 | 6 | 66 |
| Total Snowfall (") | 10.3 | 8.0 | 6.1 | 1.7 | 0.0 | 0.0 | 0.0 | 0.0 | 0.0 | 0.1 | ***2.9*** | ***10.1*** | 39.2 |
| Days ≥ 1" Snow Depth | 20 | 14 | 8 | 1 | 0 | 0 | 0 | 0 | 0 | 0 | ***2*** | 13 | 58 |

## PELLSTON EMMET CNTY *Emmet County* ELEVATION 709 ft LAT/LONG 45° 34' N / 84° 47' W

| | JAN | FEB | MAR | APR | MAY | JUN | JUL | AUG | SEP | OCT | NOV | DEC | YEAR |
|---|---|---|---|---|---|---|---|---|---|---|---|---|---|
| Maximum Temp °F | 25.4 | 27.4 | 37.2 | 51.4 | 65.6 | 74.3 | 79.3 | 76.6 | 67.4 | 55.6 | 41.6 | 30.6 | 52.7 |
| Minimum Temp °F | 7.7 | 5.9 | 16.2 | 29.4 | 39.3 | 48.1 | 53.8 | 52.4 | 45.1 | 36.0 | 27.5 | 16.3 | 31.5 |
| Mean Temp °F | 16.6 | 16.6 | 26.7 | 40.5 | 52.5 | 61.2 | 66.5 | 64.5 | 56.3 | 45.8 | 34.6 | 23.5 | 42.1 |
| Days Max Temp ≥ 90 °F | 0 | 0 | 0 | 0 | 0 | 1 | 2 | 1 | 0 | 0 | 0 | 0 | 4 |
| Days Max Temp ≤ 32 °F | 24 | 19 | 9 | 1 | 0 | 0 | 0 | 0 | 0 | 0 | 5 | 17 | 75 |
| Days Min Temp ≤ 32 °F | 31 | 28 | 29 | 20 | 8 | 1 | 0 | 0 | 3 | 12 | 22 | 29 | 183 |
| Days Min Temp ≤ 0 °F | 9 | 11 | 4 | 0 | 0 | 0 | 0 | 0 | 0 | 0 | 0 | 4 | 28 |
| Heating Degree Days | 1495 | 1360 | 1179 | 733 | 396 | 152 | 55 | 89 | 275 | 590 | 906 | 1280 | 8510 |
| Cooling Degree Days | 0 | 0 | 0 | 5 | 16 | 46 | 124 | 90 | 22 | 1 | 0 | 0 | 304 |
| Total Precipitation (") | 2.21 | 1.50 | 2.12 | 2.67 | 2.71 | 2.72 | 2.52 | 3.36 | 4.07 | 3.26 | 3.12 | 2.52 | 32.78 |
| Days ≥ 0.1" Precip | 7 | 5 | 5 | 7 | 6 | 6 | 5 | 7 | 8 | 8 | 8 | 7 | 79 |
| Total Snowfall (") | 33.7 | 20.7 | 12.8 | 5.4 | 0.3 | 0.0 | 0.0 | 0.0 | 0.0 | 0.9 | 12.9 | 27.2 | 113.9 |
| Days ≥ 1" Snow Depth | 31 | 28 | 25 | 7 | 0 | 0 | 0 | 0 | 0 | 0 | 9 | 24 | 124 |

## PETOSKEY *Emmet County* ELEVATION 689 ft LAT/LONG 45° 22' N / 84° 58' W

| | JAN | FEB | MAR | APR | MAY | JUN | JUL | AUG | SEP | OCT | NOV | DEC | YEAR |
|---|---|---|---|---|---|---|---|---|---|---|---|---|---|
| Maximum Temp °F | 27.6 | 28.9 | 38.1 | 50.3 | 62.1 | 70.7 | 76.3 | 75.5 | 68.5 | 57.3 | 43.8 | 33.0 | 52.7 |
| Minimum Temp °F | 14.5 | 12.0 | 20.7 | 32.0 | 41.8 | 51.5 | 58.3 | 57.7 | 50.9 | 41.5 | 31.6 | 21.7 | 36.2 |
| Mean Temp °F | 21.1 | 20.5 | 29.4 | 41.2 | 52.0 | 61.0 | 67.4 | 66.7 | 59.8 | 49.4 | 37.7 | 27.4 | 44.5 |
| Days Max Temp ≥ 90 °F | 0 | 0 | 0 | 0 | 0 | 1 | 1 | 1 | 0 | 0 | 0 | 0 | 3 |
| Days Max Temp ≤ 32 °F | 21 | 18 | 9 | 1 | 0 | 0 | 0 | 0 | 0 | 0 | 3 | 14 | 66 |
| Days Min Temp ≤ 32 °F | 30 | 27 | 27 | 16 | 3 | 0 | 0 | 0 | 0 | 3 | 17 | 27 | 150 |
| Days Min Temp ≤ 0 °F | 3 | 5 | 2 | 0 | 0 | 0 | 0 | 0 | 0 | 0 | 0 | 1 | 11 |
| Heating Degree Days | 1354 | 1251 | 1096 | 708 | 403 | 155 | 39 | 48 | 188 | 481 | 813 | 1159 | 7695 |
| Cooling Degree Days | 0 | 0 | 0 | 2 | 4 | 32 | 111 | 104 | 32 | 1 | 0 | 0 | 286 |
| Total Precipitation (") | 1.95 | 1.23 | 1.90 | 2.61 | 2.65 | 2.76 | 3.10 | 3.54 | 3.89 | 3.17 | 2.85 | 2.26 | 31.91 |
| Days ≥ 0.1" Precip | 7 | 4 | 5 | 7 | 6 | 7 | 5 | 7 | 8 | 8 | 8 | 7 | 79 |
| Total Snowfall (") | 38.3 | 21.4 | 10.9 | 4.1 | 0.2 | 0.0 | 0.0 | 0.0 | 0.0 | 0.4 | 10.7 | 30.1 | 116.1 |
| Days ≥ 1" Snow Depth | na | na | na | na | 0 | 0 | 0 | 0 | 0 | 0 | na | na | na |

## PONTIAC STATE HOSP *Oakland County* ELEVATION 971 ft LAT/LONG 42° 39 ' N / 83° 18 ' W

| | JAN | FEB | MAR | APR | MAY | JUN | JUL | AUG | SEP | OCT | NOV | DEC | YEAR |
|---|---|---|---|---|---|---|---|---|---|---|---|---|---|
| Maximum Temp °F | 29.6 | 33.0 | 44.4 | 58.2 | 70.5 | 79.3 | 83.4 | 81.3 | 73.7 | 60.5 | 47.0 | 34.7 | 58.0 |
| Minimum Temp °F | 15.2 | 16.6 | 25.4 | 36.2 | 47.0 | 56.1 | 61.0 | 59.5 | 52.5 | 41.4 | 32.1 | 21.6 | 38.7 |
| Mean Temp °F | 22.4 | 24.8 | 34.9 | 47.2 | 58.8 | 67.7 | 72.2 | 70.5 | 63.1 | 50.9 | 39.6 | 28.2 | 48.4 |
| Days Max Temp ≥ 90 °F | 0 | 0 | 0 | 0 | 0 | 2 | 5 | 2 | 1 | 0 | 0 | 0 | 10 |
| Days Max Temp ≤ 32 °F | 18 | 14 | 5 | 0 | 0 | 0 | 0 | 0 | 0 | 0 | 2 | 12 | 51 |
| Days Min Temp ≤ 32 °F | 30 | 26 | 25 | 11 | 1 | 0 | 0 | 0 | 0 | 5 | 17 | 28 | 143 |
| Days Min Temp ≤ 0 °F | 3 | 2 | 0 | 0 | 0 | 0 | 0 | 0 | 0 | 0 | 0 | 1 | 6 |
| Heating Degree Days | 1314 | 1128 | 925 | 533 | 231 | 49 | 7 | 18 | 124 | 436 | 756 | 1134 | 6655 |
| Cooling Degree Days | 0 | 0 | 0 | 8 | 51 | 135 | 256 | 202 | 76 | 5 | 0 | 0 | 733 |
| Total Precipitation (") | 1.53 | 1.49 | 2.33 | 2.83 | 2.75 | 3.55 | 2.94 | 3.29 | 2.99 | 2.62 | 2.77 | 2.43 | 31.52 |
| Days ≥ 0.1" Precip | 5 | 4 | 6 | 7 | 7 | 6 | 6 | 6 | 6 | 6 | 7 | 6 | 72 |
| Total Snowfall (") | 8.7 | 6.9 | 4.8 | 1.6 | 0.0 | 0.0 | 0.0 | 0.0 | 0.0 | 0.1 | 2.3 | 7.4 | 31.8 |
| Days ≥ 1" Snow Depth | 22 | 18 | 8 | 1 | 0 | 0 | 0 | 0 | 0 | 0 | 2 | 14 | 65 |

## PORT HURON *St. Clair County* ELEVATION 600 ft LAT/LONG 42° 57 ' N / 82° 27 ' W

| | JAN | FEB | MAR | APR | MAY | JUN | JUL | AUG | SEP | OCT | NOV | DEC | YEAR |
|---|---|---|---|---|---|---|---|---|---|---|---|---|---|
| Maximum Temp °F | 30.0 | 32.3 | 42.2 | 54.9 | 66.7 | 76.7 | 81.9 | 80.4 | 72.8 | 60.2 | 47.0 | 35.4 | 56.7 |
| Minimum Temp °F | 15.8 | 17.0 | 25.5 | 35.5 | 45.6 | 55.4 | 61.9 | 60.8 | 53.2 | 42.1 | 32.6 | 22.1 | 39.0 |
| Mean Temp °F | 22.9 | 24.7 | 33.9 | 45.2 | 56.2 | 66.1 | 71.9 | 70.6 | 63.0 | 51.2 | 39.8 | 28.8 | 47.9 |
| Days Max Temp ≥ 90 °F | 0 | 0 | 0 | 0 | 0 | 2 | 4 | 3 | 0 | 0 | 0 | 0 | 9 |
| Days Max Temp ≤ 32 °F | 18 | 14 | 5 | 0 | 0 | 0 | 0 | 0 | 0 | 0 | 2 | 11 | 50 |
| Days Min Temp ≤ 32 °F | 30 | 26 | 25 | 11 | 1 | 0 | 0 | 0 | 0 | 4 | 16 | 27 | 140 |
| Days Min Temp ≤ 0 °F | 3 | 2 | 0 | 0 | 0 | 0 | 0 | 0 | 0 | 0 | 0 | 1 | 6 |
| Heating Degree Days | 1298 | 1133 | 957 | 591 | 294 | 71 | 9 | 16 | 122 | 428 | 750 | 1116 | 6785 |
| Cooling Degree Days | 0 | 0 | 0 | 4 | 29 | 99 | 242 | 196 | 64 | 3 | 0 | 0 | 637 |
| Total Precipitation (") | 1.68 | 1.56 | 2.19 | 2.97 | 2.58 | 3.24 | 2.83 | 3.00 | 3.19 | 2.47 | 3.04 | 2.12 | 30.87 |
| Days ≥ 0.1" Precip | 5 | 4 | 6 | 8 | 6 | 7 | 6 | 6 | 7 | 6 | 7 | 6 | 74 |
| Total Snowfall (") | *11.3* | 9.7 | 5.5 | 1.7 | 0.0 | 0.0 | 0.0 | 0.0 | 0.0 | 0.5 | 1.7 | *8.4* | 38.8 |
| Days ≥ 1" Snow Depth | 19 | 17 | 6 | 1 | 0 | 0 | 0 | 0 | 0 | 0 | 1 | *11* | 55 |

## SAGINAW TRI CITY AP *Saginaw County* ELEVATION 662 ft LAT/LONG 43° 32 ' N / 84° 5 ' W

| | JAN | FEB | MAR | APR | MAY | JUN | JUL | AUG | SEP | OCT | NOV | DEC | YEAR |
|---|---|---|---|---|---|---|---|---|---|---|---|---|---|
| Maximum Temp °F | 27.3 | 29.7 | 40.5 | 55.0 | 67.8 | 76.9 | 81.7 | 79.0 | 70.6 | 58.3 | 44.8 | 32.9 | 55.4 |
| Minimum Temp °F | 14.4 | 15.8 | 25.3 | 36.4 | 46.5 | 55.8 | 60.5 | 58.5 | 51.0 | 40.3 | 31.4 | 21.1 | 38.1 |
| Mean Temp °F | 20.9 | 22.8 | 32.9 | 45.7 | 57.2 | 66.4 | 71.2 | 68.8 | 60.8 | 49.4 | 38.2 | 27.0 | 46.8 |
| Days Max Temp ≥ 90 °F | 0 | 0 | 0 | 0 | 0 | 2 | 4 | 2 | 0 | 0 | 0 | 0 | 8 |
| Days Max Temp ≤ 32 °F | 21 | 17 | 7 | 0 | 0 | 0 | 0 | 0 | 0 | 0 | 3 | 14 | 62 |
| Days Min Temp ≤ 32 °F | 30 | 26 | 24 | 10 | 1 | 0 | 0 | 0 | 0 | 5 | 17 | 28 | 141 |
| Days Min Temp ≤ 0 °F | 4 | 3 | 0 | 0 | 0 | 0 | 0 | 0 | 0 | 0 | 0 | 1 | 8 |
| Heating Degree Days | 1360 | 1186 | 987 | 577 | 270 | 64 | 13 | 32 | 168 | 483 | 799 | 1171 | 7110 |
| Cooling Degree Days | 0 | 0 | 0 | 6 | 38 | 99 | 213 | 149 | 44 | 2 | 0 | 0 | 551 |
| Total Precipitation (") | 1.74 | 1.36 | 2.32 | 3.03 | 2.67 | 3.02 | 2.18 | 3.33 | 3.84 | 2.62 | 2.76 | 2.28 | 31.15 |
| Days ≥ 0.1" Precip | 5 | 4 | 6 | 7 | 6 | 6 | 5 | 6 | 7 | 6 | 6 | 6 | 70 |
| Total Snowfall (") | 11.6 | 8.8 | 8.1 | 2.1 | 0.0 | 0.0 | 0.0 | 0.0 | 0.0 | 0.2 | 3.3 | 10.7 | 44.8 |
| Days ≥ 1" Snow Depth | 22 | 18 | 9 | 1 | 0 | 0 | 0 | 0 | 0 | 0 | 2 | 12 | 64 |

## SANDUSKY *Sanilac County* ELEVATION 771 ft LAT/LONG 43° 25 ' N / 82° 50 ' W

| | JAN | FEB | MAR | APR | MAY | JUN | JUL | AUG | SEP | OCT | NOV | DEC | YEAR |
|---|---|---|---|---|---|---|---|---|---|---|---|---|---|
| Maximum Temp °F | 28.1 | 30.9 | 41.3 | 54.7 | 67.8 | 76.9 | *81.6* | 79.6 | *72.1* | 59.7 | 46.2 | 34.4 | 56.1 |
| Minimum Temp °F | 12.9 | 13.9 | 22.8 | 33.7 | 43.7 | 53.3 | *57.9* | 56.6 | *49.8* | 39.3 | 30.8 | 20.8 | 36.3 |
| Mean Temp °F | 20.5 | *22.5* | 32.2 | 44.3 | 55.8 | 65.1 | *69.8* | 68.1 | *61.0* | 49.5 | 38.5 | 27.7 | 46.3 |
| Days Max Temp ≥ 90 °F | 0 | 0 | 0 | 0 | 0 | 2 | *3* | 2 | *0* | 0 | 0 | 0 | 7 |
| Days Max Temp ≤ 32 °F | 20 | 16 | 6 | 0 | 0 | 0 | *0* | 0 | *0* | 0 | 2 | 12 | 56 |
| Days Min Temp ≤ 32 °F | 30 | 27 | 27 | 15 | 3 | 0 | *0* | 0 | *0* | 6 | 18 | 28 | 154 |
| Days Min Temp ≤ 0 °F | 4 | 4 | 0 | 0 | 0 | 0 | *0* | 0 | *0* | 0 | 0 | 1 | 9 |
| Heating Degree Days | 1371 | *1194* | 1013 | 620 | 307 | 87 | *19* | 35 | *164* | 477 | 787 | 1149 | 7223 |
| Cooling Degree Days | *0* | *0* | 0 | 4 | 31 | 80 | na | 133 | na | *1* | 0 | *0* | na |
| Total Precipitation (") | 1.54 | *1.17* | 2.10 | 2.38 | 2.51 | 3.16 | *2.61* | 2.97 | *3.80* | 2.28 | 2.50 | *2.18* | 29.20 |
| Days ≥ 0.1" Precip | *5* | *3* | *5* | 5 | 6 | 7 | *5* | 6 | *7* | 6 | 6 | *6* | 67 |
| Total Snowfall (") | *13.5* | *7.9* | *8.9* | 2.1 | 0.0 | 0.0 | *0.0* | 0.0 | *0.0* | *0.0* | *3.4* | na | na |
| Days ≥ 1" Snow Depth | *23* | *19* | na | 1 | 0 | 0 | *0* | 0 | *0* | 0 | 2 | *15* | na |

## SAULT STE MARIE WSO *Chippewa County* ELEVATION 718 ft LAT/LONG 46° 28 ' N / 84° 22 ' W

| | JAN | FEB | MAR | APR | MAY | JUN | JUL | AUG | SEP | OCT | NOV | DEC | YEAR |
|---|---|---|---|---|---|---|---|---|---|---|---|---|---|
| Maximum Temp °F | 21.2 | 23.5 | 33.1 | 48.0 | 62.6 | 70.3 | 75.8 | 73.8 | 64.6 | 52.6 | 38.6 | 27.0 | 49.3 |
| Minimum Temp °F | 4.5 | 5.1 | 15.5 | 28.7 | 38.5 | 45.7 | 51.4 | 51.8 | 44.4 | 35.7 | 25.6 | 12.9 | 30.0 |
| Mean Temp °F | 12.9 | 14.3 | 24.3 | 38.4 | 50.5 | 58.1 | 63.6 | 62.8 | 54.6 | 44.2 | 32.1 | 20.0 | 39.7 |
| Days Max Temp ≥ 90 °F | 0 | 0 | 0 | 0 | 0 | 0 | 1 | 0 | 0 | 0 | 0 | 0 | 1 |
| Days Max Temp ≤ 32 °F | 26 | 22 | 14 | 2 | 0 | 0 | 0 | 0 | 0 | 0 | 7 | 21 | 92 |
| Days Min Temp ≤ 32 °F | 31 | 28 | 29 | 21 | 8 | 1 | 0 | 0 | 2 | 11 | 23 | 30 | 184 |
| Days Min Temp ≤ 0 °F | 12 | 11 | 4 | 0 | 0 | 0 | 0 | 0 | 0 | 0 | 0 | 6 | 33 |
| Heating Degree Days | 1611 | 1427 | 1254 | 792 | 448 | 220 | 92 | 111 | 318 | 637 | 979 | 1389 | 9278 |
| Cooling Degree Days | 0 | 0 | 0 | 1 | 6 | 17 | 61 | 52 | 11 | 0 | 0 | 0 | 148 |
| Total Precipitation (") | 2.48 | 1.64 | 2.28 | 2.52 | 2.69 | 3.11 | 2.99 | 3.54 | 3.81 | 3.50 | 3.56 | 2.82 | 34.94 |
| Days ≥ 0.1" Precip | 9 | 5 | 6 | 6 | 6 | 7 | 6 | 7 | 8 | 9 | 10 | 9 | 88 |
| Total Snowfall (") | 34.2 | 20.3 | 14.2 | 6.4 | 0.5 | 0.0 | 0.0 | 0.0 | 0.1 | 2.7 | 16.7 | 31.4 | 126.5 |
| Days ≥ 1" Snow Depth | 31 | 28 | 29 | 10 | 0 | 0 | 0 | 0 | 0 | 1 | 12 | 28 | 139 |

## SENEY WILDLIFE REF *Schoolcraft County* ELEVATION 712 ft LAT/LONG 46° 17 ' N / 85° 57 ' W

| | JAN | FEB | MAR | APR | MAY | JUN | JUL | AUG | SEP | OCT | NOV | DEC | YEAR |
|---|---|---|---|---|---|---|---|---|---|---|---|---|---|
| Maximum Temp °F | 25.0 | 28.2 | 37.6 | ***51.1*** | 65.6 | 74.1 | 80.1 | 77.2 | 68.0 | ***55.7*** | ***41.1*** | ***30.0*** | 52.8 |
| Minimum Temp °F | 6.3 | ***5.9*** | 14.9 | ***29.5*** | 40.0 | 48.1 | 53.9 | 52.9 | 46.4 | ***37.4*** | ***26.9*** | ***15.2*** | 31.4 |
| Mean Temp °F | 15.6 | ***17.0*** | 26.3 | ***40.3*** | 52.9 | 61.1 | 67.0 | 65.1 | 57.3 | ***46.6*** | ***34.0*** | ***22.6*** | 42.2 |
| Days Max Temp ≥ 90 °F | 0 | 0 | 0 | 0 | 0 | 1 | 2 | 1 | 0 | 0 | 0 | 0 | 4 |
| Days Max Temp ≤ 32 °F | 23 | 18 | 8 | 1 | 0 | 0 | 0 | 0 | 0 | 0 | 4 | 17 | 71 |
| Days Min Temp ≤ 32 °F | 29 | 26 | 28 | 19 | 6 | 0 | 0 | 0 | 1 | 10 | 21 | 27 | 167 |
| Days Min Temp ≤ 0 °F | 10 | 10 | 5 | 0 | 0 | 0 | 0 | 0 | 0 | 0 | 0 | 4 | 29 |
| Heating Degree Days | 1525 | ***1349*** | 1194 | ***732*** | 379 | 148 | 47 | 73 | 246 | ***565*** | ***923*** | ***1306*** | 8487 |
| Cooling Degree Days | ***0*** | na | 0 | ***1*** | ***11*** | 39 | ***129*** | ***84*** | ***18*** | na | na | na | na |
| Total Precipitation (") | 2.00 | 1.30 | 1.91 | 1.99 | 2.86 | 3.27 | 3.44 | 3.27 | 3.60 | 3.30 | 2.66 | 2.18 | 31.78 |
| Days ≥ 0.1" Precip | 6 | 4 | 5 | 5 | 6 | 7 | 6 | 7 | 7 | 7 | 7 | 6 | 73 |
| Total Snowfall (") | 33.9 | 21.4 | 14.6 | 3.9 | 0.3 | 0.0 | 0.0 | 0.0 | 0.1 | 1.3 | 14.2 | ***29.9*** | 119.6 |
| Days ≥ 1" Snow Depth | 28 | 26 | 26 | 11 | 0 | 0 | 0 | 0 | 0 | 1 | 10 | 24 | 126 |

## SOUTH HAVEN *Van Buren County* ELEVATION 591 ft LAT/LONG 42° 24 ' N / 86° 16 ' W

| | JAN | FEB | MAR | APR | MAY | JUN | JUL | AUG | SEP | OCT | NOV | DEC | YEAR |
|---|---|---|---|---|---|---|---|---|---|---|---|---|---|
| Maximum Temp °F | 31.4 | 33.8 | 44.0 | 55.1 | 65.6 | 74.9 | 79.2 | 78.1 | 72.2 | 61.4 | 48.4 | 36.8 | 56.7 |
| Minimum Temp °F | 18.6 | 20.2 | 28.2 | 37.7 | 47.0 | 56.5 | 62.0 | 61.0 | 54.2 | 44.4 | 34.7 | 24.6 | 40.8 |
| Mean Temp °F | 25.0 | 27.0 | 36.1 | 46.4 | 56.4 | 65.7 | 70.6 | 69.6 | 63.2 | 52.9 | 41.6 | 30.8 | 48.8 |
| Days Max Temp ≥ 90 °F | 0 | 0 | 0 | 0 | 0 | 0 | 1 | 1 | 0 | 0 | 0 | 0 | 2 |
| Days Max Temp ≤ 32 °F | 16 | 12 | 4 | 0 | 0 | 0 | 0 | 0 | 0 | 0 | 1 | 9 | 42 |
| Days Min Temp ≤ 32 °F | 29 | 25 | 21 | 8 | 1 | 0 | 0 | 0 | 0 | 2 | 12 | 25 | 123 |
| Days Min Temp ≤ 0 °F | 1 | 1 | 0 | 0 | 0 | 0 | 0 | 0 | 0 | 0 | 0 | 0 | 2 |
| Heating Degree Days | 1233 | 1065 | 888 | 555 | 290 | 76 | 16 | 21 | 114 | 376 | 695 | 1054 | 6383 |
| Cooling Degree Days | 0 | 0 | 0 | 5 | 31 | 108 | 221 | 192 | 76 | 5 | 0 | 0 | 638 |
| Total Precipitation (") | ***2.19*** | ***1.44*** | 2.24 | 3.39 | 3.02 | 3.63 | 3.54 | 3.71 | 4.24 | 3.03 | 3.54 | ***2.94*** | 36.91 |
| Days ≥ 0.1" Precip | ***7*** | ***5*** | 6 | 8 | 6 | 6 | 6 | 7 | 7 | 7 | 8 | ***7*** | 80 |
| Total Snowfall (") | na | na | ***5.3*** | 1.4 | 0.0 | 0.0 | 0.0 | 0.0 | 0.0 | 0.4 | ***3.2*** | na | na |
| Days ≥ 1" Snow Depth | 19 | 17 | 6 | 1 | 0 | 0 | 0 | 0 | 0 | 0 | ***2*** | na | na |

## ST CHARLES *Saginaw County* ELEVATION 600 ft LAT/LONG 43° 18 ' N / 84° 8 ' W

| | JAN | FEB | MAR | APR | MAY | JUN | JUL | AUG | SEP | OCT | NOV | DEC | YEAR |
|---|---|---|---|---|---|---|---|---|---|---|---|---|---|
| Maximum Temp °F | ***29.4*** | 32.6 | 43.1 | 57.6 | 70.7 | 79.7 | 83.9 | 81.5 | 73.3 | 61.0 | 46.7 | 34.7 | 57.9 |
| Minimum Temp °F | ***14.0*** | 14.8 | 24.0 | 35.1 | 45.1 | 54.2 | 58.4 | 56.4 | 48.9 | 39.0 | 30.6 | 20.5 | 36.8 |
| Mean Temp °F | ***21.7*** | 23.7 | 33.5 | 46.4 | 57.9 | 66.9 | 71.2 | 69.0 | 61.1 | 50.1 | 38.7 | 27.6 | 47.3 |
| Days Max Temp ≥ 90 °F | ***0*** | 0 | 0 | 0 | 0 | 3 | 6 | 4 | 1 | 0 | 0 | 0 | 14 |
| Days Max Temp ≤ 32 °F | ***18*** | ***14*** | 5 | 0 | 0 | 0 | 0 | 0 | 0 | 0 | 2 | ***13*** | 52 |
| Days Min Temp ≤ 32 °F | ***30*** | ***27*** | ***25*** | 13 | 3 | 0 | 0 | 0 | 2 | 8 | 18 | ***28*** | 154 |
| Days Min Temp ≤ 0 °F | ***5*** | ***4*** | 1 | 0 | 0 | 0 | 0 | 0 | 0 | 0 | 0 | ***2*** | 12 |
| Heating Degree Days | ***1336*** | ***1166*** | 971 | 558 | 251 | 60 | 13 | 32 | 165 | 463 | 784 | 1152 | 6951 |
| Cooling Degree Days | ***0*** | ***0*** | ***0*** | ***6*** | ***30*** | ***95*** | ***181*** | ***141*** | ***39*** | ***2*** | ***0*** | ***0*** | 494 |
| Total Precipitation (") | 1.73 | 1.18 | 2.29 | 2.78 | 2.65 | 3.22 | 2.67 | 3.54 | 3.30 | 2.63 | 2.84 | ***2.09*** | 30.92 |
| Days ≥ 0.1" Precip | ***6*** | ***4*** | 6 | ***7*** | 6 | 6 | 5 | 6 | 6 | 6 | 6 | ***6*** | 70 |
| Total Snowfall (") | ***12.8*** | ***9.5*** | 8.9 | 2.2 | 0.0 | 0.0 | 0.0 | 0.0 | 0.0 | 0.4 | 3.2 | ***10.3*** | 47.3 |
| Days ≥ 1" Snow Depth | na | ***20*** | ***9*** | 1 | 0 | 0 | 0 | 0 | 0 | 0 | 2 | ***14*** | na |

## ST JAMES 2S BEAVR IS *Charlevoix County* ELEVATION 670 ft LAT/LONG 45° 43 ' N / 85° 31 ' W

| | JAN | FEB | MAR | APR | MAY | JUN | JUL | AUG | SEP | OCT | NOV | DEC | YEAR |
|---|---|---|---|---|---|---|---|---|---|---|---|---|---|
| Maximum Temp °F | 26.2 | 26.8 | 36.0 | ***48.1*** | 61.3 | 70.2 | 75.8 | 74.3 | ***66.7*** | 55.3 | 41.9 | 32.2 | 51.2 |
| Minimum Temp °F | 13.8 | 11.1 | 20.1 | ***30.6*** | 40.1 | 49.1 | 56.1 | 56.5 | ***50.0*** | 40.5 | 31.0 | 21.6 | 35.0 |
| Mean Temp °F | 20.1 | 19.0 | 28.1 | ***39.4*** | 50.7 | 59.7 | 65.9 | 65.4 | ***58.4*** | 47.9 | 36.5 | 26.9 | 43.2 |
| Days Max Temp ≥ 90 °F | 0 | 0 | 0 | 0 | 0 | 0 | 0 | 0 | 0 | 0 | 0 | 0 | 0 |
| Days Max Temp ≤ 32 °F | 22 | 20 | 10 | 1 | 0 | 0 | 0 | 0 | 0 | 0 | 3 | 14 | 70 |
| Days Min Temp ≤ 32 °F | 30 | 28 | 29 | 18 | 5 | 0 | 0 | 0 | 0 | 4 | 17 | 28 | 159 |
| Days Min Temp ≤ 0 °F | 4 | 5 | 1 | 0 | 0 | 0 | 0 | 0 | 0 | 0 | 0 | 0 | 10 |
| Heating Degree Days | 1387 | 1293 | 1138 | ***762*** | 437 | 175 | 46 | 56 | ***209*** | 524 | 850 | 1173 | 8050 |
| Cooling Degree Days | ***0*** | 0 | 0 | ***0*** | ***2*** | ***28*** | ***91*** | ***82*** | na | 0 | ***0*** | 0 | na |
| Total Precipitation (") | 1.89 | 1.15 | 1.98 | 2.56 | 2.90 | 2.83 | 2.62 | 3.23 | 3.68 | 3.12 | 2.70 | 2.05 | 30.71 |
| Days ≥ 0.1" Precip | ***5*** | 4 | ***4*** | ***6*** | ***6*** | 6 | 5 | ***6*** | ***6*** | ***6*** | ***6*** | ***6*** | 66 |
| Total Snowfall (") | 25.2 | 15.5 | 10.7 | 3.9 | 0.2 | 0.0 | 0.0 | 0.0 | 0.0 | 0.0 | ***6.1*** | ***18.5*** | 80.1 |
| Days ≥ 1" Snow Depth | na | na | na | ***4*** | 0 | 0 | 0 | 0 | 0 | 0 | ***3*** | na | na |

## ST JOHNS *Clinton County* ELEVATION 741 ft LAT/LONG 43° 3 ' N / 84° 34 ' W

| | JAN | FEB | MAR | APR | MAY | JUN | JUL | AUG | SEP | OCT | NOV | DEC | YEAR |
|---|---|---|---|---|---|---|---|---|---|---|---|---|---|
| Maximum Temp °F | 29.7 | 32.7 | 44.1 | 58.1 | 70.9 | 79.4 | 83.8 | 81.2 | 73.7 | 61.2 | 47.1 | 35.0 | 58.1 |
| Minimum Temp °F | 13.8 | 15.1 | 24.4 | 35.4 | 46.0 | 55.2 | 59.6 | 57.4 | 50.3 | 39.9 | 30.4 | 20.3 | 37.3 |
| Mean Temp °F | 21.8 | 23.9 | 34.3 | 46.8 | 58.4 | 67.3 | 71.7 | 69.4 | 62.0 | 50.6 | 38.8 | 27.6 | 47.7 |
| Days Max Temp ≥ 90 °F | 0 | 0 | 0 | 0 | 0 | 2 | 5 | 3 | 1 | 0 | 0 | 0 | 11 |
| Days Max Temp ≤ 32 °F | 19 | 14 | 5 | 0 | 0 | 0 | 0 | 0 | 0 | 0 | 2 | 12 | 52 |
| Days Min Temp ≤ 32 °F | 30 | 27 | 25 | 13 | 2 | 0 | 0 | 0 | 1 | 7 | 19 | 28 | 152 |
| Days Min Temp ≤ 0 °F | 4 | 3 | 0 | 0 | 0 | 0 | 0 | 0 | 0 | 0 | 0 | 1 | 8 |
| Heating Degree Days | 1333 | 1152 | 946 | 547 | 238 | 54 | 11 | 26 | 146 | 448 | 780 | 1151 | 6832 |
| Cooling Degree Days | 0 | 0 | 1 | 8 | 44 | 113 | 226 | 158 | 55 | 3 | 0 | 0 | 608 |
| Total Precipitation (") | 1.58 | 1.25 | 2.24 | 3.31 | 2.82 | 3.51 | 2.94 | 3.66 | 3.95 | 2.94 | 2.75 | 2.09 | 33.04 |
| Days ≥ 0.1" Precip | 4 | 4 | 5 | 8 | 6 | 7 | 6 | 6 | 7 | 6 | 6 | 5 | 70 |
| Total Snowfall (") | 12.8 | 9.8 | 6.6 | 2.2 | 0.0 | 0.0 | 0.0 | 0.0 | 0.0 | 0.3 | 3.4 | 9.8 | 44.9 |
| Days ≥ 1" Snow Depth | 20 | 15 | 7 | 1 | 0 | 0 | 0 | 0 | 0 | 0 | 2 | 12 | 57 |

## STAMBAUGH 2 SSE *Iron County* ELEVATION 1621 ft LAT/LONG 46° 4 ' N / 88° 37 ' W

| | JAN | FEB | MAR | APR | MAY | JUN | JUL | AUG | SEP | OCT | NOV | DEC | YEAR |
|---|---|---|---|---|---|---|---|---|---|---|---|---|---|
| Maximum Temp °F | 21.7 | 26.5 | 37.4 | 52.4 | 66.2 | 73.8 | 78.3 | 75.8 | 66.3 | 54.5 | 38.0 | 25.9 | 51.4 |
| Minimum Temp °F | -2.1 | -0.2 | 12.1 | 26.4 | 37.3 | 45.9 | 50.9 | 48.9 | 41.3 | 32.0 | 20.5 | 6.8 | 26.7 |
| Mean Temp °F | 9.8 | 13.2 | 24.8 | 39.4 | 51.7 | 59.9 | 64.6 | 62.4 | 53.8 | 43.3 | 29.3 | 16.4 | 39.1 |
| Days Max Temp ≥ 90 °F | 0 | 0 | 0 | 0 | 0 | 0 | 1 | 1 | 0 | 0 | 0 | 0 | 2 |
| Days Max Temp ≤ 32 °F | 26 | 20 | 10 | 1 | 0 | 0 | 0 | 0 | 0 | 0 | 10 | 23 | 90 |
| Days Min Temp ≤ 32 °F | 31 | 28 | 29 | 22 | 11 | 3 | 0 | 1 | 6 | 18 | 27 | 31 | 207 |
| Days Min Temp ≤ 0 °F | 16 | 14 | 7 | 1 | 0 | 0 | 0 | 0 | 0 | 0 | 2 | 10 | 50 |
| Heating Degree Days | 1709 | 1459 | 1240 | 762 | 415 | 181 | 86 | 127 | 342 | 667 | 1066 | 1501 | 9555 |
| Cooling Degree Days | 0 | 0 | 0 | 2 | 9 | 31 | 71 | 50 | 9 | 0 | 0 | 0 | 172 |
| Total Precipitation (") | 1.05 | 0.75 | 1.61 | 2.37 | 3.30 | 3.73 | 3.39 | 3.84 | 4.03 | 2.91 | 2.15 | 1.42 | 30.55 |
| Days ≥ 0.1" Precip | 4 | 2 | 4 | 6 | 7 | 8 | 7 | 7 | 8 | 6 | 6 | 4 | 69 |
| Total Snowfall (") | 17.0 | 10.3 | 11.9 | 5.7 | 0.8 | 0.0 | 0.0 | 0.0 | 0.0 | 2.0 | 10.6 | 16.9 | 75.2 |
| Days ≥ 1" Snow Depth | 31 | 28 | 27 | 6 | 0 | 0 | 0 | 0 | 0 | 1 | 12 | 29 | 134 |

## STANDISH 5 SW *Arenac County* ELEVATION 645 ft LAT/LONG 43° 57 ' N / 84° 2 ' W

| | JAN | FEB | MAR | APR | MAY | JUN | JUL | AUG | SEP | OCT | NOV | DEC | YEAR |
|---|---|---|---|---|---|---|---|---|---|---|---|---|---|
| Maximum Temp °F | 27.9 | 30.5 | 40.2 | 54.3 | 67.2 | 76.3 | 81.3 | 78.8 | 70.7 | 58.2 | 44.6 | 33.2 | 55.3 |
| Minimum Temp °F | 9.5 | 10.8 | 20.6 | 32.1 | 41.4 | 50.9 | 55.5 | 53.3 | 46.3 | 35.9 | 28.2 | 18.1 | 33.6 |
| Mean Temp °F | 18.8 | 20.7 | 30.4 | 43.2 | 54.3 | 63.6 | 68.4 | 66.1 | 58.5 | 47.0 | 36.5 | 25.7 | 44.4 |
| Days Max Temp ≥ 90 °F | 0 | 0 | 0 | 0 | 0 | 2 | 3 | 2 | 0 | 0 | 0 | 0 | 7 |
| Days Max Temp ≤ 32 °F | 20 | 16 | 6 | 0 | 0 | 0 | 0 | 0 | 0 | 0 | 2 | 14 | 58 |
| Days Min Temp ≤ 32 °F | 30 | 28 | 27 | 17 | 5 | 0 | 0 | 0 | 2 | 11 | 22 | 29 | 171 |
| Days Min Temp ≤ 0 °F | 8 | 7 | 2 | 0 | 0 | 0 | 0 | 0 | 0 | 0 | 0 | 3 | 20 |
| Heating Degree Days | 1427 | 1244 | 1065 | 650 | 345 | 107 | 31 | 59 | 219 | 551 | 849 | 1212 | 7759 |
| Cooling Degree Days | 0 | 0 | 0 | 4 | 21 | 59 | 141 | 94 | 24 | 1 | 0 | 0 | 344 |
| Total Precipitation (") | 1.55 | 1.11 | 2.11 | 2.76 | 2.79 | 3.25 | 2.53 | 3.63 | 3.49 | 2.71 | 2.49 | 1.85 | 30.27 |
| Days ≥ 0.1" Precip | ***5*** | 4 | 5 | 7 | 6 | 6 | 5 | 6 | 7 | 6 | ***6*** | 5 | 68 |
| Total Snowfall (") | ***13.7*** | ***8.1*** | ***7.6*** | 1.7 | 0.0 | 0.0 | 0.0 | 0.0 | 0.0 | 0.1 | ***2.9*** | ***8.4*** | 42.5 |
| Days ≥ 1" Snow Depth | na | na | ***12*** | ***1*** | 0 | 0 | 0 | 0 | 0 | 0 | ***2*** | na | na |

## STEPHENSON 8 WNW *Menominee County* ELEVATION 690 ft LAT/LONG 45° 27' N / 87° 45' W

| | JAN | FEB | MAR | APR | MAY | JUN | JUL | AUG | SEP | OCT | NOV | DEC | YEAR |
|---|---|---|---|---|---|---|---|---|---|---|---|---|---|
| Maximum Temp °F | 24.6 | 28.5 | 39.6 | 54.0 | 67.5 | 75.9 | 80.5 | 78.0 | 68.7 | 57.0 | 41.3 | 29.3 | 53.7 |
| Minimum Temp °F | 2.7 | 5.4 | 17.8 | 30.5 | 40.2 | 49.2 | 54.5 | 52.6 | 44.9 | 34.9 | 24.2 | 11.8 | 30.7 |
| Mean Temp °F | 13.7 | 16.9 | 28.7 | 42.2 | 53.9 | 62.6 | 67.5 | 65.4 | 56.8 | 46.0 | 32.8 | 20.6 | 42.3 |
| Days Max Temp ≥ 90 °F | 0 | 0 | 0 | 0 | 0 | 1 | 3 | 1 | 0 | 0 | 0 | 0 | 5 |
| Days Max Temp ≤ 32 °F | 24 | 18 | 7 | 1 | 0 | 0 | 0 | 0 | 0 | 0 | 5 | 19 | 74 |
| Days Min Temp ≤ 32 °F | 31 | 28 | 28 | 19 | 7 | 1 | 0 | 0 | 3 | 14 | 25 | 30 | 186 |
| Days Min Temp ≤ 0 °F | 14 | 11 | 3 | 0 | 0 | 0 | 0 | 0 | 0 | 0 | 1 | 6 | 35 |
| Heating Degree Days | 1586 | 1350 | 1117 | 678 | 353 | 125 | 42 | 71 | 259 | 584 | 961 | 1369 | 8495 |
| Cooling Degree Days | 0 | 0 | 0 | 3 | 17 | 59 | 120 | 85 | 18 | 0 | 0 | 0 | 302 |
| Total Precipitation (") | 1.54 | 1.00 | 2.08 | 2.53 | 3.36 | 3.74 | 3.53 | 3.51 | 3.78 | 2.75 | 2.58 | 1.98 | 32.38 |
| Days ≥ 0.1" Precip | 5 | 3 | 5 | 6 | 7 | 8 | 7 | 7 | 7 | 6 | 6 | 6 | 73 |
| Total Snowfall (") | 17.3 | 9.4 | 11.1 | 4.0 | 0.5 | 0.0 | 0.0 | 0.0 | 0.0 | 0.5 | 6.0 | 15.9 | 64.7 |
| Days ≥ 1" Snow Depth | 30 | 27 | 20 | 4 | 0 | 0 | 0 | 0 | 0 | 0 | 6 | 21 | 108 |

## THREE RIVERS *St. Joseph County* ELEVATION 820 ft LAT/LONG 41° 57' N / 85° 38' W

| | JAN | FEB | MAR | APR | MAY | JUN | JUL | AUG | SEP | OCT | NOV | DEC | YEAR |
|---|---|---|---|---|---|---|---|---|---|---|---|---|---|
| Maximum Temp °F | 30.6 | 34.3 | 45.9 | 59.5 | 71.4 | 80.5 | 83.9 | 81.6 | 74.5 | 62.3 | 48.2 | 36.1 | 59.1 |
| Minimum Temp °F | 15.1 | 16.8 | 26.7 | 37.1 | 46.9 | 56.1 | 60.3 | 58.0 | 51.0 | 40.2 | 31.7 | 21.7 | 38.5 |
| Mean Temp °F | 22.9 | 25.6 | 36.3 | 48.3 | 59.2 | 68.3 | 72.1 | 69.8 | 62.8 | 51.2 | 40.0 | 28.9 | 48.8 |
| Days Max Temp ≥ 90 °F | 0 | 0 | 0 | 0 | 1 | 3 | 6 | 3 | 1 | 0 | 0 | 0 | 14 |
| Days Max Temp ≤ 32 °F | 17 | 13 | 4 | 0 | 0 | 0 | 0 | 0 | 0 | 0 | 2 | 11 | 47 |
| Days Min Temp ≤ 32 °F | 29 | 26 | 23 | 10 | 2 | 0 | 0 | 0 | 1 | 7 | 17 | 27 | 142 |
| Days Min Temp ≤ 0 °F | 4 | 3 | 0 | 0 | 0 | 0 | 0 | 0 | 0 | 0 | 0 | 1 | 8 |
| Heating Degree Days | 1300 | 1106 | 885 | 501 | 221 | 45 | 10 | 23 | 135 | 426 | 745 | 1111 | 6508 |
| Cooling Degree Days | 0 | 0 | 1 | 9 | 50 | 150 | 249 | 183 | 73 | 5 | 0 | 0 | 720 |
| Total Precipitation (") | 1.77 | 1.50 | 2.52 | 3.38 | 3.35 | 3.72 | 4.10 | 3.47 | 3.66 | 3.05 | 3.04 | 2.70 | 36.26 |
| Days ≥ 0.1" Precip | 5 | 4 | 6 | 8 | 7 | 7 | 7 | 6 | 7 | 6 | 7 | 7 | 77 |
| Total Snowfall (") | 7.9 | 6.1 | 5.9 | 1.6 | 0.0 | 0.0 | 0.0 | 0.0 | 0.0 | 0.4 | 3.4 | 7.6 | 32.9 |
| Days ≥ 1" Snow Depth | 21 | 17 | 8 | 1 | 0 | 0 | 0 | 0 | 0 | 0 | 3 | 14 | 64 |

## TRAVERSE CITY AP *Grand Traverse County* ELEVATION 618 ft LAT/LONG 44° 44' N / 85° 35' W

| | JAN | FEB | MAR | APR | MAY | JUN | JUL | AUG | SEP | OCT | NOV | DEC | YEAR |
|---|---|---|---|---|---|---|---|---|---|---|---|---|---|
| Maximum Temp °F | 26.6 | 28.6 | 38.8 | 53.0 | 66.7 | 76.0 | 81.2 | 78.4 | 69.7 | 57.5 | 43.4 | 32.0 | 54.3 |
| Minimum Temp °F | 13.6 | 11.8 | 20.6 | 31.9 | 41.2 | 51.3 | 57.6 | 56.4 | 49.4 | 39.5 | 30.4 | 20.5 | 35.3 |
| Mean Temp °F | 20.1 | 20.2 | 29.8 | 42.5 | 54.0 | 63.7 | 69.4 | 67.4 | 59.5 | 48.6 | 36.9 | 26.2 | 44.9 |
| Days Max Temp ≥ 90 °F | 0 | 0 | 0 | 0 | 1 | 2 | 4 | 2 | 0 | 0 | 0 | 0 | 9 |
| Days Max Temp ≤ 32 °F | 22 | 18 | 9 | 1 | 0 | 0 | 0 | 0 | 0 | 0 | 3 | 15 | 68 |
| Days Min Temp ≤ 32 °F | 30 | 27 | 27 | 17 | 6 | 0 | 0 | 0 | 1 | 7 | 19 | 28 | 162 |
| Days Min Temp ≤ 0 °F | 4 | 5 | 2 | 0 | 0 | 0 | 0 | 0 | 0 | 0 | 0 | 1 | 12 |
| Heating Degree Days | 1382 | 1260 | 1085 | 674 | 358 | 114 | 27 | 48 | 196 | 508 | 836 | 1195 | 7683 |
| Cooling Degree Days | 0 | 0 | 0 | 6 | 26 | 75 | 176 | 133 | 39 | 1 | 0 | 0 | 456 |
| Total Precipitation (") | 2.20 | 1.45 | 1.75 | 2.50 | 2.22 | 3.18 | 2.88 | 3.08 | 4.05 | 3.08 | 2.59 | 2.33 | 31.31 |
| Days ≥ 0.1" Precip | 7 | 4 | 5 | 6 | 5 | 6 | 5 | 6 | 8 | 7 | 7 | 7 | 73 |
| Total Snowfall (") | 28.2 | 17.5 | 9.7 | 2.6 | 0.1 | 0.0 | 0.0 | 0.0 | 0.0 | 0.4 | 8.3 | 22.3 | 89.1 |
| Days ≥ 1" Snow Depth | 28 | 26 | 19 | 2 | 0 | 0 | 0 | 0 | 0 | 0 | 6 | 21 | 102 |

## VANDERBILT 11 ENE *Otsego County* ELEVATION 932 ft LAT/LONG 45° 10' N / 84° 27' W

| | JAN | FEB | MAR | APR | MAY | JUN | JUL | AUG | SEP | OCT | NOV | DEC | YEAR |
|---|---|---|---|---|---|---|---|---|---|---|---|---|---|
| Maximum Temp °F | 25.3 | 27.7 | 37.7 | 52.6 | 67.6 | 76.2 | 80.7 | 77.2 | 67.6 | 55.3 | 40.8 | 29.9 | 53.2 |
| Minimum Temp °F | 4.7 | 2.8 | 13.0 | 26.8 | 36.1 | 45.1 | 49.9 | 48.6 | 41.4 | 32.6 | 24.7 | 13.2 | 28.2 |
| Mean Temp °F | 15.0 | 15.3 | 25.4 | 39.7 | 51.9 | 60.7 | 65.3 | 62.9 | 54.5 | 44.0 | 32.8 | 21.6 | 40.8 |
| Days Max Temp ≥ 90 °F | 0 | 0 | 0 | 0 | 1 | 2 | 3 | 1 | 0 | 0 | 0 | 0 | 7 |
| Days Max Temp ≤ 32 °F | ***24*** | 19 | 10 | 1 | 0 | 0 | 0 | 0 | 0 | 0 | 6 | ***19*** | 79 |
| Days Min Temp ≤ 32 °F | 31 | 28 | 29 | 22 | 12 | 3 | 0 | 1 | 7 | 18 | 24 | 30 | 205 |
| Days Min Temp ≤ 0 °F | ***12*** | 12 | 6 | 0 | 0 | 0 | 0 | 0 | 0 | 0 | 0 | 6 | 36 |
| Heating Degree Days | 1550 | 1398 | 1218 | 761 | 415 | 173 | 78 | 120 | 328 | 647 | 961 | 1341 | 8990 |
| Cooling Degree Days | 0 | 0 | 0 | 5 | 17 | 49 | 109 | 77 | 19 | 0 | 0 | 0 | 276 |
| Total Precipitation (") | 2.13 | 1.37 | 1.99 | 2.42 | 2.82 | 2.47 | 3.22 | 3.47 | 3.55 | 3.00 | 2.63 | 2.23 | 31.30 |
| Days ≥ 0.1" Precip | ***7*** | ***4*** | ***6*** | ***6*** | ***6*** | 6 | 6 | 7 | 8 | 8 | na | ***7*** | na |
| Total Snowfall (") | 29.2 | 17.1 | 12.2 | 5.3 | 0.5 | 0.0 | 0.0 | 0.0 | 0.0 | 0.7 | ***12.4*** | ***23.5*** | 100.9 |
| Days ≥ 1" Snow Depth | 30 | 28 | 29 | ***11*** | 0 | 0 | 0 | 0 | 0 | 0 | ***9*** | 26 | 133 |

## WATERSMEET *Gogebic County* ELEVATION 1611 ft LAT/LONG 46° 16 ' N / 89° 11 ' W

| | JAN | FEB | MAR | APR | MAY | JUN | JUL | AUG | SEP | OCT | NOV | DEC | YEAR |
|---|---|---|---|---|---|---|---|---|---|---|---|---|---|
| Maximum Temp °F | 21.0 | 25.0 | 36.3 | 50.5 | 64.6 | 73.5 | 78.3 | 75.7 | 66.1 | 53.7 | 36.8 | 25.4 | 50.6 |
| Minimum Temp °F | 0.9 | 2.9 | 14.3 | 27.6 | 38.3 | 47.5 | 53.3 | 51.4 | 43.6 | 34.7 | 22.4 | 9.1 | 28.8 |
| Mean Temp °F | 11.2 | 14.0 | 25.3 | 39.1 | 51.5 | 60.5 | 65.8 | 63.6 | 54.9 | 44.2 | 29.6 | 17.3 | 39.8 |
| Days Max Temp ≥ 90 °F | 0 | 0 | 0 | 0 | 0 | 1 | 2 | 1 | 0 | 0 | 0 | 0 | 4 |
| Days Max Temp ≤ 32 °F | 27 | 21 | 11 | 2 | 0 | 0 | 0 | 0 | 0 | 1 | 11 | 24 | 97 |
| Days Min Temp ≤ 32 °F | 31 | 27 | 29 | 21 | 10 | 2 | 0 | 1 | 4 | 14 | 26 | 31 | 196 |
| Days Min Temp ≤ 0 °F | 14 | 13 | 6 | 1 | 0 | 0 | 0 | 0 | 0 | 0 | 1 | 9 | 44 |
| Heating Degree Days | 1661 | 1438 | 1223 | 774 | 428 | 171 | 69 | 106 | 316 | 639 | 1055 | 1473 | 9353 |
| Cooling Degree Days | 0 | 0 | 0 | 2 | 13 | 44 | 87 | 66 | 13 | 0 | 0 | 0 | 225 |
| Total Precipitation (") | *1.65* | *1.11* | 1.90 | 2.45 | 3.25 | 3.87 | 3.16 | 3.79 | 3.87 | 3.34 | *2.49* | 2.25 | 33.13 |
| Days ≥ 0.1" Precip | na | na | *5* | *7* | *6* | na | 7 | *7* | *9* | 8 | na | na | na |
| Total Snowfall (") | na | na | na | *5.3* | 0.7 | 0.0 | 0.0 | 0.0 | 0.0 | 2.2 | na | na | na |
| Days ≥ 1" Snow Depth | na | na | na | na | *0* | 0 | 0 | 0 | 0 | *0* | na | na | na |

## WEST BRANCH 3 SE *Ogemaw County* ELEVATION 885 ft LAT/LONG 44° 15 ' N / 84° 12 ' W

| | JAN | FEB | MAR | APR | MAY | JUN | JUL | AUG | SEP | OCT | NOV | DEC | YEAR |
|---|---|---|---|---|---|---|---|---|---|---|---|---|---|
| Maximum Temp °F | 27.0 | 30.0 | 40.3 | 54.5 | 67.9 | 76.3 | 81.0 | 78.4 | 69.7 | 57.5 | 43.6 | 32.0 | 54.9 |
| Minimum Temp °F | 8.5 | 9.3 | 19.2 | 31.5 | 41.4 | 50.3 | 55.5 | 53.8 | 46.3 | 35.5 | 27.2 | 16.4 | 32.9 |
| Mean Temp °F | 17.7 | 19.7 | 29.8 | 43.0 | 54.7 | 63.4 | 68.3 | 66.1 | 58.0 | 46.5 | 35.5 | 24.2 | 43.9 |
| Days Max Temp ≥ 90 °F | 0 | 0 | 0 | 0 | 0 | 1 | 3 | 1 | 0 | 0 | 0 | 0 | 5 |
| Days Max Temp ≤ 32 °F | 22 | 17 | 7 | 1 | 0 | 0 | 0 | 0 | 0 | 0 | 4 | 16 | 67 |
| Days Min Temp ≤ 32 °F | 31 | 28 | 28 | 17 | 5 | 0 | 0 | 0 | 2 | 13 | 22 | 30 | 176 |
| Days Min Temp ≤ 0 °F | 9 | 8 | 2 | 0 | 0 | 0 | 0 | 0 | 0 | 0 | 0 | 3 | 22 |
| Heating Degree Days | 1459 | 1273 | 1086 | 656 | 333 | 110 | 32 | 60 | 228 | 568 | 880 | 1257 | 7942 |
| Cooling Degree Days | 0 | 0 | 0 | 3 | 22 | 61 | 145 | 103 | 23 | 0 | 0 | 0 | 357 |
| Total Precipitation (") | 1.50 | 1.15 | 1.97 | 2.52 | 2.99 | 3.02 | 2.90 | 3.46 | 3.72 | 2.74 | 2.58 | 1.92 | 30.47 |
| Days ≥ 0.1" Precip | 4 | 3 | 5 | 6 | 6 | 7 | 5 | 7 | 7 | 6 | 6 | 5 | 67 |
| Total Snowfall (") | 14.9 | 10.2 | 8.4 | 1.9 | 0.1 | 0.0 | 0.0 | 0.0 | 0.0 | 0.2 | 4.9 | 11.7 | 52.3 |
| Days ≥ 1" Snow Depth | 29 | 26 | 19 | 3 | 0 | 0 | 0 | 0 | 0 | 0 | 4 | 19 | 100 |

## WHITEFISH POINT *Chippewa County* ELEVATION 617 ft LAT/LONG 46° 46 ' N / 84° 57 ' W

| | JAN | FEB | MAR | APR | MAY | JUN | JUL | AUG | SEP | OCT | NOV | DEC | YEAR |
|---|---|---|---|---|---|---|---|---|---|---|---|---|---|
| Maximum Temp °F | 24.0 | 25.4 | 34.1 | 44.5 | 56.5 | 64.9 | 71.7 | 71.7 | 64.1 | 52.7 | 39.8 | 29.5 | 48.2 |
| Minimum Temp °F | 10.8 | 8.4 | 16.1 | 28.3 | 36.7 | 44.2 | 50.7 | 53.3 | 47.9 | 38.8 | 28.7 | 18.1 | 31.8 |
| Mean Temp °F | 17.4 | 16.9 | 25.1 | 36.4 | 46.6 | 54.6 | 61.2 | 62.5 | 56.0 | 45.8 | 34.3 | 23.8 | 40.1 |
| Days Max Temp ≥ 90 °F | 0 | 0 | 0 | 0 | 0 | 0 | 0 | 0 | 0 | 0 | 0 | 0 | 0 |
| Days Max Temp ≤ 32 °F | 25 | 21 | *11* | 1 | 0 | 0 | 0 | 0 | 0 | 0 | 5 | 19 | 82 |
| Days Min Temp ≤ 32 °F | 31 | 28 | 30 | 21 | 7 | 1 | 0 | 0 | 1 | *7* | 20 | 29 | 175 |
| Days Min Temp ≤ 0 °F | 5 | 8 | 3 | 0 | 0 | 0 | 0 | 0 | 0 | 0 | 0 | 1 | 17 |
| Heating Degree Days | 1469 | 1352 | 1229 | 851 | 563 | 311 | 140 | 108 | 272 | 590 | 915 | 1269 | 9069 |
| Cooling Degree Days | 0 | 0 | 0 | 0 | 0 | 5 | 35 | 51 | 10 | 0 | 0 | 0 | 101 |
| Total Precipitation (") | 3.23 | 2.10 | 1.86 | 1.95 | 2.60 | 3.12 | 3.04 | 3.56 | 3.25 | 3.27 | 3.09 | 3.23 | 34.30 |
| Days ≥ 0.1" Precip | *11* | 7 | *5* | *5* | 5 | 7 | 6 | 7 | 8 | *9* | 9 | 11 | 90 |
| Total Snowfall (") | 41.2 | 24.6 | 13.8 | 4.2 | 0.0 | 0.0 | 0.0 | 0.0 | 0.0 | 1.0 | 13.2 | 35.5 | 133.5 |
| Days ≥ 1" Snow Depth | 31 | 28 | 31 | 15 | 0 | 0 | 0 | 0 | 0 | 0 | *9* | 26 | 140 |

## YPSILANTI E MICH UNI *Washtenaw County* ELEVATION 791 ft LAT/LONG 42° 15 ' N / 83° 37 ' W

| | JAN | FEB | MAR | APR | MAY | JUN | JUL | AUG | SEP | OCT | NOV | DEC | YEAR |
|---|---|---|---|---|---|---|---|---|---|---|---|---|---|
| Maximum Temp °F | 30.6 | 34.2 | 45.7 | 59.3 | 71.8 | 80.1 | *84.2* | *81.7* | *74.3* | *62.0* | *48.1* | *35.9* | 59.0 |
| Minimum Temp °F | 17.0 | 19.1 | 27.7 | 38.4 | 49.3 | 58.1 | *63.2* | *61.2* | *53.8* | *42.6* | *33.4* | *23.4* | 40.6 |
| Mean Temp °F | 23.8 | 26.7 | 36.7 | 48.9 | 60.6 | 69.1 | *73.7* | *71.5* | *64.1* | *52.4* | *40.8* | *29.6* | 49.8 |
| Days Max Temp ≥ 90 °F | 0 | 0 | 0 | 0 | 1 | 2 | *6* | *3* | 1 | 0 | *0* | 0 | 13 |
| Days Max Temp ≤ 32 °F | *17* | 13 | 4 | 0 | 0 | 0 | *0* | *0* | 0 | 0 | *2* | *10* | 46 |
| Days Min Temp ≤ 32 °F | 29 | 25 | 22 | 8 | 0 | 0 | *0* | *0* | 0 | 3 | *15* | *26* | 128 |
| Days Min Temp ≤ 0 °F | *3* | 2 | 0 | 0 | 0 | 0 | *0* | *0* | 0 | 0 | *0* | *1* | 6 |
| Heating Degree Days | 1269 | 1075 | 869 | 487 | 186 | 34 | *3* | *11* | *107* | *391* | *720* | *1090* | 6242 |
| Cooling Degree Days | 0 | 0 | 1 | 10 | 58 | 169 | 293 | 220 | 84 | 5 | 0 | 0 | 840 |
| Total Precipitation (") | 1.63 | 1.48 | 2.54 | 3.04 | 3.06 | 3.16 | *2.93* | *3.22* | 3.45 | 2.25 | 3.01 | *2.48* | 32.25 |
| Days ≥ 0.1" Precip | *5* | *4* | *6* | 7 | *6* | *6* | *6* | *6* | *7* | 5 | 7 | *6* | 71 |
| Total Snowfall (") | *9.9* | *7.2* | 5.1 | 1.2 | 0.0 | 0.0 | 0.0 | 0.0 | 0.0 | 0.1 | *2.3* | *9.7* | 35.5 |
| Days ≥ 1" Snow Depth | *20* | 16 | 6 | *1* | 0 | 0 | 0 | 0 | 0 | 0 | *2* | *12* | 57 |

## JANUARY MINIMUM TEMPERATURE °F

### LOWEST

| Rank | Place | Value |
|---|---|---|
| 1 | Stambaugh | -2.1 |
| 2 | Bergland Dam | -1.1 |
| 3 | Ironwood | -0.6 |
| 4 | Champion | 0.3 |
| 5 | Watersmeet | 0.9 |
| 6 | Iron mountain | 1.4 |
| 7 | Stephenson | 2.7 |
| 8 | Alberta | 3.0 |
| 9 | Marquette-Co | 3.4 |
| 10 | Sault Ste. Marie | 4.5 |
| 11 | Vanderbilt | 4.7 |
| 12 | Lupton | 4.9 |
| 13 | Seney | 6.3 |
| 14 | Grayling | 6.8 |
| 15 | Manistique | 7.1 |
| | Newberry | 7.1 |
| 17 | Detour | 7.4 |
| 18 | Escanaba | 7.7 |
| | Pellston | 7.7 |
| 20 | Hancock | 8.0 |
| | Munising | 8.0 |
| 22 | Hghtn Lk-6 WSW | 8.1 |
| 23 | Mio | 8.3 |
| 24 | Lake City | 8.4 |
| 25 | West Branch | 8.5 |

### HIGHEST

| Rank | Place | Value |
|---|---|---|
| 1 | Detroit-City | 18.9 |
| 2 | Grand Haven | 18.7 |
| 3 | South Haven | 18.6 |
| 4 | Grosse Pnte Farms | 17.8 |
| 5 | Benton Harbor | 17.6 |
| 6 | Holland | 17.1 |
| | Muskegon | 17.1 |
| 8 | Mt. Clemens | 17.0 |
| | Ypsilanti | 17.0 |
| 10 | Eau Claire | 16.7 |
| 11 | Manistee | 16.5 |
| 12 | Ann Arbor | 16.4 |
| | Monroe | 16.4 |
| 14 | Dearborn | 16.3 |
| | Frankfort | 16.3 |
| 16 | Detroit-Metro | 16.1 |
| | Montague | 16.1 |
| 18 | Port Huron | 15.8 |
| 19 | Kalamazoo | 15.7 |
| | Ludington | 15.7 |
| 21 | Hart | 15.6 |
| 22 | Gull Lake | 15.2 |
| | Pontiac | 15.2 |
| 24 | Three Rivers | 15.1 |
| 25 | Grand Rapids | 15.0 |

## JULY MAXIMUM TEMPERATURE °F

### HIGHEST

| Rank | Place | Value |
|---|---|---|
| 1 | Monroe | 85.0 |
| 2 | Kalamazoo | 84.6 |
| 3 | Dearborn | 84.2 |
| | Ypsilanti | 84.2 |
| 5 | Caro | 84.0 |
| 6 | Gull Lake | 83.9 |
| | St. Charles | 83.9 |
| | Three Rivers | 83.9 |
| 9 | St. Johns | 83.8 |
| 10 | Dowagiac | 83.6 |
| | Grosse Pnte Farms | 83.6 |
| 12 | Ann Arbor | 83.4 |
| | Eau Claire | 83.4 |
| | Greenville | 83.4 |
| | Ionia | 83.4 |
| | Pontiac | 83.4 |
| 17 | Alma | 83.3 |
| | Detroit-Metro | 83.3 |
| 19 | Adrian | 83.2 |
| | Allegan | 83.2 |
| 21 | Detroit-City | 83.1 |
| | Gladwin | 83.1 |
| 23 | Hastings | 82.9 |
| | Jackson | 82.9 |
| 25 | Battle Creek | 82.8 |

### LOWEST

| Rank | Place | Value |
|---|---|---|
| 1 | Whitefish Point | 71.7 |
| 2 | Manistique | 74.0 |
| 3 | Escanaba | 74.6 |
| 4 | Fayette | 75.2 |
| 5 | Grand Marais | 75.4 |
| | Hancock | 75.4 |
| 7 | Munising | 75.6 |
| 8 | Marquette | 75.7 |
| 9 | Sault Ste. Marie | 75.8 |
| | St. James | 75.8 |
| 11 | Marquette-Co | 76.3 |
| | Newberry | 76.3 |
| | Petoskey | 76.3 |
| 14 | Detour | 76.4 |
| 15 | Frankfort | 76.7 |
| 16 | Cross Village | 76.8 |
| 17 | Alpena-Wastewtr | 77.0 |
| 18 | Ironwood | 77.5 |
| 19 | Bergland Dam | 77.9 |
| | Big Bay | 77.9 |
| 21 | Cheboygan | 78.0 |
| | Harbor Beach | 78.0 |
| 23 | Stambaugh | 78.3 |
| | Watersmeet | 78.3 |
| 25 | Alberta | 78.4 |

## ANNUAL PRECIPITATION (")

### HIGHEST

| Rank | Place | Value |
|---|---|---|
| 1 | Bloomingdale | 40.84 |
| 2 | Dowagiac | 40.38 |
| 3 | Allegan | 39.82 |
| 4 | Gull Lake | 38.43 |
| 5 | Grand Rapids | 38.27 |
| 6 | Hillsdale | 38.06 |
| 7 | Benton Harbor | 38.03 |
| 8 | Bergland Dam | 37.49 |
| 9 | Kalamazoo | 37.48 |
| 10 | South Haven | 36.91 |
| 11 | Holland | 36.77 |
| 12 | Eau Claire | 36.67 |
| 13 | Battle Creek | 36.57 |
| 14 | Gaylord | 36.46 |
| 15 | Hart | 36.39 |
| 16 | Ionia | 36.34 |
| 17 | Three Rivers | 36.26 |
| 18 | Munising | 36.25 |
| 19 | Marquette-Co | 36.00 |
| 20 | Maple City | 35.58 |
| 21 | Coldwater | 35.40 |
| 22 | Baldwin | 35.34 |
| 23 | Hastings | 35.24 |
| 24 | Greenville | 35.01 |
| 25 | Frankfort | 34.99 |

### LOWEST

| Rank | Place | Value |
|---|---|---|
| 1 | Mio | 27.68 |
| 2 | Alpena-Wastewtr | 28.50 |
| 3 | Escanaba | 28.62 |
| 4 | Cheboygan | 28.77 |
| 5 | Fayette | 28.79 |
| 6 | Hghtn Lk-Roscmn | 28.80 |
| 7 | Hale Loud Dam | 28.88 |
| 8 | Cross Village | 28.91 |
| 9 | Alpena-Phelps | 29.20 |
| | Sandusky | 29.20 |
| 11 | Iron mountain | 29.35 |
| 12 | Hghtn Lk-6 WSW | 29.49 |
| 13 | Grand Marais | 29.51 |
| 14 | Detour | 29.63 |
| 15 | Milford | 29.73 |
| 16 | Big Bay | 29.81 |
| 17 | Standish | 30.27 |
| 18 | Lupton | 30.31 |
| 19 | Manistique | 30.43 |
| 20 | West Branch | 30.47 |
| 21 | Owosso | 30.55 |
| | Stambaugh | 30.55 |
| 23 | Jackson | 30.65 |
| 24 | Marquette | 30.69 |
| 25 | St. James | 30.71 |

## ANNUAL SNOWFALL (")

### HIGHEST

| Rank | Place | Value |
|---|---|---|
| 1 | Hancock | 216.1 |
| 2 | Bergland Dam | 175.6 |
| 3 | Ironwood | 174.7 |
| 4 | Marquette-Co | 166.4 |
| 5 | Maple City | 150.3 |
| 6 | Grand Marais | 149.9 |
| 7 | Gaylord | 149.0 |
| 8 | Alberta | 145.7 |
| 9 | Munising | 145.4 |
| 10 | Champion | 137.6 |
| 11 | Whitefish Point | 133.5 |
| 12 | Sault Ste. Marie | 126.5 |
| 13 | Seney | 119.6 |
| 14 | Boyne Falls | 118.4 |
| 15 | Petoskey | 116.1 |
| 16 | Big Bay | 116.0 |
| 17 | Marquette | 115.9 |
| 18 | Frankfort | 114.6 |
| | Newberry | 114.6 |
| 20 | Pellston | 113.9 |
| 21 | Muskegon | 107.5 |
| 22 | Grayling | 105.4 |
| 23 | East Jordan | 101.1 |
| 24 | Vanderbilt | 100.9 |
| 25 | Hart | 97.9 |

### LOWEST

| Rank | Place | Value |
|---|---|---|
| 1 | Grosse Pnte Farms | 27.4 |
| 2 | Adrian | 30.1 |
| 3 | Pontiac | 31.8 |
| 4 | Three Rivers | 32.9 |
| 5 | Dearborn | 33.1 |
| 6 | Ypsilanti | 35.5 |
| 7 | Mt. Clemens | 36.2 |
| 8 | Caro | 37.1 |
| 9 | Jackson | 38.3 |
| 10 | Port Huron | 38.8 |
| 11 | Owosso | 39.2 |
| 12 | Milford | 40.2 |
| 13 | Lapeer | 40.5 |
| 14 | Alma | 40.9 |
| 15 | Standish | 42.5 |
| 16 | Detroit-Metro | 43.1 |
| 17 | Saginaw | 44.8 |
| 18 | St. Johns | 44.9 |
| 19 | Ann Arbor | 47.3 |
| | St. Charles | 47.3 |
| 21 | Hale Loud Dam | 48.4 |
| 22 | Flint | 49.9 |
| 23 | Charlotte | 50.0 |
| 24 | Ionia | 51.3 |
| 25 | Gladwin | 51.6 |

# MINNESOTA

PHYSICAL FEATURES. The State of Minnesota covers 84,068 square miles, of which 4,059 square miles is water (15,291 lakes greater than 10 acres). It extends about 400 miles south to north between latitudes 43° 30' and 49° N., and averages 275 miles east to west between longitudes 89° 30' and 97° W.

Elevations are less than 1,200 feet near each of the three major rivers--the Red, the Minnesota, and the Mississippi (except in the northern part). There are three areas at elevations greater than 1,600 feet: the Iron Range, paralleling the north shore of Lake Superior; the Coteau Des Prairies (also known as Buffalo Ridge), extending out of South Dakota across the southwest portion of the State; and a small area in the Lake Itasca region. The highest point above sea level is at 2,301 feet, Eagle Mountain, in the extreme northeast; and the lowest is at 602 feet, the surface of Lake Superior. Minnesota can be considered to have a continental divide in three directions; drainage is toward Hudson Bay to the north; toward the Atlantic Ocean to the east; and toward the Gulf of Mexico to the south.

GENERAL CLIMATE. Minnesota has a continental-type climate. The State is subject to frequent outbreaks of continental polar air throughout the year, with occasional Arctic outbreaks during the cold season. Occasional periods of prolonged heat occur during summer, particularly in the southern portion when warm air pushes northward from the Gulf of Mexico and the southwestern United States. Pacific Ocean air masses that move across the Western United States produce comparatively mild and dry weather at all seasons.

PRECIPITATION. Although the total precipitation is important, its distribution during the growing season is even more significant. For the most part, native vegetation grows for 7 months (April to October) and row crops grow for 5 months (May through September). During the latter 5-month period, approximately two-thirds of the annual precipitation occurs. Mean annual precipitation is 32 inches in extreme southeast Minnesota, an amount which gradually decreases to 19 inches in the extreme northwest portion of the State.

SNOWFALL. Seasonal snowfall averages near 70 inches along the north shore of Lake Superior in northeast Minnesota, gradually decreases to 40 inches along the Iowa border in the south, and measures 30 inches along the North Dakota and South Dakota borders in the west.

DROUGHT. Conditions of moderate droughts or worse are expected on the average at least once in 4 to 5 years, except in southwest Minnesota when they occur about once in every 3 years. Severe or extreme drought conditions occur on the average about once in every 8 to 9 years, except in the western divisions where they occur about once in every 6 years. Generally, the more severe droughts tend to persist or recur several years in succession.

STORMS. Thunderstorm winds generally cause more damage to property than any other weather factor. The annual frequency of thunderstorm days is about 45 days in southern Minnesota, decreasing to about 30 days along the Canadian border. Generally, 80 percent or more of these storms occur during the heavier rainfall months--from May through September. Damaging local windstorms, tornadoes, hail, and heavy rains generally occur with the stronger and more well-developed thunderstorms.

The month with the greatest frequency of tornadoes is June, followed by July, and then May. During these 3 months, nearly 75 percent of all tornadoes occur. Tornadoes have never been reported in December, January, and February. The southern one-half of Minnesota has three to four times as many tornadoes as does the northern one-half of the State.

The frequency of hailstorms shows a high of 4 to 5 days annually in southwestern Minnesota, decreasing to near 2 days in the northern portion of the State. The month with the most hail is July, with June next, and then August. During these 3 months, over 80 percent of the hail occurs. The size of the hail reported is generally in the marble-to-golfball category, with several reports annually of baseball-size and larger.

Heavy snowfalls of greater than 4 inches are common any time from mid-November through mid-April. Heavy snowfalls with blizzard conditions affect the State on the average about two times each winter. (Blizzard conditions involve snow, temperatures of 20° or less, and wind velocities of 35 m.p.h. and greater.)

Freezing rain and glaze storms are not numerous, but do coat the roads several times each season in Minnesota. The more severe ice storms cause extensive damage; such storms are not as common in the far north as they are in southern and southeastern portions of the State.

FLOODS. Major floods on the larger rivers occur on the average 1 or 2 years out of 10. Floods show the greatest frequency in April during the spring breakup of snow cover and frozen ground. Local flash flooding is most common in the hilly terrain and narrow valleys of southeast Minnesota, partly because of the intensive rainstorms in the southern portion of the State.

SUNSHINE. Sunshine amounts vary from a low in November of nearly 40 percent of possible sunshine hours to a high of about 70 percent in July, with an annual average of 58 percent. Hours of sunlight varies from near 8.5 hours in December to about 16 hours in June.

## COUNTY INDEX

## ELEVATION INDEX

| FEET | STATION NAME |
|---|---|
| 610 | DULUTH HARBOR STA |
| 610 | TWO HARBORS |
| 620 | GRAND MARAIS |
| 625 | CEDAR |
| 669 | STILLWATER 1 SE |
| 702 | WINONA |
| 732 | CHASKA |
| 741 | THEILMAN |
| 761 | JORDAN 1 S |
| 791 | NEW ULM 2 SE |
| 820 | HALLOCK |
| 830 | ST PETER 2 SW |
| 834 | MINNEAPOLIS INTL AP |
| 850 | ARGYLE 4 E |
| 879 | GEORGETOWN 1 E |
| 889 | CROOKSTON NW EXP STN |
| 900 | ST PAUL |
| 902 | MONTEVIDEO 1 SW |
| 906 | FOREST LAKE 5 NE |
| 912 | ADA |
| 942 | FARIBAULT |
| 942 | PRESTON |
| 951 | ROSEMOUNT AGRI EXP S |
| 960 | CAMBRIDGE STATE HOSP |
| 981 | BUFFALO |
| 981 | CAMPBELL 1 SSW |
| 981 | FARMINGTON 3 NW |
| 981 | OTTERTAIL |
| 991 | ZUMBROTA |
| 1001 | MORA |
| 1001 | RED LAKE FALLS |
| 1010 | MILAN 1 NW |

| FEET | STATION NAME |
|---|---|
| 1010 | SANTIAGO 3 E |
| 1020 | GAYLORD |
| 1020 | WHEATON |
| 1027 | REDWOOD FALLS MU AP |
| 1028 | ST CLOUD MUNI ARPT |
| 1030 | HINCKLEY |
| 1030 | SPRINGFIELD 1 NW |
| 1040 | BENSON |
| 1040 | STEWART |
| 1050 | ROSEAU 1 E |
| 1069 | WARROAD |
| 1070 | MILACA 1 ENE |
| 1075 | BAUDETTE |
| 1079 | ARTICHOKE LAKE |
| 1079 | ST JAMES FILT PLANT |
| 1089 | MOOSE LAKE 1 SSE |
| 1102 | HUTCHINSON 1 N |
| 1102 | MADISON SEWAGE PLANT |
| 1112 | WINNEBAGO |
| 1122 | LITTLE FALLS 1 N |
| 1129 | OKLEE |
| 1132 | LITCHFIELD |
| 1132 | MORRIS WC EXP STN |
| 1132 | WILLMAR STATE HSP |
| 1142 | AGASSIZ REFUGE |
| 1142 | LAMBERTON SW EXP STN |
| 1150 | OWATONNA |
| 1152 | WASECA EXP STATION |
| 1171 | GLENWOOD 2 WNW |
| 1171 | MARSHALL |
| 1175 | CALEDONIA |
| 1181 | INTERNATL FALLS ARPT |
| 1191 | FAIRMONT |
| 1191 | THORHULT 1 S |
| 1211 | BRAINERD |
| 1211 | MAHNOMEN 1 W |
| 1211 | MELROSE |
| 1211 | ROTHSAY |
| 1220 | AITKIN 2 E |
| 1220 | BIG FALLS |
| 1220 | GULL LAKE DAM |
| 1220 | NEW LONDON |
| 1220 | RED LAKE IND AGENCY |
| 1230 | ALBERT LEA 3 SE |
| 1234 | SANDY LAKE DAM LIBBY |
| 1240 | COLLEGEVILLE ST JOHN |
| 1243 | CANBY |
| 1250 | AUSTIN 3 S |
| 1250 | FERGUS FALLS |
| 1250 | PINE RIVER DAM |

| FEET | STATION NAME |
|---|---|
| 1270 | CLOQUET |
| 1280 | POKEGAMA DAM |
| 1285 | ISLE 12 N |
| 1289 | FOSSTON 1 E |
| 1289 | LONG PRAIRIE |
| 1297 | ROCHESTER MUNI AP |
| 1302 | LEECH LAKE DAM |
| 1302 | WRIGHT 4 NW |
| 1312 | GRAND RAPIDS FRS LAB |
| 1322 | CASS LAKE |
| 1322 | COOK 18 W |
| 1322 | WINNIBIGOSHISH DAM |
| 1340 | BEMIDJI |
| 1342 | GRAND MEADOW |
| 1342 | WINTON POWER PLANT |
| 1352 | WADENA 3 S |
| 1355 | HIBBING CHISHOLM AP |
| 1362 | DETROIT LAKES 1 NNE |
| 1362 | WINDOM |
| 1371 | COTTON 3 E |
| 1381 | TOWER 3 S |
| 1391 | KELLIHER |
| 1403 | TRACY |
| 1411 | WALKER AH GWAH CHING |
| 1424 | DULUTH INTL AP |
| 1427 | ALEXANDRIA CHANDLER |
| 1440 | GUNFLINT LAKE 10 NW |
| 1443 | PARK RAPIDS MUNI AP |
| 1500 | LUVERNE |
| 1503 | ITASCA UNIV OF MN |
| 1742 | PIPESTONE |

10 20 30 STATUTE MILES
MINNESOTA
STATION LEGEND
DATA PUBLISHED IN:
CLIMATOLOGICAL DATA
HOURLY PRECIPITATION DATA
CLIMATOLOGICAL DATA AND HOURLY PRECIPITATION DATA
For further information, refer to the station index and references notes.
DIVISIONS
1 NORTHWEST
2 NORTH CENTRAL
3 NORTHEAST
4 WEST CENTRAL
5 CENTRAL
6 EAST CENTRAL
7 SOUTHWEST
8 SOUTH CENTRAL
9 SOUTHEAST
US DOC - NOAA - NCDC - ASHEVILLE, NC
Updated January 1992
48.0
46.0
44.0
-98.0
-96.0
-94.0
-92.0
-90.0
CARIBOU 2 S
WARROAD
ROSEAU 1 E
HALLOCK
WANNASKA 1 S
KARLSTAD
BAUDETTE
CLEMENTSON
INDUS 3 W
INTERNL FALLS WSO ARPT
KETTLE FALLS
THIEF LAKE REFUGE
ARGYLE 4 E AGASSIZ REFUGE
THIEF RIVER FALLS 2
HIGH LANDING 2 NW
WASKISH 4 NE
THORHULT 1 S
BIG FALLS
ORR
GUNFLINT LAKE 10 NW
GRAND PORTAGE RNG STN
RED LAKE FALLS
OKLEE
KELLIHER
RED LAKE INDIAN AGENCY
COOK 18 W
WINTON POWER PLANT
TOWER 3 S
GRAND MARAIS
LUTSEN 3 NNE
ISABELLA 1 W
TOFTE RANGER STN
CROOKSTON NW EXP STN
BLACKDUCK
MARCELL
FOSSTON 1 E
EVELETH WASTE WATER PL
BEMIDJI
WINNIBIGOSHISH DAM
HIBBING FAA AIRPORT
SHELLY 4 SE
HALSTAD
ADA
MAHNOMEN 1 W
CASS LAKE
LEECH LAKE DAM
POKEGAMA DAM
GRAND RAPIDS FORESTRY
WHITEFACE RESERVOIR
BRIMSON 1 E
WALES 2 E
TWIN VALLEY 3 SW
ITASCA UNIV OF MINNESO
COTTON 3 E
GEORGETOWN 1 E
WALKER RANGER STN
REMER NO 2
WALKER AH GWAH CHING
TAMARAC WILDLIFE REF
FLOODWOOD 4 N
TWO HARBORS
ISLAND LAKE RESERVOIR
MOORHEAD
DETROIT LAKES 1 NNE
PARK RAPIDS 2 S
DEEP PORTAGE
SANDY LAKE DAM LIBBY
DULUTH WSO AP
DULUTH HARBOR STN
FRAZEE
WRIGHT 4 NW
PINE RIVER DAM
CLOQUET
PELICAN RAPIDS
AITKIN 2 E
ROTHSAY
OTTERTAIL
GULL LAKE DAM
MOOSE LAKE 1 SSE
HOLYOKE
BRAINERD
WADENA 3 S
FERGUS FALLS
GARRISON 2 N
ISLE 12 N
ORWELL DAM
BRUNO 7 ENE
CAMPBELL 1 SSW
ONAMIA RANGER STATION
HINCKLEY
LONG PRAIRIE
LITTLE FALLS 1 N
ALEXANDRIA WTR TR PLT
MORA
WHITE ROCK DAM
ROYALTON 5 W
ALEXANDRIA FAA AIRPORT
WHEATON
MILACA 1 ENE
MELROSE
GLENWOOD 2 WNW
BROWNS VALLEY
COLLEGEVILLE ST JOHN
CAMBRIDGE STATE HOSP
MORRIS WC EXP STN
ST CLOUD WSO AP
SANTIAGO 3 E
ST FRANCIS
WILD RIVER STATE PARK
ARTICHOKE LAKE
BENSON
NEW LONDON
ELK RIVER
FOREST LAKE 5 NE
CEDAR
BUFFALO
LOWER ST ANTHONY FALLS
MILAN 1 NW
LITCHFIELD
COKATO
ROCKFORD
VADNAIS LAKE
MADISON SEWAGE PLANT
WILLMAR STATE HOSPITAL
DELANO
NEW HOPE
STILLWATER 1 SE
WATSON 1 NE
WINSTED
GOLDEN VALLEY
UNIV OF MINN ST PAUL
MONTEVIDEO 1 SW
DAWSON
HUTCHINSON 1 N
MOUND
ST PAUL
MINNEAPOLIS WSFO AP
GRANITE FALLS POWER PL
CHASKA
HASTINGS DAM 2
OLIVIA 3 SE STEWART
YOUNG AMERICA
ROSEMOUNT AGRI EXP STN
CANBY
FARMINGTON 3 NW
JORDAN 1 S
RED WING DAM 3
MINNEOTA
REDWOOD FALLS FAA ARPT
GAYLORD
RED WING
LAKE CITY
MARSHALL
VESTA
REDWOOD FALLS 3 SW
LE SUEUR
NORTHFIELD 2 NNE
LE CENTER 2 SE
WABASHA
TYLER 3 N
LAMBERTON SW EXP STN
NEW ULM 2 SE
ZUMBROTA
THEILMAN
ST PETER 2 SW
FARIBAULT
MINNESOTA CITY DAM 5
TRACY
SPRINGFIELD 1 NW
MANKATO
OWATONNA
ELGIN
WINONA DAM 5 A
PIPESTONE
LAKE WILSON
WASECA EXP STATION
DODGE CENTER
WINONA
ST JAMES FILT PLANT
ROCHESTER WSO AP
AMBOY WWTP
LA CRESCENT DAM 7
WINDOM
RUSHFORD
WINNEBAGO
WELLS 1 NW
GRAND MEADOW
LANESBORO
LUVERNE
LAKEFIELD
FAIRMONT
BLUE EARTH
ALBERT LEA 3 SE
SPRING VALLEY
PRESTON
CALEDONIA
WORTHINGTON 2 NNE
SHERBURN 3 WSW
BRICELYN
ALBERT LEA 3 SE
AUSTIN 3 S
HARMONY
SPRING GROVE 1 W

## ADA Norman County — ELEVATION 912 ft — LAT/LONG 47° 18' N / 96° 31' W

| | JAN | FEB | MAR | APR | MAY | JUN | JUL | AUG | SEP | OCT | NOV | DEC | YEAR |
|---|---|---|---|---|---|---|---|---|---|---|---|---|---|
| Maximum Temp °F | 16.4 | 23.0 | 36.7 | 55.8 | 70.7 | 78.0 | 83.6 | 82.2 | 70.8 | 57.1 | 36.8 | 22.4 | 52.8 |
| Minimum Temp °F | -4.2 | 2.1 | 17.0 | 31.8 | 43.1 | 53.4 | 58.3 | 55.8 | 45.7 | 34.3 | 19.5 | 4.1 | 30.1 |
| Mean Temp °F | 6.2 | 12.6 | 26.9 | 43.8 | 56.9 | 65.7 | 71.0 | 69.0 | 58.3 | 45.7 | 28.2 | 13.3 | 41.5 |
| Days Max Temp ≥ 90 °F | 0 | 0 | 0 | 0 | 1 | 2 | 6 | 6 | 1 | 0 | 0 | 0 | 16 |
| Days Max Temp ≤ 32 °F | 26 | 19 | 10 | 1 | 0 | 0 | 0 | 0 | 0 | 1 | 10 | 23 | 90 |
| Days Min Temp ≤ 32 °F | 31 | 28 | 28 | 17 | 5 | 0 | 0 | 0 | 2 | 13 | 27 | 31 | 182 |
| Days Min Temp ≤ 0 °F | 19 | 13 | 5 | 0 | 0 | 0 | 0 | 0 | 0 | 0 | 2 | 13 | 52 |
| Heating Degree Days | 1823 | 1479 | 1173 | 629 | 276 | 70 | 13 | 40 | 232 | 594 | 1099 | 1600 | 9028 |
| Cooling Degree Days | 0 | 0 | 0 | 5 | 47 | 107 | 217 | 178 | 36 | 1 | 0 | 0 | 591 |
| Total Precipitation (") | 0.84 | 0.55 | 1.03 | 2.07 | 2.83 | 4.32 | 3.05 | 2.84 | 2.20 | 1.90 | 0.80 | 0.79 | 23.22 |
| Days ≥ 0.1" Precip | 3 | 2 | 3 | 4 | 6 | 7 | 6 | 5 | 4 | 4 | 3 | 3 | 50 |
| Total Snowfall (") | na | 5.2 | na | 1.8 | 0.0 | 0.0 | 0.0 | 0.0 | 0.0 | 0.9 | na | na | na |
| Days ≥ 1" Snow Depth | na | na | 17 | 3 | 0 | 0 | 0 | 0 | 0 | 0 | na | na | na |

## AGASSIZ REFUGE Marshall County — ELEVATION 1142 ft — LAT/LONG 48° 18' N / 95° 59' W

| | JAN | FEB | MAR | APR | MAY | JUN | JUL | AUG | SEP | OCT | NOV | DEC | YEAR |
|---|---|---|---|---|---|---|---|---|---|---|---|---|---|
| Maximum Temp °F | 13.6 | 20.3 | 34.3 | 52.9 | 68.0 | 75.2 | 79.7 | 78.2 | 67.5 | 54.5 | 33.8 | 19.4 | 49.8 |
| Minimum Temp °F | -8.5 | -2.8 | 11.9 | 29.2 | 42.2 | 51.3 | 55.7 | 52.6 | 42.7 | 31.6 | 16.4 | 0.3 | 26.9 |
| Mean Temp °F | 2.5 | 8.8 | 23.1 | 41.1 | 55.2 | 63.3 | 67.8 | 65.5 | 55.1 | 43.1 | 25.1 | 9.9 | 38.4 |
| Days Max Temp ≥ 90 °F | 0 | 0 | 0 | 0 | 1 | 1 | 2 | 2 | 0 | 0 | 0 | 0 | 6 |
| Days Max Temp ≤ 32 °F | 28 | 22 | 12 | 1 | 0 | 0 | 0 | 0 | 0 | 1 | 14 | 26 | 104 |
| Days Min Temp ≤ 32 °F | 31 | 28 | 29 | 20 | 5 | 0 | 0 | 0 | 4 | 17 | 28 | 31 | 193 |
| Days Min Temp ≤ 0 °F | 22 | 16 | 8 | 0 | 0 | 0 | 0 | 0 | 0 | 0 | 3 | 15 | 64 |
| Heating Degree Days | 1937 | 1585 | 1292 | 711 | 322 | 106 | 33 | 75 | 306 | 673 | 1190 | 1707 | 9937 |
| Cooling Degree Days | 0 | 0 | 0 | 2 | 33 | 66 | 131 | 107 | 16 | 0 | 0 | 0 | 355 |
| Total Precipitation (") | 0.62 | 0.41 | 0.78 | 1.42 | 2.53 | 3.81 | 3.46 | 2.98 | 2.57 | 1.55 | 0.92 | 0.59 | 21.64 |
| Days ≥ 0.1" Precip | 2 | 1 | 2 | 4 | 6 | 7 | 7 | 6 | 5 | 4 | 3 | 2 | 49 |
| Total Snowfall (") | 9.0 | 5.2 | 6.3 | 2.4 | 0.2 | 0.0 | 0.0 | 0.0 | 0.0 | 1.0 | 6.6 | 7.9 | 38.6 |
| Days ≥ 1" Snow Depth | 31 | 28 | 27 | 5 | 0 | 0 | 0 | 0 | 0 | 1 | 14 | 28 | 134 |

## AITKIN 2 E Aitkin County — ELEVATION 1220 ft — LAT/LONG 46° 32' N / 93° 43' W

| | JAN | FEB | MAR | APR | MAY | JUN | JUL | AUG | SEP | OCT | NOV | DEC | YEAR |
|---|---|---|---|---|---|---|---|---|---|---|---|---|---|
| Maximum Temp °F | 18.0 | 24.4 | 38.5 | 53.1 | 67.5 | 75.5 | 79.7 | 77.3 | 66.9 | 55.8 | 36.9 | 23.1 | 51.4 |
| Minimum Temp °F | -4.3 | 1.3 | 16.3 | 30.4 | 41.7 | 51.3 | 56.0 | 53.9 | 44.5 | 34.7 | 20.2 | 4.8 | 29.2 |
| Mean Temp °F | 6.8 | 12.9 | 27.4 | 41.7 | 54.6 | 63.4 | 67.9 | 65.6 | 55.7 | 45.3 | 28.6 | 13.9 | 40.3 |
| Days Max Temp ≥ 90 °F | 0 | 0 | 0 | 0 | 0 | 0 | 2 | 1 | 0 | 0 | 0 | 0 | 3 |
| Days Max Temp ≤ 32 °F | 26 | 19 | 8 | 1 | 0 | 0 | 0 | 0 | 0 | 0 | 11 | 24 | 89 |
| Days Min Temp ≤ 32 °F | 30 | 26 | 28 | 19 | 5 | 0 | 0 | 0 | 3 | 13 | 27 | 30 | 181 |
| Days Min Temp ≤ 0 °F | 17 | 13 | 4 | 0 | 0 | 0 | 0 | 0 | 0 | 0 | 1 | 12 | 47 |
| Heating Degree Days | 1794 | 1468 | 1156 | 692 | 330 | 100 | 35 | 69 | 290 | 604 | 1083 | 1580 | 9201 |
| Cooling Degree Days | 0 | 0 | 0 | 2 | 14 | 48 | 116 | 96 | 10 | 1 | 0 | 0 | 287 |
| Total Precipitation (") | 0.94 | 0.60 | 1.60 | 2.31 | 2.84 | 4.48 | 4.21 | 3.72 | 2.98 | 2.54 | 1.68 | 0.95 | 28.85 |
| Days ≥ 0.1" Precip | 3 | 2 | 4 | 5 | 6 | 8 | 6 | 6 | 6 | 5 | 4 | 3 | 58 |
| Total Snowfall (") | 11.2 | 6.0 | 8.8 | 3.0 | 0.1 | 0.0 | 0.0 | 0.0 | 0.0 | 0.5 | 7.0 | na | na |
| Days ≥ 1" Snow Depth | na | na | na | na | 0 | 0 | 0 | 0 | 0 | 0 | na | na | na |

## ALBERT LEA 3 SE Freeborn County — ELEVATION 1230 ft — LAT/LONG 43° 37' N / 93° 25' W

| | JAN | FEB | MAR | APR | MAY | JUN | JUL | AUG | SEP | OCT | NOV | DEC | YEAR |
|---|---|---|---|---|---|---|---|---|---|---|---|---|---|
| Maximum Temp °F | 21.4 | 26.7 | 39.3 | 55.7 | 69.5 | 79.0 | 82.6 | 80.0 | 71.3 | 58.8 | 40.8 | 26.5 | 54.3 |
| Minimum Temp °F | 2.6 | 8.1 | 22.0 | 35.3 | 47.2 | 57.1 | 61.6 | 58.8 | 49.3 | 37.1 | 23.9 | 9.8 | 34.4 |
| Mean Temp °F | 12.3 | 17.7 | 30.8 | 45.7 | 58.6 | 68.2 | 72.2 | 69.4 | 60.3 | 48.2 | 32.5 | 18.3 | 44.5 |
| Days Max Temp ≥ 90 °F | 0 | 0 | 0 | 0 | 1 | 3 | 5 | 3 | 1 | 0 | 0 | 0 | 13 |
| Days Max Temp ≤ 32 °F | 24 | 18 | 9 | 1 | 0 | 0 | 0 | 0 | 0 | 0 | 7 | 21 | 80 |
| Days Min Temp ≤ 32 °F | 31 | 28 | 26 | 11 | 2 | 0 | 0 | 0 | 1 | 10 | 25 | 31 | 165 |
| Days Min Temp ≤ 0 °F | 14 | 10 | 2 | 0 | 0 | 0 | 0 | 0 | 0 | 0 | 1 | 8 | 35 |
| Heating Degree Days | 1629 | 1329 | 1052 | 579 | 233 | 43 | 9 | 29 | 183 | 513 | 967 | 1441 | 8007 |
| Cooling Degree Days | 0 | 0 | 0 | 7 | 41 | 146 | 231 | 171 | 51 | 1 | 0 | 0 | 648 |
| Total Precipitation (") | 0.86 | 0.75 | 1.97 | 3.36 | 3.79 | 4.95 | 4.08 | 4.24 | 3.48 | 2.56 | 1.90 | 1.10 | 33.04 |
| Days ≥ 0.1" Precip | 3 | 3 | 4 | 7 | 7 | 8 | 7 | 6 | 6 | 5 | 4 | 3 | 63 |
| Total Snowfall (") | 10.0 | 6.2 | 6.4 | 3.3 | 0.0 | 0.0 | 0.0 | 0.0 | 0.0 | 0.4 | 4.7 | 9.7 | 40.7 |
| Days ≥ 1" Snow Depth | 25 | 22 | 12 | 2 | 0 | 0 | 0 | 0 | 0 | 0 | 5 | 19 | 85 |

## ALEXANDRIA CHANDLER *Douglas County* ELEVATION 1427 ft LAT/LONG 45° 52 ' N / 95° 23 ' W

| | JAN | FEB | MAR | APR | MAY | JUN | JUL | AUG | SEP | OCT | NOV | DEC | YEAR |
|---|---|---|---|---|---|---|---|---|---|---|---|---|---|
| Maximum Temp °F | 16.9 | 23.1 | 35.9 | 53.3 | 67.4 | 75.9 | 81.4 | 78.6 | 68.1 | 55.1 | 36.2 | 22.1 | 51.2 |
| Minimum Temp °F | -2.4 | 4.3 | 18.0 | 33.2 | 45.9 | 55.4 | 60.8 | 58.3 | 48.1 | 36.2 | 21.4 | 6.0 | 32.1 |
| Mean Temp °F | 7.2 | 13.7 | 27.0 | 43.3 | 56.7 | 65.7 | 71.1 | 68.5 | 58.1 | 45.7 | 28.8 | 14.1 | 41.7 |
| Days Max Temp ≥ 90 °F | 0 | 0 | 0 | 0 | 0 | 1 | 4 | 3 | 1 | 0 | 0 | 0 | 9 |
| Days Max Temp ≤ 32 °F | 27 | 21 | 11 | 1 | 0 | 0 | 0 | 0 | 0 | 0 | 11 | 25 | 96 |
| Days Min Temp ≤ 32 °F | 31 | 28 | 28 | 15 | 2 | 0 | 0 | 0 | 1 | 10 | 26 | 31 | 172 |
| Days Min Temp ≤ 0 °F | 17 | 12 | 4 | 0 | 0 | 0 | 0 | 0 | 0 | 0 | 2 | 11 | 46 |
| Heating Degree Days | 1791 | 1444 | 1173 | 647 | 279 | 70 | 13 | 36 | 233 | 593 | 1079 | 1574 | 8932 |
| Cooling Degree Days | 0 | 0 | 0 | 4 | 36 | 101 | 209 | 154 | 31 | 1 | 0 | 0 | 536 |
| Total Precipitation (") | 0.99 | 0.66 | 1.44 | 2.33 | 2.89 | 4.26 | 3.04 | 3.28 | 2.56 | 2.35 | 1.21 | 0.72 | 25.73 |
| Days ≥ 0.1" Precip | 3 | 2 | 3 | 5 | 6 | 8 | 6 | 5 | 5 | 4 | 3 | 2 | 52 |
| Total Snowfall (") | 11.0 | 6.5 | 9.2 | 3.1 | 0.0 | 0.0 | 0.0 | 0.0 | 0.0 | 0.6 | 6.4 | 6.6 | 43.4 |
| Days ≥ 1" Snow Depth | 29 | 26 | 21 | 4 | 0 | 0 | 0 | 0 | 0 | 0 | 9 | 24 | 113 |

## ARGYLE 4 E *Marshall County* ELEVATION 850 ft LAT/LONG 48° 20 ' N / 96° 47 ' W

| | JAN | FEB | MAR | APR | MAY | JUN | JUL | AUG | SEP | OCT | NOV | DEC | YEAR |
|---|---|---|---|---|---|---|---|---|---|---|---|---|---|
| Maximum Temp °F | 12.0 | 18.3 | 32.4 | 52.1 | 68.3 | 75.9 | 80.8 | 80.1 | 68.6 | 54.8 | 34.1 | 18.8 | 49.7 |
| Minimum Temp °F | -9.4 | -4.2 | 11.6 | 28.7 | 40.8 | 50.3 | 54.4 | 51.7 | 42.1 | 31.0 | 16.1 | -0.3 | 26.1 |
| Mean Temp °F | 1.3 | 7.0 | 22.0 | 40.4 | 54.5 | 63.1 | 67.6 | 65.9 | 55.4 | 42.9 | 25.1 | 9.3 | 37.9 |
| Days Max Temp ≥ 90 °F | 0 | 0 | 0 | 0 | 1 | 2 | 3 | 4 | 1 | 0 | 0 | 0 | 11 |
| Days Max Temp ≤ 32 °F | 28 | 23 | 14 | 2 | 0 | 0 | 0 | 0 | 0 | 1 | 13 | 26 | 107 |
| Days Min Temp ≤ 32 °F | 31 | 28 | 30 | 20 | 7 | 0 | 0 | 0 | 4 | 18 | 29 | 31 | 198 |
| Days Min Temp ≤ 0 °F | 22 | 17 | 8 | 1 | 0 | 0 | 0 | 0 | 0 | 0 | 3 | 16 | 67 |
| Heating Degree Days | 1974 | 1631 | 1325 | 733 | 341 | 113 | 38 | 72 | 297 | 680 | 1191 | 1723 | 10118 |
| Cooling Degree Days | 0 | 0 | 0 | 3 | 31 | 69 | 127 | 117 | 18 | 1 | 0 | 0 | 366 |
| Total Precipitation (") | 0.82 | 0.52 | 0.94 | 1.29 | 2.03 | 3.12 | 3.09 | 2.41 | 2.33 | 1.34 | 0.81 | 0.66 | 19.36 |
| Days ≥ 0.1" Precip | 3 | 2 | 3 | 3 | 5 | 7 | 6 | 5 | 4 | 4 | 3 | 2 | 47 |
| Total Snowfall (") | 10.5 | 6.3 | 6.0 | 1.6 | 0.1 | 0.0 | 0.0 | 0.0 | 0.1 | 1.0 | 6.0 | 7.7 | 39.3 |
| Days ≥ 1" Snow Depth | 28 | 24 | 18 | 3 | 0 | 0 | 0 | 0 | 0 | 1 | 8 | 20 | 102 |

## ARTICHOKE LAKE *Big Stone County* ELEVATION 1079 ft LAT/LONG 45° 22 ' N / 96° 8 ' W

| | JAN | FEB | MAR | APR | MAY | JUN | JUL | AUG | SEP | OCT | NOV | DEC | YEAR |
|---|---|---|---|---|---|---|---|---|---|---|---|---|---|
| Maximum Temp °F | 19.7 | 25.5 | 37.7 | 55.6 | 69.6 | 77.7 | 82.8 | 80.2 | 70.5 | 58.0 | 38.6 | 24.8 | 53.4 |
| Minimum Temp °F | -0.8 | 5.5 | 19.5 | 34.2 | 46.8 | 55.9 | 60.5 | 57.9 | 48.1 | 36.3 | 21.3 | 6.9 | 32.7 |
| Mean Temp °F | 9.6 | 15.7 | 28.7 | 44.9 | 58.2 | 66.8 | 71.7 | 69.1 | 59.3 | 47.2 | 30.0 | 16.0 | 43.1 |
| Days Max Temp ≥ 90 °F | 0 | 0 | 0 | 0 | 0 | 2 | 6 | 4 | 1 | 0 | 0 | 0 | 13 |
| Days Max Temp ≤ 32 °F | 24 | 18 | 9 | 1 | 0 | 0 | 0 | 0 | 0 | 0 | 9 | 22 | 83 |
| Days Min Temp ≤ 32 °F | 31 | 28 | 27 | 13 | 2 | 0 | 0 | 0 | 1 | 11 | 26 | 31 | 170 |
| Days Min Temp ≤ 0 °F | 17 | 11 | 3 | 0 | 0 | 0 | 0 | 0 | 0 | 0 | 2 | 11 | 44 |
| Heating Degree Days | 1715 | 1389 | 1120 | 600 | 240 | 53 | 11 | 31 | 204 | 549 | 1044 | 1515 | 8471 |
| Cooling Degree Days | 0 | 0 | 0 | 6 | 46 | 111 | 209 | 156 | 38 | 2 | 0 | 0 | 568 |
| Total Precipitation (") | 0.78 | 0.67 | 1.39 | 2.29 | 2.56 | 3.96 | 3.62 | 2.89 | 2.04 | 2.10 | 1.13 | 0.57 | 24.00 |
| Days ≥ 0.1" Precip | 3 | 2 | 4 | 5 | 6 | 7 | 6 | 5 | 5 | 4 | 3 | 2 | 52 |
| Total Snowfall (") | 9.6 | 7.5 | 7.7 | 2.4 | 0.1 | 0.0 | 0.0 | 0.0 | 0.0 | 0.8 | 6.3 | 6.4 | 40.8 |
| Days ≥ 1" Snow Depth | 26 | 22 | 15 | 2 | 0 | 0 | 0 | 0 | 0 | 0 | 7 | 19 | 91 |

## AUSTIN 3 S *Mower County* ELEVATION 1250 ft LAT/LONG 43° 37 ' N / 92° 59 ' W

| | JAN | FEB | MAR | APR | MAY | JUN | JUL | AUG | SEP | OCT | NOV | DEC | YEAR |
|---|---|---|---|---|---|---|---|---|---|---|---|---|---|
| Maximum Temp °F | 21.9 | 27.7 | 40.7 | 57.6 | 70.5 | 79.5 | 82.6 | 80.3 | 72.5 | 60.1 | 41.4 | 26.9 | 55.1 |
| Minimum Temp °F | 3.4 | 8.9 | 23.2 | 35.8 | 46.9 | 56.5 | 60.4 | 57.7 | 49.2 | 38.1 | 25.1 | 10.7 | 34.7 |
| Mean Temp °F | 12.7 | 18.3 | 32.0 | 46.7 | 58.7 | 68.0 | 71.5 | 69.0 | 60.9 | 49.1 | 33.3 | 18.8 | 44.9 |
| Days Max Temp ≥ 90 °F | 0 | 0 | 0 | 0 | 1 | 2 | 4 | 3 | 1 | 0 | 0 | 0 | 11 |
| Days Max Temp ≤ 32 °F | 24 | 18 | 8 | 0 | 0 | 0 | 0 | 0 | 0 | 0 | 7 | 21 | 78 |
| Days Min Temp ≤ 32 °F | 31 | 28 | 26 | 13 | 2 | 0 | 0 | 0 | 1 | 9 | 23 | 30 | 163 |
| Days Min Temp ≤ 0 °F | 13 | 9 | 2 | 0 | 0 | 0 | 0 | 0 | 0 | 0 | 1 | 8 | 33 |
| Heating Degree Days | 1619 | 1311 | 1015 | 546 | 226 | 40 | 12 | 30 | 171 | 489 | 946 | 1427 | 7832 |
| Cooling Degree Days | 0 | 0 | 0 | 6 | 40 | 150 | 228 | 169 | 62 | 2 | 0 | 0 | 657 |
| Total Precipitation (") | ***1.02*** | 0.64 | 1.52 | 2.88 | 3.84 | 4.28 | 4.32 | 4.13 | 3.88 | 2.43 | 1.81 | ***1.14*** | 31.89 |
| Days ≥ 0.1" Precip | na | ***2*** | 3 | 6 | 7 | 7 | 7 | 6 | 6 | 5 | 3 | ***3*** | na |
| Total Snowfall (") | na | 6.8 | ***5.3*** | 2.5 | 0.0 | 0.0 | 0.0 | 0.0 | 0.0 | 0.2 | 4.2 | ***9.0*** | na |
| Days ≥ 1" Snow Depth | ***24*** | ***20*** | ***13*** | 1 | 0 | 0 | 0 | 0 | 0 | 0 | 4 | na | na |

## BAUDETTE *Lake of the Woods County* ELEVATION 1075 ft LAT/LONG 48° 43 ' N / 94° 37 ' W

| | JAN | FEB | MAR | APR | MAY | JUN | JUL | AUG | SEP | OCT | NOV | DEC | YEAR |
|---|---|---|---|---|---|---|---|---|---|---|---|---|---|
| Maximum Temp °F | 14.6 | 22.2 | 35.7 | 53.1 | 67.8 | 75.4 | 80.0 | 78.0 | 67.5 | 54.2 | 34.9 | 20.2 | 50.3 |
| Minimum Temp °F | -9.0 | -3.8 | 10.7 | 28.5 | 41.2 | 50.3 | 55.2 | 52.8 | 43.6 | 33.2 | 18.0 | 0.4 | 26.8 |
| Mean Temp °F | 2.8 | 9.2 | 23.2 | 40.9 | 54.5 | 62.9 | 67.6 | 65.4 | 55.6 | 43.7 | 26.5 | 10.3 | 38.6 |
| Days Max Temp ≥ 90 °F | 0 | 0 | 0 | 0 | 0 | 1 | 2 | 2 | 0 | 0 | 0 | 0 | 5 |
| Days Max Temp ≤ 32 °F | 28 | 21 | 12 | 1 | 0 | 0 | 0 | 0 | 0 | 1 | 13 | 26 | 102 |
| Days Min Temp ≤ 32 °F | 31 | 28 | 29 | 20 | 6 | 0 | 0 | 0 | 2 | 15 | 28 | 31 | 190 |
| Days Min Temp ≤ 0 °F | 21 | 17 | 8 | 0 | 0 | 0 | 0 | 0 | 0 | 0 | 3 | 15 | 64 |
| Heating Degree Days | 1927 | 1573 | 1290 | 718 | 341 | 111 | 32 | 74 | 295 | 652 | 1150 | 1694 | 9857 |
| Cooling Degree Days | 0 | 0 | 0 | 1 | 26 | 60 | 128 | 107 | 15 | 0 | 0 | 0 | 337 |
| Total Precipitation (") | 0.63 | 0.37 | 0.79 | 1.37 | 2.49 | 3.92 | 3.38 | 3.42 | 2.65 | 2.13 | 0.96 | 0.61 | 22.72 |
| Days ≥ 0.1" Precip | 2 | 1 | 3 | 4 | 6 | 8 | 7 | 6 | 6 | 5 | 3 | 2 | 53 |
| Total Snowfall (") | 8.5 | *6.3* | *6.2* | *3.2* | 0.0 | 0.0 | 0.0 | 0.0 | 0.0 | 0.8 | 8.4 | *8.2* | 41.6 |
| Days ≥ 1" Snow Depth | na | na | na | na | 0 | 0 | 0 | 0 | 0 | 0 | na | *14* | na |

## BEMIDJI *Beltrami County* ELEVATION 1340 ft LAT/LONG 47° 28 ' N / 94° 53 ' W

| | JAN | FEB | MAR | APR | MAY | JUN | JUL | AUG | SEP | OCT | NOV | DEC | YEAR |
|---|---|---|---|---|---|---|---|---|---|---|---|---|---|
| Maximum Temp °F | 13.8 | 21.0 | 34.1 | 50.6 | 65.3 | 73.4 | 78.7 | 76.2 | 65.0 | 52.4 | 33.5 | 19.4 | 48.6 |
| Minimum Temp °F | -7.8 | -2.7 | 12.4 | 27.9 | 40.6 | 50.7 | 56.2 | 53.6 | 43.8 | 33.3 | 17.6 | 1.3 | 27.2 |
| Mean Temp °F | 3.0 | 9.2 | 23.3 | 39.3 | 53.0 | 62.0 | 67.4 | 65.0 | 54.4 | 42.9 | 25.6 | 10.4 | 38.0 |
| Days Max Temp ≥ 90 °F | 0 | 0 | 0 | 0 | 0 | 1 | 2 | 2 | 0 | 0 | 0 | 0 | 5 |
| Days Max Temp ≤ 32 °F | 28 | 21 | 13 | 2 | 0 | 0 | 0 | 0 | 0 | 1 | 15 | 26 | 106 |
| Days Min Temp ≤ 32 °F | 31 | 27 | 29 | 21 | 7 | 0 | 0 | 0 | 3 | 15 | 27 | 30 | 190 |
| Days Min Temp ≤ 0 °F | 21 | 16 | 7 | 0 | 0 | 0 | 0 | 0 | 0 | 0 | 3 | 14 | 61 |
| Heating Degree Days | 1921 | 1575 | 1286 | 766 | 381 | 132 | 43 | 84 | 327 | 678 | 1175 | 1691 | 10059 |
| Cooling Degree Days | 0 | *0* | *0* | *1* | *14* | *42* | 125 | 96 | 12 | *1* | 0 | 0 | 291 |
| Total Precipitation (") | 0.66 | 0.45 | 0.94 | 1.71 | 2.51 | 4.15 | 3.80 | 3.55 | 2.67 | 2.23 | 0.88 | 0.74 | 24.29 |
| Days ≥ 0.1" Precip | 2 | 2 | 3 | 4 | 6 | 8 | 7 | 6 | 6 | 5 | 3 | 3 | 55 |
| Total Snowfall (") | 8.7 | 4.7 | 7.0 | 2.9 | 0.1 | 0.0 | 0.0 | 0.0 | 0.0 | 1.0 | 6.0 | 7.6 | 38.0 |
| Days ≥ 1" Snow Depth | 30 | 26 | 25 | 7 | 0 | 0 | 0 | 0 | 0 | 1 | 11 | 27 | 127 |

## BENSON *Swift County* ELEVATION 1040 ft LAT/LONG 45° 19 ' N / 95° 36 ' W

| | JAN | FEB | MAR | APR | MAY | JUN | JUL | AUG | SEP | OCT | NOV | DEC | YEAR |
|---|---|---|---|---|---|---|---|---|---|---|---|---|---|
| Maximum Temp °F | 19.4 | 25.7 | 38.9 | 56.7 | 70.9 | 79.1 | 83.7 | 81.0 | 71.6 | 58.6 | 39.0 | 25.2 | 54.2 |
| Minimum Temp °F | 0.3 | 6.8 | 20.6 | 34.6 | 46.8 | 56.3 | 60.9 | 58.3 | 48.5 | 36.6 | 22.2 | 8.3 | 33.4 |
| Mean Temp °F | 9.9 | 16.3 | 29.7 | 45.7 | 58.8 | 67.7 | 72.3 | 69.7 | 60.1 | 47.7 | 30.7 | 16.8 | 43.8 |
| Days Max Temp ≥ 90 °F | 0 | 0 | 0 | 0 | 1 | 3 | 7 | 4 | 1 | 0 | 0 | 0 | 16 |
| Days Max Temp ≤ 32 °F | 25 | 18 | 9 | 0 | 0 | 0 | 0 | 0 | 0 | 0 | 8 | 22 | 82 |
| Days Min Temp ≤ 32 °F | 31 | 28 | 26 | 13 | 2 | 0 | 0 | 0 | 1 | 10 | 26 | 31 | 168 |
| Days Min Temp ≤ 0 °F | 16 | 10 | 3 | 0 | 0 | 0 | 0 | 0 | 0 | 0 | 1 | 9 | 39 |
| Heating Degree Days | 1707 | 1370 | 1086 | 577 | 225 | 46 | 9 | 27 | 190 | 533 | 1024 | 1489 | 8283 |
| Cooling Degree Days | 0 | 0 | 0 | 6 | 53 | 137 | 238 | 182 | 49 | 2 | 0 | 0 | 667 |
| Total Precipitation (") | 0.90 | 0.81 | 1.68 | 2.45 | 3.02 | 4.68 | 3.72 | 3.65 | 2.77 | 2.39 | 1.40 | 0.82 | 28.29 |
| Days ≥ 0.1" Precip | 3 | 2 | 4 | 5 | 6 | 8 | 6 | 6 | 5 | 4 | 3 | 2 | 54 |
| Total Snowfall (") | 10.4 | 8.3 | 8.8 | 1.7 | 0.0 | 0.0 | 0.0 | 0.0 | 0.0 | 0.5 | 6.5 | 7.8 | 44.0 |
| Days ≥ 1" Snow Depth | 28 | 25 | 18 | 2 | 0 | 0 | 0 | 0 | 0 | 0 | 9 | 23 | 105 |

## BIG FALLS *Koochiching County* ELEVATION 1220 ft LAT/LONG 48° 12 ' N / 93° 48 ' W

| | JAN | FEB | MAR | APR | MAY | JUN | JUL | AUG | SEP | OCT | NOV | DEC | YEAR |
|---|---|---|---|---|---|---|---|---|---|---|---|---|---|
| Maximum Temp °F | 15.6 | 25.1 | 38.5 | 54.6 | 68.9 | 76.1 | 80.6 | 78.5 | 67.2 | 54.7 | *34.8* | 21.0 | 51.3 |
| Minimum Temp °F | -8.8 | -1.8 | 12.1 | 27.7 | 39.5 | 48.5 | 53.0 | 50.7 | 41.8 | *32.0* | *16.4* | -0.3 | 25.9 |
| Mean Temp °F | 3.4 | 11.6 | 25.2 | 41.2 | 54.3 | 62.5 | 66.8 | 64.6 | 54.6 | *43.3* | na | 10.1 | na |
| Days Max Temp ≥ 90 °F | 0 | 0 | 0 | 0 | 1 | 1 | 3 | 2 | 0 | 0 | 0 | 0 | 7 |
| Days Max Temp ≤ 32 °F | 27 | 20 | 10 | 1 | 0 | 0 | 0 | 0 | 0 | 0 | *11* | 25 | 94 |
| Days Min Temp ≤ 32 °F | 29 | 27 | 27 | 21 | 8 | 1 | 0 | 0 | 5 | na | 24 | 29 | na |
| Days Min Temp ≤ 0 °F | *18* | 14 | 7 | 0 | 0 | 0 | 0 | 0 | 0 | 0 | 3 | 14 | 56 |
| Heating Degree Days | 1913 | 1505 | 1228 | 705 | 346 | 121 | 45 | 89 | 320 | *670* | na | *1697* | na |
| Cooling Degree Days | *0* | 0 | *0* | 2 | 25 | 58 | 110 | 87 | 11 | *0* | na | na | na |
| Total Precipitation (") | 0.95 | 0.58 | 1.15 | 1.79 | 2.79 | 4.12 | 3.81 | 3.52 | 3.11 | 2.41 | 1.30 | 1.03 | 26.56 |
| Days ≥ 0.1" Precip | 3 | 2 | 3 | 4 | 6 | 8 | 7 | 6 | 7 | 5 | 3 | 3 | 57 |
| Total Snowfall (") | 13.1 | 8.4 | 10.4 | 4.8 | 0.3 | 0.0 | 0.0 | 0.0 | 0.0 | 1.4 | *9.2* | 12.3 | 59.9 |
| Days ≥ 1" Snow Depth | na | 20 | *24* | 8 | 0 | 0 | 0 | 0 | 0 | 1 | *10* | *19* | na |

## BRAINERD *Crow Wing County* ELEVATION 1211 ft LAT/LONG 46° 22 ' N / 94° 12 ' W

| | JAN | FEB | MAR | APR | MAY | JUN | JUL | AUG | SEP | OCT | NOV | DEC | YEAR |
|---|---|---|---|---|---|---|---|---|---|---|---|---|---|
| Maximum Temp °F | ***19.3*** | ***25.8*** | 37.6 | 54.0 | 67.8 | 76.0 | ***81.0*** | 78.5 | 67.5 | ***55.3*** | 37.4 | ***23.2*** | 52.0 |
| Minimum Temp °F | ***-7.6*** | -1.7 | 14.1 | 30.2 | 42.4 | 51.2 | 56.6 | 53.5 | 43.3 | 31.9 | 18.4 | ***1.7*** | 27.8 |
| Mean Temp °F | ***6.2*** | ***12.5*** | 26.0 | 42.3 | 55.3 | 63.8 | ***68.9*** | 66.1 | 55.3 | ***43.7*** | 28.0 | ***12.4*** | 40.0 |
| Days Max Temp ≥ 90 °F | 0 | 0 | 0 | 0 | 0 | 1 | 3 | 2 | 0 | 0 | 0 | 0 | 6 |
| Days Max Temp ≤ 32 °F | ***26*** | na | ***8*** | 1 | 0 | 0 | 0 | 0 | 0 | 0 | ***10*** | na | na |
| Days Min Temp ≤ 32 °F | 31 | ***28*** | ***29*** | 19 | 4 | 0 | 0 | 0 | 4 | na | 28 | ***31*** | na |
| Days Min Temp ≤ 0 °F | ***20*** | ***14*** | ***5*** | 0 | 0 | 0 | 0 | 0 | 0 | 0 | 2 | ***14*** | 55 |
| Heating Degree Days | na | ***1457*** | ***1183*** | 673 | 314 | ***98*** | ***28*** | 68 | ***298*** | ***651*** | ***1101*** | ***1625*** | na |
| Cooling Degree Days | 0 | 0 | 0 | 3 | 23 | 71 | 150 | 105 | 16 | 1 | 0 | 0 | 369 |
| Total Precipitation (") | 0.78 | 0.57 | 1.45 | 2.25 | 3.05 | 4.35 | 3.78 | 3.34 | 3.05 | 2.47 | 1.50 | 0.80 | 27.39 |
| Days ≥ 0.1" Precip | 3 | 2 | 4 | 5 | 6 | 8 | 7 | 6 | 5 | 4 | 3 | 2 | 55 |
| Total Snowfall (") | 13.3 | 6.1 | 9.6 | 2.1 | 0.0 | 0.0 | 0.0 | 0.0 | 0.0 | 0.5 | 6.9 | 9.1 | 47.6 |
| Days ≥ 1" Snow Depth | na | na | na | na | 0 | 0 | 0 | 0 | 0 | 0 | na | na | na |

## BUFFALO *Wright County* ELEVATION 981 ft LAT/LONG 45° 10 ' N / 93° 53 ' W

| | JAN | FEB | MAR | APR | MAY | JUN | JUL | AUG | SEP | OCT | NOV | DEC | YEAR |
|---|---|---|---|---|---|---|---|---|---|---|---|---|---|
| Maximum Temp °F | 20.7 | 27.1 | 39.4 | 57.0 | 71.0 | 79.4 | 83.9 | 81.1 | 71.4 | 58.9 | 39.4 | 25.3 | 54.5 |
| Minimum Temp °F | 1.0 | 7.4 | 20.0 | 34.3 | 46.3 | 55.8 | 60.9 | 58.4 | 49.0 | 37.5 | 23.5 | 9.0 | 33.6 |
| Mean Temp °F | 10.9 | 17.3 | 29.7 | 45.7 | 58.7 | 67.6 | 72.4 | 69.8 | 60.2 | 48.3 | 31.5 | 17.1 | 44.1 |
| Days Max Temp ≥ 90 °F | 0 | 0 | 0 | 0 | 1 | 3 | 6 | 4 | 1 | 0 | 0 | 0 | 15 |
| Days Max Temp ≤ 32 °F | 24 | 17 | 8 | 0 | 0 | 0 | 0 | 0 | 0 | 0 | 8 | 22 | 79 |
| Days Min Temp ≤ 32 °F | 31 | 28 | 26 | 13 | 2 | 0 | 0 | 0 | 1 | 9 | 25 | 31 | 166 |
| Days Min Temp ≤ 0 °F | 15 | 10 | 2 | 0 | 0 | 0 | 0 | 0 | 0 | 0 | 1 | 9 | 37 |
| Heating Degree Days | 1675 | 1342 | 1086 | 577 | 230 | 46 | 9 | 26 | 185 | 515 | 997 | 1478 | 8166 |
| Cooling Degree Days | 0 | 0 | 0 | 6 | 42 | 132 | 243 | 176 | 46 | 2 | 0 | 0 | 647 |
| Total Precipitation (") | 0.87 | 0.78 | 1.58 | 2.72 | 3.07 | 4.46 | 3.56 | 3.99 | 3.16 | 2.44 | 1.70 | 0.98 | 29.31 |
| Days ≥ 0.1" Precip | 3 | 2 | 4 | 6 | 7 | 7 | 6 | 6 | 6 | 4 | 4 | 3 | 58 |
| Total Snowfall (") | 9.8 | 7.4 | 9.7 | 2.4 | 0.1 | 0.0 | 0.0 | 0.0 | 0.0 | 0.4 | 7.6 | 9.3 | 46.7 |
| Days ≥ 1" Snow Depth | 29 | 26 | 19 | 3 | 0 | 0 | 0 | 0 | 0 | 0 | 9 | 24 | 110 |

## CALEDONIA *Houston County* ELEVATION 1175 ft LAT/LONG 43° 38 ' N / 91° 30 ' W

| | JAN | FEB | MAR | APR | MAY | JUN | JUL | AUG | SEP | OCT | NOV | DEC | YEAR |
|---|---|---|---|---|---|---|---|---|---|---|---|---|---|
| Maximum Temp °F | 22.7 | 28.5 | 40.9 | 56.6 | 68.7 | 77.9 | 81.9 | 79.6 | 70.6 | 58.8 | 41.3 | 27.8 | 54.6 |
| Minimum Temp °F | 3.3 | 8.7 | 21.5 | 34.2 | 45.8 | 55.0 | 59.6 | 57.2 | 48.0 | 37.0 | 24.1 | 10.9 | 33.8 |
| Mean Temp °F | 13.0 | 18.6 | 31.3 | 45.4 | 57.3 | 66.5 | 70.8 | 68.4 | 59.3 | 48.0 | 32.7 | 19.4 | 44.2 |
| Days Max Temp ≥ 90 °F | 0 | 0 | 0 | 0 | 0 | 2 | 4 | 2 | 0 | 0 | 0 | 0 | 8 |
| Days Max Temp ≤ 32 °F | 23 | 17 | 7 | 0 | 0 | 0 | 0 | 0 | 0 | 0 | 6 | 20 | 73 |
| Days Min Temp ≤ 32 °F | 31 | 28 | 26 | 13 | 2 | 0 | 0 | 0 | 1 | 10 | 24 | 31 | 166 |
| Days Min Temp ≤ 0 °F | 13 | 9 | 2 | 0 | 0 | 0 | 0 | 0 | 0 | 0 | 1 | 7 | 32 |
| Heating Degree Days | 1607 | 1304 | 1037 | 585 | 262 | 61 | 15 | 34 | 203 | 525 | 964 | 1409 | 8006 |
| Cooling Degree Days | 0 | 0 | 0 | 7 | 30 | 120 | 202 | 151 | 42 | 1 | 0 | 0 | 553 |
| Total Precipitation (") | 1.12 | 0.89 | 2.06 | 3.85 | 3.81 | 4.56 | 4.37 | 4.30 | 4.19 | 2.48 | 2.27 | 1.52 | 35.42 |
| Days ≥ 0.1" Precip | 3 | 2 | 5 | 7 | 8 | 7 | 7 | 7 | 7 | 5 | 5 | 4 | 67 |
| Total Snowfall (") | 10.5 | 8.2 | 7.4 | 3.0 | 0.0 | 0.0 | 0.0 | 0.0 | 0.0 | 0.2 | 5.2 | 9.7 | 44.2 |
| Days ≥ 1" Snow Depth | ***23*** | ***20*** | na | na | 0 | 0 | 0 | 0 | 0 | 0 | 4 | ***21*** | na |

## CAMBRIDGE STATE HOSP *Isanti County* ELEVATION 960 ft LAT/LONG 45° 34 ' N / 93° 14 ' W

| | JAN | FEB | MAR | APR | MAY | JUN | JUL | AUG | SEP | OCT | NOV | DEC | YEAR |
|---|---|---|---|---|---|---|---|---|---|---|---|---|---|
| Maximum Temp °F | 18.4 | 25.4 | 37.5 | 54.6 | 68.4 | 76.7 | 81.5 | 78.3 | 68.2 | 56.2 | 38.1 | 24.7 | 52.3 |
| Minimum Temp °F | -1.7 | 5.2 | 18.5 | 33.0 | 44.8 | 54.0 | 59.1 | 56.3 | 46.7 | 35.4 | 21.3 | 7.2 | 31.6 |
| Mean Temp °F | 8.4 | 15.3 | 28.0 | 43.9 | 56.7 | 65.4 | 70.2 | 67.3 | 57.5 | 45.8 | 29.7 | 16.0 | 42.0 |
| Days Max Temp ≥ 90 °F | 0 | 0 | 0 | 0 | 1 | 1 | 4 | 2 | 0 | 0 | 0 | 0 | 8 |
| Days Max Temp ≤ 32 °F | 26 | 18 | 9 | 1 | 0 | 0 | 0 | 0 | 0 | 0 | 9 | 23 | 86 |
| Days Min Temp ≤ 32 °F | 31 | 28 | 28 | 15 | 3 | 0 | 0 | 0 | 2 | 12 | 26 | 31 | 176 |
| Days Min Temp ≤ 0 °F | 17 | 11 | 3 | 0 | 0 | 0 | 0 | 0 | 0 | 0 | 1 | 10 | 42 |
| Heating Degree Days | 1753 | 1398 | 1140 | 632 | 281 | 73 | 20 | 51 | 248 | 590 | 1052 | 1515 | 8753 |
| Cooling Degree Days | ***0*** | 0 | ***0*** | 5 | ***38*** | ***89*** | 188 | 133 | 29 | 1 | 0 | 0 | 483 |
| Total Precipitation (") | 0.92 | 0.65 | 1.28 | 2.30 | 3.12 | 4.59 | 3.93 | 4.05 | 3.30 | 2.58 | 1.69 | 0.96 | 29.37 |
| Days ≥ 0.1" Precip | 3 | 2 | 4 | 5 | 7 | 8 | 6 | 6 | 6 | 5 | 4 | 3 | 59 |
| Total Snowfall (") | 11.4 | 6.9 | 8.3 | 2.0 | 0.0 | 0.0 | 0.0 | 0.0 | 0.0 | 0.6 | 6.9 | 9.5 | 45.6 |
| Days ≥ 1" Snow Depth | 29 | 27 | 20 | 3 | 0 | 0 | 0 | 0 | 0 | 0 | 8 | 24 | 111 |

## CAMPBELL 1 SSW *Wilkin County* ELEVATION 981 ft LAT/LONG 46° 6' N / 96° 24' W

| | JAN | FEB | MAR | APR | MAY | JUN | JUL | AUG | SEP | OCT | NOV | DEC | YEAR |
|---|---|---|---|---|---|---|---|---|---|---|---|---|---|
| Maximum Temp °F | 16.8 | 22.8 | 36.3 | 54.9 | 69.8 | 77.5 | 83.0 | 81.5 | 70.6 | 57.7 | 37.6 | 23.7 | 52.7 |
| Minimum Temp °F | -4.1 | 2.0 | 17.3 | 32.1 | 44.3 | 53.6 | 57.7 | 55.3 | 45.0 | 33.1 | 19.1 | 4.7 | 30.0 |
| Mean Temp °F | 6.3 | 12.4 | 26.8 | 43.6 | 57.1 | 65.6 | 70.4 | 68.4 | 57.8 | 45.4 | 28.4 | 14.2 | 41.4 |
| Days Max Temp ≥ 90 °F | 0 | 0 | 0 | 0 | 1 | 2 | 6 | 5 | 1 | 0 | 0 | 0 | 15 |
| Days Max Temp ≤ 32 °F | 26 | 20 | 11 | 1 | 0 | 0 | 0 | 0 | 0 | 0 | 10 | 22 | 90 |
| Days Min Temp ≤ 32 °F | 31 | 28 | 28 | 16 | 3 | 0 | 0 | 0 | 2 | 15 | 27 | 31 | 181 |
| Days Min Temp ≤ 0 °F | 19 | 13 | 4 | 0 | 0 | 0 | 0 | 0 | 0 | 0 | 2 | 12 | 50 |
| Heating Degree Days | 1816 | 1480 | 1176 | 638 | 275 | 73 | 20 | 42 | 243 | 601 | 1092 | 1571 | 9027 |
| Cooling Degree Days | 0 | 0 | 0 | 5 | 45 | 103 | 198 | 157 | 35 | 2 | 0 | 0 | 545 |
| Total Precipitation (") | 0.60 | 0.45 | 1.11 | 2.17 | 2.39 | 3.85 | 3.31 | 2.57 | 2.31 | 1.77 | 0.86 | 0.53 | 21.92 |
| Days ≥ 0.1" Precip | *2* | *2* | 3 | 5 | 6 | 7 | 6 | 5 | 4 | 4 | 2 | *2* | 48 |
| Total Snowfall (") | *8.0* | *5.0* | *5.1* | *1.9* | 0.0 | 0.0 | 0.0 | 0.0 | 0.0 | 0.2 | *5.1* | *4.9* | 30.2 |
| Days ≥ 1" Snow Depth | na | na | na | *2* | 0 | 0 | 0 | 0 | 0 | 0 | *7* | na | na |

## CANBY *Yellow Medicine County* ELEVATION 1243 ft LAT/LONG 44° 43' N / 96° 17' W

| | JAN | FEB | MAR | APR | MAY | JUN | JUL | AUG | SEP | OCT | NOV | DEC | YEAR |
|---|---|---|---|---|---|---|---|---|---|---|---|---|---|
| Maximum Temp °F | 22.8 | 28.1 | 40.7 | 58.0 | 71.8 | 80.9 | 86.0 | 83.3 | 73.1 | 60.4 | 40.8 | 27.5 | 56.1 |
| Minimum Temp °F | 2.9 | 8.6 | 21.6 | 34.2 | 46.3 | 56.2 | 61.3 | 58.9 | 48.8 | 36.8 | 22.7 | 9.4 | 34.0 |
| Mean Temp °F | 12.9 | 18.4 | 31.2 | 46.1 | 59.1 | 68.6 | 73.7 | 71.1 | 60.9 | 48.7 | 31.8 | 18.5 | 45.1 |
| Days Max Temp ≥ 90 °F | 0 | 0 | 0 | 0 | 1 | 5 | 10 | 7 | 2 | 0 | 0 | 0 | 25 |
| Days Max Temp ≤ 32 °F | 22 | 17 | 8 | 0 | 0 | 0 | 0 | 0 | 0 | 0 | 8 | 19 | 74 |
| Days Min Temp ≤ 32 °F | 31 | 28 | 27 | 13 | 2 | 0 | 0 | 0 | 1 | 11 | 25 | 31 | 169 |
| Days Min Temp ≤ 0 °F | 14 | 9 | 2 | 0 | 0 | 0 | 0 | 0 | 0 | 0 | 1 | 8 | 34 |
| Heating Degree Days | 1613 | 1311 | 1043 | 565 | 222 | 43 | 8 | 23 | 175 | 505 | 990 | 1437 | 7935 |
| Cooling Degree Days | 0 | 0 | 0 | 9 | 52 | 152 | 264 | 203 | 56 | 3 | 0 | 0 | 739 |
| Total Precipitation (") | 0.76 | 0.83 | 1.74 | 2.37 | 2.83 | 4.34 | 3.46 | 2.86 | 2.61 | 1.95 | 1.52 | 0.81 | 26.08 |
| Days ≥ 0.1" Precip | 2 | 2 | 4 | 5 | 6 | 7 | 6 | 5 | 5 | 4 | 3 | 2 | 51 |
| Total Snowfall (") | 7.4 | 7.4 | 9.6 | 2.2 | 0.0 | 0.0 | 0.0 | 0.0 | 0.0 | 0.7 | 7.0 | 7.2 | 41.5 |
| Days ≥ 1" Snow Depth | 27 | 24 | 17 | 3 | 0 | 0 | 0 | 0 | 0 | 1 | 8 | 21 | 101 |

## CASS LAKE *Cass County* ELEVATION 1322 ft LAT/LONG 47° 23' N / 94° 38' W

| | JAN | FEB | MAR | APR | MAY | JUN | JUL | AUG | SEP | OCT | NOV | DEC | YEAR |
|---|---|---|---|---|---|---|---|---|---|---|---|---|---|
| Maximum Temp °F | 15.6 | 22.8 | 35.8 | 51.5 | 66.1 | 74.6 | 79.6 | 77.3 | 66.2 | 53.4 | 35.1 | 21.3 | 49.9 |
| Minimum Temp °F | -11.0 | -5.5 | 11.0 | 27.1 | 39.8 | 49.7 | 54.8 | 52.4 | 43.1 | 31.8 | 16.2 | -1.4 | 25.7 |
| Mean Temp °F | 2.3 | 8.7 | 23.4 | 39.3 | 53.0 | 62.1 | 67.2 | 64.9 | 54.7 | 42.6 | 25.7 | 10.0 | 37.8 |
| Days Max Temp ≥ 90 °F | 0 | 0 | 0 | 0 | 0 | 1 | 3 | 2 | 0 | 0 | 0 | 0 | 6 |
| Days Max Temp ≤ 32 °F | 28 | 21 | 12 | 1 | 0 | 0 | 0 | 0 | 0 | 1 | 13 | 25 | 101 |
| Days Min Temp ≤ 32 °F | 31 | 28 | 29 | 21 | 8 | 1 | 0 | 0 | 4 | 17 | 28 | 31 | 198 |
| Days Min Temp ≤ 0 °F | 22 | 17 | 8 | 0 | 0 | 0 | 0 | 0 | 0 | 0 | 3 | 17 | 67 |
| Heating Degree Days | 1942 | 1587 | 1284 | 765 | 385 | 136 | 49 | 90 | 321 | 687 | 1174 | 1703 | 10123 |
| Cooling Degree Days | 0 | 0 | 0 | 2 | 22 | 61 | 124 | 101 | 13 | 0 | 0 | 0 | 323 |
| Total Precipitation (") | 0.88 | 0.54 | 1.22 | 2.08 | 2.66 | 4.06 | 3.96 | 3.39 | 2.81 | 2.50 | 1.20 | 0.88 | 26.18 |
| Days ≥ 0.1" Precip | 3 | 2 | 3 | 5 | 6 | 7 | 7 | 6 | 6 | 5 | 3 | 2 | 55 |
| Total Snowfall (") | 11.9 | 6.0 | 8.5 | 3.6 | 0.1 | 0.0 | 0.0 | 0.0 | 0.0 | 0.9 | 7.3 | 10.1 | 48.4 |
| Days ≥ 1" Snow Depth | na | na | na | *5* | 0 | 0 | 0 | 0 | 0 | 1 | na | na | na |

## CEDAR *Anoka County* ELEVATION 625 ft LAT/LONG 45° 18' N / 93° 20' W

| | JAN | FEB | MAR | APR | MAY | JUN | JUL | AUG | SEP | OCT | NOV | DEC | YEAR |
|---|---|---|---|---|---|---|---|---|---|---|---|---|---|
| Maximum Temp °F | 21.3 | 27.6 | 40.7 | 58.2 | 71.2 | 79.0 | 83.4 | 80.3 | 71.2 | 59.1 | 40.0 | 26.4 | 54.9 |
| Minimum Temp °F | 0.2 | 5.8 | 19.8 | 33.9 | 45.4 | 54.4 | 59.6 | 57.1 | 48.1 | 36.8 | 22.8 | 9.1 | 32.8 |
| Mean Temp °F | 10.8 | 16.7 | 30.3 | 46.1 | 58.3 | 66.8 | 71.5 | 68.7 | 59.7 | 48.0 | 31.4 | 17.8 | 43.8 |
| Days Max Temp ≥ 90 °F | 0 | 0 | 0 | 0 | 1 | 2 | 6 | 3 | 1 | 0 | 0 | 0 | 13 |
| Days Max Temp ≤ 32 °F | 24 | 17 | 7 | 0 | 0 | 0 | 0 | 0 | 0 | 0 | 7 | 21 | 76 |
| Days Min Temp ≤ 32 °F | 31 | 28 | 26 | 14 | 3 | 0 | 0 | 0 | 2 | 11 | 25 | 31 | 171 |
| Days Min Temp ≤ 0 °F | 15 | 11 | 3 | 0 | 0 | 0 | 0 | 0 | 0 | 0 | 1 | 9 | 39 |
| Heating Degree Days | 1678 | 1358 | 1071 | 566 | 241 | 56 | 12 | 35 | 197 | 524 | 1001 | 1459 | 8198 |
| Cooling Degree Days | 0 | 0 | 0 | 8 | 46 | 117 | 218 | 149 | 41 | 2 | 0 | 0 | 581 |
| Total Precipitation (") | 1.02 | 0.80 | 1.73 | 2.57 | 3.47 | 4.24 | 3.96 | 4.49 | 3.50 | 2.62 | 1.87 | 1.01 | 31.28 |
| Days ≥ 0.1" Precip | 3 | 3 | 4 | 6 | 7 | 8 | 7 | 7 | 6 | 5 | 4 | 3 | 63 |
| Total Snowfall (") | 11.9 | 8.2 | 9.3 | 2.6 | 0.0 | 0.0 | 0.0 | 0.0 | 0.0 | 0.6 | 8.8 | 9.1 | 50.5 |
| Days ≥ 1" Snow Depth | 29 | 25 | 20 | 3 | 0 | 0 | 0 | 0 | 0 | 0 | 8 | 24 | 109 |

## CHASKA *Carver County* ELEVATION 732 ft LAT/LONG 44° 48 ' N / 93° 35 ' W

| | JAN | FEB | MAR | APR | MAY | JUN | JUL | AUG | SEP | OCT | NOV | DEC | YEAR |
|---|---|---|---|---|---|---|---|---|---|---|---|---|---|
| Maximum Temp °F | 22.9 | 28.6 | 42.0 | 59.0 | 72.0 | 81.2 | 85.2 | 82.1 | 72.5 | 60.2 | 41.3 | 27.4 | 56.2 |
| Minimum Temp °F | 2.0 | 8.0 | 22.7 | 35.9 | 47.3 | 57.0 | 61.8 | 59.2 | 49.7 | 38.0 | 24.6 | 10.4 | 34.7 |
| Mean Temp °F | 12.5 | 18.3 | 32.4 | 47.5 | 59.7 | 69.1 | 73.5 | 70.7 | 61.1 | 49.2 | 33.0 | 18.9 | 45.5 |
| Days Max Temp ≥ 90 °F | 0 | 0 | 0 | 0 | 1 | 4 | 8 | 5 | 1 | 0 | 0 | 0 | 19 |
| Days Max Temp ≤ 32 °F | 22 | 16 | 6 | 0 | 0 | 0 | 0 | 0 | 0 | 0 | 6 | 19 | 69 |
| Days Min Temp ≤ 32 °F | 31 | 27 | *25* | 11 | 1 | 0 | 0 | 0 | 1 | 9 | 24 | 31 | 160 |
| Days Min Temp ≤ 0 °F | 14 | 9 | *2* | 0 | 0 | 0 | 0 | 0 | 0 | 0 | 1 | 8 | 34 |
| Heating Degree Days | 1625 | 1314 | 1004 | 526 | 208 | 32 | 5 | 21 | 166 | 489 | 953 | 1421 | 7764 |
| Cooling Degree Days | 0 | 0 | 0 | 8 | 59 | 173 | 281 | 211 | 64 | 3 | 0 | 0 | 799 |
| Total Precipitation (") | 0.78 | 0.69 | 1.74 | 2.53 | 3.55 | 4.42 | 4.32 | 4.06 | 3.24 | 2.33 | 1.83 | 0.92 | 30.41 |
| Days ≥ 0.1" Precip | 3 | 2 | 4 | 6 | 7 | 7 | 6 | 6 | 6 | 5 | 4 | 3 | 59 |
| Total Snowfall (") | 11.6 | 7.8 | 8.8 | 2.1 | 0.0 | 0.0 | 0.0 | 0.0 | 0.0 | 0.2 | 6.8 | 9.6 | 46.9 |
| Days ≥ 1" Snow Depth | 28 | 25 | 15 | 1 | 0 | 0 | 0 | 0 | 0 | 0 | 6 | 19 | 94 |

## CLOQUET *Carlton County* ELEVATION 1270 ft LAT/LONG 46° 41 ' N / 92° 30 ' W

| | JAN | FEB | MAR | APR | MAY | JUN | JUL | AUG | SEP | OCT | NOV | DEC | YEAR |
|---|---|---|---|---|---|---|---|---|---|---|---|---|---|
| Maximum Temp °F | 18.0 | 24.8 | 36.4 | 52.3 | 67.1 | 75.1 | 80.2 | 77.1 | 66.6 | 53.9 | 35.6 | 22.7 | 50.8 |
| Minimum Temp °F | -2.2 | 2.7 | 15.4 | 28.4 | 38.6 | 47.4 | 54.1 | 52.6 | 44.3 | 34.0 | 20.6 | 5.5 | 28.5 |
| Mean Temp °F | 7.9 | 13.8 | 25.9 | 40.4 | 52.9 | 61.3 | 67.1 | 64.9 | 55.5 | 44.0 | 28.1 | 14.1 | 39.7 |
| Days Max Temp ≥ 90 °F | 0 | 0 | 0 | 0 | 0 | 1 | 3 | 1 | 0 | 0 | 0 | 0 | 5 |
| Days Max Temp ≤ 32 °F | 28 | 20 | 10 | 1 | 0 | 0 | 0 | 0 | 0 | 0 | 11 | 25 | 95 |
| Days Min Temp ≤ 32 °F | 31 | 28 | 30 | 21 | 8 | 1 | 0 | 0 | 3 | 14 | 27 | 31 | 194 |
| Days Min Temp ≤ 0 °F | 17 | 13 | 5 | 0 | 0 | 0 | 0 | 0 | 0 | 0 | 2 | 12 | 49 |
| Heating Degree Days | 1767 | 1441 | 1205 | 732 | 378 | 143 | 43 | 78 | 293 | 645 | 1099 | 1573 | 9397 |
| Cooling Degree Days | 0 | 0 | 0 | 0 | 10 | 40 | 118 | 91 | 13 | 0 | 0 | 0 | 272 |
| Total Precipitation (") | 1.24 | 0.72 | 1.78 | 2.18 | 3.26 | 4.29 | 4.02 | 4.02 | 4.05 | 2.72 | 1.95 | 1.22 | 31.45 |
| Days ≥ 0.1" Precip | 4 | 2 | 5 | 5 | 7 | 8 | 7 | 7 | 7 | 5 | 4 | 4 | 65 |
| Total Snowfall (") | 16.8 | 9.1 | 11.4 | 4.1 | 0.2 | 0.0 | 0.0 | 0.0 | 0.0 | 1.1 | 11.3 | 14.7 | 68.7 |
| Days ≥ 1" Snow Depth | 31 | 28 | 28 | 10 | 0 | 0 | 0 | 0 | 0 | 0 | 16 | 30 | 143 |

## COLLEGEVILLE ST JOHN *Stearns County* ELEVATION 1240 ft LAT/LONG 45° 35 ' N / 94° 24 ' W

| | JAN | FEB | MAR | APR | MAY | JUN | JUL | AUG | SEP | OCT | NOV | DEC | YEAR |
|---|---|---|---|---|---|---|---|---|---|---|---|---|---|
| Maximum Temp °F | 19.3 | 25.8 | 38.4 | 55.9 | 69.8 | 77.9 | 82.5 | 79.6 | 69.8 | 57.5 | 38.5 | 24.5 | 53.3 |
| Minimum Temp °F | 0.3 | 6.5 | 19.7 | 34.3 | 46.6 | 55.8 | 61.0 | 58.7 | 49.3 | 38.0 | 23.2 | 8.1 | 33.5 |
| Mean Temp °F | 9.8 | 16.2 | 29.1 | 45.1 | 58.2 | 66.9 | 71.8 | 69.2 | 59.6 | 47.8 | 30.9 | 16.3 | 43.4 |
| Days Max Temp ≥ 90 °F | 0 | 0 | 0 | 0 | 0 | 2 | 4 | 3 | 0 | 0 | 0 | 0 | 9 |
| Days Max Temp ≤ 32 °F | 25 | 18 | 9 | 0 | 0 | 0 | 0 | 0 | 0 | 0 | 9 | 23 | 84 |
| Days Min Temp ≤ 32 °F | 31 | 28 | 27 | 14 | 2 | 0 | 0 | 0 | 0 | 8 | 25 | 31 | 166 |
| Days Min Temp ≤ 0 °F | 16 | 11 | 3 | 0 | 0 | 0 | 0 | 0 | 0 | 0 | 1 | 10 | 41 |
| Heating Degree Days | 1708 | 1374 | 1105 | 593 | 240 | 51 | 9 | 28 | 198 | 530 | 1018 | 1504 | 8358 |
| Cooling Degree Days | 0 | 0 | 0 | 5 | 47 | 124 | 237 | 176 | 44 | 2 | 0 | 0 | 635 |
| Total Precipitation (") | 0.90 | 0.75 | 1.76 | 2.54 | 3.27 | 4.72 | 3.25 | 3.70 | 3.34 | 2.60 | 1.56 | 0.87 | 29.26 |
| Days ≥ 0.1" Precip | 3 | 2 | 4 | 5 | 7 | 8 | 6 | 6 | 6 | 5 | 3 | 3 | 58 |
| Total Snowfall (") | 11.6 | 8.1 | 10.3 | 3.0 | 0.1 | 0.0 | 0.0 | 0.0 | 0.0 | 0.5 | 8.6 | 9.4 | 51.6 |
| Days ≥ 1" Snow Depth | 29 | 26 | 20 | 3 | 0 | 0 | 0 | 0 | 0 | 0 | 9 | 24 | 111 |

## COOK 18 W *St. Louis County* ELEVATION 1322 ft LAT/LONG 47° 52 ' N / 93° 4 ' W

| | JAN | FEB | MAR | APR | MAY | JUN | JUL | AUG | SEP | OCT | NOV | DEC | YEAR |
|---|---|---|---|---|---|---|---|---|---|---|---|---|---|
| Maximum Temp °F | 16.6 | 24.3 | 37.0 | 53.4 | 67.1 | 73.9 | 78.5 | 76.4 | 65.7 | 53.3 | 34.2 | 21.1 | 50.1 |
| Minimum Temp °F | -7.1 | -0.7 | 12.2 | 27.2 | 38.3 | 47.4 | 53.4 | 51.1 | 42.5 | 32.0 | 17.5 | 1.7 | 26.3 |
| Mean Temp °F | 4.7 | 11.8 | 24.6 | 40.3 | 52.7 | 60.7 | 66.0 | 63.8 | 54.1 | 42.7 | 25.9 | 11.4 | 38.2 |
| Days Max Temp ≥ 90 °F | 0 | 0 | 0 | 0 | 0 | 0 | 1 | 1 | 0 | 0 | 0 | 0 | 2 |
| Days Max Temp ≤ 32 °F | 27 | 20 | *11* | 1 | 0 | 0 | 0 | 0 | 0 | 1 | 13 | 26 | 99 |
| Days Min Temp ≤ 32 °F | 31 | 28 | *29* | 23 | 10 | 1 | 0 | 1 | 5 | 17 | 28 | 31 | 204 |
| Days Min Temp ≤ 0 °F | 19 | 14 | 7 | 1 | 0 | 0 | 0 | 0 | 0 | 0 | 3 | 14 | 58 |
| Heating Degree Days | 1866 | 1500 | 1245 | 734 | 389 | 158 | 57 | 101 | 331 | 685 | 1166 | 1657 | 9889 |
| Cooling Degree Days | *0* | *0* | *0* | 1 | 18 | 34 | 101 | 78 | 9 | 0 | 0 | 0 | 241 |
| Total Precipitation (") | 1.11 | 0.58 | 1.07 | 1.58 | 2.35 | 4.13 | 3.78 | 3.92 | 3.16 | 2.36 | 1.36 | 0.96 | 26.36 |
| Days ≥ 0.1" Precip | 4 | *2* | *3* | 4 | 5 | 8 | 7 | 7 | 7 | 5 | 3 | 3 | 58 |
| Total Snowfall (") | na | *7.6* | na | na | 0.9 | 0.0 | 0.0 | 0.0 | 0.0 | 1.2 | *13.6* | *13.2* | na |
| Days ≥ 1" Snow Depth | na | na | *22* | *9* | 0 | 0 | 0 | 0 | 0 | 0 | *12* | na | na |

## COTTON 3 E *St. Louis County* ELEVATION 1371 ft LAT/LONG 47° 11' N / 92° 25' W

| | JAN | FEB | MAR | APR | MAY | JUN | JUL | AUG | SEP | OCT | NOV | DEC | YEAR |
|---|---|---|---|---|---|---|---|---|---|---|---|---|---|
| Maximum Temp °F | 17.9 | 25.0 | 37.1 | 52.2 | na | 73.6 | 78.6 | 75.7 | 65.7 | 53.4 | 35.5 | 22.1 | na |
| Minimum Temp °F | -7.9 | -2.1 | 11.8 | 26.0 | na | 46.0 | 51.7 | 49.4 | 41.4 | 31.7 | 17.3 | 1.1 | na |
| Mean Temp °F | 5.0 | 11.5 | 24.5 | 39.1 | na | 59.8 | 65.3 | 62.6 | 53.6 | 42.6 | 26.4 | 11.6 | na |
| Days Max Temp ≥ 90 °F | 0 | 0 | 0 | 0 | 0 | 0 | 1 | 1 | 0 | 0 | 0 | 0 | 2 |
| Days Max Temp ≤ 32 °F | 28 | 20 | 10 | 1 | 0 | 0 | 0 | 0 | 0 | 1 | 13 | 26 | 99 |
| Days Min Temp ≤ 32 °F | 31 | 28 | 29 | 23 | 9 | 2 | 0 | 1 | 7 | 17 | 28 | 31 | 206 |
| Days Min Temp ≤ 0 °F | 20 | 15 | 7 | 1 | 0 | 0 | 0 | 0 | 0 | 0 | 3 | 15 | 61 |
| Heating Degree Days | 1856 | 1504 | 1249 | 771 | na | 178 | 69 | 125 | 346 | 687 | 1151 | 1651 | na |
| Cooling Degree Days | 0 | 0 | 0 | 0 | na | 30 | 78 | 65 | 9 | 0 | 0 | 0 | na |
| Total Precipitation (") | 0.90 | 0.47 | 1.15 | 2.23 | *2.53* | 4.21 | 4.21 | 3.69 | 3.07 | 2.47 | 1.63 | 0.94 | 27.50 |
| Days ≥ 0.1" Precip | 3 | *2* | 3 | 6 | 6 | *8* | 8 | 7 | 6 | 5 | 5 | 4 | 63 |
| Total Snowfall (") | *11.9* | 6.1 | *8.2* | 3.0 | 0.1 | 0.0 | 0.0 | 0.0 | 0.0 | 0.8 | 11.3 | 11.4 | 52.8 |
| Days ≥ 1" Snow Depth | 31 | 28 | 25 | 7 | 0 | 0 | 0 | 0 | 0 | 0 | *14* | 30 | 135 |

## CROOKSTON NW EXP STN *Polk County* ELEVATION 889 ft LAT/LONG 47° 48' N / 96° 36' W

| | JAN | FEB | MAR | APR | MAY | JUN | JUL | AUG | SEP | OCT | NOV | DEC | YEAR |
|---|---|---|---|---|---|---|---|---|---|---|---|---|---|
| Maximum Temp °F | 13.6 | 20.2 | 33.6 | 53.3 | 69.0 | 76.7 | 81.8 | 80.6 | 69.3 | 55.6 | 34.6 | 20.1 | 50.7 |
| Minimum Temp °F | -6.2 | -0.1 | 15.2 | 31.3 | 43.4 | 52.8 | 57.3 | 55.1 | 45.2 | 33.6 | 18.1 | 2.2 | 29.0 |
| Mean Temp °F | 3.7 | 10.1 | 24.4 | 42.3 | 56.3 | 64.8 | 69.6 | 67.9 | 57.3 | 44.6 | 26.4 | 11.2 | 39.9 |
| Days Max Temp ≥ 90 °F | 0 | 0 | 0 | 0 | 1 | 2 | 4 | 4 | 1 | 0 | 0 | 0 | 12 |
| Days Max Temp ≤ 32 °F | 28 | 22 | 13 | 1 | 0 | 0 | 0 | 0 | 0 | 1 | 13 | 26 | 104 |
| Days Min Temp ≤ 32 °F | 31 | 28 | 28 | 17 | 5 | 0 | 0 | 0 | 3 | 14 | 27 | 31 | 184 |
| Days Min Temp ≤ 0 °F | 21 | 15 | 6 | 0 | 0 | 0 | 0 | 0 | 0 | 0 | 3 | 14 | 59 |
| Heating Degree Days | 1906 | 1549 | 1251 | 676 | 297 | 85 | 23 | 50 | 259 | 626 | 1153 | 1665 | 9540 |
| Cooling Degree Days | 0 | 0 | 0 | 4 | 40 | 81 | 163 | 136 | 23 | 1 | 0 | 0 | 448 |
| Total Precipitation (") | 0.51 | 0.45 | 0.77 | 1.45 | 2.44 | 3.71 | 2.78 | 2.93 | 2.27 | 1.58 | 0.74 | 0.49 | 20.12 |
| Days ≥ 0.1" Precip | 2 | 1 | 2 | 4 | 5 | 7 | 5 | 6 | 5 | 4 | 2 | 2 | 45 |
| Total Snowfall (") | 9.3 | 5.8 | 5.9 | 1.7 | 0.2 | 0.0 | 0.0 | 0.0 | 0.0 | 0.7 | 5.1 | 7.0 | 35.7 |
| Days ≥ 1" Snow Depth | 29 | 24 | 18 | 2 | 0 | 0 | 0 | 0 | 0 | 0 | 9 | 24 | 106 |

## DETROIT LAKES 1 NNE *Becker County* ELEVATION 1362 ft LAT/LONG 46° 49' N / 95° 51' W

| | JAN | FEB | MAR | APR | MAY | JUN | JUL | AUG | SEP | OCT | NOV | DEC | YEAR |
|---|---|---|---|---|---|---|---|---|---|---|---|---|---|
| Maximum Temp °F | 15.9 | 22.7 | 35.8 | 54.4 | 68.6 | 75.8 | 80.9 | 78.9 | 68.6 | 55.5 | *35.7* | 20.8 | 51.1 |
| Minimum Temp °F | -6.3 | -0.7 | 14.0 | 29.5 | 41.6 | 50.7 | 55.5 | 53.1 | 43.7 | 33.1 | *18.5* | 2.0 | 27.9 |
| Mean Temp °F | 4.8 | 11.1 | 24.9 | 41.9 | 55.1 | 63.3 | 68.2 | 66.0 | 56.2 | 44.3 | *27.1* | 11.4 | 39.5 |
| Days Max Temp ≥ 90 °F | 0 | 0 | 0 | 0 | 0 | 1 | 3 | 3 | 0 | 0 | 0 | 0 | 7 |
| Days Max Temp ≤ 32 °F | 27 | 21 | 11 | 1 | 0 | 0 | 0 | 0 | 0 | 0 | 12 | 26 | 98 |
| Days Min Temp ≤ 32 °F | 31 | 28 | 28 | 19 | 7 | 0 | 0 | 0 | 4 | 15 | 27 | 31 | 190 |
| Days Min Temp ≤ 0 °F | 20 | 14 | 6 | 0 | 0 | 0 | 0 | 0 | 0 | 0 | 3 | 14 | 57 |
| Heating Degree Days | 1863 | 1518 | 1236 | 689 | 321 | 109 | 34 | 76 | 284 | 635 | *1129* | 1656 | 9550 |
| Cooling Degree Days | 0 | *0* | *0* | 3 | 28 | 72 | 149 | *124* | 25 | 0 | 0 | 0 | 401 |
| Total Precipitation (") | 0.70 | 0.55 | 1.07 | 1.77 | 2.54 | 4.25 | 3.71 | 3.30 | 2.66 | 2.29 | 0.88 | 0.74 | 24.46 |
| Days ≥ 0.1" Precip | 2 | 2 | *3* | 5 | 6 | 8 | 6 | 6 | 5 | 5 | 3 | 3 | 54 |
| Total Snowfall (") | 11.0 | *6.2* | 7.6 | 2.7 | 0.0 | 0.0 | 0.0 | 0.0 | 0.0 | 1.3 | 6.9 | 8.9 | 44.6 |
| Days ≥ 1" Snow Depth | *31* | na | na | *3* | 0 | 0 | 0 | 0 | 0 | 1 | *9* | na | na |

## DULUTH HARBOR STA *St. Louis County* ELEVATION 610 ft LAT/LONG 46° 46' N / 92° 5' W

| | JAN | FEB | MAR | APR | MAY | JUN | JUL | AUG | SEP | OCT | NOV | DEC | YEAR |
|---|---|---|---|---|---|---|---|---|---|---|---|---|---|
| Maximum Temp °F | 19.2 | 24.4 | 33.7 | 45.3 | 55.5 | 65.3 | 73.8 | 72.4 | 63.8 | 52.4 | 37.2 | 24.6 | 47.3 |
| Minimum Temp °F | 2.6 | 8.1 | 20.1 | 31.9 | 40.6 | 49.0 | 57.9 | 58.4 | 50.3 | 40.4 | 26.5 | 11.6 | 33.1 |
| Mean Temp °F | 10.9 | 16.3 | 26.9 | 38.6 | 48.1 | 57.1 | 65.9 | 65.4 | 57.0 | 46.4 | 31.9 | 18.1 | 40.2 |
| Days Max Temp ≥ 90 °F | 0 | 0 | 0 | 0 | 0 | 0 | 0 | 0 | 0 | 0 | 0 | 0 | 0 |
| Days Max Temp ≤ 32 °F | 27 | 21 | 13 | 1 | 0 | 0 | 0 | 0 | 0 | 0 | 8 | 23 | 93 |
| Days Min Temp ≤ 32 °F | 31 | 28 | 29 | 15 | 1 | 0 | 0 | 0 | 0 | 3 | 21 | 30 | 158 |
| Days Min Temp ≤ 0 °F | 14 | 9 | 1 | 0 | 0 | 0 | 0 | 0 | 0 | 0 | 1 | 7 | 32 |
| Heating Degree Days | 1672 | 1372 | 1173 | 784 | 521 | 243 | 60 | 65 | 244 | 569 | 985 | 1448 | 9136 |
| Cooling Degree Days | 0 | 0 | 0 | 0 | 5 | *16* | *111* | 106 | 13 | 0 | 0 | 0 | 251 |
| Total Precipitation (") | 1.00 | 0.56 | 1.44 | 1.66 | 2.50 | 3.72 | 3.72 | 3.57 | 3.60 | 2.37 | 1.38 | 0.96 | 26.48 |
| Days ≥ 0.1" Precip | 3 | 2 | 4 | 4 | 6 | 6 | 6 | 6 | 6 | 4 | 3 | 3 | 53 |
| Total Snowfall (") | 11.4 | *5.5* | 7.8 | 1.9 | 0.1 | 0.0 | 0.0 | 0.0 | 0.0 | 0.2 | 4.1 | *10.2* | 41.2 |
| Days ≥ 1" Snow Depth | *29* | *24* | 21 | 4 | 0 | 0 | 0 | 0 | 0 | 0 | 4 | *21* | 103 |

## DULUTH INTL AP *St. Louis County* ELEVATION 1424 ft LAT/LONG 46° 50 ' N / 92° 11 ' W

| | JAN | FEB | MAR | APR | MAY | JUN | JUL | AUG | SEP | OCT | NOV | DEC | YEAR |
|---|---|---|---|---|---|---|---|---|---|---|---|---|---|
| Maximum Temp °F | 16.2 | 22.0 | 33.1 | 48.3 | 62.1 | 70.5 | 76.3 | 73.6 | 63.6 | 51.4 | 34.4 | 21.4 | 47.7 |
| Minimum Temp °F | -2.0 | 3.2 | 16.0 | 29.3 | 40.0 | 48.2 | 54.5 | 52.9 | 44.5 | 34.5 | 20.9 | 5.9 | 29.0 |
| Mean Temp °F | 7.1 | 12.6 | 24.5 | 38.8 | 51.1 | 59.4 | 65.4 | 63.3 | 54.1 | 43.0 | 27.6 | 13.7 | 38.4 |
| Days Max Temp ≥ 90 °F | 0 | 0 | 0 | 0 | 0 | 0 | 1 | 1 | 0 | 0 | 0 | 0 | 2 |
| Days Max Temp ≤ 32 °F | 28 | 22 | 15 | 2 | 0 | 0 | 0 | 0 | 0 | 1 | 13 | 26 | 107 |
| Days Min Temp ≤ 32 °F | 31 | 28 | 29 | 20 | 5 | 0 | 0 | 0 | 2 | 13 | 26 | 31 | 185 |
| Days Min Temp ≤ 0 °F | 17 | 13 | 4 | 0 | 0 | 0 | 0 | 0 | 0 | 0 | 2 | 12 | 48 |
| Heating Degree Days | 1792 | 1475 | 1248 | 779 | 430 | 188 | 68 | 110 | 332 | 674 | 1114 | 1587 | 9797 |
| Cooling Degree Days | 0 | 0 | 0 | 0 | 9 | 28 | 91 | 70 | 9 | 0 | 0 | 0 | 207 |
| Total Precipitation (") | 1.30 | 0.74 | 1.82 | 2.24 | 2.92 | 4.16 | 3.89 | 3.96 | 4.02 | 2.58 | 2.03 | 1.18 | 30.84 |
| Days ≥ 0.1" Precip | 3 | 2 | 5 | 6 | 7 | 8 | 7 | 7 | 7 | 5 | 5 | 4 | 66 |
| Total Snowfall (") | 18.9 | 10.0 | 13.4 | 6.2 | 0.6 | 0.0 | 0.0 | 0.0 | 0.1 | 1.6 | 15.0 | 15.6 | 81.4 |
| Days ≥ 1" Snow Depth | 31 | 28 | 29 | 11 | 0 | 0 | 0 | 0 | 0 | 1 | 14 | 29 | 143 |

## FAIRMONT *Martin County* ELEVATION 1191 ft LAT/LONG 43° 38 ' N / 94° 28 ' W

| | JAN | FEB | MAR | APR | MAY | JUN | JUL | AUG | SEP | OCT | NOV | DEC | YEAR |
|---|---|---|---|---|---|---|---|---|---|---|---|---|---|
| Maximum Temp °F | 22.0 | 27.7 | 40.4 | 56.7 | 70.7 | 79.8 | 83.2 | 80.4 | 71.9 | 59.2 | 40.8 | 26.8 | 55.0 |
| Minimum Temp °F | 3.7 | 9.6 | 22.8 | 36.3 | 48.4 | 58.0 | 61.9 | 59.3 | 50.1 | 38.4 | 24.3 | 10.4 | 35.3 |
| Mean Temp °F | 12.9 | 18.6 | 31.6 | 46.5 | 59.6 | 68.9 | 72.6 | 69.9 | 61.0 | 48.8 | 32.6 | 18.6 | 45.1 |
| Days Max Temp ≥ 90 °F | 0 | 0 | 0 | 0 | 1 | 3 | 6 | 3 | 1 | 0 | 0 | 0 | 14 |
| Days Max Temp ≤ 32 °F | 23 | 17 | 8 | 0 | 0 | 0 | 0 | 0 | 0 | 0 | 7 | 21 | 76 |
| Days Min Temp ≤ 32 °F | 31 | 28 | 26 | 11 | 1 | 0 | 0 | 0 | 1 | 8 | 24 | 31 | 161 |
| Days Min Temp ≤ 0 °F | 13 | 8 | 2 | 0 | 0 | 0 | 0 | 0 | 0 | 0 | 1 | 7 | 31 |
| Heating Degree Days | 1612 | 1303 | 1029 | 553 | 208 | 34 | 7 | 22 | 165 | 499 | 966 | 1432 | 7830 |
| Cooling Degree Days | 0 | 0 | 0 | 7 | 50 | 163 | 249 | 185 | 55 | 2 | 0 | 0 | 711 |
| Total Precipitation (") | 0.77 | 0.78 | 1.90 | 3.21 | 3.56 | 4.67 | 4.11 | 4.10 | 3.00 | 2.35 | 1.82 | 1.14 | 31.41 |
| Days ≥ 0.1" Precip | 2 | 2 | 4 | 6 | 7 | 7 | 6 | 6 | 6 | 4 | 4 | 3 | 57 |
| Total Snowfall (") | 8.8 | 7.4 | 8.6 | 3.0 | 0.0 | 0.0 | 0.0 | 0.0 | 0.0 | 0.6 | 5.3 | 9.7 | 43.4 |
| Days ≥ 1" Snow Depth | 26 | 21 | 13 | 2 | 0 | 0 | 0 | 0 | 0 | 0 | 6 | 20 | 88 |

## FARIBAULT *Rice County* ELEVATION 942 ft LAT/LONG 44° 18 ' N / 93° 16 ' W

| | JAN | FEB | MAR | APR | MAY | JUN | JUL | AUG | SEP | OCT | NOV | DEC | YEAR |
|---|---|---|---|---|---|---|---|---|---|---|---|---|---|
| Maximum Temp °F | 21.7 | 27.3 | 40.1 | 56.9 | 69.9 | 79.1 | 83.2 | 80.3 | 71.7 | 59.5 | 41.1 | 27.1 | 54.8 |
| Minimum Temp °F | 1.1 | 6.5 | 20.8 | 34.3 | 45.7 | 55.2 | 59.8 | 57.4 | 48.5 | 37.0 | 23.6 | 9.6 | 33.3 |
| Mean Temp °F | 11.4 | 17.0 | 30.4 | 45.6 | 57.9 | 67.2 | 71.6 | 68.9 | 60.1 | 48.3 | 32.4 | 18.4 | 44.1 |
| Days Max Temp ≥ 90 °F | 0 | 0 | 0 | 0 | 1 | 3 | 6 | 3 | 1 | 0 | 0 | 0 | 14 |
| Days Max Temp ≤ 32 °F | 24 | 17 | 8 | 0 | 0 | 0 | 0 | 0 | 0 | 0 | 7 | 20 | 76 |
| Days Min Temp ≤ 32 °F | 31 | 28 | 26 | 14 | 2 | 0 | 0 | 0 | 1 | 11 | 25 | 30 | 168 |
| Days Min Temp ≤ 0 °F | 15 | 11 | 3 | 0 | 0 | 0 | 0 | 0 | 0 | 0 | 1 | 9 | 39 |
| Heating Degree Days | 1657 | 1351 | 1065 | 581 | 254 | 56 | 12 | 36 | 192 | 518 | 972 | 1439 | 8133 |
| Cooling Degree Days | 0 | 0 | 0 | 7 | 38 | 132 | 218 | 161 | 50 | 2 | 0 | 0 | 608 |
| Total Precipitation (") | 1.02 | 0.80 | 1.99 | 3.03 | 3.67 | 3.81 | 4.27 | 4.15 | 3.58 | 2.34 | 1.87 | 1.13 | 31.66 |
| Days ≥ 0.1" Precip | 3 | 2 | 5 | 7 | 8 | 7 | 6 | 6 | 6 | 5 | 5 | 3 | 63 |
| Total Snowfall (") | 10.5 | 8.3 | 8.7 | 2.9 | 0.0 | 0.0 | 0.0 | 0.0 | 0.0 | 0.0 | 5.4 | 10.1 | 45.9 |
| Days ≥ 1" Snow Depth | 28 | 24 | 16 | 2 | 0 | 0 | 0 | 0 | 0 | 0 | 6 | 20 | 96 |

## FARMINGTON 3 NW *Dakota County* ELEVATION 981 ft LAT/LONG 44° 40 ' N / 93° 11 ' W

| | JAN | FEB | MAR | APR | MAY | JUN | JUL | AUG | SEP | OCT | NOV | DEC | YEAR |
|---|---|---|---|---|---|---|---|---|---|---|---|---|---|
| Maximum Temp °F | 20.9 | 26.7 | 39.7 | 57.5 | 70.9 | 79.3 | 82.8 | 80.0 | 70.9 | 59.0 | 40.2 | 26.2 | 54.5 |
| Minimum Temp °F | 2.3 | 8.3 | 21.9 | 35.7 | 47.3 | 56.6 | 60.5 | 57.7 | 49.0 | 37.8 | 24.3 | 10.0 | 34.3 |
| Mean Temp °F | 11.6 | 17.5 | 30.8 | 46.6 | 59.1 | 67.9 | 71.7 | 68.9 | 60.0 | 48.4 | 32.2 | 18.1 | 44.4 |
| Days Max Temp ≥ 90 °F | 0 | 0 | 0 | 0 | 1 | 3 | 5 | 3 | 0 | 0 | 0 | 0 | 12 |
| Days Max Temp ≤ 32 °F | 24 | 17 | 8 | 0 | 0 | 0 | 0 | 0 | 0 | 0 | 7 | 21 | 77 |
| Days Min Temp ≤ 32 °F | 31 | 27 | 26 | 11 | 1 | 0 | 0 | 0 | 1 | 10 | 24 | 30 | 161 |
| Days Min Temp ≤ 0 °F | 14 | 10 | 2 | 0 | 0 | 0 | 0 | 0 | 0 | 0 | 1 | 9 | 36 |
| Heating Degree Days | 1652 | 1336 | 1053 | 550 | 221 | 42 | 9 | 29 | 190 | 511 | 977 | 1448 | 8018 |
| Cooling Degree Days | 0 | 0 | 0 | 7 | 51 | 150 | 229 | 164 | 51 | 2 | 0 | 0 | 654 |
| Total Precipitation (") | 0.87 | 0.83 | 1.98 | 2.78 | 3.51 | 4.41 | 3.89 | 3.94 | 3.48 | 2.39 | 1.88 | 1.14 | 31.10 |
| Days ≥ 0.1" Precip | 3 | 3 | 5 | 7 | 8 | 8 | 7 | 6 | 6 | 5 | 5 | 3 | 66 |
| Total Snowfall (") | ***10.5*** | 8.6 | 9.1 | 3.1 | 0.0 | 0.0 | 0.0 | 0.0 | 0.0 | 0.2 | 7.2 | 8.3 | 47.0 |
| Days ≥ 1" Snow Depth | 29 | 23 | 16 | 2 | 0 | 0 | 0 | 0 | 0 | 0 | 8 | 21 | 99 |

## FERGUS FALLS *Otter Tail County* ELEVATION 1250 ft LAT/LONG 46° 17 ' N / 96° 6 ' W

| | JAN | FEB | MAR | APR | MAY | JUN | JUL | AUG | SEP | OCT | NOV | DEC | YEAR |
|---|---|---|---|---|---|---|---|---|---|---|---|---|---|
| Maximum Temp °F | 16.3 | 22.2 | 36.0 | 53.7 | 68.4 | 76.2 | 81.3 | 79.6 | 69.1 | 56.1 | 36.6 | 22.1 | 51.5 |
| Minimum Temp °F | -3.7 | 2.1 | 17.1 | 32.2 | 45.1 | 54.2 | 59.2 | 56.8 | 46.4 | 34.6 | 20.3 | 4.5 | 30.7 |
| Mean Temp °F | 6.3 | 12.2 | 26.6 | 43.0 | 56.8 | 65.2 | 70.3 | 68.2 | 57.8 | 45.4 | 28.4 | 13.3 | 41.1 |
| Days Max Temp ≥ 90 °F | 0 | 0 | 0 | 0 | 0 | 1 | 4 | 4 | 1 | 0 | 0 | 0 | 10 |
| Days Max Temp ≤ 32 °F | 27 | 21 | 11 | 1 | 0 | 0 | 0 | 0 | 0 | 0 | 11 | 24 | 95 |
| Days Min Temp ≤ 32 °F | 31 | 28 | 28 | 16 | 3 | 0 | 0 | 0 | 1 | 13 | 27 | 31 | 178 |
| Days Min Temp ≤ 0 °F | 18 | 13 | 4 | 0 | 0 | 0 | 0 | 0 | 0 | 0 | 2 | 12 | 49 |
| Heating Degree Days | 1818 | 1487 | 1184 | 656 | 277 | 76 | 18 | 42 | 244 | 602 | 1090 | 1598 | 9092 |
| Cooling Degree Days | 0 | 0 | 0 | 4 | 35 | 87 | 178 | 144 | 28 | 1 | 0 | 0 | 477 |
| Total Precipitation (") | 0.78 | 0.53 | 1.16 | 1.94 | 2.59 | 3.85 | 3.31 | 3.16 | 2.37 | 1.86 | 1.10 | 0.65 | 23.30 |
| Days ≥ 0.1" Precip | 2 | 2 | 3 | 5 | 6 | 7 | 6 | 5 | 5 | 4 | 3 | 3 | 51 |
| Total Snowfall (") | 12.2 | 6.0 | 7.0 | 1.9 | 0.0 | 0.0 | 0.0 | 0.0 | 0.0 | 0.3 | 6.4 | 8.6 | 42.4 |
| Days ≥ 1" Snow Depth | 28 | 23 | 15 | 2 | 0 | 0 | 0 | 0 | 0 | 0 | 8 | 20 | 96 |

## FOREST LAKE 5 NE *Chisago County* ELEVATION 906 ft LAT/LONG 45° 19 ' N / 92° 56 ' W

| | JAN | FEB | MAR | APR | MAY | JUN | JUL | AUG | SEP | OCT | NOV | DEC | YEAR |
|---|---|---|---|---|---|---|---|---|---|---|---|---|---|
| Maximum Temp °F | 21.1 | 27.4 | 39.7 | 56.6 | 69.5 | 77.3 | 81.3 | 78.6 | 69.4 | 57.9 | 40.0 | 25.9 | 53.7 |
| Minimum Temp °F | 1.2 | 7.3 | 20.6 | 34.8 | 47.1 | 56.2 | 61.2 | 59.2 | 49.8 | 38.6 | 24.3 | 9.6 | 34.2 |
| Mean Temp °F | 11.2 | 17.4 | 30.2 | 45.7 | 58.3 | 66.8 | 71.3 | 68.9 | 59.7 | 48.3 | 32.2 | 17.8 | 44.0 |
| Days Max Temp ≥ 90 °F | 0 | 0 | 0 | 0 | 0 | 1 | 2 | 2 | 0 | 0 | 0 | 0 | 5 |
| Days Max Temp ≤ 32 °F | 24 | 17 | 8 | 0 | 0 | 0 | 0 | 0 | 0 | 0 | 7 | 22 | 78 |
| Days Min Temp ≤ 32 °F | 31 | 27 | 26 | 13 | 1 | 0 | 0 | 0 | 0 | 8 | 24 | 31 | 161 |
| Days Min Temp ≤ 0 °F | 15 | 10 | 2 | 0 | 0 | 0 | 0 | 0 | 0 | 0 | 1 | 8 | 36 |
| Heating Degree Days | 1667 | 1341 | 1071 | 575 | 239 | 53 | 10 | 31 | 196 | 515 | 978 | 1458 | 8134 |
| Cooling Degree Days | 0 | 0 | 0 | 6 | 46 | 135 | 233 | 186 | 50 | 3 | 0 | 0 | 659 |
| Total Precipitation (") | 0.94 | 0.80 | 1.46 | 2.46 | 3.35 | 4.88 | 4.00 | 4.24 | 3.24 | 2.62 | 1.75 | 1.03 | 30.77 |
| Days ≥ 0.1" Precip | 3 | 3 | 4 | 6 | 7 | 8 | 7 | 7 | 6 | 6 | 4 | 3 | 64 |
| Total Snowfall (") | 10.2 | 7.3 | 8.1 | 3.2 | 0.0 | 0.0 | 0.0 | 0.0 | 0.0 | 0.4 | 8.2 | 9.4 | 46.8 |
| Days ≥ 1" Snow Depth | 30 | 27 | 21 | 3 | 0 | 0 | 0 | 0 | 0 | 0 | 9 | 25 | 115 |

## FOSSTON 1 E *Polk County* ELEVATION 1289 ft LAT/LONG 47° 34 ' N / 95° 44 ' W

| | JAN | FEB | MAR | APR | MAY | JUN | JUL | AUG | SEP | OCT | NOV | DEC | YEAR |
|---|---|---|---|---|---|---|---|---|---|---|---|---|---|
| Maximum Temp °F | 13.8 | 20.9 | 34.9 | 52.9 | 67.8 | 75.1 | 80.3 | 78.9 | 67.9 | 54.8 | 34.7 | 20.5 | 50.2 |
| Minimum Temp °F | -8.0 | -1.6 | 13.9 | 29.6 | 41.8 | 50.9 | 55.1 | 53.2 | 43.4 | 32.6 | 17.6 | 1.5 | 27.5 |
| Mean Temp °F | 2.9 | 9.7 | 24.4 | 41.3 | 54.8 | 63.1 | 67.7 | 66.1 | 55.6 | 43.8 | 26.2 | 11.0 | 38.9 |
| Days Max Temp ≥ 90 °F | 0 | 0 | 0 | 0 | 0 | 1 | 3 | 3 | 1 | 0 | 0 | 0 | 8 |
| Days Max Temp ≤ 32 °F | 28 | 21 | 12 | 1 | 0 | 0 | 0 | 0 | 0 | 1 | 13 | 25 | 101 |
| Days Min Temp ≤ 32 °F | 31 | 28 | 29 | 19 | 6 | 0 | 0 | 0 | 4 | 16 | 28 | 31 | 192 |
| Days Min Temp ≤ 0 °F | 21 | 15 | 6 | 0 | 0 | 0 | 0 | 0 | 0 | 0 | 3 | 14 | 59 |
| Heating Degree Days | 1924 | 1560 | 1251 | 707 | 333 | 112 | 40 | 68 | 298 | 653 | 1159 | 1669 | 9774 |
| Cooling Degree Days | 0 | 0 | 0 | 2 | 29 | 50 | 110 | 95 | 13 | 1 | 0 | 0 | 300 |
| Total Precipitation (") | 0.72 | 0.48 | 1.02 | 1.64 | 2.62 | 4.16 | 3.42 | 3.56 | 2.56 | 2.50 | 0.84 | 0.72 | 24.24 |
| Days ≥ 0.1" Precip | 3 | 2 | 3 | 5 | 6 | 8 | 6 | 6 | 5 | 5 | 2 | 3 | 54 |
| Total Snowfall (") | 10.0 | 5.6 | 7.7 | 2.7 | 0.3 | 0.0 | 0.0 | 0.0 | 0.0 | 1.2 | 6.3 | 8.4 | 42.2 |
| Days ≥ 1" Snow Depth | 28 | 25 | 20 | 4 | 0 | 0 | 0 | 0 | 0 | 1 | 9 | 23 | 110 |

## GAYLORD *Sibley County* ELEVATION 1020 ft LAT/LONG 44° 33 ' N / 94° 13 ' W

| | JAN | FEB | MAR | APR | MAY | JUN | JUL | AUG | SEP | OCT | NOV | DEC | YEAR |
|---|---|---|---|---|---|---|---|---|---|---|---|---|---|
| Maximum Temp °F | 21.6 | 27.6 | 40.4 | 57.8 | 71.9 | 80.9 | 84.7 | 81.4 | 72.3 | 60.2 | 40.7 | 26.8 | 55.5 |
| Minimum Temp °F | 2.4 | 8.6 | 21.8 | 35.7 | 47.7 | 57.2 | 61.7 | 58.6 | 49.1 | 37.7 | 24.1 | 9.9 | 34.5 |
| Mean Temp °F | 12.0 | 18.1 | 31.2 | 46.8 | 59.8 | 69.0 | 73.2 | 70.0 | 60.7 | 48.9 | 32.4 | 18.4 | 45.0 |
| Days Max Temp ≥ 90 °F | 0 | 0 | 0 | 0 | 1 | 4 | 8 | 4 | 1 | 0 | 0 | 0 | 18 |
| Days Max Temp ≤ 32 °F | 23 | 16 | 8 | 0 | 0 | 0 | 0 | 0 | 0 | 0 | 7 | 20 | 74 |
| Days Min Temp ≤ 32 °F | 31 | 27 | 26 | 11 | 1 | 0 | 0 | 0 | 1 | 10 | 24 | 31 | 162 |
| Days Min Temp ≤ 0 °F | 14 | 9 | 2 | 0 | 0 | 0 | 0 | 0 | 0 | 0 | 1 | 8 | 34 |
| Heating Degree Days | 1640 | 1318 | 1042 | 546 | 203 | 33 | 6 | 23 | 171 | 495 | 970 | 1440 | 7887 |
| Cooling Degree Days | 0 | 0 | 0 | 8 | 53 | 169 | 275 | 192 | 49 | 2 | 0 | 0 | 748 |
| Total Precipitation (") | 0.79 | 0.73 | 1.71 | 2.85 | 3.31 | 4.68 | 3.70 | 4.36 | 3.21 | 2.27 | 1.68 | 0.95 | 30.24 |
| Days ≥ 0.1" Precip | 3 | 2 | 4 | 6 | 7 | 7 | 6 | 6 | 5 | 5 | 3 | 2 | 56 |
| Total Snowfall (") | 8.9 | 7.1 | 9.7 | 1.7 | 0.0 | 0.0 | 0.0 | 0.0 | 0.0 | 0.3 | 5.0 | 9.2 | 41.9 |
| Days ≥ 1" Snow Depth | 26 | 21 | 15 | 2 | 0 | 0 | 0 | 0 | 0 | 0 | 5 | 17 | 86 |

## GEORGETOWN 1 E *Clay County* ELEVATION 879 ft LAT/LONG 47° 5 ' N / 96° 48 ' W

| | JAN | FEB | MAR | APR | MAY | JUN | JUL | AUG | SEP | OCT | NOV | DEC | YEAR |
|---|---|---|---|---|---|---|---|---|---|---|---|---|---|
| Maximum Temp °F | 16.1 | 22.6 | 36.3 | 56.0 | 70.9 | 77.7 | 82.3 | 81.0 | 70.8 | 57.0 | 36.2 | 21.3 | 52.4 |
| Minimum Temp °F | -3.9 | 2.8 | 17.2 | 32.1 | 44.1 | 53.3 | 58.1 | 55.8 | 46.2 | 34.6 | 19.0 | 3.4 | 30.2 |
| Mean Temp °F | 6.1 | 12.8 | 26.9 | 44.1 | 57.5 | 65.5 | 70.3 | 68.4 | 58.5 | 45.8 | 27.7 | 12.4 | 41.3 |
| Days Max Temp ≥ 90 °F | 0 | 0 | 0 | 0 | 1 | 2 | 4 | 5 | 1 | 0 | 0 | 0 | 13 |
| Days Max Temp ≤ 32 °F | 27 | 20 | 11 | 1 | 0 | 0 | 0 | 0 | 0 | 0 | 11 | 24 | 94 |
| Days Min Temp ≤ 32 °F | 31 | 28 | 28 | 17 | 4 | 0 | 0 | 0 | 2 | 12 | 27 | 30 | 179 |
| Days Min Temp ≤ 0 °F | 19 | 13 | 4 | 0 | 0 | 0 | 0 | 0 | 0 | 0 | 2 | 13 | 51 |
| Heating Degree Days | 1823 | 1473 | 1174 | 632 | 262 | 71 | 17 | 40 | 228 | 590 | 1115 | 1628 | 9053 |
| Cooling Degree Days | 0 | 0 | 0 | 5 | 47 | 99 | 186 | 155 | 35 | 1 | 0 | 0 | 528 |
| Total Precipitation (") | 0.59 | 0.40 | 0.98 | 1.92 | 2.51 | 3.51 | 3.03 | 2.83 | 2.05 | 1.91 | 0.80 | 0.61 | 21.14 |
| Days ≥ 0.1" Precip | 2 | 1 | 3 | 4 | 5 | 7 | 5 | 5 | 5 | 4 | 2 | 2 | 45 |
| Total Snowfall (") | 9.1 | 4.7 | *6.3* | 2.3 | 0.1 | 0.0 | 0.0 | 0.0 | 0.0 | 1.0 | 3.9 | *6.8* | 34.2 |
| Days ≥ 1" Snow Depth | *22* | *21* | 13 | 2 | 0 | 0 | 0 | 0 | 0 | 0 | 4 | *16* | 78 |

## GLENWOOD 2 WNW *Pope County* ELEVATION 1171 ft LAT/LONG 45° 39 ' N / 95° 24 ' W

| | JAN | FEB | MAR | APR | MAY | JUN | JUL | AUG | SEP | OCT | NOV | DEC | YEAR |
|---|---|---|---|---|---|---|---|---|---|---|---|---|---|
| Maximum Temp °F | 19.8 | 25.3 | 37.6 | 56.0 | 69.4 | 77.9 | 82.6 | 80.2 | 70.5 | 58.2 | 38.8 | 24.5 | 53.4 |
| Minimum Temp °F | -0.6 | 5.2 | 18.4 | 32.8 | 44.6 | 53.7 | 58.9 | 56.5 | 46.3 | 35.4 | 21.5 | 6.7 | 31.6 |
| Mean Temp °F | 9.6 | 15.3 | 28.0 | 44.4 | 57.0 | 65.6 | 70.7 | 68.4 | 58.4 | 46.8 | 30.2 | 15.7 | 42.5 |
| Days Max Temp ≥ 90 °F | 0 | 0 | 0 | 0 | 0 | 2 | 5 | 3 | 1 | 0 | 0 | 0 | 11 |
| Days Max Temp ≤ 32 °F | *25* | 18 | 10 | 1 | 0 | 0 | 0 | 0 | 0 | 0 | 9 | 23 | 86 |
| Days Min Temp ≤ 32 °F | 31 | 28 | 28 | 16 | 3 | 0 | 0 | 0 | 1 | *13* | 27 | 31 | 178 |
| Days Min Temp ≤ 0 °F | *16* | 12 | 4 | 0 | 0 | 0 | 0 | 0 | 0 | 0 | 1 | 11 | 44 |
| Heating Degree Days | 1716 | 1400 | 1140 | 613 | 267 | 71 | 15 | 36 | 225 | 559 | 1038 | 1524 | 8604 |
| Cooling Degree Days | 0 | 0 | 0 | 5 | 34 | 98 | 189 | 144 | 33 | 1 | 0 | 0 | 504 |
| Total Precipitation (") | 0.64 | 0.50 | 1.25 | 2.09 | 3.19 | 3.92 | 3.02 | 3.16 | 2.42 | 2.46 | 1.21 | 0.54 | 24.40 |
| Days ≥ 0.1" Precip | 2 | 2 | 3 | 5 | 6 | 8 | 6 | 5 | 4 | 5 | *3* | 2 | 51 |
| Total Snowfall (") | *7.7* | 5.9 | 7.3 | 2.2 | 0.0 | 0.0 | 0.0 | 0.0 | 0.0 | 0.3 | 5.4 | *6.0* | 34.8 |
| Days ≥ 1" Snow Depth | *29* | *25* | *17* | *2* | 0 | 0 | 0 | 0 | 0 | 0 | 9 | *24* | 106 |

## GRAND MARAIS *Cook County* ELEVATION 620 ft LAT/LONG 47° 45 ' N / 90° 20 ' W

| | JAN | FEB | MAR | APR | MAY | JUN | JUL | AUG | SEP | OCT | NOV | DEC | YEAR |
|---|---|---|---|---|---|---|---|---|---|---|---|---|---|
| Maximum Temp °F | 21.8 | 25.4 | 34.4 | 45.9 | 55.3 | 62.7 | 69.3 | 69.9 | 61.6 | 51.0 | 37.5 | 26.8 | 46.8 |
| Minimum Temp °F | 3.9 | 7.4 | 18.9 | 30.4 | 38.0 | 44.1 | 51.4 | 53.7 | 46.9 | 36.9 | 24.7 | 11.7 | 30.7 |
| Mean Temp °F | 12.9 | 16.4 | 26.6 | 38.2 | 46.7 | 53.4 | 60.4 | 61.8 | 54.3 | 44.0 | 31.1 | 19.3 | 38.8 |
| Days Max Temp ≥ 90 °F | 0 | 0 | 0 | 0 | 0 | 0 | 0 | 0 | 0 | 0 | 0 | 0 | 0 |
| Days Max Temp ≤ 32 °F | 25 | 20 | 11 | 1 | 0 | 0 | 0 | 0 | 0 | 0 | 8 | 20 | 85 |
| Days Min Temp ≤ 32 °F | 31 | 28 | 28 | 18 | 4 | 0 | 0 | 0 | 1 | 9 | 23 | 30 | 172 |
| Days Min Temp ≤ 0 °F | 13 | 9 | 2 | 0 | 0 | 0 | 0 | 0 | 0 | 0 | 1 | 7 | 32 |
| Heating Degree Days | 1611 | 1368 | 1182 | 799 | 562 | 343 | 157 | 120 | 319 | 645 | 1010 | 1411 | 9527 |
| Cooling Degree Days | 0 | 0 | 0 | 0 | 0 | 2 | 20 | 37 | 3 | 0 | 0 | 0 | 62 |
| Total Precipitation (") | 0.90 | 0.57 | 1.28 | 1.69 | 2.66 | 3.50 | 3.47 | 3.25 | 3.57 | 2.60 | 1.65 | 1.04 | 26.18 |
| Days ≥ 0.1" Precip | 3 | 2 | 4 | 4 | 6 | 7 | 7 | 7 | 7 | 6 | 4 | 3 | 60 |
| Total Snowfall (") | 16.8 | 7.7 | 8.1 | 2.0 | 0.2 | 0.0 | 0.0 | 0.0 | 0.0 | 0.4 | 4.5 | 13.1 | 52.8 |
| Days ≥ 1" Snow Depth | 31 | 28 | 29 | 9 | 0 | 0 | 0 | 0 | 0 | 0 | 7 | 24 | 128 |

## GRAND MEADOW *Mower County* ELEVATION 1342 ft LAT/LONG 43° 42 ' N / 92° 34 ' W

| | JAN | FEB | MAR | APR | MAY | JUN | JUL | AUG | SEP | OCT | NOV | DEC | YEAR |
|---|---|---|---|---|---|---|---|---|---|---|---|---|---|
| Maximum Temp °F | 20.3 | 25.8 | 37.9 | 54.2 | 68.0 | 77.6 | 81.4 | 79.0 | 70.6 | 58.2 | 40.7 | 26.0 | 53.3 |
| Minimum Temp °F | 1.7 | 7.1 | 20.7 | 34.3 | 45.7 | 55.8 | 60.0 | 57.2 | 48.2 | 36.9 | 23.8 | 9.8 | 33.4 |
| Mean Temp °F | 11.0 | 16.5 | 29.3 | 44.2 | 56.9 | 66.7 | 70.7 | 68.1 | 59.4 | 47.6 | 32.2 | 17.9 | 43.4 |
| Days Max Temp ≥ 90 °F | 0 | 0 | 0 | 0 | 0 | 2 | 3 | 2 | 0 | 0 | 0 | 0 | 7 |
| Days Max Temp ≤ 32 °F | 25 | 18 | 9 | 1 | 0 | 0 | 0 | 0 | 0 | 0 | 7 | 22 | 82 |
| Days Min Temp ≤ 32 °F | 31 | 28 | 27 | 14 | 2 | 0 | 0 | 0 | 1 | 11 | 24 | 31 | 169 |
| Days Min Temp ≤ 0 °F | 15 | 10 | 2 | 0 | 0 | 0 | 0 | 0 | 0 | 0 | 1 | 8 | 36 |
| Heating Degree Days | 1671 | 1367 | 1101 | 622 | 277 | 60 | 16 | 38 | 203 | 537 | 976 | 1453 | 8321 |
| Cooling Degree Days | 0 | 0 | 0 | 7 | 35 | 131 | 205 | 157 | 50 | 2 | 0 | 0 | 587 |
| Total Precipitation (") | 1.03 | 0.79 | 1.95 | 3.31 | 4.19 | 4.26 | 4.36 | 4.39 | 3.99 | 2.59 | 2.03 | 1.22 | 34.11 |
| Days ≥ 0.1" Precip | 3 | 2 | 4 | 7 | 8 | 8 | 6 | 7 | 6 | 5 | 4 | 3 | 63 |
| Total Snowfall (") | *12.4* | *9.0* | *7.1* | 2.3 | 0.0 | 0.0 | 0.0 | 0.0 | 0.0 | 0.3 | 4.9 | *10.4* | 46.4 |
| Days ≥ 1" Snow Depth | na | na | na | *1* | 0 | 0 | 0 | 0 | 0 | 0 | na | na | na |

## GRAND RAPIDS FRS LAB *Itasca County* ELEVATION 1312 ft LAT/LONG 47° 14 ' N / 93° 30 ' W

| | JAN | FEB | MAR | APR | MAY | JUN | JUL | AUG | SEP | OCT | NOV | DEC | YEAR |
|---|---|---|---|---|---|---|---|---|---|---|---|---|---|
| Maximum Temp °F | 16.8 | 24.5 | 37.3 | 53.5 | 67.7 | 75.6 | 80.3 | 77.6 | 66.7 | 54.3 | 35.5 | 21.8 | 51.0 |
| Minimum Temp °F | -5.3 | 0.3 | 14.5 | 29.3 | 40.9 | 49.8 | 55.0 | 52.5 | 43.4 | 33.5 | 19.3 | 3.1 | 28.0 |
| Mean Temp °F | 5.7 | 12.4 | 25.9 | 41.4 | 54.3 | 62.7 | 67.7 | 65.1 | 55.1 | 43.9 | 27.5 | 12.5 | 39.5 |
| Days Max Temp ≥ 90 °F | 0 | 0 | 0 | 0 | 0 | 1 | 2 | 1 | 0 | 0 | 0 | 0 | 4 |
| Days Max Temp ≤ 32 °F | 28 | 20 | 10 | 1 | 0 | 0 | 0 | 0 | 0 | 1 | 12 | 25 | 97 |
| Days Min Temp ≤ 32 °F | 31 | 28 | 29 | 19 | 6 | 0 | 0 | 0 | 3 | 15 | 27 | 31 | 189 |
| Days Min Temp ≤ 0 °F | 19 | 14 | 6 | 0 | 0 | 0 | 0 | 0 | 0 | 0 | 2 | 13 | 54 |
| Heating Degree Days | 1835 | 1480 | 1204 | 702 | 342 | 117 | 37 | 76 | 305 | 646 | 1120 | 1624 | 9488 |
| Cooling Degree Days | 0 | 0 | 0 | 2 | 21 | 58 | 129 | 94 | 11 | 0 | 0 | 0 | 315 |
| Total Precipitation (") | 0.97 | 0.54 | 1.32 | 2.06 | 2.86 | 4.44 | 3.99 | 3.59 | 3.10 | 2.52 | 1.41 | 1.00 | 27.80 |
| Days ≥ 0.1" Precip | 3 | 2 | 4 | 5 | 7 | 8 | 7 | 7 | 7 | 5 | 4 | 3 | 62 |
| Total Snowfall (") | 14.0 | 6.7 | 9.6 | 4.1 | 0.7 | 0.0 | 0.0 | 0.0 | 0.0 | 1.1 | 9.4 | 12.2 | 57.8 |
| Days ≥ 1" Snow Depth | ***25*** | 22 | ***21*** | 5 | 0 | 0 | 0 | 0 | 0 | 0 | na | ***24*** | na |

## GULL LAKE DAM *Cass County* ELEVATION 1220 ft LAT/LONG 46° 25 ' N / 94° 21 ' W

| | JAN | FEB | MAR | APR | MAY | JUN | JUL | AUG | SEP | OCT | NOV | DEC | YEAR |
|---|---|---|---|---|---|---|---|---|---|---|---|---|---|
| Maximum Temp °F | 18.2 | 25.1 | 37.3 | 54.1 | 68.1 | 76.1 | 80.8 | 78.4 | 68.1 | 55.9 | 36.8 | 22.8 | 51.8 |
| Minimum Temp °F | -2.9 | 2.8 | 15.8 | 30.3 | 43.1 | 53.2 | 58.6 | 56.2 | 46.6 | 35.4 | 20.7 | 4.9 | 30.4 |
| Mean Temp °F | 7.5 | 14.0 | 26.6 | 42.2 | 55.6 | 64.7 | 69.7 | 67.3 | 57.4 | 45.7 | 28.8 | 13.9 | 41.1 |
| Days Max Temp ≥ 90 °F | 0 | 0 | 0 | 0 | 0 | 1 | 3 | 2 | 0 | 0 | 0 | 0 | 6 |
| Days Max Temp ≤ 32 °F | 27 | 19 | 9 | 1 | 0 | 0 | 0 | 0 | 0 | 0 | 10 | 24 | 90 |
| Days Min Temp ≤ 32 °F | 31 | 28 | 28 | 18 | 3 | 0 | 0 | 0 | 1 | 12 | 26 | 30 | 177 |
| Days Min Temp ≤ 0 °F | 17 | 13 | 4 | 0 | 0 | 0 | 0 | 0 | 0 | 0 | 1 | 11 | 46 |
| Heating Degree Days | 1775 | 1437 | 1185 | 678 | 304 | 80 | 21 | 46 | 248 | 594 | 1081 | 1579 | 9028 |
| Cooling Degree Days | 0 | 0 | 0 | 2 | 22 | 78 | 175 | 128 | 22 | 0 | 0 | 0 | 427 |
| Total Precipitation (") | 0.82 | 0.56 | 1.53 | 2.00 | 3.04 | 4.51 | 3.68 | 3.40 | 2.82 | 2.49 | 1.25 | 0.71 | 26.81 |
| Days ≥ 0.1" Precip | 3 | 2 | 4 | 5 | 6 | 8 | 6 | 6 | 6 | 4 | 3 | 3 | 56 |
| Total Snowfall (") | na | na | ***10.4*** | 3.3 | 0.2 | 0.0 | 0.0 | 0.0 | 0.0 | 0.6 | na | na | na |
| Days ≥ 1" Snow Depth | na | na | na | 7 | 0 | 0 | 0 | 0 | 0 | 0 | na | na | na |

## GUNFLINT LAKE 10 NW *Cook County* ELEVATION 1440 ft LAT/LONG 48° 10 ' N / 90° 53 ' W

| | JAN | FEB | MAR | APR | MAY | JUN | JUL | AUG | SEP | OCT | NOV | DEC | YEAR |
|---|---|---|---|---|---|---|---|---|---|---|---|---|---|
| Maximum Temp °F | 16.2 | 22.4 | 35.9 | 50.6 | 66.7 | ***72.6*** | 78.2 | ***75.4*** | ***63.3*** | ***50.0*** | ***32.7*** | ***20.3*** | 48.7 |
| Minimum Temp °F | -6.9 | -2.9 | ***10.0*** | 24.6 | 37.4 | ***46.8*** | 53.3 | ***51.4*** | ***42.4*** | ***32.0*** | ***18.1*** | ***0.6*** | 25.6 |
| Mean Temp °F | 4.7 | 9.7 | ***22.9*** | 37.7 | 52.1 | ***59.7*** | 65.8 | ***63.4*** | ***52.9*** | ***41.0*** | ***25.8*** | ***10.4*** | 37.2 |
| Days Max Temp ≥ 90 °F | 0 | 0 | 0 | 0 | 0 | 1 | 2 | 1 | 0 | ***0*** | 0 | 0 | 4 |
| Days Max Temp ≤ 32 °F | 28 | 22 | 11 | 1 | 0 | 0 | 0 | 0 | ***0*** | ***1*** | ***13*** | ***27*** | 103 |
| Days Min Temp ≤ 32 °F | 31 | 28 | 30 | ***25*** | 11 | 0 | 0 | 0 | ***3*** | ***18*** | ***27*** | ***31*** | 204 |
| Days Min Temp ≤ 0 °F | 20 | 16 | ***9*** | 1 | 0 | 0 | 0 | 0 | ***0*** | ***0*** | ***2*** | na | na |
| Heating Degree Days | 1868 | 1562 | ***1301*** | 813 | 406 | ***181*** | 59 | ***103*** | ***364*** | ***736*** | ***1171*** | ***1690*** | 10254 |
| Cooling Degree Days | 0 | 0 | 0 | 0 | 12 | 32 | 96 | 73 | 7 | 0 | 0 | 0 | 220 |
| Total Precipitation (") | 1.07 | 0.87 | 1.41 | 1.79 | 2.41 | ***3.82*** | 3.88 | 4.23 | ***3.36*** | ***2.36*** | 1.79 | 1.03 | 28.02 |
| Days ≥ 0.1" Precip | ***4*** | ***3*** | 4 | 4 | ***6*** | 8 | ***8*** | ***8*** | ***7*** | ***6*** | ***5*** | ***4*** | 67 |
| Total Snowfall (") | ***14.6*** | ***9.7*** | ***9.7*** | ***6.3*** | 0.3 | 0.0 | 0.0 | 0.0 | 0.0 | ***2.1*** | na | na | na |
| Days ≥ 1" Snow Depth | na | ***18*** | na | na | 1 | 0 | 0 | 0 | 0 | ***1*** | na | na | na |

## HALLOCK *Kittson County* ELEVATION 820 ft LAT/LONG 48° 46 ' N / 96° 57 ' W

| | JAN | FEB | MAR | APR | MAY | JUN | JUL | AUG | SEP | OCT | NOV | DEC | YEAR |
|---|---|---|---|---|---|---|---|---|---|---|---|---|---|
| Maximum Temp °F | 10.2 | 16.8 | 31.1 | 51.1 | 66.9 | 75.5 | 80.5 | 79.4 | 67.4 | 53.3 | 32.8 | 16.9 | 48.5 |
| Minimum Temp °F | -9.4 | -3.8 | 12.2 | 29.8 | 42.4 | 52.4 | 56.4 | 53.7 | 43.6 | 31.7 | 16.2 | -0.5 | 27.1 |
| Mean Temp °F | 0.4 | 6.5 | 21.7 | 40.5 | 54.7 | 63.9 | 68.5 | 66.6 | 55.5 | 42.5 | 24.5 | 8.2 | 37.8 |
| Days Max Temp ≥ 90 °F | 0 | 0 | 0 | 0 | 1 | 2 | 3 | 4 | 1 | 0 | 0 | 0 | 11 |
| Days Max Temp ≤ 32 °F | 29 | 24 | 16 | 2 | 0 | 0 | 0 | 0 | 0 | 1 | 15 | 27 | 114 |
| Days Min Temp ≤ 32 °F | 31 | 28 | 29 | 18 | 5 | 0 | 0 | 0 | 3 | 16 | 28 | 31 | 189 |
| Days Min Temp ≤ 0 °F | 22 | 18 | 8 | 0 | 0 | 0 | 0 | 0 | 0 | 0 | 4 | 16 | 68 |
| Heating Degree Days | 2002 | 1647 | 1336 | 732 | 342 | 105 | 32 | 69 | 301 | 690 | 1208 | 1758 | 10222 |
| Cooling Degree Days | 0 | 0 | 0 | 3 | 33 | 84 | 146 | ***138*** | 18 | 0 | 0 | 0 | 422 |
| Total Precipitation (") | 0.72 | 0.42 | 0.78 | 1.32 | 2.18 | 3.30 | 2.85 | 2.31 | 2.32 | 1.40 | 0.72 | 0.57 | 18.89 |
| Days ≥ 0.1" Precip | 3 | 1 | 3 | 3 | 5 | 6 | 6 | 5 | 4 | 4 | 2 | 2 | 44 |
| Total Snowfall (") | na | na | na | na | 0.0 | 0.0 | 0.0 | 0.0 | 0.0 | 0.0 | na | na | na |
| Days ≥ 1" Snow Depth | na | na | na | na | 0 | 0 | 0 | 0 | 0 | ***0*** | na | na | na |

## HIBBING CHISHOLM AP *St. Louis County* ELEVATION 1355 ft LAT/LONG 47° 23' N / 92° 51' W

| | JAN | FEB | MAR | APR | MAY | JUN | JUL | AUG | SEP | OCT | NOV | DEC | YEAR |
|---|---|---|---|---|---|---|---|---|---|---|---|---|---|
| Maximum Temp °F | 15.1 | 22.0 | 34.6 | 50.4 | 64.5 | 72.4 | 77.2 | 74.7 | 64.0 | 51.6 | 33.4 | 20.3 | 48.3 |
| Minimum Temp °F | -6.6 | -0.8 | 13.1 | 27.6 | 38.8 | 47.8 | 53.3 | 50.6 | 41.4 | 31.4 | 17.2 | 2.1 | 26.3 |
| Mean Temp °F | 4.2 | 10.6 | 23.9 | 39.0 | 51.7 | 60.1 | 65.3 | 62.7 | 52.8 | 41.5 | 25.4 | 11.2 | 37.4 |
| Days Max Temp ≥ 90 °F | 0 | 0 | 0 | 0 | 0 | 0 | 1 | 1 | 0 | 0 | 0 | 0 | 2 |
| Days Max Temp ≤ 32 °F | 28 | 22 | 13 | 1 | 0 | 0 | 0 | 0 | 0 | 1 | 14 | 27 | 106 |
| Days Min Temp ≤ 32 °F | 30 | 28 | 29 | 22 | 8 | 0 | 0 | 0 | 5 | 18 | 28 | 31 | 199 |
| Days Min Temp ≤ 0 °F | 20 | 15 | 6 | 0 | 0 | 0 | 0 | 0 | 0 | 0 | 3 | 14 | 58 |
| Heating Degree Days | 1884 | 1532 | 1268 | 773 | 415 | 171 | 66 | 119 | 368 | 721 | 1183 | 1663 | 10163 |
| Cooling Degree Days | 0 | 0 | 0 | 0 | 8 | 30 | 77 | 56 | 5 | 0 | 0 | 0 | 176 |
| Total Precipitation (") | 0.75 | 0.48 | 1.00 | 1.65 | 2.53 | 4.05 | 4.09 | 3.53 | 3.12 | 2.32 | 1.28 | 0.78 | 25.58 |
| Days ≥ 0.1" Precip | 2 | 1 | 3 | 5 | 6 | 8 | 7 | 7 | 7 | 5 | 3 | 3 | 57 |
| Total Snowfall (") | 14.4 | 7.3 | 10.5 | 4.1 | 0.3 | 0.0 | 0.0 | 0.0 | 0.0 | 1.0 | 11.5 | 11.5 | 60.6 |
| Days ≥ 1" Snow Depth | 31 | 28 | 26 | 7 | 0 | 0 | 0 | 0 | 0 | 1 | 14 | 29 | 136 |

## HINCKLEY *Pine County* ELEVATION 1030 ft LAT/LONG 46° 1' N / 92° 56' W

| | JAN | FEB | MAR | APR | MAY | JUN | JUL | AUG | SEP | OCT | NOV | DEC | YEAR |
|---|---|---|---|---|---|---|---|---|---|---|---|---|---|
| Maximum Temp °F | 18.3 | 24.4 | 37.5 | 53.8 | 67.3 | 75.9 | 80.8 | 78.4 | 68.2 | 56.0 | 38.0 | 24.0 | 51.9 |
| Minimum Temp °F | -4.0 | 1.0 | 16.7 | 30.8 | 41.3 | 50.0 | 55.3 | 52.6 | 43.1 | 33.1 | 19.9 | 4.5 | 28.7 |
| Mean Temp °F | 7.2 | 12.7 | 27.1 | 42.3 | 54.3 | 63.0 | 68.1 | 65.5 | 55.7 | 44.6 | 28.9 | 14.3 | 40.3 |
| Days Max Temp ≥ 90 °F | 0 | 0 | 0 | 0 | 0 | 1 | 3 | 2 | 0 | 0 | 0 | 0 | 6 |
| Days Max Temp ≤ 32 °F | 27 | 19 | 10 | 1 | 0 | 0 | 0 | 0 | 0 | 0 | 9 | ***23*** | 89 |
| Days Min Temp ≤ 32 °F | 31 | 28 | 29 | 18 | 5 | 0 | 0 | 0 | 3 | 16 | 27 | 31 | 188 |
| Days Min Temp ≤ 0 °F | ***18*** | 14 | 4 | 0 | 0 | 0 | 0 | 0 | 0 | 0 | 2 | 13 | 51 |
| Heating Degree Days | 1791 | 1476 | 1167 | 678 | 340 | 112 | 33 | 70 | 289 | 628 | 1076 | 1568 | 9228 |
| Cooling Degree Days | 0 | 0 | 0 | 2 | 16 | 59 | 134 | 99 | 13 | 0 | 0 | 0 | 323 |
| Total Precipitation (") | 1.01 | 0.73 | 1.72 | 2.52 | 3.33 | 4.52 | 4.04 | 4.06 | 3.28 | 2.83 | 1.82 | 1.17 | 31.03 |
| Days ≥ 0.1" Precip | 3 | 2 | 5 | 6 | 7 | 7 | 7 | 7 | 6 | 5 | 4 | 4 | 63 |
| Total Snowfall (") | ***12.0*** | 7.7 | 9.3 | 3.1 | 0.1 | 0.0 | 0.0 | 0.0 | 0.0 | 0.8 | ***10.5*** | ***11.8*** | 55.3 |
| Days ≥ 1" Snow Depth | 23 | 20 | 18 | 5 | 0 | 0 | 0 | 0 | 0 | 1 | ***9*** | ***22*** | 98 |

## HUTCHINSON 1 N *McLeod County* ELEVATION 1102 ft LAT/LONG 44° 53' N / 94° 21' W

| | JAN | FEB | MAR | APR | MAY | JUN | JUL | AUG | SEP | OCT | NOV | DEC | YEAR |
|---|---|---|---|---|---|---|---|---|---|---|---|---|---|
| Maximum Temp °F | 21.3 | 27.0 | 40.2 | 57.3 | 71.0 | 79.7 | 83.6 | 80.9 | 72.0 | 59.8 | 40.2 | 25.9 | 54.9 |
| Minimum Temp °F | 0.8 | 6.8 | 21.4 | 35.1 | 47.0 | 56.6 | 61.0 | 58.1 | 48.7 | 37.0 | 23.2 | 8.5 | 33.7 |
| Mean Temp °F | 11.0 | 16.9 | 30.9 | 46.2 | 59.0 | 68.2 | 72.3 | 69.5 | 60.3 | 48.4 | 31.7 | 17.3 | 44.3 |
| Days Max Temp ≥ 90 °F | 0 | 0 | 0 | 0 | 1 | 3 | 6 | 4 | 1 | 0 | 0 | 0 | 15 |
| Days Max Temp ≤ 32 °F | 24 | 17 | 7 | 0 | 0 | 0 | 0 | 0 | 0 | 0 | 8 | 21 | 77 |
| Days Min Temp ≤ 32 °F | 31 | 28 | 26 | 12 | 1 | 0 | 0 | 0 | 1 | 10 | 25 | 31 | 165 |
| Days Min Temp ≤ 0 °F | 15 | 10 | 2 | 0 | 0 | 0 | 0 | 0 | 0 | 0 | 1 | 9 | 37 |
| Heating Degree Days | 1669 | 1352 | 1052 | 561 | 222 | 39 | 8 | 27 | 181 | 511 | 992 | 1475 | 8089 |
| Cooling Degree Days | 0 | 0 | 0 | 7 | 48 | 142 | 244 | 176 | 47 | 2 | 0 | 0 | 666 |
| Total Precipitation (") | 0.62 | 0.60 | 1.64 | 2.47 | 3.05 | 4.62 | 3.56 | 3.91 | 2.64 | 2.13 | 1.61 | 0.78 | 27.63 |
| Days ≥ 0.1" Precip | 2 | 2 | 4 | 5 | 7 | 8 | 6 | 6 | 5 | 4 | 3 | 2 | 54 |
| Total Snowfall (") | 8.6 | 6.0 | 8.3 | 2.2 | 0.0 | 0.0 | 0.0 | 0.0 | 0.0 | 0.3 | 6.9 | 6.4 | 38.7 |
| Days ≥ 1" Snow Depth | 28 | 24 | 16 | 1 | 0 | 0 | 0 | 0 | 0 | 0 | 7 | 21 | 97 |

## INTERNATL FALLS ARPT *Koochiching County* ELEVATION 1181 ft LAT/LONG 48° 34' N / 93° 23' W

| | JAN | FEB | MAR | APR | MAY | JUN | JUL | AUG | SEP | OCT | NOV | DEC | YEAR |
|---|---|---|---|---|---|---|---|---|---|---|---|---|---|
| Maximum Temp °F | 12.1 | 19.7 | 33.3 | 50.3 | 65.2 | 73.2 | 78.3 | 75.5 | 64.1 | 51.0 | 32.2 | 17.6 | 47.7 |
| Minimum Temp °F | -9.7 | -3.3 | 11.7 | 28.1 | 40.0 | 49.4 | 54.4 | 51.9 | 42.4 | 32.3 | 16.7 | -0.9 | 26.1 |
| Mean Temp °F | 1.2 | 8.2 | 22.5 | 39.2 | 52.6 | 61.4 | 66.4 | 63.7 | 53.3 | 41.7 | 24.5 | 8.4 | 36.9 |
| Days Max Temp ≥ 90 °F | 0 | 0 | 0 | 0 | 0 | 1 | 2 | 1 | 0 | 0 | 0 | 0 | 4 |
| Days Max Temp ≤ 32 °F | 29 | 23 | 14 | 2 | 0 | 0 | 0 | 0 | 0 | 1 | 16 | 27 | 112 |
| Days Min Temp ≤ 32 °F | 31 | 28 | 29 | 21 | 7 | 0 | 0 | 0 | 4 | 17 | 28 | 31 | 196 |
| Days Min Temp ≤ 0 °F | 22 | 17 | 7 | 0 | 0 | 0 | 0 | 0 | 0 | 0 | 3 | 17 | 66 |
| Heating Degree Days | 1978 | 1601 | 1310 | 767 | 394 | 146 | 50 | 102 | 356 | 716 | 1210 | 1753 | 10383 |
| Cooling Degree Days | 0 | 0 | 0 | 1 | 19 | 43 | 98 | 78 | 7 | 0 | 0 | 0 | 246 |
| Total Precipitation (") | 0.89 | 0.60 | 1.06 | 1.48 | 2.34 | 3.92 | 3.58 | 3.09 | 3.02 | 2.05 | 1.25 | 0.83 | 24.11 |
| Days ≥ 0.1" Precip | 2 | 2 | 3 | 4 | 5 | 8 | 7 | 7 | 6 | 5 | 4 | 3 | 56 |
| Total Snowfall (") | 14.8 | 9.7 | 9.7 | 5.5 | 0.5 | 0.0 | 0.0 | 0.0 | 0.1 | 2.2 | 13.2 | 13.9 | 69.6 |
| Days ≥ 1" Snow Depth | 31 | 28 | 27 | 10 | 0 | 0 | 0 | 0 | 0 | 1 | 16 | 30 | 143 |

## ISLE 12 N *Aitkin County* ELEVATION 1285 ft LAT/LONG 46° 18 ' N / 93° 31 ' W

| | JAN | FEB | MAR | APR | MAY | JUN | JUL | AUG | SEP | OCT | NOV | DEC | YEAR |
|---|---|---|---|---|---|---|---|---|---|---|---|---|---|
| Maximum Temp °F | 17.2 | 24.2 | 36.7 | 52.3 | 66.2 | 74.7 | 79.3 | 76.9 | 66.8 | 54.6 | 36.3 | 22.1 | 50.6 |
| Minimum Temp °F | -5.2 | 0.3 | 14.4 | 29.8 | 41.5 | 50.9 | 56.2 | 54.1 | 44.8 | 33.8 | 20.1 | 2.9 | 28.6 |
| Mean Temp °F | 6.0 | 12.3 | 25.6 | 41.0 | 53.9 | 62.8 | 67.8 | 65.5 | 55.8 | 44.2 | 28.2 | 12.5 | 39.6 |
| Days Max Temp ≥ 90 °F | 0 | 0 | 0 | 0 | 0 | 0 | 2 | 1 | 0 | 0 | 0 | 0 | 3 |
| Days Max Temp ≤ 32 °F | 28 | *20* | 10 | 1 | 0 | 0 | 0 | 0 | 0 | 0 | 11 | 25 | 95 |
| Days Min Temp ≤ 32 °F | 31 | 28 | 29 | 19 | 5 | 0 | 0 | 0 | 3 | *14* | 27 | 31 | 187 |
| Days Min Temp ≤ 0 °F | 19 | *13* | 6 | 0 | 0 | 0 | 0 | 0 | 0 | 0 | 1 | 14 | 53 |
| Heating Degree Days | 1828 | 1486 | 1215 | 713 | 353 | 114 | 39 | 72 | 287 | 637 | 1096 | 1624 | 9464 |
| Cooling Degree Days | 0 | 0 | 0 | 2 | 17 | 60 | 139 | 95 | 16 | 0 | 0 | 0 | 329 |
| Total Precipitation (") | 0.69 | 0.50 | 1.38 | 2.21 | 2.94 | 4.63 | 4.73 | 3.78 | 3.07 | 2.28 | 1.49 | 0.98 | 28.68 |
| Days ≥ 0.1" Precip | 3 | *2* | *4* | 5 | 7 | 8 | 8 | 7 | 6 | 4 | 4 | *3* | 61 |
| Total Snowfall (") | 11.8 | *6.0* | *8.6* | 2.1 | 0.2 | 0.0 | 0.0 | 0.0 | 0.0 | 0.6 | *5.4* | 8.8 | 43.5 |
| Days ≥ 1" Snow Depth | *31* | na | na | 4 | 0 | 0 | 0 | 0 | 0 | 0 | na | *26* | na |

## ITASCA UNIV OF MN *Clearwater County* ELEVATION 1503 ft LAT/LONG 47° 13 ' N / 95° 12 ' W

| | JAN | FEB | MAR | APR | MAY | JUN | JUL | AUG | SEP | OCT | NOV | DEC | YEAR |
|---|---|---|---|---|---|---|---|---|---|---|---|---|---|
| Maximum Temp °F | 16.3 | 23.9 | 36.9 | 53.2 | 67.4 | 75.3 | 80.3 | 78.3 | 66.9 | 54.2 | 34.8 | 21.1 | 50.7 |
| Minimum Temp °F | -7.6 | -2.2 | 12.5 | 27.5 | 39.8 | 49.6 | 54.6 | 52.3 | 42.7 | 32.2 | 17.6 | 1.3 | 26.7 |
| Mean Temp °F | 4.4 | 10.9 | 24.7 | 40.3 | 53.6 | 62.4 | 67.4 | 65.3 | 54.8 | 43.2 | 26.2 | 11.2 | 38.7 |
| Days Max Temp ≥ 90 °F | 0 | 0 | 0 | 0 | 0 | 1 | 3 | 2 | 0 | 0 | 0 | 0 | 6 |
| Days Max Temp ≤ 32 °F | 28 | 20 | 11 | 1 | 0 | 0 | 0 | 0 | 0 | 1 | 13 | 26 | 100 |
| Days Min Temp ≤ 32 °F | 31 | 28 | 29 | 22 | 8 | 0 | 0 | 0 | 4 | 17 | 28 | 31 | 198 |
| Days Min Temp ≤ 0 °F | 20 | 15 | 7 | 0 | 0 | 0 | 0 | 0 | 0 | 0 | 3 | 14 | 59 |
| Heating Degree Days | 1878 | 1525 | 1242 | 734 | 364 | 125 | 41 | 78 | 317 | 668 | 1158 | 1664 | 9794 |
| Cooling Degree Days | 0 | 0 | 0 | 2 | 19 | 53 | 116 | 92 | 11 | 0 | 0 | 0 | 293 |
| Total Precipitation (") | 0.94 | 0.59 | 1.45 | 2.13 | 2.59 | 4.24 | 3.71 | 3.58 | 2.95 | 2.48 | 1.22 | 0.94 | 26.82 |
| Days ≥ 0.1" Precip | 3 | 2 | 4 | 5 | 6 | 8 | 7 | 6 | 6 | 5 | 3 | 3 | 58 |
| Total Snowfall (") | 12.1 | 6.7 | 10.4 | 4.4 | 0.1 | 0.0 | 0.0 | 0.0 | 0.0 | 1.4 | 8.4 | 9.6 | 53.1 |
| Days ≥ 1" Snow Depth | 31 | 28 | 27 | 8 | 0 | 0 | 0 | 0 | 0 | 1 | 13 | 28 | 136 |

## JORDAN 1 S *Scott County* ELEVATION 761 ft LAT/LONG 44° 39 ' N / 93° 37 ' W

| | JAN | FEB | MAR | APR | MAY | JUN | JUL | AUG | SEP | OCT | NOV | DEC | YEAR |
|---|---|---|---|---|---|---|---|---|---|---|---|---|---|
| Maximum Temp °F | 21.8 | 27.3 | 40.0 | 57.5 | 70.6 | 79.0 | 82.7 | 79.5 | 70.8 | 59.0 | 40.5 | 26.7 | 54.6 |
| Minimum Temp °F | 0.5 | 6.2 | 20.1 | 33.8 | 45.5 | 54.5 | 58.6 | 55.7 | 46.9 | 35.8 | 22.5 | 8.3 | 32.4 |
| Mean Temp °F | 11.2 | 16.8 | 30.1 | 45.6 | 58.1 | 66.7 | 70.6 | 67.6 | 58.9 | 47.4 | 31.5 | 17.5 | 43.5 |
| Days Max Temp ≥ 90 °F | 0 | 0 | 0 | 0 | 1 | 2 | 5 | 2 | 1 | 0 | 0 | 0 | 11 |
| Days Max Temp ≤ 32 °F | 24 | 17 | 8 | 0 | 0 | 0 | 0 | 0 | 0 | 0 | 7 | 20 | 76 |
| Days Min Temp ≤ 32 °F | 31 | 28 | 27 | 14 | 3 | 0 | 0 | 0 | 2 | 12 | 26 | 31 | 174 |
| Days Min Temp ≤ 0 °F | 15 | 11 | 3 | 0 | 0 | 0 | 0 | 0 | 0 | 0 | 1 | 10 | 40 |
| Heating Degree Days | 1664 | 1355 | 1075 | 578 | 245 | 56 | 14 | 44 | 216 | 542 | 998 | 1467 | 8254 |
| Cooling Degree Days | 0 | 0 | 0 | 7 | 37 | 117 | 189 | 127 | 37 | 1 | 0 | 0 | 515 |
| Total Precipitation (") | 0.70 | 0.60 | 1.55 | 2.47 | 3.23 | 4.51 | 3.81 | 4.15 | 3.27 | 2.34 | 1.53 | 0.82 | 28.98 |
| Days ≥ 0.1" Precip | 2 | 2 | 4 | 6 | 7 | 8 | 6 | 7 | 6 | 5 | 4 | 3 | 60 |
| Total Snowfall (") | na | *5.1* | *5.3* | 1.6 | 0.0 | 0.0 | 0.0 | 0.0 | 0.0 | 0.1 | *2.2* | *6.7* | na |
| Days ≥ 1" Snow Depth | na | na | *11* | 1 | 0 | 0 | 0 | 0 | 0 | 0 | 4 | *14* | na |

## KELLIHER *Beltrami County* ELEVATION 1391 ft LAT/LONG 47° 56 ' N / 94° 27 ' W

| | JAN | FEB | MAR | APR | MAY | JUN | JUL | AUG | SEP | OCT | NOV | DEC | YEAR |
|---|---|---|---|---|---|---|---|---|---|---|---|---|---|
| Maximum Temp °F | na | *22.4* | *36.2* | *52.1* | *65.0* | 73.4 | *78.6* | 76.5 | *65.7* | *54.0* | na | na | na |
| Minimum Temp °F | na | *-2.2* | *13.5* | *28.8* | *40.4* | *49.7* | *54.9* | 52.1 | *42.9* | *33.9* | na | na | na |
| Mean Temp °F | na | *10.3* | *24.9* | *40.5* | *52.8* | *61.4* | *66.8* | *64.4* | *54.4* | *43.9* | na | na | na |
| Days Max Temp ≥ 90 °F | 0 | *0* | *0* | 0 | 0 | 0 | 1 | 1 | 0 | 0 | 0 | 0 | 2 |
| Days Max Temp ≤ 32 °F | 25 | *19* | *10* | *1* | 0 | 0 | 0 | 0 | 0 | 0 | *13* | *24* | 92 |
| Days Min Temp ≤ 32 °F | 28 | *27* | *28* | 20 | 7 | 0 | 0 | 0 | 3 | 13 | *26* | *27* | 179 |
| Days Min Temp ≤ 0 °F | *19* | *15* | *6* | 0 | 0 | 0 | 0 | 0 | 0 | 0 | *3* | *14* | 57 |
| Heating Degree Days | na | *1534* | *1238* | *730* | *388* | *143* | *47* | *91* | na | *650* | na | na | na |
| Cooling Degree Days | na | na | na | na | na | na | na | na | na | na | na | na | na |
| Total Precipitation (") | 0.83 | 0.51 | 1.18 | 1.84 | 2.85 | 3.99 | 3.94 | 3.64 | 3.20 | 2.41 | 1.19 | 0.82 | 26.40 |
| Days ≥ 0.1" Precip | *3* | *2* | *3* | 4 | 7 | 8 | 6 | 6 | 6 | 5 | *3* | *3* | 56 |
| Total Snowfall (") | *10.6* | *5.4* | *8.1* | *3.8* | 0.3 | 0.0 | 0.0 | 0.0 | 0.0 | 1.1 | na | *9.8* | na |
| Days ≥ 1" Snow Depth | *20* | *17* | *17* | *5* | 0 | 0 | 0 | 0 | 0 | 0 | *9* | *18* | 86 |

## LAMBERTON SW EXP STN *Redwood County* ELEVATION 1142 ft LAT/LONG 44° 15' N / 95° 19' W

| | JAN | FEB | MAR | APR | MAY | JUN | JUL | AUG | SEP | OCT | NOV | DEC | YEAR |
|---|---|---|---|---|---|---|---|---|---|---|---|---|---|
| Maximum Temp °F | 22.0 | 27.1 | 39.9 | 56.7 | 71.4 | 80.5 | 83.7 | 81.0 | 72.7 | 60.3 | 41.3 | 27.0 | 55.3 |
| Minimum Temp °F | 1.6 | 7.0 | 21.0 | 34.1 | 46.0 | 56.2 | 60.2 | 57.1 | 47.2 | 35.4 | 22.2 | 8.4 | 33.0 |
| Mean Temp °F | 11.8 | 17.1 | 30.5 | 45.4 | 58.7 | 68.3 | 72.0 | 69.1 | 60.0 | 47.9 | 31.8 | 17.7 | 44.2 |
| Days Max Temp ≥ 90 °F | 0 | 0 | 0 | 0 | 2 | 5 | 6 | 4 | 1 | 0 | 0 | 0 | 18 |
| Days Max Temp ≤ 32 °F | 23 | 17 | 9 | 1 | 0 | 0 | 0 | 0 | 0 | 0 | 7 | 19 | 76 |
| Days Min Temp ≤ 32 °F | 31 | 28 | 27 | 14 | 2 | 0 | 0 | 0 | 2 | 12 | 26 | 31 | 173 |
| Days Min Temp ≤ 0 °F | 15 | 10 | 2 | 0 | 0 | 0 | 0 | 0 | 0 | 0 | 1 | 9 | 37 |
| Heating Degree Days | 1646 | 1349 | 1064 | 586 | 238 | 46 | 12 | 36 | 195 | 529 | 990 | 1461 | 8152 |
| Cooling Degree Days | 0 | 0 | 0 | 7 | 54 | 156 | 227 | 162 | 48 | 4 | 0 | 0 | 658 |
| Total Precipitation (") | 0.65 | 0.58 | 1.63 | 2.90 | 3.10 | 3.89 | 3.39 | 3.08 | 3.05 | 2.12 | 1.29 | 0.70 | 26.38 |
| Days ≥ 0.1" Precip | 2 | 2 | 4 | 6 | 7 | 7 | 6 | 5 | 6 | 4 | 3 | 2 | 54 |
| Total Snowfall (") | 9.1 | 7.4 | 7.6 | 2.5 | 0.0 | 0.0 | 0.0 | 0.0 | 0.0 | 0.4 | 6.3 | 8.3 | 41.6 |
| Days ≥ 1" Snow Depth | 24 | 20 | 10 | 2 | 0 | 0 | 0 | 0 | 0 | 0 | 7 | 18 | 81 |

## LEECH LAKE DAM *Cass County* ELEVATION 1302 ft LAT/LONG 47° 15' N / 94° 13' W

| | JAN | FEB | MAR | APR | MAY | JUN | JUL | AUG | SEP | OCT | NOV | DEC | YEAR |
|---|---|---|---|---|---|---|---|---|---|---|---|---|---|
| Maximum Temp °F | 17.1 | 25.8 | 38.1 | 53.9 | 67.5 | 75.5 | 80.0 | 77.9 | 67.5 | 55.1 | 36.1 | 22.5 | 51.4 |
| Minimum Temp °F | -5.1 | 0.9 | 14.6 | 29.3 | 41.8 | 51.5 | 56.9 | 54.1 | 45.1 | 34.3 | 19.8 | 4.0 | 28.9 |
| Mean Temp °F | 5.9 | 13.7 | 26.4 | 41.7 | 54.6 | 63.5 | 68.5 | 66.0 | 56.3 | 44.7 | 28.0 | 13.3 | 40.2 |
| Days Max Temp ≥ 90 °F | 0 | 0 | 0 | 0 | 0 | 1 | 2 | 1 | 0 | 0 | 0 | 0 | 4 |
| Days Max Temp ≤ 32 °F | 27 | 19 | 10 | 1 | 0 | 0 | 0 | 0 | 0 | 0 | 12 | 25 | 94 |
| Days Min Temp ≤ 32 °F | 31 | 28 | 29 | 20 | 5 | 0 | 0 | 0 | 2 | 14 | 27 | 31 | 187 |
| Days Min Temp ≤ 0 °F | 19 | 14 | 6 | 0 | 0 | 0 | 0 | 0 | 0 | 0 | 2 | 13 | 54 |
| Heating Degree Days | 1830 | 1446 | 1188 | 695 | 334 | 100 | 28 | 62 | 274 | 621 | 1103 | 1599 | 9280 |
| Cooling Degree Days | 0 | 0 | 0 | 2 | 22 | 64 | 141 | 105 | 16 | 0 | 0 | 0 | 350 |
| Total Precipitation (") | 0.78 | 0.45 | 1.10 | 1.87 | 2.59 | 3.78 | 3.85 | 3.47 | 2.73 | 2.45 | 1.09 | 0.84 | 25.00 |
| Days ≥ 0.1" Precip | 3 | 1 | 4 | 5 | 6 | 8 | 7 | 6 | 6 | 4 | 3 | 3 | 56 |
| Total Snowfall (") | 13.6 | 6.4 | 9.0 | 3.7 | 0.1 | 0.0 | 0.0 | 0.0 | 0.0 | 1.0 | 6.9 | 10.9 | 51.6 |
| Days ≥ 1" Snow Depth | 30 | 25 | 23 | 6 | 0 | 0 | 0 | 0 | 0 | 1 | 12 | 27 | 124 |

## LITCHFIELD *Meeker County* ELEVATION 1132 ft LAT/LONG 45° 8' N / 94° 30' W

| | JAN | FEB | MAR | APR | MAY | JUN | JUL | AUG | SEP | OCT | NOV | DEC | YEAR |
|---|---|---|---|---|---|---|---|---|---|---|---|---|---|
| Maximum Temp °F | 20.8 | 26.9 | 39.9 | 57.6 | 71.5 | 79.9 | 84.1 | 81.3 | 72.0 | 59.8 | 39.7 | 25.7 | 54.9 |
| Minimum Temp °F | 1.0 | 7.0 | 20.7 | 34.7 | 46.8 | 56.3 | 61.1 | 58.5 | 48.6 | 37.4 | 23.0 | 8.8 | 33.7 |
| Mean Temp °F | 10.9 | 17.0 | 30.3 | 46.2 | 59.2 | 68.2 | 72.6 | 69.9 | 60.4 | 48.6 | 31.4 | 17.2 | 44.3 |
| Days Max Temp ≥ 90 °F | 0 | 0 | 0 | 0 | 1 | 3 | 7 | 4 | 1 | 0 | 0 | 0 | 16 |
| Days Max Temp ≤ 32 °F | 24 | 17 | 8 | 0 | 0 | 0 | 0 | 0 | 0 | 0 | 8 | 22 | 79 |
| Days Min Temp ≤ 32 °F | 31 | 28 | 26 | 13 | 2 | 0 | 0 | 0 | 1 | 10 | 26 | 31 | 168 |
| Days Min Temp ≤ 0 °F | 15 | 10 | 3 | 0 | 0 | 0 | 0 | 0 | 0 | 0 | 1 | 9 | 38 |
| Heating Degree Days | 1673 | 1351 | 1069 | 563 | 221 | 43 | 7 | 26 | 182 | 505 | 1002 | 1475 | 8117 |
| Cooling Degree Days | 0 | 0 | 0 | 7 | 53 | 146 | 239 | 178 | 46 | 2 | 0 | 0 | 671 |
| Total Precipitation (") | 0.79 | 0.72 | 1.48 | 2.57 | 3.09 | 5.00 | 3.87 | 3.44 | 3.01 | 2.26 | 1.36 | 0.77 | 28.36 |
| Days ≥ 0.1" Precip | 3 | 2 | 3 | 5 | 7 | 8 | 6 | 6 | 5 | 4 | 3 | 2 | 54 |
| Total Snowfall (") | 9.9 | 8.0 | 8.1 | 1.8 | 0.0 | 0.0 | 0.0 | 0.0 | 0.0 | 0.2 | *8.0* | *7.7* | 43.7 |
| Days ≥ 1" Snow Depth | na | *18* | *16* | 1 | 0 | 0 | 0 | 0 | 0 | 0 | na | na | na |

## LITTLE FALLS 1 N *Morrison County* ELEVATION 1122 ft LAT/LONG 45° 58' N / 94° 21' W

| | JAN | FEB | MAR | APR | MAY | JUN | JUL | AUG | SEP | OCT | NOV | DEC | YEAR |
|---|---|---|---|---|---|---|---|---|---|---|---|---|---|
| Maximum Temp °F | 19.6 | 26.3 | 39.3 | 56.8 | 70.8 | 78.7 | 83.4 | 80.9 | 71.0 | 58.2 | 38.5 | 24.4 | 54.0 |
| Minimum Temp °F | -2.3 | 3.7 | 17.3 | 32.1 | 44.2 | 53.7 | 58.9 | 56.4 | 46.8 | 35.7 | 21.4 | 6.0 | 31.2 |
| Mean Temp °F | 8.7 | 15.0 | 28.3 | 44.5 | 57.5 | 66.2 | 71.2 | 68.7 | 58.9 | 47.0 | 30.0 | 15.2 | 42.6 |
| Days Max Temp ≥ 90 °F | 0 | 0 | 0 | 0 | 1 | 2 | 6 | 4 | 1 | 0 | 0 | 0 | 14 |
| Days Max Temp ≤ 32 °F | 26 | 18 | 8 | 0 | 0 | 0 | 0 | 0 | 0 | 0 | 9 | 23 | 84 |
| Days Min Temp ≤ 32 °F | 31 | 28 | 28 | 16 | 3 | 0 | 0 | 0 | 2 | 12 | 26 | 31 | 177 |
| Days Min Temp ≤ 0 °F | 17 | 12 | 4 | 0 | 0 | 0 | 0 | 0 | 0 | 0 | 2 | 11 | 46 |
| Heating Degree Days | 1744 | 1407 | 1131 | 612 | 258 | 60 | 14 | 35 | 213 | 553 | 1044 | 1539 | 8610 |
| Cooling Degree Days | 0 | 0 | 0 | 4 | 41 | 113 | 223 | 167 | 37 | 1 | 0 | 0 | 586 |
| Total Precipitation (") | 0.78 | 0.54 | 1.51 | 2.19 | 2.91 | 4.34 | 3.55 | 3.40 | 2.75 | 2.56 | 1.36 | 0.75 | 26.64 |
| Days ≥ 0.1" Precip | 3 | 2 | 4 | 5 | 7 | 8 | 6 | 6 | 5 | 4 | 4 | 3 | 57 |
| Total Snowfall (") | 12.0 | 7.4 | 9.5 | 2.2 | 0.1 | 0.0 | 0.0 | 0.0 | 0.0 | 0.6 | 8.0 | 9.4 | 49.2 |
| Days ≥ 1" Snow Depth | *21* | 21 | 19 | 3 | 0 | 0 | 0 | 0 | 0 | 0 | 7 | 18 | 89 |

## LONG PRAIRIE *Todd County* ELEVATION 1289 ft LAT/LONG 45° 59' N / 94° 51' W

| | JAN | FEB | MAR | APR | MAY | JUN | JUL | AUG | SEP | OCT | NOV | DEC | YEAR |
|---|---|---|---|---|---|---|---|---|---|---|---|---|---|
| Maximum Temp °F | 18.6 | 25.5 | 38.0 | 55.8 | 69.7 | 77.7 | 82.7 | 80.1 | 70.0 | 57.7 | 37.9 | 23.7 | 53.1 |
| Minimum Temp °F | -2.2 | 3.4 | 17.2 | 32.3 | 44.3 | 53.6 | 58.3 | 55.8 | 46.2 | 35.2 | 21.0 | 5.9 | 30.9 |
| Mean Temp °F | 8.2 | 14.7 | 27.7 | 44.1 | 56.9 | 65.7 | 70.5 | 68.0 | 58.1 | 46.5 | 29.5 | 14.8 | 42.1 |
| Days Max Temp ≥ 90 °F | 0 | 0 | 0 | 0 | 0 | 2 | 5 | 3 | 0 | 0 | 0 | 0 | 10 |
| Days Max Temp ≤ 32 °F | 26 | 18 | 9 | 1 | 0 | 0 | 0 | 0 | 0 | 0 | 10 | 23 | 87 |
| Days Min Temp ≤ 32 °F | 31 | 28 | 28 | 16 | 4 | 0 | 0 | 0 | 2 | 13 | 27 | 31 | 180 |
| Days Min Temp ≤ 0 °F | 17 | 12 | 4 | 0 | 0 | 0 | 0 | 0 | 0 | 0 | 2 | 11 | 46 |
| Heating Degree Days | 1758 | 1415 | 1152 | 624 | 276 | 71 | 17 | 45 | 233 | 570 | 1059 | 1553 | 8773 |
| Cooling Degree Days | 0 | 0 | 0 | 4 | 38 | 105 | 196 | 152 | 33 | 1 | 0 | 0 | 529 |
| Total Precipitation (") | 1.17 | 0.84 | 2.02 | 2.50 | 2.98 | 4.32 | 3.74 | 3.24 | 2.88 | 2.47 | 1.60 | 1.00 | 28.76 |
| Days ≥ 0.1" Precip | 4 | 3 | 5 | 6 | 6 | 8 | 6 | 6 | 5 | 4 | 4 | 3 | 60 |
| Total Snowfall (") | 12.0 | 7.9 | 10.2 | 2.9 | 0.0 | 0.0 | 0.0 | 0.0 | 0.0 | 0.8 | 7.8 | 9.0 | 50.6 |
| Days ≥ 1" Snow Depth | 30 | 27 | 23 | 5 | 0 | 0 | 0 | 0 | 0 | 1 | 10 | 26 | 122 |

## LUVERNE *Rock County* ELEVATION 1500 ft LAT/LONG 43° 39' N / 96° 12' W

| | JAN | FEB | MAR | APR | MAY | JUN | JUL | AUG | SEP | OCT | NOV | DEC | YEAR |
|---|---|---|---|---|---|---|---|---|---|---|---|---|---|
| Maximum Temp °F | 23.6 | 29.4 | 43.1 | 59.6 | 72.2 | 81.6 | 85.3 | 82.8 | 74.1 | 61.6 | 41.7 | 28.4 | 57.0 |
| Minimum Temp °F | 3.4 | 9.0 | 22.0 | 34.4 | 45.5 | 56.1 | 60.5 | 57.6 | 48.0 | 35.9 | 22.2 | 9.5 | 33.7 |
| Mean Temp °F | 13.5 | 19.2 | 32.6 | 47.0 | 58.9 | 68.9 | 73.0 | 70.2 | 61.1 | 48.7 | 32.0 | 19.0 | 45.3 |
| Days Max Temp ≥ 90 °F | 0 | 0 | 0 | 0 | 1 | 4 | 9 | 5 | 2 | 0 | 0 | 0 | 21 |
| Days Max Temp ≤ 32 °F | 22 | 16 | 6 | 0 | 0 | 0 | 0 | 0 | 0 | 0 | 7 | 19 | 70 |
| Days Min Temp ≤ 32 °F | 31 | 28 | 26 | 13 | 2 | 0 | 0 | 0 | 2 | 12 | 26 | 31 | 171 |
| Days Min Temp ≤ 0 °F | 13 | 9 | 2 | 0 | 0 | 0 | 0 | 0 | 0 | 0 | 1 | 8 | 33 |
| Heating Degree Days | 1594 | 1287 | 997 | 536 | 221 | 34 | 8 | 24 | 169 | 500 | 984 | 1421 | 7775 |
| Cooling Degree Days | 0 | 0 | 0 | 6 | 44 | 167 | 247 | 190 | 56 | 2 | 0 | 0 | 712 |
| Total Precipitation (") | 0.62 | 0.73 | 1.94 | 2.69 | 3.11 | 4.43 | 3.66 | 3.29 | 3.07 | 2.22 | 1.46 | 0.84 | 28.06 |
| Days ≥ 0.1" Precip | 2 | 2 | 4 | 6 | 7 | 7 | 6 | 5 | 6 | 4 | 4 | 2 | 55 |
| Total Snowfall (") | 9.6 | 7.0 | 8.9 | 1.5 | 0.0 | 0.0 | 0.0 | 0.0 | 0.0 | 0.9 | 6.9 | 8.6 | 43.4 |
| Days ≥ 1" Snow Depth | 25 | 21 | 12 | 1 | 0 | 0 | 0 | 0 | 0 | 0 | 7 | 19 | 85 |

## MADISON SEWAGE PLANT *Lac qui Parle County* ELEVATION 1102 ft LAT/LONG 45° 1' N / 96° 11' W

| | JAN | FEB | MAR | APR | MAY | JUN | JUL | AUG | SEP | OCT | NOV | DEC | YEAR |
|---|---|---|---|---|---|---|---|---|---|---|---|---|---|
| Maximum Temp °F | 22.8 | 28.6 | 41.5 | 58.7 | 72.8 | 81.3 | 85.9 | 83.8 | 74.6 | 62.5 | 42.5 | 28.6 | 57.0 |
| Minimum Temp °F | 1.6 | 7.6 | 20.9 | 33.9 | 45.9 | 56.1 | 60.3 | 57.6 | 47.1 | 36.3 | 22.4 | 9.0 | 33.2 |
| Mean Temp °F | 12.2 | 18.1 | 31.2 | 46.3 | 59.4 | 68.7 | 73.1 | 70.7 | 61.0 | 49.4 | 32.4 | 18.8 | 45.1 |
| Days Max Temp ≥ 90 °F | 0 | 0 | 0 | 0 | 1 | 4 | 10 | 8 | 2 | 0 | 0 | 0 | 25 |
| Days Max Temp ≤ 32 °F | 22 | 16 | 7 | 0 | 0 | 0 | 0 | 0 | 0 | 0 | 6 | 18 | 69 |
| Days Min Temp ≤ 32 °F | 31 | 28 | 27 | 14 | 2 | 0 | 0 | 0 | 1 | 11 | 26 | 31 | 171 |
| Days Min Temp ≤ 0 °F | 15 | 10 | 2 | 0 | 0 | 0 | 0 | 0 | 0 | 0 | 1 | 9 | 37 |
| Heating Degree Days | 1631 | 1318 | 1040 | 558 | 214 | 38 | 7 | 23 | 172 | 481 | 970 | 1424 | 7876 |
| Cooling Degree Days | 0 | 0 | 0 | 7 | 58 | 167 | 267 | 219 | *61* | *6* | 0 | 0 | 785 |
| Total Precipitation (") | 0.67 | 0.65 | 1.46 | 2.36 | 2.76 | 4.37 | 3.28 | 2.78 | 2.25 | 2.35 | 1.27 | 0.55 | 24.75 |
| Days ≥ 0.1" Precip | 2 | 2 | 4 | *6* | 6 | 7 | 6 | 5 | 5 | 4 | 3 | 2 | 52 |
| Total Snowfall (") | 8.1 | 7.3 | 8.1 | 1.7 | 0.0 | 0.0 | 0.0 | 0.0 | 0.0 | 0.5 | *5.5* | *5.6* | 36.8 |
| Days ≥ 1" Snow Depth | na | na | na | *1* | 0 | 0 | 0 | 0 | 0 | 0 | na | na | na |

## MAHNOMEN 1 W *Mahnomen County* ELEVATION 1211 ft LAT/LONG 47° 19' N / 95° 58' W

| | JAN | FEB | MAR | APR | MAY | JUN | JUL | AUG | SEP | OCT | NOV | DEC | YEAR |
|---|---|---|---|---|---|---|---|---|---|---|---|---|---|
| Maximum Temp °F | 14.7 | 21.4 | 35.2 | 54.0 | 69.0 | 75.8 | 80.8 | 79.8 | 69.2 | 55.8 | 35.3 | 21.0 | 51.0 |
| Minimum Temp °F | -6.2 | 0.4 | 15.7 | 31.2 | 42.9 | 52.0 | 56.2 | 54.3 | 44.9 | 33.8 | 18.8 | 2.7 | 28.9 |
| Mean Temp °F | 4.3 | 10.9 | 25.4 | 42.6 | 56.0 | 63.9 | 68.5 | 67.1 | 57.1 | 44.9 | 27.1 | 11.9 | 40.0 |
| Days Max Temp ≥ 90 °F | 0 | 0 | 0 | 0 | 0 | 1 | 3 | 4 | 1 | 0 | 0 | 0 | 9 |
| Days Max Temp ≤ 32 °F | 28 | 21 | 12 | 1 | 0 | 0 | 0 | 0 | 0 | 1 | 12 | 25 | 100 |
| Days Min Temp ≤ 32 °F | 31 | 28 | 28 | 18 | 5 | 0 | 0 | 0 | 3 | 14 | 27 | 31 | 185 |
| Days Min Temp ≤ 0 °F | 20 | 14 | 6 | 0 | 0 | 0 | 0 | 0 | 0 | 0 | 3 | 14 | 57 |
| Heating Degree Days | 1881 | 1524 | 1220 | 667 | 300 | 97 | 30 | 57 | 264 | 620 | 1131 | 1642 | 9433 |
| Cooling Degree Days | 0 | 0 | 0 | 3 | 35 | 73 | 149 | 136 | 28 | 1 | 0 | 0 | 425 |
| Total Precipitation (") | 0.94 | 0.61 | 1.22 | 1.68 | 2.37 | 4.15 | 3.11 | 3.35 | 2.37 | 2.11 | 0.93 | 0.83 | 23.67 |
| Days ≥ 0.1" Precip | 3 | 2 | 4 | 5 | 6 | 8 | 6 | 6 | 5 | 5 | 3 | 3 | 56 |
| Total Snowfall (") | 10.4 | 6.2 | 7.5 | 2.8 | 0.2 | 0.0 | 0.0 | 0.0 | 0.0 | 1.8 | 7.1 | 8.4 | 44.4 |
| Days ≥ 1" Snow Depth | 31 | 28 | *21* | 3 | 0 | 0 | 0 | 0 | 0 | 1 | 12 | 27 | 123 |

## MARSHALL *Lyon County* ELEVATION 1171 ft LAT/LONG 44° 27 ' N / 95° 47 ' W

| | JAN | FEB | MAR | APR | MAY | JUN | JUL | AUG | SEP | OCT | NOV | DEC | YEAR |
|---|---|---|---|---|---|---|---|---|---|---|---|---|---|
| Maximum Temp °F | 22.1 | 27.3 | 40.2 | 56.7 | 70.7 | 79.9 | ***83.7*** | 81.5 | 72.0 | 59.6 | 40.9 | 26.5 | 55.1 |
| Minimum Temp °F | 2.1 | 7.8 | 21.3 | 34.9 | 46.9 | 57.0 | 61.5 | 58.6 | 48.7 | 36.5 | 23.0 | 9.3 | 34.0 |
| Mean Temp °F | 12.2 | 17.4 | 30.8 | 45.9 | 58.8 | 68.6 | ***72.6*** | 70.0 | 60.4 | 48.1 | 32.0 | 17.7 | 44.5 |
| Days Max Temp ≥ 90 °F | 0 | 0 | 0 | 0 | 1 | 4 | 7 | 5 | 1 | 0 | 0 | 0 | 18 |
| Days Max Temp ≤ 32 °F | 22 | 17 | 8 | 0 | 0 | 0 | 0 | 0 | 0 | 0 | 8 | 20 | 75 |
| Days Min Temp ≤ 32 °F | 31 | 28 | 26 | 13 | 2 | 0 | 0 | 0 | 1 | 11 | 25 | 31 | 168 |
| Days Min Temp ≤ 0 °F | 15 | 10 | 2 | 0 | 0 | 0 | 0 | 0 | 0 | 0 | 1 | 9 | 37 |
| Heating Degree Days | 1635 | 1338 | 1055 | 571 | 230 | 40 | ***10*** | 28 | 188 | 520 | 983 | 1461 | 8059 |
| Cooling Degree Days | ***0*** | ***0*** | 0 | 7 | 54 | ***169*** | ***235*** | ***177*** | 60 | 4 | 0 | 0 | 706 |
| Total Precipitation (") | 0.72 | 0.69 | 1.70 | 2.51 | 2.98 | 3.96 | 3.57 | 3.01 | 2.55 | 1.98 | 1.43 | 0.85 | 25.95 |
| Days ≥ 0.1" Precip | 2 | 2 | 4 | 6 | 6 | 7 | 6 | 5 | 5 | 4 | 3 | 2 | 52 |
| Total Snowfall (") | 8.4 | 7.4 | 8.4 | 1.7 | 0.0 | 0.0 | 0.0 | 0.0 | 0.0 | 0.7 | 6.9 | 9.3 | 42.8 |
| Days ≥ 1" Snow Depth | 16 | ***19*** | 8 | 2 | 0 | 0 | 0 | 0 | 0 | 0 | ***4*** | ***14*** | 63 |

## MELROSE *Stearns County* ELEVATION 1211 ft LAT/LONG 45° 40 ' N / 94° 49 ' W

| | JAN | FEB | MAR | APR | MAY | JUN | JUL | AUG | SEP | OCT | NOV | DEC | YEAR |
|---|---|---|---|---|---|---|---|---|---|---|---|---|---|
| Maximum Temp °F | 19.0 | 25.5 | 38.4 | 56.4 | 70.5 | 78.7 | 83.6 | 80.8 | 70.7 | 58.6 | 38.2 | 24.1 | 53.7 |
| Minimum Temp °F | -1.7 | 4.6 | 18.7 | 32.9 | 45.0 | 54.4 | 59.2 | 56.7 | 47.1 | 35.9 | 21.1 | 6.5 | 31.7 |
| Mean Temp °F | 8.7 | 15.1 | 28.5 | 44.7 | 57.8 | 66.6 | 71.4 | 68.8 | 58.9 | 47.3 | 29.7 | 15.3 | 42.7 |
| Days Max Temp ≥ 90 °F | 0 | 0 | 0 | 0 | 0 | 2 | 5 | 3 | 1 | 0 | 0 | 0 | 11 |
| Days Max Temp ≤ 32 °F | 26 | 18 | 9 | 0 | 0 | 0 | 0 | 0 | 0 | 0 | 9 | 23 | 85 |
| Days Min Temp ≤ 32 °F | 31 | 28 | 28 | 16 | 3 | 0 | 0 | 0 | 2 | 12 | 27 | 31 | 178 |
| Days Min Temp ≤ 0 °F | 17 | 11 | 3 | 0 | 0 | 0 | 0 | 0 | 0 | 0 | 2 | 11 | 44 |
| Heating Degree Days | 1743 | 1405 | 1123 | 606 | 252 | 58 | 11 | 34 | 214 | 544 | 1054 | 1537 | 8581 |
| Cooling Degree Days | 0 | 0 | 0 | 4 | 42 | 120 | 224 | 166 | 39 | 1 | 0 | 0 | 596 |
| Total Precipitation (") | 0.79 | 0.68 | 1.57 | 2.54 | 3.17 | 4.57 | 3.09 | 3.32 | 2.94 | 2.37 | 1.38 | 0.76 | 27.18 |
| Days ≥ 0.1" Precip | 3 | 2 | 4 | 6 | 6 | 7 | 6 | 6 | 6 | 4 | 3 | 3 | 56 |
| Total Snowfall (") | 10.9 | 8.2 | 8.0 | 1.7 | 0.0 | 0.0 | 0.0 | 0.0 | 0.0 | 0.4 | 6.8 | ***6.6*** | 42.6 |
| Days ≥ 1" Snow Depth | na | na | na | 1 | 0 | 0 | 0 | 0 | 0 | 0 | na | na | na |

## MILACA 1 ENE *Mille Lacs County* ELEVATION 1070 ft LAT/LONG 45° 45 ' N / 93° 39 ' W

| | JAN | FEB | MAR | APR | MAY | JUN | JUL | AUG | SEP | OCT | NOV | DEC | YEAR |
|---|---|---|---|---|---|---|---|---|---|---|---|---|---|
| Maximum Temp °F | 19.1 | 25.7 | 37.9 | 55.1 | 68.6 | 77.0 | 81.8 | 79.2 | 69.3 | 56.9 | 38.4 | 24.3 | 52.8 |
| Minimum Temp °F | -2.0 | 4.4 | 18.1 | 32.4 | 43.9 | 53.2 | 58.3 | 55.4 | 45.9 | 34.7 | 21.3 | 6.2 | 31.0 |
| Mean Temp °F | 8.5 | 15.1 | 28.0 | 43.8 | 56.3 | 65.1 | 70.1 | 67.3 | 57.6 | 45.9 | 29.8 | 15.2 | 41.9 |
| Days Max Temp ≥ 90 °F | 0 | 0 | 0 | 0 | 0 | 2 | 4 | 3 | 0 | 0 | 0 | 0 | 9 |
| Days Max Temp ≤ 32 °F | 26 | 18 | 9 | 0 | 0 | 0 | 0 | 0 | 0 | 0 | 8 | 23 | 84 |
| Days Min Temp ≤ 32 °F | 30 | 28 | 28 | 16 | 3 | 0 | 0 | 0 | 2 | 13 | 26 | 31 | 177 |
| Days Min Temp ≤ 0 °F | 16 | 12 | 3 | 0 | 0 | 0 | 0 | 0 | 0 | 0 | 1 | 11 | 43 |
| Heating Degree Days | 1750 | 1405 | 1140 | 633 | 285 | 77 | 19 | 47 | 242 | 587 | 1048 | 1537 | 8770 |
| Cooling Degree Days | 0 | 0 | 0 | 5 | 27 | 102 | 193 | 134 | 26 | 1 | 0 | 0 | 488 |
| Total Precipitation (") | 0.85 | 0.68 | 1.51 | 2.33 | 3.11 | 4.66 | 3.88 | 3.91 | 3.23 | 2.47 | 1.53 | 1.01 | 29.17 |
| Days ≥ 0.1" Precip | 3 | 2 | 4 | 5 | 7 | 8 | 7 | 6 | 6 | 5 | 4 | 3 | 60 |
| Total Snowfall (") | ***11.4*** | 6.7 | ***6.5*** | 1.5 | 0.0 | 0.0 | 0.0 | 0.0 | 0.0 | 0.4 | ***6.0*** | ***9.6*** | 42.1 |
| Days ≥ 1" Snow Depth | na | na | na | ***2*** | 0 | 0 | 0 | 0 | 0 | 0 | na | na | na |

## MILAN 1 NW *Chippewa County* ELEVATION 1010 ft LAT/LONG 45° 7 ' N / 95° 56 ' W

| | JAN | FEB | MAR | APR | MAY | JUN | JUL | AUG | SEP | OCT | NOV | DEC | YEAR |
|---|---|---|---|---|---|---|---|---|---|---|---|---|---|
| Maximum Temp °F | 20.6 | 26.6 | 39.7 | 57.2 | 71.2 | 79.6 | 84.4 | 81.8 | 72.0 | 59.2 | 39.7 | 25.7 | 54.8 |
| Minimum Temp °F | -1.1 | 5.4 | 19.6 | 33.5 | 45.5 | 54.8 | 59.1 | 56.3 | 46.3 | 34.7 | 20.4 | 6.2 | 31.7 |
| Mean Temp °F | 9.8 | 16.0 | 29.7 | 45.4 | 58.4 | 67.2 | 71.8 | 69.1 | 59.2 | 47.0 | 30.0 | 16.0 | 43.3 |
| Days Max Temp ≥ 90 °F | 0 | 0 | 0 | 0 | 1 | 4 | 8 | 5 | 1 | 0 | 0 | 0 | 19 |
| Days Max Temp ≤ 32 °F | 23 | 17 | 8 | 0 | 0 | 0 | 0 | 0 | 0 | 0 | 8 | 21 | 77 |
| Days Min Temp ≤ 32 °F | 31 | 28 | 28 | 15 | 3 | 0 | 0 | 0 | 2 | 13 | 27 | 31 | 178 |
| Days Min Temp ≤ 0 °F | 17 | 11 | 3 | 0 | 0 | 0 | 0 | 0 | 0 | 0 | 2 | 11 | 44 |
| Heating Degree Days | 1709 | 1379 | 1089 | 586 | 238 | 52 | 11 | 35 | 212 | 552 | 1042 | 1514 | 8419 |
| Cooling Degree Days | 0 | 0 | 0 | 7 | 50 | 128 | 220 | 164 | 42 | 2 | 0 | 0 | 613 |
| Total Precipitation (") | 0.69 | 0.69 | 1.38 | 2.36 | 2.70 | 4.04 | 3.34 | 2.96 | 2.43 | 2.17 | 1.09 | 0.59 | 24.44 |
| Days ≥ 0.1" Precip | 3 | 2 | 4 | 5 | 6 | 7 | 6 | 5 | 5 | 4 | 3 | 2 | 52 |
| Total Snowfall (") | 9.9 | 8.5 | 8.4 | 2.1 | 0.1 | 0.0 | 0.0 | 0.0 | 0.0 | 0.6 | 5.9 | 7.1 | 42.6 |
| Days ≥ 1" Snow Depth | 26 | 23 | 16 | 2 | 0 | 0 | 0 | 0 | 0 | 0 | 7 | 21 | 95 |

## MINNEAPOLIS INTL AP *Hennepin County* ELEVATION 834 ft LAT/LONG 44° 53' N / 93° 13' W

| | JAN | FEB | MAR | APR | MAY | JUN | JUL | AUG | SEP | OCT | NOV | DEC | YEAR |
|---|---|---|---|---|---|---|---|---|---|---|---|---|---|
| Maximum Temp °F | 20.9 | 26.7 | 39.6 | 56.6 | 69.6 | 78.6 | 83.3 | 80.3 | 70.5 | 58.0 | 40.1 | 26.4 | 54.2 |
| Minimum Temp °F | 3.1 | 9.4 | 22.8 | 36.3 | 47.9 | 57.7 | 62.9 | 60.3 | 50.3 | 38.5 | 24.9 | 11.1 | 35.4 |
| Mean Temp °F | 12.0 | 18.1 | 31.2 | 46.5 | 58.8 | 68.2 | 73.2 | 70.3 | 60.5 | 48.3 | 32.5 | 18.8 | 44.9 |
| Days Max Temp ≥ 90 °F | 0 | 0 | 0 | 0 | 1 | 3 | 6 | 4 | 1 | 0 | 0 | 0 | 15 |
| Days Max Temp ≤ 32 °F | 24 | 18 | 8 | 0 | 0 | 0 | 0 | 0 | 0 | 0 | 7 | 20 | 77 |
| Days Min Temp ≤ 32 °F | 31 | 27 | 25 | 10 | 1 | 0 | 0 | 0 | 1 | 8 | 24 | 30 | 157 |
| Days Min Temp ≤ 0 °F | 14 | 8 | 2 | 0 | 0 | 0 | 0 | 0 | 0 | 0 | 1 | 7 | 32 |
| Heating Degree Days | 1639 | 1320 | 1040 | 553 | 229 | 41 | 7 | 23 | 182 | 515 | 968 | 1427 | 7944 |
| Cooling Degree Days | 0 | 0 | 0 | 6 | 46 | 150 | 268 | 194 | 52 | 1 | 0 | 0 | 717 |
| Total Precipitation (") | 1.01 | 0.86 | 1.88 | 2.51 | 3.19 | 4.32 | 3.74 | 3.76 | 2.83 | 2.32 | 1.78 | 1.07 | 29.27 |
| Days ≥ 0.1" Precip | 3 | 3 | 5 | 6 | 7 | 7 | 6 | 6 | 6 | 5 | 4 | 3 | 61 |
| Total Snowfall (") | 13.4 | 9.0 | 10.6 | 3.1 | 0.1 | 0.0 | 0.0 | 0.0 | 0.0 | 0.7 | 9.3 | 11.0 | 57.2 |
| Days ≥ 1" Snow Depth | 28 | 25 | 16 | 2 | 0 | 0 | 0 | 0 | 0 | 0 | 7 | 21 | 99 |

## MONTEVIDEO 1 SW *Lac qui Parle County* ELEVATION 902 ft LAT/LONG 44° 57' N / 95° 43' W

| | JAN | FEB | MAR | APR | MAY | JUN | JUL | AUG | SEP | OCT | NOV | DEC | YEAR |
|---|---|---|---|---|---|---|---|---|---|---|---|---|---|
| Maximum Temp °F | 21.0 | 27.0 | 40.4 | 58.2 | 72.1 | 80.4 | 85.0 | 82.7 | 73.2 | 60.6 | 40.4 | 26.5 | 55.6 |
| Minimum Temp °F | 0.9 | 7.0 | 21.3 | 34.6 | 46.7 | 56.1 | 60.3 | 57.6 | 47.7 | 36.5 | 22.4 | 8.8 | 33.3 |
| Mean Temp °F | 11.0 | 17.0 | 30.9 | 46.4 | 59.4 | 68.3 | 72.7 | 70.1 | 60.4 | 48.6 | 31.4 | 17.7 | 44.5 |
| Days Max Temp ≥ 90 °F | 0 | 0 | 0 | 0 | 1 | 4 | 9 | 6 | 2 | 0 | 0 | 0 | 22 |
| Days Max Temp ≤ 32 °F | 23 | 17 | 8 | 0 | 0 | 0 | 0 | 0 | 0 | 0 | 8 | 20 | 76 |
| Days Min Temp ≤ 32 °F | 31 | 28 | 26 | 13 | 2 | 0 | 0 | 0 | 1 | 11 | 25 | 31 | 168 |
| Days Min Temp ≤ 0 °F | 16 | 10 | 2 | 0 | 0 | 0 | 0 | 0 | 0 | 0 | 1 | 10 | 39 |
| Heating Degree Days | 1672 | 1350 | 1050 | 557 | 216 | 41 | 10 | 28 | 182 | 507 | 1000 | 1461 | 8074 |
| Cooling Degree Days | 0 | 0 | 0 | 9 | 58 | 151 | 258 | 199 | 49 | 4 | 0 | 0 | 728 |
| Total Precipitation (") | 1.02 | 1.11 | 1.62 | 2.46 | 3.16 | 4.47 | 3.28 | 3.08 | 2.55 | 1.99 | 1.39 | 0.84 | 26.97 |
| Days ≥ 0.1" Precip | 4 | 3 | 4 | 5 | 7 | 7 | 5 | 5 | 5 | 4 | 4 | 2 | 55 |
| Total Snowfall (") | 10.6 | *10.2* | 10.1 | 1.3 | 0.0 | 0.0 | 0.0 | 0.0 | 0.0 | 0.5 | 7.7 | 7.7 | 48.1 |
| Days ≥ 1" Snow Depth | *23* | 20 | 13 | 1 | 0 | 0 | 0 | 0 | 0 | 0 | 7 | 17 | 81 |

## MOOSE LAKE 1 SSE *Carlton County* ELEVATION 1089 ft LAT/LONG 46° 27' N / 92° 45' W

| | JAN | FEB | MAR | APR | MAY | JUN | JUL | AUG | SEP | OCT | NOV | DEC | YEAR |
|---|---|---|---|---|---|---|---|---|---|---|---|---|---|
| Maximum Temp °F | 19.0 | 26.0 | 37.9 | 53.7 | 67.7 | 75.6 | 81.1 | 78.0 | 68.0 | 55.9 | 37.2 | 23.5 | 52.0 |
| Minimum Temp °F | -4.3 | 1.1 | 14.9 | 27.8 | 38.1 | 46.8 | 53.7 | 52.2 | 43.5 | 33.1 | 19.9 | 4.1 | 27.6 |
| Mean Temp °F | 7.4 | 13.6 | 26.4 | 40.8 | 52.9 | 61.2 | 67.4 | 65.1 | 55.8 | 44.5 | 28.5 | 13.8 | 39.8 |
| Days Max Temp ≥ 90 °F | 0 | 0 | 0 | 0 | 0 | 1 | 3 | 2 | 0 | 0 | 0 | 0 | 6 |
| Days Max Temp ≤ 32 °F | 27 | 19 | 9 | 1 | 0 | 0 | 0 | 0 | 0 | 0 | 10 | 25 | 91 |
| Days Min Temp ≤ 32 °F | 31 | 28 | 30 | 22 | 8 | 1 | 0 | 0 | 4 | 15 | 27 | 31 | 197 |
| Days Min Temp ≤ 0 °F | 18 | 14 | 5 | 0 | 0 | 0 | 0 | 0 | 0 | 0 | 2 | 13 | 52 |
| Heating Degree Days | 1783 | 1450 | 1190 | 721 | 377 | 143 | 37 | 75 | 287 | 629 | 1088 | 1582 | 9362 |
| Cooling Degree Days | 0 | 0 | 0 | 1 | 9 | 38 | 119 | 93 | 15 | 0 | 0 | 0 | 275 |
| Total Precipitation (") | 0.93 | 0.61 | 1.53 | 2.16 | 3.01 | 4.72 | 4.17 | 3.74 | 3.73 | 2.52 | 1.72 | 1.00 | 29.84 |
| Days ≥ 0.1" Precip | 3 | 2 | 4 | 5 | 7 | 8 | 7 | 6 | 7 | 5 | 4 | 3 | 61 |
| Total Snowfall (") | 9.2 | 5.0 | 7.1 | 3.0 | 0.1 | 0.0 | 0.0 | 0.0 | 0.0 | 0.5 | 7.2 | 8.0 | 40.1 |
| Days ≥ 1" Snow Depth | 25 | 21 | 19 | 7 | 0 | 0 | 0 | 0 | 0 | 0 | 8 | 22 | 102 |

## MORA *Kanabec County* ELEVATION 1001 ft LAT/LONG 45° 53' N / 93° 18' W

| | JAN | FEB | MAR | APR | MAY | JUN | JUL | AUG | SEP | OCT | NOV | DEC | YEAR |
|---|---|---|---|---|---|---|---|---|---|---|---|---|---|
| Maximum Temp °F | 19.8 | 26.6 | 39.2 | 55.9 | 69.4 | 77.6 | 82.3 | 79.8 | 69.7 | 57.4 | 38.8 | 24.9 | 53.5 |
| Minimum Temp °F | -5.9 | 1.3 | 15.6 | 30.5 | 41.5 | 50.1 | 55.1 | 52.5 | 43.4 | 32.3 | 18.9 | 2.8 | 28.2 |
| Mean Temp °F | 7.0 | 14.0 | 27.3 | 43.2 | 55.5 | 63.8 | 68.7 | 66.2 | 56.5 | 45.0 | 28.8 | 13.9 | 40.8 |
| Days Max Temp ≥ 90 °F | 0 | 0 | 0 | 0 | 0 | 2 | 4 | 3 | 0 | 0 | 0 | 0 | 9 |
| Days Max Temp ≤ 32 °F | 26 | 18 | 8 | 0 | 0 | 0 | 0 | 0 | 0 | 0 | 8 | 23 | 83 |
| Days Min Temp ≤ 32 °F | 31 | 28 | 29 | 18 | 6 | 0 | 0 | 0 | 4 | 16 | 27 | 31 | 190 |
| Days Min Temp ≤ 0 °F | 19 | 13 | 5 | 0 | 0 | 0 | 0 | 0 | 0 | 0 | 2 | 13 | 52 |
| Heating Degree Days | 1797 | 1434 | 1160 | 649 | 310 | 96 | 31 | 62 | 272 | 615 | 1081 | 1581 | 9088 |
| Cooling Degree Days | 0 | 0 | 0 | 3 | 21 | 62 | 131 | 93 | 18 | 0 | 0 | 0 | 328 |
| Total Precipitation (") | 0.89 | 0.67 | 1.62 | 2.33 | 3.18 | 4.16 | 3.89 | 3.68 | 3.25 | 2.62 | 1.58 | 0.96 | 28.83 |
| Days ≥ 0.1" Precip | 3 | 2 | 4 | 5 | 7 | 8 | 7 | 6 | 6 | 5 | 4 | 3 | 60 |
| Total Snowfall (") | 12.1 | 7.2 | 8.5 | 1.6 | 0.0 | 0.0 | 0.0 | 0.0 | 0.0 | 0.5 | 7.3 | 9.5 | 46.7 |
| Days ≥ 1" Snow Depth | 30 | 27 | 19 | 3 | 0 | 0 | 0 | 0 | 0 | 0 | 8 | 27 | 114 |

## MORRIS WC EXP STN *Stevens County* ELEVATION 1132 ft LAT/LONG 45° 35' N / 95° 55' W

| | JAN | FEB | MAR | APR | MAY | JUN | JUL | AUG | SEP | OCT | NOV | DEC | YEAR |
|---|---|---|---|---|---|---|---|---|---|---|---|---|---|
| Maximum Temp °F | 16.9 | 22.6 | 35.6 | 53.7 | 68.3 | 76.8 | 81.5 | 79.7 | 69.5 | 56.9 | 37.6 | 23.0 | 51.8 |
| Minimum Temp °F | -3.1 | 3.2 | 17.9 | 32.9 | 45.0 | 54.7 | 59.0 | 56.0 | 45.4 | 33.4 | 20.0 | 5.4 | 30.8 |
| Mean Temp °F | 6.9 | 12.9 | 26.8 | 43.4 | 56.7 | 65.7 | 70.3 | 67.9 | 57.5 | 45.2 | 28.8 | 14.2 | 41.4 |
| Days Max Temp ≥ 90 °F | 0 | 0 | 0 | 0 | 1 | 2 | 4 | 3 | 1 | 0 | 0 | 0 | 11 |
| Days Max Temp ≤ 32 °F | 26 | 20 | 11 | 1 | 0 | 0 | 0 | 0 | 0 | 1 | 10 | 23 | 92 |
| Days Min Temp ≤ 32 °F | 31 | 28 | 28 | 15 | 3 | 0 | 0 | 0 | 2 | 15 | 27 | 31 | 180 |
| Days Min Temp ≤ 0 °F | 18 | 13 | 4 | 0 | 0 | 0 | 0 | 0 | 0 | 0 | 2 | 12 | 49 |
| Heating Degree Days | 1799 | 1467 | 1177 | 646 | 284 | 73 | 19 | 45 | 249 | 608 | 1079 | 1570 | 9016 |
| Cooling Degree Days | 0 | 0 | 0 | 4 | 41 | 103 | 183 | 140 | 31 | 1 | 0 | 0 | 503 |
| Total Precipitation (") | 0.80 | 0.69 | 1.41 | 2.34 | 2.77 | 3.95 | 3.62 | 3.12 | 2.28 | 2.23 | 1.19 | 0.74 | 25.14 |
| Days ≥ 0.1" Precip | 3 | 2 | 4 | 5 | 6 | 7 | 6 | 5 | 5 | 4 | 3 | 2 | 52 |
| Total Snowfall (") | 10.9 | 8.4 | 9.6 | 3.3 | 0.0 | 0.0 | 0.0 | 0.0 | 0.0 | 0.8 | 6.9 | 8.0 | 47.9 |
| Days ≥ 1" Snow Depth | 30 | 27 | 20 | 3 | 0 | 0 | 0 | 0 | 0 | 0 | 10 | 23 | 113 |

## NEW LONDON *Kandiyohi County* ELEVATION 1220 ft LAT/LONG 45° 18' N / 94° 56' W

| | JAN | FEB | MAR | APR | MAY | JUN | JUL | AUG | SEP | OCT | NOV | DEC | YEAR |
|---|---|---|---|---|---|---|---|---|---|---|---|---|---|
| Maximum Temp °F | 20.3 | 26.7 | 39.7 | 57.6 | 71.4 | 79.1 | 83.5 | 80.8 | 71.3 | 58.3 | 39.0 | 25.1 | 54.4 |
| Minimum Temp °F | -0.6 | 6.1 | 19.5 | 34.2 | 46.7 | 55.9 | 61.2 | 58.4 | 49.1 | 36.4 | 22.2 | 7.6 | 33.1 |
| Mean Temp °F | 9.9 | 16.4 | 29.6 | 45.9 | 59.0 | 67.5 | 72.4 | 69.6 | 60.2 | 47.4 | 30.6 | 16.4 | 43.7 |
| Days Max Temp ≥ 90 °F | 0 | 0 | 0 | 0 | 1 | 3 | 5 | 3 | 1 | 0 | 0 | 0 | 13 |
| Days Max Temp ≤ 32 °F | 25 | 17 | 8 | 0 | 0 | 0 | 0 | 0 | 0 | 0 | 9 | 22 | 81 |
| Days Min Temp ≤ 32 °F | 31 | 28 | 27 | 13 | 2 | 0 | 0 | 0 | 1 | 11 | 25 | 31 | 169 |
| Days Min Temp ≤ 0 °F | 17 | 11 | 3 | 0 | 0 | 0 | 0 | 0 | 0 | 0 | 2 | 10 | 43 |
| Heating Degree Days | 1702 | 1368 | 1089 | 569 | 221 | 46 | 7 | 26 | 185 | 542 | 1024 | 1502 | 8281 |
| Cooling Degree Days | 0 | 0 | 0 | 6 | 51 | 137 | 248 | 181 | 44 | 1 | 0 | 0 | 668 |
| Total Precipitation (") | 0.92 | 0.75 | 1.99 | 2.74 | 3.34 | 5.44 | 3.57 | 3.91 | 3.55 | 2.39 | 1.42 | 0.97 | 30.99 |
| Days ≥ 0.1" Precip | 4 | 3 | 5 | 6 | 6 | 8 | 6 | 6 | 6 | 4 | 4 | 3 | 61 |
| Total Snowfall (") | 11.7 | 8.6 | 11.1 | 2.4 | 0.1 | 0.0 | 0.0 | 0.0 | 0.0 | 0.6 | 9.2 | 10.3 | 54.0 |
| Days ≥ 1" Snow Depth | 28 | 25 | 17 | 2 | 0 | 0 | 0 | 0 | 0 | 0 | 9 | 23 | 104 |

## NEW ULM 2 SE *Brown County* ELEVATION 791 ft LAT/LONG 44° 19' N / 94° 28' W

| | JAN | FEB | MAR | APR | MAY | JUN | JUL | AUG | SEP | OCT | NOV | DEC | YEAR |
|---|---|---|---|---|---|---|---|---|---|---|---|---|---|
| Maximum Temp °F | 22.8 | 28.8 | 41.8 | 59.1 | 72.6 | 81.2 | 84.6 | 81.8 | 73.4 | 61.5 | 41.7 | 27.5 | 56.4 |
| Minimum Temp °F | 3.1 | 9.0 | 22.8 | 35.9 | 47.4 | 57.0 | 61.5 | 58.6 | 49.6 | 38.2 | 24.7 | 10.7 | 34.9 |
| Mean Temp °F | 13.0 | 18.9 | 32.3 | 47.6 | 60.0 | 69.1 | 73.1 | 70.2 | 61.6 | 49.9 | 33.3 | 19.1 | 45.7 |
| Days Max Temp ≥ 90 °F | 0 | 0 | 0 | 0 | 1 | 4 | 8 | 5 | 1 | 0 | 0 | 0 | 19 |
| Days Max Temp ≤ 32 °F | 23 | 16 | 7 | 0 | 0 | 0 | 0 | 0 | 0 | 0 | 6 | 19 | 71 |
| Days Min Temp ≤ 32 °F | 31 | 27 | 25 | 11 | 2 | 0 | 0 | 0 | 1 | 9 | 23 | 30 | 159 |
| Days Min Temp ≤ 0 °F | 14 | 9 | 2 | 0 | 0 | 0 | 0 | 0 | 0 | 0 | 1 | 8 | 34 |
| Heating Degree Days | 1609 | 1296 | 1006 | 523 | 201 | 34 | 8 | 24 | 159 | 470 | 946 | 1417 | 7693 |
| Cooling Degree Days | 0 | 0 | 0 | 8 | 53 | 161 | 254 | 185 | 60 | 3 | 0 | 0 | 724 |
| Total Precipitation (") | 0.75 | 0.79 | 1.88 | 2.84 | 3.16 | 4.36 | 3.93 | 3.85 | 3.05 | 2.35 | 1.68 | 0.89 | 29.53 |
| Days ≥ 0.1" Precip | 3 | 2 | 4 | 6 | 7 | 7 | 6 | 6 | 6 | 4 | 4 | 3 | 58 |
| Total Snowfall (") | 10.1 | 8.0 | 9.0 | 2.0 | 0.0 | 0.0 | 0.0 | 0.0 | 0.0 | 0.4 | 6.6 | 9.0 | 45.1 |
| Days ≥ 1" Snow Depth | 25 | 21 | 12 | 2 | 0 | 0 | 0 | 0 | 0 | 0 | 6 | 19 | 85 |

## OKLEE *Red Lake County* ELEVATION 1129 ft LAT/LONG 47° 46' N / 95° 55' W

| | JAN | FEB | MAR | APR | MAY | JUN | JUL | AUG | SEP | OCT | NOV | DEC | YEAR |
|---|---|---|---|---|---|---|---|---|---|---|---|---|---|
| Maximum Temp °F | 12.9 | 21.4 | 34.5 | 52.6 | 67.9 | 75.2 | 80.4 | 79.3 | 67.8 | 54.4 | 34.1 | 19.4 | 50.0 |
| Minimum Temp °F | -8.4 | -0.8 | 13.7 | 29.6 | 41.9 | 50.8 | 54.9 | 52.9 | 42.7 | 31.9 | 16.7 | 0.6 | 27.2 |
| Mean Temp °F | 2.3 | 10.3 | 24.1 | 41.0 | 54.8 | 63.0 | 67.7 | 66.2 | 55.3 | 43.2 | 25.4 | 10.1 | 38.6 |
| Days Max Temp ≥ 90 °F | 0 | 0 | 0 | 0 | 0 | 1 | 3 | 3 | 0 | 0 | 0 | 0 | 7 |
| Days Max Temp ≤ 32 °F | 27 | 21 | 13 | 1 | 0 | 0 | 0 | 0 | 0 | 1 | 13 | 26 | 102 |
| Days Min Temp ≤ 32 °F | 30 | 28 | 29 | 20 | 6 | 0 | 0 | 0 | 4 | 17 | 28 | 31 | 193 |
| Days Min Temp ≤ 0 °F | 21 | 15 | 6 | 0 | 0 | 0 | 0 | 0 | 0 | 0 | 3 | 15 | 60 |
| Heating Degree Days | 1943 | 1539 | 1260 | 716 | 326 | 113 | 36 | 67 | 302 | 669 | 1184 | 1700 | 9855 |
| Cooling Degree Days | 0 | 0 | 0 | 1 | 22 | 55 | 130 | 117 | 14 | 1 | 0 | 0 | 340 |
| Total Precipitation (") | 0.69 | 0.45 | 1.00 | 1.66 | 2.52 | 3.68 | 3.51 | 3.56 | 2.61 | 2.00 | 0.81 | 0.68 | 23.17 |
| Days ≥ 0.1" Precip | 2 | 2 | 3 | 4 | 6 | 7 | 6 | 6 | 5 | 4 | 2 | 3 | 50 |
| Total Snowfall (") | 9.3 | 4.7 | 7.0 | 2.2 | 0.1 | 0.0 | 0.0 | 0.0 | 0.0 | 1.2 | 5.6 | 7.9 | 38.0 |
| Days ≥ 1" Snow Depth | na | 24 | 21 | 3 | 0 | 0 | 0 | 0 | 0 | 0 | 8 | na | na |

## OTTERTAIL *Otter Tail County* ELEVATION 981 ft LAT/LONG 46° 25' N / 95° 34' W

| | JAN | FEB | MAR | APR | MAY | JUN | JUL | AUG | SEP | OCT | NOV | DEC | YEAR |
|---|---|---|---|---|---|---|---|---|---|---|---|---|---|
| Maximum Temp °F | 17.4 | 24.1 | 37.2 | 54.7 | 69.1 | 77.1 | 82.1 | 79.9 | 69.6 | 56.6 | 36.8 | 23.1 | 52.3 |
| Minimum Temp °F | -4.0 | 2.2 | 16.4 | 31.9 | 45.0 | 54.6 | 59.7 | 57.6 | 47.7 | 36.1 | 20.6 | 4.5 | 31.0 |
| Mean Temp °F | 6.7 | 13.1 | 26.8 | 43.3 | 57.1 | 65.9 | 70.9 | 68.8 | 58.7 | 46.4 | 28.7 | 13.8 | 41.7 |
| Days Max Temp ≥ 90 °F | 0 | 0 | 0 | 0 | 0 | 1 | 5 | 3 | 1 | 0 | 0 | 0 | 10 |
| Days Max Temp ≤ 32 °F | 26 | 19 | 10 | 1 | 0 | 0 | 0 | 0 | 0 | 0 | 10 | 24 | 90 |
| Days Min Temp ≤ 32 °F | 31 | 28 | 28 | 16 | 3 | 0 | 0 | 0 | 1 | 11 | 27 | 31 | 176 |
| Days Min Temp ≤ 0 °F | 19 | 13 | 5 | 0 | 0 | 0 | 0 | 0 | 0 | 0 | 2 | 12 | 51 |
| Heating Degree Days | 1805 | 1460 | 1177 | 648 | 270 | 65 | 14 | 35 | 221 | 572 | 1082 | 1583 | 8932 |
| Cooling Degree Days | 0 | 0 | 0 | 3 | 40 | 109 | 217 | 168 | 36 | 1 | 0 | 0 | 574 |
| Total Precipitation (") | 0.96 | 0.62 | 1.39 | 2.37 | 2.96 | 4.49 | 3.89 | 3.22 | 2.38 | 2.33 | 1.11 | 0.78 | 26.50 |
| Days ≥ 0.1" Precip | 4 | 3 | 4 | 6 | 7 | 8 | 7 | 6 | 5 | 5 | 4 | 3 | 62 |
| Total Snowfall (") | 13.2 | 8.4 | 10.0 | 4.7 | 0.1 | 0.0 | 0.0 | 0.0 | 0.0 | 1.3 | 9.3 | 9.7 | 56.7 |
| Days ≥ 1" Snow Depth | 31 | 27 | 21 | 3 | 0 | 0 | 0 | 0 | 0 | 0 | 11 | 25 | 118 |

## OWATONNA *Steele County* ELEVATION 1150 ft LAT/LONG 44° 6' N / 93° 14' W

| | JAN | FEB | MAR | APR | MAY | JUN | JUL | AUG | SEP | OCT | NOV | DEC | YEAR |
|---|---|---|---|---|---|---|---|---|---|---|---|---|---|
| Maximum Temp °F | 22.5 | 28.4 | 41.7 | 58.6 | 72.0 | 81.1 | 84.6 | 82.1 | 73.6 | 61.0 | 41.9 | 27.6 | 56.3 |
| Minimum Temp °F | 3.2 | 8.8 | 21.9 | 35.2 | 47.0 | 56.6 | 61.1 | 58.5 | 49.7 | 38.1 | 24.5 | 10.8 | 34.6 |
| Mean Temp °F | 12.9 | 18.7 | 31.8 | 46.9 | 59.5 | 68.9 | 72.9 | 70.3 | 61.7 | 49.6 | 33.2 | 19.2 | 45.5 |
| Days Max Temp ≥ 90 °F | 0 | 0 | 0 | 0 | 1 | 4 | 7 | 5 | 1 | 0 | 0 | 0 | 18 |
| Days Max Temp ≤ 32 °F | 23 | 16 | 7 | 0 | 0 | 0 | 0 | 0 | 0 | 0 | 6 | 19 | 71 |
| Days Min Temp ≤ 32 °F | 31 | 28 | 26 | 13 | 2 | 0 | 0 | 0 | 1 | 9 | 24 | 31 | 165 |
| Days Min Temp ≤ 0 °F | 13 | 9 | 2 | 0 | 0 | 0 | 0 | 0 | 0 | 0 | 1 | 8 | 33 |
| Heating Degree Days | 1612 | 1303 | 1021 | 542 | 217 | 37 | 7 | 23 | 157 | 477 | 947 | 1413 | 7756 |
| Cooling Degree Days | 0 | 0 | 0 | 9 | 57 | 173 | 262 | 202 | 72 | 4 | 0 | 0 | 779 |
| Total Precipitation (") | 0.87 | 0.67 | 1.86 | 2.94 | 3.81 | 4.27 | 4.58 | 3.94 | 3.46 | 2.32 | 1.70 | 1.06 | 31.48 |
| Days ≥ 0.1" Precip | 3 | 2 | 4 | 7 | 8 | 8 | 6 | 6 | 6 | 5 | 4 | 3 | 62 |
| Total Snowfall (") | 9.7 | 7.0 | 7.7 | 2.8 | 0.0 | 0.0 | 0.0 | 0.0 | 0.0 | 0.5 | 4.9 | 9.2 | 41.8 |
| Days ≥ 1" Snow Depth | 28 | 24 | 15 | 2 | 0 | 0 | 0 | 0 | 0 | 0 | 6 | 21 | 96 |

## PARK RAPIDS MUNI AP *Hubbard County* ELEVATION 1443 ft LAT/LONG 46° 54' N / 95° 4' W

| | JAN | FEB | MAR | APR | MAY | JUN | JUL | AUG | SEP | OCT | NOV | DEC | YEAR |
|---|---|---|---|---|---|---|---|---|---|---|---|---|---|
| Maximum Temp °F | 16.2 | 23.9 | 37.0 | 54.3 | 68.4 | 76.1 | 81.0 | 79.0 | 68.1 | 55.3 | 35.3 | 21.3 | 51.3 |
| Minimum Temp °F | -6.2 | 0.1 | 14.7 | 29.4 | 41.6 | 51.0 | 56.1 | 53.8 | 44.3 | 32.9 | 18.0 | 2.1 | 28.2 |
| Mean Temp °F | 5.1 | 12.0 | 25.9 | 41.9 | 55.0 | 63.6 | 68.6 | 66.4 | 56.2 | 44.1 | 26.7 | 11.7 | 39.8 |
| Days Max Temp ≥ 90 °F | 0 | 0 | 0 | 0 | 0 | 1 | 4 | 3 | 1 | 0 | 0 | 0 | 9 |
| Days Max Temp ≤ 32 °F | 28 | 20 | 11 | 1 | 0 | 0 | 0 | 0 | 0 | 1 | 13 | 26 | 100 |
| Days Min Temp ≤ 32 °F | 31 | 28 | 29 | 20 | 6 | 0 | 0 | 0 | 3 | 16 | 28 | 31 | 192 |
| Days Min Temp ≤ 0 °F | 19 | 14 | 6 | 0 | 0 | 0 | 0 | 0 | 0 | 0 | 2 | 14 | 55 |
| Heating Degree Days | 1857 | 1491 | 1207 | 687 | 322 | 103 | 28 | 62 | 282 | 641 | 1142 | 1648 | 9470 |
| Cooling Degree Days | 0 | 0 | 0 | 2 | 25 | 71 | 144 | 115 | 20 | 0 | 0 | 0 | 377 |
| Total Precipitation (") | 0.69 | 0.48 | 1.24 | 2.25 | 2.63 | 4.37 | 3.80 | 3.57 | 2.81 | 2.50 | 1.08 | 0.76 | 26.18 |
| Days ≥ 0.1" Precip | 3 | 2 | 4 | 5 | 6 | 8 | 7 | 6 | 6 | 4 | 3 | 2 | 56 |
| Total Snowfall (") | 12.1 | 6.3 | 11.2 | 4.6 | 0.1 | 0.0 | 0.0 | 0.0 | 0.0 | 1.2 | 6.7 | 9.4 | 51.6 |
| Days ≥ 1" Snow Depth | 31 | 28 | 24 | 5 | 0 | 0 | 0 | 0 | 0 | 1 | 11 | 28 | 128 |

## PINE RIVER DAM *Crow Wing County* ELEVATION 1250 ft LAT/LONG 46° 41' N / 94° 7' W

| | JAN | FEB | MAR | APR | MAY | JUN | JUL | AUG | SEP | OCT | NOV | DEC | YEAR |
|---|---|---|---|---|---|---|---|---|---|---|---|---|---|
| Maximum Temp °F | 19.4 | 26.8 | 39.0 | 55.4 | 69.5 | 77.6 | 82.4 | 79.7 | 69.3 | 57.3 | 38.1 | 24.0 | 53.2 |
| Minimum Temp °F | -4.8 | 1.1 | 14.5 | 29.0 | 41.9 | 51.2 | 56.7 | 54.0 | 44.8 | 34.2 | 20.0 | 3.9 | 28.9 |
| Mean Temp °F | 7.3 | 14.0 | 26.8 | 42.2 | 55.7 | 64.4 | 69.5 | 66.9 | 57.1 | 45.7 | 29.0 | 14.0 | 41.1 |
| Days Max Temp ≥ 90 °F | 0 | 0 | 0 | 0 | 0 | 1 | 5 | 3 | 0 | 0 | 0 | 0 | 9 |
| Days Max Temp ≤ 32 °F | 26 | 18 | 9 | 0 | 0 | 0 | 0 | 0 | 0 | 0 | 10 | 23 | 86 |
| Days Min Temp ≤ 32 °F | 31 | 28 | 29 | 20 | 5 | 0 | 0 | 0 | 2 | 14 | 27 | 31 | 187 |
| Days Min Temp ≤ 0 °F | 19 | 13 | 5 | 0 | 0 | 0 | 0 | 0 | 0 | 0 | 2 | 12 | 51 |
| Heating Degree Days | 1787 | 1437 | 1178 | 677 | 305 | 86 | 22 | 52 | 253 | 591 | 1072 | 1578 | 9038 |
| Cooling Degree Days | 0 | 0 | 0 | 2 | 28 | 85 | 180 | 132 | 22 | 1 | 0 | 0 | 450 |
| Total Precipitation (") | 1.03 | 0.59 | 1.69 | 2.35 | 3.07 | 4.19 | 3.80 | 3.32 | 2.98 | 2.49 | 1.63 | 0.89 | 28.03 |
| Days ≥ 0.1" Precip | 3 | 2 | 4 | 6 | 7 | 8 | 6 | 6 | 6 | 5 | 4 | 3 | 60 |
| Total Snowfall (") | 12.9 | 6.4 | 9.1 | 4.1 | 0.3 | 0.0 | 0.0 | 0.0 | 0.0 | 0.5 | *6.3* | 9.0 | 48.6 |
| Days ≥ 1" Snow Depth | *20* | 20 | 18 | 4 | 0 | 0 | 0 | 0 | 0 | 0 | *7* | 19 | 88 |

## PIPESTONE *Pipestone County* ELEVATION 1742 ft LAT/LONG 44° 0 ' N / 96° 18 ' W

| | JAN | FEB | MAR | APR | MAY | JUN | JUL | AUG | SEP | OCT | NOV | DEC | YEAR |
|---|---|---|---|---|---|---|---|---|---|---|---|---|---|
| Maximum Temp °F | 21.5 | 27.0 | 40.8 | 57.5 | 70.7 | 79.5 | 84.2 | 82.1 | 72.6 | 60.0 | 40.4 | 26.9 | 55.3 |
| Minimum Temp °F | -0.5 | 5.6 | 19.2 | 32.2 | 43.9 | 53.3 | 57.8 | 55.0 | 44.8 | 32.4 | 19.4 | 6.2 | 30.8 |
| Mean Temp °F | 10.5 | 16.3 | 30.0 | 44.9 | 57.3 | 66.4 | 71.0 | 68.6 | 58.7 | 46.3 | 29.9 | 16.6 | 43.0 |
| Days Max Temp ≥ 90 °F | 0 | 0 | 0 | 0 | 1 | 3 | 8 | 5 | 1 | 0 | 0 | 0 | 18 |
| Days Max Temp ≤ 32 °F | 23 | 18 | 8 | 1 | 0 | 0 | 0 | 0 | 0 | 0 | 8 | 20 | 78 |
| Days Min Temp ≤ 32 °F | 31 | 28 | 27 | 16 | 4 | 0 | 0 | 0 | 4 | 17 | 27 | 31 | 185 |
| Days Min Temp ≤ 0 °F | 15 | 10 | 3 | 0 | 0 | 0 | 0 | 0 | 0 | 0 | 2 | 10 | 40 |
| Heating Degree Days | 1686 | 1370 | 1077 | 601 | 264 | 66 | 20 | 42 | 226 | 578 | 1047 | 1496 | 8473 |
| Cooling Degree Days | 0 | 0 | 0 | 6 | 34 | 115 | 194 | 147 | 41 | 2 | 0 | 0 | 539 |
| Total Precipitation (") | 0.51 | 0.63 | 1.58 | 2.39 | 3.23 | 4.23 | 3.21 | 3.15 | 3.10 | 2.11 | 1.33 | 0.73 | 26.20 |
| Days ≥ 0.1" Precip | 2 | 2 | 4 | 6 | 7 | 7 | 6 | 5 | 6 | 4 | 3 | 2 | 54 |
| Total Snowfall (") | *5.5* | *6.9* | *6.4* | 2.8 | 0.0 | 0.0 | 0.0 | 0.0 | 0.0 | 0.9 | *5.2* | 6.8 | 34.5 |
| Days ≥ 1" Snow Depth | *16* | na | *9* | 1 | 0 | 0 | 0 | 0 | 0 | 0 | na | *11* | na |

## POKEGAMA DAM *Itasca County* ELEVATION 1280 ft LAT/LONG 47° 15 ' N / 93° 35 ' W

| | JAN | FEB | MAR | APR | MAY | JUN | JUL | AUG | SEP | OCT | NOV | DEC | YEAR |
|---|---|---|---|---|---|---|---|---|---|---|---|---|---|
| Maximum Temp °F | *17.7* | *25.2* | *37.9* | *54.6* | 68.2 | 75.7 | 80.2 | 77.8 | 67.3 | *55.4* | *36.5* | *22.4* | 51.6 |
| Minimum Temp °F | *-4.4* | *1.7* | *14.6* | *29.7* | 41.6 | 50.7 | 56.0 | 54.0 | 44.8 | *34.9* | *20.2* | *3.9* | 29.0 |
| Mean Temp °F | *6.7* | *13.5* | *26.3* | *42.2* | 54.9 | 63.2 | 68.1 | 65.9 | 56.1 | *45.2* | *28.4* | *13.2* | 40.3 |
| Days Max Temp ≥ 90 °F | 0 | 0 | 0 | 0 | 0 | 1 | 2 | 1 | 0 | 0 | 0 | 0 | 4 |
| Days Max Temp ≤ 32 °F | 24 | 18 | 9 | 1 | 0 | 0 | 0 | 0 | 0 | 0 | 10 | 22 | 84 |
| Days Min Temp ≤ 32 °F | 28 | 25 | 26 | 18 | 5 | 0 | 0 | 0 | 2 | 12 | 24 | 28 | 168 |
| Days Min Temp ≤ 0 °F | 16 | 12 | 5 | 0 | 0 | 0 | 0 | 0 | 0 | 0 | 2 | 11 | 46 |
| Heating Degree Days | *1805* | *1450* | *1192* | *678* | 324 | 105 | 31 | 63 | 280 | *608* | *1091* | *1602* | 9229 |
| Cooling Degree Days | na | na | na | na | 24 | 61 | 139 | 105 | 14 | na | na | na | na |
| Total Precipitation (") | 0.88 | 0.54 | 1.30 | 1.95 | 2.86 | 4.43 | 3.97 | 3.65 | 3.27 | 2.55 | 1.35 | 0.99 | 27.74 |
| Days ≥ 0.1" Precip | 3 | 2 | 4 | 5 | 6 | 8 | 8 | 6 | 7 | 5 | 3 | 3 | 60 |
| Total Snowfall (") | 11.7 | 6.1 | 9.6 | 2.8 | 0.2 | 0.0 | 0.0 | 0.0 | 0.0 | 0.9 | 7.1 | 9.9 | 48.3 |
| Days ≥ 1" Snow Depth | 30 | 27 | 25 | 7 | 0 | 0 | 0 | 0 | 0 | 1 | 13 | 29 | 132 |

## PRESTON *Fillmore County* ELEVATION 942 ft LAT/LONG 43° 40 ' N / 92° 5 ' W

| | JAN | FEB | MAR | APR | MAY | JUN | JUL | AUG | SEP | OCT | NOV | DEC | YEAR |
|---|---|---|---|---|---|---|---|---|---|---|---|---|---|
| Maximum Temp °F | 24.0 | 29.7 | 42.0 | 57.8 | 69.8 | 79.2 | 83.2 | 80.9 | 72.7 | 60.8 | 42.9 | 29.1 | 56.0 |
| Minimum Temp °F | 1.9 | 6.8 | 20.5 | 33.2 | 43.4 | 53.0 | 57.6 | 54.5 | 46.2 | 35.3 | 22.9 | 9.8 | 32.1 |
| Mean Temp °F | 13.0 | 18.3 | 31.3 | 45.5 | 56.5 | 66.1 | 70.4 | 67.7 | 59.5 | 48.1 | 32.9 | 19.5 | 44.1 |
| Days Max Temp ≥ 90 °F | 0 | 0 | 0 | 0 | 0 | 2 | 5 | 3 | 1 | 0 | 0 | 0 | 11 |
| Days Max Temp ≤ 32 °F | 22 | 16 | 6 | 0 | 0 | 0 | 0 | 0 | 0 | 0 | 5 | 18 | 67 |
| Days Min Temp ≤ 32 °F | 31 | 27 | 26 | 14 | 5 | 0 | 0 | 0 | 3 | 13 | 25 | 30 | 174 |
| Days Min Temp ≤ 0 °F | 14 | 10 | 2 | 0 | 0 | 0 | 0 | 0 | 0 | 0 | 1 | 8 | 35 |
| Heating Degree Days | 1610 | 1307 | 1040 | 583 | 282 | 67 | 19 | 46 | 204 | 520 | 956 | 1405 | 8039 |
| Cooling Degree Days | 0 | 0 | 0 | 8 | 25 | 111 | 195 | 139 | 49 | 2 | 0 | 0 | 529 |
| Total Precipitation (") | 0.89 | 0.81 | 1.89 | 3.27 | 3.80 | 4.50 | 4.27 | 4.16 | 4.18 | 2.48 | 1.94 | 1.37 | 33.56 |
| Days ≥ 0.1" Precip | 3 | 2 | 4 | 7 | 7 | 7 | 7 | 7 | 6 | 5 | 4 | 4 | 63 |
| Total Snowfall (") | 8.9 | 6.5 | *6.2* | 1.8 | 0.0 | 0.0 | 0.0 | 0.0 | 0.0 | 0.1 | 4.0 | 9.8 | 37.3 |
| Days ≥ 1" Snow Depth | *24* | *21* | na | 1 | 0 | 0 | 0 | 0 | 0 | 0 | *3* | *18* | na |

## RED LAKE FALLS *Red Lake County* ELEVATION 1001 ft LAT/LONG 47° 53 ' N / 96° 16 ' W

| | JAN | FEB | MAR | APR | MAY | JUN | JUL | AUG | SEP | OCT | NOV | DEC | YEAR |
|---|---|---|---|---|---|---|---|---|---|---|---|---|---|
| Maximum Temp °F | 13.9 | 20.9 | 34.8 | 54.4 | 69.5 | 76.8 | 81.9 | 79.8 | 69.0 | 55.4 | 34.7 | 20.0 | 50.9 |
| Minimum Temp °F | -6.7 | 0.0 | 15.2 | 31.3 | 43.2 | 52.1 | 56.7 | 54.0 | 44.7 | 33.3 | 18.1 | 1.7 | 28.6 |
| Mean Temp °F | 3.6 | 10.4 | 25.0 | 42.9 | 56.4 | 64.5 | 69.3 | 66.8 | 56.9 | 44.4 | 26.4 | 10.9 | 39.8 |
| Days Max Temp ≥ 90 °F | 0 | 0 | 0 | 0 | 1 | 1 | 4 | *3* | 1 | 0 | 0 | 0 | 10 |
| Days Max Temp ≤ 32 °F | 28 | 21 | 12 | 1 | 0 | 0 | 0 | 0 | 0 | 1 | 13 | 25 | 101 |
| Days Min Temp ≤ 32 °F | 31 | 28 | 28 | 17 | 5 | 0 | 0 | 0 | 3 | 14 | 27 | 31 | 184 |
| Days Min Temp ≤ 0 °F | 21 | 15 | 6 | 0 | 0 | 0 | 0 | 0 | 0 | 0 | 3 | 14 | 59 |
| Heating Degree Days | 1902 | 1538 | 1234 | 658 | 293 | 86 | 20 | 61 | 268 | 634 | 1151 | 1674 | 9519 |
| Cooling Degree Days | 0 | 0 | 0 | 5 | 44 | 82 | 163 | 138 | 25 | 0 | 0 | 0 | 457 |
| Total Precipitation (") | 0.68 | 0.46 | 0.93 | 1.49 | 2.52 | 4.00 | 3.10 | 3.68 | 2.56 | 1.63 | 0.82 | 0.57 | 22.44 |
| Days ≥ 0.1" Precip | 2 | 1 | 3 | 4 | 6 | 7 | *6* | 6 | 5 | 4 | 2 | 2 | 48 |
| Total Snowfall (") | 11.5 | 7.0 | 8.4 | 2.4 | 0.1 | 0.0 | 0.0 | 0.0 | 0.0 | 0.9 | 7.9 | 9.6 | 47.8 |
| Days ≥ 1" Snow Depth | 31 | 28 | 23 | 3 | 0 | 0 | 0 | 0 | 0 | 0 | 13 | 27 | 125 |

## RED LAKE IND AGENCY *Beltrami County* ELEVATION 1220 ft LAT/LONG 47° 52 ' N / 95° 2 ' W

| | JAN | FEB | MAR | APR | MAY | JUN | JUL | AUG | SEP | OCT | NOV | DEC | YEAR |
|---|---|---|---|---|---|---|---|---|---|---|---|---|---|
| Maximum Temp °F | 13.6 | 20.4 | 33.9 | 49.8 | 64.5 | 73.2 | 78.1 | 76.3 | 65.1 | 52.6 | 34.5 | 20.1 | 48.5 |
| Minimum Temp °F | -8.3 | -3.0 | 12.1 | 28.4 | 42.2 | 51.8 | 57.2 | 54.9 | 44.7 | 33.6 | 18.0 | 0.7 | 27.7 |
| Mean Temp °F | 2.6 | 8.8 | 23.0 | 39.1 | 53.4 | 62.5 | 67.7 | 65.7 | 54.9 | 43.1 | 26.3 | 10.4 | 38.1 |
| Days Max Temp ≥ 90 °F | 0 | 0 | 0 | 0 | 0 | 1 | 2 | 2 | 0 | 0 | 0 | 0 | 5 |
| Days Max Temp ≤ 32 °F | 28 | 21 | 13 | 2 | 0 | 0 | 0 | 0 | 0 | 1 | 14 | 25 | 104 |
| Days Min Temp ≤ 32 °F | 31 | 28 | 29 | 21 | 5 | 0 | 0 | 0 | 2 | 14 | 28 | 31 | 189 |
| Days Min Temp ≤ 0 °F | 22 | 17 | 7 | 0 | 0 | 0 | 0 | 0 | 0 | 0 | 3 | 15 | 64 |
| Heating Degree Days | 1933 | 1585 | 1296 | 772 | 375 | 124 | 36 | 71 | 313 | 674 | 1156 | 1690 | 10025 |
| Cooling Degree Days | 0 | 0 | 0 | 3 | 24 | 60 | 128 | 109 | 13 | 1 | 0 | 0 | 338 |
| Total Precipitation (") | 0.58 | 0.36 | 0.88 | 1.34 | 2.38 | 3.85 | 3.92 | 3.35 | 2.63 | 2.13 | 0.81 | 0.49 | 22.72 |
| Days ≥ 0.1" Precip | 2 | 1 | 3 | 4 | 6 | 7 | 7 | 6 | 6 | 4 | 2 | 2 | 50 |
| Total Snowfall (") | 11.8 | 6.2 | 8.0 | 3.2 | 0.1 | 0.0 | 0.0 | 0.0 | 0.0 | 1.1 | 8.2 | 9.0 | 47.6 |
| Days ≥ 1" Snow Depth | 31 | 28 | 25 | 6 | 0 | 0 | 0 | 0 | 0 | 1 | 12 | 26 | 129 |

## REDWOOD FALLS MU AP *Redwood County* ELEVATION 1027 ft LAT/LONG 44° 33 ' N / 95° 5 ' W

| | JAN | FEB | MAR | APR | MAY | JUN | JUL | AUG | SEP | OCT | NOV | DEC | YEAR |
|---|---|---|---|---|---|---|---|---|---|---|---|---|---|
| Maximum Temp °F | 21.7 | 27.3 | ***39.8*** | 57.9 | 72.0 | 80.5 | 84.6 | 81.9 | 72.8 | 59.4 | 40.7 | 26.2 | 55.4 |
| Minimum Temp °F | 3.4 | 10.1 | ***23.2*** | 37.1 | 49.3 | 58.3 | 62.6 | 60.0 | 50.5 | 38.3 | 24.5 | 9.5 | 35.6 |
| Mean Temp °F | 12.6 | 18.8 | ***31.5*** | 47.5 | 60.7 | 69.4 | 73.6 | 71.0 | 61.7 | 48.9 | 32.6 | 17.9 | 45.5 |
| Days Max Temp ≥ 90 °F | 0 | 0 | ***0*** | 0 | 2 | 5 | 8 | 6 | 2 | 0 | 0 | 0 | 23 |
| Days Max Temp ≤ 32 °F | 23 | 17 | ***8*** | 0 | 0 | 0 | 0 | 0 | 0 | 0 | 8 | 20 | 76 |
| Days Min Temp ≤ 32 °F | 30 | 27 | ***23*** | 8 | 1 | 0 | 0 | 0 | 1 | 8 | 23 | 31 | 152 |
| Days Min Temp ≤ 0 °F | 14 | 8 | ***2*** | 0 | 0 | 0 | 0 | 0 | 0 | 0 | 1 | 9 | 34 |
| Heating Degree Days | 1618 | 1300 | ***1031*** | 524 | 183 | 33 | 5 | 20 | 155 | 497 | 965 | 1455 | 7786 |
| Cooling Degree Days | 0 | 0 | 0 | 7 | 56 | 164 | 269 | 199 | 55 | 3 | 0 | 0 | 753 |
| Total Precipitation (") | 0.68 | 0.66 | 1.51 | 2.76 | 2.91 | 3.98 | 3.75 | 3.42 | 2.69 | 2.05 | 1.54 | 0.71 | 26.66 |
| Days ≥ 0.1" Precip | 2 | 2 | 4 | 6 | 7 | 7 | 6 | 5 | 5 | 4 | 3 | 2 | 53 |
| Total Snowfall (") | 7.0 | 6.1 | 7.5 | 1.6 | 0.0 | 0.0 | 0.0 | 0.0 | 0.0 | 0.5 | 6.2 | 6.7 | 35.6 |
| Days ≥ 1" Snow Depth | 26 | 22 | 15 | 2 | 0 | 0 | 0 | 0 | 0 | 0 | 7 | 19 | 91 |

## ROCHESTER MUNI AP *Olmsted County* ELEVATION 1297 ft LAT/LONG 43° 55 ' N / 92° 30 ' W

| | JAN | FEB | MAR | APR | MAY | JUN | JUL | AUG | SEP | OCT | NOV | DEC | YEAR |
|---|---|---|---|---|---|---|---|---|---|---|---|---|---|
| Maximum Temp °F | 20.1 | 25.8 | 38.6 | 55.3 | 68.2 | 77.6 | 81.3 | 78.6 | 69.7 | 57.5 | 40.1 | 25.8 | 53.2 |
| Minimum Temp °F | 2.6 | 8.3 | 21.7 | 34.9 | 45.6 | 55.3 | 59.8 | 57.4 | 48.4 | 37.1 | 23.9 | 10.4 | 33.8 |
| Mean Temp °F | 11.4 | 17.1 | 30.2 | 45.1 | 56.9 | 66.5 | 70.6 | 68.0 | 59.1 | 47.3 | 32.0 | 18.1 | 43.5 |
| Days Max Temp ≥ 90 °F | 0 | 0 | 0 | 0 | 0 | 2 | 3 | 2 | 0 | 0 | 0 | 0 | 7 |
| Days Max Temp ≤ 32 °F | 25 | 19 | 9 | 1 | 0 | 0 | 0 | 0 | 0 | 0 | 8 | 22 | 84 |
| Days Min Temp ≤ 32 °F | 31 | 28 | 26 | 12 | 2 | 0 | 0 | 0 | 1 | 10 | 24 | 30 | 164 |
| Days Min Temp ≤ 0 °F | 14 | 9 | 2 | 0 | 0 | 0 | 0 | 0 | 0 | 0 | 1 | 8 | 34 |
| Heating Degree Days | 1658 | 1348 | 1072 | 594 | 273 | 61 | 14 | 41 | 211 | 545 | 983 | 1448 | 8248 |
| Cooling Degree Days | 0 | 0 | 0 | 6 | 30 | 113 | 195 | 144 | 44 | 1 | 0 | 0 | 533 |
| Total Precipitation (") | 0.87 | 0.74 | 1.84 | 3.00 | 3.42 | 3.82 | 4.32 | 4.06 | 3.44 | 2.30 | 1.84 | 1.09 | 30.74 |
| Days ≥ 0.1" Precip | 3 | 2 | 5 | 7 | 7 | 7 | 7 | 6 | 6 | 5 | 4 | 3 | 62 |
| Total Snowfall (") | 11.1 | 8.1 | 8.6 | 3.8 | 0.0 | 0.0 | 0.0 | 0.0 | 0.0 | 0.8 | 6.2 | 10.8 | 49.4 |
| Days ≥ 1" Snow Depth | 28 | 24 | 17 | 3 | 0 | 0 | 0 | 0 | 0 | 0 | 6 | 22 | 100 |

## ROSEAU 1 E *Roseau County* ELEVATION 1050 ft LAT/LONG 48° 51 ' N / 95° 46 ' W

| | JAN | FEB | MAR | APR | MAY | JUN | JUL | AUG | SEP | OCT | NOV | DEC | YEAR |
|---|---|---|---|---|---|---|---|---|---|---|---|---|---|
| Maximum Temp °F | 10.7 | 18.5 | 32.1 | 51.0 | 66.4 | 73.2 | 77.8 | 76.9 | 65.9 | 52.6 | 32.2 | 16.8 | 47.8 |
| Minimum Temp °F | -10.8 | -4.4 | 9.9 | 27.5 | 39.0 | 48.1 | 53.1 | 50.9 | 42.0 | 31.4 | 15.4 | -1.8 | 25.0 |
| Mean Temp °F | -0.1 | 7.0 | 21.0 | 39.3 | 52.7 | 60.7 | 65.5 | 63.9 | 54.0 | 42.0 | 23.9 | 7.5 | 36.4 |
| Days Max Temp ≥ 90 °F | 0 | 0 | 0 | 0 | 0 | 0 | 1 | 1 | 0 | 0 | 0 | 0 | 2 |
| Days Max Temp ≤ 32 °F | 29 | 23 | 14 | 2 | 0 | 0 | 0 | 0 | 0 | 1 | 16 | 28 | 113 |
| Days Min Temp ≤ 32 °F | 31 | 28 | 30 | 22 | 9 | 1 | 0 | 1 | 4 | 17 | 28 | 31 | 202 |
| Days Min Temp ≤ 0 °F | 22 | 17 | 9 | 0 | 0 | 0 | 0 | 0 | 0 | 0 | 4 | 17 | 69 |
| Heating Degree Days | 2018 | 1635 | 1358 | 766 | 389 | 157 | 59 | 103 | 340 | 706 | 1229 | 1781 | 10541 |
| Cooling Degree Days | 0 | ***0*** | ***0*** | 1 | ***19*** | ***29*** | ***75*** | 76 | 12 | 0 | 0 | 0 | 212 |
| Total Precipitation (") | 0.76 | 0.46 | 0.65 | 1.38 | 2.18 | 3.62 | 3.39 | 3.27 | 2.62 | 1.38 | 0.79 | 0.70 | 21.20 |
| Days ≥ 0.1" Precip | na | na | ***2*** | 3 | 5 | 7 | 7 | 6 | 5 | 4 | 3 | ***3*** | na |
| Total Snowfall (") | ***10.5*** | na | na | ***3.5*** | 0.2 | 0.0 | 0.0 | 0.0 | 0.0 | 0.7 | ***5.6*** | na | na |
| Days ≥ 1" Snow Depth | na | na | na | na | 0 | 0 | 0 | 0 | 0 | 0 | na | na | na |

## ROSEMOUNT AGRI EXP S *Dakota County* ELEVATION 951 ft LAT/LONG 44° 42 ' N / 93° 7 ' W

| | JAN | FEB | MAR | APR | MAY | JUN | JUL | AUG | SEP | OCT | NOV | DEC | YEAR |
|---|---|---|---|---|---|---|---|---|---|---|---|---|---|
| Maximum Temp °F | 21.5 | 27.5 | 40.6 | 58.1 | 71.4 | 79.9 | 83.9 | 81.2 | 72.2 | 60.0 | 40.9 | 27.1 | 55.4 |
| Minimum Temp °F | 1.1 | 6.9 | 20.9 | 34.3 | 45.9 | 55.7 | 60.2 | 57.6 | 48.8 | 37.3 | 23.6 | 10.2 | 33.5 |
| Mean Temp °F | 11.3 | 17.2 | 30.7 | 46.2 | 58.7 | 67.8 | 72.1 | 69.4 | 60.5 | 48.7 | 32.3 | 18.7 | 44.5 |
| Days Max Temp ≥ 90 °F | 0 | 0 | 0 | 0 | 1 | 3 | 6 | 3 | 1 | 0 | 0 | 0 | 14 |
| Days Max Temp ≤ 32 °F | 24 | 17 | 7 | 0 | 0 | 0 | 0 | 0 | 0 | 0 | 7 | 20 | 75 |
| Days Min Temp ≤ 32 °F | 31 | 28 | 27 | 13 | 2 | 0 | 0 | 0 | 1 | 10 | 25 | 30 | 167 |
| Days Min Temp ≤ 0 °F | 15 | 10 | 2 | 0 | 0 | 0 | 0 | 0 | 0 | 0 | 1 | 8 | 36 |
| Heating Degree Days | 1661 | 1343 | 1055 | 561 | 232 | 44 | 9 | 27 | 176 | 504 | 976 | 1429 | 8017 |
| Cooling Degree Days | 0 | 0 | 0 | 5 | 45 | 145 | 240 | 179 | 53 | 2 | 0 | 0 | 669 |
| Total Precipitation (") | 1.14 | 0.94 | 2.17 | 3.03 | 3.84 | 4.75 | 4.23 | 4.20 | 3.87 | 2.72 | 2.07 | 1.23 | 34.19 |
| Days ≥ 0.1" Precip | 4 | 3 | 5 | 7 | 8 | 8 | 7 | 7 | 7 | 6 | 4 | 4 | 70 |
| Total Snowfall (") | 11.0 | 7.5 | 8.9 | 2.2 | 0.0 | 0.0 | 0.0 | 0.0 | 0.0 | 0.2 | 7.0 | 9.5 | 46.3 |
| Days ≥ 1" Snow Depth | 29 | 25 | 18 | 2 | 0 | 0 | 0 | 0 | 0 | 0 | 8 | 24 | 106 |

## ROTHSAY *Wilkin County* ELEVATION 1211 ft LAT/LONG 46° 29 ' N / 96° 16 ' W

| | JAN | FEB | MAR | APR | MAY | JUN | JUL | AUG | SEP | OCT | NOV | DEC | YEAR |
|---|---|---|---|---|---|---|---|---|---|---|---|---|---|
| Maximum Temp °F | 16.3 | 23.4 | 36.9 | 55.8 | 70.1 | 78.0 | 82.8 | 81.2 | 70.9 | 57.5 | 36.6 | 22.2 | 52.6 |
| Minimum Temp °F | -3.7 | 3.9 | 18.0 | 33.0 | 44.9 | 54.5 | 59.1 | 57.0 | 46.9 | 35.5 | 19.6 | 4.5 | 31.1 |
| Mean Temp °F | 6.3 | 13.7 | 27.5 | 44.4 | 57.5 | 66.3 | 70.9 | 69.1 | 58.9 | 46.6 | 28.2 | 13.4 | 41.9 |
| Days Max Temp ≥ 90 °F | 0 | 0 | 0 | 0 | 1 | 2 | 5 | 5 | 1 | 0 | 0 | 0 | 14 |
| Days Max Temp ≤ 32 °F | 26 | 20 | 10 | 1 | 0 | 0 | 0 | 0 | 0 | 0 | 11 | 25 | 93 |
| Days Min Temp ≤ 32 °F | 31 | 28 | *27* | 15 | 3 | 0 | 0 | 0 | 2 | 11 | 27 | 31 | 175 |
| Days Min Temp ≤ 0 °F | 19 | 12 | 4 | 0 | 0 | 0 | 0 | 0 | 0 | 0 | 2 | 13 | 50 |
| Heating Degree Days | 1819 | 1445 | 1156 | 615 | 263 | 62 | 15 | 36 | 222 | 565 | 1098 | 1596 | 8892 |
| Cooling Degree Days | 0 | 0 | *0* | 6 | 46 | 103 | 200 | 159 | 38 | *2* | 0 | 0 | 554 |
| Total Precipitation (") | 0.67 | 0.46 | 1.17 | 1.94 | 2.60 | 3.70 | 3.83 | 2.79 | 2.18 | 1.94 | 0.91 | 0.58 | 22.77 |
| Days ≥ 0.1" Precip | 2 | 2 | 3 | 4 | 6 | 7 | 6 | 5 | 4 | 4 | 3 | 2 | 48 |
| Total Snowfall (") | 10.1 | 6.2 | *7.3* | 2.6 | 0.1 | 0.0 | 0.0 | 0.0 | 0.0 | 0.7 | *6.6* | 6.7 | 40.3 |
| Days ≥ 1" Snow Depth | 28 | 20 | 16 | 2 | 0 | 0 | 0 | 0 | 0 | 0 | *10* | 21 | 97 |

## SANDY LAKE DAM LIBBY *Aitkin County* ELEVATION 1234 ft LAT/LONG 46° 48 ' N / 93° 19 ' W

| | JAN | FEB | MAR | APR | MAY | JUN | JUL | AUG | SEP | OCT | NOV | DEC | YEAR |
|---|---|---|---|---|---|---|---|---|---|---|---|---|---|
| Maximum Temp °F | 19.2 | 26.0 | 38.4 | 54.3 | 67.9 | 75.8 | 80.3 | 77.8 | 67.9 | 56.2 | *37.7* | *23.3* | 52.1 |
| Minimum Temp °F | -3.7 | 1.4 | 15.1 | 29.7 | 42.1 | 51.0 | 56.3 | 54.3 | 45.4 | 35.0 | *20.6* | 3.9 | 29.3 |
| Mean Temp °F | 7.8 | 13.7 | 26.8 | 42.0 | 55.1 | 63.5 | 68.3 | 66.2 | 56.7 | 45.6 | *29.2* | *13.7* | 40.7 |
| Days Max Temp ≥ 90 °F | 0 | 0 | 0 | 0 | 0 | 1 | 2 | 1 | 0 | 0 | 0 | 0 | 4 |
| Days Max Temp ≤ 32 °F | 26 | 18 | 8 | 1 | 0 | 0 | 0 | 0 | 0 | 0 | 9 | 23 | 85 |
| Days Min Temp ≤ 32 °F | 30 | 27 | 28 | 19 | 4 | 0 | 0 | 0 | 2 | 12 | 24 | 29 | 175 |
| Days Min Temp ≤ 0 °F | 17 | 13 | 5 | 0 | 0 | 0 | 0 | 0 | 0 | 0 | 2 | 12 | 49 |
| Heating Degree Days | 1772 | 1442 | 1179 | 683 | 320 | 101 | 32 | 62 | 265 | 595 | *1068* | *1586* | 9105 |
| Cooling Degree Days | 0 | 0 | *0* | 2 | 21 | 67 | 141 | 114 | 19 | *0* | na | na | na |
| Total Precipitation (") | 0.92 | 0.54 | 1.33 | 1.99 | 2.95 | 4.64 | 4.17 | 3.51 | 3.27 | 2.36 | 1.30 | 0.84 | 27.82 |
| Days ≥ 0.1" Precip | 3 | 2 | 4 | 5 | 7 | 8 | 7 | 6 | 7 | 5 | 3 | 3 | 60 |
| Total Snowfall (") | 16.1 | 7.9 | 10.8 | 3.5 | 0.2 | 0.0 | 0.0 | 0.0 | 0.0 | 0.9 | 8.9 | 11.5 | 59.8 |
| Days ≥ 1" Snow Depth | 30 | 28 | 26 | 7 | 0 | 0 | 0 | 0 | 0 | 0 | 11 | 27 | 129 |

## SANTIAGO 3 E *Sherburne County* ELEVATION 1010 ft LAT/LONG 45° 32 ' N / 93° 49 ' W

| | JAN | FEB | MAR | APR | MAY | JUN | JUL | AUG | SEP | OCT | NOV | DEC | YEAR |
|---|---|---|---|---|---|---|---|---|---|---|---|---|---|
| Maximum Temp °F | 20.9 | 27.5 | 40.0 | 57.7 | 71.2 | 79.3 | 84.0 | 81.0 | 71.1 | 59.1 | 39.7 | 25.9 | 54.8 |
| Minimum Temp °F | -2.2 | 3.4 | 18.1 | 32.4 | 44.0 | 52.5 | 57.4 | 54.7 | 45.3 | 34.2 | 20.5 | 6.0 | 30.5 |
| Mean Temp °F | 9.5 | 15.5 | 29.2 | 45.1 | 57.6 | 66.0 | 70.7 | 67.9 | 58.2 | 46.7 | 30.1 | 16.0 | 42.7 |
| Days Max Temp ≥ 90 °F | 0 | 0 | 0 | 0 | 1 | 3 | 6 | 4 | 1 | 0 | 0 | 0 | 15 |
| Days Max Temp ≤ 32 °F | 24 | 17 | 7 | 0 | 0 | 0 | 0 | 0 | 0 | 0 | 8 | 22 | 78 |
| Days Min Temp ≤ 32 °F | 31 | 28 | 28 | 16 | 4 | 0 | 0 | 0 | 3 | 14 | 27 | 31 | 182 |
| Days Min Temp ≤ 0 °F | 17 | 13 | 3 | 0 | 0 | 0 | 0 | 0 | 0 | 0 | 2 | 11 | 46 |
| Heating Degree Days | 1719 | 1394 | 1105 | 595 | 257 | 66 | 16 | 44 | 231 | 565 | 1037 | 1515 | 8544 |
| Cooling Degree Days | 0 | 0 | 0 | 5 | 41 | 107 | 198 | 141 | 31 | 1 | 0 | 0 | 524 |
| Total Precipitation (") | 1.15 | 0.95 | 1.57 | 2.63 | 3.22 | 4.63 | 3.96 | 4.48 | 3.26 | 2.60 | 1.65 | 1.02 | 31.12 |
| Days ≥ 0.1" Precip | 3 | 3 | 4 | 5 | 7 | 8 | 6 | 7 | 6 | 4 | 4 | 3 | 60 |
| Total Snowfall (") | 12.7 | 8.3 | 8.3 | 1.7 | 0.0 | 0.0 | 0.0 | 0.0 | 0.0 | 0.1 | *7.9* | 9.6 | 48.6 |
| Days ≥ 1" Snow Depth | 29 | 27 | 19 | 2 | 0 | 0 | 0 | 0 | 0 | 0 | 8 | 25 | 110 |

## SPRINGFIELD 1 NW *Brown County* ELEVATION 1030 ft LAT/LONG 44° 14 ' N / 94° 58 ' W

| | JAN | FEB | MAR | APR | MAY | JUN | JUL | AUG | SEP | OCT | NOV | DEC | YEAR |
|---|---|---|---|---|---|---|---|---|---|---|---|---|---|
| Maximum Temp °F | 21.5 | 26.6 | 39.5 | 56.7 | 71.5 | 80.5 | 83.5 | 80.3 | 72.4 | 60.0 | 40.8 | 26.5 | 55.0 |
| Minimum Temp °F | 1.7 | 7.3 | 21.3 | 34.3 | 46.4 | 56.5 | 60.3 | 57.1 | 47.5 | 36.0 | 22.8 | 8.9 | 33.3 |
| Mean Temp °F | 11.6 | 17.0 | 30.4 | 45.5 | 59.0 | 68.5 | 71.9 | 68.7 | 60.0 | 48.0 | 31.8 | 17.7 | 44.2 |
| Days Max Temp ≥ 90 °F | 0 | 0 | 0 | 0 | 2 | 5 | 6 | 3 | 1 | 0 | 0 | 0 | 17 |
| Days Max Temp ≤ 32 °F | 23 | 17 | 8 | 1 | 0 | 0 | 0 | 0 | 0 | 0 | 7 | 20 | 76 |
| Days Min Temp ≤ 32 °F | 31 | 28 | 26 | 14 | 2 | 0 | 0 | 0 | 2 | 12 | 26 | 31 | 172 |
| Days Min Temp ≤ 0 °F | 15 | 10 | 2 | 0 | 0 | 0 | 0 | 0 | 0 | 0 | 1 | 9 | 37 |
| Heating Degree Days | 1652 | 1351 | 1065 | 585 | 231 | 43 | 10 | 36 | 195 | 526 | 989 | 1460 | 8143 |
| Cooling Degree Days | 0 | 0 | 0 | 8 | 54 | 155 | 227 | 156 | 48 | 4 | 0 | 0 | 652 |
| Total Precipitation (") | 0.64 | 0.72 | 1.87 | 2.94 | 3.05 | 3.84 | 3.39 | 3.29 | 2.88 | 2.17 | 1.57 | 0.77 | 27.13 |
| Days ≥ 0.1" Precip | 2 | 2 | 4 | 6 | 7 | 7 | 6 | 5 | 6 | 4 | 3 | 2 | 54 |
| Total Snowfall (") | 9.3 | 7.4 | 9.8 | 2.6 | 0.0 | 0.0 | 0.0 | 0.0 | 0.0 | 0.8 | 6.9 | 9.3 | 46.1 |
| Days ≥ 1" Snow Depth | 24 | 20 | 13 | 2 | 0 | 0 | 0 | 0 | 0 | 0 | 6 | 17 | 82 |

## ST CLOUD MUNI ARPT *Sherburne County* ELEVATION 1028 ft LAT/LONG 45° 33 ' N / 94° 4 ' W

| | JAN | FEB | MAR | APR | MAY | JUN | JUL | AUG | SEP | OCT | NOV | DEC | YEAR |
|---|---|---|---|---|---|---|---|---|---|---|---|---|---|
| Maximum Temp °F | 18.6 | 24.9 | 37.8 | 55.2 | 68.7 | 77.2 | 82.1 | 79.3 | 69.0 | 56.6 | 38.0 | 23.9 | 52.6 |
| Minimum Temp °F | -2.3 | 4.0 | 18.1 | 32.5 | 43.7 | 52.9 | 58.1 | 55.4 | 45.8 | 34.5 | 20.6 | 5.8 | 30.8 |
| Mean Temp °F | 8.2 | 14.5 | 28.0 | 43.8 | 56.3 | 65.1 | 70.2 | 67.4 | 57.4 | 45.5 | 29.3 | 14.9 | 41.7 |
| Days Max Temp ≥ 90 °F | 0 | 0 | 0 | 0 | 0 | 2 | 4 | 3 | 0 | 0 | 0 | 0 | 9 |
| Days Max Temp ≤ 32 °F | 26 | 19 | 9 | 1 | 0 | 0 | 0 | 0 | 0 | 0 | 9 | 24 | 88 |
| Days Min Temp ≤ 32 °F | 31 | 28 | 28 | 16 | 3 | 0 | 0 | 0 | 2 | 14 | 27 | 31 | 180 |
| Days Min Temp ≤ 0 °F | 18 | 12 | 4 | 0 | 0 | 0 | 0 | 0 | 0 | 0 | 2 | 11 | 47 |
| Heating Degree Days | 1760 | 1422 | 1141 | 630 | 291 | 78 | 18 | 50 | 251 | 598 | 1064 | 1549 | 8852 |
| Cooling Degree Days | 0 | 0 | 0 | 3 | 30 | 89 | 183 | 130 | 28 | 1 | 0 | 0 | 464 |
| Total Precipitation (") | 0.80 | 0.64 | 1.45 | 2.44 | 2.95 | 4.85 | 3.15 | 3.79 | 3.12 | 2.36 | 1.40 | 0.84 | 27.79 |
| Days ≥ 0.1" Precip | 3 | 2 | 4 | 5 | 7 | 8 | 6 | 6 | 6 | 4 | 4 | 3 | 58 |
| Total Snowfall (") | 11.1 | 7.2 | 8.8 | 2.6 | 0.1 | 0.0 | 0.0 | 0.0 | 0.0 | 0.6 | 8.3 | 9.0 | 47.7 |
| Days ≥ 1" Snow Depth | 28 | 25 | 18 | 2 | 0 | 0 | 0 | 0 | 0 | 0 | 7 | 23 | 103 |

## ST JAMES FILT PLANT *Watonwan County* ELEVATION 1079 ft LAT/LONG 43° 59 ' N / 94° 36 ' W

| | JAN | FEB | MAR | APR | MAY | JUN | JUL | AUG | SEP | OCT | NOV | DEC | YEAR |
|---|---|---|---|---|---|---|---|---|---|---|---|---|---|
| Maximum Temp °F | 22.8 | 28.6 | 41.5 | 58.3 | 72.3 | 81.2 | 84.3 | 81.7 | 73.4 | 61.2 | 41.9 | 27.7 | 56.2 |
| Minimum Temp °F | 3.8 | 9.7 | 22.8 | 35.6 | 47.5 | 57.4 | 61.1 | 58.8 | 49.5 | 37.8 | 24.4 | 10.8 | 34.9 |
| Mean Temp °F | 13.3 | 19.2 | 32.2 | 47.0 | 59.9 | 69.3 | 72.8 | 70.2 | 61.5 | 49.6 | 33.1 | 19.3 | 45.6 |
| Days Max Temp ≥ 90 °F | 0 | 0 | 0 | 0 | 1 | 5 | 7 | 4 | 1 | 0 | 0 | 0 | 18 |
| Days Max Temp ≤ 32 °F | 23 | 16 | 7 | 0 | 0 | 0 | 0 | 0 | 0 | 0 | 7 | 19 | 72 |
| Days Min Temp ≤ 32 °F | 31 | 27 | 25 | 11 | 1 | 0 | 0 | 0 | 1 | 9 | 24 | 31 | 160 |
| Days Min Temp ≤ 0 °F | 13 | 9 | 1 | 0 | 0 | 0 | 0 | 0 | 0 | 0 | 1 | 7 | 31 |
| Heating Degree Days | 1598 | 1289 | 1009 | 540 | 205 | 33 | 8 | 23 | 160 | 479 | 949 | 1412 | 7705 |
| Cooling Degree Days | 0 | 0 | 0 | 8 | 54 | 172 | 253 | ***194*** | 66 | 3 | 0 | 0 | 750 |
| Total Precipitation (") | 0.56 | 0.50 | 1.57 | 2.90 | 3.12 | 4.42 | 3.89 | 3.48 | 3.16 | 2.05 | 1.30 | 0.78 | 27.73 |
| Days ≥ 0.1" Precip | 2 | 2 | 3 | 6 | 7 | 7 | 6 | 6 | 6 | 4 | 3 | 2 | 54 |
| Total Snowfall (") | 8.6 | 6.3 | 8.6 | 2.3 | 0.0 | 0.0 | 0.0 | 0.0 | 0.0 | 0.6 | 5.3 | 8.4 | 40.1 |
| Days ≥ 1" Snow Depth | ***23*** | ***18*** | 12 | 1 | 0 | 0 | 0 | 0 | 0 | 0 | ***4*** | ***17*** | 75 |

## ST PAUL *Ramsey County* ELEVATION 900 ft LAT/LONG 44° 57 ' N / 92° 59 ' W

| | JAN | FEB | MAR | APR | MAY | JUN | JUL | AUG | SEP | OCT | NOV | DEC | YEAR |
|---|---|---|---|---|---|---|---|---|---|---|---|---|---|
| Maximum Temp °F | 22.2 | 28.4 | 41.2 | 58.1 | 70.9 | 79.0 | 83.5 | 80.9 | 71.5 | 59.5 | 40.9 | 26.9 | 55.3 |
| Minimum Temp °F | 4.4 | 10.0 | 22.7 | 36.0 | 47.8 | 57.3 | 62.8 | 60.3 | 51.0 | 39.7 | 25.7 | 12.0 | 35.8 |
| Mean Temp °F | 13.3 | 19.2 | 32.0 | 47.1 | 59.4 | 68.2 | 73.2 | 70.6 | 61.3 | 49.6 | 33.3 | 19.5 | 45.6 |
| Days Max Temp ≥ 90 °F | 0 | 0 | 0 | 0 | 1 | 2 | 5 | 3 | 1 | 0 | 0 | 0 | 12 |
| Days Max Temp ≤ 32 °F | 23 | 17 | 7 | 0 | 0 | 0 | 0 | 0 | 0 | 0 | 7 | 20 | 74 |
| Days Min Temp ≤ 32 °F | 31 | 27 | 25 | 11 | 1 | 0 | 0 | 0 | 0 | 7 | 23 | 30 | 155 |
| Days Min Temp ≤ 0 °F | 13 | 8 | 1 | 0 | 0 | 0 | 0 | 0 | 0 | 0 | 0 | 7 | 29 |
| Heating Degree Days | 1598 | 1289 | 1017 | 537 | 215 | 40 | 6 | 20 | 164 | 475 | 944 | 1405 | 7710 |
| Cooling Degree Days | 0 | 0 | 0 | 8 | 53 | 150 | 281 | 212 | 61 | 2 | 0 | 0 | 767 |
| Total Precipitation (") | 1.00 | 0.85 | 1.90 | 2.74 | 3.48 | 4.95 | 4.15 | 3.99 | 3.33 | 2.67 | 1.92 | 1.10 | 32.08 |
| Days ≥ 0.1" Precip | 3 | 3 | 5 | 6 | 7 | 8 | 7 | 6 | 6 | 5 | 4 | 3 | 63 |
| Total Snowfall (") | 12.7 | 8.9 | 9.8 | 3.0 | 0.0 | 0.0 | 0.0 | 0.0 | 0.0 | 0.4 | 8.4 | 10.7 | 53.9 |
| Days ≥ 1" Snow Depth | 29 | 26 | 17 | 2 | 0 | 0 | 0 | 0 | 0 | 0 | 7 | 23 | 104 |

## ST PETER 2 SW *Nicollet County* ELEVATION 830 ft LAT/LONG 44° 19 ' N / 93° 58 ' W

| | JAN | FEB | MAR | APR | MAY | JUN | JUL | AUG | SEP | OCT | NOV | DEC | YEAR |
|---|---|---|---|---|---|---|---|---|---|---|---|---|---|
| Maximum Temp °F | 22.6 | 28.1 | 41.5 | 58.2 | 71.8 | 80.3 | 84.3 | 81.4 | 72.7 | 60.6 | 41.6 | 27.8 | 55.9 |
| Minimum Temp °F | 1.9 | 7.5 | 21.4 | 34.9 | 46.7 | 56.4 | 61.2 | 58.4 | 48.9 | 37.0 | 24.2 | 10.2 | 34.1 |
| Mean Temp °F | 12.3 | 17.6 | 31.6 | 46.6 | 59.3 | 68.4 | 72.8 | 70.0 | 60.8 | 48.8 | 32.9 | 19.0 | 45.0 |
| Days Max Temp ≥ 90 °F | 0 | 0 | 0 | 0 | 1 | 4 | 7 | 4 | 1 | 0 | 0 | 0 | 17 |
| Days Max Temp ≤ 32 °F | 22 | 16 | 7 | 0 | 0 | 0 | 0 | 0 | 0 | 0 | 6 | 19 | 70 |
| Days Min Temp ≤ 32 °F | 31 | 28 | 26 | 12 | 2 | 0 | 0 | 0 | 1 | 10 | 24 | 30 | 164 |
| Days Min Temp ≤ 0 °F | 14 | 10 | 2 | 0 | 0 | 0 | 0 | 0 | 0 | 0 | 1 | 8 | 35 |
| Heating Degree Days | 1631 | 1332 | 1029 | 552 | 219 | 42 | 7 | 25 | 174 | 500 | 955 | 1421 | 7887 |
| Cooling Degree Days | 0 | 0 | 0 | 9 | 55 | 164 | 264 | 202 | 68 | 3 | 0 | 0 | 765 |
| Total Precipitation (") | 0.85 | 0.63 | 1.79 | 2.65 | 3.29 | 4.63 | 3.89 | 4.12 | 3.07 | 2.18 | 1.50 | 0.99 | 29.59 |
| Days ≥ 0.1" Precip | 3 | 2 | 4 | 6 | 7 | 7 | 6 | 6 | 6 | 4 | 3 | 2 | 56 |
| Total Snowfall (") | 8.8 | 5.4 | 6.1 | *1.5* | 0.0 | 0.0 | 0.0 | 0.0 | 0.0 | 0.1 | *3.8* | *8.6* | 34.3 |
| Days ≥ 1" Snow Depth | *21* | *17* | 11 | 1 | 0 | 0 | 0 | 0 | 0 | 0 | na | *13* | na |

## STEWART *McLeod County* ELEVATION 1040 ft LAT/LONG 44° 44 ' N / 94° 30 ' W

| | JAN | FEB | MAR | APR | MAY | JUN | JUL | AUG | SEP | OCT | NOV | DEC | YEAR |
|---|---|---|---|---|---|---|---|---|---|---|---|---|---|
| Maximum Temp °F | 20.6 | 26.8 | 40.4 | 58.1 | 71.7 | 80.5 | 84.5 | 81.5 | 72.9 | 60.5 | 41.1 | 25.9 | 55.4 |
| Minimum Temp °F | 1.0 | 7.3 | 21.8 | 35.1 | 46.5 | 56.5 | 61.0 | 58.2 | 48.8 | 37.4 | 23.9 | 8.8 | 33.9 |
| Mean Temp °F | 10.8 | 17.1 | 31.2 | 46.6 | 59.1 | 68.5 | 72.8 | 69.9 | 60.9 | 48.9 | 32.5 | 17.3 | 44.6 |
| Days Max Temp ≥ 90 °F | 0 | 0 | 0 | 0 | 1 | 4 | 7 | 4 | 1 | 0 | 0 | 0 | 17 |
| Days Max Temp ≤ 32 °F | 24 | 18 | 8 | 0 | 0 | 0 | 0 | 0 | 0 | 0 | 7 | 21 | 78 |
| Days Min Temp ≤ 32 °F | 31 | 28 | 26 | 12 | 2 | 0 | 0 | 0 | 1 | 10 | 25 | 31 | 166 |
| Days Min Temp ≤ 0 °F | 15 | 10 | 2 | 0 | 0 | 0 | 0 | 0 | 0 | 0 | 1 | 9 | 37 |
| Heating Degree Days | 1676 | 1347 | 1041 | 549 | 222 | 39 | 7 | 24 | 171 | 496 | 968 | 1473 | 8013 |
| Cooling Degree Days | 0 | 0 | 0 | 7 | 45 | 151 | 258 | 184 | 52 | 2 | 0 | 0 | 699 |
| Total Precipitation (") | 0.85 | 0.75 | 1.66 | 2.82 | 2.82 | 4.46 | 3.87 | 4.13 | 3.02 | 2.16 | 1.63 | 0.85 | 29.02 |
| Days ≥ 0.1" Precip | 3 | 3 | 4 | 7 | 7 | 7 | 6 | 6 | 6 | 5 | 4 | 3 | 61 |
| Total Snowfall (") | 11.8 | 8.7 | 9.1 | 2.8 | 0.0 | 0.0 | 0.0 | 0.0 | 0.0 | 0.5 | 7.4 | 9.3 | 49.6 |
| Days ≥ 1" Snow Depth | 26 | 22 | 17 | 2 | 0 | 0 | 0 | 0 | 0 | 0 | 6 | 21 | 94 |

## STILLWATER 1 SE *Washington County* ELEVATION 669 ft LAT/LONG 45° 3 ' N / 92° 48 ' W

| | JAN | FEB | MAR | APR | MAY | JUN | JUL | AUG | SEP | OCT | NOV | DEC | YEAR |
|---|---|---|---|---|---|---|---|---|---|---|---|---|---|
| Maximum Temp °F | 22.4 | 28.7 | 41.2 | 58.0 | 71.3 | 80.0 | 84.7 | 81.7 | 71.6 | 59.5 | 41.3 | 27.3 | 55.6 |
| Minimum Temp °F | 2.1 | 7.7 | 21.2 | 35.3 | 47.4 | 56.7 | 61.8 | 59.2 | 50.1 | 38.8 | 25.1 | 10.4 | 34.7 |
| Mean Temp °F | 12.4 | 18.3 | 31.2 | 46.7 | 59.3 | 68.4 | 73.3 | 70.5 | 60.8 | 49.1 | 33.2 | 18.9 | 45.2 |
| Days Max Temp ≥ 90 °F | 0 | 0 | 0 | 0 | 1 | 3 | 7 | 4 | 0 | 0 | 0 | 0 | 15 |
| Days Max Temp ≤ 32 °F | 23 | 16 | 7 | 0 | 0 | 0 | 0 | 0 | 0 | 0 | 6 | 20 | 72 |
| Days Min Temp ≤ 32 °F | 31 | 28 | 26 | 12 | 1 | 0 | 0 | 0 | 1 | 8 | 24 | 30 | 161 |
| Days Min Temp ≤ 0 °F | 14 | 10 | 2 | 0 | 0 | 0 | 0 | 0 | 0 | 0 | 1 | 8 | 35 |
| Heating Degree Days | 1627 | 1313 | 1042 | 549 | 214 | 38 | 5 | 17 | 169 | 489 | 947 | 1424 | 7834 |
| Cooling Degree Days | 0 | 0 | 0 | 7 | 48 | 150 | 274 | 201 | 56 | 3 | 0 | 0 | 739 |
| Total Precipitation (") | 0.98 | 0.74 | 1.77 | 2.95 | 3.24 | 5.04 | 4.46 | 4.45 | 3.85 | 2.66 | 1.97 | 1.04 | 33.15 |
| Days ≥ 0.1" Precip | 3 | 2 | 4 | 6 | 7 | 8 | 7 | 6 | 6 | 5 | 4 | 3 | 61 |
| Total Snowfall (") | *11.9* | na | na | *0.9* | 0.0 | 0.0 | 0.0 | 0.0 | 0.0 | 0.0 | na | na | na |
| Days ≥ 1" Snow Depth | na | *23* | *15* | *1* | 0 | 0 | 0 | 0 | 0 | 0 | na | na | na |

## THEILMAN *Wabasha County* ELEVATION 741 ft LAT/LONG 44° 18 ' N / 92° 12 ' W

| | JAN | FEB | MAR | APR | MAY | JUN | JUL | AUG | SEP | OCT | NOV | DEC | YEAR |
|---|---|---|---|---|---|---|---|---|---|---|---|---|---|
| Maximum Temp °F | 24.9 | 30.2 | 42.4 | 58.8 | 71.8 | 80.3 | 84.1 | 82.0 | 72.5 | 61.0 | 42.7 | 29.4 | 56.7 |
| Minimum Temp °F | 3.1 | 7.6 | 21.3 | 34.5 | 44.8 | 54.4 | 59.1 | 56.6 | 48.0 | 36.8 | 24.8 | 11.1 | 33.5 |
| Mean Temp °F | 14.0 | 18.9 | 31.9 | 46.7 | 58.3 | 67.4 | 71.6 | 69.3 | 60.3 | 48.9 | 33.8 | 20.3 | 45.1 |
| Days Max Temp ≥ 90 °F | 0 | 0 | 0 | 0 | 1 | 3 | 6 | 3 | 1 | 0 | 0 | 0 | 14 |
| Days Max Temp ≤ 32 °F | 22 | 15 | 6 | 0 | 0 | 0 | 0 | 0 | 0 | 0 | 5 | 18 | 66 |
| Days Min Temp ≤ 32 °F | 31 | 27 | 26 | 14 | 4 | 0 | 0 | 0 | 2 | 12 | 24 | 30 | 170 |
| Days Min Temp ≤ 0 °F | 13 | 9 | 2 | 0 | 0 | 0 | 0 | 0 | 0 | 0 | 1 | 8 | 33 |
| Heating Degree Days | 1573 | 1295 | 1021 | 549 | 238 | 48 | 11 | 27 | 184 | 495 | 929 | 1380 | 7750 |
| Cooling Degree Days | 0 | 0 | 0 | 8 | 41 | 139 | 233 | 180 | 54 | 2 | 0 | 0 | 657 |
| Total Precipitation (") | 0.99 | 0.68 | 1.86 | 3.11 | 3.44 | 4.22 | 4.83 | 3.91 | 3.92 | 2.43 | 1.95 | 1.09 | 32.43 |
| Days ≥ 0.1" Precip | 3 | 2 | 4 | 6 | 7 | 7 | 7 | 7 | 7 | 5 | 4 | 3 | 62 |
| Total Snowfall (") | na | na | na | *0.3* | 0.0 | 0.0 | 0.0 | 0.0 | 0.0 | 0.1 | na | na | na |
| Days ≥ 1" Snow Depth | na | na | na | na | 0 | 0 | 0 | 0 | 0 | 0 | na | na | na |

## THORHULT 1 S *Beltrami County* ELEVATION 1191 ft LAT/LONG 48° 14' N / 95° 15' W

| | JAN | FEB | MAR | APR | MAY | JUN | JUL | AUG | SEP | OCT | NOV | DEC | YEAR |
|---|---|---|---|---|---|---|---|---|---|---|---|---|---|
| Maximum Temp °F | 15.0 | 22.1 | 35.5 | 53.2 | 67.0 | 74.8 | 79.3 | 77.2 | 66.2 | 54.0 | 34.4 | 20.4 | 49.9 |
| Minimum Temp °F | -9.1 | -3.3 | 11.6 | 27.9 | 39.5 | 48.5 | 53.4 | 51.1 | 41.4 | 31.6 | 16.0 | -0.6 | 25.7 |
| Mean Temp °F | 3.2 | 9.6 | 23.6 | 40.7 | 53.2 | 61.7 | 66.4 | 64.3 | 53.9 | 42.7 | 25.0 | 10.0 | 37.9 |
| Days Max Temp ≥ 90 °F | 0 | 0 | 0 | 0 | 0 | 1 | 2 | 2 | 0 | 0 | 0 | 0 | 5 |
| Days Max Temp ≤ 32 °F | 28 | 21 | 11 | 1 | 0 | 0 | 0 | 0 | 0 | 1 | 13 | 26 | 101 |
| Days Min Temp ≤ 32 °F | 31 | 28 | 29 | 21 | 8 | 1 | 0 | 0 | 5 | 17 | 27 | 31 | 198 |
| Days Min Temp ≤ 0 °F | 21 | 16 | 8 | 0 | 0 | 0 | 0 | 0 | 0 | 0 | 4 | 15 | 64 |
| Heating Degree Days | 1916 | 1562 | 1278 | 728 | 376 | 141 | 55 | 99 | 342 | 684 | 1194 | 1702 | 10077 |
| Cooling Degree Days | 0 | 0 | 0 | 2 | 21 | 56 | 116 | 94 | 12 | 0 | 0 | 0 | 301 |
| Total Precipitation (") | 0.54 | 0.36 | 0.83 | 1.46 | 2.45 | 4.04 | 3.56 | 3.83 | 2.65 | 1.95 | 0.81 | 0.59 | 23.07 |
| Days ≥ 0.1" Precip | na | 1 | 2 | 4 | 6 | 8 | 6 | 6 | 5 | 5 | na | na | na |
| Total Snowfall (") | 13.6 | 7.5 | 9.0 | 3.3 | 0.1 | 0.0 | 0.0 | 0.0 | 0.0 | 1.7 | 7.8 | 9.3 | 52.3 |
| Days ≥ 1" Snow Depth | 30 | 26 | 23 | 6 | 0 | 0 | 0 | 0 | 0 | 0 | 11 | 25 | 121 |

## TOWER 3 S *St. Louis County* ELEVATION 1381 ft LAT/LONG 47° 48' N / 92° 17' W

| | JAN | FEB | MAR | APR | MAY | JUN | JUL | AUG | SEP | OCT | NOV | DEC | YEAR |
|---|---|---|---|---|---|---|---|---|---|---|---|---|---|
| Maximum Temp °F | 15.2 | 22.6 | 35.4 | 50.8 | 65.4 | 72.9 | 77.5 | 75.0 | 64.1 | 51.8 | 34.3 | 20.5 | 48.8 |
| Minimum Temp °F | -11.1 | -6.2 | 8.4 | 24.0 | 35.5 | 44.2 | 49.3 | 47.3 | 39.1 | 29.4 | 14.8 | -2.4 | 22.7 |
| Mean Temp °F | 2.1 | 8.2 | 21.9 | 37.4 | 50.5 | 58.6 | 63.4 | 61.1 | 51.6 | 40.6 | 24.6 | 9.1 | 35.8 |
| Days Max Temp ≥ 90 °F | 0 | 0 | 0 | 0 | 0 | 0 | 1 | 1 | 0 | 0 | 0 | 0 | 2 |
| Days Max Temp ≤ 32 °F | 29 | 22 | 12 | 1 | 0 | 0 | 0 | 0 | 0 | 1 | 14 | 26 | 105 |
| Days Min Temp ≤ 32 °F | 31 | 28 | 30 | 25 | 13 | 4 | 1 | 2 | 9 | 20 | 29 | 31 | 223 |
| Days Min Temp ≤ 0 °F | 22 | 17 | 9 | 1 | 0 | 0 | 0 | 0 | 0 | 0 | 5 | 17 | 71 |
| Heating Degree Days | 1948 | 1601 | 1329 | 821 | 452 | 212 | 104 | 159 | 403 | 748 | 1206 | 1729 | 10712 |
| Cooling Degree Days | 0 | 0 | 0 | 1 | 9 | 24 | 60 | 45 | 5 | 0 | 0 | 0 | 144 |
| Total Precipitation (") | 0.98 | 0.71 | 1.18 | 1.98 | 3.20 | 4.63 | 4.25 | 4.06 | 3.99 | 2.93 | 1.41 | 0.86 | 30.18 |
| Days ≥ 0.1" Precip | 3 | 2 | 3 | 5 | 7 | 9 | 9 | 8 | 8 | 6 | 4 | 3 | 67 |
| Total Snowfall (") | 15.4 | 9.3 | 11.3 | 5.2 | 0.4 | 0.0 | 0.0 | 0.0 | 0.0 | 2.2 | 12.9 | 12.4 | 69.1 |
| Days ≥ 1" Snow Depth | 20 | 19 | 17 | na | 0 | 0 | 0 | 0 | 0 | 1 | 11 | 21 | na |

## TRACY *Lyon County* ELEVATION 1403 ft LAT/LONG 44° 14' N / 95° 37' W

| | JAN | FEB | MAR | APR | MAY | JUN | JUL | AUG | SEP | OCT | NOV | DEC | YEAR |
|---|---|---|---|---|---|---|---|---|---|---|---|---|---|
| Maximum Temp °F | 21.3 | 26.1 | 38.7 | 55.0 | 69.6 | 78.9 | 83.2 | 80.5 | 71.4 | 58.5 | 40.0 | 26.4 | 54.1 |
| Minimum Temp °F | 1.7 | 7.4 | 20.8 | 34.5 | 46.7 | 56.8 | 61.3 | 58.3 | 48.2 | 36.4 | 22.7 | 8.9 | 33.6 |
| Mean Temp °F | 11.5 | 16.8 | 29.8 | 44.7 | 58.2 | 67.8 | 72.3 | 69.4 | 59.9 | 47.5 | 31.4 | 17.5 | 43.9 |
| Days Max Temp ≥ 90 °F | 0 | 0 | 0 | 0 | 1 | 3 | 6 | 4 | 1 | 0 | 0 | 0 | 15 |
| Days Max Temp ≤ 32 °F | 23 | 18 | 10 | 1 | 0 | 0 | 0 | 0 | 0 | 0 | 8 | 20 | 80 |
| Days Min Temp ≤ 32 °F | 31 | 28 | 27 | 13 | 2 | 0 | 0 | 0 | 1 | 11 | 25 | 31 | 169 |
| Days Min Temp ≤ 0 °F | 15 | 10 | 2 | 0 | 0 | 0 | 0 | 0 | 0 | 0 | 1 | 9 | 37 |
| Heating Degree Days | 1653 | 1356 | 1085 | 607 | 247 | 50 | 11 | 33 | 193 | 541 | 1000 | 1466 | 8242 |
| Cooling Degree Days | 0 | 0 | 0 | 8 | 47 | 147 | 238 | 176 | 48 | 3 | 0 | 0 | 667 |
| Total Precipitation (") | 0.65 | 0.62 | 1.82 | 2.79 | 3.14 | 4.08 | 3.05 | 3.03 | 3.16 | 2.10 | 1.53 | 0.78 | 26.75 |
| Days ≥ 0.1" Precip | 2 | 2 | 4 | 6 | 6 | 7 | 6 | 5 | 5 | 4 | 3 | 2 | 52 |
| Total Snowfall (") | 8.1 | 5.5 | 10.4 | 3.0 | 0.0 | 0.0 | 0.0 | 0.0 | 0.0 | 0.9 | 7.4 | 8.1 | 43.4 |
| Days ≥ 1" Snow Depth | 26 | 23 | 16 | 3 | 0 | 0 | 0 | 0 | 0 | 1 | 8 | 20 | 97 |

## TWO HARBORS *Lake County* ELEVATION 610 ft LAT/LONG 47° 1' N / 91° 41' W

| | JAN | FEB | MAR | APR | MAY | JUN | JUL | AUG | SEP | OCT | NOV | DEC | YEAR |
|---|---|---|---|---|---|---|---|---|---|---|---|---|---|
| Maximum Temp °F | 22.2 | 26.4 | 35.9 | 48.1 | 57.9 | 66.8 | 74.3 | 73.2 | 64.8 | 53.5 | 38.9 | 27.3 | 49.1 |
| Minimum Temp °F | 3.8 | 7.9 | 19.5 | 30.5 | 38.3 | 44.8 | 53.1 | 54.8 | 47.4 | 37.2 | 25.3 | 11.6 | 31.2 |
| Mean Temp °F | 13.0 | 17.2 | 27.7 | 39.3 | 48.1 | 55.8 | 63.7 | 64.0 | 56.2 | 45.3 | 32.1 | 19.5 | 40.2 |
| Days Max Temp ≥ 90 °F | 0 | 0 | 0 | 0 | 0 | 0 | 1 | 0 | 0 | 0 | 0 | 0 | 1 |
| Days Max Temp ≤ 32 °F | 25 | 19 | 10 | 1 | 0 | 0 | 0 | 0 | 0 | 0 | 7 | 20 | 82 |
| Days Min Temp ≤ 32 °F | 31 | 28 | 28 | 17 | 4 | 0 | 0 | 0 | 1 | 9 | 23 | 30 | 171 |
| Days Min Temp ≤ 0 °F | 14 | 9 | 2 | 0 | 0 | 0 | 0 | 0 | 0 | 0 | 1 | 7 | 33 |
| Heating Degree Days | 1608 | 1344 | 1149 | 764 | 518 | 275 | 93 | 82 | 268 | 603 | 980 | 1406 | 9090 |
| Cooling Degree Days | 0 | 0 | 0 | 0 | 1 | 10 | 71 | 76 | 8 | 0 | 0 | 0 | 166 |
| Total Precipitation (") | 1.09 | 0.56 | 1.72 | 2.14 | 3.01 | 4.10 | 3.75 | 3.96 | 3.84 | 2.49 | 1.96 | 1.19 | 29.81 |
| Days ≥ 0.1" Precip | 3 | 2 | 4 | 5 | 7 | 8 | 6 | 7 | 7 | 5 | 4 | 4 | 62 |
| Total Snowfall (") | na | na | 7.5 | 2.1 | 0.0 | 0.0 | 0.0 | 0.0 | 0.0 | 0.0 | na | na | na |
| Days ≥ 1" Snow Depth | 29 | 28 | 21 | 4 | 0 | 0 | 0 | 0 | 0 | 0 | 8 | na | na |

## WADENA 3 S *Wadena County* ELEVATION 1352 ft LAT/LONG 46° 26' N / 95° 8' W

| | JAN | FEB | MAR | APR | MAY | JUN | JUL | AUG | SEP | OCT | NOV | DEC | YEAR |
|---|---|---|---|---|---|---|---|---|---|---|---|---|---|
| Maximum Temp °F | 15.8 | 22.2 | 35.0 | 52.7 | 66.8 | 74.2 | 79.2 | 77.6 | 66.9 | 54.8 | 35.8 | 21.5 | 50.2 |
| Minimum Temp °F | -5.6 | 0.0 | 14.4 | 29.8 | 42.4 | 52.4 | 57.0 | 54.5 | 44.4 | 32.8 | 18.4 | 2.9 | 28.6 |
| Mean Temp °F | 5.1 | 11.1 | 24.7 | 41.3 | 54.6 | 63.3 | 68.2 | 66.1 | 55.7 | 43.8 | 27.2 | 12.2 | 39.4 |
| Days Max Temp ≥ 90 °F | 0 | 0 | 0 | 0 | 0 | 1 | 2 | 2 | 0 | 0 | 0 | 0 | 5 |
| Days Max Temp ≤ 32 °F | 27 | 21 | 12 | 1 | 0 | 0 | 0 | 0 | 0 | 0 | 12 | 25 | 98 |
| Days Min Temp ≤ 32 °F | 31 | 28 | 29 | 19 | 5 | 0 | 0 | 0 | 3 | 16 | 28 | 31 | 190 |
| Days Min Temp ≤ 0 °F | 19 | 14 | 5 | 0 | 0 | 0 | 0 | 0 | 0 | 0 | 2 | 13 | 53 |
| Heating Degree Days | 1854 | 1519 | 1242 | 706 | 334 | 106 | 33 | 66 | 295 | 651 | 1130 | 1633 | 9569 |
| Cooling Degree Days | 0 | 0 | 0 | 2 | 20 | 60 | 125 | 97 | 16 | 0 | 0 | 0 | 320 |
| Total Precipitation (") | 0.89 | 0.56 | 1.63 | 2.33 | 2.91 | 4.46 | 3.31 | 2.75 | 2.64 | 2.33 | 1.34 | 0.75 | 25.90 |
| Days ≥ 0.1" Precip | 3 | 2 | 4 | 5 | 6 | 8 | 6 | 6 | 6 | 4 | 3 | 3 | 56 |
| Total Snowfall (") | 11.6 | 6.9 | 10.0 | 4.2 | 0.2 | 0.0 | 0.0 | 0.0 | 0.0 | 1.1 | 7.8 | 7.8 | 49.6 |
| Days ≥ 1" Snow Depth | 31 | 27 | 21 | 5 | 0 | 0 | 0 | 0 | 0 | 1 | 10 | 25 | 120 |

## WALKER AH GWAH CHING *Cass County* ELEVATION 1411 ft LAT/LONG 47° 4' N / 94° 35' W

| | JAN | FEB | MAR | APR | MAY | JUN | JUL | AUG | SEP | OCT | NOV | DEC | YEAR |
|---|---|---|---|---|---|---|---|---|---|---|---|---|---|
| Maximum Temp °F | 16.7 | 24.0 | 36.5 | 52.9 | 66.9 | 74.5 | 79.3 | 76.7 | 66.0 | 54.1 | 35.2 | 21.5 | 50.4 |
| Minimum Temp °F | -3.3 | 2.9 | 16.4 | 30.9 | 43.4 | 52.4 | 57.6 | 55.4 | 45.7 | 35.5 | 20.5 | 4.5 | 30.2 |
| Mean Temp °F | 6.7 | 13.5 | 26.5 | 41.9 | 55.2 | 63.5 | 68.5 | 66.1 | 55.8 | 44.8 | 27.9 | 13.1 | 40.3 |
| Days Max Temp ≥ 90 °F | 0 | 0 | 0 | 0 | 0 | 1 | 2 | 2 | 0 | 0 | 0 | 0 | 5 |
| Days Max Temp ≤ 32 °F | 28 | 20 | 11 | 1 | 0 | 0 | 0 | 0 | 0 | 1 | 12 | 26 | 99 |
| Days Min Temp ≤ 32 °F | 31 | 28 | 28 | 18 | 4 | 0 | 0 | 0 | 2 | 12 | 27 | 31 | 181 |
| Days Min Temp ≤ 0 °F | 18 | 12 | 4 | 0 | 0 | 0 | 0 | 0 | 0 | 0 | 2 | 12 | 48 |
| Heating Degree Days | 1805 | 1451 | 1188 | 686 | 319 | 100 | 28 | 63 | 288 | 619 | 1108 | 1606 | 9261 |
| Cooling Degree Days | 0 | 0 | 0 | 2 | 26 | 69 | 151 | 110 | 16 | 0 | 0 | 0 | 374 |
| Total Precipitation (") | 0.81 | 0.51 | 1.48 | 2.37 | 2.83 | 4.14 | 3.82 | 3.35 | 2.81 | 2.73 | 1.23 | 0.88 | 26.96 |
| Days ≥ 0.1" Precip | 3 | 2 | 4 | 6 | 7 | 8 | 7 | 6 | 6 | 5 | 3 | 3 | 60 |
| Total Snowfall (") | 12.5 | 5.9 | 9.3 | 3.2 | 0.0 | 0.0 | 0.0 | 0.0 | 0.0 | 0.5 | 7.9 | 10.5 | 49.8 |
| Days ≥ 1" Snow Depth | 29 | 27 | 24 | 7 | 0 | 0 | 0 | 0 | 0 | 1 | 10 | 27 | 125 |

## WARROAD *Roseau County* ELEVATION 1069 ft LAT/LONG 48° 55' N / 95° 19' W

| | JAN | FEB | MAR | APR | MAY | JUN | JUL | AUG | SEP | OCT | NOV | DEC | YEAR |
|---|---|---|---|---|---|---|---|---|---|---|---|---|---|
| Maximum Temp °F | 12.4 | 19.8 | 32.9 | 50.1 | 64.7 | 73.4 | 78.6 | 76.7 | 65.6 | 52.7 | 33.2 | 18.4 | 48.2 |
| Minimum Temp °F | -10.1 | -4.6 | 10.3 | 27.2 | 41.1 | 51.3 | 56.1 | 53.6 | 43.7 | 32.7 | 16.5 | -0.9 | 26.4 |
| Mean Temp °F | 1.1 | 7.6 | 21.6 | 38.7 | 52.9 | 62.4 | 67.3 | 65.2 | 54.6 | 42.8 | 24.9 | 8.8 | 37.3 |
| Days Max Temp ≥ 90 °F | 0 | 0 | 0 | 0 | 0 | 1 | 2 | 1 | 0 | 0 | 0 | 0 | 4 |
| Days Max Temp ≤ 32 °F | 29 | 23 | 14 | 2 | 0 | 0 | 0 | 0 | 0 | 1 | 14 | 27 | 110 |
| Days Min Temp ≤ 32 °F | 31 | 28 | 30 | 22 | 6 | 0 | 0 | 0 | 3 | 16 | 28 | 31 | 195 |
| Days Min Temp ≤ 0 °F | 22 | 17 | 9 | 0 | 0 | 0 | 0 | 0 | 0 | 0 | 3 | 16 | 67 |
| Heating Degree Days | 1980 | 1617 | 1339 | 783 | 387 | 124 | 40 | 79 | 321 | 684 | 1198 | 1740 | 10292 |
| Cooling Degree Days | 0 | 0 | 0 | 1 | 22 | 57 | 121 | 95 | 10 | 0 | 0 | 0 | 306 |
| Total Precipitation (") | 0.61 | 0.53 | 0.81 | 1.24 | 2.19 | 3.71 | 3.54 | 2.83 | 2.65 | 1.59 | 0.98 | 0.60 | 21.28 |
| Days ≥ 0.1" Precip | 2 | 2 | 2 | 3 | 6 | 7 | 7 | 5 | 6 | 4 | 3 | 2 | 49 |
| Total Snowfall (") | ***8.6*** | ***4.7*** | na | 3.1 | 0.2 | 0.0 | 0.0 | 0.0 | 0.0 | 0.7 | na | na | na |
| Days ≥ 1" Snow Depth | ***19*** | ***20*** | na | 4 | 0 | 0 | 0 | 0 | 0 | 0 | na | ***16*** | na |

## WASECA EXP STATION *Waseca County* ELEVATION 1152 ft LAT/LONG 44° 4' N / 93° 31' W

| | JAN | FEB | MAR | APR | MAY | JUN | JUL | AUG | SEP | OCT | NOV | DEC | YEAR |
|---|---|---|---|---|---|---|---|---|---|---|---|---|---|
| Maximum Temp °F | 19.9 | 25.3 | 38.0 | 54.8 | 69.1 | 78.3 | 81.8 | 79.2 | 71.0 | 58.5 | 40.1 | 25.3 | 53.4 |
| Minimum Temp °F | 0.1 | 6.4 | 20.6 | 34.6 | 46.4 | 56.2 | 60.2 | 57.4 | 48.0 | 36.0 | 23.0 | 8.1 | 33.1 |
| Mean Temp °F | 10.0 | 15.9 | 29.3 | 44.7 | 57.8 | 67.3 | 71.0 | 68.4 | 59.5 | 47.3 | 31.6 | 16.7 | 43.3 |
| Days Max Temp ≥ 90 °F | 0 | 0 | 0 | 0 | 1 | 3 | 4 | 2 | 1 | 0 | 0 | 0 | 11 |
| Days Max Temp ≤ 32 °F | 25 | 19 | 10 | 1 | 0 | 0 | 0 | 0 | 0 | 0 | 8 | 22 | 85 |
| Days Min Temp ≤ 32 °F | 31 | 28 | 27 | 13 | 2 | 0 | 0 | 0 | 1 | 12 | 25 | 31 | 170 |
| Days Min Temp ≤ 0 °F | 15 | 11 | 3 | 0 | 0 | 0 | 0 | 0 | 0 | 0 | 1 | 10 | 40 |
| Heating Degree Days | 1702 | 1382 | 1100 | 608 | 258 | 57 | 15 | 39 | 204 | 548 | 995 | 1491 | 8399 |
| Cooling Degree Days | 0 | 0 | 0 | 6 | 46 | 146 | 215 | 164 | 52 | 3 | 0 | 0 | 632 |
| Total Precipitation (") | 1.17 | 0.97 | 2.36 | 3.26 | 3.68 | 4.30 | 4.38 | 4.35 | 3.57 | 2.55 | 2.12 | 1.39 | 34.10 |
| Days ≥ 0.1" Precip | 3 | 3 | 5 | 8 | 8 | 8 | 7 | 7 | 7 | 5 | 5 | 4 | 70 |
| Total Snowfall (") | 11.3 | 8.5 | 9.9 | 4.0 | 0.0 | 0.0 | 0.0 | 0.0 | 0.0 | 0.4 | 8.2 | 11.7 | 54.0 |
| Days ≥ 1" Snow Depth | 28 | 23 | 17 | 3 | 0 | 0 | 0 | 0 | 0 | 0 | 8 | 24 | 103 |

## WHEATON *Traverse County* ELEVATION 1020 ft LAT/LONG 45° 48 ' N / 96° 29 ' W

| | JAN | FEB | MAR | APR | MAY | JUN | JUL | AUG | SEP | OCT | NOV | DEC | YEAR |
|---|---|---|---|---|---|---|---|---|---|---|---|---|---|
| Maximum Temp °F | 20.4 | 26.6 | 39.6 | 57.7 | 71.8 | 80.0 | 85.4 | 83.7 | 73.8 | 60.8 | 40.0 | 25.9 | 55.5 |
| Minimum Temp °F | 0.3 | 6.6 | 20.2 | 33.9 | 45.7 | 55.4 | 59.9 | 57.8 | 48.0 | 36.3 | 21.8 | 7.6 | 32.8 |
| Mean Temp °F | 10.4 | 16.6 | 29.9 | 45.8 | 58.8 | 67.7 | 72.7 | 70.8 | 60.9 | 48.6 | 30.9 | 16.8 | 44.2 |
| Days Max Temp ≥ 90 °F | 0 | 0 | 0 | 0 | 1 | 3 | 9 | 7 | 2 | 0 | 0 | 0 | 22 |
| Days Max Temp ≤ 32 °F | 24 | 18 | 8 | 0 | 0 | 0 | 0 | 0 | 0 | 0 | 8 | 21 | 79 |
| Days Min Temp ≤ 32 °F | 31 | 28 | 27 | 14 | 3 | 0 | 0 | 0 | 1 | 11 | 26 | 31 | 172 |
| Days Min Temp ≤ 0 °F | 16 | 10 | 3 | 0 | 0 | 0 | 0 | 0 | 0 | 0 | 1 | 9 | 39 |
| Heating Degree Days | 1691 | 1360 | 1081 | 572 | 228 | 44 | 8 | 22 | 174 | 506 | 1015 | 1490 | 8191 |
| Cooling Degree Days | 0 | 0 | 0 | 7 | 51 | 134 | 250 | 202 | 54 | 3 | 0 | 0 | 701 |
| Total Precipitation (") | 0.71 | 0.51 | 1.32 | 2.20 | 2.49 | 3.90 | 3.02 | 2.44 | 1.99 | 1.69 | 1.06 | 0.57 | 21.90 |
| Days ≥ 0.1" Precip | 2 | 2 | 3 | 5 | 6 | 7 | 5 | 5 | 4 | 3 | 3 | 2 | 47 |
| Total Snowfall (") | 8.6 | 6.5 | *7.3* | 2.1 | 0.0 | 0.0 | 0.0 | 0.0 | 0.0 | 0.1 | 5.2 | na | na |
| Days ≥ 1" Snow Depth | na | *22* | *13* | *1* | 0 | 0 | 0 | 0 | 0 | 0 | *4* | na | na |

## WILLMAR STATE HSP *Kandiyohi County* ELEVATION 1132 ft LAT/LONG 45° 8 ' N / 95° 1 ' W

| | JAN | FEB | MAR | APR | MAY | JUN | JUL | AUG | SEP | OCT | NOV | DEC | YEAR |
|---|---|---|---|---|---|---|---|---|---|---|---|---|---|
| Maximum Temp °F | 19.6 | 25.5 | 38.1 | 55.8 | 69.8 | 78.4 | 82.5 | 80.2 | 71.1 | 58.4 | 39.0 | 24.6 | 53.6 |
| Minimum Temp °F | -0.5 | 5.8 | 20.0 | 34.4 | 46.6 | 56.2 | 60.8 | 58.0 | 48.3 | 36.6 | 22.4 | 7.2 | 33.0 |
| Mean Temp °F | 9.8 | 15.9 | 29.0 | 45.2 | 58.2 | 67.3 | 71.7 | 69.1 | 59.7 | 47.6 | 30.7 | 16.0 | 43.4 |
| Days Max Temp ≥ 90 °F | 0 | 0 | 0 | 0 | 1 | 2 | 5 | 3 | 1 | 0 | 0 | 0 | 12 |
| Days Max Temp ≤ 32 °F | 25 | 18 | 9 | 0 | 0 | 0 | 0 | 0 | 0 | 0 | 9 | 22 | 83 |
| Days Min Temp ≤ 32 °F | 31 | 28 | 27 | 13 | 2 | 0 | 0 | 0 | 1 | 10 | 26 | 31 | 169 |
| Days Min Temp ≤ 0 °F | 16 | 11 | 3 | 0 | 0 | 0 | 0 | 0 | 0 | 0 | 2 | 10 | 42 |
| Heating Degree Days | 1710 | 1382 | 1108 | 593 | 241 | 50 | 9 | 28 | 194 | 537 | 1023 | 1512 | 8387 |
| Cooling Degree Days | 0 | 0 | 0 | 6 | 44 | 125 | 213 | 154 | 36 | 1 | 0 | 0 | 579 |
| Total Precipitation (") | 0.83 | 0.75 | 1.54 | 2.54 | 3.15 | 5.33 | 3.56 | 3.81 | 3.02 | 2.29 | 1.42 | 0.82 | 29.06 |
| Days ≥ 0.1" Precip | 2 | 2 | 4 | 6 | 6 | 8 | 6 | 6 | 6 | 4 | 3 | 3 | 56 |
| Total Snowfall (") | 11.4 | 8.8 | 9.6 | 1.9 | 0.0 | 0.0 | 0.0 | 0.0 | 0.0 | 0.4 | 7.8 | 9.1 | 49.0 |
| Days ≥ 1" Snow Depth | na | *15* | *9* | *1* | 0 | 0 | 0 | 0 | 0 | 0 | na | na | na |

## WINDOM *Cottonwood County* ELEVATION 1362 ft LAT/LONG 43° 52 ' N / 95° 7 ' W

| | JAN | FEB | MAR | APR | MAY | JUN | JUL | AUG | SEP | OCT | NOV | DEC | YEAR |
|---|---|---|---|---|---|---|---|---|---|---|---|---|---|
| Maximum Temp °F | 22.7 | 28.3 | 41.7 | 57.2 | 71.3 | 80.5 | 84.6 | 81.4 | 72.6 | 60.4 | 41.3 | 27.3 | 55.8 |
| Minimum Temp °F | 2.9 | 8.5 | 22.1 | 34.5 | 46.4 | 56.4 | 61.3 | 58.4 | 48.7 | 36.2 | 23.2 | 9.9 | 34.0 |
| Mean Temp °F | 12.8 | 18.4 | 31.9 | 45.9 | 58.9 | 68.5 | 72.9 | 70.0 | 60.7 | 48.3 | 32.3 | 18.6 | 44.9 |
| Days Max Temp ≥ 90 °F | 0 | 0 | 0 | 0 | 1 | 4 | 8 | 4 | 1 | 0 | 0 | 0 | 18 |
| Days Max Temp ≤ 32 °F | 23 | 17 | 7 | 1 | 0 | 0 | 0 | 0 | 0 | 0 | 7 | 19 | 74 |
| Days Min Temp ≤ 32 °F | 31 | 28 | 26 | 13 | 2 | 0 | 0 | 0 | 1 | 11 | 25 | 31 | 168 |
| Days Min Temp ≤ 0 °F | 14 | 9 | 2 | 0 | 0 | 0 | 0 | 0 | 0 | 0 | 1 | 8 | 34 |
| Heating Degree Days | 1614 | 1310 | 1019 | 572 | 228 | 41 | 9 | 26 | 174 | 515 | 976 | 1432 | 7916 |
| Cooling Degree Days | 0 | 0 | 0 | 7 | 42 | 147 | 241 | 174 | 46 | 2 | 0 | 0 | 659 |
| Total Precipitation (") | 0.73 | 0.64 | 1.94 | 3.02 | 3.33 | 4.42 | 3.91 | 3.43 | 3.25 | 2.02 | 1.57 | 0.85 | 29.11 |
| Days ≥ 0.1" Precip | 2 | 2 | 4 | 6 | 7 | 7 | 6 | 6 | 6 | 4 | 3 | 2 | 55 |
| Total Snowfall (") | 8.4 | 6.4 | 8.7 | 3.0 | 0.0 | 0.0 | 0.0 | 0.0 | 0.0 | 0.8 | 6.2 | 8.4 | 41.9 |
| Days ≥ 1" Snow Depth | 26 | 21 | 15 | 3 | 0 | 0 | 0 | 0 | 0 | 0 | 8 | 20 | 93 |

## WINNEBAGO *Faribault County* ELEVATION 1112 ft LAT/LONG 43° 46 ' N / 94° 10 ' W

| | JAN | FEB | MAR | APR | MAY | JUN | JUL | AUG | SEP | OCT | NOV | DEC | YEAR |
|---|---|---|---|---|---|---|---|---|---|---|---|---|---|
| Maximum Temp °F | 20.4 | 26.1 | 38.9 | 55.5 | 69.5 | 78.7 | 82.4 | 79.6 | 71.1 | 58.7 | 40.4 | 25.8 | 53.9 |
| Minimum Temp °F | 1.6 | 7.5 | 20.8 | 34.0 | 46.1 | 56.5 | 60.9 | 58.1 | 48.6 | 36.5 | 23.3 | 9.3 | 33.6 |
| Mean Temp °F | 11.0 | 16.8 | 29.8 | 44.8 | 57.8 | 67.6 | 71.7 | 68.9 | 59.9 | 47.6 | 31.9 | 17.5 | 43.8 |
| Days Max Temp ≥ 90 °F | 0 | 0 | 0 | 0 | 1 | 3 | 5 | 3 | 1 | 0 | 0 | 0 | 13 |
| Days Max Temp ≤ 32 °F | 24 | 18 | 9 | 1 | 0 | 0 | 0 | 0 | 0 | 0 | 8 | 21 | 81 |
| Days Min Temp ≤ 32 °F | 31 | 28 | 27 | 13 | 2 | 0 | 0 | 0 | 1 | 11 | 25 | 31 | 169 |
| Days Min Temp ≤ 0 °F | 15 | 10 | 2 | 0 | 0 | 0 | 0 | 0 | 0 | 0 | 1 | 9 | 37 |
| Heating Degree Days | 1670 | 1356 | 1084 | 604 | 256 | 50 | 12 | 33 | 193 | 534 | 987 | 1466 | 8245 |
| Cooling Degree Days | 0 | 0 | 0 | 7 | 45 | 144 | 224 | 164 | 47 | 2 | 0 | 0 | 633 |
| Total Precipitation (") | 0.90 | 0.82 | 1.77 | 2.92 | 3.65 | 4.71 | 4.12 | 3.98 | 3.12 | 2.40 | 1.68 | 1.10 | 31.17 |
| Days ≥ 0.1" Precip | 3 | 2 | 5 | 7 | 8 | 8 | 7 | 7 | 6 | 5 | 4 | 3 | 65 |
| Total Snowfall (") | 10.7 | 7.9 | 8.4 | 3.0 | 0.0 | 0.0 | 0.0 | 0.0 | 0.0 | 0.5 | 6.0 | 10.7 | 47.2 |
| Days ≥ 1" Snow Depth | 27 | 22 | 16 | 3 | 0 | 0 | 0 | 0 | 0 | 0 | 7 | 21 | 96 |

## WINNIBIGOSHISH DAM *Itasca County* ELEVATION 1322 ft LAT/LONG 47° 26 ' N / 94° 3 ' W

| | JAN | FEB | MAR | APR | MAY | JUN | JUL | AUG | SEP | OCT | NOV | DEC | YEAR |
|---|---|---|---|---|---|---|---|---|---|---|---|---|---|
| Maximum Temp °F | *16.2* | *23.9* | *36.9* | *53.8* | 67.7 | 75.5 | 79.9 | 77.3 | 66.4 | *54.2* | *35.4* | *21.8* | 50.8 |
| Minimum Temp °F | *-5.7* | *0.5* | *13.6* | *29.1* | 41.8 | 51.5 | 57.1 | 54.8 | 45.4 | *35.4* | *20.2* | *3.8* | 29.0 |
| Mean Temp °F | *5.2* | *12.2* | *25.3* | *41.5* | 54.8 | 63.5 | 68.5 | 66.1 | 55.9 | *44.8* | *27.8* | *12.8* | 39.9 |
| Days Max Temp ≥ 90 °F | 0 | 0 | 0 | 0 | 0 | 1 | 2 | 1 | 0 | 0 | 0 | 0 | 4 |
| Days Max Temp ≤ 32 °F | 25 | 19 | 9 | 1 | 0 | 0 | 0 | 0 | 0 | 0 | 11 | 23 | 88 |
| Days Min Temp ≤ 32 °F | 28 | 26 | 26 | 19 | 5 | 0 | 0 | 0 | 1 | 11 | 25 | 28 | 169 |
| Days Min Temp ≤ 0 °F | 18 | 13 | 6 | 0 | 0 | 0 | 0 | 0 | 0 | 0 | 2 | 12 | 51 |
| Heating Degree Days | *1851* | *1485* | *1226* | *701* | 328 | 97 | 25 | 57 | 282 | *618* | *1110* | *1614* | 9394 |
| Cooling Degree Days | na | na | na | na | 21 | 61 | 132 | 101 | *10* | na | na | na | na |
| Total Precipitation (") | 0.92 | 0.56 | 1.36 | 1.89 | 2.67 | 4.03 | 4.00 | 3.53 | 2.94 | 2.64 | 1.35 | 0.99 | 26.88 |
| Days ≥ 0.1" Precip | 3 | 2 | 4 | 5 | 6 | 8 | 7 | 6 | 7 | 5 | 4 | 4 | 61 |
| Total Snowfall (") | 10.8 | 5.9 | 9.2 | 3.3 | 0.0 | 0.0 | 0.0 | 0.0 | 0.0 | 0.6 | 7.4 | 9.9 | 47.1 |
| Days ≥ 1" Snow Depth | 29 | 27 | 24 | 7 | 0 | 0 | 0 | 0 | 0 | 1 | 12 | 29 | 129 |

## WINONA *Winona County* ELEVATION 702 ft LAT/LONG 44° 3 ' N / 91° 39 ' W

| | JAN | FEB | MAR | APR | MAY | JUN | JUL | AUG | SEP | OCT | NOV | DEC | YEAR |
|---|---|---|---|---|---|---|---|---|---|---|---|---|---|
| Maximum Temp °F | 23.9 | 30.1 | 42.5 | 58.6 | 70.6 | 80.0 | 84.5 | 81.3 | 72.8 | 60.8 | 44.2 | 29.4 | 56.6 |
| Minimum Temp °F | 3.2 | 8.4 | 22.1 | 36.1 | 46.5 | 56.7 | 62.1 | 58.7 | 49.1 | 38.2 | 26.0 | 11.5 | 34.9 |
| Mean Temp °F | 13.6 | 19.3 | 32.3 | 47.4 | 58.6 | 68.3 | 73.3 | 70.0 | 61.0 | 49.5 | 35.1 | 20.5 | 45.7 |
| Days Max Temp ≥ 90 °F | 0 | 0 | 0 | 0 | 1 | 3 | 7 | 4 | 1 | 0 | 0 | 0 | 16 |
| Days Max Temp ≤ 32 °F | 22 | 15 | 6 | 0 | 0 | 0 | 0 | 0 | 0 | 0 | 4 | 17 | 64 |
| Days Min Temp ≤ 32 °F | 31 | 27 | 25 | 11 | *1* | 0 | 0 | 0 | 1 | 9 | 22 | 30 | 157 |
| Days Min Temp ≤ 0 °F | 14 | 9 | 2 | 0 | 0 | 0 | 0 | 0 | 0 | 0 | 0 | 7 | 32 |
| Heating Degree Days | 1591 | 1285 | 1009 | 532 | 238 | 41 | 6 | 22 | 166 | 478 | 889 | 1373 | 7630 |
| Cooling Degree Days | 0 | 0 | 0 | 14 | *52* | 172 | 303 | *212* | *58* | 3 | 0 | 0 | 814 |
| Total Precipitation (") | *1.29* | 0.82 | 1.82 | 3.42 | 4.03 | 4.42 | 4.48 | 4.29 | 4.13 | 2.35 | 2.10 | 1.36 | 34.51 |
| Days ≥ 0.1" Precip | 4 | 2 | 4 | 6 | 8 | 8 | 7 | *6* | 7 | 6 | 4 | 4 | 66 |
| Total Snowfall (") | 13.0 | 8.5 | *7.1* | 2.0 | 0.0 | 0.0 | 0.0 | 0.0 | 0.0 | *0.4* | *3.7* | *10.5* | 45.2 |
| Days ≥ 1" Snow Depth | na | *20* | na | *1* | 0 | 0 | 0 | 0 | 0 | 0 | *2* | na | na |

## WINTON POWER PLANT *Lake County* ELEVATION 1342 ft LAT/LONG 47° 56 ' N / 91° 46 ' W

| | JAN | FEB | MAR | APR | MAY | JUN | JUL | AUG | SEP | OCT | NOV | DEC | YEAR |
|---|---|---|---|---|---|---|---|---|---|---|---|---|---|
| Maximum Temp °F | 14.5 | 21.9 | 34.9 | 49.9 | 64.7 | 72.7 | 77.8 | 75.0 | 63.5 | 50.4 | 32.3 | 19.0 | 48.0 |
| Minimum Temp °F | -7.5 | -2.4 | 11.3 | 27.0 | 40.4 | 50.2 | 55.9 | 53.7 | 44.6 | 33.8 | 18.5 | 1.1 | 27.2 |
| Mean Temp °F | 3.6 | 9.8 | 23.1 | 38.5 | 52.6 | 61.5 | 66.9 | 64.4 | 54.1 | 42.1 | 25.4 | 10.1 | 37.7 |
| Days Max Temp ≥ 90 °F | 0 | 0 | 0 | 0 | 0 | 0 | 2 | 1 | 0 | 0 | 0 | 0 | 3 |
| Days Max Temp ≤ 32 °F | 28 | 22 | 13 | 2 | 0 | 0 | 0 | 0 | 0 | 1 | 15 | 27 | 108 |
| Days Min Temp ≤ 32 °F | 31 | 28 | 29 | 22 | 6 | 0 | 0 | 0 | 2 | 14 | 28 | 30 | 190 |
| Days Min Temp ≤ 0 °F | 21 | 16 | 8 | 0 | 0 | 0 | 0 | 0 | 0 | 0 | 2 | 15 | 62 |
| Heating Degree Days | 1904 | 1556 | 1292 | 789 | 394 | 140 | 44 | 86 | 331 | 702 | 1181 | 1699 | 10118 |
| Cooling Degree Days | 0 | 0 | 0 | 0 | 17 | 43 | 112 | 81 | 8 | 0 | 0 | 0 | 261 |
| Total Precipitation (") | 1.01 | 0.68 | 1.26 | 1.78 | 2.93 | 4.28 | 3.77 | 3.76 | 3.61 | 2.64 | 1.61 | 1.02 | 28.35 |
| Days ≥ 0.1" Precip | 4 | 2 | 4 | 5 | 7 | 9 | 8 | 8 | 8 | 6 | 4 | 4 | 69 |
| Total Snowfall (") | *12.9* | 7.0 | *7.8* | 3.4 | 0.3 | 0.0 | 0.0 | 0.0 | 0.0 | 0.6 | *8.1* | *10.3* | 50.4 |
| Days ≥ 1" Snow Depth | 31 | 28 | 30 | 13 | 0 | 0 | 0 | 0 | 0 | 1 | 16 | 30 | 149 |

## WRIGHT 4 NW *Aitkin County* ELEVATION 1302 ft LAT/LONG 46° 42 ' N / 93° 4 ' W

| | JAN | FEB | MAR | APR | MAY | JUN | JUL | AUG | SEP | OCT | NOV | DEC | YEAR |
|---|---|---|---|---|---|---|---|---|---|---|---|---|---|
| Maximum Temp °F | 17.8 | 24.9 | 36.9 | 53.0 | 67.0 | 74.6 | 79.3 | 76.8 | 66.7 | 54.5 | 36.0 | 22.5 | 50.8 |
| Minimum Temp °F | -4.7 | 1.0 | 14.5 | 28.6 | 39.3 | 47.6 | 53.5 | 51.7 | 43.3 | 33.2 | 19.2 | 3.8 | 27.6 |
| Mean Temp °F | 6.6 | 13.0 | 25.7 | 40.8 | 53.2 | 61.1 | 66.4 | 64.3 | 55.0 | 43.9 | 27.6 | 13.2 | 39.2 |
| Days Max Temp ≥ 90 °F | 0 | 0 | 0 | 0 | 0 | 0 | 2 | 1 | 0 | 0 | 0 | 0 | 3 |
| Days Max Temp ≤ 32 °F | 27 | 20 | 10 | 1 | 0 | 0 | 0 | 0 | 0 | 1 | 12 | 25 | 96 |
| Days Min Temp ≤ 32 °F | 31 | 28 | 29 | 21 | 8 | 1 | 0 | 0 | 5 | 15 | 27 | 31 | 196 |
| Days Min Temp ≤ 0 °F | 18 | 13 | 6 | 0 | 0 | 0 | 0 | 0 | 0 | 0 | 2 | 13 | 52 |
| Heating Degree Days | 1809 | 1465 | 1211 | 719 | 369 | 144 | 52 | 92 | 307 | 648 | 1114 | 1602 | 9532 |
| Cooling Degree Days | 0 | 0 | 0 | 1 | 12 | 36 | 102 | 83 | 12 | 1 | 0 | 0 | 247 |
| Total Precipitation (") | 0.94 | 0.58 | 1.45 | 2.23 | 3.25 | 4.41 | 3.84 | 3.69 | 3.36 | 2.59 | 1.63 | 1.01 | 28.98 |
| Days ≥ 0.1" Precip | 3 | 2 | 4 | 5 | 7 | 8 | 7 | 7 | 7 | 5 | 4 | 3 | 62 |
| Total Snowfall (") | 13.8 | 7.1 | 10.3 | 4.6 | 0.4 | 0.0 | 0.0 | 0.0 | 0.0 | 1.6 | 11.1 | 11.2 | 60.1 |
| Days ≥ 1" Snow Depth | 31 | 28 | 26 | 7 | 0 | 0 | 0 | 0 | 0 | 1 | 12 | 29 | 134 |

**ZUMBROTA** *Goodhue County* ELEVATION 991 ft LAT/LONG 44° 18 ' N / 92° 40 ' W

| | JAN | FEB | MAR | APR | MAY | JUN | JUL | AUG | SEP | OCT | NOV | DEC | YEAR |
|---|---|---|---|---|---|---|---|---|---|---|---|---|---|
| Maximum Temp °F | 22.8 | 28.2 | 41.4 | 58.4 | 71.1 | 80.1 | 84.1 | 81.6 | 72.7 | 60.8 | 42.0 | 28.0 | 55.9 |
| Minimum Temp °F | 1.7 | 6.4 | 20.8 | 34.1 | 44.9 | 54.5 | 59.0 | 56.6 | 48.0 | 36.6 | 24.0 | 10.2 | 33.1 |
| Mean Temp °F | 12.3 | 17.3 | 31.2 | 46.3 | 58.1 | 67.3 | 71.6 | 69.1 | 60.4 | 48.8 | 33.0 | 19.2 | 44.6 |
| Days Max Temp ≥ 90 °F | 0 | 0 | 0 | 0 | 0 | 3 | 6 | 4 | 1 | 0 | 0 | 0 | 14 |
| Days Max Temp ≤ 32 °F | 23 | 17 | 7 | 0 | 0 | 0 | 0 | 0 | 0 | 0 | 6 | 19 | 72 |
| Days Min Temp ≤ 32 °F | 31 | 28 | 26 | 14 | 3 | 0 | 0 | 0 | 2 | 12 | 24 | 30 | 170 |
| Days Min Temp ≤ 0 °F | 14 | 11 | 3 | 0 | 0 | 0 | 0 | 0 | 0 | 0 | 1 | 8 | 37 |
| Heating Degree Days | 1631 | 1342 | 1042 | 561 | 249 | 55 | 13 | 33 | 185 | 502 | 952 | 1415 | 7980 |
| Cooling Degree Days | 0 | 0 | 0 | 8 | 44 | 142 | 225 | 176 | 60 | 3 | 0 | 0 | 658 |
| Total Precipitation (") | 0.98 | 0.72 | 1.88 | 3.24 | 3.69 | 4.23 | 4.34 | 3.84 | 3.70 | 2.71 | 1.98 | 1.15 | 32.46 |
| Days ≥ 0.1" Precip | 3 | 2 | 5 | 7 | 8 | 8 | 7 | 6 | 6 | 5 | 4 | 4 | 65 |
| Total Snowfall (") | 9.5 | 7.1 | 7.6 | 2.5 | 0.0 | 0.0 | 0.0 | 0.0 | 0.0 | 0.5 | 5.4 | 8.8 | 41.4 |
| Days ≥ 1" Snow Depth | na | ***15*** | na | 1 | 0 | 0 | 0 | 0 | 0 | 0 | ***4*** | ***14*** | na |

## JANUARY MINIMUM TEMPERATURE °F

### LOWEST

| | | |
|---|---|---|
| 1 | Tower | -11.1 |
| 2 | Cass Lake | -11.0 |
| 3 | Roseau | -10.8 |
| 4 | Warroad | -10.1 |
| 5 | International Falls | -9.7 |
| 6 | Argyle | -9.4 |
| | Hallock | -9.4 |
| 8 | Thorhult | -9.1 |
| 9 | Baudette | -9.0 |
| 10 | Big Falls | -8.8 |
| 11 | Agassiz | -8.5 |
| 12 | Oklee | -8.4 |
| 13 | Red Lake | -8.3 |
| 14 | Fosston | -8.0 |
| 15 | Cotton | -7.9 |
| 16 | Bemidji | -7.8 |
| 17 | Brainerd | -7.6 |
| | Itasca | -7.6 |
| 19 | Winton | -7.5 |
| 20 | Cook | -7.1 |
| 21 | Gunflint Lake | -6.9 |
| 22 | Red Lake Falls | -6.7 |
| 23 | Hibbing | -6.6 |
| 24 | Detroit Lakes | -6.3 |
| 25 | Crookston | -6.2 |

### HIGHEST

| | | |
|---|---|---|
| 1 | St. Paul | 4.4 |
| 2 | Grand Marais | 3.9 |
| 3 | St. James | 3.8 |
| | Two Harbors | 3.8 |
| 5 | Fairmont | 3.7 |
| 6 | Austin | 3.4 |
| | Luverne | 3.4 |
| | Redwood Falls | 3.4 |
| 9 | Caledonia | 3.3 |
| 10 | Owatonna | 3.2 |
| | Winona | 3.2 |
| 12 | Minneapolis | 3.1 |
| | New Ulm | 3.1 |
| | Theilman | 3.1 |
| 15 | Canby | 2.9 |
| | Windom | 2.9 |
| 17 | Albert Lea | 2.6 |
| | Duluth-Harbor | 2.6 |
| | Rochester | 2.6 |
| 20 | Gaylord | 2.4 |
| 21 | Farmington | 2.3 |
| 22 | Marshall | 2.1 |
| | Stillwater | 2.1 |
| 24 | Chaska | 2.0 |
| 25 | Preston | 1.9 |

## JULY MAXIMUM TEMPERATURE °F

### HIGHEST

| | | |
|---|---|---|
| 1 | Canby | 86.0 |
| 2 | Madison | 85.9 |
| 3 | Wheaton | 85.4 |
| 4 | Luverne | 85.3 |
| 5 | Chaska | 85.2 |
| 6 | Montevideo | 85.0 |
| 7 | Gaylord | 84.7 |
| | Stillwater | 84.7 |
| 9 | New Ulm | 84.6 |
| | Owatonna | 84.6 |
| | Redwood Falls | 84.6 |
| | Windom | 84.6 |
| 13 | Stewart | 84.5 |
| | Winona | 84.5 |
| 15 | Milan | 84.4 |
| 16 | St. James | 84.3 |
| | St. Peter | 84.3 |
| 18 | Pipestone | 84.2 |
| 19 | Litchfield | 84.1 |
| | Theilman | 84.1 |
| | Zumbrota | 84.1 |
| 22 | Santiago | 84.0 |
| 23 | Buffalo | 83.9 |
| | Rosemount | 83.9 |
| 25 | Benson | 83.7 |

### LOWEST

| | | |
|---|---|---|
| 1 | Grand Marais | 69.3 |
| 2 | Duluth-Harbor | 73.8 |
| 3 | Two Harbors | 74.3 |
| 4 | Duluth-Intl | 76.3 |
| 5 | Hibbing | 77.2 |
| 6 | Tower | 77.5 |
| 7 | Roseau | 77.8 |
| | Winton | 77.8 |
| 9 | Red Lake | 78.1 |
| 10 | Gunflint Lake | 78.2 |
| 11 | International Falls | 78.3 |
| 12 | Cook | 78.5 |
| 13 | Cotton | 78.6 |
| | Kelliher | 78.6 |
| | Warroad | 78.6 |
| 16 | Bemidji | 78.7 |
| 17 | Wadena | 79.2 |
| 18 | Isle | 79.3 |
| | Thorhult | 79.3 |
| | Walker | 79.3 |
| | Wright | 79.3 |
| 22 | Cass Lake | 79.6 |
| 23 | Agassiz | 79.7 |
| | Aitkin | 79.7 |
| 25 | Winnibigshsh Dm | 79.9 |

## ANNUAL PRECIPITATION (")

### HIGHEST

| | | |
|---|---|---|
| 1 | Caledonia | 35.42 |
| 2 | Winona | 34.51 |
| 3 | Rosemount | 34.19 |
| 4 | Grand Meadow | 34.11 |
| 5 | Waseca | 34.10 |
| 6 | Preston | 33.56 |
| 7 | Stillwater | 33.15 |
| 8 | Albert Lea | 33.04 |
| 9 | Zumbrota | 32.46 |
| 10 | Theilman | 32.43 |
| 11 | St. Paul | 32.08 |
| 12 | Austin | 31.89 |
| 13 | Faribault | 31.66 |
| 14 | Owatonna | 31.48 |
| 15 | Cloquet | 31.45 |
| 16 | Fairmont | 31.41 |
| 17 | Cedar | 31.28 |
| 18 | Winnebago | 31.17 |
| 19 | Santiago | 31.12 |
| 20 | Farmington | 31.10 |
| 21 | Hinckley | 31.03 |
| 22 | New London | 30.99 |
| 23 | Duluth-Intl | 30.84 |
| 24 | Forest Lake | 30.77 |
| 25 | Rochester | 30.74 |

### LOWEST

| | | |
|---|---|---|
| 1 | Hallock | 18.89 |
| 2 | Argyle | 19.36 |
| 3 | Crookston | 20.12 |
| 4 | Georgetown | 21.14 |
| 5 | Roseau | 21.20 |
| 6 | Warroad | 21.28 |
| 7 | Agassiz | 21.64 |
| 8 | Wheaton | 21.90 |
| 9 | Campbell | 21.92 |
| 10 | Red Lake Falls | 22.44 |
| 11 | Baudette | 22.72 |
| | Red Lake | 22.72 |
| 13 | Rothsay | 22.77 |
| 14 | Thorhult | 23.07 |
| 15 | Oklee | 23.17 |
| 16 | Ada | 23.22 |
| 17 | Fergus Falls | 23.30 |
| 18 | Mahnomen | 23.67 |
| 19 | Artichoke Lake | 24.00 |
| 20 | International Falls | 24.11 |
| 21 | Fosston | 24.24 |
| 22 | Bemidji | 24.29 |
| 23 | Glenwood | 24.40 |
| 24 | Milan | 24.44 |
| 25 | Detroit Lakes | 24.46 |

## ANNUAL SNOWFALL (")

### HIGHEST

| | | |
|---|---|---|
| 1 | Duluth-Intl | 81.4 |
| 2 | International Falls | 69.6 |
| 3 | Tower | 69.1 |
| 4 | Cloquet | 68.7 |
| 5 | Hibbing | 60.6 |
| 6 | Wright | 60.1 |
| 7 | Big Falls | 59.9 |
| 8 | Sandy Lake Dam | 59.8 |
| 9 | Grand Rapids | 57.8 |
| 10 | Minneapolis | 57.2 |
| 11 | Ottertail | 56.7 |
| 12 | Hinckley | 55.3 |
| 13 | New London | 54.0 |
| | Waseca | 54.0 |
| 15 | St. Paul | 53.9 |
| 16 | Itasca | 53.1 |
| 17 | Cotton | 52.8 |
| | Grand Marais | 52.8 |
| 19 | Thorhult | 52.3 |
| 20 | Collegeville | 51.6 |
| | Leech Lake Dam | 51.6 |
| | Park Rapids | 51.6 |
| 23 | Long Prairie | 50.6 |
| 24 | Cedar | 50.5 |
| 25 | Winton | 50.4 |

### LOWEST

| | | |
|---|---|---|
| 1 | Campbell | 30.2 |
| 2 | Georgetown | 34.2 |
| 3 | St. Peter | 34.3 |
| 4 | Pipestone | 34.5 |
| 5 | Glenwood | 34.8 |
| 6 | Redwood Falls | 35.6 |
| 7 | Crookston | 35.7 |
| 8 | Madison | 36.8 |
| 9 | Preston | 37.3 |
| 10 | Bemidji | 38.0 |
| | Oklee | 38.0 |
| 12 | Agassiz | 38.6 |
| 13 | Hutchinson | 38.7 |
| 14 | Argyle | 39.3 |
| 15 | Moose Lake | 40.1 |
| | St. James | 40.1 |
| 17 | Rothsay | 40.3 |
| 18 | Albert Lea | 40.7 |
| 19 | Artichoke Lake | 40.8 |
| 20 | Duluth-Harbor | 41.2 |
| 21 | Zumbrota | 41.4 |
| 22 | Canby | 41.5 |
| 23 | Baudette | 41.6 |
| | Lamberton | 41.6 |
| 25 | Owatonna | 41.8 |

# MISSISSIPPI

PHYSICAL FEATURES. The State of Mississippi extends on the west from the Mississippi River to about longitude 88° W., between latitude 31° and 35° N; and from the Pearl River on the west to about 88½° W. longitude below latitude 31° N. The southern boundary of this area, a sort of "panhandle", is Mississippi Sound, which is an arm of the Gulf of Mexico. Land areas near the coast line, in contrast to those of Louisiana, are sharply defined, with the land rising to elevations of 10 to 20 feet behind the beaches. The coast is cut by numerous bays. A string of islands parallels the coast a few miles offshore. The waters of Mississippi Sound provide a natural air conditioning to ameliorate the summer heat. Thus Biloxi has an average of only 55 days with temperature 90° F. or higher, while only 40 miles inland Wiggins averages 105 such days.

A triangular area comprising nearly one-third of the State, with its apex in Rankin County and its base on the coast, is composed of rolling hills at from 200 to 500 feet above sea level. The "Delta" region in the northwest extends from the Yazoo-Tallahatchie River system westward to the Mississippi River. Between the Delta and the upland prairie the land is broken by a series of ridges and valleys which are oriented in a general southwest-northeast direction. These extend from the Tennessee border to the lower Mississippi River. From Vicksburg to Natchez these ridges stop abruptly at the river forming high bluffs along its left bank. The valleys form natural paths for the northeastward passage of tornadoes and for southward drainage of cold winter air. On clear, cold nights temperatures in these valleys are lower, sometimes as much as 20° F., than on the nearby hilltops.

GENERAL CLIMATE. In its broader aspects the climate of Mississippi is determined by the huge land mass to the north, its subtropical latitude, and the Gulf of Mexico to the south, but modifications are introduced by the varied topography.

The prevailing southerly winds provide a moist, semitropical climate, with conditions often favorable for afternoon thundershowers. When the pressure distribution is altered so as to bring westerly or northerly winds, periods of hotter and drier weather interrupt the prevailing moist condition. The high humidity, combined with hot days and nights in the interior from May to September, produces discomfort at times. The principal relief is by thunderstorms, sometimes accompanied by locally violent and destructive winds.

In the colder season the State is alternately subjected to warm tropical air and cold continental air, in periods of varying length. However, cold spells seldom last over 3 or 4 days. The ground rarely freezes, and then mostly only in the extreme north and only a few inches deep. Although slowly warmed by its southward journey, the cold air occasionally brings large and rather sudden drops in temperature. In winter the Atlantic High is also sometimes located far enough west to serve as a barrier to cold air approaching the State. Most frequently this produces a pattern of warm, clear weather over the southern part of the State with cold, rainy weather to the north of the "front", but occasionally the entire State will be under the balmy influence of this subtropical anticyclone.

Mississippi is south of the average track of winter cyclones, but occasionally one moves over the State. In some winters a succession of such cyclones will develop in the Gulf of Mexico or in Texas and move over or near the State. The State is also occasionally in the path of tropical storms or hurricanes.

FLOODS. All of the State is in the Gulf of Mexico drainage. Main rivers which flow directly into the Gulf include the Tombigbee in the northeast portion, and the Pascagoula and Pearl which forms the southwestern boundary. The Mississippi River forms most of the western boundary. The flood season in Mississippi is from November through June (the period of greatest rainfall), with March and April being the months of greatest frequency. The season of high flows in the main Mississippi River is during the first 6 months of the year. In other streams flooding sometimes occurs during the summer from persistent thundershower rains, or during the late summer and early fall from heavy rains associated with tropical storms originating in the Gulf of Mexico.

PRECIPITATION. Mean annual precipitation ranges from about 50 inches in the northwest to 65 inches in the southeast. During the freeze-free season rainfall ranges from 23 to 25 inches in the Delta districts to 36 to 38 inches in the southeast. During the winter the precipitation maximum is centered over the northern and western counties (16 to 18 inches) with the minimum (13 inches) on the coast. In summer the maximum shifts to the coastal counties (19 to 21 inches) and the minimum to the Delta counties (9 to 11 inches). The spring and fall patterns are very similar to the summer pattern. The fall months are the driest of the year, precipitation ranging from about 8 to 13 inches. Fall is the most agreeable season of the year, with cool nights and mild, clear, sunny days persisting for several days, and even weeks, at a time.

While snowfall is not of much economic importance, it is not such a rare event in Mississippi as is generally believed. Measurable snow or sleet falls on some part of the State most years.

TEMPERATURE. The normal annual temperature ranges from 62° F. in the northern border counties to 68° F. in the coastal counties. The lowest January normal is 43° F. in the north-central area, ranging upward to 54° F. in the coastal district. The highest July normal is 84° F. in the upper Delta, ranging downward to 80° to 81° in parts of central and north-central districts. Temperatures of 90° F. or higher occur on an average of 55 days per year on the immediate Gulf coast under the ameliorating effect of the relatively cooler Gulf waters. There is a rapid increase in number of days 90° F. or higher inland from the coast, reaching a maximum of 105 such days in Stone County. Temperatures of 32° F. (freezing) or lower occur on an average of 11 days a year on the immediate Gulf coast, increasing to a maximum of 60 days in Panola County. Temperatures exceed 100° F. at one or more stations each summer. They drop to zero or lower in Mississippi on an average of once in 5 years and to 32° F. or lower on the Gulf coast almost every winter.

STORMS. Thunderstorms occur on an average of 50 to 60 days a year in the northern districts and 70 to 80 days a year near the coast. Thundershowers occur more frequently in July than any other month, with the least in December. Those in late fall, winter, and early spring are more apt to be attended by high winds than in summer. However, in the interior in summer after a spell of unusually high temperatures, thunderstorms may develop with local violence.

A hazard to life and property in Mississippi is the tropical cyclone which occurs from June to November. While these storms generally move into the State on the coast, they have on occasion entered as far north as Meridian and Greenville after crossing part of Alabama or Louisiana. These latter storms are usually weakened considerably by passage over land. Hurricanes which move inland over southeast Louisiana may be as damaging on the Mississippi coast as those which cross the coast line. This is especially true of those moving from the southeast because of the usually more severe winds in the northeast quadrant and because of the high seas which move across Mississippi Sound and pile up on the shore. Those which move westward offshore often cause tide and wind damage on the coast. Those which move northeastward across or south of the Louisiana Delta and move inland between Mobile and Panama City are usually less damaging because winds are offshore and tides are subnormal. Hurricanes which move inland on the Alabama coast may affect Mississippi only slightly because of less intense and offshore winds in their western portions.

About a fourth of the tropical cyclones which affect Mississippi are of hurricane intensity with winds of 74 m.p.h. or higher at some point. One-half of all hurricanes occur in September and twice as many tropical cyclones occur in August and September as during June, July, October, and November combined.

Tornadoes occur in all months in Mississippi, but the largest number of reported tornadoes occur in March, while April, February, and May rank high. Tornadoes may occur at any place in Mississippi, but are least likely in the tier of counties within the "panhandle" below 31° N. latitude; this, however, is the area where hurricanes are most likely to occur.

## COUNTY INDEX

### *Sunflower County*
MOORHEAD

### *Tate County*
INDEPENDENCE 1 W

### *Tippah County*
RIPLEY

### *Tunica County*
TUNICA 2

### *Walthall County*
TYLERTOWN 2 WNW

### *Washington County*
GREENVILLE
STONEVILLE EXP STN

### *Wayne County*
WAYNESBORO 2 W

### *Webster County*
EUPORA 2 E

### *Wilkinson County*
WOODVILLE 4 ESE

### *Winston County*
LOUISVILLE

### *Yalobusha County*
WATER VALLEY 1 NNE

### *Yazoo County*
YAZOO CITY 5 NNE

## ELEVATION INDEX

| FEET | STATION NAME |
|---|---|
| 11 | PASCAGOULA 3 NE |
| 20 | BILOXI |
| 30 | GULFPORT NAVAL CNTR |
| 49 | MERRILL |
| 49 | PICAYUNE |
| 112 | BELZONI |
| 112 | YAZOO CITY 5 NNE |
| 121 | MOORHEAD |
| 131 | GREENVILLE |
| 131 | GREENWOOD LEFLORE AP |
| 131 | STONEVILLE EXP STN |
| 151 | COLUMBIA |
| 151 | MINTER CITY 1 NE |
| 161 | PORT GIBSON 1 NW |
| 180 | CLARKSDALE |
| 190 | GRENADA 5 NNE |
| 190 | TUNICA 2 |
| 200 | RICHTON 3 SSE |
| 210 | ABERDEEN |
| 210 | MONTICELLO |
| 210 | NATCHEZ |
| 210 | OAKLEY EXP STA |
| 220 | HATTIESBURG |
| 220 | LEXINGTON 2 NNW |
| 220 | PICKENS |
| 220 | WAYNESBORO 2 W |
| 230 | BATESVILLE 2 SW |
| 230 | QUITMAN 1 N |
| 230 | SAUCIER EXP FOREST |
| 239 | LAUREL |
| 245 | WIGGINS |
| 279 | CALHOUN CITY 2 NW |
| 289 | COLLINS |
| 291 | JACKSON THOMPSON FLD |
| 302 | WATER VALLEY 1 NNE |
| 312 | MERIDIAN KEY FLD |
| 312 | POPLARVILLE EXP STN |
| 322 | KIPLING |
| 335 | D LO 2 SW |
| 341 | FULTON 3 W |
| 341 | HOUSTON |
| 345 | LIBERTY 5 W |
| 351 | NEWTON EXP STN |
| 351 | STATE UNIVERSITY |
| 361 | INDEPENDENCE 1 W |
| 361 | PELAHATCHIE |
| 361 | TUPELO C D LEMONS AP |
| 363 | HERNANDO |
| 370 | CARTHAGE 3 SW |
| 385 | CORINTH CITY |
| 390 | TYLERTOWN 2 WNW |
| 390 | WOODVILLE 4 ESE |
| 410 | HICKORY FLAT |
| 410 | PONTOTOC EXP STN |
| 413 | PHILADELPHIA 1 WSW |
| 420 | KOSCIUSKO |
| 430 | EUPORA 2 E |
| 458 | MCCOMB PIKE CO AP |
| 469 | WINONA 5 E |
| 479 | BROOKHAVEN CITY |
| 479 | FOREST 3 S |
| 479 | HOLLY SPRINGS 4 N |
| 502 | RIPLEY |
| 502 | UNIVERSITY |
| 531 | BOONEVILLE |
| 561 | LOUISVILLE |

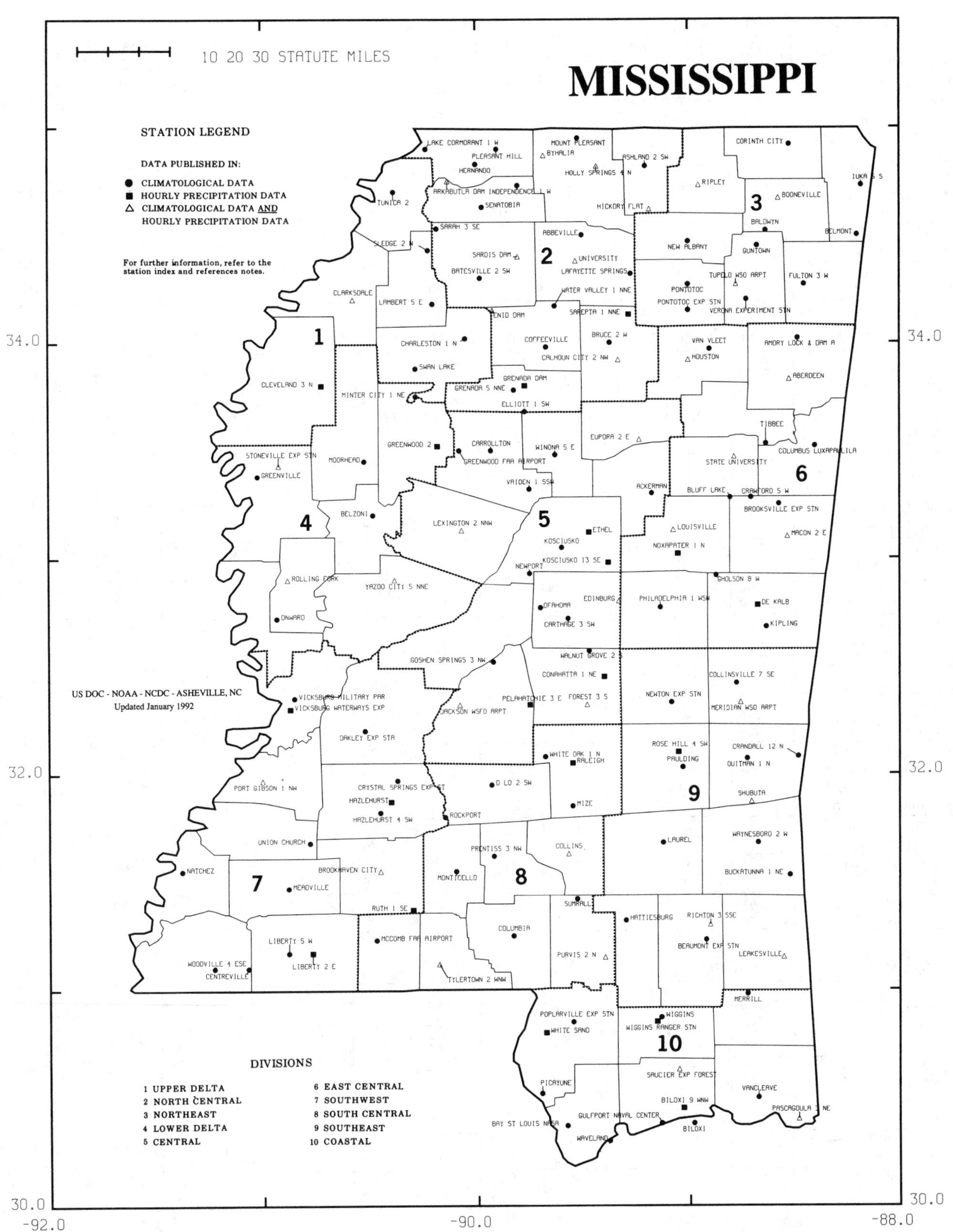

10 20 30 STATUTE MILES
MISSISSIPPI
STATION LEGEND
DATA PUBLISHED IN:
● CLIMATOLOGICAL DATA
■ HOURLY PRECIPITATION DATA
△ CLIMATOLOGICAL DATA AND HOURLY PRECIPITATION DATA
For further information, refer to the station index and references notes.
US DOC - NOAA - NCDC - ASHEVILLE, NC
Updated January 1992
DIVISIONS
1 UPPER DELTA
2 NORTH CENTRAL
3 NORTHEAST
4 LOWER DELTA
5 CENTRAL
6 EAST CENTRAL
7 SOUTHWEST
8 SOUTH CENTRAL
9 SOUTHEAST
10 COASTAL
LAKE CORMORANT 1 W
PLEASANT HILL
HERNANDO
MOUNT PLEASANT
BYHALIA
ASHLAND 2 SW
HOLLY SPRINGS 4 N
CORINTH CITY
IUKA 5 S
RIPLEY
BOONEVILLE
TUNICA 2
ARKABUTLA DAM
INDEPENDENCE 1 W
SENATOBIA
HICKORY FLAT
BALDWYN
BELMONT
SARAH 3 SE
SLEDGE 2 W
ABBEVILLE
NEW ALBANY
GUNTOWN
SARDIS DAM
UNIVERSITY
BATESVILLE 2 SW
LAFAYETTE SPRINGS
TUPELO WSO ARPT
FULTON 3 W
CLARKSDALE
LAMBERT 5 E
WATER VALLEY 1 NNE
PONTOTOC
PONTOTOC EXP STN
VERONA EXPERIMENT STN
ENID DAM
SAREPTA 1 NNE
CHARLESTON 1 N
COFFEEVILLE
BRUCE 2 W
VAN VLEET
HOUSTON
AMORY LOCK & DAM A
CALHOUN CITY 2 NW
SWAN LAKE
ABERDEEN
CLEVELAND 3 N
MINTER CITY 1 NE
GRENADA DAM
GRENADA 5 NNE
ELLIOTT 1 SW
TIBBEE
EUPORA 2 E
GREENWOOD 2
CARROLLTON
WINONA 5 E
COLUMBUS LUXAPALILA
STONEVILLE EXP STN
MOORHEAD
GREENWOOD FAA AIRPORT
STATE UNIVERSITY
GREENVILLE
VAIDEN 1 SSW
ACKERMAN
BLUFF LAKE
CRAWFORD 5 W
BELZONI
BROOKSVILLE EXP STN
LEXINGTON 2 NNW
ETHEL
LOUISVILLE
MACON 2 E
KOSCIUSKO
NOXAPATER 1 N
KOSCIUSKO 13 SE
NEWPORT
ROLLING FORK
SHOLSON 8 W
YAZOO CITY 5 NNE
EDINBURG
PHILADELPHIA 1 WSW
DE KALB
ONWARD
OFAHOMA
CARTHAGE 3 SW
KIPLING
WALNUT GROVE 2
GOSHEN SPRINGS 3 NW
CONAHATTA 1 NE
COLLINSVILLE 7 SE
VICKSBURG MILITARY PAR
VICKSBURG WATERWAYS EXP
PELAHATCHIE 3 E
FOREST 3 S
NEWTON EXP STN
JACKSON WSFO ARPT
MERIDIAN WSO ARPT
OAKLEY EXP STA
ROSE HILL 4 SW
CRANDALL 12 N
WHITE OAK 1 N
RALEIGH
PAULDING
QUITMAN 1 N
PORT GIBSON 1 NW
CRYSTAL SPRINGS EXP ST
D LO 2 SW
HAZLEHURST
SHUBUTA
MIZE
HAZLEHURST 4 SW
ROCKPORT
UNION CHURCH
LAUREL
WAYNESBORO 2 W
PRENTISS 3 NW
COLLINS
NATCHEZ
BROOKHAVEN CITY
MONTICELLO
BUCKATUNNA 1 NE
MEADVILLE
RUTH 1 SE
SUMRALL
HATTIESBURG
RICHTON 3 SSE
COLUMBIA
LIBERTY 5 W
MCCOMB FAA AIRPORT
BEAUMONT EXP STN
LEAKESVILLE
PURVIS 2 N
WOODVILLE 4 ESE
CENTREVILLE
LIBERTY 2 E
TYLERTOWN 2 WNW
MERRILL
POPLARVILLE EXP STN
WIGGINS
WHITE SAND
WIGGINS RANGER STN
SAUCIER EXP FOREST
PICAYUNE
VANCLEAVE
BILOXI 9 WNW
PASCAGOULA 3 NE
GULFPORT NAVAL CENTER
BAY ST LOUIS NASA
BILOXI
WAVELAND
1
2
3
4
5
6
7
8
9
10
34.0
32.0
30.0
-92.0
-90.0
-88.0

## ABERDEEN *Monroe County* ELEVATION 210 ft LAT/LONG 33° 50' N / 88° 33' W

| | JAN | FEB | MAR | APR | MAY | JUN | JUL | AUG | SEP | OCT | NOV | DEC | YEAR |
|---|---|---|---|---|---|---|---|---|---|---|---|---|---|
| Maximum Temp °F | 53.0 | 57.7 | 67.0 | 76.3 | 82.1 | 88.7 | 91.6 | 90.1 | 84.6 | 75.4 | 65.4 | 57.1 | 74.1 |
| Minimum Temp °F | 32.9 | 35.4 | 42.8 | 50.9 | 58.9 | 66.4 | 70.5 | 69.1 | 63.4 | 50.9 | 42.4 | 35.9 | 51.6 |
| Mean Temp °F | 43.0 | 46.6 | 54.9 | 63.6 | 70.5 | 77.6 | 81.1 | 79.6 | 74.0 | 63.2 | 53.9 | 46.5 | 62.9 |
| Days Max Temp ≥ 90 °F | 0 | 0 | 0 | 0 | 3 | 14 | 22 | 18 | 6 | 0 | 0 | 0 | 63 |
| Days Max Temp ≤ 32 °F | 1 | 0 | 0 | 0 | 0 | 0 | 0 | 0 | 0 | 0 | 0 | 0 | 1 |
| Days Min Temp ≤ 32 °F | 16 | 12 | 5 | 1 | 0 | 0 | 0 | 0 | 0 | 0 | 6 | 13 | 53 |
| Days Min Temp ≤ 0 °F | 0 | 0 | 0 | 0 | 0 | 0 | 0 | 0 | 0 | 0 | 0 | 0 | 0 |
| Heating Degree Days | 678 | 516 | 322 | 111 | 17 | 0 | 0 | 0 | 10 | 119 | 335 | 569 | 2677 |
| Cooling Degree Days | 0 | 3 | 17 | 70 | 197 | 392 | 524 | 475 | 291 | 76 | 12 | 4 | 2061 |
| Total Precipitation (") | 4.98 | 4.92 | 6.13 | 4.93 | 5.50 | 4.01 | 3.78 | 3.15 | 3.89 | 4.13 | 4.43 | 5.44 | 55.29 |
| Days ≥ 0.1" Precip | 8 | 7 | 7 | 6 | 8 | 7 | 6 | 6 | 6 | 5 | 7 | 7 | 80 |
| Total Snowfall (") | 0.9 | 0.3 | 0.5 | 0.0 | 0.0 | 0.0 | 0.0 | 0.0 | 0.0 | 0.0 | 0.0 | 0.1 | 1.8 |
| Days ≥ 1" Snow Depth | 1 | 0 | 0 | 0 | 0 | 0 | 0 | 0 | 0 | 0 | 0 | 0 | 1 |

## BATESVILLE 2 SW *Panola County* ELEVATION 230 ft LAT/LONG 34° 19' N / 89° 57' W

| | JAN | FEB | MAR | APR | MAY | JUN | JUL | AUG | SEP | OCT | NOV | DEC | YEAR |
|---|---|---|---|---|---|---|---|---|---|---|---|---|---|
| Maximum Temp °F | 49.4 | 54.3 | 63.7 | 73.5 | 80.5 | 88.1 | 91.3 | 90.3 | 85.0 | 75.1 | 63.6 | 54.1 | 72.4 |
| Minimum Temp °F | 28.5 | 31.1 | 39.6 | 48.7 | 57.1 | 65.2 | 68.4 | 65.6 | 59.4 | 47.0 | 38.7 | 32.2 | 48.5 |
| Mean Temp °F | 39.0 | 42.8 | 51.7 | 61.1 | 68.8 | 76.6 | 79.9 | 78.0 | 72.2 | 61.1 | 51.2 | 43.1 | 60.5 |
| Days Max Temp ≥ 90 °F | 0 | 0 | 0 | 0 | 2 | 13 | 21 | 19 | 9 | 1 | 0 | 0 | 65 |
| Days Max Temp ≤ 32 °F | 3 | 1 | 0 | 0 | 0 | 0 | 0 | 0 | 0 | 0 | 0 | 1 | 5 |
| Days Min Temp ≤ 32 °F | 21 | 17 | 9 | 2 | 0 | 0 | 0 | 0 | 0 | 2 | 10 | 18 | 79 |
| Days Min Temp ≤ 0 °F | 0 | 0 | 0 | 0 | 0 | 0 | 0 | 0 | 0 | 0 | 0 | 0 | 0 |
| Heating Degree Days | 799 | 622 | 416 | 168 | 39 | 1 | 0 | 0 | 22 | 172 | 416 | 673 | 3328 |
| Cooling Degree Days | 0 | 1 | 11 | 56 | 172 | 371 | 477 | 430 | 252 | 61 | 11 | 3 | 1845 |
| Total Precipitation (") | 4.39 | 4.70 | 5.80 | 5.17 | 5.96 | 4.49 | 3.91 | 3.04 | 3.50 | 3.63 | 5.54 | 5.99 | 56.12 |
| Days ≥ 0.1" Precip | 7 | 6 | 7 | 6 | 7 | 6 | 5 | 5 | 5 | 5 | 6 | 7 | 72 |
| Total Snowfall (") | ***0.7*** | 0.2 | 0.6 | 0.0 | 0.0 | 0.0 | 0.0 | 0.0 | 0.0 | 0.0 | 0.0 | 0.1 | 1.6 |
| Days ≥ 1" Snow Depth | ***0*** | 0 | 0 | 0 | 0 | 0 | 0 | 0 | 0 | 0 | 0 | 0 | 0 |

## BELZONI *Humphreys County* ELEVATION 112 ft LAT/LONG 33° 11' N / 90° 29' W

| | JAN | FEB | MAR | APR | MAY | JUN | JUL | AUG | SEP | OCT | NOV | DEC | YEAR |
|---|---|---|---|---|---|---|---|---|---|---|---|---|---|
| Maximum Temp °F | 51.4 | 56.6 | 65.8 | 75.5 | 83.0 | 90.3 | 92.5 | 91.8 | 86.9 | 77.2 | 66.0 | 56.2 | 74.4 |
| Minimum Temp °F | 33.2 | 36.3 | 44.6 | 53.1 | 61.1 | 68.5 | 71.3 | 69.1 | 63.1 | 51.1 | 43.4 | 36.3 | 52.6 |
| Mean Temp °F | 42.3 | 46.5 | 55.2 | 64.3 | 72.1 | 79.4 | 81.9 | 80.5 | 75.0 | 64.2 | 54.7 | 46.2 | 63.5 |
| Days Max Temp ≥ 90 °F | 0 | 0 | 0 | 1 | 5 | 19 | 24 | 23 | 12 | 1 | 0 | 0 | 85 |
| Days Max Temp ≤ 32 °F | 2 | 1 | 0 | 0 | 0 | 0 | 0 | 0 | 0 | 0 | 0 | 1 | 4 |
| Days Min Temp ≤ 32 °F | ***14*** | ***9*** | 3 | 0 | 0 | 0 | 0 | 0 | 0 | 0 | 4 | ***10*** | 40 |
| Days Min Temp ≤ 0 °F | 0 | 0 | 0 | 0 | 0 | 0 | 0 | 0 | 0 | 0 | 0 | 0 | 0 |
| Heating Degree Days | 696 | 518 | 316 | 105 | 13 | 0 | 0 | 0 | 10 | 113 | 317 | ***578*** | 2666 |
| Cooling Degree Days | 0 | 2 | 21 | 92 | 252 | 455 | 546 | 509 | 330 | 102 | 23 | ***4*** | 2336 |
| Total Precipitation (") | 5.21 | 4.87 | 6.04 | 5.55 | 5.75 | 3.80 | 5.09 | 3.10 | 3.01 | 3.90 | 4.80 | 6.01 | 57.13 |
| Days ≥ 0.1" Precip | 7 | 6 | 7 | 6 | 7 | 6 | 6 | 5 | 5 | 5 | 6 | 7 | 73 |
| Total Snowfall (") | 0.1 | 0.0 | 0.0 | 0.0 | 0.0 | 0.0 | 0.0 | 0.0 | 0.0 | 0.0 | 0.0 | 0.0 | 0.1 |
| Days ≥ 1" Snow Depth | 0 | 0 | 0 | 0 | 0 | 0 | 0 | 0 | 0 | 0 | 0 | 0 | 0 |

## BILOXI *Harrison County* ELEVATION 20 ft LAT/LONG 30° 24' N / 88° 54' W

| | JAN | FEB | MAR | APR | MAY | JUN | JUL | AUG | SEP | OCT | NOV | DEC | YEAR |
|---|---|---|---|---|---|---|---|---|---|---|---|---|---|
| Maximum Temp °F | ***60.2*** | 63.0 | ***68.7*** | ***76.3*** | 82.6 | ***88.3*** | ***90.1*** | ***89.6*** | 86.5 | 79.6 | 70.5 | ***64.2*** | 76.6 |
| Minimum Temp °F | ***42.9*** | ***44.8*** | ***51.3*** | ***59.6*** | ***66.1*** | ***72.7*** | ***74.5*** | ***74.0*** | 70.1 | 60.0 | 51.7 | ***46.2*** | 59.5 |
| Mean Temp °F | ***51.7*** | ***53.9*** | ***60.0*** | ***68.0*** | ***74.3*** | ***80.5*** | ***82.3*** | ***81.8*** | 78.3 | 69.8 | 61.1 | ***55.2*** | 68.1 |
| Days Max Temp ≥ 90 °F | ***0*** | 0 | ***0*** | 0 | 1 | 9 | 16 | 15 | 6 | 0 | 0 | ***0*** | 47 |
| Days Max Temp ≤ 32 °F | ***0*** | 0 | ***0*** | 0 | 0 | 0 | 0 | 0 | 0 | 0 | 0 | ***0*** | 0 |
| Days Min Temp ≤ 32 °F | ***6*** | ***3*** | ***1*** | 0 | 0 | 0 | 0 | 0 | 0 | 0 | 1 | ***3*** | 14 |
| Days Min Temp ≤ 0 °F | ***0*** | 0 | ***0*** | 0 | 0 | 0 | 0 | 0 | 0 | 0 | 0 | ***0*** | 0 |
| Heating Degree Days | ***410*** | ***312*** | ***177*** | ***37*** | ***1*** | ***0*** | ***0*** | ***0*** | 1 | 26 | 162 | ***309*** | 1435 |
| Cooling Degree Days | na | ***2*** | na | na | ***299*** | na | na | na | ***409*** | ***187*** | ***56*** | na | na |
| Total Precipitation (") | 5.87 | 5.89 | 5.42 | 4.51 | 5.11 | 4.95 | 7.02 | 6.67 | 5.35 | 3.73 | 4.20 | 5.19 | 63.91 |
| Days ≥ 0.1" Precip | 8 | 7 | ***7*** | 5 | 5 | 6 | 10 | 9 | 6 | 4 | 6 | 7 | 80 |
| Total Snowfall (") | 0.0 | 0.0 | 0.1 | 0.0 | 0.0 | 0.0 | 0.0 | 0.0 | 0.0 | 0.0 | 0.0 | 0.0 | 0.1 |
| Days ≥ 1" Snow Depth | 0 | ***0*** | 0 | 0 | 0 | 0 | 0 | 0 | 0 | 0 | 0 | 0 | 0 |

## BOONEVILLE *Prentiss County* ELEVATION 531 ft LAT/LONG 34° 39' N / 88° 34' W

| | JAN | FEB | MAR | APR | MAY | JUN | JUL | AUG | SEP | OCT | NOV | DEC | YEAR |
|---|---|---|---|---|---|---|---|---|---|---|---|---|---|
| Maximum Temp °F | 47.5 | 52.8 | 62.7 | 72.4 | 78.9 | 86.3 | 89.6 | 88.7 | 82.8 | 72.5 | 61.6 | 52.1 | 70.7 |
| Minimum Temp °F | 28.3 | 31.5 | 40.4 | 48.9 | 57.2 | 65.2 | 69.1 | 67.4 | 60.8 | 48.2 | 40.1 | 32.5 | 49.1 |
| Mean Temp °F | 37.9 | 42.2 | 51.6 | 60.7 | 68.1 | 75.8 | 79.4 | 78.1 | 71.9 | 60.4 | 50.9 | 42.4 | 60.0 |
| Days Max Temp ≥ 90 °F | 0 | 0 | 0 | 0 | 1 | 9 | 17 | 14 | 5 | 0 | 0 | 0 | 46 |
| Days Max Temp ≤ 32 °F | 3 | 1 | 0 | 0 | 0 | 0 | 0 | 0 | 0 | 0 | 0 | 2 | 6 |
| Days Min Temp ≤ 32 °F | 20 | 16 | 8 | 1 | 0 | 0 | 0 | 0 | 0 | 1 | 8 | 16 | 70 |
| Days Min Temp ≤ 0 °F | 0 | 0 | 0 | 0 | 0 | 0 | 0 | 0 | 0 | 0 | 0 | 0 | 0 |
| Heating Degree Days | 833 | 638 | 419 | 174 | 42 | 1 | 0 | 0 | 22 | 182 | 421 | 695 | 3427 |
| Cooling Degree Days | 0 | 0 | 9 | 46 | 143 | 341 | 469 | 431 | 240 | 46 | 6 | 1 | 1732 |
| Total Precipitation (") | 4.86 | 4.73 | 6.07 | 5.43 | 6.40 | 3.90 | 4.16 | 3.36 | 3.80 | 3.70 | 5.37 | 6.30 | 58.08 |
| Days ≥ 0.1" Precip | 7 | 6 | 8 | 7 | 8 | 6 | 7 | 6 | 6 | 5 | 7 | 7 | 80 |
| Total Snowfall (") | 1.5 | 0.8 | 0.1 | 0.0 | 0.0 | 0.0 | 0.0 | 0.0 | 0.0 | 0.0 | 0.0 | 0.4 | 2.8 |
| Days ≥ 1" Snow Depth | 1 | 1 | 0 | 0 | 0 | 0 | 0 | 0 | 0 | 0 | 0 | 0 | 2 |

## BROOKHAVEN CITY *Lincoln County* ELEVATION 479 ft LAT/LONG 31° 34' N / 90° 26' W

| | JAN | FEB | MAR | APR | MAY | JUN | JUL | AUG | SEP | OCT | NOV | DEC | YEAR |
|---|---|---|---|---|---|---|---|---|---|---|---|---|---|
| Maximum Temp °F | 56.7 | 61.5 | 69.3 | 77.0 | 82.2 | 88.7 | 90.7 | 90.4 | 86.1 | 77.4 | 67.9 | 60.9 | 75.7 |
| Minimum Temp °F | 35.6 | 37.9 | 45.2 | 52.6 | 59.9 | 67.0 | 69.8 | 68.6 | 63.8 | 52.0 | 44.3 | 38.9 | 53.0 |
| Mean Temp °F | 46.2 | 49.8 | 57.3 | 64.8 | 71.1 | 78.0 | 80.3 | 79.5 | 75.0 | 64.7 | 56.1 | 49.9 | 64.4 |
| Days Max Temp ≥ 90 °F | 0 | 0 | 0 | 0 | 1 | 14 | 21 | 20 | 9 | 1 | 0 | 0 | 66 |
| Days Max Temp ≤ 32 °F | 0 | 0 | 0 | 0 | 0 | 0 | 0 | 0 | 0 | 0 | 0 | 0 | 0 |
| Days Min Temp ≤ 32 °F | 14 | 9 | 4 | 0 | 0 | 0 | 0 | 0 | 0 | 0 | 5 | 11 | 43 |
| Days Min Temp ≤ 0 °F | 0 | 0 | 0 | 0 | 0 | 0 | 0 | 0 | 0 | 0 | 0 | 0 | 0 |
| Heating Degree Days | 579 | 430 | 263 | 89 | 13 | 0 | 0 | 0 | 6 | 100 | 283 | 470 | 2233 |
| Cooling Degree Days | 1 | 5 | 24 | 71 | 190 | 385 | 473 | 453 | 302 | 95 | 23 | 9 | 2031 |
| Total Precipitation (") | 5.73 | 5.79 | 6.02 | 6.15 | 5.80 | 3.86 | 4.86 | 4.94 | 3.70 | 3.39 | 4.66 | 6.21 | 61.11 |
| Days ≥ 0.1" Precip | 7 | 7 | 7 | 6 | 7 | 6 | 8 | 7 | 6 | 4 | 6 | 7 | 78 |
| Total Snowfall (") | 0.2 | 0.1 | 0.3 | 0.0 | 0.0 | 0.0 | 0.0 | 0.0 | 0.0 | 0.0 | 0.0 | 0.0 | 0.6 |
| Days ≥ 1" Snow Depth | 0 | 0 | 0 | 0 | 0 | 0 | 0 | 0 | 0 | 0 | 0 | 0 | 0 |

## CALHOUN CITY 2 NW *Calhoun County* ELEVATION 279 ft LAT/LONG 33° 55' N / 89° 20' W

| | JAN | FEB | MAR | APR | MAY | JUN | JUL | AUG | SEP | OCT | NOV | DEC | YEAR |
|---|---|---|---|---|---|---|---|---|---|---|---|---|---|
| Maximum Temp °F | 52.6 | 57.9 | 66.9 | 75.8 | 81.6 | 88.4 | 91.7 | 90.7 | 85.6 | 77.0 | 66.2 | 57.0 | 74.3 |
| Minimum Temp °F | 30.5 | 33.6 | 41.3 | 49.6 | 57.2 | 64.9 | 69.0 | 67.1 | 61.1 | 48.5 | 40.9 | 34.7 | 49.9 |
| Mean Temp °F | 41.6 | 45.8 | 54.1 | 62.6 | 69.4 | 76.7 | 80.4 | 79.0 | 73.4 | 62.8 | 53.7 | 45.9 | 62.1 |
| Days Max Temp ≥ 90 °F | 0 | 0 | 0 | 0 | 2 | 13 | 22 | 20 | 9 | 0 | 0 | 0 | 66 |
| Days Max Temp ≤ 32 °F | 2 | 0 | 0 | 0 | 0 | 0 | 0 | 0 | 0 | 0 | 0 | 1 | 3 |
| Days Min Temp ≤ 32 °F | 19 | 14 | 7 | 1 | 0 | 0 | 0 | 0 | 0 | 2 | 8 | 15 | 66 |
| Days Min Temp ≤ 0 °F | 0 | 0 | 0 | 0 | 0 | 0 | 0 | 0 | 0 | 0 | 0 | 0 | 0 |
| Heating Degree Days | 719 | 538 | 345 | 128 | 24 | 1 | 0 | 0 | 12 | 127 | 344 | 588 | 2826 |
| Cooling Degree Days | 0 | 3 | 18 | 54 | 164 | 357 | 493 | 452 | 263 | 67 | 14 | 3 | 1888 |
| Total Precipitation (") | 4.94 | 4.60 | 5.32 | 5.59 | 5.06 | 3.97 | 3.96 | 3.11 | 3.77 | 3.15 | 4.55 | 5.81 | 53.83 |
| Days ≥ 0.1" Precip | 6 | 6 | 7 | 6 | 7 | 5 | 6 | 5 | 5 | 4 | 6 | 7 | 70 |
| Total Snowfall (") | 1.1 | 0.4 | 0.3 | 0.0 | 0.0 | 0.0 | 0.0 | 0.0 | 0.0 | 0.0 | 0.0 | 0.2 | 2.0 |
| Days ≥ 1" Snow Depth | 0 | 0 | 0 | 0 | 0 | 0 | 0 | 0 | 0 | 0 | 0 | 0 | 0 |

## CARTHAGE 3 SW *Leake County* ELEVATION 370 ft LAT/LONG 32° 43' N / 89° 34' W

| | JAN | FEB | MAR | APR | MAY | JUN | JUL | AUG | SEP | OCT | NOV | DEC | YEAR |
|---|---|---|---|---|---|---|---|---|---|---|---|---|---|
| Maximum Temp °F | 54.2 | 58.8 | 67.4 | 76.2 | 82.2 | 89.3 | 91.9 | 91.3 | 86.8 | 77.0 | 67.0 | 58.4 | 75.0 |
| Minimum Temp °F | 31.7 | 34.0 | 42.0 | 50.0 | 58.1 | 65.4 | 69.2 | 67.9 | 62.2 | 48.9 | 40.6 | 34.7 | 50.4 |
| Mean Temp °F | 42.9 | 46.5 | 54.8 | 63.1 | 70.2 | 77.3 | 80.6 | 79.6 | 74.5 | 63.0 | 53.9 | 46.5 | 62.7 |
| Days Max Temp ≥ 90 °F | 0 | 0 | 0 | 0 | 3 | 16 | 23 | 22 | 12 | 1 | 0 | 0 | 77 |
| Days Max Temp ≤ 32 °F | 2 | 0 | 0 | 0 | 0 | 0 | 0 | 0 | 0 | 0 | 0 | 0 | 2 |
| Days Min Temp ≤ 32 °F | 18 | 14 | 6 | 1 | 0 | 0 | 0 | 0 | 0 | 1 | 8 | 16 | 64 |
| Days Min Temp ≤ 0 °F | 0 | 0 | 0 | 0 | 0 | 0 | 0 | 0 | 0 | 0 | 0 | 0 | 0 |
| Heating Degree Days | 679 | 522 | 332 | 122 | 23 | 0 | 0 | 0 | 10 | 132 | 344 | 571 | 2735 |
| Cooling Degree Days | 1 | 6 | 23 | 65 | 189 | 380 | 494 | 468 | 299 | 78 | 20 | 7 | 2030 |
| Total Precipitation (") | 5.49 | 5.10 | 5.92 | 5.50 | 5.54 | 3.31 | 4.66 | 3.66 | 3.53 | 3.40 | 4.78 | 5.67 | 56.56 |
| Days ≥ 0.1" Precip | 8 | 7 | 7 | 6 | 8 | 6 | 7 | 6 | 5 | 5 | 6 | 8 | 79 |
| Total Snowfall (") | 0.6 | 0.1 | 0.0 | 0.0 | 0.0 | 0.0 | 0.0 | 0.0 | 0.0 | 0.0 | 0.0 | 0.0 | 0.7 |
| Days ≥ 1" Snow Depth | 0 | 0 | 0 | 0 | 0 | 0 | 0 | 0 | 0 | 0 | 0 | 0 | 0 |

## CLARKSDALE *Coahoma County* ELEVATION 180 ft LAT/LONG 34° 12 ' N / 90° 34 ' W

| | JAN | FEB | MAR | APR | MAY | JUN | JUL | AUG | SEP | OCT | NOV | DEC | YEAR |
|---|---|---|---|---|---|---|---|---|---|---|---|---|---|
| Maximum Temp °F | 47.6 | 52.8 | 62.4 | 73.1 | 81.2 | 89.1 | 91.9 | 90.0 | 84.7 | 74.8 | 62.6 | 52.5 | 71.9 |
| Minimum Temp °F | 30.9 | 34.3 | 42.5 | 52.0 | 60.5 | 68.2 | 71.6 | 69.4 | 63.4 | 51.5 | 42.6 | 35.1 | 51.8 |
| Mean Temp °F | 39.3 | 43.6 | 52.5 | 62.6 | 70.9 | 78.7 | 81.7 | 79.7 | 74.1 | 63.2 | 52.6 | 43.8 | 61.9 |
| Days Max Temp ≥ 90 °F | 0 | 0 | 0 | 0 | 3 | 16 | 22 | 18 | 9 | 1 | 0 | 0 | 69 |
| Days Max Temp ≤ 32 °F | 4 | 2 | 0 | 0 | 0 | 0 | 0 | 0 | 0 | 0 | 0 | 1 | 7 |
| Days Min Temp ≤ 32 °F | 19 | 13 | 4 | 0 | 0 | 0 | 0 | 0 | 0 | 0 | 5 | 14 | 55 |
| Days Min Temp ≤ 0 °F | 0 | 0 | 0 | 0 | 0 | 0 | 0 | 0 | 0 | 0 | 0 | 0 | 0 |
| Heating Degree Days | 791 | 599 | 392 | 137 | 22 | 0 | 0 | 0 | 14 | 131 | 375 | 651 | 3112 |
| Cooling Degree Days | 0 | 0 | 8 | 61 | 200 | 412 | 516 | 462 | 297 | 76 | 11 | 3 | 2046 |
| Total Precipitation (") | 4.32 | 4.89 | 5.28 | 5.00 | 5.31 | 4.62 | 3.99 | 2.81 | 3.17 | 3.22 | 5.26 | 5.70 | 53.57 |
| Days ≥ 0.1" Precip | 7 | 6 | 7 | 6 | 8 | 6 | 6 | 5 | 5 | 5 | 6 | 7 | 74 |
| Total Snowfall (") | 0.9 | 0.6 | 0.6 | 0.0 | 0.0 | 0.0 | 0.0 | 0.0 | 0.0 | 0.0 | 0.0 | 0.2 | 2.3 |
| Days ≥ 1" Snow Depth | *0* | 0 | 0 | 0 | 0 | 0 | 0 | 0 | 0 | 0 | 0 | 0 | 0 |

## COLLINS *Covington County* ELEVATION 289 ft LAT/LONG 31° 38 ' N / 89° 34 ' W

| | JAN | FEB | MAR | APR | MAY | JUN | JUL | AUG | SEP | OCT | NOV | DEC | YEAR |
|---|---|---|---|---|---|---|---|---|---|---|---|---|---|
| Maximum Temp °F | 58.3 | 63.0 | 70.9 | 77.8 | 82.7 | 89.0 | 90.6 | 90.2 | 86.1 | 77.8 | 68.9 | 61.9 | 76.4 |
| Minimum Temp °F | 36.2 | 38.6 | 45.6 | 53.1 | 60.3 | 67.1 | 70.3 | 69.6 | 64.8 | 52.6 | 44.6 | 39.2 | 53.5 |
| Mean Temp °F | 47.3 | 50.8 | 58.3 | 65.5 | 71.5 | 78.0 | 80.5 | 79.9 | 75.5 | 65.2 | 56.8 | 50.6 | 65.0 |
| Days Max Temp ≥ 90 °F | 0 | 0 | 0 | 0 | 2 | 15 | 21 | 21 | 10 | 1 | 0 | 0 | 70 |
| Days Max Temp ≤ 32 °F | 0 | 0 | 0 | 0 | 0 | 0 | 0 | 0 | 0 | 0 | 0 | 0 | 0 |
| Days Min Temp ≤ 32 °F | 13 | 9 | 4 | 0 | 0 | 0 | 0 | 0 | 0 | 0 | 5 | 11 | 42 |
| Days Min Temp ≤ 0 °F | 0 | 0 | 0 | 0 | 0 | 0 | 0 | 0 | 0 | 0 | 0 | 0 | 0 |
| Heating Degree Days | 547 | 400 | 235 | 75 | 9 | 0 | 0 | 0 | 5 | 86 | 267 | 452 | 2076 |
| Cooling Degree Days | 3 | 9 | 31 | 89 | 224 | 404 | 496 | 479 | 326 | 104 | 32 | 12 | 2209 |
| Total Precipitation (") | 5.70 | 5.27 | 6.03 | 5.59 | 5.77 | 3.96 | 5.29 | 4.50 | 4.06 | 3.44 | 4.42 | 5.91 | 59.94 |
| Days ≥ 0.1" Precip | 7 | 7 | 7 | 5 | 6 | 6 | 8 | 7 | 6 | 5 | 6 | 7 | 77 |
| Total Snowfall (") | 0.1 | 0.0 | 0.1 | 0.0 | 0.0 | 0.0 | 0.0 | 0.0 | 0.0 | 0.0 | 0.0 | 0.0 | 0.2 |
| Days ≥ 1" Snow Depth | 0 | 0 | 0 | 0 | 0 | 0 | 0 | 0 | 0 | 0 | 0 | 0 | 0 |

## COLUMBIA *Marion County* ELEVATION 151 ft LAT/LONG 31° 15 ' N / 89° 50 ' W

| | JAN | FEB | MAR | APR | MAY | JUN | JUL | AUG | SEP | OCT | NOV | DEC | YEAR |
|---|---|---|---|---|---|---|---|---|---|---|---|---|---|
| Maximum Temp °F | 59.6 | 63.9 | 71.7 | 79.1 | 84.8 | 91.2 | 92.8 | 92.3 | 88.0 | 79.8 | 70.4 | 63.0 | 78.0 |
| Minimum Temp °F | 37.0 | 39.3 | 46.7 | 53.9 | 61.1 | 67.6 | 70.5 | 69.9 | 65.2 | 53.0 | 45.0 | 39.6 | 54.1 |
| Mean Temp °F | 48.3 | 51.6 | 59.2 | 66.5 | 73.0 | 79.4 | 81.7 | 81.1 | 76.6 | 66.4 | 57.7 | 51.3 | 66.1 |
| Days Max Temp ≥ 90 °F | 0 | 0 | 0 | 0 | 6 | 21 | 26 | 25 | 14 | 2 | 0 | 0 | 94 |
| Days Max Temp ≤ 32 °F | 0 | 0 | 0 | 0 | 0 | 0 | 0 | 0 | 0 | 0 | 0 | 0 | 0 |
| Days Min Temp ≤ 32 °F | 13 | 9 | 3 | 0 | 0 | 0 | 0 | 0 | 0 | 0 | 5 | 11 | 41 |
| Days Min Temp ≤ 0 °F | 0 | 0 | 0 | 0 | 0 | 0 | 0 | 0 | 0 | 0 | 0 | 0 | 0 |
| Heating Degree Days | 517 | 381 | 216 | 66 | 5 | 0 | 0 | 0 | 3 | 73 | 248 | 431 | 1940 |
| Cooling Degree Days | 5 | 9 | 41 | 103 | 258 | 442 | 528 | 522 | 362 | 132 | 37 | 15 | 2454 |
| Total Precipitation (") | 6.50 | 6.01 | 5.90 | 5.65 | 5.70 | 4.47 | 5.64 | 5.53 | 4.31 | 3.17 | 4.58 | 6.45 | 63.91 |
| Days ≥ 0.1" Precip | 8 | 7 | 7 | 6 | 7 | 7 | 9 | 8 | 6 | 4 | 6 | 8 | 83 |
| Total Snowfall (") | 0.2 | 0.0 | 0.1 | 0.0 | 0.0 | 0.0 | 0.0 | 0.0 | 0.0 | 0.0 | 0.0 | 0.0 | 0.3 |
| Days ≥ 1" Snow Depth | 0 | 0 | 0 | 0 | 0 | 0 | 0 | 0 | 0 | 0 | 0 | 0 | 0 |

## CORINTH CITY *Alcorn County* ELEVATION 385 ft LAT/LONG 34° 55 ' N / 88° 31 ' W

| | JAN | FEB | MAR | APR | MAY | JUN | JUL | AUG | SEP | OCT | NOV | DEC | YEAR |
|---|---|---|---|---|---|---|---|---|---|---|---|---|---|
| Maximum Temp °F | 49.3 | 55.1 | 64.9 | 75.0 | 81.2 | 88.7 | 92.0 | 90.4 | 84.8 | 74.9 | 63.5 | 54.1 | 72.8 |
| Minimum Temp °F | 28.6 | 31.9 | 40.1 | 49.0 | 57.1 | 64.8 | 68.6 | 66.7 | 60.5 | 47.4 | 39.7 | 32.6 | 48.9 |
| Mean Temp °F | 39.0 | 43.5 | 52.4 | 62.0 | 69.1 | 76.8 | 80.3 | 78.6 | 72.7 | 61.2 | 51.6 | 43.5 | 60.9 |
| Days Max Temp ≥ 90 °F | 0 | 0 | 0 | 0 | 3 | 14 | 22 | 19 | 8 | 0 | 0 | 0 | 66 |
| Days Max Temp ≤ 32 °F | 3 | 1 | 0 | 0 | 0 | 0 | 0 | 0 | 0 | 0 | 0 | 1 | 5 |
| Days Min Temp ≤ 32 °F | 20 | 16 | 8 | 1 | 0 | 0 | 0 | 0 | 0 | 2 | 9 | 16 | 72 |
| Days Min Temp ≤ 0 °F | 0 | 0 | 0 | 0 | 0 | 0 | 0 | 0 | 0 | 0 | 0 | 0 | 0 |
| Heating Degree Days | 800 | 601 | 394 | 146 | 31 | 1 | 0 | 0 | 19 | 163 | 402 | 659 | 3216 |
| Cooling Degree Days | 0 | 1 | 13 | 59 | 168 | 370 | 498 | 449 | 253 | 56 | 9 | 1 | 1877 |
| Total Precipitation (") | 4.36 | 4.70 | 5.80 | 5.31 | 5.87 | 3.50 | 3.97 | 3.22 | 4.15 | 3.61 | 5.22 | 5.65 | 55.36 |
| Days ≥ 0.1" Precip | 6 | 6 | 8 | 7 | 7 | 6 | 6 | 5 | 6 | 5 | 7 | 7 | 76 |
| Total Snowfall (") | 1.6 | 1.2 | 0.3 | 0.0 | 0.0 | 0.0 | 0.0 | 0.0 | 0.0 | 0.0 | 0.0 | 0.3 | 3.4 |
| Days ≥ 1" Snow Depth | 1 | 0 | 0 | 0 | 0 | 0 | 0 | 0 | 0 | 0 | 0 | 0 | 1 |

## D LO 2 SW *Simpson County* ELEVATION 335 ft LAT/LONG 31° 57 ' N / 89° 56 ' W

| | JAN | FEB | MAR | APR | MAY | JUN | JUL | AUG | SEP | OCT | NOV | DEC | YEAR |
|---|---|---|---|---|---|---|---|---|---|---|---|---|---|
| Maximum Temp °F | 55.8 | 60.2 | 68.7 | 76.8 | 82.3 | 89.1 | 91.1 | 90.6 | 86.1 | 77.1 | 68.0 | 59.9 | 75.5 |
| Minimum Temp °F | 32.1 | 34.4 | 42.1 | 50.2 | 58.0 | 65.2 | 68.6 | 67.5 | 61.9 | 48.5 | 40.3 | 34.6 | 50.3 |
| Mean Temp °F | 44.0 | 47.3 | 55.4 | 63.5 | 70.2 | 77.2 | 79.9 | 79.0 | 74.0 | 62.8 | 54.2 | 47.3 | 62.9 |
| Days Max Temp ≥ 90 °F | 0 | 0 | 0 | 0 | 2 | 15 | 22 | 20 | 10 | 1 | 0 | 0 | 70 |
| Days Max Temp ≤ 32 °F | 1 | 0 | 0 | 0 | 0 | 0 | 0 | 0 | 0 | 0 | 0 | 0 | 1 |
| Days Min Temp ≤ 32 °F | 18 | 14 | 7 | 1 | 0 | 0 | 0 | 0 | 0 | 2 | 9 | 16 | 67 |
| Days Min Temp ≤ 0 °F | 0 | 0 | 0 | 0 | 0 | 0 | 0 | 0 | 0 | 0 | 0 | 0 | 0 |
| Heating Degree Days | 648 | 497 | 314 | 116 | 21 | 0 | 0 | 0 | 12 | 137 | 338 | 550 | 2633 |
| Cooling Degree Days | 2 | 5 | 25 | 67 | 195 | 382 | 483 | 465 | 299 | 88 | 24 | 9 | 2044 |
| Total Precipitation (") | 5.30 | 5.27 | 5.95 | 5.79 | 5.68 | 4.06 | 5.06 | 4.52 | 3.61 | 3.31 | 4.82 | 6.00 | 59.37 |
| Days ≥ 0.1" Precip | 8 | 6 | 7 | 6 | 7 | 6 | 8 | 7 | 5 | 4 | 6 | 8 | 78 |
| Total Snowfall (") | 0.2 | 0.0 | 0.1 | 0.0 | 0.0 | 0.0 | 0.0 | 0.0 | 0.0 | 0.0 | 0.0 | 0.0 | 0.3 |
| Days ≥ 1" Snow Depth | 0 | 0 | 0 | 0 | 0 | 0 | 0 | 0 | 0 | 0 | 0 | 0 | 0 |

## EUPORA 2 E *Webster County* ELEVATION 430 ft LAT/LONG 33° 33 ' N / 89° 16 ' W

| | JAN | FEB | MAR | APR | MAY | JUN | JUL | AUG | SEP | OCT | NOV | DEC | YEAR |
|---|---|---|---|---|---|---|---|---|---|---|---|---|---|
| Maximum Temp °F | 53.5 | 58.8 | 67.8 | 76.4 | 82.4 | 89.0 | 91.8 | 91.0 | 85.8 | 76.4 | 66.3 | 57.5 | 74.7 |
| Minimum Temp °F | 31.7 | 34.2 | 42.0 | 49.7 | 57.6 | 64.9 | 68.3 | 66.6 | 61.2 | 48.6 | 40.7 | 35.0 | 50.0 |
| Mean Temp °F | 42.6 | 46.6 | 54.9 | 63.1 | 69.9 | 77.0 | 80.1 | 78.8 | 73.4 | 62.5 | 53.5 | 46.3 | 62.4 |
| Days Max Temp ≥ 90 °F | 0 | 0 | 0 | 0 | 3 | 15 | 23 | 20 | 9 | 0 | 0 | 0 | 70 |
| Days Max Temp ≤ 32 °F | 2 | 0 | 0 | 0 | 0 | 0 | 0 | 0 | 0 | 0 | 0 | 0 | 2 |
| Days Min Temp ≤ 32 °F | 17 | 14 | 7 | 1 | 0 | 0 | 0 | 0 | 0 | 2 | 8 | 14 | 63 |
| Days Min Temp ≤ 0 °F | 0 | 0 | 0 | 0 | 0 | 0 | 0 | 0 | 0 | 0 | 0 | 0 | 0 |
| Heating Degree Days | 688 | 517 | 328 | 120 | 23 | 1 | 0 | 0 | 12 | 135 | 349 | 575 | 2748 |
| Cooling Degree Days | 0 | 4 | 21 | 67 | 188 | 374 | 484 | 447 | 278 | 68 | 15 | 3 | 1949 |
| Total Precipitation (") | 5.21 | 4.82 | 6.45 | 5.78 | 5.05 | 3.91 | 4.24 | 3.02 | 3.84 | 4.03 | 4.66 | 6.13 | 57.14 |
| Days ≥ 0.1" Precip | 7 | 6 | 7 | 6 | 7 | 6 | 7 | 5 | 5 | 5 | 6 | 7 | 74 |
| Total Snowfall (") | *0.6* | 0.1 | 0.1 | 0.0 | 0.0 | 0.0 | 0.0 | 0.0 | 0.0 | 0.0 | 0.0 | 0.0 | 0.8 |
| Days ≥ 1" Snow Depth | 0 | 0 | 0 | 0 | 0 | 0 | 0 | 0 | 0 | 0 | 0 | 0 | 0 |

## FOREST 3 S *Scott County* ELEVATION 479 ft LAT/LONG 32° 21 ' N / 89° 29 ' W

| | JAN | FEB | MAR | APR | MAY | JUN | JUL | AUG | SEP | OCT | NOV | DEC | YEAR |
|---|---|---|---|---|---|---|---|---|---|---|---|---|---|
| Maximum Temp °F | 56.2 | 61.8 | 69.4 | 77.2 | 82.7 | 89.4 | 91.2 | 90.7 | 86.5 | 77.9 | 68.1 | 60.4 | 76.0 |
| Minimum Temp °F | 33.5 | 36.0 | 42.9 | 50.4 | 58.1 | 65.1 | 68.6 | 67.5 | 62.8 | 50.7 | 42.3 | 36.8 | 51.2 |
| Mean Temp °F | 44.9 | 49.0 | 56.2 | 63.9 | 70.4 | 77.2 | 79.9 | 79.2 | 74.6 | 64.3 | 55.3 | 48.6 | 63.6 |
| Days Max Temp ≥ 90 °F | 0 | 0 | 0 | 0 | 2 | 16 | 22 | 21 | 10 | 1 | 0 | 0 | 72 |
| Days Max Temp ≤ 32 °F | 1 | 0 | 0 | 0 | 0 | 0 | 0 | 0 | 0 | 0 | 0 | 0 | 1 |
| Days Min Temp ≤ 32 °F | 16 | 12 | 6 | 1 | 0 | 0 | 0 | 0 | 0 | 1 | 7 | 13 | 56 |
| Days Min Temp ≤ 0 °F | 0 | 0 | 0 | 0 | 0 | 0 | 0 | 0 | 0 | 0 | 0 | 0 | 0 |
| Heating Degree Days | 618 | 449 | 290 | 104 | 15 | 0 | 0 | 0 | 7 | 101 | 304 | 507 | 2395 |
| Cooling Degree Days | 1 | 5 | 23 | 68 | 191 | 380 | 481 | 466 | 306 | 90 | 23 | 7 | 2041 |
| Total Precipitation (") | 5.57 | 5.53 | 6.23 | 5.45 | 5.41 | 4.11 | 5.80 | 4.40 | 3.90 | 3.85 | 5.05 | 6.14 | 61.44 |
| Days ≥ 0.1" Precip | 8 | 7 | 7 | 6 | 7 | 6 | 9 | 6 | 6 | 5 | 6 | 8 | 81 |
| Total Snowfall (") | 0.4 | 0.0 | 0.0 | 0.0 | 0.0 | 0.0 | 0.0 | 0.0 | 0.0 | 0.0 | 0.0 | 0.0 | 0.4 |
| Days ≥ 1" Snow Depth | 0 | 0 | 0 | 0 | 0 | 0 | 0 | 0 | 0 | 0 | 0 | 0 | 0 |

## FULTON 3 W *Itawamba County* ELEVATION 341 ft LAT/LONG 34° 16 ' N / 88° 24 ' W

| | JAN | FEB | MAR | APR | MAY | JUN | JUL | AUG | SEP | OCT | NOV | DEC | YEAR |
|---|---|---|---|---|---|---|---|---|---|---|---|---|---|
| Maximum Temp °F | 51.6 | 57.2 | 66.5 | 75.9 | 81.9 | 88.9 | 92.0 | 91.0 | 85.4 | 75.6 | 64.6 | 55.8 | 73.9 |
| Minimum Temp °F | 30.7 | 33.4 | 41.1 | 48.8 | 56.6 | 64.0 | 68.0 | 66.4 | 60.7 | 48.4 | 40.8 | 34.6 | 49.5 |
| Mean Temp °F | 41.2 | 45.3 | 53.8 | 62.4 | 69.3 | 76.5 | 80.0 | 78.7 | 73.1 | 62.0 | 52.8 | 45.3 | 61.7 |
| Days Max Temp ≥ 90 °F | 0 | 0 | 0 | 0 | 2 | 14 | 22 | 20 | 8 | 0 | 0 | 0 | 66 |
| Days Max Temp ≤ 32 °F | 2 | 0 | 0 | 0 | 0 | 0 | 0 | 0 | 0 | 0 | 0 | 0 | 2 |
| Days Min Temp ≤ 32 °F | 18 | 14 | 8 | 2 | 0 | 0 | 0 | 0 | 0 | 2 | 8 | 14 | 66 |
| Days Min Temp ≤ 0 °F | 0 | 0 | 0 | 0 | 0 | 0 | 0 | 0 | 0 | 0 | 0 | 0 | 0 |
| Heating Degree Days | 733 | 551 | 356 | 134 | 27 | 1 | 0 | 0 | 14 | 146 | 369 | 607 | 2938 |
| Cooling Degree Days | 0 | 2 | 18 | 57 | 174 | 365 | 487 | 445 | 265 | 65 | 10 | 3 | 1891 |
| Total Precipitation (") | 4.95 | 5.08 | 6.19 | 5.26 | 6.43 | 3.85 | 4.27 | 3.69 | 4.26 | 3.81 | 4.89 | 6.39 | 59.07 |
| Days ≥ 0.1" Precip | 8 | 7 | 8 | 7 | 8 | 6 | 7 | 6 | 6 | 5 | 7 | 8 | 83 |
| Total Snowfall (") | 0.9 | 0.5 | 0.4 | 0.0 | 0.0 | 0.0 | 0.0 | 0.0 | 0.0 | 0.0 | 0.0 | 0.2 | 2.0 |
| Days ≥ 1" Snow Depth | 0 | 0 | 0 | 0 | 0 | 0 | 0 | 0 | 0 | 0 | 0 | 0 | 0 |

## GREENVILLE *Washington County* ELEVATION 131 ft LAT/LONG 33° 24' N / 91° 3' W

| | JAN | FEB | MAR | APR | MAY | JUN | JUL | AUG | SEP | OCT | NOV | DEC | YEAR |
|---|---|---|---|---|---|---|---|---|---|---|---|---|---|
| Maximum Temp °F | 52.1 | 57.7 | 66.4 | 76.3 | 83.4 | 90.5 | 92.9 | 91.6 | 86.5 | 77.2 | 65.8 | 56.9 | 74.8 |
| Minimum Temp °F | 33.0 | 36.3 | 44.4 | 53.2 | 61.2 | 69.0 | 71.9 | 69.9 | 63.6 | 51.8 | 43.6 | 36.9 | 52.9 |
| Mean Temp °F | 42.6 | 47.1 | 55.4 | 64.8 | 72.3 | 79.8 | 82.4 | 80.8 | 75.0 | 64.5 | 54.7 | 46.9 | 63.9 |
| Days Max Temp ≥ 90 °F | 0 | 0 | 0 | 1 | 5 | 19 | 24 | 21 | 12 | 1 | 0 | 0 | 83 |
| Days Max Temp ≤ 32 °F | 2 | 0 | 0 | 0 | 0 | 0 | 0 | 0 | 0 | 0 | 0 | 1 | 3 |
| Days Min Temp ≤ 32 °F | 16 | 11 | 3 | 0 | 0 | 0 | 0 | 0 | 0 | 0 | 4 | 11 | 45 |
| Days Min Temp ≤ 0 °F | 0 | 0 | 0 | 0 | 0 | 0 | 0 | 0 | 0 | 0 | 0 | 0 | 0 |
| Heating Degree Days | 689 | 503 | 313 | 95 | 12 | 0 | 0 | 0 | 10 | 102 | 319 | 559 | 2602 |
| Cooling Degree Days | 0 | 2 | 18 | 89 | 240 | 457 | 557 | 515 | 322 | 87 | 17 | 5 | 2309 |
| Total Precipitation (") | 4.67 | 5.15 | 5.48 | 5.10 | 5.29 | 4.24 | 3.86 | 2.37 | 2.91 | 3.61 | 5.53 | 5.28 | 53.49 |
| Days ≥ 0.1" Precip | 7 | 7 | 7 | 6 | 6 | 5 | 6 | 4 | 4 | 5 | 6 | 7 | 70 |
| Total Snowfall (") | 0.3 | 0.1 | 0.2 | 0.0 | 0.0 | 0.0 | 0.0 | 0.0 | 0.0 | 0.0 | 0.0 | 0.1 | 0.7 |
| Days ≥ 1" Snow Depth | *0* | 0 | 0 | 0 | 0 | 0 | 0 | 0 | 0 | 0 | 0 | 0 | 0 |

## GREENWOOD LEFLORE AP *Carroll County* ELEVATION 131 ft LAT/LONG 33° 30' N / 90° 12' W

| | JAN | FEB | MAR | APR | MAY | JUN | JUL | AUG | SEP | OCT | NOV | DEC | YEAR |
|---|---|---|---|---|---|---|---|---|---|---|---|---|---|
| Maximum Temp °F | 51.9 | 56.9 | 65.7 | 74.8 | 82.0 | 89.4 | 91.7 | 90.8 | 85.8 | 76.1 | 65.4 | 56.2 | 73.9 |
| Minimum Temp °F | 34.1 | 37.0 | 44.9 | 53.0 | 61.5 | 68.7 | 71.9 | 70.3 | 64.4 | 52.4 | 43.9 | 37.4 | 53.3 |
| Mean Temp °F | 43.0 | 47.0 | 55.3 | 63.9 | 71.8 | 79.1 | 81.8 | 80.5 | 75.2 | 64.3 | 54.6 | 46.9 | 63.6 |
| Days Max Temp ≥ 90 °F | 0 | 0 | 0 | 0 | 4 | 17 | 22 | 20 | 11 | 1 | 0 | 0 | 75 |
| Days Max Temp ≤ 32 °F | 2 | 1 | 0 | 0 | 0 | 0 | 0 | 0 | 0 | 0 | 0 | 1 | 4 |
| Days Min Temp ≤ 32 °F | 15 | 10 | 3 | 0 | 0 | 0 | 0 | 0 | 0 | 0 | 5 | 12 | 45 |
| Days Min Temp ≤ 0 °F | 0 | 0 | 0 | 0 | 0 | 0 | 0 | 0 | 0 | 0 | 0 | 0 | 0 |
| Heating Degree Days | 677 | 505 | 315 | 112 | 14 | 0 | 0 | 0 | 11 | 111 | 325 | 560 | 2630 |
| Cooling Degree Days | 1 | 5 | 24 | 82 | 233 | 446 | 544 | 514 | 333 | 103 | 25 | 7 | 2317 |
| Total Precipitation (") | 4.90 | 4.46 | 5.67 | 4.97 | 5.46 | 4.30 | 4.13 | 2.71 | 3.44 | 3.85 | 4.71 | 5.73 | 54.33 |
| Days ≥ 0.1" Precip | 7 | 6 | 8 | 6 | 7 | 5 | 6 | 5 | 5 | 5 | 6 | 7 | 73 |
| Total Snowfall (") | 1.0 | 0.0 | 0.2 | 0.0 | 0.0 | 0.0 | 0.0 | 0.0 | 0.0 | 0.0 | 0.0 | 0.1 | 1.3 |
| Days ≥ 1" Snow Depth | 1 | 0 | 0 | 0 | 0 | 0 | 0 | 0 | 0 | 0 | 0 | 0 | 1 |

## GRENADA 5 NNE *Grenada County* ELEVATION 190 ft LAT/LONG 33° 48' N / 89° 50' W

| | JAN | FEB | MAR | APR | MAY | JUN | JUL | AUG | SEP | OCT | NOV | DEC | YEAR |
|---|---|---|---|---|---|---|---|---|---|---|---|---|---|
| Maximum Temp °F | 51.5 | 56.5 | 65.9 | 75.4 | 81.7 | 88.9 | 91.8 | 90.5 | 85.2 | 75.6 | 65.2 | 56.3 | 73.7 |
| Minimum Temp °F | 30.2 | 33.1 | 41.4 | 50.1 | 57.9 | 65.7 | 69.4 | 67.5 | 61.4 | 49.0 | 40.7 | 34.4 | 50.1 |
| Mean Temp °F | 40.9 | 44.8 | 53.7 | 62.8 | 69.8 | 77.3 | 80.6 | 79.0 | 73.3 | 62.3 | 53.0 | 45.4 | 61.9 |
| Days Max Temp ≥ 90 °F | 0 | 0 | 0 | 0 | 3 | 15 | 23 | 20 | 9 | 1 | 0 | 0 | 71 |
| Days Max Temp ≤ 32 °F | 2 | 1 | 0 | 0 | 0 | 0 | 0 | 0 | 0 | 0 | 0 | 1 | 4 |
| Days Min Temp ≤ 32 °F | 19 | 15 | 7 | 1 | 0 | 0 | 0 | 0 | 0 | 1 | 8 | 15 | 66 |
| Days Min Temp ≤ 0 °F | 0 | 0 | 0 | 0 | 0 | 0 | 0 | 0 | 0 | 0 | 0 | 0 | 0 |
| Heating Degree Days | 742 | 566 | 361 | 131 | 26 | 1 | 0 | 0 | 14 | 144 | 366 | 604 | 2955 |
| Cooling Degree Days | 0 | 2 | 12 | 53 | 165 | 368 | 484 | 437 | 253 | 61 | 12 | 4 | 1851 |
| Total Precipitation (") | 5.14 | 4.81 | 5.85 | 5.61 | 4.93 | 4.11 | 4.64 | 3.20 | 4.14 | 3.61 | 5.12 | 6.03 | 57.19 |
| Days ≥ 0.1" Precip | 8 | 7 | 7 | 7 | 7 | 6 | 6 | 5 | 6 | 5 | 7 | 8 | 79 |
| Total Snowfall (") | 0.6 | 0.1 | 0.5 | 0.0 | 0.0 | 0.0 | 0.0 | 0.0 | 0.0 | 0.0 | 0.0 | 0.1 | 1.3 |
| Days ≥ 1" Snow Depth | 0 | 0 | 0 | 0 | 0 | 0 | 0 | 0 | 0 | 0 | 0 | 0 | 0 |

## GULFPORT NAVAL CNTR *Harrison County* ELEVATION 30 ft LAT/LONG 30° 23' N / 89° 7' W

| | JAN | FEB | MAR | APR | MAY | JUN | JUL | AUG | SEP | OCT | NOV | DEC | YEAR |
|---|---|---|---|---|---|---|---|---|---|---|---|---|---|
| Maximum Temp °F | 60.4 | 63.9 | 69.8 | 77.1 | 83.7 | 89.3 | 91.1 | 90.6 | 87.0 | 79.7 | 70.4 | 63.5 | 77.2 |
| Minimum Temp °F | 42.0 | 44.5 | 51.1 | 58.8 | 65.8 | 71.5 | 73.3 | 72.9 | 69.2 | 58.8 | 50.8 | 45.1 | 58.7 |
| Mean Temp °F | 51.2 | 54.2 | 60.5 | 68.0 | 74.8 | 80.4 | 82.2 | 81.7 | 78.1 | 69.3 | 60.7 | 54.3 | 68.0 |
| Days Max Temp ≥ 90 °F | 0 | 0 | 0 | 0 | 2 | 14 | 21 | 20 | 9 | 0 | 0 | 0 | 66 |
| Days Max Temp ≤ 32 °F | 0 | 0 | 0 | 0 | 0 | 0 | 0 | 0 | 0 | 0 | 0 | 0 | 0 |
| Days Min Temp ≤ 32 °F | 7 | 4 | 1 | 0 | 0 | 0 | 0 | 0 | 0 | 0 | 1 | 4 | 17 |
| Days Min Temp ≤ 0 °F | 0 | 0 | 0 | 0 | 0 | 0 | 0 | 0 | 0 | 0 | 0 | 0 | 0 |
| Heating Degree Days | 424 | 307 | 172 | 39 | 1 | 0 | 0 | 0 | 1 | 33 | 173 | 340 | 1490 |
| Cooling Degree Days | 2 | 7 | 41 | 117 | 313 | 464 | 546 | 533 | 395 | 170 | 49 | 12 | 2649 |
| Total Precipitation (") | 6.39 | 5.80 | 5.55 | 4.70 | 5.44 | 4.76 | 6.64 | 6.21 | 6.13 | 3.12 | 4.27 | 5.38 | 64.39 |
| Days ≥ 0.1" Precip | 8 | 7 | 7 | 5 | 6 | 7 | 9 | 9 | 7 | 3 | 6 | 6 | 80 |
| Total Snowfall (") | 0.0 | 0.0 | 0.1 | 0.0 | 0.0 | 0.0 | 0.0 | 0.0 | 0.0 | 0.0 | 0.0 | 0.0 | 0.1 |
| Days ≥ 1" Snow Depth | 0 | 0 | 0 | 0 | 0 | 0 | 0 | 0 | 0 | 0 | 0 | 0 | 0 |

## HATTIESBURG *Forrest County* ELEVATION 220 ft LAT/LONG 31° 18 ' N / 89° 17 ' W

| | JAN | FEB | MAR | APR | MAY | JUN | JUL | AUG | SEP | OCT | NOV | DEC | YEAR |
|---|---|---|---|---|---|---|---|---|---|---|---|---|---|
| Maximum Temp °F | 58.1 | 62.4 | 70.3 | 78.0 | 83.6 | 89.7 | 91.8 | 91.2 | 87.3 | 78.7 | 69.5 | 61.8 | 76.9 |
| Minimum Temp °F | 36.2 | 38.5 | 46.2 | 54.1 | 61.1 | 68.0 | 71.2 | 70.6 | 65.8 | 53.2 | 45.0 | 39.0 | 54.1 |
| Mean Temp °F | 47.1 | 50.5 | 58.3 | 66.0 | 72.4 | 78.9 | 81.5 | 81.0 | 76.6 | 65.9 | 57.2 | 50.4 | 65.5 |
| Days Max Temp ≥ 90 °F | 0 | 0 | 0 | 0 | 3 | 17 | 24 | 23 | 13 | 1 | 0 | 0 | 81 |
| Days Max Temp ≤ 32 °F | 0 | 0 | 0 | 0 | 0 | 0 | 0 | 0 | 0 | 0 | 0 | 0 | 0 |
| Days Min Temp ≤ 32 °F | 13 | 9 | 3 | 0 | 0 | 0 | 0 | 0 | 0 | 0 | 4 | 11 | 40 |
| Days Min Temp ≤ 0 °F | 0 | 0 | 0 | 0 | 0 | 0 | 0 | 0 | 0 | 0 | 0 | 0 | 0 |
| Heating Degree Days | 552 | 412 | 237 | 74 | 9 | 0 | 0 | 0 | 3 | 81 | 256 | 458 | 2082 |
| Cooling Degree Days | 3 | 8 | 37 | 99 | 251 | 427 | 519 | 509 | 352 | 122 | 33 | 12 | 2372 |
| Total Precipitation (") | 6.29 | 5.60 | 6.00 | 5.15 | 5.38 | 4.30 | 5.67 | 5.43 | 3.64 | 3.37 | 4.93 | 5.92 | 61.68 |
| Days ≥ 0.1" Precip | 8 | 6 | 7 | 5 | 6 | 6 | 8 | 8 | 6 | 4 | 6 | 7 | 77 |
| Total Snowfall (") | 0.2 | 0.2 | 0.1 | 0.0 | 0.0 | 0.0 | 0.0 | 0.0 | 0.0 | 0.0 | 0.0 | 0.0 | 0.5 |
| Days ≥ 1" Snow Depth | 0 | 0 | 0 | 0 | 0 | 0 | 0 | 0 | 0 | 0 | 0 | 0 | 0 |

## HERNANDO *DeSoto County* ELEVATION 363 ft LAT/LONG 34° 50 ' N / 90° 0 ' W

| | JAN | FEB | MAR | APR | MAY | JUN | JUL | AUG | SEP | OCT | NOV | DEC | YEAR |
|---|---|---|---|---|---|---|---|---|---|---|---|---|---|
| Maximum Temp °F | 48.4 | 53.9 | 63.6 | 73.6 | 80.3 | 88.0 | 91.1 | 89.8 | 83.9 | 74.1 | 62.3 | 52.9 | 71.8 |
| Minimum Temp °F | 30.1 | 34.0 | 42.6 | 51.3 | 59.4 | 67.2 | 70.9 | 69.0 | 62.9 | 51.6 | 42.6 | 34.7 | 51.4 |
| Mean Temp °F | 39.1 | 44.0 | 53.1 | 62.5 | 69.9 | 77.6 | 81.0 | 79.4 | 73.4 | 62.9 | 52.5 | 43.8 | 61.6 |
| Days Max Temp ≥ 90 °F | 0 | 0 | 0 | 0 | 1 | 12 | 20 | 17 | 6 | 0 | 0 | 0 | 56 |
| Days Max Temp ≤ 32 °F | 3 | 1 | 0 | 0 | 0 | 0 | 0 | 0 | 0 | 0 | 0 | 1 | 5 |
| Days Min Temp ≤ 32 °F | *18* | 13 | 5 | 1 | 0 | 0 | 0 | 0 | 0 | 0 | 6 | *13* | 56 |
| Days Min Temp ≤ 0 °F | 0 | 0 | 0 | 0 | 0 | 0 | 0 | 0 | 0 | 0 | 0 | 0 | 0 |
| Heating Degree Days | 795 | 589 | 375 | 136 | 25 | 0 | 0 | 0 | 15 | 134 | 379 | 651 | 3099 |
| Cooling Degree Days | 0 | 1 | 14 | 64 | 185 | 396 | 509 | 464 | 277 | 70 | 10 | 2 | 1992 |
| Total Precipitation (") | 4.14 | 4.58 | 5.46 | 6.02 | 5.62 | 4.43 | 3.41 | 3.31 | 3.66 | 3.33 | 5.07 | 5.78 | 54.81 |
| Days ≥ 0.1" Precip | 7 | 6 | 8 | 7 | 8 | 6 | 5 | 5 | 5 | 5 | 6 | 7 | 75 |
| Total Snowfall (") | 2.0 | 0.9 | 0.6 | 0.0 | 0.0 | 0.0 | 0.0 | 0.0 | 0.0 | 0.0 | 0.1 | 0.1 | 3.7 |
| Days ≥ 1" Snow Depth | *1* | 1 | 0 | 0 | 0 | 0 | 0 | 0 | 0 | 0 | 0 | 0 | 2 |

## HICKORY FLAT *Benton County* ELEVATION 410 ft LAT/LONG 34° 37 ' N / 89° 12 ' W

| | JAN | FEB | MAR | APR | MAY | JUN | JUL | AUG | SEP | OCT | NOV | DEC | YEAR |
|---|---|---|---|---|---|---|---|---|---|---|---|---|---|
| Maximum Temp °F | 50.3 | 55.9 | 65.3 | 75.1 | 81.2 | 88.2 | 91.6 | 90.6 | 84.9 | 75.1 | 63.5 | 54.6 | 73.0 |
| Minimum Temp °F | 29.8 | 32.4 | 40.3 | 48.8 | 56.8 | 64.5 | 68.5 | 66.6 | 60.7 | 48.3 | 40.2 | 33.6 | 49.2 |
| Mean Temp °F | 40.1 | 44.2 | 52.8 | 61.9 | 69.0 | 76.4 | 80.1 | 78.6 | 72.9 | 61.7 | 51.9 | 44.1 | 61.1 |
| Days Max Temp ≥ 90 °F | 0 | 0 | 0 | 0 | 2 | 13 | 22 | 19 | 8 | 0 | 0 | 0 | 64 |
| Days Max Temp ≤ 32 °F | 2 | 1 | 0 | 0 | 0 | 0 | 0 | 0 | 0 | 0 | 0 | 1 | 4 |
| Days Min Temp ≤ 32 °F | 19 | 15 | 9 | 2 | 0 | 0 | 0 | 0 | 0 | 2 | 9 | 15 | 71 |
| Days Min Temp ≤ 0 °F | 0 | 0 | 0 | 0 | 0 | 0 | 0 | 0 | 0 | 0 | 0 | 0 | 0 |
| Heating Degree Days | 766 | 582 | 385 | 144 | 32 | 1 | 0 | 0 | 16 | 153 | 396 | 642 | 3117 |
| Cooling Degree Days | 0 | 1 | 12 | 51 | 165 | 360 | 477 | 445 | 260 | 59 | 9 | 2 | 1841 |
| Total Precipitation (") | 4.68 | 4.99 | 6.02 | 5.61 | 5.91 | 4.38 | 4.60 | 3.76 | 3.90 | 3.49 | 5.60 | 5.99 | 58.93 |
| Days ≥ 0.1" Precip | 7 | 7 | 8 | 7 | 8 | 6 | 7 | 6 | 6 | 5 | 6 | 7 | 80 |
| Total Snowfall (") | 1.8 | 1.1 | 0.5 | 0.0 | 0.0 | 0.0 | 0.0 | 0.0 | 0.0 | 0.0 | 0.0 | 0.5 | 3.9 |
| Days ≥ 1" Snow Depth | 2 | 1 | 0 | 0 | 0 | 0 | 0 | 0 | 0 | 0 | 0 | 0 | 3 |

## HOLLY SPRINGS 4 N *Marshall County* ELEVATION 479 ft LAT/LONG 34° 49 ' N / 89° 26 ' W

| | JAN | FEB | MAR | APR | MAY | JUN | JUL | AUG | SEP | OCT | NOV | DEC | YEAR |
|---|---|---|---|---|---|---|---|---|---|---|---|---|---|
| Maximum Temp °F | 46.9 | 51.9 | 61.3 | 71.6 | 78.4 | 86.0 | 89.5 | 88.4 | 82.7 | 72.8 | 61.5 | 51.6 | 70.2 |
| Minimum Temp °F | 27.1 | 29.9 | 38.4 | 47.2 | 55.6 | 63.8 | 68.0 | 65.7 | 59.2 | 46.1 | 38.2 | 30.8 | 47.5 |
| Mean Temp °F | 37.0 | 41.0 | 49.9 | 59.4 | 67.0 | 74.9 | 78.8 | 77.1 | 71.0 | 59.5 | 49.9 | 41.2 | 58.9 |
| Days Max Temp ≥ 90 °F | 0 | 0 | 0 | 0 | 1 | 8 | 17 | 14 | 5 | 0 | 0 | 0 | 45 |
| Days Max Temp ≤ 32 °F | 4 | 2 | 0 | 0 | 0 | 0 | 0 | 0 | 0 | 0 | 0 | 2 | 8 |
| Days Min Temp ≤ 32 °F | 22 | 18 | 10 | 2 | 0 | 0 | 0 | 0 | 0 | 3 | 11 | 19 | 85 |
| Days Min Temp ≤ 0 °F | 0 | 0 | 0 | 0 | 0 | 0 | 0 | 0 | 0 | 0 | 0 | 0 | 0 |
| Heating Degree Days | 860 | 672 | 470 | 205 | 59 | 3 | 0 | 1 | 32 | 208 | 453 | 730 | 3693 |
| Cooling Degree Days | 0 | 0 | 8 | 39 | 132 | 323 | 451 | 401 | 222 | 44 | 7 | 1 | 1628 |
| Total Precipitation (") | 4.37 | 4.51 | 5.93 | 5.46 | 5.78 | 4.40 | 4.52 | 3.71 | 3.61 | 3.92 | 5.39 | 5.76 | 57.36 |
| Days ≥ 0.1" Precip | 7 | 7 | 8 | 7 | 8 | 6 | 6 | 5 | 5 | 5 | 7 | 7 | 78 |
| Total Snowfall (") | 1.4 | 0.8 | 0.5 | 0.0 | 0.0 | 0.0 | 0.0 | 0.0 | 0.0 | 0.0 | 0.0 | 0.1 | 2.8 |
| Days ≥ 1" Snow Depth | 2 | 1 | 0 | 0 | 0 | 0 | 0 | 0 | 0 | 0 | 0 | 0 | 3 |

## HOUSTON *Chickasaw County* ELEVATION 341 ft LAT/LONG 33° 54 ' N / 89° 1 ' W

| | JAN | FEB | MAR | APR | MAY | JUN | JUL | AUG | SEP | OCT | NOV | DEC | YEAR |
|---|---|---|---|---|---|---|---|---|---|---|---|---|---|
| Maximum Temp °F | 49.7 | 55.2 | 64.5 | 74.1 | 80.5 | 87.7 | 91.0 | 89.6 | 83.9 | 74.4 | 63.8 | 55.1 | 72.5 |
| Minimum Temp °F | 29.1 | 31.5 | 39.9 | 48.4 | 56.5 | 64.3 | 68.0 | 66.2 | 60.3 | 47.3 | 39.1 | 33.0 | 48.6 |
| Mean Temp °F | 39.4 | 43.4 | 52.2 | 61.3 | 68.5 | 76.0 | 79.5 | 77.9 | 72.1 | 60.9 | 51.5 | 44.1 | 60.6 |
| Days Max Temp ≥ 90 °F | 0 | 0 | 0 | 0 | 2 | 12 | 20 | 17 | 6 | 0 | 0 | 0 | 57 |
| Days Max Temp ≤ 32 °F | 2 | 1 | 0 | 0 | 0 | 0 | 0 | 0 | 0 | 0 | 0 | 1 | 4 |
| Days Min Temp ≤ 32 °F | 20 | 16 | 8 | 1 | 0 | 0 | 0 | 0 | 0 | 2 | 10 | 16 | 73 |
| Days Min Temp ≤ 0 °F | 0 | 0 | 0 | 0 | 0 | 0 | 0 | 0 | 0 | 0 | 0 | 0 | 0 |
| Heating Degree Days | 787 | 604 | 402 | 158 | 37 | 2 | 0 | 0 | 19 | 172 | 408 | 645 | 3234 |
| Cooling Degree Days | *0* | *0* | *8* | *41* | *155* | *353* | *470* | *413* | *232* | *52* | 8 | 2 | 1734 |
| Total Precipitation (") | 4.69 | 4.97 | 6.31 | 5.39 | 5.48 | 4.46 | 3.85 | 3.46 | 4.08 | 3.63 | 4.78 | 6.51 | 57.61 |
| Days ≥ 0.1" Precip | 7 | 7 | 7 | 6 | 7 | 6 | 6 | 6 | 6 | 5 | 6 | 8 | 77 |
| Total Snowfall (") | 1.8 | 0.3 | 0.4 | 0.0 | 0.0 | 0.0 | 0.0 | 0.0 | 0.0 | 0.0 | 0.1 | 0.2 | 2.8 |
| Days ≥ 1" Snow Depth | 1 | 1 | 0 | 0 | 0 | 0 | 0 | 0 | 0 | 0 | 0 | 0 | 2 |

## INDEPENDENCE 1 W *Tate County* ELEVATION 361 ft LAT/LONG 34° 44 ' N / 89° 48 ' W

| | JAN | FEB | MAR | APR | MAY | JUN | JUL | AUG | SEP | OCT | NOV | DEC | YEAR |
|---|---|---|---|---|---|---|---|---|---|---|---|---|---|
| Maximum Temp °F | 47.0 | 52.0 | 61.3 | 71.8 | 78.6 | 86.5 | 89.7 | 88.6 | 82.9 | 72.9 | 61.2 | 51.7 | 70.4 |
| Minimum Temp °F | 28.6 | 31.5 | 40.3 | 49.2 | 57.3 | 65.4 | 68.9 | 67.0 | 60.9 | 48.1 | 39.9 | 32.8 | 49.2 |
| Mean Temp °F | 37.8 | 41.8 | 50.8 | 60.5 | 68.0 | 75.9 | 79.3 | 77.8 | 71.9 | 60.5 | 50.6 | 42.3 | 59.8 |
| Days Max Temp ≥ 90 °F | 0 | 0 | 0 | 0 | 2 | 9 | 17 | 14 | 6 | 0 | 0 | 0 | 48 |
| Days Max Temp ≤ 32 °F | 4 | 2 | 0 | 0 | 0 | 0 | 0 | 0 | 0 | 0 | 0 | 2 | 8 |
| Days Min Temp ≤ 32 °F | 21 | 17 | 8 | 1 | 0 | 0 | 0 | 0 | 0 | 1 | 9 | 16 | 73 |
| Days Min Temp ≤ 0 °F | 0 | 0 | 0 | 0 | 0 | 0 | 0 | 0 | 0 | 0 | 0 | 0 | 0 |
| Heating Degree Days | 836 | 650 | 442 | 181 | 46 | 2 | 0 | 0 | 24 | 184 | 433 | 698 | 3496 |
| Cooling Degree Days | 0 | 0 | 9 | 48 | 141 | 339 | 446 | 411 | 241 | 50 | 10 | 1 | 1696 |
| Total Precipitation (") | 4.09 | 4.26 | 5.06 | 5.14 | 5.73 | 4.12 | 3.76 | 3.59 | 3.57 | 3.20 | 4.80 | 5.38 | 52.70 |
| Days ≥ 0.1" Precip | 7 | 6 | 8 | 7 | 8 | 6 | 6 | 5 | 5 | 5 | 6 | 7 | 76 |
| Total Snowfall (") | 1.4 | 1.0 | 0.1 | 0.0 | 0.0 | 0.0 | 0.0 | 0.0 | 0.0 | 0.0 | 0.1 | 0.0 | 2.6 |
| Days ≥ 1" Snow Depth | 0 | 0 | 0 | 0 | 0 | 0 | 0 | 0 | 0 | 0 | 0 | 0 | 0 |

## JACKSON THOMPSON FLD *Rankin County* ELEVATION 291 ft LAT/LONG 32° 19 ' N / 90° 5 ' W

| | JAN | FEB | MAR | APR | MAY | JUN | JUL | AUG | SEP | OCT | NOV | DEC | YEAR |
|---|---|---|---|---|---|---|---|---|---|---|---|---|---|
| Maximum Temp °F | 55.7 | 60.4 | 68.9 | 77.2 | 83.4 | 90.3 | 92.2 | 91.5 | 86.9 | 77.6 | 68.0 | 59.7 | 76.0 |
| Minimum Temp °F | 34.3 | 36.8 | 44.5 | 52.3 | 60.5 | 67.8 | 71.2 | 70.2 | 64.7 | 51.8 | 43.3 | 37.6 | 52.9 |
| Mean Temp °F | 45.0 | 48.6 | 56.7 | 64.8 | 71.9 | 79.1 | 81.7 | 80.9 | 75.9 | 64.7 | 55.6 | 48.7 | 64.5 |
| Days Max Temp ≥ 90 °F | 0 | 0 | 0 | 0 | 4 | 18 | 24 | 22 | 12 | 1 | 0 | 0 | 81 |
| Days Max Temp ≤ 32 °F | 1 | 0 | 0 | 0 | 0 | 0 | 0 | 0 | 0 | 0 | 0 | 0 | 1 |
| Days Min Temp ≤ 32 °F | 15 | 11 | 4 | 0 | 0 | 0 | 0 | 0 | 0 | 1 | 6 | 12 | 49 |
| Days Min Temp ≤ 0 °F | 0 | 0 | 0 | 0 | 0 | 0 | 0 | 0 | 0 | 0 | 0 | 0 | 0 |
| Heating Degree Days | 617 | 462 | 280 | 95 | 11 | 0 | 0 | 0 | 7 | 101 | 300 | 509 | 2382 |
| Cooling Degree Days | 2 | 8 | 32 | 90 | 240 | 443 | 537 | 518 | 342 | 104 | 30 | 11 | 2357 |
| Total Precipitation (") | 5.12 | 4.76 | 5.46 | 5.62 | 5.29 | 3.26 | 4.46 | 3.85 | 3.61 | 3.32 | 4.64 | 5.56 | 54.95 |
| Days ≥ 0.1" Precip | 7 | 7 | 7 | 6 | 7 | 5 | 7 | 6 | 5 | 4 | 6 | 8 | 75 |
| Total Snowfall (") | 0.5 | 0.2 | 0.2 | 0.0 | 0.0 | 0.0 | 0.0 | 0.0 | 0.0 | 0.0 | 0.0 | 0.0 | 0.9 |
| Days ≥ 1" Snow Depth | 0 | 0 | 0 | 0 | 0 | 0 | 0 | 0 | 0 | 0 | 0 | 0 | 0 |

## KIPLING *Kemper County* ELEVATION 322 ft LAT/LONG 32° 41 ' N / 88° 38 ' W

| | JAN | FEB | MAR | APR | MAY | JUN | JUL | AUG | SEP | OCT | NOV | DEC | YEAR |
|---|---|---|---|---|---|---|---|---|---|---|---|---|---|
| Maximum Temp °F | 55.4 | 60.3 | 68.6 | 76.6 | 82.4 | 88.8 | 90.9 | 90.3 | 86.0 | 76.7 | 67.1 | 59.0 | 75.2 |
| Minimum Temp °F | 32.9 | 35.4 | 42.3 | 49.7 | 57.6 | 64.7 | 68.5 | 67.4 | 62.1 | 49.4 | 41.3 | 35.9 | 50.6 |
| Mean Temp °F | 44.1 | 47.8 | 55.5 | 63.2 | 70.0 | 76.8 | 79.8 | 78.9 | 74.0 | 63.1 | 54.2 | 47.5 | 62.9 |
| Days Max Temp ≥ 90 °F | 0 | 0 | 0 | 0 | 2 | 14 | 21 | 20 | 10 | 1 | 0 | 0 | 68 |
| Days Max Temp ≤ 32 °F | 1 | 0 | 0 | 0 | 0 | 0 | 0 | 0 | 0 | 0 | 0 | 0 | 1 |
| Days Min Temp ≤ 32 °F | 16 | 13 | 7 | 1 | 0 | 0 | 0 | 0 | 0 | 2 | 8 | 13 | 60 |
| Days Min Temp ≤ 0 °F | 0 | 0 | 0 | 0 | 0 | 0 | 0 | 0 | 0 | 0 | 0 | 0 | 0 |
| Heating Degree Days | 642 | 482 | 310 | 115 | 19 | 0 | 0 | 0 | 10 | 125 | 332 | 542 | 2577 |
| Cooling Degree Days | 1 | 6 | 22 | 57 | 187 | 370 | 477 | 451 | 294 | 76 | 19 | 8 | 1968 |
| Total Precipitation (") | 5.43 | 5.07 | 6.29 | 5.64 | 5.30 | 3.85 | 5.01 | 3.45 | 3.37 | 3.05 | 4.13 | 5.66 | 56.25 |
| Days ≥ 0.1" Precip | 8 | 6 | 8 | 6 | 7 | 6 | 7 | 6 | 5 | 4 | 6 | 7 | 76 |
| Total Snowfall (") | 0.5 | 0.0 | 0.3 | 0.1 | 0.0 | 0.0 | 0.0 | 0.0 | 0.0 | 0.0 | 0.0 | 0.0 | 0.9 |
| Days ≥ 1" Snow Depth | 0 | 0 | 0 | 0 | 0 | 0 | 0 | 0 | 0 | 0 | 0 | 0 | 0 |

## KOSCIUSKO *Attala County* ELEVATION 420 ft LAT/LONG 33° 3 ' N / 89° 36 ' W

| | JAN | FEB | MAR | APR | MAY | JUN | JUL | AUG | SEP | OCT | NOV | DEC | YEAR |
|---|---|---|---|---|---|---|---|---|---|---|---|---|---|
| Maximum Temp °F | 52.9 | 58.2 | 67.0 | 75.7 | 81.6 | 88.8 | 91.4 | 90.7 | 85.9 | 76.4 | 65.8 | 57.1 | 74.3 |
| Minimum Temp °F | 30.4 | 32.9 | 40.5 | 48.9 | 57.4 | 64.9 | 68.4 | 67.0 | 61.3 | 48.8 | 39.9 | 34.0 | 49.5 |
| Mean Temp °F | 41.7 | 45.6 | 53.8 | 62.3 | 69.5 | 76.9 | 79.9 | 78.9 | 73.6 | 62.6 | 52.9 | 45.6 | 61.9 |
| Days Max Temp ≥ 90 °F | 0 | 0 | 0 | 0 | 3 | 15 | 23 | 20 | 10 | 1 | 0 | 0 | 72 |
| Days Max Temp ≤ 32 °F | 2 | 0 | 0 | 0 | 0 | 0 | 0 | 0 | 0 | 0 | 0 | 1 | 3 |
| Days Min Temp ≤ 32 °F | 19 | 15 | 8 | 1 | 0 | 0 | 0 | 0 | 0 | 1 | 8 | 16 | 68 |
| Days Min Temp ≤ 0 °F | 0 | 0 | 0 | 0 | 0 | 0 | 0 | 0 | 0 | 0 | 0 | 0 | 0 |
| Heating Degree Days | 716 | 543 | 357 | 135 | 27 | 1 | 0 | 0 | 14 | 138 | 367 | 599 | 2897 |
| Cooling Degree Days | 0 | 2 | 15 | 56 | 183 | 386 | 497 | 467 | 299 | 84 | 14 | 5 | 2008 |
| Total Precipitation (") | 5.70 | 5.31 | 6.35 | 6.08 | 5.64 | 3.43 | 5.19 | 3.69 | 3.86 | 3.94 | 5.01 | 6.18 | 60.38 |
| Days ≥ 0.1" Precip | 8 | 6 | 7 | 6 | 8 | 6 | 8 | 6 | 5 | 5 | 7 | 8 | 80 |
| Total Snowfall (") | 0.8 | 0.1 | 0.3 | 0.0 | 0.0 | 0.0 | 0.0 | 0.0 | 0.0 | 0.0 | 0.0 | 0.0 | 1.2 |
| Days ≥ 1" Snow Depth | *0* | 0 | 0 | 0 | 0 | 0 | 0 | 0 | 0 | 0 | 0 | 0 | 0 |

## LAUREL *Jones County* ELEVATION 239 ft LAT/LONG 31° 42 ' N / 89° 8 ' W

| | JAN | FEB | MAR | APR | MAY | JUN | JUL | AUG | SEP | OCT | NOV | DEC | YEAR |
|---|---|---|---|---|---|---|---|---|---|---|---|---|---|
| Maximum Temp °F | 57.0 | 61.2 | 69.4 | 77.5 | 82.9 | 89.5 | 91.5 | 90.7 | 86.5 | 77.7 | 68.5 | 60.5 | 76.1 |
| Minimum Temp °F | 35.2 | 37.3 | 44.5 | 52.3 | 59.7 | 66.9 | 70.1 | 69.4 | 64.4 | 51.7 | 43.3 | 37.7 | 52.7 |
| Mean Temp °F | 46.1 | 49.3 | 57.0 | 64.9 | 71.4 | 78.2 | 80.8 | 80.1 | 75.5 | 64.7 | 55.9 | 49.0 | 64.4 |
| Days Max Temp ≥ 90 °F | 0 | 0 | 0 | 0 | 3 | 16 | 22 | 21 | 10 | 1 | 0 | 0 | 73 |
| Days Max Temp ≤ 32 °F | 1 | 0 | 0 | 0 | 0 | 0 | 0 | 0 | 0 | 0 | 0 | 0 | 1 |
| Days Min Temp ≤ 32 °F | 15 | 11 | 3 | 0 | 0 | 0 | 0 | 0 | 0 | 0 | 5 | 12 | 46 |
| Days Min Temp ≤ 0 °F | 0 | 0 | 0 | 0 | 0 | 0 | 0 | 0 | 0 | 0 | 0 | 0 | 0 |
| Heating Degree Days | 583 | 443 | 269 | 88 | 12 | 0 | 0 | 0 | 6 | 98 | 288 | 497 | 2284 |
| Cooling Degree Days | 2 | 6 | 27 | 80 | 213 | 403 | 495 | 479 | 319 | 97 | 25 | 8 | 2154 |
| Total Precipitation (") | 5.66 | 4.98 | 5.77 | 5.06 | 5.38 | 3.79 | 5.75 | 4.69 | 4.18 | 3.00 | 4.46 | 5.74 | 58.46 |
| Days ≥ 0.1" Precip | 8 | 6 | 7 | 6 | 7 | 6 | 9 | 7 | 5 | 4 | 6 | 7 | 78 |
| Total Snowfall (") | 0.0 | 0.0 | 0.0 | 0.0 | 0.0 | 0.0 | 0.0 | 0.0 | 0.0 | 0.0 | 0.0 | 0.0 | 0.0 |
| Days ≥ 1" Snow Depth | 0 | 0 | 0 | 0 | 0 | 0 | 0 | 0 | 0 | 0 | 0 | 0 | 0 |

## LEXINGTON 2 NNW *Holmes County* ELEVATION 220 ft LAT/LONG 33° 7 ' N / 90° 3 ' W

| | JAN | FEB | MAR | APR | MAY | JUN | JUL | AUG | SEP | OCT | NOV | DEC | YEAR |
|---|---|---|---|---|---|---|---|---|---|---|---|---|---|
| Maximum Temp °F | 53.7 | 58.9 | 67.6 | 75.7 | 81.8 | 88.4 | 90.8 | 90.4 | 85.4 | 76.6 | 66.5 | 57.8 | 74.5 |
| Minimum Temp °F | 32.5 | 34.8 | 42.6 | 50.1 | 57.9 | 64.9 | 68.5 | 67.4 | 61.7 | 49.8 | 41.7 | 35.8 | 50.6 |
| Mean Temp °F | 43.1 | 46.9 | 55.1 | 62.9 | 69.9 | 76.7 | 79.7 | 78.9 | 73.6 | 63.2 | 54.1 | 46.7 | 62.6 |
| Days Max Temp ≥ 90 °F | 0 | 0 | 0 | 0 | 2 | 14 | 20 | 20 | 9 | 1 | 0 | 0 | 66 |
| Days Max Temp ≤ 32 °F | 1 | 0 | 0 | 0 | 0 | 0 | 0 | 0 | 0 | 0 | 0 | 0 | 1 |
| Days Min Temp ≤ 32 °F | 16 | 12 | 7 | 1 | 0 | 0 | 0 | 0 | 0 | 1 | 7 | 13 | 57 |
| Days Min Temp ≤ 0 °F | 0 | 0 | 0 | 0 | 0 | 0 | 0 | 0 | 0 | 0 | 0 | 0 | 0 |
| Heating Degree Days | 673 | 508 | 318 | 124 | 21 | 0 | 0 | 0 | 12 | 124 | 333 | 565 | 2678 |
| Cooling Degree Days | 1 | 3 | 20 | 60 | 176 | 360 | 464 | 448 | 270 | 75 | 17 | 6 | 1900 |
| Total Precipitation (") | 5.16 | 5.19 | 5.59 | 5.51 | 5.01 | 4.07 | 4.01 | 3.22 | 2.99 | 3.76 | 4.87 | 5.76 | 55.14 |
| Days ≥ 0.1" Precip | 7 | 6 | 7 | 6 | 7 | 6 | 6 | 5 | 5 | 4 | 6 | 7 | 72 |
| Total Snowfall (") | 0.6 | 0.1 | 0.4 | 0.0 | 0.0 | 0.0 | 0.0 | 0.0 | 0.0 | 0.0 | 0.0 | 0.0 | 1.1 |
| Days ≥ 1" Snow Depth | 0 | 0 | 0 | 0 | 0 | 0 | 0 | 0 | 0 | 0 | 0 | 0 | 0 |

## LIBERTY 5 W *Amite County* ELEVATION 345 ft LAT/LONG 31° 10 ' N / 90° 53 ' W

| | JAN | FEB | MAR | APR | MAY | JUN | JUL | AUG | SEP | OCT | NOV | DEC | YEAR |
|---|---|---|---|---|---|---|---|---|---|---|---|---|---|
| Maximum Temp °F | 57.2 | 61.5 | 69.4 | 77.3 | 83.3 | 89.9 | 91.5 | 91.1 | 87.2 | 78.8 | 68.8 | 60.4 | 76.4 |
| Minimum Temp °F | 34.3 | 36.4 | 43.8 | 51.5 | 58.8 | 65.8 | 68.9 | 67.7 | 63.1 | 50.6 | 42.8 | 36.3 | 51.7 |
| Mean Temp °F | 45.7 | 49.0 | 56.6 | 64.5 | 71.1 | 77.9 | 80.2 | 79.4 | 75.1 | 64.7 | 55.8 | 48.3 | 64.0 |
| Days Max Temp ≥ 90 °F | 0 | 0 | 0 | 0 | 3 | 17 | 23 | 22 | 12 | 2 | 0 | 0 | 79 |
| Days Max Temp ≤ 32 °F | 0 | 0 | 0 | 0 | 0 | 0 | 0 | 0 | 0 | 0 | 0 | 0 | 0 |
| Days Min Temp ≤ 32 °F | 15 | 11 | 5 | 1 | 0 | 0 | 0 | 0 | 0 | 0 | 6 | *14* | 52 |
| Days Min Temp ≤ 0 °F | 0 | 0 | 0 | 0 | 0 | 0 | 0 | 0 | 0 | 0 | 0 | 0 | 0 |
| Heating Degree Days | 594 | 450 | 277 | 98 | 15 | 0 | 0 | 0 | 7 | 100 | 292 | 517 | 2350 |
| Cooling Degree Days | 1 | 4 | 23 | 68 | 201 | 388 | 469 | 448 | *303* | 98 | 22 | 8 | 2033 |
| Total Precipitation (") | 6.13 | 5.67 | 6.23 | 5.42 | 5.65 | 4.69 | 5.28 | 4.65 | 4.59 | 3.08 | 4.67 | 5.90 | 61.96 |
| Days ≥ 0.1" Precip | 7 | 7 | 7 | 5 | 6 | 6 | 9 | 7 | 6 | 4 | 6 | 7 | 77 |
| Total Snowfall (") | 0.1 | 0.1 | 0.1 | 0.0 | 0.0 | 0.0 | 0.0 | 0.0 | 0.0 | 0.0 | 0.0 | 0.0 | 0.3 |
| Days ≥ 1" Snow Depth | 0 | 0 | 0 | 0 | 0 | 0 | 0 | 0 | 0 | 0 | 0 | 0 | 0 |

## LOUISVILLE *Winston County* ELEVATION 561 ft LAT/LONG 33° 7 ' N / 89° 3 ' W

| | JAN | FEB | MAR | APR | MAY | JUN | JUL | AUG | SEP | OCT | NOV | DEC | YEAR |
|---|---|---|---|---|---|---|---|---|---|---|---|---|---|
| Maximum Temp °F | 51.5 | 56.5 | 65.4 | 74.2 | 80.1 | 87.1 | 89.4 | 88.8 | 83.6 | 74.2 | 64.6 | 55.9 | 72.6 |
| Minimum Temp °F | 31.3 | 34.1 | 42.5 | 50.8 | 58.6 | 65.9 | 69.0 | 68.0 | 62.5 | 50.5 | 42.4 | 35.0 | 50.9 |
| Mean Temp °F | 41.5 | 45.3 | 53.9 | 62.5 | 69.4 | 76.6 | 79.2 | 78.4 | 73.1 | 62.3 | 53.5 | 45.5 | 61.8 |
| Days Max Temp ≥ 90 °F | 0 | 0 | 0 | 0 | 1 | 10 | 16 | 15 | 6 | 0 | 0 | 0 | 48 |
| Days Max Temp ≤ 32 °F | 2 | 1 | 0 | 0 | 0 | 0 | 0 | 0 | 0 | 0 | 0 | 1 | 4 |
| Days Min Temp ≤ 32 °F | 18 | 13 | 6 | 1 | 0 | 0 | 0 | 0 | 0 | 0 | 6 | 14 | 58 |
| Days Min Temp ≤ 0 °F | 0 | 0 | 0 | 0 | 0 | 0 | 0 | 0 | 0 | 0 | 0 | 0 | 0 |
| Heating Degree Days | 724 | 551 | 352 | 132 | 28 | 1 | 0 | 0 | 15 | 141 | 350 | 601 | 2895 |
| Cooling Degree Days | 0 | 3 | 15 | 58 | 174 | 361 | 462 | 443 | 276 | 68 | 14 | 3 | 1877 |
| Total Precipitation (") | 5.56 | 5.30 | 6.31 | 6.23 | 5.28 | 3.71 | 5.36 | 3.61 | 4.26 | 3.55 | 4.50 | 5.79 | 59.46 |
| Days ≥ 0.1" Precip | 8 | 7 | 7 | 6 | 8 | 6 | 8 | 6 | 6 | 4 | 6 | 7 | 79 |
| Total Snowfall (") | 0.3 | 0.1 | 0.0 | 0.1 | 0.0 | 0.0 | 0.0 | 0.0 | 0.0 | 0.0 | 0.0 | 0.0 | 0.5 |
| Days ≥ 1" Snow Depth | 0 | 0 | 0 | 0 | 0 | 0 | 0 | 0 | 0 | 0 | 0 | 0 | 0 |

## MCCOMB PIKE CO AP *Pike County* ELEVATION 458 ft LAT/LONG 31° 15 ' N / 90° 28 ' W

| | JAN | FEB | MAR | APR | MAY | JUN | JUL | AUG | SEP | OCT | NOV | DEC | YEAR |
|---|---|---|---|---|---|---|---|---|---|---|---|---|---|
| Maximum Temp °F | 58.3 | 62.4 | 70.3 | 77.7 | 83.3 | 89.8 | 91.4 | 91.0 | 86.7 | 78.5 | 69.1 | 61.9 | 76.7 |
| Minimum Temp °F | 36.5 | 38.8 | 45.9 | 53.4 | 60.4 | 66.7 | 69.5 | 68.9 | 64.9 | 53.3 | 44.9 | 39.4 | 53.5 |
| Mean Temp °F | 47.4 | 50.6 | 58.1 | 65.6 | 71.9 | 78.3 | 80.5 | 80.0 | 75.8 | 65.9 | 57.0 | 50.7 | 65.2 |
| Days Max Temp ≥ 90 °F | 0 | 0 | 0 | 0 | 3 | 17 | 23 | 22 | 11 | 1 | 0 | 0 | 77 |
| Days Max Temp ≤ 32 °F | 0 | 0 | 0 | 0 | 0 | 0 | 0 | 0 | 0 | 0 | 0 | 0 | 0 |
| Days Min Temp ≤ 32 °F | 13 | 9 | 3 | 0 | 0 | 0 | 0 | 0 | 0 | 0 | 5 | 10 | 40 |
| Days Min Temp ≤ 0 °F | 0 | 0 | 0 | 0 | 0 | 0 | 0 | 0 | 0 | 0 | 0 | 0 | 0 |
| Heating Degree Days | 542 | 408 | 241 | 75 | 7 | 0 | 0 | 0 | 4 | 78 | 263 | 449 | 2067 |
| Cooling Degree Days | 4 | 9 | 33 | 92 | 232 | 410 | 496 | 487 | 342 | 122 | 37 | 12 | 2276 |
| Total Precipitation (") | 5.91 | 5.83 | 6.15 | 5.70 | 5.79 | 4.31 | 5.53 | 5.36 | 4.74 | 3.24 | 4.80 | 6.23 | 63.59 |
| Days ≥ 0.1" Precip | 8 | 7 | 7 | 6 | 7 | 7 | 9 | 8 | 6 | 4 | 6 | 7 | 82 |
| Total Snowfall (") | 0.1 | 0.3 | 0.1 | 0.0 | 0.0 | 0.0 | 0.0 | 0.0 | 0.0 | 0.0 | 0.0 | 0.0 | 0.5 |
| Days ≥ 1" Snow Depth | 0 | 0 | 0 | 0 | 0 | 0 | 0 | 0 | 0 | 0 | 0 | 0 | 0 |

## MERIDIAN KEY FLD *Lauderdale County* ELEVATION 312 ft LAT/LONG 32° 20 ' N / 88° 45 ' W

| | JAN | FEB | MAR | APR | MAY | JUN | JUL | AUG | SEP | OCT | NOV | DEC | YEAR |
|---|---|---|---|---|---|---|---|---|---|---|---|---|---|
| Maximum Temp °F | 56.5 | 61.2 | 69.4 | 77.6 | 83.3 | 90.1 | 92.2 | 91.7 | 87.0 | 77.5 | 68.4 | 60.5 | 76.3 |
| Minimum Temp °F | 34.1 | 36.7 | 43.5 | 51.2 | 59.1 | 66.6 | 70.2 | 69.4 | 64.2 | 51.0 | 42.6 | 37.4 | 52.2 |
| Mean Temp °F | 45.3 | 49.0 | 56.5 | 64.4 | 71.2 | 78.3 | 81.3 | 80.6 | 75.6 | 64.3 | 55.5 | 49.0 | 64.3 |
| Days Max Temp ≥ 90 °F | 0 | 0 | 0 | 0 | 4 | 17 | 24 | 23 | 12 | 1 | 0 | 0 | 81 |
| Days Max Temp ≤ 32 °F | 0 | 0 | 0 | 0 | 0 | 0 | 0 | 0 | 0 | 0 | 0 | 0 | 0 |
| Days Min Temp ≤ 32 °F | 15 | 11 | 5 | 0 | 0 | 0 | 0 | 0 | 0 | 1 | 6 | 12 | 50 |
| Days Min Temp ≤ 0 °F | 0 | 0 | 0 | 0 | 0 | 0 | 0 | 0 | 0 | 0 | 0 | 0 | 0 |
| Heating Degree Days | 607 | 451 | 282 | 95 | 14 | 0 | 0 | 0 | 6 | 106 | 299 | 497 | 2357 |
| Cooling Degree Days | 3 | 7 | 26 | 81 | 221 | 419 | 524 | 514 | 339 | 101 | 25 | 11 | 2271 |
| Total Precipitation (") | 5.41 | 5.48 | 6.64 | 5.47 | 4.80 | 3.70 | 5.41 | 3.85 | 3.67 | 3.25 | 4.65 | 5.89 | 58.22 |
| Days ≥ 0.1" Precip | 8 | 7 | 8 | 7 | 6 | 6 | 8 | 6 | 5 | 4 | 6 | 8 | 79 |
| Total Snowfall (") | 0.3 | 0.1 | 0.2 | 0.1 | 0.0 | 0.0 | 0.0 | 0.0 | 0.0 | 0.0 | 0.0 | 0.0 | 0.7 |
| Days ≥ 1" Snow Depth | 0 | 0 | 0 | 0 | 0 | 0 | 0 | 0 | 0 | 0 | 0 | 0 | 0 |

## MERRILL *George County* ELEVATION 49 ft LAT/LONG 30° 59 ' N / 88° 43 ' W

| | JAN | FEB | MAR | APR | MAY | JUN | JUL | AUG | SEP | OCT | NOV | DEC | YEAR |
|---|---|---|---|---|---|---|---|---|---|---|---|---|---|
| Maximum Temp °F | 59.6 | 64.2 | 71.8 | 79.5 | 85.3 | 91.3 | 92.8 | 91.9 | 88.4 | 80.2 | 70.8 | 63.1 | 78.2 |
| Minimum Temp °F | 34.6 | 36.6 | 44.3 | 51.4 | 57.8 | 64.7 | 67.8 | 67.1 | 62.4 | 49.0 | 41.4 | 36.4 | 51.1 |
| Mean Temp °F | 47.1 | 50.4 | 58.1 | 65.5 | 71.6 | 78.0 | 80.3 | 79.6 | 75.5 | 64.7 | 56.1 | 49.8 | 64.7 |
| Days Max Temp ≥ 90 °F | 0 | 0 | 0 | 1 | 7 | 21 | 25 | 24 | 15 | 2 | 0 | 0 | 95 |
| Days Max Temp ≤ 32 °F | 0 | 0 | 0 | 0 | 0 | 0 | 0 | 0 | 0 | 0 | 0 | 0 | 0 |
| Days Min Temp ≤ 32 °F | 16 | 12 | 4 | 0 | 0 | 0 | 0 | 0 | 0 | 2 | 8 | 15 | 57 |
| Days Min Temp ≤ 0 °F | 0 | 0 | 0 | 0 | 0 | 0 | 0 | 0 | 0 | 0 | 0 | 0 | 0 |
| Heating Degree Days | 553 | 414 | 244 | 77 | 11 | 0 | 0 | 0 | 5 | 104 | 287 | 478 | 2173 |
| Cooling Degree Days | 2 | 9 | 37 | 86 | 221 | 396 | 479 | 459 | 309 | 101 | 28 | 13 | 2140 |
| Total Precipitation (") | 5.98 | 5.51 | 6.76 | 4.77 | 5.83 | 4.44 | 6.98 | 5.03 | 5.47 | 3.32 | 4.79 | 5.38 | 64.26 |
| Days ≥ 0.1" Precip | 8 | 7 | 7 | 5 | 6 | 7 | 10 | 8 | 6 | 4 | 6 | 7 | 81 |
| Total Snowfall (") | 0.2 | 0.0 | 0.0 | 0.0 | 0.0 | 0.0 | 0.0 | 0.0 | 0.0 | 0.0 | 0.0 | 0.0 | 0.2 |
| Days ≥ 1" Snow Depth | 0 | 0 | 0 | 0 | 0 | 0 | 0 | 0 | 0 | 0 | 0 | 0 | 0 |

## MINTER CITY 1 NE *Leflore County* ELEVATION 151 ft LAT/LONG 33° 45' N / 90° 18' W

| | JAN | FEB | MAR | APR | MAY | JUN | JUL | AUG | SEP | OCT | NOV | DEC | YEAR |
|---|---|---|---|---|---|---|---|---|---|---|---|---|---|
| Maximum Temp °F | 50.1 | 55.6 | 64.7 | 74.7 | 81.8 | 89.4 | 92.5 | 91.0 | 85.7 | 76.3 | 64.5 | 55.0 | 73.4 |
| Minimum Temp °F | 31.7 | *35.3* | 43.3 | 52.4 | 60.2 | 67.9 | 71.2 | 69.4 | 63.1 | 51.0 | 42.2 | 35.4 | 51.9 |
| Mean Temp °F | 40.9 | *45.4* | 54.0 | 63.6 | 71.0 | 78.7 | 81.9 | 80.2 | 74.4 | 63.7 | 53.4 | 45.2 | 62.7 |
| Days Max Temp ≥ 90 °F | 0 | 0 | 0 | 0 | 3 | 16 | *23* | 19 | 10 | 1 | 0 | 0 | 72 |
| Days Max Temp ≤ 32 °F | 3 | 1 | 0 | 0 | 0 | 0 | 0 | 0 | 0 | 0 | 0 | 1 | 5 |
| Days Min Temp ≤ 32 °F | *17* | *11* | 3 | 0 | 0 | 0 | 0 | 0 | 0 | 0 | *5* | na | na |
| Days Min Temp ≤ 0 °F | 0 | 0 | 0 | 0 | 0 | 0 | 0 | 0 | 0 | 0 | 0 | 0 | 0 |
| Heating Degree Days | 739 | *550* | 350 | 117 | 19 | 0 | 0 | 0 | 12 | 116 | 354 | 607 | 2864 |
| Cooling Degree Days | 0 | 1 | *14* | *80* | 215 | 428 | 544 | 495 | 312 | 82 | 15 | 3 | 2189 |
| Total Precipitation (") | 4.80 | 5.04 | 5.92 | 5.61 | 6.00 | 4.67 | 4.45 | 2.76 | 3.33 | 3.50 | 5.52 | 6.11 | 57.71 |
| Days ≥ 0.1" Precip | 6 | 5 | *7* | 6 | 7 | 6 | 6 | 4 | 5 | 4 | 6 | 7 | 69 |
| Total Snowfall (") | 0.6 | 0.2 | 0.0 | 0.0 | 0.0 | 0.0 | 0.0 | 0.0 | 0.0 | 0.0 | 0.1 | 0.2 | 1.1 |
| Days ≥ 1" Snow Depth | *0* | 0 | 0 | 0 | 0 | 0 | 0 | 0 | 0 | 0 | 0 | 0 | 0 |

## MONTICELLO *Lawrence County* ELEVATION 210 ft LAT/LONG 31° 33' N / 90° 6' W

| | JAN | FEB | MAR | APR | MAY | JUN | JUL | AUG | SEP | OCT | NOV | DEC | YEAR |
|---|---|---|---|---|---|---|---|---|---|---|---|---|---|
| Maximum Temp °F | 57.1 | 61.6 | 70.2 | 77.8 | 83.6 | 90.4 | 92.2 | 91.7 | 87.1 | 77.9 | 68.6 | 61.0 | 76.6 |
| Minimum Temp °F | 34.0 | 36.1 | 44.0 | 51.9 | 59.3 | 66.2 | 69.6 | 68.7 | 63.7 | 50.5 | 42.2 | 36.6 | 51.9 |
| Mean Temp °F | 45.6 | 48.9 | 57.1 | 64.9 | 71.5 | 78.3 | 81.0 | 80.2 | 75.4 | 64.2 | 55.4 | 48.8 | 64.3 |
| Days Max Temp ≥ 90 °F | 0 | 0 | 0 | 0 | 4 | 18 | 24 | 23 | 12 | 2 | 0 | 0 | 83 |
| Days Max Temp ≤ 32 °F | 1 | 0 | 0 | 0 | 0 | 0 | 0 | 0 | 0 | 0 | 0 | 0 | 1 |
| Days Min Temp ≤ 32 °F | 16 | 12 | 5 | 1 | 0 | 0 | 0 | 0 | 0 | 1 | 7 | 14 | 56 |
| Days Min Temp ≤ 0 °F | 0 | 0 | 0 | 0 | 0 | 0 | 0 | 0 | 0 | 0 | 0 | 0 | 0 |
| Heating Degree Days | 599 | 454 | 270 | 91 | 13 | 0 | 0 | 0 | 7 | 112 | 304 | 505 | 2355 |
| Cooling Degree Days | 2 | 6 | 32 | 83 | 224 | 417 | 510 | 495 | 326 | 100 | 25 | 10 | 2230 |
| Total Precipitation (") | 5.81 | 5.62 | 6.24 | 6.01 | 6.25 | 4.33 | 4.60 | 4.13 | 3.85 | 3.28 | 4.51 | 6.12 | 60.75 |
| Days ≥ 0.1" Precip | 8 | 7 | 7 | 6 | 7 | 7 | 8 | 7 | 6 | 4 | 6 | 8 | 81 |
| Total Snowfall (") | 0.2 | 0.1 | 0.0 | 0.0 | 0.0 | 0.0 | 0.0 | 0.0 | 0.0 | 0.0 | 0.0 | 0.0 | 0.3 |
| Days ≥ 1" Snow Depth | 0 | 0 | 0 | 0 | 0 | 0 | 0 | 0 | 0 | 0 | 0 | 0 | 0 |

## MOORHEAD *Sunflower County* ELEVATION 121 ft LAT/LONG 33° 27' N / 90° 30' W

| | JAN | FEB | MAR | APR | MAY | JUN | JUL | AUG | SEP | OCT | NOV | DEC | YEAR |
|---|---|---|---|---|---|---|---|---|---|---|---|---|---|
| Maximum Temp °F | 51.7 | 57.0 | 66.1 | 75.7 | 82.9 | 90.1 | 92.3 | 91.2 | 86.0 | 76.6 | 65.0 | 55.9 | 74.2 |
| Minimum Temp °F | 33.7 | 37.0 | 44.9 | 53.9 | 61.6 | 68.9 | 71.9 | 70.1 | 64.5 | 52.8 | 44.3 | 37.4 | 53.4 |
| Mean Temp °F | 42.7 | 47.0 | 55.5 | 64.8 | 72.2 | 79.5 | 82.1 | 80.7 | 75.3 | 64.7 | 54.7 | 46.6 | 63.8 |
| Days Max Temp ≥ 90 °F | 0 | 0 | 0 | 0 | 5 | 17 | 23 | 21 | 11 | 1 | 0 | 0 | 78 |
| Days Max Temp ≤ 32 °F | 2 | 1 | 0 | 0 | 0 | 0 | 0 | 0 | 0 | 0 | 0 | 1 | 4 |
| Days Min Temp ≤ 32 °F | 15 | 10 | 3 | 0 | 0 | 0 | 0 | 0 | 0 | 0 | 4 | 11 | 43 |
| Days Min Temp ≤ 0 °F | 0 | 0 | 0 | 0 | 0 | 0 | 0 | 0 | 0 | 0 | 0 | 0 | 0 |
| Heating Degree Days | 686 | 503 | 309 | 95 | 11 | 0 | 0 | 0 | 9 | 101 | 322 | 566 | 2602 |
| Cooling Degree Days | 0 | 3 | 20 | 89 | 241 | 451 | 552 | 519 | 340 | 102 | 19 | 5 | 2341 |
| Total Precipitation (") | 4.87 | 4.80 | 5.70 | 5.47 | 5.12 | 4.24 | 4.88 | 2.62 | 3.16 | 3.64 | 4.92 | 6.18 | 55.60 |
| Days ≥ 0.1" Precip | 7 | 6 | 7 | 6 | 7 | 6 | 6 | 5 | 4 | 4 | 6 | 7 | 71 |
| Total Snowfall (") | 0.9 | 0.2 | 0.0 | 0.0 | 0.0 | 0.0 | 0.0 | 0.0 | 0.0 | 0.0 | 0.0 | 0.0 | 1.1 |
| Days ≥ 1" Snow Depth | 1 | 0 | 0 | 0 | 0 | 0 | 0 | 0 | 0 | 0 | 0 | 0 | 1 |

## NATCHEZ *Adams County* ELEVATION 210 ft LAT/LONG 31° 34' N / 91° 24' W

| | JAN | FEB | MAR | APR | MAY | JUN | JUL | AUG | SEP | OCT | NOV | DEC | YEAR |
|---|---|---|---|---|---|---|---|---|---|---|---|---|---|
| Maximum Temp °F | 58.2 | 62.8 | 70.8 | 78.2 | 83.6 | 89.5 | 91.2 | 90.9 | 86.8 | 78.8 | 69.3 | 61.8 | 76.8 |
| Minimum Temp °F | 38.3 | 40.7 | 48.1 | 55.8 | 62.3 | 69.0 | 72.0 | 71.1 | 66.3 | 55.1 | 47.5 | 41.3 | 55.6 |
| Mean Temp °F | 48.3 | 51.8 | 59.5 | 67.0 | 73.0 | 79.3 | 81.6 | 81.0 | 76.6 | 67.0 | 58.4 | 51.6 | 66.3 |
| Days Max Temp ≥ 90 °F | 0 | 0 | 0 | 0 | 3 | 16 | 24 | 22 | 11 | 1 | 0 | 0 | 77 |
| Days Max Temp ≤ 32 °F | 0 | 0 | 0 | 0 | 0 | 0 | 0 | 0 | 0 | 0 | 0 | 0 | 0 |
| Days Min Temp ≤ 32 °F | 11 | 7 | 2 | 0 | 0 | 0 | 0 | 0 | 0 | 0 | 3 | 8 | 31 |
| Days Min Temp ≤ 0 °F | 0 | 0 | 0 | 0 | 0 | 0 | 0 | 0 | 0 | 0 | 0 | 0 | 0 |
| Heating Degree Days | 520 | 377 | 207 | 59 | 5 | 0 | 0 | 0 | 4 | 64 | 233 | 424 | 1893 |
| Cooling Degree Days | 5 | 12 | 39 | 111 | 255 | 437 | 522 | 514 | 354 | 130 | 44 | 14 | 2437 |
| Total Precipitation (") | 5.91 | 5.18 | 6.25 | 5.83 | 5.89 | 4.15 | 4.20 | 4.12 | 3.81 | 4.29 | 5.02 | 6.80 | 61.45 |
| Days ≥ 0.1" Precip | 8 | 7 | 7 | 6 | 7 | 6 | 7 | 6 | 5 | 4 | 6 | 8 | 77 |
| Total Snowfall (") | 0.5 | 0.1 | 0.1 | 0.0 | 0.0 | 0.0 | 0.0 | 0.0 | 0.0 | 0.0 | 0.0 | 0.0 | 0.7 |
| Days ≥ 1" Snow Depth | 0 | 0 | 0 | 0 | 0 | 0 | 0 | 0 | 0 | 0 | 0 | 0 | 0 |

## NEWTON EXP STN *Newton County* ELEVATION 351 ft LAT/LONG 32° 20 ' N / 89° 5 ' W

| | JAN | FEB | MAR | APR | MAY | JUN | JUL | AUG | SEP | OCT | NOV | DEC | YEAR |
|---|---|---|---|---|---|---|---|---|---|---|---|---|---|
| Maximum Temp °F | 55.1 | 59.8 | 68.2 | 76.1 | 82.0 | 89.0 | 91.5 | 90.6 | 86.1 | 76.9 | 67.2 | 59.1 | 75.1 |
| Minimum Temp °F | 32.8 | 35.2 | 42.8 | 50.4 | 58.3 | 65.4 | 69.0 | 67.9 | 62.3 | 49.4 | 41.7 | 35.9 | 50.9 |
| Mean Temp °F | 44.0 | 47.5 | 55.5 | 63.3 | 70.2 | 77.2 | 80.3 | 79.3 | 74.2 | 63.2 | 54.4 | 47.6 | 63.1 |
| Days Max Temp ≥ 90 °F | 0 | 0 | 0 | 0 | 2 | 15 | 22 | 20 | 10 | 1 | 0 | 0 | 70 |
| Days Max Temp ≤ 32 °F | 1 | 0 | 0 | 0 | 0 | 0 | 0 | 0 | 0 | 0 | 0 | 0 | 1 |
| Days Min Temp ≤ 32 °F | 17 | 13 | 6 | 1 | 0 | 0 | 0 | 0 | 0 | 1 | 7 | 14 | 59 |
| Days Min Temp ≤ 0 °F | 0 | 0 | 0 | 0 | 0 | 0 | 0 | 0 | 0 | 0 | 0 | 0 | 0 |
| Heating Degree Days | 646 | 491 | 307 | 116 | 19 | 1 | 0 | 0 | 10 | 126 | 326 | 541 | 2583 |
| Cooling Degree Days | 1 | 3 | 19 | 57 | 177 | 380 | 484 | 461 | 291 | 75 | 16 | 7 | 1971 |
| Total Precipitation (") | 5.33 | 5.37 | 6.14 | 5.54 | 4.52 | 3.24 | 4.96 | 4.05 | 3.71 | 3.44 | 4.74 | 5.58 | 56.62 |
| Days ≥ 0.1" Precip | 8 | 6 | 7 | 6 | 7 | 6 | 8 | 6 | 5 | 4 | 6 | 8 | 77 |
| Total Snowfall (") | 0.4 | 0.1 | 0.0 | 0.0 | 0.0 | 0.0 | 0.0 | 0.0 | 0.0 | 0.0 | 0.0 | 0.0 | 0.5 |
| Days ≥ 1" Snow Depth | 0 | 0 | 0 | 0 | 0 | 0 | 0 | 0 | 0 | 0 | 0 | 0 | 0 |

## OAKLEY EXP STA *Hinds County* ELEVATION 210 ft LAT/LONG 32° 12 ' N / 90° 31 ' W

| | JAN | FEB | MAR | APR | MAY | JUN | JUL | AUG | SEP | OCT | NOV | DEC | YEAR |
|---|---|---|---|---|---|---|---|---|---|---|---|---|---|
| Maximum Temp °F | 54.5 | 59.9 | 68.4 | 75.4 | 82.1 | 89.0 | 91.4 | 91.0 | 86.6 | 77.3 | 67.1 | 59.0 | 75.1 |
| Minimum Temp °F | 33.3 | 37.0 | 44.9 | 51.7 | 60.4 | 67.4 | 70.2 | 68.6 | 63.4 | 50.6 | 42.5 | 36.7 | 52.2 |
| Mean Temp °F | 43.9 | 48.5 | 56.7 | 63.6 | 71.3 | 78.2 | 80.8 | 79.9 | 75.0 | 64.0 | 54.8 | 47.9 | 63.7 |
| Days Max Temp ≥ 90 °F | 0 | 0 | 0 | 0 | 2 | 16 | 23 | 22 | 12 | 1 | 0 | 0 | 76 |
| Days Max Temp ≤ 32 °F | 1 | 0 | 0 | 0 | 0 | 0 | 0 | 0 | 0 | 0 | 0 | 1 | 2 |
| Days Min Temp ≤ 32 °F | 16 | 11 | 3 | 1 | 0 | 0 | 0 | 0 | 0 | 1 | 6 | 13 | 51 |
| Days Min Temp ≤ 0 °F | 0 | 0 | 0 | 0 | 0 | 0 | 0 | 0 | 0 | 0 | 0 | 0 | 0 |
| Heating Degree Days | 650 | 465 | 282 | 117 | 17 | 0 | 0 | 0 | 12 | 118 | 324 | 533 | 2518 |
| Cooling Degree Days | 2 | 6 | 28 | 82 | 225 | 411 | 504 | 475 | 310 | 98 | 26 | 10 | 2177 |
| Total Precipitation (") | 5.57 | 4.98 | 5.87 | 5.38 | 5.05 | 4.31 | 4.08 | 3.77 | 3.05 | 3.50 | 4.61 | 5.48 | 55.65 |
| Days ≥ 0.1" Precip | 7 | 6 | 7 | 6 | 7 | 6 | 7 | 6 | 5 | 4 | 6 | 7 | 74 |
| Total Snowfall (") | 0.5 | 0.1 | 0.2 | 0.0 | 0.0 | 0.0 | 0.0 | 0.0 | 0.0 | 0.0 | 0.0 | 0.0 | 0.8 |
| Days ≥ 1" Snow Depth | 0 | 0 | 0 | 0 | 0 | 0 | 0 | 0 | 0 | 0 | 0 | 0 | 0 |

## PASCAGOULA 3 NE *Jackson County* ELEVATION 11 ft LAT/LONG 30° 24 ' N / 88° 29 ' W

| | JAN | FEB | MAR | APR | MAY | JUN | JUL | AUG | SEP | OCT | NOV | DEC | YEAR |
|---|---|---|---|---|---|---|---|---|---|---|---|---|---|
| Maximum Temp °F | 59.5 | 63.0 | 68.9 | 76.6 | 82.9 | 88.5 | 90.0 | 89.7 | 86.9 | 79.2 | 70.0 | 63.3 | 76.5 |
| Minimum Temp °F | 41.6 | 44.1 | 51.3 | 59.2 | 65.9 | 72.4 | 74.1 | 73.5 | 70.0 | 59.0 | 51.0 | 45.0 | 58.9 |
| Mean Temp °F | 50.7 | 53.5 | 60.1 | 67.9 | 74.4 | 80.5 | 82.1 | 81.6 | 78.5 | 69.2 | 60.5 | 54.2 | 67.8 |
| Days Max Temp ≥ 90 °F | 0 | 0 | 0 | 0 | 1 | 12 | 17 | 16 | 9 | 1 | 0 | 0 | 56 |
| Days Max Temp ≤ 32 °F | 0 | 0 | 0 | 0 | 0 | 0 | 0 | 0 | 0 | 0 | 0 | 0 | 0 |
| Days Min Temp ≤ 32 °F | 7 | 4 | 1 | 0 | 0 | 0 | 0 | 0 | 0 | 0 | 1 | 4 | 17 |
| Days Min Temp ≤ 0 °F | 0 | 0 | 0 | 0 | 0 | 0 | 0 | 0 | 0 | 0 | 0 | 0 | 0 |
| Heating Degree Days | 442 | 326 | 177 | 43 | 3 | 0 | 0 | 0 | 1 | 39 | 177 | 347 | 1555 |
| Cooling Degree Days | 2 | 4 | 29 | 96 | 258 | 433 | 519 | 506 | 382 | 150 | 44 | 14 | 2437 |
| Total Precipitation (") | 5.59 | 5.73 | 5.61 | 4.17 | 6.11 | 5.16 | 6.62 | 7.31 | 6.34 | 3.71 | 4.37 | 4.63 | 65.35 |
| Days ≥ 0.1" Precip | 7 | 6 | 6 | 4 | 6 | 5 | 9 | 8 | 6 | 4 | 5 | 6 | 72 |
| Total Snowfall (") | 0.0 | 0.0 | 0.0 | 0.0 | 0.0 | 0.0 | 0.0 | 0.0 | 0.0 | 0.0 | 0.0 | 0.0 | 0.0 |
| Days ≥ 1" Snow Depth | 0 | 0 | 0 | 0 | 0 | 0 | 0 | 0 | 0 | 0 | 0 | 0 | 0 |

## PELAHATCHIE *Rankin County* ELEVATION 361 ft LAT/LONG 32° 19 ' N / 89° 47 ' W

| | JAN | FEB | MAR | APR | MAY | JUN | JUL | AUG | SEP | OCT | NOV | DEC | YEAR |
|---|---|---|---|---|---|---|---|---|---|---|---|---|---|
| Maximum Temp °F | 56.9 | 62.2 | 70.2 | 78.3 | 83.8 | 90.4 | 92.4 | 91.6 | 87.4 | 79.2 | 69.0 | 61.3 | 76.9 |
| Minimum Temp °F | 33.7 | 35.9 | 44.1 | 51.9 | 58.9 | 65.7 | 69.0 | 68.1 | 63.2 | 51.1 | 42.9 | 37.4 | 51.8 |
| Mean Temp °F | 45.3 | 49.1 | 57.2 | 65.1 | 71.3 | 78.1 | 80.7 | 79.9 | 75.3 | 65.2 | 56.0 | 49.4 | 64.4 |
| Days Max Temp ≥ 90 °F | 0 | 0 | 0 | 0 | 4 | 19 | 25 | 23 | 12 | 1 | 0 | 0 | 84 |
| Days Max Temp ≤ 32 °F | 1 | 0 | 0 | 0 | 0 | 0 | 0 | 0 | 0 | 0 | 0 | 0 | 1 |
| Days Min Temp ≤ 32 °F | 16 | 12 | 5 | 1 | 0 | 0 | 0 | 0 | 0 | 1 | 7 | 12 | 54 |
| Days Min Temp ≤ 0 °F | 0 | 0 | 0 | 0 | 0 | 0 | 0 | 0 | 0 | 0 | 0 | 0 | 0 |
| Heating Degree Days | 608 | 447 | 266 | 85 | 13 | 0 | 0 | 0 | 7 | 89 | 288 | 486 | 2289 |
| Cooling Degree Days | na | na | na | na | na | na | na | na | na | na | na | na | na |
| Total Precipitation (") | 5.48 | 4.94 | 6.20 | 5.13 | 5.63 | 3.48 | 5.34 | 3.85 | 3.41 | 3.71 | 4.68 | 6.24 | 58.09 |
| Days ≥ 0.1" Precip | 8 | 6 | 7 | 6 | 8 | 6 | 7 | 6 | 6 | 5 | 6 | 8 | 79 |
| Total Snowfall (") | 0.0 | 0.1 | 0.2 | 0.0 | 0.0 | 0.0 | 0.0 | 0.0 | 0.0 | 0.0 | 0.0 | 0.0 | 0.3 |
| Days ≥ 1" Snow Depth | 0 | 0 | 0 | 0 | 0 | 0 | 0 | 0 | 0 | 0 | 0 | 0 | 0 |

## PHILADELPHIA 1 WSW *Neshoba County* ELEVATION 413 ft LAT/LONG 32° 46 ' N / 89° 8 ' W

| | JAN | FEB | MAR | APR | MAY | JUN | JUL | AUG | SEP | OCT | NOV | DEC | YEAR |
|---|---|---|---|---|---|---|---|---|---|---|---|---|---|
| Maximum Temp °F | 53.8 | 58.6 | 67.3 | 75.6 | 81.3 | 88.3 | 90.4 | 90.0 | 85.4 | 75.9 | 66.4 | 57.8 | 74.2 |
| Minimum Temp °F | 32.1 | 34.8 | 42.8 | 50.8 | 58.4 | 65.9 | 69.6 | 68.5 | 62.9 | 49.3 | 41.4 | 35.3 | 51.0 |
| Mean Temp °F | 43.0 | 46.7 | 55.1 | 63.2 | 69.9 | 77.1 | 80.1 | 79.3 | 74.2 | 62.6 | 53.9 | 46.5 | 62.6 |
| Days Max Temp ≥ 90 °F | 0 | 0 | 0 | 0 | 2 | 13 | 19 | 18 | 9 | 1 | 0 | 0 | 62 |
| Days Max Temp ≤ 32 °F | 1 | 0 | 0 | 0 | 0 | 0 | 0 | 0 | 0 | 0 | 0 | 1 | 2 |
| Days Min Temp ≤ 32 °F | 17 | 13 | 5 | 1 | 0 | 0 | 0 | 0 | 0 | 1 | 7 | 14 | 58 |
| Days Min Temp ≤ 0 °F | 0 | 0 | 0 | 0 | 0 | 0 | 0 | 0 | 0 | 0 | 0 | 0 | 0 |
| Heating Degree Days | 677 | 512 | 321 | 119 | 24 | 1 | 0 | 0 | 11 | 136 | 340 | 569 | 2710 |
| Cooling Degree Days | 1 | 3 | 18 | 64 | 186 | 386 | 489 | 474 | 306 | 74 | 16 | 6 | 2023 |
| Total Precipitation (") | 5.61 | 5.29 | 6.15 | 5.67 | 5.47 | 3.92 | 5.04 | 3.68 | 3.63 | 3.35 | 5.02 | 5.48 | 58.31 |
| Days ≥ 0.1" Precip | 9 | 7 | 8 | 6 | 8 | 6 | 8 | 6 | 5 | 4 | 6 | 8 | 81 |
| Total Snowfall (") | 0.3 | 0.1 | 0.3 | 0.0 | 0.0 | 0.0 | 0.0 | 0.0 | 0.0 | 0.0 | 0.0 | 0.0 | 0.7 |
| Days ≥ 1" Snow Depth | 0 | 0 | 0 | 0 | 0 | 0 | 0 | 0 | 0 | 0 | 0 | 0 | 0 |

## PICAYUNE *Pearl River County* ELEVATION 49 ft LAT/LONG 30° 31 ' N / 89° 41 ' W

| | JAN | FEB | MAR | APR | MAY | JUN | JUL | AUG | SEP | OCT | NOV | DEC | YEAR |
|---|---|---|---|---|---|---|---|---|---|---|---|---|---|
| Maximum Temp °F | 61.7 | 65.2 | 71.9 | 78.7 | 84.3 | 90.1 | 91.5 | 91.0 | 87.6 | 80.1 | 71.4 | 65.0 | 78.2 |
| Minimum Temp °F | 39.0 | 41.3 | 48.5 | 54.9 | 62.3 | 68.5 | 70.9 | 70.5 | 66.2 | 54.9 | 46.9 | 41.8 | 55.5 |
| Mean Temp °F | 50.5 | 53.4 | 60.2 | 66.8 | 73.3 | 79.3 | 81.2 | 80.8 | 76.9 | 67.6 | 59.3 | 53.6 | 66.9 |
| Days Max Temp ≥ 90 °F | 0 | 0 | 0 | 0 | 3 | ***17*** | 23 | 22 | 11 | 2 | 0 | 0 | 78 |
| Days Max Temp ≤ 32 °F | 0 | 0 | 0 | 0 | 0 | 0 | 0 | 0 | 0 | 0 | 0 | 0 | 0 |
| Days Min Temp ≤ 32 °F | ***10*** | 6 | 2 | 0 | 0 | 0 | 0 | 0 | 0 | 0 | 3 | 8 | 29 |
| Days Min Temp ≤ 0 °F | 0 | 0 | 0 | 0 | 0 | 0 | 0 | 0 | 0 | 0 | 0 | 0 | 0 |
| Heating Degree Days | 451 | 333 | 189 | 56 | 4 | 1 | 0 | 0 | 4 | 58 | 210 | 368 | 1674 |
| Cooling Degree Days | 5 | 10 | 38 | 88 | 250 | 419 | 495 | 487 | 349 | 137 | 45 | 20 | 2343 |
| Total Precipitation (") | 5.48 | 5.53 | 6.14 | 4.87 | 5.00 | 4.89 | 6.23 | 5.74 | 5.15 | 3.24 | 4.10 | 5.87 | 62.24 |
| Days ≥ 0.1" Precip | 7 | 6 | 7 | 5 | 6 | 6 | 9 | 8 | 7 | 4 | 6 | 8 | 79 |
| Total Snowfall (") | 0.1 | 0.1 | 0.0 | 0.0 | 0.0 | 0.0 | 0.0 | 0.0 | 0.0 | 0.0 | 0.0 | 0.0 | 0.2 |
| Days ≥ 1" Snow Depth | 0 | 0 | 0 | 0 | 0 | 0 | 0 | 0 | 0 | 0 | 0 | 0 | 0 |

## PICKENS *Holmes County* ELEVATION 220 ft LAT/LONG 32° 53 ' N / 89° 59 ' W

| | JAN | FEB | MAR | APR | MAY | JUN | JUL | AUG | SEP | OCT | NOV | DEC | YEAR |
|---|---|---|---|---|---|---|---|---|---|---|---|---|---|
| Maximum Temp °F | 53.4 | 58.2 | 67.0 | 75.3 | 81.7 | 88.6 | 90.3 | 89.8 | 84.9 | 75.7 | 65.9 | ***57.1*** | 74.0 |
| Minimum Temp °F | 31.6 | 34.6 | 42.4 | 49.8 | 57.3 | 64.4 | 67.6 | 66.6 | 61.3 | 48.4 | 40.5 | ***34.3*** | 49.9 |
| Mean Temp °F | 42.5 | 46.4 | 54.7 | 62.6 | 69.7 | 76.5 | 79.0 | 78.2 | 73.1 | 62.1 | 53.2 | ***45.7*** | 62.0 |
| Days Max Temp ≥ 90 °F | 0 | 0 | 0 | 0 | 2 | 14 | 19 | 18 | 8 | 1 | 0 | 0 | 62 |
| Days Max Temp ≤ 32 °F | 1 | 0 | 0 | 0 | 0 | 0 | 0 | 0 | 0 | 0 | 0 | 1 | 2 |
| Days Min Temp ≤ 32 °F | 18 | 13 | 6 | 1 | 0 | 0 | 0 | 0 | 0 | 2 | 8 | 14 | 62 |
| Days Min Temp ≤ 0 °F | 0 | 0 | 0 | 0 | 0 | 0 | 0 | 0 | 0 | 0 | 0 | 0 | 0 |
| Heating Degree Days | 693 | 522 | 332 | 130 | 23 | 1 | 0 | 0 | 14 | 146 | 357 | ***595*** | 2813 |
| Cooling Degree Days | ***0*** | ***4*** | ***15*** | 41 | ***153*** | ***346*** | 428 | ***419*** | ***255*** | 58 | ***9*** | ***3*** | 1731 |
| Total Precipitation (") | 5.70 | 5.14 | 5.96 | 5.60 | 5.97 | 3.46 | 4.96 | 3.18 | 2.98 | 3.96 | 4.69 | 5.84 | 57.44 |
| Days ≥ 0.1" Precip | 7 | 7 | 7 | 6 | 7 | 5 | 7 | 5 | 5 | 5 | 6 | 7 | 74 |
| Total Snowfall (") | ***0.5*** | ***0.1*** | 0.0 | 0.0 | 0.0 | 0.0 | 0.0 | 0.0 | 0.0 | 0.0 | 0.0 | 0.1 | 0.7 |
| Days ≥ 1" Snow Depth | ***0*** | 0 | 0 | 0 | 0 | 0 | 0 | 0 | 0 | 0 | 0 | 0 | 0 |

## PONTOTOC EXP STN *Pontotoc County* ELEVATION 410 ft LAT/LONG 34° 9 ' N / 89° 0 ' W

| | JAN | FEB | MAR | APR | MAY | JUN | JUL | AUG | SEP | OCT | NOV | DEC | YEAR |
|---|---|---|---|---|---|---|---|---|---|---|---|---|---|
| Maximum Temp °F | 49.1 | 53.9 | 63.3 | 73.0 | 80.0 | 87.2 | 90.4 | 89.7 | 83.9 | 73.9 | 63.0 | 53.6 | 71.8 |
| Minimum Temp °F | 29.0 | 32.1 | 40.5 | 49.5 | 57.5 | 65.1 | 68.7 | 67.0 | 61.2 | 48.7 | 40.4 | 32.9 | 49.4 |
| Mean Temp °F | 39.1 | 43.0 | 51.9 | 61.3 | 68.8 | 76.2 | 79.6 | 78.4 | 72.5 | 61.3 | 51.7 | 43.3 | 60.6 |
| Days Max Temp ≥ 90 °F | 0 | 0 | 0 | 0 | 1 | 11 | 19 | 17 | 7 | 0 | 0 | 0 | 55 |
| Days Max Temp ≤ 32 °F | 3 | 1 | 0 | 0 | 0 | 0 | 0 | 0 | 0 | 0 | 0 | 1 | 5 |
| Days Min Temp ≤ 32 °F | 20 | 15 | 8 | 1 | 0 | 0 | 0 | 0 | 0 | 1 | 7 | 16 | 68 |
| Days Min Temp ≤ 0 °F | 0 | 0 | 0 | 0 | 0 | 0 | 0 | 0 | 0 | 0 | 0 | 0 | 0 |
| Heating Degree Days | 797 | 616 | 409 | 160 | 37 | 1 | 0 | 0 | 19 | 162 | 398 | 668 | 3267 |
| Cooling Degree Days | 0 | 0 | 10 | 51 | 164 | 356 | 475 | 441 | 257 | 57 | 8 | 1 | 1820 |
| Total Precipitation (") | 4.86 | 4.97 | 5.95 | 5.59 | 5.51 | 4.41 | 4.20 | 3.05 | 4.17 | 3.62 | 4.72 | 6.26 | 57.31 |
| Days ≥ 0.1" Precip | 7 | 7 | 8 | 7 | 8 | 7 | 6 | 5 | 6 | 5 | 7 | 8 | 81 |
| Total Snowfall (") | 0.9 | 0.5 | 0.1 | 0.0 | 0.0 | 0.0 | 0.0 | 0.0 | 0.0 | 0.0 | 0.1 | 0.1 | 1.7 |
| Days ≥ 1" Snow Depth | 0 | 0 | 0 | 0 | 0 | 0 | 0 | 0 | 0 | 0 | 0 | 0 | 0 |

## POPLARVILLE EXP STN *Pearl River County* ELEVATION 312 ft LAT/LONG 30° 51 ' N / 89° 33 ' W

| | JAN | FEB | MAR | APR | MAY | JUN | JUL | AUG | SEP | OCT | NOV | DEC | YEAR |
|---|---|---|---|---|---|---|---|---|---|---|---|---|---|
| Maximum Temp °F | 60.6 | 64.3 | 71.4 | 78.2 | 84.1 | 90.2 | 91.1 | 90.8 | 87.1 | 79.7 | 70.6 | 63.7 | 77.7 |
| Minimum Temp °F | 39.5 | 41.7 | 48.4 | 55.7 | 62.7 | 68.8 | 70.9 | 70.4 | 66.4 | 56.0 | 48.0 | 42.5 | 55.9 |
| Mean Temp °F | 50.1 | 53.0 | 59.9 | 67.0 | 73.4 | 79.5 | 81.0 | 80.6 | 76.8 | 67.8 | 59.3 | 53.1 | 66.8 |
| Days Max Temp ≥ 90 °F | 0 | 0 | 0 | 0 | 3 | 18 | 22 | 22 | 11 | 1 | 0 | 0 | 77 |
| Days Max Temp ≤ 32 °F | 0 | 0 | 0 | 0 | 0 | 0 | 0 | 0 | 0 | 0 | 0 | 0 | 0 |
| Days Min Temp ≤ 32 °F | 9 | 6 | 2 | 0 | 0 | 0 | 0 | 0 | 0 | 0 | 2 | 6 | 25 |
| Days Min Temp ≤ 0 °F | 0 | 0 | 0 | 0 | 0 | 0 | 0 | 0 | 0 | 0 | 0 | 0 | 0 |
| Heating Degree Days | 463 | 342 | 192 | 54 | 4 | 0 | 0 | 0 | 3 | 52 | 206 | 377 | 1693 |
| Cooling Degree Days | 4 | 9 | 37 | 100 | 262 | 436 | 500 | 494 | 353 | 138 | 41 | 15 | 2389 |
| Total Precipitation (") | 5.76 | 5.98 | 6.04 | 5.28 | 5.36 | 4.41 | 6.71 | 5.66 | 4.18 | 3.56 | 4.37 | 5.56 | 62.87 |
| Days ≥ 0.1" Precip | 8 | 7 | 7 | 6 | 7 | 7 | 10 | 9 | 7 | 4 | 6 | 7 | 85 |
| Total Snowfall (") | 0.1 | 0.1 | 0.2 | 0.0 | 0.0 | 0.0 | 0.0 | 0.0 | 0.0 | 0.0 | 0.0 | 0.0 | 0.4 |
| Days ≥ 1" Snow Depth | 0 | 0 | 0 | 0 | 0 | 0 | 0 | 0 | 0 | 0 | 0 | 0 | 0 |

## PORT GIBSON 1 NW *Claiborne County* ELEVATION 161 ft LAT/LONG 31° 57 ' N / 90° 59 ' W

| | JAN | FEB | MAR | APR | MAY | JUN | JUL | AUG | SEP | OCT | NOV | DEC | YEAR |
|---|---|---|---|---|---|---|---|---|---|---|---|---|---|
| Maximum Temp °F | 55.9 | 60.4 | 68.7 | 77.0 | 82.8 | 89.2 | 91.3 | 91.1 | 86.5 | 77.6 | 68.0 | 59.9 | 75.7 |
| Minimum Temp °F | 32.6 | 35.4 | 43.2 | 51.3 | 59.2 | 66.2 | 69.3 | 68.3 | 63.0 | 49.8 | 41.7 | 35.7 | 51.3 |
| Mean Temp °F | 44.3 | 47.9 | 56.0 | 64.2 | 71.1 | 77.7 | 80.4 | 79.7 | 74.8 | 63.7 | 54.9 | 47.8 | 63.5 |
| Days Max Temp ≥ 90 °F | 0 | 0 | 0 | 0 | 3 | 17 | 23 | 22 | 11 | 1 | 0 | 0 | 77 |
| Days Max Temp ≤ 32 °F | 1 | 0 | 0 | 0 | 0 | 0 | 0 | 0 | 0 | 0 | 0 | 0 | 1 |
| Days Min Temp ≤ 32 °F | 17 | 12 | 5 | 1 | 0 | 0 | 0 | 0 | 0 | 1 | 7 | 15 | 58 |
| Days Min Temp ≤ 0 °F | 0 | 0 | 0 | 0 | 0 | 0 | 0 | 0 | 0 | 0 | 0 | 0 | 0 |
| Heating Degree Days | 638 | 480 | 300 | 105 | 16 | 0 | 0 | 0 | 9 | 118 | 317 | 534 | 2517 |
| Cooling Degree Days | 1 | 6 | 30 | 81 | 215 | 397 | 487 | 473 | 311 | 91 | 23 | 7 | 2122 |
| Total Precipitation (") | 5.76 | 5.11 | 6.00 | 5.54 | 5.73 | 4.39 | 4.27 | 3.45 | 3.31 | 3.78 | 4.72 | 6.05 | 58.11 |
| Days ≥ 0.1" Precip | 7 | 6 | 7 | 6 | 6 | 6 | 7 | 6 | 5 | 4 | 6 | 7 | 73 |
| Total Snowfall (") | 0.4 | 0.1 | 0.3 | 0.0 | 0.0 | 0.0 | 0.0 | 0.0 | 0.0 | 0.0 | 0.0 | 0.0 | 0.8 |
| Days ≥ 1" Snow Depth | 0 | 0 | 0 | 0 | 0 | 0 | 0 | 0 | 0 | 0 | 0 | 0 | 0 |

## QUITMAN 1 N *Clarke County* ELEVATION 230 ft LAT/LONG 32° 2 ' N / 88° 44 ' W

| | JAN | FEB | MAR | APR | MAY | JUN | JUL | AUG | SEP | OCT | NOV | DEC | YEAR |
|---|---|---|---|---|---|---|---|---|---|---|---|---|---|
| Maximum Temp °F | 56.2 | 60.9 | 69.3 | 77.4 | 83.2 | 90.1 | 91.9 | 91.2 | 86.6 | 77.7 | 69.2 | 60.4 | 76.2 |
| Minimum Temp °F | 32.4 | 34.7 | 42.2 | 50.0 | 57.9 | 65.1 | 68.5 | 67.5 | 62.4 | 48.7 | 41.7 | 35.4 | 50.5 |
| Mean Temp °F | 44.3 | 47.8 | 55.8 | 63.8 | 70.6 | 77.7 | 80.2 | 79.4 | 74.5 | 63.2 | 55.5 | 47.9 | 63.4 |
| Days Max Temp ≥ 90 °F | 0 | 0 | 0 | 0 | 4 | 18 | 23 | 22 | 12 | 1 | 0 | 0 | 80 |
| Days Max Temp ≤ 32 °F | 1 | 0 | 0 | 0 | 0 | 0 | 0 | 0 | 0 | 0 | 0 | 0 | 1 |
| Days Min Temp ≤ 32 °F | 17 | 14 | 7 | 1 | 0 | 0 | 0 | 0 | 0 | 2 | 8 | 15 | 64 |
| Days Min Temp ≤ 0 °F | 0 | 0 | 0 | 0 | 0 | 0 | 0 | 0 | 0 | 0 | 0 | 0 | 0 |
| Heating Degree Days | 636 | 484 | 305 | 109 | 19 | 1 | 0 | 0 | 10 | 129 | 307 | 530 | 2530 |
| Cooling Degree Days | 2 | 6 | 25 | 70 | 203 | 397 | 488 | 466 | 300 | 79 | 20 | 9 | 2065 |
| Total Precipitation (") | 5.76 | 5.21 | 6.54 | 5.19 | 4.44 | 3.70 | 5.33 | 3.46 | 3.56 | 2.87 | 4.26 | 5.79 | 56.11 |
| Days ≥ 0.1" Precip | 7 | 6 | 7 | 6 | 6 | 6 | 9 | 6 | 5 | 4 | 6 | 7 | 75 |
| Total Snowfall (") | 0.3 | 0.1 | 0.2 | 0.1 | 0.0 | 0.0 | 0.0 | 0.0 | 0.0 | 0.0 | 0.0 | 0.0 | 0.7 |
| Days ≥ 1" Snow Depth | 0 | 0 | 0 | 0 | 0 | 0 | 0 | 0 | 0 | 0 | 0 | 0 | 0 |

## RICHTON 3 SSE *Perry County* ELEVATION 200 ft LAT/LONG 31° 22 ' N / 88° 56 ' W

| | JAN | FEB | MAR | APR | MAY | JUN | JUL | AUG | SEP | OCT | NOV | DEC | YEAR |
|---|---|---|---|---|---|---|---|---|---|---|---|---|---|
| Maximum Temp °F | 58.4 | 62.5 | 70.3 | 77.9 | 83.3 | 89.6 | 91.4 | 90.9 | 87.0 | 78.8 | 69.6 | 62.0 | 76.8 |
| Minimum Temp °F | 33.6 | 35.8 | 43.4 | 50.4 | 57.9 | 64.9 | 68.2 | 67.5 | 62.6 | 49.7 | 41.4 | 36.2 | 51.0 |
| Mean Temp °F | 45.8 | 49.2 | 56.8 | 64.2 | 70.7 | 77.3 | 79.8 | 79.2 | 74.8 | 64.3 | 55.5 | 49.1 | 63.9 |
| Days Max Temp ≥ 90 °F | 0 | 0 | 0 | 0 | 3 | 17 | 23 | 22 | 12 | 2 | 0 | 0 | 79 |
| Days Max Temp ≤ 32 °F | 0 | 0 | 0 | 0 | 0 | 0 | 0 | 0 | 0 | 0 | 0 | 0 | 0 |
| Days Min Temp ≤ 32 °F | 16 | 13 | 6 | 1 | 0 | 0 | 0 | 0 | 0 | 1 | 8 | 14 | 59 |
| Days Min Temp ≤ 0 °F | 0 | 0 | 0 | 0 | 0 | 0 | 0 | 0 | 0 | 0 | 0 | 0 | 0 |
| Heating Degree Days | 590 | 446 | 275 | 101 | 17 | 1 | 0 | 0 | 7 | 110 | 302 | 496 | 2345 |
| Cooling Degree Days | 3 | 7 | 29 | 72 | 206 | 385 | 472 | 458 | 309 | 103 | 28 | 11 | 2083 |
| Total Precipitation (") | 6.11 | 5.50 | 6.16 | 4.91 | 5.01 | 3.74 | 6.54 | 4.41 | 4.50 | 2.94 | 4.71 | 5.49 | 60.02 |
| Days ≥ 0.1" Precip | 8 | 7 | 7 | 6 | 6 | 7 | 10 | 8 | 6 | 4 | 6 | 7 | 82 |
| Total Snowfall (") | 0.2 | 0.1 | 0.1 | 0.1 | 0.0 | 0.0 | 0.0 | 0.0 | 0.0 | 0.0 | 0.0 | 0.0 | 0.5 |
| Days ≥ 1" Snow Depth | 0 | 0 | 0 | 0 | 0 | 0 | 0 | 0 | 0 | 0 | 0 | 0 | 0 |

## RIPLEY *Tippah County* ELEVATION 502 ft LAT/LONG 34° 43 ' N / 88° 57 ' W

| | JAN | FEB | MAR | APR | MAY | JUN | JUL | AUG | SEP | OCT | NOV | DEC | YEAR |
|---|---|---|---|---|---|---|---|---|---|---|---|---|---|
| Maximum Temp °F | 47.5 | 52.6 | 62.3 | 72.7 | 79.6 | 87.1 | 90.3 | 89.5 | 83.5 | 72.9 | 61.5 | 51.9 | 70.9 |
| Minimum Temp °F | 27.4 | 29.9 | 38.3 | 47.2 | 55.9 | 63.9 | 68.0 | 66.3 | 59.9 | 46.9 | 38.4 | 31.2 | 47.8 |
| Mean Temp °F | 37.4 | 41.2 | 50.4 | 60.0 | 67.8 | 75.5 | 79.2 | 77.9 | 71.7 | 59.9 | 50.0 | 41.6 | 59.4 |
| Days Max Temp ≥ 90 °F | 0 | 0 | 0 | 0 | 1 | 11 | 18 | 17 | 7 | 0 | 0 | 0 | 54 |
| Days Max Temp ≤ 32 °F | 3 | 1 | 0 | 0 | 0 | 0 | 0 | 0 | 0 | 0 | 0 | 1 | 5 |
| Days Min Temp ≤ 32 °F | 22 | 18 | 11 | 2 | 0 | 0 | 0 | 0 | 0 | 2 | 10 | 18 | 83 |
| Days Min Temp ≤ 0 °F | 0 | 0 | 0 | 0 | 0 | 0 | 0 | 0 | 0 | 0 | 0 | 0 | 0 |
| Heating Degree Days | 847 | 666 | 453 | 192 | 49 | 2 | 0 | 0 | 25 | 196 | 450 | 720 | 3600 |
| Cooling Degree Days | 0 | 0 | 8 | 42 | 136 | 330 | 454 | 415 | 226 | 42 | 5 | 1 | 1659 |
| Total Precipitation (") | 4.60 | 5.01 | 6.04 | 5.70 | 5.58 | 4.26 | 4.58 | 3.24 | 3.75 | 3.58 | 5.66 | 6.16 | 58.16 |
| Days ≥ 0.1" Precip | 7 | 7 | 8 | 7 | 8 | 6 | 7 | 5 | 6 | 5 | 7 | 8 | 81 |
| Total Snowfall (") | 1.2 | 1.0 | 0.3 | 0.0 | 0.0 | 0.0 | 0.0 | 0.0 | 0.0 | 0.0 | 0.0 | 0.2 | 2.7 |
| Days ≥ 1" Snow Depth | 1 | 0 | 0 | 0 | 0 | 0 | 0 | 0 | 0 | 0 | 0 | 0 | 1 |

## SAUCIER EXP FOREST *Harrison County* ELEVATION 230 ft LAT/LONG 30° 38 ' N / 89° 3 ' W

| | JAN | FEB | MAR | APR | MAY | JUN | JUL | AUG | SEP | OCT | NOV | DEC | YEAR |
|---|---|---|---|---|---|---|---|---|---|---|---|---|---|
| Maximum Temp °F | 60.8 | 64.8 | 71.7 | 78.7 | 84.3 | 89.9 | 91.2 | 90.3 | 86.7 | 79.4 | 70.3 | 63.9 | 77.7 |
| Minimum Temp °F | 40.6 | 42.8 | 49.4 | 56.2 | 62.9 | 69.0 | 71.4 | 71.1 | 67.2 | 57.0 | 49.0 | 43.5 | 56.7 |
| Mean Temp °F | 50.7 | 53.8 | 60.6 | 67.5 | 73.6 | 79.5 | 81.4 | 80.7 | 77.0 | 68.2 | 59.7 | 53.7 | 67.2 |
| Days Max Temp ≥ 90 °F | 0 | 0 | 0 | 0 | 3 | 17 | 23 | 21 | 9 | 1 | 0 | 0 | 74 |
| Days Max Temp ≤ 32 °F | 0 | 0 | 0 | 0 | 0 | 0 | 0 | 0 | 0 | 0 | 0 | 0 | 0 |
| Days Min Temp ≤ 32 °F | 8 | 5 | 1 | 0 | 0 | 0 | 0 | 0 | 0 | 0 | 1 | 5 | 20 |
| Days Min Temp ≤ 0 °F | 0 | 0 | 0 | 0 | 0 | 0 | 0 | 0 | 0 | 0 | 0 | 0 | 0 |
| Heating Degree Days | 442 | 319 | 174 | 42 | 2 | 0 | 0 | 0 | 2 | 43 | 195 | 361 | 1580 |
| Cooling Degree Days | 5 | 12 | 45 | 115 | 280 | 443 | 519 | 505 | 367 | 156 | 49 | 18 | 2514 |
| Total Precipitation (") | 6.28 | 5.87 | 6.66 | 4.98 | 5.93 | 5.33 | 7.19 | 7.20 | 5.77 | 3.19 | 4.52 | 5.75 | 68.67 |
| Days ≥ 0.1" Precip | 8 | 7 | 7 | 5 | 7 | 8 | 10 | 10 | 7 | 4 | 6 | 6 | 85 |
| Total Snowfall (") | 0.0 | 0.1 | 0.1 | 0.0 | 0.0 | 0.0 | 0.0 | 0.0 | 0.0 | 0.0 | 0.0 | 0.0 | 0.2 |
| Days ≥ 1" Snow Depth | 0 | 0 | 0 | 0 | 0 | 0 | 0 | 0 | 0 | 0 | 0 | 0 | 0 |

## STATE UNIVERSITY *Oktibbeha County* ELEVATION 351 ft LAT/LONG 33° 27 ' N / 88° 47 ' W

| | JAN | FEB | MAR | APR | MAY | JUN | JUL | AUG | SEP | OCT | NOV | DEC | YEAR |
|---|---|---|---|---|---|---|---|---|---|---|---|---|---|
| Maximum Temp °F | 51.2 | 56.1 | 65.0 | 74.4 | 81.1 | 88.4 | 91.2 | 90.3 | 84.9 | 75.2 | 64.8 | 55.7 | 73.2 |
| Minimum Temp °F | 31.4 | 34.5 | 42.8 | 51.4 | 59.5 | 67.1 | 70.7 | 68.9 | 63.2 | 50.8 | 42.6 | 35.1 | 51.5 |
| Mean Temp °F | 41.4 | 45.3 | 53.9 | 62.9 | 70.3 | 77.8 | 81.0 | 79.7 | 74.1 | 63.0 | 53.7 | 45.4 | 62.4 |
| Days Max Temp ≥ 90 °F | 0 | 0 | 0 | 0 | 2 | 14 | 21 | 19 | 8 | 1 | 0 | 0 | 65 |
| Days Max Temp ≤ 32 °F | 2 | 1 | 0 | 0 | 0 | 0 | 0 | 0 | 0 | 0 | 0 | 1 | 4 |
| Days Min Temp ≤ 32 °F | 17 | 13 | 5 | 1 | 0 | 0 | 0 | 0 | 0 | 0 | 5 | 14 | 55 |
| Days Min Temp ≤ 0 °F | 0 | 0 | 0 | 0 | 0 | 0 | 0 | 0 | 0 | 0 | 0 | 0 | 0 |
| Heating Degree Days | 726 | 550 | 353 | 125 | 23 | 1 | 0 | 0 | 11 | 127 | 344 | 604 | 2864 |
| Cooling Degree Days | 0 | 3 | 14 | 63 | 191 | 392 | 510 | 472 | 294 | 71 | 14 | 3 | 2027 |
| Total Precipitation (") | 5.20 | 5.02 | 5.91 | 5.69 | 4.95 | 3.67 | 4.67 | 3.66 | 3.96 | 3.41 | 4.11 | 5.67 | 55.92 |
| Days ≥ 0.1" Precip | 8 | 7 | 7 | 6 | 7 | 6 | 7 | 6 | 6 | 4 | 7 | 7 | 78 |
| Total Snowfall (") | 0.4 | 0.2 | 0.0 | 0.0 | 0.0 | 0.0 | 0.0 | 0.0 | 0.0 | 0.0 | 0.0 | 0.0 | 0.6 |
| Days ≥ 1" Snow Depth | 0 | 0 | 0 | 0 | 0 | 0 | 0 | 0 | 0 | 0 | 0 | 0 | 0 |

## STONEVILLE EXP STN *Washington County* ELEVATION 131 ft LAT/LONG 33° 25 ' N / 90° 55 ' W

| | JAN | FEB | MAR | APR | MAY | JUN | JUL | AUG | SEP | OCT | NOV | DEC | YEAR |
|---|---|---|---|---|---|---|---|---|---|---|---|---|---|
| Maximum Temp °F | 49.5 | 54.7 | 63.9 | 74.1 | 81.8 | 89.3 | 91.5 | 90.3 | 85.0 | 75.8 | 64.0 | 54.2 | 72.8 |
| Minimum Temp °F | 32.1 | 35.6 | 43.8 | 53.3 | 61.6 | 69.3 | 72.0 | 69.7 | 63.3 | 51.4 | 43.3 | 36.3 | 52.6 |
| Mean Temp °F | 40.9 | 45.2 | 53.9 | 63.7 | 71.7 | 79.3 | 81.8 | 80.0 | 74.2 | 63.6 | 53.7 | 45.3 | 62.8 |
| Days Max Temp ≥ 90 °F | 0 | 0 | 0 | 0 | 4 | 16 | 22 | 20 | 9 | 1 | 0 | 0 | 72 |
| Days Max Temp ≤ 32 °F | 3 | 1 | 0 | 0 | 0 | 0 | 0 | 0 | 0 | 0 | 0 | 1 | 5 |
| Days Min Temp ≤ 32 °F | 18 | 11 | 3 | 0 | 0 | 0 | 0 | 0 | 0 | 0 | 4 | 11 | 47 |
| Days Min Temp ≤ 0 °F | 0 | 0 | 0 | 0 | 0 | 0 | 0 | 0 | 0 | 0 | 0 | 0 | 0 |
| Heating Degree Days | 742 | 554 | 353 | 116 | 17 | 0 | 0 | 0 | 13 | 123 | 346 | 608 | 2872 |
| Cooling Degree Days | 0 | 1 | 14 | 81 | 230 | 442 | 529 | 484 | 293 | 83 | 14 | 4 | 2175 |
| Total Precipitation (") | 4.60 | 4.66 | 5.27 | 4.83 | 5.06 | 3.82 | 3.87 | 2.21 | 3.36 | 3.47 | 5.24 | 5.44 | 51.83 |
| Days ≥ 0.1" Precip | 7 | 6 | 7 | 6 | 7 | 6 | 6 | 4 | 5 | 5 | 6 | 7 | 72 |
| Total Snowfall (") | 1.1 | 0.3 | 0.3 | 0.0 | 0.0 | 0.0 | 0.0 | 0.0 | 0.0 | 0.0 | 0.0 | 0.1 | 1.8 |
| Days ≥ 1" Snow Depth | 1 | 0 | 0 | 0 | 0 | 0 | 0 | 0 | 0 | 0 | 0 | 0 | 1 |

## TUNICA 2 *Tunica County* ELEVATION 190 ft LAT/LONG 34° 41' N / 90° 23' W

| | JAN | FEB | MAR | APR | MAY | JUN | JUL | AUG | SEP | OCT | NOV | DEC | YEAR |
|---|---|---|---|---|---|---|---|---|---|---|---|---|---|
| Maximum Temp °F | 47.3 | 52.2 | 62.1 | 72.6 | 80.2 | 88.7 | 91.7 | 89.8 | 84.1 | 74.3 | 62.3 | 52.1 | 71.4 |
| Minimum Temp °F | 29.7 | 33.2 | 42.0 | 51.3 | 59.5 | 67.6 | 71.1 | 68.1 | 61.7 | 49.5 | 41.4 | 34.1 | 50.8 |
| Mean Temp °F | 38.5 | 42.7 | 52.1 | 62.0 | 69.9 | 78.2 | 81.4 | 79.0 | 73.0 | 61.9 | 51.9 | 43.3 | 61.2 |
| Days Max Temp ≥ 90 °F | 0 | 0 | 0 | 0 | 2 | 15 | 22 | 18 | 8 | 1 | 0 | 0 | 66 |
| Days Max Temp ≤ 32 °F | 4 | 2 | 0 | 0 | 0 | 0 | 0 | 0 | 0 | 0 | 0 | 2 | 8 |
| Days Min Temp ≤ 32 °F | 20 | 14 | 5 | 0 | 0 | 0 | 0 | 0 | 0 | 1 | 6 | 15 | 61 |
| Days Min Temp ≤ 0 °F | 0 | 0 | 0 | 0 | 0 | 0 | 0 | 0 | 0 | 0 | 0 | 0 | 0 |
| Heating Degree Days | 815 | 623 | 405 | 148 | 29 | 0 | 0 | 0 | 18 | 152 | 396 | 668 | 3254 |
| Cooling Degree Days | 0 | 0 | 11 | 63 | 192 | 417 | 524 | 456 | 274 | 64 | 11 | 2 | 2014 |
| Total Precipitation (") | 4.13 | 4.25 | 5.38 | 5.84 | 5.96 | 4.52 | 3.34 | 2.85 | 2.93 | 3.48 | 4.99 | 5.73 | 53.40 |
| Days ≥ 0.1" Precip | 7 | 6 | 8 | 7 | 7 | 6 | 5 | 4 | 5 | 5 | 6 | 7 | 73 |
| Total Snowfall (") | 1.3 | 0.6 | 0.1 | 0.0 | 0.0 | 0.0 | 0.0 | 0.0 | 0.0 | 0.0 | 0.1 | 0.0 | 2.1 |
| Days ≥ 1" Snow Depth | *0* | 0 | 0 | 0 | 0 | 0 | 0 | 0 | 0 | 0 | 0 | 0 | 0 |

## TUPELO C D LEMONS AP *Lee County* ELEVATION 361 ft LAT/LONG 34° 16' N / 88° 46' W

| | JAN | FEB | MAR | APR | MAY | JUN | JUL | AUG | SEP | OCT | NOV | DEC | YEAR |
|---|---|---|---|---|---|---|---|---|---|---|---|---|---|
| Maximum Temp °F | 50.4 | 56.3 | 65.3 | 75.2 | 81.5 | 88.8 | 91.9 | 90.9 | 85.2 | 75.2 | 63.8 | 55.0 | 73.3 |
| Minimum Temp °F | 30.1 | 32.7 | 40.6 | 48.8 | 57.2 | 65.0 | 69.2 | 67.6 | 61.5 | 48.4 | 40.0 | 33.7 | 49.6 |
| Mean Temp °F | 40.3 | 44.5 | 53.0 | 62.0 | 69.4 | 77.0 | 80.6 | 79.3 | 73.4 | 61.8 | 51.9 | 44.4 | 61.5 |
| Days Max Temp ≥ 90 °F | 0 | 0 | 0 | 0 | 3 | 14 | 22 | 20 | 8 | 0 | 0 | 0 | 67 |
| Days Max Temp ≤ 32 °F | 2 | 1 | 0 | 0 | 0 | 0 | 0 | 0 | 0 | 0 | 0 | 1 | 4 |
| Days Min Temp ≤ 32 °F | 19 | 15 | 8 | 2 | 0 | 0 | 0 | 0 | 0 | 2 | 9 | 15 | 70 |
| Days Min Temp ≤ 0 °F | 0 | 0 | 0 | 0 | 0 | 0 | 0 | 0 | 0 | 0 | 0 | 0 | 0 |
| Heating Degree Days | 761 | 574 | 377 | 142 | 29 | 1 | 0 | 0 | 14 | 150 | 395 | 634 | 3077 |
| Cooling Degree Days | 0 | 1 | 11 | 57 | 181 | 387 | 515 | 476 | 281 | 62 | 10 | 2 | 1983 |
| Total Precipitation (") | 4.59 | 4.85 | 5.98 | 4.98 | 6.20 | 3.97 | 3.75 | 2.91 | 3.55 | 3.64 | 4.67 | 6.35 | 55.44 |
| Days ≥ 0.1" Precip | 7 | 6 | 8 | 7 | 8 | 6 | 6 | 5 | 6 | 5 | 6 | 8 | 78 |
| Total Snowfall (") | 1.2 | 0.4 | 0.5 | 0.0 | 0.0 | 0.0 | 0.0 | 0.0 | 0.0 | 0.0 | 0.0 | 0.1 | 2.2 |
| Days ≥ 1" Snow Depth | *1* | 1 | 0 | 0 | 0 | 0 | 0 | 0 | 0 | 0 | 0 | 0 | 2 |

## TYLERTOWN 2 WNW *Walthall County* ELEVATION 390 ft LAT/LONG 31° 7' N / 90° 8' W

| | JAN | FEB | MAR | APR | MAY | JUN | JUL | AUG | SEP | OCT | NOV | DEC | YEAR |
|---|---|---|---|---|---|---|---|---|---|---|---|---|---|
| Maximum Temp °F | 59.7 | 63.9 | 71.3 | 78.0 | 83.5 | 89.8 | 91.1 | 90.7 | 87.0 | 79.3 | 69.8 | 62.8 | 77.2 |
| Minimum Temp °F | 38.3 | 40.4 | 47.4 | 54.4 | 61.5 | 67.6 | 70.3 | 69.7 | 65.7 | 54.7 | 46.9 | 41.4 | 54.9 |
| Mean Temp °F | 49.0 | 52.2 | 59.3 | 66.2 | 72.5 | 78.7 | 80.7 | 80.3 | 76.4 | 67.0 | 58.4 | 52.1 | 66.1 |
| Days Max Temp ≥ 90 °F | 0 | 0 | 0 | 0 | 2 | 16 | 22 | 21 | 10 | 1 | 0 | 0 | 72 |
| Days Max Temp ≤ 32 °F | 0 | 0 | 0 | 0 | 0 | 0 | 0 | 0 | 0 | 0 | 0 | 0 | 0 |
| Days Min Temp ≤ 32 °F | 11 | 8 | 3 | 0 | 0 | 0 | 0 | 0 | 0 | 0 | 4 | 9 | 35 |
| Days Min Temp ≤ 0 °F | 0 | 0 | 0 | 0 | 0 | 0 | 0 | 0 | 0 | 0 | 0 | 0 | 0 |
| Heating Degree Days | 496 | 364 | 208 | 64 | 5 | 0 | 0 | 0 | 3 | 61 | 230 | 408 | 1839 |
| Cooling Degree Days | 5 | 11 | 37 | 98 | 246 | 416 | 498 | 489 | 352 | 138 | 46 | 16 | 2352 |
| Total Precipitation (") | 6.14 | 5.79 | 5.83 | 5.78 | 6.03 | 4.32 | 6.43 | 5.23 | 4.07 | 3.48 | 4.38 | 5.99 | 63.47 |
| Days ≥ 0.1" Precip | 8 | 7 | 7 | 6 | 7 | 7 | 9 | 8 | 6 | 4 | 6 | 7 | 82 |
| Total Snowfall (") | 0.2 | 0.2 | 0.2 | 0.0 | 0.0 | 0.0 | 0.0 | 0.0 | 0.0 | 0.0 | 0.0 | 0.0 | 0.6 |
| Days ≥ 1" Snow Depth | 0 | 0 | 0 | 0 | 0 | 0 | 0 | 0 | 0 | 0 | 0 | 0 | 0 |

## UNIVERSITY *Lafayette County* ELEVATION 502 ft LAT/LONG 34° 22' N / 89° 32' W

| | JAN | FEB | MAR | APR | MAY | JUN | JUL | AUG | SEP | OCT | NOV | DEC | YEAR |
|---|---|---|---|---|---|---|---|---|---|---|---|---|---|
| Maximum Temp °F | 48.9 | 53.7 | 63.0 | 73.1 | 79.8 | 87.4 | 90.7 | 89.3 | 83.8 | 73.7 | 62.8 | 53.7 | 71.7 |
| Minimum Temp °F | 27.6 | 30.3 | 39.0 | 47.8 | 56.2 | 64.5 | 68.7 | 66.6 | 60.3 | 46.8 | 39.1 | 31.7 | 48.2 |
| Mean Temp °F | 38.3 | 42.0 | 51.0 | 60.5 | 68.0 | 76.0 | 79.7 | 78.0 | 72.1 | 60.3 | 51.0 | 42.7 | 60.0 |
| Days Max Temp ≥ 90 °F | 0 | 0 | 0 | 0 | 1 | 11 | 20 | 16 | 7 | 0 | 0 | 0 | 55 |
| Days Max Temp ≤ 32 °F | 3 | 1 | 0 | 0 | 0 | 0 | 0 | 0 | 0 | 0 | 0 | 1 | 5 |
| Days Min Temp ≤ 32 °F | 21 | 17 | 10 | 2 | 0 | 0 | 0 | 0 | 0 | 3 | 10 | 18 | 81 |
| Days Min Temp ≤ 0 °F | 0 | 0 | 0 | 0 | 0 | 0 | 0 | 0 | 0 | 0 | 0 | 0 | 0 |
| Heating Degree Days | 821 | 642 | 438 | 182 | 49 | 3 | 0 | 0 | 25 | 189 | 422 | 686 | 3457 |
| Cooling Degree Days | 0 | 1 | 11 | 51 | 163 | 370 | 493 | 445 | 260 | 53 | 10 | 2 | 1859 |
| Total Precipitation (") | 4.64 | 4.92 | 5.70 | 5.24 | 6.01 | 3.81 | 3.81 | 3.60 | 3.79 | 3.84 | 5.43 | 5.92 | 56.71 |
| Days ≥ 0.1" Precip | 7 | 7 | 8 | 7 | 8 | 6 | 6 | 6 | 6 | 5 | 7 | 7 | 80 |
| Total Snowfall (") | 1.7 | 1.0 | 0.4 | 0.0 | 0.0 | 0.0 | 0.0 | 0.0 | 0.0 | 0.0 | 0.0 | 0.2 | 3.3 |
| Days ≥ 1" Snow Depth | 1 | 1 | 0 | 0 | 0 | 0 | 0 | 0 | 0 | 0 | 0 | 0 | 2 |

## WATER VALLEY 1 NNE *Yalobusha County* ELEVATION 302 ft LAT/LONG 34° 10 ' N / 89° 38 ' W

| | JAN | FEB | MAR | APR | MAY | JUN | JUL | AUG | SEP | OCT | NOV | DEC | YEAR |
|---|---|---|---|---|---|---|---|---|---|---|---|---|---|
| Maximum Temp °F | 50.5 | 55.4 | 64.4 | 73.8 | 79.9 | 87.2 | 90.6 | 90.3 | 85.3 | 75.7 | 64.8 | 55.3 | 72.8 |
| Minimum Temp °F | 27.8 | 30.8 | 39.1 | 47.7 | 56.4 | 64.8 | 69.0 | 67.1 | 60.9 | 47.8 | 39.2 | 31.7 | 48.5 |
| Mean Temp °F | 39.2 | 43.1 | 51.8 | 60.8 | 68.2 | 76.0 | 79.8 | 78.7 | 73.1 | 61.7 | 52.0 | 43.5 | 60.7 |
| Days Max Temp ≥ 90 °F | 0 | 0 | 0 | 0 | 1 | 11 | 20 | 18 | 10 | 1 | 0 | 0 | 61 |
| Days Max Temp ≤ 32 °F | 2 | 1 | 0 | 0 | 0 | 0 | 0 | 0 | 0 | 0 | 0 | 1 | 4 |
| Days Min Temp ≤ 32 °F | 21 | 17 | 9 | 2 | 0 | 0 | 0 | 0 | 0 | 2 | 10 | 18 | 79 |
| Days Min Temp ≤ 0 °F | 0 | 0 | 0 | 0 | 0 | 0 | 0 | 0 | 0 | 0 | 0 | 0 | 0 |
| Heating Degree Days | 793 | 612 | 414 | 172 | 42 | 1 | 0 | 0 | 18 | 157 | 393 | 662 | 3264 |
| Cooling Degree Days | 0 | 1 | 11 | 49 | 148 | 347 | 471 | 443 | 272 | 66 | 13 | 3 | 1824 |
| Total Precipitation (") | 4.61 | 4.90 | 6.02 | 5.40 | 5.82 | 4.17 | 3.82 | 3.29 | 3.61 | 3.76 | 5.58 | 6.18 | 57.16 |
| Days ≥ 0.1" Precip | 7 | 7 | 8 | 7 | 8 | 6 | 7 | 5 | 6 | 5 | 6 | 7 | 79 |
| Total Snowfall (") | 1.7 | 0.8 | 0.1 | 0.0 | 0.0 | 0.0 | 0.0 | 0.0 | 0.0 | 0.0 | 0.1 | 0.3 | 3.0 |
| Days ≥ 1" Snow Depth | 1 | 1 | 0 | 0 | 0 | 0 | 0 | 0 | 0 | 0 | 0 | 0 | 2 |

## WAYNESBORO 2 W *Wayne County* ELEVATION 220 ft LAT/LONG 31° 41 ' N / 88° 37 ' W

| | JAN | FEB | MAR | APR | MAY | JUN | JUL | AUG | SEP | OCT | NOV | DEC | YEAR |
|---|---|---|---|---|---|---|---|---|---|---|---|---|---|
| Maximum Temp °F | 58.5 | 63.7 | 72.0 | 79.1 | 84.1 | 90.4 | 91.8 | 91.4 | 87.4 | 79.2 | 69.5 | 62.3 | 77.4 |
| Minimum Temp °F | 35.1 | 37.4 | 44.2 | 51.2 | 57.9 | 64.8 | 68.2 | 67.8 | 62.8 | 50.1 | 41.7 | 37.3 | 51.5 |
| Mean Temp °F | 46.8 | 50.6 | 58.1 | 65.2 | 71.0 | 77.6 | 80.1 | 79.6 | 75.1 | 64.7 | 55.6 | 49.8 | 64.5 |
| Days Max Temp ≥ 90 °F | 0 | 0 | 0 | 0 | 4 | 19 | 23 | 23 | 13 | 1 | 0 | 0 | 83 |
| Days Max Temp ≤ 32 °F | 0 | 0 | 0 | 0 | 0 | 0 | 0 | 0 | 0 | 0 | 0 | 0 | 0 |
| Days Min Temp ≤ 32 °F | 14 | 11 | 5 | 1 | 0 | 0 | 0 | 0 | 0 | 1 | 8 | 12 | 52 |
| Days Min Temp ≤ 0 °F | 0 | 0 | 0 | 0 | 0 | 0 | 0 | 0 | 0 | 0 | 0 | 0 | 0 |
| Heating Degree Days | 560 | 408 | 240 | 82 | 11 | 0 | 0 | 0 | 6 | 98 | 297 | 473 | 2175 |
| Cooling Degree Days | 3 | 9 | 31 | 81 | 201 | 389 | 477 | 479 | 316 | 99 | 23 | 7 | 2115 |
| Total Precipitation (") | 5.88 | 5.37 | 5.73 | 4.94 | 5.19 | 4.41 | 5.49 | 3.97 | 4.25 | 2.79 | 4.31 | 5.77 | 58.10 |
| Days ≥ 0.1" Precip | 7 | 6 | 6 | 5 | 7 | 7 | 9 | 7 | 6 | 4 | 5 | 7 | 76 |
| Total Snowfall (") | 0.0 | 0.1 | 0.3 | 0.0 | 0.0 | 0.0 | 0.0 | 0.0 | 0.0 | 0.0 | 0.0 | 0.0 | 0.4 |
| Days ≥ 1" Snow Depth | 0 | 0 | 0 | 0 | 0 | 0 | 0 | 0 | 0 | 0 | 0 | 0 | 0 |

## WIGGINS *Stone County* ELEVATION 245 ft LAT/LONG 30° 52 ' N / 89° 8 ' W

| | JAN | FEB | MAR | APR | MAY | JUN | JUL | AUG | SEP | OCT | NOV | DEC | YEAR |
|---|---|---|---|---|---|---|---|---|---|---|---|---|---|
| Maximum Temp °F | *60.4* | na | *71.8* | *79.4* | 84.7 | 90.8 | 92.2 | *91.4* | *88.1* | *80.7* | *71.1* | *64.4* | na |
| Minimum Temp °F | *36.9* | *38.1* | *46.5* | *54.5* | 60.7 | 67.6 | 70.1 | *69.8* | *65.3* | *54.5* | *45.7* | *40.4* | 54.2 |
| Mean Temp °F | *48.8* | na | *59.2* | *67.0* | 72.7 | 79.2 | 81.2 | *80.6* | *76.7* | *67.6* | *58.4* | *52.4* | na |
| Days Max Temp ≥ 90 °F | *0* | *0* | *0* | 0 | 5 | 20 | 25 | *24* | 13 | *1* | 0 | *0* | 88 |
| Days Max Temp ≤ 32 °F | *0* | *0* | *0* | 0 | 0 | 0 | 0 | 0 | 0 | *0* | 0 | *0* | 0 |
| Days Min Temp ≤ 32 °F | *13* | *9* | *3* | 0 | 0 | 0 | 0 | 0 | 0 | *0* | *4* | *9* | 38 |
| Days Min Temp ≤ 0 °F | *0* | *0* | *0* | 0 | 0 | 0 | 0 | 0 | 0 | *0* | 0 | *0* | 0 |
| Heating Degree Days | *502* | na | *214* | *55* | 6 | 0 | 0 | *0* | *3* | *53* | *232* | *397* | na |
| Cooling Degree Days | na | na | na | na | *250* | *435* | *514* | *500* | na | na | na | na | na |
| Total Precipitation (") | *5.75* | *5.99* | *6.43* | 5.19 | 5.19 | *4.41* | 6.50 | 5.60 | *4.47* | *3.33* | 4.43 | *5.97* | 63.26 |
| Days ≥ 0.1" Precip | *6* | *6* | *7* | 5 | 6 | *6* | *10* | *8* | 6 | *3* | *5* | *6* | 74 |
| Total Snowfall (") | *0.0* | *0.0* | *0.1* | 0.0 | 0.0 | 0.0 | 0.0 | 0.0 | 0.0 | *0.0* | 0.0 | *0.0* | 0.1 |
| Days ≥ 1" Snow Depth | *0* | *0* | *0* | 0 | 0 | 0 | 0 | *0* | 0 | *0* | 0 | *0* | 0 |

## WINONA 5 E *Montgomery County* ELEVATION 469 ft LAT/LONG 33° 29 ' N / 89° 42 ' W

| | JAN | FEB | MAR | APR | MAY | JUN | JUL | AUG | SEP | OCT | NOV | DEC | YEAR |
|---|---|---|---|---|---|---|---|---|---|---|---|---|---|
| Maximum Temp °F | 51.6 | 56.4 | 65.1 | 73.9 | 79.9 | 86.8 | 89.8 | 89.3 | 84.3 | 75.0 | 65.0 | 56.2 | 72.8 |
| Minimum Temp °F | 28.4 | 30.6 | 38.0 | 46.2 | 55.1 | 62.9 | 67.2 | 65.5 | 59.7 | 46.3 | 38.1 | 31.6 | 47.5 |
| Mean Temp °F | 40.0 | 43.5 | 51.6 | 60.1 | 67.5 | 74.9 | 78.5 | 77.4 | 72.0 | 60.7 | 51.6 | 43.9 | 60.1 |
| Days Max Temp ≥ 90 °F | 0 | 0 | 0 | 0 | 1 | 9 | 17 | 16 | 7 | 0 | 0 | 0 | 50 |
| Days Max Temp ≤ 32 °F | 2 | 1 | 0 | 0 | 0 | 0 | 0 | 0 | 0 | 0 | 0 | 1 | 4 |
| Days Min Temp ≤ 32 °F | 21 | 17 | 10 | 2 | 0 | 0 | 0 | 0 | 0 | 2 | 11 | 18 | 81 |
| Days Min Temp ≤ 0 °F | 0 | 0 | 0 | 0 | 0 | 0 | 0 | 0 | 0 | 0 | 0 | 0 | 0 |
| Heating Degree Days | 768 | 601 | 418 | 181 | 44 | 2 | 0 | 0 | 20 | 174 | 403 | 649 | 3260 |
| Cooling Degree Days | 0 | 1 | 9 | 35 | 130 | 318 | 434 | 408 | 239 | 50 | 10 | 3 | 1637 |
| Total Precipitation (") | 4.80 | 5.03 | 6.25 | 5.48 | 5.12 | 3.81 | 4.56 | 3.22 | 3.67 | 3.51 | 4.73 | 6.16 | 56.34 |
| Days ≥ 0.1" Precip | 7 | 6 | 7 | 6 | 7 | 6 | 7 | 5 | 6 | 5 | 6 | 7 | 75 |
| Total Snowfall (") | 0.8 | 0.2 | 0.6 | 0.0 | 0.0 | 0.0 | 0.0 | 0.0 | 0.0 | 0.0 | 0.0 | 0.1 | 1.7 |
| Days ≥ 1" Snow Depth | 0 | 0 | 0 | 0 | 0 | 0 | 0 | 0 | 0 | 0 | 0 | 0 | 0 |

## WOODVILLE 4 ESE *Wilkinson County* ELEVATION 390 ft LAT/LONG 31° 6 ' N / 91° 18 ' W

| | JAN | FEB | MAR | APR | MAY | JUN | JUL | AUG | SEP | OCT | NOV | DEC | YEAR |
|---|---|---|---|---|---|---|---|---|---|---|---|---|---|
| Maximum Temp °F | 58.8 | 63.0 | 70.8 | 78.3 | 83.6 | 89.1 | 90.8 | 90.5 | 86.7 | 78.7 | 69.7 | 62.6 | 76.9 |
| Minimum Temp °F | 38.3 | 40.5 | 47.6 | 55.0 | 61.9 | 67.8 | 70.6 | 70.0 | 65.5 | 55.2 | 47.1 | 41.4 | 55.1 |
| Mean Temp °F | 48.6 | 51.8 | 59.2 | 66.7 | 72.8 | 78.5 | 80.7 | 80.2 | 76.1 | 66.9 | 58.4 | 52.0 | 66.0 |
| Days Max Temp ≥ 90 °F | 0 | 0 | 0 | 0 | 2 | 15 | 22 | 21 | 9 | 1 | 0 | 0 | 70 |
| Days Max Temp ≤ 32 °F | 0 | 0 | 0 | 0 | 0 | 0 | 0 | 0 | 0 | 0 | 0 | 0 | 0 |
| Days Min Temp ≤ 32 °F | 11 | 7 | 2 | 0 | 0 | 0 | 0 | 0 | 0 | 0 | 3 | 7 | 30 |
| Days Min Temp ≤ 0 °F | 0 | 0 | 0 | 0 | 0 | 0 | 0 | 0 | 0 | 0 | 0 | 0 | 0 |
| Heating Degree Days | 509 | 374 | 212 | 59 | 4 | 0 | 0 | 0 | 4 | 63 | 230 | 406 | 1861 |
| Cooling Degree Days | 5 | 11 | 38 | 108 | 256 | 419 | 504 | 494 | 356 | 140 | 43 | 14 | 2388 |
| Total Precipitation (") | 6.18 | 5.89 | 6.76 | 6.11 | 6.26 | 5.53 | 5.98 | 4.42 | 5.00 | 3.62 | 5.06 | 6.62 | 67.43 |
| Days ≥ 0.1" Precip | 8 | 7 | 7 | 6 | 7 | 8 | 9 | 7 | 6 | 4 | 6 | 8 | 83 |
| Total Snowfall (") | 0.0 | 0.1 | 0.1 | 0.0 | 0.0 | 0.0 | 0.0 | 0.0 | 0.0 | 0.0 | 0.0 | 0.0 | 0.2 |
| Days ≥ 1" Snow Depth | 0 | 0 | 0 | 0 | 0 | 0 | 0 | 0 | 0 | 0 | 0 | 0 | 0 |

## YAZOO CITY 5 NNE *Yazoo County* ELEVATION 112 ft LAT/LONG 32° 54 ' N / 90° 23 ' W

| | JAN | FEB | MAR | APR | MAY | JUN | JUL | AUG | SEP | OCT | NOV | DEC | YEAR |
|---|---|---|---|---|---|---|---|---|---|---|---|---|---|
| Maximum Temp °F | 54.1 | 59.1 | 67.9 | 76.7 | 83.2 | 90.4 | 92.4 | 91.7 | 86.8 | 77.3 | 66.8 | 58.2 | 75.4 |
| Minimum Temp °F | 35.1 | 38.0 | 45.8 | 53.9 | 62.0 | 68.9 | 71.9 | 70.6 | 65.4 | 53.6 | 44.8 | 38.3 | 54.0 |
| Mean Temp °F | 44.6 | 48.6 | 56.8 | 65.3 | 72.6 | 79.6 | 82.1 | 81.2 | 76.1 | 65.5 | 55.8 | 48.3 | 64.7 |
| Days Max Temp ≥ 90 °F | 0 | 0 | 0 | 0 | 5 | 19 | 24 | 23 | 12 | 1 | 0 | 0 | 84 |
| Days Max Temp ≤ 32 °F | 1 | 0 | 0 | 0 | 0 | 0 | 0 | 0 | 0 | 0 | 0 | 0 | 1 |
| Days Min Temp ≤ 32 °F | 14 | 9 | 3 | 0 | 0 | 0 | 0 | 0 | 0 | 0 | 4 | 10 | 40 |
| Days Min Temp ≤ 0 °F | 0 | 0 | 0 | 0 | 0 | 0 | 0 | 0 | 0 | 0 | 0 | 0 | 0 |
| Heating Degree Days | 629 | 463 | 279 | 90 | 10 | 0 | 0 | 0 | 7 | 93 | 296 | 520 | 2387 |
| Cooling Degree Days | 2 | 7 | 31 | 101 | 259 | 454 | 548 | 525 | 356 | 118 | 30 | 10 | 2441 |
| Total Precipitation (") | 5.63 | 5.53 | 6.19 | 5.58 | 5.66 | 3.79 | 4.34 | 3.52 | 3.20 | 4.23 | 4.92 | 6.49 | 59.08 |
| Days ≥ 0.1" Precip | 8 | 7 | 7 | 7 | 7 | 6 | 6 | 5 | 5 | 5 | 6 | 7 | 76 |
| Total Snowfall (") | 0.1 | 0.0 | 0.4 | 0.0 | 0.0 | 0.0 | 0.0 | 0.0 | 0.0 | 0.0 | 0.0 | 0.0 | 0.5 |
| Days ≥ 1" Snow Depth | 0 | 0 | 0 | 0 | 0 | 0 | 0 | 0 | 0 | 0 | 0 | 0 | 0 |

## JANUARY MINIMUM TEMPERATURE °F

### LOWEST

| | | |
|---|---|---|
| 1 | Holly Springs | 27.1 |
| 2 | Ripley | 27.4 |
| 3 | University | 27.6 |
| 4 | Water Valley | 27.8 |
| 5 | Booneville | 28.3 |
| 6 | Winona | 28.4 |
| 7 | Batesville | 28.5 |
| 8 | Corinth | 28.6 |
| | Independence | 28.6 |
| 10 | Pontotoc | 29.0 |
| 11 | Houston | 29.1 |
| 12 | Tunica | 29.7 |
| 13 | Hickory Flat | 29.8 |
| 14 | Hernando | 30.1 |
| | Tupelo | 30.1 |
| 16 | Grenada | 30.2 |
| 17 | Kosciusko | 30.4 |
| 18 | Calhoun City | 30.5 |
| 19 | Fulton | 30.7 |
| 20 | Clarksdale | 30.9 |
| 21 | Louisville | 31.3 |
| 22 | State University | 31.4 |
| 23 | Pickens | 31.6 |
| 24 | Carthage | 31.7 |
| | Eupora | 31.7 |

### HIGHEST

| | | |
|---|---|---|
| 1 | Biloxi | 42.9 |
| 2 | Gulfport | 42.0 |
| 3 | Pascagoula | 41.6 |
| 4 | Saucier | 40.6 |
| 5 | Poplarville | 39.5 |
| 6 | Picayune | 39.0 |
| 7 | Natchez | 38.3 |
| | Tylertown | 38.3 |
| | Woodville | 38.3 |
| 10 | Columbia | 37.0 |
| 11 | Wiggins | 36.9 |
| 12 | McComb | 36.5 |
| 13 | Collins | 36.2 |
| | Hattiesburg | 36.2 |
| 15 | Brookhaven | 35.6 |
| 16 | Laurel | 35.2 |
| 17 | Waynesboro | 35.1 |
| | Yazoo City | 35.1 |
| 19 | Merrill | 34.6 |
| 20 | Jackson | 34.3 |
| | Liberty | 34.3 |
| 22 | Greenwood | 34.1 |
| | Meridian | 34.1 |
| 24 | Monticello | 34.0 |
| 25 | Moorhead | 33.7 |

## JULY MAXIMUM TEMPERATURE °F

### HIGHEST

| | | |
|---|---|---|
| 1 | Greenville | 92.9 |
| 2 | Columbia | 92.8 |
| | Merrill | 92.8 |
| 4 | Belzoni | 92.5 |
| | Minter City | 92.5 |
| 6 | Pelahatchie | 92.4 |
| | Yazoo City | 92.4 |
| 8 | Moorhead | 92.3 |
| 9 | Jackson | 92.2 |
| | Meridian | 92.2 |
| | Monticello | 92.2 |
| | Wiggins | 92.2 |
| 13 | Corinth | 92.0 |
| | Fulton | 92.0 |
| 15 | Carthage | 91.9 |
| | Clarksdale | 91.9 |
| | Quitman | 91.9 |
| | Tupelo | 91.9 |
| 19 | Eupora | 91.8 |
| | Grenada | 91.8 |
| | Hattiesburg | 91.8 |
| | Waynesboro | 91.8 |
| 23 | Calhoun City | 91.7 |
| | Greenwood | 91.7 |
| | Tunica | 91.7 |

### LOWEST

| | | |
|---|---|---|
| 1 | Louisville | 89.4 |
| 2 | Holly Springs | 89.5 |
| 3 | Booneville | 89.6 |
| 4 | Independence | 89.7 |
| 5 | Winona | 89.8 |
| 6 | Pascagoula | 90.0 |
| 7 | Biloxi | 90.1 |
| 8 | Pickens | 90.3 |
| | Ripley | 90.3 |
| 10 | Philadelphia | 90.4 |
| | Pontotoc | 90.4 |
| 12 | Collins | 90.6 |
| | Water Valley | 90.6 |
| 14 | Brookhaven | 90.7 |
| | University | 90.7 |
| 16 | Lexington | 90.8 |
| | Woodville | 90.8 |
| 18 | Kipling | 90.9 |
| 19 | Houston | 91.0 |
| 20 | D'Lo | 91.1 |
| | Gulfport | 91.1 |
| | Hernando | 91.1 |
| | Poplarville | 91.1 |
| | Tylertown | 91.1 |
| 25 | Forest | 91.2 |

## ANNUAL PRECIPITATION (")

### HIGHEST

| | | |
|---|---|---|
| 1 | Saucier | 68.67 |
| 2 | Woodville | 67.43 |
| 3 | Pascagoula | 65.35 |
| 4 | Gulfport | 64.39 |
| 5 | Merrill | 64.26 |
| 6 | Biloxi | 63.91 |
| | Columbia | 63.91 |
| 8 | McComb | 63.59 |
| 9 | Tylertown | 63.47 |
| 10 | Wiggins | 63.26 |
| 11 | Poplarville | 62.87 |
| 12 | Picayune | 62.24 |
| 13 | Liberty | 61.96 |
| 14 | Hattiesburg | 61.68 |
| 15 | Natchez | 61.45 |
| 16 | Forest | 61.44 |
| 17 | Brookhaven | 61.11 |
| 18 | Monticello | 60.75 |
| 19 | Kosciusko | 60.38 |
| 20 | Richton | 60.02 |
| 21 | Collins | 59.94 |
| 22 | Louisville | 59.46 |
| 23 | D'Lo | 59.37 |
| 24 | Yazoo City | 59.08 |
| 25 | Fulton | 59.07 |

### LOWEST

| | | |
|---|---|---|
| 1 | Stoneville | 51.83 |
| 2 | Independence | 52.70 |
| 3 | Tunica | 53.40 |
| 4 | Greenville | 53.49 |
| 5 | Clarksdale | 53.57 |
| 6 | Calhoun City | 53.83 |
| 7 | Greenwood | 54.33 |
| 8 | Hernando | 54.81 |
| 9 | Jackson | 54.95 |
| 10 | Lexington | 55.14 |
| 11 | Aberdeen | 55.29 |
| 12 | Corinth | 55.36 |
| 13 | Tupelo | 55.44 |
| 14 | Moorhead | 55.60 |
| 15 | Oakley | 55.65 |
| 16 | State University | 55.92 |
| 17 | Quitman | 56.11 |
| 18 | Batesville | 56.12 |
| 19 | Kipling | 56.25 |
| 20 | Winona | 56.34 |
| 21 | Carthage | 56.56 |
| 22 | Newton | 56.62 |
| 23 | University | 56.71 |
| 24 | Belzoni | 57.13 |
| 25 | Eupora | 57.14 |

## ANNUAL SNOWFALL (")

### HIGHEST

| | | |
|---|---|---|
| 1 | Hickory Flat | 3.9 |
| 2 | Hernando | 3.7 |
| 3 | Corinth | 3.4 |
| 4 | University | 3.3 |
| 5 | Water Valley | 3.0 |
| 6 | Booneville | 2.8 |
| | Holly Springs | 2.8 |
| | Houston | 2.8 |
| 9 | Ripley | 2.7 |
| 10 | Independence | 2.6 |
| 11 | Clarksdale | 2.3 |
| 12 | Tupelo | 2.2 |
| 13 | Tunica | 2.1 |
| 14 | Calhoun City | 2.0 |
| | Fulton | 2.0 |
| 16 | Aberdeen | 1.8 |
| | Stoneville | 1.8 |
| 18 | Pontotoc | 1.7 |
| | Winona | 1.7 |
| 20 | Batesville | 1.6 |
| 21 | Greenwood | 1.3 |
| | Grenada | 1.3 |
| 23 | Kosciusko | 1.2 |
| 24 | Lexington | 1.1 |
| | Minter City | 1.1 |

### LOWEST

| | | |
|---|---|---|
| 1 | Laurel | 0.0 |
| | Pascagoula | 0.0 |
| 3 | Belzoni | 0.1 |
| | Biloxi | 0.1 |
| | Gulfport | 0.1 |
| | Wiggins | 0.1 |
| 7 | Collins | 0.2 |
| | Merrill | 0.2 |
| | Picayune | 0.2 |
| | Saucier | 0.2 |
| | Woodville | 0.2 |
| 12 | Columbia | 0.3 |
| | D'Lo | 0.3 |
| | Liberty | 0.3 |
| | Monticello | 0.3 |
| | Pelahatchie | 0.3 |
| 17 | Forest | 0.4 |
| | Poplarville | 0.4 |
| | Waynesboro | 0.4 |
| 20 | Hattiesburg | 0.5 |
| | Louisville | 0.5 |
| | McComb | 0.5 |
| | Newton | 0.5 |
| | Richton | 0.5 |
| | Yazoo City | 0.5 |

# MISSOURI

PHYSICAL FEATURES. Missouri's three main terrain features are the rolling prairies of the area north of the Missouri River and in the west-central counties, the Ozarks, and the southeast lowlands, commonly called the "Bootheel." The flat lowlands of the Bootheel counties are about 250 feet above sea level. The highest elevation in the State is 1,772 feet above sea level on Taum Sauk Mountain in Iron County, in the eastern Ozarks, but the largest area of high elevation is in the central Ozarks, in Webster County, where elevations of 1,600 to 1,700 feet are common. The northwestern part of Missouri has extensive areas above 1,000 feet and some over 1,200. The northeastern part of the State slopes down to about 700 feet above sea level near the Mississippi River. The terrain varies from rugged areas bordering some of the larger streams, with deep valleys and steep hills, to broad, rolling uplands.

GENERAL CLIMATE. Missouri is an inland state, thus its climate is essentially continental. There are frequent changes in the weather, both from day to day and from season to season. Missouri is in the path of cold air moving down out of Canada, warm, moist air coming up from the Gulf of Mexico, and dry air from the west.

Annual precipitation in Missouri averages from near 50 inches in the southeastern corner to about 32 inches in the northwest. In the southeastern counties much of the precipitation comes during the fall, winter, and early spring months. In most of the western and northern counties the winter months are comparatively dry, with most of the precipitation coming in the spring, summer, and fall months.

Snow has been known to fall in Missouri as early as October, and as late as May. However, most of it falls in December, January, and February. As one would expect, the northern counties usually get the most snow. North of the Missouri River the winter snowfall averages 18 to 22 inches. This average figure tapers off to 8 to 12 inches in the southernmost counties. It is unusual for snow to stay on the ground for more than a week or two before it melts. Winter precipitation usually is in the form of rain or snow, or both. Conditions sometimes are on the borderline between rain and snow, and in these situations freezing drizzle or freezing rain occurs. This does not usually happen more than twice in a winter season..

Spring, summer, and early fall precipitation comes largely in the form of showers or thunderstorms. Thunderstorms have been observed in Missouri during the winter months, but they are most frequent from April to July. Occasionally, these produce some very heavy rains. Measurable precipitation occurs on an average of about 100 days a year. About half of these will be days with thunderstorms.

The river drainage in Missouri is wholly, either directly or indirectly, into the Mississippi River which forms the eastern boundary of the State. The northern part of the western boundary is formed by the Missouri River which then flows eastward across the State from Kansas City, entering the Mississippi just above St. Louis. Most of northern Missouri is drained by tributaries of the Missouri River, the principal ones being the Grand, Charlton, One Hundred and Two, and the Nodaway Rivers. The principal southern tributaries of the Missouri are the Osage and the Gasconade. Important tributaries which drain directly into the Mississippi within the borders of the State are the Fox, Wyaconda, Fabius, and Salt Rivers in the northeast, and the Meramec River which enters the Mississippi just below St. Louis.

Tributary flooding resulting from heavy rains and which may be expected once or twice in most years, and flash flooding along minor streams following heavy thunderstorm rains, occur most frequently in the spring and early summer, April to July, but may occur during any month. Serious flooding occurs less frequently along the main stems of the Missouri and Mississippi Rivers and usually occurs during the spring and early summer. Main stem flooding may be caused by prolonged periods of heavy rains, ice jams, or upstream flood crests synchronized with high tributary discharge.

On the average the amount of water that falls in Missouri on a square mile in a year varies from near 600 million gallons in the northwest corner to over 800 million gallons in the southeast. Some of this water runs off into the rivers and streams, some is consumed by animal life, and large amounts are evaporated back into the atmosphere, or transpired by growing vegetation. During years when precipitation comes in a fairly normal manner, moisture is stored in the top layers of the soil during the winter and early spring, when evaporation and transpiration are low. During the summer months the loss of water by evaporation and transpiration is high, and if rainfall fails to occur at frequent intervals, drought will result. Nearly every year some areas have short periods of drought in Missouri.

Tornadoes have been observed during every month of the year, with about 70 percent occurring during the 4 months, March through June. Over 25 percent of the total number were reported during May for the month of greatest frequency. Paths of the Missouri tornadoes have been short, averaging only 9½ miles. The width of tornado paths averages slightly less than 300 yards.

TEMPERATURE. Because of its inland location, Missouri is subject to frequent changes in temperature. While winters are cold and summers are hot, prolonged periods of very cold or very hot weather are unusual. Occasional periods of mild, above freezing temperatures are noted almost every winter. Conversely, during the peak of the summer season, occasional periods of dry, cool weather break up stretches of hot, humid weather.

Temperatures over 100° F. are rare, but they have occurred in every section of the State. In the summer, temperatures rise to 90° F. or higher on an average of 40 to 50 days in the west and north, and 55 to 60 days in the southeast. Temperatures below zero are infrequent, but have occurred in every county in Missouri. On the average there are 2 to 5 days a year with below zero temperatures in the northern counties, and 1 to 2 days in the southern counties, although there are some winters when temperatures do not go below zero at all.

In winter there is an average of about 110 days with temperatures below 32° F. in the northern half, the West Central Plains, and East Ozarks, about 100 days in the West Ozarks, and about 70 such days in the Bootheel counties. The average date of the first occurrence of temperatures 28° to 32° F. in the fall would come as early as mid-October in the northernmost counties and high elevations of the Ozarks, and as early as the third week in October over much of the rest of the State, except the Bootheel where the average date of the first freeze is in early November. About 1 year in 20, the first freeze would come as early as mid-September in the northern counties and late September in the rest of the State, except the Bootheel where it would occur as early as the third week in October.

# COUNTY INDEX

*Adair County*
KIRKSVILLE

*Audrain County*
MEXICO
VANDALIA

*Barry County*
CASSVILLE RANGER STN

*Barton County*
LAMAR

*Bates County*
BUTLER

*Bollinger County*
MARBLE HILL

*Buchanan County*
ST JOSEPH ROSECRANS

*Butler County*
POPLAR BLUFF
WAPPAPELLO DAM

*Caldwell County*
HAMILTON 2 W

*Callaway County*
FULTON

*Camden County*
CAMDENTON 2 NW

*Cape Girardeau County*
JACKSON

*Carroll County*
CARROLLTON

*Carter County*
VAN BUREN RS

*Chariton County*
BRUNSWICK
SALISBURY

*Christian County*
BILLINGS 2 N

*Cole County*
JEFFERSON CITY WATER

*Cooper County*
BOONVILLE

*Crawford County*
STEELVILLE 2 N

*Dade County*
LOCKWOOD

*Dallas County*
BUFFALO 3 S

*DeKalb County*
AMITY 7 WNW

*Dent County*
SALEM

*Dunklin County*
KENNETT RADIO KBOA
MALDEN MUNICIPAL AP

*Franklin County*
UNION

*Greene County*
SPRINGFIELD REG AP

*Grundy County*
SPICKARD 7 W

*Harrison County*
BETHANY

*Henry County*
CLINTON
WINDSOR

*Hickory County*
POMME DE TERRE DAM

*Howard County*
NEW FRANKLIN 1 W

*Howell County*
WEST PLAINS
WILLOW SPRGS RD KUKU

*Iron County*
ARCADIA

*Jackson County*
LEES SUMMIT WILDLIFE

*Jasper County*
JOPLIN MUNICIPAL AP

*Jefferson County*
FESTUS

*Johnson County*
WARRENSBURG 4 NW

*Laclede County*
LEBANON 2 W

*Lafayette County*
LEXINGTON 3 NE

*Lawrence County*
MT VERNON MU SW CNTR

*Lewis County*
CANTON L & D 20
STEFFENVILLE

*Lincoln County*
ELSBERRY 1 S

*Linn County*
BROOKFIELD

*McDonald County*
ANDERSON

*Madison County*
FREDERICKTOWN

*Maries County*
VICHY ROLLA NATL AP
VIENNA 2 WNW

*Marion County*
HANNIBAL WATER WORKS

***Mercer County***
PRINCETON 6 SW

***Miller County***
ELDON
LAKESIDE

***Moniteau County***
CALIFORNIA

***Morgan County***
VERSAILLES

***New Madrid County***
NEW MADRID

***Newton County***
NEOSHO

***Nodaway County***
CONCEPTION
MARYVILLE 2 E

***Osage County***
FREEDOM

***Ozark County***
DORA
WASOLA

***Pemiscot County***
CARUTHERSVILLE
PORTAGEVILLE

***Perry County***
PERRYVILLE WATR PLNT

***Pettis County***
SEDALIA WATER PLANT

***Phelps County***
ROLLA UNIV OF MO

***Pike County***
BOWLING GREEN 2 NE

***Polk County***
BOLIVAR 1 NE

***Pulaski County***
WAYNESVILLE 2 W

***Putnam County***
UNIONVILLE

***Ralls County***
SAVERTON L & D 22

***Randolph County***
MOBERLY

***Reynolds County***
BUNKER

***Ripley County***
DONIPHAN

***St. Charles County***
ST CHARLES

***St. Clair County***
APPLETON CITY
OSCEOLA

***St. Francois County***
FARMINGTON

***St. Louis County***
ST LOUIS LAMBERT AP

***Saline County***
MARSHALL
SWEET SPRINGS

***Scotland County***
MEMPHIS

***Scott County***
CAPE GIRARDEAU AP

***Shelby County***
SHELBINA

***Stoddard County***
ADVANCE 1 S

***Stone County***
GALENA 1 SW

***Taney County***
OZARK BEACH

***Texas County***
HOUSTON 3 E
LICKING 4 N
SUMMERSVILLE

***Vernon County***
NEVADA SEWAGE PLANT

***Wayne County***
CLEARWATER DAM
GREENVILLE 6 N

***Webster County***
MARSHFIELD

***Worth County***
GRANT CITY

***Wright County***
MANSFIELD
MOUNTAIN GROVE 2 N

## ELEVATION INDEX

| FEET | STATION NAME |
|---|---|
| 269 | CARUTHERSVILLE |
| 269 | KENNETT RADIO KBOA |
| 279 | PORTAGEVILLE |
| 289 | MALDEN MUNICIPAL AP |
| 302 | NEW MADRID |
| 312 | DONIPHAN |
| 331 | POPLAR BLUFF |
| 337 | CAPE GIRARDEAU AP |
| 358 | ADVANCE 1 S |
| 410 | GREENVILLE 6 N |
| 440 | MARBLE HILL |
| 449 | ELSBERRY 1 S |
| 459 | JACKSON |
| 469 | SAVERTON L & D 22 |
| 469 | WAPPAPELLO DAM |
| 480 | PERRYVILLE WATR PLNT |
| 489 | CANTON L & D 20 |
| 522 | FESTUS |
| 522 | ST CHARLES |
| 522 | UNION |
| 561 | VAN BUREN RS |
| 577 | ST LOUIS LAMBERT AP |
| 591 | LAKESIDE |
| 640 | CLEARWATER DAM |

| FEET | STATION NAME |
|---|---|
| 640 | NEW FRANKLIN 1 W |
| 650 | BRUNSWICK |
| 650 | STEFFENVILLE |
| 669 | JEFFERSON CITY WATER |
| 700 | FREDERICKTOWN |
| 700 | STEELVILLE 2 N |
| 702 | FULTON |
| 702 | OZARK BEACH |
| 702 | SWEET SPRINGS |
| 722 | SALISBURY |
| 751 | CARROLLTON |
| 751 | FREEDOM |
| 751 | HANNIBAL WATER WORKS |
| 761 | BOONVILLE |
| 770 | CLINTON |
| 770 | VIENNA 2 WNW |
| 771 | BROOKFIELD |
| 771 | OSCEOLA |
| 771 | SHELBINA |
| 771 | VANDALIA |
| 778 | MARSHALL |
| 781 | SEDALIA WATER PLANT |
| 791 | MEMPHIS |
| 801 | APPLETON CITY |
| 801 | MEXICO |
| 801 | WAYNESVILLE 2 W |
| 811 | ST JOSEPH ROSECRANS |
| 825 | LEXINGTON 3 NE |
| 840 | MOBERLY |
| 860 | BUTLER |
| 860 | NEVADA SEWAGE PLANT |
| 869 | CALIFORNIA |
| 879 | POMME DE TERRE DAM |
| 879 | SPICKARD 7 W |
| 879 | WARRENSBURG 4 NW |
| 902 | BOWLING GREEN 2 NE |
| 902 | HAMILTON 2 W |
| 902 | WINDSOR |
| 912 | BETHANY |
| 932 | ELDON |
| 932 | FARMINGTON |
| 960 | ARCADIA |
| 969 | KIRKSVILLE |
| 981 | DORA |
| 981 | LAMAR |
| 981 | PRINCETON 6 SW |
| 988 | JOPLIN MUNICIPAL AP |
| 1001 | ANDERSON |
| 1001 | LEES SUMMIT WILDLIFE |
| 1007 | WEST PLAINS |

| FEET | STATION NAME |
|---|---|
| 1010 | NEOSHO |
| 1030 | VERSAILLES |
| 1040 | CAMDENTON 2 NW |
| 1060 | UNIONVILLE |
| 1070 | BOLIVAR 1 NE |
| 1070 | LOCKWOOD |
| 1075 | AMITY 7 WNW |
| 1100 | GALENA 1 SW |
| 1112 | CONCEPTION |
| 1129 | VICHY ROLLA NATL AP |
| 1132 | GRANT CITY |
| 1152 | BUFFALO 3 S |
| 1161 | MARYVILLE 2 E |
| 1181 | LICKING 4 N |
| 1181 | SUMMERSVILLE |
| 1191 | MT VERNON MU SW CNTR |
| 1201 | ROLLA UNIV OF MO |
| 1201 | SALEM |
| 1230 | WILLOW SPRGS RD KUKU |
| 1260 | LEBANON 2 W |
| 1286 | SPRINGFIELD REG AP |
| 1289 | WASOLA |
| 1293 | HOUSTON 3 E |
| 1342 | CASSVILLE RANGER STN |
| 1362 | BILLINGS 2 N |
| 1380 | BUNKER |
| 1460 | MOUNTAIN GROVE 2 N |
| 1480 | MARSHFIELD |
| 1520 | MANSFIELD |

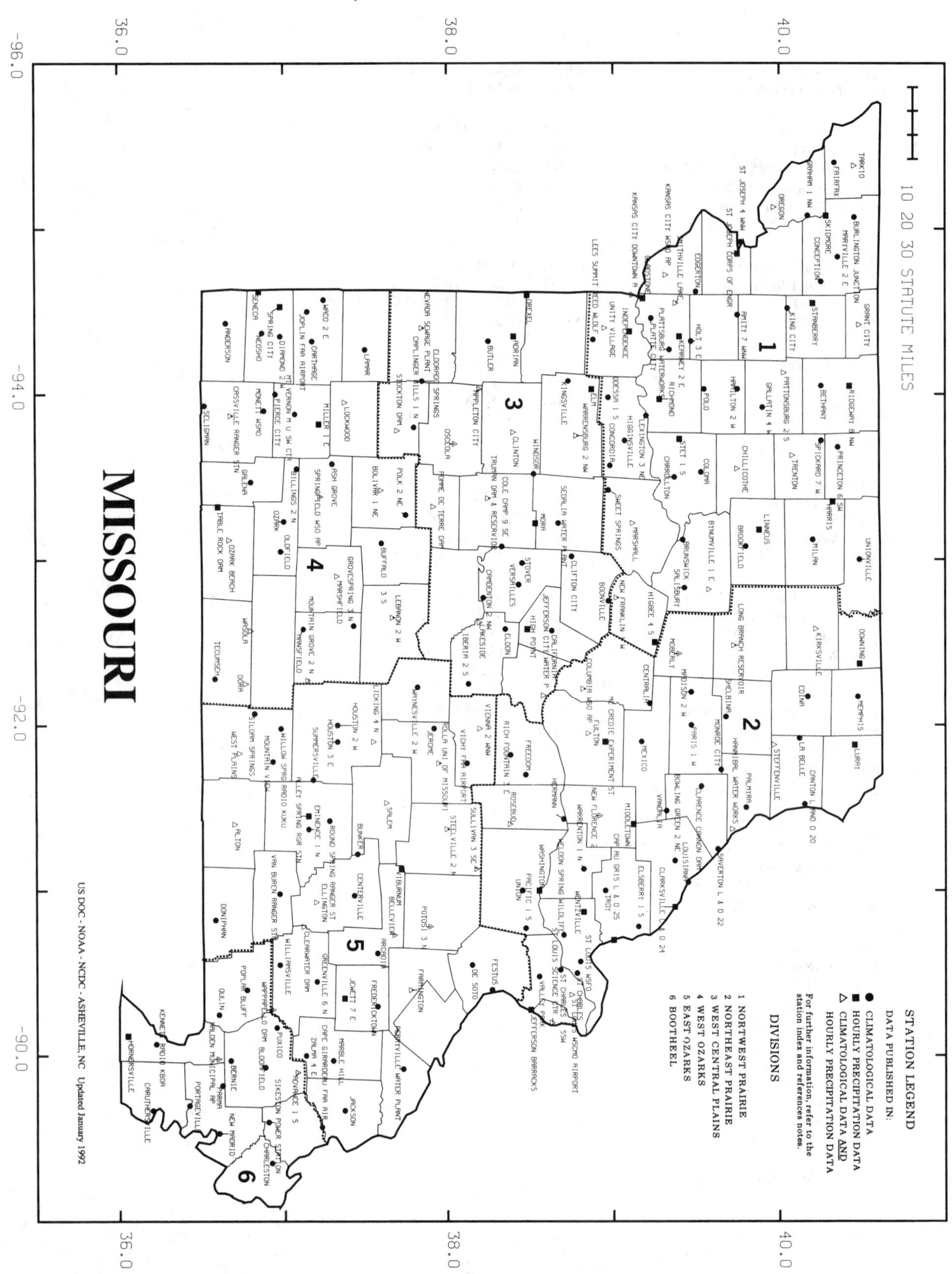
MISSOURI
10 20 30 STATUTE MILES
STATION LEGEND
DATA PUBLISHED IN:
● CLIMATOLOGICAL DATA
■ HOURLY PRECIPITATION DATA
△ CLIMATOLOGICAL DATA AND HOURLY PRECIPITATION DATA
For further information, refer to the station index and references notes.
DIVISIONS
1 NORTWEST PRAIRIE
2 NORTHEAST PRAIRIE
3 WEST CENTRAL PLAINS
4 WEST OZARKS
5 EAST OZARKS
6 BOOTHEEL
US DOC - NOAA - NCDC - ASHEVILLE, NC Updated January 1992

## ADVANCE 1 S *Stoddard County* ELEVATION 358 ft LAT/LONG 37° 6' N / 89° 55' W

| | JAN | FEB | MAR | APR | MAY | JUN | JUL | AUG | SEP | OCT | NOV | DEC | YEAR |
|---|---|---|---|---|---|---|---|---|---|---|---|---|---|
| Maximum Temp °F | 41.8 | 47.0 | 57.6 | 68.9 | 77.8 | 86.8 | 90.6 | 88.5 | 81.7 | 71.1 | 57.6 | 45.9 | 67.9 |
| Minimum Temp °F | 23.0 | 26.3 | 35.5 | 45.3 | 54.1 | 63.1 | 67.0 | 63.8 | 56.2 | 43.7 | 35.8 | 27.4 | 45.1 |
| Mean Temp °F | 32.4 | 36.7 | 46.6 | 57.1 | 66.0 | 74.9 | 78.8 | 76.2 | 69.0 | 57.4 | 46.8 | 36.7 | 56.6 |
| Days Max Temp ≥ 90 °F | 0 | 0 | 0 | 0 | 2 | 11 | 19 | 14 | 5 | 0 | 0 | 0 | 51 |
| Days Max Temp ≤ 32 °F | 7 | 4 | 0 | 0 | 0 | 0 | 0 | 0 | 0 | 0 | 0 | 3 | 14 |
| Days Min Temp ≤ 32 °F | 26 | 20 | 14 | 3 | 0 | 0 | 0 | 0 | 0 | 5 | 12 | 22 | 102 |
| Days Min Temp ≤ 0 °F | 1 | 0 | 0 | 0 | 0 | 0 | 0 | 0 | 0 | 0 | 0 | 1 | 2 |
| Heating Degree Days | 1004 | 793 | 567 | 258 | 72 | 4 | 0 | 2 | 47 | 258 | 543 | 871 | 4419 |
| Cooling Degree Days | 0 | 0 | 3 | 26 | 102 | 301 | 436 | 352 | 168 | 20 | 2 | 0 | 1410 |
| Total Precipitation (") | 3.24 | 3.25 | 4.10 | 5.16 | 4.97 | 3.46 | 3.81 | 3.15 | 3.91 | 3.48 | 4.64 | 4.05 | 47.22 |
| Days ≥ 0.1" Precip | 5 | 5 | 7 | 7 | 8 | 6 | 6 | 5 | 5 | 5 | 7 | 6 | 72 |
| Total Snowfall (") | ***3.0*** | 4.4 | 1.4 | 0.0 | 0.0 | 0.0 | 0.0 | 0.0 | 0.0 | 0.1 | 0.3 | 1.6 | 10.8 |
| Days ≥ 1" Snow Depth | na | ***4*** | ***1*** | 0 | 0 | 0 | 0 | 0 | 0 | 0 | 0 | 2 | na |

## AMITY 7 WNW *DeKalb County* ELEVATION 1075 ft LAT/LONG 39° 45' N / 94° 29' W

| | JAN | FEB | MAR | APR | MAY | JUN | JUL | AUG | SEP | OCT | NOV | DEC | YEAR |
|---|---|---|---|---|---|---|---|---|---|---|---|---|---|
| Maximum Temp °F | ***33.8*** | ***40.4*** | ***53.7*** | ***65.4*** | 74.2 | 82.7 | ***87.0*** | 85.3 | ***77.7*** | 66.1 | ***50.8*** | ***38.0*** | 62.9 |
| Minimum Temp °F | ***14.5*** | ***19.5*** | ***30.3*** | ***41.3*** | 51.3 | 60.3 | ***64.7*** | 62.4 | ***54.4*** | 42.6 | ***30.5*** | ***19.9*** | 41.0 |
| Mean Temp °F | ***24.2*** | ***29.9*** | ***42.1*** | ***53.4*** | 62.8 | 71.5 | ***75.9*** | 73.9 | ***66.0*** | 54.4 | ***40.7*** | ***28.9*** | 52.0 |
| Days Max Temp ≥ 90 °F | ***0*** | ***0*** | ***0*** | ***0*** | 0 | ***5*** | ***11*** | 9 | 3 | 0 | 0 | 0 | 28 |
| Days Max Temp ≤ 32 °F | ***13*** | ***8*** | ***2*** | ***0*** | 0 | 0 | ***0*** | 0 | 0 | 0 | ***2*** | ***10*** | 35 |
| Days Min Temp ≤ 32 °F | ***29*** | ***25*** | ***18*** | ***6*** | 0 | 0 | ***0*** | 0 | 0 | 5 | ***18*** | ***28*** | 129 |
| Days Min Temp ≤ 0 °F | ***6*** | ***3*** | ***0*** | ***0*** | 0 | 0 | ***0*** | 0 | 0 | 0 | 0 | 2 | 11 |
| Heating Degree Days | ***1259*** | ***985*** | ***705*** | ***360*** | 129 | 15 | ***1*** | 6 | ***88*** | 339 | ***722*** | ***1112*** | 5721 |
| Cooling Degree Days | 0 | 0 | ***2*** | 20 | 53 | 209 | 332 | 284 | 131 | 11 | 0 | 0 | 1042 |
| Total Precipitation (") | ***1.07*** | ***0.81*** | ***2.27*** | ***3.49*** | 4.38 | 4.43 | ***4.59*** | 3.49 | 4.79 | 3.11 | 1.72 | 1.63 | 35.78 |
| Days ≥ 0.1" Precip | ***3*** | ***3*** | ***5*** | ***7*** | 8 | 7 | ***6*** | 5 | 6 | 5 | ***4*** | ***3*** | 62 |
| Total Snowfall (") | ***5.5*** | ***3.6*** | ***2.9*** | ***0.9*** | 0.0 | 0.0 | 0.0 | 0.0 | 0.0 | 0.0 | ***1.1*** | ***3.9*** | 17.9 |
| Days ≥ 1" Snow Depth | ***10*** | ***5*** | ***2*** | 0 | 0 | 0 | 0 | 0 | 0 | 0 | ***1*** | ***4*** | 22 |

## ANDERSON *McDonald County* ELEVATION 1001 ft LAT/LONG 36° 38' N / 94° 27' W

| | JAN | FEB | MAR | APR | MAY | JUN | JUL | AUG | SEP | OCT | NOV | DEC | YEAR |
|---|---|---|---|---|---|---|---|---|---|---|---|---|---|
| Maximum Temp °F | 44.6 | 50.0 | 60.3 | 71.1 | 76.3 | 83.9 | 89.2 | 87.7 | 79.7 | 70.7 | 58.5 | 48.6 | 68.4 |
| Minimum Temp °F | 21.8 | 25.6 | 35.0 | 44.8 | 52.8 | 61.6 | 65.9 | 63.7 | 57.0 | 45.1 | 35.1 | 25.9 | 44.5 |
| Mean Temp °F | 33.2 | 37.8 | 47.7 | 58.0 | 64.6 | 72.8 | 77.6 | 75.7 | 68.4 | 57.9 | 46.8 | 37.3 | 56.5 |
| Days Max Temp ≥ 90 °F | 0 | 0 | 0 | 0 | 0 | 5 | 15 | 13 | 3 | 0 | 0 | 0 | 36 |
| Days Max Temp ≤ 32 °F | 6 | 3 | 0 | 0 | 0 | 0 | 0 | 0 | 0 | 0 | 0 | 3 | 12 |
| Days Min Temp ≤ 32 °F | 26 | 21 | 14 | 4 | 0 | 0 | 0 | 0 | 0 | 4 | 14 | 23 | 106 |
| Days Min Temp ≤ 0 °F | 2 | 1 | 0 | 0 | 0 | 0 | 0 | 0 | 0 | 0 | 0 | 1 | 4 |
| Heating Degree Days | 979 | 761 | 536 | 235 | 88 | 7 | 1 | 2 | 54 | 243 | 540 | 853 | 4299 |
| Cooling Degree Days | 0 | 0 | 5 | 32 | 79 | 254 | 401 | 357 | 171 | 26 | 3 | 0 | 1328 |
| Total Precipitation (") | 1.95 | 2.03 | 3.64 | 4.25 | 4.43 | 4.35 | 3.19 | 3.82 | 4.78 | 3.70 | 4.26 | 3.06 | 43.46 |
| Days ≥ 0.1" Precip | 4 | 4 | 6 | 7 | 8 | 7 | 4 | 5 | 6 | 5 | 5 | 5 | 66 |
| Total Snowfall (") | 3.9 | 3.8 | 2.9 | 0.0 | 0.0 | 0.0 | 0.0 | 0.0 | 0.0 | 0.0 | 0.9 | 3.0 | 14.5 |
| Days ≥ 1" Snow Depth | 6 | 4 | 2 | 0 | 0 | 0 | 0 | 0 | 0 | 0 | 0 | 3 | 15 |

## APPLETON CITY *St. Clair County* ELEVATION 801 ft LAT/LONG 38° 13' N / 94° 5' W

| | JAN | FEB | MAR | APR | MAY | JUN | JUL | AUG | SEP | OCT | NOV | DEC | YEAR |
|---|---|---|---|---|---|---|---|---|---|---|---|---|---|
| Maximum Temp °F | 40.3 | 46.1 | 58.2 | 69.3 | 77.2 | 85.5 | 90.9 | 89.7 | 81.4 | 70.3 | 55.9 | 44.6 | 67.5 |
| Minimum Temp °F | 19.5 | 23.8 | 34.1 | 44.3 | 53.4 | 62.5 | 66.6 | 64.2 | 56.8 | 45.2 | 34.6 | 24.6 | 44.1 |
| Mean Temp °F | 30.0 | 34.9 | 46.2 | 56.9 | 65.3 | 74.0 | 78.8 | 77.0 | 69.2 | 57.8 | 45.2 | 34.6 | 55.8 |
| Days Max Temp ≥ 90 °F | 0 | 0 | 0 | 0 | 1 | 8 | 19 | 16 | 5 | 0 | 0 | 0 | 49 |
| Days Max Temp ≤ 32 °F | 8 | 5 | 1 | 0 | 0 | 0 | 0 | 0 | 0 | 0 | 1 | 4 | 19 |
| Days Min Temp ≤ 32 °F | 27 | 22 | 15 | 3 | 0 | 0 | 0 | 0 | 0 | 3 | 14 | 25 | 109 |
| Days Min Temp ≤ 0 °F | 2 | 1 | 0 | 0 | 0 | 0 | 0 | 0 | 0 | 0 | 0 | 1 | 4 |
| Heating Degree Days | 1080 | 842 | 580 | 266 | 79 | 5 | 0 | 1 | 51 | 246 | 587 | 935 | 4672 |
| Cooling Degree Days | 0 | 0 | 4 | 29 | 88 | 285 | 439 | 383 | 183 | 22 | 2 | 0 | 1435 |
| Total Precipitation (") | 1.58 | 1.61 | 3.07 | 4.08 | 4.49 | 4.82 | 4.13 | 3.71 | 4.62 | 4.20 | 3.38 | 2.29 | 41.98 |
| Days ≥ 0.1" Precip | 4 | 4 | 6 | 6 | 7 | 7 | 5 | 5 | 6 | 6 | 5 | 4 | 65 |
| Total Snowfall (") | 5.3 | 4.5 | 2.3 | 0.1 | 0.0 | 0.0 | 0.0 | 0.0 | 0.0 | 0.0 | 1.0 | 3.3 | 16.5 |
| Days ≥ 1" Snow Depth | 10 | 6 | 2 | 0 | 0 | 0 | 0 | 0 | 0 | 0 | 1 | 5 | 24 |

## ARCADIA *Iron County* ELEVATION 960 ft LAT/LONG 37° 35 ' N / 90° 37 ' W

| | JAN | FEB | MAR | APR | MAY | JUN | JUL | AUG | SEP | OCT | NOV | DEC | YEAR |
|---|---|---|---|---|---|---|---|---|---|---|---|---|---|
| Maximum Temp °F | 41.8 | 47.4 | 59.3 | 71.0 | 77.8 | 84.8 | 89.6 | 87.6 | 80.2 | 70.3 | 56.6 | 46.4 | 67.7 |
| Minimum Temp °F | 21.0 | 24.4 | 33.6 | 43.9 | 51.4 | 59.8 | 64.4 | 62.2 | 55.1 | 43.7 | 35.1 | 26.8 | 43.4 |
| Mean Temp °F | 31.4 | 36.0 | 46.6 | 57.5 | 64.6 | 72.3 | 77.0 | 75.0 | 67.7 | 57.0 | 45.9 | 36.6 | 55.6 |
| Days Max Temp ≥ 90 °F | 0 | 0 | 0 | 0 | 1 | 6 | 16 | 11 | 3 | 0 | 0 | 0 | 37 |
| Days Max Temp ≤ 32 °F | 8 | 4 | 0 | 0 | 0 | 0 | 0 | 0 | 0 | 0 | 0 | 3 | 15 |
| Days Min Temp ≤ 32 °F | 26 | 21 | 15 | 5 | 1 | 0 | 0 | 0 | 0 | 5 | 13 | 22 | 108 |
| Days Min Temp ≤ 0 °F | 2 | 1 | 0 | 0 | 0 | 0 | 0 | 0 | 0 | 0 | 0 | 1 | 4 |
| Heating Degree Days | 1036 | 813 | 566 | 250 | 91 | 9 | 1 | 2 | 58 | 264 | 568 | 873 | 4531 |
| Cooling Degree Days | 0 | 0 | 2 | 30 | 81 | 230 | 380 | 326 | 139 | 17 | 1 | 0 | 1206 |
| Total Precipitation (") | 2.40 | 2.57 | 4.00 | 4.70 | 4.60 | 3.85 | 3.70 | 4.31 | 4.22 | 3.41 | 5.33 | 3.82 | 46.91 |
| Days ≥ 0.1" Precip | 5 | 5 | 7 | 8 | 8 | 6 | 5 | 6 | 6 | 6 | 7 | 6 | 75 |
| Total Snowfall (") | 4.2 | 3.9 | 2.0 | 0.5 | 0.0 | 0.0 | 0.0 | 0.0 | 0.0 | 0.1 | 0.9 | 2.3 | 13.9 |
| Days ≥ 1" Snow Depth | *6* | *4* | *1* | 0 | 0 | 0 | 0 | 0 | 0 | 0 | 0 | 1 | 12 |

## BETHANY *Harrison County* ELEVATION 912 ft LAT/LONG 40° 15 ' N / 94° 2 ' W

| | JAN | FEB | MAR | APR | MAY | JUN | JUL | AUG | SEP | OCT | NOV | DEC | YEAR |
|---|---|---|---|---|---|---|---|---|---|---|---|---|---|
| Maximum Temp °F | 33.3 | 39.3 | 52.2 | 64.8 | 74.9 | 83.9 | 88.7 | 86.3 | 78.2 | 66.6 | 50.5 | 37.9 | 63.1 |
| Minimum Temp °F | 13.8 | 18.3 | 29.5 | 41.2 | 51.2 | 60.7 | 65.2 | 62.5 | 53.8 | 42.2 | 30.6 | 20.1 | 40.8 |
| Mean Temp °F | 23.6 | 28.8 | 40.9 | 53.0 | 63.0 | 72.3 | 77.0 | 74.4 | 66.0 | 54.4 | 40.6 | 29.0 | 51.9 |
| Days Max Temp ≥ 90 °F | 0 | 0 | 0 | 0 | 0 | 6 | 15 | 10 | 3 | 0 | 0 | 0 | 34 |
| Days Max Temp ≤ 32 °F | 14 | 9 | 2 | 0 | 0 | 0 | 0 | 0 | 0 | 0 | 2 | 9 | 36 |
| Days Min Temp ≤ 32 °F | 29 | 25 | 19 | 6 | 0 | 0 | 0 | 0 | 0 | 5 | 18 | 28 | 130 |
| Days Min Temp ≤ 0 °F | 6 | 3 | 0 | 0 | 0 | 0 | 0 | 0 | 0 | 0 | 0 | 2 | 11 |
| Heating Degree Days | 1277 | 1015 | 742 | 371 | 125 | 14 | 2 | 6 | 90 | 337 | 726 | 1108 | 5813 |
| Cooling Degree Days | 0 | 0 | 2 | 22 | 59 | 230 | 375 | 307 | 134 | 11 | 0 | 0 | 1140 |
| Total Precipitation (") | 1.10 | 0.97 | 2.49 | 3.48 | 4.20 | 4.05 | 5.19 | 4.38 | 4.44 | 3.28 | 2.10 | 1.55 | 37.23 |
| Days ≥ 0.1" Precip | 3 | 3 | 5 | 7 | 7 | 7 | 6 | 6 | 7 | 5 | 5 | 3 | 64 |
| Total Snowfall (") | 8.0 | 5.8 | 3.4 | 1.0 | 0.0 | 0.0 | 0.0 | 0.0 | 0.0 | 0.1 | 2.0 | 4.9 | 25.2 |
| Days ≥ 1" Snow Depth | *13* | *8* | 2 | 0 | 0 | 0 | 0 | 0 | 0 | 0 | 1 | *5* | 29 |

## BILLINGS 2 N *Christian County* ELEVATION 1362 ft LAT/LONG 37° 4 ' N / 93° 33 ' W

| | JAN | FEB | MAR | APR | MAY | JUN | JUL | AUG | SEP | OCT | NOV | DEC | YEAR |
|---|---|---|---|---|---|---|---|---|---|---|---|---|---|
| Maximum Temp °F | 40.8 | 46.6 | 56.5 | 67.4 | 75.0 | 83.1 | *89.1* | 88.5 | na | na | 56.0 | 45.1 | na |
| Minimum Temp °F | 19.5 | 23.9 | 33.0 | 42.8 | 52.4 | 61.3 | *66.4* | 64.1 | na | na | 33.7 | 24.5 | na |
| Mean Temp °F | 30.2 | 35.3 | 44.8 | 55.1 | 63.7 | 72.2 | *77.8* | 76.3 | na | na | 44.9 | 34.8 | na |
| Days Max Temp ≥ 90 °F | 0 | 0 | 0 | 0 | 0 | 4 | 15 | *14* | *3* | 0 | 0 | 0 | 36 |
| Days Max Temp ≤ 32 °F | 8 | 5 | 1 | 0 | 0 | 0 | 0 | 0 | 0 | 0 | 1 | 5 | 20 |
| Days Min Temp ≤ 32 °F | 28 | 22 | 15 | 4 | 0 | 0 | 0 | 0 | 0 | 3 | 14 | 25 | 111 |
| Days Min Temp ≤ 0 °F | 2 | 1 | 0 | 0 | 0 | 0 | 0 | 0 | 0 | 0 | 0 | 1 | 4 |
| Heating Degree Days | 1074 | 831 | 623 | 310 | 108 | 10 | *1* | 3 | na | na | 599 | 929 | na |
| Cooling Degree Days | 0 | 0 | 3 | 21 | 70 | 242 | 409 | 377 | *164* | *19* | 1 | 0 | 1306 |
| Total Precipitation (") | 1.75 | 2.05 | 3.67 | 4.20 | 4.07 | 4.57 | 3.45 | 4.01 | 5.27 | 3.51 | 4.23 | 3.10 | 43.88 |
| Days ≥ 0.1" Precip | 4 | 4 | 6 | 7 | 7 | 7 | 5 | 5 | *6* | *6* | 6 | 5 | 68 |
| Total Snowfall (") | 5.0 | 4.8 | 3.4 | 0.5 | 0.0 | 0.0 | 0.0 | 0.0 | 0.0 | 0.1 | 1.4 | 3.2 | 18.4 |
| Days ≥ 1" Snow Depth | 8 | 5 | 2 | 0 | 0 | 0 | 0 | 0 | 0 | 0 | 1 | 3 | 19 |

## BOLIVAR 1 NE *Polk County* ELEVATION 1070 ft LAT/LONG 37° 37 ' N / 93° 25 ' W

| | JAN | FEB | MAR | APR | MAY | JUN | JUL | AUG | SEP | OCT | NOV | DEC | YEAR |
|---|---|---|---|---|---|---|---|---|---|---|---|---|---|
| Maximum Temp °F | 41.2 | 45.8 | 57.2 | 68.5 | 75.6 | 84.1 | 89.8 | 88.7 | 80.5 | 69.9 | 56.5 | 46.3 | 67.0 |
| Minimum Temp °F | 18.2 | 22.3 | 32.8 | 43.3 | 51.5 | 61.0 | 65.1 | 62.8 | 55.2 | 42.8 | 33.0 | 24.3 | 42.7 |
| Mean Temp °F | 29.7 | 34.1 | 45.0 | 55.9 | 63.6 | 72.6 | 77.5 | 75.8 | 67.9 | 56.4 | 44.8 | 35.3 | 54.9 |
| Days Max Temp ≥ 90 °F | 0 | 0 | 0 | 0 | 0 | 6 | 18 | 15 | 5 | 0 | 0 | 0 | 44 |
| Days Max Temp ≤ 32 °F | 8 | 5 | 1 | 0 | 0 | 0 | 0 | 0 | 0 | 0 | 1 | 4 | 19 |
| Days Min Temp ≤ 32 °F | 28 | 23 | 16 | 5 | 0 | 0 | 0 | 0 | 1 | 5 | 16 | 25 | 119 |
| Days Min Temp ≤ 0 °F | 2 | 2 | 0 | 0 | 0 | 0 | 0 | 0 | 0 | 0 | 0 | 1 | 5 |
| Heating Degree Days | 1087 | 865 | 616 | 292 | 111 | 11 | 2 | 4 | 70 | 284 | 602 | 912 | 4856 |
| Cooling Degree Days | 0 | 0 | 4 | *24* | 65 | 250 | 394 | 350 | 157 | 16 | 1 | 0 | 1261 |
| Total Precipitation (") | 1.77 | 2.27 | 3.54 | 4.32 | 4.77 | 4.83 | 3.65 | 3.44 | 5.01 | 4.42 | 3.95 | 2.99 | 44.96 |
| Days ≥ 0.1" Precip | 4 | 4 | 6 | 7 | 8 | 7 | 5 | 5 | 6 | 6 | 6 | 5 | 69 |
| Total Snowfall (") | na | na | *0.5* | *0.0* | 0.0 | 0.0 | 0.0 | 0.0 | 0.0 | 0.0 | 0.0 | na | na |
| Days ≥ 1" Snow Depth | na | na | *0* | *0* | 0 | 0 | 0 | 0 | 0 | 0 | 0 | na | na |

## BOONVILLE *Cooper County* ELEVATION 761 ft LAT/LONG 38° 58 ' N / 92° 45 ' W

| | JAN | FEB | MAR | APR | MAY | JUN | JUL | AUG | SEP | OCT | NOV | DEC | YEAR |
|---|---|---|---|---|---|---|---|---|---|---|---|---|---|
| Maximum Temp °F | 36.2 | 40.9 | 53.2 | 65.4 | 74.6 | 83.4 | 88.8 | 86.8 | 79.1 | 67.6 | 53.2 | 41.0 | 64.2 |
| Minimum Temp °F | 17.9 | 21.5 | 32.4 | 43.8 | 53.3 | 62.8 | 67.3 | 64.4 | 56.3 | 44.3 | 33.8 | 23.6 | 43.4 |
| Mean Temp °F | 27.1 | 31.2 | 42.8 | 54.7 | 64.0 | 73.1 | 78.1 | 75.6 | 67.8 | 56.0 | 43.6 | 32.3 | 53.9 |
| Days Max Temp ≥ 90 °F | 0 | 0 | 0 | 0 | 0 | 6 | 16 | 12 | 4 | 0 | 0 | 0 | 38 |
| Days Max Temp ≤ 32 °F | 12 | 8 | 2 | 0 | 0 | 0 | 0 | 0 | 0 | 0 | 1 | 7 | 30 |
| Days Min Temp ≤ 32 °F | 28 | 23 | 16 | 3 | 0 | 0 | 0 | 0 | 0 | 3 | 13 | 25 | 111 |
| Days Min Temp ≤ 0 °F | 3 | 2 | 0 | 0 | 0 | 0 | 0 | 0 | 0 | 0 | 0 | 1 | 6 |
| Heating Degree Days | 1170 | 949 | 684 | 329 | 111 | 11 | 1 | 3 | 67 | 297 | 640 | 1006 | 5268 |
| Cooling Degree Days | 0 | 0 | 4 | 29 | 83 | 277 | 430 | 354 | 161 | 19 | 2 | 0 | 1359 |
| Total Precipitation (") | 1.51 | 1.53 | 2.96 | 4.33 | 5.05 | 5.12 | 3.90 | 4.04 | 4.77 | 3.93 | 3.74 | 2.74 | 43.62 |
| Days ≥ 0.1" Precip | 4 | 4 | 6 | 7 | 8 | 6 | 6 | 6 | 6 | 6 | 6 | 5 | 70 |
| Total Snowfall (") | 6.1 | 6.8 | 3.5 | 0.4 | 0.0 | 0.0 | 0.0 | 0.0 | 0.0 | 0.0 | 1.5 | 4.3 | 22.6 |
| Days ≥ 1" Snow Depth | ***8*** | ***6*** | 2 | 0 | 0 | 0 | 0 | 0 | 0 | 0 | 1 | ***4*** | 21 |

## BOWLING GREEN 2 NE *Pike County* ELEVATION 902 ft LAT/LONG 39° 21 ' N / 91° 12 ' W

| | JAN | FEB | MAR | APR | MAY | JUN | JUL | AUG | SEP | OCT | NOV | DEC | YEAR |
|---|---|---|---|---|---|---|---|---|---|---|---|---|---|
| Maximum Temp °F | 34.6 | 40.4 | 52.3 | 64.9 | 74.2 | 83.1 | 88.2 | ***85.7*** | 78.9 | 67.1 | ***53.1*** | 40.3 | 63.6 |
| Minimum Temp °F | 14.7 | 18.3 | 29.6 | 40.9 | 49.8 | 59.0 | 64.3 | ***61.7*** | 53.6 | 41.6 | ***31.4*** | 20.8 | 40.5 |
| Mean Temp °F | 24.9 | 29.4 | 41.0 | 52.9 | 62.0 | 71.0 | 76.3 | ***73.7*** | ***66.2*** | 54.4 | ***42.3*** | 30.5 | 52.0 |
| Days Max Temp ≥ 90 °F | 0 | 0 | 0 | 0 | 1 | 5 | ***13*** | ***9*** | 3 | 0 | 0 | 0 | 31 |
| Days Max Temp ≤ 32 °F | ***12*** | ***8*** | 2 | 0 | 0 | 0 | 0 | 0 | 0 | 0 | 1 | ***7*** | 30 |
| Days Min Temp ≤ 32 °F | ***29*** | ***24*** | ***19*** | ***7*** | 1 | 0 | 0 | 0 | 0 | ***7*** | ***18*** | ***27*** | 132 |
| Days Min Temp ≤ 0 °F | ***5*** | 4 | 0 | 0 | 0 | 0 | 0 | 0 | 0 | 0 | 0 | 2 | 11 |
| Heating Degree Days | 1236 | 999 | 745 | 374 | 151 | 22 | 2 | ***9*** | ***91*** | 341 | ***674*** | 1061 | 5705 |
| Cooling Degree Days | 0 | 0 | 4 | 22 | 64 | 216 | 361 | 290 | 133 | 17 | 1 | 0 | 1108 |
| Total Precipitation (") | 1.50 | 1.61 | 3.05 | 4.27 | 4.28 | 3.77 | 3.41 | 3.60 | 3.85 | 2.93 | 3.47 | 2.74 | 38.48 |
| Days ≥ 0.1" Precip | ***3*** | ***3*** | ***6*** | 6 | 6 | 6 | 5 | 5 | 5 | 5 | ***6*** | 4 | 60 |
| Total Snowfall (") | na | na | ***2.0*** | 0.1 | 0.0 | 0.0 | 0.0 | 0.0 | 0.0 | 0.0 | 1.2 | ***1.3*** | na |
| Days ≥ 1" Snow Depth | na | na | na | 0 | 0 | 0 | 0 | 0 | 0 | 0 | 0 | na | na |

## BROOKFIELD *Linn County* ELEVATION 771 ft LAT/LONG 39° 47 ' N / 93° 5 ' W

| | JAN | FEB | MAR | APR | MAY | JUN | JUL | AUG | SEP | OCT | NOV | DEC | YEAR |
|---|---|---|---|---|---|---|---|---|---|---|---|---|---|
| Maximum Temp °F | 35.2 | 41.2 | 54.4 | 66.8 | 75.6 | 84.3 | 88.9 | 86.8 | 79.7 | 68.5 | 52.7 | 39.8 | 64.5 |
| Minimum Temp °F | 16.9 | 21.2 | 32.3 | 43.4 | 53.1 | 62.4 | 66.8 | 64.2 | 56.8 | 45.2 | 33.5 | 23.0 | 43.2 |
| Mean Temp °F | 26.1 | 31.3 | 43.4 | 55.1 | 64.3 | 73.3 | 77.9 | 75.6 | 68.3 | 56.9 | 43.1 | 31.4 | 53.9 |
| Days Max Temp ≥ 90 °F | 0 | 0 | 0 | 0 | 0 | 6 | 14 | 11 | 3 | 0 | 0 | 0 | 34 |
| Days Max Temp ≤ 32 °F | 12 | 7 | 1 | 0 | 0 | 0 | 0 | 0 | 0 | 0 | 1 | 7 | 28 |
| Days Min Temp ≤ 32 °F | 28 | 23 | 16 | 4 | 0 | 0 | 0 | 0 | 0 | 3 | 14 | 25 | 113 |
| Days Min Temp ≤ 0 °F | 4 | 2 | 0 | 0 | 0 | 0 | 0 | 0 | 0 | 0 | 0 | 2 | 8 |
| Heating Degree Days | 1200 | 946 | 667 | 316 | 103 | 9 | 1 | 2 | 58 | 271 | 650 | 1035 | 5258 |
| Cooling Degree Days | 0 | 0 | 4 | 29 | 82 | 276 | 421 | 354 | 180 | 22 | 1 | 0 | 1369 |
| Total Precipitation (") | 1.57 | 1.25 | 2.58 | 3.90 | 4.41 | 4.28 | 4.62 | 3.98 | 4.82 | 3.56 | 2.76 | 2.11 | 39.84 |
| Days ≥ 0.1" Precip | 3 | 3 | 6 | 7 | 8 | 7 | 6 | 6 | 7 | 6 | 5 | 4 | 68 |
| Total Snowfall (") | ***4.6*** | ***3.6*** | ***2.3*** | 0.4 | 0.0 | 0.0 | 0.0 | 0.0 | 0.0 | 0.0 | 1.1 | ***2.5*** | 14.5 |
| Days ≥ 1" Snow Depth | ***13*** | ***8*** | 3 | 0 | 0 | 0 | 0 | 0 | 0 | 0 | 1 | 6 | 31 |

## BRUNSWICK *Chariton County* ELEVATION 650 ft LAT/LONG 39° 26 ' N / 93° 8 ' W

| | JAN | FEB | MAR | APR | MAY | JUN | JUL | AUG | SEP | OCT | NOV | DEC | YEAR |
|---|---|---|---|---|---|---|---|---|---|---|---|---|---|
| Maximum Temp °F | 35.7 | 40.9 | 53.8 | 66.3 | 75.0 | 83.3 | 88.2 | 85.9 | 78.7 | 67.4 | 52.8 | 40.5 | 64.0 |
| Minimum Temp °F | 16.4 | 20.2 | 31.6 | 43.0 | 52.6 | 62.0 | 66.5 | 63.6 | 55.4 | 43.8 | 32.9 | 22.3 | 42.5 |
| Mean Temp °F | 26.1 | 30.6 | 42.7 | 54.7 | 63.8 | 72.7 | 77.4 | 74.7 | 67.1 | 55.6 | 42.9 | 31.4 | 53.3 |
| Days Max Temp ≥ 90 °F | 0 | 0 | 0 | 0 | 1 | 5 | 14 | 10 | 3 | 0 | 0 | 0 | 33 |
| Days Max Temp ≤ 32 °F | 12 | 8 | 2 | 0 | 0 | 0 | 0 | 0 | 0 | 0 | 2 | 7 | 31 |
| Days Min Temp ≤ 32 °F | 28 | 24 | 17 | 4 | 0 | 0 | 0 | 0 | 0 | 4 | 15 | 26 | 118 |
| Days Min Temp ≤ 0 °F | 4 | 3 | 0 | 0 | 0 | 0 | 0 | 0 | 0 | 0 | 0 | 1 | 8 |
| Heating Degree Days | 1200 | 965 | 686 | 326 | 112 | 10 | 1 | 5 | 74 | 304 | 658 | 1035 | 5376 |
| Cooling Degree Days | 0 | 0 | 3 | 24 | 64 | 232 | 370 | 301 | 133 | 13 | 1 | 0 | 1141 |
| Total Precipitation (") | 1.62 | 1.30 | 2.68 | 3.80 | 4.46 | 4.60 | 4.24 | 3.62 | 4.13 | 3.63 | 3.10 | 2.02 | 39.20 |
| Days ≥ 0.1" Precip | 4 | 4 | 6 | 7 | 8 | 7 | 6 | 6 | 5 | 6 | 6 | 4 | 69 |
| Total Snowfall (") | na | na | ***2.0*** | 0.1 | 0.0 | 0.0 | 0.0 | 0.0 | 0.0 | 0.0 | 1.1 | ***2.3*** | na |
| Days ≥ 1" Snow Depth | na | na | ***1*** | 0 | 0 | 0 | 0 | 0 | 0 | 0 | 1 | na | na |

## BUFFALO 3 S *Dallas County* ELEVATION 1152 ft LAT/LONG 37° 36' N / 93° 6' W

| | JAN | FEB | MAR | APR | MAY | JUN | JUL | AUG | SEP | OCT | NOV | DEC | YEAR |
|---|---|---|---|---|---|---|---|---|---|---|---|---|---|
| Maximum Temp °F | 41.5 | 47.1 | 58.3 | 69.3 | 75.7 | 83.7 | 89.4 | 88.2 | 80.1 | 69.4 | 56.0 | 45.6 | 67.0 |
| Minimum Temp °F | 20.9 | 24.9 | 34.6 | 44.8 | 52.6 | 61.3 | 65.5 | 63.6 | 56.3 | 45.3 | 35.2 | 25.6 | 44.2 |
| Mean Temp °F | 31.2 | 36.0 | 46.5 | 57.1 | 64.2 | 72.5 | 77.4 | 76.0 | 68.2 | 57.3 | 45.6 | 35.5 | 55.6 |
| Days Max Temp ≥ 90 °F | 0 | 0 | 0 | 0 | 0 | 4 | 16 | 13 | 4 | 0 | 0 | 0 | 37 |
| Days Max Temp ≤ 32 °F | 8 | 5 | 1 | 0 | 0 | 0 | 0 | 0 | 0 | 0 | 1 | 4 | 19 |
| Days Min Temp ≤ 32 °F | 26 | 20 | 14 | 4 | 0 | 0 | 0 | 0 | 0 | 4 | 13 | 23 | 104 |
| Days Min Temp ≤ 0 °F | 2 | 1 | 0 | 0 | 0 | 0 | 0 | 0 | 0 | 0 | 0 | 1 | 4 |
| Heating Degree Days | 1041 | 813 | 572 | 260 | 98 | 8 | 1 | 3 | 61 | 259 | 576 | 906 | 4598 |
| Cooling Degree Days | 0 | 0 | 3 | 29 | 66 | 231 | 381 | 348 | 158 | 19 | 1 | 0 | 1236 |
| Total Precipitation (") | 1.80 | 2.16 | 3.42 | 4.22 | 4.69 | 4.51 | 2.91 | 4.07 | 4.88 | 4.14 | 4.26 | 2.99 | 44.05 |
| Days ≥ 0.1" Precip | 4 | 4 | 6 | 7 | 7 | 7 | 5 | 6 | 6 | 6 | 6 | 5 | 69 |
| Total Snowfall (") | 4.2 | 4.6 | 3.0 | 0.2 | 0.0 | 0.0 | 0.0 | 0.0 | 0.0 | 0.0 | 1.4 | 3.6 | 17.0 |
| Days ≥ 1" Snow Depth | 9 | 7 | 2 | 0 | 0 | 0 | 0 | 0 | 0 | 0 | 1 | 4 | 23 |

## BUNKER *Reynolds County* ELEVATION 1380 ft LAT/LONG 37° 27' N / 91° 12' W

| | JAN | FEB | MAR | APR | MAY | JUN | JUL | AUG | SEP | OCT | NOV | DEC | YEAR |
|---|---|---|---|---|---|---|---|---|---|---|---|---|---|
| Maximum Temp °F | 42.1 | 47.8 | 59.1 | 69.8 | 76.1 | 83.2 | 88.2 | 86.4 | 79.6 | 69.6 | 57.2 | 46.0 | 67.1 |
| Minimum Temp °F | 21.1 | 25.0 | 34.9 | 45.2 | 53.4 | 61.3 | 65.4 | 63.0 | 56.7 | 45.3 | 35.4 | 25.3 | 44.3 |
| Mean Temp °F | 31.9 | 36.4 | 47.0 | 57.5 | 64.8 | 72.3 | 76.8 | 74.7 | 68.2 | 57.5 | 46.3 | 35.6 | 55.8 |
| Days Max Temp ≥ 90 °F | 0 | 0 | 0 | 0 | 0 | 3 | 13 | 10 | 3 | 0 | 0 | 0 | 29 |
| Days Max Temp ≤ 32 °F | 7 | 4 | 0 | 0 | 0 | 0 | 0 | 0 | 0 | 0 | 0 | 4 | 15 |
| Days Min Temp ≤ 32 °F | 26 | 21 | 14 | 3 | 0 | 0 | 0 | 0 | 0 | 3 | 13 | 23 | 103 |
| Days Min Temp ≤ 0 °F | 2 | 1 | 0 | 0 | 0 | 0 | 0 | 0 | 0 | 0 | 0 | 1 | 4 |
| Heating Degree Days | 1019 | 801 | 556 | 251 | 86 | 8 | 1 | 3 | 54 | 252 | 554 | 903 | 4488 |
| Cooling Degree Days | 0 | 0 | *3* | 32 | 72 | *223* | 370 | 316 | *158* | 20 | 2 | 0 | 1196 |
| Total Precipitation (") | 2.01 | 2.12 | 3.45 | 4.45 | 4.01 | 3.46 | 3.81 | 3.38 | 3.79 | 3.09 | 4.28 | 3.10 | 40.95 |
| Days ≥ 0.1" Precip | *3* | 3 | 5 | 6 | 6 | 5 | 5 | 5 | 5 | 4 | 5 | 4 | 56 |
| Total Snowfall (") | na | na | *0.6* | 0.0 | 0.0 | 0.0 | 0.0 | 0.0 | 0.0 | 0.0 | 0.0 | na | na |
| Days ≥ 1" Snow Depth | na | na | *0* | 0 | 0 | 0 | 0 | 0 | 0 | 0 | 0 | na | na |

## BUTLER *Bates County* ELEVATION 860 ft LAT/LONG 38° 16' N / 94° 20' W

| | JAN | FEB | MAR | APR | MAY | JUN | JUL | AUG | SEP | OCT | NOV | DEC | YEAR |
|---|---|---|---|---|---|---|---|---|---|---|---|---|---|
| Maximum Temp °F | 40.2 | 46.6 | 58.0 | 69.2 | 76.8 | 85.1 | 90.4 | 88.9 | 81.4 | 70.6 | 55.8 | 44.2 | 67.3 |
| Minimum Temp °F | 19.2 | 23.9 | 33.9 | 44.2 | 53.0 | 62.4 | 67.2 | 64.6 | 57.0 | 45.3 | 34.5 | 24.4 | 44.1 |
| Mean Temp °F | 29.7 | 35.3 | 46.0 | 56.7 | 64.9 | 73.8 | 78.9 | 76.8 | 69.2 | 57.9 | 45.2 | 34.4 | 55.7 |
| Days Max Temp ≥ 90 °F | 0 | 0 | 0 | 0 | 0 | 7 | 18 | 15 | 5 | 0 | 0 | 0 | 45 |
| Days Max Temp ≤ 32 °F | 9 | 5 | 1 | 0 | 0 | 0 | 0 | 0 | 0 | 0 | 1 | 5 | 21 |
| Days Min Temp ≤ 32 °F | 27 | 22 | 15 | 4 | 0 | 0 | 0 | 0 | 0 | 4 | 14 | 24 | 110 |
| Days Min Temp ≤ 0 °F | 3 | 1 | 0 | 0 | 0 | 0 | 0 | 0 | 0 | 0 | 0 | 1 | 5 |
| Heating Degree Days | 1086 | 832 | 587 | 270 | 90 | 6 | 1 | 1 | 51 | 243 | 590 | 942 | 4699 |
| Cooling Degree Days | 0 | 0 | 6 | 33 | 100 | 299 | 461 | 405 | 205 | 32 | 3 | 0 | 1544 |
| Total Precipitation (") | 1.76 | 1.47 | 3.07 | 4.16 | 4.48 | 5.64 | 4.28 | 4.13 | 4.67 | 3.82 | 3.20 | 2.02 | 42.70 |
| Days ≥ 0.1" Precip | 4 | 3 | 6 | 7 | 7 | 7 | 6 | 6 | 6 | 6 | 5 | 4 | 67 |
| Total Snowfall (") | *5.4* | *2.7* | 1.2 | 0.0 | 0.0 | 0.0 | 0.0 | 0.0 | 0.0 | 0.0 | 0.6 | *2.0* | 11.9 |
| Days ≥ 1" Snow Depth | na | na | *1* | 0 | 0 | 0 | 0 | 0 | 0 | 0 | 0 | na | na |

## CALIFORNIA *Moniteau County* ELEVATION 869 ft LAT/LONG 38° 38' N / 92° 34' W

| | JAN | FEB | MAR | APR | MAY | JUN | JUL | AUG | SEP | OCT | NOV | DEC | YEAR |
|---|---|---|---|---|---|---|---|---|---|---|---|---|---|
| Maximum Temp °F | *40.7* | 46.6 | 58.4 | 69.3 | *77.2* | 85.1 | 90.6 | *89.5* | 81.8 | *70.7* | *56.5* | *44.3* | 67.6 |
| Minimum Temp °F | *20.8* | 24.6 | 35.2 | 45.5 | *54.9* | 63.4 | 68.2 | *66.3* | 58.2 | *46.5* | *35.9* | *25.7* | 45.4 |
| Mean Temp °F | *30.7* | 35.6 | 47.0 | 57.4 | *66.1* | 74.3 | 79.4 | *77.9* | 70.0 | *58.5* | *46.2* | *35.1* | 56.5 |
| Days Max Temp ≥ 90 °F | 0 | 0 | 0 | 0 | 0 | 7 | 18 | *16* | 5 | 0 | 0 | 0 | 46 |
| Days Max Temp ≤ 32 °F | *8* | 5 | 1 | 0 | 0 | 0 | 0 | 0 | 0 | 0 | 1 | 4 | 19 |
| Days Min Temp ≤ 32 °F | 26 | 21 | *13* | 3 | 0 | 0 | 0 | 0 | 0 | *2* | *11* | *23* | 99 |
| Days Min Temp ≤ 0 °F | 2 | 1 | 0 | 0 | 0 | 0 | 0 | 0 | 0 | 0 | 0 | 1 | 4 |
| Heating Degree Days | *1055* | 822 | 560 | *255* | *70* | 5 | 0 | *1* | 46 | *233* | *560* | *921* | 4528 |
| Cooling Degree Days | 0 | 0 | 9 | 41 | 104 | 307 | 465 | 409 | 205 | 34 | 3 | 0 | 1577 |
| Total Precipitation (") | 1.27 | 1.59 | 3.02 | 4.15 | 5.05 | 4.33 | 3.66 | 3.34 | 4.48 | 3.47 | 3.44 | 2.39 | 40.19 |
| Days ≥ 0.1" Precip | 3 | 4 | 6 | 7 | 7 | 6 | 5 | 5 | 6 | 6 | 5 | 4 | 64 |
| Total Snowfall (") | *3.7* | *5.0* | *1.4* | 0.2 | 0.0 | 0.0 | 0.0 | 0.0 | 0.0 | 0.0 | 1.1 | *2.0* | 13.4 |
| Days ≥ 1" Snow Depth | na | na | *1* | 0 | 0 | 0 | 0 | 0 | 0 | 0 | 0 | *2* | na |

## CAMDENTON 2 NW *Camden County* ELEVATION 1040 ft LAT/LONG 38° 0 ' N / 92° 45 ' W

| | JAN | FEB | MAR | APR | MAY | JUN | JUL | AUG | SEP | OCT | NOV | DEC | YEAR |
|---|---|---|---|---|---|---|---|---|---|---|---|---|---|
| Maximum Temp °F | 42.5 | 48.6 | 59.7 | 71.2 | 77.2 | 85.1 | 90.3 | 89.0 | 81.3 | 71.2 | 57.7 | 46.3 | 68.3 |
| Minimum Temp °F | 21.4 | 25.7 | 35.2 | 45.9 | 54.2 | 63.4 | 67.8 | 65.6 | 58.2 | 46.9 | 36.2 | 26.3 | 45.6 |
| Mean Temp °F | 31.9 | 37.2 | 47.4 | 58.6 | 65.7 | 74.3 | 79.1 | 77.3 | 69.8 | 59.1 | 47.0 | 36.3 | 57.0 |
| Days Max Temp ≥ 90 °F | 0 | 0 | 0 | 1 | 0 | 7 | 18 | 15 | 5 | 0 | 0 | 0 | 46 |
| Days Max Temp ≤ 32 °F | 7 | 4 | 1 | 0 | 0 | 0 | 0 | 0 | 0 | 0 | 1 | 4 | 17 |
| Days Min Temp ≤ 32 °F | 25 | 20 | 14 | 2 | 0 | 0 | 0 | 0 | 0 | 2 | 12 | 21 | 96 |
| Days Min Temp ≤ 0 °F | 2 | 1 | 0 | 0 | 0 | 0 | 0 | 0 | 0 | 0 | 0 | 1 | 4 |
| Heating Degree Days | 1019 | 780 | 547 | 232 | 78 | 5 | 1 | 2 | 45 | 218 | 539 | 882 | 4348 |
| Cooling Degree Days | 0 | 0 | 9 | 48 | 96 | 303 | 450 | 401 | 203 | 34 | *4* | 0 | 1548 |
| Total Precipitation (") | 1.62 | 2.20 | 3.43 | 4.31 | 5.10 | 3.92 | 3.82 | 3.92 | 4.80 | 4.49 | 3.73 | 3.06 | 44.40 |
| Days ≥ 0.1" Precip | 4 | 4 | 6 | 8 | 8 | 6 | 6 | 6 | 6 | 6 | 6 | 5 | 71 |
| Total Snowfall (") | *4.3* | 4.6 | 2.8 | 0.1 | 0.0 | 0.0 | 0.0 | 0.0 | 0.0 | 0.0 | 1.0 | 3.4 | 16.2 |
| Days ≥ 1" Snow Depth | *7* | *5* | 2 | 0 | 0 | 0 | 0 | 0 | 0 | 0 | 0 | 3 | 17 |

## CANTON L & D 20 *Lewis County* ELEVATION 489 ft LAT/LONG 40° 9 ' N / 91° 31 ' W

| | JAN | FEB | MAR | APR | MAY | JUN | JUL | AUG | SEP | OCT | NOV | DEC | YEAR |
|---|---|---|---|---|---|---|---|---|---|---|---|---|---|
| Maximum Temp °F | 34.5 | 38.8 | 51.8 | 64.8 | 75.2 | 84.3 | 88.9 | 86.0 | 78.7 | 67.5 | 52.4 | 39.4 | 63.5 |
| Minimum Temp °F | 16.2 | 19.8 | 31.4 | 43.2 | 53.5 | 63.0 | 67.4 | 64.5 | 56.4 | 44.5 | 33.3 | 22.2 | 42.9 |
| Mean Temp °F | 25.4 | 29.3 | 41.6 | 54.0 | 64.4 | 73.7 | 78.2 | 75.2 | 67.6 | 56.0 | 42.9 | 30.8 | 53.3 |
| Days Max Temp ≥ 90 °F | 0 | 0 | 0 | 0 | 1 | 7 | 14 | 9 | 3 | 0 | 0 | 0 | 34 |
| Days Max Temp ≤ 32 °F | 13 | 8 | 2 | 0 | 0 | 0 | 0 | 0 | 0 | 0 | 1 | 7 | 31 |
| Days Min Temp ≤ 32 °F | 29 | 24 | 17 | 4 | 0 | 0 | 0 | 0 | 0 | 3 | 15 | 26 | 118 |
| Days Min Temp ≤ 0 °F | 4 | 3 | 0 | 0 | 0 | 0 | 0 | 0 | 0 | 0 | 0 | 2 | 9 |
| Heating Degree Days | 1223 | 1002 | 722 | 342 | 108 | 8 | 1 | 3 | 66 | 293 | 659 | 1052 | 5479 |
| Cooling Degree Days | 0 | 0 | 5 | 26 | 96 | 289 | *434* | 356 | 167 | 20 | 1 | 0 | 1394 |
| Total Precipitation (") | 1.64 | 1.46 | 2.89 | 3.63 | 4.75 | 3.77 | 4.73 | 3.61 | 4.55 | 2.97 | 3.08 | 2.20 | 39.28 |
| Days ≥ 0.1" Precip | 4 | 4 | 6 | 7 | 8 | 7 | 7 | 6 | 6 | 6 | 6 | 5 | 72 |
| Total Snowfall (") | na | na | *2.2* | 0.0 | 0.0 | 0.0 | 0.0 | 0.0 | 0.0 | 0.0 | 0.4 | na | na |
| Days ≥ 1" Snow Depth | na | na | *3* | 0 | 0 | 0 | 0 | 0 | 0 | 0 | 0 | na | na |

## CAPE GIRARDEAU AP *Scott County* ELEVATION 337 ft LAT/LONG 37° 14 ' N / 89° 34 ' W

| | JAN | FEB | MAR | APR | MAY | JUN | JUL | AUG | SEP | OCT | NOV | DEC | YEAR |
|---|---|---|---|---|---|---|---|---|---|---|---|---|---|
| Maximum Temp °F | 40.6 | 45.8 | 57.2 | 69.0 | 77.8 | 86.8 | 90.3 | 87.8 | 80.7 | 70.0 | 56.6 | 45.4 | 67.3 |
| Minimum Temp °F | 23.5 | 27.3 | 37.2 | 47.3 | 56.3 | 65.3 | 69.4 | 66.5 | 59.1 | 46.5 | 37.9 | 28.8 | 47.1 |
| Mean Temp °F | 32.1 | 36.6 | 47.2 | 58.2 | 67.1 | 76.1 | 79.9 | 77.2 | 69.9 | 58.3 | 47.3 | 37.1 | 57.3 |
| Days Max Temp ≥ 90 °F | 0 | 0 | 0 | 0 | 2 | 12 | 19 | 13 | 5 | 0 | 0 | 0 | 51 |
| Days Max Temp ≤ 32 °F | 8 | 4 | 1 | 0 | 0 | 0 | 0 | 0 | 0 | 0 | 0 | 3 | 16 |
| Days Min Temp ≤ 32 °F | 25 | 19 | 11 | 2 | 0 | 0 | 0 | 0 | 0 | 2 | 10 | 20 | 89 |
| Days Min Temp ≤ 0 °F | 1 | 0 | 0 | 0 | 0 | 0 | 0 | 0 | 0 | 0 | 0 | 0 | 1 |
| Heating Degree Days | 1013 | 796 | 548 | 233 | 61 | 2 | 0 | 1 | 39 | 237 | 527 | 858 | 4315 |
| Cooling Degree Days | 0 | 0 | 5 | 36 | 135 | 348 | 489 | 408 | 200 | 34 | 2 | 1 | 1658 |
| Total Precipitation (") | 3.14 | 3.31 | 4.49 | 4.53 | 5.04 | 3.52 | 3.53 | 3.52 | 3.79 | 3.33 | 4.38 | 4.41 | 46.99 |
| Days ≥ 0.1" Precip | 5 | 5 | 7 | 8 | 8 | 6 | 6 | 6 | 5 | 5 | 6 | 7 | 74 |
| Total Snowfall (") | 4.2 | 4.8 | 2.3 | 0.0 | 0.0 | 0.0 | 0.0 | 0.0 | 0.0 | 0.1 | 0.3 | 2.0 | 13.7 |
| Days ≥ 1" Snow Depth | 6 | 5 | 1 | 0 | 0 | 0 | 0 | 0 | 0 | 0 | 0 | 2 | 14 |

## CARROLLTON *Carroll County* ELEVATION 751 ft LAT/LONG 39° 22 ' N / 93° 30 ' W

| | JAN | FEB | MAR | APR | MAY | JUN | JUL | AUG | SEP | OCT | NOV | DEC | YEAR |
|---|---|---|---|---|---|---|---|---|---|---|---|---|---|
| Maximum Temp °F | 36.2 | 41.8 | 55.2 | 67.6 | 76.8 | 85.6 | 90.2 | 87.6 | 79.6 | 67.8 | 52.5 | 40.7 | 65.1 |
| Minimum Temp °F | 17.7 | 21.9 | 32.8 | 43.6 | 52.8 | 62.2 | 66.6 | 63.9 | 56.1 | 44.6 | 33.6 | 23.6 | 43.3 |
| Mean Temp °F | 27.0 | 31.9 | 44.0 | 55.6 | 64.9 | 73.9 | 78.4 | 75.8 | 67.9 | 56.2 | 43.1 | 32.2 | 54.2 |
| Days Max Temp ≥ 90 °F | 0 | 0 | 0 | 1 | 1 | 9 | 18 | 12 | 4 | 0 | 0 | 0 | 45 |
| Days Max Temp ≤ 32 °F | 11 | 7 | 1 | 0 | 0 | 0 | 0 | 0 | 0 | 0 | 1 | 7 | 27 |
| Days Min Temp ≤ 32 °F | 28 | 23 | 16 | 4 | 0 | 0 | 0 | 0 | 0 | 4 | 14 | 25 | 114 |
| Days Min Temp ≤ 0 °F | 3 | 2 | 0 | 0 | 0 | 0 | 0 | 0 | 0 | 0 | 0 | 1 | 6 |
| Heating Degree Days | 1172 | 929 | 647 | 301 | 93 | 6 | 1 | 2 | 62 | 287 | 652 | 1012 | 5164 |
| Cooling Degree Days | 0 | 0 | 4 | 32 | 98 | 301 | 445 | 368 | 167 | 18 | 1 | 0 | 1434 |
| Total Precipitation (") | 1.54 | 1.48 | 2.78 | 4.08 | 4.72 | 4.62 | 4.46 | 4.21 | 4.65 | 3.71 | 2.87 | 2.27 | 41.39 |
| Days ≥ 0.1" Precip | 4 | 4 | 6 | 7 | 8 | 6 | 6 | 6 | 6 | 6 | 5 | 4 | 68 |
| Total Snowfall (") | 6.8 | 6.6 | 3.6 | 0.7 | 0.0 | 0.0 | 0.0 | 0.0 | 0.0 | 0.0 | 1.5 | 4.1 | 23.3 |
| Days ≥ 1" Snow Depth | 12 | 10 | 3 | 0 | 0 | 0 | 0 | 0 | 0 | 0 | 1 | 5 | 31 |

## CARUTHERSVILLE *Pemiscot County* ELEVATION 269 ft LAT/LONG 36° 11 ' N / 89° 39 ' W

| | JAN | FEB | MAR | APR | MAY | JUN | JUL | AUG | SEP | OCT | NOV | DEC | YEAR |
|---|---|---|---|---|---|---|---|---|---|---|---|---|---|
| Maximum Temp °F | 42.8 | 47.9 | 58.5 | 69.9 | 78.6 | 87.2 | 90.6 | 88.4 | 82.1 | 72.1 | 59.4 | 48.1 | 68.8 |
| Minimum Temp °F | 25.4 | 29.3 | 38.2 | 48.1 | 57.4 | 65.6 | 69.9 | 67.3 | 60.5 | 48.0 | 39.1 | 30.6 | 48.3 |
| Mean Temp °F | 34.1 | 38.6 | 48.4 | 59.0 | 68.1 | 76.4 | 80.3 | 77.9 | 71.3 | 60.1 | 49.3 | 39.4 | 58.6 |
| Days Max Temp ≥ 90 °F | 0 | 0 | 0 | 0 | 2 | 12 | 19 | 14 | 5 | 0 | 0 | 0 | 52 |
| Days Max Temp ≤ 32 °F | 6 | 3 | 0 | 0 | 0 | 0 | 0 | 0 | 0 | 0 | 0 | 2 | 11 |
| Days Min Temp ≤ 32 °F | 23 | 18 | 9 | 1 | 0 | 0 | 0 | 0 | 0 | 1 | 8 | 19 | 79 |
| Days Min Temp ≤ 0 °F | 1 | 0 | 0 | 0 | 0 | 0 | 0 | 0 | 0 | 0 | 0 | 0 | 1 |
| Heating Degree Days | 949 | 738 | 513 | 211 | 46 | 2 | 0 | 0 | 26 | 193 | 469 | 789 | 3936 |
| Cooling Degree Days | 0 | 0 | 4 | 36 | 140 | 349 | 486 | 411 | 215 | 41 | 3 | 0 | 1685 |
| Total Precipitation (") | 3.32 | 3.83 | 4.62 | 5.05 | 4.92 | 4.47 | 3.48 | 3.16 | 3.52 | 3.34 | 4.68 | 4.95 | 49.34 |
| Days ≥ 0.1" Precip | 6 | 6 | 8 | 7 | 7 | 6 | 6 | 5 | 5 | 5 | 6 | 7 | 74 |
| Total Snowfall (") | *2.8* | 2.7 | 0.9 | 0.0 | 0.0 | 0.0 | 0.0 | 0.0 | 0.0 | 0.0 | 0.2 | 0.8 | 7.4 |
| Days ≥ 1" Snow Depth | na | *1* | 0 | 0 | 0 | 0 | 0 | 0 | 0 | 0 | 0 | *0* | na |

## CASSVILLE RANGER STN *Barry County* ELEVATION 1342 ft LAT/LONG 36° 41 ' N / 93° 52 ' W

| | JAN | FEB | MAR | APR | MAY | JUN | JUL | AUG | SEP | OCT | NOV | DEC | YEAR |
|---|---|---|---|---|---|---|---|---|---|---|---|---|---|
| Maximum Temp °F | 42.8 | 47.9 | 58.0 | 68.7 | 75.8 | 83.8 | 88.8 | 88.3 | 80.4 | 69.9 | 57.6 | *46.7* | 67.4 |
| Minimum Temp °F | 19.6 | 23.2 | 33.5 | 42.8 | 51.5 | *60.4* | 64.6 | 62.3 | 55.5 | 42.5 | 33.1 | *23.7* | 42.7 |
| Mean Temp °F | 31.0 | 35.5 | 45.8 | 55.8 | 63.7 | *72.1* | 76.8 | 75.3 | 67.9 | 56.2 | 45.4 | *35.2* | 55.1 |
| Days Max Temp ≥ 90 °F | 0 | 0 | 0 | 0 | 0 | 4 | 14 | 13 | 4 | 0 | 0 | 0 | 35 |
| Days Max Temp ≤ 32 °F | 7 | 4 | 0 | 0 | 0 | 0 | 0 | 0 | 0 | 0 | 1 | 3 | 15 |
| Days Min Temp ≤ 32 °F | 27 | 22 | 15 | 5 | 0 | 0 | 0 | 0 | 0 | 6 | 14 | 23 | 112 |
| Days Min Temp ≤ 0 °F | 2 | 1 | 0 | 0 | 0 | 0 | 0 | 0 | 0 | 0 | 0 | 1 | 4 |
| Heating Degree Days | 1047 | 826 | 594 | 290 | 109 | *11* | 2 | 3 | 62 | 287 | *592* | *917* | 4740 |
| Cooling Degree Days | *0* | *0* | 3 | 18 | *67* | *232* | 371 | 353 | 165 | *18* | *1* | na | na |
| Total Precipitation (") | 2.11 | 2.46 | 4.17 | 4.93 | 4.58 | 4.40 | 3.35 | 3.60 | 4.54 | 3.77 | 4.31 | 3.62 | 45.84 |
| Days ≥ 0.1" Precip | 4 | 4 | 6 | 7 | 7 | 6 | 5 | 5 | 6 | 5 | 5 | 5 | 65 |
| Total Snowfall (") | na | *3.2* | *2.3* | 0.3 | 0.0 | 0.0 | 0.0 | 0.0 | 0.0 | 0.0 | 0.7 | *2.2* | na |
| Days ≥ 1" Snow Depth | na | na | *1* | 0 | 0 | 0 | 0 | 0 | 0 | 0 | *0* | *2* | na |

## CLEARWATER DAM *Wayne County* ELEVATION 640 ft LAT/LONG 37° 8 ' N / 90° 47 ' W

| | JAN | FEB | MAR | APR | MAY | JUN | JUL | AUG | SEP | OCT | NOV | DEC | YEAR |
|---|---|---|---|---|---|---|---|---|---|---|---|---|---|
| Maximum Temp °F | 42.0 | 47.4 | 58.4 | 70.8 | 78.2 | 86.0 | 91.1 | 88.7 | 81.1 | 70.1 | 57.4 | *46.1* | 68.1 |
| Minimum Temp °F | 20.2 | 23.4 | 33.5 | 44.1 | 52.4 | 61.4 | 66.0 | 63.5 | 56.7 | 43.4 | 34.4 | *24.8* | 43.7 |
| Mean Temp °F | 31.1 | 35.4 | 46.0 | 57.5 | 65.3 | 73.7 | 78.6 | 76.1 | 68.9 | 56.7 | 45.9 | *35.5* | 55.9 |
| Days Max Temp ≥ 90 °F | 0 | 0 | 0 | 1 | 1 | 10 | 20 | 15 | 4 | 0 | 0 | 0 | 51 |
| Days Max Temp ≤ 32 °F | 7 | 4 | 1 | 0 | 0 | 0 | 0 | 0 | 0 | 0 | 0 | 3 | 15 |
| Days Min Temp ≤ 32 °F | 26 | 22 | 16 | 4 | 0 | 0 | 0 | 0 | 0 | 4 | 14 | 22 | 108 |
| Days Min Temp ≤ 0 °F | 2 | 1 | 0 | 0 | 0 | 0 | 0 | 0 | 0 | 0 | 0 | 1 | 4 |
| Heating Degree Days | 1042 | 830 | 588 | 253 | 81 | 6 | 1 | 1 | 46 | 273 | 568 | *908* | 4597 |
| Cooling Degree Days | 0 | 0 | 5 | 34 | 98 | 293 | 450 | 381 | 177 | 21 | 2 | *0* | 1461 |
| Total Precipitation (") | 2.95 | 2.81 | 4.24 | 4.73 | 4.35 | 3.12 | 3.62 | 3.70 | 3.61 | 3.14 | 4.70 | 4.24 | 45.21 |
| Days ≥ 0.1" Precip | 5 | 5 | 7 | 7 | 7 | 5 | 5 | 5 | 5 | 5 | 6 | 6 | 68 |
| Total Snowfall (") | na | na | *0.6* | 0.0 | 0.0 | 0.0 | 0.0 | 0.0 | 0.0 | 0.0 | 0.2 | *1.0* | na |
| Days ≥ 1" Snow Depth | *4* | 4 | 1 | 0 | 0 | 0 | 0 | 0 | 0 | 0 | 0 | 1 | 10 |

## CLINTON *Henry County* ELEVATION 770 ft LAT/LONG 38° 24 ' N / 93° 46 ' W

| | JAN | FEB | MAR | APR | MAY | JUN | JUL | AUG | SEP | OCT | NOV | DEC | YEAR |
|---|---|---|---|---|---|---|---|---|---|---|---|---|---|
| Maximum Temp °F | 37.9 | 43.8 | 55.7 | 67.3 | 75.8 | 84.4 | 90.2 | 88.7 | 80.4 | 69.5 | 54.9 | 43.3 | 66.0 |
| Minimum Temp °F | 17.0 | 20.8 | 32.1 | 42.4 | 51.9 | 61.8 | 66.4 | 63.5 | 55.3 | 42.8 | 32.9 | 22.6 | 42.5 |
| Mean Temp °F | 27.5 | 32.3 | 43.9 | 54.9 | 63.9 | 73.1 | 78.3 | 76.1 | 67.9 | 56.2 | 43.9 | 33.0 | 54.3 |
| Days Max Temp ≥ 90 °F | 0 | 0 | 0 | 0 | 0 | 7 | 17 | 15 | 5 | 0 | 0 | 0 | 44 |
| Days Max Temp ≤ 32 °F | 10 | 6 | 1 | 0 | 0 | 0 | 0 | 0 | 0 | 0 | 1 | 5 | 23 |
| Days Min Temp ≤ 32 °F | 28 | 24 | 17 | 4 | 0 | 0 | 0 | 0 | 0 | 5 | 16 | 26 | 120 |
| Days Min Temp ≤ 0 °F | 3 | 2 | 0 | 0 | 0 | 0 | 0 | 0 | 0 | 0 | 0 | 1 | 6 |
| Heating Degree Days | 1156 | 914 | 649 | 320 | 112 | 10 | 1 | 3 | 68 | 292 | 627 | 985 | 5137 |
| Cooling Degree Days | 0 | 0 | 3 | 25 | 78 | 267 | 435 | 372 | 167 | 22 | 2 | 0 | 1371 |
| Total Precipitation (") | 1.52 | 1.68 | 3.08 | 4.06 | 5.11 | 5.08 | 4.15 | 4.30 | 4.72 | 4.09 | 3.39 | 2.16 | 43.34 |
| Days ≥ 0.1" Precip | 4 | 4 | 6 | 6 | 7 | 7 | 5 | 5 | 5 | 5 | 5 | 4 | 63 |
| Total Snowfall (") | *5.7* | na | 1.5 | 0.1 | 0.0 | 0.0 | 0.0 | 0.0 | 0.0 | 0.0 | 1.3 | *2.2* | na |
| Days ≥ 1" Snow Depth | na | na | *0* | 0 | 0 | 0 | 0 | 0 | 0 | 0 | *0* | na | na |

## CONCEPTION *Nodaway County* ELEVATION 1112 ft LAT/LONG 40° 15' N / 94° 41' W

| | JAN | FEB | MAR | APR | MAY | JUN | JUL | AUG | SEP | OCT | NOV | DEC | YEAR |
|---|---|---|---|---|---|---|---|---|---|---|---|---|---|
| Maximum Temp °F | 31.8 | 37.5 | 50.6 | 63.1 | 73.3 | 82.8 | 87.5 | 85.8 | 77.7 | 65.6 | 49.7 | 37.2 | 61.9 |
| Minimum Temp °F | 12.8 | 17.5 | 28.9 | 40.7 | 51.4 | 60.8 | 65.1 | 62.4 | 53.4 | 41.9 | 29.8 | 19.0 | 40.3 |
| Mean Temp °F | 22.4 | 27.5 | 39.8 | 51.9 | 62.4 | 71.8 | 76.4 | 74.1 | 65.6 | 53.8 | 39.8 | 28.1 | 51.1 |
| Days Max Temp ≥ 90 °F | 0 | 0 | 0 | 0 | 0 | 5 | 13 | 10 | 3 | 0 | 0 | 0 | 31 |
| Days Max Temp ≤ 32 °F | 15 | 10 | 3 | 0 | 0 | 0 | 0 | 0 | 0 | 0 | 2 | 10 | 40 |
| Days Min Temp ≤ 32 °F | 29 | 25 | 20 | 6 | 0 | 0 | 0 | 0 | 0 | 5 | 19 | 28 | 132 |
| Days Min Temp ≤ 0 °F | 7 | 4 | 0 | 0 | 0 | 0 | 0 | 0 | 0 | 0 | 0 | 3 | 14 |
| Heating Degree Days | 1317 | 1053 | 777 | 401 | 143 | 16 | 2 | 6 | 95 | 357 | 751 | 1138 | 6056 |
| Cooling Degree Days | 0 | 0 | 2 | 20 | 68 | 236 | 358 | 299 | 127 | 12 | 0 | 0 | 1122 |
| Total Precipitation (") | 0.89 | 0.94 | 2.47 | 3.29 | 4.30 | 4.39 | 5.16 | 4.13 | 4.27 | 3.15 | 2.07 | 1.44 | 36.50 |
| Days ≥ 0.1" Precip | 3 | 3 | 6 | 7 | 8 | 7 | 6 | 6 | 6 | 5 | 4 | 3 | 64 |
| Total Snowfall (") | 6.0 | 5.1 | 3.7 | 0.7 | 0.0 | 0.0 | 0.0 | 0.0 | 0.0 | 0.0 | 1.2 | 3.8 | 20.5 |
| Days ≥ 1" Snow Depth | 13 | 10 | 3 | 0 | 0 | 0 | 0 | 0 | 0 | 0 | 2 | 7 | 35 |

## DONIPHAN *Ripley County* ELEVATION 312 ft LAT/LONG 36° 35' N / 90° 49' W

| | JAN | FEB | MAR | APR | MAY | JUN | JUL | AUG | SEP | OCT | NOV | DEC | YEAR |
|---|---|---|---|---|---|---|---|---|---|---|---|---|---|
| Maximum Temp °F | 44.1 | 49.9 | 59.9 | 71.3 | 79.2 | 87.1 | 91.9 | 89.8 | 82.6 | 72.4 | 59.4 | 47.9 | 69.6 |
| Minimum Temp °F | 20.4 | 23.9 | 33.4 | 43.2 | 51.8 | 60.8 | 65.3 | 62.8 | 55.3 | 42.0 | 33.6 | 24.7 | 43.1 |
| Mean Temp °F | 32.3 | 36.9 | 46.7 | 57.3 | 65.5 | 73.9 | 78.6 | 76.4 | 69.0 | 57.3 | 46.5 | 36.3 | 56.4 |
| Days Max Temp ≥ 90 °F | 0 | 0 | 0 | 0 | 2 | 12 | 22 | 17 | 6 | 1 | 0 | 0 | 60 |
| Days Max Temp ≤ 32 °F | 5 | 2 | 0 | 0 | 0 | 0 | 0 | 0 | 0 | 0 | 0 | 2 | 9 |
| Days Min Temp ≤ 32 °F | 27 | 22 | 16 | 5 | 0 | 0 | 0 | 0 | 0 | 7 | 15 | 23 | 115 |
| Days Min Temp ≤ 0 °F | 2 | 1 | 0 | 0 | 0 | 0 | 0 | 0 | 0 | 0 | 0 | 1 | 4 |
| Heating Degree Days | 1008 | 787 | 564 | 252 | 76 | 5 | 1 | 1 | 46 | 260 | 549 | 882 | 4431 |
| Cooling Degree Days | 0 | 0 | 3 | 23 | 94 | 286 | 443 | 380 | 175 | 21 | 1 | 0 | 1426 |
| Total Precipitation (") | 3.60 | 3.44 | 4.88 | 5.22 | 5.09 | 3.07 | 3.25 | 4.12 | 4.27 | 3.38 | 5.33 | 4.55 | 50.20 |
| Days ≥ 0.1" Precip | 6 | 6 | 8 | 7 | 8 | 5 | 5 | 6 | 6 | 5 | 7 | 7 | 76 |
| Total Snowfall (") | 4.3 | 4.5 | 2.0 | 0.0 | 0.0 | 0.0 | 0.0 | 0.0 | 0.0 | 0.1 | 0.8 | 1.4 | 13.1 |
| Days ≥ 1" Snow Depth | 5 | *4* | 1 | 0 | 0 | 0 | 0 | 0 | 0 | 0 | 0 | 1 | 11 |

## DORA *Ozark County* ELEVATION 981 ft LAT/LONG 36° 46' N / 92° 13' W

| | JAN | FEB | MAR | APR | MAY | JUN | JUL | AUG | SEP | OCT | NOV | DEC | YEAR |
|---|---|---|---|---|---|---|---|---|---|---|---|---|---|
| Maximum Temp °F | 43.4 | 49.8 | 59.9 | 70.3 | 76.9 | 84.3 | 90.2 | 88.9 | 81.4 | *71.3* | 58.4 | *47.9* | 68.6 |
| Minimum Temp °F | 18.9 | 23.5 | 33.0 | 42.0 | 51.5 | 60.3 | 65.4 | 63.4 | 56.1 | 42.5 | 32.9 | 23.9 | 42.8 |
| Mean Temp °F | 31.2 | 36.8 | 46.5 | 56.2 | 64.2 | 72.4 | 77.8 | 76.2 | 68.8 | *57.0* | 45.7 | *35.8* | 55.7 |
| Days Max Temp ≥ 90 °F | 0 | 0 | 0 | 0 | 0 | 6 | 18 | 15 | 5 | *0* | 0 | 0 | 44 |
| Days Max Temp ≤ 32 °F | 6 | 3 | 1 | 0 | 0 | 0 | 0 | 0 | 0 | 0 | 0 | 3 | 13 |
| Days Min Temp ≤ 32 °F | 27 | 23 | 16 | 5 | 0 | 0 | 0 | 0 | 0 | 5 | 15 | 25 | 116 |
| Days Min Temp ≤ 0 °F | 2 | 1 | 0 | 0 | 0 | 0 | 0 | 0 | 0 | 0 | 0 | 1 | 4 |
| Heating Degree Days | 1040 | 788 | 570 | 281 | 94 | 9 | 1 | 2 | 55 | *263* | 571 | *899* | 4573 |
| Cooling Degree Days | 0 | 0 | 2 | 25 | 69 | 236 | 400 | 367 | 169 | 16 | 1 | 0 | 1285 |
| Total Precipitation (") | 2.18 | 2.52 | 3.87 | 4.72 | 4.49 | 3.93 | 3.03 | 3.56 | 4.19 | 3.62 | 4.86 | 3.92 | 44.89 |
| Days ≥ 0.1" Precip | *3* | 4 | 6 | *6* | 7 | 6 | 4 | 5 | 5 | *5* | *5* | *5* | 61 |
| Total Snowfall (") | *4.0* | na | 2.3 | 0.3 | 0.0 | 0.0 | 0.0 | 0.0 | 0.0 | 0.0 | 0.7 | *2.7* | na |
| Days ≥ 1" Snow Depth | *5* | na | *1* | 0 | 0 | 0 | 0 | 0 | 0 | 0 | *0* | na | na |

## ELDON *Miller County* ELEVATION 932 ft LAT/LONG 38° 21' N / 92° 35' W

| | JAN | FEB | MAR | APR | MAY | JUN | JUL | AUG | SEP | OCT | NOV | DEC | YEAR |
|---|---|---|---|---|---|---|---|---|---|---|---|---|---|
| Maximum Temp °F | 40.2 | 46.4 | 57.2 | 68.8 | 76.8 | 84.7 | 90.0 | 88.7 | 80.2 | 69.5 | 55.8 | 44.6 | 66.9 |
| Minimum Temp °F | 19.3 | 23.5 | 33.9 | 44.8 | 53.9 | 62.7 | 67.4 | 65.0 | 57.1 | 45.6 | 35.0 | 24.6 | 44.4 |
| Mean Temp °F | 30.0 | 35.3 | 45.6 | 56.8 | 65.4 | 73.8 | 78.8 | 76.9 | 68.7 | 57.6 | 45.4 | 34.7 | 55.8 |
| Days Max Temp ≥ 90 °F | 0 | 0 | 0 | 0 | 1 | 7 | 17 | 15 | 4 | 0 | 0 | 0 | 44 |
| Days Max Temp ≤ 32 °F | 9 | 5 | 1 | 0 | 0 | 0 | 0 | 0 | 0 | 0 | 1 | 5 | 21 |
| Days Min Temp ≤ 32 °F | 27 | 23 | 15 | 3 | 0 | 0 | 0 | 0 | 0 | 2 | 13 | 24 | 107 |
| Days Min Temp ≤ 0 °F | 2 | 1 | 0 | 0 | 0 | 0 | 0 | 0 | 0 | 0 | 0 | 1 | 4 |
| Heating Degree Days | 1079 | 831 | 601 | 275 | 79 | 7 | 1 | 2 | 57 | 252 | 584 | 932 | 4700 |
| Cooling Degree Days | 0 | *0* | 5 | 30 | 84 | 276 | 440 | 376 | 167 | 22 | 2 | 0 | 1402 |
| Total Precipitation (") | 1.58 | 1.86 | 3.07 | 4.18 | 4.78 | 4.22 | 3.82 | 3.58 | 4.99 | 3.80 | 3.77 | 2.70 | 42.35 |
| Days ≥ 0.1" Precip | 3 | 4 | 6 | 7 | 7 | 7 | 6 | 5 | 6 | 6 | 6 | 5 | 68 |
| Total Snowfall (") | 3.8 | 4.1 | 2.5 | 0.2 | 0.0 | 0.0 | 0.0 | 0.0 | 0.0 | 0.0 | 1.3 | 3.9 | 15.8 |
| Days ≥ 1" Snow Depth | 8 | *5* | 1 | 0 | 0 | 0 | 0 | 0 | 0 | 0 | 1 | 4 | 19 |

## ELSBERRY 1 S *Lincoln County* ELEVATION 449 ft LAT/LONG 39° 9 ' N / 90° 47 ' W

| | JAN | FEB | MAR | APR | MAY | JUN | JUL | AUG | SEP | OCT | NOV | DEC | YEAR |
|---|---|---|---|---|---|---|---|---|---|---|---|---|---|
| Maximum Temp °F | 37.5 | 42.7 | 55.1 | 67.9 | 76.8 | 85.7 | 90.0 | 87.5 | 80.0 | 68.6 | 54.7 | 42.0 | 65.7 |
| Minimum Temp °F | 16.9 | 21.0 | 31.7 | 42.6 | 51.2 | 60.8 | 65.2 | 62.6 | 54.9 | 42.8 | 33.3 | 23.1 | 42.2 |
| Mean Temp °F | 27.2 | 31.9 | 43.4 | 55.3 | 64.0 | 73.3 | 77.6 | 75.1 | 67.5 | 55.7 | 44.0 | 32.6 | 54.0 |
| Days Max Temp ≥ 90 °F | 0 | 0 | 0 | 1 | 2 | 10 | 17 | 11 | 5 | 0 | 0 | 0 | 46 |
| Days Max Temp ≤ 32 °F | 11 | 7 | 1 | 0 | 0 | 0 | 0 | 0 | 0 | 0 | 1 | 7 | 27 |
| Days Min Temp ≤ 32 °F | 28 | 23 | 17 | 5 | 0 | 0 | 0 | 0 | 0 | 6 | 15 | 25 | 119 |
| Days Min Temp ≤ 0 °F | 4 | 2 | 0 | 0 | 0 | 0 | 0 | 0 | 0 | 0 | 0 | 1 | 7 |
| Heating Degree Days | 1166 | 929 | 668 | 314 | 113 | 11 | 2 | 3 | 66 | 303 | 625 | 999 | 5199 |
| Cooling Degree Days | 0 | 0 | 6 | 32 | 91 | 280 | 416 | 342 | 155 | 20 | 1 | 0 | 1343 |
| Total Precipitation (") | 1.76 | 1.85 | 3.28 | 4.06 | 3.91 | 3.49 | 3.48 | 3.40 | 3.79 | 3.10 | 3.37 | 3.27 | 38.76 |
| Days ≥ 0.1" Precip | 4 | 4 | 6 | 7 | 7 | 6 | 6 | 5 | 6 | 6 | 6 | 6 | 69 |
| Total Snowfall (") | 5.6 | 5.1 | 3.3 | 0.6 | 0.0 | 0.0 | 0.0 | 0.0 | 0.0 | 0.0 | 1.2 | 3.8 | 19.6 |
| Days ≥ 1" Snow Depth | na | na | *2* | 0 | 0 | 0 | 0 | 0 | 0 | 0 | *0* | na | na |

## FARMINGTON *St. Francois County* ELEVATION 932 ft LAT/LONG 37° 47 ' N / 90° 23 ' W

| | JAN | FEB | MAR | APR | MAY | JUN | JUL | AUG | SEP | OCT | NOV | DEC | YEAR |
|---|---|---|---|---|---|---|---|---|---|---|---|---|---|
| Maximum Temp °F | 39.4 | 44.6 | 55.7 | 67.9 | 75.5 | 83.7 | 88.4 | 86.6 | 79.0 | 68.2 | 55.2 | 43.9 | 65.7 |
| Minimum Temp °F | 19.1 | 22.5 | 32.7 | 43.3 | 50.7 | 59.9 | 64.3 | 61.5 | 54.3 | 41.7 | 33.5 | 24.3 | 42.3 |
| Mean Temp °F | 29.2 | 33.6 | 44.2 | 55.7 | 63.1 | 71.9 | 76.4 | 74.1 | 66.6 | 55.0 | 44.4 | 34.1 | 54.0 |
| Days Max Temp ≥ 90 °F | 0 | 0 | 0 | 0 | 0 | 6 | 14 | 10 | 3 | 0 | 0 | 0 | 33 |
| Days Max Temp ≤ 32 °F | 10 | 5 | 1 | 0 | 0 | 0 | 0 | 0 | 0 | 0 | 1 | 5 | 22 |
| Days Min Temp ≤ 32 °F | 27 | 22 | 17 | 5 | 0 | 0 | 0 | 0 | 0 | 6 | 15 | 24 | 116 |
| Days Min Temp ≤ 0 °F | 2 | 2 | 0 | 0 | 0 | 0 | 0 | 0 | 0 | 0 | 0 | 1 | 5 |
| Heating Degree Days | 1102 | 880 | 641 | 299 | 123 | 14 | 2 | 4 | 77 | 321 | 613 | 951 | 5027 |
| Cooling Degree Days | 0 | 0 | 3 | 23 | 65 | 223 | 367 | 300 | 131 | 15 | 0 | 0 | 1127 |
| Total Precipitation (") | 2.21 | 2.34 | 3.78 | 4.16 | 3.93 | 3.40 | 3.80 | 3.98 | 3.81 | 2.89 | 4.45 | 3.58 | 42.33 |
| Days ≥ 0.1" Precip | 4 | 4 | 7 | 7 | 7 | 6 | 6 | 5 | 6 | 5 | 7 | 5 | 69 |
| Total Snowfall (") | *3.2* | 5.0 | 2.3 | 0.1 | 0.0 | 0.0 | 0.0 | 0.0 | 0.0 | 0.0 | 0.7 | 2.4 | 13.7 |
| Days ≥ 1" Snow Depth | na | *5* | *2* | 0 | 0 | 0 | 0 | 0 | 0 | 0 | 0 | 2 | na |

## FESTUS *Jefferson County* ELEVATION 522 ft LAT/LONG 38° 15 ' N / 90° 25 ' W

| | JAN | FEB | MAR | APR | MAY | JUN | JUL | AUG | SEP | OCT | NOV | DEC | YEAR |
|---|---|---|---|---|---|---|---|---|---|---|---|---|---|
| Maximum Temp °F | *40.5* | *46.2* | *58.0* | *68.9* | *76.9* | *85.4* | 90.0 | *88.0* | *81.0* | *71.1* | *57.1* | *45.9* | 67.4 |
| Minimum Temp °F | *18.9* | *22.4* | *32.8* | *42.7* | *50.6* | *59.7* | 64.7 | *61.9* | *54.4* | *41.7* | *33.7* | *25.3* | 42.4 |
| Mean Temp °F | *29.7* | *34.3* | *45.4* | *55.8* | *63.7* | *72.6* | 77.4 | *75.0* | *67.7* | *56.4* | *45.4* | *35.6* | 54.9 |
| Days Max Temp ≥ 90 °F | *0* | *0* | *0* | *0* | *2* | *8* | *16* | *13* | *5* | *1* | *0* | *0* | 45 |
| Days Max Temp ≤ 32 °F | *9* | *5* | *1* | *0* | *0* | *0* | 0 | *0* | *0* | *0* | *1* | *4* | 20 |
| Days Min Temp ≤ 32 °F | *27* | *23* | *16* | *6* | *0* | *0* | 0 | *0* | *0* | *7* | *15* | *24* | 118 |
| Days Min Temp ≤ 0 °F | *3* | *1* | *0* | *0* | *0* | *0* | 0 | *0* | *0* | *0* | *0* | *1* | 5 |
| Heating Degree Days | *1088* | *860* | *607* | *293* | *108* | *10* | *1* | *3* | *63* | *283* | *584* | *905* | 4805 |
| Cooling Degree Days | na | na | na | na | na | na | *418* | na | na | na | na | na | na |
| Total Precipitation (") | *2.09* | *2.09* | *3.68* | *4.45* | *3.86* | *3.31* | 3.16 | *3.40* | *3.45* | *2.64* | *3.66* | *3.36* | 39.15 |
| Days ≥ 0.1" Precip | *4* | *4* | *7* | *7* | *7* | *5* | 5 | *5* | *5* | *5* | *6* | *6* | 66 |
| Total Snowfall (") | *4.6* | na | *2.5* | *0.5* | *0.0* | *0.0* | 0.0 | *0.0* | *0.0* | *0.0* | *1.5* | na | na |
| Days ≥ 1" Snow Depth | na | na | *1* | *0* | *0* | *0* | 0 | *0* | *0* | *0* | *1* | na | na |

## FREDERICKTOWN *Madison County* ELEVATION 700 ft LAT/LONG 37° 33 ' N / 90° 17 ' W

| | JAN | FEB | MAR | APR | MAY | JUN | JUL | AUG | SEP | OCT | NOV | DEC | YEAR |
|---|---|---|---|---|---|---|---|---|---|---|---|---|---|
| Maximum Temp °F | 40.7 | 45.8 | 56.7 | 68.6 | 76.9 | 85.2 | 90.1 | 88.1 | 80.5 | 69.4 | 56.6 | 45.1 | 67.0 |
| Minimum Temp °F | 18.0 | 21.4 | 31.8 | 41.8 | 49.9 | 59.1 | 63.6 | 60.9 | 53.4 | 40.0 | 32.6 | 23.5 | 41.3 |
| Mean Temp °F | 29.3 | 33.6 | 44.3 | 55.2 | 63.4 | 72.2 | 76.9 | 74.5 | 67.0 | 54.7 | 44.6 | 34.3 | 54.2 |
| Days Max Temp ≥ 90 °F | 0 | 0 | 0 | 0 | 1 | 8 | 18 | 13 | 4 | 0 | 0 | 0 | 44 |
| Days Max Temp ≤ 32 °F | 8 | 5 | 1 | 0 | 0 | 0 | 0 | 0 | 0 | 0 | 0 | 4 | 18 |
| Days Min Temp ≤ 32 °F | 28 | 23 | 17 | 7 | 1 | 0 | 0 | 0 | 0 | 9 | 16 | 24 | 125 |
| Days Min Temp ≤ 0 °F | 3 | 2 | 0 | 0 | 0 | 0 | 0 | 0 | 0 | 0 | 0 | 1 | 6 |
| Heating Degree Days | 1098 | 880 | 639 | 309 | 122 | 14 | 2 | 6 | 75 | 329 | 605 | 945 | 5024 |
| Cooling Degree Days | 0 | 0 | 3 | 22 | 79 | 240 | 394 | 319 | 144 | 16 | 1 | 0 | 1218 |
| Total Precipitation (") | 2.49 | 2.50 | 3.96 | 4.51 | 4.48 | 3.56 | 3.37 | 4.26 | 3.76 | 3.09 | 4.67 | 4.00 | 44.65 |
| Days ≥ 0.1" Precip | 5 | 5 | 7 | 7 | 7 | 5 | 5 | 5 | 5 | 5 | 6 | 6 | 68 |
| Total Snowfall (") | 4.8 | 4.3 | 2.2 | 0.2 | 0.0 | 0.0 | 0.0 | 0.0 | 0.0 | 0.1 | 0.8 | 2.5 | 14.9 |
| Days ≥ 1" Snow Depth | 8 | 6 | 2 | 0 | 0 | 0 | 0 | 0 | 0 | 0 | 0 | 3 | 19 |

## FREEDOM *Osage County* ELEVATION 751 ft LAT/LONG 38° 28 ' N / 91° 42 ' W

| | JAN | FEB | MAR | APR | MAY | JUN | JUL | AUG | SEP | OCT | NOV | DEC | YEAR |
|---|---|---|---|---|---|---|---|---|---|---|---|---|---|
| Maximum Temp °F | 40.5 | 46.2 | 58.0 | 69.2 | 76.7 | 85.1 | 90.3 | 88.5 | 80.7 | 69.9 | 56.3 | 45.3 | 67.2 |
| Minimum Temp °F | 19.3 | 23.3 | 33.1 | 43.5 | 51.4 | 60.7 | 65.1 | 62.5 | 55.2 | 43.6 | 34.5 | 24.8 | 43.1 |
| Mean Temp °F | 29.9 | 34.8 | 45.6 | 56.4 | 64.1 | 72.9 | 77.7 | 75.5 | 68.0 | 56.8 | 45.4 | 35.1 | 55.2 |
| Days Max Temp ≥ 90 °F | 0 | 0 | 0 | 0 | 1 | 8 | 17 | 14 | 5 | 0 | 0 | 0 | 45 |
| Days Max Temp ≤ 32 °F | 9 | 5 | 1 | 0 | 0 | 0 | 0 | 0 | 0 | 0 | 1 | 4 | 20 |
| Days Min Temp ≤ 32 °F | 27 | 22 | 16 | 4 | 0 | 0 | 0 | 0 | 0 | 5 | 13 | 23 | 110 |
| Days Min Temp ≤ 0 °F | 3 | 2 | 0 | 0 | 0 | 0 | 0 | 0 | 0 | 0 | 0 | 1 | 6 |
| Heating Degree Days | 1080 | 847 | 601 | 283 | 104 | 10 | 1 | 4 | 63 | 275 | 583 | 920 | 4771 |
| Cooling Degree Days | 0 | 0 | 6 | 28 | 74 | 258 | 410 | 346 | 157 | 22 | 3 | 0 | 1304 |
| Total Precipitation (") | 1.50 | 1.92 | 3.03 | 3.96 | 4.72 | 4.22 | 3.36 | 3.71 | 4.34 | 3.62 | 3.58 | 2.94 | 40.90 |
| Days ≥ 0.1" Precip | 4 | 4 | 6 | 7 | 7 | 6 | 6 | 5 | 6 | 6 | 6 | 5 | 68 |
| Total Snowfall (") | ***3.0*** | ***5.4*** | ***2.1*** | 0.1 | 0.0 | 0.0 | 0.0 | 0.0 | 0.0 | 0.0 | 1.3 | na | na |
| Days ≥ 1" Snow Depth | na | na | na | 0 | 0 | 0 | 0 | 0 | 0 | 0 | 0 | na | na |

## FULTON *Callaway County* ELEVATION 702 ft LAT/LONG 38° 51 ' N / 91° 57 ' W

| | JAN | FEB | MAR | APR | MAY | JUN | JUL | AUG | SEP | OCT | NOV | DEC | YEAR |
|---|---|---|---|---|---|---|---|---|---|---|---|---|---|
| Maximum Temp °F | 36.5 | 41.6 | 53.5 | 65.5 | 73.9 | 82.7 | 88.0 | 86.3 | 79.0 | 67.4 | 53.5 | 41.3 | 64.1 |
| Minimum Temp °F | 17.5 | 21.1 | 31.4 | 42.7 | 51.4 | 61.0 | 65.9 | 63.1 | 55.7 | 43.5 | 33.2 | 22.9 | 42.4 |
| Mean Temp °F | 27.0 | 31.4 | 42.5 | 54.1 | 62.7 | 71.9 | 76.9 | 74.7 | 67.4 | 55.5 | 43.4 | 32.1 | 53.3 |
| Days Max Temp ≥ 90 °F | 0 | 0 | 0 | 0 | 0 | 4 | 13 | 11 | 3 | 0 | 0 | 0 | 31 |
| Days Max Temp ≤ 32 °F | 12 | 8 | 2 | 0 | 0 | 0 | 0 | 0 | 0 | 0 | 1 | 7 | 30 |
| Days Min Temp ≤ 32 °F | 28 | 24 | 18 | 4 | 0 | 0 | 0 | 0 | 0 | 4 | 15 | 26 | 119 |
| Days Min Temp ≤ 0 °F | 4 | 2 | 0 | 0 | 0 | 0 | 0 | 0 | 0 | 0 | 0 | 1 | 7 |
| Heating Degree Days | 1171 | 944 | 695 | 340 | 131 | 13 | 2 | 4 | 69 | 307 | 643 | 1011 | 5330 |
| Cooling Degree Days | 0 | 0 | 4 | 22 | 61 | 232 | 380 | 315 | 139 | 13 | 1 | 0 | 1167 |
| Total Precipitation (") | 1.61 | 1.84 | 2.95 | 4.05 | 4.71 | 4.00 | 3.94 | 3.49 | 4.24 | 3.35 | 3.69 | 2.76 | 40.63 |
| Days ≥ 0.1" Precip | 4 | 4 | 6 | 8 | 7 | 6 | 6 | 6 | 6 | 6 | 6 | 5 | 70 |
| Total Snowfall (") | 5.5 | 6.5 | 3.6 | 0.5 | 0.0 | 0.0 | 0.0 | 0.0 | 0.0 | 0.0 | 1.6 | 4.5 | 22.2 |
| Days ≥ 1" Snow Depth | 9 | 7 | 3 | 0 | 0 | 0 | 0 | 0 | 0 | 0 | 1 | 5 | 25 |

## GALENA 1 SW *Stone County* ELEVATION 1100 ft LAT/LONG 36° 48 ' N / 93° 28 ' W

| | JAN | FEB | MAR | APR | MAY | JUN | JUL | AUG | SEP | OCT | NOV | DEC | YEAR |
|---|---|---|---|---|---|---|---|---|---|---|---|---|---|
| Maximum Temp °F | 44.4 | 50.0 | 60.5 | 71.4 | 76.9 | 84.4 | 89.5 | 88.5 | 80.5 | 71.1 | 58.6 | 48.6 | 68.7 |
| Minimum Temp °F | 20.7 | 24.0 | 33.5 | 42.7 | 51.2 | 60.0 | 64.5 | 62.8 | 56.0 | 43.4 | 33.6 | 24.8 | 43.1 |
| Mean Temp °F | 32.5 | 37.0 | 47.0 | 57.1 | 64.0 | 72.2 | 77.1 | 75.6 | 68.3 | 57.3 | 46.1 | 36.7 | 55.9 |
| Days Max Temp ≥ 90 °F | 0 | 0 | 0 | 0 | 0 | 5 | 17 | 15 | 4 | 0 | 0 | 0 | 41 |
| Days Max Temp ≤ 32 °F | 6 | 3 | 0 | 0 | 0 | 0 | 0 | 0 | 0 | 0 | 0 | 3 | 12 |
| Days Min Temp ≤ 32 °F | 27 | 22 | 16 | 6 | 0 | 0 | 0 | 0 | 0 | 5 | ***14*** | 24 | 114 |
| Days Min Temp ≤ 0 °F | 2 | 1 | 0 | 0 | 0 | 0 | 0 | 0 | 0 | 0 | 0 | 1 | 4 |
| Heating Degree Days | 1000 | 783 | 556 | 257 | 98 | 7 | 1 | 3 | 56 | 254 | 563 | 871 | 4449 |
| Cooling Degree Days | 0 | 0 | 5 | 25 | 76 | 242 | 382 | 349 | 162 | 18 | 2 | 0 | 1261 |
| Total Precipitation (") | 2.10 | 2.28 | 3.91 | 4.68 | 4.19 | 4.68 | 3.71 | 3.40 | 4.64 | 3.81 | 4.55 | 3.49 | 45.44 |
| Days ≥ 0.1" Precip | 4 | 4 | 6 | 7 | 7 | 7 | 5 | 5 | 6 | 5 | 6 | 5 | 67 |
| Total Snowfall (") | na | na | 1.4 | 0.1 | 0.0 | 0.0 | 0.0 | 0.0 | 0.0 | 0.0 | 0.5 | ***1.5*** | na |
| Days ≥ 1" Snow Depth | na | na | ***0*** | 0 | 0 | 0 | 0 | 0 | 0 | 0 | 0 | ***0*** | na |

## GRANT CITY *Worth County* ELEVATION 1132 ft LAT/LONG 40° 29 ' N / 94° 25 ' W

| | JAN | FEB | MAR | APR | MAY | JUN | JUL | AUG | SEP | OCT | NOV | DEC | YEAR |
|---|---|---|---|---|---|---|---|---|---|---|---|---|---|
| Maximum Temp °F | 32.6 | 38.6 | 51.9 | 65.4 | 74.8 | 83.8 | 88.1 | 85.8 | 77.8 | 66.2 | 49.9 | 37.3 | 62.7 |
| Minimum Temp °F | 13.7 | 18.2 | 29.7 | 41.7 | 51.9 | 60.9 | 65.5 | 62.7 | 54.6 | 42.9 | 30.9 | 19.8 | 41.0 |
| Mean Temp °F | 23.2 | 28.4 | 40.8 | 53.6 | 63.4 | 72.4 | 76.9 | 74.3 | 66.2 | 54.6 | 40.5 | 28.6 | 51.9 |
| Days Max Temp ≥ 90 °F | 0 | 0 | 0 | 0 | 0 | 6 | 13 | 9 | 3 | 0 | 0 | 0 | 31 |
| Days Max Temp ≤ 32 °F | 14 | 9 | 2 | 0 | 0 | 0 | 0 | 0 | 0 | 0 | 2 | 9 | 36 |
| Days Min Temp ≤ 32 °F | 30 | 25 | 19 | 5 | 0 | 0 | 0 | 0 | 0 | 5 | 17 | 28 | 129 |
| Days Min Temp ≤ 0 °F | 6 | 4 | 0 | 0 | 0 | 0 | 0 | 0 | 0 | 0 | 0 | 2 | 12 |
| Heating Degree Days | 1290 | 1025 | 744 | 353 | 117 | 10 | 1 | 4 | 81 | 332 | 730 | 1122 | 5809 |
| Cooling Degree Days | 0 | 0 | 3 | 24 | 72 | 237 | ***372*** | 298 | ***140*** | 10 | 0 | 0 | 1156 |
| Total Precipitation (") | 0.89 | 0.92 | 2.22 | 3.13 | 3.98 | 4.32 | 4.76 | 3.64 | 4.19 | 3.08 | 2.09 | 1.44 | 34.66 |
| Days ≥ 0.1" Precip | 3 | 3 | 5 | 6 | 8 | 7 | 7 | 6 | 6 | 5 | 4 | 3 | 63 |
| Total Snowfall (") | 6.4 | 5.0 | 4.0 | 1.1 | 0.1 | 0.0 | 0.0 | 0.0 | 0.0 | 0.2 | 2.4 | 5.0 | 24.2 |
| Days ≥ 1" Snow Depth | 15 | 11 | 4 | 1 | 0 | 0 | 0 | 0 | 0 | 0 | 2 | 8 | 41 |

## GREENVILLE 6 N *Wayne County* ELEVATION 410 ft LAT/LONG 37° 11 ' N / 90° 28 ' W

| | JAN | FEB | MAR | APR | MAY | JUN | JUL | AUG | SEP | OCT | NOV | DEC | YEAR |
|---|---|---|---|---|---|---|---|---|---|---|---|---|---|
| Maximum Temp °F | 43.2 | 48.8 | 58.9 | 70.5 | 77.9 | 85.6 | 90.2 | 88.2 | 81.0 | 70.9 | 58.4 | 46.9 | 68.4 |
| Minimum Temp °F | 19.1 | 22.7 | 32.0 | 41.8 | 50.0 | 58.8 | 63.5 | 61.1 | 54.0 | 40.2 | 32.4 | 23.2 | 41.6 |
| Mean Temp °F | 31.2 | 35.8 | 45.5 | 56.1 | 64.0 | 72.3 | 76.9 | 74.6 | 67.6 | 55.6 | 45.4 | 35.1 | 55.0 |
| Days Max Temp ≥ 90 °F | 0 | 0 | 0 | 0 | 1 | 8 | 18 | 13 | 4 | 0 | 0 | 0 | 44 |
| Days Max Temp ≤ 32 °F | 6 | 3 | 0 | 0 | 0 | 0 | 0 | 0 | 0 | 0 | 0 | 3 | 12 |
| Days Min Temp ≤ 32 °F | 26 | 22 | 17 | 7 | 1 | 0 | 0 | 0 | 0 | 9 | 17 | 24 | 123 |
| Days Min Temp ≤ 0 °F | 2 | 1 | 0 | 0 | 0 | 0 | 0 | 0 | 0 | 0 | 0 | 1 | 4 |
| Heating Degree Days | 1042 | 817 | 602 | 283 | 104 | 11 | 1 | 3 | 65 | 305 | 583 | 922 | 4738 |
| Cooling Degree Days | 0 | 0 | 3 | 19 | 70 | 231 | 382 | 319 | 148 | 16 | 2 | 0 | 1190 |
| Total Precipitation (") | 3.11 | 3.03 | 4.45 | 4.55 | 4.43 | 3.36 | 4.11 | 3.59 | 3.91 | 3.17 | 4.50 | 4.59 | 46.80 |
| Days ≥ 0.1" Precip | 5 | 6 | 7 | 7 | 7 | 6 | 6 | 6 | 6 | 5 | 6 | 7 | 74 |
| Total Snowfall (") | na | ***3.7*** | 1.5 | 0.0 | 0.0 | 0.0 | 0.0 | 0.0 | 0.0 | 0.1 | 0.7 | ***1.4*** | na |
| Days ≥ 1" Snow Depth | na | ***1*** | 1 | 0 | 0 | 0 | 0 | 0 | 0 | 0 | 0 | ***1*** | na |

## HAMILTON 2 W *Caldwell County* ELEVATION 902 ft LAT/LONG 39° 45 ' N / 94° 2 ' W

| | JAN | FEB | MAR | APR | MAY | JUN | JUL | AUG | SEP | OCT | NOV | DEC | YEAR |
|---|---|---|---|---|---|---|---|---|---|---|---|---|---|
| Maximum Temp °F | 34.1 | 39.1 | 52.6 | 64.4 | 74.1 | 83.3 | 88.3 | 86.3 | 78.3 | 66.9 | 51.5 | 39.0 | 63.2 |
| Minimum Temp °F | 12.6 | 16.7 | 28.8 | 40.1 | 50.2 | 59.5 | 64.4 | 61.6 | 52.8 | 40.0 | 29.5 | 18.3 | 39.5 |
| Mean Temp °F | 23.4 | 27.9 | 40.7 | 52.3 | 62.1 | 71.4 | 76.3 | 73.9 | 65.6 | 53.5 | 40.5 | 28.7 | 51.4 |
| Days Max Temp ≥ 90 °F | 0 | 0 | 0 | 0 | 0 | 6 | 14 | 11 | 3 | 0 | 0 | 0 | 34 |
| Days Max Temp ≤ 32 °F | 13 | 9 | 2 | 0 | 0 | 0 | 0 | 0 | 0 | 0 | 2 | 9 | 35 |
| Days Min Temp ≤ 32 °F | 30 | 26 | 20 | 8 | 1 | 0 | 0 | 0 | 1 | 7 | 19 | 28 | 140 |
| Days Min Temp ≤ 0 °F | 6 | 4 | 0 | 0 | 0 | 0 | 0 | 0 | 0 | 0 | 0 | 3 | 13 |
| Heating Degree Days | 1284 | 1041 | 748 | 390 | 144 | 17 | 3 | 7 | 96 | 364 | 728 | 1121 | 5943 |
| Cooling Degree Days | 0 | 0 | 2 | 21 | 50 | 213 | 353 | 291 | 119 | 9 | 0 | 0 | 1058 |
| Total Precipitation (") | 1.21 | 0.95 | 2.56 | 3.78 | 4.23 | 4.11 | 4.10 | 3.72 | 4.86 | 3.26 | 2.12 | 1.68 | 36.58 |
| Days ≥ 0.1" Precip | 3 | 3 | 5 | 7 | 8 | 7 | 6 | 6 | 6 | 6 | 4 | 3 | 64 |
| Total Snowfall (") | ***3.4*** | ***3.5*** | ***1.9*** | 0.1 | 0.0 | 0.0 | 0.0 | 0.0 | 0.0 | 0.0 | 0.5 | na | na |
| Days ≥ 1" Snow Depth | na | na | na | 0 | 0 | 0 | 0 | 0 | 0 | 0 | 0 | na | na |

## HANNIBAL WATER WORKS *Marion County* ELEVATION 751 ft LAT/LONG 39° 41 ' N / 91° 20 ' W

| | JAN | FEB | MAR | APR | MAY | JUN | JUL | AUG | SEP | OCT | NOV | DEC | YEAR |
|---|---|---|---|---|---|---|---|---|---|---|---|---|---|
| Maximum Temp °F | 33.5 | 38.9 | 51.3 | 64.4 | 73.9 | 82.9 | 87.1 | 84.5 | 77.4 | 66.2 | 51.5 | 38.4 | 62.5 |
| Minimum Temp °F | 16.0 | 20.1 | 31.5 | 43.5 | 53.2 | 62.1 | 66.5 | 64.0 | 56.0 | 44.2 | 33.5 | 22.0 | 42.7 |
| Mean Temp °F | 24.7 | 29.5 | 41.4 | 54.0 | 63.6 | 72.5 | 76.8 | 74.3 | 66.7 | 55.2 | 42.5 | 30.2 | 52.6 |
| Days Max Temp ≥ 90 °F | 0 | 0 | 0 | 0 | 1 | 5 | 12 | 7 | 2 | 0 | 0 | 0 | 27 |
| Days Max Temp ≤ 32 °F | 14 | 9 | 2 | 0 | 0 | 0 | 0 | 0 | 0 | 0 | 2 | 9 | 36 |
| Days Min Temp ≤ 32 °F | 28 | 24 | 18 | 4 | 0 | 0 | 0 | 0 | 0 | 3 | 14 | 26 | 117 |
| Days Min Temp ≤ 0 °F | 5 | 2 | 0 | 0 | 0 | 0 | 0 | 0 | 0 | 0 | 0 | 2 | 9 |
| Heating Degree Days | 1241 | 995 | 727 | 346 | 123 | 12 | 1 | 5 | 75 | 317 | 669 | 1072 | 5583 |
| Cooling Degree Days | 0 | 0 | 5 | 25 | 83 | 259 | 389 | 325 | 141 | 16 | 1 | 0 | 1244 |
| Total Precipitation (") | 1.66 | 1.82 | 3.12 | 3.94 | 4.56 | 3.82 | 4.64 | 4.01 | 4.04 | 3.41 | 3.31 | 2.71 | 41.04 |
| Days ≥ 0.1" Precip | 4 | 4 | 7 | 7 | 7 | 6 | 7 | 6 | 6 | 6 | 6 | 5 | 71 |
| Total Snowfall (") | 6.5 | 6.1 | 3.5 | 0.8 | 0.0 | 0.0 | 0.0 | 0.0 | 0.0 | 0.0 | 1.4 | 4.6 | 22.9 |
| Days ≥ 1" Snow Depth | 12 | 12 | 3 | 0 | 0 | 0 | 0 | 0 | 0 | 0 | 1 | 5 | 33 |

## HOUSTON 3 E *Texas County* ELEVATION 1293 ft LAT/LONG 37° 20 ' N / 91° 54 ' W

| | JAN | FEB | MAR | APR | MAY | JUN | JUL | AUG | SEP | OCT | NOV | DEC | YEAR |
|---|---|---|---|---|---|---|---|---|---|---|---|---|---|
| Maximum Temp °F | ***41.5*** | 46.3 | ***57.1*** | 69.2 | ***75.7*** | ***83.7*** | 89.5 | ***87.9*** | 80.3 | 70.0 | 57.2 | 46.8 | 67.1 |
| Minimum Temp °F | 18.6 | 22.1 | ***32.9*** | 42.8 | 50.8 | ***60.0*** | 64.1 | ***61.8*** | 54.5 | 42.1 | ***33.2*** | 23.8 | 42.2 |
| Mean Temp °F | ***30.2*** | 34.3 | ***45.0*** | 56.1 | ***63.2*** | ***71.9*** | 76.8 | ***74.8*** | 67.4 | 56.0 | ***45.1*** | 35.6 | 54.7 |
| Days Max Temp ≥ 90 °F | 0 | 0 | 0 | 0 | 0 | 4 | ***16*** | na | ***3*** | 0 | 0 | 0 | na |
| Days Max Temp ≤ 32 °F | 8 | 4 | 1 | 0 | 0 | 0 | 0 | 0 | 0 | 0 | 0 | ***3*** | 16 |
| Days Min Temp ≤ 32 °F | 27 | ***22*** | ***15*** | ***5*** | 1 | 0 | 0 | 0 | 0 | ***5*** | na | 24 | na |
| Days Min Temp ≤ 0 °F | 2 | 1 | 0 | 0 | 0 | 0 | 0 | 0 | 0 | 0 | 0 | 1 | 4 |
| Heating Degree Days | ***1079*** | 860 | ***614*** | 282 | ***123*** | ***13*** | 2 | ***4*** | 67 | ***295*** | ***595*** | ***908*** | 4842 |
| Cooling Degree Days | na | ***0*** | ***4*** | ***28*** | na | na | 391 | na | ***160*** | ***19*** | ***4*** | ***0*** | na |
| Total Precipitation (") | 2.08 | 2.34 | 3.84 | 4.85 | 4.36 | 3.44 | 3.33 | 4.02 | 4.27 | 3.68 | 4.34 | 3.37 | 43.92 |
| Days ≥ 0.1" Precip | 4 | 4 | 5 | 6 | 6 | 6 | 5 | 5 | 5 | 4 | 5 | 4 | 59 |
| Total Snowfall (") | na | na | na | 0.0 | 0.0 | 0.0 | 0.0 | 0.0 | 0.0 | 0.0 | 0.0 | na | na |
| Days ≥ 1" Snow Depth | na | na | ***0*** | 0 | 0 | 0 | 0 | 0 | 0 | 0 | 0 | na | na |

## JACKSON *Cape Girardeau County* ELEVATION 459 ft LAT/LONG 37° 23 ' N / 89° 40 ' W

| | JAN | FEB | MAR | APR | MAY | JUN | JUL | AUG | SEP | OCT | NOV | DEC | YEAR |
|---|---|---|---|---|---|---|---|---|---|---|---|---|---|
| Maximum Temp °F | 41.7 | 47.5 | 58.8 | 70.9 | 79.2 | 87.8 | 91.4 | 89.3 | 82.4 | 71.7 | 57.6 | 45.9 | 68.7 |
| Minimum Temp °F | 22.7 | 26.5 | 36.1 | 46.3 | 54.6 | 63.7 | 67.8 | 65.2 | 58.2 | 45.7 | 37.2 | 27.6 | 46.0 |
| Mean Temp °F | 32.2 | 37.0 | 47.4 | 58.6 | 66.8 | 75.8 | 79.6 | 77.2 | 70.3 | 58.7 | 47.4 | 36.8 | 57.3 |
| Days Max Temp ≥ 90 °F | 0 | 0 | 0 | 0 | 2 | 13 | 20 | 15 | 5 | 0 | 0 | 0 | 55 |
| Days Max Temp ≤ 32 °F | 8 | 3 | 0 | 0 | 0 | 0 | 0 | 0 | 0 | 0 | 0 | 3 | 14 |
| Days Min Temp ≤ 32 °F | 25 | 20 | 13 | 3 | 0 | 0 | 0 | 0 | 0 | 3 | 11 | 21 | 96 |
| Days Min Temp ≤ 0 °F | 2 | 0 | 0 | 0 | 0 | 0 | 0 | 0 | 0 | 0 | 0 | 1 | 3 |
| Heating Degree Days | 1010 | 784 | 541 | 221 | 63 | 2 | 0 | 1 | 34 | 225 | 523 | 868 | 4272 |
| Cooling Degree Days | 0 | 0 | 4 | 38 | 130 | 346 | 480 | 406 | 203 | 33 | 2 | 0 | 1642 |
| Total Precipitation (") | 2.87 | 3.22 | 4.45 | 4.67 | 5.17 | 3.70 | 3.74 | 3.64 | 3.87 | 3.38 | 4.64 | 4.00 | 47.35 |
| Days ≥ 0.1" Precip | 5 | 5 | 7 | 8 | 8 | 6 | 6 | 6 | 6 | 5 | 7 | 7 | 76 |
| Total Snowfall (") | na | *4.1* | *1.8* | 0.0 | 0.0 | 0.0 | 0.0 | 0.0 | 0.0 | 0.1 | 0.5 | na | na |
| Days ≥ 1" Snow Depth | na | *1* | *0* | 0 | 0 | 0 | 0 | 0 | 0 | 0 | 0 | *1* | na |

## JEFFERSON CITY WATER *Cole County* ELEVATION 669 ft LAT/LONG 38° 35 ' N / 92° 11 ' W

| | JAN | FEB | MAR | APR | MAY | JUN | JUL | AUG | SEP | OCT | NOV | DEC | YEAR |
|---|---|---|---|---|---|---|---|---|---|---|---|---|---|
| Maximum Temp °F | 39.7 | 45.1 | 56.6 | 68.2 | 76.4 | 84.8 | 90.2 | 88.4 | 81.0 | 69.7 | 55.8 | 44.3 | 66.7 |
| Minimum Temp °F | 17.7 | 21.2 | 31.3 | 42.3 | 51.3 | 60.8 | 65.5 | 63.0 | 54.7 | 42.5 | 32.7 | 22.8 | 42.2 |
| Mean Temp °F | 28.7 | 33.2 | 44.0 | 55.3 | 63.9 | 72.8 | 77.9 | 75.7 | 67.9 | 56.2 | 44.3 | 33.6 | 54.5 |
| Days Max Temp ≥ 90 °F | 0 | 0 | 0 | 0 | 1 | 7 | 18 | 14 | 5 | 0 | 0 | 0 | 45 |
| Days Max Temp ≤ 32 °F | 9 | 6 | 1 | 0 | 0 | 0 | 0 | 1 | 0 | 0 | 1 | 5 | 23 |
| Days Min Temp ≤ 32 °F | 28 | 24 | 18 | 5 | 0 | 0 | 0 | 0 | 0 | 5 | 16 | 25 | 121 |
| Days Min Temp ≤ 0 °F | 3 | 2 | 0 | 0 | 0 | 0 | 0 | 0 | 0 | 0 | 0 | 1 | 6 |
| Heating Degree Days | 1116 | 892 | 647 | 311 | 106 | 9 | 1 | 2 | 62 | 291 | 616 | 968 | 5021 |
| Cooling Degree Days | 0 | 0 | 4 | 26 | 71 | 257 | 414 | 345 | 152 | 18 | 1 | 0 | 1288 |
| Total Precipitation (") | 1.46 | 1.79 | 2.98 | 3.91 | 4.68 | 4.40 | 3.61 | 3.25 | 4.08 | 3.59 | 3.45 | 2.74 | 39.94 |
| Days ≥ 0.1" Precip | 4 | 4 | 6 | 7 | 8 | 7 | 6 | 5 | 6 | 6 | 6 | 5 | 70 |
| Total Snowfall (") | 4.4 | 4.4 | 2.3 | 0.2 | 0.0 | 0.0 | 0.0 | 0.0 | 0.0 | 0.0 | 1.1 | 2.7 | 15.1 |
| Days ≥ 1" Snow Depth | *5* | *4* | *1* | 0 | 0 | 0 | 0 | 0 | 0 | 0 | 0 | na | na |

## JOPLIN MUNICIPAL AP *Jasper County* ELEVATION 988 ft LAT/LONG 37° 9 ' N / 94° 30 ' W

| | JAN | FEB | MAR | APR | MAY | JUN | JUL | AUG | SEP | OCT | NOV | DEC | YEAR |
|---|---|---|---|---|---|---|---|---|---|---|---|---|---|
| Maximum Temp °F | 41.9 | 47.4 | 58.2 | 69.4 | 76.4 | 84.8 | 90.2 | 88.6 | 80.4 | 70.2 | 56.6 | 46.4 | 67.5 |
| Minimum Temp °F | 23.3 | 27.6 | 37.2 | 47.1 | 55.7 | 64.9 | 69.8 | 67.4 | 59.5 | 47.9 | 37.2 | 27.9 | 47.1 |
| Mean Temp °F | 32.6 | 37.5 | 47.8 | 58.3 | 66.1 | 74.8 | 80.0 | 78.0 | 70.0 | 59.1 | 46.9 | 37.2 | 57.4 |
| Days Max Temp ≥ 90 °F | 0 | 0 | 0 | 0 | 0 | 7 | 18 | 14 | 4 | 0 | 0 | 0 | 43 |
| Days Max Temp ≤ 32 °F | 8 | 4 | 1 | 0 | 0 | 0 | 0 | 0 | 0 | 0 | 1 | 4 | 18 |
| Days Min Temp ≤ 32 °F | 25 | 19 | 11 | 2 | 0 | 0 | 0 | 0 | 0 | 2 | 10 | 21 | 90 |
| Days Min Temp ≤ 0 °F | 1 | 0 | 0 | 0 | 0 | 0 | 0 | 0 | 0 | 0 | 0 | 1 | 2 |
| Heating Degree Days | 995 | 769 | 535 | 236 | 74 | 5 | 0 | 1 | 47 | 223 | 539 | 855 | 4279 |
| Cooling Degree Days | 0 | 0 | 8 | 44 | 118 | 327 | 489 | 439 | 218 | 42 | 4 | 0 | 1689 |
| Total Precipitation (") | 1.70 | 2.24 | 3.42 | 4.30 | 4.55 | 5.27 | 3.28 | 4.06 | 5.62 | 3.94 | 3.86 | 2.83 | 45.07 |
| Days ≥ 0.1" Precip | 4 | 4 | 6 | 7 | 8 | 7 | 5 | 6 | 6 | 6 | 5 | 5 | 69 |
| Total Snowfall (") | 3.4 | 3.6 | 2.4 | 0.0 | 0.0 | 0.0 | 0.0 | 0.0 | 0.0 | 0.0 | 0.5 | 2.7 | 12.6 |
| Days ≥ 1" Snow Depth | 6 | 4 | 1 | 0 | 0 | 0 | 0 | 0 | 0 | 0 | 0 | 3 | 14 |

## KENNETT RADIO KBOA *Dunklin County* ELEVATION 269 ft LAT/LONG 36° 13 ' N / 90° 4 ' W

| | JAN | FEB | MAR | APR | MAY | JUN | JUL | AUG | SEP | OCT | NOV | DEC | YEAR |
|---|---|---|---|---|---|---|---|---|---|---|---|---|---|
| Maximum Temp °F | 44.2 | 49.8 | 60.5 | 72.0 | 80.4 | 89.2 | 92.8 | 90.4 | 83.9 | 73.7 | 60.2 | 49.0 | 70.5 |
| Minimum Temp °F | 25.5 | 29.5 | 38.6 | 48.1 | 56.8 | 65.2 | 69.2 | 66.2 | 59.2 | 46.8 | 38.7 | 30.4 | 47.8 |
| Mean Temp °F | 34.9 | 39.7 | 49.6 | 60.1 | 68.7 | 77.3 | 81.0 | 78.3 | 71.5 | 60.3 | 49.5 | 39.7 | 59.2 |
| Days Max Temp ≥ 90 °F | 0 | 0 | 0 | 0 | 3 | 16 | 23 | 18 | 7 | 1 | 0 | 0 | 68 |
| Days Max Temp ≤ 32 °F | 5 | 2 | 0 | 0 | 0 | 0 | 0 | 0 | 0 | 0 | 0 | 2 | 9 |
| Days Min Temp ≤ 32 °F | 24 | 18 | 10 | 2 | 0 | 0 | 0 | 0 | 0 | 2 | 9 | 19 | 84 |
| Days Min Temp ≤ 0 °F | 1 | 0 | 0 | 0 | 0 | 0 | 0 | 0 | 0 | 0 | 0 | 0 | 1 |
| Heating Degree Days | 928 | 708 | 478 | 187 | 41 | 1 | 0 | 0 | 29 | 191 | 463 | 778 | 3804 |
| Cooling Degree Days | 0 | 0 | 5 | 42 | 154 | 377 | 524 | 447 | 238 | 47 | 5 | 1 | 1840 |
| Total Precipitation (") | 3.67 | 3.80 | 4.70 | 5.26 | 5.57 | 4.08 | 3.40 | 2.95 | 3.91 | 3.57 | 4.74 | 4.70 | 50.35 |
| Days ≥ 0.1" Precip | 6 | 6 | 7 | 8 | 7 | 6 | 6 | 5 | 5 | 5 | 7 | 7 | 75 |
| Total Snowfall (") | 4.1 | 3.7 | 1.0 | 0.0 | 0.0 | 0.0 | 0.0 | 0.0 | 0.0 | 0.0 | 0.3 | 1.3 | 10.4 |
| Days ≥ 1" Snow Depth | *3* | 2 | 0 | 0 | 0 | 0 | 0 | 0 | 0 | 0 | 0 | 1 | 6 |

## KIRKSVILLE *Adair County* ELEVATION 969 ft LAT/LONG 40° 12' N / 92° 34' W

| | JAN | FEB | MAR | APR | MAY | JUN | JUL | AUG | SEP | OCT | NOV | DEC | YEAR |
|---|---|---|---|---|---|---|---|---|---|---|---|---|---|
| Maximum Temp °F | 32.8 | 38.2 | 51.0 | 64.2 | 73.4 | 82.1 | 86.8 | 84.5 | 76.7 | 65.5 | 50.4 | 37.5 | 61.9 |
| Minimum Temp °F | 15.0 | 19.4 | 30.2 | 41.8 | 51.4 | 60.8 | 65.2 | 62.6 | 54.8 | 43.8 | 31.9 | 20.9 | 41.5 |
| Mean Temp °F | 24.0 | 28.8 | 40.6 | 53.0 | 62.4 | 71.5 | 76.1 | 73.6 | 65.8 | 54.7 | 41.2 | 29.2 | 51.7 |
| Days Max Temp ≥ 90 °F | 0 | 0 | 0 | 0 | 0 | 3 | 11 | 8 | 1 | 0 | 0 | 0 | 23 |
| Days Max Temp ≤ 32 °F | 14 | 10 | 3 | 0 | 0 | 0 | 0 | 0 | 0 | 0 | 2 | 10 | 39 |
| Days Min Temp ≤ 32 °F | 28 | 24 | 19 | 5 | 0 | 0 | 0 | 0 | 0 | 4 | 16 | 27 | 123 |
| Days Min Temp ≤ 0 °F | 5 | 3 | 0 | 0 | 0 | 0 | 0 | 0 | 0 | 0 | 0 | 2 | 10 |
| Heating Degree Days | 1266 | 1015 | 752 | 369 | 140 | 15 | 2 | 7 | 88 | 331 | 708 | 1102 | 5795 |
| Cooling Degree Days | 0 | 0 | 3 | 21 | 61 | 217 | 351 | 288 | 125 | 13 | 0 | 0 | 1079 |
| Total Precipitation (") | 1.21 | 0.98 | 2.43 | 3.55 | 4.39 | 4.41 | 4.67 | 3.91 | 4.39 | 3.30 | 2.62 | 1.89 | 37.75 |
| Days ≥ 0.1" Precip | 3 | 3 | 5 | 7 | 8 | 6 | 6 | 5 | 6 | 6 | 5 | 4 | 64 |
| Total Snowfall (") | 6.5 | 5.1 | 3.5 | 0.8 | 0.0 | 0.0 | 0.0 | 0.0 | 0.0 | 0.1 | 1.9 | *3.5* | 21.4 |
| Days ≥ 1" Snow Depth | na | na | *1* | 0 | 0 | 0 | 0 | 0 | 0 | 0 | 1 | na | na |

## LAKESIDE *Miller County* ELEVATION 591 ft LAT/LONG 38° 12' N / 92° 37' W

| | JAN | FEB | MAR | APR | MAY | JUN | JUL | AUG | SEP | OCT | NOV | DEC | YEAR |
|---|---|---|---|---|---|---|---|---|---|---|---|---|---|
| Maximum Temp °F | 40.7 | 45.8 | 56.6 | 68.4 | 76.3 | 84.6 | 90.2 | 88.7 | 80.5 | 70.1 | 56.8 | 45.7 | 67.0 |
| Minimum Temp °F | 18.9 | 22.3 | 32.0 | 42.5 | 51.2 | 60.8 | 65.4 | 63.4 | 56.3 | 44.3 | 35.1 | 25.4 | 43.1 |
| Mean Temp °F | 30.0 | 34.3 | 44.3 | 55.5 | 63.7 | 72.7 | 77.9 | 76.1 | 68.5 | 57.3 | 46.0 | 35.5 | 55.2 |
| Days Max Temp ≥ 90 °F | 0 | 0 | 0 | 0 | 1 | 7 | 18 | 15 | 5 | 0 | 0 | 0 | 46 |
| Days Max Temp ≤ 32 °F | 9 | 5 | 1 | 0 | 0 | 0 | 0 | 0 | 0 | 0 | 1 | 4 | 20 |
| Days Min Temp ≤ 32 °F | 28 | 23 | 18 | 4 | 0 | 0 | 0 | 0 | 0 | 3 | 13 | 24 | 113 |
| Days Min Temp ≤ 0 °F | 2 | 1 | 0 | 0 | 0 | 0 | 0 | 0 | 0 | 0 | 0 | 1 | 4 |
| Heating Degree Days | 1078 | 861 | 638 | 299 | 107 | 9 | 1 | 1 | 56 | 258 | 565 | 905 | 4778 |
| Cooling Degree Days | 0 | 0 | 4 | 24 | 70 | 256 | 421 | 369 | 172 | 21 | 3 | 0 | 1340 |
| Total Precipitation (") | 1.63 | 1.78 | 3.11 | 4.17 | 4.61 | 3.99 | 3.79 | 3.79 | 4.53 | 3.81 | 3.79 | 2.60 | 41.60 |
| Days ≥ 0.1" Precip | 4 | 4 | 6 | 8 | 7 | 7 | 5 | 5 | 6 | 6 | 6 | 5 | 69 |
| Total Snowfall (") | *2.9* | *3.7* | 1.3 | 0.1 | 0.0 | 0.0 | 0.0 | 0.0 | 0.0 | 0.0 | 0.7 | *2.2* | 10.9 |
| Days ≥ 1" Snow Depth | 6 | 5 | 1 | 0 | 0 | 0 | 0 | 0 | 0 | 0 | 0 | 3 | 15 |

## LAMAR *Barton County* ELEVATION 981 ft LAT/LONG 37° 30' N / 94° 16' W

| | JAN | FEB | MAR | APR | MAY | JUN | JUL | AUG | SEP | OCT | NOV | DEC | YEAR |
|---|---|---|---|---|---|---|---|---|---|---|---|---|---|
| Maximum Temp °F | 40.3 | 45.7 | 56.8 | 68.0 | 75.2 | 83.6 | 89.3 | 88.2 | 79.9 | 69.9 | 56.2 | 45.1 | 66.5 |
| Minimum Temp °F | 20.1 | 24.2 | 33.9 | 44.4 | 53.6 | 62.9 | 67.7 | 65.0 | 57.2 | 45.5 | 34.6 | 25.2 | 44.5 |
| Mean Temp °F | 30.2 | 35.0 | 45.4 | 56.2 | 64.5 | 73.3 | 78.5 | 76.6 | 68.6 | 57.7 | 45.5 | 35.2 | 55.6 |
| Days Max Temp ≥ 90 °F | 0 | 0 | 0 | 0 | 0 | 5 | 17 | 15 | 4 | 0 | 0 | 0 | 41 |
| Days Max Temp ≤ 32 °F | 9 | 5 | 1 | 0 | 0 | 0 | 0 | 0 | 0 | 0 | 1 | 5 | 21 |
| Days Min Temp ≤ 32 °F | 28 | 22 | 14 | 3 | 0 | 0 | 0 | 0 | 0 | 2 | 13 | 24 | 106 |
| Days Min Temp ≤ 0 °F | 2 | 1 | 0 | 0 | 0 | 0 | 0 | 0 | 0 | 0 | 0 | 1 | 4 |
| Heating Degree Days | 1072 | 841 | 605 | 282 | 97 | 9 | 1 | 3 | 60 | 251 | 583 | 917 | 4721 |
| Cooling Degree Days | 0 | 0 | 4 | 25 | 81 | 272 | 430 | 386 | 185 | 27 | 3 | 0 | 1413 |
| Total Precipitation (") | 1.78 | 2.12 | 3.86 | 4.56 | 4.93 | 5.01 | 4.21 | 3.94 | 5.42 | 4.17 | 4.10 | 2.94 | 47.04 |
| Days ≥ 0.1" Precip | 4 | 4 | 6 | 7 | 8 | 8 | 5 | 5 | 6 | 6 | 5 | 5 | 69 |
| Total Snowfall (") | *3.8* | 3.6 | 2.6 | 0.1 | 0.0 | 0.0 | 0.0 | 0.0 | 0.0 | 0.0 | 0.9 | 2.9 | 13.9 |
| Days ≥ 1" Snow Depth | na | na | *0* | 0 | 0 | 0 | 0 | 0 | 0 | 0 | 0 | *1* | na |

## LEBANON 2 W *Laclede County* ELEVATION 1260 ft LAT/LONG 37° 39' N / 92° 41' W

| | JAN | FEB | MAR | APR | MAY | JUN | JUL | AUG | SEP | OCT | NOV | DEC | YEAR |
|---|---|---|---|---|---|---|---|---|---|---|---|---|---|
| Maximum Temp °F | 41.0 | 46.7 | 58.1 | 69.4 | 76.6 | 84.5 | 89.6 | 88.3 | 80.0 | 69.6 | 55.9 | 45.6 | 67.1 |
| Minimum Temp °F | 21.3 | 25.3 | 35.0 | 45.2 | 53.3 | 62.0 | 66.6 | 64.8 | 57.4 | 46.0 | 35.8 | 26.5 | 44.9 |
| Mean Temp °F | 31.2 | 36.0 | 46.6 | 57.4 | 65.0 | 73.3 | 78.1 | 76.6 | 68.7 | 57.8 | 45.9 | 36.0 | 56.1 |
| Days Max Temp ≥ 90 °F | 0 | 0 | 0 | 0 | 0 | 6 | 17 | 14 | 4 | 0 | 0 | 0 | 41 |
| Days Max Temp ≤ 32 °F | 8 | 5 | 1 | 0 | 0 | 0 | 0 | 0 | 0 | 0 | 1 | 4 | 19 |
| Days Min Temp ≤ 32 °F | 26 | 21 | 14 | 3 | 0 | 0 | 0 | 0 | 0 | 3 | 12 | 22 | 101 |
| Days Min Temp ≤ 0 °F | 2 | 1 | 0 | 0 | 0 | 0 | 0 | 0 | 0 | 0 | 0 | 1 | 4 |
| Heating Degree Days | 1040 | 812 | 569 | 255 | 88 | 6 | 1 | 2 | 53 | 247 | 569 | 891 | 4533 |
| Cooling Degree Days | 0 | 0 | 4 | 35 | 93 | 269 | 422 | 384 | 175 | 26 | 2 | 0 | 1410 |
| Total Precipitation (") | 1.82 | 2.07 | 3.44 | 4.25 | 4.46 | 4.11 | 3.69 | 3.52 | 4.56 | 4.04 | 4.25 | 2.76 | 42.97 |
| Days ≥ 0.1" Precip | 4 | 4 | 6 | 7 | 7 | 7 | 5 | 5 | 6 | 6 | 6 | 5 | 68 |
| Total Snowfall (") | 4.2 | 4.4 | 2.3 | 0.1 | 0.0 | 0.0 | 0.0 | 0.0 | 0.0 | 0.0 | 1.2 | 2.0 | 14.2 |
| Days ≥ 1" Snow Depth | 7 | 5 | 2 | 0 | 0 | 0 | 0 | 0 | 0 | 0 | 1 | 2 | 17 |

## LEES SUMMIT WILDLIFE *Jackson County* ELEVATION 1001 ft LAT/LONG 38° 53' N / 94° 20' W

| | JAN | FEB | MAR | APR | MAY | JUN | JUL | AUG | SEP | OCT | NOV | DEC | YEAR |
|---|---|---|---|---|---|---|---|---|---|---|---|---|---|
| Maximum Temp °F | 38.0 | 44.0 | 56.4 | 67.8 | 76.1 | 84.7 | 89.8 | 88.1 | 80.1 | 69.4 | 53.8 | 42.4 | 65.9 |
| Minimum Temp °F | 17.2 | 21.7 | 32.0 | 43.2 | 52.2 | 61.1 | 65.7 | 63.3 | 55.3 | 44.0 | 32.7 | 22.6 | 42.6 |
| Mean Temp °F | 27.6 | 32.9 | 44.2 | 55.5 | 64.2 | 72.9 | 77.8 | 75.7 | 67.8 | 56.7 | 43.2 | 32.5 | 54.3 |
| Days Max Temp ≥ 90 °F | 0 | 0 | 0 | 0 | 0 | 7 | 17 | 14 | 4 | 0 | 0 | 0 | 42 |
| Days Max Temp ≤ 32 °F | 10 | 6 | 1 | 0 | 0 | 0 | 0 | 0 | 0 | 0 | 1 | 6 | 24 |
| Days Min Temp ≤ 32 °F | 29 | 23 | 17 | 4 | 0 | 0 | 0 | 0 | 0 | 4 | 15 | 26 | 118 |
| Days Min Temp ≤ 0 °F | 3 | 2 | 0 | 0 | 0 | 0 | 0 | 0 | 0 | 0 | 0 | 2 | 7 |
| Heating Degree Days | 1154 | 901 | 640 | 302 | 98 | 10 | 1 | 4 | 68 | 275 | 648 | 1000 | 5101 |
| Cooling Degree Days | 0 | 0 | 2 | 27 | 67 | 256 | 396 | 341 | 157 | 16 | 1 | 0 | 1263 |
| Total Precipitation (") | 1.29 | 1.34 | 2.61 | 3.95 | 4.95 | 5.68 | 4.12 | 4.05 | 4.68 | 3.87 | 2.65 | 1.89 | 41.08 |
| Days ≥ 0.1" Precip | 3 | 3 | 5 | 7 | 8 | 7 | 6 | 6 | 6 | 6 | 5 | 4 | 66 |
| Total Snowfall (") | 5.8 | 5.5 | 2.3 | 0.3 | 0.0 | 0.0 | 0.0 | 0.0 | 0.0 | 0.0 | 0.7 | 2.5 | 17.1 |
| Days ≥ 1" Snow Depth | 9 | 7 | 1 | 0 | 0 | 0 | 0 | 0 | 0 | 0 | 0 | 4 | 21 |

## LEXINGTON 3 NE *Lafayette County* ELEVATION 825 ft LAT/LONG 39° 12' N / 93° 52' W

| | JAN | FEB | MAR | APR | MAY | JUN | JUL | AUG | SEP | OCT | NOV | DEC | YEAR |
|---|---|---|---|---|---|---|---|---|---|---|---|---|---|
| Maximum Temp °F | 35.4 | 40.5 | 53.9 | 66.2 | 75.3 | 84.0 | 88.8 | 86.4 | 78.9 | 67.6 | 52.7 | 41.1 | 64.2 |
| Minimum Temp °F | 16.7 | 21.0 | 32.1 | 43.9 | 53.5 | 62.8 | 67.3 | 64.3 | 56.1 | 44.6 | 33.1 | 23.1 | 43.2 |
| Mean Temp °F | 26.0 | 30.8 | 43.0 | 55.1 | 64.4 | 73.5 | 78.1 | 75.4 | 67.5 | 56.1 | 42.9 | 32.1 | 53.7 |
| Days Max Temp ≥ 90 °F | 0 | 0 | 0 | 0 | 1 | 7 | 15 | 11 | 4 | 0 | 0 | 0 | 38 |
| Days Max Temp ≤ 32 °F | 12 | 8 | 2 | 0 | 0 | 0 | 0 | 0 | 0 | 0 | 1 | 7 | 30 |
| Days Min Temp ≤ 32 °F | 29 | 23 | 16 | 4 | 0 | 0 | 0 | 0 | 0 | 3 | 15 | 26 | 116 |
| Days Min Temp ≤ 0 °F | 3 | 2 | 0 | 0 | 0 | 0 | 0 | 0 | 0 | 0 | 0 | 1 | 6 |
| Heating Degree Days | 1200 | 958 | 678 | 315 | 105 | 10 | 1 | 4 | 69 | 293 | 659 | 1012 | 5304 |
| Cooling Degree Days | 0 | 0 | 3 | 27 | 74 | 260 | 395 | 313 | 144 | 16 | 1 | 0 | 1233 |
| Total Precipitation (") | 1.46 | 1.32 | 2.57 | 3.74 | 4.62 | 4.37 | 4.72 | 3.98 | 4.80 | 3.67 | 2.63 | 1.97 | 39.85 |
| Days ≥ 0.1" Precip | 3 | 4 | 5 | 6 | 7 | 7 | 6 | 6 | 6 | 6 | 5 | 4 | 65 |
| Total Snowfall (") | 5.8 | 5.3 | 3.0 | 0.7 | 0.0 | 0.0 | 0.0 | 0.0 | 0.0 | 0.0 | 1.2 | 3.2 | 19.2 |
| Days ≥ 1" Snow Depth | 9 | 8 | 2 | 0 | 0 | 0 | 0 | 0 | 0 | 0 | 1 | 3 | 23 |

## LICKING 4 N *Texas County* ELEVATION 1181 ft LAT/LONG 37° 33' N / 91° 54' W

| | JAN | FEB | MAR | APR | MAY | JUN | JUL | AUG | SEP | OCT | NOV | DEC | YEAR |
|---|---|---|---|---|---|---|---|---|---|---|---|---|---|
| Maximum Temp °F | 39.5 | 44.6 | 55.5 | 67.3 | 75.1 | 82.7 | 88.3 | 86.7 | 78.4 | 67.9 | 54.9 | 44.1 | 65.4 |
| Minimum Temp °F | 17.8 | 21.9 | 32.6 | 42.7 | 51.4 | 60.2 | 64.8 | 62.8 | 55.3 | 42.6 | 33.5 | 23.4 | 42.4 |
| Mean Temp °F | 28.7 | 33.3 | 44.1 | 55.0 | 63.3 | 71.5 | 76.6 | 74.8 | 66.9 | 55.3 | 44.2 | 33.7 | 54.0 |
| Days Max Temp ≥ 90 °F | 0 | 0 | 0 | 0 | 0 | 4 | 14 | 12 | 3 | 0 | 0 | 0 | 33 |
| Days Max Temp ≤ 32 °F | 10 | 6 | 1 | 0 | 0 | 0 | 0 | 0 | 0 | 0 | 1 | 5 | 23 |
| Days Min Temp ≤ 32 °F | 28 | 23 | 16 | 5 | 1 | 0 | 0 | 0 | 0 | 6 | 15 | 25 | 119 |
| Days Min Temp ≤ 0 °F | 3 | 2 | 0 | 0 | 0 | 0 | 0 | 0 | 0 | 0 | 0 | 1 | 6 |
| Heating Degree Days | 1118 | 889 | 645 | 317 | 122 | 17 | 3 | 6 | 79 | 314 | 618 | 963 | 5091 |
| Cooling Degree Days | 0 | 0 | 3 | 23 | 71 | 217 | 377 | 329 | 144 | 15 | 2 | 0 | 1181 |
| Total Precipitation (") | 2.21 | 2.44 | 3.59 | 4.49 | 4.58 | 4.12 | 3.44 | 3.56 | 4.57 | 3.68 | 4.33 | 3.36 | 44.37 |
| Days ≥ 0.1" Precip | 5 | 5 | 7 | 7 | 8 | 7 | 5 | 6 | 6 | 6 | 7 | 5 | 74 |
| Total Snowfall (") | 5.1 | 4.5 | 3.2 | 0.1 | 0.0 | 0.0 | 0.0 | 0.0 | 0.0 | 0.0 | 0.8 | 2.6 | 16.3 |
| Days ≥ 1" Snow Depth | 9 | 6 | 2 | 0 | 0 | 0 | 0 | 0 | 0 | 0 | 0 | 3 | 20 |

## LOCKWOOD *Dade County* ELEVATION 1070 ft LAT/LONG 37° 23' N / 93° 57' W

| | JAN | FEB | MAR | APR | MAY | JUN | JUL | AUG | SEP | OCT | NOV | DEC | YEAR |
|---|---|---|---|---|---|---|---|---|---|---|---|---|---|
| Maximum Temp °F | 42.2 | 47.8 | 58.7 | 69.5 | 76.9 | 85.3 | 90.7 | 89.7 | 81.4 | 70.8 | 56.9 | 46.4 | 68.0 |
| Minimum Temp °F | 21.9 | 26.1 | 35.7 | 45.2 | 54.2 | 63.2 | 68.0 | 65.7 | 58.3 | 47.0 | 36.2 | 26.7 | 45.7 |
| Mean Temp °F | 32.1 | 37.0 | 47.2 | 57.4 | 65.6 | 74.2 | 79.4 | 77.7 | 69.9 | 58.9 | 46.6 | 36.6 | 56.9 |
| Days Max Temp ≥ 90 °F | 0 | 0 | 0 | 0 | 0 | 8 | 19 | 17 | 5 | 0 | 0 | 0 | 49 |
| Days Max Temp ≤ 32 °F | 7 | 4 | 1 | 0 | 0 | 0 | 0 | 0 | 0 | 0 | 1 | 4 | 17 |
| Days Min Temp ≤ 32 °F | 26 | 20 | 13 | 3 | 0 | 0 | 0 | 0 | 0 | 2 | 11 | 22 | 97 |
| Days Min Temp ≤ 0 °F | 1 | 1 | 0 | 0 | 0 | 0 | 0 | 0 | 0 | 0 | 0 | 1 | 3 |
| Heating Degree Days | 1014 | 785 | 550 | 253 | 79 | 6 | 0 | 1 | 46 | 220 | 549 | 874 | 4377 |
| Cooling Degree Days | 0 | 0 | 5 | 31 | 99 | 296 | 462 | 421 | 209 | 33 | 3 | 0 | 1559 |
| Total Precipitation (") | 1.75 | 2.26 | 3.54 | 4.29 | 4.54 | 4.93 | 3.85 | 4.30 | 4.94 | 4.09 | 4.19 | 2.90 | 45.58 |
| Days ≥ 0.1" Precip | 4 | 4 | 6 | 7 | 8 | 7 | 5 | 5 | 6 | 6 | 5 | 5 | 68 |
| Total Snowfall (") | 4.5 | 4.5 | 3.8 | 0.1 | 0.0 | 0.0 | 0.0 | 0.0 | 0.0 | 0.0 | 0.9 | 3.3 | 17.1 |
| Days ≥ 1" Snow Depth | 8 | 5 | 2 | 0 | 0 | 0 | 0 | 0 | 0 | 0 | 1 | 4 | 20 |

## MALDEN MUNICIPAL AP *Dunklin County* ELEVATION 289 ft LAT/LONG 36° 36' N / 89° 59' W

| | JAN | FEB | MAR | APR | MAY | JUN | JUL | AUG | SEP | OCT | NOV | DEC | YEAR |
|---|---|---|---|---|---|---|---|---|---|---|---|---|---|
| Maximum Temp °F | 42.2 | 47.0 | 58.7 | 70.8 | 79.6 | 88.3 | 92.1 | 89.3 | 82.7 | 71.5 | ***58.6*** | ***47.5*** | 69.0 |
| Minimum Temp °F | 24.4 | 28.0 | 37.3 | 48.1 | 56.7 | 65.5 | 69.5 | 66.8 | 60.1 | 46.4 | ***38.4*** | ***30.3*** | 47.6 |
| Mean Temp °F | 33.4 | 37.5 | 48.0 | 59.5 | 68.1 | 76.9 | 80.9 | 78.1 | 71.4 | 59.0 | ***48.6*** | ***38.9*** | 58.4 |
| Days Max Temp ≥ 90 °F | 0 | 0 | 0 | 0 | 3 | 14 | 22 | 16 | 6 | 0 | ***0*** | ***0*** | 61 |
| Days Max Temp ≤ 32 °F | 7 | 4 | 0 | 0 | 0 | 0 | 0 | 0 | 0 | 0 | ***0*** | ***2*** | 13 |
| Days Min Temp ≤ 32 °F | 24 | 19 | 10 | 1 | 0 | 0 | 0 | 0 | 0 | 1 | ***9*** | ***19*** | 83 |
| Days Min Temp ≤ 0 °F | 1 | 0 | 0 | 0 | 0 | 0 | 0 | 0 | 0 | 0 | ***0*** | ***0*** | 1 |
| Heating Degree Days | 974 | 769 | 524 | 200 | 48 | 2 | 0 | 0 | 26 | 219 | ***490*** | ***802*** | 4054 |
| Cooling Degree Days | ***0*** | ***0*** | ***4*** | ***45*** | 151 | 381 | ***523*** | 433 | 234 | 39 | ***3*** | ***1*** | 1814 |
| Total Precipitation (") | 3.15 | 3.23 | 4.63 | 4.73 | 4.49 | 3.75 | 3.57 | 2.75 | 3.90 | 2.76 | ***4.02*** | ***4.48*** | 45.46 |
| Days ≥ 0.1" Precip | ***5*** | 5 | ***7*** | 7 | 7 | 6 | 5 | 4 | 5 | 5 | ***6*** | ***6*** | 68 |
| Total Snowfall (") | na | na | na | 0.0 | 0.0 | 0.0 | 0.0 | 0.0 | 0.0 | 0.0 | ***0.1*** | na | na |
| Days ≥ 1" Snow Depth | na | na | na | 0 | 0 | 0 | 0 | 0 | 0 | 0 | ***0*** | na | na |

## MANSFIELD *Wright County* ELEVATION 1520 ft LAT/LONG 37° 6' N / 92° 35' W

| | JAN | FEB | MAR | APR | MAY | JUN | JUL | AUG | SEP | OCT | NOV | DEC | YEAR |
|---|---|---|---|---|---|---|---|---|---|---|---|---|---|
| Maximum Temp °F | 42.2 | ***47.1*** | 58.2 | 69.3 | 76.2 | 84.3 | 89.4 | 88.1 | 80.0 | 69.7 | 55.4 | 46.1 | 67.2 |
| Minimum Temp °F | 18.6 | 22.7 | 31.7 | 42.1 | 50.7 | 59.0 | 63.7 | 61.4 | 54.5 | 43.4 | 33.0 | 24.3 | 42.1 |
| Mean Temp °F | 30.6 | ***35.0*** | 45.0 | 55.7 | 63.5 | 71.8 | 76.6 | 74.8 | 67.3 | 56.6 | 44.2 | ***35.2*** | 54.7 |
| Days Max Temp ≥ 90 °F | 0 | 0 | 0 | 0 | 0 | 6 | 16 | 13 | 3 | 0 | 0 | 0 | 38 |
| Days Max Temp ≤ 32 °F | 7 | ***4*** | 1 | 0 | 0 | 0 | 0 | 0 | 0 | 0 | 1 | 4 | 17 |
| Days Min Temp ≤ 32 °F | 27 | 22 | 17 | 5 | 0 | 0 | 0 | 0 | 0 | 4 | 15 | 24 | 114 |
| Days Min Temp ≤ 0 °F | 2 | 1 | 0 | 0 | 0 | 0 | 0 | 0 | 0 | 0 | 0 | 1 | 4 |
| Heating Degree Days | 1059 | ***839*** | 616 | 290 | 106 | 10 | 1 | 4 | 64 | 276 | 618 | ***915*** | 4798 |
| Cooling Degree Days | ***0*** | ***0*** | ***1*** | ***12*** | ***46*** | ***190*** | ***349*** | 301 | 121 | 16 | ***0*** | ***0*** | 1036 |
| Total Precipitation (") | 1.92 | 2.12 | 3.63 | 4.27 | 4.28 | 4.65 | 3.51 | 3.47 | 4.81 | 3.92 | 4.46 | 3.38 | 44.42 |
| Days ≥ 0.1" Precip | 4 | 4 | 6 | 7 | 7 | 7 | 5 | 6 | 6 | 6 | 6 | 5 | 69 |
| Total Snowfall (") | ***5.0*** | ***4.7*** | ***2.5*** | 0.3 | 0.0 | 0.0 | 0.0 | 0.0 | 0.0 | 0.0 | 0.7 | ***2.7*** | 15.9 |
| Days ≥ 1" Snow Depth | na | na | na | 0 | 0 | 0 | 0 | 0 | 0 | 0 | ***0*** | na | na |

## MARBLE HILL *Bollinger County* ELEVATION 440 ft LAT/LONG 37° 18' N / 89° 58' W

| | JAN | FEB | MAR | APR | MAY | JUN | JUL | AUG | SEP | OCT | NOV | DEC | YEAR |
|---|---|---|---|---|---|---|---|---|---|---|---|---|---|
| Maximum Temp °F | 42.2 | 47.6 | 58.9 | 69.9 | 77.9 | 85.9 | 89.8 | 88.0 | 81.2 | 71.3 | 57.6 | 46.4 | 68.1 |
| Minimum Temp °F | 21.1 | 24.7 | 34.3 | 43.9 | 52.2 | 60.8 | 65.2 | 62.8 | 55.6 | 43.2 | 34.7 | 26.0 | 43.7 |
| Mean Temp °F | 31.7 | 36.2 | 46.5 | 57.0 | 65.1 | 73.4 | 77.5 | 75.4 | 68.5 | 57.3 | 46.2 | 36.2 | 55.9 |
| Days Max Temp ≥ 90 °F | 0 | 0 | 0 | 0 | 1 | 9 | 17 | 13 | 4 | 0 | 0 | 0 | 44 |
| Days Max Temp ≤ 32 °F | 7 | 3 | 0 | 0 | 0 | 0 | 0 | 0 | 0 | 0 | 0 | 3 | 13 |
| Days Min Temp ≤ 32 °F | 26 | 21 | 14 | 5 | 0 | 0 | 0 | 0 | 0 | 6 | 14 | 22 | 108 |
| Days Min Temp ≤ 0 °F | 2 | 1 | 0 | 0 | 0 | 0 | 0 | 0 | 0 | 0 | 0 | 1 | 4 |
| Heating Degree Days | 1027 | 808 | 570 | 259 | 84 | 7 | 1 | 2 | 49 | 260 | 562 | 885 | 4514 |
| Cooling Degree Days | 0 | 0 | 4 | 19 | 87 | 256 | 386 | 331 | 157 | 22 | 1 | 0 | 1263 |
| Total Precipitation (") | 3.07 | 3.00 | 4.74 | 4.48 | 4.46 | 3.62 | 3.97 | 4.12 | 4.02 | 3.50 | 4.66 | 4.27 | 47.91 |
| Days ≥ 0.1" Precip | 5 | 5 | 7 | 7 | 7 | 6 | 5 | 6 | 5 | 6 | 7 | 7 | 73 |
| Total Snowfall (") | 4.3 | ***5.2*** | 2.1 | 0.0 | 0.0 | 0.0 | 0.0 | 0.0 | 0.0 | 0.1 | 0.6 | 1.7 | 14.0 |
| Days ≥ 1" Snow Depth | ***2*** | na | 1 | 0 | 0 | 0 | 0 | 0 | 0 | 0 | 0 | ***1*** | na |

## MARSHALL *Saline County* ELEVATION 778 ft LAT/LONG 39° 6' N / 93° 12' W

| | JAN | FEB | MAR | APR | MAY | JUN | JUL | AUG | SEP | OCT | NOV | DEC | YEAR |
|---|---|---|---|---|---|---|---|---|---|---|---|---|---|
| Maximum Temp °F | ***37.5*** | 43.1 | 55.9 | 68.1 | 76.7 | 85.5 | 90.2 | 87.8 | 80.1 | ***69.1*** | ***53.8*** | ***42.3*** | 65.8 |
| Minimum Temp °F | ***18.4*** | 22.8 | 33.5 | 44.6 | 54.1 | 63.3 | 67.6 | 64.8 | 56.9 | ***45.4*** | ***34.1*** | ***24.4*** | 44.2 |
| Mean Temp °F | ***28.0*** | 33.0 | 44.7 | 56.4 | 65.5 | 74.4 | 78.9 | 76.3 | 68.5 | ***57.3*** | ***44.0*** | ***33.4*** | 55.0 |
| Days Max Temp ≥ 90 °F | ***0*** | 0 | 0 | 0 | 1 | 9 | ***18*** | 13 | 5 | ***0*** | ***0*** | ***0*** | 46 |
| Days Max Temp ≤ 32 °F | ***11*** | 7 | 1 | 0 | 0 | 0 | 0 | 0 | 0 | ***0*** | ***1*** | ***6*** | 26 |
| Days Min Temp ≤ 32 °F | ***27*** | 23 | 15 | 3 | 0 | 0 | 0 | 0 | 0 | ***3*** | ***13*** | ***25*** | 109 |
| Days Min Temp ≤ 0 °F | ***3*** | 1 | 0 | 0 | 0 | 0 | 0 | 0 | 0 | ***0*** | ***0*** | ***1*** | 5 |
| Heating Degree Days | ***1142*** | 898 | 628 | 282 | 85 | 6 | 1 | 2 | 62 | ***263*** | ***626*** | ***972*** | 4967 |
| Cooling Degree Days | ***0*** | 0 | 5 | 32 | 94 | 296 | ***416*** | ***348*** | ***156*** | ***20*** | na | na | na |
| Total Precipitation (") | na | ***1.34*** | ***2.70*** | 4.01 | 4.50 | 4.09 | 3.88 | ***3.09*** | ***4.31*** | ***3.24*** | ***2.89*** | ***1.83*** | na |
| Days ≥ 0.1" Precip | na | ***3*** | ***5*** | ***7*** | ***7*** | ***6*** | ***6*** | ***5*** | ***5*** | ***5*** | ***5*** | ***4*** | na |
| Total Snowfall (") | ***4.0*** | ***4.3*** | ***1.1*** | 0.3 | 0.0 | 0.0 | 0.0 | 0.0 | 0.0 | ***0.0*** | ***0.9*** | ***3.1*** | 13.7 |
| Days ≥ 1" Snow Depth | na | na | na | 0 | 0 | 0 | 0 | 0 | 0 | ***0*** | na | na | na |

## MARSHFIELD *Webster County* ELEVATION 1480 ft LAT/LONG 37° 20 ' N / 92° 54 ' W

| | JAN | FEB | MAR | APR | MAY | JUN | JUL | AUG | SEP | OCT | NOV | DEC | YEAR |
|---|---|---|---|---|---|---|---|---|---|---|---|---|---|
| Maximum Temp °F | 41.0 | 46.1 | 56.9 | 67.8 | 75.4 | 83.4 | 88.5 | 87.2 | 79.1 | 68.4 | 55.2 | 45.1 | 66.2 |
| Minimum Temp °F | 21.0 | 24.9 | 34.7 | 44.9 | 53.7 | 62.5 | 67.2 | 65.6 | 58.1 | 46.5 | 35.5 | 25.9 | 45.0 |
| Mean Temp °F | 31.0 | 35.5 | 45.8 | 56.4 | 64.6 | 73.0 | 77.9 | 76.4 | 68.6 | 57.5 | 45.4 | 35.5 | 55.6 |
| Days Max Temp ≥ 90 °F | 0 | 0 | 0 | 0 | 0 | 4 | 14 | 12 | 3 | 0 | 0 | 0 | 33 |
| Days Max Temp ≤ 32 °F | 8 | 5 | 1 | 0 | 0 | 0 | 0 | 0 | 0 | 0 | 1 | 4 | 19 |
| Days Min Temp ≤ 32 °F | 26 | 21 | 14 | 3 | 0 | 0 | 0 | 0 | 0 | 2 | 13 | 23 | 102 |
| Days Min Temp ≤ 0 °F | 2 | 1 | 0 | 0 | 0 | 0 | 0 | 0 | 0 | 0 | 0 | 1 | 4 |
| Heating Degree Days | 1046 | 826 | 591 | 277 | 93 | 8 | 1 | 2 | 55 | 254 | 583 | 907 | 4643 |
| Cooling Degree Days | 0 | 0 | 3 | 27 | 83 | 258 | 412 | 377 | 174 | 23 | 1 | 0 | 1358 |
| Total Precipitation (") | 2.15 | 2.12 | 3.62 | 4.10 | 4.43 | 4.27 | 3.64 | 3.54 | 4.77 | 3.95 | 4.14 | 3.11 | 43.84 |
| Days ≥ 0.1" Precip | 4 | 4 | 7 | 7 | 7 | 6 | 5 | 6 | 6 | 6 | 6 | 5 | 69 |
| Total Snowfall (") | na | *4.7* | 1.6 | 0.1 | 0.0 | 0.0 | 0.0 | 0.0 | 0.0 | 0.0 | 0.7 | *2.6* | na |
| Days ≥ 1" Snow Depth | na | na | 1 | 0 | 0 | 0 | 0 | 0 | 0 | 0 | 0 | *1* | na |

## MARYVILLE 2 E *Nodaway County* ELEVATION 1161 ft LAT/LONG 40° 21 ' N / 94° 52 ' W

| | JAN | FEB | MAR | APR | MAY | JUN | JUL | AUG | SEP | OCT | NOV | DEC | YEAR |
|---|---|---|---|---|---|---|---|---|---|---|---|---|---|
| Maximum Temp °F | 32.0 | 37.7 | 50.0 | 63.4 | 73.8 | 82.9 | 87.4 | 85.0 | 77.4 | 66.0 | 49.9 | 36.6 | 61.8 |
| Minimum Temp °F | 11.0 | 15.1 | 26.6 | 38.2 | 49.2 | 58.7 | 63.2 | 60.0 | 51.1 | 39.1 | 27.8 | 17.1 | 38.1 |
| Mean Temp °F | 21.5 | 26.6 | 38.3 | 50.8 | 61.5 | 70.8 | 75.4 | 72.5 | 64.3 | 52.6 | 38.9 | 26.8 | 50.0 |
| Days Max Temp ≥ 90 °F | 0 | 0 | 0 | 0 | 1 | 6 | 13 | 9 | 3 | 0 | 0 | 0 | 32 |
| Days Max Temp ≤ 32 °F | 15 | 10 | 3 | 0 | 0 | 0 | 0 | 0 | 0 | 0 | 2 | 10 | 40 |
| Days Min Temp ≤ 32 °F | 30 | 27 | 22 | 9 | 1 | 0 | 0 | 0 | 1 | 8 | 22 | 29 | 149 |
| Days Min Temp ≤ 0 °F | 8 | 5 | 1 | 0 | 0 | 0 | 0 | 0 | 0 | 0 | 0 | 4 | 18 |
| Heating Degree Days | 1342 | 1079 | 821 | 432 | 159 | 23 | 3 | 9 | 114 | 392 | 780 | 1179 | 6333 |
| Cooling Degree Days | 0 | 0 | 1 | 19 | 56 | 215 | 341 | 257 | 113 | 10 | 0 | 0 | 1012 |
| Total Precipitation (") | 0.82 | 0.81 | 2.22 | 3.12 | 4.14 | 4.44 | 5.21 | 3.85 | 4.18 | 3.24 | 1.93 | 1.39 | 35.35 |
| Days ≥ 0.1" Precip | 3 | 3 | 5 | 6 | 7 | 7 | 7 | 6 | 6 | 5 | 4 | 3 | 62 |
| Total Snowfall (") | *5.5* | 3.4 | 2.4 | 0.7 | 0.0 | 0.0 | 0.0 | 0.0 | 0.0 | 0.0 | 0.9 | 3.5 | 16.4 |
| Days ≥ 1" Snow Depth | na | na | *1* | 0 | 0 | 0 | 0 | 0 | 0 | 0 | 0 | na | na |

## MEMPHIS *Scotland County* ELEVATION 791 ft LAT/LONG 40° 27 ' N / 92° 10 ' W

| | JAN | FEB | MAR | APR | MAY | JUN | JUL | AUG | SEP | OCT | NOV | DEC | YEAR |
|---|---|---|---|---|---|---|---|---|---|---|---|---|---|
| Maximum Temp °F | 32.8 | *38.5* | *50.9* | *64.6* | *74.5* | *83.7* | *88.2* | *85.6* | *77.9* | *66.1* | 50.6 | 37.9 | 62.6 |
| Minimum Temp °F | 13.7 | *18.1* | *28.9* | *40.5* | *49.6* | *59.0* | *63.9* | *61.4* | *53.4* | *42.2* | 30.9 | 20.4 | 40.2 |
| Mean Temp °F | 23.3 | *28.3* | *40.0* | *52.6* | *62.1* | *71.3* | *76.1* | *73.5* | *65.7* | *54.2* | 40.8 | 29.2 | 51.4 |
| Days Max Temp ≥ 90 °F | 0 | 0 | 0 | 0 | 0 | *6* | *13* | na | *2* | 0 | 0 | 0 | na |
| Days Max Temp ≤ 32 °F | 13 | *9* | 2 | 0 | 0 | 0 | 0 | 0 | 0 | 0 | 1 | 9 | 34 |
| Days Min Temp ≤ 32 °F | 28 | 24 | 19 | *6* | 1 | 0 | 0 | 0 | 0 | *5* | *17* | 27 | 127 |
| Days Min Temp ≤ 0 °F | 6 | 3 | 0 | 0 | 0 | 0 | 0 | 0 | 0 | 0 | 0 | 2 | 11 |
| Heating Degree Days | 1287 | *1031* | *770* | *377* | *145* | *16* | *2* | *6* | *82* | *345* | 720 | 1104 | 5885 |
| Cooling Degree Days | 0 | na | *1* | *15* | *63* | na | na | *298* | *117* | *12* | 0 | 0 | na |
| Total Precipitation (") | *1.35* | 0.84 | 2.07 | *3.92* | *3.98* | *3.60* | 4.63 | 3.85 | 4.68 | 2.58 | 2.18 | 1.83 | 35.51 |
| Days ≥ 0.1" Precip | 3 | 2 | 5 | na | *7* | *6* | *5* | *5* | *5* | *5* | *4* | *4* | na |
| Total Snowfall (") | *5.8* | 4.8 | 2.4 | 1.1 | 0.0 | 0.0 | 0.0 | 0.0 | 0.0 | 0.1 | 1.4 | *2.6* | 18.2 |
| Days ≥ 1" Snow Depth | *11* | *7* | 3 | 0 | 0 | 0 | 0 | 0 | 0 | 0 | 1 | *5* | 27 |

## MEXICO *Audrain County* ELEVATION 801 ft LAT/LONG 39° 10 ' N / 91° 53 ' W

| | JAN | FEB | MAR | APR | MAY | JUN | JUL | AUG | SEP | OCT | NOV | DEC | YEAR |
|---|---|---|---|---|---|---|---|---|---|---|---|---|---|
| Maximum Temp °F | 35.2 | 40.2 | 52.6 | 65.2 | 74.5 | 83.7 | 89.0 | 86.7 | 79.0 | 67.2 | 52.8 | 40.4 | 63.9 |
| Minimum Temp °F | 15.3 | 18.9 | 29.9 | 41.5 | 51.4 | 60.8 | 65.3 | 62.6 | 54.5 | 42.4 | 32.1 | 21.4 | 41.3 |
| Mean Temp °F | 25.3 | 29.6 | 41.3 | 53.4 | 62.9 | 72.3 | 77.2 | 74.7 | 66.7 | 54.8 | 42.5 | 30.9 | 52.6 |
| Days Max Temp ≥ 90 °F | 0 | 0 | 0 | 0 | 0 | 6 | 15 | 11 | 4 | 0 | 0 | 0 | 36 |
| Days Max Temp ≤ 32 °F | 12 | 9 | 2 | 0 | 0 | 0 | 0 | 0 | 0 | 0 | 1 | 8 | 32 |
| Days Min Temp ≤ 32 °F | 29 | 25 | 20 | 6 | 0 | 0 | 0 | 0 | 0 | 5 | 16 | 27 | 128 |
| Days Min Temp ≤ 0 °F | 5 | 3 | 0 | 0 | 0 | 0 | 0 | 0 | 0 | 0 | 0 | 2 | 10 |
| Heating Degree Days | 1224 | 994 | 731 | 363 | 132 | 15 | 2 | 5 | 79 | 327 | 671 | 1052 | 5595 |
| Cooling Degree Days | 0 | 0 | 3 | 23 | 71 | 253 | 399 | 332 | 147 | 15 | 1 | 0 | 1244 |
| Total Precipitation (") | 1.59 | 1.70 | 3.06 | 4.01 | 4.79 | 4.58 | 4.23 | 3.32 | 4.24 | 3.36 | 3.47 | 2.61 | 40.96 |
| Days ≥ 0.1" Precip | 4 | 4 | 7 | 8 | 9 | 7 | 6 | 6 | 6 | 6 | 7 | 5 | 75 |
| Total Snowfall (") | 5.8 | 5.9 | 3.8 | 0.3 | 0.0 | 0.0 | 0.0 | 0.0 | 0.0 | 0.0 | 1.5 | 4.4 | 21.7 |
| Days ≥ 1" Snow Depth | 7 | 5 | 2 | 0 | 0 | 0 | 0 | 0 | 0 | 0 | 1 | 4 | 19 |

## MOBERLY *Randolph County* ELEVATION 840 ft LAT/LONG 39° 25 ' N / 92° 26 ' W

| | JAN | FEB | MAR | APR | MAY | JUN | JUL | AUG | SEP | OCT | NOV | DEC | YEAR |
|---|---|---|---|---|---|---|---|---|---|---|---|---|---|
| Maximum Temp °F | 36.1 | 41.4 | 54.4 | 67.0 | 75.9 | 84.5 | 89.4 | 87.1 | 79.3 | 68.1 | 53.2 | 41.0 | 64.8 |
| Minimum Temp °F | 18.0 | 22.2 | 33.1 | 44.3 | 53.8 | 62.8 | 67.2 | 64.7 | 57.2 | 45.9 | 34.5 | 24.1 | 44.0 |
| Mean Temp °F | 27.1 | 31.8 | 43.8 | 55.6 | 64.9 | 73.7 | 78.3 | 75.9 | 68.2 | 57.0 | 43.9 | 32.6 | 54.4 |
| Days Max Temp ≥ 90 °F | 0 | 0 | 0 | 0 | 0 | 6 | 16 | 12 | 3 | 0 | 0 | 0 | 37 |
| Days Max Temp ≤ 32 °F | 11 | 7 | 1 | 0 | 0 | 0 | 0 | 0 | 0 | 0 | 1 | 6 | 26 |
| Days Min Temp ≤ 32 °F | 27 | 23 | 15 | 4 | 0 | 0 | 0 | 0 | 0 | 2 | 13 | 25 | 109 |
| Days Min Temp ≤ 0 °F | 3 | 1 | 0 | 0 | 0 | 0 | 0 | 0 | 0 | 0 | 0 | 1 | 5 |
| Heating Degree Days | 1168 | 930 | 656 | 304 | 92 | 7 | 1 | 2 | 58 | 268 | 628 | 999 | 5113 |
| Cooling Degree Days | 0 | 0 | 7 | 39 | 103 | 299 | 457 | 386 | 183 | 27 | 2 | 0 | 1503 |
| Total Precipitation (") | 1.60 | 1.61 | 2.94 | 4.08 | 4.46 | 4.36 | 4.43 | 3.67 | 4.46 | 3.58 | 2.98 | 2.32 | 40.49 |
| Days ≥ 0.1" Precip | 4 | 4 | 6 | 7 | 8 | 6 | 5 | 6 | 6 | 6 | 5 | 5 | 68 |
| Total Snowfall (") | *6.0* | *5.9* | *2.1* | 0.3 | 0.0 | 0.0 | 0.0 | 0.0 | 0.0 | 0.0 | 1.1 | *2.9* | 18.3 |
| Days ≥ 1" Snow Depth | na | na | na | 0 | 0 | 0 | 0 | 0 | 0 | 0 | 0 | na | na |

## MOUNTAIN GROVE 2 N *Wright County* ELEVATION 1460 ft LAT/LONG 37° 9 ' N / 92° 16 ' W

| | JAN | FEB | MAR | APR | MAY | JUN | JUL | AUG | SEP | OCT | NOV | DEC | YEAR |
|---|---|---|---|---|---|---|---|---|---|---|---|---|---|
| Maximum Temp °F | 40.6 | 46.2 | 56.7 | 67.7 | 75.0 | 82.8 | 88.1 | 87.0 | 79.0 | 68.7 | 55.2 | 44.7 | 66.0 |
| Minimum Temp °F | 21.0 | 25.0 | 34.4 | 44.5 | 52.9 | 61.3 | 65.9 | 63.4 | 56.5 | 45.1 | 35.2 | 25.7 | 44.2 |
| Mean Temp °F | 30.8 | 35.6 | 45.6 | 56.1 | 63.9 | 72.1 | 77.0 | 75.2 | 67.8 | 56.9 | 45.2 | 35.2 | 55.1 |
| Days Max Temp ≥ 90 °F | 0 | 0 | 0 | 0 | 0 | 3 | 13 | 11 | 2 | 0 | 0 | 0 | 29 |
| Days Max Temp ≤ 32 °F | 9 | 5 | 1 | 0 | 0 | 0 | 0 | 0 | 0 | 0 | 1 | 5 | 21 |
| Days Min Temp ≤ 32 °F | 26 | 21 | 14 | 3 | 0 | 0 | 0 | 0 | 0 | 3 | 13 | 23 | 103 |
| Days Min Temp ≤ 0 °F | 2 | 1 | 0 | 0 | 0 | 0 | 0 | 0 | 0 | 0 | 0 | 1 | 4 |
| Heating Degree Days | 1052 | 823 | 599 | 282 | 101 | 10 | 1 | 4 | 61 | 266 | 587 | 916 | 4702 |
| Cooling Degree Days | 0 | 0 | 2 | 21 | 64 | 220 | 368 | 328 | 145 | 15 | 1 | 0 | 1164 |
| Total Precipitation (") | 2.18 | 2.44 | 3.82 | 4.54 | 4.54 | 4.18 | 3.58 | 3.83 | 4.43 | 3.62 | 4.56 | 3.66 | 45.38 |
| Days ≥ 0.1" Precip | 5 | 4 | 7 | 7 | 7 | 7 | 5 | 6 | 7 | 6 | 6 | 5 | 72 |
| Total Snowfall (") | 4.1 | 3.7 | 2.8 | 0.5 | 0.0 | 0.0 | 0.0 | 0.0 | 0.0 | 0.0 | 1.2 | 2.4 | 14.7 |
| Days ≥ 1" Snow Depth | 9 | 7 | 3 | 0 | 0 | 0 | 0 | 0 | 0 | 0 | 1 | 4 | 24 |

## MT VERNON MU SW CNTR *Lawrence County* ELEVATION 1191 ft LAT/LONG 37° 4 ' N / 93° 53 ' W

| | JAN | FEB | MAR | APR | MAY | JUN | JUL | AUG | SEP | OCT | NOV | DEC | YEAR |
|---|---|---|---|---|---|---|---|---|---|---|---|---|---|
| Maximum Temp °F | 41.1 | 46.1 | 56.5 | 67.1 | 74.5 | 83.1 | 88.9 | 87.8 | 79.5 | 68.9 | 55.9 | 45.5 | 66.2 |
| Minimum Temp °F | 20.0 | 24.7 | 34.4 | 44.3 | 53.0 | 61.8 | 66.5 | 64.6 | 57.2 | 45.2 | 35.1 | 25.4 | 44.3 |
| Mean Temp °F | 30.6 | 35.4 | 45.5 | 55.7 | 63.8 | 72.5 | 77.6 | 76.2 | 68.3 | 57.1 | 45.6 | 35.5 | 55.3 |
| Days Max Temp ≥ 90 °F | 0 | 0 | 0 | 0 | 0 | 4 | 15 | 13 | 3 | 0 | 0 | 0 | 35 |
| Days Max Temp ≤ 32 °F | 8 | 5 | 1 | 0 | 0 | 0 | 0 | 0 | 0 | 0 | 1 | 5 | 20 |
| Days Min Temp ≤ 32 °F | 27 | 21 | 14 | 4 | 0 | 0 | 0 | 0 | 0 | 3 | 13 | 24 | 106 |
| Days Min Temp ≤ 0 °F | 2 | 1 | 0 | 0 | 0 | 0 | 0 | 0 | 0 | 0 | 0 | 1 | 4 |
| Heating Degree Days | 1061 | 829 | 602 | 296 | 109 | 12 | 2 | 4 | 63 | 268 | 579 | 908 | 4733 |
| Cooling Degree Days | 0 | 0 | 4 | 25 | 82 | 257 | 413 | 383 | 179 | 25 | 2 | 0 | 1370 |
| Total Precipitation (") | 1.82 | 1.99 | 3.70 | 4.15 | 4.49 | 5.19 | 3.14 | 4.21 | 5.39 | 3.83 | 4.38 | 2.88 | 45.17 |
| Days ≥ 0.1" Precip | 4 | 4 | 6 | 6 | 7 | 7 | 5 | 6 | 7 | 6 | 5 | 5 | 68 |
| Total Snowfall (") | na | *4.0* | *1.9* | 0.0 | 0.0 | 0.0 | 0.0 | 0.0 | 0.0 | 0.0 | 0.9 | *2.6* | na |
| Days ≥ 1" Snow Depth | na | na | *1* | 0 | 0 | 0 | 0 | 0 | 0 | 0 | *0* | *2* | na |

## NEOSHO *Newton County* ELEVATION 1010 ft LAT/LONG 36° 52 ' N / 94° 22 ' W

| | JAN | FEB | MAR | APR | MAY | JUN | JUL | AUG | SEP | OCT | NOV | DEC | YEAR |
|---|---|---|---|---|---|---|---|---|---|---|---|---|---|
| Maximum Temp °F | 44.7 | 50.5 | 60.9 | 71.7 | 78.0 | 85.7 | 91.0 | 90.1 | 82.0 | 72.2 | 58.8 | 48.9 | 69.5 |
| Minimum Temp °F | 22.1 | 26.2 | 35.5 | 45.1 | 53.6 | 62.1 | 66.5 | 64.0 | 57.0 | 45.5 | 35.6 | 26.8 | 45.0 |
| Mean Temp °F | 33.4 | 38.4 | 48.2 | 58.4 | 65.8 | 73.9 | 78.8 | 77.0 | 69.5 | 58.9 | 47.3 | 37.9 | 57.3 |
| Days Max Temp ≥ 90 °F | 0 | 0 | 0 | 0 | 0 | 7 | 20 | 18 | 5 | 0 | 0 | 0 | 50 |
| Days Max Temp ≤ 32 °F | 5 | 3 | 0 | 0 | 0 | 0 | 0 | 0 | 0 | 0 | 0 | 3 | 11 |
| Days Min Temp ≤ 32 °F | 26 | 20 | 14 | 4 | 0 | 0 | 0 | 0 | 0 | 4 | 12 | 22 | 102 |
| Days Min Temp ≤ 0 °F | 1 | 1 | 0 | 0 | 0 | 0 | 0 | 0 | 0 | 0 | 0 | 1 | 3 |
| Heating Degree Days | 972 | 745 | 522 | 230 | 74 | 6 | 1 | 2 | 49 | 222 | 528 | 835 | 4186 |
| Cooling Degree Days | 0 | 0 | 7 | 37 | 101 | 284 | 437 | 391 | 191 | 32 | 5 | 0 | 1485 |
| Total Precipitation (") | 1.89 | 2.23 | 3.74 | 4.51 | 4.55 | 4.57 | 3.18 | 3.79 | 4.95 | 4.27 | 4.22 | 2.96 | 44.86 |
| Days ≥ 0.1" Precip | 4 | 4 | 6 | 7 | 8 | 6 | 5 | 5 | 6 | 6 | 5 | 5 | 67 |
| Total Snowfall (") | 3.7 | 3.2 | 2.7 | 0.0 | 0.0 | 0.0 | 0.0 | 0.0 | 0.0 | 0.0 | 0.6 | 2.3 | 12.5 |
| Days ≥ 1" Snow Depth | na | na | *0* | 0 | 0 | 0 | 0 | 0 | 0 | 0 | *0* | na | na |

## NEVADA SEWAGE PLANT *Vernon County* ELEVATION 860 ft LAT/LONG 37° 50 ' N / 94° 21 ' W

| | JAN | FEB | MAR | APR | MAY | JUN | JUL | AUG | SEP | OCT | NOV | DEC | YEAR |
|---|---|---|---|---|---|---|---|---|---|---|---|---|---|
| Maximum Temp °F | 41.8 | 47.5 | 59.1 | 70.3 | 77.6 | 86.0 | 91.7 | 90.2 | 81.9 | 71.7 | 57.3 | 46.0 | 68.4 |
| Minimum Temp °F | 19.3 | 23.7 | 34.3 | 44.1 | 53.1 | 62.4 | 66.3 | 63.6 | 56.3 | 44.6 | 34.1 | 24.7 | 43.9 |
| Mean Temp °F | 30.6 | 35.6 | 46.7 | 57.3 | 65.4 | 74.2 | 79.0 | 76.9 | 69.1 | 58.2 | 45.7 | 35.4 | 56.2 |
| Days Max Temp ≥ 90 °F | 0 | 0 | 0 | 0 | 1 | 9 | 21 | 17 | 6 | 0 | 0 | 0 | 54 |
| Days Max Temp ≤ 32 °F | 8 | 5 | 1 | 0 | 0 | 0 | 0 | 0 | 0 | 0 | 1 | 4 | 19 |
| Days Min Temp ≤ 32 °F | 27 | 22 | 15 | 4 | 0 | 0 | 0 | 0 | 0 | 5 | 14 | 24 | 111 |
| Days Min Temp ≤ 0 °F | 2 | 1 | 0 | 0 | 0 | 0 | 0 | 0 | 0 | 0 | 0 | 1 | 4 |
| Heating Degree Days | 1060 | 823 | 565 | 259 | 82 | 5 | 1 | 1 | 54 | 243 | 574 | 912 | 4579 |
| Cooling Degree Days | 0 | 0 | 5 | 31 | 96 | 293 | 446 | 384 | 184 | 26 | 3 | 0 | 1468 |
| Total Precipitation (") | 1.63 | 1.85 | 3.42 | 4.35 | 5.01 | 5.49 | 3.62 | 3.98 | 4.36 | 4.33 | 3.39 | 2.39 | 43.82 |
| Days ≥ 0.1" Precip | 4 | 4 | 6 | 6 | 8 | 7 | 5 | 5 | 6 | 6 | 5 | 4 | 66 |
| Total Snowfall (") | 4.3 | 3.6 | 2.2 | 0.1 | 0.0 | 0.0 | 0.0 | 0.0 | 0.0 | 0.0 | 0.8 | 2.9 | 13.9 |
| Days ≥ 1" Snow Depth | 9 | 5 | 1 | 0 | 0 | 0 | 0 | 0 | 0 | 0 | 0 | 3 | 18 |

## NEW FRANKLIN 1 W *Howard County* ELEVATION 640 ft LAT/LONG 39° 1 ' N / 92° 46 ' W

| | JAN | FEB | MAR | APR | MAY | JUN | JUL | AUG | SEP | OCT | NOV | DEC | YEAR |
|---|---|---|---|---|---|---|---|---|---|---|---|---|---|
| Maximum Temp °F | 38.0 | 43.4 | 55.7 | 67.5 | 76.2 | 84.8 | 89.7 | 87.7 | 80.3 | 69.4 | 54.8 | 42.7 | 65.9 |
| Minimum Temp °F | 17.9 | 21.8 | 33.0 | 44.0 | 53.4 | 62.8 | 66.9 | 64.2 | 56.4 | 44.5 | 34.3 | 23.8 | 43.6 |
| Mean Temp °F | 28.0 | 32.6 | 44.4 | 55.7 | 64.8 | 73.8 | 78.3 | 75.9 | 68.4 | 57.0 | 44.5 | 33.3 | 54.7 |
| Days Max Temp ≥ 90 °F | 0 | 0 | 0 | 0 | 0 | 7 | 17 | 13 | 5 | 0 | 0 | 0 | 42 |
| Days Max Temp ≤ 32 °F | 10 | 6 | 1 | 0 | 0 | 0 | 0 | 0 | 0 | 0 | 1 | 5 | 23 |
| Days Min Temp ≤ 32 °F | 28 | 24 | 16 | 3 | 0 | 0 | 0 | 0 | 0 | 3 | 13 | 25 | 112 |
| Days Min Temp ≤ 0 °F | 3 | 2 | 0 | 0 | 0 | 0 | 0 | 0 | 0 | 0 | 0 | 1 | 6 |
| Heating Degree Days | 1140 | 907 | 637 | 298 | 92 | 6 | 1 | 3 | 60 | 268 | 609 | 977 | 4998 |
| Cooling Degree Days | 0 | 0 | 4 | 30 | 89 | 286 | 424 | 350 | 167 | 20 | 2 | 0 | 1372 |
| Total Precipitation (") | 1.40 | 1.42 | 2.74 | 3.86 | 4.49 | 4.35 | 3.49 | 3.89 | 4.14 | 3.46 | 2.81 | 2.20 | 38.25 |
| Days ≥ 0.1" Precip | 4 | 4 | 5 | 7 | 7 | 7 | 6 | 6 | 6 | 6 | 5 | 4 | 67 |
| Total Snowfall (") | na | *4.8* | *1.5* | 0.1 | 0.0 | 0.0 | 0.0 | 0.0 | 0.0 | 0.0 | 1.1 | *2.5* | na |
| Days ≥ 1" Snow Depth | 8 | 6 | 2 | 0 | 0 | 0 | 0 | 0 | 0 | 0 | 1 | 4 | 21 |

## NEW MADRID *New Madrid County* ELEVATION 302 ft LAT/LONG 36° 35 ' N / 89° 32 ' W

| | JAN | FEB | MAR | APR | MAY | JUN | JUL | AUG | SEP | OCT | NOV | DEC | YEAR |
|---|---|---|---|---|---|---|---|---|---|---|---|---|---|
| Maximum Temp °F | 41.0 | 45.9 | 56.7 | 68.7 | 78.0 | 87.0 | 90.9 | 88.9 | 82.2 | 71.1 | 57.6 | 46.1 | 67.8 |
| Minimum Temp °F | 24.5 | 28.1 | 37.7 | 47.8 | 57.0 | 65.4 | 69.5 | 66.7 | 59.4 | 47.0 | 38.9 | 29.3 | 47.6 |
| Mean Temp °F | 32.8 | 37.0 | 47.2 | 58.3 | 67.5 | 76.2 | 80.3 | 77.8 | 70.8 | 59.1 | 48.3 | 37.8 | 57.8 |
| Days Max Temp ≥ 90 °F | 0 | 0 | 0 | 0 | 1 | 11 | 20 | 15 | 5 | 0 | 0 | 0 | 52 |
| Days Max Temp ≤ 32 °F | 7 | 4 | 0 | 0 | 0 | 0 | 0 | 0 | 0 | 0 | 0 | 3 | 14 |
| Days Min Temp ≤ 32 °F | 24 | 19 | 11 | 1 | 0 | 0 | 0 | 0 | 0 | 2 | 8 | 19 | 84 |
| Days Min Temp ≤ 0 °F | 1 | 0 | 0 | 0 | 0 | 0 | 0 | 0 | 0 | 0 | 0 | 0 | 1 |
| Heating Degree Days | 992 | 783 | 547 | 227 | 57 | 2 | 0 | 1 | 33 | 214 | 496 | 837 | 4189 |
| Cooling Degree Days | 0 | 0 | 4 | 35 | 139 | 352 | 496 | 423 | 216 | 36 | 2 | 0 | 1703 |
| Total Precipitation (") | 3.36 | 3.77 | 4.56 | 5.44 | 5.22 | 3.94 | 3.78 | 2.80 | 4.13 | 3.58 | 4.76 | 4.90 | 50.24 |
| Days ≥ 0.1" Precip | 5 | 6 | 8 | 8 | 7 | 6 | 6 | 5 | 5 | 5 | 7 | 7 | 75 |
| Total Snowfall (") | na | *2.5* | 0.5 | 0.0 | 0.0 | 0.0 | 0.0 | 0.0 | 0.0 | 0.0 | 0.0 | *0.6* | na |
| Days ≥ 1" Snow Depth | na | na | 0 | 0 | 0 | 0 | 0 | 0 | 0 | 0 | 0 | *0* | na |

## OSCEOLA *St. Clair County* ELEVATION 771 ft LAT/LONG 38° 3 ' N / 93° 42 ' W

| | JAN | FEB | MAR | APR | MAY | JUN | JUL | AUG | SEP | OCT | NOV | DEC | YEAR |
|---|---|---|---|---|---|---|---|---|---|---|---|---|---|
| Maximum Temp °F | 40.8 | 46.4 | 57.9 | 69.2 | 76.6 | 85.2 | 90.8 | 89.4 | 80.7 | 70.1 | 56.2 | 45.0 | 67.4 |
| Minimum Temp °F | 20.5 | 24.7 | 34.9 | 45.2 | 53.6 | 62.8 | 67.4 | 65.3 | 57.7 | 45.9 | 35.6 | 25.8 | 44.9 |
| Mean Temp °F | 30.6 | 35.5 | 46.4 | 57.2 | 65.1 | 74.0 | 79.1 | 77.4 | 69.2 | 58.1 | 45.9 | 35.4 | 56.2 |
| Days Max Temp ≥ 90 °F | 0 | 0 | 0 | 0 | 1 | 8 | 19 | 16 | 5 | 0 | 0 | 0 | 49 |
| Days Max Temp ≤ 32 °F | 9 | 5 | 1 | 0 | 0 | 0 | 0 | 0 | 0 | 0 | 1 | 4 | 20 |
| Days Min Temp ≤ 32 °F | 26 | 21 | 14 | 3 | 0 | 0 | 0 | 0 | 0 | 3 | 13 | 23 | 103 |
| Days Min Temp ≤ 0 °F | 2 | 1 | 0 | 0 | 0 | 0 | 0 | 0 | 0 | 0 | 0 | 1 | 4 |
| Heating Degree Days | 1058 | 824 | 574 | 260 | 84 | 5 | 1 | 1 | 51 | 241 | 569 | 911 | 4579 |
| Cooling Degree Days | 0 | 0 | 7 | 36 | 96 | 305 | 468 | 413 | 199 | 31 | 4 | 0 | 1559 |
| Total Precipitation (") | 1.65 | 1.79 | 2.94 | 4.23 | 4.45 | 4.38 | 3.53 | 3.97 | 4.32 | 4.09 | 3.48 | 2.54 | 41.37 |
| Days ≥ 0.1" Precip | 4 | 4 | 6 | 7 | 7 | 7 | 5 | 6 | 6 | 6 | 5 | 5 | 68 |
| Total Snowfall (") | 5.3 | 4.5 | 2.2 | 0.3 | 0.0 | 0.0 | 0.0 | 0.0 | 0.0 | 0.0 | 1.1 | 3.1 | 16.5 |
| Days ≥ 1" Snow Depth | 7 | 5 | 1 | 0 | 0 | 0 | 0 | 0 | 0 | 0 | 0 | 3 | 16 |

## OZARK BEACH *Taney County* ELEVATION 702 ft LAT/LONG 36° 40 ' N / 93° 7 ' W

| | JAN | FEB | MAR | APR | MAY | JUN | JUL | AUG | SEP | OCT | NOV | DEC | YEAR |
|---|---|---|---|---|---|---|---|---|---|---|---|---|---|
| Maximum Temp °F | 44.8 | 49.9 | 60.1 | 71.0 | 78.0 | 86.3 | 91.6 | 90.0 | 82.4 | 72.4 | 59.6 | 49.1 | 69.6 |
| Minimum Temp °F | 19.9 | 23.2 | 32.1 | 41.7 | 50.0 | 59.3 | 63.4 | 61.4 | 55.0 | 42.4 | 33.5 | 25.1 | 42.2 |
| Mean Temp °F | 32.4 | 36.5 | 46.1 | 56.4 | 64.1 | 72.8 | 77.4 | 75.7 | 68.7 | 57.5 | 46.6 | 37.1 | 55.9 |
| Days Max Temp ≥ 90 °F | 0 | 0 | 0 | 0 | 1 | 10 | 21 | 17 | 6 | 1 | 0 | 0 | 56 |
| Days Max Temp ≤ 32 °F | 5 | 3 | 0 | 0 | 0 | 0 | 0 | 0 | 0 | 0 | 0 | 3 | 11 |
| Days Min Temp ≤ 32 °F | 27 | 24 | 17 | 5 | 0 | 0 | 0 | 0 | 0 | 4 | 15 | 25 | 117 |
| Days Min Temp ≤ 0 °F | 2 | 0 | 0 | 0 | 0 | 0 | 0 | 0 | 0 | 0 | 0 | 1 | 3 |
| Heating Degree Days | 1006 | 797 | 581 | 270 | 95 | 8 | 1 | 2 | 47 | 249 | 548 | 858 | 4462 |
| Cooling Degree Days | 0 | 0 | 2 | 18 | 70 | 245 | 387 | 351 | 164 | 17 | 2 | 0 | 1256 |
| Total Precipitation (") | 2.10 | 2.41 | 3.85 | 4.28 | 4.27 | 4.32 | 3.41 | 3.45 | 4.36 | 3.44 | 4.46 | 3.36 | 43.71 |
| Days ≥ 0.1" Precip | 4 | 4 | 6 | 7 | 7 | 7 | 5 | 5 | 6 | 5 | 6 | 5 | 67 |
| Total Snowfall (") | 3.0 | 3.1 | 2.8 | 0.1 | 0.0 | 0.0 | 0.0 | 0.0 | 0.0 | 0.0 | 0.7 | 1.7 | 11.4 |
| Days ≥ 1" Snow Depth | 4 | 3 | 1 | 0 | 0 | 0 | 0 | 0 | 0 | 0 | 0 | 2 | 10 |

## PERRYVILLE WATR PLNT *Perry County* ELEVATION 480 ft LAT/LONG 37° 44 ' N / 89° 51 ' W

| | JAN | FEB | MAR | APR | MAY | JUN | JUL | AUG | SEP | OCT | NOV | DEC | YEAR |
|---|---|---|---|---|---|---|---|---|---|---|---|---|---|
| Maximum Temp °F | ***40.5*** | 46.0 | 56.5 | 68.5 | ***77.2*** | ***85.8*** | ***90.4*** | ***88.6*** | 80.9 | 70.6 | 57.2 | ***44.9*** | 67.3 |
| Minimum Temp °F | 20.1 | 23.7 | 33.1 | ***43.2*** | 51.8 | ***61.6*** | ***65.6*** | ***62.9*** | 55.2 | 42.4 | 33.9 | ***24.7*** | 43.2 |
| Mean Temp °F | ***30.4*** | 34.9 | 44.8 | ***55.8*** | ***64.5*** | ***73.7*** | ***78.1*** | ***75.8*** | 68.1 | 56.5 | 45.6 | ***34.8*** | 55.2 |
| Days Max Temp ≥ 90 °F | 0 | 0 | 0 | 0 | 1 | ***9*** | ***18*** | ***14*** | 5 | 0 | 0 | 0 | 47 |
| Days Max Temp ≤ 32 °F | ***8*** | 5 | 1 | 0 | 0 | ***0*** | 0 | 0 | 0 | 0 | 0 | 3 | 17 |
| Days Min Temp ≤ 32 °F | 27 | 21 | 16 | 5 | 0 | 0 | 0 | 0 | 0 | ***6*** | ***14*** | 23 | 112 |
| Days Min Temp ≤ 0 °F | 2 | 1 | 0 | 0 | 0 | 0 | 0 | 0 | 0 | 0 | 0 | 1 | 4 |
| Heating Degree Days | ***1066*** | 846 | 623 | ***297*** | ***104*** | ***8*** | ***1*** | ***4*** | 64 | 282 | 578 | ***927*** | 4800 |
| Cooling Degree Days | 0 | 0 | 5 | 30 | 99 | ***307*** | 442 | ***370*** | 175 | 24 | 3 | ***0*** | 1455 |
| Total Precipitation (") | 2.23 | 2.03 | 3.47 | 3.99 | 4.36 | 3.13 | 3.55 | 3.42 | 3.50 | 3.02 | 4.60 | 2.95 | 40.25 |
| Days ≥ 0.1" Precip | ***4*** | 3 | 5 | 6 | 6 | ***5*** | ***5*** | 5 | 5 | 5 | ***5*** | 4 | 58 |
| Total Snowfall (") | na | na | ***0.3*** | 0.0 | 0.0 | 0.0 | 0.0 | 0.0 | 0.0 | 0.0 | ***0.1*** | na | na |
| Days ≥ 1" Snow Depth | na | na | ***0*** | 0 | 0 | 0 | 0 | 0 | 0 | 0 | ***0*** | na | na |

## POMME DE TERRE DAM *Hickory County* ELEVATION 879 ft LAT/LONG 37° 55 ' N / 93° 19 ' W

| | JAN | FEB | MAR | APR | MAY | JUN | JUL | AUG | SEP | OCT | NOV | DEC | YEAR |
|---|---|---|---|---|---|---|---|---|---|---|---|---|---|
| Maximum Temp °F | ***39.4*** | ***44.7*** | 55.9 | 67.6 | 75.8 | 84.5 | 90.4 | 89.1 | 80.3 | 69.3 | 55.5 | ***44.5*** | 66.4 |
| Minimum Temp °F | ***17.9*** | 22.9 | 33.1 | 44.1 | 53.3 | 63.0 | 67.7 | 65.5 | 57.4 | 45.4 | 35.3 | ***25.0*** | 44.2 |
| Mean Temp °F | ***28.3*** | ***33.7*** | 44.5 | 55.8 | 64.6 | 73.7 | 79.1 | 77.3 | 68.9 | 57.4 | 45.4 | ***34.8*** | 55.3 |
| Days Max Temp ≥ 90 °F | 0 | 0 | 0 | 0 | 0 | 7 | 18 | 16 | 5 | 0 | 0 | 0 | 46 |
| Days Max Temp ≤ 32 °F | 8 | 5 | 1 | 0 | 0 | 0 | 0 | 0 | 0 | 0 | 1 | 4 | 19 |
| Days Min Temp ≤ 32 °F | 25 | 21 | 16 | 3 | 0 | 0 | 0 | 0 | 0 | 3 | 13 | 22 | 103 |
| Days Min Temp ≤ 0 °F | 2 | 1 | 0 | 0 | 0 | 0 | 0 | 0 | 0 | 0 | 0 | 1 | 4 |
| Heating Degree Days | ***1129*** | ***875*** | 631 | 294 | 98 | 10 | 1 | 2 | 61 | 261 | 583 | ***931*** | 4876 |
| Cooling Degree Days | na | na | 3 | 29 | 88 | 289 | 462 | 407 | 196 | 27 | 2 | na | na |
| Total Precipitation (") | 1.59 | 1.85 | 3.00 | 3.99 | 4.62 | 4.05 | 3.77 | 3.29 | 4.74 | 4.03 | 3.45 | 2.44 | 40.82 |
| Days ≥ 0.1" Precip | 3 | 4 | 6 | 7 | 8 | 7 | 6 | 5 | 7 | 6 | 5 | 4 | 68 |
| Total Snowfall (") | na | na | 1.2 | 0.1 | 0.0 | 0.0 | 0.0 | 0.0 | 0.0 | 0.0 | 0.5 | ***2.1*** | na |
| Days ≥ 1" Snow Depth | na | na | 1 | 0 | 0 | 0 | 0 | 0 | 0 | 0 | ***0*** | na | na |

## POPLAR BLUFF *Butler County* ELEVATION 331 ft LAT/LONG 36° 45 ' N / 90° 24 ' W

| | JAN | FEB | MAR | APR | MAY | JUN | JUL | AUG | SEP | OCT | NOV | DEC | YEAR |
|---|---|---|---|---|---|---|---|---|---|---|---|---|---|
| Maximum Temp °F | ***42.6*** | 47.7 | 58.6 | 69.9 | 79.0 | 87.1 | 91.4 | 88.9 | 81.8 | 71.0 | 57.8 | 47.4 | 68.6 |
| Minimum Temp °F | 23.7 | 27.3 | 36.5 | 46.8 | 55.9 | 64.5 | 68.8 | 66.0 | 59.1 | 45.7 | 37.2 | 28.5 | 46.7 |
| Mean Temp °F | ***33.1*** | 37.5 | 47.6 | 58.4 | 67.5 | 75.8 | 80.1 | 77.5 | 70.4 | 58.4 | 47.5 | 38.0 | 57.7 |
| Days Max Temp ≥ 90 °F | 0 | 0 | 0 | 0 | 2 | 12 | 21 | 15 | 5 | 0 | 0 | 0 | 55 |
| Days Max Temp ≤ 32 °F | 6 | 3 | 0 | 0 | 0 | 0 | 0 | 0 | 0 | 0 | 0 | 2 | 11 |
| Days Min Temp ≤ 32 °F | 25 | 20 | 12 | 2 | 0 | 0 | 0 | 0 | 0 | 2 | 11 | 21 | 93 |
| Days Min Temp ≤ 0 °F | 1 | 0 | 0 | 0 | 0 | 0 | 0 | 0 | 0 | 0 | 0 | 0 | 1 |
| Heating Degree Days | ***983*** | 757 | 537 | 224 | ***54*** | 3 | 0 | 1 | 35 | 233 | 525 | ***831*** | 4183 |
| Cooling Degree Days | 0 | 0 | ***5*** | 31 | ***144*** | 359 | 494 | 423 | 216 | 32 | 2 | 0 | 1706 |
| Total Precipitation (") | 3.01 | 3.50 | 4.63 | 5.00 | 4.45 | 3.43 | 3.46 | 3.39 | 3.90 | 3.29 | 4.75 | 4.34 | 47.15 |
| Days ≥ 0.1" Precip | 5 | 5 | 7 | 7 | ***6*** | 5 | 5 | 5 | 5 | 5 | 6 | 7 | 68 |
| Total Snowfall (") | 3.2 | ***2.6*** | 2.0 | 0.0 | 0.0 | 0.0 | 0.0 | 0.0 | 0.0 | 0.0 | 0.3 | ***1.0*** | 9.1 |
| Days ≥ 1" Snow Depth | na | na | ***0*** | 0 | 0 | 0 | 0 | 0 | 0 | 0 | 0 | ***0*** | na |

## PORTAGEVILLE *Pemiscot County* ELEVATION 279 ft LAT/LONG 36° 25 ' N / 89° 42 ' W

| | JAN | FEB | MAR | APR | MAY | JUN | JUL | AUG | SEP | OCT | NOV | DEC | YEAR |
|---|---|---|---|---|---|---|---|---|---|---|---|---|---|
| Maximum Temp °F | 41.3 | 46.6 | 57.7 | 69.3 | 77.9 | 87.1 | 90.3 | 87.8 | 81.2 | 71.3 | 58.2 | 46.8 | 68.0 |
| Minimum Temp °F | 25.5 | 29.2 | 39.1 | 48.7 | 57.4 | 66.0 | 69.8 | 66.8 | 59.9 | 47.5 | 39.5 | 30.6 | 48.3 |
| Mean Temp °F | 33.4 | 37.9 | 48.4 | 59.0 | 67.7 | 76.6 | 80.1 | 77.3 | 70.6 | 59.4 | 48.9 | 38.8 | 58.2 |
| Days Max Temp ≥ 90 °F | 0 | 0 | 0 | 0 | 2 | 12 | 19 | 12 | 4 | 0 | 0 | 0 | 49 |
| Days Max Temp ≤ 32 °F | 8 | 4 | 1 | 0 | 0 | 0 | 0 | 0 | 0 | 0 | 0 | 3 | 16 |
| Days Min Temp ≤ 32 °F | 23 | 18 | 9 | 1 | 0 | 0 | 0 | 0 | 0 | 1 | 7 | 19 | 78 |
| Days Min Temp ≤ 0 °F | 1 | 0 | 0 | 0 | 0 | 0 | 0 | 0 | 0 | 0 | 0 | 0 | 1 |
| Heating Degree Days | 973 | 759 | 513 | 213 | 53 | 2 | 0 | 0 | 33 | 209 | 482 | 807 | 4044 |
| Cooling Degree Days | 0 | 0 | 6 | *41* | 147 | 370 | 497 | 419 | 217 | 40 | *5* | 1 | 1743 |
| Total Precipitation (") | 3.36 | 3.35 | 4.28 | 5.19 | 5.05 | 3.95 | 3.52 | 3.01 | 3.66 | 3.30 | 4.40 | 4.60 | 47.67 |
| Days ≥ 0.1" Precip | 6 | 5 | 7 | 8 | 7 | 6 | 6 | 5 | 6 | 5 | 6 | 7 | 74 |
| Total Snowfall (") | 3.9 | 3.0 | 1.5 | *0.0* | 0.0 | 0.0 | 0.0 | 0.0 | 0.0 | 0.1 | 0.4 | 1.5 | 10.4 |
| Days ≥ 1" Snow Depth | *5* | *4* | 1 | *0* | 0 | 0 | 0 | 0 | 0 | 0 | 0 | 2 | 12 |

## PRINCETON 6 SW *Mercer County* ELEVATION 981 ft LAT/LONG 40° 21 ' N / 93° 41 ' W

| | JAN | FEB | MAR | APR | MAY | JUN | JUL | AUG | SEP | OCT | NOV | DEC | YEAR |
|---|---|---|---|---|---|---|---|---|---|---|---|---|---|
| Maximum Temp °F | 33.8 | 39.0 | 52.3 | 65.1 | 74.3 | 83.2 | 87.7 | 85.7 | 77.7 | 66.4 | 50.6 | 37.8 | 62.8 |
| Minimum Temp °F | 14.1 | 17.9 | 29.7 | 41.2 | 50.9 | 60.2 | 64.5 | 61.6 | 53.6 | 41.9 | 30.0 | 19.1 | 40.4 |
| Mean Temp °F | 24.0 | 28.5 | 41.0 | 53.2 | 62.6 | 71.7 | 76.1 | 73.7 | 65.7 | 54.2 | 40.3 | 28.5 | 51.6 |
| Days Max Temp ≥ 90 °F | 0 | 0 | 0 | 0 | 0 | 5 | 13 | 9 | 3 | 0 | 0 | 0 | 30 |
| Days Max Temp ≤ 32 °F | 13 | 9 | 2 | 0 | 0 | 0 | 0 | 0 | 0 | 0 | 2 | 9 | 35 |
| Days Min Temp ≤ 32 °F | 29 | 25 | 19 | 6 | 0 | 0 | 0 | 0 | 0 | 6 | 18 | 28 | 131 |
| Days Min Temp ≤ 0 °F | 6 | 3 | 0 | 0 | 0 | 0 | 0 | 0 | 0 | 0 | 0 | 3 | 12 |
| Heating Degree Days | 1265 | 1024 | 738 | 363 | 133 | 13 | 1 | 7 | 88 | 343 | 734 | 1126 | 5835 |
| Cooling Degree Days | 0 | 0 | 2 | 22 | 57 | 222 | 343 | 283 | 119 | 10 | 0 | 0 | 1058 |
| Total Precipitation (") | 1.02 | 0.91 | 2.27 | 3.64 | 4.08 | 4.15 | 5.12 | 3.70 | 4.19 | 3.06 | 2.05 | 1.63 | 35.82 |
| Days ≥ 0.1" Precip | 3 | 3 | 6 | 7 | 8 | 7 | 6 | 6 | 7 | 6 | 5 | 4 | 68 |
| Total Snowfall (") | na | na | *1.0* | 0.8 | 0.0 | 0.0 | 0.0 | 0.0 | 0.0 | 0.0 | *1.1* | na | na |
| Days ≥ 1" Snow Depth | na | na | na | *0* | 0 | 0 | 0 | 0 | 0 | 0 | *0* | na | na |

## ROLLA UNIV OF MO *Phelps County* ELEVATION 1201 ft LAT/LONG 37° 57 ' N / 91° 46 ' W

| | JAN | FEB | MAR | APR | MAY | JUN | JUL | AUG | SEP | OCT | NOV | DEC | YEAR |
|---|---|---|---|---|---|---|---|---|---|---|---|---|---|
| Maximum Temp °F | *40.2* | 44.7 | *54.8* | *66.6* | *75.0* | *83.6* | *88.9* | *87.4* | *78.6* | *68.1* | *55.2* | *43.5* | 65.6 |
| Minimum Temp °F | *21.4* | 24.5 | *34.5* | 45.3 | 54.3 | 63.2 | 68.0 | *65.7* | 57.9 | 46.3 | *36.7* | *25.9* | 45.3 |
| Mean Temp °F | *31.1* | 34.6 | *44.7* | *56.0* | *64.7* | *73.5* | *78.5* | *76.6* | *68.2* | *57.3* | *45.9* | *34.7* | 55.5 |
| Days Max Temp ≥ 90 °F | 0 | 0 | 0 | 0 | 0 | *6* | *15* | *12* | 4 | 0 | 0 | 0 | 37 |
| Days Max Temp ≤ 32 °F | *9* | *6* | *1* | 0 | 0 | 0 | 0 | 0 | 0 | 0 | 1 | *5* | 22 |
| Days Min Temp ≤ 32 °F | *26* | 21 | *15* | *3* | 0 | 0 | 0 | 0 | 0 | *2* | *11* | *23* | 101 |
| Days Min Temp ≤ 0 °F | *2* | 1 | 0 | 0 | 0 | 0 | 0 | 0 | 0 | 0 | 0 | 1 | 4 |
| Heating Degree Days | *1044* | 849 | *628* | *299* | *95* | *8* | 1 | *3* | *64* | *263* | *568* | *932* | 4754 |
| Cooling Degree Days | 0 | 0 | 5 | 37 | 94 | 273 | 438 | 376 | 173 | 26 | 3 | 0 | 1425 |
| Total Precipitation (") | 1.88 | 2.28 | 3.17 | 4.20 | 4.70 | 3.78 | 4.22 | 3.84 | 4.13 | 3.64 | 4.31 | 3.38 | 43.53 |
| Days ≥ 0.1" Precip | 4 | 4 | *7* | 8 | 8 | 6 | 6 | 6 | 6 | 6 | 6 | *5* | 72 |
| Total Snowfall (") | *4.7* | *5.1* | na | 0.5 | 0.0 | 0.0 | 0.0 | 0.0 | 0.0 | 0.1 | 1.4 | *3.9* | na |
| Days ≥ 1" Snow Depth | *8* | *6* | *2* | 0 | 0 | 0 | 0 | 0 | 0 | 0 | 1 | *4* | 21 |

## SALEM *Dent County* ELEVATION 1201 ft LAT/LONG 37° 38 ' N / 91° 33 ' W

| | JAN | FEB | MAR | APR | MAY | JUN | JUL | AUG | SEP | OCT | NOV | DEC | YEAR |
|---|---|---|---|---|---|---|---|---|---|---|---|---|---|
| Maximum Temp °F | 42.1 | 46.5 | 58.1 | 69.3 | 76.7 | 84.0 | 89.2 | 87.7 | 80.1 | 70.2 | 57.0 | 46.1 | 67.3 |
| Minimum Temp °F | 20.9 | 24.7 | 34.4 | 44.6 | 53.0 | 61.2 | 66.0 | 63.8 | 56.4 | 45.3 | 35.8 | 25.9 | 44.3 |
| Mean Temp °F | 31.5 | 35.6 | 46.3 | 57.0 | 64.9 | 72.6 | 77.6 | 75.8 | 68.3 | 57.8 | 46.5 | 36.0 | 55.8 |
| Days Max Temp ≥ 90 °F | 0 | 0 | 0 | 0 | 0 | 6 | 15 | 12 | 3 | 0 | 0 | 0 | 36 |
| Days Max Temp ≤ 32 °F | 8 | 5 | 1 | 0 | 0 | 0 | 0 | 0 | 0 | 0 | 0 | 4 | 18 |
| Days Min Temp ≤ 32 °F | 26 | 21 | 15 | 4 | 0 | 0 | 0 | 0 | 0 | 4 | 12 | 23 | 105 |
| Days Min Temp ≤ 0 °F | 2 | 1 | 0 | 0 | 0 | 0 | 0 | 0 | 0 | 0 | 0 | 1 | 4 |
| Heating Degree Days | 1031 | 823 | 579 | 266 | 93 | 11 | 2 | 3 | 57 | 248 | 552 | 892 | 4557 |
| Cooling Degree Days | 0 | 0 | 6 | 35 | 101 | *264* | 423 | 383 | 172 | 28 | 3 | 0 | 1415 |
| Total Precipitation (") | 2.19 | 2.23 | 3.70 | 4.52 | 4.36 | 3.41 | 3.29 | 4.20 | 4.40 | 3.50 | 4.33 | 3.43 | 43.56 |
| Days ≥ 0.1" Precip | 4 | 4 | 7 | 8 | 7 | *6* | 5 | 6 | 6 | 5 | 6 | 6 | 70 |
| Total Snowfall (") | na | na | *1.0* | 0.0 | 0.0 | 0.0 | 0.0 | 0.0 | 0.0 | 0.0 | 0.2 | *1.5* | na |
| Days ≥ 1" Snow Depth | na | na | *1* | 0 | 0 | 0 | 0 | 0 | 0 | 0 | 0 | *1* | na |

## SALISBURY *Chariton County* ELEVATION 722 ft LAT/LONG 39° 26 ' N / 92° 48 ' W

| | JAN | FEB | MAR | APR | MAY | JUN | JUL | AUG | SEP | OCT | NOV | DEC | YEAR |
|---|---|---|---|---|---|---|---|---|---|---|---|---|---|
| Maximum Temp °F | 35.5 | 41.2 | 54.0 | 66.0 | 75.3 | 83.8 | 88.5 | 86.5 | 79.0 | 67.8 | 52.9 | 40.7 | 64.3 |
| Minimum Temp °F | 16.4 | 20.7 | 31.9 | 43.0 | 52.3 | 61.6 | 66.0 | 62.9 | 55.1 | 43.6 | 33.1 | 22.7 | 42.4 |
| Mean Temp °F | 25.9 | 31.0 | 43.0 | 54.5 | 63.9 | 72.7 | 77.3 | 74.7 | 67.1 | 55.7 | 43.0 | 31.7 | 53.4 |
| Days Max Temp ≥ 90 °F | 0 | 0 | 0 | 0 | 0 | 6 | 14 | 10 | 4 | 0 | 0 | 0 | 34 |
| Days Max Temp ≤ 32 °F | 12 | 7 | 1 | 0 | 0 | 0 | 0 | 0 | 0 | 0 | 1 | 7 | 28 |
| Days Min Temp ≤ 32 °F | 29 | 24 | 17 | 4 | 0 | 0 | 0 | 0 | 0 | 4 | 14 | 25 | 117 |
| Days Min Temp ≤ 0 °F | 4 | 2 | 0 | 0 | 0 | 0 | 0 | 0 | 0 | 0 | 0 | 2 | 8 |
| Heating Degree Days | 1204 | 954 | 679 | 332 | 111 | 11 | 1 | 4 | 73 | 303 | 653 | 1024 | 5349 |
| Cooling Degree Days | 0 | 0 | 4 | 27 | 79 | 253 | 388 | 316 | 141 | 18 | 1 | 0 | 1227 |
| Total Precipitation (") | 1.55 | 1.54 | 2.79 | 3.96 | 4.90 | 4.81 | 4.46 | 3.85 | 4.87 | 3.62 | 3.11 | 2.36 | 41.82 |
| Days ≥ 0.1" Precip | 4 | 4 | 6 | 7 | 8 | 7 | 6 | 6 | 6 | 5 | 6 | 5 | 70 |
| Total Snowfall (") | 5.9 | 5.5 | 3.2 | 0.6 | 0.0 | 0.0 | 0.0 | 0.0 | 0.0 | 0.0 | 1.5 | 4.3 | 21.0 |
| Days ≥ 1" Snow Depth | 11 | 8 | 2 | 0 | 0 | 0 | 0 | 0 | 0 | 0 | 1 | 4 | 26 |

## SAVERTON L & D 22 *Ralls County* ELEVATION 469 ft LAT/LONG 39° 38 ' N / 91° 15 ' W

| | JAN | FEB | MAR | APR | MAY | JUN | JUL | AUG | SEP | OCT | NOV | DEC | YEAR |
|---|---|---|---|---|---|---|---|---|---|---|---|---|---|
| Maximum Temp °F | 35.4 | 40.2 | 52.5 | 65.3 | 74.9 | 84.2 | 88.7 | 86.1 | 78.5 | 67.3 | 52.9 | 40.5 | 63.9 |
| Minimum Temp °F | 17.6 | 21.6 | 32.3 | 44.1 | 53.7 | 63.2 | 67.8 | 65.5 | 57.8 | 46.1 | 34.9 | 23.7 | 44.0 |
| Mean Temp °F | 26.5 | 30.9 | 42.4 | 54.7 | 64.4 | 73.7 | 78.3 | 75.8 | 68.2 | 56.7 | 43.9 | 32.1 | 54.0 |
| Days Max Temp ≥ 90 °F | 0 | 0 | 0 | 0 | 1 | 7 | 15 | 10 | 3 | 0 | 0 | 0 | 36 |
| Days Max Temp ≤ 32 °F | 12 | 8 | 2 | 0 | 0 | 0 | 0 | 0 | 0 | 0 | 1 | 7 | 30 |
| Days Min Temp ≤ 32 °F | 28 | 23 | 17 | 3 | 0 | 0 | 0 | 0 | 0 | 2 | 13 | 25 | 111 |
| Days Min Temp ≤ 0 °F | 4 | 2 | 0 | 0 | 0 | 0 | 0 | 0 | 0 | 0 | 0 | 2 | 8 |
| Heating Degree Days | 1188 | 958 | 698 | 327 | 104 | 8 | 1 | 2 | 57 | 275 | 629 | 1014 | 5261 |
| Cooling Degree Days | 0 | 0 | 5 | 26 | 84 | 278 | 429 | 364 | 167 | 19 | 1 | 0 | 1373 |
| Total Precipitation (") | 1.32 | 1.45 | 2.82 | 3.85 | 4.42 | 3.67 | 4.14 | 3.65 | 4.22 | 3.18 | 3.15 | 2.38 | 38.25 |
| Days ≥ 0.1" Precip | 4 | 4 | 6 | 7 | 7 | 6 | 6 | 6 | 6 | 6 | 6 | 4 | 68 |
| Total Snowfall (") | na | 4.0 | 2.1 | 0.1 | 0.0 | 0.0 | 0.0 | 0.0 | 0.0 | 0.0 | 0.4 | 3.0 | na |
| Days ≥ 1" Snow Depth | 10 | 7 | 3 | 0 | 0 | 0 | 0 | 0 | 0 | 0 | 0 | 4 | 24 |

## SEDALIA WATER PLANT *Pettis County* ELEVATION 781 ft LAT/LONG 38° 40 ' N / 93° 13 ' W

| | JAN | FEB | MAR | APR | MAY | JUN | JUL | AUG | SEP | OCT | NOV | DEC | YEAR |
|---|---|---|---|---|---|---|---|---|---|---|---|---|---|
| Maximum Temp °F | 37.7 | 42.6 | 54.9 | 66.6 | 75.1 | 83.9 | 89.2 | 87.5 | 79.7 | 68.3 | 54.1 | 42.4 | 65.2 |
| Minimum Temp °F | 16.7 | 20.4 | 31.7 | 42.3 | 51.6 | 61.2 | 65.6 | 62.6 | 54.7 | 42.7 | 32.3 | 22.8 | 42.1 |
| Mean Temp °F | 27.2 | 31.6 | 43.3 | 54.5 | 63.4 | 72.6 | 77.4 | 75.1 | 67.2 | 55.5 | 43.3 | 32.6 | 53.6 |
| Days Max Temp ≥ 90 °F | 0 | 0 | 0 | 0 | 0 | 6 | 16 | 13 | 4 | 0 | 0 | 0 | 39 |
| Days Max Temp ≤ 32 °F | 10 | 7 | 1 | 0 | 0 | 0 | 0 | 0 | 0 | 0 | 1 | 6 | 25 |
| Days Min Temp ≤ 32 °F | 28 | 24 | 18 | 5 | 0 | 0 | 0 | 0 | 0 | 5 | 16 | 26 | 122 |
| Days Min Temp ≤ 0 °F | 4 | 2 | 0 | 0 | 0 | 0 | 0 | 0 | 0 | 0 | 0 | 1 | 7 |
| Heating Degree Days | 1166 | 937 | 670 | 333 | 121 | 12 | 2 | 5 | 77 | 311 | 647 | 996 | 5277 |
| Cooling Degree Days | 0 | 0 | 3 | 22 | 60 | 239 | 382 | 309 | 136 | 15 | 1 | 0 | 1167 |
| Total Precipitation (") | 1.45 | 1.45 | 2.84 | 4.05 | 4.67 | 4.99 | 4.21 | 3.60 | 4.32 | 3.88 | 3.39 | 2.38 | 41.23 |
| Days ≥ 0.1" Precip | 4 | 4 | 6 | 7 | 8 | 7 | 6 | 5 | 6 | 6 | 6 | 5 | 70 |
| Total Snowfall (") | 3.8 | 4.3 | 1.6 | 0.5 | 0.0 | 0.0 | 0.0 | 0.0 | 0.0 | 0.0 | 1.2 | 2.2 | 13.6 |
| Days ≥ 1" Snow Depth | 7 | 5 | 1 | 0 | 0 | 0 | 0 | 0 | 0 | 0 | 1 | 3 | 17 |

## SHELBINA *Shelby County* ELEVATION 771 ft LAT/LONG 39° 42 ' N / 92° 3 ' W

| | JAN | FEB | MAR | APR | MAY | JUN | JUL | AUG | SEP | OCT | NOV | DEC | YEAR |
|---|---|---|---|---|---|---|---|---|---|---|---|---|---|
| Maximum Temp °F | 36.0 | 42.0 | 54.3 | 67.1 | 75.4 | 84.4 | 88.9 | 87.0 | 79.4 | 68.6 | 53.1 | 40.6 | 64.7 |
| Minimum Temp °F | 16.3 | 20.7 | 31.4 | 42.2 | 51.6 | 61.0 | 65.3 | 62.2 | 54.3 | 42.9 | 32.1 | 22.1 | 41.8 |
| Mean Temp °F | 26.2 | 31.5 | 43.0 | 54.9 | 63.5 | 72.8 | 77.2 | 74.7 | 66.9 | 55.8 | 42.6 | 31.2 | 53.4 |
| Days Max Temp ≥ 90 °F | 0 | 0 | 0 | 0 | 1 | 7 | 14 | 12 | 3 | 0 | 0 | 0 | 37 |
| Days Max Temp ≤ 32 °F | 12 | 7 | 1 | 0 | 0 | 0 | 0 | 0 | 0 | 0 | 1 | 7 | 28 |
| Days Min Temp ≤ 32 °F | 28 | 23 | 17 | 5 | 1 | 0 | 0 | 0 | 0 | 5 | 17 | 25 | 121 |
| Days Min Temp ≤ 0 °F | 4 | 3 | 0 | 0 | 0 | 0 | 0 | 0 | 0 | 0 | 0 | 2 | 9 |
| Heating Degree Days | 1197 | 940 | 679 | 319 | 123 | 12 | 2 | 4 | 73 | 301 | 665 | 1040 | 5355 |
| Cooling Degree Days | 0 | na | 3 | 24 | 77 | 269 | 389 | 328 | 135 | 17 | 0 | 0 | na |
| Total Precipitation (") | 1.54 | 1.38 | 2.91 | 3.82 | 4.45 | 4.15 | 4.14 | 3.92 | 4.23 | 3.38 | 3.11 | 2.32 | 39.35 |
| Days ≥ 0.1" Precip | 3 | 3 | 6 | 7 | 7 | 6 | 6 | 6 | 6 | 6 | 6 | 5 | 67 |
| Total Snowfall (") | 5.6 | 5.2 | 3.4 | 0.4 | 0.0 | 0.0 | 0.0 | 0.0 | 0.0 | 0.0 | 1.2 | 3.4 | 19.2 |
| Days ≥ 1" Snow Depth | 10 | 9 | 3 | 0 | 0 | 0 | 0 | 0 | 0 | 0 | 1 | 5 | 28 |

## SPICKARD 7 W *Grundy County* ELEVATION 879 ft LAT/LONG 40° 15 ' N / 93° 43 ' W

| | JAN | FEB | MAR | APR | MAY | JUN | JUL | AUG | SEP | OCT | NOV | DEC | YEAR |
|---|---|---|---|---|---|---|---|---|---|---|---|---|---|
| Maximum Temp °F | 33.0 | 38.5 | 51.9 | 64.4 | 73.8 | 83.1 | 88.0 | 85.8 | 77.8 | 66.6 | 50.3 | 37.9 | 62.6 |
| Minimum Temp °F | 13.6 | 18.4 | 29.7 | 41.2 | 51.1 | 60.7 | 64.9 | 62.2 | 53.5 | 42.0 | 30.4 | 19.1 | 40.6 |
| Mean Temp °F | 23.3 | 28.4 | 40.9 | 52.8 | 62.5 | 71.9 | 76.5 | 74.0 | 65.7 | 54.3 | 40.4 | 28.5 | 51.6 |
| Days Max Temp ≥ 90 °F | 0 | 0 | 0 | 0 | 0 | 5 | 13 | 9 | 3 | 0 | 0 | 0 | 30 |
| Days Max Temp ≤ 32 °F | 14 | 9 | 2 | 0 | 0 | 0 | 0 | 0 | 0 | 0 | 2 | 9 | 36 |
| Days Min Temp ≤ 32 °F | 30 | 25 | 19 | 6 | 0 | 0 | 0 | 0 | 1 | 6 | 18 | 28 | 133 |
| Days Min Temp ≤ 0 °F | 6 | 3 | 0 | 0 | 0 | 0 | 0 | 0 | 0 | 0 | 0 | 3 | 12 |
| Heating Degree Days | 1285 | 1026 | 743 | 375 | 136 | 15 | 1 | 7 | 92 | 342 | 733 | 1124 | 5879 |
| Cooling Degree Days | 0 | 0 | 2 | 19 | 59 | 229 | 366 | 301 | 125 | 12 | 0 | 0 | 1113 |
| Total Precipitation (") | 1.14 | 0.85 | 2.48 | 3.54 | 4.41 | 4.03 | 4.97 | 3.66 | 4.19 | 3.30 | 2.21 | 1.72 | 36.50 |
| Days ≥ 0.1" Precip | 3 | 3 | 5 | 7 | 8 | 7 | 6 | 6 | 7 | 6 | 5 | 4 | 67 |
| Total Snowfall (") | ***5.8*** | ***4.6*** | 3.3 | 0.3 | 0.0 | 0.0 | 0.0 | 0.0 | 0.0 | 0.0 | 1.1 | ***3.4*** | 18.5 |
| Days ≥ 1" Snow Depth | na | na | ***3*** | 0 | 0 | 0 | 0 | 0 | 0 | 0 | ***1*** | ***5*** | na |

## SPRINGFIELD REG AP *Greene County* ELEVATION 1286 ft LAT/LONG 37° 14 ' N / 93° 23 ' W

| | JAN | FEB | MAR | APR | MAY | JUN | JUL | AUG | SEP | OCT | NOV | DEC | YEAR |
|---|---|---|---|---|---|---|---|---|---|---|---|---|---|
| Maximum Temp °F | 41.6 | 46.6 | 57.4 | 67.9 | 75.5 | 84.4 | 89.6 | 88.4 | 80.0 | 69.0 | 55.9 | 45.9 | 66.9 |
| Minimum Temp °F | 21.3 | 25.4 | 34.8 | 44.4 | 53.1 | 62.2 | 66.8 | 65.1 | 57.6 | 45.8 | 35.5 | 26.2 | 44.9 |
| Mean Temp °F | 31.5 | 36.0 | 46.1 | 56.2 | 64.4 | 73.3 | 78.2 | 76.8 | 68.8 | 57.5 | 45.7 | 36.1 | 55.9 |
| Days Max Temp ≥ 90 °F | 0 | 0 | 0 | 0 | 0 | 6 | 17 | 15 | 4 | 0 | 0 | 0 | 42 |
| Days Max Temp ≤ 32 °F | 7 | 5 | 1 | 0 | 0 | 0 | 0 | 0 | 0 | 0 | 1 | 4 | 18 |
| Days Min Temp ≤ 32 °F | 26 | 20 | 14 | 3 | 0 | 0 | 0 | 0 | 0 | 2 | 12 | 23 | 100 |
| Days Min Temp ≤ 0 °F | 2 | 1 | 0 | 0 | 0 | 0 | 0 | 0 | 0 | 0 | 0 | 1 | 4 |
| Heating Degree Days | 1032 | 812 | 581 | 282 | 97 | 7 | 1 | 1 | 57 | 256 | 573 | 890 | 4589 |
| Cooling Degree Days | 0 | 0 | 3 | 27 | 89 | 277 | 436 | 398 | 186 | 25 | 2 | 0 | 1443 |
| Total Precipitation (") | 1.95 | 2.23 | 3.70 | 4.38 | 4.16 | 5.04 | 3.30 | 3.69 | 4.99 | 3.75 | 4.31 | 3.30 | 44.80 |
| Days ≥ 0.1" Precip | 4 | 5 | 7 | 7 | 8 | 7 | 5 | 6 | 6 | 6 | 6 | 5 | 72 |
| Total Snowfall (") | 5.1 | 5.0 | 3.7 | 0.4 | 0.0 | 0.0 | 0.0 | 0.0 | 0.0 | 0.0 | 1.6 | 3.8 | 19.6 |
| Days ≥ 1" Snow Depth | 8 | 6 | 2 | 0 | 0 | 0 | 0 | 0 | 0 | 0 | 1 | 4 | 21 |

## ST CHARLES *St. Charles County* ELEVATION 522 ft LAT/LONG 38° 47 ' N / 90° 29 ' W

| | JAN | FEB | MAR | APR | MAY | JUN | JUL | AUG | SEP | OCT | NOV | DEC | YEAR |
|---|---|---|---|---|---|---|---|---|---|---|---|---|---|
| Maximum Temp °F | 37.5 | 42.7 | 54.3 | ***67.1*** | 75.8 | 84.7 | 89.0 | 86.8 | 79.7 | 68.4 | ***54.7*** | 42.9 | 65.3 |
| Minimum Temp °F | 17.8 | 22.0 | 32.5 | 43.3 | 52.1 | 61.8 | 66.2 | 62.9 | 54.9 | 42.9 | 33.8 | 23.6 | 42.8 |
| Mean Temp °F | 27.6 | 32.5 | 43.5 | ***55.3*** | 64.0 | 73.3 | 77.6 | 74.8 | 67.4 | 55.7 | ***44.2*** | 33.3 | 54.1 |
| Days Max Temp ≥ 90 °F | 0 | 0 | 0 | ***0*** | 1 | 8 | 15 | 11 | 4 | 0 | ***0*** | 0 | 39 |
| Days Max Temp ≤ 32 °F | 11 | 7 | 1 | 0 | 0 | 0 | 0 | 0 | 0 | 0 | ***1*** | 6 | 26 |
| Days Min Temp ≤ 32 °F | 28 | 24 | 17 | ***4*** | 0 | 0 | 0 | 0 | 0 | 4 | 15 | 25 | 117 |
| Days Min Temp ≤ 0 °F | 3 | 1 | 0 | 0 | 0 | 0 | 0 | 0 | 0 | 0 | 0 | 1 | 5 |
| Heating Degree Days | 1152 | 910 | 665 | ***313*** | 113 | 11 | 1 | 3 | 67 | 305 | ***620*** | 973 | 5133 |
| Cooling Degree Days | 0 | 0 | 6 | 32 | 87 | 274 | ***408*** | 321 | 147 | 20 | ***3*** | 0 | 1298 |
| Total Precipitation (") | 1.99 | 2.10 | 3.07 | 4.09 | 4.05 | 3.35 | 4.06 | 3.09 | 3.32 | 2.99 | 3.37 | 3.18 | 38.66 |
| Days ≥ 0.1" Precip | 4 | 4 | 7 | 8 | 7 | 6 | 6 | 5 | 5 | 5 | 6 | 6 | 69 |
| Total Snowfall (") | na | na | ***2.1*** | 0.1 | 0.0 | 0.0 | 0.0 | 0.0 | 0.0 | 0.0 | ***0.9*** | ***3.3*** | na |
| Days ≥ 1" Snow Depth | na | na | na | 0 | 0 | 0 | 0 | 0 | 0 | 0 | ***0*** | na | na |

## ST JOSEPH ROSECRANS *Buchanan County* ELEVATION 811 ft LAT/LONG 39° 46 ' N / 94° 55 ' W

| | JAN | FEB | MAR | APR | MAY | JUN | JUL | AUG | SEP | OCT | NOV | DEC | YEAR |
|---|---|---|---|---|---|---|---|---|---|---|---|---|---|
| Maximum Temp °F | 35.2 | 41.5 | 54.5 | 67.1 | 76.5 | 85.7 | 89.5 | 86.3 | 79.2 | 68.7 | 52.6 | 39.3 | 64.7 |
| Minimum Temp °F | 15.7 | 20.7 | 32.0 | 43.0 | 53.6 | 63.3 | 67.4 | 63.6 | 55.2 | 43.2 | 31.9 | 21.3 | 42.6 |
| Mean Temp °F | 25.4 | 31.1 | 43.3 | 55.1 | 65.1 | 74.5 | 78.5 | 75.0 | 67.2 | 56.0 | 42.3 | 30.3 | 53.7 |
| Days Max Temp ≥ 90 °F | 0 | 0 | 0 | 1 | 2 | 9 | 16 | 11 | 4 | 1 | 0 | 0 | 44 |
| Days Max Temp ≤ 32 °F | 12 | 7 | 1 | 0 | 0 | 0 | 0 | 0 | 0 | 0 | 1 | 8 | 29 |
| Days Min Temp ≤ 32 °F | 29 | 24 | 16 | 5 | 0 | 0 | 0 | 0 | 0 | 5 | ***16*** | 27 | 122 |
| Days Min Temp ≤ 0 °F | 4 | 3 | 0 | 0 | 0 | 0 | 0 | 0 | 0 | 0 | 0 | 2 | 9 |
| Heating Degree Days | 1220 | 950 | 669 | 318 | 95 | 7 | 1 | 4 | 72 | 299 | 677 | 1069 | 5381 |
| Cooling Degree Days | ***0*** | 0 | ***3*** | ***31*** | ***104*** | ***310*** | ***446*** | ***318*** | ***159*** | ***21*** | ***1*** | ***0*** | 1393 |
| Total Precipitation (") | 0.97 | 0.93 | 2.29 | 3.27 | 4.64 | 4.16 | 3.70 | 3.96 | 4.22 | 3.31 | 1.81 | 1.58 | 34.84 |
| Days ≥ 0.1" Precip | 3 | 2 | 5 | 6 | 7 | 6 | 5 | 6 | 5 | 5 | 3 | 4 | 57 |
| Total Snowfall (") | na | na | ***2.2*** | ***0.3*** | ***0.0*** | ***0.0*** | ***0.0*** | ***0.0*** | ***0.0*** | na | ***0.5*** | na | na |
| Days ≥ 1" Snow Depth | ***7*** | ***7*** | ***2*** | ***0*** | ***0*** | ***0*** | ***0*** | ***0*** | ***0*** | ***0*** | ***0*** | na | na |

## ST LOUIS LAMBERT AP *St. Louis County* ELEVATION 577 ft LAT/LONG 38° 45 ' N / 90° 23 ' W

| | JAN | FEB | MAR | APR | MAY | JUN | JUL | AUG | SEP | OCT | NOV | DEC | YEAR |
|---|---|---|---|---|---|---|---|---|---|---|---|---|---|
| Maximum Temp °F | 37.8 | 43.0 | 54.9 | 67.1 | 76.1 | 85.2 | 89.3 | 87.0 | 79.8 | 68.1 | 54.4 | 42.7 | 65.5 |
| Minimum Temp °F | 20.7 | 24.7 | 35.0 | 45.9 | 55.3 | 65.2 | 69.9 | 67.3 | 59.6 | 47.2 | 36.8 | 26.3 | 46.2 |
| Mean Temp °F | 29.3 | 33.9 | 45.0 | 56.5 | 65.8 | 75.2 | 79.6 | 77.2 | 69.7 | 57.7 | 45.6 | 34.6 | 55.8 |
| Days Max Temp ≥ 90 °F | 0 | 0 | 0 | 0 | 1 | 9 | 16 | 12 | 4 | 0 | 0 | 0 | 42 |
| Days Max Temp ≤ 32 °F | 11 | 7 | 1 | 0 | 0 | 0 | 0 | 0 | 0 | 0 | 1 | 6 | 26 |
| Days Min Temp ≤ 32 °F | 26 | 21 | 13 | 3 | 0 | 0 | 0 | 0 | 0 | 2 | 11 | 22 | 98 |
| Days Min Temp ≤ 0 °F | 2 | 1 | 0 | 0 | 0 | 0 | 0 | 0 | 0 | 0 | 0 | 1 | 4 |
| Heating Degree Days | 1100 | 873 | 622 | 285 | 86 | 5 | 0 | 1 | 45 | 255 | 579 | 938 | 4789 |
| Cooling Degree Days | 0 | 0 | 8 | 42 | 124 | 344 | 493 | 423 | 214 | 38 | 5 | 0 | 1691 |
| Total Precipitation (") | 1.87 | 2.12 | 3.38 | 3.81 | 3.77 | 3.58 | 3.81 | 2.98 | 3.35 | 2.79 | 3.58 | 3.13 | 38.17 |
| Days ≥ 0.1" Precip | 4 | 4 | 7 | 7 | 7 | 6 | 6 | 5 | 6 | 5 | 6 | 5 | 68 |
| Total Snowfall (") | 6.1 | 5.2 | 3.6 | 0.7 | 0.0 | 0.0 | 0.0 | 0.0 | 0.0 | 0.0 | 1.4 | 4.3 | 21.3 |
| Days ≥ 1" Snow Depth | 8 | 6 | 2 | 0 | 0 | 0 | 0 | 0 | 0 | 0 | 1 | 4 | 21 |

## STEELVILLE 2 N *Crawford County* ELEVATION 700 ft LAT/LONG 37° 59 ' N / 91° 22 ' W

| | JAN | FEB | MAR | APR | MAY | JUN | JUL | AUG | SEP | OCT | NOV | DEC | YEAR |
|---|---|---|---|---|---|---|---|---|---|---|---|---|---|
| Maximum Temp °F | 40.7 | 46.2 | 57.3 | 69.5 | 77.3 | 84.7 | 89.9 | 87.8 | 80.0 | 69.0 | 56.7 | 44.9 | 67.0 |
| Minimum Temp °F | 16.3 | 19.8 | 29.1 | 39.2 | 47.7 | 57.2 | 62.5 | 60.1 | 52.1 | 39.1 | 30.6 | 21.3 | 39.6 |
| Mean Temp °F | 28.5 | 33.0 | 43.2 | 54.4 | 62.5 | 71.0 | 76.2 | 74.0 | 66.1 | 54.1 | 43.6 | 33.1 | 53.3 |
| Days Max Temp ≥ 90 °F | 0 | 0 | 0 | 1 | 1 | 7 | 17 | 13 | 4 | 0 | 0 | 0 | 43 |
| Days Max Temp ≤ 32 °F | 9 | 5 | 1 | 0 | 0 | 0 | 0 | 0 | 0 | 0 | 1 | 5 | 21 |
| Days Min Temp ≤ 32 °F | 28 | 24 | 20 | 9 | 2 | 0 | 0 | 0 | 1 | 10 | 18 | 25 | 137 |
| Days Min Temp ≤ 0 °F | 4 | 2 | 0 | 0 | 0 | 0 | 0 | 0 | 0 | 0 | 0 | 2 | 8 |
| Heating Degree Days | 1125 | 895 | 674 | 337 | 137 | 20 | 3 | 7 | 91 | 346 | 635 | 983 | 5253 |
| Cooling Degree Days | 0 | 0 | 4 | 20 | 61 | 209 | 354 | 299 | 125 | 12 | 1 | 0 | 1085 |
| Total Precipitation (") | 1.96 | 2.21 | 3.16 | 4.31 | 4.24 | 3.62 | 3.75 | 3.75 | 3.92 | 3.40 | 3.92 | 2.88 | 41.12 |
| Days ≥ 0.1" Precip | 4 | 4 | 7 | 7 | 7 | 6 | 6 | 6 | 6 | 6 | 7 | 5 | 71 |
| Total Snowfall (") | ***4.9*** | 4.2 | 2.0 | 0.1 | 0.0 | 0.0 | 0.0 | 0.0 | 0.0 | 0.0 | 0.7 | 2.7 | 14.6 |
| Days ≥ 1" Snow Depth | 8 | 6 | 2 | 0 | 0 | 0 | 0 | 0 | 0 | 0 | 1 | 3 | 20 |

## STEFFENVILLE *Lewis County* ELEVATION 650 ft LAT/LONG 39° 58 ' N / 91° 53 ' W

| | JAN | FEB | MAR | APR | MAY | JUN | JUL | AUG | SEP | OCT | NOV | DEC | YEAR |
|---|---|---|---|---|---|---|---|---|---|---|---|---|---|
| Maximum Temp °F | 34.3 | 39.3 | 52.8 | 65.2 | 74.3 | 83.0 | 87.9 | 85.6 | 78.5 | 66.8 | 52.1 | 39.0 | 63.2 |
| Minimum Temp °F | 15.3 | 19.4 | 30.9 | 42.1 | 51.4 | 61.1 | 65.4 | 62.5 | 54.7 | 43.3 | 32.5 | 21.6 | 41.7 |
| Mean Temp °F | 24.8 | 29.4 | 41.8 | 53.7 | 62.9 | 72.1 | 76.7 | 74.1 | 66.6 | 55.1 | 42.4 | 30.4 | 52.5 |
| Days Max Temp ≥ 90 °F | 0 | 0 | 0 | 0 | 0 | 5 | 12 | 9 | 3 | 0 | 0 | 0 | 29 |
| Days Max Temp ≤ 32 °F | 13 | 8 | 2 | 0 | 0 | 0 | 0 | 0 | 0 | 0 | 1 | 8 | 32 |
| Days Min Temp ≤ 32 °F | 28 | 24 | 18 | 5 | 0 | 0 | 0 | 0 | 0 | 5 | 16 | 26 | 122 |
| Days Min Temp ≤ 0 °F | 5 | 3 | 0 | 0 | 0 | 0 | 0 | 0 | 0 | 0 | 0 | 2 | 10 |
| Heating Degree Days | 1238 | 997 | 714 | 355 | 130 | 12 | 1 | 5 | 78 | 320 | 673 | 1068 | 5591 |
| Cooling Degree Days | 0 | 0 | 4 | 24 | 67 | 243 | 377 | 318 | 147 | 16 | 1 | 0 | 1197 |
| Total Precipitation (") | 1.36 | 1.24 | 2.66 | 3.38 | 4.48 | 3.50 | 4.54 | 3.44 | 3.97 | 3.28 | 2.95 | 2.07 | 36.87 |
| Days ≥ 0.1" Precip | 3 | 3 | 5 | 7 | 8 | 6 | 7 | 5 | 6 | 5 | 5 | 4 | 64 |
| Total Snowfall (") | 6.4 | 5.4 | 3.1 | 0.6 | 0.0 | 0.0 | 0.0 | 0.0 | 0.0 | 0.1 | 1.5 | 4.1 | 21.2 |
| Days ≥ 1" Snow Depth | 14 | 10 | 3 | 0 | 0 | 0 | 0 | 0 | 0 | 0 | 1 | 6 | 34 |

## SUMMERSVILLE *Texas County* ELEVATION 1181 ft LAT/LONG 37° 11 ' N / 91° 40 ' W

| | JAN | FEB | MAR | APR | MAY | JUN | JUL | AUG | SEP | OCT | NOV | DEC | YEAR |
|---|---|---|---|---|---|---|---|---|---|---|---|---|---|
| Maximum Temp °F | 42.8 | 47.8 | 59.1 | 70.2 | 76.1 | 84.2 | 89.3 | 87.7 | 80.0 | 69.8 | 56.6 | ***46.4*** | 67.5 |
| Minimum Temp °F | 21.4 | 25.6 | 34.8 | 44.8 | 52.2 | 60.9 | 65.6 | 63.3 | 56.6 | 45.1 | 35.6 | ***27.1*** | 44.4 |
| Mean Temp °F | 32.1 | 36.8 | 47.0 | 57.5 | 64.1 | 72.6 | 77.5 | 75.5 | 68.3 | 57.5 | 46.1 | ***36.8*** | 56.0 |
| Days Max Temp ≥ 90 °F | 0 | 0 | 0 | 0 | 0 | 4 | 15 | 12 | 3 | 0 | 0 | 0 | 34 |
| Days Max Temp ≤ 32 °F | 7 | 4 | 0 | 0 | 0 | 0 | 0 | 0 | 0 | 0 | 1 | 3 | 15 |
| Days Min Temp ≤ 32 °F | 25 | 20 | 13 | 3 | 0 | 0 | 0 | 0 | 0 | 3 | 12 | 20 | 96 |
| Days Min Temp ≤ 0 °F | 2 | 1 | 0 | 0 | 0 | 0 | 0 | 0 | 0 | 0 | 0 | 0 | 3 |
| Heating Degree Days | 1012 | 791 | 556 | 243 | 96 | 6 | 1 | 2 | 52 | 251 | 560 | ***868*** | 4438 |
| Cooling Degree Days | ***0*** | ***0*** | ***4*** | ***30*** | 68 | 254 | 416 | 359 | 165 | 23 | 1 | na | na |
| Total Precipitation (") | 1.96 | 2.29 | 3.83 | 4.21 | 4.64 | 3.96 | 3.43 | 3.39 | 4.24 | 3.35 | 3.81 | ***3.49*** | 42.60 |
| Days ≥ 0.1" Precip | 4 | 4 | 6 | 6 | 7 | 6 | 4 | 4 | 6 | 5 | 5 | 4 | 61 |
| Total Snowfall (") | 5.2 | ***4.6*** | 3.2 | 0.1 | 0.0 | 0.0 | 0.0 | 0.0 | 0.0 | 0.0 | 0.8 | ***2.4*** | 16.3 |
| Days ≥ 1" Snow Depth | ***5*** | ***6*** | ***1*** | 0 | 0 | 0 | 0 | 0 | 0 | 0 | 0 | 1 | 13 |

## SWEET SPRINGS *Saline County* ELEVATION 702 ft LAT/LONG 38° 58 ' N / 93° 25 ' W

| | JAN | FEB | MAR | APR | MAY | JUN | JUL | AUG | SEP | OCT | NOV | DEC | YEAR |
|---|---|---|---|---|---|---|---|---|---|---|---|---|---|
| Maximum Temp °F | 38.1 | 44.0 | 56.7 | 68.6 | 77.1 | 85.6 | 90.8 | 88.7 | 80.8 | 69.3 | 54.4 | 42.5 | 66.4 |
| Minimum Temp °F | 18.6 | 22.6 | 33.6 | 43.9 | 52.9 | 62.3 | 66.7 | 63.9 | 56.1 | 44.7 | 34.0 | 24.2 | 43.6 |
| Mean Temp °F | 28.4 | 33.3 | 45.2 | 56.3 | 65.1 | 74.0 | 78.7 | 76.3 | 68.5 | 57.0 | 44.2 | 33.4 | 55.0 |
| Days Max Temp ≥ 90 °F | 0 | 0 | 0 | 0 | 1 | 8 | 19 | 14 | 5 | 0 | 0 | 0 | 47 |
| Days Max Temp ≤ 32 °F | 10 | 6 | 1 | 0 | 0 | 0 | 0 | 0 | 0 | 0 | 1 | 6 | 24 |
| Days Min Temp ≤ 32 °F | 27 | 23 | 15 | 4 | 0 | 0 | 0 | 0 | 0 | 4 | 14 | 24 | 111 |
| Days Min Temp ≤ 0 °F | 3 | 2 | 0 | 0 | 0 | 0 | 0 | 0 | 0 | 0 | 0 | 1 | 6 |
| Heating Degree Days | 1130 | 888 | 611 | 285 | 87 | 6 | 1 | 2 | 58 | 268 | 618 | 974 | 4928 |
| Cooling Degree Days | 0 | 0 | 5 | 34 | 96 | 296 | 451 | 379 | 180 | 23 | 2 | 0 | 1466 |
| Total Precipitation (") | 1.41 | 1.58 | 2.98 | 4.35 | 4.65 | 4.19 | 3.97 | 3.86 | 4.57 | 3.60 | 3.08 | 2.32 | 40.56 |
| Days ≥ 0.1" Precip | 3 | 4 | 6 | 7 | 7 | 7 | 6 | 6 | 6 | 6 | 5 | 4 | 67 |
| Total Snowfall (") | 6.4 | 6.4 | 4.2 | 0.6 | 0.0 | 0.0 | 0.0 | 0.0 | 0.0 | 0.1 | 2.1 | 4.5 | 24.3 |
| Days ≥ 1" Snow Depth | 10 | 7 | 2 | 0 | 0 | 0 | 0 | 0 | 0 | 0 | 1 | 4 | 24 |

## UNION *Franklin County* ELEVATION 522 ft LAT/LONG 38° 27 ' N / 91° 0 ' W

| | JAN | FEB | MAR | APR | MAY | JUN | JUL | AUG | SEP | OCT | NOV | DEC | YEAR |
|---|---|---|---|---|---|---|---|---|---|---|---|---|---|
| Maximum Temp °F | 40.9 | 46.4 | 58.0 | 69.9 | 77.6 | 85.9 | 90.5 | 88.6 | 81.0 | 70.4 | 56.6 | 45.2 | 67.6 |
| Minimum Temp °F | 19.6 | 23.3 | 33.3 | 43.5 | 51.8 | 61.0 | 65.8 | 63.4 | 55.5 | 43.5 | 34.3 | 25.2 | 43.4 |
| Mean Temp °F | 30.3 | 34.9 | 45.7 | 56.7 | 64.7 | 73.5 | 78.2 | 76.0 | 68.3 | 57.0 | 45.5 | 35.2 | 55.5 |
| Days Max Temp ≥ 90 °F | 0 | 0 | 0 | 1 | 1 | 9 | 18 | 14 | 5 | 0 | 0 | 0 | 48 |
| Days Max Temp ≤ 32 °F | 9 | 5 | 1 | 0 | 0 | 0 | 0 | 0 | 0 | 0 | 0 | 4 | 19 |
| Days Min Temp ≤ 32 °F | 27 | 22 | 16 | 5 | 0 | 0 | 0 | 0 | 0 | 5 | 14 | 23 | 112 |
| Days Min Temp ≤ 0 °F | 2 | 1 | 0 | 0 | 0 | 0 | 0 | 0 | 0 | 0 | 0 | 1 | 4 |
| Heating Degree Days | 1070 | 844 | 598 | 274 | 93 | 7 | 1 | 2 | 55 | 267 | 582 | 917 | 4710 |
| Cooling Degree Days | 0 | 0 | 6 | 35 | 94 | 289 | 443 | 378 | 177 | 25 | 4 | 0 | 1451 |
| Total Precipitation (") | 2.03 | 2.14 | 3.49 | 4.07 | 4.06 | 3.81 | 3.71 | 3.48 | 4.17 | 3.34 | 4.00 | 3.21 | 41.51 |
| Days ≥ 0.1" Precip | 5 | 4 | 7 | 7 | 7 | 6 | 6 | 6 | 6 | 6 | 7 | 6 | 73 |
| Total Snowfall (") | 5.2 | 5.0 | 3.1 | 0.5 | 0.0 | 0.0 | 0.0 | 0.0 | 0.0 | 0.0 | 1.2 | 3.5 | 18.5 |
| Days ≥ 1" Snow Depth | 8 | 6 | 2 | 0 | 0 | 0 | 0 | 0 | 0 | 0 | 1 | 4 | 21 |

## UNIONVILLE *Putnam County* ELEVATION 1060 ft LAT/LONG 40° 28 ' N / 93° 0 ' W

| | JAN | FEB | MAR | APR | MAY | JUN | JUL | AUG | SEP | OCT | NOV | DEC | YEAR |
|---|---|---|---|---|---|---|---|---|---|---|---|---|---|
| Maximum Temp °F | 32.4 | 37.7 | 51.0 | 64.1 | 73.6 | 82.7 | 87.4 | 84.7 | ***77.0*** | 65.3 | 49.6 | ***37.0*** | 61.9 |
| Minimum Temp °F | 13.0 | 17.1 | 28.5 | 40.2 | 50.0 | 59.5 | 64.1 | 61.3 | ***53.8*** | 41.7 | 29.9 | ***18.8*** | 39.8 |
| Mean Temp °F | 22.8 | 27.4 | 39.7 | 52.2 | 61.8 | 71.1 | 75.8 | 73.0 | ***65.4*** | 53.5 | 39.8 | ***28.0*** | 50.9 |
| Days Max Temp ≥ 90 °F | 0 | 0 | 0 | 0 | 0 | 4 | 12 | 8 | ***2*** | 0 | 0 | ***0*** | 26 |
| Days Max Temp ≤ 32 °F | 15 | 10 | 2 | 0 | 0 | 0 | 0 | 0 | ***0*** | 0 | 2 | ***10*** | 39 |
| Days Min Temp ≤ 32 °F | 30 | 25 | 20 | 7 | 1 | 0 | 0 | 0 | ***0*** | 5 | ***19*** | ***28*** | 135 |
| Days Min Temp ≤ 0 °F | 6 | 4 | 0 | 0 | 0 | 0 | 0 | 0 | ***0*** | 0 | 0 | ***3*** | 13 |
| Heating Degree Days | 1303 | 1054 | 778 | 391 | 153 | 15 | 2 | 8 | ***92*** | 362 | 750 | ***1142*** | 6050 |
| Cooling Degree Days | ***0*** | ***0*** | ***1*** | ***15*** | ***44*** | ***190*** | ***317*** | ***245*** | na | ***5*** | ***0*** | ***0*** | na |
| Total Precipitation (") | 1.13 | 0.93 | 2.49 | 3.55 | 4.37 | 4.24 | 4.90 | 4.04 | ***4.52*** | 3.21 | 2.17 | 1.60 | 37.15 |
| Days ≥ 0.1" Precip | 3 | 3 | 5 | 7 | 8 | 7 | 6 | 7 | ***7*** | 6 | 4 | ***4*** | 67 |
| Total Snowfall (") | 7.0 | 5.0 | 3.0 | 1.6 | 0.0 | 0.0 | 0.0 | 0.0 | ***0.0*** | 0.1 | 2.2 | ***4.3*** | 23.2 |
| Days ≥ 1" Snow Depth | 14 | 11 | 3 | 0 | 0 | 0 | 0 | 0 | ***0*** | 0 | ***1*** | 7 | 36 |

## VAN BUREN RS *Carter County* ELEVATION 561 ft LAT/LONG 37° 0 ' N / 91° 1 ' W

| | JAN | FEB | MAR | APR | MAY | JUN | JUL | AUG | SEP | OCT | NOV | DEC | YEAR |
|---|---|---|---|---|---|---|---|---|---|---|---|---|---|
| Maximum Temp °F | 43.4 | 48.9 | 59.9 | 71.7 | 78.5 | 86.0 | 90.8 | 89.2 | 82.0 | 71.8 | 59.4 | 49.0 | 69.2 |
| Minimum Temp °F | 17.6 | 21.6 | 31.8 | 41.4 | 49.8 | 58.3 | 63.3 | 61.5 | 54.2 | 40.7 | 32.1 | 23.8 | 41.3 |
| Mean Temp °F | 30.5 | 35.0 | 45.9 | 56.6 | 64.2 | 72.2 | 77.1 | 75.4 | 68.0 | 56.3 | 45.8 | 36.5 | 55.3 |
| Days Max Temp ≥ 90 °F | 0 | 0 | 0 | 1 | 1 | 9 | 19 | 15 | 5 | 1 | 0 | 0 | 51 |
| Days Max Temp ≤ 32 °F | 6 | 3 | 0 | 0 | 0 | 0 | 0 | 0 | 0 | 0 | 0 | 2 | 11 |
| Days Min Temp ≤ 32 °F | 27 | 23 | 17 | 7 | 1 | 0 | 0 | 0 | 0 | 8 | ***17*** | 23 | 123 |
| Days Min Temp ≤ 0 °F | 3 | 1 | 0 | 0 | 0 | 0 | 0 | 0 | 0 | 0 | 0 | 1 | 5 |
| Heating Degree Days | 1062 | 840 | 590 | 274 | 100 | 11 | 2 | 2 | 57 | 287 | 572 | 877 | 4674 |
| Cooling Degree Days | 0 | 0 | 4 | 29 | 83 | 234 | ***384*** | 348 | 157 | 23 | 2 | 0 | 1264 |
| Total Precipitation (") | 3.11 | 3.06 | 4.52 | 4.67 | 4.63 | 3.68 | 4.39 | 3.87 | 4.01 | 3.10 | 4.83 | 4.68 | 48.55 |
| Days ≥ 0.1" Precip | 5 | 5 | 7 | 7 | 7 | 6 | 6 | 6 | 6 | 5 | 6 | 6 | 72 |
| Total Snowfall (") | ***3.2*** | ***2.9*** | 1.5 | 0.0 | 0.0 | 0.0 | 0.0 | 0.0 | 0.0 | 0.0 | 0.6 | ***1.8*** | 10.0 |
| Days ≥ 1" Snow Depth | na | na | ***0*** | 0 | 0 | 0 | 0 | 0 | 0 | 0 | 0 | ***1*** | na |

## VANDALIA *Audrain County* ELEVATION 771 ft LAT/LONG 39° 18 ' N / 91° 30 ' W

| | JAN | FEB | MAR | APR | MAY | JUN | JUL | AUG | SEP | OCT | NOV | DEC | YEAR |
|---|---|---|---|---|---|---|---|---|---|---|---|---|---|
| Maximum Temp °F | 36.0 | 41.5 | 54.1 | 66.1 | 75.4 | 84.5 | 89.3 | 87.1 | 79.3 | 68.1 | 53.7 | 41.8 | 64.7 |
| Minimum Temp °F | 17.6 | 21.3 | 32.0 | 42.8 | 52.3 | 61.5 | 65.9 | 63.5 | 55.5 | 43.7 | 34.0 | 24.0 | 42.8 |
| Mean Temp °F | 26.8 | 31.4 | 43.1 | 54.5 | 63.8 | 73.0 | 77.6 | 75.3 | 67.4 | 55.9 | 43.9 | 32.9 | 53.8 |
| Days Max Temp ≥ 90 °F | 0 | 0 | 0 | 0 | 1 | 7 | 15 | 11 | 3 | 0 | 0 | 0 | 37 |
| Days Max Temp ≤ 32 °F | 12 | 7 | 1 | 0 | 0 | 0 | 0 | 0 | 0 | 0 | 1 | 6 | 27 |
| Days Min Temp ≤ 32 °F | 28 | 24 | 17 | 5 | 0 | 0 | 0 | 0 | 0 | 5 | 14 | 25 | 118 |
| Days Min Temp ≤ 0 °F | 3 | 2 | 0 | 0 | 0 | 0 | 0 | 0 | 0 | 0 | 0 | 1 | 6 |
| Heating Degree Days | 1176 | 942 | 675 | 331 | 112 | 11 | 1 | 3 | 66 | 297 | 628 | 987 | 5229 |
| Cooling Degree Days | *0* | *0* | *3* | *21* | *72* | *261* | *410* | *346* | *143* | *16* | *1* | *0* | 1273 |
| Total Precipitation (") | 1.63 | 1.57 | 2.97 | 4.02 | 4.86 | 4.37 | 4.55 | 3.94 | 4.06 | 3.18 | 3.19 | 2.54 | 40.88 |
| Days ≥ 0.1" Precip | 4 | *4* | 6 | 7 | 8 | 6 | 6 | 6 | 5 | 6 | 6 | 5 | 69 |
| Total Snowfall (") | 5.4 | 5.1 | 3.1 | 0.6 | 0.0 | 0.0 | 0.0 | 0.0 | 0.0 | 0.0 | 1.8 | *4.5* | 20.5 |
| Days ≥ 1" Snow Depth | *9* | 7 | 2 | 0 | 0 | 0 | 0 | 0 | 0 | 0 | 1 | *4* | 23 |

## VERSAILLES *Morgan County* ELEVATION 1030 ft LAT/LONG 38° 26 ' N / 92° 51 ' W

| | JAN | FEB | MAR | APR | MAY | JUN | JUL | AUG | SEP | OCT | NOV | DEC | YEAR |
|---|---|---|---|---|---|---|---|---|---|---|---|---|---|
| Maximum Temp °F | 40.4 | 45.9 | 57.8 | 68.7 | 75.7 | 82.8 | 88.7 | 86.7 | 79.3 | 69.1 | 55.8 | 44.8 | 66.3 |
| Minimum Temp °F | 19.8 | 23.5 | 33.8 | 44.2 | 52.9 | 62.0 | 66.9 | 64.6 | 56.8 | 45.3 | 34.7 | 25.1 | 44.1 |
| Mean Temp °F | 30.1 | 34.7 | 45.8 | 56.5 | 64.3 | 72.4 | 77.8 | 75.7 | 68.1 | 57.2 | 45.3 | 35.0 | 55.2 |
| Days Max Temp ≥ 90 °F | 0 | 0 | 0 | 0 | 0 | 3 | 13 | 11 | 3 | 0 | 0 | 0 | 30 |
| Days Max Temp ≤ 32 °F | 8 | 5 | 1 | 0 | 0 | 0 | 0 | 0 | 0 | 0 | 1 | 4 | 19 |
| Days Min Temp ≤ 32 °F | 27 | 22 | 15 | 4 | 0 | 0 | 0 | 0 | 0 | 3 | 13 | 24 | 108 |
| Days Min Temp ≤ 0 °F | 2 | 1 | 0 | 0 | 0 | 0 | 0 | 0 | 0 | 0 | 0 | 1 | 4 |
| Heating Degree Days | 1074 | 848 | 592 | 279 | 93 | 9 | 1 | 2 | 60 | 260 | 588 | 925 | 4731 |
| Cooling Degree Days | 0 | 0 | 5 | 34 | 71 | 240 | 404 | 342 | 158 | 20 | 2 | 0 | 1276 |
| Total Precipitation (") | 1.64 | 1.91 | 3.21 | 4.43 | 4.78 | 3.91 | 3.85 | 3.98 | 4.57 | 4.30 | 3.73 | 2.75 | 43.06 |
| Days ≥ 0.1" Precip | 4 | 4 | 6 | 7 | 7 | 6 | 5 | 6 | 6 | 6 | 6 | 5 | 68 |
| Total Snowfall (") | na | na | *0.6* | 0.0 | 0.0 | 0.0 | 0.0 | 0.0 | 0.0 | 0.0 | 1.2 | *1.9* | na |
| Days ≥ 1" Snow Depth | 10 | 8 | 2 | 0 | 0 | 0 | 0 | 0 | 0 | 0 | 1 | 5 | 26 |

## VICHY ROLLA NATL AP *Maries County* ELEVATION 1129 ft LAT/LONG 38° 7 ' N / 91° 46 ' W

| | JAN | FEB | MAR | APR | MAY | JUN | JUL | AUG | SEP | OCT | NOV | DEC | YEAR |
|---|---|---|---|---|---|---|---|---|---|---|---|---|---|
| Maximum Temp °F | *38.3* | *44.1* | *54.9* | *67.0* | *74.5* | *82.8* | *87.8* | *85.7* | *77.9* | *67.6* | *54.0* | *44.4* | 64.9 |
| Minimum Temp °F | *20.6* | *24.8* | *34.6* | *45.8* | *54.6* | *63.2* | *68.2* | *65.7* | *58.6* | *47.2* | *36.5* | *27.4* | 45.6 |
| Mean Temp °F | *29.5* | *34.4* | *44.8* | *56.5* | *64.6* | *73.0* | *78.0* | *75.7* | *68.3* | *57.4* | *45.3* | *35.9* | 55.3 |
| Days Max Temp ≥ 90 °F | *0* | *0* | *0* | *0* | *0* | *5* | *14* | *10* | *3* | *0* | *0* | *0* | 32 |
| Days Max Temp ≤ 32 °F | *11* | *6* | *1* | *0* | *0* | *0* | *0* | *0* | *0* | *0* | *1* | *5* | 24 |
| Days Min Temp ≤ 32 °F | *25* | *21* | *15* | *3* | *0* | *0* | *0* | *0* | *0* | *2* | *11* | *22* | 99 |
| Days Min Temp ≤ 0 °F | *2* | *1* | *0* | *0* | *0* | *0* | *0* | *0* | *0* | *0* | *0* | *0* | 3 |
| Heating Degree Days | *1093* | *856* | *625* | *280* | *101* | *9* | *1* | *2* | *57* | *261* | *586* | *895* | 4766 |
| Cooling Degree Days | na | na | na | na | na | na | na | na | na | na | na | na | na |
| Total Precipitation (") | *1.65* | *1.55* | *3.05* | *4.23* | *4.45* | *4.07* | *3.88* | *4.07* | *4.14* | *3.29* | *3.72* | *3.00* | 41.10 |
| Days ≥ 0.1" Precip | *4* | *3* | *6* | *7* | *7* | *6* | *5* | *5* | *6* | *5* | *6* | *5* | 65 |
| Total Snowfall (") | *4.6* | *4.0* | *2.7* | *0.6* | *0.0* | *0.0* | *0.0* | *0.0* | *0.0* | *0.0* | *1.7* | *2.6* | 16.2 |
| Days ≥ 1" Snow Depth | *9* | *7* | *2* | *0* | *0* | *0* | *0* | *0* | *0* | *0* | *1* | *3* | 22 |

## VIENNA 2 WNW *Maries County* ELEVATION 770 ft LAT/LONG 38° 13 ' N / 92° 0 ' W

| | JAN | FEB | MAR | APR | MAY | JUN | JUL | AUG | SEP | OCT | NOV | DEC | YEAR |
|---|---|---|---|---|---|---|---|---|---|---|---|---|---|
| Maximum Temp °F | 39.3 | 44.8 | 55.6 | 67.7 | 75.9 | 84.0 | 89.3 | 88.1 | 80.1 | 69.2 | 55.7 | 44.3 | 66.2 |
| Minimum Temp °F | 17.7 | 21.8 | 31.8 | 42.3 | 51.2 | 60.3 | 64.8 | 62.7 | 54.5 | 42.0 | 33.0 | 23.6 | 42.1 |
| Mean Temp °F | 28.5 | 33.3 | 43.7 | 55.0 | 63.6 | 72.2 | 77.1 | 75.4 | 67.3 | 55.6 | 44.3 | 34.0 | 54.2 |
| Days Max Temp ≥ 90 °F | 0 | 0 | 0 | 0 | 0 | 6 | 17 | 14 | 5 | 0 | 0 | 0 | 42 |
| Days Max Temp ≤ 32 °F | 10 | 6 | 1 | 0 | 0 | 0 | 0 | 0 | 0 | 0 | 1 | 5 | 23 |
| Days Min Temp ≤ 32 °F | 28 | 23 | 17 | 6 | 1 | 0 | 0 | 0 | 0 | 7 | 15 | 25 | 122 |
| Days Min Temp ≤ 0 °F | 3 | 2 | 0 | 0 | 0 | 0 | 0 | 0 | 0 | 0 | 0 | 1 | 6 |
| Heating Degree Days | 1125 | 888 | 657 | 319 | 116 | 14 | 2 | 3 | 76 | 307 | 615 | 955 | 5077 |
| Cooling Degree Days | 0 | 0 | 5 | 27 | 77 | 242 | 398 | 343 | 153 | 17 | 3 | 0 | 1265 |
| Total Precipitation (") | 1.78 | 2.12 | 3.53 | 4.08 | 4.66 | 4.25 | 3.73 | 3.93 | 4.33 | 3.96 | 4.15 | 3.17 | 43.69 |
| Days ≥ 0.1" Precip | 4 | 4 | 7 | 7 | 7 | 6 | 6 | 6 | 6 | 6 | 6 | 5 | 70 |
| Total Snowfall (") | 5.7 | 4.4 | 3.0 | 0.3 | 0.0 | 0.0 | 0.0 | 0.0 | 0.0 | 0.0 | 1.6 | *3.5* | 18.5 |
| Days ≥ 1" Snow Depth | 10 | 7 | 2 | 0 | 0 | 0 | 0 | 0 | 0 | 0 | 1 | 4 | 24 |

## WAPPAPELLO DAM *Butler County* ELEVATION 469 ft LAT/LONG 36° 55' N / 90° 17' W

| | JAN | FEB | MAR | APR | MAY | JUN | JUL | AUG | SEP | OCT | NOV | DEC | YEAR |
|---|---|---|---|---|---|---|---|---|---|---|---|---|---|
| Maximum Temp °F | *41.0* | 46.9 | 57.7 | 70.1 | 77.9 | 86.6 | 91.3 | 89.0 | 81.4 | 70.6 | 57.8 | 46.7 | 68.1 |
| Minimum Temp °F | *21.4* | 25.9 | 35.9 | 47.1 | 55.5 | 64.5 | 68.7 | 66.1 | 59.3 | 46.3 | 37.1 | *27.9* | 46.3 |
| Mean Temp °F | *31.2* | 36.5 | 46.6 | 58.6 | 66.7 | 75.6 | 80.0 | 77.5 | 70.4 | 58.5 | 47.5 | *37.4* | 57.2 |
| Days Max Temp ≥ 90 °F | 0 | 0 | 0 | 0 | 2 | 11 | 20 | 15 | 5 | 0 | 0 | 0 | 53 |
| Days Max Temp ≤ 32 °F | 7 | 4 | 1 | 0 | 0 | 0 | 0 | 0 | 0 | 0 | 0 | 2 | 14 |
| Days Min Temp ≤ 32 °F | 25 | 20 | 12 | 2 | 0 | 0 | 0 | 0 | 0 | 1 | 10 | 21 | 91 |
| Days Min Temp ≤ 0 °F | 1 | 0 | 0 | 0 | 0 | 0 | 0 | 0 | 0 | 0 | 0 | 0 | 1 |
| Heating Degree Days | *1041* | 799 | 568 | 221 | 63 | 3 | 0 | 0 | 32 | 228 | 520 | *850* | 4325 |
| Cooling Degree Days | na | *0* | *4* | 38 | 121 | 344 | 500 | 427 | 214 | 30 | 2 | na | na |
| Total Precipitation (") | 3.53 | 3.32 | 4.61 | 4.87 | 4.43 | 3.29 | 3.47 | 3.33 | 3.98 | 3.45 | 4.99 | 4.13 | 47.40 |
| Days ≥ 0.1" Precip | 5 | 5 | 6 | 7 | 7 | 6 | 5 | 5 | 6 | 5 | 6 | 6 | 69 |
| Total Snowfall (") | *3.1* | *2.5* | 1.3 | 0.0 | 0.0 | 0.0 | 0.0 | 0.0 | 0.0 | 0.0 | 0.3 | 1.4 | 8.6 |
| Days ≥ 1" Snow Depth | na | *3* | 1 | 0 | 0 | 0 | 0 | 0 | 0 | 0 | 0 | *1* | na |

## WARRENSBURG 4 NW *Johnson County* ELEVATION 879 ft LAT/LONG 38° 46' N / 93° 44' W

| | JAN | FEB | MAR | APR | MAY | JUN | JUL | AUG | SEP | OCT | NOV | DEC | YEAR |
|---|---|---|---|---|---|---|---|---|---|---|---|---|---|
| Maximum Temp °F | 36.8 | *43.7* | 55.4 | 67.3 | 75.9 | 84.0 | 89.6 | 87.6 | 80.4 | 69.2 | 54.3 | 42.5 | 65.6 |
| Minimum Temp °F | 16.9 | *22.5* | 32.3 | 43.6 | 53.0 | 62.1 | 66.5 | 63.8 | 55.8 | 44.3 | 34.0 | 23.7 | 43.2 |
| Mean Temp °F | 26.9 | *33.1* | 43.9 | 55.5 | 64.4 | 73.1 | 78.1 | 75.7 | 68.1 | 56.8 | 44.2 | 33.2 | 54.4 |
| Days Max Temp ≥ 90 °F | 0 | *0* | 0 | 0 | 0 | 6 | 16 | 13 | 5 | 0 | 0 | 0 | 40 |
| Days Max Temp ≤ 32 °F | 11 | *6* | 1 | 0 | 0 | 0 | 0 | 0 | 0 | 0 | 1 | 6 | 25 |
| Days Min Temp ≤ 32 °F | 28 | *23* | 17 | 4 | 0 | 0 | 0 | 0 | 0 | 4 | 13 | 25 | 114 |
| Days Min Temp ≤ 0 °F | 3 | *2* | 0 | 0 | 0 | 0 | 0 | 0 | 0 | 0 | 0 | 1 | 6 |
| Heating Degree Days | 1175 | *894* | 651 | 304 | 102 | 11 | 1 | 4 | 65 | 277 | 619 | 980 | 5083 |
| Cooling Degree Days | *0* | na | *1* | *20* | *79* | *255* | *413* | *333* | *162* | *14* | *2* | *0* | na |
| Total Precipitation (") | 1.33 | *1.34* | 3.05 | 4.29 | 4.44 | 5.14 | 4.63 | 3.79 | 4.51 | 3.83 | 3.34 | 2.11 | 41.80 |
| Days ≥ 0.1" Precip | *3* | *3* | 6 | 7 | 7 | 7 | 6 | 5 | 5 | 6 | 5 | 4 | 64 |
| Total Snowfall (") | *5.1* | *4.8* | *2.7* | *0.3* | 0.0 | 0.0 | 0.0 | 0.0 | 0.0 | 0.0 | 0.9 | *2.7* | 16.5 |
| Days ≥ 1" Snow Depth | na | na | na | *0* | 0 | 0 | 0 | 0 | 0 | 0 | 0 | na | na |

## WASOLA *Ozark County* ELEVATION 1289 ft LAT/LONG 36° 47' N / 92° 34' W

| | JAN | FEB | MAR | APR | MAY | JUN | JUL | AUG | SEP | OCT | NOV | DEC | YEAR |
|---|---|---|---|---|---|---|---|---|---|---|---|---|---|
| Maximum Temp °F | *44.2* | *48.7* | *59.2* | *70.7* | *77.3* | na | *89.8* | *87.7* | *80.0* | *70.6* | *57.2* | *47.4* | na |
| Minimum Temp °F | *23.0* | *26.0* | *35.1* | *45.4* | *53.7* | na | *66.4* | *64.2* | *57.7* | *46.0* | *36.3* | *26.6* | na |
| Mean Temp °F | *33.6* | *37.4* | *47.1* | *58.1* | *65.6* | na | *78.1* | *76.0* | *68.8* | *58.3* | *46.8* | *37.0* | na |
| Days Max Temp ≥ 90 °F | *0* | *0* | *0* | *0* | *0* | 5 | *16* | *13* | *3* | 0 | 0 | 0 | 37 |
| Days Max Temp ≤ 32 °F | *7* | *3* | *1* | *0* | *0* | *0* | *0* | *0* | *0* | 0 | 0 | *3* | 14 |
| Days Min Temp ≤ 32 °F | *24* | *20* | *13* | *3* | *0* | *0* | *0* | *0* | *0* | *2* | 10 | *21* | 93 |
| Days Min Temp ≤ 0 °F | *1* | *0* | *0* | *0* | *0* | *0* | *0* | *0* | *0* | *0* | 0 | *0* | 1 |
| Heating Degree Days | *967* | *776* | *552* | *230* | *71* | na | *1* | *2* | *45* | *230* | *540* | *861* | na |
| Cooling Degree Days | na | na | na | na | na | na | na | na | na | na | na | na | na |
| Total Precipitation (") | 2.10 | 2.38 | 3.76 | 4.13 | 4.24 | 3.52 | 3.20 | 3.74 | 4.54 | 3.58 | 4.49 | 3.25 | 42.93 |
| Days ≥ 0.1" Precip | 3 | 3 | *5* | *6* | 6 | 5 | 4 | 5 | 5 | 4 | 5 | *4* | 55 |
| Total Snowfall (") | *1.8* | *2.6* | *1.3* | 0.4 | 0.0 | 0.0 | 0.0 | 0.0 | 0.0 | 0.0 | 0.9 | *1.3* | 8.3 |
| Days ≥ 1" Snow Depth | na | na | *0* | 0 | 0 | 0 | 0 | 0 | 0 | 0 | *0* | na | na |

## WAYNESVILLE 2 W *Pulaski County* ELEVATION 801 ft LAT/LONG 37° 50' N / 92° 11' W

| | JAN | FEB | MAR | APR | MAY | JUN | JUL | AUG | SEP | OCT | NOV | DEC | YEAR |
|---|---|---|---|---|---|---|---|---|---|---|---|---|---|
| Maximum Temp °F | 45.1 | 50.0 | 61.2 | 71.9 | 77.5 | 84.5 | 89.7 | 88.4 | 81.0 | 72.0 | 59.1 | 48.8 | 69.1 |
| Minimum Temp °F | 18.6 | 22.1 | 31.8 | 41.7 | 50.2 | 59.4 | 63.9 | 61.6 | 54.3 | 42.2 | 32.8 | 23.7 | 41.9 |
| Mean Temp °F | 31.9 | 36.1 | 46.5 | 56.9 | 63.9 | 72.0 | 76.8 | 75.0 | 67.7 | 57.1 | 46.0 | 36.3 | 55.5 |
| Days Max Temp ≥ 90 °F | 0 | 0 | 0 | 1 | 0 | 6 | 17 | 14 | 5 | 1 | 0 | 0 | 44 |
| Days Max Temp ≤ 32 °F | 6 | 3 | 0 | 0 | 0 | 0 | 0 | 0 | 0 | 0 | 0 | 3 | 12 |
| Days Min Temp ≤ 32 °F | 27 | 23 | 17 | 7 | 1 | 0 | 0 | 0 | 0 | 7 | 16 | 24 | 122 |
| Days Min Temp ≤ 0 °F | 3 | 2 | 0 | 0 | 0 | 0 | 0 | 0 | 0 | 0 | 0 | 1 | 6 |
| Heating Degree Days | 1020 | 810 | 573 | 269 | 105 | 12 | 1 | 3 | 65 | 264 | 566 | 884 | 4572 |
| Cooling Degree Days | 0 | 0 | 6 | 29 | 72 | 227 | 375 | 328 | 149 | 19 | 2 | 0 | 1207 |
| Total Precipitation (") | 2.07 | 2.40 | 3.68 | 3.97 | 4.70 | 4.20 | 3.70 | 3.75 | 4.43 | 4.12 | 4.25 | 3.30 | 44.57 |
| Days ≥ 0.1" Precip | 5 | 5 | 7 | 8 | 8 | 7 | 5 | 6 | 6 | 7 | 6 | 6 | 76 |
| Total Snowfall (") | 7.6 | 6.2 | 4.6 | 0.4 | 0.0 | 0.0 | 0.0 | 0.0 | 0.0 | 0.0 | 1.9 | 3.6 | 24.3 |
| Days ≥ 1" Snow Depth | 8 | 5 | 2 | 0 | 0 | 0 | 0 | 0 | 0 | 0 | 1 | 3 | 19 |

## WEST PLAINS *Howell County* ELEVATION 1007 ft LAT/LONG 36° 44' N / 91° 51' W

| | JAN | FEB | MAR | APR | MAY | JUN | JUL | AUG | SEP | OCT | NOV | DEC | YEAR |
|---|---|---|---|---|---|---|---|---|---|---|---|---|---|
| Maximum Temp °F | 43.1 | 48.2 | 58.6 | 69.5 | 76.5 | 84.3 | 89.6 | 88.0 | 80.1 | 70.3 | 57.3 | 46.9 | 67.7 |
| Minimum Temp °F | 21.1 | 24.6 | 33.9 | 43.6 | 52.1 | 60.9 | 65.7 | 63.6 | 56.6 | 43.8 | 34.1 | 25.5 | 43.8 |
| Mean Temp °F | 32.1 | 36.4 | 46.3 | 56.6 | 64.4 | 72.6 | 77.7 | 75.8 | 68.4 | 57.1 | 45.7 | 36.2 | 55.8 |
| Days Max Temp ≥ 90 °F | 0 | 0 | 0 | 0 | 0 | 5 | 16 | 13 | 3 | 0 | 0 | 0 | 37 |
| Days Max Temp ≤ 32 °F | 6 | 3 | 1 | 0 | 0 | 0 | 0 | 0 | 0 | 0 | 0 | 3 | 13 |
| Days Min Temp ≤ 32 °F | 27 | 22 | 15 | 4 | 0 | 0 | 0 | 0 | 0 | 4 | 14 | 24 | 110 |
| Days Min Temp ≤ 0 °F | 1 | 0 | 0 | 0 | 0 | 0 | 0 | 0 | 0 | 0 | 0 | 1 | 2 |
| Heating Degree Days | 1012 | 800 | 576 | 268 | 90 | 6 | 1 | 1 | 51 | 259 | 573 | 886 | 4523 |
| Cooling Degree Days | 0 | 0 | 3 | 22 | 74 | 240 | 407 | 362 | 167 | 18 | 1 | 0 | 1294 |
| Total Precipitation (") | 2.56 | 2.91 | 4.38 | 4.92 | 4.29 | 4.32 | 3.12 | 3.59 | 4.19 | 3.48 | 4.74 | 4.13 | 46.63 |
| Days ≥ 0.1" Precip | 5 | 5 | 7 | 7 | 7 | 7 | 5 | 5 | 6 | 5 | 6 | 6 | 71 |
| Total Snowfall (") | 3.5 | 4.1 | 2.5 | 0.2 | 0.0 | 0.0 | 0.0 | 0.0 | 0.0 | 0.1 | 0.7 | 2.5 | 13.6 |
| Days ≥ 1" Snow Depth | 6 | 4 | 1 | 0 | 0 | 0 | 0 | 0 | 0 | 0 | 0 | 2 | 13 |

## WILLOW SPRGS RD KUKU *Howell County* ELEVATION 1230 ft LAT/LONG 36° 59' N / 91° 59' W

| | JAN | FEB | MAR | APR | MAY | JUN | JUL | AUG | SEP | OCT | NOV | DEC | YEAR |
|---|---|---|---|---|---|---|---|---|---|---|---|---|---|
| Maximum Temp °F | 42.1 | 47.4 | 57.9 | 68.9 | 76.0 | 83.5 | 88.9 | 87.6 | 79.5 | 69.5 | 56.0 | 45.8 | 66.9 |
| Minimum Temp °F | 21.3 | 25.2 | 34.1 | 43.7 | 51.9 | 60.6 | 65.1 | 63.4 | 56.6 | 44.6 | 34.5 | 25.8 | 43.9 |
| Mean Temp °F | 31.7 | 36.3 | 46.0 | 56.3 | 64.0 | 72.1 | 77.1 | 75.5 | 68.1 | 57.0 | 45.3 | 35.8 | 55.4 |
| Days Max Temp ≥ 90 °F | 0 | 0 | 0 | 0 | 0 | 4 | 14 | 12 | 3 | 0 | 0 | 0 | 33 |
| Days Max Temp ≤ 32 °F | 7 | 4 | 1 | 0 | 0 | 0 | 0 | 0 | 0 | 0 | 1 | 4 | 17 |
| Days Min Temp ≤ 32 °F | 26 | 20 | 15 | 5 | 1 | 0 | 0 | 0 | 0 | 4 | 13 | 23 | 107 |
| Days Min Temp ≤ 0 °F | 2 | 1 | 0 | 0 | 0 | 0 | 0 | 0 | 0 | 0 | 0 | 1 | 4 |
| Heating Degree Days | 1024 | 803 | 584 | 276 | 100 | 7 | 1 | 1 | 55 | 263 | 584 | 898 | 4596 |
| Cooling Degree Days | 0 | 0 | 4 | 24 | 79 | 241 | 400 | 364 | 167 | 23 | 1 | 0 | 1303 |
| Total Precipitation (") | 2.33 | 2.50 | 3.95 | 5.00 | 4.57 | 3.97 | 3.30 | 3.46 | 4.09 | 3.64 | 4.34 | 3.77 | 44.92 |
| Days ≥ 0.1" Precip | 4 | 4 | 7 | 7 | 8 | 6 | 5 | 6 | 6 | 5 | 6 | 5 | 69 |
| Total Snowfall (") | *3.4* | 2.5 | 2.9 | 0.3 | 0.0 | 0.0 | 0.0 | 0.0 | 0.0 | 0.0 | 0.6 | 2.2 | 11.9 |
| Days ≥ 1" Snow Depth | *5* | 4 | 1 | 0 | 0 | 0 | 0 | 0 | 0 | 0 | 0 | 1 | 11 |

## WINDSOR *Henry County* ELEVATION 902 ft LAT/LONG 38° 32' N / 93° 32' W

| | JAN | FEB | MAR | APR | MAY | JUN | JUL | AUG | SEP | OCT | NOV | DEC | YEAR |
|---|---|---|---|---|---|---|---|---|---|---|---|---|---|
| Maximum Temp °F | 36.8 | 43.3 | 54.9 | 66.7 | 75.7 | 83.6 | 89.7 | 87.7 | 79.8 | 68.7 | 54.1 | 42.8 | 65.3 |
| Minimum Temp °F | 16.1 | 21.3 | 31.4 | 42.0 | 51.8 | 61.0 | 65.5 | 62.5 | 54.9 | 42.7 | 32.6 | 22.7 | 42.0 |
| Mean Temp °F | 26.5 | 32.3 | 43.2 | 54.4 | 63.8 | 72.3 | 77.6 | 75.1 | 67.4 | 55.7 | 43.4 | 32.8 | 53.7 |
| Days Max Temp ≥ 90 °F | 0 | 0 | 0 | 0 | 0 | 5 | *17* | 13 | 4 | 0 | 0 | 0 | 39 |
| Days Max Temp ≤ 32 °F | 11 | 7 | 1 | 0 | 0 | 0 | 0 | 0 | 0 | 0 | 1 | 6 | 26 |
| Days Min Temp ≤ 32 °F | 29 | 24 | 18 | 4 | 0 | 0 | 0 | 0 | 0 | 5 | 16 | 26 | 122 |
| Days Min Temp ≤ 0 °F | 4 | 2 | 0 | 0 | 0 | 0 | 0 | 0 | 0 | 0 | 0 | 1 | 7 |
| Heating Degree Days | 1187 | 916 | 672 | 330 | 109 | 12 | 1 | 4 | 72 | 301 | 643 | 994 | 5241 |
| Cooling Degree Days | 0 | 0 | 3 | 17 | 70 | *248* | *418* | *328* | 154 | 16 | 1 | 0 | 1255 |
| Total Precipitation (") | *1.65* | 1.54 | 2.96 | 4.19 | 4.38 | 4.88 | 4.03 | 3.79 | 4.50 | 4.10 | 3.46 | 2.20 | 41.68 |
| Days ≥ 0.1" Precip | na | *3* | 5 | 6 | 7 | 7 | 5 | 5 | 6 | 6 | *5* | 4 | na |
| Total Snowfall (") | *5.8* | *5.3* | *2.4* | 0.1 | 0.0 | 0.0 | 0.0 | 0.0 | 0.0 | 0.0 | 1.1 | 3.7 | 18.4 |
| Days ≥ 1" Snow Depth | na | na | na | 0 | 0 | 0 | 0 | 0 | 0 | 0 | *0* | na | na |

## JANUARY MINIMUM TEMPERATURE °F

### LOWEST

| Rank | Place | Value |
|---|---|---|
| 1 | Maryville | 11.0 |
| 2 | Hamilton | 12.6 |
| 3 | Conception | 12.8 |
| 4 | Unionville | 13.0 |
| 5 | Spickard | 13.6 |
| 6 | Grant City | 13.7 |
| | Memphis | 13.7 |
| 8 | Bethany | 13.8 |
| 9 | Princeton | 14.1 |
| 10 | Amity | 14.5 |
| 11 | Bowling Green | 14.7 |
| 12 | Kirksville | 15.0 |
| 13 | Mexico | 15.3 |
| | Steffenville | 15.3 |
| 15 | St. Joseph | 15.7 |
| 16 | Hannibal | 16.0 |
| 17 | Windsor | 16.1 |
| 18 | Canton L & D | 16.2 |
| 19 | Shelbina | 16.3 |
| | Steelville | 16.3 |
| 21 | Brunswick | 16.4 |
| | Salisbury | 16.4 |
| 23 | Lexington | 16.7 |
| | Sedalia | 16.7 |
| 25 | Brookfield | 16.9 |

### HIGHEST

| Rank | Place | Value |
|---|---|---|
| 1 | Kennett | 25.5 |
| | Portageville | 25.5 |
| 3 | Caruthersville | 25.4 |
| 4 | New Madrid | 24.5 |
| 5 | Malden | 24.4 |
| 6 | Poplar Bluff | 23.7 |
| 7 | Cape Girardeau | 23.5 |
| 8 | Joplin | 23.3 |
| 9 | Advance | 23.0 |
| | Wasola | 23.0 |
| 11 | Jackson | 22.7 |
| 12 | Neosho | 22.1 |
| 13 | Lockwood | 21.9 |
| 14 | Anderson | 21.8 |
| 15 | Camdenton | 21.4 |
| | Rolla | 21.4 |
| | Summersville | 21.4 |
| | Wappapello Dam | 21.4 |
| 19 | Lebanon | 21.3 |
| | Springfield | 21.3 |
| | Willow Springs | 21.3 |
| 22 | Bunker | 21.1 |
| | Marble Hill | 21.1 |
| | West Plains | 21.1 |
| 25 | Arcadia | 21.0 |

## JULY MAXIMUM TEMPERATURE °F

### HIGHEST

| Rank | Place | Value |
|---|---|---|
| 1 | Kennett | 92.8 |
| 2 | Malden | 92.1 |
| 3 | Doniphan | 91.9 |
| 4 | Nevada | 91.7 |
| 5 | Ozark Beach | 91.6 |
| 6 | Jackson | 91.4 |
| | Poplar Bluff | 91.4 |
| 8 | Wappapello Dam | 91.3 |
| 9 | Clearwater Dam | 91.1 |
| 10 | Neosho | 91.0 |
| 11 | Appleton City | 90.9 |
| | New Madrid | 90.9 |
| 13 | Osceola | 90.8 |
| | Sweet Springs | 90.8 |
| | Van Buren | 90.8 |
| 16 | Lockwood | 90.7 |
| 17 | Advance | 90.6 |
| | California | 90.6 |
| | Caruthersville | 90.6 |
| 20 | Union | 90.5 |
| 21 | Butler | 90.4 |
| | Perryville | 90.4 |
| | Pomme d Terr Dm | 90.4 |
| 24 | Camdenton | 90.3 |
| | Cape Girardeau | 90.3 |

### LOWEST

| Rank | Place | Value |
|---|---|---|
| 1 | Kirksville | 86.8 |
| 2 | Amity | 87.0 |
| 3 | Hannibal | 87.1 |
| 4 | Maryville | 87.4 |
| | Unionville | 87.4 |
| 6 | Conception | 87.5 |
| 7 | Princeton | 87.7 |
| 8 | Vichy | 87.8 |
| 9 | Steffenville | 87.9 |
| 10 | Fulton | 88.0 |
| | Spickard | 88.0 |
| 12 | Grant City | 88.1 |
| | Mountain Grove | 88.1 |
| 14 | Bowling Green | 88.2 |
| | Brunswick | 88.2 |
| | Bunker | 88.2 |
| | Memphis | 88.2 |
| 18 | Hamilton | 88.3 |
| | Licking | 88.3 |
| 20 | Farmington | 88.4 |
| 21 | Marshfield | 88.5 |
| | Salisbury | 88.5 |
| 23 | Bethany | 88.7 |
| | Saverton L & D | 88.7 |
| | Versailles | 88.7 |

## ANNUAL PRECIPITATION (")

### HIGHEST

| Rank | Place | Value |
|---|---|---|
| 1 | Kennett | 50.35 |
| 2 | New Madrid | 50.24 |
| 3 | Doniphan | 50.20 |
| 4 | Caruthersville | 49.34 |
| 5 | Van Buren | 48.55 |
| 6 | Marble Hill | 47.91 |
| 7 | Portageville | 47.67 |
| 8 | Wappapello Dam | 47.40 |
| 9 | Jackson | 47.35 |
| 10 | Advance | 47.22 |
| 11 | Poplar Bluff | 47.15 |
| 12 | Lamar | 47.04 |
| 13 | Cape Girardeau | 46.99 |
| 14 | Arcadia | 46.91 |
| 15 | Greenville | 46.80 |
| 16 | West Plains | 46.63 |
| 17 | Cassville | 45.84 |
| 18 | Lockwood | 45.58 |
| 19 | Malden | 45.46 |
| 20 | Galena | 45.44 |
| 21 | Mountain Grove | 45.38 |
| 22 | Clearwater Dam | 45.21 |
| 23 | Mt. Vernon | 45.17 |
| 24 | Joplin | 45.07 |
| 25 | Bolivar | 44.96 |

### LOWEST

| Rank | Place | Value |
|---|---|---|
| 1 | Grant City | 34.66 |
| 2 | St. Joseph | 34.84 |
| 3 | Maryville | 35.35 |
| 4 | Memphis | 35.51 |
| 5 | Amity | 35.78 |
| 6 | Princeton | 35.82 |
| 7 | Conception | 36.50 |
| | Spickard | 36.50 |
| 9 | Hamilton | 36.58 |
| 10 | Steffenville | 36.87 |
| 11 | Unionville | 37.15 |
| 12 | Bethany | 37.23 |
| 13 | Kirksville | 37.75 |
| 14 | St. Louis | 38.17 |
| 15 | New Franklin | 38.25 |
| | Saverton L & D | 38.25 |
| 17 | Bowling Green | 38.48 |
| 18 | St. Charles | 38.66 |
| 19 | Elsberry | 38.76 |
| 20 | Festus | 39.15 |
| 21 | Brunswick | 39.20 |
| 22 | Canton L & D | 39.28 |
| 23 | Shelbina | 39.35 |
| 24 | Brookfield | 39.84 |
| 25 | Lexington | 39.85 |

## ANNUAL SNOWFALL (")

### HIGHEST

| Rank | Place | Value |
|---|---|---|
| 1 | Bethany | 25.2 |
| 2 | Sweet Springs | 24.3 |
| | Waynesville | 24.3 |
| 4 | Grant City | 24.2 |
| 5 | Carrollton | 23.3 |
| 6 | Unionville | 23.2 |
| 7 | Hannibal | 22.9 |
| 8 | Boonville | 22.6 |
| 9 | Fulton | 22.2 |
| 10 | Mexico | 21.7 |
| 11 | Kirksville | 21.4 |
| 12 | St. Louis | 21.3 |
| 13 | Steffenville | 21.2 |
| 14 | Salisbury | 21.0 |
| 15 | Conception | 20.5 |
| | Vandalia | 20.5 |
| 17 | Elsberry | 19.6 |
| | Springfield | 19.6 |
| 19 | Lexington | 19.2 |
| | Shelbina | 19.2 |
| 21 | Spickard | 18.5 |
| | Union | 18.5 |
| | Vienna | 18.5 |
| 24 | Billings | 18.4 |
| | Windsor | 18.4 |

### LOWEST

| Rank | Place | Value |
|---|---|---|
| 1 | Caruthersville | 7.4 |
| 2 | Wasola | 8.3 |
| 3 | Wappapello Dam | 8.6 |
| 4 | Poplar Bluff | 9.1 |
| 5 | Van Buren | 10.0 |
| 6 | Kennett | 10.4 |
| | Portageville | 10.4 |
| 8 | Advance | 10.8 |
| 9 | Lakeside | 10.9 |
| 10 | Ozark Beach | 11.4 |
| 11 | Butler | 11.9 |
| | Willow Springs | 11.9 |
| 13 | Neosho | 12.5 |
| 14 | Joplin | 12.6 |
| 15 | Doniphan | 13.1 |
| 16 | California | 13.4 |
| 17 | Sedalia | 13.6 |
| | West Plains | 13.6 |
| 19 | Cape Girardeau | 13.7 |
| | Farmington | 13.7 |
| | Marshall | 13.7 |
| 22 | Arcadia | 13.9 |
| | Lamar | 13.9 |
| | Nevada | 13.9 |
| 25 | Marble Hill | 14.0 |

# MONTANA

PHYSICAL FEATURES. Montana, with an area of 146,316 square miles, is the fourth largest State of the Union. Climatic variations are large. The half of the State southwest of a line from the southeastern corner to the Canadian Border north of Cut Bank in Glacier County is very mountainous, while the northeastern half is very much like Great Plains country, broken occasionally by wide valleys and isolated groups of hills. The extent of the climatic variations one should expect is indicated by the range in elevation of from 1,800 feet above sea level where the Kootenai River enters Idaho to 12,850 feet at Granite Peak near Yellowstone Park. Half the State lies over 4,000 feet above sea level.

The Continental Divide traverses the western half of the State in roughly a north-south direction. To the west of the Divide, Montana is drained by the Kootenai, Clark Fork, and Flathead Rivers into the Pacific Ocean through the Columbia River. Many of the tributary streams in this region have their origin in the high elevations of the western slopes of the Rockies. Most streams traverse narrow canyons, at least through parts of their length, affording many valuable waterpower sites. A relatively small area located between the Hudson Bay Divide and the Rocky Mountains is drained by the St. Mary River, which finds its way to the Hudson Bay through the Saskatchewan River. The remainder of the State is drained by the Missouri River, which is formed by the confluence of the Gallatin, Madison, and Jefferson Rivers at Three Forks, travels northward through deep canyons in the Big Belt Mountains, and flows through the lower lying northeastern portion of the State. The Yellowstone River, the principal tributary of the Missouri in Montana and which has its source in Wyoming, drains the southeastern section of the State and has its confluence with the Missouri just east of the Montana - North Dakota line.

GENERAL CLIMATE. The Continental Divide exerts a marked influence on the climate of adjacent areas. West of the Divide the climate might be termed a modified north Pacific coast type, while to the east, climatic characteristics are decidedly continental. On the west of the mountain barrier winters are milder, precipitation is more evenly distributed throughout the year, summers are cooler in general, and winds are lighter than on the eastern side. There is more cloudiness in the west in all seasons, humidity runs a bit higher, and the growing season is shorter than in the eastern plains areas.

Cold waves, which cover parts of Montana on the average of 6 to 12 times a winter, are confined mostly to the sections northeast of a Glacier Park - Miles City line. A few of these cold waves cover the entire area east of the Divide, and 1 or 2 a season will cover the State all the way from the Dakotas to Idaho. With temperatures well below 0° and sometimes strong winds with blowing snow, these cold waves can be very inconvenient and even dangerous. In small areas ideally situated for radiation cooling, low temperatures can fall to -50° or lower.

During the summer months hot weather occurs fairly often in the eastern parts of the State. Temperatures of over 100° sometimes occur in the lower elevation areas west of the Divide during the summer, but hot spells are less frequent and of shorter duration than in the plains sections. Hot spells nowhere become oppressive, however, because summer nights almost invariably are cool and pleasant. In the areas with elevations above 4,000 feet, extremely hot weather is almost unknown. Summer days, however, are usually warm enough for light summer clothing.

Winters, while usually cold, have few extended cold spells. Between cold waves there are periods, sometimes longer than 10 days, of mild but often windy weather. These warm, windy winter periods occur almost entirely along the eastern slopes of the Divide and are popularly known as "chinook" weather. The so-called "chinook" belt extends from the Browning-Shelby area southeastward to the Yellowstone Valley above Billings. Through this belt, "chinook" winds frequently reach speeds of 25 to 50 m.p.h. or more and can persist, with little interruptions, for several days.

Most Montana lakes freeze over every winter, but Flathead Lake, between Polson and Kalispell, freezes over completely only during the coldest winters, and 1 year in 10. All rivers carry floating ice during the late winter or early spring. Few streams freeze solid; water generally continues to flow beneath the ice. During coldest winters "anchor" ice, which builds from the bottom of shallow streams, on rare occasions causes some flooding.

PRECIPITATION. Precipitation varies widely and depends largely upon topographic influences. Areas adjacent to mountain ranges in general are the wettest, although there are a few exceptions where the "rain-shadow" effect appears. Generally, nearly half the annual long-term average total falls in the 3 months, May through July.

SNOWFALL. Annual snowfall varies from quite heavy, 300 inches, in some parts of the mountains in the western half of the State, to around 20 inches at some stations in the two northern Divisions east of the Continental Divide. Most snow falls during the November-March period, but heavy snowstorms can occur as early as mid-September or as late as May 1 in the higher southwestern half of the State. In eastern sections early or late season snows are not very common. Mountain snowpacks in the wetter areas often exceed 100 inches in depth as the annual snow season approaches its end around April 1 to 15.

The greatest volume of flow of Montana's rivers occurs during the spring and early summer months with the melting of the winter snowpack. Heavy rains falling during the spring thaw constitute a serious flood threat. Ice jams, which occur during the spring breakup, usually in March, cause backwater flooding. Flash-floods, although restricted in scope, are probably the most numerous and result from locally heavy rainstorms in the spring and summer.

STORMS. Severe storms of several types can occur, but the most troublesome are hailstorms. Their occurrence is limited mainly to July and August, infrequently in June and September. Tornadoes develop infrequently and occur almost entirely east of the Divide, largely in the eastern third of the State. Severe windstorms of a general nature are rare but can occur locally, mainly east of the Divide, from a few to several times a year. Drought in its most severe form is practically unknown, but dry years do occur in some sections. All parts of the State rarely suffer from dryness at the same time.

# COUNTY INDEX

### Missoula County
LINDBERGH LAKE
MISSOULA JOHNSN-BELL
POTOMAC
SEELEY LAKE RS

### Musselshell County
MELSTONE
ROUNDUP

### Park County
GARDINER
LIVINGSTON MISSION
WILSALL 8 ENE

### Petroleum County
FLATWILLOW 4 ENE

### Phillips County
CONTENT 3 SSE
FORKS 4 NNE
MALTA 35 S

### Pondera County
CONRAD
VALIER

### Powder River County
BIDDLE 8 SW
BROADUS
MOORHEAD 9 NE
SONNETTE 2 WNW

### Powell County
DEER LODGE 3 W

### Prairie County
TERRY
TERRY 21 NNW

### Ravalli County
DARBY
HAMILTON
STEVENSVILLE
SULA 3 ENE
WESTERN AG RESEARCH

### Richland County
SAVAGE
SIDNEY

### Roosevelt County
BREDETTE
CULBERTSON

### Rosebud County
BIRNEY
BRANDENBERG
COLSTRIP
INGOMAR 14 NE
ROCK SPRINGS

### Sanders County
HERON 2 NW
THOMPSON FALLS P H
TROUT CREEK RS

### Sheridan County
MEDICINE LAKE 3 SE
RAYMOND BORDER STN
REDSTONE
WESTBY

### Silver Bow County
BUTTE BERT MOONEY AP
DIVIDE 2 NW

### Stillwater County
COLUMBUS
MYSTIC LAKE
RAPELJE 4 S

### Sweet Grass County
BIG TIMBER
MELVILLE 4 W

### Teton County
BLACKLEAF
CHOTEAU AP
FAIRFIELD

### Toole County
GOLDBUTTE 7 N
SWEETGRASS

### Treasure County
HYSHAM
HYSHAM 25 SSE

### Valley County
GLASGOW INTL AP
OPHEIM 10 N
OPHEIM 12 SSE

### Wheatland County
HARLOWTON
JUDITH GAP 13 E

### Wibaux County
CARLYLE 12 NW
WIBAUX 2 E

### Yellowstone County
BILLINGS LOGAN AP
BILLINGS WATER PLANT
HUNTLEY EXP STN

## ELEVATION INDEX

| FEET | STATION NAME |
|---|---|
| 1923 | CULBERTSON |
| 1932 | TROY |
| 1952 | MEDICINE LAKE 3 SE |
| 1952 | SIDNEY |
| 1991 | SAVAGE |
| 2070 | FORT PECK POWER PLAN |
| 2080 | GLENDIVE |
| 2096 | LIBBY 1 NE RS |
| 2113 | REDSTONE |
| 2113 | WESTBY |
| 2251 | TERRY |
| 2280 | HERON 2 NW |
| 2284 | GLASGOW INTL AP |
| 2341 | CONTENT 3 SSE |
| 2352 | RAYMOND BORDER STN |
| 2362 | TROUT CREEK RS |
| 2363 | HARLEM 4 W |
| 2382 | THOMPSON FALLS P H |
| 2401 | VIDA 6 NE |
| 2411 | CHINOOK |
| 2441 | CIRCLE |
| 2480 | MIZPAH 4 NNW |
| 2532 | EUREKA RS |
| 2572 | LOMA 1 WNW |
| 2585 | HAVRE CITY-COUNTY AP |
| 2592 | JORDAN |
| 2602 | FORKS 4 NNE |
| 2629 | MILES CITY MUNI AP |
| 2631 | BROCKWAY 3 WSW |
| 2641 | FORT BENTON |
| 2651 | HYSHAM |
| 2651 | MALTA 35 S |
| 2671 | WIBAUX 2 E |
| 2680 | SAINT REGIS 1 NE |
| 2690 | BREDETTE |
| 2690 | FORT ASSINNIBOINE |

| FEET | STATION NAME |
|---|---|
| 2700 | SUPERIOR |
| 2703 | BIG SANDY |
| 2723 | COHAGEN |
| 2723 | TROY 18 N |
| 2733 | POLSON KERR DAM |
| 2743 | SIMPSON 6 NW |
| 2753 | MOSBY 2 ENE |
| 2772 | BRANDENBERG |
| 2772 | PLEVNA |
| 2782 | INGOMAR 14 NE |
| 2802 | POWDERVILLE 8 NNE |
| 2831 | GILDFORD |
| 2851 | TIBER DAM |
| 2890 | HARDIN |
| 2890 | MELSTONE |
| 2904 | ST IGNATIUS |
| 2943 | CRESTON |
| 2953 | ILIAD |
| 2953 | OPHEIM 12 SSE |
| 2965 | KALISPELL GLACIER AP |
| 2982 | OPHEIM 10 N |
| 2982 | RUDYARD 27 N |
| 2992 | HUNTLEY EXP STN |
| 3002 | FORTINE 1 N |
| 3022 | ROCK SPRINGS |
| 3031 | BROADUS |
| 3031 | CARLYLE 12 NW |
| 3061 | BIGFORK 13 S |
| 3104 | BILLINGS WATER PLANT |
| 3104 | CHESTER |
| 3104 | HYSHAM 25 SSE |
| 3104 | SWAN LAKE |
| 3143 | FLATWILLOW 4 ENE |
| 3143 | GERALDINE |
| 3153 | HUNGRY HORSE DAM |
| 3153 | WEST GLACIER |
| 3182 | BIRNEY |
| 3182 | OLNEY |
| 3202 | MOORHEAD 9 NE |
| 3202 | YELLOWTAIL DAM |
| 3205 | MISSOULA JOHNSN-BELL |
| 3222 | COLSTRIP |
| 3232 | ROUNDUP |
| 3251 | WINIFRED |
| 3261 | TERRY 21 NNW |
| 3300 | JOPLIN |
| 3304 | RIDGEWAY 1 S |
| 3373 | STEVENSVILLE |
| 3390 | CASCADE 5 S |
| 3432 | EKALAKA |
| 3451 | ROY 8 NE |
| 3471 | SWEETGRASS |
| 3491 | GRASS RANGE |
| 3491 | HOLTER DAM |
| 3504 | BUSBY |
| 3504 | GOLDBUTTE 7 N |
| 3524 | CONRAD |
| 3533 | HAMILTON |
| 3563 | SUN RIVER 4 S |
| 3567 | BILLINGS LOGAN AP |
| 3592 | COLUMBUS |
| 3602 | BIDDLE 8 SW |
| 3602 | LIBBY 32 SSE |
| 3602 | POTOMAC |
| 3602 | WESTERN AG RESEARCH |
| 3612 | DENTON 1 NNE |
| 3671 | CANYON FERRY DAM |
| 3681 | BRIDGER 3 S |
| 3688 | GREAT FALLS INTL AP |
| 3691 | POLEBRIDGE |
| 3704 | JOLIET |
| 3734 | WYOLA 1 SW |
| 3802 | BARBER |
| 3812 | CHOTEAU AP |
| 3812 | VALIER |
| 3822 | DARBY |
| 3828 | HELENA ARPT |
| 3832 | TOWNSEND |
| 3839 | CUT BANK MUNI AP |
| 3904 | SONNETTE 2 WNW |
| 3944 | DRUMMOND AVIATION |
| 3983 | FAIRFIELD |
| 4032 | SEELEY LAKE RS |
| 4042 | TRIDENT |
| 4070 | AUGUSTA |
| 4104 | BIG TIMBER |
| 4132 | LEWISTOWN MUNI AP |
| 4134 | RAPELJE 4 S |
| 4167 | HARLOWTON |
| 4300 | ROGERS PASS 9 NNE |
| 4304 | MOCCASIN EXP STN |
| 4304 | SHONKIN 7 S |
| 4314 | STANFORD |
| 4324 | BLACKLEAF |
| 4344 | DEL BONITA |
| 4403 | SULA 3 ENE |
| 4459 | BOZEMAN GALLATIN FLD |
| 4462 | BABB 6 NE |
| 4482 | RYEGATE 18 NNW |
| 4505 | LINCOLN RS |
| 4505 | LINDBERGH LAKE |
| 4593 | GIBSON DAM |
| 4603 | CASCADE 20 SSE |
| 4659 | LIVINGSTON MISSION |
| 4662 | TWIN BRIDGES |
| 4734 | NORRIS MADISON P H |
| 4783 | BOZEMAN 6 W EXP FARM |
| 4823 | MARTINSDALE 3 NNW |
| 4833 | DEER LODGE 3 W |
| 4862 | BOZEMAN MONTANA SU |
| 4882 | BOULDER |
| 4900 | AUSTIN 1 W |
| 4954 | ENNIS |
| 5052 | GLEN 4 N |
| 5105 | JUDITH GAP 13 E |
| 5233 | DILLON WMCE |
| 5282 | PHILIPSBURG RS |
| 5305 | GARDINER |
| 5374 | MELVILLE 4 W |
| 5413 | DIVIDE 2 NW |
| 5505 | PONY |
| 5535 | BUTTE BERT MOONEY AP |
| 5799 | ALDER 17 S |
| 5843 | WILSALL 8 ENE |
| 5850 | RED LODGE |
| 5853 | VIRGINIA CITY |
| 5882 | LENNEP 5 SW |
| 5915 | BOZEMAN 12 NE |
| 6063 | WISDOM |
| 6273 | LIMA |
| 6493 | HEBGEN DAM |
| 6565 | MYSTIC LAKE |
| 6667 | WEST YELLOWSTONE |
| 6706 | LAKEVIEW |

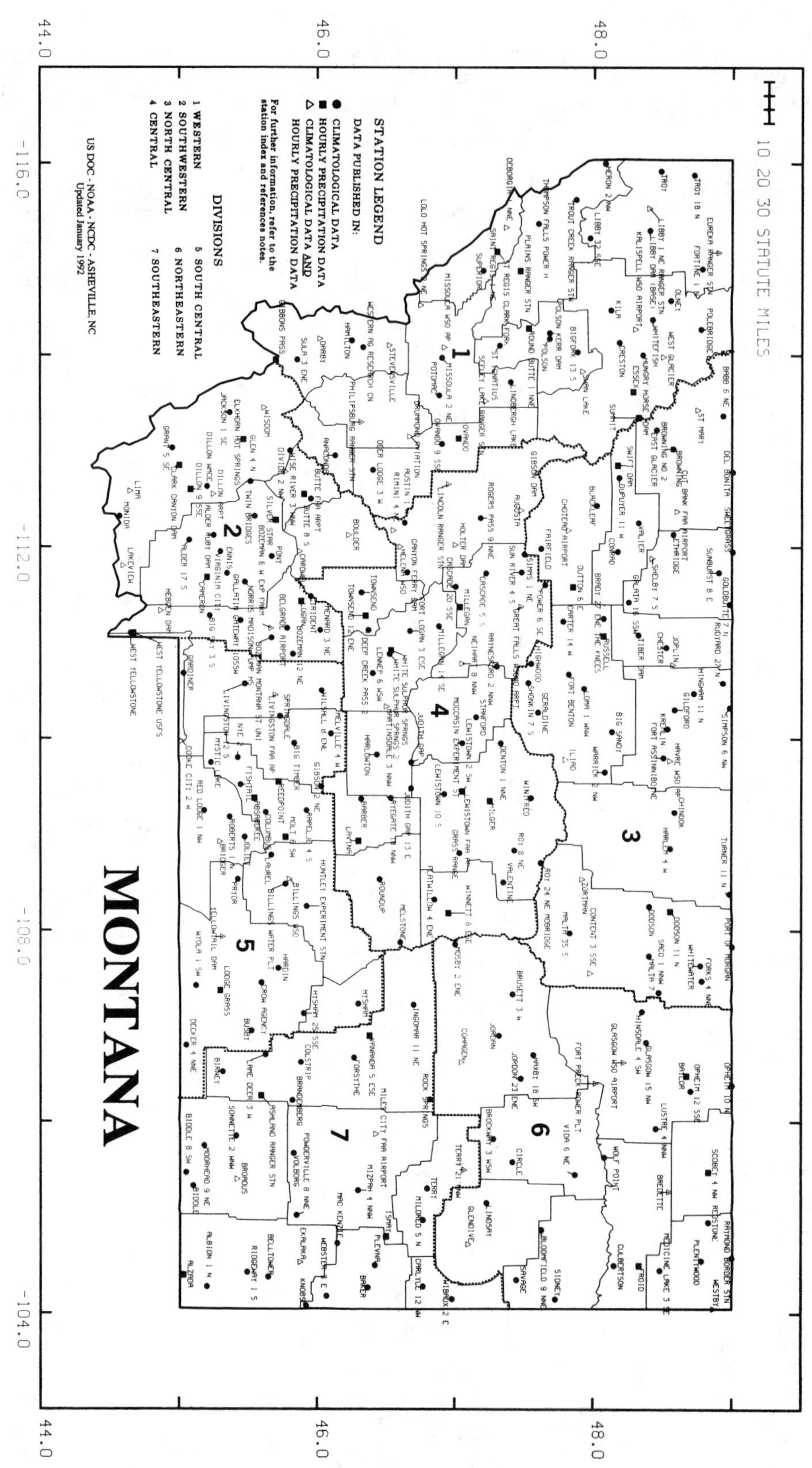
MONTANA
10 20 30 STATUTE MILES
STATION LEGEND
DATA PUBLISHED IN:
● CLIMATOLOGICAL DATA
■ HOURLY PRECIPITATION DATA
△ CLIMATOLOGICAL DATA AND HOURLY PRECIPITATION DATA
For further information, refer to the station index and reference notes.
DIVISIONS
1 WESTERN
2 SOUTHWESTERN
3 NORTH CENTRAL
4 CENTRAL
5 SOUTH CENTRAL
6 NORTHEASTERN
7 SOUTHEASTERN
US DOC - NOAA - NCDC - ASHEVILLE, NC
Updated January 1992
44.0
46.0
48.0
-116.0
-112.0
-108.0
-104.0

## ALDER 17 S *Madison County* ELEVATION 5799 ft LAT/LONG 45° 4 ' N / 112° 4 ' W

| | JAN | FEB | MAR | APR | MAY | JUN | JUL | AUG | SEP | OCT | NOV | DEC | YEAR |
|---|---|---|---|---|---|---|---|---|---|---|---|---|---|
| Maximum Temp °F | 32.6 | 37.1 | 43.1 | 52.1 | 61.7 | 70.6 | 79.4 | 78.3 | 68.2 | 57.1 | 40.6 | 32.5 | 54.4 |
| Minimum Temp °F | 12.6 | 15.7 | 20.8 | 27.2 | 34.6 | 41.3 | 45.6 | 44.4 | 36.4 | 28.7 | 19.2 | 12.8 | 28.3 |
| Mean Temp °F | 22.6 | 26.4 | 32.0 | 39.7 | 48.2 | 56.0 | 62.5 | 61.4 | 52.3 | 42.9 | 29.9 | 22.7 | 41.4 |
| Days Max Temp ≥ 90 °F | 0 | 0 | 0 | 0 | 0 | 0 | 1 | 1 | 0 | 0 | 0 | 0 | 2 |
| Days Max Temp ≤ 32 °F | 13 | 7 | 4 | 1 | 0 | 0 | 0 | 0 | 0 | 1 | 7 | 14 | 47 |
| Days Min Temp ≤ 32 °F | 30 | 27 | 28 | 22 | 12 | 3 | 0 | 2 | 10 | 21 | 27 | 30 | 212 |
| Days Min Temp ≤ 0 °F | 6 | 3 | 1 | 0 | 0 | 0 | 0 | 0 | 0 | 0 | 2 | 5 | 17 |
| Heating Degree Days | 1307 | 1083 | 1017 | 754 | 516 | 275 | 105 | 136 | 377 | 677 | 1046 | 1306 | 8599 |
| Cooling Degree Days | 0 | 0 | 0 | 0 | 0 | 11 | 33 | 28 | 2 | 0 | 0 | 0 | 74 |
| Total Precipitation (") | 0.30 | 0.28 | 0.68 | 1.03 | 2.28 | 2.41 | 1.60 | 1.46 | 1.39 | 0.96 | 0.66 | 0.38 | 13.43 |
| Days ≥ 0.1" Precip | 1 | 1 | 2 | 3 | 7 | 7 | 4 | 5 | 4 | 3 | 2 | 1 | 40 |
| Total Snowfall (") | 6.2 | 5.1 | 9.9 | ***8.6*** | ***3.6*** | 0.3 | 0.0 | 0.0 | ***0.7*** | na | ***7.7*** | 6.8 | na |
| Days ≥ 1" Snow Depth | na | na | na | ***2*** | 0 | 0 | 0 | 0 | 0 | na | na | na | na |

## AUGUSTA *Lewis and Clark County* ELEVATION 4070 ft LAT/LONG 47° 29 ' N / 112° 23 ' W

| | JAN | FEB | MAR | APR | MAY | JUN | JUL | AUG | SEP | OCT | NOV | DEC | YEAR |
|---|---|---|---|---|---|---|---|---|---|---|---|---|---|
| Maximum Temp °F | 35.2 | 41.1 | 47.7 | 57.3 | 66.4 | 74.6 | 82.4 | 81.8 | 72.5 | 61.2 | 45.0 | 37.0 | 58.5 |
| Minimum Temp °F | 11.3 | 15.4 | 21.5 | 29.4 | 37.8 | 45.4 | 49.3 | 47.7 | 40.0 | 31.8 | 21.8 | 14.2 | 30.5 |
| Mean Temp °F | 23.3 | 28.3 | 34.6 | 43.4 | 52.1 | 60.0 | 65.8 | 64.8 | 56.3 | 46.5 | 33.4 | 25.6 | 44.5 |
| Days Max Temp ≥ 90 °F | 0 | 0 | 0 | 0 | 0 | 2 | 6 | 6 | 1 | 0 | 0 | 0 | 15 |
| Days Max Temp ≤ 32 °F | 10 | 6 | 3 | 1 | 0 | 0 | 0 | 0 | 0 | 1 | 4 | 8 | 33 |
| Days Min Temp ≤ 32 °F | 27 | 25 | 26 | 20 | 7 | 0 | 0 | 0 | 5 | 16 | 24 | 27 | 177 |
| Days Min Temp ≤ 0 °F | ***9*** | 5 | 2 | 0 | 0 | 0 | 0 | 0 | 0 | 0 | 3 | 6 | 25 |
| Heating Degree Days | 1289 | 1031 | 933 | 643 | 396 | 176 | 54 | 73 | 271 | 568 | 942 | 1216 | 7592 |
| Cooling Degree Days | 0 | 0 | 0 | 0 | 5 | 32 | 75 | 68 | 14 | 2 | 0 | 0 | 196 |
| Total Precipitation (") | 0.53 | 0.44 | 0.65 | 1.16 | 2.32 | 2.51 | 1.39 | 1.50 | 1.29 | 0.79 | 0.48 | 0.54 | 13.60 |
| Days ≥ 0.1" Precip | 2 | 2 | 2 | 3 | 6 | 5 | 3 | 4 | 3 | 2 | 2 | 2 | 36 |
| Total Snowfall (") | na | na | na | na | 0.0 | 0.0 | 0.0 | 0.0 | 0.3 | na | na | na | na |
| Days ≥ 1" Snow Depth | na | na | na | na | 0 | 0 | 0 | 0 | 0 | ***0*** | na | na | na |

## AUSTIN 1 W *Lewis and Clark County* ELEVATION 4900 ft LAT/LONG 46° 39 ' N / 112° 16 ' W

| | JAN | FEB | MAR | APR | MAY | JUN | JUL | AUG | SEP | OCT | NOV | DEC | YEAR |
|---|---|---|---|---|---|---|---|---|---|---|---|---|---|
| Maximum Temp °F | 30.5 | 35.4 | 42.8 | 52.4 | 62.3 | 70.6 | 79.5 | 78.6 | 67.2 | 55.0 | 38.9 | 31.1 | 53.7 |
| Minimum Temp °F | 11.7 | 16.2 | 21.1 | 28.2 | 36.1 | 43.0 | 48.1 | 47.4 | 39.2 | 31.5 | 20.8 | 13.1 | 29.7 |
| Mean Temp °F | 21.1 | 25.8 | 31.9 | 40.3 | 49.2 | 56.8 | 63.8 | 63.0 | 53.2 | 43.3 | 29.9 | 22.1 | 41.7 |
| Days Max Temp ≥ 90 °F | 0 | 0 | 0 | 0 | 0 | 0 | 2 | 2 | 0 | 0 | 0 | 0 | 4 |
| Days Max Temp ≤ 32 °F | 14 | 9 | 4 | 1 | 0 | 0 | 0 | 0 | 0 | 1 | 7 | 15 | 51 |
| Days Min Temp ≤ 32 °F | 30 | 27 | 28 | 22 | 10 | 1 | 0 | 0 | 6 | 16 | 27 | 30 | 197 |
| Days Min Temp ≤ 0 °F | 8 | 4 | 2 | 0 | 0 | 0 | 0 | 0 | 0 | 0 | 2 | 5 | 21 |
| Heating Degree Days | 1355 | 1100 | 1018 | 733 | 483 | 253 | 88 | 106 | 353 | 667 | 1047 | 1324 | 8527 |
| Cooling Degree Days | 0 | 0 | 0 | 0 | 2 | 14 | 47 | 46 | 5 | 0 | 0 | 0 | 114 |
| Total Precipitation (") | 1.10 | 0.72 | 1.01 | 1.39 | 2.21 | 2.22 | 1.53 | 1.40 | 1.54 | 1.03 | 0.98 | 1.07 | 16.20 |
| Days ≥ 0.1" Precip | 4 | 3 | 4 | 5 | 7 | 6 | 4 | 4 | 4 | 4 | 4 | 4 | 53 |
| Total Snowfall (") | na | na | na | na | 0.7 | 0.0 | 0.0 | 0.3 | 0.7 | ***2.5*** | na | na | na |
| Days ≥ 1" Snow Depth | na | na | na | ***2*** | 0 | 0 | 0 | 0 | 0 | 2 | 7 | 15 | na |

## BABB 6 NE *Glacier County* ELEVATION 4462 ft LAT/LONG 48° 56 ' N / 113° 21 ' W

| | JAN | FEB | MAR | APR | MAY | JUN | JUL | AUG | SEP | OCT | NOV | DEC | YEAR |
|---|---|---|---|---|---|---|---|---|---|---|---|---|---|
| Maximum Temp °F | 31.7 | 36.6 | 42.4 | 51.4 | 61.6 | 68.8 | 75.7 | 75.9 | 65.9 | 55.9 | 41.1 | ***32.2*** | 53.3 |
| Minimum Temp °F | 7.8 | 12.4 | 19.4 | 27.4 | 35.1 | 41.3 | 44.2 | 43.7 | 36.1 | 30.0 | 19.7 | ***10.4*** | 27.3 |
| Mean Temp °F | 19.8 | 24.5 | 31.0 | 39.4 | 48.3 | 55.1 | 60.0 | 59.9 | 51.0 | 43.0 | 30.5 | ***21.3*** | 40.3 |
| Days Max Temp ≥ 90 °F | 0 | 0 | 0 | 0 | 0 | 0 | 1 | 1 | 0 | 0 | 0 | 0 | 2 |
| Days Max Temp ≤ 32 °F | 12 | 7 | 5 | 1 | 0 | 0 | 0 | 0 | 0 | 1 | 5 | ***11*** | 42 |
| Days Min Temp ≤ 32 °F | 29 | 26 | 27 | 22 | 12 | 2 | 0 | 1 | 9 | ***19*** | ***25*** | 27 | 199 |
| Days Min Temp ≤ 0 °F | 11 | 7 | 3 | 0 | 0 | 0 | 0 | 0 | 0 | 0 | 3 | 8 | 32 |
| Heating Degree Days | 1397 | 1136 | 1047 | 762 | 510 | 295 | 167 | 165 | 415 | 677 | 1041 | ***1335*** | 8947 |
| Cooling Degree Days | 0 | 0 | 0 | 0 | 1 | 6 | 20 | 23 | 2 | 1 | 0 | ***0*** | 53 |
| Total Precipitation (") | 0.78 | 0.60 | 0.85 | 1.26 | 2.55 | 3.18 | 1.96 | 2.03 | 1.83 | 1.01 | 0.79 | 0.71 | 17.55 |
| Days ≥ 0.1" Precip | 3 | 2 | 3 | 4 | 5 | 6 | 4 | 5 | 4 | 3 | 3 | 3 | 45 |
| Total Snowfall (") | na | na | na | na | na | 0.0 | 0.0 | 0.0 | na | na | na | na | na |
| Days ≥ 1" Snow Depth | na | na | na | na | na | ***0*** | ***0*** | ***0*** | ***0*** | na | na | na | na |

## BARBER *Golden Valley County* ELEVATION 3802 ft LAT/LONG 46° 19 ' N / 109° 23 ' W

| | JAN | FEB | MAR | APR | MAY | JUN | JUL | AUG | SEP | OCT | NOV | DEC | YEAR |
|---|---|---|---|---|---|---|---|---|---|---|---|---|---|
| Maximum Temp °F | 35.1 | 40.9 | 47.8 | 58.1 | 67.6 | 75.7 | 82.3 | 80.5 | 70.5 | 61.4 | 45.0 | 37.3 | 58.5 |
| Minimum Temp °F | 9.7 | 14.2 | 20.8 | 29.5 | 38.7 | 46.8 | 51.6 | 49.7 | 40.4 | 31.0 | 19.2 | 12.0 | 30.3 |
| Mean Temp °F | 22.4 | 27.6 | 34.3 | 43.8 | 53.1 | 61.3 | 67.0 | 65.2 | 55.5 | 46.3 | 32.1 | 24.7 | 44.4 |
| Days Max Temp ≥ 90 °F | 0 | 0 | 0 | 0 | 0 | 2 | 4 | 3 | 0 | 0 | 0 | 0 | 9 |
| Days Max Temp ≤ 32 °F | 11 | 7 | 3 | 0 | 0 | 0 | 0 | 0 | 0 | 0 | 4 | 9 | 34 |
| Days Min Temp ≤ 32 °F | 28 | 26 | 28 | 20 | 6 | 0 | 0 | 0 | 4 | 17 | ***26*** | 30 | 185 |
| Days Min Temp ≤ 0 °F | 8 | 4 | 2 | 0 | 0 | 0 | 0 | 0 | 0 | 0 | 2 | 6 | 22 |
| Heating Degree Days | 1315 | 1050 | 938 | 628 | 364 | 143 | 35 | 63 | 287 | 574 | 983 | 1245 | 7625 |
| Cooling Degree Days | 0 | 0 | 0 | 0 | 3 | 44 | 101 | 77 | 9 | 1 | 0 | 0 | 235 |
| Total Precipitation (") | 0.54 | 0.32 | 0.51 | 1.09 | 2.14 | 2.28 | 1.48 | 1.28 | 1.12 | 0.92 | 0.47 | 0.43 | 12.58 |
| Days ≥ 0.1" Precip | 2 | 1 | 2 | 4 | 5 | 5 | 4 | 3 | 3 | 2 | 2 | 2 | 35 |
| Total Snowfall (") | na | na | na | na | 0.0 | 0.0 | 0.0 | 0.0 | 0.2 | na | na | na | na |
| Days ≥ 1" Snow Depth | na | na | na | na | 0 | 0 | 0 | 0 | 0 | ***0*** | na | na | na |

## BIDDLE 8 SW *Powder River County* ELEVATION 3602 ft LAT/LONG 45° 3 ' N / 105° 28 ' W

| | JAN | FEB | MAR | APR | MAY | JUN | JUL | AUG | SEP | OCT | NOV | DEC | YEAR |
|---|---|---|---|---|---|---|---|---|---|---|---|---|---|
| Maximum Temp °F | 33.1 | 38.7 | 47.4 | 58.4 | 68.5 | 78.2 | 86.8 | 86.2 | 74.4 | 61.5 | 45.2 | 35.2 | 59.5 |
| Minimum Temp °F | 8.2 | 13.1 | 21.4 | 30.7 | 40.2 | 49.1 | 55.0 | 53.0 | 42.9 | 32.0 | 20.1 | 10.5 | 31.4 |
| Mean Temp °F | 20.7 | 25.9 | 34.4 | 44.6 | 54.4 | 63.7 | 70.9 | 69.6 | 58.7 | 46.8 | 32.7 | 22.8 | 45.4 |
| Days Max Temp ≥ 90 °F | 0 | 0 | 0 | 0 | 1 | 4 | 12 | 12 | 3 | 0 | 0 | 0 | 32 |
| Days Max Temp ≤ 32 °F | 13 | 8 | 4 | 0 | 0 | 0 | 0 | 0 | 0 | 0 | 5 | 11 | 41 |
| Days Min Temp ≤ 32 °F | 31 | 27 | 28 | ***17*** | 5 | 0 | 0 | 0 | 3 | ***16*** | 28 | 29 | 184 |
| Days Min Temp ≤ 0 °F | 10 | 5 | 1 | 0 | 0 | 0 | 0 | 0 | 0 | 0 | 2 | 7 | 25 |
| Heating Degree Days | 1369 | 1098 | 943 | 609 | 335 | 109 | 18 | 29 | 220 | 560 | 963 | 1302 | 7555 |
| Cooling Degree Days | 0 | 0 | 0 | 1 | 15 | 87 | 218 | 191 | 36 | 1 | 0 | 0 | 549 |
| Total Precipitation (") | 0.44 | 0.35 | 0.76 | 1.54 | 2.42 | 2.70 | 1.52 | 0.91 | 1.52 | 1.02 | 0.55 | 0.52 | 14.25 |
| Days ≥ 0.1" Precip | 1 | 1 | 3 | 4 | 6 | 6 | 4 | 3 | 3 | 3 | 2 | 2 | 38 |
| Total Snowfall (") | 7.9 | ***5.8*** | ***7.6*** | na | ***0.8*** | 0.2 | 0.0 | 0.0 | 0.9 | 2.3 | ***5.5*** | ***8.0*** | na |
| Days ≥ 1" Snow Depth | 26 | 15 | 9 | ***2*** | 0 | 0 | 0 | 0 | 0 | 1 | 7 | 18 | 78 |

## BIG SANDY *Chouteau County* ELEVATION 2703 ft LAT/LONG 48° 10 ' N / 110° 7 ' W

| | JAN | FEB | MAR | APR | MAY | JUN | JUL | AUG | SEP | OCT | NOV | DEC | YEAR |
|---|---|---|---|---|---|---|---|---|---|---|---|---|---|
| Maximum Temp °F | 27.8 | 35.0 | 47.0 | 60.2 | 71.1 | 79.5 | 87.0 | 86.2 | 74.1 | 62.3 | 43.3 | ***32.3*** | 58.8 |
| Minimum Temp °F | 4.3 | 9.7 | 21.1 | 31.1 | 40.8 | 48.7 | 52.5 | 50.9 | 40.8 | 31.3 | 18.5 | 8.6 | 29.9 |
| Mean Temp °F | 16.1 | 22.3 | 34.1 | 45.7 | 56.0 | 64.1 | 69.8 | 68.6 | 57.5 | 46.9 | 30.9 | ***20.5*** | 44.4 |
| Days Max Temp ≥ 90 °F | 0 | 0 | 0 | 0 | 1 | 5 | 13 | 12 | 3 | 0 | 0 | 0 | 34 |
| Days Max Temp ≤ 32 °F | 16 | 10 | 4 | 1 | 0 | 0 | 0 | 0 | 0 | 1 | 6 | 12 | 50 |
| Days Min Temp ≤ 32 °F | 30 | 27 | 27 | 17 | 5 | 1 | 0 | 0 | 5 | 16 | 26 | 29 | 183 |
| Days Min Temp ≤ 0 °F | 13 | 8 | 2 | 0 | 0 | 0 | 0 | 0 | 0 | 0 | 3 | 8 | 34 |
| Heating Degree Days | 1497 | 1199 | 952 | 574 | 290 | 99 | 27 | 40 | 247 | 554 | 1016 | ***1376*** | 7871 |
| Cooling Degree Days | 0 | 0 | 0 | 3 | 24 | 89 | 175 | 147 | 20 | 1 | 0 | ***0*** | 459 |
| Total Precipitation (") | 0.64 | 0.41 | 0.59 | 1.09 | 2.35 | 2.50 | 1.59 | 1.51 | 1.51 | 0.78 | 0.49 | 0.50 | 13.96 |
| Days ≥ 0.1" Precip | 2 | 1 | ***2*** | 3 | 5 | 6 | 4 | 4 | 4 | 2 | 2 | 1 | 36 |
| Total Snowfall (") | na | na | na | na | na | ***0.0*** | ***0.0*** | ***0.0*** | ***0.1*** | na | na | na | na |
| Days ≥ 1" Snow Depth | na | na | na | na | na | ***0*** | ***0*** | ***0*** | ***0*** | na | na | na | na |

## BIG TIMBER *Sweet Grass County* ELEVATION 4104 ft LAT/LONG 45° 50 ' N / 109° 57 ' W

| | JAN | FEB | MAR | APR | MAY | JUN | JUL | AUG | SEP | OCT | NOV | DEC | YEAR |
|---|---|---|---|---|---|---|---|---|---|---|---|---|---|
| Maximum Temp °F | 36.5 | 41.7 | 48.9 | 59.1 | 68.6 | 78.0 | 86.4 | 85.5 | 73.6 | 61.4 | 45.5 | 38.0 | 60.3 |
| Minimum Temp °F | 15.5 | 19.0 | 24.0 | 31.7 | 39.6 | 47.3 | 52.2 | 50.7 | 41.4 | 33.9 | 24.8 | 17.9 | 33.2 |
| Mean Temp °F | 26.0 | 30.4 | 36.5 | 45.4 | 54.2 | 62.7 | 69.3 | 68.1 | 57.5 | 47.7 | 35.2 | 28.0 | 46.8 |
| Days Max Temp ≥ 90 °F | 0 | 0 | 0 | 0 | 1 | 4 | 12 | 11 | 2 | 0 | 0 | 0 | 30 |
| Days Max Temp ≤ 32 °F | 10 | 6 | 3 | 0 | 0 | 0 | 0 | 0 | 0 | 1 | 4 | 8 | 32 |
| Days Min Temp ≤ 32 °F | 26 | 23 | 24 | 17 | 4 | 0 | 0 | 0 | 3 | 14 | 22 | 26 | 159 |
| Days Min Temp ≤ 0 °F | 7 | 3 | 2 | 0 | 0 | 0 | 0 | 0 | 0 | 0 | 1 | 4 | 17 |
| Heating Degree Days | 1205 | 971 | 876 | 582 | 334 | 120 | 23 | 35 | 237 | 532 | 888 | 1140 | 6943 |
| Cooling Degree Days | 0 | 0 | 0 | 1 | 7 | 64 | 163 | 148 | 20 | 1 | 0 | 0 | 404 |
| Total Precipitation (") | 0.61 | 0.44 | 0.93 | 1.69 | 2.82 | 2.82 | 1.45 | 1.27 | 1.38 | 1.45 | 0.73 | 0.61 | 16.20 |
| Days ≥ 0.1" Precip | 2 | 2 | 3 | 5 | 6 | 6 | 4 | 3 | 4 | 3 | 3 | 2 | 43 |
| Total Snowfall (") | na | na | na | na | ***0.3*** | 0.0 | ***0.0*** | ***0.1*** | ***0.3*** | na | na | na | na |
| Days ≥ 1" Snow Depth | na | na | na | na | na | ***0*** | ***0*** | ***0*** | ***0*** | na | na | na | na |

## BIGFORK 13 S *Lake County* ELEVATION 3061 ft LAT/LONG 47° 53 ' N / 114° 2 ' W

| | JAN | FEB | MAR | APR | MAY | JUN | JUL | AUG | SEP | OCT | NOV | DEC | YEAR |
|---|---|---|---|---|---|---|---|---|---|---|---|---|---|
| Maximum Temp °F | 32.9 | 37.1 | 45.0 | 55.0 | 64.2 | 71.2 | ***79.3*** | ***79.0*** | 67.2 | 55.5 | 41.2 | 34.7 | 55.2 |
| Minimum Temp °F | 21.4 | 23.8 | 28.0 | 34.0 | 40.7 | 47.4 | ***52.2*** | na | 43.4 | 36.0 | 29.1 | 23.9 | na |
| Mean Temp °F | 27.2 | 30.5 | 36.5 | 44.4 | 52.5 | 59.3 | ***65.8*** | na | ***55.3*** | 45.8 | 35.1 | 29.3 | na |
| Days Max Temp ≥ 90 °F | 0 | 0 | 0 | 0 | 0 | 0 | na | na | 0 | 0 | 0 | 0 | na |
| Days Max Temp ≤ 32 °F | 13 | 7 | 2 | 0 | 0 | 0 | 0 | 0 | 0 | 0 | ***4*** | 12 | 38 |
| Days Min Temp ≤ 32 °F | 27 | 24 | 23 | ***13*** | 2 | 0 | 0 | 0 | ***1*** | ***8*** | ***19*** | 26 | 143 |
| Days Min Temp ≤ 0 °F | 1 | 1 | 0 | 0 | 0 | 0 | 0 | 0 | 0 | 0 | 0 | 1 | 3 |
| Heating Degree Days | 1165 | 966 | 876 | 611 | 385 | 188 | na | na | ***290*** | 590 | 890 | 1099 | na |
| Cooling Degree Days | 0 | 0 | 0 | 0 | 3 | 23 | ***78*** | 78 | 4 | 0 | ***0*** | 0 | 186 |
| Total Precipitation (") | 1.85 | 1.15 | 1.32 | 1.54 | 2.67 | 3.00 | 1.58 | 1.83 | 1.88 | 1.45 | 1.72 | 1.90 | 21.89 |
| Days ≥ 0.1" Precip | 6 | 4 | 5 | 4 | 6 | 7 | 4 | ***4*** | ***5*** | 4 | 6 | ***6*** | 61 |
| Total Snowfall (") | na | na | na | na | ***0.0*** | ***0.0*** | ***0.0*** | ***0.0*** | ***0.0*** | ***0.1*** | na | na | na |
| Days ≥ 1" Snow Depth | na | na | na | na | ***0*** | ***0*** | ***0*** | ***0*** | ***0*** | ***0*** | na | na | na |

## BILLINGS LOGAN AP *Yellowstone County* ELEVATION 3567 ft LAT/LONG 45° 48 ' N / 108° 32 ' W

| | JAN | FEB | MAR | APR | MAY | JUN | JUL | AUG | SEP | OCT | NOV | DEC | YEAR |
|---|---|---|---|---|---|---|---|---|---|---|---|---|---|
| Maximum Temp °F | 32.0 | 38.5 | 46.6 | 57.2 | 67.3 | 77.5 | 85.9 | 84.7 | 72.1 | 60.0 | 44.1 | 35.1 | 58.4 |
| Minimum Temp °F | 14.0 | 19.1 | 25.8 | 34.2 | 43.5 | 51.8 | 57.9 | 56.5 | 46.6 | 37.1 | 25.6 | 17.3 | 35.8 |
| Mean Temp °F | 23.0 | 28.8 | 36.3 | 45.7 | 55.4 | 64.7 | 71.9 | 70.6 | 59.4 | 48.5 | 34.9 | 26.2 | 47.1 |
| Days Max Temp ≥ 90 °F | 0 | 0 | 0 | 0 | 0 | 4 | 12 | 11 | 2 | 0 | 0 | 0 | 29 |
| Days Max Temp ≤ 32 °F | 14 | 8 | 5 | 1 | 0 | 0 | 0 | 0 | 0 | 1 | 5 | 11 | 45 |
| Days Min Temp ≤ 32 °F | 28 | 24 | 23 | 13 | 2 | 0 | 0 | 0 | 1 | 9 | 22 | 27 | 149 |
| Days Min Temp ≤ 0 °F | 8 | 4 | 1 | 0 | 0 | 0 | 0 | 0 | 0 | 0 | 1 | 4 | 18 |
| Heating Degree Days | 1297 | 1015 | 885 | 574 | 306 | 94 | 17 | 26 | 207 | 507 | 897 | 1196 | 7021 |
| Cooling Degree Days | 0 | 0 | 0 | 3 | 21 | 106 | 241 | 221 | 47 | 5 | 0 | 0 | 644 |
| Total Precipitation (") | 0.81 | 0.61 | 1.11 | 1.77 | 2.37 | 2.06 | 1.19 | 0.93 | 1.38 | 1.26 | 0.82 | 0.76 | 15.07 |
| Days ≥ 0.1" Precip | 3 | 2 | 4 | 5 | 6 | 5 | 3 | 3 | 3 | 3 | 3 | 2 | 42 |
| Total Snowfall (") | 10.0 | 6.5 | 10.2 | 8.1 | 2.0 | 0.0 | 0.0 | 0.0 | 1.4 | 4.1 | 7.4 | 8.9 | 58.6 |
| Days ≥ 1" Snow Depth | 18 | 12 | 8 | 3 | 0 | 0 | 0 | 0 | 0 | 2 | 7 | 16 | 66 |

## BILLINGS WATER PLANT *Yellowstone County* ELEVATION 3104 ft LAT/LONG 45° 46 ' N / 108° 29 ' W

| | JAN | FEB | MAR | APR | MAY | JUN | JUL | AUG | SEP | OCT | NOV | DEC | YEAR |
|---|---|---|---|---|---|---|---|---|---|---|---|---|---|
| Maximum Temp °F | 36.9 | 43.4 | 52.6 | 62.7 | 71.9 | 80.8 | 88.4 | 87.5 | 76.6 | 65.3 | 48.6 | 38.9 | 62.8 |
| Minimum Temp °F | 12.1 | 16.9 | 24.3 | 32.8 | 42.2 | 49.9 | 55.1 | 53.2 | 43.4 | 34.1 | 23.5 | 14.8 | 33.5 |
| Mean Temp °F | 24.6 | 30.2 | 38.4 | 47.8 | 57.1 | 65.4 | 71.8 | 70.4 | 60.0 | 49.7 | 36.1 | 26.9 | 48.2 |
| Days Max Temp ≥ 90 °F | 0 | 0 | 0 | 0 | 1 | 6 | 15 | 14 | 3 | 0 | 0 | 0 | 39 |
| Days Max Temp ≤ 32 °F | 10 | 6 | 2 | 0 | 0 | 0 | 0 | 0 | 0 | 0 | 3 | 8 | 29 |
| Days Min Temp ≤ 32 °F | 29 | 26 | 25 | 15 | 3 | 0 | 0 | 0 | 2 | 12 | 26 | 29 | 167 |
| Days Min Temp ≤ 0 °F | 8 | 4 | 1 | 0 | 0 | 0 | 0 | 0 | 0 | 0 | 1 | 5 | 19 |
| Heating Degree Days | 1247 | 977 | 816 | 511 | 256 | 77 | 10 | 19 | 180 | 467 | 862 | 1175 | 6597 |
| Cooling Degree Days | 0 | 0 | 0 | 2 | 22 | 112 | 226 | 205 | 38 | 2 | 0 | 0 | 607 |
| Total Precipitation (") | 0.63 | 0.50 | 0.76 | 1.49 | 2.31 | 2.29 | 1.09 | 1.00 | 1.40 | 1.27 | 0.69 | 0.57 | 14.00 |
| Days ≥ 0.1" Precip | 2 | 2 | 3 | 4 | 5 | 5 | 3 | 3 | 4 | 3 | 2 | 2 | 38 |
| Total Snowfall (") | na | na | na | na | na | na | na | na | na | na | na | na | na |
| Days ≥ 1" Snow Depth | na | na | na | na | na | na | na | na | na | na | na | na | na |

## BIRNEY *Rosebud County* ELEVATION 3182 ft LAT/LONG 45° 19 ' N / 106° 31 ' W

| | JAN | FEB | MAR | APR | MAY | JUN | JUL | AUG | SEP | OCT | NOV | DEC | YEAR |
|---|---|---|---|---|---|---|---|---|---|---|---|---|---|
| Maximum Temp °F | 33.2 | 40.5 | 50.9 | 62.6 | 72.4 | 81.3 | 89.4 | 87.9 | 76.2 | 64.0 | 45.3 | 35.5 | 61.6 |
| Minimum Temp °F | 5.1 | 11.3 | 21.3 | 30.6 | 40.1 | 49.0 | 53.6 | 51.2 | 41.0 | 30.5 | 18.3 | 8.2 | 30.0 |
| Mean Temp °F | 19.1 | 25.9 | 36.1 | 46.7 | 56.3 | 65.2 | 71.6 | 69.6 | 58.6 | 47.3 | 31.9 | 21.9 | 45.9 |
| Days Max Temp ≥ 90 °F | 0 | 0 | 0 | 0 | 2 | 6 | 17 | 14 | 3 | 0 | 0 | 0 | 42 |
| Days Max Temp ≤ 32 °F | 13 | 6 | 2 | 0 | 0 | 0 | 0 | 0 | 0 | 0 | 4 | ***11*** | 36 |
| Days Min Temp ≤ 32 °F | 31 | 28 | 28 | 18 | 5 | 0 | 0 | 0 | 4 | 19 | 28 | 31 | 192 |
| Days Min Temp ≤ 0 °F | 11 | 6 | 2 | 0 | 0 | 0 | 0 | 0 | 0 | 0 | 2 | 7 | 28 |
| Heating Degree Days | 1415 | 1098 | 889 | 546 | 280 | 81 | 11 | 22 | 211 | 542 | 988 | 1329 | 7412 |
| Cooling Degree Days | 0 | 0 | 0 | 2 | 20 | 118 | 233 | 199 | 29 | 1 | 0 | 0 | 602 |
| Total Precipitation (") | 0.54 | 0.43 | 0.78 | 1.42 | 1.96 | 2.75 | 1.31 | 0.91 | 1.22 | 1.06 | 0.71 | 0.58 | 13.67 |
| Days ≥ 0.1" Precip | 2 | 2 | 3 | 4 | 5 | 6 | 4 | 2 | 3 | 3 | 2 | 2 | 38 |
| Total Snowfall (") | ***6.8*** | na | na | 3.3 | 0.3 | 0.0 | 0.0 | 0.0 | 0.5 | ***0.5*** | ***3.7*** | na | na |
| Days ≥ 1" Snow Depth | na | na | 7 | 1 | 0 | 0 | 0 | 0 | 0 | ***0*** | na | ***14*** | na |

## BLACKLEAF *Teton County* ELEVATION 4324 ft LAT/LONG 48° 1' N / 112° 29' W

| | JAN | FEB | MAR | APR | MAY | JUN | JUL | AUG | SEP | OCT | NOV | DEC | YEAR |
|---|---|---|---|---|---|---|---|---|---|---|---|---|---|
| Maximum Temp °F | 32.8 | 37.9 | 43.4 | 53.3 | 62.5 | 70.5 | 78.4 | 78.1 | 68.4 | 58.0 | ***42.7*** | 35.0 | 55.1 |
| Minimum Temp °F | 8.3 | 12.6 | 19.2 | 27.9 | 36.3 | 43.7 | 47.2 | 46.0 | 38.1 | 29.3 | ***18.5*** | 10.9 | 28.2 |
| Mean Temp °F | 20.6 | 25.3 | 31.3 | 40.6 | 49.4 | 57.1 | 62.9 | 62.0 | 53.3 | 43.7 | ***30.6*** | 23.0 | 41.7 |
| Days Max Temp ≥ 90 °F | 0 | 0 | 0 | 0 | 0 | 0 | 2 | 2 | 0 | 0 | 0 | 0 | 4 |
| Days Max Temp ≤ 32 °F | 13 | 9 | 5 | 1 | 0 | 0 | 0 | 0 | 0 | 1 | 5 | 11 | 45 |
| Days Min Temp ≤ 32 °F | 28 | 26 | 28 | 21 | 9 | 1 | 0 | 1 | 7 | ***19*** | ***27*** | ***29*** | 196 |
| Days Min Temp ≤ 0 °F | 11 | 7 | 3 | 0 | 0 | 0 | 0 | 0 | 0 | 0 | 3 | 8 | 32 |
| Heating Degree Days | 1376 | 1117 | 1037 | 727 | 479 | 245 | 107 | 132 | 351 | 657 | ***1024*** | 1304 | 8556 |
| Cooling Degree Days | 0 | 0 | 0 | 0 | 2 | 17 | 51 | 47 | 6 | 1 | 0 | 0 | 124 |
| Total Precipitation (") | 0.57 | 0.48 | 0.71 | 1.15 | 2.21 | 2.62 | 1.43 | 1.81 | 1.17 | 0.58 | 0.47 | 0.50 | 13.70 |
| Days ≥ 0.1" Precip | 2 | 2 | 2 | 3 | 5 | 6 | 3 | 4 | 3 | 2 | 2 | ***2*** | 36 |
| Total Snowfall (") | na | na | na | na | 1.1 | 0.1 | 0.0 | 0.2 | ***1.2*** | ***2.7*** | ***6.3*** | na | na |
| Days ≥ 1" Snow Depth | na | na | na | na | 0 | 0 | 0 | 0 | 0 | na | na | na | na |

## BOULDER *Jefferson County* ELEVATION 4882 ft LAT/LONG 46° 14' N / 112° 6' W

| | JAN | FEB | MAR | APR | MAY | JUN | JUL | AUG | SEP | OCT | NOV | DEC | YEAR |
|---|---|---|---|---|---|---|---|---|---|---|---|---|---|
| Maximum Temp °F | 33.9 | 39.1 | 45.7 | 55.2 | 64.8 | 73.6 | 82.2 | 81.6 | 70.8 | 59.4 | 42.2 | 34.5 | 56.9 |
| Minimum Temp °F | 9.9 | 14.1 | 20.0 | 27.2 | 35.1 | 42.7 | 47.8 | 46.1 | 36.7 | 27.7 | 18.3 | 10.7 | 28.0 |
| Mean Temp °F | 21.9 | 26.6 | 32.9 | 41.2 | 50.0 | 58.2 | 65.0 | 63.8 | 53.8 | 43.5 | 30.3 | 22.6 | 42.5 |
| Days Max Temp ≥ 90 °F | 0 | 0 | 0 | 0 | 0 | 1 | 5 | 5 | 0 | 0 | 0 | 0 | 11 |
| Days Max Temp ≤ 32 °F | 11 | 6 | 4 | 1 | 0 | 0 | 0 | 0 | 0 | 0 | 5 | 12 | 39 |
| Days Min Temp ≤ 32 °F | 30 | 27 | 29 | 23 | 10 | 2 | 0 | 1 | 8 | 22 | 28 | 30 | 210 |
| Days Min Temp ≤ 0 °F | 8 | 5 | 2 | 0 | 0 | 0 | 0 | 0 | 0 | 0 | 3 | 6 | 24 |
| Heating Degree Days | 1330 | 1078 | 989 | 706 | 459 | 219 | 69 | 89 | 336 | 657 | 1036 | 1308 | 8276 |
| Cooling Degree Days | 0 | 0 | 0 | 0 | 1 | 22 | 72 | 58 | 5 | 0 | 0 | 0 | 158 |
| Total Precipitation (") | 0.55 | 0.31 | 0.60 | 0.83 | 1.90 | 2.00 | 1.47 | 1.40 | 1.14 | 0.60 | 0.51 | 0.46 | 11.77 |
| Days ≥ 0.1" Precip | 2 | 1 | 2 | 3 | 6 | 6 | 5 | 4 | 3 | 2 | 2 | 2 | 38 |
| Total Snowfall (") | na | na | na | na | ***0.5*** | 0.0 | 0.0 | 0.0 | ***0.0*** | na | na | na | na |
| Days ≥ 1" Snow Depth | na | na | na | na | ***0*** | 0 | 0 | 0 | 0 | na | na | na | na |

## BOZEMAN 12 NE *Gallatin County* ELEVATION 5915 ft LAT/LONG 45° 49' N / 110° 53' W

| | JAN | FEB | MAR | APR | MAY | JUN | JUL | AUG | SEP | OCT | NOV | DEC | YEAR |
|---|---|---|---|---|---|---|---|---|---|---|---|---|---|
| Maximum Temp °F | 32.4 | 36.2 | 41.1 | 48.7 | 58.2 | 66.7 | 74.8 | 74.6 | 64.1 | 53.7 | 39.3 | 32.8 | 51.9 |
| Minimum Temp °F | 8.4 | 10.4 | 15.2 | 22.7 | 30.2 | 36.5 | 39.5 | 38.1 | 31.9 | 25.7 | 16.5 | 8.7 | 23.7 |
| Mean Temp °F | 20.5 | 23.3 | 28.2 | 35.7 | 44.2 | 51.6 | 57.2 | 56.4 | 48.0 | 39.7 | 27.9 | 20.8 | 37.8 |
| Days Max Temp ≥ 90 °F | 0 | 0 | 0 | 0 | 0 | 0 | 0 | 0 | 0 | 0 | 0 | 0 | 0 |
| Days Max Temp ≤ 32 °F | 14 | 9 | 6 | 1 | 0 | 0 | 0 | 0 | 0 | 1 | 8 | 14 | 53 |
| Days Min Temp ≤ 32 °F | 30 | 26 | 30 | 28 | 21 | 8 | 3 | 4 | 17 | 26 | 28 | 30 | 251 |
| Days Min Temp ≤ 0 °F | 9 | 7 | 4 | 1 | 0 | 0 | 0 | 0 | 0 | 0 | 4 | 9 | 34 |
| Heating Degree Days | 1376 | 1170 | 1134 | 871 | 637 | 394 | 236 | 260 | 504 | 777 | 1105 | 1365 | 9829 |
| Cooling Degree Days | 0 | 0 | 0 | 0 | 0 | 0 | 2 | 1 | 0 | 0 | 0 | 0 | 3 |
| Total Precipitation (") | 2.75 | 1.83 | 2.62 | 3.52 | 4.54 | 4.51 | 2.23 | 2.45 | 3.09 | 2.66 | 2.59 | 2.31 | 35.10 |
| Days ≥ 0.1" Precip | 8 | 6 | 8 | 9 | 11 | 9 | 6 | 5 | 7 | 6 | 8 | 8 | 91 |
| Total Snowfall (") | 38.5 | 28.4 | 38.7 | 31.1 | 12.2 | 1.5 | 0.1 | 0.2 | 3.8 | 13.4 | 29.3 | 35.2 | 232.4 |
| Days ≥ 1" Snow Depth | 30 | 28 | 29 | 21 | 5 | 0 | 0 | 0 | 1 | 6 | 19 | 29 | 168 |

## BOZEMAN 6 W EXP FARM *Gallatin County* ELEVATION 4783 ft LAT/LONG 45° 40' N / 111° 9' W

| | JAN | FEB | MAR | APR | MAY | JUN | JUL | AUG | SEP | OCT | NOV | DEC | YEAR |
|---|---|---|---|---|---|---|---|---|---|---|---|---|---|
| Maximum Temp °F | 32.7 | 37.8 | 45.3 | 55.2 | 64.7 | 73.3 | 81.0 | 80.9 | 70.4 | 58.2 | 41.6 | 33.4 | 56.2 |
| Minimum Temp °F | 11.5 | 16.1 | 22.4 | 29.8 | 37.6 | 44.2 | 48.5 | 47.4 | 39.5 | 31.2 | 20.4 | 12.4 | 30.1 |
| Mean Temp °F | 22.2 | 27.0 | 33.9 | 42.5 | 51.2 | 58.8 | 64.8 | 64.1 | 55.0 | 44.7 | 31.0 | 22.9 | 43.2 |
| Days Max Temp ≥ 90 °F | 0 | 0 | 0 | 0 | 0 | 1 | 3 | 3 | 0 | 0 | 0 | 0 | 7 |
| Days Max Temp ≤ 32 °F | 13 | 7 | 3 | 0 | 0 | 0 | 0 | 0 | 0 | 0 | 6 | 13 | 42 |
| Days Min Temp ≤ 32 °F | 29 | 26 | 27 | 20 | 7 | 1 | 0 | 0 | 5 | 17 | 26 | 29 | 187 |
| Days Min Temp ≤ 0 °F | 7 | 4 | 1 | 0 | 0 | 0 | 0 | 0 | 0 | 0 | 2 | 6 | 20 |
| Heating Degree Days | 1323 | 1068 | 959 | 668 | 423 | 202 | 66 | 77 | 302 | 623 | 1014 | 1298 | 8023 |
| Cooling Degree Days | 0 | 0 | 0 | 0 | 2 | 25 | 69 | 59 | 7 | 0 | 0 | 0 | 162 |
| Total Precipitation (") | 0.53 | 0.48 | 1.11 | 1.59 | 2.76 | 2.78 | 1.54 | 1.31 | 1.66 | 1.46 | 0.91 | 0.57 | 16.70 |
| Days ≥ 0.1" Precip | 2 | 2 | 4 | 5 | 7 | 7 | 4 | 4 | 4 | 4 | 3 | 2 | 48 |
| Total Snowfall (") | 9.4 | 7.3 | 12.7 | 9.4 | 1.8 | 0.2 | 0.0 | 0.1 | 0.5 | ***4.7*** | 9.3 | 9.9 | 65.3 |
| Days ≥ 1" Snow Depth | 25 | 19 | 16 | 4 | 1 | 0 | 0 | 0 | 0 | 3 | 12 | 22 | 102 |

## BOZEMAN GALLATIN FLD *Gallatin County* ELEVATION 4459 ft LAT/LONG 45° 47' N / 111° 9' W

| | JAN | FEB | MAR | APR | MAY | JUN | JUL | AUG | SEP | OCT | NOV | DEC | YEAR |
|---|---|---|---|---|---|---|---|---|---|---|---|---|---|
| Maximum Temp °F | 29.5 | 35.4 | 43.3 | 54.6 | 64.6 | 73.9 | 83.6 | 82.7 | 70.3 | 57.9 | 41.0 | 30.9 | 55.6 |
| Minimum Temp °F | 7.2 | 12.7 | 20.0 | 28.9 | 37.2 | 44.3 | 49.2 | 47.9 | 38.8 | 29.1 | 17.9 | 7.8 | 28.4 |
| Mean Temp °F | 18.4 | 24.1 | 31.7 | 41.8 | 50.9 | 59.2 | 66.4 | 65.4 | 54.6 | 43.5 | 29.4 | 19.4 | 42.1 |
| Days Max Temp ≥ 90 °F | 0 | 0 | 0 | 0 | 0 | 2 | 8 | 8 | 1 | 0 | 0 | 0 | 19 |
| Days Max Temp ≤ 32 °F | 15 | 9 | 5 | 1 | 0 | 0 | 0 | 0 | 0 | 1 | 7 | 16 | 54 |
| Days Min Temp ≤ 32 °F | 30 | 27 | 29 | 21 | 7 | 1 | 0 | 0 | 6 | 21 | 28 | 30 | 200 |
| Days Min Temp ≤ 0 °F | 10 | 6 | 2 | 0 | 0 | 0 | 0 | 0 | 0 | 0 | 3 | 9 | 30 |
| Heating Degree Days | 1436 | 1150 | 1026 | 690 | 431 | 193 | 52 | 66 | 315 | 658 | 1061 | 1409 | 8487 |
| Cooling Degree Days | 0 | 0 | 0 | 0 | 2 | 27 | 99 | 83 | 8 | 0 | 0 | 0 | 219 |
| Total Precipitation (") | 0.58 | 0.51 | 0.97 | 1.36 | 2.48 | 2.57 | 1.20 | 1.20 | 1.53 | 1.21 | 0.80 | 0.58 | 14.99 |
| Days ≥ 0.1" Precip | 2 | 2 | 3 | 4 | 7 | 7 | 4 | 4 | 4 | 4 | 3 | 2 | 46 |
| Total Snowfall (") | 7.1 | 5.2 | 8.6 | 6.3 | 2.4 | 0.1 | 0.0 | 0.0 | 0.5 | 2.2 | 5.6 | 6.8 | 44.8 |
| Days ≥ 1" Snow Depth | 24 | 16 | 14 | 3 | 1 | 0 | 0 | 0 | 0 | 1 | 9 | 20 | 88 |

## BOZEMAN MONTANA SU *Gallatin County* ELEVATION 4862 ft LAT/LONG 45° 40' N / 111° 3' W

| | JAN | FEB | MAR | APR | MAY | JUN | JUL | AUG | SEP | OCT | NOV | DEC | YEAR |
|---|---|---|---|---|---|---|---|---|---|---|---|---|---|
| Maximum Temp °F | 33.4 | 38.1 | 45.0 | 54.9 | 64.5 | 73.4 | 81.7 | 81.1 | 70.4 | 58.8 | 42.0 | 34.0 | 56.4 |
| Minimum Temp °F | 13.9 | 17.9 | 23.4 | 31.1 | 39.4 | 46.4 | 52.0 | 50.7 | 41.9 | 33.1 | 22.5 | 14.7 | 32.3 |
| Mean Temp °F | 23.7 | 28.0 | 34.3 | 43.0 | 52.0 | 60.0 | 66.8 | 65.9 | 56.2 | 46.0 | 32.3 | 24.4 | 44.4 |
| Days Max Temp ≥ 90 °F | 0 | 0 | 0 | 0 | 0 | 1 | 4 | 3 | 1 | 0 | 0 | 0 | 9 |
| Days Max Temp ≤ 32 °F | 12 | 7 | 3 | 1 | 0 | 0 | 0 | 0 | 0 | 1 | 6 | 12 | 42 |
| Days Min Temp ≤ 32 °F | 29 | 26 | 26 | 17 | 6 | 0 | 0 | 0 | 4 | 14 | 25 | 29 | 176 |
| Days Min Temp ≤ 0 °F | 6 | 3 | 1 | 0 | 0 | 0 | 0 | 0 | 0 | 0 | 1 | 4 | 15 |
| Heating Degree Days | 1276 | 1037 | 946 | 653 | 401 | 181 | 47 | 60 | 275 | 583 | 975 | 1253 | 7687 |
| Cooling Degree Days | 0 | 0 | 0 | 0 | 5 | 43 | 114 | 104 | 17 | 1 | 0 | 0 | 284 |
| Total Precipitation (") | 0.82 | 0.64 | 1.42 | 1.99 | 3.22 | 3.06 | 1.49 | 1.42 | 1.88 | 1.67 | 1.14 | 0.75 | 19.50 |
| Days ≥ 0.1" Precip | 3 | 2 | 5 | 6 | 8 | 8 | 4 | 4 | 5 | 5 | 3 | 3 | 56 |
| Total Snowfall (") | 12.9 | 10.0 | 17.9 | 14.0 | 4.8 | 0.4 | 0.0 | 0.1 | 1.3 | 5.6 | 12.0 | 12.3 | 91.3 |
| Days ≥ 1" Snow Depth | 29 | 22 | 18 | 5 | 1 | 0 | 0 | 0 | 0 | 3 | 15 | 26 | 119 |

## BRANDENBERG *Rosebud County* ELEVATION 2772 ft LAT/LONG 45° 49' N / 106° 13' W

| | JAN | FEB | MAR | APR | MAY | JUN | JUL | AUG | SEP | OCT | NOV | DEC | YEAR |
|---|---|---|---|---|---|---|---|---|---|---|---|---|---|
| Maximum Temp °F | 31.1 | 38.2 | 48.2 | 60.1 | 70.6 | 80.2 | 88.9 | 87.2 | 74.7 | 61.3 | 44.4 | 33.9 | 59.9 |
| Minimum Temp °F | 7.4 | 13.1 | 22.2 | 31.8 | 41.0 | 49.7 | 54.2 | 52.0 | 42.1 | 32.3 | 20.6 | 10.9 | 31.4 |
| Mean Temp °F | 19.3 | 25.7 | 35.2 | 46.0 | 55.8 | 65.0 | 71.6 | 69.7 | 58.4 | 46.9 | 32.5 | 22.4 | 45.7 |
| Days Max Temp ≥ 90 °F | 0 | 0 | 0 | 0 | 1 | 5 | 15 | 14 | 3 | 0 | 0 | 0 | 38 |
| Days Max Temp ≤ 32 °F | 14 | 9 | 4 | 0 | 0 | 0 | 0 | 0 | 0 | 0 | 5 | 12 | 44 |
| Days Min Temp ≤ 32 °F | 30 | 27 | 27 | 16 | 5 | 0 | 0 | 0 | 4 | 15 | 27 | 30 | 181 |
| Days Min Temp ≤ 0 °F | 10 | 6 | 2 | 0 | 0 | 0 | 0 | 0 | 0 | 0 | 2 | 6 | 26 |
| Heating Degree Days | 1413 | 1103 | 917 | 566 | 293 | 84 | 13 | 26 | 220 | 557 | 969 | 1314 | 7475 |
| Cooling Degree Days | 0 | 0 | 0 | 2 | 22 | 113 | 232 | 201 | 30 | 1 | 0 | 0 | 601 |
| Total Precipitation (") | 0.63 | 0.46 | 0.77 | 1.64 | 2.24 | 2.52 | 1.08 | 1.08 | 1.32 | 1.26 | 0.72 | 0.65 | 14.37 |
| Days ≥ 0.1" Precip | 2 | 2 | 2 | 4 | 6 | 6 | 3 | 3 | 3 | 3 | 2 | 2 | 38 |
| Total Snowfall (") | 8.6 | 6.0 | 6.4 | 4.7 | 0.7 | 0.0 | 0.0 | 0.0 | 0.8 | 1.5 | 5.9 | 9.1 | 43.7 |
| Days ≥ 1" Snow Depth | 24 | 17 | 9 | 1 | 0 | 0 | 0 | 0 | 0 | 1 | 7 | 20 | 79 |

## BREDETTE *Roosevelt County* ELEVATION 2690 ft LAT/LONG 48° 33' N / 105° 16' W

| | JAN | FEB | MAR | APR | MAY | JUN | JUL | AUG | SEP | OCT | NOV | DEC | YEAR |
|---|---|---|---|---|---|---|---|---|---|---|---|---|---|
| Maximum Temp °F | 18.8 | 25.6 | 38.8 | 54.6 | 67.8 | 76.9 | 83.4 | 82.4 | 69.7 | 56.2 | 36.3 | 23.6 | 52.8 |
| Minimum Temp °F | 0.5 | 6.8 | 18.4 | 30.6 | 41.6 | 49.9 | 54.7 | 53.3 | 43.2 | 32.5 | 17.9 | 5.8 | 29.6 |
| Mean Temp °F | 9.6 | 16.2 | 28.6 | 42.6 | 54.7 | 63.4 | 69.1 | 67.9 | 56.5 | 44.4 | 27.1 | 14.7 | 41.2 |
| Days Max Temp ≥ 90 °F | 0 | 0 | 0 | 0 | 1 | 3 | 7 | 8 | 2 | 0 | 0 | 0 | 21 |
| Days Max Temp ≤ 32 °F | 23 | 17 | 9 | 1 | 0 | 0 | 0 | 0 | 0 | 1 | 10 | 21 | 82 |
| Days Min Temp ≤ 32 °F | 31 | 27 | 29 | 18 | 4 | 0 | 0 | 0 | 3 | 14 | 27 | 31 | 184 |
| Days Min Temp ≤ 0 °F | 16 | 10 | 3 | 0 | 0 | 0 | 0 | 0 | 0 | 0 | 3 | 11 | 43 |
| Heating Degree Days | 1714 | 1373 | 1121 | 665 | 330 | 110 | 28 | 55 | 278 | 627 | 1131 | 1555 | 8987 |
| Cooling Degree Days | 0 | 0 | 0 | 2 | 26 | 84 | 169 | 164 | 32 | 2 | 0 | 0 | 479 |
| Total Precipitation (") | 0.34 | 0.22 | 0.45 | 0.88 | 1.93 | 2.75 | 2.20 | 1.59 | 1.26 | 0.66 | 0.31 | 0.34 | 12.93 |
| Days ≥ 0.1" Precip | 1 | 1 | 1 | 2 | 5 | 6 | 4 | 3 | 3 | 2 | 1 | 1 | 30 |
| Total Snowfall (") | 5.7 | 4.0 | 5.4 | 3.6 | 0.9 | 0.0 | 0.0 | 0.0 | 0.2 | 1.9 | 4.2 | 6.2 | 32.1 |
| Days ≥ 1" Snow Depth | 30 | 23 | 13 | 4 | 0 | 0 | 0 | 0 | 0 | 1 | 8 | 24 | 103 |

## BRIDGER 3 S *Carbon County* ELEVATION 3681 ft LAT/LONG 45° 18 ' N / 108° 54 ' W

| | JAN | FEB | MAR | APR | MAY | JUN | JUL | AUG | SEP | OCT | NOV | DEC | YEAR |
|---|---|---|---|---|---|---|---|---|---|---|---|---|---|
| Maximum Temp °F | 33.2 | 40.6 | 49.9 | 59.7 | 69.8 | 79.0 | 87.0 | 86.0 | 74.8 | 62.8 | 45.7 | 35.7 | 60.4 |
| Minimum Temp °F | 11.6 | 17.1 | 23.7 | 31.8 | 40.6 | 48.2 | 53.3 | 51.6 | 42.3 | 33.7 | 23.1 | 14.7 | 32.6 |
| Mean Temp °F | 22.4 | 28.8 | 36.8 | 45.7 | 55.2 | 63.6 | 70.2 | 68.8 | 58.6 | 48.3 | 34.5 | 25.2 | 46.5 |
| Days Max Temp ≥ 90 °F | 0 | 0 | 0 | 0 | 1 | 5 | 12 | 11 | 3 | 0 | 0 | 0 | 32 |
| Days Max Temp ≤ 32 °F | 12 | 7 | 3 | 0 | 0 | 0 | 0 | 0 | 0 | 0 | 4 | 10 | 36 |
| Days Min Temp ≤ 32 °F | 29 | 26 | 26 | 16 | 4 | 0 | 0 | 0 | 3 | 13 | 25 | 28 | 170 |
| Days Min Temp ≤ 0 °F | 8 | 3 | 1 | 0 | 0 | 0 | 0 | 0 | 0 | 0 | 1 | 5 | 18 |
| Heating Degree Days | 1314 | 1015 | 866 | 572 | 310 | 104 | 17 | 27 | 216 | 512 | 908 | 1226 | 7087 |
| Cooling Degree Days | 0 | 0 | 0 | 1 | 15 | 80 | 186 | 160 | 29 | 1 | 0 | 0 | 472 |
| Total Precipitation (") | 0.78 | 0.54 | 1.07 | 1.73 | 2.34 | 1.84 | 0.78 | 0.80 | 1.33 | 1.18 | 0.71 | 0.63 | 13.73 |
| Days ≥ 0.1" Precip | 2 | 2 | 3 | 4 | 5 | 4 | 2 | 2 | 3 | 3 | 2 | 2 | 34 |
| Total Snowfall (") | 8.8 | 6.2 | 9.7 | 6.8 | 1.1 | 0.0 | 0.0 | 0.0 | 1.1 | *4.2* | *5.7* | 7.3 | 50.9 |
| Days ≥ 1" Snow Depth | 22 | 14 | 7 | 2 | 0 | 0 | 0 | 0 | 0 | *1* | *7* | 16 | 69 |

## BROADUS *Powder River County* ELEVATION 3031 ft LAT/LONG 45° 26 ' N / 105° 24 ' W

| | JAN | FEB | MAR | APR | MAY | JUN | JUL | AUG | SEP | OCT | NOV | DEC | YEAR |
|---|---|---|---|---|---|---|---|---|---|---|---|---|---|
| Maximum Temp °F | 31.7 | 38.5 | 47.9 | 59.3 | 69.4 | 78.7 | 86.7 | 85.7 | 74.1 | 61.7 | 45.3 | 34.7 | 59.5 |
| Minimum Temp °F | 5.4 | 11.5 | 20.6 | 30.6 | 40.6 | 49.9 | 54.8 | 52.0 | 41.0 | 29.6 | 18.2 | 8.2 | 30.2 |
| Mean Temp °F | 18.6 | 25.1 | 34.3 | 45.0 | 55.0 | 64.3 | 70.8 | 68.9 | 57.6 | 45.7 | 31.8 | 21.4 | 44.9 |
| Days Max Temp ≥ 90 °F | 0 | 0 | 0 | 0 | 1 | 4 | 13 | 11 | 3 | 0 | 0 | 0 | 32 |
| Days Max Temp ≤ 32 °F | 14 | 9 | 4 | 0 | 0 | 0 | 0 | 0 | 0 | 1 | 5 | 12 | 45 |
| Days Min Temp ≤ 32 °F | 31 | 27 | 29 | 18 | 5 | 0 | 0 | 0 | 5 | 20 | 29 | 31 | 195 |
| Days Min Temp ≤ 0 °F | 11 | 6 | 2 | 0 | 0 | 0 | 0 | 0 | 0 | 0 | 2 | 8 | 29 |
| Heating Degree Days | 1436 | 1120 | 946 | 595 | 318 | 94 | 18 | 32 | 244 | 592 | 992 | 1344 | 7731 |
| Cooling Degree Days | 0 | 0 | 0 | 2 | 22 | 96 | 223 | 186 | 29 | 1 | 0 | 0 | 559 |
| Total Precipitation (") | 0.50 | 0.41 | 0.85 | 1.62 | 2.29 | 2.30 | 1.37 | 0.86 | 1.23 | 1.05 | 0.59 | 0.56 | 13.63 |
| Days ≥ 0.1" Precip | 2 | 1 | 3 | 4 | 6 | 6 | 3 | 2 | 3 | 3 | 2 | 2 | 37 |
| Total Snowfall (") | 8.1 | 6.6 | 8.7 | 6.2 | 1.5 | 0.1 | 0.0 | 0.0 | 0.9 | 2.6 | 6.1 | 9.2 | 50.0 |
| Days ≥ 1" Snow Depth | 24 | 17 | 9 | 1 | 0 | 0 | 0 | 0 | 0 | 1 | 6 | 18 | 76 |

## BROCKWAY 3 WSW *McCone County* ELEVATION 2631 ft LAT/LONG 47° 17 ' N / 105° 49 ' W

| | JAN | FEB | MAR | APR | MAY | JUN | JUL | AUG | SEP | OCT | NOV | DEC | YEAR |
|---|---|---|---|---|---|---|---|---|---|---|---|---|---|
| Maximum Temp °F | 25.5 | 32.0 | 43.7 | 57.5 | 69.0 | 78.3 | 85.7 | 84.9 | 73.1 | 60.1 | 41.5 | 30.0 | 56.8 |
| Minimum Temp °F | 3.5 | 9.5 | 19.7 | 30.1 | 40.3 | 48.9 | 53.3 | 52.0 | 41.4 | 30.5 | 17.8 | 7.3 | 29.5 |
| Mean Temp °F | 14.5 | 20.8 | 31.8 | 43.8 | 54.7 | 63.6 | 69.5 | 68.5 | 57.3 | 45.3 | 29.7 | 18.7 | 43.2 |
| Days Max Temp ≥ 90 °F | 0 | 0 | 0 | 0 | 1 | 4 | 11 | 10 | 2 | 0 | 0 | 0 | 28 |
| Days Max Temp ≤ 32 °F | *19* | 12 | *6* | 1 | 0 | 0 | 0 | 0 | 0 | 1 | 6 | *17* | 62 |
| Days Min Temp ≤ 32 °F | 31 | 28 | 28 | 19 | 5 | 0 | 0 | 0 | 4 | *18* | 27 | 31 | 191 |
| Days Min Temp ≤ 0 °F | *12* | 8 | 3 | 0 | 0 | 0 | 0 | 0 | 0 | 0 | 3 | 9 | 35 |
| Heating Degree Days | 1564 | 1245 | 1022 | 629 | 333 | 107 | 30 | 46 | 256 | 604 | 1048 | 1431 | 8315 |
| Cooling Degree Days | 0 | 0 | 0 | 1 | 22 | 83 | 176 | 178 | 30 | 1 | 0 | 0 | 491 |
| Total Precipitation (") | 0.26 | 0.15 | 0.38 | 1.10 | 1.93 | 2.59 | 1.59 | 1.30 | 1.32 | 0.67 | 0.26 | 0.27 | 11.82 |
| Days ≥ 0.1" Precip | 1 | 0 | 1 | 3 | 5 | 5 | 4 | 3 | 3 | 2 | 1 | 1 | 29 |
| Total Snowfall (") | na | na | na | na | na | *0.0* | *0.0* | *0.0* | *0.0* | na | na | na | na |
| Days ≥ 1" Snow Depth | na | na | na | na | na | *0* | *0* | *0* | *0* | na | na | na | na |

## BUSBY *Big Horn County* ELEVATION 3504 ft LAT/LONG 45° 32 ' N / 106° 57 ' W

| | JAN | FEB | MAR | APR | MAY | JUN | JUL | AUG | SEP | OCT | NOV | DEC | YEAR |
|---|---|---|---|---|---|---|---|---|---|---|---|---|---|
| Maximum Temp °F | 31.4 | 37.5 | 47.1 | 59.4 | 69.2 | 78.5 | 87.5 | 86.8 | 75.2 | 61.4 | 44.9 | 33.6 | 59.4 |
| Minimum Temp °F | 5.1 | 10.0 | 19.9 | 29.3 | 38.3 | 46.4 | 50.9 | *48.9* | 39.4 | 29.1 | 17.4 | 6.9 | 28.5 |
| Mean Temp °F | 18.3 | 23.8 | 33.5 | 44.3 | 53.8 | 62.5 | 69.2 | *67.8* | 57.3 | 45.3 | 31.2 | 20.3 | 43.9 |
| Days Max Temp ≥ 90 °F | 0 | 0 | 0 | 0 | 1 | 4 | *13* | *12* | 3 | 0 | 0 | 0 | 33 |
| Days Max Temp ≤ 32 °F | 13 | 8 | 3 | 0 | 0 | 0 | 0 | 0 | 0 | 0 | 5 | *12* | 41 |
| Days Min Temp ≤ 32 °F | 31 | 28 | 29 | 20 | 7 | 1 | 0 | 0 | 5 | 21 | 28 | 31 | 201 |
| Days Min Temp ≤ 0 °F | 12 | 7 | 2 | 0 | 0 | 0 | 0 | 0 | 0 | 0 | 3 | 8 | 32 |
| Heating Degree Days | 1443 | 1158 | 969 | 614 | 348 | 124 | 23 | *39* | 241 | 605 | 1008 | 1380 | 7952 |
| Cooling Degree Days | 0 | 0 | 0 | 0 | 7 | 67 | 161 | 146 | 18 | 0 | 0 | 0 | 399 |
| Total Precipitation (") | 0.76 | 0.55 | 0.77 | 1.40 | 2.29 | 2.67 | 1.18 | 0.90 | 1.58 | 1.33 | 0.86 | 0.73 | 15.02 |
| Days ≥ 0.1" Precip | 3 | 2 | 3 | 4 | 6 | 6 | 3 | 3 | 4 | 4 | 3 | 2 | 43 |
| Total Snowfall (") | *10.7* | *7.6* | 8.0 | 5.9 | 1.3 | 0.0 | 0.0 | 0.0 | 0.8 | *2.6* | 8.7 | 11.2 | 56.8 |
| Days ≥ 1" Snow Depth | 27 | 23 | 15 | 2 | 0 | 0 | 0 | 0 | 0 | 1 | 9 | 24 | 101 |

## BUTTE BERT MOONEY AP *Silver Bow County* ELEVATION 5535 ft LAT/LONG 45° 57' N / 112° 30' W

| | JAN | FEB | MAR | APR | MAY | JUN | JUL | AUG | SEP | OCT | NOV | DEC | YEAR |
|---|---|---|---|---|---|---|---|---|---|---|---|---|---|
| Maximum Temp °F | 29.2 | 34.0 | 40.9 | 50.9 | 60.8 | 70.2 | 79.4 | 78.5 | 67.0 | 55.1 | 38.8 | 29.6 | 52.9 |
| Minimum Temp °F | 5.8 | 9.2 | 17.4 | 26.1 | 34.1 | 41.4 | 45.6 | 44.0 | 35.1 | 26.2 | 15.5 | 5.4 | 25.5 |
| Mean Temp °F | 17.5 | 21.6 | 29.1 | 38.5 | 47.5 | 55.8 | 62.5 | 61.2 | 51.1 | 40.7 | 27.2 | 17.5 | 39.2 |
| Days Max Temp ≥ 90 °F | 0 | 0 | 0 | 0 | 0 | 1 | 2 | 2 | 0 | 0 | 0 | 0 | 5 |
| Days Max Temp ≤ 32 °F | 17 | 11 | 6 | 1 | 0 | 0 | 0 | 0 | 0 | 1 | 8 | 18 | 62 |
| Days Min Temp ≤ 32 °F | 30 | 27 | 30 | 25 | 12 | 2 | 0 | 1 | 11 | 26 | 28 | 30 | 222 |
| Days Min Temp ≤ 0 °F | 11 | 7 | 3 | 0 | 0 | 0 | 0 | 0 | 0 | 0 | 4 | 11 | 36 |
| Heating Degree Days | 1463 | 1219 | 1105 | 788 | 536 | 277 | 105 | 136 | 413 | 747 | 1128 | 1467 | 9384 |
| Cooling Degree Days | 0 | 0 | 0 | 0 | 0 | 10 | 37 | 28 | 1 | 0 | 0 | 0 | 76 |
| Total Precipitation (") | 0.52 | 0.38 | 0.76 | 0.94 | 1.85 | 2.20 | 1.36 | 1.33 | 1.20 | 0.77 | 0.56 | 0.44 | 12.31 |
| Days ≥ 0.1" Precip | 1 | 1 | 3 | 3 | 6 | 6 | 4 | 4 | 4 | 3 | 2 | 1 | 38 |
| Total Snowfall (") | 8.1 | 6.5 | 10.6 | 8.4 | 3.5 | 0.3 | 0.0 | 0.3 | 1.3 | 4.0 | 6.8 | 7.6 | 57.4 |
| Days ≥ 1" Snow Depth | 26 | 21 | 15 | 5 | 1 | 0 | 0 | 0 | 0 | 2 | 11 | 24 | 105 |

## CANYON FERRY DAM *Lewis and Clark County* ELEVATION 3671 ft LAT/LONG 46° 39' N / 111° 44' W

| | JAN | FEB | MAR | APR | MAY | JUN | JUL | AUG | SEP | OCT | NOV | DEC | YEAR |
|---|---|---|---|---|---|---|---|---|---|---|---|---|---|
| Maximum Temp °F | 31.5 | 37.2 | 46.5 | 57.2 | 67.1 | 75.4 | 83.7 | 83.3 | 71.9 | 59.2 | 43.1 | 33.6 | 57.5 |
| Minimum Temp °F | 13.9 | 17.5 | 24.5 | 32.4 | 40.2 | 47.6 | 52.2 | 51.5 | 43.4 | 35.1 | 25.7 | 17.2 | 33.4 |
| Mean Temp °F | 22.7 | 27.4 | 35.5 | 44.8 | 53.6 | 61.5 | 68.0 | 67.4 | 57.6 | 47.1 | 34.4 | 25.4 | 45.4 |
| Days Max Temp ≥ 90 °F | 0 | 0 | 0 | 0 | 0 | 1 | 7 | 6 | 0 | 0 | 0 | 0 | 14 |
| Days Max Temp ≤ 32 °F | 13 | 8 | 3 | 0 | 0 | 0 | 0 | 0 | 0 | 0 | 4 | 12 | 40 |
| Days Min Temp ≤ 32 °F | 28 | 26 | 25 | 15 | 3 | 0 | 0 | 0 | 2 | 11 | 22 | 28 | 160 |
| Days Min Temp ≤ 0 °F | 7 | 3 | 1 | 0 | 0 | 0 | 0 | 0 | 0 | 0 | 1 | 4 | 16 |
| Heating Degree Days | 1306 | 1056 | 907 | 600 | 348 | 138 | 32 | 39 | 232 | 546 | 910 | 1220 | 7334 |
| Cooling Degree Days | 0 | 0 | 0 | 0 | 3 | 42 | 117 | 119 | 16 | 0 | 0 | 0 | 297 |
| Total Precipitation (") | 0.50 | 0.31 | 0.58 | 1.00 | 1.84 | 1.89 | 1.39 | 1.29 | 1.18 | 0.67 | 0.47 | 0.49 | 11.61 |
| Days ≥ 0.1" Precip | 2 | 1 | 2 | 4 | 5 | 6 | 4 | 3 | 3 | 2 | 2 | 2 | 36 |
| Total Snowfall (") | na | na | na | na | 0.0 | 0.0 | 0.0 | 0.0 | 0.0 | na | na | na | na |
| Days ≥ 1" Snow Depth | na | na | na | na | 0 | 0 | 0 | 0 | 0 | *0* | na | na | na |

## CARLYLE 12 NW *Wibaux County* ELEVATION 3031 ft LAT/LONG 46° 46' N / 104° 17' W

| | JAN | FEB | MAR | APR | MAY | JUN | JUL | AUG | SEP | OCT | NOV | DEC | YEAR |
|---|---|---|---|---|---|---|---|---|---|---|---|---|---|
| Maximum Temp °F | 25.2 | 31.1 | 42.1 | 55.4 | 66.9 | 75.7 | 83.4 | 82.6 | 71.3 | 58.7 | 40.6 | 29.6 | 55.2 |
| Minimum Temp °F | 5.0 | 10.5 | 20.0 | 30.2 | 40.8 | 50.0 | 55.2 | 53.7 | 44.0 | 33.2 | 20.1 | 9.8 | 31.0 |
| Mean Temp °F | 15.1 | 20.8 | 31.1 | 42.8 | 53.9 | 62.8 | 69.3 | 68.2 | 57.7 | 46.0 | 30.4 | 19.7 | 43.2 |
| Days Max Temp ≥ 90 °F | 0 | 0 | 0 | 0 | 0 | 2 | 7 | 8 | 2 | 0 | 0 | 0 | 19 |
| Days Max Temp ≤ 32 °F | 19 | 13 | 7 | 1 | 0 | 0 | 0 | 0 | 0 | 1 | 8 | 17 | 66 |
| Days Min Temp ≤ 32 °F | 30 | 27 | 28 | 19 | 4 | 0 | 0 | 0 | 2 | *14* | 26 | 30 | 180 |
| Days Min Temp ≤ 0 °F | 12 | 8 | 2 | 0 | 0 | 0 | 0 | 0 | 0 | 0 | 2 | 8 | 32 |
| Heating Degree Days | 1544 | 1244 | 1046 | 659 | 351 | 119 | 29 | 47 | 249 | 585 | 1033 | 1400 | 8306 |
| Cooling Degree Days | 0 | 0 | 0 | 1 | 16 | 73 | 181 | 161 | 35 | 1 | 0 | 0 | 468 |
| Total Precipitation (") | 0.51 | 0.39 | 0.65 | 1.54 | 2.30 | 2.74 | 1.78 | 1.55 | 1.59 | 1.09 | 0.58 | 0.52 | 15.24 |
| Days ≥ 0.1" Precip | 2 | 1 | 2 | 4 | 5 | 6 | 4 | 3 | 4 | 3 | 2 | 2 | 38 |
| Total Snowfall (") | 10.4 | 7.3 | 7.3 | 6.2 | 1.4 | 0.0 | 0.0 | 0.0 | 0.7 | 1.8 | 6.0 | 9.0 | 50.1 |
| Days ≥ 1" Snow Depth | na | na | na | na | 0 | 0 | 0 | *0* | 0 | *0* | na | na | na |

## CASCADE 20 SSE *Cascade County* ELEVATION 4603 ft LAT/LONG 47° 0' N / 111° 35' W

| | JAN | FEB | MAR | APR | MAY | JUN | JUL | AUG | SEP | OCT | NOV | DEC | YEAR |
|---|---|---|---|---|---|---|---|---|---|---|---|---|---|
| Maximum Temp °F | 35.1 | 39.8 | 45.5 | 54.4 | 63.6 | 71.8 | 80.0 | 79.6 | 69.2 | 58.4 | 43.7 | 36.4 | 56.5 |
| Minimum Temp °F | 11.1 | 15.1 | 20.0 | 27.3 | 34.9 | 41.4 | 43.7 | 42.9 | 35.8 | 29.4 | 20.8 | 13.6 | 28.0 |
| Mean Temp °F | 23.1 | 27.5 | 32.8 | 40.9 | 49.3 | 56.6 | 61.9 | 61.3 | 52.5 | 43.9 | 32.3 | 25.0 | 42.3 |
| Days Max Temp ≥ 90 °F | 0 | 0 | 0 | 0 | 0 | 1 | 3 | 4 | 0 | 0 | 0 | 0 | 8 |
| Days Max Temp ≤ 32 °F | 11 | 6 | 3 | 1 | 0 | 0 | 0 | 0 | 0 | 1 | 4 | 9 | 35 |
| Days Min Temp ≤ 32 °F | 28 | 25 | 27 | 22 | 11 | 3 | 1 | 2 | 10 | 19 | 24 | 27 | 199 |
| Days Min Temp ≤ 0 °F | 9 | 5 | 3 | 0 | 0 | 0 | 0 | 0 | 0 | 0 | 3 | 6 | 26 |
| Heating Degree Days | 1293 | 1053 | 992 | 718 | 481 | 255 | 120 | 138 | 371 | 647 | 976 | 1233 | 8277 |
| Cooling Degree Days | 0 | 0 | 0 | 0 | 1 | 12 | 30 | 34 | 3 | 0 | 0 | 0 | 80 |
| Total Precipitation (") | 0.58 | 0.37 | 0.82 | 1.31 | 2.66 | 2.69 | 1.52 | 1.43 | 1.58 | 1.01 | 0.56 | 0.53 | 15.06 |
| Days ≥ 0.1" Precip | 2 | 1 | 3 | 4 | 6 | 6 | 4 | 4 | 4 | 3 | 2 | 2 | 41 |
| Total Snowfall (") | na | na | na | na | na | na | *0.0* | *0.0* | na | na | na | na | na |
| Days ≥ 1" Snow Depth | na | na | na | na | na | *0* | *0* | *0* | na | na | na | na | na |

## CASCADE 5 S *Cascade County* ELEVATION 3390 ft LAT/LONG 47° 13 ' N / 111° 43 ' W

| | JAN | FEB | MAR | APR | MAY | JUN | JUL | AUG | SEP | OCT | NOV | DEC | YEAR |
|---|---|---|---|---|---|---|---|---|---|---|---|---|---|
| Maximum Temp °F | 34.5 | 40.4 | 47.6 | 57.2 | 66.7 | 75.0 | 83.3 | 82.6 | 71.8 | 61.0 | 45.3 | 37.2 | 58.5 |
| Minimum Temp °F | 12.9 | 17.8 | 23.2 | 31.6 | 39.8 | 47.1 | 50.1 | 48.8 | 40.8 | 34.3 | 24.7 | 16.9 | 32.3 |
| Mean Temp °F | 23.7 | 29.1 | 35.4 | 44.4 | 53.3 | 61.1 | 66.7 | 65.8 | 56.3 | 47.7 | 35.0 | 27.1 | 45.5 |
| Days Max Temp ≥ 90 °F | 0 | 0 | 0 | 0 | 0 | 3 | 8 | 8 | 2 | 0 | 0 | 0 | 21 |
| Days Max Temp ≤ 32 °F | 11 | 7 | 4 | 1 | 0 | 0 | 0 | 0 | 0 | 1 | 4 | 9 | 37 |
| Days Min Temp ≤ 32 °F | 26 | 22 | 24 | 17 | 5 | 0 | 0 | 0 | 5 | 13 | 20 | 25 | 157 |
| Days Min Temp ≤ 0 °F | 9 | 5 | 2 | 0 | 0 | 0 | 0 | 0 | 0 | 0 | 3 | 6 | 25 |
| Heating Degree Days | 1275 | 1007 | 911 | 611 | 361 | 151 | 44 | 65 | 274 | 531 | 892 | 1171 | 7293 |
| Cooling Degree Days | 0 | 0 | 0 | 1 | 6 | 43 | 96 | 97 | 18 | 3 | 0 | 0 | 264 |
| Total Precipitation (") | 0.66 | 0.46 | 1.08 | 1.79 | 2.75 | 2.51 | 1.58 | 1.63 | 1.62 | 1.16 | 0.64 | 0.63 | 16.51 |
| Days ≥ 0.1" Precip | 2 | 1 | 4 | 5 | 6 | 6 | 4 | 4 | 4 | 3 | 2 | 2 | 43 |
| Total Snowfall (") | 12.7 | 9.0 | 13.5 | 10.2 | 0.7 | 0.1 | 0.0 | 0.2 | 1.3 | 4.0 | 10.1 | 13.0 | 74.8 |
| Days ≥ 1" Snow Depth | 13 | 11 | 9 | 3 | 0 | 0 | 0 | 0 | 0 | 2 | 7 | 12 | 57 |

## CHESTER *Liberty County* ELEVATION 3104 ft LAT/LONG 48° 30 ' N / 110° 57 ' W

| | JAN | FEB | MAR | APR | MAY | JUN | JUL | AUG | SEP | OCT | NOV | DEC | YEAR |
|---|---|---|---|---|---|---|---|---|---|---|---|---|---|
| Maximum Temp °F | 27.4 | 34.7 | 45.2 | 58.1 | 68.6 | 76.4 | 83.3 | 82.7 | 72.0 | 60.4 | 41.8 | 31.1 | 56.8 |
| Minimum Temp °F | 2.4 | 7.3 | 17.8 | 28.0 | 38.4 | 46.1 | 50.0 | 48.6 | 38.4 | 27.8 | 15.1 | 5.5 | 27.1 |
| Mean Temp °F | 14.9 | 21.1 | 31.5 | 43.0 | 53.5 | 61.3 | 66.7 | 65.7 | 55.2 | 44.1 | 28.5 | 18.3 | 42.0 |
| Days Max Temp ≥ 90 °F | 0 | 0 | 0 | 0 | 0 | 3 | 8 | 7 | 1 | 0 | 0 | 0 | 19 |
| Days Max Temp ≤ 32 °F | 16 | 11 | 5 | 1 | 0 | 0 | 0 | 0 | 0 | 1 | 6 | 13 | 53 |
| Days Min Temp ≤ 32 °F | 30 | 28 | 30 | 21 | 7 | 1 | 0 | 0 | 7 | ***21*** | 28 | 31 | 204 |
| Days Min Temp ≤ 0 °F | 13 | 9 | 3 | 0 | 0 | 0 | 0 | 0 | 0 | 0 | 4 | 10 | 39 |
| Heating Degree Days | 1550 | 1236 | 1031 | 652 | 356 | 149 | 50 | 66 | 300 | 643 | 1089 | 1443 | 8565 |
| Cooling Degree Days | 0 | 0 | 0 | 1 | 12 | 52 | 110 | 104 | 11 | 0 | 0 | 0 | 290 |
| Total Precipitation (") | 0.44 | 0.26 | 0.53 | 0.79 | 1.77 | 2.29 | 1.51 | 1.30 | 0.82 | 0.44 | 0.35 | 0.42 | 10.92 |
| Days ≥ 0.1" Precip | 2 | 1 | 2 | 3 | 5 | 6 | 4 | 3 | 2 | 2 | 1 | 2 | 33 |
| Total Snowfall (") | na | na | na | na | na | na | ***0.0*** | ***0.0*** | na | na | na | na | na |
| Days ≥ 1" Snow Depth | na | na | na | na | na | na | ***0*** | ***0*** | na | na | na | na | na |

## CHINOOK *Blaine County* ELEVATION 2411 ft LAT/LONG 48° 35 ' N / 109° 12 ' W

| | JAN | FEB | MAR | APR | MAY | JUN | JUL | AUG | SEP | OCT | NOV | DEC | YEAR |
|---|---|---|---|---|---|---|---|---|---|---|---|---|---|
| Maximum Temp °F | 25.4 | 32.8 | 45.5 | 59.3 | 70.4 | 78.7 | 85.1 | 84.3 | 72.5 | 60.1 | 41.6 | 30.1 | 57.2 |
| Minimum Temp °F | 2.2 | 7.8 | 19.0 | 30.3 | 40.7 | 48.9 | 52.6 | 50.6 | 39.8 | 30.0 | 16.5 | 6.5 | 28.7 |
| Mean Temp °F | 13.6 | 20.5 | 32.2 | 44.8 | 55.6 | 63.8 | 68.9 | 67.5 | 56.2 | 45.1 | 29.1 | 18.3 | 43.0 |
| Days Max Temp ≥ 90 °F | 0 | 0 | 0 | 0 | 1 | 4 | 9 | 9 | 2 | 0 | 0 | 0 | 25 |
| Days Max Temp ≤ 32 °F | 18 | 12 | 5 | 0 | 0 | 0 | 0 | 0 | 0 | 1 | 7 | 15 | 58 |
| Days Min Temp ≤ 32 °F | 30 | 27 | 29 | 19 | 4 | 0 | 0 | 0 | 5 | ***18*** | 28 | 30 | 190 |
| Days Min Temp ≤ 0 °F | 14 | 8 | 3 | 0 | 0 | 0 | 0 | 0 | 0 | 0 | 3 | 10 | 38 |
| Heating Degree Days | 1604 | 1250 | 1010 | 600 | 297 | 97 | 23 | 47 | 274 | 611 | 1072 | 1432 | 8317 |
| Cooling Degree Days | ***0*** | 0 | 0 | 1 | 13 | 73 | 135 | 121 | 8 | 0 | 0 | 0 | 351 |
| Total Precipitation (") | 0.58 | 0.31 | 0.51 | 1.12 | 2.11 | 2.21 | 1.72 | 1.36 | 1.53 | 0.62 | 0.40 | 0.49 | 12.96 |
| Days ≥ 0.1" Precip | 2 | 1 | 2 | 3 | 4 | 5 | 4 | 3 | 3 | 2 | 2 | 2 | 33 |
| Total Snowfall (") | na | na | na | na | 0.3 | 0.0 | 0.0 | 0.0 | 0.0 | na | na | na | na |
| Days ≥ 1" Snow Depth | na | na | na | na | 0 | 0 | 0 | 0 | 0 | ***0*** | na | na | na |

## CHOTEAU AP *Teton County* ELEVATION 3812 ft LAT/LONG 47° 49 ' N / 112° 10 ' W

| | JAN | FEB | MAR | APR | MAY | JUN | JUL | AUG | SEP | OCT | NOV | DEC | YEAR |
|---|---|---|---|---|---|---|---|---|---|---|---|---|---|
| Maximum Temp °F | 33.3 | 39.3 | 46.3 | 56.3 | 66.2 | 74.2 | 81.9 | 81.3 | 71.0 | 60.4 | 43.8 | 36.1 | 57.5 |
| Minimum Temp °F | 12.2 | 16.8 | 23.3 | 31.5 | 40.4 | 48.0 | 52.2 | 51.1 | 43.0 | 35.2 | 23.6 | 16.4 | 32.8 |
| Mean Temp °F | 22.8 | 28.1 | 34.8 | 43.9 | 53.3 | 61.1 | 67.1 | 66.2 | 57.0 | 47.8 | 33.7 | 26.3 | 45.2 |
| Days Max Temp ≥ 90 °F | 0 | 0 | 0 | 0 | 0 | 1 | 5 | 6 | 1 | 0 | 0 | 0 | 13 |
| Days Max Temp ≤ 32 °F | 12 | 8 | 4 | 1 | 0 | 0 | 0 | 0 | 0 | 1 | 4 | 9 | 39 |
| Days Min Temp ≤ 32 °F | 27 | 24 | 25 | 16 | 4 | 0 | 0 | 0 | 3 | 12 | 23 | 26 | 160 |
| Days Min Temp ≤ 0 °F | 9 | 5 | 2 | 0 | 0 | 0 | 0 | 0 | 0 | 0 | 2 | 5 | 23 |
| Heating Degree Days | 1306 | 1037 | 929 | 626 | 362 | 154 | 44 | 63 | 258 | 529 | 931 | 1196 | 7435 |
| Cooling Degree Days | 0 | 0 | 0 | 1 | 11 | 49 | 112 | 116 | 30 | 5 | 0 | 0 | 324 |
| Total Precipitation (") | 0.28 | 0.21 | 0.37 | 0.83 | 1.81 | 2.25 | 1.32 | 1.37 | 1.01 | 0.47 | 0.28 | 0.28 | 10.48 |
| Days ≥ 0.1" Precip | 1 | 1 | 1 | 2 | 5 | 5 | 3 | 4 | 2 | 2 | 1 | 1 | 28 |
| Total Snowfall (") | 8.2 | 5.9 | 6.7 | 5.7 | 1.0 | 0.0 | 0.0 | 0.2 | 1.4 | ***2.8*** | 5.8 | 7.4 | 45.1 |
| Days ≥ 1" Snow Depth | ***14*** | ***9*** | na | 2 | 0 | 0 | 0 | 0 | 0 | 1 | ***5*** | ***12*** | na |

## CIRCLE *McCone County* ELEVATION 2441 ft LAT/LONG 47° 26 ' N / 105° 35 ' W

| | JAN | FEB | MAR | APR | MAY | JUN | JUL | AUG | SEP | OCT | NOV | DEC | YEAR |
|---|---|---|---|---|---|---|---|---|---|---|---|---|---|
| Maximum Temp °F | 24.5 | 31.9 | 42.9 | 57.1 | 69.1 | 78.6 | 85.9 | 84.9 | 72.4 | 59.4 | 41.2 | 29.5 | 56.5 |
| Minimum Temp °F | 2.2 | 9.1 | 19.5 | 31.2 | 41.6 | 50.4 | 55.1 | 53.5 | 42.2 | 31.4 | 18.7 | 7.5 | 30.2 |
| Mean Temp °F | 13.4 | 20.6 | 31.2 | 44.1 | 55.4 | 64.5 | 70.5 | 69.3 | 57.3 | 45.4 | 30.0 | 18.5 | 43.4 |
| Days Max Temp ≥ 90 °F | 0 | 0 | 0 | 0 | 1 | 4 | 12 | 11 | 3 | 0 | 0 | 0 | 31 |
| Days Max Temp ≤ 32 °F | 19 | 13 | 6 | 1 | 0 | 0 | 0 | 0 | 0 | 1 | 7 | 16 | 63 |
| Days Min Temp ≤ 32 °F | 31 | 28 | 28 | 17 | 4 | 0 | 0 | 0 | 4 | 17 | 28 | 30 | 187 |
| Days Min Temp ≤ 0 °F | 14 | 8 | 3 | 0 | 0 | 0 | 0 | 0 | 0 | 0 | 2 | 10 | 37 |
| Heating Degree Days | 1598 | 1251 | 1042 | 620 | 315 | 97 | 26 | 43 | 257 | 600 | 1045 | 1436 | 8330 |
| Cooling Degree Days | 0 | 0 | 0 | 2 | 27 | 97 | 209 | 196 | 34 | 1 | 0 | 0 | 566 |
| Total Precipitation (") | 0.53 | 0.32 | 0.61 | 1.38 | 2.01 | 2.68 | 1.86 | 1.39 | 1.37 | 0.77 | 0.42 | 0.50 | 13.84 |
| Days ≥ 0.1" Precip | 2 | 1 | 2 | 4 | 5 | 6 | 4 | 3 | 3 | 2 | 2 | 2 | 36 |
| Total Snowfall (") | ***7.7*** | ***4.8*** | ***5.0*** | 2.6 | 0.6 | 0.0 | 0.0 | 0.0 | 0.2 | 1.4 | ***3.6*** | 5.9 | 31.8 |
| Days ≥ 1" Snow Depth | ***20*** | ***16*** | ***9*** | 2 | 0 | 0 | 0 | 0 | 0 | 1 | 5 | 14 | 67 |

## COHAGEN *Garfield County* ELEVATION 2723 ft LAT/LONG 47° 4 ' N / 106° 37 ' W

| | JAN | FEB | MAR | APR | MAY | JUN | JUL | AUG | SEP | OCT | NOV | DEC | YEAR |
|---|---|---|---|---|---|---|---|---|---|---|---|---|---|
| Maximum Temp °F | 29.0 | 36.0 | 47.8 | 60.5 | 70.4 | 79.5 | 87.1 | 86.7 | 75.1 | 61.9 | 44.1 | 33.0 | 59.3 |
| Minimum Temp °F | 4.3 | 10.4 | 20.7 | 30.2 | 40.1 | 48.8 | 53.0 | 51.6 | 40.8 | 30.4 | 18.3 | 7.6 | 29.7 |
| Mean Temp °F | 16.7 | 23.2 | 34.2 | 45.4 | 55.3 | 64.2 | 70.1 | 69.2 | 58.0 | 46.2 | 31.2 | 20.3 | 44.5 |
| Days Max Temp ≥ 90 °F | 0 | 0 | 0 | 0 | 1 | 5 | 13 | 13 | 3 | 0 | 0 | 0 | 35 |
| Days Max Temp ≤ 32 °F | 15 | 10 | 4 | 0 | 0 | 0 | 0 | 0 | 0 | 1 | 5 | 13 | 48 |
| Days Min Temp ≤ 32 °F | 30 | 27 | 28 | 18 | 6 | 0 | 0 | 0 | 5 | 17 | 27 | 30 | 188 |
| Days Min Temp ≤ 0 °F | 13 | 7 | 2 | 0 | 0 | 0 | 0 | 0 | 0 | 0 | 3 | 9 | 34 |
| Heating Degree Days | 1493 | 1174 | 947 | 580 | 307 | 96 | 21 | 35 | 232 | 577 | 1007 | 1381 | 7850 |
| Cooling Degree Days | 0 | 0 | 0 | 1 | 18 | 90 | 195 | 180 | 28 | 1 | 0 | 0 | 513 |
| Total Precipitation (") | 0.49 | 0.27 | 0.45 | 1.12 | 2.09 | 2.24 | 1.61 | 0.98 | 1.36 | 0.82 | 0.35 | 0.49 | 12.27 |
| Days ≥ 0.1" Precip | 2 | 1 | 2 | 3 | 5 | 6 | 3 | 2 | 3 | 2 | 1 | 2 | 32 |
| Total Snowfall (") | 6.3 | 4.3 | 4.2 | 3.9 | 1.0 | 0.0 | 0.0 | 0.0 | 0.2 | 1.2 | 3.5 | 6.7 | 31.3 |
| Days ≥ 1" Snow Depth | 22 | 15 | 9 | 2 | 0 | 0 | 0 | 0 | 0 | 1 | 7 | 18 | 74 |

## COLSTRIP *Rosebud County* ELEVATION 3222 ft LAT/LONG 45° 53 ' N / 106° 36 ' W

| | JAN | FEB | MAR | APR | MAY | JUN | JUL | AUG | SEP | OCT | NOV | DEC | YEAR |
|---|---|---|---|---|---|---|---|---|---|---|---|---|---|
| Maximum Temp °F | 34.0 | 40.3 | 48.5 | 59.5 | 69.9 | 79.0 | 87.1 | 86.1 | 74.2 | 61.3 | 45.8 | 36.1 | 60.2 |
| Minimum Temp °F | 8.7 | 14.6 | 22.7 | 31.7 | 41.3 | 49.9 | 54.6 | 52.9 | 43.0 | 32.9 | 21.9 | 11.9 | 32.2 |
| Mean Temp °F | 21.4 | 27.5 | 35.2 | 45.6 | 55.6 | 64.5 | 70.9 | 69.5 | 58.7 | 47.1 | 33.9 | 24.1 | 46.2 |
| Days Max Temp ≥ 90 °F | 0 | 0 | 0 | 0 | 1 | 4 | 14 | 13 | 3 | 0 | 0 | 0 | 35 |
| Days Max Temp ≤ 32 °F | 11 | 8 | 3 | 0 | 0 | 0 | 0 | 0 | 0 | 1 | ***4*** | 10 | 37 |
| Days Min Temp ≤ 32 °F | 29 | 27 | 26 | 16 | 4 | 0 | 0 | 0 | 3 | 14 | 26 | 29 | 174 |
| Days Min Temp ≤ 0 °F | 10 | 5 | 2 | 0 | 0 | 0 | 0 | 0 | 0 | 0 | 2 | 6 | 25 |
| Heating Degree Days | 1342 | 1053 | 896 | 576 | 300 | 96 | 19 | 32 | 217 | 549 | 926 | 1263 | 7269 |
| Cooling Degree Days | 0 | 0 | 0 | 2 | 17 | 95 | 197 | 177 | 34 | 2 | 0 | 0 | 524 |
| Total Precipitation (") | 0.60 | 0.45 | 0.80 | 1.53 | 2.49 | 2.65 | 1.25 | 1.12 | 1.43 | 1.27 | 0.59 | 0.54 | 14.72 |
| Days ≥ 0.1" Precip | 2 | 2 | 3 | 4 | 6 | 7 | 3 | 3 | 3 | 3 | 2 | 2 | 40 |
| Total Snowfall (") | na | na | na | na | ***0.7*** | 0.0 | 0.0 | 0.0 | 0.6 | ***1.6*** | na | na | na |
| Days ≥ 1" Snow Depth | ***21*** | ***15*** | ***8*** | ***2*** | 0 | 0 | 0 | 0 | ***0*** | ***1*** | ***8*** | ***17*** | 72 |

## COLUMBUS *Stillwater County* ELEVATION 3592 ft LAT/LONG 45° 38 ' N / 109° 15 ' W

| | JAN | FEB | MAR | APR | MAY | JUN | JUL | AUG | SEP | OCT | NOV | DEC | YEAR |
|---|---|---|---|---|---|---|---|---|---|---|---|---|---|
| Maximum Temp °F | 35.9 | 42.5 | 50.6 | 60.3 | 69.9 | 78.7 | 86.6 | 85.3 | 74.3 | 63.1 | 46.4 | 37.2 | 60.9 |
| Minimum Temp °F | 9.6 | 14.6 | 22.7 | 30.8 | 39.6 | 47.2 | 52.5 | 50.4 | 41.0 | 31.3 | 21.0 | 11.9 | 31.1 |
| Mean Temp °F | 22.8 | 28.6 | 36.7 | 45.6 | 54.8 | 63.0 | 69.6 | 68.0 | 57.7 | 47.2 | 33.7 | 24.6 | 46.0 |
| Days Max Temp ≥ 90 °F | 0 | 0 | 0 | 0 | 1 | 4 | 11 | 10 | 2 | 0 | 0 | 0 | 28 |
| Days Max Temp ≤ 32 °F | 11 | 6 | 2 | 0 | 0 | 0 | 0 | 0 | 0 | 0 | 4 | 10 | 33 |
| Days Min Temp ≤ 32 °F | 30 | 26 | 27 | 17 | 4 | 0 | 0 | 0 | 4 | 17 | 27 | 30 | 182 |
| Days Min Temp ≤ 0 °F | 9 | 4 | 1 | 0 | 0 | 0 | 0 | 0 | 0 | 0 | 1 | 6 | 21 |
| Heating Degree Days | 1304 | 1024 | 871 | 576 | 316 | 110 | 19 | 30 | 228 | 543 | 934 | 1245 | 7200 |
| Cooling Degree Days | 0 | 0 | 0 | 0 | 9 | 63 | 157 | 130 | 15 | 0 | 0 | 0 | 374 |
| Total Precipitation (") | 0.61 | 0.54 | 1.01 | 1.73 | 2.97 | 2.14 | 1.28 | 1.07 | 1.56 | 1.27 | 0.66 | 0.60 | 15.44 |
| Days ≥ 0.1" Precip | ***2*** | 2 | 4 | 5 | 6 | 5 | 3 | 3 | 4 | 3 | 2 | 3 | 42 |
| Total Snowfall (") | na | na | na | na | 1.4 | 0.0 | 0.0 | 0.0 | 0.5 | ***2.7*** | na | na | na |
| Days ≥ 1" Snow Depth | na | na | na | na | 0 | 0 | 0 | 0 | 0 | ***1*** | na | na | na |

## CONRAD *Pondera County* ELEVATION 3524 ft LAT/LONG 48° 10 ' N / 111° 58 ' W

| | JAN | FEB | MAR | APR | MAY | JUN | JUL | AUG | SEP | OCT | NOV | DEC | YEAR |
|---|---|---|---|---|---|---|---|---|---|---|---|---|---|
| Maximum Temp °F | 31.0 | 37.3 | 45.6 | 57.1 | 67.0 | 74.2 | 81.0 | 80.7 | 70.4 | 60.0 | 43.3 | 34.3 | 56.8 |
| Minimum Temp °F | 6.0 | 11.3 | 19.2 | 27.9 | 37.6 | 45.2 | 48.7 | 47.5 | 38.3 | 29.3 | 17.3 | 9.6 | 28.2 |
| Mean Temp °F | 18.6 | 24.3 | 32.4 | 42.5 | 52.4 | 59.7 | 64.9 | 64.1 | 54.4 | 44.7 | 30.3 | 22.0 | 42.5 |
| Days Max Temp ≥ 90 °F | 0 | 0 | 0 | 0 | 0 | 1 | 4 | 5 | 1 | 0 | 0 | 0 | 11 |
| Days Max Temp ≤ 32 °F | 13 | 9 | 4 | 1 | 0 | 0 | 0 | 0 | 0 | 1 | 5 | 11 | 44 |
| Days Min Temp ≤ 32 °F | 30 | 27 | 28 | 22 | 8 | 1 | 0 | 0 | 7 | 19 | 27 | 29 | 198 |
| Days Min Temp ≤ 0 °F | 12 | 7 | 3 | 0 | 0 | 0 | 0 | 0 | 0 | 0 | 3 | 8 | 33 |
| Heating Degree Days | 1436 | 1143 | 1004 | 668 | 389 | 179 | 72 | 88 | 324 | 623 | 1034 | 1328 | 8288 |
| Cooling Degree Days | 0 | 0 | 0 | 0 | 4 | 26 | 61 | 61 | 5 | 1 | 0 | 0 | 158 |
| Total Precipitation (") | 0.49 | 0.33 | 0.65 | 1.09 | 2.04 | 2.34 | 1.29 | 1.51 | 1.05 | 0.53 | 0.46 | 0.47 | 12.25 |
| Days ≥ 0.1" Precip | 1 | 1 | 2 | 3 | 5 | 6 | 4 | 4 | 3 | 2 | 2 | 2 | 35 |
| Total Snowfall (") | *8.2* | 5.1 | *5.8* | *4.7* | 0.3 | 0.1 | 0.0 | 0.3 | *0.6* | na | *5.3* | *6.7* | na |
| Days ≥ 1" Snow Depth | na | na | na | na | *0* | *0* | *0* | *0* | *0* | *1* | na | na | na |

## CONTENT 3 SSE *Phillips County* ELEVATION 2341 ft LAT/LONG 47° 59 ' N / 107° 34 ' W

| | JAN | FEB | MAR | APR | MAY | JUN | JUL | AUG | SEP | OCT | NOV | DEC | YEAR |
|---|---|---|---|---|---|---|---|---|---|---|---|---|---|
| Maximum Temp °F | 26.8 | 33.1 | 45.2 | *58.6* | 69.5 | 78.0 | 85.2 | 84.3 | 72.7 | 60.0 | 42.2 | 30.8 | 57.2 |
| Minimum Temp °F | 4.2 | 9.6 | 20.4 | *30.3* | 40.9 | 49.2 | 53.8 | 52.0 | 41.4 | 31.1 | 18.2 | 7.3 | 29.9 |
| Mean Temp °F | 15.5 | 21.4 | 32.8 | *44.5* | 55.2 | 63.6 | 69.5 | 68.2 | 57.0 | 45.5 | 30.2 | 19.1 | 43.5 |
| Days Max Temp ≥ 90 °F | 0 | 0 | 0 | 0 | 1 | 4 | 10 | 10 | 2 | 0 | 0 | 0 | 27 |
| Days Max Temp ≤ 32 °F | 17 | 12 | 5 | 1 | 0 | 0 | 0 | 0 | 0 | 1 | 6 | 14 | 56 |
| Days Min Temp ≤ 32 °F | 30 | 27 | *27* | *18* | *5* | 0 | 0 | 0 | 4 | *16* | 27 | 30 | 184 |
| Days Min Temp ≤ 0 °F | *13* | 8 | 3 | 0 | 0 | 0 | 0 | 0 | 0 | 0 | 3 | 9 | 36 |
| Heating Degree Days | 1531 | 1226 | 993 | *610* | 313 | 108 | 26 | 44 | 255 | 597 | 1037 | 1417 | 8157 |
| Cooling Degree Days | 0 | 0 | 0 | 1 | 20 | 84 | 175 | 155 | 21 | 1 | 0 | 0 | 457 |
| Total Precipitation (") | 0.40 | 0.22 | 0.50 | 0.98 | 2.04 | 2.33 | 1.65 | 1.25 | 1.10 | 0.70 | 0.31 | 0.34 | 11.82 |
| Days ≥ 0.1" Precip | 2 | 1 | na | na | *5* | 5 | 4 | 3 | 2 | *2* | 2 | 1 | na |
| Total Snowfall (") | na | na | na | na | *0.1* | 0.0 | 0.0 | 0.0 | 0.0 | *0.5* | na | na | na |
| Days ≥ 1" Snow Depth | na | na | na | na | *0* | 0 | 0 | 0 | 0 | na | na | na | na |

## CRESTON *Flathead County* ELEVATION 2943 ft LAT/LONG 48° 11 ' N / 114° 8 ' W

| | JAN | FEB | MAR | APR | MAY | JUN | JUL | AUG | SEP | OCT | NOV | DEC | YEAR |
|---|---|---|---|---|---|---|---|---|---|---|---|---|---|
| Maximum Temp °F | 30.4 | 36.0 | 44.7 | 55.2 | 64.7 | 71.6 | 79.1 | 79.6 | 68.1 | 55.6 | 40.0 | 31.8 | 54.7 |
| Minimum Temp °F | 15.4 | 18.8 | 24.8 | 31.6 | 38.7 | 45.2 | 48.3 | 46.8 | 38.2 | 30.2 | 24.4 | 17.7 | 31.7 |
| Mean Temp °F | 22.9 | 27.4 | 34.8 | 43.4 | 51.7 | 58.5 | 63.7 | 63.2 | 53.2 | 42.9 | 32.2 | 24.7 | 43.2 |
| Days Max Temp ≥ 90 °F | 0 | 0 | 0 | 0 | 0 | 0 | 3 | 3 | 0 | 0 | 0 | 0 | 6 |
| Days Max Temp ≤ 32 °F | 15 | 8 | 2 | 0 | 0 | 0 | 0 | 0 | 0 | 0 | 5 | 15 | 45 |
| Days Min Temp ≤ 32 °F | 29 | 26 | 27 | 17 | 5 | 0 | 0 | 0 | 6 | 20 | 25 | 29 | 184 |
| Days Min Temp ≤ 0 °F | 5 | 3 | 1 | 0 | 0 | 0 | 0 | 0 | 0 | 0 | 1 | 3 | 13 |
| Heating Degree Days | 1297 | 1054 | 931 | 641 | 407 | 210 | 91 | 99 | 351 | 678 | 978 | 1241 | 7978 |
| Cooling Degree Days | 0 | 0 | 0 | 0 | 5 | 21 | 52 | 52 | 3 | 0 | 0 | 0 | 133 |
| Total Precipitation (") | 1.45 | 1.12 | 1.15 | 1.55 | 2.35 | 2.96 | 1.83 | 1.58 | 1.65 | 1.24 | 1.52 | 1.56 | 19.96 |
| Days ≥ 0.1" Precip | 5 | 4 | 4 | 5 | 6 | 7 | 5 | 4 | 5 | 4 | 5 | 5 | 59 |
| Total Snowfall (") | *13.7* | 8.4 | 4.6 | 0.9 | 0.1 | 0.0 | 0.0 | 0.0 | 0.0 | 1.0 | 5.5 | *15.5* | 49.7 |
| Days ≥ 1" Snow Depth | 24 | 19 | 9 | 1 | 0 | 0 | 0 | 0 | 0 | 1 | 8 | 21 | 83 |

## CULBERTSON *Roosevelt County* ELEVATION 1923 ft LAT/LONG 48° 9 ' N / 104° 31 ' W

| | JAN | FEB | MAR | APR | MAY | JUN | JUL | AUG | SEP | OCT | NOV | DEC | YEAR |
|---|---|---|---|---|---|---|---|---|---|---|---|---|---|
| Maximum Temp °F | 20.8 | 28.9 | 42.9 | 59.0 | 71.1 | 79.6 | 85.9 | 85.2 | 72.8 | 60.0 | 39.4 | 26.2 | 56.0 |
| Minimum Temp °F | -2.5 | 4.7 | 16.9 | 28.4 | 40.0 | 48.4 | 52.5 | 50.5 | 39.8 | 29.0 | 15.1 | 3.0 | 27.2 |
| Mean Temp °F | 9.2 | 16.8 | 29.9 | 43.7 | 55.6 | 64.0 | 69.2 | 67.9 | 56.3 | 44.5 | 27.3 | 14.6 | 41.6 |
| Days Max Temp ≥ 90 °F | 0 | 0 | 0 | 0 | 1 | 4 | 10 | 10 | 2 | 0 | 0 | 0 | 27 |
| Days Max Temp ≤ 32 °F | 22 | 15 | 7 | 0 | 0 | 0 | 0 | 0 | 0 | 1 | 8 | 19 | 72 |
| Days Min Temp ≤ 32 °F | 31 | 28 | 30 | 21 | 7 | 0 | 0 | 0 | 6 | 20 | 29 | 31 | 203 |
| Days Min Temp ≤ 0 °F | 17 | 11 | 4 | 0 | 0 | 0 | 0 | 0 | 0 | 0 | 4 | 12 | 48 |
| Heating Degree Days | 1730 | 1354 | 1080 | 632 | 307 | 94 | 26 | 51 | 275 | 629 | 1126 | 1558 | 8862 |
| Cooling Degree Days | 0 | 0 | 0 | 2 | 27 | 86 | 173 | 168 | 21 | 0 | 0 | 0 | 477 |
| Total Precipitation (") | 0.41 | 0.23 | 0.42 | 1.15 | 2.31 | 2.90 | 2.01 | 1.36 | 1.56 | 0.77 | 0.35 | 0.40 | 13.87 |
| Days ≥ 0.1" Precip | 2 | 1 | 1 | 3 | 6 | 6 | 5 | 3 | 4 | 2 | 1 | 1 | 35 |
| Total Snowfall (") | na | na | na | na | na | *0.0* | *0.0* | *0.0* | *0.2* | *0.7* | na | na | na |
| Days ≥ 1" Snow Depth | 28 | 21 | 12 | 1 | 0 | 0 | 0 | 0 | 0 | 1 | 8 | 20 | 91 |

## CUT BANK MUNI AP *Glacier County* ELEVATION 3839 ft LAT/LONG 48° 36 ' N / 112° 22 ' W

| | JAN | FEB | MAR | APR | MAY | JUN | JUL | AUG | SEP | OCT | NOV | DEC | YEAR |
|---|---|---|---|---|---|---|---|---|---|---|---|---|---|
| Maximum Temp °F | 27.6 | 33.5 | 41.3 | 52.4 | 62.6 | 70.5 | 78.2 | 77.6 | 67.1 | 56.0 | 39.4 | 30.8 | 53.1 |
| Minimum Temp °F | 7.6 | 12.0 | 19.7 | 28.7 | 37.7 | 45.3 | 49.3 | 48.4 | 39.5 | 30.9 | 19.2 | 11.2 | 29.1 |
| Mean Temp °F | 17.7 | 22.8 | 30.5 | 40.6 | 50.1 | 57.9 | 63.7 | 63.0 | 53.4 | 43.5 | 29.3 | 21.0 | 41.1 |
| Days Max Temp ≥ 90 °F | 0 | 0 | 0 | 0 | 0 | 1 | 2 | 3 | 1 | 0 | 0 | 0 | 7 |
| Days Max Temp ≤ 32 °F | 15 | 11 | 7 | 2 | 0 | 0 | 0 | 0 | 0 | 2 | 8 | 14 | 59 |
| Days Min Temp ≤ 32 °F | 28 | 26 | 28 | 20 | 7 | 0 | 0 | 0 | 5 | 18 | 26 | 28 | 186 |
| Days Min Temp ≤ 0 °F | 12 | 7 | 3 | 0 | 0 | 0 | 0 | 0 | 0 | 0 | 3 | 8 | 33 |
| Heating Degree Days | 1460 | 1186 | 1062 | 727 | 456 | 225 | 91 | 114 | 350 | 661 | 1066 | 1358 | 8756 |
| Cooling Degree Days | 0 | 0 | 0 | 0 | 3 | 23 | 61 | 65 | 8 | 1 | 0 | 0 | 161 |
| Total Precipitation (") | 0.42 | 0.30 | 0.52 | 0.91 | 2.09 | 2.60 | 1.56 | 1.75 | 1.13 | 0.47 | 0.38 | 0.32 | 12.45 |
| Days ≥ 0.1" Precip | 1 | 1 | 2 | 3 | 5 | 6 | 4 | 4 | 3 | 1 | 1 | 1 | 32 |
| Total Snowfall (") | 5.7 | 4.1 | 5.7 | 4.6 | 1.2 | 0.1 | 0.0 | 0.0 | 0.7 | 2.5 | 4.6 | 4.4 | 33.6 |
| Days ≥ 1" Snow Depth | 17 | 14 | 10 | 4 | 1 | 0 | 0 | 0 | 0 | 2 | 8 | 13 | 69 |

## DARBY *Ravalli County* ELEVATION 3822 ft LAT/LONG 46° 2 ' N / 114° 11 ' W

| | JAN | FEB | MAR | APR | MAY | JUN | JUL | AUG | SEP | OCT | NOV | DEC | YEAR |
|---|---|---|---|---|---|---|---|---|---|---|---|---|---|
| Maximum Temp °F | 36.1 | 42.5 | 50.1 | 58.4 | 67.4 | 75.3 | 83.9 | 83.5 | 72.6 | 61.5 | 45.0 | 36.8 | 59.4 |
| Minimum Temp °F | 18.6 | 21.5 | 26.0 | 31.2 | 38.3 | 44.4 | 47.6 | 47.0 | 40.0 | 32.7 | 24.9 | 18.7 | 32.6 |
| Mean Temp °F | 27.4 | 32.0 | 38.1 | 44.9 | 52.9 | 59.9 | 65.8 | 65.3 | 56.3 | 47.1 | 35.0 | 27.8 | 46.0 |
| Days Max Temp ≥ 90 °F | 0 | 0 | 0 | 0 | 0 | 2 | 8 | 8 | 1 | 0 | 0 | 0 | 19 |
| Days Max Temp ≤ 32 °F | 10 | 4 | 1 | 0 | 0 | 0 | 0 | 0 | 0 | 0 | 3 | 8 | 26 |
| Days Min Temp ≤ 32 °F | 27 | 25 | 25 | 17 | 6 | 1 | 0 | 0 | 4 | 15 | 24 | 28 | 172 |
| Days Min Temp ≤ 0 °F | 3 | 1 | 0 | 0 | 0 | 0 | 0 | 0 | 0 | 0 | 1 | 2 | 7 |
| Heating Degree Days | 1161 | 924 | 829 | 598 | 373 | 176 | 58 | 67 | 266 | 548 | 898 | 1149 | 7047 |
| Cooling Degree Days | 0 | 0 | 0 | 0 | 7 | 34 | 90 | 85 | 12 | 0 | 0 | 0 | 228 |
| Total Precipitation (") | 1.66 | 0.99 | 1.00 | 1.07 | 2.02 | 1.96 | 1.00 | 1.24 | 1.28 | 1.04 | 1.34 | 1.30 | 15.90 |
| Days ≥ 0.1" Precip | 5 | 4 | 3 | 4 | 5 | 5 | 3 | 4 | 4 | 3 | 5 | 4 | 49 |
| Total Snowfall (") | na | na | na | na | na | *0.0* | *0.0* | *0.0* | *0.3* | na | na | na | na |
| Days ≥ 1" Snow Depth | na | na | na | na | na | *0* | *0* | *0* | *0* | na | na | na | na |

## DEER LODGE 3 W *Powell County* ELEVATION 4833 ft LAT/LONG 46° 23 ' N / 112° 48 ' W

| | JAN | FEB | MAR | APR | MAY | JUN | JUL | AUG | SEP | OCT | NOV | DEC | YEAR |
|---|---|---|---|---|---|---|---|---|---|---|---|---|---|
| Maximum Temp °F | 32.5 | 37.8 | 45.1 | 54.9 | 63.5 | 71.6 | 79.9 | 80.0 | 69.7 | 57.9 | 41.6 | 32.9 | 55.6 |
| Minimum Temp °F | 9.7 | 13.9 | 19.6 | 25.7 | 32.8 | 39.7 | 43.0 | 41.8 | 33.7 | 25.7 | 17.1 | 10.0 | 26.1 |
| Mean Temp °F | 21.1 | 25.9 | 32.4 | 40.3 | 48.2 | 55.7 | 61.5 | 60.9 | 51.7 | 41.9 | 29.4 | 21.5 | 40.9 |
| Days Max Temp ≥ 90 °F | 0 | 0 | 0 | 0 | 0 | 0 | 3 | 4 | 1 | 0 | 0 | 0 | 8 |
| Days Max Temp ≤ 32 °F | 13 | 7 | 3 | 0 | 0 | 0 | 0 | 0 | 0 | 0 | 6 | 14 | 43 |
| Days Min Temp ≤ 32 °F | 30 | 27 | 30 | 25 | 14 | 4 | 1 | 2 | 13 | 25 | 28 | 30 | 229 |
| Days Min Temp ≤ 0 °F | 8 | 4 | 2 | 0 | 0 | 0 | 0 | 0 | 0 | 0 | 3 | 7 | 24 |
| Heating Degree Days | 1355 | 1099 | 1005 | 733 | 513 | 281 | 131 | 146 | 396 | 711 | 1062 | 1343 | 8775 |
| Cooling Degree Days | 0 | 0 | 0 | 0 | 0 | 7 | 23 | 21 | 1 | 0 | 0 | 0 | 52 |
| Total Precipitation (") | 0.40 | 0.29 | 0.48 | 0.74 | 1.83 | 1.80 | 1.25 | 1.32 | 1.07 | 0.61 | 0.40 | 0.39 | 10.58 |
| Days ≥ 0.1" Precip | 1 | 1 | 1 | 2 | 5 | 6 | 3 | 4 | 3 | 2 | 1 | 1 | 30 |
| Total Snowfall (") | *9.6* | *5.3* | na | *4.7* | 0.2 | 0.2 | 0.0 | 0.0 | 0.0 | *1.6* | na | na | na |
| Days ≥ 1" Snow Depth | *20* | na | na | *2* | 0 | 0 | 0 | 0 | 0 | *0* | *6* | na | na |

## DEL BONITA *Glacier County* ELEVATION 4344 ft LAT/LONG 49° 0 ' N / 112° 47 ' W

| | JAN | FEB | MAR | APR | MAY | JUN | JUL | AUG | SEP | OCT | NOV | DEC | YEAR |
|---|---|---|---|---|---|---|---|---|---|---|---|---|---|
| Maximum Temp °F | 27.3 | 33.0 | 40.0 | 51.4 | 62.3 | 69.5 | 76.3 | 76.3 | 66.6 | 55.4 | 38.4 | 29.8 | 52.2 |
| Minimum Temp °F | 7.0 | 12.0 | 19.2 | 27.5 | 36.6 | 43.7 | 47.3 | 46.8 | 38.7 | 30.4 | 18.5 | 10.7 | 28.2 |
| Mean Temp °F | 17.2 | 22.6 | 29.6 | 39.5 | 49.5 | 56.6 | 61.8 | 61.6 | 52.7 | 42.9 | 28.5 | 20.3 | 40.2 |
| Days Max Temp ≥ 90 °F | 0 | 0 | 0 | 0 | 0 | 0 | 1 | 1 | 0 | 0 | 0 | 0 | 2 |
| Days Max Temp ≤ 32 °F | 16 | 11 | 7 | 2 | 0 | 0 | 0 | 0 | 0 | 2 | 8 | 15 | 61 |
| Days Min Temp ≤ 32 °F | 29 | 26 | 28 | 22 | 10 | 1 | 0 | 0 | 7 | 18 | 27 | 29 | 197 |
| Days Min Temp ≤ 0 °F | 11 | 7 | 3 | 0 | 0 | 0 | 0 | 0 | 0 | 0 | 3 | 8 | 32 |
| Heating Degree Days | 1478 | 1193 | 1092 | 759 | 475 | 257 | 130 | 141 | 370 | 679 | 1089 | 1380 | 9043 |
| Cooling Degree Days | 0 | 0 | 0 | 0 | 3 | 16 | 38 | 43 | 8 | 1 | 0 | 0 | 109 |
| Total Precipitation (") | 0.57 | 0.40 | 0.79 | 1.31 | 2.45 | 3.19 | 1.50 | 1.84 | 1.54 | 0.59 | 0.53 | 0.42 | 15.13 |
| Days ≥ 0.1" Precip | 2 | 1 | 3 | 4 | 6 | 7 | 4 | 4 | 4 | 2 | 2 | 2 | 41 |
| Total Snowfall (") | 9.1 | 6.5 | 11.0 | 11.3 | *2.5* | 0.0 | 0.0 | 0.2 | 3.5 | *5.8* | 8.0 | *9.0* | 66.9 |
| Days ≥ 1" Snow Depth | *19* | na | na | *5* | 1 | 0 | 0 | 0 | *1* | *3* | *10* | na | na |

## DENTON 1 NNE *Fergus County* ELEVATION 3612 ft LAT/LONG 47° 19 ' N / 109° 57 ' W

| | JAN | FEB | MAR | APR | MAY | JUN | JUL | AUG | SEP | OCT | NOV | DEC | YEAR |
|---|---|---|---|---|---|---|---|---|---|---|---|---|---|
| Maximum Temp °F | 32.4 | 38.6 | 46.7 | 57.5 | 66.2 | 74.9 | ***82.7*** | 82.3 | 71.2 | 60.4 | 44.7 | 36.0 | 57.8 |
| Minimum Temp °F | 6.1 | 10.9 | 18.8 | 27.9 | 36.6 | 43.9 | ***46.8*** | 45.7 | 36.5 | 28.0 | 16.9 | 9.0 | 27.3 |
| Mean Temp °F | 19.3 | 24.8 | 32.8 | 42.9 | 51.4 | 59.4 | ***64.7*** | 64.0 | 53.9 | 44.2 | 30.8 | 22.5 | 42.6 |
| Days Max Temp ≥ 90 °F | 0 | 0 | 0 | 0 | 0 | 2 | ***6*** | ***7*** | 1 | 0 | 0 | 0 | 16 |
| Days Max Temp ≤ 32 °F | 12 | 8 | 4 | 1 | 0 | 0 | 0 | 0 | 0 | 1 | 4 | 10 | 40 |
| Days Min Temp ≤ 32 °F | 29 | 27 | 28 | ***21*** | 9 | 2 | 0 | 1 | 9 | na | 27 | 29 | na |
| Days Min Temp ≤ 0 °F | 12 | 7 | 3 | 0 | 0 | 0 | 0 | 0 | 0 | 0 | 3 | 8 | 33 |
| Heating Degree Days | 1413 | 1131 | 993 | 657 | 415 | 184 | ***69*** | 87 | 334 | 637 | 1020 | 1309 | 8249 |
| Cooling Degree Days | 0 | 0 | 0 | 0 | 3 | 27 | ***61*** | 64 | 5 | 0 | 0 | 0 | 160 |
| Total Precipitation (") | 0.61 | 0.36 | 0.67 | 1.12 | 2.69 | 2.73 | 1.83 | 1.66 | 1.52 | 0.97 | 0.43 | 0.56 | 15.15 |
| Days ≥ 0.1" Precip | 2 | 1 | 3 | 3 | 7 | 6 | 4 | 4 | 4 | 3 | 1 | 2 | 40 |
| Total Snowfall (") | ***12.8*** | ***6.3*** | na | ***3.7*** | 1.5 | 0.1 | 0.0 | 0.0 | 0.5 | ***2.1*** | na | na | na |
| Days ≥ 1" Snow Depth | na | na | na | ***2*** | 0 | 0 | 0 | 0 | 0 | na | na | na | na |

## DILLON WMCE *Beaverhead County* ELEVATION 5233 ft LAT/LONG 45° 12 ' N / 112° 38 ' W

| | JAN | FEB | MAR | APR | MAY | JUN | JUL | AUG | SEP | OCT | NOV | DEC | YEAR |
|---|---|---|---|---|---|---|---|---|---|---|---|---|---|
| Maximum Temp °F | 34.9 | 40.8 | 47.7 | 57.1 | 66.9 | 75.2 | 83.6 | 82.0 | 72.1 | 60.9 | 43.6 | 34.7 | 58.3 |
| Minimum Temp °F | 13.4 | 16.6 | 22.1 | 28.7 | 36.2 | 43.2 | 47.1 | 45.6 | 38.2 | 30.8 | 21.1 | 13.5 | 29.7 |
| Mean Temp °F | 24.1 | 28.7 | 34.9 | 42.9 | 51.6 | 59.3 | 65.4 | 63.8 | 55.1 | 46.0 | 32.4 | 24.1 | 44.0 |
| Days Max Temp ≥ 90 °F | 0 | 0 | 0 | 0 | 0 | 1 | 5 | 3 | 0 | 0 | 0 | 0 | 9 |
| Days Max Temp ≤ 32 °F | 11 | 5 | 2 | 0 | 0 | 0 | 0 | 0 | 0 | 0 | 4 | 11 | 33 |
| Days Min Temp ≤ 32 °F | 29 | 27 | 28 | 21 | 9 | 1 | 0 | 0 | 6 | 17 | 26 | 30 | 194 |
| Days Min Temp ≤ 0 °F | 6 | 3 | 1 | 0 | 0 | 0 | 0 | 0 | 0 | 0 | 2 | 5 | 17 |
| Heating Degree Days | 1261 | 1016 | 925 | 656 | 411 | 185 | 52 | 79 | 293 | 583 | 973 | 1261 | 7695 |
| Cooling Degree Days | 0 | 0 | 0 | 0 | 2 | 21 | 69 | 46 | 4 | 0 | 0 | 0 | 142 |
| Total Precipitation (") | 0.32 | 0.22 | 0.66 | 1.13 | 2.06 | 1.90 | 1.31 | 1.22 | 1.14 | 0.74 | 0.43 | 0.31 | 11.44 |
| Days ≥ 0.1" Precip | 1 | 1 | 2 | 3 | 6 | 5 | 4 | 3 | 3 | 2 | 1 | 1 | 32 |
| Total Snowfall (") | na | na | na | na | na | na | ***0.0*** | ***0.0*** | na | na | na | na | na |
| Days ≥ 1" Snow Depth | na | na | na | na | na | ***0*** | ***0*** | ***0*** | na | na | na | na | na |

## DIVIDE 2 NW *Silver Bow County* ELEVATION 5413 ft LAT/LONG 45° 46 ' N / 112° 47 ' W

| | JAN | FEB | MAR | APR | MAY | JUN | JUL | AUG | SEP | OCT | NOV | DEC | YEAR |
|---|---|---|---|---|---|---|---|---|---|---|---|---|---|
| Maximum Temp °F | 31.3 | 36.5 | 43.2 | 52.0 | 61.4 | 70.1 | 78.5 | 77.3 | 66.9 | 55.8 | 39.7 | 30.8 | 53.6 |
| Minimum Temp °F | 10.4 | 13.5 | 20.1 | 27.4 | 35.3 | 42.4 | 47.3 | 46.2 | 37.8 | 29.3 | 19.1 | 10.6 | 28.3 |
| Mean Temp °F | 20.9 | 25.0 | 31.6 | 39.7 | 48.4 | 56.3 | 62.9 | 61.8 | 52.4 | 42.6 | 29.4 | 20.7 | 41.0 |
| Days Max Temp ≥ 90 °F | 0 | 0 | 0 | 0 | 0 | 0 | 1 | 1 | 0 | 0 | 0 | 0 | 2 |
| Days Max Temp ≤ 32 °F | 15 | 8 | 4 | 1 | 0 | 0 | 0 | 0 | 0 | 1 | 7 | 16 | 52 |
| Days Min Temp ≤ 32 °F | 30 | 28 | 30 | 24 | 10 | 2 | 0 | 0 | 6 | 21 | 28 | 31 | 210 |
| Days Min Temp ≤ 0 °F | 7 | 4 | 1 | 0 | 0 | 0 | 0 | 0 | 0 | 0 | 2 | 5 | 19 |
| Heating Degree Days | 1362 | 1122 | 1026 | 752 | 509 | 265 | 98 | 123 | 373 | 688 | 1062 | 1366 | 8746 |
| Cooling Degree Days | 0 | 0 | 0 | 0 | 1 | 12 | 42 | 28 | 2 | 0 | 0 | 0 | 85 |
| Total Precipitation (") | 0.54 | 0.40 | 0.80 | 1.05 | 1.84 | 2.10 | 1.35 | 1.45 | 1.27 | 0.77 | 0.62 | 0.54 | 12.73 |
| Days ≥ 0.1" Precip | 2 | 1 | 3 | 4 | 6 | 6 | 4 | 5 | 4 | 2 | 2 | 2 | 41 |
| Total Snowfall (") | 7.4 | 5.5 | 7.3 | 4.9 | 0.9 | 0.0 | 0.0 | 0.1 | 0.8 | 1.6 | 4.6 | 6.7 | 39.8 |
| Days ≥ 1" Snow Depth | 21 | 12 | 6 | 2 | 0 | 0 | 0 | 0 | 0 | 1 | 6 | 18 | 66 |

## DRUMMOND AVIATION *Granite County* ELEVATION 3944 ft LAT/LONG 46° 40 ' N / 113° 9 ' W

| | JAN | FEB | MAR | APR | MAY | JUN | JUL | AUG | SEP | OCT | NOV | DEC | YEAR |
|---|---|---|---|---|---|---|---|---|---|---|---|---|---|
| Maximum Temp °F | 31.1 | 38.3 | 47.7 | 58.0 | 67.0 | 75.0 | 83.6 | 83.0 | 71.9 | 59.0 | 41.3 | 31.3 | 57.3 |
| Minimum Temp °F | 11.5 | 15.9 | 22.1 | 28.4 | 35.6 | 42.9 | 45.3 | 44.1 | 36.2 | 28.2 | 20.2 | 12.1 | 28.5 |
| Mean Temp °F | 21.4 | 27.1 | 34.9 | 43.2 | 51.3 | 59.0 | 64.5 | 63.6 | 54.1 | 43.7 | 30.8 | 21.7 | 42.9 |
| Days Max Temp ≥ 90 °F | 0 | 0 | 0 | 0 | 0 | 2 | 8 | 8 | 1 | 0 | 0 | 0 | 19 |
| Days Max Temp ≤ 32 °F | 14 | 6 | 2 | 0 | 0 | 0 | 0 | 0 | 0 | 0 | 5 | 15 | 42 |
| Days Min Temp ≤ 32 °F | 30 | 27 | 28 | 21 | 10 | 2 | 0 | 1 | 10 | 22 | 27 | 30 | 208 |
| Days Min Temp ≤ 0 °F | 7 | 3 | 1 | 0 | 0 | 0 | 0 | 0 | 0 | 0 | 2 | 6 | 19 |
| Heating Degree Days | 1347 | 1062 | 926 | 648 | 419 | 194 | 69 | 85 | 325 | 655 | 1021 | 1336 | 8087 |
| Cooling Degree Days | 0 | 0 | 0 | 0 | 3 | 24 | 58 | 52 | 4 | 0 | 0 | 0 | 141 |
| Total Precipitation (") | 0.94 | 0.56 | 0.81 | 1.03 | 1.86 | 1.91 | 1.22 | 1.37 | 1.11 | 0.80 | 0.69 | 0.79 | 13.09 |
| Days ≥ 0.1" Precip | 3 | 2 | 3 | 3 | 6 | 5 | 4 | 4 | 3 | 3 | 3 | 3 | 42 |
| Total Snowfall (") | 8.9 | 5.4 | 6.1 | 5.0 | 2.3 | 0.1 | 0.0 | 0.1 | 0.9 | 1.6 | 5.3 | 7.5 | 43.2 |
| Days ≥ 1" Snow Depth | 24 | 15 | 6 | 1 | 0 | 0 | 0 | 0 | 0 | 0 | 7 | 19 | 72 |

## EKALAKA *Carter County* ELEVATION 3432 ft LAT/LONG 45° 53 ' N / 104° 32 ' W

| | JAN | FEB | MAR | APR | MAY | JUN | JUL | AUG | SEP | OCT | NOV | DEC | YEAR |
|---|---|---|---|---|---|---|---|---|---|---|---|---|---|
| Maximum Temp °F | 29.3 | 34.8 | 44.5 | 57.1 | 68.3 | 77.5 | 85.5 | 84.1 | 72.0 | 59.1 | 42.1 | 32.3 | 57.2 |
| Minimum Temp °F | 6.6 | 12.4 | 20.8 | 30.9 | 40.8 | 49.7 | 55.3 | 53.3 | 42.8 | 31.9 | 19.8 | 10.4 | 31.2 |
| Mean Temp °F | 18.0 | 23.6 | 32.6 | 44.0 | 54.6 | 63.6 | 70.4 | 68.7 | 57.4 | 45.6 | 31.0 | 21.4 | 44.2 |
| Days Max Temp ≥ 90 °F | 0 | 0 | 0 | 0 | 0 | 3 | 10 | 9 | 2 | 0 | 0 | 0 | 24 |
| Days Max Temp ≤ 32 °F | 16 | 10 | 6 | 1 | 0 | 0 | 0 | 0 | 0 | 1 | 6 | 14 | 54 |
| Days Min Temp ≤ 32 °F | 30 | 27 | 27 | 18 | 5 | 0 | 0 | 0 | 4 | 15 | 27 | 30 | 183 |
| Days Min Temp ≤ 0 °F | 11 | 6 | 2 | 0 | 0 | 0 | 0 | 0 | 0 | 0 | 2 | 7 | 28 |
| Heating Degree Days | 1454 | 1162 | 996 | 623 | 331 | 110 | 25 | 43 | 258 | 597 | 1015 | 1347 | 7961 |
| Cooling Degree Days | 0 | 0 | 0 | 1 | 20 | 92 | 211 | 174 | 33 | 0 | 0 | 0 | 531 |
| Total Precipitation (") | 0.53 | 0.42 | 0.74 | 1.78 | 2.57 | 3.58 | 1.79 | 1.20 | 1.61 | 1.34 | 0.67 | 0.60 | 16.83 |
| Days ≥ 0.1" Precip | 2 | 1 | 2 | 4 | 6 | 7 | 4 | 3 | 4 | 3 | 2 | 2 | 40 |
| Total Snowfall (") | na | na | na | na | na | na | na | na | na | na | na | na | na |
| Days ≥ 1" Snow Depth | na | na | na | na | na | na | na | na | na | na | na | na | na |

## ENNIS *Madison County* ELEVATION 4954 ft LAT/LONG 45° 21 ' N / 111° 43 ' W

| | JAN | FEB | MAR | APR | MAY | JUN | JUL | AUG | SEP | OCT | NOV | DEC | YEAR |
|---|---|---|---|---|---|---|---|---|---|---|---|---|---|
| Maximum Temp °F | 33.5 | 38.8 | 45.8 | 55.5 | 65.6 | 74.4 | 82.7 | 81.4 | 70.9 | 59.3 | 42.3 | 33.6 | 57.0 |
| Minimum Temp °F | 14.7 | 17.7 | 23.1 | 29.2 | 36.4 | 43.1 | 47.5 | 45.8 | 37.9 | 31.0 | 22.7 | 15.7 | 30.4 |
| Mean Temp °F | 24.1 | 28.3 | 34.5 | 42.4 | 51.0 | 58.8 | 65.1 | 63.6 | 54.4 | 45.2 | 32.5 | 24.6 | 43.7 |
| Days Max Temp ≥ 90 °F | 0 | 0 | 0 | 0 | 0 | 1 | 4 | 4 | 0 | 0 | 0 | 0 | 9 |
| Days Max Temp ≤ 32 °F | 12 | 7 | 3 | 0 | 0 | 0 | 0 | 0 | 0 | 1 | 5 | 13 | 41 |
| Days Min Temp ≤ 32 °F | 29 | 25 | 26 | 20 | 9 | 1 | 0 | 0 | 7 | 18 | 24 | 28 | 187 |
| Days Min Temp ≤ 0 °F | 5 | 3 | 1 | 0 | 0 | 0 | 0 | 0 | 0 | 0 | 1 | 4 | 14 |
| Heating Degree Days | 1261 | 1032 | 940 | 671 | 428 | 196 | 57 | 79 | 315 | 608 | 963 | 1247 | 7797 |
| Cooling Degree Days | 0 | 0 | 0 | 0 | 0 | 18 | 66 | 44 | 4 | 0 | 0 | 0 | 132 |
| Total Precipitation (") | 0.37 | 0.40 | 0.80 | 1.30 | 2.12 | 2.37 | 1.34 | 1.38 | 1.31 | 0.97 | 0.65 | 0.41 | 13.42 |
| Days ≥ 0.1" Precip | 1 | 2 | 3 | 4 | 6 | 6 | 4 | 4 | 4 | 3 | 2 | 1 | 40 |
| Total Snowfall (") | na | na | na | na | na | *0.0* | *0.0* | *0.0* | *0.2* | na | na | na | na |
| Days ≥ 1" Snow Depth | na | na | na | na | na | *0* | *0* | *0* | *0* | na | na | na | na |

## EUREKA RS *Lincoln County* ELEVATION 2532 ft LAT/LONG 48° 54 ' N / 115° 4 ' W

| | JAN | FEB | MAR | APR | MAY | JUN | JUL | AUG | SEP | OCT | NOV | DEC | YEAR |
|---|---|---|---|---|---|---|---|---|---|---|---|---|---|
| Maximum Temp °F | 29.7 | 37.9 | 48.8 | 59.3 | 69.0 | 76.1 | 84.1 | 83.8 | 72.1 | 57.1 | 39.7 | 30.5 | 57.3 |
| Minimum Temp °F | 15.0 | 19.9 | 26.4 | 33.0 | 40.1 | 46.2 | 49.3 | 48.2 | 40.2 | 32.1 | 25.1 | 17.5 | 32.8 |
| Mean Temp °F | 22.4 | 28.9 | 37.6 | 46.2 | 54.6 | 61.2 | 66.7 | 66.0 | 56.2 | 44.6 | 32.4 | 24.0 | 45.1 |
| Days Max Temp ≥ 90 °F | 0 | 0 | 0 | 0 | 0 | 3 | 10 | 10 | 2 | 0 | 0 | 0 | 25 |
| Days Max Temp ≤ 32 °F | 15 | 6 | 1 | 0 | 0 | 0 | 0 | 0 | 0 | 0 | 5 | 16 | 43 |
| Days Min Temp ≤ 32 °F | 28 | 24 | 23 | 15 | 4 | 0 | 0 | 0 | 5 | 16 | 23 | 28 | 166 |
| Days Min Temp ≤ 0 °F | 6 | 3 | 1 | 0 | 0 | 0 | 0 | 0 | 0 | 0 | 1 | 4 | 15 |
| Heating Degree Days | 1316 | 1013 | 842 | 558 | 323 | 150 | 52 | 62 | 270 | 625 | 970 | 1263 | 7444 |
| Cooling Degree Days | 0 | 0 | 0 | 1 | 11 | 49 | 109 | 105 | 13 | 0 | 0 | 0 | 288 |
| Total Precipitation (") | 1.25 | 0.78 | 0.74 | 0.91 | 1.65 | 2.07 | 1.41 | 1.29 | 1.13 | 0.89 | 1.21 | 1.08 | 14.41 |
| Days ≥ 0.1" Precip | 4 | 3 | 3 | 3 | 5 | 5 | 4 | 4 | 4 | 3 | 4 | 4 | 46 |
| Total Snowfall (") | *14.2* | *7.5* | *4.9* | *1.2* | 0.0 | 0.0 | 0.0 | 0.0 | 0.1 | *0.4* | na | na | na |
| Days ≥ 1" Snow Depth | na | na | na | na | *0* | *0* | *0* | *0* | *0* | *0* | na | na | na |

## FAIRFIELD *Teton County* ELEVATION 3983 ft LAT/LONG 47° 37 ' N / 111° 59 ' W

| | JAN | FEB | MAR | APR | MAY | JUN | JUL | AUG | SEP | OCT | NOV | DEC | YEAR |
|---|---|---|---|---|---|---|---|---|---|---|---|---|---|
| Maximum Temp °F | 32.4 | 39.0 | 46.6 | 56.4 | 66.0 | 73.3 | 80.4 | 80.6 | 70.9 | 60.1 | 43.4 | 34.9 | 57.0 |
| Minimum Temp °F | 11.6 | 16.1 | 22.3 | 30.6 | 39.6 | 46.9 | 50.8 | 50.3 | 42.1 | 34.2 | 22.8 | 15.0 | 31.9 |
| Mean Temp °F | 22.0 | 27.6 | 34.4 | 43.6 | 52.8 | 60.2 | 65.6 | 65.4 | 56.5 | 47.2 | 33.1 | 25.0 | 44.4 |
| Days Max Temp ≥ 90 °F | 0 | 0 | 0 | 0 | 0 | 1 | 3 | 5 | 1 | 0 | 0 | 0 | 10 |
| Days Max Temp ≤ 32 °F | 13 | 7 | 4 | 1 | 0 | 0 | 0 | 0 | 0 | 1 | 4 | 10 | 40 |
| Days Min Temp ≤ 32 °F | 28 | 25 | 26 | 18 | 5 | 0 | 0 | 0 | 3 | 13 | 24 | 27 | 169 |
| Days Min Temp ≤ 0 °F | 9 | 5 | 2 | 0 | 0 | 0 | 0 | 0 | 0 | 0 | 2 | 6 | 24 |
| Heating Degree Days | 1327 | 1050 | 941 | 637 | 378 | 172 | 62 | 73 | 272 | 546 | 950 | 1235 | 7643 |
| Cooling Degree Days | 0 | 0 | 0 | 1 | 10 | 35 | 84 | 91 | 23 | 2 | 0 | 0 | 246 |
| Total Precipitation (") | 0.42 | 0.27 | 0.61 | 1.13 | 2.20 | 2.29 | 1.46 | 1.58 | 1.17 | 0.60 | 0.34 | 0.38 | 12.45 |
| Days ≥ 0.1" Precip | 1 | 1 | 2 | 3 | 5 | 5 | 4 | 4 | 3 | 2 | 1 | 1 | 32 |
| Total Snowfall (") | 8.4 | 5.8 | *9.9* | *7.1* | 1.4 | 0.2 | 0.0 | 0.1 | 1.9 | *2.4* | 6.5 | *7.8* | 51.5 |
| Days ≥ 1" Snow Depth | 17 | 13 | *9* | *4* | 0 | 0 | 0 | 0 | 1 | 2 | *8* | 13 | 67 |

## FLATWILLOW 4 ENE *Petroleum County* ELEVATION 3143 ft LAT/LONG 46° 51 ' N / 108° 19 ' W

| | JAN | FEB | MAR | APR | MAY | JUN | JUL | AUG | SEP | OCT | NOV | DEC | YEAR |
|---|---|---|---|---|---|---|---|---|---|---|---|---|---|
| Maximum Temp °F | 33.5 | 39.8 | 48.3 | 59.2 | 69.2 | 78.0 | 85.6 | 85.1 | 73.7 | 62.5 | 46.0 | 36.9 | 59.8 |
| Minimum Temp °F | 9.3 | 14.2 | 22.3 | 31.1 | 40.0 | 48.6 | 53.1 | 51.5 | 42.4 | 33.4 | 21.3 | 12.5 | 31.6 |
| Mean Temp °F | 21.4 | 27.0 | 35.3 | 45.2 | 54.6 | 63.3 | 69.4 | 68.3 | 58.1 | 48.0 | 33.7 | 24.7 | 45.8 |
| Days Max Temp ≥ 90 °F | 0 | 0 | 0 | 0 | 1 | 4 | 11 | 11 | 2 | 0 | 0 | 0 | 29 |
| Days Max Temp ≤ 32 °F | 12 | 8 | 4 | 1 | 0 | 0 | 0 | 0 | 0 | 1 | 4 | 10 | 40 |
| Days Min Temp ≤ 32 °F | 29 | 26 | 27 | 17 | 5 | 0 | 0 | 0 | 3 | 14 | 25 | 29 | 175 |
| Days Min Temp ≤ 0 °F | 10 | 6 | 2 | 0 | 0 | 0 | 0 | 0 | 0 | 0 | 2 | 6 | 26 |
| Heating Degree Days | 1345 | 1066 | 914 | 589 | 325 | 106 | 23 | 35 | 229 | 522 | 933 | 1243 | 7330 |
| Cooling Degree Days | 0 | 0 | 0 | 1 | 12 | 74 | 164 | 155 | 27 | 2 | 0 | 0 | 435 |
| Total Precipitation (") | 0.57 | 0.34 | 0.70 | 1.29 | 2.66 | 2.45 | 1.50 | 1.24 | 1.11 | 0.88 | 0.44 | 0.49 | 13.67 |
| Days ≥ 0.1" Precip | 2 | 1 | 2 | 4 | 6 | 6 | 4 | 3 | 3 | 3 | 2 | 2 | 38 |
| Total Snowfall (") | 8.0 | 4.4 | 5.4 | 3.9 | 0.8 | 0.0 | 0.0 | 0.0 | 0.6 | 1.8 | 3.9 | 7.0 | 35.8 |
| Days ≥ 1" Snow Depth | na | na | na | *1* | 0 | 0 | 0 | 0 | 0 | 1 | *4* | na | na |

## FORKS 4 NNE *Phillips County* ELEVATION 2602 ft LAT/LONG 48° 47 ' N / 107° 26 ' W

| | JAN | FEB | MAR | APR | MAY | JUN | JUL | AUG | SEP | OCT | NOV | DEC | YEAR |
|---|---|---|---|---|---|---|---|---|---|---|---|---|---|
| Maximum Temp °F | 20.4 | 26.9 | 40.2 | 56.0 | 68.1 | 76.7 | 83.7 | 83.1 | 71.0 | 57.7 | 38.1 | 25.7 | 54.0 |
| Minimum Temp °F | -0.4 | 5.2 | 17.4 | 29.5 | 40.1 | 48.0 | 52.8 | 51.7 | 41.4 | 30.7 | 16.7 | 5.0 | 28.2 |
| Mean Temp °F | 10.0 | 16.1 | 28.8 | 42.8 | 54.1 | 62.4 | 68.2 | 67.4 | 56.2 | 44.3 | 27.4 | 15.4 | 41.1 |
| Days Max Temp ≥ 90 °F | 0 | 0 | 0 | 0 | 1 | 3 | 7 | 9 | 2 | 0 | 0 | 0 | 22 |
| Days Max Temp ≤ 32 °F | 22 | 16 | 9 | 1 | 0 | 0 | 0 | 0 | 0 | 1 | 9 | 19 | 77 |
| Days Min Temp ≤ 32 °F | 30 | 28 | 29 | 19 | 6 | 0 | 0 | 0 | 4 | 17 | 28 | 30 | 191 |
| Days Min Temp ≤ 0 °F | 16 | 11 | 4 | 0 | 0 | 0 | 0 | 0 | 0 | 0 | 3 | 10 | 44 |
| Heating Degree Days | 1704 | 1376 | 1116 | 661 | 344 | 130 | 35 | 58 | 283 | 638 | 1122 | 1534 | 9001 |
| Cooling Degree Days | 0 | 0 | 0 | 1 | 20 | 75 | 152 | 156 | 28 | 0 | 0 | 0 | 432 |
| Total Precipitation (") | 0.40 | 0.30 | 0.55 | 0.94 | 2.05 | 2.55 | 2.06 | 1.22 | 1.15 | 0.62 | 0.42 | 0.37 | 12.63 |
| Days ≥ 0.1" Precip | 1 | 1 | 2 | 2 | 5 | 6 | 5 | 3 | 3 | 2 | 2 | 1 | 33 |
| Total Snowfall (") | 6.6 | 4.4 | 5.3 | *2.6* | 0.8 | 0.0 | 0.0 | 0.0 | 0.1 | 1.7 | *4.3* | 5.0 | 30.8 |
| Days ≥ 1" Snow Depth | 27 | 23 | 16 | 3 | 0 | 0 | 0 | 0 | 0 | 2 | 9 | 20 | 100 |

## FORT ASSINNIBOINE *Hill County* ELEVATION 2690 ft LAT/LONG 48° 30 ' N / 109° 48 ' W

| | JAN | FEB | MAR | APR | MAY | JUN | JUL | AUG | SEP | OCT | NOV | DEC | YEAR |
|---|---|---|---|---|---|---|---|---|---|---|---|---|---|
| Maximum Temp °F | 26.6 | 33.6 | 46.1 | 59.8 | 70.8 | 79.3 | 86.4 | 86.1 | 73.9 | 61.3 | 42.4 | 31.3 | 58.1 |
| Minimum Temp °F | 4.9 | 10.1 | 20.5 | 30.6 | 40.6 | 48.5 | 52.7 | 51.3 | 41.5 | 32.2 | 19.1 | 9.1 | 30.1 |
| Mean Temp °F | 15.7 | 21.9 | 33.3 | 45.2 | 55.7 | 64.0 | 69.5 | 68.7 | 57.8 | 46.9 | 30.8 | 20.2 | 44.1 |
| Days Max Temp ≥ 90 °F | 0 | 0 | 0 | 0 | 1 | 4 | 12 | 12 | 3 | 0 | 0 | 0 | 32 |
| Days Max Temp ≤ 32 °F | 16 | 12 | 5 | 1 | 0 | 0 | 0 | 0 | 0 | 1 | 6 | 14 | 55 |
| Days Min Temp ≤ 32 °F | 29 | 27 | 27 | 17 | 4 | 0 | 0 | 0 | 4 | 15 | 26 | 29 | 178 |
| Days Min Temp ≤ 0 °F | 13 | 9 | 3 | 0 | 0 | 0 | 0 | 0 | 0 | 0 | 3 | 9 | 37 |
| Heating Degree Days | 1524 | 1212 | 975 | 588 | 295 | 99 | 23 | 38 | 238 | 556 | 1020 | 1383 | 7951 |
| Cooling Degree Days | 0 | 0 | 0 | 2 | 20 | 85 | 169 | 164 | 26 | 2 | 0 | 0 | 468 |
| Total Precipitation (") | 0.53 | 0.34 | 0.66 | 1.05 | 2.01 | 2.09 | 1.71 | 1.43 | 1.30 | 0.66 | 0.43 | 0.52 | 12.73 |
| Days ≥ 0.1" Precip | 2 | 1 | 2 | 3 | 4 | 5 | 4 | 3 | 3 | 2 | 2 | 2 | 33 |
| Total Snowfall (") | na | na | na | na | na | *0.0* | *0.0* | *0.0* | *0.0* | na | na | na | na |
| Days ≥ 1" Snow Depth | *23* | *18* | na | *1* | 0 | 0 | 0 | 0 | 0 | *1* | *6* | na | na |

## FORT BENTON *Chouteau County* ELEVATION 2641 ft LAT/LONG 47° 49 ' N / 110° 40 ' W

| | JAN | FEB | MAR | APR | MAY | JUN | JUL | AUG | SEP | OCT | NOV | DEC | YEAR |
|---|---|---|---|---|---|---|---|---|---|---|---|---|---|
| Maximum Temp °F | 32.0 | 39.3 | 49.2 | 60.7 | 70.4 | 78.4 | 85.5 | 84.6 | 73.6 | 62.7 | 45.7 | 35.7 | 59.8 |
| Minimum Temp °F | 8.4 | 13.4 | 22.2 | 31.4 | 40.8 | 48.5 | 52.2 | 50.9 | 40.9 | 31.7 | 20.3 | 12.0 | 31.1 |
| Mean Temp °F | 20.2 | 26.4 | 35.7 | 46.1 | 55.6 | 63.5 | 68.9 | 67.7 | 57.3 | 47.2 | 33.0 | 23.9 | 45.5 |
| Days Max Temp ≥ 90 °F | 0 | 0 | 0 | 0 | 1 | 4 | 10 | 9 | 2 | 0 | 0 | 0 | 26 |
| Days Max Temp ≤ 32 °F | 13 | 9 | 4 | 0 | 0 | 0 | 0 | 0 | 0 | 1 | 5 | 11 | 43 |
| Days Min Temp ≤ 32 °F | 28 | 26 | 25 | 16 | 4 | 0 | 0 | 0 | 5 | 16 | 25 | 28 | 173 |
| Days Min Temp ≤ 0 °F | 11 | 6 | 2 | 0 | 0 | 0 | 0 | 0 | 0 | 0 | 3 | 7 | 29 |
| Heating Degree Days | 1384 | 1085 | 900 | 561 | 293 | 100 | 23 | 40 | 241 | 546 | 953 | 1270 | 7396 |
| Cooling Degree Days | 0 | 0 | 0 | 1 | 15 | 75 | 153 | 146 | 16 | 1 | 0 | 0 | 407 |
| Total Precipitation (") | 0.73 | 0.43 | 0.89 | 1.28 | 2.32 | 2.58 | 1.42 | 1.51 | 1.27 | 0.84 | 0.58 | 0.62 | 14.47 |
| Days ≥ 0.1" Precip | 3 | 2 | 3 | 4 | 5 | 6 | 3 | 4 | 3 | 3 | 2 | 2 | 40 |
| Total Snowfall (") | 15.2 | 9.5 | 9.6 | 5.0 | 0.6 | 0.0 | 0.0 | 0.0 | 0.3 | *2.1* | 8.1 | 12.7 | 63.1 |
| Days ≥ 1" Snow Depth | 17 | 12 | 7 | 1 | 0 | 0 | 0 | 0 | 0 | 1 | 7 | 14 | 59 |

## FORT PECK POWER PLAN *McCone County* ELEVATION 2070 ft LAT/LONG 48° 1 ' N / 106° 24 ' W

| | JAN | FEB | MAR | APR | MAY | JUN | JUL | AUG | SEP | OCT | NOV | DEC | YEAR |
|---|---|---|---|---|---|---|---|---|---|---|---|---|---|
| Maximum Temp °F | 23.7 | 30.6 | 43.3 | 58.0 | 70.4 | 79.8 | 86.6 | 85.7 | 73.4 | 60.6 | 42.2 | 30.2 | 57.0 |
| Minimum Temp °F | 3.7 | 9.8 | 21.2 | 33.0 | 43.7 | 52.4 | 56.9 | 55.7 | 45.3 | 36.2 | 22.4 | 10.0 | 32.5 |
| Mean Temp °F | 13.7 | 20.3 | 32.3 | 45.5 | 57.1 | 66.1 | 71.8 | 70.7 | 59.4 | 48.4 | 32.4 | 20.1 | 44.8 |
| Days Max Temp ≥ 90 °F | 0 | 0 | 0 | 0 | 1 | 5 | 12 | 11 | 2 | 0 | 0 | 0 | 31 |
| Days Max Temp ≤ 32 °F | 20 | 14 | 6 | 1 | 0 | 0 | 0 | 0 | 0 | 0 | 6 | 15 | 62 |
| Days Min Temp ≤ 32 °F | 30 | 27 | 27 | 14 | 2 | 0 | 0 | 0 | 2 | 10 | 24 | 30 | 166 |
| Days Min Temp ≤ 0 °F | 14 | 8 | 2 | 0 | 0 | 0 | 0 | 0 | 0 | 0 | 2 | 8 | 34 |
| Heating Degree Days | 1586 | 1257 | 1007 | 578 | 260 | 64 | 10 | 24 | 200 | 509 | 973 | 1387 | 7855 |
| Cooling Degree Days | 0 | 0 | 0 | 1 | 29 | 120 | 230 | 223 | 38 | 2 | 0 | 0 | 643 |
| Total Precipitation (") | 0.33 | 0.18 | 0.39 | 1.02 | 1.88 | 2.21 | 2.04 | 1.39 | 1.14 | 0.74 | 0.26 | 0.28 | 11.86 |
| Days ≥ 0.1" Precip | 1 | 1 | 1 | 2 | 4 | 5 | 4 | 3 | 3 | 2 | 1 | 1 | 28 |
| Total Snowfall (") | na | na | na | na | ***0.0*** | ***0.0*** | ***0.0*** | ***0.0*** | ***0.0*** | na | na | na | na |
| Days ≥ 1" Snow Depth | na | na | na | na | ***0*** | ***0*** | ***0*** | ***0*** | ***0*** | na | na | na | na |

## FORTINE 1 N *Lincoln County* ELEVATION 3002 ft LAT/LONG 48° 47 ' N / 114° 54 ' W

| | JAN | FEB | MAR | APR | MAY | JUN | JUL | AUG | SEP | OCT | NOV | DEC | YEAR |
|---|---|---|---|---|---|---|---|---|---|---|---|---|---|
| Maximum Temp °F | 29.0 | 37.3 | 47.2 | 57.2 | 66.7 | 74.1 | 81.8 | 81.7 | 70.7 | 56.1 | 38.2 | 30.1 | 55.8 |
| Minimum Temp °F | 13.7 | 18.7 | 24.2 | 30.4 | 36.6 | 43.3 | 46.5 | 45.6 | 37.3 | 29.6 | 23.0 | 16.2 | 30.4 |
| Mean Temp °F | 21.4 | 28.1 | 35.7 | 43.8 | 51.7 | 58.7 | 64.1 | 63.7 | 54.0 | 42.9 | 30.6 | 23.2 | 43.2 |
| Days Max Temp ≥ 90 °F | 0 | 0 | 0 | 0 | 0 | 2 | 7 | 7 | 1 | 0 | 0 | 0 | 17 |
| Days Max Temp ≤ 32 °F | 16 | 7 | 1 | 0 | 0 | 0 | 0 | 0 | 0 | 0 | 7 | 17 | 48 |
| Days Min Temp ≤ 32 °F | 29 | 26 | 26 | 19 | 9 | 1 | 0 | 0 | 8 | 19 | 25 | 29 | 191 |
| Days Min Temp ≤ 0 °F | 7 | 3 | 1 | 0 | 0 | 0 | 0 | 0 | 0 | 0 | 1 | 4 | 16 |
| Heating Degree Days | 1347 | 1038 | 901 | 630 | 409 | 200 | 84 | 91 | 328 | 680 | 1025 | 1290 | 8023 |
| Cooling Degree Days | 0 | 0 | 0 | 0 | ***4*** | 19 | 61 | 61 | 4 | 0 | 0 | 0 | 149 |
| Total Precipitation (") | 1.29 | 0.80 | 0.82 | 1.08 | 1.84 | 2.27 | 1.61 | 1.36 | 1.25 | 0.92 | 1.25 | 1.25 | 15.74 |
| Days ≥ 0.1" Precip | 5 | 3 | 3 | 3 | 5 | 7 | 5 | 4 | 4 | 4 | 5 | 5 | 53 |
| Total Snowfall (") | 12.1 | 6.6 | 3.7 | 1.2 | 0.0 | 0.0 | 0.0 | 0.0 | 0.1 | 1.4 | 6.3 | 10.1 | 41.5 |
| Days ≥ 1" Snow Depth | 23 | 17 | 6 | 0 | 0 | 0 | 0 | 0 | 0 | 1 | 8 | 21 | 76 |

## GARDINER *Park County* ELEVATION 5305 ft LAT/LONG 45° 2 ' N / 110° 42 ' W

| | JAN | FEB | MAR | APR | MAY | JUN | JUL | AUG | SEP | OCT | NOV | DEC | YEAR |
|---|---|---|---|---|---|---|---|---|---|---|---|---|---|
| Maximum Temp °F | 33.0 | 38.5 | 46.6 | 56.1 | 66.4 | 76.0 | 85.1 | 84.0 | 73.7 | 60.9 | 42.2 | 32.7 | 57.9 |
| Minimum Temp °F | 13.7 | 17.2 | 23.4 | 30.0 | 38.0 | 45.4 | 51.3 | 50.3 | 41.2 | 32.7 | 22.3 | 14.6 | 31.7 |
| Mean Temp °F | 23.4 | 27.9 | 35.0 | 43.0 | 52.2 | 60.8 | 68.2 | 67.2 | 57.5 | 46.8 | 32.3 | 23.7 | 44.8 |
| Days Max Temp ≥ 90 °F | 0 | 0 | 0 | 0 | 0 | 2 | 9 | 8 | 1 | 0 | 0 | 0 | 20 |
| Days Max Temp ≤ 32 °F | ***13*** | ***6*** | ***2*** | 0 | 0 | 0 | 0 | 0 | 0 | 0 | ***4*** | 13 | 38 |
| Days Min Temp ≤ 32 °F | ***30*** | 26 | ***26*** | ***19*** | 7 | 1 | 0 | 0 | 4 | ***14*** | ***26*** | 29 | 182 |
| Days Min Temp ≤ 0 °F | ***5*** | ***2*** | 1 | 0 | 0 | 0 | 0 | 0 | 0 | 0 | 1 | 3 | 12 |
| Heating Degree Days | 1282 | 1042 | 923 | 652 | 391 | 165 | 31 | 41 | 238 | 558 | 976 | 1277 | 7576 |
| Cooling Degree Days | ***0*** | ***0*** | ***0*** | ***0*** | 5 | ***56*** | 140 | 124 | ***22*** | ***0*** | ***0*** | 0 | 347 |
| Total Precipitation (") | 0.36 | ***0.27*** | ***0.52*** | 0.64 | 1.80 | 1.50 | 1.21 | 0.96 | 0.97 | ***0.78*** | ***0.64*** | ***0.54*** | 10.19 |
| Days ≥ 0.1" Precip | 1 | ***1*** | ***1*** | 2 | ***5*** | ***5*** | 4 | 3 | ***3*** | na | na | ***2*** | na |
| Total Snowfall (") | na | na | na | na | ***0.5*** | ***0.0*** | ***0.0*** | ***0.0*** | ***0.1*** | na | na | na | na |
| Days ≥ 1" Snow Depth | na | na | na | na | 0 | 0 | 0 | 0 | 0 | na | na | na | na |

## GERALDINE *Chouteau County* ELEVATION 3143 ft LAT/LONG 47° 36 ' N / 110° 16 ' W

| | JAN | FEB | MAR | APR | MAY | JUN | JUL | AUG | SEP | OCT | NOV | DEC | YEAR |
|---|---|---|---|---|---|---|---|---|---|---|---|---|---|
| Maximum Temp °F | 32.9 | 39.1 | 47.2 | 58.1 | 68.1 | 76.6 | 84.5 | 84.3 | 72.8 | 61.4 | 45.0 | 36.5 | 58.9 |
| Minimum Temp °F | 8.8 | 14.2 | 22.2 | 31.0 | 40.3 | 47.9 | 51.7 | 50.8 | 41.5 | 32.5 | 20.7 | 12.5 | 31.2 |
| Mean Temp °F | 20.9 | 26.7 | 34.8 | 44.6 | 54.3 | 62.3 | 68.2 | 67.6 | 57.2 | 47.0 | 32.9 | 24.5 | 45.1 |
| Days Max Temp ≥ 90 °F | 0 | 0 | 0 | 0 | 0 | 3 | 9 | 10 | 2 | 0 | 0 | 0 | 24 |
| Days Max Temp ≤ 32 °F | 12 | 8 | 4 | 1 | 0 | 0 | 0 | 0 | 0 | 1 | 5 | 10 | 41 |
| Days Min Temp ≤ 32 °F | 28 | 25 | 25 | 17 | 5 | 0 | 0 | 0 | 4 | 15 | 24 | 28 | 171 |
| Days Min Temp ≤ 0 °F | 12 | 6 | 2 | 0 | 0 | 0 | 0 | 0 | 0 | 0 | 3 | 7 | 30 |
| Heating Degree Days | 1365 | 1076 | 931 | 608 | 335 | 126 | 32 | 46 | 252 | 554 | 956 | 1250 | 7531 |
| Cooling Degree Days | 0 | 0 | 0 | 1 | 14 | 64 | 147 | 149 | 25 | 1 | 0 | 0 | 401 |
| Total Precipitation (") | 0.80 | 0.48 | 0.89 | 1.31 | 2.74 | 2.63 | 1.74 | 1.60 | 1.41 | 0.90 | 0.59 | 0.69 | 15.78 |
| Days ≥ 0.1" Precip | 3 | 2 | 2 | 4 | 6 | 6 | 4 | 4 | 4 | 3 | 2 | 2 | 42 |
| Total Snowfall (") | 13.5 | 7.5 | 8.7 | 6.4 | 1.8 | 0.1 | 0.0 | 0.0 | 0.6 | 2.8 | 7.6 | 11.5 | 60.5 |
| Days ≥ 1" Snow Depth | 14 | 9 | 6 | 1 | 0 | 0 | 0 | 0 | 0 | 1 | 6 | 11 | 48 |

## GIBSON DAM *Lewis and Clark County* ELEVATION 4593 ft LAT/LONG 47° 36 ' N / 112° 46 ' W

| | JAN | FEB | MAR | APR | MAY | JUN | JUL | AUG | SEP | OCT | NOV | DEC | YEAR |
|---|---|---|---|---|---|---|---|---|---|---|---|---|---|
| Maximum Temp °F | 33.6 | 38.4 | 43.4 | 51.9 | 61.2 | 69.1 | 77.3 | 76.9 | 67.2 | 56.7 | 41.5 | 34.4 | 54.3 |
| Minimum Temp °F | 13.1 | 16.7 | 21.4 | 28.6 | 35.8 | 42.7 | 46.6 | 45.6 | 38.4 | 32.3 | 22.9 | 15.8 | 30.0 |
| Mean Temp °F | 23.4 | 27.6 | 32.4 | 40.3 | 48.5 | 55.9 | 62.0 | 61.3 | 52.8 | 44.5 | 32.2 | 25.1 | 42.2 |
| Days Max Temp ≥ 90 °F | 0 | 0 | 0 | 0 | 0 | 0 | 1 | 1 | 0 | 0 | 0 | 0 | 2 |
| Days Max Temp ≤ 32 °F | 12 | 7 | 4 | 1 | 0 | 0 | 0 | 0 | 0 | 1 | 5 | 12 | 42 |
| Days Min Temp ≤ 32 °F | 28 | 25 | 27 | 22 | 9 | 1 | 0 | 0 | 6 | 16 | 24 | 27 | 185 |
| Days Min Temp ≤ 0 °F | 8 | 5 | 2 | 0 | 0 | 0 | 0 | 0 | 0 | 0 | 2 | 5 | 22 |
| Heating Degree Days | 1284 | 1050 | 1003 | 735 | 504 | 274 | 115 | 136 | 363 | 627 | 977 | 1231 | 8299 |
| Cooling Degree Days | 0 | 0 | 0 | 0 | 1 | 9 | 29 | 26 | 3 | 0 | 0 | 0 | 68 |
| Total Precipitation (") | 1.04 | 0.70 | 0.89 | 1.49 | 2.87 | 2.85 | 1.46 | 1.76 | 1.47 | 0.99 | 0.88 | 0.85 | 17.25 |
| Days ≥ 0.1" Precip | 4 | 2 | 3 | 4 | 6 | 7 | 4 | 4 | 4 | 3 | 3 | 3 | 47 |
| Total Snowfall (") | 13.0 | 10.4 | 11.0 | 8.5 | 1.6 | 0.1 | 0.0 | 0.2 | 1.3 | *4.9* | 7.9 | 10.6 | 69.5 |
| Days ≥ 1" Snow Depth | 18 | 14 | 12 | 4 | 1 | 0 | 0 | 0 | 1 | 2 | 8 | 14 | 74 |

## GILDFORD *Hill County* ELEVATION 2831 ft LAT/LONG 48° 35 ' N / 110° 18 ' W

| | JAN | FEB | MAR | APR | MAY | JUN | JUL | AUG | SEP | OCT | NOV | DEC | YEAR |
|---|---|---|---|---|---|---|---|---|---|---|---|---|---|
| Maximum Temp °F | 25.7 | 32.9 | 43.9 | 57.1 | 68.5 | 76.5 | 83.7 | 83.0 | 71.2 | 59.1 | 41.0 | 30.3 | 56.1 |
| Minimum Temp °F | 3.8 | 9.3 | 19.5 | 29.7 | 39.7 | 47.5 | 51.4 | 50.2 | 40.4 | 30.5 | 17.5 | 8.0 | 29.0 |
| Mean Temp °F | 14.7 | 21.1 | 31.7 | 43.4 | 54.1 | 62.0 | 67.6 | 66.6 | 55.8 | 44.8 | 29.3 | 19.2 | 42.5 |
| Days Max Temp ≥ 90 °F | 0 | 0 | 0 | 0 | 0 | 3 | 8 | 8 | 1 | 0 | 0 | 0 | 20 |
| Days Max Temp ≤ 32 °F | 17 | 12 | 6 | 1 | 0 | 0 | 0 | 0 | 0 | 1 | 7 | 14 | 58 |
| Days Min Temp ≤ 32 °F | 30 | 27 | 28 | 19 | 5 | 0 | 0 | 0 | 5 | 17 | 27 | 30 | 188 |
| Days Min Temp ≤ 0 °F | 14 | 9 | 3 | 0 | 0 | 0 | 0 | 0 | 0 | 0 | 4 | 9 | 39 |
| Heating Degree Days | 1556 | 1235 | 1026 | 641 | 339 | 133 | 38 | 60 | 286 | 619 | 1066 | 1417 | 8416 |
| Cooling Degree Days | 0 | 0 | 0 | 1 | 10 | 56 | 118 | 113 | 13 | 1 | 0 | 0 | 312 |
| Total Precipitation (") | 0.40 | 0.23 | 0.50 | 0.86 | 1.79 | 2.25 | 1.42 | 1.37 | 1.15 | 0.53 | 0.31 | 0.37 | 11.18 |
| Days ≥ 0.1" Precip | 2 | 1 | 1 | 3 | 4 | 5 | 4 | 4 | 3 | 2 | 1 | 1 | 31 |
| Total Snowfall (") | na | na | na | na | *0.2* | 0.0 | 0.0 | 0.1 | *0.0* | na | na | na | na |
| Days ≥ 1" Snow Depth | na | na | na | na | 0 | 0 | 0 | 0 | 0 | *1* | na | na | na |

## GLASGOW INTL AP *Valley County* ELEVATION 2284 ft LAT/LONG 48° 13 ' N / 106° 37 ' W

| | JAN | FEB | MAR | APR | MAY | JUN | JUL | AUG | SEP | OCT | NOV | DEC | YEAR |
|---|---|---|---|---|---|---|---|---|---|---|---|---|---|
| Maximum Temp °F | 19.8 | 26.9 | 40.8 | 56.5 | 68.2 | 77.2 | 84.1 | 83.2 | 70.6 | 57.9 | 38.9 | 25.9 | 54.2 |
| Minimum Temp °F | 1.0 | 7.7 | 19.6 | 32.0 | 42.8 | 51.3 | 56.5 | 55.4 | 44.0 | 32.9 | 18.7 | 6.6 | 30.7 |
| Mean Temp °F | 10.4 | 17.3 | 30.2 | 44.3 | 55.5 | 64.2 | 70.3 | 69.3 | 57.4 | 45.4 | 28.8 | 16.3 | 42.4 |
| Days Max Temp ≥ 90 °F | 0 | 0 | 0 | 0 | 1 | 3 | 9 | 9 | 2 | 0 | 0 | 0 | 24 |
| Days Max Temp ≤ 32 °F | 22 | 16 | 8 | 1 | 0 | 0 | 0 | 0 | 0 | 1 | 9 | 19 | 76 |
| Days Min Temp ≤ 32 °F | 31 | 28 | 28 | 15 | 2 | 0 | 0 | 0 | 2 | 13 | 27 | 31 | 177 |
| Days Min Temp ≤ 0 °F | 15 | 10 | 3 | 0 | 0 | 0 | 0 | 0 | 0 | 0 | 2 | 9 | 39 |
| Heating Degree Days | 1690 | 1341 | 1072 | 617 | 305 | 94 | 20 | 40 | 251 | 600 | 1079 | 1506 | 8615 |
| Cooling Degree Days | 0 | 0 | 0 | 2 | 25 | 95 | 200 | 191 | 29 | 1 | 0 | 0 | 543 |
| Total Precipitation (") | 0.39 | 0.25 | 0.42 | 0.70 | 1.67 | 2.04 | 1.70 | 1.36 | 0.95 | 0.64 | 0.30 | 0.36 | 10.78 |
| Days ≥ 0.1" Precip | 1 | 1 | 1 | 2 | 4 | 5 | 4 | 3 | 2 | 2 | 1 | 1 | 27 |
| Total Snowfall (") | 7.3 | 4.2 | 4.1 | 2.6 | 0.6 | 0.0 | 0.0 | 0.0 | 0.2 | 1.1 | 3.2 | 5.8 | 29.1 |
| Days ≥ 1" Snow Depth | 26 | 19 | 13 | 3 | 0 | 0 | 0 | 0 | 0 | 1 | 6 | 19 | 87 |

## GLEN 4 N *Madison County* ELEVATION 5052 ft LAT/LONG 45° 31 ' N / 112° 41 ' W

| | JAN | FEB | MAR | APR | MAY | JUN | JUL | AUG | SEP | OCT | NOV | DEC | YEAR |
|---|---|---|---|---|---|---|---|---|---|---|---|---|---|
| Maximum Temp °F | 34.7 | 40.4 | 47.3 | 56.6 | 65.7 | 74.4 | 82.8 | 81.0 | 70.3 | 60.1 | 43.8 | 34.6 | 57.6 |
| Minimum Temp °F | 8.7 | 12.2 | 19.7 | 27.6 | 36.2 | 43.3 | 47.9 | 45.6 | 36.3 | 27.9 | 18.1 | 9.0 | 27.7 |
| Mean Temp °F | 21.7 | *26.4* | 33.5 | 42.1 | 51.0 | 59.0 | 65.3 | 63.2 | 53.3 | 44.0 | 30.9 | 21.8 | 42.7 |
| Days Max Temp ≥ 90 °F | 0 | 0 | 0 | 0 | 0 | 1 | 6 | 5 | 0 | 0 | 0 | 0 | 12 |
| Days Max Temp ≤ 32 °F | 11 | *5* | 2 | 0 | 0 | 0 | 0 | 0 | 0 | 0 | na | *12* | na |
| Days Min Temp ≤ 32 °F | 30 | 27 | 29 | 22 | 9 | 1 | 0 | 1 | 8 | *22* | *27* | 30 | 206 |
| Days Min Temp ≤ 0 °F | 7 | 4 | 1 | 0 | 0 | 0 | 0 | 0 | 0 | 0 | 2 | 7 | 21 |
| Heating Degree Days | 1336 | *1082* | 969 | 680 | 429 | 197 | 57 | 95 | 346 | 642 | 1019 | 1332 | 8184 |
| Cooling Degree Days | 0 | *0* | *0* | 0 | 3 | 27 | 75 | *43* | *2* | 0 | 0 | 0 | 150 |
| Total Precipitation (") | 0.20 | 0.15 | 0.38 | 0.65 | 1.54 | 1.97 | 1.18 | 1.22 | 0.99 | 0.42 | 0.33 | 0.17 | 9.20 |
| Days ≥ 0.1" Precip | *0* | 0 | 1 | 2 | 5 | 5 | 4 | 4 | 3 | 2 | 1 | 1 | 28 |
| Total Snowfall (") | na | na | na | na | *0.1* | 0.0 | 0.0 | 0.0 | *0.4* | *0.3* | na | na | na |
| Days ≥ 1" Snow Depth | na | na | na | na | 0 | 0 | 0 | 0 | *0* | *0* | na | na | na |

## GLENDIVE *Dawson County* ELEVATION 2080 ft LAT/LONG 47° 6 ' N / 104° 43 ' W

| | JAN | FEB | MAR | APR | MAY | JUN | JUL | AUG | SEP | OCT | NOV | DEC | YEAR |
|---|---|---|---|---|---|---|---|---|---|---|---|---|---|
| Maximum Temp °F | 25.7 | 33.2 | 45.7 | 60.2 | 72.0 | 81.1 | 88.5 | 87.9 | 75.5 | 62.3 | 42.7 | 30.4 | 58.8 |
| Minimum Temp °F | 3.6 | 9.8 | 21.4 | 33.4 | 44.7 | 54.0 | 58.9 | 56.8 | 45.5 | 34.6 | 21.2 | 9.3 | 32.8 |
| Mean Temp °F | 14.7 | 21.5 | 33.6 | 46.8 | 58.4 | 67.5 | 73.7 | 72.4 | 60.5 | 48.5 | 31.9 | 19.8 | 45.8 |
| Days Max Temp ≥ 90 °F | 0 | 0 | 0 | 0 | 2 | 6 | 15 | 14 | 4 | 0 | 0 | 0 | 41 |
| Days Max Temp ≤ 32 °F | 18 | 12 | 5 | 0 | 0 | 0 | 0 | 0 | 0 | 0 | 6 | 15 | 56 |
| Days Min Temp ≤ 32 °F | 31 | 27 | 27 | 14 | 2 | 0 | 0 | 0 | 2 | 12 | 27 | 31 | 173 |
| Days Min Temp ≤ 0 °F | 13 | 8 | 2 | 0 | 0 | 0 | 0 | 0 | 0 | 0 | 1 | 8 | 32 |
| Heating Degree Days | 1556 | 1222 | 968 | 541 | 236 | 51 | 9 | 18 | 184 | 507 | 985 | 1394 | 7671 |
| Cooling Degree Days | 0 | 0 | 0 | 3 | 40 | 133 | 274 | 250 | 47 | 2 | 0 | 0 | 749 |
| Total Precipitation (") | 0.46 | 0.30 | 0.53 | 1.25 | 2.06 | 2.89 | 1.70 | 1.38 | 1.42 | 0.85 | 0.40 | 0.45 | 13.69 |
| Days ≥ 0.1" Precip | 2 | 1 | 2 | 3 | 5 | 6 | 4 | 3 | 3 | 2 | 2 | 1 | 34 |
| Total Snowfall (") | ***7.3*** | ***4.8*** | ***3.8*** | ***1.6*** | 0.3 | 0.0 | 0.0 | 0.0 | 0.1 | 0.7 | ***2.8*** | ***5.9*** | 27.3 |
| Days ≥ 1" Snow Depth | 23 | 18 | 8 | 1 | 0 | 0 | 0 | 0 | 0 | 0 | 6 | 19 | 75 |

## GOLDBUTTE 7 N *Toole County* ELEVATION 3504 ft LAT/LONG 48° 59 ' N / 111° 24 ' W

| | JAN | FEB | MAR | APR | MAY | JUN | JUL | AUG | SEP | OCT | NOV | DEC | YEAR |
|---|---|---|---|---|---|---|---|---|---|---|---|---|---|
| Maximum Temp °F | 30.3 | 35.6 | 43.7 | 55.2 | 65.4 | 72.9 | 79.9 | 79.3 | 68.6 | 57.5 | 41.4 | 33.0 | 55.2 |
| Minimum Temp °F | 7.7 | 12.5 | 20.4 | 29.6 | 38.4 | 45.4 | 48.9 | 47.9 | 39.6 | 31.6 | 19.6 | 11.9 | 29.5 |
| Mean Temp °F | 19.0 | 24.1 | 32.1 | 42.4 | 51.9 | 59.2 | 64.4 | 63.6 | 54.1 | 44.6 | 30.5 | 22.4 | 42.4 |
| Days Max Temp ≥ 90 °F | 0 | 0 | 0 | 0 | 0 | 1 | 3 | 4 | 1 | 0 | 0 | 0 | 9 |
| Days Max Temp ≤ 32 °F | 14 | 10 | 6 | 1 | 0 | 0 | 0 | 0 | 0 | 1 | 6 | 12 | 50 |
| Days Min Temp ≤ 32 °F | 28 | 26 | 27 | 19 | 7 | 1 | 0 | 1 | 6 | 16 | 25 | 28 | 184 |
| Days Min Temp ≤ 0 °F | 11 | 7 | 3 | 0 | 0 | 0 | 0 | 0 | 0 | 0 | 3 | 8 | 32 |
| Heating Degree Days | 1421 | 1148 | 1014 | 670 | 403 | 196 | 83 | 103 | 333 | 628 | 1028 | 1314 | 8341 |
| Cooling Degree Days | 0 | 0 | 0 | 0 | 6 | 29 | 68 | 66 | 13 | 1 | 0 | 0 | 183 |
| Total Precipitation (") | 0.42 | 0.32 | 0.64 | 1.07 | 2.13 | 2.77 | 1.37 | 1.79 | 1.51 | 0.77 | 0.48 | 0.40 | 13.67 |
| Days ≥ 0.1" Precip | 1 | 1 | 2 | 3 | 5 | 6 | 3 | 4 | 4 | 2 | 2 | 1 | 34 |
| Total Snowfall (") | 10.2 | 7.8 | 11.6 | 8.7 | 2.0 | 0.0 | 0.0 | 0.3 | 0.8 | 5.7 | 7.2 | 9.3 | 63.6 |
| Days ≥ 1" Snow Depth | ***9*** | ***6*** | ***7*** | ***3*** | 1 | 0 | 0 | 0 | 0 | 2 | 6 | ***8*** | 42 |

## GRASS RANGE *Fergus County* ELEVATION 3491 ft LAT/LONG 47° 1 ' N / 108° 48 ' W

| | JAN | FEB | MAR | APR | MAY | JUN | JUL | AUG | SEP | OCT | NOV | DEC | YEAR |
|---|---|---|---|---|---|---|---|---|---|---|---|---|---|
| Maximum Temp °F | 36.3 | 41.8 | 49.2 | 58.9 | 69.0 | 77.9 | 85.0 | 84.4 | 73.5 | 63.2 | 47.0 | 39.4 | 60.5 |
| Minimum Temp °F | 9.8 | 14.3 | 21.7 | 30.0 | 38.9 | 46.4 | 50.7 | 49.5 | 40.5 | 32.1 | 20.7 | 13.2 | 30.7 |
| Mean Temp °F | 23.1 | 28.1 | 35.5 | 44.5 | 54.0 | 62.1 | 67.9 | 67.0 | 57.0 | 47.7 | 33.8 | 26.3 | 45.6 |
| Days Max Temp ≥ 90 °F | 0 | 0 | 0 | 0 | 1 | 3 | 10 | 10 | 2 | 0 | 0 | 0 | 26 |
| Days Max Temp ≤ 32 °F | 11 | 7 | 3 | 0 | 0 | 0 | 0 | 0 | 0 | 0 | 4 | 8 | 33 |
| Days Min Temp ≤ 32 °F | 29 | 26 | 27 | 19 | 5 | 0 | 0 | 0 | 4 | 15 | 26 | 29 | 180 |
| Days Min Temp ≤ 0 °F | 10 | 6 | 2 | 0 | 0 | 0 | 0 | 0 | 0 | 0 | 2 | 6 | 26 |
| Heating Degree Days | 1294 | 1038 | 908 | 609 | 342 | 132 | 37 | 48 | 251 | 529 | 929 | 1192 | 7309 |
| Cooling Degree Days | 0 | 0 | 0 | 1 | 13 | 69 | 139 | 122 | 18 | 1 | 0 | 0 | 363 |
| Total Precipitation (") | 0.87 | 0.33 | 0.90 | 1.46 | 3.00 | 2.88 | 2.01 | 1.60 | 1.39 | 0.93 | 0.54 | 0.63 | 16.54 |
| Days ≥ 0.1" Precip | 3 | 1 | 3 | 4 | 6 | 6 | 5 | 4 | 4 | 3 | 2 | 2 | 43 |
| Total Snowfall (") | 12.0 | ***5.6*** | ***10.4*** | 6.8 | 1.3 | 0.2 | 0.0 | 0.0 | 0.4 | 2.7 | ***6.3*** | ***7.4*** | 53.1 |
| Days ≥ 1" Snow Depth | na | na | na | na | ***0*** | ***0*** | ***0*** | ***0*** | ***0*** | na | na | na | na |

## GREAT FALLS INTL AP *Cascade County* ELEVATION 3688 ft LAT/LONG 47° 29 ' N / 111° 21 ' W

| | JAN | FEB | MAR | APR | MAY | JUN | JUL | AUG | SEP | OCT | NOV | DEC | YEAR |
|---|---|---|---|---|---|---|---|---|---|---|---|---|---|
| Maximum Temp °F | 30.8 | 37.2 | 44.7 | 55.4 | 65.5 | 74.7 | 83.1 | 81.9 | 69.9 | 58.4 | 42.9 | 34.0 | 56.5 |
| Minimum Temp °F | 11.8 | 16.6 | 23.4 | 32.1 | 40.9 | 48.8 | 53.4 | 52.5 | 43.4 | 35.0 | 23.9 | 15.6 | 33.1 |
| Mean Temp °F | 21.3 | 26.9 | 34.0 | 43.8 | 53.2 | 61.8 | 68.3 | 67.2 | 56.7 | 46.7 | 33.4 | 24.8 | 44.8 |
| Days Max Temp ≥ 90 °F | 0 | 0 | 0 | 0 | 0 | 2 | 8 | 7 | 1 | 0 | 0 | 0 | 18 |
| Days Max Temp ≤ 32 °F | 13 | 9 | 5 | 1 | 0 | 0 | 0 | 0 | 0 | 1 | 5 | 11 | 45 |
| Days Min Temp ≤ 32 °F | 27 | 24 | 25 | 16 | 3 | 0 | 0 | 0 | 3 | 12 | 22 | 26 | 158 |
| Days Min Temp ≤ 0 °F | 10 | 5 | 2 | 0 | 0 | 0 | 0 | 0 | 0 | 0 | 2 | 6 | 25 |
| Heating Degree Days | 1350 | 1068 | 953 | 631 | 368 | 146 | 38 | 59 | 270 | 561 | 941 | 1241 | 7626 |
| Cooling Degree Days | 0 | 0 | 0 | 2 | 11 | 51 | 121 | 121 | 19 | 3 | 0 | 0 | 328 |
| Total Precipitation (") | 0.87 | 0.56 | 1.07 | 1.51 | 2.40 | 2.43 | 1.36 | 1.60 | 1.25 | 0.92 | 0.65 | 0.78 | 15.40 |
| Days ≥ 0.1" Precip | 3 | 2 | 3 | 4 | 6 | 5 | 3 | 4 | 3 | 3 | 2 | 3 | 41 |
| Total Snowfall (") | 10.8 | 7.4 | 11.1 | 9.3 | 2.1 | 0.2 | 0.0 | 0.3 | 1.7 | 4.2 | 7.2 | 9.4 | 63.7 |
| Days ≥ 1" Snow Depth | 17 | 12 | 9 | 3 | 0 | 0 | 0 | 0 | 0 | 2 | 7 | 14 | 64 |

## HAMILTON *Ravalli County* ELEVATION 3533 ft LAT/LONG 46° 15 ' N / 114° 9 ' W

| | JAN | FEB | MAR | APR | MAY | JUN | JUL | AUG | SEP | OCT | NOV | DEC | YEAR |
|---|---|---|---|---|---|---|---|---|---|---|---|---|---|
| Maximum Temp °F | 35.4 | 41.9 | 49.9 | 58.8 | 67.4 | 75.2 | 83.5 | 82.7 | 71.6 | 59.7 | 43.8 | 35.0 | 58.7 |
| Minimum Temp °F | 17.6 | 21.0 | 26.2 | 32.4 | 39.4 | 45.8 | 49.3 | 48.2 | 40.2 | 31.5 | 24.1 | 17.5 | 32.8 |
| Mean Temp °F | 26.5 | 31.4 | 38.1 | 45.6 | 53.4 | 60.5 | 66.4 | 65.5 | 55.9 | 45.7 | 34.0 | 26.3 | 45.8 |
| Days Max Temp ≥ 90 °F | 0 | 0 | 0 | 0 | 0 | 2 | 8 | 6 | 0 | 0 | 0 | 0 | 16 |
| Days Max Temp ≤ 32 °F | 10 | 5 | 1 | 0 | 0 | 0 | 0 | 0 | 0 | 0 | 4 | 11 | 31 |
| Days Min Temp ≤ 32 °F | 28 | 25 | 25 | 16 | 5 | 0 | 0 | 0 | 4 | 17 | 24 | 29 | 173 |
| Days Min Temp ≤ 0 °F | 4 | 2 | 0 | 0 | 0 | 0 | 0 | 0 | 0 | 0 | 1 | 3 | 10 |
| Heating Degree Days | 1188 | 942 | 828 | 575 | 357 | 164 | 50 | 62 | 275 | 593 | 924 | 1193 | 7151 |
| Cooling Degree Days | 0 | 0 | 0 | 0 | 8 | 39 | 99 | 85 | 8 | 0 | 0 | 0 | 239 |
| Total Precipitation (") | 1.20 | 0.75 | 0.87 | 0.98 | 1.70 | 1.65 | 0.94 | 1.19 | 1.15 | 0.81 | 0.92 | 0.93 | 13.09 |
| Days ≥ 0.1" Precip | 4 | 3 | 3 | 3 | 5 | 5 | 3 | 4 | 4 | 2 | 3 | 3 | 42 |
| Total Snowfall (") | na | na | na | na | na | ***0.0*** | ***0.0*** | ***0.0*** | na | na | na | na | na |
| Days ≥ 1" Snow Depth | na | na | na | na | ***0*** | ***0*** | ***0*** | ***0*** | ***0*** | na | na | na | na |

## HARDIN *Big Horn County* ELEVATION 2890 ft LAT/LONG 45° 45 ' N / 107° 36 ' W

| | JAN | FEB | MAR | APR | MAY | JUN | JUL | AUG | SEP | OCT | NOV | DEC | YEAR |
|---|---|---|---|---|---|---|---|---|---|---|---|---|---|
| Maximum Temp °F | 33.2 | 40.4 | 50.4 | 61.4 | 72.6 | 81.6 | 90.2 | 89.3 | ***76.5*** | ***64.7*** | ***47.2*** | 35.7 | 61.9 |
| Minimum Temp °F | 7.4 | 12.7 | 22.3 | 31.7 | 41.4 | 49.6 | 54.5 | 52.6 | ***41.8*** | ***31.5*** | ***20.1*** | 9.8 | 31.3 |
| Mean Temp °F | 20.3 | 26.6 | 36.4 | 46.6 | 57.0 | 65.6 | 72.4 | 71.2 | ***59.2*** | ***48.1*** | ***33.7*** | 22.8 | 46.7 |
| Days Max Temp ≥ 90 °F | 0 | 0 | 0 | 0 | 2 | 7 | ***18*** | ***16*** | ***4*** | 0 | 0 | 0 | 47 |
| Days Max Temp ≤ 32 °F | 12 | 8 | 4 | 0 | 0 | 0 | 0 | 0 | 0 | 1 | 4 | ***10*** | 39 |
| Days Min Temp ≤ 32 °F | 30 | 27 | 27 | ***16*** | ***3*** | 0 | 0 | 0 | 3 | ***17*** | ***26*** | 30 | 179 |
| Days Min Temp ≤ 0 °F | 11 | 5 | 2 | 0 | 0 | 0 | 0 | 0 | 0 | 0 | ***2*** | 7 | 27 |
| Heating Degree Days | 1380 | 1079 | 880 | 548 | 264 | 80 | 12 | 23 | ***204*** | ***516*** | ***935*** | 1305 | 7226 |
| Cooling Degree Days | 0 | 0 | ***0*** | ***3*** | ***25*** | 133 | 265 | 251 | ***45*** | ***2*** | ***0*** | ***0*** | 724 |
| Total Precipitation (") | 0.66 | 0.37 | 0.56 | 1.48 | 1.98 | 2.00 | 1.05 | 0.77 | 1.41 | 1.19 | 0.62 | ***0.52*** | 12.61 |
| Days ≥ 0.1" Precip | ***2*** | 1 | 2 | 4 | ***5*** | 5 | 3 | 2 | 3 | ***3*** | ***2*** | ***2*** | 34 |
| Total Snowfall (") | na | na | na | na | ***0.0*** | 0.0 | 0.0 | 0.0 | ***0.2*** | na | na | na | na |
| Days ≥ 1" Snow Depth | na | na | na | na | ***0*** | 0 | 0 | 0 | 0 | na | na | na | na |

## HARLEM 4 W *Blaine County* ELEVATION 2363 ft LAT/LONG 48° 33 ' N / 108° 51 ' W

| | JAN | FEB | MAR | APR | MAY | JUN | JUL | AUG | SEP | OCT | NOV | DEC | YEAR |
|---|---|---|---|---|---|---|---|---|---|---|---|---|---|
| Maximum Temp °F | 26.4 | 33.8 | 45.6 | 59.7 | 70.3 | 78.6 | 85.3 | 84.5 | 73.3 | 61.0 | 42.2 | 29.9 | 57.6 |
| Minimum Temp °F | 2.6 | 7.6 | 18.7 | 29.7 | ***39.8*** | 48.5 | 52.4 | 50.8 | 40.0 | 29.8 | 16.2 | 6.0 | 28.5 |
| Mean Temp °F | 14.5 | 20.7 | 32.2 | 44.7 | ***55.1*** | 63.6 | 68.9 | 67.7 | 56.7 | 45.4 | 29.2 | 18.2 | 43.1 |
| Days Max Temp ≥ 90 °F | 0 | 0 | 0 | 0 | 1 | 3 | 10 | 10 | 2 | 0 | 0 | 0 | 26 |
| Days Max Temp ≤ 32 °F | 17 | ***12*** | 5 | 0 | 0 | 0 | 0 | 0 | 0 | 1 | 7 | 15 | 57 |
| Days Min Temp ≤ 32 °F | 31 | 28 | 29 | 19 | ***6*** | 0 | 0 | 0 | 5 | 19 | 28 | 31 | 196 |
| Days Min Temp ≤ 0 °F | 13 | ***9*** | 3 | 0 | 0 | 0 | 0 | 0 | 0 | 0 | 3 | 9 | 37 |
| Heating Degree Days | 1561 | 1250 | 1011 | 604 | ***318*** | 101 | 25 | 47 | 261 | 601 | 1075 | 1446 | 8300 |
| Cooling Degree Days | 0 | 0 | 0 | 1 | ***14*** | 71 | 146 | 137 | 15 | 0 | 0 | 0 | 384 |
| Total Precipitation (") | 0.49 | 0.33 | 0.35 | 0.84 | 1.77 | 2.06 | 1.78 | 1.09 | 1.36 | 0.55 | 0.33 | 0.38 | 11.33 |
| Days ≥ 0.1" Precip | 2 | 1 | 1 | 2 | ***4*** | 5 | 4 | 3 | 3 | 2 | 1 | 1 | 29 |
| Total Snowfall (") | ***6.1*** | na | na | ***1.1*** | ***0.0*** | 0.0 | 0.0 | 0.0 | 0.0 | na | na | na | na |
| Days ≥ 1" Snow Depth | na | na | na | na | ***0*** | 0 | 0 | 0 | 0 | na | na | na | na |

## HARLOWTON *Wheatland County* ELEVATION 4167 ft LAT/LONG 46° 27 ' N / 109° 50 ' W

| | JAN | FEB | MAR | APR | MAY | JUN | JUL | AUG | SEP | OCT | NOV | DEC | YEAR |
|---|---|---|---|---|---|---|---|---|---|---|---|---|---|
| Maximum Temp °F | 34.8 | 40.5 | 47.1 | 56.8 | 66.6 | 75.3 | 82.8 | 82.1 | 71.0 | 60.2 | 44.0 | 36.2 | 58.1 |
| Minimum Temp °F | 12.4 | 15.8 | 20.9 | 28.7 | 37.3 | 45.2 | 49.6 | 48.1 | 39.1 | 31.3 | 21.5 | 14.7 | 30.4 |
| Mean Temp °F | 23.6 | 28.2 | 34.1 | 42.8 | 52.0 | 60.3 | 66.2 | 65.2 | 55.1 | 45.8 | 32.8 | 25.5 | 44.3 |
| Days Max Temp ≥ 90 °F | 0 | 0 | 0 | 0 | 0 | 2 | 5 | 6 | 1 | 0 | 0 | 0 | 14 |
| Days Max Temp ≤ 32 °F | 11 | 6 | 3 | 0 | 0 | 0 | 0 | 0 | 0 | 0 | 4 | 10 | 34 |
| Days Min Temp ≤ 32 °F | 28 | 25 | 28 | 21 | 8 | 1 | 0 | 0 | 5 | 16 | 26 | 28 | 186 |
| Days Min Temp ≤ 0 °F | 8 | 4 | 2 | 0 | 0 | 0 | 0 | 0 | 0 | 0 | 2 | 5 | 21 |
| Heating Degree Days | 1278 | 1034 | 952 | 660 | 398 | 169 | 46 | 67 | 298 | 589 | 961 | 1220 | 7672 |
| Cooling Degree Days | 0 | 0 | 0 | 0 | 4 | 41 | 94 | 81 | 9 | 1 | 0 | 0 | 230 |
| Total Precipitation (") | 0.51 | 0.41 | 0.71 | 1.23 | 2.28 | 2.67 | 1.61 | 1.53 | 1.27 | 0.83 | 0.52 | 0.43 | 14.00 |
| Days ≥ 0.1" Precip | 2 | 1 | 2 | 3 | 6 | 6 | 4 | 4 | 3 | 2 | 2 | 1 | 36 |
| Total Snowfall (") | 7.8 | 6.0 | 7.6 | 5.2 | 1.1 | 0.1 | 0.0 | 0.0 | 1.3 | 2.5 | 5.2 | 6.0 | 42.8 |
| Days ≥ 1" Snow Depth | 13 | 10 | 6 | 1 | 0 | 0 | 0 | 0 | 0 | 1 | 6 | 11 | 48 |

## HAVRE CITY-COUNTY AP *Hill County* ELEVATION 2585 ft LAT/LONG 48° 33 ' N / 109° 46 ' W

| | JAN | FEB | MAR | APR | MAY | JUN | JUL | AUG | SEP | OCT | NOV | DEC | YEAR |
|---|---|---|---|---|---|---|---|---|---|---|---|---|---|
| Maximum Temp °F | 24.3 | 31.5 | 43.4 | 57.0 | 68.2 | 76.8 | 84.3 | 83.4 | 71.2 | 59.0 | 40.7 | 29.6 | 55.8 |
| Minimum Temp °F | 3.5 | 9.2 | 20.4 | 31.1 | 41.5 | 49.3 | 53.6 | 52.4 | 41.9 | 31.0 | 17.7 | 7.9 | 30.0 |
| Mean Temp °F | 13.9 | 20.4 | 32.0 | 44.1 | 54.9 | 63.1 | 69.0 | 67.9 | 56.5 | 45.0 | 29.3 | 18.8 | 42.9 |
| Days Max Temp ≥ 90 °F | 0 | 0 | 0 | 0 | 0 | 4 | 9 | 9 | 2 | 0 | 0 | 0 | 24 |
| Days Max Temp ≤ 32 °F | 17 | 13 | 6 | 1 | 0 | 0 | 0 | 0 | 0 | 1 | 7 | 15 | 60 |
| Days Min Temp ≤ 32 °F | 30 | 27 | 27 | 16 | 3 | 0 | 0 | 0 | 3 | 17 | 27 | 30 | 180 |
| Days Min Temp ≤ 0 °F | 14 | 9 | 3 | 0 | 0 | 0 | 0 | 0 | 0 | 0 | 4 | 9 | 39 |
| Heating Degree Days | 1580 | 1254 | 1018 | 621 | 318 | 115 | 27 | 48 | 267 | 614 | 1066 | 1430 | 8358 |
| Cooling Degree Days | 0 | 0 | 0 | 1 | 14 | 64 | 139 | 133 | 16 | 1 | 0 | 0 | 368 |
| Total Precipitation (") | 0.54 | 0.37 | 0.68 | 0.96 | 1.63 | 1.86 | 1.57 | 1.29 | 1.11 | 0.55 | 0.43 | 0.52 | 11.51 |
| Days ≥ 0.1" Precip | 2 | 1 | 2 | 3 | 4 | 5 | 4 | 3 | 3 | 2 | 2 | 2 | 33 |
| Total Snowfall (") | 9.3 | 6.4 | 7.6 | 6.3 | 1.4 | 0.0 | 0.0 | 0.0 | 0.4 | 2.3 | 5.3 | 7.9 | 46.9 |
| Days ≥ 1" Snow Depth | 21 | 17 | 11 | 3 | 0 | 0 | 0 | 0 | 0 | 1 | 7 | 16 | 76 |

## HEBGEN DAM *Gallatin County* ELEVATION 6493 ft LAT/LONG 44° 52 ' N / 111° 20 ' W

| | JAN | FEB | MAR | APR | MAY | JUN | JUL | AUG | SEP | OCT | NOV | DEC | YEAR |
|---|---|---|---|---|---|---|---|---|---|---|---|---|---|
| Maximum Temp °F | 21.9 | 27.6 | 36.8 | 46.7 | 59.3 | 69.0 | 78.0 | 76.9 | 66.5 | 52.1 | 33.1 | 22.1 | 49.2 |
| Minimum Temp °F | 2.7 | 4.6 | 12.0 | 22.3 | 31.5 | 38.6 | 43.5 | 42.4 | 35.2 | 27.6 | 17.1 | 4.2 | 23.5 |
| Mean Temp °F | 12.3 | 16.1 | 24.4 | 34.5 | 45.4 | 53.8 | 60.8 | 59.7 | 50.9 | 39.9 | 25.1 | 13.2 | 36.3 |
| Days Max Temp ≥ 90 °F | 0 | 0 | 0 | 0 | 0 | 0 | 0 | 0 | 0 | 0 | 0 | 0 | 0 |
| Days Max Temp ≤ 32 °F | 27 | 20 | 8 | 1 | 0 | 0 | 0 | 0 | 0 | 1 | 14 | 27 | 98 |
| Days Min Temp ≤ 32 °F | 31 | 28 | 31 | 27 | 17 | 4 | 0 | 1 | 9 | 24 | 29 | 31 | 232 |
| Days Min Temp ≤ 0 °F | 13 | 11 | 6 | 0 | 0 | 0 | 0 | 0 | 0 | 0 | 2 | 12 | 44 |
| Heating Degree Days | 1628 | 1374 | 1251 | 907 | 601 | 330 | 134 | 164 | 417 | 773 | 1191 | 1601 | 10371 |
| Cooling Degree Days | 0 | 0 | 0 | 0 | 0 | 2 | 9 | 5 | 0 | 0 | 0 | 0 | 16 |
| Total Precipitation (") | 3.25 | 2.32 | 2.51 | 1.91 | 2.67 | 3.14 | 2.01 | 1.93 | 1.98 | 1.64 | 2.70 | 3.32 | 29.38 |
| Days ≥ 0.1" Precip | 10 | 8 | 8 | 6 | 8 | 8 | 6 | 5 | 5 | 4 | 8 | 10 | 86 |
| Total Snowfall (") | 47.8 | 33.7 | 28.9 | 10.3 | 3.0 | 0.4 | 0.0 | 0.0 | 0.5 | 4.9 | 28.9 | 49.6 | 208.0 |
| Days ≥ 1" Snow Depth | 31 | 28 | 31 | 24 | 4 | 0 | 0 | 0 | 0 | 3 | 22 | 31 | 174 |

## HELENA ARPT *Lewis and Clark County* ELEVATION 3828 ft LAT/LONG 46° 36 ' N / 112° 0 ' W

| | JAN | FEB | MAR | APR | MAY | JUN | JUL | AUG | SEP | OCT | NOV | DEC | YEAR |
|---|---|---|---|---|---|---|---|---|---|---|---|---|---|
| Maximum Temp °F | 30.1 | 36.8 | 45.7 | 56.3 | 66.0 | 74.7 | 83.3 | 82.1 | 70.3 | 58.1 | 41.9 | 31.4 | 56.4 |
| Minimum Temp °F | 10.2 | 15.3 | 22.9 | 31.0 | 39.9 | 47.6 | 52.6 | 51.0 | 41.2 | 31.3 | 20.5 | 11.4 | 31.2 |
| Mean Temp °F | 20.2 | 26.0 | 34.3 | 43.7 | 53.0 | 61.2 | 68.0 | 66.6 | 55.8 | 44.7 | 31.2 | 21.5 | 43.9 |
| Days Max Temp ≥ 90 °F | 0 | 0 | 0 | 0 | 0 | 2 | 8 | 7 | 1 | 0 | 0 | 0 | 18 |
| Days Max Temp ≤ 32 °F | 14 | 8 | 4 | 0 | 0 | 0 | 0 | 0 | 0 | 1 | 5 | 15 | 47 |
| Days Min Temp ≤ 32 °F | 29 | 27 | 27 | 18 | 3 | 0 | 0 | 0 | 3 | 17 | 27 | 30 | 181 |
| Days Min Temp ≤ 0 °F | 9 | 5 | 1 | 0 | 0 | 0 | 0 | 0 | 0 | 0 | 2 | 6 | 23 |
| Heating Degree Days | 1385 | 1094 | 944 | 633 | 369 | 150 | 36 | 53 | 283 | 622 | 1007 | 1344 | 7920 |
| Cooling Degree Days | 0 | 0 | 0 | 0 | 6 | 52 | 141 | 124 | 13 | 0 | 0 | 0 | 336 |
| Total Precipitation (") | 0.63 | 0.43 | 0.72 | 0.98 | 1.62 | 1.90 | 1.21 | 1.29 | 1.15 | 0.66 | 0.49 | 0.54 | 11.62 |
| Days ≥ 0.1" Precip | 2 | 2 | 3 | 3 | 5 | 5 | 3 | 4 | 3 | 2 | 2 | 2 | 36 |
| Total Snowfall (") | 9.0 | 6.0 | 7.1 | 5.2 | 1.4 | 0.1 | 0.0 | 0.2 | 1.5 | 2.9 | 5.3 | 8.1 | 46.8 |
| Days ≥ 1" Snow Depth | 19 | 12 | 8 | 2 | 0 | 0 | 0 | 0 | 0 | 1 | 6 | 14 | 62 |

## HERON 2 NW *Sanders County* ELEVATION 2280 ft LAT/LONG 48° 4 ' N / 115° 57 ' W

| | JAN | FEB | MAR | APR | MAY | JUN | JUL | AUG | SEP | OCT | NOV | DEC | YEAR |
|---|---|---|---|---|---|---|---|---|---|---|---|---|---|
| Maximum Temp °F | 31.6 | 37.5 | 46.0 | 57.0 | 66.9 | 73.5 | 81.0 | 80.5 | 69.3 | 55.3 | 39.4 | 32.0 | 55.8 |
| Minimum Temp °F | 19.4 | 22.1 | 26.0 | 31.8 | 37.9 | 43.8 | 46.2 | 45.4 | 39.4 | 32.4 | 27.8 | 21.3 | 32.8 |
| Mean Temp °F | 25.5 | 29.9 | 36.0 | 44.4 | 52.4 | 58.7 | 63.6 | 63.0 | 54.4 | 43.9 | 33.6 | 26.7 | 44.3 |
| Days Max Temp ≥ 90 °F | 0 | 0 | 0 | 0 | 0 | 1 | 6 | 5 | 0 | 0 | 0 | 0 | 12 |
| Days Max Temp ≤ 32 °F | 14 | 5 | 1 | 0 | 0 | 0 | 0 | 0 | 0 | 0 | 4 | 14 | 38 |
| Days Min Temp ≤ 32 °F | 30 | 26 | 26 | 17 | 7 | 1 | 0 | 1 | 4 | 16 | 22 | 29 | 179 |
| Days Min Temp ≤ 0 °F | 3 | 1 | 0 | 0 | 0 | 0 | 0 | 0 | 0 | 0 | 0 | 2 | 6 |
| Heating Degree Days | 1217 | 985 | 892 | 612 | 386 | 198 | 84 | 101 | 314 | 649 | 935 | 1181 | 7554 |
| Cooling Degree Days | 0 | 0 | 0 | 0 | 4 | 18 | 46 | 45 | 2 | 0 | 0 | 0 | 115 |
| Total Precipitation (") | 4.54 | 3.24 | 2.47 | 2.12 | 2.47 | 2.74 | 1.30 | 1.59 | 1.86 | 2.14 | 4.24 | 4.34 | 33.05 |
| Days ≥ 0.1" Precip | 11 | 9 | 8 | 7 | 7 | 7 | 4 | 4 | 5 | 6 | 11 | 12 | 91 |
| Total Snowfall (") | 29.0 | 16.4 | 6.2 | 1.4 | 0.1 | 0.0 | 0.0 | 0.0 | 0.0 | 0.5 | 9.2 | 24.9 | 87.7 |
| Days ≥ 1" Snow Depth | 30 | 27 | 20 | 1 | 0 | 0 | 0 | 0 | 0 | 0 | 9 | 26 | 113 |

## HOLTER DAM *Lewis and Clark County* ELEVATION 3491 ft LAT/LONG 47° 0 ' N / 112° 1 ' W

| | JAN | FEB | MAR | APR | MAY | JUN | JUL | AUG | SEP | OCT | NOV | DEC | YEAR |
|---|---|---|---|---|---|---|---|---|---|---|---|---|---|
| Maximum Temp °F | 34.9 | 40.1 | 47.0 | 56.9 | 67.1 | 75.9 | 84.3 | 83.7 | 71.6 | 59.8 | 45.1 | 37.1 | 58.6 |
| Minimum Temp °F | 18.5 | 22.2 | 27.4 | 34.8 | 42.7 | 50.4 | 54.7 | 53.7 | 46.1 | 40.3 | 30.4 | 22.4 | 37.0 |
| Mean Temp °F | 26.7 | 31.2 | 37.2 | 45.8 | 54.9 | 63.1 | 69.5 | 68.7 | 58.9 | 50.1 | 37.8 | 29.8 | 47.8 |
| Days Max Temp ≥ 90 °F | 0 | 0 | 0 | 0 | 0 | 2 | 8 | 9 | 1 | 0 | 0 | 0 | 20 |
| Days Max Temp ≤ 32 °F | 10 | 6 | 3 | 1 | 0 | 0 | 0 | 0 | 0 | 1 | 3 | 8 | 32 |
| Days Min Temp ≤ 32 °F | 24 | 20 | 20 | 11 | 2 | 0 | 0 | 0 | 1 | 6 | 15 | 22 | 121 |
| Days Min Temp ≤ 0 °F | 6 | 3 | 1 | 0 | 0 | 0 | 0 | 0 | 0 | 0 | 1 | 4 | 15 |
| Heating Degree Days | 1181 | 949 | 855 | 569 | 314 | 111 | 21 | 30 | 207 | 459 | 810 | 1086 | 6592 |
| Cooling Degree Days | 0 | 0 | 0 | 1 | 11 | 71 | 164 | 166 | 31 | 3 | 0 | 0 | 447 |
| Total Precipitation (") | 0.43 | 0.27 | 0.53 | 1.27 | 2.19 | 2.01 | 1.44 | 1.36 | 1.25 | 0.66 | 0.35 | 0.38 | 12.14 |
| Days ≥ 0.1" Precip | 2 | 1 | 2 | 4 | 5 | 5 | 3 | 4 | 4 | 2 | 1 | 1 | 34 |
| Total Snowfall (") | na | na | na | na | ***0.8*** | na | ***0.0*** | ***0.0*** | ***0.5*** | na | na | na | na |
| Days ≥ 1" Snow Depth | na | na | na | na | ***0*** | ***0*** | ***0*** | ***0*** | ***0*** | ***1*** | na | na | na |

## HUNGRY HORSE DAM *Flathead County* ELEVATION 3153 ft LAT/LONG 48° 23 ' N / 114° 3 ' W

| | JAN | FEB | MAR | APR | MAY | JUN | JUL | AUG | SEP | OCT | NOV | DEC | YEAR |
|---|---|---|---|---|---|---|---|---|---|---|---|---|---|
| Maximum Temp °F | 29.4 | 34.6 | 42.9 | 53.3 | 64.1 | 72.2 | 79.8 | 79.7 | 66.1 | 52.1 | 37.4 | 30.8 | 53.5 |
| Minimum Temp °F | 16.2 | 18.8 | 24.1 | 32.1 | 39.8 | 46.3 | 50.4 | 49.7 | 40.9 | 32.9 | 25.5 | 18.9 | 33.0 |
| Mean Temp °F | 22.8 | 26.7 | 33.5 | 42.8 | 52.0 | 59.2 | 65.1 | 64.7 | 53.5 | 42.5 | 31.4 | 24.8 | 43.3 |
| Days Max Temp ≥ 90 °F | 0 | 0 | 0 | 0 | 0 | 1 | 5 | 5 | 0 | 0 | 0 | 0 | 11 |
| Days Max Temp ≤ 32 °F | 16 | 9 | 3 | 0 | 0 | 0 | 0 | 0 | 0 | 1 | 7 | 17 | 53 |
| Days Min Temp ≤ 32 °F | 29 | 26 | 26 | 16 | 3 | 0 | 0 | 0 | 3 | 15 | 24 | 29 | 171 |
| Days Min Temp ≤ 0 °F | 5 | 3 | 1 | 0 | 0 | 0 | 0 | 0 | 0 | 0 | 1 | 2 | 12 |
| Heating Degree Days | 1301 | 1076 | 970 | 662 | 400 | 198 | 80 | 87 | 345 | 692 | 1000 | 1238 | 8049 |
| Cooling Degree Days | 0 | 0 | 0 | 0 | 8 | 38 | 92 | 90 | 5 | 0 | 0 | 0 | 233 |
| Total Precipitation (") | 3.63 | 2.46 | 2.23 | 2.26 | 3.01 | 3.53 | 2.12 | 1.93 | 2.76 | 2.75 | 3.82 | 3.59 | 34.09 |
| Days ≥ 0.1" Precip | 8 | 6 | 7 | 6 | 7 | 7 | 5 | 4 | 5 | 6 | 8 | 9 | 78 |
| Total Snowfall (") | ***20.9*** | ***13.1*** | ***8.0*** | na | 0.0 | 0.2 | 0.0 | 0.0 | 0.0 | na | na | na | na |
| Days ≥ 1" Snow Depth | 25 | 22 | 20 | ***4*** | 0 | 0 | 0 | 0 | 0 | na | ***10*** | 23 | na |

## HUNTLEY EXP STN *Yellowstone County* ELEVATION 2992 ft LAT/LONG 45° 55 ' N / 108° 15 ' W

| | JAN | FEB | MAR | APR | MAY | JUN | JUL | AUG | SEP | OCT | NOV | DEC | YEAR |
|---|---|---|---|---|---|---|---|---|---|---|---|---|---|
| Maximum Temp °F | 32.6 | 39.9 | 49.3 | 60.4 | 69.9 | 78.7 | 86.2 | 85.6 | 74.1 | 62.6 | 46.0 | 35.9 | 60.1 |
| Minimum Temp °F | 7.9 | 13.6 | 21.7 | 30.6 | 40.1 | 48.1 | 52.5 | 50.7 | 40.7 | 30.6 | 19.5 | 10.5 | 30.5 |
| Mean Temp °F | 20.3 | 26.8 | 35.5 | 45.6 | 55.0 | 63.4 | 69.4 | 68.2 | 57.4 | 46.6 | 32.8 | 23.2 | 45.4 |
| Days Max Temp ≥ 90 °F | 0 | 0 | 0 | 0 | 1 | 4 | 12 | 11 | 2 | 0 | 0 | 0 | 30 |
| Days Max Temp ≤ 32 °F | 12 | 8 | 4 | 0 | 0 | 0 | 0 | 0 | 0 | 0 | 4 | 11 | 39 |
| Days Min Temp ≤ 32 °F | 30 | 27 | 27 | 18 | 4 | 0 | 0 | 0 | 4 | 19 | 28 | 30 | 187 |
| Days Min Temp ≤ 0 °F | 10 | 5 | 2 | 0 | 0 | 0 | 0 | 0 | 0 | 0 | 2 | 7 | 26 |
| Heating Degree Days | 1380 | 1073 | 908 | 578 | 311 | 107 | 25 | 34 | 241 | 565 | 959 | 1290 | 7471 |
| Cooling Degree Days | 0 | 0 | 0 | 1 | 11 | 73 | 150 | 139 | 20 | 1 | 0 | 0 | 395 |
| Total Precipitation (") | 0.63 | 0.47 | 0.84 | 1.59 | 2.28 | 2.28 | 1.30 | 1.17 | 1.50 | 1.11 | 0.67 | 0.62 | 14.46 |
| Days ≥ 0.1" Precip | 2 | 2 | 3 | 5 | 6 | 5 | 3 | 3 | 4 | 3 | 2 | 2 | 40 |
| Total Snowfall (") | 9.8 | ***6.4*** | ***8.2*** | ***3.1*** | 0.5 | 0.0 | 0.0 | 0.0 | 0.6 | na | 6.7 | 9.5 | na |
| Days ≥ 1" Snow Depth | ***21*** | ***14*** | ***9*** | ***2*** | 0 | 0 | 0 | 0 | 0 | ***1*** | ***8*** | ***18*** | 73 |

## HYSHAM *Treasure County* ELEVATION 2651 ft LAT/LONG 46° 18 ' N / 107° 13 ' W

| | JAN | FEB | MAR | APR | MAY | JUN | JUL | AUG | SEP | OCT | NOV | DEC | YEAR |
|---|---|---|---|---|---|---|---|---|---|---|---|---|---|
| Maximum Temp °F | 32.3 | 39.5 | 50.5 | 62.2 | 72.2 | 81.1 | 88.4 | 87.5 | 76.2 | 64.1 | 45.9 | 35.4 | 61.3 |
| Minimum Temp °F | 8.2 | 13.4 | 22.8 | 32.3 | 41.9 | 50.3 | 55.2 | 53.3 | 43.0 | 32.7 | 21.1 | 11.7 | 32.2 |
| Mean Temp °F | 20.2 | 26.5 | 36.7 | 47.3 | 57.1 | 65.7 | 71.8 | 70.4 | 59.6 | 48.4 | 33.5 | 23.6 | 46.7 |
| Days Max Temp ≥ 90 °F | 0 | 0 | 0 | 0 | 2 | 6 | 16 | 14 | 4 | 0 | 0 | 0 | 42 |
| Days Max Temp ≤ 32 °F | 13 | 8 | 3 | 0 | 0 | 0 | 0 | 0 | 0 | 0 | 4 | 11 | 39 |
| Days Min Temp ≤ 32 °F | 30 | 26 | 26 | 15 | 3 | 0 | 0 | 0 | 3 | 15 | 26 | 30 | 174 |
| Days Min Temp ≤ 0 °F | 11 | 6 | 2 | 0 | 0 | 0 | 0 | 0 | 0 | 0 | 2 | 7 | 28 |
| Heating Degree Days | 1383 | 1082 | 871 | 526 | 256 | 69 | 10 | 20 | 188 | 508 | 939 | 1279 | 7131 |
| Cooling Degree Days | 0 | 0 | 0 | 2 | 23 | 112 | 223 | 197 | 30 | 2 | 0 | 0 | 589 |
| Total Precipitation (") | 0.54 | 0.34 | 0.75 | 1.38 | 2.38 | 2.16 | 1.44 | 0.88 | 1.42 | 1.10 | 0.53 | 0.54 | 13.46 |
| Days ≥ 0.1" Precip | 2 | 1 | 3 | 4 | 5 | 6 | 3 | 3 | 3 | 3 | 2 | 2 | 37 |
| Total Snowfall (") | ***9.4*** | na | na | ***3.4*** | ***1.0*** | 0.0 | 0.0 | 0.0 | ***0.4*** | ***1.6*** | ***6.3*** | 7.6 | na |
| Days ≥ 1" Snow Depth | na | na | na | na | ***0*** | ***0*** | 0 | ***0*** | ***0*** | ***0*** | na | na | na |

## HYSHAM 25 SSE *Treasure County* ELEVATION 3104 ft LAT/LONG 45° 54 ' N / 107° 8 ' W

| | JAN | FEB | MAR | APR | MAY | JUN | JUL | AUG | SEP | OCT | NOV | DEC | YEAR |
|---|---|---|---|---|---|---|---|---|---|---|---|---|---|
| Maximum Temp °F | 30.6 | 37.7 | 46.3 | 57.6 | 68.2 | 77.8 | 87.4 | ***86.1*** | 73.7 | 61.8 | 44.5 | 34.5 | 58.9 |
| Minimum Temp °F | 4.4 | 11.1 | 20.8 | 30.1 | 38.5 | 47.0 | 52.5 | ***49.5*** | 39.5 | 29.2 | 17.8 | 8.4 | 29.1 |
| Mean Temp °F | 17.5 | 24.4 | 33.6 | 43.9 | 53.4 | 62.4 | 70.0 | ***67.8*** | 56.6 | 45.5 | 31.2 | 21.4 | 44.0 |
| Days Max Temp ≥ 90 °F | 0 | 0 | 0 | 0 | 1 | 4 | 13 | ***12*** | 3 | 0 | 0 | 0 | 33 |
| Days Max Temp ≤ 32 °F | 14 | 8 | 4 | 1 | 0 | 0 | 0 | 0 | 0 | 1 | ***6*** | 11 | 45 |
| Days Min Temp ≤ 32 °F | 30 | 27 | 28 | ***18*** | 7 | 1 | 0 | ***0*** | 6 | 20 | 28 | 30 | 195 |
| Days Min Temp ≤ 0 °F | 12 | 7 | 2 | 0 | 0 | 0 | 0 | 0 | 0 | 0 | 3 | 8 | 32 |
| Heating Degree Days | 1467 | 1139 | 966 | 628 | 362 | 133 | 26 | ***54*** | 272 | 598 | 1009 | 1345 | 7999 |
| Cooling Degree Days | 0 | 0 | 0 | 0 | 12 | 81 | 178 | 169 | 25 | 1 | 0 | 0 | 466 |
| Total Precipitation (") | 0.60 | 0.44 | 0.81 | 1.61 | 2.24 | 2.70 | 1.33 | 0.84 | 1.46 | 1.11 | 0.70 | 0.59 | 14.43 |
| Days ≥ 0.1" Precip | 2 | 2 | 2 | 4 | 6 | 7 | 3 | ***2*** | ***4*** | 3 | 3 | 2 | 40 |
| Total Snowfall (") | 11.1 | 7.2 | 8.5 | ***6.8*** | 1.5 | 0.0 | 0.0 | 0.0 | ***1.0*** | 2.4 | 7.1 | 11.0 | 56.6 |
| Days ≥ 1" Snow Depth | 23 | 16 | 9 | ***3*** | 0 | 0 | 0 | 0 | 0 | 1 | 7 | 18 | 77 |

## ILIAD *Chouteau County* ELEVATION 2953 ft LAT/LONG 47° 48 ' N / 109° 47 ' W

| | JAN | FEB | MAR | APR | MAY | JUN | JUL | AUG | SEP | OCT | NOV | DEC | YEAR |
|---|---|---|---|---|---|---|---|---|---|---|---|---|---|
| Maximum Temp °F | 28.5 | 35.8 | 46.9 | 59.7 | ***69.6*** | 77.7 | 85.8 | 85.0 | 73.2 | 61.4 | 43.1 | 32.4 | 58.3 |
| Minimum Temp °F | 5.8 | 11.1 | 21.5 | 31.2 | ***40.1*** | 47.5 | 51.9 | 50.3 | 39.9 | 30.1 | 17.7 | 8.4 | 29.6 |
| Mean Temp °F | 17.4 | 23.5 | 34.2 | 45.3 | ***54.8*** | 62.6 | 69.0 | 67.7 | 56.6 | 45.8 | 30.4 | 20.5 | 44.0 |
| Days Max Temp ≥ 90 °F | 0 | 0 | 0 | 0 | 0 | 3 | ***10*** | 10 | 2 | 0 | 0 | 0 | 25 |
| Days Max Temp ≤ 32 °F | 15 | 10 | 4 | 0 | 0 | 0 | 0 | 0 | 0 | 1 | 6 | 13 | 49 |
| Days Min Temp ≤ 32 °F | 30 | 27 | 27 | 17 | ***4*** | 0 | 0 | 0 | 5 | ***17*** | 27 | 30 | 184 |
| Days Min Temp ≤ 0 °F | 11 | 7 | 2 | 0 | 0 | 0 | 0 | 0 | 0 | 0 | 3 | 8 | 31 |
| Heating Degree Days | 1471 | 1168 | 947 | 585 | ***321*** | 122 | 29 | 47 | 264 | 591 | 1031 | 1380 | 7956 |
| Cooling Degree Days | 0 | 0 | 0 | 1 | 12 | 66 | 150 | 139 | 20 | 1 | 0 | 0 | 389 |
| Total Precipitation (") | 0.66 | ***0.44*** | 0.64 | 1.12 | 2.13 | 2.31 | 1.37 | 1.42 | 1.15 | 0.70 | 0.46 | 0.44 | 12.84 |
| Days ≥ 0.1" Precip | na | na | ***2*** | ***3*** | 5 | 5 | 3 | 3 | 3 | 2 | ***2*** | ***1*** | na |
| Total Snowfall (") | na | na | na | na | ***0.5*** | ***0.0*** | ***0.0*** | ***0.0*** | ***0.0*** | na | na | na | na |
| Days ≥ 1" Snow Depth | na | na | na | na | ***0*** | ***0*** | ***0*** | ***0*** | ***0*** | na | na | na | na |

## INGOMAR 14 NE *Rosebud County* ELEVATION 2782 ft LAT/LONG 46° 42 ' N / 107° 13 ' W

| | JAN | FEB | MAR | APR | MAY | JUN | JUL | AUG | SEP | OCT | NOV | DEC | YEAR |
|---|---|---|---|---|---|---|---|---|---|---|---|---|---|
| Maximum Temp °F | 29.6 | 36.9 | 48.0 | 60.4 | 71.1 | 80.0 | 88.1 | 87.5 | 75.6 | 63.0 | 44.7 | 33.8 | 59.9 |
| Minimum Temp °F | 3.3 | 9.6 | 20.2 | 30.2 | 40.4 | 49.3 | 54.5 | 52.2 | 40.9 | 29.9 | 17.2 | 6.5 | 29.5 |
| Mean Temp °F | 16.5 | 23.2 | 34.1 | 45.3 | 55.8 | 64.6 | 71.4 | 69.9 | 58.3 | 46.5 | 31.0 | 20.2 | 44.7 |
| Days Max Temp ≥ 90 °F | 0 | 0 | 0 | 0 | 1 | 5 | 15 | 14 | 3 | 0 | 0 | 0 | 38 |
| Days Max Temp ≤ 32 °F | 15 | 10 | 4 | 0 | 0 | 0 | 0 | 0 | 0 | 0 | 5 | 13 | 47 |
| Days Min Temp ≤ 32 °F | 30 | 27 | 28 | 18 | 4 | 0 | 0 | 0 | 4 | 19 | 28 | 30 | 188 |
| Days Min Temp ≤ 0 °F | 13 | 8 | 2 | 0 | 0 | 0 | 0 | 0 | 0 | 0 | 3 | 10 | 36 |
| Heating Degree Days | 1501 | 1174 | 951 | 586 | 295 | 89 | 13 | 27 | 222 | 568 | 1016 | 1385 | 7827 |
| Cooling Degree Days | 0 | 0 | 0 | 1 | 17 | 98 | 210 | 188 | 27 | 0 | 0 | 0 | 541 |
| Total Precipitation (") | 0.38 | 0.23 | 0.40 | 0.98 | 2.07 | 2.33 | 1.53 | 0.90 | 1.27 | 0.71 | 0.35 | 0.45 | 11.60 |
| Days ≥ 0.1" Precip | 1 | 1 | 1 | 3 | 5 | 6 | 3 | 2 | 3 | 2 | 1 | 2 | 30 |
| Total Snowfall (") | 8.8 | 4.8 | 5.2 | 3.6 | 1.2 | 0.0 | 0.0 | 0.0 | 0.5 | 1.5 | 5.2 | 8.2 | 39.0 |
| Days ≥ 1" Snow Depth | 19 | 15 | 7 | 1 | 0 | 0 | 0 | 0 | 0 | 1 | 6 | 15 | 64 |

## JOLIET *Carbon County* ELEVATION 3704 ft LAT/LONG 45° 29 ' N / 108° 58 ' W

| | JAN | FEB | MAR | APR | MAY | JUN | JUL | AUG | SEP | OCT | NOV | DEC | YEAR |
|---|---|---|---|---|---|---|---|---|---|---|---|---|---|
| Maximum Temp °F | 35.2 | 41.8 | 50.0 | 59.9 | 69.7 | 78.8 | 86.6 | 85.1 | 74.1 | 62.4 | 45.7 | 37.5 | 60.6 |
| Minimum Temp °F | 11.1 | 16.4 | 22.6 | 30.7 | 39.1 | 46.6 | 52.0 | 50.0 | 41.0 | 32.6 | 21.7 | 14.0 | 31.5 |
| Mean Temp °F | 23.2 | 29.2 | 36.3 | 45.3 | 54.4 | 62.7 | 69.3 | 67.6 | 57.6 | 47.6 | 33.7 | 25.8 | 46.1 |
| Days Max Temp ≥ 90 °F | 0 | 0 | 0 | 0 | 1 | 4 | 12 | 10 | 2 | 0 | 0 | 0 | 29 |
| Days Max Temp ≤ 32 °F | 11 | 6 | 2 | 0 | 0 | 0 | 0 | 0 | 0 | 0 | 4 | 9 | 32 |
| Days Min Temp ≤ 32 °F | 29 | 26 | 27 | 18 | 5 | 0 | 0 | 0 | 4 | 14 | 25 | 29 | 177 |
| Days Min Temp ≤ 0 °F | 8 | 4 | 1 | 0 | 0 | 0 | 0 | 0 | 0 | 0 | 2 | 5 | 20 |
| Heating Degree Days | 1291 | 1004 | 883 | 585 | 328 | 118 | 20 | 37 | 236 | 535 | 934 | 1210 | 7181 |
| Cooling Degree Days | 0 | 0 | 0 | 1 | 10 | 69 | 165 | 134 | 20 | 0 | 0 | 0 | 399 |
| Total Precipitation (") | 0.73 | 0.61 | 1.23 | 1.96 | 3.07 | 2.04 | 1.07 | 1.14 | 1.53 | 1.46 | 0.74 | 0.67 | 16.25 |
| Days ≥ 0.1" Precip | 3 | 2 | 4 | 5 | 6 | 5 | 3 | 3 | 4 | 3 | 3 | 3 | 44 |
| Total Snowfall (") | 11.6 | 7.6 | 10.0 | ***6.0*** | 0.6 | 0.0 | 0.0 | 0.0 | 1.0 | 3.6 | 8.3 | 10.5 | 59.2 |
| Days ≥ 1" Snow Depth | 24 | 15 | 9 | 2 | 0 | 0 | 0 | 0 | 0 | 1 | 8 | 19 | 78 |

## JOPLIN *Liberty County* ELEVATION 3300 ft LAT/LONG 48° 34 ' N / 110° 46 ' W

| | JAN | FEB | MAR | APR | MAY | JUN | JUL | AUG | SEP | OCT | NOV | DEC | YEAR |
|---|---|---|---|---|---|---|---|---|---|---|---|---|---|
| Maximum Temp °F | 25.9 | 32.3 | 42.8 | 55.3 | 66.7 | 74.3 | 81.9 | 81.4 | 69.6 | 58.2 | 39.5 | 29.8 | 54.8 |
| Minimum Temp °F | 4.4 | 9.5 | 19.4 | 29.0 | 38.8 | 46.2 | 50.5 | 49.7 | 39.7 | 30.6 | 17.2 | 8.8 | 28.7 |
| Mean Temp °F | 15.2 | 20.9 | 31.1 | 42.2 | 52.8 | 60.3 | 66.2 | 65.6 | 54.7 | 44.4 | 28.4 | 19.3 | 41.8 |
| Days Max Temp ≥ 90 °F | 0 | 0 | 0 | 0 | 0 | 2 | na | 6 | 1 | 0 | 0 | 0 | na |
| Days Max Temp ≤ 32 °F | 17 | 12 | 5 | 1 | 0 | 0 | 0 | 0 | 0 | 1 | 7 | 15 | 58 |
| Days Min Temp ≤ 32 °F | 30 | 28 | 29 | 20 | 6 | 0 | 0 | 0 | 6 | *17* | 27 | 30 | 193 |
| Days Min Temp ≤ 0 °F | 13 | 8 | 2 | 0 | 0 | 0 | 0 | 0 | 0 | 0 | 3 | 8 | 34 |
| Heating Degree Days | 1541 | 1239 | 1043 | 679 | 379 | 173 | 56 | 75 | 316 | 633 | 1086 | 1412 | 8632 |
| Cooling Degree Days | 0 | 0 | 0 | 1 | 11 | 46 | 95 | 103 | 12 | 1 | 0 | 0 | 269 |
| Total Precipitation (") | 0.29 | 0.19 | 0.42 | 0.81 | 1.96 | 2.10 | 1.40 | 1.31 | 0.92 | 0.41 | 0.27 | 0.24 | 10.32 |
| Days ≥ 0.1" Precip | 1 | 1 | 1 | 2 | 5 | 6 | 4 | 4 | 3 | 2 | 1 | 1 | 31 |
| Total Snowfall (") | *4.3* | *2.8* | *2.5* | *2.5* | 0.3 | 0.0 | 0.0 | 0.0 | 0.2 | *0.6* | na | 3.9 | na |
| Days ≥ 1" Snow Depth | *17* | na | na | 2 | 0 | 0 | 0 | 0 | 0 | 1 | *4* | *13* | na |

## JORDAN *Garfield County* ELEVATION 2592 ft LAT/LONG 47° 19 ' N / 106° 54 ' W

| | JAN | FEB | MAR | APR | MAY | JUN | JUL | AUG | SEP | OCT | NOV | DEC | YEAR |
|---|---|---|---|---|---|---|---|---|---|---|---|---|---|
| Maximum Temp °F | 28.2 | 35.9 | 47.7 | 60.5 | 71.8 | 81.4 | 88.8 | 88.2 | 75.8 | 62.7 | 44.3 | 32.6 | 59.8 |
| Minimum Temp °F | 3.0 | 9.2 | 19.8 | 30.3 | 40.5 | 49.6 | 54.4 | 52.4 | 41.4 | 30.6 | 18.2 | 7.3 | 29.7 |
| Mean Temp °F | 15.6 | 22.6 | 33.8 | 45.4 | 56.2 | 65.5 | 71.6 | 70.3 | 58.6 | 46.7 | 31.3 | 20.0 | 44.8 |
| Days Max Temp ≥ 90 °F | 0 | 0 | 0 | 0 | 2 | 6 | 16 | 15 | 4 | 0 | 0 | 0 | 43 |
| Days Max Temp ≤ 32 °F | 16 | 11 | 4 | 0 | 0 | 0 | 0 | 0 | 0 | 0 | 5 | 13 | 49 |
| Days Min Temp ≤ 32 °F | 30 | 27 | 27 | 18 | 5 | 0 | 0 | 0 | 4 | 18 | 27 | 30 | 186 |
| Days Min Temp ≤ 0 °F | 13 | 8 | 2 | 0 | 0 | 0 | 0 | 0 | 0 | 0 | 3 | 9 | 35 |
| Heating Degree Days | 1527 | 1192 | 962 | 582 | 285 | 76 | 13 | 27 | 219 | 562 | 1004 | 1390 | 7839 |
| Cooling Degree Days | 0 | 0 | 0 | 1 | 27 | 117 | 229 | 215 | 34 | 1 | 0 | 0 | 624 |
| Total Precipitation (") | 0.54 | 0.34 | 0.57 | 1.19 | 2.12 | 2.33 | 1.89 | 1.15 | 1.28 | 0.85 | 0.37 | 0.51 | 13.14 |
| Days ≥ 0.1" Precip | 2 | 1 | 2 | 4 | 5 | 5 | 4 | 3 | 3 | 2 | 1 | 2 | 34 |
| Total Snowfall (") | na | na | na | na | *0.0* | *0.0* | *0.0* | *0.0* | *0.0* | na | na | na | na |
| Days ≥ 1" Snow Depth | na | na | na | na | *0* | *0* | *0* | *0* | *0* | na | na | na | na |

## JUDITH GAP 13 E *Wheatland County* ELEVATION 5105 ft LAT/LONG 46° 41 ' N / 109° 29 ' W

| | JAN | FEB | MAR | APR | MAY | JUN | JUL | AUG | SEP | OCT | NOV | DEC | YEAR |
|---|---|---|---|---|---|---|---|---|---|---|---|---|---|
| Maximum Temp °F | 31.8 | 36.8 | 42.8 | 52.4 | 62.1 | 70.1 | 77.2 | 76.7 | 65.6 | 56.0 | 41.1 | 33.5 | 53.8 |
| Minimum Temp °F | 8.8 | 13.2 | 18.6 | 26.3 | 34.8 | 42.0 | 46.8 | 46.1 | 37.6 | 29.1 | 18.2 | 11.3 | 27.7 |
| Mean Temp °F | 20.3 | 25.0 | 30.7 | 39.4 | 48.5 | 56.1 | 62.0 | 61.4 | 51.6 | 42.6 | 29.7 | 22.5 | 40.8 |
| Days Max Temp ≥ 90 °F | 0 | 0 | 0 | 0 | 0 | 0 | 1 | 1 | 0 | 0 | 0 | 0 | 2 |
| Days Max Temp ≤ 32 °F | *13* | 8 | 5 | 1 | 0 | 0 | 0 | 0 | 0 | 1 | *6* | 13 | 47 |
| Days Min Temp ≤ 32 °F | 30 | 27 | 30 | 23 | 12 | 2 | 0 | 1 | 7 | *20* | 27 | 30 | 209 |
| Days Min Temp ≤ 0 °F | 9 | 5 | 2 | 0 | 0 | 0 | 0 | 0 | 0 | 0 | 2 | 6 | 24 |
| Heating Degree Days | 1380 | 1123 | 1056 | 762 | 506 | 272 | 120 | 140 | 397 | 687 | 1053 | 1314 | 8810 |
| Cooling Degree Days | 0 | 0 | 0 | 0 | 0 | 14 | 34 | 31 | 3 | 0 | 0 | 0 | 82 |
| Total Precipitation (") | 0.76 | 0.46 | 0.80 | 1.36 | 2.46 | 2.92 | 2.04 | 1.63 | 1.38 | 0.84 | 0.51 | 0.63 | 15.79 |
| Days ≥ 0.1" Precip | 3 | 2 | 3 | 4 | 6 | 7 | 5 | 4 | 4 | 3 | 2 | 3 | 46 |
| Total Snowfall (") | na | na | na | na | na | na | *0.0* | *0.0* | na | na | na | na | na |
| Days ≥ 1" Snow Depth | na | na | na | na | na | *0* | *0* | *0* | na | na | na | na | na |

## KALISPELL GLACIER AP *Flathead County* ELEVATION 2965 ft LAT/LONG 48° 18 ' N / 114° 16 ' W

| | JAN | FEB | MAR | APR | MAY | JUN | JUL | AUG | SEP | OCT | NOV | DEC | YEAR |
|---|---|---|---|---|---|---|---|---|---|---|---|---|---|
| Maximum Temp °F | 28.7 | 34.8 | 44.4 | 55.7 | 65.0 | 72.2 | 80.6 | 80.6 | 68.8 | 55.3 | 38.6 | 30.1 | 54.6 |
| Minimum Temp °F | 13.4 | 17.7 | 24.3 | 31.4 | 38.7 | 45.0 | 48.2 | 47.4 | 38.5 | 29.2 | 23.5 | 15.8 | 31.1 |
| Mean Temp °F | 21.0 | 26.3 | 34.4 | 43.6 | 51.9 | 58.6 | 64.5 | 64.1 | 53.6 | 42.3 | 31.1 | 23.0 | 42.9 |
| Days Max Temp ≥ 90 °F | 0 | 0 | 0 | 0 | 0 | 1 | 6 | 6 | 0 | 0 | 0 | 0 | 13 |
| Days Max Temp ≤ 32 °F | 17 | 9 | 3 | 0 | 0 | 0 | 0 | 0 | 0 | 1 | 6 | 17 | 53 |
| Days Min Temp ≤ 32 °F | 29 | 26 | 27 | 18 | 6 | 1 | 0 | 0 | 5 | 21 | 26 | 29 | 188 |
| Days Min Temp ≤ 0 °F | 6 | 3 | 1 | 0 | 0 | 0 | 0 | 0 | 0 | 0 | 1 | 4 | 15 |
| Heating Degree Days | 1357 | 1087 | 943 | 636 | 403 | 204 | 78 | 89 | 338 | 697 | 1011 | 1295 | 8138 |
| Cooling Degree Days | 0 | 0 | 0 | 0 | 6 | 20 | 58 | 64 | 3 | 0 | 0 | 0 | 151 |
| Total Precipitation (") | 1.57 | 1.08 | 0.94 | 1.13 | 1.84 | 2.34 | 1.29 | 1.36 | 1.20 | 0.91 | 1.34 | 1.62 | 16.62 |
| Days ≥ 0.1" Precip | 6 | 4 | 4 | 3 | 6 | 6 | 4 | 4 | 4 | 3 | 5 | 6 | 55 |
| Total Snowfall (") | 17.0 | 10.0 | 5.6 | 2.5 | 0.4 | 0.0 | 0.0 | 0.0 | 0.1 | 1.4 | 8.1 | 16.9 | 62.0 |
| Days ≥ 1" Snow Depth | 25 | 21 | 10 | 1 | 0 | 0 | 0 | 0 | 0 | 1 | 7 | 19 | 84 |

## LAKEVIEW *Beaverhead County* ELEVATION 6706 ft LAT/LONG 44° 36 ' N / 111° 48 ' W

| | JAN | FEB | MAR | APR | MAY | JUN | JUL | AUG | SEP | OCT | NOV | DEC | YEAR |
|---|---|---|---|---|---|---|---|---|---|---|---|---|---|
| Maximum Temp °F | 23.4 | 28.1 | 35.3 | 46.4 | 57.9 | 67.5 | 76.0 | 75.3 | 65.8 | 53.0 | 35.4 | 24.5 | 49.1 |
| Minimum Temp °F | 0.5 | 3.2 | 10.2 | 21.5 | 30.0 | 36.3 | 40.8 | 39.5 | 31.8 | 23.4 | 13.5 | 2.1 | 21.1 |
| Mean Temp °F | 12.0 | 15.7 | 22.8 | 34.0 | 44.0 | 51.9 | 58.4 | 57.4 | 48.7 | 38.2 | 24.5 | 13.3 | 35.1 |
| Days Max Temp ≥ 90 °F | 0 | 0 | 0 | 0 | 0 | 0 | 0 | 0 | 0 | 0 | 0 | 0 | 0 |
| Days Max Temp ≤ 32 °F | 24 | 17 | 9 | 2 | 0 | 0 | 0 | 0 | 0 | 1 | 11 | 21 | 85 |
| Days Min Temp ≤ 32 °F | 30 | 27 | 30 | 27 | 19 | 8 | 2 | 3 | 15 | 26 | 28 | 29 | 244 |
| Days Min Temp ≤ 0 °F | 14 | 11 | 7 | 1 | 0 | 0 | 0 | 0 | 0 | 0 | 5 | 12 | 50 |
| Heating Degree Days | 1640 | 1388 | 1301 | 923 | 645 | 387 | 200 | 231 | 481 | 824 | 1208 | 1597 | 10825 |
| Cooling Degree Days | 0 | *0* | 0 | 0 | *0* | 2 | 4 | *2* | *0* | *0* | *0* | *0* | 8 |
| Total Precipitation (") | 1.33 | 0.87 | 1.78 | 1.63 | 2.37 | 2.88 | 1.84 | 1.57 | 1.78 | 1.32 | 1.36 | 1.24 | 19.97 |
| Days ≥ 0.1" Precip | 3 | 3 | 5 | 5 | 6 | 7 | 5 | 4 | 4 | 3 | 4 | 4 | 53 |
| Total Snowfall (") | na | *10.0* | 16.4 | *9.7* | 4.5 | 0.9 | 0.0 | 0.0 | 2.2 | *3.6* | na | na | na |
| Days ≥ 1" Snow Depth | 29 | 26 | 29 | 20 | 3 | 0 | 0 | 0 | 0 | na | *17* | 27 | na |

## LENNEP 5 SW *Meagher County* ELEVATION 5882 ft LAT/LONG 46° 24 ' N / 110° 41 ' W

| | JAN | FEB | MAR | APR | MAY | JUN | JUL | AUG | SEP | OCT | NOV | DEC | YEAR |
|---|---|---|---|---|---|---|---|---|---|---|---|---|---|
| Maximum Temp °F | 29.7 | 33.1 | 39.2 | 49.0 | 59.2 | 67.9 | 75.8 | 75.9 | 65.1 | 53.5 | 37.4 | 29.9 | 51.3 |
| Minimum Temp °F | 11.5 | 13.7 | 18.5 | 25.6 | 32.9 | 40.3 | 44.4 | 43.7 | 36.2 | 28.6 | 18.8 | 11.9 | 27.2 |
| Mean Temp °F | 20.6 | 23.4 | 28.9 | 37.4 | 46.1 | 54.1 | 60.1 | 59.8 | 50.7 | 41.1 | 28.2 | 21.0 | 39.3 |
| Days Max Temp ≥ 90 °F | 0 | 0 | 0 | 0 | 0 | 0 | 0 | 0 | 0 | 0 | 0 | 0 | 0 |
| Days Max Temp ≤ 32 °F | 17 | 12 | 7 | 2 | 0 | 0 | 0 | 0 | 0 | 1 | 9 | 17 | 65 |
| Days Min Temp ≤ 32 °F | 30 | 28 | 29 | 25 | 15 | 3 | 1 | 1 | 9 | 21 | 28 | 30 | 220 |
| Days Min Temp ≤ 0 °F | 7 | 4 | 2 | 0 | 0 | 0 | 0 | 0 | 0 | 0 | 2 | 5 | 20 |
| Heating Degree Days | 1371 | 1169 | 1112 | 823 | 580 | 324 | 162 | 173 | 423 | 736 | 1100 | 1361 | 9334 |
| Cooling Degree Days | 0 | 0 | 0 | 0 | 0 | 5 | 21 | 16 | 1 | 0 | 0 | 0 | 43 |
| Total Precipitation (") | 0.99 | 0.67 | 0.96 | 1.25 | 2.30 | 2.62 | 1.73 | 1.51 | 1.46 | 1.15 | 0.87 | 0.82 | 16.33 |
| Days ≥ 0.1" Precip | 3 | 3 | 4 | 5 | 7 | 8 | 5 | 5 | 5 | 4 | 3 | 4 | 56 |
| Total Snowfall (") | na | na | na | na | na | *0.2* | *0.0* | *0.0* | *0.4* | na | na | na | na |
| Days ≥ 1" Snow Depth | na | na | na | na | na | *0* | *0* | *0* | *0* | na | na | na | na |

## LEWISTOWN MUNI AP *Fergus County* ELEVATION 4132 ft LAT/LONG 47° 3 ' N / 109° 27 ' W

| | JAN | FEB | MAR | APR | MAY | JUN | JUL | AUG | SEP | OCT | NOV | DEC | YEAR |
|---|---|---|---|---|---|---|---|---|---|---|---|---|---|
| Maximum Temp °F | 30.6 | 35.5 | 42.0 | 53.1 | 63.0 | 71.6 | 80.1 | 80.2 | 68.6 | 57.6 | 42.7 | 34.2 | 54.9 |
| Minimum Temp °F | 8.9 | 13.8 | 20.4 | 29.1 | 37.5 | 45.0 | 49.4 | 49.0 | 40.1 | 31.3 | 20.3 | 12.4 | 29.8 |
| Mean Temp °F | 19.8 | 24.7 | 31.2 | 41.1 | 50.3 | 58.3 | 64.8 | 64.6 | 54.4 | 44.5 | 31.5 | 23.3 | 42.4 |
| Days Max Temp ≥ 90 °F | 0 | 0 | 0 | 0 | 0 | 1 | 4 | 5 | 1 | 0 | 0 | 0 | 11 |
| Days Max Temp ≤ 32 °F | 14 | 10 | 6 | 2 | 0 | 0 | 0 | 0 | 0 | 1 | 6 | 12 | 51 |
| Days Min Temp ≤ 32 °F | 29 | 26 | 28 | 20 | 7 | 1 | 0 | 0 | 4 | 16 | 26 | 28 | 185 |
| Days Min Temp ≤ 0 °F | 10 | 6 | 2 | 0 | 0 | 0 | 0 | 0 | 0 | 0 | 3 | 7 | 28 |
| Heating Degree Days | 1397 | 1132 | 1040 | 709 | 451 | 214 | 74 | 88 | 323 | 630 | 997 | 1286 | 8341 |
| Cooling Degree Days | 0 | 0 | 0 | 0 | 2 | 29 | 80 | 85 | 11 | 1 | 0 | 0 | 208 |
| Total Precipitation (") | 1.01 | 0.63 | 1.04 | 1.34 | 2.95 | 2.97 | 1.89 | 1.81 | 1.51 | 1.12 | 0.76 | 0.85 | 17.88 |
| Days ≥ 0.1" Precip | 4 | 2 | 4 | 4 | 7 | 7 | 5 | 4 | 4 | 3 | 3 | 3 | 50 |
| Total Snowfall (") | 11.9 | 7.1 | 10.1 | 9.4 | 5.0 | 0.1 | 0.0 | 0.0 | 1.1 | 4.8 | 6.6 | 10.3 | 66.4 |
| Days ≥ 1" Snow Depth | 25 | 20 | 17 | 6 | 2 | 0 | 0 | 0 | 0 | 3 | 9 | 19 | 101 |

## LIBBY 1 NE RS *Lincoln County* ELEVATION 2096 ft LAT/LONG 48° 24 ' N / 115° 32 ' W

| | JAN | FEB | MAR | APR | MAY | JUN | JUL | AUG | SEP | OCT | NOV | DEC | YEAR |
|---|---|---|---|---|---|---|---|---|---|---|---|---|---|
| Maximum Temp °F | 31.9 | 40.6 | 50.7 | 61.2 | 71.2 | 78.6 | 86.2 | 86.6 | 74.3 | 58.4 | 40.6 | 31.8 | 59.3 |
| Minimum Temp °F | 17.6 | 22.1 | 25.7 | 31.0 | 37.8 | 44.3 | 47.2 | 46.3 | 39.2 | 32.8 | 26.9 | 19.9 | 32.6 |
| Mean Temp °F | 24.8 | 31.4 | 38.2 | 46.1 | 54.5 | 61.5 | 66.7 | 66.5 | 56.7 | 45.6 | 33.8 | 25.8 | 46.0 |
| Days Max Temp ≥ 90 °F | 0 | 0 | 0 | 0 | 1 | 5 | 13 | 13 | 2 | 0 | 0 | 0 | 34 |
| Days Max Temp ≤ 32 °F | 13 | 4 | 1 | 0 | 0 | 0 | 0 | 0 | 0 | 0 | 4 | 15 | 37 |
| Days Min Temp ≤ 32 °F | 30 | 26 | 26 | 18 | 8 | 1 | 0 | 0 | 5 | 15 | 23 | 30 | 182 |
| Days Min Temp ≤ 0 °F | 4 | 1 | 0 | 0 | 0 | 0 | 0 | 0 | 0 | 0 | 0 | 2 | 7 |
| Heating Degree Days | 1241 | 942 | 822 | 560 | 325 | 142 | 49 | 53 | 252 | 594 | 929 | 1208 | 7117 |
| Cooling Degree Days | 0 | 0 | 0 | 0 | 11 | 51 | 107 | 111 | 9 | 0 | 0 | 0 | 289 |
| Total Precipitation (") | 2.14 | 1.38 | 1.22 | 1.02 | 1.48 | 1.75 | 1.19 | 1.03 | 1.11 | 1.31 | 2.22 | 2.21 | 18.06 |
| Days ≥ 0.1" Precip | 6 | 4 | 5 | 4 | 5 | 5 | 3 | 3 | 4 | 4 | 7 | 7 | 57 |
| Total Snowfall (") | 16.4 | *7.4* | na | na | 0.0 | 0.0 | 0.0 | 0.0 | 0.0 | na | na | *16.9* | na |
| Days ≥ 1" Snow Depth | 29 | 24 | 12 | *1* | 0 | 0 | 0 | 0 | 0 | *0* | na | na | na |

## LIBBY 32 SSE *Lincoln County* ELEVATION 3602 ft LAT/LONG 47° 58' N / 115° 14' W

| | JAN | FEB | MAR | APR | MAY | JUN | JUL | AUG | SEP | OCT | NOV | DEC | YEAR |
|---|---|---|---|---|---|---|---|---|---|---|---|---|---|
| Maximum Temp °F | 29.7 | 36.1 | 43.7 | 53.2 | 63.1 | 70.3 | 77.9 | 78.3 | 67.5 | 54.2 | 36.8 | 29.4 | 53.4 |
| Minimum Temp °F | 13.7 | 16.6 | 21.6 | 27.4 | 33.5 | 39.8 | 42.3 | 41.5 | 34.2 | 27.5 | 21.5 | 14.5 | 27.8 |
| Mean Temp °F | 21.7 | 26.4 | 32.7 | 40.3 | 48.3 | 55.0 | 60.1 | 60.0 | 50.9 | 40.9 | 29.2 | 22.0 | 40.6 |
| Days Max Temp ≥ 90 °F | 0 | 0 | 0 | 0 | 0 | 0 | 2 | 3 | 0 | 0 | 0 | 0 | 5 |
| Days Max Temp ≤ 32 °F | 16 | 7 | 2 | 0 | 0 | 0 | 0 | 0 | 0 | 0 | 7 | 17 | 49 |
| Days Min Temp ≤ 32 °F | 30 | 28 | 30 | 24 | 14 | 5 | 2 | 3 | 13 | 23 | 28 | 30 | 230 |
| Days Min Temp ≤ 0 °F | 6 | 3 | 1 | 0 | 0 | 0 | 0 | 0 | 0 | 0 | 1 | 5 | 16 |
| Heating Degree Days | 1335 | 1085 | 995 | 734 | 511 | 299 | 167 | 169 | 418 | 741 | 1068 | 1327 | 8849 |
| Cooling Degree Days | 0 | 0 | 0 | 0 | 1 | 2 | 7 | 8 | 0 | 0 | 0 | 0 | 18 |
| Total Precipitation (") | 3.19 | 2.19 | 1.79 | 1.48 | 1.95 | 2.32 | 1.23 | 1.33 | 1.47 | 1.81 | 2.86 | 2.87 | 24.49 |
| Days ≥ 0.1" Precip | 9 | 6 | 6 | 5 | 7 | 6 | 3 | 4 | 4 | 6 | 9 | 9 | 74 |
| Total Snowfall (") | 25.8 | 16.7 | 10.5 | 5.2 | 1.0 | 0.0 | 0.0 | 0.2 | 0.2 | 2.9 | 15.2 | 23.2 | 100.9 |
| Days ≥ 1" Snow Depth | 30 | 28 | 27 | 9 | 0 | 0 | 0 | 0 | 0 | 1 | 16 | 30 | 141 |

## LIMA *Beaverhead County* ELEVATION 6273 ft LAT/LONG 44° 39' N / 112° 35' W

| | JAN | FEB | MAR | APR | MAY | JUN | JUL | AUG | SEP | OCT | NOV | DEC | YEAR |
|---|---|---|---|---|---|---|---|---|---|---|---|---|---|
| Maximum Temp °F | 28.2 | 33.2 | 41.4 | 51.9 | 62.0 | 71.0 | 79.8 | 78.4 | 68.7 | 56.3 | 38.3 | 28.4 | 53.1 |
| Minimum Temp °F | 7.1 | 10.3 | 17.3 | 24.9 | 32.4 | 39.7 | 44.2 | 43.1 | 35.1 | 26.7 | 16.2 | 7.8 | 25.4 |
| Mean Temp °F | 17.7 | 21.8 | 29.4 | 38.4 | 47.2 | 55.4 | 62.0 | 60.8 | 51.9 | 41.5 | 27.3 | 18.1 | 39.3 |
| Days Max Temp ≥ 90 °F | 0 | 0 | 0 | 0 | 0 | 0 | 1 | 1 | 0 | 0 | 0 | 0 | 2 |
| Days Max Temp ≤ 32 °F | 19 | 11 | 4 | 1 | 0 | 0 | 0 | 0 | 0 | 1 | 9 | 20 | 65 |
| Days Min Temp ≤ 32 °F | 31 | 28 | 30 | 25 | 16 | 4 | 1 | 1 | 11 | 24 | 29 | 31 | 231 |
| Days Min Temp ≤ 0 °F | 9 | 5 | 2 | 0 | 0 | 0 | 0 | 0 | 0 | 0 | 3 | 7 | 26 |
| Heating Degree Days | 1461 | 1215 | 1097 | 791 | 546 | 288 | 111 | 141 | 388 | 722 | 1126 | 1449 | 9335 |
| Cooling Degree Days | 0 | 0 | 0 | 0 | 0 | 6 | 27 | 16 | 1 | 0 | 0 | 0 | 50 |
| Total Precipitation (") | 0.37 | 0.32 | 0.65 | 1.07 | 2.00 | 2.00 | 1.54 | 1.42 | 1.16 | 0.84 | 0.49 | 0.43 | 12.29 |
| Days ≥ 0.1" Precip | 1 | 1 | 2 | 3 | 6 | 6 | 5 | 4 | 4 | 3 | 2 | 2 | 39 |
| Total Snowfall (") | 7.6 | 5.9 | 10.6 | *11.2* | 7.6 | 0.2 | 0.0 | 0.0 | *1.9* | *6.5* | 7.0 | 8.1 | 66.6 |
| Days ≥ 1" Snow Depth | 26 | 21 | 11 | *4* | 1 | 0 | 0 | 0 | 0 | *2* | *10* | *24* | 99 |

## LINCOLN RS *Lewis and Clark County* ELEVATION 4505 ft LAT/LONG 46° 57' N / 112° 39' W

| | JAN | FEB | MAR | APR | MAY | JUN | JUL | AUG | SEP | OCT | NOV | DEC | YEAR |
|---|---|---|---|---|---|---|---|---|---|---|---|---|---|
| Maximum Temp °F | 30.3 | 36.5 | 44.7 | 54.1 | 64.0 | 72.2 | 80.6 | 80.5 | 69.2 | 56.0 | 38.8 | 31.0 | 54.8 |
| Minimum Temp °F | 11.3 | 14.6 | 20.1 | 26.3 | 33.3 | 39.6 | 42.0 | 40.6 | 32.9 | 26.8 | 19.7 | 12.0 | 26.6 |
| Mean Temp °F | 20.8 | 25.6 | 32.5 | 40.2 | 48.7 | 55.9 | 61.4 | 60.6 | 51.0 | 41.4 | 29.3 | 21.5 | 40.7 |
| Days Max Temp ≥ 90 °F | 0 | 0 | 0 | 0 | 0 | 1 | 4 | 4 | 1 | 0 | 0 | 0 | 10 |
| Days Max Temp ≤ 32 °F | 15 | 8 | 2 | 0 | 0 | 0 | 0 | 0 | 0 | 1 | 7 | 16 | 49 |
| Days Min Temp ≤ 32 °F | 30 | 27 | 29 | 25 | 15 | 5 | 2 | 3 | 15 | 24 | 28 | 30 | 233 |
| Days Min Temp ≤ 0 °F | 7 | 5 | 2 | 0 | 0 | 0 | 0 | 0 | 0 | 0 | 3 | 6 | 23 |
| Heating Degree Days | 1364 | 1107 | 1001 | 738 | 500 | 275 | 133 | 152 | 414 | 726 | 1065 | 1342 | 8817 |
| Cooling Degree Days | 0 | 0 | 0 | 0 | 1 | 10 | 25 | 21 | 1 | 0 | 0 | 0 | 58 |
| Total Precipitation (") | 2.09 | 1.29 | 1.18 | 1.40 | 2.18 | 2.04 | 1.22 | 1.46 | 1.31 | 1.21 | 1.33 | 1.86 | 18.57 |
| Days ≥ 0.1" Precip | 7 | 5 | 4 | 4 | 6 | 6 | 3 | 4 | 4 | 4 | 5 | 7 | 59 |
| Total Snowfall (") | 22.1 | 14.1 | 11.2 | *8.2* | *1.5* | 0.0 | 0.0 | 0.0 | *0.5* | *2.9* | 9.5 | 18.9 | 88.9 |
| Days ≥ 1" Snow Depth | 29 | 27 | 27 | *8* | *1* | 0 | 0 | 0 | *0* | 2 | *13* | 29 | 136 |

## LINDBERGH LAKE *Missoula County* ELEVATION 4505 ft LAT/LONG 47° 24' N / 113° 43' W

| | JAN | FEB | MAR | APR | MAY | JUN | JUL | AUG | SEP | OCT | NOV | DEC | YEAR |
|---|---|---|---|---|---|---|---|---|---|---|---|---|---|
| Maximum Temp °F | 29.6 | 35.6 | 42.8 | 52.1 | 62.3 | 70.2 | 78.4 | 78.5 | 67.2 | 53.5 | 36.2 | 29.4 | 53.0 |
| Minimum Temp °F | 13.7 | 16.7 | 21.5 | 28.4 | 36.0 | 43.0 | 46.9 | 46.2 | 38.2 | 31.1 | 22.4 | 14.8 | 29.9 |
| Mean Temp °F | 21.6 | 26.2 | 32.0 | 40.3 | 49.2 | 56.6 | 62.7 | 62.4 | 52.7 | 42.3 | 29.3 | 22.1 | 41.5 |
| Days Max Temp ≥ 90 °F | 0 | 0 | 0 | 0 | 0 | 1 | 2 | 3 | 0 | 0 | 0 | 0 | 6 |
| Days Max Temp ≤ 32 °F | 17 | 8 | 3 | 0 | 0 | 0 | 0 | 0 | 0 | 1 | 9 | 19 | 57 |
| Days Min Temp ≤ 32 °F | 30 | 28 | 30 | 23 | 9 | 1 | 0 | 0 | 6 | 18 | 28 | 30 | 203 |
| Days Min Temp ≤ 0 °F | 6 | 3 | 1 | 0 | 0 | 0 | 0 | 0 | 0 | 0 | 1 | 4 | 15 |
| Heating Degree Days | 1338 | 1090 | 1015 | 737 | 484 | 258 | 112 | 119 | 366 | 697 | 1063 | 1323 | 8602 |
| Cooling Degree Days | 0 | 0 | 0 | 0 | 2 | 12 | 41 | 39 | 3 | 0 | 0 | 0 | 97 |
| Total Precipitation (") | 3.38 | 2.23 | 2.00 | 1.74 | 2.19 | 2.51 | 1.32 | 1.44 | 1.72 | 1.84 | 3.01 | 3.16 | 26.54 |
| Days ≥ 0.1" Precip | 10 | 8 | *8* | 6 | 7 | 7 | 4 | 4 | 5 | 6 | 9 | 10 | 84 |
| Total Snowfall (") | 37.3 | 24.4 | 21.4 | 8.1 | 2.1 | 0.0 | 0.0 | 0.0 | 0.4 | 4.6 | 22.4 | 33.0 | 153.7 |
| Days ≥ 1" Snow Depth | 31 | 28 | 31 | 21 | 2 | 0 | 0 | 0 | 0 | 2 | 18 | 30 | 163 |

## LIVINGSTON MISSION *Park County* ELEVATION 4659 ft LAT/LONG 45° 42 ' N / 110° 27 ' W

| | JAN | FEB | MAR | APR | MAY | JUN | JUL | AUG | SEP | OCT | NOV | DEC | YEAR |
|---|---|---|---|---|---|---|---|---|---|---|---|---|---|
| Maximum Temp °F | 34.7 | 39.4 | 45.7 | 54.8 | 64.3 | 73.5 | 82.8 | 82.1 | 70.7 | 59.0 | 43.0 | 35.9 | 57.2 |
| Minimum Temp °F | 17.6 | 20.5 | 24.5 | 31.1 | 38.6 | 46.2 | 51.1 | 49.5 | 41.0 | 34.3 | 25.9 | 19.2 | 33.3 |
| Mean Temp °F | 26.4 | 30.0 | 35.1 | 43.0 | 51.5 | 59.8 | 67.0 | 65.8 | 55.9 | 46.7 | 34.5 | 27.7 | 45.3 |
| Days Max Temp ≥ 90 °F | 0 | 0 | 0 | 0 | 0 | 1 | 7 | 7 | 1 | 0 | 0 | 0 | 16 |
| Days Max Temp ≤ 32 °F | 11 | 7 | 4 | 1 | 0 | 0 | 0 | 0 | 0 | 1 | 5 | 11 | 40 |
| Days Min Temp ≤ 32 °F | 26 | 22 | 24 | 18 | 6 | 0 | 0 | 0 | 4 | 13 | 21 | 26 | 160 |
| Days Min Temp ≤ 0 °F | 5 | 3 | 1 | 0 | 0 | 0 | 0 | 0 | 0 | 0 | 1 | 4 | 14 |
| Heating Degree Days | 1190 | 983 | 919 | 654 | 414 | 179 | 45 | 60 | 282 | 563 | 907 | 1149 | 7345 |
| Cooling Degree Days | 0 | 0 | 0 | 1 | 3 | 35 | 117 | 95 | 17 | 1 | 0 | 0 | 269 |
| Total Precipitation (") | 0.61 | 0.42 | 0.92 | 1.60 | 2.90 | 2.63 | 1.60 | 1.36 | 1.64 | 1.29 | 0.74 | 0.47 | 16.18 |
| Days ≥ 0.1" Precip | 2 | 1 | 3 | 5 | 6 | 6 | 4 | 4 | 4 | 3 | 2 | 2 | 42 |
| Total Snowfall (") | 10.5 | 6.5 | 13.2 | 9.3 | 1.8 | 0.1 | 0.0 | 0.2 | 1.3 | 3.6 | 7.9 | 8.6 | 63.0 |
| Days ≥ 1" Snow Depth | 11 | 8 | 9 | 4 | 1 | 0 | 0 | 0 | 0 | 2 | 7 | 10 | 52 |

## LOMA 1 WNW *Chouteau County* ELEVATION 2572 ft LAT/LONG 47° 56 ' N / 110° 30 ' W

| | JAN | FEB | MAR | APR | MAY | JUN | JUL | AUG | SEP | OCT | NOV | DEC | YEAR |
|---|---|---|---|---|---|---|---|---|---|---|---|---|---|
| Maximum Temp °F | 30.7 | 38.5 | 48.6 | 60.5 | 71.1 | 79.5 | 87.5 | 86.3 | 74.5 | 62.9 | 45.4 | 35.1 | 60.1 |
| Minimum Temp °F | 4.7 | 10.4 | 20.9 | 30.7 | 40.4 | 48.7 | 52.8 | 50.8 | 40.7 | 30.6 | 18.2 | 8.7 | 29.8 |
| Mean Temp °F | 17.7 | 24.5 | 34.8 | 45.6 | 55.7 | 64.1 | 70.2 | 68.6 | 57.6 | 46.7 | 31.8 | 21.9 | 44.9 |
| Days Max Temp ≥ 90 °F | 0 | 0 | 0 | 0 | 1 | 5 | 14 | 12 | 3 | 0 | 0 | 0 | 35 |
| Days Max Temp ≤ 32 °F | 14 | 9 | 4 | 0 | 0 | 0 | 0 | 0 | 0 | 1 | 5 | 11 | 44 |
| Days Min Temp ≤ 32 °F | 30 | 27 | 27 | 18 | 4 | 0 | 0 | 0 | 5 | 18 | 27 | 30 | 186 |
| Days Min Temp ≤ 0 °F | 13 | 8 | 2 | 0 | 0 | 0 | 0 | 0 | 0 | 0 | 3 | 9 | 35 |
| Heating Degree Days | 1462 | 1138 | 931 | 576 | 294 | 94 | 20 | 39 | 239 | 559 | 989 | 1331 | 7672 |
| Cooling Degree Days | 0 | 0 | 0 | 2 | 18 | 81 | 176 | 159 | 20 | 1 | 0 | 0 | 457 |
| Total Precipitation (") | 0.71 | 0.44 | 0.79 | 1.17 | 2.11 | 2.56 | 1.31 | 1.56 | 1.19 | 0.74 | 0.55 | 0.56 | 13.69 |
| Days ≥ 0.1" Precip | 3 | 2 | 2 | 3 | 5 | 6 | 3 | 3 | 3 | 2 | 2 | 2 | 36 |
| Total Snowfall (") | 13.3 | 8.5 | 7.0 | ***4.2*** | 0.9 | 0.0 | 0.0 | 0.0 | 0.5 | ***1.5*** | 7.1 | 10.8 | 53.8 |
| Days ≥ 1" Snow Depth | 19 | 16 | 9 | 2 | 0 | 0 | 0 | 0 | 0 | 1 | 7 | 16 | 70 |

## MALTA 35 S *Phillips County* ELEVATION 2651 ft LAT/LONG 47° 51 ' N / 108° 0 ' W

| | JAN | FEB | MAR | APR | MAY | JUN | JUL | AUG | SEP | OCT | NOV | DEC | YEAR |
|---|---|---|---|---|---|---|---|---|---|---|---|---|---|
| Maximum Temp °F | 26.8 | 33.4 | 45.3 | 57.8 | 69.2 | 77.7 | 85.9 | 85.3 | 72.8 | ***60.4*** | 41.3 | 31.6 | 57.3 |
| Minimum Temp °F | 3.5 | 8.7 | 20.4 | 29.8 | 40.4 | 49.0 | 53.5 | 52.5 | 40.9 | ***30.9*** | 17.9 | 7.6 | 29.6 |
| Mean Temp °F | 15.1 | 21.1 | 32.9 | 43.8 | 54.8 | 63.4 | 69.7 | 68.9 | 56.9 | ***45.7*** | 29.6 | 19.6 | 43.5 |
| Days Max Temp ≥ 90 °F | 0 | 0 | 0 | 0 | 1 | 4 | ***11*** | ***12*** | 2 | 0 | 0 | 0 | 30 |
| Days Max Temp ≤ 32 °F | 16 | 12 | 6 | 1 | 0 | 0 | 0 | 0 | 0 | 0 | ***7*** | ***14*** | 56 |
| Days Min Temp ≤ 32 °F | 30 | 27 | 27 | 19 | ***5*** | 0 | 0 | 0 | 4 | na | ***28*** | 30 | na |
| Days Min Temp ≤ 0 °F | 13 | 9 | 3 | 0 | 0 | 0 | 0 | 0 | 0 | 0 | 3 | 9 | 37 |
| Heating Degree Days | 1542 | 1232 | 987 | 630 | 319 | 112 | 23 | 41 | 262 | ***591*** | 1056 | 1402 | 8197 |
| Cooling Degree Days | 0 | 0 | 0 | 1 | 11 | 82 | 173 | 176 | 21 | 1 | 0 | 0 | 465 |
| Total Precipitation (") | 0.42 | 0.31 | 0.60 | 1.16 | 2.19 | 2.27 | 1.67 | 1.36 | 1.21 | 0.54 | 0.35 | 0.42 | 12.50 |
| Days ≥ 0.1" Precip | 1 | 1 | 2 | 3 | 5 | 5 | 4 | 3 | 2 | 2 | 1 | 1 | 30 |
| Total Snowfall (") | 8.4 | 5.2 | 6.4 | ***6.5*** | 1.1 | 0.0 | 0.0 | 0.0 | 0.1 | ***1.4*** | ***4.8*** | 6.0 | 39.9 |
| Days ≥ 1" Snow Depth | 22 | 17 | 11 | ***3*** | 0 | 0 | 0 | 0 | 0 | 1 | ***7*** | ***17*** | 78 |

## MARTINSDALE 3 NNW *Meagher County* ELEVATION 4823 ft LAT/LONG 46° 27 ' N / 110° 19 ' W

| | JAN | FEB | MAR | APR | MAY | JUN | JUL | AUG | SEP | OCT | NOV | DEC | YEAR |
|---|---|---|---|---|---|---|---|---|---|---|---|---|---|
| Maximum Temp °F | 34.9 | 39.9 | 45.9 | 56.2 | 65.8 | 73.3 | ***79.8*** | 80.0 | 70.4 | ***60.0*** | 43.6 | 35.9 | 57.1 |
| Minimum Temp °F | 12.1 | 15.1 | 19.6 | 26.5 | 34.4 | 41.2 | ***45.0*** | 43.1 | 35.8 | ***29.0*** | 20.0 | 13.7 | 28.0 |
| Mean Temp °F | 23.5 | 27.5 | 32.8 | 41.4 | 50.1 | 57.3 | ***62.4*** | 61.6 | 53.2 | ***44.5*** | 31.8 | 24.8 | 42.6 |
| Days Max Temp ≥ 90 °F | 0 | 0 | 0 | 0 | 0 | 1 | 2 | 2 | 0 | 0 | 0 | 0 | 5 |
| Days Max Temp ≤ 32 °F | 11 | 6 | 3 | 1 | 0 | 0 | 0 | 0 | 0 | 0 | 4 | 10 | 35 |
| Days Min Temp ≤ 32 °F | 29 | 26 | 28 | 24 | 12 | 2 | 0 | 1 | ***10*** | ***20*** | 27 | 29 | 208 |
| Days Min Temp ≤ 0 °F | 8 | 4 | 2 | 0 | 0 | 0 | 0 | 0 | 0 | 0 | 2 | 5 | 21 |
| Heating Degree Days | 1282 | 1051 | 992 | 701 | 458 | 233 | ***103*** | 123 | 352 | ***628*** | 993 | 1240 | 8156 |
| Cooling Degree Days | 0 | 0 | 0 | 0 | 0 | 18 | na | 28 | 3 | 0 | 0 | 0 | na |
| Total Precipitation (") | 0.59 | 0.31 | 0.70 | 1.14 | 2.20 | 2.21 | 1.80 | 1.49 | 1.31 | 0.76 | 0.51 | 0.43 | 13.45 |
| Days ≥ 0.1" Precip | 2 | 1 | 2 | 3 | 6 | 6 | 5 | 4 | 3 | 3 | 2 | 2 | 39 |
| Total Snowfall (") | 11.6 | 7.6 | 11.9 | 6.7 | 0.9 | 0.0 | 0.0 | 0.0 | 1.0 | 4.1 | 6.7 | 8.6 | 59.1 |
| Days ≥ 1" Snow Depth | 10 | 7 | 6 | 1 | 0 | 0 | 0 | 0 | 0 | 2 | 5 | 9 | 40 |

## MEDICINE LAKE 3 SE *Sheridan County* ELEVATION 1952 ft LAT/LONG 48° 29 ' N / 104° 27 ' W

| | JAN | FEB | MAR | APR | MAY | JUN | JUL | AUG | SEP | OCT | NOV | DEC | YEAR |
|---|---|---|---|---|---|---|---|---|---|---|---|---|---|
| Maximum Temp °F | 20.2 | 28.2 | 41.8 | 57.5 | 70.2 | 78.9 | 85.5 | 84.5 | 72.7 | 59.6 | 38.7 | 25.8 | 55.3 |
| Minimum Temp °F | -1.4 | 5.4 | 17.5 | 30.4 | 42.2 | 51.0 | 55.3 | 53.5 | 43.0 | 31.9 | 17.2 | 4.4 | 29.2 |
| Mean Temp °F | 9.4 | 16.8 | 29.7 | 44.0 | 56.3 | 65.0 | 70.4 | 69.0 | 57.8 | 45.8 | 27.9 | 15.1 | 42.3 |
| Days Max Temp ≥ 90 °F | 0 | 0 | 0 | 0 | 1 | 4 | 9 | 10 | 2 | 0 | 0 | 0 | 26 |
| Days Max Temp ≤ 32 °F | 22 | 15 | 7 | 1 | 0 | 0 | 0 | 0 | 0 | 0 | 9 | 19 | 73 |
| Days Min Temp ≤ 32 °F | 31 | 28 | 29 | 18 | 4 | 0 | 0 | 0 | 3 | 16 | 28 | 31 | 188 |
| Days Min Temp ≤ 0 °F | 16 | 11 | 4 | 0 | 0 | 0 | 0 | 0 | 0 | 0 | 3 | 11 | 45 |
| Heating Degree Days | 1721 | 1356 | 1088 | 625 | 289 | 83 | 19 | 43 | 241 | 589 | 1107 | 1542 | 8703 |
| Cooling Degree Days | 0 | 0 | 0 | 2 | 31 | 105 | 199 | 191 | 32 | 1 | 0 | 0 | 561 |
| Total Precipitation (") | 0.41 | 0.28 | 0.49 | 1.18 | 2.11 | 2.52 | 1.88 | 1.46 | 1.23 | 0.67 | 0.30 | 0.38 | 12.91 |
| Days ≥ 0.1" Precip | 1 | 1 | 2 | 3 | 6 | 6 | 4 | 4 | 4 | 2 | 1 | 1 | 35 |
| Total Snowfall (") | 6.4 | 4.6 | 4.5 | 3.9 | 0.1 | 0.0 | 0.0 | 0.0 | 0.1 | 0.4 | 2.8 | 5.4 | 28.2 |
| Days ≥ 1" Snow Depth | 26 | 19 | 11 | 3 | 0 | 0 | 0 | 0 | 0 | 0 | 8 | 20 | 87 |

## MELSTONE *Musselshell County* ELEVATION 2890 ft LAT/LONG 46° 36 ' N / 107° 52 ' W

| | JAN | FEB | MAR | APR | MAY | JUN | JUL | AUG | SEP | OCT | NOV | DEC | YEAR |
|---|---|---|---|---|---|---|---|---|---|---|---|---|---|
| Maximum Temp °F | 32.7 | 39.7 | 48.9 | 60.0 | 70.3 | 79.7 | 87.7 | 86.9 | 75.2 | 62.8 | 45.3 | 35.7 | 60.4 |
| Minimum Temp °F | 10.0 | 15.2 | 23.5 | 32.4 | 41.7 | 50.3 | 55.2 | 53.9 | 43.7 | 33.8 | 21.8 | 12.9 | 32.9 |
| Mean Temp °F | 21.4 | 27.5 | 36.2 | 46.2 | 56.0 | 65.1 | 71.5 | 70.4 | 59.4 | 48.3 | 33.6 | 24.3 | 46.7 |
| Days Max Temp ≥ 90 °F | 0 | 0 | 0 | 0 | 1 | 5 | ***14*** | 12 | 3 | 0 | 0 | 0 | 35 |
| Days Max Temp ≤ 32 °F | 13 | 8 | 3 | 0 | 0 | 0 | 0 | 0 | 0 | 0 | 4 | 10 | 38 |
| Days Min Temp ≤ 32 °F | 29 | 26 | 26 | 15 | 4 | 0 | 0 | 0 | 2 | ***13*** | 24 | 29 | 168 |
| Days Min Temp ≤ 0 °F | 10 | 5 | 1 | 0 | 0 | 0 | 0 | 0 | 0 | 0 | 2 | 6 | 24 |
| Heating Degree Days | 1347 | 1051 | 886 | 558 | 287 | 84 | 14 | 22 | 195 | 515 | 941 | 1245 | 7145 |
| Cooling Degree Days | 0 | 0 | 0 | 2 | 20 | 106 | 220 | 202 | 37 | 3 | 0 | 0 | 590 |
| Total Precipitation (") | 0.62 | 0.42 | 0.78 | 1.49 | 2.69 | 2.69 | 1.60 | 1.27 | 1.52 | 1.02 | 0.56 | 0.60 | 15.26 |
| Days ≥ 0.1" Precip | 3 | 2 | 2 | 4 | 6 | 6 | 4 | 3 | 3 | 3 | 2 | 2 | 40 |
| Total Snowfall (") | 10.3 | 6.7 | 7.9 | 6.1 | 1.4 | 0.0 | 0.0 | 0.0 | 0.5 | 2.7 | 6.3 | 9.3 | 51.2 |
| Days ≥ 1" Snow Depth | 19 | 13 | 7 | 2 | 0 | 0 | 0 | 0 | 0 | 1 | 6 | 14 | 62 |

## MELVILLE 4 W *Sweet Grass County* ELEVATION 5374 ft LAT/LONG 46° 6 ' N / 110° 3 ' W

| | JAN | FEB | MAR | APR | MAY | JUN | JUL | AUG | SEP | OCT | NOV | DEC | YEAR |
|---|---|---|---|---|---|---|---|---|---|---|---|---|---|
| Maximum Temp °F | 33.9 | 37.4 | 42.0 | 50.8 | 60.3 | 68.3 | 75.8 | 75.2 | 65.4 | 55.5 | 41.4 | 35.1 | 53.4 |
| Minimum Temp °F | 10.8 | 13.6 | 18.5 | 26.3 | 35.0 | 42.4 | 47.4 | 46.4 | 38.3 | 29.9 | 18.7 | 12.1 | 28.3 |
| Mean Temp °F | 22.4 | 25.5 | 30.3 | 38.6 | 47.6 | 55.4 | 61.6 | 60.9 | 51.9 | 42.7 | 30.1 | 23.7 | 40.9 |
| Days Max Temp ≥ 90 °F | 0 | 0 | 0 | 0 | 0 | 0 | 0 | 0 | 0 | 0 | 0 | 0 | 0 |
| Days Max Temp ≤ 32 °F | 12 | 8 | 6 | 1 | 0 | 0 | 0 | 0 | 0 | 1 | 6 | 11 | 45 |
| Days Min Temp ≤ 32 °F | 29 | 27 | 29 | 23 | 11 | 1 | 0 | 1 | 7 | 18 | 27 | 29 | 202 |
| Days Min Temp ≤ 0 °F | 8 | 5 | 3 | 0 | 0 | 0 | 0 | 0 | 0 | 0 | 2 | 6 | 24 |
| Heating Degree Days | 1315 | 1108 | 1069 | 786 | 530 | 289 | 128 | 147 | 391 | 683 | 1041 | 1276 | 8763 |
| Cooling Degree Days | 0 | 0 | 0 | 0 | 0 | 9 | 30 | 21 | 3 | 0 | 0 | 0 | 63 |
| Total Precipitation (") | 0.65 | 0.49 | 1.18 | 1.80 | 2.82 | 3.06 | 1.99 | 1.69 | 1.48 | 1.14 | 0.76 | 0.50 | 17.56 |
| Days ≥ 0.1" Precip | 2 | 2 | 4 | 5 | 7 | 8 | 5 | 4 | 4 | 3 | 3 | 2 | 49 |
| Total Snowfall (") | na | na | na | na | na | ***0.0*** | ***0.0*** | 0.0 | ***1.9*** | na | na | na | na |
| Days ≥ 1" Snow Depth | na | na | na | na | na | ***0*** | 0 | 0 | ***0*** | na | na | na | na |

## MILES CITY MUNI AP *Custer County* ELEVATION 2629 ft LAT/LONG 46° 26 ' N / 105° 52 ' W

| | JAN | FEB | MAR | APR | MAY | JUN | JUL | AUG | SEP | OCT | NOV | DEC | YEAR |
|---|---|---|---|---|---|---|---|---|---|---|---|---|---|
| Maximum Temp °F | 25.7 | 33.2 | 45.1 | 58.3 | 69.6 | 79.5 | 87.8 | 86.2 | 73.5 | 59.8 | 42.0 | 30.3 | 57.6 |
| Minimum Temp °F | 6.3 | 12.4 | 22.9 | 34.1 | 44.8 | 54.0 | 60.2 | 58.5 | 46.6 | 35.0 | 21.6 | 10.4 | 33.9 |
| Mean Temp °F | 16.0 | 22.9 | 34.0 | 46.3 | 57.2 | 66.8 | 74.0 | 72.3 | 60.1 | 47.4 | 31.9 | 20.4 | 45.8 |
| Days Max Temp ≥ 90 °F | 0 | 0 | 0 | 0 | 1 | 5 | 14 | 12 | 3 | 0 | 0 | 0 | 35 |
| Days Max Temp ≤ 32 °F | 18 | 12 | 6 | 1 | 0 | 0 | 0 | 0 | 0 | 0 | 7 | 15 | 59 |
| Days Min Temp ≤ 32 °F | 30 | 27 | 26 | 12 | 2 | 0 | 0 | 0 | 2 | 11 | 26 | 30 | 166 |
| Days Min Temp ≤ 0 °F | 12 | 7 | 2 | 0 | 0 | 0 | 0 | 0 | 0 | 0 | 2 | 7 | 30 |
| Heating Degree Days | 1516 | 1184 | 954 | 558 | 265 | 67 | 10 | 21 | 197 | 539 | 988 | 1378 | 7677 |
| Cooling Degree Days | 0 | 0 | 0 | 4 | 35 | 139 | 294 | 263 | 53 | 2 | 0 | 0 | 790 |
| Total Precipitation (") | 0.55 | 0.40 | 0.60 | 1.51 | 2.13 | 2.76 | 1.66 | 1.14 | 1.29 | 1.02 | 0.51 | 0.57 | 14.14 |
| Days ≥ 0.1" Precip | 2 | 1 | 2 | 4 | 6 | 6 | 4 | 3 | 3 | 2 | 2 | 2 | 37 |
| Total Snowfall (") | 6.1 | 4.3 | 4.6 | 4.1 | 0.9 | 0.0 | 0.0 | 0.0 | 0.4 | 1.1 | 4.0 | 5.8 | 31.3 |
| Days ≥ 1" Snow Depth | 23 | 17 | 10 | 2 | 0 | 0 | 0 | 0 | 0 | 1 | 6 | 17 | 76 |

## MISSOULA JOHNSN-BELL *Missoula County* ELEVATION 3205 ft LAT/LONG 46° 55 ' N / 114° 5 ' W

| | JAN | FEB | MAR | APR | MAY | JUN | JUL | AUG | SEP | OCT | NOV | DEC | YEAR |
|---|---|---|---|---|---|---|---|---|---|---|---|---|---|
| Maximum Temp °F | 30.7 | 37.1 | 47.3 | 57.6 | 66.4 | 74.7 | 83.8 | 83.3 | 71.3 | 57.3 | 40.1 | 30.3 | 56.7 |
| Minimum Temp °F | 16.2 | 20.2 | 26.2 | 32.2 | 39.1 | 45.9 | 50.1 | 49.3 | 40.5 | 31.3 | 23.9 | 16.3 | 32.6 |
| Mean Temp °F | 23.5 | 28.7 | 36.8 | 44.9 | 52.8 | 60.3 | 67.0 | 66.3 | 55.9 | 44.3 | 32.0 | 23.3 | 44.7 |
| Days Max Temp ≥ 90 °F | 0 | 0 | 0 | 0 | 0 | 3 | 9 | 9 | 1 | 0 | 0 | 0 | 22 |
| Days Max Temp ≤ 32 °F | 15 | 7 | 2 | 0 | 0 | 0 | 0 | 0 | 0 | 0 | 6 | 17 | 47 |
| Days Min Temp ≤ 32 °F | 29 | 26 | 26 | 17 | 5 | 0 | 0 | 0 | 4 | 17 | 26 | 30 | 180 |
| Days Min Temp ≤ 0 °F | 4 | 2 | 0 | 0 | 0 | 0 | 0 | 0 | 0 | 0 | 1 | 3 | 10 |
| Heating Degree Days | 1282 | 1019 | 868 | 595 | 376 | 171 | 49 | 58 | 277 | 634 | 982 | 1286 | 7597 |
| Cooling Degree Days | 0 | 0 | 0 | 0 | 6 | 41 | 108 | 105 | 9 | 0 | 0 | 0 | 269 |
| Total Precipitation (") | 1.16 | 0.72 | 0.96 | 1.04 | 1.80 | 1.77 | 0.98 | 1.14 | 1.03 | 0.76 | 0.82 | 1.04 | 13.22 |
| Days ≥ 0.1" Precip | 4 | 2 | 3 | 3 | 5 | 5 | 3 | 3 | 3 | 2 | 3 | 3 | 39 |
| Total Snowfall (") | 12.1 | 7.4 | 5.7 | 2.1 | 0.5 | 0.0 | 0.0 | 0.0 | 0.0 | 0.9 | 5.5 | 10.6 | 44.8 |
| Days ≥ 1" Snow Depth | 25 | 15 | 6 | 0 | 0 | 0 | 0 | 0 | 0 | 0 | 5 | 19 | 70 |

## MIZPAH 4 NNW *Custer County* ELEVATION 2480 ft LAT/LONG 46° 15 ' N / 105° 18 ' W

| | JAN | FEB | MAR | APR | MAY | JUN | JUL | AUG | SEP | OCT | NOV | DEC | YEAR |
|---|---|---|---|---|---|---|---|---|---|---|---|---|---|
| Maximum Temp °F | 27.6 | 34.9 | 46.8 | 60.1 | 71.0 | 80.3 | 89.0 | 87.8 | 75.7 | 62.1 | 43.6 | 31.8 | 59.2 |
| Minimum Temp °F | 1.9 | 8.9 | 20.8 | 31.6 | 41.8 | 50.6 | 55.5 | 53.0 | 42.0 | 30.5 | 18.3 | 6.2 | 30.1 |
| Mean Temp °F | 14.8 | 22.0 | 33.8 | 45.9 | 56.4 | 65.5 | 72.2 | 70.4 | 58.9 | 46.4 | 31.0 | 19.1 | 44.7 |
| Days Max Temp ≥ 90 °F | 0 | 0 | 0 | 0 | 1 | 5 | 15 | 14 | 4 | 0 | 0 | 0 | 39 |
| Days Max Temp ≤ 32 °F | 16 | 11 | 4 | 0 | 0 | 0 | 0 | 0 | 0 | 0 | 6 | 14 | 51 |
| Days Min Temp ≤ 32 °F | 31 | 27 | 28 | 16 | 4 | 0 | 0 | 0 | 5 | 19 | 28 | 31 | 189 |
| Days Min Temp ≤ 0 °F | 14 | 8 | 2 | 0 | 0 | 0 | 0 | 0 | 0 | 0 | 2 | 9 | 35 |
| Heating Degree Days | 1553 | 1209 | 961 | 569 | 280 | 77 | 12 | 25 | 214 | 571 | 1014 | 1420 | 7905 |
| Cooling Degree Days | 0 | 0 | 0 | 2 | 30 | 121 | 261 | 227 | 37 | 0 | 0 | 0 | 678 |
| Total Precipitation (") | 0.39 | 0.24 | 0.55 | 1.50 | 2.20 | 2.52 | 1.59 | 1.12 | 1.32 | 0.99 | 0.47 | 0.43 | 13.32 |
| Days ≥ 0.1" Precip | 1 | 1 | 1 | 4 | 5 | 6 | 3 | 3 | 3 | 3 | 2 | 1 | 33 |
| Total Snowfall (") | 5.8 | 3.0 | 4.8 | 3.7 | 0.7 | 0.0 | 0.0 | 0.0 | 0.5 | 1.0 | 4.2 | 6.6 | 30.3 |
| Days ≥ 1" Snow Depth | 24 | 18 | 9 | 1 | 0 | 0 | 0 | 0 | 0 | 0 | 6 | 19 | 77 |

## MOCCASIN EXP STN *Judith Basin County* ELEVATION 4304 ft LAT/LONG 47° 3 ' N / 109° 57 ' W

| | JAN | FEB | MAR | APR | MAY | JUN | JUL | AUG | SEP | OCT | NOV | DEC | YEAR |
|---|---|---|---|---|---|---|---|---|---|---|---|---|---|
| Maximum Temp °F | 33.2 | 36.9 | 42.9 | 52.7 | 62.5 | 70.8 | 79.5 | 79.7 | 68.1 | 57.6 | 43.5 | 35.9 | 55.3 |
| Minimum Temp °F | 9.6 | 14.1 | 20.5 | 29.4 | 38.2 | 45.8 | 50.5 | 50.0 | 41.0 | 31.8 | 20.8 | 12.9 | 30.4 |
| Mean Temp °F | 21.4 | 25.5 | 31.7 | 41.0 | 50.4 | 58.3 | 65.0 | 64.9 | 54.6 | 44.7 | 32.2 | 24.4 | 42.8 |
| Days Max Temp ≥ 90 °F | 0 | 0 | 0 | 0 | 0 | 1 | 3 | 4 | 1 | 0 | 0 | 0 | 9 |
| Days Max Temp ≤ 32 °F | 12 | 9 | 6 | 2 | 0 | 0 | 0 | 0 | 0 | 1 | 5 | 10 | 45 |
| Days Min Temp ≤ 32 °F | 29 | 26 | 28 | 20 | 7 | 0 | 0 | 0 | 4 | 16 | 25 | 28 | 183 |
| Days Min Temp ≤ 0 °F | 9 | 6 | 2 | 0 | 0 | 0 | 0 | 0 | 0 | 0 | 2 | 6 | 25 |
| Heating Degree Days | 1345 | 1109 | 1025 | 712 | 449 | 220 | 77 | 87 | 325 | 622 | 977 | 1252 | 8200 |
| Cooling Degree Days | 0 | 0 | 0 | 0 | 4 | 32 | 87 | 94 | 19 | 1 | 0 | 0 | 237 |
| Total Precipitation (") | 0.66 | 0.47 | 0.88 | 1.30 | 2.88 | 2.96 | 1.94 | 1.78 | 1.52 | 0.99 | 0.57 | 0.59 | 16.54 |
| Days ≥ 0.1" Precip | 2 | 2 | 3 | 4 | 7 | 7 | 5 | 4 | 4 | 3 | 2 | 2 | 45 |
| Total Snowfall (") | na | ***8.2*** | na | na | na | ***0.1*** | ***0.0*** | ***0.0*** | 0.7 | ***5.7*** | ***8.2*** | ***11.9*** | na |
| Days ≥ 1" Snow Depth | na | na | na | na | na | ***0*** | ***0*** | ***0*** | ***1*** | ***3*** | na | na | na |

## MOORHEAD 9 NE *Powder River County* ELEVATION 3202 ft LAT/LONG 45° 11 ' N / 105° 44 ' W

| | JAN | FEB | MAR | APR | MAY | JUN | JUL | AUG | SEP | OCT | NOV | DEC | YEAR |
|---|---|---|---|---|---|---|---|---|---|---|---|---|---|
| Maximum Temp °F | 32.2 | 39.0 | ***48.7*** | 60.2 | ***69.7*** | 79.4 | 87.9 | 87.5 | 75.8 | 62.5 | 44.9 | 34.0 | 60.2 |
| Minimum Temp °F | 7.2 | 13.0 | ***21.9*** | 31.3 | ***39.9*** | 49.3 | 53.5 | 51.6 | 41.8 | 31.4 | 19.9 | 9.5 | 30.9 |
| Mean Temp °F | 19.7 | 26.0 | ***35.4*** | 45.8 | ***54.8*** | 64.4 | 70.7 | 69.6 | 58.8 | 46.9 | 32.4 | 21.8 | 45.5 |
| Days Max Temp ≥ 90 °F | 0 | 0 | 0 | 0 | ***1*** | 4 | na | ***13*** | 4 | 0 | 0 | 0 | na |
| Days Max Temp ≤ 32 °F | 13 | 8 | ***3*** | 0 | 0 | 0 | 0 | 0 | 0 | 0 | 5 | ***12*** | 41 |
| Days Min Temp ≤ 32 °F | 31 | 27 | ***28*** | ***17*** | ***5*** | 0 | 0 | 0 | 4 | 17 | 28 | 31 | 188 |
| Days Min Temp ≤ 0 °F | 10 | 5 | ***1*** | 0 | 0 | 0 | 0 | 0 | 0 | 0 | 2 | 8 | 26 |
| Heating Degree Days | 1400 | 1094 | ***909*** | 572 | ***322*** | 92 | 17 | 25 | 212 | 556 | 970 | 1335 | 7504 |
| Cooling Degree Days | 0 | 0 | ***0*** | 2 | 17 | 97 | 205 | 188 | 34 | 1 | 0 | 0 | 544 |
| Total Precipitation (") | 0.41 | 0.29 | 0.64 | 1.45 | 2.15 | 2.57 | 1.51 | 0.89 | 1.21 | 0.99 | 0.48 | 0.45 | 13.04 |
| Days ≥ 0.1" Precip | 2 | 1 | ***2*** | 4 | ***6*** | 6 | 3 | 3 | 3 | 3 | 2 | 2 | 37 |
| Total Snowfall (") | 6.3 | 4.5 | na | ***4.2*** | ***0.7*** | 0.0 | 0.0 | 0.0 | ***0.3*** | ***0.7*** | na | 7.0 | na |
| Days ≥ 1" Snow Depth | 23 | ***18*** | 7 | ***1*** | 0 | 0 | 0 | 0 | 0 | 1 | ***7*** | 19 | 76 |

## MOSBY 2 ENE *Garfield County* ELEVATION 2753 ft LAT/LONG 47° 0 ' N / 107° 51 ' W

| | JAN | FEB | MAR | APR | MAY | JUN | JUL | AUG | SEP | OCT | NOV | DEC | YEAR |
|---|---|---|---|---|---|---|---|---|---|---|---|---|---|
| Maximum Temp °F | 30.8 | 38.0 | 47.9 | 59.3 | 69.9 | 79.5 | 88.0 | 87.2 | 74.7 | 63.1 | 45.9 | 35.5 | 60.0 |
| Minimum Temp °F | 7.1 | 13.2 | 23.2 | 32.8 | 42.6 | 51.0 | 56.5 | 54.9 | 44.3 | 34.2 | 21.6 | 11.7 | 32.8 |
| Mean Temp °F | 19.0 | 25.6 | 35.6 | 46.1 | 56.3 | 65.3 | 72.3 | 71.0 | 59.5 | 48.7 | 33.8 | 23.6 | 46.4 |
| Days Max Temp ≥ 90 °F | 0 | 0 | 0 | 0 | 1 | 5 | 14 | 13 | 3 | 0 | 0 | 0 | 36 |
| Days Max Temp ≤ 32 °F | 13 | 9 | 4 | 0 | 0 | 0 | 0 | 0 | 0 | 0 | 5 | 11 | 42 |
| Days Min Temp ≤ 32 °F | 28 | 26 | 26 | 14 | 3 | 0 | 0 | 0 | 2 | 12 | 25 | 29 | 165 |
| Days Min Temp ≤ 0 °F | 11 | 6 | 1 | 0 | 0 | 0 | 0 | 0 | 0 | 0 | 2 | 7 | 27 |
| Heating Degree Days | 1423 | 1106 | 905 | 563 | 283 | 82 | 14 | 21 | 204 | 501 | 929 | 1277 | 7308 |
| Cooling Degree Days | *0* | 0 | 0 | 1 | 21 | 102 | 227 | *203* | *37* | *2* | 0 | 0 | 593 |
| Total Precipitation (") | 0.58 | 0.34 | 0.60 | 1.28 | 2.53 | 2.28 | 1.75 | 1.18 | 1.36 | 0.84 | 0.41 | 0.46 | 13.61 |
| Days ≥ 0.1" Precip | 2 | 1 | 2 | 4 | 6 | 5 | 4 | 3 | 3 | 2 | 2 | 2 | 36 |
| Total Snowfall (") | *7.4* | 3.9 | 4.1 | 3.8 | 1.4 | 0.1 | 0.0 | 0.0 | 0.1 | 1.8 | 3.3 | 6.1 | 32.0 |
| Days ≥ 1" Snow Depth | na | *14* | *7* | 1 | 0 | 0 | 0 | 0 | 0 | 0 | 4 | *12* | na |

## MYSTIC LAKE *Stillwater County* ELEVATION 6565 ft LAT/LONG 45° 14 ' N / 109° 45 ' W

| | JAN | FEB | MAR | APR | MAY | JUN | JUL | AUG | SEP | OCT | NOV | DEC | YEAR |
|---|---|---|---|---|---|---|---|---|---|---|---|---|---|
| Maximum Temp °F | 33.9 | 37.0 | 41.8 | 49.4 | 58.7 | 67.4 | 74.7 | 74.1 | 64.7 | 54.7 | 40.2 | 33.8 | 52.5 |
| Minimum Temp °F | 16.6 | 18.3 | 21.5 | 27.9 | 36.4 | 44.3 | 50.1 | 50.0 | 41.5 | 34.0 | 24.0 | 18.1 | 31.9 |
| Mean Temp °F | 25.3 | 27.7 | 31.7 | 38.7 | 47.6 | 55.9 | 62.4 | 62.0 | 53.1 | 44.4 | 32.1 | 25.9 | 42.2 |
| Days Max Temp ≥ 90 °F | 0 | 0 | 0 | 0 | 0 | 0 | 0 | 0 | 0 | 0 | 0 | 0 | 0 |
| Days Max Temp ≤ 32 °F | 12 | 8 | 5 | 2 | 0 | 0 | 0 | 0 | 0 | 1 | 6 | 13 | 47 |
| Days Min Temp ≤ 32 °F | 27 | 25 | 26 | 21 | 10 | 1 | 0 | 0 | 5 | 14 | 23 | 27 | 179 |
| Days Min Temp ≤ 0 °F | 5 | 3 | 2 | 0 | 0 | 0 | 0 | 0 | 0 | 0 | 1 | 4 | 15 |
| Heating Degree Days | 1224 | 1048 | 1026 | 783 | 534 | 280 | 116 | 126 | 359 | 633 | 981 | 1204 | 8314 |
| Cooling Degree Days | 0 | 0 | 0 | 0 | 0 | 16 | 50 | 42 | 10 | 0 | 0 | 0 | 118 |
| Total Precipitation (") | 1.37 | 1.12 | 2.20 | 2.81 | 3.79 | 3.21 | 2.31 | 1.83 | 2.22 | 1.97 | 1.52 | 1.30 | 25.65 |
| Days ≥ 0.1" Precip | 5 | 4 | 7 | 7 | 9 | 9 | 7 | 6 | 6 | 5 | 5 | 5 | 75 |
| Total Snowfall (") | 19.2 | 16.4 | 31.7 | 31.7 | 13.3 | 1.5 | 0.1 | 0.0 | 7.0 | 15.5 | 20.5 | 19.8 | 176.7 |
| Days ≥ 1" Snow Depth | 24 | 22 | 25 | 16 | 5 | 0 | 0 | 0 | 2 | 7 | 17 | 20 | 138 |

## NORRIS MADISON P H *Madison County* ELEVATION 4734 ft LAT/LONG 45° 29 ' N / 111° 38 ' W

| | JAN | FEB | MAR | APR | MAY | JUN | JUL | AUG | SEP | OCT | NOV | DEC | YEAR |
|---|---|---|---|---|---|---|---|---|---|---|---|---|---|
| Maximum Temp °F | 35.6 | 40.3 | 46.7 | 55.9 | 65.8 | 74.8 | 83.9 | 83.2 | 72.4 | 60.1 | 43.6 | 35.8 | 58.2 |
| Minimum Temp °F | 19.2 | 22.4 | 26.8 | 33.5 | 41.3 | 48.5 | 54.1 | 53.3 | 44.7 | 36.8 | 27.6 | 20.5 | 35.7 |
| Mean Temp °F | 27.4 | 31.4 | 36.8 | 44.7 | 53.6 | 61.7 | 69.0 | 68.2 | 58.6 | 48.5 | 35.6 | 28.2 | 47.0 |
| Days Max Temp ≥ 90 °F | 0 | 0 | 0 | 0 | 0 | 1 | 7 | 7 | 1 | 0 | 0 | 0 | 16 |
| Days Max Temp ≤ 32 °F | 10 | 5 | 2 | 0 | 0 | 0 | 0 | 0 | 0 | 0 | 4 | 10 | 31 |
| Days Min Temp ≤ 32 °F | 27 | 23 | 22 | 14 | 3 | 0 | 0 | 0 | 2 | 8 | 20 | 26 | 145 |
| Days Min Temp ≤ 0 °F | 4 | 2 | 1 | 0 | 0 | 0 | 0 | 0 | 0 | 0 | 1 | 2 | 10 |
| Heating Degree Days | 1159 | 943 | 866 | 601 | 353 | 141 | 28 | 35 | 217 | 506 | 874 | 1135 | 6858 |
| Cooling Degree Days | 0 | 0 | 0 | 0 | 7 | 52 | 154 | 151 | 32 | 1 | 0 | 0 | 397 |
| Total Precipitation (") | 0.52 | 0.42 | 1.20 | 1.75 | 3.08 | 2.80 | 1.60 | 1.47 | 1.75 | 1.36 | 0.80 | 0.54 | 17.29 |
| Days ≥ 0.1" Precip | 2 | 2 | 4 | 5 | 7 | 7 | 5 | 4 | 5 | 4 | 3 | 2 | 50 |
| Total Snowfall (") | na | *6.2* | na | na | 2.5 | 0.0 | 0.0 | 0.0 | 0.5 | na | na | na | na |
| Days ≥ 1" Snow Depth | na | na | na | *3* | 1 | 0 | 0 | 0 | 0 | *2* | na | na | na |

## OLNEY *Flathead County* ELEVATION 3182 ft LAT/LONG 48° 32 ' N / 114° 34 ' W

| | JAN | FEB | MAR | APR | MAY | JUN | JUL | AUG | SEP | OCT | NOV | DEC | YEAR |
|---|---|---|---|---|---|---|---|---|---|---|---|---|---|
| Maximum Temp °F | 29.3 | 36.3 | *45.3* | 56.3 | 66.7 | 73.8 | 80.9 | 81.0 | 69.4 | 54.9 | 36.7 | 28.2 | 54.9 |
| Minimum Temp °F | 12.7 | 14.5 | *21.2* | 27.3 | 34.2 | 40.5 | 43.1 | 42.0 | 34.3 | 27.0 | 21.3 | 13.6 | 27.6 |
| Mean Temp °F | 21.0 | 25.4 | *33.2* | 41.9 | 50.5 | 57.2 | 62.0 | 61.5 | 51.9 | 41.0 | 29.0 | 20.9 | 41.3 |
| Days Max Temp ≥ 90 °F | 0 | 0 | 0 | 0 | 0 | 1 | 5 | 6 | 0 | 0 | 0 | 0 | 12 |
| Days Max Temp ≤ 32 °F | *17* | 7 | 1 | 0 | 0 | 0 | 0 | 0 | 0 | 0 | 7 | 19 | 51 |
| Days Min Temp ≤ 32 °F | *30* | 28 | *29* | 24 | 13 | 4 | 1 | 2 | 13 | 24 | 28 | 31 | 227 |
| Days Min Temp ≤ 0 °F | 6 | 4 | 1 | 0 | 0 | 0 | 0 | 0 | 0 | 0 | 2 | 5 | 18 |
| Heating Degree Days | 1360 | 1111 | *977* | 688 | 445 | 239 | 118 | 132 | 390 | 738 | 1073 | 1361 | 8632 |
| Cooling Degree Days | 0 | 0 | 0 | 0 | 2 | 12 | 35 | 36 | 2 | 0 | 0 | 0 | 87 |
| Total Precipitation (") | 2.80 | 2.02 | 1.31 | 1.29 | 2.12 | 2.91 | 1.75 | 1.45 | 1.35 | 1.35 | 2.40 | 2.27 | 23.02 |
| Days ≥ 0.1" Precip | 8 | 6 | *5* | 4 | 6 | 7 | 5 | 4 | 4 | 5 | 7 | 7 | 68 |
| Total Snowfall (") | na | na | na | na | *0.0* | *0.0* | *0.0* | *0.0* | *0.1* | na | na | na | na |
| Days ≥ 1" Snow Depth | *23* | *19* | *20* | *5* | 0 | 0 | 0 | 0 | 0 | 1 | *12* | 20 | 100 |

## OPHEIM 10 N *Valley County* ELEVATION 2982 ft LAT/LONG 49° 0 ' N / 106° 23 ' W

| | JAN | FEB | MAR | APR | MAY | JUN | JUL | AUG | SEP | OCT | NOV | DEC | YEAR |
|---|---|---|---|---|---|---|---|---|---|---|---|---|---|
| Maximum Temp °F | 18.4 | 25.1 | 38.2 | 54.6 | 67.3 | 75.7 | 81.8 | 81.6 | 69.5 | 56.3 | 36.5 | 24.4 | 52.5 |
| Minimum Temp °F | -3.8 | 2.6 | 14.8 | 26.6 | 37.0 | 45.6 | 49.3 | 47.5 | 37.4 | 27.6 | 13.4 | 1.6 | 25.0 |
| Mean Temp °F | 7.3 | 13.9 | 26.4 | 40.6 | 52.2 | 60.7 | 65.5 | 64.6 | 53.4 | 42.0 | 25.0 | 13.0 | 38.7 |
| Days Max Temp ≥ 90 °F | 0 | 0 | 0 | 0 | 0 | 2 | 5 | 7 | 1 | 0 | 0 | 0 | 15 |
| Days Max Temp ≤ 32 °F | 24 | 17 | 10 | 1 | 0 | 0 | 0 | 0 | 0 | 1 | 11 | 20 | 84 |
| Days Min Temp ≤ 32 °F | 31 | 28 | 30 | 23 | 10 | 1 | 0 | 1 | 9 | 21 | 29 | 31 | 214 |
| Days Min Temp ≤ 0 °F | 17 | 13 | 5 | 0 | 0 | 0 | 0 | 0 | 0 | 0 | 5 | 13 | 53 |
| Heating Degree Days | 1787 | 1440 | 1190 | 724 | 400 | 163 | 66 | 95 | 351 | 706 | 1192 | 1607 | 9721 |
| Cooling Degree Days | 0 | 0 | 0 | 0 | 12 | 52 | 93 | 95 | 10 | 0 | 0 | 0 | 262 |
| Total Precipitation (") | 0.35 | 0.29 | 0.39 | 0.56 | 1.98 | 2.60 | 2.22 | 1.26 | 1.27 | 0.55 | 0.27 | 0.29 | 12.03 |
| Days ≥ 0.1" Precip | *1* | 1 | 1 | 2 | 5 | 6 | 5 | 3 | 3 | 2 | 1 | 1 | 31 |
| Total Snowfall (") | na | na | na | na | na | na | *0.0* | *0.0* | na | na | na | na | na |
| Days ≥ 1" Snow Depth | na | na | na | na | na | *0* | *0* | *0* | *0* | na | na | na | na |

## OPHEIM 12 SSE *Valley County* ELEVATION 2953 ft LAT/LONG 48° 41 ' N / 106° 19 ' W

| | JAN | FEB | MAR | APR | MAY | JUN | JUL | AUG | SEP | OCT | NOV | DEC | YEAR |
|---|---|---|---|---|---|---|---|---|---|---|---|---|---|
| Maximum Temp °F | 17.9 | 24.6 | 37.7 | 52.8 | 65.8 | 74.1 | 80.7 | 79.9 | 67.5 | 54.7 | 35.8 | 23.0 | 51.2 |
| Minimum Temp °F | -2.9 | 3.4 | 15.4 | 27.3 | 37.8 | 46.4 | 50.7 | 49.0 | 38.4 | 28.0 | 13.9 | 1.3 | 25.7 |
| Mean Temp °F | 7.5 | 14.0 | 26.6 | 40.1 | 51.8 | 60.3 | 65.7 | 64.5 | 53.0 | 41.4 | 24.9 | 12.2 | 38.5 |
| Days Max Temp ≥ 90 °F | 0 | 0 | 0 | 0 | 0 | 2 | 5 | 6 | 1 | 0 | 0 | 0 | 14 |
| Days Max Temp ≤ 32 °F | 24 | 17 | 10 | 2 | 0 | 0 | 0 | 0 | 0 | 2 | 11 | 22 | 88 |
| Days Min Temp ≤ 32 °F | 31 | 28 | 30 | 23 | 8 | 1 | 0 | 0 | 6 | 22 | 29 | 31 | 209 |
| Days Min Temp ≤ 0 °F | 16 | 12 | 4 | 0 | 0 | 0 | 0 | 0 | 0 | 0 | 5 | 14 | 51 |
| Heating Degree Days | 1779 | 1434 | 1185 | 742 | 409 | 170 | 64 | 97 | 363 | 725 | 1197 | 1634 | 9799 |
| Cooling Degree Days | 0 | 0 | 0 | 0 | 9 | 46 | 90 | 99 | 9 | 0 | 0 | 0 | 253 |
| Total Precipitation (") | 0.26 | 0.20 | 0.34 | 0.85 | 1.92 | 2.47 | 2.10 | 1.32 | 1.28 | 0.62 | 0.27 | 0.25 | 11.88 |
| Days ≥ 0.1" Precip | 1 | 0 | 1 | 2 | 5 | 6 | 5 | 3 | 3 | 2 | 1 | 1 | 30 |
| Total Snowfall (") | 8.5 | 6.1 | 6.1 | 4.4 | 1.7 | 0.0 | 0.0 | 0.0 | 0.5 | 2.1 | 5.8 | 7.8 | 43.0 |
| Days ≥ 1" Snow Depth | 29 | 24 | 18 | 5 | 1 | 0 | 0 | 0 | 0 | 2 | 10 | 24 | 113 |

## PHILIPSBURG RS *Granite County* ELEVATION 5282 ft LAT/LONG 46° 19 ' N / 113° 18 ' W

| | JAN | FEB | MAR | APR | MAY | JUN | JUL | AUG | SEP | OCT | NOV | DEC | YEAR |
|---|---|---|---|---|---|---|---|---|---|---|---|---|---|
| Maximum Temp °F | 32.7 | 37.7 | 44.5 | 53.4 | 62.2 | 70.9 | 79.7 | 79.6 | 69.3 | 58.2 | 41.5 | 33.3 | 55.2 |
| Minimum Temp °F | 13.7 | 16.3 | 20.9 | 26.6 | 33.2 | 39.7 | 42.4 | 41.2 | 34.1 | 27.8 | 20.5 | 13.9 | 27.5 |
| Mean Temp °F | 23.2 | 27.0 | 32.7 | 40.0 | 47.7 | 55.4 | 61.1 | 60.4 | 51.7 | 43.0 | 31.1 | 23.6 | 41.4 |
| Days Max Temp ≥ 90 °F | 0 | 0 | 0 | 0 | 0 | 0 | 3 | 3 | 0 | 0 | 0 | 0 | 6 |
| Days Max Temp ≤ 32 °F | 13 | 7 | 3 | 0 | 0 | 0 | 0 | 0 | 0 | 0 | 6 | 13 | 42 |
| Days Min Temp ≤ 32 °F | 29 | 26 | 28 | 24 | 15 | 4 | 1 | 2 | 13 | 23 | 26 | 29 | 220 |
| Days Min Temp ≤ 0 °F | 6 | 4 | 2 | 0 | 0 | 0 | 0 | 0 | 0 | 0 | 2 | 5 | 19 |
| Heating Degree Days | 1289 | 1067 | 995 | 743 | 530 | 291 | 137 | 154 | 394 | 675 | 1012 | 1271 | 8558 |
| Cooling Degree Days | 0 | 0 | 0 | 0 | 1 | 10 | 26 | 24 | 1 | 0 | 0 | 0 | 62 |
| Total Precipitation (") | 0.66 | 0.46 | 0.83 | 1.45 | 2.42 | 2.27 | 1.30 | 1.65 | 1.43 | 1.03 | 0.64 | 0.58 | 14.72 |
| Days ≥ 0.1" Precip | 2 | 2 | 3 | 5 | 7 | 7 | 4 | 5 | 4 | 3 | 2 | 2 | 46 |
| Total Snowfall (") | na | na | na | na | 1.0 | 0.0 | 0.0 | 0.1 | *0.3* | na | na | na | na |
| Days ≥ 1" Snow Depth | *25* | *18* | *11* | 2 | 1 | 0 | 0 | 0 | 0 | *1* | na | na | na |

## PLEVNA *Fallon County* ELEVATION 2772 ft LAT/LONG 46° 25 ' N / 104° 30 ' W

| | JAN | FEB | MAR | APR | MAY | JUN | JUL | AUG | SEP | OCT | NOV | DEC | YEAR |
|---|---|---|---|---|---|---|---|---|---|---|---|---|---|
| Maximum Temp °F | 27.9 | 33.9 | 45.4 | 58.9 | 70.3 | 79.9 | 88.5 | 87.2 | 75.3 | 61.9 | 43.2 | 31.5 | 58.7 |
| Minimum Temp °F | 2.6 | 9.4 | 19.6 | 30.5 | 40.7 | 49.8 | 54.9 | 52.4 | 41.8 | 30.7 | 18.2 | 7.0 | 29.8 |
| Mean Temp °F | 15.2 | 21.8 | 32.5 | 44.7 | 55.5 | 64.9 | 71.7 | 69.8 | 58.6 | 46.3 | 30.6 | 19.3 | 44.2 |
| Days Max Temp ≥ 90 °F | 0 | 0 | 0 | 0 | 1 | 5 | 14 | 13 | 3 | 0 | 0 | 0 | 36 |
| Days Max Temp ≤ 32 °F | 16 | 11 | 5 | 0 | 0 | 0 | 0 | 0 | 0 | 0 | 6 | *14* | 52 |
| Days Min Temp ≤ 32 °F | 30 | 28 | 29 | 18 | 5 | 0 | 0 | 0 | 4 | 18 | 28 | 31 | 191 |
| Days Min Temp ≤ 0 °F | *11* | 7 | 3 | 0 | 0 | 0 | 0 | 0 | 0 | 0 | 2 | *8* | 31 |
| Heating Degree Days | 1541 | 1216 | 999 | 603 | 308 | 88 | 16 | 33 | 224 | 581 | 1027 | 1411 | 8047 |
| Cooling Degree Days | 0 | 0 | 0 | 2 | 27 | 112 | 237 | 192 | 40 | 1 | 0 | 0 | 611 |
| Total Precipitation (") | 0.54 | 0.36 | 0.69 | 1.48 | 2.23 | 2.63 | 1.81 | 1.35 | 1.49 | 1.01 | 0.52 | 0.48 | 14.59 |
| Days ≥ 0.1" Precip | 2 | 1 | 2 | 4 | 5 | 6 | 4 | 3 | 4 | 2 | 2 | 2 | 37 |
| Total Snowfall (") | *6.7* | 3.3 | *6.3* | *3.1* | 0.7 | 0.0 | 0.0 | 0.0 | 0.4 | *1.0* | *3.9* | *5.7* | 31.1 |
| Days ≥ 1" Snow Depth | na | na | na | na | na | *0* | *0* | *0* | *0* | na | na | na | na |

## POLEBRIDGE *Flathead County* ELEVATION 3691 ft LAT/LONG 48° 47 ' N / 114° 16 ' W

| | JAN | FEB | MAR | APR | MAY | JUN | JUL | AUG | SEP | OCT | NOV | DEC | YEAR |
|---|---|---|---|---|---|---|---|---|---|---|---|---|---|
| Maximum Temp °F | 29.2 | 36.0 | 43.5 | *53.9* | 63.4 | 71.0 | 78.9 | 79.2 | 68.0 | 55.3 | 37.7 | 29.5 | 53.8 |
| Minimum Temp °F | 7.2 | 10.5 | 17.2 | *25.3* | 32.4 | 38.8 | *40.5* | 39.2 | 31.7 | 24.0 | 17.8 | 9.3 | 24.5 |
| Mean Temp °F | 18.2 | 23.2 | 30.4 | *39.6* | 48.0 | 54.9 | *59.6* | 59.2 | 49.8 | 39.7 | 27.8 | 19.4 | 39.2 |
| Days Max Temp ≥ 90 °F | 0 | 0 | 0 | 0 | 0 | 0 | 3 | *4* | 1 | 0 | 0 | 0 | 8 |
| Days Max Temp ≤ 32 °F | 17 | 7 | 2 | 0 | 0 | 0 | 0 | 0 | 0 | 1 | 7 | 18 | 52 |
| Days Min Temp ≤ 32 °F | 31 | 28 | 30 | *26* | 16 | 5 | 2 | 4 | 17 | *28* | 28 | 30 | 245 |
| Days Min Temp ≤ 0 °F | 10 | 6 | 3 | 0 | 0 | 0 | 0 | 0 | 0 | 0 | 3 | 8 | 30 |
| Heating Degree Days | 1446 | 1172 | 1066 | *755* | 522 | 302 | *177* | 187 | 451 | 780 | 1111 | 1406 | 9375 |
| Cooling Degree Days | 0 | 0 | 0 | 0 | 1 | 5 | 16 | 12 | 0 | 0 | 0 | 0 | 34 |
| Total Precipitation (") | 2.55 | 1.76 | 1.42 | 1.18 | 1.71 | 2.28 | 1.45 | 1.28 | 1.24 | 1.36 | 2.28 | 2.44 | 20.95 |
| Days ≥ 0.1" Precip | 8 | 6 | 6 | *4* | 5 | 6 | 5 | 4 | 4 | 5 | 7 | 8 | 68 |
| Total Snowfall (") | *31.6* | *18.5* | 9.0 | 3.2 | 0.2 | 0.4 | 0.0 | 0.0 | 0.3 | 2.5 | *14.7* | *22.6* | 103.0 |
| Days ≥ 1" Snow Depth | 31 | 28 | 28 | 10 | 0 | 0 | 0 | 0 | 0 | *2* | 16 | 29 | 144 |

## POLSON KERR DAM *Lake County* ELEVATION 2733 ft LAT/LONG 47° 41 ' N / 114° 15 ' W

| | JAN | FEB | MAR | APR | MAY | JUN | JUL | AUG | SEP | OCT | NOV | DEC | YEAR |
|---|---|---|---|---|---|---|---|---|---|---|---|---|---|
| Maximum Temp °F | 32.5 | 38.3 | 47.1 | 57.4 | 65.9 | 72.9 | 81.8 | 82.0 | 70.4 | 57.6 | 41.2 | 33.2 | 56.7 |
| Minimum Temp °F | 19.8 | 22.8 | 27.4 | 33.9 | 41.1 | 47.9 | 52.1 | 51.9 | 43.4 | 35.0 | 27.5 | 21.3 | 35.3 |
| Mean Temp °F | 26.2 | 30.6 | 37.3 | 45.7 | 53.5 | 60.4 | 67.0 | 67.0 | 56.9 | 46.3 | 34.4 | 27.3 | 46.1 |
| Days Max Temp ≥ 90 °F | 0 | 0 | 0 | 0 | 0 | 1 | 6 | 6 | 0 | 0 | 0 | 0 | 13 |
| Days Max Temp ≤ 32 °F | 13 | 7 | 2 | 0 | 0 | 0 | 0 | 0 | 0 | 0 | 4 | 13 | 39 |
| Days Min Temp ≤ 32 °F | 28 | 24 | 24 | 13 | 3 | 0 | 0 | 0 | 2 | 11 | 22 | 28 | 155 |
| Days Min Temp ≤ 0 °F | 3 | 1 | 0 | 0 | 0 | 0 | 0 | 0 | 0 | 0 | 0 | 1 | 5 |
| Heating Degree Days | 1197 | 966 | 853 | 572 | 352 | 161 | 45 | 48 | 249 | 571 | 912 | 1163 | 7089 |
| Cooling Degree Days | 0 | 0 | 0 | 0 | 7 | 40 | 118 | 128 | 14 | 0 | 0 | 0 | 307 |
| Total Precipitation (") | 1.04 | 0.76 | 0.81 | 1.14 | 2.03 | 2.21 | 1.25 | 1.27 | 1.20 | 0.91 | 0.98 | 1.01 | 14.61 |
| Days ≥ 0.1" Precip | 4 | 3 | 3 | 3 | 6 | 6 | 4 | 4 | 4 | 3 | 3 | 4 | 47 |
| Total Snowfall (") | na | na | na | na | *0.0* | *0.0* | *0.0* | *0.0* | *0.0* | *0.1* | na | na | na |
| Days ≥ 1" Snow Depth | *23* | na | *7* | *0* | 0 | 0 | 0 | 0 | 0 | *0* | na | na | na |

## PONY *Madison County* ELEVATION 5505 ft LAT/LONG 45° 40 ' N / 111° 54 ' W

| | JAN | FEB | MAR | APR | MAY | JUN | JUL | AUG | SEP | OCT | NOV | DEC | YEAR |
|---|---|---|---|---|---|---|---|---|---|---|---|---|---|
| Maximum Temp °F | 33.7 | 37.3 | 42.9 | 52.3 | 61.1 | 69.5 | 77.6 | 77.1 | 66.8 | 56.0 | 41.1 | 34.1 | 54.1 |
| Minimum Temp °F | 12.5 | 15.2 | 20.6 | 28.2 | 36.2 | 42.9 | 48.2 | 46.9 | 39.0 | 30.7 | 20.0 | 13.5 | 29.5 |
| Mean Temp °F | 23.1 | 26.3 | 31.8 | 40.3 | 48.7 | 56.2 | 62.9 | 62.0 | 52.9 | 43.4 | 30.6 | 23.8 | 41.8 |
| Days Max Temp ≥ 90 °F | 0 | 0 | 0 | 0 | 0 | 0 | 1 | 1 | 0 | 0 | 0 | 0 | 2 |
| Days Max Temp ≤ 32 °F | 12 | 8 | 4 | 1 | 0 | 0 | 0 | 0 | 0 | 1 | 6 | 13 | 45 |
| Days Min Temp ≤ 32 °F | 30 | 27 | 28 | 22 | 10 | 2 | 0 | 0 | 6 | 18 | 27 | 30 | 200 |
| Days Min Temp ≤ 0 °F | *6* | 3 | 2 | 0 | 0 | 0 | 0 | 0 | 0 | 0 | 2 | 4 | 17 |
| Heating Degree Days | 1294 | 1087 | 1023 | 734 | 500 | 268 | 103 | 124 | 359 | 663 | 1026 | 1272 | 8453 |
| Cooling Degree Days | 0 | 0 | 0 | 0 | 1 | 14 | 55 | 38 | 4 | 0 | 0 | 0 | 112 |
| Total Precipitation (") | 0.65 | 0.60 | 1.37 | 1.90 | 3.10 | 2.51 | 1.51 | 1.43 | 1.82 | 1.32 | 0.91 | 0.61 | 17.73 |
| Days ≥ 0.1" Precip | 2 | 2 | 4 | 5 | 8 | 7 | 4 | 4 | 5 | 4 | 3 | 3 | 51 |
| Total Snowfall (") | na | *9.1* | *17.7* | *12.2* | *4.0* | 0.1 | 0.0 | 0.1 | 3.8 | na | na | na | na |
| Days ≥ 1" Snow Depth | na | 18 | na | *4* | 1 | 0 | 0 | 0 | 1 | na | na | *23* | na |

## POTOMAC *Missoula County* ELEVATION 3602 ft LAT/LONG 46° 53 ' N / 113° 36 ' W

| | JAN | FEB | MAR | APR | MAY | JUN | JUL | AUG | SEP | OCT | NOV | DEC | YEAR |
|---|---|---|---|---|---|---|---|---|---|---|---|---|---|
| Maximum Temp °F | 30.6 | 38.0 | 46.7 | 57.1 | 66.5 | 73.8 | 82.4 | 83.1 | 71.3 | 58.0 | 39.7 | 29.3 | 56.4 |
| Minimum Temp °F | 9.2 | 12.8 | 19.9 | 26.1 | 32.9 | 39.6 | 41.2 | 39.9 | 32.7 | 24.9 | 18.3 | 8.4 | 25.5 |
| Mean Temp °F | 19.9 | 25.4 | 33.4 | 41.6 | 49.7 | 56.7 | 61.8 | 61.5 | 52.0 | 41.5 | 29.0 | 18.8 | 40.9 |
| Days Max Temp ≥ 90 °F | 0 | 0 | 0 | 0 | 0 | 1 | *6* | 7 | 1 | 0 | 0 | 0 | 15 |
| Days Max Temp ≤ 32 °F | *14* | 6 | 2 | 0 | 0 | 0 | 0 | 0 | 0 | 0 | 6 | 18 | 46 |
| Days Min Temp ≤ 32 °F | 30 | 27 | 30 | *25* | *16* | 4 | 2 | 3 | 16 | *26* | 28 | 30 | 237 |
| Days Min Temp ≤ 0 °F | 8 | 5 | 1 | 0 | 0 | 0 | 0 | 0 | 0 | 0 | 2 | 7 | 23 |
| Heating Degree Days | 1391 | 1112 | 973 | 697 | 469 | 251 | 122 | 128 | 384 | 722 | 1074 | 1425 | 8748 |
| Cooling Degree Days | 0 | 0 | 0 | 0 | 3 | 10 | 29 | 27 | 1 | 0 | 0 | 0 | 70 |
| Total Precipitation (") | 1.56 | 1.02 | 0.76 | 0.97 | 1.63 | 1.91 | 0.90 | 1.04 | 1.12 | 0.93 | 1.22 | 1.53 | 14.59 |
| Days ≥ 0.1" Precip | 5 | 3 | 3 | 3 | 5 | 5 | 3 | 3 | 3 | 3 | 5 | 5 | 46 |
| Total Snowfall (") | 16.8 | *9.3* | *4.9* | na | 0.1 | 0.0 | 0.0 | 0.0 | 0.1 | *0.6* | *8.1* | *14.6* | na |
| Days ≥ 1" Snow Depth | 30 | 22 | *11* | 0 | 0 | 0 | 0 | 0 | 0 | 1 | 10 | 25 | 99 |

## POWDERVILLE 8 NNE *Custer County* ELEVATION 2802 ft LAT/LONG 45° 51 ' N / 105° 1 ' W

| | JAN | FEB | MAR | APR | MAY | JUN | JUL | AUG | SEP | OCT | NOV | DEC | YEAR |
|---|---|---|---|---|---|---|---|---|---|---|---|---|---|
| Maximum Temp °F | 29.6 | 36.5 | 48.1 | 60.2 | 71.3 | 81.1 | 89.8 | 88.7 | 76.5 | 62.0 | 44.3 | 33.0 | 60.1 |
| Minimum Temp °F | 4.2 | 11.0 | 21.0 | 31.5 | 41.4 | 50.8 | 56.3 | 54.0 | 42.3 | 30.5 | 18.8 | 7.9 | 30.8 |
| Mean Temp °F | 16.9 | 23.5 | 34.4 | 45.9 | 56.4 | 66.0 | 73.1 | 71.6 | 59.4 | 46.3 | 31.5 | 20.5 | 45.5 |
| Days Max Temp ≥ 90 °F | 0 | 0 | 0 | 0 | 1 | 6 | 17 | 16 | 5 | 0 | 0 | 0 | 45 |
| Days Max Temp ≤ 32 °F | 15 | 10 | 3 | 0 | 0 | 0 | 0 | 0 | 0 | 0 | *5* | 13 | 46 |
| Days Min Temp ≤ 32 °F | 31 | 27 | *28* | *16* | 4 | 0 | 0 | 0 | 4 | *17* | 28 | 31 | 186 |
| Days Min Temp ≤ 0 °F | 12 | 7 | 2 | 0 | 0 | 0 | 0 | 0 | 0 | 0 | 2 | 8 | 31 |
| Heating Degree Days | 1486 | 1164 | 942 | 567 | 281 | 71 | 11 | 19 | 204 | 575 | 997 | 1375 | 7692 |
| Cooling Degree Days | 0 | 0 | 0 | 2 | 27 | 129 | 280 | 249 | 45 | 0 | 0 | 0 | 732 |
| Total Precipitation (") | 0.44 | 0.29 | 0.47 | 1.65 | 2.14 | 2.82 | 1.63 | 0.97 | 1.44 | 1.18 | 0.54 | 0.47 | 14.04 |
| Days ≥ 0.1" Precip | na | *1* | na | *4* | *5* | 6 | 3 | 2 | 3 | *3* | na | *2* | na |
| Total Snowfall (") | na | na | na | na | na | *0.0* | *0.0* | *0.0* | 0.2 | *0.6* | na | na | na |
| Days ≥ 1" Snow Depth | na | na | na | na | *0* | 0 | 0 | 0 | 0 | *0* | na | na | na |

## RAPELJE 4 S *Stillwater County* ELEVATION 4134 ft LAT/LONG 45° 55 ' N / 109° 15 ' W

| | JAN | FEB | MAR | APR | MAY | JUN | JUL | AUG | SEP | OCT | NOV | DEC | YEAR |
|---|---|---|---|---|---|---|---|---|---|---|---|---|---|
| Maximum Temp °F | 35.2 | 40.6 | 47.4 | 57.6 | 67.5 | 77.3 | 85.9 | 85.5 | 73.7 | 61.7 | 44.9 | 36.9 | 59.5 |
| Minimum Temp °F | 11.6 | 15.4 | 21.6 | 29.6 | 38.1 | 46.0 | 51.4 | 50.3 | 41.1 | 32.1 | 21.6 | 13.9 | 31.1 |
| Mean Temp °F | 23.4 | 28.0 | 34.5 | 43.6 | 52.8 | 61.7 | 68.7 | 67.9 | 57.5 | 46.9 | 33.3 | 25.4 | 45.3 |
| Days Max Temp ≥ 90 °F | 0 | 0 | 0 | 0 | 0 | 4 | 12 | 12 | 3 | 0 | 0 | 0 | 31 |
| Days Max Temp ≤ 32 °F | 11 | 7 | 4 | 1 | 0 | 0 | 0 | 0 | 0 | 1 | 4 | 10 | 38 |
| Days Min Temp ≤ 32 °F | 29 | 26 | 28 | 20 | 7 | 0 | 0 | 0 | 4 | 16 | 26 | 28 | 184 |
| Days Min Temp ≤ 0 °F | 9 | 5 | 2 | 0 | 0 | 0 | 0 | 0 | 0 | 0 | 2 | 6 | 24 |
| Heating Degree Days | 1284 | 1038 | 938 | 635 | 377 | 145 | 32 | 43 | 245 | 556 | 946 | 1220 | 7459 |
| Cooling Degree Days | 0 | 0 | 0 | 0 | 7 | 60 | 151 | 139 | 26 | 3 | 0 | 0 | 386 |
| Total Precipitation (") | 0.65 | 0.58 | 1.05 | 1.59 | 2.65 | 2.30 | 1.65 | 1.35 | 1.46 | 1.33 | 0.73 | 0.56 | 15.90 |
| Days ≥ 0.1" Precip | 2 | 2 | 3 | 4 | 6 | 6 | 4 | 3 | 3 | 3 | 2 | 2 | 40 |
| Total Snowfall (") | 11.0 | 9.1 | 13.8 | *7.8* | 1.7 | 0.0 | 0.0 | 0.0 | 1.6 | *5.6* | *9.4* | 10.0 | 70.0 |
| Days ≥ 1" Snow Depth | na | na | na | *2* | 0 | 0 | 0 | 0 | 0 | *1* | na | na | na |

## RAYMOND BORDER STN *Sheridan County* ELEVATION 2352 ft LAT/LONG 49° 0 ' N / 104° 35 ' W

| | JAN | FEB | MAR | APR | MAY | JUN | JUL | AUG | SEP | OCT | NOV | DEC | YEAR |
|---|---|---|---|---|---|---|---|---|---|---|---|---|---|
| Maximum Temp °F | 18.7 | 26.1 | 39.1 | 56.0 | 68.9 | 77.2 | 83.4 | 83.0 | 71.2 | 58.5 | 36.8 | 24.3 | 53.6 |
| Minimum Temp °F | -1.9 | 4.7 | 16.7 | 29.3 | 40.5 | 49.3 | 53.4 | 51.8 | 41.8 | 31.3 | 16.2 | 3.5 | 28.1 |
| Mean Temp °F | 8.4 | 15.3 | 28.0 | 42.7 | 54.7 | 63.3 | 68.4 | 67.4 | 56.5 | 44.9 | 26.5 | 13.9 | 40.8 |
| Days Max Temp ≥ 90 °F | 0 | 0 | 0 | 0 | 1 | 3 | 7 | 9 | 2 | 0 | 0 | 0 | 22 |
| Days Max Temp ≤ 32 °F | 23 | 16 | 9 | 1 | 0 | 0 | 0 | 0 | 0 | 1 | 11 | 20 | 81 |
| Days Min Temp ≤ 32 °F | 30 | 27 | 29 | 19 | 5 | 0 | 0 | 0 | 4 | 17 | 28 | 31 | 190 |
| Days Min Temp ≤ 0 °F | 17 | 11 | 4 | 0 | 0 | 0 | 0 | 0 | 0 | 0 | 4 | 12 | 48 |
| Heating Degree Days | 1755 | 1396 | 1143 | 663 | 330 | 111 | 33 | 60 | 274 | 616 | 1149 | 1580 | 9110 |
| Cooling Degree Days | 0 | 0 | 0 | 1 | 22 | 80 | 154 | 153 | 26 | 1 | 0 | 0 | 437 |
| Total Precipitation (") | 0.30 | 0.19 | 0.37 | 0.93 | 2.01 | 2.61 | 2.21 | 1.69 | 1.51 | 0.71 | 0.23 | 0.33 | 13.09 |
| Days ≥ 0.1" Precip | 1 | *1* | *1* | 2 | 5 | 6 | 5 | 4 | 4 | 2 | 1 | *1* | 33 |
| Total Snowfall (") | na | na | na | na | 0.4 | 0.0 | 0.0 | 0.0 | 0.0 | *0.7* | na | na | na |
| Days ≥ 1" Snow Depth | na | na | na | na | 0 | 0 | 0 | 0 | 0 | *0* | na | na | na |

## RED LODGE *Carbon County* ELEVATION 5850 ft LAT/LONG 45° 11 ' N / 109° 15 ' W

| | JAN | FEB | MAR | APR | MAY | JUN | JUL | AUG | SEP | OCT | NOV | DEC | YEAR |
|---|---|---|---|---|---|---|---|---|---|---|---|---|---|
| Maximum Temp °F | 33.1 | 36.5 | 42.7 | 52.1 | 61.8 | 70.9 | 78.6 | 77.6 | 66.7 | 55.3 | 40.5 | 34.0 | 54.2 |
| Minimum Temp °F | 12.3 | 15.5 | 20.6 | 28.5 | 37.1 | 44.1 | 49.9 | 49.1 | 40.3 | 32.1 | 21.1 | 14.3 | 30.4 |
| Mean Temp °F | 22.7 | 26.1 | 31.7 | 40.3 | 49.5 | 57.5 | 64.3 | 63.4 | 53.5 | 43.7 | 30.8 | 24.2 | 42.3 |
| Days Max Temp ≥ 90 °F | 0 | 0 | 0 | 0 | 0 | 0 | 2 | 1 | 0 | 0 | 0 | 0 | 3 |
| Days Max Temp ≤ 32 °F | 13 | 8 | 5 | 2 | 0 | 0 | 0 | 0 | 0 | 1 | 6 | 13 | 48 |
| Days Min Temp ≤ 32 °F | 29 | 27 | 28 | 20 | 9 | 1 | 0 | 0 | 5 | 16 | 25 | 29 | 189 |
| Days Min Temp ≤ 0 °F | 7 | 4 | 1 | 0 | 0 | 0 | 0 | 0 | 0 | 0 | 2 | 5 | 19 |
| Heating Degree Days | 1305 | 1093 | 1027 | 733 | 476 | 238 | 84 | 102 | 348 | 653 | 1019 | 1259 | 8337 |
| Cooling Degree Days | 0 | 0 | 0 | 0 | 3 | 24 | 68 | 52 | 8 | 0 | 0 | 0 | 155 |
| Total Precipitation (") | 1.54 | 1.22 | 2.64 | 3.36 | 3.86 | 2.88 | 1.59 | 1.44 | 2.35 | 2.07 | 1.64 | 1.32 | 25.91 |
| Days ≥ 0.1" Precip | 5 | 4 | 6 | 7 | 7 | 7 | 5 | 4 | 5 | 5 | 4 | 4 | 63 |
| Total Snowfall (") | 24.3 | 18.7 | 31.5 | 29.3 | 10.4 | 0.2 | 0.0 | 0.0 | 7.5 | 14.3 | 20.6 | 22.4 | 179.2 |
| Days ≥ 1" Snow Depth | 28 | 23 | 21 | 10 | 2 | 0 | 0 | 0 | 2 | 5 | 16 | 24 | 131 |

## REDSTONE *Sheridan County* ELEVATION 2113 ft LAT/LONG 48° 50 ' N / 104° 57 ' W

| | JAN | FEB | MAR | APR | MAY | JUN | JUL | AUG | SEP | OCT | NOV | DEC | YEAR |
|---|---|---|---|---|---|---|---|---|---|---|---|---|---|
| Maximum Temp °F | 20.8 | 28.0 | 40.8 | 56.6 | 69.6 | 78.3 | 84.7 | 83.7 | 71.4 | 58.1 | 38.1 | 26.0 | 54.7 |
| Minimum Temp °F | -3.2 | 3.7 | 15.8 | 27.2 | 38.7 | 47.4 | 51.2 | 48.9 | 38.3 | 28.5 | 14.2 | 2.1 | 26.1 |
| Mean Temp °F | 8.8 | 15.9 | 28.3 | 41.9 | 54.1 | 62.9 | 68.0 | 66.3 | 54.7 | 43.1 | 26.1 | 14.1 | 40.4 |
| Days Max Temp ≥ 90 °F | 0 | 0 | 0 | 0 | 1 | 4 | 9 | 9 | 2 | 0 | 0 | 0 | 25 |
| Days Max Temp ≤ 32 °F | 21 | 15 | 8 | 1 | 0 | 0 | 0 | 0 | 0 | 1 | 9 | 19 | 74 |
| Days Min Temp ≤ 32 °F | 31 | 28 | 30 | 22 | 8 | 1 | 0 | 1 | 8 | ***21*** | 29 | 31 | 210 |
| Days Min Temp ≤ 0 °F | 18 | 12 | 4 | 0 | 0 | 0 | 0 | 0 | 0 | 0 | 5 | 14 | 53 |
| Heating Degree Days | 1741 | 1382 | 1131 | 687 | 344 | 118 | 36 | 73 | 316 | 674 | 1161 | 1574 | 9237 |
| Cooling Degree Days | 0 | 0 | 0 | 1 | 16 | 72 | 130 | 129 | 13 | 0 | 0 | 0 | 361 |
| Total Precipitation (") | 0.33 | 0.22 | 0.53 | 0.95 | 1.98 | 2.31 | 2.07 | 1.39 | 1.36 | 0.66 | 0.31 | 0.34 | 12.45 |
| Days ≥ 0.1" Precip | 1 | 1 | 2 | 3 | 6 | 5 | 5 | 3 | 3 | 2 | 1 | 1 | 33 |
| Total Snowfall (") | na | na | na | na | ***0.1*** | ***0.0*** | ***0.0*** | ***0.0*** | ***0.0*** | na | na | na | na |
| Days ≥ 1" Snow Depth | 23 | 16 | 10 | ***2*** | 0 | 0 | 0 | 0 | 0 | ***1*** | ***6*** | ***18*** | 76 |

## RIDGEWAY 1 S *Carter County* ELEVATION 3304 ft LAT/LONG 45° 30 ' N / 104° 26 ' W

| | JAN | FEB | MAR | APR | MAY | JUN | JUL | AUG | SEP | OCT | NOV | DEC | YEAR |
|---|---|---|---|---|---|---|---|---|---|---|---|---|---|
| Maximum Temp °F | 28.4 | 33.9 | 43.6 | 56.9 | 67.8 | 76.6 | 84.7 | 84.3 | 72.6 | 60.0 | 42.5 | 32.2 | 57.0 |
| Minimum Temp °F | 2.9 | 8.8 | 19.0 | 30.5 | 40.5 | 49.6 | 55.2 | 52.6 | 41.3 | 29.9 | 17.1 | 6.6 | 29.5 |
| Mean Temp °F | 15.7 | 21.4 | 31.3 | 43.7 | 54.1 | 63.1 | 70.0 | 68.5 | 57.0 | 45.0 | 29.9 | 19.4 | 43.3 |
| Days Max Temp ≥ 90 °F | 0 | 0 | 0 | 0 | 0 | 3 | 9 | 9 | 2 | 0 | 0 | 0 | 23 |
| Days Max Temp ≤ 32 °F | 16 | 11 | 6 | 1 | 0 | 0 | 0 | 0 | 0 | 0 | ***6*** | 14 | 54 |
| Days Min Temp ≤ 32 °F | 31 | 28 | 29 | 18 | 5 | 0 | 0 | 0 | 5 | ***19*** | 29 | 31 | 195 |
| Days Min Temp ≤ 0 °F | 12 | ***7*** | 2 | 0 | 0 | 0 | 0 | 0 | 0 | 0 | 2 | 8 | 31 |
| Heating Degree Days | 1526 | 1222 | 1035 | 633 | 342 | 119 | 28 | 41 | 264 | 614 | 1049 | 1406 | 8279 |
| Cooling Degree Days | 0 | 0 | 0 | 1 | 15 | 80 | 194 | 162 | 27 | 0 | 0 | 0 | 479 |
| Total Precipitation (") | 0.37 | 0.35 | 0.69 | 1.59 | 2.23 | 2.61 | 1.71 | 1.03 | 1.25 | 1.03 | 0.50 | 0.48 | 13.84 |
| Days ≥ 0.1" Precip | ***1*** | 1 | 2 | 4 | 5 | 6 | 4 | 3 | 3 | 2 | 2 | 1 | 34 |
| Total Snowfall (") | ***5.1*** | 4.3 | 7.5 | na | 0.4 | 0.1 | 0.0 | 0.0 | 0.8 | 1.7 | 4.4 | 5.9 | na |
| Days ≥ 1" Snow Depth | na | na | na | na | ***0*** | 0 | 0 | 0 | 0 | na | na | na | na |

## ROCK SPRINGS *Rosebud County* ELEVATION 3022 ft LAT/LONG 46° 49 ' N / 106° 14 ' W

| | JAN | FEB | MAR | APR | MAY | JUN | JUL | AUG | SEP | OCT | NOV | DEC | YEAR |
|---|---|---|---|---|---|---|---|---|---|---|---|---|---|
| Maximum Temp °F | 26.3 | 32.2 | 44.2 | 56.9 | 67.6 | 76.9 | 85.0 | 84.3 | 72.1 | 59.7 | 41.4 | 30.0 | 56.4 |
| Minimum Temp °F | 5.0 | 10.5 | 20.5 | 30.6 | 40.7 | 49.5 | 54.5 | 53.3 | 42.4 | 32.0 | 19.0 | 8.1 | 30.5 |
| Mean Temp °F | 15.7 | 21.4 | 32.4 | 43.8 | 54.1 | 63.2 | 69.7 | 68.9 | 57.3 | 45.9 | 30.3 | 19.0 | 43.5 |
| Days Max Temp ≥ 90 °F | 0 | 0 | 0 | 0 | 0 | 3 | ***9*** | 9 | 2 | 0 | 0 | 0 | 23 |
| Days Max Temp ≤ 32 °F | 17 | ***12*** | 6 | 1 | 0 | 0 | 0 | 0 | 0 | 1 | ***7*** | ***15*** | 59 |
| Days Min Temp ≤ 32 °F | 30 | 27 | 28 | ***18*** | 4 | 0 | 0 | 0 | 3 | ***15*** | 27 | 30 | 182 |
| Days Min Temp ≤ 0 °F | 12 | 8 | 2 | 0 | 0 | 0 | 0 | 0 | 0 | 0 | 2 | 8 | 32 |
| Heating Degree Days | 1527 | 1227 | 1003 | 629 | 343 | 114 | 25 | 40 | 251 | 582 | 1036 | 1421 | 8198 |
| Cooling Degree Days | 0 | 0 | 0 | 1 | 16 | 79 | 178 | 173 | 24 | 1 | 0 | 0 | 472 |
| Total Precipitation (") | 0.32 | 0.18 | 0.31 | 1.08 | 2.09 | 2.55 | 1.46 | 0.85 | 1.24 | 0.75 | 0.33 | 0.28 | 11.44 |
| Days ≥ 0.1" Precip | 1 | 0 | 1 | 3 | 5 | 6 | 4 | 3 | 3 | 2 | 1 | 1 | 30 |
| Total Snowfall (") | na | na | na | na | ***0.0*** | ***0.0*** | ***0.0*** | ***0.0*** | ***0.0*** | na | na | na | na |
| Days ≥ 1" Snow Depth | na | na | na | na | ***0*** | ***0*** | ***0*** | ***0*** | ***0*** | na | na | na | na |

## ROGERS PASS 9 NNE *Lewis and Clark County* ELEVATION 4300 ft LAT/LONG 47° 11 ' N / 112° 18 ' W

| | JAN | FEB | MAR | APR | MAY | JUN | JUL | AUG | SEP | OCT | NOV | DEC | YEAR |
|---|---|---|---|---|---|---|---|---|---|---|---|---|---|
| Maximum Temp °F | 32.4 | 38.5 | 44.2 | 54.0 | 63.4 | 72.0 | 81.2 | 80.8 | 69.1 | 58.0 | 41.4 | 33.8 | 55.7 |
| Minimum Temp °F | 12.2 | 17.8 | 22.4 | 29.8 | 37.5 | 44.4 | 49.4 | 48.5 | 38.8 | 32.6 | 22.1 | 14.8 | 30.9 |
| Mean Temp °F | 22.3 | 28.2 | 33.3 | 41.9 | 50.5 | 58.2 | 65.3 | 64.7 | 54.0 | 45.3 | 31.8 | 24.3 | 43.3 |
| Days Max Temp ≥ 90 °F | 0 | 0 | 0 | 0 | 0 | 1 | 5 | 4 | 0 | 0 | 0 | 0 | 10 |
| Days Max Temp ≤ 32 °F | 12 | 7 | 4 | 1 | 0 | 0 | 0 | 0 | 0 | 1 | 5 | 11 | 41 |
| Days Min Temp ≤ 32 °F | 28 | 24 | 26 | 19 | 7 | 1 | 0 | 0 | 6 | 14 | 24 | 29 | 178 |
| Days Min Temp ≤ 0 °F | 9 | 5 | 2 | 0 | 0 | 0 | 0 | 0 | 0 | 0 | 3 | 5 | 24 |
| Heating Degree Days | 1319 | 1033 | 976 | 687 | 445 | 220 | 72 | 94 | 336 | 605 | 991 | 1255 | 8033 |
| Cooling Degree Days | 0 | 0 | 0 | 1 | 3 | 26 | 77 | 83 | 14 | 1 | 0 | 0 | 205 |
| Total Precipitation (") | 0.91 | 0.65 | 1.30 | 1.75 | 3.20 | 3.09 | 1.62 | 1.78 | 1.78 | 1.11 | 0.69 | 0.99 | 18.87 |
| Days ≥ 0.1" Precip | 3 | 2 | 4 | 5 | 7 | 6 | 4 | 4 | 4 | 3 | 3 | 4 | 49 |
| Total Snowfall (") | ***14.3*** | ***12.4*** | na | na | ***5.5*** | ***0.0*** | 0.0 | 0.0 | ***3.6*** | na | na | na | na |
| Days ≥ 1" Snow Depth | ***17*** | ***14*** | 11 | 5 | 1 | 0 | 0 | 0 | 1 | 2 | ***7*** | 13 | 71 |

## ROUNDUP *Musselshell County* ELEVATION 3232 ft LAT/LONG 46° 28' N / 108° 34' W

| | JAN | FEB | MAR | APR | MAY | JUN | JUL | AUG | SEP | OCT | NOV | DEC | YEAR |
|---|---|---|---|---|---|---|---|---|---|---|---|---|---|
| Maximum Temp °F | 36.1 | 42.7 | 50.6 | 60.8 | 71.0 | 80.1 | ***87.6*** | 87.0 | 75.6 | 63.9 | 46.8 | 37.8 | 61.7 |
| Minimum Temp °F | 11.6 | 16.4 | 23.4 | 32.4 | 41.6 | 49.7 | 54.6 | 52.5 | 42.5 | 33.4 | 22.5 | 14.5 | 32.9 |
| Mean Temp °F | 23.9 | 29.6 | 37.1 | 46.6 | 56.4 | 65.0 | ***71.1*** | 69.7 | 59.1 | 48.7 | 34.7 | 26.1 | 47.3 |
| Days Max Temp ≥ 90 °F | 0 | 0 | 0 | 0 | 1 | 5 | ***14*** | 13 | 4 | 0 | 0 | 0 | 37 |
| Days Max Temp ≤ 32 °F | 10 | 6 | 3 | 0 | 0 | 0 | 0 | 0 | 0 | 1 | 3 | 8 | 31 |
| Days Min Temp ≤ 32 °F | 28 | 25 | 26 | 16 | 3 | 0 | 0 | 0 | 3 | 14 | 24 | 28 | 167 |
| Days Min Temp ≤ 0 °F | 8 | 4 | 2 | 0 | 0 | 0 | 0 | 0 | 0 | 0 | 1 | 5 | 20 |
| Heating Degree Days | 1268 | 994 | 859 | 546 | 277 | 85 | ***16*** | 27 | 206 | 502 | 905 | 1201 | 6886 |
| Cooling Degree Days | 0 | 0 | 0 | 1 | 15 | 95 | 203 | 169 | 29 | 2 | 0 | 0 | 514 |
| Total Precipitation (") | 0.45 | 0.35 | 0.60 | 1.17 | 2.24 | 2.39 | 1.59 | 1.28 | 1.33 | 1.03 | 0.32 | 0.44 | 13.19 |
| Days ≥ 0.1" Precip | 2 | 1 | 2 | 4 | 6 | 6 | 4 | 3 | 3 | 3 | 1 | 2 | 37 |
| Total Snowfall (") | na | na | na | na | na | na | na | na | na | na | na | na | na |
| Days ≥ 1" Snow Depth | na | na | na | na | na | na | na | na | na | na | na | na | na |

## ROY 8 NE *Fergus County* ELEVATION 3451 ft LAT/LONG 47° 25' N / 108° 50' W

| | JAN | FEB | MAR | APR | MAY | JUN | JUL | AUG | SEP | OCT | NOV | DEC | YEAR |
|---|---|---|---|---|---|---|---|---|---|---|---|---|---|
| Maximum Temp °F | 30.3 | 36.5 | 45.6 | 57.0 | 67.4 | 76.6 | 84.6 | 84.3 | 72.4 | 61.0 | 44.1 | 34.0 | 57.8 |
| Minimum Temp °F | 6.6 | 11.6 | 21.2 | 30.8 | 40.5 | 48.3 | 53.1 | 51.9 | 41.7 | 32.0 | 19.3 | 9.8 | 30.6 |
| Mean Temp °F | 18.4 | 24.1 | 33.4 | 43.9 | 53.9 | 62.5 | 68.9 | 68.1 | 57.1 | 46.5 | 31.8 | 22.0 | 44.2 |
| Days Max Temp ≥ 90 °F | 0 | 0 | 0 | 0 | 0 | 3 | 10 | 10 | 2 | 0 | 0 | 0 | 25 |
| Days Max Temp ≤ 32 °F | 14 | 10 | 5 | 1 | 0 | 0 | 0 | 0 | 0 | 1 | 5 | 12 | 48 |
| Days Min Temp ≤ 32 °F | 30 | 26 | 27 | 18 | 4 | 0 | 0 | 0 | 3 | 15 | 26 | 30 | 179 |
| Days Min Temp ≤ 0 °F | 12 | 7 | 2 | 0 | 0 | 0 | 0 | 0 | 0 | 0 | 3 | 8 | 32 |
| Heating Degree Days | 1439 | 1150 | 974 | 626 | 343 | 126 | 28 | 43 | 253 | 568 | 991 | 1330 | 7871 |
| Cooling Degree Days | 0 | 0 | 0 | 0 | 11 | 66 | 155 | 143 | 22 | 2 | 0 | 0 | 399 |
| Total Precipitation (") | 0.55 | 0.38 | 0.73 | 1.24 | 2.72 | 2.46 | 1.82 | 1.51 | 1.31 | 0.73 | 0.41 | 0.49 | 14.35 |
| Days ≥ 0.1" Precip | 2 | 2 | 3 | 3 | 6 | 6 | 4 | 4 | 3 | 2 | 2 | 2 | 39 |
| Total Snowfall (") | 9.5 | 6.5 | 7.8 | 6.8 | 1.3 | 0.1 | 0.0 | 0.0 | 0.2 | 2.2 | 5.1 | 8.0 | 47.5 |
| Days ≥ 1" Snow Depth | 19 | 15 | 11 | 3 | 0 | 0 | 0 | 0 | 0 | 1 | 7 | 16 | 72 |

## RUDYARD 27 N *Hill County* ELEVATION 2982 ft LAT/LONG 48° 59' N / 110° 35' W

| | JAN | FEB | MAR | APR | MAY | JUN | JUL | AUG | SEP | OCT | NOV | DEC | YEAR |
|---|---|---|---|---|---|---|---|---|---|---|---|---|---|
| Maximum Temp °F | 24.9 | 31.7 | 42.7 | 55.6 | ***67.6*** | 75.3 | 82.4 | 82.1 | 70.0 | 58.7 | 39.8 | 30.5 | 55.1 |
| Minimum Temp °F | 3.4 | 8.9 | 19.9 | 30.2 | ***40.5*** | 48.0 | 51.8 | 51.4 | 40.7 | 31.5 | 17.3 | 7.4 | 29.3 |
| Mean Temp °F | 14.0 | 20.3 | 31.4 | 42.9 | ***54.0*** | 61.7 | 67.1 | 66.9 | 55.3 | 45.2 | 28.7 | 19.0 | 42.2 |
| Days Max Temp ≥ 90 °F | 0 | 0 | 0 | 0 | 0 | 2 | 7 | 8 | 1 | 0 | 0 | 0 | 18 |
| Days Max Temp ≤ 32 °F | 16 | 13 | 7 | 1 | 0 | 0 | 0 | 0 | 0 | 1 | ***8*** | 15 | 61 |
| Days Min Temp ≤ 32 °F | 29 | 27 | 27 | ***18*** | ***4*** | 0 | 0 | 0 | ***5*** | ***16*** | 27 | 30 | 183 |
| Days Min Temp ≤ 0 °F | 13 | 9 | 3 | 0 | 0 | 0 | 0 | 0 | 0 | 0 | 3 | 9 | 37 |
| Heating Degree Days | 1567 | 1257 | 1036 | 655 | ***341*** | 142 | 42 | 59 | 298 | ***612*** | 1082 | 1421 | 8512 |
| Cooling Degree Days | 0 | 0 | 0 | 1 | ***10*** | 55 | 116 | 126 | 16 | ***0*** | ***0*** | 0 | 324 |
| Total Precipitation (") | 0.35 | 0.26 | 0.41 | 0.89 | 1.54 | 2.23 | 1.48 | 1.22 | 1.06 | 0.45 | 0.32 | 0.35 | 10.56 |
| Days ≥ 0.1" Precip | ***1*** | ***1*** | ***1*** | 2 | 4 | 5 | 3 | 3 | 2 | 1 | ***1*** | ***1*** | 25 |
| Total Snowfall (") | na | na | na | na | ***0.3*** | 0.0 | 0.0 | 0.0 | 0.5 | ***0.8*** | na | na | na |
| Days ≥ 1" Snow Depth | na | na | na | na | 0 | 0 | 0 | 0 | 0 | na | na | na | na |

## RYEGATE 18 NNW *Golden Valley County* ELEVATION 4482 ft LAT/LONG 46° 32' N / 109° 23' W

| | JAN | FEB | MAR | APR | MAY | JUN | JUL | AUG | SEP | OCT | NOV | DEC | YEAR |
|---|---|---|---|---|---|---|---|---|---|---|---|---|---|
| Maximum Temp °F | 32.6 | 38.0 | 44.8 | 55.4 | 65.0 | 73.8 | 81.5 | 81.2 | 70.0 | 58.7 | 43.1 | 34.8 | 56.6 |
| Minimum Temp °F | 9.7 | 14.0 | 19.8 | 28.0 | 36.6 | 44.2 | 49.5 | 48.3 | 38.7 | 29.8 | 19.0 | 11.8 | 29.1 |
| Mean Temp °F | 21.2 | 26.0 | 32.3 | 41.7 | 50.8 | 59.0 | 65.5 | 64.8 | 54.4 | 44.3 | 31.1 | 23.3 | 42.9 |
| Days Max Temp ≥ 90 °F | 0 | 0 | 0 | 0 | 0 | 2 | 5 | 5 | 1 | 0 | 0 | 0 | 13 |
| Days Max Temp ≤ 32 °F | 13 | 8 | 4 | 1 | 0 | 0 | 0 | 0 | 0 | 1 | 5 | 12 | 44 |
| Days Min Temp ≤ 32 °F | 30 | 27 | 29 | 21 | 9 | 1 | 0 | 0 | 6 | 18 | 27 | 30 | 198 |
| Days Min Temp ≤ 0 °F | 9 | 5 | 2 | 0 | 0 | 0 | 0 | 0 | 0 | 0 | 2 | 6 | 24 |
| Heating Degree Days | 1354 | 1095 | 1007 | 692 | 434 | 201 | 61 | 79 | 322 | 636 | 1012 | 1286 | 8179 |
| Cooling Degree Days | 0 | 0 | 0 | 0 | 3 | 38 | 89 | 86 | 11 | 1 | 0 | 0 | 228 |
| Total Precipitation (") | 0.56 | 0.34 | 0.55 | 1.05 | 1.90 | 2.43 | 1.72 | 1.44 | 1.12 | 0.77 | 0.46 | 0.47 | 12.81 |
| Days ≥ 0.1" Precip | 2 | 1 | 2 | 3 | 5 | 6 | 5 | 4 | 3 | 3 | 2 | 2 | 38 |
| Total Snowfall (") | 8.6 | 4.8 | 6.5 | 4.7 | 1.2 | 0.0 | 0.0 | 0.0 | 0.8 | 3.2 | 5.2 | 7.3 | 42.3 |
| Days ≥ 1" Snow Depth | 16 | 11 | 8 | 2 | 0 | 0 | 0 | 0 | 0 | 1 | 7 | 12 | 57 |

## SAINT REGIS 1 NE *Mineral County* ELEVATION 2680 ft LAT/LONG 47° 18' N / 115° 6' W

| | JAN | FEB | MAR | APR | MAY | JUN | JUL | AUG | SEP | OCT | NOV | DEC | YEAR |
|---|---|---|---|---|---|---|---|---|---|---|---|---|---|
| Maximum Temp °F | 33.3 | 39.8 | 49.6 | 59.3 | 69.0 | 77.3 | 85.6 | 86.0 | 74.4 | 59.7 | 41.6 | 33.1 | 59.1 |
| Minimum Temp °F | 18.0 | 20.8 | 24.9 | 30.6 | 35.9 | 42.9 | 45.2 | 44.3 | 36.9 | 30.1 | 24.9 | 19.2 | 31.1 |
| Mean Temp °F | 25.7 | 30.3 | 37.3 | 45.0 | 52.5 | 60.1 | 65.4 | 65.1 | 55.7 | 44.9 | 33.3 | 26.2 | 45.1 |
| Days Max Temp ≥ 90 °F | 0 | 0 | 0 | 0 | 1 | 3 | 12 | 12 | 2 | 0 | 0 | 0 | 30 |
| Days Max Temp ≤ 32 °F | 11 | 4 | 1 | 0 | 0 | 0 | 0 | 0 | 0 | 0 | 4 | 12 | 32 |
| Days Min Temp ≤ 32 °F | 30 | 27 | 26 | 18 | 11 | 2 | 0 | 1 | 8 | 20 | 25 | 30 | 198 |
| Days Min Temp ≤ 0 °F | 4 | 2 | 0 | 0 | 0 | 0 | 0 | 0 | 0 | 0 | 1 | 2 | 9 |
| Heating Degree Days | 1213 | 973 | 854 | 595 | 387 | 172 | 67 | 67 | 283 | 617 | 947 | 1198 | 7373 |
| Cooling Degree Days | 0 | 0 | 0 | 0 | 4 | 33 | 85 | 82 | ***11*** | 0 | 0 | 0 | 215 |
| Total Precipitation (") | 2.59 | 1.66 | 1.36 | 1.41 | 1.54 | 1.69 | 1.08 | 1.34 | 1.38 | 1.47 | 2.09 | 2.15 | 19.76 |
| Days ≥ 0.1" Precip | na | na | na | na | ***4*** | 6 | 3 | 4 | 4 | 5 | na | na | na |
| Total Snowfall (") | na | na | na | na | na | na | na | na | na | na | na | na | na |
| Days ≥ 1" Snow Depth | na | na | na | na | na | na | na | na | na | na | na | na | na |

## SAVAGE *Richland County* ELEVATION 1991 ft LAT/LONG 47° 27' N / 104° 21' W

| | JAN | FEB | MAR | APR | MAY | JUN | JUL | AUG | SEP | OCT | NOV | DEC | YEAR |
|---|---|---|---|---|---|---|---|---|---|---|---|---|---|
| Maximum Temp °F | 23.8 | 31.4 | 44.5 | 59.2 | 71.4 | 79.6 | 86.9 | 85.9 | 74.0 | 60.6 | 41.4 | 28.6 | 57.3 |
| Minimum Temp °F | 2.4 | 9.2 | 20.2 | 31.6 | 43.1 | 52.0 | 56.7 | 54.6 | 44.1 | 33.6 | 20.3 | 7.8 | 31.3 |
| Mean Temp °F | 13.1 | 20.3 | 32.4 | 45.4 | 57.3 | 65.9 | 71.8 | 70.3 | 59.1 | 47.1 | 30.9 | 18.2 | 44.3 |
| Days Max Temp ≥ 90 °F | 0 | 0 | 0 | 0 | 1 | 4 | 12 | 12 | 3 | 0 | 0 | 0 | 32 |
| Days Max Temp ≤ 32 °F | 19 | 14 | 6 | 1 | 0 | 0 | 0 | 0 | 0 | 0 | 7 | 17 | 64 |
| Days Min Temp ≤ 32 °F | 30 | 28 | 28 | 16 | 3 | 0 | 0 | 0 | 2 | 13 | 27 | 31 | 178 |
| Days Min Temp ≤ 0 °F | 14 | 9 | 2 | 0 | 0 | 0 | 0 | 0 | 0 | 0 | 2 | 9 | 36 |
| Heating Degree Days | 1605 | 1256 | 1004 | 583 | 259 | 67 | 11 | 27 | 208 | 546 | 1016 | 1444 | 8026 |
| Cooling Degree Days | 0 | 0 | 0 | 3 | 38 | 120 | 247 | 223 | 40 | 1 | 0 | 0 | 672 |
| Total Precipitation (") | 0.36 | 0.23 | 0.48 | 1.31 | 2.06 | 3.06 | 2.00 | 1.29 | 1.49 | 0.83 | 0.38 | 0.40 | 13.89 |
| Days ≥ 0.1" Precip | 1 | 1 | 2 | 4 | 5 | 6 | 4 | 3 | 3 | 2 | 2 | 1 | 34 |
| Total Snowfall (") | 7.3 | 5.4 | 5.6 | 3.5 | 0.9 | 0.0 | 0.0 | 0.0 | 0.3 | 1.3 | 4.6 | 8.8 | 37.7 |
| Days ≥ 1" Snow Depth | 24 | 18 | 8 | 2 | 0 | 0 | 0 | 0 | 0 | 0 | 6 | 18 | 76 |

## SEELEY LAKE RS *Missoula County* ELEVATION 4032 ft LAT/LONG 47° 13' N / 113° 31' W

| | JAN | FEB | MAR | APR | MAY | JUN | JUL | AUG | SEP | OCT | NOV | DEC | YEAR |
|---|---|---|---|---|---|---|---|---|---|---|---|---|---|
| Maximum Temp °F | 30.2 | 37.2 | 44.5 | 53.8 | 63.6 | 71.9 | 80.8 | 81.3 | 70.1 | 56.9 | 38.7 | 29.8 | 54.9 |
| Minimum Temp °F | 10.8 | 13.5 | 20.0 | 27.3 | 34.6 | 41.2 | 43.4 | 42.4 | 35.6 | 29.3 | 21.8 | 12.4 | 27.7 |
| Mean Temp °F | 20.5 | 25.3 | 32.3 | 40.6 | 49.1 | 56.6 | 62.1 | 61.9 | 52.9 | 43.1 | 30.3 | 21.1 | 41.3 |
| Days Max Temp ≥ 90 °F | 0 | 0 | 0 | 0 | 0 | 1 | 4 | 6 | 1 | 0 | 0 | 0 | 12 |
| Days Max Temp ≤ 32 °F | 17 | 7 | 2 | 0 | 0 | 0 | 0 | 0 | 0 | 0 | 7 | 19 | 52 |
| Days Min Temp ≤ 32 °F | 30 | 28 | 29 | 24 | 13 | 2 | 1 | 1 | 10 | ***22*** | 27 | 30 | 217 |
| Days Min Temp ≤ 0 °F | 8 | 5 | 2 | 0 | 0 | 0 | 0 | 0 | 0 | 0 | 2 | 6 | 23 |
| Heating Degree Days | 1373 | 1112 | 1009 | 727 | 487 | 257 | 118 | 123 | 361 | 673 | 1035 | 1355 | 8630 |
| Cooling Degree Days | 0 | 0 | 0 | 0 | 3 | 11 | 34 | 38 | 2 | 0 | 0 | 0 | 88 |
| Total Precipitation (") | 2.82 | 1.68 | 1.37 | 1.17 | 1.90 | 2.22 | 1.11 | 1.31 | 1.41 | 1.31 | 2.08 | 2.54 | 20.92 |
| Days ≥ 0.1" Precip | 8 | 5 | 5 | 4 | 6 | 6 | 4 | 4 | 4 | 4 | 7 | 8 | 65 |
| Total Snowfall (") | 34.0 | 17.2 | 12.6 | 3.9 | 0.7 | 0.0 | 0.0 | 0.0 | 0.0 | 1.3 | 15.7 | 30.4 | 115.8 |
| Days ≥ 1" Snow Depth | 30 | 27 | 27 | 8 | 0 | 0 | 0 | 0 | 0 | 1 | 13 | 28 | 134 |

## SHONKIN 7 S *Chouteau County* ELEVATION 4304 ft LAT/LONG 47° 32' N / 110° 35' W

| | JAN | FEB | MAR | APR | MAY | JUN | JUL | AUG | SEP | OCT | NOV | DEC | YEAR |
|---|---|---|---|---|---|---|---|---|---|---|---|---|---|
| Maximum Temp °F | 32.9 | 38.4 | 43.5 | ***52.7*** | ***62.1*** | ***70.7*** | 78.9 | ***79.3*** | ***67.3*** | ***57.1*** | ***42.4*** | 35.3 | 55.1 |
| Minimum Temp °F | 11.1 | 17.2 | 22.2 | ***30.1*** | ***38.3*** | ***44.7*** | 49.0 | 48.5 | ***40.5*** | ***33.2*** | ***22.5*** | ***16.1*** | 31.1 |
| Mean Temp °F | 22.0 | 27.8 | 32.9 | ***41.4*** | ***50.3*** | ***57.7*** | 64.0 | ***63.7*** | ***54.0*** | ***45.1*** | ***32.5*** | ***26.0*** | 43.1 |
| Days Max Temp ≥ 90 °F | 0 | 0 | 0 | 0 | 0 | 0 | 2 | 3 | 1 | 0 | 0 | 0 | 6 |
| Days Max Temp ≤ 32 °F | 11 | 7 | 4 | 1 | 0 | 0 | 0 | 0 | 0 | 1 | 4 | 10 | 38 |
| Days Min Temp ≤ 32 °F | 25 | 22 | 23 | 16 | 6 | 0 | 0 | 0 | 4 | 12 | 21 | 25 | 154 |
| Days Min Temp ≤ 0 °F | 9 | 5 | 2 | 0 | 0 | 0 | 0 | 0 | 0 | 0 | 2 | 6 | 24 |
| Heating Degree Days | 1331 | 1043 | 989 | ***701*** | ***454*** | ***227*** | 90 | ***100*** | ***339*** | ***611*** | ***973*** | ***1205*** | 8063 |
| Cooling Degree Days | ***0*** | ***0*** | ***0*** | na | na | na | ***65*** | na | na | na | na | na | na |
| Total Precipitation (") | 1.47 | 1.04 | 2.01 | 2.75 | 4.74 | 4.43 | 2.19 | 2.12 | 2.92 | 1.96 | 1.35 | 1.47 | 28.45 |
| Days ≥ 0.1" Precip | 5 | 3 | 5 | 6 | 7 | 7 | 4 | 4 | 5 | 4 | 4 | 4 | 58 |
| Total Snowfall (") | ***22.1*** | 15.8 | 23.8 | 20.4 | 8.8 | 0.0 | 0.0 | 0.2 | 2.0 | ***10.9*** | 15.4 | 23.9 | 143.3 |
| Days ≥ 1" Snow Depth | 20 | 16 | 13 | 7 | 2 | 0 | 0 | 0 | 1 | 4 | 9 | 17 | 89 |

## SIDNEY *Richland County* ELEVATION 1952 ft LAT/LONG 47° 43' N / 104° 8' W

| | JAN | FEB | MAR | APR | MAY | JUN | JUL | AUG | SEP | OCT | NOV | DEC | YEAR |
|---|---|---|---|---|---|---|---|---|---|---|---|---|---|
| Maximum Temp °F | 23.1 | 30.7 | 44.2 | 59.4 | 71.6 | 79.3 | 85.4 | 84.1 | 72.4 | 59.5 | 40.0 | 27.8 | 56.5 |
| Minimum Temp °F | 1.7 | 8.2 | 19.2 | 30.5 | 41.9 | 50.8 | 55.0 | 52.7 | 42.6 | 32.3 | 18.7 | 6.8 | 30.0 |
| Mean Temp °F | 12.4 | 19.5 | 31.7 | 45.0 | 56.8 | 65.1 | 70.2 | 68.4 | 57.5 | 45.9 | 29.4 | 17.3 | 43.3 |
| Days Max Temp ≥ 90 °F | 0 | 0 | 0 | 0 | 2 | 4 | 9 | 9 | 2 | 0 | 0 | 0 | 26 |
| Days Max Temp ≤ 32 °F | 20 | 14 | 6 | 0 | 0 | 0 | 0 | 0 | 0 | 1 | 8 | 18 | 67 |
| Days Min Temp ≤ 32 °F | 31 | 28 | 28 | 18 | 5 | 0 | 0 | 0 | 3 | 15 | 28 | 31 | 187 |
| Days Min Temp ≤ 0 °F | 15 | 9 | 3 | 0 | 0 | 0 | 0 | 0 | 0 | 0 | 3 | 10 | 40 |
| Heating Degree Days | 1628 | 1281 | 1024 | 596 | 274 | 75 | 15 | 40 | 243 | 586 | 1063 | 1473 | 8298 |
| Cooling Degree Days | 0 | 0 | 0 | 2 | 36 | 104 | 209 | 185 | 29 | 0 | 0 | 0 | 565 |
| Total Precipitation (") | 0.43 | 0.31 | 0.52 | 1.25 | 2.05 | 2.88 | 1.90 | 1.32 | 1.52 | 0.87 | 0.50 | 0.49 | 14.04 |
| Days ≥ 0.1" Precip | 1 | 1 | 2 | 3 | 5 | 6 | 5 | 3 | 3 | 2 | 2 | 1 | 34 |
| Total Snowfall (") | 7.4 | 6.2 | 5.0 | 3.3 | 0.9 | 0.0 | 0.0 | 0.0 | 0.5 | 1.9 | 6.0 | 8.1 | 39.3 |
| Days ≥ 1" Snow Depth | 25 | 19 | 9 | 2 | 0 | 0 | 0 | 0 | 0 | 1 | 8 | 20 | 84 |

## SIMPSON 6 NW *Hill County* ELEVATION 2743 ft LAT/LONG 48° 59' N / 110° 19' W

| | JAN | FEB | MAR | APR | MAY | JUN | JUL | AUG | SEP | OCT | NOV | DEC | YEAR |
|---|---|---|---|---|---|---|---|---|---|---|---|---|---|
| Maximum Temp °F | 22.3 | 30.0 | 42.5 | 57.1 | 69.6 | 77.4 | 84.6 | 84.0 | 72.5 | ***58.8*** | 39.3 | 28.0 | 55.5 |
| Minimum Temp °F | -0.2 | 5.4 | 17.5 | 28.6 | 38.4 | 46.1 | 50.3 | 49.0 | 38.5 | 28.4 | 14.8 | 4.7 | 26.8 |
| Mean Temp °F | 11.1 | 17.5 | 30.1 | 42.9 | 54.0 | 61.8 | 67.5 | 66.5 | 55.5 | ***43.6*** | 27.0 | 16.3 | 41.2 |
| Days Max Temp ≥ 90 °F | 0 | 0 | 0 | 0 | 1 | 4 | 9 | 10 | 2 | 0 | 0 | 0 | 26 |
| Days Max Temp ≤ 32 °F | 18 | 14 | 7 | 1 | 0 | 0 | 0 | 0 | 0 | 1 | 8 | 16 | 65 |
| Days Min Temp ≤ 32 °F | 29 | 26 | 28 | 21 | 7 | 1 | 0 | 0 | 6 | ***20*** | 28 | 30 | 196 |
| Days Min Temp ≤ 0 °F | 15 | 10 | 3 | 0 | 0 | 0 | 0 | 0 | 0 | 0 | 4 | 11 | 43 |
| Heating Degree Days | 1672 | 1335 | 1076 | 657 | 341 | 140 | 39 | 61 | 293 | ***662*** | 1136 | 1505 | 8917 |
| Cooling Degree Days | ***0*** | 0 | 0 | 0 | 11 | 62 | 130 | 123 | 14 | 0 | 0 | 0 | 340 |
| Total Precipitation (") | 0.32 | 0.23 | 0.43 | 0.66 | 1.54 | 2.17 | 1.51 | 1.39 | 1.11 | 0.45 | 0.34 | 0.30 | 10.45 |
| Days ≥ 0.1" Precip | 1 | 1 | 2 | 2 | 4 | 5 | 4 | 3 | 2 | 1 | 1 | 1 | 27 |
| Total Snowfall (") | 5.5 | 4.0 | 4.9 | 3.6 | 0.3 | 0.0 | 0.0 | 0.0 | 0.4 | 1.6 | 4.2 | 4.6 | 29.1 |
| Days ≥ 1" Snow Depth | 20 | 17 | 11 | 3 | 0 | 0 | 0 | 0 | 0 | 1 | 7 | 15 | 74 |

## SONNETTE 2 WNW *Powder River County* ELEVATION 3904 ft LAT/LONG 45° 25' N / 105° 51' W

| | JAN | FEB | MAR | APR | MAY | JUN | JUL | AUG | SEP | OCT | NOV | DEC | YEAR |
|---|---|---|---|---|---|---|---|---|---|---|---|---|---|
| Maximum Temp °F | 32.0 | 37.5 | 45.7 | 56.0 | 66.5 | 75.8 | 84.7 | 84.1 | 72.8 | 59.3 | 43.5 | 34.5 | 57.7 |
| Minimum Temp °F | 5.7 | 11.3 | 19.6 | 28.5 | 37.9 | 46.1 | 51.0 | 49.2 | 39.7 | 29.4 | 17.7 | 8.7 | 28.7 |
| Mean Temp °F | 18.8 | 24.5 | 32.7 | 42.3 | 52.3 | 61.0 | 67.9 | 66.7 | 56.3 | 44.3 | 30.7 | 21.6 | 43.3 |
| Days Max Temp ≥ 90 °F | 0 | 0 | 0 | 0 | 0 | 2 | 9 | 9 | 2 | 0 | 0 | 0 | 22 |
| Days Max Temp ≤ 32 °F | 14 | 8 | 5 | 1 | 0 | 0 | 0 | 0 | 0 | 1 | 6 | 12 | 47 |
| Days Min Temp ≤ 32 °F | 31 | 27 | 29 | ***22*** | 8 | 1 | 0 | 0 | 5 | ***19*** | 29 | 31 | 202 |
| Days Min Temp ≤ 0 °F | 11 | 6 | 2 | 0 | 0 | 0 | 0 | 0 | 0 | 0 | 2 | 8 | 29 |
| Heating Degree Days | 1426 | 1137 | 996 | 677 | 395 | 159 | 41 | 56 | 277 | 635 | 1024 | 1347 | 8170 |
| Cooling Degree Days | 0 | 0 | 0 | 1 | 8 | 54 | 138 | 120 | 20 | 0 | 0 | 0 | 341 |
| Total Precipitation (") | 0.57 | 0.39 | 0.77 | 1.67 | 2.60 | 2.58 | 1.49 | 1.06 | 1.46 | 1.14 | 0.68 | 0.55 | 14.96 |
| Days ≥ 0.1" Precip | ***2*** | ***2*** | 3 | 4 | 6 | 6 | 4 | 3 | 4 | 3 | 2 | 2 | 41 |
| Total Snowfall (") | na | na | na | na | ***0.2*** | 0.1 | 0.0 | 0.0 | 0.0 | na | na | na | na |
| Days ≥ 1" Snow Depth | na | na | na | na | 1 | 0 | 0 | 0 | 0 | ***0*** | na | na | na |

## ST IGNATIUS *Lake County* ELEVATION 2904 ft LAT/LONG 47° 19' N / 114° 6' W

| | JAN | FEB | MAR | APR | MAY | JUN | JUL | AUG | SEP | OCT | NOV | DEC | YEAR |
|---|---|---|---|---|---|---|---|---|---|---|---|---|---|
| Maximum Temp °F | 33.6 | 39.4 | 48.9 | 59.1 | 68.2 | 75.9 | 83.9 | 83.6 | 71.6 | 58.4 | 42.2 | 34.3 | 58.3 |
| Minimum Temp °F | 18.5 | 21.7 | 26.6 | 32.7 | 39.5 | 46.1 | 49.1 | 48.2 | 40.7 | 32.8 | 25.8 | 19.6 | 33.4 |
| Mean Temp °F | 26.1 | 30.6 | 37.8 | 45.9 | 53.9 | 61.0 | 66.5 | 65.9 | 56.2 | 45.6 | 34.1 | 27.0 | 45.9 |
| Days Max Temp ≥ 90 °F | 0 | 0 | 0 | 0 | 0 | 2 | 9 | 8 | 1 | 0 | 0 | 0 | 20 |
| Days Max Temp ≤ 32 °F | 12 | 6 | 2 | 0 | 0 | 0 | 0 | 0 | 0 | 0 | 4 | 12 | 36 |
| Days Min Temp ≤ 32 °F | 27 | 24 | 24 | 16 | 5 | 0 | 0 | 0 | 3 | 15 | 23 | 27 | 164 |
| Days Min Temp ≤ 0 °F | 4 | 2 | 1 | 0 | 0 | 0 | 0 | 0 | 0 | 0 | 1 | 2 | 10 |
| Heating Degree Days | 1200 | 966 | 837 | 567 | 343 | 153 | 51 | 56 | 268 | 594 | 922 | 1172 | 7129 |
| Cooling Degree Days | 0 | 0 | 0 | 0 | 6 | 43 | 100 | 92 | 10 | 0 | 0 | 0 | 251 |
| Total Precipitation (") | 1.10 | 0.70 | 1.18 | 1.45 | 2.43 | 2.52 | 1.33 | 1.38 | 1.46 | 1.10 | 1.00 | 1.01 | 16.66 |
| Days ≥ 0.1" Precip | 4 | 2 | 4 | 4 | 6 | 6 | 4 | 4 | 4 | 4 | 3 | 3 | 48 |
| Total Snowfall (") | 10.7 | 7.4 | 6.8 | 1.7 | 0.2 | 0.0 | 0.0 | 0.0 | 0.1 | 0.7 | 5.6 | 11.3 | 44.5 |
| Days ≥ 1" Snow Depth | 20 | 13 | 7 | 0 | 0 | 0 | 0 | 0 | 0 | 0 | 6 | 16 | 62 |

## STANFORD *Judith Basin County* ELEVATION 4314 ft LAT/LONG 47° 10 ' N / 110° 15 ' W

| | JAN | FEB | MAR | APR | MAY | JUN | JUL | AUG | SEP | OCT | NOV | DEC | YEAR |
|---|---|---|---|---|---|---|---|---|---|---|---|---|---|
| Maximum Temp °F | ***35.4*** | 39.7 | 45.6 | 55.7 | 64.9 | 72.8 | 80.2 | 80.4 | 70.0 | 59.4 | ***44.9*** | ***38.0*** | 57.2 |
| Minimum Temp °F | ***11.8*** | 15.6 | 21.4 | 29.3 | 37.7 | 45.0 | 49.4 | 48.7 | 40.4 | 32.0 | ***21.4*** | ***14.3*** | 30.6 |
| Mean Temp °F | ***23.6*** | 27.7 | 33.5 | 42.5 | 51.3 | 58.9 | 64.9 | 64.6 | 55.2 | 45.7 | ***33.2*** | ***26.1*** | 43.9 |
| Days Max Temp ≥ 90 °F | 0 | 0 | 0 | 0 | 0 | 1 | 3 | 4 | 1 | 0 | 0 | 0 | 9 |
| Days Max Temp ≤ 32 °F | ***10*** | 7 | 5 | 1 | 0 | 0 | 0 | 0 | 0 | 1 | ***4*** | ***8*** | 36 |
| Days Min Temp ≤ 32 °F | ***28*** | 25 | 27 | 20 | 8 | 1 | 0 | 0 | 6 | 16 | 24 | ***26*** | 181 |
| Days Min Temp ≤ 0 °F | ***9*** | 5 | 2 | 0 | 0 | 0 | 0 | 0 | 0 | 0 | ***2*** | ***5*** | 23 |
| Heating Degree Days | ***1277*** | 1047 | 970 | 667 | 421 | 202 | 72 | 88 | 308 | 591 | ***947*** | ***1198*** | 7788 |
| Cooling Degree Days | 0 | 0 | 0 | 1 | 6 | 39 | 93 | 100 | 21 | 2 | 0 | ***0*** | 262 |
| Total Precipitation (") | 0.77 | 0.49 | 0.87 | 1.47 | 2.97 | 2.95 | 1.94 | 1.69 | 1.47 | 0.99 | 0.62 | 0.63 | 16.86 |
| Days ≥ 0.1" Precip | ***3*** | 1 | 3 | 4 | 7 | 7 | 5 | 4 | 4 | 3 | 2 | ***2*** | 45 |
| Total Snowfall (") | na | na | 9.0 | ***10.0*** | 3.7 | 0.2 | 0.0 | 0.1 | 1.6 | 4.5 | na | na | na |
| Days ≥ 1" Snow Depth | ***19*** | ***15*** | 11 | 3 | 1 | 0 | 0 | 0 | 0 | 2 | ***7*** | ***14*** | 72 |

## STEVENSVILLE *Ravalli County* ELEVATION 3373 ft LAT/LONG 46° 31 ' N / 114° 6 ' W

| | JAN | FEB | MAR | APR | MAY | JUN | JUL | AUG | SEP | OCT | NOV | DEC | YEAR |
|---|---|---|---|---|---|---|---|---|---|---|---|---|---|
| Maximum Temp °F | 33.7 | 41.0 | 50.0 | 59.7 | 68.4 | 76.0 | 84.0 | 83.5 | 72.3 | 59.5 | 42.9 | 33.8 | 58.7 |
| Minimum Temp °F | 17.0 | 20.4 | 25.1 | 30.6 | 37.6 | 44.8 | 47.2 | 45.9 | 38.5 | 30.2 | 23.7 | 16.9 | 31.5 |
| Mean Temp °F | 25.4 | 30.7 | 37.6 | 45.2 | 53.1 | 60.4 | 65.6 | 64.7 | 55.4 | 44.8 | 33.3 | 25.4 | 45.1 |
| Days Max Temp ≥ 90 °F | 0 | 0 | 0 | 0 | 0 | 2 | 8 | 8 | 1 | 0 | 0 | 0 | 19 |
| Days Max Temp ≤ 32 °F | 12 | 5 | 1 | 0 | 0 | 0 | 0 | 0 | 0 | 0 | 4 | 13 | 35 |
| Days Min Temp ≤ 32 °F | 28 | 25 | 26 | 18 | 7 | 1 | 0 | 0 | 6 | 19 | 25 | 29 | 184 |
| Days Min Temp ≤ 0 °F | 4 | 2 | 0 | 0 | 0 | 0 | 0 | 0 | 0 | 0 | 1 | 3 | 10 |
| Heating Degree Days | 1222 | 962 | 844 | 588 | 367 | 161 | 53 | 66 | 287 | 618 | 944 | 1223 | 7335 |
| Cooling Degree Days | 0 | 0 | 0 | 0 | 6 | 34 | 86 | 72 | 7 | 0 | 0 | 0 | 205 |
| Total Precipitation (") | 1.22 | 0.77 | 0.78 | 0.79 | 1.46 | 1.57 | 0.81 | 1.02 | 0.97 | 0.62 | 0.90 | 0.94 | 11.85 |
| Days ≥ 0.1" Precip | 3 | 3 | 3 | 3 | 4 | 5 | 2 | 3 | 3 | 2 | 3 | 3 | 37 |
| Total Snowfall (") | na | na | na | na | ***0.2*** | na | na | na | na | na | na | na | na |
| Days ≥ 1" Snow Depth | na | na | na | na | ***0*** | na | na | na | na | na | na | na | na |

## SULA 3 ENE *Ravalli County* ELEVATION 4403 ft LAT/LONG 45° 50 ' N / 113° 59 ' W

| | JAN | FEB | MAR | APR | MAY | JUN | JUL | AUG | SEP | OCT | NOV | DEC | YEAR |
|---|---|---|---|---|---|---|---|---|---|---|---|---|---|
| Maximum Temp °F | 33.6 | 40.9 | 48.3 | 57.0 | 65.3 | 73.2 | 82.0 | 81.6 | 71.9 | 60.3 | 43.2 | 33.6 | 57.6 |
| Minimum Temp °F | 9.5 | 13.7 | 19.7 | 25.9 | 31.9 | 37.9 | 39.8 | 38.7 | 31.1 | 24.2 | 18.4 | 10.7 | 25.1 |
| Mean Temp °F | 21.6 | 27.3 | 34.0 | 41.5 | 48.7 | 55.6 | 60.9 | 60.5 | 51.5 | 42.3 | 30.8 | 22.1 | 41.4 |
| Days Max Temp ≥ 90 °F | 0 | 0 | 0 | 0 | 0 | 1 | 5 | 6 | 1 | 0 | 0 | 0 | 13 |
| Days Max Temp ≤ 32 °F | 12 | 4 | 1 | 0 | 0 | 0 | 0 | 0 | 0 | 0 | 4 | 13 | 34 |
| Days Min Temp ≤ 32 °F | 30 | 27 | 29 | 24 | 16 | 6 | 4 | 6 | 18 | 26 | 28 | 30 | 244 |
| Days Min Temp ≤ 0 °F | 8 | 5 | 2 | 0 | 0 | 0 | 0 | 0 | 0 | 0 | 2 | 7 | 24 |
| Heating Degree Days | 1340 | 1058 | 954 | 699 | 499 | 281 | 139 | 157 | 399 | 700 | 1020 | 1323 | 8569 |
| Cooling Degree Days | 0 | 0 | 0 | 0 | 0 | 5 | 19 | 16 | 0 | 0 | 0 | 0 | 40 |
| Total Precipitation (") | 1.02 | 0.72 | 0.96 | 1.36 | 2.25 | 2.23 | 1.28 | 1.47 | 1.32 | 1.14 | 1.07 | 0.93 | 15.75 |
| Days ≥ 0.1" Precip | 4 | 3 | 3 | 5 | 7 | 7 | 4 | 4 | 4 | 4 | 4 | 3 | 52 |
| Total Snowfall (") | na | na | na | na | na | ***0.0*** | ***0.0*** | ***0.0*** | ***0.6*** | na | na | na | na |
| Days ≥ 1" Snow Depth | na | na | na | na | na | ***0*** | ***0*** | ***0*** | ***0*** | na | na | na | na |

## SUN RIVER 4 S *Cascade County* ELEVATION 3563 ft LAT/LONG 47° 28 ' N / 111° 46 ' W

| | JAN | FEB | MAR | APR | MAY | JUN | JUL | AUG | SEP | OCT | NOV | DEC | YEAR |
|---|---|---|---|---|---|---|---|---|---|---|---|---|---|
| Maximum Temp °F | 34.0 | 40.2 | 47.1 | 57.3 | 66.5 | 74.4 | 81.8 | 80.7 | 70.9 | 60.2 | 44.6 | 36.7 | 57.9 |
| Minimum Temp °F | 11.4 | 16.0 | 21.2 | 30.1 | 38.6 | 46.3 | 49.9 | 48.4 | 40.3 | 33.2 | 22.7 | 14.6 | 31.1 |
| Mean Temp °F | 22.7 | 28.1 | 34.2 | 43.7 | 52.6 | 60.3 | 65.9 | 64.5 | 55.6 | 46.7 | 33.7 | 25.7 | 44.5 |
| Days Max Temp ≥ 90 °F | 0 | 0 | 0 | 0 | 0 | 2 | 6 | 5 | 1 | 0 | 0 | 0 | 14 |
| Days Max Temp ≤ 32 °F | 11 | 7 | 4 | 1 | 0 | 0 | 0 | 0 | 0 | 1 | 4 | 9 | 37 |
| Days Min Temp ≤ 32 °F | 26 | 23 | 26 | 18 | 6 | 1 | 0 | 0 | 5 | 14 | 22 | 26 | 167 |
| Days Min Temp ≤ 0 °F | 9 | 5 | 3 | 0 | 0 | 0 | 0 | 0 | 0 | 0 | 3 | 7 | 27 |
| Heating Degree Days | 1305 | 1035 | 950 | 632 | 383 | 168 | 53 | 79 | 288 | 561 | 933 | 1215 | 7602 |
| Cooling Degree Days | 0 | 0 | 0 | 1 | 5 | 39 | 82 | 82 | 12 | 2 | 0 | 0 | 223 |
| Total Precipitation (") | 0.47 | 0.33 | 0.69 | 1.28 | 2.03 | 2.31 | 1.54 | 1.50 | 1.05 | 0.74 | 0.47 | 0.46 | 12.87 |
| Days ≥ 0.1" Precip | 1 | 1 | 2 | 3 | 5 | 5 | 4 | 4 | 3 | 3 | 2 | 2 | 35 |
| Total Snowfall (") | 7.3 | 5.0 | ***7.6*** | 5.7 | 0.8 | 0.1 | 0.0 | 0.1 | 0.6 | 2.3 | 5.2 | ***7.8*** | 42.5 |
| Days ≥ 1" Snow Depth | 14 | 10 | 7 | 2 | 0 | 0 | 0 | 0 | 0 | 1 | 6 | ***11*** | 51 |

## SUPERIOR *Mineral County* ELEVATION 2700 ft LAT/LONG 47° 11 ' N / 114° 52 ' W

| | JAN | FEB | MAR | APR | MAY | JUN | JUL | AUG | SEP | OCT | NOV | DEC | YEAR |
|---|---|---|---|---|---|---|---|---|---|---|---|---|---|
| Maximum Temp °F | 34.4 | 42.3 | 51.5 | 60.5 | 69.6 | 76.7 | 85.0 | 84.9 | 74.2 | 60.5 | 42.6 | 33.7 | 59.7 |
| Minimum Temp °F | 19.6 | 22.6 | 26.9 | 32.4 | 39.0 | 45.9 | 49.2 | 48.8 | 41.0 | 33.0 | 26.9 | 20.6 | 33.8 |
| Mean Temp °F | 27.0 | 32.5 | 39.2 | 46.4 | 54.3 | 61.3 | 67.1 | 66.9 | 57.7 | 46.8 | 34.8 | 27.3 | 46.8 |
| Days Max Temp ≥ 90 °F | 0 | 0 | 0 | 0 | 1 | 4 | 11 | 10 | 2 | 0 | 0 | 0 | 28 |
| Days Max Temp ≤ 32 °F | 10 | 3 | 1 | 0 | 0 | 0 | 0 | 0 | 0 | 0 | 3 | 11 | 28 |
| Days Min Temp ≤ 32 °F | 27 | 25 | 25 | 16 | 5 | 0 | 0 | 0 | 3 | 15 | 22 | 28 | 166 |
| Days Min Temp ≤ 0 °F | 3 | 1 | 0 | 0 | 0 | 0 | 0 | 0 | 0 | 0 | 0 | 1 | 5 |
| Heating Degree Days | 1171 | 912 | 792 | 550 | 330 | 144 | 42 | 44 | 228 | 558 | 900 | 1162 | 6833 |
| Cooling Degree Days | 0 | 0 | 0 | 0 | 8 | 43 | 103 | 110 | 13 | 0 | 0 | 0 | 277 |
| Total Precipitation (") | 1.69 | 0.98 | 1.27 | 1.22 | 1.65 | 1.86 | 0.95 | 1.37 | 1.10 | 1.09 | 1.53 | 1.47 | 16.18 |
| Days ≥ 0.1" Precip | 5 | 3 | 5 | 4 | 5 | 5 | 3 | 4 | 3 | 3 | 5 | 4 | 49 |
| Total Snowfall (") | na | *5.8* | *3.6* | *0.5* | 0.1 | 0.0 | 0.0 | 0.0 | 0.0 | 0.1 | na | na | na |
| Days ≥ 1" Snow Depth | *20* | *13* | *5* | 0 | 0 | 0 | 0 | 0 | 0 | 0 | *4* | na | na |

## SWAN LAKE *Lake County* ELEVATION 3104 ft LAT/LONG 47° 55 ' N / 113° 50 ' W

| | JAN | FEB | MAR | APR | MAY | JUN | JUL | AUG | SEP | OCT | NOV | DEC | YEAR |
|---|---|---|---|---|---|---|---|---|---|---|---|---|---|
| Maximum Temp °F | 30.2 | 38.0 | 46.0 | 56.3 | 66.0 | 72.7 | 80.1 | 79.4 | *67.8* | 55.7 | 38.7 | 30.3 | 55.1 |
| Minimum Temp °F | 14.6 | 18.9 | 23.0 | 28.8 | 35.2 | 41.1 | 43.3 | 42.5 | *35.9* | 29.3 | 23.5 | 16.3 | 29.4 |
| Mean Temp °F | 22.4 | 28.5 | 34.5 | 42.6 | 50.6 | 57.0 | 61.7 | 61.0 | *51.9* | 42.5 | 31.1 | 23.3 | 42.3 |
| Days Max Temp ≥ 90 °F | 0 | 0 | 0 | 0 | 0 | 1 | 3 | 3 | 0 | 0 | 0 | 0 | 7 |
| Days Max Temp ≤ 32 °F | *15* | 6 | 2 | 0 | 0 | 0 | 0 | 0 | 0 | 0 | *6* | 17 | 46 |
| Days Min Temp ≤ 32 °F | 31 | 28 | 29 | *22* | 10 | 2 | 0 | 1 | 9 | na | *27* | 30 | na |
| Days Min Temp ≤ 0 °F | 5 | 2 | 1 | 0 | 0 | 0 | 0 | 0 | 0 | 0 | 1 | 3 | 12 |
| Heating Degree Days | 1316 | 1025 | 936 | 667 | 442 | 243 | 116 | 138 | *388* | 689 | 1011 | 1286 | 8257 |
| Cooling Degree Days | 0 | 0 | 0 | *0* | *3* | *10* | 19 | *17* | na | 0 | *0* | 0 | na |
| Total Precipitation (") | 3.25 | 2.37 | 1.93 | 1.66 | 2.43 | 2.93 | 1.42 | 1.96 | 2.13 | 2.16 | 3.35 | 3.36 | 28.95 |
| Days ≥ 0.1" Precip | *8* | *5* | *5* | 4 | 6 | 6 | 3 | 4 | 5 | *5* | na | *8* | na |
| Total Snowfall (") | 35.3 | 20.8 | 11.4 | *3.7* | 0.7 | 0.0 | 0.0 | 0.0 | 0.3 | *2.8* | *21.2* | 36.8 | 133.0 |
| Days ≥ 1" Snow Depth | 27 | 24 | 24 | *9* | 0 | 0 | 0 | 0 | 0 | *1* | *14* | 26 | 125 |

## SWEETGRASS *Toole County* ELEVATION 3471 ft LAT/LONG 49° 0 ' N / 111° 57 ' W

| | JAN | FEB | MAR | APR | MAY | JUN | JUL | AUG | SEP | OCT | NOV | DEC | YEAR |
|---|---|---|---|---|---|---|---|---|---|---|---|---|---|
| Maximum Temp °F | 28.6 | 34.6 | 43.8 | 55.6 | 66.4 | 73.8 | 80.9 | 80.3 | 69.8 | 58.9 | 41.4 | 31.9 | 55.5 |
| Minimum Temp °F | 7.4 | 12.3 | 21.2 | 30.9 | 40.3 | 47.9 | 51.8 | 50.9 | 42.0 | 32.8 | 19.8 | 11.1 | 30.7 |
| Mean Temp °F | 18.1 | 23.5 | 32.5 | 43.3 | 53.4 | 60.8 | 66.5 | 65.6 | 55.9 | 45.9 | 30.6 | 21.5 | 43.1 |
| Days Max Temp ≥ 90 °F | 0 | 0 | 0 | 0 | 0 | 1 | 4 | 5 | 1 | 0 | 0 | 0 | 11 |
| Days Max Temp ≤ 32 °F | 14 | 10 | 6 | 1 | 0 | 0 | 0 | 0 | 0 | 1 | 6 | 12 | 50 |
| Days Min Temp ≤ 32 °F | 28 | 24 | 26 | 17 | 5 | 0 | 0 | 0 | 4 | 14 | 25 | 28 | 171 |
| Days Min Temp ≤ 0 °F | 12 | 6 | 2 | 0 | 0 | 0 | 0 | 0 | 0 | 0 | 3 | 8 | 31 |
| Heating Degree Days | 1449 | 1167 | 999 | 645 | 363 | 159 | 53 | 74 | 288 | 586 | 1026 | 1342 | 8151 |
| Cooling Degree Days | 0 | 0 | 0 | 1 | 10 | 42 | 94 | 99 | *20* | 2 | 0 | 0 | 268 |
| Total Precipitation (") | 0.37 | 0.24 | 0.56 | 1.10 | 2.38 | 3.27 | 1.79 | 2.04 | 1.46 | 0.65 | 0.41 | 0.36 | 14.63 |
| Days ≥ 0.1" Precip | 2 | 1 | 2 | 3 | 5 | 7 | 4 | 5 | 4 | 2 | 1 | 1 | 37 |
| Total Snowfall (") | na | na | na | na | *0.0* | 0.0 | 0.0 | 0.0 | *0.3* | na | na | na | na |
| Days ≥ 1" Snow Depth | na | na | na | na | 0 | 0 | 0 | 0 | 0 | na | na | na | na |

## TERRY *Prairie County* ELEVATION 2251 ft LAT/LONG 46° 47 ' N / 105° 19 ' W

| | JAN | FEB | MAR | APR | MAY | JUN | JUL | AUG | SEP | OCT | NOV | DEC | YEAR |
|---|---|---|---|---|---|---|---|---|---|---|---|---|---|
| Maximum Temp °F | 25.2 | 32.9 | 44.9 | 58.5 | 70.1 | 79.6 | 87.2 | 86.7 | 74.1 | 60.8 | 43.0 | 31.3 | 57.9 |
| Minimum Temp °F | -0.3 | 6.2 | 17.9 | 29.7 | 41.0 | 50.5 | 54.9 | 52.0 | 40.1 | 28.8 | 16.5 | 4.7 | 28.5 |
| Mean Temp °F | 12.5 | 19.6 | 31.2 | 44.1 | 55.7 | 65.1 | 71.0 | 69.3 | 57.2 | 44.8 | 29.6 | 17.9 | 43.2 |
| Days Max Temp ≥ 90 °F | 0 | 0 | 0 | 0 | 1 | 5 | 13 | 13 | 3 | 0 | 0 | 0 | 35 |
| Days Max Temp ≤ 32 °F | 17 | 13 | 5 | 1 | 0 | 0 | 0 | 0 | 0 | 1 | 6 | *16* | 59 |
| Days Min Temp ≤ 32 °F | 30 | 27 | 29 | 19 | 5 | 0 | 0 | 0 | 5 | 21 | 29 | 30 | 195 |
| Days Min Temp ≤ 0 °F | 14 | 10 | 3 | 0 | 0 | 0 | 0 | 0 | 0 | 0 | 2 | *10* | 39 |
| Heating Degree Days | 1625 | 1276 | 1040 | 620 | 304 | 87 | 21 | 38 | 259 | 620 | 1060 | 1455 | 8405 |
| Cooling Degree Days | 0 | 0 | 0 | 1 | 23 | 109 | 203 | 190 | 23 | 0 | 0 | 0 | 549 |
| Total Precipitation (") | na | *0.19* | 0.35 | 1.09 | 1.87 | 2.38 | 1.43 | 1.22 | 1.27 | 0.80 | 0.37 | 0.24 | na |
| Days ≥ 0.1" Precip | na | na | 1 | 3 | 5 | 6 | 4 | 3 | 3 | 2 | *1* | *1* | na |
| Total Snowfall (") | na | na | na | na | 0.5 | 0.0 | 0.0 | 0.0 | 0.3 | na | na | na | na |
| Days ≥ 1" Snow Depth | na | na | na | na | na | *0* | *0* | *0* | *0* | na | na | na | na |

## TERRY 21 NNW *Prairie County* ELEVATION 3261 ft LAT/LONG 47° 4 ' N / 105° 30 ' W

| | JAN | FEB | MAR | APR | MAY | JUN | JUL | AUG | SEP | OCT | NOV | DEC | YEAR |
|---|---|---|---|---|---|---|---|---|---|---|---|---|---|
| Maximum Temp °F | 25.3 | 31.0 | 41.9 | 54.9 | 66.1 | 74.9 | 82.7 | 82.0 | 69.9 | 57.1 | 39.9 | 29.2 | 54.6 |
| Minimum Temp °F | 4.1 | 9.5 | 19.1 | 29.2 | 39.9 | 48.7 | 53.8 | 52.5 | 41.6 | 30.9 | 18.1 | 8.1 | 29.6 |
| Mean Temp °F | 14.7 | 20.3 | 30.5 | 42.1 | 53.0 | 61.8 | 68.3 | 67.3 | 55.8 | 44.0 | 29.0 | 18.6 | 42.1 |
| Days Max Temp ≥ 90 °F | 0 | 0 | 0 | 0 | 0 | 2 | 6 | 7 | 1 | 0 | 0 | 0 | 16 |
| Days Max Temp ≤ 32 °F | 19 | 13 | 7 | 1 | 0 | 0 | 0 | 0 | 0 | 1 | 8 | 16 | 65 |
| Days Min Temp ≤ 32 °F | 31 | 28 | 29 | 20 | 6 | 0 | 0 | 0 | 4 | 17 | 28 | 30 | 193 |
| Days Min Temp ≤ 0 °F | 13 | 9 | 3 | 0 | 0 | 0 | 0 | 0 | 0 | 0 | 2 | 9 | 36 |
| Heating Degree Days | 1556 | 1256 | 1063 | 681 | 373 | 137 | 35 | 53 | 289 | 645 | 1073 | 1433 | 8594 |
| Cooling Degree Days | 0 | 0 | 0 | 0 | 12 | 58 | 147 | 136 | 20 | 0 | 0 | 0 | 373 |
| Total Precipitation (") | 0.48 | 0.24 | 0.47 | 1.40 | 2.19 | 2.86 | 1.74 | 1.28 | 1.48 | 0.91 | 0.43 | 0.46 | 13.94 |
| Days ≥ 0.1" Precip | 2 | 1 | 2 | 3 | 5 | 6 | 4 | 3 | 3 | 3 | 1 | *2* | 35 |
| Total Snowfall (") | na | na | na | na | na | *0.0* | *0.0* | *0.0* | *0.4* | na | na | na | na |
| Days ≥ 1" Snow Depth | na | na | na | na | 0 | 0 | 0 | 0 | 0 | *1* | na | na | na |

## THOMPSON FALLS P H *Sanders County* ELEVATION 2382 ft LAT/LONG 47° 36 ' N / 115° 22 ' W

| | JAN | FEB | MAR | APR | MAY | JUN | JUL | AUG | SEP | OCT | NOV | DEC | YEAR |
|---|---|---|---|---|---|---|---|---|---|---|---|---|---|
| Maximum Temp °F | 34.6 | 42.4 | 52.2 | 62.3 | 71.4 | 78.4 | 86.9 | 87.1 | 75.7 | 60.9 | 42.8 | 34.6 | 60.8 |
| Minimum Temp °F | 20.8 | 23.8 | 27.7 | 33.2 | 39.8 | 46.3 | 49.6 | 49.2 | 41.8 | 34.1 | 28.4 | 22.5 | 34.8 |
| Mean Temp °F | 27.7 | 33.1 | 39.9 | 47.8 | 55.6 | 62.4 | 68.3 | 68.2 | 58.8 | 47.5 | 35.6 | 28.6 | 47.8 |
| Days Max Temp ≥ 90 °F | 0 | 0 | 0 | 0 | 1 | 5 | 14 | 14 | 3 | 0 | 0 | 0 | 37 |
| Days Max Temp ≤ 32 °F | 9 | 3 | 0 | 0 | 0 | 0 | 0 | 0 | 0 | 0 | 2 | 10 | 24 |
| Days Min Temp ≤ 32 °F | 28 | 25 | 24 | 15 | 4 | 0 | 0 | 0 | 3 | 13 | 21 | 28 | 161 |
| Days Min Temp ≤ 0 °F | 2 | 1 | 0 | 0 | 0 | 0 | 0 | 0 | 0 | 0 | 0 | 1 | 4 |
| Heating Degree Days | 1150 | 894 | 770 | 511 | 295 | 122 | 34 | 35 | 201 | 536 | 876 | 1122 | 6546 |
| Cooling Degree Days | 0 | 0 | 0 | 1 | 16 | 60 | 141 | 151 | 25 | 0 | 0 | 0 | 394 |
| Total Precipitation (") | 2.83 | 1.91 | 1.76 | 1.51 | 2.04 | 2.18 | 1.14 | 1.31 | 1.26 | 1.61 | 2.46 | 2.53 | 22.54 |
| Days ≥ 0.1" Precip | 8 | 6 | 6 | 5 | 6 | 6 | 3 | 4 | 4 | 5 | 8 | 8 | 69 |
| Total Snowfall (") | na | na | na | na | 0.0 | 0.0 | 0.0 | 0.0 | 0.0 | *0.0* | na | na | na |
| Days ≥ 1" Snow Depth | *26* | 18 | *7* | *0* | 0 | 0 | 0 | 0 | 0 | *0* | na | na | na |

## TIBER DAM *Liberty County* ELEVATION 2851 ft LAT/LONG 48° 19 ' N / 111° 5 ' W

| | JAN | FEB | MAR | APR | MAY | JUN | JUL | AUG | SEP | OCT | NOV | DEC | YEAR |
|---|---|---|---|---|---|---|---|---|---|---|---|---|---|
| Maximum Temp °F | 28.9 | 37.6 | 46.7 | 59.6 | 70.3 | 78.4 | 85.3 | 85.4 | 74.2 | 63.1 | 44.9 | 33.4 | 59.0 |
| Minimum Temp °F | 5.9 | 11.5 | 20.6 | 30.1 | 40.3 | 48.1 | 51.9 | 50.7 | 41.0 | 31.5 | 19.2 | 9.5 | 30.0 |
| Mean Temp °F | 17.2 | 24.6 | 33.7 | 44.9 | 55.3 | 63.3 | 68.6 | 68.1 | 57.6 | 47.3 | 32.1 | 21.5 | 44.5 |
| Days Max Temp ≥ 90 °F | 0 | 0 | 0 | 0 | 1 | 4 | 11 | 11 | 3 | 0 | 0 | 0 | 30 |
| Days Max Temp ≤ 32 °F | 14 | 9 | 4 | 1 | 0 | 0 | 0 | 0 | 0 | 0 | 4 | 11 | 43 |
| Days Min Temp ≤ 32 °F | 28 | 27 | 26 | 18 | 4 | 0 | 0 | 0 | 4 | 15 | 25 | 28 | 175 |
| Days Min Temp ≤ 0 °F | 12 | 7 | 2 | 0 | 0 | 0 | 0 | 0 | 0 | 0 | 3 | 8 | 32 |
| Heating Degree Days | 1479 | 1136 | 964 | 598 | 303 | 108 | 29 | 39 | 238 | 542 | 981 | 1343 | 7760 |
| Cooling Degree Days | *0* | 0 | *0* | 1 | 13 | 69 | 131 | 139 | 21 | 1 | 0 | *0* | 375 |
| Total Precipitation (") | 0.39 | 0.24 | 0.54 | 0.93 | 1.66 | 2.09 | 1.30 | 1.39 | 0.91 | 0.57 | 0.33 | 0.33 | 10.68 |
| Days ≥ 0.1" Precip | 1 | 1 | 2 | 3 | 5 | 5 | 3 | 3 | 2 | 2 | 1 | 1 | 29 |
| Total Snowfall (") | na | na | na | na | *0.0* | *0.0* | *0.0* | *0.0* | *0.0* | na | na | na | na |
| Days ≥ 1" Snow Depth | na | na | na | na | *0* | 0 | 0 | 0 | 0 | na | na | na | na |

## TOWNSEND *Broadwater County* ELEVATION 3832 ft LAT/LONG 46° 19 ' N / 111° 31 ' W

| | JAN | FEB | MAR | APR | MAY | JUN | JUL | AUG | SEP | OCT | NOV | DEC | YEAR |
|---|---|---|---|---|---|---|---|---|---|---|---|---|---|
| Maximum Temp °F | 33.8 | 39.6 | 48.1 | 58.2 | 67.4 | 75.0 | 82.3 | 81.9 | 71.2 | 60.0 | 44.3 | 34.8 | 58.1 |
| Minimum Temp °F | 11.6 | 15.3 | 22.4 | 30.4 | 38.5 | 46.0 | 50.4 | 48.3 | 39.4 | 30.8 | 21.7 | 13.0 | 30.6 |
| Mean Temp °F | 22.7 | 27.5 | 35.3 | 44.3 | 53.0 | 60.6 | 66.3 | 65.1 | 55.3 | 45.5 | 33.0 | 23.9 | 44.4 |
| Days Max Temp ≥ 90 °F | 0 | 0 | 0 | 0 | 0 | 2 | 6 | 6 | 1 | 0 | 0 | 0 | 15 |
| Days Max Temp ≤ 32 °F | 12 | 7 | 3 | 0 | 0 | 0 | 0 | 0 | 0 | 0 | 4 | 12 | 38 |
| Days Min Temp ≤ 32 °F | 29 | 26 | 27 | 18 | 6 | 0 | 0 | 0 | 5 | 18 | 26 | 29 | 184 |
| Days Min Temp ≤ 0 °F | 7 | 4 | 2 | 0 | 0 | 0 | 0 | 0 | 0 | 0 | 2 | 5 | 20 |
| Heating Degree Days | 1304 | 1053 | 912 | 613 | 369 | 163 | 51 | 68 | 294 | 600 | 952 | 1267 | 7646 |
| Cooling Degree Days | 0 | 0 | 0 | 0 | 5 | 39 | 91 | 82 | 9 | 0 | 0 | 0 | 226 |
| Total Precipitation (") | 0.40 | 0.25 | 0.64 | 0.80 | 1.67 | 2.04 | 1.39 | 1.29 | 1.16 | 0.64 | 0.39 | 0.37 | 11.04 |
| Days ≥ 0.1" Precip | 1 | 1 | 2 | 3 | 5 | 6 | 4 | 4 | 4 | 2 | 1 | 1 | 34 |
| Total Snowfall (") | *5.1* | 3.2 | *4.1* | na | 0.1 | 0.0 | 0.0 | 0.0 | 0.1 | *0.9* | *3.4* | *5.2* | na |
| Days ≥ 1" Snow Depth | na | na | na | na | *0* | *0* | *0* | *0* | *0* | na | na | na | na |

## TRIDENT *Gallatin County* ELEVATION 4042 ft LAT/LONG 45° 57 ' N / 111° 29 ' W

| | JAN | FEB | MAR | APR | MAY | JUN | JUL | AUG | SEP | OCT | NOV | DEC | YEAR |
|---|---|---|---|---|---|---|---|---|---|---|---|---|---|
| Maximum Temp °F | 34.5 | 40.8 | 48.5 | 58.7 | 68.9 | 77.7 | 86.0 | 85.2 | 73.4 | 61.1 | 44.3 | 35.5 | 59.6 |
| Minimum Temp °F | 11.8 | 16.8 | 23.7 | 31.2 | 39.7 | 47.1 | 51.5 | 49.7 | 40.4 | 31.7 | 21.5 | 13.2 | 31.5 |
| Mean Temp °F | 23.2 | 28.8 | 36.1 | 45.0 | 54.3 | 62.4 | 68.8 | 67.5 | 56.9 | 46.4 | 33.0 | 24.4 | 45.6 |
| Days Max Temp ≥ 90 °F | 0 | 0 | 0 | 0 | 1 | 4 | 12 | 11 | 2 | 0 | 0 | 0 | 30 |
| Days Max Temp ≤ 32 °F | 11 | 6 | 2 | 0 | 0 | 0 | 0 | 0 | 0 | 0 | 5 | 11 | 35 |
| Days Min Temp ≤ 32 °F | 29 | 26 | 26 | 18 | 4 | 0 | 0 | 0 | 4 | 17 | 25 | 29 | 178 |
| Days Min Temp ≤ 0 °F | 7 | 3 | 1 | 0 | 0 | 0 | 0 | 0 | 0 | 0 | 2 | 5 | 18 |
| Heating Degree Days | 1292 | 1014 | 889 | 595 | 331 | 125 | 27 | 40 | 251 | 570 | 956 | 1253 | 7343 |
| Cooling Degree Days | 0 | 0 | 0 | 0 | 8 | 62 | 150 | 130 | 15 | 0 | 0 | 0 | 365 |
| Total Precipitation (") | 0.38 | 0.27 | 0.68 | 1.09 | 2.17 | 2.21 | 1.45 | 1.18 | 1.44 | 0.83 | 0.54 | 0.26 | 12.50 |
| Days ≥ 0.1" Precip | 1 | 1 | 2 | 4 | 6 | 6 | 4 | 4 | 4 | 3 | 2 | 1 | 38 |
| Total Snowfall (") | ***5.5*** | ***3.4*** | ***7.5*** | ***3.1*** | 0.3 | 0.0 | 0.0 | 0.0 | 0.4 | ***1.9*** | na | ***5.1*** | na |
| Days ≥ 1" Snow Depth | ***16*** | ***7*** | na | ***1*** | 0 | 0 | 0 | 0 | 0 | ***1*** | ***5*** | ***14*** | na |

## TROUT CREEK RS *Sanders County* ELEVATION 2362 ft LAT/LONG 47° 52 ' N / 115° 37 ' W

| | JAN | FEB | MAR | APR | MAY | JUN | JUL | AUG | SEP | OCT | NOV | DEC | YEAR |
|---|---|---|---|---|---|---|---|---|---|---|---|---|---|
| Maximum Temp °F | 33.8 | 40.8 | 50.0 | 60.0 | 69.4 | 76.1 | 84.3 | 85.4 | 74.5 | 59.9 | 41.9 | 33.7 | 59.2 |
| Minimum Temp °F | 19.3 | 22.1 | 26.2 | 31.1 | 36.7 | 42.6 | 45.1 | 44.7 | 38.6 | 32.0 | 27.7 | 21.1 | 32.3 |
| Mean Temp °F | 26.6 | 31.5 | 38.1 | 45.6 | 53.1 | 59.4 | 64.7 | 65.1 | 56.5 | 45.9 | 34.8 | 27.4 | 45.7 |
| Days Max Temp ≥ 90 °F | 0 | 0 | 0 | 0 | 1 | 3 | 10 | 12 | 2 | 0 | 0 | 0 | 28 |
| Days Max Temp ≤ 32 °F | 11 | 3 | 0 | 0 | 0 | 0 | 0 | 0 | 0 | 0 | 3 | 11 | 28 |
| Days Min Temp ≤ 32 °F | 28 | 25 | 25 | 18 | 9 | 1 | 0 | 1 | 5 | 17 | 22 | 28 | 179 |
| Days Min Temp ≤ 0 °F | 3 | 2 | 0 | 0 | 0 | 0 | 0 | 0 | 0 | 0 | 0 | 2 | 7 |
| Heating Degree Days | 1185 | 940 | 827 | 577 | 368 | 187 | 75 | 72 | 258 | 585 | 899 | 1158 | 7131 |
| Cooling Degree Days | 0 | 0 | 0 | 0 | 8 | 26 | 68 | 80 | 9 | 0 | 0 | 0 | 191 |
| Total Precipitation (") | 4.08 | 2.75 | 2.23 | 1.81 | 2.13 | 2.31 | 1.35 | 1.38 | 1.43 | 1.87 | 3.71 | 3.72 | 28.77 |
| Days ≥ 0.1" Precip | 10 | 8 | 7 | 6 | 7 | 6 | 4 | 4 | 4 | 5 | 10 | 11 | 82 |
| Total Snowfall (") | 22.1 | 10.6 | na | ***0.3*** | 0.0 | 0.0 | 0.0 | 0.0 | 0.0 | ***0.0*** | ***6.9*** | 19.7 | na |
| Days ≥ 1" Snow Depth | 27 | 22 | 10 | 0 | 0 | 0 | 0 | 0 | 0 | 0 | ***6*** | 19 | 84 |

## TROY *Lincoln County* ELEVATION 1932 ft LAT/LONG 48° 29 ' N / 115° 55 ' W

| | JAN | FEB | MAR | APR | MAY | JUN | JUL | AUG | SEP | OCT | NOV | DEC | YEAR |
|---|---|---|---|---|---|---|---|---|---|---|---|---|---|
| Maximum Temp °F | 32.3 | 39.4 | 49.6 | 60.1 | ***69.0*** | 76.3 | 85.2 | 85.3 | 73.3 | ***58.0*** | ***40.4*** | ***32.0*** | 58.4 |
| Minimum Temp °F | 19.1 | 22.1 | 26.2 | 31.8 | ***38.8*** | 45.0 | 47.8 | 47.1 | 40.4 | ***33.2*** | ***28.6*** | ***21.0*** | 33.4 |
| Mean Temp °F | 25.8 | 30.8 | 37.9 | 46.0 | ***53.9*** | 60.7 | 66.5 | 66.2 | 56.9 | ***45.5*** | ***34.6*** | ***26.5*** | 45.9 |
| Days Max Temp ≥ 90 °F | 0 | 0 | 0 | 0 | 1 | 3 | 11 | 11 | 2 | 0 | 0 | 0 | 28 |
| Days Max Temp ≤ 32 °F | ***13*** | ***4*** | 1 | 0 | 0 | 0 | 0 | 0 | 0 | 0 | na | ***15*** | na |
| Days Min Temp ≤ 32 °F | 29 | 26 | ***25*** | ***16*** | na | 0 | 0 | 0 | 4 | ***14*** | ***20*** | ***28*** | na |
| Days Min Temp ≤ 0 °F | ***3*** | 1 | 0 | 0 | 0 | 0 | 0 | 0 | 0 | 0 | 0 | 1 | 5 |
| Heating Degree Days | ***1211*** | 959 | 831 | ***566*** | ***345*** | 156 | 51 | 55 | 248 | ***597*** | ***903*** | ***1186*** | 7108 |
| Cooling Degree Days | 0 | 0 | 0 | 0 | 8 | ***34*** | 102 | 100 | 10 | 0 | 0 | 0 | 254 |
| Total Precipitation (") | 3.16 | 2.08 | 1.83 | 1.61 | 1.83 | 1.97 | 1.08 | 1.29 | 1.37 | 1.75 | 3.41 | 3.26 | 24.64 |
| Days ≥ 0.1" Precip | na | na | ***6*** | ***5*** | ***6*** | ***5*** | 3 | 4 | 4 | na | na | na | na |
| Total Snowfall (") | na | na | na | na | na | na | na | ***0.0*** | na | na | na | na | na |
| Days ≥ 1" Snow Depth | na | na | na | na | na | na | na | ***0*** | na | ***0*** | na | na | na |

## TROY 18 N *Lincoln County* ELEVATION 2723 ft LAT/LONG 48° 44 ' N / 115° 53 ' W

| | JAN | FEB | MAR | APR | MAY | JUN | JUL | AUG | SEP | OCT | NOV | DEC | YEAR |
|---|---|---|---|---|---|---|---|---|---|---|---|---|---|
| Maximum Temp °F | 30.3 | ***37.6*** | ***46.1*** | 56.7 | 67.1 | 73.9 | 81.7 | 82.5 | 70.6 | ***55.7*** | ***38.2*** | ***30.1*** | 55.9 |
| Minimum Temp °F | 16.5 | ***19.9*** | ***24.2*** | 29.6 | 36.6 | 43.1 | 46.3 | 45.9 | 38.8 | ***31.5*** | ***25.5*** | ***18.4*** | 31.4 |
| Mean Temp °F | 23.4 | ***28.8*** | ***35.2*** | 43.2 | 51.9 | 58.5 | 63.9 | 64.2 | 54.7 | ***43.6*** | ***31.9*** | ***24.3*** | 43.6 |
| Days Max Temp ≥ 90 °F | 0 | 0 | 0 | 0 | 0 | 1 | 6 | 7 | 1 | 0 | 0 | 0 | 15 |
| Days Max Temp ≤ 32 °F | 15 | ***5*** | 1 | 0 | 0 | 0 | 0 | 0 | 0 | 0 | ***5*** | ***17*** | 43 |
| Days Min Temp ≤ 32 °F | 30 | ***27*** | 28 | 22 | 8 | 1 | 0 | 0 | 5 | ***17*** | 25 | ***30*** | 193 |
| Days Min Temp ≤ 0 °F | 4 | 2 | 0 | 0 | 0 | 0 | 0 | 0 | 0 | 0 | 1 | 3 | 10 |
| Heating Degree Days | 1283 | ***1017*** | ***919*** | 650 | 404 | 206 | 86 | 81 | 307 | ***656*** | ***987*** | ***1258*** | 7854 |
| Cooling Degree Days | 0 | 0 | 0 | 0 | 6 | ***16*** | 55 | ***60*** | 4 | ***0*** | ***0*** | na | na |
| Total Precipitation (") | 4.25 | 2.88 | 2.59 | 2.31 | 2.41 | 2.68 | 1.55 | 1.54 | 2.08 | 2.87 | 4.65 | 4.56 | 34.37 |
| Days ≥ 0.1" Precip | ***9*** | 7 | 7 | 7 | 7 | 7 | 4 | 4 | 5 | 7 | ***10*** | ***9*** | 83 |
| Total Snowfall (") | na | ***15.8*** | na | na | 0.0 | 0.0 | 0.0 | 0.0 | 0.0 | ***1.4*** | ***12.7*** | na | na |
| Days ≥ 1" Snow Depth | ***30*** | ***27*** | na | na | 0 | 0 | 0 | 0 | 0 | ***1*** | na | ***29*** | na |

## TWIN BRIDGES *Madison County* ELEVATION 4662 ft LAT/LONG 45° 32' N / 112° 20' W

| | JAN | FEB | MAR | APR | MAY | JUN | JUL | AUG | SEP | OCT | NOV | DEC | YEAR |
|---|---|---|---|---|---|---|---|---|---|---|---|---|---|
| Maximum Temp °F | 34.4 | 40.5 | 47.7 | 57.3 | 67.0 | 75.6 | 83.6 | 81.8 | 71.4 | 60.2 | 43.5 | 34.3 | 58.1 |
| Minimum Temp °F | 11.5 | 14.8 | 21.0 | 27.7 | 35.4 | 42.5 | 45.9 | 43.5 | 35.1 | 27.2 | 19.3 | 11.7 | 28.0 |
| Mean Temp °F | 23.0 | 27.7 | 34.4 | 42.5 | 51.2 | 59.1 | 64.8 | 62.7 | 53.3 | 43.7 | 31.4 | 23.0 | 43.1 |
| Days Max Temp ≥ 90 °F | 0 | 0 | 0 | 0 | 0 | 2 | 6 | 5 | 0 | 0 | 0 | 0 | 13 |
| Days Max Temp ≤ 32 °F | 11 | 5 | 3 | 0 | 0 | 0 | 0 | 0 | 0 | 0 | 4 | 12 | 35 |
| Days Min Temp ≤ 32 °F | 30 | 27 | 29 | 22 | 11 | 2 | 0 | 1 | 11 | 23 | 27 | 30 | 213 |
| Days Min Temp ≤ 0 °F | 7 | 4 | 1 | 0 | 0 | 0 | 0 | 0 | 0 | 0 | 2 | 5 | 19 |
| Heating Degree Days | 1297 | 1048 | 943 | 668 | 420 | 188 | 58 | 96 | 347 | 652 | 1000 | 1295 | 8012 |
| Cooling Degree Days | 0 | 0 | 0 | 0 | 1 | 18 | 57 | 31 | 1 | 0 | 0 | 0 | 108 |
| Total Precipitation (") | 0.20 | 0.16 | 0.48 | 0.82 | 1.67 | 1.79 | 1.15 | 1.07 | 1.11 | 0.53 | 0.41 | 0.23 | 9.62 |
| Days ≥ 0.1" Precip | 0 | 1 | 2 | 3 | 5 | 5 | 3 | 3 | 3 | 2 | 1 | 1 | 29 |
| Total Snowfall (") | na | na | na | na | ***0.0*** | 0.0 | 0.0 | 0.1 | 0.0 | ***0.1*** | na | na | na |
| Days ≥ 1" Snow Depth | na | na | na | na | 0 | 0 | 0 | 0 | 0 | ***0*** | na | na | na |

## VALIER *Pondera County* ELEVATION 3812 ft LAT/LONG 48° 19' N / 112° 15' W

| | JAN | FEB | MAR | APR | MAY | JUN | JUL | AUG | SEP | OCT | NOV | DEC | YEAR |
|---|---|---|---|---|---|---|---|---|---|---|---|---|---|
| Maximum Temp °F | 31.0 | 36.8 | 44.3 | 55.3 | 65.3 | 73.0 | 80.1 | 79.9 | 69.7 | 59.1 | 42.1 | 33.7 | 55.9 |
| Minimum Temp °F | 9.2 | 13.6 | 20.8 | 29.6 | 38.9 | 46.5 | 50.7 | 49.7 | 41.3 | 33.1 | 21.2 | 13.1 | 30.6 |
| Mean Temp °F | 20.1 | 25.2 | 32.6 | 42.5 | 52.1 | 59.8 | 65.4 | 64.8 | 55.5 | 46.1 | 31.7 | 23.4 | 43.3 |
| Days Max Temp ≥ 90 °F | 0 | 0 | 0 | 0 | 0 | 1 | 3 | 4 | 0 | 0 | 0 | 0 | 8 |
| Days Max Temp ≤ 32 °F | 13 | 9 | 5 | 1 | 0 | 0 | 0 | 0 | 0 | 1 | 5 | 11 | 45 |
| Days Min Temp ≤ 32 °F | 29 | 26 | 27 | 19 | 5 | 0 | 0 | 0 | 4 | 14 | 25 | 28 | 177 |
| Days Min Temp ≤ 0 °F | 11 | 6 | 2 | 0 | 0 | 0 | 0 | 0 | 0 | 0 | 2 | 7 | 28 |
| Heating Degree Days | 1388 | 1117 | 999 | 670 | 395 | 179 | 63 | 77 | 292 | 579 | 994 | 1283 | 8036 |
| Cooling Degree Days | 0 | 0 | 0 | 1 | 6 | 32 | 81 | 85 | 15 | 2 | 0 | 0 | 222 |
| Total Precipitation (") | 0.37 | 0.26 | 0.50 | 0.95 | 2.05 | 2.70 | 1.43 | 1.70 | 1.16 | 0.60 | 0.35 | 0.31 | 12.38 |
| Days ≥ 0.1" Precip | 1 | 1 | 2 | 3 | 5 | 6 | 4 | 4 | 3 | 2 | 1 | 1 | 33 |
| Total Snowfall (") | na | na | ***4.2*** | na | 0.3 | 0.0 | 0.0 | 0.0 | 0.4 | ***0.7*** | na | na | na |
| Days ≥ 1" Snow Depth | ***14*** | ***13*** | ***9*** | ***2*** | 0 | 0 | 0 | 0 | 0 | 1 | ***7*** | ***12*** | 58 |

## VIDA 6 NE *McCone County* ELEVATION 2401 ft LAT/LONG 47° 52' N / 105° 27' W

| | JAN | FEB | MAR | APR | MAY | JUN | JUL | AUG | SEP | OCT | NOV | DEC | YEAR |
|---|---|---|---|---|---|---|---|---|---|---|---|---|---|
| Maximum Temp °F | 23.7 | 30.9 | 42.8 | 57.6 | 69.1 | 78.1 | 84.8 | 83.9 | 72.2 | 59.0 | 40.7 | 28.7 | 56.0 |
| Minimum Temp °F | 1.6 | 8.5 | 19.7 | 30.7 | 41.5 | 50.3 | 54.7 | 52.9 | 42.5 | 32.1 | 18.6 | 7.4 | 30.0 |
| Mean Temp °F | 12.3 | 19.5 | 31.4 | 44.0 | 55.3 | 64.3 | 69.8 | 68.4 | 57.4 | 45.7 | 29.7 | 18.1 | 43.0 |
| Days Max Temp ≥ 90 °F | 0 | 0 | 0 | 0 | 1 | 4 | 9 | 9 | 2 | 0 | 0 | 0 | 25 |
| Days Max Temp ≤ 32 °F | 19 | 13 | 7 | 1 | 0 | 0 | 0 | 0 | 0 | 1 | 8 | 17 | 66 |
| Days Min Temp ≤ 32 °F | 30 | 27 | 26 | 17 | ***4*** | 0 | 0 | 0 | 3 | 15 | 27 | 30 | 179 |
| Days Min Temp ≤ 0 °F | 14 | 9 | 3 | 0 | 0 | 0 | 0 | 0 | 0 | 0 | 3 | 9 | 38 |
| Heating Degree Days | 1630 | 1277 | 1036 | 625 | 309 | 93 | 23 | 47 | 255 | 593 | 1052 | ***1447*** | 8387 |
| Cooling Degree Days | 0 | 0 | 0 | 2 | ***22*** | ***96*** | ***179*** | 169 | 26 | 1 | 0 | ***0*** | 495 |
| Total Precipitation (") | 0.58 | 0.32 | 0.68 | 1.48 | 2.25 | 2.78 | 1.96 | 1.48 | 1.40 | 0.84 | 0.46 | 0.52 | 14.75 |
| Days ≥ 0.1" Precip | 2 | 1 | 2 | 3 | 5 | 5 | 4 | 3 | 3 | 2 | 2 | 2 | 34 |
| Total Snowfall (") | ***6.8*** | ***4.4*** | 5.0 | 4.8 | 0.7 | 0.0 | 0.0 | 0.0 | 0.3 | ***0.8*** | ***3.7*** | ***6.6*** | 33.1 |
| Days ≥ 1" Snow Depth | ***19*** | ***13*** | 8 | 2 | 0 | 0 | 0 | 0 | 0 | ***1*** | ***5*** | ***16*** | 64 |

## VIRGINIA CITY *Madison County* ELEVATION 5853 ft LAT/LONG 45° 18' N / 111° 56' W

| | JAN | FEB | MAR | APR | MAY | JUN | JUL | AUG | SEP | OCT | NOV | DEC | YEAR |
|---|---|---|---|---|---|---|---|---|---|---|---|---|---|
| Maximum Temp °F | 33.6 | 37.9 | 43.6 | 52.8 | 62.2 | 71.3 | 79.9 | 79.0 | 68.5 | 57.4 | 41.1 | 33.5 | 55.1 |
| Minimum Temp °F | 12.5 | 15.2 | 20.6 | 27.6 | 35.9 | 42.9 | 48.6 | 47.0 | 38.2 | 29.8 | 19.8 | 12.6 | 29.2 |
| Mean Temp °F | 23.1 | 26.6 | 32.1 | 40.2 | 49.1 | 57.1 | 64.3 | 63.0 | 53.4 | 43.6 | 30.5 | 23.1 | 42.2 |
| Days Max Temp ≥ 90 °F | 0 | 0 | 0 | 0 | 0 | 0 | 2 | 1 | 0 | 0 | 0 | 0 | 3 |
| Days Max Temp ≤ 32 °F | 12 | 7 | 3 | 1 | 0 | 0 | 0 | 0 | 0 | 1 | 6 | 13 | 43 |
| Days Min Temp ≤ 32 °F | 30 | 27 | 29 | 23 | 10 | 2 | 0 | 1 | 7 | 19 | 27 | 30 | 205 |
| Days Min Temp ≤ 0 °F | 6 | 3 | 1 | 0 | 0 | 0 | 0 | 0 | 0 | 0 | 2 | 5 | 17 |
| Heating Degree Days | 1295 | 1079 | 1013 | 738 | 489 | 246 | 80 | 105 | 348 | 656 | 1029 | 1295 | 8373 |
| Cooling Degree Days | 0 | 0 | 0 | 0 | 0 | 13 | 53 | 41 | 4 | 0 | 0 | 0 | 111 |
| Total Precipitation (") | 0.62 | 0.52 | 1.02 | 1.42 | 2.40 | 2.54 | 1.74 | 1.52 | 1.40 | 1.04 | 0.98 | 0.65 | 15.85 |
| Days ≥ 0.1" Precip | 2 | 2 | 3 | 4 | 7 | 7 | 5 | 5 | 4 | 3 | 3 | 2 | 47 |
| Total Snowfall (") | na | na | na | na | ***4.4*** | 0.7 | 0.0 | 0.0 | 1.1 | na | na | na | na |
| Days ≥ 1" Snow Depth | na | na | na | na | ***0*** | 0 | 0 | 0 | 0 | ***1*** | na | na | na |

## WEST GLACIER *Flathead County* ELEVATION 3153 ft LAT/LONG 48° 30 ' N / 113° 59 ' W

| | JAN | FEB | MAR | APR | MAY | JUN | JUL | AUG | SEP | OCT | NOV | DEC | YEAR |
|---|---|---|---|---|---|---|---|---|---|---|---|---|---|
| Maximum Temp °F | 29.0 | 34.7 | 42.9 | 53.6 | 64.5 | 71.9 | 78.9 | 78.6 | 66.7 | 53.1 | 36.9 | 29.6 | 53.4 |
| Minimum Temp °F | 15.3 | 18.4 | 23.7 | 30.3 | 37.3 | 43.8 | 47.2 | 46.5 | 38.7 | 31.4 | 24.8 | 17.7 | 31.3 |
| Mean Temp °F | 22.2 | 26.6 | 33.3 | 42.0 | 50.9 | 57.9 | 63.1 | 62.6 | 52.7 | 42.3 | 30.9 | 23.7 | 42.4 |
| Days Max Temp ≥ 90 °F | 0 | 0 | 0 | 0 | 0 | 0 | 2 | 2 | 0 | 0 | 0 | 0 | 4 |
| Days Max Temp ≤ 32 °F | 17 | 8 | 2 | 0 | 0 | 0 | 0 | 0 | 0 | 0 | 7 | 17 | 51 |
| Days Min Temp ≤ 32 °F | 30 | 27 | 28 | 20 | 7 | 1 | 0 | 0 | 5 | 19 | 25 | 30 | 192 |
| Days Min Temp ≤ 0 °F | 5 | 3 | 1 | 0 | 0 | 0 | 0 | 0 | 0 | 0 | 1 | 3 | 13 |
| Heating Degree Days | 1321 | 1079 | 975 | 683 | 432 | 221 | 98 | 110 | 366 | 697 | 1018 | 1275 | 8275 |
| Cooling Degree Days | 0 | 0 | 0 | 0 | 2 | 16 | 45 | 45 | 2 | 0 | 0 | 0 | 110 |
| Total Precipitation (") | 3.34 | 2.12 | 1.63 | 1.69 | 2.58 | 3.25 | 1.85 | 1.74 | 2.05 | 1.99 | 2.98 | 3.17 | 28.39 |
| Days ≥ 0.1" Precip | 9 | 7 | 6 | 5 | 7 | 8 | 5 | 5 | 6 | 6 | 8 | 9 | 81 |
| Total Snowfall (") | 38.4 | 20.8 | 10.9 | *1.7* | 0.1 | 0.3 | 0.0 | 0.0 | 0.1 | 1.7 | 17.4 | 38.8 | 130.2 |
| Days ≥ 1" Snow Depth | 31 | 28 | 30 | 12 | 0 | 0 | 0 | 0 | 0 | 1 | 16 | 30 | 148 |

## WEST YELLOWSTONE *Gallatin County* ELEVATION 6667 ft LAT/LONG 44° 39 ' N / 111° 6 ' W

| | JAN | FEB | MAR | APR | MAY | JUN | JUL | AUG | SEP | OCT | NOV | DEC | YEAR |
|---|---|---|---|---|---|---|---|---|---|---|---|---|---|
| Maximum Temp °F | 24.3 | 30.2 | 37.9 | 46.2 | 58.6 | 68.5 | 77.8 | 76.4 | 65.2 | 51.2 | 33.0 | 23.4 | 49.4 |
| Minimum Temp °F | 2.2 | 3.7 | 10.9 | 20.5 | 29.4 | 36.6 | 41.2 | 38.9 | 30.4 | 22.4 | 11.7 | 1.6 | 20.8 |
| Mean Temp °F | 13.2 | 16.9 | 24.4 | 33.4 | 44.0 | 52.6 | 59.5 | 57.7 | 47.9 | 36.8 | 22.4 | 12.5 | 35.1 |
| Days Max Temp ≥ 90 °F | 0 | 0 | 0 | 0 | 0 | 0 | 0 | 1 | 0 | 0 | 0 | 0 | 1 |
| Days Max Temp ≤ 32 °F | 26 | 16 | 6 | 1 | 0 | 0 | 0 | 0 | 0 | 1 | 14 | 26 | 90 |
| Days Min Temp ≤ 32 °F | 31 | 28 | 30 | 28 | 22 | 8 | 2 | 4 | 19 | 28 | 29 | 31 | 260 |
| Days Min Temp ≤ 0 °F | 14 | 11 | 7 | 1 | 0 | 0 | 0 | 0 | 0 | 1 | 6 | 14 | 54 |
| Heating Degree Days | 1600 | 1351 | 1252 | 941 | 643 | 367 | 168 | 223 | 508 | 867 | 1273 | 1621 | 10814 |
| Cooling Degree Days | 0 | *0* | 0 | 0 | 0 | 2 | 4 | 3 | 0 | 0 | 0 | 0 | 9 |
| Total Precipitation (") | 2.11 | 1.63 | 1.69 | 1.49 | 2.00 | 2.32 | 1.75 | 1.47 | 1.58 | 1.32 | 2.09 | 2.34 | 21.79 |
| Days ≥ 0.1" Precip | 7 | 5 | 5 | 5 | 7 | 7 | 5 | 4 | 4 | 4 | 7 | 7 | 67 |
| Total Snowfall (") | *32.6* | na | *20.9* | *12.0* | 3.4 | 0.3 | 0.0 | 0.0 | 0.8 | *7.9* | *27.0* | *36.3* | na |
| Days ≥ 1" Snow Depth | 30 | 28 | 31 | 25 | 5 | 0 | 0 | 0 | 0 | 3 | *23* | 30 | 175 |

## WESTBY *Sheridan County* ELEVATION 2113 ft LAT/LONG 48° 52 ' N / 104° 3 ' W

| | JAN | FEB | MAR | APR | MAY | JUN | JUL | AUG | SEP | OCT | NOV | DEC | YEAR |
|---|---|---|---|---|---|---|---|---|---|---|---|---|---|
| Maximum Temp °F | 16.5 | 24.4 | 38.0 | 54.6 | 68.2 | 76.8 | *82.9* | 82.6 | 70.6 | 56.6 | 36.6 | 22.8 | 52.6 |
| Minimum Temp °F | -4.9 | 2.2 | 15.1 | 28.4 | 40.6 | 50.0 | *54.1* | 52.3 | 41.5 | 30.1 | 16.1 | 2.2 | 27.3 |
| Mean Temp °F | 5.8 | 13.3 | 26.5 | 41.5 | 54.4 | 63.4 | *68.5* | 67.5 | 56.1 | 43.4 | 26.4 | 12.5 | 39.9 |
| Days Max Temp ≥ 90 °F | 0 | 0 | 0 | 0 | 1 | 3 | *7* | 7 | 2 | 0 | 0 | 0 | 20 |
| Days Max Temp ≤ 32 °F | 25 | 18 | 10 | 1 | 0 | 0 | 0 | 0 | 0 | 1 | 11 | 21 | 87 |
| Days Min Temp ≤ 32 °F | 31 | 28 | 30 | 21 | 5 | 0 | 0 | 0 | 4 | na | 28 | 31 | na |
| Days Min Temp ≤ 0 °F | 18 | 13 | 5 | 0 | 0 | 0 | 0 | 0 | 0 | 0 | 3 | 13 | 52 |
| Heating Degree Days | 1834 | 1456 | 1184 | 698 | 339 | 107 | *33* | 59 | 284 | 664 | 1152 | 1623 | 9433 |
| Cooling Degree Days | 0 | 0 | 0 | *0* | 18 | *62* | na | 140 | 21 | 0 | 0 | 0 | na |
| Total Precipitation (") | 0.45 | 0.29 | 0.50 | 1.10 | 1.99 | 2.69 | 2.31 | 1.79 | 1.29 | 0.69 | 0.30 | 0.44 | 13.84 |
| Days ≥ 0.1" Precip | 2 | 1 | 2 | 3 | 5 | 6 | 5 | 4 | 3 | 2 | 1 | 2 | 36 |
| Total Snowfall (") | 6.3 | 4.5 | *4.6* | *3.8* | 0.6 | 0.0 | 0.0 | 0.0 | *0.0* | na | *3.5* | *5.7* | na |
| Days ≥ 1" Snow Depth | *26* | *21* | *15* | *4* | 0 | 0 | 0 | 0 | 0 | na | *6* | *20* | na |

## WESTERN AG RESEARCH *Ravalli County* ELEVATION 3602 ft LAT/LONG 46° 20 ' N / 114° 4 ' W

| | JAN | FEB | MAR | APR | MAY | JUN | JUL | AUG | SEP | OCT | NOV | DEC | YEAR |
|---|---|---|---|---|---|---|---|---|---|---|---|---|---|
| Maximum Temp °F | 34.9 | 41.9 | 50.3 | 59.3 | 67.9 | 75.7 | 83.7 | 82.9 | 71.7 | 59.9 | 43.6 | 34.8 | 58.9 |
| Minimum Temp °F | 17.3 | 21.0 | 26.0 | 31.2 | 38.0 | 44.7 | 48.3 | 47.3 | 39.8 | 31.6 | 23.8 | 17.1 | 32.2 |
| Mean Temp °F | 26.1 | 31.5 | 38.2 | 45.3 | 53.0 | 60.2 | 66.0 | 65.1 | 55.8 | 45.8 | 33.7 | 26.0 | 45.6 |
| Days Max Temp ≥ 90 °F | 0 | 0 | 0 | 0 | 0 | 3 | 7 | 6 | 0 | 0 | 0 | 0 | 16 |
| Days Max Temp ≤ 32 °F | 11 | 4 | 1 | 0 | 0 | 0 | 0 | 0 | 0 | 0 | 3 | 12 | 31 |
| Days Min Temp ≤ 32 °F | 28 | 25 | 25 | 18 | 6 | 1 | 0 | 0 | 4 | 17 | 24 | 28 | 176 |
| Days Min Temp ≤ 0 °F | 4 | 2 | 0 | 0 | 0 | 0 | 0 | 0 | 0 | 0 | 1 | 3 | 10 |
| Heating Degree Days | 1199 | 941 | 824 | 586 | 369 | 169 | 50 | 63 | 278 | 590 | 931 | 1205 | 7205 |
| Cooling Degree Days | 0 | 0 | 0 | 0 | 6 | 40 | 96 | 87 | 8 | 0 | 0 | 0 | 237 |
| Total Precipitation (") | 0.75 | 0.46 | 0.66 | 0.90 | 1.71 | 1.63 | 0.93 | 1.17 | 1.02 | 0.73 | 0.68 | 0.60 | 11.24 |
| Days ≥ 0.1" Precip | 3 | 2 | 2 | 3 | 5 | 5 | 3 | 4 | 3 | 2 | 2 | 2 | 36 |
| Total Snowfall (") | na | na | na | na | 0.0 | 0.0 | 0.0 | 0.0 | 0.1 | na | na | na | na |
| Days ≥ 1" Snow Depth | na | na | na | na | 0 | 0 | 0 | 0 | 0 | na | na | na | na |

## WIBAUX 2 E *Wibaux County* ELEVATION 2671 ft LAT/LONG 46° 59 ' N / 104° 9 ' W

| | JAN | FEB | MAR | APR | MAY | JUN | JUL | AUG | SEP | OCT | NOV | DEC | YEAR |
|---|---|---|---|---|---|---|---|---|---|---|---|---|---|
| Maximum Temp °F | 24.2 | 30.7 | 42.3 | 56.3 | 68.5 | 77.3 | 85.2 | 84.6 | 72.3 | 58.8 | 39.9 | 28.6 | 55.7 |
| Minimum Temp °F | 1.1 | 7.4 | 17.9 | 28.8 | 39.3 | 48.2 | 52.2 | 50.9 | 40.3 | 29.9 | 16.8 | 5.9 | 28.2 |
| Mean Temp °F | 12.7 | 19.1 | 30.2 | 42.5 | 54.0 | 62.8 | 68.7 | 67.8 | 56.3 | 44.4 | 28.4 | 17.3 | 42.0 |
| Days Max Temp ≥ 90 °F | 0 | 0 | 0 | 0 | 1 | 3 | 10 | 10 | 2 | 0 | 0 | 0 | 26 |
| Days Max Temp ≤ 32 °F | 20 | 14 | 7 | 1 | 0 | 0 | 0 | 0 | 0 | 1 | 8 | 17 | 68 |
| Days Min Temp ≤ 32 °F | 31 | 28 | 29 | 21 | 6 | 1 | 0 | 0 | 5 | 19 | 28 | 31 | 199 |
| Days Min Temp ≤ 0 °F | 15 | 10 | 3 | 0 | 0 | 0 | 0 | 0 | 0 | 0 | 3 | 11 | 42 |
| Heating Degree Days | 1620 | 1292 | 1074 | 668 | 348 | 121 | 34 | 51 | 280 | 633 | 1092 | 1474 | 8687 |
| Cooling Degree Days | 0 | 0 | 0 | 1 | 16 | 70 | 158 | 147 | 27 | 0 | 0 | 0 | 419 |
| Total Precipitation (") | 0.31 | 0.20 | 0.55 | 1.36 | 2.38 | 2.75 | 1.89 | 1.46 | 1.58 | 1.04 | 0.43 | 0.29 | 14.24 |
| Days ≥ 0.1" Precip | 1 | 1 | 2 | 4 | 6 | 6 | 5 | 3 | 4 | 2 | 1 | 1 | 36 |
| Total Snowfall (") | 7.0 | 4.0 | 5.4 | na | 0.9 | 0.0 | 0.0 | 0.0 | 0.3 | 1.6 | *5.4* | 6.0 | na |
| Days ≥ 1" Snow Depth | na | na | na | na | 0 | 0 | 0 | 0 | 0 | *1* | *6* | *11* | na |

## WILSALL 8 ENE *Park County* ELEVATION 5843 ft LAT/LONG 46° 2 ' N / 110° 30 ' W

| | JAN | FEB | MAR | APR | MAY | JUN | JUL | AUG | SEP | OCT | NOV | DEC | YEAR |
|---|---|---|---|---|---|---|---|---|---|---|---|---|---|
| Maximum Temp °F | 33.1 | 37.1 | 42.4 | 50.9 | 60.9 | 69.8 | 78.2 | 77.9 | 67.0 | 55.6 | 40.6 | 33.8 | 53.9 |
| Minimum Temp °F | 11.5 | 14.7 | 19.2 | 26.3 | 34.4 | 41.3 | 46.0 | 45.2 | 37.6 | 29.9 | 20.3 | 13.2 | 28.3 |
| Mean Temp °F | 22.4 | 25.9 | 30.8 | 38.6 | 47.7 | 55.6 | 62.1 | 61.5 | 52.3 | 42.8 | 30.5 | 23.5 | 41.1 |
| Days Max Temp ≥ 90 °F | 0 | 0 | 0 | 0 | 0 | 0 | 1 | 1 | 0 | 0 | 0 | 0 | 2 |
| Days Max Temp ≤ 32 °F | 13 | 7 | 5 | 1 | 0 | 0 | 0 | 0 | 0 | 1 | 6 | 13 | 46 |
| Days Min Temp ≤ 32 °F | 30 | 27 | 28 | 24 | 12 | 2 | 0 | 0 | 6 | 18 | 26 | 30 | 203 |
| Days Min Temp ≤ 0 °F | 7 | 3 | 2 | 0 | 0 | 0 | 0 | 0 | 0 | 0 | 2 | 4 | 18 |
| Heating Degree Days | 1316 | 1098 | 1053 | 785 | 530 | 284 | 112 | 126 | 378 | 681 | 1030 | 1280 | 8673 |
| Cooling Degree Days | 0 | 0 | 0 | 0 | 0 | 10 | 36 | 27 | 3 | 0 | 0 | 0 | 76 |
| Total Precipitation (") | 0.90 | 0.75 | 1.51 | 2.04 | 3.49 | 3.37 | 1.93 | 1.79 | 2.00 | 1.52 | 1.03 | 0.87 | 21.20 |
| Days ≥ 0.1" Precip | 3 | 3 | 5 | 7 | 9 | 8 | 6 | 5 | 6 | 4 | 4 | 4 | 64 |
| Total Snowfall (") | 15.4 | 12.4 | 18.9 | 14.9 | 7.2 | 0.7 | 0.0 | 0.1 | 2.8 | 6.0 | 10.6 | 14.7 | 103.7 |
| Days ≥ 1" Snow Depth | 23 | 19 | 21 | 10 | 1 | 0 | 0 | 0 | 0 | 2 | 9 | 20 | 105 |

## WINIFRED *Fergus County* ELEVATION 3251 ft LAT/LONG 47° 33 ' N / 109° 23 ' W

| | JAN | FEB | MAR | APR | MAY | JUN | JUL | AUG | SEP | OCT | NOV | DEC | YEAR |
|---|---|---|---|---|---|---|---|---|---|---|---|---|---|
| Maximum Temp °F | 30.4 | 37.0 | 45.9 | 57.1 | 67.2 | 76.1 | 84.2 | 84.2 | 72.2 | 60.6 | 43.9 | 34.4 | 57.8 |
| Minimum Temp °F | 6.9 | 12.1 | 21.0 | 30.1 | 39.4 | 47.1 | 51.0 | 49.4 | 39.8 | 30.3 | 18.9 | 9.8 | 29.7 |
| Mean Temp °F | 18.7 | 24.6 | 33.5 | 43.6 | 53.3 | 61.6 | 67.7 | 66.8 | 56.0 | 45.5 | 31.5 | 22.1 | 43.7 |
| Days Max Temp ≥ 90 °F | 0 | 0 | 0 | 0 | 0 | 3 | 10 | 10 | 2 | 0 | 0 | 0 | 25 |
| Days Max Temp ≤ 32 °F | 14 | 9 | 5 | 1 | 0 | 0 | 0 | 0 | 0 | 1 | 5 | 11 | 46 |
| Days Min Temp ≤ 32 °F | 29 | 27 | 27 | 19 | 5 | 0 | 0 | 0 | 5 | 18 | 27 | 29 | 186 |
| Days Min Temp ≤ 0 °F | 12 | 7 | 2 | 0 | 0 | 0 | 0 | 0 | 0 | 0 | 3 | 8 | 32 |
| Heating Degree Days | 1432 | 1136 | 970 | 636 | 362 | 143 | 40 | 57 | 282 | 598 | 999 | 1324 | 7979 |
| Cooling Degree Days | 0 | 0 | 0 | 1 | 10 | 58 | 128 | 123 | 18 | 1 | 0 | 0 | 339 |
| Total Precipitation (") | 0.77 | 0.43 | 0.77 | 1.36 | 2.70 | 2.86 | 1.60 | 1.75 | 1.38 | 0.83 | 0.58 | 0.72 | 15.75 |
| Days ≥ 0.1" Precip | 3 | 1 | 3 | 4 | 6 | 6 | 4 | 4 | 3 | 3 | 2 | 3 | 42 |
| Total Snowfall (") | na | na | na | na | na | na | na | na | na | na | na | na | na |
| Days ≥ 1" Snow Depth | na | na | na | na | na | na | na | na | na | na | na | na | na |

## WISDOM *Beaverhead County* ELEVATION 6063 ft LAT/LONG 45° 37 ' N / 113° 27 ' W

| | JAN | FEB | MAR | APR | MAY | JUN | JUL | AUG | SEP | OCT | NOV | DEC | YEAR |
|---|---|---|---|---|---|---|---|---|---|---|---|---|---|
| Maximum Temp °F | 27.1 | 31.7 | 39.2 | 49.0 | 59.6 | 68.7 | 77.9 | 77.3 | 66.9 | 54.8 | 37.0 | 27.5 | 51.4 |
| Minimum Temp °F | 1.9 | 3.5 | 11.3 | 20.6 | 28.3 | 35.7 | 37.3 | 34.5 | 27.2 | 20.0 | 12.1 | 2.5 | 19.6 |
| Mean Temp °F | 14.5 | 17.6 | 25.3 | 34.8 | 44.0 | 52.3 | 57.6 | 55.9 | 47.0 | 37.5 | 24.6 | 15.0 | 35.5 |
| Days Max Temp ≥ 90 °F | 0 | 0 | 0 | 0 | 0 | 0 | 0 | 1 | 0 | 0 | 0 | 0 | 1 |
| Days Max Temp ≤ 32 °F | 19 | 13 | 6 | 1 | 0 | 0 | 0 | 0 | 0 | 1 | 9 | 20 | 69 |
| Days Min Temp ≤ 32 °F | 30 | 28 | 31 | 29 | 23 | 9 | 5 | 12 | 22 | 29 | 29 | 31 | 278 |
| Days Min Temp ≤ 0 °F | 13 | 11 | 6 | 1 | 0 | 0 | 0 | 0 | 0 | 0 | 5 | 13 | 49 |
| Heating Degree Days | 1560 | 1333 | 1225 | 899 | 645 | 376 | 224 | 276 | 532 | 847 | 1207 | 1544 | 10668 |
| Cooling Degree Days | 0 | 0 | 0 | 0 | 0 | 1 | 3 | 2 | 0 | 0 | 0 | 0 | 6 |
| Total Precipitation (") | 0.58 | 0.44 | 0.63 | 0.91 | 1.60 | 1.79 | 1.18 | 1.16 | 1.12 | 0.71 | 0.70 | 0.63 | 11.45 |
| Days ≥ 0.1" Precip | 2 | 1 | 2 | 3 | 5 | 6 | 4 | 4 | 3 | 2 | 3 | 2 | 37 |
| Total Snowfall (") | na | na | na | na | na | *0.0* | 0.0 | 0.0 | 0.2 | na | na | na | na |
| Days ≥ 1" Snow Depth | na | na | na | na | na | na | *0* | *0* | na | na | na | na | na |

## WYOLA 1 SW *Big Horn County* ELEVATION 3734 ft LAT/LONG 45° 8 ' N / 107° 23 ' W

| | JAN | FEB | MAR | APR | MAY | JUN | JUL | AUG | SEP | OCT | NOV | DEC | YEAR |
|---|---|---|---|---|---|---|---|---|---|---|---|---|---|
| Maximum Temp °F | 35.6 | 40.7 | 49.5 | 60.4 | 69.6 | 78.3 | 86.4 | 86.0 | 74.6 | 63.1 | 46.5 | 37.5 | 60.7 |
| Minimum Temp °F | 7.8 | 12.3 | 20.7 | 29.4 | 37.5 | 45.5 | 49.8 | 48.0 | 38.7 | 29.5 | 18.9 | 9.8 | 29.0 |
| Mean Temp °F | 21.8 | 26.5 | 35.1 | 44.8 | 53.6 | 61.9 | 68.2 | 67.2 | 56.7 | 46.3 | 32.9 | 23.7 | 44.9 |
| Days Max Temp ≥ 90 °F | 0 | 0 | 0 | 0 | 0 | 4 | 12 | 11 | 2 | 0 | 0 | 0 | 29 |
| Days Max Temp ≤ 32 °F | 10 | 6 | 2 | 0 | 0 | 0 | 0 | 0 | 0 | 0 | 4 | 10 | 32 |
| Days Min Temp ≤ 32 °F | 30 | 27 | 28 | 20 | 8 | 1 | 0 | 0 | 6 | 20 | 28 | 30 | 198 |
| Days Min Temp ≤ 0 °F | 9 | 5 | 2 | 0 | 0 | 0 | 0 | 0 | 0 | 0 | 2 | 7 | 25 |
| Heating Degree Days | 1332 | 1079 | 919 | 601 | 350 | 129 | 28 | 41 | 260 | 574 | 956 | 1273 | 7542 |
| Cooling Degree Days | 0 | 0 | 0 | 0 | 3 | 49 | 119 | 102 | 19 | 1 | 0 | 0 | 293 |
| Total Precipitation (") | 0.87 | 0.72 | 1.15 | 2.04 | 2.68 | 2.65 | 1.33 | 0.80 | 1.72 | 1.60 | 1.03 | 0.80 | 17.39 |
| Days ≥ 0.1" Precip | 3 | 3 | 4 | 6 | 6 | 6 | 3 | 2 | 5 | 4 | 3 | 3 | 48 |
| Total Snowfall (") | *15.3* | *11.8* | *13.8* | 9.7 | 1.7 | 0.1 | 0.0 | 0.0 | 1.7 | 3.4 | 8.0 | *14.9* | 80.4 |
| Days ≥ 1" Snow Depth | na | na | na | *1* | 0 | 0 | 0 | 0 | 0 | 1 | *5* | na | na |

## YELLOWTAIL DAM *Big Horn County* ELEVATION 3202 ft LAT/LONG 45° 19 ' N / 107° 55 ' W

| | JAN | FEB | MAR | APR | MAY | JUN | JUL | AUG | SEP | OCT | NOV | DEC | YEAR |
|---|---|---|---|---|---|---|---|---|---|---|---|---|---|
| Maximum Temp °F | 38.5 | 43.7 | 51.9 | 62.0 | 71.7 | 81.1 | 89.9 | 89.7 | 77.7 | 65.4 | 48.8 | 40.8 | 63.4 |
| Minimum Temp °F | 17.0 | 21.4 | 27.9 | 36.7 | 45.0 | 52.6 | 58.0 | 56.7 | 47.7 | 39.1 | 28.2 | 20.5 | 37.6 |
| Mean Temp °F | 27.8 | 32.6 | 39.9 | 49.4 | 58.4 | 66.9 | 73.9 | 73.2 | 62.7 | 52.3 | 38.5 | 30.7 | 50.5 |
| Days Max Temp ≥ 90 °F | 0 | 0 | 0 | 0 | 1 | 7 | 17 | 17 | 5 | 0 | 0 | 0 | 47 |
| Days Max Temp ≤ 32 °F | 9 | 6 | 2 | 0 | 0 | 0 | 0 | 0 | 0 | 0 | 3 | 7 | 27 |
| Days Min Temp ≤ 32 °F | 25 | 22 | 20 | 9 | 1 | 0 | 0 | 0 | 1 | 7 | 18 | 24 | 127 |
| Days Min Temp ≤ 0 °F | 6 | 3 | 1 | 0 | 0 | 0 | 0 | 0 | 0 | 0 | 1 | 3 | 14 |
| Heating Degree Days | 1150 | 908 | 772 | 467 | 228 | 61 | 6 | 11 | 139 | 399 | 790 | 1057 | 5988 |
| Cooling Degree Days | 0 | 0 | 1 | 8 | 34 | 142 | 290 | 283 | 82 | 12 | 0 | 0 | 852 |
| Total Precipitation (") | 0.92 | 0.74 | 1.40 | 2.19 | 3.12 | 2.54 | 1.40 | 1.09 | 1.91 | 1.78 | 1.02 | 0.82 | 18.93 |
| Days ≥ 0.1" Precip | 3 | 2 | 5 | 6 | 7 | 6 | 3 | 3 | 5 | 4 | 3 | 3 | 50 |
| Total Snowfall (") | na | na | na | na | *0.1* | 0.0 | 0.0 | 0.0 | *0.2* | na | na | na | na |
| Days ≥ 1" Snow Depth | na | na | na | na | *0* | 0 | 0 | 0 | *0* | na | na | na | na |

## JANUARY MINIMUM TEMPERATURE °F

### LOWEST

| | | |
|---|---|---|
| 1 | Westby | -4.9 |
| 2 | Opheim-10 N | -3.8 |
| 3 | Redstone | -3.2 |
| 4 | Opheim-12 SSE | -2.9 |
| 5 | Culbertson | -2.5 |
| 6 | Raymond | -1.9 |
| 7 | Medicine Lake | -1.4 |
| 8 | Forks | -0.4 |
| 9 | Terry | -0.3 |
| 10 | Simpson | -0.2 |
| 11 | Bredette | 0.5 |
| | Lakeview | 0.5 |
| 13 | Glasgow | 1.0 |
| 14 | Wibaux | 1.1 |
| 15 | Vida | 1.6 |
| 16 | Sidney | 1.7 |
| 17 | Mizpah | 1.9 |
| | Wisdom | 1.9 |
| 19 | Chinook | 2.2 |
| | Circle | 2.2 |
| | West Yellowstone | 2.2 |
| 22 | Chester | 2.4 |
| | Savage | 2.4 |
| 24 | Harlem | 2.6 |
| | Plevna | 2.6 |

### HIGHEST

| | | |
|---|---|---|
| 1 | Bigfork | 21.4 |
| 2 | Thompson Falls | 20.8 |
| 3 | Polson Kerr Dam | 19.8 |
| 4 | Superior | 19.6 |
| 5 | Heron | 19.4 |
| 6 | Trout Creek | 19.3 |
| 7 | Norris | 19.2 |
| 8 | Troy | 19.1 |
| 9 | Darby | 18.6 |
| 10 | Holter Dam | 18.5 |
| | St. Ignatius | 18.5 |
| 12 | St. Regis | 18.0 |
| 13 | Hamilton | 17.6 |
| | Libby-1 NE | 17.6 |
| | Livingston | 17.6 |
| 16 | Western | 17.3 |
| 17 | Stevensville | 17.0 |
| | Yellowtail Dam | 17.0 |
| 19 | Mystic Lake | 16.6 |
| 20 | Troy-18 N | 16.5 |
| 21 | Hungry Horse Dm | 16.2 |
| | Missoula | 16.2 |
| 23 | Big Timber | 15.5 |
| 24 | Creston | 15.4 |
| 25 | West Glacier | 15.3 |

## JULY MAXIMUM TEMPERATURE °F

### HIGHEST

| | | |
|---|---|---|
| 1 | Hardin | 90.2 |
| 2 | Yellowtail Dam | 89.9 |
| 3 | Powderville | 89.8 |
| 4 | Birney | 89.4 |
| 5 | Mizpah | 89.0 |
| 6 | Brandenberg | 88.9 |
| 7 | Jordan | 88.8 |
| 8 | Glendive | 88.5 |
| | Plevna | 88.5 |
| 10 | Billings-Water | 88.4 |
| | Hysham | 88.4 |
| 12 | Ingomar | 88.1 |
| 13 | Mosby | 88.0 |
| 14 | Moorhead | 87.9 |
| 15 | Miles City | 87.8 |
| 16 | Melstone | 87.7 |
| 17 | Roundup | 87.6 |
| 18 | Busby | 87.5 |
| | Loma | 87.5 |
| 20 | Hysham-25 SSE | 87.4 |
| 21 | Terry | 87.2 |
| 22 | Cohagen | 87.1 |
| | Colstrip | 87.1 |
| 24 | Big Sandy | 87.0 |
| | Bridger | 87.0 |

### LOWEST

| | | |
|---|---|---|
| 1 | Mystic Lake | 74.7 |
| 2 | Bozeman-12 NE | 74.8 |
| 3 | Babb | 75.7 |
| 4 | Lennep | 75.8 |
| | Melville | 75.8 |
| 6 | Lakeview | 76.0 |
| 7 | Del Bonita | 76.3 |
| 8 | Judith Gap | 77.2 |
| 9 | Gibson Dam | 77.3 |
| 10 | Pony | 77.6 |
| 11 | West Yellowstone | 77.8 |
| 12 | Libby-32 SSE | 77.9 |
| | Wisdom | 77.9 |
| 14 | Hebgen Dam | 78.0 |
| 15 | Cut Bank | 78.2 |
| | Wilsall | 78.2 |
| 17 | Blackleaf | 78.4 |
| | Lindbergh Lake | 78.4 |
| 19 | Divide | 78.5 |
| 20 | Red Lodge | 78.6 |
| 21 | Polebridge | 78.9 |
| | Shonkin | 78.9 |
| | West Glacier | 78.9 |
| 24 | Creston | 79.1 |
| 25 | Bigfork | 79.3 |

## ANNUAL PRECIPITATION (")

### HIGHEST

| | | |
|---|---|---|
| 1 | Bozeman-12 NE | 35.10 |
| 2 | Troy-18 N | 34.37 |
| 3 | Hungry Horse Dm | 34.09 |
| 4 | Heron | 33.05 |
| 5 | Hebgen Dam | 29.38 |
| 6 | Swan Lake | 28.95 |
| 7 | Trout Creek | 28.77 |
| 8 | Shonkin | 28.45 |
| 9 | West Glacier | 28.39 |
| 10 | Lindbergh Lake | 26.54 |
| 11 | Red Lodge | 25.91 |
| 12 | Mystic Lake | 25.65 |
| 13 | Troy | 24.64 |
| 14 | Libby-32 SSE | 24.49 |
| 15 | Olney | 23.02 |
| 16 | Thompson Falls | 22.54 |
| 17 | Bigfork | 21.89 |
| 18 | West Yellowstone | 21.79 |
| 19 | Wilsall | 21.20 |
| 20 | Polebridge | 20.95 |
| 21 | Seeley Lake | 20.92 |
| 22 | Lakeview | 19.97 |
| 23 | Creston | 19.96 |
| 24 | St. Regis | 19.76 |
| 25 | Bozeman-MSU | 19.50 |

### LOWEST

| | | |
|---|---|---|
| 1 | Glen | 9.20 |
| 2 | Twin Bridges | 9.62 |
| 3 | Gardiner | 10.19 |
| 4 | Joplin | 10.32 |
| 5 | Simpson | 10.45 |
| 6 | Choteau | 10.48 |
| 7 | Rudyard | 10.56 |
| 8 | Deer Lodge | 10.58 |
| 9 | Tiber Dam | 10.68 |
| 10 | Glasgow | 10.78 |
| 11 | Chester | 10.92 |
| 12 | Townsend | 11.04 |
| 13 | Gildford | 11.18 |
| 14 | Western | 11.24 |
| 15 | Harlem | 11.33 |
| 16 | Dillon | 11.44 |
| | Rock Springs | 11.44 |
| 18 | Wisdom | 11.45 |
| 19 | Havre | 11.51 |
| 20 | Ingomar | 11.60 |
| 21 | Canyon Ferry Dm | 11.61 |
| 22 | Helena | 11.62 |
| 23 | Boulder | 11.77 |
| 24 | Brockway | 11.82 |
| | Content | 11.82 |

## ANNUAL SNOWFALL (")

### HIGHEST

| | | |
|---|---|---|
| 1 | Bozeman-12 NE | 232.4 |
| 2 | Hebgen Dam | 208.0 |
| 3 | Red Lodge | 179.2 |
| 4 | Mystic Lake | 176.7 |
| 5 | Lindbergh Lake | 153.7 |
| 6 | Shonkin | 143.3 |
| 7 | Swan Lake | 133.0 |
| 8 | West Glacier | 130.2 |
| 9 | Seeley Lake | 115.8 |
| 10 | Wilsall | 103.7 |
| 11 | Polebridge | 103.0 |
| 12 | Libby-32 SSE | 100.9 |
| 13 | Bozeman-MSU | 91.3 |
| 14 | Lincoln | 88.9 |
| 15 | Heron | 87.7 |
| 16 | Wyola | 80.4 |
| 17 | Cascade-5 S | 74.8 |
| 18 | Rapelje | 70.0 |
| 19 | Gibson Dam | 69.5 |
| 20 | Del Bonita | 66.9 |
| 21 | Lima | 66.6 |
| 22 | Lewistown | 66.4 |
| 23 | Bozeman-6 W | 65.3 |
| 24 | Great Falls | 63.7 |
| 25 | Goldbutte | 63.6 |

### LOWEST

| | | |
|---|---|---|
| 1 | Glendive | 27.3 |
| 2 | Medicine Lake | 28.2 |
| 3 | Glasgow | 29.1 |
| | Simpson | 29.1 |
| 5 | Mizpah | 30.3 |
| 6 | Forks | 30.8 |
| 7 | Plevna | 31.1 |
| 8 | Cohagen | 31.3 |
| | Miles City | 31.3 |
| 10 | Circle | 31.8 |
| 11 | Mosby | 32.0 |
| 12 | Bredette | 32.1 |
| 13 | Vida | 33.1 |
| 14 | Cut Bank | 33.6 |
| 15 | Flatwillow | 35.8 |
| 16 | Savage | 37.7 |
| 17 | Ingomar | 39.0 |
| 18 | Sidney | 39.3 |
| 19 | Divide | 39.8 |
| 20 | Malta | 39.9 |
| 21 | Fortine | 41.5 |
| 22 | Ryegate | 42.3 |
| 23 | Sun River | 42.5 |
| 24 | Harlowton | 42.8 |
| 25 | Opheim-12 SSE | 43.0 |

# NEBRASKA

PHYSICAL FEATURES. Nebraska, one of the Great Plains States, is located in the north-central portion of the United States. The area of the State is 76,653 square miles, of which about 600 are water. On the eastern boundary, along the Missouri River, the elevation rises from less than 900 feet in the southeast to 1,200 feet in the northeast. The elevation also increases westward to about 3,000 feet in the southwest and 5,000 feet in the northwest. The landscape changes from level or gently rolling prairie in the east, to rounded sandhills in the north-central part, and thence westward to high plains.

All of Nebraska is drained by the Missouri River System. The direction of flow is mostly west to east, but in the southeastern section the flow is from northwest to southeast. The Missouri River forms the eastern boundary of the State and a part of the northern boundary. The major tributary is the Platte River, with its two main branches which rise in the high elevations of Colorado. Other important tributaries are the Niobrara River in the north and the Republican and Big Blue Rivers in the south.

Greatest volume of flow occurs during May, June, and July, the months of heaviest rainfall. Although the heaviest snowfall occurs in February and March, it usually does not accumulate to any considerable depth and so the resultant runoff does not materially affect river stages.

GENERAL CLIMATE. The climate is typical of the interior of large continents in middle latitudes; that is, rather light rainfall, low humidity, hot summers, cold winters, great variations in temperature and rainfall from year to year, and frequent changes in weather from day to day. The rapid changes in weather are brought about by invasion of large masses of air of different characteristics, such as warm, moist air from the Gulf of Mexico; hot, dry air from the Southwest; cool, dry air from the north Pacific Ocean; and cold, dry air from northwestern Canada.

The Rocky Mountains to the west have a profound influence on the climate of Nebraska. Air crossing the mountains from the west loses much of its moisture on the windward side and becomes warmer and drier as it descends on the eastern slopes; therefore, no significant amount of moisture which falls as rain or snow reaches the State from the Pacific Ocean. The moisture supply for precipitation comes from the Gulf of Mexico. The remoteness from the source of supply is one of the reasons for the wide variation in rainfall from year to year. Moist air from the Gulf is often deflected eastward before it reaches Nebraska. Downslope winds from the Rocky Mountains occasionally cause large, rapid changes to higher temperatures, particularly during the winter.

Although hot nights in summer occur rather frequently in the east, they are almost unknown in the higher elevations of the western, less humid, part of the State where rapid cooling after sunset generally occurs.

TEMPERATURE. The mean annual temperature varies from about 53° F. along the eastern half of the southern border to about 45° F. in the northwest corner. Maximum temperatures above 100° F. have occurred throughout the State in the months of June, July, August, and September. Temperatures of 110° F. or higher have been recorded over most of the State, except in parts of the northwest. Minimum temperatures of zero or below occur on an average about 10 days a year in the southeast and 25 days in the northwest. Minima below -40° F. have been recorded a few times at northern and western stations. Although the winter climate is classed as cold, there are frequent periods of mild, pleasant weather.

The average date of the last freeze (32° F.) in spring ranges from about April 25 in the extreme southeast to about May 21 in a small area in the northwest portion, while the first in the fall varies from about October 6 in the southeast to about September 20 in the extreme northwest. Hence the average length of the growing season (freeze-free season) ranges from 164 days in the southeast to 122 in the northwest.

PRECIPITATION. The average annual precipitation in the eastern third of the State is about 27 inches; in the central third, about 22 inches; and in the western third, about 18 inches. The amount decreases rather uniformly from 33 inches in the southeast corner to about 14 inches in a small area near the western border. On the average nearly 80 percent of the yearly total falls in the 6 months from April to September. During July and August, rainfall normally diminishes slowly in the east portion. In the west it decreases more rapidly, so that the August average is only a little over one-half the June average in many localities.

Excessive rates of rainfall for short periods occur frequently in summer thundershowers. In some seasons thundershowers are numerous and well distributed, but sometimes they are scattered and infrequent. The result is great variability in the monthly amounts of rainfall in different years and also in the annual amounts from year to year. In dry years, periods of 15 to 20 days without appreciable rain may occur in June, July, and August; and under such conditions, hot, dry winds often cause serious and extensive damage to crops. The precipitation records show successions of wet and dry epochs.

Floods may be expected once or twice in most years in smaller streams in the eastern third of the State, but less frequently over the west and central portions, and are generally caused by short duration, high intensity rainfall. Severe flooding occurs infrequently on the Missouri and is usually caused by rapid melting of heavy snowpacks in the upper portion of the basin, attended by moderate to heavy rains.

The average seasonal snowfall is approximately 29 inches. Snowfall usually increases during the late winter and reaches a maximum in March over most of the State. The higher regions in the west portion frequently have heavy snows in April, and occasionally in May.

OTHER CLIMATIC ELEMENTS. Sunshine for the year averages about 65 percent of the possible amount, ranging from about 55 percent in December to nearly 80 percent in July.

There are frequent changes in wind direction at all seasons of the year, but the prevailing direction is from the south or southeast from May to September, and from the northwest or north during the remainder of the year, except that westerly winds predominate in the southwest portion during the autumn and winter months. The average is about 9 m.p.h.

A few tornadoes occur within the State nearly every year; the average is about 10 per year. Although tornadoes are usually very small, both in width and in length of path, there is almost total destruction where the whirling funnel cloud touches the ground.

The number of hailstorms averages between 20 and 25 per year, occurring mostly in June, July, and August.

## COUNTY INDEX

***Adams County***
HASTINGS 4 N

***Antelope County***
OAKDALE

***Arthur County***
ARTHUR

***Banner County***
HARRISBURG 10 NW

***Blaine County***
BREWSTER
PURDUM

***Boone County***
ALBION 6 WSW

***Box Butte County***
ALLIANCE 1 WNW
HEMINGFORD

***Boyd County***
BUTTE

***Brown County***
AINSWORTH

***Buffalo County***
KEARNEY 4 NE
RAVENNA

***Burt County***
TEKAMAH

***Butler County***
DAVID CITY

***Cass County***
WEEPING WATER

***Cedar County***
GAVINS POINT DAM
HARTINGTON

***Chase County***
ENDERS LAKE
IMPERIAL

***Cherry County***
BROWNLEE
MERRIMAN
MULLEN 21 NW
VALENTINE LAKES GAME
VALENTINE MILLER FLD

***Cheyenne County***
DALTON
LODGEPOLE
SIDNEY 6 NNW

***Colfax County***
CLARKSON

***Cuming County***
WEST POINT

***Custer County***
ANSELMO 2 SE
BROKEN BOW 2 W
OCONTO

***Dawes County***
CHADRON

***Dawson County***
GOTHENBURG

***Deuel County***
BIG SPRINGS

***Dixon County***
NE NEBRASKA EXP STN
WAKEFIELD

***Dodge County***
FREMONT

***Douglas County***
OMAHA EPPLEY AIRFLD
OMAHA WSFO

***Dundy County***
BENKELMAN

***Fillmore County***
FAIRMONT
GENEVA

***Frontier County***
CURTIS 3 NNE
MEDICINE CREEK DAM

***Furnas County***
BEAVER CITY
CAMBRIDGE

***Garden County***
CRESCENT LAKE NWR
OSHKOSH 1 W

***Garfield County***
BURWELL 4 SE

***Gosper County***
CANADAY STEAM PLANT

***Grant County***
HYANNIS

***Greeley County***
GREELEY

***Hall County***
GRAND ISLAND ARPT

***Hamilton County***
AURORA

***Harlan County***
HARLAN COUNTY LAKE

***Hayes County***
HAYES CENTER

***Hitchcock County***
CULBERTSON
TRENTON DAM

***Holt County***
ATKINSON
EWING
O'NEILL

***Hooker County***
MULLEN

***Howard County***
SAINT PAUL 4 N

***Jefferson County***
FAIRBURY

***Johnson County***
TECUMSEH

***Kearney County***
MINDEN

***Keith County***
KINGSLEY DAM
OGALLALA

***Keya Paha County***
SPRINGVIEW

***Kimball County***
KIMBALL 2 N

***Knox County***
CREIGHTON
NIOBRARA

***Lancaster County***
LINCOLN
LINCOLN MUNI AP

***Lincoln County***
HERSHEY 5 SSE
NORTH PLATTE BRD FLD
NORTH PLATTE EXP FRM
WALLACE 2 W

***Madison County***
MADISON
NORFOLK STEFAN AP

***Merrick County***
CENTRAL CITY

***Morrill County***
BRIDGEPORT

***Nance County***
GENOA 2 W

***Nemaha County***
AUBURN 5 ESE

***Nuckolls County***
SUPERIOR

***Otoe County***
NEBRASKA CITY
SYRACUSE

***Pawnee County***
PAWNEE CITY

***Perkins County***
MADRID

***Phelps County***
HOLDREGE

***Pierce County***
OSMOND

***Platte County***
COLUMBUS 3 NE

***Polk County***
OSCEOLA

***Red Willow County***
MCCOOK
RED WILLOW DAM

***Richardson County***
FALLS CITY 2 NE

***Rock County***
NEWPORT
ROSE 7 WNW

***Saline County***
CRETE

***Saunders County***
ASHLAND 2
WAHOO

***Scotts Bluff County***
MITCHELL 5 E
SCOTTSBLUFF CNTY AP

***Seward County***
SEWARD

***Sheridan County***
ELLSWORTH 15 NNE
HAY SPRINGS
HAY SPRINGS 12 S

***Sherman County***
LOUP CITY

***Sioux County***
HARRISON

***Stanton County***
STANTON

***Thayer County***
HEBRON

***Thomas County***
HALSEY 2 W

***Thurston County***
WALTHILL

***Valley County***
NORTH LOUP

***Washington County***
BLAIR

***Webster County***
RED CLOUD

***York County***
YORK

## ELEVATION INDEX

| FEET | STATION NAME |
|---|---|
| 930 | AUBURN 5 ESE |
| 971 | NEBRASKA CITY |
| 980 | FALLS CITY 2 NE |
| 997 | OMAHA EPPLEY AIRFLD |
| 1050 | SYRACUSE |
| 1060 | TEKAMAH |
| 1070 | ASHLAND 2 |
| 1122 | BLAIR |
| 1142 | WAHOO |
| 1165 | LINCOLN |
| 1170 | TECUMSEH |
| 1181 | PAWNEE CITY |
| 1194 | LINCOLN MUNI AP |
| 1201 | FREMONT |
| 1211 | WALTHILL |

| FEET | STATION NAME |
|---|---|
| 1230 | NIOBRARA |
| 1240 | GAVINS POINT DAM |
| 1240 | WEEPING WATER |
| 1312 | FAIRBURY |
| 1312 | WEST POINT |
| 1319 | OMAHA WSFO |
| 1362 | CRETE |
| 1381 | HARTINGTON |
| 1411 | WAKEFIELD |
| 1440 | COLUMBUS 3 NE |
| 1440 | SEWARD |
| 1460 | HEBRON |
| 1472 | STANTON |
| 1480 | NE NEBRASKA EXP STN |
| 1544 | NORFOLK STEFAN AP |
| 1572 | CLARKSON |
| 1572 | SUPERIOR |
| 1581 | GENOA 2 W |
| 1581 | MADISON |
| 1611 | YORK |
| 1621 | DAVID CITY |
| 1631 | GENEVA |
| 1640 | FAIRMONT |
| 1640 | OSCEOLA |
| 1650 | OSMOND |
| 1660 | CREIGHTON |
| 1690 | CENTRAL CITY |
| 1713 | OAKDALE |
| 1722 | RED CLOUD |
| 1762 | ALBION 6 WSW |
| 1791 | AURORA |
| 1801 | SAINT PAUL 4 N |
| 1850 | EWING |
| 1860 | GRAND ISLAND ARPT |
| 1903 | BUTTE |
| 1932 | HASTINGS 4 N |
| 1962 | NORTH LOUP |
| 1982 | O'NEILL |
| 2001 | HARLAN COUNTY LAKE |
| 2001 | RAVENNA |
| 2021 | GREELEY |
| 2060 | LOUP CITY |
| 2113 | ATKINSON |
| 2129 | KEARNEY 4 NE |
| 2170 | BEAVER CITY |
| 2172 | MINDEN |
| 2180 | BURWELL 4 SE |
| 2231 | NEWPORT |
| 2260 | CAMBRIDGE |
| 2342 | HOLDREGE |

| FEET | STATION NAME |
|---|---|
| 2362 | CANADAY STEAM PLANT |
| 2392 | MEDICINE CREEK DAM |
| 2441 | SPRINGVIEW |
| 2470 | BROKEN BOW 2 W |
| 2503 | BREWSTER |
| 2513 | ROSE 7 WNW |
| 2523 | AINSWORTH |
| 2530 | MCCOOK |
| 2562 | RED WILLOW DAM |
| 2572 | GOTHENBURG |
| 2581 | OCONTO |
| 2582 | VALENTINE MILLER FLD |
| 2592 | CULBERTSON |
| 2602 | ANSELMO 2 SE |
| 2661 | TRENTON DAM |
| 2690 | PURDUM |
| 2705 | HALSEY 2 W |
| 2722 | CURTIS 3 NNE |
| 2789 | NORTH PLATTE BRD FLD |
| 2821 | BROWNLEE |
| 2933 | VALENTINE LAKES GAME |
| 2952 | HERSHEY 5 SSE |
| 2963 | BENKELMAN |
| 3031 | HAYES CENTER |
| 3031 | NORTH PLATTE EXP FRM |
| 3081 | ENDERS LAKE |
| 3100 | WALLACE 2 W |
| 3202 | MADRID |
| 3222 | MULLEN |
| 3222 | OGALLALA |
| 3251 | MERRIMAN |
| 3281 | IMPERIAL |
| 3304 | KINGSLEY DAM |
| 3312 | CHADRON |
| 3363 | BIG SPRINGS |
| 3392 | OSHKOSH 1 W |
| 3450 | MULLEN 21 NW |
| 3504 | ARTHUR |
| 3666 | BRIDGEPORT |
| 3734 | HYANNIS |
| 3812 | HAY SPRINGS 12 S |
| 3822 | CRESCENT LAKE NWR |
| 3832 | LODGEPOLE |
| 3871 | HAY SPRINGS |
| 3966 | SCOTTSBLUFF CNTY AP |
| 3983 | ALLIANCE 1 WNW |
| 3983 | ELLSWORTH 15 NNE |
| 4081 | MITCHELL 5 E |
| 4272 | DALTON |
| 4272 | HEMINGFORD |

| FEET | STATION NAME |
|---|---|
| 4324 | SIDNEY 6 NNW |
| 4462 | HARRISBURG 10 NW |
| 4734 | KIMBALL 2 N |
| 4852 | HARRISON |

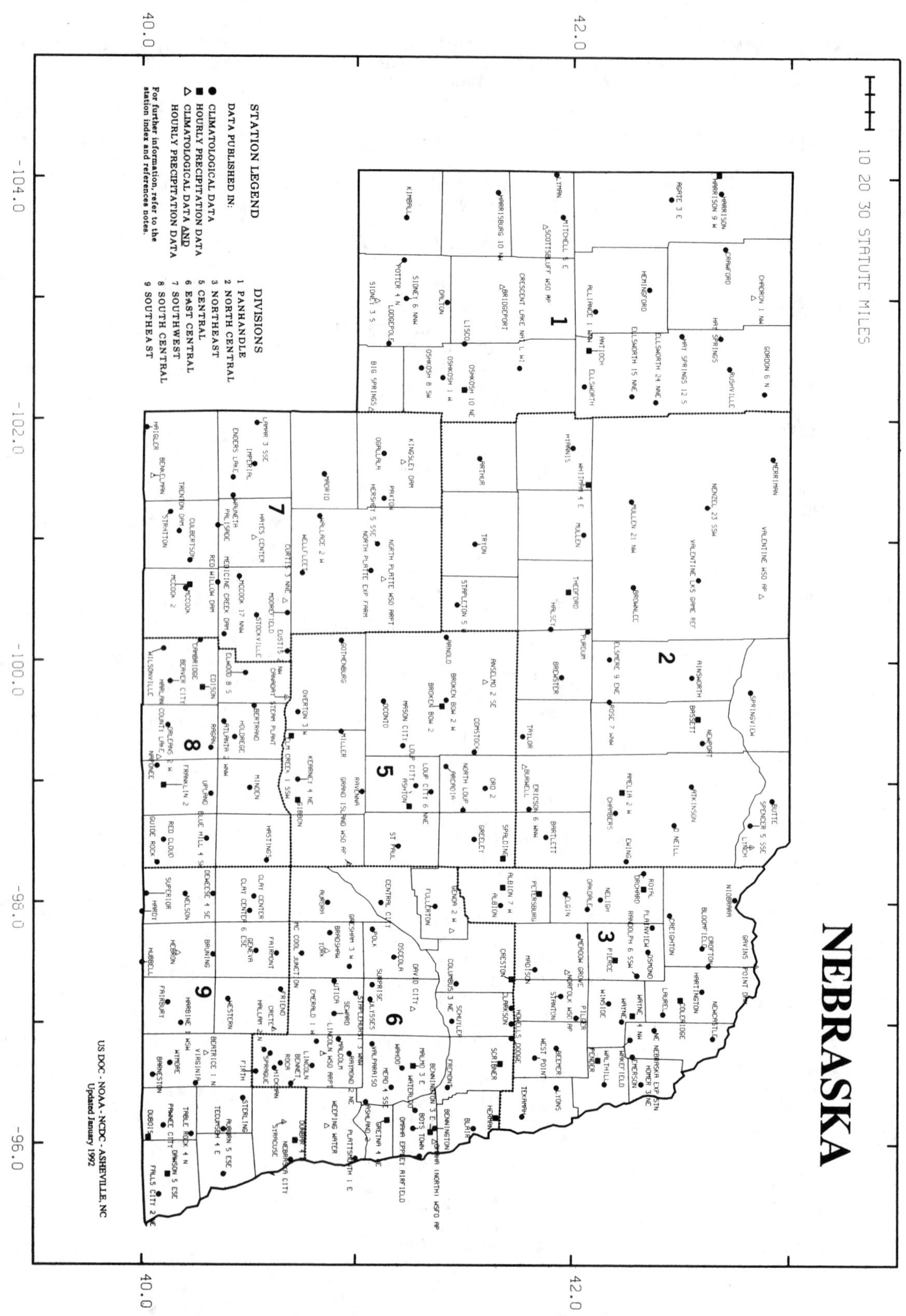
NEBRASKA
10 20 30 STATUTE MILES
STATION LEGEND
DATA PUBLISHED IN:
● CLIMATOLOGICAL DATA
■ HOURLY PRECIPITATION DATA
△ CLIMATOLOGICAL DATA AND HOURLY PRECIPITATION DATA
For further information, refer to the station index and reference notes.
DIVISIONS
1 PANHANDLE
2 NORTH CENTRAL
3 NORTHEAST
5 CENTRAL
6 EAST CENTRAL
7 SOUTHWEST
8 SOUTH CENTRAL
9 SOUTHEAST
US DOC - NOAA - NCDC - ASHEVILLE, NC
Updated January 1992

## AINSWORTH *Brown County* ELEVATION 2523 ft LAT/LONG 42° 33 ' N / 99° 52 ' W

| | JAN | FEB | MAR | APR | MAY | JUN | JUL | AUG | SEP | OCT | NOV | DEC | YEAR |
|---|---|---|---|---|---|---|---|---|---|---|---|---|---|
| Maximum Temp °F | 33.1 | 38.1 | 48.4 | 61.2 | 72.0 | 81.7 | 87.5 | 85.7 | 76.4 | 64.3 | 46.3 | 36.2 | 60.9 |
| Minimum Temp °F | 11.9 | 16.3 | 25.1 | 36.0 | 46.9 | 56.6 | 62.0 | 60.1 | 50.3 | 38.8 | 25.5 | 15.7 | 37.1 |
| Mean Temp °F | 22.5 | 27.2 | 36.8 | 48.6 | 59.5 | 69.2 | 74.8 | 72.9 | 63.4 | 51.6 | 35.9 | 26.0 | 49.0 |
| Days Max Temp ≥ 90 °F | 0 | 0 | 0 | 0 | 1 | 5 | 13 | 10 | 3 | 0 | 0 | 0 | 32 |
| Days Max Temp ≤ 32 °F | 13 | 10 | 5 | 0 | 0 | 0 | 0 | 0 | 0 | 0 | 4 | 12 | 44 |
| Days Min Temp ≤ 32 °F | 29 | 26 | 23 | 11 | 1 | 0 | 0 | 0 | 1 | 7 | 22 | 29 | 149 |
| Days Min Temp ≤ 0 °F | 8 | 4 | 1 | 0 | 0 | 0 | 0 | 0 | 0 | 0 | 1 | 4 | 18 |
| Heating Degree Days | 1311 | 1060 | 869 | 492 | 203 | 34 | 6 | 12 | 130 | 417 | 866 | 1203 | 6603 |
| Cooling Degree Days | 0 | 0 | 0 | 13 | 41 | 168 | 303 | 258 | 90 | 7 | 0 | 0 | 880 |
| Total Precipitation (") | 0.44 | 0.57 | 1.35 | 2.22 | 3.22 | 3.29 | 3.21 | 2.69 | 2.42 | 1.38 | 0.96 | 0.51 | 22.26 |
| Days ≥ 0.1" Precip | 1 | 2 | 3 | 5 | 6 | 6 | 6 | 5 | 4 | 3 | 2 | 2 | 45 |
| Total Snowfall (") | 6.0 | 5.9 | 8.1 | 4.3 | 0.0 | 0.0 | 0.0 | 0.0 | 0.3 | 1.7 | 7.3 | 6.6 | 40.2 |
| Days ≥ 1" Snow Depth | 18 | 13 | 8 | 2 | 0 | 0 | 0 | 0 | 0 | 1 | 6 | 14 | 62 |

## ALBION 6 WSW *Boone County* ELEVATION 1762 ft LAT/LONG 41° 41 ' N / 98° 0 ' W

| | JAN | FEB | MAR | APR | MAY | JUN | JUL | AUG | SEP | OCT | NOV | DEC | YEAR |
|---|---|---|---|---|---|---|---|---|---|---|---|---|---|
| Maximum Temp °F | 31.2 | 36.1 | 47.6 | 61.5 | 72.1 | 82.0 | 86.9 | 84.6 | 76.2 | 64.3 | 46.9 | 35.1 | 60.4 |
| Minimum Temp °F | 7.9 | 12.8 | 23.4 | 35.0 | 46.2 | 56.5 | 61.5 | 59.2 | 48.5 | 35.6 | 23.5 | 12.8 | 35.2 |
| Mean Temp °F | 19.6 | 24.5 | 35.5 | 48.3 | 59.2 | 69.3 | 74.2 | 71.9 | 62.4 | 50.0 | 35.2 | 24.0 | 47.8 |
| Days Max Temp ≥ 90 °F | 0 | 0 | 0 | 0 | 1 | 6 | 12 | 9 | 3 | 0 | 0 | 0 | 31 |
| Days Max Temp ≤ 32 °F | 16 | 11 | 5 | 0 | 0 | 0 | 0 | 0 | 0 | 0 | 4 | 12 | 48 |
| Days Min Temp ≤ 32 °F | 31 | 27 | 26 | 12 | 2 | 0 | 0 | 0 | 1 | 11 | 26 | 31 | 167 |
| Days Min Temp ≤ 0 °F | 10 | 5 | 1 | 0 | 0 | 0 | 0 | 0 | 0 | 0 | 1 | 5 | 22 |
| Heating Degree Days | 1403 | 1138 | 907 | 501 | 214 | 37 | 7 | 15 | 147 | 462 | 888 | 1266 | 6985 |
| Cooling Degree Days | 0 | 0 | 0 | 11 | 38 | 182 | 286 | 233 | 81 | 3 | 0 | 0 | 834 |
| Total Precipitation (") | 0.47 | 0.75 | 2.06 | 2.64 | 4.01 | 4.41 | 3.81 | 3.53 | 2.71 | 1.78 | 1.27 | 0.65 | 28.09 |
| Days ≥ 0.1" Precip | 1 | 2 | 4 | 5 | 7 | 7 | 6 | 5 | 5 | 4 | 3 | 2 | 51 |
| Total Snowfall (") | 5.1 | 6.8 | 5.4 | 1.5 | 0.0 | 0.0 | 0.0 | 0.0 | 0.0 | 0.5 | 4.3 | 6.9 | 30.5 |
| Days ≥ 1" Snow Depth | *16* | 13 | *6* | 1 | 0 | 0 | 0 | 0 | 0 | 0 | 4 | 12 | 52 |

## ALLIANCE 1 WNW *Box Butte County* ELEVATION 3983 ft LAT/LONG 42° 6 ' N / 102° 52 ' W

| | JAN | FEB | MAR | APR | MAY | JUN | JUL | AUG | SEP | OCT | NOV | DEC | YEAR |
|---|---|---|---|---|---|---|---|---|---|---|---|---|---|
| Maximum Temp °F | 36.2 | 41.8 | 48.7 | 58.9 | 69.2 | 79.7 | 87.2 | 85.1 | 75.1 | 62.6 | 47.2 | 38.1 | 60.8 |
| Minimum Temp °F | 10.7 | 15.4 | 22.8 | 32.2 | 42.5 | 52.1 | 57.8 | 55.6 | 44.9 | 32.7 | 21.2 | 12.8 | 33.4 |
| Mean Temp °F | 23.5 | 28.6 | 35.8 | 45.6 | 55.9 | 66.0 | 72.5 | 70.5 | 60.0 | 47.7 | 34.3 | 25.4 | 47.2 |
| Days Max Temp ≥ 90 °F | 0 | 0 | 0 | 0 | 0 | 5 | 13 | 10 | 2 | 0 | 0 | 0 | 30 |
| Days Max Temp ≤ 32 °F | 10 | 7 | 4 | 1 | 0 | 0 | 0 | 0 | 0 | 0 | 4 | 10 | 36 |
| Days Min Temp ≤ 32 °F | 30 | 27 | 28 | 15 | 3 | 0 | 0 | 0 | 2 | 15 | 27 | 30 | 177 |
| Days Min Temp ≤ 0 °F | 7 | 4 | 1 | 0 | 0 | 0 | 0 | 0 | 0 | 0 | 1 | 4 | 17 |
| Heating Degree Days | 1282 | 1023 | 898 | 578 | 294 | 73 | 10 | 21 | 190 | 531 | 915 | 1222 | 7037 |
| Cooling Degree Days | *0* | 0 | 0 | 1 | 19 | 114 | 254 | 212 | 52 | 1 | 0 | 0 | 653 |
| Total Precipitation (") | 0.37 | 0.32 | 0.77 | 1.68 | 2.89 | 3.12 | 2.21 | 1.65 | 1.29 | 0.87 | 0.49 | 0.39 | 16.05 |
| Days ≥ 0.1" Precip | 1 | 1 | 2 | 4 | 7 | 7 | 5 | 4 | 3 | 3 | 2 | 1 | 40 |
| Total Snowfall (") | *6.4* | 5.0 | *7.6* | 3.8 | 0.5 | 0.0 | 0.0 | 0.0 | 0.3 | 2.4 | *5.2* | na | na |
| Days ≥ 1" Snow Depth | *13* | 8 | 5 | 2 | 0 | 0 | 0 | 0 | 0 | 1 | *7* | *10* | 46 |

## ANSELMO 2 SE *Custer County* ELEVATION 2602 ft LAT/LONG 41° 37 ' N / 99° 52 ' W

| | JAN | FEB | MAR | APR | MAY | JUN | JUL | AUG | SEP | OCT | NOV | DEC | YEAR |
|---|---|---|---|---|---|---|---|---|---|---|---|---|---|
| Maximum Temp °F | 34.6 | 40.1 | 51.1 | 62.8 | 72.8 | 82.1 | 87.5 | 85.5 | 76.7 | 65.0 | 48.0 | 37.1 | 61.9 |
| Minimum Temp °F | 9.1 | 14.2 | 23.2 | 33.2 | 44.0 | 54.1 | 59.7 | 57.4 | 46.9 | 34.4 | 21.6 | 12.3 | 34.2 |
| Mean Temp °F | 21.8 | 27.2 | 37.2 | 48.1 | 58.4 | 68.1 | 73.7 | 71.5 | 61.8 | 49.8 | 34.8 | 24.6 | 48.1 |
| Days Max Temp ≥ 90 °F | 0 | 0 | 0 | 0 | 1 | 6 | 13 | 10 | 3 | 0 | 0 | 0 | 33 |
| Days Max Temp ≤ 32 °F | 12 | 8 | 3 | 0 | 0 | 0 | 0 | 0 | 0 | 0 | 4 | 11 | 38 |
| Days Min Temp ≤ 32 °F | 31 | 27 | 26 | 15 | 4 | 0 | 0 | 0 | 2 | 13 | 26 | 31 | 175 |
| Days Min Temp ≤ 0 °F | 8 | 4 | 1 | 0 | 0 | 0 | 0 | 0 | 0 | 0 | 1 | 5 | 19 |
| Heating Degree Days | 1334 | 1059 | 856 | 507 | 230 | 49 | 7 | 16 | 158 | 467 | 899 | 1246 | 6828 |
| Cooling Degree Days | 0 | 0 | 0 | 9 | 37 | 160 | 278 | 235 | 78 | 3 | 0 | 0 | 800 |
| Total Precipitation (") | 0.48 | 0.59 | 1.50 | 2.62 | 3.43 | 3.92 | 3.58 | 2.92 | 2.16 | 1.50 | 1.16 | 0.55 | 24.41 |
| Days ≥ 0.1" Precip | 1 | 2 | 3 | 5 | 6 | 6 | 6 | 5 | 4 | 3 | 2 | 1 | 44 |
| Total Snowfall (") | 6.6 | 5.8 | 7.8 | 3.6 | 0.0 | 0.0 | 0.0 | 0.0 | 0.1 | 1.7 | 6.8 | 6.8 | 39.2 |
| Days ≥ 1" Snow Depth | 17 | 12 | 6 | 2 | 0 | 0 | 0 | 0 | 0 | 1 | 6 | 13 | 57 |

## ARTHUR *Arthur County* ELEVATION 3504 ft LAT/LONG 41° 34' N / 101° 41' W

| | JAN | FEB | MAR | APR | MAY | JUN | JUL | AUG | SEP | OCT | NOV | DEC | YEAR |
|---|---|---|---|---|---|---|---|---|---|---|---|---|---|
| Maximum Temp °F | 35.9 | 41.1 | 49.6 | 60.9 | 70.6 | 80.2 | 86.5 | 84.9 | 75.8 | 64.1 | 48.0 | 38.6 | 61.4 |
| Minimum Temp °F | 9.8 | 14.4 | 22.2 | 32.6 | 42.8 | 52.5 | 58.5 | 56.4 | 45.9 | 32.9 | 21.2 | 12.2 | 33.4 |
| Mean Temp °F | 23.1 | 27.8 | 36.0 | 46.7 | 56.7 | 66.3 | 72.5 | 70.7 | 60.9 | 48.6 | 34.7 | 25.6 | 47.5 |
| Days Max Temp ≥ 90 °F | 0 | 0 | 0 | 0 | 1 | 5 | 12 | 9 | 3 | 0 | 0 | 0 | 30 |
| Days Max Temp ≤ 32 °F | 11 | 8 | 4 | 0 | 0 | 0 | 0 | 0 | 0 | 0 | 4 | 10 | 37 |
| Days Min Temp ≤ 32 °F | 31 | 28 | 28 | 15 | 3 | 0 | 0 | 0 | 2 | 15 | 27 | 31 | 180 |
| Days Min Temp ≤ 0 °F | 8 | 4 | 1 | 0 | 0 | 0 | 0 | 0 | 0 | 0 | 1 | 5 | 19 |
| Heating Degree Days | 1292 | 1045 | 893 | 548 | 269 | 66 | 10 | 19 | 172 | 505 | 903 | 1216 | 6938 |
| Cooling Degree Days | 0 | 0 | 0 | 3 | 21 | 123 | 254 | 220 | 63 | 1 | 0 | 0 | 685 |
| Total Precipitation (") | 0.38 | 0.39 | 1.12 | 1.72 | 3.25 | 2.98 | 3.55 | 1.95 | 1.54 | 0.93 | 0.67 | 0.42 | 18.90 |
| Days ≥ 0.1" Precip | 1 | 1 | 3 | 4 | 6 | 6 | 6 | 4 | 4 | 2 | 2 | 2 | 41 |
| Total Snowfall (") | 6.3 | 5.0 | 8.0 | 4.2 | 0.3 | 0.0 | 0.0 | 0.0 | 0.3 | 1.7 | 5.4 | 5.7 | 36.9 |
| Days ≥ 1" Snow Depth | 18 | 12 | 8 | 2 | 0 | 0 | 0 | 0 | 0 | 1 | 6 | 14 | 61 |

## ASHLAND 2 *Saunders County* ELEVATION 1070 ft LAT/LONG 41° 4' N / 96° 20' W

| | JAN | FEB | MAR | APR | MAY | JUN | JUL | AUG | SEP | OCT | NOV | DEC | YEAR |
|---|---|---|---|---|---|---|---|---|---|---|---|---|---|
| Maximum Temp °F | 31.9 | 37.4 | 49.9 | 63.4 | 73.3 | 82.9 | 87.5 | 85.5 | 77.1 | 65.7 | 49.1 | 36.1 | 61.7 |
| Minimum Temp °F | 9.6 | 14.5 | 26.2 | 38.6 | 49.6 | 60.0 | 65.0 | 62.3 | 52.3 | 39.6 | 27.3 | 15.9 | 38.4 |
| Mean Temp °F | 20.7 | 26.0 | 38.1 | 51.1 | 61.5 | 71.5 | 76.3 | 73.9 | 64.7 | 52.7 | 38.2 | 26.0 | 50.1 |
| Days Max Temp ≥ 90 °F | 0 | 0 | 0 | 1 | 1 | 6 | 12 | 10 | 4 | 0 | 0 | 0 | 34 |
| Days Max Temp ≤ 32 °F | 15 | 11 | 4 | 0 | 0 | 0 | 0 | 0 | 0 | 0 | 3 | 12 | 45 |
| Days Min Temp ≤ 32 °F | 31 | 27 | 22 | 8 | 1 | 0 | 0 | 0 | 1 | 7 | 22 | 30 | 149 |
| Days Min Temp ≤ 0 °F | 8 | 5 | 1 | 0 | 0 | 0 | 0 | 0 | 0 | 0 | 0 | 4 | 18 |
| Heating Degree Days | 1366 | 1093 | 825 | 427 | 167 | 21 | 3 | 9 | 114 | 386 | 797 | 1201 | 6409 |
| Cooling Degree Days | 0 | 0 | 1 | 21 | 62 | 230 | 356 | 286 | 123 | 7 | 0 | 0 | 1086 |
| Total Precipitation (") | 0.67 | 0.65 | 1.90 | 2.84 | 4.45 | 4.08 | 3.71 | 4.08 | 3.16 | 2.29 | 1.48 | 0.99 | 30.30 |
| Days ≥ 0.1" Precip | 2 | 2 | 4 | 6 | 8 | 7 | 6 | 5 | 5 | 4 | 3 | 3 | 55 |
| Total Snowfall (") | 5.5 | 5.7 | 3.3 | 0.8 | 0.1 | 0.0 | 0.0 | 0.0 | 0.0 | 0.2 | 2.8 | 5.1 | 23.5 |
| Days ≥ 1" Snow Depth | na | na | na | 0 | 0 | 0 | 0 | 0 | 0 | 0 | *0* | na | na |

## ATKINSON *Holt County* ELEVATION 2113 ft LAT/LONG 42° 32' N / 98° 59' W

| | JAN | FEB | MAR | APR | MAY | JUN | JUL | AUG | SEP | OCT | NOV | DEC | YEAR |
|---|---|---|---|---|---|---|---|---|---|---|---|---|---|
| Maximum Temp °F | 32.1 | 36.8 | 48.2 | 62.0 | 72.5 | 81.6 | 86.8 | 85.5 | 76.6 | 64.4 | 46.0 | 34.8 | 60.6 |
| Minimum Temp °F | 10.1 | 14.7 | 24.5 | 35.8 | 46.9 | 56.4 | 61.6 | 59.4 | 49.3 | 37.4 | 24.6 | 14.2 | 36.2 |
| Mean Temp °F | 21.3 | 25.8 | 36.4 | 49.0 | 59.7 | 69.1 | 74.2 | 72.5 | 63.0 | 50.9 | 35.4 | 24.5 | 48.5 |
| Days Max Temp ≥ 90 °F | 0 | 0 | 0 | 0 | 1 | 5 | 11 | 10 | 4 | 0 | 0 | 0 | 31 |
| Days Max Temp ≤ 32 °F | 14 | 11 | 4 | 0 | 0 | 0 | 0 | 0 | 0 | 0 | 5 | 13 | 47 |
| Days Min Temp ≤ 32 °F | 29 | 27 | 24 | 11 | 1 | 0 | 0 | 0 | 1 | 9 | 23 | 30 | 155 |
| Days Min Temp ≤ 0 °F | 8 | 5 | 1 | 0 | 0 | 0 | 0 | 0 | 0 | 0 | 1 | 5 | 20 |
| Heating Degree Days | 1349 | 1100 | 880 | 482 | 200 | 34 | 7 | 14 | 137 | 437 | 883 | 1249 | 6772 |
| Cooling Degree Days | 0 | 0 | 0 | 14 | 48 | 181 | 308 | 266 | 94 | 7 | 0 | 0 | 918 |
| Total Precipitation (") | 0.50 | 0.66 | 1.53 | 2.43 | 3.22 | 3.83 | 3.29 | 2.62 | 2.43 | 1.63 | 1.07 | 0.63 | 23.84 |
| Days ≥ 0.1" Precip | 1 | 2 | 4 | 6 | 7 | 7 | 6 | 5 | 5 | 4 | 3 | 2 | 52 |
| Total Snowfall (") | 5.8 | 5.3 | 7.0 | 3.2 | 0.0 | 0.0 | 0.0 | 0.0 | 0.1 | 1.2 | 6.3 | 7.4 | 36.3 |
| Days ≥ 1" Snow Depth | na | *10* | na | 1 | 0 | 0 | 0 | 0 | 0 | 0 | *4* | na | na |

## AUBURN 5 ESE *Nemaha County* ELEVATION 930 ft LAT/LONG 40° 22' N / 95° 45' W

| | JAN | FEB | MAR | APR | MAY | JUN | JUL | AUG | SEP | OCT | NOV | DEC | YEAR |
|---|---|---|---|---|---|---|---|---|---|---|---|---|---|
| Maximum Temp °F | 34.1 | 39.9 | 53.1 | 66.2 | 75.8 | 85.3 | 89.1 | 86.9 | 79.3 | 67.8 | 50.9 | 38.3 | 63.9 |
| Minimum Temp °F | 13.7 | 18.3 | 29.7 | 41.2 | 52.0 | 61.7 | 66.1 | 63.4 | 54.6 | 42.7 | 30.1 | 18.9 | 41.0 |
| Mean Temp °F | 23.9 | 29.1 | 41.4 | 53.7 | 63.9 | 73.5 | 77.6 | 75.2 | 67.0 | 55.3 | 40.5 | 28.6 | 52.5 |
| Days Max Temp ≥ 90 °F | 0 | 0 | 0 | 1 | 2 | 8 | 15 | 12 | 5 | 0 | 0 | 0 | 43 |
| Days Max Temp ≤ 32 °F | 14 | 9 | 2 | 0 | 0 | 0 | 0 | 0 | 0 | 0 | 2 | 9 | 36 |
| Days Min Temp ≤ 32 °F | 30 | 25 | 20 | 6 | 1 | 0 | 0 | 0 | 0 | 6 | 19 | 29 | 136 |
| Days Min Temp ≤ 0 °F | 6 | 3 | 0 | 0 | 0 | 0 | 0 | 0 | 0 | 0 | 0 | 3 | 12 |
| Heating Degree Days | 1267 | 1006 | 727 | 357 | 115 | 9 | 1 | 4 | 79 | 317 | 728 | 1122 | 5732 |
| Cooling Degree Days | 0 | 0 | 3 | 31 | 89 | 284 | 405 | 330 | 162 | 20 | 1 | 0 | 1325 |
| Total Precipitation (") | 0.82 | 0.98 | 2.34 | 2.85 | 4.00 | 3.74 | 4.44 | 3.68 | 3.82 | 2.78 | 1.71 | 1.18 | 32.34 |
| Days ≥ 0.1" Precip | 2 | 3 | 5 | 6 | 7 | 6 | 7 | 6 | 6 | 4 | 4 | 3 | 59 |
| Total Snowfall (") | 6.7 | 6.1 | 4.9 | 1.7 | 0.0 | 0.0 | 0.0 | 0.0 | 0.0 | 0.2 | 2.4 | 5.3 | 27.3 |
| Days ≥ 1" Snow Depth | 15 | 12 | 4 | 1 | 0 | 0 | 0 | 0 | 0 | 0 | 2 | 8 | 42 |

## AURORA *Hamilton County* ELEVATION 1791 ft LAT/LONG 40° 52 ' N / 98° 1 ' W

| | JAN | FEB | MAR | APR | MAY | JUN | JUL | AUG | SEP | OCT | NOV | DEC | YEAR |
|---|---|---|---|---|---|---|---|---|---|---|---|---|---|
| Maximum Temp °F | 33.2 | 38.5 | 50.4 | 64.2 | 74.1 | 84.2 | 88.2 | 86.0 | 77.7 | 65.8 | 48.6 | 36.6 | 62.3 |
| Minimum Temp °F | 12.2 | 16.8 | 27.1 | 38.7 | 50.0 | 59.6 | 64.2 | 61.9 | 52.0 | 39.8 | 27.0 | 16.8 | 38.8 |
| Mean Temp °F | 22.7 | 27.7 | 38.8 | 51.5 | 62.1 | 72.0 | 76.2 | 73.9 | 64.9 | 52.9 | 37.8 | 26.7 | 50.6 |
| Days Max Temp ≥ 90 °F | 0 | 0 | 0 | 1 | 2 | 8 | 14 | 11 | 4 | 0 | 0 | 0 | 40 |
| Days Max Temp ≤ 32 °F | 14 | 10 | 3 | 0 | 0 | 0 | 0 | 0 | 0 | 0 | 3 | 11 | 41 |
| Days Min Temp ≤ 32 °F | 30 | 26 | 22 | 7 | 1 | 0 | 0 | 0 | 1 | 6 | 22 | 30 | 145 |
| Days Min Temp ≤ 0 °F | 7 | 4 | 0 | 0 | 0 | 0 | 0 | 0 | 0 | 0 | 0 | 3 | 14 |
| Heating Degree Days | 1304 | 1046 | 807 | 412 | 148 | 20 | 3 | 6 | 105 | 381 | 809 | 1179 | 6220 |
| Cooling Degree Days | 0 | 0 | 1 | 17 | 59 | 222 | 330 | 271 | 107 | 5 | 0 | 0 | 1012 |
| Total Precipitation (") | 0.65 | 0.72 | 2.20 | 2.83 | 4.36 | 4.24 | 3.32 | 2.97 | 2.85 | 2.02 | 1.36 | 0.88 | 28.40 |
| Days ≥ 0.1" Precip | 2 | 2 | 4 | 5 | 7 | 6 | 6 | 5 | 5 | 4 | 3 | 2 | 51 |
| Total Snowfall (") | 6.1 | 6.5 | 5.8 | 1.4 | 0.1 | 0.0 | 0.0 | 0.0 | 0.2 | 0.5 | 4.1 | 7.3 | 32.0 |
| Days ≥ 1" Snow Depth | na | *14* | 5 | 1 | 0 | 0 | 0 | 0 | 0 | 0 | 3 | *11* | na |

## BEAVER CITY *Furnas County* ELEVATION 2170 ft LAT/LONG 40° 8 ' N / 99° 50 ' W

| | JAN | FEB | MAR | APR | MAY | JUN | JUL | AUG | SEP | OCT | NOV | DEC | YEAR |
|---|---|---|---|---|---|---|---|---|---|---|---|---|---|
| Maximum Temp °F | 40.3 | 46.7 | 57.6 | 69.3 | 77.7 | 88.5 | 93.8 | 92.0 | 83.6 | 71.8 | 53.7 | 43.0 | 68.2 |
| Minimum Temp °F | 12.5 | 17.2 | 26.5 | 37.3 | 47.8 | 57.8 | 63.1 | 60.4 | 50.1 | 37.1 | 24.8 | 15.8 | 37.5 |
| Mean Temp °F | 26.4 | 32.0 | 42.1 | 53.3 | 62.8 | 73.2 | 78.5 | 76.2 | 66.8 | 54.5 | 39.3 | 29.4 | 52.9 |
| Days Max Temp ≥ 90 °F | 0 | 0 | 0 | 1 | 3 | 14 | 22 | 20 | 9 | 2 | 0 | 0 | 71 |
| Days Max Temp ≤ 32 °F | 9 | 5 | 1 | 0 | 0 | 0 | 0 | 0 | 0 | 0 | 2 | 6 | 23 |
| Days Min Temp ≤ 32 °F | 31 | 27 | 23 | 10 | 2 | 0 | 0 | 0 | 1 | 10 | 25 | 30 | 159 |
| Days Min Temp ≤ 0 °F | 6 | 3 | 0 | 0 | 0 | 0 | 0 | 0 | 0 | 0 | 0 | 3 | 12 |
| Heating Degree Days | 1190 | 925 | 704 | 360 | 134 | 15 | 1 | 4 | 79 | 332 | 765 | 1096 | 5605 |
| Cooling Degree Days | 0 | 0 | 1 | 21 | 71 | 277 | 420 | 369 | 160 | 11 | 0 | 0 | 1330 |
| Total Precipitation (") | 0.67 | 0.67 | 1.79 | 2.03 | 3.46 | 3.82 | 3.22 | 2.67 | 2.08 | 1.64 | 1.09 | 0.71 | 23.85 |
| Days ≥ 0.1" Precip | 2 | 2 | 4 | 5 | 7 | 6 | 6 | 4 | 4 | 3 | 2 | 2 | 47 |
| Total Snowfall (") | 6.7 | 5.6 | 6.0 | 1.4 | 0.0 | 0.0 | 0.0 | 0.0 | 0.4 | 0.4 | 3.8 | 6.2 | 30.5 |
| Days ≥ 1" Snow Depth | 14 | 9 | 4 | 1 | 0 | 0 | 0 | 0 | 0 | 0 | 3 | 9 | 40 |

## BENKELMAN *Dundy County* ELEVATION 2963 ft LAT/LONG 40° 3 ' N / 101° 32 ' W

| | JAN | FEB | MAR | APR | MAY | JUN | JUL | AUG | SEP | OCT | NOV | DEC | YEAR |
|---|---|---|---|---|---|---|---|---|---|---|---|---|---|
| Maximum Temp °F | 40.0 | 45.1 | 54.2 | 65.3 | 74.4 | 85.1 | 91.1 | 89.1 | 80.1 | 68.9 | 52.5 | 43.0 | 65.7 |
| Minimum Temp °F | 11.4 | 16.1 | 24.6 | 35.1 | 45.7 | 55.9 | 61.7 | 58.9 | 48.0 | 34.5 | 22.6 | 14.0 | 35.7 |
| Mean Temp °F | 25.8 | 30.6 | 39.4 | 50.3 | 60.1 | 70.6 | 76.4 | 74.0 | 64.1 | 51.7 | 37.6 | 28.6 | 50.8 |
| Days Max Temp ≥ 90 °F | 0 | 0 | 0 | 1 | 2 | 11 | 19 | 17 | 8 | 1 | 0 | 0 | 59 |
| Days Max Temp ≤ 32 °F | 9 | 6 | 3 | 0 | 0 | 0 | 0 | 0 | 0 | 0 | 3 | 7 | 28 |
| Days Min Temp ≤ 32 °F | 31 | 28 | 25 | 12 | 2 | 0 | 0 | 0 | 2 | 12 | 27 | 31 | 170 |
| Days Min Temp ≤ 0 °F | 6 | 3 | 1 | 0 | 0 | 0 | 0 | 0 | 0 | 0 | 0 | 3 | 13 |
| Heating Degree Days | 1210 | 964 | 787 | 444 | 193 | 32 | 4 | 9 | 121 | 410 | 816 | 1123 | 6113 |
| Cooling Degree Days | 0 | 0 | 0 | 10 | 45 | 217 | 362 | 312 | 110 | 4 | 0 | 0 | 1060 |
| Total Precipitation (") | 0.57 | 0.47 | 1.32 | 1.57 | 3.00 | 2.94 | 3.00 | 2.06 | 1.40 | 1.14 | 0.75 | 0.51 | 18.73 |
| Days ≥ 0.1" Precip | 2 | 1 | 3 | 3 | 6 | 6 | 5 | 4 | 3 | 3 | 2 | 2 | 40 |
| Total Snowfall (") | 7.2 | 3.7 | 6.2 | 1.1 | 0.0 | 0.0 | 0.0 | 0.0 | 0.2 | 0.9 | 4.0 | 5.3 | 28.6 |
| Days ≥ 1" Snow Depth | 12 | 8 | *4* | 1 | 0 | 0 | 0 | 0 | 0 | 1 | 2 | 8 | 36 |

## BIG SPRINGS *Deuel County* ELEVATION 3363 ft LAT/LONG 41° 4 ' N / 102° 5 ' W

| | JAN | FEB | MAR | APR | MAY | JUN | JUL | AUG | SEP | OCT | NOV | DEC | YEAR |
|---|---|---|---|---|---|---|---|---|---|---|---|---|---|
| Maximum Temp °F | 37.5 | 43.5 | 52.2 | 63.1 | 72.0 | 81.4 | 87.9 | 85.9 | 77.2 | 65.4 | 50.0 | 40.1 | 63.0 |
| Minimum Temp °F | 11.3 | 15.6 | 24.3 | 34.1 | 44.6 | 54.0 | 60.0 | 57.7 | 47.0 | 34.4 | 22.6 | 13.3 | 34.9 |
| Mean Temp °F | 24.4 | 29.6 | 38.3 | 48.6 | 58.3 | 67.7 | 73.9 | 71.8 | 62.1 | 49.9 | 36.3 | 26.7 | 49.0 |
| Days Max Temp ≥ 90 °F | 0 | 0 | 0 | 0 | 1 | *5* | 13 | *11* | 3 | 0 | 0 | 0 | 33 |
| Days Max Temp ≤ 32 °F | 10 | 6 | 3 | 0 | 0 | 0 | 0 | 0 | 0 | 0 | 3 | 8 | 30 |
| Days Min Temp ≤ 32 °F | 30 | 28 | 26 | 13 | 2 | 0 | 0 | 0 | 1 | 12 | 27 | 31 | 170 |
| Days Min Temp ≤ 0 °F | 6 | 3 | 1 | 0 | 0 | 0 | 0 | 0 | 0 | 0 | 1 | 4 | 15 |
| Heating Degree Days | 1250 | 993 | 820 | 487 | 228 | 51 | 5 | 13 | 141 | 462 | 852 | 1180 | 6482 |
| Cooling Degree Days | 0 | 0 | 0 | 3 | 24 | 133 | 265 | 221 | 58 | 2 | 0 | 0 | 706 |
| Total Precipitation (") | 0.44 | 0.40 | 1.25 | 1.77 | 3.24 | 2.89 | 2.15 | 1.80 | 1.02 | 0.79 | 0.63 | 0.50 | 16.88 |
| Days ≥ 0.1" Precip | 2 | 2 | 3 | 4 | 6 | 6 | 5 | 4 | 2 | 2 | 2 | 1 | 39 |
| Total Snowfall (") | na | na | na | *0.6* | 0.1 | 0.0 | 0.0 | 0.0 | 0.0 | 0.1 | na | na | na |
| Days ≥ 1" Snow Depth | na | na | na | *0* | 0 | 0 | 0 | 0 | 0 | 0 | na | na | na |

## BLAIR *Washington County* ELEVATION 1122 ft LAT/LONG 41° 33 ' N / 96° 8 ' W

| | JAN | FEB | MAR | APR | MAY | JUN | JUL | AUG | SEP | OCT | NOV | DEC | YEAR |
|---|---|---|---|---|---|---|---|---|---|---|---|---|---|
| Maximum Temp °F | 30.4 | 35.0 | 48.2 | 62.7 | 73.5 | 82.4 | 86.3 | 84.0 | 75.7 | 64.4 | 47.6 | 34.6 | 60.4 |
| Minimum Temp °F | 10.4 | 15.1 | 26.8 | 39.1 | 49.9 | 60.1 | 64.5 | 61.7 | 52.3 | 40.1 | 27.4 | 16.1 | 38.6 |
| Mean Temp °F | 20.4 | 25.1 | 37.6 | 50.9 | 61.7 | 71.2 | 75.4 | 72.9 | 64.0 | 52.2 | 37.5 | 25.4 | 49.5 |
| Days Max Temp ≥ 90 °F | 0 | 0 | 0 | 0 | 1 | 6 | 10 | 8 | 2 | 0 | 0 | 0 | 27 |
| Days Max Temp ≤ 32 °F | 16 | 13 | 4 | 0 | 0 | 0 | 0 | 0 | 0 | 0 | 4 | 13 | 50 |
| Days Min Temp ≤ 32 °F | 30 | 26 | 22 | 7 | 1 | 0 | 0 | 0 | 0 | 7 | 21 | 30 | 144 |
| Days Min Temp ≤ 0 °F | 9 | 5 | 0 | 0 | 0 | 0 | 0 | 0 | 0 | 0 | 0 | 3 | 17 |
| Heating Degree Days | 1377 | 1120 | 844 | 431 | 158 | 21 | 3 | 10 | 118 | 399 | 818 | 1222 | 6521 |
| Cooling Degree Days | 0 | 0 | 0 | 20 | 60 | 215 | 325 | 258 | 105 | 6 | 0 | 0 | 989 |
| Total Precipitation (") | 0.68 | 0.78 | 2.56 | 2.87 | 3.91 | 4.30 | 3.54 | 3.00 | 3.63 | 2.35 | 1.40 | 1.04 | 30.06 |
| Days ≥ 0.1" Precip | 2 | 2 | 5 | 6 | 7 | 7 | 6 | 5 | 5 | 4 | 3 | 2 | 54 |
| Total Snowfall (") | 6.0 | 5.7 | 3.9 | 1.0 | 0.0 | 0.0 | 0.0 | 0.0 | 0.0 | 0.4 | 2.2 | *5.4* | 24.6 |
| Days ≥ 1" Snow Depth | na | na | na | 0 | 0 | 0 | 0 | 0 | 0 | 0 | *1* | na | na |

## BREWSTER *Blaine County* ELEVATION 2503 ft LAT/LONG 41° 57 ' N / 99° 52 ' W

| | JAN | FEB | MAR | APR | MAY | JUN | JUL | AUG | SEP | OCT | NOV | DEC | YEAR |
|---|---|---|---|---|---|---|---|---|---|---|---|---|---|
| Maximum Temp °F | 32.8 | 37.5 | 48.3 | 60.5 | 71.2 | 81.1 | *87.6* | *85.6* | 75.8 | 64.4 | 47.1 | 36.0 | 60.7 |
| Minimum Temp °F | 7.8 | 12.5 | 22.4 | 33.6 | 44.2 | 54.2 | *60.2* | *57.6* | 46.3 | 33.9 | 21.5 | 11.4 | 33.8 |
| Mean Temp °F | 20.3 | 25.0 | 35.4 | 47.1 | 57.7 | 67.7 | *74.0* | *71.6* | 61.1 | 49.1 | 34.4 | 23.6 | 47.3 |
| Days Max Temp ≥ 90 °F | 0 | 0 | 0 | 0 | 1 | 5 | 12 | 10 | 4 | 0 | 0 | 0 | 32 |
| Days Max Temp ≤ 32 °F | 14 | 10 | 5 | 0 | 0 | 0 | 0 | 0 | 0 | 0 | 5 | 12 | 46 |
| Days Min Temp ≤ 32 °F | 31 | 27 | 26 | 14 | 3 | 0 | 0 | 0 | 2 | 13 | 26 | 31 | 173 |
| Days Min Temp ≤ 0 °F | 9 | 5 | 1 | 0 | 0 | 0 | 0 | 0 | 0 | 0 | 1 | 5 | 21 |
| Heating Degree Days | 1381 | 1123 | 908 | 534 | 249 | 53 | *9* | *20* | 174 | 487 | 913 | 1276 | 7127 |
| Cooling Degree Days | 0 | *0* | 0 | na | *27* | *147* | na | na | *67* | *2* | 0 | 0 | na |
| Total Precipitation (") | 0.48 | 0.64 | 1.12 | 1.99 | 3.16 | 3.66 | 3.18 | 3.13 | 1.56 | 1.23 | 0.93 | 0.67 | 21.75 |
| Days ≥ 0.1" Precip | 1 | 1 | 3 | 5 | 5 | 6 | 5 | *4* | 4 | 3 | 2 | 2 | 41 |
| Total Snowfall (") | na | *4.9* | na | *1.8* | 0.0 | 0.0 | 0.0 | 0.0 | 0.0 | 0.5 | *5.5* | *5.4* | na |
| Days ≥ 1" Snow Depth | na | *6* | na | 1 | 0 | 0 | 0 | 0 | 0 | 1 | *4* | na | na |

## BRIDGEPORT *Morrill County* ELEVATION 3666 ft LAT/LONG 41° 40 ' N / 103° 6 ' W

| | JAN | FEB | MAR | APR | MAY | JUN | JUL | AUG | SEP | OCT | NOV | DEC | YEAR |
|---|---|---|---|---|---|---|---|---|---|---|---|---|---|
| Maximum Temp °F | 39.6 | 46.0 | 54.1 | 64.4 | 74.2 | 84.1 | 90.8 | 88.8 | 79.9 | 68.0 | 50.8 | 41.4 | 65.2 |
| Minimum Temp °F | 12.0 | 16.3 | 24.1 | 33.1 | 43.7 | 53.3 | 59.0 | 56.6 | 45.9 | 33.2 | 21.9 | 13.4 | 34.4 |
| Mean Temp °F | 25.8 | 31.2 | 39.1 | 48.8 | 59.0 | 68.7 | 74.9 | 72.8 | 63.0 | 50.6 | 36.4 | 27.4 | 49.8 |
| Days Max Temp ≥ 90 °F | 0 | 0 | 0 | 0 | 2 | 9 | 19 | 16 | 5 | 0 | 0 | 0 | 51 |
| Days Max Temp ≤ 32 °F | 9 | 4 | 2 | 0 | 0 | 0 | 0 | 0 | 0 | 0 | 3 | 7 | 25 |
| Days Min Temp ≤ 32 °F | 30 | 27 | 26 | 14 | 3 | 0 | 0 | 0 | 2 | 14 | 27 | 30 | 173 |
| Days Min Temp ≤ 0 °F | 6 | 3 | 1 | 0 | 0 | 0 | 0 | 0 | 0 | 0 | 1 | 4 | 15 |
| Heating Degree Days | 1208 | 952 | 796 | 482 | 212 | 40 | 3 | 8 | 127 | 441 | 852 | 1158 | 6279 |
| Cooling Degree Days | 0 | 0 | 0 | 4 | 34 | 170 | 318 | 273 | 77 | 1 | 0 | 0 | 877 |
| Total Precipitation (") | 0.38 | 0.33 | 0.87 | 1.53 | 2.69 | 2.87 | 2.44 | 1.55 | 1.32 | 0.92 | 0.56 | 0.38 | 15.84 |
| Days ≥ 0.1" Precip | 1 | 1 | 2 | 4 | 6 | 6 | 5 | 4 | 3 | 3 | 2 | 1 | 38 |
| Total Snowfall (") | 7.0 | 6.0 | 7.6 | 3.5 | 0.3 | 0.0 | 0.0 | 0.0 | 0.5 | 1.8 | 5.7 | 7.4 | 39.8 |
| Days ≥ 1" Snow Depth | 12 | 6 | 4 | 1 | 0 | 0 | 0 | 0 | 0 | 1 | 4 | 11 | 39 |

## BROKEN BOW 2 W *Custer County* ELEVATION 2470 ft LAT/LONG 41° 24 ' N / 99° 38 ' W

| | JAN | FEB | MAR | APR | MAY | JUN | JUL | AUG | SEP | OCT | NOV | DEC | YEAR |
|---|---|---|---|---|---|---|---|---|---|---|---|---|---|
| Maximum Temp °F | 35.4 | 40.4 | 50.5 | 62.7 | 72.2 | 82.0 | 87.0 | 85.4 | 77.1 | 65.8 | 48.4 | 38.1 | 62.1 |
| Minimum Temp °F | 9.8 | 14.4 | 23.2 | 33.6 | 44.9 | 54.7 | 60.3 | 57.9 | 47.4 | 35.0 | 21.9 | 12.7 | 34.7 |
| Mean Temp °F | 22.6 | 27.4 | 36.8 | 48.2 | 58.6 | 68.4 | 73.7 | 71.7 | 62.3 | 50.4 | 35.2 | 25.4 | 48.4 |
| Days Max Temp ≥ 90 °F | 0 | 0 | 0 | 0 | 1 | 6 | 12 | 10 | 4 | 0 | 0 | 0 | 33 |
| Days Max Temp ≤ 32 °F | 11 | 8 | 4 | 0 | 0 | 0 | 0 | 0 | 0 | 0 | 4 | 10 | 37 |
| Days Min Temp ≤ 32 °F | 31 | 27 | 26 | 14 | 3 | 0 | 0 | 0 | 2 | 12 | 27 | 30 | 172 |
| Days Min Temp ≤ 0 °F | 8 | 4 | 1 | 0 | 0 | 0 | 0 | 0 | 0 | 0 | 1 | 5 | 19 |
| Heating Degree Days | 1308 | 1056 | 867 | 503 | 223 | 44 | 8 | 17 | 151 | 449 | 888 | 1221 | 6735 |
| Cooling Degree Days | 0 | 0 | 0 | 7 | 28 | 144 | 265 | 218 | 74 | 3 | 0 | 0 | 739 |
| Total Precipitation (") | 0.44 | 0.51 | 1.34 | 2.24 | 3.19 | 3.93 | 3.74 | 2.48 | 2.01 | 1.40 | 0.86 | 0.47 | 22.61 |
| Days ≥ 0.1" Precip | 1 | 2 | 3 | 5 | 7 | 6 | 6 | 5 | 4 | 3 | 2 | 2 | 46 |
| Total Snowfall (") | 5.5 | 5.0 | 5.9 | 2.3 | 0.2 | 0.0 | 0.0 | 0.0 | 0.0 | 0.9 | 4.5 | 6.0 | 30.3 |
| Days ≥ 1" Snow Depth | *12* | 7 | *4* | 1 | 0 | 0 | 0 | 0 | 0 | 0 | 3 | *9* | 36 |

## BROWNLEE *Cherry County* ELEVATION 2821 ft LAT/LONG 42° 17 ' N / 100° 37 ' W

| | JAN | FEB | MAR | APR | MAY | JUN | JUL | AUG | SEP | OCT | NOV | DEC | YEAR |
|---|---|---|---|---|---|---|---|---|---|---|---|---|---|
| Maximum Temp °F | 34.7 | 39.1 | 46.7 | na | 69.7 | 80.4 | 86.2 | 85.2 | 75.0 | 64.2 | 47.3 | 37.6 | na |
| Minimum Temp °F | 8.0 | 12.4 | 20.4 | 31.6 | 41.5 | 51.6 | 56.9 | 54.5 | 42.1 | 31.3 | 19.3 | 10.7 | 31.7 |
| Mean Temp °F | 21.5 | 25.8 | 33.6 | na | 55.6 | 66.0 | 71.7 | 69.8 | 58.6 | 47.8 | 33.3 | 24.1 | na |
| Days Max Temp ≥ 90 °F | 0 | 0 | 0 | 0 | 0 | 4 | 11 | 9 | 3 | 0 | 0 | 0 | 27 |
| Days Max Temp ≤ 32 °F | 13 | 9 | 5 | 0 | 0 | 0 | 0 | 0 | 0 | 0 | 5 | 11 | 43 |
| Days Min Temp ≤ 32 °F | 31 | 28 | 28 | 15 | 4 | 0 | 0 | 0 | 4 | 17 | 27 | 30 | 184 |
| Days Min Temp ≤ 0 °F | 9 | 5 | 1 | 0 | 0 | 0 | 0 | 0 | 0 | 0 | 1 | 6 | 22 |
| Heating Degree Days | 1341 | 1104 | 968 | na | na | 75 | 17 | 27 | 225 | 529 | 942 | 1262 | na |
| Cooling Degree Days | 0 | 0 | 0 | 4 | 16 | 105 | 227 | 180 | 45 | 2 | 0 | 0 | 579 |
| Total Precipitation (") | 0.31 | 0.44 | 1.38 | 2.09 | 3.11 | 2.80 | 3.09 | 2.13 | 1.73 | 1.07 | 0.80 | 0.51 | 19.46 |
| Days ≥ 0.1" Precip | 1 | 1 | 2 | na | na | 5 | 6 | 4 | 4 | 2 | 1 | 2 | na |
| Total Snowfall (") | 4.5 | 5.1 | 6.5 | 3.4 | 0.1 | 0.0 | 0.0 | 0.0 | 0.3 | 0.7 | 5.3 | 6.7 | 32.6 |
| Days ≥ 1" Snow Depth | 16 | 11 | 8 | 2 | 0 | 0 | 0 | 0 | 0 | 0 | 7 | 13 | 57 |

## BURWELL 4 SE *Garfield County* ELEVATION 2180 ft LAT/LONG 41° 47 ' N / 99° 8 ' W

| | JAN | FEB | MAR | APR | MAY | JUN | JUL | AUG | SEP | OCT | NOV | DEC | YEAR |
|---|---|---|---|---|---|---|---|---|---|---|---|---|---|
| Maximum Temp °F | 34.7 | 39.2 | 49.7 | 62.6 | 72.9 | 82.6 | 87.4 | 85.9 | 76.3 | 64.9 | 47.0 | 36.5 | 61.6 |
| Minimum Temp °F | 10.5 | 14.9 | 23.9 | 34.8 | 46.1 | 56.1 | 61.1 | 59.1 | 48.1 | 35.8 | 22.6 | 12.9 | 35.5 |
| Mean Temp °F | 22.6 | 27.1 | 36.9 | 48.7 | 59.5 | 69.3 | 74.3 | 72.5 | 62.2 | 50.4 | 34.9 | 24.7 | 48.6 |
| Days Max Temp ≥ 90 °F | 0 | 0 | 0 | 0 | 1 | 6 | 13 | 11 | 4 | 0 | 0 | 0 | 35 |
| Days Max Temp ≤ 32 °F | 13 | 9 | 4 | 0 | 0 | 0 | 0 | 0 | 0 | 0 | 4 | 11 | 41 |
| Days Min Temp ≤ 32 °F | 31 | 27 | 25 | 12 | 2 | 0 | 0 | 0 | 2 | 11 | 26 | 31 | 167 |
| Days Min Temp ≤ 0 °F | 7 | 4 | 1 | 0 | 0 | 0 | 0 | 0 | 0 | 0 | 1 | 5 | 18 |
| Heating Degree Days | 1308 | 1065 | 866 | 488 | 203 | 33 | 6 | 13 | 150 | 451 | 897 | 1241 | 6721 |
| Cooling Degree Days | 0 | 0 | 0 | 12 | 46 | 192 | 310 | 267 | 92 | 4 | 0 | 0 | 923 |
| Total Precipitation (") | 0.42 | 0.60 | 1.16 | 2.20 | 3.09 | 4.00 | 3.07 | 3.01 | 2.23 | 1.63 | 0.95 | 0.50 | 22.86 |
| Days ≥ 0.1" Precip | 1 | 2 | 3 | 5 | 6 | 7 | 6 | 5 | 4 | 4 | 2 | 2 | 47 |
| Total Snowfall (") | 4.7 | 4.4 | 4.6 | 2.2 | 0.0 | 0.0 | 0.0 | 0.0 | 0.0 | 0.8 | 4.4 | 5.3 | 26.4 |
| Days ≥ 1" Snow Depth | 17 | 11 | 5 | 1 | 0 | 0 | 0 | 0 | 0 | 0 | 5 | 12 | 51 |

## BUTTE *Boyd County* ELEVATION 1903 ft LAT/LONG 42° 55 ' N / 98° 50 ' W

| | JAN | FEB | MAR | APR | MAY | JUN | JUL | AUG | SEP | OCT | NOV | DEC | YEAR |
|---|---|---|---|---|---|---|---|---|---|---|---|---|---|
| Maximum Temp °F | 31.0 | 36.5 | 48.4 | 62.5 | 73.2 | 82.7 | 88.3 | 86.8 | 77.1 | 64.5 | 45.9 | 34.3 | 60.9 |
| Minimum Temp °F | 9.6 | 14.7 | 24.8 | 36.6 | 47.6 | 57.3 | 62.8 | 60.8 | 50.5 | 38.4 | 25.0 | 13.9 | 36.8 |
| Mean Temp °F | 20.3 | 25.7 | 36.6 | 49.6 | 60.4 | 70.0 | 75.6 | 73.8 | 63.8 | 51.5 | 35.5 | 24.1 | 48.9 |
| Days Max Temp ≥ 90 °F | 0 | 0 | 0 | 1 | 1 | 5 | 13 | 11 | 4 | 0 | 0 | 0 | 35 |
| Days Max Temp ≤ 32 °F | 15 | 11 | 4 | 0 | 0 | 0 | 0 | 0 | 0 | 0 | 5 | 13 | 48 |
| Days Min Temp ≤ 32 °F | 30 | 27 | 24 | 10 | 1 | 0 | 0 | 0 | 1 | 8 | 23 | 30 | 154 |
| Days Min Temp ≤ 0 °F | 9 | 5 | 1 | 0 | 0 | 0 | 0 | 0 | 0 | 0 | 1 | 5 | 21 |
| Heating Degree Days | 1379 | 1104 | 872 | 467 | 185 | 27 | 5 | 9 | 124 | 419 | 877 | 1260 | 6728 |
| Cooling Degree Days | 0 | 0 | 0 | 16 | 53 | 191 | 328 | 288 | 104 | 7 | 0 | 0 | 987 |
| Total Precipitation (") | 0.46 | 0.59 | 1.59 | 2.40 | 3.43 | 3.75 | 3.09 | 2.78 | 2.30 | 1.65 | 1.05 | 0.54 | 23.63 |
| Days ≥ 0.1" Precip | 1 | 2 | 3 | 5 | 6 | 6 | 5 | 5 | 4 | 3 | 3 | 2 | 45 |
| Total Snowfall (") | 5.7 | 5.6 | 6.2 | 2.0 | 0.0 | 0.0 | 0.0 | 0.0 | 0.1 | 0.8 | 5.8 | 6.4 | 32.6 |
| Days ≥ 1" Snow Depth | na | na | na | 0 | 0 | 0 | 0 | 0 | 0 | 0 | na | na | na |

## CAMBRIDGE *Furnas County* ELEVATION 2260 ft LAT/LONG 40° 16 ' N / 100° 10 ' W

| | JAN | FEB | MAR | APR | MAY | JUN | JUL | AUG | SEP | OCT | NOV | DEC | YEAR |
|---|---|---|---|---|---|---|---|---|---|---|---|---|---|
| Maximum Temp °F | 39.1 | 45.2 | 55.4 | 67.0 | 75.2 | 85.7 | 90.2 | 88.6 | 80.9 | 69.4 | 52.2 | 41.8 | 65.9 |
| Minimum Temp °F | 12.4 | 17.3 | 26.2 | 36.7 | 47.4 | 57.6 | 62.9 | 60.5 | 49.7 | 36.3 | 24.2 | 15.3 | 37.2 |
| Mean Temp °F | 25.8 | 31.3 | 40.8 | 51.9 | 61.3 | 71.7 | 76.6 | 74.6 | 65.3 | 52.9 | 38.2 | 28.6 | 51.6 |
| Days Max Temp ≥ 90 °F | 0 | 0 | 0 | 1 | 2 | 10 | 17 | 15 | 7 | 1 | 0 | 0 | 53 |
| Days Max Temp ≤ 32 °F | 10 | 6 | 2 | 0 | 0 | 0 | 0 | 0 | 0 | 0 | 2 | 7 | 27 |
| Days Min Temp ≤ 32 °F | 31 | 27 | 24 | 10 | 1 | 0 | 0 | 0 | 1 | 11 | 25 | 31 | 161 |
| Days Min Temp ≤ 0 °F | 5 | 3 | 1 | 0 | 0 | 0 | 0 | 0 | 0 | 0 | 0 | 3 | 12 |
| Heating Degree Days | 1210 | 946 | 743 | 397 | 158 | 20 | 2 | 6 | 99 | 376 | 796 | 1123 | 5876 |
| Cooling Degree Days | 0 | 0 | 0 | 12 | 45 | 228 | 360 | 308 | 120 | 4 | 0 | 0 | 1077 |
| Total Precipitation (") | 0.48 | 0.55 | 1.41 | 1.98 | 3.69 | 3.66 | 3.49 | 2.68 | 1.59 | 1.18 | 1.05 | 0.57 | 22.33 |
| Days ≥ 0.1" Precip | 2 | 2 | 3 | 4 | 7 | 6 | 6 | 5 | 4 | 3 | 2 | 2 | 46 |
| Total Snowfall (") | 5.9 | 5.3 | 4.7 | 1.3 | 0.0 | 0.0 | 0.0 | 0.0 | 0.2 | 0.5 | 4.4 | 6.9 | 29.2 |
| Days ≥ 1" Snow Depth | 13 | 11 | 3 | 0 | 0 | 0 | 0 | 0 | 0 | 0 | 3 | 9 | 39 |

## CANADAY STEAM PLANT *Gosper County* ELEVATION 2362 ft LAT/LONG 40° 42 ' N / 99° 42 ' W

| | JAN | FEB | MAR | APR | MAY | JUN | JUL | AUG | SEP | OCT | NOV | DEC | YEAR |
|---|---|---|---|---|---|---|---|---|---|---|---|---|---|
| Maximum Temp °F | 34.6 | 40.0 | 50.5 | 62.8 | 72.5 | 82.8 | 87.3 | 85.0 | 76.9 | 65.2 | 48.9 | 38.0 | 62.0 |
| Minimum Temp °F | 11.6 | 16.1 | 25.5 | 36.5 | 47.7 | 57.5 | 62.8 | 60.3 | 49.9 | 37.3 | 24.9 | 15.4 | 37.1 |
| Mean Temp °F | 23.1 | 28.1 | 38.0 | 49.7 | 60.1 | 70.2 | 75.1 | 72.7 | 63.4 | 51.3 | 36.9 | 26.7 | 49.6 |
| Days Max Temp ≥ 90 °F | 0 | 0 | 0 | 0 | 1 | 7 | 12 | 9 | 4 | 0 | 0 | 0 | 33 |
| Days Max Temp ≤ 32 °F | 13 | 9 | 4 | 0 | 0 | 0 | 0 | 0 | 0 | 0 | 4 | 10 | 40 |
| Days Min Temp ≤ 32 °F | 31 | 27 | 24 | 9 | 1 | 0 | 0 | 0 | 1 | 8 | 25 | 30 | 156 |
| Days Min Temp ≤ 0 °F | 7 | 4 | 1 | 0 | 0 | 0 | 0 | 0 | 0 | 0 | 0 | 3 | 15 |
| Heating Degree Days | 1292 | 1035 | 831 | 462 | 190 | 31 | 4 | 12 | 126 | 423 | 836 | 1180 | 6422 |
| Cooling Degree Days | 0 | 0 | 0 | 11 | 42 | 190 | 305 | 252 | 94 | 4 | 0 | 0 | 898 |
| Total Precipitation (") | 0.43 | 0.43 | 1.28 | 1.91 | 3.65 | 3.63 | 3.77 | 2.69 | 1.99 | 1.44 | 0.87 | 0.50 | 22.59 |
| Days ≥ 0.1" Precip | 2 | 1 | 3 | 4 | 7 | 6 | 6 | 5 | 4 | 3 | 2 | 1 | 44 |
| Total Snowfall (") | 5.3 | 3.5 | 2.8 | 0.7 | 0.2 | 0.0 | 0.0 | 0.0 | 0.0 | 0.8 | ***2.6*** | 5.3 | 21.2 |
| Days ≥ 1" Snow Depth | 12 | 8 | 3 | 1 | 0 | 0 | 0 | 0 | 0 | 0 | 3 | 7 | 34 |

## CENTRAL CITY *Merrick County* ELEVATION 1690 ft LAT/LONG 41° 7 ' N / 98° 0 ' W

| | JAN | FEB | MAR | APR | MAY | JUN | JUL | AUG | SEP | OCT | NOV | DEC | YEAR |
|---|---|---|---|---|---|---|---|---|---|---|---|---|---|
| Maximum Temp °F | 34.1 | 39.9 | 52.1 | 65.7 | 74.9 | 84.3 | 87.8 | 85.9 | 78.2 | 67.1 | 49.2 | 37.4 | 63.1 |
| Minimum Temp °F | 13.4 | 18.0 | 28.4 | 39.7 | 50.9 | 60.7 | 65.2 | 63.1 | 53.7 | 41.3 | 28.0 | 17.8 | 40.0 |
| Mean Temp °F | 23.7 | 29.0 | 40.3 | 52.8 | 62.9 | 72.5 | 76.5 | 74.5 | 66.0 | 54.3 | 38.6 | 27.6 | 51.6 |
| Days Max Temp ≥ 90 °F | 0 | 0 | 0 | 1 | 2 | 8 | 13 | 10 | 3 | 0 | 0 | 0 | 37 |
| Days Max Temp ≤ 32 °F | 13 | 9 | 3 | 0 | 0 | 0 | 0 | 0 | 0 | 0 | 3 | 10 | 38 |
| Days Min Temp ≤ 32 °F | 30 | 26 | 21 | 7 | 0 | 0 | 0 | 0 | 0 | 5 | 20 | 30 | 139 |
| Days Min Temp ≤ 0 °F | 6 | 3 | 0 | 0 | 0 | 0 | 0 | 0 | 0 | 0 | 0 | 3 | 12 |
| Heating Degree Days | 1272 | 1011 | 761 | 378 | 130 | 14 | 2 | 5 | 88 | 339 | 785 | 1152 | 5937 |
| Cooling Degree Days | 0 | 0 | 1 | 25 | 73 | 247 | 351 | 303 | 140 | 9 | 0 | 0 | 1149 |
| Total Precipitation (") | 0.57 | 0.69 | 1.89 | 2.60 | 4.19 | 4.02 | 3.46 | 2.60 | 3.16 | 1.84 | 1.26 | 0.76 | 27.04 |
| Days ≥ 0.1" Precip | 1 | 2 | 4 | 5 | 7 | 6 | 6 | 5 | 5 | 3 | 3 | 2 | 49 |
| Total Snowfall (") | 3.9 | 4.4 | 3.7 | 0.4 | 0.2 | 0.0 | 0.0 | 0.0 | 0.0 | 0.3 | 3.0 | 5.6 | 21.5 |
| Days ≥ 1" Snow Depth | 17 | 12 | 6 | 0 | 0 | 0 | 0 | 0 | 0 | 0 | 3 | 11 | 49 |

## CHADRON *Dawes County* ELEVATION 3312 ft LAT/LONG 42° 50 ' N / 103° 0 ' W

| | JAN | FEB | MAR | APR | MAY | JUN | JUL | AUG | SEP | OCT | NOV | DEC | YEAR |
|---|---|---|---|---|---|---|---|---|---|---|---|---|---|
| Maximum Temp °F | 34.8 | 40.3 | 49.6 | 59.7 | 70.4 | 81.2 | 89.3 | 88.1 | 77.6 | 64.2 | 47.4 | 37.8 | 61.7 |
| Minimum Temp °F | 9.9 | 14.2 | 23.0 | 33.1 | 43.0 | 52.8 | 59.1 | 57.3 | 46.3 | 33.7 | 21.8 | 12.6 | 33.9 |
| Mean Temp °F | 22.4 | 27.3 | 36.4 | 46.4 | 56.8 | 67.1 | 74.2 | 72.7 | 62.0 | 49.0 | 34.6 | 25.2 | 47.8 |
| Days Max Temp ≥ 90 °F | 0 | 0 | 0 | 0 | 1 | 6 | 16 | 15 | 5 | 0 | 0 | 0 | 43 |
| Days Max Temp ≤ 32 °F | 12 | 8 | 4 | 1 | 0 | 0 | 0 | 0 | 0 | 0 | 5 | 10 | 40 |
| Days Min Temp ≤ 32 °F | 30 | 28 | 26 | 15 | 3 | 0 | 0 | 0 | 2 | 13 | 25 | 29 | 171 |
| Days Min Temp ≤ 0 °F | 8 | 5 | 1 | 0 | 0 | 0 | 0 | 0 | 0 | 0 | 1 | 5 | 20 |
| Heating Degree Days | 1315 | 1059 | 881 | 552 | 274 | 62 | 8 | 14 | 163 | 492 | 906 | 1227 | 6953 |
| Cooling Degree Days | 0 | 0 | 0 | 3 | 28 | 139 | ***312*** | ***280*** | ***86*** | 2 | ***0*** | 0 | 850 |
| Total Precipitation (") | 0.42 | 0.40 | 0.78 | 1.80 | 2.80 | 2.95 | 2.15 | 1.48 | 1.41 | 0.91 | 0.47 | 0.42 | 15.99 |
| Days ≥ 0.1" Precip | 1 | 1 | 2 | 4 | 6 | 6 | 5 | 3 | 4 | 2 | 2 | 1 | 37 |
| Total Snowfall (") | 6.4 | 6.1 | 8.2 | 4.4 | 0.7 | 0.0 | 0.0 | 0.0 | 0.5 | 2.6 | 4.8 | 7.7 | 41.4 |
| Days ≥ 1" Snow Depth | 16 | ***11*** | 7 | 2 | 0 | 0 | 0 | 0 | 0 | 1 | 5 | 12 | 54 |

## CLARKSON *Colfax County* ELEVATION 1572 ft LAT/LONG 41° 43 ' N / 97° 7 ' W

| | JAN | FEB | MAR | APR | MAY | JUN | JUL | AUG | SEP | OCT | NOV | DEC | YEAR |
|---|---|---|---|---|---|---|---|---|---|---|---|---|---|
| Maximum Temp °F | 30.4 | 36.4 | 48.9 | 63.4 | 74.0 | 83.6 | 87.2 | 84.8 | 76.0 | 64.1 | 46.7 | 34.6 | 60.8 |
| Minimum Temp °F | 9.2 | 14.6 | 25.5 | 37.1 | 48.6 | 58.5 | 63.2 | 60.7 | 50.6 | 38.2 | 25.4 | 14.7 | 37.2 |
| Mean Temp °F | 19.8 | 25.5 | 37.2 | 50.2 | 61.3 | 71.0 | 75.2 | 72.8 | 63.3 | 51.2 | 36.0 | 24.6 | 49.0 |
| Days Max Temp ≥ 90 °F | 0 | 0 | 0 | 1 | 1 | 7 | 12 | 8 | 3 | 0 | 0 | 0 | 32 |
| Days Max Temp ≤ 32 °F | 16 | 11 | 4 | 0 | 0 | 0 | 0 | 0 | 0 | 0 | 4 | 12 | 47 |
| Days Min Temp ≤ 32 °F | 31 | 27 | 24 | 10 | 1 | 0 | 0 | 0 | 1 | 9 | 23 | 30 | 156 |
| Days Min Temp ≤ 0 °F | 9 | 4 | 1 | 0 | 0 | 0 | 0 | 0 | 0 | 0 | 1 | 4 | 19 |
| Heating Degree Days | 1394 | 1108 | 854 | 448 | 166 | 21 | 4 | 11 | 129 | 428 | 862 | 1245 | 6670 |
| Cooling Degree Days | 0 | 0 | 0 | 18 | 57 | 212 | 318 | 255 | 96 | 3 | 0 | 0 | 959 |
| Total Precipitation (") | 0.59 | 0.80 | 2.13 | 2.56 | 4.45 | 4.59 | 3.44 | 2.95 | 3.06 | 2.42 | 1.42 | 0.94 | 29.35 |
| Days ≥ 0.1" Precip | 2 | 2 | 4 | 5 | 7 | 7 | 5 | 5 | 5 | 4 | 3 | 2 | 51 |
| Total Snowfall (") | ***5.9*** | ***5.0*** | ***4.7*** | 1.5 | 0.0 | 0.0 | 0.0 | 0.0 | 0.1 | 0.9 | 3.9 | ***6.1*** | 28.1 |
| Days ≥ 1" Snow Depth | na | na | ***3*** | 1 | 0 | 0 | 0 | 0 | 0 | 0 | 3 | 10 | na |

## COLUMBUS 3 NE *Platte County* ELEVATION 1440 ft LAT/LONG 41° 26' N / 97° 21' W

| | JAN | FEB | MAR | APR | MAY | JUN | JUL | AUG | SEP | OCT | NOV | DEC | YEAR |
|---|---|---|---|---|---|---|---|---|---|---|---|---|---|
| Maximum Temp °F | 30.6 | 36.2 | 48.8 | 62.9 | 73.7 | 84.0 | 88.0 | 85.4 | 76.5 | 63.9 | 46.3 | 34.4 | 60.9 |
| Minimum Temp °F | 11.0 | 15.9 | 27.1 | 39.1 | 50.7 | 60.5 | 65.2 | 62.9 | 53.0 | 40.3 | 26.7 | 15.8 | 39.0 |
| Mean Temp °F | 20.8 | 26.0 | 38.0 | 51.0 | 62.2 | 72.3 | 76.6 | 74.2 | 64.8 | 52.1 | 36.5 | 25.1 | 50.0 |
| Days Max Temp ≥ 90 °F | 0 | 0 | 0 | 1 | 1 | 8 | 14 | 10 | 3 | 0 | 0 | 0 | 37 |
| Days Max Temp ≤ 32 °F | 16 | 11 | 4 | 0 | 0 | 0 | 0 | 0 | 0 | 0 | 4 | 13 | 48 |
| Days Min Temp ≤ 32 °F | 31 | 27 | 22 | 7 | 1 | 0 | 0 | 0 | 0 | 6 | 23 | 30 | 147 |
| Days Min Temp ≤ 0 °F | 7 | 4 | 0 | 0 | 0 | 0 | 0 | 0 | 0 | 0 | 0 | 4 | 15 |
| Heating Degree Days | 1364 | 1093 | 832 | 426 | 149 | 17 | 2 | 6 | 109 | 402 | 848 | 1230 | 6478 |
| Cooling Degree Days | 0 | 0 | 1 | 22 | 75 | 260 | 380 | 311 | 129 | 8 | 0 | 0 | 1186 |
| Total Precipitation (") | 0.48 | 0.77 | 1.87 | 2.63 | 4.08 | 4.26 | 3.49 | 3.19 | 3.12 | 2.30 | 1.30 | 0.86 | 28.35 |
| Days ≥ 0.1" Precip | 2 | 2 | 4 | 5 | 7 | 6 | 6 | 5 | 5 | 4 | 3 | 2 | 51 |
| Total Snowfall (") | 4.7 | 5.8 | 4.2 | 1.0 | 0.0 | 0.0 | 0.0 | 0.0 | 0.0 | 0.9 | 3.4 | 5.7 | 25.7 |
| Days ≥ 1" Snow Depth | 17 | 14 | 6 | 1 | 0 | 0 | 0 | 0 | 0 | 0 | 3 | 12 | 53 |

## CREIGHTON *Knox County* ELEVATION 1660 ft LAT/LONG 42° 27' N / 97° 54' W

| | JAN | FEB | MAR | APR | MAY | JUN | JUL | AUG | SEP | OCT | NOV | DEC | YEAR |
|---|---|---|---|---|---|---|---|---|---|---|---|---|---|
| Maximum Temp °F | 30.9 | 36.8 | 49.2 | 63.7 | 74.3 | 84.2 | 88.4 | 86.3 | 77.5 | 65.1 | 46.6 | 34.6 | 61.5 |
| Minimum Temp °F | 9.4 | 15.0 | 25.7 | 37.3 | 48.6 | 58.1 | 62.8 | 60.3 | 50.5 | 38.6 | 25.5 | 14.0 | 37.2 |
| Mean Temp °F | 20.2 | 25.9 | 37.5 | 50.6 | 61.5 | 71.2 | 75.8 | 73.3 | 64.0 | 51.9 | 36.0 | 24.3 | 49.3 |
| Days Max Temp ≥ 90 °F | 0 | 0 | 0 | 1 | 1 | 7 | 14 | 11 | 4 | 0 | 0 | 0 | 38 |
| Days Max Temp ≤ 32 °F | 16 | 11 | 4 | 0 | 0 | 0 | 0 | 0 | 0 | 0 | 4 | 13 | 48 |
| Days Min Temp ≤ 32 °F | 30 | 26 | 23 | 9 | 1 | 0 | 0 | 0 | 1 | 9 | 23 | 30 | 152 |
| Days Min Temp ≤ 0 °F | 9 | 5 | 1 | 0 | 0 | 0 | 0 | 0 | 0 | 0 | 1 | 5 | 21 |
| Heating Degree Days | 1383 | 1097 | 847 | 439 | 165 | 23 | 4 | 9 | 121 | 407 | 862 | 1255 | 6612 |
| Cooling Degree Days | 0 | 0 | 1 | 20 | 63 | 229 | 345 | 287 | 109 | 9 | 0 | 0 | 1063 |
| Total Precipitation (") | 0.51 | 0.61 | 1.63 | 2.32 | 3.45 | 3.79 | 3.25 | 2.93 | 2.26 | 1.67 | 1.20 | 0.66 | 24.28 |
| Days ≥ 0.1" Precip | 2 | 2 | 4 | 5 | 7 | 6 | 6 | 5 | 5 | 4 | 3 | 2 | 51 |
| Total Snowfall (") | na | na | na | *1.0* | 0.0 | 0.0 | 0.0 | 0.0 | 0.0 | 0.2 | na | na | na |
| Days ≥ 1" Snow Depth | na | na | na | 0 | 0 | 0 | 0 | 0 | 0 | 0 | na | na | na |

## CRESCENT LAKE NWR *Garden County* ELEVATION 3822 ft LAT/LONG 41° 45' N / 102° 26' W

| | JAN | FEB | MAR | APR | MAY | JUN | JUL | AUG | SEP | OCT | NOV | DEC | YEAR |
|---|---|---|---|---|---|---|---|---|---|---|---|---|---|
| Maximum Temp °F | 36.5 | 42.3 | 50.3 | 61.1 | 70.4 | 79.9 | 86.6 | 85.1 | 76.0 | 64.7 | 48.8 | 39.4 | 61.8 |
| Minimum Temp °F | 9.2 | 14.0 | 22.3 | 32.4 | 42.6 | 51.4 | 57.0 | 54.6 | 44.0 | 31.8 | 20.7 | 11.6 | 32.6 |
| Mean Temp °F | 22.8 | 28.2 | 36.3 | 46.8 | 56.5 | 65.7 | 71.8 | 69.9 | 60.0 | 48.3 | 34.8 | 25.5 | 47.2 |
| Days Max Temp ≥ 90 °F | 0 | 0 | 0 | 0 | 0 | 4 | 12 | 9 | 3 | 0 | 0 | 0 | 28 |
| Days Max Temp ≤ 32 °F | 11 | 7 | 4 | 1 | 0 | 0 | 0 | 0 | 0 | 0 | 4 | 9 | 36 |
| Days Min Temp ≤ 32 °F | 31 | 28 | 27 | 15 | 3 | 0 | 0 | 0 | 3 | 17 | 28 | 31 | 183 |
| Days Min Temp ≤ 0 °F | 8 | 4 | 1 | 0 | 0 | 0 | 0 | 0 | 0 | 0 | 1 | 5 | 19 |
| Heating Degree Days | 1303 | 1033 | 881 | 541 | 272 | 71 | 10 | 21 | 183 | 514 | 900 | 1218 | 6947 |
| Cooling Degree Days | 0 | 0 | 0 | 2 | 16 | 104 | 225 | 186 | 43 | 1 | 0 | 0 | 577 |
| Total Precipitation (") | 0.32 | 0.36 | 0.92 | 1.61 | 3.21 | 3.03 | 2.40 | 1.65 | 1.63 | 0.98 | 0.59 | 0.33 | 17.03 |
| Days ≥ 0.1" Precip | 1 | 1 | 3 | 4 | 6 | 6 | 5 | 4 | 3 | 3 | 2 | 1 | 39 |
| Total Snowfall (") | 5.7 | 5.3 | 7.9 | 4.0 | 0.7 | 0.0 | 0.0 | 0.0 | 0.3 | 2.0 | 5.2 | 5.8 | 36.9 |
| Days ≥ 1" Snow Depth | *14* | 9 | 6 | 2 | 0 | 0 | 0 | 0 | 0 | 1 | 4 | 10 | 46 |

## CRETE *Saline County* ELEVATION 1362 ft LAT/LONG 40° 37' N / 96° 57' W

| | JAN | FEB | MAR | APR | MAY | JUN | JUL | AUG | SEP | OCT | NOV | DEC | YEAR |
|---|---|---|---|---|---|---|---|---|---|---|---|---|---|
| Maximum Temp °F | 34.1 | 40.2 | 52.6 | 65.5 | 75.0 | 85.0 | 88.9 | 86.6 | 78.5 | 67.6 | 50.2 | 38.1 | 63.5 |
| Minimum Temp °F | 13.7 | 18.1 | 29.2 | 40.5 | 51.1 | 60.8 | 65.3 | 63.1 | 54.0 | 42.2 | 29.3 | 18.7 | 40.5 |
| Mean Temp °F | 23.9 | 29.2 | 40.9 | 53.0 | 63.1 | 72.9 | 77.2 | 74.9 | 66.3 | 54.9 | 39.8 | 28.4 | 52.0 |
| Days Max Temp ≥ 90 °F | 0 | 0 | 0 | 0 | 1 | 8 | 15 | 11 | 3 | 0 | 0 | 0 | 38 |
| Days Max Temp ≤ 32 °F | 13 | 9 | 2 | 0 | 0 | 0 | 0 | 0 | 0 | 0 | 3 | 10 | 37 |
| Days Min Temp ≤ 32 °F | 30 | 26 | 20 | 6 | 1 | 0 | 0 | 0 | 0 | 5 | 19 | 29 | 136 |
| Days Min Temp ≤ 0 °F | 6 | 3 | 0 | 0 | 0 | 0 | 0 | 0 | 0 | 0 | 0 | 3 | 12 |
| Heating Degree Days | 1268 | 1005 | 741 | 370 | 126 | 12 | 1 | 5 | 84 | 323 | 751 | 1128 | 5814 |
| Cooling Degree Days | 0 | 0 | 2 | 23 | 68 | 258 | 378 | 315 | 142 | 12 | 0 | 0 | 1198 |
| Total Precipitation (") | 0.61 | 0.70 | 2.11 | 2.77 | 4.13 | 4.11 | 3.74 | 3.35 | 3.84 | 2.33 | 1.31 | 0.94 | 29.94 |
| Days ≥ 0.1" Precip | 2 | 2 | 4 | 6 | 7 | 6 | 6 | 5 | 5 | 4 | 3 | 2 | 52 |
| Total Snowfall (") | 5.0 | 5.7 | 4.0 | 0.6 | 0.1 | 0.0 | 0.0 | 0.0 | 0.0 | 0.2 | 2.5 | 5.3 | 23.4 |
| Days ≥ 1" Snow Depth | 17 | 11 | 4 | 0 | 0 | 0 | 0 | 0 | 0 | 0 | 3 | 9 | 44 |

## CULBERTSON *Hitchcock County* ELEVATION 2592 ft LAT/LONG 40° 13 ' N / 100° 50 ' W

| | JAN | FEB | MAR | APR | MAY | JUN | JUL | AUG | SEP | OCT | NOV | DEC | YEAR |
|---|---|---|---|---|---|---|---|---|---|---|---|---|---|
| Maximum Temp °F | 37.4 | 43.3 | 53.4 | 64.7 | 73.9 | 84.9 | 90.3 | 88.5 | 80.0 | 67.6 | 51.0 | 40.7 | 64.6 |
| Minimum Temp °F | 10.4 | 15.1 | 23.5 | 33.9 | 44.8 | 55.1 | 60.6 | 58.4 | 47.7 | 34.0 | 22.2 | 13.7 | 34.9 |
| Mean Temp °F | 24.1 | 29.2 | 38.5 | 49.4 | 59.4 | 70.0 | 75.5 | 73.4 | 63.9 | 50.8 | 36.6 | 27.2 | 49.8 |
| Days Max Temp ≥ 90 °F | 0 | 0 | 0 | 1 | 2 | 10 | 17 | 16 | 7 | 1 | 0 | 0 | 54 |
| Days Max Temp ≤ 32 °F | 10 | 7 | 3 | 0 | 0 | 0 | 0 | 0 | 0 | 0 | 2 | 8 | 30 |
| Days Min Temp ≤ 32 °F | 31 | 28 | 27 | 14 | 3 | 0 | 0 | 0 | 2 | 14 | 28 | 31 | 178 |
| Days Min Temp ≤ 0 °F | 6 | 4 | 1 | 0 | 0 | 0 | 0 | 0 | 0 | 0 | 1 | 3 | 15 |
| Heating Degree Days | 1263 | 1005 | 815 | 469 | 208 | 32 | 4 | 10 | 121 | 438 | 846 | 1165 | 6376 |
| Cooling Degree Days | 0 | 0 | 0 | 8 | 41 | 195 | 334 | 285 | 98 | 3 | 0 | 0 | 964 |
| Total Precipitation (") | 0.58 | 0.51 | 1.28 | 1.97 | 3.45 | 3.45 | 3.40 | 2.69 | 1.56 | 1.32 | 0.96 | 0.52 | 21.69 |
| Days ≥ 0.1" Precip | 2 | 2 | 3 | 4 | 7 | 6 | 6 | 5 | 3 | 3 | 2 | 1 | 44 |
| Total Snowfall (") | 7.4 | 4.8 | 6.2 | 2.0 | 0.0 | 0.0 | 0.0 | 0.0 | 0.3 | 0.4 | 4.3 | 5.4 | 30.8 |
| Days ≥ 1" Snow Depth | 14 | 10 | 5 | 1 | 0 | 0 | 0 | 0 | 0 | 0 | 4 | 9 | 43 |

## CURTIS 3 NNE *Frontier County* ELEVATION 2722 ft LAT/LONG 40° 40 ' N / 100° 30 ' W

| | JAN | FEB | MAR | APR | MAY | JUN | JUL | AUG | SEP | OCT | NOV | DEC | YEAR |
|---|---|---|---|---|---|---|---|---|---|---|---|---|---|
| Maximum Temp °F | 38.8 | 44.4 | 54.0 | 66.3 | 74.5 | 84.7 | 90.2 | 88.5 | 80.1 | 68.5 | 50.5 | 41.7 | 65.2 |
| Minimum Temp °F | 10.1 | 15.0 | 23.6 | 34.3 | 44.9 | 55.0 | 60.3 | 57.6 | 46.8 | 33.6 | 21.4 | 13.2 | 34.7 |
| Mean Temp °F | 24.7 | 29.8 | 38.8 | 50.3 | 59.8 | 69.9 | 75.2 | 73.0 | 63.5 | 51.1 | 36.0 | 27.5 | 50.0 |
| Days Max Temp ≥ 90 °F | 0 | 0 | 0 | 1 | 2 | 9 | 17 | 15 | 5 | 1 | 0 | 0 | 50 |
| Days Max Temp ≤ 32 °F | 10 | 6 | 2 | 0 | 0 | 0 | 0 | 0 | 0 | 0 | 3 | 8 | 29 |
| Days Min Temp ≤ 32 °F | 31 | 27 | 27 | 13 | 3 | 0 | 0 | 0 | 2 | 14 | 27 | 31 | 175 |
| Days Min Temp ≤ 0 °F | 7 | 4 | 1 | 0 | 0 | 0 | 0 | 0 | 0 | 0 | 1 | 4 | 17 |
| Heating Degree Days | 1243 | 986 | 806 | 439 | 193 | 30 | 4 | 9 | 124 | 429 | 864 | 1155 | 6282 |
| Cooling Degree Days | *0* | 0 | 0 | 8 | 35 | 182 | 312 | 261 | 88 | 2 | 0 | 0 | 888 |
| Total Precipitation (") | 0.45 | 0.49 | 1.26 | 1.86 | 3.27 | 3.66 | 3.35 | 2.35 | 1.63 | 1.15 | 0.73 | 0.47 | 20.67 |
| Days ≥ 0.1" Precip | 1 | 1 | 3 | 4 | 6 | 6 | 6 | 4 | 4 | 3 | 2 | 1 | 41 |
| Total Snowfall (") | *6.7* | *4.9* | *4.8* | 1.3 | 0.0 | 0.0 | 0.0 | 0.0 | 0.2 | 0.4 | *3.5* | *5.9* | 27.7 |
| Days ≥ 1" Snow Depth | 12 | 8 | 4 | 1 | 0 | 0 | 0 | 0 | 0 | 0 | 3 | 8 | 36 |

## DALTON *Cheyenne County* ELEVATION 4272 ft LAT/LONG 41° 25 ' N / 102° 58 ' W

| | JAN | FEB | MAR | APR | MAY | JUN | JUL | AUG | SEP | OCT | NOV | DEC | YEAR |
|---|---|---|---|---|---|---|---|---|---|---|---|---|---|
| Maximum Temp °F | 38.9 | 43.9 | 50.8 | 61.1 | 70.6 | 80.7 | 87.7 | 85.9 | 76.9 | 65.0 | 49.2 | 40.8 | 62.6 |
| Minimum Temp °F | 13.5 | 17.0 | 23.2 | 31.8 | 41.3 | 50.6 | 56.5 | 54.5 | 44.8 | 33.7 | 22.9 | 15.4 | 33.8 |
| Mean Temp °F | 26.2 | 30.5 | 37.0 | 46.5 | 56.0 | 65.7 | 72.1 | 70.2 | 60.9 | 49.4 | 36.1 | 28.2 | 48.2 |
| Days Max Temp ≥ 90 °F | 0 | 0 | 0 | 0 | 0 | 5 | 15 | 11 | 3 | 0 | 0 | 0 | 34 |
| Days Max Temp ≤ 32 °F | 9 | 6 | 4 | 1 | 0 | 0 | 0 | 0 | 0 | 0 | 4 | 7 | 31 |
| Days Min Temp ≤ 32 °F | 30 | 27 | 27 | 16 | 4 | 0 | 0 | 0 | 2 | 13 | 26 | 29 | 174 |
| Days Min Temp ≤ 0 °F | 5 | 3 | 1 | 0 | 0 | 0 | 0 | 0 | 0 | 0 | 1 | 4 | 14 |
| Heating Degree Days | 1196 | 969 | 860 | 550 | 285 | 72 | 9 | 18 | 162 | 480 | 860 | 1136 | 6597 |
| Cooling Degree Days | 0 | 0 | 0 | 1 | 12 | 98 | 216 | 176 | 39 | 1 | 0 | 0 | 543 |
| Total Precipitation (") | 0.49 | 0.55 | 1.32 | 2.00 | 3.27 | 3.38 | 2.57 | 1.91 | 1.29 | 1.17 | 0.80 | 0.54 | 19.29 |
| Days ≥ 0.1" Precip | 2 | 2 | 3 | 4 | 6 | 6 | 5 | 4 | 3 | 3 | 2 | 2 | 42 |
| Total Snowfall (") | 8.1 | 7.8 | 10.8 | 6.3 | 0.6 | 0.0 | 0.0 | 0.0 | 0.2 | 2.8 | 7.2 | 8.0 | 51.8 |
| Days ≥ 1" Snow Depth | *11* | *6* | *4* | 1 | 0 | 0 | 0 | 0 | 0 | 1 | 5 | *8* | 36 |

## DAVID CITY *Butler County* ELEVATION 1621 ft LAT/LONG 41° 15 ' N / 97° 8 ' W

| | JAN | FEB | MAR | APR | MAY | JUN | JUL | AUG | SEP | OCT | NOV | DEC | YEAR |
|---|---|---|---|---|---|---|---|---|---|---|---|---|---|
| Maximum Temp °F | 30.7 | 36.2 | 48.6 | 62.6 | 73.4 | 83.4 | 88.0 | 85.5 | 76.9 | 64.8 | 47.4 | 34.9 | 61.0 |
| Minimum Temp °F | 10.7 | 15.2 | 26.1 | 38.2 | 49.1 | 59.4 | 64.5 | 61.8 | 52.3 | 40.2 | 27.0 | 15.9 | 38.4 |
| Mean Temp °F | 20.7 | 25.7 | 37.4 | 50.4 | 61.2 | 71.4 | 76.2 | 73.7 | 64.6 | 52.5 | 37.2 | 25.4 | 49.7 |
| Days Max Temp ≥ 90 °F | 0 | 0 | 0 | 1 | 1 | 7 | 13 | 10 | 3 | 0 | 0 | 0 | 35 |
| Days Max Temp ≤ 32 °F | 16 | 11 | 4 | 0 | 0 | 0 | 0 | 0 | 0 | 0 | 4 | 12 | 47 |
| Days Min Temp ≤ 32 °F | 31 | 27 | 22 | 8 | 1 | 0 | 0 | 0 | 1 | 6 | 21 | 30 | 147 |
| Days Min Temp ≤ 0 °F | 8 | 5 | 0 | 0 | 0 | 0 | 0 | 0 | 0 | 0 | 0 | 4 | 17 |
| Heating Degree Days | 1368 | 1102 | 849 | 443 | 169 | 24 | 3 | 8 | 111 | 391 | 827 | 1221 | 6516 |
| Cooling Degree Days | 0 | 0 | 1 | 18 | 54 | 220 | 342 | 269 | 112 | 7 | 0 | 0 | 1023 |
| Total Precipitation (") | 0.61 | 0.77 | 2.25 | 2.84 | 4.48 | 4.88 | 3.08 | 3.32 | 3.44 | 2.24 | 1.42 | 0.98 | 30.31 |
| Days ≥ 0.1" Precip | 2 | 2 | 4 | 5 | 8 | 7 | 6 | 5 | 5 | 4 | 3 | 2 | 53 |
| Total Snowfall (") | *6.0* | 6.3 | *5.5* | 1.4 | 0.0 | 0.0 | 0.0 | 0.0 | 0.2 | 0.7 | 4.1 | 5.9 | 30.1 |
| Days ≥ 1" Snow Depth | na | na | *6* | *0* | 0 | 0 | 0 | 0 | 0 | 0 | *2* | *9* | na |

## ELLSWORTH 15 NNE *Sheridan County* ELEVATION 3983 ft LAT/LONG 42° 16 ' N / 102° 12 ' W

| | JAN | FEB | MAR | APR | MAY | JUN | JUL | AUG | SEP | OCT | NOV | DEC | YEAR |
|---|---|---|---|---|---|---|---|---|---|---|---|---|---|
| Maximum Temp °F | 34.0 | 39.1 | 47.9 | 58.8 | 69.0 | 78.7 | 85.8 | 84.8 | 75.0 | 63.0 | 46.7 | 36.9 | 60.0 |
| Minimum Temp °F | 9.0 | 13.3 | 21.6 | 31.7 | 41.5 | 51.0 | 57.2 | 55.4 | 44.8 | 32.8 | 20.7 | 11.3 | 32.5 |
| Mean Temp °F | 21.6 | 26.2 | 34.8 | 45.3 | 55.3 | 64.9 | 71.5 | 70.1 | 59.9 | 48.0 | 33.7 | 24.1 | 46.3 |
| Days Max Temp ≥ 90 °F | 0 | 0 | 0 | 0 | 0 | 3 | 11 | 10 | 3 | 0 | 0 | 0 | 27 |
| Days Max Temp ≤ 32 °F | 13 | 9 | 5 | 1 | 0 | 0 | 0 | 0 | 0 | 0 | 5 | 10 | 43 |
| Days Min Temp ≤ 32 °F | 31 | 28 | 28 | 16 | 4 | 0 | 0 | 0 | 3 | 14 | 27 | 31 | 182 |
| Days Min Temp ≤ 0 °F | 8 | 5 | 1 | 0 | 0 | 0 | 0 | 0 | 0 | 0 | 1 | 5 | 20 |
| Heating Degree Days | 1341 | 1089 | 930 | 585 | 310 | 86 | 16 | 25 | 192 | 523 | 932 | 1260 | 7289 |
| Cooling Degree Days | 0 | 0 | 0 | 2 | 14 | 89 | 214 | 191 | 47 | 1 | 0 | 0 | 558 |
| Total Precipitation (") | 0.31 | 0.45 | 0.81 | 1.78 | 2.80 | 3.15 | 2.83 | 2.08 | 1.53 | 1.13 | 0.45 | 0.31 | 17.63 |
| Days ≥ 0.1" Precip | 1 | 1 | 2 | 4 | 6 | 6 | 6 | 4 | 4 | 3 | 2 | 1 | 40 |
| Total Snowfall (") | 6.6 | 5.7 | 8.1 | 4.5 | 0.2 | 0.0 | 0.0 | 0.0 | 0.4 | 1.9 | 5.6 | 5.0 | 38.0 |
| Days ≥ 1" Snow Depth | na | na | na | 1 | 0 | 0 | 0 | 0 | 0 | na | *2* | na | na |

## ENDERS LAKE *Chase County* ELEVATION 3081 ft LAT/LONG 40° 25 ' N / 101° 31 ' W

| | JAN | FEB | MAR | APR | MAY | JUN | JUL | AUG | SEP | OCT | NOV | DEC | YEAR |
|---|---|---|---|---|---|---|---|---|---|---|---|---|---|
| Maximum Temp °F | 37.7 | 42.9 | 51.7 | 62.9 | 72.4 | 82.9 | 89.3 | 87.7 | 78.7 | 66.8 | 50.2 | 40.7 | 63.7 |
| Minimum Temp °F | 10.7 | 15.0 | 23.2 | 33.7 | 43.9 | 53.9 | 59.6 | 57.3 | 46.5 | 33.2 | 21.9 | 12.9 | 34.3 |
| Mean Temp °F | 24.2 | 29.0 | 37.4 | 48.3 | 58.2 | 68.4 | 74.4 | 72.5 | 62.6 | 50.0 | 36.1 | 26.8 | 49.0 |
| Days Max Temp ≥ 90 °F | 0 | 0 | 0 | 0 | 1 | 8 | 16 | 14 | 6 | 1 | 0 | 0 | 46 |
| Days Max Temp ≤ 32 °F | 10 | 7 | 4 | 0 | 0 | 0 | 0 | 0 | 0 | 0 | 3 | 8 | 32 |
| Days Min Temp ≤ 32 °F | 31 | 28 | 27 | 13 | 2 | 0 | 0 | 0 | 2 | 15 | 27 | 31 | 176 |
| Days Min Temp ≤ 0 °F | 6 | 4 | 1 | 0 | 0 | 0 | 0 | 0 | 0 | 0 | 0 | 4 | 15 |
| Heating Degree Days | 1259 | 1010 | 848 | 497 | 233 | 43 | 7 | 12 | 142 | 460 | 860 | 1177 | 6548 |
| Cooling Degree Days | 0 | 0 | 0 | 4 | 24 | 150 | 295 | 252 | 83 | 2 | 0 | 0 | 810 |
| Total Precipitation (") | 0.47 | 0.46 | 1.20 | 1.92 | 3.12 | 3.26 | 3.20 | 2.30 | 1.36 | 1.15 | 0.69 | 0.46 | 19.59 |
| Days ≥ 0.1" Precip | 1 | 1 | 3 | 4 | 6 | 6 | 6 | 4 | 3 | 2 | 2 | 1 | 39 |
| Total Snowfall (") | 6.7 | 3.9 | 6.1 | 1.8 | 0.1 | 0.0 | 0.0 | 0.0 | 0.0 | 1.1 | 4.1 | 5.4 | 29.2 |
| Days ≥ 1" Snow Depth | 14 | 9 | 4 | 1 | 0 | 0 | 0 | 0 | 0 | 1 | 4 | 11 | 44 |

## EWING *Holt County* ELEVATION 1850 ft LAT/LONG 42° 15 ' N / 98° 21 ' W

| | JAN | FEB | MAR | APR | MAY | JUN | JUL | AUG | SEP | OCT | NOV | DEC | YEAR |
|---|---|---|---|---|---|---|---|---|---|---|---|---|---|
| Maximum Temp °F | 32.0 | 37.5 | 49.2 | 63.2 | 73.2 | 82.5 | 87.0 | 85.2 | 76.8 | 65.1 | 46.8 | 35.4 | 61.2 |
| Minimum Temp °F | 8.7 | 14.0 | 24.7 | 36.5 | 47.7 | 57.3 | 62.5 | 60.2 | 49.8 | 37.5 | 24.4 | 13.2 | 36.4 |
| Mean Temp °F | 20.4 | 25.8 | 37.0 | 49.9 | 60.5 | 69.9 | 74.7 | 72.7 | 63.4 | 51.3 | 35.6 | 24.3 | 48.8 |
| Days Max Temp ≥ 90 °F | 0 | 0 | 0 | 1 | 1 | 5 | 12 | 9 | 3 | 0 | 0 | 0 | 31 |
| Days Max Temp ≤ 32 °F | 15 | 11 | 4 | 0 | 0 | 0 | 0 | 0 | 0 | 0 | 4 | 12 | 46 |
| Days Min Temp ≤ 32 °F | 31 | 27 | 24 | 11 | 2 | 0 | 0 | 0 | 1 | 10 | 24 | 30 | 160 |
| Days Min Temp ≤ 0 °F | 9 | 5 | 1 | 0 | 0 | 0 | 0 | 0 | 0 | 0 | 1 | 5 | 21 |
| Heating Degree Days | 1376 | 1103 | 863 | 457 | 181 | 29 | 6 | 12 | 129 | 423 | 876 | 1256 | 6711 |
| Cooling Degree Days | 0 | 0 | 0 | 16 | 51 | 193 | 309 | 261 | 93 | 6 | 0 | 0 | 929 |
| Total Precipitation (") | 0.54 | 0.59 | 1.54 | 2.35 | 3.32 | 3.98 | 3.21 | 3.16 | 2.20 | 1.70 | 1.06 | 0.63 | 24.28 |
| Days ≥ 0.1" Precip | 2 | 2 | 3 | 5 | 7 | 6 | 6 | 6 | 5 | 4 | 3 | 2 | 51 |
| Total Snowfall (") | 5.5 | 5.6 | 5.0 | 2.5 | 0.0 | 0.0 | 0.0 | 0.0 | 0.0 | 0.7 | 4.5 | 6.4 | 30.2 |
| Days ≥ 1" Snow Depth | 18 | 12 | 6 | 1 | 0 | 0 | 0 | 0 | 0 | 0 | 4 | 13 | 54 |

## FAIRBURY *Jefferson County* ELEVATION 1312 ft LAT/LONG 40° 8 ' N / 97° 11 ' W

| | JAN | FEB | MAR | APR | MAY | JUN | JUL | AUG | SEP | OCT | NOV | DEC | YEAR |
|---|---|---|---|---|---|---|---|---|---|---|---|---|---|
| Maximum Temp °F | 33.8 | 39.4 | 51.8 | 64.5 | 73.7 | 83.4 | 88.9 | 86.5 | 78.2 | 66.8 | 50.3 | 38.3 | 63.0 |
| Minimum Temp °F | 12.0 | 15.7 | 26.8 | 38.2 | 49.1 | 59.2 | 64.7 | 62.5 | 52.6 | 39.8 | 27.2 | 17.3 | 38.8 |
| Mean Temp °F | 22.9 | 27.6 | 39.3 | 51.4 | 61.4 | 71.3 | 76.8 | 74.5 | 65.4 | 53.4 | 38.8 | 27.8 | 50.9 |
| Days Max Temp ≥ 90 °F | 0 | 0 | 0 | 0 | 1 | 7 | 14 | 11 | 5 | 0 | 0 | 0 | 38 |
| Days Max Temp ≤ 32 °F | 14 | 10 | 3 | 0 | 0 | 0 | 0 | 0 | 0 | 0 | 3 | 10 | 40 |
| Days Min Temp ≤ 32 °F | 30 | 27 | 22 | 9 | 1 | 0 | 0 | 0 | 0 | 7 | 22 | 30 | 148 |
| Days Min Temp ≤ 0 °F | 7 | 4 | 0 | 0 | 0 | 0 | 0 | 0 | 0 | 0 | 0 | 3 | 14 |
| Heating Degree Days | 1298 | 1050 | 790 | 417 | 160 | 22 | 2 | 6 | 104 | 368 | 779 | 1146 | 6142 |
| Cooling Degree Days | 0 | 0 | 1 | 20 | 53 | 225 | 372 | 305 | 135 | 9 | 0 | 0 | 1120 |
| Total Precipitation (") | 0.71 | 0.82 | 2.16 | 2.63 | 3.91 | 4.18 | 4.36 | 4.09 | 3.35 | 2.15 | 1.42 | 0.99 | 30.77 |
| Days ≥ 0.1" Precip | 2 | 2 | 4 | 6 | 8 | 6 | 6 | 6 | 5 | 4 | 3 | 2 | 54 |
| Total Snowfall (") | 6.3 | 7.1 | 4.4 | 0.6 | 0.0 | 0.0 | 0.0 | 0.0 | 0.0 | 0.1 | 2.8 | 5.2 | 26.5 |
| Days ≥ 1" Snow Depth | 15 | 11 | 4 | 0 | 0 | 0 | 0 | 0 | 0 | 0 | 3 | 8 | 41 |

## FAIRMONT *Fillmore County* ELEVATION 1640 ft LAT/LONG 40° 38 ' N / 97° 35 ' W

| | JAN | FEB | MAR | APR | MAY | JUN | JUL | AUG | SEP | OCT | NOV | DEC | YEAR |
|---|---|---|---|---|---|---|---|---|---|---|---|---|---|
| Maximum Temp °F | 34.2 | 40.0 | 52.1 | 65.1 | 74.7 | 84.5 | 88.5 | 86.4 | 78.6 | 67.1 | 50.0 | 38.3 | 63.3 |
| Minimum Temp °F | 12.4 | 17.1 | 27.7 | 38.7 | 50.0 | 59.7 | 64.4 | 62.1 | 52.8 | 40.7 | 27.5 | 17.2 | 39.2 |
| Mean Temp °F | 23.4 | 28.6 | 39.9 | 51.9 | 62.4 | 72.1 | 76.5 | 74.3 | 65.7 | 53.9 | 38.8 | 27.8 | 51.3 |
| Days Max Temp ≥ 90 °F | 0 | 0 | 0 | 0 | 2 | 8 | 14 | 12 | 4 | 0 | 0 | 0 | 40 |
| Days Max Temp ≤ 32 °F | 13 | 9 | 3 | 0 | 0 | 0 | 0 | 0 | 0 | 0 | 3 | 10 | 38 |
| Days Min Temp ≤ 32 °F | 31 | 26 | 21 | 7 | 0 | 0 | 0 | 0 | 0 | 6 | 22 | 30 | 143 |
| Days Min Temp ≤ 0 °F | 6 | 3 | 0 | 0 | 0 | 0 | 0 | 0 | 0 | 0 | 0 | 3 | 12 |
| Heating Degree Days | 1284 | 1022 | 772 | 398 | 138 | 16 | 2 | 6 | 92 | 348 | 780 | 1148 | 6006 |
| Cooling Degree Days | 0 | 0 | 0 | 17 | 62 | 235 | 358 | 291 | 127 | 7 | 0 | 0 | 1097 |
| Total Precipitation (") | 0.63 | 0.71 | 2.19 | 2.75 | 4.15 | 3.80 | 3.30 | 3.07 | 3.11 | 2.14 | 1.29 | 0.86 | 28.00 |
| Days ≥ 0.1" Precip | 2 | 2 | 4 | 5 | 7 | 6 | 6 | 5 | 5 | 3 | 3 | 2 | 50 |
| Total Snowfall (") | 6.8 | 6.2 | 6.0 | 1.3 | 0.2 | 0.0 | 0.0 | 0.0 | 0.1 | 0.4 | 3.6 | 6.5 | 31.1 |
| Days ≥ 1" Snow Depth | na | *12* | 5 | 1 | 0 | 0 | 0 | 0 | 0 | 0 | 3 | *9* | na |

## FALLS CITY 2 NE *Richardson County* ELEVATION 980 ft LAT/LONG 40° 5 ' N / 95° 35 ' W

| | JAN | FEB | MAR | APR | MAY | JUN | JUL | AUG | SEP | OCT | NOV | DEC | YEAR |
|---|---|---|---|---|---|---|---|---|---|---|---|---|---|
| Maximum Temp °F | 35.3 | 40.9 | 53.8 | 66.4 | 76.1 | 85.2 | 89.5 | 87.3 | 79.3 | 68.0 | 51.7 | 39.6 | 64.4 |
| Minimum Temp °F | 15.0 | 19.0 | 29.9 | 41.2 | 51.8 | 61.2 | 65.3 | 62.6 | 54.1 | 42.7 | 30.4 | 19.9 | 41.1 |
| Mean Temp °F | 25.1 | 30.0 | 41.9 | 53.8 | 64.0 | 73.2 | 77.4 | 75.0 | 66.7 | 55.3 | 41.1 | 29.8 | 52.8 |
| Days Max Temp ≥ 90 °F | 0 | 0 | 0 | 0 | 1 | 9 | 16 | 12 | 5 | 0 | 0 | 0 | 43 |
| Days Max Temp ≤ 32 °F | 13 | 8 | 2 | 0 | 0 | 0 | 0 | 0 | 0 | 0 | 2 | 9 | 34 |
| Days Min Temp ≤ 32 °F | 29 | 25 | 19 | 6 | 0 | 0 | 0 | 0 | 0 | 5 | 18 | 28 | 130 |
| Days Min Temp ≤ 0 °F | 5 | 3 | 0 | 0 | 0 | 0 | 0 | 0 | 0 | 0 | 0 | 2 | 10 |
| Heating Degree Days | 1228 | 981 | 713 | 352 | 111 | 11 | 2 | 5 | 82 | 315 | 711 | 1086 | 5597 |
| Cooling Degree Days | 0 | 0 | 2 | 25 | 75 | 254 | 386 | 309 | 145 | 15 | 0 | 0 | 1211 |
| Total Precipitation (") | 0.89 | 0.83 | 2.29 | 3.01 | 4.11 | 3.93 | 5.46 | 4.15 | 4.42 | 2.79 | 1.96 | 1.12 | 34.96 |
| Days ≥ 0.1" Precip | 3 | 3 | 4 | 6 | 7 | 6 | 7 | 6 | 6 | 4 | 4 | 3 | 59 |
| Total Snowfall (") | 6.1 | 5.0 | 3.5 | 0.8 | 0.0 | 0.0 | 0.0 | 0.0 | 0.0 | 0.0 | 1.2 | 3.6 | 20.2 |
| Days ≥ 1" Snow Depth | 14 | 11 | 3 | 0 | 0 | 0 | 0 | 0 | 0 | 0 | 2 | 6 | 36 |

## FREMONT *Dodge County* ELEVATION 1201 ft LAT/LONG 41° 26 ' N / 96° 29 ' W

| | JAN | FEB | MAR | APR | MAY | JUN | JUL | AUG | SEP | OCT | NOV | DEC | YEAR |
|---|---|---|---|---|---|---|---|---|---|---|---|---|---|
| Maximum Temp °F | 32.2 | 38.0 | 50.9 | 65.1 | 75.5 | 84.7 | 88.1 | 85.8 | 77.9 | 65.9 | 48.7 | 36.1 | 62.4 |
| Minimum Temp °F | 12.0 | 17.0 | 28.1 | 39.7 | 50.6 | 60.1 | 64.7 | 62.1 | 52.9 | 40.9 | 28.5 | 17.4 | 39.5 |
| Mean Temp °F | 22.1 | 27.5 | 39.5 | 52.5 | 63.1 | 72.5 | 76.4 | 73.9 | 65.4 | 53.4 | 38.6 | 26.8 | 51.0 |
| Days Max Temp ≥ 90 °F | 0 | 0 | 0 | 1 | 2 | 8 | 13 | 10 | 3 | 0 | 0 | 0 | 37 |
| Days Max Temp ≤ 32 °F | 15 | 10 | 3 | 0 | 0 | 0 | 0 | 0 | 0 | 0 | 3 | 11 | 42 |
| Days Min Temp ≤ 32 °F | 30 | 26 | 21 | 7 | 1 | 0 | 0 | 0 | 0 | 6 | 20 | 29 | 140 |
| Days Min Temp ≤ 0 °F | 7 | 3 | 0 | 0 | 0 | 0 | 0 | 0 | 0 | 0 | 0 | 3 | 13 |
| Heating Degree Days | 1323 | 1051 | 785 | 386 | 130 | 13 | 2 | 6 | 93 | 364 | 785 | 1179 | 6117 |
| Cooling Degree Days | 0 | 0 | 1 | 22 | 77 | 255 | 366 | 299 | 131 | 9 | 0 | 0 | 1160 |
| Total Precipitation (") | 0.74 | 0.74 | 2.20 | 2.69 | 4.32 | 4.66 | 3.23 | 3.30 | 3.57 | 2.43 | 1.51 | 1.04 | 30.43 |
| Days ≥ 0.1" Precip | 2 | 2 | 5 | 6 | 8 | 7 | 6 | 5 | 6 | 4 | 3 | 3 | 57 |
| Total Snowfall (") | 6.5 | 6.6 | 5.7 | 1.6 | 0.0 | 0.0 | 0.0 | 0.0 | 0.0 | 0.7 | 4.2 | 7.0 | 32.3 |
| Days ≥ 1" Snow Depth | 18 | 13 | 6 | 1 | 0 | 0 | 0 | 0 | 0 | 0 | 4 | 12 | 54 |

## GAVINS POINT DAM *Cedar County* ELEVATION 1240 ft LAT/LONG 42° 51 ' N / 97° 29 ' W

| | JAN | FEB | MAR | APR | MAY | JUN | JUL | AUG | SEP | OCT | NOV | DEC | YEAR |
|---|---|---|---|---|---|---|---|---|---|---|---|---|---|
| Maximum Temp °F | 27.3 | 32.2 | 44.5 | 59.6 | 71.3 | 80.8 | 86.3 | 84.3 | 74.8 | 62.1 | 44.3 | 31.6 | 58.3 |
| Minimum Temp °F | 6.9 | 12.3 | 24.1 | 37.0 | 48.5 | 58.1 | 63.5 | 61.5 | 51.1 | 38.5 | 25.6 | 12.8 | 36.7 |
| Mean Temp °F | 17.1 | 22.3 | 34.3 | 48.3 | 59.9 | 69.5 | 74.9 | 72.9 | 62.9 | 50.3 | 35.0 | 22.2 | 47.5 |
| Days Max Temp ≥ 90 °F | 0 | 0 | 0 | 0 | 1 | 5 | 11 | 9 | 3 | 0 | 0 | 0 | 29 |
| Days Max Temp ≤ 32 °F | 18 | 14 | 6 | 0 | 0 | 0 | 0 | 0 | 0 | 0 | 5 | 16 | 59 |
| Days Min Temp ≤ 32 °F | 31 | 27 | 25 | 9 | 1 | 0 | 0 | 0 | 1 | 7 | 23 | 30 | 154 |
| Days Min Temp ≤ 0 °F | 11 | 7 | 1 | 0 | 0 | 0 | 0 | 0 | 0 | 0 | 1 | 5 | 25 |
| Heating Degree Days | 1479 | 1201 | 945 | 503 | 200 | 35 | 6 | 11 | 140 | 453 | 894 | 1320 | 7187 |
| Cooling Degree Days | 0 | 0 | 0 | 14 | 47 | 172 | 293 | 248 | 84 | 4 | 0 | 0 | 862 |
| Total Precipitation (") | 0.46 | 0.58 | 1.66 | 2.30 | 3.63 | 3.89 | 2.99 | 2.87 | 2.42 | 1.81 | 1.10 | 0.73 | 24.44 |
| Days ≥ 0.1" Precip | 1 | 2 | 4 | 5 | 6 | 6 | 5 | 5 | 6 | 3 | 2 | 2 | 47 |
| Total Snowfall (") | na | na | na | *0.9* | 0.0 | 0.0 | 0.0 | 0.0 | 0.0 | 0.1 | na | na | na |
| Days ≥ 1" Snow Depth | na | na | na | *0* | 0 | 0 | 0 | 0 | 0 | 0 | na | na | na |

## GENEVA *Fillmore County* ELEVATION 1631 ft LAT/LONG 40° 32 ' N / 97° 36 ' W

| | JAN | FEB | MAR | APR | MAY | JUN | JUL | AUG | SEP | OCT | NOV | DEC | YEAR |
|---|---|---|---|---|---|---|---|---|---|---|---|---|---|
| Maximum Temp °F | 34.2 | 40.0 | 52.1 | 65.0 | 74.3 | 83.9 | 88.0 | 85.7 | 77.7 | 66.2 | 49.7 | 37.9 | 62.9 |
| Minimum Temp °F | 13.6 | 18.2 | 28.6 | 40.0 | 51.0 | 60.6 | 65.2 | 63.1 | 53.7 | 41.9 | 28.6 | 18.5 | 40.3 |
| Mean Temp °F | 23.9 | 29.1 | 40.3 | 52.6 | 62.7 | 72.2 | 76.6 | 74.4 | 65.8 | 54.1 | 39.2 | 28.2 | 51.6 |
| Days Max Temp ≥ 90 °F | 0 | 0 | 0 | 0 | 1 | 7 | 13 | 10 | 4 | 0 | 0 | 0 | 35 |
| Days Max Temp ≤ 32 °F | 13 | 9 | 2 | 0 | 0 | 0 | 0 | 0 | 0 | 0 | 3 | 10 | 37 |
| Days Min Temp ≤ 32 °F | 30 | 26 | 20 | 6 | 1 | 0 | 0 | 0 | 0 | 4 | 20 | 29 | 136 |
| Days Min Temp ≤ 0 °F | 6 | 3 | 0 | 0 | 0 | 0 | 0 | 0 | 0 | 0 | 0 | 2 | 11 |
| Heating Degree Days | 1267 | 1007 | 758 | 381 | 132 | 15 | 2 | 5 | 92 | 345 | 769 | 1134 | 5907 |
| Cooling Degree Days | 0 | 0 | 1 | 22 | 67 | 240 | 363 | 304 | 134 | 10 | 0 | 0 | 1141 |
| Total Precipitation (") | 0.62 | 0.71 | 2.18 | 3.02 | 4.40 | 4.40 | 3.52 | 3.35 | 3.40 | 2.22 | 1.30 | 0.76 | 29.88 |
| Days ≥ 0.1" Precip | 2 | *2* | 4 | 6 | 7 | 6 | 6 | 5 | 5 | 4 | 2 | 2 | 51 |
| Total Snowfall (") | 5.9 | 6.4 | 5.1 | 1.1 | 0.1 | 0.0 | 0.0 | 0.0 | 0.1 | 0.4 | 2.4 | 5.0 | 26.5 |
| Days ≥ 1" Snow Depth | na | na | *3* | 1 | 0 | 0 | 0 | 0 | 0 | 0 | *2* | *7* | na |

## GENOA 2 W *Nance County* ELEVATION 1581 ft LAT/LONG 41° 27 ' N / 97° 43 ' W

| | JAN | FEB | MAR | APR | MAY | JUN | JUL | AUG | SEP | OCT | NOV | DEC | YEAR |
|---|---|---|---|---|---|---|---|---|---|---|---|---|---|
| Maximum Temp °F | 32.6 | 38.2 | 50.4 | 64.5 | 74.1 | 83.5 | 87.2 | 85.2 | 77.2 | 65.6 | 48.0 | 36.0 | 61.9 |
| Minimum Temp °F | 10.2 | 15.1 | 25.9 | 37.3 | 48.4 | 58.1 | 62.7 | 60.4 | 50.7 | 38.0 | 25.3 | 14.9 | 37.2 |
| Mean Temp °F | 21.4 | 26.7 | 38.2 | 51.0 | 61.3 | 70.9 | 75.0 | 72.8 | 64.0 | 51.8 | 36.7 | 25.5 | 49.6 |
| Days Max Temp ≥ 90 °F | 0 | 0 | 0 | 1 | 2 | 7 | 12 | 9 | 3 | 0 | 0 | 0 | 34 |
| Days Max Temp ≤ 32 °F | 15 | 10 | 3 | 0 | 0 | 0 | 0 | 0 | 0 | 0 | 3 | 11 | 42 |
| Days Min Temp ≤ 32 °F | 31 | 27 | 24 | 10 | 2 | 0 | 0 | 0 | 1 | 9 | 24 | 30 | 158 |
| Days Min Temp ≤ 0 °F | 8 | 4 | 0 | 0 | 0 | 0 | 0 | 0 | 0 | 0 | 1 | 4 | 17 |
| Heating Degree Days | 1345 | 1076 | 827 | 426 | 164 | 21 | 3 | 10 | 117 | 408 | 843 | 1220 | 6460 |
| Cooling Degree Days | 0 | 0 | 1 | 18 | 57 | 211 | 307 | 258 | 103 | 5 | 0 | 0 | 960 |
| Total Precipitation (") | 0.58 | 0.88 | 2.06 | 2.37 | 4.16 | 4.65 | 3.63 | 3.22 | 2.76 | 1.95 | 1.39 | 0.92 | 28.57 |
| Days ≥ 0.1" Precip | 2 | 2 | 4 | 5 | 7 | 6 | 6 | 5 | 5 | 4 | 3 | 3 | 52 |
| Total Snowfall (") | 5.3 | 6.5 | 4.8 | 1.5 | 0.1 | 0.0 | 0.0 | 0.0 | 0.1 | 0.9 | 5.2 | 7.5 | 31.9 |
| Days ≥ 1" Snow Depth | 18 | 13 | 7 | 1 | 0 | 0 | 0 | 0 | 0 | 0 | 4 | 14 | 57 |

## GOTHENBURG *Dawson County* ELEVATION 2572 ft LAT/LONG 40° 56 ' N / 100° 10 ' W

| | JAN | FEB | MAR | APR | MAY | JUN | JUL | AUG | SEP | OCT | NOV | DEC | YEAR |
|---|---|---|---|---|---|---|---|---|---|---|---|---|---|
| Maximum Temp °F | 36.1 | 42.1 | 52.3 | 64.3 | 73.5 | 84.0 | 88.5 | 86.6 | 77.7 | 65.9 | 49.0 | 39.2 | 63.3 |
| Minimum Temp °F | 12.0 | 16.7 | 25.6 | 36.5 | 47.4 | 57.2 | 62.2 | 59.9 | 49.5 | 36.9 | 23.9 | 14.9 | 36.9 |
| Mean Temp °F | 24.1 | 29.4 | 39.0 | 50.4 | 60.5 | 70.7 | 75.4 | 73.3 | 63.6 | 51.4 | 36.5 | 27.1 | 50.1 |
| Days Max Temp ≥ 90 °F | 0 | 0 | 0 | 0 | 1 | 8 | 14 | 12 | 4 | 0 | 0 | 0 | 39 |
| Days Max Temp ≤ 32 °F | 11 | 7 | 3 | 0 | 0 | 0 | 0 | 0 | 0 | 0 | 3 | 9 | 33 |
| Days Min Temp ≤ 32 °F | 31 | 27 | 24 | 10 | 2 | 0 | 0 | 0 | 1 | 9 | 26 | 30 | 160 |
| Days Min Temp ≤ 0 °F | 6 | 3 | 0 | 0 | 0 | 0 | 0 | 0 | 0 | 0 | 1 | 3 | 13 |
| Heating Degree Days | 1261 | 998 | 800 | 439 | 178 | 24 | 3 | 8 | 121 | 417 | 849 | 1169 | 6267 |
| Cooling Degree Days | 0 | 0 | 0 | 10 | 42 | 202 | 315 | 267 | 86 | 1 | 0 | 0 | 923 |
| Total Precipitation (") | 0.46 | 0.51 | 1.34 | 2.13 | 3.45 | 4.04 | 3.48 | 2.54 | 1.55 | 1.32 | 0.74 | 0.53 | 22.09 |
| Days ≥ 0.1" Precip | 2 | 2 | 3 | 4 | 7 | 6 | 6 | 4 | 4 | 3 | 2 | 2 | 45 |
| Total Snowfall (") | na | na | *4.3* | 1.1 | 0.1 | 0.0 | 0.0 | 0.0 | 0.0 | 0.7 | *2.1* | na | na |
| Days ≥ 1" Snow Depth | na | *8* | *3* | 0 | 0 | 0 | 0 | 0 | 0 | 0 | *2* | na | na |

## GRAND ISLAND ARPT *Hall County* ELEVATION 1860 ft LAT/LONG 40° 58 ' N / 98° 19 ' W

| | JAN | FEB | MAR | APR | MAY | JUN | JUL | AUG | SEP | OCT | NOV | DEC | YEAR |
|---|---|---|---|---|---|---|---|---|---|---|---|---|---|
| Maximum Temp °F | 32.6 | 38.0 | 50.1 | 63.4 | 73.2 | 83.8 | 88.1 | 85.9 | 77.2 | 65.2 | 48.3 | 36.6 | 61.9 |
| Minimum Temp °F | 11.7 | 16.5 | 26.8 | 38.2 | 49.4 | 59.2 | 64.4 | 61.9 | 51.7 | 39.0 | 26.1 | 16.0 | 38.4 |
| Mean Temp °F | 22.2 | 27.3 | 38.4 | 50.8 | 61.3 | 71.5 | 76.3 | 73.9 | 64.5 | 52.1 | 37.2 | 26.3 | 50.2 |
| Days Max Temp ≥ 90 °F | 0 | 0 | 0 | 1 | 1 | 8 | 13 | 11 | 4 | 0 | 0 | 0 | 38 |
| Days Max Temp ≤ 32 °F | 14 | 11 | 4 | 0 | 0 | 0 | 0 | 0 | 0 | 0 | 3 | 12 | 44 |
| Days Min Temp ≤ 32 °F | 31 | 27 | 22 | 8 | 1 | 0 | 0 | 0 | 0 | 7 | 23 | 30 | 149 |
| Days Min Temp ≤ 0 °F | 7 | 4 | 0 | 0 | 0 | 0 | 0 | 0 | 0 | 0 | 0 | 3 | 14 |
| Heating Degree Days | 1321 | 1057 | 818 | 431 | 165 | 21 | 2 | 8 | 113 | 400 | 826 | 1193 | 6355 |
| Cooling Degree Days | 0 | 0 | 1 | 17 | 57 | 227 | 349 | 288 | 116 | 7 | 0 | 0 | 1062 |
| Total Precipitation (") | 0.55 | 0.75 | 1.88 | 2.49 | 3.77 | 4.07 | 3.08 | 2.94 | 2.71 | 1.51 | 1.18 | 0.77 | 25.70 |
| Days ≥ 0.1" Precip | 2 | 2 | 4 | 5 | 7 | 6 | 6 | 5 | 5 | 3 | 3 | 2 | 50 |
| Total Snowfall (") | 5.9 | 6.8 | 6.1 | 1.5 | 0.1 | 0.0 | 0.0 | 0.0 | 0.1 | 0.8 | 4.2 | 7.2 | 32.7 |
| Days ≥ 1" Snow Depth | 16 | 12 | 5 | 1 | 0 | 0 | 0 | 0 | 0 | 0 | 4 | 10 | 48 |

## GREELEY *Greeley County* ELEVATION 2021 ft LAT/LONG 41° 33 ' N / 98° 32 ' W

| | JAN | FEB | MAR | APR | MAY | JUN | JUL | AUG | SEP | OCT | NOV | DEC | YEAR |
|---|---|---|---|---|---|---|---|---|---|---|---|---|---|
| Maximum Temp °F | 33.1 | 39.0 | 50.4 | 64.0 | 73.6 | 83.3 | 88.0 | 86.4 | 78.1 | 66.5 | 48.2 | 37.0 | 62.3 |
| Minimum Temp °F | 8.7 | 14.4 | 24.3 | 35.4 | 46.2 | 56.0 | 61.1 | 58.6 | 48.6 | 36.0 | 23.5 | 13.3 | 35.5 |
| Mean Temp °F | 20.9 | 26.7 | 37.4 | 49.7 | 59.9 | 69.7 | 74.5 | 72.5 | 63.4 | 51.3 | 35.8 | 25.2 | 48.9 |
| Days Max Temp ≥ 90 °F | 0 | 0 | 0 | 1 | 1 | 7 | 13 | 11 | 4 | 0 | 0 | 0 | 37 |
| Days Max Temp ≤ 32 °F | 14 | 10 | 3 | 0 | 0 | 0 | 0 | 0 | 0 | 0 | 3 | 11 | 41 |
| Days Min Temp ≤ 32 °F | 31 | 27 | 26 | 11 | 2 | 0 | 0 | 0 | 1 | 11 | 25 | 30 | 164 |
| Days Min Temp ≤ 0 °F | 9 | 5 | 1 | 0 | 0 | 0 | 0 | 0 | 0 | 0 | 1 | 4 | 20 |
| Heating Degree Days | 1360 | 1074 | 850 | 459 | 190 | 32 | 5 | 11 | 128 | 424 | 869 | 1227 | 6629 |
| Cooling Degree Days | 0 | 0 | 0 | 11 | 33 | 177 | 289 | 241 | 91 | 2 | 0 | 0 | 844 |
| Total Precipitation (") | 0.45 | 0.61 | 1.81 | 2.52 | 3.59 | 3.99 | 3.82 | 3.06 | 2.65 | 1.69 | 1.26 | 0.66 | 26.11 |
| Days ≥ 0.1" Precip | 1 | 2 | 3 | 5 | 7 | 6 | 6 | 5 | 5 | 4 | 2 | 2 | 48 |
| Total Snowfall (") | na | na | na | 0.1 | 0.0 | 0.0 | 0.0 | 0.0 | 0.0 | 0.3 | na | na | na |
| Days ≥ 1" Snow Depth | na | na | na | 0 | 0 | 0 | 0 | 0 | 0 | 0 | na | na | na |

## HALSEY 2 W *Thomas County* ELEVATION 2705 ft LAT/LONG 41° 54 ' N / 100° 19 ' W

| | JAN | FEB | MAR | APR | MAY | JUN | JUL | AUG | SEP | OCT | NOV | DEC | YEAR |
|---|---|---|---|---|---|---|---|---|---|---|---|---|---|
| Maximum Temp °F | 34.9 | 41.3 | 50.7 | 63.5 | 73.6 | 83.4 | 89.7 | 87.7 | 78.1 | 66.1 | 48.1 | 37.9 | 62.9 |
| Minimum Temp °F | 9.3 | 14.2 | 22.8 | 33.6 | 44.1 | 53.2 | 59.3 | 56.8 | 46.3 | 33.7 | 22.0 | 12.6 | 34.0 |
| Mean Temp °F | 22.1 | 27.8 | 36.8 | 48.6 | 58.9 | 68.3 | 74.5 | 72.2 | 62.2 | 50.0 | 35.1 | 25.3 | 48.5 |
| Days Max Temp ≥ 90 °F | 0 | 0 | 0 | 0 | 1 | 7 | 16 | 13 | 5 | 0 | 0 | 0 | 42 |
| Days Max Temp ≤ 32 °F | 12 | 8 | 4 | 0 | 0 | 0 | 0 | 0 | 0 | 0 | 4 | 10 | 38 |
| Days Min Temp ≤ 32 °F | 30 | 27 | 26 | 14 | 3 | 0 | 0 | 0 | 2 | 15 | 26 | 30 | 173 |
| Days Min Temp ≤ 0 °F | 9 | 4 | 1 | 0 | 0 | 0 | 0 | 0 | 0 | 0 | 1 | 5 | 20 |
| Heating Degree Days | 1325 | 1045 | 868 | 491 | 216 | 42 | 5 | 12 | 146 | 462 | 891 | 1227 | 6730 |
| Cooling Degree Days | na | na | na | na | na | na | na | na | na | na | na | na | na |
| Total Precipitation (") | 0.54 | 0.71 | 1.34 | 2.60 | 3.39 | 3.59 | 2.96 | 2.62 | 1.92 | 1.21 | 0.98 | 0.62 | 22.48 |
| Days ≥ 0.1" Precip | 2 | 2 | na | 5 | 7 | 6 | 5 | 4 | 4 | 3 | 2 | na | na |
| Total Snowfall (") | na | na | na | na | 0.2 | 0.0 | 0.0 | 0.0 | 0.2 | 0.9 | na | na | na |
| Days ≥ 1" Snow Depth | na | na | na | 1 | 0 | 0 | 0 | 0 | 0 | 0 | na | na | na |

## HARLAN COUNTY LAKE *Harlan County* ELEVATION 2001 ft LAT/LONG 40° 5 ' N / 99° 12 ' W

| | JAN | FEB | MAR | APR | MAY | JUN | JUL | AUG | SEP | OCT | NOV | DEC | YEAR |
|---|---|---|---|---|---|---|---|---|---|---|---|---|---|
| Maximum Temp °F | 35.7 | 40.3 | 51.9 | 64.0 | 72.8 | 83.2 | 89.2 | 87.0 | 78.3 | 66.9 | 50.1 | 39.4 | 63.2 |
| Minimum Temp °F | 10.5 | 14.8 | 25.4 | 37.0 | 47.5 | 57.7 | 63.1 | 60.5 | 50.3 | 37.7 | 24.7 | 15.2 | 37.0 |
| Mean Temp °F | 23.1 | 27.6 | 38.6 | 50.5 | 60.2 | 70.5 | 76.2 | 73.8 | 64.3 | 52.3 | 37.4 | 27.4 | 50.2 |
| Days Max Temp ≥ 90 °F | 0 | 0 | 0 | 0 | 1 | 7 | 15 | 12 | 5 | 0 | 0 | 0 | 40 |
| Days Max Temp ≤ 32 °F | 12 | 9 | 3 | 0 | 0 | 0 | 0 | 0 | 0 | 0 | 2 | 9 | 35 |
| Days Min Temp ≤ 32 °F | 31 | 27 | 24 | 9 | 1 | 0 | 0 | 0 | 1 | 8 | 25 | 31 | 157 |
| Days Min Temp ≤ 0 °F | 7 | 4 | 0 | 0 | 0 | 0 | 0 | 0 | 0 | 0 | 0 | 3 | 14 |
| Heating Degree Days | 1292 | 1050 | 811 | 436 | 186 | 29 | 3 | 8 | 116 | 394 | 821 | 1160 | 6306 |
| Cooling Degree Days | 0 | 0 | 1 | 13 | 44 | 218 | 366 | 303 | 118 | 7 | 0 | 0 | 1070 |
| Total Precipitation (") | 0.38 | 0.50 | 1.62 | 1.99 | 3.77 | 3.45 | 3.81 | 3.08 | 2.34 | 1.59 | 0.80 | 0.44 | 23.77 |
| Days ≥ 0.1" Precip | 1 | 2 | 3 | 4 | 7 | 6 | 6 | 5 | 4 | 3 | 2 | 1 | 44 |
| Total Snowfall (") | 5.1 | 4.5 | 3.3 | 0.4 | 0.1 | 0.0 | 0.0 | 0.0 | 0.2 | 0.4 | 2.0 | 4.1 | 20.1 |
| Days ≥ 1" Snow Depth | 11 | 8 | 3 | 0 | 0 | 0 | 0 | 0 | 0 | 0 | 2 | 6 | 30 |

## HARRISBURG 10 NW *Banner County* ELEVATION 4462 ft LAT/LONG 41° 39 ' N / 103° 53 ' W

| | JAN | FEB | MAR | APR | MAY | JUN | JUL | AUG | SEP | OCT | NOV | DEC | YEAR |
|---|---|---|---|---|---|---|---|---|---|---|---|---|---|
| Maximum Temp °F | 38.4 | 42.8 | 49.3 | 59.2 | 68.9 | 79.0 | 86.2 | 84.7 | 75.1 | 63.2 | 48.7 | 40.4 | 61.3 |
| Minimum Temp °F | 12.0 | 15.8 | 22.3 | 30.3 | 40.0 | 49.3 | 55.0 | 52.5 | 42.4 | 30.6 | 20.2 | 13.0 | 32.0 |
| Mean Temp °F | 25.2 | 29.3 | 35.8 | 44.8 | 54.5 | 64.2 | 70.6 | 68.6 | 58.8 | 46.9 | 34.5 | 26.8 | 46.7 |
| Days Max Temp ≥ 90 °F | 0 | 0 | 0 | 0 | 0 | 4 | 11 | 8 | 2 | 0 | 0 | 0 | 25 |
| Days Max Temp ≤ 32 °F | 9 | 6 | 4 | 1 | 0 | 0 | 0 | 0 | 0 | 1 | 4 | 8 | 33 |
| Days Min Temp ≤ 32 °F | 30 | 27 | 28 | 18 | 5 | 0 | 0 | 0 | 4 | 18 | 27 | 30 | 187 |
| Days Min Temp ≤ 0 °F | 6 | 4 | 1 | 0 | 0 | 0 | 0 | 0 | 0 | 0 | 1 | 5 | 17 |
| Heating Degree Days | 1227 | 1000 | 898 | 601 | 329 | 97 | 16 | 29 | 212 | 553 | 910 | 1180 | 7052 |
| Cooling Degree Days | 0 | 0 | 0 | 0 | 12 | 89 | 197 | 161 | 35 | 1 | 0 | 0 | 495 |
| Total Precipitation (") | 0.39 | 0.40 | 0.98 | 1.55 | 2.52 | 2.47 | 1.97 | 1.34 | 1.11 | 0.89 | 0.58 | 0.36 | 14.56 |
| Days ≥ 0.1" Precip | 1 | 1 | 3 | 4 | 6 | 5 | 5 | 4 | 3 | 3 | 2 | 1 | 38 |
| Total Snowfall (") | 7.7 | 6.3 | 9.5 | 6.0 | 0.8 | 0.0 | 0.0 | 0.0 | 0.3 | 2.8 | 6.8 | 7.8 | 48.0 |
| Days ≥ 1" Snow Depth | 16 | 12 | 9 | 4 | 0 | 0 | 0 | 0 | 0 | 2 | 8 | 15 | 66 |

## HARRISON *Sioux County* ELEVATION 4852 ft LAT/LONG 42° 41 ' N / 103° 53 ' W

| | JAN | FEB | MAR | APR | MAY | JUN | JUL | AUG | SEP | OCT | NOV | DEC | YEAR |
|---|---|---|---|---|---|---|---|---|---|---|---|---|---|
| Maximum Temp °F | 32.4 | 36.7 | 44.4 | 54.6 | 65.1 | 75.9 | 84.3 | 83.1 | 72.6 | 59.7 | 43.7 | 34.8 | 57.3 |
| Minimum Temp °F | 9.0 | 13.4 | 20.4 | 29.3 | 38.9 | 48.6 | 54.8 | 52.8 | 42.3 | 30.9 | 19.4 | 11.2 | 30.9 |
| Mean Temp °F | 20.7 | 25.1 | 32.4 | 42.0 | 52.0 | 62.3 | 69.6 | 68.0 | 57.4 | 45.3 | 31.6 | 23.0 | 44.1 |
| Days Max Temp ≥ 90 °F | 0 | 0 | 0 | 0 | 0 | 3 | 9 | 7 | 2 | 0 | 0 | 0 | 21 |
| Days Max Temp ≤ 32 °F | 14 | 10 | 6 | 1 | 0 | 0 | 0 | 0 | 0 | 1 | 7 | 13 | 52 |
| Days Min Temp ≤ 32 °F | 31 | 28 | 29 | 20 | 7 | 1 | 0 | 0 | 4 | 17 | 27 | 30 | 194 |
| Days Min Temp ≤ 0 °F | 8 | 5 | 2 | 0 | 0 | 0 | 0 | 0 | 0 | 0 | 2 | 6 | 23 |
| Heating Degree Days | 1366 | 1122 | 1003 | 684 | 402 | 134 | 28 | 40 | 248 | 604 | 996 | 1295 | 7922 |
| Cooling Degree Days | 0 | 0 | 0 | 0 | 6 | 67 | 179 | 148 | 29 | 0 | 0 | 0 | 429 |
| Total Precipitation (") | 0.41 | 0.46 | 1.14 | 2.05 | 3.21 | 2.64 | 2.18 | 1.31 | 1.37 | 1.14 | 0.62 | 0.50 | 17.03 |
| Days ≥ 0.1" Precip | 2 | 2 | 3 | 5 | 6 | 6 | 5 | 3 | 3 | 3 | 2 | 2 | 42 |
| Total Snowfall (") | 8.1 | 8.4 | 11.4 | 8.8 | 1.6 | 0.0 | 0.0 | 0.0 | 1.0 | 4.1 | 7.0 | 8.7 | 59.1 |
| Days ≥ 1" Snow Depth | *12* | 9 | 9 | 4 | 1 | 0 | 0 | 0 | 0 | 2 | 6 | *11* | 54 |

## HARTINGTON *Cedar County* ELEVATION 1381 ft LAT/LONG 42° 37 ' N / 97° 16 ' W

| | JAN | FEB | MAR | APR | MAY | JUN | JUL | AUG | SEP | OCT | NOV | DEC | YEAR |
|---|---|---|---|---|---|---|---|---|---|---|---|---|---|
| Maximum Temp °F | 29.9 | 35.8 | 48.5 | 63.4 | 74.6 | 84.1 | 88.4 | 86.3 | 77.6 | 65.1 | 46.1 | 33.6 | 61.1 |
| Minimum Temp °F | 9.1 | 14.5 | 25.5 | 37.4 | 48.7 | 58.5 | 63.4 | 61.3 | 51.6 | 39.7 | 24.7 | 14.2 | 37.4 |
| Mean Temp °F | 19.6 | 25.2 | 37.0 | 50.4 | 61.7 | 71.3 | 75.9 | 73.8 | 64.6 | 52.4 | 35.4 | 23.9 | 49.3 |
| Days Max Temp ≥ 90 °F | 0 | 0 | 0 | 1 | 2 | 8 | 14 | 11 | 4 | 0 | 0 | 0 | 40 |
| Days Max Temp ≤ 32 °F | 16 | 11 | 4 | 0 | 0 | 0 | 0 | 0 | 0 | 0 | 4 | 13 | 48 |
| Days Min Temp ≤ 32 °F | 30 | 27 | 23 | 9 | 1 | 0 | 0 | 0 | 0 | 7 | 23 | 30 | 150 |
| Days Min Temp ≤ 0 °F | 9 | 5 | 1 | 0 | 0 | 0 | 0 | 0 | 0 | 0 | 1 | 5 | 21 |
| Heating Degree Days | 1402 | 1118 | 861 | 443 | 159 | 20 | 3 | 8 | 111 | 394 | 880 | 1267 | 6666 |
| Cooling Degree Days | 0 | 0 | 0 | 17 | 60 | 213 | 323 | 270 | 110 | 9 | 0 | 0 | 1002 |
| Total Precipitation (") | 0.53 | 0.58 | 1.92 | 2.40 | 3.45 | 4.35 | 2.64 | 2.86 | 2.52 | 1.94 | 1.19 | 0.75 | 25.13 |
| Days ≥ 0.1" Precip | 2 | 2 | 4 | 5 | 7 | 6 | 5 | 5 | 5 | 3 | 3 | 2 | 49 |
| Total Snowfall (") | 6.8 | 6.0 | 5.5 | 2.3 | 0.0 | 0.0 | 0.0 | 0.0 | 0.0 | 0.6 | 5.3 | 6.8 | 33.3 |
| Days ≥ 1" Snow Depth | na | na | na | 0 | 0 | 0 | 0 | 0 | 0 | 0 | na | na | na |

## HASTINGS 4 N *Adams County* ELEVATION 1932 ft LAT/LONG 40° 35 ' N / 98° 23 ' W

| | JAN | FEB | MAR | APR | MAY | JUN | JUL | AUG | SEP | OCT | NOV | DEC | YEAR |
|---|---|---|---|---|---|---|---|---|---|---|---|---|---|
| Maximum Temp °F | 34.1 | 39.7 | 51.0 | 64.2 | 74.1 | 84.6 | 89.0 | 86.7 | 78.0 | 65.8 | 48.9 | 37.4 | 62.8 |
| Minimum Temp °F | 12.9 | 17.6 | 27.3 | 38.8 | 50.0 | 59.6 | 64.1 | 61.5 | 52.0 | 40.3 | 27.1 | 17.2 | 39.0 |
| Mean Temp °F | 23.5 | 28.7 | 39.2 | 51.5 | 62.1 | 72.1 | 76.6 | 74.2 | 65.1 | 53.1 | 38.0 | 27.3 | 50.9 |
| Days Max Temp ≥ 90 °F | 0 | 0 | 0 | 1 | 1 | 9 | 15 | 12 | 5 | 0 | 0 | 0 | 43 |
| Days Max Temp ≤ 32 °F | 13 | 9 | 3 | 0 | 0 | 0 | 0 | 0 | 0 | 0 | 3 | 10 | 38 |
| Days Min Temp ≤ 32 °F | 30 | 26 | 22 | 7 | 0 | 0 | 0 | 0 | 0 | 6 | 22 | 30 | 143 |
| Days Min Temp ≤ 0 °F | 6 | 3 | 0 | 0 | 0 | 0 | 0 | 0 | 0 | 0 | 0 | 3 | 12 |
| Heating Degree Days | 1278 | 1019 | 795 | 409 | 144 | 16 | 2 | 7 | 101 | 374 | 803 | 1163 | 6111 |
| Cooling Degree Days | 0 | 0 | 1 | 17 | 61 | 243 | 366 | 300 | 124 | 10 | 0 | 0 | 1122 |
| Total Precipitation (") | 0.60 | 0.75 | 2.01 | 2.67 | 4.59 | 4.02 | 3.77 | 3.24 | 2.99 | 1.78 | 1.15 | 0.89 | 28.46 |
| Days ≥ 0.1" Precip | 2 | 2 | 4 | 5 | 8 | 6 | 6 | 5 | 5 | 4 | 3 | 2 | 52 |
| Total Snowfall (") | *5.1* | *6.1* | 5.5 | 0.9 | 0.2 | 0.0 | 0.0 | 0.0 | 0.2 | 0.4 | *3.4* | 6.2 | 28.0 |
| Days ≥ 1" Snow Depth | *13* | *11* | 4 | 0 | 0 | 0 | 0 | 0 | 0 | 0 | *4* | *9* | 41 |

## HAY SPRINGS *Sheridan County* ELEVATION 3871 ft LAT/LONG 42° 41 ' N / 102° 41 ' W

| | JAN | FEB | MAR | APR | MAY | JUN | JUL | AUG | SEP | OCT | NOV | DEC | YEAR |
|---|---|---|---|---|---|---|---|---|---|---|---|---|---|
| Maximum Temp °F | 34.9 | 41.2 | 49.4 | 59.6 | 70.5 | 79.7 | 87.0 | 85.9 | 76.7 | 63.7 | 46.6 | 37.9 | 61.1 |
| Minimum Temp °F | 9.9 | 15.2 | 22.4 | 31.8 | 41.6 | 50.9 | 57.0 | 55.3 | 44.3 | 32.4 | 21.0 | 12.6 | 32.9 |
| Mean Temp °F | 22.4 | 28.2 | 36.0 | 45.7 | 56.1 | 65.3 | 72.0 | 70.6 | 60.5 | 48.1 | 33.8 | 25.3 | 47.0 |
| Days Max Temp ≥ 90 °F | 0 | 0 | 0 | 0 | 0 | 4 | 13 | 12 | 4 | 0 | 0 | 0 | 33 |
| Days Max Temp ≤ 32 °F | 12 | 7 | 4 | 0 | 0 | 0 | 0 | 0 | 0 | 0 | 5 | 10 | 38 |
| Days Min Temp ≤ 32 °F | 30 | 27 | 28 | 17 | 4 | 0 | 0 | 0 | 3 | 16 | 27 | 30 | 182 |
| Days Min Temp ≤ 0 °F | 9 | 4 | 1 | 0 | 0 | 0 | 0 | 0 | 0 | 0 | 1 | 5 | 20 |
| Heating Degree Days | 1314 | 1033 | 893 | 573 | 285 | 76 | 13 | 19 | 178 | 518 | 928 | 1226 | 7056 |
| Cooling Degree Days | *0* | *0* | *0* | *2* | *16* | *95* | 225 | 204 | 51 | 0 | 0 | 0 | 593 |
| Total Precipitation (") | 0.52 | 0.57 | 1.33 | 2.54 | 2.98 | 3.07 | 3.10 | 1.76 | 1.39 | 1.28 | 0.81 | 0.61 | 19.96 |
| Days ≥ 0.1" Precip | 2 | 2 | 4 | 6 | 7 | 7 | 6 | 4 | 4 | 3 | 2 | 2 | 49 |
| Total Snowfall (") | *9.5* | na | *10.2* | 6.9 | 0.4 | 0.1 | 0.0 | 0.0 | 0.5 | *3.4* | 8.0 | *10.2* | na |
| Days ≥ 1" Snow Depth | na | na | na | *1* | 0 | 0 | 0 | 0 | 0 | *1* | *4* | na | na |

## HAY SPRINGS 12 S *Sheridan County* ELEVATION 3812 ft LAT/LONG 42° 30 ' N / 102° 42 ' W

| | JAN | FEB | MAR | APR | MAY | JUN | JUL | AUG | SEP | OCT | NOV | DEC | YEAR |
|---|---|---|---|---|---|---|---|---|---|---|---|---|---|
| Maximum Temp °F | 35.7 | 40.9 | 49.8 | 60.7 | 71.3 | 80.8 | 87.0 | 85.5 | 76.5 | 64.0 | 47.1 | 37.6 | 61.4 |
| Minimum Temp °F | 9.0 | 13.3 | 21.3 | 30.5 | 41.0 | 50.1 | 56.0 | 54.0 | 43.1 | 31.3 | 19.6 | 10.8 | 31.7 |
| Mean Temp °F | 22.4 | 27.2 | 35.6 | 45.6 | 56.2 | 65.5 | 71.5 | 69.8 | 59.8 | 47.6 | 33.4 | 24.3 | 46.6 |
| Days Max Temp ≥ 90 °F | 0 | 0 | 0 | 0 | 1 | 6 | 12 | 10 | 3 | 0 | 0 | 0 | 32 |
| Days Max Temp ≤ 32 °F | 11 | 7 | 4 | 0 | 0 | 0 | 0 | 0 | 0 | 0 | 4 | 10 | 36 |
| Days Min Temp ≤ 32 °F | 31 | 28 | 29 | 18 | 5 | 0 | 0 | 0 | 3 | 18 | 28 | 31 | 191 |
| Days Min Temp ≤ 0 °F | 8 | 5 | 1 | 0 | 0 | 0 | 0 | 0 | 0 | 0 | 1 | 6 | 21 |
| Heating Degree Days | 1317 | 1062 | 904 | 576 | 281 | 74 | 10 | 20 | 188 | 532 | 943 | 1257 | 7164 |
| Cooling Degree Days | 0 | 0 | 0 | 1 | 18 | 102 | 217 | 189 | 40 | 1 | 0 | 0 | 568 |
| Total Precipitation (") | 0.30 | 0.28 | 0.70 | 1.69 | 2.83 | 3.09 | 2.69 | 1.74 | 1.34 | 0.90 | 0.47 | 0.27 | 16.30 |
| Days ≥ 0.1" Precip | 1 | 1 | 2 | 4 | 6 | 6 | 5 | 4 | 4 | 2 | 1 | 1 | 37 |
| Total Snowfall (") | 6.1 | 6.1 | 7.6 | 5.8 | 0.8 | 0.0 | 0.0 | 0.0 | 0.3 | 2.6 | 6.4 | 6.5 | 42.2 |
| Days ≥ 1" Snow Depth | 15 | 11 | 6 | 2 | 0 | 0 | 0 | 0 | 0 | 1 | 5 | 12 | 52 |

## HAYES CENTER *Hayes County* ELEVATION 3031 ft LAT/LONG 40° 31 ' N / 100° 58 ' W

| | JAN | FEB | MAR | APR | MAY | JUN | JUL | AUG | SEP | OCT | NOV | DEC | YEAR |
|---|---|---|---|---|---|---|---|---|---|---|---|---|---|
| Maximum Temp °F | 37.4 | 42.5 | 51.5 | 62.8 | 72.0 | 82.2 | 88.1 | 86.3 | 77.4 | 65.8 | 49.9 | 40.4 | 63.0 |
| Minimum Temp °F | 13.5 | 17.6 | 25.4 | 35.6 | 45.9 | 55.6 | 61.1 | 59.0 | 49.2 | 37.3 | 25.1 | 16.5 | 36.8 |
| Mean Temp °F | 25.7 | 30.1 | 38.5 | 49.2 | 59.0 | 68.9 | 74.6 | 72.7 | 63.3 | 51.6 | 37.5 | 28.6 | 50.0 |
| Days Max Temp ≥ 90 °F | 0 | 0 | 0 | 0 | 1 | 7 | 14 | 12 | 4 | 0 | 0 | 0 | 38 |
| Days Max Temp ≤ 32 °F | 11 | 7 | 3 | 0 | 0 | 0 | 0 | 0 | 0 | 0 | 3 | 9 | 33 |
| Days Min Temp ≤ 32 °F | 30 | 27 | 24 | 11 | 1 | 0 | 0 | 0 | 1 | 8 | 24 | 30 | 156 |
| Days Min Temp ≤ 0 °F | 5 | 3 | 1 | 0 | 0 | 0 | 0 | 0 | 0 | 0 | 0 | 3 | 12 |
| Heating Degree Days | 1212 | 979 | 816 | 473 | 212 | 40 | 6 | 12 | 129 | 414 | 818 | 1123 | 6234 |
| Cooling Degree Days | 0 | 0 | 0 | 8 | 31 | 174 | 307 | 266 | 95 | 5 | 0 | 0 | 886 |
| Total Precipitation (") | 0.50 | 0.56 | 1.45 | 2.07 | 3.19 | 3.60 | 3.55 | 2.28 | 1.46 | 1.34 | 0.82 | 0.53 | 21.35 |
| Days ≥ 0.1" Precip | 1 | 2 | 4 | 4 | 7 | 6 | 6 | 5 | 3 | 3 | 2 | 1 | 44 |
| Total Snowfall (") | 7.2 | *6.6* | *7.7* | 3.4 | 0.1 | 0.0 | 0.0 | 0.0 | 0.0 | 1.1 | 4.5 | 5.5 | 36.1 |
| Days ≥ 1" Snow Depth | 14 | 11 | 6 | 1 | 0 | 0 | 0 | 0 | 0 | 1 | 5 | 10 | 48 |

## HEBRON *Thayer County* ELEVATION 1460 ft LAT/LONG 40° 10 ' N / 97° 35 ' W

| | JAN | FEB | MAR | APR | MAY | JUN | JUL | AUG | SEP | OCT | NOV | DEC | YEAR |
|---|---|---|---|---|---|---|---|---|---|---|---|---|---|
| Maximum Temp °F | 33.6 | 39.5 | 52.0 | 64.5 | 73.9 | 84.1 | 89.5 | 87.2 | 78.8 | 67.2 | 50.2 | 38.1 | 63.2 |
| Minimum Temp °F | 12.6 | 16.7 | 28.0 | 39.8 | 50.6 | 60.7 | 65.9 | 63.3 | 53.2 | 40.3 | 27.9 | 17.8 | 39.7 |
| Mean Temp °F | 23.1 | 28.1 | 40.0 | 52.2 | 62.3 | 72.4 | 77.7 | 75.3 | 66.1 | 53.8 | 39.1 | 28.0 | 51.5 |
| Days Max Temp ≥ 90 °F | 0 | 0 | 0 | 0 | 1 | 8 | 15 | 12 | 5 | 0 | 0 | 0 | 41 |
| Days Max Temp ≤ 32 °F | 14 | 10 | 3 | 0 | 0 | 0 | 0 | 0 | 0 | 0 | 2 | 10 | 39 |
| Days Min Temp ≤ 32 °F | 30 | 26 | 21 | 7 | 1 | 0 | 0 | 0 | 0 | 6 | 22 | 30 | 143 |
| Days Min Temp ≤ 0 °F | 6 | 3 | 0 | 0 | 0 | 0 | 0 | 0 | 0 | 0 | 0 | 2 | 11 |
| Heating Degree Days | 1293 | 1034 | 769 | 394 | 147 | 19 | 2 | 6 | 96 | 354 | 772 | 1141 | 6027 |
| Cooling Degree Days | 0 | 0 | 2 | 21 | 64 | 250 | 397 | 324 | 143 | 9 | 0 | 0 | 1210 |
| Total Precipitation (") | 0.66 | 0.82 | 2.14 | 2.61 | 4.17 | 4.07 | 4.04 | 3.49 | 3.16 | 2.04 | 1.31 | 1.00 | 29.51 |
| Days ≥ 0.1" Precip | 2 | 2 | 4 | 6 | 8 | 6 | 6 | 6 | 5 | 4 | 3 | 2 | 54 |
| Total Snowfall (") | 5.4 | 6.0 | 5.1 | 1.1 | 0.1 | 0.0 | 0.0 | 0.0 | 0.1 | 0.4 | 2.9 | 5.5 | 26.6 |
| Days ≥ 1" Snow Depth | 17 | 12 | 4 | 1 | 0 | 0 | 0 | 0 | 0 | 0 | 3 | 9 | 46 |

## HEMINGFORD *Box Butte County* ELEVATION 4272 ft LAT/LONG 42° 21 ' N / 103° 5 ' W

| | JAN | FEB | MAR | APR | MAY | JUN | JUL | AUG | SEP | OCT | NOV | DEC | YEAR |
|---|---|---|---|---|---|---|---|---|---|---|---|---|---|
| Maximum Temp °F | 35.5 | 40.2 | 47.6 | 57.4 | 67.8 | 78.5 | 86.4 | 84.6 | 74.6 | 62.4 | 46.5 | 38.2 | 60.0 |
| Minimum Temp °F | 13.2 | 17.0 | 23.9 | 32.6 | 42.6 | 52.1 | 57.9 | 56.1 | 46.3 | 35.4 | 23.7 | 15.6 | 34.7 |
| Mean Temp °F | 24.4 | 28.7 | 35.8 | 45.0 | 55.2 | 65.3 | 72.2 | 70.4 | 60.5 | 48.9 | 35.1 | 26.9 | 47.4 |
| Days Max Temp ≥ 90 °F | 0 | 0 | 0 | 0 | 0 | 4 | 12 | 10 | 2 | 0 | 0 | 0 | 28 |
| Days Max Temp ≤ 32 °F | 12 | 7 | 5 | 1 | 0 | 0 | 0 | 0 | 0 | 1 | 5 | 10 | 41 |
| Days Min Temp ≤ 32 °F | 30 | 27 | 26 | 15 | 3 | 0 | 0 | 0 | 2 | 11 | 25 | 30 | 169 |
| Days Min Temp ≤ 0 °F | 6 | 3 | 1 | 0 | 0 | 0 | 0 | 0 | 0 | 0 | 1 | 4 | 15 |
| Heating Degree Days | 1255 | 1020 | 899 | 594 | 313 | 84 | 14 | 24 | 182 | 495 | 891 | 1175 | 6946 |
| Cooling Degree Days | 0 | 0 | 0 | 3 | 17 | 107 | 239 | 206 | 59 | 3 | 0 | 0 | 634 |
| Total Precipitation (") | 0.49 | 0.50 | 1.07 | 1.87 | 3.27 | 2.82 | 2.40 | 1.54 | 1.29 | 0.97 | 0.57 | 0.43 | 17.22 |
| Days ≥ 0.1" Precip | 1 | 2 | 3 | 5 | 6 | 6 | 5 | 4 | 3 | 3 | 2 | 2 | 42 |
| Total Snowfall (") | 7.6 | 7.0 | 10.1 | 8.1 | 1.4 | 0.0 | 0.0 | 0.0 | 0.6 | 3.9 | 6.9 | 7.4 | 53.0 |
| Days ≥ 1" Snow Depth | 20 | 15 | 11 | 4 | 1 | 0 | 0 | 0 | 0 | 2 | 9 | 17 | 79 |

## HERSHEY 5 SSE *Lincoln County* ELEVATION 2952 ft LAT/LONG 41° 6 ' N / 100° 58 ' W

| | JAN | FEB | MAR | APR | MAY | JUN | JUL | AUG | SEP | OCT | NOV | DEC | YEAR |
|---|---|---|---|---|---|---|---|---|---|---|---|---|---|
| Maximum Temp °F | 35.3 | 42.1 | 52.1 | 63.3 | 72.9 | 82.9 | *87.9* | *86.6* | 77.5 | 66.4 | 49.1 | 39.1 | 62.9 |
| Minimum Temp °F | 11.2 | 15.7 | 24.5 | 34.8 | 45.5 | 55.4 | 60.3 | *58.0* | 47.2 | 35.3 | 22.9 | 14.1 | 35.4 |
| Mean Temp °F | 23.3 | 28.9 | 38.3 | 49.1 | 59.2 | 69.2 | *74.1* | *72.3* | 62.4 | 50.8 | 36.0 | 26.6 | 49.2 |
| Days Max Temp ≥ 90 °F | 0 | 0 | 0 | 0 | 1 | *6* | *13* | 11 | 5 | 0 | 0 | 0 | 36 |
| Days Max Temp ≤ 32 °F | 12 | 7 | 3 | 0 | 0 | 0 | 0 | 0 | 0 | 0 | 4 | 9 | 35 |
| Days Min Temp ≤ 32 °F | 31 | 28 | 26 | *13* | 2 | 0 | 0 | 0 | 2 | 11 | 26 | 31 | 170 |
| Days Min Temp ≤ 0 °F | 7 | 3 | 0 | 0 | 0 | 0 | 0 | 0 | 0 | 0 | 1 | 3 | 14 |
| Heating Degree Days | 1288 | 1014 | 821 | 476 | 206 | 34 | *6* | *11* | 143 | 435 | 863 | 1183 | 6480 |
| Cooling Degree Days | 0 | 0 | 0 | 7 | 31 | *168* | na | 254 | 84 | 4 | 0 | 0 | na |
| Total Precipitation (") | 0.58 | 0.44 | 1.09 | 1.89 | 3.13 | 3.41 | 3.30 | 2.06 | 1.37 | 1.23 | 0.71 | 0.45 | 19.66 |
| Days ≥ 0.1" Precip | 2 | 1 | 3 | 4 | 7 | 6 | 6 | 4 | 3 | 3 | 2 | 1 | 42 |
| Total Snowfall (") | 5.4 | 4.5 | 5.2 | 1.9 | 0.0 | 0.0 | 0.0 | 0.0 | 0.1 | 1.7 | 4.7 | 5.4 | 28.9 |
| Days ≥ 1" Snow Depth | 16 | 9 | 5 | 1 | 0 | 0 | 0 | 0 | 0 | 1 | 5 | 11 | 48 |

## HOLDREGE *Phelps County* ELEVATION 2342 ft LAT/LONG 40° 26 ' N / 99° 23 ' W

| | JAN | FEB | MAR | APR | MAY | JUN | JUL | AUG | SEP | OCT | NOV | DEC | YEAR |
|---|---|---|---|---|---|---|---|---|---|---|---|---|---|
| Maximum Temp °F | 34.7 | 40.7 | 51.4 | 64.0 | 73.2 | 83.8 | 88.1 | 85.9 | 77.9 | 66.0 | 48.8 | 38.0 | 62.7 |
| Minimum Temp °F | 12.9 | 17.4 | 26.3 | 37.0 | 48.0 | 57.7 | 62.7 | 60.5 | 51.0 | 39.1 | 26.0 | 16.7 | 37.9 |
| Mean Temp °F | 23.9 | 29.1 | 38.9 | 50.5 | 60.6 | 70.8 | 75.4 | 73.2 | 64.5 | 52.6 | 37.4 | 27.4 | 50.4 |
| Days Max Temp ≥ 90 °F | 0 | 0 | 0 | 0 | 1 | 8 | 13 | 10 | 4 | 0 | 0 | 0 | 36 |
| Days Max Temp ≤ 32 °F | 13 | 9 | 3 | 0 | 0 | 0 | 0 | 0 | 0 | 0 | 3 | 10 | 38 |
| Days Min Temp ≤ 32 °F | 30 | 27 | 23 | 9 | 1 | 0 | 0 | 0 | 1 | 7 | 23 | 30 | 151 |
| Days Min Temp ≤ 0 °F | 6 | 3 | 0 | 0 | 0 | 0 | 0 | 0 | 0 | 0 | 0 | 3 | 12 |
| Heating Degree Days | 1269 | 1008 | 804 | 436 | 174 | 26 | 3 | 9 | 110 | 387 | 821 | 1160 | 6207 |
| Cooling Degree Days | 0 | 0 | 1 | 13 | 43 | 199 | 320 | 264 | 105 | 6 | 0 | 0 | 951 |
| Total Precipitation (") | 0.53 | 0.57 | 1.80 | 2.14 | 4.21 | 4.03 | 4.10 | 3.00 | 2.22 | 1.68 | 1.08 | 0.61 | 25.97 |
| Days ≥ 0.1" Precip | 1 | 2 | 4 | 5 | 7 | 6 | 6 | 5 | 5 | 3 | 2 | 2 | 48 |
| Total Snowfall (") | 6.0 | 5.1 | 5.8 | 2.0 | 0.1 | 0.0 | 0.0 | 0.0 | 0.3 | 1.0 | 4.2 | 5.8 | 30.3 |
| Days ≥ 1" Snow Depth | 18 | 11 | 6 | 1 | 0 | 0 | 0 | 0 | 0 | 0 | 4 | 11 | 51 |

## HYANNIS *Grant County* ELEVATION 3734 ft LAT/LONG 41° 59 ' N / 101° 48 ' W

| | JAN | FEB | MAR | APR | MAY | JUN | JUL | AUG | SEP | OCT | NOV | DEC | YEAR |
|---|---|---|---|---|---|---|---|---|---|---|---|---|---|
| Maximum Temp °F | *35.3* | *40.4* | *49.1* | *60.5* | *70.9* | *80.3* | 86.6 | *84.9* | *75.7* | *63.6* | *47.5* | *39.3* | 61.2 |
| Minimum Temp °F | *11.4* | *14.8* | *23.4* | *32.8* | *43.6* | *52.6* | 58.2 | *56.3* | na | *33.4* | *21.8* | *14.4* | na |
| Mean Temp °F | *23.4* | *27.6* | *36.3* | *46.7* | *57.3* | *66.5* | 72.4 | *70.6* | na | *48.5* | *34.8* | *26.9* | na |
| Days Max Temp ≥ 90 °F | *0* | *0* | *0* | 0 | *0* | 4 | 11 | *9* | *2* | *0* | 0 | 0 | 26 |
| Days Max Temp ≤ 32 °F | *11* | *8* | *4* | 0 | *0* | 0 | 0 | *0* | *0* | *0* | *4* | *9* | 36 |
| Days Min Temp ≤ 32 °F | *29* | *27* | *26* | *15* | *3* | 0 | 0 | *0* | *2* | *13* | *25* | *30* | 170 |
| Days Min Temp ≤ 0 °F | *6* | *4* | *1* | 0 | *0* | 0 | 0 | *0* | *0* | *0* | 1 | *4* | 16 |
| Heating Degree Days | *1284* | *1049* | *884* | *546* | *252* | *61* | 9 | *19* | na | *506* | *901* | *1175* | na |
| Cooling Degree Days | 0 | 0 | 0 | 4 | 21 | 122 | 246 | 210 | *56* | 1 | *0* | 0 | 660 |
| Total Precipitation (") | *0.37* | *0.50* | *1.34* | *1.80* | 3.07 | 3.02 | 3.21 | *2.08* | *1.50* | *1.16* | 0.72 | 0.36 | 19.13 |
| Days ≥ 0.1" Precip | *1* | *2* | *3* | *4* | *5* | 6 | 6 | *4* | *4* | *3* | *2* | *1* | 41 |
| Total Snowfall (") | *4.9* | *5.5* | *8.1* | *3.7* | *0.2* | 0.0 | 0.0 | *0.0* | *0.4* | *0.7* | *6.2* | *5.1* | 34.8 |
| Days ≥ 1" Snow Depth | na | na | na | *2* | *0* | 0 | 0 | *0* | *0* | na | *6* | *13* | na |

## IMPERIAL *Chase County* ELEVATION 3281 ft LAT/LONG 40° 31 ' N / 101° 38 ' W

| | JAN | FEB | MAR | APR | MAY | JUN | JUL | AUG | SEP | OCT | NOV | DEC | YEAR |
|---|---|---|---|---|---|---|---|---|---|---|---|---|---|
| Maximum Temp °F | 39.2 | 45.2 | 53.9 | 65.0 | 74.3 | 84.2 | 89.9 | 87.9 | 79.6 | 67.6 | 50.7 | 41.4 | 64.9 |
| Minimum Temp °F | 14.4 | 18.6 | 26.0 | 35.8 | 46.6 | 56.4 | 61.8 | 59.8 | 49.8 | 37.1 | 24.8 | 16.4 | 37.3 |
| Mean Temp °F | 26.8 | 31.9 | 39.9 | 50.4 | 60.5 | 70.3 | 75.9 | 73.9 | 64.7 | 52.4 | 37.8 | 29.0 | 51.1 |
| Days Max Temp ≥ 90 °F | 0 | 0 | 0 | 0 | 1 | 9 | 18 | 14 | 5 | 0 | 0 | 0 | 47 |
| Days Max Temp ≤ 32 °F | 9 | 6 | 2 | 0 | 0 | 0 | 0 | 0 | 0 | 0 | 3 | 7 | 27 |
| Days Min Temp ≤ 32 °F | 30 | 27 | 24 | 11 | 1 | 0 | 0 | 0 | 1 | 9 | 25 | 30 | 158 |
| Days Min Temp ≤ 0 °F | 4 | 2 | 0 | 0 | 0 | 0 | 0 | 0 | 0 | 0 | 0 | 3 | 9 |
| Heating Degree Days | 1177 | 927 | 770 | 436 | 175 | 26 | 2 | 5 | 100 | 388 | 810 | 1112 | 5928 |
| Cooling Degree Days | 0 | 0 | 0 | 7 | 44 | 200 | 346 | 300 | 108 | 3 | 0 | 0 | 1008 |
| Total Precipitation (") | 0.54 | 0.47 | 1.32 | 1.78 | 3.19 | 3.33 | 2.91 | 2.36 | 1.35 | 1.17 | 0.70 | 0.46 | 19.58 |
| Days ≥ 0.1" Precip | 2 | 1 | 3 | 4 | 6 | 6 | 6 | 4 | 3 | 3 | 2 | 1 | 41 |
| Total Snowfall (") | 5.9 | 4.6 | 7.5 | 3.0 | 0.2 | 0.0 | 0.0 | 0.0 | 0.2 | 2.0 | 4.9 | 5.2 | 33.5 |
| Days ≥ 1" Snow Depth | 14 | 8 | 4 | 1 | 0 | 0 | 0 | 0 | 0 | 1 | 5 | 11 | 44 |

## KEARNEY 4 NE *Buffalo County* ELEVATION 2129 ft LAT/LONG 40° 44' N / 99° 1' W

| | JAN | FEB | MAR | APR | MAY | JUN | JUL | AUG | SEP | OCT | NOV | DEC | YEAR |
|---|---|---|---|---|---|---|---|---|---|---|---|---|---|
| Maximum Temp °F | 34.1 | 39.4 | 50.3 | 63.1 | 72.8 | 83.2 | 88.0 | 85.8 | 77.4 | 65.6 | 48.6 | 37.5 | 62.2 |
| Minimum Temp °F | 10.7 | 15.0 | 25.1 | 36.4 | 47.7 | 57.8 | 62.5 | 59.9 | 49.6 | 37.2 | 24.5 | 14.9 | 36.8 |
| Mean Temp °F | 22.4 | 27.3 | 37.7 | 49.8 | 60.3 | 70.5 | 75.3 | 72.9 | 63.5 | 51.4 | 36.6 | 26.2 | 49.5 |
| Days Max Temp ≥ 90 °F | 0 | 0 | 0 | 1 | 1 | 7 | 14 | 10 | 4 | 0 | 0 | 0 | 37 |
| Days Max Temp ≤ 32 °F | 13 | 10 | 4 | 0 | 0 | 0 | 0 | 0 | 0 | 0 | 3 | 11 | 41 |
| Days Min Temp ≤ 32 °F | 31 | 27 | 25 | 10 | 1 | 0 | 0 | 0 | 1 | 9 | 25 | 31 | 160 |
| Days Min Temp ≤ 0 °F | 8 | 4 | 1 | 0 | 0 | 0 | 0 | 0 | 0 | 0 | 0 | 3 | 16 |
| Heating Degree Days | 1313 | 1058 | 838 | 459 | 186 | 30 | 4 | 12 | 128 | 419 | 847 | 1195 | 6489 |
| Cooling Degree Days | 0 | 0 | 1 | 13 | 48 | 209 | 325 | 267 | 105 | 4 | 0 | 0 | 972 |
| Total Precipitation (") | 0.53 | 0.63 | 1.81 | 2.28 | 3.88 | 4.03 | 3.35 | 2.73 | 2.37 | 1.75 | 1.03 | 0.72 | 25.11 |
| Days ≥ 0.1" Precip | 2 | 2 | 4 | 5 | 7 | 6 | 6 | 5 | 4 | 3 | 3 | 2 | 49 |
| Total Snowfall (") | 6.3 | 5.4 | 5.9 | 1.8 | 0.2 | 0.0 | 0.0 | 0.0 | 0.2 | 0.7 | 4.1 | 7.7 | 32.3 |
| Days ≥ 1" Snow Depth | na | *6* | *4* | 0 | 0 | 0 | 0 | 0 | 0 | 0 | *2* | na | na |

## KIMBALL 2 N *Kimball County* ELEVATION 4734 ft LAT/LONG 41° 14' N / 103° 40' W

| | JAN | FEB | MAR | APR | MAY | JUN | JUL | AUG | SEP | OCT | NOV | DEC | YEAR |
|---|---|---|---|---|---|---|---|---|---|---|---|---|---|
| Maximum Temp °F | 38.9 | 43.2 | 49.5 | 59.3 | 69.0 | 79.6 | 86.8 | 84.8 | 75.4 | 63.3 | 48.4 | 40.4 | 61.5 |
| Minimum Temp °F | 13.2 | 16.6 | 22.7 | 31.3 | 41.1 | 50.7 | 56.5 | 54.0 | 43.8 | 32.4 | 21.9 | 14.2 | 33.2 |
| Mean Temp °F | 26.0 | 29.9 | 36.1 | 45.3 | 55.1 | 65.2 | 71.7 | 69.4 | 59.6 | 47.8 | 35.2 | 27.4 | 47.4 |
| Days Max Temp ≥ 90 °F | 0 | 0 | 0 | 0 | 0 | 5 | 12 | 9 | 3 | 0 | 0 | 0 | 29 |
| Days Max Temp ≤ 32 °F | 9 | 6 | 4 | 1 | 0 | 0 | 0 | 0 | 0 | 1 | 4 | 8 | 33 |
| Days Min Temp ≤ 32 °F | 30 | 27 | 28 | 17 | 3 | 0 | 0 | 0 | 2 | 15 | 27 | 30 | 179 |
| Days Min Temp ≤ 0 °F | 5 | 2 | 1 | 0 | 0 | 0 | 0 | 0 | 0 | 0 | 1 | 3 | 12 |
| Heating Degree Days | 1201 | 983 | 888 | 583 | 311 | 82 | 11 | 21 | 191 | 526 | 888 | 1161 | 6846 |
| Cooling Degree Days | 0 | 0 | 0 | 0 | 12 | 107 | 223 | 176 | 40 | 1 | 0 | 0 | 559 |
| Total Precipitation (") | 0.41 | 0.30 | 1.14 | 1.58 | 2.71 | 2.91 | 2.72 | 1.89 | 1.27 | 0.94 | 0.60 | 0.46 | 16.93 |
| Days ≥ 0.1" Precip | 1 | 1 | 3 | 4 | 6 | 6 | 6 | 4 | 3 | 3 | 2 | 2 | 41 |
| Total Snowfall (") | 7.0 | 4.5 | 10.7 | 5.6 | 0.9 | 0.0 | 0.0 | 0.0 | 0.3 | 3.0 | 6.9 | 7.6 | 46.5 |
| Days ≥ 1" Snow Depth | 11 | 7 | 6 | 2 | 0 | 0 | 0 | 0 | 0 | 1 | 5 | 10 | 42 |

## KINGSLEY DAM *Keith County* ELEVATION 3304 ft LAT/LONG 41° 13' N / 101° 39' W

| | JAN | FEB | MAR | APR | MAY | JUN | JUL | AUG | SEP | OCT | NOV | DEC | YEAR |
|---|---|---|---|---|---|---|---|---|---|---|---|---|---|
| Maximum Temp °F | 36.7 | 43.1 | 52.0 | 63.5 | 72.8 | 82.7 | 89.2 | 87.4 | 78.4 | 66.5 | 49.6 | 39.3 | 63.4 |
| Minimum Temp °F | 14.6 | 18.1 | 25.9 | 36.3 | 46.7 | 56.2 | 61.8 | 59.9 | 50.4 | 39.0 | 27.3 | 18.4 | 37.9 |
| Mean Temp °F | 25.7 | 30.6 | 39.0 | 49.9 | 59.8 | 69.5 | 75.5 | 73.7 | 64.4 | 52.8 | 38.5 | 28.8 | 50.7 |
| Days Max Temp ≥ 90 °F | 0 | 0 | 0 | 0 | 1 | 7 | 16 | 14 | 5 | 0 | 0 | 0 | 43 |
| Days Max Temp ≤ 32 °F | 10 | 6 | 2 | 0 | 0 | 0 | 0 | 0 | 0 | 0 | 3 | 8 | 29 |
| Days Min Temp ≤ 32 °F | 30 | 27 | 24 | 9 | 1 | 0 | 0 | 0 | 1 | 6 | 21 | 29 | 148 |
| Days Min Temp ≤ 0 °F | 5 | 3 | 1 | 0 | 0 | 0 | 0 | 0 | 0 | 0 | 0 | 2 | 11 |
| Heating Degree Days | 1212 | 964 | 799 | 450 | 193 | 31 | 3 | 7 | 104 | 375 | 789 | 1115 | 6042 |
| Cooling Degree Days | 0 | 0 | 0 | 7 | 38 | 177 | 330 | 293 | 102 | 5 | 0 | 0 | 952 |
| Total Precipitation (") | 0.54 | 0.57 | 1.41 | 1.78 | 3.16 | 2.99 | 2.62 | 1.86 | 1.38 | 0.98 | 0.76 | 0.50 | 18.55 |
| Days ≥ 0.1" Precip | 2 | 2 | 3 | 4 | 6 | 6 | 6 | 4 | 3 | 3 | 2 | 2 | 43 |
| Total Snowfall (") | 7.3 | 6.1 | 9.0 | 3.0 | 0.4 | 0.0 | 0.0 | 0.0 | 0.1 | 1.2 | 6.0 | 6.3 | 39.4 |
| Days ≥ 1" Snow Depth | 15 | 9 | 5 | 1 | 0 | 0 | 0 | 0 | 0 | 0 | 4 | 11 | 45 |

## LINCOLN *Lancaster County* ELEVATION 1165 ft LAT/LONG 40° 49' N / 96° 42' W

| | JAN | FEB | MAR | APR | MAY | JUN | JUL | AUG | SEP | OCT | NOV | DEC | YEAR |
|---|---|---|---|---|---|---|---|---|---|---|---|---|---|
| Maximum Temp °F | 32.6 | 37.5 | 50.4 | 63.6 | 74.0 | 84.3 | 88.8 | 86.3 | 77.5 | 66.0 | 49.2 | 37.2 | 62.3 |
| Minimum Temp °F | 13.9 | 17.9 | 29.2 | 41.1 | 52.2 | 62.7 | 67.9 | 65.4 | 55.3 | 43.0 | 29.9 | 19.5 | 41.5 |
| Mean Temp °F | 23.3 | 27.7 | 39.8 | 52.4 | 63.1 | 73.5 | 75.7 | 75.8 | 66.5 | 54.5 | 39.6 | 28.4 | 51.7 |
| Days Max Temp ≥ 90 °F | 0 | 0 | 0 | 1 | 2 | 8 | 14 | 11 | 3 | 0 | 0 | 0 | 39 |
| Days Max Temp ≤ 32 °F | 14 | 11 | 3 | 0 | 0 | 0 | 0 | 0 | 0 | 0 | 3 | 10 | 41 |
| Days Min Temp ≤ 32 °F | 29 | 25 | 19 | 6 | 0 | 0 | 0 | 0 | 0 | 4 | 18 | 28 | 129 |
| Days Min Temp ≤ 0 °F | 6 | 3 | 0 | 0 | 0 | 0 | 0 | 0 | 0 | 0 | 0 | 2 | 11 |
| Heating Degree Days | 1287 | 1046 | 775 | 392 | 131 | 14 | 1 | 4 | 84 | 336 | 755 | 1130 | 5955 |
| Cooling Degree Days | 0 | 0 | 2 | 27 | 76 | 281 | 425 | 346 | 144 | 9 | 0 | 0 | 1310 |
| Total Precipitation (") | 0.57 | 0.76 | 2.01 | 2.78 | 4.14 | 4.20 | 3.89 | 3.45 | 3.32 | 2.24 | 1.33 | 0.98 | 29.67 |
| Days ≥ 0.1" Precip | 2 | 2 | 4 | 5 | 7 | 6 | 5 | 5 | 5 | 4 | 3 | 2 | 50 |
| Total Snowfall (") | 5.0 | 5.7 | 3.7 | 0.8 | 0.1 | 0.0 | 0.0 | 0.0 | 0.0 | 0.4 | 2.7 | 5.2 | 23.6 |
| Days ≥ 1" Snow Depth | 14 | 10 | 4 | 0 | 0 | 0 | 0 | 0 | 0 | 0 | 2 | 7 | 37 |

## LINCOLN MUNI AP *Lancaster County* ELEVATION 1194 ft LAT/LONG 40° 51 ' N / 96° 46 ' W

| | JAN | FEB | MAR | APR | MAY | JUN | JUL | AUG | SEP | OCT | NOV | DEC | YEAR |
|---|---|---|---|---|---|---|---|---|---|---|---|---|---|
| Maximum Temp °F | 33.0 | 38.1 | 51.6 | 64.2 | 74.3 | 84.8 | 89.6 | 86.6 | 78.1 | 65.8 | 48.8 | 36.4 | 62.6 |
| Minimum Temp °F | 11.4 | 16.1 | 28.0 | 39.3 | 50.3 | 60.4 | 66.1 | 63.5 | 53.2 | 40.0 | 27.1 | 15.9 | 39.3 |
| Mean Temp °F | 22.2 | 27.1 | 39.8 | 51.8 | 62.3 | 72.7 | 77.8 | 75.1 | 65.7 | 52.9 | 38.0 | 26.1 | 51.0 |
| Days Max Temp ≥ 90 °F | 0 | 0 | 0 | 1 | 1 | 9 | 16 | 11 | 4 | 0 | 0 | 0 | 42 |
| Days Max Temp ≤ 32 °F | 14 | 11 | 3 | 0 | 0 | 0 | 0 | 0 | 0 | 0 | 3 | 11 | 42 |
| Days Min Temp ≤ 32 °F | 30 | 26 | 21 | 7 | 1 | 0 | 0 | 0 | 0 | 6 | 22 | 30 | 143 |
| Days Min Temp ≤ 0 °F | 7 | 4 | 0 | 0 | 0 | 0 | 0 | 0 | 0 | 0 | 0 | 4 | 15 |
| Heating Degree Days | 1319 | 1062 | 775 | 407 | 139 | 14 | 1 | 5 | 98 | 377 | 805 | 1198 | 6200 |
| Cooling Degree Days | 0 | 0 | 1 | 20 | 66 | 258 | 399 | 323 | 133 | 10 | 0 | 0 | 1210 |
| Total Precipitation (") | 0.71 | 0.58 | 2.51 | 2.87 | 3.82 | 3.43 | 3.72 | 3.47 | 3.21 | 2.02 | 1.43 | 0.97 | 28.74 |
| Days ≥ 0.1" Precip | 2 | 2 | 5 | 6 | 7 | 6 | 6 | 5 | 5 | 4 | 3 | 2 | 53 |
| Total Snowfall (") | 5.7 | 5.0 | 4.8 | 1.6 | 0.0 | 0.0 | 0.0 | 0.0 | 0.0 | 0.3 | 3.1 | 5.5 | 26.0 |
| Days ≥ 1" Snow Depth | 16 | 12 | 5 | 1 | 0 | 0 | 0 | 0 | 0 | 0 | 3 | 9 | 46 |

## LODGEPOLE *Cheyenne County* ELEVATION 3832 ft LAT/LONG 41° 9 ' N / 102° 38 ' W

| | JAN | FEB | MAR | APR | MAY | JUN | JUL | AUG | SEP | OCT | NOV | DEC | YEAR |
|---|---|---|---|---|---|---|---|---|---|---|---|---|---|
| Maximum Temp °F | 40.8 | 46.6 | 54.4 | 64.7 | 74.1 | 84.6 | 91.4 | 89.5 | 80.5 | 68.6 | 51.1 | 41.9 | 65.7 |
| Minimum Temp °F | 13.4 | 17.3 | 24.1 | 33.2 | 43.2 | 52.6 | 58.8 | 56.6 | 46.1 | 34.5 | 22.9 | 14.8 | 34.8 |
| Mean Temp °F | 27.1 | 32.0 | 39.3 | 49.0 | 58.7 | 68.6 | 75.1 | 73.1 | 63.3 | 51.6 | 37.2 | 28.3 | 50.3 |
| Days Max Temp ≥ 90 °F | 0 | 0 | 0 | 0 | 1 | 10 | 20 | 17 | 6 | 0 | 0 | 0 | 54 |
| Days Max Temp ≤ 32 °F | 7 | 4 | 2 | 0 | 0 | 0 | 0 | 0 | 0 | 0 | 2 | 6 | 21 |
| Days Min Temp ≤ 32 °F | 31 | 27 | 26 | 14 | 3 | 0 | 0 | 0 | 2 | 12 | 27 | 30 | 172 |
| Days Min Temp ≤ 0 °F | 5 | 3 | 1 | 0 | 0 | 0 | 0 | 0 | 0 | 0 | 1 | 3 | 13 |
| Heating Degree Days | 1168 | 923 | 789 | 476 | 219 | 41 | 3 | 8 | 122 | 409 | 829 | 1132 | 6119 |
| Cooling Degree Days | 0 | 0 | 0 | 4 | 34 | 172 | 328 | 286 | 87 | 2 | 0 | 0 | 913 |
| Total Precipitation (") | 0.42 | 0.48 | 1.30 | 1.77 | 3.19 | 3.06 | 2.68 | 2.17 | 1.25 | 0.93 | 0.73 | 0.49 | 18.47 |
| Days ≥ 0.1" Precip | 1 | 2 | 3 | 4 | 6 | 6 | 5 | 4 | 3 | 3 | 2 | 2 | 41 |
| Total Snowfall (") | 5.6 | 5.7 | 7.9 | 3.8 | 0.3 | 0.0 | 0.0 | 0.0 | 0.2 | 1.0 | 6.3 | 6.7 | 37.5 |
| Days ≥ 1" Snow Depth | 13 | na | 4 | 1 | 0 | 0 | 0 | 0 | 0 | 0 | na | na | na |

## LOUP CITY *Sherman County* ELEVATION 2060 ft LAT/LONG 41° 17 ' N / 98° 58 ' W

| | JAN | FEB | MAR | APR | MAY | JUN | JUL | AUG | SEP | OCT | NOV | DEC | YEAR |
|---|---|---|---|---|---|---|---|---|---|---|---|---|---|
| Maximum Temp °F | 33.6 | 38.6 | 49.7 | 62.7 | 72.4 | 81.6 | 87.1 | 85.4 | 76.6 | 65.6 | 48.6 | 37.7 | 61.6 |
| Minimum Temp °F | 8.7 | 13.6 | 24.2 | 35.5 | 46.9 | 56.5 | 62.0 | 59.6 | 48.6 | 36.0 | 23.5 | 13.2 | 35.7 |
| Mean Temp °F | 21.2 | 26.1 | 37.0 | 49.1 | 59.6 | 69.1 | 74.6 | 72.5 | 62.6 | 50.8 | 36.1 | 25.5 | 48.7 |
| Days Max Temp ≥ 90 °F | 0 | 0 | 0 | 0 | 1 | 6 | 12 | 10 | 3 | 0 | 0 | 0 | 32 |
| Days Max Temp ≤ 32 °F | 14 | 11 | 4 | 0 | 0 | 0 | 0 | 0 | 0 | 0 | 3 | 10 | 42 |
| Days Min Temp ≤ 32 °F | 31 | 27 | 25 | 11 | 2 | 0 | 0 | 0 | 1 | 10 | 25 | 31 | 163 |
| Days Min Temp ≤ 0 °F | 8 | 5 | 1 | 0 | 0 | 0 | 0 | 0 | 0 | 0 | 1 | 4 | 19 |
| Heating Degree Days | 1353 | 1090 | 862 | 476 | 201 | 39 | 5 | 13 | 142 | 438 | 860 | 1219 | 6698 |
| Cooling Degree Days | 0 | 0 | 0 | 11 | 36 | 159 | 291 | 247 | 84 | 3 | 0 | 0 | 831 |
| Total Precipitation (") | 0.53 | 0.69 | 2.03 | 2.76 | 3.74 | 4.11 | 3.68 | 2.67 | 2.38 | 1.70 | 1.37 | 0.81 | 26.47 |
| Days ≥ 0.1" Precip | 2 | 2 | 4 | 5 | 7 | 6 | 6 | 5 | 5 | 4 | 3 | 2 | 51 |
| Total Snowfall (") | 7.1 | 6.1 | 6.6 | 1.7 | 0.2 | 0.0 | 0.0 | 0.0 | 0.1 | 0.8 | 5.4 | 8.3 | 36.3 |
| Days ≥ 1" Snow Depth | na | na | na | 0 | 0 | 0 | 0 | 0 | 0 | 0 | na | na | na |

## MADISON *Madison County* ELEVATION 1581 ft LAT/LONG 41° 50 ' N / 97° 27 ' W

| | JAN | FEB | MAR | APR | MAY | JUN | JUL | AUG | SEP | OCT | NOV | DEC | YEAR |
|---|---|---|---|---|---|---|---|---|---|---|---|---|---|
| Maximum Temp °F | 30.6 | 36.1 | 48.3 | 62.8 | 73.5 | 83.4 | 87.4 | 85.0 | 76.8 | 65.0 | 46.9 | 34.8 | 60.9 |
| Minimum Temp °F | 8.5 | 13.8 | 24.6 | 36.4 | 47.8 | 57.9 | 62.5 | 59.9 | 49.8 | 37.0 | 24.3 | 13.6 | 36.3 |
| Mean Temp °F | 19.5 | 25.0 | 36.5 | 49.6 | 60.6 | 70.7 | 75.0 | 72.5 | 63.3 | 51.0 | 35.6 | 24.2 | 48.6 |
| Days Max Temp ≥ 90 °F | 0 | 0 | 0 | 1 | 2 | 7 | 12 | 10 | 3 | 0 | 0 | 0 | 35 |
| Days Max Temp ≤ 32 °F | 16 | 11 | 4 | 0 | 0 | 0 | 0 | 0 | 0 | 0 | 4 | 12 | 47 |
| Days Min Temp ≤ 32 °F | 30 | 27 | 25 | 10 | 2 | 0 | 0 | 0 | 1 | 10 | 24 | 30 | 159 |
| Days Min Temp ≤ 0 °F | 9 | 5 | 0 | 0 | 0 | 0 | 0 | 0 | 0 | 0 | 1 | 4 | 19 |
| Heating Degree Days | 1402 | 1124 | 879 | 467 | 182 | 28 | 4 | 13 | 133 | 433 | 875 | 1259 | 6799 |
| Cooling Degree Days | 0 | 0 | 0 | 13 | 45 | 188 | 283 | 226 | 85 | 3 | 0 | 0 | 843 |
| Total Precipitation (") | 0.47 | 0.67 | 1.81 | 2.34 | 4.11 | 4.25 | 3.77 | 2.81 | 2.53 | 1.94 | 1.18 | 0.77 | 26.65 |
| Days ≥ 0.1" Precip | 1 | 2 | 4 | 5 | 7 | 6 | 6 | 5 | 5 | 4 | 2 | 2 | 49 |
| Total Snowfall (") | 4.4 | 4.9 | 4.3 | 0.9 | 0.0 | 0.0 | 0.0 | 0.0 | 0.0 | 0.6 | 3.7 | 6.7 | 25.5 |
| Days ≥ 1" Snow Depth | 12 | 9 | 4 | 0 | 0 | 0 | 0 | 0 | 0 | 0 | 2 | 10 | 37 |

## MADRID *Perkins County* ELEVATION 3202 ft LAT/LONG 40° 51 ' N / 101° 33 ' W

| | JAN | FEB | MAR | APR | MAY | JUN | JUL | AUG | SEP | OCT | NOV | DEC | YEAR |
|---|---|---|---|---|---|---|---|---|---|---|---|---|---|
| Maximum Temp °F | 38.1 | 44.5 | 53.7 | 65.2 | 75.0 | 85.9 | 91.8 | 89.6 | 81.0 | 68.6 | 50.2 | 40.5 | 65.3 |
| Minimum Temp °F | 12.9 | 17.3 | 25.2 | 35.1 | 45.7 | 55.6 | 60.9 | 58.8 | 48.6 | 36.0 | 23.9 | 15.1 | 36.3 |
| Mean Temp °F | 25.5 | 30.9 | 39.5 | 50.2 | 60.4 | 70.8 | 76.3 | 74.3 | 64.8 | 52.3 | 37.1 | 27.8 | 50.8 |
| Days Max Temp ≥ 90 °F | 0 | 0 | 0 | 0 | 2 | 11 | 20 | 18 | 7 | 0 | 0 | 0 | 58 |
| Days Max Temp ≤ 32 °F | 10 | 6 | 3 | 0 | 0 | 0 | 0 | 0 | 0 | 0 | 3 | 8 | 30 |
| Days Min Temp ≤ 32 °F | 30 | 27 | 25 | 12 | 2 | 0 | 0 | 0 | 1 | 10 | 26 | 30 | 163 |
| Days Min Temp ≤ 0 °F | 5 | 3 | 1 | 0 | 0 | 0 | 0 | 0 | 0 | 0 | 0 | 3 | 12 |
| Heating Degree Days | 1217 | 955 | 784 | 443 | 181 | 25 | 2 | 5 | 100 | 391 | 832 | 1145 | 6080 |
| Cooling Degree Days | 0 | 0 | 0 | 8 | 49 | 208 | 352 | 307 | 109 | 3 | 0 | 0 | 1036 |
| Total Precipitation (") | 0.48 | 0.50 | 1.33 | 1.74 | 3.23 | 3.37 | 3.05 | 2.13 | 1.30 | 1.04 | 0.74 | 0.43 | 19.34 |
| Days ≥ 0.1" Precip | 2 | 1 | 3 | 4 | 6 | 6 | 6 | 4 | 3 | 2 | 2 | 2 | 41 |
| Total Snowfall (") | 7.5 | 6.1 | 9.2 | 3.3 | 0.1 | 0.0 | 0.0 | 0.0 | 0.1 | 1.7 | 5.8 | 7.0 | 40.8 |
| Days ≥ 1" Snow Depth | 12 | 7 | 5 | 1 | 0 | 0 | 0 | 0 | 0 | 1 | 4 | 9 | 39 |

## MCCOOK *Red Willow County* ELEVATION 2530 ft LAT/LONG 40° 12 ' N / 100° 36 ' W

| | JAN | FEB | MAR | APR | MAY | JUN | JUL | AUG | SEP | OCT | NOV | DEC | YEAR |
|---|---|---|---|---|---|---|---|---|---|---|---|---|---|
| Maximum Temp °F | 38.5 | 43.8 | 53.6 | 65.0 | 73.8 | 84.4 | 90.3 | 87.8 | 79.3 | 67.6 | 51.3 | 41.1 | 64.7 |
| Minimum Temp °F | 13.8 | 18.0 | 26.2 | 36.5 | 47.2 | 57.1 | 62.9 | 60.7 | 50.5 | 37.3 | 24.8 | 16.6 | 37.6 |
| Mean Temp °F | 26.2 | 30.9 | 39.9 | 50.8 | 60.5 | 70.8 | 76.6 | 74.2 | 64.9 | 52.5 | 38.1 | 28.9 | 51.2 |
| Days Max Temp ≥ 90 °F | 0 | 0 | 0 | 1 | 2 | 9 | 17 | 14 | 6 | 1 | 0 | 0 | 50 |
| Days Max Temp ≤ 32 °F | 11 | 7 | 3 | 0 | 0 | 0 | 0 | 0 | 0 | 0 | 3 | 8 | 32 |
| Days Min Temp ≤ 32 °F | 31 | 27 | 24 | 10 | 1 | 0 | 0 | 0 | 1 | 8 | 25 | 30 | 157 |
| Days Min Temp ≤ 0 °F | 4 | 3 | 0 | 0 | 0 | 0 | 0 | 0 | 0 | 0 | 0 | 3 | 10 |
| Heating Degree Days | 1197 | 955 | 771 | 428 | 181 | 29 | 2 | 7 | 105 | 390 | 801 | 1113 | 5979 |
| Cooling Degree Days | 0 | 0 | 0 | 11 | 43 | 207 | 343 | 293 | 108 | 5 | 0 | 0 | 1010 |
| Total Precipitation (") | 0.52 | 0.60 | 1.29 | 2.06 | 3.26 | 3.12 | 3.46 | 2.67 | 1.42 | 1.22 | 0.94 | 0.59 | 21.15 |
| Days ≥ 0.1" Precip | ***2*** | 2 | ***3*** | 4 | 6 | 5 | 6 | 5 | 3 | 3 | ***2*** | ***2*** | 43 |
| Total Snowfall (") | na | ***5.3*** | ***6.6*** | ***2.5*** | 0.0 | 0.0 | 0.0 | 0.0 | 0.1 | 0.3 | ***4.7*** | ***6.1*** | na |
| Days ≥ 1" Snow Depth | na | ***9*** | na | ***1*** | 0 | 0 | 0 | 0 | 0 | 0 | ***3*** | na | na |

## MEDICINE CREEK DAM *Frontier County* ELEVATION 2392 ft LAT/LONG 40° 23 ' N / 100° 13 ' W

| | JAN | FEB | MAR | APR | MAY | JUN | JUL | AUG | SEP | OCT | NOV | DEC | YEAR |
|---|---|---|---|---|---|---|---|---|---|---|---|---|---|
| Maximum Temp °F | 36.6 | 41.9 | 52.2 | 64.1 | 72.9 | 83.3 | 88.8 | 87.1 | 78.6 | 67.2 | 50.6 | 39.9 | 63.6 |
| Minimum Temp °F | 10.1 | 14.9 | 24.4 | 35.6 | 46.5 | 56.5 | 61.9 | 59.2 | 48.5 | 35.2 | 23.2 | 13.7 | 35.8 |
| Mean Temp °F | 23.4 | 28.4 | 38.3 | 49.9 | 59.7 | 70.0 | 75.4 | 73.2 | 63.6 | 51.2 | 37.0 | 26.8 | 49.7 |
| Days Max Temp ≥ 90 °F | 0 | 0 | 0 | 1 | 1 | 8 | 15 | 13 | 5 | 1 | 0 | 0 | 44 |
| Days Max Temp ≤ 32 °F | 12 | 8 | 3 | 0 | 0 | 0 | 0 | 0 | 0 | 0 | 3 | 9 | 35 |
| Days Min Temp ≤ 32 °F | 31 | 28 | 25 | 11 | 2 | 0 | 0 | 0 | 2 | 12 | 26 | 31 | 168 |
| Days Min Temp ≤ 0 °F | 7 | 4 | 0 | 0 | 0 | 0 | 0 | 0 | 0 | 0 | 1 | 4 | 16 |
| Heating Degree Days | 1279 | 1025 | 821 | 456 | 197 | 33 | 4 | 9 | 126 | 423 | 836 | 1177 | 6386 |
| Cooling Degree Days | 0 | 0 | 0 | 11 | 41 | 195 | 325 | 277 | 102 | 4 | 0 | 0 | 955 |
| Total Precipitation (") | 0.41 | 0.47 | 1.31 | 1.91 | 3.38 | 3.71 | 3.53 | 2.73 | 1.69 | 1.22 | 0.92 | 0.45 | 21.73 |
| Days ≥ 0.1" Precip | 1 | 1 | 3 | 4 | 7 | 6 | 6 | 5 | 3 | 3 | 2 | 1 | 42 |
| Total Snowfall (") | ***4.4*** | ***4.3*** | ***2.9*** | 1.2 | 0.0 | 0.0 | 0.0 | 0.0 | 0.0 | 0.0 | ***1.9*** | ***4.6*** | 19.3 |
| Days ≥ 1" Snow Depth | 12 | 8 | 4 | 0 | 0 | 0 | 0 | 0 | 0 | 0 | 3 | 9 | 36 |

## MERRIMAN *Cherry County* ELEVATION 3251 ft LAT/LONG 42° 55 ' N / 101° 41 ' W

| | JAN | FEB | MAR | APR | MAY | JUN | JUL | AUG | SEP | OCT | NOV | DEC | YEAR |
|---|---|---|---|---|---|---|---|---|---|---|---|---|---|
| Maximum Temp °F | 33.4 | 39.5 | 48.4 | 60.2 | 71.3 | 80.8 | 87.9 | 86.6 | 77.2 | 65.0 | 47.0 | 37.0 | 61.2 |
| Minimum Temp °F | 9.8 | 14.5 | 23.0 | 33.3 | 43.5 | 52.9 | 58.7 | 56.4 | 46.0 | 34.4 | 22.0 | 13.4 | 34.0 |
| Mean Temp °F | 21.6 | 27.0 | 35.7 | 46.8 | 57.4 | 66.9 | 73.3 | 71.5 | 61.6 | 49.7 | 34.5 | 25.3 | 47.6 |
| Days Max Temp ≥ 90 °F | 0 | 0 | 0 | 0 | 1 | 6 | 14 | 13 | 5 | 1 | 0 | 0 | 40 |
| Days Max Temp ≤ 32 °F | 13 | 9 | 4 | 0 | 0 | 0 | 0 | 0 | 0 | 0 | 5 | 11 | 42 |
| Days Min Temp ≤ 32 °F | 30 | 27 | 26 | 14 | 3 | 0 | 0 | 0 | 2 | 12 | 25 | 30 | 169 |
| Days Min Temp ≤ 0 °F | 8 | 5 | 1 | 0 | 0 | 0 | 0 | 0 | 0 | 0 | 1 | 5 | 20 |
| Heating Degree Days | 1339 | 1065 | 900 | 544 | 259 | 66 | 11 | 22 | 169 | 471 | 908 | 1225 | 6979 |
| Cooling Degree Days | 0 | 0 | 0 | 6 | 34 | 135 | 271 | 232 | 75 | 5 | 0 | 0 | 758 |
| Total Precipitation (") | 0.38 | 0.45 | 0.91 | 1.78 | 2.60 | 3.20 | 2.77 | 1.89 | 1.45 | 0.97 | 0.61 | 0.43 | 17.44 |
| Days ≥ 0.1" Precip | 1 | 2 | 3 | 4 | 6 | 6 | 5 | 4 | 3 | 2 | 2 | 1 | 39 |
| Total Snowfall (") | 4.5 | ***5.0*** | ***8.3*** | 4.2 | 0.2 | 0.0 | 0.0 | 0.0 | 0.2 | 2.2 | 5.9 | 5.2 | 35.7 |
| Days ≥ 1" Snow Depth | ***17*** | ***13*** | ***7*** | 2 | 0 | 0 | 0 | 0 | 0 | 1 | 7 | 15 | 62 |

## MINDEN *Kearney County* ELEVATION 2172 ft LAT/LONG 40° 30 ' N / 98° 57 ' W

| | JAN | FEB | MAR | APR | MAY | JUN | JUL | AUG | SEP | OCT | NOV | DEC | YEAR |
|---|---|---|---|---|---|---|---|---|---|---|---|---|---|
| Maximum Temp °F | 36.0 | 42.3 | 53.4 | 65.8 | 74.6 | 84.8 | 89.2 | 87.1 | 79.5 | 67.9 | 50.1 | 39.2 | 64.2 |
| Minimum Temp °F | 12.9 | 17.4 | 27.2 | 38.0 | 49.0 | 58.6 | 63.7 | 61.6 | 51.7 | 39.5 | 26.4 | 17.0 | 38.6 |
| Mean Temp °F | 24.5 | 29.9 | 40.3 | 51.9 | 61.8 | 71.8 | 76.5 | 74.4 | 65.6 | 53.7 | 38.3 | 28.1 | 51.4 |
| Days Max Temp ≥ 90 °F | 0 | 0 | 0 | 1 | 1 | 8 | 15 | 13 | 5 | 0 | 0 | 0 | 43 |
| Days Max Temp ≤ 32 °F | 11 | 8 | 2 | 0 | 0 | 0 | 0 | 0 | 0 | 0 | 2 | 9 | 32 |
| Days Min Temp ≤ 32 °F | 30 | 26 | 22 | 9 | 1 | 0 | 0 | 0 | 1 | 7 | 23 | 30 | 149 |
| Days Min Temp ≤ 0 °F | 6 | 3 | 0 | 0 | 0 | 0 | 0 | 0 | 0 | 0 | 0 | 3 | 12 |
| Heating Degree Days | 1250 | 984 | 760 | 398 | 149 | 18 | 2 | 6 | 93 | 355 | 795 | 1136 | 5946 |
| Cooling Degree Days | 0 | 0 | 1 | 18 | 61 | 236 | 363 | 306 | 132 | 10 | 0 | 0 | 1127 |
| Total Precipitation (") | 0.42 | 0.55 | 1.85 | 2.17 | 3.89 | 3.97 | 3.91 | 2.90 | 2.58 | 1.61 | 1.03 | 0.56 | 25.44 |
| Days ≥ 0.1" Precip | 1 | 2 | 3 | 5 | 7 | 6 | 6 | 5 | 4 | 3 | 2 | 2 | 46 |
| Total Snowfall (") | 5.6 | 5.0 | 5.6 | 2.2 | 0.0 | 0.0 | 0.0 | 0.0 | 0.3 | 0.6 | ***3.5*** | 4.8 | 27.6 |
| Days ≥ 1" Snow Depth | 15 | 11 | 5 | 1 | 0 | 0 | 0 | 0 | 0 | 0 | 3 | 10 | 45 |

## MITCHELL 5 E *Scotts Bluff County* ELEVATION 4081 ft LAT/LONG 41° 57 ' N / 103° 41 ' W

| | JAN | FEB | MAR | APR | MAY | JUN | JUL | AUG | SEP | OCT | NOV | DEC | YEAR |
|---|---|---|---|---|---|---|---|---|---|---|---|---|---|
| Maximum Temp °F | 37.0 | 42.3 | 49.7 | 60.2 | 70.4 | 80.8 | 86.6 | 84.1 | 75.7 | 64.2 | ***48.6*** | 39.3 | 61.6 |
| Minimum Temp °F | 12.1 | 15.8 | 23.5 | 32.0 | 42.2 | 52.4 | 57.4 | 54.4 | 44.1 | 32.4 | 22.1 | 13.3 | 33.5 |
| Mean Temp °F | 24.6 | 29.1 | 36.6 | 46.1 | 56.3 | 66.7 | 72.0 | 69.3 | 59.9 | 48.3 | ***35.4*** | 26.3 | 47.5 |
| Days Max Temp ≥ 90 °F | 0 | 0 | 0 | 0 | 1 | 6 | 11 | 7 | 2 | 0 | 0 | 0 | 27 |
| Days Max Temp ≤ 32 °F | 10 | 7 | 4 | 1 | 0 | 0 | 0 | 0 | 0 | 1 | 3 | 9 | 35 |
| Days Min Temp ≤ 32 °F | 30 | 27 | 27 | 16 | 3 | 0 | 0 | 0 | 2 | 15 | 26 | 30 | 176 |
| Days Min Temp ≤ 0 °F | 6 | 3 | 1 | 0 | 0 | 0 | 0 | 0 | 0 | 0 | 1 | 4 | 15 |
| Heating Degree Days | 1245 | 1007 | 873 | 561 | 282 | 67 | 9 | 21 | 184 | 510 | ***880*** | 1192 | 6831 |
| Cooling Degree Days | 0 | 0 | 0 | 1 | 23 | 139 | 236 | 180 | 43 | 1 | ***0*** | 0 | 623 |
| Total Precipitation (") | 0.29 | 0.29 | 0.63 | 1.27 | 2.62 | 2.61 | 2.00 | 1.26 | 1.13 | 0.84 | 0.41 | 0.28 | 13.63 |
| Days ≥ 0.1" Precip | 1 | 1 | 2 | 4 | 5 | 5 | 5 | 3 | 3 | 2 | 2 | 1 | 34 |
| Total Snowfall (") | ***4.8*** | ***4.3*** | ***5.9*** | ***2.8*** | 0.3 | 0.0 | 0.0 | 0.0 | 0.3 | 1.7 | na | na | na |
| Days ≥ 1" Snow Depth | na | na | ***5*** | ***1*** | 0 | 0 | 0 | 0 | 0 | 1 | na | na | na |

## MULLEN *Hooker County* ELEVATION 3222 ft LAT/LONG 42° 3 ' N / 101° 3 ' W

| | JAN | FEB | MAR | APR | MAY | JUN | JUL | AUG | SEP | OCT | NOV | DEC | YEAR |
|---|---|---|---|---|---|---|---|---|---|---|---|---|---|
| Maximum Temp °F | 35.8 | 41.1 | 50.7 | 62.7 | 73.3 | 83.1 | 88.7 | 86.7 | 76.8 | 64.8 | 47.2 | ***37.5*** | 62.4 |
| Minimum Temp °F | 11.9 | 15.6 | 23.5 | 33.9 | 44.4 | 53.8 | 59.4 | 57.7 | 46.9 | 35.5 | 23.5 | ***14.1*** | 35.0 |
| Mean Temp °F | 23.8 | 28.4 | 37.2 | 48.3 | 58.9 | 68.5 | 74.1 | 72.2 | 61.9 | 50.2 | 35.4 | ***25.9*** | 48.7 |
| Days Max Temp ≥ 90 °F | 0 | 0 | 0 | 0 | 1 | 7 | 15 | 12 | 3 | 0 | 0 | 0 | 38 |
| Days Max Temp ≤ 32 °F | 11 | 7 | 4 | 0 | 0 | 0 | 0 | 0 | 0 | 0 | 4 | ***10*** | 36 |
| Days Min Temp ≤ 32 °F | 30 | 27 | 26 | 14 | 3 | 0 | 0 | 0 | 2 | 11 | 25 | 30 | 168 |
| Days Min Temp ≤ 0 °F | 7 | 4 | 1 | 0 | 0 | 0 | 0 | 0 | 0 | 0 | 1 | ***4*** | 17 |
| Heating Degree Days | 1268 | 1026 | 857 | 498 | 216 | 41 | 6 | 12 | 156 | 456 | 883 | ***1208*** | 6627 |
| Cooling Degree Days | 0 | ***0*** | 0 | 6 | 35 | 155 | 285 | 245 | 70 | 3 | 0 | 0 | 799 |
| Total Precipitation (") | 0.43 | 0.51 | 1.27 | 2.22 | 3.24 | 3.24 | 3.62 | 2.60 | 1.77 | 1.02 | 0.93 | 0.51 | 21.36 |
| Days ≥ 0.1" Precip | 1 | 1 | 3 | 4 | 6 | 6 | 6 | 5 | 3 | 3 | 2 | 2 | 42 |
| Total Snowfall (") | 7.9 | 5.8 | 8.2 | 5.0 | 0.1 | 0.0 | 0.0 | 0.0 | 0.2 | 1.5 | 7.9 | ***8.2*** | 44.8 |
| Days ≥ 1" Snow Depth | 15 | 10 | 7 | 2 | 0 | 0 | 0 | 0 | 0 | 1 | 6 | ***12*** | 53 |

## MULLEN 21 NW *Cherry County* ELEVATION 3450 ft LAT/LONG 42° 16 ' N / 101° 21 ' W

| | JAN | FEB | MAR | APR | MAY | JUN | JUL | AUG | SEP | OCT | NOV | DEC | YEAR |
|---|---|---|---|---|---|---|---|---|---|---|---|---|---|
| Maximum Temp °F | 32.3 | 37.8 | 46.8 | 58.4 | 69.7 | 79.7 | 86.2 | 85.0 | 75.5 | 62.9 | 46.2 | ***36.7*** | 59.8 |
| Minimum Temp °F | 9.0 | 13.9 | 21.5 | 31.6 | 42.0 | 51.2 | 56.8 | 55.3 | 44.0 | 32.6 | 20.8 | ***12.6*** | 32.6 |
| Mean Temp °F | 20.7 | 25.9 | 34.2 | 45.0 | 55.9 | 65.5 | 71.5 | 70.1 | 59.8 | 47.8 | 33.5 | ***24.6*** | 46.2 |
| Days Max Temp ≥ 90 °F | 0 | 0 | 0 | 0 | 1 | 4 | 12 | 10 | 3 | 0 | 0 | 0 | 30 |
| Days Max Temp ≤ 32 °F | 14 | 10 | 6 | 1 | 0 | 0 | 0 | 0 | 0 | 0 | 5 | 11 | 47 |
| Days Min Temp ≤ 32 °F | 31 | 28 | 28 | 16 | 4 | 0 | 0 | 0 | 3 | 14 | 27 | 31 | 182 |
| Days Min Temp ≤ 0 °F | 9 | 4 | 1 | 0 | 0 | 0 | 0 | 0 | 0 | 0 | 1 | 4 | 19 |
| Heating Degree Days | 1369 | 1097 | 950 | 595 | 294 | 82 | 17 | 26 | 201 | 548 | 920 | ***1246*** | 7345 |
| Cooling Degree Days | ***0*** | 0 | 0 | 4 | ***18*** | 109 | 219 | 205 | 59 | 3 | 0 | 0 | 617 |
| Total Precipitation (") | 0.72 | 0.65 | 1.53 | 2.36 | 3.11 | 3.73 | 3.40 | 2.54 | 1.72 | 1.23 | 1.21 | 0.70 | 22.90 |
| Days ≥ 0.1" Precip | ***2*** | ***2*** | 4 | 5 | 6 | 6 | 6 | 5 | 4 | 3 | 3 | ***3*** | 49 |
| Total Snowfall (") | ***8.6*** | ***6.3*** | ***11.7*** | 7.0 | 0.3 | 0.0 | 0.0 | 0.0 | 0.8 | 2.5 | ***9.3*** | ***8.2*** | 54.7 |
| Days ≥ 1" Snow Depth | 21 | ***14*** | ***11*** | 4 | 0 | 0 | 0 | 0 | 0 | ***2*** | ***9*** | ***16*** | 77 |

## NE NEBRASKA EXP STN *Dixon County* ELEVATION 1480 ft LAT/LONG 42° 23 ' N / 96° 58 ' W

| | JAN | FEB | MAR | APR | MAY | JUN | JUL | AUG | SEP | OCT | NOV | DEC | YEAR |
|---|---|---|---|---|---|---|---|---|---|---|---|---|---|
| Maximum Temp °F | 27.6 | 32.6 | 45.5 | 60.3 | 71.7 | 81.0 | 85.4 | 83.0 | 74.6 | 62.7 | 45.0 | 32.2 | 58.5 |
| Minimum Temp °F | 5.5 | 11.0 | 23.1 | 35.2 | 46.7 | 56.9 | 61.4 | 58.6 | 48.2 | 35.6 | 23.5 | 11.5 | 34.8 |
| Mean Temp °F | 16.6 | 21.8 | 34.3 | 47.8 | 59.2 | 69.0 | 73.4 | 70.8 | 61.4 | 49.2 | 34.3 | 21.9 | 46.6 |
| Days Max Temp ≥ 90 °F | 0 | 0 | 0 | 0 | 1 | 5 | 9 | 6 | 2 | 0 | 0 | 0 | 23 |
| Days Max Temp ≤ 32 °F | 18 | 14 | 6 | 0 | 0 | 0 | 0 | 0 | 0 | 0 | 5 | 15 | 58 |
| Days Min Temp ≤ 32 °F | 31 | 28 | 26 | 12 | 2 | 0 | 0 | 0 | 2 | 11 | 25 | 31 | 168 |
| Days Min Temp ≤ 0 °F | 12 | 7 | 1 | 0 | 0 | 0 | 0 | 0 | 0 | 0 | 1 | 6 | 27 |
| Heating Degree Days | 1496 | 1213 | 944 | 518 | 218 | 40 | 8 | 21 | 167 | 488 | 916 | 1331 | 7360 |
| Cooling Degree Days | 0 | 0 | 0 | 11 | 43 | 168 | 263 | 207 | 72 | 3 | 0 | 0 | 767 |
| Total Precipitation (") | 0.50 | 0.74 | 1.83 | 2.54 | 3.96 | 4.41 | 2.89 | 2.63 | 2.63 | 1.99 | 1.28 | 0.79 | 26.19 |
| Days ≥ 0.1" Precip | 1 | 2 | 4 | 6 | 7 | 6 | 5 | 5 | 5 | 4 | 3 | 2 | 50 |
| Total Snowfall (") | 4.8 | 5.1 | 5.0 | 1.3 | 0.0 | 0.0 | 0.0 | 0.0 | 0.0 | 0.7 | 4.2 | 5.4 | 26.5 |
| Days ≥ 1" Snow Depth | na | na | *7* | 1 | 0 | 0 | 0 | 0 | 0 | 0 | *5* | *14* | na |

## NEBRASKA CITY *Otoe County* ELEVATION 971 ft LAT/LONG 40° 41 ' N / 95° 50 ' W

| | JAN | FEB | MAR | APR | MAY | JUN | JUL | AUG | SEP | OCT | NOV | DEC | YEAR |
|---|---|---|---|---|---|---|---|---|---|---|---|---|---|
| Maximum Temp °F | 32.1 | 37.6 | 50.4 | 63.5 | 73.9 | 83.4 | 87.3 | 85.0 | 77.2 | 65.7 | 49.4 | 36.7 | 61.9 |
| Minimum Temp °F | 12.0 | 16.6 | 27.7 | 39.3 | 50.2 | 59.9 | 64.9 | 62.1 | 52.9 | 40.9 | 28.7 | 17.9 | 39.4 |
| Mean Temp °F | 22.1 | 27.1 | 39.0 | 51.4 | 62.1 | 71.7 | 76.1 | 73.6 | 65.1 | 53.3 | 39.1 | 27.3 | 50.7 |
| Days Max Temp ≥ 90 °F | 0 | 0 | 0 | 0 | 1 | 6 | 12 | 9 | 3 | 0 | 0 | 0 | 31 |
| Days Max Temp ≤ 32 °F | 15 | 10 | 3 | 0 | 0 | 0 | 0 | 0 | 0 | 0 | 3 | 11 | 42 |
| Days Min Temp ≤ 32 °F | 30 | 26 | 22 | 7 | 0 | 0 | 0 | 0 | 0 | 6 | 21 | 29 | 141 |
| Days Min Temp ≤ 0 °F | 7 | 4 | 0 | 0 | 0 | 0 | 0 | 0 | 0 | 0 | 0 | 3 | 14 |
| Heating Degree Days | 1324 | 1062 | 798 | 415 | 150 | 18 | 3 | 7 | 103 | 368 | 772 | 1161 | 6181 |
| Cooling Degree Days | 0 | 0 | 2 | 19 | 58 | 227 | 346 | 274 | 122 | 9 | 0 | 0 | 1057 |
| Total Precipitation (") | 0.91 | 0.93 | 2.35 | 3.19 | 3.98 | 3.89 | 5.37 | 3.82 | 3.61 | 2.80 | 1.66 | 1.35 | 33.86 |
| Days ≥ 0.1" Precip | 2 | 3 | 5 | 6 | 8 | 6 | 6 | 6 | 6 | 5 | 4 | 3 | 60 |
| Total Snowfall (") | 7.3 | 6.9 | 3.9 | 1.0 | 0.1 | 0.0 | 0.0 | 0.0 | 0.0 | 0.2 | 1.7 | 5.2 | 26.3 |
| Days ≥ 1" Snow Depth | 14 | 11 | *4* | 1 | 0 | 0 | 0 | 0 | 0 | 0 | 2 | 7 | 39 |

## NEWPORT *Rock County* ELEVATION 2231 ft LAT/LONG 42° 36 ' N / 99° 20 ' W

| | JAN | FEB | MAR | APR | MAY | JUN | JUL | AUG | SEP | OCT | NOV | DEC | YEAR |
|---|---|---|---|---|---|---|---|---|---|---|---|---|---|
| Maximum Temp °F | 31.0 | 36.1 | 46.7 | 59.9 | 71.4 | 81.6 | 87.8 | 86.2 | 76.2 | 63.7 | 46.0 | 35.1 | 60.1 |
| Minimum Temp °F | 8.9 | 13.8 | 24.0 | 35.8 | 46.6 | 56.1 | 61.3 | 59.2 | 49.0 | 36.8 | 24.0 | 13.2 | 35.7 |
| Mean Temp °F | 20.0 | 25.0 | 35.4 | 47.8 | 59.0 | 68.9 | 74.6 | 72.7 | 62.6 | 50.3 | 35.0 | 24.2 | 48.0 |
| Days Max Temp ≥ 90 °F | 0 | 0 | 0 | 0 | 1 | 6 | 14 | 12 | 4 | 1 | 0 | 0 | 38 |
| Days Max Temp ≤ 32 °F | 16 | 12 | 6 | 0 | 0 | 0 | 0 | 0 | 0 | 0 | 6 | 13 | 53 |
| Days Min Temp ≤ 32 °F | 30 | 27 | 25 | 11 | 1 | 0 | 0 | 0 | 1 | 9 | 24 | 30 | 158 |
| Days Min Temp ≤ 0 °F | 10 | 5 | 1 | 0 | 0 | 0 | 0 | 0 | 0 | 0 | 1 | 5 | 22 |
| Heating Degree Days | 1390 | 1124 | 912 | 516 | 221 | 44 | 9 | 17 | 152 | 458 | 894 | 1259 | 6996 |
| Cooling Degree Days | 0 | 0 | 0 | 12 | 45 | 175 | 307 | 264 | 91 | 7 | 0 | 0 | 901 |
| Total Precipitation (") | 0.51 | 0.74 | 1.49 | 2.27 | 3.45 | 3.91 | 3.48 | 2.48 | 2.39 | 1.67 | 1.13 | 0.73 | 24.25 |
| Days ≥ 0.1" Precip | 2 | 2 | 3 | 5 | 6 | 6 | 6 | 4 | 5 | 4 | 3 | 2 | 48 |
| Total Snowfall (") | 5.9 | 6.6 | 8.4 | 5.1 | 0.0 | 0.0 | 0.0 | 0.0 | 0.3 | 2.3 | 7.6 | 7.9 | 44.1 |
| Days ≥ 1" Snow Depth | 20 | 16 | 10 | 2 | 0 | 0 | 0 | 0 | 0 | 1 | 8 | 16 | 73 |

## NIOBRARA *Knox County* ELEVATION 1230 ft LAT/LONG 42° 45 ' N / 98° 1 ' W

| | JAN | FEB | MAR | APR | MAY | JUN | JUL | AUG | SEP | OCT | NOV | DEC | YEAR |
|---|---|---|---|---|---|---|---|---|---|---|---|---|---|
| Maximum Temp °F | 30.8 | 36.4 | 48.9 | 63.7 | 74.4 | 83.9 | 88.4 | 86.7 | 77.5 | 65.1 | 45.7 | 34.1 | 61.3 |
| Minimum Temp °F | 9.2 | 14.8 | 25.7 | 37.3 | 48.2 | 57.7 | 63.2 | 61.1 | 51.1 | 38.5 | 25.4 | 14.2 | 37.2 |
| Mean Temp °F | 20.0 | 25.6 | 37.3 | 50.5 | 61.3 | 70.8 | 75.8 | 74.0 | 64.3 | 51.8 | 35.6 | 24.2 | 49.3 |
| Days Max Temp ≥ 90 °F | 0 | 0 | 0 | 1 | 2 | 8 | 14 | 12 | 4 | 0 | 0 | 0 | 41 |
| Days Max Temp ≤ 32 °F | 16 | 11 | 4 | 0 | 0 | 0 | 0 | 0 | 0 | 0 | 4 | 13 | 48 |
| Days Min Temp ≤ 32 °F | 30 | 26 | 24 | 9 | 1 | 0 | 0 | 0 | 1 | 8 | 23 | 30 | 152 |
| Days Min Temp ≤ 0 °F | 9 | 5 | 1 | 0 | 0 | 0 | 0 | 0 | 0 | 0 | 1 | 5 | 21 |
| Heating Degree Days | 1388 | 1106 | 853 | 441 | 168 | 24 | 3 | 7 | 116 | 409 | 877 | 1259 | 6651 |
| Cooling Degree Days | 0 | 0 | 0 | 20 | 67 | 215 | 350 | 299 | 116 | 9 | 0 | 0 | 1076 |
| Total Precipitation (") | 0.38 | 0.49 | 1.34 | 2.29 | 3.27 | 3.59 | 3.16 | 2.53 | 2.28 | 1.64 | 0.95 | 0.48 | 22.40 |
| Days ≥ 0.1" Precip | 1 | 1 | 3 | 5 | 6 | 6 | 5 | 4 | 5 | 3 | 3 | 2 | 44 |
| Total Snowfall (") | 5.1 | 5.1 | 4.9 | 1.4 | 0.0 | 0.0 | 0.0 | 0.0 | 0.0 | 0.5 | 4.3 | 6.4 | 27.7 |
| Days ≥ 1" Snow Depth | 19 | 13 | 7 | 1 | 0 | 0 | 0 | 0 | 0 | 0 | 4 | 14 | 58 |

## NORFOLK STEFAN AP *Madison County* ELEVATION 1544 ft LAT/LONG 41° 59 ' N / 97° 26 ' W

| | JAN | FEB | MAR | APR | MAY | JUN | JUL | AUG | SEP | OCT | NOV | DEC | YEAR |
|---|---|---|---|---|---|---|---|---|---|---|---|---|---|
| Maximum Temp °F | 29.8 | 34.9 | 47.3 | 61.8 | 72.8 | 82.7 | 86.9 | 84.4 | 75.6 | 63.5 | 46.0 | 33.6 | 59.9 |
| Minimum Temp °F | 9.0 | 14.3 | 25.6 | 37.8 | 49.1 | 58.9 | 64.0 | 61.5 | 51.1 | 38.3 | 25.3 | 14.0 | 37.4 |
| Mean Temp °F | 19.4 | 24.7 | 36.5 | 49.8 | 60.9 | 70.8 | 75.5 | 73.0 | 63.3 | 50.9 | 35.7 | 23.8 | 48.7 |
| Days Max Temp ≥ 90 °F | 0 | 0 | 0 | 1 | 1 | 7 | 12 | 9 | 3 | 0 | 0 | 0 | 33 |
| Days Max Temp ≤ 32 °F | 16 | 12 | 5 | 0 | 0 | 0 | 0 | 0 | 0 | 0 | 4 | 14 | 51 |
| Days Min Temp ≤ 32 °F | 31 | 27 | 24 | 9 | 1 | 0 | 0 | 0 | 1 | 8 | 24 | 30 | 155 |
| Days Min Temp ≤ 0 °F | 9 | 5 | 1 | 0 | 0 | 0 | 0 | 0 | 0 | 0 | 1 | 4 | 20 |
| Heating Degree Days | 1408 | 1133 | 878 | 462 | 174 | 25 | 3 | 11 | 131 | 435 | 873 | 1272 | 6805 |
| Cooling Degree Days | 0 | 0 | 0 | 19 | 58 | 213 | 326 | 267 | 100 | 5 | 0 | 0 | 988 |
| Total Precipitation (") | 0.60 | 0.80 | 1.83 | 2.34 | 3.70 | 4.54 | 3.45 | 2.64 | 2.56 | 1.78 | 1.17 | 0.76 | 26.17 |
| Days ≥ 0.1" Precip | 2 | 2 | 4 | 5 | 7 | 6 | 6 | 5 | 5 | 4 | 3 | 2 | 51 |
| Total Snowfall (") | 6.1 | 6.0 | 5.5 | 2.1 | 0.0 | 0.0 | 0.0 | 0.0 | 0.0 | 0.7 | 4.3 | 6.7 | 31.4 |
| Days ≥ 1" Snow Depth | 19 | 13 | 8 | 1 | 0 | 0 | 0 | 0 | 0 | 0 | 4 | 15 | 60 |

## NORTH LOUP *Valley County* ELEVATION 1962 ft LAT/LONG 41° 30 ' N / 98° 46 ' W

| | JAN | FEB | MAR | APR | MAY | JUN | JUL | AUG | SEP | OCT | NOV | DEC | YEAR |
|---|---|---|---|---|---|---|---|---|---|---|---|---|---|
| Maximum Temp °F | 33.6 | 39.0 | 50.0 | 64.0 | 72.9 | 82.5 | 86.1 | 84.0 | 76.1 | 65.1 | 47.8 | 36.9 | 61.5 |
| Minimum Temp °F | 10.3 | 15.3 | 25.3 | 36.4 | 47.8 | 57.6 | 62.6 | 60.2 | 50.3 | 37.5 | 24.4 | 14.3 | 36.8 |
| Mean Temp °F | 22.0 | 27.2 | 37.7 | 50.2 | 60.4 | 70.1 | 74.3 | 72.1 | 63.2 | 51.3 | 36.1 | 25.6 | 49.2 |
| Days Max Temp ≥ 90 °F | 0 | 0 | 0 | 1 | 1 | 5 | 10 | 7 | 2 | 0 | 0 | 0 | 26 |
| Days Max Temp ≤ 32 °F | 14 | 9 | 3 | 0 | 0 | 0 | 0 | 0 | 0 | 0 | 3 | 11 | 40 |
| Days Min Temp ≤ 32 °F | 31 | 27 | 25 | 11 | 1 | 0 | 0 | 0 | 1 | 9 | 25 | 30 | 160 |
| Days Min Temp ≤ 0 °F | 8 | 4 | 1 | 0 | 0 | 0 | 0 | 0 | 0 | 0 | 1 | 4 | 18 |
| Heating Degree Days | 1327 | 1060 | 841 | 445 | 179 | 27 | 4 | 11 | 127 | 421 | 860 | 1215 | 6517 |
| Cooling Degree Days | 0 | 0 | 0 | 12 | 44 | 191 | 286 | 240 | 86 | 3 | 0 | 0 | 862 |
| Total Precipitation (") | 0.46 | 0.63 | 1.62 | 2.32 | 3.22 | 3.89 | 3.55 | 3.33 | 2.42 | 1.65 | 1.11 | 0.69 | 24.89 |
| Days ≥ 0.1" Precip | 2 | 2 | 4 | 5 | 6 | 7 | 6 | 5 | 5 | 3 | 3 | 2 | 50 |
| Total Snowfall (") | 5.0 | 5.3 | 5.1 | 1.2 | 0.2 | 0.0 | 0.0 | 0.0 | 0.0 | 0.8 | 4.6 | 7.7 | 29.9 |
| Days ≥ 1" Snow Depth | 16 | 12 | 5 | 1 | 0 | 0 | 0 | 0 | 0 | 0 | 3 | 12 | 49 |

## NORTH PLATTE BRD FLD *Lincoln County* ELEVATION 2789 ft LAT/LONG 41° 8 ' N / 100° 42 ' W

| | JAN | FEB | MAR | APR | MAY | JUN | JUL | AUG | SEP | OCT | NOV | DEC | YEAR |
|---|---|---|---|---|---|---|---|---|---|---|---|---|---|
| Maximum Temp °F | 34.9 | 40.8 | 50.7 | 62.5 | 71.7 | 81.6 | 87.3 | 85.8 | 77.2 | 65.4 | 48.7 | 38.2 | 62.1 |
| Minimum Temp °F | 9.3 | 14.3 | 23.4 | 34.2 | 44.8 | 54.4 | 60.1 | 57.7 | 46.3 | 33.4 | 21.1 | 11.9 | 34.2 |
| Mean Temp °F | 22.1 | 27.5 | 37.1 | 48.4 | 58.3 | 68.0 | 73.7 | 71.8 | 61.8 | 49.4 | 34.9 | 25.1 | 48.2 |
| Days Max Temp ≥ 90 °F | 0 | 0 | 0 | 0 | 1 | 6 | 13 | 11 | 4 | 0 | 0 | 0 | 35 |
| Days Max Temp ≤ 32 °F | 12 | 8 | 4 | 0 | 0 | 0 | 0 | 0 | 0 | 0 | 4 | 10 | 38 |
| Days Min Temp ≤ 32 °F | 31 | 28 | 27 | 13 | 2 | 0 | 0 | 0 | 2 | 14 | 28 | 31 | 176 |
| Days Min Temp ≤ 0 °F | 8 | 4 | 1 | 0 | 0 | 0 | 0 | 0 | 0 | 0 | 1 | 5 | 19 |
| Heating Degree Days | 1325 | 1051 | 859 | 497 | 229 | 44 | 6 | 15 | 157 | 478 | 895 | 1232 | 6788 |
| Cooling Degree Days | 0 | 0 | 0 | 8 | 28 | 155 | 291 | 248 | 76 | 1 | 0 | 0 | 807 |
| Total Precipitation (") | 0.43 | 0.49 | 1.17 | 1.94 | 3.20 | 3.43 | 3.15 | 1.94 | 1.31 | 1.17 | 0.74 | 0.49 | 19.46 |
| Days ≥ 0.1" Precip | 1 | 1 | 3 | 4 | 6 | 6 | 6 | 4 | 3 | 3 | 2 | 2 | 41 |
| Total Snowfall (") | 5.9 | 4.4 | 5.7 | 2.5 | 0.1 | 0.0 | 0.0 | 0.0 | 0.1 | 1.6 | 4.3 | 5.1 | 29.7 |
| Days ≥ 1" Snow Depth | 14 | 9 | 4 | 1 | 0 | 0 | 0 | 0 | 0 | 1 | 4 | 9 | 42 |

## NORTH PLATTE EXP FRM *Lincoln County* ELEVATION 3031 ft LAT/LONG 41° 4 ' N / 100° 45 ' W

| | JAN | FEB | MAR | APR | MAY | JUN | JUL | AUG | SEP | OCT | NOV | DEC | YEAR |
|---|---|---|---|---|---|---|---|---|---|---|---|---|---|
| Maximum Temp °F | 33.7 | 39.5 | 49.3 | 61.1 | 70.7 | 81.0 | 87.7 | 86.1 | 77.0 | 64.7 | 47.9 | 37.7 | 61.4 |
| Minimum Temp °F | 11.2 | 15.3 | 24.4 | 35.0 | 45.3 | 55.0 | 60.6 | 58.5 | 47.9 | 36.0 | 23.8 | 14.6 | 35.6 |
| Mean Temp °F | 22.5 | 27.4 | 36.9 | 48.1 | 58.0 | 68.0 | 74.2 | 72.3 | 62.5 | 50.4 | 35.8 | 26.2 | 48.5 |
| Days Max Temp ≥ 90 °F | 0 | 0 | 0 | 0 | 1 | 5 | 13 | 12 | 4 | 0 | 0 | 0 | 35 |
| Days Max Temp ≤ 32 °F | 13 | 9 | 4 | 0 | 0 | 0 | 0 | 0 | 0 | 0 | 5 | 11 | 42 |
| Days Min Temp ≤ 32 °F | 31 | 27 | 25 | 12 | 2 | 0 | 0 | 0 | 1 | 10 | 25 | 30 | 163 |
| Days Min Temp ≤ 0 °F | 7 | 4 | 1 | 0 | 0 | 0 | 0 | 0 | 0 | 0 | 1 | 3 | 16 |
| Heating Degree Days | 1312 | 1054 | 865 | 506 | 236 | 47 | 6 | 15 | 145 | 451 | 868 | 1197 | 6702 |
| Cooling Degree Days | 0 | 0 | 0 | 8 | 23 | 150 | 285 | 244 | 76 | 3 | 0 | 0 | 789 |
| Total Precipitation (") | 0.39 | 0.49 | 1.08 | 1.95 | 3.40 | 3.46 | 2.89 | 1.99 | 1.44 | 1.31 | 0.70 | 0.44 | 19.54 |
| Days ≥ 0.1" Precip | 1 | 1 | 3 | 4 | 7 | 6 | 6 | 4 | 3 | 3 | 2 | 1 | 41 |
| Total Snowfall (") | 6.9 | 5.2 | 6.3 | 2.8 | 0.0 | 0.0 | 0.0 | 0.0 | 0.2 | 1.8 | 5.0 | 5.9 | 34.1 |
| Days ≥ 1" Snow Depth | na | na | na | *0* | 0 | 0 | 0 | 0 | 0 | 0 | na | na | na |

## O'NEILL *Holt County* ELEVATION 1982 ft LAT/LONG 42° 27 ' N / 98° 38 ' W

| | JAN | FEB | MAR | APR | MAY | JUN | JUL | AUG | SEP | OCT | NOV | DEC | YEAR |
|---|---|---|---|---|---|---|---|---|---|---|---|---|---|
| Maximum Temp °F | 30.0 | 35.8 | 47.5 | 61.7 | 72.8 | 82.6 | 88.1 | 86.4 | 76.4 | 63.9 | 45.1 | 33.7 | 60.3 |
| Minimum Temp °F | 8.8 | 13.9 | 23.7 | 35.5 | 46.6 | 56.2 | 61.6 | 59.6 | 49.2 | 37.2 | 23.9 | 13.2 | 35.8 |
| Mean Temp °F | 19.5 | 24.9 | 35.6 | 48.6 | 59.7 | 69.4 | 74.9 | 73.0 | 62.9 | 50.6 | 34.5 | 23.4 | 48.1 |
| Days Max Temp ≥ 90 °F | 0 | 0 | 0 | 0 | 1 | 6 | 14 | 12 | 4 | 0 | 0 | 0 | 37 |
| Days Max Temp ≤ 32 °F | 16 | 11 | 5 | 0 | 0 | 0 | 0 | 0 | 0 | 0 | 5 | 14 | 51 |
| Days Min Temp ≤ 32 °F | 30 | 27 | 25 | 12 | 2 | 0 | 0 | 0 | 1 | 10 | 24 | 30 | 161 |
| Days Min Temp ≤ 0 °F | 9 | 5 | 1 | 0 | 0 | 0 | 0 | 0 | 0 | 0 | 1 | 5 | 21 |
| Heating Degree Days | 1407 | 1127 | 904 | 493 | 201 | 34 | 8 | 13 | 143 | 445 | 907 | 1281 | 6963 |
| Cooling Degree Days | 0 | 0 | 0 | 12 | 45 | 185 | 319 | 272 | 91 | 5 | 0 | 0 | 929 |
| Total Precipitation (") | 0.55 | 0.56 | 1.56 | 2.34 | 3.29 | 3.59 | 3.30 | 2.68 | 2.24 | 1.63 | 1.08 | 0.71 | 23.53 |
| Days ≥ 0.1" Precip | 2 | 2 | 3 | 5 | 7 | 6 | 6 | 5 | 5 | 4 | 3 | 2 | 50 |
| Total Snowfall (") | 5.2 | 5.6 | 5.7 | 3.3 | 0.0 | 0.0 | 0.0 | 0.0 | 0.0 | 1.0 | 4.4 | 6.9 | 32.1 |
| Days ≥ 1" Snow Depth | 17 | 13 | *9* | 1 | 0 | 0 | 0 | 0 | 0 | 0 | 4 | 14 | 58 |

## OAKDALE *Antelope County* ELEVATION 1713 ft LAT/LONG 42° 4 ' N / 97° 58 ' W

| | JAN | FEB | MAR | APR | MAY | JUN | JUL | AUG | SEP | OCT | NOV | DEC | YEAR |
|---|---|---|---|---|---|---|---|---|---|---|---|---|---|
| Maximum Temp °F | 30.7 | 35.5 | 47.1 | 60.9 | 71.5 | 81.2 | 86.2 | 83.9 | 75.1 | 63.6 | 46.4 | 34.7 | 59.7 |
| Minimum Temp °F | 7.1 | 12.3 | 23.8 | 35.9 | 47.1 | 56.9 | 62.0 | 59.3 | 48.4 | 35.3 | 23.1 | 12.0 | 35.3 |
| Mean Temp °F | 18.9 | 23.9 | 35.5 | 48.5 | 59.3 | 69.1 | 74.1 | 71.6 | 61.8 | 49.5 | 34.8 | 23.4 | 47.5 |
| Days Max Temp ≥ 90 °F | 0 | 0 | 0 | 1 | 1 | 5 | 11 | 8 | 3 | 0 | 0 | 0 | 29 |
| Days Max Temp ≤ 32 °F | 16 | 12 | 5 | 0 | 0 | 0 | 0 | 0 | 0 | 0 | 5 | 13 | 51 |
| Days Min Temp ≤ 32 °F | 31 | 27 | 25 | 11 | 2 | 0 | 0 | 0 | 2 | 12 | 26 | 31 | 167 |
| Days Min Temp ≤ 0 °F | 10 | 6 | 1 | 0 | 0 | 0 | 0 | 0 | 0 | 0 | 1 | 6 | 24 |
| Heating Degree Days | 1424 | 1153 | 909 | 499 | 213 | 41 | 8 | 18 | 162 | 477 | 900 | 1285 | 7089 |
| Cooling Degree Days | 0 | 0 | 0 | 13 | 42 | 171 | 276 | 229 | 79 | 3 | 0 | 0 | 813 |
| Total Precipitation (") | 0.52 | 0.68 | 1.67 | 2.52 | 3.69 | 4.30 | 3.36 | 2.81 | 2.37 | 1.81 | 1.14 | 0.73 | 25.60 |
| Days ≥ 0.1" Precip | 2 | 2 | 4 | 5 | 7 | 6 | 6 | 5 | 5 | 4 | 3 | 2 | 51 |
| Total Snowfall (") | 5.3 | 5.8 | 5.3 | 2.2 | 0.0 | 0.0 | 0.0 | 0.0 | 0.0 | 0.8 | 5.0 | 6.7 | 31.1 |
| Days ≥ 1" Snow Depth | 18 | 13 | 6 | 1 | 0 | 0 | 0 | 0 | 0 | 0 | 4 | 14 | 56 |

## OCONTO *Custer County* ELEVATION 2581 ft LAT/LONG 41° 9 ' N / 99° 46 ' W

| | JAN | FEB | MAR | APR | MAY | JUN | JUL | AUG | SEP | OCT | NOV | DEC | YEAR |
|---|---|---|---|---|---|---|---|---|---|---|---|---|---|
| Maximum Temp °F | 35.5 | 40.9 | 51.3 | 63.6 | 72.4 | 82.1 | 87.0 | 85.6 | 77.6 | 65.6 | 48.6 | 39.1 | 62.4 |
| Minimum Temp °F | 11.4 | 16.0 | 24.8 | 35.3 | 46.2 | 55.8 | 60.8 | 59.0 | 49.1 | 36.9 | 24.0 | 15.4 | 36.2 |
| Mean Temp °F | 23.5 | 28.5 | 38.1 | 49.4 | 59.3 | 69.0 | 73.9 | 72.3 | 63.4 | 51.3 | 36.3 | 27.3 | 49.4 |
| Days Max Temp ≥ 90 °F | 0 | 0 | 0 | 0 | 1 | 6 | 12 | 10 | 3 | 0 | 0 | 0 | 32 |
| Days Max Temp ≤ 32 °F | 12 | 8 | 3 | 0 | 0 | 0 | 0 | 0 | 0 | 0 | 3 | 9 | 35 |
| Days Min Temp ≤ 32 °F | 31 | 27 | 25 | 11 | 2 | 0 | 0 | 0 | 1 | 10 | 25 | 30 | 162 |
| Days Min Temp ≤ 0 °F | 7 | 4 | 1 | 0 | 0 | 0 | 0 | 0 | 0 | 0 | 1 | 3 | 16 |
| Heating Degree Days | 1281 | 1023 | 828 | 467 | 203 | 34 | 6 | 12 | 124 | 423 | 854 | 1163 | 6418 |
| Cooling Degree Days | 0 | 0 | 0 | *9* | 36 | 172 | 288 | 245 | 83 | 4 | 0 | 0 | 837 |
| Total Precipitation (") | 0.64 | 0.54 | 1.62 | 1.97 | 3.46 | 3.97 | 3.51 | 2.43 | 1.94 | 1.38 | 0.92 | 0.61 | 22.99 |
| Days ≥ 0.1" Precip | 2 | 2 | 4 | 5 | 7 | 6 | 6 | 5 | 4 | 3 | 2 | 2 | 48 |
| Total Snowfall (") | na | na | na | *1.7* | 0.0 | 0.0 | 0.0 | 0.0 | 0.0 | 0.7 | *4.9* | na | na |
| Days ≥ 1" Snow Depth | na | na | na | *0* | 0 | 0 | 0 | 0 | 0 | 0 | na | na | na |

## OGALLALA *Keith County* ELEVATION 3222 ft LAT/LONG 41° 8 ' N / 101° 43 ' W

| | JAN | FEB | MAR | APR | MAY | JUN | JUL | AUG | SEP | OCT | NOV | DEC | YEAR |
|---|---|---|---|---|---|---|---|---|---|---|---|---|---|
| Maximum Temp °F | 35.7 | 42.0 | 50.9 | 62.6 | 72.3 | 82.5 | 89.6 | 87.4 | 77.9 | 65.7 | 49.7 | 39.5 | 63.0 |
| Minimum Temp °F | 10.8 | 15.4 | 23.9 | 34.1 | 44.8 | 54.7 | 60.1 | 57.4 | 46.5 | 33.6 | 22.3 | 13.2 | 34.7 |
| Mean Temp °F | 23.3 | 28.7 | 37.5 | 48.4 | 58.6 | 68.7 | 74.9 | 72.4 | 62.2 | 49.7 | 36.0 | 26.4 | 48.9 |
| Days Max Temp ≥ 90 °F | 0 | 0 | 0 | 0 | 1 | 8 | 16 | 14 | 5 | 0 | 0 | 0 | 44 |
| Days Max Temp ≤ 32 °F | 11 | 8 | 3 | 0 | 0 | 0 | 0 | 0 | 0 | 0 | 3 | 8 | 33 |
| Days Min Temp ≤ 32 °F | 31 | 28 | 26 | 13 | 2 | 0 | 0 | 0 | 2 | 13 | 27 | 31 | 173 |
| Days Min Temp ≤ 0 °F | 6 | 4 | 1 | 0 | 0 | 0 | 0 | 0 | 0 | 0 | 0 | 4 | 15 |
| Heating Degree Days | 1288 | 1017 | 847 | 494 | 224 | 44 | 6 | 12 | 149 | 470 | 863 | 1192 | 6606 |
| Cooling Degree Days | 0 | 0 | 0 | 7 | 34 | 177 | 326 | 261 | 82 | 2 | 0 | 0 | 889 |
| Total Precipitation (") | 0.51 | 0.39 | 1.33 | 1.66 | 3.32 | 2.91 | 2.74 | 1.86 | 1.15 | 0.95 | 0.67 | 0.57 | 18.06 |
| Days ≥ 0.1" Precip | 2 | 1 | 3 | 4 | 6 | 6 | 5 | 4 | 3 | 2 | 2 | 2 | 40 |
| Total Snowfall (") | *6.9* | *4.3* | *7.9* | *1.6* | 0.3 | 0.0 | 0.0 | 0.0 | 0.2 | *0.3* | *4.0* | *3.8* | 29.3 |
| Days ≥ 1" Snow Depth | na | na | na | *0* | 0 | 0 | 0 | 0 | 0 | *0* | na | na | na |

## OMAHA EPPLEY AIRFLD *Douglas County* ELEVATION 997 ft LAT/LONG 41° 18 ' N / 95° 54 ' W

| | JAN | FEB | MAR | APR | MAY | JUN | JUL | AUG | SEP | OCT | NOV | DEC | YEAR |
|---|---|---|---|---|---|---|---|---|---|---|---|---|---|
| Maximum Temp °F | 31.4 | 37.0 | 50.4 | 64.0 | 74.1 | 83.8 | 87.5 | 85.3 | 76.7 | 65.0 | 48.4 | 35.5 | 61.6 |
| Minimum Temp °F | 11.1 | 16.7 | 28.1 | 39.9 | 50.8 | 60.6 | 65.8 | 63.2 | 53.4 | 40.8 | 28.2 | 16.7 | 39.6 |
| Mean Temp °F | 21.3 | 26.9 | 39.2 | 52.0 | 62.5 | 72.2 | 76.7 | 74.3 | 65.1 | 52.9 | 38.3 | 26.1 | 50.6 |
| Days Max Temp ≥ 90 °F | 0 | 0 | 0 | 1 | 1 | 7 | 12 | 9 | 3 | 0 | 0 | 0 | 33 |
| Days Max Temp ≤ 32 °F | 15 | 11 | 3 | 0 | 0 | 0 | 0 | 0 | 0 | 0 | 3 | 11 | 43 |
| Days Min Temp ≤ 32 °F | 30 | 26 | 21 | 7 | 0 | 0 | 0 | 0 | 0 | 6 | 21 | 29 | 140 |
| Days Min Temp ≤ 0 °F | 7 | 4 | 0 | 0 | 0 | 0 | 0 | 0 | 0 | 0 | 0 | 3 | 14 |
| Heating Degree Days | 1349 | 1070 | 794 | 401 | 145 | 15 | 1 | 6 | 102 | 381 | 794 | 1198 | 6256 |
| Cooling Degree Days | 0 | 0 | 2 | 20 | 64 | 224 | 353 | 291 | 114 | 8 | 0 | 0 | 1076 |
| Total Precipitation (") | 0.81 | 0.77 | 2.00 | 2.70 | 4.29 | 4.13 | 3.74 | 2.98 | 3.74 | 2.29 | 1.59 | 1.06 | 30.10 |
| Days ≥ 0.1" Precip | 2 | 2 | 4 | 6 | 7 | 7 | 6 | 5 | 6 | 4 | 3 | 3 | 55 |
| Total Snowfall (") | 6.4 | 6.0 | 5.0 | 1.1 | 0.0 | 0.0 | 0.0 | 0.0 | 0.0 | 0.4 | 3.1 | 5.5 | 27.5 |
| Days ≥ 1" Snow Depth | 16 | 12 | 5 | 1 | 0 | 0 | 0 | 0 | 0 | 0 | 3 | 10 | 47 |

## OMAHA WSFO *Douglas County* ELEVATION 1319 ft LAT/LONG 41° 22 ' N / 96° 1 ' W

| | JAN | FEB | MAR | APR | MAY | JUN | JUL | AUG | SEP | OCT | NOV | DEC | YEAR |
|---|---|---|---|---|---|---|---|---|---|---|---|---|---|
| Maximum Temp °F | 29.9 | 35.3 | 48.0 | 61.9 | 72.4 | 81.9 | 85.7 | 83.5 | 75.0 | 63.2 | 46.5 | 34.0 | 59.8 |
| Minimum Temp °F | 11.8 | 17.1 | 28.1 | 40.4 | 51.7 | 61.3 | 66.2 | 63.7 | 54.5 | 42.4 | 29.0 | 17.6 | 40.3 |
| Mean Temp °F | 20.9 | 26.3 | 38.1 | 51.2 | 62.1 | 71.6 | 76.0 | 73.6 | 64.8 | 52.8 | 37.8 | 25.8 | 50.1 |
| Days Max Temp ≥ 90 °F | 0 | 0 | 0 | 0 | 1 | 5 | 9 | 7 | 2 | 0 | 0 | 0 | 24 |
| Days Max Temp ≤ 32 °F | 16 | 12 | 4 | 0 | 0 | 0 | 0 | 0 | 0 | 0 | 4 | 13 | 49 |
| Days Min Temp ≤ 32 °F | 30 | 26 | 20 | 6 | 0 | 0 | 0 | 0 | 0 | 4 | 19 | 29 | 134 |
| Days Min Temp ≤ 0 °F | 7 | 4 | 0 | 0 | 0 | 0 | 0 | 0 | 0 | 0 | 0 | 3 | 14 |
| Heating Degree Days | 1363 | 1088 | 830 | 423 | 152 | 18 | 2 | 7 | 107 | 383 | 809 | 1209 | 6391 |
| Cooling Degree Days | 0 | 0 | 2 | 22 | 69 | 233 | 363 | 293 | 124 | 10 | 0 | 0 | 1116 |
| Total Precipitation (") | 0.73 | 0.78 | 2.22 | 2.86 | 4.36 | 3.92 | 3.63 | 3.16 | 3.65 | 2.51 | 1.42 | 1.01 | 30.25 |
| Days ≥ 0.1" Precip | 2 | 2 | 5 | 6 | 8 | 7 | 6 | 5 | 6 | 4 | 3 | 3 | 57 |
| Total Snowfall (") | 7.1 | 6.6 | 5.6 | 1.7 | 0.0 | 0.0 | 0.0 | 0.0 | 0.0 | 0.7 | 3.5 | 5.5 | 30.7 |
| Days ≥ 1" Snow Depth | 17 | 13 | 6 | 1 | 0 | 0 | 0 | 0 | 0 | 0 | 3 | 10 | 50 |

## OSCEOLA *Polk County* ELEVATION 1640 ft LAT/LONG 41° 11 ' N / 97° 33 ' W

| | JAN | FEB | MAR | APR | MAY | JUN | JUL | AUG | SEP | OCT | NOV | DEC | YEAR |
|---|---|---|---|---|---|---|---|---|---|---|---|---|---|
| Maximum Temp °F | 32.7 | 38.6 | 51.1 | 64.9 | 74.8 | 84.8 | 88.4 | 86.4 | 78.4 | 66.3 | 48.5 | 36.3 | 62.6 |
| Minimum Temp °F | 11.5 | 16.1 | 27.0 | 38.3 | 49.8 | 59.4 | 63.9 | 61.5 | 51.8 | 39.8 | 26.5 | 16.1 | 38.5 |
| Mean Temp °F | 22.1 | 27.4 | 39.1 | 51.6 | 62.3 | 72.1 | 76.2 | 74.0 | 65.1 | 53.0 | 37.5 | 26.2 | 50.6 |
| Days Max Temp ≥ 90 °F | 0 | 0 | 0 | 1 | 1 | 8 | 14 | 11 | 4 | 0 | 0 | 0 | 39 |
| Days Max Temp ≤ 32 °F | 14 | 10 | 3 | 0 | 0 | 0 | 0 | 0 | 0 | 0 | 3 | 11 | 41 |
| Days Min Temp ≤ 32 °F | 31 | 26 | 22 | 9 | 1 | 0 | 0 | 0 | 1 | 7 | 22 | 30 | 149 |
| Days Min Temp ≤ 0 °F | 7 | 4 | 0 | 0 | 0 | 0 | 0 | 0 | 0 | 0 | 1 | 3 | 15 |
| Heating Degree Days | 1323 | 1056 | 797 | 407 | 143 | 16 | 2 | 6 | 99 | 373 | 817 | 1195 | 6234 |
| Cooling Degree Days | 0 | 0 | 1 | 20 | 65 | 241 | 347 | 290 | 120 | 6 | 0 | 0 | 1090 |
| Total Precipitation (") | 0.62 | 0.77 | 2.21 | 2.73 | 4.43 | 4.42 | 3.47 | 3.15 | 3.01 | 2.14 | 1.48 | 0.96 | 29.39 |
| Days ≥ 0.1" Precip | 2 | 2 | 4 | 5 | 8 | 6 | 6 | 5 | 5 | 4 | 3 | 2 | 52 |
| Total Snowfall (") | 5.9 | 6.8 | 5.4 | 1.4 | 0.4 | 0.0 | 0.0 | 0.0 | 0.1 | 0.7 | 4.3 | 6.8 | 31.8 |
| Days ≥ 1" Snow Depth | *13* | *10* | *2* | 0 | 0 | 0 | 0 | 0 | 0 | 0 | 4 | *9* | 38 |

## OSHKOSH 1 W *Garden County* ELEVATION 3392 ft LAT/LONG 41° 24 ' N / 102° 20 ' W

| | JAN | FEB | MAR | APR | MAY | JUN | JUL | AUG | SEP | OCT | NOV | DEC | YEAR |
|---|---|---|---|---|---|---|---|---|---|---|---|---|---|
| Maximum Temp °F | 37.5 | 43.9 | 52.5 | 63.9 | 72.8 | 82.4 | 88.6 | 86.9 | 78.3 | 66.8 | 49.8 | 40.1 | 63.6 |
| Minimum Temp °F | 10.4 | 15.6 | 23.4 | 32.9 | 43.7 | 52.9 | 58.6 | 56.2 | 45.6 | 32.9 | 21.2 | 12.0 | 33.8 |
| Mean Temp °F | 23.9 | 29.8 | 38.0 | 48.4 | 58.3 | 67.7 | 73.6 | 71.6 | 61.9 | 49.8 | 35.5 | 26.1 | 48.7 |
| Days Max Temp ≥ 90 °F | 0 | 0 | 0 | 0 | 1 | 7 | 14 | 12 | 4 | 0 | 0 | 0 | 38 |
| Days Max Temp ≤ 32 °F | 10 | 5 | 2 | 0 | 0 | 0 | 0 | 0 | 0 | 0 | 3 | 8 | 28 |
| Days Min Temp ≤ 32 °F | 31 | 28 | 26 | 14 | 2 | 0 | 0 | 0 | 2 | 15 | 28 | 31 | 177 |
| Days Min Temp ≤ 0 °F | 7 | 3 | 1 | 0 | 0 | 0 | 0 | 0 | 0 | 0 | 1 | 5 | 17 |
| Heating Degree Days | 1267 | 987 | 830 | 493 | 228 | 46 | 4 | 11 | 144 | 464 | 878 | 1200 | 6552 |
| Cooling Degree Days | 0 | 0 | 0 | 3 | 26 | 143 | 283 | 244 | 68 | 1 | 0 | 0 | 768 |
| Total Precipitation (") | 0.31 | 0.38 | 1.04 | 1.67 | 3.18 | 2.88 | 2.73 | 1.88 | 1.43 | 0.96 | 0.62 | 0.41 | 17.49 |
| Days ≥ 0.1" Precip | 1 | 1 | 3 | 4 | 6 | 6 | 5 | 4 | 3 | 2 | 2 | 1 | 38 |
| Total Snowfall (") | 5.0 | 4.2 | 5.5 | 2.4 | 0.4 | 0.0 | 0.0 | 0.0 | 0.2 | 1.2 | 4.9 | 5.3 | 29.1 |
| Days ≥ 1" Snow Depth | 14 | 9 | 4 | 1 | 0 | 0 | 0 | 0 | 0 | 0 | 5 | 11 | 44 |

## OSMOND *Pierce County* ELEVATION 1650 ft LAT/LONG 42° 21 ' N / 97° 36 ' W

| | JAN | FEB | MAR | APR | MAY | JUN | JUL | AUG | SEP | OCT | NOV | DEC | YEAR |
|---|---|---|---|---|---|---|---|---|---|---|---|---|---|
| Maximum Temp °F | 30.3 | 36.1 | 48.4 | 63.3 | 74.5 | 84.5 | 88.4 | 86.1 | 77.4 | 65.0 | 46.1 | 34.0 | 61.2 |
| Minimum Temp °F | 7.9 | 13.7 | 25.0 | 36.6 | 48.3 | 58.0 | 62.8 | 60.2 | 50.1 | 37.5 | 24.2 | 13.0 | 36.4 |
| Mean Temp °F | 18.9 | 24.9 | 36.7 | 49.8 | 61.5 | 71.3 | 75.6 | 73.2 | 63.8 | 51.3 | 35.2 | 23.5 | 48.8 |
| Days Max Temp ≥ 90 °F | 0 | 0 | 0 | 1 | 1 | 8 | 14 | 10 | 4 | 0 | 0 | 0 | 38 |
| Days Max Temp ≤ 32 °F | 16 | 11 | 4 | 0 | 0 | 0 | 0 | 0 | 0 | 0 | 4 | 13 | 48 |
| Days Min Temp ≤ 32 °F | 31 | 27 | 25 | 11 | 1 | 0 | 0 | 0 | 1 | 10 | 25 | 31 | 162 |
| Days Min Temp ≤ 0 °F | 10 | 5 | 1 | 0 | 0 | 0 | 0 | 0 | 0 | 0 | 1 | 5 | 22 |
| Heating Degree Days | 1423 | 1124 | 870 | 459 | 164 | 22 | 3 | 10 | 124 | 427 | 887 | 1278 | 6791 |
| Cooling Degree Days | 0 | 0 | 0 | 16 | 58 | 227 | 331 | 269 | 98 | 5 | 0 | 0 | 1004 |
| Total Precipitation (") | 0.45 | 0.77 | 1.69 | 2.61 | 3.41 | 3.98 | 3.35 | 2.94 | 2.58 | 1.65 | 1.16 | 0.62 | 25.21 |
| Days ≥ 0.1" Precip | 2 | 2 | 4 | 5 | 6 | 6 | 5 | 5 | 5 | 3 | 2 | 2 | 47 |
| Total Snowfall (") | na | na | *3.9* | 1.4 | 0.0 | 0.0 | 0.0 | 0.0 | 0.0 | 0.5 | *4.5* | *5.7* | na |
| Days ≥ 1" Snow Depth | na | na | *5* | 0 | 0 | 0 | 0 | 0 | 0 | 0 | *3* | na | na |

## PAWNEE CITY *Pawnee County* ELEVATION 1181 ft LAT/LONG 40° 6 ' N / 96° 9 ' W

| | JAN | FEB | MAR | APR | MAY | JUN | JUL | AUG | SEP | OCT | NOV | DEC | YEAR |
|---|---|---|---|---|---|---|---|---|---|---|---|---|---|
| Maximum Temp °F | 35.9 | 42.3 | 54.8 | 67.9 | 77.1 | 86.6 | 91.2 | 89.2 | 80.3 | 69.1 | 52.2 | 39.8 | 65.5 |
| Minimum Temp °F | 15.5 | 20.1 | 31.4 | 43.0 | 53.2 | 62.3 | 66.9 | 64.6 | 55.3 | 43.7 | 31.4 | 20.5 | 42.3 |
| Mean Temp °F | 25.7 | 31.2 | 43.1 | 55.5 | 65.1 | 74.5 | 79.1 | 77.0 | 67.8 | 56.4 | 41.8 | 30.2 | 54.0 |
| Days Max Temp ≥ 90 °F | 0 | 0 | 0 | 1 | 2 | 11 | 18 | 15 | 5 | 0 | 0 | 0 | 52 |
| Days Max Temp ≤ 32 °F | 12 | 8 | 1 | 0 | 0 | 0 | 0 | 0 | 0 | 0 | 2 | 8 | 31 |
| Days Min Temp ≤ 32 °F | 28 | 23 | 17 | 4 | 0 | 0 | 0 | 0 | 0 | 4 | 17 | 27 | 120 |
| Days Min Temp ≤ 0 °F | 5 | 3 | 0 | 0 | 0 | 0 | 0 | 0 | 0 | 0 | 0 | 2 | 10 |
| Heating Degree Days | 1213 | 948 | 676 | 306 | 96 | 7 | 1 | 2 | 69 | 285 | 690 | 1072 | 5365 |
| Cooling Degree Days | 0 | 0 | 4 | 35 | 98 | 297 | 445 | 383 | 167 | 18 | 1 | 0 | 1448 |
| Total Precipitation (") | 0.77 | 0.78 | 2.17 | 2.76 | 3.84 | 4.58 | 4.50 | 3.91 | 3.96 | 2.49 | 1.49 | 1.13 | 32.38 |
| Days ≥ 0.1" Precip | 2 | 2 | 4 | 5 | 7 | 6 | 6 | 6 | 6 | 4 | 3 | 2 | 53 |
| Total Snowfall (") | *6.0* | *5.4* | *2.7* | 0.3 | 0.0 | 0.0 | 0.0 | 0.0 | 0.0 | 0.0 | 1.3 | 4.5 | 20.2 |
| Days ≥ 1" Snow Depth | *9* | *5* | 2 | 0 | 0 | 0 | 0 | 0 | 0 | 0 | *0* | na | na |

## PURDUM *Blaine County* ELEVATION 2690 ft LAT/LONG 42° 4 ' N / 100° 15 ' W

| | JAN | FEB | MAR | APR | MAY | JUN | JUL | AUG | SEP | OCT | NOV | DEC | YEAR |
|---|---|---|---|---|---|---|---|---|---|---|---|---|---|
| Maximum Temp °F | 35.0 | 41.1 | 50.1 | 62.6 | 72.6 | 82.3 | 87.9 | 86.3 | 77.5 | 65.9 | *49.0* | *38.5* | 62.4 |
| Minimum Temp °F | 9.1 | 14.4 | 22.9 | 33.5 | 44.3 | 54.1 | 59.8 | 57.3 | 47.0 | 34.6 | 22.5 | 12.7 | 34.4 |
| Mean Temp °F | 22.1 | 27.8 | 36.5 | 48.1 | 58.5 | 68.3 | 73.9 | 71.9 | 62.3 | 50.3 | *35.8* | *25.5* | 48.4 |
| Days Max Temp ≥ 90 °F | 0 | 0 | 0 | 0 | 1 | 6 | 13 | 11 | 4 | 0 | 0 | 0 | 35 |
| Days Max Temp ≤ 32 °F | 12 | 8 | 4 | 0 | 0 | 0 | 0 | 0 | 0 | 0 | 3 | 10 | 37 |
| Days Min Temp ≤ 32 °F | 31 | 27 | *26* | 14 | 3 | 0 | 0 | 0 | 2 | 13 | 26 | 30 | 172 |
| Days Min Temp ≤ 0 °F | 8 | 4 | 1 | 0 | 0 | 0 | 0 | 0 | 0 | 0 | 1 | 4 | 18 |
| Heating Degree Days | 1327 | 1046 | 879 | 507 | 224 | 42 | 6 | 14 | 144 | 453 | *870* | *1218* | 6730 |
| Cooling Degree Days | 0 | *0* | *0* | *7* | *28* | *142* | *279* | *234* | 72 | 4 | *0* | *0* | 766 |
| Total Precipitation (") | 0.49 | 0.73 | 1.38 | 2.41 | 3.71 | 3.23 | 3.01 | 2.43 | 1.93 | 1.24 | 0.95 | 0.56 | 22.07 |
| Days ≥ 0.1" Precip | 2 | 2 | 3 | 5 | 6 | 6 | 5 | 4 | 4 | 3 | 2 | 2 | 44 |
| Total Snowfall (") | 5.0 | *4.7* | 6.4 | 2.8 | 0.0 | 0.0 | 0.0 | 0.0 | 0.0 | 0.8 | *5.8* | *5.8* | 31.3 |
| Days ≥ 1" Snow Depth | 12 | *10* | *5* | 1 | 0 | 0 | 0 | 0 | 0 | 0 | 4 | na | na |

## RAVENNA *Buffalo County* ELEVATION 2001 ft LAT/LONG 41° 2 ' N / 98° 55 ' W

| | JAN | FEB | MAR | APR | MAY | JUN | JUL | AUG | SEP | OCT | NOV | DEC | YEAR |
|---|---|---|---|---|---|---|---|---|---|---|---|---|---|
| Maximum Temp °F | 34.9 | 40.4 | 51.8 | 64.6 | 74.3 | 84.2 | 88.5 | 86.5 | 78.0 | 66.3 | 49.0 | 37.9 | 63.0 |
| Minimum Temp °F | 10.1 | 15.0 | 24.9 | 36.3 | 47.2 | 57.2 | 62.6 | 60.1 | 49.3 | 36.6 | 23.7 | 13.8 | 36.4 |
| Mean Temp °F | 22.5 | 27.7 | 38.4 | 50.5 | 60.8 | 70.7 | 75.6 | 73.3 | 63.7 | 51.5 | 36.4 | 25.9 | 49.7 |
| Days Max Temp ≥ 90 °F | 0 | 0 | 0 | 1 | 1 | 8 | 14 | 12 | 4 | 0 | 0 | 0 | 40 |
| Days Max Temp ≤ 32 °F | 12 | 9 | 3 | 0 | 0 | 0 | 0 | 0 | 0 | 0 | 3 | 10 | 37 |
| Days Min Temp ≤ 32 °F | 31 | 27 | 24 | 11 | 2 | 0 | 0 | 0 | 2 | 11 | 25 | 30 | 163 |
| Days Min Temp ≤ 0 °F | 8 | 4 | 1 | 0 | 0 | 0 | 0 | 0 | 0 | 0 | 1 | 4 | 18 |
| Heating Degree Days | 1311 | 1045 | 820 | 439 | 175 | 27 | 3 | 9 | 125 | 419 | 852 | 1205 | 6430 |
| Cooling Degree Days | 0 | 0 | 0 | 12 | 45 | 200 | 322 | 270 | 95 | 4 | 0 | 0 | 948 |
| Total Precipitation (") | 0.49 | 0.61 | 1.88 | 2.44 | 3.95 | 4.58 | 3.48 | 2.91 | 2.26 | 1.78 | 1.21 | 0.67 | 26.26 |
| Days ≥ 0.1" Precip | 2 | 2 | 4 | 5 | 7 | 6 | 6 | 5 | 4 | 4 | 3 | 2 | 50 |
| Total Snowfall (") | 5.5 | 5.2 | 4.5 | *0.6* | 0.2 | 0.0 | 0.0 | 0.0 | 0.1 | 0.9 | 3.7 | *6.3* | 27.0 |
| Days ≥ 1" Snow Depth | na | na | na | 0 | 0 | 0 | 0 | 0 | 0 | 0 | *1* | na | na |

## RED CLOUD *Webster County* ELEVATION 1722 ft LAT/LONG 40° 6 ' N / 98° 31 ' W

| | JAN | FEB | MAR | APR | MAY | JUN | JUL | AUG | SEP | OCT | NOV | DEC | YEAR |
|---|---|---|---|---|---|---|---|---|---|---|---|---|---|
| Maximum Temp °F | 35.9 | 41.6 | 53.5 | 65.9 | 74.9 | 85.4 | 90.7 | 88.3 | 79.7 | 67.9 | 51.2 | 40.1 | 64.6 |
| Minimum Temp °F | 10.4 | 15.0 | 25.2 | 36.9 | 47.7 | 57.6 | 62.8 | 60.0 | 49.4 | 36.5 | 24.0 | 14.6 | 36.7 |
| Mean Temp °F | 23.2 | 28.2 | 39.4 | 51.4 | 61.4 | 71.5 | 76.8 | 74.2 | 64.6 | 52.2 | 37.6 | 27.4 | 50.7 |
| Days Max Temp ≥ 90 °F | 0 | 0 | 0 | 1 | 2 | 10 | 17 | 14 | 6 | 1 | 0 | 0 | 51 |
| Days Max Temp ≤ 32 °F | 12 | 8 | 2 | 0 | 0 | 0 | 0 | 0 | 0 | 0 | 2 | 8 | 32 |
| Days Min Temp ≤ 32 °F | 31 | 27 | 24 | 10 | 2 | 0 | 0 | 0 | 1 | 11 | 25 | 30 | 161 |
| Days Min Temp ≤ 0 °F | 7 | 4 | 0 | 0 | 0 | 0 | 0 | 0 | 0 | 0 | 0 | 3 | 14 |
| Heating Degree Days | 1291 | 1033 | 789 | 413 | 163 | 23 | 2 | 8 | 113 | 399 | 815 | 1159 | 6208 |
| Cooling Degree Days | 0 | 0 | 0 | 12 | 44 | 218 | 358 | 291 | 112 | 6 | 0 | 0 | 1041 |
| Total Precipitation (") | 0.51 | 0.61 | 1.92 | 2.28 | 4.13 | 3.64 | 4.05 | 3.19 | 2.75 | 1.66 | 1.14 | 0.73 | 26.61 |
| Days ≥ 0.1" Precip | 2 | 2 | 3 | 5 | 8 | 6 | 6 | 5 | 5 | 3 | 3 | 2 | 50 |
| Total Snowfall (") | na | na | *3.0* | 0.3 | 0.0 | 0.0 | 0.0 | 0.0 | 0.2 | 0.2 | *2.1* | na | na |
| Days ≥ 1" Snow Depth | na | na | *3* | 0 | 0 | 0 | 0 | 0 | 0 | 0 | *3* | na | na |

## RED WILLOW DAM *Red Willow County* ELEVATION 2562 ft LAT/LONG 40° 21 ' N / 100° 39 ' W

| | JAN | FEB | MAR | APR | MAY | JUN | JUL | AUG | SEP | OCT | NOV | DEC | YEAR |
|---|---|---|---|---|---|---|---|---|---|---|---|---|---|
| Maximum Temp °F | 37.7 | 42.7 | 52.5 | 64.1 | 73.1 | 83.0 | 88.7 | *87.7* | *77.6* | *67.2* | 50.6 | 40.9 | 63.8 |
| Minimum Temp °F | 10.6 | 14.6 | 23.4 | 34.4 | 44.7 | 55.1 | 60.7 | 58.1 | *47.2* | 34.5 | 22.4 | 13.9 | 35.0 |
| Mean Temp °F | 24.2 | 28.7 | 38.0 | 49.3 | 59.0 | 69.1 | 74.6 | *72.9* | *62.4* | *50.9* | 36.5 | 27.5 | 49.4 |
| Days Max Temp ≥ 90 °F | 0 | 0 | 0 | 0 | 1 | 7 | 15 | 13 | 4 | 1 | 0 | 0 | 41 |
| Days Max Temp ≤ 32 °F | 10 | 8 | 3 | 0 | 0 | 0 | 0 | 0 | 0 | 0 | 2 | 8 | 31 |
| Days Min Temp ≤ 32 °F | 31 | 28 | 27 | 13 | 2 | 0 | 0 | 0 | 1 | 12 | 26 | 30 | 170 |
| Days Min Temp ≤ 0 °F | 6 | 3 | 1 | 0 | 0 | 0 | 0 | 0 | 0 | 0 | 0 | 3 | 13 |
| Heating Degree Days | 1258 | 1021 | 831 | 472 | 212 | 39 | 5 | *8* | *145* | *432* | 846 | 1158 | 6427 |
| Cooling Degree Days | 0 | 0 | 0 | 8 | 34 | 174 | 299 | *276* | na | *3* | *0* | 0 | na |
| Total Precipitation (") | 0.39 | 0.46 | 1.12 | 1.89 | 3.12 | 3.45 | 3.26 | 2.54 | 1.57 | 1.26 | 0.74 | 0.49 | 20.29 |
| Days ≥ 0.1" Precip | 1 | 1 | 3 | 4 | 6 | 6 | 6 | 4 | 3 | 2 | 2 | 1 | 39 |
| Total Snowfall (") | 4.7 | *3.0* | *2.5* | 0.9 | 0.0 | 0.0 | 0.0 | 0.0 | 0.0 | 0.4 | 1.7 | 4.7 | 17.9 |
| Days ≥ 1" Snow Depth | 12 | 9 | 4 | 1 | 0 | 0 | 0 | 0 | 0 | 0 | 2 | *9* | 37 |

## ROSE 7 WNW *Rock County* ELEVATION 2513 ft LAT/LONG 42° 10 ' N / 99° 40 ' W

| | JAN | FEB | MAR | APR | MAY | JUN | JUL | AUG | SEP | OCT | NOV | DEC | YEAR |
|---|---|---|---|---|---|---|---|---|---|---|---|---|---|
| Maximum Temp °F | *31.1* | *36.2* | *46.3* | *60.0* | *70.8* | *79.8* | *86.0* | *83.8* | *74.9* | *63.7* | *45.6* | *35.2* | 59.5 |
| Minimum Temp °F | *7.5* | *11.5* | *21.5* | *33.0* | *43.9* | *53.4* | *58.9* | *56.1* | *45.1* | *33.3* | *20.2* | *11.2* | 33.0 |
| Mean Temp °F | *19.2* | *23.9* | *33.9* | *46.5* | *57.4* | *66.6* | *72.5* | *70.0* | *60.1* | *48.5* | *32.9* | *23.2* | 46.2 |
| Days Max Temp ≥ 90 °F | *0* | *0* | *0* | *1* | *0* | *4* | na | *8* | *3* | *0* | *0* | *0* | na |
| Days Max Temp ≤ 32 °F | *15* | *11* | *5* | *1* | *0* | *0* | *0* | *0* | *0* | *0* | *6* | *12* | 50 |
| Days Min Temp ≤ 32 °F | *31* | *27* | *28* | na | *3* | *0* | *0* | *0* | *3* | *14* | *27* | *31* | na |
| Days Min Temp ≤ 0 °F | *9* | *6* | *1* | *0* | *0* | *0* | *0* | *0* | *0* | *0* | *2* | *6* | 24 |
| Heating Degree Days | *1416* | *1155* | *957* | *552* | *250* | *62* | *12* | *28* | *195* | *505* | *956* | *1288* | 7376 |
| Cooling Degree Days | 0 | 0 | 0 | 6 | 23 | 120 | 230 | 195 | 55 | 2 | 0 | 0 | 631 |
| Total Precipitation (") | *0.59* | *0.66* | *1.30* | *1.94* | *3.42* | *3.37* | *3.09* | *2.96* | *1.71* | *1.31* | *1.02* | *0.57* | 21.94 |
| Days ≥ 0.1" Precip | *2* | *2* | na | *5* | *6* | *6* | *6* | *5* | *4* | *3* | *2* | *1* | na |
| Total Snowfall (") | *6.4* | *5.5* | na | *2.5* | *0.0* | *0.0* | *0.0* | *0.0* | *0.2* | *1.1* | *7.1* | *7.2* | na |
| Days ≥ 1" Snow Depth | na | na | na | na | *0* | *0* | *0* | *0* | *0* | *0* | na | na | na |

## SAINT PAUL 4 N *Howard County* ELEVATION 1801 ft LAT/LONG 41° 12 ' N / 98° 27 ' W

| | JAN | FEB | MAR | APR | MAY | JUN | JUL | AUG | SEP | OCT | NOV | DEC | YEAR |
|---|---|---|---|---|---|---|---|---|---|---|---|---|---|
| Maximum Temp °F | 33.1 | 38.7 | 49.7 | 63.1 | 72.2 | 81.8 | 85.9 | 84.0 | 75.9 | 64.3 | 47.5 | 36.3 | 61.0 |
| Minimum Temp °F | 10.9 | 15.8 | 26.3 | 38.0 | 49.1 | 58.9 | 63.7 | 61.4 | 51.2 | 38.6 | 25.3 | 14.9 | 37.8 |
| Mean Temp °F | 22.1 | 27.2 | 38.0 | 50.6 | 60.7 | 70.4 | 74.8 | 72.7 | 63.6 | 51.4 | 36.4 | 25.7 | 49.5 |
| Days Max Temp ≥ 90 °F | 0 | 0 | 0 | 1 | 1 | 5 | 10 | 8 | 3 | 0 | 0 | 0 | 28 |
| Days Max Temp ≤ 32 °F | 14 | 10 | 3 | 0 | 0 | 0 | 0 | 0 | 0 | 0 | 4 | 11 | 42 |
| Days Min Temp ≤ 32 °F | 31 | 26 | 23 | 8 | 1 | 0 | 0 | 0 | 1 | 8 | 24 | 30 | 152 |
| Days Min Temp ≤ 0 °F | 7 | 4 | 0 | 0 | 0 | 0 | 0 | 0 | 0 | 0 | 1 | 4 | 16 |
| Heating Degree Days | 1326 | 1059 | 830 | 436 | 175 | 26 | 4 | 10 | 124 | 418 | 851 | 1214 | 6473 |
| Cooling Degree Days | 0 | 0 | 1 | 17 | 50 | 207 | 313 | 270 | 104 | 4 | 0 | 0 | 966 |
| Total Precipitation (") | 0.40 | 0.63 | 1.78 | 2.37 | 4.13 | 4.20 | 3.18 | 2.71 | 2.56 | 1.55 | 0.99 | 0.62 | 25.12 |
| Days ≥ 0.1" Precip | 1 | 2 | 4 | 4 | 7 | 6 | 6 | 5 | 4 | 3 | 2 | 2 | 46 |
| Total Snowfall (") | na | na | na | *0.6* | 0.2 | 0.0 | 0.0 | 0.0 | 0.1 | 0.5 | *3.3* | na | na |
| Days ≥ 1" Snow Depth | na | na | na | *0* | 0 | 0 | 0 | 0 | 0 | 0 | na | na | na |

## SCOTTSBLUFF CNTY AP *Scotts Bluff County* ELEVATION 3966 ft LAT/LONG 41° 52 ' N / 103° 36 ' W

| | JAN | FEB | MAR | APR | MAY | JUN | JUL | AUG | SEP | OCT | NOV | DEC | YEAR |
|---|---|---|---|---|---|---|---|---|---|---|---|---|---|
| Maximum Temp °F | 38.0 | 43.6 | 51.1 | 61.6 | 71.4 | 81.9 | 89.2 | 87.1 | 77.5 | 64.8 | 49.0 | 39.7 | 62.9 |
| Minimum Temp °F | 12.4 | 16.5 | 24.0 | 33.1 | 43.3 | 53.1 | 58.7 | 56.3 | 45.6 | 33.5 | 22.1 | 13.1 | 34.3 |
| Mean Temp °F | 25.2 | 30.1 | 37.6 | 47.3 | 57.4 | 67.5 | 74.0 | 71.7 | 61.6 | 49.2 | 35.6 | 26.4 | 48.6 |
| Days Max Temp ≥ 90 °F | 0 | 0 | 0 | 0 | 1 | 7 | 16 | 13 | 4 | 0 | 0 | 0 | 41 |
| Days Max Temp ≤ 32 °F | 9 | 6 | 3 | 0 | 0 | 0 | 0 | 0 | 0 | 0 | 3 | 9 | 30 |
| Days Min Temp ≤ 32 °F | 30 | 27 | 27 | 14 | 2 | 0 | 0 | 0 | 2 | 13 | 27 | 30 | 172 |
| Days Min Temp ≤ 0 °F | 6 | 3 | 1 | 0 | 0 | 0 | 0 | 0 | 0 | 0 | 1 | 5 | 16 |
| Heating Degree Days | 1228 | 979 | 842 | 525 | 251 | 55 | 6 | 12 | 156 | 485 | 876 | 1189 | 6604 |
| Cooling Degree Days | 0 | 0 | 0 | 3 | 24 | 144 | 283 | 243 | 65 | 1 | 0 | 0 | 763 |
| Total Precipitation (") | 0.53 | 0.54 | 1.09 | 1.60 | 2.56 | 2.82 | 2.09 | 1.17 | 1.11 | 0.98 | 0.74 | 0.58 | 15.81 |
| Days ≥ 0.1" Precip | 2 | 2 | 3 | 4 | 6 | 6 | 5 | 3 | 3 | 3 | 2 | 2 | 41 |
| Total Snowfall (") | 6.8 | 5.9 | 8.6 | 4.9 | 0.9 | 0.0 | 0.0 | 0.0 | 0.5 | 3.0 | 5.9 | 7.3 | 43.8 |
| Days ≥ 1" Snow Depth | 13 | 8 | 5 | 2 | 0 | 0 | 0 | 0 | 0 | 1 | 5 | 12 | 46 |

## SEWARD *Seward County* ELEVATION 1440 ft LAT/LONG 40° 55 ' N / 97° 6 ' W

| | JAN | FEB | MAR | APR | MAY | JUN | JUL | AUG | SEP | OCT | NOV | DEC | YEAR |
|---|---|---|---|---|---|---|---|---|---|---|---|---|---|
| Maximum Temp °F | 32.8 | 38.7 | 51.2 | 65.3 | 75.0 | 85.2 | 89.2 | 86.4 | 77.6 | 66.0 | 48.6 | 36.8 | 62.7 |
| Minimum Temp °F | 13.3 | 17.9 | 28.8 | 40.4 | 51.3 | 61.1 | 65.8 | 63.4 | 54.4 | 42.4 | 28.8 | 18.5 | 40.5 |
| Mean Temp °F | 23.1 | 28.3 | 40.1 | 52.9 | 63.2 | 73.2 | 77.5 | 74.9 | 66.0 | 54.2 | 38.7 | 27.7 | 51.7 |
| Days Max Temp ≥ 90 °F | 0 | 0 | 0 | 1 | 1 | 9 | 15 | 11 | 3 | 0 | 0 | 0 | 40 |
| Days Max Temp ≤ 32 °F | 14 | 10 | 3 | 0 | 0 | 0 | 0 | 0 | 0 | 0 | 3 | 10 | 40 |
| Days Min Temp ≤ 32 °F | 30 | 25 | 20 | 6 | 0 | 0 | 0 | 0 | 0 | 5 | 20 | 29 | 135 |
| Days Min Temp ≤ 0 °F | 6 | 3 | 0 | 0 | 0 | 0 | 0 | 0 | 0 | 0 | 0 | 3 | 12 |
| Heating Degree Days | 1293 | 1029 | 768 | 374 | 124 | 12 | 1 | 5 | 86 | 340 | 782 | 1150 | 5964 |
| Cooling Degree Days | 0 | 0 | 2 | 25 | 76 | 269 | 397 | 327 | 142 | 11 | 0 | 0 | 1249 |
| Total Precipitation (") | 0.60 | 0.62 | 1.93 | 2.62 | 3.98 | 3.78 | 3.54 | 3.25 | 3.21 | 2.14 | 1.33 | 0.81 | 27.81 |
| Days ≥ 0.1" Precip | 2 | 2 | 4 | 6 | 8 | 6 | 6 | 5 | 5 | 4 | 3 | 2 | 53 |
| Total Snowfall (") | 5.0 | 5.5 | 4.2 | 1.0 | 0.2 | 0.0 | 0.0 | 0.0 | 0.0 | 0.2 | 2.7 | 5.2 | 24.0 |
| Days ≥ 1" Snow Depth | 18 | 13 | 6 | 1 | 0 | 0 | 0 | 0 | 0 | 0 | 3 | 9 | 50 |

## SIDNEY 6 NNW *Cheyenne County* ELEVATION 4324 ft LAT/LONG 41° 14 ' N / 103° 0 ' W

| | JAN | FEB | MAR | APR | MAY | JUN | JUL | AUG | SEP | OCT | NOV | DEC | YEAR |
|---|---|---|---|---|---|---|---|---|---|---|---|---|---|
| Maximum Temp °F | 37.7 | 43.1 | 49.8 | 59.9 | 69.4 | 79.9 | 87.4 | 85.4 | 76.0 | 63.6 | 48.4 | 40.0 | 61.7 |
| Minimum Temp °F | 11.2 | 15.2 | 21.7 | 30.4 | 40.5 | 50.1 | 56.1 | 53.9 | 43.8 | 32.1 | 20.8 | 13.1 | 32.4 |
| Mean Temp °F | 24.4 | 29.2 | 35.7 | 45.2 | 55.0 | 65.0 | 71.8 | 69.6 | 60.0 | 47.8 | 34.6 | 26.5 | 47.1 |
| Days Max Temp ≥ 90 °F | 0 | 0 | 0 | 0 | 0 | 5 | 14 | 10 | 3 | 0 | 0 | 0 | 32 |
| Days Max Temp ≤ 32 °F | 10 | 6 | 4 | 1 | 0 | 0 | 0 | 0 | 0 | 1 | 5 | 8 | 35 |
| Days Min Temp ≤ 32 °F | 31 | 28 | 28 | 18 | 5 | 0 | 0 | 0 | 3 | 16 | 28 | 30 | 187 |
| Days Min Temp ≤ 0 °F | 6 | 3 | 1 | 0 | 0 | 0 | 0 | 0 | 0 | 0 | 1 | 4 | 15 |
| Heating Degree Days | 1251 | 1005 | 901 | 590 | 315 | 85 | 14 | 24 | 187 | 527 | 904 | 1186 | 6989 |
| Cooling Degree Days | 0 | 0 | 0 | 1 | 9 | 88 | 215 | 176 | 48 | 1 | 0 | 0 | 538 |
| Total Precipitation (") | 0.29 | 0.34 | 0.96 | 1.45 | 2.76 | 3.21 | 2.41 | 1.96 | 1.12 | 0.89 | 0.55 | 0.32 | 16.26 |
| Days ≥ 0.1" Precip | 1 | 1 | 3 | 4 | 6 | 6 | 6 | 4 | 3 | 2 | 1 | 1 | 38 |
| Total Snowfall (") | *4.2* | 4.2 | *7.1* | 2.4 | 0.3 | 0.0 | 0.0 | 0.0 | 0.2 | 1.3 | *3.5* | *4.3* | 27.5 |
| Days ≥ 1" Snow Depth | na | na | na | 1 | 0 | 0 | 0 | 0 | 0 | 1 | *2* | na | na |

## SPRINGVIEW *Keya Paha County* ELEVATION 2441 ft LAT/LONG 42° 49 ' N / 99° 44 ' W

| | JAN | FEB | MAR | APR | MAY | JUN | JUL | AUG | SEP | OCT | NOV | DEC | YEAR |
|---|---|---|---|---|---|---|---|---|---|---|---|---|---|
| Maximum Temp °F | 30.5 | 35.4 | 45.6 | 58.8 | 70.2 | 80.3 | 87.0 | 85.2 | 75.2 | 62.6 | 44.8 | 34.3 | 59.2 |
| Minimum Temp °F | 8.9 | 13.4 | 22.9 | 35.0 | 45.9 | 55.8 | 61.3 | 59.0 | 48.4 | 36.5 | 23.6 | 13.2 | 35.3 |
| Mean Temp °F | 19.7 | 24.4 | 34.3 | 46.9 | 58.1 | 68.1 | 74.1 | 72.2 | 61.8 | 49.6 | 34.2 | 23.8 | 47.3 |
| Days Max Temp ≥ 90 °F | 0 | 0 | 0 | 0 | 1 | 5 | 12 | 10 | 4 | 1 | 0 | 0 | 33 |
| Days Max Temp ≤ 32 °F | 16 | 12 | 7 | 1 | 0 | 0 | 0 | 0 | 0 | 0 | 6 | 14 | 56 |
| Days Min Temp ≤ 32 °F | 30 | 27 | 26 | 12 | 2 | 0 | 0 | 0 | 1 | 10 | 24 | 30 | 162 |
| Days Min Temp ≤ 0 °F | 9 | 6 | 1 | 0 | 0 | 0 | 0 | 0 | 0 | 0 | 1 | 5 | 22 |
| Heating Degree Days | 1399 | 1139 | 945 | 544 | 244 | 51 | 9 | 19 | 165 | 478 | 918 | 1273 | 7184 |
| Cooling Degree Days | 0 | 0 | 0 | 9 | 34 | 145 | 274 | 231 | 72 | 5 | 0 | 0 | 770 |
| Total Precipitation (") | 0.35 | 0.52 | 1.32 | 2.19 | 3.32 | 3.39 | 3.08 | 2.38 | 2.27 | 1.34 | 0.78 | 0.45 | 21.39 |
| Days ≥ 0.1" Precip | 1 | 1 | 3 | 5 | 6 | 6 | 6 | 5 | 4 | 3 | 2 | 2 | 44 |
| Total Snowfall (") | 4.8 | 5.9 | 7.0 | 4.1 | 0.0 | 0.0 | 0.0 | 0.0 | 0.1 | 1.4 | 5.5 | 6.1 | 34.9 |
| Days ≥ 1" Snow Depth | 11 | 9 | 7 | 1 | 0 | 0 | 0 | 0 | 0 | 1 | *3* | 9 | 41 |

## STANTON *Stanton County* ELEVATION 1472 ft LAT/LONG 41° 57' N / 97° 14' W

| | JAN | FEB | MAR | APR | MAY | JUN | JUL | AUG | SEP | OCT | NOV | DEC | YEAR |
|---|---|---|---|---|---|---|---|---|---|---|---|---|---|
| Maximum Temp °F | 30.9 | 36.8 | 49.5 | 64.1 | 74.3 | 83.7 | 87.4 | 85.3 | 76.9 | 64.9 | 46.9 | 34.6 | 61.3 |
| Minimum Temp °F | 9.6 | 14.8 | 25.9 | 37.5 | 48.6 | 58.3 | 63.2 | 61.0 | 51.2 | 39.1 | 25.7 | 14.7 | 37.5 |
| Mean Temp °F | 20.3 | 25.8 | 37.7 | 50.8 | 61.4 | 71.0 | 75.3 | 73.1 | 64.1 | 52.0 | 36.3 | 24.7 | 49.4 |
| Days Max Temp ≥ 90 °F | 0 | 0 | 0 | 1 | 1 | 7 | 13 | 9 | 3 | 0 | 0 | 0 | 34 |
| Days Max Temp ≤ 32 °F | 16 | 11 | 4 | 0 | 0 | 0 | 0 | 0 | 0 | 0 | 4 | 13 | 48 |
| Days Min Temp ≤ 32 °F | 31 | 27 | 23 | 10 | 1 | 0 | 0 | 0 | 1 | 8 | 23 | 30 | 154 |
| Days Min Temp ≤ 0 °F | 9 | 5 | 1 | 0 | 0 | 0 | 0 | 0 | 0 | 0 | 1 | 4 | 20 |
| Heating Degree Days | 1380 | 1099 | 839 | 432 | 162 | 21 | 3 | 10 | 118 | 403 | 854 | 1244 | 6565 |
| Cooling Degree Days | 0 | 0 | 1 | 19 | 59 | 215 | 322 | 271 | 108 | 6 | 0 | 0 | 1001 |
| Total Precipitation (") | 0.60 | 0.85 | 2.13 | 2.60 | 4.31 | 4.60 | 3.74 | 2.81 | 2.65 | 2.07 | 1.41 | 0.95 | 28.72 |
| Days ≥ 0.1" Precip | 2 | 3 | 4 | 6 | 7 | 7 | 6 | 5 | 5 | 4 | 3 | 3 | 55 |
| Total Snowfall (") | 5.7 | 6.1 | 5.2 | 1.5 | 0.0 | 0.0 | 0.0 | 0.0 | 0.0 | 0.7 | 3.3 | 7.0 | 29.5 |
| Days ≥ 1" Snow Depth | na | na | na | 0 | 0 | 0 | 0 | 0 | 0 | 0 | *3* | na | na |

## SUPERIOR *Nuckolls County* ELEVATION 1572 ft LAT/LONG 40° 2' N / 98° 4' W

| | JAN | FEB | MAR | APR | MAY | JUN | JUL | AUG | SEP | OCT | NOV | DEC | YEAR |
|---|---|---|---|---|---|---|---|---|---|---|---|---|---|
| Maximum Temp °F | 36.3 | 42.5 | 54.8 | 66.6 | 75.5 | 85.7 | 90.4 | 88.1 | 79.4 | 67.4 | 51.1 | 40.1 | 64.8 |
| Minimum Temp °F | 14.6 | 19.1 | 29.8 | 40.6 | 51.1 | 61.0 | 65.8 | 63.6 | 54.0 | 41.6 | 28.7 | 18.9 | 40.7 |
| Mean Temp °F | 25.4 | 30.8 | 42.3 | 53.7 | 63.3 | 73.3 | 78.1 | 75.8 | 66.7 | 54.5 | 40.0 | 29.6 | 52.8 |
| Days Max Temp ≥ 90 °F | 0 | 0 | 0 | 1 | 1 | 10 | 17 | 13 | 5 | 0 | 0 | 0 | 47 |
| Days Max Temp ≤ 32 °F | 12 | 7 | 2 | 0 | 0 | 0 | 0 | 0 | 0 | 0 | 2 | 8 | 31 |
| Days Min Temp ≤ 32 °F | 30 | 26 | 19 | 6 | 0 | 0 | 0 | 0 | 0 | 5 | 20 | 30 | 136 |
| Days Min Temp ≤ 0 °F | 5 | 2 | 0 | 0 | 0 | 0 | 0 | 0 | 0 | 0 | 0 | 2 | 9 |
| Heating Degree Days | 1220 | 958 | 697 | 351 | 119 | 12 | 1 | 3 | 81 | 332 | 745 | 1092 | 5611 |
| Cooling Degree Days | 0 | 0 | 1 | 24 | 79 | 280 | 420 | 357 | 160 | 12 | 0 | 0 | 1333 |
| Total Precipitation (") | 0.65 | 0.77 | 2.15 | 2.54 | 4.11 | 3.60 | 3.70 | 3.27 | 3.19 | 2.01 | 1.36 | 0.99 | 28.34 |
| Days ≥ 0.1" Precip | 2 | 2 | 4 | 5 | 8 | 6 | 6 | 6 | 5 | 4 | 3 | 3 | 54 |
| Total Snowfall (") | 6.1 | 6.1 | 4.1 | 0.9 | 0.1 | 0.0 | 0.0 | 0.0 | 0.1 | 0.2 | 3.1 | 5.9 | 26.6 |
| Days ≥ 1" Snow Depth | 16 | 11 | 4 | 1 | 0 | 0 | 0 | 0 | 0 | 0 | 3 | 9 | 44 |

## SYRACUSE *Otoe County* ELEVATION 1050 ft LAT/LONG 40° 39' N / 96° 11' W

| | JAN | FEB | MAR | APR | MAY | JUN | JUL | AUG | SEP | OCT | NOV | DEC | YEAR |
|---|---|---|---|---|---|---|---|---|---|---|---|---|---|
| Maximum Temp °F | 32.9 | 38.3 | 51.4 | 64.7 | 74.9 | 84.9 | 89.2 | 86.8 | 78.7 | 66.9 | 50.1 | 37.4 | 63.0 |
| Minimum Temp °F | 11.1 | 15.6 | 27.2 | 39.3 | 50.2 | 60.3 | 65.0 | 62.2 | 52.4 | 39.7 | 27.7 | 16.9 | 39.0 |
| Mean Temp °F | 22.1 | 27.0 | 39.4 | 52.0 | 62.6 | 72.6 | 77.1 | 74.5 | 65.6 | 53.3 | 38.9 | 27.2 | 51.0 |
| Days Max Temp ≥ 90 °F | 0 | 0 | 0 | 0 | 1 | 9 | 15 | 12 | 5 | 0 | 0 | 0 | 42 |
| Days Max Temp ≤ 32 °F | 14 | 10 | 3 | 0 | 0 | 0 | 0 | 0 | 0 | 0 | 3 | 10 | 40 |
| Days Min Temp ≤ 32 °F | 30 | 26 | 22 | 8 | 1 | 0 | 0 | 0 | 1 | 8 | 21 | 30 | 147 |
| Days Min Temp ≤ 0 °F | 7 | 4 | 0 | 0 | 0 | 0 | 0 | 0 | 0 | 0 | 0 | 3 | 14 |
| Heating Degree Days | 1325 | 1066 | 790 | 401 | 142 | 16 | 2 | 5 | 101 | 371 | 776 | 1167 | 6162 |
| Cooling Degree Days | 0 | 0 | 2 | 23 | 70 | 254 | 383 | 308 | 134 | 9 | 0 | 0 | 1183 |
| Total Precipitation (") | 0.70 | 0.70 | 2.28 | 2.77 | 4.01 | 3.79 | 4.43 | 3.68 | 3.54 | 2.54 | 1.50 | 0.97 | 30.91 |
| Days ≥ 0.1" Precip | 2 | 2 | 5 | 6 | 8 | 6 | 6 | 6 | 5 | 5 | 3 | 2 | 56 |
| Total Snowfall (") | 5.6 | 5.9 | 4.4 | 0.7 | 0.1 | 0.0 | 0.0 | 0.0 | 0.0 | 0.1 | 1.9 | 4.3 | 23.0 |
| Days ≥ 1" Snow Depth | 13 | 8 | 3 | 0 | 0 | 0 | 0 | 0 | 0 | 0 | 1 | *6* | 31 |

## TECUMSEH *Johnson County* ELEVATION 1170 ft LAT/LONG 40° 23' N / 96° 7' W

| | JAN | FEB | MAR | APR | MAY | JUN | JUL | AUG | SEP | OCT | NOV | DEC | YEAR |
|---|---|---|---|---|---|---|---|---|---|---|---|---|---|
| Maximum Temp °F | 33.4 | 39.1 | 51.6 | 64.4 | 74.7 | 84.5 | 89.1 | 86.9 | 78.5 | 67.0 | 50.5 | 38.0 | 63.1 |
| Minimum Temp °F | 11.5 | 15.9 | 27.8 | 39.3 | 50.1 | 60.1 | 65.0 | 62.2 | 52.3 | 39.6 | 28.0 | 17.2 | 39.1 |
| Mean Temp °F | 22.5 | 27.5 | 39.7 | 51.8 | 62.4 | 72.3 | 77.1 | 74.6 | 65.5 | 53.3 | 39.3 | 27.6 | 51.1 |
| Days Max Temp ≥ 90 °F | 0 | 0 | 0 | 1 | 1 | 8 | 15 | 12 | 5 | 0 | 0 | 0 | 42 |
| Days Max Temp ≤ 32 °F | 14 | 9 | 3 | 0 | 0 | 0 | 0 | 0 | 0 | 0 | 2 | 10 | 38 |
| Days Min Temp ≤ 32 °F | 30 | 26 | 21 | 8 | 1 | 0 | 0 | 0 | 1 | 8 | 21 | 30 | 146 |
| Days Min Temp ≤ 0 °F | 7 | 4 | 0 | 0 | 0 | 0 | 0 | 0 | 0 | 0 | 0 | 3 | 14 |
| Heating Degree Days | 1312 | 1052 | 779 | 407 | 147 | 17 | 2 | 7 | 102 | 370 | 765 | 1152 | 6112 |
| Cooling Degree Days | 0 | 0 | 2 | 25 | 68 | 248 | 384 | 313 | 134 | 11 | 0 | 0 | 1185 |
| Total Precipitation (") | 0.88 | 0.92 | 2.52 | 2.74 | 4.24 | 3.84 | 4.57 | 3.76 | 3.69 | 2.42 | 1.71 | 1.23 | 32.52 |
| Days ≥ 0.1" Precip | 3 | 3 | 5 | 6 | 7 | 6 | 7 | 6 | 6 | 5 | 4 | 3 | 61 |
| Total Snowfall (") | 7.2 | 7.4 | 4.8 | 1.1 | 0.0 | 0.0 | 0.0 | 0.0 | 0.0 | 0.2 | 2.1 | 5.0 | 27.8 |
| Days ≥ 1" Snow Depth | 14 | 11 | 4 | 1 | 0 | 0 | 0 | 0 | 0 | 0 | 2 | 7 | 39 |

## TEKAMAH *Burt County* ELEVATION 1060 ft LAT/LONG 41° 47 ' N / 96° 13 ' W

| | JAN | FEB | MAR | APR | MAY | JUN | JUL | AUG | SEP | OCT | NOV | DEC | YEAR |
|---|---|---|---|---|---|---|---|---|---|---|---|---|---|
| Maximum Temp °F | 30.7 | 36.3 | 49.1 | 63.9 | 75.1 | 84.7 | 88.1 | 85.7 | 77.7 | 65.8 | 48.0 | 34.6 | 61.6 |
| Minimum Temp °F | 10.0 | 15.9 | 26.7 | 38.7 | 50.4 | 60.1 | 64.4 | 61.5 | 51.8 | 39.8 | 27.4 | 15.3 | 38.5 |
| Mean Temp °F | 20.4 | 26.4 | 37.9 | 51.4 | 62.8 | 72.3 | 76.3 | 73.7 | 64.8 | 52.8 | 37.7 | 25.0 | 50.1 |
| Days Max Temp ≥ 90 °F | 0 | 0 | 0 | 1 | 2 | 8 | 13 | 9 | 3 | 0 | 0 | 0 | 36 |
| Days Max Temp ≤ 32 °F | 16 | 11 | 4 | 0 | 0 | 0 | 0 | 0 | 0 | 0 | 3 | 12 | 46 |
| Days Min Temp ≤ 32 °F | 30 | 26 | 22 | 7 | 1 | 0 | 0 | 0 | 0 | 7 | 22 | 30 | 145 |
| Days Min Temp ≤ 0 °F | 8 | 4 | 0 | 0 | 0 | 0 | 0 | 0 | 0 | 0 | 1 | 4 | 17 |
| Heating Degree Days | 1378 | 1085 | 831 | 419 | 141 | 14 | 2 | 7 | 104 | 383 | 812 | 1234 | 6410 |
| Cooling Degree Days | 0 | 0 | 1 | 21 | 74 | 236 | 336 | 268 | 109 | 7 | 0 | 0 | 1052 |
| Total Precipitation (") | 0.69 | 0.73 | 2.19 | 2.79 | 4.05 | 4.14 | 3.44 | 3.33 | 3.52 | 2.34 | 1.41 | 0.92 | 29.55 |
| Days ≥ 0.1" Precip | 2 | 2 | 4 | 6 | 7 | 7 | 7 | 5 | 6 | 4 | 3 | 2 | 55 |
| Total Snowfall (") | 6.6 | 6.1 | 4.3 | 0.7 | 0.0 | 0.0 | 0.0 | 0.0 | 0.0 | 0.5 | 2.6 | 6.2 | 27.0 |
| Days ≥ 1" Snow Depth | na | na | *4* | 0 | 0 | 0 | 0 | 0 | 0 | 0 | *1* | na | na |

## TRENTON DAM *Hitchcock County* ELEVATION 2661 ft LAT/LONG 40° 11 ' N / 101° 1 ' W

| | JAN | FEB | MAR | APR | MAY | JUN | JUL | AUG | SEP | OCT | NOV | DEC | YEAR |
|---|---|---|---|---|---|---|---|---|---|---|---|---|---|
| Maximum Temp °F | 38.2 | 43.8 | 53.4 | 65.0 | 73.6 | 83.8 | 89.9 | 88.5 | 79.5 | 68.0 | 51.2 | 40.9 | 64.7 |
| Minimum Temp °F | 11.9 | 16.0 | 24.7 | 35.9 | 45.8 | 56.3 | 62.1 | 59.7 | 49.4 | 36.2 | 24.3 | 15.0 | 36.4 |
| Mean Temp °F | 25.1 | 29.9 | 39.0 | 50.4 | 59.7 | 70.0 | 76.0 | 74.2 | 64.4 | 52.1 | 37.8 | 27.8 | 50.5 |
| Days Max Temp ≥ 90 °F | 0 | 0 | 0 | 0 | 2 | 8 | 17 | 16 | 6 | 1 | 0 | 0 | 50 |
| Days Max Temp ≤ 32 °F | 10 | 7 | 3 | 0 | 0 | 0 | 0 | 0 | 0 | 0 | 3 | 8 | 31 |
| Days Min Temp ≤ 32 °F | 31 | 28 | 25 | 11 | 2 | 0 | 0 | 0 | 1 | 9 | 26 | 31 | 164 |
| Days Min Temp ≤ 0 °F | 5 | 3 | 1 | 0 | 0 | 0 | 0 | 0 | 0 | 0 | 0 | 3 | 12 |
| Heating Degree Days | 1229 | 984 | 798 | 437 | 198 | 32 | 4 | 7 | 115 | 398 | 810 | 1146 | 6158 |
| Cooling Degree Days | 0 | 0 | 0 | 9 | 44 | 209 | 359 | 314 | 119 | 5 | 0 | 0 | 1059 |
| Total Precipitation (") | 0.53 | 0.47 | 1.30 | 1.90 | 3.40 | 3.28 | 3.47 | 2.26 | 1.56 | 1.33 | 0.97 | 0.52 | 20.99 |
| Days ≥ 0.1" Precip | 2 | 2 | 3 | 4 | 7 | 6 | 6 | 5 | 3 | 3 | 2 | 1 | 44 |
| Total Snowfall (") | 7.0 | 3.9 | 6.1 | 2.0 | 0.0 | 0.0 | 0.0 | 0.0 | 0.2 | 1.3 | 4.4 | 5.6 | 30.5 |
| Days ≥ 1" Snow Depth | 13 | 7 | 4 | 1 | 0 | 0 | 0 | 0 | 0 | 0 | 4 | 9 | 38 |

## VALENTINE LAKES GAME *Cherry County* ELEVATION 2933 ft LAT/LONG 42° 35 ' N / 100° 41 ' W

| | JAN | FEB | MAR | APR | MAY | JUN | JUL | AUG | SEP | OCT | NOV | DEC | YEAR |
|---|---|---|---|---|---|---|---|---|---|---|---|---|---|
| Maximum Temp °F | 34.8 | 40.0 | 48.9 | 61.0 | 71.9 | 81.6 | 87.5 | 85.6 | 76.6 | 64.5 | 47.1 | 37.1 | 61.4 |
| Minimum Temp °F | 12.1 | 16.1 | 24.4 | 35.6 | 46.1 | 55.3 | 60.7 | 58.7 | 48.6 | 37.5 | 24.8 | 15.3 | 36.3 |
| Mean Temp °F | 23.5 | 28.1 | 36.7 | 48.3 | 59.0 | 68.5 | 74.2 | 72.2 | 62.6 | 51.0 | 36.0 | 26.2 | 48.9 |
| Days Max Temp ≥ 90 °F | 0 | 0 | 0 | 0 | 1 | 5 | 13 | 10 | 3 | 0 | 0 | 0 | 32 |
| Days Max Temp ≤ 32 °F | 12 | 8 | 4 | 0 | 0 | 0 | 0 | 0 | 0 | 0 | 4 | 11 | 39 |
| Days Min Temp ≤ 32 °F | 29 | 26 | 24 | 12 | 2 | 0 | 0 | 0 | 1 | 9 | 23 | 29 | 155 |
| Days Min Temp ≤ 0 °F | 7 | 4 | 1 | 0 | 0 | 0 | 0 | 0 | 0 | 0 | 1 | 4 | 17 |
| Heating Degree Days | 1281 | 1035 | 871 | 500 | 213 | 43 | 9 | 14 | 143 | 432 | 864 | 1196 | 6601 |
| Cooling Degree Days | 0 | 0 | 0 | 9 | 35 | 153 | 283 | 240 | 85 | 6 | 0 | 0 | 811 |
| Total Precipitation (") | 0.38 | 0.46 | 1.16 | 2.24 | 3.26 | 3.23 | 3.80 | 2.66 | 1.70 | 1.20 | 0.87 | 0.51 | 21.47 |
| Days ≥ 0.1" Precip | 1 | 1 | 3 | 5 | 6 | 5 | 6 | 5 | 4 | 3 | 2 | 2 | 43 |
| Total Snowfall (") | 5.6 | 6.6 | 8.1 | 4.5 | 0.2 | 0.0 | 0.0 | 0.0 | 0.6 | 1.2 | 7.3 | 6.0 | 40.1 |
| Days ≥ 1" Snow Depth | 14 | 12 | 7 | 2 | 0 | 0 | 0 | 0 | 0 | 1 | 6 | 10 | 52 |

## VALENTINE MILLER FLD *Cherry County* ELEVATION 2582 ft LAT/LONG 42° 53 ' N / 100° 33 ' W

| | JAN | FEB | MAR | APR | MAY | JUN | JUL | AUG | SEP | OCT | NOV | DEC | YEAR |
|---|---|---|---|---|---|---|---|---|---|---|---|---|---|
| Maximum Temp °F | 32.6 | 37.6 | 47.4 | 60.1 | 71.5 | 81.7 | 88.2 | 86.4 | 76.4 | 63.4 | 46.1 | 35.8 | 60.6 |
| Minimum Temp °F | 6.8 | 11.7 | 21.3 | 33.0 | 44.0 | 53.9 | 59.8 | 57.4 | 46.1 | 33.4 | 20.6 | 10.0 | 33.2 |
| Mean Temp °F | 19.8 | 24.7 | 34.4 | 46.6 | 57.7 | 67.8 | 74.0 | 71.9 | 61.3 | 48.4 | 33.4 | 22.9 | 46.9 |
| Days Max Temp ≥ 90 °F | 0 | 0 | 0 | 0 | 1 | 7 | 14 | 13 | 5 | 0 | 0 | 0 | 40 |
| Days Max Temp ≤ 32 °F | 14 | 11 | 5 | 1 | 0 | 0 | 0 | 0 | 0 | 0 | 5 | 12 | 48 |
| Days Min Temp ≤ 32 °F | 31 | 28 | 27 | 15 | 3 | 0 | 0 | 0 | 2 | 14 | 27 | 31 | 178 |
| Days Min Temp ≤ 0 °F | 10 | 6 | 2 | 0 | 0 | 0 | 0 | 0 | 0 | 0 | 1 | 7 | 26 |
| Heating Degree Days | 1398 | 1132 | 943 | 551 | 251 | 53 | 10 | 21 | 177 | 509 | 942 | 1298 | 7285 |
| Cooling Degree Days | 0 | 0 | 0 | 7 | 36 | 148 | 288 | 241 | 74 | 2 | 0 | 0 | 796 |
| Total Precipitation (") | 0.31 | 0.43 | 1.08 | 1.80 | 2.87 | 3.02 | 3.07 | 2.47 | 1.46 | 0.98 | 0.69 | 0.39 | 18.57 |
| Days ≥ 0.1" Precip | 1 | 1 | 3 | 4 | 6 | 6 | 5 | 4 | 4 | 3 | 2 | 1 | 40 |
| Total Snowfall (") | 4.9 | 5.8 | 7.9 | 4.0 | 0.1 | 0.0 | 0.0 | 0.0 | 0.6 | 1.2 | 5.7 | 5.4 | 35.6 |
| Days ≥ 1" Snow Depth | 15 | 13 | 8 | 2 | 0 | 0 | 0 | 0 | 0 | 1 | 5 | 13 | 57 |

## WAHOO *Saunders County* ELEVATION 1142 ft LAT/LONG 41° 12 ' N / 96° 38 ' W

| | JAN | FEB | MAR | APR | MAY | JUN | JUL | AUG | SEP | OCT | NOV | DEC | YEAR |
|---|---|---|---|---|---|---|---|---|---|---|---|---|---|
| Maximum Temp °F | 32.6 | 38.3 | 51.1 | 64.7 | 75.2 | 84.6 | 88.4 | 85.9 | 78.1 | 66.7 | 49.1 | 37.6 | 62.7 |
| Minimum Temp °F | 11.5 | 15.9 | 27.4 | 38.8 | 49.6 | 59.7 | 64.1 | 61.0 | 51.6 | 39.7 | 27.3 | 17.2 | 38.7 |
| Mean Temp °F | 22.1 | 27.1 | 39.2 | 51.8 | 62.4 | 72.2 | 76.2 | 73.3 | 64.9 | 53.2 | 38.3 | 27.4 | 50.7 |
| Days Max Temp ≥ 90 °F | 0 | 0 | 0 | 0 | 1 | 8 | 14 | 10 | 3 | 0 | 0 | 0 | 36 |
| Days Max Temp ≤ 32 °F | 15 | 10 | 3 | 0 | 0 | 0 | 0 | 0 | 0 | 0 | 3 | 9 | 40 |
| Days Min Temp ≤ 32 °F | 30 | 26 | 22 | 8 | 1 | 0 | 0 | 0 | 1 | 7 | 21 | 29 | 145 |
| Days Min Temp ≤ 0 °F | 7 | 4 | 0 | 0 | 0 | 0 | 0 | 0 | 0 | 0 | 0 | 2 | 13 |
| Heating Degree Days | 1326 | 1060 | 793 | 404 | 140 | 16 | 2 | 8 | 103 | 369 | 796 | 1159 | 6176 |
| Cooling Degree Days | 0 | 0 | 2 | 17 | 61 | 244 | 344 | 260 | 118 | 6 | 0 | 0 | 1052 |
| Total Precipitation (") | 0.67 | 0.89 | 2.41 | 3.41 | 4.99 | 4.70 | 3.90 | 4.04 | 4.24 | 2.86 | 1.75 | 1.04 | 34.90 |
| Days ≥ 0.1" Precip | 3 | 3 | 6 | 8 | 9 | 8 | 7 | 7 | 7 | 5 | 5 | 3 | 71 |
| Total Snowfall (") | 7.7 | 8.1 | 6.5 | 1.7 | 0.2 | 0.0 | 0.0 | 0.0 | 0.0 | 0.6 | 3.8 | 8.6 | 37.2 |
| Days ≥ 1" Snow Depth | na | na | na | 0 | 0 | 0 | 0 | 0 | 0 | 0 | na | na | na |

## WAKEFIELD *Dixon County* ELEVATION 1411 ft LAT/LONG 42° 16 ' N / 96° 52 ' W

| | JAN | FEB | MAR | APR | MAY | JUN | JUL | AUG | SEP | OCT | NOV | DEC | YEAR |
|---|---|---|---|---|---|---|---|---|---|---|---|---|---|
| Maximum Temp °F | 29.5 | 35.5 | 48.4 | 63.5 | 74.1 | 83.7 | 87.1 | 84.9 | 76.7 | 64.8 | 46.4 | 33.4 | 60.7 |
| Minimum Temp °F | 7.8 | 13.5 | 24.8 | 36.7 | 48.3 | 58.3 | 62.7 | 60.0 | 50.4 | 38.1 | 25.0 | 13.3 | 36.6 |
| Mean Temp °F | 18.7 | 24.5 | 36.6 | 50.1 | 61.3 | 71.0 | 74.9 | 72.5 | 63.6 | 51.5 | 35.7 | 23.4 | 48.7 |
| Days Max Temp ≥ 90 °F | 0 | 0 | 0 | 1 | 1 | 7 | 12 | 8 | 3 | 0 | 0 | 0 | 32 |
| Days Max Temp ≤ 32 °F | 17 | 12 | 4 | 0 | 0 | 0 | 0 | 0 | 0 | 0 | 4 | 13 | 50 |
| Days Min Temp ≤ 32 °F | 31 | 27 | 24 | 10 | 2 | 0 | 0 | 0 | 1 | 9 | 24 | 30 | 158 |
| Days Min Temp ≤ 0 °F | 10 | 6 | 1 | 0 | 0 | 0 | 0 | 0 | 0 | 0 | 1 | 5 | 23 |
| Heating Degree Days | 1431 | 1136 | 873 | 450 | 169 | 21 | 4 | 10 | 123 | 420 | 872 | 1285 | 6794 |
| Cooling Degree Days | 0 | 0 | 0 | 19 | 65 | 226 | 322 | 260 | 103 | 7 | 0 | 0 | 1002 |
| Total Precipitation (") | 0.68 | 0.85 | 2.28 | 2.84 | 3.81 | 4.16 | 3.37 | 2.61 | 2.93 | 2.23 | 1.50 | 0.98 | 28.24 |
| Days ≥ 0.1" Precip | 2 | 2 | 4 | 6 | 7 | 7 | 5 | 5 | 5 | 4 | 3 | 2 | 52 |
| Total Snowfall (") | 7.2 | 6.4 | 6.8 | 1.9 | 0.0 | 0.0 | 0.0 | 0.0 | 0.0 | 1.1 | 5.0 | 8.6 | 37.0 |
| Days ≥ 1" Snow Depth | 19 | 11 | 7 | 0 | 0 | 0 | 0 | 0 | 0 | 0 | 4 | 15 | 56 |

## WALLACE 2 W *Lincoln County* ELEVATION 3100 ft LAT/LONG 40° 50 ' N / 101° 12 ' W

| | JAN | FEB | MAR | APR | MAY | JUN | JUL | AUG | SEP | OCT | NOV | DEC | YEAR |
|---|---|---|---|---|---|---|---|---|---|---|---|---|---|
| Maximum Temp °F | 36.0 | 42.1 | 51.8 | 63.6 | 72.8 | 82.7 | 88.4 | 86.7 | 78.4 | 66.3 | 48.8 | 38.7 | 63.0 |
| Minimum Temp °F | 10.8 | 15.2 | 23.2 | 33.1 | 43.7 | 53.7 | 59.1 | 57.0 | 46.4 | 33.9 | 21.5 | 12.9 | 34.2 |
| Mean Temp °F | 23.5 | 28.7 | 37.5 | 48.4 | 58.3 | 68.2 | 73.8 | 71.9 | 62.4 | 50.1 | 35.2 | 25.8 | 48.7 |
| Days Max Temp ≥ 90 °F | 0 | 0 | 0 | 0 | 1 | 7 | 14 | 11 | 4 | 0 | 0 | 0 | 37 |
| Days Max Temp ≤ 32 °F | 11 | 7 | 3 | 0 | 0 | 0 | 0 | 0 | 0 | 0 | 4 | 10 | 35 |
| Days Min Temp ≤ 32 °F | 31 | 28 | 27 | 14 | 3 | 0 | 0 | 0 | 2 | 13 | 27 | 31 | 176 |
| Days Min Temp ≤ 0 °F | 6 | 3 | 1 | 0 | 0 | 0 | 0 | 0 | 0 | 0 | 1 | 4 | 15 |
| Heating Degree Days | 1282 | 1019 | 843 | 494 | 227 | 42 | 5 | 11 | 142 | 458 | 887 | 1207 | 6617 |
| Cooling Degree Days | 0 | 0 | 0 | 3 | 23 | 146 | 278 | 242 | 77 | 2 | 0 | 0 | 771 |
| Total Precipitation (") | 0.34 | 0.40 | 1.11 | 1.57 | 3.17 | 3.07 | 2.88 | 2.03 | 1.25 | 1.13 | 0.56 | 0.42 | 17.93 |
| Days ≥ 0.1" Precip | 1 | 1 | 3 | 4 | 6 | 6 | 6 | 4 | 3 | 3 | 2 | 1 | 40 |
| Total Snowfall (") | 5.0 | 4.4 | 6.8 | 1.9 | 0.1 | 0.0 | 0.0 | 0.0 | 0.2 | 1.7 | 4.2 | 5.2 | 29.5 |
| Days ≥ 1" Snow Depth | 14 | 9 | 5 | 1 | 0 | 0 | 0 | 0 | 0 | 1 | 5 | 11 | 46 |

## WALTHILL *Thurston County* ELEVATION 1211 ft LAT/LONG 42° 9 ' N / 96° 30 ' W

| | JAN | FEB | MAR | APR | MAY | JUN | JUL | AUG | SEP | OCT | NOV | DEC | YEAR |
|---|---|---|---|---|---|---|---|---|---|---|---|---|---|
| Maximum Temp °F | 30.7 | 36.1 | *48.5* | *64.5* | 75.0 | 84.6 | *87.9* | *85.9* | *77.4* | 65.8 | *46.4* | *33.6* | 61.4 |
| Minimum Temp °F | 9.3 | 14.3 | *25.0* | *36.5* | 47.9 | 58.0 | *62.2* | *60.3* | *50.7* | 38.3 | *25.2* | *13.8* | 36.8 |
| Mean Temp °F | 20.0 | 25.2 | *36.8* | *50.5* | 61.5 | 71.3 | *75.1* | *73.1* | *64.0* | 52.1 | *35.8* | *23.7* | 49.1 |
| Days Max Temp ≥ 90 °F | 0 | 0 | 0 | 1 | 1 | 8 | *13* | *9* | *3* | 0 | 0 | 0 | 35 |
| Days Max Temp ≤ 32 °F | *16* | 11 | *4* | 0 | 0 | 0 | 0 | 0 | 0 | 0 | *3* | *13* | 47 |
| Days Min Temp ≤ 32 °F | 30 | 27 | *24* | *10* | 1 | 0 | 0 | 0 | *1* | *9* | *23* | 30 | 155 |
| Days Min Temp ≤ 0 °F | 8 | 5 | *1* | 0 | 0 | 0 | 0 | 0 | 0 | 0 | 0 | 5 | 19 |
| Heating Degree Days | 1389 | *1115* | *869* | *442* | 165 | 24 | *3* | *10* | *120* | 401 | *868* | *1273* | 6679 |
| Cooling Degree Days | 0 | 0 | 0 | 17 | 69 | 226 | 323 | 276 | *112* | 6 | 0 | 0 | 1029 |
| Total Precipitation (") | 0.60 | 0.73 | 2.05 | 2.59 | 4.07 | 4.16 | 3.75 | 2.87 | 3.28 | 2.29 | 1.37 | 0.93 | 28.69 |
| Days ≥ 0.1" Precip | *2* | *2* | *4* | 5 | 7 | 7 | 5 | *5* | *6* | *4* | *3* | *2* | 52 |
| Total Snowfall (") | na | na | na | *1.8* | 0.0 | 0.0 | 0.0 | 0.0 | 0.0 | *0.6* | na | na | na |
| Days ≥ 1" Snow Depth | na | na | na | *0* | 0 | 0 | 0 | 0 | 0 | *0* | na | na | na |

## WEEPING WATER *Cass County* ELEVATION 1240 ft LAT/LONG 40° 53 ' N / 96° 8 ' W

| | JAN | FEB | MAR | APR | MAY | JUN | JUL | AUG | SEP | OCT | NOV | DEC | YEAR |
|---|---|---|---|---|---|---|---|---|---|---|---|---|---|
| Maximum Temp °F | 33.1 | 38.9 | 51.4 | 64.9 | 74.8 | 84.0 | 87.7 | 85.3 | 77.3 | 66.3 | 49.3 | 36.8 | 62.5 |
| Minimum Temp °F | 11.8 | 16.6 | 28.0 | 39.8 | 50.3 | 59.9 | 64.7 | 62.3 | 53.4 | 40.9 | 28.1 | 16.8 | 39.4 |
| Mean Temp °F | 22.5 | 27.8 | 39.7 | 52.4 | 62.5 | 72.1 | 76.2 | 73.8 | 65.4 | 53.6 | 38.8 | 26.8 | 51.0 |
| Days Max Temp ≥ 90 °F | 0 | 0 | 0 | 1 | 1 | 7 | 13 | 9 | 3 | 0 | 0 | 0 | 34 |
| Days Max Temp ≤ 32 °F | 14 | 9 | 2 | 0 | 0 | 0 | 0 | 0 | 0 | 0 | 2 | 11 | 38 |
| Days Min Temp ≤ 32 °F | 30 | 26 | 21 | 7 | 1 | 0 | 0 | 0 | 0 | 6 | 21 | 29 | 141 |
| Days Min Temp ≤ 0 °F | 7 | 4 | 0 | 0 | 0 | 0 | 0 | 0 | 0 | 0 | 0 | 3 | 14 |
| Heating Degree Days | 1311 | 1043 | 779 | 389 | 137 | 12 | 1 | 7 | 95 | 360 | 781 | 1179 | 6094 |
| Cooling Degree Days | 0 | 0 | 2 | 21 | 60 | 225 | 340 | 272 | 120 | 8 | 0 | 0 | 1048 |
| Total Precipitation (") | 0.85 | 0.87 | 2.24 | 2.92 | 4.51 | 4.31 | 4.60 | 4.07 | 3.81 | 2.69 | 1.57 | 1.18 | 33.62 |
| Days ≥ 0.1" Precip | 2 | 2 | 5 | 6 | 7 | 6 | 6 | 6 | 6 | 5 | 4 | 3 | 58 |
| Total Snowfall (") | *7.2* | *6.8* | *3.4* | 0.9 | 0.1 | 0.0 | 0.0 | 0.0 | 0.0 | 0.2 | *2.0* | na | na |
| Days ≥ 1" Snow Depth | na | na | *4* | 0 | 0 | 0 | 0 | 0 | 0 | 0 | *1* | na | na |

## WEST POINT *Cuming County* ELEVATION 1312 ft LAT/LONG 41° 50 ' N / 96° 43 ' W

| | JAN | FEB | MAR | APR | MAY | JUN | JUL | AUG | SEP | OCT | NOV | DEC | YEAR |
|---|---|---|---|---|---|---|---|---|---|---|---|---|---|
| Maximum Temp °F | 29.4 | 34.9 | 47.6 | 62.3 | 73.4 | 83.0 | 87.1 | 84.6 | 76.0 | 64.1 | 46.5 | 33.5 | 60.2 |
| Minimum Temp °F | 8.2 | 13.4 | 25.1 | 37.4 | 48.6 | 58.8 | 63.6 | 60.8 | 50.5 | 38.0 | 25.5 | 13.9 | 37.0 |
| Mean Temp °F | 18.8 | 24.1 | 36.4 | 49.9 | 61.0 | 70.9 | 75.4 | 72.7 | 63.3 | 51.1 | 36.0 | 23.7 | 48.6 |
| Days Max Temp ≥ 90 °F | 0 | 0 | 0 | 0 | 1 | 7 | 12 | 8 | 3 | 0 | 0 | 0 | 31 |
| Days Max Temp ≤ 32 °F | 17 | 12 | 5 | 0 | 0 | 0 | 0 | 0 | 0 | 0 | 4 | 14 | 52 |
| Days Min Temp ≤ 32 °F | 31 | 27 | 24 | 9 | 1 | 0 | 0 | 0 | 1 | 9 | 23 | 30 | 155 |
| Days Min Temp ≤ 0 °F | 10 | 6 | 1 | 0 | 0 | 0 | 0 | 0 | 0 | 0 | 1 | 4 | 22 |
| Heating Degree Days | 1425 | 1147 | 881 | 459 | 174 | 25 | 4 | 11 | 131 | 432 | 862 | 1274 | 6825 |
| Cooling Degree Days | 0 | 0 | 0 | 17 | 51 | 209 | 317 | 247 | 88 | 4 | 0 | 0 | 933 |
| Total Precipitation (") | 0.61 | 0.81 | 2.12 | 2.68 | 4.15 | 4.43 | 3.52 | 3.16 | 3.08 | 2.33 | 1.49 | 0.97 | 29.35 |
| Days ≥ 0.1" Precip | 2 | 2 | 4 | 6 | 7 | 7 | 6 | 5 | 5 | 4 | 3 | 2 | 53 |
| Total Snowfall (") | 6.5 | *6.6* | 5.2 | 1.5 | 0.0 | 0.0 | 0.0 | 0.0 | 0.1 | 0.5 | *4.7* | 7.7 | 32.8 |
| Days ≥ 1" Snow Depth | 18 | 15 | 7 | 1 | 0 | 0 | 0 | 0 | 0 | 0 | 4 | 13 | 58 |

## YORK *York County* ELEVATION 1611 ft LAT/LONG 40° 52 ' N / 97° 35 ' W

| | JAN | FEB | MAR | APR | MAY | JUN | JUL | AUG | SEP | OCT | NOV | DEC | YEAR |
|---|---|---|---|---|---|---|---|---|---|---|---|---|---|
| Maximum Temp °F | 33.0 | 38.5 | 50.3 | 63.8 | 74.2 | 84.9 | 88.7 | 86.3 | 77.6 | 65.9 | 48.9 | 37.0 | 62.4 |
| Minimum Temp °F | 12.5 | 16.9 | 27.6 | 39.2 | 50.7 | 60.8 | 65.6 | 63.1 | 53.3 | 40.9 | 27.6 | 17.4 | 39.6 |
| Mean Temp °F | 22.8 | 27.7 | 39.0 | 51.5 | 62.5 | 72.8 | 77.2 | 74.8 | 65.5 | 53.4 | 38.2 | 27.2 | 51.1 |
| Days Max Temp ≥ 90 °F | 0 | 0 | 0 | 1 | 2 | 9 | 14 | 11 | 4 | 0 | 0 | 0 | 41 |
| Days Max Temp ≤ 32 °F | 14 | 10 | 3 | 0 | 0 | 0 | 0 | 0 | 0 | 0 | 3 | 11 | 41 |
| Days Min Temp ≤ 32 °F | 31 | 26 | 21 | 7 | 0 | 0 | 0 | 0 | 0 | 5 | 21 | 30 | 141 |
| Days Min Temp ≤ 0 °F | 6 | 3 | 0 | 0 | 0 | 0 | 0 | 0 | 0 | 0 | 0 | 3 | 12 |
| Heating Degree Days | 1301 | 1046 | 801 | 410 | 141 | 15 | 2 | 5 | 97 | 364 | 797 | 1166 | 6145 |
| Cooling Degree Days | 0 | 0 | 1 | 18 | 68 | 258 | 378 | 308 | 130 | 8 | 0 | 0 | 1169 |
| Total Precipitation (") | 0.74 | 0.76 | 2.20 | 2.72 | 4.23 | 4.15 | 3.84 | 3.16 | 2.88 | 2.05 | 1.47 | 0.97 | 29.17 |
| Days ≥ 0.1" Precip | 2 | 2 | 4 | 5 | 8 | 6 | 6 | 5 | 5 | 4 | 3 | 2 | 52 |
| Total Snowfall (") | 7.0 | 6.3 | 6.0 | 1.0 | 0.1 | 0.0 | 0.0 | 0.0 | 0.1 | 0.4 | 3.9 | 7.6 | 32.4 |
| Days ≥ 1" Snow Depth | na | na | na | 0 | 0 | 0 | 0 | 0 | 0 | 0 | *2* | na | na |

## JANUARY MINIMUM TEMPERATURE °F

| LOWEST | | | HIGHEST | | |
|---|---|---|---|---|---|
| 1 | Northeast Nbraska | 5.5 | 1 | Pawnee City | 15.5 |
| 2 | Valentine-Miller | 6.8 | 2 | Falls City | 15.0 |
| 3 | Gavins Point Dam | 6.9 | 3 | Kingsley Dam | 14.6 |
| 4 | Oakdale | 7.1 | | Superior | 14.6 |
| 5 | Rose | 7.5 | 5 | Imperial | 14.4 |
| 6 | Brewster | 7.8 | 6 | Lincoln | 13.9 |
| | Wakefield | 7.8 | 7 | McCook | 13.8 |
| 8 | Albion | 7.9 | 8 | Auburn | 13.7 |
| | Osmond | 7.9 | | Crete | 13.7 |
| 10 | Brownlee | 8.0 | 10 | Geneva | 13.6 |
| 11 | West Point | 8.2 | 11 | Dalton | 13.5 |
| 12 | Madison | 8.5 | | Hayes Center | 13.5 |
| 13 | Ewing | 8.7 | 13 | Central City | 13.4 |
| | Greeley | 8.7 | | Lodgepole | 13.4 |
| | Loup City | 8.7 | 15 | Seward | 13.3 |
| 16 | O'Neill | 8.8 | 16 | Hemingford | 13.2 |
| 17 | Newport | 8.9 | | Kimball | 13.2 |
| | Springview | 8.9 | 18 | Hastings | 12.9 |
| 19 | Ellsworth | 9.0 | | Holdrege | 12.9 |
| | Harrison | 9.0 | | Madrid | 12.9 |
| | Hay Springs-12 S | 9.0 | | Minden | 12.9 |
| | Mullen-21 NW | 9.0 | 22 | Hebron | 12.6 |
| | Norfolk | 9.0 | 23 | Beaver City | 12.5 |
| 24 | Anselmo | 9.1 | | York | 12.5 |
| | Hartington | 9.1 | 25 | Cambridge | 12.4 |

## JULY MAXIMUM TEMPERATURE °F

| HIGHEST | | | LOWEST | | |
|---|---|---|---|---|---|
| 1 | Beaver City | 93.8 | 1 | Harrison | 84.3 |
| 2 | Madrid | 91.8 | 2 | Nrtheast Nebraska | 85.4 |
| 3 | Lodgepole | 91.4 | 3 | Omaha-WSFO | 85.7 |
| 4 | Pawnee City | 91.2 | 4 | Ellsworth | 85.8 |
| 5 | Benkelman | 91.1 | 5 | St. Paul | 85.9 |
| 6 | Bridgeport | 90.8 | 6 | Rose | 86.0 |
| 7 | Red Cloud | 90.7 | 7 | North Loup | 86.1 |
| 8 | Superior | 90.4 | 8 | Brownlee | 86.2 |
| 9 | Culbertson | 90.3 | | Harrisburg | 86.2 |
| | McCook | 90.3 | | Mullen-21 NW | 86.2 |
| 11 | Cambridge | 90.2 | | Oakdale | 86.2 |
| | Curtis | 90.2 | 12 | Blair | 86.3 |
| 13 | Imperial | 89.9 | | Gavins Point Dam | 86.3 |
| | Trenton Dam | 89.9 | 14 | Hemingford | 86.4 |
| 15 | Halsey | 89.7 | 15 | Arthur | 86.5 |
| 16 | Lincoln-Muni | 89.6 | 16 | Crescent Lake | 86.6 |
| | Ogallala | 89.6 | | Hyannis | 86.6 |
| 18 | Falls City | 89.5 | | Mitchell | 86.6 |
| | Hebron | 89.5 | 19 | Atkinson | 86.8 |
| 20 | Chadron | 89.3 | | Kimball | 86.8 |
| | Enders Lake | 89.3 | 21 | Albion | 86.9 |
| 22 | Harlan | 89.2 | | Norfolk | 86.9 |
| | Kingsley Dam | 89.2 | 23 | Broken Bow | 87.0 |
| | Minden | 89.2 | | Ewing | 87.0 |
| | Scottsbluff | 89.2 | | Hay Springs | 87.0 |

## ANNUAL PRECIPITATION (")

| HIGHEST | | | LOWEST | | |
|---|---|---|---|---|---|
| 1 | Falls City | 34.96 | 1 | Mitchell | 13.63 |
| 2 | Wahoo | 34.90 | 2 | Harrisburg | 14.56 |
| 3 | Nebraska City | 33.86 | 3 | Scottsbluff | 15.81 |
| 4 | Weeping Water | 33.62 | 4 | Bridgeport | 15.84 |
| 5 | Tecumseh | 32.52 | 5 | Chadron | 15.99 |
| 6 | Pawnee City | 32.38 | 6 | Alliance | 16.05 |
| 7 | Auburn | 32.34 | 7 | Sidney | 16.26 |
| 8 | Syracuse | 30.91 | 8 | Hay Springs-12 S | 16.30 |
| 9 | Fairbury | 30.77 | 9 | Big Springs | 16.88 |
| 10 | Fremont | 30.43 | 10 | Kimball | 16.93 |
| 11 | David City | 30.31 | 11 | Crescent Lake | 17.03 |
| 12 | Ashland | 30.30 | | Harrison | 17.03 |
| 13 | Omaha-WSFO | 30.25 | 13 | Hemingford | 17.22 |
| 14 | Omaha-Eppley | 30.10 | 14 | Merriman | 17.44 |
| 15 | Blair | 30.06 | 15 | Oshkosh | 17.49 |
| 16 | Crete | 29.94 | 16 | Ellsworth | 17.63 |
| 17 | Geneva | 29.88 | 17 | Wallace | 17.93 |
| 18 | Lincoln | 29.67 | 18 | Ogallala | 18.06 |
| 19 | Tekamah | 29.55 | 19 | Lodgepole | 18.47 |
| 20 | Hebron | 29.51 | 20 | Kingsley Dam | 18.55 |
| 21 | Osceola | 29.39 | 21 | Valentine-Miller | 18.57 |
| 22 | Clarkson | 29.35 | 22 | Benkelman | 18.73 |
| | West Point | 29.35 | 23 | Arthur | 18.90 |
| 24 | York | 29.17 | 24 | Hyannis | 19.13 |
| 25 | Lincoln-Muni | 28.74 | 25 | Dalton | 19.29 |

## ANNUAL SNOWFALL (")

| HIGHEST | | | LOWEST | | |
|---|---|---|---|---|---|
| 1 | Harrison | 59.1 | 1 | Red Willow Dam | 17.9 |
| 2 | Mullen-21 NW | 54.7 | 2 | Medicine Crk Dm | 19.3 |
| 3 | Hemingford | 53.0 | 3 | Harlan | 20.1 |
| 4 | Dalton | 51.8 | 4 | Falls City | 20.2 |
| 5 | Harrisburg | 48.0 | | Pawnee City | 20.2 |
| 6 | Kimball | 46.5 | 6 | Canaday | 21.2 |
| 7 | Mullen | 44.8 | 7 | Central City | 21.5 |
| 8 | Newport | 44.1 | 8 | Syracuse | 23.0 |
| 9 | Scottsbluff | 43.8 | 9 | Crete | 23.4 |
| 10 | Hay Springs-12 S | 42.2 | 10 | Ashland | 23.5 |
| 11 | Chadron | 41.4 | 11 | Lincoln | 23.6 |
| 12 | Madrid | 40.8 | 12 | Seward | 24.0 |
| 13 | Ainsworth | 40.2 | 13 | Blair | 24.6 |
| 14 | Valentine-Lakes | 40.1 | 14 | Madison | 25.5 |
| 15 | Bridgeport | 39.8 | 15 | Columbus | 25.7 |
| 16 | Kingsley Dam | 39.4 | 16 | Lincoln-Muni | 26.0 |
| 17 | Anselmo | 39.2 | 17 | Nebraska City | 26.3 |
| 18 | Ellsworth | 38.0 | 18 | Burwell | 26.4 |
| 19 | Lodgepole | 37.5 | 19 | Fairbury | 26.5 |
| 20 | Wahoo | 37.2 | | Geneva | 26.5 |
| 21 | Wakefield | 37.0 | | Northeast Nbraska | 26.5 |
| 22 | Arthur | 36.9 | 22 | Hebron | 26.6 |
| | Crescent Lake | 36.9 | | Superior | 26.6 |
| 24 | Atkinson | 36.3 | 24 | Ravenna | 27.0 |
| | Loup City | 36.3 | | Tekamah | 27.0 |

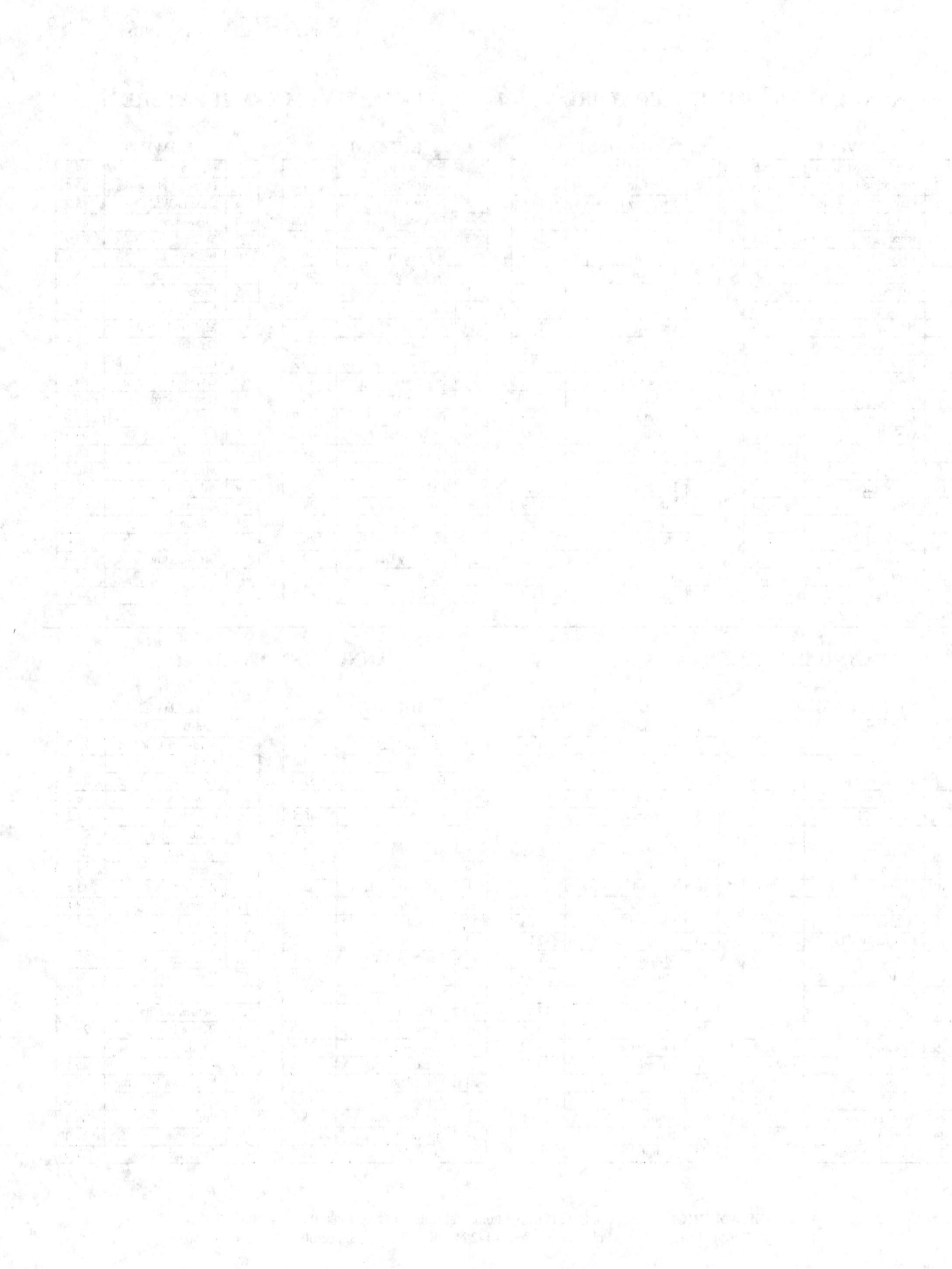

# NEVADA

PHYSICAL FEATURES. Nevada is primarily a plateau area. The eastern part has an average elevation of between 5,000 and 6,000 feet above sea level; the western portion between 3,800 and 5,000 feet, the lower limit being in the vicinity of Pyramid Lake and Carson Sink; and the southern part generally between 2,000 and 3,000 feet. From the lower elevations of the west portion there is a fairly rapid rise westward to the summits of the eastern ranges of the Sierra Nevada. The southwestern part slopes down toward Death Valley, California, and the southern portion toward the channel of the Colorado River, the elevation of which is less than 1,000 feet above sea level. The extreme northeastern part slopes northerly, draining into the Snake River and thence into the Columbia.

On the Nevada plateau there are many mountain ranges, most of them 50 to 100 miles long, running generally north and south. The only east-west range is in the northeast. It forms the southern limit of the Columbia River Basin. With the exception of this small drainage area and another limited region in the southeast which drains into the Colorado River, the State lies within the confines of the Great Basin, and the waters of its streams disappear into sinks or flow into lakes with no outlets.

GENERAL CLIMATE. Nevada lies just east and to the leeward of the Sierra Nevada Range, a massive mountain barrier which has a marked influence on the climate of the State. One of the greatest contrasts in precipitation found within a short distance in the United States occurs between the western, or California, slopes of the Sierras and the valleys just to the east of this range. The prevailing winds are from the west, and as moist air associated with storms from the Pacific Ocean ascends the western slopes of the Sierras, a large portion of the original moisture falls as precipitation. As the air descends the eastern slope, it is warmed by compression, so that very little precipitation occurs. The effects of this mountain barrier are felt not only in the extreme western part, but generally throughout the State, with the result that the lowlands of Nevada are largely desert or semidesert.

With its varied and rugged topography--its mountain ranges, narrow valleys and low, sage-covered deserts, ranging in elevation from about 1,500 to more than 10,000 feet--Nevada presents wide local variations of temperature and rainfall. The most striking climatic features are bright sunshine, small annual precipitation in the valleys and deserts, heavy snowfall in the higher mountains, dryness and purity of air, and phenomenally large daily ranges of temperature.

TEMPERATURE. The mean annual temperatures vary from the middle 40s in the northeastern part to around 50° F. in the west, and to the middle 60s in the south. In the northeastern portion summers are short and hot, winters long and cold. In the west, the summers are also short and hot, but the winters are only moderately cold; while in the south the summers are long and hot and the winters short and mild. Prolonged periods of extremely cold weather are rare, due primarily to the mountains east and north of the State which act as a barrier to the intensely cold continental Arctic air masses.

In Nevada there is relatively strong insolation of heat during the day and rapid nighttime cooling, because of the clear air, resulting in wide daily ranges in temperature. Even after the hottest days, the nights are usually cool. At Reno the average range between the highest and the lowest daily temperatures is 29° F. in January, increasing month by month to 45° F. in July. In summer temperatures above 100° F. occur rather frequently in the extreme southern portion and occasionally over the remainder of the State. However, the humidity is normally low so that corresponding temperatures are less disagreeable in Nevada than in more humid climates.

PRECIPITATION. Nevada's precipitation mostly occurs during the winter season and on the average is less than in any other State. Precipitation is lightest over the lower parts of the western plateau, a series of long valleys extending from the State border opposite Death Valley in California northward to the Idaho line. Over the more southerly of those valleys the average annual precipitation is less than 5 inches. From this low average it ranges upward to 18 inches in Lamoille Canyon on the western side of the Ruby Mountains of northeast Nevada, and up to about 28 inches at Marlette Lake high in the most easterly range of the Sierras. Variations in precipitation are due mainly to differences in elevation and exposure to precipitation-bearing winds. The average annual number of days with measurable precipitation varies considerably.

Snowfall is usually heavy in the mountains, particularly in the north. Mountain snowfall forms the main source of water for streamflow. In years when winter and spring snowfall is light, the result is a shortage of water. Melting of the mountain snowpack in the spring usually causes some flooding in northern and extreme western streams during the period April to June, but damaging floods of this type are infrequent. Rain floods in which snowmelt is also a factor usually occur from November to March. Heavy summer thunderstorms cause flooding in local streams, but they usually occur over sparsely settled mountainous areas and, therefore, are seldom destructive.

OTHER CLIMATIC ELEMENTS. The State has a generous supply of sunshine, the average percentage of the possible amount at northern and central locations being generally between 65 and 75 percent and at southern locations above 80 percent.

The low humidity and abundant sunshine produce rapid evaporation. Annual amounts in the extreme southern portion of the State, as measured in evaporation pans, average over 100 inches. In northern and central sections amounts average roughly half as much.

Winds are generally light. Storms with high winds rarely occur, and still more rarely cause appreciable damage, except locally along the east slope of the Sierras. The prevailing wind direction is west, although at a few stations, because of local topography, it is south or southwest.

Dust or sandstorms occur occasionally, particularly over the southern part during the spring months when storms are moving through the region more frequently than at other seasons of the year.

Thunderstorms are infrequent, the average annual number being 12 at Winnemucca, 13 at Reno and Las Vegas, 22 at Elko, and 30 at Ely. Summer thunderstorms develop occasionally into heavy local downpours of rain. These storms, locally termed cloudbursts, may bring to a locality as much rain in a few hours as would normally fall in several months. Tornadoes are extremely rare.

Over the northern and central portions of the State, freezes continue until late in spring and being early in autumn. The shortest freeze-free season is in the extreme northeast, and the longest in the extreme south, the range being from less than 100 days at several stations in the northeast to around 140 in the west, and to over 225 in the extreme south.

# COUNTY INDEX



# ELEVATION INDEX

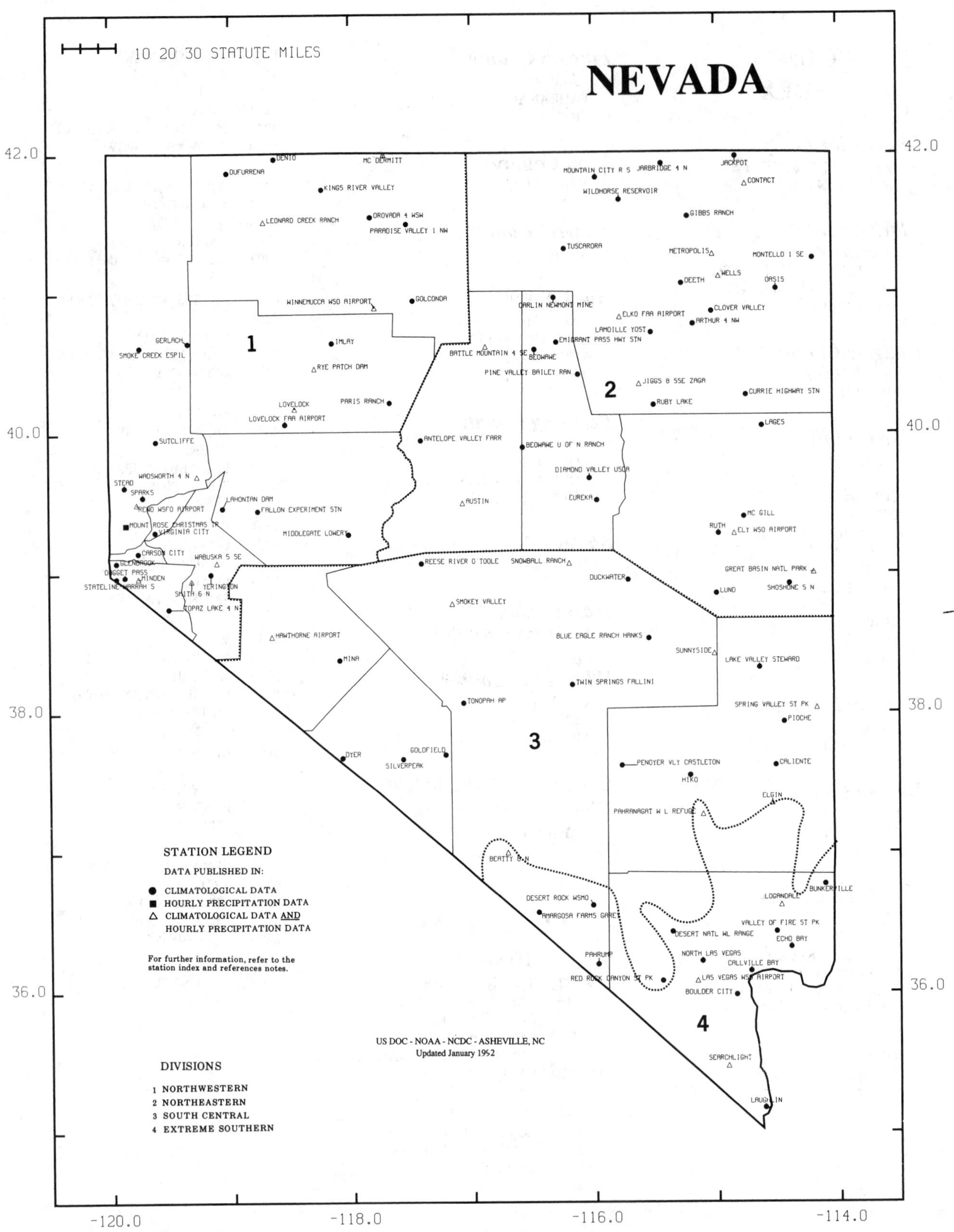

10 20 30 STATUTE MILES
NEVADA
42.0
40.0
38.0
36.0
-120.0
-118.0
-116.0
-114.0
DENIO
MC DERMITT
DUFURRENA
KINGS RIVER VALLEY
LEONARD CREEK RANCH
OROVADA 4 WSW
PARADISE VALLEY 1 NW
WINNEMUCCA WSO AIRPORT
GOLCONDA
GERLACH
SMOKE CREEK ESPIL
IMLAY
RYE PATCH DAM
LOVELOCK
LOVELOCK FAA AIRPORT
PARIS RANCH
SUTCLIFFE
WADSWORTH 4 N
STEAD
SPARKS
RENO WSFO AIRPORT
MOUNT ROSE CHRISTMAS TR
VIRGINIA CITY
LAHONTAN DAM
FALLON EXPERIMENT STN
MIDDLEGATE LOWERY
CARSON CITY
GLENBROOK
DAGGET PASS
STATELINE HARRAH S
MINDEN
WABUSKA 5 SE
YERINGTON
SMITH 6 N
TOPAZ LAKE 4 N
HAWTHORNE AIRPORT
MINA
DYER
GOLDFIELD
SILVERPEAK
MOUNTAIN CITY R S
JARBRIDGE 4 N
JACKPOT
CONTACT
WILDHORSE RESERVOIR
GIBBS RANCH
TUSCARORA
METROPOLIS
MONTELLO 1 SE
WELLS
DEETH
OASIS
CARLIN NEWMONT MINE
ELKO FAA AIRPORT
CLOVER VALLEY
ARTHUR 4 NW
LAMOILLE YOST
EMIGRANT PASS HWY STN
BATTLE MOUNTAIN 4 SE
BEOWAWE
PINE VALLEY BAILEY RAN
JIGGS 8 SSE ZAGA
RUBY LAKE
CURRIE HIGHWAY STN
LAGES
ANTELOPE VALLEY FARR
BEOWAWE U OF N RANCH
DIAMOND VALLEY USDA
AUSTIN
EUREKA
MC GILL
RUTH
ELY WSO AIRPORT
REESE RIVER O TOOLE
SNOWBALL RANCH
DUCKWATER
GREAT BASIN NATL PARK
LUND
SHOSHONE 5 N
SMOKEY VALLEY
BLUE EAGLE RANCH HANKS
SUNNYSIDE
LAKE VALLEY STEWARD
TWIN SPRINGS FALLINI
TONOPAH AP
SPRING VALLEY ST PK
PIOCHE
PENOYER VLY CASTLETON
HIKO
CALIENTE
ELGIN
PAHRANAGAT W L REFUGE
BEATTY 8 N
DESERT ROCK WSMO
AMARGOSA FARMS GAREY
LOGANDALE
BUNKERVILLE
VALLEY OF FIRE ST PK
DESERT NATL WL RANGE
ECHO BAY
NORTH LAS VEGAS
CALLVILLE BAY
PAHRUMP
RED ROCK CANYON ST PK
LAS VEGAS WSO AIRPORT
BOULDER CITY
SEARCHLIGHT
LAUGHLIN
1
2
3
4
STATION LEGEND
DATA PUBLISHED IN:
CLIMATOLOGICAL DATA
HOURLY PRECIPITATION DATA
CLIMATOLOGICAL DATA AND HOURLY PRECIPITATION DATA
For further information, refer to the station index and references notes.
US DOC - NOAA - NCDC - ASHEVILLE, NC
Updated January 1992
DIVISIONS
1 NORTHWESTERN
2 NORTHEASTERN
3 SOUTH CENTRAL
4 EXTREME SOUTHERN

## ARTHUR 4 NW *Elko County* ELEVATION 6404 ft LAT/LONG 40° 48 ' N / 115° 11 ' W

| | JAN | FEB | MAR | APR | MAY | JUN | JUL | AUG | SEP | OCT | NOV | DEC | YEAR |
|---|---|---|---|---|---|---|---|---|---|---|---|---|---|
| Maximum Temp °F | 35.6 | 39.4 | 45.3 | 54.1 | 64.2 | 73.2 | 82.4 | 81.4 | 72.2 | 60.5 | 44.7 | 35.7 | 57.4 |
| Minimum Temp °F | 13.9 | 17.4 | 23.1 | 28.6 | 35.8 | 42.6 | 49.2 | 48.3 | 40.0 | 31.3 | 22.2 | 14.4 | 30.6 |
| Mean Temp °F | 24.8 | 28.5 | 34.3 | 41.4 | 50.0 | 57.9 | 65.9 | 64.9 | 56.2 | 45.9 | 33.5 | 25.1 | 44.0 |
| Days Max Temp ≥ 90 °F | 0 | 0 | 0 | 0 | 0 | 0 | 3 | 2 | 0 | 0 | 0 | 0 | 5 |
| Days Max Temp ≤ 32 °F | 11 | 5 | 2 | 0 | 0 | 0 | 0 | 0 | 0 | 0 | 4 | 10 | 32 |
| Days Min Temp ≤ 32 °F | 30 | 27 | 28 | 21 | 10 | *1* | 0 | 0 | 4 | 17 | 27 | 31 | 196 |
| Days Min Temp ≤ 0 °F | 4 | 2 | 0 | 0 | 0 | 0 | 0 | 0 | 0 | 0 | 1 | 3 | 10 |
| Heating Degree Days | 1239 | 1024 | 944 | 700 | 457 | 218 | 46 | 63 | 264 | 585 | 940 | 1231 | 7711 |
| Cooling Degree Days | 0 | 0 | 0 | 0 | 0 | 13 | 82 | 70 | 7 | 0 | 0 | 0 | 172 |
| Total Precipitation (") | 1.53 | 1.44 | 1.42 | 1.23 | 1.34 | 1.05 | 0.64 | 0.80 | 1.01 | 1.11 | 1.70 | 1.58 | 14.85 |
| Days ≥ 0.1" Precip | 5 | 5 | 5 | 4 | 4 | 4 | 2 | 2 | 3 | 3 | 5 | 5 | 47 |
| Total Snowfall (") | *11.6* | *9.9* | *8.0* | 2.8 | 1.1 | 0.0 | 0.0 | 0.0 | 0.0 | 1.3 | *7.1* | 12.4 | 54.2 |
| Days ≥ 1" Snow Depth | na | na | *14* | 3 | 1 | 0 | 0 | 0 | 0 | 1 | *9* | *19* | na |

## AUSTIN *Lander County* ELEVATION 6611 ft LAT/LONG 39° 30 ' N / 117° 5 ' W

| | JAN | FEB | MAR | APR | MAY | JUN | JUL | AUG | SEP | OCT | NOV | DEC | YEAR |
|---|---|---|---|---|---|---|---|---|---|---|---|---|---|
| Maximum Temp °F | 41.9 | 45.1 | 49.3 | 56.0 | 66.1 | 77.3 | 87.1 | 85.2 | 76.0 | 63.8 | 49.3 | 42.2 | 61.6 |
| Minimum Temp °F | 19.9 | 22.5 | 25.7 | 30.1 | 38.0 | 45.9 | 53.8 | 52.6 | 44.8 | 35.7 | 26.3 | 20.3 | 34.6 |
| Mean Temp °F | 30.9 | 33.9 | 37.5 | 43.1 | 52.1 | 61.6 | 70.5 | 68.9 | 60.4 | 49.8 | 37.8 | 31.3 | 48.2 |
| Days Max Temp ≥ 90 °F | 0 | 0 | 0 | 0 | 0 | 3 | 12 | 9 | 1 | 0 | 0 | 0 | 25 |
| Days Max Temp ≤ 32 °F | 5 | 2 | 1 | 0 | 0 | 0 | 0 | 0 | 0 | 0 | 2 | 5 | 15 |
| Days Min Temp ≤ 32 °F | 29 | 25 | 25 | 19 | 8 | 1 | 0 | 0 | 2 | 10 | 22 | 28 | 169 |
| Days Min Temp ≤ 0 °F | 1 | 0 | 0 | 0 | 0 | 0 | 0 | 0 | 0 | 0 | 0 | 1 | 2 |
| Heating Degree Days | 1049 | 873 | 846 | 650 | 398 | 152 | 15 | 31 | 168 | 466 | 808 | 1038 | 6494 |
| Cooling Degree Days | 0 | 0 | 0 | 0 | 7 | 66 | 191 | 172 | 42 | 2 | 0 | 0 | 480 |
| Total Precipitation (") | 1.25 | 1.16 | 1.76 | 1.68 | 1.67 | 1.07 | 0.59 | 0.80 | 0.80 | 1.01 | 1.28 | 1.22 | 14.29 |
| Days ≥ 0.1" Precip | 4 | 4 | 5 | 4 | 4 | 3 | 2 | 2 | 2 | 3 | 4 | 4 | 41 |
| Total Snowfall (") | 13.4 | 11.0 | 16.1 | 12.4 | 6.5 | 0.4 | 0.0 | 0.0 | 0.4 | 3.1 | 8.0 | 10.9 | 82.2 |
| Days ≥ 1" Snow Depth | 16 | 11 | 8 | 4 | 1 | 0 | 0 | 0 | 0 | 1 | 5 | 13 | 59 |

## BATTLE MOUNTAIN 4 SE *Lander County* ELEVATION 4529 ft LAT/LONG 40° 37 ' N / 116° 52 ' W

| | JAN | FEB | MAR | APR | MAY | JUN | JUL | AUG | SEP | OCT | NOV | DEC | YEAR |
|---|---|---|---|---|---|---|---|---|---|---|---|---|---|
| Maximum Temp °F | 41.1 | 48.5 | 55.5 | 63.4 | 73.7 | 83.6 | 93.1 | 91.1 | 81.3 | 68.1 | 51.6 | 41.5 | 66.0 |
| Minimum Temp °F | 16.6 | 21.9 | 25.9 | 30.3 | 38.7 | 45.7 | 51.4 | 48.9 | 39.9 | 29.7 | 23.1 | 15.8 | 32.3 |
| Mean Temp °F | 28.9 | 35.2 | 40.7 | 46.9 | 56.2 | 64.7 | 72.3 | 70.0 | 60.6 | 48.9 | 37.4 | 28.6 | 49.2 |
| Days Max Temp ≥ 90 °F | 0 | 0 | 0 | 0 | 2 | 10 | 24 | 20 | 6 | 0 | 0 | 0 | 62 |
| Days Max Temp ≤ 32 °F | 6 | 1 | 0 | 0 | 0 | 0 | 0 | 0 | 0 | 0 | 1 | 5 | 13 |
| Days Min Temp ≤ 32 °F | 28 | 25 | 25 | 19 | 7 | 1 | 0 | 0 | 5 | 20 | 24 | 29 | 183 |
| Days Min Temp ≤ 0 °F | 4 | 1 | 0 | 0 | 0 | 0 | 0 | 0 | 0 | 0 | 0 | 3 | 8 |
| Heating Degree Days | 1112 | 834 | 747 | 539 | 280 | 92 | 8 | 22 | 160 | 492 | 823 | 1122 | 6231 |
| Cooling Degree Days | 0 | 0 | 0 | 1 | 19 | 97 | 239 | 195 | 41 | 1 | 0 | 0 | 593 |
| Total Precipitation (") | 0.63 | 0.59 | 0.70 | 0.88 | 1.04 | 0.88 | 0.31 | 0.43 | 0.66 | 0.64 | 0.72 | 0.74 | 8.22 |
| Days ≥ 0.1" Precip | 2 | 2 | 3 | 3 | 3 | 3 | 1 | 1 | 2 | 2 | 2 | 2 | 26 |
| Total Snowfall (") | 4.8 | 4.1 | 2.6 | 2.3 | 0.3 | 0.0 | 0.0 | 0.0 | 0.0 | 0.3 | 1.8 | 6.3 | 22.5 |
| Days ≥ 1" Snow Depth | 9 | 3 | 1 | 0 | 0 | 0 | 0 | 0 | 0 | 0 | 1 | 7 | 21 |

## BEOWAWE *Eureka County* ELEVATION 4701 ft LAT/LONG 40° 36 ' N / 116° 29 ' W

| | JAN | FEB | MAR | APR | MAY | JUN | JUL | AUG | SEP | OCT | NOV | DEC | YEAR |
|---|---|---|---|---|---|---|---|---|---|---|---|---|---|
| Maximum Temp °F | 39.0 | 46.0 | 53.6 | 61.8 | 71.9 | 81.8 | 91.4 | 89.2 | 79.8 | 66.9 | 49.6 | 39.7 | 64.2 |
| Minimum Temp °F | *13.3* | 20.4 | 25.5 | 30.0 | 37.6 | 44.6 | 50.8 | 48.6 | 39.0 | 29.5 | 22.6 | 14.6 | 31.4 |
| Mean Temp °F | *26.0* | 33.3 | 39.6 | 45.9 | 54.8 | 63.2 | 71.1 | 68.9 | 59.4 | 48.2 | 36.2 | 27.2 | 47.8 |
| Days Max Temp ≥ 90 °F | 0 | 0 | 0 | 0 | 1 | 7 | *18* | 16 | *4* | 0 | 0 | 0 | 46 |
| Days Max Temp ≤ 32 °F | *9* | 2 | 0 | 0 | 0 | 0 | 0 | 0 | 0 | 0 | 1 | *7* | 19 |
| Days Min Temp ≤ 32 °F | *30* | *26* | na | *17* | *8* | 1 | 0 | 0 | 5 | *20* | *25* | 30 | na |
| Days Min Temp ≤ 0 °F | 6 | 1 | 0 | 0 | 0 | 0 | 0 | 0 | 0 | 0 | 0 | 3 | 10 |
| Heating Degree Days | *1203* | 889 | *778* | *568* | *321* | 112 | 14 | 37 | 186 | *522* | *873* | 1166 | 6669 |
| Cooling Degree Days | 0 | 0 | 0 | 1 | 15 | 62 | 165 | 148 | 28 | 0 | 0 | 0 | 419 |
| Total Precipitation (") | 0.67 | 0.64 | 0.75 | 0.86 | 1.13 | 0.84 | 0.31 | 0.57 | 0.57 | 0.61 | 0.87 | 0.85 | 8.67 |
| Days ≥ 0.1" Precip | *2* | 2 | 2 | 2 | *2* | 2 | 1 | 1 | 2 | 2 | 3 | 2 | 23 |
| Total Snowfall (") | na | na | na | *0.8* | 0.1 | 0.0 | 0.0 | 0.0 | 0.0 | 0.3 | *1.2* | *3.8* | na |
| Days ≥ 1" Snow Depth | na | na | na | *0* | 0 | 0 | 0 | 0 | 0 | 0 | *1* | na | na |

## BOULDER CITY *Clark County* ELEVATION 2525 ft LAT/LONG 35° 59 ' N / 114° 51 ' W

| | JAN | FEB | MAR | APR | MAY | JUN | JUL | AUG | SEP | OCT | NOV | DEC | YEAR |
|---|---|---|---|---|---|---|---|---|---|---|---|---|---|
| Maximum Temp °F | 55.1 | 61.0 | 68.1 | 76.7 | 86.3 | 96.7 | 101.7 | 99.6 | 92.2 | 80.0 | 64.8 | 55.3 | 78.1 |
| Minimum Temp °F | 38.5 | 42.6 | 46.9 | 53.0 | 61.4 | 70.5 | 76.1 | 74.7 | 68.1 | 57.7 | 46.0 | 38.5 | 56.2 |
| Mean Temp °F | 46.8 | 51.8 | 57.5 | 65.0 | 73.9 | 83.6 | 88.9 | 87.2 | 80.2 | 68.9 | 55.4 | 46.9 | 67.2 |
| Days Max Temp ≥ 90 °F | 0 | 0 | 0 | 2 | 12 | 25 | 30 | 30 | 20 | 4 | 0 | 0 | 123 |
| Days Max Temp ≤ 32 °F | 0 | 0 | 0 | 0 | 0 | 0 | 0 | 0 | 0 | 0 | 0 | 0 | 0 |
| Days Min Temp ≤ 32 °F | 5 | 2 | 1 | 0 | 0 | 0 | 0 | 0 | 0 | 0 | 0 | 5 | 13 |
| Days Min Temp ≤ 0 °F | 0 | 0 | 0 | 0 | 0 | 0 | 0 | 0 | 0 | 0 | 0 | 0 | 0 |
| Heating Degree Days | 558 | 367 | 247 | 95 | 20 | 1 | 0 | 0 | 2 | 47 | 287 | 554 | 2178 |
| Cooling Degree Days | 0 | 3 | 20 | 126 | 316 | 582 | 747 | 700 | 473 | 180 | 9 | 0 | 3156 |
| Total Precipitation (") | 0.69 | 0.70 | 0.94 | 0.30 | 0.23 | 0.12 | 0.47 | 0.92 | 0.58 | 0.31 | 0.60 | 0.58 | 6.44 |
| Days ≥ 0.1" Precip | 2 | 2 | 2 | 1 | 1 | 0 | 1 | 2 | 1 | 1 | 1 | 2 | 16 |
| Total Snowfall (") | 0.0 | 0.0 | 0.0 | 0.0 | 0.0 | 0.0 | 0.0 | 0.0 | 0.0 | 0.0 | 0.0 | 0.0 | 0.0 |
| Days ≥ 1" Snow Depth | 0 | 0 | 0 | 0 | 0 | 0 | 0 | 0 | 0 | 0 | 0 | 0 | 0 |

## CALIENTE *Lincoln County* ELEVATION 4403 ft LAT/LONG 37° 37 ' N / 114° 31 ' W

| | JAN | FEB | MAR | APR | MAY | JUN | JUL | AUG | SEP | OCT | NOV | DEC | YEAR |
|---|---|---|---|---|---|---|---|---|---|---|---|---|---|
| Maximum Temp °F | 46.8 | 53.3 | 59.9 | 68.4 | 78.2 | 89.0 | 95.3 | 93.0 | 84.8 | 73.4 | 57.6 | 47.3 | 70.6 |
| Minimum Temp °F | 18.5 | 23.3 | 28.8 | 34.0 | 42.1 | 50.0 | 56.8 | 55.5 | 46.0 | 35.3 | 25.9 | 18.4 | 36.2 |
| Mean Temp °F | 32.6 | 38.3 | 44.4 | 51.2 | 60.2 | 69.5 | 76.0 | 74.3 | 65.4 | 54.4 | 41.8 | 32.9 | 53.4 |
| Days Max Temp ≥ 90 °F | 0 | 0 | 0 | 0 | 3 | 15 | 26 | 23 | 9 | 1 | 0 | 0 | 77 |
| Days Max Temp ≤ 32 °F | 1 | 1 | 0 | 0 | 0 | 0 | 0 | 0 | 0 | 0 | 0 | 1 | 3 |
| Days Min Temp ≤ 32 °F | 30 | 25 | 22 | *13* | 3 | 0 | 0 | 0 | 1 | *10* | *25* | *30* | 159 |
| Days Min Temp ≤ 0 °F | 1 | 0 | 0 | 0 | 0 | 0 | 0 | 0 | 0 | 0 | 0 | 1 | 2 |
| Heating Degree Days | 996 | 747 | 632 | 408 | 172 | 24 | 1 | 1 | 63 | 327 | 689 | 987 | 5047 |
| Cooling Degree Days | 0 | 0 | 0 | 2 | 29 | 179 | 356 | 308 | 91 | 3 | 0 | 0 | 968 |
| Total Precipitation (") | 0.94 | 0.93 | 1.32 | 0.75 | 0.69 | 0.30 | 0.92 | 1.03 | 0.76 | 0.80 | 0.84 | 0.64 | 9.92 |
| Days ≥ 0.1" Precip | 3 | 2 | 4 | 2 | 2 | 1 | 3 | 3 | 2 | 2 | 2 | *2* | 28 |
| Total Snowfall (") | na | na | *0.7* | *0.0* | 0.0 | 0.0 | 0.0 | 0.0 | 0.1 | 0.0 | *0.8* | na | na |
| Days ≥ 1" Snow Depth | na | na | *0* | 0 | 0 | 0 | 0 | 0 | 0 | *0* | *1* | na | na |

## CARSON CITY *Carson City Independent City* ELEVATION 4652 ft LAT/LONG 39° 9 ' N / 119° 46 ' W

| | JAN | FEB | MAR | APR | MAY | JUN | JUL | AUG | SEP | OCT | NOV | DEC | YEAR |
|---|---|---|---|---|---|---|---|---|---|---|---|---|---|
| Maximum Temp °F | 45.4 | 50.9 | 56.5 | 62.9 | 72.1 | 81.0 | 89.2 | 87.7 | 79.7 | 69.0 | 54.6 | 45.6 | 66.2 |
| Minimum Temp °F | 20.4 | 24.0 | 28.5 | 32.2 | 38.9 | 45.6 | 50.2 | 48.5 | 41.0 | 32.0 | 25.5 | 19.8 | 33.9 |
| Mean Temp °F | 33.0 | 37.5 | 42.5 | 47.6 | 55.5 | 63.3 | 69.7 | 68.1 | 60.4 | 50.5 | 40.0 | 32.7 | 50.1 |
| Days Max Temp ≥ 90 °F | 0 | 0 | 0 | 0 | 0 | 5 | 17 | 14 | 2 | 0 | 0 | 0 | 38 |
| Days Max Temp ≤ 32 °F | 3 | 1 | 0 | 0 | 0 | 0 | 0 | 0 | 0 | 0 | 0 | 3 | 7 |
| Days Min Temp ≤ 32 °F | 27 | 24 | 21 | 16 | 6 | 1 | 0 | 0 | 4 | 16 | 23 | 27 | 165 |
| Days Min Temp ≤ 0 °F | 1 | 0 | 0 | 0 | 0 | 0 | 0 | 0 | 0 | 0 | 0 | 1 | 2 |
| Heating Degree Days | 986 | 771 | 690 | 515 | 295 | 100 | 15 | 26 | 158 | 442 | 742 | 994 | 5734 |
| Cooling Degree Days | 0 | 0 | 0 | 0 | 11 | 63 | 171 | 144 | 34 | 0 | 0 | 0 | 423 |
| Total Precipitation (") | 1.79 | 1.41 | 1.06 | 0.40 | 0.46 | 0.38 | 0.25 | 0.33 | 0.51 | 0.69 | 1.36 | 1.46 | 10.10 |
| Days ≥ 0.1" Precip | 3 | 3 | 3 | 1 | 1 | 1 | 1 | 1 | 1 | 2 | 3 | 3 | 23 |
| Total Snowfall (") | *6.9* | 4.0 | 2.2 | 0.7 | 0.3 | 0.0 | 0.0 | 0.0 | 0.1 | 0.1 | *1.1* | 4.5 | 19.9 |
| Days ≥ 1" Snow Depth | *5* | 3 | 1 | 0 | 0 | 0 | 0 | 0 | 0 | 0 | *1* | 3 | 13 |

## CLOVER VALLEY *Elko County* ELEVATION 5804 ft LAT/LONG 40° 44 ' N / 115° 2 ' W

| | JAN | FEB | MAR | APR | MAY | JUN | JUL | AUG | SEP | OCT | NOV | DEC | YEAR |
|---|---|---|---|---|---|---|---|---|---|---|---|---|---|
| Maximum Temp °F | 36.9 | 42.5 | 49.6 | 58.0 | 67.0 | 76.6 | *85.2* | 84.1 | *75.7* | 63.9 | 47.8 | 37.7 | 60.4 |
| Minimum Temp °F | 12.8 | 17.7 | 23.8 | 28.2 | 35.4 | 43.0 | *49.0* | 47.6 | *39.3* | 30.0 | 21.5 | 13.4 | 30.1 |
| Mean Temp °F | 24.9 | 30.1 | 36.7 | 43.1 | 51.2 | 59.8 | *67.1* | 65.9 | *57.5* | 47.2 | 34.7 | 25.6 | 45.3 |
| Days Max Temp ≥ 90 °F | 0 | 0 | 0 | 0 | 0 | 1 | na | *5* | *0* | 0 | 0 | 0 | na |
| Days Max Temp ≤ 32 °F | 10 | *4* | 1 | 0 | 0 | 0 | *0* | 0 | 0 | 0 | 2 | 8 | 25 |
| Days Min Temp ≤ 32 °F | 29 | *26* | 27 | *21* | na | 1 | *0* | 0 | *5* | *18* | *27* | 30 | na |
| Days Min Temp ≤ 0 °F | 6 | 3 | 0 | 0 | 0 | 0 | 0 | 0 | 0 | 0 | 1 | 4 | 14 |
| Heating Degree Days | 1237 | 978 | 869 | 648 | 421 | 179 | *30* | 50 | *229* | 545 | 903 | 1215 | 7304 |
| Cooling Degree Days | 0 | 0 | 0 | 0 | 1 | 32 | 111 | 88 | 14 | 0 | 0 | 0 | 246 |
| Total Precipitation (") | 1.33 | 1.24 | 1.29 | 1.10 | 1.24 | 0.99 | 0.49 | 0.69 | 0.91 | 0.93 | 1.74 | 1.61 | 13.56 |
| Days ≥ 0.1" Precip | 4 | 4 | 5 | 4 | 4 | 3 | *1* | 2 | 3 | 3 | 5 | 4 | 42 |
| Total Snowfall (") | na | na | na | na | *0.1* | 0.0 | 0.0 | 0.0 | 0.1 | 0.4 | na | na | na |
| Days ≥ 1" Snow Depth | na | na | na | na | *0* | 0 | 0 | 0 | 0 | 0 | na | na | na |

## CONTACT *Elko County* ELEVATION 5374 ft LAT/LONG 41° 47 ' N / 114° 45 ' W

| | JAN | FEB | MAR | APR | MAY | JUN | JUL | AUG | SEP | OCT | NOV | DEC | YEAR |
|---|---|---|---|---|---|---|---|---|---|---|---|---|---|
| Maximum Temp °F | 39.6 | 44.2 | 50.5 | 58.9 | 68.6 | 78.9 | 88.8 | 86.5 | ***77.2*** | 64.5 | 48.1 | 39.3 | 62.1 |
| Minimum Temp °F | 15.1 | 19.0 | 24.4 | 28.1 | 35.4 | 42.7 | 47.8 | 46.1 | ***37.8*** | 28.4 | 21.6 | 14.2 | 30.1 |
| Mean Temp °F | 27.4 | 31.6 | 37.5 | 43.5 | 52.1 | 60.8 | 68.4 | 66.3 | ***57.5*** | 46.5 | 34.9 | 26.7 | 46.1 |
| Days Max Temp ≥ 90 °F | 0 | 0 | 0 | 0 | 0 | 5 | 16 | 11 | ***2*** | 0 | 0 | 0 | 34 |
| Days Max Temp ≤ 32 °F | 6 | 3 | 0 | 0 | 0 | 0 | 0 | 0 | 0 | 0 | 2 | 7 | 18 |
| Days Min Temp ≤ 32 °F | 29 | 27 | 27 | 23 | 11 | 2 | 0 | 0 | 6 | 22 | 27 | ***30*** | 204 |
| Days Min Temp ≤ 0 °F | 4 | 2 | 0 | 0 | 0 | 0 | 0 | 0 | 0 | 0 | 1 | 4 | 11 |
| Heating Degree Days | 1156 | 936 | 847 | 638 | 397 | 162 | 22 | 49 | ***232*** | 566 | 896 | 1179 | 7080 |
| Cooling Degree Days | 0 | 0 | 0 | 0 | 5 | 47 | 132 | ***114*** | ***15*** | 0 | 0 | ***0*** | 313 |
| Total Precipitation (") | 0.66 | 0.44 | 0.84 | 0.79 | 1.47 | 1.24 | 0.64 | 0.73 | 0.84 | 0.84 | 0.86 | 0.82 | 10.17 |
| Days ≥ 0.1" Precip | 3 | 2 | 3 | 3 | 5 | 4 | 2 | 2 | 3 | 3 | 3 | 4 | 37 |
| Total Snowfall (") | ***9.8*** | na | ***5.8*** | ***3.1*** | 1.8 | 0.1 | 0.0 | 0.0 | 0.1 | 1.1 | ***5.6*** | ***11.7*** | na |
| Days ≥ 1" Snow Depth | na | na | na | 0 | 0 | 0 | 0 | 0 | 0 | 0 | na | na | na |

## DEETH *Elko County* ELEVATION 5344 ft LAT/LONG 41° 4 ' N / 115° 17 ' W

| | JAN | FEB | MAR | APR | MAY | JUN | JUL | AUG | SEP | OCT | NOV | DEC | YEAR |
|---|---|---|---|---|---|---|---|---|---|---|---|---|---|
| Maximum Temp °F | 34.4 | 40.3 | 48.8 | 58.2 | 68.6 | ***77.2*** | ***87.7*** | 85.8 | 77.0 | 63.8 | 46.5 | 35.2 | 60.3 |
| Minimum Temp °F | 8.0 | 13.7 | 21.1 | 26.4 | 34.2 | ***40.0*** | ***43.9*** | 41.3 | 32.7 | 23.5 | 18.1 | 9.0 | 26.0 |
| Mean Temp °F | 21.2 | 27.0 | 35.0 | 42.4 | 51.5 | ***58.6*** | na | ***63.4*** | 54.8 | 43.7 | 32.4 | 22.1 | na |
| Days Max Temp ≥ 90 °F | 0 | 0 | 0 | 0 | 0 | 2 | ***11*** | ***10*** | 1 | 0 | 0 | 0 | 24 |
| Days Max Temp ≤ 32 °F | 12 | ***6*** | 1 | 0 | 0 | 0 | 0 | 0 | 0 | 0 | 3 | ***11*** | 33 |
| Days Min Temp ≤ 32 °F | 30 | 27 | ***29*** | ***24*** | ***11*** | 3 | 1 | 3 | ***16*** | ***26*** | 28 | 30 | 228 |
| Days Min Temp ≤ 0 °F | 9 | 5 | 1 | 0 | 0 | 0 | 0 | 0 | 0 | 0 | 2 | 8 | 25 |
| Heating Degree Days | 1350 | 1065 | 924 | 671 | 411 | ***197*** | na | ***86*** | 301 | 655 | 973 | 1333 | na |
| Cooling Degree Days | 0 | 0 | 0 | 0 | 1 | ***11*** | na | ***40*** | 3 | 0 | 0 | 0 | na |
| Total Precipitation (") | 0.95 | 0.83 | 1.26 | 1.24 | 1.52 | 1.08 | 0.51 | 0.58 | 0.80 | 0.89 | 1.25 | 1.02 | 11.93 |
| Days ≥ 0.1" Precip | 3 | 3 | 4 | 4 | 4 | 3 | 1 | 2 | 2 | 3 | 4 | 3 | 36 |
| Total Snowfall (") | na | na | na | na | ***0.0*** | 0.0 | 0.0 | 0.0 | 0.0 | ***0.0*** | na | na | na |
| Days ≥ 1" Snow Depth | na | na | na | ***0*** | 0 | 0 | 0 | 0 | 0 | ***0*** | na | na | na |

## DENIO *Humboldt County* ELEVATION 4193 ft LAT/LONG 41° 59 ' N / 118° 38 ' W

| | JAN | FEB | MAR | APR | MAY | JUN | JUL | AUG | SEP | OCT | NOV | DEC | YEAR |
|---|---|---|---|---|---|---|---|---|---|---|---|---|---|
| Maximum Temp °F | 41.1 | 47.6 | 54.2 | 62.5 | 72.4 | 81.4 | 91.3 | 89.5 | 79.4 | 66.8 | 50.7 | 41.5 | 64.9 |
| Minimum Temp °F | 20.1 | 24.4 | 27.6 | 31.2 | 38.0 | 45.9 | 52.2 | 51.2 | 41.5 | 33.0 | 25.7 | 20.1 | 34.2 |
| Mean Temp °F | 30.6 | 36.0 | 40.9 | 46.9 | 55.2 | 63.7 | 71.8 | 70.4 | 60.5 | 49.8 | 38.1 | 30.9 | 49.6 |
| Days Max Temp ≥ 90 °F | 0 | 0 | 0 | 0 | 1 | 8 | 20 | 18 | 4 | 0 | 0 | 0 | 51 |
| Days Max Temp ≤ 32 °F | 5 | 2 | 0 | 0 | 0 | 0 | 0 | 0 | 0 | 0 | 1 | 4 | 12 |
| Days Min Temp ≤ 32 °F | 28 | 23 | ***22*** | 17 | 8 | 1 | 0 | 0 | 4 | 15 | 23 | 28 | 169 |
| Days Min Temp ≤ 0 °F | ***2*** | 0 | 0 | 0 | 0 | 0 | 0 | 0 | 0 | 0 | 0 | 1 | 3 |
| Heating Degree Days | 1059 | 811 | 740 | 537 | 307 | 111 | 14 | 22 | 170 | 467 | 800 | 1050 | 6088 |
| Cooling Degree Days | 0 | 0 | 0 | 1 | 13 | 83 | 221 | 192 | 41 | 2 | 0 | 0 | 553 |
| Total Precipitation (") | 0.79 | 0.78 | 1.12 | 0.94 | 0.95 | 0.85 | 0.27 | 0.52 | 0.55 | 0.58 | 1.24 | 0.87 | 9.46 |
| Days ≥ 0.1" Precip | 2 | 3 | 4 | 3 | 3 | 3 | 1 | 1 | 2 | 2 | 4 | 3 | 31 |
| Total Snowfall (") | ***5.3*** | 3.3 | ***3.0*** | na | 0.2 | 0.2 | 0.0 | 0.0 | 0.0 | 0.5 | 4.0 | ***5.6*** | na |
| Days ≥ 1" Snow Depth | ***8*** | ***2*** | ***1*** | ***0*** | 0 | 0 | 0 | 0 | 0 | 0 | ***2*** | na | na |

## DESERT NATL WL RANGE *Clark County* ELEVATION 2923 ft LAT/LONG 36° 26 ' N / 115° 22 ' W

| | JAN | FEB | MAR | APR | MAY | JUN | JUL | AUG | SEP | OCT | NOV | DEC | YEAR |
|---|---|---|---|---|---|---|---|---|---|---|---|---|---|
| Maximum Temp °F | 57.2 | 62.3 | 68.5 | 76.4 | 85.9 | 96.5 | 101.7 | 99.6 | 92.0 | 80.4 | 65.7 | 56.9 | 78.6 |
| Minimum Temp °F | 29.1 | 32.8 | 38.1 | 44.0 | 52.3 | 60.6 | 66.9 | 65.5 | 57.6 | 47.0 | 35.9 | 29.1 | 46.6 |
| Mean Temp °F | 43.2 | 47.6 | 53.3 | 60.2 | 69.1 | 78.7 | 84.4 | 82.6 | 74.8 | 63.7 | 50.8 | 43.0 | 62.6 |
| Days Max Temp ≥ 90 °F | 0 | 0 | 0 | 2 | 12 | 24 | 30 | 30 | 19 | 4 | 0 | 0 | 121 |
| Days Max Temp ≤ 32 °F | 0 | 0 | 0 | 0 | 0 | 0 | 0 | 0 | 0 | 0 | 0 | 0 | 0 |
| Days Min Temp ≤ 32 °F | ***22*** | 13 | 6 | 1 | 0 | 0 | 0 | 0 | 0 | 1 | ***10*** | 22 | 75 |
| Days Min Temp ≤ 0 °F | 0 | 0 | 0 | 0 | 0 | 0 | 0 | 0 | 0 | 0 | 0 | 0 | 0 |
| Heating Degree Days | 669 | 485 | 358 | 167 | 41 | 2 | 0 | 0 | 5 | 103 | 418 | 675 | 2923 |
| Cooling Degree Days | 0 | 0 | 3 | 43 | 187 | 431 | 607 | 560 | 309 | 67 | 1 | 0 | 2208 |
| Total Precipitation (") | 0.44 | 0.51 | 0.75 | 0.32 | 0.25 | 0.17 | 0.44 | 0.46 | 0.33 | 0.27 | 0.36 | 0.45 | 4.75 |
| Days ≥ 0.1" Precip | 1 | 1 | 2 | 1 | 1 | 0 | 1 | 1 | 1 | 1 | 1 | 1 | 12 |
| Total Snowfall (") | 0.5 | 0.0 | 0.0 | 0.0 | 0.0 | 0.0 | 0.0 | 0.0 | 0.0 | 0.0 | 0.0 | 0.2 | 0.7 |
| Days ≥ 1" Snow Depth | 0 | 0 | 0 | 0 | 0 | 0 | 0 | 0 | 0 | 0 | 0 | 0 | 0 |

## DYER *Esmeralda County* ELEVATION 4899 ft LAT/LONG 37° 41 ' N / 118° 5 ' W

| | JAN | FEB | MAR | APR | MAY | JUN | JUL | AUG | SEP | OCT | NOV | DEC | YEAR |
|---|---|---|---|---|---|---|---|---|---|---|---|---|---|
| Maximum Temp °F | 46.3 | 53.2 | 59.3 | 66.9 | 76.5 | 86.4 | 93.1 | 90.7 | 82.7 | 71.5 | 56.7 | 46.6 | 69.2 |
| Minimum Temp °F | 16.6 | 22.0 | 27.2 | 32.0 | 40.0 | 48.0 | 54.1 | 51.9 | 43.3 | 33.0 | 23.5 | 16.5 | 34.0 |
| Mean Temp °F | 31.5 | 37.6 | 43.3 | 49.4 | 58.3 | 67.2 | 73.6 | 71.3 | 63.0 | 52.2 | 40.1 | 31.5 | 51.6 |
| Days Max Temp ≥ 90 °F | 0 | 0 | 0 | 0 | 2 | 12 | 24 | 20 | 6 | 0 | 0 | 0 | 64 |
| Days Max Temp ≤ 32 °F | 2 | 1 | 0 | 0 | 0 | 0 | 0 | 0 | 0 | 0 | 0 | 2 | 5 |
| Days Min Temp ≤ 32 °F | 29 | 25 | 24 | 16 | 5 | 0 | 0 | 0 | 3 | 15 | 25 | 30 | 172 |
| Days Min Temp ≤ 0 °F | 2 | 0 | 0 | 0 | 0 | 0 | 0 | 0 | 0 | 0 | 0 | 1 | 3 |
| Heating Degree Days | 1033 | 767 | 666 | 460 | 217 | 46 | 1 | 6 | 101 | 390 | 740 | 1030 | 5457 |
| Cooling Degree Days | 0 | 0 | 0 | 1 | 24 | 132 | 278 | 213 | 54 | 1 | 0 | 0 | 703 |
| Total Precipitation (") | 0.31 | 0.43 | 0.56 | 0.40 | 0.53 | 0.27 | 0.52 | 0.52 | 0.49 | 0.35 | 0.47 | 0.33 | 5.18 |
| Days ≥ 0.1" Precip | 1 | 1 | 2 | 1 | 1 | 1 | 2 | 1 | 1 | 1 | 1 | 1 | 14 |
| Total Snowfall (") | 2.2 | 2.1 | 2.2 | 1.7 | 0.2 | 0.0 | 0.0 | 0.0 | 0.1 | 0.1 | 1.3 | 1.3 | 11.2 |
| Days ≥ 1" Snow Depth | 3 | 2 | 0 | 0 | 0 | 0 | 0 | 0 | 0 | 0 | 0 | 1 | 6 |

## ELKO MUNICIPAL AP *Elko County* ELEVATION 5074 ft LAT/LONG 40° 50 ' N / 115° 47 ' W

| | JAN | FEB | MAR | APR | MAY | JUN | JUL | AUG | SEP | OCT | NOV | DEC | YEAR |
|---|---|---|---|---|---|---|---|---|---|---|---|---|---|
| Maximum Temp °F | 36.7 | 43.0 | 50.9 | 59.2 | 69.4 | 80.2 | 90.4 | 88.4 | 78.5 | 65.4 | 48.3 | 37.2 | 62.3 |
| Minimum Temp °F | 13.3 | 19.5 | 25.3 | 29.5 | 36.7 | 44.0 | 49.8 | 48.1 | 38.8 | 29.1 | 21.8 | 13.6 | 30.8 |
| Mean Temp °F | 25.0 | 31.3 | 38.1 | 44.4 | 53.1 | 62.1 | 70.1 | 68.3 | 58.6 | 47.3 | 35.1 | 25.4 | 46.6 |
| Days Max Temp ≥ 90 °F | 0 | 0 | 0 | 0 | 0 | 6 | 20 | 15 | 4 | 0 | 0 | 0 | 45 |
| Days Max Temp ≤ 32 °F | 9 | 4 | 1 | 0 | 0 | 0 | 0 | 0 | 0 | 0 | 2 | 9 | 25 |
| Days Min Temp ≤ 32 °F | 29 | 26 | 26 | 20 | 8 | 1 | 0 | 0 | 7 | 21 | 25 | 29 | 192 |
| Days Min Temp ≤ 0 °F | 6 | 2 | 0 | 0 | 0 | 0 | 0 | 0 | 0 | 0 | 1 | 4 | 13 |
| Heating Degree Days | 1233 | 946 | 826 | 612 | 366 | 131 | 18 | 33 | 207 | 543 | 891 | 1221 | 7027 |
| Cooling Degree Days | 0 | 0 | 0 | 0 | 3 | 49 | 171 | 136 | 23 | 0 | 0 | 0 | 382 |
| Total Precipitation (") | 0.95 | 0.81 | 0.95 | 0.77 | 0.94 | 0.77 | 0.33 | 0.59 | 0.64 | 0.66 | 1.11 | 1.03 | 9.55 |
| Days ≥ 0.1" Precip | 3 | 3 | 3 | 3 | 3 | 3 | 1 | 1 | 2 | 2 | 4 | 4 | 32 |
| Total Snowfall (") | 8.3 | 5.8 | 4.7 | 2.8 | 1.2 | 0.0 | 0.0 | 0.0 | 0.1 | 0.9 | 5.3 | 8.9 | 38.0 |
| Days ≥ 1" Snow Depth | 14 | 6 | 2 | 1 | 0 | 0 | 0 | 0 | 0 | 0 | 3 | 10 | 36 |

## ELY YELLAND FIELD *White Pine County* ELEVATION 6263 ft LAT/LONG 39° 17 ' N / 114° 51 ' W

| | JAN | FEB | MAR | APR | MAY | JUN | JUL | AUG | SEP | OCT | NOV | DEC | YEAR |
|---|---|---|---|---|---|---|---|---|---|---|---|---|---|
| Maximum Temp °F | 39.9 | 43.7 | 49.3 | 57.6 | 67.6 | 78.6 | 87.0 | 84.6 | 75.4 | 63.1 | 48.8 | 40.3 | 61.3 |
| Minimum Temp °F | 9.7 | 14.9 | 21.4 | 26.0 | 33.5 | 40.5 | 47.7 | 46.4 | 37.2 | 27.8 | 18.6 | 10.3 | 27.8 |
| Mean Temp °F | 24.8 | 29.3 | 35.4 | 41.8 | 50.6 | 59.6 | 67.4 | 65.5 | 56.3 | 45.5 | 33.7 | 25.3 | 44.6 |
| Days Max Temp ≥ 90 °F | 0 | 0 | 0 | 0 | 0 | 3 | 12 | 7 | 0 | 0 | 0 | 0 | 22 |
| Days Max Temp ≤ 32 °F | 7 | 3 | 1 | 0 | 0 | 0 | 0 | 0 | 0 | 0 | 2 | 7 | 20 |
| Days Min Temp ≤ 32 °F | 31 | 28 | 29 | 25 | 14 | 3 | 0 | 0 | 8 | 24 | 28 | 31 | 221 |
| Days Min Temp ≤ 0 °F | 7 | 3 | 1 | 0 | 0 | 0 | 0 | 0 | 0 | 0 | 1 | 6 | 18 |
| Heating Degree Days | 1239 | 1000 | 912 | 689 | 440 | 179 | 23 | 50 | 260 | 598 | 932 | 1223 | 7545 |
| Cooling Degree Days | 0 | 0 | 0 | 0 | 0 | 27 | 102 | 76 | 8 | 0 | 0 | 0 | 213 |
| Total Precipitation (") | 0.73 | 0.68 | 0.98 | 0.89 | 1.20 | 0.73 | 0.71 | 0.90 | 1.01 | 0.94 | 0.68 | 0.69 | 10.14 |
| Days ≥ 0.1" Precip | 2 | 2 | 4 | 3 | 3 | 2 | 2 | 3 | 2 | 3 | 2 | 2 | 30 |
| Total Snowfall (") | 9.0 | 7.2 | 8.7 | 6.1 | 2.8 | 0.1 | 0.0 | 0.0 | 0.5 | 2.8 | 6.0 | 7.9 | 51.1 |
| Days ≥ 1" Snow Depth | 18 | 13 | 5 | 2 | 0 | 0 | 0 | 0 | 0 | 1 | 4 | 13 | 56 |

## EMIGRANT PASS HWY *Eureka County* ELEVATION 5764 ft LAT/LONG 40° 39 ' N / 116° 18 ' W

| | JAN | FEB | MAR | APR | MAY | JUN | JUL | AUG | SEP | OCT | NOV | DEC | YEAR |
|---|---|---|---|---|---|---|---|---|---|---|---|---|---|
| Maximum Temp °F | 37.8 | 41.9 | 48.7 | 56.9 | 66.9 | 78.6 | 89.1 | 86.4 | 76.8 | 63.9 | 47.8 | 38.1 | 61.1 |
| Minimum Temp °F | 18.3 | 22.6 | 26.7 | 31.8 | 38.7 | 46.9 | 55.7 | 53.4 | 44.6 | 34.0 | ***26.0*** | 18.9 | 34.8 |
| Mean Temp °F | 28.1 | 32.3 | 37.7 | 44.4 | 52.8 | 62.8 | 72.4 | 69.9 | 60.7 | 48.9 | ***36.8*** | 28.5 | 47.9 |
| Days Max Temp ≥ 90 °F | 0 | 0 | 0 | 0 | 0 | 4 | 18 | 12 | 1 | 0 | 0 | 0 | 35 |
| Days Max Temp ≤ 32 °F | 8 | 4 | 0 | 0 | 0 | 0 | 0 | 0 | 0 | 0 | 1 | 7 | 20 |
| Days Min Temp ≤ 32 °F | 29 | 25 | 23 | 17 | 6 | 0 | 0 | 0 | 1 | 12 | ***24*** | 29 | 166 |
| Days Min Temp ≤ 0 °F | 2 | 1 | 0 | 0 | 0 | 0 | 0 | 0 | 0 | 0 | 0 | 1 | 4 |
| Heating Degree Days | 1143 | 918 | 839 | 614 | 379 | 132 | 12 | 33 | 169 | 495 | ***842*** | 1123 | 6699 |
| Cooling Degree Days | 0 | 0 | 0 | 1 | 9 | 80 | 238 | 220 | 64 | 2 | 0 | 0 | 614 |
| Total Precipitation (") | 1.07 | 1.07 | 1.30 | 1.22 | 1.49 | 1.02 | 0.36 | 0.61 | 0.76 | 0.97 | 1.35 | 1.09 | 12.31 |
| Days ≥ 0.1" Precip | 4 | 4 | 5 | 4 | 4 | 3 | 1 | 2 | 2 | 3 | 5 | 4 | 41 |
| Total Snowfall (") | na | ***8.4*** | 6.8 | 3.7 | 1.8 | 0.1 | 0.0 | 0.0 | 0.0 | 0.9 | ***6.0*** | na | na |
| Days ≥ 1" Snow Depth | ***11*** | ***7*** | ***2*** | 1 | 0 | 0 | 0 | 0 | 0 | 0 | ***2*** | ***10*** | 33 |

## EUREKA *Eureka County* ELEVATION 6545 ft LAT/LONG 39° 31' N / 115° 58' W

| | JAN | FEB | MAR | APR | MAY | JUN | JUL | AUG | SEP | OCT | NOV | DEC | YEAR |
|---|---|---|---|---|---|---|---|---|---|---|---|---|---|
| Maximum Temp °F | 37.7 | 41.6 | 47.4 | 56.0 | 65.9 | 76.4 | 85.4 | 83.0 | 74.2 | 62.3 | 46.6 | 38.4 | 59.6 |
| Minimum Temp °F | 16.9 | 20.2 | 24.5 | 29.2 | 37.0 | 44.7 | 52.8 | 52.1 | 43.9 | 34.1 | 24.2 | 17.3 | 33.1 |
| Mean Temp °F | 27.3 | 30.9 | 36.0 | 42.6 | 51.5 | 60.6 | 69.1 | 67.6 | 59.1 | 48.2 | 35.4 | 27.9 | 46.4 |
| Days Max Temp ≥ 90 °F | 0 | 0 | 0 | 0 | 0 | 2 | 7 | 5 | 0 | 0 | 0 | 0 | 14 |
| Days Max Temp ≤ 32 °F | 9 | 4 | 2 | 0 | 0 | 0 | 0 | 0 | 0 | 0 | 3 | 7 | 25 |
| Days Min Temp ≤ 32 °F | 30 | 26 | 25 | 20 | 10 | 2 | 0 | 0 | 3 | 13 | 24 | 29 | 182 |
| Days Min Temp ≤ 0 °F | 2 | 1 | 0 | 0 | 0 | 0 | 0 | 0 | 0 | 0 | 0 | 2 | 5 |
| Heating Degree Days | 1161 | 955 | 892 | 667 | 416 | 168 | 19 | 36 | 196 | 513 | 881 | 1144 | 7048 |
| Cooling Degree Days | 0 | 0 | 0 | 0 | 4 | 39 | 137 | 121 | 26 | 0 | 0 | 0 | 327 |
| Total Precipitation (") | 0.97 | 0.88 | 1.39 | 1.23 | 1.55 | 0.97 | 0.74 | 1.00 | 1.01 | 1.00 | 1.02 | 1.09 | 12.85 |
| Days ≥ 0.1" Precip | 3 | 3 | 4 | 4 | 4 | 3 | 2 | 3 | 2 | 2 | 3 | 3 | 36 |
| Total Snowfall (") | na | ***5.6*** | ***11.3*** | 6.3 | 3.8 | 0.1 | 0.0 | 0.0 | 0.7 | 2.2 | ***6.4*** | ***9.9*** | na |
| Days ≥ 1" Snow Depth | na | na | na | ***2*** | 1 | 0 | 0 | 0 | 0 | 1 | na | na | na |

## FALLON EXPERIMENT ST *Churchill County* ELEVATION 3970 ft LAT/LONG 39° 27' N / 118° 47' W

| | JAN | FEB | MAR | APR | MAY | JUN | JUL | AUG | SEP | OCT | NOV | DEC | YEAR |
|---|---|---|---|---|---|---|---|---|---|---|---|---|---|
| Maximum Temp °F | 44.2 | 52.4 | 59.1 | 65.1 | 73.9 | 82.9 | 91.0 | 89.2 | 80.4 | 68.9 | 54.7 | 45.2 | 67.3 |
| Minimum Temp °F | 19.0 | 23.5 | 28.5 | 33.7 | 41.2 | 48.0 | 53.6 | 51.5 | 43.0 | 33.3 | 25.7 | 18.6 | 35.0 |
| Mean Temp °F | 31.6 | 38.0 | 43.8 | 49.4 | 57.6 | 65.5 | 72.3 | 70.3 | 61.8 | 51.1 | 40.2 | 31.9 | 51.1 |
| Days Max Temp ≥ 90 °F | 0 | 0 | 0 | 0 | 1 | 8 | 21 | 16 | 4 | 0 | 0 | 0 | 50 |
| Days Max Temp ≤ 32 °F | 4 | 1 | 0 | 0 | 0 | 0 | 0 | 0 | 0 | 0 | 0 | 3 | 8 |
| Days Min Temp ≤ 32 °F | 29 | 24 | 22 | 13 | 3 | 0 | 0 | 0 | 2 | 14 | 24 | 28 | 159 |
| Days Min Temp ≤ 0 °F | 1 | 0 | 0 | 0 | 0 | 0 | 0 | 0 | 0 | 0 | 0 | 1 | 2 |
| Heating Degree Days | 1027 | 755 | 647 | 461 | 239 | 72 | 7 | 17 | 132 | 424 | 738 | 1019 | 5538 |
| Cooling Degree Days | 0 | 0 | 0 | 1 | 16 | 89 | 223 | 186 | 42 | 1 | 0 | 0 | 558 |
| Total Precipitation (") | 0.50 | 0.44 | 0.44 | 0.56 | 0.57 | 0.56 | 0.21 | 0.29 | 0.36 | 0.42 | 0.46 | 0.39 | 5.20 |
| Days ≥ 0.1" Precip | 2 | 1 | 1 | 2 | 2 | 1 | 1 | 1 | 1 | 1 | 2 | 1 | 16 |
| Total Snowfall (") | ***3.0*** | ***1.5*** | ***1.3*** | 0.3 | 0.0 | 0.0 | 0.0 | 0.0 | 0.0 | 0.0 | 0.5 | na | na |
| Days ≥ 1" Snow Depth | na | ***1*** | 0 | 0 | 0 | 0 | 0 | 0 | 0 | 0 | ***0*** | ***3*** | na |

## GIBBS RANCH *Elko County* ELEVATION 6004 ft LAT/LONG 41° 33' N / 115° 13' W

| | JAN | FEB | MAR | APR | MAY | JUN | JUL | AUG | SEP | OCT | NOV | DEC | YEAR |
|---|---|---|---|---|---|---|---|---|---|---|---|---|---|
| Maximum Temp °F | 35.2 | 39.8 | 46.3 | 55.1 | 65.0 | 74.1 | ***84.1*** | ***83.2*** | ***74.5*** | ***62.2*** | ***44.6*** | 36.3 | 58.4 |
| Minimum Temp °F | 7.0 | 11.8 | 19.5 | 24.6 | 31.9 | 39.0 | ***45.4*** | ***43.7*** | ***35.1*** | ***26.1*** | ***18.0*** | 9.2 | 25.9 |
| Mean Temp °F | 21.1 | 26.0 | 32.9 | 39.8 | 48.5 | 56.6 | ***64.8*** | ***63.4*** | ***54.8*** | ***44.3*** | ***31.3*** | 22.8 | 42.2 |
| Days Max Temp ≥ 90 °F | 0 | 0 | 0 | 0 | 0 | 1 | ***6*** | ***4*** | 0 | 0 | 0 | 0 | 11 |
| Days Max Temp ≤ 32 °F | ***10*** | ***6*** | 1 | 0 | 0 | 0 | 0 | 0 | 0 | 0 | ***3*** | na | na |
| Days Min Temp ≤ 32 °F | 30 | ***27*** | ***29*** | na | na | ***4*** | 1 | 1 | na | na | na | 29 | na |
| Days Min Temp ≤ 0 °F | ***8*** | ***5*** | 1 | 0 | 0 | 0 | 0 | 0 | 0 | 0 | ***2*** | ***6*** | 22 |
| Heating Degree Days | 1355 | 1095 | 986 | ***745*** | ***502*** | 258 | ***64*** | ***86*** | ***301*** | ***634*** | ***1001*** | 1302 | 8329 |
| Cooling Degree Days | 0 | 0 | 0 | 0 | 1 | 12 | ***68*** | 45 | 6 | ***0*** | 0 | ***0*** | 132 |
| Total Precipitation (") | 0.98 | 0.71 | 0.85 | 0.72 | 1.27 | 1.07 | 0.64 | 0.65 | 0.66 | 0.69 | 1.20 | 1.01 | 10.45 |
| Days ≥ 0.1" Precip | 3 | 3 | 3 | 2 | 4 | 3 | 2 | 2 | 2 | 3 | 4 | 3 | 34 |
| Total Snowfall (") | na | na | na | ***0.7*** | ***0.1*** | 0.1 | 0.0 | 0.0 | 0.1 | 0.1 | na | na | na |
| Days ≥ 1" Snow Depth | na | na | na | ***0*** | ***0*** | 0 | 0 | 0 | 0 | ***0*** | na | na | na |

## GLENBROOK *Douglas County* ELEVATION 6404 ft LAT/LONG 39° 5' N / 119° 56' W

| | JAN | FEB | MAR | APR | MAY | JUN | JUL | AUG | SEP | OCT | NOV | DEC | YEAR |
|---|---|---|---|---|---|---|---|---|---|---|---|---|---|
| Maximum Temp °F | 42.0 | 43.9 | 47.4 | 53.5 | 62.4 | 71.5 | 79.1 | 78.6 | 72.4 | 61.5 | 48.5 | 42.1 | 58.6 |
| Minimum Temp °F | 23.8 | 24.9 | 26.7 | 29.5 | 35.8 | 42.7 | 48.5 | 48.9 | 43.8 | 36.2 | 29.3 | 24.3 | 34.5 |
| Mean Temp °F | 32.9 | 34.4 | 37.1 | 41.5 | 49.1 | 57.1 | 63.9 | 63.8 | 58.1 | 48.9 | 38.8 | 33.3 | 46.6 |
| Days Max Temp ≥ 90 °F | 0 | 0 | 0 | 0 | 0 | 0 | 1 | 1 | 0 | 0 | 0 | 0 | 2 |
| Days Max Temp ≤ 32 °F | 3 | 2 | 1 | 0 | 0 | 0 | 0 | 0 | 0 | 0 | 1 | 4 | 11 |
| Days Min Temp ≤ 32 °F | 29 | 26 | 26 | 21 | 9 | 2 | 0 | 0 | 1 | 8 | 20 | 28 | 170 |
| Days Min Temp ≤ 0 °F | 0 | 0 | 0 | 0 | 0 | 0 | 0 | 0 | 0 | 0 | 0 | 0 | 0 |
| Heating Degree Days | 988 | 858 | 859 | 697 | 486 | 237 | 75 | 79 | 206 | 493 | 781 | 976 | 6735 |
| Cooling Degree Days | 0 | 0 | 0 | 0 | 0 | 7 | 37 | 40 | 6 | 0 | 0 | 0 | 90 |
| Total Precipitation (") | 2.89 | 2.34 | 2.22 | 0.98 | 0.67 | 0.51 | 0.38 | 0.55 | 0.76 | 1.13 | 2.41 | 2.43 | 17.27 |
| Days ≥ 0.1" Precip | 5 | 4 | 5 | 3 | 2 | 2 | 1 | 1 | 2 | 3 | 5 | 5 | 38 |
| Total Snowfall (") | 14.9 | ***15.0*** | 14.6 | 6.3 | 1.4 | 0.2 | 0.0 | 0.0 | 0.0 | 0.8 | 7.6 | ***15.5*** | 76.3 |
| Days ≥ 1" Snow Depth | na | na | na | na | ***0*** | 0 | 0 | 0 | 0 | ***0*** | na | na | na |

## GOLCONDA *Humboldt County* ELEVATION 4393 ft LAT/LONG 40° 57 ' N / 117° 29 ' W

| | JAN | FEB | MAR | APR | MAY | JUN | JUL | AUG | SEP | OCT | NOV | DEC | YEAR |
|---|---|---|---|---|---|---|---|---|---|---|---|---|---|
| Maximum Temp °F | 41.2 | 48.3 | 55.2 | 62.9 | 72.9 | 83.4 | 92.7 | 90.7 | 81.0 | 68.1 | 51.2 | 40.8 | 65.7 |
| Minimum Temp °F | 19.4 | 24.1 | 28.1 | 32.1 | 39.8 | 47.6 | 54.2 | 51.8 | 42.3 | 32.9 | 26.0 | 19.0 | 34.8 |
| Mean Temp °F | 30.3 | 36.2 | 41.7 | 47.6 | 56.4 | 65.5 | 73.5 | 71.3 | 61.7 | 50.5 | 38.6 | 29.9 | 50.3 |
| Days Max Temp ≥ 90 °F | 0 | 0 | 0 | 0 | 2 | 9 | 23 | 20 | 6 | 0 | 0 | 0 | 60 |
| Days Max Temp ≤ 32 °F | 6 | 1 | 0 | 0 | 0 | 0 | 0 | 0 | 0 | 0 | 1 | 5 | 13 |
| Days Min Temp ≤ 32 °F | 27 | 23 | 23 | 15 | 5 | 1 | 0 | 0 | 3 | 14 | 23 | 28 | 162 |
| Days Min Temp ≤ 0 °F | 2 | 0 | 0 | 0 | 0 | 0 | 0 | 0 | 0 | 0 | 0 | 2 | 4 |
| Heating Degree Days | 1068 | 805 | 716 | 516 | 277 | 83 | 8 | 18 | 141 | 443 | 784 | 1080 | 5939 |
| Cooling Degree Days | 0 | 0 | 0 | 2 | 18 | 107 | 259 | 231 | 56 | 2 | 0 | 0 | 675 |
| Total Precipitation (") | 0.60 | 0.56 | 0.71 | 0.58 | 0.82 | 0.75 | 0.25 | 0.44 | 0.46 | 0.53 | 0.90 | 0.77 | 7.37 |
| Days ≥ 0.1" Precip | 2 | 2 | 2 | 2 | 3 | 3 | 1 | 1 | 1 | 2 | 3 | 3 | 25 |
| Total Snowfall (") | na | ***2.3*** | ***0.9*** | 0.0 | 0.0 | 0.0 | 0.0 | 0.0 | 0.0 | 0.3 | na | na | na |
| Days ≥ 1" Snow Depth | na | na | ***0*** | 0 | 0 | 0 | 0 | 0 | 0 | 0 | na | ***4*** | na |

## GOLDFIELD *Esmeralda County* ELEVATION 5705 ft LAT/LONG 37° 43 ' N / 117° 13 ' W

| | JAN | FEB | MAR | APR | MAY | JUN | JUL | AUG | SEP | OCT | NOV | DEC | YEAR |
|---|---|---|---|---|---|---|---|---|---|---|---|---|---|
| Maximum Temp °F | 42.5 | 47.3 | 52.8 | 60.7 | 70.2 | 80.7 | 88.3 | 86.3 | 77.5 | 66.4 | 51.7 | 42.2 | 63.9 |
| Minimum Temp °F | 21.8 | 26.0 | 29.9 | 35.1 | 43.6 | 52.1 | 60.0 | 58.3 | 49.9 | 39.5 | 28.9 | 21.2 | 38.9 |
| Mean Temp °F | 32.2 | 36.7 | 41.4 | 47.9 | 56.9 | 66.5 | 74.2 | 72.3 | 63.7 | 52.9 | 40.3 | 31.7 | 51.4 |
| Days Max Temp ≥ 90 °F | 0 | 0 | 0 | 0 | 0 | 4 | ***14*** | 9 | 1 | 0 | 0 | 0 | 28 |
| Days Max Temp ≤ 32 °F | 4 | 2 | 1 | 0 | 0 | 0 | 0 | 0 | 0 | 0 | 1 | 5 | 13 |
| Days Min Temp ≤ 32 °F | 29 | 23 | 19 | 11 | 3 | 0 | 0 | 0 | 1 | 6 | 19 | 29 | 140 |
| Days Min Temp ≤ 0 °F | 0 | 0 | 0 | 0 | 0 | 0 | 0 | 0 | 0 | 0 | 0 | 0 | 0 |
| Heating Degree Days | 1012 | 794 | 725 | 505 | 259 | 68 | 3 | 9 | 101 | 369 | 735 | 1024 | 5604 |
| Cooling Degree Days | 0 | 0 | 0 | 1 | 17 | 125 | 295 | 256 | 71 | 4 | ***0*** | 0 | 769 |
| Total Precipitation (") | 0.67 | 0.87 | 0.79 | 0.59 | 0.59 | 0.45 | 0.44 | 0.63 | 0.63 | 0.50 | 0.39 | 0.30 | 6.85 |
| Days ≥ 0.1" Precip | 1 | 2 | 2 | 2 | 1 | 1 | 1 | 2 | 1 | 1 | 1 | 1 | 16 |
| Total Snowfall (") | ***2.4*** | ***4.8*** | ***4.3*** | ***2.2*** | 0.5 | 0.0 | 0.0 | 0.0 | 0.1 | 0.4 | ***0.9*** | ***1.8*** | 17.4 |
| Days ≥ 1" Snow Depth | na | na | ***1*** | ***0*** | 0 | 0 | 0 | 0 | 0 | 0 | ***0*** | na | na |

## IMLAY *Pershing County* ELEVATION 4213 ft LAT/LONG 40° 40 ' N / 118° 9 ' W

| | JAN | FEB | MAR | APR | MAY | JUN | JUL | AUG | SEP | OCT | NOV | DEC | YEAR |
|---|---|---|---|---|---|---|---|---|---|---|---|---|---|
| Maximum Temp °F | 43.0 | 50.3 | 57.3 | ***64.0*** | 73.5 | 83.1 | 92.3 | 90.3 | 81.1 | ***69.1*** | 53.0 | 43.4 | 66.7 |
| Minimum Temp °F | 19.1 | 24.3 | 28.8 | ***33.8*** | 42.2 | 50.9 | 57.4 | 54.7 | 45.2 | ***34.4*** | 26.0 | 18.7 | 36.3 |
| Mean Temp °F | 31.0 | 37.3 | 43.1 | ***48.9*** | 58.0 | 67.0 | 74.9 | 72.6 | 63.2 | ***51.8*** | 39.5 | 31.1 | 51.5 |
| Days Max Temp ≥ 90 °F | 0 | 0 | 0 | 0 | 2 | 8 | ***21*** | ***18*** | ***6*** | 0 | 0 | 0 | 55 |
| Days Max Temp ≤ 32 °F | 5 | 1 | 0 | 0 | 0 | 0 | 0 | 0 | 0 | 0 | 0 | ***3*** | 9 |
| Days Min Temp ≤ 32 °F | ***28*** | ***25*** | na | na | 3 | 0 | 0 | 0 | 2 | na | na | 29 | na |
| Days Min Temp ≤ 0 °F | 1 | 0 | 0 | 0 | 0 | 0 | 0 | 0 | 0 | 0 | 0 | 1 | 2 |
| Heating Degree Days | 1044 | 776 | 672 | ***477*** | 243 | 69 | 6 | 14 | 115 | ***406*** | 759 | 1044 | 5625 |
| Cooling Degree Days | 0 | 0 | 0 | 1 | 30 | 125 | 294 | 239 | 67 | 6 | 0 | 0 | 762 |
| Total Precipitation (") | 0.65 | 0.65 | 0.80 | 0.76 | 0.88 | 0.75 | 0.24 | 0.46 | 0.44 | 0.50 | 0.84 | 0.75 | 7.72 |
| Days ≥ 0.1" Precip | ***3*** | ***2*** | ***3*** | ***3*** | 2 | 2 | 1 | ***1*** | 1 | 1 | 3 | ***3*** | 25 |
| Total Snowfall (") | na | na | na | ***0.1*** | 0.0 | 0.0 | 0.0 | 0.0 | 0.0 | 0.3 | ***0.5*** | na | na |
| Days ≥ 1" Snow Depth | na | na | na | ***0*** | 0 | 0 | 0 | 0 | 0 | 0 | ***0*** | na | na |

## LAHONTAN DAM *Churchill County* ELEVATION 4163 ft LAT/LONG 39° 28 ' N / 119° 4 ' W

| | JAN | FEB | MAR | APR | MAY | JUN | JUL | AUG | SEP | OCT | NOV | DEC | YEAR |
|---|---|---|---|---|---|---|---|---|---|---|---|---|---|
| Maximum Temp °F | 44.1 | 51.5 | 58.2 | 65.2 | 75.0 | 84.3 | 93.5 | 91.8 | 82.3 | 70.2 | 55.1 | 45.2 | 68.0 |
| Minimum Temp °F | 22.7 | 27.4 | 33.4 | 39.0 | 47.3 | 55.0 | 62.8 | 61.4 | 52.3 | 42.0 | 31.3 | 23.1 | 41.5 |
| Mean Temp °F | 33.4 | 39.5 | 45.8 | 52.1 | 61.2 | 69.7 | 78.2 | 76.7 | 67.3 | 56.1 | 43.3 | 34.2 | 54.8 |
| Days Max Temp ≥ 90 °F | 0 | 0 | 0 | 0 | 2 | 10 | 24 | 21 | 7 | 0 | 0 | 0 | 64 |
| Days Max Temp ≤ 32 °F | 4 | 1 | 0 | 0 | 0 | 0 | 0 | 0 | 0 | 0 | 0 | 3 | 8 |
| Days Min Temp ≤ 32 °F | 28 | 21 | 14 | 6 | 1 | 0 | 0 | 0 | 0 | ***3*** | ***17*** | ***28*** | 118 |
| Days Min Temp ≤ 0 °F | 1 | 0 | 0 | 0 | 0 | 0 | 0 | 0 | 0 | 0 | 0 | 1 | 2 |
| Heating Degree Days | 973 | 713 | 589 | 386 | 175 | 45 | 2 | 5 | 65 | 286 | 645 | 949 | 4833 |
| Cooling Degree Days | 0 | 0 | 0 | 9 | 70 | 207 | 424 | 392 | 152 | 16 | 0 | 0 | 1270 |
| Total Precipitation (") | 0.46 | 0.47 | 0.52 | 0.38 | 0.42 | 0.51 | 0.34 | 0.51 | 0.39 | 0.30 | 0.58 | 0.47 | 5.35 |
| Days ≥ 0.1" Precip | 2 | 1 | 1 | 1 | 1 | 1 | 1 | 1 | 1 | 1 | 2 | 1 | 14 |
| Total Snowfall (") | na | na | ***0.8*** | 0.1 | 0.1 | 0.0 | 0.0 | 0.0 | 0.0 | 0.0 | ***0.1*** | ***1.4*** | na |
| Days ≥ 1" Snow Depth | na | na | ***0*** | 0 | 0 | 0 | 0 | 0 | 0 | ***0*** | ***0*** | na | na |

## LAS VEGAS MCCRN INTL *Clark County* ELEVATION 2169 ft LAT/LONG 36° 5 ' N / 115° 10 ' W

| | JAN | FEB | MAR | APR | MAY | JUN | JUL | AUG | SEP | OCT | NOV | DEC | YEAR |
|---|---|---|---|---|---|---|---|---|---|---|---|---|---|
| Maximum Temp °F | 56.7 | 62.7 | 69.1 | 77.7 | 87.9 | 98.7 | 104.1 | 101.7 | 93.8 | 80.9 | 66.0 | 56.6 | 79.7 |
| Minimum Temp °F | 34.5 | 39.2 | 44.8 | 51.4 | 60.8 | 70.2 | 76.7 | 74.8 | 66.6 | 54.4 | 42.3 | 34.5 | 54.2 |
| Mean Temp °F | 45.6 | 51.0 | 57.0 | 64.6 | 74.4 | 84.5 | 90.4 | 88.3 | 80.2 | 67.7 | 54.2 | 45.6 | 67.0 |
| Days Max Temp ≥ 90 °F | 0 | 0 | 0 | 3 | 16 | 26 | 30 | 30 | 22 | 6 | 0 | 0 | 133 |
| Days Max Temp ≤ 32 °F | 0 | 0 | 0 | 0 | 0 | 0 | 0 | 0 | 0 | 0 | 0 | 0 | 0 |
| Days Min Temp ≤ 32 °F | 11 | 4 | 1 | 0 | 0 | 0 | 0 | 0 | 0 | 0 | 2 | 11 | 29 |
| Days Min Temp ≤ 0 °F | 0 | 0 | 0 | 0 | 0 | 0 | 0 | 0 | 0 | 0 | 0 | 0 | 0 |
| Heating Degree Days | 595 | 390 | 258 | 95 | 16 | 0 | 0 | 0 | 2 | 54 | 322 | 596 | 2328 |
| Cooling Degree Days | 0 | 2 | 18 | 116 | 323 | 611 | 791 | 735 | 477 | 151 | 6 | 0 | 3230 |
| Total Precipitation (") | 0.54 | 0.62 | 0.60 | 0.21 | 0.28 | 0.13 | 0.34 | 0.49 | 0.23 | 0.21 | 0.42 | 0.46 | 4.53 |
| Days ≥ 0.1" Precip | 2 | 2 | 1 | 1 | 1 | 0 | 1 | 1 | 1 | 1 | 1 | 1 | 13 |
| Total Snowfall (") | 0.8 | 0.1 | 0.0 | 0.0 | 0.0 | 0.0 | 0.0 | 0.0 | 0.0 | 0.0 | 0.0 | 0.1 | 1.0 |
| Days ≥ 1" Snow Depth | 0 | 0 | 0 | 0 | 0 | 0 | 0 | 0 | 0 | 0 | 0 | 0 | 0 |

## LEONARD CREEK RANCH *Humboldt County* ELEVATION 4232 ft LAT/LONG 41° 31 ' N / 118° 43 ' W

| | JAN | FEB | MAR | APR | MAY | JUN | JUL | AUG | SEP | OCT | NOV | DEC | YEAR |
|---|---|---|---|---|---|---|---|---|---|---|---|---|---|
| Maximum Temp °F | 39.5 | 47.2 | 54.9 | 63.5 | 73.6 | 82.7 | 92.1 | 89.7 | 80.2 | 67.1 | 50.2 | 39.9 | 65.1 |
| Minimum Temp °F | 20.0 | 25.7 | 30.2 | 34.6 | 42.5 | 49.5 | 55.8 | 54.5 | 46.5 | 37.4 | 28.2 | 20.6 | 37.1 |
| Mean Temp °F | 29.8 | 36.5 | 42.6 | 49.1 | 58.1 | 66.1 | 74.0 | 72.1 | 63.4 | 52.2 | 39.2 | 30.3 | 51.1 |
| Days Max Temp ≥ 90 °F | 0 | 0 | 0 | 0 | 2 | 8 | 21 | 18 | 5 | 0 | 0 | 0 | 54 |
| Days Max Temp ≤ 32 °F | 6 | 1 | 0 | 0 | 0 | 0 | 0 | 0 | 0 | 0 | 1 | 5 | 13 |
| Days Min Temp ≤ 32 °F | 28 | 23 | 19 | 12 | 3 | 0 | 0 | 0 | 1 | 8 | 21 | 28 | 143 |
| Days Min Temp ≤ 0 °F | 1 | 0 | 0 | 0 | 0 | 0 | 0 | 0 | 0 | 0 | 0 | 1 | 2 |
| Heating Degree Days | 1085 | 798 | 688 | 473 | 239 | 78 | 8 | 17 | 115 | 391 | 767 | 1070 | 5729 |
| Cooling Degree Days | 0 | 0 | 0 | 3 | 32 | 115 | 276 | 240 | 82 | 4 | 0 | 0 | 752 |
| Total Precipitation (") | 1.04 | 0.86 | 0.91 | 0.69 | 0.68 | 0.66 | 0.31 | 0.46 | 0.47 | 0.52 | 1.06 | 1.07 | 8.73 |
| Days ≥ 0.1" Precip | 3 | 3 | 3 | 2 | 2 | 2 | 1 | 1 | 2 | 2 | 4 | 4 | 29 |
| Total Snowfall (") | *5.6* | na | 1.3 | 0.8 | 0.2 | 0.1 | 0.0 | 0.0 | 0.0 | 0.8 | *1.8* | *6.9* | na |
| Days ≥ 1" Snow Depth | *12* | *4* | *0* | 0 | 0 | 0 | 0 | 0 | 0 | 0 | *1* | *8* | 25 |

## LOVELOCK *Pershing County* ELEVATION 3983 ft LAT/LONG 40° 11 ' N / 118° 28 ' W

| | JAN | FEB | MAR | APR | MAY | JUN | JUL | AUG | SEP | OCT | NOV | DEC | YEAR |
|---|---|---|---|---|---|---|---|---|---|---|---|---|---|
| Maximum Temp °F | 42.6 | 51.0 | 58.6 | 66.4 | 75.7 | 84.1 | 92.6 | 90.7 | 82.0 | 69.6 | 53.4 | 43.7 | 67.5 |
| Minimum Temp °F | 18.7 | 24.0 | 28.8 | 34.3 | 42.7 | 49.7 | 55.8 | 53.2 | 44.9 | 35.1 | 25.7 | 19.1 | 36.0 |
| Mean Temp °F | 30.7 | 37.5 | 43.7 | 50.3 | 59.2 | 66.9 | 74.2 | 72.0 | 63.5 | 52.4 | 39.6 | 31.4 | 51.8 |
| Days Max Temp ≥ 90 °F | 0 | 0 | 0 | 0 | 2 | 9 | 22 | 20 | *5* | 0 | 0 | 0 | 58 |
| Days Max Temp ≤ 32 °F | 5 | 1 | 0 | 0 | 0 | 0 | 0 | 0 | 0 | 0 | 0 | 3 | 9 |
| Days Min Temp ≤ 32 °F | 29 | 24 | 21 | *12* | 2 | 0 | 0 | 0 | 1 | 11 | 24 | 29 | 153 |
| Days Min Temp ≤ 0 °F | 1 | 0 | 0 | 0 | 0 | 0 | 0 | 0 | 0 | 0 | 0 | 1 | 2 |
| Heating Degree Days | 1057 | 769 | 652 | 433 | 202 | 60 | 5 | 11 | 99 | 389 | 756 | 1035 | 5468 |
| Cooling Degree Days | 0 | 0 | 0 | 1 | 26 | 117 | 264 | 221 | 54 | 3 | 0 | 0 | 686 |
| Total Precipitation (") | 0.50 | 0.47 | 0.45 | 0.59 | 0.47 | 0.49 | 0.15 | 0.34 | 0.43 | 0.44 | 0.55 | 0.58 | 5.46 |
| Days ≥ 0.1" Precip | *2* | *2* | 2 | 2 | 1 | 2 | 0 | 1 | *1* | 2 | 2 | 2 | 19 |
| Total Snowfall (") | na | na | na | 0.0 | 0.0 | 0.0 | 0.0 | 0.0 | 0.0 | 0.1 | *0.7* | na | na |
| Days ≥ 1" Snow Depth | na | na | na | 0 | 0 | 0 | 0 | 0 | 0 | 0 | 1 | na | na |

## LOVELOCK DERBY FLD *Pershing County* ELEVATION 3904 ft LAT/LONG 40° 4 ' N / 118° 33 ' W

| | JAN | FEB | MAR | APR | MAY | JUN | JUL | AUG | SEP | OCT | NOV | DEC | YEAR |
|---|---|---|---|---|---|---|---|---|---|---|---|---|---|
| Maximum Temp °F | 42.4 | 50.4 | 57.6 | 64.5 | 75.0 | 84.4 | *94.4* | 91.7 | *82.3* | *69.3* | *53.2* | 42.8 | 67.3 |
| Minimum Temp °F | 16.5 | 22.1 | 27.2 | 33.0 | 42.2 | 49.6 | *56.0* | 53.3 | *44.4* | *33.1* | *23.4* | 16.3 | 34.8 |
| Mean Temp °F | 29.5 | 36.3 | 42.4 | 48.8 | 58.6 | 67.1 | *75.2* | 72.5 | *63.4* | *51.3* | *38.4* | 29.5 | 51.1 |
| Days Max Temp ≥ 90 °F | 0 | 0 | 0 | 0 | 2 | 10 | 25 | 21 | *7* | 0 | 0 | 0 | 65 |
| Days Max Temp ≤ 32 °F | 5 | 1 | 0 | 0 | 0 | 0 | 0 | 0 | *0* | 0 | 0 | 4 | 10 |
| Days Min Temp ≤ 32 °F | 29 | 25 | 23 | 14 | 3 | 0 | 0 | 0 | *2* | 14 | 25 | 30 | 165 |
| Days Min Temp ≤ 0 °F | 2 | 0 | 0 | 0 | 0 | 0 | 0 | 0 | *0* | 0 | 0 | 2 | 4 |
| Heating Degree Days | 1095 | 803 | 693 | 480 | 220 | 60 | *3* | 12 | *104* | *421* | *792* | 1093 | 5776 |
| Cooling Degree Days | *0* | *0* | *0* | *1* | *30* | *123* | na | *252* | na | na | na | *0* | na |
| Total Precipitation (") | 0.49 | 0.41 | 0.50 | 0.54 | 0.42 | 0.55 | 0.18 | 0.39 | *0.39* | 0.28 | 0.51 | 0.56 | 5.22 |
| Days ≥ 0.1" Precip | 2 | 2 | 2 | 2 | 1 | 1 | 1 | 1 | *1* | *1* | 1 | 2 | 17 |
| Total Snowfall (") | 2.1 | 1.3 | *0.2* | 0.4 | 0.0 | 0.0 | 0.0 | 0.0 | *0.0* | 0.0 | 0.3 | 2.6 | 6.9 |
| Days ≥ 1" Snow Depth | 5 | 2 | *0* | 0 | 0 | 0 | 0 | 0 | *0* | 0 | 0 | 3 | 10 |

## LUND *White Pine County* ELEVATION 5554 ft LAT/LONG 38° 51 ' N / 115° 0 ' W

| | JAN | FEB | MAR | APR | MAY | JUN | JUL | AUG | SEP | OCT | NOV | DEC | YEAR |
|---|---|---|---|---|---|---|---|---|---|---|---|---|---|
| Maximum Temp °F | 42.2 | 47.2 | 53.5 | 61.5 | 71.2 | 81.0 | 88.4 | 86.5 | 78.8 | 67.4 | 52.7 | 43.3 | 64.5 |
| Minimum Temp °F | 14.0 | 18.9 | 23.7 | 28.8 | 36.7 | 43.7 | 49.9 | 48.8 | 40.9 | 31.6 | 21.8 | 14.9 | 31.1 |
| Mean Temp °F | 28.1 | 33.1 | 38.7 | 45.2 | 53.9 | 62.4 | 69.2 | 67.7 | 59.9 | 49.5 | 37.2 | 29.2 | 47.8 |
| Days Max Temp ≥ 90 °F | 0 | 0 | 0 | 0 | 0 | 5 | 14 | 9 | 1 | 0 | 0 | 0 | 29 |
| Days Max Temp ≤ 32 °F | 5 | 2 | 0 | 0 | 0 | 0 | 0 | 0 | 0 | 0 | 1 | 4 | 12 |
| Days Min Temp ≤ 32 °F | 31 | 27 | 27 | 22 | *8* | 1 | 0 | 0 | 3 | *17* | *28* | 31 | 195 |
| Days Min Temp ≤ 0 °F | 3 | 1 | 0 | 0 | 0 | 0 | 0 | 0 | 0 | 0 | 0 | 2 | 6 |
| Heating Degree Days | 1137 | 893 | 809 | 588 | 339 | 117 | 13 | 28 | 165 | 473 | 828 | 1104 | 6494 |
| Cooling Degree Days | 0 | 0 | 0 | 0 | 3 | 51 | 148 | 123 | 19 | 0 | 0 | 0 | 344 |
| Total Precipitation (") | 0.86 | 0.74 | 1.12 | 0.89 | 1.01 | 0.90 | 0.81 | 1.12 | 0.94 | 0.93 | 0.76 | 0.78 | 10.86 |
| Days ≥ 0.1" Precip | 2 | 2 | 3 | 3 | 2 | 2 | 2 | 3 | 2 | 2 | 2 | 2 | 27 |
| Total Snowfall (") | na | *3.6* | *4.0* | *2.7* | 0.7 | 0.0 | 0.0 | 0.0 | 0.0 | 0.3 | *2.3* | *4.1* | na |
| Days ≥ 1" Snow Depth | na | na | na | *1* | 0 | 0 | 0 | 0 | 0 | 0 | *0* | na | na |

## MC GILL *White Pine County* ELEVATION 6345 ft LAT/LONG 39° 24 ' N / 114° 46 ' W

| | JAN | FEB | MAR | APR | MAY | JUN | JUL | AUG | SEP | OCT | NOV | DEC | YEAR |
|---|---|---|---|---|---|---|---|---|---|---|---|---|---|
| Maximum Temp °F | 38.8 | 42.4 | 48.3 | 56.3 | 66.4 | 77.0 | 85.6 | 83.3 | 74.2 | 62.5 | 48.1 | 39.7 | 60.2 |
| Minimum Temp °F | 15.5 | 19.3 | 24.2 | 29.6 | 37.6 | 46.1 | 54.2 | 52.4 | 43.0 | 33.1 | 23.6 | 16.4 | 32.9 |
| Mean Temp °F | 27.2 | 30.9 | 36.3 | 43.0 | 52.1 | 61.6 | 69.9 | 67.9 | 58.6 | 47.8 | 35.8 | 28.1 | 46.6 |
| Days Max Temp ≥ 90 °F | 0 | 0 | 0 | 0 | 0 | 2 | 7 | 5 | 0 | 0 | 0 | 0 | 14 |
| Days Max Temp ≤ 32 °F | 8 | 5 | 2 | 0 | 0 | 0 | 0 | 0 | 0 | 0 | 3 | 7 | 25 |
| Days Min Temp ≤ 32 °F | 30 | 26 | 27 | 20 | 8 | 1 | 0 | 0 | 2 | 14 | 25 | 30 | 183 |
| Days Min Temp ≤ 0 °F | 3 | 2 | 0 | 0 | 0 | 0 | 0 | 0 | 0 | 0 | 0 | 2 | 7 |
| Heating Degree Days | 1165 | 955 | 886 | 655 | 396 | 148 | 13 | 31 | 205 | 527 | 869 | 1137 | 6987 |
| Cooling Degree Days | 0 | 0 | 0 | 0 | 2 | 54 | 160 | 134 | 22 | 0 | 0 | 0 | 372 |
| Total Precipitation (") | 0.45 | 0.44 | 0.61 | 0.78 | 1.08 | 0.85 | 0.78 | 0.83 | 0.94 | 0.92 | 0.50 | 0.50 | 8.68 |
| Days ≥ 0.1" Precip | 2 | 1 | 2 | 2 | 3 | 3 | 2 | 3 | 2 | 2 | 2 | 1 | 25 |
| Total Snowfall (") | na | na | na | na | *0.2* | 0.0 | 0.0 | 0.0 | 0.0 | 0.4 | 1.0 | na | na |
| Days ≥ 1" Snow Depth | na | na | *3* | *1* | *0* | 0 | 0 | 0 | 0 | 0 | 3 | *6* | na |

## METROPOLIS *Elko County* ELEVATION 5799 ft LAT/LONG 41° 17 ' N / 115° 1 ' W

| | JAN | FEB | MAR | APR | MAY | JUN | JUL | AUG | SEP | OCT | NOV | DEC | YEAR |
|---|---|---|---|---|---|---|---|---|---|---|---|---|---|
| Maximum Temp °F | 35.5 | 41.0 | 48.0 | 57.1 | 67.7 | 77.0 | 87.0 | 86.3 | 75.9 | 63.4 | *45.8* | *35.7* | 60.0 |
| Minimum Temp °F | 17.3 | 21.2 | 26.2 | 30.6 | 39.2 | 46.4 | 54.4 | 53.9 | 44.5 | 35.9 | *25.9* | 17.5 | 34.4 |
| Mean Temp °F | 26.4 | 31.1 | 37.1 | 43.9 | 53.5 | 61.7 | 70.7 | 70.1 | 60.2 | 49.6 | *35.8* | *26.6* | 47.2 |
| Days Max Temp ≥ 90 °F | 0 | 0 | 0 | 0 | 0 | 3 | *13* | 10 | 1 | 0 | 0 | 0 | 27 |
| Days Max Temp ≤ 32 °F | 11 | 5 | 1 | 0 | 0 | 0 | 0 | 0 | 0 | 0 | 3 | *11* | 31 |
| Days Min Temp ≤ 32 °F | 30 | 26 | 24 | 18 | *7* | 1 | 0 | 0 | 2 | 10 | 23 | 30 | 171 |
| Days Min Temp ≤ 0 °F | 2 | 1 | 0 | 0 | 0 | 0 | 0 | 0 | 0 | 0 | 0 | 2 | 5 |
| Heating Degree Days | 1189 | 950 | 857 | 625 | 359 | 150 | 18 | 23 | 176 | 468 | *873* | *1184* | 6872 |
| Cooling Degree Days | 0 | 0 | 0 | 0 | 9 | 60 | 192 | 183 | 42 | 1 | 0 | 0 | 487 |
| Total Precipitation (") | 1.17 | 1.05 | 1.20 | 1.18 | 1.62 | 1.30 | 0.67 | 0.71 | 0.88 | 1.16 | 1.61 | 1.24 | 13.79 |
| Days ≥ 0.1" Precip | 3 | 3 | *3* | 4 | 5 | 4 | *2* | 2 | 2 | 3 | 4 | 4 | 39 |
| Total Snowfall (") | 8.3 | *8.0* | *7.6* | *3.7* | *2.4* | 0.1 | 0.0 | 0.0 | 0.1 | *1.0* | *4.4* | na | na |
| Days ≥ 1" Snow Depth | 17 | *14* | na | *1* | 1 | 0 | 0 | 0 | 0 | *0* | na | na | na |

## MINA *Mineral County* ELEVATION 4554 ft LAT/LONG 38° 23 ' N / 118° 6 ' W

| | JAN | FEB | MAR | APR | MAY | JUN | JUL | AUG | SEP | OCT | NOV | DEC | YEAR |
|---|---|---|---|---|---|---|---|---|---|---|---|---|---|
| Maximum Temp °F | 45.9 | 52.6 | 58.7 | 65.7 | 75.9 | 85.9 | 94.5 | 92.3 | 83.5 | 71.1 | 56.1 | 46.2 | 69.0 |
| Minimum Temp °F | 22.0 | 26.3 | 31.5 | 36.8 | 45.8 | 54.8 | 61.9 | 59.5 | 49.9 | 38.7 | 29.8 | 22.1 | 39.9 |
| Mean Temp °F | 34.0 | 39.4 | 45.1 | 51.3 | 60.9 | 70.4 | 78.2 | 75.9 | 66.7 | 54.9 | 42.9 | 34.2 | 54.5 |
| Days Max Temp ≥ 90 °F | 0 | 0 | 0 | 0 | 2 | 12 | 26 | 22 | 8 | 0 | 0 | 0 | 70 |
| Days Max Temp ≤ 32 °F | 2 | 1 | 0 | 0 | 0 | 0 | 0 | 0 | 0 | 0 | 0 | 2 | 5 |
| Days Min Temp ≤ 32 °F | 28 | 23 | 17 | 8 | 1 | 0 | 0 | 0 | 0 | 6 | 19 | 27 | 129 |
| Days Min Temp ≤ 0 °F | 0 | 0 | 0 | 0 | 0 | 0 | 0 | 0 | 0 | 0 | 0 | 0 | 0 |
| Heating Degree Days | 956 | 714 | 608 | 408 | 174 | 36 | 1 | 4 | 62 | 313 | 655 | 949 | 4880 |
| Cooling Degree Days | 0 | 0 | 0 | 7 | 61 | 218 | 413 | 367 | 130 | 9 | 0 | 0 | 1205 |
| Total Precipitation (") | 0.41 | 0.43 | 0.59 | 0.64 | 0.59 | 0.50 | 0.47 | 0.48 | 0.41 | 0.44 | 0.41 | 0.35 | 5.72 |
| Days ≥ 0.1" Precip | 1 | 1 | 2 | 2 | 2 | 1 | 1 | 1 | 1 | 1 | 1 | 1 | 15 |
| Total Snowfall (") | 2.8 | 2.4 | 2.3 | 1.3 | 0.3 | 0.0 | 0.0 | 0.0 | 0.0 | 0.5 | 0.9 | 1.7 | 12.2 |
| Days ≥ 1" Snow Depth | na | na | *0* | 0 | 0 | 0 | 0 | 0 | 0 | 0 | *0* | na | na |

## MINDEN *Douglas County* ELEVATION 4709 ft LAT/LONG 38° 58' N / 119° 46' W

| | JAN | FEB | MAR | APR | MAY | JUN | JUL | AUG | SEP | OCT | NOV | DEC | YEAR |
|---|---|---|---|---|---|---|---|---|---|---|---|---|---|
| Maximum Temp °F | 45.1 | 51.1 | 56.5 | 63.1 | 72.4 | 81.5 | 89.8 | 88.5 | 80.9 | 69.9 | 55.3 | 46.2 | 66.7 |
| Minimum Temp °F | 16.8 | 20.8 | 25.2 | 29.4 | 36.4 | 43.1 | 47.6 | 45.9 | 38.6 | 29.3 | 22.9 | 16.8 | 31.1 |
| Mean Temp °F | 31.0 | 36.0 | 40.9 | 46.3 | 54.4 | 62.3 | 68.7 | 67.3 | 59.8 | 49.6 | 39.1 | 31.5 | 48.9 |
| Days Max Temp ≥ 90 °F | 0 | 0 | 0 | 0 | 0 | 5 | 17 | 15 | 4 | 0 | 0 | 0 | 41 |
| Days Max Temp ≤ 32 °F | 3 | 1 | 0 | 0 | 0 | 0 | 0 | 0 | 0 | 0 | 0 | 3 | 7 |
| Days Min Temp ≤ 32 °F | 29 | 26 | 26 | 20 | 9 | 2 | 0 | 0 | 6 | 21 | 27 | 29 | 195 |
| Days Min Temp ≤ 0 °F | 2 | 1 | 0 | 0 | 0 | 0 | 0 | 0 | 0 | 0 | 0 | 2 | 5 |
| Heating Degree Days | 1048 | 813 | 741 | 556 | 325 | 120 | 22 | 36 | 172 | 470 | 770 | 1032 | 6105 |
| Cooling Degree Days | 0 | 0 | 0 | 0 | 5 | 49 | 136 | 115 | 27 | 0 | 0 | 0 | 332 |
| Total Precipitation (") | 1.28 | 1.01 | 0.87 | 0.33 | 0.36 | 0.33 | 0.27 | 0.37 | 0.40 | 0.54 | 1.00 | 1.04 | 7.80 |
| Days ≥ 0.1" Precip | 3 | 3 | 2 | 1 | 1 | 1 | 1 | 1 | 1 | 2 | 3 | 3 | 22 |
| Total Snowfall (") | na | ***2.6*** | ***2.6*** | 0.8 | 0.2 | 0.0 | 0.0 | 0.0 | 0.0 | 0.2 | na | na | na |
| Days ≥ 1" Snow Depth | ***5*** | ***2*** | ***1*** | 0 | 0 | 0 | 0 | 0 | 0 | 0 | na | na | na |

## MONTELLO 1 SE *Elko County* ELEVATION 4882 ft LAT/LONG 41° 16' N / 114° 12' W

| | JAN | FEB | MAR | APR | MAY | JUN | JUL | AUG | SEP | OCT | NOV | DEC | YEAR |
|---|---|---|---|---|---|---|---|---|---|---|---|---|---|
| Maximum Temp °F | 36.1 | 42.1 | ***51.9*** | ***61.4*** | 71.7 | 81.9 | 92.4 | 90.0 | 79.3 | 65.3 | 47.9 | 37.2 | 63.1 |
| Minimum Temp °F | 9.5 | 15.9 | ***23.4*** | ***27.9*** | 36.1 | 43.9 | 49.8 | 48.1 | 37.9 | 27.3 | 19.0 | 9.8 | 29.1 |
| Mean Temp °F | 23.0 | 29.0 | ***37.7*** | ***44.7*** | 54.0 | 62.9 | 71.1 | 69.1 | 58.6 | 46.3 | 33.5 | 23.5 | 46.1 |
| Days Max Temp ≥ 90 °F | 0 | 0 | 0 | 0 | 1 | 8 | ***20*** | ***17*** | 4 | 0 | 0 | 0 | 50 |
| Days Max Temp ≤ 32 °F | 11 | 4 | 0 | 0 | 0 | 0 | 0 | 0 | 0 | 0 | 1 | ***8*** | 24 |
| Days Min Temp ≤ 32 °F | 30 | 27 | ***28*** | ***21*** | ***9*** | 1 | 0 | 0 | 7 | 22 | 27 | 31 | 203 |
| Days Min Temp ≤ 0 °F | 7 | 3 | ***0*** | 0 | 0 | 0 | 0 | 0 | 0 | 0 | 1 | 6 | 17 |
| Heating Degree Days | 1295 | 1012 | ***849*** | ***601*** | ***342*** | 122 | 14 | 26 | 209 | 572 | 944 | 1281 | 7267 |
| Cooling Degree Days | 0 | 0 | ***0*** | 0 | 3 | 65 | 190 | ***159*** | 20 | 0 | 0 | 0 | 437 |
| Total Precipitation (") | 0.49 | 0.47 | 0.64 | 0.66 | 0.97 | 0.92 | 0.77 | 0.67 | 0.65 | 0.54 | 0.67 | 0.48 | 7.93 |
| Days ≥ 0.1" Precip | 1 | 2 | ***2*** | 2 | 3 | 3 | 2 | 2 | 2 | 2 | 2 | 2 | 25 |
| Total Snowfall (") | na | ***2.4*** | na | ***0.0*** | 0.0 | 0.0 | 0.0 | 0.0 | 0.0 | 0.2 | ***1.1*** | na | na |
| Days ≥ 1" Snow Depth | na | na | na | ***0*** | 0 | 0 | 0 | 0 | 0 | 0 | ***0*** | na | na |

## MOUNTAIN CITY R S *Elko County* ELEVATION 5640 ft LAT/LONG 41° 50' N / 115° 58' W

| | JAN | FEB | MAR | APR | MAY | JUN | JUL | AUG | SEP | OCT | NOV | DEC | YEAR |
|---|---|---|---|---|---|---|---|---|---|---|---|---|---|
| Maximum Temp °F | 38.0 | 42.1 | 47.3 | 55.6 | 64.8 | 74.4 | 84.4 | 83.9 | 75.0 | 63.3 | 46.5 | 38.4 | 59.5 |
| Minimum Temp °F | 9.9 | 13.4 | 19.3 | 23.8 | 30.3 | 36.0 | 40.0 | 38.3 | 30.3 | 22.7 | 18.5 | 10.5 | 24.4 |
| Mean Temp °F | 24.0 | 27.8 | 33.2 | 39.7 | 47.5 | 55.2 | 62.2 | 61.2 | 52.6 | 43.0 | 32.5 | 24.5 | 41.9 |
| Days Max Temp ≥ 90 °F | 0 | 0 | 0 | 0 | 0 | 1 | 6 | 6 | 0 | 0 | 0 | 0 | 13 |
| Days Max Temp ≤ 32 °F | 7 | 4 | 1 | 0 | 0 | 0 | 0 | 0 | 0 | 0 | 3 | 8 | 23 |
| Days Min Temp ≤ 32 °F | 30 | 27 | 29 | 27 | 19 | 9 | 4 | 6 | 19 | 27 | ***28*** | 29 | 254 |
| Days Min Temp ≤ 0 °F | ***8*** | 4 | 1 | 0 | 0 | 0 | 0 | 0 | 0 | 0 | 2 | 6 | 21 |
| Heating Degree Days | 1265 | 1049 | 979 | 753 | 530 | 289 | 106 | 131 | 366 | 676 | 969 | 1249 | 8362 |
| Cooling Degree Days | 0 | 0 | 0 | 0 | 0 | 2 | 24 | 19 | 0 | 0 | 0 | 0 | 45 |
| Total Precipitation (") | 1.22 | 1.00 | 1.14 | 1.12 | 1.51 | 1.03 | 0.60 | 0.65 | 0.84 | 1.01 | 1.47 | 1.28 | 12.87 |
| Days ≥ 0.1" Precip | 4 | 4 | 4 | 4 | 4 | 3 | 2 | 2 | 3 | 3 | 5 | ***4*** | 42 |
| Total Snowfall (") | na | na | na | na | ***0.3*** | 0.0 | 0.0 | 0.0 | 0.0 | 1.0 | na | na | na |
| Days ≥ 1" Snow Depth | na | na | na | na | ***0*** | 0 | 0 | 0 | 0 | 0 | na | na | na |

## OROVADA 4 WSW *Humboldt County* ELEVATION 4290 ft LAT/LONG 41° 33' N / 117° 50' W

| | JAN | FEB | MAR | APR | MAY | JUN | JUL | AUG | SEP | OCT | NOV | DEC | YEAR |
|---|---|---|---|---|---|---|---|---|---|---|---|---|---|
| Maximum Temp °F | 39.7 | 47.1 | 53.2 | 60.6 | 71.1 | 81.3 | 90.7 | 89.3 | 79.3 | 66.3 | 49.7 | 39.8 | 64.0 |
| Minimum Temp °F | 19.7 | 24.0 | 27.9 | 31.8 | 38.8 | 46.4 | 52.0 | 50.4 | 41.2 | 32.9 | 25.4 | 19.2 | 34.1 |
| Mean Temp °F | 29.7 | 35.6 | 40.6 | 46.3 | 55.0 | 63.9 | 71.4 | 69.9 | 60.3 | 49.6 | 37.6 | 29.5 | 49.1 |
| Days Max Temp ≥ 90 °F | 0 | 0 | 0 | 0 | 1 | 7 | 20 | 17 | 5 | 0 | 0 | 0 | 50 |
| Days Max Temp ≤ 32 °F | 7 | 2 | 0 | 0 | 0 | 0 | 0 | 0 | 0 | 0 | 1 | 6 | 16 |
| Days Min Temp ≤ 32 °F | 28 | 24 | 22 | ***16*** | 6 | 1 | 0 | 0 | 4 | 14 | 23 | 28 | 166 |
| Days Min Temp ≤ 0 °F | 2 | 1 | 0 | 0 | 0 | 0 | 0 | 0 | 0 | 0 | 0 | 2 | 5 |
| Heating Degree Days | 1087 | 825 | 750 | 556 | 315 | 109 | 19 | 26 | 170 | 470 | 816 | 1092 | 6235 |
| Cooling Degree Days | 0 | 0 | 0 | 0 | 12 | 75 | 193 | 174 | 35 | 1 | 0 | 0 | 490 |
| Total Precipitation (") | 0.89 | 0.71 | 1.03 | 1.22 | 1.08 | 1.06 | 0.29 | 0.57 | 0.63 | 0.76 | 1.14 | 0.90 | 10.28 |
| Days ≥ 0.1" Precip | 3 | 2 | 3 | 4 | 3 | 3 | 1 | 1 | 2 | 2 | 4 | 4 | 32 |
| Total Snowfall (") | ***3.5*** | ***2.0*** | 2.8 | ***1.9*** | 0.0 | 0.0 | 0.0 | 0.0 | 0.0 | 0.9 | ***2.4*** | ***4.9*** | 18.4 |
| Days ≥ 1" Snow Depth | na | ***1*** | na | ***0*** | 0 | 0 | 0 | 0 | 0 | 0 | na | na | na |

## PAHRANAGAT W L REF *Lincoln County* ELEVATION 3402 ft LAT/LONG 37° 16 ' N / 115° 7 ' W

| | JAN | FEB | MAR | APR | MAY | JUN | JUL | AUG | SEP | OCT | NOV | DEC | YEAR |
|---|---|---|---|---|---|---|---|---|---|---|---|---|---|
| Maximum Temp °F | 53.0 | 58.4 | 64.6 | 72.3 | 82.1 | 92.0 | 97.8 | 95.9 | 88.8 | 77.1 | 62.3 | 52.8 | 74.8 |
| Minimum Temp °F | 26.8 | 30.9 | 35.4 | 40.9 | 49.2 | 56.9 | 63.6 | 62.1 | 53.5 | 43.6 | 33.0 | 26.3 | 43.5 |
| Mean Temp °F | 39.9 | 44.7 | 50.0 | 56.6 | 65.7 | 74.4 | 80.8 | 79.0 | 71.2 | 60.4 | 47.7 | 39.6 | 59.2 |
| Days Max Temp ≥ 90 °F | 0 | 0 | 0 | 1 | 5 | 19 | 29 | 27 | 15 | 3 | 0 | 0 | 99 |
| Days Max Temp ≤ 32 °F | 0 | 0 | 0 | 0 | 0 | 0 | 0 | 0 | 0 | 0 | 0 | 0 | 0 |
| Days Min Temp ≤ 32 °F | 26 | 17 | 9 | 3 | 0 | 0 | 0 | 0 | 0 | 2 | 14 | 26 | 97 |
| Days Min Temp ≤ 0 °F | 0 | 0 | 0 | 0 | 0 | 0 | 0 | 0 | 0 | 0 | 0 | 0 | 0 |
| Heating Degree Days | 768 | 568 | 458 | 257 | 72 | 6 | 0 | 0 | 15 | 170 | 513 | 782 | 3609 |
| Cooling Degree Days | 0 | 0 | 0 | 20 | 113 | 340 | 517 | 474 | 231 | 41 | 0 | 0 | 1736 |
| Total Precipitation (") | 0.67 | 0.64 | 0.78 | 0.62 | 0.45 | 0.24 | 0.56 | 0.76 | 0.40 | 0.55 | 0.55 | 0.51 | 6.73 |
| Days ≥ 0.1" Precip | 2 | 2 | 3 | 2 | 1 | 1 | 1 | 2 | 1 | 1 | 2 | 2 | 20 |
| Total Snowfall (") | 0.4 | 0.2 | 0.1 | 0.1 | 0.0 | 0.0 | 0.0 | 0.0 | 0.0 | 0.0 | 0.2 | 0.6 | 1.6 |
| Days ≥ 1" Snow Depth | 0 | 0 | 0 | 0 | 0 | 0 | 0 | 0 | 0 | 0 | 0 | 0 | 0 |

## PAHRUMP *Nye County* ELEVATION 2831 ft LAT/LONG 36° 13 ' N / 116° 0 ' W

| | JAN | FEB | MAR | APR | MAY | JUN | JUL | AUG | SEP | OCT | NOV | DEC | YEAR |
|---|---|---|---|---|---|---|---|---|---|---|---|---|---|
| Maximum Temp °F | 57.4 | 62.4 | 67.5 | 75.2 | 84.5 | 95.0 | 100.6 | 99.0 | 91.8 | 80.5 | 66.5 | 57.8 | 78.2 |
| Minimum Temp °F | 26.9 | 31.9 | 37.3 | 43.0 | 51.5 | 60.1 | 67.0 | 65.2 | 56.1 | 44.2 | 33.1 | 26.3 | 45.2 |
| Mean Temp °F | 42.2 | 47.2 | 52.4 | 59.1 | 68.0 | 77.5 | 83.8 | 82.1 | 74.0 | 62.4 | 49.8 | 42.1 | 61.7 |
| Days Max Temp ≥ 90 °F | 0 | 0 | 0 | 2 | 10 | 23 | 29 | 28 | 19 | 5 | 0 | 0 | 116 |
| Days Max Temp ≤ 32 °F | 0 | 0 | 0 | 0 | 0 | 0 | 0 | 0 | 0 | 0 | 0 | 0 | 0 |
| Days Min Temp ≤ 32 °F | *24* | *15* | 8 | 1 | 0 | 0 | 0 | 0 | 0 | 2 | 14 | 25 | 89 |
| Days Min Temp ≤ 0 °F | 0 | 0 | 0 | 0 | 0 | 0 | 0 | 0 | 0 | 0 | 0 | 0 | 0 |
| Heating Degree Days | 701 | 496 | 383 | 196 | 56 | 2 | 0 | 0 | 9 | 128 | 450 | 709 | 3130 |
| Cooling Degree Days | 0 | 0 | 2 | 35 | 150 | 378 | 568 | 522 | 283 | 51 | 0 | 0 | 1989 |
| Total Precipitation (") | 0.63 | 0.80 | 0.77 | 0.36 | 0.26 | 0.09 | 0.39 | 0.45 | 0.23 | 0.18 | 0.44 | 0.56 | 5.16 |
| Days ≥ 0.1" Precip | 2 | 2 | 2 | 1 | 1 | 0 | 1 | 1 | 1 | 1 | 1 | 2 | 15 |
| Total Snowfall (") | *0.0* | 0.1 | 0.0 | 0.0 | 0.0 | 0.0 | 0.0 | 0.0 | 0.0 | 0.0 | 0.0 | 0.0 | 0.1 |
| Days ≥ 1" Snow Depth | *0* | 0 | 0 | 0 | 0 | 0 | 0 | 0 | 0 | 0 | 0 | 0 | 0 |

## PARADISE VALLEY 1 NW *Humboldt County* ELEVATION 4682 ft LAT/LONG 41° 30 ' N / 117° 32 ' W

| | JAN | FEB | MAR | APR | MAY | JUN | JUL | AUG | SEP | OCT | NOV | DEC | YEAR |
|---|---|---|---|---|---|---|---|---|---|---|---|---|---|
| Maximum Temp °F | 39.7 | 46.2 | 53.3 | 61.1 | 70.8 | 80.6 | 90.1 | 88.5 | 79.3 | 67.1 | 49.8 | 40.2 | 63.9 |
| Minimum Temp °F | 16.7 | 21.6 | 25.7 | 29.3 | 35.7 | 42.5 | 47.8 | 46.2 | 38.2 | 29.9 | 23.2 | 16.4 | 31.1 |
| Mean Temp °F | 28.3 | 33.9 | 39.5 | 45.2 | 53.3 | 61.6 | 69.0 | 67.4 | 58.8 | 48.5 | 36.5 | 28.3 | 47.5 |
| Days Max Temp ≥ 90 °F | 0 | 0 | 0 | 0 | 1 | 6 | 19 | 16 | 4 | 0 | 0 | 0 | 46 |
| Days Max Temp ≤ 32 °F | 6 | 1 | 0 | 0 | 0 | 0 | 0 | 0 | 0 | 0 | 1 | 5 | 13 |
| Days Min Temp ≤ 32 °F | 29 | 26 | 26 | 21 | 11 | 2 | 0 | 1 | 6 | 19 | 26 | 30 | 197 |
| Days Min Temp ≤ 0 °F | 3 | 1 | 0 | 0 | 0 | 0 | 0 | 0 | 0 | 0 | 0 | 2 | 6 |
| Heating Degree Days | 1132 | 871 | 784 | 587 | 359 | 139 | 23 | 42 | 198 | 503 | 849 | 1130 | 6617 |
| Cooling Degree Days | 0 | 0 | 0 | 0 | 5 | 42 | 141 | 120 | 21 | 0 | 0 | 0 | 329 |
| Total Precipitation (") | 1.22 | 1.01 | 0.95 | 0.65 | 0.73 | 0.76 | 0.32 | 0.39 | 0.50 | 0.67 | 1.43 | 1.40 | 10.03 |
| Days ≥ 0.1" Precip | 4 | 3 | 3 | 2 | 3 | 2 | 1 | 1 | 2 | 2 | 4 | 5 | 32 |
| Total Snowfall (") | 7.1 | 4.7 | 2.4 | 0.8 | 0.1 | 0.0 | 0.0 | 0.0 | 0.0 | 0.6 | 3.6 | 8.3 | 27.6 |
| Days ≥ 1" Snow Depth | 15 | 8 | 1 | 0 | 0 | 0 | 0 | 0 | 0 | 0 | 4 | 12 | 40 |

## PIOCHE *Lincoln County* ELEVATION 6115 ft LAT/LONG 37° 56 ' N / 114° 27 ' W

| | JAN | FEB | MAR | APR | MAY | JUN | JUL | AUG | SEP | OCT | NOV | DEC | YEAR |
|---|---|---|---|---|---|---|---|---|---|---|---|---|---|
| Maximum Temp °F | 41.6 | 45.5 | 51.2 | 59.3 | 69.4 | 80.3 | 87.2 | 85.1 | 76.5 | 64.8 | 50.1 | 42.2 | 62.8 |
| Minimum Temp °F | 22.0 | 24.8 | 29.1 | 34.8 | 43.6 | 51.9 | 58.4 | 57.5 | 49.7 | 39.5 | 29.1 | 22.5 | 38.6 |
| Mean Temp °F | 31.9 | 35.2 | 40.2 | 47.1 | 56.5 | 66.1 | 72.8 | 71.3 | 63.1 | 52.0 | 39.6 | 32.4 | 50.7 |
| Days Max Temp ≥ 90 °F | 0 | 0 | 0 | 0 | 0 | 4 | 12 | 8 | 0 | 0 | 0 | 0 | 24 |
| Days Max Temp ≤ 32 °F | 5 | 2 | 0 | 0 | 0 | 0 | 0 | 0 | 0 | 0 | 1 | 4 | 12 |
| Days Min Temp ≤ 32 °F | 29 | 24 | 21 | 12 | 3 | 0 | 0 | 0 | 1 | 6 | 19 | 28 | 143 |
| Days Min Temp ≤ 0 °F | 1 | 0 | 0 | 0 | 0 | 0 | 0 | 0 | 0 | 0 | 0 | 0 | 1 |
| Heating Degree Days | 1021 | 834 | 763 | 530 | 269 | 68 | 4 | 9 | 104 | 398 | 755 | 1006 | 5761 |
| Cooling Degree Days | 0 | 0 | 0 | 1 | 15 | *106* | *239* | *219* | 59 | 5 | 0 | 0 | 644 |
| Total Precipitation (") | 1.47 | 1.47 | 1.78 | 1.01 | 1.03 | 0.45 | 1.18 | 1.30 | 0.92 | 1.05 | 1.05 | 1.30 | 14.01 |
| Days ≥ 0.1" Precip | 3 | 3 | 4 | 3 | 3 | 1 | 3 | 3 | 2 | 3 | 2 | 3 | 33 |
| Total Snowfall (") | na | na | *3.4* | 2.3 | 0.0 | 0.1 | 0.0 | 0.0 | 0.0 | 0.7 | na | na | na |
| Days ≥ 1" Snow Depth | na | na | na | 1 | 0 | 0 | 0 | 0 | 0 | 0 | *2* | na | na |

## RENO CANNON INTL AP *Washoe County* ELEVATION 4403 ft LAT/LONG 39° 30 ' N / 119° 47 ' W

| | JAN | FEB | MAR | APR | MAY | JUN | JUL | AUG | SEP | OCT | NOV | DEC | YEAR |
|---|---|---|---|---|---|---|---|---|---|---|---|---|---|
| Maximum Temp °F | 44.9 | 51.6 | 56.9 | 63.8 | 73.4 | 82.7 | 91.3 | 89.5 | 81.1 | 69.6 | 54.7 | 45.2 | 67.1 |
| Minimum Temp °F | 21.0 | 24.6 | 28.4 | 32.3 | 39.1 | 45.3 | 50.0 | 48.4 | 41.5 | 32.8 | 25.9 | 20.0 | 34.1 |
| Mean Temp °F | 33.0 | 38.1 | 42.7 | 48.1 | 56.3 | 64.0 | 70.7 | 69.0 | 61.3 | 51.2 | 40.3 | 32.6 | 50.6 |
| Days Max Temp ≥ 90 °F | 0 | 0 | 0 | 0 | 1 | 7 | 20 | 17 | 5 | 0 | 0 | 0 | 50 |
| Days Max Temp ≤ 32 °F | 3 | 1 | 0 | 0 | 0 | 0 | 0 | 0 | 0 | 0 | 0 | 3 | 7 |
| Days Min Temp ≤ 32 °F | 28 | 24 | 22 | 16 | 5 | 1 | 0 | 0 | 3 | 16 | 24 | 28 | 167 |
| Days Min Temp ≤ 0 °F | 1 | 0 | 0 | 0 | 0 | 0 | 0 | 0 | 0 | 0 | 0 | 1 | 2 |
| Heating Degree Days | 985 | 753 | 686 | 502 | 275 | 94 | 13 | 24 | 141 | 422 | 734 | 997 | 5626 |
| Cooling Degree Days | 0 | 0 | 0 | 1 | 17 | 91 | 221 | 189 | 54 | 2 | 0 | 0 | 575 |
| Total Precipitation (") | 1.01 | 0.90 | 0.75 | 0.34 | 0.53 | 0.45 | 0.27 | 0.31 | 0.39 | 0.39 | 0.87 | 0.94 | 7.15 |
| Days ≥ 0.1" Precip | 3 | 3 | 3 | 1 | 2 | 2 | 1 | 1 | 1 | 1 | 3 | 3 | 24 |
| Total Snowfall (") | 5.5 | 5.4 | 3.6 | 1.1 | 0.6 | 0.0 | 0.0 | 0.0 | 0.1 | 0.4 | 2.7 | 5.0 | 24.4 |
| Days ≥ 1" Snow Depth | 6 | 3 | 1 | 0 | 0 | 0 | 0 | 0 | 0 | 0 | 1 | 4 | 15 |

## RUBY LAKE *Elko County* ELEVATION 6014 ft LAT/LONG 40° 12 ' N / 115° 30 ' W

| | JAN | FEB | MAR | APR | MAY | JUN | JUL | AUG | SEP | OCT | NOV | DEC | YEAR |
|---|---|---|---|---|---|---|---|---|---|---|---|---|---|
| Maximum Temp °F | 39.1 | 43.3 | 49.8 | 57.6 | 67.7 | 78.3 | 87.0 | 85.5 | 75.9 | ***64.0*** | 48.5 | 39.0 | 61.3 |
| Minimum Temp °F | 14.4 | 18.6 | 25.5 | 30.8 | 38.4 | 45.4 | 52.0 | 49.9 | 41.1 | 31.4 | 23.5 | 14.8 | 32.2 |
| Mean Temp °F | 26.8 | 31.0 | 37.7 | 44.2 | 53.1 | 61.9 | 69.6 | 67.7 | 58.5 | ***47.7*** | 36.0 | 27.0 | 46.8 |
| Days Max Temp ≥ 90 °F | 0 | 0 | 0 | 0 | 0 | 3 | na | ***8*** | 0 | 0 | 0 | 0 | na |
| Days Max Temp ≤ 32 °F | 7 | 3 | 1 | 0 | 0 | 0 | 0 | 0 | 0 | 0 | 2 | 6 | 19 |
| Days Min Temp ≤ 32 °F | 29 | ***26*** | ***26*** | na | 7 | 1 | 0 | 0 | 4 | ***17*** | ***25*** | 30 | na |
| Days Min Temp ≤ 0 °F | 5 | 2 | 0 | 0 | 0 | 0 | 0 | 0 | 0 | 0 | 0 | ***3*** | 10 |
| Heating Degree Days | 1178 | 953 | 837 | 616 | 366 | 135 | 12 | 28 | 203 | ***528*** | 869 | 1173 | 6898 |
| Cooling Degree Days | 0 | 0 | 0 | 0 | 3 | 55 | 153 | 113 | 15 | 0 | 0 | 0 | 339 |
| Total Precipitation (") | 1.26 | 1.10 | 1.27 | 0.97 | 1.34 | 0.90 | 0.54 | 0.82 | 0.88 | 1.15 | 1.55 | 1.30 | 13.08 |
| Days ≥ 0.1" Precip | 4 | 3 | 4 | 3 | 4 | 3 | 2 | 2 | 2 | 3 | 4 | 4 | 38 |
| Total Snowfall (") | 11.4 | ***8.4*** | ***5.8*** | 2.6 | 0.9 | 0.0 | 0.0 | 0.0 | 0.2 | 1.0 | 5.2 | ***10.8*** | 46.3 |
| Days ≥ 1" Snow Depth | na | na | na | ***0*** | 0 | 0 | 0 | 0 | 0 | 0 | ***4*** | na | na |

## RYE PATCH DAM *Pershing County* ELEVATION 4163 ft LAT/LONG 40° 28 ' N / 118° 18 ' W

| | JAN | FEB | MAR | APR | MAY | JUN | JUL | AUG | SEP | OCT | NOV | DEC | YEAR |
|---|---|---|---|---|---|---|---|---|---|---|---|---|---|
| Maximum Temp °F | 42.6 | 50.6 | 57.9 | 65.4 | 75.5 | 85.0 | 94.2 | 92.3 | 83.3 | 70.6 | 54.0 | 43.8 | 67.9 |
| Minimum Temp °F | 17.3 | 22.4 | 26.5 | 31.0 | 39.2 | 46.3 | 51.7 | 49.7 | 41.3 | 31.3 | 24.4 | 17.3 | 33.2 |
| Mean Temp °F | 30.0 | 36.5 | 42.3 | 48.3 | 57.4 | 65.7 | 73.0 | 71.1 | 62.3 | 51.0 | 39.1 | 30.6 | 50.6 |
| Days Max Temp ≥ 90 °F | 0 | 0 | 0 | 0 | 2 | 11 | 25 | 22 | 8 | 1 | 0 | 0 | 69 |
| Days Max Temp ≤ 32 °F | 5 | 1 | 0 | 0 | 0 | 0 | 0 | 0 | 0 | 0 | 0 | 3 | 9 |
| Days Min Temp ≤ 32 °F | 29 | 25 | 24 | 18 | 6 | 1 | 0 | 0 | 4 | 17 | 23 | 29 | 176 |
| Days Min Temp ≤ 0 °F | 2 | 0 | 0 | 0 | 0 | 0 | 0 | 0 | 0 | 0 | 0 | 2 | 4 |
| Heating Degree Days | 1078 | 796 | 698 | 496 | 250 | 76 | 7 | 16 | 123 | 431 | 769 | 1061 | 5801 |
| Cooling Degree Days | 0 | 0 | 0 | 1 | 26 | 108 | 258 | 216 | 57 | 2 | ***0*** | 0 | 668 |
| Total Precipitation (") | 0.71 | 0.63 | 0.83 | 0.95 | 0.87 | 0.85 | 0.33 | 0.45 | 0.45 | 0.66 | 0.83 | 0.77 | 8.33 |
| Days ≥ 0.1" Precip | 3 | 2 | 3 | 3 | 3 | 2 | 1 | 1 | 1 | 2 | 3 | 3 | 27 |
| Total Snowfall (") | na | na | na | ***0.0*** | 0.0 | 0.0 | 0.0 | 0.0 | 0.0 | 0.0 | ***1.0*** | na | na |
| Days ≥ 1" Snow Depth | na | na | na | 0 | 0 | 0 | 0 | 0 | 0 | 0 | ***0*** | na | na |

## SEARCHLIGHT *Clark County* ELEVATION 3540 ft LAT/LONG 35° 28 ' N / 114° 55 ' W

| | JAN | FEB | MAR | APR | MAY | JUN | JUL | AUG | SEP | OCT | NOV | DEC | YEAR |
|---|---|---|---|---|---|---|---|---|---|---|---|---|---|
| Maximum Temp °F | 53.9 | 58.9 | 64.8 | 73.1 | 82.7 | 93.1 | 97.7 | 95.5 | 88.5 | 77.1 | 62.9 | 53.7 | 75.2 |
| Minimum Temp °F | 35.7 | 38.3 | 41.6 | 47.1 | 55.6 | 65.1 | 70.9 | 69.4 | 62.7 | 53.2 | 42.0 | 35.6 | 51.4 |
| Mean Temp °F | 44.8 | 48.7 | 53.2 | 60.1 | 69.2 | 79.1 | 84.3 | 82.5 | 75.6 | 65.2 | 52.5 | 44.7 | 63.3 |
| Days Max Temp ≥ 90 °F | 0 | 0 | 0 | 0 | 6 | 21 | 30 | 27 | 14 | 2 | 0 | 0 | 100 |
| Days Max Temp ≤ 32 °F | 0 | 0 | 0 | 0 | 0 | 0 | 0 | 0 | 0 | 0 | 0 | 0 | 0 |
| Days Min Temp ≤ 32 °F | 10 | 5 | 3 | 1 | 0 | 0 | 0 | 0 | 0 | 0 | 3 | 10 | 32 |
| Days Min Temp ≤ 0 °F | 0 | 0 | 0 | 0 | 0 | 0 | 0 | 0 | 0 | 0 | 0 | 0 | 0 |
| Heating Degree Days | 619 | 454 | 364 | 179 | 43 | 3 | 0 | 0 | 7 | 91 | 371 | 623 | 2754 |
| Cooling Degree Days | 0 | 1 | 5 | 60 | 182 | 440 | 590 | 547 | 338 | 103 | 3 | 0 | 2269 |
| Total Precipitation (") | 0.92 | 0.98 | 1.04 | 0.40 | 0.25 | 0.10 | 0.88 | 1.27 | 0.58 | 0.48 | 0.57 | 0.87 | 8.34 |
| Days ≥ 0.1" Precip | 2 | 3 | 3 | 1 | 1 | 0 | 2 | 2 | 1 | 1 | 1 | 2 | 19 |
| Total Snowfall (") | 0.3 | 0.0 | 0.1 | 0.0 | 0.0 | 0.0 | 0.0 | 0.0 | 0.0 | 0.0 | 0.0 | 0.5 | 0.9 |
| Days ≥ 1" Snow Depth | 0 | 0 | 0 | 0 | 0 | 0 | 0 | 0 | 0 | 0 | 0 | ***0*** | 0 |

## SMOKEY VALLEY *Nye County* ELEVATION 5625 ft LAT/LONG 38° 47 ' N / 117° 10 ' W

| | JAN | FEB | MAR | APR | MAY | JUN | JUL | AUG | SEP | OCT | NOV | DEC | YEAR |
|---|---|---|---|---|---|---|---|---|---|---|---|---|---|
| Maximum Temp °F | 43.6 | 50.0 | 56.1 | 64.0 | 73.8 | 83.6 | 91.2 | 88.7 | 80.6 | 69.1 | 53.7 | 43.8 | 66.5 |
| Minimum Temp °F | 15.1 | 20.2 | 25.6 | 30.3 | 38.4 | 46.5 | 53.0 | 51.4 | 42.5 | 32.5 | 22.7 | 15.3 | 32.8 |
| Mean Temp °F | 29.4 | 35.3 | 40.9 | 47.2 | 56.2 | 65.1 | 72.2 | 70.1 | 61.5 | 50.8 | 38.2 | 29.6 | 49.7 |
| Days Max Temp ≥ 90 °F | 0 | 0 | 0 | 0 | 0 | 8 | ***21*** | 15 | 3 | 0 | 0 | 0 | 47 |
| Days Max Temp ≤ 32 °F | 3 | 1 | 0 | 0 | 0 | 0 | 0 | 0 | 0 | 0 | 1 | 4 | 9 |
| Days Min Temp ≤ 32 °F | 29 | 25 | 25 | ***18*** | ***6*** | 1 | 0 | 0 | 3 | ***15*** | 26 | 30 | 178 |
| Days Min Temp ≤ 0 °F | 2 | 1 | 0 | 0 | 0 | 0 | 0 | 0 | 0 | 0 | 0 | 2 | 5 |
| Heating Degree Days | 1097 | ***840*** | 741 | 529 | 277 | 77 | 3 | 14 | 133 | 433 | 799 | 1092 | 6035 |
| Cooling Degree Days | 0 | 0 | 0 | 0 | 13 | 98 | 238 | ***192*** | 44 | 0 | 0 | 0 | 585 |
| Total Precipitation (") | 0.57 | 0.70 | 0.81 | 0.55 | 0.59 | 0.48 | 0.61 | 0.87 | 0.62 | 0.54 | 0.60 | 0.41 | 7.35 |
| Days ≥ 0.1" Precip | 2 | 2 | 2 | 2 | 1 | 1 | 2 | 2 | 2 | 2 | 2 | 1 | 21 |
| Total Snowfall (") | na | na | na | 0.1 | 0.0 | 0.0 | 0.0 | 0.0 | 0.0 | 0.0 | ***0.9*** | na | na |
| Days ≥ 1" Snow Depth | na | na | na | 0 | 0 | 0 | 0 | 0 | 0 | 0 | ***0*** | na | na |

## SUNNYSIDE *Nye County* ELEVATION 5335 ft LAT/LONG 38° 26 ' N / 115° 1 ' W

| | JAN | FEB | MAR | APR | MAY | JUN | JUL | AUG | SEP | OCT | NOV | DEC | YEAR |
|---|---|---|---|---|---|---|---|---|---|---|---|---|---|
| Maximum Temp °F | 43.8 | 49.3 | 55.3 | 63.9 | 73.7 | 84.3 | 91.1 | 88.7 | ***80.6*** | 68.7 | 54.2 | 45.3 | 66.6 |
| Minimum Temp °F | 14.8 | 21.1 | 26.5 | 31.2 | 38.9 | 47.2 | 53.8 | 51.5 | 42.7 | 32.0 | 22.8 | 15.5 | 33.2 |
| Mean Temp °F | 29.3 | 35.2 | 40.9 | 47.7 | 56.4 | 65.7 | 72.5 | 70.1 | ***61.7*** | 50.5 | 38.5 | 30.4 | 49.9 |
| Days Max Temp ≥ 90 °F | 0 | 0 | 0 | 0 | 1 | 9 | na | ***14*** | 3 | 0 | 0 | 0 | na |
| Days Max Temp ≤ 32 °F | ***3*** | 1 | 0 | 0 | 0 | 0 | 0 | 0 | 0 | 0 | 1 | ***2*** | 7 |
| Days Min Temp ≤ 32 °F | 30 | 26 | na | na | na | 0 | 0 | 0 | ***2*** | na | ***26*** | ***30*** | na |
| Days Min Temp ≤ 0 °F | 3 | 1 | 0 | 0 | 0 | 0 | 0 | 0 | 0 | 0 | 0 | ***2*** | 6 |
| Heating Degree Days | 1100 | 834 | 739 | 515 | 269 | 65 | 3 | 11 | ***126*** | 439 | 787 | 1064 | 5952 |
| Cooling Degree Days | 0 | 0 | 0 | 0 | 8 | 99 | 235 | 184 | 34 | 0 | 0 | 0 | 560 |
| Total Precipitation (") | 0.76 | 0.72 | 1.02 | 0.73 | 0.84 | 0.45 | 0.91 | 1.02 | 0.90 | 0.96 | 0.66 | 0.64 | 9.61 |
| Days ≥ 0.1" Precip | 3 | 2 | 3 | 2 | 2 | 1 | 2 | 2 | 2 | 2 | 2 | 2 | 25 |
| Total Snowfall (") | na | ***2.3*** | ***3.2*** | ***1.0*** | 0.1 | 0.0 | 0.0 | 0.0 | 0.0 | 0.2 | 2.5 | ***3.5*** | na |
| Days ≥ 1" Snow Depth | na | na | na | ***0*** | 0 | 0 | 0 | 0 | 0 | 0 | ***1*** | na | na |

## TONOPAH AP *Nye County* ELEVATION 5425 ft LAT/LONG 38° 4 ' N / 117° 7 ' W

| | JAN | FEB | MAR | APR | MAY | JUN | JUL | AUG | SEP | OCT | NOV | DEC | YEAR |
|---|---|---|---|---|---|---|---|---|---|---|---|---|---|
| Maximum Temp °F | 43.8 | 49.4 | 55.3 | 63.2 | 73.5 | 83.9 | 91.1 | 88.7 | 79.9 | 68.1 | 53.1 | 44.1 | 66.2 |
| Minimum Temp °F | 18.9 | 23.7 | 28.3 | 33.5 | 42.2 | 50.8 | 56.6 | 54.9 | 47.2 | 37.0 | 26.1 | 19.0 | 36.5 |
| Mean Temp °F | 31.4 | 36.6 | 41.8 | 48.4 | 57.9 | 67.4 | 73.9 | 71.8 | 63.6 | 52.6 | 39.6 | 31.6 | 51.4 |
| Days Max Temp ≥ 90 °F | 0 | 0 | 0 | 0 | 1 | 9 | 21 | 15 | 3 | 0 | 0 | 0 | 49 |
| Days Max Temp ≤ 32 °F | 3 | 1 | 0 | 0 | 0 | 0 | 0 | 0 | 0 | 0 | 1 | 4 | 9 |
| Days Min Temp ≤ 32 °F | 30 | 26 | 23 | 14 | 3 | 0 | 0 | 0 | 1 | 8 | 24 | 30 | 159 |
| Days Min Temp ≤ 0 °F | 1 | 0 | 0 | 0 | 0 | 0 | 0 | 0 | 0 | 0 | 0 | 1 | 2 |
| Heating Degree Days | 1034 | 796 | 711 | 492 | 233 | 50 | 2 | 8 | 96 | 380 | 755 | 1029 | 5586 |
| Cooling Degree Days | 0 | 0 | 0 | 1 | 21 | 140 | 287 | 232 | 65 | 4 | 0 | 0 | 750 |
| Total Precipitation (") | 0.39 | 0.51 | 0.61 | 0.43 | 0.55 | 0.32 | 0.59 | 0.59 | 0.48 | 0.41 | 0.48 | 0.31 | 5.67 |
| Days ≥ 0.1" Precip | 1 | 2 | 2 | 1 | 2 | 1 | 1 | 2 | 2 | 1 | 1 | 1 | 17 |
| Total Snowfall (") | 3.1 | 3.3 | 3.0 | 1.3 | 0.4 | 0.0 | 0.0 | 0.0 | 0.0 | 0.1 | 1.7 | 2.5 | 15.4 |
| Days ≥ 1" Snow Depth | 4 | ***2*** | ***1*** | ***0*** | 0 | 0 | 0 | 0 | 0 | 0 | 1 | 3 | 11 |

## TUSCARORA *Elko County* ELEVATION 6184 ft LAT/LONG 41° 19 ' N / 116° 14 ' W

| | JAN | FEB | MAR | APR | MAY | JUN | JUL | AUG | SEP | OCT | NOV | DEC | YEAR |
|---|---|---|---|---|---|---|---|---|---|---|---|---|---|
| Maximum Temp °F | 36.5 | 39.8 | 45.5 | 52.9 | 63.5 | 73.8 | 84.0 | 83.1 | 72.5 | ***61.3*** | 44.2 | ***37.0*** | 57.8 |
| Minimum Temp °F | 15.5 | 18.8 | 23.5 | 27.8 | 34.8 | 41.8 | 49.0 | 47.9 | 39.0 | ***31.1*** | 22.7 | ***17.5*** | 30.8 |
| Mean Temp °F | 26.1 | 29.3 | 34.5 | 40.4 | 49.2 | 57.8 | 66.5 | 65.5 | 55.8 | ***46.3*** | 33.5 | ***27.3*** | 44.4 |
| Days Max Temp ≥ 90 °F | 0 | 0 | 0 | 0 | 0 | 1 | 6 | 5 | 0 | 0 | 0 | 0 | 12 |
| Days Max Temp ≤ 32 °F | ***10*** | ***5*** | 2 | 0 | 0 | 0 | 0 | 0 | 0 | 0 | 4 | ***9*** | 30 |
| Days Min Temp ≤ 32 °F | 30 | 27 | 27 | 23 | ***13*** | 3 | 0 | 0 | 7 | ***17*** | ***27*** | ***30*** | 204 |
| Days Min Temp ≤ 0 °F | ***4*** | ***1*** | 0 | 0 | 0 | 0 | 0 | 0 | 0 | 0 | 0 | ***2*** | 7 |
| Heating Degree Days | 1201 | 1002 | 939 | 732 | 485 | 229 | 47 | 64 | 282 | ***575*** | 940 | ***1162*** | 7658 |
| Cooling Degree Days | 0 | ***0*** | ***0*** | ***0*** | ***1*** | ***24*** | ***95*** | ***94*** | ***13*** | ***0*** | ***0*** | ***0*** | 227 |
| Total Precipitation (") | 1.13 | 0.84 | 1.05 | 0.84 | 1.23 | 1.11 | 0.70 | 0.51 | 0.84 | 0.96 | 1.50 | 1.48 | 12.19 |
| Days ≥ 0.1" Precip | ***4*** | 4 | 4 | 3 | 4 | 4 | 2 | 2 | 3 | 3 | 5 | 5 | 43 |
| Total Snowfall (") | na | na | na | ***2.3*** | ***1.9*** | ***0.0*** | 0.0 | 0.0 | 0.2 | ***1.4*** | na | na | na |
| Days ≥ 1" Snow Depth | ***21*** | ***16*** | na | ***1*** | 0 | 0 | 0 | 0 | 0 | ***0*** | ***6*** | ***19*** | na |

## VIRGINIA CITY *Storey County* ELEVATION 6345 ft LAT/LONG 39° 18' N / 119° 39' W

| | JAN | FEB | MAR | APR | MAY | JUN | JUL | AUG | SEP | OCT | NOV | DEC | YEAR |
|---|---|---|---|---|---|---|---|---|---|---|---|---|---|
| Maximum Temp °F | 40.6 | 43.5 | 48.3 | 55.3 | 64.6 | 73.6 | 82.7 | 81.2 | 72.8 | 61.6 | 48.3 | 41.6 | 59.5 |
| Minimum Temp °F | 23.2 | 25.8 | 29.4 | 34.3 | 42.2 | 50.3 | 57.9 | 56.8 | 49.4 | 40.0 | 29.8 | 24.3 | 38.6 |
| Mean Temp °F | 32.0 | 34.7 | 38.9 | 44.8 | 53.4 | 62.0 | 70.3 | 69.0 | 61.1 | 50.8 | 39.1 | 33.0 | 49.1 |
| Days Max Temp ≥ 90 °F | 0 | 0 | 0 | 0 | 0 | 0 | 5 | 3 | 0 | 0 | 0 | 0 | 8 |
| Days Max Temp ≤ 32 °F | 6 | 3 | 1 | 0 | 0 | 0 | 0 | 0 | 0 | 0 | 2 | 5 | 17 |
| Days Min Temp ≤ 32 °F | 26 | 22 | *20* | 13 | 5 | 1 | 0 | 0 | 1 | 7 | 18 | 26 | 139 |
| Days Min Temp ≤ 0 °F | 0 | 0 | 0 | 0 | 0 | 0 | 0 | 0 | 0 | 0 | 0 | 0 | 0 |
| Heating Degree Days | 1018 | 849 | 803 | 598 | 361 | 150 | 25 | 37 | 160 | 436 | 772 | 986 | 6195 |
| Cooling Degree Days | 0 | 0 | 0 | 0 | 13 | 67 | 199 | 174 | 55 | 6 | 0 | 0 | 514 |
| Total Precipitation (") | 2.12 | 1.99 | 1.74 | 0.67 | 0.82 | 0.70 | 0.33 | 0.55 | 0.63 | 0.87 | 1.81 | 1.96 | 14.19 |
| Days ≥ 0.1" Precip | 5 | 5 | 4 | 2 | 2 | 2 | 1 | 1 | 2 | 2 | 4 | 4 | 34 |
| Total Snowfall (") | *11.9* | 14.6 | 8.5 | 2.0 | 1.4 | 0.0 | 0.0 | 0.0 | 0.3 | 0.8 | 7.7 | 13.0 | 60.2 |
| Days ≥ 1" Snow Depth | *16* | 14 | 6 | 1 | 1 | 0 | 0 | 0 | 0 | 1 | 6 | 11 | 56 |

## WELLS *Elko County* ELEVATION 5633 ft LAT/LONG 41° 7' N / 114° 58' W

| | JAN | FEB | MAR | APR | MAY | JUN | JUL | AUG | SEP | OCT | NOV | DEC | YEAR |
|---|---|---|---|---|---|---|---|---|---|---|---|---|---|
| Maximum Temp °F | 35.2 | 40.3 | 47.7 | 56.5 | 66.7 | 77.1 | 86.8 | 84.7 | 74.8 | 61.9 | 45.6 | 35.8 | 59.4 |
| Minimum Temp °F | 10.9 | 16.3 | 22.7 | 27.5 | 34.5 | 41.7 | 47.9 | 46.0 | 36.7 | 27.2 | 20.0 | 11.0 | 28.5 |
| Mean Temp °F | 23.1 | 28.3 | 35.2 | 42.1 | 50.6 | 59.4 | 67.4 | 65.4 | 55.8 | 44.6 | 32.8 | 23.4 | 44.0 |
| Days Max Temp ≥ 90 °F | 0 | 0 | 0 | 0 | 0 | 3 | 12 | 8 | 0 | 0 | 0 | 0 | 23 |
| Days Max Temp ≤ 32 °F | 12 | 5 | 1 | 0 | 0 | 0 | 0 | 0 | 0 | 0 | 3 | 10 | 31 |
| Days Min Temp ≤ 32 °F | 30 | 27 | 28 | 23 | 12 | 3 | 0 | 1 | 9 | 23 | 27 | 30 | 213 |
| Days Min Temp ≤ 0 °F | 7 | 4 | 1 | 0 | 0 | 0 | 0 | 0 | 0 | 0 | 1 | 6 | 19 |
| Heating Degree Days | 1293 | 1029 | 916 | 681 | 439 | 186 | 30 | 58 | 275 | 626 | 958 | 1283 | 7774 |
| Cooling Degree Days | 0 | 0 | 0 | 0 | 1 | 28 | 105 | 80 | 8 | 0 | 0 | 0 | 222 |
| Total Precipitation (") | 0.76 | 0.76 | 0.95 | 0.91 | 1.10 | 0.97 | 0.49 | 0.57 | 0.87 | 0.79 | 1.10 | 0.95 | 10.22 |
| Days ≥ 0.1" Precip | 3 | 3 | 3 | 4 | 3 | 3 | 2 | 2 | 2 | 3 | 4 | 3 | 35 |
| Total Snowfall (") | 8.8 | 7.9 | 7.9 | 4.4 | 1.8 | 0.1 | 0.0 | 0.0 | 0.1 | 1.3 | 6.5 | 11.2 | 50.0 |
| Days ≥ 1" Snow Depth | 21 | 14 | 6 | 1 | 0 | 0 | 0 | 0 | 0 | 1 | 6 | 19 | 68 |

## WINNEMUCCA MUNI AP *Humboldt County* ELEVATION 4301 ft LAT/LONG 40° 54' N / 117° 48' W

| | JAN | FEB | MAR | APR | MAY | JUN | JUL | AUG | SEP | OCT | NOV | DEC | YEAR |
|---|---|---|---|---|---|---|---|---|---|---|---|---|---|
| Maximum Temp °F | 41.7 | 48.9 | 55.3 | 62.6 | 73.0 | 83.1 | 92.7 | 90.7 | 80.6 | 67.8 | 51.7 | 42.1 | 65.9 |
| Minimum Temp °F | 17.6 | 22.8 | 25.9 | 30.0 | 38.0 | 45.7 | 51.6 | 48.9 | 39.2 | 29.2 | 23.0 | 16.8 | 32.4 |
| Mean Temp °F | 29.7 | 35.9 | 40.6 | 46.3 | 55.5 | 64.4 | 72.2 | 69.8 | 59.9 | 48.5 | 37.4 | 29.5 | 49.1 |
| Days Max Temp ≥ 90 °F | 0 | 0 | 0 | 0 | 2 | 10 | 22 | 20 | 6 | 0 | 0 | 0 | 60 |
| Days Max Temp ≤ 32 °F | 6 | 1 | 0 | 0 | 0 | 0 | 0 | 0 | 0 | 0 | 1 | 5 | 13 |
| Days Min Temp ≤ 32 °F | 28 | 24 | 25 | 19 | 7 | 1 | 0 | 0 | 6 | 20 | 25 | 29 | 184 |
| Days Min Temp ≤ 0 °F | 3 | 1 | 0 | 0 | 0 | 0 | 0 | 0 | 0 | 0 | 0 | 3 | 7 |
| Heating Degree Days | 1088 | 815 | 748 | 553 | 298 | 99 | 11 | 25 | 173 | 503 | 821 | 1095 | 6229 |
| Cooling Degree Days | 0 | 0 | 0 | 1 | 14 | 87 | 228 | 183 | 32 | 0 | 0 | 0 | 545 |
| Total Precipitation (") | 0.74 | 0.56 | 0.80 | 0.82 | 0.83 | 0.74 | 0.27 | 0.38 | 0.40 | 0.61 | 0.93 | 0.83 | 7.91 |
| Days ≥ 0.1" Precip | 2 | 2 | 3 | 3 | 3 | 2 | 1 | 1 | 1 | 2 | 3 | 3 | 26 |
| Total Snowfall (") | 4.7 | 3.9 | 3.7 | 2.3 | 0.7 | 0.0 | 0.0 | 0.0 | 0.0 | 0.6 | 2.5 | 5.4 | 23.8 |
| Days ≥ 1" Snow Depth | 10 | 4 | 1 | 0 | 0 | 0 | 0 | 0 | 0 | 0 | 2 | 7 | 24 |

## YERINGTON *Lyon County* ELEVATION 4383 ft LAT/LONG 38° 59' N / 119° 10' W

| | JAN | FEB | MAR | APR | MAY | JUN | JUL | AUG | SEP | OCT | NOV | DEC | YEAR |
|---|---|---|---|---|---|---|---|---|---|---|---|---|---|
| Maximum Temp °F | 46.2 | 53.3 | 59.6 | 66.5 | 75.6 | 84.6 | 92.2 | 90.6 | 82.4 | 70.7 | 55.8 | 46.5 | 68.7 |
| Minimum Temp °F | 19.4 | 23.4 | 28.9 | 33.9 | 41.7 | 48.9 | 54.2 | 52.2 | 44.4 | 34.8 | 25.8 | 18.7 | 35.5 |
| Mean Temp °F | 32.8 | 38.3 | 44.3 | 50.2 | 58.7 | 66.8 | 73.2 | 71.4 | 63.4 | 52.8 | 40.8 | 32.6 | 52.1 |
| Days Max Temp ≥ 90 °F | 0 | 0 | 0 | 0 | 1 | 9 | 22 | 20 | 6 | 0 | 0 | 0 | 58 |
| Days Max Temp ≤ 32 °F | 3 | 0 | 0 | 0 | 0 | 0 | 0 | 0 | 0 | 0 | 0 | 2 | 5 |
| Days Min Temp ≤ 32 °F | 29 | 25 | 21 | 13 | 3 | 0 | 0 | 0 | 2 | 11 | 24 | 29 | 157 |
| Days Min Temp ≤ 0 °F | 1 | 0 | 0 | 0 | 0 | 0 | 0 | 0 | 0 | 0 | 0 | 1 | 2 |
| Heating Degree Days | 991 | 747 | 637 | 437 | 212 | 54 | 4 | 12 | 100 | 373 | 716 | 997 | 5280 |
| Cooling Degree Days | 0 | 0 | 0 | 1 | 29 | 119 | 262 | 229 | 69 | 2 | 0 | 0 | 711 |
| Total Precipitation (") | 0.54 | 0.50 | 0.57 | 0.43 | 0.59 | 0.51 | 0.35 | 0.39 | 0.27 | 0.42 | 0.47 | 0.45 | 5.49 |
| Days ≥ 0.1" Precip | 2 | 1 | 2 | 1 | 2 | 1 | 1 | 1 | 1 | 1 | 2 | 2 | 17 |
| Total Snowfall (") | *1.9* | *1.6* | *1.3* | 0.3 | 0.1 | 0.0 | 0.0 | 0.0 | 0.1 | 0.2 | *0.4* | na | na |
| Days ≥ 1" Snow Depth | na | na | *0* | 0 | 0 | 0 | 0 | 0 | 0 | 0 | *1* | na | na |

## JANUARY MINIMUM TEMPERATURE °F

### LOWEST

| | | |
|---|---|---|
| 1 | Gibbs | 7.0 |
| 2 | Deeth | 8.0 |
| 3 | Montello | 9.5 |
| 4 | Ely | 9.7 |
| 5 | Mountain City | 9.9 |
| 6 | Wells | 10.9 |
| 7 | Clover Valley | 12.8 |
| 8 | Beowawe | 13.3 |
| | Elko | 13.3 |
| 10 | Arthur | 13.9 |
| 11 | Lund | 14.0 |
| 12 | Ruby Lake | 14.4 |
| 13 | Sunnyside | 14.8 |
| 14 | Contact | 15.1 |
| | Smokey Valley | 15.1 |
| 16 | McGill | 15.5 |
| | Tuscarora | 15.5 |
| 18 | Lovelock-Derby | 16.5 |
| 19 | Battle Mountain | 16.6 |
| | Dyer | 16.6 |
| 21 | Paradise Valley | 16.7 |
| 22 | Minden | 16.8 |
| 23 | Eureka | 16.9 |
| 24 | Metropolis | 17.3 |
| | Rye Patch Dam | 17.3 |

### HIGHEST

| | | |
|---|---|---|
| 1 | Boulder City | 38.5 |
| 2 | Searchlight | 35.7 |
| 3 | Las Vegas | 34.5 |
| 4 | Desert | 29.1 |
| 5 | Pahrump | 26.9 |
| 6 | Pahranagat | 26.8 |
| 7 | Glenbrook | 23.8 |
| 8 | Virginia City | 23.2 |
| 9 | Lahontan Dam | 22.7 |
| 10 | Mina | 22.0 |
| | Pioche | 22.0 |
| 12 | Goldfield | 21.8 |
| 13 | Reno | 21.0 |
| 14 | Carson City | 20.4 |
| 15 | Denio | 20.1 |
| 16 | Leonard Creek | 20.0 |
| 17 | Austin | 19.9 |
| 18 | Orovada | 19.7 |
| 19 | Golconda | 19.4 |
| | Yerington | 19.4 |
| 21 | Imlay | 19.1 |
| 22 | Fallon | 19.0 |
| 23 | Tonopah | 18.9 |
| 24 | Lovelock | 18.7 |
| 25 | Caliente | 18.5 |

## JULY MAXIMUM TEMPERATURE °F

### HIGHEST

| | | |
|---|---|---|
| 1 | Las Vegas | 104.1 |
| 2 | Boulder City | 101.7 |
| | Desert | 101.7 |
| 4 | Pahrump | 100.6 |
| 5 | Pahranagat | 97.8 |
| 6 | Searchlight | 97.7 |
| 7 | Caliente | 95.3 |
| 8 | Mina | 94.5 |
| 9 | Lovelock-Derby | 94.4 |
| 10 | Rye Patch Dam | 94.2 |
| 11 | Lahontan Dam | 93.5 |
| 12 | Battle Mountain | 93.1 |
| | Dyer | 93.1 |
| 14 | Golconda | 92.7 |
| | Winnemucca | 92.7 |
| 16 | Lovelock | 92.6 |
| 17 | Montello | 92.4 |
| 18 | Imlay | 92.3 |
| 19 | Yerington | 92.2 |
| 20 | Leonard Creek | 92.1 |
| 21 | Beowawe | 91.4 |
| 22 | Denio | 91.3 |
| | Reno | 91.3 |
| 24 | Smokey Valley | 91.2 |
| 25 | Sunnyside | 91.1 |

### LOWEST

| | | |
|---|---|---|
| 1 | Glenbrook | 79.1 |
| 2 | Arthur | 82.4 |
| 3 | Virginia City | 82.7 |
| 4 | Tuscarora | 84.0 |
| 5 | Gibbs | 84.1 |
| 6 | Mountain City | 84.4 |
| 7 | Clover Valley | 85.2 |
| 8 | Eureka | 85.4 |
| 9 | McGill | 85.6 |
| 10 | Wells | 86.8 |
| 11 | Ely | 87.0 |
| | Metropolis | 87.0 |
| | Ruby Lake | 87.0 |
| 14 | Austin | 87.1 |
| 15 | Pioche | 87.2 |
| 16 | Deeth | 87.7 |
| 17 | Goldfield | 88.3 |
| 18 | Lund | 88.4 |
| 19 | Contact | 88.8 |
| 20 | Emigrant Pass | 89.1 |
| 21 | Carson City | 89.2 |
| 22 | Minden | 89.8 |
| 23 | Paradise Valley | 90.1 |
| 24 | Elko | 90.4 |
| 25 | Orovada | 90.7 |

## ANNUAL PRECIPITATION (")

### HIGHEST

| | | |
|---|---|---|
| 1 | Glenbrook | 17.27 |
| 2 | Arthur | 14.85 |
| 3 | Austin | 14.29 |
| 4 | Virginia City | 14.19 |
| 5 | Pioche | 14.01 |
| 6 | Metropolis | 13.79 |
| 7 | Clover Valley | 13.56 |
| 8 | Ruby Lake | 13.08 |
| 9 | Mountain City | 12.87 |
| 10 | Eureka | 12.85 |
| 11 | Emigrant Pass | 12.31 |
| 12 | Tuscarora | 12.19 |
| 13 | Deeth | 11.93 |
| 14 | Lund | 10.86 |
| 15 | Gibbs | 10.45 |
| 16 | Orovada | 10.28 |
| 17 | Wells | 10.22 |
| 18 | Contact | 10.17 |
| 19 | Ely | 10.14 |
| 20 | Carson City | 10.10 |
| 21 | Paradise Valley | 10.03 |
| 22 | Caliente | 9.92 |
| 23 | Sunnyside | 9.61 |
| 24 | Elko | 9.55 |
| 25 | Denio | 9.46 |

### LOWEST

| | | |
|---|---|---|
| 1 | Las Vegas | 4.53 |
| 2 | Desert | 4.75 |
| 3 | Pahrump | 5.16 |
| 4 | Dyer | 5.18 |
| 5 | Fallon | 5.20 |
| 6 | Lovelock-Derby | 5.22 |
| 7 | Lahontan Dam | 5.35 |
| 8 | Lovelock | 5.46 |
| 9 | Yerington | 5.49 |
| 10 | Tonopah | 5.67 |
| 11 | Mina | 5.72 |
| 12 | Boulder City | 6.44 |
| 13 | Pahranagat | 6.73 |
| 14 | Goldfield | 6.85 |
| 15 | Reno | 7.15 |
| 16 | Smokey Valley | 7.35 |
| 17 | Golconda | 7.37 |
| 18 | Imlay | 7.72 |
| 19 | Minden | 7.80 |
| 20 | Winnemucca | 7.91 |
| 21 | Montello | 7.93 |
| 22 | Battle Mountain | 8.22 |
| 23 | Rye Patch Dam | 8.33 |
| 24 | Searchlight | 8.34 |
| 25 | Beowawe | 8.67 |

## ANNUAL SNOWFALL (")

### HIGHEST

| | | |
|---|---|---|
| 1 | Austin | 82.2 |
| 2 | Glenbrook | 76.3 |
| 3 | Virginia City | 60.2 |
| 4 | Arthur | 54.2 |
| 5 | Ely | 51.1 |
| 6 | Wells | 50.0 |
| 7 | Ruby Lake | 46.3 |
| 8 | Elko | 38.0 |
| 9 | Paradise Valley | 27.6 |
| 10 | Reno | 24.4 |
| 11 | Winnemucca | 23.8 |
| 12 | Battle Mountain | 22.5 |
| 13 | Carson City | 19.9 |
| 14 | Orovada | 18.4 |
| 15 | Goldfield | 17.4 |
| 16 | Tonopah | 15.4 |
| 17 | Mina | 12.2 |
| 18 | Dyer | 11.2 |
| 19 | Lovelock-Derby | 6.9 |
| 20 | Pahranagat | 1.6 |
| 21 | Las Vegas | 1.0 |
| 22 | Searchlight | 0.9 |
| 23 | Desert | 0.7 |
| 24 | Pahrump | 0.1 |
| 25 | Boulder City | 0.0 |

### LOWEST

| | | |
|---|---|---|
| 1 | Boulder City | 0.0 |
| 2 | Pahrump | 0.1 |
| 3 | Desert | 0.7 |
| 4 | Searchlight | 0.9 |
| 5 | Las Vegas | 1.0 |
| 6 | Pahranagat | 1.6 |
| 7 | Lovelock-Derby | 6.9 |
| 8 | Dyer | 11.2 |
| 9 | Mina | 12.2 |
| 10 | Tonopah | 15.4 |
| 11 | Goldfield | 17.4 |
| 12 | Orovada | 18.4 |
| 13 | Carson City | 19.9 |
| 14 | Battle Mountain | 22.5 |
| 15 | Winnemucca | 23.8 |
| 16 | Reno | 24.4 |
| 17 | Paradise Valley | 27.6 |
| 18 | Elko | 38.0 |
| 19 | Ruby Lake | 46.3 |
| 20 | Wells | 50.0 |
| 21 | Ely | 51.1 |
| 22 | Arthur | 54.2 |
| 23 | Virginia City | 60.2 |
| 24 | Glenbrook | 76.3 |
| 25 | Austin | 82.2 |

# NEW HAMPSHIRE

PHYSICAL FEATURES. New Hampshire occupies 9,304 square miles. From below the 43d parallel of latitude it extends nearly 200 miles northward to beyond the 45th parallel. At its southern border, New Hampshire extends westward from the Atlantic coastline for nearly 100 miles. It narrows to less than 20 miles in width at its northern tip. The eastern border lies near 71° W. longitude. Its western border is the Connecticut River, except in the extreme north.

The terrain is hilly to mountainous. Elevations of less than 500 feet above sea level are found only in the coastal area of the southeast, the Merrimac River Valley, and the central and southern portions of the Connecticut River Valley. Elsewhere the general elevation is from 500 to 1,500 feet, excepting up to near 2,500 feet in the extreme north. Numerous hills and mountains extend to heights of 2,000 to 4,000 feet above sea level over most of the State except in the southeast. Many White Mountain peaks rise above 4,000 feet; Mt. Washington reaches 6,288 feet above sea level. This is the highest mountain in the northeastern United States.

The glacier of the great Ice Age accounts for much of the topography, including many of the 1,300 lakes and ponds. The largest is Lake Winnepesaukee which covers an area of 71 square miles in the central part of the State. Inland waters cover about 280 square miles. The two principal rivers in the State are the Connecticut and the Merrimack Rivers, both of which flow in a southerly direction.

GENERAL CLIMATE. Characteristics of New Hampshire climate are: (1) changeableness of the weather, (2) large range of temperature, both daily and annual, (3) great differences between the same seasons in different years, (4) equable distribution of precipitation, and (5) considerable diversity from place to place. The regional climatic influences are modified in New Hampshire by varying distances from the ocean, elevations, and types of terrain. The State has been divided into two climatological divisions (Northern and Southern) which take into account the main features of these modifying factors.

New Hampshire lies in the "prevailing westerlies", the belt of generally eastward air movement which encircles the globe in middle latitudes. Embedded in this circulation are extensive masses of air originating in higher or lower latitudes and interacting to produce low-pressure storm systems. Relative to most other sections of the country, a large number of such storms pass over or near New Hampshire. The majority of air masses affecting this State belong to three types: (1) cold, dry air pouring down from subarctic North America, (2) warm, moist air streaming up on a long overland journey from the Gulf of Mexico and eastward, and (3) cool, damp air moving in from the North Atlantic. Because the atmospheric flow is usually offshore, New Hampshire is more influenced by the first two types than it is by the third.

The procession of contrasting air masses and the relatively frequent passage of storms bring about approximately twice-weekly alternation from fair to cloudy or stormy conditions, often attended by abrupt changes in temperature, moisture, sunshine, wind direction and speed. There is no regular or persistent rhythm to this sequence, and it is interrupted by periods during which the weather patterns continue the same for several days, infrequently for several weeks. New Hampshire weather, however, is cited for variety rather than monotony.

The Northern Division is the area least affected by the ocean influences and most affected by higher elevations as well as by its more northerly latitude. In the Southern Division, lower elevation and latitude tend to cause higher temperatures, though this is modified seasonally by ocean influences.

TEMPERATURE. The annual temperature averages near 41° F. in the Northern Division and near 46° F. in the Southern. Summer temperatures are comfortable for the most part. They are reasonably uniform over the State, excepting topographical extremes. Hot days with maxima of 90° F. or higher average from only a few per year in the extreme north to 5 to 15 per year over most of the rest of the State. Average temperatures vary from place to place more in the winter than in summer. Days with subzero readings are relatively few along the immediate coast but are common inland. They average from 25 to 50 in number per year in most of the Northern Division and from 10 to 25 in the Southern Division. The average date of the last freezing temperature in spring ranges from early in June at the colder locations to late in April at a few southern stations. For most of the State the growing season begins in May and usually ends in the latter part of September.

PRECIPITATION. New Hampshire is fortunate in having its precipitation rather evenly distributed through the year. Low pressure, or frontal, storm systems are the principal year-round moisture producers. This activity ebbs somewhat in summer, but thunderstorms are of increased activity at this time, tending to make up the difference. Though brief and often of small extent, the thunderstorms produce the heaviest local rainfall intensities. Rains of 1 to 2 inches in 1 hour can be expected at least once in a 10-year period. Prolonged droughts are infrequent; shorter dry spells in summer are fairly common. Widespread floods are infrequent. Floods occur most often in the spring when they are caused by a combination of rain and melting snow.

Total annual precipitation averages near 44 inches in the Northern Division and 41 inches in the Southern. The distribution is quite uniform over the Southern Division. The mountainous character of much of central and northern New Hampshire, and the generally higher elevations there, account for the greater annual totals and variability from place to place. Considerable rain or wet snow falls along the coast in winter, while farther inland snow is more generally the rule. Occasionally freezing rain occurs, coating exposed surfaces with troublesome ice. This problem is less frequent in northern New Hampshire. Most areas can expect at least one occurrence of glaze in the season. Measurable amounts of precipitation fall on an average of 1 day in 3. Frequency is higher at higher elevations and in extreme northern New Hampshire, up to 140 to 150 days per year.

SNOWFALL. Average annual amounts of snowfall in the Southern Division increase from around 50 inches near the coast to 60 to 80 inches inland. Totals vary greatly in the Northern Division. Along the Connecticut River in the southern portion, totals average near 60 inches but increase to over 100 inches at the higher elevations of the northern and western portions. The summit of Mt. Washington receives nearly 185 inches. Bethlehem, only about 20 miles to the west, receives only about 70 inches per year. The number of days with 1 inch or more of snowfall varies from near 20 per season over much of the Southern Division up to 30 to 40 in the Northern Division and even to 50 or more at the highest elevations.

Snow cover is continuous through the whole winter season as a rule. Most frequent exceptions are found along the immediate coast and sometimes in extreme southern New Hampshire. Snow cover reaches its maximum depth, on the average, during the latter half of February in the Southern Division. In the Northern Division, the greatest depth comes in early March. Water stored in the snow makes an important contribution to a continuous water supply. The spring melting is usually too gradual to produce serious flooding.

OTHER CLIMATIC ELEMENTS. Sunshine averages over 50 percent of the possible amount in the Southern Division. and the lower elevations of the Northern Division. Higher elevations and peaks are cloudier, especially in winter, reducing the percentage to less than 50 percent generally. Mt. Washington reports an average of only 33 percent. Persistent fogs are sometimes experienced along the coast and on the higher elevations inland. Duration of fogs diminishes inland over flat and valley locations. But the shorter duration heavy ground fogs of early morning occur frequently at susceptible places in these areas. The number of days with fog probably varies from about 20 to 90 per year over the State.

The prevailing wind, on a yearly basis, comes from a westerly direction. It is predominantly from the northwest in winter and from the southwest in summer. Along the coast in spring and summer the sea breeze is important. These onshore winds, from the cool ocean, may come inland for 10 miles or so.

Coastal storms or "northeasters" can be a serious weather hazard in southeastern New Hampshire, decreasing in importance northward. They generate very strong winds and heavy rain or snow. They can produce abnormally high wind-driven tides. Occasionally in summer or fall storms of tropical origin affect New Hampshire. These may be similar (except for snow) to the northeasters. Only a very few retain near or full hurricane force. Tornadoes are not common phenomena, yet many years may have one or more. Most tornadoes are small, affecting a very localized area. About 80 percent of tornadoes occur between May 15 and September 15. Thunder and hailstorms have a similar frequency maximum from mid-spring to early fall. Thunderstorms occur on 15 to 30 days per year. The most severe are attended by hail.

# COUNTY INDEX

### *Cheshire County*
KEENE
SURRY MOUNTAIN LAKE

### *Coos County*
BERLIN
COLEBROOK
FIRST CONN LAKE
LANCASTER
MOUNT WASHINGTON
NORTH STRATFORD
PINKHAM NOTCH

### *Grafton County*
BENTON 5 SW
GRAFTON
HANOVER
MONROE 5 NNE
PLYMOUTH

### *Hillsborough County*
MASSABESIC LAKE
NASHUA 2 NNW
PETERBORO 2 S

### *Merrimack County*
CONCORD MUNI AP
MOUNT SUNAPEE

### *Rockingham County*
EPPING

### *Strafford County*
DURHAM

# ELEVATION INDEX

| FEET | STATION NAME |
|---|---|
| 69 | DURHAM |
| 141 | NASHUA 2 NNW |
| 200 | EPPING |
| 249 | MASSABESIC LAKE |
| 354 | CONCORD MUNI AP |
| 489 | KEENE |
| 541 | SURRY MOUNTAIN LAKE |
| 600 | HANOVER |
| 660 | PLYMOUTH |
| 679 | MONROE 5 NNE |
| 840 | GRAFTON |
| 879 | LANCASTER |
| 902 | NORTH STRATFORD |
| 1001 | PETERBORO 2 S |
| 1112 | BERLIN |
| 1122 | COLEBROOK |
| 1201 | BENTON 5 SW |
| 1312 | MOUNT SUNAPEE |
| 1660 | FIRST CONN LAKE |
| 2001 | PINKHAM NOTCH |
| 6270 | MOUNT WASHINGTON |

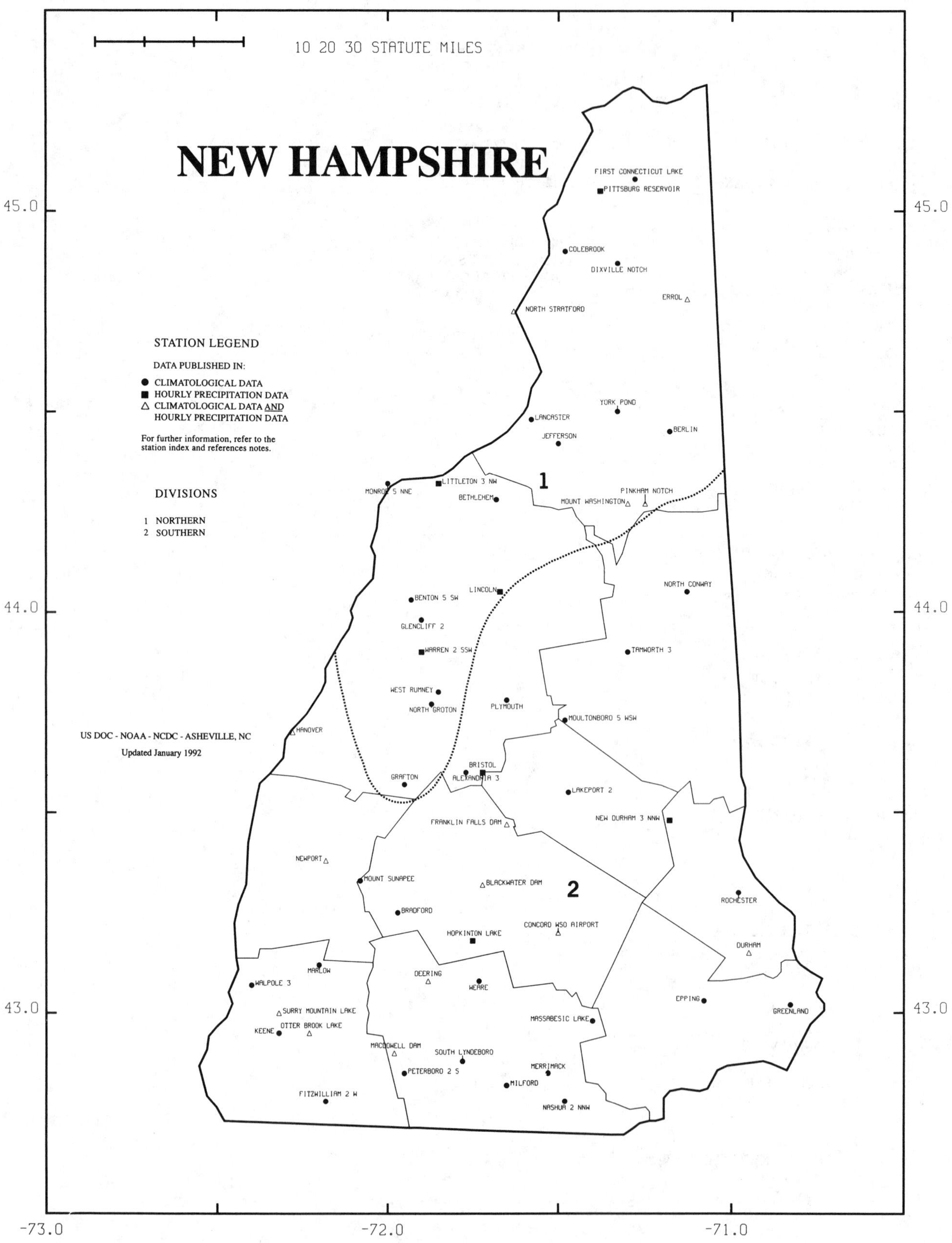
10 20 30 STATUTE MILES
NEW HAMPSHIRE
STATION LEGEND
DATA PUBLISHED IN:
CLIMATOLOGICAL DATA
HOURLY PRECIPITATION DATA
CLIMATOLOGICAL DATA AND HOURLY PRECIPITATION DATA
For further information, refer to the station index and references notes.
DIVISIONS
1 NORTHERN
2 SOUTHERN
US DOC - NOAA - NCDC - ASHEVILLE, NC
Updated January 1992
FIRST CONNECTICUT LAKE
PITTSBURG RESERVOIR
COLEBROOK
DIXVILLE NOTCH
ERROL
NORTH STRATFORD
YORK POND
LANCASTER
BERLIN
JEFFERSON
1
LITTLETON 3 NW
MONROE 5 NNE
BETHLEHEM
PINKHAM NOTCH
MOUNT WASHINGTON
LINCOLN
NORTH CONWAY
BENTON 5 SW
GLENCLIFF 2
WARREN 2 SSW
TAMWORTH 3
WEST RUMNEY
NORTH GROTON
PLYMOUTH
MOULTONBORO 5 WSW
HANOVER
BRISTOL
GRAFTON
ALEXANDRIA 3
LAKEPORT 2
FRANKLIN FALLS DAM
NEW DURHAM 3 NNW
NEWPORT
MOUNT SUNAPEE
BLACKWATER DAM
2
ROCHESTER
BRADFORD
CONCORD WSO AIRPORT
HOPKINTON LAKE
DURHAM
MARLOW
DEERING
WEARE
WALPOLE 3
EPPING
GREENLAND
SURRY MOUNTAIN LAKE
OTTER BROOK LAKE
KEENE
MASSABESIC LAKE
MACDOWELL DAM
SOUTH LYNDEBORO
MERRIMACK
PETERBORO 2 S
MILFORD
FITZWILLIAM 2 W
NASHUA 2 NNW
45.0
44.0
43.0
-73.0
-72.0
-71.0

## BENTON 5 SW *Grafton County* ELEVATION 1201 ft LAT/LONG 44° 2 ' N / 71° 56 ' W

| | JAN | FEB | MAR | APR | MAY | JUN | JUL | AUG | SEP | OCT | NOV | DEC | YEAR |
|---|---|---|---|---|---|---|---|---|---|---|---|---|---|
| Maximum Temp °F | 26.7 | 29.5 | 39.0 | 52.4 | 65.6 | 73.7 | 78.0 | 75.7 | 66.9 | 56.0 | 43.1 | 31.2 | 53.2 |
| Minimum Temp °F | 7.0 | 8.8 | 19.4 | 31.1 | 41.7 | 50.3 | 54.9 | 53.3 | 45.5 | 35.9 | 26.9 | 13.7 | 32.4 |
| Mean Temp °F | 16.9 | 19.2 | 29.3 | 41.8 | 53.7 | 62.0 | 66.5 | 64.5 | 56.2 | 46.0 | 35.0 | 22.5 | 42.8 |
| Days Max Temp ≥ 90 °F | 0 | 0 | 0 | 0 | 0 | 0 | 1 | 0 | 0 | 0 | 0 | 0 | 1 |
| Days Max Temp ≤ 32 °F | 21 | 18 | 8 | 0 | 0 | 0 | 0 | 0 | 0 | 0 | 5 | 17 | 69 |
| Days Min Temp ≤ 32 °F | 31 | 27 | 28 | 18 | 4 | 0 | 0 | 0 | 2 | 12 | 22 | 30 | 174 |
| Days Min Temp ≤ 0 °F | 11 | 9 | 2 | 0 | 0 | 0 | 0 | 0 | 0 | 0 | 0 | 6 | 28 |
| Heating Degree Days | 1486 | 1289 | 1101 | 691 | 352 | 129 | 46 | 80 | 271 | 583 | 893 | 1312 | 8233 |
| Cooling Degree Days | 0 | 0 | 0 | 1 | 9 | 44 | 101 | 74 | 15 | 0 | 0 | 0 | 244 |
| Total Precipitation (") | 2.18 | 2.02 | 2.42 | 2.69 | 3.37 | 3.77 | 3.65 | 4.22 | 3.53 | 3.26 | 3.28 | 2.72 | 37.11 |
| Days ≥ 0.1" Precip | 5 | 5 | 6 | 7 | 8 | 8 | 8 | 8 | 7 | 7 | 7 | 7 | 83 |
| Total Snowfall (") | 16.8 | 15.2 | 13.2 | 5.6 | 0.3 | 0.0 | 0.0 | 0.0 | 0.0 | 0.5 | 5.9 | 16.7 | 74.2 |
| Days ≥ 1" Snow Depth | 29 | 26 | 21 | 5 | 0 | 0 | 0 | 0 | 0 | 0 | 7 | 22 | 110 |

## BERLIN *Coos County* ELEVATION 1112 ft LAT/LONG 44° 29 ' N / 71° 10 ' W

| | JAN | FEB | MAR | APR | MAY | JUN | JUL | AUG | SEP | OCT | NOV | DEC | YEAR |
|---|---|---|---|---|---|---|---|---|---|---|---|---|---|
| Maximum Temp °F | 25.9 | 28.9 | 38.7 | 51.5 | 65.7 | ***74.1*** | 79.0 | ***76.9*** | 68.2 | 56.3 | 42.9 | 30.6 | 53.2 |
| Minimum Temp °F | 2.9 | 4.8 | 16.2 | 29.7 | 40.3 | ***49.9*** | 54.3 | ***52.4*** | 43.4 | 33.8 | 25.1 | 10.7 | 30.3 |
| Mean Temp °F | 14.4 | 16.9 | 27.5 | 40.7 | 53.0 | ***62.0*** | 66.7 | ***64.6*** | 55.8 | 45.0 | 34.0 | 20.7 | 41.8 |
| Days Max Temp ≥ 90 °F | 0 | 0 | 0 | 0 | 0 | 1 | 2 | 1 | 0 | 0 | 0 | 0 | 4 |
| Days Max Temp ≤ 32 °F | 22 | 17 | 9 | 1 | 0 | 0 | 0 | 0 | 0 | 0 | 5 | 17 | 71 |
| Days Min Temp ≤ 32 °F | 31 | 28 | 28 | 19 | 6 | 0 | 0 | 0 | 3 | 14 | 24 | 30 | 183 |
| Days Min Temp ≤ 0 °F | 14 | 12 | 4 | 0 | 0 | 0 | 0 | 0 | 0 | 0 | 0 | 7 | 37 |
| Heating Degree Days | 1563 | 1352 | 1156 | 724 | 375 | ***135*** | 50 | ***85*** | 281 | 612 | 922 | 1368 | 8623 |
| Cooling Degree Days | 0 | 0 | 0 | 0 | 9 | 49 | 114 | 87 | 14 | 1 | 0 | 0 | 274 |
| Total Precipitation (") | 2.44 | 2.15 | 2.62 | 3.04 | 3.25 | 3.89 | 3.28 | 4.27 | 3.29 | 3.55 | 3.57 | 3.08 | 38.43 |
| Days ≥ 0.1" Precip | 5 | 5 | 7 | ***7*** | ***8*** | ***9*** | 8 | ***8*** | 6 | 7 | 7 | 7 | 84 |
| Total Snowfall (") | 20.0 | ***16.8*** | 14.0 | 5.5 | 0.2 | 0.0 | 0.0 | 0.0 | 0.0 | 0.1 | 6.8 | 19.6 | 83.0 |
| Days ≥ 1" Snow Depth | ***30*** | ***27*** | 24 | 7 | 0 | 0 | 0 | 0 | 0 | 0 | ***6*** | 22 | 116 |

## COLEBROOK *Coos County* ELEVATION 1122 ft LAT/LONG 44° 54 ' N / 71° 28 ' W

| | JAN | FEB | MAR | APR | MAY | JUN | JUL | AUG | SEP | OCT | NOV | DEC | YEAR |
|---|---|---|---|---|---|---|---|---|---|---|---|---|---|
| Maximum Temp °F | 24.1 | 27.9 | 37.9 | 51.4 | 65.8 | 72.7 | 77.3 | 75.0 | 66.6 | 55.1 | 41.6 | 28.7 | 52.0 |
| Minimum Temp °F | 0.2 | 2.3 | 14.3 | 27.0 | 37.7 | 47.2 | 51.7 | 50.4 | 42.9 | 33.1 | 23.9 | 9.0 | 28.3 |
| Mean Temp °F | 12.2 | 15.1 | 26.2 | 39.2 | 51.8 | 60.0 | 64.5 | 62.8 | 54.8 | 44.1 | 32.9 | 18.9 | 40.2 |
| Days Max Temp ≥ 90 °F | 0 | 0 | 0 | 0 | 0 | 0 | 0 | 0 | 0 | 0 | 0 | 0 | 0 |
| Days Max Temp ≤ 32 °F | 23 | 18 | 10 | 1 | 0 | 0 | 0 | 0 | 0 | 0 | 6 | 19 | 77 |
| Days Min Temp ≤ 32 °F | 31 | 28 | 29 | 22 | 10 | 2 | 0 | 1 | 5 | 15 | 24 | 30 | 197 |
| Days Min Temp ≤ 0 °F | 16 | 13 | 6 | 0 | 0 | 0 | 0 | 0 | 0 | 0 | 1 | 9 | 45 |
| Heating Degree Days | 1631 | 1404 | 1197 | 767 | 409 | 176 | 81 | 118 | 311 | 640 | 958 | ***1430*** | 9122 |
| Cooling Degree Days | 0 | 0 | 0 | 0 | 8 | 30 | 82 | 58 | 10 | 0 | 0 | 0 | 188 |
| Total Precipitation (") | 2.56 | 2.03 | 2.59 | 2.62 | 3.50 | 4.35 | 4.04 | 4.72 | 3.71 | 3.34 | 3.48 | 3.14 | 40.08 |
| Days ≥ 0.1" Precip | 6 | 6 | 6 | 7 | 8 | ***9*** | 8 | 9 | 7 | 7 | 8 | 8 | 89 |
| Total Snowfall (") | ***21.7*** | ***18.6*** | 14.1 | ***5.2*** | 0.0 | 0.0 | 0.0 | 0.0 | 0.0 | 0.6 | 10.0 | ***23.9*** | 94.1 |
| Days ≥ 1" Snow Depth | 30 | 28 | ***26*** | ***8*** | 0 | 0 | 0 | 0 | 0 | 0 | 10 | 27 | 129 |

## CONCORD MUNI AP *Merrimack County* ELEVATION 354 ft LAT/LONG 43° 12 ' N / 71° 31 ' W

| | JAN | FEB | MAR | APR | MAY | JUN | JUL | AUG | SEP | OCT | NOV | DEC | YEAR |
|---|---|---|---|---|---|---|---|---|---|---|---|---|---|
| Maximum Temp °F | 30.0 | 33.1 | 42.7 | 56.7 | 69.1 | 77.4 | 82.5 | 80.2 | 71.5 | 60.2 | 47.2 | 34.8 | 57.1 |
| Minimum Temp °F | 7.9 | 10.9 | 22.0 | 31.7 | 41.8 | 51.4 | 56.7 | 55.1 | 46.2 | 34.9 | 27.1 | 15.1 | 33.4 |
| Mean Temp °F | 19.0 | 22.0 | 32.4 | 44.2 | 55.5 | 64.4 | 69.6 | 67.7 | 58.9 | 47.6 | 37.2 | 25.0 | 45.3 |
| Days Max Temp ≥ 90 °F | 0 | 0 | 0 | 0 | 1 | 2 | 5 | 3 | 0 | 0 | 0 | 0 | 11 |
| Days Max Temp ≤ 32 °F | 18 | 14 | 4 | 0 | 0 | 0 | 0 | 0 | 0 | 0 | 2 | 13 | 51 |
| Days Min Temp ≤ 32 °F | 30 | 27 | 26 | 17 | 5 | 0 | 0 | 0 | 2 | 14 | 22 | 29 | 172 |
| Days Min Temp ≤ 0 °F | 10 | 7 | 1 | 0 | 0 | 0 | 0 | 0 | 0 | 0 | 0 | 5 | 23 |
| Heating Degree Days | 1421 | 1207 | 1003 | 619 | 308 | 92 | 23 | 46 | 210 | 535 | 827 | 1234 | 7525 |
| Cooling Degree Days | 0 | 0 | 0 | 1 | 20 | 75 | 180 | 138 | 35 | 1 | 0 | 0 | 450 |
| Total Precipitation (") | 2.49 | 2.39 | 2.90 | 2.97 | 3.19 | 3.20 | 3.20 | 3.40 | 3.07 | 3.10 | 3.60 | 3.25 | 36.76 |
| Days ≥ 0.1" Precip | 6 | 5 | 6 | 6 | 7 | 7 | 6 | 6 | 6 | 6 | 7 | 6 | 74 |
| Total Snowfall (") | 18.2 | 14.9 | 10.8 | 2.8 | 0.0 | 0.0 | 0.0 | 0.0 | 0.0 | 0.1 | 4.4 | 13.8 | 65.0 |
| Days ≥ 1" Snow Depth | 26 | 23 | 17 | 2 | 0 | 0 | 0 | 0 | 0 | 0 | 4 | 18 | 90 |

## DURHAM *Strafford County* ELEVATION 69 ft LAT/LONG 43° 8 ' N / 70° 56 ' W

| | JAN | FEB | MAR | APR | MAY | JUN | JUL | AUG | SEP | OCT | NOV | DEC | YEAR |
|---|---|---|---|---|---|---|---|---|---|---|---|---|---|
| Maximum Temp °F | 33.1 | 36.3 | 45.2 | 57.5 | 68.9 | 77.9 | 83.2 | 81.2 | 73.1 | 62.1 | 49.3 | 37.4 | 58.8 |
| Minimum Temp °F | 11.3 | 14.1 | 23.6 | 32.6 | 42.2 | 51.6 | 57.0 | 55.4 | 47.2 | 36.8 | 29.1 | 18.0 | 34.9 |
| Mean Temp °F | 22.2 | 25.2 | 34.4 | 45.1 | 55.6 | 64.8 | 70.1 | 68.4 | 60.1 | 49.5 | 39.2 | 27.7 | 46.9 |
| Days Max Temp ≥ 90 °F | 0 | 0 | 0 | 0 | 1 | 2 | 4 | 3 | 1 | 0 | 0 | 0 | 11 |
| Days Max Temp ≤ 32 °F | 14 | 9 | 2 | 0 | 0 | 0 | 0 | 0 | 0 | 0 | 1 | 9 | 35 |
| Days Min Temp ≤ 32 °F | 30 | 27 | 25 | 16 | 4 | 0 | 0 | 0 | 2 | 11 | 20 | 29 | 164 |
| Days Min Temp ≤ 0 °F | 7 | 4 | 1 | 0 | 0 | 0 | 0 | 0 | 0 | 0 | 0 | 2 | 14 |
| Heating Degree Days | 1319 | 1117 | 941 | 592 | 301 | 82 | 16 | 35 | 177 | 476 | 767 | 1149 | 6972 |
| Cooling Degree Days | 0 | 0 | 0 | 1 | 17 | 81 | 185 | 143 | 41 | 2 | 0 | 0 | 470 |
| Total Precipitation (") | 2.84 | 2.88 | 3.29 | 3.90 | 3.60 | 3.49 | 3.00 | 3.52 | 3.38 | 3.58 | 4.64 | 4.00 | 42.12 |
| Days ≥ 0.1" Precip | 6 | 5 | 7 | 7 | 8 | 7 | 6 | 6 | 6 | 6 | 8 | 7 | 79 |
| Total Snowfall (") | 15.8 | 12.3 | 8.7 | 2.0 | 0.0 | 0.0 | 0.0 | 0.0 | 0.0 | 0.2 | 2.6 | 13.1 | 54.7 |
| Days ≥ 1" Snow Depth | 24 | 21 | 13 | 1 | 0 | 0 | 0 | 0 | 0 | 0 | 2 | 16 | 77 |

## EPPING *Rockingham County* ELEVATION 200 ft LAT/LONG 43° 3 ' N / 71° 4 ' W

| | JAN | FEB | MAR | APR | MAY | JUN | JUL | AUG | SEP | OCT | NOV | DEC | YEAR |
|---|---|---|---|---|---|---|---|---|---|---|---|---|---|
| Maximum Temp °F | 32.4 | 35.8 | 44.8 | 57.2 | 69.1 | 77.8 | 82.8 | 80.5 | 72.3 | 61.4 | 48.7 | 36.9 | 58.3 |
| Minimum Temp °F | 11.4 | 14.1 | 23.6 | 32.8 | 42.4 | 51.9 | 57.1 | 55.6 | 46.9 | 36.3 | 28.9 | 18.0 | 34.9 |
| Mean Temp °F | 21.9 | 25.0 | 34.2 | 45.0 | 55.8 | 64.9 | 70.0 | 68.1 | 59.6 | 48.9 | 38.9 | 27.5 | 46.7 |
| Days Max Temp ≥ 90 °F | 0 | 0 | 0 | 0 | 1 | 2 | 4 | 2 | 1 | 0 | 0 | 0 | 10 |
| Days Max Temp ≤ 32 °F | 16 | 11 | 3 | 0 | 0 | 0 | 0 | 0 | 0 | 0 | 1 | 10 | 41 |
| Days Min Temp ≤ 32 °F | 30 | 27 | 26 | 16 | 4 | 0 | 0 | 0 | 2 | 12 | 20 | 29 | 166 |
| Days Min Temp ≤ 0 °F | 7 | 4 | 1 | 0 | 0 | 0 | 0 | 0 | 0 | 0 | 0 | 2 | 14 |
| Heating Degree Days | 1328 | 1124 | 946 | 594 | 296 | 81 | 18 | 38 | 191 | 495 | 778 | 1157 | 7046 |
| Cooling Degree Days | 0 | 0 | 0 | 1 | 20 | 87 | 186 | 140 | 39 | 2 | 0 | 0 | 475 |
| Total Precipitation (") | 3.18 | 3.27 | 3.62 | 3.94 | 3.62 | 3.74 | 3.37 | 3.62 | 3.62 | 3.54 | 4.53 | 4.22 | 44.27 |
| Days ≥ 0.1" Precip | 6 | 6 | 7 | 7 | 7 | 7 | 7 | 6 | 6 | 6 | 8 | 7 | 80 |
| Total Snowfall (") | 16.1 | 14.7 | 10.2 | 2.4 | 0.0 | 0.0 | 0.0 | 0.0 | 0.0 | 0.1 | 3.2 | 13.0 | 59.7 |
| Days ≥ 1" Snow Depth | na | na | na | *1* | 0 | 0 | 0 | 0 | 0 | 0 | *1* | na | na |

## FIRST CONN LAKE *Coos County* ELEVATION 1660 ft LAT/LONG 45° 5 ' N / 71° 17 ' W

| | JAN | FEB | MAR | APR | MAY | JUN | JUL | AUG | SEP | OCT | NOV | DEC | YEAR |
|---|---|---|---|---|---|---|---|---|---|---|---|---|---|
| Maximum Temp °F | 20.3 | 23.5 | 33.6 | 46.0 | 60.4 | 69.1 | 73.8 | 71.6 | 63.2 | 51.5 | 38.0 | 25.5 | 48.0 |
| Minimum Temp °F | -3.6 | -3.2 | 8.5 | 24.3 | 36.2 | 46.3 | 51.0 | 49.0 | 40.8 | 31.2 | 21.5 | 5.8 | 25.7 |
| Mean Temp °F | 8.4 | 10.2 | 21.1 | 35.2 | 48.3 | 57.7 | 62.4 | 60.3 | 52.0 | 41.4 | 29.8 | 15.7 | 36.9 |
| Days Max Temp ≥ 90 °F | 0 | 0 | 0 | 0 | 0 | 0 | 0 | 0 | 0 | 0 | 0 | 0 | 0 |
| Days Max Temp ≤ 32 °F | 26 | 22 | 15 | 3 | 0 | 0 | 0 | 0 | 0 | 1 | 9 | 23 | 99 |
| Days Min Temp ≤ 32 °F | 31 | 28 | 30 | 25 | 10 | 1 | 0 | 0 | 6 | 18 | 26 | 31 | 206 |
| Days Min Temp ≤ 0 °F | 18 | 17 | 10 | 1 | 0 | 0 | 0 | 0 | 0 | 0 | 1 | 11 | 58 |
| Heating Degree Days | 1742 | 1543 | 1353 | 889 | 511 | 230 | 119 | 166 | 389 | 731 | 1045 | 1521 | 10239 |
| Cooling Degree Days | 0 | 0 | 0 | 0 | 2 | 16 | 44 | 28 | 4 | 0 | 0 | 0 | 94 |
| Total Precipitation (") | 2.77 | 2.30 | 2.81 | 3.02 | 3.92 | 4.66 | 4.33 | 5.00 | 4.24 | 3.80 | 3.94 | 3.51 | 44.30 |
| Days ≥ 0.1" Precip | na | *6* | *8* | *8* | *9* | 10 | *9* | 10 | *9* | *9* | *10* | *8* | na |
| Total Snowfall (") | 33.9 | 27.1 | 26.6 | 11.4 | 1.2 | 0.0 | 0.0 | 0.0 | 0.0 | 1.9 | 17.7 | 35.2 | 155.0 |
| Days ≥ 1" Snow Depth | 31 | 28 | 31 | 21 | 2 | 0 | 0 | 0 | 0 | 2 | 14 | 30 | 159 |

## GRAFTON *Grafton County* ELEVATION 840 ft LAT/LONG 43° 34 ' N / 71° 57 ' W

| | JAN | FEB | MAR | APR | MAY | JUN | JUL | AUG | SEP | OCT | NOV | DEC | YEAR |
|---|---|---|---|---|---|---|---|---|---|---|---|---|---|
| Maximum Temp °F | 27.8 | 31.4 | 40.9 | 54.1 | 67.4 | 75.1 | 79.8 | 77.4 | 68.8 | 57.7 | 44.4 | 32.2 | 54.8 |
| Minimum Temp °F | 4.1 | 6.1 | 17.5 | 28.1 | 38.3 | 47.7 | 52.5 | 51.3 | 42.8 | 32.8 | 25.1 | 11.9 | 29.9 |
| Mean Temp °F | 16.0 | 18.8 | 29.2 | 41.1 | 52.9 | 61.4 | 66.2 | 64.4 | 55.9 | 45.3 | 34.8 | 22.1 | 42.3 |
| Days Max Temp ≥ 90 °F | 0 | 0 | 0 | 0 | 0 | 1 | 2 | 1 | 0 | 0 | 0 | 0 | 4 |
| Days Max Temp ≤ 32 °F | 20 | 15 | 6 | 0 | 0 | 0 | 0 | 0 | 0 | 0 | 3 | 16 | 60 |
| Days Min Temp ≤ 32 °F | 31 | 27 | 29 | 21 | 10 | 2 | 0 | 1 | 6 | 17 | 24 | 30 | 198 |
| Days Min Temp ≤ 0 °F | 13 | 11 | 3 | 0 | 0 | 0 | 0 | 0 | 0 | 0 | 0 | 7 | 34 |
| Heating Degree Days | 1513 | 1301 | 1102 | 710 | 375 | 143 | 56 | 90 | 283 | 606 | 901 | 1324 | 8404 |
| Cooling Degree Days | 0 | 0 | 0 | 0 | 7 | 37 | 99 | 75 | 15 | 0 | 0 | 0 | 233 |
| Total Precipitation (") | 2.43 | 2.47 | 2.83 | 3.10 | 3.81 | 3.68 | 3.83 | 3.88 | 3.43 | 3.65 | 3.52 | 3.32 | 39.95 |
| Days ≥ 0.1" Precip | 6 | 5 | 7 | 7 | 7 | 8 | 7 | 7 | 6 | 6 | 7 | 7 | 80 |
| Total Snowfall (") | 21.6 | 17.6 | 13.3 | 5.0 | 0.4 | 0.0 | 0.0 | 0.0 | 0.0 | 0.6 | 6.7 | 17.9 | 83.1 |
| Days ≥ 1" Snow Depth | na | na | na | *6* | 0 | 0 | 0 | 0 | 0 | 0 | *4* | na | na |

## HANOVER *Grafton County* ELEVATION 600 ft LAT/LONG 43° 42 ' N / 72° 17 ' W

| | JAN | FEB | MAR | APR | MAY | JUN | JUL | AUG | SEP | OCT | NOV | DEC | YEAR |
|---|---|---|---|---|---|---|---|---|---|---|---|---|---|
| Maximum Temp °F | 28.3 | 32.7 | 42.3 | 56.4 | 69.8 | 77.9 | 82.5 | 80.2 | 70.7 | 58.1 | 45.0 | 32.6 | 56.4 |
| Minimum Temp °F | 7.3 | 9.9 | 21.2 | 32.1 | 43.0 | 52.4 | 57.6 | 56.3 | 48.4 | 37.1 | 28.5 | 15.1 | 34.1 |
| Mean Temp °F | 17.8 | 21.3 | 31.7 | 44.3 | 56.4 | 65.1 | 70.0 | 68.3 | 59.6 | 47.6 | 36.8 | 23.9 | 45.2 |
| Days Max Temp ≥ 90 °F | 0 | 0 | 0 | 0 | 1 | 2 | 4 | 3 | 0 | 0 | 0 | 0 | 10 |
| Days Max Temp ≤ 32 °F | 19 | 14 | 4 | 0 | 0 | 0 | 0 | 0 | 0 | 0 | 2 | 15 | 54 |
| Days Min Temp ≤ 32 °F | 31 | 27 | 27 | 17 | 4 | 0 | 0 | 0 | 1 | 10 | 21 | 29 | 167 |
| Days Min Temp ≤ 0 °F | 10 | 8 | 2 | 0 | 0 | 0 | 0 | 0 | 0 | 0 | 0 | 5 | 25 |
| Heating Degree Days | 1456 | 1226 | 1026 | 617 | 280 | 76 | 15 | 33 | 188 | 532 | 840 | 1274 | 7563 |
| Cooling Degree Days | 0 | 0 | 0 | 3 | 21 | 83 | 195 | 150 | 33 | 1 | 0 | 0 | 486 |
| Total Precipitation (") | 2.55 | 2.24 | 2.65 | 2.91 | 3.51 | 3.40 | 3.53 | 3.77 | 3.57 | 3.24 | 3.45 | 3.18 | 38.00 |
| Days ≥ 0.1" Precip | 6 | 5 | 6 | 6 | 8 | 8 | 7 | 7 | 7 | 6 | 8 | 7 | 81 |
| Total Snowfall (") | 19.0 | 13.9 | 9.5 | 2.2 | 0.2 | 0.0 | 0.0 | 0.0 | 0.0 | 0.2 | 4.4 | 17.7 | 67.1 |
| Days ≥ 1" Snow Depth | 26 | 26 | 19 | 2 | 0 | 0 | 0 | 0 | 0 | 0 | 5 | 20 | 98 |

## KEENE *Cheshire County* ELEVATION 489 ft LAT/LONG 42° 55 ' N / 72° 17 ' W

| | JAN | FEB | MAR | APR | MAY | JUN | JUL | AUG | SEP | OCT | NOV | DEC | YEAR |
|---|---|---|---|---|---|---|---|---|---|---|---|---|---|
| Maximum Temp °F | 31.6 | 35.0 | 44.5 | 58.3 | 70.9 | 78.7 | 83.3 | 81.1 | 72.7 | 61.4 | 48.2 | 35.8 | 58.5 |
| Minimum Temp °F | 10.1 | 12.2 | 22.5 | 32.9 | 43.3 | 52.4 | 57.3 | 56.2 | 47.8 | 36.8 | 29.3 | 17.2 | 34.8 |
| Mean Temp °F | 20.9 | 23.6 | 33.6 | 45.6 | 57.1 | 65.6 | 70.4 | 68.7 | 60.2 | 49.1 | 38.8 | 26.5 | 46.7 |
| Days Max Temp ≥ 90 °F | 0 | 0 | 0 | 0 | 1 | 2 | 4 | 3 | 1 | 0 | 0 | 0 | 11 |
| Days Max Temp ≤ 32 °F | 16 | 11 | 3 | 0 | 0 | 0 | 0 | 0 | 0 | 0 | 1 | 11 | 42 |
| Days Min Temp ≤ 32 °F | 30 | 27 | 26 | 15 | 4 | 0 | 0 | 0 | 1 | 12 | 20 | 29 | 164 |
| Days Min Temp ≤ 0 °F | 8 | 6 | 1 | 0 | 0 | 0 | 0 | 0 | 0 | 0 | 0 | 4 | 19 |
| Heating Degree Days | 1362 | 1162 | 969 | 578 | 260 | 69 | 16 | 32 | 176 | 488 | 780 | 1187 | 7079 |
| Cooling Degree Days | 0 | 0 | 0 | 2 | 23 | 84 | 189 | 153 | 41 | 3 | 0 | 0 | 495 |
| Total Precipitation (") | 2.82 | 2.52 | 3.15 | 3.17 | 3.84 | 3.72 | 3.78 | 3.95 | 3.31 | 3.20 | 3.53 | 3.36 | 40.35 |
| Days ≥ 0.1" Precip | 6 | 5 | 7 | 7 | 7 | 7 | 7 | 7 | 6 | 6 | 8 | 7 | 80 |
| Total Snowfall (") | 15.8 | 12.9 | 10.6 | 2.6 | 0.1 | 0.0 | 0.0 | 0.0 | 0.0 | 0.0 | 3.6 | 13.0 | 58.6 |
| Days ≥ 1" Snow Depth | 27 | 25 | 18 | 2 | 0 | 0 | 0 | 0 | 0 | 0 | 4 | 18 | 94 |

## LANCASTER *Coos County* ELEVATION 879 ft LAT/LONG 44° 29 ' N / 71° 34 ' W

| | JAN | FEB | MAR | APR | MAY | JUN | JUL | AUG | SEP | OCT | NOV | DEC | YEAR |
|---|---|---|---|---|---|---|---|---|---|---|---|---|---|
| Maximum Temp °F | 25.4 | 29.9 | 40.2 | 54.0 | 68.2 | 75.6 | 79.9 | *77.8* | 68.9 | 57.0 | 42.6 | 29.8 | 54.1 |
| Minimum Temp °F | 1.5 | 3.7 | 16.2 | 28.4 | 39.1 | 48.0 | 52.8 | *52.0* | 43.6 | 33.4 | 24.6 | 10.2 | 29.5 |
| Mean Temp °F | 13.5 | 16.8 | 28.2 | 41.3 | 53.7 | 61.8 | 66.4 | *64.9* | 56.3 | 45.2 | 33.6 | 20.0 | 41.8 |
| Days Max Temp ≥ 90 °F | 0 | 0 | 0 | 0 | 0 | 1 | 1 | 0 | 0 | 0 | 0 | 0 | 2 |
| Days Max Temp ≤ 32 °F | 22 | 17 | 7 | 1 | 0 | 0 | 0 | 0 | 0 | 0 | 5 | 18 | 70 |
| Days Min Temp ≤ 32 °F | 31 | 28 | 29 | 21 | 8 | 1 | 0 | 0 | 3 | 15 | 24 | 30 | 190 |
| Days Min Temp ≤ 0 °F | 15 | 12 | 5 | 0 | 0 | 0 | 0 | 0 | 0 | 0 | 0 | 7 | 39 |
| Heating Degree Days | 1593 | 1357 | 1134 | 706 | 353 | 133 | 51 | *77* | 269 | 606 | 935 | 1389 | 8603 |
| Cooling Degree Days | 0 | 0 | 0 | 0 | 8 | 41 | 105 | 81 | 13 | 0 | 0 | 0 | 248 |
| Total Precipitation (") | 2.28 | 1.89 | 2.34 | 2.67 | 3.34 | 4.04 | 3.78 | 4.28 | 3.45 | 3.10 | 3.29 | 2.92 | 37.38 |
| Days ≥ 0.1" Precip | 6 | 5 | 7 | 7 | 8 | 8 | 8 | *8* | 7 | 7 | 9 | 8 | 88 |
| Total Snowfall (") | 17.5 | 15.4 | 12.1 | 3.9 | 0.0 | 0.0 | 0.0 | 0.0 | 0.0 | 0.4 | 5.7 | 17.6 | 72.6 |
| Days ≥ 1" Snow Depth | 29 | 27 | 25 | 6 | 0 | 0 | 0 | 0 | 0 | 0 | 6 | 24 | 117 |

## MASSABESIC LAKE *Hillsborough County* ELEVATION 249 ft LAT/LONG 42° 59 ' N / 71° 24 ' W

| | JAN | FEB | MAR | APR | MAY | JUN | JUL | AUG | SEP | OCT | NOV | DEC | YEAR |
|---|---|---|---|---|---|---|---|---|---|---|---|---|---|
| Maximum Temp °F | 31.9 | 34.7 | 43.6 | 55.9 | 68.0 | 77.0 | 82.0 | 80.2 | 71.9 | 61.1 | 49.3 | 36.7 | 57.7 |
| Minimum Temp °F | 9.6 | 12.5 | 23.2 | 33.7 | 44.1 | 53.4 | 58.5 | 57.2 | 48.1 | 37.1 | 29.5 | 17.6 | 35.4 |
| Mean Temp °F | 20.8 | 23.7 | 33.4 | 44.8 | 56.1 | 65.2 | 70.3 | 68.7 | 60.1 | 49.1 | 39.4 | 27.2 | 46.6 |
| Days Max Temp ≥ 90 °F | 0 | 0 | 0 | 0 | 0 | 2 | 4 | 3 | 1 | 0 | 0 | 0 | 10 |
| Days Max Temp ≤ 32 °F | 16 | 12 | 4 | 0 | 0 | 0 | 0 | 0 | 0 | 0 | 1 | 11 | 44 |
| Days Min Temp ≤ 32 °F | 30 | 27 | 26 | 14 | 2 | 0 | 0 | 0 | 1 | 10 | 20 | 29 | 159 |
| Days Min Temp ≤ 0 °F | 7 | 5 | 1 | 0 | 0 | 0 | 0 | 0 | 0 | 0 | 0 | 2 | 15 |
| Heating Degree Days | 1365 | 1159 | 972 | 601 | 288 | 79 | 17 | 32 | 180 | 489 | 760 | 1165 | 7107 |
| Cooling Degree Days | 0 | 0 | 0 | 1 | 20 | 92 | 193 | *153* | 39 | 2 | 0 | 0 | 500 |
| Total Precipitation (") | 2.70 | 2.37 | 2.87 | 3.15 | 3.53 | 3.61 | 3.59 | 3.90 | 3.30 | 3.30 | 3.79 | 3.47 | 39.58 |
| Days ≥ 0.1" Precip | 6 | 5 | 6 | 7 | 8 | 7 | 7 | 6 | 6 | 6 | 8 | 7 | 79 |
| Total Snowfall (") | 15.5 | 13.6 | 8.8 | *2.0* | 0.0 | 0.0 | 0.0 | 0.0 | 0.1 | 0.0 | 3.1 | 11.8 | 54.9 |
| Days ≥ 1" Snow Depth | *24* | na | *15* | *1* | 0 | 0 | 0 | 0 | 0 | 0 | 2 | *15* | na |

## MONROE 5 NNE *Grafton County* ELEVATION 679 ft LAT/LONG 44° 19 ' N / 72° 0 ' W

| | JAN | FEB | MAR | APR | MAY | JUN | JUL | AUG | SEP | OCT | NOV | DEC | YEAR |
|---|---|---|---|---|---|---|---|---|---|---|---|---|---|
| Maximum Temp °F | 24.4 | 28.4 | 38.9 | 52.3 | 66.3 | 74.3 | 79.4 | 77.1 | 67.5 | 55.1 | 42.5 | 29.2 | 53.0 |
| Minimum Temp °F | 1.9 | 2.8 | 16.4 | 30.3 | 41.4 | 50.4 | 55.6 | 54.3 | 45.7 | 34.8 | 26.4 | 11.0 | 30.9 |
| Mean Temp °F | 13.2 | 15.6 | 27.6 | 41.3 | 53.9 | 62.4 | 67.6 | 65.7 | 56.6 | 45.0 | 34.5 | 20.2 | 42.0 |
| Days Max Temp ≥ 90 °F | 0 | 0 | 0 | 0 | 0 | 1 | 2 | 1 | 0 | 0 | 0 | 0 | 4 |
| Days Max Temp ≤ 32 °F | 23 | 18 | 8 | 1 | 0 | 0 | 0 | 0 | 0 | 0 | 4 | 18 | 72 |
| Days Min Temp ≤ 32 °F | 31 | 28 | 29 | 19 | 5 | 0 | 0 | 0 | 1 | 14 | 23 | 30 | 180 |
| Days Min Temp ≤ 0 °F | 15 | 13 | 4 | 0 | 0 | 0 | 0 | 0 | 0 | 0 | 0 | 7 | 39 |
| Heating Degree Days | 1602 | 1387 | 1152 | 706 | 351 | 126 | 41 | 70 | 262 | 615 | 909 | 1383 | 8604 |
| Cooling Degree Days | 0 | 0 | 0 | 0 | 10 | 48 | 122 | 97 | 17 | 0 | 0 | 0 | 294 |
| Total Precipitation (") | 2.41 | 1.81 | 2.32 | 2.57 | 2.93 | 3.96 | 3.32 | 4.06 | 3.55 | 3.29 | 3.21 | 2.62 | 36.05 |
| Days ≥ 0.1" Precip | 5 | 5 | 6 | 7 | na | 9 | 8 | 8 | 7 | 7 | na | 7 | na |
| Total Snowfall (") | 18.0 | 13.6 | 10.4 | 3.2 | 0.0 | 0.0 | 0.0 | 0.0 | 0.0 | na | 3.6 | na | na |
| Days ≥ 1" Snow Depth | na | na | na | na | 0 | 0 | 0 | 0 | 0 | na | na | na | na |

## MOUNT SUNAPEE *Merrimack County* ELEVATION 1312 ft LAT/LONG 43° 20 ' N / 72° 5 ' W

| | JAN | FEB | MAR | APR | MAY | JUN | JUL | AUG | SEP | OCT | NOV | DEC | YEAR |
|---|---|---|---|---|---|---|---|---|---|---|---|---|---|
| Maximum Temp °F | 29.1 | 31.8 | 41.1 | 54.0 | 67.4 | 74.7 | 79.2 | 77.2 | 68.5 | 58.1 | 45.1 | 33.2 | 55.0 |
| Minimum Temp °F | 11.0 | 12.7 | 22.0 | 32.4 | 43.5 | 52.7 | 57.6 | 56.2 | 48.2 | 38.1 | 28.7 | 17.1 | 35.0 |
| Mean Temp °F | 20.0 | 22.2 | 31.6 | 43.2 | 55.6 | 63.7 | 68.4 | 66.9 | 58.4 | 48.1 | 36.9 | 25.2 | 45.0 |
| Days Max Temp ≥ 90 °F | 0 | 0 | 0 | 0 | 0 | 0 | 0 | 0 | 0 | 0 | 0 | 0 | 0 |
| Days Max Temp ≤ 32 °F | 19 | 15 | 6 | 0 | 0 | 0 | 0 | 0 | 0 | 0 | 2 | 15 | 57 |
| Days Min Temp ≤ 32 °F | 30 | 27 | 27 | 16 | 2 | 0 | 0 | 0 | 1 | 9 | 20 | 29 | 161 |
| Days Min Temp ≤ 0 °F | 7 | 4 | 1 | 0 | 0 | 0 | 0 | 0 | 0 | 0 | 0 | 2 | 14 |
| Heating Degree Days | 1387 | 1201 | 1029 | 648 | 301 | 101 | 26 | 48 | 218 | 519 | 836 | 1228 | 7542 |
| Cooling Degree Days | 0 | 0 | 0 | 1 | 20 | 64 | 143 | 121 | 27 | 2 | 0 | 0 | 378 |
| Total Precipitation (") | 2.51 | 2.71 | 3.08 | 3.62 | 4.13 | 3.84 | 3.62 | 4.08 | 3.60 | 3.67 | 3.98 | 3.43 | 42.27 |
| Days ≥ 0.1" Precip | 5 | 4 | 6 | 7 | 8 | 8 | 7 | 7 | 7 | 6 | 7 | 6 | 78 |
| Total Snowfall (") | 20.7 | 18.8 | 14.1 | 5.3 | 0.2 | 0.0 | 0.0 | 0.0 | 0.0 | 0.0 | 4.4 | 16.1 | 79.6 |
| Days ≥ 1" Snow Depth | na | na | na | na | 0 | 0 | 0 | 0 | 0 | 0 | na | na | na |

## MOUNT WASHINGTON *Coos County* ELEVATION 6270 ft LAT/LONG 44° 16 ' N / 71° 18 ' W

| | JAN | FEB | MAR | APR | MAY | JUN | JUL | AUG | SEP | OCT | NOV | DEC | YEAR |
|---|---|---|---|---|---|---|---|---|---|---|---|---|---|
| Maximum Temp °F | 12.6 | 13.1 | 20.2 | 29.4 | 41.1 | 50.0 | 54.2 | 52.7 | 45.8 | 36.1 | 27.3 | 17.9 | 33.4 |
| Minimum Temp °F | -4.6 | -3.2 | 5.3 | 16.5 | 28.8 | 38.1 | 43.2 | 41.8 | 34.6 | 24.1 | 13.7 | 1.6 | 20.0 |
| Mean Temp °F | 4.0 | 5.0 | 12.8 | 23.0 | 35.0 | 44.1 | 48.7 | 47.3 | 40.2 | 30.1 | 20.6 | 9.7 | 26.7 |
| Days Max Temp ≥ 90 °F | 0 | 0 | 0 | 0 | 0 | 0 | 0 | 0 | 0 | 0 | 0 | 0 | 0 |
| Days Max Temp ≤ 32 °F | 29 | 26 | 25 | 19 | 7 | 1 | 0 | 0 | 2 | 11 | 20 | 27 | 167 |
| Days Min Temp ≤ 32 °F | 31 | 28 | 31 | 28 | 20 | 7 | 2 | 3 | 12 | 24 | 28 | 31 | 245 |
| Days Min Temp ≤ 0 °F | 19 | 17 | 11 | 2 | 0 | 0 | 0 | 0 | 0 | 0 | 4 | 14 | 67 |
| Heating Degree Days | 1889 | 1693 | 1614 | 1254 | 923 | 621 | 499 | 541 | 735 | 1075 | 1327 | 1710 | 13881 |
| Cooling Degree Days | 0 | 0 | 0 | 0 | 0 | 0 | 0 | 0 | 0 | 0 | 0 | 0 | 0 |
| Total Precipitation (") | 7.99 | 8.48 | 9.28 | 8.09 | 7.61 | 8.03 | 7.11 | 8.40 | 8.06 | 7.13 | 10.17 | 9.60 | 99.95 |
| Days ≥ 0.1" Precip | 15 | 13 | 15 | 13 | 13 | 13 | 12 | 12 | 12 | 12 | 15 | 16 | 161 |
| Total Snowfall (") | 51.5 | 48.7 | 54.0 | 40.0 | 11.9 | 1.3 | 0.0 | 0.2 | 2.4 | 13.3 | 39.0 | 54.0 | 316.3 |
| Days ≥ 1" Snow Depth | 29 | 28 | 30 | 27 | 11 | 0 | 0 | 0 | 1 | 9 | 19 | 28 | 182 |

## NASHUA 2 NNW *Hillsborough County* ELEVATION 141 ft LAT/LONG 42° 47 ' N / 71° 29 ' W

| | JAN | FEB | MAR | APR | MAY | JUN | JUL | AUG | SEP | OCT | NOV | DEC | YEAR |
|---|---|---|---|---|---|---|---|---|---|---|---|---|---|
| Maximum Temp °F | 32.8 | 35.9 | 44.9 | 57.3 | 69.0 | 77.2 | 82.5 | 80.4 | 72.0 | 61.2 | 49.2 | 37.1 | 58.3 |
| Minimum Temp °F | 11.5 | 13.7 | 23.6 | 33.2 | 43.5 | 52.8 | 58.0 | 56.7 | 48.2 | 36.7 | 28.9 | 18.1 | 35.4 |
| Mean Temp °F | 22.2 | 24.8 | 34.3 | 45.3 | 56.3 | 65.0 | 70.3 | 68.6 | 60.1 | 49.0 | 39.1 | 27.6 | 46.9 |
| Days Max Temp ≥ 90 °F | 0 | 0 | 0 | 0 | 1 | 2 | 4 | 2 | 0 | 0 | 0 | 0 | 9 |
| Days Max Temp ≤ 32 °F | 15 | 10 | 3 | 0 | 0 | 0 | 0 | 0 | 0 | 0 | 1 | 9 | 38 |
| Days Min Temp ≤ 32 °F | 30 | 27 | 27 | 15 | 3 | 0 | 0 | 0 | 1 | 11 | 21 | 29 | 164 |
| Days Min Temp ≤ 0 °F | 6 | 4 | 0 | 0 | 0 | 0 | 0 | 0 | 0 | 0 | 0 | 2 | 12 |
| Heating Degree Days | 1321 | 1128 | 946 | 587 | 281 | 79 | 15 | 33 | 178 | 491 | 771 | 1151 | 6981 |
| Cooling Degree Days | 0 | 0 | 0 | 1 | 21 | 84 | 195 | 162 | 42 | 2 | 0 | 0 | 507 |
| Total Precipitation (") | 3.30 | 3.17 | 3.79 | 3.61 | 3.51 | 3.78 | 3.36 | 3.78 | 3.51 | 3.43 | 4.26 | 4.02 | 43.52 |
| Days ≥ 0.1" Precip | 6 | 6 | 7 | 7 | 7 | 7 | 6 | 6 | 6 | 6 | 8 | 7 | 79 |
| Total Snowfall (") | 15.5 | 14.9 | 10.9 | 1.8 | 0.0 | 0.0 | 0.0 | 0.0 | 0.0 | 0.0 | 3.4 | 13.1 | 59.6 |
| Days ≥ 1" Snow Depth | na | na | na | 1 | 0 | 0 | 0 | 0 | 0 | 0 | na | na | na |

## NORTH STRATFORD *Coos County* ELEVATION 902 ft LAT/LONG 44° 45' N / 71° 38' W

| | JAN | FEB | MAR | APR | MAY | JUN | JUL | AUG | SEP | OCT | NOV | DEC | YEAR |
|---|---|---|---|---|---|---|---|---|---|---|---|---|---|
| Maximum Temp °F | 25.2 | 29.7 | 39.6 | 52.7 | 68.2 | 75.8 | 79.9 | 77.9 | 68.8 | 56.6 | 42.3 | 30.5 | 53.9 |
| Minimum Temp °F | 1.2 | 3.5 | 15.4 | 28.4 | 39.7 | 48.8 | 53.3 | 52.2 | 43.7 | 33.6 | 25.1 | 11.6 | 29.7 |
| Mean Temp °F | 13.5 | 16.5 | 27.6 | 40.7 | 54.0 | 62.3 | 66.6 | 65.0 | 56.3 | 45.1 | 33.7 | 21.1 | 41.9 |
| Days Max Temp ≥ 90 °F | 0 | 0 | 0 | 0 | 0 | 1 | 1 | 1 | 0 | 0 | 0 | 0 | 3 |
| Days Max Temp ≤ 32 °F | 23 | 17 | 8 | 1 | 0 | 0 | 0 | 0 | 0 | 0 | 5 | 17 | 71 |
| Days Min Temp ≤ 32 °F | 31 | 28 | 28 | 21 | 7 | 1 | 0 | 0 | 4 | 15 | 23 | 30 | 188 |
| Days Min Temp ≤ 0 °F | 15 | 13 | 5 | 0 | 0 | 0 | 0 | 0 | 0 | 0 | 0 | 7 | 40 |
| Heating Degree Days | 1593 | 1366 | 1152 | 722 | 344 | 125 | 49 | 76 | 274 | 610 | 932 | 1357 | 8600 |
| Cooling Degree Days | 0 | 0 | 0 | 0 | 8 | 43 | 106 | 85 | 14 | 1 | 0 | 0 | 257 |
| Total Precipitation (") | 2.51 | 2.13 | 2.58 | 2.75 | 3.29 | 3.83 | 3.84 | 4.53 | 3.52 | 3.32 | 3.37 | 3.11 | 38.78 |
| Days ≥ 0.1" Precip | 6 | 5 | 7 | 7 | 8 | 8 | 7 | na | 7 | 8 | 8 | 8 | na |
| Total Snowfall (") | 21.0 | 18.6 | 15.2 | 6.3 | 0.0 | 0.0 | 0.0 | 0.0 | 0.0 | 0.3 | 6.9 | 22.7 | 91.0 |
| Days ≥ 1" Snow Depth | 29 | 26 | 26 | 8 | 0 | 0 | 0 | 0 | 0 | 0 | 8 | 25 | 122 |

## PETERBORO 2 S *Hillsborough County* ELEVATION 1001 ft LAT/LONG 42° 51' N / 71° 57' W

| | JAN | FEB | MAR | APR | MAY | JUN | JUL | AUG | SEP | OCT | NOV | DEC | YEAR |
|---|---|---|---|---|---|---|---|---|---|---|---|---|---|
| Maximum Temp °F | 30.6 | 33.8 | 43.0 | 55.7 | 67.7 | 75.0 | 79.7 | 77.5 | 69.5 | 59.1 | 46.8 | 34.9 | 56.1 |
| Minimum Temp °F | 11.0 | 13.2 | 22.4 | 32.2 | 41.9 | 50.8 | 56.4 | 54.7 | 46.8 | 36.5 | 28.5 | 16.9 | 34.3 |
| Mean Temp °F | 20.8 | 23.5 | 32.7 | 44.0 | 54.8 | 62.9 | 68.1 | 66.1 | 58.2 | 47.8 | 37.7 | 25.9 | 45.2 |
| Days Max Temp ≥ 90 °F | 0 | 0 | 0 | 0 | 0 | 0 | 1 | 0 | 0 | 0 | 0 | 0 | 1 |
| Days Max Temp ≤ 32 °F | 18 | 13 | 4 | 0 | 0 | 0 | 0 | 0 | 0 | 0 | 1 | 13 | 49 |
| Days Min Temp ≤ 32 °F | 30 | 27 | 26 | 16 | 4 | 0 | 0 | 0 | 2 | 12 | 20 | 29 | 166 |
| Days Min Temp ≤ 0 °F | 7 | 4 | 1 | 0 | 0 | 0 | 0 | 0 | 0 | 0 | 0 | 3 | 15 |
| Heating Degree Days | 1364 | 1165 | 994 | 625 | 325 | 114 | 33 | 58 | 225 | 526 | 813 | 1206 | 7448 |
| Cooling Degree Days | 0 | 0 | na | 0 | na | na | 148 | 94 | 27 | 1 | 0 | 0 | na |
| Total Precipitation (") | 3.16 | 3.11 | 3.47 | 3.66 | 4.09 | 3.84 | 3.92 | 4.10 | 3.26 | 3.58 | 4.23 | 4.10 | 44.52 |
| Days ≥ 0.1" Precip | 6 | 5 | 6 | 6 | 7 | 7 | 7 | 7 | 6 | 6 | 7 | 7 | 77 |
| Total Snowfall (") | 18.6 | 16.4 | 12.6 | 3.6 | 0.4 | 0.0 | 0.0 | 0.0 | 0.0 | 0.2 | 6.4 | 15.9 | 74.1 |
| Days ≥ 1" Snow Depth | 26 | 25 | 20 | 2 | 0 | 0 | 0 | 0 | 0 | 0 | 6 | 20 | 99 |

## PINKHAM NOTCH *Coos County* ELEVATION 2001 ft LAT/LONG 44° 16' N / 71° 15' W

| | JAN | FEB | MAR | APR | MAY | JUN | JUL | AUG | SEP | OCT | NOV | DEC | YEAR |
|---|---|---|---|---|---|---|---|---|---|---|---|---|---|
| Maximum Temp °F | 24.6 | 26.7 | 35.3 | 47.0 | 60.9 | 69.1 | 73.8 | 71.5 | 63.2 | 52.8 | 40.3 | 29.4 | 49.5 |
| Minimum Temp °F | 3.8 | 5.5 | 15.2 | 27.5 | 38.7 | 48.1 | 52.7 | 50.6 | 42.3 | 32.2 | 23.3 | 10.6 | 29.2 |
| Mean Temp °F | 14.2 | 16.1 | 25.2 | 37.3 | 49.8 | 58.6 | 63.3 | 61.1 | 52.8 | 42.5 | 31.8 | 20.0 | 39.4 |
| Days Max Temp ≥ 90 °F | 0 | 0 | 0 | 0 | 0 | 0 | 0 | 0 | 0 | 0 | 0 | 0 | 0 |
| Days Max Temp ≤ 32 °F | 23 | 20 | 12 | 2 | 0 | 0 | 0 | 0 | 0 | 0 | 7 | 19 | 83 |
| Days Min Temp ≤ 32 °F | 31 | 28 | 30 | 22 | 7 | 0 | 0 | 0 | 3 | 17 | 25 | 30 | 193 |
| Days Min Temp ≤ 0 °F | 13 | 10 | 4 | 0 | 0 | 0 | 0 | 0 | 0 | 0 | 0 | 6 | 33 |
| Heating Degree Days | 1569 | 1375 | 1226 | 827 | 467 | 205 | 98 | 145 | 365 | 689 | 989 | 1388 | 9343 |
| Cooling Degree Days | 0 | 0 | 0 | 0 | 4 | 21 | 54 | 29 | 4 | 0 | 0 | 0 | 112 |
| Total Precipitation (") | 4.26 | 3.86 | 4.72 | 4.90 | 4.49 | 4.99 | 4.34 | 5.19 | 4.73 | 4.97 | 5.65 | 5.11 | 57.21 |
| Days ≥ 0.1" Precip | 7 | 7 | 9 | 8 | 9 | 10 | 8 | 8 | 8 | 8 | 9 | 9 | 100 |
| Total Snowfall (") | 29.4 | 26.2 | 27.1 | 13.3 | 1.0 | 0.0 | 0.0 | 0.0 | 0.1 | 1.3 | 13.1 | 30.9 | 142.4 |
| Days ≥ 1" Snow Depth | 30 | 28 | 30 | 19 | 2 | 0 | 0 | 0 | 0 | 0 | 10 | 27 | 146 |

## PLYMOUTH *Grafton County* ELEVATION 660 ft LAT/LONG 43° 47' N / 71° 39' W

| | JAN | FEB | MAR | APR | MAY | JUN | JUL | AUG | SEP | OCT | NOV | DEC | YEAR |
|---|---|---|---|---|---|---|---|---|---|---|---|---|---|
| Maximum Temp °F | 26.7 | 30.4 | 40.0 | 53.2 | 66.8 | 75.1 | 80.1 | 78.0 | 68.4 | 57.5 | 44.0 | 31.4 | 54.3 |
| Minimum Temp °F | 4.1 | 5.9 | 17.5 | 29.2 | 38.7 | 48.3 | 53.2 | 51.3 | 42.8 | 32.2 | 24.7 | 12.1 | 30.0 |
| Mean Temp °F | 15.5 | 18.2 | 28.8 | 41.2 | 52.8 | 61.7 | 66.7 | 64.7 | 55.6 | 44.9 | 34.4 | 21.8 | 42.2 |
| Days Max Temp ≥ 90 °F | 0 | 0 | 0 | 0 | 0 | 1 | 2 | 1 | 0 | 0 | 0 | 0 | 4 |
| Days Max Temp ≤ 32 °F | 21 | 16 | 7 | 0 | 0 | 0 | 0 | 0 | 0 | 0 | 3 | 16 | 63 |
| Days Min Temp ≤ 32 °F | 31 | 28 | 30 | 21 | 7 | 1 | 0 | 0 | 4 | 18 | 25 | 30 | 195 |
| Days Min Temp ≤ 0 °F | 13 | 11 | 3 | 0 | 0 | 0 | 0 | 0 | 0 | 0 | 0 | 6 | 33 |
| Heating Degree Days | 1530 | 1317 | 1115 | 707 | 379 | 138 | 48 | 81 | 290 | 618 | 911 | 1333 | 8467 |
| Cooling Degree Days | 0 | 0 | 0 | 0 | 6 | 38 | 107 | 76 | 13 | 0 | 0 | 0 | 240 |
| Total Precipitation (") | 3.19 | 2.97 | 3.39 | 3.31 | 3.84 | 3.78 | 3.95 | 3.97 | 3.41 | 3.68 | 4.19 | 3.81 | 43.49 |
| Days ≥ 0.1" Precip | 7 | 6 | 7 | 7 | 8 | 8 | 7 | 7 | 7 | 7 | 8 | 7 | 86 |
| Total Snowfall (") | 20.9 | 17.4 | 12.6 | 4.6 | 0.2 | 0.0 | 0.0 | 0.0 | 0.0 | 0.3 | 4.6 | 18.2 | 78.8 |
| Days ≥ 1" Snow Depth | 23 | 22 | 21 | 8 | 0 | 0 | 0 | 0 | 0 | 0 | 5 | 18 | 97 |

## SURRY MOUNTAIN LAKE *Cheshire County* ELEVATION 541 ft LAT/LONG 43° 0 ' N / 72° 19 ' W

| | JAN | FEB | MAR | APR | MAY | JUN | JUL | AUG | SEP | OCT | NOV | DEC | YEAR |
|---|---|---|---|---|---|---|---|---|---|---|---|---|---|
| Maximum Temp °F | 29.2 | 32.3 | 41.6 | 54.4 | 67.1 | 75.3 | 80.3 | 78.1 | 69.9 | ***59.1*** | ***46.3*** | ***33.3*** | 55.6 |
| Minimum Temp °F | 5.4 | 8.2 | 20.0 | 31.0 | 41.7 | 50.8 | 55.7 | 54.2 | 45.3 | ***33.6*** | ***26.8*** | ***13.4*** | 32.2 |
| Mean Temp °F | 17.4 | 20.2 | 30.9 | 42.7 | 54.4 | 63.1 | 68.0 | 66.2 | 57.7 | ***46.4*** | ***36.6*** | ***23.4*** | 43.9 |
| Days Max Temp ≥ 90 °F | 0 | 0 | 0 | 0 | 0 | 1 | 2 | 1 | 0 | 0 | 0 | 0 | 4 |
| Days Max Temp ≤ 32 °F | 18 | 14 | 5 | 0 | 0 | 0 | 0 | 0 | 0 | 0 | 2 | 14 | 53 |
| Days Min Temp ≤ 32 °F | 30 | 27 | 28 | 18 | 5 | 0 | 0 | 0 | 2 | 15 | 23 | 29 | 177 |
| Days Min Temp ≤ 0 °F | 11 | 8 | 2 | 0 | 0 | 0 | 0 | 0 | 0 | 0 | 0 | 5 | 26 |
| Heating Degree Days | 1471 | 1257 | 1052 | 662 | 335 | 112 | 33 | 59 | 237 | ***571*** | ***845*** | ***1284*** | 7918 |
| Cooling Degree Days | ***0*** | ***0*** | ***0*** | ***1*** | ***13*** | ***50*** | ***136*** | ***104*** | ***22*** | na | na | na | na |
| Total Precipitation (") | 2.82 | 2.55 | 2.97 | 2.97 | 4.05 | 3.48 | 3.78 | 3.89 | 3.07 | 3.31 | 3.35 | 3.24 | 39.48 |
| Days ≥ 0.1" Precip | 6 | 6 | 6 | 7 | 8 | 7 | 6 | 7 | 6 | 6 | 8 | 7 | 80 |
| Total Snowfall (") | 16.7 | 13.4 | 9.5 | 2.6 | 0.1 | 0.0 | 0.0 | 0.0 | 0.0 | 0.0 | 4.3 | 14.7 | 61.3 |
| Days ≥ 1" Snow Depth | 28 | 26 | 20 | 3 | 0 | 0 | 0 | 0 | 0 | 0 | 5 | 20 | 102 |

## JANUARY MINIMUM TEMPERATURE °F

### LOWEST

| | | |
|---|---|---|
| 1 | Mt Washington | -4.6 |
| 2 | First Conn Lake | -3.6 |
| 3 | Colebrook | 0.2 |
| 4 | North Stratford | 1.2 |
| 5 | Lancaster | 1.5 |
| 6 | Monroe | 1.9 |
| 7 | Berlin | 2.9 |
| 8 | Pinkham Notch | 3.8 |
| 9 | Grafton | 4.1 |
| | Plymouth | 4.1 |
| 11 | Surry Mntain Lk | 5.4 |
| 12 | Benton | 7.0 |
| 13 | Hanover | 7.3 |
| 14 | Concord | 7.9 |
| 15 | Massabesic Lake | 9.6 |
| 16 | Keene | 10.1 |
| 17 | Mount Sunapee | 11.0 |
| | Peterboro | 11.0 |
| 19 | Durham | 11.3 |
| 20 | Epping | 11.4 |
| 21 | Nashua | 11.5 |

### HIGHEST

| | | |
|---|---|---|
| 1 | Nashua | 11.5 |
| 2 | Epping | 11.4 |
| 3 | Durham | 11.3 |
| 4 | Mount Sunapee | 11.0 |
| | Peterboro | 11.0 |
| 6 | Keene | 10.1 |
| 7 | Massabesic Lake | 9.6 |
| 8 | Concord | 7.9 |
| 9 | Hanover | 7.3 |
| 10 | Benton | 7.0 |
| 11 | Surry Mntain Lk | 5.4 |
| 12 | Grafton | 4.1 |
| | Plymouth | 4.1 |
| 14 | Pinkham Notch | 3.8 |
| 15 | Berlin | 2.9 |
| 16 | Monroe | 1.9 |
| 17 | Lancaster | 1.5 |
| 18 | North Stratford | 1.2 |
| 19 | Colebrook | 0.2 |
| 20 | First Conn Lake | -3.6 |
| 21 | Mt Washington | -4.6 |

## JULY MAXIMUM TEMPERATURE °F

### HIGHEST

| | | |
|---|---|---|
| 1 | Keene | 83.3 |
| 2 | Durham | 83.2 |
| 3 | Epping | 82.8 |
| 4 | Concord | 82.5 |
| | Hanover | 82.5 |
| | Nashua | 82.5 |
| 7 | Massabesic Lake | 82.0 |
| 8 | Surry Mntain Lk | 80.3 |
| 9 | Plymouth | 80.1 |
| 10 | Lancaster | 79.9 |
| | North Stratford | 79.9 |
| 12 | Grafton | 79.8 |
| 13 | Peterboro | 79.7 |
| 14 | Monroe | 79.4 |
| 15 | Mount Sunapee | 79.2 |
| 16 | Berlin | 79.0 |
| 17 | Benton | 78.0 |
| 18 | Colebrook | 77.3 |
| 19 | First Conn Lake | 73.8 |
| | Pinkham Notch | 73.8 |
| 21 | Mt Washington | 54.2 |

### LOWEST

| | | |
|---|---|---|
| 1 | Mt Washington | 54.2 |
| 2 | First Conn Lake | 73.8 |
| | Pinkham Notch | 73.8 |
| 4 | Colebrook | 77.3 |
| 5 | Benton | 78.0 |
| 6 | Berlin | 79.0 |
| 7 | Mount Sunapee | 79.2 |
| 8 | Monroe | 79.4 |
| 9 | Peterboro | 79.7 |
| 10 | Grafton | 79.8 |
| 11 | Lancaster | 79.9 |
| | North Stratford | 79.9 |
| 13 | Plymouth | 80.1 |
| 14 | Surry Mntain Lk | 80.3 |
| 15 | Massabesic Lake | 82.0 |
| 16 | Concord | 82.5 |
| | Hanover | 82.5 |
| | Nashua | 82.5 |
| 19 | Epping | 82.8 |
| 20 | Durham | 83.2 |
| 21 | Keene | 83.3 |

## ANNUAL PRECIPITATION (")

### HIGHEST

| | | |
|---|---|---|
| 1 | Mt Washington | 99.95 |
| 2 | Pinkham Notch | 57.21 |
| 3 | Peterboro | 44.52 |
| 4 | First Conn Lake | 44.30 |
| 5 | Epping | 44.27 |
| 6 | Nashua | 43.52 |
| 7 | Plymouth | 43.49 |
| 8 | Mount Sunapee | 42.27 |
| 9 | Durham | 42.12 |
| 10 | Keene | 40.35 |
| 11 | Colebrook | 40.08 |
| 12 | Grafton | 39.95 |
| 13 | Massabesic Lake | 39.58 |
| 14 | Surry Mntain Lk | 39.48 |
| 15 | North Stratford | 38.78 |
| 16 | Berlin | 38.43 |
| 17 | Hanover | 38.00 |
| 18 | Lancaster | 37.38 |
| 19 | Benton | 37.11 |
| 20 | Concord | 36.76 |
| 21 | Monroe | 36.05 |

### LOWEST

| | | |
|---|---|---|
| 1 | Monroe | 36.05 |
| 2 | Concord | 36.76 |
| 3 | Benton | 37.11 |
| 4 | Lancaster | 37.38 |
| 5 | Hanover | 38.00 |
| 6 | Berlin | 38.43 |
| 7 | North Stratford | 38.78 |
| 8 | Surry Mntain Lk | 39.48 |
| 9 | Massabesic Lake | 39.58 |
| 10 | Grafton | 39.95 |
| 11 | Colebrook | 40.08 |
| 12 | Keene | 40.35 |
| 13 | Durham | 42.12 |
| 14 | Mount Sunapee | 42.27 |
| 15 | Plymouth | 43.49 |
| 16 | Nashua | 43.52 |
| 17 | Epping | 44.27 |
| 18 | First Conn Lake | 44.30 |
| 19 | Peterboro | 44.52 |
| 20 | Pinkham Notch | 57.21 |
| 21 | Mt Washington | 99.95 |

## ANNUAL SNOWFALL (")

### HIGHEST

| | | |
|---|---|---|
| 1 | Mt Washington | 316.3 |
| 2 | First Conn Lake | 155.0 |
| 3 | Pinkham Notch | 142.4 |
| 4 | Colebrook | 94.1 |
| 5 | North Stratford | 91.0 |
| 6 | Grafton | 83.1 |
| 7 | Berlin | 83.0 |
| 8 | Mount Sunapee | 79.6 |
| 9 | Plymouth | 78.8 |
| 10 | Benton | 74.2 |
| 11 | Peterboro | 74.1 |
| 12 | Lancaster | 72.6 |
| 13 | Hanover | 67.1 |
| 14 | Concord | 65.0 |
| 15 | Surry Mntain Lk | 61.3 |
| 16 | Epping | 59.7 |
| 17 | Nashua | 59.6 |
| 18 | Keene | 58.6 |
| 19 | Massabesic Lake | 54.9 |
| 20 | Durham | 54.7 |

### LOWEST

| | | |
|---|---|---|
| 1 | Durham | 54.7 |
| 2 | Massabesic Lake | 54.9 |
| 3 | Keene | 58.6 |
| 4 | Nashua | 59.6 |
| 5 | Epping | 59.7 |
| 6 | Surry Mntain Lk | 61.3 |
| 7 | Concord | 65.0 |
| 8 | Hanover | 67.1 |
| 9 | Lancaster | 72.6 |
| 10 | Peterboro | 74.1 |
| 11 | Benton | 74.2 |
| 12 | Plymouth | 78.8 |
| 13 | Mount Sunapee | 79.6 |
| 14 | Berlin | 83.0 |
| 15 | Grafton | 83.1 |
| 16 | North Stratford | 91.0 |
| 17 | Colebrook | 94.1 |
| 18 | Pinkham Notch | 142.4 |
| 19 | First Conn Lake | 155.0 |
| 20 | Mt Washington | 316.3 |

# NEW JERSEY

PHYSICAL FEATURES. New Jersey, though one of the smaller states, has a varied topography. In the northwestern part a section comprising about one-fifth of the area of the State is known as the Highlands and Kittatinny Valley. This region is traversed by several low mountain ridges extending northeasterly across the State with valleys and rolling hills between. The highest of these ranges is the Kittatinny, which rises from the banks of the Delaware River at the famous Delaware Water Gap. To the eastward the region is studded with numerous lakes, some of the largest of which are Lakes Hopatcong, Mohawk, and Greenwood. Elevations up to 1,800 feet above sea level are found in the Kittatinny Mountains near the New York State line.

South and east of the Highlands is a region of about equal area known as the Red Sandstone Plain, or the Piedmont of New Jersey. It is generally hilly in its northwestern part, becoming rolling and then flat toward the south and southeast. At its northeastern corner are the Palisades, cliffs which rise abruptly from the Hudson River to heights of 200 to 500 feet. The seacoast section extends from Sandy Hook to Cape May, or about 125 miles. This area is characterized by long stretches of sandy beaches. Tidewater marshes become numerous toward the south.

In the southern interior a region known as the Pines is covered with scrubby forests of pine and some oak. The land is low and some of it is swampy. In fact, most of the State that lies south of a line connecting Jersey City and Trenton is low and flat with few elevations higher than 100 feet above mean sea level, these being mainly in Monmouth County.

About 30 percent of the area of New Jersey drains into the Delaware River and Delaware Bay, which form the western boundary. Nearly half of Sussex County, in the northwest, drains northward through the Wallkill River into the Hudson River of New York. The remainder of the State drains directly into the Atlantic Ocean through the Passaic, Hackensack, and Raritan Rivers in the north, and a number of small rivers and streams in the south.

GENERAL CLIMATE. The extreme length of the State is 166 miles and its greatest width only about 65. The difference in climate is quite marked between the southern tip at Cape May and the northern extremity in the Kittatinny Mountains. The former locality is almost surrounded by water and is fairly well removed from the influence of the frequent storms that cross the Great Lakes region and move out the St. Lawrence Valley. The northern extremity is well within the zone of influence of these storms and, in addition, lies at elevations varying from 800 to 1,800 feet. The influence of these high elevations on the temperature is considerable. The differences between these two localities are particularly marked in the winter, Cape May having a normal January temperature about the same as that of southwestern Virginia, while that of Layton, in the extreme northwest, is similar to that of the northern area of Ohio. Since the prevailing winds are mostly offshore, the ocean influence does not have full effect.

TEMPERATURE. Temperature differences between the northern and southern parts of the State are greatest in winter and least in summer. Nearly every station has registered readings of 100° F. or higher at some time, and all of them have records of zero or below. In the northern Highland area, the average date of last freeze (32° F) in spring is about May 2, and that of the first in fall, October 12. On the seacoast corresponding dates are April 6 and November 9, while in the central and southern interior the dates are April 23 and October 19. Freeze-free days in the northern Highlands average 163, with 217 along the seacoast and 179 in the central and southern interior.

PRECIPITATION. Northern New Jersey is near enough to the paths of the storms which cross the Great Lakes region and pass down the St. Lawrence Valley to receive part of its precipitation from that source. However, the heaviest general rains are produced by coastal storms of tropical origin. The centers of these storms usually pass some distance offshore, with heaviest rainfall and strongest wind near the coast. On several occasions tropical storms have moved inland along the south Atlantic coast, and then moved northward either through or to the west of New Jersey. The damage by high tides to coastal installations during the passage of a tropical storm is often severe, whether the storm passes offshore or inland.

The average annual precipitation ranges from about 40 inches along the southeast coast to 51 inches in north-central parts of the State. In other sections the annual averages are mostly between 43 and 47 inches. Rainfall is well distributed during the warm months. Heavy 24-hour falls of 7 or 8 inches are occasionally recorded. Brief periods of drought during the growing season are not uncommon, but prolonged droughts are relatively rare, occurring on the average once in 15 years. Flooding in New Jersey is usually caused by heavy general rains, at times associated with storms of tropical origin. Local flooding results from ice gorging.

The season during which measurable quantities of snow are likely to fall extends from about October 15 to April 20 in the Highlands, and from about November 15 to March 15 in the vicinity of Cape May. Average seasonal amounts range from about 13 inches at Cape May to nearly 50 inches in the Highlands. Snowfalls of 10 or more inches in a single storm are occasional occurrences.

The number of days a month with measurable precipitation averages 8 for each of the fall months (September, October, and November) and 9 to 12 for the other months of the year; the average yearly number is 120. Midday relative humidity averages 68 percent along the seacoast and 57 percent or less at inland locations.

Normally, sunshine varies from slightly over one-half of the possible amount in the northern counties to about 60 percent in the south. The prevailing wind is from the northwest from October to April, inclusive, and from the southwest for the other months of the year.

## COUNTY INDEX

### *Atlantic County*
ATLANTIC CITY INT AP
ATLANTIC CITY MARINA

### *Burlington County*
INDIAN MILLS 2 W
MOORESTOWN
PEMBERTON 3 S

### *Cape May County*
BELLEPLAIN ST FOREST
CAPE MAY 2 NW

### *Cumberland County*
MILLVILLE MUNI AP
SEABROOK FARMS

### *Essex County*
CANOE BROOK
ESSEX FELLS SERV BLD
NEWARK INTL ARPT

### *Gloucester County*
GLASSBORO

### *Hunterdon County*
FLEMINGTON 5 NNW
LAMBERTVILLE

### *Mercer County*
HIGHTSTOWN 2 W

### *Monmouth County*
LONG BRANCH OAKHURST

### *Morris County*
BOONTON 1 SE
LONG VALLEY

### *Ocean County*
TOMS RIVER
TUCKERTON

### *Passaic County*
CHARLOTTEBURG RESERV
LITTLE FALLS

### *Salem County*
WOODSTOWN

### *Somerset County*
SOMERVILLE 3 NW

### *Sussex County*
NEWTON ST PAUL ABBEY
SUSSEX

### *Union County*
PLAINFIELD

## ELEVATION INDEX

| FEET | STATION NAME |
|---|---|
| 7 | NEWARK INTL ARPT |
| 10 | TOMS RIVER |
| 16 | ATLANTIC CITY MARINA |
| 20 | CAPE MAY 2 NW |
| 20 | TUCKERTON |
| 30 | LONG BRANCH OAKHURST |
| 49 | BELLEPLAIN ST FOREST |
| 49 | PEMBERTON 3 S |
| 49 | WOODSTOWN |
| 56 | MOORESTOWN |
| 59 | LAMBERTVILLE |
| 59 | SOMERVILLE 3 NW |
| 66 | ATLANTIC CITY INT AP |
| 72 | MILLVILLE MUNI AP |
| 89 | PLAINFIELD |
| 102 | HIGHTSTOWN 2 W |
| 102 | INDIAN MILLS 2 W |
| 102 | SEABROOK FARMS |
| 151 | GLASSBORO |
| 161 | LITTLE FALLS |
| 180 | CANOE BROOK |
| 249 | ESSEX FELLS SERV BLD |
| 260 | FLEMINGTON 5 NNW |
| 280 | BOONTON 1 SE |
| 449 | SUSSEX |
| 541 | LONG VALLEY |
| 571 | NEWTON ST PAUL ABBEY |
| 761 | CHARLOTTEBURG RESERV |

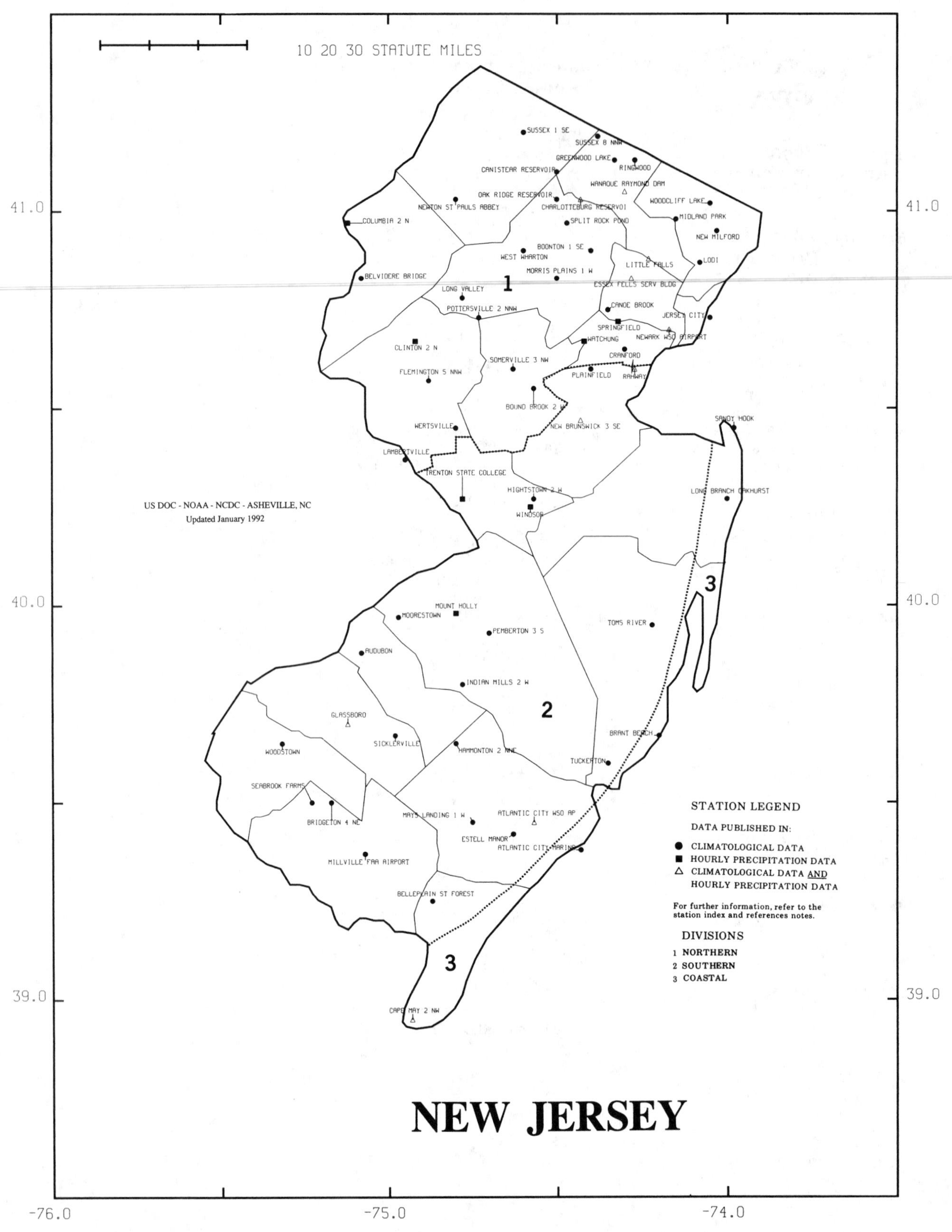
10 20 30 STATUTE MILES
41.0
40.0
39.0
-76.0
-75.0
-74.0
US DOC - NOAA - NCDC - ASHEVILLE, NC
Updated January 1992
STATION LEGEND
DATA PUBLISHED IN:
CLIMATOLOGICAL DATA
HOURLY PRECIPITATION DATA
CLIMATOLOGICAL DATA AND HOURLY PRECIPITATION DATA
For further information, refer to the station index and references notes.
DIVISIONS
1 NORTHERN
2 SOUTHERN
3 COASTAL
NEW JERSEY

## ATLANTIC CITY INT AP *Atlantic County* ELEVATION 66 ft LAT/LONG 39° 27 ' N / 74° 35 ' W

| | JAN | FEB | MAR | APR | MAY | JUN | JUL | AUG | SEP | OCT | NOV | DEC | YEAR |
|---|---|---|---|---|---|---|---|---|---|---|---|---|---|
| Maximum Temp °F | 40.9 | 43.1 | 51.3 | 61.4 | 71.1 | 80.4 | 84.7 | 83.3 | 76.6 | 66.0 | 56.0 | 46.2 | 63.4 |
| Minimum Temp °F | 22.1 | 23.8 | 31.2 | 39.7 | 49.5 | 58.8 | 65.2 | 63.5 | 55.4 | 43.6 | 35.7 | 26.9 | 42.9 |
| Mean Temp °F | 31.6 | 33.5 | 41.2 | 50.5 | 60.4 | 69.6 | 75.0 | 73.4 | 66.0 | 54.8 | 45.9 | 36.6 | 53.2 |
| Days Max Temp ≥ 90 °F | 0 | 0 | 0 | 0 | 1 | 4 | 7 | 5 | 1 | 0 | 0 | 0 | 18 |
| Days Max Temp ≤ 32 °F | 7 | 5 | 1 | 0 | 0 | 0 | 0 | 0 | 0 | 0 | 0 | 3 | 16 |
| Days Min Temp ≤ 32 °F | 25 | 22 | 17 | 6 | 0 | 0 | 0 | 0 | 0 | 4 | 13 | 22 | 109 |
| Days Min Temp ≤ 0 °F | 1 | 0 | 0 | 0 | 0 | 0 | 0 | 0 | 0 | 0 | 0 | 0 | 1 |
| Heating Degree Days | 1029 | 884 | 730 | 433 | 184 | 29 | 1 | 7 | 73 | 325 | 568 | 875 | 5138 |
| Cooling Degree Days | 0 | 0 | 1 | 7 | 56 | 188 | 349 | 275 | 115 | 19 | 1 | 0 | 1011 |
| Total Precipitation (") | 3.18 | 2.96 | 3.83 | 3.41 | 3.26 | 2.54 | 4.01 | 4.41 | 2.84 | 2.71 | 3.20 | 3.27 | 39.62 |
| Days ≥ 0.1" Precip | 7 | 6 | 7 | 7 | 6 | 5 | 6 | 6 | 5 | 5 | 6 | 6 | 72 |
| Total Snowfall (") | 5.2 | 6.4 | 1.8 | 0.4 | 0.0 | 0.0 | 0.0 | 0.0 | 0.0 | 0.0 | 0.5 | 1.9 | 16.2 |
| Days ≥ 1" Snow Depth | 4 | 4 | 1 | 0 | 0 | 0 | 0 | 0 | 0 | 0 | 0 | 2 | 11 |

## ATLANTIC CITY MARINA *Atlantic County* ELEVATION 16 ft LAT/LONG 39° 22 ' N / 74° 25 ' W

| | JAN | FEB | MAR | APR | MAY | JUN | JUL | AUG | SEP | OCT | NOV | DEC | YEAR |
|---|---|---|---|---|---|---|---|---|---|---|---|---|---|
| Maximum Temp °F | 39.9 | 42.0 | 48.7 | 57.4 | 66.0 | 74.7 | 80.4 | 79.9 | 73.9 | 64.1 | 54.9 | 46.3 | 60.7 |
| Minimum Temp °F | 27.1 | 29.1 | 36.0 | 44.4 | 53.8 | 62.6 | 68.6 | 68.1 | 61.9 | 51.0 | 41.8 | 33.2 | 48.1 |
| Mean Temp °F | 33.5 | 35.5 | 42.4 | 50.9 | 59.9 | 68.6 | 74.5 | 74.0 | 67.9 | 57.5 | 48.4 | 39.8 | 54.4 |
| Days Max Temp ≥ 90 °F | 0 | 0 | 0 | 0 | 0 | 0 | 2 | 1 | 0 | 0 | 0 | 0 | 3 |
| Days Max Temp ≤ 32 °F | 7 | 4 | 1 | 0 | 0 | 0 | 0 | 0 | 0 | 0 | 0 | 2 | 14 |
| Days Min Temp ≤ 32 °F | 22 | 18 | 10 | 1 | 0 | 0 | 0 | 0 | 0 | 0 | 4 | 14 | 69 |
| Days Min Temp ≤ 0 °F | 0 | 0 | 0 | 0 | 0 | 0 | 0 | 0 | 0 | 0 | 0 | 0 | 0 |
| Heating Degree Days | 969 | 826 | 694 | 418 | 178 | 20 | 1 | 2 | 33 | 241 | 492 | 775 | 4649 |
| Cooling Degree Days | 0 | 0 | 0 | 2 | 38 | 152 | 326 | 293 | 133 | 21 | 1 | 0 | 966 |
| Total Precipitation (") | 3.16 | 2.87 | 3.60 | 3.11 | 3.00 | 2.44 | 3.41 | 4.15 | 2.54 | 2.43 | 2.97 | 3.48 | 37.16 |
| Days ≥ 0.1" Precip | 7 | 6 | 6 | 6 | 6 | 5 | 6 | 6 | 5 | 4 | 6 | 6 | 69 |
| Total Snowfall (") | na | na | na | na | na | na | na | na | na | na | na | na | na |
| Days ≥ 1" Snow Depth | na | na | na | na | na | na | na | na | na | na | na | na | na |

## BELLEPLAIN ST FOREST *Cape May County* ELEVATION 49 ft LAT/LONG 39° 16 ' N / 74° 52 ' W

| | JAN | FEB | MAR | APR | MAY | JUN | JUL | AUG | SEP | OCT | NOV | DEC | YEAR |
|---|---|---|---|---|---|---|---|---|---|---|---|---|---|
| Maximum Temp °F | 42.9 | 45.6 | 54.4 | 65.2 | 74.7 | 82.4 | 86.3 | 84.8 | 78.9 | 68.5 | 58.5 | 48.3 | 65.9 |
| Minimum Temp °F | 22.4 | 23.9 | 31.3 | 39.3 | 49.1 | 58.2 | 64.2 | 62.7 | 55.5 | 44.0 | 36.1 | 27.5 | 42.9 |
| Mean Temp °F | 32.7 | 34.8 | 42.9 | 52.3 | 61.9 | 70.3 | 75.3 | 73.8 | 67.3 | 56.3 | 47.3 | 37.9 | 54.4 |
| Days Max Temp ≥ 90 °F | 0 | 0 | 0 | 0 | 1 | 4 | 9 | 6 | 2 | 0 | 0 | 0 | 22 |
| Days Max Temp ≤ 32 °F | 5 | 3 | 0 | 0 | 0 | 0 | 0 | 0 | 0 | 0 | 0 | 2 | 10 |
| Days Min Temp ≤ 32 °F | 25 | 22 | 18 | 8 | 1 | 0 | 0 | 0 | 0 | 5 | 12 | 22 | 113 |
| Days Min Temp ≤ 0 °F | 1 | 0 | 0 | 0 | 0 | 0 | 0 | 0 | 0 | 0 | 0 | 0 | 1 |
| Heating Degree Days | 995 | 847 | 680 | 384 | 145 | 24 | 1 | 5 | 57 | 283 | 526 | 832 | 4779 |
| Cooling Degree Days | 0 | 0 | 1 | 8 | 60 | 195 | 345 | 267 | 125 | 21 | 2 | 0 | 1024 |
| Total Precipitation (") | 3.36 | 3.01 | 3.87 | 3.57 | 3.55 | 2.88 | 3.70 | 4.99 | 3.37 | 3.42 | 3.16 | 3.55 | 42.43 |
| Days ≥ 0.1" Precip | 6 | 6 | 7 | 7 | 7 | 5 | 6 | 6 | 5 | 5 | 6 | 6 | 72 |
| Total Snowfall (") | 4.0 | *5.4* | 1.6 | 0.1 | 0.0 | 0.0 | 0.0 | 0.0 | 0.0 | 0.0 | 0.4 | *1.4* | 12.9 |
| Days ≥ 1" Snow Depth | *5* | *4* | *1* | 0 | 0 | 0 | 0 | 0 | 0 | 0 | 0 | *2* | 12 |

## BOONTON 1 SE *Morris County* ELEVATION 280 ft LAT/LONG 40° 54 ' N / 74° 24 ' W

| | JAN | FEB | MAR | APR | MAY | JUN | JUL | AUG | SEP | OCT | NOV | DEC | YEAR |
|---|---|---|---|---|---|---|---|---|---|---|---|---|---|
| Maximum Temp °F | 35.4 | 38.3 | 47.7 | 59.6 | 70.6 | 78.9 | 83.8 | 82.1 | 74.8 | 63.5 | 52.4 | 40.9 | 60.7 |
| Minimum Temp °F | 18.0 | 19.6 | 28.8 | 38.7 | 48.2 | 57.0 | 62.1 | 60.6 | 52.4 | 40.5 | 33.7 | 24.3 | 40.3 |
| Mean Temp °F | 26.7 | 29.0 | 38.3 | 49.2 | 59.4 | 68.0 | 73.0 | 71.4 | 63.7 | 52.0 | 43.1 | 32.7 | 50.5 |
| Days Max Temp ≥ 90 °F | 0 | 0 | 0 | 0 | 0 | 2 | 5 | 3 | 1 | 0 | 0 | 0 | 11 |
| Days Max Temp ≤ 32 °F | 12 | 8 | 1 | 0 | 0 | 0 | 0 | 0 | 0 | 0 | 0 | 6 | 27 |
| Days Min Temp ≤ 32 °F | 29 | 26 | 21 | 6 | 0 | 0 | 0 | 0 | 0 | 6 | 14 | 26 | 128 |
| Days Min Temp ≤ 0 °F | 2 | 1 | 0 | 0 | 0 | 0 | 0 | 0 | 0 | 0 | 0 | 0 | 3 |
| Heating Degree Days | 1180 | 1010 | 822 | 472 | 200 | 36 | 4 | 10 | 104 | 399 | 652 | 995 | 5884 |
| Cooling Degree Days | 0 | 0 | 0 | 4 | 41 | 142 | 280 | 217 | 72 | 5 | 0 | 0 | 761 |
| Total Precipitation (") | 3.52 | 3.01 | 4.15 | 4.16 | 4.85 | 4.34 | 4.56 | 4.34 | 4.74 | 3.68 | 4.30 | 3.84 | 49.49 |
| Days ≥ 0.1" Precip | 6 | 6 | 7 | 7 | 8 | 7 | 7 | 7 | 6 | 6 | 7 | 6 | 80 |
| Total Snowfall (") | 8.4 | 9.6 | 5.3 | 0.7 | 0.0 | 0.0 | 0.0 | 0.0 | 0.0 | 0.0 | 0.8 | 4.4 | 29.2 |
| Days ≥ 1" Snow Depth | na | na | *3* | 0 | 0 | 0 | 0 | 0 | 0 | 0 | 0 | na | na |

## CANOE BROOK *Essex County* ELEVATION 180 ft LAT/LONG 40° 45 ' N / 74° 21 ' W

| | JAN | FEB | MAR | APR | MAY | JUN | JUL | AUG | SEP | OCT | NOV | DEC | YEAR |
|---|---|---|---|---|---|---|---|---|---|---|---|---|---|
| Maximum Temp °F | 36.8 | 39.8 | 49.1 | 60.8 | 71.6 | 80.2 | 85.2 | 83.4 | 76.2 | 65.1 | 53.7 | 42.4 | 62.0 |
| Minimum Temp °F | 16.9 | 18.6 | 28.0 | 37.3 | 47.1 | 56.6 | 61.8 | 60.6 | 52.5 | 40.0 | 32.8 | 23.4 | 39.6 |
| Mean Temp °F | 26.9 | 29.2 | 38.6 | 49.1 | 59.4 | 68.4 | 73.6 | 72.1 | 64.4 | 52.6 | 43.2 | 32.9 | 50.9 |
| Days Max Temp ≥ 90 °F | 0 | 0 | 0 | 0 | 1 | 3 | 8 | 5 | 1 | 0 | 0 | 0 | 18 |
| Days Max Temp ≤ 32 °F | 10 | 7 | 1 | 0 | 0 | 0 | 0 | 0 | 0 | 0 | 0 | 5 | 23 |
| Days Min Temp ≤ 32 °F | 29 | 26 | 22 | 9 | 1 | 0 | 0 | 0 | 0 | 6 | 16 | 27 | 136 |
| Days Min Temp ≤ 0 °F | 3 | 1 | 0 | 0 | 0 | 0 | 0 | 0 | 0 | 0 | 0 | 0 | 4 |
| Heating Degree Days | 1179 | 1009 | 818 | 479 | 209 | 39 | 4 | 9 | 97 | 388 | 651 | 993 | 5875 |
| Cooling Degree Days | 0 | 0 | 0 | 5 | 44 | 154 | 291 | 235 | 85 | 8 | 0 | 0 | 822 |
| Total Precipitation (") | 3.70 | 3.11 | 4.37 | 4.08 | 4.79 | 4.25 | 4.65 | 4.74 | 4.75 | 3.95 | 4.44 | 4.05 | 50.88 |
| Days ≥ 0.1" Precip | 6 | 6 | 7 | 7 | 8 | 7 | 7 | 6 | 6 | 5 | 7 | 6 | 78 |
| Total Snowfall (") | 9.2 | 8.7 | 4.7 | 0.7 | 0.0 | 0.0 | 0.0 | 0.0 | 0.0 | 0.0 | 0.5 | 3.4 | 27.2 |
| Days ≥ 1" Snow Depth | 15 | 12 | 4 | 0 | 0 | 0 | 0 | 0 | 0 | 0 | 0 | 4 | 35 |

## CAPE MAY 2 NW *Cape May County* ELEVATION 20 ft LAT/LONG 38° 57 ' N / 74° 57 ' W

| | JAN | FEB | MAR | APR | MAY | JUN | JUL | AUG | SEP | OCT | NOV | DEC | YEAR |
|---|---|---|---|---|---|---|---|---|---|---|---|---|---|
| Maximum Temp °F | 40.7 | 42.4 | 49.8 | 59.5 | 68.9 | 78.0 | 83.5 | 82.5 | 76.9 | 66.2 | 56.4 | 46.5 | 62.6 |
| Minimum Temp °F | 26.5 | 27.7 | 34.7 | 43.0 | 52.4 | 61.5 | 67.4 | 66.6 | 60.6 | 50.0 | 41.3 | 31.7 | 47.0 |
| Mean Temp °F | 33.6 | 35.1 | 42.2 | 51.3 | 60.7 | 69.8 | 75.4 | 74.6 | 68.8 | 58.1 | 48.9 | 39.1 | 54.8 |
| Days Max Temp ≥ 90 °F | 0 | 0 | 0 | 0 | 0 | 1 | 4 | 2 | 1 | 0 | 0 | 0 | 8 |
| Days Max Temp ≤ 32 °F | 6 | 4 | 1 | 0 | 0 | 0 | 0 | 0 | 0 | 0 | 0 | 2 | 13 |
| Days Min Temp ≤ 32 °F | 23 | 20 | 11 | 2 | 0 | 0 | 0 | 0 | 0 | 1 | 5 | 16 | 78 |
| Days Min Temp ≤ 0 °F | 0 | 0 | 0 | 0 | 0 | 0 | 0 | 0 | 0 | 0 | 0 | 0 | 0 |
| Heating Degree Days | 965 | 838 | 699 | 408 | 160 | 17 | 0 | 2 | 32 | 229 | 477 | 796 | 4623 |
| Cooling Degree Days | 0 | 0 | 0 | 4 | 45 | 184 | 351 | 298 | 146 | 24 | 1 | 0 | 1053 |
| Total Precipitation (") | 3.45 | 3.00 | 4.02 | 3.43 | 3.54 | 3.12 | 3.44 | 3.97 | 2.99 | 3.15 | 3.09 | 3.54 | 40.74 |
| Days ≥ 0.1" Precip | 7 | 7 | 7 | 7 | 6 | 5 | 5 | 6 | 5 | 5 | 6 | 6 | 72 |
| Total Snowfall (") | *4.1* | *6.6* | 1.9 | 0.1 | 0.0 | 0.0 | 0.0 | 0.0 | 0.0 | 0.0 | 0.4 | 1.4 | 14.5 |
| Days ≥ 1" Snow Depth | 3 | *5* | *1* | 0 | 0 | 0 | 0 | 0 | 0 | 0 | 0 | 1 | 10 |

## CHARLOTTEBURG RESERV *Passaic County* ELEVATION 761 ft LAT/LONG 41° 2 ' N / 74° 26 ' W

| | JAN | FEB | MAR | APR | MAY | JUN | JUL | AUG | SEP | OCT | NOV | DEC | YEAR |
|---|---|---|---|---|---|---|---|---|---|---|---|---|---|
| Maximum Temp °F | 34.3 | 36.9 | 46.1 | 58.0 | 69.1 | 77.1 | 82.2 | 80.6 | 73.3 | 62.5 | 51.3 | 39.6 | 59.2 |
| Minimum Temp °F | 15.3 | 16.4 | 25.3 | 35.5 | 45.2 | 54.1 | 59.0 | 57.3 | 49.4 | 38.1 | 31.3 | 21.5 | 37.4 |
| Mean Temp °F | 24.9 | 26.7 | 35.7 | 46.8 | 57.2 | 65.6 | 70.7 | 68.9 | 61.4 | 50.3 | 41.4 | 30.6 | 48.4 |
| Days Max Temp ≥ 90 °F | 0 | 0 | 0 | 0 | 0 | 1 | 4 | 1 | 1 | 0 | 0 | 0 | 7 |
| Days Max Temp ≤ 32 °F | 13 | 10 | 3 | 0 | 0 | 0 | 0 | 0 | 0 | 0 | 0 | 8 | 34 |
| Days Min Temp ≤ 32 °F | 29 | 26 | 25 | 11 | 1 | 0 | 0 | 0 | 1 | 9 | 18 | 28 | 148 |
| Days Min Temp ≤ 0 °F | 3 | 2 | 0 | 0 | 0 | 0 | 0 | 0 | 0 | 0 | 0 | 1 | 6 |
| Heating Degree Days | 1239 | 1076 | 902 | 543 | 259 | 66 | 11 | 25 | 150 | 450 | 702 | 1061 | 6484 |
| Cooling Degree Days | 0 | 0 | 0 | 3 | 27 | 96 | 208 | 152 | 46 | 4 | 0 | 0 | 536 |
| Total Precipitation (") | 3.64 | 3.28 | 4.47 | 4.36 | 4.78 | 4.33 | 4.48 | 4.73 | 4.76 | 4.07 | 4.75 | 4.10 | 51.75 |
| Days ≥ 0.1" Precip | 7 | 6 | 7 | 7 | 7 | 7 | 7 | 7 | 6 | 6 | 7 | 6 | 80 |
| Total Snowfall (") | 10.0 | 10.7 | 7.4 | 1.3 | 0.0 | 0.0 | 0.0 | 0.0 | 0.0 | 0.1 | 1.1 | 5.6 | 36.2 |
| Days ≥ 1" Snow Depth | 16 | 14 | 7 | 1 | 0 | 0 | 0 | 0 | 0 | 0 | 1 | 5 | 44 |

## ESSEX FELLS SERV BLD *Essex County* ELEVATION 249 ft LAT/LONG 40° 50 ' N / 74° 18 ' W

| | JAN | FEB | MAR | APR | MAY | JUN | JUL | AUG | SEP | OCT | NOV | DEC | YEAR |
|---|---|---|---|---|---|---|---|---|---|---|---|---|---|
| Maximum Temp °F | 35.9 | 39.4 | 48.9 | 60.7 | 71.7 | 79.8 | 84.8 | 83.0 | 75.5 | 64.3 | 53.2 | 41.1 | 61.5 |
| Minimum Temp °F | 17.9 | 20.0 | 27.9 | 37.6 | 47.4 | 56.5 | 61.9 | 60.1 | 52.2 | 40.6 | 33.6 | 23.7 | 40.0 |
| Mean Temp °F | 27.0 | 30.1 | 38.4 | 49.2 | 59.6 | 68.3 | 73.4 | 71.6 | 63.9 | 52.5 | 43.4 | 32.4 | 50.8 |
| Days Max Temp ≥ 90 °F | 0 | 0 | 0 | 0 | 1 | 3 | 7 | 4 | 1 | 0 | 0 | 0 | 16 |
| Days Max Temp ≤ 32 °F | 11 | 7 | 1 | 0 | 0 | 0 | 0 | 0 | 0 | 0 | 0 | 6 | 25 |
| Days Min Temp ≤ 32 °F | 29 | 26 | 22 | 8 | 1 | 0 | 0 | 0 | 0 | 6 | 15 | 27 | 134 |
| Days Min Temp ≤ 0 °F | 2 | 0 | 0 | 0 | 0 | 0 | 0 | 0 | 0 | 0 | 0 | 0 | 2 |
| Heating Degree Days | 1172 | 980 | 817 | 475 | 199 | 38 | 3 | 9 | 103 | 388 | 642 | 1004 | 5830 |
| Cooling Degree Days | 0 | 0 | 1 | 5 | 40 | 136 | 279 | 211 | 72 | 6 | 0 | 0 | 750 |
| Total Precipitation (") | 3.84 | 3.11 | 4.08 | 4.37 | 4.80 | 4.24 | 4.73 | 4.51 | 4.77 | 3.84 | 4.25 | 4.19 | 50.73 |
| Days ≥ 0.1" Precip | 5 | 5 | 6 | 7 | 8 | 7 | 7 | 7 | 6 | 5 | 6 | 6 | 75 |
| Total Snowfall (") | *6.1* | *5.2* | 4.1 | 0.7 | 0.0 | 0.0 | 0.0 | 0.0 | 0.0 | 0.1 | 0.5 | *2.8* | 19.5 |
| Days ≥ 1" Snow Depth | na | na | *2* | 0 | 0 | 0 | 0 | 0 | 0 | 0 | 0 | *2* | na |

## FLEMINGTON 5 NNW *Hunterdon County* ELEVATION 260 ft LAT/LONG 40° 34 ' N / 74° 53 ' W

| | JAN | FEB | MAR | APR | MAY | JUN | JUL | AUG | SEP | OCT | NOV | DEC | YEAR |
|---|---|---|---|---|---|---|---|---|---|---|---|---|---|
| Maximum Temp °F | 36.6 | 39.6 | 49.2 | 61.5 | 72.2 | 80.8 | 85.8 | 83.8 | 76.4 | 65.2 | 53.7 | 42.4 | 62.3 |
| Minimum Temp °F | 17.2 | 18.9 | 27.4 | 36.6 | 46.2 | 55.7 | 61.6 | 59.7 | 52.0 | 39.8 | 32.2 | 23.6 | 39.2 |
| Mean Temp °F | 26.9 | 29.3 | 38.3 | 49.1 | 59.2 | 68.3 | 73.7 | 71.8 | 64.2 | 52.5 | 43.0 | 33.1 | 50.8 |
| Days Max Temp ≥ 90 °F | 0 | 0 | 0 | 0 | 1 | 4 | 9 | 5 | 2 | 0 | 0 | 0 | 21 |
| Days Max Temp ≤ 32 °F | 10 | 7 | 1 | 0 | 0 | 0 | 0 | 0 | 0 | 0 | 0 | 5 | 23 |
| Days Min Temp ≤ 32 °F | 29 | 26 | 23 | 10 | 1 | 0 | 0 | 0 | 0 | 8 | 17 | 26 | 140 |
| Days Min Temp ≤ 0 °F | 3 | 1 | 0 | 0 | 0 | 0 | 0 | 0 | 0 | 0 | 0 | 0 | 4 |
| Heating Degree Days | 1174 | 1003 | 822 | 476 | 206 | 37 | 3 | 12 | 100 | 387 | 654 | 983 | 5857 |
| Cooling Degree Days | 0 | 0 | 0 | 5 | 41 | 140 | 301 | 229 | 83 | 9 | 0 | 0 | 808 |
| Total Precipitation (") | 3.78 | 3.01 | 4.19 | 3.90 | 4.71 | 4.37 | 4.68 | 4.18 | 4.14 | 3.55 | 3.90 | 4.00 | 48.41 |
| Days ≥ 0.1" Precip | 7 | 6 | 7 | 7 | 8 | 7 | 7 | 7 | 6 | 5 | 7 | 7 | 81 |
| Total Snowfall (") | 10.7 | 9.8 | 6.0 | 0.7 | 0.0 | 0.0 | 0.0 | 0.0 | 0.0 | 0.1 | 0.6 | 4.7 | 32.6 |
| Days ≥ 1" Snow Depth | 15 | 11 | 4 | 0 | 0 | 0 | 0 | 0 | 0 | 0 | 0 | 4 | 34 |

## GLASSBORO *Gloucester County* ELEVATION 151 ft LAT/LONG 39° 42 ' N / 75° 7 ' W

| | JAN | FEB | MAR | APR | MAY | JUN | JUL | AUG | SEP | OCT | NOV | DEC | YEAR |
|---|---|---|---|---|---|---|---|---|---|---|---|---|---|
| Maximum Temp °F | 38.9 | 41.7 | 50.9 | 62.0 | 72.4 | 81.2 | 85.7 | 84.3 | 77.6 | 65.9 | 55.4 | 44.5 | 63.4 |
| Minimum Temp °F | 22.7 | 24.4 | 32.7 | 41.8 | 51.4 | 60.6 | 66.2 | 64.7 | 57.3 | 45.0 | 37.4 | 28.3 | 44.4 |
| Mean Temp °F | 30.8 | 33.1 | 41.9 | 51.9 | 61.9 | 70.9 | 76.0 | 74.6 | 67.4 | 55.5 | 46.5 | 36.4 | 53.9 |
| Days Max Temp ≥ 90 °F | 0 | 0 | 0 | 0 | 1 | 4 | 8 | 5 | 2 | 0 | 0 | 0 | 20 |
| Days Max Temp ≤ 32 °F | 8 | 6 | 1 | 0 | 0 | 0 | 0 | 0 | 0 | 0 | 0 | 3 | 18 |
| Days Min Temp ≤ 32 °F | 26 | 23 | 15 | 3 | 0 | 0 | 0 | 0 | 0 | 2 | 9 | 22 | 100 |
| Days Min Temp ≤ 0 °F | 1 | 0 | 0 | 0 | 0 | 0 | 0 | 0 | 0 | 0 | 0 | 0 | 1 |
| Heating Degree Days | 1052 | 895 | 712 | 397 | 147 | 19 | 1 | 3 | 53 | 303 | 551 | 879 | 5012 |
| Cooling Degree Days | 0 | 0 | 1 | 9 | 66 | 214 | 370 | 297 | 129 | 17 | 1 | 0 | 1104 |
| Total Precipitation (") | 3.40 | 2.77 | 3.94 | 3.72 | 4.09 | 3.76 | 4.33 | 4.38 | 3.60 | 3.33 | 3.41 | 3.77 | 44.50 |
| Days ≥ 0.1" Precip | 7 | 6 | 7 | 7 | 7 | 6 | 7 | 6 | 5 | 5 | 6 | 7 | 76 |
| Total Snowfall (") | na | na | na | 0.0 | 0.0 | 0.0 | 0.0 | 0.0 | 0.0 | 0.0 | 0.0 | *1.5* | na |
| Days ≥ 1" Snow Depth | na | na | na | 0 | 0 | 0 | 0 | 0 | 0 | 0 | 0 | *2* | na |

## HIGHTSTOWN 2 W *Mercer County* ELEVATION 102 ft LAT/LONG 40° 17 ' N / 74° 31 ' W

| | JAN | FEB | MAR | APR | MAY | JUN | JUL | AUG | SEP | OCT | NOV | DEC | YEAR |
|---|---|---|---|---|---|---|---|---|---|---|---|---|---|
| Maximum Temp °F | 37.8 | 40.7 | 50.0 | 61.4 | 72.1 | 80.9 | 85.3 | 83.6 | 76.5 | 65.3 | 54.6 | 43.4 | 62.6 |
| Minimum Temp °F | 21.0 | 22.9 | 30.9 | 39.5 | 49.2 | 58.3 | 63.6 | 62.2 | 54.5 | 42.9 | 35.6 | 27.0 | 42.3 |
| Mean Temp °F | 29.4 | 31.8 | 40.4 | 50.5 | 60.7 | 69.6 | 74.5 | 72.9 | 65.5 | 54.1 | 45.1 | 35.2 | 52.5 |
| Days Max Temp ≥ 90 °F | 0 | 0 | 0 | 0 | 1 | 4 | 8 | 5 | 1 | 0 | 0 | 0 | 19 |
| Days Max Temp ≤ 32 °F | 9 | 6 | 1 | 0 | 0 | 0 | 0 | 0 | 0 | 0 | 0 | 4 | 20 |
| Days Min Temp ≤ 32 °F | 27 | 23 | 18 | 6 | 0 | 0 | 0 | 0 | 0 | 5 | 12 | 23 | 114 |
| Days Min Temp ≤ 0 °F | 1 | 1 | 0 | 0 | 0 | 0 | 0 | 0 | 0 | 0 | 0 | 0 | 2 |
| Heating Degree Days | 1095 | 930 | 755 | 435 | 173 | 28 | 2 | 8 | 81 | 341 | 591 | 916 | 5355 |
| Cooling Degree Days | 0 | 0 | 1 | 6 | 53 | 173 | 317 | 246 | 98 | 13 | 1 | 0 | 908 |
| Total Precipitation (") | 3.37 | 2.78 | 3.92 | 3.85 | 4.27 | 3.79 | 4.93 | 4.72 | 4.10 | 3.25 | 3.71 | 3.81 | 46.50 |
| Days ≥ 0.1" Precip | 7 | 6 | 7 | 7 | 8 | 7 | 7 | 7 | 6 | 5 | 6 | 7 | 80 |
| Total Snowfall (") | 7.7 | 8.4 | 4.4 | 0.6 | 0.0 | 0.0 | 0.0 | 0.0 | 0.0 | 0.1 | 0.5 | 3.3 | 25.0 |
| Days ≥ 1" Snow Depth | 10 | 8 | 3 | 0 | 0 | 0 | 0 | 0 | 0 | 0 | 0 | 3 | 24 |

## INDIAN MILLS 2 W *Burlington County* ELEVATION 102 ft LAT/LONG 39° 48 ' N / 74° 47 ' W

| | JAN | FEB | MAR | APR | MAY | JUN | JUL | AUG | SEP | OCT | NOV | DEC | YEAR |
|---|---|---|---|---|---|---|---|---|---|---|---|---|---|
| Maximum Temp °F | 40.2 | 43.5 | 52.6 | 64.1 | 74.4 | 82.8 | 86.9 | 85.3 | 78.4 | 67.3 | 56.7 | 45.7 | 64.8 |
| Minimum Temp °F | 20.8 | 22.8 | 30.5 | 38.8 | 48.6 | 57.6 | 63.0 | 61.5 | 53.9 | 42.3 | 34.8 | 26.3 | 41.7 |
| Mean Temp °F | 30.5 | 33.2 | 41.6 | 51.4 | 61.5 | 70.2 | 75.0 | 73.4 | 66.1 | 54.8 | 45.8 | 36.1 | 53.3 |
| Days Max Temp ≥ 90 °F | 0 | 0 | 0 | 0 | 2 | 5 | 10 | 7 | 2 | 0 | 0 | 0 | 26 |
| Days Max Temp ≤ 32 °F | 7 | 4 | 0 | 0 | 0 | 0 | 0 | 0 | 0 | 0 | 0 | 3 | 14 |
| Days Min Temp ≤ 32 °F | 27 | 23 | 18 | 8 | 1 | 0 | 0 | 0 | 0 | 7 | 13 | 22 | 119 |
| Days Min Temp ≤ 0 °F | 1 | 1 | 0 | 0 | 0 | 0 | 0 | 0 | 0 | 0 | 0 | 0 | 2 |
| Heating Degree Days | 1062 | 893 | 721 | 410 | 160 | 22 | 2 | 7 | 73 | 325 | 572 | 890 | 5137 |
| Cooling Degree Days | 0 | 0 | 1 | 11 | 70 | 198 | 345 | 274 | 117 | 20 | 2 | 0 | 1038 |
| Total Precipitation (") | 3.71 | 3.00 | 4.12 | 3.86 | 3.78 | 3.54 | 4.48 | 4.92 | 3.55 | 3.16 | 3.60 | 4.17 | 45.89 |
| Days ≥ 0.1" Precip | 7 | 5 | 7 | 7 | 7 | 6 | 6 | 7 | 5 | 5 | 6 | 7 | 75 |
| Total Snowfall (") | 7.2 | 6.4 | 3.3 | 0.3 | 0.0 | 0.0 | 0.0 | 0.0 | 0.0 | 0.0 | 0.5 | 2.8 | 20.5 |
| Days ≥ 1" Snow Depth | 8 | 5 | 2 | 0 | 0 | 0 | 0 | 0 | 0 | 0 | 0 | 2 | 17 |

## LAMBERTVILLE *Hunterdon County* ELEVATION 59 ft LAT/LONG 40° 22 ' N / 74° 57 ' W

| | JAN | FEB | MAR | APR | MAY | JUN | JUL | AUG | SEP | OCT | NOV | DEC | YEAR |
|---|---|---|---|---|---|---|---|---|---|---|---|---|---|
| Maximum Temp °F | 38.9 | 42.4 | 52.1 | 63.9 | 75.1 | 83.1 | 87.4 | 85.6 | 78.6 | 67.1 | 55.3 | 43.8 | 64.4 |
| Minimum Temp °F | 20.8 | 22.7 | 30.5 | 39.1 | 49.2 | 58.5 | 63.7 | 62.4 | 54.8 | 42.8 | 34.9 | 26.2 | 42.1 |
| Mean Temp °F | 29.8 | 32.6 | 41.4 | 51.6 | 62.2 | 70.8 | 75.6 | 74.0 | 66.7 | 55.0 | 45.1 | 35.0 | 53.3 |
| Days Max Temp ≥ 90 °F | 0 | 0 | 0 | 0 | 1 | 5 | 11 | 8 | 2 | 0 | 0 | 0 | 27 |
| Days Max Temp ≤ 32 °F | 8 | 4 | 0 | 0 | 0 | 0 | 0 | 0 | 0 | 0 | 0 | 3 | 15 |
| Days Min Temp ≤ 32 °F | 27 | 23 | 19 | 7 | 0 | 0 | 0 | 0 | 0 | 5 | 13 | 24 | 118 |
| Days Min Temp ≤ 0 °F | 1 | 1 | 0 | 0 | 0 | 0 | 0 | 0 | 0 | 0 | 0 | 0 | 2 |
| Heating Degree Days | 1085 | 909 | 727 | 404 | 140 | 16 | 1 | 5 | 63 | 318 | 591 | 922 | 5181 |
| Cooling Degree Days | 0 | 0 | 1 | 8 | 64 | 201 | 349 | 281 | 118 | 17 | 0 | 0 | 1039 |
| Total Precipitation (") | 3.47 | 2.86 | 4.10 | 3.90 | 4.46 | 3.93 | 4.81 | 4.26 | 4.10 | 3.18 | 3.84 | 3.87 | 46.78 |
| Days ≥ 0.1" Precip | 7 | 6 | 8 | 7 | 8 | 7 | 7 | 7 | 6 | 5 | 7 | 7 | 82 |
| Total Snowfall (") | 7.6 | 7.1 | 3.9 | 0.7 | 0.0 | 0.0 | 0.0 | 0.0 | 0.0 | 0.1 | 0.6 | 3.5 | 23.5 |
| Days ≥ 1" Snow Depth | 12 | 8 | 2 | 0 | 0 | 0 | 0 | 0 | 0 | 0 | 0 | 3 | 25 |

## LITTLE FALLS *Passaic County* ELEVATION 161 ft LAT/LONG 40° 53 ' N / 74° 14 ' W

| | JAN | FEB | MAR | APR | MAY | JUN | JUL | AUG | SEP | OCT | NOV | DEC | YEAR |
|---|---|---|---|---|---|---|---|---|---|---|---|---|---|
| Maximum Temp °F | 37.0 | 40.0 | 49.3 | 61.2 | 71.9 | 80.8 | 85.9 | 83.7 | 76.1 | 64.8 | 53.9 | 42.4 | 62.2 |
| Minimum Temp °F | 19.7 | 21.7 | 30.1 | 40.0 | 49.8 | 59.1 | 64.3 | 63.0 | 54.7 | 42.4 | 35.3 | 25.6 | 42.1 |
| Mean Temp °F | 28.4 | 30.9 | 39.7 | 50.6 | 60.8 | 70.0 | 75.2 | 73.4 | 65.4 | 53.6 | 44.6 | 34.0 | 52.2 |
| Days Max Temp ≥ 90 °F | 0 | 0 | 0 | 0 | 1 | 4 | 9 | 5 | 2 | 0 | 0 | 0 | 21 |
| Days Max Temp ≤ 32 °F | 10 | 6 | 1 | 0 | 0 | 0 | 0 | 0 | 0 | 0 | 0 | 4 | 21 |
| Days Min Temp ≤ 32 °F | 28 | 24 | 19 | 5 | 0 | 0 | 0 | 0 | 0 | 4 | 12 | 24 | 116 |
| Days Min Temp ≤ 0 °F | 1 | 0 | 0 | 0 | 0 | 0 | 0 | 0 | 0 | 0 | 0 | 0 | 1 |
| Heating Degree Days | 1129 | 957 | 778 | 434 | 174 | 27 | 2 | 6 | 79 | 354 | 606 | 956 | 5502 |
| Cooling Degree Days | 0 | 0 | 1 | 7 | 58 | 187 | 337 | 269 | 94 | 8 | 0 | 0 | 961 |
| Total Precipitation (") | 3.54 | 3.06 | 4.34 | 4.29 | 4.98 | 4.16 | 4.38 | 4.62 | 4.99 | 3.82 | 4.36 | 4.02 | 50.56 |
| Days ≥ 0.1" Precip | 6 | 6 | 7 | 7 | 7 | 7 | 7 | 6 | 6 | 5 | 7 | 6 | 77 |
| Total Snowfall (") | na | na | ***3.4*** | 0.1 | 0.0 | 0.0 | 0.0 | 0.0 | 0.0 | 0.0 | 0.1 | na | na |
| Days ≥ 1" Snow Depth | na | ***9*** | ***3*** | 0 | 0 | 0 | 0 | 0 | 0 | 0 | 0 | ***5*** | na |

## LONG BRANCH OAKHURST *Monmouth County* ELEVATION 30 ft LAT/LONG 40° 16 ' N / 74° 0 ' W

| | JAN | FEB | MAR | APR | MAY | JUN | JUL | AUG | SEP | OCT | NOV | DEC | YEAR |
|---|---|---|---|---|---|---|---|---|---|---|---|---|---|
| Maximum Temp °F | 39.5 | 41.7 | 48.8 | 59.0 | 68.1 | 77.6 | 82.7 | 81.5 | 75.2 | 64.7 | 55.4 | 45.2 | 61.6 |
| Minimum Temp °F | 22.8 | 24.9 | 31.8 | 40.3 | 49.9 | 59.5 | 65.4 | 64.3 | 57.5 | 46.3 | 38.1 | 28.6 | 44.1 |
| Mean Temp °F | 31.2 | 33.3 | 40.3 | 49.6 | 59.0 | 68.5 | 74.1 | 72.9 | 66.4 | 55.5 | 46.8 | 36.9 | 52.9 |
| Days Max Temp ≥ 90 °F | 0 | 0 | 0 | 0 | 1 | 2 | 5 | 2 | 1 | 0 | 0 | 0 | 11 |
| Days Max Temp ≤ 32 °F | 8 | 5 | 1 | 0 | 0 | 0 | 0 | 0 | 0 | 0 | 0 | 3 | 17 |
| Days Min Temp ≤ 32 °F | 25 | 22 | 16 | 4 | 0 | 0 | 0 | 0 | 0 | 2 | 9 | 21 | 99 |
| Days Min Temp ≤ 0 °F | 1 | 0 | 0 | 0 | 0 | 0 | 0 | 0 | 0 | 0 | 0 | 0 | 1 |
| Heating Degree Days | 1041 | 888 | 758 | 459 | 210 | 30 | 1 | 5 | 59 | 299 | 540 | 864 | 5154 |
| Cooling Degree Days | 0 | 0 | 1 | 2 | 36 | 145 | 296 | 241 | 100 | 13 | 1 | 0 | 835 |
| Total Precipitation (") | 3.73 | 3.25 | 4.24 | 3.98 | 4.30 | 3.42 | 4.37 | 5.09 | 3.56 | 3.49 | 4.10 | 4.09 | 47.62 |
| Days ≥ 0.1" Precip | 6 | 6 | 7 | 6 | 7 | 6 | 6 | 7 | 6 | 5 | 7 | 7 | 76 |
| Total Snowfall (") | 4.9 | 5.9 | 3.0 | 0.3 | 0.0 | 0.0 | 0.0 | 0.0 | 0.0 | 0.0 | 0.1 | 1.5 | 15.7 |
| Days ≥ 1" Snow Depth | 6 | 5 | 2 | 0 | 0 | 0 | 0 | 0 | 0 | 0 | 0 | 1 | 14 |

## LONG VALLEY *Morris County* ELEVATION 541 ft LAT/LONG 40° 47 ' N / 74° 47 ' W

| | JAN | FEB | MAR | APR | MAY | JUN | JUL | AUG | SEP | OCT | NOV | DEC | YEAR |
|---|---|---|---|---|---|---|---|---|---|---|---|---|---|
| Maximum Temp °F | 35.4 | 38.3 | 47.7 | 59.8 | 69.7 | 77.5 | 81.9 | 80.0 | 72.4 | 62.4 | 52.0 | 40.6 | 59.8 |
| Minimum Temp °F | 15.7 | 17.2 | 25.1 | 34.8 | 44.5 | 53.3 | 58.6 | 57.3 | 49.5 | 37.7 | 30.6 | 21.6 | 37.2 |
| Mean Temp °F | 25.6 | 27.8 | 36.5 | 47.3 | 57.2 | 65.4 | 70.3 | 68.7 | 61.0 | 50.1 | 41.3 | 31.2 | 48.5 |
| Days Max Temp ≥ 90 °F | 0 | 0 | 0 | 0 | 0 | 1 | 3 | 1 | 0 | 0 | 0 | 0 | 5 |
| Days Max Temp ≤ 32 °F | 11 | 8 | 2 | 0 | 0 | 0 | 0 | 0 | 0 | 0 | 0 | 6 | 27 |
| Days Min Temp ≤ 32 °F | 29 | 26 | 25 | 12 | 2 | 0 | 0 | 0 | 0 | 10 | 19 | 28 | 151 |
| Days Min Temp ≤ 0 °F | 3 | 2 | 0 | 0 | 0 | 0 | 0 | 0 | 0 | 0 | 0 | 1 | 6 |
| Heating Degree Days | 1216 | 1044 | 877 | 528 | 256 | 66 | 13 | 25 | 156 | 458 | 703 | 1042 | 6384 |
| Cooling Degree Days | 0 | 0 | 0 | 3 | 23 | 87 | 198 | 143 | 41 | 3 | 0 | 0 | 498 |
| Total Precipitation (") | 3.75 | 3.34 | 4.25 | 4.44 | 4.79 | 4.58 | 5.03 | 5.12 | 4.73 | 3.91 | 4.49 | 4.15 | 52.58 |
| Days ≥ 0.1" Precip | 7 | 6 | 7 | 7 | 8 | 7 | 7 | 7 | 7 | 6 | 7 | 7 | 83 |
| Total Snowfall (") | 10.6 | 10.0 | 6.8 | 1.2 | 0.0 | 0.0 | 0.0 | 0.0 | 0.0 | 0.2 | 1.0 | 4.8 | 34.6 |
| Days ≥ 1" Snow Depth | na | na | na | 0 | 0 | 0 | 0 | 0 | 0 | 0 | 0 | na | na |

## MILLVILLE MUNI AP *Cumberland County* ELEVATION 72 ft LAT/LONG 39° 22' N / 75° 4' W

| | JAN | FEB | MAR | APR | MAY | JUN | JUL | AUG | SEP | OCT | NOV | DEC | YEAR |
|---|---|---|---|---|---|---|---|---|---|---|---|---|---|
| Maximum Temp °F | 40.1 | 42.9 | 51.7 | 62.6 | 72.5 | 81.2 | 85.6 | 84.1 | 77.4 | 66.4 | 56.2 | 45.7 | 63.9 |
| Minimum Temp °F | 23.2 | 24.6 | 32.6 | 41.1 | 51.1 | 60.6 | 66.7 | 65.3 | 57.7 | 45.4 | 37.1 | 28.3 | 44.5 |
| Mean Temp °F | 31.7 | 33.7 | 42.2 | 51.9 | 61.8 | 71.0 | 76.2 | 74.7 | 67.6 | 55.9 | 46.7 | 37.0 | 54.2 |
| Days Max Temp ≥ 90 °F | 0 | 0 | 0 | 0 | 1 | 3 | 8 | 5 | 2 | 0 | 0 | 0 | 19 |
| Days Max Temp ≤ 32 °F | 7 | 5 | 1 | 0 | 0 | 0 | 0 | 0 | 0 | 0 | 0 | 3 | 16 |
| Days Min Temp ≤ 32 °F | 25 | 22 | 16 | 5 | 0 | 0 | 0 | 0 | 0 | 2 | 10 | 21 | 101 |
| Days Min Temp ≤ 0 °F | 1 | 0 | 0 | 0 | 0 | 0 | 0 | 0 | 0 | 0 | 0 | 0 | 1 |
| Heating Degree Days | 1026 | 876 | 702 | 395 | 148 | 17 | 1 | 2 | 50 | 294 | 544 | 860 | 4915 |
| Cooling Degree Days | 0 | 0 | 1 | 9 | 65 | 212 | 380 | 304 | 132 | 22 | 1 | 0 | 1126 |
| Total Precipitation (") | 3.29 | 3.22 | 4.10 | 3.58 | 3.82 | 3.10 | 3.78 | 4.11 | 3.35 | 3.09 | 3.16 | 3.75 | 42.35 |
| Days ≥ 0.1" Precip | 6 | 6 | 7 | 7 | 7 | 5 | 6 | 6 | 5 | 5 | 6 | 6 | 72 |
| Total Snowfall (") | 4.9 | 4.7 | 1.8 | 0.2 | 0.0 | 0.0 | 0.0 | 0.0 | 0.0 | 0.0 | 0.4 | 1.9 | 13.9 |
| Days ≥ 1" Snow Depth | 6 | 5 | 1 | 0 | 0 | 0 | 0 | 0 | 0 | 0 | 0 | 2 | 14 |

## MOORESTOWN *Burlington County* ELEVATION 56 ft LAT/LONG 39° 58' N / 74° 58' W

| | JAN | FEB | MAR | APR | MAY | JUN | JUL | AUG | SEP | OCT | NOV | DEC | YEAR |
|---|---|---|---|---|---|---|---|---|---|---|---|---|---|
| Maximum Temp °F | 40.1 | 43.6 | 53.1 | 64.5 | 74.7 | 82.9 | 86.9 | 84.9 | 78.3 | 67.0 | 55.5 | ***44.9*** | 64.7 |
| Minimum Temp °F | 22.6 | 24.1 | 31.9 | 40.4 | 50.1 | 59.5 | 64.6 | 63.1 | 55.5 | 43.9 | 36.0 | 27.2 | 43.2 |
| Mean Temp °F | 31.4 | 33.8 | 42.5 | 52.5 | 62.4 | 71.2 | 75.7 | 74.1 | 66.9 | 55.5 | 45.8 | ***36.1*** | 54.0 |
| Days Max Temp ≥ 90 °F | 0 | 0 | 0 | 1 | 2 | 6 | 10 | 7 | 2 | 0 | 0 | ***0*** | 28 |
| Days Max Temp ≤ 32 °F | 8 | 5 | 1 | 0 | 0 | 0 | 0 | 0 | 0 | 0 | 0 | ***3*** | 17 |
| Days Min Temp ≤ 32 °F | 27 | 23 | 17 | 5 | 0 | 0 | 0 | 0 | 0 | 3 | 12 | ***22*** | 109 |
| Days Min Temp ≤ 0 °F | 1 | 0 | 0 | 0 | 0 | 0 | 0 | 0 | 0 | 0 | 0 | 0 | 1 |
| Heating Degree Days | 1038 | 874 | 695 | 378 | 141 | 19 | 1 | 5 | 62 | 303 | 572 | ***884*** | 4972 |
| Cooling Degree Days | 0 | 0 | 2 | 12 | 84 | 238 | 384 | 300 | 134 | 19 | 1 | 0 | 1174 |
| Total Precipitation (") | 3.50 | 2.84 | 3.98 | 3.70 | 4.33 | 3.92 | 5.21 | 5.30 | 3.74 | 3.31 | 3.46 | 3.83 | 47.12 |
| Days ≥ 0.1" Precip | ***6*** | 5 | 7 | 6 | 7 | 6 | 7 | 7 | 6 | 5 | 6 | ***6*** | 74 |
| Total Snowfall (") | na | na | na | 0.1 | 0.0 | 0.0 | 0.0 | 0.0 | 0.0 | 0.0 | 0.0 | na | na |
| Days ≥ 1" Snow Depth | na | na | na | 0 | 0 | 0 | 0 | 0 | 0 | 0 | 0 | na | na |

## NEWARK INTL ARPT *Essex County* ELEVATION 7 ft LAT/LONG 40° 42' N / 74° 10' W

| | JAN | FEB | MAR | APR | MAY | JUN | JUL | AUG | SEP | OCT | NOV | DEC | YEAR |
|---|---|---|---|---|---|---|---|---|---|---|---|---|---|
| Maximum Temp °F | 38.1 | 41.0 | 50.3 | 61.9 | 72.4 | 81.5 | 86.2 | 84.5 | 76.9 | 65.7 | 54.8 | 43.7 | 63.1 |
| Minimum Temp °F | 23.8 | 25.7 | 33.7 | 43.6 | 53.9 | 63.4 | 69.2 | 68.0 | 60.0 | 48.3 | 39.5 | 29.8 | 46.6 |
| Mean Temp °F | 31.0 | 33.3 | 42.0 | 52.7 | 63.2 | 72.5 | 77.8 | 76.3 | 68.5 | 57.0 | 47.1 | 36.7 | 54.8 |
| Days Max Temp ≥ 90 °F | 0 | 0 | 0 | 0 | 2 | 5 | 9 | 7 | 2 | 0 | 0 | 0 | 25 |
| Days Max Temp ≤ 32 °F | 9 | 6 | 1 | 0 | 0 | 0 | 0 | 0 | 0 | 0 | 0 | 4 | 20 |
| Days Min Temp ≤ 32 °F | 24 | 21 | 13 | 1 | 0 | 0 | 0 | 0 | 0 | 1 | 6 | 19 | 85 |
| Days Min Temp ≤ 0 °F | 1 | 0 | 0 | 0 | 0 | 0 | 0 | 0 | 0 | 0 | 0 | 0 | 1 |
| Heating Degree Days | 1048 | 887 | 707 | 371 | 121 | 11 | 0 | 2 | 40 | 259 | 531 | 869 | 4846 |
| Cooling Degree Days | 0 | 0 | 2 | 11 | 86 | 265 | 430 | 363 | 158 | 22 | 2 | 0 | 1339 |
| Total Precipitation (") | 3.42 | 2.97 | 4.15 | 3.76 | 4.33 | 3.25 | 4.44 | 4.06 | 3.81 | 3.03 | 3.86 | 3.59 | 44.67 |
| Days ≥ 0.1" Precip | 6 | 6 | 7 | 7 | 7 | 6 | 7 | 6 | 6 | 5 | 6 | 7 | 76 |
| Total Snowfall (") | 8.1 | 8.8 | 4.5 | 0.7 | 0.0 | 0.0 | 0.0 | 0.0 | 0.0 | 0.0 | 0.6 | 3.0 | 25.7 |
| Days ≥ 1" Snow Depth | 11 | 8 | 2 | 0 | 0 | 0 | 0 | 0 | 0 | 0 | 0 | 2 | 23 |

## NEWTON ST PAUL ABBEY *Sussex County* ELEVATION 571 ft LAT/LONG 41° 3' N / 74° 45' W

| | JAN | FEB | MAR | APR | MAY | JUN | JUL | AUG | SEP | OCT | NOV | DEC | YEAR |
|---|---|---|---|---|---|---|---|---|---|---|---|---|---|
| Maximum Temp °F | 33.5 | 36.7 | 46.3 | 58.9 | 70.3 | 78.4 | 83.3 | 81.2 | 73.5 | 62.1 | 50.5 | 38.8 | 59.5 |
| Minimum Temp °F | 13.6 | 15.5 | 25.0 | 35.2 | 44.8 | 53.8 | 58.7 | 56.9 | 48.7 | 36.8 | 30.0 | 20.5 | 36.6 |
| Mean Temp °F | 23.6 | 26.1 | 35.7 | 47.1 | 57.6 | 66.2 | 71.0 | 69.0 | 61.2 | 49.5 | 40.3 | 29.7 | 48.1 |
| Days Max Temp ≥ 90 °F | 0 | 0 | 0 | 0 | 0 | 2 | 5 | 3 | 1 | 0 | 0 | 0 | 11 |
| Days Max Temp ≤ 32 °F | 13 | 9 | 2 | 0 | 0 | 0 | 0 | 0 | 0 | 0 | 1 | 8 | 33 |
| Days Min Temp ≤ 32 °F | 30 | 27 | 25 | 12 | 2 | 0 | 0 | 0 | 1 | 11 | 20 | 28 | 156 |
| Days Min Temp ≤ 0 °F | 4 | 3 | 0 | 0 | 0 | 0 | 0 | 0 | 0 | 0 | 0 | 1 | 8 |
| Heating Degree Days | 1277 | 1091 | 902 | 533 | 248 | 64 | 13 | 27 | 157 | 477 | 735 | 1088 | 6612 |
| Cooling Degree Days | 0 | 0 | 1 | 3 | 27 | 101 | 215 | 155 | 47 | 4 | 0 | 0 | 553 |
| Total Precipitation (") | 3.04 | 2.67 | 3.50 | 3.93 | 4.29 | 4.47 | 4.32 | 4.59 | 4.13 | 3.46 | 3.92 | 3.43 | 45.75 |
| Days ≥ 0.1" Precip | 6 | 6 | 7 | 8 | 8 | 8 | 7 | 7 | 6 | 6 | 7 | 6 | 82 |
| Total Snowfall (") | 10.4 | 10.9 | 8.2 | 1.5 | 0.0 | 0.0 | 0.0 | 0.0 | 0.0 | 0.1 | 1.9 | 5.8 | 38.8 |
| Days ≥ 1" Snow Depth | 19 | 14 | 7 | 1 | 0 | 0 | 0 | 0 | 0 | 0 | 1 | 8 | 50 |

## PEMBERTON 3 S *Burlington County* ELEVATION 49 ft LAT/LONG 39° 56 ' N / 74° 42 ' W

| | JAN | FEB | MAR | APR | MAY | JUN | JUL | AUG | SEP | OCT | NOV | DEC | YEAR |
|---|---|---|---|---|---|---|---|---|---|---|---|---|---|
| Maximum Temp °F | 40.6 | 43.9 | 53.0 | 64.0 | 74.6 | 82.5 | 86.6 | 85.4 | 78.8 | 68.0 | 56.9 | 45.7 | 65.0 |
| Minimum Temp °F | 21.4 | 23.0 | 30.5 | 38.1 | 48.2 | 57.1 | 62.4 | 61.6 | 54.1 | 42.8 | 35.4 | 26.8 | 41.8 |
| Mean Temp °F | 31.0 | 33.5 | 41.8 | 51.0 | 61.4 | 69.8 | 74.5 | 73.5 | 66.5 | 55.4 | 46.2 | 36.3 | 53.4 |
| Days Max Temp ≥ 90 °F | 0 | 0 | 0 | 0 | 1 | 5 | 9 | 8 | 2 | 0 | 0 | 0 | 25 |
| Days Max Temp ≤ 32 °F | 7 | 4 | 0 | 0 | 0 | 0 | 0 | 0 | 0 | 0 | 0 | 3 | 14 |
| Days Min Temp ≤ 32 °F | 26 | 22 | 18 | 9 | 1 | 0 | 0 | 0 | 0 | 6 | 13 | 22 | 117 |
| Days Min Temp ≤ 0 °F | 1 | 1 | 0 | 0 | 0 | 0 | 0 | 0 | 0 | 0 | 0 | 0 | 2 |
| Heating Degree Days | 1046 | 884 | 714 | 419 | 156 | 24 | 2 | 5 | 67 | 307 | 560 | 883 | 5067 |
| Cooling Degree Days | 0 | 0 | 2 | 6 | 60 | 178 | 321 | 269 | 119 | 19 | 2 | 0 | 976 |
| Total Precipitation (") | 3.57 | 2.88 | 4.13 | 3.61 | 4.16 | 4.03 | 4.86 | 4.95 | 3.75 | 3.38 | 3.64 | 3.98 | 46.94 |
| Days ≥ 0.1" Precip | 7 | 6 | 7 | 6 | 8 | 7 | 6 | 6 | 6 | 5 | 6 | 7 | 77 |
| Total Snowfall (") | 6.5 | 6.7 | 3.5 | 0.3 | 0.0 | 0.0 | 0.0 | 0.0 | 0.0 | 0.0 | 0.5 | 2.5 | 20.0 |
| Days ≥ 1" Snow Depth | 8 | 6 | 2 | 0 | 0 | 0 | 0 | 0 | 0 | 0 | 0 | 2 | 18 |

## PLAINFIELD *Union County* ELEVATION 89 ft LAT/LONG 40° 36 ' N / 74° 24 ' W

| | JAN | FEB | MAR | APR | MAY | JUN | JUL | AUG | SEP | OCT | NOV | DEC | YEAR |
|---|---|---|---|---|---|---|---|---|---|---|---|---|---|
| Maximum Temp °F | 37.1 | 41.5 | 51.2 | 62.7 | 73.5 | 81.9 | 86.4 | 84.8 | 77.2 | 65.9 | 53.8 | 42.2 | 63.2 |
| Minimum Temp °F | 21.7 | 23.6 | 31.3 | 40.1 | 49.9 | 58.8 | 63.9 | 62.6 | 55.1 | 43.5 | 35.8 | 27.5 | 42.8 |
| Mean Temp °F | 29.4 | 32.5 | 41.3 | 51.4 | 61.7 | 70.4 | 75.2 | 73.7 | 66.2 | 54.7 | 44.8 | 34.9 | 53.0 |
| Days Max Temp ≥ 90 °F | 0 | 0 | 0 | 0 | 1 | 5 | 10 | 7 | 2 | 0 | 0 | 0 | 25 |
| Days Max Temp ≤ 32 °F | 10 | 5 | 1 | 0 | 0 | 0 | 0 | 0 | 0 | 0 | 0 | 4 | 20 |
| Days Min Temp ≤ 32 °F | 26 | 23 | 18 | 5 | 0 | 0 | 0 | 0 | 0 | 4 | 12 | 22 | 110 |
| Days Min Temp ≤ 0 °F | 1 | 0 | 0 | 0 | 0 | 0 | 0 | 0 | 0 | 0 | 0 | 0 | 1 |
| Heating Degree Days | 1095 | 910 | 729 | 409 | 150 | 20 | 1 | 4 | 66 | 324 | 599 | 926 | 5233 |
| Cooling Degree Days | 0 | 0 | 1 | 8 | 62 | 195 | 340 | 275 | 107 | 12 | 0 | 0 | 1000 |
| Total Precipitation (") | 3.57 | 2.98 | 4.14 | 3.91 | 4.72 | 3.88 | 5.17 | 4.74 | 4.26 | 3.71 | 4.09 | 3.88 | 49.05 |
| Days ≥ 0.1" Precip | 7 | 6 | 7 | 7 | 8 | 7 | 7 | 7 | 6 | 5 | 7 | 7 | 81 |
| Total Snowfall (") | 9.5 | 9.7 | 5.0 | 0.6 | 0.0 | 0.0 | 0.0 | 0.0 | 0.0 | 0.0 | 0.7 | 4.0 | 29.5 |
| Days ≥ 1" Snow Depth | 14 | 11 | 3 | 0 | 0 | 0 | 0 | 0 | 0 | 0 | 0 | 3 | 31 |

## SEABROOK FARMS *Cumberland County* ELEVATION 102 ft LAT/LONG 39° 30 ' N / 75° 14 ' W

| | JAN | FEB | MAR | APR | MAY | JUN | JUL | AUG | SEP | OCT | NOV | DEC | YEAR |
|---|---|---|---|---|---|---|---|---|---|---|---|---|---|
| Maximum Temp °F | 39.8 | 42.0 | 51.3 | 62.2 | 72.2 | 81.6 | 85.7 | 84.2 | 77.8 | 66.6 | 55.7 | 45.1 | 63.7 |
| Minimum Temp °F | 23.0 | 24.4 | 32.9 | 41.3 | 51.1 | 61.0 | 66.1 | 64.5 | 56.7 | 44.8 | 37.1 | 28.2 | 44.3 |
| Mean Temp °F | 31.4 | 33.2 | 42.1 | 51.8 | 61.7 | 71.3 | 75.9 | 74.4 | 67.2 | 55.7 | 46.4 | 36.7 | 54.0 |
| Days Max Temp ≥ 90 °F | 0 | 0 | 0 | 0 | 1 | 4 | 8 | 5 | 2 | 0 | 0 | 0 | 20 |
| Days Max Temp ≤ 32 °F | 8 | 6 | 1 | 0 | 0 | 0 | 0 | 0 | 0 | 0 | 0 | 3 | 18 |
| Days Min Temp ≤ 32 °F | 26 | 23 | 16 | 4 | 0 | 0 | 0 | 0 | 0 | 2 | 10 | 22 | 103 |
| Days Min Temp ≤ 0 °F | 0 | 0 | 0 | 0 | 0 | 0 | 0 | 0 | 0 | 0 | 0 | 0 | 0 |
| Heating Degree Days | 1032 | 890 | 704 | 399 | 152 | 17 | 1 | 3 | 55 | 296 | 552 | 871 | 4972 |
| Cooling Degree Days | na | na | na | na | na | na | 378 | 292 | 122 | 18 | 0 | 0 | na |
| Total Precipitation (") | 3.56 | 2.66 | 3.88 | 3.22 | 3.83 | 3.19 | 4.45 | 4.50 | 3.48 | 3.33 | 3.25 | 4.00 | 43.35 |
| Days ≥ 0.1" Precip | 7 | 5 | 7 | 6 | 7 | 6 | 7 | 7 | 6 | 5 | 6 | 6 | 75 |
| Total Snowfall (") | na | na | na | 0.2 | 0.0 | 0.0 | 0.0 | 0.0 | 0.0 | 0.0 | 0.2 | na | na |
| Days ≥ 1" Snow Depth | na | na | na | 0 | 0 | 0 | 0 | 0 | 0 | 0 | 0 | na | na |

## SOMERVILLE 3 NW *Somerset County* ELEVATION 59 ft LAT/LONG 40° 35 ' N / 74° 37 ' W

| | JAN | FEB | MAR | APR | MAY | JUN | JUL | AUG | SEP | OCT | NOV | DEC | YEAR |
|---|---|---|---|---|---|---|---|---|---|---|---|---|---|
| Maximum Temp °F | 36.7 | 39.6 | 49.3 | 61.3 | 72.0 | 80.5 | 85.4 | 83.2 | 75.7 | 64.2 | 53.4 | 42.0 | 61.9 |
| Minimum Temp °F | 18.3 | 19.8 | 27.9 | 37.4 | 47.3 | 56.4 | 61.7 | 60.4 | 52.5 | 40.6 | 33.1 | 24.2 | 40.0 |
| Mean Temp °F | 27.5 | 29.7 | 38.6 | 49.4 | 59.7 | 68.5 | 73.6 | 71.8 | 64.1 | 52.4 | 43.3 | 33.1 | 51.0 |
| Days Max Temp ≥ 90 °F | 0 | 0 | 0 | 0 | 1 | 3 | 8 | 4 | 1 | 0 | 0 | 0 | 17 |
| Days Max Temp ≤ 32 °F | 10 | 7 | 1 | 0 | 0 | 0 | 0 | 0 | 0 | 0 | 0 | 5 | 23 |
| Days Min Temp ≤ 32 °F | 28 | 25 | 22 | 8 | 1 | 0 | 0 | 0 | 0 | 6 | 16 | 26 | 132 |
| Days Min Temp ≤ 0 °F | 2 | 1 | 0 | 0 | 0 | 0 | 0 | 0 | 0 | 0 | 0 | 0 | 3 |
| Heating Degree Days | 1155 | 990 | 810 | 467 | 193 | 33 | 3 | 9 | 99 | 389 | 645 | 982 | 5775 |
| Cooling Degree Days | 0 | 0 | 0 | 4 | 39 | 139 | 282 | 216 | 76 | 7 | 0 | 0 | 763 |
| Total Precipitation (") | 3.48 | 2.81 | 3.90 | 3.88 | 4.37 | 4.09 | 4.68 | 4.66 | 4.24 | 3.42 | 3.73 | 3.74 | 47.00 |
| Days ≥ 0.1" Precip | 7 | 6 | 7 | 7 | 7 | 7 | 7 | 7 | 6 | 5 | 6 | 7 | 79 |
| Total Snowfall (") | 8.4 | 8.4 | 5.4 | 0.9 | 0.0 | 0.0 | 0.0 | 0.0 | 0.0 | 0.0 | 0.7 | 3.6 | 27.4 |
| Days ≥ 1" Snow Depth | 13 | 11 | 4 | 0 | 0 | 0 | 0 | 0 | 0 | 0 | 0 | 4 | 32 |

## SUSSEX *Sussex County* ELEVATION 449 ft LAT/LONG 41° 13' N / 74° 36' W

| | JAN | FEB | MAR | APR | MAY | JUN | JUL | AUG | SEP | OCT | NOV | DEC | YEAR |
|---|---|---|---|---|---|---|---|---|---|---|---|---|---|
| Maximum Temp °F | 33.6 | 36.9 | 46.2 | 59.0 | 70.2 | 78.3 | 83.2 | 81.5 | 73.7 | 62.5 | 50.7 | 38.5 | 59.5 |
| Minimum Temp °F | 12.8 | 14.8 | 24.4 | 34.7 | 44.3 | 53.5 | 58.5 | 56.7 | 48.3 | 36.7 | 29.2 | 19.7 | 36.1 |
| Mean Temp °F | 23.2 | 25.9 | 35.4 | 46.9 | 57.3 | 65.9 | 70.9 | 69.1 | 61.0 | 49.6 | 39.9 | 29.1 | 47.9 |
| Days Max Temp ≥ 90 °F | 0 | 0 | 0 | 0 | 0 | 2 | 5 | 3 | 1 | 0 | 0 | 0 | 11 |
| Days Max Temp ≤ 32 °F | 13 | 9 | 2 | 0 | 0 | 0 | 0 | 0 | 0 | 0 | 0 | 8 | 32 |
| Days Min Temp ≤ 32 °F | 30 | 27 | 26 | 12 | 2 | 0 | 0 | 0 | 1 | 12 | 20 | 28 | 158 |
| Days Min Temp ≤ 0 °F | 5 | 3 | 0 | 0 | 0 | 0 | 0 | 0 | 0 | 0 | 0 | 2 | 10 |
| Heating Degree Days | 1291 | 1099 | 912 | 541 | 257 | 64 | 11 | 25 | 165 | 473 | 745 | 1105 | 6688 |
| Cooling Degree Days | 0 | 0 | 0 | 4 | 29 | 104 | 221 | *171* | 54 | 4 | 0 | 0 | 587 |
| Total Precipitation (") | 3.41 | 2.95 | 3.80 | 4.19 | 4.43 | 4.63 | 4.30 | 4.40 | 3.92 | 3.57 | 3.97 | 3.64 | 47.21 |
| Days ≥ 0.1" Precip | 7 | 6 | 7 | 8 | 8 | 8 | 7 | 7 | 6 | 6 | 7 | 7 | 84 |
| Total Snowfall (") | na | na | *7.9* | 1.5 | 0.0 | 0.0 | 0.0 | 0.0 | 0.0 | 0.0 | 2.1 | na | na |
| Days ≥ 1" Snow Depth | na | na | na | 0 | 0 | 0 | 0 | 0 | 0 | 0 | 1 | na | na |

## TOMS RIVER *Ocean County* ELEVATION 10 ft LAT/LONG 39° 57' N / 74° 13' W

| | JAN | FEB | MAR | APR | MAY | JUN | JUL | AUG | SEP | OCT | NOV | DEC | YEAR |
|---|---|---|---|---|---|---|---|---|---|---|---|---|---|
| Maximum Temp °F | 40.7 | 42.8 | 51.3 | 61.7 | 72.0 | 80.9 | 85.8 | 84.3 | 77.9 | 66.9 | 57.0 | 46.3 | 64.0 |
| Minimum Temp °F | 20.0 | 21.8 | 29.9 | 38.3 | 48.7 | 57.6 | 63.0 | 61.6 | 53.8 | 41.7 | 34.4 | 25.4 | 41.3 |
| Mean Temp °F | 30.4 | 32.3 | 40.6 | 50.0 | 60.3 | 69.3 | 74.4 | 72.9 | 65.9 | 54.3 | 45.7 | 35.9 | 52.7 |
| Days Max Temp ≥ 90 °F | 0 | 0 | 0 | 0 | 1 | 4 | 9 | 6 | 2 | 0 | 0 | 0 | 22 |
| Days Max Temp ≤ 32 °F | 7 | 5 | 1 | 0 | 0 | 0 | 0 | 0 | 0 | 0 | 0 | 2 | 15 |
| Days Min Temp ≤ 32 °F | 27 | 24 | 19 | 8 | 0 | 0 | 0 | 0 | 0 | 5 | 14 | 24 | 121 |
| Days Min Temp ≤ 0 °F | 1 | 1 | 0 | 0 | 0 | 0 | 0 | 0 | 0 | 0 | 0 | 0 | 2 |
| Heating Degree Days | 1066 | 916 | 749 | 447 | 184 | 30 | 3 | 7 | 72 | 336 | 572 | 896 | 5278 |
| Cooling Degree Days | 0 | 0 | 1 | 4 | *61* | 174 | 324 | 248 | 104 | 9 | 2 | 0 | 927 |
| Total Precipitation (") | 3.70 | 3.20 | 4.25 | 3.95 | 3.94 | 3.42 | 4.78 | 4.95 | 3.60 | 3.50 | 3.89 | 4.16 | 47.34 |
| Days ≥ 0.1" Precip | 7 | 6 | 7 | 7 | 7 | 7 | 7 | 7 | 6 | 5 | 7 | 7 | 80 |
| Total Snowfall (") | na | na | *1.8* | 0.1 | 0.0 | 0.0 | 0.0 | 0.0 | 0.0 | 0.0 | 0.0 | na | na |
| Days ≥ 1" Snow Depth | na | na | na | 0 | 0 | 0 | 0 | 0 | 0 | 0 | 0 | na | na |

## TUCKERTON *Ocean County* ELEVATION 20 ft LAT/LONG 39° 36' N / 74° 20' W

| | JAN | FEB | MAR | APR | MAY | JUN | JUL | AUG | SEP | OCT | NOV | DEC | YEAR |
|---|---|---|---|---|---|---|---|---|---|---|---|---|---|
| Maximum Temp °F | 41.1 | 43.3 | 51.3 | 61.4 | 71.2 | 80.4 | 85.3 | 84.0 | 77.4 | 66.8 | 56.7 | 46.4 | 63.8 |
| Minimum Temp °F | 22.7 | 24.1 | 31.6 | 40.0 | 50.0 | 59.7 | 65.6 | 64.5 | 57.1 | 45.4 | 36.8 | 27.9 | 43.8 |
| Mean Temp °F | 31.9 | 33.7 | 41.4 | 50.7 | 60.6 | 70.1 | 75.5 | 74.3 | 67.3 | 56.1 | 46.8 | 37.1 | 53.8 |
| Days Max Temp ≥ 90 °F | 0 | 0 | 0 | 0 | 1 | 4 | 9 | 6 | 2 | 0 | 0 | 0 | 22 |
| Days Max Temp ≤ 32 °F | 6 | 4 | 0 | 0 | 0 | 0 | 0 | 0 | 0 | 0 | 0 | 2 | 12 |
| Days Min Temp ≤ 32 °F | 25 | 22 | 17 | 6 | 0 | 0 | 0 | 0 | 0 | 3 | 11 | 21 | 105 |
| Days Min Temp ≤ 0 °F | 1 | 0 | 0 | 0 | 0 | 0 | 0 | 0 | 0 | 0 | 0 | 0 | 1 |
| Heating Degree Days | 1019 | 878 | 724 | 428 | 175 | 22 | 1 | 4 | 55 | 288 | 541 | 857 | 4992 |
| Cooling Degree Days | 0 | 0 | 0 | 6 | 57 | 202 | 362 | 297 | 135 | 20 | 1 | 0 | 1080 |
| Total Precipitation (") | 3.68 | 3.24 | 4.48 | 3.99 | 3.60 | 3.00 | 4.31 | 4.86 | 3.18 | 3.23 | 3.77 | 4.00 | 45.34 |
| Days ≥ 0.1" Precip | 7 | 6 | 7 | 6 | 6 | 6 | 6 | 7 | 5 | 5 | 6 | 7 | 74 |
| Total Snowfall (") | 6.1 | 7.9 | 2.8 | 0.3 | 0.0 | 0.0 | 0.0 | 0.0 | 0.0 | 0.0 | 0.6 | 2.8 | 20.5 |
| Days ≥ 1" Snow Depth | 6 | 6 | 2 | 0 | 0 | 0 | 0 | 0 | 0 | 0 | 0 | 2 | 16 |

## WOODSTOWN *Salem County* ELEVATION 49 ft LAT/LONG 39° 39' N / 75° 19' W

| | JAN | FEB | MAR | APR | MAY | JUN | JUL | AUG | SEP | OCT | NOV | DEC | YEAR |
|---|---|---|---|---|---|---|---|---|---|---|---|---|---|
| Maximum Temp °F | 40.5 | 43.7 | 53.4 | 65.1 | 75.5 | 83.9 | 87.9 | 86.0 | 79.2 | 67.6 | 56.8 | 45.8 | 65.5 |
| Minimum Temp °F | 22.8 | 24.4 | 32.3 | 40.5 | 50.6 | 59.9 | 65.2 | 63.5 | 56.4 | 45.0 | 37.0 | 28.1 | 43.8 |
| Mean Temp °F | 31.7 | 34.1 | 42.9 | 52.8 | 63.1 | 71.9 | 76.5 | 74.7 | 67.9 | 56.3 | 47.0 | 37.0 | 54.7 |
| Days Max Temp ≥ 90 °F | 0 | 0 | 0 | 0 | 1 | 6 | 12 | 8 | 2 | 0 | 0 | 0 | 29 |
| Days Max Temp ≤ 32 °F | 7 | 4 | 0 | 0 | 0 | 0 | 0 | 0 | 0 | 0 | 0 | 3 | 14 |
| Days Min Temp ≤ 32 °F | 26 | 23 | 16 | 5 | 0 | 0 | 0 | 0 | 0 | 3 | 10 | 22 | 105 |
| Days Min Temp ≤ 0 °F | 1 | 0 | 0 | 0 | 0 | 0 | 0 | 0 | 0 | 0 | 0 | 0 | 1 |
| Heating Degree Days | 1026 | 866 | 681 | 367 | 120 | 12 | 1 | 2 | 50 | 279 | 537 | 862 | 4803 |
| Cooling Degree Days | 0 | 0 | 2 | 10 | 78 | 238 | 390 | 309 | 142 | 23 | 1 | 0 | 1193 |
| Total Precipitation (") | 3.34 | 2.75 | 3.88 | 3.70 | 3.87 | 3.79 | 4.54 | 4.21 | 3.58 | 3.36 | 3.52 | 3.85 | 44.39 |
| Days ≥ 0.1" Precip | 6 | 5 | 7 | 7 | 7 | 6 | 7 | 6 | 5 | 5 | 6 | 7 | 74 |
| Total Snowfall (") | 5.5 | 5.5 | 2.6 | 0.5 | 0.0 | 0.0 | 0.0 | 0.0 | 0.0 | 0.1 | 0.4 | 3.1 | 17.7 |
| Days ≥ 1" Snow Depth | 8 | 6 | 1 | 0 | 0 | 0 | 0 | 0 | 0 | 0 | 0 | *3* | 18 |

## JANUARY MINIMUM TEMPERATURE °F

### LOWEST

| | | |
|---|---|---|
| 1 | Sussex | 12.8 |
| 2 | Newton | 13.6 |
| 3 | Charlottebrg Rsrvr | 15.3 |
| 4 | Long Valley | 15.7 |
| 5 | Canoe Brook | 16.9 |
| 6 | Flemington | 17.2 |
| 7 | Essex Fells | 17.9 |
| 8 | Boonton | 18.0 |
| 9 | Somerville | 18.3 |
| 10 | Little Falls | 19.7 |
| 11 | Toms River | 20.0 |
| 12 | Indian Mills | 20.8 |
| | Lambertville | 20.8 |
| 14 | Hightstown | 21.0 |
| 15 | Pemberton | 21.4 |
| 16 | Plainfield | 21.7 |
| 17 | Atlantic City-Int | 22.1 |
| 18 | Belleplain | 22.4 |
| 19 | Moorestown | 22.6 |
| 20 | Glassboro | 22.7 |
| | Tuckerton | 22.7 |
| 22 | Long Branch | 22.8 |
| | Woodstown | 22.8 |
| 24 | Seabrook Farms | 23.0 |
| 25 | Millville | 23.2 |

### HIGHEST

| | | |
|---|---|---|
| 1 | Atlantic City-Mar | 27.1 |
| 2 | Cape May | 26.5 |
| 3 | Newark | 23.8 |
| 4 | Millville | 23.2 |
| 5 | Seabrook Farms | 23.0 |
| 6 | Long Branch | 22.8 |
| | Woodstown | 22.8 |
| 8 | Glassboro | 22.7 |
| | Tuckerton | 22.7 |
| 10 | Moorestown | 22.6 |
| 11 | Belleplain | 22.4 |
| 12 | Atlantic City-Int | 22.1 |
| 13 | Plainfield | 21.7 |
| 14 | Pemberton | 21.4 |
| 15 | Hightstown | 21.0 |
| 16 | Indian Mills | 20.8 |
| | Lambertville | 20.8 |
| 18 | Toms River | 20.0 |
| 19 | Little Falls | 19.7 |
| 20 | Somerville | 18.3 |
| 21 | Boonton | 18.0 |
| 22 | Essex Fells | 17.9 |
| 23 | Flemington | 17.2 |
| 24 | Canoe Brook | 16.9 |
| 25 | Long Valley | 15.7 |

## JULY MAXIMUM TEMPERATURE °F

### HIGHEST

| | | |
|---|---|---|
| 1 | Woodstown | 87.9 |
| 2 | Lambertville | 87.4 |
| 3 | Indian Mills | 86.9 |
| | Moorestown | 86.9 |
| 5 | Pemberton | 86.6 |
| 6 | Plainfield | 86.4 |
| 7 | Belleplain | 86.3 |
| 8 | Newark | 86.2 |
| 9 | Little Falls | 85.9 |
| 10 | Flemington | 85.8 |
| | Toms River | 85.8 |
| 12 | Glassboro | 85.7 |
| | Seabrook Farms | 85.7 |
| 14 | Millville | 85.6 |
| 15 | Somerville | 85.4 |
| 16 | Hightstown | 85.3 |
| | Tuckerton | 85.3 |
| 18 | Canoe Brook | 85.2 |
| 19 | Essex Fells | 84.8 |
| 20 | Atlantic City-Int | 84.7 |
| 21 | Boonton | 83.8 |
| 22 | Cape May | 83.5 |
| 23 | Newton | 83.3 |
| 24 | Sussex | 83.2 |
| 25 | Long Branch | 82.7 |

### LOWEST

| | | |
|---|---|---|
| 1 | Atlantic City-Mar | 80.4 |
| 2 | Long Valley | 81.9 |
| 3 | Charlottebrg Rsrvr | 82.2 |
| 4 | Long Branch | 82.7 |
| 5 | Sussex | 83.2 |
| 6 | Newton | 83.3 |
| 7 | Cape May | 83.5 |
| 8 | Boonton | 83.8 |
| 9 | Atlantic City-Int | 84.7 |
| 10 | Essex Fells | 84.8 |
| 11 | Canoe Brook | 85.2 |
| 12 | Hightstown | 85.3 |
| | Tuckerton | 85.3 |
| 14 | Somerville | 85.4 |
| 15 | Millville | 85.6 |
| 16 | Glassboro | 85.7 |
| | Seabrook Farms | 85.7 |
| 18 | Flemington | 85.8 |
| | Toms River | 85.8 |
| 20 | Little Falls | 85.9 |
| 21 | Newark | 86.2 |
| 22 | Belleplain | 86.3 |
| 23 | Plainfield | 86.4 |
| 24 | Pemberton | 86.6 |
| 25 | Indian Mills | 86.9 |

## ANNUAL PRECIPITATION (")

### HIGHEST

| | | |
|---|---|---|
| 1 | Long Valley | 52.58 |
| 2 | Charlottebrg Rsrvr | 51.75 |
| 3 | Canoe Brook | 50.88 |
| 4 | Essex Fells | 50.73 |
| 5 | Little Falls | 50.56 |
| 6 | Boonton | 49.49 |
| 7 | Plainfield | 49.05 |
| 8 | Flemington | 48.41 |
| 9 | Long Branch | 47.62 |
| 10 | Toms River | 47.34 |
| 11 | Sussex | 47.21 |
| 12 | Moorestown | 47.12 |
| 13 | Somerville | 47.00 |
| 14 | Pemberton | 46.94 |
| 15 | Lambertville | 46.78 |
| 16 | Hightstown | 46.50 |
| 17 | Indian Mills | 45.89 |
| 18 | Newton | 45.75 |
| 19 | Tuckerton | 45.34 |
| 20 | Newark | 44.67 |
| 21 | Glassboro | 44.50 |
| 22 | Woodstown | 44.39 |
| 23 | Seabrook Farms | 43.35 |
| 24 | Belleplain | 42.43 |
| 25 | Millville | 42.35 |

### LOWEST

| | | |
|---|---|---|
| 1 | Atlantic City-Mar | 37.16 |
| 2 | Atlantic City-Int | 39.62 |
| 3 | Cape May | 40.74 |
| 4 | Millville | 42.35 |
| 5 | Belleplain | 42.43 |
| 6 | Seabrook Farms | 43.35 |
| 7 | Woodstown | 44.39 |
| 8 | Glassboro | 44.50 |
| 9 | Newark | 44.67 |
| 10 | Tuckerton | 45.34 |
| 11 | Newton | 45.75 |
| 12 | Indian Mills | 45.89 |
| 13 | Hightstown | 46.50 |
| 14 | Lambertville | 46.78 |
| 15 | Pemberton | 46.94 |
| 16 | Somerville | 47.00 |
| 17 | Moorestown | 47.12 |
| 18 | Sussex | 47.21 |
| 19 | Toms River | 47.34 |
| 20 | Long Branch | 47.62 |
| 21 | Flemington | 48.41 |
| 22 | Plainfield | 49.05 |
| 23 | Boonton | 49.49 |
| 24 | Little Falls | 50.56 |
| 25 | Essex Fells | 50.73 |

## ANNUAL SNOWFALL (")

### HIGHEST

| | | |
|---|---|---|
| 1 | Newton | 38.8 |
| 2 | Charlottebrg Rsrvr | 36.2 |
| 3 | Long Valley | 34.6 |
| 4 | Flemington | 32.6 |
| 5 | Plainfield | 29.5 |
| 6 | Boonton | 29.2 |
| 7 | Somerville | 27.4 |
| 8 | Canoe Brook | 27.2 |
| 9 | Newark | 25.7 |
| 10 | Hightstown | 25.0 |
| 11 | Lambertville | 23.5 |
| 12 | Indian Mills | 20.5 |
| | Tuckerton | 20.5 |
| 14 | Pemberton | 20.0 |
| 15 | Essex Fells | 19.5 |
| 16 | Woodstown | 17.7 |
| 17 | Atlantic City-Int | 16.2 |
| 18 | Long Branch | 15.7 |
| 19 | Cape May | 14.5 |
| 20 | Millville | 13.9 |
| 21 | Belleplain | 12.9 |

### LOWEST

| | | |
|---|---|---|
| 1 | Belleplain | 12.9 |
| 2 | Millville | 13.9 |
| 3 | Cape May | 14.5 |
| 4 | Long Branch | 15.7 |
| 5 | Atlantic City-Int | 16.2 |
| 6 | Woodstown | 17.7 |
| 7 | Essex Fells | 19.5 |
| 8 | Pemberton | 20.0 |
| 9 | Indian Mills | 20.5 |
| | Tuckerton | 20.5 |
| 11 | Lambertville | 23.5 |
| 12 | Hightstown | 25.0 |
| 13 | Newark | 25.7 |
| 14 | Canoe Brook | 27.2 |
| 15 | Somerville | 27.4 |
| 16 | Boonton | 29.2 |
| 17 | Plainfield | 29.5 |
| 18 | Flemington | 32.6 |
| 19 | Long Valley | 34.6 |
| 20 | Charlottebrg Rsrvr | 36.2 |
| 21 | Newton | 38.8 |

# NEW MEXICO

PHYSICAL FEATURES. New Mexico, with a total area of 121,666 square miles, is in the southwestern part of the country. The State, approximately 350 miles square, lies mostly between latitudes 32° and 37° N. and longitudes 103° and 109° W. The State's topography consists mainly of high plateaus or mesas, with numerous mountain ranges, canyons, valleys, and normally dry arroyos. Average elevation is about 5,700 feet above sea level. The lowest point is upstream from the Red Bluff Reservoir at 2,817 feet where the Pecos River flows into Texas. The highest point is Wheeler Peak at 13,161 feet above sea level. The principal sources of moisture for the scant rains and snows that fall on the State are the Pacific Ocean, 500 miles to the west, and the Gulf of Mexico, 500 miles to the southeast.

New Mexico is divided into three major areas by mountain ranges and highlands, oriented in a general north-south direction, which merge in the north. The Northern Mountains and Central Highlands, between longitudes 105° and 106° W., are the western boundary of the Northeastern and Southeastern Plains which slope gradually eastward and southeastward. The northern part of these eastern plains lies within the Arkansas River Basin and is drained mostly by the Canadian River and the Cimarron River. West of the mountain ranges that form the Continental Divide, whose height decreases to a markedly lower elevation in southern New Mexico, rivers drain into the Gulf of California through the Colorado River system. Between the Northern Mountains and the Central Highland system and the Continental Divide system is the Rio Grande Valley which widens toward the south.

GENERAL CLIMATE. New Mexico has a mild, arid or semiarid, continental climate characterized by light precipitation totals, abundant sunshine, low relative humidities, and a relatively large annual and diurnal temperature range. The highest mountains have climate characteristics common to the Rocky Mountains.

Location and topography play major roles in determining the climate of New Mexico, particularly true for any specific locality. Both the ruggedness of the terrain and its direction of slope are important. The eastern plains open to the Great Plains of Texas and Oklahoma and to their northward extension into central Canada. At times during winter months, cold continental air masses move southward out of central Canada and invade this area, producing blizzard and cold-wave conditions. These air masses occasionally cross the Central Highlands, which greatly modify and warm the air masses before they reach the Rio Grande Valley.

PRECIPITATION. Average annual precipitation ranges from less than 10 inches over much of the southern desert and the Rio Grande and San Juan Valleys, to more than 20 inches at higher elevations in the State. A wide variation in annual totals is characteristic of arid and semiarid climates.

Summer rains fall almost entirely during brief, but frequently intense, thunderstorms. The general southeasterly circulation from the Gulf of Mexico brings moisture for these storms into the State, and strong surface heating combined with orographic lifting as the air moves over higher terrain causes convective air currents and condensation. July and August are the rainiest months over most of the State, with from 30 to 40 percent of the year's total moisture falling at that time. The San Juan Valley area is least affected by this summer circulation, receiving about 25 percent of its annual rainfall during July and August. During the warmest 6 months of the year, May through October, total precipitation averages from 60 percent of the annual total in the Northwestern Plateau to 80 percent of the annual total in the eastern plains.

Winter precipitation is caused mainly by frontal activity associated with the general movement of Pacific Ocean storms across the country from west to east. As these storms move inland, much of the moisture is precipitated over the coastal and inland mountain ranges of California, Nevada, Arizona, and Utah. Much of the remaining moisture falls on the western slope of the Continental Divide and over northern and high central mountain ranges. Winter is the driest season in New Mexico except for the portion west of the Continental Divide. This dryness is most noticeable in the Central Valley and on eastern slopes of the mountains.

Much of the winter precipitation falls as snow in the mountain areas, but it may occur as either rain or snow in the valleys. Average annual snowfall ranges from about 3 inches at the Southern Desert and Southeastern Plains stations to well over 100 inches at Northern Mountain stations. It may exceed 300 inches in the highest mountains of the north.

FLOODS. General floods are seldom widespread in New Mexico. Heavy summer thunderstorms may bring several inches of rain to small areas in a short time. Because of the rough terrain and sparse vegetation in many areas, run-offs from these storms frequently cause local flash floods. Normally dry arroyos may overflow their banks for several hours, halting traffic where water crosses highways, and damaging bridges, culverts, and roadways. Snowmelt during April to June, especially in combination with a warm rain, and heavy general rains during August to October may occasionally cause flooding of the larger rivers.

TEMPERATURE. Elevation is a greater factor in determining the temperature of any specific locality than its latitude. During the summer months, individual daytime temperatures quite often exceed 100° at elevations below 5,000 feet; but the average monthly maximum temperatures during July, the warmest month, range from slightly above 90° at lower elevations to the upper 70s at high elevations. Warmest days quite often occur in June before the thunderstorm season sets in; during July and August, afternoon convective storms tend to shut off afternoon solar insolation, lowering temperatures before they reach their potential daily high. A preponderance of clear skies and low relative humidity permits rapid cooling by radiation from the earth after sundown; consequently, nights are usually comfortable in summer.

In January, the coldest month, average daytime temperatures range from the middle 50s in the southern and central valleys to the middle 30s in the higher elevations of the north. Minimum temperatures below freezing are common in all sections of the State during the winter, but subzero temperatures are rare except in the mountains. The freeze-free season ranges from more than 200 days in the southern valleys to less than 80 days in the northern mountains where some high mountain valleys have freezes in summer months.

SEVERE STORMS. On rare occasions, a tropical hurricane may cause heavy rain in eastern and central New Mexico as it moves inland from the western part of the Gulf of Mexico. Also on rare occasions, a tropical storm moving inland from the Gulf of California area may cause heavy rain to fall in southwestern New Mexico. Tornadoes are occasionally reported in New Mexico, most frequently during afternoon and early evening hours from May through August.

Thunderstorms are relatively frequent in summer, averaging in numbers from 40 in the south to more than 70 in the northeast, the latter area having the second greatest thunderstorm frequency in the country. Occasionally, these heavy thunderstorms are accompanied by hail, with the greatest hail frequency occurring near and to the east of Los Alamos.

OTHER CLIMATIC ELEMENTS. Plentiful sunshine occurs in New Mexico, with from 75 to 80 percent of the possible sunshine being received. In winter, this prevalence is particularly noticeable with from 70 to 75 percent of the possible sunshine being received. It is not uncommon for as much as 90 percent of the possible sunshine to occur in November and in some of the spring months. The average number of hours of annual sunshine ranges from near 3,700 in the southwest to 2,800 in the north-central portions.

Average relative humidities are lower in the valleys but higher in the mountains because of the lower mountain temperatures. Relative humidity ranges from an average of near 65 percent at about sunrise to near 30 percent in midafternoon; however, afternoon humidities in warmer months are often less than 20 percent and occasionally may go as low as 4 percent. The prevalent low relative humidities during periods of extreme temperatures ease the effect of summer and winter temperatures on comfort.

Wind speeds over the State are usually moderate, although relatively strong winds often accompany occasional frontal activity during late winter and spring months and sometimes occur just in advance of thunderstorms. Frontal winds may exceed 30 m.p.h. for several hours and reach peak speeds of more than 50 m.p.h. Spring is the windy season. Blowing dust and serious soil erosion of unprotected fields may be a problem during dry spells. Winds are generally stronger in the eastern plains than in other parts of the State. Winds generally predominate from the southeast in summer and from the west in winter, but local surface wind directions will vary greatly because of local topography and mountain and valley breezes.

Potential evaporation in New Mexico is much greater than average annual precipitation.

## COUNTY INDEX

### *Bernalillo County*
ALBUQUERQUE INTL AP
SANDIA PARK

### *Catron County*
BEAVERHEAD R S
LUNA R S
RESERVE RANGER STN

### *Chaves County*
BITTER LAKES WL REF
ELK 2 E

### *Cibola County*
EL MORRO NATL MON
GRANTS MILAN MUNI AP
LAGUNA

### *Colfax County*
CIMARRON 4 SW
EAGLE NEST
LAKE MALOYA
MAXWELL 3 NW
RATON FILTER PLANT
SPRINGER

### *Curry County*
CLOVIS
CLOVIS 13 N
MELROSE

### *DeBaca County*
FORT SUMNER
YESO 2 S

### *Dona Ana County*
HATCH 2 W
JORNADA EXP RANGE

### *Eddy County*
ARTESIA 6 S
CARLSBAD
CARLSBAD CAVERN CITY
CARLSBAD CAVERNS
HOPE

### *Grant County*
CLIFF 11 SE
FAYWOOD
FORT BAYARD
GILA HOT SPRINGS
HACHITA
MIMBRES RS
REDROCK 1 NNE
WHITE SIGNAL

### *Guadalupe County*
DILIA
NEWKIRK
SANTA ROSA

### *Harding County*
MOSQUERO 1 NE
ROY

### *Hidalgo County*
ANIMAS
LORDSBURG 4 SE

### *Lea County*
CROSSROADS 2
HOBBS
JAL
MALJAMAR 4 SE
PEARL
TATUM

### *Lincoln County*
CARRIZOZO 1 SW
RAMON 8 SW
RUIDOSO

### *Los Alamos County*
LOS ALAMOS

### *Luna County*
COLUMBUS
GAGE 4 ESE

### *McKinley County*
MCGAFFEY 5 SE
STAR LAKE
ZUNI

### *Mora County*
GASCON
OCATE 1 N
VALMORA

### *Otero County*
ALAMOGORDO
MOUNTAIN PARK
OROGRANDE
TULAROSA
WHITE SANDS NATL MON

### *Quay County*
CAMERON
RAGLAND 3 SSW
SAN JON
TUCUMCARI 4 NE

### *Rio Arriba County*
ABIQUIU DAM
ALCALDE
CHAMA
DULCE
EL RITO
EL VADO DAM
LYBROOK
TIERRA AMARILLA 4 N

### *Roosevelt County*
ELIDA
PORTALES

### *Sandoval County*
JEMEZ SPRINGS
TORREON NAVAJO MISSI
WOLF CANYON

### *San Juan County*
AZTEC RUINS NATL MON
BLOOMFIELD 3 SE
CHACO CANYN NATL MON
FRUITLAND 2 E
NAVAJO DAM
OTIS
SHIPROCK

### *San Miguel County*
BELL RANCH
CONCHAS DAM
LAS VEGAS MUNI AP
PECOS RS

### *Santa Fe County*
ESPANOLA
STANLEY 1 NNE

### *Sierra County*
ALEMAN RANCH
CABALLO DAM
ELEPHANT BUTTE DAM
HILLSBORO
WINSTON

### *Socorro County*

AUGUSTINE 2 E
BERNARDO
BOSQUE DEL APACHE
GRAN QUIVIRA NATL MO
SOCORRO

### *Taos County*

CERRO
RED RIVER
TAOS

### *Torrance County*

ESTANCIA
MOUNTAINAIR
PEDERNAL 4 E

### *Union County*

AMISTAD 5 SSW
CLAYTON MUNI ARPK
DES MOINES
GRENVILLE
PASAMONTE

### *Valencia County*

FENCE LAKE
LOS LUNAS 3 SSW

## ELEVATION INDEX

| FEET | STATION NAME |
|---|---|
| 3120 | CARLSBAD |
| 3153 | JAL |
| 3235 | CARLSBAD CAVERN CITY |
| 3320 | ARTESIA 6 S |
| 3622 | HOBBS |
| 3671 | BITTER LAKES WL REF |
| 3802 | PEARL |
| 4003 | PORTALES |
| 4003 | WHITE SANDS NATL MON |
| 4032 | COLUMBUS |
| 4032 | FORT SUMNER |
| 4042 | HATCH 2 W |
| 4091 | HOPE |
| 4104 | CROSSROADS 2 |
| 4104 | TATUM |
| 4104 | TUCUMCARI 4 NE |
| 4153 | MALJAMAR 4 SE |
| 4153 | REDROCK 1 NNE |
| 4183 | OROGRANDE |
| 4190 | CABALLO DAM |

| FEET | STATION NAME |
|---|---|
| 4232 | SAN JON |
| 4242 | CONCHAS DAM |
| 4252 | LORDSBURG 4 SE |
| 4272 | JORNADA EXP RANGE |
| 4290 | CLOVIS |
| 4350 | ALAMOGORDO |
| 4354 | ELIDA |
| 4413 | ANIMAS |
| 4442 | CARLSBAD CAVERNS |
| 4442 | CLOVIS 13 N |
| 4442 | TULAROSA |
| 4482 | GAGE 4 ESE |
| 4495 | AMISTAD 5 SSW |
| 4505 | BELL RANCH |
| 4505 | HACHITA |
| 4524 | BOSQUE DEL APACHE |
| 4534 | ALEMAN RANCH |
| 4573 | NEWKIRK |
| 4583 | ELEPHANT BUTTE DAM |
| 4603 | CAMERON |
| 4603 | MELROSE |
| 4603 | SANTA ROSA |
| 4623 | SOCORRO |
| 4734 | BERNARDO |
| 4803 | CLIFF 11 SE |
| 4842 | LOS LUNAS 3 SSW |
| 4852 | YESO 2 S |
| 4954 | SHIPROCK |
| 5000 | CLAYTON MUNI ARPK |
| 5060 | RAGLAND 3 SSW |
| 5092 | FAYWOOD |
| 5144 | DILIA |
| 5174 | FRUITLAND 2 E |
| 5311 | ALBUQUERQUE INTL AP |
| 5328 | RAMON 8 SW |
| 5335 | HILLSBORO |
| 5426 | CARRIZOZO 1 SW |
| 5554 | MOSQUERO 1 NE |
| 5594 | ESPANOLA |
| 5604 | GILA HOT SPRINGS |
| 5643 | AZTEC RUINS NATL MON |
| 5653 | PASAMONTE |
| 5682 | ALCALDE |
| 5705 | ELK 2 E |
| 5774 | NAVAJO DAM |
| 5794 | BLOOMFIELD 3 SE |
| 5833 | RESERVE RANGER STN |
| 5843 | LAGUNA |
| 5880 | SPRINGER |
| 5882 | ROY |

| FEET | STATION NAME |
|---|---|
| 5915 | MAXWELL 3 NW |
| 5994 | GRENVILLE |
| 6073 | WHITE SIGNAL |
| 6106 | JEMEZ SPRINGS |
| 6115 | ESTANCIA |
| 6135 | CHACO CANYN NATL MON |
| 6155 | FORT BAYARD |
| 6204 | PEDERNAL 4 E |
| 6204 | WINSTON |
| 6253 | MIMBRES RS |
| 6306 | VALMORA |
| 6380 | STANLEY 1 NNE |
| 6404 | ABIQUIU DAM |
| 6434 | CIMARRON 4 SW |
| 6452 | ZUNI |
| 6499 | MOUNTAINAIR |
| 6526 | GRANTS MILAN MUNI AP |
| 6624 | DES MOINES |
| 6624 | GRAN QUIVIRA NATL MO |
| 6644 | STAR LAKE |
| 6703 | TORREON NAVAJO MISSI |
| 6726 | BEAVERHEAD R S |
| 6726 | MOUNTAIN PARK |
| 6760 | RUIDOSO |
| 6775 | DULCE |
| 6804 | EL VADO DAM |
| 6863 | LAS VEGAS MUNI AP |
| 6873 | EL RITO |
| 6883 | OTIS |
| 6906 | PECOS RS |
| 6936 | RATON FILTER PLANT |
| 6985 | TAOS |
| 7014 | SANDIA PARK |
| 7024 | AUGUSTINE 2 E |
| 7054 | LUNA R S |
| 7073 | FENCE LAKE |
| 7205 | LYBROOK |
| 7224 | EL MORRO NATL MON |
| 7365 | LOS ALAMOS |
| 7405 | LAKE MALOYA |
| 7427 | TIERRA AMARILLA 4 N |
| 7664 | CERRO |
| 7674 | OCATE 1 N |
| 7854 | CHAMA |
| 7999 | MCGAFFEY 5 SE |
| 8156 | WOLF CANYON |
| 8254 | EAGLE NEST |
| 8254 | GASCON |
| 8675 | RED RIVER |

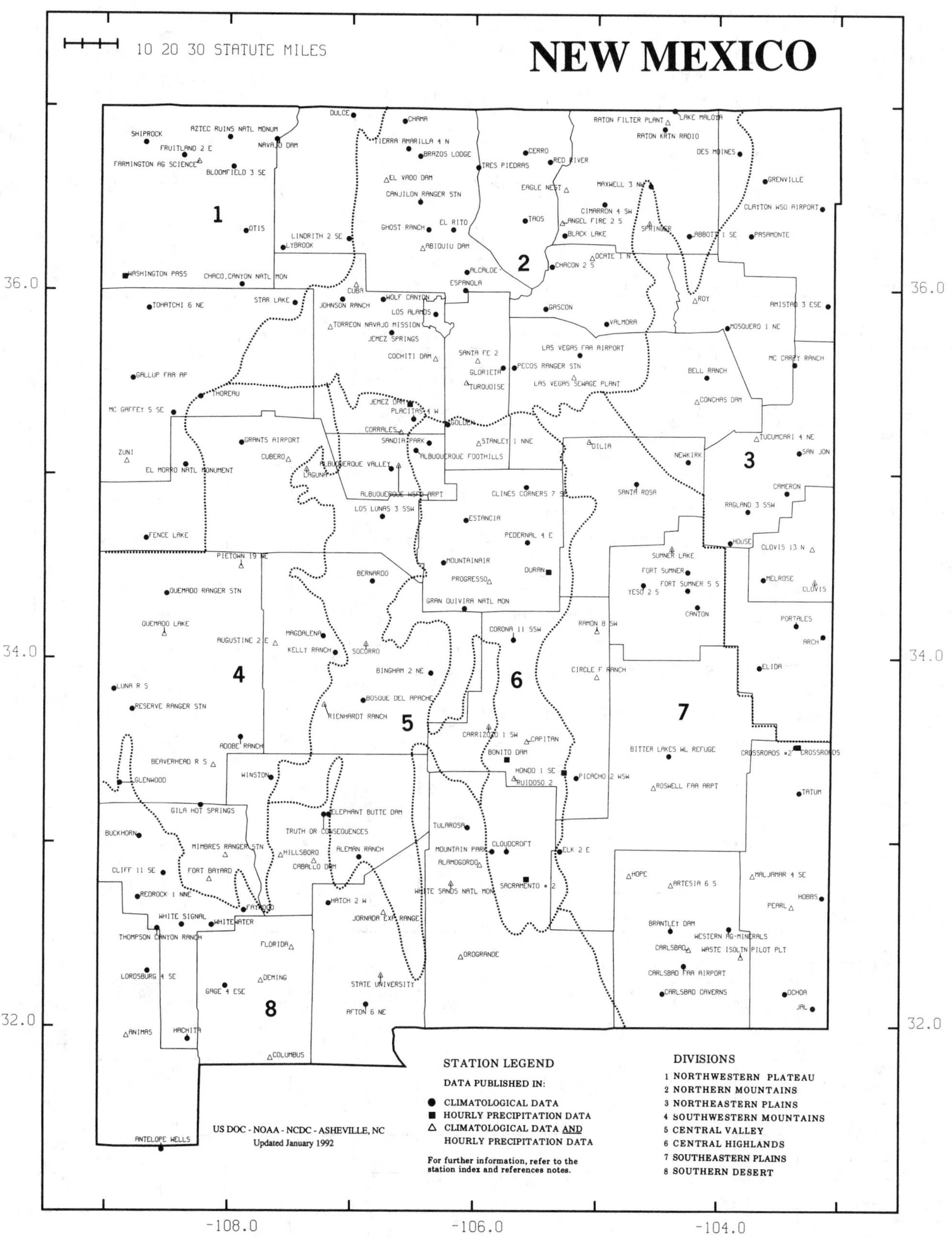
10 20 30 STATUTE MILES
NEW MEXICO
36.0
34.0
32.0
-108.0
-106.0
-104.0
1
2
3
4
5
6
7
8
SHIPROCK
AZTEC RUINS NATL MONUM
FRUITLAND 2 E
NAVAJO DAM
FARMINGTON AG SCIENCE
BLOOMFIELD 3 SE
DULCE
CHAMA
TIERRA AMARILLA 4 N
BRAZOS LODGE
EL VADO DAM
CANJILON RANGER STN
GHOST RANCH
EL RITO
ABIQUIU DAM
TRES PIEDRAS
CERRO
RED RIVER
EAGLE NEST
TAOS
ANGEL FIRE 2 S
BLACK LAKE
CIMARRON 4 SW
MAXWELL 3 NW
RATON FILTER PLANT
LAKE MALOYA
RATON KRTN RADIO
DES MOINES
GRENVILLE
CLAYTON WSO AIRPORT
SPRINGER
ABBOTT 1 SE
PASAMONTE
OCATE 1 N
CHACON 2 S
OTIS
LINDRITH 2 SE
LYBROOK
WASHINGTON PASS
CHACO CANYON NATL MON
ALCALDE
ESPANOLA
CUBA
WOLF CANYON
JOHNSON RANCH
LOS ALAMOS
TORREON NAVAJO MISSION
JEMEZ SPRINGS
COCHITI DAM
SANTA FE 2
GLORIETA
PECOS RANGER STN
TURQUOISE
GASCON
VALMORA
LAS VEGAS FAA AIRPORT
LAS VEGAS SEWAGE PLANT
ROY
AMISTAD 3 ESE
MOSQUERO 1 NE
MC CARTY RANCH
BELL RANCH
CONCHAS DAM
TOHATCHI 6 NE
STAR LAKE
GALLUP FAA AP
THOREAU
MC GAFFEY 5 SE
GRANTS AIRPORT
ZUNI
CUBERO
EL MORRO NATL MONUMENT
LAGUNA
JEMEZ DAM
PLACITAS 4 W
GOLDEN
CORRALES
SANDIA PARK
ALBUQUERQUE FOOTHILLS
ALBUQUERQUE VALLEY
ALBUQUERQUE WSFO ARPT
STANLEY 1 NNE
LOS LUNAS 3 SSW
CLINES CORNERS 7 SE
ESTANCIA
PEDERNAL 4 E
MOUNTAINAIR
PROGRESSO
DURAN
GRAN QUIVIRA NATL MON
DILIA
NEWKIRK
SANTA ROSA
TUCUMCARI 4 NE
SAN JON
CAMERON
RAGLAND 3 SSW
HOUSE
CLOVIS 13 N
SUMNER LAKE
FORT SUMNER
FORT SUMNER 5 S
YESO 2 S
MELROSE
CLOVIS
CANTON
PORTALES
ARCH
ELIDA
FENCE LAKE
PIETOWN 19 NE
BERNARDO
QUEMADO RANGER STN
QUEMADO LAKE
AUGUSTINE 2 E
MAGDALENA
KELLY RANCH
SOCORRO
BINGHAM 2 NE
BOSQUE DEL APACHE
RIENHARDT RANCH
CORONA 11 SSW
RAMON 8 SW
CIRCLE F RANCH
LUNA R S
RESERVE RANGER STN
ADOBE RANCH
BEAVERHEAD R S
GLENWOOD
WINSTON
CARRIZOZO 1 SW
CAPITAN
BONITO DAM
HONDO 1 SE
RUIDOSO 2
PICACHO 2 WSW
BITTER LAKES WL REFUGE
CROSSROADS #2
CROSSROADS
ROSWELL FAA ARPT
TATUM
GILA HOT SPRINGS
BUCKHORN
MIMBRES RANGER STN
ELEPHANT BUTTE DAM
TRUTH OR CONSEQUENCES
HILLSBORO
CABALLO DAM
ALEMAN RANCH
TULAROSA
MOUNTAIN PARK
CLOUDCROFT
ALAMOGORDO
ELK 2 E
CLIFF 11 SE
FORT BAYARD
REDROCK 1 NNE
FAYWOOD
WHITE SIGNAL
WHITEWATER
THOMPSON CANYON RANCH
HATCH 2 W
JORNADA EXP RANGE
WHITE SANDS NATL MON
SACRAMENTO #2
HOPE
ARTESIA 6 S
MALJAMAR 4 SE
HOBBS
PEARL
BRANTLEY DAM
WESTERN AG-MINERALS
CARLSBAD
WASTE ISOLTN PILOT PLT
CARLSBAD FAA AIRPORT
CARLSBAD CAVERNS
OCHOA
JAL
FLORIDA
OROGRANDE
LORDSBURG 4 SE
DEMING
GAGE 4 ESE
STATE UNIVERSITY
AFTON 6 NE
ANIMAS
HACHITA
COLUMBUS
ANTELOPE WELLS
US DOC - NOAA - NCDC - ASHEVILLE, NC
Updated January 1992
STATION LEGEND
DATA PUBLISHED IN:
● CLIMATOLOGICAL DATA
■ HOURLY PRECIPITATION DATA
△ CLIMATOLOGICAL DATA AND HOURLY PRECIPITATION DATA
For further information, refer to the station index and references notes.
DIVISIONS
1 NORTHWESTERN PLATEAU
2 NORTHERN MOUNTAINS
3 NORTHEASTERN PLAINS
4 SOUTHWESTERN MOUNTAINS
5 CENTRAL VALLEY
6 CENTRAL HIGHLANDS
7 SOUTHEASTERN PLAINS
8 SOUTHERN DESERT

## ABIQUIU DAM *Rio Arriba County* ELEVATION 6404 ft LAT/LONG 36° 14' N / 106° 26' W

| | JAN | FEB | MAR | APR | MAY | JUN | JUL | AUG | SEP | OCT | NOV | DEC | YEAR |
|---|---|---|---|---|---|---|---|---|---|---|---|---|---|
| Maximum Temp °F | na | 46.3 | 54.4 | 63.0 | 72.0 | 82.5 | 86.5 | 83.7 | 77.1 | 66.6 | 52.7 | 43.3 | na |
| Minimum Temp °F | na | 20.1 | 27.5 | 33.9 | 42.7 | 51.5 | 57.2 | 55.5 | 48.0 | 37.3 | 26.2 | 17.5 | na |
| Mean Temp °F | na | 33.2 | 40.9 | 48.5 | 57.3 | 67.0 | 71.9 | 69.6 | 62.5 | 52.0 | 39.5 | 30.3 | na |
| Days Max Temp ≥ 90 °F | 0 | 0 | 0 | 0 | 0 | 5 | 8 | 4 | 0 | 0 | 0 | 0 | 17 |
| Days Max Temp ≤ 32 °F | *4* | *2* | 0 | 0 | 0 | 0 | 0 | 0 | 0 | 0 | 1 | *3* | 10 |
| Days Min Temp ≤ 32 °F | *30* | 27 | *24* | *12* | 2 | 0 | 0 | 0 | 0 | 7 | 25 | 30 | 157 |
| Days Min Temp ≤ 0 °F | *1* | 0 | 0 | 0 | 0 | 0 | 0 | 0 | 0 | 0 | 0 | 1 | 2 |
| Heating Degree Days | na | *889* | *736* | 489 | 238 | 44 | 2 | 7 | 100 | 397 | 758 | 1068 | na |
| Cooling Degree Days | 0 | 0 | 0 | 0 | 7 | 121 | 222 | 166 | 36 | 0 | 0 | 0 | 552 |
| Total Precipitation (") | 0.37 | 0.24 | 0.54 | 0.51 | 1.00 | 0.76 | 1.79 | 2.09 | 1.13 | 0.84 | 0.53 | 0.34 | 10.14 |
| Days ≥ 0.1" Precip | 1 | 1 | 1 | 2 | 3 | 2 | 5 | 6 | 3 | 3 | 2 | 1 | 30 |
| Total Snowfall (") | na | na | *0.7* | 0.6 | 0.0 | 0.0 | 0.0 | 0.0 | 0.0 | 0.1 | 0.3 | na | na |
| Days ≥ 1" Snow Depth | na | na | *0* | 0 | 0 | 0 | 0 | 0 | 0 | 0 | 0 | na | na |

## ALAMOGORDO *Otero County* ELEVATION 4350 ft LAT/LONG 32° 54' N / 105° 58' W

| | JAN | FEB | MAR | APR | MAY | JUN | JUL | AUG | SEP | OCT | NOV | DEC | YEAR |
|---|---|---|---|---|---|---|---|---|---|---|---|---|---|
| Maximum Temp °F | 56.2 | 61.2 | 68.5 | 77.4 | 85.4 | 94.6 | 94.0 | 90.8 | 85.8 | 77.0 | 64.9 | 56.7 | 76.0 |
| Minimum Temp °F | 29.0 | 32.8 | 38.5 | 45.6 | 53.9 | 62.6 | 66.1 | 64.0 | 58.0 | 47.2 | 36.2 | 29.4 | 46.9 |
| Mean Temp °F | 42.6 | 47.0 | 53.5 | 61.5 | 69.7 | 78.7 | 80.1 | 77.4 | 71.9 | 62.1 | 50.6 | 43.0 | 61.5 |
| Days Max Temp ≥ 90 °F | 0 | 0 | 0 | 1 | 9 | 25 | 24 | 20 | 9 | 1 | 0 | 0 | 89 |
| Days Max Temp ≤ 32 °F | 0 | 0 | 0 | 0 | 0 | 0 | 0 | 0 | 0 | 0 | 0 | 0 | 0 |
| Days Min Temp ≤ 32 °F | 22 | 14 | 7 | 1 | 0 | 0 | 0 | 0 | 0 | 1 | 10 | 21 | 76 |
| Days Min Temp ≤ 0 °F | 0 | 0 | 0 | 0 | 0 | 0 | 0 | 0 | 0 | 0 | 0 | 0 | 0 |
| Heating Degree Days | 687 | 500 | 351 | 134 | 20 | 0 | 0 | 0 | 10 | 122 | 426 | 674 | 2924 |
| Cooling Degree Days | 0 | 0 | 2 | 39 | 171 | 427 | 478 | 401 | 221 | 37 | 1 | 0 | 1777 |
| Total Precipitation (") | 0.78 | 0.56 | 0.50 | 0.30 | 0.66 | 0.90 | 2.28 | 2.46 | 1.83 | 1.31 | 0.76 | 1.03 | 13.37 |
| Days ≥ 0.1" Precip | 2 | 2 | 2 | 1 | 2 | 2 | 5 | 6 | 5 | 3 | 2 | 3 | 35 |
| Total Snowfall (") | 2.2 | 0.7 | 0.4 | 0.1 | 0.0 | 0.0 | 0.0 | 0.0 | 0.0 | 0.1 | 0.3 | 1.4 | 5.2 |
| Days ≥ 1" Snow Depth | *0* | 0 | 0 | 0 | 0 | 0 | 0 | 0 | 0 | 0 | 0 | 0 | 0 |

## ALBUQUERQUE INTL AP *Bernalillo County* ELEVATION 5311 ft LAT/LONG 35° 3' N / 106° 37' W

| | JAN | FEB | MAR | APR | MAY | JUN | JUL | AUG | SEP | OCT | NOV | DEC | YEAR |
|---|---|---|---|---|---|---|---|---|---|---|---|---|---|
| Maximum Temp °F | 47.2 | 53.8 | 61.9 | 71.0 | 79.5 | 90.1 | 92.6 | 89.0 | 82.1 | 70.9 | 57.1 | 47.6 | 70.2 |
| Minimum Temp °F | 22.8 | 27.2 | 32.9 | 40.1 | 49.0 | 58.7 | 64.6 | 62.7 | 55.2 | 43.2 | 31.2 | 23.6 | 42.6 |
| Mean Temp °F | 35.1 | 40.6 | 47.4 | 55.6 | 64.3 | 74.4 | 78.6 | 75.9 | 68.7 | 57.1 | 44.2 | 35.6 | 56.5 |
| Days Max Temp ≥ 90 °F | 0 | 0 | 0 | 0 | 2 | 17 | 23 | 16 | 4 | 0 | 0 | 0 | 62 |
| Days Max Temp ≤ 32 °F | 2 | 1 | 0 | 0 | 0 | 0 | 0 | 0 | 0 | 0 | 0 | 2 | 5 |
| Days Min Temp ≤ 32 °F | 29 | 22 | 15 | 4 | 0 | 0 | 0 | 0 | 0 | 2 | 17 | 28 | 117 |
| Days Min Temp ≤ 0 °F | 0 | 0 | 0 | 0 | 0 | 0 | 0 | 0 | 0 | 0 | 0 | 0 | 0 |
| Heating Degree Days | 921 | 683 | 539 | 282 | 87 | 4 | 0 | 0 | 27 | 248 | 618 | 905 | 4314 |
| Cooling Degree Days | 0 | 0 | 0 | 10 | 76 | 312 | 427 | 353 | 153 | 8 | 0 | 0 | 1339 |
| Total Precipitation (") | 0.46 | 0.48 | 0.54 | 0.48 | 0.65 | 0.64 | 1.32 | 1.76 | 1.00 | 0.93 | 0.55 | 0.56 | 9.37 |
| Days ≥ 0.1" Precip | 2 | 2 | 2 | 1 | 2 | 2 | 4 | 4 | 3 | 2 | 2 | 2 | 28 |
| Total Snowfall (") | 2.9 | 2.4 | 1.9 | 0.7 | 0.0 | 0.0 | 0.0 | 0.0 | 0.0 | 0.3 | 1.0 | 2.4 | 11.6 |
| Days ≥ 1" Snow Depth | 2 | 1 | 0 | 0 | 0 | 0 | 0 | 0 | 0 | 0 | 0 | 2 | 5 |

## ALCALDE *Rio Arriba County* ELEVATION 5682 ft LAT/LONG 36° 6' N / 106° 4' W

| | JAN | FEB | MAR | APR | MAY | JUN | JUL | AUG | SEP | OCT | NOV | DEC | YEAR |
|---|---|---|---|---|---|---|---|---|---|---|---|---|---|
| Maximum Temp °F | 44.9 | 51.4 | 59.3 | 67.9 | 76.5 | 86.2 | 89.5 | 86.3 | 80.0 | 70.4 | 56.3 | 46.1 | 67.9 |
| Minimum Temp °F | 16.2 | 21.0 | 26.5 | 32.9 | 40.5 | 48.7 | 55.4 | 54.0 | 45.1 | 33.5 | 23.7 | 16.7 | 34.5 |
| Mean Temp °F | 30.6 | 36.2 | 43.0 | 50.4 | 58.6 | 67.5 | 72.5 | 70.2 | 62.6 | 52.0 | 40.0 | 31.4 | 51.3 |
| Days Max Temp ≥ 90 °F | 0 | 0 | 0 | 0 | 1 | 10 | 16 | 9 | 2 | 0 | 0 | 0 | 38 |
| Days Max Temp ≤ 32 °F | 2 | 0 | 0 | 0 | 0 | 0 | 0 | 0 | 0 | 0 | 0 | 2 | 4 |
| Days Min Temp ≤ 32 °F | 30 | 27 | 25 | 15 | 3 | 0 | 0 | 0 | 1 | 14 | 26 | 29 | 170 |
| Days Min Temp ≤ 0 °F | 2 | 0 | 0 | 0 | 0 | 0 | 0 | 0 | 0 | 0 | 0 | 1 | 3 |
| Heating Degree Days | 1061 | 807 | 677 | 431 | 201 | 29 | 0 | 6 | 99 | 398 | 743 | 1034 | 5486 |
| Cooling Degree Days | 0 | 0 | 0 | 0 | 10 | 127 | 236 | 186 | 35 | 0 | 0 | 0 | 594 |
| Total Precipitation (") | 0.40 | 0.35 | 0.53 | 0.58 | 0.89 | 0.89 | 1.38 | 1.94 | 1.38 | 0.90 | 0.78 | 0.45 | 10.47 |
| Days ≥ 0.1" Precip | 1 | 1 | 2 | 2 | 3 | 2 | 4 | 4 | 3 | 2 | 2 | 1 | 27 |
| Total Snowfall (") | 2.7 | 2.5 | 1.9 | 0.2 | 0.1 | 0.0 | 0.0 | 0.0 | 0.0 | 0.2 | 1.1 | 1.8 | 10.5 |
| Days ≥ 1" Snow Depth | *2* | *1* | 0 | 0 | 0 | 0 | 0 | 0 | 0 | 0 | *0* | *2* | 5 |

## ALEMAN RANCH *Sierra County* ELEVATION 4534 ft LAT/LONG 32° 55 ' N / 106° 56 ' W

| | JAN | FEB | MAR | APR | MAY | JUN | JUL | AUG | SEP | OCT | NOV | DEC | YEAR |
|---|---|---|---|---|---|---|---|---|---|---|---|---|---|
| Maximum Temp °F | 54.4 | 59.8 | 66.9 | 75.4 | 83.1 | 92.7 | 92.9 | 89.8 | 84.2 | 75.0 | 62.8 | 54.3 | 74.3 |
| Minimum Temp °F | 24.2 | 27.1 | 31.9 | 38.8 | 46.9 | 56.2 | 62.6 | 61.0 | 53.8 | 42.2 | 31.1 | 24.9 | 41.7 |
| Mean Temp °F | 39.4 | 43.5 | 49.4 | 57.1 | 65.0 | 74.5 | 77.8 | 75.4 | 69.0 | 58.6 | 47.0 | 39.6 | 58.0 |
| Days Max Temp ≥ 90 °F | 0 | 0 | 0 | 0 | 5 | 21 | 23 | 18 | 6 | 0 | 0 | 0 | 73 |
| Days Max Temp ≤ 32 °F | 0 | 0 | 0 | 0 | 0 | 0 | 0 | 0 | 0 | 0 | 0 | 0 | 0 |
| Days Min Temp ≤ 32 °F | 27 | 21 | 17 | 6 | 0 | 0 | 0 | 0 | 0 | 3 | 18 | 26 | 118 |
| Days Min Temp ≤ 0 °F | 0 | 0 | 0 | 0 | 0 | 0 | 0 | 0 | 0 | 0 | 0 | 0 | 0 |
| Heating Degree Days | 788 | 602 | 477 | 237 | 62 | 2 | 0 | 0 | 19 | 202 | 533 | 780 | 3702 |
| Cooling Degree Days | 0 | 0 | 0 | 9 | 71 | 303 | 405 | 333 | 148 | 8 | 0 | 0 | 1277 |
| Total Precipitation (") | 0.47 | 0.38 | 0.28 | 0.19 | 0.52 | 0.73 | 2.11 | 2.40 | 1.43 | 1.01 | 0.65 | 0.92 | 11.09 |
| Days ≥ 0.1" Precip | 2 | 1 | 1 | 1 | 1 | 2 | 5 | 6 | 4 | 2 | 2 | 3 | 30 |
| Total Snowfall (") | 2.2 | 0.9 | 0.6 | 0.1 | 0.0 | 0.0 | 0.0 | 0.0 | 0.0 | 0.0 | 0.6 | 2.9 | 7.3 |
| Days ≥ 1" Snow Depth | 1 | 0 | 0 | 0 | 0 | 0 | 0 | 0 | 0 | 0 | 0 | 2 | 3 |

## AMISTAD 5 SSW *Union County* ELEVATION 4495 ft LAT/LONG 35° 55 ' N / 103° 6 ' W

| | JAN | FEB | MAR | APR | MAY | JUN | JUL | AUG | SEP | OCT | NOV | DEC | YEAR |
|---|---|---|---|---|---|---|---|---|---|---|---|---|---|
| Maximum Temp °F | 49.7 | 54.4 | 62.4 | 71.4 | 78.8 | 88.1 | 91.5 | 88.9 | 82.1 | 72.5 | 59.7 | 50.8 | 70.9 |
| Minimum Temp °F | 19.2 | 22.3 | 28.8 | 37.3 | 46.7 | 56.2 | 61.4 | 59.7 | 51.8 | 40.0 | 28.4 | 20.8 | 39.4 |
| Mean Temp °F | 34.5 | 38.3 | 45.6 | 54.4 | 62.8 | 72.2 | 76.5 | 74.3 | 67.0 | 56.3 | 44.1 | 35.8 | 55.2 |
| Days Max Temp ≥ 90 °F | 0 | 0 | 0 | 0 | 3 | 14 | 21 | 16 | 5 | 1 | 0 | 0 | 60 |
| Days Max Temp ≤ 32 °F | 4 | 2 | 1 | 0 | 0 | 0 | 0 | 0 | 0 | 0 | 0 | 3 | 10 |
| Days Min Temp ≤ 32 °F | 29 | 25 | 21 | 8 | 1 | 0 | 0 | 0 | 0 | 5 | 21 | 28 | 138 |
| Days Min Temp ≤ 0 °F | 1 | 1 | 0 | 0 | 0 | 0 | 0 | 0 | 0 | 0 | 0 | 1 | 3 |
| Heating Degree Days | 939 | 746 | 593 | 317 | 118 | 11 | 1 | 2 | 55 | 275 | 621 | 897 | 4575 |
| Cooling Degree Days | 0 | 0 | 0 | 6 | 47 | 229 | 359 | 308 | 124 | 6 | 0 | 0 | 1079 |
| Total Precipitation (") | 0.29 | 0.32 | 0.66 | 0.83 | 2.19 | 2.06 | 2.85 | 2.63 | 1.71 | 0.98 | 0.62 | 0.41 | 15.55 |
| Days ≥ 0.1" Precip | 1 | 1 | 2 | 2 | 4 | 4 | 5 | 5 | 3 | 2 | 2 | 2 | 33 |
| Total Snowfall (") | *2.5* | *2.7* | na | 0.6 | 0.2 | 0.0 | 0.0 | 0.0 | 0.0 | 0.2 | 0.6 | *3.4* | na |
| Days ≥ 1" Snow Depth | na | *0* | na | 0 | 0 | 0 | 0 | 0 | 0 | 0 | *0* | na | na |

## ANIMAS *Hidalgo County* ELEVATION 4413 ft LAT/LONG 31° 57 ' N / 108° 49 ' W

| | JAN | FEB | MAR | APR | MAY | JUN | JUL | AUG | SEP | OCT | NOV | DEC | YEAR |
|---|---|---|---|---|---|---|---|---|---|---|---|---|---|
| Maximum Temp °F | 57.5 | 62.6 | 69.3 | 77.7 | 86.1 | 95.3 | 94.4 | 91.3 | 86.8 | 78.1 | 66.0 | 57.4 | 76.9 |
| Minimum Temp °F | 26.3 | 29.1 | 33.7 | 39.5 | 47.6 | 56.9 | 63.2 | 61.5 | 55.4 | 43.9 | 32.6 | 26.7 | 43.0 |
| Mean Temp °F | 41.9 | 45.9 | 51.5 | 58.6 | 66.9 | 76.1 | 78.8 | 76.4 | 71.2 | 61.1 | 49.3 | 42.1 | 60.0 |
| Days Max Temp ≥ 90 °F | 0 | 0 | 0 | 1 | *9* | 25 | 26 | 21 | *10* | 2 | 0 | 0 | 94 |
| Days Max Temp ≤ 32 °F | 0 | 0 | 0 | 0 | 0 | 0 | 0 | 0 | 0 | 0 | 0 | 0 | 0 |
| Days Min Temp ≤ 32 °F | 24 | 19 | 14 | 6 | 1 | 0 | 0 | 0 | 0 | 2 | 15 | 24 | 105 |
| Days Min Temp ≤ 0 °F | 0 | 0 | 0 | 0 | 0 | 0 | 0 | 0 | 0 | 0 | 0 | 0 | 0 |
| Heating Degree Days | 708 | 533 | 411 | 196 | 44 | 1 | 0 | 0 | 10 | 144 | 464 | 703 | 3214 |
| Cooling Degree Days | 0 | 0 | 0 | 15 | 107 | 350 | 431 | 361 | 211 | 24 | 0 | 0 | 1499 |
| Total Precipitation (") | 0.69 | 0.56 | 0.48 | 0.21 | 0.24 | 0.43 | 2.21 | 2.34 | 1.51 | 1.17 | 0.71 | 1.32 | 11.87 |
| Days ≥ 0.1" Precip | 2 | 2 | 2 | 1 | 1 | 1 | 5 | 5 | 3 | 3 | 2 | 3 | 30 |
| Total Snowfall (") | 1.7 | 1.0 | 0.6 | 0.3 | 0.0 | 0.0 | 0.0 | 0.0 | 0.0 | 0.0 | 0.6 | 2.4 | 6.6 |
| Days ≥ 1" Snow Depth | 1 | 0 | 0 | 0 | 0 | 0 | 0 | 0 | 0 | 0 | 0 | 1 | 2 |

## ARTESIA 6 S *Eddy County* ELEVATION 3320 ft LAT/LONG 32° 46 ' N / 104° 23 ' W

| | JAN | FEB | MAR | APR | MAY | JUN | JUL | AUG | SEP | OCT | NOV | DEC | YEAR |
|---|---|---|---|---|---|---|---|---|---|---|---|---|---|
| Maximum Temp °F | 55.9 | 61.5 | 69.2 | 78.1 | 85.2 | 92.8 | 93.7 | 91.4 | 84.9 | 76.9 | 65.4 | 57.5 | 76.0 |
| Minimum Temp °F | 22.0 | 25.6 | 32.5 | 41.3 | 50.7 | 59.7 | 64.1 | 62.0 | 54.5 | 41.9 | 30.5 | 22.5 | 42.3 |
| Mean Temp °F | 39.0 | 43.6 | 50.9 | 59.7 | 67.9 | 76.3 | 78.9 | 76.8 | 69.7 | 59.4 | 48.0 | 40.0 | 59.2 |
| Days Max Temp ≥ 90 °F | 0 | 0 | 0 | 2 | 10 | 22 | 25 | 22 | 10 | 2 | 0 | 0 | 93 |
| Days Max Temp ≤ 32 °F | 1 | 1 | 0 | 0 | 0 | 0 | 0 | 0 | 0 | 0 | 0 | 1 | 3 |
| Days Min Temp ≤ 32 °F | 28 | 23 | 16 | 4 | 0 | 0 | 0 | 0 | 0 | 3 | 18 | 28 | 120 |
| Days Min Temp ≤ 0 °F | 1 | 0 | 0 | 0 | 0 | 0 | 0 | 0 | 0 | 0 | 0 | 0 | 1 |
| Heating Degree Days | 800 | 599 | 433 | 182 | 42 | 2 | 0 | 1 | 32 | 190 | 505 | 767 | 3553 |
| Cooling Degree Days | 0 | 0 | 3 | 32 | 144 | 360 | 440 | 389 | 182 | 20 | 1 | 0 | 1571 |
| Total Precipitation (") | 0.41 | 0.45 | 0.31 | 0.48 | 1.22 | 1.52 | 1.45 | 2.41 | 2.46 | 1.19 | 0.61 | 0.55 | 13.06 |
| Days ≥ 0.1" Precip | 1 | 2 | 1 | 1 | 3 | 3 | 3 | 5 | 4 | 3 | 2 | 2 | 30 |
| Total Snowfall (") | 2.1 | 1.6 | 0.8 | 0.6 | 0.0 | 0.0 | 0.0 | 0.0 | 0.0 | 0.1 | 0.9 | 2.3 | 8.4 |
| Days ≥ 1" Snow Depth | 3 | *1* | 0 | 0 | 0 | 0 | 0 | 0 | 0 | 0 | 1 | *1* | 6 |

## AUGUSTINE 2 E *Socorro County* ELEVATION 7024 ft LAT/LONG 34° 5' N / 107° 41' W

| | JAN | FEB | MAR | APR | MAY | JUN | JUL | AUG | SEP | OCT | NOV | DEC | YEAR |
|---|---|---|---|---|---|---|---|---|---|---|---|---|---|
| Maximum Temp °F | *46.7* | 50.7 | *57.1* | *65.6* | *72.9* | *83.2* | 84.1 | 80.4 | *76.0* | 67.8 | *55.8* | *47.5* | 65.6 |
| Minimum Temp °F | *13.4* | *17.2* | *20.6* | *26.1* | *34.6* | *43.5* | 51.5 | 49.5 | 41.4 | 29.4 | *20.0* | *13.0* | 30.0 |
| Mean Temp °F | *30.1* | *33.9* | *38.9* | *45.9* | *53.8* | *63.4* | 67.8 | 65.0 | *58.7* | 48.6 | *37.9* | *30.2* | 47.9 |
| Days Max Temp ≥ 90 °F | 0 | 0 | 0 | 0 | 0 | na | 4 | 1 | 0 | 0 | 0 | 0 | na |
| Days Max Temp ≤ 32 °F | 2 | *1* | 0 | 0 | 0 | 0 | 0 | 0 | 0 | 0 | 0 | 2 | 5 |
| Days Min Temp ≤ 32 °F | 30 | *27* | 28 | *23* | *11* | *2* | 0 | 0 | *3* | *19* | *27* | 30 | 200 |
| Days Min Temp ≤ 0 °F | *3* | *1* | 0 | 0 | 0 | 0 | 0 | 0 | 0 | 0 | 0 | *2* | 6 |
| Heating Degree Days | *1075* | *871* | *803* | *567* | *342* | *89* | 13 | 40 | *187* | 494 | *804* | *1061* | 6346 |
| Cooling Degree Days | 0 | 0 | 0 | 0 | 3 | 60 | 105 | 57 | 8 | 0 | 0 | 0 | 233 |
| Total Precipitation (") | 0.47 | 0.42 | 0.39 | 0.21 | 0.71 | 0.52 | 2.32 | 2.87 | 1.73 | 1.27 | 0.47 | 0.62 | 12.00 |
| Days ≥ 0.1" Precip | 1 | 1 | 1 | 1 | 2 | 1 | 6 | 7 | 4 | 2 | 2 | *2* | 30 |
| Total Snowfall (") | *2.0* | *2.7* | 1.8 | 0.2 | 0.0 | 0.0 | 0.0 | 0.0 | 0.0 | 0.2 | *0.1* | *2.8* | 9.8 |
| Days ≥ 1" Snow Depth | *3* | *1* | 1 | 0 | 0 | 0 | 0 | 0 | 0 | 0 | 0 | *1* | 6 |

## AZTEC RUINS NATL MON *San Juan County* ELEVATION 5643 ft LAT/LONG 36° 50' N / 108° 0' W

| | JAN | FEB | MAR | APR | MAY | JUN | JUL | AUG | SEP | OCT | NOV | DEC | YEAR |
|---|---|---|---|---|---|---|---|---|---|---|---|---|---|
| Maximum Temp °F | 42.6 | 49.9 | 58.6 | 67.9 | 77.1 | 87.4 | 92.1 | 89.3 | 81.8 | 69.7 | 54.5 | 43.8 | 67.9 |
| Minimum Temp °F | 14.9 | 20.5 | 25.7 | 31.5 | 39.6 | 48.1 | 56.0 | 54.9 | 46.1 | 35.4 | 24.7 | 16.8 | 34.5 |
| Mean Temp °F | 28.8 | 35.2 | 42.2 | 49.7 | 58.4 | 67.8 | 74.1 | 72.1 | 64.0 | 52.6 | 39.6 | 30.3 | 51.2 |
| Days Max Temp ≥ 90 °F | 0 | 0 | 0 | 0 | 1 | 12 | 23 | 17 | 3 | 0 | 0 | 0 | 56 |
| Days Max Temp ≤ 32 °F | 3 | 1 | 0 | 0 | 0 | 0 | 0 | 0 | 0 | 0 | 0 | 3 | 7 |
| Days Min Temp ≤ 32 °F | 30 | 27 | 26 | 16 | 5 | 0 | 0 | 0 | 1 | 10 | 25 | 30 | 170 |
| Days Min Temp ≤ 0 °F | 3 | 0 | 0 | 0 | 0 | 0 | 0 | 0 | 0 | 0 | 0 | 2 | 5 |
| Heating Degree Days | 1116 | 834 | 700 | 453 | 209 | 33 | 0 | 2 | 79 | 378 | 755 | 1067 | 5626 |
| Cooling Degree Days | 0 | 0 | 0 | 0 | 13 | 143 | 281 | 243 | 61 | 0 | 0 | 0 | 741 |
| Total Precipitation (") | 0.95 | 0.81 | 0.93 | 0.67 | 0.74 | 0.39 | 0.96 | 1.16 | 1.02 | 1.12 | 0.94 | 0.89 | 10.58 |
| Days ≥ 0.1" Precip | 3 | 3 | 3 | 2 | 2 | 1 | 3 | 3 | 3 | 3 | 3 | 3 | 32 |
| Total Snowfall (") | 4.8 | 3.0 | 1.0 | 0.1 | 0.0 | 0.0 | 0.0 | 0.0 | 0.0 | 0.3 | 0.7 | 4.8 | 14.7 |
| Days ≥ 1" Snow Depth | na | *2* | 0 | 0 | 0 | 0 | 0 | 0 | 0 | 0 | 0 | 4 | na |

## BEAVERHEAD R S *Catron County* ELEVATION 6726 ft LAT/LONG 33° 27' N / 108° 7' W

| | JAN | FEB | MAR | APR | MAY | JUN | JUL | AUG | SEP | OCT | NOV | DEC | YEAR |
|---|---|---|---|---|---|---|---|---|---|---|---|---|---|
| Maximum Temp °F | *48.1* | 52.2 | 58.3 | 66.5 | 74.1 | 83.6 | 84.4 | 81.4 | 77.1 | 68.2 | *57.4* | *48.7* | 66.7 |
| Minimum Temp °F | *12.5* | *16.4* | 21.0 | 24.8 | 32.4 | 40.6 | 49.8 | 48.8 | 41.1 | 29.3 | *19.1* | 13.2 | 29.1 |
| Mean Temp °F | *30.3* | *34.2* | 39.7 | 45.7 | 53.3 | 62.1 | 67.1 | 65.1 | 59.1 | 48.8 | *38.3* | *31.0* | 47.9 |
| Days Max Temp ≥ 90 °F | 0 | 0 | 0 | 0 | 0 | 5 | 6 | 1 | 0 | 0 | 0 | 0 | 12 |
| Days Max Temp ≤ 32 °F | 2 | 0 | 0 | 0 | 0 | 0 | 0 | 0 | 0 | 0 | 0 | 1 | 3 |
| Days Min Temp ≤ 32 °F | 29 | 26 | 30 | 26 | 16 | 3 | 0 | 0 | 3 | *22* | 28 | 30 | 213 |
| Days Min Temp ≤ 0 °F | 3 | 1 | 0 | 0 | 0 | 0 | 0 | 0 | 0 | 0 | 0 | 2 | 6 |
| Heating Degree Days | *1069* | *864* | 778 | 573 | 356 | 111 | 13 | 33 | 175 | 495 | *793* | *1050* | 6310 |
| Cooling Degree Days | na | *0* | 0 | 0 | 1 | 39 | 86 | 52 | 7 | 0 | 0 | 0 | na |
| Total Precipitation (") | 0.88 | 0.92 | 0.76 | 0.38 | 0.71 | 0.70 | 2.87 | 3.37 | 2.13 | 1.54 | 0.86 | 1.36 | 16.48 |
| Days ≥ 0.1" Precip | 2 | 2 | 2 | 1 | 2 | 2 | 7 | 8 | 4 | 3 | 2 | 3 | 38 |
| Total Snowfall (") | *5.6* | *5.1* | *2.5* | 0.3 | 0.0 | 0.0 | 0.0 | 0.0 | 0.0 | 0.4 | 1.6 | *8.7* | 24.2 |
| Days ≥ 1" Snow Depth | na | *3* | na | 0 | 0 | 0 | 0 | 0 | 0 | 0 | *1* | na | na |

## BELL RANCH *San Miguel County* ELEVATION 4505 ft LAT/LONG 35° 32' N / 104° 6' W

| | JAN | FEB | MAR | APR | MAY | JUN | JUL | AUG | SEP | OCT | NOV | DEC | YEAR |
|---|---|---|---|---|---|---|---|---|---|---|---|---|---|
| Maximum Temp °F | 52.8 | 57.8 | 65.3 | 73.7 | 80.6 | 89.8 | 91.9 | 89.5 | 82.8 | 74.1 | 62.5 | *53.2* | 72.8 |
| Minimum Temp °F | 17.7 | 21.9 | 28.5 | 37.8 | 47.5 | 57.1 | 63.4 | 61.4 | 53.1 | 39.9 | 27.9 | *18.7* | 39.6 |
| Mean Temp °F | 35.3 | 39.9 | 46.9 | 55.8 | 64.1 | 73.5 | 77.7 | 75.5 | 68.0 | 57.1 | 45.3 | *35.9* | 56.3 |
| Days Max Temp ≥ 90 °F | 0 | 0 | 0 | 0 | 4 | 17 | 21 | 18 | 6 | 0 | 0 | 0 | 66 |
| Days Max Temp ≤ 32 °F | 2 | 1 | 0 | 0 | 0 | 0 | 0 | 0 | 0 | 0 | 0 | 1 | 4 |
| Days Min Temp ≤ 32 °F | 28 | 25 | 22 | 8 | 1 | 0 | 0 | 0 | 0 | 4 | 22 | 27 | 137 |
| Days Min Temp ≤ 0 °F | 1 | 0 | 0 | 0 | 0 | 0 | 0 | 0 | 0 | 0 | 0 | 1 | 2 |
| Heating Degree Days | 914 | 702 | 555 | 278 | 88 | 6 | 0 | 1 | 36 | 248 | 586 | *894* | 4308 |
| Cooling Degree Days | *0* | 0 | 0 | 8 | 66 | 282 | *396* | 348 | *140* | 7 | 0 | na | na |
| Total Precipitation (") | 0.28 | 0.31 | 0.55 | 0.60 | 1.56 | 2.18 | 2.51 | 2.68 | 1.62 | 1.00 | 0.72 | 0.37 | 14.38 |
| Days ≥ 0.1" Precip | 1 | 1 | 2 | 2 | 3 | 4 | 5 | 5 | 3 | 2 | 1 | 1 | 30 |
| Total Snowfall (") | na | 2.6 | *2.0* | 0.3 | 0.0 | 0.0 | 0.0 | 0.0 | 0.0 | 0.4 | 0.9 | *2.6* | na |
| Days ≥ 1" Snow Depth | na | *0* | *0* | 0 | 0 | 0 | 0 | 0 | 0 | 0 | *0* | na | na |

## BERNARDO *Socorro County* ELEVATION 4734 ft LAT/LONG 34° 26' N / 106° 49' W

| | JAN | FEB | MAR | APR | MAY | JUN | JUL | AUG | SEP | OCT | NOV | DEC | YEAR |
|---|---|---|---|---|---|---|---|---|---|---|---|---|---|
| Maximum Temp °F | 51.6 | 57.9 | 66.1 | 75.0 | 83.1 | 92.7 | 94.7 | 91.4 | 84.7 | 75.1 | 61.9 | 52.1 | 73.9 |
| Minimum Temp °F | ***18.1*** | 22.5 | 28.2 | 34.5 | 43.9 | 50.4 | 59.0 | 57.3 | 49.0 | 36.7 | 26.2 | 19.7 | 37.1 |
| Mean Temp °F | ***34.6*** | 40.1 | 47.2 | 54.7 | 63.5 | 71.4 | 76.7 | 74.2 | 66.8 | 55.9 | 44.0 | 35.6 | 55.4 |
| Days Max Temp ≥ 90 °F | 0 | 0 | 0 | 1 | 6 | 21 | 26 | 20 | 7 | 1 | 0 | 0 | 82 |
| Days Max Temp ≤ 32 °F | 1 | 0 | 0 | 0 | 0 | 0 | 0 | 0 | 0 | 0 | 0 | 1 | 2 |
| Days Min Temp ≤ 32 °F | ***28*** | 23 | 20 | 12 | 4 | 0 | 0 | 0 | 1 | 9 | 22 | 27 | 146 |
| Days Min Temp ≤ 0 °F | ***1*** | 0 | 0 | 0 | 0 | 0 | 0 | 0 | 0 | 0 | 0 | 1 | 2 |
| Heating Degree Days | ***936*** | 696 | 545 | 310 | 112 | 15 | 0 | 2 | 47 | 281 | 623 | 905 | 4472 |
| Cooling Degree Days | ***0*** | ***0*** | ***0*** | 11 | 116 | ***303*** | 444 | 372 | 153 | 7 | 0 | 0 | 1406 |
| Total Precipitation (") | 0.31 | 0.30 | 0.26 | 0.33 | 0.54 | 0.46 | 1.30 | 1.82 | 1.30 | 1.02 | 0.41 | 0.49 | 8.54 |
| Days ≥ 0.1" Precip | 1 | 1 | 1 | 1 | 2 | 1 | 3 | 5 | 3 | 2 | 1 | 2 | 23 |
| Total Snowfall (") | ***1.2*** | 0.7 | 0.2 | 0.2 | 0.0 | 0.0 | 0.0 | 0.0 | 0.0 | 0.2 | 0.2 | ***1.2*** | 3.9 |
| Days ≥ 1" Snow Depth | ***0*** | 0 | 0 | 0 | 0 | 0 | 0 | 0 | 0 | 0 | 0 | ***0*** | 0 |

## BITTER LAKES WL REF *Chaves County* ELEVATION 3671 ft LAT/LONG 33° 29' N / 104° 24' W

| | JAN | FEB | MAR | APR | MAY | JUN | JUL | AUG | SEP | OCT | NOV | DEC | YEAR |
|---|---|---|---|---|---|---|---|---|---|---|---|---|---|
| Maximum Temp °F | 55.4 | 61.1 | 69.5 | 78.5 | 85.8 | 94.2 | 95.1 | 92.9 | 85.6 | 77.2 | ***64.5*** | 56.8 | 76.4 |
| Minimum Temp °F | 20.8 | 24.9 | 31.6 | 40.5 | 49.7 | 59.0 | 63.4 | 61.8 | ***54.3*** | 40.4 | ***28.7*** | 21.0 | 41.3 |
| Mean Temp °F | 38.1 | 43.0 | 50.6 | 59.5 | 67.8 | 76.6 | 79.3 | 77.4 | ***69.9*** | 58.9 | ***46.6*** | 39.0 | 58.9 |
| Days Max Temp ≥ 90 °F | 0 | 0 | 0 | 3 | ***10*** | 22 | 26 | 23 | ***11*** | 2 | 0 | 0 | 97 |
| Days Max Temp ≤ 32 °F | 2 | 1 | 0 | 0 | 0 | 0 | 0 | 0 | 0 | 0 | 0 | 2 | 5 |
| Days Min Temp ≤ 32 °F | ***29*** | 23 | ***16*** | 5 | 0 | 0 | 0 | 0 | 0 | 5 | ***20*** | ***28*** | 126 |
| Days Min Temp ≤ 0 °F | 1 | 0 | 0 | 0 | 0 | 0 | 0 | 0 | 0 | 0 | 0 | 1 | 2 |
| Heating Degree Days | 825 | 614 | 440 | 187 | 48 | 2 | 0 | 0 | ***32*** | 208 | ***545*** | 800 | 3701 |
| Cooling Degree Days | 0 | 0 | 2 | 31 | 144 | 376 | 451 | 411 | 201 | 16 | 0 | 0 | 1632 |
| Total Precipitation (") | 0.43 | 0.43 | 0.34 | 0.50 | 1.25 | 1.69 | 2.18 | 2.62 | 1.90 | 1.13 | 0.59 | 0.57 | 13.63 |
| Days ≥ 0.1" Precip | 1 | 1 | 1 | 1 | 2 | 3 | 3 | 4 | 4 | 2 | 1 | 2 | 25 |
| Total Snowfall (") | 2.5 | 2.2 | 0.6 | 0.0 | 0.0 | 0.0 | 0.0 | 0.0 | 0.0 | 0.2 | 0.7 | 2.0 | 8.2 |
| Days ≥ 1" Snow Depth | 1 | 1 | 0 | 0 | 0 | 0 | 0 | 0 | 0 | 0 | 0 | 1 | 3 |

## BLOOMFIELD 3 SE *San Juan County* ELEVATION 5794 ft LAT/LONG 36° 40' N / 107° 58' W

| | JAN | FEB | MAR | APR | MAY | JUN | JUL | AUG | SEP | OCT | NOV | DEC | YEAR |
|---|---|---|---|---|---|---|---|---|---|---|---|---|---|
| Maximum Temp °F | 40.7 | 48.4 | 57.4 | 66.9 | 76.6 | 87.7 | 91.6 | 88.6 | 80.6 | 68.5 | 53.4 | 42.8 | 66.9 |
| Minimum Temp °F | 17.7 | 23.6 | 29.4 | 35.4 | 44.4 | 53.3 | 60.2 | 58.5 | 50.7 | 39.1 | 27.9 | 20.0 | 38.4 |
| Mean Temp °F | 29.1 | 36.0 | 43.4 | 51.2 | 60.6 | 70.5 | 75.9 | 73.6 | 65.7 | 53.9 | 40.7 | 31.3 | 52.7 |
| Days Max Temp ≥ 90 °F | 0 | 0 | 0 | 0 | 1 | 13 | 22 | 14 | 2 | 0 | 0 | 0 | 52 |
| Days Max Temp ≤ 32 °F | 5 | 1 | 0 | 0 | 0 | 0 | 0 | 0 | 0 | 0 | 0 | 3 | 9 |
| Days Min Temp ≤ 32 °F | 29 | 26 | 21 | 11 | 1 | 0 | 0 | 0 | 0 | 5 | 22 | 29 | 144 |
| Days Min Temp ≤ 0 °F | 1 | 0 | 0 | 0 | 0 | 0 | 0 | 0 | 0 | 0 | 0 | 1 | 2 |
| Heating Degree Days | 1106 | 812 | 662 | 408 | 159 | 17 | 0 | 2 | 51 | 341 | 724 | 1039 | 5321 |
| Cooling Degree Days | ***0*** | 0 | 0 | 1 | 28 | 192 | 331 | 272 | 76 | 1 | 0 | 0 | 901 |
| Total Precipitation (") | 0.60 | 0.61 | 0.81 | 0.60 | 0.60 | 0.41 | 1.06 | 1.23 | 0.98 | 1.05 | 0.84 | 0.64 | 9.43 |
| Days ≥ 0.1" Precip | 2 | 2 | 3 | 2 | 2 | 1 | 3 | 3 | 3 | 3 | 2 | 2 | 28 |
| Total Snowfall (") | ***3.3*** | 2.0 | 1.3 | 0.6 | 0.0 | 0.0 | 0.0 | 0.0 | 0.0 | 0.2 | 0.7 | 3.8 | 11.9 |
| Days ≥ 1" Snow Depth | ***5*** | ***2*** | ***0*** | 0 | 0 | 0 | 0 | 0 | 0 | 0 | 0 | na | na |

## BOSQUE DEL APACHE *Socorro County* ELEVATION 4524 ft LAT/LONG 33° 46' N / 106° 54' W

| | JAN | FEB | MAR | APR | MAY | JUN | JUL | AUG | SEP | OCT | NOV | DEC | YEAR |
|---|---|---|---|---|---|---|---|---|---|---|---|---|---|
| Maximum Temp °F | 54.9 | 61.5 | 69.2 | 77.7 | 85.5 | 94.5 | 95.4 | 92.3 | 86.5 | 77.9 | 65.2 | 54.6 | 76.3 |
| Minimum Temp °F | 20.2 | 24.2 | 30.2 | 36.7 | 45.1 | 53.0 | 59.9 | 58.7 | 50.5 | 38.5 | 27.0 | 20.0 | 38.7 |
| Mean Temp °F | 37.6 | 42.9 | 49.8 | 57.1 | 65.3 | 73.7 | 77.7 | 75.5 | 68.5 | 58.2 | 46.1 | 37.3 | 57.5 |
| Days Max Temp ≥ 90 °F | 0 | 0 | 0 | 1 | 9 | 24 | 27 | 23 | 10 | 1 | 0 | 0 | 95 |
| Days Max Temp ≤ 32 °F | 0 | 0 | 0 | 0 | 0 | 0 | 0 | 0 | 0 | 0 | 0 | 0 | 0 |
| Days Min Temp ≤ 32 °F | 27 | 23 | 18 | ***8*** | 1 | 0 | 0 | 0 | 0 | 7 | 22 | 29 | 135 |
| Days Min Temp ≤ 0 °F | 0 | 0 | 0 | 0 | 0 | 0 | 0 | 0 | 0 | 0 | 0 | 0 | 0 |
| Heating Degree Days | 844 | 609 | 466 | 235 | 61 | 4 | 0 | 0 | 22 | 215 | ***554*** | ***846*** | 3856 |
| Cooling Degree Days | 0 | 0 | 0 | 12 | 87 | 287 | 399 | 348 | 149 | 8 | ***0*** | 0 | 1290 |
| Total Precipitation (") | 0.38 | 0.42 | 0.31 | 0.25 | 0.59 | 0.68 | 1.33 | 2.03 | 1.46 | 1.19 | 0.55 | 0.75 | 9.94 |
| Days ≥ 0.1" Precip | 1 | 1 | 1 | 1 | 2 | 2 | 3 | 5 | 4 | 2 | 1 | 2 | 25 |
| Total Snowfall (") | ***1.0*** | 1.0 | 0.1 | 0.1 | 0.0 | 0.0 | 0.0 | 0.0 | 0.0 | 0.2 | 0.2 | ***2.8*** | 5.4 |
| Days ≥ 1" Snow Depth | ***0*** | 0 | 0 | 0 | 0 | 0 | 0 | 0 | 0 | 0 | 0 | ***1*** | 1 |

## CABALLO DAM *Sierra County* ELEVATION 4190 ft LAT/LONG 32° 54 ' N / 107° 18 ' W

| | JAN | FEB | MAR | APR | MAY | JUN | JUL | AUG | SEP | OCT | NOV | DEC | YEAR |
|---|---|---|---|---|---|---|---|---|---|---|---|---|---|
| Maximum Temp °F | 56.2 | 61.5 | 68.5 | 76.8 | 84.7 | 94.4 | 95.1 | 92.1 | 86.4 | 77.5 | 65.8 | 56.2 | 76.3 |
| Minimum Temp °F | 25.8 | 29.4 | 35.0 | 41.8 | 49.9 | 59.3 | 65.5 | 63.5 | 56.4 | 44.2 | 33.5 | 26.6 | 44.2 |
| Mean Temp °F | 41.0 | 45.5 | 51.8 | 59.3 | 67.3 | 76.9 | 80.3 | 77.8 | 71.4 | 60.9 | 49.6 | 41.4 | 60.3 |
| Days Max Temp ≥ 90 °F | 0 | 0 | 0 | 1 | 8 | 23 | 26 | 23 | 11 | 1 | 0 | 0 | 93 |
| Days Max Temp ≤ 32 °F | 0 | 0 | 0 | 0 | 0 | 0 | 0 | 0 | 0 | 0 | 0 | 0 | 0 |
| Days Min Temp ≤ 32 °F | 26 | 19 | 11 | 2 | 0 | 0 | 0 | 0 | 0 | 2 | 14 | 26 | 100 |
| Days Min Temp ≤ 0 °F | 0 | 0 | 0 | 0 | 0 | 0 | 0 | 0 | 0 | 0 | 0 | 0 | 0 |
| Heating Degree Days | 736 | 544 | 403 | 181 | 41 | 1 | 0 | 0 | 12 | 149 | 453 | 724 | 3244 |
| Cooling Degree Days | 0 | 0 | 1 | 21 | 125 | 382 | 480 | 401 | 213 | 26 | 0 | 0 | 1649 |
| Total Precipitation (") | 0.48 | 0.36 | 0.21 | 0.13 | 0.44 | 0.66 | 1.91 | 2.27 | 1.55 | 0.96 | 0.61 | 0.86 | 10.44 |
| Days ≥ 0.1" Precip | 2 | 1 | 1 | 1 | 1 | 1 | 4 | 5 | 4 | 2 | 2 | 2 | 26 |
| Total Snowfall (") | 0.4 | 0.0 | 0.1 | 0.0 | 0.0 | 0.0 | 0.0 | 0.0 | 0.0 | 0.0 | 0.3 | 0.6 | 1.4 |
| Days ≥ 1" Snow Depth | 0 | 0 | 0 | 0 | 0 | 0 | 0 | 0 | 0 | 0 | 0 | 0 | 0 |

## CAMERON *Quay County* ELEVATION 4603 ft LAT/LONG 34° 54 ' N / 103° 23 ' W

| | JAN | FEB | MAR | APR | MAY | JUN | JUL | AUG | SEP | OCT | NOV | DEC | YEAR |
|---|---|---|---|---|---|---|---|---|---|---|---|---|---|
| Maximum Temp °F | 49.5 | 54.1 | 62.3 | 71.1 | 78.6 | 87.3 | 89.7 | 86.8 | 80.4 | 71.1 | 58.4 | 50.3 | 70.0 |
| Minimum Temp °F | 22.0 | 25.0 | 30.7 | 38.5 | 47.3 | 56.4 | 60.6 | 59.1 | 52.4 | 41.5 | 30.9 | 23.5 | 40.7 |
| Mean Temp °F | 35.8 | 39.6 | 46.5 | 54.9 | 63.0 | 71.9 | 75.1 | 73.0 | 66.4 | 56.3 | 44.6 | 36.9 | 55.3 |
| Days Max Temp ≥ 90 °F | 0 | 0 | 0 | 0 | 2 | 13 | 18 | 11 | 4 | 0 | 0 | 0 | 48 |
| Days Max Temp ≤ 32 °F | 3 | 1 | 1 | 0 | 0 | 0 | 0 | 0 | 0 | 0 | 1 | 2 | 8 |
| Days Min Temp ≤ 32 °F | 28 | 22 | 17 | 6 | 1 | 0 | 0 | 0 | 0 | 4 | 17 | 26 | 121 |
| Days Min Temp ≤ 0 °F | 1 | 0 | 0 | 0 | 0 | 0 | 0 | 0 | 0 | 0 | 0 | 1 | 2 |
| Heating Degree Days | 899 | 710 | 567 | 303 | 112 | 12 | 1 | 3 | 55 | 273 | 605 | 863 | 4403 |
| Cooling Degree Days | 0 | 0 | 0 | 5 | 54 | 234 | 323 | 267 | 107 | 8 | 0 | 0 | 998 |
| Total Precipitation (") | 0.55 | 0.65 | 0.89 | 1.06 | 2.11 | 2.77 | 3.00 | 3.31 | 2.16 | 1.42 | 0.83 | 0.64 | 19.39 |
| Days ≥ 0.1" Precip | 1 | 2 | 2 | 3 | 4 | 5 | 6 | 6 | 4 | 3 | 2 | 2 | 40 |
| Total Snowfall (") | na | na | 1.8 | 0.2 | 0.0 | 0.0 | 0.0 | 0.0 | 0.0 | 0.6 | 1.5 | *3.4* | na |
| Days ≥ 1" Snow Depth | na | *1* | 1 | 0 | 0 | 0 | 0 | 0 | 0 | 0 | 0 | *1* | na |

## CARLSBAD *Eddy County* ELEVATION 3120 ft LAT/LONG 32° 25 ' N / 104° 14 ' W

| | JAN | FEB | MAR | APR | MAY | JUN | JUL | AUG | SEP | OCT | NOV | DEC | YEAR |
|---|---|---|---|---|---|---|---|---|---|---|---|---|---|
| Maximum Temp °F | 57.4 | 62.9 | 71.2 | 79.6 | 86.4 | 94.2 | 94.8 | 92.8 | 86.5 | 78.9 | 66.9 | 59.1 | 77.6 |
| Minimum Temp °F | 27.6 | 31.3 | 38.2 | 46.9 | 55.7 | 63.9 | 67.6 | 66.2 | 59.1 | 46.5 | 36.1 | 28.8 | 47.3 |
| Mean Temp °F | 42.5 | 47.1 | 54.7 | 63.3 | 71.1 | 79.1 | 81.2 | 79.5 | 72.8 | 62.8 | 51.5 | 44.0 | 62.5 |
| Days Max Temp ≥ 90 °F | 0 | 0 | 0 | 3 | *12* | 23 | *25* | *23* | *12* | 3 | 0 | 0 | 101 |
| Days Max Temp ≤ 32 °F | 1 | 0 | 0 | 0 | 0 | 0 | 0 | 0 | 0 | 0 | 0 | 1 | 2 |
| Days Min Temp ≤ 32 °F | 24 | 16 | 7 | 1 | 0 | 0 | 0 | 0 | 0 | 1 | *9* | *20* | 78 |
| Days Min Temp ≤ 0 °F | 0 | 0 | 0 | 0 | 0 | 0 | 0 | 0 | 0 | 0 | 0 | 0 | 0 |
| Heating Degree Days | 691 | 499 | 325 | 111 | 21 | 1 | *0* | 0 | 20 | *122* | 400 | *644* | 2834 |
| Cooling Degree Days | 0 | 1 | 9 | 67 | 224 | 451 | 522 | 479 | 269 | *62* | 4 | 0 | 2088 |
| Total Precipitation (") | 0.43 | 0.50 | 0.31 | 0.58 | 1.30 | 1.60 | 1.91 | 2.07 | 2.88 | 1.18 | 0.65 | 0.58 | 13.99 |
| Days ≥ 0.1" Precip | 1 | 2 | 1 | 1 | 2 | 3 | 3 | 4 | 4 | 2 | 1 | 1 | 25 |
| Total Snowfall (") | *0.7* | *1.0* | 0.3 | 0.0 | 0.0 | 0.0 | 0.0 | 0.0 | 0.0 | 0.0 | 0.6 | na | na |
| Days ≥ 1" Snow Depth | *0* | *0* | 0 | 0 | 0 | 0 | 0 | 0 | 0 | 0 | 0 | *0* | 0 |

## CARLSBAD CAVERN CITY *Eddy County* ELEVATION 3235 ft LAT/LONG 32° 20 ' N / 104° 16 ' W

| | JAN | FEB | MAR | APR | MAY | JUN | JUL | AUG | SEP | OCT | NOV | DEC | YEAR |
|---|---|---|---|---|---|---|---|---|---|---|---|---|---|
| Maximum Temp °F | 56.1 | 62.1 | 70.4 | 79.4 | 86.9 | 94.9 | 95.1 | 92.5 | 85.8 | 77.1 | 65.3 | 57.8 | 76.9 |
| Minimum Temp °F | 28.3 | 32.1 | 38.9 | 47.7 | 56.5 | 64.8 | 68.6 | 66.7 | 59.9 | 48.1 | 36.8 | 29.4 | 48.2 |
| Mean Temp °F | 42.3 | 47.1 | 54.7 | 63.6 | 71.7 | 79.8 | 81.8 | 79.7 | 72.9 | 62.6 | 51.1 | 43.6 | 62.6 |
| Days Max Temp ≥ 90 °F | 0 | 0 | 0 | 3 | 13 | 24 | 26 | 23 | 12 | 3 | 0 | 0 | 104 |
| Days Max Temp ≤ 32 °F | 1 | 0 | 0 | 0 | 0 | 0 | 0 | 0 | 0 | 0 | 0 | 1 | 2 |
| Days Min Temp ≤ 32 °F | 22 | 15 | 7 | 1 | 0 | 0 | 0 | 0 | 0 | 1 | 9 | 21 | 76 |
| Days Min Temp ≤ 0 °F | 0 | 0 | 0 | 0 | 0 | 0 | 0 | 0 | 0 | 0 | 0 | 0 | 0 |
| Heating Degree Days | 698 | 499 | 325 | 108 | 17 | 0 | 0 | 0 | 19 | 125 | 414 | 656 | 2861 |
| Cooling Degree Days | 0 | 1 | 8 | 68 | 225 | 451 | 514 | 463 | 258 | 57 | 3 | 0 | 2048 |
| Total Precipitation (") | 0.42 | 0.42 | 0.26 | 0.51 | 1.38 | 1.48 | 1.92 | 2.47 | 2.88 | 1.03 | 0.63 | 0.56 | 13.96 |
| Days ≥ 0.1" Precip | 1 | 1 | 1 | 1 | 3 | 3 | 3 | 4 | 4 | 2 | 2 | 2 | 27 |
| Total Snowfall (") | 1.7 | 1.8 | 0.8 | 0.5 | 0.0 | 0.0 | 0.0 | 0.0 | 0.0 | 0.1 | 1.0 | 2.0 | 7.9 |
| Days ≥ 1" Snow Depth | 1 | 1 | 0 | 0 | 0 | 0 | 0 | 0 | 0 | 0 | 0 | 1 | 3 |

## CARLSBAD CAVERNS *Eddy County* ELEVATION 4442 ft LAT/LONG 32° 11' N / 104° 27' W

| | JAN | FEB | MAR | APR | MAY | JUN | JUL | AUG | SEP | OCT | NOV | DEC | YEAR |
|---|---|---|---|---|---|---|---|---|---|---|---|---|---|
| Maximum Temp °F | 55.1 | 59.0 | 66.4 | 75.0 | 81.8 | 89.5 | 89.9 | 87.6 | 81.5 | 73.9 | 63.7 | 56.8 | 73.4 |
| Minimum Temp °F | 32.7 | 35.6 | 41.8 | 49.3 | 56.9 | 63.6 | 65.6 | 64.6 | 59.2 | 51.1 | 41.2 | 34.5 | 49.7 |
| Mean Temp °F | 43.8 | 47.2 | 54.1 | 62.2 | 69.3 | 76.5 | 77.8 | 76.1 | 70.4 | 62.5 | 52.2 | 45.6 | 61.5 |
| Days Max Temp ≥ 90 °F | 0 | 0 | 0 | 1 | 5 | 16 | 18 | 13 | 5 | 1 | 0 | 0 | 59 |
| Days Max Temp ≤ 32 °F | 2 | 1 | 0 | 0 | 0 | 0 | 0 | 0 | 0 | 0 | 0 | 1 | 4 |
| Days Min Temp ≤ 32 °F | 15 | 10 | 5 | 1 | 0 | 0 | 0 | 0 | 0 | 0 | *6* | 13 | 50 |
| Days Min Temp ≤ 0 °F | 0 | 0 | 0 | 0 | 0 | 0 | 0 | 0 | 0 | 0 | 0 | 0 | 0 |
| Heating Degree Days | 650 | 499 | 342 | 135 | 36 | 3 | 0 | 2 | 34 | 138 | 381 | 595 | 2815 |
| Cooling Degree Days | 0 | 2 | 10 | 60 | 174 | 364 | 408 | 363 | 206 | 62 | 5 | 0 | 1654 |
| Total Precipitation (") | 0.44 | 0.51 | 0.33 | 0.51 | 1.62 | 1.95 | 2.11 | 2.74 | 3.62 | 1.26 | 0.57 | 0.57 | 16.23 |
| Days ≥ 0.1" Precip | 2 | 2 | 1 | 1 | 3 | 3 | 4 | 5 | 5 | 3 | 1 | 1 | 31 |
| Total Snowfall (") | 1.4 | 1.3 | 1.1 | 0.0 | 0.0 | 0.0 | 0.0 | 0.0 | 0.0 | 0.1 | 0.7 | 2.2 | 6.8 |
| Days ≥ 1" Snow Depth | *1* | 1 | 0 | 0 | 0 | 0 | 0 | 0 | 0 | 0 | 0 | *0* | 2 |

## CARRIZOZO 1 SW *Lincoln County* ELEVATION 5426 ft LAT/LONG 33° 39' N / 105° 53' W

| | JAN | FEB | MAR | APR | MAY | JUN | JUL | AUG | SEP | OCT | NOV | DEC | YEAR |
|---|---|---|---|---|---|---|---|---|---|---|---|---|---|
| Maximum Temp °F | 51.5 | 56.3 | 63.6 | 72.6 | 80.5 | 90.1 | 90.8 | 87.6 | 81.9 | 73.0 | 60.6 | 52.3 | 71.7 |
| Minimum Temp °F | 22.3 | 25.7 | 31.4 | 38.1 | 46.6 | 55.4 | 60.3 | 58.3 | 51.8 | 40.5 | 29.5 | 22.7 | 40.2 |
| Mean Temp °F | 36.9 | 41.0 | 47.5 | 55.3 | 63.6 | 72.8 | 75.6 | 73.0 | 66.9 | 56.8 | 45.1 | 37.5 | 56.0 |
| Days Max Temp ≥ 90 °F | 0 | 0 | 0 | 0 | 3 | 17 | 20 | 13 | 3 | 0 | 0 | 0 | 56 |
| Days Max Temp ≤ 32 °F | 1 | 0 | 0 | 0 | 0 | 0 | 0 | 0 | 0 | 0 | 0 | 1 | 2 |
| Days Min Temp ≤ 32 °F | 28 | 22 | 18 | 8 | 1 | 0 | 0 | 0 | 0 | 5 | 20 | 27 | 129 |
| Days Min Temp ≤ 0 °F | 0 | 0 | 0 | 0 | 0 | 0 | 0 | 0 | 0 | 0 | 0 | 0 | 0 |
| Heating Degree Days | 863 | 672 | 536 | 290 | 96 | 6 | 0 | 1 | 38 | 255 | 591 | 845 | 4193 |
| Cooling Degree Days | 0 | 0 | 0 | 7 | 61 | 261 | 331 | 268 | 108 | 9 | 0 | 0 | 1045 |
| Total Precipitation (") | 0.69 | 0.63 | 0.59 | 0.34 | 0.75 | 0.84 | 2.20 | 2.72 | 1.63 | 1.24 | 0.81 | 0.90 | 13.34 |
| Days ≥ 0.1" Precip | 2 | 2 | 2 | 1 | 2 | 3 | 5 | 6 | 4 | 2 | 2 | 3 | 34 |
| Total Snowfall (") | 2.5 | 2.0 | 1.0 | 0.5 | 0.0 | 0.0 | 0.0 | 0.0 | 0.0 | 0.2 | 0.9 | *1.8* | 8.9 |
| Days ≥ 1" Snow Depth | na | na | 0 | 0 | 0 | 0 | 0 | 0 | 0 | 0 | 0 | na | na |

## CERRO *Taos County* ELEVATION 7664 ft LAT/LONG 36° 44' N / 105° 36' W

| | JAN | FEB | MAR | APR | MAY | JUN | JUL | AUG | SEP | OCT | NOV | DEC | YEAR |
|---|---|---|---|---|---|---|---|---|---|---|---|---|---|
| Maximum Temp °F | 35.3 | 40.5 | 49.4 | 59.5 | 68.2 | 77.9 | 81.9 | *79.0* | *73.5* | 63.1 | *48.0* | 37.5 | 59.5 |
| Minimum Temp °F | 6.2 | 12.1 | 21.1 | 27.3 | 34.8 | 43.0 | 49.0 | *47.8* | *41.3* | 30.5 | *18.9* | 8.7 | 28.4 |
| Mean Temp °F | 20.8 | 26.4 | 35.1 | 43.4 | 51.6 | 60.5 | 65.5 | *63.4* | *57.4* | 46.8 | *33.5* | 23.1 | 44.0 |
| Days Max Temp ≥ 90 °F | 0 | 0 | 0 | 0 | 0 | 1 | 1 | 0 | 0 | 0 | 0 | 0 | 2 |
| Days Max Temp ≤ 32 °F | 11 | 5 | 1 | 0 | 0 | 0 | 0 | 0 | 0 | 0 | 2 | 9 | 28 |
| Days Min Temp ≤ 32 °F | 31 | 28 | 29 | 22 | 10 | 1 | 0 | 0 | 3 | 19 | 26 | 30 | 199 |
| Days Min Temp ≤ 0 °F | 10 | 5 | 1 | 0 | 0 | 0 | 0 | 0 | 0 | 0 | 1 | 7 | 24 |
| Heating Degree Days | 1363 | 1084 | 918 | 642 | 409 | 146 | 30 | *65* | *223* | 558 | *939* | 1291 | 7668 |
| Cooling Degree Days | 0 | 0 | 0 | 0 | *0* | 16 | 47 | *17* | *3* | 0 | na | 0 | na |
| Total Precipitation (") | 0.63 | 0.56 | 0.92 | 0.77 | 1.08 | 1.06 | 1.93 | 2.23 | 1.45 | 1.07 | 0.95 | 0.82 | 13.47 |
| Days ≥ 0.1" Precip | 2 | 2 | 3 | 3 | 3 | 3 | 5 | 6 | 4 | 3 | 2 | 3 | 39 |
| Total Snowfall (") | 11.3 | 9.2 | 11.4 | 3.3 | 0.4 | 0.0 | 0.0 | 0.0 | 0.1 | 2.2 | 8.7 | *14.0* | 60.6 |
| Days ≥ 1" Snow Depth | na | na | na | *0* | 0 | 0 | 0 | 0 | 0 | *0* | na | na | na |

## CHACO CANYN NATL MON *San Juan County* ELEVATION 6135 ft LAT/LONG 36° 4' N / 107° 58' W

| | JAN | FEB | MAR | APR | MAY | JUN | JUL | AUG | SEP | OCT | NOV | DEC | YEAR |
|---|---|---|---|---|---|---|---|---|---|---|---|---|---|
| Maximum Temp °F | 42.4 | 48.5 | 57.1 | 66.5 | 75.7 | 86.1 | 89.9 | 87.1 | 80.0 | 68.7 | 54.1 | 43.7 | 66.7 |
| Minimum Temp °F | 11.8 | 18.0 | 22.7 | 28.3 | 37.0 | 46.0 | 54.3 | 52.6 | 43.8 | 31.3 | 21.2 | 13.0 | 31.7 |
| Mean Temp °F | 27.1 | 33.2 | 39.9 | 47.4 | 56.4 | 66.1 | 72.1 | 69.8 | 61.9 | 50.0 | 37.6 | 28.4 | 49.2 |
| Days Max Temp ≥ 90 °F | 0 | 0 | 0 | 0 | 0 | 10 | 17 | 10 | 1 | 0 | 0 | 0 | 38 |
| Days Max Temp ≤ 32 °F | 4 | 1 | 0 | 0 | 0 | 0 | 0 | 0 | 0 | 0 | 0 | 4 | 9 |
| Days Min Temp ≤ 32 °F | 30 | 27 | 28 | 20 | 9 | 1 | 0 | 0 | 3 | 17 | 26 | 30 | 191 |
| Days Min Temp ≤ 0 °F | 5 | 1 | 0 | 0 | 0 | 0 | 0 | 0 | 0 | 0 | 0 | 4 | 10 |
| Heating Degree Days | 1168 | 891 | 769 | 521 | 266 | 54 | 1 | 7 | 116 | 457 | 815 | 1126 | 6191 |
| Cooling Degree Days | 0 | 0 | 0 | 0 | 8 | 105 | 223 | *171* | 29 | 0 | 0 | 0 | 536 |
| Total Precipitation (") | 0.55 | 0.51 | 0.61 | 0.46 | 0.69 | 0.54 | 1.26 | 1.26 | 1.19 | 1.04 | 0.73 | 0.57 | 9.41 |
| Days ≥ 0.1" Precip | 2 | 2 | 2 | 1 | 2 | 2 | 4 | 4 | 3 | 3 | 2 | 2 | 29 |
| Total Snowfall (") | *3.5* | 3.0 | 1.8 | 0.5 | 0.0 | 0.0 | 0.0 | 0.0 | 0.0 | 0.6 | 1.5 | *3.8* | 14.7 |
| Days ≥ 1" Snow Depth | na | *1* | 1 | 0 | 0 | 0 | 0 | 0 | 0 | 0 | 1 | *4* | na |

## CHAMA *Rio Arriba County* ELEVATION 7854 ft LAT/LONG 36° 54 ' N / 106° 35 ' W

| | JAN | FEB | MAR | APR | MAY | JUN | JUL | AUG | SEP | OCT | NOV | DEC | YEAR |
|---|---|---|---|---|---|---|---|---|---|---|---|---|---|
| Maximum Temp °F | 36.8 | 40.6 | 46.5 | 56.0 | 65.3 | 75.8 | 80.4 | 77.4 | 71.0 | 60.8 | 46.3 | 38.2 | 57.9 |
| Minimum Temp °F | 4.8 | 9.0 | 16.1 | 23.1 | 30.7 | 37.6 | 44.9 | 44.2 | 36.7 | 26.9 | 16.2 | 8.0 | 24.8 |
| Mean Temp °F | 20.9 | 24.8 | 31.3 | 39.6 | 48.0 | 56.7 | 62.7 | 60.8 | 53.9 | 43.9 | 31.3 | 23.1 | 41.4 |
| Days Max Temp ≥ 90 °F | 0 | 0 | 0 | 0 | 0 | 0 | 1 | 0 | 0 | 0 | 0 | 0 | 1 |
| Days Max Temp ≤ 32 °F | 9 | 4 | 1 | 0 | 0 | 0 | 0 | 0 | 0 | 0 | 3 | 8 | 25 |
| Days Min Temp ≤ 32 °F | 31 | 28 | 31 | 27 | 19 | 6 | 1 | 0 | 8 | 25 | 29 | 31 | 236 |
| Days Min Temp ≤ 0 °F | 10 | 6 | 2 | 0 | 0 | 0 | 0 | 0 | 0 | 0 | 2 | 7 | 27 |
| Heating Degree Days | 1361 | 1127 | 1033 | 749 | 520 | 246 | 79 | 127 | 327 | 647 | 1004 | 1292 | 8512 |
| Cooling Degree Days | 0 | 0 | 0 | 0 | 0 | 3 | 10 | 6 | 0 | 0 | 0 | 0 | 19 |
| Total Precipitation (") | 1.93 | 1.76 | 2.13 | 1.28 | 1.39 | 1.22 | 2.19 | 2.90 | 2.17 | 2.02 | 1.89 | 1.94 | 22.82 |
| Days ≥ 0.1" Precip | 5 | 5 | 6 | 4 | 4 | 4 | 6 | 8 | 5 | 5 | 5 | 5 | 62 |
| Total Snowfall (") | 24.6 | 20.9 | 19.0 | 5.6 | 0.8 | 0.0 | 0.0 | 0.0 | 0.1 | 4.4 | 12.5 | 22.2 | 110.1 |
| Days ≥ 1" Snow Depth | 29 | 26 | 23 | *5* | 0 | 0 | 0 | 0 | 0 | 2 | 9 | 25 | 119 |

## CIMARRON 4 SW *Colfax County* ELEVATION 6434 ft LAT/LONG 36° 31 ' N / 104° 55 ' W

| | JAN | FEB | MAR | APR | MAY | JUN | JUL | AUG | SEP | OCT | NOV | DEC | YEAR |
|---|---|---|---|---|---|---|---|---|---|---|---|---|---|
| Maximum Temp °F | 46.4 | 49.8 | 55.4 | 63.7 | 71.5 | 80.9 | 83.7 | 80.6 | 75.8 | 68.0 | 55.3 | 47.9 | 64.9 |
| Minimum Temp °F | 16.1 | 18.8 | 24.3 | 31.4 | 39.5 | 47.6 | 52.4 | 50.9 | 44.0 | 33.8 | 23.4 | 16.9 | 33.3 |
| Mean Temp °F | 31.3 | 34.3 | 39.9 | 47.6 | 55.5 | 64.3 | 68.0 | 65.8 | 59.9 | 50.9 | 39.4 | 32.4 | 49.1 |
| Days Max Temp ≥ 90 °F | 0 | 0 | 0 | 0 | 0 | 3 | 5 | 1 | 0 | 0 | 0 | 0 | 9 |
| Days Max Temp ≤ 32 °F | 3 | 2 | 1 | 0 | 0 | 0 | 0 | 0 | 0 | 0 | 1 | 3 | 10 |
| Days Min Temp ≤ 32 °F | 30 | 26 | 26 | 16 | 4 | 0 | 0 | 0 | 1 | 13 | 25 | 30 | 171 |
| Days Min Temp ≤ 0 °F | 2 | 1 | 0 | 0 | 0 | 0 | 0 | 0 | 0 | 0 | 0 | 2 | 5 |
| Heating Degree Days | 1039 | 860 | 771 | 516 | 291 | 70 | 12 | 31 | 156 | 431 | 762 | 1002 | 5941 |
| Cooling Degree Days | 0 | 0 | 0 | 0 | 4 | 52 | 99 | 56 | 8 | 0 | 0 | 0 | 219 |
| Total Precipitation (") | 0.42 | 0.52 | 0.87 | 0.95 | 2.10 | 2.14 | 2.89 | 3.42 | 1.80 | 0.88 | 0.67 | 0.46 | 17.12 |
| Days ≥ 0.1" Precip | 1 | 2 | 3 | 3 | 4 | 5 | 7 | 7 | 4 | 2 | 1 | 1 | 40 |
| Total Snowfall (") | 5.3 | *7.7* | *6.0* | 3.9 | 0.9 | 0.0 | 0.0 | 0.0 | 0.2 | 1.9 | 5.0 | 6.8 | 37.7 |
| Days ≥ 1" Snow Depth | *4* | na | *2* | 1 | 0 | 0 | 0 | 0 | 0 | 0 | *2* | na | na |

## CLAYTON MUNI ARPK *Union County* ELEVATION 5000 ft LAT/LONG 36° 27 ' N / 103° 9 ' W

| | JAN | FEB | MAR | APR | MAY | JUN | JUL | AUG | SEP | OCT | NOV | DEC | YEAR |
|---|---|---|---|---|---|---|---|---|---|---|---|---|---|
| Maximum Temp °F | 47.4 | 50.8 | 57.4 | 66.5 | 73.8 | 83.4 | 87.3 | 84.7 | 77.4 | 68.3 | 56.0 | 47.7 | 66.7 |
| Minimum Temp °F | 19.6 | 22.6 | 28.2 | 37.2 | 46.0 | 55.3 | 60.4 | 58.7 | 51.3 | 40.1 | 28.4 | 20.9 | 39.1 |
| Mean Temp °F | 33.5 | 36.7 | 42.8 | 51.9 | 59.9 | 69.4 | 73.8 | 71.7 | 64.3 | 54.2 | 42.2 | 34.3 | 52.9 |
| Days Max Temp ≥ 90 °F | 0 | 0 | 0 | 0 | 1 | 7 | 12 | 8 | 2 | 0 | 0 | 0 | 30 |
| Days Max Temp ≤ 32 °F | 5 | 3 | 2 | 0 | 0 | 0 | 0 | 0 | 0 | 0 | 1 | 4 | 15 |
| Days Min Temp ≤ 32 °F | 29 | 25 | 22 | 8 | 1 | 0 | 0 | 0 | 0 | 6 | 20 | 28 | 139 |
| Days Min Temp ≤ 0 °F | 1 | 1 | 0 | 0 | 0 | 0 | 0 | 0 | 0 | 0 | 0 | 1 | 3 |
| Heating Degree Days | 968 | 792 | 681 | 392 | 181 | 27 | 4 | 7 | 94 | 334 | 676 | 944 | 5100 |
| Cooling Degree Days | 0 | 0 | *1* | 6 | 30 | 173 | 288 | 238 | 89 | 7 | 0 | 0 | 832 |
| Total Precipitation (") | 0.26 | 0.29 | 0.60 | 0.95 | 2.10 | 2.40 | 2.75 | 2.73 | 1.84 | 0.92 | 0.50 | 0.34 | 15.68 |
| Days ≥ 0.1" Precip | 1 | 1 | 2 | 2 | 5 | 5 | 6 | 6 | 4 | 2 | 2 | 1 | 37 |
| Total Snowfall (") | 4.1 | 3.4 | 5.3 | 1.9 | 0.6 | 0.0 | 0.0 | 0.0 | 0.2 | 1.1 | 3.0 | 4.5 | 24.1 |
| Days ≥ 1" Snow Depth | 6 | 3 | 3 | 1 | 0 | 0 | 0 | 0 | 0 | 0 | 3 | 5 | 21 |

## CLIFF 11 SE *Grant County* ELEVATION 4803 ft LAT/LONG 32° 52 ' N / 108° 31 ' W

| | JAN | FEB | MAR | APR | MAY | JUN | JUL | AUG | SEP | OCT | NOV | DEC | YEAR |
|---|---|---|---|---|---|---|---|---|---|---|---|---|---|
| Maximum Temp °F | 55.6 | 60.0 | 65.9 | 74.1 | 82.2 | 92.1 | 92.6 | 89.5 | 84.9 | 75.9 | 64.1 | 55.6 | 74.4 |
| Minimum Temp °F | 22.0 | 24.2 | 28.7 | 33.6 | 41.2 | 50.4 | 58.9 | 57.5 | 50.1 | 38.0 | 27.2 | 21.6 | 37.8 |
| Mean Temp °F | 38.8 | 42.1 | 47.3 | 53.9 | 61.7 | 71.3 | 75.7 | 73.5 | 67.5 | 56.9 | 45.7 | 38.6 | 56.1 |
| Days Max Temp ≥ 90 °F | 0 | 0 | 0 | 0 | 4 | 20 | 22 | 16 | 7 | 1 | 0 | 0 | 70 |
| Days Max Temp ≤ 32 °F | 0 | 0 | 0 | 0 | 0 | 0 | 0 | 0 | 0 | 0 | 0 | 0 | 0 |
| Days Min Temp ≤ 32 °F | 28 | 24 | 22 | 14 | 3 | 0 | 0 | 0 | 0 | 8 | 23 | 27 | 149 |
| Days Min Temp ≤ 0 °F | 0 | 0 | 0 | 0 | 0 | 0 | 0 | 0 | 0 | 0 | 0 | 0 | 0 |
| Heating Degree Days | 804 | 640 | 541 | 329 | 123 | 9 | 0 | 0 | 25 | 247 | 571 | 810 | 4099 |
| Cooling Degree Days | 0 | 0 | 0 | 2 | 27 | 203 | 315 | 264 | 105 | 1 | 0 | 0 | 917 |
| Total Precipitation (") | 1.18 | 1.10 | 0.94 | 0.35 | 0.44 | 0.40 | 2.85 | 3.02 | 1.80 | 1.42 | 0.89 | 1.49 | 15.88 |
| Days ≥ 0.1" Precip | 3 | 3 | 3 | 1 | 2 | 1 | 6 | 7 | 4 | 3 | 2 | 4 | 39 |
| Total Snowfall (") | 1.7 | 1.2 | 0.9 | 0.3 | 0.0 | 0.0 | 0.0 | 0.0 | 0.0 | 0.0 | 0.5 | 1.5 | 6.1 |
| Days ≥ 1" Snow Depth | 1 | 0 | 0 | 0 | 0 | 0 | 0 | 0 | 0 | 0 | 0 | 0 | 1 |

## CLOVIS *Curry County* ELEVATION 4290 ft LAT/LONG 34° 25 ' N / 103° 12 ' W

| | JAN | FEB | MAR | APR | MAY | JUN | JUL | AUG | SEP | OCT | NOV | DEC | YEAR |
|---|---|---|---|---|---|---|---|---|---|---|---|---|---|
| Maximum Temp °F | 50.6 | 55.3 | 63.3 | 72.2 | 79.9 | 88.2 | 90.4 | 87.9 | 81.6 | 72.3 | 60.5 | 52.3 | 71.2 |
| Minimum Temp °F | 22.9 | 26.3 | 32.3 | 40.7 | 49.8 | 58.8 | 63.0 | 61.5 | 54.3 | 43.0 | 32.4 | 24.8 | 42.5 |
| Mean Temp °F | 36.7 | 40.8 | 47.8 | 56.5 | 64.9 | 73.5 | 76.7 | 74.7 | 68.0 | 57.7 | 46.5 | 38.6 | 56.9 |
| Days Max Temp ≥ 90 °F | 0 | 0 | 0 | 1 | 4 | 15 | 19 | 14 | 5 | 1 | 0 | 0 | 59 |
| Days Max Temp ≤ 32 °F | 3 | 2 | 1 | 0 | 0 | 0 | 0 | 0 | 0 | 0 | 1 | 3 | 10 |
| Days Min Temp ≤ 32 °F | 28 | 22 | 16 | 4 | 0 | 0 | 0 | 0 | 0 | 2 | 15 | 26 | 113 |
| Days Min Temp ≤ 0 °F | 0 | 0 | 0 | 0 | 0 | 0 | 0 | 0 | 0 | 0 | 0 | 1 | 1 |
| Heating Degree Days | 869 | 676 | 527 | 261 | 86 | 9 | 1 | 2 | 48 | 241 | 549 | 813 | 4082 |
| Cooling Degree Days | 0 | 0 | 1 | 14 | 89 | 281 | 373 | 327 | 152 | 16 | 0 | 0 | 1253 |
| Total Precipitation (") | 0.45 | 0.43 | 0.60 | 0.86 | 1.96 | 2.74 | 2.67 | 3.27 | 2.19 | 1.36 | 0.68 | 0.64 | 17.85 |
| Days ≥ 0.1" Precip | 1 | 2 | 2 | 2 | 4 | 5 | 5 | 6 | 4 | 3 | 2 | 2 | 38 |
| Total Snowfall (") | 2.9 | 3.5 | 1.7 | 0.2 | 0.0 | 0.0 | 0.0 | 0.0 | 0.0 | 0.2 | 1.7 | 3.8 | 14.0 |
| Days ≥ 1" Snow Depth | 3 | 2 | *1* | 0 | 0 | 0 | 0 | 0 | 0 | 0 | 1 | 3 | 10 |

## CLOVIS 13 N *Curry County* ELEVATION 4442 ft LAT/LONG 34° 36 ' N / 103° 13 ' W

| | JAN | FEB | MAR | APR | MAY | JUN | JUL | AUG | SEP | OCT | NOV | DEC | YEAR |
|---|---|---|---|---|---|---|---|---|---|---|---|---|---|
| Maximum Temp °F | 52.5 | 57.2 | 65.2 | 73.7 | 80.7 | 89.4 | 91.0 | 88.5 | 82.8 | 74.1 | 61.6 | 53.9 | 72.6 |
| Minimum Temp °F | 22.6 | 25.3 | 30.9 | 39.0 | 48.1 | 56.9 | 61.4 | 60.0 | 53.2 | 42.1 | 31.6 | 24.4 | 41.3 |
| Mean Temp °F | 37.5 | 41.3 | 48.1 | 56.4 | 64.4 | 73.2 | 76.2 | 74.2 | 68.0 | 58.2 | 46.6 | 39.2 | 56.9 |
| Days Max Temp ≥ 90 °F | 0 | 0 | 0 | 0 | 4 | 16 | 20 | 15 | 6 | 0 | 0 | 0 | 61 |
| Days Max Temp ≤ 32 °F | 2 | 1 | 0 | 0 | 0 | 0 | 0 | 0 | 0 | 0 | 0 | 2 | 5 |
| Days Min Temp ≤ 32 °F | 28 | 23 | 17 | 6 | 1 | 0 | 0 | 0 | 0 | 3 | 16 | 25 | 119 |
| Days Min Temp ≤ 0 °F | 0 | 0 | 0 | 0 | 0 | 0 | 0 | 0 | 0 | 0 | 0 | 0 | 0 |
| Heating Degree Days | 844 | 665 | 517 | 262 | 83 | 7 | 0 | 1 | 39 | 220 | 545 | 792 | 3975 |
| Cooling Degree Days | 0 | 0 | 1 | 9 | 74 | 271 | 360 | 306 | 146 | 11 | 0 | 0 | 1178 |
| Total Precipitation (") | 0.30 | 0.35 | 0.57 | 0.78 | 2.37 | 2.42 | 2.53 | 3.09 | 2.02 | 1.31 | 0.58 | 0.47 | 16.79 |
| Days ≥ 0.1" Precip | 1 | 1 | 2 | 2 | 4 | 4 | 5 | 5 | 4 | 2 | 1 | 1 | 32 |
| Total Snowfall (") | na | na | *0.2* | 0.2 | 0.0 | 0.0 | 0.0 | 0.0 | 0.0 | 0.1 | *0.3* | na | na |
| Days ≥ 1" Snow Depth | na | *0* | 0 | 0 | 0 | 0 | 0 | 0 | 0 | 0 | *0* | na | na |

## COLUMBUS *Luna County* ELEVATION 4032 ft LAT/LONG 31° 49 ' N / 107° 38 ' W

| | JAN | FEB | MAR | APR | MAY | JUN | JUL | AUG | SEP | OCT | NOV | DEC | YEAR |
|---|---|---|---|---|---|---|---|---|---|---|---|---|---|
| Maximum Temp °F | 58.0 | 63.7 | 70.9 | 79.3 | 87.2 | 96.4 | 95.6 | 92.6 | 87.7 | 78.8 | 67.0 | 58.3 | 78.0 |
| Minimum Temp °F | 29.1 | 32.5 | 37.9 | 44.7 | 53.1 | 62.7 | 66.8 | 64.7 | 58.6 | 47.2 | 35.9 | 29.5 | 46.9 |
| Mean Temp °F | 43.6 | 48.1 | 54.4 | 62.0 | 70.1 | 79.5 | 81.2 | 78.7 | 73.2 | 63.1 | 51.5 | 43.9 | 62.4 |
| Days Max Temp ≥ 90 °F | 0 | 0 | 0 | 2 | 11 | 27 | 27 | 24 | 13 | 2 | 0 | 0 | 106 |
| Days Max Temp ≤ 32 °F | 0 | 0 | 0 | 0 | 0 | 0 | 0 | 0 | 0 | 0 | 0 | 0 | 0 |
| Days Min Temp ≤ 32 °F | 20 | 14 | 7 | 2 | 0 | 0 | 0 | 0 | 0 | 1 | 10 | 20 | 74 |
| Days Min Temp ≤ 0 °F | 0 | 0 | 0 | 0 | 0 | 0 | 0 | 0 | 0 | 0 | 0 | 0 | 0 |
| Heating Degree Days | 657 | 471 | 324 | 125 | 17 | 0 | 0 | 0 | 7 | 108 | 400 | 647 | 2756 |
| Cooling Degree Days | 0 | 0 | 4 | 60 | 212 | 483 | 533 | 451 | 287 | 61 | 2 | 0 | 2093 |
| Total Precipitation (") | 0.56 | 0.47 | 0.28 | 0.26 | 0.30 | 0.31 | 2.10 | 2.06 | 1.52 | 0.99 | 0.55 | 0.94 | 10.34 |
| Days ≥ 0.1" Precip | 2 | 2 | 1 | 1 | 1 | 1 | 5 | 5 | 4 | 2 | 2 | 3 | 29 |
| Total Snowfall (") | *0.8* | 0.3 | 0.0 | 0.2 | 0.0 | 0.0 | 0.0 | 0.0 | 0.0 | 0.0 | 0.4 | 1.2 | 2.9 |
| Days ≥ 1" Snow Depth | 0 | 0 | 0 | 0 | 0 | 0 | 0 | 0 | 0 | 0 | 0 | *0* | 0 |

## CONCHAS DAM *San Miguel County* ELEVATION 4242 ft LAT/LONG 35° 24 ' N / 104° 11 ' W

| | JAN | FEB | MAR | APR | MAY | JUN | JUL | AUG | SEP | OCT | NOV | DEC | YEAR |
|---|---|---|---|---|---|---|---|---|---|---|---|---|---|
| Maximum Temp °F | 52.7 | 57.2 | 64.2 | 72.6 | 80.4 | 89.9 | 93.2 | 90.8 | 83.8 | 74.5 | 62.0 | 53.8 | 72.9 |
| Minimum Temp °F | 23.3 | 27.1 | 34.3 | 43.2 | 51.8 | 60.9 | 65.8 | 63.8 | 56.6 | 45.0 | 33.3 | 25.4 | 44.2 |
| Mean Temp °F | 38.0 | 42.2 | 49.3 | 57.9 | 66.1 | 75.5 | 79.5 | 77.3 | 70.2 | 59.8 | 47.7 | 39.6 | 58.6 |
| Days Max Temp ≥ 90 °F | 0 | 0 | 0 | 1 | 4 | 17 | 24 | 20 | 8 | 1 | 0 | 0 | 75 |
| Days Max Temp ≤ 32 °F | 2 | 1 | 0 | 0 | 0 | 0 | 0 | 0 | 0 | 0 | 0 | 2 | 5 |
| Days Min Temp ≤ 32 °F | 27 | 21 | 12 | 3 | 0 | 0 | 0 | 0 | 0 | 2 | 14 | 26 | 105 |
| Days Min Temp ≤ 0 °F | 0 | 0 | 0 | 0 | 0 | 0 | 0 | 0 | 0 | 0 | 0 | 0 | 0 |
| Heating Degree Days | 822 | 636 | 480 | 223 | 68 | 6 | 1 | 1 | 33 | 187 | 513 | 779 | 3749 |
| Cooling Degree Days | 0 | 0 | 1 | 21 | 106 | 332 | 462 | 400 | 202 | 27 | 1 | 0 | 1552 |
| Total Precipitation (") | 0.41 | 0.40 | 0.66 | 0.74 | 1.50 | 1.92 | 2.51 | 2.57 | 1.61 | 1.07 | 0.66 | 0.41 | 14.46 |
| Days ≥ 0.1" Precip | 1 | 1 | 2 | 2 | 3 | 4 | 5 | 6 | 3 | 2 | 1 | 1 | 31 |
| Total Snowfall (") | *3.3* | 3.5 | 1.1 | 0.3 | 0.2 | 0.0 | 0.0 | 0.0 | 0.0 | 0.4 | 0.9 | *2.9* | 12.6 |
| Days ≥ 1" Snow Depth | *2* | 1 | 1 | 0 | 0 | 0 | 0 | 0 | 0 | 0 | 0 | 2 | 6 |

## CROSSROADS 2 *Lea County* ELEVATION 4104 ft LAT/LONG 33° 29 ' N / 103° 21 ' W

| | JAN | FEB | MAR | APR | MAY | JUN | JUL | AUG | SEP | OCT | NOV | DEC | YEAR |
|---|---|---|---|---|---|---|---|---|---|---|---|---|---|
| Maximum Temp °F | 54.6 | 59.2 | 67.1 | 75.4 | 82.4 | 90.2 | 91.3 | 88.7 | 82.7 | 74.7 | 63.2 | 56.0 | 73.8 |
| Minimum Temp °F | 23.4 | 26.4 | 32.7 | 40.9 | 49.9 | 58.8 | 62.6 | 60.8 | 54.3 | 42.8 | 32.0 | 24.7 | 42.4 |
| Mean Temp °F | 39.0 | 42.8 | 49.9 | 58.2 | 66.2 | 74.5 | 76.9 | 74.8 | 68.5 | 58.8 | 47.6 | 40.4 | 58.1 |
| Days Max Temp ≥ 90 °F | 0 | 0 | 0 | 1 | 5 | 16 | 20 | 15 | 6 | 1 | 0 | 0 | 64 |
| Days Max Temp ≤ 32 °F | 2 | 1 | 0 | 0 | 0 | 0 | 0 | 0 | 0 | 0 | 0 | 2 | 5 |
| Days Min Temp ≤ 32 °F | 27 | 22 | 14 | 4 | 0 | 0 | 0 | 0 | 0 | 2 | 16 | 26 | 111 |
| Days Min Temp ≤ 0 °F | 0 | 0 | 0 | 0 | 0 | 0 | 0 | 0 | 0 | 0 | 0 | 0 | 0 |
| Heating Degree Days | 798 | 619 | 463 | 216 | 58 | 3 | 0 | 1 | 38 | 205 | 515 | 757 | 3673 |
| Cooling Degree Days | 0 | 0 | 0 | 16 | 105 | ***310*** | 396 | 330 | 157 | 14 | 1 | 0 | 1329 |
| Total Precipitation (") | 0.35 | 0.42 | 0.44 | 0.65 | 1.87 | 2.15 | 2.72 | 3.00 | 2.23 | 1.35 | 0.56 | 0.44 | 16.18 |
| Days ≥ 0.1" Precip | 1 | 1 | 1 | 2 | 3 | 4 | 4 | 5 | 4 | 2 | 1 | 1 | 29 |
| Total Snowfall (") | na | ***0.7*** | 0.3 | 0.0 | 0.0 | 0.0 | 0.0 | 0.0 | 0.0 | 0.1 | 0.6 | ***1.5*** | na |
| Days ≥ 1" Snow Depth | na | ***0*** | 0 | 0 | 0 | 0 | 0 | 0 | 0 | 0 | 0 | ***0*** | na |

## DES MOINES *Union County* ELEVATION 6624 ft LAT/LONG 36° 46 ' N / 103° 50 ' W

| | JAN | FEB | MAR | APR | MAY | JUN | JUL | AUG | SEP | OCT | NOV | DEC | YEAR |
|---|---|---|---|---|---|---|---|---|---|---|---|---|---|
| Maximum Temp °F | 44.9 | 48.0 | 54.0 | 62.8 | 70.1 | 79.5 | 83.2 | 80.6 | 74.9 | 65.8 | 52.7 | 45.3 | 63.5 |
| Minimum Temp °F | 17.1 | 19.6 | 25.0 | 32.7 | 41.5 | 51.2 | 56.6 | 54.9 | 47.5 | 36.4 | 25.2 | 18.5 | 35.5 |
| Mean Temp °F | 31.1 | 33.8 | 39.5 | 47.8 | 55.9 | 65.4 | 69.9 | 67.8 | 61.2 | 51.1 | 39.0 | 31.9 | 49.5 |
| Days Max Temp ≥ 90 °F | 0 | 0 | 0 | 0 | 0 | 2 | 3 | 1 | 0 | 0 | 0 | 0 | 6 |
| Days Max Temp ≤ 32 °F | 4 | 3 | 2 | 0 | 0 | 0 | 0 | 0 | 0 | 0 | 2 | 4 | 15 |
| Days Min Temp ≤ 32 °F | 30 | 26 | 25 | 15 | 4 | 0 | 0 | 0 | 1 | 10 | 23 | 29 | 163 |
| Days Min Temp ≤ 0 °F | 2 | 1 | 0 | 0 | 0 | 0 | 0 | 0 | 0 | 0 | 0 | 2 | 5 |
| Heating Degree Days | 1045 | 874 | 782 | 511 | 282 | 71 | 9 | 21 | 139 | 424 | 774 | 1018 | 5950 |
| Cooling Degree Days | 0 | 0 | 0 | 1 | 6 | 94 | 171 | 120 | 36 | 0 | 0 | 0 | 428 |
| Total Precipitation (") | 0.40 | 0.47 | 1.08 | 1.01 | 2.29 | 2.24 | 3.29 | 2.87 | 2.14 | 0.99 | 0.79 | 0.54 | 18.11 |
| Days ≥ 0.1" Precip | 1 | 1 | 3 | 3 | 5 | 5 | 7 | 6 | 4 | 2 | 2 | 2 | 41 |
| Total Snowfall (") | 3.9 | 4.3 | 8.2 | 3.6 | 2.0 | 0.0 | 0.0 | 0.0 | 0.4 | 2.3 | 5.5 | 5.6 | 35.8 |
| Days ≥ 1" Snow Depth | 9 | 6 | 6 | 3 | 1 | 0 | 0 | 0 | 0 | 1 | 5 | 8 | 39 |

## DILIA *Guadalupe County* ELEVATION 5144 ft LAT/LONG 35° 11 ' N / 105° 3 ' W

| | JAN | FEB | MAR | APR | MAY | JUN | JUL | AUG | SEP | OCT | NOV | DEC | YEAR |
|---|---|---|---|---|---|---|---|---|---|---|---|---|---|
| Maximum Temp °F | 50.8 | 55.3 | 62.0 | 70.7 | 78.7 | 88.3 | 90.8 | 88.0 | 81.4 | 72.5 | 59.9 | 51.9 | 70.9 |
| Minimum Temp °F | 21.3 | 24.0 | 29.5 | 36.6 | 45.2 | 54.6 | 59.9 | 58.1 | 50.9 | 39.6 | 29.6 | 22.4 | 39.3 |
| Mean Temp °F | 36.1 | 39.7 | 45.6 | 53.6 | 62.0 | 71.5 | 75.4 | 73.0 | 66.2 | 56.0 | 44.8 | 37.2 | 55.1 |
| Days Max Temp ≥ 90 °F | 0 | 0 | 0 | 0 | 2 | 14 | 20 | 14 | 4 | 0 | 0 | 0 | 54 |
| Days Max Temp ≤ 32 °F | 2 | 1 | 0 | 0 | 0 | 0 | 0 | 0 | 0 | 0 | 0 | 1 | 4 |
| Days Min Temp ≤ 32 °F | 28 | 24 | 20 | 9 | 1 | 0 | 0 | 0 | 0 | 5 | 19 | 27 | 133 |
| Days Min Temp ≤ 0 °F | 1 | 0 | 0 | 0 | 0 | 0 | 0 | 0 | 0 | 0 | 0 | 0 | 1 |
| Heating Degree Days | 890 | 709 | 594 | 339 | 125 | 10 | 1 | 2 | 52 | 276 | 600 | 856 | 4454 |
| Cooling Degree Days | 0 | 0 | 0 | 5 | 40 | 230 | 339 | 282 | 99 | 5 | 0 | 0 | 1000 |
| Total Precipitation (") | 0.56 | 0.51 | 0.65 | 0.76 | 1.38 | 1.33 | 2.84 | 2.72 | 1.82 | 1.08 | 0.74 | 0.64 | 15.03 |
| Days ≥ 0.1" Precip | 2 | 1 | 2 | 2 | 3 | 3 | 6 | 6 | 4 | 2 | 2 | 2 | 35 |
| Total Snowfall (") | 6.0 | ***4.5*** | 5.1 | 0.8 | 0.5 | 0.0 | 0.0 | 0.0 | 0.0 | 0.9 | 2.5 | na | na |
| Days ≥ 1" Snow Depth | na | na | ***1*** | 0 | 0 | 0 | 0 | 0 | 0 | 0 | 0 | na | na |

## DULCE *Rio Arriba County* ELEVATION 6775 ft LAT/LONG 36° 56 ' N / 107° 0 ' W

| | JAN | FEB | MAR | APR | MAY | JUN | JUL | AUG | SEP | OCT | NOV | DEC | YEAR |
|---|---|---|---|---|---|---|---|---|---|---|---|---|---|
| Maximum Temp °F | 40.3 | 45.0 | 52.3 | 61.5 | 69.9 | 80.0 | 84.5 | 81.9 | 75.3 | 65.3 | 50.7 | 41.7 | 62.4 |
| Minimum Temp °F | 5.7 | 11.4 | 19.6 | 24.6 | 31.3 | 37.9 | 46.4 | 46.4 | 37.5 | 27.2 | 17.6 | 9.0 | 26.2 |
| Mean Temp °F | 23.0 | 28.3 | 36.0 | 43.1 | 50.8 | 58.9 | 65.5 | 64.2 | 56.4 | 46.3 | 34.1 | 25.4 | 44.3 |
| Days Max Temp ≥ 90 °F | 0 | 0 | 0 | 0 | 0 | 2 | 4 | 1 | 0 | 0 | 0 | 0 | 7 |
| Days Max Temp ≤ 32 °F | 4 | 1 | 0 | 0 | 0 | 0 | 0 | 0 | 0 | 0 | 1 | 4 | 10 |
| Days Min Temp ≤ 32 °F | 31 | 27 | 30 | 26 | 19 | 6 | 1 | 0 | 8 | 24 | 29 | 31 | 232 |
| Days Min Temp ≤ 0 °F | 11 | 4 | 0 | 0 | 0 | 0 | 0 | 0 | 0 | 0 | 1 | 6 | 22 |
| Heating Degree Days | 1296 | 1028 | 893 | 652 | 434 | 185 | 35 | 57 | 252 | 574 | 917 | 1222 | 7545 |
| Cooling Degree Days | 0 | 0 | 0 | 0 | 0 | 14 | 52 | 49 | 2 | 0 | 0 | 0 | 117 |
| Total Precipitation (") | 1.42 | 1.25 | 1.62 | 0.98 | 1.45 | 0.85 | 1.81 | 2.70 | 1.69 | 1.59 | 1.43 | 1.51 | 18.30 |
| Days ≥ 0.1" Precip | 4 | 4 | 5 | 3 | 4 | 3 | 5 | 6 | 4 | 4 | 4 | 4 | 50 |
| Total Snowfall (") | 14.1 | 12.1 | 7.9 | ***1.2*** | 0.0 | 0.0 | 0.0 | 0.0 | 0.0 | 1.2 | 5.6 | ***15.0*** | 57.1 |
| Days ≥ 1" Snow Depth | 22 | ***17*** | 7 | ***0*** | 0 | 0 | 0 | 0 | 0 | 0 | 5 | 16 | 67 |

## EAGLE NEST *Colfax County* ELEVATION 8254 ft LAT/LONG 36° 34 ' N / 105° 16 ' W

| | JAN | FEB | MAR | APR | MAY | JUN | JUL | AUG | SEP | OCT | NOV | DEC | YEAR |
|---|---|---|---|---|---|---|---|---|---|---|---|---|---|
| Maximum Temp °F | 36.9 | 40.5 | 46.4 | 55.3 | 64.0 | 73.8 | 77.5 | 74.7 | 69.5 | 60.7 | 47.6 | 39.0 | 57.2 |
| Minimum Temp °F | 0.5 | 5.6 | 15.3 | 22.6 | 29.7 | 36.3 | 42.6 | 41.7 | 33.8 | 23.6 | 14.7 | 4.5 | 22.6 |
| Mean Temp °F | 18.7 | 23.1 | 30.9 | 39.0 | 46.9 | 55.1 | 60.0 | 58.3 | 51.7 | 42.2 | 31.2 | 21.8 | 39.9 |
| Days Max Temp ≥ 90 °F | 0 | 0 | 0 | 0 | 0 | 0 | 0 | 0 | 0 | 0 | 0 | 0 | 0 |
| Days Max Temp ≤ 32 °F | 8 | 5 | 3 | 0 | 0 | 0 | 0 | 0 | 0 | 0 | 3 | 8 | 27 |
| Days Min Temp ≤ 32 °F | 30 | 28 | 29 | 27 | 20 | 7 | 0 | 1 | 12 | 27 | 29 | 31 | 241 |
| Days Min Temp ≤ 0 °F | 15 | 10 | 3 | 0 | 0 | 0 | 0 | 0 | 0 | 0 | 3 | 11 | 42 |
| Heating Degree Days | 1427 | 1177 | 1049 | 775 | 553 | 291 | 148 | 203 | 392 | 700 | 1007 | 1334 | 9056 |
| Cooling Degree Days | 0 | 0 | 0 | 0 | 0 | 1 | 2 | 1 | 0 | 0 | 0 | 0 | 4 |
| Total Precipitation (") | 0.68 | 0.66 | 1.13 | 0.84 | 1.31 | 1.41 | 2.69 | 3.06 | 1.29 | 0.85 | 0.78 | 0.80 | 15.50 |
| Days ≥ 0.1" Precip | 2 | 2 | 3 | 3 | 4 | 4 | 7 | 9 | 4 | 2 | 2 | 3 | 45 |
| Total Snowfall (") | 10.1 | 10.4 | 13.0 | 5.5 | 2.2 | 0.0 | 0.0 | 0.0 | 0.3 | *2.2* | 7.2 | 10.5 | 61.4 |
| Days ≥ 1" Snow Depth | *14* | na | *8* | *2* | 0 | 0 | 0 | 0 | 0 | *1* | *5* | na | na |

## EL MORRO NATL MON *Cibola County* ELEVATION 7224 ft LAT/LONG 35° 3 ' N / 108° 21 ' W

| | JAN | FEB | MAR | APR | MAY | JUN | JUL | AUG | SEP | OCT | NOV | DEC | YEAR |
|---|---|---|---|---|---|---|---|---|---|---|---|---|---|
| Maximum Temp °F | 43.2 | 47.1 | 53.5 | 62.9 | 71.9 | 82.4 | 84.9 | 81.8 | 76.3 | 66.4 | 52.9 | 44.8 | 64.0 |
| Minimum Temp °F | 13.0 | 18.0 | 23.3 | 27.9 | 34.8 | 42.7 | 51.1 | 49.8 | 42.3 | 31.3 | 21.5 | 14.8 | 30.9 |
| Mean Temp °F | 28.1 | 32.5 | 38.4 | 45.4 | 53.4 | 62.6 | 68.1 | 65.8 | 59.3 | 48.9 | 37.3 | 29.8 | 47.5 |
| Days Max Temp ≥ 90 °F | 0 | 0 | 0 | 0 | 0 | 4 | 6 | 2 | 0 | 0 | 0 | 0 | 12 |
| Days Max Temp ≤ 32 °F | 3 | 1 | 0 | 0 | 0 | 0 | 0 | 0 | 0 | 0 | 1 | 3 | 8 |
| Days Min Temp ≤ 32 °F | 31 | 28 | 29 | 23 | 11 | 2 | 0 | 0 | 2 | 18 | 27 | 30 | 201 |
| Days Min Temp ≤ 0 °F | 5 | 1 | 0 | 0 | 0 | 0 | 0 | 0 | 0 | 0 | 1 | 3 | 10 |
| Heating Degree Days | 1137 | 910 | 816 | 580 | 354 | 105 | 8 | 29 | 171 | 494 | 825 | 1084 | 6513 |
| Cooling Degree Days | 0 | 0 | 0 | 0 | 1 | 40 | 97 | 67 | 7 | 0 | 0 | 0 | 212 |
| Total Precipitation (") | 1.14 | 0.94 | 1.17 | 0.75 | 0.82 | 0.62 | 1.99 | 2.67 | 1.43 | 1.18 | 0.96 | 1.12 | 14.79 |
| Days ≥ 0.1" Precip | 3 | 3 | 4 | 2 | 3 | 2 | 5 | 7 | 4 | 3 | 3 | 4 | 43 |
| Total Snowfall (") | 12.1 | 8.9 | 8.7 | 3.5 | 0.7 | 0.0 | 0.0 | 0.0 | 0.0 | 1.8 | 5.6 | 12.1 | 53.4 |
| Days ≥ 1" Snow Depth | 20 | 11 | *3* | 0 | 0 | 0 | 0 | 0 | 0 | 0 | 2 | 10 | 46 |

## EL RITO *Rio Arriba County* ELEVATION 6873 ft LAT/LONG 36° 20 ' N / 106° 11 ' W

| | JAN | FEB | MAR | APR | MAY | JUN | JUL | AUG | SEP | OCT | NOV | DEC | YEAR |
|---|---|---|---|---|---|---|---|---|---|---|---|---|---|
| Maximum Temp °F | 41.2 | 46.1 | 53.4 | 62.4 | 70.9 | 80.5 | 84.4 | 81.5 | 76.2 | 66.0 | 51.6 | 42.4 | 63.1 |
| Minimum Temp °F | 15.7 | 20.1 | 25.4 | 32.2 | 39.4 | 47.1 | 53.5 | 52.1 | 45.6 | 36.0 | 25.4 | 17.6 | 34.2 |
| Mean Temp °F | 28.5 | 33.1 | 39.4 | 47.3 | 55.2 | 63.8 | 69.0 | 66.8 | 60.9 | 51.0 | 38.6 | 30.0 | 48.6 |
| Days Max Temp ≥ 90 °F | 0 | 0 | 0 | 0 | 0 | 2 | 4 | 1 | 0 | 0 | 0 | 0 | 7 |
| Days Max Temp ≤ 32 °F | 4 | 2 | 0 | 0 | 0 | 0 | 0 | 0 | 0 | 0 | 1 | 3 | 10 |
| Days Min Temp ≤ 32 °F | 30 | 27 | 26 | 16 | 5 | 1 | 0 | 0 | 1 | 8 | 24 | 30 | 168 |
| Days Min Temp ≤ 0 °F | 2 | 1 | 0 | 0 | 0 | 0 | 0 | 0 | 0 | 0 | 0 | 1 | 4 |
| Heating Degree Days | 1125 | 892 | 786 | 524 | 299 | 82 | 5 | 18 | 132 | 427 | 786 | 1077 | 6153 |
| Cooling Degree Days | 0 | 0 | 0 | 0 | 2 | 58 | 134 | 95 | 19 | 1 | 0 | 0 | 309 |
| Total Precipitation (") | 0.74 | 0.59 | 0.85 | 0.70 | 0.92 | 0.96 | 1.79 | 2.31 | 1.26 | 1.11 | 0.93 | 0.65 | 12.81 |
| Days ≥ 0.1" Precip | 2 | 2 | 3 | 2 | 3 | 3 | 5 | 7 | 4 | 3 | 2 | 2 | 38 |
| Total Snowfall (") | *7.4* | *8.1* | 5.5 | 1.4 | 0.0 | 0.0 | 0.0 | 0.0 | 0.0 | 1.2 | 3.6 | *7.5* | 34.7 |
| Days ≥ 1" Snow Depth | na | na | *1* | 0 | 0 | 0 | 0 | 0 | 0 | 0 | 1 | na | na |

## EL VADO DAM *Rio Arriba County* ELEVATION 6804 ft LAT/LONG 36° 36 ' N / 106° 44 ' W

| | JAN | FEB | MAR | APR | MAY | JUN | JUL | AUG | SEP | OCT | NOV | DEC | YEAR |
|---|---|---|---|---|---|---|---|---|---|---|---|---|---|
| Maximum Temp °F | 39.2 | 43.9 | 50.4 | 59.8 | 69.1 | 79.9 | 84.7 | 81.6 | 74.9 | 64.8 | 50.5 | 41.3 | 61.7 |
| Minimum Temp °F | 4.8 | 11.8 | 20.8 | 25.6 | 33.0 | 40.1 | 48.2 | 47.5 | 38.3 | 27.7 | 18.5 | 10.0 | 27.2 |
| Mean Temp °F | 22.0 | 27.9 | 35.6 | 42.7 | 51.1 | 60.0 | 66.5 | 64.5 | 56.6 | 46.3 | 34.5 | 25.7 | 44.4 |
| Days Max Temp ≥ 90 °F | 0 | 0 | 0 | 0 | 0 | 2 | 5 | 1 | 0 | 0 | 0 | 0 | 8 |
| Days Max Temp ≤ 32 °F | 6 | 2 | 1 | 0 | 0 | 0 | 0 | 0 | 0 | 0 | 1 | 5 | 15 |
| Days Min Temp ≤ 32 °F | 31 | 28 | 29 | 25 | 14 | 3 | 0 | 0 | 6 | 24 | 28 | 31 | 219 |
| Days Min Temp ≤ 0 °F | 12 | 5 | 0 | 0 | 0 | 0 | 0 | 0 | 0 | 0 | 0 | 6 | 23 |
| Heating Degree Days | 1327 | 1042 | 904 | 661 | 425 | 162 | 24 | 54 | 246 | 574 | 907 | 1212 | 7538 |
| Cooling Degree Days | 0 | 0 | 0 | 0 | 1 | 24 | 72 | 53 | 3 | 0 | 0 | 0 | 153 |
| Total Precipitation (") | 0.98 | 0.79 | 1.04 | 0.82 | 1.23 | 0.91 | 1.86 | 2.78 | 1.61 | 1.33 | 1.04 | 0.94 | 15.33 |
| Days ≥ 0.1" Precip | 4 | 3 | 4 | 3 | 4 | 3 | 5 | 7 | 5 | 4 | 4 | 3 | 49 |
| Total Snowfall (") | 13.1 | 8.5 | 6.1 | *1.5* | 0.0 | 0.0 | 0.0 | 0.0 | 0.0 | 0.9 | 4.9 | 10.1 | 45.1 |
| Days ≥ 1" Snow Depth | 23 | 18 | 8 | 1 | 0 | 0 | 0 | 0 | 0 | 0 | 4 | 12 | 66 |

## ELEPHANT BUTTE DAM *Sierra County* ELEVATION 4583 ft LAT/LONG 33° 9 ' N / 107° 11 ' W

| | JAN | FEB | MAR | APR | MAY | JUN | JUL | AUG | SEP | OCT | NOV | DEC | YEAR |
|---|---|---|---|---|---|---|---|---|---|---|---|---|---|
| Maximum Temp °F | 54.1 | 59.7 | 67.1 | 75.5 | 83.2 | 92.7 | 93.4 | 90.3 | 84.8 | 75.5 | 63.3 | 53.8 | 74.4 |
| Minimum Temp °F | 28.7 | 32.4 | 38.4 | 45.4 | 53.7 | 62.8 | 67.3 | 65.2 | 58.9 | 47.8 | 36.7 | 29.2 | 47.2 |
| Mean Temp °F | 41.4 | 46.0 | 52.8 | 60.5 | 68.5 | 77.8 | 80.4 | 77.8 | 71.9 | 61.7 | 50.0 | 41.5 | 60.9 |
| Days Max Temp ≥ 90 °F | 0 | 0 | 0 | 1 | 5 | 21 | 25 | 18 | 7 | 0 | 0 | 0 | 77 |
| Days Max Temp ≤ 32 °F | 0 | 0 | 0 | 0 | 0 | 0 | 0 | 0 | 0 | 0 | 0 | 1 | 1 |
| Days Min Temp ≤ 32 °F | 24 | 15 | 6 | 1 | 0 | 0 | 0 | 0 | 0 | 0 | 8 | 23 | 77 |
| Days Min Temp ≤ 0 °F | 0 | 0 | 0 | 0 | 0 | 0 | 0 | 0 | 0 | 0 | 0 | 0 | 0 |
| Heating Degree Days | 724 | 530 | 374 | 156 | 31 | 1 | 0 | 0 | 10 | 132 | 443 | 721 | 3122 |
| Cooling Degree Days | 0 | 0 | 2 | 38 | 159 | 414 | 488 | 405 | 223 | 32 | 0 | 0 | 1761 |
| Total Precipitation (") | 0.43 | 0.30 | 0.24 | 0.13 | 0.51 | 0.67 | 1.46 | 2.33 | 1.50 | 1.07 | 0.64 | 0.84 | 10.12 |
| Days ≥ 0.1" Precip | 2 | 1 | 1 | 0 | 1 | 2 | 4 | 5 | 3 | 2 | 1 | 2 | 24 |
| Total Snowfall (") | ***1.1*** | 0.2 | 0.2 | 0.1 | 0.0 | 0.0 | 0.0 | 0.0 | 0.0 | 0.3 | 0.3 | 0.8 | 3.0 |
| Days ≥ 1" Snow Depth | 1 | 0 | 0 | 0 | 0 | 0 | 0 | 0 | 0 | 0 | 0 | 0 | 1 |

## ELIDA *Roosevelt County* ELEVATION 4354 ft LAT/LONG 33° 57 ' N / 103° 39 ' W

| | JAN | FEB | MAR | APR | MAY | JUN | JUL | AUG | SEP | OCT | NOV | DEC | YEAR |
|---|---|---|---|---|---|---|---|---|---|---|---|---|---|
| Maximum Temp °F | 51.9 | 56.5 | 64.5 | 73.8 | 81.0 | 89.6 | 91.0 | 88.9 | 82.8 | 73.6 | 61.5 | 53.3 | 72.4 |
| Minimum Temp °F | 23.1 | 26.3 | 32.5 | 40.8 | 49.6 | 58.8 | 62.9 | 61.0 | 54.1 | 42.8 | 31.9 | 24.3 | 42.3 |
| Mean Temp °F | 37.5 | 41.5 | 48.5 | 57.3 | 65.3 | 74.2 | 77.0 | 75.0 | 68.5 | 58.2 | 46.7 | 38.8 | 57.4 |
| Days Max Temp ≥ 90 °F | 0 | 0 | 0 | 1 | 5 | 16 | 20 | 16 | 6 | 1 | 0 | 0 | 65 |
| Days Max Temp ≤ 32 °F | 3 | 1 | 1 | 0 | 0 | 0 | 0 | 0 | 0 | 0 | 0 | 2 | 7 |
| Days Min Temp ≤ 32 °F | 28 | 21 | 15 | 5 | 1 | 0 | 0 | 0 | 0 | 4 | 16 | 25 | 115 |
| Days Min Temp ≤ 0 °F | 1 | 0 | 0 | 0 | 0 | 0 | 0 | 0 | 0 | 0 | 0 | 1 | 2 |
| Heating Degree Days | 844 | 658 | 505 | 242 | 79 | 8 | 1 | 2 | 43 | 224 | 542 | 805 | 3953 |
| Cooling Degree Days | 0 | 0 | 1 | 18 | 100 | 309 | 381 | 334 | 159 | 20 | 0 | 0 | 1322 |
| Total Precipitation (") | 0.49 | 0.36 | 0.39 | 0.66 | 1.50 | 2.14 | 2.93 | 2.83 | 2.03 | 1.09 | 0.71 | 0.49 | 15.62 |
| Days ≥ 0.1" Precip | 1 | 1 | 1 | 1 | 3 | 4 | 5 | 5 | 4 | 2 | 2 | 1 | 30 |
| Total Snowfall (") | ***2.0*** | na | ***0.6*** | 0.4 | 0.0 | 0.0 | 0.0 | 0.0 | 0.0 | 0.3 | ***1.4*** | ***2.6*** | na |
| Days ≥ 1" Snow Depth | na | na | ***0*** | 0 | 0 | 0 | 0 | 0 | 0 | 0 | ***0*** | na | na |

## ELK 2 E *Chaves County* ELEVATION 5705 ft LAT/LONG 32° 56 ' N / 105° 17 ' W

| | JAN | FEB | MAR | APR | MAY | JUN | JUL | AUG | SEP | OCT | NOV | DEC | YEAR |
|---|---|---|---|---|---|---|---|---|---|---|---|---|---|
| Maximum Temp °F | 54.5 | 57.5 | ***63.5*** | 70.6 | 76.8 | ***84.3*** | 83.8 | ***81.3*** | 77.0 | 71.2 | 62.1 | ***56.2*** | 69.9 |
| Minimum Temp °F | 21.3 | 23.7 | ***28.5*** | 35.0 | 42.8 | ***51.2*** | 55.8 | ***54.6*** | 47.9 | 37.1 | 27.7 | ***21.8*** | 37.3 |
| Mean Temp °F | 37.9 | 40.6 | ***46.0*** | 52.8 | 59.8 | ***67.7*** | 69.9 | ***68.0*** | 62.5 | 54.2 | 44.9 | ***39.0*** | 53.6 |
| Days Max Temp ≥ 90 °F | 0 | 0 | 0 | 0 | 1 | 6 | 5 | 2 | 0 | 0 | 0 | 0 | 14 |
| Days Max Temp ≤ 32 °F | 1 | 0 | 0 | 0 | 0 | 0 | 0 | 0 | 0 | 0 | 0 | 1 | 2 |
| Days Min Temp ≤ 32 °F | 28 | 24 | 21 | 10 | 2 | 0 | 0 | 0 | 0 | 7 | 23 | 27 | 142 |
| Days Min Temp ≤ 0 °F | 1 | 0 | 0 | 0 | 0 | 0 | 0 | 0 | 0 | 0 | 0 | 0 | 1 |
| Heating Degree Days | 834 | 684 | ***582*** | 360 | ***172*** | ***25*** | 3 | ***10*** | 98 | 329 | 596 | ***799*** | 4492 |
| Cooling Degree Days | 0 | 0 | ***0*** | ***1*** | na | na | ***143*** | na | 26 | 0 | 0 | ***0*** | na |
| Total Precipitation (") | 0.55 | 0.56 | 0.36 | 0.56 | 1.22 | 1.94 | 2.47 | 3.86 | 2.91 | 1.24 | 0.71 | 0.68 | 17.06 |
| Days ≥ 0.1" Precip | 2 | 2 | 1 | 1 | 3 | 4 | 6 | 8 | 6 | 3 | 2 | 2 | 40 |
| Total Snowfall (") | 4.5 | 4.0 | 1.9 | 0.4 | 0.1 | 0.0 | 0.0 | 0.0 | 0.0 | 1.7 | 1.7 | ***6.3*** | 20.6 |
| Days ≥ 1" Snow Depth | ***2*** | ***1*** | 0 | 0 | 0 | 0 | 0 | 0 | 0 | 0 | 0 | ***1*** | 4 |

## ESPANOLA *Santa Fe County* ELEVATION 5594 ft LAT/LONG 35° 59 ' N / 106° 5 ' W

| | JAN | FEB | MAR | APR | MAY | JUN | JUL | AUG | SEP | OCT | NOV | DEC | YEAR |
|---|---|---|---|---|---|---|---|---|---|---|---|---|---|
| Maximum Temp °F | 45.2 | 52.0 | 59.9 | 69.3 | ***77.6*** | ***87.5*** | ***90.5*** | 86.9 | 80.8 | 71.5 | 57.6 | 47.0 | 68.8 |
| Minimum Temp °F | 15.0 | 20.5 | 26.4 | 33.0 | ***41.8*** | ***50.2*** | ***56.9*** | 55.2 | 46.6 | 34.6 | 23.4 | 15.6 | 34.9 |
| Mean Temp °F | 30.1 | 36.3 | 43.2 | 51.2 | ***59.7*** | ***68.9*** | ***73.7*** | 71.1 | 63.8 | 53.1 | 40.5 | 31.3 | 51.9 |
| Days Max Temp ≥ 90 °F | 0 | 0 | 0 | 0 | 1 | 11 | 17 | 11 | 3 | 0 | 0 | 0 | 43 |
| Days Max Temp ≤ 32 °F | 2 | 0 | 0 | 0 | 0 | 0 | 0 | 0 | 0 | 0 | 0 | 2 | 4 |
| Days Min Temp ≤ 32 °F | 30 | 26 | 26 | 15 | 2 | 0 | 0 | 0 | 0 | 12 | 27 | 30 | 168 |
| Days Min Temp ≤ 0 °F | 2 | 0 | 0 | 0 | 0 | 0 | 0 | 0 | 0 | 0 | 0 | 1 | 3 |
| Heating Degree Days | 1075 | 806 | 669 | 408 | ***170*** | ***21*** | ***0*** | 3 | 78 | 364 | 728 | 1038 | 5360 |
| Cooling Degree Days | 0 | 0 | 0 | 1 | 14 | 153 | ***282*** | 212 | 57 | 0 | 0 | 0 | 719 |
| Total Precipitation (") | 0.45 | 0.41 | 0.70 | 0.65 | 0.93 | 0.80 | 1.48 | 2.05 | 1.19 | 0.91 | 0.74 | 0.55 | 10.86 |
| Days ≥ 0.1" Precip | 1 | 1 | 2 | 2 | 2 | 2 | 3 | 5 | 3 | 2 | 2 | 2 | 27 |
| Total Snowfall (") | ***3.1*** | na | 1.3 | 0.4 | 0.0 | 0.0 | 0.0 | 0.0 | 0.0 | 0.2 | 0.7 | na | na |
| Days ≥ 1" Snow Depth | na | na | ***0*** | 0 | 0 | 0 | 0 | 0 | 0 | 0 | 0 | na | na |

## ESTANCIA *Torrance County* ELEVATION 6115 ft LAT/LONG 34° 45 ' N / 106° 4 ' W

| | JAN | FEB | MAR | APR | MAY | JUN | JUL | AUG | SEP | OCT | NOV | DEC | YEAR |
|---|---|---|---|---|---|---|---|---|---|---|---|---|---|
| Maximum Temp °F | 46.9 | 52.3 | 60.0 | 68.8 | 77.0 | 87.2 | 89.7 | 86.1 | 80.0 | 70.5 | 57.2 | 47.9 | 68.6 |
| Minimum Temp °F | 17.2 | 20.7 | 25.5 | 31.5 | 39.6 | 47.6 | 53.8 | 52.3 | 44.7 | 33.4 | 24.2 | 16.8 | 33.9 |
| Mean Temp °F | 32.1 | 36.6 | 42.8 | 50.2 | 58.4 | 67.4 | 71.8 | 69.2 | 62.4 | 52.0 | 40.7 | 32.4 | 51.3 |
| Days Max Temp ≥ 90 °F | 0 | 0 | 0 | 0 | 1 | 11 | 18 | 9 | 1 | 0 | 0 | 0 | 40 |
| Days Max Temp ≤ 32 °F | 3 | 1 | 0 | 0 | 0 | 0 | 0 | 0 | 0 | 0 | 0 | 2 | 6 |
| Days Min Temp ≤ 32 °F | 29 | 26 | 25 | 17 | 5 | 0 | 0 | 0 | 2 | 14 | 25 | 30 | 173 |
| Days Min Temp ≤ 0 °F | 2 | 0 | 0 | 0 | 0 | 0 | 0 | 0 | 0 | 0 | 0 | 2 | 4 |
| Heating Degree Days | 1012 | 797 | 682 | 438 | 210 | 30 | 1 | 6 | 101 | 397 | 723 | 1006 | 5403 |
| Cooling Degree Days | 0 | 0 | 0 | 1 | 13 | 133 | 229 | 175 | 37 | 0 | 0 | 0 | 588 |
| Total Precipitation (") | 0.58 | 0.58 | 0.65 | 0.56 | 1.28 | 0.90 | 2.23 | 2.56 | 1.63 | 1.36 | 0.85 | 1.02 | 14.20 |
| Days ≥ 0.1" Precip | 2 | 2 | 2 | 2 | 3 | 2 | 6 | 6 | 4 | 3 | 2 | 3 | 37 |
| Total Snowfall (") | na | na | na | 0.5 | 0.0 | 0.0 | 0.0 | 0.0 | 0.0 | 0.9 | *1.9* | na | na |
| Days ≥ 1" Snow Depth | na | *3* | na | 0 | 0 | 0 | 0 | 0 | 0 | 0 | 1 | *8* | na |

## FAYWOOD *Grant County* ELEVATION 5092 ft LAT/LONG 32° 36 ' N / 107° 53 ' W

| | JAN | FEB | MAR | APR | MAY | JUN | JUL | AUG | SEP | OCT | NOV | DEC | YEAR |
|---|---|---|---|---|---|---|---|---|---|---|---|---|---|
| Maximum Temp °F | 55.3 | 59.4 | 65.4 | 73.9 | 81.7 | 91.2 | 91.5 | 88.5 | 83.6 | 74.5 | 63.3 | 55.6 | 73.7 |
| Minimum Temp °F | 25.7 | 28.0 | 32.6 | 38.5 | 45.9 | 54.9 | 60.6 | 58.9 | 53.1 | 42.2 | 31.2 | 26.0 | 41.5 |
| Mean Temp °F | 40.5 | 43.7 | 49.0 | 56.2 | 63.9 | 73.1 | 76.1 | 73.7 | 68.4 | 58.4 | 47.3 | 40.9 | 57.6 |
| Days Max Temp ≥ 90 °F | 0 | 0 | 0 | 0 | 3 | 18 | 20 | 13 | 4 | 0 | 0 | 0 | 58 |
| Days Max Temp ≤ 32 °F | 0 | 0 | 0 | 0 | 0 | 0 | 0 | 0 | 0 | 0 | 0 | 0 | 0 |
| Days Min Temp ≤ 32 °F | 26 | 21 | 15 | 5 | 1 | 0 | 0 | 0 | 0 | *2* | 17 | 25 | 112 |
| Days Min Temp ≤ 0 °F | 0 | 0 | 0 | 0 | 0 | 0 | 0 | 0 | 0 | 0 | 0 | 0 | 0 |
| Heating Degree Days | 753 | 595 | 488 | 261 | 82 | 5 | 0 | 0 | 21 | 209 | 525 | 742 | 3681 |
| Cooling Degree Days | 0 | 0 | 0 | 6 | 55 | 253 | 330 | 261 | 122 | 10 | 0 | 0 | 1037 |
| Total Precipitation (") | 0.80 | 0.59 | 0.47 | 0.17 | 0.33 | 0.82 | 2.49 | 2.88 | 1.62 | 1.36 | 0.91 | 1.27 | 13.71 |
| Days ≥ 0.1" Precip | 2 | 2 | 2 | 1 | 1 | 2 | 6 | 7 | 4 | 3 | 2 | 3 | 35 |
| Total Snowfall (") | 1.1 | 0.7 | 0.3 | 0.0 | 0.0 | 0.0 | 0.0 | 0.0 | 0.0 | 0.1 | 0.2 | 1.7 | 4.1 |
| Days ≥ 1" Snow Depth | *0* | *0* | 0 | 0 | 0 | 0 | 0 | 0 | 0 | 0 | 0 | *0* | 0 |

## FENCE LAKE *Valencia County* ELEVATION 7073 ft LAT/LONG 34° 39 ' N / 108° 40 ' W

| | JAN | FEB | MAR | APR | MAY | JUN | JUL | AUG | SEP | OCT | NOV | DEC | YEAR |
|---|---|---|---|---|---|---|---|---|---|---|---|---|---|
| Maximum Temp °F | 44.3 | 48.7 | 55.1 | 63.9 | 73.5 | 84.1 | 86.2 | 82.8 | 77.4 | 67.1 | 54.1 | 46.3 | 65.3 |
| Minimum Temp °F | 14.0 | 17.9 | 22.6 | 27.1 | 34.8 | 43.3 | 51.7 | 50.1 | 42.6 | 31.5 | 21.1 | 15.3 | 31.0 |
| Mean Temp °F | 29.2 | 33.3 | 38.9 | 45.5 | 54.2 | 63.7 | 69.0 | 66.4 | 60.0 | 49.3 | 37.6 | 30.8 | 48.2 |
| Days Max Temp ≥ 90 °F | 0 | 0 | 0 | 0 | 0 | 7 | 9 | 2 | 0 | 0 | 0 | 0 | 18 |
| Days Max Temp ≤ 32 °F | 3 | 1 | 0 | 0 | 0 | 0 | 0 | 0 | 0 | 0 | 1 | 2 | 7 |
| Days Min Temp ≤ 32 °F | 31 | 28 | 29 | 23 | 11 | 2 | 0 | 0 | 2 | *17* | 28 | 30 | 201 |
| Days Min Temp ≤ 0 °F | 3 | 1 | 0 | 0 | 0 | 0 | 0 | 0 | 0 | 0 | 0 | 3 | 7 |
| Heating Degree Days | 1104 | 888 | 803 | 578 | 331 | 84 | 5 | 21 | 152 | 479 | 814 | 1053 | 6312 |
| Cooling Degree Days | 0 | 0 | 0 | 0 | 1 | 55 | 115 | 74 | 8 | 0 | 0 | 0 | 253 |
| Total Precipitation (") | 0.96 | 0.89 | 1.17 | 0.64 | 0.56 | 0.53 | 2.32 | 2.55 | 1.48 | 1.29 | 0.96 | 1.01 | 14.36 |
| Days ≥ 0.1" Precip | 3 | 3 | 4 | 2 | 2 | 2 | 6 | 7 | 4 | 3 | 3 | 3 | 42 |
| Total Snowfall (") | na | *5.9* | na | 2.8 | 0.2 | 0.0 | 0.0 | 0.0 | 0.0 | 0.8 | 3.9 | na | na |
| Days ≥ 1" Snow Depth | na | na | na | *0* | 0 | 0 | 0 | 0 | 0 | 0 | *1* | na | na |

## FORT BAYARD *Grant County* ELEVATION 6155 ft LAT/LONG 32° 48 ' N / 108° 9 ' W

| | JAN | FEB | MAR | APR | MAY | JUN | JUL | AUG | SEP | OCT | NOV | DEC | YEAR |
|---|---|---|---|---|---|---|---|---|---|---|---|---|---|
| Maximum Temp °F | 51.7 | 55.3 | 60.9 | 69.3 | 77.3 | 87.3 | 87.3 | 84.2 | 79.7 | 71.2 | 59.6 | 52.2 | 69.7 |
| Minimum Temp °F | 25.3 | 27.4 | 31.2 | 36.7 | 44.1 | 54.0 | 58.7 | 56.8 | 51.4 | 41.9 | 31.4 | 25.9 | 40.4 |
| Mean Temp °F | 38.6 | 41.4 | 46.1 | 53.0 | 60.7 | 70.7 | 73.0 | 70.5 | 65.5 | 56.6 | 45.5 | 39.1 | 55.1 |
| Days Max Temp ≥ 90 °F | 0 | 0 | 0 | 0 | 1 | 11 | 11 | 3 | 1 | 0 | 0 | 0 | 27 |
| Days Max Temp ≤ 32 °F | 0 | 0 | 0 | 0 | 0 | 0 | 0 | 0 | 0 | 0 | 0 | 0 | 0 |
| Days Min Temp ≤ 32 °F | 26 | 21 | 18 | 9 | 1 | 0 | 0 | 0 | 0 | 3 | 17 | 26 | 121 |
| Days Min Temp ≤ 0 °F | 0 | 0 | 0 | 0 | 0 | 0 | 0 | 0 | 0 | 0 | 0 | 0 | 0 |
| Heating Degree Days | 812 | 660 | 579 | 356 | 152 | 13 | 0 | 3 | 46 | 258 | 578 | 796 | 4253 |
| Cooling Degree Days | 0 | 0 | 0 | 3 | 29 | 210 | 252 | 189 | 74 | 2 | 0 | 0 | 759 |
| Total Precipitation (") | 0.90 | 0.73 | 0.56 | 0.19 | 0.53 | 0.86 | 3.36 | 3.40 | 1.94 | 1.45 | 0.94 | 1.35 | 16.21 |
| Days ≥ 0.1" Precip | 2 | 2 | 2 | 1 | 1 | 2 | 8 | 8 | 5 | 3 | 2 | 4 | 40 |
| Total Snowfall (") | *1.6* | 1.1 | 0.5 | 0.1 | 0.0 | 0.0 | 0.0 | 0.0 | 0.0 | 0.0 | 0.4 | na | na |
| Days ≥ 1" Snow Depth | na | *0* | 0 | 0 | 0 | 0 | 0 | 0 | 0 | 0 | 0 | na | na |

## FORT SUMNER *DeBaca County* ELEVATION 4032 ft LAT/LONG 34° 28 ' N / 104° 15 ' W

| | JAN | FEB | MAR | APR | MAY | JUN | JUL | AUG | SEP | OCT | NOV | DEC | YEAR |
|---|---|---|---|---|---|---|---|---|---|---|---|---|---|
| Maximum Temp °F | 53.1 | 57.6 | 66.2 | 74.0 | 81.2 | 89.2 | 91.3 | 88.6 | 82.4 | 73.5 | 62.3 | 53.7 | 72.8 |
| Minimum Temp °F | 22.2 | 25.7 | 32.4 | 40.6 | 49.6 | 58.6 | 63.1 | 61.4 | 53.7 | 41.3 | 30.8 | 22.8 | 41.9 |
| Mean Temp °F | 37.6 | 41.7 | 49.3 | 57.4 | 65.4 | 73.9 | 77.2 | 75.0 | 68.0 | 57.5 | 46.6 | 38.3 | 57.3 |
| Days Max Temp ≥ 90 °F | 0 | 0 | 0 | 0 | 4 | 16 | 20 | 15 | 5 | 1 | 0 | 0 | 61 |
| Days Max Temp ≤ 32 °F | 2 | 1 | 0 | 0 | 0 | 0 | 0 | 0 | 0 | 0 | 0 | 2 | 5 |
| Days Min Temp ≤ 32 °F | 28 | 22 | *16* | 4 | 0 | 0 | 0 | 0 | 0 | 3 | 18 | 27 | 118 |
| Days Min Temp ≤ 0 °F | 1 | 0 | 0 | 0 | 0 | 0 | 0 | 0 | 0 | 0 | 0 | 0 | 1 |
| Heating Degree Days | 840 | 652 | 480 | 235 | 68 | 5 | 0 | 1 | 40 | 240 | 547 | 821 | 3929 |
| Cooling Degree Days | 0 | 0 | 1 | 11 | 82 | 283 | 380 | 331 | 146 | 10 | 0 | 0 | 1244 |
| Total Precipitation (") | 0.45 | 0.42 | 0.49 | 0.55 | 1.39 | 1.63 | 2.22 | 3.37 | 2.09 | 1.40 | 0.65 | 0.53 | 15.19 |
| Days ≥ 0.1" Precip | 2 | 1 | 1 | 1 | 3 | 3 | 4 | 5 | 4 | 3 | 2 | 2 | 31 |
| Total Snowfall (") | *3.2* | *2.6* | *0.7* | 0.2 | 0.0 | 0.0 | 0.0 | 0.0 | 0.0 | 0.2 | 1.2 | *2.7* | 10.8 |
| Days ≥ 1" Snow Depth | *3* | *1* | *0* | 0 | 0 | 0 | 0 | 0 | 0 | 0 | 0 | *2* | 6 |

## FRUITLAND 2 E *San Juan County* ELEVATION 5174 ft LAT/LONG 36° 44 ' N / 108° 24 ' W

| | JAN | FEB | MAR | APR | MAY | JUN | JUL | AUG | SEP | OCT | NOV | DEC | YEAR |
|---|---|---|---|---|---|---|---|---|---|---|---|---|---|
| Maximum Temp °F | 41.9 | 50.5 | 59.7 | 68.8 | 77.8 | 87.7 | 91.6 | 89.0 | 81.7 | 69.8 | 54.6 | 43.4 | 68.0 |
| Minimum Temp °F | 17.1 | 22.0 | 27.6 | 32.8 | 40.7 | 49.4 | 57.5 | 56.2 | 46.9 | 35.4 | 25.7 | 17.9 | 35.8 |
| Mean Temp °F | 29.5 | 36.3 | 43.7 | 50.9 | 59.3 | 68.6 | 74.5 | 72.6 | 64.4 | 52.6 | 40.2 | 30.7 | 51.9 |
| Days Max Temp ≥ 90 °F | 0 | 0 | 0 | 0 | 1 | 12 | 22 | 15 | 2 | 0 | 0 | 0 | 52 |
| Days Max Temp ≤ 32 °F | 5 | 1 | 0 | 0 | 0 | 0 | 0 | 0 | 0 | 0 | 0 | 3 | 9 |
| Days Min Temp ≤ 32 °F | 29 | 25 | 23 | 15 | 4 | 0 | 0 | 0 | 1 | 11 | 24 | 30 | 162 |
| Days Min Temp ≤ 0 °F | 1 | 0 | 0 | 0 | 0 | 0 | 0 | 0 | 0 | 0 | 0 | 1 | 2 |
| Heating Degree Days | 1092 | 804 | 654 | 418 | 185 | 25 | 0 | 2 | 73 | 377 | 738 | 1057 | 5425 |
| Cooling Degree Days | 0 | 0 | 0 | 1 | 16 | 147 | 309 | 261 | 62 | 0 | 0 | 0 | 796 |
| Total Precipitation (") | 0.69 | 0.62 | 0.74 | 0.52 | 0.50 | 0.28 | 0.77 | 0.98 | 0.97 | 0.96 | 0.78 | 0.70 | 8.51 |
| Days ≥ 0.1" Precip | 2 | 2 | 3 | 2 | 2 | 1 | 2 | 3 | 3 | 3 | 3 | 2 | 28 |
| Total Snowfall (") | 4.1 | *3.4* | *1.4* | 0.3 | 0.0 | 0.0 | 0.0 | 0.0 | 0.0 | 0.3 | 1.2 | na | na |
| Days ≥ 1" Snow Depth | na | na | *1* | 0 | 0 | 0 | 0 | 0 | 0 | 0 | 0 | na | na |

## GAGE 4 ESE *Luna County* ELEVATION 4482 ft LAT/LONG 32° 13 ' N / 108° 5 ' W

| | JAN | FEB | MAR | APR | MAY | JUN | JUL | AUG | SEP | OCT | NOV | DEC | YEAR |
|---|---|---|---|---|---|---|---|---|---|---|---|---|---|
| Maximum Temp °F | *56.2* | 61.0 | 68.7 | *76.9* | 85.4 | 94.9 | 94.7 | 91.9 | 86.5 | *77.6* | *65.2* | 55.7 | 76.2 |
| Minimum Temp °F | *25.1* | 28.3 | 32.8 | *38.4* | 47.4 | 57.0 | 63.8 | 62.5 | 55.3 | *43.3* | *32.0* | 25.3 | 42.6 |
| Mean Temp °F | *40.7* | 44.6 | 50.8 | *57.7* | 66.5 | 76.0 | 79.3 | 77.2 | 70.9 | *60.5* | *48.7* | 40.5 | 59.5 |
| Days Max Temp ≥ 90 °F | *0* | 0 | 0 | *1* | 10 | 25 | 26 | 22 | 11 | *1* | *0* | 0 | 96 |
| Days Max Temp ≤ 32 °F | *0* | 0 | 0 | 0 | 0 | 0 | 0 | 0 | 0 | *0* | *0* | 0 | 0 |
| Days Min Temp ≤ 32 °F | *26* | 21 | 14 | *6* | 1 | 0 | 0 | 0 | 0 | *1* | *15* | 26 | 110 |
| Days Min Temp ≤ 0 °F | *0* | 0 | 0 | 0 | 0 | 0 | 0 | 0 | 0 | *0* | *0* | 0 | 0 |
| Heating Degree Days | *747* | 568 | 435 | *223* | 52 | 1 | 0 | 0 | 13 | *156* | *484* | 752 | 3431 |
| Cooling Degree Days | *0* | *0* | *1* | *18* | *135* | 368 | 459 | 404 | 224 | *23* | *0* | 0 | 1632 |
| Total Precipitation (") | 0.70 | 0.59 | 0.44 | 0.13 | 0.25 | 0.55 | 2.22 | 2.20 | 1.27 | *1.09* | 0.80 | 1.45 | 11.69 |
| Days ≥ 0.1" Precip | *2* | 2 | 2 | 1 | 1 | 1 | 5 | 5 | 3 | *2* | 2 | *3* | 29 |
| Total Snowfall (") | na | na | *0.0* | *0.1* | 0.0 | 0.0 | 0.0 | 0.0 | 0.0 | *0.0* | *0.0* | na | na |
| Days ≥ 1" Snow Depth | na | *0* | 0 | *0* | 0 | 0 | 0 | 0 | 0 | *0* | *0* | na | na |

## GASCON *Mora County* ELEVATION 8254 ft LAT/LONG 35° 54 ' N / 105° 26 ' W

| | JAN | FEB | MAR | APR | MAY | JUN | JUL | AUG | SEP | OCT | NOV | DEC | YEAR |
|---|---|---|---|---|---|---|---|---|---|---|---|---|---|
| Maximum Temp °F | 43.5 | 44.5 | 48.5 | 55.6 | 63.2 | 72.8 | 75.5 | 72.8 | 68.0 | 60.5 | 50.2 | 44.4 | 58.3 |
| Minimum Temp °F | 14.8 | 16.5 | 21.2 | 26.3 | 33.2 | 40.7 | 45.1 | 43.8 | 38.3 | 29.6 | 21.4 | 16.1 | 28.9 |
| Mean Temp °F | 29.2 | 30.6 | 34.9 | 41.0 | 48.2 | 56.8 | 60.3 | 58.4 | 53.2 | 45.1 | 35.8 | 30.3 | 43.7 |
| Days Max Temp ≥ 90 °F | 0 | 0 | 0 | 0 | 0 | 0 | 0 | 0 | 0 | 0 | 0 | 0 | 0 |
| Days Max Temp ≤ 32 °F | 4 | 3 | 2 | 0 | 0 | 0 | 0 | 0 | 0 | 0 | 1 | 4 | 14 |
| Days Min Temp ≤ 32 °F | 30 | 28 | 29 | 24 | 14 | 2 | 0 | 0 | 5 | 21 | 28 | 30 | 211 |
| Days Min Temp ≤ 0 °F | 2 | 1 | 1 | 0 | 0 | 0 | 0 | 0 | 0 | 0 | 0 | 2 | 6 |
| Heating Degree Days | 1103 | 966 | 927 | 715 | 515 | 245 | 143 | 199 | 347 | 610 | 869 | 1072 | 7711 |
| Cooling Degree Days | 0 | 0 | 0 | 0 | 0 | 6 | 4 | 0 | 0 | 0 | 0 | 0 | 10 |
| Total Precipitation (") | 0.98 | 1.17 | 1.86 | 1.49 | 2.08 | 1.99 | 4.17 | 4.34 | 2.20 | 1.51 | 1.24 | 1.09 | 24.12 |
| Days ≥ 0.1" Precip | 3 | 3 | 4 | 3 | 5 | 5 | 9 | 10 | 5 | 3 | 2 | 3 | 55 |
| Total Snowfall (") | 20.1 | 21.2 | 29.8 | 16.7 | 4.7 | 0.0 | 0.0 | 0.0 | 0.4 | 7.9 | 14.5 | 20.9 | 136.2 |
| Days ≥ 1" Snow Depth | 13 | *9* | *8* | *3* | 0 | 0 | 0 | 0 | 0 | 2 | 6 | 11 | 52 |

## GILA HOT SPRINGS *Grant County* ELEVATION 5604 ft LAT/LONG 33° 12 ' N / 108° 13 ' W

| | JAN | FEB | MAR | APR | MAY | JUN | JUL | AUG | SEP | OCT | NOV | DEC | YEAR |
|---|---|---|---|---|---|---|---|---|---|---|---|---|---|
| Maximum Temp °F | 54.7 | 58.8 | 64.3 | 72.4 | 79.6 | 89.1 | 88.9 | 85.8 | 81.8 | 73.8 | 63.0 | 54.7 | 72.2 |
| Minimum Temp °F | 19.1 | 21.5 | 25.4 | 30.1 | 36.7 | 44.1 | 53.7 | 53.1 | 45.4 | 33.9 | 23.7 | 19.2 | 33.8 |
| Mean Temp °F | 36.9 | 40.2 | 44.9 | 51.3 | 58.2 | 66.6 | 71.3 | 69.5 | 63.6 | 53.9 | 43.4 | 36.9 | 53.1 |
| Days Max Temp ≥ 90 °F | 0 | 0 | 0 | 0 | 2 | 13 | 15 | 6 | 2 | 0 | 0 | 0 | 38 |
| Days Max Temp ≤ 32 °F | 0 | 0 | 0 | 0 | 0 | 0 | 0 | 0 | 0 | 0 | 0 | 0 | 0 |
| Days Min Temp ≤ 32 °F | 29 | 26 | 27 | 20 | 7 | 1 | 0 | 0 | 1 | 15 | 26 | 29 | 181 |
| Days Min Temp ≤ 0 °F | 1 | 0 | 0 | 0 | 0 | 0 | 0 | 0 | 0 | 0 | 0 | 0 | 1 |
| Heating Degree Days | 864 | 694 | 617 | 406 | 208 | 35 | 1 | 2 | 69 | 338 | 643 | 864 | 4741 |
| Cooling Degree Days | 0 | 0 | 0 | 0 | 5 | 97 | 192 | 157 | 34 | 0 | 0 | 0 | 485 |
| Total Precipitation (") | 1.07 | 1.04 | 0.90 | 0.39 | 0.69 | 0.62 | 2.88 | 3.21 | 1.98 | 1.54 | 1.04 | 1.57 | 16.93 |
| Days ≥ 0.1" Precip | 3 | 3 | 3 | 1 | 2 | 2 | 7 | 8 | 5 | 3 | 2 | 4 | 43 |
| Total Snowfall (") | 3.0 | 1.1 | 0.4 | 0.0 | 0.0 | 0.0 | 0.0 | 0.0 | 0.0 | 0.2 | 0.3 | *1.7* | 6.7 |
| Days ≥ 1" Snow Depth | *2* | *1* | 0 | 0 | 0 | 0 | 0 | 0 | 0 | 0 | 0 | *1* | 4 |

## GRAN QUIVIRA NATL MO *Socorro County* ELEVATION 6624 ft LAT/LONG 34° 16 ' N / 106° 5 ' W

| | JAN | FEB | MAR | APR | MAY | JUN | JUL | AUG | SEP | OCT | NOV | DEC | YEAR |
|---|---|---|---|---|---|---|---|---|---|---|---|---|---|
| Maximum Temp °F | 48.2 | 53.3 | 60.5 | 69.4 | 77.6 | 87.4 | 88.8 | 85.3 | 79.7 | 70.8 | 57.3 | 49.3 | 69.0 |
| Minimum Temp °F | 21.1 | 24.2 | 28.7 | 35.0 | 43.0 | 51.7 | 56.0 | 54.7 | 48.4 | 38.5 | 28.0 | 21.8 | 37.6 |
| Mean Temp °F | 34.7 | 38.8 | 44.6 | 52.2 | 60.3 | 69.5 | 72.4 | 70.0 | 64.1 | 54.7 | 42.5 | 35.6 | 53.3 |
| Days Max Temp ≥ 90 °F | 0 | 0 | 0 | 0 | 1 | 12 | 15 | 7 | 1 | 0 | 0 | 0 | 36 |
| Days Max Temp ≤ 32 °F | 2 | 1 | 0 | 0 | 0 | 0 | 0 | 0 | 0 | 0 | 0 | 1 | 4 |
| Days Min Temp ≤ 32 °F | 29 | 25 | 22 | 11 | 2 | 0 | 0 | 0 | 0 | 6 | 21 | 29 | 145 |
| Days Min Temp ≤ 0 °F | 0 | 0 | 0 | 0 | 0 | 0 | 0 | 0 | 0 | 0 | 0 | 0 | 0 |
| Heating Degree Days | 933 | 735 | 624 | 378 | 159 | 18 | 1 | 6 | 70 | 314 | 668 | 905 | 4811 |
| Cooling Degree Days | 0 | 0 | 0 | 2 | 25 | 181 | 250 | 186 | 58 | 2 | 0 | 0 | 704 |
| Total Precipitation (") | 0.71 | 0.75 | 0.70 | 0.65 | 1.04 | 1.05 | 2.82 | 3.54 | 2.02 | 1.41 | 0.95 | 1.15 | 16.79 |
| Days ≥ 0.1" Precip | 2 | 3 | 2 | 2 | 3 | 3 | 7 | 8 | 5 | 3 | 2 | 3 | 43 |
| Total Snowfall (") | *5.4* | *5.4* | 2.6 | 0.4 | 0.0 | 0.0 | 0.0 | 0.0 | 0.0 | 0.5 | 2.1 | *6.1* | 22.5 |
| Days ≥ 1" Snow Depth | na | na | *1* | 0 | 0 | 0 | 0 | 0 | 0 | 0 | 0 | na | na |

## GRANTS MILAN MUNI AP *Cibola County* ELEVATION 6526 ft LAT/LONG 35° 10 ' N / 107° 54 ' W

| | JAN | FEB | MAR | APR | MAY | JUN | JUL | AUG | SEP | OCT | NOV | DEC | YEAR |
|---|---|---|---|---|---|---|---|---|---|---|---|---|---|
| Maximum Temp °F | 45.3 | 51.1 | 58.0 | 67.5 | 75.4 | 85.8 | 87.6 | 84.4 | 78.7 | 68.7 | 55.4 | 46.4 | 67.0 |
| Minimum Temp °F | 14.1 | 18.7 | 24.1 | 29.8 | 38.4 | 46.7 | 54.7 | 52.6 | 44.5 | 32.8 | 22.7 | 14.9 | 32.8 |
| Mean Temp °F | 29.7 | 34.9 | 41.1 | 48.7 | 56.9 | 66.3 | 71.2 | 68.5 | 61.6 | 50.8 | 39.1 | 30.7 | 50.0 |
| Days Max Temp ≥ 90 °F | 0 | 0 | 0 | 0 | 0 | 10 | 13 | 5 | 1 | 0 | 0 | 0 | 29 |
| Days Max Temp ≤ 32 °F | 3 | 1 | 0 | 0 | 0 | 0 | 0 | 0 | 0 | 0 | 0 | 3 | 7 |
| Days Min Temp ≤ 32 °F | 30 | 27 | 27 | 19 | 6 | 0 | 0 | 0 | 1 | 15 | 26 | 30 | 181 |
| Days Min Temp ≤ 0 °F | 3 | 1 | 0 | 0 | 0 | 0 | 0 | 0 | 0 | 0 | 0 | 2 | 6 |
| Heating Degree Days | 1087 | 844 | 735 | 484 | 249 | 46 | 1 | 8 | 117 | 435 | 770 | 1057 | 5833 |
| Cooling Degree Days | 0 | 0 | 0 | 0 | 8 | 105 | 211 | 151 | 29 | 0 | 0 | 0 | 504 |
| Total Precipitation (") | 0.53 | 0.49 | 0.56 | 0.49 | 0.71 | 0.56 | 1.72 | 2.19 | 1.52 | 1.08 | 0.65 | 0.73 | 11.23 |
| Days ≥ 0.1" Precip | 2 | 2 | 2 | 1 | 2 | 2 | 5 | 6 | 4 | 3 | 2 | 2 | 33 |
| Total Snowfall (") | na | *2.4* | 1.2 | 0.3 | 0.0 | 0.0 | 0.0 | 0.0 | 0.0 | 0.4 | 0.6 | *4.4* | na |
| Days ≥ 1" Snow Depth | na | *1* | 1 | 0 | 0 | 0 | 0 | 0 | 0 | 0 | 0 | *3* | na |

## GRENVILLE *Union County* ELEVATION 5994 ft LAT/LONG 36° 36 ' N / 103° 37 ' W

| | JAN | FEB | MAR | APR | MAY | JUN | JUL | AUG | SEP | OCT | NOV | DEC | YEAR |
|---|---|---|---|---|---|---|---|---|---|---|---|---|---|
| Maximum Temp °F | 46.2 | 49.8 | 56.2 | 65.2 | 72.3 | 82.0 | 86.1 | 83.5 | 77.2 | 68.2 | 55.2 | 47.3 | 65.8 |
| Minimum Temp °F | 16.7 | 19.2 | 24.8 | 32.4 | 41.6 | 50.6 | 55.6 | 54.2 | 46.9 | 35.7 | 25.2 | 18.3 | 35.1 |
| Mean Temp °F | 31.5 | 34.5 | 40.5 | 48.8 | 57.0 | 66.3 | 70.8 | 68.9 | 62.1 | 51.9 | 40.3 | 32.8 | 50.4 |
| Days Max Temp ≥ 90 °F | 0 | 0 | 0 | 0 | 1 | 6 | 10 | 5 | 1 | 0 | 0 | 0 | 23 |
| Days Max Temp ≤ 32 °F | 4 | 3 | 2 | 0 | 0 | 0 | 0 | 0 | 0 | 0 | 1 | 4 | 14 |
| Days Min Temp ≤ 32 °F | 30 | 27 | 26 | 15 | 3 | 0 | 0 | 0 | 1 | 10 | 24 | 30 | 166 |
| Days Min Temp ≤ 0 °F | 2 | 1 | 0 | 0 | 0 | 0 | 0 | 0 | 0 | 0 | 0 | 1 | 4 |
| Heating Degree Days | 1032 | 853 | 753 | 479 | 252 | 55 | 7 | 12 | 117 | 399 | 735 | 991 | 5685 |
| Cooling Degree Days | 0 | 0 | 0 | 1 | 8 | 104 | 192 | 149 | 40 | 1 | 0 | 0 | 495 |
| Total Precipitation (") | 0.25 | 0.31 | 0.71 | 1.00 | 2.45 | 2.36 | 3.22 | 3.01 | 1.94 | 0.80 | 0.58 | 0.32 | 16.95 |
| Days ≥ 0.1" Precip | 1 | 1 | 2 | 3 | 5 | 5 | 7 | 6 | 4 | 2 | 2 | 1 | 39 |
| Total Snowfall (") | *1.8* | *2.0* | 3.5 | 0.7 | 0.1 | 0.0 | 0.0 | 0.0 | 0.2 | 0.4 | 1.0 | *2.3* | 12.0 |
| Days ≥ 1" Snow Depth | na | na | na | *0* | 0 | 0 | 0 | 0 | 0 | 0 | *0* | na | na |

## HACHITA *Grant County* ELEVATION 4505 ft LAT/LONG 31° 55' N / 108° 19' W

| | JAN | FEB | MAR | APR | MAY | JUN | JUL | AUG | SEP | OCT | NOV | DEC | YEAR |
|---|---|---|---|---|---|---|---|---|---|---|---|---|---|
| Maximum Temp °F | 58.1 | 62.5 | 69.2 | 77.6 | 85.4 | 94.6 | 94.2 | 91.3 | 87.1 | 77.9 | 66.1 | 57.9 | 76.8 |
| Minimum Temp °F | 26.4 | 28.5 | 33.2 | 39.0 | 47.2 | 56.9 | 63.5 | 61.8 | 55.2 | 44.0 | 32.7 | 26.9 | 42.9 |
| Mean Temp °F | 42.2 | 45.6 | 51.2 | 58.3 | 66.4 | 75.8 | 78.9 | 76.5 | 71.2 | 61.0 | 49.4 | 42.4 | 59.9 |
| Days Max Temp ≥ 90 °F | 0 | 0 | 0 | 1 | 8 | 25 | 26 | 22 | 12 | 1 | 0 | 0 | 95 |
| Days Max Temp ≤ 32 °F | 0 | 0 | 0 | 0 | 0 | 0 | 0 | 0 | 0 | 0 | 0 | 0 | 0 |
| Days Min Temp ≤ 32 °F | 25 | 20 | 15 | 7 | 1 | 0 | 0 | 0 | 0 | 2 | 15 | 24 | 109 |
| Days Min Temp ≤ 0 °F | 0 | 0 | 0 | 0 | 0 | 0 | 0 | 0 | 0 | 0 | 0 | 0 | 0 |
| Heating Degree Days | 699 | 540 | 420 | 206 | 50 | 1 | 0 | 0 | 10 | 146 | 460 | 692 | 3224 |
| Cooling Degree Days | 0 | 0 | 1 | 18 | 112 | 352 | 443 | 375 | 218 | 25 | 0 | 0 | 1544 |
| Total Precipitation (") | 0.68 | 0.52 | 0.46 | 0.17 | 0.22 | 0.47 | 2.37 | 2.22 | 1.23 | 0.99 | 0.65 | 1.19 | 11.17 |
| Days ≥ 0.1" Precip | 2 | 2 | 2 | 1 | 1 | 1 | 6 | 5 | 3 | 3 | 2 | 3 | 31 |
| Total Snowfall (") | 1.2 | 0.9 | 0.3 | 0.5 | 0.0 | 0.0 | 0.0 | 0.0 | 0.0 | 0.0 | 0.4 | ***1.7*** | 5.0 |
| Days ≥ 1" Snow Depth | ***0*** | 0 | 0 | 0 | 0 | 0 | 0 | 0 | 0 | 0 | 0 | ***1*** | 1 |

## HATCH 2 W *Dona Ana County* ELEVATION 4042 ft LAT/LONG 32° 40' N / 107° 9' W

| | JAN | FEB | MAR | APR | MAY | JUN | JUL | AUG | SEP | OCT | NOV | DEC | YEAR |
|---|---|---|---|---|---|---|---|---|---|---|---|---|---|
| Maximum Temp °F | 59.3 | 63.8 | 70.4 | 78.4 | 86.2 | 95.1 | 95.4 | 92.9 | 87.8 | 79.2 | 68.2 | 59.6 | 78.0 |
| Minimum Temp °F | 23.6 | 27.1 | 33.2 | 39.7 | 47.6 | 55.9 | 62.4 | 60.6 | 53.4 | 40.7 | 29.8 | 24.0 | 41.5 |
| Mean Temp °F | 41.5 | 45.4 | 51.8 | 59.1 | 66.9 | 75.5 | 78.8 | 76.7 | 70.6 | 60.0 | 49.0 | 41.8 | 59.8 |
| Days Max Temp ≥ 90 °F | 0 | 0 | 0 | 1 | 10 | 25 | 27 | 24 | 14 | 2 | 0 | 0 | 103 |
| Days Max Temp ≤ 32 °F | 0 | 0 | 0 | 0 | 0 | 0 | 0 | 0 | 0 | 0 | 0 | 0 | 0 |
| Days Min Temp ≤ 32 °F | 27 | 22 | 14 | 5 | 0 | 0 | 0 | 0 | 0 | 4 | 20 | 27 | 119 |
| Days Min Temp ≤ 0 °F | 0 | 0 | 0 | 0 | 0 | 0 | 0 | 0 | 0 | 0 | 0 | 0 | 0 |
| Heating Degree Days | 722 | 545 | 404 | 184 | 39 | 1 | 0 | 0 | 13 | 170 | 473 | 712 | 3263 |
| Cooling Degree Days | 0 | 0 | 1 | 13 | 107 | 325 | 428 | 377 | 192 | 17 | 0 | 0 | 1460 |
| Total Precipitation (") | 0.59 | 0.40 | 0.21 | 0.17 | 0.46 | 0.59 | 2.05 | 2.38 | 1.48 | 1.04 | 0.58 | 0.92 | 10.87 |
| Days ≥ 0.1" Precip | 2 | 1 | 1 | 1 | 1 | 1 | 5 | 5 | 4 | 2 | 1 | 3 | 27 |
| Total Snowfall (") | ***1.4*** | 0.3 | 0.3 | 0.2 | 0.0 | 0.0 | 0.0 | 0.0 | 0.0 | 0.0 | 0.6 | 0.9 | 3.7 |
| Days ≥ 1" Snow Depth | na | 0 | 0 | 0 | 0 | 0 | 0 | 0 | 0 | 0 | 0 | ***0*** | na |

## HILLSBORO *Sierra County* ELEVATION 5335 ft LAT/LONG 32° 54' N / 107° 34' W

| | JAN | FEB | MAR | APR | MAY | JUN | JUL | AUG | SEP | OCT | NOV | DEC | YEAR |
|---|---|---|---|---|---|---|---|---|---|---|---|---|---|
| Maximum Temp °F | 54.4 | 59.4 | 65.9 | 73.9 | 81.6 | 90.9 | 90.7 | 87.6 | 82.4 | 73.9 | 62.6 | 54.2 | 73.1 |
| Minimum Temp °F | 24.3 | 27.5 | 32.8 | 39.1 | 46.4 | 54.7 | 60.6 | 58.6 | 51.8 | 41.0 | 30.8 | 24.7 | 41.0 |
| Mean Temp °F | 39.4 | 43.5 | 49.4 | 56.5 | 64.0 | 72.8 | 75.7 | 73.1 | 67.1 | 57.5 | 46.7 | 39.5 | 57.1 |
| Days Max Temp ≥ 90 °F | 0 | 0 | 0 | 0 | 3 | 18 | 19 | 12 | 4 | 0 | 0 | 0 | 56 |
| Days Max Temp ≤ 32 °F | 0 | 0 | 0 | 0 | 0 | 0 | 0 | 0 | 0 | 0 | 0 | 0 | 0 |
| Days Min Temp ≤ 32 °F | 26 | 21 | 15 | 6 | 0 | 0 | 0 | 0 | 0 | 3 | 18 | 26 | 115 |
| Days Min Temp ≤ 0 °F | 0 | 0 | 0 | 0 | 0 | 0 | 0 | 0 | 0 | 0 | 0 | 0 | 0 |
| Heating Degree Days | 786 | 602 | 476 | 252 | 78 | 4 | 0 | 0 | 29 | 232 | 541 | 784 | 3784 |
| Cooling Degree Days | 0 | 0 | 0 | 5 | 49 | 242 | 326 | 257 | 93 | 3 | 0 | 0 | 975 |
| Total Precipitation (") | 0.63 | 0.47 | 0.26 | 0.23 | 0.66 | 0.76 | 2.56 | 3.08 | 2.26 | 1.38 | 0.73 | 1.19 | 14.21 |
| Days ≥ 0.1" Precip | 2 | 1 | 1 | 1 | 2 | 2 | 6 | 7 | 4 | 3 | 1 | 2 | 32 |
| Total Snowfall (") | 2.5 | 1.5 | 0.5 | 0.1 | 0.0 | 0.0 | 0.0 | 0.0 | 0.0 | 0.8 | 1.0 | 2.4 | 8.8 |
| Days ≥ 1" Snow Depth | 2 | 0 | 0 | 0 | 0 | 0 | 0 | 0 | 0 | 0 | 0 | 2 | 4 |

## HOBBS *Lea County* ELEVATION 3622 ft LAT/LONG 32° 42' N / 103° 8' W

| | JAN | FEB | MAR | APR | MAY | JUN | JUL | AUG | SEP | OCT | NOV | DEC | YEAR |
|---|---|---|---|---|---|---|---|---|---|---|---|---|---|
| Maximum Temp °F | 56.9 | 62.1 | 70.1 | 78.6 | 85.2 | 92.3 | 92.9 | 90.8 | 85.0 | 77.2 | 65.8 | 58.9 | 76.3 |
| Minimum Temp °F | 28.9 | 32.3 | 38.8 | 47.3 | 55.2 | 63.3 | 66.7 | 65.3 | 59.0 | 48.7 | 37.7 | 30.4 | 47.8 |
| Mean Temp °F | 43.0 | 47.2 | 54.5 | 63.0 | 70.2 | 77.8 | 79.8 | 78.1 | 72.0 | 63.0 | 51.8 | 44.7 | 62.1 |
| Days Max Temp ≥ 90 °F | 0 | 0 | 0 | 1 | 9 | 21 | 24 | 20 | 9 | 2 | 0 | 0 | 86 |
| Days Max Temp ≤ 32 °F | 1 | 1 | 0 | 0 | 0 | 0 | 0 | 0 | 0 | 0 | 0 | 1 | 3 |
| Days Min Temp ≤ 32 °F | 21 | 15 | 7 | 1 | 0 | 0 | 0 | 0 | 0 | 1 | 8 | 18 | 71 |
| Days Min Temp ≤ 0 °F | 0 | 0 | 0 | 0 | 0 | 0 | 0 | 0 | 0 | 0 | 0 | 0 | 0 |
| Heating Degree Days | 676 | 495 | 329 | 114 | 24 | 1 | 0 | 1 | 21 | 118 | 390 | 624 | 2793 |
| Cooling Degree Days | 0 | 1 | 7 | 59 | 197 | 413 | 479 | 435 | 242 | 61 | 2 | 0 | 1896 |
| Total Precipitation (") | 0.46 | 0.65 | 0.52 | 0.77 | 2.47 | 1.82 | 2.63 | 2.69 | 2.88 | 1.46 | 0.85 | 0.75 | 17.95 |
| Days ≥ 0.1" Precip | 2 | 2 | 1 | 2 | 4 | 3 | 4 | 4 | 4 | 2 | 2 | 2 | 32 |
| Total Snowfall (") | 1.8 | 1.6 | 0.8 | 0.4 | 0.0 | 0.0 | 0.0 | 0.0 | 0.0 | 0.2 | 1.0 | 1.2 | 7.0 |
| Days ≥ 1" Snow Depth | ***0*** | 0 | 0 | 0 | 0 | 0 | 0 | 0 | 0 | 0 | 0 | ***0*** | 0 |

## HOPE *Eddy County* ELEVATION 4091 ft LAT/LONG 32° 49 ' N / 104° 44 ' W

| | JAN | FEB | MAR | APR | MAY | JUN | JUL | AUG | SEP | OCT | NOV | DEC | YEAR |
|---|---|---|---|---|---|---|---|---|---|---|---|---|---|
| Maximum Temp °F | 56.9 | 62.8 | *69.5* | 77.6 | 84.9 | 92.7 | 93.1 | 90.8 | 84.7 | 77.5 | 66.1 | 58.9 | 76.3 |
| Minimum Temp °F | 27.2 | 29.8 | 35.0 | 41.9 | 50.6 | 59.1 | 63.1 | 62.2 | 55.5 | 45.8 | 34.5 | 27.6 | 44.4 |
| Mean Temp °F | 42.2 | 46.3 | *52.3* | *60.0* | 67.8 | 75.9 | 78.1 | 76.5 | 70.1 | 61.7 | 50.3 | 43.4 | 60.4 |
| Days Max Temp ≥ 90 °F | 0 | 0 | *0* | 1 | 9 | 21 | 23 | 20 | 9 | 2 | 0 | 0 | 85 |
| Days Max Temp ≤ 32 °F | 1 | 0 | *0* | 0 | 0 | 0 | 0 | 0 | 0 | 0 | 0 | 1 | 2 |
| Days Min Temp ≤ 32 °F | 23 | 18 | 12 | 3 | 0 | 0 | 0 | 0 | 0 | 2 | 13 | 23 | 94 |
| Days Min Temp ≤ 0 °F | 0 | 0 | 0 | 0 | 0 | 0 | 0 | 0 | 0 | 0 | 0 | 0 | 0 |
| Heating Degree Days | 698 | 522 | *389* | *167* | 37 | 1 | 0 | 1 | 26 | 139 | 433 | 664 | 3077 |
| Cooling Degree Days | 0 | 1 | *4* | *34* | 133 | 346 | 417 | 388 | 200 | 40 | 1 | 0 | 1564 |
| Total Precipitation (") | 0.43 | 0.48 | 0.27 | 0.65 | 1.25 | 1.90 | 2.03 | 2.52 | 2.66 | 1.16 | 0.62 | 0.60 | 14.57 |
| Days ≥ 0.1" Precip | 1 | 1 | 1 | 1 | 3 | 3 | 4 | 5 | 4 | 2 | 1 | 1 | 27 |
| Total Snowfall (") | *3.2* | *2.1* | *0.6* | 1.5 | 0.0 | 0.0 | 0.0 | 0.0 | 0.0 | 0.5 | 1.7 | *3.7* | 13.3 |
| Days ≥ 1" Snow Depth | na | na | *0* | 0 | 0 | 0 | 0 | 0 | 0 | *0* | 0 | na | na |

## JAL *Lea County* ELEVATION 3153 ft LAT/LONG 32° 7 ' N / 103° 12 ' W

| | JAN | FEB | MAR | APR | MAY | JUN | JUL | AUG | SEP | OCT | NOV | DEC | YEAR |
|---|---|---|---|---|---|---|---|---|---|---|---|---|---|
| Maximum Temp °F | 59.4 | 65.3 | 73.5 | 82.0 | 88.2 | 94.6 | 95.3 | 93.3 | 87.3 | 79.5 | 68.2 | 61.3 | 79.0 |
| Minimum Temp °F | 28.5 | 32.4 | 39.3 | 48.2 | 56.7 | 64.9 | 68.2 | 66.7 | 60.3 | 48.9 | 37.3 | 30.0 | 48.4 |
| Mean Temp °F | 44.0 | 48.9 | 56.4 | 65.1 | 72.4 | 79.8 | 81.8 | 80.0 | 73.8 | 64.3 | 52.8 | 45.7 | 63.8 |
| Days Max Temp ≥ 90 °F | 0 | 0 | 1 | 5 | 15 | 24 | 26 | 24 | 13 | 4 | 0 | 0 | 112 |
| Days Max Temp ≤ 32 °F | 1 | 0 | 0 | 0 | 0 | 0 | 0 | 0 | 0 | 0 | 0 | 0 | 1 |
| Days Min Temp ≤ 32 °F | 22 | 15 | 6 | 1 | 0 | 0 | 0 | 0 | 0 | 1 | 9 | 19 | 73 |
| Days Min Temp ≤ 0 °F | 0 | 0 | 0 | 0 | 0 | 0 | 0 | 0 | 0 | 0 | 0 | 0 | 0 |
| Heating Degree Days | 645 | 450 | 274 | 79 | 13 | 0 | 0 | 0 | 14 | 94 | 364 | 592 | 2525 |
| Cooling Degree Days | 0 | 1 | 12 | 83 | 243 | 438 | 511 | 472 | 279 | 75 | 3 | 0 | 2117 |
| Total Precipitation (") | 0.43 | 0.54 | 0.43 | 0.65 | 1.54 | 1.68 | 1.87 | 2.08 | 2.35 | 1.25 | 0.74 | 0.54 | 14.10 |
| Days ≥ 0.1" Precip | 1 | 2 | 1 | 1 | 3 | 3 | 3 | 4 | 4 | 2 | 1 | 1 | 26 |
| Total Snowfall (") | 1.7 | 0.5 | 0.5 | 0.0 | 0.0 | 0.0 | 0.0 | 0.0 | 0.0 | 0.1 | 1.0 | 1.1 | 4.9 |
| Days ≥ 1" Snow Depth | 1 | 0 | 0 | 0 | 0 | 0 | 0 | 0 | 0 | 0 | 0 | 0 | 1 |

## JEMEZ SPRINGS *Sandoval County* ELEVATION 6106 ft LAT/LONG 35° 47 ' N / 106° 41 ' W

| | JAN | FEB | MAR | APR | MAY | JUN | JUL | AUG | SEP | OCT | NOV | DEC | YEAR |
|---|---|---|---|---|---|---|---|---|---|---|---|---|---|
| Maximum Temp °F | 45.7 | 50.7 | 57.2 | 66.4 | 74.7 | 85.0 | 87.9 | 84.5 | 78.6 | 68.7 | 55.2 | 46.5 | 66.8 |
| Minimum Temp °F | 19.4 | 23.4 | 28.7 | 34.7 | 42.1 | 50.5 | 56.0 | 54.4 | 47.6 | 37.8 | 28.1 | 20.9 | 37.0 |
| Mean Temp °F | 32.6 | 37.1 | 42.9 | 50.6 | 58.4 | 67.8 | 72.0 | 69.5 | 63.1 | 53.3 | 41.7 | 33.7 | 51.9 |
| Days Max Temp ≥ 90 °F | 0 | 0 | 0 | 0 | 0 | 8 | 12 | 5 | 1 | 0 | 0 | 0 | 26 |
| Days Max Temp ≤ 32 °F | 2 | 1 | 0 | 0 | 0 | 0 | 0 | 0 | 0 | 0 | 0 | 1 | 4 |
| Days Min Temp ≤ 32 °F | 30 | 26 | 23 | 11 | 2 | 0 | 0 | 0 | 0 | 6 | 22 | 30 | 150 |
| Days Min Temp ≤ 0 °F | 1 | 0 | 0 | 0 | 0 | 0 | 0 | 0 | 0 | 0 | 0 | 0 | 1 |
| Heating Degree Days | 998 | 781 | 677 | 426 | 206 | 32 | 1 | 5 | 83 | 356 | 694 | 963 | 5222 |
| Cooling Degree Days | 0 | 0 | 0 | 0 | 11 | 136 | 228 | 162 | 41 | 0 | 0 | 0 | 578 |
| Total Precipitation (") | 1.20 | 0.89 | 1.19 | 0.86 | 1.14 | 1.13 | 2.61 | 3.14 | 1.85 | 1.41 | 1.16 | 1.09 | 17.67 |
| Days ≥ 0.1" Precip | 3 | 3 | 3 | 2 | 3 | 3 | 7 | 8 | 5 | 3 | 3 | 3 | 46 |
| Total Snowfall (") | 8.3 | 6.4 | 5.0 | 1.6 | 0.2 | 0.0 | 0.0 | 0.0 | 0.0 | 0.2 | 2.4 | 7.6 | 31.7 |
| Days ≥ 1" Snow Depth | *11* | 4 | 1 | 0 | 0 | 0 | 0 | 0 | 0 | 0 | 1 | 7 | 24 |

## JORNADA EXP RANGE *Dona Ana County* ELEVATION 4272 ft LAT/LONG 32° 37 ' N / 106° 44 ' W

| | JAN | FEB | MAR | APR | MAY | JUN | JUL | AUG | SEP | OCT | NOV | DEC | YEAR |
|---|---|---|---|---|---|---|---|---|---|---|---|---|---|
| Maximum Temp °F | 56.5 | 61.3 | 68.5 | 76.9 | 84.9 | 94.2 | 95.2 | 91.8 | 86.3 | 77.5 | 65.5 | 56.5 | 76.3 |
| Minimum Temp °F | 21.1 | 24.8 | 30.2 | 37.3 | 45.6 | 55.1 | 62.8 | 60.8 | 53.3 | 40.2 | 28.1 | 21.8 | 40.1 |
| Mean Temp °F | 38.9 | 43.1 | 49.4 | 57.1 | 65.3 | 74.7 | 79.0 | 76.4 | 69.8 | 58.9 | 46.8 | 39.2 | 58.2 |
| Days Max Temp ≥ 90 °F | 0 | 0 | 0 | 1 | 8 | 24 | 25 | 22 | 10 | 1 | 0 | 0 | 91 |
| Days Max Temp ≤ 32 °F | 0 | 0 | 0 | 0 | 0 | 0 | 0 | 0 | 0 | 0 | 0 | 0 | 0 |
| Days Min Temp ≤ 32 °F | 27 | 22 | 19 | 9 | 1 | 0 | 0 | 0 | 0 | 5 | 21 | 26 | 130 |
| Days Min Temp ≤ 0 °F | 0 | 0 | 0 | 0 | 0 | 0 | 0 | 0 | 0 | 0 | 0 | 0 | 0 |
| Heating Degree Days | 802 | 612 | 478 | 239 | 67 | 3 | 0 | 0 | 17 | 200 | 538 | 793 | 3749 |
| Cooling Degree Days | 0 | 0 | 0 | 9 | 80 | 298 | 423 | *357* | 163 | 11 | 0 | 0 | 1341 |
| Total Precipitation (") | 0.58 | 0.39 | 0.29 | 0.21 | 0.47 | 0.70 | 1.93 | 2.33 | 1.32 | 0.99 | 0.61 | 0.95 | 10.77 |
| Days ≥ 0.1" Precip | 2 | 2 | 1 | 1 | 1 | 1 | 5 | 5 | 3 | 3 | 2 | 3 | 29 |
| Total Snowfall (") | 1.1 | 0.3 | 0.1 | 0.0 | 0.0 | 0.0 | 0.0 | 0.0 | 0.0 | 0.0 | 0.3 | *1.5* | 3.3 |
| Days ≥ 1" Snow Depth | 0 | 0 | 0 | 0 | 0 | 0 | 0 | 0 | 0 | 0 | 0 | 1 | 1 |

## LAGUNA *Cibola County* ELEVATION 5843 ft LAT/LONG 35° 2 ' N / 107° 24 ' W

| | JAN | FEB | MAR | APR | MAY | JUN | JUL | AUG | SEP | OCT | NOV | DEC | YEAR |
|---|---|---|---|---|---|---|---|---|---|---|---|---|---|
| Maximum Temp °F | 47.2 | 52.8 | 60.0 | 68.9 | 77.2 | 88.0 | 90.2 | 86.9 | 80.5 | 70.4 | 57.8 | 48.4 | 69.0 |
| Minimum Temp °F | 19.1 | 23.1 | 28.3 | 34.4 | 43.2 | 52.4 | 59.0 | 57.7 | 49.5 | 37.8 | 27.4 | 19.9 | 37.7 |
| Mean Temp °F | 33.1 | 38.0 | 44.1 | 51.7 | 60.2 | 70.2 | 74.6 | 72.3 | 65.0 | 54.1 | 42.6 | 34.2 | 53.3 |
| Days Max Temp ≥ 90 °F | 0 | 0 | 0 | 0 | 1 | 13 | 18 | 10 | 2 | 0 | 0 | 0 | 44 |
| Days Max Temp ≤ 32 °F | 1 | 0 | 0 | 0 | 0 | 0 | 0 | 0 | 0 | 0 | 0 | 1 | 2 |
| Days Min Temp ≤ 32 °F | 29 | 25 | 22 | 12 | 2 | 0 | 0 | 0 | 0 | 7 | 22 | 29 | 148 |
| Days Min Temp ≤ 0 °F | 1 | 0 | 0 | 0 | 0 | 0 | 0 | 0 | 0 | 0 | 0 | 0 | 1 |
| Heating Degree Days | 982 | 756 | 640 | 394 | 166 | 17 | 0 | 2 | 61 | 334 | 664 | 949 | 4965 |
| Cooling Degree Days | 0 | 0 | 0 | 3 | 25 | 194 | 306 | 260 | 76 | 2 | 0 | 0 | 866 |
| Total Precipitation (") | 0.51 | 0.44 | 0.42 | 0.40 | 0.81 | 0.55 | 1.76 | 2.13 | 1.49 | 1.19 | 0.42 | 0.59 | 10.71 |
| Days ≥ 0.1" Precip | 2 | 1 | 1 | 1 | 2 | 1 | 5 | 5 | 4 | 2 | 1 | 2 | 27 |
| Total Snowfall (") | 4.1 | 2.2 | 1.5 | 0.8 | 0.2 | 0.0 | 0.0 | 0.0 | 0.0 | 0.7 | 0.6 | 3.6 | 13.7 |
| Days ≥ 1" Snow Depth | 3 | 1 | 1 | 0 | 0 | 0 | 0 | 0 | 0 | 0 | 1 | 4 | 10 |

## LAKE MALOYA *Colfax County* ELEVATION 7405 ft LAT/LONG 36° 59 ' N / 104° 22 ' W

| | JAN | FEB | MAR | APR | MAY | JUN | JUL | AUG | SEP | OCT | NOV | DEC | YEAR |
|---|---|---|---|---|---|---|---|---|---|---|---|---|---|
| Maximum Temp °F | 41.8 | 43.9 | 49.4 | 57.4 | 65.9 | 75.1 | 78.9 | 76.1 | 70.2 | 61.3 | 48.7 | 42.0 | 59.2 |
| Minimum Temp °F | 8.8 | 11.2 | 17.8 | 25.8 | 34.8 | 42.8 | 47.8 | 46.5 | 39.4 | 29.9 | 19.7 | 11.4 | 28.0 |
| Mean Temp °F | 25.5 | 27.6 | 33.8 | 41.6 | 50.4 | 59.0 | 63.4 | 61.3 | 54.8 | 45.6 | 34.2 | 26.7 | 43.7 |
| Days Max Temp ≥ 90 °F | 0 | 0 | 0 | 0 | 0 | 0 | 0 | 0 | 0 | 0 | 0 | 0 | 0 |
| Days Max Temp ≤ 32 °F | 6 | 4 | 2 | 1 | 0 | 0 | 0 | 0 | 0 | 0 | 3 | 6 | 22 |
| Days Min Temp ≤ 32 °F | 31 | 28 | 31 | 26 | 10 | 1 | 0 | 0 | 3 | 20 | 29 | 31 | 210 |
| Days Min Temp ≤ 0 °F | 6 | 4 | 1 | 0 | 0 | 0 | 0 | 0 | 0 | 0 | 1 | 4 | 16 |
| Heating Degree Days | 1218 | 1050 | 962 | 695 | 447 | 183 | 64 | 113 | 299 | 594 | 917 | 1180 | 7722 |
| Cooling Degree Days | 0 | 0 | 0 | 0 | 0 | 10 | 18 | 6 | 0 | 0 | 0 | 0 | 34 |
| Total Precipitation (") | 0.89 | 1.12 | 1.79 | 1.71 | 2.99 | 2.73 | 3.77 | 3.73 | 2.19 | 1.38 | 1.47 | 0.85 | 24.62 |
| Days ≥ 0.1" Precip | 3 | 3 | 5 | 4 | 7 | 6 | 8 | 8 | 5 | 4 | 3 | 3 | 59 |
| Total Snowfall (") | 15.2 | 15.9 | 22.7 | 12.4 | 1.9 | 0.0 | 0.0 | 0.0 | 0.5 | 5.2 | 15.4 | 15.5 | 104.7 |
| Days ≥ 1" Snow Depth | na | na | *7* | na | 0 | 0 | 0 | 0 | 0 | *1* | *5* | na | na |

## LAS VEGAS MUNI AP *San Miguel County* ELEVATION 6863 ft LAT/LONG 35° 39 ' N / 105° 9 ' W

| | JAN | FEB | MAR | APR | MAY | JUN | JUL | AUG | SEP | OCT | NOV | DEC | YEAR |
|---|---|---|---|---|---|---|---|---|---|---|---|---|---|
| Maximum Temp °F | 45.3 | 48.8 | 54.6 | 62.8 | 70.7 | 80.3 | 82.9 | 80.2 | 74.3 | 65.6 | 53.9 | 46.5 | 63.8 |
| Minimum Temp °F | 18.1 | 20.8 | 25.7 | 32.6 | 40.6 | 49.2 | 53.8 | 52.3 | 46.0 | 35.8 | 25.9 | 19.3 | 35.0 |
| Mean Temp °F | 31.7 | 34.8 | 40.2 | 47.7 | 55.7 | 64.8 | 68.4 | 66.3 | 60.2 | 50.7 | 39.9 | 32.9 | 49.4 |
| Days Max Temp ≥ 90 °F | 0 | 0 | 0 | 0 | 0 | 3 | 3 | 1 | 0 | 0 | 0 | 0 | 7 |
| Days Max Temp ≤ 32 °F | 4 | 2 | 1 | 0 | 0 | 0 | 0 | 0 | 0 | 0 | 1 | 4 | 12 |
| Days Min Temp ≤ 32 °F | 29 | 26 | 25 | 14 | 3 | 0 | 0 | 0 | 1 | 10 | 23 | 29 | 160 |
| Days Min Temp ≤ 0 °F | 1 | 1 | 0 | 0 | 0 | 0 | 0 | 0 | 0 | 0 | 0 | 1 | 3 |
| Heating Degree Days | 1025 | 845 | 762 | 512 | 287 | 67 | 12 | 29 | 150 | 437 | 745 | 987 | 5858 |
| Cooling Degree Days | 0 | 0 | 0 | 0 | 5 | 77 | 126 | 85 | 17 | 1 | 0 | 0 | 311 |
| Total Precipitation (") | 0.36 | 0.38 | 0.66 | 0.90 | 1.94 | 1.99 | 3.51 | 3.70 | 1.96 | 1.04 | 0.75 | 0.53 | 17.72 |
| Days ≥ 0.1" Precip | 1 | 1 | 2 | 2 | 4 | 5 | 7 | 8 | 4 | 3 | 2 | 2 | 41 |
| Total Snowfall (") | 6.4 | 6.5 | 7.9 | 3.5 | 0.7 | 0.1 | 0.0 | 0.0 | 0.0 | 1.8 | 4.1 | 8.0 | 39.0 |
| Days ≥ 1" Snow Depth | 11 | 6 | 4 | 2 | 0 | 0 | 0 | 0 | 0 | 1 | 3 | 9 | 36 |

## LORDSBURG 4 SE *Hidalgo County* ELEVATION 4252 ft LAT/LONG 32° 21 ' N / 108° 42 ' W

| | JAN | FEB | MAR | APR | MAY | JUN | JUL | AUG | SEP | OCT | NOV | DEC | YEAR |
|---|---|---|---|---|---|---|---|---|---|---|---|---|---|
| Maximum Temp °F | 58.1 | 62.9 | 69.5 | 78.3 | 86.4 | 95.8 | 96.0 | 93.2 | 88.0 | 78.6 | 66.2 | 58.0 | 77.6 |
| Minimum Temp °F | 25.2 | 27.2 | 32.5 | 38.2 | 46.6 | 57.0 | 64.1 | 62.4 | 55.0 | 42.6 | 30.2 | 25.2 | 42.2 |
| Mean Temp °F | 41.6 | 45.0 | 51.1 | 58.3 | 66.5 | 76.4 | 80.0 | 77.8 | 71.5 | 60.6 | 48.2 | 41.6 | 59.9 |
| Days Max Temp ≥ 90 °F | 0 | 0 | 0 | 1 | 11 | 26 | 28 | 25 | 13 | 2 | 0 | 0 | 106 |
| Days Max Temp ≤ 32 °F | 0 | 0 | 0 | 0 | 0 | 0 | 0 | 0 | 0 | 0 | 0 | 0 | 0 |
| Days Min Temp ≤ 32 °F | 26 | 22 | 16 | 7 | 1 | 0 | 0 | 0 | 0 | 3 | 18 | 25 | 118 |
| Days Min Temp ≤ 0 °F | 0 | 0 | 0 | 0 | 0 | 0 | 0 | 0 | 0 | 0 | 0 | 0 | 0 |
| Heating Degree Days | 720 | 556 | 425 | 208 | 47 | 1 | 0 | 0 | 11 | 155 | 498 | 718 | 3339 |
| Cooling Degree Days | 0 | 0 | 0 | 14 | 110 | 363 | 466 | 400 | 215 | 23 | 0 | 0 | 1591 |
| Total Precipitation (") | 0.91 | 0.74 | 0.79 | 0.24 | 0.38 | 0.53 | 1.88 | 2.09 | 1.32 | 1.20 | 0.80 | 1.41 | 12.29 |
| Days ≥ 0.1" Precip | 3 | 2 | 3 | 1 | 1 | 1 | 5 | 5 | 3 | 3 | 2 | 3 | 32 |
| Total Snowfall (") | 0.9 | 0.9 | 0.6 | 0.1 | 0.0 | 0.0 | 0.0 | 0.0 | 0.0 | 0.0 | 0.3 | 1.2 | 4.0 |
| Days ≥ 1" Snow Depth | *0* | 0 | 0 | 0 | 0 | 0 | 0 | 0 | 0 | 0 | 0 | 0 | 0 |

## LOS ALAMOS *Los Alamos County* ELEVATION 7365 ft LAT/LONG 35° 52 ' N / 106° 19 ' W

| | JAN | FEB | MAR | APR | MAY | JUN | JUL | AUG | SEP | OCT | NOV | DEC | YEAR |
|---|---|---|---|---|---|---|---|---|---|---|---|---|---|
| Maximum Temp °F | 39.4 | 43.5 | 50.0 | 58.7 | 67.3 | 77.9 | 80.5 | 77.2 | 71.1 | 61.2 | 48.2 | 40.5 | 59.6 |
| Minimum Temp °F | 17.6 | 21.3 | 26.8 | 33.4 | 41.6 | 51.0 | 55.0 | 53.2 | 46.9 | 37.0 | 26.4 | 19.2 | 35.8 |
| Mean Temp °F | 28.5 | 32.4 | 38.5 | 46.1 | 54.5 | 64.5 | 67.8 | 65.2 | 59.0 | 49.1 | 37.3 | 29.9 | 47.7 |
| Days Max Temp ≥ 90 °F | 0 | 0 | 0 | 0 | 0 | 1 | 1 | 0 | 0 | 0 | 0 | 0 | 2 |
| Days Max Temp ≤ 32 °F | 6 | 3 | 1 | 0 | 0 | 0 | 0 | 0 | 0 | 0 | 2 | 6 | 18 |
| Days Min Temp ≤ 32 °F | 31 | 27 | 24 | 13 | 3 | 0 | 0 | 0 | 0 | 8 | 24 | 30 | 160 |
| Days Min Temp ≤ 0 °F | 1 | 0 | 0 | 0 | 0 | 0 | 0 | 0 | 0 | 0 | 0 | 0 | 1 |
| Heating Degree Days | 1123 | 913 | 817 | 561 | 323 | 80 | 16 | 42 | 182 | 486 | 824 | 1082 | 6449 |
| Cooling Degree Days | 0 | 0 | 0 | 0 | 5 | 76 | 110 | 55 | 11 | 0 | 0 | 0 | 257 |
| Total Precipitation (") | 0.90 | 0.76 | 1.26 | 0.95 | 1.43 | 1.44 | 3.27 | 3.74 | 2.09 | 1.33 | 1.13 | 1.07 | 19.37 |
| Days ≥ 0.1" Precip | 2 | 2 | 3 | 3 | 4 | 3 | 8 | 9 | 5 | 3 | 3 | 3 | 48 |
| Total Snowfall (") | 13.0 | 9.4 | 11.2 | 4.2 | 0.9 | 0.0 | 0.0 | 0.0 | 0.1 | 2.4 | 4.4 | 12.3 | 57.9 |
| Days ≥ 1" Snow Depth | *21* | 11 | *4* | 1 | 0 | 0 | 0 | 0 | 0 | 1 | *3* | 13 | 54 |

## LOS LUNAS 3 SSW *Valencia County* ELEVATION 4842 ft LAT/LONG 34° 46 ' N / 106° 45 ' W

| | JAN | FEB | MAR | APR | MAY | JUN | JUL | AUG | SEP | OCT | NOV | DEC | YEAR |
|---|---|---|---|---|---|---|---|---|---|---|---|---|---|
| Maximum Temp °F | 50.8 | 57.5 | 65.3 | 73.4 | 80.7 | 90.1 | 92.2 | 89.2 | 83.3 | 73.6 | 60.4 | 50.8 | 72.3 |
| Minimum Temp °F | 18.6 | 23.0 | 29.3 | 35.7 | 44.4 | 53.0 | 60.4 | 59.0 | 50.3 | 37.8 | 26.4 | 19.3 | 38.1 |
| Mean Temp °F | 34.7 | 40.3 | 47.3 | 54.6 | 62.6 | 71.6 | 76.3 | 74.1 | 66.8 | 55.7 | 43.4 | 35.1 | 55.2 |
| Days Max Temp ≥ 90 °F | 0 | 0 | 0 | 0 | 2 | 17 | 23 | 17 | 4 | 0 | 0 | 0 | 63 |
| Days Max Temp ≤ 32 °F | 1 | 0 | 0 | 0 | 0 | 0 | 0 | 0 | 0 | 0 | 0 | 1 | 2 |
| Days Min Temp ≤ 32 °F | 29 | 24 | 21 | 10 | 1 | 0 | 0 | 0 | 0 | 7 | 24 | 29 | 145 |
| Days Min Temp ≤ 0 °F | 1 | 0 | 0 | 0 | 0 | 0 | 0 | 0 | 0 | 0 | 0 | 0 | 1 |
| Heating Degree Days | 931 | 691 | 542 | 307 | 106 | 7 | 0 | 1 | 35 | 283 | 641 | 920 | 4464 |
| Cooling Degree Days | 0 | 0 | 0 | 3 | 49 | 238 | 368 | 314 | 110 | 2 | 0 | 0 | 1084 |
| Total Precipitation (") | 0.40 | 0.43 | 0.42 | 0.45 | 0.58 | 0.54 | 1.34 | 1.94 | 1.35 | 1.00 | 0.53 | 0.54 | 9.52 |
| Days ≥ 0.1" Precip | 1 | 2 | 1 | 1 | 2 | 1 | 4 | 4 | 3 | 2 | 1 | 2 | 24 |
| Total Snowfall (") | *1.4* | *0.8* | 0.0 | 0.3 | 0.0 | 0.0 | 0.0 | 0.0 | 0.0 | 0.3 | 0.4 | 1.7 | 4.9 |
| Days ≥ 1" Snow Depth | *0* | 0 | 0 | 0 | 0 | 0 | 0 | 0 | 0 | 0 | 0 | *1* | 1 |

## LUNA R S *Catron County* ELEVATION 7054 ft LAT/LONG 33° 50 ' N / 108° 56 ' W

| | JAN | FEB | MAR | APR | MAY | JUN | JUL | AUG | SEP | OCT | NOV | DEC | YEAR |
|---|---|---|---|---|---|---|---|---|---|---|---|---|---|
| Maximum Temp °F | 48.5 | 51.2 | 56.2 | 64.3 | 72.0 | 81.8 | 82.9 | 79.8 | 75.7 | 67.7 | 56.4 | *49.8* | 65.5 |
| Minimum Temp °F | 11.9 | 14.8 | 19.3 | 21.9 | 29.0 | 36.2 | 46.5 | 45.7 | 38.0 | 27.0 | 17.0 | *12.1* | 26.6 |
| Mean Temp °F | 30.2 | 33.0 | 37.8 | 43.1 | 50.5 | 59.0 | 64.8 | 62.8 | 56.8 | 47.3 | 36.7 | *31.0* | 46.1 |
| Days Max Temp ≥ 90 °F | 0 | 0 | 0 | 0 | 0 | 3 | 3 | 0 | 0 | 0 | 0 | 0 | 6 |
| Days Max Temp ≤ 32 °F | 1 | 1 | 0 | 0 | 0 | 0 | 0 | 0 | 0 | 0 | 0 | 1 | 3 |
| Days Min Temp ≤ 32 °F | 31 | 27 | 29 | 28 | 23 | 10 | 0 | 0 | 7 | 24 | 29 | 30 | 238 |
| Days Min Temp ≤ 0 °F | 4 | 2 | 0 | 0 | 0 | 0 | 0 | 0 | 0 | 0 | 0 | 4 | 10 |
| Heating Degree Days | 1073 | 898 | 837 | 650 | 442 | 184 | 37 | 77 | 240 | 540 | 842 | *1047* | 6867 |
| Cooling Degree Days | 0 | 0 | 0 | 0 | 0 | 12 | 28 | 14 | 1 | 0 | 0 | 0 | 55 |
| Total Precipitation (") | 0.98 | 0.89 | 0.86 | 0.37 | 0.70 | 0.61 | 3.01 | 3.34 | 2.02 | 1.59 | 1.07 | 1.41 | 16.85 |
| Days ≥ 0.1" Precip | 3 | 3 | 2 | 1 | 2 | 2 | 8 | 9 | 5 | 4 | 2 | 3 | 44 |
| Total Snowfall (") | *3.5* | na | *1.8* | *0.0* | 0.0 | 0.0 | 0.0 | 0.0 | 0.0 | 0.1 | *0.9* | na | na |
| Days ≥ 1" Snow Depth | *3* | na | *1* | *0* | 0 | 0 | 0 | 0 | 0 | 0 | *0* | na | na |

## LYBROOK *Rio Arriba County* ELEVATION 7205 ft LAT/LONG 36° 14 ' N / 107° 34 ' W

| | JAN | FEB | MAR | APR | MAY | JUN | JUL | AUG | SEP | OCT | NOV | DEC | YEAR |
|---|---|---|---|---|---|---|---|---|---|---|---|---|---|
| Maximum Temp °F | 38.0 | 42.3 | 50.6 | 59.0 | 69.4 | *80.1* | *84.1* | *80.9* | *73.6* | *62.7* | 48.6 | 39.2 | 60.7 |
| Minimum Temp °F | 14.4 | 19.2 | 25.9 | 32.5 | 40.0 | 50.5 | 55.7 | 53.7 | *46.6* | *36.1* | 24.7 | 16.4 | 34.6 |
| Mean Temp °F | 26.1 | 30.7 | 38.3 | 45.9 | 54.5 | *65.3* | *70.0* | *67.3* | *60.0* | *49.4* | 36.7 | 27.7 | 47.7 |
| Days Max Temp ≥ 90 °F | 0 | 0 | 0 | 0 | 0 | 3 | *4* | 1 | 0 | 0 | 0 | 0 | 8 |
| Days Max Temp ≤ 32 °F | *8* | 4 | 1 | 0 | 0 | 0 | 0 | 0 | 0 | 0 | *2* | 6 | 21 |
| Days Min Temp ≤ 32 °F | 31 | 27 | *25* | *14* | 4 | 0 | 0 | 0 | 1 | *9* | 25 | 30 | 166 |
| Days Min Temp ≤ 0 °F | 2 | 1 | 0 | 0 | 0 | 0 | 0 | 0 | 0 | 0 | 0 | 1 | 4 |
| Heating Degree Days | 1201 | 962 | 822 | 568 | 323 | *74* | *4* | *21* | *160* | *478* | 843 | 1148 | 6604 |
| Cooling Degree Days | 0 | 0 | 0 | *0* | 4 | *110* | na | 101 | 21 | 0 | 0 | 0 | na |
| Total Precipitation (") | *0.70* | 0.54 | 0.63 | 0.35 | 0.56 | 0.55 | 1.16 | 2.08 | 1.29 | *1.01* | 0.72 | 0.72 | 10.31 |
| Days ≥ 0.1" Precip | *2* | *2* | 2 | 1 | 2 | 1 | 3 | 5 | *3* | *3* | *2* | *2* | 28 |
| Total Snowfall (") | na | na | *3.5* | *1.0* | 0.0 | 0.0 | 0.0 | 0.0 | 0.0 | *0.8* | *2.4* | na | na |
| Days ≥ 1" Snow Depth | na | na | *1* | *0* | 0 | 0 | 0 | 0 | 0 | *0* | *1* | na | na |

## MALJAMAR 4 SE *Lea County* ELEVATION 4153 ft LAT/LONG 32° 50 ' N / 103° 44 ' W

| | JAN | FEB | MAR | APR | MAY | JUN | JUL | AUG | SEP | OCT | NOV | DEC | YEAR |
|---|---|---|---|---|---|---|---|---|---|---|---|---|---|
| Maximum Temp °F | 55.7 | 61.1 | 69.2 | 78.2 | 85.6 | 93.3 | 94.3 | 91.9 | 85.7 | 77.0 | 64.9 | 57.4 | 76.2 |
| Minimum Temp °F | 25.1 | 28.3 | 35.0 | 43.1 | 51.7 | 59.8 | 63.7 | 62.1 | 55.8 | 44.4 | 33.2 | 26.1 | 44.0 |
| Mean Temp °F | 40.4 | 44.7 | 52.1 | 60.7 | 68.7 | 76.6 | 79.1 | 77.0 | 70.7 | 60.7 | 49.0 | 41.8 | 60.1 |
| Days Max Temp ≥ 90 °F | 0 | 0 | 0 | 2 | 9 | 22 | 25 | 21 | 11 | 2 | 0 | 0 | 92 |
| Days Max Temp ≤ 32 °F | 1 | 0 | 0 | 0 | 0 | 0 | 0 | 0 | 0 | 0 | 0 | 1 | 2 |
| Days Min Temp ≤ 32 °F | 25 | 19 | 11 | 2 | 0 | 0 | 0 | 0 | 0 | 2 | 14 | 24 | 97 |
| Days Min Temp ≤ 0 °F | 0 | 0 | 0 | 0 | 0 | 0 | 0 | 0 | 0 | 0 | 0 | 0 | 0 |
| Heating Degree Days | 755 | 566 | 396 | 157 | 31 | 2 | 0 | 1 | 26 | 161 | 472 | 713 | 3280 |
| Cooling Degree Days | 0 | 0 | 3 | 38 | 163 | 395 | 470 | 429 | 230 | 44 | 1 | 0 | 1773 |
| Total Precipitation (") | 0.39 | 0.49 | 0.40 | 0.50 | 1.86 | 1.69 | 2.35 | 2.62 | 2.94 | 1.09 | 0.75 | 0.64 | 15.72 |
| Days ≥ 0.1" Precip | 1 | 2 | 1 | 1 | 3 | 3 | 5 | 5 | 4 | 2 | 2 | 2 | 31 |
| Total Snowfall (") | ***2.3*** | 1.9 | 1.3 | 0.3 | 0.0 | 0.0 | 0.0 | 0.0 | 0.0 | 0.2 | 0.8 | ***1.8*** | 8.6 |
| Days ≥ 1" Snow Depth | ***1*** | ***0*** | 0 | 0 | 0 | 0 | 0 | 0 | 0 | 0 | 0 | ***1*** | 2 |

## MAXWELL 3 NW *Colfax County* ELEVATION 5915 ft LAT/LONG 36° 33 ' N / 104° 33 ' W

| | JAN | FEB | MAR | APR | MAY | JUN | JUL | AUG | SEP | OCT | NOV | DEC | YEAR |
|---|---|---|---|---|---|---|---|---|---|---|---|---|---|
| Maximum Temp °F | 45.8 | 50.9 | 57.1 | 65.8 | 73.1 | 82.1 | 86.2 | 83.3 | 77.4 | 68.3 | 55.4 | 46.9 | 66.0 |
| Minimum Temp °F | 10.8 | 14.0 | 21.6 | 29.1 | 38.2 | 46.6 | 52.6 | 51.1 | 43.7 | 31.7 | 20.7 | 12.4 | 31.0 |
| Mean Temp °F | 28.3 | 32.5 | 39.4 | 47.3 | 55.7 | 64.4 | 69.4 | 67.2 | 60.6 | 50.0 | 38.1 | 29.7 | 48.6 |
| Days Max Temp ≥ 90 °F | 0 | 0 | 0 | 0 | 0 | 5 | 9 | 4 | 1 | 0 | 0 | 0 | 19 |
| Days Max Temp ≤ 32 °F | 5 | 2 | 1 | 0 | 0 | 0 | 0 | 0 | 0 | 0 | 1 | 4 | 13 |
| Days Min Temp ≤ 32 °F | 30 | 27 | 28 | 20 | 6 | 0 | 0 | 0 | 2 | 17 | 28 | 30 | 188 |
| Days Min Temp ≤ 0 °F | 4 | 2 | 0 | 0 | 0 | 0 | 0 | 0 | 0 | 0 | 0 | 4 | 10 |
| Heating Degree Days | 1131 | 909 | 788 | 524 | 285 | 71 | 6 | 19 | 143 | 458 | 801 | 1086 | 6221 |
| Cooling Degree Days | 0 | 0 | 0 | 0 | 4 | 62 | 143 | 103 | 18 | 0 | 0 | 0 | 330 |
| Total Precipitation (") | 0.27 | 0.25 | 0.47 | 0.61 | 1.95 | 2.20 | 2.60 | 3.39 | 1.82 | 0.88 | 0.56 | 0.27 | 15.27 |
| Days ≥ 0.1" Precip | 1 | 1 | 2 | 2 | 4 | 5 | 6 | 7 | 3 | 2 | 1 | 1 | 35 |
| Total Snowfall (") | ***3.9*** | ***3.7*** | ***3.1*** | ***0.7*** | 0.8 | 0.0 | 0.0 | 0.0 | 0.2 | 1.3 | 3.0 | na | na |
| Days ≥ 1" Snow Depth | na | ***1*** | na | 0 | 0 | 0 | 0 | 0 | 0 | 0 | ***1*** | na | na |

## MCGAFFEY 5 SE *McKinley County* ELEVATION 7999 ft LAT/LONG 35° 20 ' N / 108° 27 ' W

| | JAN | FEB | MAR | APR | MAY | JUN | JUL | AUG | SEP | OCT | NOV | DEC | YEAR |
|---|---|---|---|---|---|---|---|---|---|---|---|---|---|
| Maximum Temp °F | 39.2 | 41.9 | 47.2 | 55.6 | 65.0 | 76.6 | 79.9 | 77.0 | 71.6 | 62.1 | 48.9 | 40.6 | 58.8 |
| Minimum Temp °F | 8.1 | 11.4 | 17.8 | 23.8 | 31.0 | 38.9 | 46.3 | 45.1 | 37.7 | 27.3 | 17.3 | 10.0 | 26.2 |
| Mean Temp °F | 23.6 | 26.7 | 32.6 | 39.7 | 48.1 | 57.8 | 63.2 | 61.1 | 54.7 | 44.8 | 33.1 | 25.3 | 42.6 |
| Days Max Temp ≥ 90 °F | 0 | 0 | 0 | 0 | 0 | 1 | 1 | 0 | 0 | 0 | 0 | 0 | 2 |
| Days Max Temp ≤ 32 °F | 7 | 4 | 2 | 0 | 0 | 0 | 0 | 0 | 0 | 0 | 2 | 5 | 20 |
| Days Min Temp ≤ 32 °F | 31 | 28 | 31 | 27 | 17 | 4 | 0 | 0 | 6 | ***24*** | 29 | 31 | 228 |
| Days Min Temp ≤ 0 °F | 8 | 4 | 2 | 0 | 0 | 0 | 0 | 0 | 0 | 0 | 2 | 6 | 22 |
| Heating Degree Days | 1276 | 1075 | 999 | 752 | 518 | 218 | 71 | 121 | 303 | 621 | 952 | 1223 | 8129 |
| Cooling Degree Days | 0 | 0 | 0 | 0 | 0 | 8 | 18 | 6 | 0 | 0 | 0 | 0 | 32 |
| Total Precipitation (") | 1.80 | 1.50 | 2.05 | 1.08 | 0.99 | 0.68 | 2.37 | 2.52 | 1.69 | 1.49 | 1.68 | 1.72 | 19.57 |
| Days ≥ 0.1" Precip | 4 | 4 | 6 | 3 | 3 | 2 | 7 | 7 | 4 | 3 | 3 | 4 | 50 |
| Total Snowfall (") | na | na | na | na | 0.5 | 0.0 | 0.0 | 0.0 | 0.1 | 2.2 | ***8.2*** | na | na |
| Days ≥ 1" Snow Depth | na | na | na | ***2*** | 0 | 0 | 0 | 0 | 0 | 0 | na | na | na |

## MELROSE *Curry County* ELEVATION 4603 ft LAT/LONG 34° 26 ' N / 103° 38 ' W

| | JAN | FEB | MAR | APR | MAY | JUN | JUL | AUG | SEP | OCT | NOV | DEC | YEAR |
|---|---|---|---|---|---|---|---|---|---|---|---|---|---|
| Maximum Temp °F | 51.8 | 56.7 | 64.4 | 73.3 | 80.9 | 89.1 | 90.6 | 87.9 | 82.0 | 72.8 | 60.7 | 52.8 | 71.9 |
| Minimum Temp °F | 22.8 | 25.8 | 31.6 | 39.8 | 49.1 | 58.3 | 62.8 | 61.2 | 54.1 | 42.5 | 31.6 | 24.2 | 42.0 |
| Mean Temp °F | 37.3 | 41.3 | 48.0 | 56.6 | 65.0 | 73.7 | 76.7 | 74.5 | 68.1 | 57.7 | 46.2 | 38.5 | 57.0 |
| Days Max Temp ≥ 90 °F | 0 | 0 | 0 | 0 | 4 | 16 | 19 | 13 | 4 | 0 | 0 | 0 | 56 |
| Days Max Temp ≤ 32 °F | 2 | 1 | 0 | 0 | 0 | 0 | 0 | 0 | 0 | 0 | 0 | 2 | 5 |
| Days Min Temp ≤ 32 °F | 28 | 22 | 16 | 5 | 0 | 0 | 0 | 0 | 0 | 3 | 16 | 26 | 116 |
| Days Min Temp ≤ 0 °F | 0 | 0 | 0 | 0 | 0 | 0 | 0 | 0 | 0 | 0 | 0 | 0 | 0 |
| Heating Degree Days | 851 | 662 | 519 | 254 | 75 | 6 | 0 | 1 | 38 | 233 | 559 | 814 | 4012 |
| Cooling Degree Days | 0 | 0 | 0 | 10 | 88 | 293 | 376 | 326 | 149 | 12 | 0 | 0 | 1254 |
| Total Precipitation (") | 0.40 | 0.48 | 0.64 | 0.74 | 1.69 | 2.10 | 3.11 | 3.31 | 2.22 | 1.28 | 0.62 | 0.57 | 17.16 |
| Days ≥ 0.1" Precip | 1 | 1 | 2 | 2 | 4 | 4 | 6 | 6 | 4 | 3 | 2 | 2 | 37 |
| Total Snowfall (") | 3.2 | 3.4 | 1.9 | 0.6 | 0.1 | 0.0 | 0.0 | 0.0 | 0.0 | 0.4 | 1.8 | 3.9 | 15.3 |
| Days ≥ 1" Snow Depth | 3 | 2 | 0 | 0 | 0 | 0 | 0 | 0 | 0 | 0 | 1 | 3 | 9 |

## MIMBRES RS *Grant County* ELEVATION 6253 ft LAT/LONG 32° 56' N / 108° 1' W

| | JAN | FEB | MAR | APR | MAY | JUN | JUL | AUG | SEP | OCT | NOV | DEC | YEAR |
|---|---|---|---|---|---|---|---|---|---|---|---|---|---|
| Maximum Temp °F | 52.3 | 55.7 | 61.3 | 69.6 | 77.5 | 87.2 | 86.4 | 83.4 | 79.6 | 71.3 | 60.3 | 53.0 | 69.8 |
| Minimum Temp °F | 20.0 | 21.7 | 24.8 | 29.0 | 36.0 | 44.9 | 52.7 | 51.4 | 44.5 | 34.0 | 24.4 | 20.1 | 33.6 |
| Mean Temp °F | 36.2 | 38.7 | 43.1 | 49.3 | 56.8 | 66.1 | 69.6 | 67.4 | 62.1 | 52.7 | 42.4 | 36.6 | 51.8 |
| Days Max Temp ≥ 90 °F | 0 | 0 | 0 | 0 | 1 | 11 | 10 | 3 | 1 | 0 | 0 | 0 | 26 |
| Days Max Temp ≤ 32 °F | 0 | 0 | 0 | 0 | 0 | 0 | 0 | 0 | 0 | 0 | 0 | 0 | 0 |
| Days Min Temp ≤ 32 °F | 30 | 26 | 28 | 22 | 9 | 1 | 0 | 0 | 1 | 13 | 27 | 30 | 187 |
| Days Min Temp ≤ 0 °F | 0 | 0 | 0 | 0 | 0 | 0 | 0 | 0 | 0 | 0 | 0 | 0 | 0 |
| Heating Degree Days | 886 | 736 | 673 | 465 | 253 | 43 | 3 | 11 | 102 | 375 | 673 | 874 | 5094 |
| Cooling Degree Days | 0 | 0 | 0 | 0 | 7 | 96 | 153 | 107 | 24 | 0 | 0 | 0 | 387 |
| Total Precipitation (") | 1.00 | 0.76 | 0.78 | 0.32 | 0.68 | 0.85 | 3.59 | 4.17 | 2.23 | 1.62 | 1.12 | 1.40 | 18.52 |
| Days ≥ 0.1" Precip | 3 | 2 | 3 | 1 | 2 | 2 | 9 | 9 | 5 | 3 | 2 | 3 | 44 |
| Total Snowfall (") | *2.2* | *2.4* | 2.6 | 0.3 | 0.0 | 0.0 | 0.0 | 0.0 | 0.0 | 0.5 | 0.9 | *5.1* | 14.0 |
| Days ≥ 1" Snow Depth | *3* | *1* | *1* | 0 | 0 | 0 | 0 | 0 | 0 | 0 | 0 | *3* | 8 |

## MOSQUERO 1 NE *Harding County* ELEVATION 5554 ft LAT/LONG 35° 49' N / 103° 55' W

| | JAN | FEB | MAR | APR | MAY | JUN | JUL | AUG | SEP | OCT | NOV | DEC | YEAR |
|---|---|---|---|---|---|---|---|---|---|---|---|---|---|
| Maximum Temp °F | 46.2 | 50.2 | 58.4 | 67.3 | 74.7 | 84.1 | 86.9 | 83.9 | 77.7 | 68.6 | 55.4 | 46.6 | 66.7 |
| Minimum Temp °F | 18.7 | 21.7 | 27.7 | 35.1 | 44.1 | 53.5 | 58.7 | 57.1 | 49.7 | 38.2 | 27.0 | 19.9 | 37.6 |
| Mean Temp °F | 32.5 | 36.0 | 43.1 | 51.3 | 59.4 | 68.8 | 72.8 | 70.6 | 63.8 | 53.5 | 41.2 | 33.3 | 52.2 |
| Days Max Temp ≥ 90 °F | 0 | 0 | 0 | 0 | 1 | 8 | 11 | 5 | 1 | 0 | 0 | 0 | 26 |
| Days Max Temp ≤ 32 °F | 4 | 2 | 1 | 0 | 0 | 0 | 0 | 0 | 0 | 0 | 1 | 4 | 12 |
| Days Min Temp ≤ 32 °F | 30 | 26 | 22 | 10 | 2 | 0 | 0 | 0 | 0 | 7 | 23 | 29 | 149 |
| Days Min Temp ≤ 0 °F | 1 | 1 | 0 | 0 | 0 | 0 | 0 | 0 | 0 | 0 | 0 | 1 | 3 |
| Heating Degree Days | 1002 | 813 | 673 | 408 | 186 | 30 | 3 | 6 | 90 | 353 | 706 | 976 | 5246 |
| Cooling Degree Days | 0 | 0 | 0 | 3 | 23 | 163 | 264 | 194 | 66 | 2 | 0 | 0 | 715 |
| Total Precipitation (") | 0.40 | 0.35 | 0.73 | 0.87 | 2.08 | 2.21 | 2.89 | 3.11 | 1.91 | 1.08 | 0.73 | 0.51 | 16.87 |
| Days ≥ 0.1" Precip | 1 | 1 | 2 | 2 | 4 | 5 | 6 | 6 | 3 | 3 | 2 | 2 | 37 |
| Total Snowfall (") | 4.0 | 3.1 | 3.8 | 1.7 | 0.8 | 0.0 | 0.0 | 0.0 | 0.0 | 1.5 | 2.8 | 4.5 | 22.2 |
| Days ≥ 1" Snow Depth | 6 | 3 | 1 | 0 | 0 | 0 | 0 | 0 | 0 | 0 | 2 | *6* | 18 |

## MOUNTAIN PARK *Otero County* ELEVATION 6726 ft LAT/LONG 32° 57' N / 105° 50' W

| | JAN | FEB | MAR | APR | MAY | JUN | JUL | AUG | SEP | OCT | NOV | DEC | YEAR |
|---|---|---|---|---|---|---|---|---|---|---|---|---|---|
| Maximum Temp °F | 48.7 | 51.7 | 58.0 | 66.1 | 73.1 | 81.9 | 81.4 | 78.6 | 74.6 | 67.4 | 56.3 | 50.0 | 65.6 |
| Minimum Temp °F | 25.4 | 27.5 | 30.9 | 37.5 | 44.1 | 53.2 | 56.1 | 54.8 | 49.8 | 40.9 | 31.2 | 26.4 | 39.8 |
| Mean Temp °F | 37.1 | 39.6 | 44.5 | 51.8 | 58.6 | 67.6 | 68.8 | 66.7 | 62.2 | 54.2 | 43.8 | 38.2 | 52.8 |
| Days Max Temp ≥ 90 °F | 0 | 0 | 0 | 0 | 0 | 3 | 2 | 1 | 0 | 0 | 0 | 0 | 6 |
| Days Max Temp ≤ 32 °F | 1 | 1 | 0 | 0 | 0 | 0 | 0 | 0 | 0 | 0 | 0 | 1 | 3 |
| Days Min Temp ≤ 32 °F | 26 | 22 | 18 | 7 | 1 | 0 | 0 | 0 | 0 | 3 | 17 | 26 | 120 |
| Days Min Temp ≤ 0 °F | 0 | 0 | 0 | 0 | 0 | 0 | 0 | 0 | 0 | 0 | 0 | 0 | 0 |
| Heating Degree Days | 859 | 711 | 629 | 389 | 201 | 30 | 9 | 22 | 98 | 330 | 628 | 823 | 4729 |
| Cooling Degree Days | 0 | 0 | 0 | 1 | 10 | 123 | 127 | 87 | 21 | 0 | 0 | 0 | 369 |
| Total Precipitation (") | 1.20 | 1.22 | 0.97 | 0.52 | 1.07 | 1.35 | 3.23 | 4.24 | 2.70 | 1.71 | 1.29 | 1.64 | 21.14 |
| Days ≥ 0.1" Precip | 3 | 4 | 3 | 1 | 3 | 3 | 7 | 10 | 6 | 3 | 3 | 3 | 49 |
| Total Snowfall (") | 3.6 | *3.5* | 3.0 | 0.9 | 0.1 | 0.0 | 0.0 | 0.0 | 0.0 | 1.0 | 1.6 | *3.9* | 17.6 |
| Days ≥ 1" Snow Depth | *3* | *2* | *0* | 0 | 0 | 0 | 0 | 0 | 0 | 0 | 0 | na | na |

## MOUNTAINAIR *Torrance County* ELEVATION 6499 ft LAT/LONG 34° 32' N / 106° 15' W

| | JAN | FEB | MAR | APR | MAY | JUN | JUL | AUG | SEP | OCT | NOV | DEC | YEAR |
|---|---|---|---|---|---|---|---|---|---|---|---|---|---|
| Maximum Temp °F | 45.9 | 51.4 | 58.4 | 67.1 | 74.9 | 84.9 | 87.2 | 83.6 | 77.9 | 68.5 | 56.2 | 46.6 | 66.9 |
| Minimum Temp °F | 18.6 | 22.5 | 26.9 | 31.8 | 40.1 | 49.0 | 54.4 | 52.9 | 46.4 | 35.7 | 25.8 | 19.2 | 35.3 |
| Mean Temp °F | 32.3 | 36.9 | 42.7 | 49.5 | 57.6 | 67.0 | 70.8 | 68.2 | 62.2 | 52.1 | 41.1 | 32.9 | 51.1 |
| Days Max Temp ≥ 90 °F | 0 | 0 | 0 | 0 | 0 | 8 | 11 | 3 | 0 | 0 | 0 | 0 | 22 |
| Days Max Temp ≤ 32 °F | 2 | 1 | 0 | 0 | 0 | 0 | 0 | 0 | 0 | 0 | 0 | 2 | 5 |
| Days Min Temp ≤ 32 °F | 29 | 26 | 24 | 16 | 4 | 0 | 0 | 0 | 1 | 10 | 24 | 28 | 162 |
| Days Min Temp ≤ 0 °F | 1 | 0 | 0 | 0 | 0 | 0 | 0 | 0 | 0 | 0 | 0 | 1 | 2 |
| Heating Degree Days | 1007 | 786 | 683 | 460 | 232 | 40 | 2 | 10 | 103 | 393 | 711 | 988 | 5415 |
| Cooling Degree Days | 0 | 0 | 0 | 0 | 10 | 119 | 196 | 124 | 32 | *0* | *0* | *0* | 481 |
| Total Precipitation (") | 0.72 | 0.72 | 0.71 | 0.58 | 0.91 | 0.73 | 2.51 | 2.78 | 1.88 | 1.38 | 0.78 | 1.07 | 14.77 |
| Days ≥ 0.1" Precip | 2 | 3 | 2 | 2 | 2 | 2 | 5 | 6 | 4 | 3 | 2 | 2 | 35 |
| Total Snowfall (") | 6.6 | *5.3* | 4.2 | 1.2 | 0.4 | 0.0 | 0.0 | 0.0 | 0.0 | 0.5 | 1.2 | *5.9* | 25.3 |
| Days ≥ 1" Snow Depth | na | *2* | na | 0 | 0 | 0 | 0 | 0 | 0 | 0 | *0* | na | na |

## NAVAJO DAM *San Juan County* ELEVATION 5774 ft LAT/LONG 36° 49 ' N / 107° 37 ' W

| | JAN | FEB | MAR | APR | MAY | JUN | JUL | AUG | SEP | OCT | NOV | DEC | YEAR |
|---|---|---|---|---|---|---|---|---|---|---|---|---|---|
| Maximum Temp °F | 39.1 | 46.1 | 55.1 | 64.4 | 74.1 | 85.0 | 90.5 | 87.5 | 79.2 | 67.5 | 51.5 | 40.5 | 65.0 |
| Minimum Temp °F | 17.7 | 22.3 | 28.7 | 34.9 | 43.7 | 52.8 | 60.0 | 58.4 | 50.3 | 39.1 | 28.2 | 20.1 | 38.0 |
| Mean Temp °F | 28.5 | 34.2 | 41.9 | 49.6 | 58.9 | 68.9 | 75.3 | 73.0 | 64.8 | 53.3 | 39.9 | 30.3 | 51.5 |
| Days Max Temp ≥ 90 °F | 0 | 0 | 0 | 0 | 0 | 9 | 20 | 12 | 1 | 0 | 0 | 0 | 42 |
| Days Max Temp ≤ 32 °F | 6 | 1 | 0 | 0 | 0 | 0 | 0 | 0 | 0 | 0 | 1 | 4 | 12 |
| Days Min Temp ≤ 32 °F | 31 | 27 | 23 | 11 | 2 | 0 | 0 | 0 | 0 | 5 | 22 | 30 | 151 |
| Days Min Temp ≤ 0 °F | 1 | 0 | 0 | 0 | 0 | 0 | 0 | 0 | 0 | 0 | 0 | 1 | 2 |
| Heating Degree Days | 1127 | 862 | 705 | 456 | 197 | 31 | 0 | 3 | 66 | 362 | 747 | 1068 | 5624 |
| Cooling Degree Days | 0 | 0 | 0 | 1 | 22 | 177 | 335 | 287 | 82 | 2 | 0 | 0 | 906 |
| Total Precipitation (") | 1.13 | 0.99 | 1.28 | 0.92 | 0.96 | 0.57 | 1.11 | 1.74 | 1.12 | 1.28 | 1.21 | 1.23 | 13.54 |
| Days ≥ 0.1" Precip | 3 | 3 | 4 | 3 | 3 | 2 | 3 | 5 | 3 | 3 | 3 | 3 | 38 |
| Total Snowfall (") | na | ***3.6*** | ***1.0*** | 0.1 | 0.0 | 0.0 | 0.0 | 0.0 | 0.0 | 0.3 | ***0.7*** | na | na |
| Days ≥ 1" Snow Depth | ***10*** | ***5*** | 1 | 0 | 0 | 0 | 0 | 0 | 0 | 0 | 0 | ***6*** | 22 |

## NEWKIRK *Guadalupe County* ELEVATION 4573 ft LAT/LONG 35° 4 ' N / 104° 16 ' W

| | JAN | FEB | MAR | APR | MAY | JUN | JUL | AUG | SEP | OCT | NOV | DEC | YEAR |
|---|---|---|---|---|---|---|---|---|---|---|---|---|---|
| Maximum Temp °F | 52.2 | 57.5 | ***65.0*** | 73.4 | 81.6 | 90.7 | 92.6 | 90.1 | 83.9 | 74.4 | 61.5 | 53.1 | 73.0 |
| Minimum Temp °F | 22.1 | 25.4 | ***31.6*** | 39.4 | 48.6 | 58.0 | 63.2 | 61.6 | 54.4 | 42.4 | 31.3 | 23.5 | 41.8 |
| Mean Temp °F | 37.2 | 41.5 | ***48.3*** | 56.5 | 65.1 | 74.4 | 77.9 | 75.9 | 69.2 | 58.4 | 46.4 | 38.3 | 57.4 |
| Days Max Temp ≥ 90 °F | 0 | 0 | 0 | 0 | 4 | 18 | 23 | 18 | 6 | 0 | 0 | 0 | 69 |
| Days Max Temp ≤ 32 °F | 2 | 1 | 0 | 0 | 0 | 0 | 0 | 0 | 0 | 0 | 0 | 2 | 5 |
| Days Min Temp ≤ 32 °F | 27 | 21 | 16 | 6 | 1 | 0 | 0 | 0 | 0 | 4 | 17 | 25 | 117 |
| Days Min Temp ≤ 0 °F | 1 | 0 | 0 | 0 | 0 | 0 | 0 | 0 | 0 | 0 | 0 | 0 | 1 |
| Heating Degree Days | 857 | 659 | ***509*** | 260 | 77 | 6 | 0 | 1 | 33 | 216 | 550 | 822 | 3990 |
| Cooling Degree Days | 0 | 0 | ***1*** | 14 | 81 | 296 | 414 | 351 | ***177*** | 15 | 0 | 0 | 1349 |
| Total Precipitation (") | 0.36 | 0.41 | 0.59 | 0.61 | 1.44 | 1.79 | 2.66 | 3.03 | 1.50 | 1.26 | 0.64 | 0.42 | 14.71 |
| Days ≥ 0.1" Precip | 1 | 1 | 2 | 2 | 3 | 3 | 5 | 6 | 4 | 2 | 2 | 1 | 32 |
| Total Snowfall (") | 4.7 | 4.7 | ***1.9*** | 1.4 | 0.2 | 0.0 | 0.0 | 0.0 | 0.0 | 0.7 | 1.8 | 4.9 | 20.3 |
| Days ≥ 1" Snow Depth | ***3*** | ***1*** | 0 | 0 | 0 | 0 | 0 | 0 | 0 | 0 | 0 | 2 | 6 |

## OCATE 1 N *Mora County* ELEVATION 7674 ft LAT/LONG 36° 11 ' N / 105° 3 ' W

| | JAN | FEB | MAR | APR | MAY | JUN | JUL | AUG | SEP | OCT | NOV | DEC | YEAR |
|---|---|---|---|---|---|---|---|---|---|---|---|---|---|
| Maximum Temp °F | 45.5 | 47.9 | 52.6 | 61.0 | 68.8 | 78.0 | 80.9 | 78.0 | 73.1 | 64.7 | 52.9 | 46.3 | 62.5 |
| Minimum Temp °F | 13.4 | 16.2 | 20.4 | 27.2 | 35.5 | 43.1 | 48.5 | 47.5 | 40.6 | 29.6 | 20.6 | 14.0 | 29.7 |
| Mean Temp °F | 29.6 | 32.1 | 36.5 | 44.1 | 52.2 | 60.6 | 64.7 | 62.8 | 56.9 | 47.2 | 36.8 | 30.2 | 46.1 |
| Days Max Temp ≥ 90 °F | 0 | 0 | 0 | 0 | 0 | 1 | 1 | 0 | 0 | 0 | 0 | 0 | 2 |
| Days Max Temp ≤ 32 °F | 3 | 2 | 1 | 0 | 0 | 0 | 0 | 0 | 0 | 0 | 1 | 3 | 10 |
| Days Min Temp ≤ 32 °F | 30 | 27 | 28 | 22 | 10 | 1 | 0 | 0 | 3 | 20 | 26 | 30 | 197 |
| Days Min Temp ≤ 0 °F | 3 | 3 | 1 | 0 | 0 | 0 | 0 | 0 | 0 | 0 | 1 | 3 | 11 |
| Heating Degree Days | 1092 | 923 | 876 | 617 | 390 | 139 | 39 | 80 | 240 | 546 | 838 | 1074 | 6854 |
| Cooling Degree Days | 0 | 0 | 0 | 0 | 1 | 15 | 31 | 22 | 5 | 0 | 0 | 0 | 74 |
| Total Precipitation (") | 0.32 | 0.33 | 0.67 | 0.81 | 2.01 | 2.39 | 3.66 | 3.80 | 1.74 | 0.89 | 0.65 | 0.47 | 17.74 |
| Days ≥ 0.1" Precip | 1 | 1 | 2 | 2 | 5 | 5 | 8 | 9 | 4 | 2 | 2 | 2 | 43 |
| Total Snowfall (") | 5.5 | 5.2 | 7.6 | 3.9 | 0.9 | 0.0 | 0.0 | 0.0 | 0.1 | 2.4 | 3.8 | 6.5 | 35.9 |
| Days ≥ 1" Snow Depth | na | ***1*** | na | 0 | 0 | 0 | 0 | 0 | 0 | 0 | ***1*** | na | na |

## OROGRANDE *Otero County* ELEVATION 4183 ft LAT/LONG 32° 22 ' N / 106° 5 ' W

| | JAN | FEB | MAR | APR | MAY | JUN | JUL | AUG | SEP | OCT | NOV | DEC | YEAR |
|---|---|---|---|---|---|---|---|---|---|---|---|---|---|
| Maximum Temp °F | 56.9 | 62.4 | 69.6 | 78.0 | 85.9 | 95.3 | 95.5 | 92.3 | 86.9 | 78.0 | 66.1 | 57.4 | 77.0 |
| Minimum Temp °F | 27.3 | 31.2 | 36.0 | 43.9 | 51.3 | 61.1 | 65.3 | 63.4 | 57.5 | 45.9 | 34.6 | 27.3 | 45.4 |
| Mean Temp °F | 42.1 | 46.7 | 52.8 | 60.9 | 68.6 | 78.2 | 80.4 | 77.9 | 72.2 | 62.0 | 50.4 | 42.4 | 61.2 |
| Days Max Temp ≥ 90 °F | 0 | 0 | 0 | 1 | 10 | 25 | 27 | 23 | 11 | 1 | 0 | 0 | 98 |
| Days Max Temp ≤ 32 °F | 0 | 0 | 0 | 0 | 0 | 0 | 0 | 0 | 0 | 0 | 0 | 0 | 0 |
| Days Min Temp ≤ 32 °F | 23 | 17 | 9 | 2 | 0 | 0 | 0 | 0 | 0 | 1 | 12 | 23 | 87 |
| Days Min Temp ≤ 0 °F | 0 | 0 | 0 | 0 | 0 | 0 | 0 | 0 | 0 | 0 | 0 | 0 | 0 |
| Heating Degree Days | 704 | 509 | 374 | 147 | 29 | 1 | 0 | 0 | 12 | 127 | 432 | 694 | 3029 |
| Cooling Degree Days | 0 | 0 | 2 | 35 | 149 | 423 | 483 | 410 | 233 | 36 | 1 | 0 | 1772 |
| Total Precipitation (") | 0.52 | 0.32 | 0.20 | 0.27 | 0.62 | 1.11 | 1.92 | 2.65 | 1.73 | 1.19 | 0.56 | 0.87 | 11.96 |
| Days ≥ 0.1" Precip | 2 | 1 | 1 | 1 | 2 | 2 | 4 | 5 | 4 | ***2*** | 2 | 2 | 28 |
| Total Snowfall (") | 1.5 | 0.5 | 0.2 | 0.4 | 0.0 | 0.0 | 0.0 | 0.0 | 0.0 | 0.1 | 0.5 | 1.5 | 4.7 |
| Days ≥ 1" Snow Depth | ***0*** | 0 | 0 | 0 | 0 | 0 | 0 | 0 | 0 | 0 | 0 | 1 | 1 |

## OTIS *San Juan County* ELEVATION 6883 ft LAT/LONG 36° 19' N / 107° 52' W

| | JAN | FEB | MAR | APR | MAY | JUN | JUL | AUG | SEP | OCT | NOV | DEC | YEAR |
|---|---|---|---|---|---|---|---|---|---|---|---|---|---|
| Maximum Temp °F | 37.4 | 42.8 | 51.2 | 60.6 | 70.4 | 81.2 | 85.5 | 82.4 | 74.9 | 63.2 | 48.5 | 39.3 | 61.5 |
| Minimum Temp °F | 18.0 | 22.6 | 27.6 | 33.2 | 41.8 | 51.5 | 57.1 | 55.2 | 48.8 | 38.8 | 27.3 | 19.8 | 36.8 |
| Mean Temp °F | 27.7 | 32.7 | 39.5 | 47.0 | 56.2 | 66.4 | 71.3 | 68.8 | 61.8 | 51.1 | 38.0 | 29.6 | 49.2 |
| Days Max Temp ≥ 90 °F | 0 | 0 | 0 | 0 | 0 | 4 | 6 | 2 | 0 | 0 | 0 | 0 | 12 |
| Days Max Temp ≤ 32 °F | 9 | 3 | 1 | 0 | 0 | 0 | 0 | 0 | 0 | 0 | 2 | 7 | 22 |
| Days Min Temp ≤ 32 °F | 30 | 26 | 23 | ***13*** | 3 | 0 | 0 | 0 | 0 | 6 | 21 | 29 | 151 |
| Days Min Temp ≤ 0 °F | 1 | 0 | 0 | 0 | 0 | 0 | 0 | 0 | 0 | 0 | 0 | 1 | 2 |
| Heating Degree Days | 1147 | 903 | 785 | 535 | 274 | 55 | 2 | 10 | 116 | 426 | 805 | 1092 | 6150 |
| Cooling Degree Days | 0 | 0 | 0 | 0 | 7 | 114 | 191 | 137 | 28 | 0 | 0 | 0 | 477 |
| Total Precipitation (") | 0.64 | 0.49 | 0.70 | 0.47 | 0.66 | 0.77 | 1.51 | 1.77 | 1.02 | 0.96 | 0.74 | 0.63 | 10.36 |
| Days ≥ 0.1" Precip | 2 | 2 | 2 | 1 | 2 | 2 | 4 | 4 | 3 | 3 | 2 | 2 | 29 |
| Total Snowfall (") | 7.3 | 5.4 | 4.4 | 1.2 | 0.7 | 0.0 | 0.0 | 0.0 | 0.0 | 0.9 | 2.8 | 6.3 | 29.0 |
| Days ≥ 1" Snow Depth | ***14*** | 6 | 2 | 0 | 0 | 0 | 0 | 0 | 0 | 0 | 2 | 7 | 31 |

## PASAMONTE *Union County* ELEVATION 5653 ft LAT/LONG 36° 18' N / 103° 44' W

| | JAN | FEB | MAR | APR | MAY | JUN | JUL | AUG | SEP | OCT | NOV | DEC | YEAR |
|---|---|---|---|---|---|---|---|---|---|---|---|---|---|
| Maximum Temp °F | 46.2 | 49.7 | 56.8 | 65.9 | 73.9 | 83.4 | 87.9 | 84.9 | 77.5 | 67.9 | 55.5 | 47.2 | 66.4 |
| Minimum Temp °F | 15.9 | 18.8 | 25.0 | 32.9 | 41.8 | 51.6 | 56.3 | 54.8 | 47.2 | 35.8 | 24.6 | 17.2 | 35.2 |
| Mean Temp °F | 31.1 | 34.3 | 40.9 | 49.5 | 57.9 | 67.3 | 72.1 | 69.9 | 62.4 | 51.9 | 40.0 | 32.2 | 50.8 |
| Days Max Temp ≥ 90 °F | 0 | 0 | 0 | 0 | 1 | 8 | 13 | 8 | 2 | 0 | 0 | 0 | 32 |
| Days Max Temp ≤ 32 °F | 5 | 3 | 2 | 0 | 0 | 0 | 0 | 0 | 0 | 0 | 2 | 4 | 16 |
| Days Min Temp ≤ 32 °F | 31 | 27 | 26 | 14 | 3 | 0 | 0 | 0 | 0 | 10 | 25 | 30 | 166 |
| Days Min Temp ≤ 0 °F | 2 | 1 | 0 | 0 | 0 | 0 | 0 | 0 | 0 | 0 | 0 | 2 | 5 |
| Heating Degree Days | 1045 | 860 | 739 | 461 | 229 | 45 | 5 | 11 | 115 | 400 | 743 | 1010 | 5663 |
| Cooling Degree Days | 0 | 0 | 0 | 2 | 12 | 121 | 221 | 170 | 44 | 1 | 0 | 0 | 571 |
| Total Precipitation (") | 0.26 | 0.32 | 0.70 | 0.81 | 2.22 | 2.28 | 3.11 | 3.18 | 1.82 | 0.88 | 0.57 | 0.30 | 16.45 |
| Days ≥ 0.1" Precip | 1 | 1 | 2 | 2 | 4 | 5 | 6 | 7 | 4 | 2 | 2 | 1 | 37 |
| Total Snowfall (") | 3.5 | 3.4 | 4.1 | 2.0 | 0.5 | 0.0 | 0.0 | 0.0 | 0.0 | 0.9 | 2.8 | 3.4 | 20.6 |
| Days ≥ 1" Snow Depth | ***4*** | 3 | 2 | 1 | 0 | 0 | 0 | 0 | 0 | 0 | 2 | 5 | 17 |

## PEARL *Lea County* ELEVATION 3802 ft LAT/LONG 32° 39' N / 103° 23' W

| | JAN | FEB | MAR | APR | MAY | JUN | JUL | AUG | SEP | OCT | NOV | DEC | YEAR |
|---|---|---|---|---|---|---|---|---|---|---|---|---|---|
| Maximum Temp °F | 55.7 | 60.6 | 68.4 | 76.9 | 83.6 | 90.5 | 91.9 | 89.7 | 83.3 | ***75.7*** | 65.1 | 57.9 | 74.9 |
| Minimum Temp °F | 25.9 | 29.0 | 35.3 | 44.0 | 52.6 | 60.7 | 64.6 | 62.9 | 56.4 | 45.5 | 34.7 | 27.4 | 44.9 |
| Mean Temp °F | 40.8 | 44.8 | 51.8 | 60.5 | 68.1 | 75.6 | 78.3 | 76.3 | 69.9 | ***60.7*** | 49.9 | 42.7 | 60.0 |
| Days Max Temp ≥ 90 °F | 0 | 0 | 0 | 1 | 7 | 17 | 22 | 18 | 7 | 1 | 0 | 0 | 73 |
| Days Max Temp ≤ 32 °F | 1 | 1 | 0 | 0 | 0 | 0 | 0 | 0 | 0 | 0 | 0 | 1 | 3 |
| Days Min Temp ≤ 32 °F | 25 | 19 | 11 | 3 | 0 | 0 | 0 | 0 | 0 | ***2*** | 12 | 22 | 94 |
| Days Min Temp ≤ 0 °F | 0 | 0 | 0 | 0 | 0 | 0 | 0 | 0 | 0 | 0 | 0 | 0 | 0 |
| Heating Degree Days | 742 | 563 | 404 | 163 | 38 | 3 | 0 | 1 | 31 | ***162*** | 445 | 686 | 3238 |
| Cooling Degree Days | 0 | 0 | 2 | 29 | 142 | ***345*** | 422 | 377 | 193 | 27 | 1 | 0 | 1538 |
| Total Precipitation (") | 0.33 | 0.48 | 0.40 | 0.57 | 1.87 | 2.41 | 2.08 | 2.43 | 2.82 | 1.23 | 0.61 | 0.65 | 15.88 |
| Days ≥ 0.1" Precip | 1 | 2 | 1 | 2 | 3 | 3 | 3 | 4 | 4 | 2 | 2 | 2 | 29 |
| Total Snowfall (") | 1.5 | 1.1 | 0.2 | 0.3 | 0.0 | 0.0 | 0.0 | 0.0 | 0.0 | 0.2 | 1.2 | 1.2 | 5.7 |
| Days ≥ 1" Snow Depth | 1 | 0 | 0 | 0 | 0 | 0 | 0 | 0 | 0 | 0 | 0 | 0 | 1 |

## PECOS RS *San Miguel County* ELEVATION 6906 ft LAT/LONG 35° 35' N / 105° 41' W

| | JAN | FEB | MAR | APR | MAY | JUN | JUL | AUG | SEP | OCT | NOV | DEC | YEAR |
|---|---|---|---|---|---|---|---|---|---|---|---|---|---|
| Maximum Temp °F | 46.4 | 48.9 | 54.2 | 62.6 | 71.7 | 81.9 | 84.4 | 81.6 | ***75.3*** | ***66.4*** | 53.8 | 47.8 | 64.6 |
| Minimum Temp °F | 14.4 | 18.9 | 23.9 | 30.1 | 38.4 | 47.7 | 53.1 | 51.4 | 44.2 | 33.8 | 23.2 | 16.5 | 33.0 |
| Mean Temp °F | 30.5 | 33.9 | 38.9 | 46.4 | 55.1 | 64.9 | 68.7 | 66.5 | ***59.9*** | ***50.2*** | 38.5 | 32.1 | 48.8 |
| Days Max Temp ≥ 90 °F | 0 | 0 | 0 | 0 | 0 | 4 | 5 | 2 | 0 | 0 | 0 | 0 | 11 |
| Days Max Temp ≤ 32 °F | 2 | 1 | 0 | 0 | 0 | 0 | 0 | 0 | 0 | 0 | 1 | 2 | 6 |
| Days Min Temp ≤ 32 °F | 31 | 27 | 28 | ***19*** | 6 | 0 | 0 | 0 | 1 | ***13*** | ***27*** | 30 | 182 |
| Days Min Temp ≤ 0 °F | ***2*** | 1 | 0 | 0 | 0 | 0 | 0 | 0 | 0 | 0 | 0 | 2 | 5 |
| Heating Degree Days | ***1057*** | 871 | 803 | 552 | 306 | 68 | 12 | 26 | ***160*** | ***454*** | 789 | 1005 | 6103 |
| Cooling Degree Days | 0 | 0 | 0 | 0 | 3 | 71 | 132 | 84 | 12 | 0 | 0 | 0 | 302 |
| Total Precipitation (") | 0.71 | 0.70 | 0.97 | 0.78 | 1.30 | 1.32 | 3.09 | 3.63 | 2.06 | 1.12 | 0.98 | 0.65 | 17.31 |
| Days ≥ 0.1" Precip | 2 | 2 | 3 | 2 | 3 | 4 | 8 | 9 | 5 | 3 | 3 | 2 | 46 |
| Total Snowfall (") | ***4.3*** | ***3.8*** | ***3.6*** | 1.7 | 0.0 | 0.0 | 0.0 | 0.0 | 0.0 | 0.3 | ***2.0*** | ***4.5*** | 20.2 |
| Days ≥ 1" Snow Depth | na | na | ***1*** | 0 | 0 | 0 | 0 | 0 | 0 | 0 | na | na | na |

## PEDERNAL 4 E *Torrance County* ELEVATION 6204 ft LAT/LONG 34° 38 ' N / 105° 34 ' W

| | JAN | FEB | MAR | APR | MAY | JUN | JUL | AUG | SEP | OCT | NOV | DEC | YEAR |
|---|---|---|---|---|---|---|---|---|---|---|---|---|---|
| Maximum Temp °F | 43.0 | 48.7 | 56.8 | 65.8 | 73.8 | 84.2 | 86.9 | 83.8 | 77.5 | 67.9 | 54.5 | 44.5 | 65.6 |
| Minimum Temp °F | 17.9 | 20.8 | 25.5 | 31.7 | 40.2 | 49.7 | 55.0 | 53.6 | 46.7 | 35.7 | 25.6 | 18.6 | 35.1 |
| Mean Temp °F | 30.5 | 34.8 | 41.1 | 48.8 | 57.0 | 67.0 | 71.0 | 68.7 | 62.1 | 51.8 | 40.1 | 31.5 | 50.4 |
| Days Max Temp ≥ 90 °F | 0 | 0 | 0 | 0 | 0 | 8 | 12 | 5 | 1 | 0 | 0 | 0 | 26 |
| Days Max Temp ≤ 32 °F | 6 | 2 | 1 | 0 | 0 | 0 | 0 | 0 | 0 | 0 | 1 | 5 | 15 |
| Days Min Temp ≤ 32 °F | 30 | 27 | 26 | 16 | 4 | 0 | 0 | 0 | 1 | 10 | 24 | 30 | 168 |
| Days Min Temp ≤ 0 °F | 1 | 1 | 0 | 0 | 0 | 0 | 0 | 0 | 0 | 0 | 0 | 1 | 3 |
| Heating Degree Days | 1064 | 848 | 733 | 480 | 250 | 43 | 4 | 13 | 113 | 403 | 740 | 1031 | 5722 |
| Cooling Degree Days | 0 | 0 | 0 | 0 | 10 | 122 | 205 | 147 | 37 | 0 | 0 | 0 | 521 |
| Total Precipitation (") | 0.32 | 0.39 | 0.38 | 0.38 | 1.00 | 1.22 | 2.07 | 2.87 | 1.28 | 0.97 | 0.55 | 0.58 | 12.01 |
| Days ≥ 0.1" Precip | 1 | 1 | 1 | 1 | 2 | 3 | 5 | 6 | 3 | 2 | 2 | 1 | 28 |
| Total Snowfall (") | *3.6* | na | na | 0.7 | 0.3 | 0.0 | 0.0 | 0.0 | 0.0 | 0.2 | *1.8* | *6.3* | na |
| Days ≥ 1" Snow Depth | *5* | *3* | na | 0 | 0 | 0 | 0 | 0 | 0 | 0 | *1* | *4* | na |

## PORTALES *Roosevelt County* ELEVATION 4003 ft LAT/LONG 34° 11 ' N / 103° 20 ' W

| | JAN | FEB | MAR | APR | MAY | JUN | JUL | AUG | SEP | OCT | NOV | DEC | YEAR |
|---|---|---|---|---|---|---|---|---|---|---|---|---|---|
| Maximum Temp °F | 54.2 | 58.9 | 67.3 | 75.9 | 83.0 | 90.5 | 91.7 | 89.4 | 84.1 | 75.2 | 62.8 | 54.7 | 74.0 |
| Minimum Temp °F | 22.3 | 25.6 | 32.1 | 40.8 | 50.1 | 59.6 | 63.7 | 62.1 | 54.9 | 42.7 | 31.3 | 23.6 | 42.4 |
| Mean Temp °F | 38.3 | 42.3 | 49.7 | 58.4 | 66.5 | 75.1 | 77.7 | 75.8 | 69.5 | 59.0 | 47.1 | 39.2 | 58.2 |
| Days Max Temp ≥ 90 °F | 0 | 0 | 0 | 1 | 6 | 18 | 21 | 17 | 7 | 1 | 0 | 0 | 71 |
| Days Max Temp ≤ 32 °F | 2 | 1 | 0 | 0 | 0 | 0 | 0 | 0 | 0 | 0 | 0 | 1 | 4 |
| Days Min Temp ≤ 32 °F | 28 | 22 | 15 | 5 | 0 | 0 | 0 | 0 | 0 | 3 | 17 | 26 | 116 |
| Days Min Temp ≤ 0 °F | 1 | 0 | 0 | 0 | 0 | 0 | 0 | 0 | 0 | 0 | 0 | 0 | 1 |
| Heating Degree Days | 821 | 635 | 469 | 210 | 56 | 3 | 0 | 1 | 27 | 200 | 531 | 793 | 3746 |
| Cooling Degree Days | 0 | 0 | 0 | 16 | 104 | 322 | 398 | 348 | 175 | 15 | 0 | 0 | 1378 |
| Total Precipitation (") | 0.51 | 0.42 | 0.49 | 0.68 | 1.70 | 2.57 | 2.70 | 3.03 | 2.03 | 1.31 | 0.72 | 0.59 | 16.75 |
| Days ≥ 0.1" Precip | 1 | 1 | 1 | 1 | 3 | 4 | 5 | 6 | 4 | 3 | 2 | 2 | 33 |
| Total Snowfall (") | 1.9 | 2.5 | 0.7 | 0.1 | 0.0 | 0.0 | 0.0 | 0.0 | 0.0 | 0.2 | 0.9 | 2.5 | 8.8 |
| Days ≥ 1" Snow Depth | 2 | *1* | 0 | 0 | 0 | 0 | 0 | 0 | 0 | 0 | 0 | *1* | 4 |

## RAGLAND 3 SSW *Quay County* ELEVATION 5060 ft LAT/LONG 34° 48 ' N / 103° 45 ' W

| | JAN | FEB | MAR | APR | MAY | JUN | JUL | AUG | SEP | OCT | NOV | DEC | YEAR |
|---|---|---|---|---|---|---|---|---|---|---|---|---|---|
| Maximum Temp °F | 49.3 | 54.3 | 62.5 | 71.5 | 78.8 | 88.1 | 89.8 | 86.9 | 80.2 | 71.3 | 58.5 | 50.1 | 70.1 |
| Minimum Temp °F | 21.8 | 25.0 | 30.3 | 38.7 | 47.1 | 56.2 | 60.7 | 59.3 | 52.3 | 41.5 | 31.1 | 23.3 | 40.6 |
| Mean Temp °F | 35.6 | 39.7 | 46.4 | 55.1 | 63.0 | 72.2 | 75.3 | 73.1 | 66.3 | 56.4 | 44.8 | 36.7 | 55.4 |
| Days Max Temp ≥ 90 °F | 0 | 0 | 0 | 0 | 2 | 14 | 17 | 11 | 3 | 0 | 0 | 0 | 47 |
| Days Max Temp ≤ 32 °F | 3 | 2 | 1 | 0 | 0 | 0 | 0 | 0 | 0 | 0 | 1 | 3 | 10 |
| Days Min Temp ≤ 32 °F | 28 | 23 | 19 | 6 | 1 | 0 | 0 | 0 | 0 | 3 | 17 | 27 | 124 |
| Days Min Temp ≤ 0 °F | 1 | 0 | 0 | 0 | 0 | 0 | 0 | 0 | 0 | 0 | 0 | 1 | 2 |
| Heating Degree Days | 906 | 708 | 569 | 295 | 109 | 11 | 1 | 3 | 55 | 267 | 600 | 870 | 4394 |
| Cooling Degree Days | 0 | 0 | 0 | 5 | 49 | 247 | 336 | 273 | 105 | 5 | 0 | 0 | 1020 |
| Total Precipitation (") | 0.46 | 0.58 | 0.75 | 0.93 | 1.83 | 2.10 | 3.09 | 3.29 | 2.29 | 1.24 | 0.75 | 0.54 | 17.85 |
| Days ≥ 0.1" Precip | 1 | 2 | 2 | 2 | 4 | 4 | 5 | 6 | 4 | 3 | 2 | 2 | 37 |
| Total Snowfall (") | 4.1 | 4.9 | 2.8 | 2.0 | 0.0 | 0.0 | 0.0 | 0.0 | 0.0 | 0.2 | 1.8 | 4.0 | 19.8 |
| Days ≥ 1" Snow Depth | 4 | 2 | 1 | 0 | 0 | 0 | 0 | 0 | 0 | 0 | 1 | 3 | 11 |

## RAMON 8 SW *Lincoln County* ELEVATION 5328 ft LAT/LONG 34° 9 ' N / 105° 0 ' W

| | JAN | FEB | MAR | APR | MAY | JUN | JUL | AUG | SEP | OCT | NOV | DEC | YEAR |
|---|---|---|---|---|---|---|---|---|---|---|---|---|---|
| Maximum Temp °F | 51.4 | 56.6 | 63.5 | 72.0 | 79.6 | 88.5 | *90.2* | *87.3* | 81.0 | 72.1 | 59.9 | 52.3 | 71.2 |
| Minimum Temp °F | 21.7 | 24.2 | 29.1 | 36.7 | 45.3 | 54.1 | *59.4* | *57.8* | 50.6 | 39.7 | 28.8 | 22.6 | 39.2 |
| Mean Temp °F | 36.6 | 40.4 | 46.3 | 54.4 | 62.5 | 71.3 | *74.8* | *72.6* | 65.8 | 55.9 | 44.3 | 37.5 | 55.2 |
| Days Max Temp ≥ 90 °F | 0 | 0 | 0 | 0 | *3* | *13* | *16* | na | *3* | 0 | 0 | 0 | na |
| Days Max Temp ≤ 32 °F | 2 | 1 | 0 | 0 | 0 | 0 | 0 | 0 | 0 | 0 | 0 | 2 | 5 |
| Days Min Temp ≤ 32 °F | *28* | na | *20* | *9* | 1 | 0 | 0 | 0 | 0 | *5* | na | *26* | na |
| Days Min Temp ≤ 0 °F | 1 | 0 | 0 | 0 | 0 | 0 | 0 | 0 | 0 | 0 | 0 | 1 | 2 |
| Heating Degree Days | 875 | 683 | 572 | *314* | 117 | 9 | *1* | *2* | 52 | 281 | *612* | *846* | 4364 |
| Cooling Degree Days | 0 | 0 | 0 | 5 | 45 | 216 | 301 | 247 | 85 | 6 | 0 | 0 | 905 |
| Total Precipitation (") | 0.42 | 0.41 | 0.48 | 0.61 | 1.36 | 1.46 | 1.97 | 2.34 | 1.82 | 1.07 | 0.60 | 0.52 | 13.06 |
| Days ≥ 0.1" Precip | 2 | 1 | 2 | 2 | 3 | 3 | 5 | *5* | 4 | 2 | *1* | 1 | 31 |
| Total Snowfall (") | *4.7* | na | 2.4 | 1.0 | 0.1 | 0.0 | 0.0 | 0.0 | 0.0 | 0.7 | 1.3 | *5.1* | na |
| Days ≥ 1" Snow Depth | *3* | na | 1 | 0 | 0 | 0 | 0 | 0 | 0 | 0 | *0* | *2* | na |

## RATON FILTER PLANT *Colfax County* ELEVATION 6936 ft LAT/LONG 36° 55 ' N / 104° 26 ' W

| | JAN | FEB | MAR | APR | MAY | JUN | JUL | AUG | SEP | OCT | NOV | DEC | YEAR |
|---|---|---|---|---|---|---|---|---|---|---|---|---|---|
| Maximum Temp °F | 44.4 | 46.9 | 52.4 | 60.4 | 68.3 | 77.9 | 82.4 | 79.4 | 73.5 | 64.8 | 52.7 | 45.4 | 62.4 |
| Minimum Temp °F | 18.6 | 20.8 | 25.5 | 32.8 | 41.4 | 50.5 | 54.9 | 53.6 | 46.6 | 36.7 | 27.0 | 20.3 | 35.7 |
| Mean Temp °F | 31.5 | 33.9 | 39.0 | 46.6 | 54.9 | 64.2 | 68.7 | 66.5 | 60.1 | 50.8 | 39.9 | 32.9 | 49.1 |
| Days Max Temp ≥ 90 °F | 0 | 0 | 0 | 0 | 0 | 1 | 2 | 0 | 0 | 0 | 0 | 0 | 3 |
| Days Max Temp ≤ 32 °F | 5 | 3 | 2 | 0 | 0 | 0 | 0 | 0 | 0 | 0 | 1 | 4 | 15 |
| Days Min Temp ≤ 32 °F | 30 | 26 | 25 | 14 | 3 | 0 | 0 | 0 | 1 | 8 | 21 | 28 | 156 |
| Days Min Temp ≤ 0 °F | 1 | 1 | 0 | 0 | 0 | 0 | 0 | 0 | 0 | 0 | 0 | 1 | 3 |
| Heating Degree Days | 1030 | 871 | 798 | 545 | 311 | 83 | 11 | 29 | 159 | 435 | 747 | 990 | 6009 |
| Cooling Degree Days | 0 | 0 | 0 | 0 | 3 | 71 | 133 | 86 | 18 | 0 | 0 | 0 | 311 |
| Total Precipitation (") | 0.47 | 0.46 | 1.00 | 1.06 | 2.49 | 2.15 | 2.84 | 3.55 | 1.56 | 1.00 | 0.64 | 0.58 | 17.80 |
| Days ≥ 0.1" Precip | 2 | 1 | 3 | 3 | 6 | 5 | 7 | 8 | 4 | 3 | 2 | 2 | 46 |
| Total Snowfall (") | 6.9 | 5.2 | 7.2 | 3.2 | 0.7 | 0.0 | 0.0 | 0.0 | 0.2 | 1.4 | 4.0 | 7.0 | 35.8 |
| Days ≥ 1" Snow Depth | ***3*** | ***2*** | ***2*** | 1 | 0 | 0 | 0 | 0 | 0 | 0 | 1 | ***4*** | 13 |

## RED RIVER *Taos County* ELEVATION 8675 ft LAT/LONG 36° 42 ' N / 105° 24 ' W

| | JAN | FEB | MAR | APR | MAY | JUN | JUL | AUG | SEP | OCT | NOV | DEC | YEAR |
|---|---|---|---|---|---|---|---|---|---|---|---|---|---|
| Maximum Temp °F | 36.9 | 39.9 | 45.3 | 54.0 | 62.9 | 72.8 | 76.3 | 73.3 | 68.0 | 58.2 | 44.6 | 37.9 | 55.8 |
| Minimum Temp °F | 4.9 | 8.5 | 16.2 | 22.4 | 29.5 | 35.9 | 41.6 | 40.6 | 34.1 | 25.3 | 15.6 | 7.6 | 23.5 |
| Mean Temp °F | 20.9 | 24.2 | 30.8 | 38.2 | 46.2 | 54.4 | 59.0 | 57.0 | 51.1 | 41.8 | 30.1 | 22.8 | 39.7 |
| Days Max Temp ≥ 90 °F | 0 | 0 | 0 | 0 | 0 | 0 | 0 | 0 | 0 | 0 | 0 | 0 | 0 |
| Days Max Temp ≤ 32 °F | 8 | 5 | 2 | 0 | 0 | 0 | 0 | 0 | 0 | 0 | 3 | 8 | 26 |
| Days Min Temp ≤ 32 °F | 31 | 28 | 30 | 28 | 23 | 8 | 0 | 1 | 12 | 27 | 29 | 31 | 248 |
| Days Min Temp ≤ 0 °F | 11 | 7 | 2 | 1 | 0 | 0 | 0 | 0 | 0 | 0 | 3 | 9 | 33 |
| Heating Degree Days | 1360 | 1145 | 1054 | 796 | 576 | 311 | 180 | 241 | 411 | 713 | 1039 | 1303 | 9129 |
| Cooling Degree Days | 0 | 0 | 0 | 0 | 0 | 0 | 0 | 0 | 0 | 0 | 0 | 0 | 0 |
| Total Precipitation (") | 1.07 | 1.10 | 2.01 | 1.70 | 1.85 | 1.65 | 2.85 | 3.38 | 1.76 | 1.64 | 1.62 | 1.36 | 21.99 |
| Days ≥ 0.1" Precip | 4 | 3 | 6 | 5 | 5 | 5 | 8 | 10 | 5 | 4 | 4 | 4 | 63 |
| Total Snowfall (") | 19.4 | 20.3 | 36.1 | 23.9 | 7.6 | 0.2 | 0.0 | 0.0 | 0.9 | 9.4 | 20.6 | 22.1 | 160.5 |
| Days ≥ 1" Snow Depth | 26 | 23 | 19 | 7 | 1 | 0 | 0 | 0 | 0 | 3 | 12 | 22 | 113 |

## REDROCK 1 NNE *Grant County* ELEVATION 4153 ft LAT/LONG 32° 42 ' N / 108° 44 ' W

| | JAN | FEB | MAR | APR | MAY | JUN | JUL | AUG | SEP | OCT | NOV | DEC | YEAR |
|---|---|---|---|---|---|---|---|---|---|---|---|---|---|
| Maximum Temp °F | 58.3 | 62.9 | 69.2 | 77.5 | 85.4 | 94.9 | 95.1 | 92.1 | 87.5 | 78.1 | 66.5 | 58.2 | 77.1 |
| Minimum Temp °F | 25.2 | 28.0 | 32.0 | 36.6 | 43.9 | 52.7 | 62.4 | 61.7 | 54.1 | 41.6 | 29.9 | 25.1 | 41.1 |
| Mean Temp °F | 41.7 | 45.5 | 50.6 | 57.1 | 64.6 | 73.8 | 78.7 | 76.9 | 70.8 | 59.9 | 48.2 | 41.7 | 59.1 |
| Days Max Temp ≥ 90 °F | 0 | 0 | 0 | 1 | 8 | 25 | 26 | 23 | 12 | 2 | 0 | 0 | 97 |
| Days Max Temp ≤ 32 °F | 0 | 0 | 0 | 0 | 0 | 0 | 0 | 0 | 0 | 0 | 0 | 0 | 0 |
| Days Min Temp ≤ 32 °F | 26 | 21 | 17 | 9 | 1 | 0 | 0 | 0 | 0 | 4 | 20 | 26 | 124 |
| Days Min Temp ≤ 0 °F | 0 | 0 | 0 | 0 | 0 | 0 | 0 | 0 | 0 | 0 | 0 | 0 | 0 |
| Heating Degree Days | 715 | 546 | 440 | 237 | 67 | 3 | 0 | 0 | 9 | 169 | 497 | 716 | 3399 |
| Cooling Degree Days | 0 | 0 | 0 | 8 | 69 | 288 | 423 | 376 | 194 | 14 | 0 | 0 | 1372 |
| Total Precipitation (") | 1.00 | 0.88 | 0.77 | 0.23 | 0.41 | 0.41 | 2.63 | 2.52 | 1.67 | 1.25 | 0.76 | 1.38 | 13.91 |
| Days ≥ 0.1" Precip | 3 | 3 | 2 | 1 | 1 | 2 | 6 | 6 | 4 | 3 | 2 | 3 | 36 |
| Total Snowfall (") | 1.1 | 0.7 | 0.3 | 0.1 | 0.0 | 0.0 | 0.0 | 0.0 | 0.0 | 0.0 | 0.2 | 1.7 | 4.1 |
| Days ≥ 1" Snow Depth | 0 | 0 | 0 | 0 | 0 | 0 | 0 | 0 | 0 | 0 | 0 | 1 | 1 |

## RESERVE RANGER STN *Catron County* ELEVATION 5833 ft LAT/LONG 33° 43 ' N / 108° 47 ' W

| | JAN | FEB | MAR | APR | MAY | JUN | JUL | AUG | SEP | OCT | NOV | DEC | YEAR |
|---|---|---|---|---|---|---|---|---|---|---|---|---|---|
| Maximum Temp °F | 52.1 | ***55.5*** | 61.0 | 69.2 | 76.9 | 86.9 | ***88.2*** | 85.6 | ***80.7*** | ***72.4*** | 60.8 | 53.3 | 70.2 |
| Minimum Temp °F | 16.8 | 18.5 | 22.5 | 27.0 | 34.0 | 42.4 | ***52.3*** | 50.9 | ***43.5*** | ***31.6*** | 22.2 | 15.9 | 31.5 |
| Mean Temp °F | 34.5 | ***37.1*** | 41.8 | 48.1 | 55.5 | 64.7 | ***70.3*** | 68.3 | ***62.1*** | ***52.1*** | 41.5 | 34.6 | 50.9 |
| Days Max Temp ≥ 90 °F | 0 | 0 | 0 | 0 | 1 | 11 | 14 | 6 | 1 | 0 | 0 | 0 | 33 |
| Days Max Temp ≤ 32 °F | 0 | 0 | 0 | 0 | 0 | 0 | 0 | 0 | 0 | 0 | 0 | 1 | 1 |
| Days Min Temp ≤ 32 °F | 29 | 25 | 28 | 24 | 12 | 2 | 0 | 0 | 2 | 17 | 26 | 29 | 194 |
| Days Min Temp ≤ 0 °F | 1 | 0 | 0 | 0 | 0 | 0 | 0 | 0 | 0 | 0 | 0 | 1 | 2 |
| Heating Degree Days | 940 | ***783*** | 713 | 500 | 291 | 70 | ***4*** | 9 | ***105*** | ***395*** | ***697*** | 938 | 5445 |
| Cooling Degree Days | 0 | ***0*** | 0 | 0 | 3 | 75 | ***160*** | 125 | 27 | ***0*** | 0 | ***0*** | 390 |
| Total Precipitation (") | 1.06 | 1.12 | 1.07 | 0.41 | 0.63 | 0.69 | 2.51 | 2.94 | 1.96 | 1.66 | 1.25 | 1.57 | 16.87 |
| Days ≥ 0.1" Precip | 3 | 2 | 3 | 1 | 2 | 2 | 6 | 7 | 5 | 3 | 2 | 3 | 39 |
| Total Snowfall (") | ***2.3*** | 1.7 | 1.0 | 0.2 | 0.0 | 0.0 | 0.0 | 0.0 | 0.0 | 0.0 | 0.2 | na | na |
| Days ≥ 1" Snow Depth | ***2*** | 1 | 0 | 0 | 0 | 0 | 0 | 0 | 0 | 0 | 0 | ***1*** | 4 |

## ROY *Harding County* ELEVATION 5882 ft LAT/LONG 35° 57 ' N / 104° 12 ' W

| | JAN | FEB | MAR | APR | MAY | JUN | JUL | AUG | SEP | OCT | NOV | DEC | YEAR |
|---|---|---|---|---|---|---|---|---|---|---|---|---|---|
| Maximum Temp °F | 46.9 | 50.9 | 57.4 | 66.1 | 73.7 | 82.6 | 85.8 | 83.5 | 77.1 | 67.8 | 55.8 | 47.4 | 66.3 |
| Minimum Temp °F | 17.9 | 21.1 | 26.6 | 34.1 | 43.2 | 52.0 | 57.3 | 55.7 | 48.8 | 37.4 | 26.9 | 19.3 | 36.7 |
| Mean Temp °F | 32.4 | 36.0 | 42.0 | 50.1 | 58.5 | 67.3 | 71.6 | 69.6 | 63.0 | 52.6 | 41.4 | 33.4 | 51.5 |
| Days Max Temp ≥ 90 °F | 0 | 0 | 0 | 0 | 0 | 5 | 8 | 4 | 1 | 0 | 0 | 0 | 18 |
| Days Max Temp ≤ 32 °F | 4 | 2 | 1 | 0 | 0 | 0 | 0 | 0 | 0 | 0 | 1 | 3 | 11 |
| Days Min Temp ≤ 32 °F | 30 | 26 | 23 | 12 | 2 | 0 | 0 | 0 | 0 | 7 | 22 | 29 | 151 |
| Days Min Temp ≤ 0 °F | 1 | 1 | 0 | 0 | 0 | 0 | 0 | 0 | 0 | 0 | 0 | 1 | 3 |
| Heating Degree Days | 1004 | 811 | 706 | 440 | 210 | 42 | 4 | 10 | 96 | 380 | 702 | 973 | 5378 |
| Cooling Degree Days | 0 | 0 | 0 | 1 | 11 | 114 | 210 | 163 | 43 | 0 | 0 | 0 | 542 |
| Total Precipitation (") | 0.34 | 0.40 | 0.67 | 0.76 | 1.88 | 2.24 | 2.95 | 3.22 | 1.79 | 1.04 | 0.50 | 0.47 | 16.26 |
| Days ≥ 0.1" Precip | 1 | 1 | 2 | 2 | 4 | 4 | 7 | 6 | 4 | 2 | 1 | 2 | 36 |
| Total Snowfall (") | na | na | *0.3* | *0.1* | 0.0 | 0.0 | 0.0 | 0.0 | 0.0 | 0.2 | 1.2 | na | na |
| Days ≥ 1" Snow Depth | na | na | 1 | 0 | 0 | 0 | 0 | 0 | 0 | 0 | 1 | na | na |

## RUIDOSO *Lincoln County* ELEVATION 6760 ft LAT/LONG 33° 20 ' N / 105° 41 ' W

| | JAN | FEB | MAR | APR | MAY | JUN | JUL | AUG | SEP | OCT | NOV | DEC | YEAR |
|---|---|---|---|---|---|---|---|---|---|---|---|---|---|
| Maximum Temp °F | *49.9* | *51.7* | *57.2* | *64.5* | *72.2* | *80.7* | *80.7* | *78.5* | *74.2* | *66.0* | *56.9* | *50.8* | 65.3 |
| Minimum Temp °F | *17.1* | *18.6* | *22.5* | *27.2* | *33.5* | na | *47.0* | na | *40.7* | *30.8* | *23.0* | *18.1* | na |
| Mean Temp °F | *33.5* | *35.2* | *39.8* | *45.9* | *52.9* | na | *63.9* | na | *57.4* | *48.4* | *40.0* | *34.5* | na |
| Days Max Temp ≥ 90 °F | *0* | *0* | *0* | *0* | *0* | *2* | *1* | *0* | *0* | *0* | *0* | *0* | 3 |
| Days Max Temp ≤ 32 °F | *1* | *1* | *0* | *0* | *0* | *0* | *0* | *0* | *0* | *0* | *0* | *1* | 3 |
| Days Min Temp ≤ 32 °F | *29* | *26* | *27* | *22* | *15* | *3* | *0* | *0* | *3* | *19* | *25* | *28* | 197 |
| Days Min Temp ≤ 0 °F | *2* | *1* | *0* | *0* | *0* | *0* | *0* | *0* | *0* | *0* | *0* | *1* | 4 |
| Heating Degree Days | *968* | *834* | *773* | *569* | *370* | na | *53* | na | *221* | *506* | *744* | *937* | na |
| Cooling Degree Days | na | na | na | na | na | na | na | na | na | na | na | na | na |
| Total Precipitation (") | *1.22* | *1.27* | *1.09* | *0.70* | *1.10* | *2.12* | *3.47* | *4.19* | *2.63* | *1.92* | *1.09* | *1.85* | 22.65 |
| Days ≥ 0.1" Precip | *4* | *4* | *4* | *2* | *3* | *5* | *9* | *10* | *7* | *4* | *3* | *4* | 59 |
| Total Snowfall (") | *10.1* | na | *6.9* | *2.9* | *0.0* | *0.0* | *0.0* | *0.0* | *0.0* | *2.3* | *3.1* | na | na |
| Days ≥ 1" Snow Depth | na | na | na | na | *0* | *0* | *0* | *0* | *0* | na | na | na | na |

## SAN JON *Quay County* ELEVATION 4232 ft LAT/LONG 35° 7 ' N / 103° 20 ' W

| | JAN | FEB | MAR | APR | MAY | JUN | JUL | AUG | SEP | OCT | NOV | DEC | YEAR |
|---|---|---|---|---|---|---|---|---|---|---|---|---|---|
| Maximum Temp °F | 52.7 | 57.1 | 65.4 | 74.2 | 82.0 | 91.2 | 93.4 | 90.8 | 84.2 | 74.6 | 62.1 | 53.5 | 73.4 |
| Minimum Temp °F | 23.0 | 26.7 | 33.7 | 42.2 | 51.3 | 60.8 | 65.5 | 63.7 | 56.2 | 44.4 | 32.6 | 24.7 | 43.7 |
| Mean Temp °F | 37.9 | 41.9 | 49.6 | 58.2 | 66.7 | 76.0 | 79.5 | 77.3 | 70.2 | 59.5 | 47.4 | 39.1 | 58.6 |
| Days Max Temp ≥ 90 °F | 0 | 0 | 0 | 1 | 6 | 19 | 24 | 20 | 9 | 1 | 0 | 0 | 80 |
| Days Max Temp ≤ 32 °F | 2 | 1 | 0 | 0 | 0 | 0 | 0 | 0 | 0 | 0 | 0 | 2 | 5 |
| Days Min Temp ≤ 32 °F | 26 | 21 | 14 | 5 | 0 | 0 | 0 | 0 | 0 | 3 | 15 | 25 | 109 |
| Days Min Temp ≤ 0 °F | 1 | 0 | 0 | 0 | 0 | 0 | 0 | 0 | 0 | 0 | 0 | 1 | 2 |
| Heating Degree Days | 834 | 645 | 474 | 224 | 63 | 6 | 0 | 1 | 33 | 199 | 523 | 796 | 3798 |
| Cooling Degree Days | 0 | 0 | 3 | 28 | 115 | 340 | 459 | 392 | 201 | 31 | 0 | 0 | 1569 |
| Total Precipitation (") | 0.51 | 0.58 | 0.84 | 1.03 | 2.12 | 2.38 | 2.86 | 3.24 | 1.95 | 1.30 | 0.79 | 0.46 | 18.06 |
| Days ≥ 0.1" Precip | 2 | 1 | 2 | 2 | 4 | 4 | 5 | 5 | 4 | 3 | 2 | 1 | 35 |
| Total Snowfall (") | 4.2 | 3.9 | 2.5 | 1.0 | 0.0 | 0.0 | 0.0 | 0.0 | 0.0 | 0.7 | 1.3 | 3.0 | 16.6 |
| Days ≥ 1" Snow Depth | *2* | *1* | 1 | 0 | 0 | 0 | 0 | 0 | 0 | 0 | 0 | *1* | 5 |

## SANDIA PARK *Bernalillo County* ELEVATION 7014 ft LAT/LONG 35° 10 ' N / 106° 22 ' W

| | JAN | FEB | MAR | APR | MAY | JUN | JUL | AUG | SEP | OCT | NOV | DEC | YEAR |
|---|---|---|---|---|---|---|---|---|---|---|---|---|---|
| Maximum Temp °F | 43.0 | 47.1 | 53.4 | 62.3 | 71.2 | 82.1 | 84.5 | 80.9 | 74.5 | 64.9 | 51.8 | 43.7 | 63.3 |
| Minimum Temp °F | 17.9 | 21.9 | 27.1 | 33.3 | 40.6 | 48.9 | 53.6 | 52.2 | 46.0 | 35.6 | 26.4 | 19.3 | 35.2 |
| Mean Temp °F | 30.5 | 34.5 | 40.3 | 47.9 | 56.0 | 65.5 | 69.1 | 66.6 | 60.3 | 50.3 | 39.1 | 31.5 | 49.3 |
| Days Max Temp ≥ 90 °F | 0 | 0 | 0 | 0 | 0 | 5 | 6 | 1 | 0 | 0 | 0 | 0 | 12 |
| Days Max Temp ≤ 32 °F | 3 | 1 | 1 | 0 | 0 | 0 | 0 | 0 | 0 | 0 | 1 | 3 | 9 |
| Days Min Temp ≤ 32 °F | 30 | 26 | 25 | 14 | 4 | 0 | 0 | 0 | 1 | 11 | 24 | 29 | 164 |
| Days Min Temp ≤ 0 °F | 2 | 0 | 0 | 0 | 0 | 0 | 0 | 0 | 0 | 0 | 0 | 1 | 3 |
| Heating Degree Days | 1063 | 854 | 760 | 507 | 277 | 59 | 8 | 24 | 151 | 450 | 770 | 1032 | 5955 |
| Cooling Degree Days | 0 | 0 | 0 | 0 | 4 | 71 | 122 | 71 | 14 | 0 | 0 | 0 | 282 |
| Total Precipitation (") | 1.32 | 1.27 | 1.53 | 0.94 | 1.32 | 1.10 | 3.08 | 3.03 | 1.96 | 1.63 | 1.45 | 1.42 | 20.05 |
| Days ≥ 0.1" Precip | 4 | 3 | 4 | 2 | 3 | 3 | 7 | 7 | 5 | 3 | 3 | 3 | 47 |
| Total Snowfall (") | *13.9* | 12.3 | 10.9 | 4.7 | 0.8 | 0.0 | 0.0 | 0.0 | 0.0 | *1.7* | 7.8 | 12.6 | 64.7 |
| Days ≥ 1" Snow Depth | na | na | na | 1 | 0 | 0 | 0 | 0 | 0 | 0 | *3* | na | na |

## SANTA ROSA *Guadalupe County* ELEVATION 4603 ft LAT/LONG 34° 56 ' N / 104° 40 ' W

| | JAN | FEB | MAR | APR | MAY | JUN | JUL | AUG | SEP | OCT | NOV | DEC | YEAR |
|---|---|---|---|---|---|---|---|---|---|---|---|---|---|
| Maximum Temp °F | 54.4 | 59.1 | 66.1 | 74.3 | 82.1 | 90.8 | 92.6 | 90.0 | 84.2 | 75.5 | 62.8 | 54.7 | 73.9 |
| Minimum Temp °F | 24.9 | 27.6 | 33.3 | 41.0 | 49.4 | 58.3 | 63.2 | 61.1 | 54.1 | 42.5 | 32.9 | 25.9 | 42.9 |
| Mean Temp °F | 39.7 | 43.4 | 49.7 | 57.7 | 65.8 | 74.5 | 77.9 | 75.6 | 69.2 | 59.0 | 47.9 | 40.3 | 58.4 |
| Days Max Temp ≥ 90 °F | 0 | 0 | 0 | 0 | 5 | 18 | 23 | 18 | 6 | 1 | 0 | 0 | 71 |
| Days Max Temp ≤ 32 °F | 1 | 1 | 0 | 0 | 0 | 0 | 0 | 0 | 0 | 0 | 0 | 1 | 3 |
| Days Min Temp ≤ 32 °F | 24 | 20 | 14 | 4 | 0 | 0 | 0 | 0 | 0 | 3 | 15 | 23 | 103 |
| Days Min Temp ≤ 0 °F | 0 | 0 | 0 | 0 | 0 | 0 | 0 | 0 | 0 | 0 | 0 | 0 | 0 |
| Heating Degree Days | 778 | 604 | 468 | 222 | 60 | 3 | 0 | 1 | 25 | 196 | 508 | 759 | 3624 |
| Cooling Degree Days | 0 | 0 | 0 | 12 | 92 | 307 | 407 | 346 | 164 | 15 | 1 | 0 | 1344 |
| Total Precipitation (") | 0.39 | 0.43 | 0.59 | 0.66 | 1.32 | 1.79 | 2.42 | 3.07 | 1.75 | 1.13 | 0.80 | 0.61 | 14.96 |
| Days ≥ 0.1" Precip | 1 | 1 | 2 | 2 | 3 | 4 | 5 | 6 | 4 | 3 | 2 | 2 | 35 |
| Total Snowfall (") | 2.6 | 2.6 | 0.9 | 0.8 | 0.1 | 0.0 | 0.0 | 0.0 | 0.0 | 0.5 | 1.5 | 3.8 | 12.8 |
| Days ≥ 1" Snow Depth | 2 | 1 | 0 | 0 | 0 | 0 | 0 | 0 | 0 | 0 | 0 | 3 | 6 |

## SHIPROCK *San Juan County* ELEVATION 4954 ft LAT/LONG 36° 46 ' N / 108° 44 ' W

| | JAN | FEB | MAR | APR | MAY | JUN | JUL | AUG | SEP | OCT | NOV | DEC | YEAR |
|---|---|---|---|---|---|---|---|---|---|---|---|---|---|
| Maximum Temp °F | 43.1 | 51.1 | 60.5 | 70.1 | *79.7* | 89.5 | *94.2* | 91.4 | 84.2 | 71.7 | 56.2 | 43.9 | 69.6 |
| Minimum Temp °F | 16.0 | 21.3 | 27.2 | 33.6 | *42.7* | 50.1 | *57.3* | 56.1 | 46.8 | 34.9 | 25.1 | 17.2 | 35.7 |
| Mean Temp °F | 29.6 | 36.2 | 43.8 | 51.8 | *61.2* | 69.8 | *75.7* | 73.8 | 65.5 | 53.4 | 40.7 | 30.6 | 52.7 |
| Days Max Temp ≥ 90 °F | 0 | 0 | 0 | 0 | 2 | 17 | *26* | 22 | 6 | 0 | 0 | 0 | 73 |
| Days Max Temp ≤ 32 °F | 4 | 1 | 0 | 0 | 0 | 0 | 0 | 0 | 0 | 0 | 0 | 3 | 8 |
| Days Min Temp ≤ 32 °F | 30 | 26 | 22 | 13 | *3* | 0 | 0 | 0 | 1 | 11 | 24 | 30 | 160 |
| Days Min Temp ≤ 0 °F | 1 | 0 | 0 | 0 | 0 | 0 | 0 | 0 | 0 | 0 | 0 | 1 | 2 |
| Heating Degree Days | 1091 | 806 | 649 | 392 | *145* | 23 | *0* | 3 | 55 | 356 | 724 | 1060 | 5304 |
| Cooling Degree Days | 0 | 0 | 0 | 3 | 42 | 204 | 348 | 302 | 96 | 2 | 0 | 0 | 997 |
| Total Precipitation (") | 0.67 | 0.51 | 0.47 | 0.39 | 0.65 | 0.28 | 0.64 | 1.06 | 0.74 | 0.95 | 0.72 | 0.71 | 7.79 |
| Days ≥ 0.1" Precip | 2 | 2 | 2 | 1 | *2* | 1 | 1 | 3 | 2 | 2 | 2 | 2 | 22 |
| Total Snowfall (") | na | na | 0.3 | 0.0 | 0.0 | 0.0 | 0.0 | 0.0 | 0.0 | 0.0 | 0.3 | na | na |
| Days ≥ 1" Snow Depth | na | *0* | 0 | 0 | 0 | 0 | 0 | 0 | 0 | 0 | 0 | na | na |

## SOCORRO *Socorro County* ELEVATION 4623 ft LAT/LONG 34° 4 ' N / 106° 54 ' W

| | JAN | FEB | MAR | APR | MAY | JUN | JUL | AUG | SEP | OCT | NOV | DEC | YEAR |
|---|---|---|---|---|---|---|---|---|---|---|---|---|---|
| Maximum Temp °F | 52.3 | 59.3 | 67.0 | 75.5 | 82.6 | 91.4 | 92.8 | 89.8 | 84.0 | 75.1 | 62.0 | 52.1 | 73.7 |
| Minimum Temp °F | 21.0 | 25.1 | 31.1 | 37.7 | 45.9 | 53.8 | 60.4 | 58.6 | 50.6 | 39.2 | 27.9 | 22.3 | 39.5 |
| Mean Temp °F | 36.7 | 42.2 | 49.1 | 56.6 | 64.2 | 72.6 | 76.7 | 74.2 | 67.3 | 57.2 | 45.0 | 37.2 | 56.6 |
| Days Max Temp ≥ 90 °F | 0 | 0 | 0 | 0 | 4 | 20 | 24 | 17 | 5 | 0 | 0 | 0 | 70 |
| Days Max Temp ≤ 32 °F | 1 | 0 | 0 | 0 | 0 | 0 | 0 | 0 | 0 | 0 | 0 | 1 | 2 |
| Days Min Temp ≤ 32 °F | 27 | 23 | 18 | 8 | 1 | 0 | 0 | 0 | 0 | 6 | 22 | 27 | 132 |
| Days Min Temp ≤ 0 °F | 0 | 0 | 0 | 0 | 0 | 0 | 0 | 0 | 0 | 0 | 0 | 0 | 0 |
| Heating Degree Days | 872 | 636 | 486 | 252 | 78 | 4 | 0 | 0 | 31 | 241 | 594 | 856 | 4050 |
| Cooling Degree Days | 0 | 0 | 0 | 11 | 65 | 259 | 375 | 310 | 118 | *4* | 0 | 0 | 1142 |
| Total Precipitation (") | 0.37 | 0.38 | 0.31 | 0.31 | 0.53 | 0.57 | 1.41 | 2.18 | 1.57 | 1.14 | 0.49 | 0.62 | 9.88 |
| Days ≥ 0.1" Precip | 1 | 1 | 1 | 1 | 2 | 1 | 4 | 5 | 4 | 2 | 1 | 2 | 25 |
| Total Snowfall (") | *1.0* | 1.8 | 0.4 | 0.0 | 0.0 | 0.0 | 0.0 | 0.0 | 0.0 | 0.3 | 0.7 | *3.1* | 7.3 |
| Days ≥ 1" Snow Depth | *1* | *0* | 0 | 0 | 0 | 0 | 0 | 0 | 0 | 0 | 0 | *1* | 2 |

## SPRINGER *Colfax County* ELEVATION 5880 ft LAT/LONG 36° 21 ' N / 104° 35 ' W

| | JAN | FEB | MAR | APR | MAY | JUN | JUL | AUG | SEP | OCT | NOV | DEC | YEAR |
|---|---|---|---|---|---|---|---|---|---|---|---|---|---|
| Maximum Temp °F | 47.4 | 52.9 | 59.9 | 67.8 | 75.7 | 85.1 | 88.2 | 85.6 | 79.9 | 71.0 | 57.2 | 47.7 | 68.2 |
| Minimum Temp °F | 12.6 | 16.6 | 23.5 | 30.8 | 40.1 | 48.9 | 54.0 | 52.6 | 44.9 | 32.9 | 21.9 | 13.7 | 32.7 |
| Mean Temp °F | 30.0 | 34.8 | 41.7 | 49.3 | 57.9 | 67.0 | 71.2 | 69.1 | 62.4 | 51.9 | 39.6 | 30.7 | 50.5 |
| Days Max Temp ≥ 90 °F | 0 | 0 | 0 | 0 | 1 | 9 | 14 | 8 | 2 | 0 | 0 | 0 | 34 |
| Days Max Temp ≤ 32 °F | 4 | 1 | 0 | 0 | 0 | 0 | 0 | 0 | 0 | 0 | 1 | 4 | 10 |
| Days Min Temp ≤ 32 °F | 30 | 27 | 27 | 17 | 4 | 0 | 0 | 0 | 1 | 16 | 26 | 30 | 178 |
| Days Min Temp ≤ 0 °F | 4 | 2 | 0 | 0 | 0 | 0 | 0 | 0 | 0 | 0 | 0 | 3 | 9 |
| Heating Degree Days | 1077 | 846 | 714 | 464 | 223 | 35 | 3 | 7 | 101 | 399 | 756 | 1055 | 5680 |
| Cooling Degree Days | 0 | 0 | 0 | 0 | 9 | 95 | 186 | 137 | 33 | 0 | 0 | 0 | 460 |
| Total Precipitation (") | 0.34 | 0.33 | 0.74 | 0.83 | 2.00 | 2.00 | 2.70 | 3.58 | 1.80 | 0.99 | 0.63 | 0.36 | 16.30 |
| Days ≥ 0.1" Precip | 1 | 1 | 2 | 2 | 4 | 4 | 6 | 7 | 4 | 3 | 2 | 1 | 37 |
| Total Snowfall (") | 4.6 | 4.1 | 6.1 | 2.2 | 0.8 | 0.0 | 0.0 | 0.0 | 0.1 | 1.5 | 2.7 | 5.2 | 27.3 |
| Days ≥ 1" Snow Depth | 6 | 3 | 2 | 1 | 0 | 0 | 0 | 0 | 0 | 0 | 2 | 6 | 20 |

## STANLEY 1 NNE *Santa Fe County* ELEVATION 6380 ft LAT/LONG 35° 10' N / 105° 58' W

| | JAN | FEB | MAR | APR | MAY | JUN | JUL | AUG | SEP | OCT | NOV | DEC | YEAR |
|---|---|---|---|---|---|---|---|---|---|---|---|---|---|
| Maximum Temp °F | 42.6 | 48.4 | 56.1 | 64.9 | 73.8 | 83.9 | 87.0 | 83.8 | 77.6 | 67.7 | 54.1 | 44.3 | 65.4 |
| Minimum Temp °F | 15.5 | 19.4 | 24.6 | 30.7 | 39.2 | 48.1 | 53.7 | 52.1 | 44.4 | 33.1 | 23.6 | 15.9 | 33.4 |
| Mean Temp °F | 29.1 | 33.9 | 40.4 | 47.8 | 56.5 | 66.0 | 70.4 | 68.0 | 61.0 | 50.4 | 38.9 | 30.1 | 49.4 |
| Days Max Temp ≥ 90 °F | 0 | 0 | 0 | 0 | 0 | 7 | 11 | 4 | 0 | 0 | 0 | 0 | 22 |
| Days Max Temp ≤ 32 °F | 4 | 2 | 1 | 0 | 0 | 0 | 0 | 0 | 0 | 0 | 1 | 4 | 12 |
| Days Min Temp ≤ 32 °F | 31 | 27 | 27 | 18 | 5 | 0 | 0 | 0 | 1 | 14 | 26 | 31 | 180 |
| Days Min Temp ≤ 0 °F | 2 | 1 | 0 | 0 | 0 | 0 | 0 | 0 | 0 | 0 | 0 | 1 | 4 |
| Heating Degree Days | 1107 | 872 | 755 | 511 | 262 | 49 | 3 | 15 | 133 | 445 | 776 | 1074 | 6002 |
| Cooling Degree Days | 0 | 0 | 0 | 0 | 7 | 89 | 164 | 115 | 20 | 0 | 0 | 0 | 395 |
| Total Precipitation (") | 0.44 | 0.40 | 0.51 | 0.52 | 1.12 | 1.17 | 2.21 | 2.49 | 1.55 | 1.26 | 0.65 | 0.51 | 12.83 |
| Days ≥ 0.1" Precip | 2 | 2 | 2 | 2 | 3 | 3 | 6 | 6 | 4 | 3 | 2 | 2 | 37 |
| Total Snowfall (") | 4.1 | 4.3 | 2.6 | 1.3 | 0.1 | 0.0 | 0.0 | 0.0 | 0.0 | 0.7 | *2.5* | 4.5 | 20.1 |
| Days ≥ 1" Snow Depth | 9 | 4 | 1 | 1 | 0 | 0 | 0 | 0 | 0 | 0 | 2 | 8 | 25 |

## STAR LAKE *McKinley County* ELEVATION 6644 ft LAT/LONG 35° 56' N / 107° 28' W

| | JAN | FEB | MAR | APR | MAY | JUN | JUL | AUG | SEP | OCT | NOV | DEC | YEAR |
|---|---|---|---|---|---|---|---|---|---|---|---|---|---|
| Maximum Temp °F | 39.4 | 45.4 | 54.4 | 63.9 | 72.8 | 83.3 | 86.9 | 83.9 | 77.1 | 66.2 | 51.7 | 41.6 | 63.9 |
| Minimum Temp °F | 9.6 | 16.7 | 22.4 | 27.3 | 35.8 | 44.6 | 52.3 | 51.0 | 42.7 | 30.6 | 20.4 | 11.3 | 30.4 |
| Mean Temp °F | 24.5 | 31.1 | 38.4 | 45.6 | 54.3 | 64.0 | 69.6 | 67.5 | 59.9 | 48.5 | 36.1 | 26.5 | 47.2 |
| Days Max Temp ≥ 90 °F | 0 | 0 | 0 | 0 | 0 | 5 | 9 | 4 | 0 | 0 | 0 | 0 | 18 |
| Days Max Temp ≤ 32 °F | 7 | 2 | 0 | 0 | 0 | 0 | 0 | 0 | 0 | 0 | 1 | 6 | 16 |
| Days Min Temp ≤ 32 °F | 31 | 28 | 28 | 22 | 10 | 1 | 0 | 0 | 2 | 18 | 28 | 30 | 198 |
| Days Min Temp ≤ 0 °F | 7 | 2 | 0 | 0 | 0 | 0 | 0 | 0 | 0 | 0 | 1 | 5 | 15 |
| Heating Degree Days | 1248 | 951 | 817 | 575 | 326 | 81 | 4 | 18 | 158 | 507 | 862 | 1187 | 6734 |
| Cooling Degree Days | 0 | 0 | 0 | 0 | 2 | 65 | 148 | 112 | *12* | 0 | 0 | 0 | 339 |
| Total Precipitation (") | 0.48 | 0.38 | 0.50 | 0.48 | 0.67 | 0.66 | 1.41 | 1.90 | 1.18 | 0.88 | 0.67 | 0.58 | 9.79 |
| Days ≥ 0.1" Precip | 2 | 1 | 2 | 1 | 2 | 2 | 4 | 5 | 3 | 3 | 2 | 2 | 29 |
| Total Snowfall (") | 5.1 | 3.8 | 3.4 | 0.8 | 0.7 | 0.0 | 0.0 | 0.0 | 0.0 | 0.5 | 1.8 | 4.6 | 20.7 |
| Days ≥ 1" Snow Depth | 14 | 6 | 1 | 0 | 0 | 0 | 0 | 0 | 0 | 0 | 1 | 7 | 29 |

## TAOS *Taos County* ELEVATION 6985 ft LAT/LONG 36° 25' N / 105° 34' W

| | JAN | FEB | MAR | APR | MAY | JUN | JUL | AUG | SEP | OCT | NOV | DEC | YEAR |
|---|---|---|---|---|---|---|---|---|---|---|---|---|---|
| Maximum Temp °F | 39.9 | 44.7 | 52.7 | 62.0 | 70.5 | 80.9 | 85.0 | 82.0 | 75.8 | 65.2 | 51.1 | 41.7 | 62.6 |
| Minimum Temp °F | 8.4 | 15.6 | 23.1 | 29.1 | 37.2 | 45.5 | 51.2 | 49.6 | 42.2 | 31.6 | 20.7 | 11.1 | 30.4 |
| Mean Temp °F | 24.2 | 30.1 | 37.9 | 45.6 | 53.9 | 63.2 | 68.1 | 65.9 | 59.0 | 48.4 | 36.0 | 26.4 | 46.6 |
| Days Max Temp ≥ 90 °F | 0 | 0 | 0 | 0 | 0 | 3 | 5 | 1 | 0 | 0 | 0 | 0 | 9 |
| Days Max Temp ≤ 32 °F | 5 | 2 | 0 | 0 | 0 | 0 | 0 | 0 | 0 | 0 | 1 | 4 | 12 |
| Days Min Temp ≤ 32 °F | 31 | 28 | 29 | 21 | 7 | 0 | 0 | 0 | 2 | 18 | 28 | 31 | 195 |
| Days Min Temp ≤ 0 °F | 8 | 2 | 0 | 0 | 0 | 0 | 0 | 0 | 0 | 0 | 1 | 4 | 15 |
| Heating Degree Days | 1260 | 977 | 834 | 577 | 339 | 90 | 8 | 29 | 178 | 508 | 865 | 1189 | 6854 |
| Cooling Degree Days | 0 | 0 | 0 | 0 | 1 | 51 | 111 | 69 | 7 | 0 | 0 | 0 | 239 |
| Total Precipitation (") | 0.59 | 0.52 | 0.82 | 0.80 | 1.17 | 1.05 | 1.55 | 2.15 | 1.42 | 1.11 | 0.90 | 0.74 | 12.82 |
| Days ≥ 0.1" Precip | 2 | 2 | 3 | 2 | 4 | 3 | 5 | 6 | 3 | 3 | 3 | 2 | 38 |
| Total Snowfall (") | 8.6 | 6.5 | 5.7 | 2.4 | 0.5 | 0.0 | 0.0 | 0.0 | 0.0 | 0.7 | 4.1 | 9.0 | 37.5 |
| Days ≥ 1" Snow Depth | *15* | 9 | 3 | 1 | 0 | 0 | 0 | 0 | 0 | 0 | 3 | 10 | 41 |

## TATUM *Lea County* ELEVATION 4104 ft LAT/LONG 33° 15' N / 103° 20' W

| | JAN | FEB | MAR | APR | MAY | JUN | JUL | AUG | SEP | OCT | NOV | DEC | YEAR |
|---|---|---|---|---|---|---|---|---|---|---|---|---|---|
| Maximum Temp °F | 55.0 | 60.1 | 68.0 | 75.8 | 82.9 | *90.9* | 91.3 | 89.6 | 83.5 | 75.2 | 63.2 | 55.6 | 74.3 |
| Minimum Temp °F | 22.7 | 25.2 | 31.4 | 39.6 | 48.6 | *57.4* | 61.3 | 59.6 | 52.9 | 41.2 | 30.1 | 23.1 | 41.1 |
| Mean Temp °F | 38.8 | 42.7 | 49.7 | 57.7 | 65.8 | *74.1* | 76.3 | 74.6 | 68.2 | 58.3 | 46.7 | 39.4 | 57.7 |
| Days Max Temp ≥ 90 °F | 0 | 0 | 0 | 1 | 6 | 18 | 19 | 17 | 7 | 1 | 0 | 0 | 69 |
| Days Max Temp ≤ 32 °F | 2 | 1 | 0 | 0 | 0 | 0 | 0 | 0 | 0 | 0 | 0 | 2 | 5 |
| Days Min Temp ≤ 32 °F | 28 | 23 | 18 | 6 | 1 | 0 | 0 | 0 | 0 | 4 | 18 | 27 | 125 |
| Days Min Temp ≤ 0 °F | 0 | 0 | 0 | 0 | 0 | 0 | 0 | 0 | 0 | 0 | 0 | 0 | 0 |
| Heating Degree Days | 805 | 624 | 469 | 230 | 66 | *5* | 0 | 1 | 41 | 220 | 544 | 788 | 3793 |
| Cooling Degree Days | 0 | 0 | 1 | 16 | 94 | *298* | 376 | 333 | 158 | 12 | 1 | 0 | 1289 |
| Total Precipitation (") | 0.37 | 0.44 | 0.53 | 0.50 | 2.22 | 2.24 | 2.75 | 2.54 | 2.62 | 1.52 | 0.70 | 0.50 | 16.93 |
| Days ≥ 0.1" Precip | 1 | 1 | 1 | 1 | 4 | 4 | 5 | 5 | 4 | 2 | 2 | *1* | 31 |
| Total Snowfall (") | *0.8* | *1.6* | *1.0* | 0.6 | 0.0 | 0.0 | 0.0 | 0.0 | 0.0 | 0.1 | 0.3 | na | na |
| Days ≥ 1" Snow Depth | na | na | *0* | 0 | 0 | 0 | 0 | 0 | 0 | 0 | 0 | na | na |

## TIERRA AMARILLA 4 N *Rio Arriba County* ELEVATION 7427 ft LAT/LONG 36° 45' N / 106° 34' W

| | JAN | FEB | MAR | APR | MAY | JUN | JUL | AUG | SEP | OCT | NOV | DEC | YEAR |
|---|---|---|---|---|---|---|---|---|---|---|---|---|---|
| Maximum Temp °F | *37.9* | 42.6 | *49.6* | *59.2* | *68.0* | 77.5 | 82.1 | *79.5* | *73.0* | *63.2* | *48.2* | *39.5* | 60.0 |
| Minimum Temp °F | *3.2* | 9.0 | *17.9* | *24.2* | *31.2* | 37.8 | *45.4* | *44.7* | *36.6* | *27.2* | *16.4* | *7.5* | 25.1 |
| Mean Temp °F | *20.6* | *25.7* | *33.8* | *41.7* | *49.6* | 57.7 | *63.8* | *62.1* | *54.8* | *45.2* | *32.3* | *23.5* | 42.6 |
| Days Max Temp ≥ 90 °F | 0 | 0 | 0 | 0 | 0 | 1 | 1 | 0 | 0 | 0 | 0 | 0 | 2 |
| Days Max Temp ≤ 32 °F | 7 | 2 | 1 | 0 | 0 | 0 | 0 | 0 | 0 | 0 | 2 | 6 | 18 |
| Days Min Temp ≤ 32 °F | 30 | 28 | 30 | 24 | 18 | 5 | 0 | 0 | 7 | 23 | *27* | 30 | 222 |
| Days Min Temp ≤ 0 °F | 12 | 6 | 1 | 0 | 0 | 0 | 0 | 0 | 0 | 0 | 2 | *8* | 29 |
| Heating Degree Days | *1371* | *1103* | *961* | *693* | *470* | 217 | *57* | *95* | *299* | *606* | *980* | *1279* | 8131 |
| Cooling Degree Days | *0* | 0 | *0* | *0* | na | *4* | *19* | na | na | *0* | na | *0* | na |
| Total Precipitation (") | 1.02 | 1.07 | 1.13 | 0.80 | 1.27 | 0.91 | 2.00 | 2.77 | 1.84 | 1.42 | 1.20 | 0.97 | 16.40 |
| Days ≥ 0.1" Precip | 3 | 3 | 4 | 2 | *3* | 3 | 6 | 7 | 4 | 4 | *3* | 3 | 45 |
| Total Snowfall (") | *11.9* | *11.4* | *10.5* | 2.6 | 0.2 | 0.0 | 0.0 | 0.0 | 0.0 | *2.2* | na | *12.8* | na |
| Days ≥ 1" Snow Depth | *22* | 16 | *11* | 1 | 0 | 0 | 0 | 0 | 0 | 1 | *5* | *14* | 70 |

## TORREON NAVAJO MISSI *Sandoval County* ELEVATION 6703 ft LAT/LONG 35° 48' N / 107° 11' W

| | JAN | FEB | MAR | APR | MAY | JUN | JUL | AUG | SEP | OCT | NOV | DEC | YEAR |
|---|---|---|---|---|---|---|---|---|---|---|---|---|---|
| Maximum Temp °F | 41.2 | 46.9 | 56.0 | 65.7 | 74.7 | 85.7 | 88.8 | 85.5 | 78.4 | 67.4 | 52.5 | 42.7 | 65.5 |
| Minimum Temp °F | 14.8 | 20.2 | 25.3 | 31.1 | 39.8 | 48.9 | 55.9 | 54.2 | 46.6 | 35.0 | 23.9 | 16.2 | 34.3 |
| Mean Temp °F | 28.0 | 33.6 | 40.6 | 48.4 | 57.3 | 67.3 | 72.4 | 69.9 | 62.5 | 51.2 | 38.2 | 29.5 | 49.9 |
| Days Max Temp ≥ 90 °F | 0 | 0 | 0 | 0 | 0 | 10 | 15 | 7 | 1 | 0 | 0 | 0 | 33 |
| Days Max Temp ≤ 32 °F | 5 | 2 | 0 | 0 | 0 | 0 | 0 | 0 | 0 | 0 | 1 | 4 | 12 |
| Days Min Temp ≤ 32 °F | 30 | 27 | 26 | 17 | 5 | 0 | 0 | 0 | 1 | 11 | 26 | 30 | 173 |
| Days Min Temp ≤ 0 °F | 3 | 1 | 0 | 0 | 0 | 0 | 0 | 0 | 0 | 0 | 0 | 2 | 6 |
| Heating Degree Days | 1138 | 881 | 748 | 491 | 240 | 38 | 1 | 7 | 100 | 420 | 797 | 1095 | 5956 |
| Cooling Degree Days | 0 | 0 | 0 | 0 | 8 | 124 | 234 | 179 | 35 | 0 | 0 | 0 | 580 |
| Total Precipitation (") | 0.54 | 0.43 | 0.63 | 0.55 | 0.81 | 0.64 | 1.52 | 1.88 | 1.22 | 0.98 | 0.72 | 0.55 | 10.47 |
| Days ≥ 0.1" Precip | 2 | 2 | 2 | 2 | 2 | 2 | 4 | 5 | 4 | 3 | 2 | 2 | 32 |
| Total Snowfall (") | *5.2* | 4.0 | 2.8 | 0.9 | 0.3 | 0.0 | 0.0 | 0.0 | 0.0 | 0.7 | 1.8 | 4.5 | 20.2 |
| Days ≥ 1" Snow Depth | 9 | 3 | 1 | 0 | 0 | 0 | 0 | 0 | 0 | 0 | 1 | 6 | 20 |

## TUCUMCARI 4 NE *Quay County* ELEVATION 4104 ft LAT/LONG 35° 12' N / 103° 41' W

| | JAN | FEB | MAR | APR | MAY | JUN | JUL | AUG | SEP | OCT | NOV | DEC | YEAR |
|---|---|---|---|---|---|---|---|---|---|---|---|---|---|
| Maximum Temp °F | 53.7 | 58.0 | 65.9 | 74.2 | 81.6 | 90.5 | 93.3 | 90.7 | 84.2 | 75.1 | 62.7 | 54.3 | 73.7 |
| Minimum Temp °F | 23.6 | 27.2 | 33.8 | 42.0 | 50.9 | 59.8 | 64.4 | 62.3 | 55.4 | 44.0 | 33.4 | 25.2 | 43.5 |
| Mean Temp °F | 38.6 | 42.6 | 49.8 | 58.1 | 66.3 | 75.2 | 78.9 | 76.5 | 69.8 | 59.6 | 48.0 | 39.8 | 58.6 |
| Days Max Temp ≥ 90 °F | 0 | 0 | 0 | 1 | 5 | 18 | 25 | 20 | 8 | 1 | 0 | 0 | 78 |
| Days Max Temp ≤ 32 °F | 2 | 1 | 0 | 0 | 0 | 0 | 0 | 0 | 0 | 0 | 0 | 2 | 5 |
| Days Min Temp ≤ 32 °F | 26 | 21 | 13 | 4 | 0 | 0 | 0 | 0 | 0 | 2 | 14 | 25 | 105 |
| Days Min Temp ≤ 0 °F | 0 | 0 | 0 | 0 | 0 | 0 | 0 | 0 | 0 | 0 | 0 | 0 | 0 |
| Heating Degree Days | 810 | 625 | 465 | 222 | 65 | 6 | 0 | 1 | 32 | 195 | 503 | 775 | 3699 |
| Cooling Degree Days | 0 | 0 | 2 | 22 | 99 | 312 | 434 | 369 | 181 | 24 | 1 | 0 | 1444 |
| Total Precipitation (") | 0.39 | 0.47 | 0.71 | 0.98 | 1.77 | 2.06 | 2.64 | 2.77 | 1.78 | 1.21 | 0.72 | 0.43 | 15.93 |
| Days ≥ 0.1" Precip | 1 | 1 | 2 | 2 | 4 | 4 | 5 | 5 | 3 | 2 | 2 | 1 | 32 |
| Total Snowfall (") | na | *4.0* | 1.9 | 1.0 | 0.0 | 0.0 | 0.0 | 0.0 | 0.0 | 0.4 | 0.9 | *3.5* | na |
| Days ≥ 1" Snow Depth | na | *2* | 1 | 0 | 0 | 0 | 0 | 0 | 0 | 0 | 1 | na | na |

## TULAROSA *Otero County* ELEVATION 4442 ft LAT/LONG 33° 4' N / 106° 2' W

| | JAN | FEB | MAR | APR | MAY | JUN | JUL | AUG | SEP | OCT | NOV | DEC | YEAR |
|---|---|---|---|---|---|---|---|---|---|---|---|---|---|
| Maximum Temp °F | *56.5* | 60.7 | 68.3 | 76.2 | 83.6 | 92.9 | 93.1 | *90.5* | *85.0* | *75.5* | *63.5* | *55.4* | 75.1 |
| Minimum Temp °F | *28.3* | 31.7 | 36.8 | 43.5 | *51.6* | 59.5 | 63.4 | *61.9* | *55.9* | *45.8* | *34.9* | 28.4 | 45.1 |
| Mean Temp °F | *42.5* | 46.3 | 52.5 | 60.0 | *67.6* | 76.2 | 78.2 | *76.2* | *70.4* | *60.6* | *49.2* | *41.8* | 60.1 |
| Days Max Temp ≥ 90 °F | 0 | 0 | 0 | 1 | 5 | *21* | *25* | *20* | *7* | *0* | 0 | 0 | 79 |
| Days Max Temp ≤ 32 °F | 0 | 0 | 0 | 0 | 0 | 0 | 0 | 0 | 0 | 0 | 0 | 0 | 0 |
| Days Min Temp ≤ 32 °F | *23* | *15* | *9* | 3 | 0 | 0 | 0 | 0 | 0 | *1* | *11* | *23* | 85 |
| Days Min Temp ≤ 0 °F | 0 | 0 | 0 | 0 | 0 | 0 | 0 | 0 | 0 | 0 | 0 | 0 | 0 |
| Heating Degree Days | *691* | 523 | 383 | 167 | *37* | 1 | 0 | *0* | *14* | *147* | *456* | *711* | 3130 |
| Cooling Degree Days | *0* | *0* | 2 | 26 | 126 | 345 | 410 | 352 | 177 | 19 | 0 | 0 | 1457 |
| Total Precipitation (") | 0.54 | 0.53 | 0.44 | 0.27 | 0.62 | 0.71 | 1.95 | 2.06 | 1.28 | 1.24 | 0.72 | 0.95 | 11.31 |
| Days ≥ 0.1" Precip | 2 | 2 | 1 | 1 | 2 | 2 | 5 | 5 | *3* | *2* | *2* | 3 | 30 |
| Total Snowfall (") | *0.3* | *0.4* | 0.1 | 0.0 | 0.0 | 0.0 | 0.0 | 0.0 | 0.0 | 0.0 | 0.4 | *0.3* | 1.5 |
| Days ≥ 1" Snow Depth | *0* | *0* | 0 | 0 | 0 | 0 | 0 | 0 | 0 | 0 | *0* | 0 | 0 |

## VALMORA *Mora County* ELEVATION 6306 ft LAT/LONG 35° 49 ' N / 104° 56 ' W

| | JAN | FEB | MAR | APR | MAY | JUN | JUL | AUG | SEP | OCT | NOV | DEC | YEAR |
|---|---|---|---|---|---|---|---|---|---|---|---|---|---|
| Maximum Temp °F | 48.5 | 51.8 | 57.2 | 64.7 | 72.4 | 81.7 | 85.0 | 82.4 | 76.9 | 68.8 | 57.6 | 50.1 | 66.4 |
| Minimum Temp °F | 13.8 | 17.3 | 23.2 | 29.9 | 38.3 | 46.5 | 52.4 | 50.9 | 43.5 | 32.0 | 22.2 | 14.8 | 32.1 |
| Mean Temp °F | 31.1 | 34.6 | 40.2 | 47.3 | 55.4 | 64.1 | 68.7 | 66.6 | 60.2 | 50.4 | 40.0 | 32.4 | 49.3 |
| Days Max Temp ≥ 90 °F | 0 | 0 | 0 | 0 | 0 | 5 | 7 | 3 | 1 | 0 | 0 | 0 | 16 |
| Days Max Temp ≤ 32 °F | 3 | 2 | 1 | 0 | 0 | 0 | 0 | 0 | 0 | 0 | 1 | 2 | 9 |
| Days Min Temp ≤ 32 °F | 30 | 27 | 27 | 19 | 6 | 0 | 0 | 0 | 1 | 17 | 27 | 30 | 184 |
| Days Min Temp ≤ 0 °F | 3 | 1 | 0 | 0 | 0 | 0 | 0 | 0 | 0 | 0 | 0 | 2 | 6 |
| Heating Degree Days | 1043 | 850 | 762 | 523 | 294 | 75 | 8 | 24 | 149 | 445 | 745 | 1002 | 5920 |
| Cooling Degree Days | 0 | 0 | 0 | 0 | 4 | 67 | 137 | 95 | 16 | 0 | 0 | 0 | 319 |
| Total Precipitation (") | 0.39 | 0.42 | 0.71 | 0.78 | 1.95 | 2.03 | 3.56 | 3.33 | 2.11 | 0.98 | 0.68 | 0.58 | 17.52 |
| Days ≥ 0.1" Precip | 1 | 2 | 2 | 2 | 4 | 4 | 7 | 7 | 4 | 2 | 1 | 2 | 38 |
| Total Snowfall (") | *5.2* | *5.4* | *5.9* | 2.3 | 0.0 | 0.0 | 0.0 | 0.0 | 0.0 | 1.9 | *2.1* | 6.9 | 29.7 |
| Days ≥ 1" Snow Depth | *4* | 2 | 2 | 1 | 0 | 0 | 0 | 0 | 0 | 0 | 1 | 4 | 14 |

## WHITE SANDS NATL MON *Otero County* ELEVATION 4003 ft LAT/LONG 32° 47 ' N / 106° 10 ' W

| | JAN | FEB | MAR | APR | MAY | JUN | JUL | AUG | SEP | OCT | NOV | DEC | YEAR |
|---|---|---|---|---|---|---|---|---|---|---|---|---|---|
| Maximum Temp °F | 56.9 | 62.7 | 70.8 | 79.6 | 87.5 | 96.5 | 97.0 | 93.8 | 88.2 | 78.7 | 65.5 | 56.6 | 77.8 |
| Minimum Temp °F | 22.7 | 25.9 | 31.3 | 39.1 | 48.0 | 57.9 | 63.5 | 60.8 | 53.6 | 40.4 | 28.6 | 22.5 | 41.2 |
| Mean Temp °F | 39.8 | 44.3 | 51.0 | 59.4 | 67.8 | 77.2 | 80.3 | 77.3 | 71.0 | 59.6 | 47.2 | 39.6 | 59.5 |
| Days Max Temp ≥ 90 °F | 0 | 0 | 0 | 2 | 12 | 27 | 28 | 25 | 13 | 2 | 0 | 0 | 109 |
| Days Max Temp ≤ 32 °F | 0 | 0 | 0 | 0 | 0 | 0 | 0 | 0 | 0 | 0 | 0 | 0 | 0 |
| Days Min Temp ≤ 32 °F | 27 | 21 | 17 | 7 | 1 | 0 | 0 | 0 | 0 | 6 | 21 | 27 | 127 |
| Days Min Temp ≤ 0 °F | 0 | 0 | 0 | 0 | 0 | 0 | 0 | 0 | 0 | 0 | 0 | 0 | 0 |
| Heating Degree Days | 773 | 577 | 427 | 184 | 39 | 1 | 0 | 0 | 12 | 183 | 526 | 782 | 3504 |
| Cooling Degree Days | 0 | 0 | 0 | 27 | 130 | 382 | 470 | 392 | 194 | 17 | 0 | 0 | 1612 |
| Total Precipitation (") | 0.56 | 0.39 | 0.29 | 0.28 | 0.49 | 0.87 | 1.44 | 2.17 | 1.25 | 0.97 | 0.60 | 0.87 | 10.18 |
| Days ≥ 0.1" Precip | 2 | 1 | 1 | 1 | 2 | 2 | 4 | 5 | 3 | 2 | 2 | 2 | 27 |
| Total Snowfall (") | 1.2 | 0.4 | 0.1 | 0.0 | 0.0 | 0.0 | 0.0 | 0.0 | 0.0 | 0.1 | 0.3 | 1.7 | 3.8 |
| Days ≥ 1" Snow Depth | *0* | 0 | 0 | 0 | 0 | 0 | 0 | 0 | 0 | 0 | 0 | 1 | 1 |

## WHITE SIGNAL *Grant County* ELEVATION 6073 ft LAT/LONG 32° 33 ' N / 108° 22 ' W

| | JAN | FEB | MAR | APR | MAY | JUN | JUL | AUG | SEP | OCT | NOV | DEC | YEAR |
|---|---|---|---|---|---|---|---|---|---|---|---|---|---|
| Maximum Temp °F | 51.7 | 55.0 | 61.0 | 69.3 | 77.7 | 87.2 | 87.0 | 83.8 | 79.1 | 71.0 | 59.3 | 51.9 | 69.5 |
| Minimum Temp °F | 23.8 | 26.5 | 30.1 | 35.5 | 43.3 | 53.0 | 59.0 | 56.7 | 49.8 | 39.1 | 29.4 | 24.2 | 39.2 |
| Mean Temp °F | 37.7 | 40.7 | 45.4 | 52.4 | 60.6 | 70.1 | 73.0 | 70.4 | 64.5 | 55.1 | 44.4 | 38.1 | 54.4 |
| Days Max Temp ≥ 90 °F | 0 | 0 | 0 | 0 | 1 | 11 | 10 | 4 | 1 | 0 | 0 | 0 | 27 |
| Days Max Temp ≤ 32 °F | 0 | 0 | 0 | 0 | 0 | 0 | 0 | 0 | 0 | 0 | 0 | 1 | 1 |
| Days Min Temp ≤ 32 °F | 27 | 23 | 20 | 12 | 2 | 0 | 0 | 0 | 0 | 5 | 21 | 27 | 137 |
| Days Min Temp ≤ 0 °F | 0 | 0 | 0 | 0 | 0 | 0 | 0 | 0 | 0 | 0 | 0 | 0 | 0 |
| Heating Degree Days | 838 | 678 | 600 | 372 | 157 | 16 | 1 | 5 | 64 | 303 | 612 | 828 | 4474 |
| Cooling Degree Days | 0 | 0 | 0 | 2 | 26 | 179 | 229 | 164 | 48 | 2 | 0 | 0 | 650 |
| Total Precipitation (") | 1.25 | 1.07 | 0.88 | 0.31 | 0.45 | 0.58 | 2.77 | 2.91 | 1.64 | 1.35 | 1.07 | 1.84 | 16.12 |
| Days ≥ 0.1" Precip | 3 | 3 | 3 | 1 | 1 | 2 | 7 | 6 | 4 | 3 | 2 | 4 | 39 |
| Total Snowfall (") | 3.6 | *3.6* | 2.6 | 0.5 | 0.0 | 0.0 | 0.0 | 0.0 | 0.0 | 0.4 | 1.5 | 4.3 | 16.5 |
| Days ≥ 1" Snow Depth | na | na | *0* | 0 | 0 | 0 | 0 | 0 | 0 | 0 | 0 | *1* | na |

## WINSTON *Sierra County* ELEVATION 6204 ft LAT/LONG 33° 21 ' N / 107° 39 ' W

| | JAN | FEB | MAR | APR | MAY | JUN | JUL | AUG | SEP | OCT | NOV | DEC | YEAR |
|---|---|---|---|---|---|---|---|---|---|---|---|---|---|
| Maximum Temp °F | 52.9 | 57.0 | 62.1 | 70.3 | 77.2 | 87.0 | 87.1 | 83.7 | 79.1 | 72.0 | 61.1 | 52.8 | 70.2 |
| Minimum Temp °F | 18.5 | 22.4 | 26.1 | 31.5 | 38.5 | 48.0 | 54.3 | 52.1 | 44.7 | 34.2 | 24.1 | 18.2 | 34.4 |
| Mean Temp °F | 35.7 | 39.7 | 44.1 | 50.9 | 57.9 | 67.5 | 70.7 | 67.9 | 61.9 | 53.1 | 42.6 | 35.7 | 52.3 |
| Days Max Temp ≥ 90 °F | 0 | 0 | 0 | 0 | 1 | na | na | 3 | 1 | 0 | 0 | 0 | na |
| Days Max Temp ≤ 32 °F | 1 | 0 | 0 | 0 | 0 | 0 | 0 | 0 | 0 | 0 | 0 | 0 | 1 |
| Days Min Temp ≤ 32 °F | 30 | 24 | *25* | na | na | 0 | 0 | 0 | 1 | na | *27* | 30 | na |
| Days Min Temp ≤ 0 °F | 1 | 0 | 0 | 0 | 0 | 0 | 0 | 0 | 0 | 0 | 0 | 0 | 1 |
| Heating Degree Days | 904 | 708 | 642 | 417 | 220 | 32 | 1 | 8 | 106 | 364 | 666 | 904 | 4972 |
| Cooling Degree Days | 0 | 0 | 0 | 0 | 7 | 130 | 172 | 102 | 19 | 0 | 0 | 0 | 430 |
| Total Precipitation (") | 0.49 | 0.41 | 0.27 | 0.22 | 0.82 | 0.77 | 2.78 | 3.39 | 2.30 | 1.16 | 0.52 | 0.77 | 13.90 |
| Days ≥ 0.1" Precip | *2* | 1 | 1 | 1 | 2 | 2 | 7 | 7 | 4 | 2 | 1 | 2 | 32 |
| Total Snowfall (") | *3.8* | 2.5 | *1.4* | 0.1 | 0.0 | 0.0 | 0.0 | 0.0 | 0.0 | 0.9 | 1.4 | 4.9 | 15.0 |
| Days ≥ 1" Snow Depth | na | na | *0* | 0 | 0 | 0 | 0 | 0 | 0 | 0 | 0 | na | na |

## WOLF CANYON *Sandoval County* ELEVATION 8156 ft LAT/LONG 35° 58 ' N / 106° 46 ' W

| | JAN | FEB | MAR | APR | MAY | JUN | JUL | AUG | SEP | OCT | NOV | DEC | YEAR |
|---|---|---|---|---|---|---|---|---|---|---|---|---|---|
| Maximum Temp °F | 37.9 | 40.2 | 45.3 | 54.3 | 63.1 | 73.2 | 76.3 | 73.1 | 67.4 | 58.2 | 45.5 | 38.9 | 56.1 |
| Minimum Temp °F | 7.2 | 10.4 | 16.8 | 22.7 | 28.6 | 35.0 | 42.1 | 41.4 | 34.4 | 25.1 | 16.4 | 9.2 | 24.1 |
| Mean Temp °F | 22.6 | 25.3 | 31.1 | 38.5 | 45.9 | 54.2 | 59.2 | 57.3 | 51.0 | 41.7 | 31.0 | 24.1 | 40.2 |
| Days Max Temp ≥ 90 °F | 0 | 0 | 0 | 0 | 0 | 0 | 0 | 0 | 0 | 0 | 0 | 0 | 0 |
| Days Max Temp ≤ 32 °F | 8 | 5 | 2 | 0 | 0 | 0 | 0 | 0 | 0 | 0 | 3 | 7 | 25 |
| Days Min Temp ≤ 32 °F | 31 | 28 | 31 | 29 | 24 | 10 | 0 | 1 | 11 | 27 | 29 | 31 | 252 |
| Days Min Temp ≤ 0 °F | 8 | 4 | 1 | 0 | 0 | 0 | 0 | 0 | 0 | 0 | 2 | 6 | 21 |
| Heating Degree Days | 1309 | 1114 | 1045 | 788 | 586 | 319 | 174 | 233 | 415 | 716 | 1014 | 1262 | 8975 |
| Cooling Degree Days | 0 | 0 | 0 | 0 | 0 | 0 | 1 | 0 | 0 | 0 | 0 | 0 | 1 |
| Total Precipitation (") | 1.86 | 1.66 | 2.14 | 1.23 | 1.48 | 1.34 | 3.25 | 3.82 | 2.06 | 1.78 | 1.73 | 1.75 | 24.10 |
| Days ≥ 0.1" Precip | 5 | 5 | 6 | 3 | 4 | 4 | 8 | 9 | 5 | 4 | 4 | 5 | 62 |
| Total Snowfall (") | 23.8 | 23.2 | 26.0 | 11.4 | 3.2 | 0.0 | 0.0 | 0.0 | 0.2 | 5.1 | 13.4 | 21.8 | 128.1 |
| Days ≥ 1" Snow Depth | 25 | 22 | 18 | 6 | 1 | 0 | 0 | 0 | 0 | 2 | 9 | 21 | 104 |

## YESO 2 S *DeBaca County* ELEVATION 4852 ft LAT/LONG 34° 24 ' N / 104° 37 ' W

| | JAN | FEB | MAR | APR | MAY | JUN | JUL | AUG | SEP | OCT | NOV | DEC | YEAR |
|---|---|---|---|---|---|---|---|---|---|---|---|---|---|
| Maximum Temp °F | *53.1* | *57.6* | *65.2* | 73.4 | 81.4 | 89.7 | 91.6 | *88.2* | 82.3 | 74.2 | 62.1 | *53.0* | 72.7 |
| Minimum Temp °F | *23.2* | *25.9* | *31.4* | 38.8 | 47.5 | 56.0 | 61.1 | *59.3* | 52.4 | 41.5 | 31.3 | *23.9* | 41.0 |
| Mean Temp °F | *38.2* | *41.8* | *48.3* | 56.2 | 64.5 | 72.9 | 76.4 | *73.8* | 67.4 | 57.9 | 46.7 | *38.4* | 56.9 |
| Days Max Temp ≥ 90 °F | 0 | 0 | 0 | 0 | *4* | *16* | *20* | *15* | *4* | 0 | 0 | 0 | 59 |
| Days Max Temp ≤ 32 °F | 2 | 1 | 0 | 0 | 0 | 0 | 0 | 0 | 0 | 0 | 0 | 1 | 4 |
| Days Min Temp ≤ 32 °F | na | na | *15* | *6* | 1 | 0 | 0 | 0 | 0 | 3 | *17* | na | na |
| Days Min Temp ≤ 0 °F | 0 | 0 | 0 | 0 | 0 | 0 | 0 | 0 | 0 | 0 | 0 | 0 | 0 |
| Heating Degree Days | *826* | *649* | *511* | 267 | 84 | 7 | *0* | *2* | 40 | 229 | 543 | *817* | 3975 |
| Cooling Degree Days | *0* | 0 | *1* | 8 | 74 | 249 | 360 | 294 | 119 | 12 | 0 | 0 | 1117 |
| Total Precipitation (") | 0.53 | 0.47 | 0.55 | 0.63 | 1.35 | 1.46 | 2.17 | 2.92 | 1.93 | 1.29 | 0.61 | 0.56 | 14.47 |
| Days ≥ 0.1" Precip | 2 | 1 | 2 | 2 | 3 | 3 | 5 | 5 | 4 | 2 | 1 | 2 | 32 |
| Total Snowfall (") | 4.9 | *4.3* | 2.0 | 1.0 | 0.2 | 0.0 | 0.0 | 0.0 | 0.0 | 0.9 | 1.8 | 4.8 | 19.9 |
| Days ≥ 1" Snow Depth | na | *1* | 1 | 0 | 0 | 0 | 0 | 0 | 0 | 0 | 0 | 2 | na |

## ZUNI *McKinley County* ELEVATION 6452 ft LAT/LONG 35° 6 ' N / 108° 47 ' W

| | JAN | FEB | MAR | APR | MAY | JUN | JUL | AUG | SEP | OCT | NOV | DEC | YEAR |
|---|---|---|---|---|---|---|---|---|---|---|---|---|---|
| Maximum Temp °F | 46.1 | 50.7 | 56.6 | 65.7 | 74.2 | 84.5 | 88.0 | 84.6 | 78.9 | 69.1 | 55.8 | 47.3 | 66.8 |
| Minimum Temp °F | *16.0* | 20.5 | 25.4 | 30.4 | 37.8 | 46.0 | 53.8 | 53.2 | 45.5 | 34.5 | *24.4* | *17.2* | 33.7 |
| Mean Temp °F | *30.9* | 35.5 | 41.0 | 48.2 | 56.0 | 65.3 | 71.0 | 69.0 | 62.2 | 51.8 | *40.0* | *32.2* | 50.3 |
| Days Max Temp ≥ 90 °F | 0 | 0 | 0 | 0 | 0 | 7 | 13 | 6 | 1 | 0 | 0 | 0 | 27 |
| Days Max Temp ≤ 32 °F | 2 | 1 | 0 | 0 | 0 | 0 | 0 | 0 | 0 | 0 | 0 | 2 | 5 |
| Days Min Temp ≤ 32 °F | *30* | 27 | 26 | 20 | 6 | 0 | 0 | 0 | 1 | 13 | 25 | *30* | 178 |
| Days Min Temp ≤ 0 °F | *2* | 0 | 0 | 0 | 0 | 0 | 0 | 0 | 0 | 0 | 0 | 2 | 4 |
| Heating Degree Days | *1049* | 826 | 736 | 504 | 276 | 56 | 1 | 9 | 103 | 402 | *744* | *1009* | 5715 |
| Cooling Degree Days | 0 | 0 | 0 | 0 | 2 | 71 | 175 | 141 | 23 | 0 | 0 | 0 | 412 |
| Total Precipitation (") | 0.93 | 0.73 | 1.08 | 0.62 | 0.60 | 0.42 | 2.08 | 2.39 | 1.41 | 1.14 | 0.87 | 1.08 | 13.35 |
| Days ≥ 0.1" Precip | 3 | 2 | 3 | 2 | 2 | 1 | 5 | 6 | 3 | 3 | 2 | 3 | 35 |
| Total Snowfall (") | *2.8* | *2.9* | *2.1* | 0.7 | 0.1 | 0.0 | 0.0 | 0.0 | 0.0 | 0.3 | *1.1* | 4.9 | 14.9 |
| Days ≥ 1" Snow Depth | *4* | na | *1* | 0 | 0 | 0 | 0 | 0 | 0 | 0 | *1* | *4* | na |

## JANUARY MINIMUM TEMPERATURE °F

### LOWEST

| | | |
|---|---|---|
| 1 | Eagle Nest | 0.5 |
| 2 | Tierra Amarilla | 3.2 |
| 3 | Chama | 4.8 |
| | El Vado Dam | 4.8 |
| 5 | Red River | 4.9 |
| 6 | Dulce | 5.7 |
| 7 | Cerro | 6.2 |
| 8 | Wolf Canyon | 7.2 |
| 9 | McGaffey | 8.1 |
| 10 | Taos | 8.4 |
| 11 | Lake Maloya | 8.8 |
| 12 | Star Lake | 9.6 |
| 13 | Maxwell | 10.8 |
| 14 | Chaco Canyon | 11.8 |
| 15 | Luna | 11.9 |
| 16 | Beaverhead | 12.5 |
| 17 | Springer | 12.6 |
| 18 | El Morro | 13.0 |
| 19 | Augustine | 13.4 |
| | Ocate | 13.4 |
| 21 | Valmora | 13.8 |
| 22 | Fence Lake | 14.0 |
| 23 | Grants | 14.1 |
| 24 | Lybrook | 14.4 |
| | Pecos | 14.4 |

### HIGHEST

| | | |
|---|---|---|
| 1 | Carlsbad-Caverns | 32.7 |
| 2 | Columbus | 29.1 |
| 3 | Alamogordo | 29.0 |
| 4 | Hobbs | 28.9 |
| 5 | Elephant Btte Dm | 28.7 |
| 6 | Jal | 28.5 |
| 7 | Carlsbad-Cvrn Cy | 28.3 |
| | Tularosa | 28.3 |
| 9 | Carlsbad | 27.6 |
| 10 | Orogrande | 27.3 |
| 11 | Hope | 27.2 |
| 12 | Hachita | 26.4 |
| 13 | Animas | 26.3 |
| 14 | Pearl | 25.9 |
| 15 | Caballo Dam | 25.8 |
| 16 | Faywood | 25.7 |
| 17 | Mountain Park | 25.4 |
| 18 | Fort Bayard | 25.3 |
| 19 | Lordsburg | 25.2 |
| | Redrock | 25.2 |
| 21 | Gage | 25.1 |
| | Maljamar | 25.1 |
| 23 | Santa Rosa | 24.9 |
| 24 | Hillsboro | 24.3 |
| 25 | Aleman | 24.2 |

## JULY MAXIMUM TEMPERATURE °F

### HIGHEST

| | | |
|---|---|---|
| 1 | White Sands | 97.0 |
| 2 | Lordsburg | 96.0 |
| 3 | Columbus | 95.6 |
| 4 | Orogrande | 95.5 |
| 5 | Bosque del Apche | 95.4 |
| | Hatch | 95.4 |
| 7 | Jal | 95.3 |
| 8 | Jornada | 95.2 |
| 9 | Bitter Lakes | 95.1 |
| | Caballo Dam | 95.1 |
| | Carlsbad-Cvrn Cy | 95.1 |
| | Redrock | 95.1 |
| 13 | Carlsbad | 94.8 |
| 14 | Bernardo | 94.7 |
| | Gage | 94.7 |
| 16 | Animas | 94.4 |
| 17 | Maljamar | 94.3 |
| 18 | Hachita | 94.2 |
| | Shiprock | 94.2 |
| 20 | Alamogordo | 94.0 |
| 21 | Artesia | 93.7 |
| 22 | Elephant Btte Dm | 93.4 |
| | San Jon | 93.4 |
| 24 | Tucumcari | 93.3 |
| 25 | Conchas Dam | 93.2 |

### LOWEST

| | | |
|---|---|---|
| 1 | Gascon | 75.5 |
| 2 | Red River | 76.3 |
| | Wolf Canyon | 76.3 |
| 4 | Eagle Nest | 77.5 |
| 5 | Lake Maloya | 78.9 |
| 6 | McGaffey | 79.9 |
| 7 | Chama | 80.4 |
| 8 | Los Alamos | 80.5 |
| 9 | Ruidoso | 80.7 |
| 10 | Ocate | 80.9 |
| 11 | Mountain Park | 81.4 |
| 12 | Cerro | 81.9 |
| 13 | Tierra Amarilla | 82.1 |
| 14 | Raton | 82.4 |
| 15 | Las Vegas | 82.9 |
| | Luna | 82.9 |
| 17 | Des Moines | 83.2 |
| 18 | Cimarron | 83.7 |
| 19 | Elk | 83.8 |
| 20 | Augustine | 84.1 |
| | Lybrook | 84.1 |
| 22 | Beaverhead | 84.4 |
| | El Rito | 84.4 |
| | Pecos | 84.4 |
| 25 | Dulce | 84.5 |

## ANNUAL PRECIPITATION (")

### HIGHEST

| | | |
|---|---|---|
| 1 | Lake Maloya | 24.62 |
| 2 | Gascon | 24.12 |
| 3 | Wolf Canyon | 24.10 |
| 4 | Chama | 22.82 |
| 5 | Ruidoso | 22.65 |
| 6 | Red River | 21.99 |
| 7 | Mountain Park | 21.14 |
| 8 | Sandia Park | 20.05 |
| 9 | McGaffey | 19.57 |
| 10 | Cameron | 19.39 |
| 11 | Los Alamos | 19.37 |
| 12 | Mimbres | 18.52 |
| 13 | Dulce | 18.30 |
| 14 | Des Moines | 18.11 |
| 15 | San Jon | 18.06 |
| 16 | Hobbs | 17.95 |
| 17 | Clovis | 17.85 |
| | Ragland | 17.85 |
| 19 | Raton | 17.80 |
| 20 | Ocate | 17.74 |
| 21 | Las Vegas | 17.72 |
| 22 | Jemez Springs | 17.67 |
| 23 | Valmora | 17.52 |
| 24 | Pecos | 17.31 |
| 25 | Melrose | 17.16 |

### LOWEST

| | | |
|---|---|---|
| 1 | Shiprock | 7.79 |
| 2 | Fruitland | 8.51 |
| 3 | Bernardo | 8.54 |
| 4 | Albuquerque | 9.37 |
| 5 | Chaco Canyon | 9.41 |
| 6 | Bloomfield | 9.43 |
| 7 | Los Lunas | 9.52 |
| 8 | Star Lake | 9.79 |
| 9 | Socorro | 9.88 |
| 10 | Bosque del Apche | 9.94 |
| 11 | Elephant Btte Dm | 10.12 |
| 12 | Abiquiu Dam | 10.14 |
| 13 | White Sands | 10.18 |
| 14 | Lybrook | 10.31 |
| 15 | Columbus | 10.34 |
| 16 | Otis | 10.36 |
| 17 | Caballo Dam | 10.44 |
| 18 | Alcalde | 10.47 |
| | Torreon | 10.47 |
| 20 | Aztec Ruins | 10.58 |
| 21 | Laguna | 10.71 |
| 22 | Jornada | 10.77 |
| 23 | Espanola | 10.86 |
| 24 | Hatch | 10.87 |
| 25 | Aleman | 11.09 |

## ANNUAL SNOWFALL (")

### HIGHEST

| | | |
|---|---|---|
| 1 | Red River | 160.5 |
| 2 | Gascon | 136.2 |
| 3 | Wolf Canyon | 128.1 |
| 4 | Chama | 110.1 |
| 5 | Lake Maloya | 104.7 |
| 6 | Sandia Park | 64.7 |
| 7 | Eagle Nest | 61.4 |
| 8 | Cerro | 60.6 |
| 9 | Los Alamos | 57.9 |
| 10 | Dulce | 57.1 |
| 11 | El Morro | 53.4 |
| 12 | El Vado Dam | 45.1 |
| 13 | Las Vegas | 39.0 |
| 14 | Cimarron | 37.7 |
| 15 | Taos | 37.5 |
| 16 | Ocate | 35.9 |
| 17 | Des Moines | 35.8 |
| | Raton | 35.8 |
| 19 | El Rito | 34.7 |
| 20 | Jemez Springs | 31.7 |
| 21 | Valmora | 29.7 |
| 22 | Otis | 29.0 |
| 23 | Springer | 27.3 |
| 24 | Mountainair | 25.3 |
| 25 | Beaverhead | 24.2 |

### LOWEST

| | | |
|---|---|---|
| 1 | Caballo Dam | 1.4 |
| 2 | Tularosa | 1.5 |
| 3 | Columbus | 2.9 |
| 4 | Elephant Btte Dm | 3.0 |
| 5 | Jornada | 3.3 |
| 6 | Hatch | 3.7 |
| 7 | White Sands | 3.8 |
| 8 | Bernardo | 3.9 |
| 9 | Lordsburg | 4.0 |
| 10 | Faywood | 4.1 |
| | Redrock | 4.1 |
| 12 | Orogrande | 4.7 |
| 13 | Jal | 4.9 |
| | Los Lunas | 4.9 |
| 15 | Hachita | 5.0 |
| 16 | Alamogordo | 5.2 |
| 17 | Bosque del Apche | 5.4 |
| 18 | Pearl | 5.7 |
| 19 | Cliff | 6.1 |
| 20 | Animas | 6.6 |
| 21 | Gila Hot Springs | 6.7 |
| 22 | Carlsbad-Caverns | 6.8 |
| 23 | Hobbs | 7.0 |
| 24 | Aleman | 7.3 |
| | Socorro | 7.3 |

# NEW YORK

PHYSICAL FEATURES. New York State contains 49,576 square miles, inclusive of 1,637 square miles of inland water, but exclusive of the boundary-water areas of Long Island Sound, New York Harbor, Lake Ontario, and Lake Erie. The major portion of the State lies generally between latitudes 42° and 45° N. and between longitudes 73° 30' and 79° 45' W. However, in the extreme southeast, a triangular portion extends southward to about latitude 40° 30' N., while Long Island lies eastward to about longitude 72° W.

The principal highland regions of the State are the Adirondacks in the northeast and the Appalachian Plateau (Southern Plateau) in the south. A minor highland region occurs in southeastern New York where the Hudson River has cut a valley between the Palisades on the west, near the New Jersey border, and the Taconic Mountains on the east, along the Connecticut and Massachusetts border. Just west of the Adirondacks and the upper Black River Valley in Lewis County is another minor highland known as Tug Hill. Much of the eastern border of the State consists of a long, narrow lowland region which is occupied by Lake Champlain, Lake George, and the middle and lower portions of the Hudson Valley.

Approximately 40 percent of New York State has an elevation of more than 1,000 feet above sea level. In northwestern Essex County are a number of peaks with an elevation of between 4,000 to 5,000 feet. The highest point, Mount Marcy, reaches a height of 5,344 feet above sea level. The Appalachian Plateau merges variously into the Great Lakes Plain of western New York with gradual- to steep-sloping terrain. This Plateau is penetrated by the valleys of the Finger Lakes which extend southward from the Great Lakes Plain. Other prominent lakes plus innumerable smaller lakes and ponds dot the landscape, with more than 1,500 in the Adirondack region alone.

GENERAL CLIMATE. The climate of New York State is broadly representative of the humid continental type which prevails in the Northeastern United States, but its diversity is not usually encountered within an area of comparable size. The geographical position of the State and the usual course of air masses, governed by the large-scale patterns of atmospheric circulation, provide general climatic controls. Differences in latitude, character of the topography, and proximity to large bodies of water have pronounced effects on the climate.

Lengthy periods of either abnormally cold or warm weather result from the movement of great high pressure (anticyclonic) systems into and through the Eastern United States. Cold winter temperatures prevail over New York whenever Arctic air masses, under high barometric pressure, flow southward from central Canada or from Hudson Bay. High pressure systems often move just off the Atlantic coast, become more or less stagnant for several days, and then a persistent air flow from the southwest or south affects the State. This circulation brings the very warm, often humid weather of the summer season and the mild, more pleasant temperatures during the fall, winter, and spring seasons.

TEMPERATURE. Many atmospheric and physiographic controls on the climate result in a considerable variation of temperature conditions over New York State. The average annual mean temperature ranges from about 40° in the Adirondacks to near 55° in the New York City area. The winters are long and cold in the Plateau Divisions of the State. Winter temperatures are moderated considerably in the Great Lakes Plain of western New York. The moderating influence of Lakes Erie and Ontario is comparable to that produced by the Atlantic Ocean in the southern portion of the Hudson Valley.

The summer climate is cool in the Adirondacks, Catskills, and higher elevations of the Southern Plateau. The New York City area and lower portions of the Hudson Valley have rather warm summers by comparison, with some periods of high, uncomfortable humidity. The remainder of New York State enjoys pleasantly warm summers, marred by only occasional, brief intervals of sultry conditions. Summer daytime temperatures usually range from the upper 70s to mid-80s over much of the State. The moderating effect of Lakes Erie and Ontario on temperatures assumes practical importance during the spring and fall seasons. The lake waters warm slowly in the spring, the effect of which is to reduce the warming of the atmosphere over adjacent land areas. In the fall season, the lake waters cool more slowly than the land areas and thus serve as a heat source.

PRECIPITATION. Moisture for precipitation in New York State is transported primarily from the Gulf of Mexico and Atlantic Ocean through circulation patterns and storm systems of the atmosphere. Distribution of precipitation within the State is greatly influenced by topography and proximity to the Great Lakes or Atlantic Ocean. Average annual amounts in excess of 50 inches occur in the western Adirondacks, Tug Hill area, and the Catskills, while slightly less than that amount is noted in the higher elevations of the Western Plateau southeast of Lake Erie. Areas of least rainfall, with average accumulations of about 30 inches, occur near Lake Ontario in the extreme western counties, in the lower half of the Genesee River Valley, and in the vicinity of Lake Champlain.

New York State has a fairly uniform distribution of precipitation during the year. There are no distinctly dry or wet seasons which are regularly repeated on an annual basis. Minimum precipitation occurs in the winter season. Maximum amounts are noted in the summer season throughout the State except along the Great Lakes where slight peaks of similar magnitude occur in both the spring and fall seasons.

SNOWFALL. The climate of New York State is marked by abundant snowfall. With the exception of the Coastal Division, the State receives an average seasonal amount of 40 inches or more. The average snowfall is greater than 70 inches over some 60 percent of New York's area. The moderating influence of the Atlantic Ocean reduces the snow accumulation to 25 to 35 inches in the New York City area and on Long Island. About one-third of the winter season precipitation in the Coastal Division occurs from storms which also yield at least 1 inch of snow. The great bulk of the winter precipitation in upstate New York comes as snow.

A durable snow cover generally begins to develop in the Adirondacks and northern lowlands by late November and remains on the ground until various times in April, depending upon late winter snowfall and early spring temperatures. The Southern Plateau, Great Lakes Plain in southern portions of western upstate New York, and the Hudson Valley experience a continuous snow cover from about mid-December to mid-March, with maximum depths usually occurring in February. Bare ground may occur briefly in the lower elevations of these regions during some winters. From late December or early January through February, the Atlantic coastal region of the State experiences alternating periods of measurable snow cover and bare ground.

FLOODS. Although major floods are relatively infrequent, the greatest potential and frequency for floods occur in the early spring when substantial rains combine with rapid snowmelting to produce a heavy runoff. Damaging floods are caused at other times of the year by prolonged periods of heavy rainfall.

WINDS AND STORMS. The prevailing wind is generally from the west in New York State. A southwest component becomes evident in winds during the warmer months while a northwest component is characteristic of the colder one-half of the year. Thunderstorms occur on an average of about 30 days in a year throughout the State. Destructive winds and lightning strikes in local areas are common with the more vigorous warm-season thunderstorms. Locally, hail occurs with more severe thunderstorms. Tornadoes are not common. About 3 or 4 of these storms strike limited, localized areas of New York State in most years. Tornadoes occur generally between late May and late August. Storms of freezing rain occur on one or more occasions during the winter season and often affect a wide area of the State in any one incident. Such storms are usually limited to a thin but dangerous coating of ice on exposed surfaces. Hurricanes and tropical storms periodically cause serious and heavy losses in the vicinity of Long Island and southeastern upstate New York. The greatest storm hazard in terms of area affected is heavy snow. Coastal northeaster storms occur with some frequency in most winters. Blizzard conditions of heavy snow, high winds, and rapidly falling temperature occur occasionally, but are much less characteristic of New York's climate than in the plains of Midwestern United States.

OTHER CLIMATIC ELEMENTS. The climate of the State features much cloudy weather during the months of November, December, and January in upstate New York. From June through September, however, about 60 to 70 percent of the possible sunshine hours is received. In the Atlantic coastal region, the sunshine hours increases from 50 percent of possible in the winter to about 65 percent of possible in the summer. The occurrence of heavy dense fog is variable over the State. The valleys and ridges of the Southern Plateau are most subject to periods of fog, with occurrences averaging about 50 days in a year. In the Great Lakes Plain and northern valleys, the frequency decreases to only 10 to 20 days annually. In those portions of the State with greater maritime influence, the frequency of dense fog in a year ranges from about 35 days on the south shore of Long Island to 25 days in the Hudson Valley.

## COUNTY INDEX

***Albany County***
ALBANY COUNTY AP
ALCOVE DAM

***Allegany County***
ALFRED
ANGELICA
BOLIVAR

***Broome County***
BINGHAMTON LINK FLD

***Cattaraugus County***
ALLEGANY STATE PARK
FRANKLINVILLE 1 SSW
LITTLE VALLEY

***Cayuga County***
AURORA RESEARCH FARM

***Chautauqua County***
FREDONIA
WESTFIELD 3 SW

***Chemung County***
ELMIRA

***Chenango County***
NORWICH

***Clinton County***
CHAZY
DANNEMORA
PERU 2 WSW
PLATTSBURGH AFB

***Columbia County***
HUDSON CORRECTIONAL

***Cortland County***
CORTLAND

***Delaware County***
DEPOSIT
WALTON

***Dutchess County***
GLENHAM
MILLBROOK
POUGHKEEPSIE

***Erie County***
BUFFALO GR BUFFLO AP
COLDEN 1 N
GOWANDA PSYCHIATRIC

***Essex County***
LAKE PLACID 2 S

***Franklin County***
CHASM FALLS
TUPPER LAKE SUNMOUNT

***Fulton County***
GLOVERSVILLE

***Genesee County***
BATAVIA

***Hamilton County***
INDIAN LAKE 2 SW

***Herkimer County***
LITTLE FALLS CITY RE
OLD FORGE
STILLWATER RESERVOIR

***Jefferson County***
WATERTOWN
WATERTOWN AP

***Kings County***
NEW YORK

***Lewis County***
LOWVILLE

***Livingston County***
DANSVILLE
HEMLOCK
MOUNT MORRIS 2 W

***Monroe County***
ROCHESTER INTL AP

***Nassau County***
MINEOLA

***New York County***
NEW YORK CNTRL PARK

***Niagara County***
LOCKPORT 2 NE
LOCKPORT 4 NE

***Oneida County***
BOONVILLE 2 SSW
CAMDEN 2 NW
UTICA ONEIDA CNTY AP

***Onondaga County***
SYRACUSE HANCOCK AP

***Ontario County***
CANANDAIGUA 3 S

***Orange County***
MIDDLETOWN 2 NW
PORT JERVIS
WEST POINT

***Orleans County***
ALBION 2 NE

***Oswego County***
OSWEGO EAST

***Otsego County***
CHERRY VALLEY 2 NNE
COOPERSTOWN

***Queens County***
NEW YORK J F KENNEDY
NEW YORK LAGUARDIA

***Rensselaer County***
GRAFTON
TROY L & D

***Rockland County***
SUFFERN

***St. Lawrence County***
CANTON 4 SE
GOUVERNEUR 3 NW
LAWRENCEVILLE
MASSENA AP
OGDENSBURG 4 NE
WANAKENA RANGER SCHO

### *Saratoga County*
SARATOGA SPRINGS 4 S

### *Steuben County*
BATH

### *Suffolk County*
BRIDGEHAMPTON
GREENPORT POWER HOUS
PATCHOGUE 2 N
RIVERHEAD RES FARM
SETAUKET STRONG

### *Sullivan County*
LIBERTY 1 NE

### *Tioga County*
SPENCER 2 N

### *Tompkins County*
ITHACA CORNELL UNIV

### *Ulster County*
MOHONK LAKE
SLIDE MOUNTAIN

### *Warren County*
GLENS FALLS AP
GLENS FALLS FARM

### *Washington County*
SALEM
WHITEHALL

### *Wayne County*
SODUS CENTER

### *Westchester County*
DOBBS FERRY ARDSLEY
YORKTOWN HEIGHTS 1 W

### *Wyoming County*
ARCADE
WARSAW 6 SW

## ELEVATION INDEX

| FEET | STATION NAME |
|---|---|
| 20 | GREENPORT POWER HOUS |
| 20 | NEW YORK |
| 20 | NEW YORK J F KENNEDY |
| 20 | TROY L & D |
| 39 | SETAUKET STRONG |
| 52 | NEW YORK LAGUARDIA |
| 59 | BRIDGEHAMPTON |
| 59 | HUDSON CORRECTIONAL |
| 59 | PATCHOGUE 2 N |
| 100 | RIVERHEAD RES FARM |
| 121 | CHAZY |
| 121 | WHITEHALL |
| 131 | MINEOLA |
| 144 | NEW YORK CNTRL PARK |
| 161 | POUGHKEEPSIE |
| 171 | PLATTSBURGH AFB |
| 207 | MASSENA AP |
| 230 | GLENHAM |
| 239 | DOBBS FERRY ARDSLEY |
| 269 | SUFFERN |
| 279 | OGDENSBURG 4 NE |
| 289 | OSWEGO EAST |
| 292 | ALBANY COUNTY AP |
| 312 | SARATOGA SPRINGS 4 S |
| 322 | WEST POINT |
| 331 | WATERTOWN AP |
| 341 | GLENS FALLS AP |
| 413 | CANTON 4 SE |
| 420 | GOUVERNEUR 3 NW |
| 420 | SODUS CENTER |
| 427 | SYRACUSE HANCOCK AP |
| 440 | LOCKPORT 4 NE |
| 469 | ALBION 2 NE |
| 469 | PORT JERVIS |
| 479 | LAWRENCEVILLE |
| 489 | SALEM |
| 502 | GLENS FALLS FARM |
| 502 | WATERTOWN |
| 512 | PERU 2 WSW |
| 522 | LOCKPORT 2 NE |
| 541 | CAMDEN 2 NW |
| 554 | ROCHESTER INTL AP |
| 600 | ALCOVE DAM |
| 620 | MIDDLETOWN 2 NW |
| 669 | YORKTOWN HEIGHTS 1 W |
| 702 | DANSVILLE |
| 712 | BUFFALO GR BUFFLO AP |
| 722 | CANANDAIGUA 3 S |
| 732 | UTICA ONEIDA CNTY AP |
| 761 | FREDONIA |
| 771 | GLOVERSVILLE |
| 820 | MILLBROOK |
| 830 | AURORA RESEARCH FARM |
| 840 | ELMIRA |
| 860 | GOWANDA PSYCHIATRIC |
| 879 | MOUNT MORRIS 2 W |
| 900 | BATAVIA |
| 902 | HEMLOCK |
| 902 | LITTLE FALLS CITY RE |
| 922 | LOWVILLE |
| 951 | ITHACA CORNELL UNIV |
| 1001 | DEPOSIT |
| 1025 | COLDEN 1 N |
| 1030 | CHASM FALLS |
| 1050 | WESTFIELD 3 SW |
| 1070 | NORWICH |
| 1100 | SPENCER 2 N |
| 1112 | BATH |
| 1132 | CORTLAND |
| 1240 | COOPERSTOWN |
| 1240 | WALTON |
| 1250 | MOHONK LAKE |
| 1342 | DANNEMORA |
| 1362 | CHERRY VALLEY 2 NNE |
| 1421 | ANGELICA |
| 1480 | ARCADE |
| 1503 | ALLEGANY STATE PARK |
| 1512 | WANAKENA RANGER SCHO |
| 1562 | BOLIVAR |
| 1562 | GRAFTON |
| 1581 | BOONVILLE 2 SSW |
| 1581 | LIBERTY 1 NE |
| 1581 | LITTLE VALLEY |
| 1590 | BINGHAMTON LINK FLD |
| 1591 | FRANKLINVILLE 1 SSW |
| 1660 | INDIAN LAKE 2 SW |
| 1680 | TUPPER LAKE SUNMOUNT |
| 1703 | STILLWATER RESERVOIR |
| 1722 | WARSAW 6 SW |
| 1732 | OLD FORGE |
| 1762 | ALFRED |
| 1860 | LAKE PLACID 2 S |
| 2651 | SLIDE MOUNTAIN |

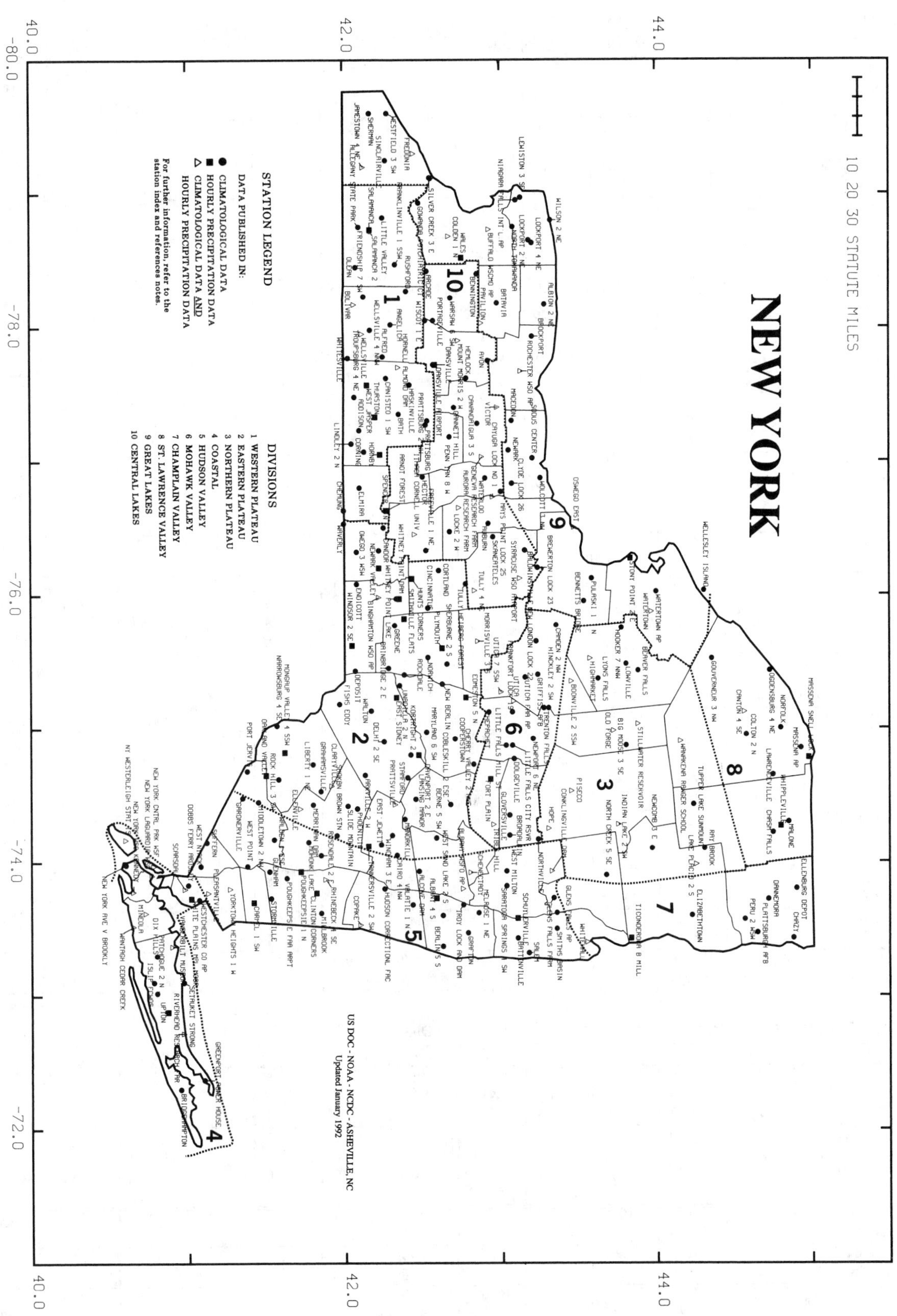
NEW YORK
10 20 30 STATUTE MILES
STATION LEGEND
DATA PUBLISHED IN:
● CLIMATOLOGICAL DATA
■ HOURLY PRECIPITATION DATA
△ CLIMATOLOGICAL DATA AND HOURLY PRECIPITATION DATA
For further information, refer to the station index and reference notes.
DIVISIONS
1 WESTERN PLATEAU
2 EASTERN PLATEAU
3 NORTHERN PLATEAU
4 COASTAL
5 HUDSON VALLEY
6 MOHAWK VALLEY
7 CHAMPLAIN VALLEY
8 ST. LAWRENCE VALLEY
9 GREAT LAKES
10 CENTRAL LAKES
US DOC - NOAA - NCDC - ASHEVILLE, NC
Updated January 1992
40.0
42.0
44.0
-80.0
-78.0
-76.0
-74.0
-72.0

## ALBANY COUNTY AP *Albany County* ELEVATION 292 ft LAT/LONG 42° 45 ' N / 73° 48 ' W

| | JAN | FEB | MAR | APR | MAY | JUN | JUL | AUG | SEP | OCT | NOV | DEC | YEAR |
|---|---|---|---|---|---|---|---|---|---|---|---|---|---|
| Maximum Temp °F | 30.2 | 33.4 | 43.8 | 57.9 | 69.8 | 77.9 | 82.8 | 80.5 | 72.3 | 60.7 | 48.2 | 35.5 | 57.8 |
| Minimum Temp °F | 11.6 | 14.1 | 24.5 | 35.5 | 45.7 | 54.5 | 59.6 | 57.8 | 49.6 | 38.6 | 30.8 | 19.0 | 36.8 |
| Mean Temp °F | 20.9 | 23.8 | 34.2 | 46.7 | 57.8 | 66.2 | 71.2 | 69.2 | 61.0 | 49.7 | 39.5 | 27.3 | 47.3 |
| Days Max Temp ≥ 90 °F | 0 | 0 | 0 | 0 | 0 | 2 | 4 | 2 | 1 | 0 | 0 | 0 | 9 |
| Days Max Temp ≤ 32 °F | 17 | 13 | 4 | 0 | 0 | 0 | 0 | 0 | 0 | 0 | 1 | 11 | 46 |
| Days Min Temp ≤ 32 °F | 30 | 26 | 25 | 12 | 2 | 0 | 0 | 0 | 1 | 9 | 18 | 27 | 150 |
| Days Min Temp ≤ 0 °F | 7 | 4 | 0 | 0 | 0 | 0 | 0 | 0 | 0 | 0 | 0 | 2 | 13 |
| Heating Degree Days | 1360 | 1157 | 948 | 546 | 244 | 61 | 9 | 27 | 162 | 471 | 759 | 1162 | 6906 |
| Cooling Degree Days | 0 | 0 | 1 | 4 | 31 | 101 | 220 | 168 | 48 | 4 | 0 | 0 | 577 |
| Total Precipitation (") | 2.36 | 2.20 | 3.01 | 3.19 | 3.48 | 3.69 | 3.43 | 3.46 | 3.02 | 3.00 | 3.33 | 3.00 | 37.17 |
| Days ≥ 0.1" Precip | 5 | 5 | 6 | 7 | 8 | 7 | 6 | 7 | 6 | 5 | 7 | 7 | 76 |
| Total Snowfall (") | 16.8 | 13.3 | 10.6 | 2.6 | 0.1 | 0.0 | 0.0 | 0.0 | 0.0 | 0.2 | 5.1 | 15.9 | 64.6 |
| Days ≥ 1" Snow Depth | 22 | 18 | 9 | 1 | 0 | 0 | 0 | 0 | 0 | 0 | 3 | 13 | 66 |

## ALBION 2 NE *Orleans County* ELEVATION 469 ft LAT/LONG 43° 16 ' N / 78° 8 ' W

| | JAN | FEB | MAR | APR | MAY | JUN | JUL | AUG | SEP | OCT | NOV | DEC | YEAR |
|---|---|---|---|---|---|---|---|---|---|---|---|---|---|
| Maximum Temp °F | ***30.8*** | ***33.0*** | ***42.8*** | 56.2 | 68.7 | 77.4 | ***82.3*** | ***80.2*** | ***72.5*** | ***60.6*** | ***48.0*** | ***36.3*** | 57.4 |
| Minimum Temp °F | ***16.5*** | ***17.6*** | ***25.6*** | 36.2 | 46.4 | 55.4 | ***60.9*** | ***59.6*** | ***52.6*** | ***41.6*** | ***33.3*** | ***22.9*** | 39.1 |
| Mean Temp °F | ***23.7*** | ***25.3*** | ***34.2*** | 46.2 | 57.6 | 66.4 | ***71.6*** | ***69.9*** | ***62.6*** | ***51.1*** | ***40.6*** | ***29.6*** | 48.2 |
| Days Max Temp ≥ 90 °F | ***0*** | ***0*** | ***0*** | 0 | 0 | 1 | ***4*** | ***2*** | ***0*** | ***0*** | ***0*** | ***0*** | 7 |
| Days Max Temp ≤ 32 °F | ***17*** | ***14*** | ***6*** | 0 | 0 | 0 | ***0*** | ***0*** | ***0*** | ***0*** | ***1*** | ***10*** | 48 |
| Days Min Temp ≤ 32 °F | ***29*** | ***25*** | ***24*** | 11 | 1 | 0 | ***0*** | ***0*** | ***0*** | ***4*** | ***15*** | ***26*** | 135 |
| Days Min Temp ≤ 0 °F | ***2*** | ***2*** | ***0*** | 0 | 0 | 0 | ***0*** | ***0*** | ***0*** | ***0*** | ***0*** | ***1*** | 5 |
| Heating Degree Days | ***1276*** | ***1114*** | ***947*** | 562 | 257 | 63 | ***8*** | ***19*** | ***131*** | ***428*** | ***726*** | ***1090*** | 6621 |
| Cooling Degree Days | 0 | 0 | 0 | 5 | 34 | 111 | 234 | 184 | 67 | 4 | 0 | 0 | 639 |
| Total Precipitation (") | ***2.28*** | ***2.12*** | ***2.81*** | 3.21 | 2.99 | 3.68 | ***2.75*** | ***3.11*** | ***3.63*** | ***2.69*** | ***3.29*** | ***3.33*** | 35.89 |
| Days ≥ 0.1" Precip | na | ***6*** | ***7*** | 8 | 7 | 7 | ***6*** | ***6*** | ***7*** | ***7*** | ***9*** | na | na |
| Total Snowfall (") | ***18.4*** | ***16.4*** | ***9.3*** | ***2.5*** | ***0.3*** | 0.0 | ***0.0*** | ***0.0*** | ***0.0*** | ***0.1*** | ***5.4*** | ***16.4*** | 68.8 |
| Days ≥ 1" Snow Depth | na | ***16*** | na | ***1*** | ***0*** | 0 | 0 | 0 | ***0*** | ***0*** | ***4*** | na | na |

## ALCOVE DAM *Albany County* ELEVATION 600 ft LAT/LONG 42° 28 ' N / 73° 56 ' W

| | JAN | FEB | MAR | APR | MAY | JUN | JUL | AUG | SEP | OCT | NOV | DEC | YEAR |
|---|---|---|---|---|---|---|---|---|---|---|---|---|---|
| Maximum Temp °F | 30.5 | 32.9 | 42.3 | 55.6 | 68.1 | 76.3 | 81.3 | 79.4 | 71.1 | 59.8 | 47.5 | 35.6 | 56.7 |
| Minimum Temp °F | 9.3 | 10.9 | 21.5 | 32.7 | 42.7 | 51.8 | 56.8 | 54.7 | 46.9 | 35.1 | 28.1 | 16.5 | 33.9 |
| Mean Temp °F | 20.0 | 22.0 | 32.0 | 44.2 | 55.4 | 64.1 | 69.1 | 67.1 | 59.0 | 47.5 | 37.9 | 26.0 | 45.4 |
| Days Max Temp ≥ 90 °F | 0 | 0 | 0 | 0 | 0 | 1 | 3 | 2 | 0 | 0 | 0 | 0 | 6 |
| Days Max Temp ≤ 32 °F | 18 | 14 | 5 | 0 | 0 | 0 | 0 | 0 | 0 | 0 | 1 | 11 | 49 |
| Days Min Temp ≤ 32 °F | 30 | 27 | 27 | 16 | 4 | 0 | 0 | 0 | 1 | 13 | 22 | 29 | 169 |
| Days Min Temp ≤ 0 °F | 7 | 6 | 1 | 0 | 0 | 0 | 0 | 0 | 0 | 0 | 0 | 3 | 17 |
| Heating Degree Days | 1390 | 1207 | 1018 | 618 | 306 | 94 | 21 | 48 | 205 | 537 | 807 | 1201 | 7452 |
| Cooling Degree Days | 0 | 0 | 0 | 1 | 18 | 66 | 162 | 119 | 31 | 1 | 0 | 0 | 398 |
| Total Precipitation (") | 2.26 | 2.22 | 3.12 | 3.43 | 3.98 | 3.85 | 3.82 | 3.30 | 3.47 | 3.06 | 3.71 | 2.98 | 39.20 |
| Days ≥ 0.1" Precip | 4 | 5 | 6 | 6 | 7 | 7 | 6 | 7 | 6 | 5 | 7 | 6 | 72 |
| Total Snowfall (") | na | na | na | na | na | ***0.0*** | 0.0 | 0.0 | 0.0 | ***0.1*** | na | na | na |
| Days ≥ 1" Snow Depth | na | na | na | na | na | ***0*** | 0 | 0 | 0 | ***0*** | na | na | na |

## ALFRED *Allegany County* ELEVATION 1762 ft LAT/LONG 42° 15 ' N / 77° 47 ' W

| | JAN | FEB | MAR | APR | MAY | JUN | JUL | AUG | SEP | OCT | NOV | DEC | YEAR |
|---|---|---|---|---|---|---|---|---|---|---|---|---|---|
| Maximum Temp °F | 29.9 | 32.9 | 42.8 | 55.9 | 67.4 | 75.5 | 79.1 | 77.3 | 70.0 | 58.8 | 46.3 | 35.3 | 55.9 |
| Minimum Temp °F | 11.6 | 12.5 | 21.3 | 31.5 | 41.3 | 50.0 | 54.8 | 53.5 | 47.0 | 36.2 | 28.7 | 18.9 | 33.9 |
| Mean Temp °F | 20.8 | 22.7 | 32.0 | 43.8 | 54.4 | 62.8 | 67.0 | 65.4 | 58.5 | 47.5 | 37.5 | 27.1 | 45.0 |
| Days Max Temp ≥ 90 °F | 0 | 0 | 0 | 0 | 0 | 0 | 1 | 1 | 0 | 0 | 0 | 0 | 2 |
| Days Max Temp ≤ 32 °F | 18 | 14 | 6 | 0 | 0 | 0 | 0 | 0 | 0 | 0 | 2 | 12 | 52 |
| Days Min Temp ≤ 32 °F | 30 | 27 | 26 | 17 | 6 | 0 | 0 | 0 | 2 | 11 | 20 | 28 | 167 |
| Days Min Temp ≤ 0 °F | 6 | 6 | 2 | 0 | 0 | 0 | 0 | 0 | 0 | 0 | 0 | 2 | 16 |
| Heating Degree Days | 1364 | 1189 | 1016 | 633 | 336 | 116 | 39 | 61 | 216 | 537 | 817 | 1168 | 7492 |
| Cooling Degree Days | 0 | 0 | 0 | 1 | 13 | 50 | 109 | 74 | 26 | 1 | 0 | 0 | 274 |
| Total Precipitation (") | 2.03 | 1.95 | 2.67 | 2.99 | 3.37 | 4.40 | 3.72 | 3.22 | 3.75 | 3.15 | 3.41 | 2.97 | 37.63 |
| Days ≥ 0.1" Precip | 6 | 6 | 7 | 8 | 9 | 8 | 8 | 7 | 7 | 7 | 9 | 8 | 90 |
| Total Snowfall (") | 19.1 | 16.4 | 14.0 | 4.1 | 0.6 | 0.0 | 0.0 | 0.0 | 0.0 | 0.5 | 8.4 | 18.8 | 81.9 |
| Days ≥ 1" Snow Depth | 26 | 24 | 16 | 3 | 0 | 0 | 0 | 0 | 0 | 0 | 6 | ***20*** | 95 |

## ALLEGANY STATE PARK *Cattaraugus County* ELEVATION 1503 ft LAT/LONG 42° 6 ' N / 78° 45 ' W

| | JAN | FEB | MAR | APR | MAY | JUN | JUL | AUG | SEP | OCT | NOV | DEC | YEAR |
|---|---|---|---|---|---|---|---|---|---|---|---|---|---|
| Maximum Temp °F | 29.9 | 32.3 | 42.0 | 55.1 | 66.6 | 74.9 | 78.7 | 76.9 | 69.4 | 58.1 | 45.9 | 34.6 | 55.4 |
| Minimum Temp °F | 12.0 | 12.3 | 20.5 | 31.2 | 40.0 | 48.9 | 53.4 | 52.4 | 46.1 | 35.9 | 28.9 | 19.1 | 33.4 |
| Mean Temp °F | 21.0 | 22.3 | 31.3 | 43.1 | 53.3 | 61.9 | 66.1 | 64.7 | 57.8 | 47.0 | 37.4 | 26.9 | 44.4 |
| Days Max Temp ≥ 90 °F | 0 | 0 | 0 | 0 | 0 | 0 | 1 | 1 | 0 | 0 | 0 | 0 | 2 |
| Days Max Temp ≤ 32 °F | 18 | 15 | 7 | 1 | 0 | 0 | 0 | 0 | 0 | 0 | 3 | 13 | 57 |
| Days Min Temp ≤ 32 °F | 30 | 27 | 27 | 18 | 8 | 1 | 0 | 0 | 2 | 12 | 21 | 29 | 175 |
| Days Min Temp ≤ 0 °F | 6 | 5 | 2 | 0 | 0 | 0 | 0 | 0 | 0 | 0 | 0 | 2 | 15 |
| Heating Degree Days | 1359 | 1200 | 1039 | 650 | 367 | 134 | 53 | 75 | 235 | 551 | 821 | 1176 | 7660 |
| Cooling Degree Days | 0 | 0 | 0 | 2 | 12 | 42 | 94 | 68 | 23 | 1 | 0 | 0 | 242 |
| Total Precipitation (") | 2.67 | 2.42 | 3.12 | 3.48 | 3.88 | 4.88 | 4.27 | 4.18 | 4.46 | 3.78 | 4.30 | 3.88 | 45.32 |
| Days ≥ 0.1" Precip | 9 | 7 | 9 | 9 | 9 | 9 | 9 | 8 | 8 | 10 | 10 | 11 | 108 |
| Total Snowfall (") | na | na | na | na | 0.0 | 0.0 | 0.0 | 0.0 | 0.0 | 0.0 | na | na | na |
| Days ≥ 1" Snow Depth | 27 | 25 | 19 | 2 | 0 | 0 | 0 | 0 | 0 | 0 | *7* | 20 | 100 |

## ANGELICA *Allegany County* ELEVATION 1421 ft LAT/LONG 42° 18 ' N / 78° 2 ' W

| | JAN | FEB | MAR | APR | MAY | JUN | JUL | AUG | SEP | OCT | NOV | DEC | YEAR |
|---|---|---|---|---|---|---|---|---|---|---|---|---|---|
| Maximum Temp °F | 31.2 | 33.4 | 43.2 | 56.4 | 68.0 | 75.8 | 79.7 | 77.6 | 70.6 | 59.5 | 46.9 | 35.8 | 56.5 |
| Minimum Temp °F | 12.7 | 12.6 | 21.3 | 31.4 | 40.4 | 49.0 | 54.0 | 52.9 | 46.5 | 36.1 | 29.2 | 19.4 | 33.8 |
| Mean Temp °F | 22.0 | 23.0 | 32.3 | 43.9 | 54.3 | 62.4 | 66.9 | 65.3 | 58.6 | 47.8 | 38.1 | 27.6 | 45.2 |
| Days Max Temp ≥ 90 °F | 0 | 0 | 0 | 0 | 0 | 1 | 1 | 1 | 0 | 0 | 0 | 0 | 3 |
| Days Max Temp ≤ 32 °F | 17 | 13 | 6 | 0 | 0 | 0 | 0 | 0 | 0 | 0 | 2 | 12 | 50 |
| Days Min Temp ≤ 32 °F | 30 | 26 | 26 | 17 | 8 | 1 | 0 | 0 | 2 | 11 | 20 | 28 | 169 |
| Days Min Temp ≤ 0 °F | 6 | 6 | 2 | 0 | 0 | 0 | 0 | 0 | 0 | 0 | 0 | 2 | 16 |
| Heating Degree Days | 1328 | 1179 | 1008 | 628 | 339 | 125 | 43 | 64 | 213 | 526 | 801 | 1154 | 7408 |
| Cooling Degree Days | 0 | 0 | 0 | 2 | 10 | 44 | 103 | 72 | 25 | 1 | 0 | 0 | 257 |
| Total Precipitation (") | 1.85 | 2.04 | 2.29 | 2.87 | 3.06 | 4.59 | 3.75 | 3.68 | 3.70 | 3.01 | 3.16 | 2.62 | 36.62 |
| Days ≥ 0.1" Precip | 6 | 6 | 7 | 7 | 8 | 9 | 8 | 7 | 8 | 7 | 9 | 8 | 90 |
| Total Snowfall (") | 15.3 | 11.5 | 9.0 | 2.9 | 0.3 | 0.0 | 0.0 | 0.0 | 0.0 | 0.3 | 5.9 | 14.0 | 59.2 |
| Days ≥ 1" Snow Depth | 24 | 22 | 13 | 2 | 0 | 0 | 0 | 0 | 0 | 0 | 5 | 16 | 82 |

## ARCADE *Wyoming County* ELEVATION 1480 ft LAT/LONG 42° 32 ' N / 78° 25 ' W

| | JAN | FEB | MAR | APR | MAY | JUN | JUL | AUG | SEP | OCT | NOV | DEC | YEAR |
|---|---|---|---|---|---|---|---|---|---|---|---|---|---|
| Maximum Temp °F | 28.5 | 31.6 | 41.4 | 55.7 | 66.9 | 75.3 | 78.9 | 77.0 | 69.9 | 58.7 | 45.9 | 34.4 | 55.4 |
| Minimum Temp °F | 11.8 | 12.5 | 21.0 | 32.1 | 41.4 | 50.5 | 55.2 | 53.7 | 47.4 | 37.7 | 29.9 | 19.1 | 34.4 |
| Mean Temp °F | 20.1 | 22.1 | 31.2 | 43.9 | 54.2 | 62.9 | 67.1 | 65.4 | 58.7 | 48.2 | 37.9 | 26.8 | 44.9 |
| Days Max Temp ≥ 90 °F | 0 | 0 | 0 | 0 | 0 | 0 | 1 | 0 | 0 | 0 | 0 | 0 | 1 |
| Days Max Temp ≤ 32 °F | 19 | 15 | 7 | 1 | 0 | 0 | 0 | 0 | 0 | 0 | 3 | 13 | 58 |
| Days Min Temp ≤ 32 °F | 30 | 26 | 26 | 16 | 7 | 1 | 0 | 0 | 2 | 10 | 20 | 28 | 166 |
| Days Min Temp ≤ 0 °F | 6 | 6 | 2 | 0 | 0 | 0 | 0 | 0 | 0 | 0 | 0 | 3 | 17 |
| Heating Degree Days | 1386 | 1207 | 1039 | 627 | 343 | 115 | 42 | 65 | 213 | 515 | 805 | 1177 | 7534 |
| Cooling Degree Days | 0 | 0 | 0 | 2 | 15 | 58 | 126 | 88 | 34 | 1 | 0 | 0 | 324 |
| Total Precipitation (") | 2.60 | 2.20 | 2.65 | 3.29 | 3.54 | 4.53 | 3.91 | 3.95 | 4.56 | 3.44 | 4.10 | 3.64 | 42.41 |
| Days ≥ 0.1" Precip | 8 | 6 | *8* | 8 | 8 | 7 | 7 | 8 | 9 | 9 | 11 | 10 | 99 |
| Total Snowfall (") | 29.9 | 21.4 | *12.3* | *4.7* | *0.4* | 0.0 | 0.0 | 0.0 | 0.0 | 0.8 | 13.3 | 25.2 | 108.0 |
| Days ≥ 1" Snow Depth | 27 | 23 | 16 | 3 | 0 | 0 | 0 | 0 | 0 | 0 | 7 | 20 | 96 |

## AURORA RESEARCH FARM *Cayuga County* ELEVATION 830 ft LAT/LONG 42° 44 ' N / 76° 39 ' W

| | JAN | FEB | MAR | APR | MAY | JUN | JUL | AUG | SEP | OCT | NOV | DEC | YEAR |
|---|---|---|---|---|---|---|---|---|---|---|---|---|---|
| Maximum Temp °F | 30.4 | 31.9 | 41.5 | 54.9 | 66.9 | 76.1 | 81.1 | 79.5 | 71.6 | 59.5 | 47.6 | 35.9 | 56.4 |
| Minimum Temp °F | 15.0 | 15.9 | 24.6 | 35.5 | 45.9 | 55.3 | 59.8 | 58.4 | 51.3 | 40.7 | 32.6 | 21.9 | 38.1 |
| Mean Temp °F | 22.8 | 24.0 | 33.1 | 45.2 | 56.4 | 65.8 | 70.5 | 69.0 | 61.4 | 50.1 | 40.1 | 28.9 | 47.3 |
| Days Max Temp ≥ 90 °F | 0 | 0 | 0 | 0 | 1 | 1 | 3 | 2 | 1 | 0 | 0 | 0 | 8 |
| Days Max Temp ≤ 32 °F | 17 | 15 | 7 | 1 | 0 | 0 | 0 | 0 | 0 | 0 | 2 | 11 | 53 |
| Days Min Temp ≤ 32 °F | 29 | 26 | 24 | 13 | 1 | 0 | 0 | 0 | 0 | 5 | 16 | 26 | 140 |
| Days Min Temp ≤ 0 °F | 4 | 2 | 0 | 0 | 0 | 0 | 0 | 0 | 0 | 0 | 0 | 1 | 7 |
| Heating Degree Days | 1303 | 1153 | 983 | 592 | 292 | 81 | 19 | 32 | 156 | 459 | 740 | 1111 | 6921 |
| Cooling Degree Days | 0 | 0 | 1 | 7 | 38 | 115 | 213 | 169 | 59 | 5 | 1 | 0 | 608 |
| Total Precipitation (") | 1.70 | 1.75 | 2.41 | 3.15 | 3.16 | 4.04 | 3.42 | 3.65 | 3.98 | 3.17 | 3.39 | 2.59 | 36.41 |
| Days ≥ 0.1" Precip | 5 | 5 | 7 | 8 | 8 | 8 | 8 | 7 | 7 | 7 | 8 | 6 | 84 |
| Total Snowfall (") | 14.2 | 12.4 | 10.6 | 4.1 | 0.3 | 0.0 | 0.0 | 0.0 | 0.0 | 0.4 | 5.0 | 12.8 | 59.8 |
| Days ≥ 1" Snow Depth | 23 | 22 | 14 | 3 | 0 | 0 | 0 | 0 | 0 | 0 | 4 | 16 | 82 |

## BATAVIA *Genesee County* ELEVATION 900 ft LAT/LONG 43° 0 ' N / 78° 11 ' W

| | JAN | FEB | MAR | APR | MAY | JUN | JUL | AUG | SEP | OCT | NOV | DEC | YEAR |
|---|---|---|---|---|---|---|---|---|---|---|---|---|---|
| Maximum Temp °F | 31.5 | 33.2 | 43.5 | 57.0 | 68.6 | 77.6 | 81.6 | 79.5 | 72.2 | 61.0 | 48.2 | 36.3 | 57.5 |
| Minimum Temp °F | 15.4 | 15.6 | 24.6 | 35.4 | 45.8 | 55.0 | 59.8 | 58.2 | 51.3 | 40.6 | 32.4 | 21.7 | 38.0 |
| Mean Temp °F | 23.4 | 24.4 | 34.1 | 46.2 | 57.3 | 66.3 | 70.7 | 68.8 | 61.8 | 50.8 | 40.3 | 29.0 | 47.8 |
| Days Max Temp ≥ 90 °F | 0 | 0 | 0 | 0 | 0 | 1 | 2 | 1 | 0 | 0 | 0 | 0 | 4 |
| Days Max Temp ≤ 32 °F | 16 | 14 | 6 | 0 | 0 | 0 | 0 | 0 | 0 | 0 | 2 | 11 | 49 |
| Days Min Temp ≤ 32 °F | 29 | 26 | 24 | 13 | 2 | 0 | 0 | 0 | 0 | 6 | 16 | 27 | 143 |
| Days Min Temp ≤ 0 °F | 4 | 3 | 1 | 0 | 0 | 0 | 0 | 0 | 0 | 0 | 0 | 1 | 9 |
| Heating Degree Days | 1282 | 1139 | 953 | 562 | 265 | 65 | 12 | 28 | 146 | 436 | 733 | 1108 | 6729 |
| Cooling Degree Days | 0 | 0 | 0 | 7 | 32 | 107 | 209 | 157 | 57 | 4 | 0 | 0 | 573 |
| Total Precipitation (") | 1.66 | 1.81 | 2.13 | 3.08 | 3.25 | 3.72 | 3.16 | 3.78 | 3.88 | 3.00 | 2.93 | 2.55 | 34.95 |
| Days ≥ 0.1" Precip | 5 | 4 | 6 | 7 | 8 | 7 | 6 | 7 | 8 | 8 | 8 | 8 | 82 |
| Total Snowfall (") | 22.8 | 19.1 | 12.3 | 4.3 | 0.5 | 0.0 | 0.0 | 0.0 | 0.0 | 0.3 | 8.6 | 20.7 | 88.6 |
| Days ≥ 1" Snow Depth | 22 | 20 | 10 | 2 | 0 | 0 | 0 | 0 | 0 | 0 | 4 | 16 | 74 |

## BATH *Steuben County* ELEVATION 1112 ft LAT/LONG 42° 20 ' N / 77° 20 ' W

| | JAN | FEB | MAR | APR | MAY | JUN | JUL | AUG | SEP | OCT | NOV | DEC | YEAR |
|---|---|---|---|---|---|---|---|---|---|---|---|---|---|
| Maximum Temp °F | 31.8 | 33.4 | 42.9 | 56.5 | 67.9 | 76.5 | 81.0 | 79.3 | 71.9 | 60.0 | 47.2 | 35.9 | 57.0 |
| Minimum Temp °F | 12.8 | 12.8 | 21.6 | 32.5 | 41.6 | 50.3 | 55.1 | 53.4 | 46.7 | 36.1 | 29.5 | 19.5 | 34.3 |
| Mean Temp °F | 22.5 | 23.1 | 32.3 | 44.6 | 54.8 | 63.4 | 68.1 | 66.3 | 59.3 | 48.2 | 38.5 | 27.8 | 45.7 |
| Days Max Temp ≥ 90 °F | 0 | 0 | 0 | 0 | 0 | 1 | 3 | 1 | 0 | 0 | 0 | 0 | 5 |
| Days Max Temp ≤ 32 °F | 16 | 13 | 5 | 0 | 0 | 0 | 0 | 0 | 0 | 0 | 2 | 11 | 47 |
| Days Min Temp ≤ 32 °F | 30 | 27 | 26 | 16 | 5 | 0 | 0 | 0 | 2 | 12 | 20 | 28 | 166 |
| Days Min Temp ≤ 0 °F | 5 | 5 | 1 | 0 | 0 | 0 | 0 | 0 | 0 | 0 | 0 | 2 | 13 |
| Heating Degree Days | 1303 | 1176 | 1007 | 608 | 323 | 107 | 34 | 54 | 200 | 516 | 790 | 1149 | 7267 |
| Cooling Degree Days | 0 | 0 | 0 | 2 | 15 | 61 | 137 | 93 | 29 | 1 | 0 | 0 | 338 |
| Total Precipitation (") | 1.50 | 1.63 | 2.08 | 2.65 | 2.92 | 3.67 | 3.31 | 2.76 | 3.19 | 2.55 | 2.94 | 2.27 | 31.47 |
| Days ≥ 0.1" Precip | 4 | 4 | 6 | 7 | 7 | 7 | 7 | 6 | 6 | 6 | 7 | 6 | 73 |
| Total Snowfall (") | 10.3 | 10.5 | 8.2 | 2.0 | 0.1 | 0.0 | 0.0 | 0.0 | 0.0 | 0.1 | 3.3 | 10.7 | 45.2 |
| Days ≥ 1" Snow Depth | 21 | 22 | 12 | 2 | 0 | 0 | 0 | 0 | 0 | 0 | 4 | 16 | 77 |

## BINGHAMTON LINK FLD *Broome County* ELEVATION 1590 ft LAT/LONG 42° 13 ' N / 75° 59 ' W

| | JAN | FEB | MAR | APR | MAY | JUN | JUL | AUG | SEP | OCT | NOV | DEC | YEAR |
|---|---|---|---|---|---|---|---|---|---|---|---|---|---|
| Maximum Temp °F | 28.2 | 30.6 | 40.4 | 54.0 | 65.6 | 73.9 | 78.7 | 76.4 | 68.2 | 56.8 | 44.9 | 33.4 | 54.3 |
| Minimum Temp °F | 14.5 | 15.8 | 24.5 | 35.5 | 46.2 | 54.7 | 59.8 | 58.1 | 50.5 | 39.9 | 31.5 | 20.8 | 37.7 |
| Mean Temp °F | 21.4 | 23.2 | 32.5 | 44.8 | 55.9 | 64.3 | 69.3 | 67.3 | 59.4 | 48.4 | 38.2 | 27.2 | 46.0 |
| Days Max Temp ≥ 90 °F | 0 | 0 | 0 | 0 | 0 | 0 | 1 | 0 | 0 | 0 | 0 | 0 | 1 |
| Days Max Temp ≤ 32 °F | 20 | 16 | 8 | 1 | 0 | 0 | 0 | 0 | 0 | 0 | 4 | 14 | 63 |
| Days Min Temp ≤ 32 °F | 30 | 26 | 24 | 12 | 1 | 0 | 0 | 0 | 0 | 6 | 17 | 27 | 143 |
| Days Min Temp ≤ 0 °F | 4 | 2 | 0 | 0 | 0 | 0 | 0 | 0 | 0 | 0 | 0 | 1 | 7 |
| Heating Degree Days | 1346 | 1173 | 1002 | 605 | 298 | 91 | 21 | 43 | 195 | 510 | 797 | 1166 | 7247 |
| Cooling Degree Days | 0 | 0 | 0 | 4 | 26 | 74 | 169 | 124 | 36 | 2 | 1 | 0 | 436 |
| Total Precipitation (") | 2.43 | 2.30 | 2.92 | 3.20 | 3.44 | 3.67 | 3.54 | 3.48 | 3.52 | 2.88 | 3.39 | 3.11 | 37.88 |
| Days ≥ 0.1" Precip | 6 | 6 | 7 | 7 | 8 | 8 | 7 | 7 | 6 | 7 | 8 | 7 | 84 |
| Total Snowfall (") | 19.6 | 15.5 | 13.5 | 4.5 | 0.3 | 0.0 | 0.0 | 0.0 | 0.0 | 0.9 | 7.4 | 18.2 | 79.9 |
| Days ≥ 1" Snow Depth | 24 | 22 | 14 | 2 | 0 | 0 | 0 | 0 | 0 | 0 | 5 | 17 | 84 |

## BOLIVAR *Allegany County* ELEVATION 1562 ft LAT/LONG 42° 4 ' N / 78° 10 ' W

| | JAN | FEB | MAR | APR | MAY | JUN | JUL | AUG | SEP | OCT | NOV | DEC | YEAR |
|---|---|---|---|---|---|---|---|---|---|---|---|---|---|
| Maximum Temp °F | 29.9 | 32.9 | 42.3 | 54.9 | 66.8 | 74.2 | 78.4 | 76.7 | 69.0 | 58.4 | 46.4 | 34.8 | 55.4 |
| Minimum Temp °F | 10.3 | 11.0 | 20.7 | 31.1 | 40.4 | 48.5 | 53.6 | 52.6 | 45.8 | 34.5 | 28.4 | 17.6 | 32.9 |
| Mean Temp °F | 20.1 | 22.0 | 31.5 | 43.0 | 53.6 | 61.3 | 66.0 | 64.7 | 57.4 | 46.4 | 37.4 | 26.2 | 44.1 |
| Days Max Temp ≥ 90 °F | 0 | 0 | 0 | 0 | 0 | 0 | 1 | 0 | 0 | 0 | 0 | 0 | 1 |
| Days Max Temp ≤ 32 °F | 19 | 15 | 6 | 1 | 0 | 0 | 0 | 0 | 0 | 0 | 3 | 13 | 57 |
| Days Min Temp ≤ 32 °F | 30 | 27 | 26 | 17 | 8 | 1 | 0 | 0 | 2 | 14 | 21 | 29 | 175 |
| Days Min Temp ≤ 0 °F | 8 | 6 | 2 | 0 | 0 | 0 | 0 | 0 | 0 | 0 | 0 | 3 | 19 |
| Heating Degree Days | 1386 | 1209 | 1031 | 654 | 358 | 145 | 56 | 74 | 241 | 569 | 821 | 1196 | 7740 |
| Cooling Degree Days | 0 | 0 | 0 | 2 | 11 | 44 | 102 | 73 | 18 | 1 | 0 | 0 | 251 |
| Total Precipitation (") | 2.06 | 2.02 | 2.83 | 3.13 | 3.39 | 5.17 | 4.47 | 3.73 | 4.22 | 3.15 | 3.32 | 3.04 | 40.53 |
| Days ≥ 0.1" Precip | 6 | 6 | 8 | 8 | 8 | 9 | 9 | 8 | 9 | 8 | 8 | 9 | 96 |
| Total Snowfall (") | 19.3 | 16.2 | 12.4 | 3.1 | 0.2 | 0.0 | 0.0 | 0.0 | 0.0 | 0.4 | 6.3 | 20.4 | 78.3 |
| Days ≥ 1" Snow Depth | 26 | 25 | 16 | 3 | 0 | 0 | 0 | 0 | 0 | 0 | 6 | 20 | 96 |

## BOONVILLE 2 SSW *Oneida County* ELEVATION 1581 ft LAT/LONG 43° 27 ' N / 75° 21 ' W

| | JAN | FEB | MAR | APR | MAY | JUN | JUL | AUG | SEP | OCT | NOV | DEC | YEAR |
|---|---|---|---|---|---|---|---|---|---|---|---|---|---|
| Maximum Temp °F | 23.9 | 26.4 | 35.9 | 50.1 | 63.2 | 71.4 | 76.0 | 73.8 | 65.6 | 54.1 | 41.2 | 29.1 | 50.9 |
| Minimum Temp °F | 7.2 | 9.1 | 19.1 | 31.6 | 42.8 | 51.2 | 56.3 | 54.9 | 47.1 | 36.6 | 27.2 | 14.3 | 33.1 |
| Mean Temp °F | 15.6 | 17.8 | 27.6 | 40.9 | 53.0 | 61.3 | 66.2 | 64.4 | 56.4 | 45.4 | 34.2 | 21.8 | 42.0 |
| Days Max Temp ≥ 90 °F | 0 | 0 | 0 | 0 | 0 | 0 | 0 | 0 | 0 | 0 | 0 | 0 | 0 |
| Days Max Temp ≤ 32 °F | 25 | 21 | 12 | 2 | 0 | 0 | 0 | 0 | 0 | 0 | 6 | 19 | 85 |
| Days Min Temp ≤ 32 °F | 31 | 27 | 28 | 18 | 3 | 0 | 0 | 0 | 1 | 11 | 22 | 30 | 171 |
| Days Min Temp ≤ 0 °F | 10 | 8 | 2 | 0 | 0 | 0 | 0 | 0 | 0 | 0 | 0 | 5 | 25 |
| Heating Degree Days | 1527 | 1327 | 1154 | 718 | 376 | 147 | 50 | 81 | 268 | 601 | 916 | 1334 | 8499 |
| Cooling Degree Days | 0 | 0 | 0 | 2 | 11 | 39 | 97 | 66 | 14 | 0 | 0 | 0 | 229 |
| Total Precipitation (") | 5.36 | 4.71 | 4.88 | 4.75 | 4.49 | 4.68 | 3.93 | 4.97 | 5.86 | 4.76 | 6.17 | 6.03 | 60.59 |
| Days ≥ 0.1" Precip | 13 | 11 | 11 | 9 | 9 | 9 | 7 | 8 | 9 | 10 | 12 | 13 | 121 |
| Total Snowfall (") | 59.5 | 49.0 | 33.7 | 10.9 | 1.5 | 0.0 | 0.0 | 0.0 | 0.0 | 2.5 | 23.3 | 49.4 | 229.8 |
| Days ≥ 1" Snow Depth | 30 | 28 | 30 | 16 | 1 | 0 | 0 | 0 | 0 | 1 | 13 | 28 | 147 |

## BRIDGEHAMPTON *Suffolk County* ELEVATION 59 ft LAT/LONG 40° 57 ' N / 72° 18 ' W

| | JAN | FEB | MAR | APR | MAY | JUN | JUL | AUG | SEP | OCT | NOV | DEC | YEAR |
|---|---|---|---|---|---|---|---|---|---|---|---|---|---|
| Maximum Temp °F | 37.8 | 38.6 | 45.9 | 55.4 | 65.1 | 74.2 | 80.2 | 79.5 | 72.8 | 62.7 | 53.0 | 43.3 | 59.0 |
| Minimum Temp °F | 22.3 | 23.8 | 30.4 | 38.5 | 47.7 | 57.2 | 63.5 | 62.9 | 55.4 | 44.8 | 36.9 | 28.1 | 42.6 |
| Mean Temp °F | 30.0 | 31.2 | 38.2 | 46.9 | 56.4 | 65.7 | 71.9 | 71.2 | 64.1 | 53.7 | 45.0 | 35.7 | 50.8 |
| Days Max Temp ≥ 90 °F | 0 | 0 | 0 | 0 | 0 | 0 | 2 | 1 | 0 | 0 | 0 | 0 | 3 |
| Days Max Temp ≤ 32 °F | 9 | 7 | 1 | 0 | 0 | 0 | 0 | 0 | 0 | 0 | 0 | 4 | 21 |
| Days Min Temp ≤ 32 °F | 27 | 23 | 18 | 6 | 0 | 0 | 0 | 0 | 0 | 3 | 10 | 21 | 108 |
| Days Min Temp ≤ 0 °F | 0 | 0 | 0 | 0 | 0 | 0 | 0 | 0 | 0 | 0 | 0 | 0 | 0 |
| Heating Degree Days | 1076 | 947 | 825 | 536 | 269 | 52 | 4 | 8 | 89 | 347 | 594 | 901 | 5648 |
| Cooling Degree Days | 0 | 0 | 0 | 0 | 12 | 90 | 236 | 202 | 65 | 6 | 0 | 0 | 611 |
| Total Precipitation (") | 4.21 | 3.69 | 4.49 | 3.78 | 3.75 | 3.52 | 2.99 | 3.80 | 3.47 | 3.32 | 4.50 | 4.55 | 46.07 |
| Days ≥ 0.1" Precip | 7 | 6 | 8 | 6 | 7 | 6 | 5 | 5 | 5 | 6 | 7 | 8 | 76 |
| Total Snowfall (") | 8.5 | 8.5 | 5.5 | 0.5 | 0.0 | 0.0 | 0.0 | 0.0 | 0.0 | 0.0 | 0.8 | 2.9 | 26.7 |
| Days ≥ 1" Snow Depth | 10 | 9 | 4 | 0 | 0 | 0 | 0 | 0 | 0 | 0 | 0 | 2 | 25 |

## BUFFALO GR BUFFLO AP *Erie County* ELEVATION 712 ft LAT/LONG 42° 56 ' N / 78° 44 ' W

| | JAN | FEB | MAR | APR | MAY | JUN | JUL | AUG | SEP | OCT | NOV | DEC | YEAR |
|---|---|---|---|---|---|---|---|---|---|---|---|---|---|
| Maximum Temp °F | 30.4 | 31.9 | 41.6 | 54.7 | 66.4 | 75.4 | 80.3 | 78.2 | 70.6 | 59.1 | 47.2 | 36.0 | 56.0 |
| Minimum Temp °F | 17.2 | 17.7 | 25.8 | 36.5 | 47.1 | 56.6 | 62.1 | 60.3 | 53.0 | 42.5 | 33.9 | 23.4 | 39.7 |
| Mean Temp °F | 23.8 | 24.8 | 33.7 | 45.6 | 56.8 | 66.0 | 71.2 | 69.3 | 61.8 | 50.8 | 40.6 | 29.8 | 47.8 |
| Days Max Temp ≥ 90 °F | 0 | 0 | 0 | 0 | 0 | 1 | 2 | 1 | 0 | 0 | 0 | 0 | 4 |
| Days Max Temp ≤ 32 °F | 17 | 15 | 7 | 1 | 0 | 0 | 0 | 0 | 0 | 0 | 2 | 11 | 53 |
| Days Min Temp ≤ 32 °F | 28 | 26 | 24 | 10 | 0 | 0 | 0 | 0 | 0 | 3 | 14 | 25 | 130 |
| Days Min Temp ≤ 0 °F | 2 | 1 | 0 | 0 | 0 | 0 | 0 | 0 | 0 | 0 | 0 | 1 | 4 |
| Heating Degree Days | 1269 | 1128 | 962 | 579 | 276 | 66 | 8 | 22 | 144 | 438 | 726 | 1086 | 6704 |
| Cooling Degree Days | 0 | 0 | 0 | 5 | 32 | 96 | 225 | 170 | 55 | 4 | 0 | 0 | 587 |
| Total Precipitation (") | 2.82 | 2.33 | 2.84 | 2.96 | 3.19 | 3.67 | 3.15 | 3.98 | 3.82 | 3.25 | 3.98 | 3.81 | 39.80 |
| Days ≥ 0.1" Precip | 8 | 7 | 8 | 8 | 7 | 7 | 6 | 7 | 7 | 8 | 10 | 10 | 93 |
| Total Snowfall (") | 25.4 | 18.2 | 11.7 | 3.5 | 0.3 | 0.0 | 0.0 | 0.0 | 0.0 | 0.3 | 10.6 | 22.9 | 92.9 |
| Days ≥ 1" Snow Depth | 21 | 21 | 10 | 2 | 0 | 0 | 0 | 0 | 0 | 0 | 4 | 16 | 74 |

## CAMDEN 2 NW *Oneida County* ELEVATION 541 ft LAT/LONG 43° 22 ' N / 75° 47 ' W

| | JAN | FEB | MAR | APR | MAY | JUN | JUL | AUG | SEP | OCT | NOV | DEC | YEAR |
|---|---|---|---|---|---|---|---|---|---|---|---|---|---|
| Maximum Temp °F | 28.7 | 30.3 | 40.8 | 54.5 | 67.5 | *75.6* | na | *78.5* | *70.1* | 58.1 | *46.1* | 34.0 | na |
| Minimum Temp °F | 8.7 | 8.1 | 19.9 | 32.0 | 41.8 | *50.1* | na | *54.1* | *46.3* | 36.0 | *29.6* | 17.1 | na |
| Mean Temp °F | 18.7 | 19.3 | 30.4 | 43.3 | 54.6 | *62.9* | na | *66.3* | *58.2* | 47.1 | *37.9* | 25.5 | na |
| Days Max Temp ≥ 90 °F | 0 | 0 | 0 | 0 | 0 | *1* | *2* | *1* | *0* | 0 | 0 | 0 | 4 |
| Days Max Temp ≤ 32 °F | 19 | 17 | 7 | 1 | 0 | *0* | *0* | *0* | *0* | 0 | 2 | 12 | 58 |
| Days Min Temp ≤ 32 °F | 30 | 27 | 25 | *17* | 6 | *1* | *0* | *0* | *3* | 12 | 19 | 28 | 168 |
| Days Min Temp ≤ 0 °F | 9 | 9 | 3 | 0 | 0 | *0* | *0* | *0* | *0* | 0 | 0 | 4 | 25 |
| Heating Degree Days | 1428 | 1287 | 1066 | 647 | 330 | *123* | na | *59* | *224* | 549 | *807* | 1219 | na |
| Cooling Degree Days | 0 | 0 | 0 | 1 | 16 | 57 | na | 107 | 28 | 1 | 0 | 0 | na |
| Total Precipitation (") | 4.15 | 3.32 | 3.78 | 4.06 | 4.01 | *4.33* | *4.28* | *4.68* | *5.22* | 4.36 | *5.25* | 4.44 | 51.88 |
| Days ≥ 0.1" Precip | 11 | 8 | 10 | 9 | 9 | *9* | *7* | *8* | *9* | 9 | 11 | 11 | 111 |
| Total Snowfall (") | 39.1 | 31.8 | 17.2 | *3.0* | 0.0 | 0.0 | *0.0* | *0.0* | *0.0* | 0.1 | *7.6* | 28.5 | 127.3 |
| Days ≥ 1" Snow Depth | na | na | na | na | *0* | *0* | *0* | *0* | *0* | *0* | na | *20* | na |

## CANANDAIGUA 3 S *Ontario County* ELEVATION 722 ft LAT/LONG 42° 51' N / 77° 17' W

| | JAN | FEB | MAR | APR | MAY | JUN | JUL | AUG | SEP | OCT | NOV | DEC | YEAR |
|---|---|---|---|---|---|---|---|---|---|---|---|---|---|
| Maximum Temp °F | 32.0 | 33.5 | 42.3 | 55.3 | 67.3 | 76.5 | 81.7 | 79.7 | 72.1 | 60.4 | 48.8 | 37.4 | 57.2 |
| Minimum Temp °F | 15.9 | 16.4 | 24.8 | 35.3 | 45.5 | 55.3 | 60.7 | 59.4 | 52.6 | 41.7 | 33.5 | 23.0 | 38.7 |
| Mean Temp °F | 24.0 | 25.0 | 33.6 | 45.3 | 56.4 | 65.9 | 71.2 | 69.6 | 62.4 | 51.1 | 41.2 | 30.2 | 48.0 |
| Days Max Temp ≥ 90 °F | 0 | 0 | 0 | 0 | 0 | 1 | 3 | 2 | 1 | 0 | 0 | 0 | 7 |
| Days Max Temp ≤ 32 °F | 16 | 13 | 6 | 0 | 0 | 0 | 0 | 0 | 0 | 0 | 1 | 9 | 45 |
| Days Min Temp ≤ 32 °F | 29 | 26 | 24 | 12 | 1 | 0 | 0 | 0 | 0 | 4 | 14 | 25 | 135 |
| Days Min Temp ≤ 0 °F | 3 | 2 | 0 | 0 | 0 | 0 | 0 | 0 | 0 | 0 | 0 | 0 | 5 |
| Heating Degree Days | 1265 | 1122 | 967 | 587 | 288 | 72 | 12 | 25 | 136 | 430 | 707 | 1071 | 6682 |
| Cooling Degree Days | 0 | 0 | 0 | 3 | 31 | 102 | 226 | 170 | 61 | 6 | 0 | 0 | 599 |
| Total Precipitation (") | 1.61 | 1.60 | 2.18 | 3.07 | 2.83 | 3.72 | 3.08 | 3.30 | 3.34 | 2.84 | 3.03 | 2.34 | 32.94 |
| Days ≥ 0.1" Precip | 4 | 5 | 6 | 7 | 8 | 7 | 7 | 7 | 7 | 7 | 8 | 7 | 80 |
| Total Snowfall (") | na | na | na | na | *0.0* | 0.0 | 0.0 | 0.0 | 0.0 | 0.0 | na | na | na |
| Days ≥ 1" Snow Depth | na | na | na | na | 0 | 0 | 0 | 0 | 0 | 0 | na | na | na |

## CANTON 4 SE *St. Lawrence County* ELEVATION 413 ft LAT/LONG 44° 35' N / 75° 10' W

| | JAN | FEB | MAR | APR | MAY | JUN | JUL | AUG | SEP | OCT | NOV | DEC | YEAR |
|---|---|---|---|---|---|---|---|---|---|---|---|---|---|
| Maximum Temp °F | 25.2 | 27.4 | 38.2 | 52.5 | 65.4 | 73.9 | 79.3 | 77.0 | 68.7 | 56.7 | 44.3 | 30.9 | 53.3 |
| Minimum Temp °F | 3.6 | 5.4 | 18.1 | 31.9 | 43.3 | 52.4 | 57.4 | 55.6 | 47.0 | 36.1 | 27.4 | 12.0 | 32.5 |
| Mean Temp °F | 14.4 | 16.5 | 28.2 | 42.2 | 54.4 | 63.2 | 68.3 | 66.3 | 57.9 | 46.4 | 35.9 | 21.5 | 42.9 |
| Days Max Temp ≥ 90 °F | 0 | 0 | 0 | 0 | 0 | 0 | 1 | 1 | 0 | 0 | 0 | 0 | 2 |
| Days Max Temp ≤ 32 °F | 21 | 18 | 10 | 1 | 0 | 0 | 0 | 0 | 0 | 0 | 4 | 16 | 70 |
| Days Min Temp ≤ 32 °F | 30 | 27 | 27 | 17 | 4 | 0 | 0 | 0 | 2 | 12 | 21 | 29 | 169 |
| Days Min Temp ≤ 0 °F | 13 | 11 | 3 | 0 | 0 | 0 | 0 | 0 | 0 | 0 | 0 | 7 | 34 |
| Heating Degree Days | 1564 | 1366 | 1134 | 678 | 340 | 117 | 39 | 68 | 238 | 572 | 867 | 1343 | 8326 |
| Cooling Degree Days | 0 | 0 | 0 | 2 | 18 | 64 | 155 | 115 | 29 | 2 | 0 | 0 | 385 |
| Total Precipitation (") | 2.02 | 1.92 | 2.25 | 2.89 | 2.95 | 3.27 | 3.28 | 4.02 | 3.99 | 3.21 | 3.53 | 2.81 | 36.14 |
| Days ≥ 0.1" Precip | 6 | 5 | 6 | 7 | 8 | 8 | 7 | 8 | 8 | 8 | 9 | 7 | 87 |
| Total Snowfall (") | 22.9 | 19.9 | 13.5 | 4.4 | 0.2 | 0.0 | 0.0 | 0.0 | 0.0 | 0.5 | 6.5 | 21.8 | 89.7 |
| Days ≥ 1" Snow Depth | 27 | 25 | 19 | 4 | 0 | 0 | 0 | 0 | 0 | 0 | 7 | 21 | 103 |

## CHASM FALLS *Franklin County* ELEVATION 1030 ft LAT/LONG 44° 45' N / 74° 13' W

| | JAN | FEB | MAR | APR | MAY | JUN | JUL | AUG | SEP | OCT | NOV | DEC | YEAR |
|---|---|---|---|---|---|---|---|---|---|---|---|---|---|
| Maximum Temp °F | 25.6 | 28.4 | 38.7 | 52.4 | 66.5 | 74.8 | 78.8 | 75.9 | 67.5 | 56.4 | 43.0 | 30.3 | 53.2 |
| Minimum Temp °F | 4.7 | 6.2 | 17.3 | 30.0 | 41.0 | 50.3 | 55.0 | 53.4 | 45.8 | 35.9 | 26.5 | 12.2 | 31.5 |
| Mean Temp °F | 15.2 | 17.3 | 28.0 | 41.2 | 53.8 | 62.6 | 66.9 | 64.7 | 56.7 | 46.2 | 34.8 | 21.3 | 42.4 |
| Days Max Temp ≥ 90 °F | 0 | 0 | 0 | 0 | 0 | 1 | 1 | 0 | 0 | 0 | 0 | 0 | 2 |
| Days Max Temp ≤ 32 °F | 21 | 18 | 9 | 1 | 0 | 0 | 0 | 0 | 0 | 0 | 5 | 17 | 71 |
| Days Min Temp ≤ 32 °F | 30 | 27 | 28 | 19 | 6 | 1 | 0 | 0 | 2 | 12 | 23 | 29 | 177 |
| Days Min Temp ≤ 0 °F | 12 | 11 | 4 | 0 | 0 | 0 | 0 | 0 | 0 | 0 | 0 | 7 | 34 |
| Heating Degree Days | 1540 | 1341 | 1140 | 708 | 356 | 124 | 48 | 83 | 261 | 577 | 901 | 1349 | 8428 |
| Cooling Degree Days | 0 | 0 | 0 | 0 | 11 | 48 | 108 | 70 | 14 | 0 | 0 | 0 | 251 |
| Total Precipitation (") | 2.59 | 2.44 | 2.74 | 3.02 | 3.35 | 4.00 | 4.22 | 5.37 | 4.18 | 3.56 | 3.89 | 3.47 | 42.83 |
| Days ≥ 0.1" Precip | 9 | 7 | 8 | 9 | 9 | 9 | 8 | 10 | 9 | 9 | 10 | 9 | 106 |
| Total Snowfall (") | 27.6 | 25.9 | 19.5 | 7.3 | 0.4 | 0.0 | 0.0 | 0.0 | 0.0 | 0.9 | 13.9 | 27.1 | 122.6 |
| Days ≥ 1" Snow Depth | 30 | 28 | 29 | 11 | 1 | 0 | 0 | 0 | 0 | 1 | 11 | 26 | 137 |

## CHAZY *Clinton County* ELEVATION 121 ft LAT/LONG 44° 53' N / 73° 23' W

| | JAN | FEB | MAR | APR | MAY | JUN | JUL | AUG | SEP | OCT | NOV | DEC | YEAR |
|---|---|---|---|---|---|---|---|---|---|---|---|---|---|
| Maximum Temp °F | 26.4 | 28.8 | 39.3 | 54.5 | 67.5 | 75.9 | 80.7 | 78.2 | 68.9 | 56.8 | 44.4 | 31.7 | 54.4 |
| Minimum Temp °F | 5.6 | 7.4 | 19.4 | 33.0 | 43.7 | 53.1 | 58.0 | 55.9 | 47.4 | 37.5 | 28.1 | 13.8 | 33.6 |
| Mean Temp °F | 16.1 | 18.2 | 29.4 | 43.8 | 55.6 | 64.5 | 69.3 | 67.1 | 58.2 | 47.2 | 36.3 | 22.8 | 44.0 |
| Days Max Temp ≥ 90 °F | 0 | 0 | 0 | 0 | 0 | 1 | 2 | 1 | 0 | 0 | 0 | 0 | 4 |
| Days Max Temp ≤ 32 °F | 20 | 17 | 8 | 0 | 0 | 0 | 0 | 0 | 0 | 0 | 3 | 15 | 63 |
| Days Min Temp ≤ 32 °F | 30 | 27 | 27 | 15 | 3 | 0 | 0 | 0 | 1 | 10 | 21 | 28 | 162 |
| Days Min Temp ≤ 0 °F | 12 | 9 | 3 | 0 | 0 | 0 | 0 | 0 | 0 | 0 | 0 | 6 | 30 |
| Heating Degree Days | 1512 | 1319 | 1096 | 632 | 300 | 82 | 20 | 48 | 222 | 547 | 855 | 1304 | 7937 |
| Cooling Degree Days | 0 | 0 | 0 | 2 | 14 | 67 | 162 | 118 | 22 | 1 | 0 | 0 | 386 |
| Total Precipitation (") | *1.22* | na | *1.51* | 2.55 | 2.96 | 3.09 | 3.30 | 3.91 | 3.23 | 2.93 | 2.81 | *1.96* | na |
| Days ≥ 0.1" Precip | na | na | na | na | *7* | 7 | 7 | 8 | 7 | 7 | na | na | na |
| Total Snowfall (") | na | na | *7.6* | *2.9* | 0.1 | 0.0 | 0.0 | 0.0 | 0.0 | 0.4 | *5.0* | *13.5* | na |
| Days ≥ 1" Snow Depth | na | *22* | na | *2* | 0 | 0 | 0 | 0 | 0 | 0 | na | *17* | na |

## CHERRY VALLEY 2 NNE *Otsego County* ELEVATION 1362 ft LAT/LONG 42° 49 ' N / 74° 44 ' W

| | JAN | FEB | MAR | APR | MAY | JUN | JUL | AUG | SEP | OCT | NOV | DEC | YEAR |
|---|---|---|---|---|---|---|---|---|---|---|---|---|---|
| Maximum Temp °F | 27.4 | 29.8 | 39.6 | 53.5 | 65.9 | 73.9 | 78.3 | 76.0 | 68.0 | 57.0 | 44.4 | 32.2 | 53.8 |
| Minimum Temp °F | 10.6 | 12.1 | 21.8 | 33.0 | 43.6 | 52.5 | 57.5 | 56.0 | 48.8 | 38.3 | 29.3 | 17.1 | 35.1 |
| Mean Temp °F | 19.0 | 21.0 | 30.7 | 43.3 | 54.7 | 63.2 | 67.9 | 66.0 | 58.4 | 47.7 | 36.9 | 24.6 | 44.4 |
| Days Max Temp ≥ 90 °F | 0 | 0 | 0 | 0 | 0 | 0 | 0 | 0 | 0 | 0 | 0 | 0 | 0 |
| Days Max Temp ≤ 32 °F | 20 | 17 | 8 | 1 | 0 | 0 | 0 | 0 | 0 | 0 | 4 | 16 | 66 |
| Days Min Temp ≤ 32 °F | 30 | 27 | 27 | 15 | 3 | 0 | 0 | 0 | 1 | 9 | 20 | 29 | 161 |
| Days Min Temp ≤ 0 °F | 7 | 5 | 1 | 0 | 0 | 0 | 0 | 0 | 0 | 0 | 0 | 3 | 16 |
| Heating Degree Days | 1419 | 1237 | 1057 | 646 | 328 | 110 | 31 | 59 | 220 | 533 | 837 | 1244 | 7721 |
| Cooling Degree Days | 0 | 0 | 0 | 1 | 15 | 55 | 128 | 92 | 26 | 1 | 0 | 0 | 318 |
| Total Precipitation (") | 2.68 | 2.52 | 3.48 | 3.82 | 4.24 | 4.45 | 4.21 | 3.91 | 4.05 | 3.49 | 4.04 | 3.60 | 44.49 |
| Days ≥ 0.1" Precip | 8 | 6 | 8 | 9 | 9 | 9 | 7 | 8 | 8 | 8 | 9 | 9 | 98 |
| Total Snowfall (") | 29.2 | 20.3 | 20.9 | 6.4 | 1.1 | 0.0 | 0.0 | 0.0 | 0.0 | 0.9 | 12.5 | 28.7 | 120.0 |
| Days ≥ 1" Snow Depth | 25 | 24 | 17 | 5 | 0 | 0 | 0 | 0 | 0 | 0 | 6 | 20 | 97 |

## COLDEN 1 N *Erie County* ELEVATION 1025 ft LAT/LONG 42° 40 ' N / 78° 41 ' W

| | JAN | FEB | MAR | APR | MAY | JUN | JUL | AUG | SEP | OCT | NOV | DEC | YEAR |
|---|---|---|---|---|---|---|---|---|---|---|---|---|---|
| Maximum Temp °F | 29.8 | 32.0 | 41.7 | 54.9 | 66.9 | 75.4 | 79.5 | 77.7 | 70.4 | 59.1 | 46.7 | 35.0 | 55.8 |
| Minimum Temp °F | 12.8 | 12.6 | 21.4 | 32.4 | 41.9 | 50.9 | 55.7 | 54.6 | 48.0 | 37.6 | 30.4 | 20.3 | 34.9 |
| Mean Temp °F | 21.4 | 22.3 | 31.6 | 43.7 | 54.4 | 63.2 | 67.6 | 66.2 | 59.2 | 48.4 | 38.6 | 27.7 | 45.4 |
| Days Max Temp ≥ 90 °F | 0 | 0 | 0 | 0 | 0 | 0 | 1 | 1 | 0 | 0 | 0 | 0 | 2 |
| Days Max Temp ≤ 32 °F | 18 | 15 | 7 | 1 | 0 | 0 | 0 | 0 | 0 | 0 | 3 | 12 | 56 |
| Days Min Temp ≤ 32 °F | 30 | 27 | 26 | 16 | 5 | 0 | 0 | 0 | 1 | 9 | 19 | 28 | 161 |
| Days Min Temp ≤ 0 °F | 6 | 5 | 2 | 0 | 0 | 0 | 0 | 0 | 0 | 0 | 0 | 2 | 15 |
| Heating Degree Days | 1346 | 1199 | 1028 | 635 | 338 | 113 | 33 | 54 | 201 | 510 | 787 | 1149 | 7393 |
| Cooling Degree Days | 0 | 0 | 0 | 3 | 16 | 55 | 119 | 90 | 31 | 1 | 0 | 0 | 315 |
| Total Precipitation (") | 3.56 | 2.85 | 3.46 | 3.71 | 3.52 | 4.16 | 3.88 | 4.13 | 4.87 | 3.85 | 5.11 | 4.62 | 47.72 |
| Days ≥ 0.1" Precip | 11 | 9 | 10 | 9 | 9 | 8 | 7 | 8 | 9 | 10 | 12 | 13 | 115 |
| Total Snowfall (") | 46.9 | 31.1 | 18.7 | 6.5 | 0.4 | 0.0 | 0.0 | 0.0 | 0.0 | 0.7 | 19.5 | 38.8 | 162.6 |
| Days ≥ 1" Snow Depth | 28 | 26 | 20 | 3 | 0 | 0 | 0 | 0 | 0 | 0 | 7 | 23 | 107 |

## COOPERSTOWN *Otsego County* ELEVATION 1240 ft LAT/LONG 42° 42 ' N / 74° 55 ' W

| | JAN | FEB | MAR | APR | MAY | JUN | JUL | AUG | SEP | OCT | NOV | DEC | YEAR |
|---|---|---|---|---|---|---|---|---|---|---|---|---|---|
| Maximum Temp °F | 30.0 | 32.8 | 42.4 | 55.9 | 67.9 | 75.5 | 79.8 | 77.8 | 69.9 | 59.2 | 46.7 | 34.7 | 56.1 |
| Minimum Temp °F | 10.6 | 11.6 | 21.4 | 31.8 | 42.0 | 50.9 | 55.7 | 54.8 | 47.5 | 36.9 | 29.0 | 17.6 | 34.2 |
| Mean Temp °F | 20.3 | 22.2 | 32.0 | 43.9 | 55.0 | 63.2 | 67.8 | 66.3 | 58.7 | 48.1 | 37.9 | 26.2 | 45.1 |
| Days Max Temp ≥ 90 °F | 0 | 0 | 0 | 0 | 0 | 0 | 1 | 1 | 0 | 0 | 0 | 0 | 2 |
| Days Max Temp ≤ 32 °F | 18 | 14 | 6 | 0 | 0 | 0 | 0 | 0 | 0 | 0 | 2 | 13 | 53 |
| Days Min Temp ≤ 32 °F | 30 | 27 | 27 | 16 | 6 | 0 | 0 | 0 | 2 | 11 | 20 | 28 | 167 |
| Days Min Temp ≤ 0 °F | 7 | 6 | 2 | 0 | 0 | 0 | 0 | 0 | 0 | 0 | 0 | 3 | 18 |
| Heating Degree Days | 1379 | 1202 | 1018 | 628 | 318 | 106 | 32 | 51 | 210 | 519 | 807 | 1197 | 7467 |
| Cooling Degree Days | 0 | 0 | 0 | 1 | 11 | 55 | 132 | 95 | 27 | 1 | 0 | 0 | 322 |
| Total Precipitation (") | 2.53 | 2.18 | 3.19 | 3.42 | 3.66 | 4.29 | 3.65 | 3.64 | 3.92 | 3.18 | 3.50 | 3.15 | 40.31 |
| Days ≥ 0.1" Precip | 7 | 6 | 8 | 8 | 8 | 8 | 7 | 7 | 8 | 7 | 8 | 8 | 90 |
| Total Snowfall (") | 22.0 | 16.7 | 15.1 | 5.1 | 0.9 | 0.0 | 0.0 | 0.0 | 0.0 | 0.2 | 7.5 | 19.9 | 87.4 |
| Days ≥ 1" Snow Depth | 27 | 26 | 20 | 4 | 0 | 0 | 0 | 0 | 0 | 0 | 6 | 20 | 103 |

## CORTLAND *Cortland County* ELEVATION 1132 ft LAT/LONG 42° 36 ' N / 76° 11 ' W

| | JAN | FEB | MAR | APR | MAY | JUN | JUL | AUG | SEP | OCT | NOV | DEC | YEAR |
|---|---|---|---|---|---|---|---|---|---|---|---|---|---|
| Maximum Temp °F | 29.9 | 31.6 | 41.3 | 54.5 | 67.1 | 75.9 | 80.7 | 79.0 | 70.6 | 58.8 | 46.4 | 34.8 | 55.9 |
| Minimum Temp °F | 14.4 | 14.7 | 23.7 | 34.5 | 44.4 | 53.4 | 58.3 | 56.7 | 49.3 | 39.2 | 31.6 | 21.2 | 36.8 |
| Mean Temp °F | 22.2 | 23.2 | 32.5 | 44.5 | 55.7 | 64.7 | 69.5 | 67.9 | 60.0 | 49.0 | 39.0 | 28.0 | 46.4 |
| Days Max Temp ≥ 90 °F | 0 | 0 | 0 | 0 | 0 | 1 | 3 | 2 | 0 | 0 | 0 | 0 | 6 |
| Days Max Temp ≤ 32 °F | 18 | 15 | 8 | 1 | 0 | 0 | 0 | 0 | 0 | 0 | 3 | 12 | 57 |
| Days Min Temp ≤ 32 °F | 29 | 26 | 24 | 14 | 2 | 0 | 0 | 0 | 0 | 7 | 16 | 26 | 144 |
| Days Min Temp ≤ 0 °F | 5 | 4 | 1 | 0 | 0 | 0 | 0 | 0 | 0 | 0 | 0 | 2 | 12 |
| Heating Degree Days | 1321 | 1174 | 1001 | 611 | 305 | 93 | 23 | 41 | 186 | 491 | 772 | 1140 | 7158 |
| Cooling Degree Days | 0 | 0 | 0 | 3 | 26 | 92 | 188 | 143 | 40 | 2 | 0 | 0 | 494 |
| Total Precipitation (") | 2.56 | 2.45 | 3.01 | 3.24 | 3.31 | 3.87 | 3.48 | 3.32 | 3.86 | 3.27 | 3.67 | 3.70 | 39.74 |
| Days ≥ 0.1" Precip | 7 | 6 | 7 | 7 | 8 | 8 | 7 | 7 | 7 | 8 | 9 | 9 | 90 |
| Total Snowfall (") | 22.4 | 19.0 | 13.9 | 3.6 | 0.1 | 0.0 | 0.0 | 0.0 | 0.0 | 0.6 | 7.5 | 23.1 | 90.2 |
| Days ≥ 1" Snow Depth | 26 | 25 | 17 | 3 | 0 | 0 | 0 | 0 | 0 | 0 | 5 | 19 | 95 |

## DANNEMORA *Clinton County* ELEVATION 1342 ft LAT/LONG 44° 43 ' N / 73° 43 ' W

| | JAN | FEB | MAR | APR | MAY | JUN | JUL | AUG | SEP | OCT | NOV | DEC | YEAR |
|---|---|---|---|---|---|---|---|---|---|---|---|---|---|
| Maximum Temp °F | 25.7 | 28.3 | 38.7 | 52.3 | 65.9 | 74.5 | 79.1 | 76.6 | 67.9 | 56.1 | 42.2 | 30.5 | 53.2 |
| Minimum Temp °F | 7.0 | 9.5 | 20.2 | 32.9 | 44.1 | 53.1 | 58.2 | 56.3 | 47.8 | 37.5 | 27.4 | 13.9 | 34.0 |
| Mean Temp °F | 16.4 | 18.9 | 29.5 | 42.6 | 55.0 | 63.8 | 68.7 | 66.5 | 57.8 | 46.8 | 34.8 | 22.2 | 43.6 |
| Days Max Temp ≥ 90 °F | 0 | 0 | 0 | 0 | 0 | 1 | 2 | 0 | 0 | 0 | 0 | 0 | 3 |
| Days Max Temp ≤ 32 °F | 22 | 18 | 9 | 1 | 0 | 0 | 0 | 0 | 0 | 0 | 5 | 17 | 72 |
| Days Min Temp ≤ 32 °F | 30 | 27 | 27 | 15 | 2 | 0 | 0 | 0 | 1 | 10 | 21 | 29 | 162 |
| Days Min Temp ≤ 0 °F | 10 | 7 | 2 | 0 | 0 | 0 | 0 | 0 | 0 | 0 | 0 | 5 | 24 |
| Heating Degree Days | 1502 | 1296 | 1095 | 668 | 323 | 99 | 24 | 54 | 233 | 558 | 898 | 1321 | 8071 |
| Cooling Degree Days | 0 | 0 | 0 | 2 | 17 | 62 | 143 | 96 | 20 | 1 | 0 | 0 | 341 |
| Total Precipitation (") | 1.99 | 1.90 | 2.17 | 2.83 | 3.06 | 3.50 | 3.43 | 4.26 | 3.54 | 3.05 | 3.32 | 2.91 | 35.96 |
| Days ≥ 0.1" Precip | 6 | 5 | 6 | 7 | 8 | 8 | 7 | 8 | 7 | 7 | 8 | 7 | 84 |
| Total Snowfall (") | na | na | na | *4.2* | 0.2 | 0.0 | 0.0 | 0.0 | 0.0 | 0.6 | na | na | na |
| Days ≥ 1" Snow Depth | na | na | na | na | 0 | 0 | 0 | 0 | 0 | 0 | na | na | na |

## DANSVILLE *Livingston County* ELEVATION 702 ft LAT/LONG 42° 34 ' N / 77° 42 ' W

| | JAN | FEB | MAR | APR | MAY | JUN | JUL | AUG | SEP | OCT | NOV | DEC | YEAR |
|---|---|---|---|---|---|---|---|---|---|---|---|---|---|
| Maximum Temp °F | 32.6 | 34.4 | 44.0 | 56.9 | 68.9 | 78.0 | 82.6 | 80.8 | 73.0 | 61.4 | 49.2 | 37.7 | 58.3 |
| Minimum Temp °F | 15.3 | 15.7 | 24.0 | 34.3 | 43.8 | 53.0 | 57.9 | 56.4 | 49.3 | 39.0 | 32.1 | 22.3 | 36.9 |
| Mean Temp °F | 24.0 | 25.1 | 34.0 | 45.6 | 56.4 | 65.5 | 70.3 | 68.6 | 61.2 | 50.2 | 40.7 | 30.0 | 47.6 |
| Days Max Temp ≥ 90 °F | 0 | 0 | 0 | 0 | 0 | 2 | 5 | 3 | 1 | 0 | 0 | 0 | 11 |
| Days Max Temp ≤ 32 °F | 15 | 13 | 5 | 0 | 0 | 0 | 0 | 0 | 0 | 0 | 1 | 9 | 43 |
| Days Min Temp ≤ 32 °F | 29 | 26 | 24 | 14 | 3 | 0 | 0 | 0 | 1 | 7 | 16 | 26 | 146 |
| Days Min Temp ≤ 0 °F | 4 | 3 | 1 | 0 | 0 | 0 | 0 | 0 | 0 | 0 | 0 | 1 | 9 |
| Heating Degree Days | 1265 | 1120 | 954 | 581 | 290 | 84 | 19 | 36 | 163 | 456 | 723 | 1078 | 6769 |
| Cooling Degree Days | 0 | 0 | 1 | 6 | 29 | 100 | 200 | 155 | 55 | 3 | 0 | 0 | 549 |
| Total Precipitation (") | 1.39 | 1.33 | 1.82 | 2.67 | 2.87 | 3.65 | 3.03 | 3.26 | 3.42 | 2.65 | 2.80 | 2.13 | 31.02 |
| Days ≥ 0.1" Precip | 4 | 4 | 5 | 8 | 8 | 8 | 7 | 7 | 7 | 7 | 7 | 6 | 78 |
| Total Snowfall (") | 14.1 | 11.5 | 7.9 | 2.7 | 0.3 | 0.0 | 0.0 | 0.0 | 0.0 | 0.1 | 4.7 | 12.1 | 53.4 |
| Days ≥ 1" Snow Depth | 18 | 15 | 6 | 1 | 0 | 0 | 0 | 0 | 0 | 0 | 2 | 10 | 52 |

## DEPOSIT *Delaware County* ELEVATION 1001 ft LAT/LONG 42° 4 ' N / 75° 26 ' W

| | JAN | FEB | MAR | APR | MAY | JUN | JUL | AUG | SEP | OCT | NOV | DEC | YEAR |
|---|---|---|---|---|---|---|---|---|---|---|---|---|---|
| Maximum Temp °F | 31.0 | 34.2 | 44.3 | 58.4 | 69.9 | 77.0 | 80.8 | 79.0 | 71.1 | 60.3 | 47.7 | 35.3 | 57.4 |
| Minimum Temp °F | 12.2 | 13.5 | 23.1 | 33.3 | 43.2 | 52.5 | 57.2 | 56.5 | 49.6 | 38.0 | 30.3 | 19.3 | 35.7 |
| Mean Temp °F | 21.6 | 23.9 | 33.7 | 45.9 | 56.6 | 64.8 | 69.0 | 67.8 | 60.3 | 49.1 | 39.0 | 27.4 | 46.6 |
| Days Max Temp ≥ 90 °F | 0 | 0 | 0 | 0 | 0 | 1 | 2 | 1 | 0 | 0 | 0 | 0 | 4 |
| Days Max Temp ≤ 32 °F | 17 | 13 | 4 | 0 | 0 | 0 | 0 | 0 | 0 | 0 | 2 | 11 | 47 |
| Days Min Temp ≤ 32 °F | 30 | 26 | 25 | 15 | 4 | 0 | 0 | 0 | 1 | 10 | 19 | 28 | 158 |
| Days Min Temp ≤ 0 °F | 7 | 6 | 1 | 0 | 0 | 0 | 0 | 0 | 0 | 0 | 0 | 2 | 16 |
| Heating Degree Days | 1342 | 1156 | 965 | 570 | 275 | 79 | 23 | 36 | 172 | 488 | 774 | 1160 | 7040 |
| Cooling Degree Days | 0 | 0 | 0 | 2 | 16 | 64 | 149 | 114 | 34 | 2 | 0 | 0 | 381 |
| Total Precipitation (") | 2.71 | 2.60 | 3.37 | 3.73 | 3.91 | 3.98 | 3.96 | 4.24 | 3.53 | 3.44 | 4.05 | 3.50 | 43.02 |
| Days ≥ 0.1" Precip | 7 | 7 | 8 | 9 | 9 | 8 | 8 | 8 | 7 | 7 | 9 | 8 | 95 |
| Total Snowfall (") | *17.3* | na | na | *3.1* | 0.1 | 0.0 | 0.0 | 0.0 | 0.0 | 0.0 | *5.0* | *17.8* | na |
| Days ≥ 1" Snow Depth | *21* | na | na | 1 | 0 | 0 | 0 | 0 | 0 | 0 | *3* | na | na |

## DOBBS FERRY ARDSLEY *Westchester County* ELEVATION 239 ft LAT/LONG 41° 1 ' N / 73° 52 ' W

| | JAN | FEB | MAR | APR | MAY | JUN | JUL | AUG | SEP | OCT | NOV | DEC | YEAR |
|---|---|---|---|---|---|---|---|---|---|---|---|---|---|
| Maximum Temp °F | 37.1 | 40.1 | 49.6 | 61.8 | 72.3 | 80.4 | 85.3 | 83.4 | 75.7 | 64.4 | 53.5 | 42.1 | 62.1 |
| Minimum Temp °F | 22.5 | 24.0 | 31.6 | 40.9 | 50.7 | 59.8 | 65.3 | 64.3 | 56.9 | 46.2 | 37.8 | 28.2 | 44.0 |
| Mean Temp °F | 29.8 | 32.1 | 40.6 | 51.4 | 61.5 | 70.1 | 75.3 | 73.9 | 66.3 | 55.3 | 45.7 | 35.2 | 53.1 |
| Days Max Temp ≥ 90 °F | 0 | 0 | 0 | 0 | 1 | 3 | 7 | 4 | 1 | 0 | 0 | 0 | 16 |
| Days Max Temp ≤ 32 °F | 10 | 6 | 1 | 0 | 0 | 0 | 0 | 0 | 0 | 0 | 0 | 4 | 21 |
| Days Min Temp ≤ 32 °F | 26 | 23 | 16 | 4 | 0 | 0 | 0 | 0 | 0 | 2 | 8 | 21 | 100 |
| Days Min Temp ≤ 0 °F | 1 | 0 | 0 | 0 | 0 | 0 | 0 | 0 | 0 | 0 | 0 | 0 | 1 |
| Heating Degree Days | 1083 | 923 | 750 | 408 | 152 | 19 | 1 | 3 | 63 | 305 | 575 | 918 | 5200 |
| Cooling Degree Days | 0 | 0 | 1 | 5 | 52 | 181 | 327 | 273 | 109 | 13 | 1 | 0 | 962 |
| Total Precipitation (") | 3.84 | 3.42 | 4.57 | 4.33 | 4.85 | 3.81 | 4.23 | 4.35 | 4.38 | 3.84 | 4.55 | 4.39 | 50.56 |
| Days ≥ 0.1" Precip | 7 | 6 | 8 | 7 | 8 | 7 | 7 | 7 | 6 | 6 | 7 | 7 | 83 |
| Total Snowfall (") | 9.8 | 10.5 | 6.5 | 0.7 | 0.0 | 0.0 | 0.0 | 0.0 | 0.0 | 0.0 | 0.9 | 5.2 | 33.6 |
| Days ≥ 1" Snow Depth | 15 | 13 | 5 | 0 | 0 | 0 | 0 | 0 | 0 | 0 | 0 | 6 | 39 |

## ELMIRA *Chemung County* ELEVATION 840 ft LAT/LONG 42° 5 ' N / 76° 47 ' W

| | JAN | FEB | MAR | APR | MAY | JUN | JUL | AUG | SEP | OCT | NOV | DEC | YEAR |
|---|---|---|---|---|---|---|---|---|---|---|---|---|---|
| Maximum Temp °F | 32.1 | 34.4 | 43.8 | 57.2 | 69.3 | 78.0 | 82.6 | 80.6 | 72.6 | 60.6 | 48.7 | 37.0 | 58.1 |
| Minimum Temp °F | 14.1 | 14.5 | 23.0 | 33.4 | 42.7 | 51.8 | 56.9 | 55.2 | 48.1 | 37.0 | 30.5 | 20.9 | 35.7 |
| Mean Temp °F | 23.1 | 24.5 | 33.4 | 45.3 | 56.1 | 64.9 | 69.8 | 68.0 | 60.4 | 48.8 | 39.6 | 29.0 | 46.9 |
| Days Max Temp ≥ 90 °F | 0 | 0 | 0 | 0 | 1 | 2 | 5 | 3 | 1 | 0 | 0 | 0 | 12 |
| Days Max Temp ≤ 32 °F | 16 | 13 | 5 | 0 | 0 | 0 | 0 | 0 | 0 | 0 | 1 | 10 | 45 |
| Days Min Temp ≤ 32 °F | 29 | 26 | 25 | 16 | 4 | 0 | 0 | 0 | 1 | 10 | 18 | 27 | 156 |
| Days Min Temp ≤ 0 °F | 5 | 4 | 0 | 0 | 0 | 0 | 0 | 0 | 0 | 0 | 0 | 2 | 11 |
| Heating Degree Days | 1292 | 1138 | 974 | 586 | 294 | 87 | 19 | 38 | 175 | 496 | 756 | 1109 | 6964 |
| Cooling Degree Days | 0 | 0 | 0 | 4 | 28 | 97 | 200 | 145 | 47 | 2 | 0 | 0 | 523 |
| Total Precipitation (") | 1.80 | 1.87 | 2.50 | 2.73 | 3.11 | 3.95 | 3.55 | 3.30 | 3.24 | 2.79 | 3.06 | 2.38 | 34.28 |
| Days ≥ 0.1" Precip | 5 | 5 | 6 | 7 | 8 | 8 | 7 | 6 | 7 | 6 | 7 | 6 | 78 |
| Total Snowfall (") | 10.4 | 9.8 | 8.9 | 1.6 | 0.3 | 0.0 | 0.0 | 0.0 | 0.0 | 0.2 | 2.9 | 9.6 | 43.7 |
| Days ≥ 1" Snow Depth | 19 | 17 | 8 | 1 | 0 | 0 | 0 | 0 | 0 | 0 | 1 | 10 | 56 |

## FRANKLINVILLE 1 SSW *Cattaraugus County* ELEVATION 1591 ft LAT/LONG 42° 20 ' N / 78° 27 ' W

| | JAN | FEB | MAR | APR | MAY | JUN | JUL | AUG | SEP | OCT | NOV | DEC | YEAR |
|---|---|---|---|---|---|---|---|---|---|---|---|---|---|
| Maximum Temp °F | 28.8 | 30.8 | 40.5 | 53.4 | 65.4 | 73.9 | 77.9 | 76.1 | 68.8 | 57.9 | 45.3 | 33.9 | 54.4 |
| Minimum Temp °F | 9.9 | 10.1 | 18.6 | 30.1 | 38.9 | 48.3 | 53.0 | 51.6 | 45.3 | 35.0 | 28.0 | 17.6 | 32.2 |
| Mean Temp °F | 19.3 | 20.5 | 29.6 | 41.8 | 52.2 | 61.3 | 65.5 | 63.8 | 57.1 | 46.5 | 36.7 | 25.8 | 43.3 |
| Days Max Temp ≥ 90 °F | 0 | 0 | 0 | 0 | 0 | 0 | 1 | 0 | 0 | 0 | 0 | 0 | 1 |
| Days Max Temp ≤ 32 °F | 19 | 16 | 8 | 1 | 0 | 0 | 0 | 0 | 0 | 0 | 3 | 14 | 61 |
| Days Min Temp ≤ 32 °F | 30 | 27 | 28 | 19 | 9 | 1 | 0 | 0 | 3 | 13 | 21 | 29 | 180 |
| Days Min Temp ≤ 0 °F | 8 | 8 | 3 | 0 | 0 | 0 | 0 | 0 | 0 | 0 | 0 | 3 | 22 |
| Heating Degree Days | 1411 | 1251 | 1092 | 691 | 398 | 149 | 63 | 88 | 249 | 568 | 842 | 1208 | 8010 |
| Cooling Degree Days | 0 | 0 | 0 | 1 | 9 | 39 | 85 | 58 | 18 | 1 | 0 | 0 | 211 |
| Total Precipitation (") | 2.33 | 1.95 | 2.66 | 3.14 | 3.50 | 4.26 | 3.97 | 3.96 | 4.21 | 3.66 | 3.73 | 3.14 | 40.51 |
| Days ≥ 0.1" Precip | 8 | 6 | 8 | 8 | 9 | 9 | 8 | 8 | 9 | 9 | 10 | 10 | 102 |
| Total Snowfall (") | ***28.7*** | 17.3 | 14.1 | 4.8 | 0.2 | 0.0 | 0.0 | 0.0 | 0.0 | 0.6 | ***11.2*** | ***26.1*** | 103.0 |
| Days ≥ 1" Snow Depth | 25 | 24 | 16 | 3 | 0 | 0 | 0 | 0 | 0 | 0 | 7 | 21 | 96 |

## FREDONIA *Chautauqua County* ELEVATION 761 ft LAT/LONG 42° 27 ' N / 79° 18 ' W

| | JAN | FEB | MAR | APR | MAY | JUN | JUL | AUG | SEP | OCT | NOV | DEC | YEAR |
|---|---|---|---|---|---|---|---|---|---|---|---|---|---|
| Maximum Temp °F | 32.1 | 33.8 | 43.7 | 56.2 | 67.6 | 76.5 | 80.5 | 78.6 | 72.4 | 61.3 | 49.1 | 37.7 | 57.5 |
| Minimum Temp °F | 18.4 | 18.0 | 26.3 | 36.8 | 46.9 | 56.4 | 61.4 | 60.3 | 54.0 | 43.8 | 35.3 | 25.3 | 40.2 |
| Mean Temp °F | 25.3 | 26.0 | 35.1 | 46.6 | 57.3 | 66.5 | 70.9 | 69.5 | 63.2 | 52.6 | 42.2 | 31.5 | 48.9 |
| Days Max Temp ≥ 90 °F | 0 | 0 | 0 | 0 | 0 | 0 | 1 | 1 | 0 | 0 | 0 | 0 | 2 |
| Days Max Temp ≤ 32 °F | 16 | 13 | 6 | 0 | 0 | 0 | 0 | 0 | 0 | 0 | 1 | 9 | 45 |
| Days Min Temp ≤ 32 °F | 28 | 25 | 23 | 10 | 1 | 0 | 0 | 0 | 0 | 2 | 12 | 25 | 126 |
| Days Min Temp ≤ 0 °F | 2 | 2 | 0 | 0 | 0 | 0 | 0 | 0 | 0 | 0 | 0 | 0 | 4 |
| Heating Degree Days | 1225 | 1096 | 922 | 553 | 264 | 64 | 10 | 20 | 115 | 385 | 676 | 1033 | 6363 |
| Cooling Degree Days | 0 | 0 | 1 | 7 | 35 | 107 | 204 | 169 | 71 | 6 | 0 | 0 | 600 |
| Total Precipitation (") | 2.36 | 2.06 | 2.72 | 3.21 | 3.04 | 3.84 | 3.77 | 3.88 | 4.92 | 3.99 | 4.31 | 3.39 | 41.49 |
| Days ≥ 0.1" Precip | 8 | 6 | 8 | 8 | 7 | 7 | 7 | 7 | 8 | 9 | 11 | 9 | 95 |
| Total Snowfall (") | 25.5 | 16.7 | 10.7 | 2.7 | 0.3 | 0.0 | 0.0 | 0.0 | 0.0 | 0.3 | 7.9 | 20.5 | 84.6 |
| Days ≥ 1" Snow Depth | 24 | 21 | 11 | 1 | 0 | 0 | 0 | 0 | 0 | 0 | 4 | 17 | 78 |

## GLENHAM *Dutchess County* ELEVATION 230 ft LAT/LONG 41° 31 ' N / 73° 56 ' W

| | JAN | FEB | MAR | APR | MAY | JUN | JUL | AUG | SEP | OCT | NOV | DEC | YEAR |
|---|---|---|---|---|---|---|---|---|---|---|---|---|---|
| Maximum Temp °F | 35.5 | 38.6 | 48.1 | 60.4 | 72.3 | 80.9 | 86.0 | 84.3 | 76.2 | 64.4 | 52.4 | 40.5 | 61.6 |
| Minimum Temp °F | 16.3 | 18.2 | 27.9 | 38.9 | 49.0 | 58.0 | 63.1 | 61.5 | 53.4 | 41.8 | 33.5 | 22.8 | 40.4 |
| Mean Temp °F | 25.9 | 28.4 | 38.0 | 49.7 | 60.7 | 69.4 | 74.6 | 72.9 | 64.9 | 53.1 | 42.9 | 31.7 | 51.0 |
| Days Max Temp ≥ 90 °F | 0 | 0 | 0 | 0 | 1 | 4 | 9 | 6 | 2 | 0 | 0 | 0 | 22 |
| Days Max Temp ≤ 32 °F | 12 | 8 | 2 | 0 | 0 | 0 | 0 | 0 | 0 | 0 | 0 | 6 | 28 |
| Days Min Temp ≤ 32 °F | 28 | 25 | 22 | 7 | 0 | 0 | 0 | 0 | 0 | 5 | 15 | 26 | 128 |
| Days Min Temp ≤ 0 °F | 3 | 1 | 0 | 0 | 0 | 0 | 0 | 0 | 0 | 0 | 0 | 1 | 5 |
| Heating Degree Days | 1205 | 1027 | 829 | 462 | 180 | 35 | 4 | 10 | 93 | 371 | 655 | 1027 | 5898 |
| Cooling Degree Days | 0 | 0 | 1 | 8 | 60 | 179 | 326 | 268 | 99 | 11 | 1 | 0 | 953 |
| Total Precipitation (") | 2.93 | 2.84 | 3.50 | 3.71 | 4.37 | 4.11 | 4.26 | 4.04 | 3.96 | 3.31 | 3.71 | 3.61 | 44.35 |
| Days ≥ 0.1" Precip | 6 | 6 | 7 | 7 | 7 | 7 | 7 | 6 | 6 | 6 | 6 | 7 | 78 |
| Total Snowfall (") | 11.1 | 10.0 | 6.7 | 1.0 | 0.0 | 0.0 | 0.0 | 0.0 | 0.0 | 0.1 | 2.0 | 8.1 | 39.0 |
| Days ≥ 1" Snow Depth | 15 | 13 | 6 | 1 | 0 | 0 | 0 | 0 | 0 | 0 | 1 | 7 | 43 |

## GLENS FALLS AP *Warren County* ELEVATION 341 ft LAT/LONG 43° 21 ' N / 73° 37 ' W

| | JAN | FEB | MAR | APR | MAY | JUN | JUL | AUG | SEP | OCT | NOV | DEC | YEAR |
|---|---|---|---|---|---|---|---|---|---|---|---|---|---|
| Maximum Temp °F | 27.5 | 31.0 | 41.6 | 56.1 | 67.7 | 76.6 | 81.5 | 79.0 | 70.2 | 58.2 | 45.6 | 33.2 | 55.7 |
| Minimum Temp °F | 6.3 | 9.5 | 21.8 | 33.5 | 43.7 | 52.8 | 57.9 | 56.3 | 47.5 | 36.3 | 28.3 | 15.9 | 34.2 |
| Mean Temp °F | 16.9 | 20.3 | 31.7 | 44.8 | 55.8 | 64.7 | 69.7 | 67.7 | 58.9 | 47.3 | 37.0 | 24.6 | 44.9 |
| Days Max Temp ≥ 90 °F | 0 | 0 | 0 | 0 | 0 | 1 | 3 | 1 | 0 | 0 | 0 | 0 | 5 |
| Days Max Temp ≤ 32 °F | 20 | 15 | 5 | 0 | 0 | 0 | 0 | 0 | 0 | 0 | 2 | 14 | 56 |
| Days Min Temp ≤ 32 °F | 30 | 27 | 26 | 15 | 3 | 0 | 0 | 0 | 1 | 11 | 21 | 28 | 162 |
| Days Min Temp ≤ 0 °F | 10 | 8 | 1 | 0 | 0 | 0 | 0 | 0 | 0 | 0 | 0 | 5 | 24 |
| Heating Degree Days | 1484 | 1256 | 1024 | 600 | 296 | 79 | 18 | 40 | 209 | 544 | 834 | 1246 | 7630 |
| Cooling Degree Days | 0 | 0 | 0 | 1 | 18 | 74 | 179 | 131 | 29 | 2 | 0 | 0 | 434 |
| Total Precipitation (") | 2.48 | 2.22 | 3.04 | 3.10 | 3.72 | 3.36 | 3.18 | 3.61 | 3.23 | 2.93 | 3.25 | 2.96 | 37.08 |
| Days ≥ 0.1" Precip | 5 | 5 | 6 | 7 | 8 | 7 | 6 | 7 | 6 | 6 | 7 | 6 | 76 |
| Total Snowfall (") | 17.9 | 13.8 | 11.1 | 2.3 | 0.1 | 0.0 | 0.0 | 0.0 | 0.0 | 0.0 | 3.8 | 16.1 | 65.1 |
| Days ≥ 1" Snow Depth | 26 | 24 | 14 | 1 | 0 | 0 | 0 | 0 | 0 | 0 | 3 | 17 | 85 |

## GLENS FALLS FARM *Warren County* ELEVATION 502 ft LAT/LONG 43° 20 ' N / 73° 44 ' W

| | JAN | FEB | MAR | APR | MAY | JUN | JUL | AUG | SEP | OCT | NOV | DEC | YEAR |
|---|---|---|---|---|---|---|---|---|---|---|---|---|---|
| Maximum Temp °F | 30.0 | 33.6 | 43.8 | 58.2 | 70.5 | 78.7 | 83.0 | 80.7 | 72.2 | 60.6 | 46.6 | 34.4 | 57.7 |
| Minimum Temp °F | 7.8 | 10.3 | 21.2 | 32.6 | 43.3 | 51.9 | 57.0 | 55.2 | 47.2 | 36.1 | 28.1 | 15.5 | 33.9 |
| Mean Temp °F | 18.9 | 22.0 | 32.5 | 45.4 | 56.9 | 65.3 | 70.0 | 67.9 | 59.7 | 48.6 | 37.4 | 24.9 | 45.8 |
| Days Max Temp ≥ 90 °F | 0 | 0 | 0 | 0 | 0 | 2 | 4 | 2 | 0 | 0 | 0 | 0 | 8 |
| Days Max Temp ≤ 32 °F | 17 | 12 | 3 | 0 | 0 | 0 | 0 | 0 | 0 | 0 | 1 | 12 | 45 |
| Days Min Temp ≤ 32 °F | 31 | 27 | 27 | 15 | 3 | 0 | 0 | 0 | 2 | 12 | 21 | 29 | 167 |
| Days Min Temp ≤ 0 °F | 9 | 8 | 1 | 0 | 0 | 0 | 0 | 0 | 0 | 0 | 0 | 5 | 23 |
| Heating Degree Days | 1423 | 1208 | 999 | 584 | 266 | 70 | 13 | 34 | 187 | 511 | 821 | 1236 | 7352 |
| Cooling Degree Days | 0 | 0 | 0 | 3 | 23 | 93 | 197 | 146 | 37 | 2 | 0 | 0 | 501 |
| Total Precipitation (") | 2.82 | 2.59 | 3.57 | 3.79 | 4.56 | 4.14 | 3.86 | 4.39 | 4.01 | 3.49 | 4.26 | 3.58 | 45.06 |
| Days ≥ 0.1" Precip | 6 | 5 | 7 | 8 | 8 | 8 | 7 | 7 | 7 | 7 | 8 | 8 | 86 |
| Total Snowfall (") | 19.3 | 13.1 | 11.4 | 2.2 | 0.0 | 0.0 | 0.0 | 0.0 | 0.0 | 0.0 | 4.0 | 16.8 | 66.8 |
| Days ≥ 1" Snow Depth | 28 | 26 | 19 | 3 | 0 | 0 | 0 | 0 | 0 | 0 | 4 | 20 | 100 |

## GLOVERSVILLE *Fulton County* ELEVATION 771 ft LAT/LONG 43° 3 ' N / 74° 21 ' W

| | JAN | FEB | MAR | APR | MAY | JUN | JUL | AUG | SEP | OCT | NOV | DEC | YEAR |
|---|---|---|---|---|---|---|---|---|---|---|---|---|---|
| Maximum Temp °F | 27.8 | 31.1 | 41.1 | 56.0 | 68.4 | 76.1 | 80.6 | 78.4 | 70.2 | 58.6 | 45.3 | 33.0 | 55.6 |
| Minimum Temp °F | 10.0 | 11.7 | 22.0 | 33.5 | 44.4 | 53.5 | 58.3 | 56.7 | 49.1 | 37.6 | 29.8 | 17.3 | 35.3 |
| Mean Temp °F | 18.9 | 21.4 | 31.6 | 44.8 | 56.4 | 64.8 | 69.5 | 67.6 | 59.7 | 48.2 | 37.6 | 25.2 | 45.5 |
| Days Max Temp ≥ 90 °F | 0 | 0 | 0 | 0 | 0 | 1 | 2 | 1 | 0 | 0 | 0 | 0 | 4 |
| Days Max Temp ≤ 32 °F | 20 | 16 | 6 | 0 | 0 | 0 | 0 | 0 | 0 | 0 | 2 | 14 | 58 |
| Days Min Temp ≤ 32 °F | 30 | 27 | 26 | 14 | 2 | 0 | 0 | 0 | 1 | 10 | 19 | 29 | 158 |
| Days Min Temp ≤ 0 °F | 8 | 6 | 1 | 0 | 0 | 0 | 0 | 0 | 0 | 0 | 0 | 3 | 18 |
| Heating Degree Days | 1421 | 1225 | 1029 | 600 | 277 | 79 | 17 | 38 | 185 | 515 | 817 | 1227 | 7430 |
| Cooling Degree Days | 0 | 0 | 0 | 1 | 19 | 78 | 167 | 121 | 29 | 1 | 0 | 0 | 416 |
| Total Precipitation (") | 2.90 | 2.70 | 3.47 | 3.64 | 4.08 | 4.22 | 3.87 | 3.94 | 3.98 | 3.43 | 3.87 | 3.60 | 43.70 |
| Days ≥ 0.1" Precip | 7 | 6 | 7 | 8 | 8 | 8 | 7 | 7 | 7 | 7 | 8 | 8 | 88 |
| Total Snowfall (") | 20.1 | 17.0 | 13.8 | 3.0 | 0.2 | 0.0 | 0.0 | 0.0 | 0.0 | 0.0 | 5.0 | 18.5 | 77.6 |
| Days ≥ 1" Snow Depth | 28 | 27 | 23 | 4 | 0 | 0 | 0 | 0 | 0 | 0 | 5 | 21 | 108 |

## GOUVERNEUR 3 NW *St. Lawrence County* ELEVATION 420 ft LAT/LONG 44° 21 ' N / 75° 31 ' W

| | JAN | FEB | MAR | APR | MAY | JUN | JUL | AUG | SEP | OCT | NOV | DEC | YEAR |
|---|---|---|---|---|---|---|---|---|---|---|---|---|---|
| Maximum Temp °F | 26.8 | 29.6 | 40.5 | 55.0 | 67.7 | 75.8 | 81.1 | 78.8 | 70.3 | 58.3 | 44.9 | 32.0 | 55.1 |
| Minimum Temp °F | 5.5 | 7.7 | 19.2 | 32.3 | 42.3 | 51.3 | 56.2 | 54.6 | 46.8 | 36.7 | 28.3 | 13.7 | 32.9 |
| Mean Temp °F | 16.2 | 18.6 | 29.9 | 43.7 | 55.0 | 63.6 | 68.7 | 66.7 | 58.6 | 47.5 | 36.7 | 22.9 | 44.0 |
| Days Max Temp ≥ 90 °F | 0 | 0 | 0 | 0 | 0 | 0 | 2 | 1 | 0 | 0 | 0 | 0 | 3 |
| Days Max Temp ≤ 32 °F | 20 | 17 | 7 | 0 | 0 | 0 | 0 | 0 | 0 | 0 | 3 | 15 | 62 |
| Days Min Temp ≤ 32 °F | 30 | 27 | 26 | 16 | 4 | 0 | 0 | 0 | 2 | 10 | 20 | 28 | 163 |
| Days Min Temp ≤ 0 °F | 12 | 9 | 3 | 0 | 0 | 0 | 0 | 0 | 0 | 0 | 0 | 7 | 31 |
| Heating Degree Days | 1509 | 1303 | 1082 | 636 | 316 | 101 | 27 | 53 | 215 | 536 | 844 | 1301 | 7923 |
| Cooling Degree Days | 0 | 0 | 0 | 2 | 10 | 53 | 143 | 109 | 25 | 1 | 0 | 0 | 343 |
| Total Precipitation (") | 2.19 | 2.01 | 2.39 | 3.13 | 3.15 | 3.17 | 2.79 | 3.79 | 4.11 | 3.32 | 3.88 | 3.14 | 37.07 |
| Days ≥ 0.1" Precip | 6 | 5 | 7 | 8 | 8 | 7 | 7 | 7 | 8 | 8 | 10 | 8 | 89 |
| Total Snowfall (") | 22.1 | 16.6 | 13.0 | 3.4 | 0.2 | 0.0 | 0.0 | 0.0 | 0.0 | 0.5 | 8.4 | 20.8 | 85.0 |
| Days ≥ 1" Snow Depth | 28 | 25 | 19 | 3 | 0 | 0 | 0 | 0 | 0 | 0 | 6 | 21 | 102 |

## GOWANDA PSYCHIATRIC *Erie County* ELEVATION 860 ft LAT/LONG 42° 29 ' N / 78° 56 ' W

| | JAN | FEB | MAR | APR | MAY | JUN | JUL | AUG | SEP | OCT | NOV | DEC | YEAR |
|---|---|---|---|---|---|---|---|---|---|---|---|---|---|
| Maximum Temp °F | 32.4 | 34.4 | 44.9 | 58.0 | 69.5 | 77.7 | 81.6 | 79.6 | 72.4 | 61.2 | 49.1 | 37.5 | 58.2 |
| Minimum Temp °F | 16.4 | 16.7 | 25.3 | 35.4 | 45.7 | 55.1 | 59.7 | 58.3 | 51.9 | 41.7 | 33.6 | 23.3 | 38.6 |
| Mean Temp °F | 24.4 | 25.5 | 35.1 | 46.8 | 57.7 | 66.4 | 70.6 | 69.0 | 62.2 | 51.5 | 41.4 | 30.5 | 48.4 |
| Days Max Temp ≥ 90 °F | 0 | 0 | 0 | 0 | 0 | 1 | 2 | 1 | 0 | 0 | 0 | 0 | 4 |
| Days Max Temp ≤ 32 °F | 15 | 13 | 5 | 0 | 0 | 0 | 0 | 0 | 0 | 0 | 1 | 10 | 44 |
| Days Min Temp ≤ 32 °F | 29 | 26 | 24 | 12 | 2 | 0 | 0 | 0 | 0 | 4 | 14 | 26 | 137 |
| Days Min Temp ≤ 0 °F | 3 | 3 | 1 | 0 | 0 | 0 | 0 | 0 | 0 | 0 | 0 | 1 | 8 |
| Heating Degree Days | 1253 | 1108 | 922 | 546 | 256 | 66 | 13 | 25 | 138 | 418 | 702 | 1062 | 6509 |
| Cooling Degree Days | 0 | 0 | 1 | 7 | 38 | 112 | 215 | 167 | 61 | 5 | 0 | 0 | 606 |
| Total Precipitation (") | 2.53 | 1.97 | *2.69* | *2.83* | 3.03 | 3.62 | 3.90 | 3.52 | 4.32 | 3.10 | 3.68 | 3.24 | 38.43 |
| Days ≥ 0.1" Precip | 8 | 6 | *7* | 6 | 7 | 6 | 6 | 6 | 7 | 7 | 9 | 9 | 84 |
| Total Snowfall (") | na | na | na | na | *0.0* | *0.0* | 0.0 | 0.0 | *0.0* | 0.0 | na | na | na |
| Days ≥ 1" Snow Depth | na | na | na | na | *0* | *0* | 0 | 0 | *0* | 0 | na | na | na |

## GRAFTON *Rensselaer County* ELEVATION 1562 ft LAT/LONG 42° 47 ' N / 73° 28 ' W

| | JAN | FEB | MAR | APR | MAY | JUN | JUL | AUG | SEP | OCT | NOV | DEC | YEAR |
|---|---|---|---|---|---|---|---|---|---|---|---|---|---|
| Maximum Temp °F | 28.0 | 30.6 | 40.5 | 54.2 | 66.4 | 73.8 | 78.3 | 75.9 | 67.6 | 57.0 | 44.6 | 33.0 | 54.2 |
| Minimum Temp °F | 11.0 | 12.7 | 22.3 | 33.6 | 44.7 | 53.2 | 58.0 | 56.6 | 49.3 | 39.2 | 29.5 | 17.5 | 35.6 |
| Mean Temp °F | 19.5 | 21.7 | 31.4 | 43.9 | 55.6 | 63.5 | 68.2 | 66.3 | 58.5 | 48.1 | 37.1 | 25.3 | 44.9 |
| Days Max Temp ≥ 90 °F | 0 | 0 | 0 | 0 | 0 | 0 | 0 | 0 | 0 | 0 | 0 | 0 | 0 |
| Days Max Temp ≤ 32 °F | 21 | 16 | 8 | 1 | 0 | 0 | 0 | 0 | 0 | 0 | 4 | 15 | 65 |
| Days Min Temp ≤ 32 °F | 30 | 27 | 26 | 14 | 2 | 0 | 0 | 0 | 1 | 8 | 20 | 28 | 156 |
| Days Min Temp ≤ 0 °F | 7 | 5 | 1 | 0 | 0 | 0 | 0 | 0 | 0 | 0 | 0 | 3 | 16 |
| Heating Degree Days | 1404 | 1216 | 1036 | 628 | 301 | 100 | 26 | 53 | 216 | 518 | 831 | 1225 | 7554 |
| Cooling Degree Days | 0 | 0 | 0 | 2 | 14 | 56 | 134 | 95 | 25 | 1 | 0 | 0 | 327 |
| Total Precipitation (") | 2.62 | 2.39 | 3.26 | 3.89 | 4.48 | 4.86 | 4.35 | 4.70 | 4.10 | 3.89 | 4.11 | 3.16 | 45.81 |
| Days ≥ 0.1" Precip | 7 | 6 | 8 | 9 | 9 | 9 | 8 | 8 | 7 | 8 | 9 | 8 | 96 |
| Total Snowfall (") | 19.8 | 16.6 | 14.0 | 6.0 | 0.7 | 0.0 | 0.0 | 0.0 | 0.0 | 0.9 | 7.8 | 17.4 | 83.2 |
| Days ≥ 1" Snow Depth | 26 | 25 | 19 | 4 | 0 | 0 | 0 | 0 | 0 | 0 | 6 | 19 | 99 |

## GREENPORT POWER HOUS *Suffolk County* ELEVATION 20 ft LAT/LONG 41° 6 ' N / 72° 22 ' W

| | JAN | FEB | MAR | APR | MAY | JUN | JUL | AUG | SEP | OCT | NOV | DEC | YEAR |
|---|---|---|---|---|---|---|---|---|---|---|---|---|---|
| Maximum Temp °F | *37.3* | *38.3* | 45.9 | 55.4 | 65.8 | 74.5 | 80.6 | 80.0 | 73.0 | 63.0 | 53.8 | *43.5* | 59.3 |
| Minimum Temp °F | *22.7* | 23.7 | 30.5 | 39.2 | 48.6 | 58.3 | 64.9 | 64.6 | 57.6 | 47.3 | 39.0 | *28.7* | 43.8 |
| Mean Temp °F | *30.1* | *31.1* | 38.2 | 47.4 | 57.2 | 66.4 | 72.8 | 72.3 | 65.4 | 55.2 | 46.4 | *36.1* | 51.6 |
| Days Max Temp ≥ 90 °F | 0 | 0 | 0 | 0 | 0 | 0 | 2 | 1 | 0 | 0 | 0 | 0 | 3 |
| Days Max Temp ≤ 32 °F | *9* | *7* | 2 | 0 | 0 | 0 | 0 | 0 | 0 | 0 | 0 | *3* | 21 |
| Days Min Temp ≤ 32 °F | *26* | *23* | 18 | 4 | 0 | 0 | 0 | 0 | 0 | 1 | 7 | *21* | 100 |
| Days Min Temp ≤ 0 °F | 0 | 0 | 0 | 0 | 0 | 0 | 0 | 0 | 0 | 0 | 0 | 0 | 0 |
| Heating Degree Days | *1076* | *952* | 823 | 524 | 247 | 44 | 2 | 4 | 65 | 305 | 551 | *888* | 5481 |
| Cooling Degree Days | 0 | 0 | 0 | 0 | 15 | 102 | 256 | 235 | 88 | 8 | 0 | 0 | 704 |
| Total Precipitation (") | 3.90 | 3.25 | 4.00 | 3.85 | 3.57 | 3.79 | 3.29 | 4.36 | 3.40 | 3.54 | 4.15 | 4.17 | 45.27 |
| Days ≥ 0.1" Precip | *7* | *6* | 7 | 7 | 7 | 6 | 5 | 6 | 6 | 6 | 7 | *7* | 77 |
| Total Snowfall (") | na | na | na | *0.1* | *0.0* | *0.0* | *0.0* | *0.0* | *0.0* | *0.0* | na | na | na |
| Days ≥ 1" Snow Depth | na | na | na | *0* | *0* | 0 | *0* | *0* | *0* | *0* | na | na | na |

## HEMLOCK *Livingston County* ELEVATION 902 ft LAT/LONG 42° 47 ' N / 77° 37 ' W

| | JAN | FEB | MAR | APR | MAY | JUN | JUL | AUG | SEP | OCT | NOV | DEC | YEAR |
|---|---|---|---|---|---|---|---|---|---|---|---|---|---|
| Maximum Temp °F | 30.8 | 32.4 | 42.2 | 55.9 | 67.4 | 76.5 | 80.6 | 78.8 | 71.1 | 59.5 | 47.5 | 36.8 | 56.6 |
| Minimum Temp °F | 13.3 | 12.9 | 22.3 | 33.6 | 43.5 | 53.1 | 58.5 | 56.8 | 50.5 | 39.9 | 31.4 | 21.3 | 36.4 |
| Mean Temp °F | 22.1 | 22.7 | 32.3 | 44.8 | 55.5 | 64.8 | 69.6 | 67.8 | 60.8 | 49.7 | 39.5 | 29.1 | 46.6 |
| Days Max Temp ≥ 90 °F | 0 | 0 | 0 | 0 | 0 | 1 | 2 | 1 | 0 | 0 | 0 | 0 | 4 |
| Days Max Temp ≤ 32 °F | 17 | 14 | 7 | 0 | 0 | 0 | 0 | 0 | 0 | 0 | 2 | 10 | 50 |
| Days Min Temp ≤ 32 °F | 30 | 27 | 26 | 14 | 3 | 0 | 0 | 0 | 1 | 6 | 17 | 26 | 150 |
| Days Min Temp ≤ 0 °F | 5 | 5 | 1 | 0 | 0 | 0 | 0 | 0 | 0 | 0 | 0 | 2 | 13 |
| Heating Degree Days | 1323 | 1188 | 1008 | 600 | 305 | 83 | 19 | 37 | 166 | 471 | 758 | 1105 | 7063 |
| Cooling Degree Days | 0 | 0 | 0 | 3 | 16 | 80 | 173 | *129* | 48 | 4 | 0 | 0 | 453 |
| Total Precipitation (") | 1.51 | 1.48 | 2.19 | 2.89 | 3.12 | 3.77 | 3.29 | 3.59 | 3.61 | 2.98 | 2.99 | 2.32 | 33.74 |
| Days ≥ 0.1" Precip | 4 | 4 | 6 | 7 | 8 | 8 | 7 | 7 | 8 | 8 | 8 | 6 | 81 |
| Total Snowfall (") | na | na | na | na | 0.0 | 0.0 | 0.0 | 0.0 | 0.0 | *0.0* | na | na | na |
| Days ≥ 1" Snow Depth | na | na | na | na | 0 | 0 | 0 | 0 | 0 | 0 | na | na | na |

## HUDSON CORRECTIONAL *Columbia County* ELEVATION 59 ft LAT/LONG 42° 15 ' N / 73° 48 ' W

| | JAN | FEB | MAR | APR | MAY | JUN | JUL | AUG | SEP | OCT | NOV | DEC | YEAR |
|---|---|---|---|---|---|---|---|---|---|---|---|---|---|
| Maximum Temp °F | 33.0 | 36.4 | 46.8 | 61.4 | 72.9 | 80.5 | 85.1 | 82.3 | 73.9 | 62.6 | 50.2 | 38.0 | 60.3 |
| Minimum Temp °F | 13.7 | 15.9 | 25.0 | 35.5 | 45.9 | 55.1 | 60.0 | 58.7 | 51.2 | 40.0 | 32.1 | 21.5 | 37.9 |
| Mean Temp °F | 23.4 | 26.1 | 36.0 | 48.5 | 59.4 | 67.8 | 72.6 | 70.4 | 62.6 | 51.3 | 41.2 | 29.8 | 49.1 |
| Days Max Temp ≥ 90 °F | 0 | 0 | 0 | 0 | 1 | 3 | 7 | 4 | 1 | 0 | 0 | 0 | 16 |
| Days Max Temp ≤ 32 °F | 14 | 9 | 2 | 0 | 0 | 0 | 0 | 0 | 0 | 0 | 0 | 8 | 33 |
| Days Min Temp ≤ 32 °F | 29 | 26 | 24 | 12 | 1 | 0 | 0 | 0 | 0 | 7 | 17 | 28 | 144 |
| Days Min Temp ≤ 0 °F | 5 | 3 | 0 | 0 | 0 | 0 | 0 | 0 | 0 | 0 | 0 | 1 | 9 |
| Heating Degree Days | 1283 | 1090 | 894 | 493 | 202 | 39 | 4 | 18 | 126 | 422 | 709 | 1085 | 6365 |
| Cooling Degree Days | 0 | 0 | 1 | 4 | 37 | 125 | 244 | 191 | 57 | 4 | 0 | 0 | 663 |
| Total Precipitation (") | 2.64 | 2.40 | 3.14 | 3.41 | 4.10 | 3.66 | 3.64 | 3.84 | 3.66 | 3.37 | 3.40 | 3.21 | 40.47 |
| Days ≥ 0.1" Precip | 5 | 5 | *5* | 7 | 8 | 7 | 6 | 7 | 7 | 6 | 7 | 6 | 76 |
| Total Snowfall (") | na | na | *6.2* | *1.7* | 0.0 | 0.0 | 0.0 | 0.0 | 0.0 | 0.1 | *1.3* | na | na |
| Days ≥ 1" Snow Depth | 19 | *16* | 8 | 1 | 0 | 0 | 0 | 0 | 0 | 0 | 1 | 11 | 56 |

## INDIAN LAKE 2 SW *Hamilton County* ELEVATION 1660 ft LAT/LONG 43° 45 ' N / 74° 17 ' W

| | JAN | FEB | MAR | APR | MAY | JUN | JUL | AUG | SEP | OCT | NOV | DEC | YEAR |
|---|---|---|---|---|---|---|---|---|---|---|---|---|---|
| Maximum Temp °F | 25.0 | 27.6 | 37.0 | 49.5 | 63.0 | 71.1 | 75.7 | 73.7 | 65.5 | 54.1 | 41.6 | 29.9 | 51.1 |
| Minimum Temp °F | 2.0 | 3.3 | 14.0 | 27.0 | 37.7 | 46.9 | 51.8 | 50.4 | 43.0 | 32.4 | 24.3 | 10.5 | 28.6 |
| Mean Temp °F | 13.6 | 15.5 | 25.5 | 38.3 | 50.4 | 59.0 | 63.7 | 62.1 | 54.3 | 43.2 | 33.0 | 20.2 | 39.9 |
| Days Max Temp ≥ 90 °F | 0 | 0 | 0 | 0 | 0 | 0 | 0 | 0 | 0 | 0 | 0 | 0 | 0 |
| Days Max Temp ≤ 32 °F | 23 | 19 | 10 | 1 | 0 | 0 | 0 | 0 | 0 | 0 | 5 | 18 | 76 |
| Days Min Temp ≤ 32 °F | 30 | 28 | 29 | 23 | 10 | 1 | 0 | 0 | 4 | 17 | 24 | 30 | 196 |
| Days Min Temp ≤ 0 °F | 15 | 13 | 5 | 0 | 0 | 0 | 0 | 0 | 0 | 0 | 0 | 7 | 40 |
| Heating Degree Days | 1589 | 1394 | 1217 | 794 | 450 | 197 | 90 | 122 | 324 | 668 | 954 | 1382 | 9181 |
| Cooling Degree Days | 0 | 0 | 0 | 0 | 4 | 22 | 64 | 41 | 7 | 0 | 0 | 0 | 138 |
| Total Precipitation (") | 2.62 | 2.27 | 2.97 | 2.90 | 3.55 | 3.52 | 3.28 | 4.15 | 4.19 | 3.55 | 3.66 | 2.86 | 39.52 |
| Days ≥ 0.1" Precip | 6 | 6 | 7 | 7 | 8 | 8 | 7 | 8 | 8 | 8 | 8 | 7 | 88 |
| Total Snowfall (") | na | na | na | na | *0.0* | 0.0 | 0.0 | 0.0 | 0.0 | *0.6* | na | na | na |
| Days ≥ 1" Snow Depth | na | na | na | na | *0* | 0 | 0 | 0 | 0 | *0* | na | na | na |

## ITHACA CORNELL UNIV *Tompkins County* ELEVATION 951 ft LAT/LONG 42° 27 ' N / 76° 28 ' W

| | JAN | FEB | MAR | APR | MAY | JUN | JUL | AUG | SEP | OCT | NOV | DEC | YEAR |
|---|---|---|---|---|---|---|---|---|---|---|---|---|---|
| Maximum Temp °F | 30.1 | 31.7 | 41.2 | 54.3 | 66.3 | 74.9 | 79.6 | 78.1 | 70.2 | 58.5 | 46.7 | 35.3 | 55.6 |
| Minimum Temp °F | 13.1 | 13.6 | 23.2 | 34.0 | 43.5 | 52.7 | 57.3 | 56.0 | 48.8 | 38.4 | 31.3 | 20.6 | 36.0 |
| Mean Temp °F | 21.6 | 22.7 | 32.2 | 44.2 | 54.9 | 63.9 | 68.5 | 67.1 | 59.5 | 48.5 | 39.0 | 28.0 | 45.8 |
| Days Max Temp ≥ 90 °F | 0 | 0 | 0 | 0 | 0 | 1 | 2 | 1 | 0 | 0 | 0 | 0 | 4 |
| Days Max Temp ≤ 32 °F | 18 | 15 | 7 | 1 | 0 | 0 | 0 | 0 | 0 | 0 | 2 | 12 | 55 |
| Days Min Temp ≤ 32 °F | 29 | 26 | 25 | 14 | 4 | 0 | 0 | 0 | 1 | 9 | 17 | 27 | 152 |
| Days Min Temp ≤ 0 °F | 5 | 5 | 1 | 0 | 0 | 0 | 0 | 0 | 0 | 0 | 0 | 2 | 13 |
| Heating Degree Days | 1340 | 1188 | 1009 | 622 | 326 | 108 | 34 | 51 | 195 | 508 | 772 | 1140 | 7293 |
| Cooling Degree Days | 0 | 0 | 1 | 3 | 21 | 73 | 153 | 123 | 38 | 2 | 0 | 0 | 414 |
| Total Precipitation (") | 1.87 | 1.91 | 2.46 | 3.04 | 3.20 | 3.74 | 3.59 | 3.58 | 3.74 | 3.21 | 3.17 | 2.62 | 36.13 |
| Days ≥ 0.1" Precip | 5 | 5 | 6 | 7 | 8 | 8 | 8 | 7 | 7 | 8 | 8 | 7 | 84 |
| Total Snowfall (") | 17.4 | 13.9 | 11.7 | 3.6 | 0.2 | 0.0 | 0.0 | 0.0 | 0.0 | 0.6 | 5.5 | 14.8 | 67.7 |
| Days ≥ 1" Snow Depth | 23 | 22 | 12 | 2 | 0 | 0 | 0 | 0 | 0 | 0 | 4 | 15 | 78 |

## LAKE PLACID 2 S *Essex County* ELEVATION 1860 ft LAT/LONG 44° 17 ' N / 73° 59 ' W

| | JAN | FEB | MAR | APR | MAY | JUN | JUL | AUG | SEP | OCT | NOV | DEC | YEAR |
|---|---|---|---|---|---|---|---|---|---|---|---|---|---|
| Maximum Temp °F | 24.9 | 28.0 | 37.9 | 50.6 | 63.8 | 71.9 | 76.3 | 73.8 | 65.8 | 54.6 | 42.0 | 30.1 | 51.6 |
| Minimum Temp °F | 3.2 | 4.8 | 15.3 | 27.9 | 38.3 | 47.3 | 52.0 | 50.6 | 43.1 | 33.3 | 23.9 | 10.2 | 29.2 |
| Mean Temp °F | 14.0 | 16.5 | 26.6 | 39.2 | 51.1 | 59.6 | 64.2 | 62.2 | 54.5 | 44.0 | 33.0 | 20.2 | 40.4 |
| Days Max Temp ≥ 90 °F | 0 | 0 | 0 | 0 | 0 | 0 | 0 | 0 | 0 | 0 | 0 | 0 | 0 |
| Days Max Temp ≤ 32 °F | 23 | 19 | 10 | 1 | 0 | 0 | 0 | 0 | 0 | 0 | 6 | 18 | 77 |
| Days Min Temp ≤ 32 °F | 31 | 27 | 29 | 21 | 10 | 1 | 0 | 0 | 4 | 15 | 24 | 30 | 192 |
| Days Min Temp ≤ 0 °F | 13 | 11 | 5 | 0 | 0 | 0 | 0 | 0 | 0 | 0 | 1 | 9 | 39 |
| Heating Degree Days | 1576 | 1365 | 1182 | 766 | 432 | 187 | 83 | 123 | 317 | 645 | 954 | 1385 | 9015 |
| Cooling Degree Days | 0 | 0 | 0 | 0 | 5 | 26 | 67 | 44 | 8 | 1 | 0 | 0 | 151 |
| Total Precipitation (") | 2.26 | 1.98 | 2.53 | 2.90 | 3.28 | 3.61 | 3.61 | 4.56 | 4.19 | 3.31 | 3.40 | 2.70 | 38.33 |
| Days ≥ 0.1" Precip | 6 | 5 | 7 | 8 | 9 | 9 | 9 | 10 | 8 | 8 | 8 | 8 | 95 |
| Total Snowfall (") | na | na | na | na | *0.8* | *0.0* | *0.0* | 0.0 | 0.0 | *1.4* | na | na | na |
| Days ≥ 1" Snow Depth | na | na | na | na | na | *0* | na | *0* | *0* | na | na | na | na |

## LAWRENCEVILLE *St. Lawrence County* ELEVATION 479 ft LAT/LONG 44° 45 ' N / 74° 40 ' W

| | JAN | FEB | MAR | APR | MAY | JUN | JUL | AUG | SEP | OCT | NOV | DEC | YEAR |
|---|---|---|---|---|---|---|---|---|---|---|---|---|---|
| Maximum Temp °F | 25.2 | 28.1 | 39.0 | 53.9 | 67.3 | 75.4 | 80.1 | 77.6 | 69.1 | 57.4 | 44.0 | 30.3 | 54.0 |
| Minimum Temp °F | 5.8 | 8.1 | 19.2 | 32.8 | 43.8 | 53.3 | 58.6 | 57.0 | 48.8 | 38.3 | 28.2 | 13.4 | 33.9 |
| Mean Temp °F | 15.5 | 18.2 | 29.1 | 43.5 | 55.5 | 64.4 | 69.4 | 67.3 | 59.0 | 47.8 | 36.1 | 21.9 | 44.0 |
| Days Max Temp ≥ 90 °F | 0 | 0 | 0 | 0 | 0 | 0 | 1 | 1 | 0 | 0 | 0 | 0 | 2 |
| Days Max Temp ≤ 32 °F | 21 | 18 | 9 | 1 | 0 | 0 | 0 | 0 | 0 | 0 | 5 | 17 | 71 |
| Days Min Temp ≤ 32 °F | 30 | 27 | 27 | 16 | 4 | 0 | 0 | 0 | 1 | 9 | 21 | 29 | 164 |
| Days Min Temp ≤ 0 °F | 12 | 9 | 3 | 0 | 0 | 0 | 0 | 0 | 0 | 0 | 0 | 6 | 30 |
| Heating Degree Days | 1529 | 1318 | 1104 | 641 | 310 | 94 | 23 | 50 | 207 | 528 | 862 | 1333 | 7999 |
| Cooling Degree Days | 0 | 0 | 0 | 3 | 22 | 79 | 177 | 134 | 33 | 3 | 0 | 0 | 451 |
| Total Precipitation (") | 1.83 | 1.84 | 2.07 | 2.80 | 2.81 | 3.58 | 3.41 | 4.19 | 3.73 | 3.03 | 3.24 | 2.64 | 35.17 |
| Days ≥ 0.1" Precip | 6 | 5 | 7 | 8 | 8 | 8 | 7 | 8 | 8 | 8 | 9 | 7 | 89 |
| Total Snowfall (") | 17.1 | 14.8 | 11.9 | 4.7 | 0.4 | 0.0 | 0.0 | 0.0 | 0.0 | 0.8 | 8.0 | 16.6 | 74.3 |
| Days ≥ 1" Snow Depth | 27 | 25 | 18 | 4 | 0 | 0 | 0 | 0 | 0 | 0 | 7 | 21 | 102 |

## LIBERTY 1 NE *Sullivan County* ELEVATION 1581 ft LAT/LONG 41° 49 ' N / 74° 45 ' W

| | JAN | FEB | MAR | APR | MAY | JUN | JUL | AUG | SEP | OCT | NOV | DEC | YEAR |
|---|---|---|---|---|---|---|---|---|---|---|---|---|---|
| Maximum Temp °F | 29.7 | 32.9 | 41.9 | 54.7 | 66.5 | 74.5 | 79.3 | 77.8 | 69.6 | 58.6 | 46.3 | 34.4 | 55.5 |
| Minimum Temp °F | 11.2 | 12.8 | 22.0 | 33.1 | 42.7 | 50.9 | 56.3 | 54.4 | 46.7 | 36.0 | 28.5 | 17.9 | 34.4 |
| Mean Temp °F | 20.5 | 22.9 | 32.0 | 43.9 | 54.6 | 62.7 | 67.8 | 66.1 | 58.2 | 47.4 | 37.4 | 26.2 | 45.0 |
| Days Max Temp ≥ 90 °F | 0 | 0 | 0 | 0 | 0 | 0 | 2 | 1 | 0 | 0 | 0 | 0 | 3 |
| Days Max Temp ≤ 32 °F | 19 | 14 | 6 | 0 | 0 | 0 | 0 | 0 | 0 | 0 | 2 | 13 | 54 |
| Days Min Temp ≤ 32 °F | 30 | 27 | 27 | 15 | 3 | 0 | 0 | 0 | 1 | 12 | 21 | 29 | 165 |
| Days Min Temp ≤ 0 °F | 6 | 4 | 1 | 0 | 0 | 0 | 0 | 0 | 0 | 0 | 0 | 2 | 13 |
| Heating Degree Days | 1372 | 1184 | 1017 | 627 | 328 | 116 | 32 | 52 | 222 | 541 | 821 | 1197 | 7509 |
| Cooling Degree Days | 0 | 0 | 0 | 2 | 15 | 52 | 142 | 97 | 23 | 1 | 0 | 0 | 332 |
| Total Precipitation (") | 3.34 | 2.93 | 3.68 | 4.30 | 4.67 | 4.54 | 4.40 | 4.54 | 4.23 | 3.85 | 4.23 | 3.98 | 48.69 |
| Days ≥ 0.1" Precip | 7 | 6 | 8 | 8 | 9 | 8 | 8 | 8 | 7 | 7 | 8 | 8 | 92 |
| Total Snowfall (") | 16.7 | 12.6 | 13.4 | 2.8 | 0.2 | 0.0 | 0.0 | 0.0 | 0.0 | 0.3 | 5.8 | 15.6 | 67.4 |
| Days ≥ 1" Snow Depth | 27 | 25 | 19 | 3 | 0 | 0 | 0 | 0 | 0 | 0 | 4 | 18 | 96 |

## LITTLE FALLS CITY RE *Herkimer County* ELEVATION 902 ft LAT/LONG 43° 4 ' N / 74° 52 ' W

| | JAN | FEB | MAR | APR | MAY | JUN | JUL | AUG | SEP | OCT | NOV | DEC | YEAR |
|---|---|---|---|---|---|---|---|---|---|---|---|---|---|
| Maximum Temp °F | 27.6 | 30.5 | 40.3 | 54.6 | 67.5 | 75.8 | ***80.5*** | ***78.2*** | 70.1 | 58.3 | 45.4 | 33.0 | 55.2 |
| Minimum Temp °F | 9.1 | 11.1 | 20.6 | 32.3 | 43.0 | 51.7 | ***56.7*** | 54.8 | 47.1 | 36.5 | 28.3 | 16.1 | 33.9 |
| Mean Temp °F | 18.4 | 20.8 | 30.5 | 43.5 | 55.2 | 63.8 | ***68.6*** | ***66.5*** | 58.6 | 47.5 | 36.9 | 24.6 | 44.6 |
| Days Max Temp ≥ 90 °F | 0 | 0 | 0 | 0 | 0 | 1 | 2 | 1 | 0 | 0 | 0 | 0 | 4 |
| Days Max Temp ≤ 32 °F | 20 | 16 | 6 | 0 | 0 | 0 | 0 | 0 | 0 | 0 | 2 | 14 | 58 |
| Days Min Temp ≤ 32 °F | 30 | 27 | 27 | ***16*** | 3 | 0 | 0 | 0 | 1 | 11 | 20 | 29 | 164 |
| Days Min Temp ≤ 0 °F | 8 | 6 | 2 | 0 | 0 | 0 | 0 | 0 | 0 | 0 | 0 | 3 | 19 |
| Heating Degree Days | 1441 | 1242 | 1064 | 640 | 313 | 101 | ***27*** | ***49*** | 213 | 540 | 835 | 1250 | 7715 |
| Cooling Degree Days | 0 | 0 | 0 | 1 | 14 | 63 | ***143*** | na | 24 | 1 | 0 | 0 | na |
| Total Precipitation (") | 2.47 | 2.20 | 2.74 | 3.71 | 3.81 | 4.50 | 4.02 | 3.78 | 4.31 | 3.39 | 4.21 | 3.14 | 42.28 |
| Days ≥ 0.1" Precip | 7 | 6 | 7 | 8 | 8 | ***9*** | ***7*** | ***7*** | 8 | 8 | ***8*** | 8 | 91 |
| Total Snowfall (") | 24.1 | 16.3 | 12.2 | 3.2 | 0.3 | 0.0 | 0.0 | 0.0 | 0.0 | 0.1 | 5.8 | 17.6 | 79.6 |
| Days ≥ 1" Snow Depth | 28 | 25 | 19 | 2 | 0 | 0 | 0 | 0 | 0 | 0 | 5 | 21 | 100 |

## LITTLE VALLEY *Cattaraugus County* ELEVATION 1581 ft LAT/LONG 42° 15 ' N / 78° 48 ' W

| | JAN | FEB | MAR | APR | MAY | JUN | JUL | AUG | SEP | OCT | NOV | DEC | YEAR |
|---|---|---|---|---|---|---|---|---|---|---|---|---|---|
| Maximum Temp °F | 29.6 | 31.5 | 41.1 | 54.0 | 66.0 | 74.6 | 79.0 | 77.1 | 69.3 | 57.9 | 45.9 | 34.4 | 55.0 |
| Minimum Temp °F | 12.2 | 12.1 | 20.5 | 31.5 | 40.5 | 49.6 | 54.4 | 53.3 | 47.3 | 37.1 | 29.7 | 19.6 | 34.0 |
| Mean Temp °F | 20.9 | 21.8 | 30.8 | 42.8 | 53.3 | 62.1 | 66.7 | 65.2 | 58.4 | 47.5 | 37.8 | 27.1 | 44.5 |
| Days Max Temp ≥ 90 °F | 0 | 0 | 0 | 0 | 0 | 0 | 1 | 0 | 0 | 0 | 0 | 0 | 1 |
| Days Max Temp ≤ 32 °F | 19 | 15 | 8 | 1 | 0 | 0 | 0 | 0 | 0 | 0 | 3 | 14 | 60 |
| Days Min Temp ≤ 32 °F | 30 | 27 | 26 | 18 | 7 | 0 | 0 | 0 | 1 | 10 | 20 | 28 | 167 |
| Days Min Temp ≤ 0 °F | 6 | 6 | 2 | 0 | 0 | 0 | 0 | 0 | 0 | 0 | 0 | 2 | 16 |
| Heating Degree Days | 1360 | 1213 | 1053 | 660 | 369 | 133 | 47 | 66 | 222 | 537 | 809 | 1168 | 7637 |
| Cooling Degree Days | 0 | 0 | 0 | 2 | 13 | 47 | 115 | 81 | 26 | 1 | 0 | 0 | 285 |
| Total Precipitation (") | 3.52 | 3.07 | 3.58 | 3.74 | 3.77 | 4.78 | 4.36 | 4.57 | 4.84 | 4.16 | 4.99 | 4.61 | 49.99 |
| Days ≥ 0.1" Precip | 12 | 9 | 10 | 10 | 9 | 10 | 9 | 9 | 9 | 10 | 13 | 14 | 124 |
| Total Snowfall (") | 34.1 | 24.2 | 18.9 | 6.6 | 0.5 | 0.0 | 0.0 | 0.0 | 0.0 | 1.0 | 16.4 | 38.8 | 140.5 |
| Days ≥ 1" Snow Depth | 26 | 24 | 16 | 3 | 0 | 0 | 0 | 0 | 0 | 0 | 8 | 21 | 98 |

## LOCKPORT 2 NE *Niagara County* ELEVATION 522 ft LAT/LONG 43° 11 ' N / 78° 39 ' W

| | JAN | FEB | MAR | APR | MAY | JUN | JUL | AUG | SEP | OCT | NOV | DEC | YEAR |
|---|---|---|---|---|---|---|---|---|---|---|---|---|---|
| Maximum Temp °F | 31.0 | 33.2 | 42.6 | 56.5 | 68.2 | 76.8 | 81.2 | 78.9 | 71.2 | 59.7 | 47.8 | 36.3 | 57.0 |
| Minimum Temp °F | 16.3 | 17.1 | 24.8 | 35.5 | 45.9 | 55.1 | 60.7 | 59.2 | 52.1 | 41.7 | 33.0 | 22.7 | 38.7 |
| Mean Temp °F | 23.7 | 25.2 | 33.7 | 46.0 | 57.0 | 66.0 | 70.9 | 69.1 | 61.7 | 50.8 | 40.4 | 29.5 | 47.8 |
| Days Max Temp ≥ 90 °F | 0 | 0 | 0 | 0 | 0 | 1 | 1 | 1 | 0 | 0 | 0 | 0 | 3 |
| Days Max Temp ≤ 32 °F | 17 | 14 | 6 | 0 | 0 | 0 | 0 | 0 | 0 | 0 | 1 | 11 | 49 |
| Days Min Temp ≤ 32 °F | 29 | 26 | 25 | 12 | 2 | 0 | 0 | 0 | 0 | 4 | 16 | 26 | 140 |
| Days Min Temp ≤ 0 °F | 3 | 2 | 0 | 0 | 0 | 0 | 0 | 0 | 0 | 0 | 0 | 1 | 6 |
| Heating Degree Days | 1274 | 1118 | 963 | 567 | 269 | 66 | 10 | 23 | 146 | 439 | 731 | 1093 | 6699 |
| Cooling Degree Days | 0 | 0 | 0 | 5 | 31 | 95 | 213 | 158 | 54 | 4 | 0 | 0 | 560 |
| Total Precipitation (") | 2.49 | 2.26 | 2.76 | 3.20 | 3.00 | 3.41 | 2.90 | 3.86 | 3.91 | 2.94 | 3.68 | 3.38 | 37.79 |
| Days ≥ 0.1" Precip | 8 | 7 | 7 | 8 | 7 | 7 | 5 | 7 | 7 | 8 | 9 | 9 | 89 |
| Total Snowfall (") | 25.3 | 19.4 | 12.0 | 4.7 | 0.5 | 0.0 | 0.0 | 0.0 | 0.0 | 0.2 | 8.0 | 19.6 | 89.7 |
| Days ≥ 1" Snow Depth | 23 | 22 | 12 | 1 | 0 | 0 | 0 | 0 | 0 | 0 | 4 | 17 | 79 |

## LOCKPORT 4 NE *Niagara County* ELEVATION 440 ft LAT/LONG 43° 12 ' N / 78° 38 ' W

| | JAN | FEB | MAR | APR | MAY | JUN | JUL | AUG | SEP | OCT | NOV | DEC | YEAR |
|---|---|---|---|---|---|---|---|---|---|---|---|---|---|
| Maximum Temp °F | 30.7 | 32.1 | 41.2 | 54.4 | 66.5 | 75.9 | 80.8 | 79.1 | 71.6 | 59.7 | 47.6 | 35.9 | 56.3 |
| Minimum Temp °F | 16.3 | 16.8 | 24.9 | 35.7 | 45.5 | 55.0 | 60.1 | 58.6 | 51.4 | 41.0 | 32.9 | 22.6 | 38.4 |
| Mean Temp °F | 23.5 | 24.5 | 33.1 | 45.1 | 56.0 | 65.5 | 70.5 | 68.9 | 61.5 | 50.4 | 40.3 | 29.3 | 47.4 |
| Days Max Temp ≥ 90 °F | 0 | 0 | 0 | 0 | 0 | 1 | 2 | 1 | 0 | 0 | 0 | 0 | 4 |
| Days Max Temp ≤ 32 °F | 17 | 15 | 7 | 1 | 0 | 0 | 0 | 0 | 0 | 0 | 2 | 11 | 53 |
| Days Min Temp ≤ 32 °F | 29 | 26 | 25 | 12 | 1 | 0 | 0 | 0 | 0 | 4 | 15 | 26 | 138 |
| Days Min Temp ≤ 0 °F | 2 | 1 | 0 | 0 | 0 | 0 | 0 | 0 | 0 | 0 | 0 | 0 | 3 |
| Heating Degree Days | 1279 | 1138 | 984 | 595 | 297 | 79 | 15 | 28 | 153 | 452 | 735 | 1099 | 6854 |
| Cooling Degree Days | 0 | 0 | 0 | 5 | 29 | 97 | 208 | 163 | 56 | 4 | 0 | 0 | 562 |
| Total Precipitation (") | 2.27 | 2.16 | 2.63 | 3.17 | 2.87 | 3.42 | 2.92 | 3.74 | 3.93 | 2.91 | 3.27 | 3.27 | 36.56 |
| Days ≥ 0.1" Precip | 7 | 6 | 7 | 8 | 7 | 7 | 6 | 7 | 7 | 8 | 9 | 8 | 87 |
| Total Snowfall (") | 17.1 | 14.9 | 8.9 | 2.8 | 0.3 | 0.0 | 0.0 | 0.0 | 0.0 | 0.1 | 4.3 | 15.2 | 63.6 |
| Days ≥ 1" Snow Depth | 23 | 21 | 12 | 2 | 0 | 0 | 0 | 0 | 0 | 0 | 4 | 15 | 77 |

## LOWVILLE *Lewis County* ELEVATION 922 ft LAT/LONG 43° 48 ' N / 75° 29 ' W

| | JAN | FEB | MAR | APR | MAY | JUN | JUL | AUG | SEP | OCT | NOV | DEC | YEAR |
|---|---|---|---|---|---|---|---|---|---|---|---|---|---|
| Maximum Temp °F | 26.0 | 28.9 | 38.5 | 52.8 | 65.8 | 74.5 | 79.2 | 77.0 | 68.3 | 56.6 | 43.9 | 31.3 | 53.6 |
| Minimum Temp °F | 6.6 | 8.5 | 19.8 | 32.6 | 42.6 | 51.4 | 56.0 | 54.6 | 46.4 | 36.6 | 28.3 | 14.5 | 33.2 |
| Mean Temp °F | 16.3 | 18.8 | 29.2 | 42.7 | 54.2 | 63.0 | 67.7 | 65.8 | 57.4 | 46.6 | 36.1 | 22.9 | 43.4 |
| Days Max Temp ≥ 90 °F | 0 | 0 | 0 | 0 | 0 | 0 | 1 | 1 | 0 | 0 | 0 | 0 | 2 |
| Days Max Temp ≤ 32 °F | 21 | 17 | 9 | 1 | 0 | 0 | 0 | 0 | 0 | 0 | 4 | 16 | 68 |
| Days Min Temp ≤ 32 °F | 30 | 27 | 27 | 16 | 4 | 0 | 0 | 0 | 2 | 10 | 20 | 29 | 165 |
| Days Min Temp ≤ 0 °F | 11 | 9 | 2 | 0 | 0 | 0 | 0 | 0 | 0 | 0 | 0 | 5 | 27 |
| Heating Degree Days | 1505 | 1297 | 1104 | 664 | 342 | 117 | 37 | 64 | 245 | 564 | 859 | 1298 | 8096 |
| Cooling Degree Days | 0 | 0 | 0 | 2 | 12 | 49 | 120 | 90 | 20 | 1 | 0 | 0 | 294 |
| Total Precipitation (") | 3.08 | 2.48 | 2.84 | 3.24 | 3.12 | 3.55 | 3.34 | 3.78 | 4.06 | 3.50 | 4.18 | 3.65 | 40.82 |
| Days ≥ 0.1" Precip | 8 | 6 | 7 | 7 | 8 | 8 | 7 | 8 | 8 | 8 | 10 | 9 | 94 |
| Total Snowfall (") | 35.2 | 22.6 | 16.2 | 4.7 | 0.4 | 0.0 | 0.0 | 0.0 | 0.0 | 0.8 | 11.1 | 30.0 | 121.0 |
| Days ≥ 1" Snow Depth | *29* | *26* | *22* | 5 | 0 | 0 | 0 | 0 | 0 | 0 | *7* | 23 | 112 |

## MASSENA AP *St. Lawrence County* ELEVATION 207 ft LAT/LONG 44° 56 ' N / 74° 51 ' W

| | JAN | FEB | MAR | APR | MAY | JUN | JUL | AUG | SEP | OCT | NOV | DEC | YEAR |
|---|---|---|---|---|---|---|---|---|---|---|---|---|---|
| Maximum Temp °F | 24.0 | 26.6 | 37.5 | 53.1 | 66.9 | 75.6 | 80.9 | 78.1 | 68.9 | 56.7 | 43.2 | 29.4 | 53.4 |
| Minimum Temp °F | 3.8 | 6.1 | 18.8 | 32.9 | 43.9 | 52.8 | 57.9 | 56.0 | 47.3 | 36.7 | 27.7 | 12.3 | 33.0 |
| Mean Temp °F | 14.0 | 16.4 | 28.1 | 43.0 | 55.4 | 64.2 | 69.4 | 67.0 | 58.1 | 46.8 | 35.5 | 20.9 | 43.2 |
| Days Max Temp ≥ 90 °F | 0 | 0 | 0 | 0 | 0 | 1 | 3 | 1 | 0 | 0 | 0 | 0 | 5 |
| Days Max Temp ≤ 32 °F | 21 | 19 | 10 | 1 | 0 | 0 | 0 | 0 | 0 | 0 | 5 | 17 | 73 |
| Days Min Temp ≤ 32 °F | 30 | 27 | 27 | 15 | 3 | 0 | 0 | 0 | 1 | 11 | 21 | 29 | 164 |
| Days Min Temp ≤ 0 °F | 12 | 10 | 3 | 0 | 0 | 0 | 0 | 0 | 0 | 0 | 0 | 7 | 32 |
| Heating Degree Days | 1578 | 1368 | 1137 | 654 | 310 | 92 | 22 | 53 | 227 | 561 | 880 | 1361 | 8243 |
| Cooling Degree Days | 0 | 0 | 0 | 2 | 19 | 72 | 178 | 124 | 26 | 2 | 0 | 0 | 423 |
| Total Precipitation (") | 2.35 | 2.07 | 2.35 | 2.92 | 2.67 | 3.31 | 3.09 | 3.71 | 3.64 | 2.86 | 3.16 | 3.21 | 35.34 |
| Days ≥ 0.1" Precip | 7 | 6 | 7 | 7 | 7 | 7 | 7 | 7 | 7 | 7 | 8 | 8 | 85 |
| Total Snowfall (") | 17.6 | 15.5 | 10.7 | 3.9 | 0.1 | 0.0 | 0.0 | 0.0 | 0.0 | 1.0 | 6.3 | 19.4 | 74.5 |
| Days ≥ 1" Snow Depth | 27 | 25 | 18 | 3 | 0 | 0 | 0 | 0 | 0 | 0 | 5 | 20 | 98 |

## MIDDLETOWN 2 NW *Orange County* ELEVATION 620 ft LAT/LONG 41° 27 ' N / 74° 26 ' W

| | JAN | FEB | MAR | APR | MAY | JUN | JUL | AUG | SEP | OCT | NOV | DEC | YEAR |
|---|---|---|---|---|---|---|---|---|---|---|---|---|---|
| Maximum Temp °F | 34.2 | 37.8 | 47.7 | 60.7 | 71.8 | 79.4 | 83.9 | 82.0 | 74.5 | 63.6 | 51.1 | 39.0 | 60.5 |
| Minimum Temp °F | 17.2 | 18.8 | 27.6 | 38.7 | 49.1 | 57.8 | 62.9 | 61.3 | 54.0 | 43.0 | 34.6 | 23.7 | 40.7 |
| Mean Temp °F | 25.7 | 28.3 | 37.7 | 49.7 | 60.5 | 68.6 | 73.4 | 71.7 | 64.3 | 53.3 | 42.8 | 31.4 | 50.6 |
| Days Max Temp ≥ 90 °F | 0 | 0 | 0 | 0 | 0 | 2 | 4 | 2 | 1 | 0 | 0 | 0 | 9 |
| Days Max Temp ≤ 32 °F | 13 | 8 | 2 | 0 | 0 | 0 | 0 | 0 | 0 | 0 | 0 | 7 | 30 |
| Days Min Temp ≤ 32 °F | 29 | 26 | 22 | 7 | 0 | 0 | 0 | 0 | 0 | 3 | 13 | 26 | 126 |
| Days Min Temp ≤ 0 °F | 2 | 1 | 0 | 0 | 0 | 0 | 0 | 0 | 0 | 0 | 0 | 0 | 3 |
| Heating Degree Days | 1212 | 1028 | 841 | 456 | 175 | 29 | 2 | 8 | 95 | 361 | 659 | 1034 | 5900 |
| Cooling Degree Days | 0 | 0 | 0 | 6 | 46 | 147 | 279 | 220 | 81 | 7 | 0 | 0 | 786 |
| Total Precipitation (") | 2.70 | 2.51 | 3.22 | 3.73 | 4.51 | 4.23 | 3.89 | 4.00 | 3.78 | 3.40 | 3.74 | 3.22 | 42.93 |
| Days ≥ 0.1" Precip | 6 | 5 | 7 | 7 | 8 | 8 | 7 | 6 | 6 | 6 | 7 | 6 | 79 |
| Total Snowfall (") | na | na | na | na | na | 0.0 | 0.0 | 0.0 | 0.0 | *0.0* | na | na | na |
| Days ≥ 1" Snow Depth | na | na | na | na | na | 0 | 0 | 0 | 0 | *0* | na | na | na |

## MILLBROOK *Dutchess County* ELEVATION 820 ft LAT/LONG 41° 51 ' N / 73° 37 ' W

| | JAN | FEB | MAR | APR | MAY | JUN | JUL | AUG | SEP | OCT | NOV | DEC | YEAR |
|---|---|---|---|---|---|---|---|---|---|---|---|---|---|
| Maximum Temp °F | 33.4 | 36.7 | 46.2 | 58.4 | 69.9 | 77.2 | 82.0 | 80.1 | 72.2 | 61.8 | 49.8 | 38.0 | 58.8 |
| Minimum Temp °F | 12.5 | 14.6 | 24.4 | 34.4 | 44.3 | 52.6 | 57.3 | 56.4 | 48.4 | 37.0 | 29.4 | 19.0 | 35.9 |
| Mean Temp °F | 22.9 | 25.6 | 35.4 | 46.5 | 57.1 | 64.9 | 69.7 | 68.3 | 60.3 | 49.5 | 39.6 | 28.6 | 47.4 |
| Days Max Temp ≥ 90 °F | 0 | 0 | 0 | 0 | 0 | 1 | 3 | 1 | 0 | 0 | 0 | 0 | 5 |
| Days Max Temp ≤ 32 °F | 14 | 9 | 2 | 0 | 0 | 0 | 0 | 0 | 0 | 0 | 1 | 8 | 34 |
| Days Min Temp ≤ 32 °F | 29 | 26 | 24 | 12 | 2 | 0 | 0 | 0 | 1 | 11 | 20 | 27 | 152 |
| Days Min Temp ≤ 0 °F | 5 | 4 | 0 | 0 | 0 | 0 | 0 | 0 | 0 | 0 | 0 | 2 | 11 |
| Heating Degree Days | 1301 | 1104 | 916 | 556 | 258 | 71 | 14 | 27 | 171 | 476 | 755 | 1124 | 6773 |
| Cooling Degree Days | *0* | 0 | *0* | 1 | 20 | 75 | 176 | 135 | 34 | 3 | 0 | 0 | 444 |
| Total Precipitation (") | 2.74 | 2.56 | 2.95 | 3.21 | 4.05 | 4.09 | 4.06 | 4.26 | 3.58 | 3.28 | 3.18 | 3.41 | 41.37 |
| Days ≥ 0.1" Precip | 6 | 5 | 6 | 6 | 8 | 7 | 7 | 7 | 6 | 6 | 6 | 7 | 77 |
| Total Snowfall (") | *12.4* | 9.9 | *8.0* | *1.8* | 0.2 | 0.0 | 0.0 | 0.0 | 0.0 | 0.0 | *2.2* | 9.7 | 44.2 |
| Days ≥ 1" Snow Depth | *18* | *14* | *8* | *1* | 0 | 0 | 0 | 0 | 0 | 0 | *1* | na | na |

## MINEOLA *Nassau County* ELEVATION 131 ft LAT/LONG 40° 44 ' N / 73° 38 ' W

| | JAN | FEB | MAR | APR | MAY | JUN | JUL | AUG | SEP | OCT | NOV | DEC | YEAR |
|---|---|---|---|---|---|---|---|---|---|---|---|---|---|
| Maximum Temp °F | 37.1 | 39.1 | 47.6 | 57.6 | 67.8 | 77.6 | 82.6 | 81.0 | 73.5 | 62.8 | 52.9 | 42.5 | 60.2 |
| Minimum Temp °F | 24.7 | 25.9 | 33.1 | 41.1 | 50.3 | 60.1 | 66.0 | 64.9 | 57.7 | 47.1 | 39.7 | 30.2 | 45.1 |
| Mean Temp °F | 31.0 | 32.5 | 40.4 | 49.4 | 59.1 | 68.8 | 74.4 | 73.0 | 65.6 | 55.0 | 46.3 | 36.4 | 52.7 |
| Days Max Temp ≥ 90 °F | 0 | 0 | 0 | 0 | 0 | 2 | 4 | 3 | 1 | 0 | 0 | 0 | 10 |
| Days Max Temp ≤ 32 °F | 10 | 7 | 1 | 0 | 0 | 0 | 0 | 0 | 0 | 0 | 0 | 4 | 22 |
| Days Min Temp ≤ 32 °F | 24 | 21 | 14 | 3 | 0 | 0 | 0 | 0 | 0 | 1 | 5 | 18 | 86 |
| Days Min Temp ≤ 0 °F | 0 | 0 | 0 | 0 | 0 | 0 | 0 | 0 | 0 | 0 | 0 | 0 | 0 |
| Heating Degree Days | 1048 | 911 | 756 | 465 | 208 | 27 | 2 | 6 | 70 | 311 | 555 | 880 | 5239 |
| Cooling Degree Days | 0 | 0 | 0 | 3 | 35 | 160 | 312 | 247 | 91 | *10* | *0* | 0 | 858 |
| Total Precipitation (") | 3.42 | 2.99 | 4.19 | 4.04 | 4.25 | 3.44 | 3.79 | 3.80 | 3.67 | 3.36 | 3.87 | 3.91 | 44.73 |
| Days ≥ 0.1" Precip | 6 | 6 | 7 | 6 | 7 | 6 | 6 | 6 | 6 | 5 | 6 | 7 | 74 |
| Total Snowfall (") | 7.0 | 9.2 | 4.0 | 0.5 | 0.0 | 0.0 | 0.0 | 0.0 | 0.0 | 0.0 | 0.2 | 2.9 | 23.8 |
| Days ≥ 1" Snow Depth | 9 | 8 | 2 | 0 | 0 | 0 | 0 | 0 | 0 | 0 | 0 | 2 | 21 |

## MOHONK LAKE *Ulster County* ELEVATION 1250 ft LAT/LONG 41° 46 ' N / 74° 9 ' W

| | JAN | FEB | MAR | APR | MAY | JUN | JUL | AUG | SEP | OCT | NOV | DEC | YEAR |
|---|---|---|---|---|---|---|---|---|---|---|---|---|---|
| Maximum Temp °F | 30.8 | 34.1 | 43.8 | 57.2 | 67.9 | 75.1 | 79.6 | 77.6 | 69.9 | 59.0 | 47.1 | 35.7 | 56.5 |
| Minimum Temp °F | 16.4 | 17.8 | 26.3 | 37.4 | 48.6 | 57.5 | 62.9 | 61.6 | 54.1 | 43.6 | 33.7 | 22.5 | 40.2 |
| Mean Temp °F | 23.6 | 26.0 | 35.0 | 47.3 | 58.3 | 66.3 | 71.3 | 69.6 | 62.0 | 51.3 | 40.4 | 29.1 | 48.4 |
| Days Max Temp ≥ 90 °F | 0 | 0 | 0 | 0 | 0 | 0 | 2 | 0 | 0 | 0 | 0 | 0 | 2 |
| Days Max Temp ≤ 32 °F | 17 | 13 | 4 | 0 | 0 | 0 | 0 | 0 | 0 | 0 | 2 | 12 | 48 |
| Days Min Temp ≤ 32 °F | 29 | 26 | 23 | 9 | 0 | 0 | 0 | 0 | 0 | 3 | 14 | 27 | 131 |
| Days Min Temp ≤ 0 °F | 3 | 1 | 0 | 0 | 0 | 0 | 0 | 0 | 0 | 0 | 0 | 1 | 5 |
| Heating Degree Days | 1275 | 1096 | 922 | 529 | 232 | 56 | 7 | 18 | 134 | 421 | 730 | 1106 | 6526 |
| Cooling Degree Days | 0 | 0 | 0 | 4 | 35 | 112 | 228 | 170 | 53 | 3 | 0 | 0 | 605 |
| Total Precipitation (") | 3.30 | 3.12 | 4.10 | 4.13 | 5.11 | 4.16 | 4.16 | 4.44 | 4.38 | 3.77 | 4.15 | 3.94 | 48.76 |
| Days ≥ 0.1" Precip | 7 | 6 | 7 | 7 | 8 | 7 | 7 | 7 | 6 | 6 | 7 | 7 | 82 |
| Total Snowfall (") | 15.5 | 14.6 | 13.7 | 3.3 | 0.5 | 0.0 | 0.0 | 0.0 | 0.0 | 0.2 | 4.3 | 13.7 | 65.8 |
| Days ≥ 1" Snow Depth | 25 | 23 | 18 | 2 | 0 | 0 | 0 | 0 | 0 | 0 | 3 | 14 | 85 |

### MOUNT MORRIS 2 W *Livingston County* ELEVATION 879 ft LAT/LONG 42° 44 ' N / 77° 54 ' W

| | JAN | FEB | MAR | APR | MAY | JUN | JUL | AUG | SEP | OCT | NOV | DEC | YEAR |
|---|---|---|---|---|---|---|---|---|---|---|---|---|---|
| Maximum Temp °F | 31.1 | 32.5 | 42.1 | 55.3 | 67.3 | 76.6 | 81.1 | 79.4 | 71.6 | 59.7 | 47.2 | 36.0 | 56.7 |
| Minimum Temp °F | 14.6 | 14.8 | 23.5 | 34.7 | 44.6 | 54.0 | 58.9 | 57.1 | 49.6 | 39.0 | 31.2 | 21.0 | 36.9 |
| Mean Temp °F | 22.8 | 23.7 | 32.8 | 45.0 | 56.0 | 65.3 | 70.0 | 68.3 | 60.6 | 49.3 | 39.2 | 28.5 | 46.8 |
| Days Max Temp ≥ 90 °F | 0 | 0 | 0 | 0 | 0 | 1 | 3 | 1 | 0 | 0 | 0 | 0 | 5 |
| Days Max Temp ≤ 32 °F | 17 | 14 | 7 | 1 | 0 | 0 | 0 | 0 | 0 | 0 | 2 | 11 | 52 |
| Days Min Temp ≤ 32 °F | 29 | 26 | 25 | 13 | 2 | 0 | 0 | 0 | 0 | 7 | 17 | 27 | 146 |
| Days Min Temp ≤ 0 °F | 4 | 3 | 0 | 0 | 0 | 0 | 0 | 0 | 0 | 0 | 0 | 1 | 8 |
| Heating Degree Days | 1300 | 1160 | 992 | 598 | 298 | 80 | 18 | 35 | 170 | 482 | 766 | 1124 | 7023 |
| Cooling Degree Days | 0 | 0 | 0 | 5 | 25 | 91 | 189 | 141 | 40 | 2 | 0 | 0 | 493 |
| Total Precipitation (") | 1.38 | 1.42 | 1.72 | 2.44 | 2.67 | 3.56 | 3.11 | 3.22 | 3.30 | 2.51 | 2.70 | 2.15 | 30.18 |
| Days ≥ 0.1" Precip | 4 | 4 | 5 | 6 | 7 | 7 | 6 | 7 | 7 | 6 | 7 | 6 | 72 |
| Total Snowfall (") | na | na | na | 1.3 | 0.2 | 0.0 | 0.0 | 0.0 | 0.0 | 0.0 | 4.2 | na | na |
| Days ≥ 1" Snow Depth | 21 | 19 | 11 | 2 | 0 | 0 | 0 | 0 | 0 | 0 | *3* | 15 | 71 |

### NEW YORK *Kings County* ELEVATION 20 ft LAT/LONG 40° 36 ' N / 73° 59 ' W

| | JAN | FEB | MAR | APR | MAY | JUN | JUL | AUG | SEP | OCT | NOV | DEC | YEAR |
|---|---|---|---|---|---|---|---|---|---|---|---|---|---|
| Maximum Temp °F | 38.3 | 41.1 | 49.4 | 60.2 | 70.6 | 79.5 | 84.8 | 83.4 | 76.1 | 65.0 | 54.3 | 43.3 | 62.2 |
| Minimum Temp °F | 25.5 | 27.1 | 34.4 | 43.7 | 53.9 | 63.4 | 69.3 | 68.3 | 61.0 | 49.9 | 41.1 | 31.0 | 47.4 |
| Mean Temp °F | 31.9 | 34.1 | 41.9 | 52.0 | 62.3 | 71.5 | 77.1 | 75.9 | 68.6 | 57.5 | 47.7 | 37.1 | 54.8 |
| Days Max Temp ≥ 90 °F | 0 | 0 | 0 | 0 | 1 | 3 | 6 | 4 | 1 | 0 | 0 | 0 | 15 |
| Days Max Temp ≤ 32 °F | 9 | 5 | 1 | 0 | 0 | 0 | 0 | 0 | 0 | 0 | 0 | 4 | 19 |
| Days Min Temp ≤ 32 °F | 23 | 20 | 11 | 1 | 0 | 0 | 0 | 0 | 0 | 0 | 4 | 16 | 75 |
| Days Min Temp ≤ 0 °F | 0 | 0 | 0 | 0 | 0 | 0 | 0 | 0 | 0 | 0 | 0 | 0 | 0 |
| Heating Degree Days | 1018 | 865 | 709 | 389 | 132 | 11 | 0 | 1 | 33 | 243 | 514 | 857 | 4772 |
| Cooling Degree Days | 0 | 0 | 0 | 4 | 61 | 221 | 393 | 342 | 147 | 17 | 1 | 0 | 1186 |
| Total Precipitation (") | 3.33 | 3.01 | 3.97 | 3.89 | 4.37 | 3.38 | 4.12 | 4.26 | 3.92 | 3.16 | 3.89 | 3.69 | 44.99 |
| Days ≥ 0.1" Precip | 6 | 6 | 7 | 6 | 7 | 7 | 6 | 7 | 6 | 5 | 6 | 6 | 75 |
| Total Snowfall (") | 7.2 | 8.6 | 3.9 | 0.4 | 0.0 | 0.0 | 0.0 | 0.0 | 0.0 | 0.0 | 0.4 | 2.6 | 23.1 |
| Days ≥ 1" Snow Depth | 8 | 8 | 3 | 0 | 0 | 0 | 0 | 0 | 0 | 0 | 0 | 2 | 21 |

### NEW YORK CNTRL PARK *New York County* ELEVATION 144 ft LAT/LONG 40° 47 ' N / 73° 58 ' W

| | JAN | FEB | MAR | APR | MAY | JUN | JUL | AUG | SEP | OCT | NOV | DEC | YEAR |
|---|---|---|---|---|---|---|---|---|---|---|---|---|---|
| Maximum Temp °F | 37.9 | 40.5 | 49.8 | 61.5 | 72.0 | 80.3 | 85.4 | 84.0 | 76.1 | 64.9 | 54.2 | 43.3 | 62.5 |
| Minimum Temp °F | 25.5 | 26.9 | 34.6 | 44.0 | 53.9 | 63.1 | 68.6 | 67.6 | 60.2 | 49.5 | 41.1 | 31.3 | 47.2 |
| Mean Temp °F | 31.7 | 33.7 | 42.2 | 52.8 | 63.0 | 71.8 | 77.1 | 75.8 | 68.2 | 57.2 | 47.7 | 37.4 | 54.9 |
| Days Max Temp ≥ 90 °F | 0 | 0 | 0 | 0 | 1 | 3 | 8 | 5 | 1 | 0 | 0 | 0 | 18 |
| Days Max Temp ≤ 32 °F | 10 | 6 | 1 | 0 | 0 | 0 | 0 | 0 | 0 | 0 | 0 | 4 | 21 |
| Days Min Temp ≤ 32 °F | 23 | 20 | 11 | 1 | 0 | 0 | 0 | 0 | 0 | 0 | 4 | 16 | 75 |
| Days Min Temp ≤ 0 °F | 0 | 0 | 0 | 0 | 0 | 0 | 0 | 0 | 0 | 0 | 0 | 0 | 0 |
| Heating Degree Days | 1025 | 878 | 700 | 369 | 124 | 14 | 1 | 2 | 41 | 253 | 514 | 850 | 4771 |
| Cooling Degree Days | 0 | 0 | 1 | 10 | 74 | 233 | 394 | 341 | 144 | 19 | 2 | 0 | 1218 |
| Total Precipitation (") | 3.55 | 3.23 | 4.22 | 3.90 | 4.63 | 3.77 | 4.44 | 4.40 | 4.16 | 3.63 | 4.34 | 4.10 | 48.37 |
| Days ≥ 0.1" Precip | 6 | 6 | 7 | 6 | 7 | 7 | 7 | 7 | 6 | 5 | 6 | 7 | 77 |
| Total Snowfall (") | 7.2 | 8.5 | 3.9 | 0.4 | 0.0 | 0.0 | 0.0 | 0.0 | 0.0 | 0.0 | 0.4 | 2.7 | 23.1 |
| Days ≥ 1" Snow Depth | 10 | 9 | 3 | 0 | 0 | 0 | 0 | 0 | 0 | 0 | 0 | 2 | 24 |

### NEW YORK J F KENNEDY *Queens County* ELEVATION 20 ft LAT/LONG 40° 39 ' N / 73° 47 ' W

| | JAN | FEB | MAR | APR | MAY | JUN | JUL | AUG | SEP | OCT | NOV | DEC | YEAR |
|---|---|---|---|---|---|---|---|---|---|---|---|---|---|
| Maximum Temp °F | 37.9 | 39.9 | 48.1 | 58.4 | 67.9 | 77.2 | 83.0 | 82.0 | 74.9 | 64.2 | 53.9 | 43.5 | 60.9 |
| Minimum Temp °F | 25.3 | 26.7 | 34.0 | 43.0 | 52.7 | 62.1 | 68.3 | 67.6 | 60.2 | 49.3 | 40.7 | 31.2 | 46.8 |
| Mean Temp °F | 31.6 | 33.3 | 41.1 | 50.7 | 60.3 | 69.6 | 75.7 | 74.8 | 67.6 | 56.8 | 47.3 | 37.4 | 53.9 |
| Days Max Temp ≥ 90 °F | 0 | 0 | 0 | 0 | 0 | 2 | 4 | 3 | 1 | 0 | 0 | 0 | 10 |
| Days Max Temp ≤ 32 °F | 9 | 6 | 1 | 0 | 0 | 0 | 0 | 0 | 0 | 0 | 0 | 4 | 20 |
| Days Min Temp ≤ 32 °F | 23 | 20 | 12 | 1 | 0 | 0 | 0 | 0 | 0 | 0 | 4 | 17 | 77 |
| Days Min Temp ≤ 0 °F | 0 | 0 | 0 | 0 | 0 | 0 | 0 | 0 | 0 | 0 | 0 | 0 | 0 |
| Heating Degree Days | 1028 | 888 | 735 | 423 | 170 | 16 | 0 | 2 | 42 | 260 | 525 | 849 | 4938 |
| Cooling Degree Days | 0 | 0 | 0 | 1 | 38 | 181 | 353 | 313 | 129 | 14 | 0 | 0 | 1029 |
| Total Precipitation (") | 3.15 | 2.81 | 3.74 | 3.70 | 4.01 | 3.44 | 3.74 | 3.72 | 3.37 | 2.91 | 3.48 | 3.52 | 41.59 |
| Days ≥ 0.1" Precip | 6 | 6 | 7 | 6 | 7 | 6 | 6 | 6 | 6 | 5 | 6 | 7 | 74 |
| Total Snowfall (") | 7.0 | 7.8 | 3.6 | 0.5 | 0.0 | 0.0 | 0.0 | 0.0 | 0.0 | 0.0 | 0.3 | 2.6 | 21.8 |
| Days ≥ 1" Snow Depth | 7 | 6 | 2 | 0 | 0 | 0 | 0 | 0 | 0 | 0 | 0 | 2 | 17 |

## NEW YORK LAGUARDIA *Queens County* ELEVATION 52 ft LAT/LONG 40° 46 ' N / 73° 52 ' W

| | JAN | FEB | MAR | APR | MAY | JUN | JUL | AUG | SEP | OCT | NOV | DEC | YEAR |
|---|---|---|---|---|---|---|---|---|---|---|---|---|---|
| Maximum Temp °F | 37.5 | 39.8 | 48.4 | 59.6 | 70.2 | 79.2 | 84.4 | 82.7 | 75.2 | 64.0 | 53.5 | 43.1 | 61.5 |
| Minimum Temp °F | 25.8 | 27.4 | 34.5 | 44.1 | 54.0 | 63.5 | 69.3 | 68.6 | 61.4 | 50.5 | 41.5 | 31.7 | 47.7 |
| Mean Temp °F | 31.7 | 33.6 | 41.5 | 51.9 | 62.1 | 71.4 | 76.9 | 75.7 | 68.3 | 57.3 | 47.5 | 37.4 | 54.6 |
| Days Max Temp ≥ 90 °F | 0 | 0 | 0 | 0 | 1 | 3 | 6 | 4 | 1 | 0 | 0 | 0 | 15 |
| Days Max Temp ≤ 32 °F | 10 | 6 | 1 | 0 | 0 | 0 | 0 | 0 | 0 | 0 | 0 | 4 | 21 |
| Days Min Temp ≤ 32 °F | 23 | 20 | 11 | 1 | 0 | 0 | 0 | 0 | 0 | 0 | 3 | 15 | 73 |
| Days Min Temp ≤ 0 °F | 0 | 0 | 0 | 0 | 0 | 0 | 0 | 0 | 0 | 0 | 0 | 0 | 0 |
| Heating Degree Days | 1027 | 880 | 723 | 394 | 137 | 14 | 1 | 1 | 39 | 250 | 519 | 849 | 4834 |
| Cooling Degree Days | 0 | 0 | 0 | 6 | 66 | 231 | 393 | 344 | 149 | 21 | 1 | 0 | 1211 |
| Total Precipitation (") | 3.04 | 2.74 | 3.85 | 3.58 | 4.00 | 3.56 | 4.10 | 4.06 | 3.51 | 3.07 | 3.69 | 3.54 | 42.74 |
| Days ≥ 0.1" Precip | 6 | 6 | 7 | 6 | 7 | 6 | 7 | 7 | 6 | 5 | 6 | 7 | 76 |
| Total Snowfall (") | 7.1 | 8.4 | 3.8 | 0.4 | 0.0 | 0.0 | 0.0 | 0.0 | 0.0 | 0.0 | 0.4 | 2.8 | 22.9 |
| Days ≥ 1" Snow Depth | 8 | 7 | 2 | 0 | 0 | 0 | 0 | 0 | 0 | 0 | 0 | 2 | 19 |

## NORWICH *Chenango County* ELEVATION 1070 ft LAT/LONG 42° 33 ' N / 75° 32 ' W

| | JAN | FEB | MAR | APR | MAY | JUN | JUL | AUG | SEP | OCT | NOV | DEC | YEAR |
|---|---|---|---|---|---|---|---|---|---|---|---|---|---|
| Maximum Temp °F | 30.6 | 33.1 | 43.0 | 56.3 | 68.3 | 76.2 | 80.7 | 78.8 | 70.8 | 59.5 | 47.1 | 35.3 | 56.6 |
| Minimum Temp °F | 10.2 | 11.0 | 21.0 | 32.0 | 41.7 | 50.5 | 55.3 | 53.9 | 46.6 | 35.7 | 28.4 | 18.2 | 33.7 |
| Mean Temp °F | 20.4 | 22.0 | 32.0 | 44.1 | 55.0 | 63.4 | 68.1 | 66.4 | 58.7 | 47.6 | 37.8 | 26.8 | 45.2 |
| Days Max Temp ≥ 90 °F | 0 | 0 | 0 | 0 | 0 | 1 | 2 | 1 | 0 | 0 | 0 | 0 | 4 |
| Days Max Temp ≤ 32 °F | 17 | 13 | 5 | 0 | 0 | 0 | 0 | 0 | 0 | 0 | 2 | 11 | 48 |
| Days Min Temp ≤ 32 °F | 30 | 27 | 26 | 17 | 5 | 0 | 0 | 0 | 2 | 12 | 21 | 28 | 168 |
| Days Min Temp ≤ 0 °F | 7 | 7 | 1 | 0 | 0 | 0 | 0 | 0 | 0 | 0 | 0 | 3 | 18 |
| Heating Degree Days | 1375 | 1206 | 1015 | 620 | 319 | 108 | 32 | 52 | 212 | 533 | 811 | 1179 | 7462 |
| Cooling Degree Days | 0 | 0 | 0 | 2 | 16 | 67 | 153 | 111 | 30 | 1 | 0 | 0 | 380 |
| Total Precipitation (") | 2.46 | 2.23 | 2.95 | 3.34 | 3.75 | 4.10 | 3.58 | 3.51 | 4.06 | 3.32 | 3.91 | 3.36 | 40.57 |
| Days ≥ 0.1" Precip | 7 | 6 | 7 | 8 | 9 | 8 | 7 | 7 | 7 | 8 | 9 | 8 | 91 |
| Total Snowfall (") | 17.6 | 13.9 | 10.9 | 3.2 | 0.3 | 0.0 | 0.0 | 0.0 | 0.0 | 0.6 | 5.9 | 16.9 | 69.3 |
| Days ≥ 1" Snow Depth | 26 | 26 | 16 | 2 | 0 | 0 | 0 | 0 | 0 | 0 | 5 | 16 | 91 |

## OGDENSBURG 4 NE *St. Lawrence County* ELEVATION 279 ft LAT/LONG 44° 44 ' N / 75° 27 ' W

| | JAN | FEB | MAR | APR | MAY | JUN | JUL | AUG | SEP | OCT | NOV | DEC | YEAR |
|---|---|---|---|---|---|---|---|---|---|---|---|---|---|
| Maximum Temp °F | 25.6 | 28.4 | 39.3 | 53.6 | 66.9 | 75.8 | 81.2 | 79.0 | 70.0 | 58.0 | 44.6 | 31.5 | 54.5 |
| Minimum Temp °F | 6.8 | 8.6 | 20.6 | 33.8 | 44.8 | 53.9 | 59.6 | 57.3 | 48.8 | 38.9 | 29.6 | 15.6 | 34.9 |
| Mean Temp °F | 16.3 | 18.5 | 30.0 | 43.7 | 55.9 | 64.9 | 70.4 | 68.2 | 59.4 | 48.5 | 37.2 | 23.6 | 44.7 |
| Days Max Temp ≥ 90 °F | 0 | 0 | 0 | 0 | 0 | 1 | 1 | 1 | 0 | 0 | 0 | 0 | 3 |
| Days Max Temp ≤ 32 °F | 21 | 18 | 7 | 0 | 0 | 0 | 0 | 0 | 0 | 0 | 3 | 15 | 64 |
| Days Min Temp ≤ 32 °F | 30 | 27 | 27 | 14 | 2 | 0 | 0 | 0 | 1 | 8 | 19 | 28 | 156 |
| Days Min Temp ≤ 0 °F | 11 | 8 | 2 | 0 | 0 | 0 | 0 | 0 | 0 | 0 | 0 | 5 | 26 |
| Heating Degree Days | 1506 | 1306 | 1080 | 633 | 293 | 79 | 13 | 34 | 190 | 507 | 829 | 1277 | 7747 |
| Cooling Degree Days | 0 | 0 | 0 | 1 | 20 | 84 | 212 | 149 | 24 | 2 | 0 | 0 | 492 |
| Total Precipitation (") | 2.09 | 1.92 | 1.96 | 2.62 | 2.80 | 3.13 | 3.02 | 3.60 | 3.64 | 2.94 | 3.14 | 2.83 | 33.69 |
| Days ≥ 0.1" Precip | 6 | 5 | 5 | 8 | 7 | 7 | 7 | 7 | 7 | 7 | 8 | *7* | 81 |
| Total Snowfall (") | na | ***12.0*** | na | ***2.5*** | 0.0 | 0.0 | 0.0 | 0.0 | 0.0 | 0.2 | na | na | na |
| Days ≥ 1" Snow Depth | 26 | 26 | 19 | ***2*** | 0 | 0 | 0 | 0 | 0 | 0 | ***4*** | ***19*** | 96 |

## OLD FORGE *Herkimer County* ELEVATION 1732 ft LAT/LONG 43° 43 ' N / 74° 58 ' W

| | JAN | FEB | MAR | APR | MAY | JUN | JUL | AUG | SEP | OCT | NOV | DEC | YEAR |
|---|---|---|---|---|---|---|---|---|---|---|---|---|---|
| Maximum Temp °F | 24.1 | 27.9 | 37.2 | 50.5 | 64.5 | 71.7 | 76.4 | 73.9 | 65.6 | 54.3 | 41.2 | 29.5 | 51.4 |
| Minimum Temp °F | 1.0 | 3.0 | 13.7 | 26.0 | 37.7 | 46.5 | 51.4 | 50.4 | 43.1 | 32.4 | 23.6 | 9.6 | 28.2 |
| Mean Temp °F | 12.7 | 15.5 | 25.4 | 38.2 | 51.1 | 59.1 | 63.9 | 62.2 | 54.4 | 43.4 | 32.5 | 19.6 | 39.8 |
| Days Max Temp ≥ 90 °F | 0 | 0 | 0 | 0 | 0 | 0 | 0 | 0 | 0 | 0 | 0 | 0 | 0 |
| Days Max Temp ≤ 32 °F | 24 | 19 | 10 | 2 | 0 | 0 | 0 | 0 | 0 | 0 | 7 | 19 | 81 |
| Days Min Temp ≤ 32 °F | 31 | 28 | 29 | 23 | 10 | 2 | 0 | 0 | 4 | 17 | 24 | 30 | 198 |
| Days Min Temp ≤ 0 °F | 15 | 13 | 6 | 0 | 0 | 0 | 0 | 0 | 0 | 0 | 1 | 9 | 44 |
| Heating Degree Days | 1620 | 1394 | 1220 | 796 | 430 | 194 | 84 | 117 | 320 | 663 | 970 | 1403 | 9211 |
| Cooling Degree Days | 0 | 0 | 0 | 1 | 5 | 26 | 60 | 38 | 6 | 0 | 0 | 0 | 136 |
| Total Precipitation (") | 3.70 | 2.99 | 3.64 | 3.84 | 4.20 | 4.21 | 4.04 | 4.86 | 5.21 | 4.30 | 5.00 | 4.35 | 50.34 |
| Days ≥ 0.1" Precip | 11 | 8 | 9 | 9 | 9 | 9 | 8 | 9 | 9 | 10 | 12 | 11 | 114 |
| Total Snowfall (") | 55.3 | 41.4 | 37.2 | 12.9 | 2.2 | 0.0 | 0.0 | 0.0 | 0.0 | 3.6 | 24.9 | 48.0 | 225.5 |
| Days ≥ 1" Snow Depth | ***30*** | ***27*** | 27 | 13 | 1 | 0 | 0 | 0 | 0 | 1 | 13 | 27 | 139 |

## OSWEGO EAST *Oswego County* ELEVATION 289 ft LAT/LONG 43° 27 ' N / 76° 31 ' W

| | JAN | FEB | MAR | APR | MAY | JUN | JUL | AUG | SEP | OCT | NOV | DEC | YEAR |
|---|---|---|---|---|---|---|---|---|---|---|---|---|---|
| Maximum Temp °F | 29.9 | 31.6 | 40.6 | 53.1 | 64.9 | 74.5 | 79.7 | 78.0 | 70.3 | 58.8 | 47.0 | 35.2 | 55.3 |
| Minimum Temp °F | 16.1 | 17.3 | 26.0 | 36.4 | 45.6 | 55.0 | 61.6 | 60.8 | 53.7 | 43.5 | 34.9 | 23.3 | 39.5 |
| Mean Temp °F | 23.0 | 24.5 | 33.3 | 44.8 | 55.3 | 64.8 | 70.7 | 69.4 | 62.1 | 51.2 | 41.0 | 29.3 | 47.4 |
| Days Max Temp ≥ 90 °F | 0 | 0 | 0 | 0 | 0 | 0 | 1 | 1 | 0 | 0 | 0 | 0 | 2 |
| Days Max Temp ≤ 32 °F | 18 | 15 | 7 | 0 | 0 | 0 | 0 | 0 | 0 | 0 | 1 | 11 | 52 |
| Days Min Temp ≤ 32 °F | 29 | 26 | 24 | 10 | 1 | 0 | 0 | 0 | 0 | 3 | 12 | 25 | 130 |
| Days Min Temp ≤ 0 °F | 4 | 2 | 0 | 0 | 0 | 0 | 0 | 0 | 0 | 0 | 0 | 1 | 7 |
| Heating Degree Days | 1294 | 1137 | 975 | 602 | 313 | 87 | 10 | 17 | 134 | 426 | 715 | 1099 | 6809 |
| Cooling Degree Days | 0 | 0 | 0 | 3 | 19 | 83 | 203 | 167 | 51 | 4 | 0 | 0 | 530 |
| Total Precipitation (") | 3.61 | 3.05 | 3.16 | 3.37 | 3.22 | 3.45 | 2.85 | 3.67 | 4.19 | 3.75 | 4.44 | 3.92 | 42.68 |
| Days ≥ 0.1" Precip | 10 | 9 | 9 | 8 | 8 | 8 | 6 | 7 | 8 | 9 | 10 | 11 | 103 |
| Total Snowfall (") | 51.2 | 38.4 | 18.0 | 4.1 | 0.0 | 0.0 | 0.0 | 0.0 | 0.0 | 0.3 | 8.0 | 33.9 | 153.9 |
| Days ≥ 1" Snow Depth | 27 | 25 | 17 | 2 | 0 | 0 | 0 | 0 | 0 | 0 | 4 | 18 | 93 |

## PATCHOGUE 2 N *Suffolk County* ELEVATION 59 ft LAT/LONG 40° 48 ' N / 73° 1 ' W

| | JAN | FEB | MAR | APR | MAY | JUN | JUL | AUG | SEP | OCT | NOV | DEC | YEAR |
|---|---|---|---|---|---|---|---|---|---|---|---|---|---|
| Maximum Temp °F | 38.7 | 40.6 | 48.8 | 59.2 | 69.5 | 78.1 | 83.3 | 82.4 | 75.5 | 64.8 | 54.3 | 44.1 | 61.6 |
| Minimum Temp °F | 20.9 | 22.4 | 29.5 | 37.9 | 47.7 | 57.2 | 63.6 | 62.8 | 55.1 | 44.0 | 36.1 | 27.1 | 42.0 |
| Mean Temp °F | 29.9 | 31.5 | 39.2 | 48.6 | 58.6 | 67.7 | 73.4 | 72.6 | 65.3 | 54.4 | 45.2 | 35.6 | 51.8 |
| Days Max Temp ≥ 90 °F | 0 | 0 | 0 | 0 | 0 | 1 | 4 | 3 | 1 | 0 | 0 | 0 | 9 |
| Days Max Temp ≤ 32 °F | 8 | 5 | 1 | 0 | 0 | 0 | 0 | 0 | 0 | 0 | 0 | 3 | 17 |
| Days Min Temp ≤ 32 °F | 27 | 24 | 20 | 8 | 1 | 0 | 0 | 0 | 0 | 4 | 12 | 22 | 118 |
| Days Min Temp ≤ 0 °F | 1 | 1 | 0 | 0 | 0 | 0 | 0 | 0 | 0 | 0 | 0 | 0 | 2 |
| Heating Degree Days | 1082 | 938 | 792 | 487 | 213 | 35 | 2 | 6 | 72 | 329 | 586 | 903 | 5445 |
| Cooling Degree Days | 0 | 0 | 0 | 1 | 27 | 146 | 296 | 251 | 93 | 11 | 0 | 0 | 825 |
| Total Precipitation (") | 4.01 | 3.60 | 4.52 | 4.25 | 4.04 | 4.18 | 3.57 | 4.73 | 3.53 | 3.73 | 4.61 | 4.63 | 49.40 |
| Days ≥ 0.1" Precip | 8 | 7 | 8 | 8 | 7 | 7 | 6 | 7 | 6 | 6 | 8 | 8 | 86 |
| Total Snowfall (") | 9.5 | 10.0 | 6.1 | 0.6 | 0.0 | 0.0 | 0.0 | 0.0 | 0.0 | 0.0 | 0.7 | 4.2 | 31.1 |
| Days ≥ 1" Snow Depth | 10 | 7 | 3 | 0 | 0 | 0 | 0 | 0 | 0 | 0 | 0 | 3 | 23 |

## PERU 2 WSW *Clinton County* ELEVATION 512 ft LAT/LONG 44° 34 ' N / 73° 34 ' W

| | JAN | FEB | MAR | APR | MAY | JUN | JUL | AUG | SEP | OCT | NOV | DEC | YEAR |
|---|---|---|---|---|---|---|---|---|---|---|---|---|---|
| Maximum Temp °F | 27.7 | 30.2 | 40.9 | 54.8 | 68.0 | 76.4 | 81.6 | 78.9 | 69.7 | 58.0 | 45.0 | 32.3 | 55.3 |
| Minimum Temp °F | 7.3 | 9.7 | 20.7 | 32.6 | 43.4 | 53.0 | 58.1 | 55.9 | 47.6 | 37.2 | 28.3 | 14.8 | 34.1 |
| Mean Temp °F | 17.6 | 20.0 | 30.8 | 43.7 | 55.7 | 64.7 | 69.9 | 67.4 | 58.7 | 47.6 | 36.7 | 23.6 | 44.7 |
| Days Max Temp ≥ 90 °F | 0 | 0 | 0 | 0 | 0 | 1 | 3 | 1 | 0 | 0 | 0 | 0 | 5 |
| Days Max Temp ≤ 32 °F | 20 | 16 | 7 | 0 | 0 | 0 | 0 | 0 | 0 | 0 | 3 | 15 | 61 |
| Days Min Temp ≤ 32 °F | 30 | 27 | 27 | 16 | 3 | 0 | 0 | 0 | 1 | 10 | 20 | 29 | 163 |
| Days Min Temp ≤ 0 °F | 11 | 8 | 2 | 0 | 0 | 0 | 0 | 0 | 0 | 0 | 0 | 5 | 26 |
| Heating Degree Days | 1463 | 1267 | 1052 | 635 | 301 | 81 | 16 | 45 | 212 | 533 | 843 | 1276 | 7724 |
| Cooling Degree Days | 0 | 0 | 0 | 3 | 22 | 77 | 188 | 130 | 28 | 2 | 0 | 0 | 450 |
| Total Precipitation (") | 1.40 | 1.51 | 1.71 | 2.67 | 2.64 | 3.15 | 2.93 | 3.45 | 2.87 | 2.56 | 2.71 | 2.15 | 29.75 |
| Days ≥ 0.1" Precip | *3* | *3* | 4 | 6 | 7 | 7 | 6 | 7 | 6 | 6 | 6 | *4* | 65 |
| Total Snowfall (") | *13.9* | 12.3 | 9.1 | 4.0 | 0.0 | 0.0 | 0.0 | 0.0 | 0.0 | 0.5 | 4.9 | 14.3 | 59.0 |
| Days ≥ 1" Snow Depth | na | na | na | *1* | 0 | 0 | 0 | 0 | 0 | 0 | *2* | na | na |

## PLATTSBURGH AFB *Clinton County* ELEVATION 171 ft LAT/LONG 44° 42 ' N / 73° 28 ' W

| | JAN | FEB | MAR | APR | MAY | JUN | JUL | AUG | SEP | OCT | NOV | DEC | YEAR |
|---|---|---|---|---|---|---|---|---|---|---|---|---|---|
| Maximum Temp °F | 26.5 | 28.8 | 39.4 | 53.8 | 67.0 | 76.1 | 80.9 | 78.0 | 69.3 | 57.1 | 44.5 | 32.3 | 54.5 |
| Minimum Temp °F | 7.7 | 9.1 | 21.4 | 34.2 | 45.1 | 54.8 | 59.9 | 58.1 | 50.0 | 38.8 | 29.8 | 15.9 | 35.4 |
| Mean Temp °F | 17.1 | 19.0 | 30.4 | 44.0 | 56.1 | 65.4 | 70.4 | 68.0 | 59.7 | 48.0 | 37.2 | 24.1 | 44.9 |
| Days Max Temp ≥ 90 °F | 0 | 0 | 0 | 0 | 0 | 1 | 3 | 1 | 0 | 0 | 0 | 0 | 5 |
| Days Max Temp ≤ 32 °F | 20 | 17 | 7 | 0 | 0 | 0 | 0 | 0 | 0 | 0 | 2 | 14 | 60 |
| Days Min Temp ≤ 32 °F | 30 | 27 | 26 | 14 | 2 | 0 | 0 | 0 | 0 | 8 | 19 | 28 | 154 |
| Days Min Temp ≤ 0 °F | 10 | 8 | 2 | 0 | 0 | 0 | 0 | 0 | 0 | 0 | 0 | 4 | 24 |
| Heating Degree Days | 1478 | 1294 | 1064 | 625 | 289 | 69 | 13 | 37 | 188 | 523 | 829 | 1261 | 7670 |
| Cooling Degree Days | 0 | 0 | 0 | 2 | 11 | 63 | 158 | 117 | 29 | 1 | 0 | 0 | 381 |
| Total Precipitation (") | 1.96 | 1.68 | 2.14 | 2.80 | 2.88 | 3.21 | 2.95 | 3.98 | 3.30 | 2.69 | 3.01 | 2.46 | 33.06 |
| Days ≥ 0.1" Precip | 6 | 5 | 6 | 7 | 7 | 8 | 8 | 9 | 7 | 7 | 8 | 7 | 85 |
| Total Snowfall (") | 17.2 | 13.3 | 9.7 | 3.9 | 0.0 | 0.0 | 0.0 | 0.0 | 0.0 | 0.3 | 5.8 | 14.6 | 64.8 |
| Days ≥ 1" Snow Depth | 24 | *24* | 15 | 2 | 0 | 0 | 0 | 0 | 0 | 0 | *3* | 15 | 83 |

## PORT JERVIS *Orange County* ELEVATION 469 ft LAT/LONG 41° 23 ' N / 74° 41 ' W

| | JAN | FEB | MAR | APR | MAY | JUN | JUL | AUG | SEP | OCT | NOV | DEC | YEAR |
|---|---|---|---|---|---|---|---|---|---|---|---|---|---|
| Maximum Temp °F | 34.2 | 38.1 | 48.6 | 62.3 | 72.9 | 80.4 | 84.5 | 81.9 | 73.2 | 61.8 | 50.5 | 38.7 | 60.6 |
| Minimum Temp °F | 16.2 | 18.1 | 26.5 | 36.4 | 46.8 | 55.5 | 60.6 | 59.5 | 51.9 | 40.0 | 32.3 | 22.3 | 38.8 |
| Mean Temp °F | 25.2 | 28.1 | 37.6 | 49.4 | 59.9 | 67.9 | 72.6 | 70.7 | 62.6 | 50.9 | 41.4 | 30.5 | 49.7 |
| Days Max Temp ≥ 90 °F | 0 | 0 | 0 | 0 | 1 | 3 | 6 | 3 | 1 | 0 | 0 | 0 | 14 |
| Days Max Temp ≤ 32 °F | 13 | 8 | 2 | 0 | 0 | 0 | 0 | 0 | 0 | 0 | 1 | 7 | 31 |
| Days Min Temp ≤ 32 °F | 29 | 26 | 23 | 11 | 1 | 0 | 0 | 0 | 0 | 7 | 16 | 27 | 140 |
| Days Min Temp ≤ 0 °F | 3 | 2 | 0 | 0 | 0 | 0 | 0 | 0 | 0 | 0 | 0 | 1 | 6 |
| Heating Degree Days | 1227 | 1035 | 844 | 466 | 189 | 36 | 4 | 14 | 127 | 432 | 700 | 1061 | 6135 |
| Cooling Degree Days | 0 | 0 | 0 | 6 | 42 | 131 | 262 | 201 | 66 | 5 | 0 | 0 | 713 |
| Total Precipitation (") | 3.01 | 2.85 | 3.73 | 3.83 | 4.40 | 3.95 | 4.16 | 3.78 | 4.02 | 3.18 | 3.78 | 3.55 | 44.24 |
| Days ≥ 0.1" Precip | 6 | 5 | 7 | 7 | 8 | 7 | 7 | 6 | 6 | 5 | 6 | 7 | 77 |
| Total Snowfall (") | 12.1 | 10.5 | 9.2 | 1.4 | 0.0 | 0.0 | 0.0 | 0.0 | 0.0 | 0.0 | 2.8 | 9.3 | 45.3 |
| Days ≥ 1" Snow Depth | 22 | 19 | 9 | 0 | 0 | 0 | 0 | 0 | 0 | 0 | 1 | 10 | 61 |

## POUGHKEEPSIE *Dutchess County* ELEVATION 161 ft LAT/LONG 41° 38 ' N / 73° 53 ' W

| | JAN | FEB | MAR | APR | MAY | JUN | JUL | AUG | SEP | OCT | NOV | DEC | YEAR |
|---|---|---|---|---|---|---|---|---|---|---|---|---|---|
| Maximum Temp °F | 33.3 | 36.9 | 46.9 | 59.4 | 70.4 | 78.9 | 83.8 | 82.0 | 73.9 | 62.4 | 50.6 | 38.7 | 59.8 |
| Minimum Temp °F | 14.5 | 17.1 | 26.4 | 36.4 | 46.6 | 55.5 | 60.9 | 59.7 | 51.1 | 39.1 | 31.5 | 21.3 | 38.3 |
| Mean Temp °F | 23.9 | 27.1 | 36.7 | 47.9 | 58.5 | 67.2 | 72.4 | 70.9 | 62.5 | 50.8 | 41.0 | 30.0 | 49.1 |
| Days Max Temp ≥ 90 °F | 0 | 0 | 0 | 0 | 1 | 3 | 5 | 4 | 1 | 0 | 0 | 0 | 14 |
| Days Max Temp ≤ 32 °F | 14 | 9 | 2 | 0 | 0 | 0 | 0 | 0 | 0 | 0 | 0 | 8 | 33 |
| Days Min Temp ≤ 32 °F | 29 | 26 | 23 | 11 | 1 | 0 | 0 | 0 | 0 | 9 | 17 | 27 | 143 |
| Days Min Temp ≤ 0 °F | 4 | 2 | 0 | 0 | 0 | 0 | 0 | 0 | 0 | 0 | 0 | 1 | 7 |
| Heating Degree Days | 1265 | 1066 | 870 | 509 | 225 | 48 | 6 | 15 | 131 | 437 | 712 | 1077 | 6361 |
| Cooling Degree Days | 0 | 0 | 0 | 3 | 34 | 106 | 243 | 195 | 64 | 4 | 0 | 0 | 649 |
| Total Precipitation (") | 2.43 | 2.51 | 3.29 | 3.61 | 4.67 | 3.79 | 4.28 | 3.94 | 3.47 | 3.19 | 3.55 | 3.14 | 41.87 |
| Days ≥ 0.1" Precip | 5 | 6 | 7 | 7 | 8 | 7 | 7 | 6 | 5 | 5 | 7 | 7 | 77 |
| Total Snowfall (") | 10.1 | 9.5 | 6.3 | 1.4 | 0.0 | 0.0 | 0.0 | 0.0 | 0.0 | 0.0 | 2.0 | 8.1 | 37.4 |
| Days ≥ 1" Snow Depth | 19 | 14 | 6 | 1 | 0 | 0 | 0 | 0 | 0 | 0 | 1 | 8 | 49 |

## RIVERHEAD RES FARM *Suffolk County* ELEVATION 100 ft LAT/LONG 40° 58 ' N / 72° 43 ' W

| | JAN | FEB | MAR | APR | MAY | JUN | JUL | AUG | SEP | OCT | NOV | DEC | YEAR |
|---|---|---|---|---|---|---|---|---|---|---|---|---|---|
| Maximum Temp °F | 38.1 | 39.4 | 47.7 | 58.7 | 70.0 | 78.7 | 83.5 | 82.1 | 75.0 | 64.3 | 53.8 | 43.7 | 61.3 |
| Minimum Temp °F | 23.4 | 24.6 | 31.6 | 39.5 | 49.3 | 58.6 | 64.4 | 63.8 | 57.1 | 46.6 | 38.3 | 29.1 | 43.9 |
| Mean Temp °F | 30.8 | 32.0 | 39.7 | 49.1 | 59.7 | 68.7 | 74.0 | 73.0 | 66.1 | 55.5 | 46.1 | 36.4 | 52.6 |
| Days Max Temp ≥ 90 °F | 0 | 0 | 0 | 0 | 1 | 2 | 4 | 2 | 1 | 0 | 0 | 0 | 10 |
| Days Max Temp ≤ 32 °F | 8 | 6 | 1 | 0 | 0 | 0 | 0 | 0 | 0 | 0 | 0 | 3 | 18 |
| Days Min Temp ≤ 32 °F | 26 | 23 | 18 | 4 | 0 | 0 | 0 | 0 | 0 | 1 | 7 | 20 | 99 |
| Days Min Temp ≤ 0 °F | 0 | 0 | 0 | 0 | 0 | 0 | 0 | 0 | 0 | 0 | 0 | 0 | 0 |
| Heating Degree Days | 1054 | 924 | 778 | 473 | 189 | 26 | 1 | 3 | 57 | 296 | 561 | 879 | 5241 |
| Cooling Degree Days | 0 | 0 | 0 | 2 | 38 | 164 | 314 | 263 | 101 | 10 | 0 | 0 | 892 |
| Total Precipitation (") | 3.91 | 3.38 | 4.17 | 3.91 | 3.77 | 3.59 | 3.33 | 4.10 | 3.40 | 3.47 | 4.29 | 4.28 | 45.60 |
| Days ≥ 0.1" Precip | 7 | 6 | 7 | 7 | 7 | 6 | 6 | 6 | 5 | 6 | 7 | 7 | 77 |
| Total Snowfall (") | 9.8 | 9.1 | 4.7 | 0.5 | 0.0 | 0.0 | 0.0 | 0.0 | 0.0 | 0.0 | 0.4 | 3.5 | 28.0 |
| Days ≥ 1" Snow Depth | 10 | 9 | 3 | 0 | 0 | 0 | 0 | 0 | 0 | 0 | 0 | 3 | 25 |

## ROCHESTER INTL AP *Monroe County* ELEVATION 554 ft LAT/LONG 43° 7 ' N / 77° 40 ' W

| | JAN | FEB | MAR | APR | MAY | JUN | JUL | AUG | SEP | OCT | NOV | DEC | YEAR |
|---|---|---|---|---|---|---|---|---|---|---|---|---|---|
| Maximum Temp °F | 30.8 | 32.4 | 42.3 | 56.1 | 67.9 | 77.2 | 82.2 | 79.8 | 71.9 | 60.2 | 48.0 | 36.3 | 57.1 |
| Minimum Temp °F | 16.3 | 16.8 | 25.4 | 36.1 | 46.2 | 55.4 | 60.7 | 59.1 | 51.7 | 41.3 | 33.2 | 23.0 | 38.8 |
| Mean Temp °F | 23.6 | 24.6 | 33.9 | 46.1 | 57.1 | 66.3 | 71.4 | 69.5 | 61.8 | 50.8 | 40.6 | 29.7 | 48.0 |
| Days Max Temp ≥ 90 °F | 0 | 0 | 0 | 0 | 0 | 2 | 4 | 2 | 0 | 0 | 0 | 0 | 8 |
| Days Max Temp ≤ 32 °F | 17 | 14 | 7 | 0 | 0 | 0 | 0 | 0 | 0 | 0 | 1 | 10 | 49 |
| Days Min Temp ≤ 32 °F | 29 | 26 | 23 | 11 | 1 | 0 | 0 | 0 | 0 | 4 | 15 | 25 | 134 |
| Days Min Temp ≤ 0 °F | 3 | 2 | 0 | 0 | 0 | 0 | 0 | 0 | 0 | 0 | 0 | 1 | 6 |
| Heating Degree Days | 1278 | 1133 | 957 | 565 | 271 | 65 | 9 | 24 | 147 | 439 | 725 | 1088 | 6701 |
| Cooling Degree Days | 0 | 0 | 0 | 5 | 33 | 95 | 218 | 168 | 54 | 5 | 0 | 0 | 578 |
| Total Precipitation (") | 2.14 | 2.03 | 2.38 | 2.66 | 2.66 | 3.01 | 2.78 | 3.39 | 3.20 | 2.55 | 2.91 | 2.84 | 32.55 |
| Days ≥ 0.1" Precip | 6 | 6 | 7 | 7 | 7 | 6 | 6 | 7 | 7 | 6 | 8 | 8 | 81 |
| Total Snowfall (") | 26.1 | 23.2 | 13.7 | 4.6 | 0.4 | 0.0 | 0.0 | 0.0 | 0.0 | 0.2 | 6.7 | 21.9 | 96.8 |
| Days ≥ 1" Snow Depth | 23 | 22 | 12 | 2 | 0 | 0 | 0 | 0 | 0 | 0 | 3 | 16 | 78 |

## SALEM *Washington County* ELEVATION 489 ft LAT/LONG 43° 10 ' N / 73° 19 ' W

| | JAN | FEB | MAR | APR | MAY | JUN | JUL | AUG | SEP | OCT | NOV | DEC | YEAR |
|---|---|---|---|---|---|---|---|---|---|---|---|---|---|
| Maximum Temp °F | 30.7 | 33.3 | 43.6 | 57.6 | 69.4 | 77.6 | 82.4 | 80.4 | 72.3 | 60.6 | 47.5 | 35.7 | 57.6 |
| Minimum Temp °F | 6.4 | 8.3 | 21.0 | 32.5 | 42.4 | 51.5 | 55.7 | 53.6 | 45.1 | 34.6 | 27.2 | 15.2 | 32.8 |
| Mean Temp °F | 18.6 | 20.8 | 32.3 | 45.1 | 55.9 | 64.5 | 69.1 | 67.1 | 58.7 | 47.7 | 37.4 | 25.5 | 45.2 |
| Days Max Temp ≥ 90 °F | 0 | 0 | 0 | 0 | 0 | 1 | 4 | 2 | 0 | 0 | 0 | 0 | 7 |
| Days Max Temp ≤ 32 °F | 17 | 12 | 4 | 0 | 0 | 0 | 0 | 0 | 0 | 0 | 1 | 10 | 44 |
| Days Min Temp ≤ 32 °F | 30 | 27 | 27 | 16 | 4 | 0 | 0 | 0 | 3 | 13 | 21 | 29 | 170 |
| Days Min Temp ≤ 0 °F | 11 | 9 | 1 | 0 | 0 | 0 | 0 | 0 | 0 | 0 | 0 | 4 | 25 |
| Heating Degree Days | 1435 | 1242 | 1007 | 593 | 293 | 87 | 23 | 47 | 214 | 531 | 822 | 1219 | 7513 |
| Cooling Degree Days | 0 | 0 | 0 | 2 | 18 | 79 | 166 | 122 | 29 | 2 | 0 | 0 | 418 |
| Total Precipitation (") | 2.40 | 1.70 | 2.43 | 2.80 | 3.46 | 3.46 | 3.02 | 3.65 | 3.46 | 2.85 | 3.24 | 2.74 | 35.21 |
| Days ≥ 0.1" Precip | 5 | 4 | 6 | 6 | 7 | 7 | 6 | 6 | 6 | 6 | 7 | 6 | 72 |
| Total Snowfall (") | *13.5* | *9.4* | *7.6* | 3.0 | 0.1 | 0.0 | 0.0 | 0.0 | 0.0 | 0.0 | *4.1* | *11.5* | 49.2 |
| Days ≥ 1" Snow Depth | *24* | *19* | *12* | 1 | 0 | 0 | 0 | 0 | 0 | 0 | *3* | *13* | 72 |

## SARATOGA SPRINGS 4 S *Saratoga County* ELEVATION 312 ft LAT/LONG 43° 2 ' N / 73° 49 ' W

| | JAN | FEB | MAR | APR | MAY | JUN | JUL | AUG | SEP | OCT | NOV | DEC | YEAR |
|---|---|---|---|---|---|---|---|---|---|---|---|---|---|
| Maximum Temp °F | 30.3 | 33.6 | 44.4 | 59.0 | 71.5 | 79.3 | 83.8 | 81.3 | 72.7 | 60.9 | 47.9 | 35.2 | 58.3 |
| Minimum Temp °F | 9.4 | 11.3 | 22.6 | 33.7 | 44.3 | 53.4 | 58.4 | 56.7 | 48.1 | 37.2 | 29.3 | 17.3 | 35.1 |
| Mean Temp °F | 19.9 | 22.5 | 33.5 | 46.3 | 57.9 | 66.3 | 71.1 | 69.0 | 60.5 | 49.1 | 38.6 | 26.3 | 46.8 |
| Days Max Temp ≥ 90 °F | 0 | 0 | 0 | 0 | 0 | 2 | 5 | 2 | 0 | 0 | 0 | 0 | 9 |
| Days Max Temp ≤ 32 °F | 17 | 13 | 3 | 0 | 0 | 0 | 0 | 0 | 0 | 0 | 1 | 10 | 44 |
| Days Min Temp ≤ 32 °F | 30 | 26 | 26 | 15 | 3 | 0 | 0 | 0 | 1 | 11 | 20 | 28 | 160 |
| Days Min Temp ≤ 0 °F | 9 | 7 | 1 | 0 | 0 | 0 | 0 | 0 | 0 | 0 | 0 | 4 | 21 |
| Heating Degree Days | 1393 | 1195 | 969 | 555 | 241 | 58 | 9 | 26 | 175 | 490 | 787 | 1194 | 7092 |
| Cooling Degree Days | 0 | 0 | 0 | 4 | 33 | 108 | 225 | 169 | 48 | 4 | 0 | 0 | 591 |
| Total Precipitation (") | 2.82 | 2.49 | 3.32 | 3.62 | 4.14 | 4.07 | 3.54 | 3.93 | 3.47 | 3.41 | 3.83 | 3.47 | 42.11 |
| Days ≥ 0.1" Precip | 6 | 5 | 7 | 7 | 8 | 7 | 6 | 7 | 6 | 6 | 8 | 7 | 80 |
| Total Snowfall (") | 18.2 | 13.3 | 10.5 | 2.6 | 0.0 | 0.0 | 0.0 | 0.0 | 0.0 | 0.0 | 4.0 | 15.2 | 63.8 |
| Days ≥ 1" Snow Depth | 28 | 26 | 16 | 2 | 0 | 0 | 0 | 0 | 0 | 0 | 3 | 19 | 94 |

## SETAUKET STRONG *Suffolk County* ELEVATION 39 ft LAT/LONG 40° 57 ' N / 73° 6 ' W

| | JAN | FEB | MAR | APR | MAY | JUN | JUL | AUG | SEP | OCT | NOV | DEC | YEAR |
|---|---|---|---|---|---|---|---|---|---|---|---|---|---|
| Maximum Temp °F | 38.2 | 40.1 | 48.3 | 59.5 | 69.7 | 78.1 | 82.9 | 81.3 | 74.4 | 64.2 | 54.1 | 43.7 | 61.2 |
| Minimum Temp °F | 23.6 | 24.5 | 31.1 | 39.6 | 49.0 | 58.6 | 64.6 | 64.3 | 57.9 | 47.4 | 38.9 | 29.6 | 44.1 |
| Mean Temp °F | 30.9 | 32.3 | 39.8 | 49.6 | 59.4 | 68.3 | 73.8 | 72.8 | 66.2 | 55.8 | 46.6 | 36.7 | 52.7 |
| Days Max Temp ≥ 90 °F | 0 | 0 | 0 | 0 | 0 | 1 | 4 | 1 | 0 | 0 | 0 | 0 | 6 |
| Days Max Temp ≤ 32 °F | 8 | 5 | 1 | 0 | 0 | 0 | 0 | 0 | 0 | 0 | 0 | 3 | 17 |
| Days Min Temp ≤ 32 °F | 26 | 23 | 18 | 5 | 0 | 0 | 0 | 0 | 0 | 1 | 7 | 19 | 99 |
| Days Min Temp ≤ 0 °F | 0 | 0 | 0 | 0 | 0 | 0 | 0 | 0 | 0 | 0 | 0 | 0 | 0 |
| Heating Degree Days | 1050 | 915 | 776 | 458 | 196 | 29 | 1 | 3 | 55 | 286 | 547 | 871 | 5187 |
| Cooling Degree Days | 0 | 0 | 0 | 2 | 33 | 142 | 287 | 245 | 94 | 10 | 0 | 0 | 813 |
| Total Precipitation (") | 3.55 | 3.05 | 4.24 | 3.95 | 3.89 | 3.66 | 3.46 | 4.00 | 3.56 | 3.49 | 4.15 | 4.16 | 45.16 |
| Days ≥ 0.1" Precip | *5* | *6* | *7* | 7 | 7 | 6 | 6 | 6 | 5 | 6 | *7* | *7* | 75 |
| Total Snowfall (") | 5.7 | 5.5 | 2.9 | 0.3 | 0.0 | 0.0 | 0.0 | 0.0 | 0.0 | 0.0 | 0.2 | 2.2 | 16.8 |
| Days ≥ 1" Snow Depth | na | na | *1* | 0 | 0 | 0 | 0 | 0 | 0 | 0 | 0 | *1* | na |

## SLIDE MOUNTAIN *Ulster County* ELEVATION 2651 ft LAT/LONG 42° 1 ' N / 74° 25 ' W

| | JAN | FEB | MAR | APR | MAY | JUN | JUL | AUG | SEP | OCT | NOV | DEC | YEAR |
|---|---|---|---|---|---|---|---|---|---|---|---|---|---|
| Maximum Temp °F | 25.7 | 28.0 | 36.6 | 48.6 | 60.9 | 67.5 | 72.0 | 70.3 | 63.1 | 53.2 | 41.8 | 30.4 | 49.8 |
| Minimum Temp °F | 8.3 | 9.0 | 17.5 | 29.0 | 39.6 | 47.8 | 52.7 | 51.7 | 44.7 | 34.1 | 25.9 | 15.0 | 31.3 |
| Mean Temp °F | 17.0 | 18.6 | 27.0 | 38.8 | 50.2 | 57.6 | 62.4 | 61.1 | 54.0 | 43.7 | 33.9 | 22.7 | 40.6 |
| Days Max Temp ≥ 90 °F | 0 | 0 | 0 | 0 | 0 | 0 | 0 | 0 | 0 | 0 | 0 | 0 | 0 |
| Days Max Temp ≤ 32 °F | 23 | 18 | 11 | 2 | 0 | 0 | 0 | 0 | 0 | 0 | 6 | 18 | 78 |
| Days Min Temp ≤ 32 °F | 30 | 27 | 29 | 20 | 7 | 0 | 0 | 0 | 2 | 14 | 24 | 30 | 183 |
| Days Min Temp ≤ 0 °F | 8 | 7 | 2 | 0 | 0 | 0 | 0 | 0 | 0 | 0 | 0 | 3 | 20 |
| Heating Degree Days | 1482 | 1305 | 1170 | 778 | 453 | 225 | 110 | 138 | 332 | 655 | 927 | 1304 | 8879 |
| Cooling Degree Days | 0 | 0 | 0 | 0 | 4 | 12 | 41 | 25 | 6 | 0 | 0 | 0 | 88 |
| Total Precipitation (") | 4.54 | 4.18 | 5.40 | 5.28 | 5.73 | 5.17 | 4.67 | 5.02 | 5.04 | 5.03 | 5.81 | 5.10 | 60.97 |
| Days ≥ 0.1" Precip | 8 | 7 | 8 | 9 | 9 | 8 | 8 | 8 | 8 | 7 | 9 | 9 | 98 |
| Total Snowfall (") | 23.0 | 20.8 | 20.5 | 5.9 | 1.7 | 0.0 | 0.0 | 0.0 | 0.0 | 0.6 | 8.9 | 20.2 | 101.6 |
| Days ≥ 1" Snow Depth | 29 | 27 | 26 | 9 | 1 | 0 | 0 | 0 | 0 | 0 | 9 | 23 | 124 |

## SODUS CENTER *Wayne County* ELEVATION 420 ft LAT/LONG 43° 12 ' N / 77° 1 ' W

| | JAN | FEB | MAR | APR | MAY | JUN | JUL | AUG | SEP | OCT | NOV | DEC | YEAR |
|---|---|---|---|---|---|---|---|---|---|---|---|---|---|
| Maximum Temp °F | 31.6 | 33.5 | 42.9 | 56.1 | 68.2 | 77.5 | 81.9 | 80.0 | 72.5 | 60.6 | 48.6 | 36.4 | 57.5 |
| Minimum Temp °F | 16.3 | 16.7 | 25.6 | 36.3 | 45.8 | 54.6 | 60.1 | 59.0 | 52.3 | 41.6 | 33.6 | 22.8 | 38.7 |
| Mean Temp °F | 24.0 | 25.1 | 34.3 | 46.2 | 57.0 | 66.1 | 71.0 | 69.5 | 62.4 | 51.1 | 41.2 | 29.6 | 48.1 |
| Days Max Temp ≥ 90 °F | 0 | 0 | 0 | 0 | 0 | 2 | 4 | 2 | 1 | 0 | 0 | 0 | 9 |
| Days Max Temp ≤ 32 °F | 16 | 14 | 5 | 0 | 0 | 0 | 0 | 0 | 0 | 0 | 1 | 10 | 46 |
| Days Min Temp ≤ 32 °F | 28 | 26 | 24 | 11 | 2 | 0 | 0 | 0 | 0 | 5 | 14 | 26 | 136 |
| Days Min Temp ≤ 0 °F | 3 | 2 | 0 | 0 | 0 | 0 | 0 | 0 | 0 | 0 | 0 | 1 | 6 |
| Heating Degree Days | 1265 | 1119 | 946 | 564 | 268 | 69 | 13 | 23 | 131 | 429 | 709 | 1090 | 6626 |
| Cooling Degree Days | 0 | 0 | 1 | 5 | 27 | 96 | 207 | 169 | 58 | 4 | 0 | 0 | 567 |
| Total Precipitation (") | 2.31 | 1.94 | 2.42 | 3.25 | 3.12 | 3.62 | 3.05 | 3.44 | 4.01 | 3.83 | 4.12 | 2.87 | 37.98 |
| Days ≥ 0.1" Precip | 7 | 6 | 7 | 8 | 7 | 7 | 6 | 7 | 8 | 9 | 10 | 8 | 90 |
| Total Snowfall (") | 25.5 | 17.9 | 12.4 | 2.9 | 0.0 | 0.0 | 0.0 | 0.0 | 0.0 | 0.1 | 5.2 | 22.0 | 86.0 |
| Days ≥ 1" Snow Depth | 25 | 23 | 14 | 1 | 0 | 0 | 0 | 0 | 0 | 0 | 4 | 17 | 84 |

## SPENCER 2 N *Tioga County* ELEVATION 1100 ft LAT/LONG 42° 11 ' N / 76° 30 ' W

| | JAN | FEB | MAR | APR | MAY | JUN | JUL | AUG | SEP | OCT | NOV | DEC | YEAR |
|---|---|---|---|---|---|---|---|---|---|---|---|---|---|
| Maximum Temp °F | 30.9 | 33.7 | 43.3 | 56.4 | 68.3 | 76.9 | 81.5 | 79.6 | 71.6 | 59.7 | 46.8 | 35.5 | 57.0 |
| Minimum Temp °F | 10.5 | 11.5 | 20.7 | 30.6 | 40.3 | 49.4 | 53.8 | 52.7 | 45.5 | 34.7 | 27.8 | 17.4 | 32.9 |
| Mean Temp °F | 20.7 | 22.6 | 32.0 | 43.5 | 54.4 | 63.2 | 67.7 | 66.2 | 58.6 | 47.2 | 37.3 | 26.4 | 45.0 |
| Days Max Temp ≥ 90 °F | 0 | 0 | 0 | 0 | 0 | 1 | 3 | 2 | 0 | 0 | 0 | 0 | 6 |
| Days Max Temp ≤ 32 °F | 17 | 13 | 5 | 0 | 0 | 0 | 0 | 0 | 0 | 0 | 2 | 12 | 49 |
| Days Min Temp ≤ 32 °F | 30 | 27 | 27 | 19 | 7 | 1 | 0 | 0 | 2 | 14 | 21 | 29 | 177 |
| Days Min Temp ≤ 0 °F | 7 | 7 | 1 | 0 | 0 | 0 | 0 | 0 | 0 | 0 | 0 | 4 | 19 |
| Heating Degree Days | 1366 | 1190 | 1016 | 638 | 336 | 111 | 37 | 56 | 217 | 545 | 823 | 1188 | 7523 |
| Cooling Degree Days | 0 | 0 | 0 | *1* | *14* | *56* | *129* | 93 | 29 | 1 | *0* | 0 | 323 |
| Total Precipitation (") | 2.24 | 2.11 | 2.76 | 3.26 | 3.29 | 4.00 | 3.53 | 3.43 | 3.57 | 3.34 | 3.48 | 3.06 | 38.07 |
| Days ≥ 0.1" Precip | 5 | 5 | 6 | 7 | 8 | 8 | 7 | 7 | 7 | 7 | 7 | 7 | 81 |
| Total Snowfall (") | 14.6 | 12.7 | 8.9 | 2.2 | 0.2 | 0.0 | 0.0 | 0.0 | 0.0 | 0.2 | 4.8 | 13.2 | 56.8 |
| Days ≥ 1" Snow Depth | 21 | *22* | 9 | 1 | 0 | 0 | 0 | 0 | 0 | 0 | 4 | 14 | 71 |

## STILLWATER RESERVOIR *Herkimer County* ELEVATION 1703 ft LAT/LONG 43° 53 ' N / 75° 2 ' W

| | JAN | FEB | MAR | APR | MAY | JUN | JUL | AUG | SEP | OCT | NOV | DEC | YEAR |
|---|---|---|---|---|---|---|---|---|---|---|---|---|---|
| Maximum Temp °F | 24.0 | 26.8 | 36.4 | 49.1 | 63.0 | 70.5 | 75.3 | 73.5 | 65.4 | 54.3 | 41.3 | 29.2 | 50.7 |
| Minimum Temp °F | 0.7 | 1.9 | 13.1 | 28.0 | 40.3 | 49.6 | 54.7 | 53.6 | 45.9 | 34.7 | 25.7 | 10.2 | 29.9 |
| Mean Temp °F | 12.4 | 14.3 | 24.8 | 38.6 | 51.6 | 60.1 | 65.0 | 63.6 | 55.7 | 44.6 | 33.5 | 19.7 | 40.3 |
| Days Max Temp ≥ 90 °F | 0 | 0 | 0 | 0 | 0 | 0 | 0 | 0 | 0 | 0 | 0 | 0 | 0 |
| Days Max Temp ≤ 32 °F | 24 | 20 | 11 | 2 | 0 | 0 | 0 | 0 | 0 | 0 | 7 | 19 | 83 |
| Days Min Temp ≤ 32 °F | 31 | 27 | 29 | 21 | 6 | 0 | 0 | 0 | 2 | 14 | 23 | 30 | 183 |
| Days Min Temp ≤ 0 °F | 15 | 14 | 7 | 0 | 0 | 0 | 0 | 0 | 0 | 0 | 0 | 8 | 44 |
| Heating Degree Days | 1628 | 1426 | 1239 | 786 | 415 | 173 | 68 | 93 | 284 | 628 | 937 | 1397 | 9074 |
| Cooling Degree Days | 0 | 0 | 0 | 0 | 7 | 27 | 76 | 54 | 10 | 1 | 0 | 0 | 175 |
| Total Precipitation (") | 3.18 | 2.60 | 3.16 | 3.76 | 3.94 | 4.26 | 4.04 | 4.81 | 4.86 | 4.13 | 4.77 | 4.26 | 47.77 |
| Days ≥ 0.1" Precip | 10 | 8 | 10 | 9 | 9 | 9 | 9 | 10 | 9 | 10 | 11 | 11 | 115 |
| Total Snowfall (") | 49.9 | 36.9 | 28.4 | 10.4 | 1.5 | 0.0 | 0.0 | 0.0 | 0.0 | 2.2 | 20.0 | 42.9 | 192.2 |
| Days ≥ 1" Snow Depth | na | na | na | na | *0* | *0* | 0 | 0 | 0 | 1 | na | na | na |

## SUFFERN *Rockland County* ELEVATION 269 ft LAT/LONG 41° 7 ' N / 74° 9 ' W

| | JAN | FEB | MAR | APR | MAY | JUN | JUL | AUG | SEP | OCT | NOV | DEC | YEAR |
|---|---|---|---|---|---|---|---|---|---|---|---|---|---|
| Maximum Temp °F | *35.9* | 38.2 | 47.7 | *59.6* | *70.8* | 79.0 | *84.0* | *82.5* | *74.7* | *63.4* | *52.3* | *40.7* | 60.7 |
| Minimum Temp °F | *17.0* | 18.0 | 27.6 | *38.0* | *47.5* | 56.4 | *61.7* | *59.8* | *51.8* | *39.5* | *32.5* | *22.9* | 39.4 |
| Mean Temp °F | *26.5* | 28.1 | 37.7 | *48.9* | *59.2* | 67.7 | *72.9* | *71.2* | *63.3* | *51.5* | *42.4* | *31.8* | 50.1 |
| Days Max Temp ≥ 90 °F | *0* | 0 | 0 | *0* | *1* | 2 | *6* | *4* | *1* | *0* | *0* | *0* | 14 |
| Days Max Temp ≤ 32 °F | *11* | 8 | *2* | 0 | *0* | 0 | *0* | *0* | *0* | *0* | *0* | *6* | 27 |
| Days Min Temp ≤ 32 °F | *29* | 26 | 23 | *7* | *1* | 0 | *0* | *0* | *0* | *7* | *16* | *27* | 136 |
| Days Min Temp ≤ 0 °F | *2* | 1 | 0 | *0* | *0* | 0 | *0* | *0* | *0* | *0* | *0* | *0* | 3 |
| Heating Degree Days | *1186* | 1034 | 841 | *480* | *210* | 40 | *4* | *13* | *112* | *415* | *672* | *1022* | 6029 |
| Cooling Degree Days | na | na | *0* | *5* | na | *147* | na | na | na | na | na | na | na |
| Total Precipitation (") | *3.28* | 3.00 | 3.99 | 4.24 | *5.04* | 4.31 | *4.20* | *4.72* | *5.07* | *4.27* | *4.38* | *3.65* | 50.15 |
| Days ≥ 0.1" Precip | *6* | *5* | *6* | *8* | *8* | 7 | *7* | na | *6* | *6* | *6* | na | na |
| Total Snowfall (") | na | na | na | *0.7* | *0.0* | 0.0 | *0.0* | *0.0* | *0.0* | *0.0* | *0.3* | na | na |
| Days ≥ 1" Snow Depth | na | na | na | na | na | *0* | na | na | na | na | na | na | na |

## SYRACUSE HANCOCK AP *Onondaga County* ELEVATION 427 ft LAT/LONG 43° 7 ' N / 76° 7 ' W

| | JAN | FEB | MAR | APR | MAY | JUN | JUL | AUG | SEP | OCT | NOV | DEC | YEAR |
|---|---|---|---|---|---|---|---|---|---|---|---|---|---|
| Maximum Temp °F | 30.6 | 32.6 | 42.4 | 56.4 | 68.3 | 76.8 | 81.6 | 79.4 | 71.3 | 59.7 | 47.8 | 36.0 | 56.9 |
| Minimum Temp °F | 14.4 | 15.6 | 24.9 | 35.8 | 45.9 | 54.6 | 60.0 | 58.8 | 51.4 | 40.8 | 32.8 | 21.6 | 38.1 |
| Mean Temp °F | 22.6 | 24.1 | 33.7 | 46.1 | 57.2 | 65.8 | 70.8 | 69.1 | 61.4 | 50.3 | 40.4 | 28.8 | 47.5 |
| Days Max Temp ≥ 90 °F | 0 | 0 | 0 | 0 | 0 | 1 | 4 | 2 | 0 | 0 | 0 | 0 | 7 |
| Days Max Temp ≤ 32 °F | 17 | 14 | 6 | 0 | 0 | 0 | 0 | 0 | 0 | 0 | 1 | 11 | 49 |
| Days Min Temp ≤ 32 °F | 29 | 25 | 24 | 12 | 1 | 0 | 0 | 0 | 0 | 5 | 15 | 26 | 137 |
| Days Min Temp ≤ 0 °F | 5 | 3 | 1 | 0 | 0 | 0 | 0 | 0 | 0 | 0 | 0 | 2 | 11 |
| Heating Degree Days | 1309 | 1148 | 964 | 564 | 265 | 71 | 11 | 27 | 153 | 454 | 732 | 1115 | 6813 |
| Cooling Degree Days | 0 | 0 | 1 | 5 | 32 | 95 | 216 | 164 | 48 | 3 | 0 | 0 | 564 |
| Total Precipitation (") | 2.42 | 2.09 | 2.90 | 3.48 | 3.47 | 3.80 | 4.05 | 3.63 | 4.13 | 3.25 | 3.72 | 3.25 | 40.19 |
| Days ≥ 0.1" Precip | 7 | 6 | 7 | 8 | 8 | 8 | 7 | 7 | 7 | 8 | 9 | 9 | 91 |
| Total Snowfall (") | 32.7 | 25.0 | 17.0 | 4.6 | 0.1 | 0.0 | 0.0 | 0.0 | 0.0 | 0.6 | 9.3 | 27.6 | 116.9 |
| Days ≥ 1" Snow Depth | 25 | 23 | 14 | 2 | 0 | 0 | 0 | 0 | 0 | 0 | 5 | 17 | 86 |

## TROY L & D *Rensselaer County* ELEVATION 20 ft LAT/LONG 42° 45 ' N / 73° 41 ' W

| | JAN | FEB | MAR | APR | MAY | JUN | JUL | AUG | SEP | OCT | NOV | DEC | YEAR |
|---|---|---|---|---|---|---|---|---|---|---|---|---|---|
| Maximum Temp °F | 30.9 | 33.5 | 43.6 | 57.3 | 70.0 | 78.5 | 83.7 | 81.7 | 73.3 | 61.3 | 49.0 | 36.4 | 58.3 |
| Minimum Temp °F | 12.6 | 14.5 | 25.0 | 37.0 | 47.7 | 57.1 | 62.4 | 60.4 | 51.8 | 40.3 | 32.5 | 20.8 | 38.5 |
| Mean Temp °F | 21.8 | 24.0 | 34.3 | 47.2 | 58.9 | 67.8 | 73.1 | 71.1 | 62.5 | 50.9 | 40.7 | 28.6 | 48.4 |
| Days Max Temp ≥ 90 °F | 0 | 0 | 0 | 0 | 1 | 2 | 6 | 3 | 1 | 0 | 0 | 0 | 13 |
| Days Max Temp ≤ 32 °F | 16 | 12 | 4 | 0 | 0 | 0 | 0 | 0 | 0 | 0 | 1 | 10 | 43 |
| Days Min Temp ≤ 32 °F | 30 | 26 | 24 | 9 | 0 | 0 | 0 | 0 | 0 | 5 | 16 | 27 | 137 |
| Days Min Temp ≤ 0 °F | 6 | 3 | 0 | 0 | 0 | 0 | 0 | 0 | 0 | 0 | 0 | 1 | 10 |
| Heating Degree Days | 1333 | 1151 | 943 | 532 | 215 | 45 | 4 | 15 | 129 | 435 | 721 | 1120 | 6643 |
| Cooling Degree Days | 0 | 0 | 0 | 4 | 35 | 131 | 270 | 211 | 63 | 4 | 1 | 0 | 719 |
| Total Precipitation (") | 1.99 | 1.91 | 2.75 | 3.23 | 3.61 | 3.79 | 3.96 | 3.86 | 3.18 | 3.07 | 3.17 | 2.47 | 36.99 |
| Days ≥ 0.1" Precip | 5 | 5 | 6 | 7 | 8 | 7 | 7 | 7 | 6 | 6 | 7 | 6 | 77 |
| Total Snowfall (") | ***11.8*** | ***10.7*** | 7.8 | ***1.7*** | 0.0 | 0.0 | 0.0 | 0.0 | 0.0 | 0.1 | 3.4 | ***10.4*** | 45.9 |
| Days ≥ 1" Snow Depth | 22 | 18 | 9 | 1 | 0 | 0 | 0 | 0 | 0 | 0 | 2 | 12 | 64 |

## TUPPER LAKE SUNMOUNT *Franklin County* ELEVATION 1680 ft LAT/LONG 44° 14 ' N / 74° 26 ' W

| | JAN | FEB | MAR | APR | MAY | JUN | JUL | AUG | SEP | OCT | NOV | DEC | YEAR |
|---|---|---|---|---|---|---|---|---|---|---|---|---|---|
| Maximum Temp °F | 24.3 | 27.0 | 37.0 | 49.6 | 63.6 | 71.9 | 76.7 | 74.4 | 66.2 | 54.8 | 41.5 | 29.5 | 51.4 |
| Minimum Temp °F | 1.9 | 3.0 | 14.2 | 27.9 | 39.1 | 48.3 | 53.1 | 51.3 | 43.3 | 33.2 | 24.0 | 9.9 | 29.1 |
| Mean Temp °F | 13.1 | 15.0 | 25.6 | 38.7 | 51.4 | 60.1 | 64.9 | 62.9 | 54.8 | 44.0 | 32.8 | 19.7 | 40.3 |
| Days Max Temp ≥ 90 °F | 0 | 0 | 0 | 0 | 0 | 0 | 1 | 0 | 0 | 0 | 0 | 0 | 1 |
| Days Max Temp ≤ 32 °F | 23 | 20 | 10 | 2 | 0 | 0 | 0 | 0 | 0 | 0 | 6 | 18 | 79 |
| Days Min Temp ≤ 32 °F | 31 | 27 | 29 | 22 | 8 | 1 | 0 | 0 | 4 | 15 | 24 | 30 | 191 |
| Days Min Temp ≤ 0 °F | 15 | 13 | 5 | 0 | 0 | 0 | 0 | 0 | 0 | 0 | 0 | 8 | 41 |
| Heating Degree Days | 1605 | 1406 | 1212 | 782 | 423 | 180 | 75 | 114 | 311 | 642 | 960 | 1397 | 9107 |
| Cooling Degree Days | 0 | 0 | 0 | 1 | 5 | 30 | 78 | 52 | 8 | 0 | 0 | 0 | 174 |
| Total Precipitation (") | 2.73 | 2.49 | 2.70 | 3.19 | 3.67 | 3.63 | 4.24 | 4.58 | 4.18 | 3.36 | 3.69 | 3.14 | 41.60 |
| Days ≥ 0.1" Precip | 8 | 7 | 8 | 8 | 9 | 8 | 8 | 9 | 8 | 8 | 10 | 9 | 100 |
| Total Snowfall (") | 23.8 | 19.6 | 15.5 | 6.0 | 0.8 | 0.0 | 0.0 | 0.0 | 0.0 | 0.8 | 10.7 | 25.3 | 102.5 |
| Days ≥ 1" Snow Depth | 30 | 27 | 26 | 8 | 0 | 0 | 0 | 0 | 0 | 1 | 10 | 24 | 126 |

## UTICA ONEIDA CNTY AP *Oneida County* ELEVATION 732 ft LAT/LONG 43° 9 ' N / 75° 23 ' W

| | JAN | FEB | MAR | APR | MAY | JUN | JUL | AUG | SEP | OCT | NOV | DEC | YEAR |
|---|---|---|---|---|---|---|---|---|---|---|---|---|---|
| Maximum Temp °F | 27.9 | 30.1 | 39.9 | 54.4 | 66.8 | 75.4 | 80.2 | 78.1 | 69.5 | 57.5 | 45.2 | 33.1 | 54.8 |
| Minimum Temp °F | 13.0 | 14.4 | 24.1 | 35.7 | 45.9 | 54.8 | 60.1 | 58.8 | 50.8 | 40.1 | 31.7 | 19.6 | 37.4 |
| Mean Temp °F | 20.4 | 22.3 | 32.0 | 45.1 | 56.4 | 65.1 | 70.2 | 68.5 | 60.2 | 48.8 | 38.5 | 26.4 | 46.2 |
| Days Max Temp ≥ 90 °F | 0 | 0 | 0 | 0 | 0 | 1 | 2 | 1 | 0 | 0 | 0 | 0 | 4 |
| Days Max Temp ≤ 32 °F | 19 | 16 | 8 | 0 | 0 | 0 | 0 | 0 | 0 | 0 | 3 | 14 | 60 |
| Days Min Temp ≤ 32 °F | 29 | 26 | 24 | 11 | 1 | 0 | 0 | 0 | 0 | 6 | 16 | 27 | 140 |
| Days Min Temp ≤ 0 °F | 6 | 4 | 1 | 0 | 0 | 0 | 0 | 0 | 0 | 0 | 0 | 2 | 13 |
| Heating Degree Days | 1373 | 1200 | 1016 | 594 | 283 | 82 | 15 | 34 | 181 | 498 | 789 | 1190 | 7255 |
| Cooling Degree Days | 0 | 0 | 0 | 2 | 25 | 79 | 186 | 139 | 39 | 2 | 0 | 0 | 472 |
| Total Precipitation (") | 3.45 | 3.03 | 3.58 | 3.72 | 3.83 | 4.24 | 3.84 | 3.78 | 4.48 | 3.40 | 4.15 | 4.24 | 45.74 |
| Days ≥ 0.1" Precip | 10 | 8 | 9 | 8 | 9 | 8 | 6 | 7 | 7 | 8 | 10 | 11 | 101 |
| Total Snowfall (") | 25.7 | 19.3 | 16.4 | 3.6 | 0.2 | 0.0 | 0.0 | 0.0 | 0.0 | 0.5 | 8.7 | 22.3 | 96.7 |
| Days ≥ 1" Snow Depth | 26 | 25 | 17 | 2 | 0 | 0 | 0 | 0 | 0 | 0 | 5 | 19 | 94 |

## WALTON *Delaware County* ELEVATION 1240 ft LAT/LONG 42° 10' N / 75° 8' W

| | JAN | FEB | MAR | APR | MAY | JUN | JUL | AUG | SEP | OCT | NOV | DEC | YEAR |
|---|---|---|---|---|---|---|---|---|---|---|---|---|---|
| Maximum Temp °F | 30.9 | 34.0 | 43.7 | 57.3 | 69.3 | 77.0 | 81.5 | 79.3 | 71.0 | 60.0 | 47.0 | 35.6 | 57.2 |
| Minimum Temp °F | 11.4 | 12.9 | 22.6 | 32.5 | 42.1 | 50.7 | 55.4 | 54.6 | 47.9 | 36.9 | 29.5 | 18.7 | 34.6 |
| Mean Temp °F | 21.2 | 23.4 | 33.2 | 44.9 | 55.7 | 63.9 | 68.4 | 67.0 | 59.5 | 48.5 | 38.3 | 27.2 | 45.9 |
| Days Max Temp ≥ 90 °F | 0 | 0 | 0 | 0 | 0 | 1 | 2 | 1 | 0 | 0 | 0 | 0 | 4 |
| Days Max Temp ≤ 32 °F | 17 | 13 | 5 | 0 | 0 | 0 | 0 | 0 | 0 | 0 | 2 | 12 | 49 |
| Days Min Temp ≤ 32 °F | 30 | 26 | 25 | 16 | 6 | 0 | 0 | 0 | 2 | 12 | 19 | 28 | 164 |
| Days Min Temp ≤ 0 °F | 7 | 6 | 1 | 0 | 0 | 0 | 0 | 0 | 0 | 0 | 0 | 3 | 17 |
| Heating Degree Days | 1354 | 1167 | 980 | 597 | 296 | 95 | 27 | 44 | 193 | 508 | 794 | 1166 | 7221 |
| Cooling Degree Days | 0 | 0 | 0 | 1 | 14 | 62 | 143 | 107 | 32 | 2 | 0 | 0 | 361 |
| Total Precipitation (") | 2.79 | 2.66 | 3.57 | 3.80 | 4.30 | 4.19 | 4.08 | 4.25 | 3.84 | 3.64 | 4.24 | 3.59 | 44.95 |
| Days ≥ 0.1" Precip | 6 | 6 | 8 | 9 | 9 | 9 | 8 | 8 | 7 | 7 | 8 | 8 | 93 |
| Total Snowfall (") | 23.1 | 20.4 | 17.1 | 6.1 | 0.5 | 0.0 | 0.0 | 0.0 | 0.0 | 0.6 | 8.9 | 20.5 | 97.2 |
| Days ≥ 1" Snow Depth | 27 | 25 | 18 | 3 | 0 | 0 | 0 | 0 | 0 | 0 | 7 | 20 | 100 |

## WANAKENA RANGER SCHO *St. Lawrence County* ELEVATION 1512 ft LAT/LONG 44° 9' N / 74° 54' W

| | JAN | FEB | MAR | APR | MAY | JUN | JUL | AUG | SEP | OCT | NOV | DEC | YEAR |
|---|---|---|---|---|---|---|---|---|---|---|---|---|---|
| Maximum Temp °F | 25.9 | 28.5 | 38.6 | 51.8 | 65.8 | 73.4 | 77.7 | 75.3 | 66.9 | 55.5 | 42.2 | 30.6 | 52.7 |
| Minimum Temp °F | 2.8 | 3.8 | 14.9 | 28.7 | 40.0 | 48.9 | 53.7 | 52.3 | 44.5 | 34.5 | 25.3 | 11.4 | 30.1 |
| Mean Temp °F | 14.3 | 16.2 | 26.7 | 40.3 | 52.9 | 61.2 | 65.7 | 63.8 | 55.7 | 45.0 | 33.8 | 21.0 | 41.4 |
| Days Max Temp ≥ 90 °F | 0 | 0 | 0 | 0 | 0 | 0 | 0 | 0 | 0 | 0 | 0 | 0 | 0 |
| Days Max Temp ≤ 32 °F | 22 | 18 | 9 | 1 | 0 | 0 | 0 | 0 | 0 | 0 | 5 | 17 | 72 |
| Days Min Temp ≤ 32 °F | 30 | 27 | 28 | 21 | 8 | 1 | 0 | 0 | 3 | 14 | 23 | 29 | 184 |
| Days Min Temp ≤ 0 °F | 13 | 12 | 5 | 0 | 0 | 0 | 0 | 0 | 0 | 0 | 0 | 7 | 37 |
| Heating Degree Days | 1564 | 1375 | 1179 | 735 | 378 | 150 | 61 | 91 | 285 | 614 | 930 | 1353 | 8715 |
| Cooling Degree Days | 0 | 0 | 0 | 1 | 10 | 40 | 96 | 65 | 12 | 0 | 0 | 0 | 224 |
| Total Precipitation (") | 2.75 | 2.36 | 2.80 | 3.19 | 3.77 | 3.94 | 4.13 | 4.48 | 4.49 | 3.74 | 4.16 | 3.42 | 43.23 |
| Days ≥ 0.1" Precip | 9 | 7 | 8 | 8 | 9 | 9 | 8 | 9 | 8 | 8 | 11 | 9 | 103 |
| Total Snowfall (") | 32.6 | 26.7 | 21.1 | 7.4 | 1.4 | 0.0 | 0.0 | 0.0 | 0.0 | 1.5 | 14.6 | 29.5 | 134.8 |
| Days ≥ 1" Snow Depth | 30 | 28 | 28 | 9 | 0 | 0 | 0 | 0 | 0 | 1 | 12 | 26 | 134 |

## WARSAW 6 SW *Wyoming County* ELEVATION 1722 ft LAT/LONG 42° 41' N / 78° 12' W

| | JAN | FEB | MAR | APR | MAY | JUN | JUL | AUG | SEP | OCT | NOV | DEC | YEAR |
|---|---|---|---|---|---|---|---|---|---|---|---|---|---|
| Maximum Temp °F | 27.3 | 29.5 | 39.0 | 52.8 | 64.5 | 73.3 | ***77.5*** | 75.4 | 68.2 | 57.0 | 44.2 | 32.6 | 53.4 |
| Minimum Temp °F | 11.3 | 11.9 | 20.3 | 32.4 | 42.4 | 51.7 | ***55.9*** | 54.5 | 47.8 | 37.8 | 29.0 | 18.1 | 34.4 |
| Mean Temp °F | 19.3 | 20.7 | 29.8 | 42.7 | 53.5 | 62.5 | ***66.7*** | 65.0 | 58.0 | 47.4 | 36.6 | 25.3 | 44.0 |
| Days Max Temp ≥ 90 °F | 0 | 0 | 0 | 0 | 0 | 0 | 1 | 0 | 0 | 0 | 0 | 0 | 1 |
| Days Max Temp ≤ 32 °F | 21 | 17 | 11 | 1 | 0 | 0 | 0 | 0 | 0 | 0 | 5 | 16 | 71 |
| Days Min Temp ≤ 32 °F | 30 | 27 | 27 | 17 | ***5*** | 0 | 0 | 0 | 1 | 9 | 21 | 29 | 166 |
| Days Min Temp ≤ 0 °F | 7 | 5 | 2 | 0 | 0 | 0 | 0 | 0 | 0 | 0 | 0 | 2 | 16 |
| Heating Degree Days | 1409 | 1245 | 1085 | 665 | 365 | 126 | ***47*** | 74 | 231 | 540 | 847 | 1224 | 7858 |
| Cooling Degree Days | 0 | 0 | 0 | 3 | 17 | 50 | 115 | 76 | 25 | 1 | 0 | 0 | 287 |
| Total Precipitation (") | 2.69 | 2.25 | 2.93 | 3.32 | 3.51 | 4.20 | 3.76 | 3.95 | 4.31 | 3.54 | 4.04 | 3.50 | 42.00 |
| Days ≥ 0.1" Precip | 9 | 8 | 9 | 9 | 9 | 9 | 8 | 8 | 9 | 9 | 10 | 12 | 109 |
| Total Snowfall (") | ***25.0*** | 20.4 | ***15.2*** | 4.3 | 0.4 | 0.0 | 0.0 | 0.0 | 0.0 | 0.7 | ***12.3*** | ***21.6*** | 99.9 |
| Days ≥ 1" Snow Depth | ***27*** | 24 | 18 | 4 | 0 | 0 | 0 | 0 | 0 | 0 | 8 | ***22*** | 103 |

## WATERTOWN *Jefferson County* ELEVATION 502 ft LAT/LONG 43° 58' N / 75° 52' W

| | JAN | FEB | MAR | APR | MAY | JUN | JUL | AUG | SEP | OCT | NOV | DEC | YEAR |
|---|---|---|---|---|---|---|---|---|---|---|---|---|---|
| Maximum Temp °F | 27.9 | 29.8 | 39.8 | 53.3 | 65.6 | 74.6 | 79.8 | 77.9 | 69.7 | 57.7 | 45.9 | 33.5 | 54.6 |
| Minimum Temp °F | 8.4 | 10.0 | 21.7 | 34.9 | 46.0 | 55.6 | 60.8 | 59.4 | 51.0 | 39.6 | 30.7 | 16.8 | 36.2 |
| Mean Temp °F | 18.2 | 20.0 | 30.8 | 44.1 | 55.8 | 65.1 | 70.4 | 68.6 | 60.4 | 48.7 | 38.3 | 25.2 | 45.5 |
| Days Max Temp ≥ 90 °F | 0 | 0 | 0 | 0 | 0 | 0 | 2 | 1 | 0 | 0 | 0 | 0 | 3 |
| Days Max Temp ≤ 32 °F | 18 | 16 | 8 | 1 | 0 | 0 | 0 | 0 | 0 | 0 | 3 | 13 | 59 |
| Days Min Temp ≤ 32 °F | 29 | 26 | 25 | 13 | 2 | 0 | 0 | 0 | 0 | 7 | 17 | 28 | 147 |
| Days Min Temp ≤ 0 °F | 10 | 8 | 2 | 0 | 0 | 0 | 0 | 0 | 0 | 0 | 0 | 4 | 24 |
| Heating Degree Days | 1445 | 1266 | 1055 | 623 | 302 | 85 | 19 | 35 | 176 | 502 | 793 | 1228 | 7529 |
| Cooling Degree Days | 0 | 0 | 0 | 5 | 25 | 87 | 196 | 155 | 43 | 3 | 0 | 0 | 514 |
| Total Precipitation (") | 3.15 | 2.52 | 2.66 | 3.07 | 3.32 | 3.63 | 2.81 | 4.07 | 4.41 | 3.71 | 4.59 | 3.91 | 41.85 |
| Days ≥ 0.1" Precip | 9 | 7 | 7 | 8 | 8 | 7 | 6 | 7 | 8 | 9 | 11 | 10 | 97 |
| Total Snowfall (") | 31.7 | 22.4 | 13.2 | 3.3 | 0.2 | 0.0 | 0.0 | 0.0 | 0.0 | 0.5 | 7.9 | 27.2 | 106.4 |
| Days ≥ 1" Snow Depth | 26 | 24 | 17 | 3 | 0 | 0 | 0 | 0 | 0 | 0 | 5 | 20 | 95 |

## WATERTOWN AP *Jefferson County* ELEVATION 331 ft LAT/LONG 44° 0 ' N / 76° 1 ' W

| | JAN | FEB | MAR | APR | MAY | JUN | JUL | AUG | SEP | OCT | NOV | DEC | YEAR |
|---|---|---|---|---|---|---|---|---|---|---|---|---|---|
| Maximum Temp °F | 28.0 | 29.8 | 39.9 | 53.7 | 65.1 | 73.9 | 79.5 | 77.8 | 69.7 | 58.0 | 45.7 | 33.5 | 54.6 |
| Minimum Temp °F | 8.6 | 9.9 | 21.0 | 33.3 | 43.3 | 52.2 | 57.7 | 56.4 | 48.2 | 38.3 | 29.9 | 16.3 | 34.6 |
| Mean Temp °F | 18.4 | 19.9 | 30.5 | 43.5 | 54.2 | 63.1 | 68.6 | 67.1 | 59.0 | 48.2 | 37.8 | 25.0 | 44.6 |
| Days Max Temp ≥ 90 °F | 0 | 0 | 0 | 0 | 0 | 0 | 1 | 1 | 0 | 0 | 0 | 0 | 2 |
| Days Max Temp ≤ 32 °F | 18 | 16 | 8 | 0 | 0 | 0 | 0 | 0 | 0 | 0 | 2 | 13 | 57 |
| Days Min Temp ≤ 32 °F | 29 | 26 | 26 | 15 | 3 | 0 | 0 | 0 | 2 | 10 | 18 | 27 | 156 |
| Days Min Temp ≤ 0 °F | 9 | 8 | 2 | 0 | 0 | 0 | 0 | 0 | 0 | 0 | 0 | 5 | 24 |
| Heating Degree Days | 1438 | 1268 | 1063 | 639 | 339 | 116 | 27 | 50 | 210 | 520 | 809 | 1235 | 7714 |
| Cooling Degree Days | 0 | 0 | 0 | 2 | 13 | 48 | 143 | 113 | 33 | 3 | 0 | 0 | 355 |
| Total Precipitation (") | 2.13 | 2.14 | 2.06 | 2.65 | 2.73 | 2.80 | 2.00 | 3.02 | 3.51 | 2.80 | 3.33 | 2.75 | 31.92 |
| Days ≥ 0.1" Precip | 7 | 6 | 6 | 7 | 7 | 6 | 5 | 6 | 7 | 7 | 8 | 8 | 80 |
| Total Snowfall (") | 24.1 | 18.4 | 10.0 | 2.5 | 0.0 | 0.0 | 0.0 | 0.0 | 0.0 | 0.4 | 6.1 | 20.6 | 82.1 |
| Days ≥ 1" Snow Depth | 24 | 23 | 14 | 2 | 0 | 0 | 0 | 0 | 0 | 0 | 4 | 17 | 84 |

## WEST POINT *Orange County* ELEVATION 322 ft LAT/LONG 41° 23 ' N / 73° 58 ' W

| | JAN | FEB | MAR | APR | MAY | JUN | JUL | AUG | SEP | OCT | NOV | DEC | YEAR |
|---|---|---|---|---|---|---|---|---|---|---|---|---|---|
| Maximum Temp °F | 35.3 | 38.6 | 48.7 | 62.0 | 72.9 | 81.3 | 86.2 | 83.8 | 75.4 | 63.5 | 51.8 | 39.9 | 61.6 |
| Minimum Temp °F | 19.6 | 21.1 | 29.6 | 39.7 | 49.9 | 58.8 | 63.9 | 62.6 | 55.4 | 44.7 | 36.2 | 25.6 | 42.3 |
| Mean Temp °F | 27.5 | 29.9 | 39.2 | 50.9 | 61.4 | 70.1 | 75.1 | 73.2 | 65.4 | 54.1 | 44.0 | 32.8 | 52.0 |
| Days Max Temp ≥ 90 °F | 0 | 0 | 0 | 0 | 1 | 4 | 9 | 5 | 1 | 0 | 0 | 0 | 20 |
| Days Max Temp ≤ 32 °F | 11 | 7 | 1 | 0 | 0 | 0 | 0 | 0 | 0 | 0 | 0 | 6 | 25 |
| Days Min Temp ≤ 32 °F | 27 | 25 | 19 | 5 | 0 | 0 | 0 | 0 | 0 | 2 | 11 | 24 | 113 |
| Days Min Temp ≤ 0 °F | 1 | 0 | 0 | 0 | 0 | 0 | 0 | 0 | 0 | 0 | 0 | 0 | 1 |
| Heating Degree Days | 1157 | 985 | 795 | 425 | 155 | 20 | 1 | 5 | 75 | 338 | 624 | 993 | 5573 |
| Cooling Degree Days | 0 | 0 | 1 | 8 | 54 | 176 | 324 | 257 | 90 | 6 | 0 | 0 | 916 |
| Total Precipitation (") | 3.30 | 3.17 | 3.95 | 4.01 | 4.83 | 4.10 | 4.10 | 4.36 | 4.20 | 3.92 | 4.43 | 3.99 | 48.36 |
| Days ≥ 0.1" Precip | 7 | 6 | 7 | 7 | 8 | 7 | 7 | 7 | 6 | 6 | 7 | 7 | 82 |
| Total Snowfall (") | 10.2 | ***10.7*** | 6.7 | 0.4 | 0.0 | 0.0 | 0.0 | 0.0 | 0.0 | 0.0 | 0.9 | 6.6 | 35.5 |
| Days ≥ 1" Snow Depth | 14 | 13 | 4 | 0 | 0 | 0 | 0 | 0 | 0 | 0 | ***0*** | 6 | 37 |

## WESTFIELD 3 SW *Chautauqua County* ELEVATION 1050 ft LAT/LONG 42° 17 ' N / 79° 36 ' W

| | JAN | FEB | MAR | APR | MAY | JUN | JUL | AUG | SEP | OCT | NOV | DEC | YEAR |
|---|---|---|---|---|---|---|---|---|---|---|---|---|---|
| Maximum Temp °F | 31.0 | 32.8 | 42.1 | 54.4 | 65.8 | 74.7 | 78.8 | 76.9 | 70.0 | 58.8 | 47.5 | 36.7 | 55.8 |
| Minimum Temp °F | 18.0 | 17.8 | 26.1 | 37.0 | 47.9 | 57.2 | 62.3 | 61.2 | 54.8 | 44.1 | 35.0 | 24.8 | 40.5 |
| Mean Temp °F | 24.5 | 25.4 | 34.1 | 45.7 | 56.8 | 66.0 | 70.6 | 69.1 | 62.5 | 51.5 | 41.3 | 30.8 | 48.2 |
| Days Max Temp ≥ 90 °F | 0 | 0 | 0 | 0 | 0 | 0 | 1 | 0 | 0 | 0 | 0 | 0 | 1 |
| Days Max Temp ≤ 32 °F | 17 | 14 | 7 | 1 | 0 | 0 | 0 | 0 | 0 | 0 | 2 | 11 | 52 |
| Days Min Temp ≤ 32 °F | 28 | 25 | 24 | 10 | 1 | 0 | 0 | 0 | 0 | 2 | 12 | 25 | 127 |
| Days Min Temp ≤ 0 °F | 2 | 2 | 0 | 0 | 0 | 0 | 0 | 0 | 0 | 0 | 0 | 0 | 4 |
| Heating Degree Days | 1248 | 1114 | 951 | 578 | 278 | 72 | 12 | 23 | 130 | 418 | 706 | 1055 | 6585 |
| Cooling Degree Days | 0 | 0 | 1 | 7 | 37 | 101 | 208 | 159 | 61 | 4 | 0 | 0 | 578 |
| Total Precipitation (") | 2.39 | 2.21 | 3.04 | 3.52 | 3.64 | 4.36 | 4.11 | 4.54 | 5.59 | 4.74 | 4.80 | 3.62 | 46.56 |
| Days ≥ 0.1" Precip | 7 | 6 | 8 | 8 | 8 | 8 | 7 | 8 | 8 | 10 | 11 | 10 | 99 |
| Total Snowfall (") | 22.2 | 16.5 | 12.0 | 3.2 | 0.3 | 0.0 | 0.0 | 0.0 | 0.0 | 0.5 | 9.0 | 24.9 | 88.6 |
| Days ≥ 1" Snow Depth | 25 | 22 | 14 | 2 | 0 | 0 | 0 | 0 | 0 | 0 | 5 | 17 | 85 |

## WHITEHALL *Washington County* ELEVATION 121 ft LAT/LONG 43° 33 ' N / 73° 24 ' W

| | JAN | FEB | MAR | APR | MAY | JUN | JUL | AUG | SEP | OCT | NOV | DEC | YEAR |
|---|---|---|---|---|---|---|---|---|---|---|---|---|---|
| Maximum Temp °F | 29.0 | 32.4 | 43.0 | 57.9 | 71.3 | 80.1 | 84.8 | 81.6 | 72.2 | 59.9 | 46.9 | 34.2 | 57.8 |
| Minimum Temp °F | 8.9 | 10.9 | 22.9 | 35.2 | 46.5 | 55.4 | 59.9 | 58.3 | 50.6 | 40.0 | 31.3 | 18.0 | 36.5 |
| Mean Temp °F | 18.9 | 21.6 | 33.0 | 46.6 | 59.0 | 67.8 | 72.4 | 70.0 | 61.4 | 49.9 | 39.1 | 26.2 | 47.2 |
| Days Max Temp ≥ 90 °F | 0 | 0 | 0 | 0 | 1 | 3 | 6 | 2 | 0 | 0 | 0 | 0 | 12 |
| Days Max Temp ≤ 32 °F | 18 | 14 | 4 | 0 | 0 | 0 | 0 | 0 | 0 | 0 | 1 | 12 | 49 |
| Days Min Temp ≤ 32 °F | 30 | 27 | 25 | 13 | 1 | 0 | 0 | 0 | 0 | 7 | 17 | 28 | 148 |
| Days Min Temp ≤ 0 °F | 9 | 6 | 1 | 0 | 0 | 0 | 0 | 0 | 0 | 0 | 0 | 3 | 19 |
| Heating Degree Days | 1423 | 1218 | 986 | 550 | 212 | 43 | 5 | 19 | 149 | 463 | 770 | 1197 | 7035 |
| Cooling Degree Days | ***0*** | 0 | 0 | 3 | 30 | 122 | 239 | 178 | 52 | 3 | 0 | 0 | 627 |
| Total Precipitation (") | 2.64 | 2.32 | 2.88 | 3.09 | 3.84 | 3.16 | 3.50 | 4.04 | 3.79 | 3.30 | 3.63 | 2.96 | 39.15 |
| Days ≥ 0.1" Precip | 6 | 5 | 7 | 7 | 8 | 7 | 7 | 7 | 6 | 6 | 8 | 7 | 81 |
| Total Snowfall (") | 16.7 | 13.9 | 11.2 | 2.2 | 0.0 | 0.0 | 0.0 | 0.0 | 0.0 | 0.0 | ***4.1*** | 16.5 | 64.6 |
| Days ≥ 1" Snow Depth | na | na | na | na | 0 | 0 | 0 | 0 | 0 | 0 | na | na | na |

## YORKTOWN HEIGHTS 1 W *Westchester County* ELEVATION 669 ft LAT/LONG 41° 16 ' N / 73° 48 ' W

| | JAN | FEB | MAR | APR | MAY | JUN | JUL | AUG | SEP | OCT | NOV | DEC | YEAR |
|---|---|---|---|---|---|---|---|---|---|---|---|---|---|
| Maximum Temp °F | 33.5 | 36.7 | 46.2 | 58.6 | 69.5 | 77.1 | 82.0 | 80.3 | 73.1 | 62.1 | 50.6 | 38.7 | 59.0 |
| Minimum Temp °F | 17.3 | 19.4 | 27.9 | 38.6 | 48.6 | 57.2 | 62.5 | 61.0 | 53.3 | 42.4 | 34.3 | 23.6 | 40.5 |
| Mean Temp °F | 25.4 | 28.1 | 37.1 | 48.6 | 59.1 | 67.2 | 72.3 | 70.7 | 63.2 | 52.3 | 42.4 | 31.1 | 49.8 |
| Days Max Temp ≥ 90 °F | 0 | 0 | 0 | 0 | 0 | 1 | 3 | 1 | 1 | 0 | 0 | 0 | 6 |
| Days Max Temp ≤ 32 °F | 14 | 10 | 2 | 0 | 0 | 0 | 0 | 0 | 0 | 0 | 1 | 8 | 35 |
| Days Min Temp ≤ 32 °F | 29 | 25 | 22 | 6 | 0 | 0 | 0 | 0 | 0 | 4 | 14 | 26 | 126 |
| Days Min Temp ≤ 0 °F | 2 | 1 | 0 | 0 | 0 | 0 | 0 | 0 | 0 | 0 | 0 | 0 | 3 |
| Heating Degree Days | 1221 | 1037 | 859 | 492 | 208 | 46 | 6 | 14 | 110 | 392 | 670 | 1043 | 6098 |
| Cooling Degree Days | 0 | 0 | 1 | 4 | 36 | 125 | 249 | 195 | 62 | 5 | 0 | 0 | 677 |
| Total Precipitation (") | 3.63 | 3.17 | 4.20 | 4.42 | 4.90 | 4.35 | 4.46 | 4.77 | 4.27 | 3.82 | 4.58 | 4.03 | 50.60 |
| Days ≥ 0.1" Precip | 7 | 6 | 7 | 8 | 8 | 7 | 7 | 7 | 6 | 5 | 7 | 7 | 82 |
| Total Snowfall (") | 10.3 | 10.4 | 7.2 | 1.3 | 0.1 | 0.0 | 0.0 | 0.0 | 0.0 | 0.3 | 1.8 | 6.7 | 38.1 |
| Days ≥ 1" Snow Depth | 19 | 17 | 9 | 1 | 0 | 0 | 0 | 0 | 0 | 0 | 2 | 10 | 58 |

## JANUARY MINIMUM TEMPERATURE °F

### LOWEST

| | | |
|---|---|---|
| 1 | Stillwater Rservr | 0.7 |
| 2 | Old Forge | 1.0 |
| 3 | Tupper Lake | 1.9 |
| 4 | Indian Lake | 2.0 |
| 5 | Wanakena | 2.8 |
| 6 | Lake Placid | 3.2 |
| 7 | Canton | 3.6 |
| 8 | Massena | 3.8 |
| 9 | Chasm Falls | 4.7 |
| 10 | Gouverneur | 5.5 |
| 11 | Chazy | 5.6 |
| 12 | Lawrenceville | 5.8 |
| 13 | Glens Falls-Ap | 6.3 |
| 14 | Salem | 6.4 |
| 15 | Lowville | 6.6 |
| 16 | Ogdensburg | 6.8 |
| 17 | Dannemora | 7.0 |
| 18 | Boonville | 7.2 |
| 19 | Peru | 7.3 |
| 20 | Plattsburgh | 7.7 |
| 21 | Glens Falls-Farm | 7.8 |
| 22 | Slide Mountain | 8.3 |
| 23 | Watertown | 8.4 |
| 24 | Watertown-Ap | 8.6 |
| 25 | Camden | 8.7 |

### HIGHEST

| | | |
|---|---|---|
| 1 | New Yrk-LaGrdia | 25.8 |
| 2 | New York | 25.5 |
| | New York-Ctrl Pk | 25.5 |
| 4 | New York-JFK | 25.3 |
| 5 | Mineola | 24.7 |
| 6 | Setauket | 23.6 |
| 7 | Riverhead | 23.4 |
| 8 | Greenport | 22.7 |
| 9 | Dobbs Ferry | 22.5 |
| 10 | Bridgehampton | 22.3 |
| 11 | Patchogue | 20.9 |
| 12 | West Point | 19.6 |
| 13 | Fredonia | 18.4 |
| 14 | Westfield | 18.0 |
| 15 | Yorktown Heights | 17.3 |
| 16 | Buffalo | 17.2 |
| | Middletown | 17.2 |
| 18 | Suffern | 17.0 |
| 19 | Albion | 16.5 |
| 20 | Gowanda | 16.4 |
| | Mohonk Lake | 16.4 |
| 22 | Glenham | 16.3 |
| | Lockport-2 NE | 16.3 |
| | Lockport-4 NE | 16.3 |
| | Rochester | 16.3 |

## JULY MAXIMUM TEMPERATURE °F

### HIGHEST

| | | |
|---|---|---|
| 1 | West Point | 86.2 |
| 2 | Glenham | 86.0 |
| 3 | New York-Ctrl Pk | 85.4 |
| 4 | Dobbs Ferry | 85.3 |
| 5 | Hudson | 85.1 |
| 6 | New York | 84.8 |
| | Whitehall | 84.8 |
| 8 | Port Jervis | 84.5 |
| 9 | New Yrk-LaGrdia | 84.4 |
| 10 | Suffern | 84.0 |
| 11 | Middletown | 83.9 |
| 12 | Poughkeepsie | 83.8 |
| | Saratoga Springs | 83.8 |
| 14 | Troy L & D | 83.7 |
| 15 | Riverhead | 83.5 |
| 16 | Patchogue | 83.3 |
| 17 | Glens Falls-Farm | 83.0 |
| | New York-JFK | 83.0 |
| 19 | Setauket | 82.9 |
| 20 | Albany | 82.8 |
| 21 | Dansville | 82.6 |
| | Elmira | 82.6 |
| | Mineola | 82.6 |
| 24 | Salem | 82.4 |
| 25 | Albion | 82.3 |

### LOWEST

| | | |
|---|---|---|
| 1 | Slide Mountain | 72.0 |
| 2 | Stillwater Rservr | 75.3 |
| 3 | Indian Lake | 75.7 |
| 4 | Boonville | 76.0 |
| 5 | Lake Placid | 76.3 |
| 6 | Old Forge | 76.4 |
| 7 | Tupper Lake | 76.7 |
| 8 | Warsaw | 77.5 |
| 9 | Wanakena | 77.7 |
| 10 | Franklinville | 77.9 |
| 11 | Cherry Valley | 78.3 |
| | Grafton | 78.3 |
| 13 | Bolivar | 78.4 |
| 14 | Allegany | 78.7 |
| | Binghamton | 78.7 |
| 16 | Chasm Falls | 78.8 |
| | Westfield | 78.8 |
| 18 | Arcade | 78.9 |
| 19 | Little Valley | 79.0 |
| 20 | Alfred | 79.1 |
| | Dannemora | 79.1 |
| 22 | Lowville | 79.2 |
| 23 | Canton | 79.3 |
| | Liberty | 79.3 |
| 25 | Colden | 79.5 |

## ANNUAL PRECIPITATION (")

### HIGHEST

| | | |
|---|---|---|
| 1 | Slide Mountain | 60.97 |
| 2 | Boonville | 60.59 |
| 3 | Camden | 51.88 |
| 4 | Yorktown Heights | 50.60 |
| 5 | Dobbs Ferry | 50.56 |
| 6 | Old Forge | 50.34 |
| 7 | Suffern | 50.15 |
| 8 | Little Valley | 49.99 |
| 9 | Patchogue | 49.40 |
| 10 | Mohonk Lake | 48.76 |
| 11 | Liberty | 48.69 |
| 12 | New York-Ctrl Pk | 48.37 |
| 13 | West Point | 48.36 |
| 14 | Stillwater Rservr | 47.77 |
| 15 | Colden | 47.72 |
| 16 | Westfield | 46.56 |
| 17 | Bridgehampton | 46.07 |
| 18 | Grafton | 45.81 |
| 19 | Utica | 45.74 |
| 20 | Riverhead | 45.60 |
| 21 | Allegany | 45.32 |
| 22 | Greenport | 45.27 |
| 23 | Setauket | 45.16 |
| 24 | Glens Falls-Farm | 45.06 |
| 25 | New York | 44.99 |

### LOWEST

| | | |
|---|---|---|
| 1 | Peru | 29.75 |
| 2 | Mount Morris | 30.18 |
| 3 | Dansville | 31.02 |
| 4 | Bath | 31.47 |
| 5 | Watertown-Ap | 31.92 |
| 6 | Rochester | 32.55 |
| 7 | Canandaigua | 32.94 |
| 8 | Plattsburgh | 33.06 |
| 9 | Ogdensburg | 33.69 |
| 10 | Hemlock | 33.74 |
| 11 | Elmira | 34.28 |
| 12 | Batavia | 34.95 |
| 13 | Lawrenceville | 35.17 |
| 14 | Salem | 35.21 |
| 15 | Massena | 35.34 |
| 16 | Albion | 35.89 |
| 17 | Dannemora | 35.96 |
| 18 | Ithaca | 36.13 |
| 19 | Canton | 36.14 |
| 20 | Aurora | 36.41 |
| 21 | Lockport-4 NE | 36.56 |
| 22 | Angelica | 36.62 |
| 23 | Troy L & D | 36.99 |
| 24 | Gouverneur | 37.07 |
| 25 | Glens Falls-Ap | 37.08 |

## ANNUAL SNOWFALL (")

### HIGHEST

| | | |
|---|---|---|
| 1 | Boonville | 229.8 |
| 2 | Old Forge | 225.5 |
| 3 | Stillwater Rservr | 192.2 |
| 4 | Colden | 162.6 |
| 5 | Oswego | 153.9 |
| 6 | Little Valley | 140.5 |
| 7 | Wanakena | 134.8 |
| 8 | Camden | 127.3 |
| 9 | Chasm Falls | 122.6 |
| 10 | Lowville | 121.0 |
| 11 | Cherry Valley | 120.0 |
| 12 | Syracuse | 116.9 |
| 13 | Arcade | 108.0 |
| 14 | Watertown | 106.4 |
| 15 | Franklinville | 103.0 |
| 16 | Tupper Lake | 102.5 |
| 17 | Slide Mountain | 101.6 |
| 18 | Warsaw | 99.9 |
| 19 | Walton | 97.2 |
| 20 | Rochester | 96.8 |
| 21 | Utica | 96.7 |
| 22 | Buffalo | 92.9 |
| 23 | Cortland | 90.2 |
| 24 | Canton | 89.7 |
| | Lockport-2 NE | 89.7 |

### LOWEST

| | | |
|---|---|---|
| 1 | Setauket | 16.8 |
| 2 | New York-JFK | 21.8 |
| 3 | New Yrk-LaGrdia | 22.9 |
| 4 | New York | 23.1 |
| | New York-Ctrl Pk | 23.1 |
| 6 | Mineola | 23.8 |
| 7 | Bridgehampton | 26.7 |
| 8 | Riverhead | 28.0 |
| 9 | Patchogue | 31.1 |
| 10 | Dobbs Ferry | 33.6 |
| 11 | West Point | 35.5 |
| 12 | Poughkeepsie | 37.4 |
| 13 | Yorktown Heights | 38.1 |
| 14 | Glenham | 39.0 |
| 15 | Elmira | 43.7 |
| 16 | Millbrook | 44.2 |
| 17 | Bath | 45.2 |
| 18 | Port Jervis | 45.3 |
| 19 | Troy L & D | 45.9 |
| 20 | Salem | 49.2 |
| 21 | Dansville | 53.4 |
| 22 | Spencer | 56.8 |
| 23 | Peru | 59.0 |
| 24 | Angelica | 59.2 |
| 25 | Aurora | 59.8 |

# NORTH CAROLINA

PHYSICAL FEATURES. North Carolina lies between 33½° and 37° N. latitude and between 75° and 84½° W. longitude. The span of longitude is greater than that of any other state east of the Mississippi River. The greatest length from east to west is 503 miles. The greatest breadth from north to south is 187 miles. The total area is 52,712 square miles: 49,142 square miles of land and 3,570 square miles of water.

The range of altitude is also the greatest of any eastern state. North Carolina rises from sea level along the Atlantic coast to 6,684 feet at the summit of Mount Mitchell, the highest peak in the eastern United States. Mount Mitchell is in the heart of the Blue Ridge Range. This Range, along with the Great Smokies, lies partly in North Carolina and partly in Tennessee and forms the highest part of the Appalachian Mountains.

The three principal physiographic divisions of the eastern United States are particularly well developed in North Carolina. Beginning in the east, they are: the Coastal Plain, the Piedmont, and the Mountains. The land and water areas of the Coastal Plain Division comprise nearly half the area of the State. The tidewater portion is generally flat and swampy, while the interior is gently sloping and, for the most part, naturally well drained. The Piedmont Division rises gently from about 200 feet at the fall line to near 1,500 feet at the base of the mountains; its area is about one-third of the State. The land is mostly gently rolling. There are several ranges of steeper hills. The Mountain Division is the smallest of the three, little more than one-fifth of the State's area. In elevation it ranges downward from Mount Mitchell's peak to about 1,000 feet above mean sea level in the lowest valleys. There are more than 40 peaks higher than 6,000 feet and about 80 others over 5,000 feet high.

North Carolina rivers fall into two groups: those that flow into the Atlantic Ocean and those that drain westward into the Mississippi River system. The two are separated by a ridge averaging 2,200 feet above mean sea level. A second chain of mountains ranging up to 6,000 feet marks the western boundary of the State. Most of the State, including the Coastal Plain, the Piedmont, and the eastern and southern slopes, drains into the Atlantic Ocean. The principal rivers involved are the Roanoke, Tar, Neuse, Cape Fear, Yadkin, and Catawba. The main stream draining the extreme western part of North Carolina is the French Broad River. The northern mountains are drained by streams flowing into the Ohio River system. All eventually reach the Mississippi.

GENERAL CLIMATE. North Carolina has the most varied climate of any eastern state. This is due mainly to its wide range in elevation and distance from the ocean. In all seasons of the year the average temperature varies more than 20° from the lower coast to the highest mountain elevations. Altitude also has an important effect on rainfall. The rainiest part of the eastern United States, with an annual average of more than 80 inches, is in southwestern North Carolina where moist southerly winds are forced upward in passing over the mountain barrier.

In winter the greater part of North Carolina is partially protected by the mountain ranges from the frequent outbreaks of cold which move southeastward across the Central States. Such outbreaks often spread southward all the way to the Gulf of Mexico without attaining strength and depth to cross the Appalachian Range. When cold waves do break across they are usually modified by the crossing and the descent on the eastern slopes. The temperature drops to around 10° over central North Carolina once or twice during an average winter. Near the coast a comparable figure is some 10 degrees higher, and in the upper mountains 10 degrees lower. Temperatures as low as 0° are rare outside the mountains, but have occurred at one time or another throughout the western part of the State.

Winter temperatures in the eastern Coastal Plain are modified by the proximity of the Atlantic Ocean. This effect raises the average winter temperature and reduces the average day-to-night range. The Gulf Stream, contrary to popular opinion, has little direct effect on North Carolina temperatures, even on the immediate coast. The Stream lies some 50 miles offshore at its nearest point. The southern reaches of the cold Labrador Current pass between the Gulf Stream and the North Carolina coast. This offsets any warming effect the Stream might otherwise have on coastal temperatures. The meeting of the two opposing currents does provide a breeding ground for rough weather. Not infrequently low pressure storms having their origin there develop major proportions, causing rain on the North Carolina coast and over states to the north.

In spring the storm systems that bring cold weather southward reach North Carolina less forcefully than in winter, and temperatures begin to modify. Day-to-day variations in temperature are less pronounced, and warm weather is more likely to occur in conjunction with fair weather. During the summer, when the drying of the air is sufficient to keep cloudiness at a minimum for several days, temperatures may occasionally reach 100° or a little higher in interior sections at elevations below 1,500 feet. Ordinarily, however, summer cloudiness develops to limit the sun's heating while temperatures are still in the 90° range. Autumn is the season of most rapidly changing temperature, the daily downward trend being greater than the corresponding rise in spring. The dropoff is most rapid in October and continues almost as fast in November.

PRECIPITATION. There are no distinct wet and dry seasons in North Carolina. There is some seasonal variation in average precipitation. Summer rainfall is normally the greatest, and July the wettest month. Since the rain at this time of year comes mostly with thunderstorms and convective showers, it is also more variable than at other seasons. Daily showers are not uncommon, nor are periods of one or two weeks without rain. Autumn is the driest season, and October the driest month. Precipitation in winter and spring occurs mostly with migratory low pressure storms. It appears with greater regularity and more even distribution than summer showers.

Winter precipitation usually occurs with southerly through easterly winds, and is seldom associated with very cold weather. Snow and sleet occur on an average of once or twice a year near the coast, and not much more often over the southeastern half of the State. Over the Mountains and western Piedmont frozen precipitation sometimes occurs with interior low pressure storms. In the extreme west it can happen with a cold front passage from northwest. Average winter snowfall ranges from about 1 inch per year on the Outer Banks and the lower coast, to about 9 inches in the northern Piedmont and southern Mountains. Some of the higher mountain peaks and upper slopes receive an average of nearly 50 inches a year.

OTHER CLIMATIC ELEMENTS. Relative humidity may vary greatly from day to day and even from hour to hour, especially in winter. The average relative humidity, however, does not vary greatly from season to season, there being a slight tendency for highest averages in winter and lowest in spring. The lowest relative humidities are found over the southern Piedmont; the highest are along the immediate coast.

Sunshine is abundant, the average annual percentage of possible ranging from 60 to 65 percent at most recording points. Measurable rain falls on about 120 days. Prevailing winds blow from southwest 10 months of the year, and from northeast during September and October. The average wind speed for interior locations is about 8 m.p.h., for coastal points about 12.

STORMS AND FLOODS. Intense rainstorms occur in the precipitous mountain terrain, especially in the southern portion. Streams here rise quickly to flood, and almost as quickly subside when rain ends. Floods occur frequently, affecting some part of North Carolina each year. Floods may occur at any season, but are most frequent in early spring, summer, and early fall. Rains associated with West Indian hurricanes are the main cause of summer and fall floods. The greatest economic loss entailed in North Carolina because of stormy weather is that due to summer thunderstorms. These usually affect only limited areas. In any given locality, 40 to 50 days with thunderstorms may be expected in a year.

North Carolina is outside the principal tornado area of the United States, experiencing an average of less than 4 per year. Tropical hurricanes come close enough to influence North Carolina weather about twice in an average year. Only about once in 10 years, on the average, does this type storm strike the State with sufficient force to do much damage.

## COUNTY INDEX

*New Hanover County*
WILMINGTON 7 N
WILMINGTON NEW HANVR

*Northampton County*
JACKSON

*Orange County*
CHAPEL HILL 2 W

*Pasquotank County*
ELIZABETH CITY

*Pender County*
WILLARD 4 SW

*Person County*
ROXBORO 7 ESE

*Pitt County*
GREENVILLE

*Polk County*
TRYON

*Randolph County*
ASHEBORO 2 W

*Richmond County*
HAMLET

*Robeson County*
LUMBERTON 3 SE

*Rockingham County*
REIDSVILLE 2 NW

*Rowan County*
SALISBURY

*Scotland County*
LAURINBURG

*Stanly County*
ALBEMARLE

*Stokes County*
DANBURY 1 NW

*Surry County*
MOUNT AIRY

*Swain County*
OCONALUFTEE

*Transylvania County*
PISGAH FOREST 1 N

*Union County*
MONROE 4 SE

*Vance County*
HENDERSON 2 NNW

*Wake County*
RALEIGH 4 SW
RALEIGH DURHAM AP
RALEIGH STATE UNIV

*Warren County*
ARCOLA

*Washington County*
PLYMOUTH 5 E

*Watauga County*
BLOWING ROCK 1 NW

*Wayne County*
GOLDSBORO S-J AFB

*Wilkes County*
NORTH WILKESBORO
W KERR SCOTT RESERVO

*Wilson County*
WILSON 3 SW

*Yadkin County*
YADKINVILLE 6 E

*Yancey County*
CELO 2 S

## ELEVATION INDEX

| FEET | STATION NAME |
|---|---|
| 0 | NEW HOLLAND |
| 8 | CAPE HATTERAS NWS |
| 10 | BELHAVEN |
| 10 | CEDAR ISLAND |
| 10 | ELIZABETH CITY |
| 10 | HATTERAS |
| 10 | MOREHEAD CITY 2 WNW |
| 20 | EDENTON |
| 20 | NEW BERN CRAVEN CO |
| 20 | PLYMOUTH 5 E |
| 20 | SOUTHPORT 5 N |
| 20 | WILLIAMSTON 1 E |
| 30 | GREENVILLE |
| 36 | WILMINGTON NEW HANVR |
| 39 | TARBORO 1 S |
| 39 | WILMINGTON 7 N |
| 49 | KINSTON 5 SE |
| 49 | LEWISTON |
| 59 | ELIZABETHTOWN LOCK 2 |
| 59 | WILLARD 4 SW |
| 79 | WHITEVILLE 7 NW |
| 102 | FAYETTEVILLE |
| 109 | GOLDSBORO S-J AFB |
| 110 | ROCKY MOUNT 8 ESE |
| 131 | JACKSON |
| 131 | LUMBERTON 3 SE |
| 151 | SMITHFIELD |
| 151 | WILSON 3 SW |
| 200 | DUNN 4 NW |
| 210 | ROANOKE RAPIDS |
| 230 | LAURINBURG |
| 330 | ARCOLA |
| 331 | CLAYTON 3 W |
| 351 | HAMLET |
| 381 | LOUISBURG |
| 400 | RALEIGH STATE UNIV |
| 410 | DURHAM |
| 420 | RALEIGH 4 SW |
| 420 | WADESBORO |
| 443 | RALEIGH DURHAM AP |
| 479 | HENDERSON 2 NNW |
| 502 | CHAPEL HILL 2 W |
| 502 | OXFORD 1 E |
| 586 | MONROE 4 SE |
| 600 | ALBEMARLE |
| 630 | SILER CITY 2 S |
| 640 | BURLINGTON FIRE STN |
| 700 | GASTONIA |
| 702 | ROXBORO 7 ESE |
| 712 | CONCORD |
| 732 | JACKSON SPRINGS 5 WN |
| 760 | LEXINGTON |
| 761 | SALISBURY |

| FEET | STATION NAME |
|---|---|
| 768 | CHARLOTTE DOUGLAS AP |
| 801 | DANBURY 1 NW |
| 860 | ASHEBORO 2 W |
| 860 | YADKINVILLE 6 E |
| 889 | REIDSVILLE 2 NW |
| 902 | GRNSBR,HGH-PT,AP |
| 902 | SHELBY 2 NNE |
| 912 | HIGH POINT |
| 951 | STATESVILLE 2 NNE |
| 971 | NORTH WILKESBORO |
| 981 | LINCOLNTON 4 W |
| 1070 | MOUNT AIRY |
| 1079 | TRYON |
| 1102 | W KERR SCOTT RESERVO |
| 1112 | TAPOCO |
| 1142 | MORGANTON |
| 1145 | HICKORY REGIONAL AP |
| 1289 | LENOIR |
| 1430 | MARION |
| 1581 | MURPHY 2 NE |
| 1821 | ANDREWS |
| 1952 | MARSHALL |
| 2041 | OCONALUFTEE |
| 2070 | FLETCHER 3 W |
| 2103 | CULLOWHEE |
| 2113 | BENT CREEK |
| 2113 | FRANKLIN 3 W |
| 2113 | PISGAH FOREST 1 N |
| 2152 | HENDERSONVILLE 1 NE |
| 2160 | FLETCHER 2 NE |
| 2162 | ASHEVILLE REGIONL AP |
| 2251 | COWEETA EXP STN |
| 2267 | ASHEVILLE |
| 2402 | BLACK MOUNTAIN 2 W |
| 2621 | CATALOOCHEE |
| 2641 | WAYNESVILLE 1 E |
| 2703 | CELO 2 S |
| 2875 | TRANSOU |
| 2904 | JEFFERSON 2 ESE |
| 3602 | BLOWING ROCK 1 NW |
| 3743 | BANNER ELK |
| 3763 | HIGHLANDS |
| 5305 | GRANDFATHER MOUNTAIN |

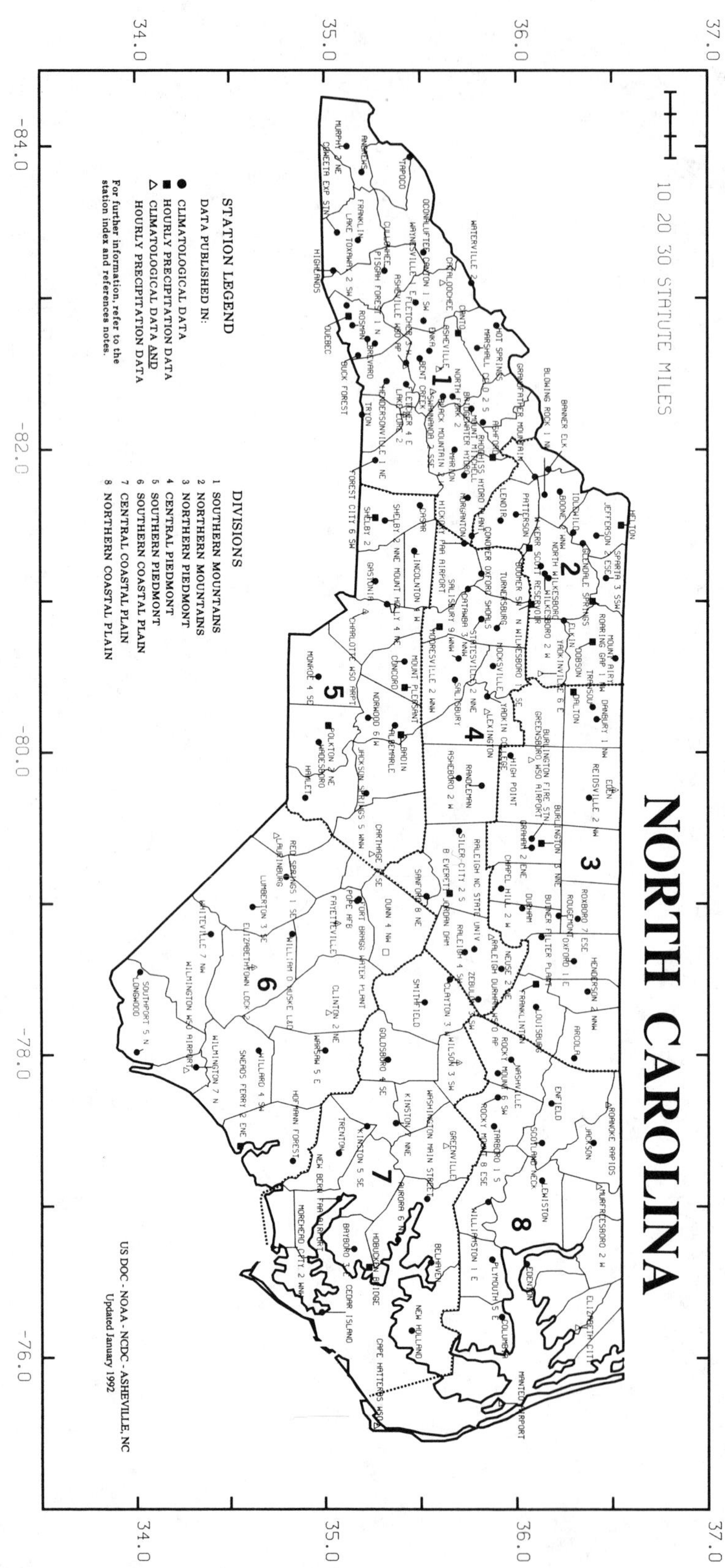
NORTH CAROLINA
10 20 30 STATUTE MILES
STATION LEGEND
DATA PUBLISHED IN:
● CLIMATOLOGICAL DATA
■ HOURLY PRECIPITATION DATA
△ CLIMATOLOGICAL DATA AND HOURLY PRECIPITATION DATA
For further information, refer to the station index and references notes.
DIVISIONS
1 SOUTHERN MOUNTAINS
2 NORTHERN MOUNTAINS
3 NORTHERN PIEDMONT
4 CENTRAL PIEDMONT
5 SOUTHERN PIEDMONT
6 SOUTHERN COASTAL PLAIN
7 CENTRAL COASTAL PLAIN
8 NORTHERN COASTAL PLAIN
US DOC - NOAA - NCDC - ASHEVILLE, NC
Updated January 1992
34.0
35.0
36.0
37.0
-84.0
-82.0
-80.0
-78.0
-76.0

## ALBEMARLE *Stanly County* ELEVATION 600 ft LAT/LONG 35° 25' N / 80° 12' W

| | JAN | FEB | MAR | APR | MAY | JUN | JUL | AUG | SEP | OCT | NOV | DEC | YEAR |
|---|---|---|---|---|---|---|---|---|---|---|---|---|---|
| Maximum Temp °F | 51.7 | 55.9 | 64.5 | 73.7 | 79.5 | 85.9 | 89.0 | 87.4 | 82.1 | 72.8 | 63.9 | 54.9 | 71.8 |
| Minimum Temp °F | 29.8 | 31.9 | 38.8 | 46.0 | 54.9 | 63.1 | 67.5 | 66.3 | 60.0 | 47.6 | 39.0 | 32.3 | 48.1 |
| Mean Temp °F | 40.8 | 43.9 | 51.7 | 59.9 | 67.2 | 74.5 | 78.3 | 76.9 | 71.1 | 60.3 | 51.5 | 43.6 | 60.0 |
| Days Max Temp ≥ 90 °F | 0 | 0 | 0 | 1 | 1 | 8 | 14 | 12 | 4 | 0 | 0 | 0 | 40 |
| Days Max Temp ≤ 32 °F | 1 | 0 | 0 | 0 | 0 | 0 | 0 | 0 | 0 | 0 | 0 | 0 | 1 |
| Days Min Temp ≤ 32 °F | 19 | 16 | 10 | 3 | 0 | 0 | 0 | 0 | 0 | 2 | 9 | 17 | 76 |
| Days Min Temp ≤ 0 °F | 0 | 0 | 0 | 0 | 0 | 0 | 0 | 0 | 0 | 0 | 0 | 0 | 0 |
| Heating Degree Days | 745 | 590 | 414 | 184 | 50 | 4 | 0 | 0 | 19 | 178 | 404 | 657 | 3245 |
| Cooling Degree Days | 0 | 1 | 8 | 38 | 140 | 335 | 459 | 396 | 221 | 46 | 4 | 2 | 1650 |
| Total Precipitation (") | 3.88 | 3.86 | 4.82 | 3.05 | 4.35 | 4.62 | 4.85 | 4.38 | 3.56 | 3.43 | 3.28 | 3.43 | 47.51 |
| Days ≥ 0.1" Precip | 7 | 6 | 7 | 6 | 7 | 6 | 8 | 6 | 5 | 5 | 6 | 6 | 75 |
| Total Snowfall (") | ***1.6*** | ***2.3*** | 1.1 | 0.0 | 0.0 | 0.0 | 0.0 | 0.0 | 0.0 | 0.0 | 0.0 | 0.7 | 5.7 |
| Days ≥ 1" Snow Depth | na | ***1*** | ***0*** | 0 | 0 | 0 | 0 | 0 | 0 | 0 | 0 | 0 | na |

## ANDREWS *Cherokee County* ELEVATION 1821 ft LAT/LONG 35° 12' N / 83° 49' W

| | JAN | FEB | MAR | APR | MAY | JUN | JUL | AUG | SEP | OCT | NOV | DEC | YEAR |
|---|---|---|---|---|---|---|---|---|---|---|---|---|---|
| Maximum Temp °F | 47.9 | 52.0 | 60.5 | 69.7 | 76.2 | 82.6 | 85.2 | 84.3 | 79.4 | 70.1 | 60.8 | 52.0 | 68.4 |
| Minimum Temp °F | 23.0 | 24.9 | 32.4 | 39.6 | 48.5 | 56.6 | 61.0 | 60.0 | 53.9 | 40.9 | 33.0 | 26.6 | 41.7 |
| Mean Temp °F | 35.4 | 38.5 | 46.4 | 54.7 | 62.4 | 69.6 | 73.1 | 72.2 | 66.7 | 55.5 | 46.9 | 39.3 | 55.1 |
| Days Max Temp ≥ 90 °F | 0 | 0 | 0 | 0 | 0 | 2 | 6 | 4 | 1 | 0 | 0 | 0 | 13 |
| Days Max Temp ≤ 32 °F | 2 | 1 | 0 | 0 | 0 | 0 | 0 | 0 | 0 | 0 | 0 | 1 | 4 |
| Days Min Temp ≤ 32 °F | 24 | 22 | 17 | 8 | 1 | 0 | 0 | 0 | 0 | 7 | 17 | 23 | 119 |
| Days Min Temp ≤ 0 °F | 1 | 0 | 0 | 0 | 0 | 0 | 0 | 0 | 0 | 0 | 0 | 0 | 1 |
| Heating Degree Days | 909 | 743 | 570 | 310 | 120 | 17 | 1 | 2 | 49 | 295 | 537 | 789 | 4342 |
| Cooling Degree Days | 0 | 0 | 1 | 7 | 48 | 179 | 283 | 237 | 109 | 10 | 1 | 0 | 875 |
| Total Precipitation (") | 6.09 | 5.99 | 6.66 | 4.88 | 5.12 | 4.95 | 5.34 | 5.50 | 4.18 | 3.64 | 5.05 | 6.00 | 63.40 |
| Days ≥ 0.1" Precip | 10 | 8 | 10 | 8 | 9 | 9 | 10 | 9 | 7 | 6 | 7 | 9 | 102 |
| Total Snowfall (") | ***3.0*** | 3.4 | 1.7 | 0.4 | 0.0 | 0.0 | 0.0 | 0.0 | 0.0 | 0.0 | 0.1 | 0.4 | 9.0 |
| Days ≥ 1" Snow Depth | na | ***2*** | ***0*** | 0 | 0 | 0 | 0 | 0 | 0 | 0 | 0 | 0 | na |

## ARCOLA *Warren County* ELEVATION 330 ft LAT/LONG 36° 17' N / 77° 59' W

| | JAN | FEB | MAR | APR | MAY | JUN | JUL | AUG | SEP | OCT | NOV | DEC | YEAR |
|---|---|---|---|---|---|---|---|---|---|---|---|---|---|
| Maximum Temp °F | 48.0 | 51.7 | 60.9 | 71.4 | 78.0 | 85.5 | 89.0 | 87.2 | 82.0 | 71.4 | 63.1 | 53.1 | 70.1 |
| Minimum Temp °F | 27.0 | 29.1 | 36.4 | 44.5 | 53.6 | 61.3 | 66.2 | 64.8 | 58.1 | 45.6 | 37.9 | 31.2 | 46.3 |
| Mean Temp °F | 37.5 | 40.5 | 48.7 | 58.0 | 65.8 | 73.4 | 77.7 | 76.0 | 70.1 | 58.6 | 50.5 | 42.2 | 58.3 |
| Days Max Temp ≥ 90 °F | 0 | 0 | 0 | 0 | 2 | 8 | 15 | 11 | 4 | 0 | 0 | 0 | 40 |
| Days Max Temp ≤ 32 °F | 3 | 1 | 0 | 0 | 0 | 0 | 0 | 0 | 0 | 0 | 0 | 1 | 5 |
| Days Min Temp ≤ 32 °F | 23 | 19 | 11 | 3 | 0 | 0 | 0 | 0 | 0 | 3 | 10 | 18 | 87 |
| Days Min Temp ≤ 0 °F | 0 | 0 | 0 | 0 | 0 | 0 | 0 | 0 | 0 | 0 | 0 | 0 | 0 |
| Heating Degree Days | 845 | 687 | 503 | 234 | 73 | 7 | 0 | 2 | 28 | 219 | 433 | 700 | 3731 |
| Cooling Degree Days | 0 | ***0*** | 4 | 27 | 105 | 274 | 413 | 329 | 173 | 29 | 3 | 1 | 1358 |
| Total Precipitation (") | 3.52 | 3.53 | 4.12 | 2.89 | 4.06 | 3.74 | 4.79 | 4.88 | 3.18 | 3.34 | 3.11 | 3.20 | 44.36 |
| Days ≥ 0.1" Precip | 7 | 6 | 7 | 6 | 7 | 6 | 7 | 7 | 5 | 5 | 5 | 6 | 74 |
| Total Snowfall (") | 1.8 | 2.8 | 1.3 | 0.0 | 0.0 | 0.0 | 0.0 | 0.0 | 0.0 | 0.0 | 0.1 | 0.6 | 6.6 |
| Days ≥ 1" Snow Depth | 2 | 2 | 1 | 0 | 0 | 0 | 0 | 0 | 0 | 0 | 0 | 1 | 6 |

## ASHEBORO 2 W *Randolph County* ELEVATION 860 ft LAT/LONG 35° 42' N / 79° 47' W

| | JAN | FEB | MAR | APR | MAY | JUN | JUL | AUG | SEP | OCT | NOV | DEC | YEAR |
|---|---|---|---|---|---|---|---|---|---|---|---|---|---|
| Maximum Temp °F | 49.7 | 54.0 | 63.3 | 73.0 | 78.4 | 84.8 | 88.4 | 86.7 | 81.4 | 71.7 | 62.9 | 53.6 | 70.7 |
| Minimum Temp °F | 30.1 | 32.4 | 40.0 | 47.9 | 56.1 | 63.9 | 68.2 | 67.0 | 61.0 | 49.2 | 40.9 | 33.7 | 49.2 |
| Mean Temp °F | 39.9 | 43.2 | 51.7 | 60.4 | 67.3 | 74.4 | 78.3 | 76.9 | 71.2 | 60.5 | 51.9 | 43.7 | 60.0 |
| Days Max Temp ≥ 90 °F | 0 | 0 | 0 | 0 | 1 | 6 | 13 | 11 | 4 | 0 | 0 | 0 | 35 |
| Days Max Temp ≤ 32 °F | 2 | 0 | 0 | 0 | 0 | 0 | 0 | 0 | 0 | 0 | 0 | 1 | 3 |
| Days Min Temp ≤ 32 °F | 19 | 15 | 8 | 1 | 0 | 0 | 0 | 0 | 0 | 1 | 6 | 15 | 65 |
| Days Min Temp ≤ 0 °F | 0 | 0 | 0 | 0 | 0 | 0 | 0 | 0 | 0 | 0 | 0 | 0 | 0 |
| Heating Degree Days | 772 | 609 | 415 | 175 | 50 | 4 | 0 | 0 | 19 | 174 | 392 | 654 | 3264 |
| Cooling Degree Days | 0 | 1 | 10 | 45 | 135 | 316 | 445 | 380 | 213 | 46 | 5 | 1 | 1597 |
| Total Precipitation (") | 3.83 | 3.69 | 4.25 | 3.27 | 4.39 | 3.90 | 4.54 | 4.82 | 3.52 | 3.49 | 2.94 | 3.32 | 45.96 |
| Days ≥ 0.1" Precip | 7 | 6 | 7 | 6 | 7 | 6 | 8 | 7 | 5 | 5 | 5 | 7 | 76 |
| Total Snowfall (") | 2.2 | 2.7 | 1.3 | 0.0 | 0.0 | 0.0 | 0.0 | 0.0 | 0.0 | 0.0 | 0.1 | 0.6 | 6.9 |
| Days ≥ 1" Snow Depth | 2 | 1 | 0 | 0 | 0 | 0 | 0 | 0 | 0 | 0 | 0 | 0 | 3 |

## ASHEVILLE *Buncombe County* ELEVATION 2267 ft LAT/LONG 35° 36 ' N / 82° 32 ' W

| | JAN | FEB | MAR | APR | MAY | JUN | JUL | AUG | SEP | OCT | NOV | DEC | YEAR |
|---|---|---|---|---|---|---|---|---|---|---|---|---|---|
| Maximum Temp °F | 46.0 | 49.8 | 58.4 | 67.7 | 74.4 | 81.2 | 84.5 | 83.0 | 77.1 | 67.4 | 58.1 | 49.9 | 66.5 |
| Minimum Temp °F | 27.1 | 29.4 | 37.0 | 45.2 | 52.6 | 60.0 | 64.1 | 63.1 | 57.1 | 45.7 | 37.6 | 31.0 | 45.8 |
| Mean Temp °F | 36.6 | 39.6 | 47.7 | 56.5 | 63.5 | 70.6 | 74.3 | 73.1 | 67.1 | 56.6 | 47.9 | 40.5 | 56.2 |
| Days Max Temp ≥ 90 °F | 0 | 0 | 0 | 0 | 0 | 1 | 5 | 3 | 0 | 0 | 0 | 0 | 9 |
| Days Max Temp ≤ 32 °F | 4 | 2 | 0 | 0 | 0 | 0 | 0 | 0 | 0 | 0 | 0 | 2 | 8 |
| Days Min Temp ≤ 32 °F | 21 | 18 | 11 | 2 | 0 | 0 | 0 | 0 | 0 | 2 | 10 | 19 | 83 |
| Days Min Temp ≤ 0 °F | 1 | 0 | 0 | 0 | 0 | 0 | 0 | 0 | 0 | 0 | 0 | 0 | 1 |
| Heating Degree Days | 874 | 711 | 532 | 263 | 98 | 11 | 1 | 2 | 47 | 265 | 508 | 753 | 4065 |
| Cooling Degree Days | 0 | 0 | 3 | 14 | 62 | 205 | 321 | 263 | 117 | 15 | 1 | 0 | 1001 |
| Total Precipitation (") | 2.51 | 3.10 | 3.92 | 3.06 | 3.62 | 3.19 | 3.03 | 3.66 | 3.05 | 2.62 | 2.85 | 2.74 | 37.35 |
| Days ≥ 0.1" Precip | 6 | 6 | 7 | 6 | 7 | 7 | 7 | 7 | 6 | 5 | 5 | 6 | 75 |
| Total Snowfall (") | 4.5 | 4.2 | 3.0 | 0.8 | 0.0 | 0.0 | 0.0 | 0.0 | 0.0 | 0.0 | 0.8 | 2.3 | 15.6 |
| Days ≥ 1" Snow Depth | *4* | *3* | *1* | *0* | *0* | *0* | *0* | 0 | 0 | 0 | 0 | *1* | 9 |

## ASHEVILLE REGIONL AP *Buncombe County* ELEVATION 2162 ft LAT/LONG 35° 26 ' N / 82° 33 ' W

| | JAN | FEB | MAR | APR | MAY | JUN | JUL | AUG | SEP | OCT | NOV | DEC | YEAR |
|---|---|---|---|---|---|---|---|---|---|---|---|---|---|
| Maximum Temp °F | 46.8 | 50.4 | 59.1 | 68.2 | 74.6 | 81.3 | 84.2 | 82.7 | 77.2 | 68.1 | 58.9 | 50.7 | 66.9 |
| Minimum Temp °F | 25.5 | 27.6 | 35.0 | 42.5 | 50.6 | 58.5 | 63.2 | 62.2 | 55.8 | 43.5 | 35.4 | 29.3 | 44.1 |
| Mean Temp °F | 36.2 | 39.1 | 47.1 | 55.4 | 62.7 | 69.9 | 73.7 | 72.5 | 66.5 | 55.8 | 47.2 | 40.0 | 55.5 |
| Days Max Temp ≥ 90 °F | 0 | 0 | 0 | 0 | 0 | 2 | 5 | 2 | 0 | 0 | 0 | 0 | 9 |
| Days Max Temp ≤ 32 °F | 3 | 1 | 0 | 0 | 0 | 0 | 0 | 0 | 0 | 0 | 0 | 1 | 5 |
| Days Min Temp ≤ 32 °F | 23 | 20 | 13 | 4 | 0 | 0 | 0 | 0 | 0 | 4 | 13 | 21 | 98 |
| Days Min Temp ≤ 0 °F | 1 | 0 | 0 | 0 | 0 | 0 | 0 | 0 | 0 | 0 | 0 | 0 | 1 |
| Heating Degree Days | 888 | 726 | 548 | 288 | 111 | 14 | 1 | 3 | 54 | 286 | 529 | 767 | 4215 |
| Cooling Degree Days | 0 | 0 | 1 | 6 | 50 | 189 | 304 | 250 | 108 | 10 | 0 | 0 | 918 |
| Total Precipitation (") | 3.28 | 3.78 | 4.67 | 3.29 | 4.59 | 4.09 | 4.35 | 4.57 | 3.70 | 3.44 | 3.61 | 3.51 | 46.88 |
| Days ≥ 0.1" Precip | 6 | 6 | 7 | 6 | 8 | 7 | 8 | 8 | 6 | 6 | 6 | 6 | 80 |
| Total Snowfall (") | 4.6 | 4.4 | 3.0 | 0.6 | 0.0 | 0.0 | 0.0 | 0.0 | 0.0 | 0.0 | 0.7 | 2.0 | 15.3 |
| Days ≥ 1" Snow Depth | 5 | 3 | 1 | 0 | 0 | 0 | 0 | 0 | 0 | 0 | 0 | 1 | 10 |

## BANNER ELK *Avery County* ELEVATION 3743 ft LAT/LONG 36° 9 ' N / 81° 52 ' W

| | JAN | FEB | MAR | APR | MAY | JUN | JUL | AUG | SEP | OCT | NOV | DEC | YEAR |
|---|---|---|---|---|---|---|---|---|---|---|---|---|---|
| Maximum Temp °F | 41.7 | 44.1 | 52.4 | 61.1 | 67.8 | 74.1 | 77.2 | 76.2 | 71.2 | 62.7 | 53.7 | 45.7 | 60.7 |
| Minimum Temp °F | 20.5 | 22.4 | 29.9 | 37.5 | 45.2 | 52.2 | 56.3 | 55.2 | 49.9 | 39.2 | 31.2 | 24.8 | 38.7 |
| Mean Temp °F | 31.1 | 33.3 | 41.1 | 49.3 | 56.5 | 63.2 | 66.7 | 65.8 | 60.6 | 51.0 | 42.5 | 35.3 | 49.7 |
| Days Max Temp ≥ 90 °F | 0 | 0 | 0 | 0 | 0 | 0 | 0 | 0 | 0 | 0 | 0 | 0 | 0 |
| Days Max Temp ≤ 32 °F | 7 | 4 | 2 | 0 | 0 | 0 | 0 | 0 | 0 | 0 | 1 | 4 | 18 |
| Days Min Temp ≤ 32 °F | 26 | 23 | 18 | 10 | 2 | 0 | 0 | 0 | 1 | 8 | 17 | 24 | 129 |
| Days Min Temp ≤ 0 °F | 2 | 1 | 0 | 0 | 0 | 0 | 0 | 0 | 0 | 0 | 0 | 1 | 4 |
| Heating Degree Days | 1043 | 890 | 733 | 465 | 258 | 79 | 22 | 33 | 146 | 429 | 669 | 915 | 5682 |
| Cooling Degree Days | 0 | 0 | 0 | 0 | 4 | 41 | 109 | 71 | 24 | 1 | 0 | 0 | 250 |
| Total Precipitation (") | 3.62 | 3.86 | 4.97 | 4.24 | 4.76 | 4.52 | 4.66 | 4.61 | 4.07 | 3.92 | 3.78 | 3.15 | 50.16 |
| Days ≥ 0.1" Precip | 9 | 8 | 9 | 8 | 10 | 9 | 9 | 8 | 7 | 7 | 7 | 7 | 98 |
| Total Snowfall (") | 12.3 | 12.7 | 8.3 | 2.7 | 0.2 | 0.0 | 0.0 | 0.0 | 0.0 | 0.4 | 3.1 | 6.6 | 46.3 |
| Days ≥ 1" Snow Depth | *8* | *8* | 4 | 1 | 0 | 0 | 0 | 0 | 0 | 0 | 2 | 5 | 28 |

## BELHAVEN *Beaufort County* ELEVATION 10 ft LAT/LONG 35° 33 ' N / 76° 38 ' W

| | JAN | FEB | MAR | APR | MAY | JUN | JUL | AUG | SEP | OCT | NOV | DEC | YEAR |
|---|---|---|---|---|---|---|---|---|---|---|---|---|---|
| Maximum Temp °F | 51.3 | 54.7 | 62.3 | *71.4* | 78.4 | 85.3 | *88.6* | 87.0 | 82.5 | 72.9 | 65.1 | 55.5 | 71.3 |
| Minimum Temp °F | 31.9 | 33.9 | 40.9 | 49.5 | 58.3 | 66.7 | 71.0 | 69.6 | 63.7 | 51.8 | 44.1 | 35.4 | 51.4 |
| Mean Temp °F | 41.6 | 44.3 | 51.6 | *60.4* | 68.4 | 76.0 | *79.8* | 78.3 | 73.1 | 62.4 | 54.8 | 45.5 | 61.4 |
| Days Max Temp ≥ 90 °F | 0 | 0 | 0 | 0 | 2 | 7 | *13* | 9 | 3 | 0 | 0 | 0 | 34 |
| Days Max Temp ≤ 32 °F | 1 | 0 | 0 | 0 | 0 | 0 | 0 | 0 | 0 | 0 | 0 | 0 | 1 |
| Days Min Temp ≤ 32 °F | 17 | 14 | 6 | 0 | 0 | 0 | 0 | 0 | 0 | 0 | 4 | 13 | 54 |
| Days Min Temp ≤ 0 °F | 0 | 0 | 0 | 0 | 0 | 0 | 0 | 0 | 0 | 0 | 0 | 0 | 0 |
| Heating Degree Days | 718 | 579 | 418 | *177* | 40 | 2 | *0* | 0 | 7 | 134 | 314 | 600 | 2989 |
| Cooling Degree Days | 0 | 1 | 9 | 43 | 158 | 349 | *486* | 407 | 255 | 64 | 13 | 2 | 1787 |
| Total Precipitation (") | 4.17 | 3.08 | 4.21 | 3.08 | 4.69 | 4.56 | 6.07 | 5.65 | 4.12 | 3.32 | 3.09 | 3.18 | 49.22 |
| Days ≥ 0.1" Precip | 9 | 6 | 7 | 6 | 7 | 7 | 9 | 8 | 6 | 5 | 5 | 6 | 81 |
| Total Snowfall (") | 1.3 | 1.8 | 1.2 | 0.1 | 0.0 | 0.0 | 0.0 | 0.0 | 0.0 | 0.0 | 0.0 | 0.4 | 4.8 |
| Days ≥ 1" Snow Depth | 1 | 1 | 0 | 0 | 0 | 0 | 0 | 0 | 0 | 0 | 0 | 0 | 2 |

## BENT CREEK *Buncombe County* ELEVATION 2113 ft LAT/LONG 35° 30 ' N / 82° 36 ' W

| | JAN | FEB | MAR | APR | MAY | JUN | JUL | AUG | SEP | OCT | NOV | DEC | YEAR |
|---|---|---|---|---|---|---|---|---|---|---|---|---|---|
| Maximum Temp °F | 47.4 | 51.2 | 59.9 | 69.4 | 75.2 | 81.3 | 84.3 | 82.8 | 77.7 | 68.9 | 59.6 | 51.1 | 67.4 |
| Minimum Temp °F | 25.4 | 27.1 | 34.0 | 41.0 | 49.3 | 56.9 | 61.6 | 60.8 | 54.8 | 43.0 | 35.0 | 28.8 | 43.1 |
| Mean Temp °F | 36.5 | 39.2 | 47.0 | 55.2 | 62.3 | 69.1 | 73.0 | 71.9 | 66.2 | 56.0 | 47.3 | 40.0 | 55.3 |
| Days Max Temp ≥ 90 °F | 0 | 0 | 0 | 0 | 0 | 1 | 5 | 2 | 0 | 0 | 0 | 0 | 8 |
| Days Max Temp ≤ 32 °F | 2 | 1 | 0 | 0 | 0 | 0 | 0 | 0 | 0 | 0 | 0 | 1 | 4 |
| Days Min Temp ≤ 32 °F | 23 | 20 | 15 | 6 | 1 | 0 | 0 | 0 | 0 | 5 | 14 | 21 | 105 |
| Days Min Temp ≤ 0 °F | 1 | 0 | 0 | 0 | 0 | 0 | 0 | 0 | 0 | 0 | 0 | 0 | 1 |
| Heating Degree Days | 877 | 723 | 553 | 293 | 115 | 18 | 1 | 3 | 55 | 282 | 526 | 769 | 4215 |
| Cooling Degree Days | 0 | 0 | 1 | 6 | 46 | 169 | 281 | 232 | 104 | 12 | 1 | 0 | 852 |
| Total Precipitation (") | 3.23 | 3.77 | 5.06 | 3.62 | 4.51 | 3.48 | 3.95 | 4.26 | 3.90 | 3.60 | 3.86 | 3.47 | 46.71 |
| Days ≥ 0.1" Precip | 7 | 6 | 8 | 6 | 8 | 7 | 8 | 7 | 6 | 6 | 6 | 6 | 81 |
| Total Snowfall (") | *2.9* | *3.1* | *1.5* | 0.5 | 0.0 | 0.0 | 0.0 | 0.0 | 0.0 | 0.0 | *0.0* | *1.2* | 9.2 |
| Days ≥ 1" Snow Depth | *2* | *1* | *1* | 0 | 0 | 0 | 0 | 0 | 0 | 0 | 0 | *1* | 5 |

## BLACK MOUNTAIN 2 W *Buncombe County* ELEVATION 2402 ft LAT/LONG 35° 37 ' N / 82° 19 ' W

| | JAN | FEB | MAR | APR | MAY | JUN | JUL | AUG | SEP | OCT | NOV | DEC | YEAR |
|---|---|---|---|---|---|---|---|---|---|---|---|---|---|
| Maximum Temp °F | 48.9 | 52.3 | 60.7 | 69.8 | 75.5 | 81.5 | 84.6 | 83.1 | 77.8 | 69.7 | 60.5 | 52.7 | 68.1 |
| Minimum Temp °F | 25.5 | 27.8 | 34.6 | 42.0 | 49.4 | 56.9 | 61.1 | 60.5 | 54.4 | 43.0 | 35.1 | 29.0 | 43.3 |
| Mean Temp °F | 37.2 | 40.1 | 47.7 | 55.9 | 62.4 | 69.2 | 72.9 | 71.8 | 66.1 | 56.4 | 47.8 | 40.8 | 55.7 |
| Days Max Temp ≥ 90 °F | 0 | 0 | 0 | 0 | 0 | 2 | 4 | 3 | 0 | 0 | 0 | 0 | 9 |
| Days Max Temp ≤ 32 °F | 2 | 1 | 0 | 0 | 0 | 0 | 0 | 0 | 0 | 0 | 0 | 1 | 4 |
| Days Min Temp ≤ 32 °F | 23 | 20 | 14 | 5 | 0 | 0 | 0 | 0 | 0 | 5 | 13 | 20 | 100 |
| Days Min Temp ≤ 0 °F | 0 | 0 | 0 | 0 | 0 | 0 | 0 | 0 | 0 | 0 | 0 | 0 | 0 |
| Heating Degree Days | 851 | 692 | 531 | 271 | 110 | 17 | 2 | 3 | 56 | 269 | 511 | 744 | 4057 |
| Cooling Degree Days | 0 | 0 | 1 | 7 | 40 | 153 | 263 | 220 | 95 | 12 | 0 | 0 | 791 |
| Total Precipitation (") | 3.44 | 3.67 | 4.88 | 3.89 | 5.07 | 4.33 | 4.33 | 4.43 | 3.64 | 4.07 | 4.05 | 3.50 | 49.30 |
| Days ≥ 0.1" Precip | 7 | 6 | 8 | 7 | 9 | 7 | 8 | 8 | 6 | 6 | 7 | 6 | 85 |
| Total Snowfall (") | *3.8* | *3.0* | 1.8 | 0.3 | 0.0 | 0.0 | 0.0 | 0.0 | 0.0 | 0.0 | 0.4 | 1.4 | 10.7 |
| Days ≥ 1" Snow Depth | na | na | 0 | 0 | 0 | 0 | 0 | 0 | 0 | 0 | 0 | *0* | na |

## BLOWING ROCK 1 NW *Watauga County* ELEVATION 3602 ft LAT/LONG 36° 8 ' N / 81° 41 ' W

| | JAN | FEB | MAR | APR | MAY | JUN | JUL | AUG | SEP | OCT | NOV | DEC | YEAR |
|---|---|---|---|---|---|---|---|---|---|---|---|---|---|
| Maximum Temp °F | 38.3 | 41.3 | 49.5 | 58.9 | 66.1 | 72.5 | 76.2 | 74.7 | 68.9 | 59.7 | 51.2 | 42.4 | 58.3 |
| Minimum Temp °F | 20.7 | 22.3 | 30.1 | 38.7 | 47.1 | 54.2 | 58.5 | 57.2 | 51.4 | 40.6 | 33.0 | 25.0 | 39.9 |
| Mean Temp °F | 29.5 | 31.8 | 39.8 | 48.8 | 56.6 | 63.4 | 67.4 | 66.0 | 60.2 | 50.2 | 42.1 | 33.8 | 49.1 |
| Days Max Temp ≥ 90 °F | 0 | 0 | 0 | 0 | 0 | 0 | 0 | 0 | 0 | 0 | 0 | 0 | 0 |
| Days Max Temp ≤ 32 °F | 9 | 6 | 2 | 0 | 0 | 0 | 0 | 0 | 0 | 0 | 2 | 6 | 25 |
| Days Min Temp ≤ 32 °F | 26 | 23 | 18 | 9 | 1 | 0 | 0 | 0 | 0 | 6 | 15 | 23 | 121 |
| Days Min Temp ≤ 0 °F | 2 | 1 | 0 | 0 | 0 | 0 | 0 | 0 | 0 | 0 | 0 | 1 | 4 |
| Heating Degree Days | 1092 | 931 | 773 | 480 | 260 | 82 | 22 | 37 | 158 | 453 | 681 | 961 | 5930 |
| Cooling Degree Days | 0 | 0 | 0 | 2 | 7 | 53 | 121 | 77 | 22 | 1 | 0 | 0 | 283 |
| Total Precipitation (") | 4.60 | 4.79 | 6.52 | 5.77 | 6.58 | 6.06 | 5.91 | 6.20 | 5.75 | 5.36 | 5.75 | 4.58 | 67.87 |
| Days ≥ 0.1" Precip | 8 | 8 | 9 | 8 | 10 | 9 | 10 | 10 | 7 | 6 | 7 | 8 | 100 |
| Total Snowfall (") | 10.2 | 9.9 | 5.8 | 1.0 | 0.0 | 0.0 | 0.0 | 0.0 | 0.0 | 0.1 | 2.0 | 6.1 | 35.1 |
| Days ≥ 1" Snow Depth | 9 | 8 | 4 | 1 | 0 | 0 | 0 | 0 | 0 | 0 | 1 | 5 | 28 |

## BURLINGTON FIRE STN *Alamance County* ELEVATION 640 ft LAT/LONG 36° 5 ' N / 79° 25 ' W

| | JAN | FEB | MAR | APR | MAY | JUN | JUL | AUG | SEP | OCT | NOV | DEC | YEAR |
|---|---|---|---|---|---|---|---|---|---|---|---|---|---|
| Maximum Temp °F | 48.5 | 52.6 | 61.9 | 72.4 | 78.7 | 86.0 | 89.8 | 88.2 | 81.8 | 71.6 | 62.5 | 52.8 | 70.6 |
| Minimum Temp °F | 27.4 | 29.7 | 37.2 | 45.7 | 54.4 | 63.3 | 67.6 | 65.9 | 59.1 | 46.5 | 38.2 | 30.8 | 47.2 |
| Mean Temp °F | 38.0 | 41.1 | 49.6 | 59.1 | 66.5 | 74.7 | 78.7 | 77.1 | 70.5 | 59.1 | 50.4 | 41.8 | 58.9 |
| Days Max Temp ≥ 90 °F | 0 | 0 | 0 | 1 | 2 | 9 | 17 | 14 | 5 | 0 | 0 | 0 | 48 |
| Days Max Temp ≤ 32 °F | 2 | 1 | 0 | 0 | 0 | 0 | 0 | 0 | 0 | 0 | 0 | 1 | 4 |
| Days Min Temp ≤ 32 °F | 23 | 19 | 10 | 1 | 0 | 0 | 0 | 0 | 0 | 1 | 9 | 18 | 81 |
| Days Min Temp ≤ 0 °F | 0 | 0 | 0 | 0 | 0 | 0 | 0 | 0 | 0 | 0 | 0 | 0 | 0 |
| Heating Degree Days | 830 | 667 | 476 | 207 | 64 | 4 | 0 | 1 | 25 | 208 | 436 | 711 | 3629 |
| Cooling Degree Days | 0 | 0 | 4 | 33 | 116 | 322 | 454 | 381 | 188 | 33 | 2 | 0 | 1533 |
| Total Precipitation (") | 3.57 | 3.40 | 4.03 | 3.12 | 4.32 | 4.09 | 4.78 | 4.14 | 3.41 | 3.30 | 2.92 | 3.25 | 44.33 |
| Days ≥ 0.1" Precip | 7 | 6 | 7 | 6 | 8 | 6 | 7 | 7 | 5 | 5 | 5 | 6 | 75 |
| Total Snowfall (") | *1.8* | *1.3* | *0.2* | 0.0 | 0.0 | 0.0 | 0.0 | 0.0 | 0.0 | 0.0 | 0.0 | 0.5 | 3.8 |
| Days ≥ 1" Snow Depth | 2 | *1* | 0 | 0 | 0 | 0 | 0 | 0 | 0 | 0 | 0 | 0 | 3 |

## CAPE HATTERAS NWS *Dare County* ELEVATION 8 ft LAT/LONG 35° 16 ' N / 75° 33 ' W

| | JAN | FEB | MAR | APR | MAY | JUN | JUL | AUG | SEP | OCT | NOV | DEC | YEAR |
|---|---|---|---|---|---|---|---|---|---|---|---|---|---|
| Maximum Temp °F | 52.8 | 53.9 | 59.8 | 67.6 | 74.8 | 81.3 | 85.3 | 84.9 | 81.1 | 72.5 | 65.0 | 57.5 | 69.7 |
| Minimum Temp °F | 37.6 | 38.0 | 43.7 | 51.0 | 59.7 | 67.6 | 72.4 | 72.1 | 67.8 | 58.6 | 49.8 | 42.0 | 55.0 |
| Mean Temp °F | 45.3 | 46.0 | 51.8 | 59.3 | 67.3 | 74.5 | 78.9 | 78.6 | 74.5 | 65.6 | 57.4 | 49.8 | 62.4 |
| Days Max Temp ≥ 90 °F | 0 | 0 | 0 | 0 | 0 | 1 | 3 | 2 | 1 | 0 | 0 | 0 | 7 |
| Days Max Temp ≤ 32 °F | 1 | 0 | 0 | 0 | 0 | 0 | 0 | 0 | 0 | 0 | 0 | 0 | 1 |
| Days Min Temp ≤ 32 °F | 10 | 8 | 3 | 0 | 0 | 0 | 0 | 0 | 0 | 0 | 1 | 5 | 27 |
| Days Min Temp ≤ 0 °F | 0 | 0 | 0 | 0 | 0 | 0 | 0 | 0 | 0 | 0 | 0 | 0 | 0 |
| Heating Degree Days | 605 | 532 | 408 | 191 | 46 | 3 | 0 | 0 | 2 | 73 | 244 | 469 | 2573 |
| Cooling Degree Days | 0 | 2 | 7 | 36 | 138 | 313 | 465 | 433 | 298 | 102 | 31 | 4 | 1829 |
| Total Precipitation (") | 5.60 | 3.97 | 4.83 | 3.09 | 3.93 | 3.68 | 5.15 | 6.16 | 5.30 | 5.33 | 4.74 | 4.57 | 56.35 |
| Days ≥ 0.1" Precip | 8 | 6 | 8 | 5 | 6 | 6 | 8 | 8 | 6 | 6 | 6 | 6 | 79 |
| Total Snowfall (") | 0.4 | 0.5 | 0.2 | 0.0 | 0.0 | 0.0 | 0.0 | 0.0 | 0.0 | 0.0 | 0.0 | 0.7 | 1.8 |
| Days ≥ 1" Snow Depth | 0 | 0 | 0 | 0 | 0 | 0 | 0 | 0 | 0 | 0 | 0 | 0 | 0 |

## CATALOOCHEE *Haywood County* ELEVATION 2621 ft LAT/LONG 35° 38 ' N / 83° 5 ' W

| | JAN | FEB | MAR | APR | MAY | JUN | JUL | AUG | SEP | OCT | NOV | DEC | YEAR |
|---|---|---|---|---|---|---|---|---|---|---|---|---|---|
| Maximum Temp °F | 44.8 | 49.0 | 57.2 | 66.4 | 72.0 | 78.3 | 80.9 | 79.7 | 75.1 | 66.8 | 57.7 | 48.5 | 64.7 |
| Minimum Temp °F | 21.0 | 23.2 | 29.7 | 37.0 | 45.0 | 52.6 | 57.0 | 56.1 | 50.4 | 37.8 | 30.6 | 24.3 | 38.7 |
| Mean Temp °F | 32.9 | 36.2 | 43.5 | 51.7 | 58.5 | 65.5 | 69.0 | 68.0 | 62.8 | 52.3 | 44.2 | 36.4 | 51.8 |
| Days Max Temp ≥ 90 °F | 0 | 0 | 0 | 0 | 0 | 0 | 0 | 0 | 0 | 0 | 0 | 0 | 0 |
| Days Max Temp ≤ 32 °F | 4 | 3 | 0 | 0 | 0 | 0 | 0 | 0 | 0 | 0 | 0 | 2 | 9 |
| Days Min Temp ≤ 32 °F | 26 | 23 | 20 | 10 | 2 | 0 | 0 | 0 | 1 | 11 | 19 | 25 | 137 |
| Days Min Temp ≤ 0 °F | 2 | 1 | 0 | 0 | 0 | 0 | 0 | 0 | 0 | 0 | 0 | 0 | 3 |
| Heating Degree Days | 988 | 808 | 660 | 393 | 204 | 46 | 9 | 13 | 99 | 389 | 617 | 880 | 5106 |
| Cooling Degree Days | 0 | 0 | 0 | 1 | 11 | 77 | 155 | 110 | 38 | 3 | 0 | 0 | 395 |
| Total Precipitation (") | 4.40 | 4.43 | 5.32 | 4.08 | 4.71 | 4.03 | 4.62 | 5.14 | 3.56 | 2.94 | 3.56 | 4.28 | 51.07 |
| Days ≥ 0.1" Precip | 8 | 7 | 8 | 8 | 9 | 8 | 9 | 9 | 7 | 6 | 7 | 7 | 93 |
| Total Snowfall (") | na | na | na | na | ***0.0*** | ***0.0*** | ***0.0*** | ***0.0*** | ***0.0*** | na | na | na | na |
| Days ≥ 1" Snow Depth | na | na | na | ***0*** | ***0*** | ***0*** | ***0*** | ***0*** | ***0*** | ***0*** | ***0*** | na | na |

## CEDAR ISLAND *Carteret County* ELEVATION 10 ft LAT/LONG 34° 59 ' N / 76° 18 ' W

| | JAN | FEB | MAR | APR | MAY | JUN | JUL | AUG | SEP | OCT | NOV | DEC | YEAR |
|---|---|---|---|---|---|---|---|---|---|---|---|---|---|
| Maximum Temp °F | 52.5 | 55.0 | 62.8 | 72.3 | 79.6 | 85.7 | 89.0 | 87.3 | 82.1 | 72.7 | 64.3 | 56.4 | 71.6 |
| Minimum Temp °F | 35.6 | 37.0 | 43.6 | 51.4 | 59.9 | 67.3 | 71.4 | 70.9 | 66.8 | 56.9 | 47.6 | 39.5 | 54.0 |
| Mean Temp °F | 44.1 | 46.1 | 53.2 | 61.9 | 69.8 | 76.5 | 80.2 | 79.1 | 74.5 | 64.8 | 56.0 | 47.9 | 62.8 |
| Days Max Temp ≥ 90 °F | 0 | 0 | 0 | 1 | 2 | 8 | 15 | 10 | 3 | 0 | 0 | 0 | 39 |
| Days Max Temp ≤ 32 °F | 1 | 0 | 0 | 0 | 0 | 0 | 0 | 0 | 0 | 0 | 0 | 0 | 1 |
| Days Min Temp ≤ 32 °F | 12 | 10 | 3 | 0 | 0 | 0 | 0 | 0 | 0 | 0 | 2 | 8 | 35 |
| Days Min Temp ≤ 0 °F | 0 | 0 | 0 | 0 | 0 | 0 | 0 | 0 | 0 | 0 | 0 | 0 | 0 |
| Heating Degree Days | 642 | 530 | 367 | 139 | 23 | 1 | 0 | 0 | 2 | 84 | 279 | 526 | 2593 |
| Cooling Degree Days | 0 | 1 | 10 | 56 | 185 | 374 | 506 | 446 | 294 | 86 | 18 | 3 | 1979 |
| Total Precipitation (") | 5.11 | 3.77 | 4.62 | 2.98 | 4.16 | 4.37 | 6.90 | 6.49 | 5.99 | 4.51 | 3.80 | 4.42 | 57.12 |
| Days ≥ 0.1" Precip | 9 | 7 | 7 | 5 | 8 | 6 | 10 | 9 | 7 | 5 | 6 | 7 | 86 |
| Total Snowfall (") | 1.2 | 0.8 | 0.6 | 0.0 | 0.0 | 0.0 | 0.0 | 0.0 | 0.0 | 0.0 | 0.0 | 0.6 | 3.2 |
| Days ≥ 1" Snow Depth | 0 | 0 | 0 | 0 | 0 | 0 | 0 | 0 | 0 | 0 | 0 | 0 | 0 |

## CELO 2 S *Yancey County* ELEVATION 2703 ft LAT/LONG 35° 50 ' N / 82° 11 ' W

| | JAN | FEB | MAR | APR | MAY | JUN | JUL | AUG | SEP | OCT | NOV | DEC | YEAR |
|---|---|---|---|---|---|---|---|---|---|---|---|---|---|
| Maximum Temp °F | 45.4 | 48.2 | 56.2 | 65.3 | 71.6 | 77.4 | 80.5 | 79.1 | 73.9 | 65.5 | 57.2 | 49.1 | 64.1 |
| Minimum Temp °F | 20.3 | 22.3 | 29.9 | 37.5 | 45.3 | 52.6 | 57.5 | 56.6 | 50.0 | 37.5 | 30.0 | 23.5 | 38.6 |
| Mean Temp °F | 32.9 | 35.3 | 43.1 | 51.4 | 58.5 | 65.1 | 69.1 | 67.9 | 61.9 | 51.5 | 43.6 | 36.3 | 51.4 |
| Days Max Temp ≥ 90 °F | 0 | 0 | 0 | 0 | 0 | 0 | 1 | 0 | 0 | 0 | 0 | 0 | 1 |
| Days Max Temp ≤ 32 °F | 4 | 3 | 1 | 0 | 0 | 0 | 0 | 0 | 0 | 0 | 0 | 2 | 10 |
| Days Min Temp ≤ 32 °F | 27 | 23 | 19 | 9 | 3 | 0 | 0 | 0 | 1 | 11 | 19 | 25 | 137 |
| Days Min Temp ≤ 0 °F | 2 | 1 | 0 | 0 | 0 | 0 | 0 | 0 | 0 | 0 | 0 | 0 | 3 |
| Heating Degree Days | 989 | 832 | 672 | 402 | 208 | 58 | 11 | 18 | 121 | 412 | 634 | 882 | 5239 |
| Cooling Degree Days | 0 | 0 | 1 | 1 | 14 | 82 | 170 | 118 | 42 | 2 | 0 | 0 | 430 |
| Total Precipitation (") | 4.43 | 4.87 | 6.28 | 4.58 | 5.59 | 4.44 | 4.71 | 5.17 | 4.70 | 5.08 | 5.04 | 4.24 | 59.13 |
| Days ≥ 0.1" Precip | 7 | 7 | 9 | 8 | 9 | 8 | 9 | 9 | 7 | 6 | 7 | 7 | 93 |
| Total Snowfall (") | 6.2 | 5.2 | 3.2 | 0.8 | 0.0 | 0.0 | 0.0 | 0.0 | 0.0 | 0.0 | 0.7 | 2.5 | 18.6 |
| Days ≥ 1" Snow Depth | 6 | 4 | 2 | 0 | 0 | 0 | 0 | 0 | 0 | 0 | 1 | 2 | 15 |

## CHAPEL HILL 2 W *Orange County* ELEVATION 502 ft LAT/LONG 35° 55 ' N / 79° 4 ' W

| | JAN | FEB | MAR | APR | MAY | JUN | JUL | AUG | SEP | OCT | NOV | DEC | YEAR |
|---|---|---|---|---|---|---|---|---|---|---|---|---|---|
| Maximum Temp °F | 48.8 | 52.6 | 61.4 | 71.5 | 78.1 | 85.1 | 89.1 | 87.3 | 81.8 | 71.4 | 62.9 | 53.1 | 70.3 |
| Minimum Temp °F | 26.6 | 28.7 | 36.5 | 44.8 | 53.4 | 61.6 | 65.8 | 64.6 | 57.8 | 44.9 | 37.1 | 30.1 | 46.0 |
| Mean Temp °F | 37.7 | 40.7 | 49.0 | 58.2 | 65.8 | 73.4 | 77.5 | 76.0 | 69.8 | 58.2 | 50.0 | 41.6 | 58.2 |
| Days Max Temp ≥ 90 °F | 0 | 0 | 0 | 0 | 2 | 7 | 15 | 11 | 4 | 0 | 0 | 0 | 39 |
| Days Max Temp ≤ 32 °F | 2 | 1 | 0 | 0 | 0 | 0 | 0 | 0 | 0 | 0 | 0 | 1 | 4 |
| Days Min Temp ≤ 32 °F | 23 | 20 | 12 | 3 | 0 | 0 | 0 | 0 | 0 | 3 | 11 | 20 | 92 |
| Days Min Temp ≤ 0 °F | 0 | 0 | 0 | 0 | 0 | 0 | 0 | 0 | 0 | 0 | 0 | 0 | 0 |
| Heating Degree Days | 839 | 681 | 496 | 230 | 76 | 8 | 0 | 2 | 30 | 230 | 448 | 718 | 3758 |
| Cooling Degree Days | 0 | 1 | 7 | 32 | 117 | 289 | 414 | 354 | 184 | 27 | 3 | 1 | 1429 |
| Total Precipitation (") | 3.77 | 3.61 | 4.50 | 3.02 | 4.59 | 4.12 | 3.87 | 4.40 | 3.21 | 3.67 | 3.33 | 3.29 | 45.38 |
| Days ≥ 0.1" Precip | 7 | 6 | 8 | 6 | 7 | 6 | 7 | 7 | 5 | 5 | 5 | 6 | 75 |
| Total Snowfall (") | 1.6 | 2.4 | 1.4 | 0.0 | 0.0 | 0.0 | 0.0 | 0.0 | 0.0 | 0.0 | 0.0 | 0.7 | 6.1 |
| Days ≥ 1" Snow Depth | 1 | 1 | 0 | 0 | 0 | 0 | 0 | 0 | 0 | 0 | 0 | 0 | 2 |

## CHARLOTTE DOUGLAS AP *Mecklenburg County* ELEVATION 768 ft LAT/LONG 35° 13 ' N / 80° 56 ' W

| | JAN | FEB | MAR | APR | MAY | JUN | JUL | AUG | SEP | OCT | NOV | DEC | YEAR |
|---|---|---|---|---|---|---|---|---|---|---|---|---|---|
| Maximum Temp °F | 49.6 | 53.9 | 62.8 | 72.2 | 78.6 | 85.4 | 88.8 | 87.1 | 81.4 | 71.5 | 62.1 | 53.2 | 70.6 |
| Minimum Temp °F | 30.5 | 32.8 | 40.3 | 48.5 | 57.1 | 65.4 | 69.7 | 68.6 | 62.4 | 50.2 | 40.9 | 34.0 | 50.0 |
| Mean Temp °F | 40.0 | 43.4 | 51.5 | 60.4 | 67.9 | 75.4 | 79.3 | 77.9 | 71.9 | 60.8 | 51.5 | 43.6 | 60.3 |
| Days Max Temp ≥ 90 °F | 0 | 0 | 0 | 0 | 1 | 8 | 15 | 11 | 4 | 0 | 0 | 0 | 39 |
| Days Max Temp ≤ 32 °F | 2 | 0 | 0 | 0 | 0 | 0 | 0 | 0 | 0 | 0 | 0 | 0 | 2 |
| Days Min Temp ≤ 32 °F | 18 | 15 | 7 | 1 | 0 | 0 | 0 | 0 | 0 | 1 | 6 | 15 | 63 |
| Days Min Temp ≤ 0 °F | 0 | 0 | 0 | 0 | 0 | 0 | 0 | 0 | 0 | 0 | 0 | 0 | 0 |
| Heating Degree Days | 766 | 605 | 417 | 174 | 45 | 3 | 0 | 0 | 15 | 166 | 403 | 658 | 3252 |
| Cooling Degree Days | 0 | 1 | 8 | 48 | 158 | 361 | 490 | 419 | 241 | 53 | 5 | 2 | 1786 |
| Total Precipitation (") | 3.74 | 3.60 | 4.60 | 2.55 | 3.88 | 3.30 | 3.87 | 3.93 | 3.52 | 3.46 | 3.24 | 3.34 | 43.03 |
| Days ≥ 0.1" Precip | 7 | 7 | 8 | 5 | 7 | 6 | 7 | 6 | 5 | 5 | 5 | 6 | 74 |
| Total Snowfall (") | 2.2 | 2.3 | 1.3 | 0.0 | 0.0 | 0.0 | 0.0 | 0.0 | 0.0 | 0.0 | 0.1 | 0.6 | 6.5 |
| Days ≥ 1" Snow Depth | 2 | 1 | 0 | 0 | 0 | 0 | 0 | 0 | 0 | 0 | 0 | 0 | 3 |

## CLAYTON 3 W *Johnston County* ELEVATION 331 ft LAT/LONG 35° 39 ' N / 78° 30 ' W

| | JAN | FEB | MAR | APR | MAY | JUN | JUL | AUG | SEP | OCT | NOV | DEC | YEAR |
|---|---|---|---|---|---|---|---|---|---|---|---|---|---|
| Maximum Temp °F | 50.8 | ***54.6*** | 63.9 | 73.7 | ***80.5*** | ***86.8*** | ***89.5*** | 88.0 | 83.1 | 73.2 | ***64.7*** | ***55.0*** | 72.0 |
| Minimum Temp °F | 30.3 | ***32.1*** | 39.9 | 47.8 | ***56.6*** | ***64.2*** | ***68.6*** | 67.9 | 61.4 | 49.9 | ***41.7*** | ***33.7*** | 49.5 |
| Mean Temp °F | 40.8 | ***43.4*** | 51.9 | 60.8 | ***68.6*** | ***75.6*** | ***79.1*** | 78.0 | 72.3 | 61.6 | ***53.2*** | ***44.4*** | 60.8 |
| Days Max Temp ≥ 90 °F | 0 | 0 | 0 | 1 | 3 | 9 | 15 | 12 | 5 | 0 | 0 | 0 | 45 |
| Days Max Temp ≤ 32 °F | 1 | 1 | 0 | 0 | 0 | 0 | 0 | 0 | 0 | 0 | 0 | 1 | 3 |
| Days Min Temp ≤ 32 °F | 18 | 16 | 7 | 1 | 0 | 0 | 0 | 0 | 0 | 1 | 6 | 15 | 64 |
| Days Min Temp ≤ 0 °F | 0 | 0 | 0 | 0 | 0 | 0 | 0 | 0 | 0 | 0 | 0 | 0 | 0 |
| Heating Degree Days | 745 | ***607*** | 410 | 169 | ***37*** | ***3*** | ***0*** | 1 | 14 | 152 | ***355*** | ***634*** | 3127 |
| Cooling Degree Days | ***0*** | na | ***14*** | ***51*** | ***169*** | na | na | ***425*** | ***236*** | ***63*** | ***8*** | ***2*** | na |
| Total Precipitation (") | 3.56 | 3.67 | 4.10 | 2.96 | 3.91 | 3.81 | 4.88 | 4.29 | 3.37 | 2.92 | 3.12 | 3.17 | 43.76 |
| Days ≥ 0.1" Precip | 7 | 6 | 7 | 5 | 7 | 6 | 7 | 6 | 5 | 4 | 5 | 6 | 71 |
| Total Snowfall (") | 1.3 | 1.8 | 0.6 | 0.0 | 0.0 | 0.0 | 0.0 | 0.0 | 0.0 | 0.0 | 0.0 | 0.3 | 4.0 |
| Days ≥ 1" Snow Depth | ***1*** | ***0*** | 0 | 0 | 0 | 0 | 0 | 0 | 0 | 0 | 0 | 0 | 1 |

## CONCORD *Cabarrus County* ELEVATION 712 ft LAT/LONG 35° 25 ' N / 80° 35 ' W

| | JAN | FEB | MAR | APR | MAY | JUN | JUL | AUG | SEP | OCT | NOV | DEC | YEAR |
|---|---|---|---|---|---|---|---|---|---|---|---|---|---|
| Maximum Temp °F | 50.1 | 54.6 | 63.3 | 73.3 | 79.9 | 86.8 | 90.3 | 88.7 | 83.3 | 73.1 | 63.8 | 54.3 | 71.8 |
| Minimum Temp °F | 27.7 | 29.7 | 37.3 | 45.7 | 54.8 | 63.5 | 68.1 | 66.8 | 60.0 | 46.9 | 38.1 | 31.0 | 47.5 |
| Mean Temp °F | 38.9 | 42.1 | 50.3 | 59.6 | 67.4 | 75.1 | 79.2 | 77.8 | 71.7 | 60.0 | 51.0 | 42.7 | 59.7 |
| Days Max Temp ≥ 90 °F | 0 | 0 | 0 | 1 | 3 | 11 | 19 | 15 | 6 | 0 | 0 | 0 | 55 |
| Days Max Temp ≤ 32 °F | 1 | 0 | 0 | 0 | 0 | 0 | 0 | 0 | 0 | 0 | 0 | 1 | 2 |
| Days Min Temp ≤ 32 °F | 22 | 19 | 10 | 2 | 0 | 0 | 0 | 0 | 0 | 2 | 10 | 19 | 84 |
| Days Min Temp ≤ 0 °F | 0 | 0 | 0 | 0 | 0 | 0 | 0 | 0 | 0 | 0 | 0 | 0 | 0 |
| Heating Degree Days | 802 | 640 | 454 | 196 | 55 | 4 | 0 | 1 | 19 | 186 | 419 | 686 | 3462 |
| Cooling Degree Days | 0 | 1 | 6 | 38 | 136 | 339 | 472 | 406 | 224 | 43 | 4 | 1 | 1670 |
| Total Precipitation (") | 3.89 | 3.47 | 4.61 | 3.04 | 4.04 | 4.43 | 4.74 | 3.82 | 3.64 | 3.84 | 3.21 | 3.40 | 46.13 |
| Days ≥ 0.1" Precip | 7 | 6 | 8 | 6 | 7 | 7 | 7 | 6 | 5 | 5 | 6 | 6 | 76 |
| Total Snowfall (") | 1.7 | 1.8 | 1.0 | 0.0 | 0.0 | 0.0 | 0.0 | 0.0 | 0.0 | 0.0 | 0.0 | 0.5 | 5.0 |
| Days ≥ 1" Snow Depth | 2 | 1 | 0 | 0 | 0 | 0 | 0 | 0 | 0 | 0 | 0 | 0 | 3 |

## COWEETA EXP STN *Macon County* ELEVATION 2251 ft LAT/LONG 35° 4 ' N / 83° 26 ' W

| | JAN | FEB | MAR | APR | MAY | JUN | JUL | AUG | SEP | OCT | NOV | DEC | YEAR |
|---|---|---|---|---|---|---|---|---|---|---|---|---|---|
| Maximum Temp °F | 48.3 | 52.1 | 59.8 | 68.6 | 74.4 | 80.5 | 83.3 | 82.1 | 77.2 | 68.9 | 60.2 | 51.8 | 67.3 |
| Minimum Temp °F | 23.6 | 25.5 | 33.0 | 40.5 | 48.1 | 55.0 | 59.1 | 58.6 | 53.2 | 40.9 | 33.0 | 26.6 | 41.4 |
| Mean Temp °F | 36.0 | 38.8 | 46.4 | 54.6 | 61.3 | 67.8 | 71.2 | 70.4 | 65.2 | 54.9 | 46.6 | 39.3 | 54.4 |
| Days Max Temp ≥ 90 °F | 0 | 0 | 0 | 0 | 0 | 1 | 3 | 1 | 0 | 0 | 0 | 0 | 5 |
| Days Max Temp ≤ 32 °F | 2 | 1 | 0 | 0 | 0 | 0 | 0 | 0 | 0 | 0 | 0 | 1 | 4 |
| Days Min Temp ≤ 32 °F | 25 | 22 | 16 | 7 | 1 | 0 | 0 | 0 | 0 | 7 | 17 | 23 | 118 |
| Days Min Temp ≤ 0 °F | 1 | 0 | 0 | 0 | 0 | 0 | 0 | 0 | 0 | 0 | 0 | 0 | 1 |
| Heating Degree Days | 893 | 733 | 570 | 312 | 135 | 25 | 3 | 4 | 63 | 309 | 545 | 792 | 4384 |
| Cooling Degree Days | 0 | 0 | 1 | 5 | 31 | 133 | 227 | 186 | 78 | 5 | 0 | 0 | 666 |
| Total Precipitation (") | 6.60 | 6.89 | 7.90 | 5.56 | 6.37 | 5.06 | 4.83 | 5.70 | 5.45 | 4.82 | 6.27 | 6.71 | 72.16 |
| Days ≥ 0.1" Precip | 9 | 8 | 9 | 8 | 10 | 8 | 9 | 9 | 8 | 6 | 8 | 8 | 100 |
| Total Snowfall (") | na | na | *1.4* | 0.3 | 0.0 | 0.0 | 0.0 | 0.0 | 0.0 | 0.0 | *0.0* | *0.3* | na |
| Days ≥ 1" Snow Depth | na | na | *1* | 0 | 0 | 0 | 0 | 0 | 0 | 0 | *0* | *0* | na |

## CULLOWHEE *Jackson County* ELEVATION 2103 ft LAT/LONG 35° 19 ' N / 83° 11 ' W

| | JAN | FEB | MAR | APR | MAY | JUN | JUL | AUG | SEP | OCT | NOV | DEC | YEAR |
|---|---|---|---|---|---|---|---|---|---|---|---|---|---|
| Maximum Temp °F | 48.7 | 52.9 | 61.8 | 70.5 | 76.4 | 82.2 | 84.8 | 83.5 | 78.3 | 70.0 | 60.5 | 52.1 | 68.5 |
| Minimum Temp °F | 25.7 | 27.5 | 34.2 | 41.1 | 49.4 | 56.8 | 61.3 | 60.7 | 55.5 | 43.2 | 34.6 | 28.8 | 43.2 |
| Mean Temp °F | 37.2 | 40.3 | 48.0 | 55.8 | 63.0 | 69.5 | 73.1 | 72.1 | 66.9 | 56.6 | 47.6 | 40.5 | 55.9 |
| Days Max Temp ≥ 90 °F | 0 | 0 | 0 | 0 | 0 | 1 | 5 | 2 | 0 | 0 | 0 | 0 | 8 |
| Days Max Temp ≤ 32 °F | 1 | 0 | 0 | 0 | 0 | 0 | 0 | 0 | 0 | 0 | 0 | 1 | 2 |
| Days Min Temp ≤ 32 °F | 22 | 19 | 14 | 6 | 1 | 0 | 0 | 0 | 0 | 5 | 15 | 20 | 102 |
| Days Min Temp ≤ 0 °F | 1 | 0 | 0 | 0 | 0 | 0 | 0 | 0 | 0 | 0 | 0 | 0 | 1 |
| Heating Degree Days | 854 | 693 | 520 | 275 | 101 | 14 | 1 | 2 | 42 | 262 | 516 | 754 | 4034 |
| Cooling Degree Days | 0 | 0 | 0 | 9 | 52 | 180 | 283 | 245 | 114 | 12 | 1 | 0 | 896 |
| Total Precipitation (") | 4.29 | 4.60 | 5.50 | 3.74 | 4.64 | 4.38 | 4.40 | 4.40 | 3.55 | 3.27 | 4.01 | 4.41 | 51.19 |
| Days ≥ 0.1" Precip | 8 | 8 | 9 | 7 | 9 | 8 | 9 | 8 | 7 | 6 | 7 | 8 | 94 |
| Total Snowfall (") | 3.5 | 3.2 | 2.3 | 0.0 | 0.0 | 0.0 | 0.0 | 0.0 | 0.0 | 0.1 | 0.5 | 1.7 | 11.3 |
| Days ≥ 1" Snow Depth | *2* | 1 | 0 | 0 | 0 | 0 | 0 | 0 | 0 | 0 | 0 | 1 | 4 |

## DANBURY 1 NW *Stokes County* ELEVATION 801 ft LAT/LONG 36° 25 ' N / 80° 12 ' W

| | JAN | FEB | MAR | APR | MAY | JUN | JUL | AUG | SEP | OCT | NOV | DEC | YEAR |
|---|---|---|---|---|---|---|---|---|---|---|---|---|---|
| Maximum Temp °F | 46.6 | 50.5 | 59.1 | 69.4 | 76.1 | 83.1 | 86.9 | 85.4 | 79.7 | 69.9 | 60.5 | 50.8 | 68.2 |
| Minimum Temp °F | 24.5 | 26.8 | 34.2 | 42.4 | 50.7 | 59.8 | 64.5 | 63.1 | 55.9 | 42.6 | 34.8 | 28.0 | 43.9 |
| Mean Temp °F | 35.6 | 38.6 | 46.7 | 55.9 | 63.5 | 71.5 | 75.7 | 74.2 | 67.8 | 56.3 | 47.7 | 39.4 | 56.1 |
| Days Max Temp ≥ 90 °F | 0 | 0 | 0 | 0 | 0 | 4 | 9 | 7 | 3 | 0 | 0 | 0 | 23 |
| Days Max Temp ≤ 32 °F | 3 | 1 | 0 | 0 | 0 | 0 | 0 | 0 | 0 | 0 | 0 | 1 | 5 |
| Days Min Temp ≤ 32 °F | 24 | 21 | 13 | 4 | 0 | 0 | 0 | 0 | 0 | 5 | 13 | 21 | 101 |
| Days Min Temp ≤ 0 °F | 0 | 0 | 0 | 0 | 0 | 0 | 0 | 0 | 0 | 0 | 0 | 0 | 0 |
| Heating Degree Days | 906 | 737 | 564 | 283 | 108 | 11 | 1 | 3 | 47 | 279 | 513 | 787 | 4239 |
| Cooling Degree Days | 0 | 0 | 3 | 18 | 71 | 234 | 359 | 298 | 144 | 17 | 1 | 0 | 1145 |
| Total Precipitation (") | 3.55 | 3.31 | 4.75 | 3.67 | 4.59 | 3.70 | 5.06 | 3.99 | 3.99 | 4.25 | 3.29 | 3.54 | 47.69 |
| Days ≥ 0.1" Precip | 6 | 6 | 7 | 7 | 7 | 6 | 8 | 7 | 6 | 5 | 6 | 6 | 77 |
| Total Snowfall (") | *2.2* | na | 1.8 | 0.1 | 0.0 | 0.0 | 0.0 | 0.0 | 0.0 | 0.0 | 0.0 | 0.9 | na |
| Days ≥ 1" Snow Depth | 4 | *2* | 0 | 0 | 0 | 0 | 0 | 0 | 0 | 0 | 0 | 1 | 7 |

## DUNN 4 NW *Harnett County* ELEVATION 200 ft LAT/LONG 35° 19 ' N / 78° 41 ' W

| | JAN | FEB | MAR | APR | MAY | JUN | JUL | AUG | SEP | OCT | NOV | DEC | YEAR |
|---|---|---|---|---|---|---|---|---|---|---|---|---|---|
| Maximum Temp °F | 51.4 | 55.9 | 64.6 | 74.2 | 80.3 | 86.5 | 89.5 | 87.8 | 83.0 | 73.4 | 64.6 | 55.7 | 72.2 |
| Minimum Temp °F | 29.4 | 32.2 | 39.8 | 47.3 | 55.9 | 63.9 | 68.1 | 67.0 | 60.4 | 47.8 | 39.4 | 32.4 | 48.6 |
| Mean Temp °F | 40.5 | 44.1 | 52.2 | 60.8 | 68.1 | 75.2 | 78.8 | 77.3 | 71.7 | 60.6 | 52.0 | 44.2 | 60.5 |
| Days Max Temp ≥ 90 °F | 0 | 0 | 0 | 1 | 3 | 9 | 16 | 12 | 5 | 0 | 0 | 0 | 46 |
| Days Max Temp ≤ 32 °F | 1 | 0 | 0 | 0 | 0 | 0 | 0 | 0 | 0 | 0 | 0 | 0 | 1 |
| Days Min Temp ≤ 32 °F | 19 | 16 | 9 | 1 | 0 | 0 | 0 | 0 | 0 | 1 | 9 | 17 | 72 |
| Days Min Temp ≤ 0 °F | 0 | 0 | 0 | 0 | 0 | 0 | 0 | 0 | 0 | 0 | 0 | 0 | 0 |
| Heating Degree Days | 756 | 583 | 397 | 166 | 43 | 3 | 0 | 0 | 15 | 175 | 391 | 641 | 3170 |
| Cooling Degree Days | 0 | 1 | 9 | 44 | 143 | 338 | 463 | 393 | 226 | 48 | 8 | 2 | 1675 |
| Total Precipitation (") | 3.66 | 3.69 | 4.43 | 3.06 | 3.72 | 4.53 | 5.77 | 5.09 | 3.50 | 2.91 | 2.97 | 3.72 | 47.05 |
| Days ≥ 0.1" Precip | 7 | 7 | 7 | 5 | 7 | 7 | 8 | 8 | 6 | 4 | 5 | 6 | 77 |
| Total Snowfall (") | *1.3* | 1.2 | 0.4 | 0.0 | 0.0 | 0.0 | 0.0 | 0.0 | 0.0 | 0.0 | 0.0 | 0.2 | 3.1 |
| Days ≥ 1" Snow Depth | *0* | 0 | 0 | 0 | 0 | 0 | 0 | 0 | 0 | 0 | 0 | 0 | 0 |

## DURHAM *Durham County* ELEVATION 410 ft LAT/LONG 36° 2 ' N / 78° 58 ' W

| | JAN | FEB | MAR | APR | MAY | JUN | JUL | AUG | SEP | OCT | NOV | DEC | YEAR |
|---|---|---|---|---|---|---|---|---|---|---|---|---|---|
| Maximum Temp °F | na | *52.7* | *62.1* | *72.8* | *79.3* | *85.7* | *89.6* | *88.2* | *82.8* | *72.1* | *63.1* | *53.7* | na |
| Minimum Temp °F | *26.5* | *27.5* | *35.5* | *44.6* | *53.3* | 62.0 | *66.7* | 65.3 | 58.4 | 45.0 | 36.7 | *30.2* | 46.0 |
| Mean Temp °F | na | *40.2* | *48.9* | *58.7* | *66.4* | *73.8* | *78.2* | *76.9* | *70.8* | *58.6* | *49.8* | *42.1* | na |
| Days Max Temp ≥ 90 °F | na | *0* | *0* | *0* | *2* | *8* | *17* | *13* | 5 | *0* | *0* | *0* | na |
| Days Max Temp ≤ 32 °F | na | *0* | *0* | *0* | *0* | *0* | *0* | 0 | 0 | *0* | *0* | *0* | na |
| Days Min Temp ≤ 32 °F | *23* | *20* | *13* | *3* | *0* | 0 | *0* | 0 | 0 | 3 | 11 | 19 | 92 |
| Days Min Temp ≤ 0 °F | *0* | *0* | *0* | *0* | *0* | 0 | *0* | 0 | 0 | 0 | 0 | 0 | 0 |
| Heating Degree Days | na | *694* | *497* | *213* | *64* | *6* | *0* | *1* | *21* | *217* | *451* | *705* | na |
| Cooling Degree Days | na | na | na | na | na | na | na | na | na | na | na | na | na |
| Total Precipitation (") | 3.87 | 3.75 | 4.53 | 3.23 | 4.74 | 3.88 | 4.21 | 4.57 | 3.58 | 3.80 | 3.27 | 3.53 | 46.96 |
| Days ≥ 0.1" Precip | 7 | 6 | 8 | 6 | 8 | 6 | 7 | 7 | 5 | 5 | 5 | 7 | 77 |
| Total Snowfall (") | 2.2 | 2.2 | 1.1 | 0.0 | 0.0 | 0.0 | 0.0 | 0.0 | 0.0 | 0.0 | 0.1 | 0.4 | 6.0 |
| Days ≥ 1" Snow Depth | 1 | *1* | 0 | 0 | 0 | 0 | 0 | 0 | 0 | 0 | 0 | 0 | 2 |

## EDENTON *Chowan County* ELEVATION 20 ft LAT/LONG 36° 3 ' N / 76° 37 ' W

| | JAN | FEB | MAR | APR | MAY | JUN | JUL | AUG | SEP | OCT | NOV | DEC | YEAR |
|---|---|---|---|---|---|---|---|---|---|---|---|---|---|
| Maximum Temp °F | 51.8 | 55.0 | 63.2 | 72.3 | 78.8 | 85.2 | 88.4 | 86.6 | 81.8 | 72.6 | 64.8 | 56.2 | 71.4 |
| Minimum Temp °F | 32.8 | 34.3 | 41.1 | 48.9 | 57.9 | 66.1 | 70.8 | 69.5 | 63.9 | 52.5 | 44.5 | 36.6 | 51.6 |
| Mean Temp °F | 42.3 | 44.7 | 52.2 | 60.6 | 68.4 | 75.7 | 79.6 | 78.1 | 72.9 | 62.6 | 54.6 | 46.4 | 61.5 |
| Days Max Temp ≥ 90 °F | 0 | 0 | 0 | 1 | 1 | 6 | 12 | 9 | 2 | 0 | 0 | 0 | 31 |
| Days Max Temp ≤ 32 °F | 1 | 0 | 0 | 0 | 0 | 0 | 0 | 0 | 0 | 0 | 0 | 0 | 1 |
| Days Min Temp ≤ 32 °F | 15 | 13 | 6 | 0 | 0 | 0 | 0 | 0 | 0 | 0 | 3 | 11 | 48 |
| Days Min Temp ≤ 0 °F | 0 | 0 | 0 | 0 | 0 | 0 | 0 | 0 | 0 | 0 | 0 | 0 | 0 |
| Heating Degree Days | 696 | 567 | 400 | 170 | 39 | 3 | 0 | 0 | 8 | 126 | 315 | 570 | 2894 |
| Cooling Degree Days | 0 | 1 | 10 | 40 | 158 | 350 | 486 | 398 | 239 | 58 | 11 | 1 | 1752 |
| Total Precipitation (") | 4.23 | 3.42 | 4.30 | 3.34 | 4.45 | 4.19 | 5.66 | 5.44 | 4.01 | 3.21 | 2.73 | 3.12 | 48.10 |
| Days ≥ 0.1" Precip | 8 | 7 | 7 | 6 | 7 | 6 | 8 | 8 | 5 | 5 | 5 | 6 | 78 |
| Total Snowfall (") | 1.3 | 1.6 | 0.9 | 0.1 | 0.0 | 0.0 | 0.0 | 0.0 | 0.0 | 0.0 | 0.0 | 0.2 | 4.1 |
| Days ≥ 1" Snow Depth | *0* | 0 | 0 | 0 | 0 | 0 | 0 | 0 | 0 | 0 | 0 | 0 | 0 |

## ELIZABETH CITY *Pasquotank County* ELEVATION 10 ft LAT/LONG 36° 18 ' N / 76° 13 ' W

| | JAN | FEB | MAR | APR | MAY | JUN | JUL | AUG | SEP | OCT | NOV | DEC | YEAR |
|---|---|---|---|---|---|---|---|---|---|---|---|---|---|
| Maximum Temp °F | 51.0 | 53.7 | 62.1 | 71.1 | 78.0 | 84.7 | 88.5 | 87.1 | 82.5 | 72.8 | 64.9 | 55.8 | 71.0 |
| Minimum Temp °F | 31.1 | 33.0 | 40.0 | 47.6 | 56.9 | 65.2 | 70.1 | 68.8 | 62.9 | 51.3 | 43.0 | 35.1 | 50.4 |
| Mean Temp °F | 41.1 | 43.2 | 51.1 | 59.2 | 67.5 | 75.0 | 79.3 | 78.0 | 72.7 | 62.1 | 54.0 | 45.5 | 60.7 |
| Days Max Temp ≥ 90 °F | 0 | 0 | 0 | 0 | 2 | 7 | 13 | 10 | 3 | 0 | 0 | 0 | 35 |
| Days Max Temp ≤ 32 °F | 1 | 0 | 0 | 0 | 0 | 0 | 0 | 0 | 0 | 0 | 0 | 0 | 1 |
| Days Min Temp ≤ 32 °F | 17 | 14 | 8 | 1 | 0 | 0 | 0 | 0 | 0 | 1 | 5 | 14 | 60 |
| Days Min Temp ≤ 0 °F | 0 | 0 | 0 | 0 | 0 | 0 | 0 | 0 | 0 | 0 | 0 | 0 | 0 |
| Heating Degree Days | 735 | 610 | 433 | 205 | 55 | 5 | 0 | 0 | 9 | 142 | 338 | 600 | 3132 |
| Cooling Degree Days | 0 | 1 | 10 | 38 | 154 | 344 | 495 | 419 | 260 | 67 | 13 | 1 | 1802 |
| Total Precipitation (") | 4.34 | 3.27 | 4.25 | 3.10 | 4.20 | 4.29 | 5.86 | 5.50 | 4.19 | 3.09 | 2.90 | 3.09 | 48.08 |
| Days ≥ 0.1" Precip | 7 | 6 | 7 | 6 | 7 | 6 | 8 | 7 | 5 | 5 | 5 | 6 | 75 |
| Total Snowfall (") | *0.3* | *0.1* | 0.1 | 0.0 | 0.0 | 0.0 | 0.0 | 0.0 | 0.0 | 0.0 | 0.0 | 0.1 | 0.6 |
| Days ≥ 1" Snow Depth | 0 | *0* | 0 | 0 | 0 | 0 | 0 | 0 | 0 | 0 | 0 | 0 | 0 |

## ELIZABETHTOWN LOCK 2 *Bladen County* ELEVATION 59 ft LAT/LONG 34° 38 ' N / 78° 35 ' W

| | JAN | FEB | MAR | APR | MAY | JUN | JUL | AUG | SEP | OCT | NOV | DEC | YEAR |
|---|---|---|---|---|---|---|---|---|---|---|---|---|---|
| Maximum Temp °F | 54.7 | 58.7 | 67.4 | 75.4 | 81.1 | 86.0 | 89.2 | 88.3 | 83.8 | 75.2 | 67.2 | 59.0 | 73.8 |
| Minimum Temp °F | 32.1 | 34.0 | 41.3 | 48.7 | 57.2 | 64.4 | 69.0 | 68.2 | 62.5 | 50.8 | 41.7 | 34.6 | 50.4 |
| Mean Temp °F | 43.4 | 46.4 | 54.4 | 62.1 | 69.2 | 75.2 | 79.1 | 78.3 | 73.2 | 63.0 | 54.4 | 46.7 | 62.1 |
| Days Max Temp ≥ 90 °F | 0 | 0 | 0 | 1 | 2 | 7 | 16 | 13 | 5 | 0 | 0 | 0 | 44 |
| Days Max Temp ≤ 32 °F | 1 | 0 | 0 | 0 | 0 | 0 | 0 | 0 | 0 | 0 | 0 | 0 | 1 |
| Days Min Temp ≤ 32 °F | 16 | 13 | 6 | 1 | 0 | 0 | 0 | 0 | 0 | 1 | 6 | 14 | 57 |
| Days Min Temp ≤ 0 °F | 0 | 0 | 0 | 0 | 0 | 0 | 0 | 0 | 0 | 0 | 0 | 0 | 0 |
| Heating Degree Days | 662 | 521 | 335 | 142 | 30 | 2 | 0 | 0 | 7 | 128 | 325 | 562 | 2714 |
| Cooling Degree Days | 0 | 2 | *12* | 49 | *165* | 334 | 462 | *427* | *258* | *78* | *15* | *3* | 1805 |
| Total Precipitation (") | 3.92 | 3.30 | 4.43 | 2.90 | 3.55 | 4.24 | 6.16 | 5.86 | 3.97 | 2.72 | 2.63 | 3.23 | 46.91 |
| Days ≥ 0.1" Precip | 7 | 6 | 6 | 5 | 6 | 6 | 8 | 7 | 5 | 4 | 4 | 5 | 69 |
| Total Snowfall (") | *0.3* | 0.7 | 0.0 | 0.0 | 0.0 | 0.0 | 0.0 | 0.0 | 0.0 | 0.0 | 0.0 | 0.1 | 1.1 |
| Days ≥ 1" Snow Depth | *0* | 0 | 0 | 0 | 0 | 0 | 0 | 0 | 0 | 0 | 0 | 0 | 0 |

## FAYETTEVILLE *Cumberland County* ELEVATION 102 ft LAT/LONG 35° 3 ' N / 78° 51 ' W

| | JAN | FEB | MAR | APR | MAY | JUN | JUL | AUG | SEP | OCT | NOV | DEC | YEAR |
|---|---|---|---|---|---|---|---|---|---|---|---|---|---|
| Maximum Temp °F | 51.7 | 55.4 | 64.0 | 73.7 | 80.2 | 87.0 | 90.3 | 88.3 | 83.5 | 73.9 | 65.3 | 55.9 | 72.4 |
| Minimum Temp °F | 29.7 | 31.8 | 39.2 | 47.6 | 56.4 | 64.7 | 69.7 | 68.3 | 61.8 | 48.7 | 40.0 | 32.8 | 49.2 |
| Mean Temp °F | 40.7 | 43.6 | 51.6 | 60.7 | 68.3 | 75.9 | 80.0 | 78.3 | 72.7 | 61.3 | 52.7 | 44.4 | 60.9 |
| Days Max Temp ≥ 90 °F | 0 | 0 | 0 | 1 | 3 | 11 | 18 | 14 | 6 | 0 | 0 | 0 | 53 |
| Days Max Temp ≤ 32 °F | 1 | 0 | 0 | 0 | 0 | 0 | 0 | 0 | 0 | 0 | 0 | 0 | 1 |
| Days Min Temp ≤ 32 °F | 19 | 17 | 8 | 1 | 0 | 0 | 0 | 0 | 0 | 1 | 8 | 17 | 71 |
| Days Min Temp ≤ 0 °F | 0 | 0 | 0 | 0 | 0 | 0 | 0 | 0 | 0 | 0 | 0 | 0 | 0 |
| Heating Degree Days | 746 | 597 | 417 | 174 | 45 | 4 | 0 | 0 | 14 | 161 | 374 | 635 | 3167 |
| Cooling Degree Days | 0 | 1 | 9 | 47 | 155 | 360 | 497 | 420 | 249 | 56 | 8 | 3 | 1805 |
| Total Precipitation (") | 3.70 | 3.51 | 4.14 | 2.95 | 3.42 | 4.27 | 5.41 | 5.47 | 4.06 | 2.80 | 2.88 | 3.23 | 45.84 |
| Days ≥ 0.1" Precip | 7 | 6 | 7 | 5 | 7 | 6 | 8 | 8 | 5 | 5 | 5 | 6 | 75 |
| Total Snowfall (") | *0.7* | 0.6 | 0.3 | 0.0 | 0.0 | 0.0 | 0.0 | 0.0 | 0.0 | 0.0 | 0.0 | 0.4 | 2.0 |
| Days ≥ 1" Snow Depth | *1* | 0 | 0 | 0 | 0 | 0 | 0 | 0 | 0 | 0 | 0 | 0 | 1 |

## FLETCHER 2 NE *Buncombe County* ELEVATION 2160 ft LAT/LONG 35° 26 ' N / 82° 26 ' W

| | JAN | FEB | MAR | APR | MAY | JUN | JUL | AUG | SEP | OCT | NOV | DEC | YEAR |
|---|---|---|---|---|---|---|---|---|---|---|---|---|---|
| Maximum Temp °F | 46.1 | 49.9 | 57.9 | 67.2 | 74.2 | 80.2 | 83.4 | 82.2 | 77.0 | 67.9 | 59.0 | 50.8 | 66.3 |
| Minimum Temp °F | 21.2 | 23.4 | 30.4 | 37.8 | 46.6 | 55.0 | 59.4 | 58.3 | 51.8 | 39.4 | 31.0 | 25.2 | 40.0 |
| Mean Temp °F | 33.6 | 36.6 | 44.2 | 52.5 | 60.5 | 67.6 | 71.5 | 70.3 | 64.4 | 53.7 | 45.0 | 38.0 | 53.2 |
| Days Max Temp ≥ 90 °F | 0 | 0 | 0 | 0 | 0 | 1 | 3 | 2 | 0 | 0 | 0 | 0 | 6 |
| Days Max Temp ≤ 32 °F | 4 | 2 | 0 | 0 | 0 | 0 | 0 | 0 | 0 | 0 | 0 | 1 | 7 |
| Days Min Temp ≤ 32 °F | 26 | 23 | 19 | 9 | 2 | 0 | 0 | 0 | 0 | 9 | 18 | 24 | 130 |
| Days Min Temp ≤ 0 °F | 1 | 0 | 0 | 0 | 0 | 0 | 0 | 0 | 0 | 0 | 0 | 0 | 1 |
| Heating Degree Days | 965 | 794 | 641 | 370 | 159 | 29 | 3 | 8 | 83 | 347 | 593 | 830 | 4822 |
| Cooling Degree Days | 0 | 0 | 1 | 3 | 31 | 134 | *239* | *189* | 79 | 6 | *0* | 0 | 682 |
| Total Precipitation (") | 3.21 | 3.50 | 4.58 | 3.17 | 4.36 | 3.77 | 4.24 | 4.59 | 3.60 | 3.45 | 3.35 | 3.08 | 44.90 |
| Days ≥ 0.1" Precip | 6 | 6 | 7 | 6 | 7 | 6 | 8 | 7 | 6 | 5 | 5 | 6 | 75 |
| Total Snowfall (") | *3.9* | 3.5 | 2.4 | 0.5 | 0.0 | 0.0 | 0.0 | 0.0 | 0.0 | 0.0 | 0.5 | *0.9* | 11.7 |
| Days ≥ 1" Snow Depth | 5 | 3 | 1 | 0 | 0 | 0 | 0 | 0 | 0 | 0 | 0 | 1 | 10 |

## FLETCHER 3 W *Henderson County* ELEVATION 2070 ft LAT/LONG 35° 26 ' N / 82° 34 ' W

| | JAN | FEB | MAR | APR | MAY | JUN | JUL | AUG | SEP | OCT | NOV | DEC | YEAR |
|---|---|---|---|---|---|---|---|---|---|---|---|---|---|
| Maximum Temp °F | 46.3 | 50.2 | 58.7 | 67.8 | 74.1 | 80.6 | 83.7 | 82.2 | 76.9 | 67.9 | 58.7 | 50.1 | 66.4 |
| Minimum Temp °F | 24.4 | 26.3 | 33.6 | 40.7 | 49.2 | 56.9 | 61.4 | 60.6 | 54.4 | 41.8 | 34.1 | 27.8 | 42.6 |
| Mean Temp °F | 35.4 | 38.3 | 46.1 | 54.3 | 61.7 | 68.8 | 72.6 | 71.4 | 65.7 | 54.9 | 46.4 | 38.9 | 54.5 |
| Days Max Temp ≥ 90 °F | 0 | 0 | 0 | 0 | 0 | 1 | 4 | 2 | 0 | 0 | 0 | 0 | 7 |
| Days Max Temp ≤ 32 °F | 3 | 2 | 0 | 0 | 0 | 0 | 0 | 0 | 0 | 0 | 0 | 2 | 7 |
| Days Min Temp ≤ 32 °F | 24 | 22 | 15 | 6 | 1 | 0 | 0 | 0 | 0 | 6 | 15 | 22 | 111 |
| Days Min Temp ≤ 0 °F | 1 | 0 | 0 | 0 | 0 | 0 | 0 | 0 | 0 | 0 | 0 | 0 | 1 |
| Heating Degree Days | 912 | 749 | 578 | 320 | 130 | 20 | 2 | 4 | 63 | 312 | 551 | 801 | 4442 |
| Cooling Degree Days | 0 | 0 | 1 | 3 | 34 | 148 | 256 | 209 | 84 | 6 | 0 | 0 | 741 |
| Total Precipitation (") | 3.71 | 4.07 | 5.33 | 3.55 | 5.04 | 4.51 | 4.78 | 5.52 | 4.01 | 3.87 | 3.98 | 3.98 | 52.35 |
| Days ≥ 0.1" Precip | 7 | 7 | 8 | 6 | 9 | 8 | 9 | 9 | 7 | 6 | 7 | 7 | 90 |
| Total Snowfall (") | 3.8 | 3.4 | 2.6 | 0.5 | 0.0 | 0.0 | 0.0 | 0.0 | 0.0 | 0.0 | 0.3 | 1.6 | 12.2 |
| Days ≥ 1" Snow Depth | *1* | na | 1 | 0 | 0 | 0 | 0 | 0 | 0 | 0 | 0 | *0* | na |

## FRANKLIN 3 W *Macon County* ELEVATION 2113 ft LAT/LONG 35° 11 ' N / 83° 23 ' W

| | JAN | FEB | MAR | APR | MAY | JUN | JUL | AUG | SEP | OCT | NOV | DEC | YEAR |
|---|---|---|---|---|---|---|---|---|---|---|---|---|---|
| Maximum Temp °F | 49.6 | 53.6 | 61.7 | 70.6 | 76.7 | 82.8 | 85.8 | 84.5 | 79.3 | 70.9 | 61.4 | 52.8 | 69.1 |
| Minimum Temp °F | 25.0 | 26.8 | 34.0 | 40.7 | 49.1 | 57.1 | 62.0 | 61.6 | 55.7 | 42.5 | 33.8 | 27.9 | 43.0 |
| Mean Temp °F | 37.3 | 40.3 | 47.8 | 55.6 | 62.9 | 70.0 | 74.0 | 73.1 | 67.5 | 56.7 | 47.7 | 40.4 | 56.1 |
| Days Max Temp ≥ 90 °F | 0 | 0 | 0 | 0 | 0 | 2 | 8 | 4 | 1 | 0 | 0 | 0 | 15 |
| Days Max Temp ≤ 32 °F | 1 | 1 | 0 | 0 | 0 | 0 | 0 | 0 | 0 | 0 | 0 | 1 | 3 |
| Days Min Temp ≤ 32 °F | 23 | 20 | 15 | 7 | 1 | 0 | 0 | 0 | 0 | 6 | 15 | 21 | 108 |
| Days Min Temp ≤ 0 °F | 1 | 0 | 0 | 0 | 0 | 0 | 0 | 0 | 0 | 0 | 0 | 0 | 1 |
| Heating Degree Days | 849 | 693 | 525 | 281 | 104 | 12 | 0 | 1 | 40 | 260 | 514 | 757 | 4036 |
| Cooling Degree Days | 0 | 0 | 0 | 7 | 46 | 184 | 308 | 262 | 119 | 12 | 0 | 0 | 938 |
| Total Precipitation (") | 4.59 | 4.76 | 5.60 | 4.07 | 4.82 | 4.15 | 4.09 | 4.77 | 4.13 | 3.33 | 4.31 | 4.71 | 53.33 |
| Days ≥ 0.1" Precip | 8 | 7 | 9 | 7 | 9 | 8 | 9 | 8 | 7 | 6 | 7 | 8 | 93 |
| Total Snowfall (") | 2.9 | *2.5* | 0.9 | 0.5 | 0.0 | 0.0 | 0.0 | 0.0 | 0.0 | 0.0 | 0.2 | 1.0 | 8.0 |
| Days ≥ 1" Snow Depth | 2 | 1 | 0 | 0 | 0 | 0 | 0 | 0 | 0 | 0 | 0 | 1 | 4 |

## GASTONIA *Gaston County* ELEVATION 700 ft LAT/LONG 35° 16' N / 81° 8' W

| | JAN | FEB | MAR | APR | MAY | JUN | JUL | AUG | SEP | OCT | NOV | DEC | YEAR |
|---|---|---|---|---|---|---|---|---|---|---|---|---|---|
| Maximum Temp °F | 51.1 | 55.7 | 64.7 | 73.7 | 80.0 | 86.4 | 89.7 | 88.0 | 82.5 | 72.9 | 63.2 | 54.5 | 71.9 |
| Minimum Temp °F | 29.9 | 31.9 | 39.5 | 47.3 | 55.7 | 63.8 | 68.2 | 67.0 | 60.8 | 48.7 | 39.5 | 33.0 | 48.8 |
| Mean Temp °F | 40.5 | 43.8 | 52.1 | 60.5 | 67.9 | 75.1 | 79.0 | 77.6 | 71.7 | 60.8 | 51.4 | 43.7 | 60.3 |
| Days Max Temp ≥ 90 °F | 0 | 0 | 0 | 0 | 2 | 8 | 16 | 13 | 5 | 0 | 0 | 0 | 44 |
| Days Max Temp ≤ 32 °F | 1 | 0 | 0 | 0 | 0 | 0 | 0 | 0 | 0 | 0 | 0 | 0 | 1 |
| Days Min Temp ≤ 32 °F | 19 | 16 | 8 | 2 | 0 | 0 | 0 | 0 | 0 | 2 | 9 | 16 | 72 |
| Days Min Temp ≤ 0 °F | 0 | 0 | 0 | 0 | 0 | 0 | 0 | 0 | 0 | 0 | 0 | 0 | 0 |
| Heating Degree Days | 752 | 592 | 398 | 169 | 42 | 3 | 0 | 0 | 17 | 166 | 406 | 652 | 3197 |
| Cooling Degree Days | 0 | 1 | 7 | 42 | 146 | 347 | 473 | 412 | 233 | 48 | 3 | 1 | 1713 |
| Total Precipitation (") | 3.99 | 3.75 | 4.58 | 2.75 | 4.08 | 3.71 | 3.84 | 4.66 | 3.99 | 3.66 | 3.13 | 3.77 | 45.91 |
| Days ≥ 0.1" Precip | 7 | 6 | 7 | 5 | 7 | 7 | 7 | 7 | 5 | 5 | 6 | 7 | 76 |
| Total Snowfall (") | na | *0.3* | 0.0 | 0.0 | 0.0 | 0.0 | 0.0 | 0.0 | 0.0 | 0.0 | 0.0 | *0.0* | na |
| Days ≥ 1" Snow Depth | na | *0* | 0 | 0 | 0 | 0 | 0 | 0 | 0 | 0 | 0 | *0* | na |

## GOLDSBORO S-J AFB *Wayne County* ELEVATION 109 ft LAT/LONG 35° 20' N / 77° 58' W

| | JAN | FEB | MAR | APR | MAY | JUN | JUL | AUG | SEP | OCT | NOV | DEC | YEAR |
|---|---|---|---|---|---|---|---|---|---|---|---|---|---|
| Maximum Temp °F | 51.6 | 54.7 | 63.6 | 73.3 | 80.3 | 86.9 | 90.3 | 88.5 | 83.7 | 73.8 | 65.2 | 56.1 | 72.3 |
| Minimum Temp °F | 31.3 | 33.1 | 40.8 | 49.1 | 58.0 | 65.7 | 70.5 | 69.3 | 62.7 | 49.7 | 40.8 | 34.2 | 50.4 |
| Mean Temp °F | 41.4 | 43.9 | 52.2 | 61.2 | 69.2 | 76.3 | 80.4 | 78.9 | 73.2 | 61.8 | 53.1 | 45.2 | 61.4 |
| Days Max Temp ≥ 90 °F | 0 | 0 | 0 | 1 | 3 | 11 | 18 | 14 | 6 | 0 | 0 | 0 | 53 |
| Days Max Temp ≤ 32 °F | 2 | 0 | 0 | 0 | 0 | 0 | 0 | 0 | 0 | 0 | 0 | 0 | 2 |
| Days Min Temp ≤ 32 °F | 17 | 15 | 6 | 0 | 0 | 0 | 0 | 0 | 0 | 1 | 7 | 15 | 61 |
| Days Min Temp ≤ 0 °F | 0 | 0 | 0 | 0 | 0 | 0 | 0 | 0 | 0 | 0 | 0 | 0 | 0 |
| Heating Degree Days | 724 | 591 | 401 | 166 | 37 | 4 | 0 | 0 | 12 | 150 | 365 | 611 | 3061 |
| Cooling Degree Days | 0 | 4 | 13 | 61 | 190 | 379 | 519 | 448 | 267 | 57 | 11 | 4 | 1953 |
| Total Precipitation (") | 4.28 | 3.60 | 4.44 | 3.25 | 3.97 | 3.87 | 5.92 | 5.85 | 4.07 | 2.91 | 3.05 | 3.42 | 48.63 |
| Days ≥ 0.1" Precip | 8 | 7 | 7 | 6 | 7 | 6 | 8 | 8 | 5 | 5 | 5 | 6 | 78 |
| Total Snowfall (") | 1.4 | 1.7 | 0.9 | 0.0 | 0.0 | 0.0 | 0.0 | 0.0 | 0.0 | 0.0 | 0.0 | 0.6 | 4.6 |
| Days ≥ 1" Snow Depth | 1 | 1 | 0 | 0 | 0 | 0 | 0 | 0 | 0 | 0 | 0 | 0 | 2 |

## GRANDFATHER MOUNTAIN *Avery County* ELEVATION 5305 ft LAT/LONG 36° 6' N / 81° 49' W

| | JAN | FEB | MAR | APR | MAY | JUN | JUL | AUG | SEP | OCT | NOV | DEC | YEAR |
|---|---|---|---|---|---|---|---|---|---|---|---|---|---|
| Maximum Temp °F | 35.8 | 37.7 | 45.1 | 54.3 | 60.6 | 66.4 | 69.6 | 68.2 | 62.9 | 54.7 | 46.5 | 39.6 | 53.5 |
| Minimum Temp °F | 19.4 | 21.2 | 28.2 | 36.8 | 45.3 | 52.6 | 56.5 | 55.8 | 50.4 | 40.2 | 31.5 | 24.0 | 38.5 |
| Mean Temp °F | 27.6 | 29.4 | 36.7 | 45.6 | 53.0 | 59.5 | 63.0 | 62.0 | 56.7 | 47.5 | 39.1 | 31.8 | 46.0 |
| Days Max Temp ≥ 90 °F | 0 | 0 | 0 | 0 | 0 | 0 | 0 | 0 | 0 | 0 | 0 | 0 | 0 |
| Days Max Temp ≤ 32 °F | 11 | 9 | 4 | 1 | 0 | 0 | 0 | 0 | 0 | 0 | 3 | 7 | 35 |
| Days Min Temp ≤ 32 °F | 26 | 23 | 19 | 10 | 2 | 0 | 0 | 0 | 0 | 7 | 15 | 23 | 125 |
| Days Min Temp ≤ 0 °F | 3 | 2 | 0 | 0 | 0 | 0 | 0 | 0 | 0 | 0 | 0 | 1 | 6 |
| Heating Degree Days | 1149 | 996 | 869 | 575 | 365 | 164 | 80 | 104 | 245 | 536 | 771 | 1021 | 6875 |
| Cooling Degree Days | 0 | 0 | 0 | 0 | 0 | 8 | 35 | 21 | 3 | 0 | 0 | 0 | 67 |
| Total Precipitation (") | 4.05 | 4.50 | 5.55 | 5.09 | 6.26 | 6.15 | 5.39 | 5.78 | 5.81 | 5.38 | 4.72 | 4.01 | 62.69 |
| Days ≥ 0.1" Precip | 10 | 9 | 10 | 8 | 11 | 10 | 10 | 9 | 8 | 6 | 7 | 8 | 106 |
| Total Snowfall (") | 14.4 | 13.9 | 8.8 | 3.9 | 0.5 | 0.0 | 0.0 | 0.0 | 0.0 | 0.5 | 3.4 | 9.0 | 54.4 |
| Days ≥ 1" Snow Depth | 11 | 12 | 6 | 2 | 0 | 0 | 0 | 0 | 0 | 0 | 2 | 7 | 40 |

## GREENVILLE *Pitt County* ELEVATION 30 ft LAT/LONG 35° 37' N / 77° 22' W

| | JAN | FEB | MAR | APR | MAY | JUN | JUL | AUG | SEP | OCT | NOV | DEC | YEAR |
|---|---|---|---|---|---|---|---|---|---|---|---|---|---|
| Maximum Temp °F | 51.5 | 55.0 | 63.8 | 73.3 | 80.0 | 86.3 | 89.9 | 88.1 | 83.3 | 73.7 | 65.3 | 56.1 | 72.2 |
| Minimum Temp °F | 30.0 | 32.2 | 39.7 | 47.5 | 56.4 | 64.1 | 69.0 | 67.6 | 61.1 | 48.7 | 40.1 | 33.2 | 49.1 |
| Mean Temp °F | 40.8 | 43.6 | 51.8 | 60.4 | 68.2 | 75.2 | 79.5 | 77.9 | 72.2 | 61.2 | 52.7 | 44.7 | 60.7 |
| Days Max Temp ≥ 90 °F | 0 | 0 | 0 | 1 | 3 | 10 | 17 | 13 | 5 | 0 | 0 | 0 | 49 |
| Days Max Temp ≤ 32 °F | 1 | 0 | 0 | 0 | 0 | 0 | 0 | 0 | 0 | 0 | 0 | 0 | 1 |
| Days Min Temp ≤ 32 °F | 19 | 16 | 8 | 1 | 0 | 0 | 0 | 0 | 0 | 1 | 8 | 16 | 69 |
| Days Min Temp ≤ 0 °F | 0 | 0 | 0 | 0 | 0 | 0 | 0 | 0 | 0 | 0 | 0 | 0 | 0 |
| Heating Degree Days | 744 | 600 | 416 | 180 | 46 | 3 | 0 | 0 | 13 | 163 | 372 | 623 | 3160 |
| Cooling Degree Days | 0 | 2 | 12 | 49 | 161 | 342 | 487 | 408 | 241 | 56 | 11 | 3 | 1772 |
| Total Precipitation (") | 4.25 | 3.40 | 4.02 | 3.17 | 4.03 | 4.59 | 5.38 | 6.10 | 4.08 | 3.13 | 2.73 | 3.18 | 48.06 |
| Days ≥ 0.1" Precip | 7 | 6 | 7 | 6 | 7 | 7 | 8 | 8 | 5 | 5 | 5 | 6 | 77 |
| Total Snowfall (") | *0.7* | 1.5 | 0.9 | 0.0 | 0.0 | 0.0 | 0.0 | 0.0 | 0.0 | 0.0 | 0.0 | 0.2 | 3.3 |
| Days ≥ 1" Snow Depth | *0* | 0 | 0 | 0 | 0 | 0 | 0 | 0 | 0 | 0 | 0 | 0 | 0 |

## GRNSBR,HGH PT,W-S AP *Guilford County* ELEVATION 902 ft LAT/LONG 36° 5 ' N / 79° 57 ' W

| | JAN | FEB | MAR | APR | MAY | JUN | JUL | AUG | SEP | OCT | NOV | DEC | YEAR |
|---|---|---|---|---|---|---|---|---|---|---|---|---|---|
| Maximum Temp °F | 46.9 | 51.1 | 60.4 | 70.4 | 77.1 | 84.1 | 87.7 | 85.8 | 79.9 | 69.6 | 60.3 | 51.0 | 68.7 |
| Minimum Temp °F | 27.2 | 29.6 | 37.5 | 45.9 | 54.7 | 63.3 | 67.9 | 66.5 | 59.6 | 47.0 | 38.1 | 31.1 | 47.4 |
| Mean Temp °F | 37.1 | 40.4 | 49.0 | 58.2 | 66.0 | 73.7 | 77.8 | 76.1 | 69.8 | 58.3 | 49.2 | 41.1 | 58.1 |
| Days Max Temp ≥ 90 °F | 0 | 0 | 0 | 0 | 1 | 6 | 12 | 9 | 2 | 0 | 0 | 0 | 30 |
| Days Max Temp ≤ 32 °F | 3 | 1 | 0 | 0 | 0 | 0 | 0 | 0 | 0 | 0 | 0 | 1 | 5 |
| Days Min Temp ≤ 32 °F | 22 | 19 | 10 | 2 | 0 | 0 | 0 | 0 | 0 | 1 | 9 | 18 | 81 |
| Days Min Temp ≤ 0 °F | 0 | 0 | 0 | 0 | 0 | 0 | 0 | 0 | 0 | 0 | 0 | 0 | 0 |
| Heating Degree Days | 859 | 689 | 495 | 228 | 71 | 6 | 0 | 1 | 29 | 226 | 469 | 735 | 3808 |
| Cooling Degree Days | 0 | 0 | 5 | 28 | 109 | 288 | 422 | 349 | 181 | 29 | 2 | 0 | 1413 |
| Total Precipitation (") | 3.18 | 3.15 | 3.86 | 3.09 | 4.05 | 3.55 | 4.54 | 3.74 | 3.53 | 3.43 | 2.92 | 3.16 | 42.20 |
| Days ≥ 0.1" Precip | 7 | 6 | 7 | 6 | 7 | 6 | 8 | 6 | 4 | 5 | 5 | 6 | 73 |
| Total Snowfall (") | 3.5 | 3.0 | 1.7 | 0.0 | 0.0 | 0.0 | 0.0 | 0.0 | 0.0 | 0.0 | 0.2 | 0.9 | 9.3 |
| Days ≥ 1" Snow Depth | 3 | 3 | 1 | 0 | 0 | 0 | 0 | 0 | 0 | 0 | 0 | 1 | 8 |

## HAMLET *Richmond County* ELEVATION 351 ft LAT/LONG 34° 53 ' N / 79° 43 ' W

| | JAN | FEB | MAR | APR | MAY | JUN | JUL | AUG | SEP | OCT | NOV | DEC | YEAR |
|---|---|---|---|---|---|---|---|---|---|---|---|---|---|
| Maximum Temp °F | 52.1 | 56.9 | 65.9 | 75.5 | 81.9 | 88.2 | 91.2 | 89.1 | 84.4 | 74.6 | 65.3 | 56.2 | 73.4 |
| Minimum Temp °F | 28.3 | 30.5 | 38.0 | 46.3 | 55.3 | 63.1 | 67.6 | 66.4 | 60.1 | 47.5 | 37.8 | 31.2 | 47.7 |
| Mean Temp °F | 40.2 | 43.7 | 52.0 | 60.9 | 68.6 | 75.7 | 79.4 | 77.8 | 72.3 | 61.1 | 51.6 | 43.7 | 60.6 |
| Days Max Temp ≥ 90 °F | 0 | 0 | 0 | 1 | 4 | 13 | 21 | 16 | 7 | 1 | 0 | 0 | 63 |
| Days Max Temp ≤ 32 °F | 1 | 0 | 0 | 0 | 0 | 0 | 0 | 0 | 0 | 0 | 0 | 0 | 1 |
| Days Min Temp ≤ 32 °F | 21 | 17 | 11 | 2 | 0 | 0 | 0 | 0 | 0 | 2 | 11 | 19 | 83 |
| Days Min Temp ≤ 0 °F | 0 | 0 | 0 | 0 | 0 | 0 | 0 | 0 | 0 | 0 | 0 | 0 | 0 |
| Heating Degree Days | 763 | 594 | 406 | 166 | 37 | 3 | 0 | 0 | 15 | 165 | 403 | 654 | 3206 |
| Cooling Degree Days | 0 | 1 | 9 | 46 | 150 | 342 | 471 | 388 | 226 | 52 | 5 | 1 | 1691 |
| Total Precipitation (") | 3.87 | 3.60 | 4.25 | 2.86 | 4.09 | 4.56 | 6.45 | 4.39 | 3.63 | 3.82 | 3.20 | 3.23 | 47.95 |
| Days ≥ 0.1" Precip | 7 | 7 | 8 | 5 | 7 | 7 | 9 | 7 | 5 | 5 | 6 | 7 | 80 |
| Total Snowfall (") | *0.7* | 1.1 | 0.6 | 0.0 | 0.0 | 0.0 | 0.0 | 0.0 | 0.0 | 0.0 | 0.0 | 0.2 | 2.6 |
| Days ≥ 1" Snow Depth | *1* | *0* | 0 | 0 | 0 | 0 | 0 | 0 | 0 | 0 | 0 | 0 | 1 |

## HATTERAS *Dare County* ELEVATION 10 ft LAT/LONG 35° 13 ' N / 75° 41 ' W

| | JAN | FEB | MAR | APR | MAY | JUN | JUL | AUG | SEP | OCT | NOV | DEC | YEAR |
|---|---|---|---|---|---|---|---|---|---|---|---|---|---|
| Maximum Temp °F | 52.2 | 54.0 | 60.4 | 68.2 | 75.5 | 81.8 | 85.7 | 85.7 | 82.3 | 73.5 | 65.6 | 58.2 | 70.3 |
| Minimum Temp °F | 36.9 | 38.9 | 44.5 | 52.9 | 61.3 | 68.7 | 73.1 | 73.1 | 68.7 | 60.0 | 51.1 | 43.4 | 56.1 |
| Mean Temp °F | 44.6 | 46.5 | 52.4 | 60.6 | 68.4 | 75.3 | 79.4 | 79.4 | 75.5 | 66.8 | 58.4 | 50.8 | 63.2 |
| Days Max Temp ≥ 90 °F | 0 | 0 | 0 | 0 | 0 | 1 | 3 | 3 | 0 | 0 | 0 | 0 | 7 |
| Days Max Temp ≤ 32 °F | 1 | 0 | 0 | 0 | 0 | 0 | 0 | 0 | 0 | 0 | 0 | 0 | 1 |
| Days Min Temp ≤ 32 °F | 10 | 7 | 2 | 0 | 0 | 0 | 0 | 0 | 0 | 0 | 0 | 3 | 22 |
| Days Min Temp ≤ 0 °F | 0 | 0 | 0 | 0 | 0 | 0 | 0 | 0 | 0 | 0 | 0 | 0 | 0 |
| Heating Degree Days | 626 | 517 | 384 | 151 | 25 | 2 | 0 | 0 | 0 | 51 | 214 | 436 | 2406 |
| Cooling Degree Days | na | 0 | 2 | 29 | 151 | 333 | 477 | na | na | na | na | na | na |
| Total Precipitation (") | 5.37 | 3.96 | 4.48 | 2.92 | 3.75 | 4.10 | 5.07 | 5.09 | 4.51 | 4.61 | 4.65 | 4.48 | 52.99 |
| Days ≥ 0.1" Precip | 8 | 7 | 7 | 5 | 6 | 6 | 7 | 7 | 5 | 6 | 6 | 7 | 77 |
| Total Snowfall (") | 0.3 | 0.3 | 0.4 | 0.0 | 0.0 | 0.0 | 0.0 | 0.0 | 0.0 | 0.0 | 0.0 | 0.1 | 1.1 |
| Days ≥ 1" Snow Depth | 0 | 0 | 0 | 0 | 0 | 0 | 0 | 0 | 0 | 0 | 0 | 0 | 0 |

## HENDERSON 2 NNW *Vance County* ELEVATION 479 ft LAT/LONG 36° 22 ' N / 78° 25 ' W

| | JAN | FEB | MAR | APR | MAY | JUN | JUL | AUG | SEP | OCT | NOV | DEC | YEAR |
|---|---|---|---|---|---|---|---|---|---|---|---|---|---|
| Maximum Temp °F | 47.8 | 51.6 | 60.6 | 71.0 | 78.3 | 85.2 | 88.8 | 87.2 | 81.7 | 71.0 | 62.1 | 52.2 | 69.8 |
| Minimum Temp °F | 24.2 | 26.0 | 34.4 | 42.0 | 52.1 | 61.0 | 65.6 | 64.0 | 57.7 | 43.2 | 35.6 | 27.8 | 44.5 |
| Mean Temp °F | 36.0 | 38.9 | 47.5 | 56.5 | 65.2 | na | 77.2 | 75.7 | 69.8 | 57.1 | 48.9 | 40.1 | na |
| Days Max Temp ≥ 90 °F | 0 | 0 | 0 | 0 | 1 | 7 | 15 | 11 | 4 | 0 | 0 | 0 | 38 |
| Days Max Temp ≤ 32 °F | 2 | 1 | 0 | 0 | 0 | 0 | 0 | 0 | 0 | 0 | 0 | 1 | 4 |
| Days Min Temp ≤ 32 °F | 24 | 22 | 15 | 6 | 0 | 0 | 0 | 0 | 0 | 5 | 13 | 22 | 107 |
| Days Min Temp ≤ 0 °F | 0 | 0 | 0 | 0 | 0 | 0 | 0 | 0 | 0 | 0 | 0 | 0 | 0 |
| Heating Degree Days | 891 | 731 | 541 | 271 | 87 | 11 | 1 | 3 | 32 | 261 | 482 | 765 | 4076 |
| Cooling Degree Days | 0 | 0 | 6 | 21 | 103 | na | 415 | 349 | 185 | 30 | 3 | 1 | na |
| Total Precipitation (") | 3.54 | 3.27 | 3.93 | 2.89 | 3.95 | 3.46 | 4.22 | 4.45 | 3.49 | 3.51 | 3.12 | 3.05 | 42.88 |
| Days ≥ 0.1" Precip | 6 | 5 | 7 | 6 | 6 | 6 | 6 | 7 | 5 | 5 | 5 | 6 | 70 |
| Total Snowfall (") | na | 2.8 | na | 0.0 | 0.0 | 0.0 | 0.0 | 0.0 | 0.0 | 0.0 | 0.0 | 0.2 | na |
| Days ≥ 1" Snow Depth | 1 | na | 0 | 0 | 0 | 0 | 0 | 0 | 0 | 0 | 0 | 0 | na |

## HENDERSONVILLE 1 NE *Henderson County* ELEVATION 2152 ft LAT/LONG 35° 20 ' N / 82° 28 ' W

| | JAN | FEB | MAR | APR | MAY | JUN | JUL | AUG | SEP | OCT | NOV | DEC | YEAR |
|---|---|---|---|---|---|---|---|---|---|---|---|---|---|
| Maximum Temp °F | 47.8 | 51.5 | 60.3 | 69.3 | 75.4 | 81.5 | 84.7 | 82.9 | 77.5 | 68.5 | 59.4 | 51.2 | 67.5 |
| Minimum Temp °F | 26.2 | 28.3 | 35.4 | 42.6 | 51.0 | 58.5 | 63.0 | 61.8 | 55.6 | 43.7 | 35.3 | 29.4 | 44.2 |
| Mean Temp °F | 37.0 | 39.9 | 47.9 | 56.0 | 63.2 | 70.0 | 73.9 | 72.4 | 66.6 | 56.1 | 47.4 | 40.3 | 55.9 |
| Days Max Temp ≥ 90 °F | 0 | 0 | 0 | 0 | 0 | 2 | 6 | 3 | 0 | 0 | 0 | 0 | 11 |
| Days Max Temp ≤ 32 °F | 3 | 1 | 0 | 0 | 0 | 0 | 0 | 0 | 0 | 0 | 0 | 1 | 5 |
| Days Min Temp ≤ 32 °F | 22 | 19 | 13 | 4 | 0 | 0 | 0 | 0 | 0 | 4 | 13 | 20 | 95 |
| Days Min Temp ≤ 0 °F | 1 | 0 | 0 | 0 | 0 | 0 | 0 | 0 | 0 | 0 | 0 | 0 | 1 |
| Heating Degree Days | 862 | 703 | 524 | 272 | 97 | 13 | 1 | 3 | 52 | 279 | 523 | 759 | 4088 |
| Cooling Degree Days | 0 | 0 | 2 | 11 | 56 | 205 | 329 | 270 | 124 | 17 | 0 | 0 | 1014 |
| Total Precipitation (") | 4.23 | 4.35 | 5.74 | 3.91 | 5.22 | 4.77 | 4.72 | 5.96 | 4.30 | 4.36 | 4.39 | 4.35 | 56.30 |
| Days ≥ 0.1" Precip | 7 | 7 | 8 | 6 | 9 | 8 | 8 | 9 | 6 | 6 | 6 | 7 | 87 |
| Total Snowfall (") | 3.8 | 3.5 | 2.5 | 0.3 | 0.0 | 0.0 | 0.0 | 0.0 | 0.0 | 0.0 | 0.6 | 0.8 | 11.5 |
| Days ≥ 1" Snow Depth | na | *1* | 0 | 0 | 0 | 0 | 0 | 0 | 0 | 0 | 0 | 0 | na |

## HICKORY REGIONAL AP *Burke County* ELEVATION 1145 ft LAT/LONG 35° 44 ' N / 81° 23 ' W

| | JAN | FEB | MAR | APR | MAY | JUN | JUL | AUG | SEP | OCT | NOV | DEC | YEAR |
|---|---|---|---|---|---|---|---|---|---|---|---|---|---|
| Maximum Temp °F | 47.8 | 51.7 | 60.6 | 70.2 | 76.9 | 83.8 | 87.1 | 85.4 | 79.5 | 70.0 | 60.4 | 51.4 | 68.7 |
| Minimum Temp °F | 28.4 | 30.6 | 38.3 | 46.6 | 55.0 | 63.1 | 67.5 | 66.2 | 59.8 | 47.6 | 38.9 | 31.9 | 47.8 |
| Mean Temp °F | 38.1 | 41.2 | 49.5 | 58.4 | 66.0 | 73.5 | 77.3 | 75.8 | 69.7 | 58.8 | 49.7 | 41.7 | 58.3 |
| Days Max Temp ≥ 90 °F | 0 | 0 | 0 | 0 | 1 | 5 | 11 | 7 | 2 | 0 | 0 | 0 | 26 |
| Days Max Temp ≤ 32 °F | 2 | 1 | 0 | 0 | 0 | 0 | 0 | 0 | 0 | 0 | 0 | 1 | 4 |
| Days Min Temp ≤ 32 °F | 20 | 18 | 9 | 2 | 0 | 0 | 0 | 0 | 0 | 1 | 8 | 17 | 75 |
| Days Min Temp ≤ 0 °F | 0 | 0 | 0 | 0 | 0 | 0 | 0 | 0 | 0 | 0 | 0 | 0 | 0 |
| Heating Degree Days | 826 | 667 | 477 | 217 | 64 | 6 | 0 | 1 | 26 | 212 | 456 | 716 | 3668 |
| Cooling Degree Days | 0 | 0 | 5 | 29 | 105 | 298 | 428 | 355 | 181 | 34 | 2 | 0 | 1437 |
| Total Precipitation (") | 3.78 | 4.06 | 4.95 | 3.34 | 4.57 | 4.68 | 4.45 | 4.14 | 4.21 | 3.72 | 3.58 | 3.74 | 49.22 |
| Days ≥ 0.1" Precip | 6 | 7 | 8 | 6 | 8 | 7 | 7 | 7 | 6 | 5 | 6 | 6 | 79 |
| Total Snowfall (") | 4.3 | 3.0 | 1.7 | 0.1 | 0.0 | 0.0 | 0.0 | 0.0 | 0.0 | 0.0 | 0.1 | 0.8 | 10.0 |
| Days ≥ 1" Snow Depth | 3 | 2 | 1 | 0 | 0 | 0 | 0 | 0 | 0 | 0 | 0 | 0 | 6 |

## HIGH POINT *Guilford County* ELEVATION 912 ft LAT/LONG 35° 57 ' N / 80° 0 ' W

| | JAN | FEB | MAR | APR | MAY | JUN | JUL | AUG | SEP | OCT | NOV | DEC | YEAR |
|---|---|---|---|---|---|---|---|---|---|---|---|---|---|
| Maximum Temp °F | 49.1 | 53.5 | 62.6 | 72.4 | 78.8 | 85.3 | 88.7 | 86.9 | 81.5 | 71.6 | 61.9 | 52.7 | 70.4 |
| Minimum Temp °F | 29.4 | 31.4 | 39.0 | 47.2 | 55.4 | 63.2 | 67.6 | 66.2 | 60.0 | 48.4 | 40.3 | 33.0 | 48.4 |
| Mean Temp °F | 39.2 | 42.5 | 50.8 | 59.8 | 67.1 | 74.3 | 78.2 | 76.6 | 70.7 | 60.0 | 51.1 | 42.8 | 59.4 |
| Days Max Temp ≥ 90 °F | 0 | 0 | 0 | 0 | 1 | 7 | 14 | 10 | 3 | 0 | 0 | 0 | 35 |
| Days Max Temp ≤ 32 °F | 2 | 0 | 0 | 0 | 0 | 0 | 0 | 0 | 0 | 0 | 0 | 1 | 3 |
| Days Min Temp ≤ 32 °F | 20 | 17 | 8 | 2 | 0 | 0 | 0 | 0 | 0 | 1 | 7 | 16 | 71 |
| Days Min Temp ≤ 0 °F | 0 | 0 | 0 | 0 | 0 | 0 | 0 | 0 | 0 | 0 | 0 | 0 | 0 |
| Heating Degree Days | 792 | 630 | 440 | 187 | 51 | 5 | 0 | 1 | 22 | 184 | 413 | 681 | 3406 |
| Cooling Degree Days | 0 | 1 | 7 | 37 | 125 | 307 | 430 | 364 | 197 | 41 | 3 | 0 | 1512 |
| Total Precipitation (") | 3.57 | 3.52 | 4.23 | 3.27 | 4.35 | 3.74 | 4.11 | 4.31 | 3.29 | 3.56 | 3.16 | 3.42 | 44.53 |
| Days ≥ 0.1" Precip | 7 | 6 | 7 | 6 | 7 | 7 | 7 | 7 | 5 | 5 | 5 | 6 | 75 |
| Total Snowfall (") | 2.9 | 2.2 | 0.9 | 0.0 | 0.0 | 0.0 | 0.0 | 0.0 | 0.0 | 0.0 | 0.1 | 0.8 | 6.9 |
| Days ≥ 1" Snow Depth | 2 | 1 | 0 | 0 | 0 | 0 | 0 | 0 | 0 | 0 | 0 | 0 | 3 |

## HIGHLANDS *Macon County* ELEVATION 3763 ft LAT/LONG 35° 3 ' N / 83° 12 ' W

| | JAN | FEB | MAR | APR | MAY | JUN | JUL | AUG | SEP | OCT | NOV | DEC | YEAR |
|---|---|---|---|---|---|---|---|---|---|---|---|---|---|
| Maximum Temp °F | 43.0 | 45.9 | 54.4 | 63.0 | 69.1 | 74.7 | 77.5 | 76.0 | 70.7 | 61.9 | 53.1 | 45.7 | 61.3 |
| Minimum Temp °F | 24.5 | 25.8 | 33.0 | 40.2 | 48.1 | 54.9 | 59.0 | 58.3 | 52.9 | 41.9 | 33.7 | 28.0 | 41.7 |
| Mean Temp °F | 33.8 | 35.9 | 43.7 | 51.6 | 58.7 | 64.8 | 68.3 | 67.2 | 61.8 | 52.0 | 43.4 | 36.9 | 51.5 |
| Days Max Temp ≥ 90 °F | 0 | 0 | 0 | 0 | 0 | 0 | 0 | 0 | 0 | 0 | 0 | 0 | 0 |
| Days Max Temp ≤ 32 °F | 3 | 2 | 1 | 0 | 0 | 0 | 0 | 0 | 0 | 0 | 0 | 3 | 9 |
| Days Min Temp ≤ 32 °F | 23 | 21 | 15 | 7 | 1 | 0 | 0 | 0 | 0 | 6 | 14 | 22 | 109 |
| Days Min Temp ≤ 0 °F | 1 | 0 | 0 | 0 | 0 | 0 | 0 | 0 | 0 | 0 | 0 | 0 | 1 |
| Heating Degree Days | 961 | 816 | 654 | 396 | 197 | 50 | 12 | 18 | 112 | 399 | 641 | 867 | 5123 |
| Cooling Degree Days | 0 | 0 | 0 | 0 | 5 | 52 | 121 | 77 | 21 | 1 | 0 | 0 | 277 |
| Total Precipitation (") | 6.65 | 6.73 | 8.50 | 6.13 | 8.09 | 7.23 | 6.66 | 7.14 | 6.84 | 6.29 | 7.80 | 7.53 | 85.59 |
| Days ≥ 0.1" Precip | 9 | 8 | 9 | 8 | 10 | 9 | 11 | 10 | 8 | 7 | 8 | 9 | 106 |
| Total Snowfall (") | 5.2 | 4.7 | 3.0 | 0.9 | 0.2 | 0.0 | 0.0 | 0.0 | 0.0 | 0.0 | 0.7 | 1.9 | 16.6 |
| Days ≥ 1" Snow Depth | *5* | 3 | 1 | 0 | 0 | 0 | 0 | 0 | 0 | 0 | 0 | 2 | 11 |

## JACKSON *Northampton County* ELEVATION 131 ft LAT/LONG 36° 24' N / 77° 25' W

| | JAN | FEB | MAR | APR | MAY | JUN | JUL | AUG | SEP | OCT | NOV | DEC | YEAR |
|---|---|---|---|---|---|---|---|---|---|---|---|---|---|
| Maximum Temp °F | 49.6 | 53.7 | 62.5 | 72.8 | 79.6 | 86.6 | 90.1 | 88.4 | 83.2 | 72.7 | 63.8 | 54.0 | 71.4 |
| Minimum Temp °F | 28.8 | 30.4 | 37.8 | 45.5 | 54.8 | 63.0 | 67.9 | 66.4 | 60.0 | 47.9 | 39.4 | 32.1 | 47.8 |
| Mean Temp °F | 39.2 | 42.1 | 50.1 | 59.2 | 67.3 | 74.8 | 79.0 | 77.4 | 71.6 | 60.3 | 51.6 | 43.1 | 59.6 |
| Days Max Temp ≥ 90 °F | 0 | 0 | 0 | 1 | 2 | 11 | 17 | 14 | 6 | 0 | 0 | 0 | 51 |
| Days Max Temp ≤ 32 °F | 2 | 1 | 0 | 0 | 0 | 0 | 0 | 0 | 0 | 0 | 0 | 1 | 4 |
| Days Min Temp ≤ 32 °F | 21 | 17 | 10 | 2 | 0 | 0 | 0 | 0 | 0 | 2 | 9 | 17 | 78 |
| Days Min Temp ≤ 0 °F | 0 | 0 | 0 | 0 | 0 | 0 | 0 | 0 | 0 | 0 | 0 | 0 | 0 |
| Heating Degree Days | 792 | 642 | 462 | 208 | 55 | 5 | 0 | 1 | 16 | 182 | 403 | 674 | 3440 |
| Cooling Degree Days | 0 | 0 | 8 | 40 | 136 | 333 | 477 | 400 | 227 | 45 | 6 | 2 | 1674 |
| Total Precipitation (") | 3.80 | 3.39 | 4.12 | 3.08 | 4.27 | 3.81 | 4.93 | 4.63 | 3.37 | 3.17 | 2.90 | 3.29 | 44.76 |
| Days ≥ 0.1" Precip | 8 | 7 | 7 | 6 | 7 | 7 | 8 | 6 | 5 | 4 | 5 | 6 | 76 |
| Total Snowfall (") | *1.5* | *2.6* | 1.6 | 0.2 | 0.0 | 0.0 | 0.0 | 0.0 | 0.0 | 0.0 | 0.0 | 0.7 | 6.6 |
| Days ≥ 1" Snow Depth | 2 | 2 | 1 | 0 | 0 | 0 | 0 | 0 | 0 | 0 | 0 | 1 | 6 |

## JACKSON SPRINGS 5 WN *Montgomery County* ELEVATION 732 ft LAT/LONG 35° 13' N / 79° 44' W

| | JAN | FEB | MAR | APR | MAY | JUN | JUL | AUG | SEP | OCT | NOV | DEC | YEAR |
|---|---|---|---|---|---|---|---|---|---|---|---|---|---|
| Maximum Temp °F | 49.7 | 53.8 | 62.6 | 72.6 | 79.1 | 85.8 | 89.3 | 87.5 | 82.8 | 72.6 | 63.4 | 53.9 | 71.1 |
| Minimum Temp °F | 29.3 | 31.3 | 38.9 | 47.5 | 55.8 | 63.4 | 67.7 | 66.5 | 60.4 | 48.8 | 40.9 | 33.0 | 48.6 |
| Mean Temp °F | 39.6 | 42.5 | 50.8 | 60.1 | 67.5 | 74.7 | 78.5 | 77.0 | 71.6 | 60.7 | 52.2 | 43.5 | 59.9 |
| Days Max Temp ≥ 90 °F | 0 | 0 | 0 | 1 | 2 | 9 | 16 | 12 | 5 | 0 | 0 | 0 | 45 |
| Days Max Temp ≤ 32 °F | 2 | 1 | 0 | 0 | 0 | 0 | 0 | 0 | 0 | 0 | 0 | 1 | 4 |
| Days Min Temp ≤ 32 °F | 19 | 16 | 8 | 1 | 0 | 0 | 0 | 0 | 0 | 1 | 6 | 16 | 67 |
| Days Min Temp ≤ 0 °F | 0 | 0 | 0 | 0 | 0 | 0 | 0 | 0 | 0 | 0 | 0 | 0 | 0 |
| Heating Degree Days | 781 | 628 | 441 | 186 | 51 | 5 | 0 | 1 | 17 | 169 | 384 | 661 | 3324 |
| Cooling Degree Days | 0 | 1 | 8 | 43 | 143 | 325 | 451 | 379 | 220 | 50 | 6 | 1 | 1627 |
| Total Precipitation (") | 3.99 | 3.58 | 4.49 | 2.93 | 3.68 | 4.21 | 5.00 | 4.44 | 3.53 | 3.70 | 3.13 | 3.39 | 46.07 |
| Days ≥ 0.1" Precip | 7 | 6 | 8 | 6 | 7 | 6 | 8 | 7 | 5 | 5 | 5 | 7 | 77 |
| Total Snowfall (") | na | 1.5 | 0.8 | 0.0 | 0.0 | 0.0 | 0.0 | 0.0 | 0.0 | 0.0 | 0.0 | 0.5 | na |
| Days ≥ 1" Snow Depth | na | 0 | 0 | 0 | 0 | 0 | 0 | 0 | 0 | 0 | 0 | 0 | na |

## JEFFERSON 2 ESE *Ashe County* ELEVATION 2904 ft LAT/LONG 36° 25' N / 81° 29' W

| | JAN | FEB | MAR | APR | MAY | JUN | JUL | AUG | SEP | OCT | NOV | DEC | YEAR |
|---|---|---|---|---|---|---|---|---|---|---|---|---|---|
| Maximum Temp °F | 42.5 | 45.5 | 54.7 | 64.0 | 70.6 | 77.2 | 80.5 | 79.2 | 73.2 | 63.9 | 54.9 | 46.8 | 62.7 |
| Minimum Temp °F | 22.3 | 24.1 | 31.7 | 38.9 | 46.9 | 54.3 | 58.9 | 57.6 | 51.2 | 39.5 | 32.0 | 25.6 | 40.3 |
| Mean Temp °F | 32.5 | 34.9 | 43.3 | 51.4 | 58.8 | 65.8 | 69.7 | 68.4 | 62.2 | 51.7 | 43.5 | 36.2 | 51.5 |
| Days Max Temp ≥ 90 °F | 0 | 0 | 0 | 0 | 0 | 0 | 1 | 0 | 0 | 0 | 0 | 0 | 1 |
| Days Max Temp ≤ 32 °F | 6 | 4 | 1 | 0 | 0 | 0 | 0 | 0 | 0 | 0 | 0 | 3 | 14 |
| Days Min Temp ≤ 32 °F | 25 | 22 | 17 | 8 | 2 | 0 | 0 | 0 | 0 | 8 | 17 | 24 | 123 |
| Days Min Temp ≤ 0 °F | 1 | 0 | 0 | 0 | 0 | 0 | 0 | 0 | 0 | 0 | 0 | 0 | 1 |
| Heating Degree Days | 1001 | 844 | 668 | 402 | 199 | 47 | 8 | 15 | 119 | 407 | 638 | 884 | 5232 |
| Cooling Degree Days | 0 | 0 | 0 | 2 | 15 | 83 | 182 | 133 | 47 | 2 | 0 | 0 | 464 |
| Total Precipitation (") | 3.36 | 3.69 | 4.50 | 3.95 | 4.86 | 4.13 | 4.63 | 4.22 | 4.05 | 4.06 | 3.73 | 3.37 | 48.55 |
| Days ≥ 0.1" Precip | 7 | 6 | 8 | 7 | 9 | 7 | 9 | 8 | 6 | 5 | 6 | 6 | 84 |
| Total Snowfall (") | 6.4 | 6.6 | 3.3 | 0.5 | 0.0 | 0.0 | 0.0 | 0.0 | 0.0 | 0.1 | 1.1 | 3.7 | 21.7 |
| Days ≥ 1" Snow Depth | 6 | 6 | 2 | 0 | 0 | 0 | 0 | 0 | 0 | 0 | 1 | 4 | 19 |

## KINSTON 5 SE *Lenoir County* ELEVATION 49 ft LAT/LONG 35° 16' N / 77° 35' W

| | JAN | FEB | MAR | APR | MAY | JUN | JUL | AUG | SEP | OCT | NOV | DEC | YEAR |
|---|---|---|---|---|---|---|---|---|---|---|---|---|---|
| Maximum Temp °F | 51.8 | 55.2 | 63.7 | 73.2 | 79.8 | 86.2 | 89.6 | 88.1 | 83.4 | 74.0 | 65.7 | 56.1 | 72.2 |
| Minimum Temp °F | 30.0 | 31.6 | 38.9 | 46.3 | 54.8 | 62.6 | 67.3 | 66.0 | 59.7 | 47.7 | 40.0 | 32.6 | 48.1 |
| Mean Temp °F | 40.9 | 43.4 | 51.3 | 59.8 | 67.3 | 74.4 | 78.5 | 77.1 | 71.6 | 60.9 | 52.9 | 44.4 | 60.2 |
| Days Max Temp ≥ 90 °F | 0 | 0 | 0 | 1 | 3 | 9 | 17 | 14 | 5 | 0 | 0 | 0 | 49 |
| Days Max Temp ≤ 32 °F | 2 | 0 | 0 | 0 | 0 | 0 | 0 | 0 | 0 | 0 | 0 | 0 | 2 |
| Days Min Temp ≤ 32 °F | 19 | 17 | 9 | 1 | 0 | 0 | 0 | 0 | 0 | 2 | 8 | 17 | 73 |
| Days Min Temp ≤ 0 °F | 0 | 0 | 0 | 0 | 0 | 0 | 0 | 0 | 0 | 0 | 0 | 0 | 0 |
| Heating Degree Days | 741 | 605 | 427 | 193 | 57 | 6 | 0 | 1 | 16 | 169 | 368 | 636 | 3219 |
| Cooling Degree Days | 0 | 2 | 10 | 41 | 142 | 316 | 448 | 376 | 216 | 50 | 11 | 3 | 1615 |
| Total Precipitation (") | 4.28 | 3.40 | 4.21 | 3.45 | 4.32 | 4.48 | 6.00 | 5.88 | 4.38 | 3.31 | 2.73 | 3.55 | 49.99 |
| Days ≥ 0.1" Precip | 8 | 6 | 7 | 6 | 7 | 7 | 9 | 8 | 6 | 5 | 4 | 6 | 79 |
| Total Snowfall (") | 0.7 | 0.9 | 0.8 | 0.0 | 0.0 | 0.0 | 0.0 | 0.0 | 0.0 | 0.0 | 0.0 | 0.4 | 2.8 |
| Days ≥ 1" Snow Depth | 1 | 0 | 0 | 0 | 0 | 0 | 0 | 0 | 0 | 0 | 0 | 0 | 1 |

## LAURINBURG *Scotland County* ELEVATION 230 ft LAT/LONG 34° 47 ' N / 79° 27 ' W

| | JAN | FEB | MAR | APR | MAY | JUN | JUL | AUG | SEP | OCT | NOV | DEC | YEAR |
|---|---|---|---|---|---|---|---|---|---|---|---|---|---|
| Maximum Temp °F | 54.0 | 58.7 | 67.2 | 76.7 | 83.0 | 89.0 | 91.6 | 89.7 | 85.2 | 75.9 | 66.8 | 57.8 | 74.6 |
| Minimum Temp °F | 31.8 | 34.0 | 41.1 | 48.6 | 57.2 | 64.8 | 69.2 | 68.1 | 62.0 | 50.1 | 41.7 | 34.9 | 50.3 |
| Mean Temp °F | 43.0 | 46.4 | 54.2 | 62.7 | 70.1 | 77.0 | 80.4 | 78.9 | 73.7 | 63.0 | 54.3 | 46.4 | 62.5 |
| Days Max Temp ≥ 90 °F | 0 | 0 | 0 | 2 | 5 | 15 | 21 | 17 | 8 | 1 | 0 | 0 | 69 |
| Days Max Temp ≤ 32 °F | 1 | 0 | 0 | 0 | 0 | 0 | 0 | 0 | 0 | 0 | 0 | 0 | 1 |
| Days Min Temp ≤ 32 °F | 17 | 14 | 7 | 1 | 0 | 0 | 0 | 0 | 0 | 1 | 7 | 14 | 61 |
| Days Min Temp ≤ 0 °F | 0 | 0 | 0 | 0 | 0 | 0 | 0 | 0 | 0 | 0 | 0 | 0 | 0 |
| Heating Degree Days | 677 | 522 | 342 | 131 | 25 | 1 | 0 | 0 | 9 | 125 | 328 | 574 | 2734 |
| Cooling Degree Days | 0 | 2 | 14 | 59 | 192 | 385 | 505 | 434 | 270 | 72 | 13 | 3 | 1949 |
| Total Precipitation (") | 3.97 | 3.64 | 4.37 | 2.68 | 3.60 | 5.14 | 6.13 | 5.05 | 4.06 | 3.25 | 3.01 | 3.26 | 48.16 |
| Days ≥ 0.1" Precip | 7 | 6 | 7 | 5 | 6 | 7 | 9 | 7 | 5 | 5 | 5 | 6 | 75 |
| Total Snowfall (") | 0.6 | 1.1 | 0.6 | 0.0 | 0.0 | 0.0 | 0.0 | 0.0 | 0.0 | 0.0 | 0.0 | 0.3 | 2.6 |
| Days ≥ 1" Snow Depth | 1 | 1 | 0 | 0 | 0 | 0 | 0 | 0 | 0 | 0 | 0 | 0 | 2 |

## LENOIR *Caldwell County* ELEVATION 1289 ft LAT/LONG 35° 55 ' N / 81° 32 ' W

| | JAN | FEB | MAR | APR | MAY | JUN | JUL | AUG | SEP | OCT | NOV | DEC | YEAR |
|---|---|---|---|---|---|---|---|---|---|---|---|---|---|
| Maximum Temp °F | 49.7 | 53.9 | 63.0 | 72.7 | 78.8 | 84.9 | 88.1 | 86.5 | 80.9 | 71.9 | 62.3 | 53.3 | 70.5 |
| Minimum Temp °F | 27.0 | 29.3 | 36.6 | 44.4 | 52.9 | 60.8 | 65.1 | 63.9 | 57.7 | 45.7 | 37.0 | 30.2 | 45.9 |
| Mean Temp °F | 38.4 | 41.6 | 49.8 | 58.5 | 65.9 | 72.9 | 76.6 | 75.2 | 69.3 | 58.8 | 49.7 | 41.8 | 58.2 |
| Days Max Temp ≥ 90 °F | 0 | 0 | 0 | 0 | 1 | 6 | 12 | 9 | 3 | 0 | 0 | 0 | 31 |
| Days Max Temp ≤ 32 °F | 1 | 0 | 0 | 0 | 0 | 0 | 0 | 0 | 0 | 0 | 0 | 0 | 1 |
| Days Min Temp ≤ 32 °F | 22 | 19 | 11 | 3 | 0 | 0 | 0 | 0 | 0 | 2 | 11 | 20 | 88 |
| Days Min Temp ≤ 0 °F | 0 | 0 | 0 | 0 | 0 | 0 | 0 | 0 | 0 | 0 | 0 | 0 | 0 |
| Heating Degree Days | 817 | 654 | 467 | 212 | 63 | 5 | 0 | 1 | 28 | 210 | 455 | 712 | 3624 |
| Cooling Degree Days | 0 | 0 | 4 | 24 | 107 | 287 | 409 | 353 | 180 | 33 | 1 | 0 | 1398 |
| Total Precipitation (") | 3.56 | 3.85 | 4.72 | 3.81 | 4.67 | 4.53 | 4.81 | 4.19 | 4.25 | 3.90 | 3.64 | 3.59 | 49.52 |
| Days ≥ 0.1" Precip | 7 | 6 | 7 | 6 | 8 | 7 | 8 | 7 | 6 | 5 | 6 | 6 | 79 |
| Total Snowfall (") | *3.4* | *3.2* | 1.8 | 0.1 | 0.0 | 0.0 | 0.0 | 0.0 | 0.0 | 0.0 | 0.0 | 0.8 | 9.3 |
| Days ≥ 1" Snow Depth | *1* | *1* | 0 | 0 | 0 | 0 | 0 | 0 | 0 | 0 | 0 | *0* | 2 |

## LEWISTON *Bertie County* ELEVATION 49 ft LAT/LONG 36° 8 ' N / 77° 11 ' W

| | JAN | FEB | MAR | APR | MAY | JUN | JUL | AUG | SEP | OCT | NOV | DEC | YEAR |
|---|---|---|---|---|---|---|---|---|---|---|---|---|---|
| Maximum Temp °F | 51.3 | 54.6 | 63.6 | 73.2 | 80.0 | 86.2 | 89.9 | 88.3 | 83.1 | 74.2 | 65.0 | 56.5 | 72.2 |
| Minimum Temp °F | 29.3 | 30.9 | 37.9 | 45.7 | 54.5 | 62.0 | 67.1 | 65.5 | 59.0 | 48.2 | 39.6 | 33.4 | 47.8 |
| Mean Temp °F | 40.3 | 42.8 | 50.8 | 59.5 | 67.3 | 74.1 | 78.5 | 76.9 | 71.1 | 61.2 | 52.3 | 45.0 | 60.0 |
| Days Max Temp ≥ 90 °F | 0 | 0 | 0 | 1 | 3 | 8 | 17 | 13 | 5 | 0 | 0 | 0 | 47 |
| Days Max Temp ≤ 32 °F | 1 | 0 | 0 | 0 | 0 | 0 | 0 | 0 | 0 | 0 | 0 | 0 | 1 |
| Days Min Temp ≤ 32 °F | 19 | 17 | 10 | 3 | 0 | 0 | 0 | 0 | 0 | 2 | 9 | 15 | 75 |
| Days Min Temp ≤ 0 °F | 0 | 0 | 0 | 0 | 0 | 0 | 0 | 0 | 0 | 0 | 0 | 0 | 0 |
| Heating Degree Days | 757 | 623 | 444 | 199 | 57 | 6 | 0 | 1 | 19 | 163 | 384 | 616 | 3269 |
| Cooling Degree Days | 1 | 3 | 11 | 45 | 158 | 328 | 483 | 403 | 218 | 64 | 12 | 3 | 1729 |
| Total Precipitation (") | 3.82 | 3.39 | 4.04 | 2.97 | 4.35 | 4.02 | 5.74 | 5.05 | 3.72 | 3.09 | 2.53 | 3.37 | 46.09 |
| Days ≥ 0.1" Precip | 7 | 6 | 7 | 6 | 7 | 6 | 8 | 7 | 5 | 4 | 5 | 6 | 74 |
| Total Snowfall (") | 1.3 | 2.1 | 1.0 | 0.0 | 0.0 | 0.0 | 0.0 | 0.0 | 0.0 | 0.0 | 0.0 | 0.1 | 4.5 |
| Days ≥ 1" Snow Depth | 1 | 1 | 0 | 0 | 0 | 0 | 0 | 0 | 0 | 0 | 0 | 0 | 2 |

## LEXINGTON *Davidson County* ELEVATION 760 ft LAT/LONG 35° 51 ' N / 80° 16 ' W

| | JAN | FEB | MAR | APR | MAY | JUN | JUL | AUG | SEP | OCT | NOV | DEC | YEAR |
|---|---|---|---|---|---|---|---|---|---|---|---|---|---|
| Maximum Temp °F | 49.5 | 53.9 | 63.6 | 73.6 | 80.1 | 86.5 | 90.0 | 88.2 | 82.4 | 72.5 | 62.5 | 53.2 | 71.3 |
| Minimum Temp °F | 28.3 | 30.6 | 38.1 | 46.0 | 54.8 | 63.0 | 67.2 | 65.8 | 59.2 | 47.2 | 38.5 | 31.7 | 47.5 |
| Mean Temp °F | 38.9 | 42.3 | 50.9 | 59.8 | 67.5 | 74.8 | 78.6 | 77.0 | 70.8 | 59.9 | 50.5 | 42.5 | 59.5 |
| Days Max Temp ≥ 90 °F | 0 | 0 | 0 | 0 | 2 | 9 | 17 | 13 | 4 | 0 | 0 | 0 | 45 |
| Days Max Temp ≤ 32 °F | 1 | 0 | 0 | 0 | 0 | 0 | 0 | 0 | 0 | 0 | 0 | 0 | 1 |
| Days Min Temp ≤ 32 °F | 20 | 18 | 10 | 2 | 0 | 0 | 0 | 0 | 0 | 2 | 10 | 18 | 80 |
| Days Min Temp ≤ 0 °F | 0 | 0 | 0 | 0 | 0 | 0 | 0 | 0 | 0 | 0 | 0 | 0 | 0 |
| Heating Degree Days | 802 | 635 | 436 | 185 | 50 | 4 | 0 | 1 | 23 | 189 | 433 | 693 | 3451 |
| Cooling Degree Days | 0 | 0 | 5 | 25 | 115 | 295 | 423 | 356 | 183 | 34 | 2 | 1 | 1439 |
| Total Precipitation (") | 3.66 | 3.82 | 4.43 | 3.27 | 4.04 | 4.17 | 3.81 | 3.91 | 3.45 | 3.52 | 3.34 | 3.55 | 44.97 |
| Days ≥ 0.1" Precip | 7 | 6 | 7 | 6 | 7 | 6 | 7 | 6 | 5 | 5 | 6 | 6 | 74 |
| Total Snowfall (") | 3.0 | 2.6 | 1.2 | 0.0 | 0.0 | 0.0 | 0.0 | 0.0 | 0.0 | 0.0 | 0.1 | 0.6 | 7.5 |
| Days ≥ 1" Snow Depth | 2 | 1 | 0 | 0 | 0 | 0 | 0 | 0 | 0 | 0 | 0 | 0 | 3 |

## LINCOLNTON 4 W *Lincoln County* ELEVATION 981 ft LAT/LONG 35° 28' N / 81° 19' W

| | JAN | FEB | MAR | APR | MAY | JUN | JUL | AUG | SEP | OCT | NOV | DEC | YEAR |
|---|---|---|---|---|---|---|---|---|---|---|---|---|---|
| Maximum Temp °F | 49.5 | 53.9 | 62.8 | 72.0 | 78.2 | 84.9 | 88.2 | 86.6 | 81.0 | 71.6 | 61.8 | 52.9 | 70.3 |
| Minimum Temp °F | 28.6 | 30.8 | 38.5 | 46.1 | 54.9 | 62.7 | 66.8 | 65.5 | 59.3 | 47.2 | 38.6 | 31.9 | 47.6 |
| Mean Temp °F | 39.0 | 42.4 | 50.6 | 59.0 | 66.6 | 73.8 | 77.5 | 76.1 | 70.2 | 59.4 | 50.2 | 42.4 | 58.9 |
| Days Max Temp ≥ 90 °F | 0 | 0 | 0 | 0 | 0 | 6 | 12 | 8 | 3 | 0 | 0 | 0 | 29 |
| Days Max Temp ≤ 32 °F | 1 | 0 | 0 | 0 | 0 | 0 | 0 | 0 | 0 | 0 | 0 | 0 | 1 |
| Days Min Temp ≤ 32 °F | 20 | 17 | 9 | 2 | 0 | 0 | 0 | 0 | 0 | 2 | 9 | 17 | 76 |
| Days Min Temp ≤ 0 °F | 0 | 0 | 0 | 0 | 0 | 0 | 0 | 0 | 0 | 0 | 0 | 0 | 0 |
| Heating Degree Days | 798 | 633 | 442 | 201 | 55 | 4 | 0 | 1 | 23 | 199 | 439 | 693 | 3488 |
| Cooling Degree Days | 0 | 0 | 4 | 29 | 112 | 302 | 421 | 358 | 182 | 36 | 2 | 1 | 1447 |
| Total Precipitation (") | 3.81 | 3.70 | 4.79 | 3.22 | 4.66 | 3.87 | 4.18 | 4.24 | 3.62 | 4.00 | 3.40 | 3.83 | 47.32 |
| Days ≥ 0.1" Precip | 6 | 6 | 8 | 6 | 8 | 6 | 7 | 7 | 5 | 5 | 6 | 6 | 76 |
| Total Snowfall (") | *3.3* | 2.3 | 1.6 | 0.0 | 0.0 | 0.0 | 0.0 | 0.0 | 0.0 | 0.0 | 0.0 | 0.7 | 7.9 |
| Days ≥ 1" Snow Depth | 3 | 1 | 1 | 0 | 0 | 0 | 0 | 0 | 0 | 0 | 0 | 0 | 5 |

## LOUISBURG *Franklin County* ELEVATION 381 ft LAT/LONG 36° 6' N / 78° 18' W

| | JAN | FEB | MAR | APR | MAY | JUN | JUL | AUG | SEP | OCT | NOV | DEC | YEAR |
|---|---|---|---|---|---|---|---|---|---|---|---|---|---|
| Maximum Temp °F | 49.3 | 52.8 | 62.5 | 72.6 | 79.3 | 86.4 | 89.7 | 88.1 | 82.6 | 72.4 | 63.0 | 53.4 | 71.0 |
| Minimum Temp °F | 23.9 | 25.1 | 33.4 | 41.3 | 50.5 | 59.5 | 64.8 | 62.9 | 55.5 | 42.5 | 33.6 | 26.5 | 43.3 |
| Mean Temp °F | 36.7 | 38.8 | 48.0 | 57.0 | 64.8 | 73.0 | 77.3 | 75.5 | 69.2 | 57.5 | 48.1 | 40.0 | 57.2 |
| Days Max Temp ≥ 90 °F | 0 | 0 | 0 | 1 | 3 | 11 | 18 | 13 | 5 | 0 | 0 | 0 | 51 |
| Days Max Temp ≤ 32 °F | 2 | 1 | 0 | 0 | 0 | 0 | 0 | 0 | 0 | 0 | 0 | 1 | 4 |
| Days Min Temp ≤ 32 °F | 24 | 21 | 16 | 6 | 1 | 0 | 0 | 0 | 0 | 6 | 16 | 23 | 113 |
| Days Min Temp ≤ 0 °F | 0 | 0 | 0 | 0 | 0 | 0 | 0 | 0 | 0 | 0 | 0 | 0 | 0 |
| Heating Degree Days | 869 | 735 | 525 | 260 | 90 | 11 | 1 | 2 | 37 | 247 | 503 | 768 | 4048 |
| Cooling Degree Days | 0 | *0* | *6* | 25 | *87* | 275 | 401 | *311* | *161* | 23 | 2 | 1 | 1292 |
| Total Precipitation (") | 3.76 | 3.61 | 4.23 | 2.94 | 4.51 | 3.47 | 4.47 | 5.08 | 3.35 | 3.37 | 2.99 | 3.10 | 44.88 |
| Days ≥ 0.1" Precip | 7 | 6 | 7 | 5 | 7 | 5 | 7 | 6 | 5 | 5 | 5 | 6 | 71 |
| Total Snowfall (") | 1.5 | 1.4 | 0.7 | 0.0 | 0.0 | 0.0 | 0.0 | 0.0 | 0.0 | 0.0 | 0.0 | 0.5 | 4.1 |
| Days ≥ 1" Snow Depth | 1 | 1 | 0 | 0 | 0 | 0 | 0 | 0 | 0 | 0 | 0 | 0 | 2 |

## LUMBERTON 3 SE *Robeson County* ELEVATION 131 ft LAT/LONG 34° 42' N / 79° 4' W

| | JAN | FEB | MAR | APR | MAY | JUN | JUL | AUG | SEP | OCT | NOV | DEC | YEAR |
|---|---|---|---|---|---|---|---|---|---|---|---|---|---|
| Maximum Temp °F | 52.0 | 55.9 | 64.6 | 74.3 | 80.9 | 86.7 | 89.8 | 88.2 | 83.7 | 74.5 | 66.0 | *57.0* | 72.8 |
| Minimum Temp °F | 28.8 | 30.8 | 38.4 | 46.5 | 54.8 | 63.1 | 67.6 | 66.1 | 59.7 | 46.6 | 38.8 | 32.1 | 47.8 |
| Mean Temp °F | 40.4 | 43.4 | 51.5 | 60.4 | 67.8 | 74.9 | 78.7 | 77.2 | 71.7 | 60.6 | 52.5 | *44.7* | 60.3 |
| Days Max Temp ≥ 90 °F | 0 | 0 | 0 | 1 | 3 | 10 | 16 | 14 | 6 | 0 | 0 | 0 | 50 |
| Days Max Temp ≤ 32 °F | 1 | 0 | 0 | 0 | 0 | 0 | 0 | 0 | 0 | 0 | 0 | 0 | 1 |
| Days Min Temp ≤ 32 °F | 20 | 17 | 9 | 2 | 0 | 0 | 0 | 0 | 0 | 3 | 9 | 17 | 77 |
| Days Min Temp ≤ 0 °F | 0 | 0 | 0 | 0 | 0 | 0 | 0 | 0 | 0 | 0 | 0 | 0 | 0 |
| Heating Degree Days | 754 | 605 | 416 | 179 | 49 | 5 | 0 | 1 | 17 | 176 | 381 | *627* | 3210 |
| Cooling Degree Days | *0* | *0* | *5* | *38* | 136 | *332* | *455* | *374* | *219* | *42* | *10* | na | na |
| Total Precipitation (") | 3.75 | 3.35 | 4.18 | 2.73 | 4.01 | 4.93 | 5.64 | 5.36 | 3.88 | 2.94 | 2.78 | 3.21 | 46.76 |
| Days ≥ 0.1" Precip | 7 | 6 | 7 | 5 | 6 | 7 | 8 | 7 | 5 | 5 | 5 | 6 | 74 |
| Total Snowfall (") | *0.4* | 1.1 | 0.9 | 0.0 | 0.0 | 0.0 | 0.0 | 0.0 | 0.0 | 0.0 | 0.0 | 0.2 | 2.6 |
| Days ≥ 1" Snow Depth | 1 | 0 | 0 | 0 | 0 | 0 | 0 | 0 | 0 | 0 | 0 | *0* | 1 |

## MARION *McDowell County* ELEVATION 1430 ft LAT/LONG 35° 41' N / 82° 1' W

| | JAN | FEB | MAR | APR | MAY | JUN | JUL | AUG | SEP | OCT | NOV | DEC | YEAR |
|---|---|---|---|---|---|---|---|---|---|---|---|---|---|
| Maximum Temp °F | 48.8 | 53.4 | 62.9 | 72.2 | 78.4 | 84.5 | 87.5 | 85.5 | 79.5 | 70.4 | 60.6 | 52.4 | 69.7 |
| Minimum Temp °F | 27.7 | 30.2 | 37.4 | 45.4 | 53.7 | 61.3 | 65.7 | 64.7 | 58.8 | 47.0 | 38.1 | 31.5 | 46.8 |
| Mean Temp °F | 38.3 | 41.8 | 50.2 | 58.9 | 66.1 | 72.9 | 76.6 | 75.1 | 69.2 | 58.8 | 49.4 | 42.0 | 58.3 |
| Days Max Temp ≥ 90 °F | 0 | 0 | 0 | 0 | 1 | 5 | 10 | 6 | 1 | 0 | 0 | 0 | 23 |
| Days Max Temp ≤ 32 °F | 1 | 0 | 0 | 0 | 0 | 0 | 0 | 0 | 0 | 0 | 0 | 0 | 1 |
| Days Min Temp ≤ 32 °F | 21 | 18 | 10 | 2 | 0 | 0 | 0 | 0 | 0 | 1 | 9 | 18 | 79 |
| Days Min Temp ≤ 0 °F | 0 | 0 | 0 | 0 | 0 | 0 | 0 | 0 | 0 | 0 | 0 | 0 | 0 |
| Heating Degree Days | 821 | 648 | 456 | 202 | 52 | 6 | 0 | 1 | 27 | 209 | 464 | 707 | 3593 |
| Cooling Degree Days | 0 | 0 | 3 | 24 | 95 | 274 | 400 | 336 | 164 | 26 | 0 | 0 | 1322 |
| Total Precipitation (") | 3.81 | 4.19 | 5.46 | 4.13 | 5.39 | 4.82 | 4.53 | 5.02 | 4.58 | 4.61 | 4.57 | 4.26 | 55.37 |
| Days ≥ 0.1" Precip | 7 | 6 | 8 | 6 | 8 | 7 | 9 | 8 | 6 | 6 | 7 | 6 | 84 |
| Total Snowfall (") | *4.8* | *3.4* | 2.4 | 0.3 | 0.0 | 0.0 | 0.0 | 0.0 | 0.0 | 0.0 | 0.2 | 1.2 | 12.3 |
| Days ≥ 1" Snow Depth | na | *1* | *0* | 0 | 0 | 0 | 0 | 0 | 0 | 0 | 0 | *0* | na |

## MARSHALL *Madison County* ELEVATION 1952 ft LAT/LONG 35° 48 ' N / 82° 41 ' W

| | JAN | FEB | MAR | APR | MAY | JUN | JUL | AUG | SEP | OCT | NOV | DEC | YEAR |
|---|---|---|---|---|---|---|---|---|---|---|---|---|---|
| Maximum Temp °F | 45.9 | 50.2 | 59.5 | 68.8 | 75.0 | 81.2 | 84.3 | 83.1 | 78.2 | 68.8 | 59.0 | 50.7 | 67.1 |
| Minimum Temp °F | 24.7 | 26.8 | 34.0 | 41.5 | 49.6 | 57.2 | 61.8 | 60.8 | 54.9 | 42.6 | 34.9 | 28.9 | 43.1 |
| Mean Temp °F | 35.3 | 38.5 | 46.8 | 55.2 | 62.3 | 69.2 | 73.1 | 72.0 | 66.6 | 55.7 | 46.9 | 39.8 | 55.1 |
| Days Max Temp ≥ 90 °F | 0 | 0 | 0 | 0 | 0 | 1 | 4 | 2 | 1 | 0 | 0 | 0 | 8 |
| Days Max Temp ≤ 32 °F | 4 | 2 | 1 | 0 | 0 | 0 | 0 | 0 | 0 | 0 | 0 | 1 | 8 |
| Days Min Temp ≤ 32 °F | 23 | 20 | 15 | 6 | 0 | 0 | 0 | 0 | 0 | 5 | 14 | 20 | 103 |
| Days Min Temp ≤ 0 °F | 1 | 0 | 0 | 0 | 0 | 0 | 0 | 0 | 0 | 0 | 0 | 0 | 1 |
| Heating Degree Days | 914 | 742 | 558 | 296 | 118 | 18 | 1 | 3 | 52 | 291 | 536 | 774 | 4303 |
| Cooling Degree Days | 0 | 0 | 2 | 7 | 48 | 165 | 280 | 237 | 109 | 12 | 0 | 0 | 860 |
| Total Precipitation (") | 2.90 | 3.18 | 4.00 | 3.33 | 3.72 | 3.39 | 3.96 | 3.96 | 2.82 | 2.49 | 2.83 | 2.89 | 39.47 |
| Days ≥ 0.1" Precip | 7 | 7 | 8 | 7 | 8 | 8 | 8 | 8 | 6 | 5 | 7 | 7 | 86 |
| Total Snowfall (") | 4.7 | 3.5 | *2.2* | 0.9 | 0.0 | 0.0 | 0.0 | 0.0 | 0.0 | 0.0 | 0.8 | 1.8 | 13.9 |
| Days ≥ 1" Snow Depth | 4 | 3 | 1 | 0 | 0 | 0 | 0 | 0 | 0 | 0 | 0 | *1* | 9 |

## MONROE 4 SE *Union County* ELEVATION 586 ft LAT/LONG 34° 56 ' N / 80° 31 ' W

| | JAN | FEB | MAR | APR | MAY | JUN | JUL | AUG | SEP | OCT | NOV | DEC | YEAR |
|---|---|---|---|---|---|---|---|---|---|---|---|---|---|
| Maximum Temp °F | 51.5 | 55.9 | 64.6 | 73.7 | 79.9 | 86.4 | 89.3 | 87.6 | 82.7 | 73.2 | 63.9 | 55.2 | 72.0 |
| Minimum Temp °F | 30.6 | 32.7 | 40.1 | 47.6 | 56.1 | 63.9 | 68.2 | 66.8 | 60.6 | 48.2 | 39.9 | 33.6 | 49.0 |
| Mean Temp °F | 41.1 | 44.3 | 52.4 | 60.7 | 68.0 | 75.2 | 78.8 | 77.3 | 71.7 | 60.7 | 51.9 | 44.4 | 60.5 |
| Days Max Temp ≥ 90 °F | 0 | 0 | 0 | 0 | 1 | 8 | 15 | 12 | 4 | 0 | 0 | 0 | 40 |
| Days Max Temp ≤ 32 °F | 1 | 0 | 0 | 0 | 0 | 0 | 0 | 0 | 0 | 0 | 0 | 0 | 1 |
| Days Min Temp ≤ 32 °F | 18 | 15 | 9 | 2 | 0 | 0 | 0 | 0 | 0 | 2 | 9 | 16 | 71 |
| Days Min Temp ≤ 0 °F | 0 | 0 | 0 | 0 | 0 | 0 | 0 | 0 | 0 | 0 | 0 | 0 | 0 |
| Heating Degree Days | 735 | 579 | 393 | 167 | 42 | 3 | 0 | 0 | 18 | 170 | 391 | 633 | 3131 |
| Cooling Degree Days | 0 | 1 | 9 | 41 | 143 | 342 | 458 | 390 | 222 | 49 | 5 | 2 | 1662 |
| Total Precipitation (") | 4.25 | 3.82 | 4.84 | 2.84 | 3.75 | 4.18 | 4.77 | 5.03 | 3.81 | 3.91 | 3.18 | 3.64 | 48.02 |
| Days ≥ 0.1" Precip | 7 | 7 | 8 | 5 | 7 | 7 | 8 | 6 | 5 | 5 | 5 | 6 | 76 |
| Total Snowfall (") | *1.4* | 1.7 | 0.8 | 0.0 | 0.0 | 0.0 | 0.0 | 0.0 | 0.0 | 0.0 | 0.1 | 0.6 | 4.6 |
| Days ≥ 1" Snow Depth | 2 | 1 | 0 | 0 | 0 | 0 | 0 | 0 | 0 | 0 | 0 | 0 | 3 |

## MOREHEAD CITY 2 WNW *Carteret County* ELEVATION 10 ft LAT/LONG 34° 42 ' N / 76° 44 ' W

| | JAN | FEB | MAR | APR | MAY | JUN | JUL | AUG | SEP | OCT | NOV | DEC | YEAR |
|---|---|---|---|---|---|---|---|---|---|---|---|---|---|
| Maximum Temp °F | 55.1 | 57.3 | 63.5 | 71.1 | 77.8 | 83.7 | 87.2 | 86.7 | 83.5 | 75.5 | 67.6 | 59.7 | 72.4 |
| Minimum Temp °F | 35.3 | 36.4 | 43.0 | 51.0 | 60.2 | 68.2 | 72.3 | 71.5 | 66.2 | 55.6 | 46.6 | 38.9 | 53.8 |
| Mean Temp °F | 45.2 | 46.9 | 53.3 | 61.0 | 69.0 | 76.0 | 79.8 | 79.1 | 74.9 | 65.6 | 57.1 | 49.4 | 63.1 |
| Days Max Temp ≥ 90 °F | 0 | 0 | 0 | 0 | 0 | 2 | 6 | 5 | 2 | 0 | 0 | 0 | 15 |
| Days Max Temp ≤ 32 °F | 0 | 0 | 0 | 0 | 0 | 0 | 0 | 0 | 0 | 0 | 0 | 0 | 0 |
| Days Min Temp ≤ 32 °F | 13 | 11 | 5 | 1 | 0 | 0 | 0 | 0 | 0 | 0 | 3 | 10 | 43 |
| Days Min Temp ≤ 0 °F | 0 | 0 | 0 | 0 | 0 | 0 | 0 | 0 | 0 | 0 | 0 | 0 | 0 |
| Heating Degree Days | 608 | 507 | 361 | 149 | 27 | 1 | 0 | 0 | 2 | 78 | 253 | 477 | 2463 |
| Cooling Degree Days | 0 | 0 | 4 | 42 | 167 | 359 | 498 | 449 | 308 | 107 | 26 | 2 | 1962 |
| Total Precipitation (") | 5.08 | 3.93 | 4.37 | 2.56 | 4.66 | 4.18 | 6.46 | 6.60 | 5.33 | 4.45 | 3.78 | 4.42 | 55.82 |
| Days ≥ 0.1" Precip | 8 | 6 | 7 | 5 | 7 | 6 | 9 | 9 | 7 | 5 | 6 | 6 | 81 |
| Total Snowfall (") | 0.3 | 0.8 | 0.8 | 0.0 | 0.0 | 0.0 | 0.0 | 0.0 | 0.0 | 0.0 | 0.0 | 0.1 | 2.0 |
| Days ≥ 1" Snow Depth | 0 | 0 | 0 | 0 | 0 | 0 | 0 | 0 | 0 | 0 | 0 | 0 | 0 |

## MORGANTON *Burke County* ELEVATION 1142 ft LAT/LONG 35° 45 ' N / 81° 41 ' W

| | JAN | FEB | MAR | APR | MAY | JUN | JUL | AUG | SEP | OCT | NOV | DEC | YEAR |
|---|---|---|---|---|---|---|---|---|---|---|---|---|---|
| Maximum Temp °F | 50.3 | 54.5 | 63.7 | 73.1 | 79.4 | 85.8 | 89.0 | 87.3 | 81.4 | 72.1 | 62.3 | 53.5 | 71.0 |
| Minimum Temp °F | 26.7 | 28.9 | 36.1 | 44.0 | 52.3 | 60.2 | 64.7 | 63.6 | 57.1 | 44.9 | 36.1 | 29.7 | 45.4 |
| Mean Temp °F | 38.5 | 41.8 | 50.0 | 58.6 | 65.9 | 73.1 | 76.8 | 75.5 | 69.3 | 58.6 | 49.3 | 41.6 | 58.3 |
| Days Max Temp ≥ 90 °F | 0 | 0 | 0 | 0 | 1 | 8 | 15 | 11 | 3 | 0 | 0 | 0 | 38 |
| Days Max Temp ≤ 32 °F | 1 | 0 | 0 | 0 | 0 | 0 | 0 | 0 | 0 | 0 | 0 | 0 | 1 |
| Days Min Temp ≤ 32 °F | 22 | 19 | 12 | 4 | 0 | 0 | 0 | 0 | 0 | 4 | 11 | 20 | 92 |
| Days Min Temp ≤ 0 °F | 0 | 0 | 0 | 0 | 0 | 0 | 0 | 0 | 0 | 0 | 0 | 0 | 0 |
| Heating Degree Days | 814 | 650 | 462 | 209 | 63 | 6 | 0 | 1 | 31 | 218 | 467 | 719 | 3640 |
| Cooling Degree Days | 0 | 0 | 3 | 22 | 95 | 271 | 392 | 332 | 161 | 24 | 1 | 0 | 1301 |
| Total Precipitation (") | 3.92 | 4.11 | 4.85 | 3.62 | 4.81 | 4.74 | 4.12 | 4.05 | 4.28 | 4.12 | 3.75 | 3.78 | 50.15 |
| Days ≥ 0.1" Precip | 7 | 6 | 8 | 6 | 7 | 7 | 8 | 7 | 6 | 5 | 6 | 6 | 79 |
| Total Snowfall (") | 3.4 | 2.4 | 1.5 | 0.1 | 0.0 | 0.0 | 0.0 | 0.0 | 0.0 | 0.0 | 0.2 | 0.7 | 8.3 |
| Days ≥ 1" Snow Depth | 4 | 2 | 1 | 0 | 0 | 0 | 0 | 0 | 0 | 0 | 0 | 1 | 8 |

## MOUNT AIRY *Surry County* ELEVATION 1070 ft LAT/LONG 36° 31 ' N / 80° 37 ' W

| | JAN | FEB | MAR | APR | MAY | JUN | JUL | AUG | SEP | OCT | NOV | DEC | YEAR |
|---|---|---|---|---|---|---|---|---|---|---|---|---|---|
| Maximum Temp °F | 47.1 | 51.6 | 61.3 | 71.2 | 78.2 | 84.7 | 87.8 | 86.3 | 80.6 | 70.8 | 60.5 | 50.8 | 69.2 |
| Minimum Temp °F | 25.6 | 27.5 | 35.0 | 42.6 | 51.7 | 59.9 | 64.3 | 63.1 | 56.6 | 44.1 | 35.8 | 28.9 | 44.6 |
| Mean Temp °F | 36.4 | 39.6 | 48.2 | 57.0 | 65.0 | 72.3 | 76.1 | 74.7 | 68.6 | 57.5 | 48.2 | 39.9 | 57.0 |
| Days Max Temp ≥ 90 °F | 0 | 0 | 0 | 0 | 1 | 6 | 12 | 8 | 2 | 0 | 0 | 0 | 29 |
| Days Max Temp ≤ 32 °F | 2 | 1 | 0 | 0 | 0 | 0 | 0 | 0 | 0 | 0 | 0 | 1 | 4 |
| Days Min Temp ≤ 32 °F | 23 | 21 | 13 | 4 | 0 | 0 | 0 | 0 | 0 | 4 | 13 | 20 | 98 |
| Days Min Temp ≤ 0 °F | 0 | 0 | 0 | 0 | 0 | 0 | 0 | 0 | 0 | 0 | 0 | 0 | 0 |
| Heating Degree Days | 880 | 710 | 516 | 249 | 78 | 7 | 0 | 1 | 35 | 245 | 500 | 772 | 3993 |
| Cooling Degree Days | 0 | 0 | 2 | 15 | 88 | 256 | 378 | 316 | 157 | 23 | 1 | 0 | 1236 |
| Total Precipitation (") | 3.48 | 3.43 | 4.40 | 3.80 | 4.50 | 3.64 | 4.51 | 3.93 | 4.15 | 3.72 | 3.26 | 3.35 | 46.17 |
| Days ≥ 0.1" Precip | 7 | 6 | 8 | 7 | 8 | 7 | 8 | 7 | 6 | 5 | 6 | 6 | 81 |
| Total Snowfall (") | 3.9 | 4.0 | 1.4 | 0.0 | 0.0 | 0.0 | 0.0 | 0.0 | 0.0 | 0.0 | 0.3 | 1.6 | 11.2 |
| Days ≥ 1" Snow Depth | 3 | 2 | 0 | 0 | 0 | 0 | 0 | 0 | 0 | 0 | 0 | 1 | 6 |

## MURPHY 2 NE *Cherokee County* ELEVATION 1581 ft LAT/LONG 35° 6 ' N / 84° 2 ' W

| | JAN | FEB | MAR | APR | MAY | JUN | JUL | AUG | SEP | OCT | NOV | DEC | YEAR |
|---|---|---|---|---|---|---|---|---|---|---|---|---|---|
| Maximum Temp °F | 48.0 | 52.8 | 60.7 | 70.0 | 76.5 | 83.3 | ***86.2*** | 85.4 | 80.3 | ***71.0*** | 61.2 | ***51.8*** | 68.9 |
| Minimum Temp °F | 25.0 | 26.9 | 33.2 | 40.0 | 48.8 | 57.4 | ***62.1*** | 61.3 | 55.5 | ***42.1*** | 34.2 | ***27.5*** | 42.8 |
| Mean Temp °F | 36.5 | 39.9 | 47.0 | 55.0 | 62.7 | 70.4 | ***74.2*** | 73.4 | 67.9 | ***56.5*** | 47.7 | ***39.7*** | 55.9 |
| Days Max Temp ≥ 90 °F | 0 | 0 | 0 | 0 | 0 | 3 | ***9*** | 5 | 2 | ***0*** | 0 | ***0*** | 19 |
| Days Max Temp ≤ 32 °F | 3 | 1 | 0 | 0 | 0 | 0 | ***0*** | 0 | 0 | ***0*** | 0 | ***1*** | 5 |
| Days Min Temp ≤ 32 °F | 22 | 20 | 16 | 8 | 1 | 0 | ***0*** | 0 | 0 | ***7*** | 16 | ***21*** | 111 |
| Days Min Temp ≤ 0 °F | 1 | 0 | 0 | 0 | 0 | 0 | ***0*** | 0 | 0 | ***0*** | 0 | ***0*** | 1 |
| Heating Degree Days | 876 | 703 | 553 | 302 | 117 | 14 | ***1*** | 1 | 37 | ***268*** | 513 | ***777*** | 4162 |
| Cooling Degree Days | 0 | 0 | 1 | 9 | 51 | 190 | 309 | 269 | 132 | 16 | 1 | ***0*** | 978 |
| Total Precipitation (") | 5.12 | 5.03 | 5.62 | 4.43 | 4.66 | 4.33 | 5.38 | 5.09 | 3.59 | 3.11 | 4.33 | 4.49 | 55.18 |
| Days ≥ 0.1" Precip | 9 | 8 | 9 | 8 | 8 | 8 | 9 | 8 | 7 | 5 | 7 | 8 | 94 |
| Total Snowfall (") | na | na | na | ***0.0*** | ***0.0*** | ***0.0*** | na | ***0.0*** | ***0.0*** | na | na | na | na |
| Days ≥ 1" Snow Depth | na | na | na | na | na | na | na | ***0*** | ***0*** | na | na | na | na |

## NEW BERN CRAVEN CO *Craven County* ELEVATION 20 ft LAT/LONG 35° 5 ' N / 77° 2 ' W

| | JAN | FEB | MAR | APR | MAY | JUN | JUL | AUG | SEP | OCT | NOV | DEC | YEAR |
|---|---|---|---|---|---|---|---|---|---|---|---|---|---|
| Maximum Temp °F | 53.8 | 56.9 | 64.7 | 73.6 | 79.9 | 85.7 | 88.9 | 87.5 | 83.3 | 74.8 | 66.6 | 58.0 | 72.8 |
| Minimum Temp °F | 33.7 | 35.3 | 42.1 | 49.9 | 58.8 | 66.4 | 71.1 | 70.1 | 64.7 | 53.2 | 44.0 | 36.5 | 52.2 |
| Mean Temp °F | 43.8 | 46.1 | 53.4 | 61.7 | 69.4 | 76.1 | 80.0 | 78.8 | 74.0 | 64.0 | 55.3 | 47.3 | 62.5 |
| Days Max Temp ≥ 90 °F | 0 | 0 | 0 | 1 | 2 | 8 | 14 | 11 | 4 | 0 | 0 | 0 | 40 |
| Days Max Temp ≤ 32 °F | 1 | 0 | 0 | 0 | 0 | 0 | 0 | 0 | 0 | 0 | 0 | 0 | 1 |
| Days Min Temp ≤ 32 °F | 15 | 12 | 5 | 0 | 0 | 0 | 0 | 0 | 0 | 0 | 4 | 12 | 48 |
| Days Min Temp ≤ 0 °F | 0 | 0 | 0 | 0 | 0 | 0 | 0 | 0 | 0 | 0 | 0 | 0 | 0 |
| Heating Degree Days | 653 | 529 | 365 | 147 | 31 | 2 | 0 | 0 | 5 | 106 | 302 | 546 | 2686 |
| Cooling Degree Days | 1 | 4 | 14 | 59 | 189 | 377 | 518 | 453 | 293 | 89 | 20 | 4 | 2021 |
| Total Precipitation (") | 4.62 | 3.90 | 4.29 | 3.04 | 4.55 | 4.78 | 6.58 | 6.57 | 4.84 | 3.07 | 3.09 | 3.76 | 53.09 |
| Days ≥ 0.1" Precip | 8 | 6 | 7 | 6 | 7 | 7 | 10 | 9 | 6 | 5 | 5 | 6 | 82 |
| Total Snowfall (") | 1.1 | 0.9 | 0.4 | 0.0 | 0.0 | 0.0 | 0.0 | 0.0 | 0.0 | 0.0 | 0.0 | 0.5 | 2.9 |
| Days ≥ 1" Snow Depth | 1 | 0 | 0 | 0 | 0 | 0 | 0 | 0 | 0 | 0 | 0 | 0 | 1 |

## NEW HOLLAND *Hyde County* ELEVATION 0 ft LAT/LONG 35° 27 ' N / 76° 11 ' W

| | JAN | FEB | MAR | APR | MAY | JUN | JUL | AUG | SEP | OCT | NOV | DEC | YEAR |
|---|---|---|---|---|---|---|---|---|---|---|---|---|---|
| Maximum Temp °F | na | na | na | ***72.3*** | na | na | na | na | na | na | na | na | na |
| Minimum Temp °F | na | na | na | ***48.9*** | na | na | na | na | na | na | na | na | na |
| Mean Temp °F | na | na | na | ***60.6*** | na | na | na | na | na | na | na | na | na |
| Days Max Temp ≥ 90 °F | 0 | 0 | 0 | 0 | 1 | 5 | ***11*** | 9 | ***2*** | 0 | 0 | 0 | 28 |
| Days Max Temp ≤ 32 °F | 1 | 0 | 0 | 0 | 0 | 0 | 0 | 0 | 0 | 0 | 0 | 0 | 1 |
| Days Min Temp ≤ 32 °F | 13 | 11 | 5 | 1 | 0 | 0 | 0 | 0 | 0 | 0 | 3 | 9 | 42 |
| Days Min Temp ≤ 0 °F | 0 | 0 | 0 | 0 | 0 | 0 | 0 | 0 | 0 | 0 | 0 | 0 | 0 |
| Heating Degree Days | na | na | na | ***164*** | na | na | na | na | na | na | na | na | na |
| Cooling Degree Days | na | na | na | na | na | na | na | na | na | na | na | na | na |
| Total Precipitation (") | 4.49 | 3.37 | 4.00 | 3.23 | 4.16 | 4.32 | 5.57 | 6.34 | 4.72 | 4.27 | 3.46 | 3.49 | 51.42 |
| Days ≥ 0.1" Precip | 7 | ***5*** | 6 | 4 | 6 | 5 | 8 | 6 | 4 | 4 | 4 | 5 | 64 |
| Total Snowfall (") | ***0.6*** | 0.8 | 0.1 | 0.0 | 0.0 | 0.0 | 0.0 | 0.0 | 0.0 | 0.0 | 0.0 | 0.4 | 1.9 |
| Days ≥ 1" Snow Depth | ***0*** | 0 | 0 | 0 | 0 | 0 | 0 | 0 | 0 | 0 | 0 | 0 | 0 |

## NORTH WILKESBORO *Wilkes County* ELEVATION 971 ft LAT/LONG 36° 10 ' N / 81° 9 ' W

| | JAN | FEB | MAR | APR | MAY | JUN | JUL | AUG | SEP | OCT | NOV | DEC | YEAR |
|---|---|---|---|---|---|---|---|---|---|---|---|---|---|
| Maximum Temp °F | 47.0 | 51.1 | 60.2 | 70.3 | 77.1 | 83.9 | 87.2 | 85.7 | 79.6 | 70.1 | 60.6 | 50.9 | 68.6 |
| Minimum Temp °F | 22.8 | 24.8 | 32.7 | 41.2 | 50.2 | 59.1 | 63.5 | 62.1 | 55.0 | 41.5 | 33.2 | 26.2 | 42.7 |
| Mean Temp °F | 34.9 | 38.0 | 46.5 | 55.7 | 63.7 | 71.5 | 75.4 | 73.9 | 67.3 | 55.8 | 46.9 | 38.6 | 55.7 |
| Days Max Temp ≥ 90 °F | 0 | 0 | 0 | 0 | 1 | 5 | 11 | 8 | 2 | 0 | 0 | 0 | 27 |
| Days Max Temp ≤ 32 °F | 2 | 1 | 0 | 0 | 0 | 0 | 0 | 0 | 0 | 0 | 0 | 1 | 4 |
| Days Min Temp ≤ 32 °F | 26 | 23 | 16 | 6 | 1 | 0 | 0 | 0 | 0 | 6 | 16 | 24 | 118 |
| Days Min Temp ≤ 0 °F | 0 | 0 | 0 | 0 | 0 | 0 | 0 | 0 | 0 | 0 | 0 | 0 | 0 |
| Heating Degree Days | 925 | 756 | 569 | 287 | 105 | 12 | 1 | 3 | 50 | 290 | 537 | 812 | 4347 |
| Cooling Degree Days | 0 | 0 | 3 | 18 | 78 | 247 | 363 | 301 | 141 | 16 | 1 | 0 | 1168 |
| Total Precipitation (") | 3.89 | 3.92 | 4.83 | 4.16 | 4.66 | 4.70 | 4.37 | 4.58 | 4.18 | 4.01 | 3.50 | 3.70 | 50.50 |
| Days ≥ 0.1" Precip | 7 | 7 | 8 | 7 | 8 | 7 | 8 | 8 | 6 | 5 | 7 | 7 | 85 |
| Total Snowfall (") | 5.0 | 3.7 | 2.4 | 0.0 | 0.0 | 0.0 | 0.0 | 0.0 | 0.0 | 0.0 | 0.2 | 1.6 | 12.9 |
| Days ≥ 1" Snow Depth | 4 | 3 | 1 | 0 | 0 | 0 | 0 | 0 | 0 | 0 | 0 | 1 | 9 |

## OCONALUFTEE *Swain County* ELEVATION 2041 ft LAT/LONG 35° 31 ' N / 83° 18 ' W

| | JAN | FEB | MAR | APR | MAY | JUN | JUL | AUG | SEP | OCT | NOV | DEC | YEAR |
|---|---|---|---|---|---|---|---|---|---|---|---|---|---|
| Maximum Temp °F | 47.5 | 51.6 | 59.9 | 69.0 | 75.5 | 81.8 | 84.9 | 84.0 | 79.0 | 70.0 | 60.5 | 51.5 | 67.9 |
| Minimum Temp °F | 20.9 | 22.7 | 29.5 | 37.4 | 45.5 | 53.6 | 57.9 | 57.1 | 50.7 | 38.2 | 30.2 | 23.8 | 39.0 |
| Mean Temp °F | 34.2 | 37.2 | 44.7 | 53.2 | 60.5 | 67.8 | 71.4 | 70.6 | 64.8 | 54.1 | 45.4 | 37.7 | 53.5 |
| Days Max Temp ≥ 90 °F | 0 | 0 | 0 | 0 | 0 | 1 | 5 | 3 | 0 | 0 | 0 | 0 | 9 |
| Days Max Temp ≤ 32 °F | 2 | 1 | 0 | 0 | 0 | 0 | 0 | 0 | 0 | 0 | 0 | 1 | 4 |
| Days Min Temp ≤ 32 °F | 26 | 23 | 20 | 10 | 2 | 0 | 0 | 0 | 0 | 11 | 19 | 25 | 136 |
| Days Min Temp ≤ 0 °F | 1 | 1 | 0 | 0 | 0 | 0 | 0 | 0 | 0 | 0 | 0 | 0 | 2 |
| Heating Degree Days | 947 | 779 | 620 | 349 | 155 | 25 | 3 | 4 | 72 | 335 | 583 | 839 | 4711 |
| Cooling Degree Days | 0 | 0 | 0 | 3 | 33 | 136 | 238 | 206 | 86 | 7 | 0 | 0 | 709 |
| Total Precipitation (") | 5.14 | 4.81 | 6.02 | 4.41 | 5.22 | 4.52 | 4.80 | 4.53 | 3.96 | 3.55 | 4.54 | 5.20 | 56.70 |
| Days ≥ 0.1" Precip | 8 | 8 | 9 | 8 | 9 | 8 | 9 | 9 | 7 | 6 | 7 | 8 | 96 |
| Total Snowfall (") | ***3.7*** | na | ***1.0*** | 0.9 | 0.0 | 0.0 | 0.0 | 0.0 | 0.0 | 0.0 | 0.3 | na | na |
| Days ≥ 1" Snow Depth | na | na | na | 0 | 0 | 0 | 0 | 0 | 0 | 0 | ***0*** | na | na |

## OXFORD 1 E *Granville County* ELEVATION 502 ft LAT/LONG 36° 17 ' N / 78° 37 ' W

| | JAN | FEB | MAR | APR | MAY | JUN | JUL | AUG | SEP | OCT | NOV | DEC | YEAR |
|---|---|---|---|---|---|---|---|---|---|---|---|---|---|
| Maximum Temp °F | 49.0 | 53.0 | 62.4 | 72.2 | 79.3 | 85.7 | 89.5 | 87.5 | 82.2 | 71.8 | 62.3 | 53.0 | 70.7 |
| Minimum Temp °F | 28.9 | 30.7 | 38.0 | 45.9 | 55.0 | 62.7 | 67.1 | 66.0 | 59.6 | 47.7 | 39.3 | 32.1 | 47.8 |
| Mean Temp °F | 39.0 | 41.9 | 50.2 | 59.1 | 67.2 | 74.2 | 78.3 | 76.8 | 70.9 | 59.7 | 50.8 | 42.6 | 59.2 |
| Days Max Temp ≥ 90 °F | 0 | 0 | 0 | 0 | 2 | 8 | 15 | 12 | 4 | 0 | 0 | 0 | 41 |
| Days Max Temp ≤ 32 °F | 2 | 1 | 0 | 0 | 0 | 0 | 0 | 0 | 0 | 0 | 0 | 1 | 4 |
| Days Min Temp ≤ 32 °F | 20 | 18 | 10 | 2 | 0 | 0 | 0 | 0 | 0 | 2 | 8 | 17 | 77 |
| Days Min Temp ≤ 0 °F | 0 | 0 | 0 | 0 | 0 | 0 | 0 | 0 | 0 | 0 | 0 | 0 | 0 |
| Heating Degree Days | 800 | 647 | 458 | 203 | 53 | 4 | 0 | 1 | 20 | 194 | 424 | 689 | 3493 |
| Cooling Degree Days | 0 | 1 | 8 | 35 | 138 | 308 | 448 | 376 | 209 | 43 | 4 | 1 | 1571 |
| Total Precipitation (") | 3.54 | 3.42 | 4.14 | 2.99 | 4.16 | 3.35 | 4.58 | 4.39 | 3.53 | 3.57 | 3.05 | 3.21 | 43.93 |
| Days ≥ 0.1" Precip | 7 | 6 | 7 | 6 | 7 | 6 | 6 | 7 | 5 | 5 | 6 | 6 | 74 |
| Total Snowfall (") | 2.3 | 2.4 | 1.4 | 0.0 | 0.0 | 0.0 | 0.0 | 0.0 | 0.0 | 0.0 | 0.0 | 0.7 | 6.8 |
| Days ≥ 1" Snow Depth | 2 | 2 | 0 | 0 | 0 | 0 | 0 | 0 | 0 | 0 | 0 | 0 | 4 |

## PISGAH FOREST 1 N *Transylvania County* ELEVATION 2113 ft LAT/LONG 35° 16 ' N / 82° 42 ' W

| | JAN | FEB | MAR | APR | MAY | JUN | JUL | AUG | SEP | OCT | NOV | DEC | YEAR |
|---|---|---|---|---|---|---|---|---|---|---|---|---|---|
| Maximum Temp °F | 47.4 | 50.9 | 58.7 | 68.1 | 74.2 | 80.6 | 83.6 | 82.3 | 77.1 | 68.7 | 59.8 | 51.3 | 66.9 |
| Minimum Temp °F | 22.7 | 24.5 | 30.9 | 39.1 | 47.2 | 55.3 | 59.8 | 58.9 | 53.4 | 40.5 | 32.0 | 25.8 | 40.8 |
| Mean Temp °F | 35.1 | 37.7 | 44.8 | 53.6 | 60.7 | 68.0 | 71.7 | 70.6 | 65.2 | 54.6 | 45.9 | 38.6 | 53.9 |
| Days Max Temp ≥ 90 °F | 0 | 0 | 0 | 0 | 0 | 1 | 4 | 2 | 0 | 0 | 0 | 0 | 7 |
| Days Max Temp ≤ 32 °F | 2 | 1 | 0 | 0 | 0 | 0 | 0 | 0 | 0 | 0 | 0 | 1 | 4 |
| Days Min Temp ≤ 32 °F | 25 | 23 | 19 | 8 | 1 | 0 | 0 | 0 | 0 | 8 | 18 | 24 | 126 |
| Days Min Temp ≤ 0 °F | 1 | 0 | 0 | 0 | 0 | 0 | 0 | 0 | 0 | 0 | 0 | 0 | 1 |
| Heating Degree Days | 921 | 765 | 619 | 338 | 152 | 26 | 3 | 5 | 66 | 322 | 566 | 809 | 4592 |
| Cooling Degree Days | 0 | 0 | 0 | 3 | 25 | 133 | 237 | 185 | 77 | 7 | 0 | 0 | 667 |
| Total Precipitation (") | 5.32 | 5.16 | 6.56 | 4.40 | 6.06 | 5.14 | 5.85 | 6.06 | 4.89 | 5.02 | 5.42 | 5.33 | 65.21 |
| Days ≥ 0.1" Precip | 7 | 7 | 9 | 7 | 10 | 8 | 9 | 9 | 7 | 6 | 7 | 8 | 94 |
| Total Snowfall (") | ***4.8*** | 2.7 | 1.5 | 0.4 | 0.0 | 0.0 | 0.0 | 0.0 | 0.0 | 0.0 | 0.3 | ***1.6*** | 11.3 |
| Days ≥ 1" Snow Depth | 3 | 2 | 1 | 0 | 0 | 0 | 0 | 0 | 0 | 0 | 0 | ***0*** | 6 |

## PLYMOUTH 5 E *Washington County* ELEVATION 20 ft LAT/LONG 35° 52 ' N / 76° 39 ' W

| | JAN | FEB | MAR | APR | MAY | JUN | JUL | AUG | SEP | OCT | NOV | DEC | YEAR |
|---|---|---|---|---|---|---|---|---|---|---|---|---|---|
| Maximum Temp °F | 53.0 | 56.3 | 64.4 | 73.7 | 80.5 | 86.8 | 89.8 | 88.2 | 83.6 | 74.2 | 66.1 | 57.4 | 72.8 |
| Minimum Temp °F | 30.9 | 32.3 | 38.7 | 46.1 | 55.2 | 63.2 | 67.9 | 66.6 | 60.9 | 49.8 | 41.4 | 34.3 | 48.9 |
| Mean Temp °F | 42.0 | 44.3 | 51.6 | 60.0 | 67.9 | 75.0 | 78.9 | 77.5 | 72.3 | 62.1 | 53.8 | 45.9 | 60.9 |
| Days Max Temp ≥ 90 °F | 0 | 0 | 0 | 1 | 3 | 10 | 17 | 13 | 5 | 0 | 0 | 0 | 49 |
| Days Max Temp ≤ 32 °F | 1 | 0 | 0 | 0 | 0 | 0 | 0 | 0 | 0 | 0 | 0 | 0 | 1 |
| Days Min Temp ≤ 32 °F | 18 | 16 | 10 | 2 | 0 | 0 | 0 | 0 | 0 | 1 | 7 | 15 | 69 |
| Days Min Temp ≤ 0 °F | 0 | 0 | 0 | 0 | 0 | 0 | 0 | 0 | 0 | 0 | 0 | 0 | 0 |
| Heating Degree Days | 708 | 580 | 418 | 186 | 47 | 3 | 0 | 0 | 9 | 140 | 344 | 587 | 3022 |
| Cooling Degree Days | 0 | 2 | 10 | 43 | 159 | 340 | 473 | 400 | 243 | 61 | 12 | 3 | 1746 |
| Total Precipitation (") | 4.28 | 3.40 | 4.63 | 3.38 | 4.62 | 4.47 | 5.92 | 5.63 | 4.33 | 3.41 | 3.01 | 3.22 | 50.30 |
| Days ≥ 0.1" Precip | 8 | 6 | 7 | 6 | 8 | 7 | 9 | 7 | 6 | 5 | 5 | 6 | 80 |
| Total Snowfall (") | 1.2 | 1.6 | 0.8 | 0.0 | 0.0 | 0.0 | 0.0 | 0.0 | 0.0 | 0.0 | 0.0 | 0.3 | 3.9 |
| Days ≥ 1" Snow Depth | 1 | 1 | 0 | 0 | 0 | 0 | 0 | 0 | 0 | 0 | 0 | 0 | 2 |

## RALEIGH 4 SW *Wake County* ELEVATION 420 ft LAT/LONG 35° 44 ' N / 78° 41 ' W

| | JAN | FEB | MAR | APR | MAY | JUN | JUL | AUG | SEP | OCT | NOV | DEC | YEAR |
|---|---|---|---|---|---|---|---|---|---|---|---|---|---|
| Maximum Temp °F | 51.0 | 55.0 | 63.7 | 73.1 | 79.6 | 86.0 | 89.0 | 87.6 | 82.7 | 72.7 | 63.8 | 54.9 | 71.6 |
| Minimum Temp °F | 30.4 | 32.4 | 39.8 | 47.5 | 56.0 | 63.7 | 68.0 | 67.0 | 60.9 | 48.9 | 40.8 | 33.9 | 49.1 |
| Mean Temp °F | 40.7 | 43.7 | 51.8 | 60.3 | 67.8 | 74.8 | 78.5 | 77.3 | 71.9 | 60.8 | 52.3 | 44.4 | 60.4 |
| Days Max Temp ≥ 90 °F | 0 | 0 | 0 | 1 | 1 | 7 | 14 | 11 | 4 | 0 | 0 | 0 | 38 |
| Days Max Temp ≤ 32 °F | 2 | 0 | 0 | 0 | 0 | 0 | 0 | 0 | 0 | 0 | 0 | 0 | 2 |
| Days Min Temp ≤ 32 °F | 18 | 16 | 8 | 2 | 0 | 0 | 0 | 0 | 0 | 1 | 7 | 15 | 67 |
| Days Min Temp ≤ 0 °F | 0 | 0 | 0 | 0 | 0 | 0 | 0 | 0 | 0 | 0 | 0 | 0 | 0 |
| Heating Degree Days | 746 | 595 | 412 | 179 | 45 | 4 | 0 | 0 | 15 | 167 | 382 | 633 | 3178 |
| Cooling Degree Days | 0 | 1 | 9 | 46 | 148 | 331 | 453 | 389 | 224 | 49 | 7 | 2 | 1659 |
| Total Precipitation (") | 3.80 | 3.55 | 4.21 | 2.76 | 4.39 | 4.03 | 4.48 | 4.42 | 3.25 | 3.44 | 3.08 | 3.24 | 44.65 |
| Days ≥ 0.1" Precip | 7 | 6 | 8 | 5 | 7 | 6 | 7 | 6 | 5 | 4 | 5 | 6 | 72 |
| Total Snowfall (") | 1.3 | 2.2 | 1.0 | 0.0 | 0.0 | 0.0 | 0.0 | 0.0 | 0.0 | 0.0 | 0.0 | 0.4 | 4.9 |
| Days ≥ 1" Snow Depth | *1* | *0* | 0 | 0 | 0 | 0 | 0 | 0 | 0 | 0 | 0 | 0 | 1 |

## RALEIGH DURHAM AP *Wake County* ELEVATION 443 ft LAT/LONG 35° 52 ' N / 78° 47 ' W

| | JAN | FEB | MAR | APR | MAY | JUN | JUL | AUG | SEP | OCT | NOV | DEC | YEAR |
|---|---|---|---|---|---|---|---|---|---|---|---|---|---|
| Maximum Temp °F | 49.2 | 53.1 | 62.1 | 72.1 | 78.4 | 85.2 | 88.5 | 86.8 | 81.4 | 71.5 | 62.6 | 53.3 | 70.4 |
| Minimum Temp °F | 28.7 | 31.0 | 38.3 | 46.2 | 55.0 | 63.3 | 68.2 | 67.1 | 60.8 | 48.1 | 39.3 | 32.5 | 48.2 |
| Mean Temp °F | 39.0 | 42.0 | 50.3 | 59.2 | 66.7 | 74.3 | 78.4 | 77.0 | 71.1 | 59.8 | 51.0 | 42.9 | 59.3 |
| Days Max Temp ≥ 90 °F | 0 | 0 | 0 | 1 | 1 | 7 | 13 | 10 | 3 | 0 | 0 | 0 | 35 |
| Days Max Temp ≤ 32 °F | 2 | 1 | 0 | 0 | 0 | 0 | 0 | 0 | 0 | 0 | 0 | 1 | 4 |
| Days Min Temp ≤ 32 °F | 20 | 17 | 10 | 2 | 0 | 0 | 0 | 0 | 0 | 1 | 9 | 17 | 76 |
| Days Min Temp ≤ 0 °F | 0 | 0 | 0 | 0 | 0 | 0 | 0 | 0 | 0 | 0 | 0 | 0 | 0 |
| Heating Degree Days | 800 | 642 | 460 | 208 | 61 | 4 | 0 | 1 | 20 | 192 | 420 | 679 | 3487 |
| Cooling Degree Days | 0 | 1 | 10 | 42 | 132 | 321 | 457 | 385 | 215 | 46 | 6 | 2 | 1617 |
| Total Precipitation (") | 3.48 | 3.43 | 3.89 | 2.53 | 4.00 | 3.48 | 4.10 | 3.99 | 3.12 | 3.07 | 2.71 | 3.10 | 40.90 |
| Days ≥ 0.1" Precip | 6 | 6 | 7 | 5 | 7 | 6 | 8 | 6 | 5 | 5 | 5 | 6 | 72 |
| Total Snowfall (") | 2.0 | 2.9 | 1.4 | 0.1 | 0.0 | 0.0 | 0.0 | 0.0 | 0.0 | 0.0 | 0.1 | 0.6 | 7.1 |
| Days ≥ 1" Snow Depth | 2 | 2 | 1 | 0 | 0 | 0 | 0 | 0 | 0 | 0 | 0 | 0 | 5 |

## RALEIGH STATE UNIV *Wake County* ELEVATION 400 ft LAT/LONG 35° 47 ' N / 78° 40 ' W

| | JAN | FEB | MAR | APR | MAY | JUN | JUL | AUG | SEP | OCT | NOV | DEC | YEAR |
|---|---|---|---|---|---|---|---|---|---|---|---|---|---|
| Maximum Temp °F | 48.8 | 52.8 | 61.5 | 71.4 | 77.9 | 84.9 | 88.7 | 87.0 | 81.7 | 71.1 | 62.8 | 52.8 | 70.1 |
| Minimum Temp °F | 29.3 | 31.4 | 39.2 | 47.8 | 56.5 | 64.7 | 69.2 | 68.0 | 61.5 | 49.1 | 41.4 | 33.0 | 49.3 |
| Mean Temp °F | 39.0 | 42.1 | 50.4 | 59.6 | 67.2 | 74.8 | 79.0 | 77.5 | 71.6 | 60.2 | 52.1 | 42.9 | 59.7 |
| Days Max Temp ≥ 90 °F | 0 | 0 | 0 | 0 | 1 | 7 | 13 | 11 | 4 | 0 | 0 | 0 | 36 |
| Days Max Temp ≤ 32 °F | 2 | 1 | 0 | 0 | 0 | 0 | 0 | 0 | 0 | 0 | 0 | 1 | 4 |
| Days Min Temp ≤ 32 °F | 19 | 16 | 8 | 1 | 0 | 0 | 0 | 0 | 0 | 1 | 6 | 16 | 67 |
| Days Min Temp ≤ 0 °F | 0 | 0 | 0 | 0 | 0 | 0 | 0 | 0 | 0 | 0 | 0 | 0 | 0 |
| Heating Degree Days | 798 | 640 | 454 | 197 | 55 | 5 | 0 | 1 | 18 | 183 | 389 | 678 | 3418 |
| Cooling Degree Days | 0 | 1 | 10 | 45 | 141 | 335 | 464 | 391 | 224 | 44 | 6 | 2 | 1663 |
| Total Precipitation (") | 3.91 | 3.49 | 4.33 | 2.79 | 4.37 | 4.09 | 4.47 | 4.27 | 3.11 | 3.55 | 2.89 | 3.33 | 44.60 |
| Days ≥ 0.1" Precip | 7 | 7 | 8 | 6 | 7 | 6 | 8 | 7 | 5 | 5 | 5 | 6 | 77 |
| Total Snowfall (") | 1.6 | 2.2 | 1.1 | 0.0 | 0.0 | 0.0 | 0.0 | 0.0 | 0.0 | 0.0 | 0.0 | 0.6 | 5.5 |
| Days ≥ 1" Snow Depth | 1 | 1 | 0 | 0 | 0 | 0 | 0 | 0 | 0 | 0 | 0 | 0 | 2 |

## REIDSVILLE 2 NW *Rockingham County* ELEVATION 889 ft LAT/LONG 36° 23 ' N / 79° 42 ' W

| | JAN | FEB | MAR | APR | MAY | JUN | JUL | AUG | SEP | OCT | NOV | DEC | YEAR |
|---|---|---|---|---|---|---|---|---|---|---|---|---|---|
| Maximum Temp °F | 45.6 | 49.6 | 58.6 | 68.9 | 76.1 | 83.4 | 87.0 | 85.5 | 79.9 | 69.3 | 59.9 | 50.1 | 67.8 |
| Minimum Temp °F | 26.5 | 28.7 | 36.5 | 45.7 | 54.2 | 62.2 | 66.1 | 64.3 | 57.7 | 45.6 | 38.3 | 30.4 | 46.3 |
| Mean Temp °F | 36.1 | 39.2 | 47.6 | 57.4 | 65.2 | 72.8 | 76.6 | 74.9 | 68.8 | 57.5 | 49.1 | 40.3 | 57.1 |
| Days Max Temp ≥ 90 °F | 0 | 0 | 0 | 0 | 1 | 5 | 10 | 8 | 3 | 0 | 0 | 0 | 27 |
| Days Max Temp ≤ 32 °F | 4 | 2 | 0 | 0 | 0 | 0 | 0 | 0 | 0 | 0 | 0 | 1 | 7 |
| Days Min Temp ≤ 32 °F | 23 | 19 | 11 | 2 | 0 | 0 | 0 | 0 | 0 | 2 | 9 | 19 | 85 |
| Days Min Temp ≤ 0 °F | 0 | 0 | 0 | 0 | 0 | 0 | 0 | 0 | 0 | 0 | 0 | 0 | 0 |
| Heating Degree Days | 889 | 723 | 537 | 249 | 83 | 10 | 1 | 2 | 39 | 249 | 473 | 759 | 4014 |
| Cooling Degree Days | 0 | 0 | 5 | 29 | 104 | 282 | 397 | 324 | 167 | 27 | 2 | 0 | 1337 |
| Total Precipitation (") | 3.67 | 3.28 | 4.42 | 3.42 | 4.15 | 3.80 | 4.89 | 3.71 | 3.47 | 3.68 | 3.14 | 3.25 | 44.88 |
| Days ≥ 0.1" Precip | 7 | 7 | 7 | 6 | 7 | 6 | 7 | 6 | 5 | 5 | 5 | 6 | 74 |
| Total Snowfall (") | 5.0 | 4.1 | 2.2 | 0.0 | 0.0 | 0.0 | 0.0 | 0.0 | 0.0 | 0.0 | 0.1 | 1.1 | 12.5 |
| Days ≥ 1" Snow Depth | 5 | 3 | 1 | 0 | 0 | 0 | 0 | 0 | 0 | 0 | 0 | 1 | 10 |

## ROANOKE RAPIDS *Halifax County* ELEVATION 210 ft LAT/LONG 36° 29 ' N / 77° 40 ' W

| | JAN | FEB | MAR | APR | MAY | JUN | JUL | AUG | SEP | OCT | NOV | DEC | YEAR |
|---|---|---|---|---|---|---|---|---|---|---|---|---|---|
| Maximum Temp °F | na | *51.2* | *60.0* | *70.2* | *77.5* | *85.3* | na | *87.6* | *81.5* | *70.8* | *62.5* | *51.9* | na |
| Minimum Temp °F | *27.3* | *29.3* | *36.7* | *45.0* | *54.3* | *62.7* | *67.6* | *65.8* | *59.0* | *46.1* | *39.0* | *30.7* | 47.0 |
| Mean Temp °F | na | *40.3* | *48.5* | *57.6* | *65.9* | *74.0* | na | *76.7* | *70.3* | *58.4* | *50.8* | *41.2* | na |
| Days Max Temp ≥ 90 °F | *0* | *0* | *0* | *0* | *2* | *8* | *15* | *12* | *4* | *0* | *0* | *0* | 41 |
| Days Max Temp ≤ 32 °F | *2* | *1* | *0* | *0* | *0* | *0* | *0* | *0* | *0* | *0* | *0* | *1* | 4 |
| Days Min Temp ≤ 32 °F | *22* | *19* | *10* | *2* | *0* | *0* | *0* | *0* | *0* | *2* | *8* | *19* | 82 |
| Days Min Temp ≤ 0 °F | *0* | *0* | *0* | *0* | *0* | *0* | *0* | *0* | *0* | *0* | *0* | *0* | 0 |
| Heating Degree Days | na | *693* | *511* | *242* | 73 | *5* | na | *1* | *27* | *224* | *426* | *732* | na |
| Cooling Degree Days | 0 | 0 | 5 | 25 | 114 | 298 | 447 | 365 | 187 | 30 | 3 | 1 | 1475 |
| Total Precipitation (") | *3.95* | *3.38* | *4.23* | *3.32* | *4.09* | *3.63* | *4.02* | *4.43* | *3.91* | *3.37* | *3.50* | *3.31* | 45.14 |
| Days ≥ 0.1" Precip | *7* | *6* | *7* | *6* | *8* | *6* | *7* | *7* | *5* | *5* | *5* | *7* | 76 |
| Total Snowfall (") | na | na | na | *0.0* | *0.0* | *0.0* | *0.0* | *0.0* | *0.0* | *0.0* | *0.0* | *0.0* | na |
| Days ≥ 1" Snow Depth | na | na | na | *0* | *0* | *0* | *0* | *0* | *0* | *0* | *0* | *0* | na |

## ROCKY MOUNT 8 ESE *Edgecombe County* ELEVATION 110 ft LAT/LONG 35° 54 ' N / 77° 43 ' W

| | JAN | FEB | MAR | APR | MAY | JUN | JUL | AUG | SEP | OCT | NOV | DEC | YEAR |
|---|---|---|---|---|---|---|---|---|---|---|---|---|---|
| Maximum Temp °F | 50.7 | 54.7 | 63.4 | 73.4 | 79.9 | 86.1 | 89.6 | 88.1 | *83.4* | 73.8 | 65.1 | 55.6 | 72.0 |
| Minimum Temp °F | 30.0 | 32.1 | 39.2 | 47.1 | 56.1 | 63.7 | 68.3 | 67.0 | *60.4* | 48.8 | 40.2 | 33.5 | 48.9 |
| Mean Temp °F | 40.4 | 43.4 | 51.3 | 60.3 | 68.0 | 74.9 | 79.0 | 77.6 | *71.9* | 61.3 | 52.7 | 44.6 | 60.5 |
| Days Max Temp ≥ 90 °F | 0 | 0 | 0 | 1 | 2 | 8 | 16 | 14 | *5* | 0 | 0 | 0 | 46 |
| Days Max Temp ≤ 32 °F | 2 | 0 | 0 | 0 | 0 | 0 | 0 | 0 | *0* | 0 | 0 | 1 | 3 |
| Days Min Temp ≤ 32 °F | 19 | 16 | 9 | 1 | 0 | 0 | 0 | 0 | *0* | 1 | 8 | 16 | 70 |
| Days Min Temp ≤ 0 °F | 0 | 0 | 0 | 0 | 0 | 0 | 0 | 0 | *0* | 0 | 0 | 0 | 0 |
| Heating Degree Days | 757 | 605 | 427 | 182 | 47 | 4 | 0 | 1 | *18* | 160 | 374 | 629 | 3204 |
| Cooling Degree Days | *0* | *1* | 12 | *46* | *155* | *336* | *470* | *399* | na | *48* | 7 | *2* | na |
| Total Precipitation (") | 3.86 | 3.52 | 4.32 | 3.16 | 3.88 | 4.25 | 4.91 | 4.82 | 3.81 | 2.91 | 2.71 | 3.22 | 45.37 |
| Days ≥ 0.1" Precip | 8 | 7 | 7 | 6 | 7 | 6 | 7 | 7 | 5 | 4 | 5 | 6 | 75 |
| Total Snowfall (") | 1.3 | 2.2 | 0.9 | 0.0 | 0.0 | 0.0 | 0.0 | 0.0 | 0.0 | 0.0 | 0.0 | 0.1 | 4.5 |
| Days ≥ 1" Snow Depth | 1 | 2 | 0 | 0 | 0 | 0 | 0 | 0 | 0 | 0 | 0 | 0 | 3 |

## ROXBORO 7 ESE *Person County* ELEVATION 702 ft LAT/LONG 36° 23 ' N / 78° 59 ' W

| | JAN | FEB | MAR | APR | MAY | JUN | JUL | AUG | SEP | OCT | NOV | DEC | YEAR |
|---|---|---|---|---|---|---|---|---|---|---|---|---|---|
| Maximum Temp °F | 47.3 | 51.4 | 60.7 | 71.0 | 77.6 | 84.7 | 88.3 | 86.8 | 81.0 | 70.6 | 61.6 | 51.4 | 69.4 |
| Minimum Temp °F | 26.0 | 28.4 | 35.8 | 44.5 | 53.2 | 60.9 | 65.2 | 63.4 | 56.4 | 44.8 | 37.2 | 29.4 | 45.4 |
| Mean Temp °F | 36.7 | 39.9 | 48.2 | 57.8 | 65.3 | 72.9 | 76.8 | 75.1 | 68.8 | 57.8 | 49.4 | 40.4 | 57.4 |
| Days Max Temp ≥ 90 °F | 0 | 0 | 0 | 0 | 1 | 7 | 13 | 10 | 3 | 0 | 0 | 0 | 34 |
| Days Max Temp ≤ 32 °F | 3 | 1 | 0 | 0 | 0 | 0 | 0 | 0 | 0 | 0 | 0 | 1 | 5 |
| Days Min Temp ≤ 32 °F | 23 | 20 | 12 | 3 | 0 | 0 | 0 | 0 | 0 | 3 | 10 | 20 | 91 |
| Days Min Temp ≤ 0 °F | 0 | 0 | 0 | 0 | 0 | 0 | 0 | 0 | 0 | 0 | 0 | 0 | 0 |
| Heating Degree Days | 872 | 701 | 517 | 241 | 80 | 9 | 0 | 2 | 41 | 238 | 464 | 756 | 3921 |
| Cooling Degree Days | 0 | 0 | 5 | 27 | 93 | 258 | 381 | 323 | 160 | 26 | 3 | 1 | 1277 |
| Total Precipitation (") | 3.80 | 3.30 | 4.22 | 2.99 | 3.81 | 3.32 | 4.70 | 4.06 | 3.60 | 3.78 | 3.14 | 3.35 | 44.07 |
| Days ≥ 0.1" Precip | 7 | 6 | 8 | 6 | 7 | 6 | 7 | 6 | 5 | 5 | 6 | 6 | 75 |
| Total Snowfall (") | 4.3 | 3.6 | 2.0 | 0.0 | 0.0 | 0.0 | 0.0 | 0.0 | 0.0 | 0.0 | 0.3 | 1.3 | 11.5 |
| Days ≥ 1" Snow Depth | 3 | 3 | 1 | 0 | 0 | 0 | 0 | 0 | 0 | 0 | 0 | 1 | 8 |

## SALISBURY *Rowan County* ELEVATION 761 ft LAT/LONG 35° 40 ' N / 80° 29 ' W

| | JAN | FEB | MAR | APR | MAY | JUN | JUL | AUG | SEP | OCT | NOV | DEC | YEAR |
|---|---|---|---|---|---|---|---|---|---|---|---|---|---|
| Maximum Temp °F | 50.0 | 54.3 | 63.5 | 73.1 | 79.5 | 85.9 | 89.4 | 87.8 | 81.9 | 72.0 | 62.7 | 53.3 | 71.1 |
| Minimum Temp °F | 29.0 | 31.1 | 38.8 | 46.6 | 55.2 | 63.2 | 67.5 | 66.4 | 59.7 | 47.2 | 38.8 | 31.9 | 47.9 |
| Mean Temp °F | 39.5 | 42.7 | 51.2 | 59.9 | 67.4 | 74.6 | 78.5 | 77.2 | 70.8 | 59.6 | 50.8 | 42.6 | 59.6 |
| Days Max Temp ≥ 90 °F | 0 | 0 | 0 | 0 | 2 | 8 | 16 | 12 | 4 | 0 | 0 | 0 | 42 |
| Days Max Temp ≤ 32 °F | 2 | 0 | 0 | 0 | 0 | 0 | 0 | 0 | 0 | 0 | 0 | 0 | 2 |
| Days Min Temp ≤ 32 °F | 20 | 17 | 9 | 2 | 0 | 0 | 0 | 0 | 0 | 2 | 9 | 17 | 76 |
| Days Min Temp ≤ 0 °F | 0 | 0 | 0 | 0 | 0 | 0 | 0 | 0 | 0 | 0 | 0 | 0 | 0 |
| Heating Degree Days | 784 | 622 | 429 | 183 | 49 | 4 | 0 | 0 | 22 | 193 | 424 | 688 | 3398 |
| Cooling Degree Days | 0 | 0 | 6 | 36 | 133 | 321 | 448 | 397 | 206 | 39 | 3 | 1 | 1590 |
| Total Precipitation (") | 3.32 | 3.73 | 4.34 | 3.11 | 3.90 | 4.07 | 3.78 | 3.44 | 3.32 | 3.70 | 3.03 | 3.33 | 43.07 |
| Days ≥ 0.1" Precip | 6 | 6 | 7 | 6 | 7 | 6 | 6 | 6 | 4 | 5 | 5 | 6 | 70 |
| Total Snowfall (") | *2.7* | *2.6* | 0.9 | 0.0 | 0.0 | 0.0 | 0.0 | 0.0 | 0.0 | 0.0 | 0.0 | 0.5 | 6.7 |
| Days ≥ 1" Snow Depth | na | *0* | 0 | 0 | 0 | 0 | 0 | 0 | 0 | 0 | 0 | 0 | na |

## SHELBY 2 NNE *Cleveland County* ELEVATION 902 ft LAT/LONG 35° 18 ' N / 81° 32 ' W

| | JAN | FEB | MAR | APR | MAY | JUN | JUL | AUG | SEP | OCT | NOV | DEC | YEAR |
|---|---|---|---|---|---|---|---|---|---|---|---|---|---|
| Maximum Temp °F | 50.7 | 55.3 | 64.1 | 72.9 | 79.2 | 85.8 | 88.9 | 87.3 | 81.7 | 72.1 | 63.1 | 53.8 | 71.2 |
| Minimum Temp °F | 28.6 | 31.2 | 38.3 | 46.0 | 54.2 | 62.0 | 66.1 | 65.2 | 58.6 | 46.2 | 38.3 | 31.6 | 47.2 |
| Mean Temp °F | 39.7 | 43.2 | 51.2 | 59.5 | 66.7 | 73.9 | 77.5 | 76.3 | 70.2 | 59.1 | 50.7 | 42.7 | 59.2 |
| Days Max Temp ≥ 90 °F | 0 | 0 | 0 | 0 | 1 | 7 | 14 | 10 | 3 | 0 | 0 | 0 | 35 |
| Days Max Temp ≤ 32 °F | 1 | 0 | 0 | 0 | 0 | 0 | 0 | 0 | 0 | 0 | 0 | 0 | 1 |
| Days Min Temp ≤ 32 °F | 20 | 17 | 10 | 2 | 0 | 0 | 0 | 0 | 0 | 2 | 10 | 17 | 78 |
| Days Min Temp ≤ 0 °F | 0 | 0 | 0 | 0 | 0 | 0 | 0 | 0 | 0 | 0 | 0 | 0 | 0 |
| Heating Degree Days | 777 | 608 | 425 | 189 | 51 | 4 | 0 | 0 | 24 | 204 | 425 | 685 | 3392 |
| Cooling Degree Days | 0 | 0 | 6 | 30 | 119 | 303 | 424 | 369 | 187 | 31 | 1 | 1 | 1471 |
| Total Precipitation (") | 3.87 | 3.91 | 4.86 | 3.15 | 4.66 | 4.22 | 4.07 | 4.81 | 3.51 | 3.89 | 3.49 | 3.94 | 48.38 |
| Days ≥ 0.1" Precip | 6 | 6 | 7 | 6 | 7 | 6 | 7 | 7 | 5 | 5 | 6 | *6* | 74 |
| Total Snowfall (") | 3.2 | 2.5 | 1.2 | 0.0 | 0.0 | 0.0 | 0.0 | 0.0 | 0.0 | 0.0 | 0.0 | 0.6 | 7.5 |
| Days ≥ 1" Snow Depth | na | *0* | *0* | 0 | 0 | 0 | 0 | 0 | 0 | 0 | 0 | *0* | na |

## SILER CITY 2 S *Chatham County* ELEVATION 630 ft LAT/LONG 35° 43 ' N / 79° 27 ' W

| | JAN | FEB | MAR | APR | MAY | JUN | JUL | AUG | SEP | OCT | NOV | DEC | YEAR |
|---|---|---|---|---|---|---|---|---|---|---|---|---|---|
| Maximum Temp °F | 48.9 | 52.7 | 61.5 | 71.1 | 77.7 | 84.6 | 88.4 | 86.7 | 81.5 | 71.5 | 62.6 | 53.2 | 70.0 |
| Minimum Temp °F | 25.2 | 27.0 | 35.2 | 43.2 | 52.3 | 60.7 | 65.0 | 63.5 | 56.7 | 43.8 | 35.6 | 28.4 | 44.7 |
| Mean Temp °F | 37.1 | 39.9 | 48.4 | 57.2 | 65.0 | 72.7 | 76.7 | 75.1 | 69.2 | 57.6 | 49.1 | 40.9 | 57.4 |
| Days Max Temp ≥ 90 °F | 0 | 0 | 0 | 0 | 1 | 6 | 13 | 10 | 3 | 0 | 0 | 0 | 33 |
| Days Max Temp ≤ 32 °F | 2 | 1 | 0 | 0 | 0 | 0 | 0 | 0 | 0 | 0 | 0 | 1 | 4 |
| Days Min Temp ≤ 32 °F | 23 | 20 | 12 | 5 | 1 | 0 | 0 | 0 | 0 | 4 | 13 | 21 | 99 |
| Days Min Temp ≤ 0 °F | 0 | 0 | 0 | 0 | 0 | 0 | 0 | 0 | 0 | 0 | 0 | 0 | 0 |
| Heating Degree Days | 859 | 703 | 513 | 253 | 86 | 11 | 0 | 3 | 36 | 245 | 472 | 742 | 3923 |
| Cooling Degree Days | 0 | 1 | 7 | 32 | 118 | 295 | 425 | 356 | 188 | 32 | 3 | 1 | 1458 |
| Total Precipitation (") | 3.98 | 3.62 | 4.62 | 3.10 | 4.53 | 3.97 | 4.93 | 4.37 | 3.50 | 3.78 | 3.08 | 3.29 | 46.77 |
| Days ≥ 0.1" Precip | 7 | 6 | 7 | 6 | 7 | 6 | 8 | 6 | 5 | 5 | 5 | 6 | 74 |
| Total Snowfall (") | 2.3 | 2.4 | 0.7 | 0.0 | 0.0 | 0.0 | 0.0 | 0.0 | 0.0 | 0.0 | 0.0 | 0.4 | 5.8 |
| Days ≥ 1" Snow Depth | *1* | 1 | 0 | 0 | 0 | 0 | 0 | 0 | 0 | 0 | 0 | 0 | 2 |

## SMITHFIELD *Johnston County* ELEVATION 151 ft LAT/LONG 35° 31 ' N / 78° 21 ' W

| | JAN | FEB | MAR | APR | MAY | JUN | JUL | AUG | SEP | OCT | NOV | DEC | YEAR |
|---|---|---|---|---|---|---|---|---|---|---|---|---|---|
| Maximum Temp °F | 51.7 | 56.0 | 64.7 | 74.6 | 81.0 | 87.4 | 90.5 | 88.7 | 83.8 | 73.4 | 64.8 | 55.8 | 72.7 |
| Minimum Temp °F | 30.0 | 32.1 | 39.3 | 46.7 | 55.1 | 62.5 | 67.1 | 65.9 | 59.1 | 46.9 | 38.6 | 32.5 | 48.0 |
| Mean Temp °F | 40.9 | 44.1 | 52.0 | 60.7 | 68.1 | 75.0 | 78.8 | 77.3 | 71.5 | 60.1 | 51.7 | 44.2 | 60.4 |
| Days Max Temp ≥ 90 °F | 0 | 0 | 0 | 1 | 3 | 11 | 18 | 14 | 6 | 0 | 0 | 0 | 53 |
| Days Max Temp ≤ 32 °F | 1 | 0 | 0 | 0 | 0 | 0 | 0 | 0 | 0 | 0 | 0 | 0 | 1 |
| Days Min Temp ≤ 32 °F | 19 | 16 | 9 | 2 | 0 | 0 | 0 | 0 | 0 | 3 | 10 | 17 | 76 |
| Days Min Temp ≤ 0 °F | 0 | 0 | 0 | 0 | 0 | 0 | 0 | 0 | 0 | 0 | 0 | 0 | 0 |
| Heating Degree Days | 740 | 586 | 405 | 170 | 42 | 3 | 0 | 0 | 17 | 185 | 399 | 641 | 3188 |
| Cooling Degree Days | 0 | 1 | 10 | 43 | 147 | 329 | 456 | 383 | 213 | 41 | 6 | 2 | 1631 |
| Total Precipitation (") | 3.89 | 3.67 | 4.42 | 3.01 | 4.10 | 4.03 | 5.35 | 4.48 | 3.86 | 2.97 | 2.86 | 3.01 | 45.65 |
| Days ≥ 0.1" Precip | 8 | 6 | 7 | 5 | 7 | 6 | 8 | 7 | 5 | 5 | 5 | 6 | 75 |
| Total Snowfall (") | *0.9* | 1.6 | 0.7 | 0.0 | 0.0 | 0.0 | 0.0 | 0.0 | 0.0 | 0.0 | 0.0 | 0.0 | 3.2 |
| Days ≥ 1" Snow Depth | *0* | 0 | 0 | 0 | 0 | 0 | 0 | 0 | 0 | 0 | 0 | 0 | 0 |

## SOUTHPORT 5 N *Brunswick County* ELEVATION 20 ft LAT/LONG 34° 0 ' N / 78° 1 ' W

| | JAN | FEB | MAR | APR | MAY | JUN | JUL | AUG | SEP | OCT | NOV | DEC | YEAR |
|---|---|---|---|---|---|---|---|---|---|---|---|---|---|
| Maximum Temp °F | 55.3 | 57.9 | 64.9 | 72.8 | ***79.2*** | 85.5 | 88.7 | ***87.7*** | 83.6 | 75.5 | ***68.4*** | 59.6 | 73.3 |
| Minimum Temp °F | 32.2 | 34.6 | 41.7 | 49.6 | ***57.9*** | 66.5 | 71.0 | ***69.5*** | 63.4 | 51.3 | ***43.0*** | 35.8 | 51.4 |
| Mean Temp °F | 43.9 | 46.3 | 53.5 | 61.2 | ***68.6*** | 76.0 | 79.9 | ***78.7*** | 73.5 | 63.4 | ***55.6*** | 47.7 | 62.4 |
| Days Max Temp ≥ 90 °F | 0 | 0 | 0 | 0 | 1 | 5 | 12 | 9 | 2 | 0 | 0 | 0 | 29 |
| Days Max Temp ≤ 32 °F | 1 | 0 | 0 | 0 | 0 | 0 | 0 | 0 | 0 | 0 | 0 | 0 | 1 |
| Days Min Temp ≤ 32 °F | 16 | 13 | 6 | 1 | 0 | 0 | 0 | 0 | 0 | 1 | ***6*** | 13 | 56 |
| Days Min Temp ≤ 0 °F | 0 | 0 | 0 | 0 | 0 | 0 | 0 | 0 | 0 | 0 | 0 | 0 | 0 |
| Heating Degree Days | 648 | 522 | 357 | 151 | ***30*** | 1 | 0 | ***0*** | 5 | 116 | ***292*** | 534 | 2656 |
| Cooling Degree Days | 0 | 2 | 7 | 40 | ***136*** | 355 | 486 | ***418*** | 248 | 73 | 20 | 4 | 1789 |
| Total Precipitation (") | 4.99 | 4.25 | 4.57 | 2.87 | 4.08 | 4.73 | 6.44 | 7.57 | 7.26 | 3.69 | 3.36 | 4.32 | 58.13 |
| Days ≥ 0.1" Precip | 8 | 6 | 6 | 4 | 7 | 6 | 8 | 10 | 8 | 5 | 5 | 7 | 80 |
| Total Snowfall (") | ***0.0*** | 0.4 | 0.3 | 0.0 | 0.0 | 0.0 | 0.0 | 0.0 | 0.0 | 0.0 | 0.0 | 0.5 | 1.2 |
| Days ≥ 1" Snow Depth | ***0*** | ***0*** | 0 | 0 | 0 | 0 | 0 | 0 | 0 | 0 | ***0*** | 0 | 0 |

## STATESVILLE 2 NNE *Iredell County* ELEVATION 951 ft LAT/LONG 35° 49 ' N / 80° 53 ' W

| | JAN | FEB | MAR | APR | MAY | JUN | JUL | AUG | SEP | OCT | NOV | DEC | YEAR |
|---|---|---|---|---|---|---|---|---|---|---|---|---|---|
| Maximum Temp °F | 49.4 | 53.9 | 63.2 | 73.0 | 79.2 | 85.5 | 88.7 | 86.9 | 81.3 | 71.8 | 62.0 | 52.6 | 70.6 |
| Minimum Temp °F | 25.7 | 27.6 | 35.1 | 42.8 | 52.4 | 60.4 | 64.9 | 63.8 | 56.8 | 44.3 | 35.3 | 28.8 | 44.8 |
| Mean Temp °F | 37.6 | 40.8 | 49.2 | 57.9 | 65.9 | 73.0 | 76.9 | 75.4 | 69.1 | 58.1 | 48.6 | 40.6 | 57.8 |
| Days Max Temp ≥ 90 °F | 0 | 0 | 0 | 0 | 1 | 7 | 14 | 10 | 3 | 0 | 0 | 0 | 35 |
| Days Max Temp ≤ 32 °F | 2 | 0 | 0 | 0 | 0 | 0 | 0 | 0 | 0 | 0 | 0 | 0 | 2 |
| Days Min Temp ≤ 32 °F | 23 | 20 | 13 | 5 | 0 | 0 | 0 | 0 | 0 | 4 | 14 | 20 | 99 |
| Days Min Temp ≤ 0 °F | 0 | 0 | 0 | 0 | 0 | 0 | 0 | 0 | 0 | 0 | 0 | 0 | 0 |
| Heating Degree Days | 844 | 677 | 487 | 229 | 66 | 7 | 0 | 1 | 33 | 233 | 486 | 750 | 3813 |
| Cooling Degree Days | 0 | 0 | 5 | 26 | 115 | 287 | 411 | 343 | 172 | 32 | 2 | 1 | 1394 |
| Total Precipitation (") | 3.44 | 3.62 | 4.61 | 3.00 | 4.39 | 4.39 | 3.95 | 3.99 | 3.80 | 3.55 | 3.24 | 3.69 | 45.67 |
| Days ≥ 0.1" Precip | 7 | 6 | 8 | 6 | 7 | 7 | 7 | 7 | 5 | 5 | 6 | 7 | 78 |
| Total Snowfall (") | 3.3 | 2.2 | 1.4 | 0.0 | 0.0 | 0.0 | 0.0 | 0.0 | 0.0 | 0.0 | 0.2 | 0.4 | 7.5 |
| Days ≥ 1" Snow Depth | 2 | 1 | 0 | 0 | 0 | 0 | 0 | 0 | 0 | 0 | 0 | 0 | 3 |

## TAPOCO *Graham County* ELEVATION 1112 ft LAT/LONG 35° 27 ' N / 83° 56 ' W

| | JAN | FEB | MAR | APR | MAY | JUN | JUL | AUG | SEP | OCT | NOV | DEC | YEAR |
|---|---|---|---|---|---|---|---|---|---|---|---|---|---|
| Maximum Temp °F | 49.4 | 53.5 | 62.9 | 72.3 | 77.9 | 84.0 | 86.5 | 85.4 | 79.9 | 71.1 | 61.7 | 53.3 | 69.8 |
| Minimum Temp °F | 28.4 | 30.1 | 37.5 | 44.4 | 51.7 | 58.7 | 62.1 | 61.7 | 57.2 | 46.3 | 38.0 | 32.1 | 45.7 |
| Mean Temp °F | 38.9 | 41.8 | 50.2 | 58.4 | 64.8 | 71.4 | 74.3 | 73.6 | 68.6 | 58.7 | 49.9 | 42.8 | 57.8 |
| Days Max Temp ≥ 90 °F | 0 | 0 | 0 | 0 | 0 | 4 | 9 | 6 | 1 | 0 | 0 | 0 | 20 |
| Days Max Temp ≤ 32 °F | 2 | 1 | 0 | 0 | 0 | 0 | 0 | 0 | 0 | 0 | 0 | 1 | 4 |
| Days Min Temp ≤ 32 °F | 20 | 17 | 11 | 3 | 0 | 0 | 0 | 0 | 0 | 2 | 10 | 17 | 80 |
| Days Min Temp ≤ 0 °F | 0 | 0 | 0 | 0 | 0 | 0 | 0 | 0 | 0 | 0 | 0 | 0 | 0 |
| Heating Degree Days | 802 | 648 | 456 | 213 | 70 | 7 | 0 | 1 | 28 | 210 | 449 | 683 | 3567 |
| Cooling Degree Days | 0 | 0 | 4 | 20 | 74 | 221 | 318 | 281 | 142 | 22 | 2 | 1 | 1085 |
| Total Precipitation (") | 5.25 | 5.09 | 6.19 | 4.56 | 5.51 | 5.16 | 6.24 | 4.85 | 4.12 | 3.27 | 4.34 | 5.17 | 59.75 |
| Days ≥ 0.1" Precip | 9 | 8 | 9 | 8 | 9 | 9 | 9 | 8 | 7 | 6 | 7 | 8 | 97 |
| Total Snowfall (") | na | ***2.9*** | 0.6 | 0.3 | 0.0 | 0.0 | 0.0 | 0.0 | 0.0 | 0.0 | 0.0 | ***0.3*** | na |
| Days ≥ 1" Snow Depth | na | na | ***0*** | 0 | 0 | 0 | 0 | 0 | 0 | 0 | 0 | ***0*** | na |

## TARBORO 1 S *Edgecombe County* ELEVATION 39 ft LAT/LONG 35° 53 ' N / 77° 32 ' W

| | JAN | FEB | MAR | APR | MAY | JUN | JUL | AUG | SEP | OCT | NOV | DEC | YEAR |
|---|---|---|---|---|---|---|---|---|---|---|---|---|---|
| Maximum Temp °F | 52.0 | 55.1 | 64.1 | 73.9 | 81.0 | 87.4 | 90.8 | 89.1 | 84.2 | 74.2 | 65.2 | 55.8 | 72.7 |
| Minimum Temp °F | 29.0 | 30.9 | 38.0 | 45.9 | 55.1 | 63.0 | 67.8 | 66.4 | 60.1 | 47.8 | 39.1 | 32.1 | 47.9 |
| Mean Temp °F | 40.6 | 43.1 | 51.1 | 59.9 | 68.0 | 75.1 | 79.3 | 77.8 | 72.1 | 61.1 | 52.2 | 43.9 | 60.4 |
| Days Max Temp ≥ 90 °F | 0 | 0 | 0 | 1 | 4 | 11 | 20 | 16 | 6 | 0 | 0 | 0 | 58 |
| Days Max Temp ≤ 32 °F | 1 | 0 | 0 | 0 | 0 | 0 | 0 | 0 | 0 | 0 | 0 | 0 | 1 |
| Days Min Temp ≤ 32 °F | 20 | 17 | 10 | 2 | 0 | 0 | 0 | 0 | 0 | 2 | 9 | 17 | 77 |
| Days Min Temp ≤ 0 °F | 0 | 0 | 0 | 0 | 0 | 0 | 0 | 0 | 0 | 0 | 0 | 0 | 0 |
| Heating Degree Days | 750 | 615 | 431 | 188 | 46 | 4 | 0 | 0 | 13 | 164 | 386 | 648 | 3245 |
| Cooling Degree Days | 0 | 1 | 9 | 40 | 151 | 338 | 472 | 397 | 228 | 50 | 7 | 1 | 1694 |
| Total Precipitation (") | 4.04 | 3.58 | 4.20 | 3.02 | 3.81 | 3.94 | 4.70 | 5.31 | 4.01 | 2.96 | 2.56 | 3.10 | 45.23 |
| Days ≥ 0.1" Precip | 7 | 7 | 7 | 6 | 7 | 6 | 7 | 7 | 5 | 4 | 5 | 6 | 74 |
| Total Snowfall (") | 1.7 | 2.8 | 1.5 | 0.0 | 0.0 | 0.0 | 0.0 | 0.0 | 0.0 | 0.0 | 0.0 | 0.3 | 6.3 |
| Days ≥ 1" Snow Depth | 1 | ***1*** | 0 | 0 | 0 | 0 | 0 | 0 | 0 | 0 | 0 | 0 | 2 |

## TRANSOU *Ashe County* ELEVATION 2875 ft LAT/LONG 36° 24 ' N / 81° 18 ' W

| | JAN | FEB | MAR | APR | MAY | JUN | JUL | AUG | SEP | OCT | NOV | DEC | YEAR |
|---|---|---|---|---|---|---|---|---|---|---|---|---|---|
| Maximum Temp °F | 42.1 | 45.2 | 54.2 | 63.6 | 70.1 | 76.6 | 80.1 | 78.6 | 72.8 | 63.6 | 54.2 | 45.9 | 62.3 |
| Minimum Temp °F | 20.9 | 22.8 | 30.3 | 37.4 | 45.3 | 52.6 | 57.0 | 55.6 | 49.5 | 38.1 | 30.7 | 24.1 | 38.7 |
| Mean Temp °F | 31.6 | 34.0 | 42.2 | 50.5 | 57.7 | 64.6 | 68.6 | 67.1 | 61.2 | 50.9 | 42.5 | 35.0 | 50.5 |
| Days Max Temp ≥ 90 °F | 0 | 0 | 0 | 0 | 0 | 0 | 0 | 0 | 0 | 0 | 0 | 0 | 0 |
| Days Max Temp ≤ 32 °F | 6 | 3 | 1 | 0 | 0 | 0 | 0 | 0 | 0 | 0 | 1 | 3 | 14 |
| Days Min Temp ≤ 32 °F | 26 | 23 | 19 | 10 | 3 | 0 | 0 | 0 | 1 | 11 | 18 | 24 | 135 |
| Days Min Temp ≤ 0 °F | 2 | 0 | 0 | 0 | 0 | 0 | 0 | 0 | 0 | 0 | 0 | 1 | 3 |
| Heating Degree Days | 1029 | 868 | 699 | 430 | 227 | 63 | 12 | 25 | 139 | 431 | 669 | 923 | 5515 |
| Cooling Degree Days | 0 | 0 | 0 | 1 | 11 | 66 | 149 | 100 | 33 | 2 | 0 | 0 | 362 |
| Total Precipitation (") | 3.55 | 4.00 | 5.49 | 4.71 | 5.66 | 4.77 | 5.00 | 4.48 | 4.63 | 5.14 | 4.76 | 3.80 | 55.99 |
| Days ≥ 0.1" Precip | 7 | 6 | 8 | 8 | 9 | 8 | 9 | 7 | 6 | 6 | 7 | 7 | 88 |
| Total Snowfall (") | 8.4 | 7.2 | 3.6 | 1.4 | 0.0 | 0.0 | 0.0 | 0.0 | 0.0 | 0.1 | 0.9 | 4.2 | 25.8 |
| Days ≥ 1" Snow Depth | 8 | 6 | 1 | 0 | 0 | 0 | 0 | 0 | 0 | 0 | 0 | 3 | 18 |

## TRYON *Polk County* ELEVATION 1079 ft LAT/LONG 35° 12 ' N / 82° 14 ' W

| | JAN | FEB | MAR | APR | MAY | JUN | JUL | AUG | SEP | OCT | NOV | DEC | YEAR |
|---|---|---|---|---|---|---|---|---|---|---|---|---|---|
| Maximum Temp °F | 51.6 | 56.1 | 65.1 | 74.0 | 80.0 | 85.9 | 88.7 | 87.0 | 81.3 | 72.5 | 63.1 | 54.8 | 71.7 |
| Minimum Temp °F | 30.5 | 32.3 | 39.4 | 46.3 | 54.5 | 61.9 | 66.5 | 65.4 | 59.6 | 48.2 | 39.9 | 33.4 | 48.2 |
| Mean Temp °F | 41.1 | 44.2 | 52.3 | 60.2 | 67.3 | 73.9 | 77.6 | 76.2 | 70.5 | 60.4 | 51.5 | 44.2 | 60.0 |
| Days Max Temp ≥ 90 °F | 0 | 0 | 0 | 0 | 1 | 8 | 14 | 9 | 3 | 0 | 0 | 0 | 35 |
| Days Max Temp ≤ 32 °F | 1 | 0 | 0 | 0 | 0 | 0 | 0 | 0 | 0 | 0 | 0 | 0 | 1 |
| Days Min Temp ≤ 32 °F | 18 | 15 | 8 | 2 | 0 | 0 | 0 | 0 | 0 | 1 | 7 | 16 | 67 |
| Days Min Temp ≤ 0 °F | 0 | 0 | 0 | 0 | 0 | 0 | 0 | 0 | 0 | 0 | 0 | 0 | 0 |
| Heating Degree Days | 735 | 581 | 393 | 170 | 42 | 3 | 0 | 0 | 17 | 170 | 399 | 639 | 3149 |
| Cooling Degree Days | 0 | 0 | 6 | 33 | 124 | 304 | 429 | 370 | 194 | 37 | 2 | 0 | 1499 |
| Total Precipitation (") | 5.11 | 5.16 | 6.59 | 4.47 | 6.15 | 5.52 | 5.45 | 6.00 | 5.27 | 5.33 | 4.86 | 4.96 | 64.87 |
| Days ≥ 0.1" Precip | 7 | 7 | 9 | 7 | 8 | 8 | 8 | 8 | 7 | 6 | 7 | 7 | 89 |
| Total Snowfall (") | 3.4 | 2.8 | 1.5 | 0.0 | 0.0 | 0.0 | 0.0 | 0.0 | 0.0 | 0.0 | 0.1 | 0.9 | 8.7 |
| Days ≥ 1" Snow Depth | 3 | 2 | 1 | 0 | 0 | 0 | 0 | 0 | 0 | 0 | 0 | 0 | 6 |

## W KERR SCOTT RESERVO *Wilkes County* ELEVATION 1102 ft LAT/LONG 36° 8 ' N / 81° 14 ' W

| | JAN | FEB | MAR | APR | MAY | JUN | JUL | AUG | SEP | OCT | NOV | DEC | YEAR |
|---|---|---|---|---|---|---|---|---|---|---|---|---|---|
| Maximum Temp °F | 47.1 | 51.2 | 60.1 | 69.9 | 77.0 | 84.0 | 87.5 | 85.8 | 79.8 | 70.3 | 60.6 | 51.2 | 68.7 |
| Minimum Temp °F | 23.2 | 25.4 | 33.3 | 41.3 | 50.7 | 59.3 | 63.8 | 62.5 | 55.9 | 42.8 | 34.3 | 26.9 | 43.3 |
| Mean Temp °F | 35.2 | 38.4 | 46.7 | 55.7 | 63.9 | 71.7 | 75.6 | 74.2 | 67.9 | 56.5 | 47.5 | 39.1 | 56.0 |
| Days Max Temp ≥ 90 °F | 0 | 0 | 0 | 0 | 1 | 6 | 12 | 8 | 2 | 0 | 0 | 0 | 29 |
| Days Max Temp ≤ 32 °F | 3 | 1 | 0 | 0 | 0 | 0 | 0 | 0 | 0 | 0 | 0 | 1 | 5 |
| Days Min Temp ≤ 32 °F | 26 | 22 | 15 | 6 | 0 | 0 | 0 | 0 | 0 | 4 | 14 | 23 | 110 |
| Days Min Temp ≤ 0 °F | 1 | 0 | 0 | 0 | 0 | 0 | 0 | 0 | 0 | 0 | 0 | 0 | 1 |
| Heating Degree Days | 918 | 745 | 561 | 289 | 99 | 12 | 1 | 2 | 44 | 269 | 519 | 796 | 4255 |
| Cooling Degree Days | 0 | 0 | 4 | 16 | 80 | 248 | 366 | 303 | 143 | 18 | 0 | 0 | 1178 |
| Total Precipitation (") | 3.93 | 3.91 | 5.05 | 4.19 | 4.88 | 4.70 | 4.67 | 5.15 | 4.75 | 3.99 | 3.69 | 3.82 | 52.73 |
| Days ≥ 0.1" Precip | 7 | 7 | 8 | 7 | 8 | 7 | 8 | 8 | 6 | 5 | 6 | 7 | 84 |
| Total Snowfall (") | 4.6 | *4.4* | 2.1 | 0.0 | 0.0 | 0.0 | 0.0 | 0.0 | 0.0 | 0.0 | 0.0 | 0.9 | 12.0 |
| Days ≥ 1" Snow Depth | *3* | *3* | 1 | 0 | 0 | 0 | 0 | 0 | 0 | 0 | 0 | 1 | 8 |

## WADESBORO *Anson County* ELEVATION 420 ft LAT/LONG 34° 57 ' N / 80° 4 ' W

| | JAN | FEB | MAR | APR | MAY | JUN | JUL | AUG | SEP | OCT | NOV | DEC | YEAR |
|---|---|---|---|---|---|---|---|---|---|---|---|---|---|
| Maximum Temp °F | 50.8 | 55.3 | 63.9 | 73.7 | 80.1 | 86.8 | 90.2 | 88.4 | 83.3 | 73.5 | 64.5 | 54.9 | 72.1 |
| Minimum Temp °F | 30.5 | 32.8 | 40.6 | 49.1 | 57.3 | 65.2 | 69.5 | 68.2 | 62.2 | 49.5 | 41.6 | 33.9 | 50.0 |
| Mean Temp °F | 40.7 | 44.0 | 52.3 | 61.4 | 68.8 | 76.0 | 79.9 | 78.3 | 72.8 | 61.5 | 53.1 | 44.4 | 61.1 |
| Days Max Temp ≥ 90 °F | 0 | 0 | 0 | 1 | 2 | 10 | 18 | 14 | 5 | 0 | 0 | 0 | 50 |
| Days Max Temp ≤ 32 °F | 1 | 0 | 0 | 0 | 0 | 0 | 0 | 0 | 0 | 0 | 0 | 0 | 1 |
| Days Min Temp ≤ 32 °F | 18 | 15 | 7 | 1 | 0 | 0 | 0 | 0 | 0 | 1 | 6 | 15 | 63 |
| Days Min Temp ≤ 0 °F | 0 | 0 | 0 | 0 | 0 | 0 | 0 | 0 | 0 | 0 | 0 | 0 | 0 |
| Heating Degree Days | 748 | 587 | 398 | 158 | 37 | 3 | 0 | 0 | 13 | 153 | 360 | 632 | 3089 |
| Cooling Degree Days | 0 | 1 | 11 | 55 | 167 | 370 | 497 | 425 | 254 | 57 | 8 | 1 | 1846 |
| Total Precipitation (") | 4.16 | 3.67 | 4.63 | 2.66 | 3.70 | 4.40 | 5.21 | 4.78 | 3.77 | 3.51 | 2.90 | 3.35 | 46.74 |
| Days ≥ 0.1" Precip | 8 | 6 | 8 | 5 | 6 | 6 | 8 | 7 | 5 | 5 | 5 | 6 | 75 |
| Total Snowfall (") | 1.4 | 1.9 | 0.8 | 0.0 | 0.0 | 0.0 | 0.0 | 0.0 | 0.0 | 0.0 | 0.1 | 0.6 | 4.8 |
| Days ≥ 1" Snow Depth | 1 | 1 | 0 | 0 | 0 | 0 | 0 | 0 | 0 | 0 | 0 | 0 | 2 |

## WAYNESVILLE 1 E *Haywood County* ELEVATION 2641 ft LAT/LONG 35° 29' N / 82° 57' W

| | JAN | FEB | MAR | APR | MAY | JUN | JUL | AUG | SEP | OCT | NOV | DEC | YEAR |
|---|---|---|---|---|---|---|---|---|---|---|---|---|---|
| Maximum Temp °F | 47.6 | 51.0 | 59.9 | 68.3 | 74.1 | 80.3 | 82.9 | 81.6 | 76.3 | 67.8 | 58.7 | 51.2 | 66.6 |
| Minimum Temp °F | 23.8 | 25.5 | 32.9 | 39.9 | 47.2 | 54.6 | 58.9 | 58.3 | 52.6 | 40.5 | 32.7 | 27.3 | 41.2 |
| Mean Temp °F | 35.7 | 38.3 | 46.4 | 54.1 | 60.7 | 67.5 | 70.9 | 70.0 | 64.5 | 54.2 | 45.7 | 39.3 | 53.9 |
| Days Max Temp ≥ 90 °F | 0 | 0 | 0 | 0 | 0 | 0 | 2 | 1 | 0 | 0 | 0 | 0 | 3 |
| Days Max Temp ≤ 32 °F | 3 | 2 | 0 | 0 | 0 | 0 | 0 | 0 | 0 | 0 | 0 | 1 | 6 |
| Days Min Temp ≤ 32 °F | 23 | 21 | 16 | 7 | 1 | 0 | 0 | 0 | 0 | 8 | 16 | 21 | 113 |
| Days Min Temp ≤ 0 °F | 1 | 0 | 0 | 0 | 0 | 0 | 0 | 0 | 0 | 0 | 0 | 0 | 1 |
| Heating Degree Days | 901 | 748 | 571 | 325 | 149 | 26 | 2 | 5 | 73 | 332 | 572 | 791 | 4495 |
| Cooling Degree Days | 0 | 0 | 1 | 3 | 27 | 122 | 218 | 178 | 74 | 5 | 0 | 0 | 628 |
| Total Precipitation (") | 3.81 | 4.36 | 5.26 | 3.57 | 4.35 | 3.81 | 3.96 | 4.46 | 3.45 | 2.89 | 3.69 | 4.14 | 47.75 |
| Days ≥ 0.1" Precip | 7 | 7 | 8 | 7 | 9 | 7 | 8 | 8 | 6 | 6 | 6 | 7 | 86 |
| Total Snowfall (") | 4.6 | 4.5 | 2.8 | 1.0 | 0.2 | 0.0 | 0.0 | 0.0 | 0.0 | 0.0 | 0.4 | 2.2 | 15.7 |
| Days ≥ 1" Snow Depth | 3 | *2* | 1 | 0 | 0 | 0 | 0 | 0 | 0 | 0 | 0 | 1 | 7 |

## WHITEVILLE 7 NW *Columbus County* ELEVATION 79 ft LAT/LONG 34° 24' N / 78° 44' W

| | JAN | FEB | MAR | APR | MAY | JUN | JUL | AUG | SEP | OCT | NOV | DEC | YEAR |
|---|---|---|---|---|---|---|---|---|---|---|---|---|---|
| Maximum Temp °F | 54.3 | 58.1 | 66.4 | 75.1 | 81.4 | 87.0 | 90.1 | 88.6 | 84.3 | 75.2 | 67.3 | 58.2 | 73.8 |
| Minimum Temp °F | 32.0 | 33.9 | 40.9 | 48.1 | 56.7 | 64.3 | 68.8 | 67.7 | 61.9 | 49.8 | 41.5 | 34.6 | 50.0 |
| Mean Temp °F | 43.2 | 46.0 | 53.7 | 61.6 | 69.1 | 75.7 | 79.5 | 78.2 | 73.1 | 62.5 | 54.4 | 46.4 | 62.0 |
| Days Max Temp ≥ 90 °F | 0 | 0 | 0 | 0 | 3 | 10 | 18 | 15 | 6 | 0 | 0 | 0 | 52 |
| Days Max Temp ≤ 32 °F | 1 | 0 | 0 | 0 | 0 | 0 | 0 | 0 | 0 | 0 | 0 | 0 | 1 |
| Days Min Temp ≤ 32 °F | 17 | 14 | 7 | 1 | 0 | 0 | 0 | 0 | 0 | 1 | 7 | 15 | 62 |
| Days Min Temp ≤ 0 °F | 0 | 0 | 0 | 0 | 0 | 0 | 0 | 0 | 0 | 0 | 0 | 0 | 0 |
| Heating Degree Days | 671 | 531 | 358 | 149 | 33 | 2 | 0 | 0 | 9 | 136 | 326 | 573 | 2788 |
| Cooling Degree Days | 0 | 2 | 13 | 47 | 163 | 349 | 479 | 411 | 256 | 64 | 14 | 4 | 1802 |
| Total Precipitation (") | 3.91 | 3.48 | 4.51 | 2.87 | 4.46 | 4.64 | 5.92 | 5.46 | 4.39 | 2.84 | 2.71 | 3.23 | 48.42 |
| Days ≥ 0.1" Precip | 8 | 6 | 8 | 5 | 7 | 7 | 9 | 8 | 6 | 4 | 5 | 6 | 79 |
| Total Snowfall (") | 0.5 | 1.2 | 0.5 | 0.0 | 0.0 | 0.0 | 0.0 | 0.0 | 0.0 | 0.0 | 0.0 | 0.6 | 2.8 |
| Days ≥ 1" Snow Depth | 0 | 1 | 0 | 0 | 0 | 0 | 0 | 0 | 0 | 0 | 0 | 0 | 1 |

## WILLARD 4 SW *Pender County* ELEVATION 59 ft LAT/LONG 34° 39' N / 78° 2' W

| | JAN | FEB | MAR | APR | MAY | JUN | JUL | AUG | SEP | OCT | NOV | DEC | YEAR |
|---|---|---|---|---|---|---|---|---|---|---|---|---|---|
| Maximum Temp °F | 55.6 | 59.5 | 67.9 | 76.5 | 82.1 | 87.1 | 90.0 | 88.2 | 83.7 | 75.1 | 67.5 | 59.5 | 74.4 |
| Minimum Temp °F | 32.6 | 34.5 | 41.0 | 48.2 | 57.1 | 64.4 | 68.7 | 67.9 | 62.2 | 51.1 | 42.6 | 35.2 | 50.5 |
| Mean Temp °F | 44.1 | 47.0 | 54.5 | 62.3 | 69.6 | 75.8 | 79.3 | 78.1 | 73.0 | 63.1 | 55.0 | 47.4 | 62.4 |
| Days Max Temp ≥ 90 °F | 0 | 0 | 0 | 1 | 4 | 10 | 18 | 12 | 5 | 0 | 0 | 0 | 50 |
| Days Max Temp ≤ 32 °F | 1 | 0 | 0 | 0 | 0 | 0 | 0 | 0 | 0 | 0 | 0 | 0 | 1 |
| Days Min Temp ≤ 32 °F | 16 | 13 | 7 | 1 | 0 | 0 | 0 | 0 | 0 | 1 | 6 | 14 | 58 |
| Days Min Temp ≤ 0 °F | 0 | 0 | 0 | 0 | 0 | 0 | 0 | 0 | 0 | 0 | 0 | 0 | 0 |
| Heating Degree Days | 641 | 504 | 334 | 134 | 26 | 2 | 0 | 0 | 7 | 121 | 309 | 544 | 2622 |
| Cooling Degree Days | 0 | 3 | 17 | 62 | 184 | 354 | 473 | 409 | 251 | 71 | 18 | 5 | 1847 |
| Total Precipitation (") | 4.33 | 3.45 | 4.48 | 2.93 | 4.46 | 4.89 | 7.70 | 6.71 | 4.76 | 2.78 | 2.83 | 3.41 | 52.73 |
| Days ≥ 0.1" Precip | 8 | 6 | 8 | 5 | 7 | 7 | 10 | 9 | 6 | 5 | 5 | 6 | 82 |
| Total Snowfall (") | 0.8 | 1.1 | 0.6 | 0.0 | 0.0 | 0.0 | 0.0 | 0.0 | 0.0 | 0.0 | 0.0 | 0.9 | 3.4 |
| Days ≥ 1" Snow Depth | 0 | 0 | 0 | 0 | 0 | 0 | 0 | 0 | 0 | 0 | 0 | 0 | 0 |

## WILLIAMSTON 1 E *Martin County* ELEVATION 20 ft LAT/LONG 35° 51' N / 77° 2' W

| | JAN | FEB | MAR | APR | MAY | JUN | JUL | AUG | SEP | OCT | NOV | DEC | YEAR |
|---|---|---|---|---|---|---|---|---|---|---|---|---|---|
| Maximum Temp °F | 52.3 | 55.1 | 63.5 | 73.1 | 79.5 | 85.9 | 89.3 | 87.8 | 83.3 | 73.6 | 65.5 | 55.8 | 72.1 |
| Minimum Temp °F | 31.5 | 32.9 | 40.3 | 47.6 | 56.4 | 64.5 | 69.3 | 68.0 | 62.0 | 49.6 | 41.8 | 34.0 | 49.8 |
| Mean Temp °F | 41.9 | 44.0 | 51.9 | 60.4 | 68.0 | 75.2 | 79.3 | 77.9 | 72.7 | 61.6 | 53.7 | 44.9 | 61.0 |
| Days Max Temp ≥ 90 °F | 0 | 0 | 0 | 1 | 2 | 8 | 16 | 12 | 4 | 0 | 0 | 0 | 43 |
| Days Max Temp ≤ 32 °F | 1 | 0 | 0 | 0 | 0 | 0 | 0 | 0 | 0 | 0 | 0 | 0 | 1 |
| Days Min Temp ≤ 32 °F | 17 | 15 | 7 | 1 | 0 | 0 | 0 | 0 | 0 | 1 | 6 | 15 | 62 |
| Days Min Temp ≤ 0 °F | 0 | 0 | 0 | 0 | 0 | 0 | 0 | 0 | 0 | 0 | 0 | 0 | 0 |
| Heating Degree Days | 709 | 588 | 410 | 176 | 45 | 4 | 0 | 0 | 10 | 153 | 346 | 618 | 3059 |
| Cooling Degree Days | ***0*** | ***2*** | ***10*** | ***36*** | ***144*** | 330 | ***469*** | ***381*** | ***233*** | ***49*** | ***12*** | ***2*** | 1668 |
| Total Precipitation (") | 4.24 | 3.08 | 4.15 | 3.09 | 4.30 | 4.25 | 5.37 | 5.69 | 4.17 | 3.54 | 2.68 | 3.02 | 47.58 |
| Days ≥ 0.1" Precip | 8 | 6 | 7 | 6 | 7 | 6 | 8 | 7 | 5 | 5 | 5 | 6 | 76 |
| Total Snowfall (") | 1.6 | ***1.8*** | 1.2 | 0.0 | 0.0 | 0.0 | 0.0 | 0.0 | 0.0 | 0.0 | 0.0 | 0.4 | 5.0 |
| Days ≥ 1" Snow Depth | 1 | 1 | 0 | 0 | 0 | 0 | 0 | 0 | 0 | 0 | 0 | 0 | 2 |

## WILMINGTON 7 N *New Hanover County* ELEVATION 39 ft LAT/LONG 34° 19 ' N / 77° 55 ' W

| | JAN | FEB | MAR | APR | MAY | JUN | JUL | AUG | SEP | OCT | NOV | DEC | YEAR |
|---|---|---|---|---|---|---|---|---|---|---|---|---|---|
| Maximum Temp °F | 55.4 | 58.4 | 66.1 | 74.6 | 80.8 | 86.6 | 89.6 | 88.0 | 83.7 | 74.9 | 67.8 | 59.3 | 73.8 |
| Minimum Temp °F | 33.0 | 34.9 | 42.1 | 49.2 | 58.1 | 65.9 | 70.3 | 69.1 | 63.6 | 51.9 | 43.6 | 36.2 | 51.5 |
| Mean Temp °F | 44.2 | 46.7 | 54.1 | 61.9 | 69.5 | 76.2 | 80.0 | 78.6 | 73.7 | 63.5 | 55.7 | 47.8 | 62.7 |
| Days Max Temp ≥ 90 °F | 0 | 0 | 0 | 1 | 3 | 9 | 16 | 12 | 4 | 0 | 0 | 0 | 45 |
| Days Max Temp ≤ 32 °F | 1 | 0 | 0 | 0 | 0 | 0 | 0 | 0 | 0 | 0 | 0 | 0 | 1 |
| Days Min Temp ≤ 32 °F | 16 | 13 | 7 | 1 | 0 | 0 | 0 | 0 | 0 | 1 | 6 | 13 | 57 |
| Days Min Temp ≤ 0 °F | 0 | 0 | 0 | 0 | 0 | 0 | 0 | 0 | 0 | 0 | 0 | 0 | 0 |
| Heating Degree Days | 639 | 514 | 346 | 147 | 33 | 2 | 0 | 0 | 7 | 118 | 292 | 531 | 2629 |
| Cooling Degree Days | 0 | 3 | 14 | 57 | 180 | 376 | 494 | 431 | 271 | 77 | 21 | 6 | 1930 |
| Total Precipitation (") | 4.49 | 3.68 | 4.55 | 3.03 | 4.37 | 4.96 | 8.36 | 7.54 | 5.16 | 2.86 | 3.17 | 4.05 | 56.22 |
| Days ≥ 0.1" Precip | 8 | 6 | 7 | 5 | 7 | 7 | 10 | 10 | 6 | 5 | 5 | 6 | 82 |
| Total Snowfall (") | 0.6 | 0.8 | 0.6 | 0.0 | 0.0 | 0.0 | 0.0 | 0.0 | 0.0 | 0.0 | 0.0 | 0.8 | 2.8 |
| Days ≥ 1" Snow Depth | 0 | 0 | 0 | 0 | 0 | 0 | 0 | 0 | 0 | 0 | 0 | 0 | 0 |

## WILMINGTON NEW HANVR *New Hanover County* ELEVATION 36 ft LAT/LONG 34° 16 ' N / 77° 55 ' W

| | JAN | FEB | MAR | APR | MAY | JUN | JUL | AUG | SEP | OCT | NOV | DEC | YEAR |
|---|---|---|---|---|---|---|---|---|---|---|---|---|---|
| Maximum Temp °F | 55.7 | 58.7 | 65.9 | 74.3 | 80.6 | 86.4 | 89.9 | 88.4 | 84.2 | 75.8 | 68.0 | 59.9 | 74.0 |
| Minimum Temp °F | 35.1 | 36.7 | 43.5 | 51.2 | 59.8 | 67.4 | 72.1 | 71.1 | 65.7 | 54.0 | 45.1 | 38.0 | 53.3 |
| Mean Temp °F | 45.4 | 47.7 | 54.7 | 62.8 | 70.2 | 77.0 | 81.0 | 79.8 | 75.0 | 64.9 | 56.6 | 49.0 | 63.7 |
| Days Max Temp ≥ 90 °F | 0 | 0 | 0 | 1 | 2 | 8 | 17 | 13 | 5 | 0 | 0 | 0 | 46 |
| Days Max Temp ≤ 32 °F | 1 | 0 | 0 | 0 | 0 | 0 | 0 | 0 | 0 | 0 | 0 | 0 | 1 |
| Days Min Temp ≤ 32 °F | 13 | 11 | 4 | 0 | 0 | 0 | 0 | 0 | 0 | 0 | 3 | 10 | 41 |
| Days Min Temp ≤ 0 °F | 0 | 0 | 0 | 0 | 0 | 0 | 0 | 0 | 0 | 0 | 0 | 0 | 0 |
| Heating Degree Days | 602 | 485 | 329 | 128 | 25 | 1 | 0 | 0 | 3 | 92 | 269 | 495 | 2429 |
| Cooling Degree Days | 1 | 5 | 16 | 63 | 189 | 383 | 522 | 457 | 304 | 95 | 27 | 6 | 2068 |
| Total Precipitation (") | 4.24 | 3.56 | 4.32 | 2.78 | 4.26 | 5.40 | 8.19 | 7.39 | 5.15 | 2.97 | 3.21 | 3.83 | 55.30 |
| Days ≥ 0.1" Precip | 8 | 6 | 7 | 5 | 6 | 7 | 10 | 9 | 6 | 4 | 5 | 6 | 79 |
| Total Snowfall (") | 0.5 | 0.6 | 0.4 | 0.0 | 0.0 | 0.0 | 0.0 | 0.0 | 0.0 | 0.0 | 0.0 | 0.7 | 2.2 |
| Days ≥ 1" Snow Depth | 0 | 0 | 0 | 0 | 0 | 0 | 0 | 0 | 0 | 0 | 0 | 0 | 0 |

## WILSON 3 SW *Wilson County* ELEVATION 151 ft LAT/LONG 35° 43 ' N / 77° 55 ' W

| | JAN | FEB | MAR | APR | MAY | JUN | JUL | AUG | SEP | OCT | NOV | DEC | YEAR |
|---|---|---|---|---|---|---|---|---|---|---|---|---|---|
| Maximum Temp °F | 50.6 | 54.4 | 62.9 | 73.0 | 79.7 | 86.9 | 90.2 | 88.5 | 83.6 | 73.4 | 64.6 | 55.1 | 71.9 |
| Minimum Temp °F | 28.8 | 31.1 | 38.0 | 46.4 | 55.3 | 63.2 | 67.8 | 66.5 | 59.8 | 47.5 | 39.1 | 32.0 | 48.0 |
| Mean Temp °F | 39.7 | 42.8 | 50.5 | 59.7 | 67.6 | 75.1 | 79.0 | 77.6 | 71.7 | 60.5 | 51.9 | 43.6 | 60.0 |
| Days Max Temp ≥ 90 °F | 0 | 0 | 0 | 1 | 3 | 11 | 18 | 15 | 6 | 0 | 0 | 0 | 54 |
| Days Max Temp ≤ 32 °F | 2 | 0 | 0 | 0 | 0 | 0 | 0 | 0 | 0 | 0 | 0 | 1 | 3 |
| Days Min Temp ≤ 32 °F | 21 | 17 | 9 | 1 | 0 | 0 | 0 | 0 | 0 | 2 | 9 | 18 | 77 |
| Days Min Temp ≤ 0 °F | 0 | 0 | 0 | 0 | 0 | 0 | 0 | 0 | 0 | 0 | 0 | 0 | 0 |
| Heating Degree Days | 777 | 623 | 451 | 194 | 52 | 4 | 0 | 1 | 18 | 179 | 395 | 659 | 3353 |
| Cooling Degree Days | 0 | 2 | 9 | 39 | 140 | 334 | 462 | 392 | 217 | 44 | 7 | 2 | 1648 |
| Total Precipitation (") | 3.95 | 3.50 | 4.27 | 3.22 | 4.19 | 4.10 | 5.33 | 5.06 | 3.68 | 2.90 | 2.97 | 3.33 | 46.50 |
| Days ≥ 0.1" Precip | 7 | 7 | 7 | 6 | 7 | 6 | 7 | 7 | 5 | 5 | 5 | 6 | 75 |
| Total Snowfall (") | 1.3 | 1.5 | 1.0 | 0.0 | 0.0 | 0.0 | 0.0 | 0.0 | 0.0 | 0.0 | 0.0 | 0.3 | 4.1 |
| Days ≥ 1" Snow Depth | 1 | 1 | 0 | 0 | 0 | 0 | 0 | 0 | 0 | 0 | 0 | 0 | 2 |

## YADKINVILLE 6 E *Yadkin County* ELEVATION 860 ft LAT/LONG 36° 8 ' N / 80° 33 ' W

| | JAN | FEB | MAR | APR | MAY | JUN | JUL | AUG | SEP | OCT | NOV | DEC | YEAR |
|---|---|---|---|---|---|---|---|---|---|---|---|---|---|
| Maximum Temp °F | 48.8 | 53.4 | 63.0 | 72.8 | 79.1 | 85.6 | 89.1 | 87.6 | 82.0 | 72.2 | 62.0 | 52.3 | 70.7 |
| Minimum Temp °F | 26.8 | 29.0 | 36.5 | 44.1 | 53.0 | 60.8 | 65.0 | 63.8 | 57.4 | 45.3 | 37.0 | 30.3 | 45.8 |
| Mean Temp °F | 37.8 | 41.3 | 49.8 | 58.5 | 66.1 | 73.2 | 77.0 | 75.7 | 69.7 | 58.8 | 49.5 | 41.3 | 58.2 |
| Days Max Temp ≥ 90 °F | 0 | 0 | 0 | 1 | 2 | 8 | 15 | 12 | 5 | 0 | 0 | 0 | 43 |
| Days Max Temp ≤ 32 °F | 1 | 0 | 0 | 0 | 0 | 0 | 0 | 0 | 0 | 0 | 0 | 0 | 1 |
| Days Min Temp ≤ 32 °F | 22 | 19 | 12 | 4 | 0 | 0 | 0 | 0 | 0 | 3 | 11 | 19 | 90 |
| Days Min Temp ≤ 0 °F | 0 | 0 | 0 | 0 | 0 | 0 | 0 | 0 | 0 | 0 | 0 | 0 | 0 |
| Heating Degree Days | 835 | 664 | 470 | 214 | 62 | 5 | 0 | 1 | 27 | 215 | 461 | 729 | 3683 |
| Cooling Degree Days | 0 | 0 | 5 | 27 | 111 | 288 | 408 | 350 | 184 | 35 | 2 | 0 | 1410 |
| Total Precipitation (") | 3.57 | 3.48 | 4.62 | 3.24 | 4.46 | 3.89 | 4.21 | 3.64 | 3.68 | 3.87 | 3.08 | 3.52 | 45.26 |
| Days ≥ 0.1" Precip | 7 | 6 | 8 | 7 | 8 | 7 | 8 | 6 | 5 | 5 | 6 | 6 | 79 |
| Total Snowfall (") | 4.1 | 3.3 | 1.5 | 0.0 | 0.0 | 0.0 | 0.0 | 0.0 | 0.0 | 0.0 | 0.2 | 0.8 | 9.9 |
| Days ≥ 1" Snow Depth | 4 | 2 | 1 | 0 | 0 | 0 | 0 | 0 | 0 | 0 | 0 | 1 | 8 |

## JANUARY MINIMUM TEMPERATURE °F

### LOWEST

| | | |
|---|---|---|
| 1 | Grandfather Mntn | 19.4 |
| 2 | Celo | 20.3 |
| 3 | Banner Elk | 20.5 |
| 4 | Blowing Rock | 20.7 |
| 5 | Oconaluftee | 20.9 |
| | Transou | 20.9 |
| 7 | Cataloochee | 21.0 |
| 8 | Fletcher-2 NE | 21.2 |
| 9 | Jefferson | 22.3 |
| 10 | Pisgah Forest | 22.7 |
| 11 | North Wilkesboro | 22.8 |
| 12 | Andrews | 23.0 |
| 13 | West Kerr Scott | 23.2 |
| 14 | Coweeta | 23.6 |
| 15 | Waynesville | 23.8 |
| 16 | Louisburg | 23.9 |
| 17 | Henderson | 24.2 |
| 18 | Fletcher-3 W | 24.4 |
| 19 | Danbury | 24.5 |
| | Highlands | 24.5 |
| 21 | Marshall | 24.7 |
| 22 | Franklin | 25.0 |
| | Murphy | 25.0 |
| 24 | Siler City | 25.2 |
| 25 | Bent Creek | 25.4 |

### HIGHEST

| | | |
|---|---|---|
| 1 | Cape Hatteras | 37.6 |
| 2 | Hatteras | 36.9 |
| 3 | Cedar Island | 35.6 |
| 4 | Morehead City | 35.3 |
| 5 | Wilmington-N H | 35.1 |
| 6 | New Bern | 33.7 |
| 7 | Wilmington-7 N | 33.0 |
| 8 | Edenton | 32.8 |
| 9 | Willard | 32.6 |
| 10 | Southport | 32.2 |
| 11 | Elizabethtown Lk | 32.1 |
| 12 | Whiteville | 32.0 |
| 13 | Belhaven | 31.9 |
| 14 | Laurinburg | 31.8 |
| 15 | Williamston | 31.5 |
| 16 | Goldsboro | 31.3 |
| 17 | Elizabeth City | 31.1 |
| 18 | Plymouth | 30.9 |
| 19 | Monroe | 30.6 |
| 20 | Charlotte | 30.5 |
| | Tryon | 30.5 |
| | Wadesboro | 30.5 |
| 23 | Raleigh-4 SW | 30.4 |
| 24 | Clayton | 30.3 |
| 25 | Asheboro | 30.1 |

## JULY MAXIMUM TEMPERATURE °F

### HIGHEST

| | | |
|---|---|---|
| 1 | Laurinburg | 91.6 |
| 2 | Hamlet | 91.2 |
| 3 | Tarboro | 90.8 |
| 4 | Smithfield | 90.5 |
| 5 | Concord | 90.3 |
| | Fayetteville | 90.3 |
| | Goldsboro | 90.3 |
| 8 | Wadesboro | 90.2 |
| | Wilson | 90.2 |
| 10 | Jackson | 90.1 |
| | Whiteville | 90.1 |
| 12 | Lexington | 90.0 |
| | Willard | 90.0 |
| 14 | Greenville | 89.9 |
| | Lewiston | 89.9 |
| | Wilmington-N H | 89.9 |
| 17 | Burlington | 89.8 |
| | Lumberton | 89.8 |
| | Plymouth | 89.8 |
| 20 | Gastonia | 89.7 |
| | Louisburg | 89.7 |
| 22 | Durham | 89.6 |
| | Kinston | 89.6 |
| | Rocky Mount | 89.6 |
| | Wilmington-7 N | 89.6 |

### LOWEST

| | | |
|---|---|---|
| 1 | Grandfather Mntn | 69.6 |
| 2 | Blowing Rock | 76.2 |
| 3 | Banner Elk | 77.2 |
| 4 | Highlands | 77.5 |
| 5 | Transou | 80.1 |
| 6 | Celo | 80.5 |
| | Jefferson | 80.5 |
| 8 | Cataloochee | 80.9 |
| 9 | Waynesville | 82.9 |
| 10 | Coweeta | 83.3 |
| 11 | Fletcher-2 NE | 83.4 |
| 12 | Pisgah Forest | 83.6 |
| 13 | Fletcher-3 W | 83.7 |
| 14 | Asheville-Reg | 84.2 |
| 15 | Bent Creek | 84.3 |
| | Marshall | 84.3 |
| 17 | Asheville | 84.5 |
| 18 | Black Mountain | 84.6 |
| 19 | Hendersonville | 84.7 |
| 20 | Cullowhee | 84.8 |
| 21 | Oconaluftee | 84.9 |
| 22 | Andrews | 85.2 |
| 23 | Cape Hatteras | 85.3 |
| 24 | Hatteras | 85.7 |
| 25 | Franklin | 85.8 |

## ANNUAL PRECIPITATION (")

### HIGHEST

| | | |
|---|---|---|
| 1 | Highlands | 85.59 |
| 2 | Coweeta | 72.16 |
| 3 | Blowing Rock | 67.87 |
| 4 | Pisgah Forest | 65.21 |
| 5 | Tryon | 64.87 |
| 6 | Andrews | 63.40 |
| 7 | Grandfather Mntn | 62.69 |
| 8 | Tapoco | 59.75 |
| 9 | Celo | 59.13 |
| 10 | Southport | 58.13 |
| 11 | Cedar Island | 57.12 |
| 12 | Oconaluftee | 56.70 |
| 13 | Cape Hatteras | 56.35 |
| 14 | Hendersonville | 56.30 |
| 15 | Wilmington-7 N | 56.22 |
| 16 | Transou | 55.99 |
| 17 | Morehead City | 55.82 |
| 18 | Marion | 55.37 |
| 19 | Wilmington-N H | 55.30 |
| 20 | Murphy | 55.18 |
| 21 | Franklin | 53.33 |
| 22 | New Bern | 53.09 |
| 23 | Hatteras | 52.99 |
| 24 | West Kerr Scott | 52.73 |
| | Willard | 52.73 |

### LOWEST

| | | |
|---|---|---|
| 1 | Asheville | 37.35 |
| 2 | Marshall | 39.47 |
| 3 | Raleigh-Durham | 40.90 |
| 4 | Greensboro | 42.20 |
| 5 | Henderson | 42.88 |
| 6 | Charlotte | 43.03 |
| 7 | Salisbury | 43.07 |
| 8 | Clayton | 43.76 |
| 9 | Oxford | 43.93 |
| 10 | Roxboro | 44.07 |
| 11 | Burlington | 44.33 |
| 12 | Arcola | 44.36 |
| 13 | High Point | 44.53 |
| 14 | Raleigh-St Univ | 44.60 |
| 15 | Raleigh-4 SW | 44.65 |
| 16 | Jackson | 44.76 |
| 17 | Louisburg | 44.88 |
| | Reidsville | 44.88 |
| 19 | Fletcher-2 NE | 44.90 |
| 20 | Lexington | 44.97 |
| 21 | Roanoke Rapids | 45.14 |
| 22 | Tarboro | 45.23 |
| 23 | Yadkinville | 45.26 |
| 24 | Rocky Mount | 45.37 |
| 25 | Chapel Hill | 45.38 |

## ANNUAL SNOWFALL (")

### HIGHEST

| | | |
|---|---|---|
| 1 | Grandfather Mntn | 54.4 |
| 2 | Banner Elk | 46.3 |
| 3 | Blowing Rock | 35.1 |
| 4 | Transou | 25.8 |
| 5 | Jefferson | 21.7 |
| 6 | Celo | 18.6 |
| 7 | Highlands | 16.6 |
| 8 | Waynesville | 15.7 |
| 9 | Asheville | 15.6 |
| 10 | Asheville-Reg | 15.3 |
| 11 | Marshall | 13.9 |
| 12 | North Wilkesboro | 12.9 |
| 13 | Reidsville | 12.5 |
| 14 | Marion | 12.3 |
| 15 | Fletcher-3 W | 12.2 |
| 16 | West Kerr Scott | 12.0 |
| 17 | Fletcher-2 NE | 11.7 |
| 18 | Hendersonville | 11.5 |
| | Roxboro | 11.5 |
| 20 | Cullowhee | 11.3 |
| | Pisgah Forest | 11.3 |
| 22 | Mount Airy | 11.2 |
| 23 | Black Mountain | 10.7 |
| 24 | Hickory | 10.0 |
| 25 | Yadkinville | 9.9 |

### LOWEST

| | | |
|---|---|---|
| 1 | Elizabeth City | 0.6 |
| 2 | Elizabethtown Lk | 1.1 |
| | Hatteras | 1.1 |
| 4 | Southport | 1.2 |
| 5 | Cape Hatteras | 1.8 |
| 6 | New Holland | 1.9 |
| 7 | Fayetteville | 2.0 |
| | Morehead City | 2.0 |
| 9 | Wilmington-N H | 2.2 |
| 10 | Hamlet | 2.6 |
| | Laurinburg | 2.6 |
| | Lumberton | 2.6 |
| 13 | Kinston | 2.8 |
| | Whiteville | 2.8 |
| | Wilmington-7 N | 2.8 |
| 16 | New Bern | 2.9 |
| 17 | Dunn | 3.1 |
| 18 | Cedar Island | 3.2 |
| | Smithfield | 3.2 |
| 20 | Greenville | 3.3 |
| 21 | Willard | 3.4 |
| 22 | Burlington | 3.8 |
| 23 | Plymouth | 3.9 |
| 24 | Clayton | 4.0 |
| 25 | Edenton | 4.1 |

# NORTH DAKOTA

PHYSICAL FEATURES. North Dakota is typically plains country located near the center of the North American Continent. The eastern part of the State is flat, with an elevation in the Red River Valley of 780 feet at Pembina in the north to 962 feet above sea level at Wahpeton in the south. To the westward there is a gradual rise of terrain until an elevation of 3,468 feet is reached at Black Butte in the southwestern part of the State. The Turtle Mountains in the north-central part of the State are only about 500 feet higher than the surrounding area, with the highest elevation about 2,300 feet above sea level.

GENERAL CLIMATE. Summers are usually very pleasant, but hot winds and periods of prolonged high temperatures occur occasionally. However, minimum temperatures are seldom above 70° F., so it is unusual to have uncomfortable nights. Winters are usually cold with occasional ones that are open and mild.

TEMPERATURE. The annual mean temperature for North Dakota ranges from about 36° F. in the northeast to 43° F. in the extreme south. Temperatures above 100° F. are occasionally recorded, and zero readings are common in winter. The average number of days a year when the temperature reaches 90° F. or higher is 14, and the average number with zero or lower is 53. The average growing season is about 121 days, ranging from 110 days in the northeast and north-central to 135 in the extreme south. For the State, the average date of the last freeze in spring is May 19, and the first in fall is September 18. Freezing temperatures have occurred, however, as late as the first part of June and as early in the fall as the first few days of September.

PRECIPITATION. Precipitation in the eastern third of the State averages about 19 inches, in the middle third about 16 inches, and in the western third about 15 inches. On an average, about 77 percent of the annual precipitation occurs during the crop-growing freeze-free season, April to September, and almost 50 percent falls during May, June, and July. The normal precipitation for the driest months, November to February, is about one-half an inch a month. The greatest amount falls between 5 p.m. and 8 p.m. and again about midnight. In North Dakota, precipitation is considered the most important climatic factor.

Most of the rain in the summer months occurs in storms accompanied by thunder and lightning, often with heavy falls for a short time. The average number of thunderstorm days in 30, mostly in June, July, and August. In most years at least some part of the State is visited by a storm that brings a rainfall of 2 or 3 inches in 24 hours, and occasionally 5 or 6 inches falls in 1 day. On an average, rain falls about 1 day in 4 during the summer months. The annual number of days with measurable precipitation averages 66, ranging from about 50 in the west to 90 in the east.

The first light snow in autumn occasionally falls in September, but usually very little occurs until after October. The average number of days with 0.1 inch or more of snow is 23. The average annual snowfall is 32 inches with the greatest amount in the northeast and least in the southwest. Occasionally there is heavy snowfall in winter, and the amount of snow on the ground accumulates to a considerable depth.

RIVERS AND FLOODS. The streams of North Dakota fall into two main groups -- those in the west and south-central portions draining into the Missouri Basin, and those in the east and north-central portions draining into the Red River of the North.

Some of the important tributaries which drain into the Missouri in North Dakota are: the Cannonball, Grand, Heart, Knife, Little Missouri, and James. Local floods occur occasionally on all the tributaries, mainly associated with ice breakup, notably on the Heart River where serious floods have occurred from ice jams. Floods along the main stem of the Missouri in the past have been caused primarily by snowmelt in the high plains. The resulting flooding has been almost invariably aggravated by ice jams.

The streams draining the east and north-central portions of North Dakota flow into the Red River of the North, which flows in a northerly direction between Minnesota and North Dakota into Canada. The most important tributaries in the eastern portion of North Dakota are the Sheyenne and the Pembina. The latter rises in the province of Manitoba,

Canada. In the north-central portion the Souris River originates in the province of Saskatchewan, Canada, flows southeastward into North Dakota, and then curves back into Canada and flows in a northerly direction into the Assiniboine River which empties into the Red River of the North above the International Boundary.

Floods in the Red River of the North Basin occur primarily during the spring season (April and May) and are caused chiefly by melting snow. Ice conditions, particularly on the northward flowing streams, increase flood crests and occasionally cause extremely high flood stages due to jams. Early freeze-up in the fall before snow occurs is also a contributing factor in producing flood conditions in the spring. Considerably higher crests result along the tributaries and the main stem of the river if the snowmelt is accompanied by a period of prolonged heavy rains. Major rainstorms of sufficient magnitude to cause more than local flooding (without snowmelt) are extremely rare.

OTHER CLIMATIC ELEMENTS. The prevailing direction of the wind in all months of the year is from the northwest, unless it is influenced by local conditions. More southerly winds are observed during the summer than during the winter. The average annual wind speed is about 11 m.p.h. The highest speeds are in spring and the lowest in late summer. High winds frequently accompany severe thunderstorms. Tornadoes are reported in North Dakota.

The average relative humidity is about 68 percent, slightly higher in the east than in the west. Humidity is frequently low during the afternoon in summer, sometimes below 20 percent. Dense fogs are experienced, on an average, on only 8 days of the year.

The average number of clear days is 160, partly cloudy 100, and cloudy 105. On a clear day the sun shines for more than 15 hours from the middle of May to the end of July. The yearly average amount of sunshine is 59 percent of the possible amount, with 74 percent in July and 72 percent in August.

# COUNTY INDEX

***Adams County***
HETTINGER

***Barnes County***
VALLEY CITY 3 NNW

***Benson County***
LEEDS

***Billings County***
FAIRFIELD
MEDORA

***Bottineau County***
BOTTINEAU
WESTHOPE
WILLOW CITY

***Bowman County***
BOWMAN

***Burke County***
BOWBELLS
POWERS LAKE 1 N

***Burleigh County***
BISMARCK MUNI AP
MOFFIT 3 SE

***Cass County***
FARGO HECTOR FIELD

***Cavalier County***
LANGDON EXPERIMENT F

***Dickey County***
ELLENDALE
FULLERTON 1 ESE
OAKES 2 S

***Divide County***
CROSBY
FORTUNA 1 W

***Dunn County***
DUNN CENTER 2 SW

***Emmons County***
LINTON

***Foster County***
CARRINGTON
MC HENRY 3 W

***Golden Valley County***
TROTTERS 3 SSE

***Grand Forks County***
GRAND FORKS INTL AP
GRAND FORKS UNIV

***Grant County***
CARSON
PRETTY ROCK

***Griggs County***
COOPERSTOWN

***Hettinger County***
MOTT
NEW ENGLAND

***Kidder County***
PETTIBONE
STEELE
TUTTLE

***LaMoure County***
LA MOURE

***Logan County***
GACKLE
NAPOLEON

***McHenry County***
DRAKE 9 NE
GRANVILLE
TOWNER 2 NE
UPHAM 3 N
VELVA

***McIntosh County***
ASHLEY
WISHEK

***McKenzie County***
KEENE 3 S
WATFORD CITY 14 S

***McLean County***
BUTTE
GARRISON 1 NNW
MAX
TURTLE LAKE
UNDERWOOD
WILTON

***Mercer County***
BEULAH 1 W

***Morton County***
BREIEN
HEBRON
MANDAN EXP STN
NEW SALEM

***Mountrail County***
STANLEY 3 NNW

***Nelson County***
MC VILLE
PETERSBURG 2 N

***Oliver County***
CENTER 4 SE

***Pembina County***
CAVALIER 7 NW
PEMBINA

***Pierce County***
RUGBY

***Ramsey County***
DEVILS LAKE KDLR

***Ransom County***
LISBON

***Renville County***
MOHALL

***Richland County***
MC LEOD 3 E
WAHPETON 3 N

***Rolette County***
BELCOURT KEYA RADIO
ROLLA 3 NW

***Sargent County***
FORMAN 5 SSE

***Sheridan County***
MC CLUSKY

***Sioux County***
FORT YATES 4 SW

***Slope County***
AMIDON

***Stark County***
DICKINSON EXP STN
DICKINSON MUNI AP
RICHARDTON ABBEY

***Steele County***
COLGATE
SHARON

***Stutsman County***
JAMESTOWN MUNI AP
JAMESTOWN STATE HOSP

***Towner County***
HANSBORO 4 NNE

***Traill County***
HILLSBORO 3 N

***Walsh County***
GRAFTON
PARK RIVER

***Ward County***
FOXHOLM 7 N
KENMARE 1 WSW
MINOT EXPERIMENT STN
MINOT FAA AP

***Wells County***
FESSENDEN
HURDSFIELD 8 SW

***Williams County***
TIOGA 1 E
WILDROSE
WILLISTON EXP FARM
WILLISTON SLOULIN AP

## ELEVATION INDEX

| FEET | STATION NAME |
|---|---|
| 791 | PEMBINA |
| 830 | GRAFTON |
| 830 | GRAND FORKS UNIV |
| 852 | GRAND FORKS INTL AP |
| 889 | CAVALIER 7 NW |
| 896 | FARGO HECTOR FIELD |
| 909 | HILLSBORO 3 N |
| 961 | WAHPETON 3 N |
| 991 | PARK RIVER |
| 1079 | MC LEOD 3 E |
| 1152 | LISBON |
| 1181 | COLGATE |
| 1230 | VALLEY CITY 3 NNW |
| 1250 | FORMAN 5 SSE |
| 1381 | LA MOURE |
| 1430 | COOPERSTOWN |
| 1430 | UPHAM 3 N |
| 1440 | FULLERTON 1 ESE |
| 1460 | ELLENDALE |
| 1467 | JAMESTOWN STATE HOSP |
| 1470 | DEVILS LAKE KDLR |
| 1470 | MC VILLE |
| 1470 | WILLOW CITY |
| 1480 | TOWNER 2 NE |
| 1496 | JAMESTOWN MUNI AP |
| 1503 | GRANVILLE |
| 1512 | MC HENRY 3 W |
| 1512 | VELVA |
| 1512 | WESTHOPE |
| 1522 | LEEDS |
| 1522 | PETERSBURG 2 N |
| 1522 | SHARON |
| 1542 | HANSBORO 4 NNE |
| 1542 | RUGBY |
| 1552 | DRAKE 9 NE |
| 1581 | CARRINGTON |
| 1601 | OAKES 2 S |
| 1611 | FESSENDEN |
| 1611 | FOXHOLM 7 N |
| 1621 | LANGDON EXPERIMENT F |
| 1640 | BOTTINEAU |
| 1650 | MOHALL |
| 1663 | BISMARCK MUNI AP |
| 1670 | FORT YATES 4 SW |
| 1713 | LINTON |
| 1726 | MINOT FAA AP |

| FEET | STATION NAME |
|---|---|
| 1752 | MANDAN EXP STN |
| 1762 | MOFFIT 3 SE |
| 1772 | MINOT EXPERIMENT STN |
| 1781 | BEULAH 1 W |
| 1801 | KENMARE 1 WSW |
| 1831 | BREIEN |
| 1860 | PETTIBONE |
| 1860 | STEELE |
| 1880 | BUTTE |
| 1880 | TUTTLE |
| 1900 | WILLISTON SLOULIN AP |
| 1903 | TURTLE LAKE |
| 1923 | ROLLA 3 NW |
| 1926 | GARRISON 1 NNW |
| 1939 | HURDSFIELD 8 SW |
| 1942 | MC CLUSKY |
| 1952 | CROSBY |
| 1952 | GACKLE |
| 1962 | BELCOURT KEYA RADIO |
| 1962 | BOWBELLS |
| 1962 | NAPOLEON |
| 1972 | WATFORD CITY 14 S |
| 1990 | CENTER 4 SE |
| 2011 | WISHEK |
| 2031 | ASHLEY |
| 2031 | UNDERWOOD |
| 2090 | MAX |
| 2113 | WILLISTON EXP FARM |
| 2129 | NEW SALEM |
| 2162 | HEBRON |
| 2162 | WILTON |
| 2192 | DUNN CENTER 2 SW |
| 2211 | POWERS LAKE 1 N |
| 2251 | MEDORA |
| 2260 | STANLEY 3 NNW |
| 2260 | WILDROSE |
| 2280 | TIOGA 1 E |
| 2333 | CARSON |
| 2349 | FORTUNA 1 W |
| 2421 | TROTTERS 3 SSE |
| 2441 | MOTT |
| 2461 | DICKINSON EXP STN |
| 2470 | KEENE 3 S |
| 2470 | RICHARDTON ABBEY |
| 2480 | PRETTY ROCK |
| 2602 | DICKINSON MUNI AP |
| 2621 | NEW ENGLAND |
| 2680 | HETTINGER |
| 2753 | FAIRFIELD |
| 2913 | AMIDON |

| FEET | STATION NAME |
| --- | --- |
| 2980 | BOWMAN |

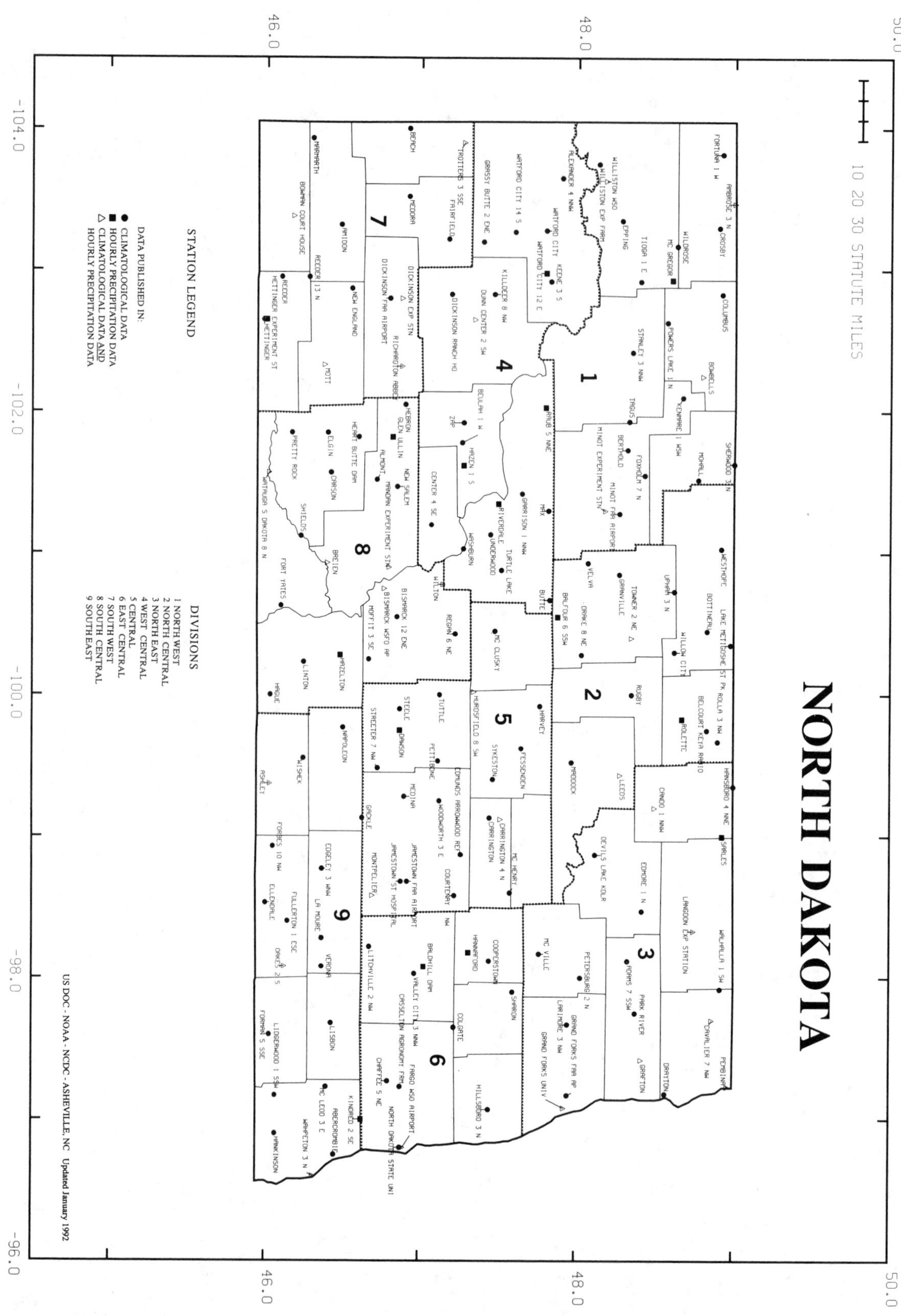
NORTH DAKOTA
10 20 30 STATUTE MILES
STATION LEGEND
DATA PUBLISHED IN:
● CLIMATOLOGICAL DATA
■ HOURLY PRECIPITATION DATA
△ CLIMATOLOGICAL DATA AND HOURLY PRECIPITATION DATA
DIVISIONS
1 NORTH WEST
2 NORTH CENTRAL
3 NORTH EAST
4 WEST CENTRAL
5 CENTRAL
6 EAST CENTRAL
7 SOUTH WEST
8 SOUTH CENTRAL
9 SOUTHEAST
US DOC - NOAA - NCDC - ASHEVILLE, NC Updated January 1992

## AMIDON *Slope County* ELEVATION 2913 ft LAT/LONG 46° 29 ' N / 103° 19 ' W

| | JAN | FEB | MAR | APR | MAY | JUN | JUL | AUG | SEP | OCT | NOV | DEC | YEAR |
|---|---|---|---|---|---|---|---|---|---|---|---|---|---|
| Maximum Temp °F | 24.3 | 30.0 | 40.8 | 53.5 | 66.2 | 75.2 | 82.8 | 82.6 | 70.6 | 57.5 | 39.7 | 29.2 | 54.4 |
| Minimum Temp °F | 3.6 | 9.2 | 19.5 | 30.5 | 41.5 | 50.7 | 56.0 | 54.1 | 43.6 | 32.4 | 19.4 | 8.5 | 30.8 |
| Mean Temp °F | 14.0 | 19.6 | 30.2 | 42.1 | 53.8 | 63.0 | 69.5 | 68.4 | 57.1 | 45.0 | 29.6 | 18.8 | 42.6 |
| Days Max Temp ≥ 90 °F | 0 | 0 | 0 | 0 | 1 | 2 | 7 | 9 | 2 | 0 | 0 | 0 | 21 |
| Days Max Temp ≤ 32 °F | 19 | 14 | 8 | 1 | 0 | 0 | 0 | 0 | 0 | 1 | 9 | 17 | 69 |
| Days Min Temp ≤ 32 °F | 30 | 28 | 28 | 18 | 5 | 0 | 0 | 0 | 3 | 15 | 27 | 31 | 185 |
| Days Min Temp ≤ 0 °F | 13 | 8 | 3 | 0 | 0 | 0 | 0 | 0 | 0 | 0 | 2 | 8 | 34 |
| Heating Degree Days | 1580 | 1275 | 1075 | 682 | 356 | 121 | 32 | 49 | 266 | 614 | 1058 | 1428 | 8536 |
| Cooling Degree Days | 0 | 0 | 0 | 2 | 22 | 80 | 188 | 173 | 37 | 2 | 0 | 0 | 504 |
| Total Precipitation (") | 0.39 | 0.32 | 0.57 | 1.44 | 2.40 | 3.29 | 2.28 | 1.39 | 1.39 | 1.05 | 0.49 | 0.39 | 15.40 |
| Days ≥ 0.1" Precip | 1 | 1 | 2 | 4 | 5 | 7 | 5 | 4 | 4 | 2 | 2 | 1 | 38 |
| Total Snowfall (") | *5.3* | *3.8* | na | *3.9* | 0.7 | 0.0 | 0.0 | 0.0 | 0.2 | 1.2 | na | na | na |
| Days ≥ 1" Snow Depth | na | *13* | *10* | *3* | 1 | 0 | 0 | 0 | 0 | 1 | na | na | na |

## ASHLEY *McIntosh County* ELEVATION 2031 ft LAT/LONG 46° 2 ' N / 99° 22 ' W

| | JAN | FEB | MAR | APR | MAY | JUN | JUL | AUG | SEP | OCT | NOV | DEC | YEAR |
|---|---|---|---|---|---|---|---|---|---|---|---|---|---|
| Maximum Temp °F | 19.0 | 25.3 | 38.0 | 54.9 | 68.5 | 76.8 | 83.6 | 82.2 | 71.3 | 58.1 | 38.0 | 24.4 | 53.3 |
| Minimum Temp °F | -2.0 | 4.3 | 16.7 | 30.0 | 42.0 | 51.6 | 56.4 | 53.6 | 43.0 | 31.6 | 17.5 | 4.5 | 29.1 |
| Mean Temp °F | 8.5 | 14.8 | 27.4 | 42.5 | 55.2 | 64.2 | 70.0 | 67.9 | 57.2 | 44.9 | 27.8 | 14.5 | 41.2 |
| Days Max Temp ≥ 90 °F | 0 | 0 | 0 | 0 | 1 | 2 | 7 | 6 | 2 | 0 | 0 | 0 | 18 |
| Days Max Temp ≤ 32 °F | 24 | 19 | 10 | 1 | 0 | 0 | 0 | 0 | 0 | 1 | 10 | 21 | 86 |
| Days Min Temp ≤ 32 °F | 31 | 28 | 29 | 19 | 5 | 0 | 0 | 0 | 3 | 16 | 29 | 31 | 191 |
| Days Min Temp ≤ 0 °F | 18 | 11 | 4 | 0 | 0 | 0 | 0 | 0 | 0 | 0 | 2 | 12 | 47 |
| Heating Degree Days | 1747 | 1412 | 1160 | 670 | 315 | 93 | 26 | 51 | 262 | 619 | 1111 | 1561 | 9027 |
| Cooling Degree Days | 0 | 0 | 0 | 3 | 25 | 85 | 193 | 150 | 32 | 1 | 0 | 0 | 489 |
| Total Precipitation (") | 0.35 | 0.33 | 0.87 | 1.63 | 2.57 | 3.47 | 2.53 | 2.16 | 1.41 | 1.29 | 0.53 | 0.34 | 17.48 |
| Days ≥ 0.1" Precip | 1 | 1 | 2 | 4 | 6 | 6 | 5 | 4 | 3 | 3 | 2 | 1 | 38 |
| Total Snowfall (") | 5.2 | 4.2 | 5.2 | 3.6 | 0.2 | 0.0 | 0.0 | 0.0 | 0.0 | 0.7 | 4.6 | 4.3 | 28.0 |
| Days ≥ 1" Snow Depth | 24 | 21 | 15 | 2 | 0 | 0 | 0 | 0 | 0 | 0 | 9 | 18 | 89 |

## BELCOURT KEYA RADIO *Rolette County* ELEVATION 1962 ft LAT/LONG 48° 50 ' N / 99° 45 ' W

| | JAN | FEB | MAR | APR | MAY | JUN | JUL | AUG | SEP | OCT | NOV | DEC | YEAR |
|---|---|---|---|---|---|---|---|---|---|---|---|---|---|
| Maximum Temp °F | *13.7* | 20.3 | 32.3 | 50.2 | *65.0* | 72.9 | 78.7 | 77.5 | *65.9* | *53.7* | *33.0* | *19.8* | 48.6 |
| Minimum Temp °F | -9.6 | *-3.5* | 10.5 | *25.0* | *37.5* | 47.2 | *52.0* | 49.6 | *39.2* | *28.0* | *13.1* | *-2.3* | 23.9 |
| Mean Temp °F | *2.4* | *8.3* | 21.5 | *37.3* | *50.9* | 60.1 | *65.2* | 63.6 | *52.5* | *40.9* | *23.0* | *8.9* | 36.2 |
| Days Max Temp ≥ 90 °F | 0 | 0 | 0 | 0 | 0 | 1 | 2 | 2 | 0 | 0 | 0 | 0 | 5 |
| Days Max Temp ≤ 32 °F | 26 | *22* | *15* | 3 | 0 | 0 | 0 | 0 | 0 | 1 | 15 | *24* | 106 |
| Days Min Temp ≤ 32 °F | 30 | 27 | 30 | *24* | *9* | 1 | 0 | 0 | 7 | *21* | 28 | 30 | 207 |
| Days Min Temp ≤ 0 °F | 21 | *15* | *8* | 0 | 0 | 0 | 0 | 0 | 0 | 0 | 5 | *16* | 65 |
| Heating Degree Days | *1942* | *1596* | 1344 | *822* | *438* | 175 | *66* | 108 | *380* | *740* | *1262* | *1753* | 10626 |
| Cooling Degree Days | na | na | *0* | *0* | *9* | 35 | na | 71 | na | *0* | na | *0* | na |
| Total Precipitation (") | 0.53 | 0.39 | 0.60 | 1.52 | 2.18 | 3.31 | 3.10 | 2.59 | 2.04 | 1.20 | 0.53 | 0.47 | 18.46 |
| Days ≥ 0.1" Precip | 2 | *1* | 2 | 3 | 5 | 6 | 5 | 5 | 4 | 3 | 2 | *1* | 39 |
| Total Snowfall (") | 9.1 | *7.9* | *7.0* | 4.4 | 1.0 | 0.0 | 0.0 | 0.0 | 0.1 | 2.1 | *6.3* | *7.1* | 45.0 |
| Days ≥ 1" Snow Depth | *25* | *22* | *16* | *5* | 1 | 0 | 0 | 0 | 0 | 1 | *10* | *22* | 102 |

## BEULAH 1 W *Mercer County* ELEVATION 1781 ft LAT/LONG 47° 16 ' N / 101° 47 ' W

| | JAN | FEB | MAR | APR | MAY | JUN | JUL | AUG | SEP | OCT | NOV | DEC | YEAR |
|---|---|---|---|---|---|---|---|---|---|---|---|---|---|
| Maximum Temp °F | 19.8 | 26.5 | 39.5 | 55.6 | 69.2 | 77.5 | 84.2 | 83.0 | 71.4 | 58.6 | 38.4 | 25.4 | 54.1 |
| Minimum Temp °F | -2.1 | 4.5 | 16.7 | 28.8 | 40.3 | 49.9 | 54.1 | 52.0 | 41.5 | 30.3 | 16.8 | 4.1 | 28.1 |
| Mean Temp °F | 8.8 | 15.5 | 28.1 | 42.2 | 54.8 | 63.7 | 69.2 | 67.5 | 56.4 | 44.5 | 27.7 | 14.8 | 41.1 |
| Days Max Temp ≥ 90 °F | 0 | 0 | 0 | 0 | 1 | 4 | 8 | 8 | 2 | 0 | 0 | 0 | 23 |
| Days Max Temp ≤ 32 °F | 22 | 17 | 10 | 1 | 0 | 0 | 0 | 0 | 0 | 1 | 10 | 20 | 81 |
| Days Min Temp ≤ 32 °F | 31 | 28 | 29 | 21 | 7 | 0 | 0 | 0 | 4 | 19 | 29 | 31 | 199 |
| Days Min Temp ≤ 0 °F | 16 | 11 | 4 | 0 | 0 | 0 | 0 | 0 | 0 | 0 | 2 | 11 | 44 |
| Heating Degree Days | 1739 | 1393 | 1136 | 679 | 333 | 109 | 29 | 58 | 279 | 631 | 1114 | 1554 | 9054 |
| Cooling Degree Days | 0 | 0 | 0 | 4 | 30 | 94 | 172 | 154 | 27 | 1 | 0 | 0 | 482 |
| Total Precipitation (") | 0.39 | 0.32 | 0.69 | 1.82 | 2.25 | 3.36 | 2.29 | 1.67 | 1.62 | 1.18 | 0.60 | 0.42 | 16.61 |
| Days ≥ 0.1" Precip | 1 | 1 | 2 | 4 | 5 | 6 | 5 | 4 | 4 | 3 | 2 | 2 | 39 |
| Total Snowfall (") | *5.7* | *4.8* | *4.3* | *2.7* | 0.1 | 0.0 | 0.0 | 0.0 | 0.1 | *1.5* | 5.1 | *5.2* | 29.5 |
| Days ≥ 1" Snow Depth | *28* | *21* | *12* | *3* | 0 | 0 | 0 | 0 | 0 | *1* | 9 | *23* | 97 |

## BISMARCK MUNI AP *Burleigh County* ELEVATION 1663 ft LAT/LONG 46° 46 ' N / 100° 45 ' W

| | JAN | FEB | MAR | APR | MAY | JUN | JUL | AUG | SEP | OCT | NOV | DEC | YEAR |
|---|---|---|---|---|---|---|---|---|---|---|---|---|---|
| Maximum Temp °F | 19.8 | 26.2 | 39.1 | 55.0 | 68.3 | 76.9 | 84.2 | 82.6 | 70.9 | 57.7 | 38.3 | 25.2 | 53.7 |
| Minimum Temp °F | -1.9 | 5.3 | 18.4 | 30.8 | 42.4 | 51.3 | 55.9 | 53.9 | 43.3 | 31.8 | 17.9 | 4.2 | 29.4 |
| Mean Temp °F | 9.0 | 15.8 | 28.8 | 42.9 | 55.3 | 64.1 | 70.1 | 68.2 | 57.1 | 44.8 | 28.1 | 14.7 | 41.6 |
| Days Max Temp ≥ 90 °F | 0 | 0 | 0 | 0 | 1 | 3 | 8 | 8 | 2 | 0 | 0 | 0 | 22 |
| Days Max Temp ≤ 32 °F | 23 | 17 | 9 | 1 | 0 | 0 | 0 | 0 | 0 | 1 | 9 | 20 | 80 |
| Days Min Temp ≤ 32 °F | 31 | 28 | 28 | 18 | 4 | 0 | 0 | 0 | 3 | 16 | 28 | 31 | 187 |
| Days Min Temp ≤ 0 °F | 17 | 11 | 3 | 0 | 0 | 0 | 0 | 0 | 0 | 0 | 2 | 12 | 45 |
| Heating Degree Days | 1735 | 1385 | 1116 | 657 | 312 | 94 | 19 | 48 | 260 | 619 | 1100 | 1554 | 8899 |
| Cooling Degree Days | 0 | 0 | 0 | 3 | 25 | 87 | 190 | 163 | 28 | 0 | 0 | 0 | 496 |
| Total Precipitation (") | 0.45 | 0.43 | 0.80 | 1.57 | 2.13 | 2.66 | 2.46 | 1.76 | 1.63 | 1.04 | 0.60 | 0.48 | 16.01 |
| Days ≥ 0.1" Precip | 1 | 1 | 2 | 4 | 5 | 6 | 5 | 4 | 3 | 2 | 2 | 1 | 36 |
| Total Snowfall (") | 8.0 | 7.7 | 8.5 | 4.4 | 0.9 | 0.0 | 0.0 | 0.0 | 0.3 | 2.1 | 7.7 | 8.2 | 47.8 |
| Days ≥ 1" Snow Depth | 26 | 21 | 12 | 3 | 0 | 0 | 0 | 0 | 0 | 1 | 9 | 21 | 93 |

## BOTTINEAU *Bottineau County* ELEVATION 1640 ft LAT/LONG 48° 50 ' N / 100° 27 ' W

| | JAN | FEB | MAR | APR | MAY | JUN | JUL | AUG | SEP | OCT | NOV | DEC | YEAR |
|---|---|---|---|---|---|---|---|---|---|---|---|---|---|
| Maximum Temp °F | 12.6 | 19.3 | 32.7 | 51.7 | 66.8 | 74.8 | 79.7 | 79.0 | 67.1 | 53.6 | 32.9 | 18.3 | 49.0 |
| Minimum Temp °F | -8.1 | -1.3 | 13.2 | 28.9 | 41.4 | 50.4 | 54.3 | 52.1 | 41.5 | 29.4 | 14.2 | -1.3 | 26.2 |
| Mean Temp °F | 2.3 | 9.0 | 23.0 | 40.3 | 54.2 | 62.6 | 67.0 | 65.6 | 54.3 | 41.5 | 23.6 | 8.5 | 37.7 |
| Days Max Temp ≥ 90 °F | 0 | 0 | 0 | 0 | 1 | 2 | 3 | 4 | 1 | 0 | 0 | 0 | 11 |
| Days Max Temp ≤ 32 °F | 28 | 23 | 14 | 2 | 0 | 0 | 0 | 0 | 0 | 1 | 15 | 27 | 110 |
| Days Min Temp ≤ 32 °F | 31 | 28 | 30 | 20 | 6 | 0 | 0 | 0 | 4 | 20 | 29 | 31 | 199 |
| Days Min Temp ≤ 0 °F | 21 | 15 | 7 | 0 | 0 | 0 | 0 | 0 | 0 | 0 | 5 | 16 | 64 |
| Heating Degree Days | 1944 | 1578 | 1297 | 736 | 349 | 126 | 47 | 82 | 332 | 722 | 1237 | 1748 | 10198 |
| Cooling Degree Days | 0 | 0 | 0 | 2 | 25 | 66 | 113 | 108 | 15 | 0 | 0 | 0 | 329 |
| Total Precipitation (") | 0.51 | 0.37 | 0.68 | 1.32 | 1.99 | 3.26 | 3.11 | 2.72 | 2.01 | 1.23 | 0.56 | 0.56 | 18.32 |
| Days ≥ 0.1" Precip | 2 | 1 | 2 | 3 | 5 | 7 | 6 | 5 | 4 | 3 | 2 | 2 | 42 |
| Total Snowfall (") | 8.5 | 5.3 | 6.0 | 3.2 | 0.3 | 0.0 | 0.0 | 0.0 | 0.1 | 1.8 | 6.3 | 7.6 | 39.1 |
| Days ≥ 1" Snow Depth | 30 | 25 | 20 | 5 | 0 | 0 | 0 | 0 | 0 | 2 | 13 | 28 | 123 |

## BOWBELLS *Burke County* ELEVATION 1962 ft LAT/LONG 48° 48 ' N / 102° 15 ' W

| | JAN | FEB | MAR | APR | MAY | JUN | JUL | AUG | SEP | OCT | NOV | DEC | YEAR |
|---|---|---|---|---|---|---|---|---|---|---|---|---|---|
| Maximum Temp °F | 15.1 | 21.3 | 34.4 | 52.1 | 66.7 | 75.4 | 80.9 | 79.8 | 67.6 | 54.6 | 34.2 | 21.4 | 50.3 |
| Minimum Temp °F | -5.1 | 1.2 | 13.8 | 27.6 | 39.2 | 48.6 | 53.1 | 50.4 | 40.5 | 29.2 | 15.4 | 1.9 | 26.3 |
| Mean Temp °F | 5.0 | 11.2 | 24.1 | 39.9 | 53.0 | 62.0 | 67.0 | 65.0 | 54.1 | 41.9 | 24.8 | 11.7 | 38.3 |
| Days Max Temp ≥ 90 °F | 0 | 0 | 0 | 0 | 1 | 2 | 5 | 5 | 1 | 0 | 0 | 0 | 14 |
| Days Max Temp ≤ 32 °F | 25 | 20 | 13 | 2 | 0 | 0 | 0 | 0 | 0 | 1 | 13 | 23 | 97 |
| Days Min Temp ≤ 32 °F | 31 | 28 | 30 | 21 | 7 | 0 | 0 | 0 | 5 | 20 | 29 | 31 | 202 |
| Days Min Temp ≤ 0 °F | 19 | 14 | 6 | 0 | 0 | 0 | 0 | 0 | 0 | 0 | 4 | 14 | 57 |
| Heating Degree Days | 1859 | 1515 | 1262 | 747 | 380 | 134 | 49 | 87 | 343 | 707 | 1201 | 1652 | 9936 |
| Cooling Degree Days | 0 | 0 | 0 | 1 | 19 | 53 | 111 | 95 | 16 | 0 | 0 | 0 | 295 |
| Total Precipitation (") | 0.50 | 0.43 | 0.58 | 1.50 | 2.15 | 2.96 | 2.79 | 2.12 | 2.05 | 1.05 | 0.36 | 0.32 | 16.81 |
| Days ≥ 0.1" Precip | 2 | 1 | 2 | 3 | 5 | 6 | 6 | 5 | 4 | 3 | 2 | 1 | 40 |
| Total Snowfall (") | 9.0 | 5.6 | 4.8 | 4.0 | 0.2 | 0.0 | 0.0 | 0.0 | 0.1 | 1.4 | ***4.5*** | 5.7 | 35.3 |
| Days ≥ 1" Snow Depth | na | na | na | ***4*** | 0 | 0 | 0 | 0 | 0 | ***0*** | na | na | na |

## BOWMAN *Bowman County* ELEVATION 2980 ft LAT/LONG 46° 11 ' N / 103° 23 ' W

| | JAN | FEB | MAR | APR | MAY | JUN | JUL | AUG | SEP | OCT | NOV | DEC | YEAR |
|---|---|---|---|---|---|---|---|---|---|---|---|---|---|
| Maximum Temp °F | 24.3 | 29.9 | 40.8 | 54.0 | 65.9 | 74.8 | 82.6 | 82.2 | 70.1 | 57.7 | 40.2 | 29.2 | 54.3 |
| Minimum Temp °F | 3.7 | 8.5 | 19.2 | 30.3 | 41.3 | 50.7 | 55.9 | 53.7 | 42.7 | 31.6 | 18.6 | 8.2 | 30.4 |
| Mean Temp °F | 14.0 | 19.2 | 30.0 | 42.2 | 53.6 | 62.8 | 69.3 | 68.0 | 56.4 | 44.7 | 29.4 | 18.7 | 42.4 |
| Days Max Temp ≥ 90 °F | 0 | 0 | 0 | 0 | 0 | 2 | 7 | 8 | 2 | 0 | 0 | 0 | 19 |
| Days Max Temp ≤ 32 °F | 19 | 14 | 8 | 1 | 0 | 0 | 0 | 0 | 0 | 1 | 8 | 17 | 68 |
| Days Min Temp ≤ 32 °F | 31 | 28 | 29 | 19 | 5 | 0 | 0 | 0 | 4 | 16 | 28 | 31 | 191 |
| Days Min Temp ≤ 0 °F | 13 | 8 | 3 | 0 | 0 | 0 | 0 | 0 | 0 | 0 | 2 | 8 | 34 |
| Heating Degree Days | 1576 | 1287 | 1077 | 679 | 359 | 122 | 31 | 50 | 281 | 624 | 1062 | 1428 | 8576 |
| Cooling Degree Days | 0 | 0 | 0 | 1 | 16 | 67 | 171 | 148 | 28 | 1 | 0 | 0 | 432 |
| Total Precipitation (") | 0.42 | 0.37 | 0.66 | 1.39 | 2.44 | 3.31 | 2.20 | 1.31 | 1.35 | 1.13 | 0.47 | 0.42 | 15.47 |
| Days ≥ 0.1" Precip | 1 | 1 | 2 | 4 | 6 | 7 | 4 | 4 | 4 | 2 | 2 | 2 | 39 |
| Total Snowfall (") | ***8.8*** | 7.6 | ***8.2*** | 5.6 | 0.6 | 0.0 | 0.0 | 0.0 | 0.7 | 2.3 | 8.3 | ***8.9*** | 51.0 |
| Days ≥ 1" Snow Depth | 26 | 20 | 16 | 6 | 1 | 0 | 0 | 0 | 0 | 1 | 11 | 23 | 104 |

## BREIEN *Morton County* ELEVATION 1831 ft LAT/LONG 46° 24' N / 100° 56' W

| | JAN | FEB | MAR | APR | MAY | JUN | JUL | AUG | SEP | OCT | NOV | DEC | YEAR |
|---|---|---|---|---|---|---|---|---|---|---|---|---|---|
| Maximum Temp °F | 22.2 | 29.3 | 42.3 | 58.2 | 71.3 | 79.5 | 86.3 | 85.4 | ***73.8*** | ***60.4*** | 40.9 | 28.3 | 56.5 |
| Minimum Temp °F | -2.8 | 4.9 | 17.5 | 29.3 | 40.9 | 50.4 | 55.0 | 52.8 | ***41.7*** | ***29.3*** | 16.5 | ***3.9*** | 28.3 |
| Mean Temp °F | 10.0 | 17.1 | 29.7 | 43.6 | 56.1 | 64.9 | 70.7 | 69.1 | ***57.8*** | ***44.9*** | 28.5 | ***16.0*** | 42.4 |
| Days Max Temp ≥ 90 °F | 0 | 0 | 0 | 0 | 1 | 4 | ***11*** | 10 | 3 | 0 | 0 | 0 | 29 |
| Days Max Temp ≤ 32 °F | 21 | ***15*** | 7 | 1 | 0 | 0 | 0 | 0 | 0 | 0 | 8 | ***18*** | 70 |
| Days Min Temp ≤ 32 °F | 31 | 28 | 29 | 19 | 6 | 0 | 0 | 0 | ***4*** | ***19*** | ***28*** | 30 | 194 |
| Days Min Temp ≤ 0 °F | 18 | ***11*** | 3 | 0 | 0 | 0 | 0 | 0 | 0 | 0 | 3 | ***11*** | 46 |
| Heating Degree Days | 1702 | 1350 | 1086 | 637 | 291 | 81 | 18 | 38 | ***243*** | ***617*** | 1091 | ***1514*** | 8668 |
| Cooling Degree Days | 0 | 0 | 0 | 2 | 26 | 94 | 207 | 177 | 34 | 0 | 0 | 0 | 540 |
| Total Precipitation (") | 0.36 | 0.38 | 0.65 | 1.69 | 2.37 | 3.22 | 2.58 | 1.83 | 1.54 | 1.00 | 0.46 | 0.37 | 16.45 |
| Days ≥ 0.1" Precip | 1 | 1 | 2 | 4 | ***5*** | 6 | 4 | 4 | ***3*** | 2 | 2 | ***1*** | 35 |
| Total Snowfall (") | ***6.7*** | ***6.3*** | ***7.7*** | ***2.4*** | 0.8 | 0.0 | 0.0 | 0.0 | 0.0 | ***1.2*** | na | ***6.3*** | na |
| Days ≥ 1" Snow Depth | na | na | na | ***1*** | 0 | 0 | 0 | 0 | 0 | 0 | na | na | na |

## BUTTE *McLean County* ELEVATION 1880 ft LAT/LONG 47° 50' N / 100° 40' W

| | JAN | FEB | MAR | APR | MAY | JUN | JUL | AUG | SEP | OCT | NOV | DEC | YEAR |
|---|---|---|---|---|---|---|---|---|---|---|---|---|---|
| Maximum Temp °F | 17.9 | 24.5 | 37.1 | 55.6 | 70.1 | 78.6 | 84.4 | 83.4 | 71.1 | 57.1 | 36.9 | 23.0 | 53.3 |
| Minimum Temp °F | -0.7 | 5.6 | 17.8 | 31.1 | 43.0 | 52.5 | 57.2 | 55.3 | 45.4 | 34.0 | 19.8 | 5.5 | 30.5 |
| Mean Temp °F | 8.6 | 15.1 | 27.5 | 43.4 | 56.6 | 65.5 | 70.8 | 69.4 | 58.2 | 45.5 | 28.3 | 14.3 | 41.9 |
| Days Max Temp ≥ 90 °F | 0 | 0 | 0 | 0 | 1 | 3 | 8 | 9 | 2 | 0 | 0 | 0 | 23 |
| Days Max Temp ≤ 32 °F | 23 | ***18*** | 11 | 1 | 0 | 0 | 0 | 0 | 0 | 1 | 10 | 22 | 86 |
| Days Min Temp ≤ 32 °F | 30 | 28 | 29 | 17 | 4 | 0 | 0 | 0 | 2 | 13 | 26 | 30 | 179 |
| Days Min Temp ≤ 0 °F | 16 | 11 | 3 | 0 | 0 | 0 | 0 | 0 | 0 | 0 | 2 | 11 | 43 |
| Heating Degree Days | 1745 | 1407 | 1157 | 644 | 287 | 75 | 16 | 38 | 238 | 598 | 1093 | 1568 | 8866 |
| Cooling Degree Days | 0 | 0 | 0 | 4 | 39 | 107 | 205 | 188 | 40 | 1 | 0 | 0 | 584 |
| Total Precipitation (") | 0.47 | 0.35 | 0.72 | 1.60 | 2.13 | 2.75 | 2.69 | 1.74 | 1.68 | 1.24 | 0.55 | 0.44 | 16.36 |
| Days ≥ 0.1" Precip | 2 | 1 | 2 | 4 | 5 | 6 | 5 | 4 | 4 | 2 | 2 | 1 | 38 |
| Total Snowfall (") | ***6.9*** | ***5.1*** | ***3.8*** | 2.7 | 0.1 | 0.0 | 0.0 | 0.0 | 0.0 | 1.9 | 5.3 | 5.9 | 31.7 |
| Days ≥ 1" Snow Depth | ***22*** | ***16*** | 13 | 3 | 0 | 0 | 0 | 0 | 0 | 1 | 7 | 15 | 77 |

## CARRINGTON *Foster County* ELEVATION 1581 ft LAT/LONG 47° 27' N / 99° 8' W

| | JAN | FEB | MAR | APR | MAY | JUN | JUL | AUG | SEP | OCT | NOV | DEC | YEAR |
|---|---|---|---|---|---|---|---|---|---|---|---|---|---|
| Maximum Temp °F | 15.6 | 21.7 | 34.4 | 51.9 | 66.8 | 74.7 | 80.4 | 79.3 | 68.1 | 55.2 | 35.0 | 21.3 | 50.4 |
| Minimum Temp °F | -3.9 | 2.0 | 15.3 | 29.5 | 41.6 | 51.4 | 56.3 | 53.5 | 43.2 | 31.9 | 16.9 | 2.9 | 28.4 |
| Mean Temp °F | 5.9 | 11.9 | 24.9 | 40.7 | 54.2 | 63.1 | 68.4 | 66.4 | 55.7 | 43.6 | 26.0 | 12.1 | 39.4 |
| Days Max Temp ≥ 90 °F | 0 | 0 | 0 | 0 | 0 | 2 | 4 | 4 | 1 | 0 | 0 | 0 | 11 |
| Days Max Temp ≤ 32 °F | 26 | 21 | 13 | 2 | 0 | 0 | 0 | 0 | 0 | 1 | 13 | 24 | 100 |
| Days Min Temp ≤ 32 °F | 31 | 28 | 30 | 20 | 6 | 0 | 0 | 0 | 2 | 16 | 29 | 31 | 193 |
| Days Min Temp ≤ 0 °F | 19 | 14 | 5 | 0 | 0 | 0 | 0 | 0 | 0 | 0 | 3 | 14 | 55 |
| Heating Degree Days | 1831 | 1496 | 1237 | 724 | 347 | 113 | 32 | 66 | 295 | 657 | 1165 | 1635 | 9598 |
| Cooling Degree Days | 0 | 0 | 0 | 3 | 24 | 70 | 151 | 123 | 22 | 1 | 0 | 0 | 394 |
| Total Precipitation (") | 0.64 | 0.51 | 0.91 | 1.53 | 2.11 | 3.48 | 2.83 | 1.95 | 1.73 | 1.17 | 0.73 | 0.52 | 18.11 |
| Days ≥ 0.1" Precip | 2 | 2 | 3 | 4 | 5 | 7 | 6 | 5 | 4 | 3 | 2 | 2 | 45 |
| Total Snowfall (") | 9.2 | 7.2 | 7.2 | 3.8 | 0.7 | 0.0 | 0.0 | 0.0 | 0.0 | 1.5 | 6.9 | 7.4 | 43.9 |
| Days ≥ 1" Snow Depth | 29 | 24 | 18 | 4 | 0 | 0 | 0 | 0 | 0 | 1 | ***10*** | 21 | 107 |

## CARSON *Grant County* ELEVATION 2333 ft LAT/LONG 46° 25' N / 101° 34' W

| | JAN | FEB | MAR | APR | MAY | JUN | JUL | AUG | SEP | OCT | NOV | DEC | YEAR |
|---|---|---|---|---|---|---|---|---|---|---|---|---|---|
| Maximum Temp °F | 20.1 | 26.2 | 38.0 | 53.3 | 66.8 | 74.7 | 81.8 | 80.8 | 69.5 | 57.2 | 38.3 | 25.5 | 52.7 |
| Minimum Temp °F | -0.6 | 5.2 | 17.0 | 29.6 | 41.5 | 50.9 | 56.0 | 53.7 | 42.8 | 31.1 | 17.4 | 4.9 | 29.1 |
| Mean Temp °F | 9.8 | 15.7 | 27.5 | 41.5 | 54.2 | 62.8 | 68.9 | 67.3 | 56.2 | 44.2 | 27.9 | 15.2 | 40.9 |
| Days Max Temp ≥ 90 °F | 0 | 0 | 0 | 0 | 0 | 2 | 6 | 6 | 2 | 0 | 0 | 0 | 16 |
| Days Max Temp ≤ 32 °F | 22 | 18 | 11 | 2 | 0 | 0 | 0 | 0 | 0 | 1 | 10 | 20 | 84 |
| Days Min Temp ≤ 32 °F | 31 | 28 | 29 | 20 | 5 | 0 | 0 | 0 | 4 | 17 | 28 | 31 | 193 |
| Days Min Temp ≤ 0 °F | 16 | 11 | 3 | 0 | 0 | 0 | 0 | 0 | 0 | 0 | 3 | 11 | 44 |
| Heating Degree Days | 1710 | 1388 | 1156 | 700 | 347 | 121 | 34 | 58 | 288 | 640 | 1107 | 1538 | 9087 |
| Cooling Degree Days | 0 | 0 | 0 | 3 | 21 | 68 | 162 | 142 | 28 | 1 | 0 | 0 | 425 |
| Total Precipitation (") | 0.28 | 0.39 | 0.79 | 1.86 | 2.46 | 3.30 | 2.45 | 1.81 | 1.34 | 1.18 | 0.46 | 0.34 | 16.66 |
| Days ≥ 0.1" Precip | 1 | 1 | 2 | 5 | 5 | 7 | 5 | 4 | 3 | 3 | 2 | 1 | 39 |
| Total Snowfall (") | 4.7 | 5.2 | 6.5 | 4.4 | 0.6 | 0.0 | 0.0 | 0.0 | 0.2 | 1.9 | 4.2 | 5.4 | 33.1 |
| Days ≥ 1" Snow Depth | ***19*** | ***14*** | ***11*** | 4 | 0 | 0 | 0 | 0 | 0 | 1 | 5 | na | na |

## CAVALIER 7 NW *Pembina County* ELEVATION 889 ft LAT/LONG 48° 52 ' N / 97° 42 ' W

| | JAN | FEB | MAR | APR | MAY | JUN | JUL | AUG | SEP | OCT | NOV | DEC | YEAR |
|---|---|---|---|---|---|---|---|---|---|---|---|---|---|
| Maximum Temp °F | 12.3 | 18.9 | 32.1 | 51.5 | 68.0 | 75.4 | 79.5 | 79.1 | 68.0 | 54.1 | 33.4 | 18.4 | 49.2 |
| Minimum Temp °F | -7.2 | -1.5 | 13.8 | 29.3 | 41.3 | 51.0 | 55.4 | 52.8 | 43.1 | 31.8 | 16.3 | 0.5 | 27.2 |
| Mean Temp °F | 2.6 | 8.7 | 23.0 | 40.4 | 54.7 | 63.2 | 67.5 | 65.9 | 55.6 | 43.0 | 24.9 | 9.5 | 38.3 |
| Days Max Temp ≥ 90 °F | 0 | 0 | 0 | 0 | 1 | 2 | 2 | 4 | 1 | 0 | 0 | 0 | 10 |
| Days Max Temp ≤ 32 °F | 27 | 23 | 15 | 2 | 0 | 0 | 0 | 0 | 0 | 1 | 14 | 25 | 107 |
| Days Min Temp ≤ 32 °F | 31 | 28 | 30 | 20 | 6 | 0 | 0 | 0 | 3 | 16 | 29 | 31 | 194 |
| Days Min Temp ≤ 0 °F | 22 | 16 | 6 | 0 | 0 | 0 | 0 | 0 | 0 | 0 | 4 | 16 | 64 |
| Heating Degree Days | 1934 | 1586 | 1296 | 732 | 339 | 112 | 35 | 71 | 297 | 676 | 1196 | 1718 | 9992 |
| Cooling Degree Days | 0 | 0 | 0 | 2 | 30 | 60 | 110 | 103 | 15 | 1 | 0 | 0 | 321 |
| Total Precipitation (") | 0.42 | 0.33 | 0.70 | 1.34 | 2.21 | 3.06 | 3.22 | 2.56 | 1.88 | 1.37 | 0.58 | 0.44 | 18.11 |
| Days ≥ 0.1" Precip | 1 | 1 | 2 | 3 | 5 | 6 | 6 | 5 | 4 | 3 | 2 | 2 | 40 |
| Total Snowfall (") | *7.3* | *5.2* | *5.3* | 2.3 | 0.6 | 0.0 | 0.0 | 0.0 | 0.1 | 2.2 | *7.3* | 6.5 | 36.8 |
| Days ≥ 1" Snow Depth | 31 | 27 | 24 | 5 | 0 | 0 | 0 | 0 | 0 | 1 | *14* | 26 | 128 |

## CENTER 4 SE *Oliver County* ELEVATION 1990 ft LAT/LONG 47° 4 ' N / 101° 12 ' W

| | JAN | FEB | MAR | APR | MAY | JUN | JUL | AUG | SEP | OCT | NOV | DEC | YEAR |
|---|---|---|---|---|---|---|---|---|---|---|---|---|---|
| Maximum Temp °F | 20.6 | 26.2 | 39.2 | 54.9 | 68.4 | 76.4 | 82.9 | 82.1 | 70.5 | 57.7 | 38.5 | 25.7 | 53.6 |
| Minimum Temp °F | -0.5 | 5.5 | 17.5 | 29.9 | 41.3 | 50.6 | 55.4 | 53.0 | 42.9 | 31.7 | 17.5 | 5.4 | 29.2 |
| Mean Temp °F | 10.0 | 15.9 | 28.4 | 42.4 | 54.8 | 63.5 | 69.2 | 67.5 | 56.7 | 44.7 | 28.0 | 15.6 | 41.4 |
| Days Max Temp ≥ 90 °F | 0 | 0 | 0 | 0 | 1 | 2 | 6 | 7 | 1 | 0 | 0 | 0 | 17 |
| Days Max Temp ≤ 32 °F | 22 | 18 | 9 | 1 | 0 | 0 | 0 | 0 | 0 | 1 | 9 | 20 | 80 |
| Days Min Temp ≤ 32 °F | 31 | 28 | 29 | 19 | 5 | 0 | 0 | 0 | 4 | 16 | 28 | 31 | 191 |
| Days Min Temp ≤ 0 °F | 17 | 11 | 4 | 0 | 0 | 0 | 0 | 0 | 0 | 0 | 2 | 11 | 45 |
| Heating Degree Days | 1702 | 1381 | 1130 | 673 | 325 | 103 | 26 | 53 | 269 | 622 | 1105 | 1528 | 8917 |
| Cooling Degree Days | 0 | 0 | 0 | 3 | 20 | 70 | 166 | 148 | 26 | 1 | 0 | 0 | 434 |
| Total Precipitation (") | 0.42 | 0.42 | 0.67 | 1.75 | 2.15 | 3.09 | 2.59 | 1.83 | 1.93 | 1.38 | 0.59 | 0.47 | 17.29 |
| Days ≥ 0.1" Precip | 2 | 2 | 2 | 4 | 5 | 7 | 5 | 4 | 4 | 3 | 2 | 2 | 42 |
| Total Snowfall (") | 5.2 | 4.7 | 6.0 | 3.0 | 0.6 | 0.0 | 0.0 | 0.0 | 0.1 | 1.5 | 4.4 | 5.4 | 30.9 |
| Days ≥ 1" Snow Depth | *26* | *22* | *14* | 3 | 0 | 0 | 0 | 0 | 0 | 1 | *7* | 18 | 91 |

## COLGATE *Steele County* ELEVATION 1181 ft LAT/LONG 47° 14 ' N / 97° 39 ' W

| | JAN | FEB | MAR | APR | MAY | JUN | JUL | AUG | SEP | OCT | NOV | DEC | YEAR |
|---|---|---|---|---|---|---|---|---|---|---|---|---|---|
| Maximum Temp °F | 15.3 | 21.9 | 35.7 | 55.1 | 70.4 | 77.9 | 83.3 | 82.5 | 71.1 | 56.9 | 35.5 | 21.1 | 52.2 |
| Minimum Temp °F | -4.8 | 1.1 | 16.3 | 30.2 | 42.1 | 51.3 | 55.5 | 53.5 | 43.8 | 32.4 | 17.2 | 2.8 | 28.5 |
| Mean Temp °F | 5.3 | 11.5 | 26.0 | 42.6 | 56.3 | 64.6 | 69.5 | 68.0 | 57.5 | 44.6 | 26.4 | 12.1 | 40.4 |
| Days Max Temp ≥ 90 °F | 0 | 0 | 0 | 0 | 1 | 2 | 6 | 7 | 2 | 0 | 0 | 0 | 18 |
| Days Max Temp ≤ 32 °F | 26 | 20 | 11 | 1 | 0 | 0 | 0 | 0 | 0 | 1 | 13 | 24 | 96 |
| Days Min Temp ≤ 32 °F | 31 | 28 | 29 | 19 | 5 | 0 | 0 | 0 | 3 | 15 | 28 | 31 | 189 |
| Days Min Temp ≤ 0 °F | 19 | 14 | 5 | 0 | 0 | 0 | 0 | 0 | 0 | 0 | 3 | 14 | 55 |
| Heating Degree Days | 1851 | 1504 | 1205 | 666 | 293 | 84 | 21 | 48 | 252 | 627 | 1153 | 1636 | 9340 |
| Cooling Degree Days | 0 | 0 | 0 | 4 | 36 | 81 | 171 | 152 | 28 | 1 | 0 | 0 | 473 |
| Total Precipitation (") | 0.47 | 0.27 | 0.81 | 1.53 | 2.39 | 3.34 | 2.59 | 2.49 | 2.04 | 1.47 | 0.60 | 0.43 | 18.43 |
| Days ≥ 0.1" Precip | 2 | 1 | 2 | 4 | 6 | 6 | 5 | 5 | 4 | 3 | 2 | 2 | 42 |
| Total Snowfall (") | 7.5 | 4.0 | 4.3 | 2.1 | 0.3 | 0.0 | 0.0 | 0.0 | 0.0 | 0.7 | *5.8* | 5.3 | 30.0 |
| Days ≥ 1" Snow Depth | 24 | 21 | 15 | 2 | 0 | 0 | 0 | 0 | 0 | 0 | *7* | 21 | 90 |

## COOPERSTOWN *Griggs County* ELEVATION 1430 ft LAT/LONG 47° 26 ' N / 98° 7 ' W

| | JAN | FEB | MAR | APR | MAY | JUN | JUL | AUG | SEP | OCT | NOV | DEC | YEAR |
|---|---|---|---|---|---|---|---|---|---|---|---|---|---|
| Maximum Temp °F | 16.2 | 23.1 | 36.1 | 54.9 | 69.9 | 77.7 | 83.1 | 81.9 | 70.8 | 57.2 | 35.6 | 21.4 | 52.3 |
| Minimum Temp °F | -4.1 | 2.4 | 16.1 | 30.5 | 42.7 | 52.0 | 56.6 | 54.2 | 44.1 | 32.6 | 17.3 | 2.6 | 28.9 |
| Mean Temp °F | 6.0 | 12.8 | 26.2 | 42.7 | 56.3 | 64.9 | 69.9 | 68.0 | 57.5 | 44.9 | 26.5 | 12.0 | 40.6 |
| Days Max Temp ≥ 90 °F | 0 | 0 | 0 | 0 | 1 | 2 | 6 | 6 | 2 | 0 | 0 | 0 | 17 |
| Days Max Temp ≤ 32 °F | 26 | 20 | 11 | 1 | 0 | 0 | 0 | 0 | 0 | 1 | 12 | 24 | 95 |
| Days Min Temp ≤ 32 °F | 31 | 28 | 29 | 19 | 5 | 0 | 0 | 0 | 3 | 15 | 28 | 31 | 189 |
| Days Min Temp ≤ 0 °F | 19 | 13 | 5 | 0 | 0 | 0 | 0 | 0 | 0 | 0 | 3 | 14 | 54 |
| Heating Degree Days | 1826 | 1471 | 1196 | 662 | 292 | 81 | 19 | 47 | 254 | 616 | 1150 | 1639 | 9253 |
| Cooling Degree Days | 0 | 0 | 0 | 4 | 34 | 83 | 171 | 146 | 28 | 0 | 0 | 0 | 466 |
| Total Precipitation (") | 0.61 | 0.43 | 0.96 | 1.51 | 2.50 | 3.50 | 3.38 | 2.54 | 2.28 | 1.38 | 0.82 | 0.55 | 20.46 |
| Days ≥ 0.1" Precip | 2 | 2 | 3 | 4 | 6 | 7 | 6 | 5 | 4 | 3 | 2 | 2 | 46 |
| Total Snowfall (") | *8.0* | *4.9* | *5.7* | 2.6 | 0.2 | 0.0 | 0.0 | 0.0 | 0.0 | *0.8* | na | *6.1* | na |
| Days ≥ 1" Snow Depth | *29* | *25* | 17 | 2 | 0 | 0 | 0 | 0 | 0 | 1 | *11* | 23 | 108 |

## CROSBY *Divide County* ELEVATION 1952 ft LAT/LONG 48° 54' N / 103° 18' W

| | JAN | FEB | MAR | APR | MAY | JUN | JUL | AUG | SEP | OCT | NOV | DEC | YEAR |
|---|---|---|---|---|---|---|---|---|---|---|---|---|---|
| Maximum Temp °F | 17.1 | 24.3 | 38.2 | 55.7 | 69.5 | 77.7 | 83.6 | 82.3 | 70.3 | 56.4 | 35.6 | 22.5 | 52.8 |
| Minimum Temp °F | -2.8 | 4.0 | 16.7 | 29.6 | 41.4 | 50.5 | 54.8 | 52.4 | 42.5 | 31.7 | 16.7 | 3.4 | 28.4 |
| Mean Temp °F | 7.2 | 14.2 | 27.4 | 42.7 | 55.5 | 64.1 | 69.2 | 67.4 | 56.5 | 44.1 | 26.2 | 13.0 | 40.6 |
| Days Max Temp ≥ 90 °F | 0 | 0 | 0 | 0 | 1 | 3 | 7 | 7 | 1 | 0 | 0 | 0 | 19 |
| Days Max Temp ≤ 32 °F | 25 | 18 | 10 | 1 | 0 | 0 | 0 | 0 | 0 | 1 | 12 | 22 | 89 |
| Days Min Temp ≤ 32 °F | 31 | 28 | 29 | 19 | 5 | 0 | 0 | 0 | 3 | 16 | 28 | 31 | 190 |
| Days Min Temp ≤ 0 °F | 18 | 12 | 4 | 0 | 0 | 0 | 0 | 0 | 0 | 0 | 3 | 12 | 49 |
| Heating Degree Days | 1791 | 1429 | 1159 | 665 | 310 | 94 | 24 | 55 | 274 | 642 | 1159 | 1611 | 9213 |
| Cooling Degree Days | 0 | 0 | 0 | 2 | 29 | 87 | 168 | 153 | 22 | 0 | 0 | 0 | 461 |
| Total Precipitation (") | 0.50 | 0.31 | 0.54 | 1.22 | 2.08 | 2.56 | 2.56 | 1.69 | 1.63 | 0.88 | 0.37 | 0.47 | 14.81 |
| Days ≥ 0.1" Precip | 2 | 1 | 2 | 3 | 5 | 6 | 5 | 4 | 4 | 3 | 1 | 2 | 38 |
| Total Snowfall (") | 9.1 | 4.8 | 5.0 | 4.8 | 0.5 | 0.0 | 0.0 | 0.0 | 0.0 | *1.9* | 5.1 | 7.8 | 39.0 |
| Days ≥ 1" Snow Depth | 25 | 20 | 14 | 4 | 0 | 0 | 0 | 0 | 0 | 1 | 8 | 20 | 92 |

## DEVILS LAKE KDLR *Ramsey County* ELEVATION 1470 ft LAT/LONG 48° 7' N / 98° 52' W

| | JAN | FEB | MAR | APR | MAY | JUN | JUL | AUG | SEP | OCT | NOV | DEC | YEAR |
|---|---|---|---|---|---|---|---|---|---|---|---|---|---|
| Maximum Temp °F | 14.6 | 21.4 | 33.2 | 51.7 | 66.9 | 75.0 | 80.5 | 79.2 | 67.4 | 54.0 | 34.0 | 20.2 | 49.8 |
| Minimum Temp °F | -4.2 | 2.4 | 15.4 | 30.3 | 43.0 | 52.9 | 57.4 | 54.9 | 44.5 | 33.4 | 18.5 | 3.4 | 29.3 |
| Mean Temp °F | 5.1 | 11.9 | 24.3 | 41.0 | 55.0 | 64.0 | 69.0 | 67.1 | 56.0 | 43.7 | 26.2 | 11.8 | 39.6 |
| Days Max Temp ≥ 90 °F | 0 | 0 | 0 | 0 | 1 | 2 | 4 | 4 | 1 | 0 | 0 | 0 | 12 |
| Days Max Temp ≤ 32 °F | 26 | 21 | 13 | 2 | 0 | 0 | 0 | 0 | 0 | 1 | 14 | 25 | 102 |
| Days Min Temp ≤ 32 °F | 31 | 28 | 29 | 19 | 4 | 0 | 0 | 0 | 2 | 14 | 28 | 31 | 186 |
| Days Min Temp ≤ 0 °F | *19* | 13 | 5 | 0 | 0 | 0 | 0 | 0 | 0 | 0 | 2 | 13 | 52 |
| Heating Degree Days | 1859 | 1496 | 1254 | 713 | 329 | 99 | 27 | 59 | 288 | 653 | 1155 | 1646 | 9578 |
| Cooling Degree Days | 0 | 0 | 0 | 2 | 30 | 77 | 153 | 133 | 22 | 1 | 0 | 0 | 418 |
| Total Precipitation (") | 0.57 | 0.47 | 0.84 | 1.01 | 2.10 | 3.73 | 3.11 | 2.10 | 1.89 | 1.29 | 0.77 | 0.56 | 18.44 |
| Days ≥ 0.1" Precip | 2 | 2 | 3 | 3 | 5 | 7 | 6 | 5 | 4 | 3 | 2 | 2 | 44 |
| Total Snowfall (") | na | *4.5* | na | 2.5 | 0.4 | 0.0 | 0.0 | 0.0 | 0.0 | 1.7 | 5.1 | na | na |
| Days ≥ 1" Snow Depth | *30* | *25* | *20* | 3 | 0 | 0 | 0 | 0 | 0 | 1 | 10 | *25* | 114 |

## DICKINSON EXP STN *Stark County* ELEVATION 2461 ft LAT/LONG 46° 53' N / 102° 48' W

| | JAN | FEB | MAR | APR | MAY | JUN | JUL | AUG | SEP | OCT | NOV | DEC | YEAR |
|---|---|---|---|---|---|---|---|---|---|---|---|---|---|
| Maximum Temp °F | 22.4 | 28.4 | 39.8 | 53.9 | 66.8 | 75.5 | 82.6 | 82.4 | 70.5 | 57.8 | 39.4 | 28.0 | 54.0 |
| Minimum Temp °F | -0.9 | 4.7 | 16.2 | 28.2 | 39.5 | 48.8 | 53.4 | 51.1 | 39.9 | 28.8 | 15.5 | 4.5 | 27.5 |
| Mean Temp °F | 10.8 | 16.6 | 28.0 | 41.0 | 53.2 | 62.1 | 68.0 | 66.8 | 55.2 | 43.3 | 27.5 | 16.3 | 40.7 |
| Days Max Temp ≥ 90 °F | 0 | 0 | 0 | 0 | 1 | 2 | 7 | 8 | 2 | 0 | 0 | 0 | 20 |
| Days Max Temp ≤ 32 °F | 20 | 16 | 9 | 1 | 0 | 0 | 0 | 0 | 0 | 1 | 9 | 18 | 74 |
| Days Min Temp ≤ 32 °F | 31 | 28 | 30 | 21 | 6 | 0 | 0 | 0 | 5 | 21 | 29 | 31 | 202 |
| Days Min Temp ≤ 0 °F | 16 | 11 | 4 | 0 | 0 | 0 | 0 | 0 | 0 | 0 | 3 | 11 | 45 |
| Heating Degree Days | 1678 | 1363 | 1140 | 712 | 372 | 135 | 43 | 67 | 311 | 667 | 1120 | 1505 | 9113 |
| Cooling Degree Days | 0 | 0 | 0 | 1 | 16 | 66 | 142 | 130 | 21 | 0 | 0 | 0 | 376 |
| Total Precipitation (") | 0.36 | 0.29 | 0.61 | 1.78 | 2.35 | 3.70 | 2.08 | 1.54 | 1.59 | 1.18 | 0.51 | 0.42 | 16.41 |
| Days ≥ 0.1" Precip | 1 | 1 | 2 | 5 | 5 | 7 | 5 | 4 | 4 | 3 | 2 | 1 | 40 |
| Total Snowfall (") | na | na | na | na | 0.4 | 0.0 | 0.0 | 0.0 | 0.4 | *1.7* | na | na | na |
| Days ≥ 1" Snow Depth | na | na | *10* | 2 | 0 | 0 | 0 | 0 | 0 | 1 | *5* | na | na |

## DICKINSON MUNI AP *Stark County* ELEVATION 2602 ft LAT/LONG 46° 48' N / 102° 48' W

| | JAN | FEB | MAR | APR | MAY | JUN | JUL | AUG | SEP | OCT | NOV | DEC | YEAR |
|---|---|---|---|---|---|---|---|---|---|---|---|---|---|
| Maximum Temp °F | 22.8 | 28.7 | 40.3 | 54.5 | 67.1 | 75.6 | 83.3 | 82.2 | 70.3 | 57.4 | 39.0 | 27.6 | 54.1 |
| Minimum Temp °F | 3.6 | 9.3 | 19.5 | 30.8 | 41.6 | 50.6 | 55.5 | 54.1 | 43.5 | 32.7 | 19.3 | 8.2 | 30.7 |
| Mean Temp °F | 13.2 | 19.0 | 29.9 | 42.7 | 54.4 | 63.1 | 69.5 | 68.2 | 56.9 | 45.1 | 29.2 | 17.9 | 42.4 |
| Days Max Temp ≥ 90 °F | 0 | 0 | 0 | 0 | 1 | 2 | 7 | 8 | 2 | 0 | 0 | 0 | 20 |
| Days Max Temp ≤ 32 °F | 20 | 15 | 9 | 1 | 0 | 0 | 0 | 0 | 0 | 1 | 10 | 18 | 74 |
| Days Min Temp ≤ 32 °F | 31 | 27 | 28 | 17 | 4 | 0 | 0 | 0 | 3 | 15 | 27 | 30 | 182 |
| Days Min Temp ≤ 0 °F | 13 | 9 | 3 | 0 | 0 | 0 | 0 | 0 | 0 | 0 | 2 | 9 | 36 |
| Heating Degree Days | 1603 | 1293 | 1080 | 665 | 339 | 110 | 30 | 51 | 269 | 611 | 1069 | 1455 | 8575 |
| Cooling Degree Days | 0 | 0 | 0 | 2 | 22 | 77 | 190 | 167 | 33 | 2 | 0 | 0 | 493 |
| Total Precipitation (") | 0.39 | 0.34 | 0.69 | 1.86 | 2.32 | 3.31 | 2.01 | 1.33 | 1.57 | 1.15 | 0.49 | 0.39 | 15.85 |
| Days ≥ 0.1" Precip | 1 | 1 | 2 | 4 | 5 | 7 | 4 | 3 | 4 | 2 | 1 | 1 | 35 |
| Total Snowfall (") | 5.5 | 4.1 | 6.4 | 5.5 | 0.9 | 0.0 | 0.0 | 0.0 | 0.4 | 2.0 | 4.9 | 5.3 | 35.0 |
| Days ≥ 1" Snow Depth | 23 | 17 | 12 | 4 | 1 | 0 | 0 | 0 | 0 | 1 | 9 | 17 | 84 |

## DRAKE 9 NE *McHenry County* ELEVATION 1552 ft LAT/LONG 48° 2 ' N / 100° 17 ' W

| | JAN | FEB | MAR | APR | MAY | JUN | JUL | AUG | SEP | OCT | NOV | DEC | YEAR |
|---|---|---|---|---|---|---|---|---|---|---|---|---|---|
| Maximum Temp °F | 16.3 | 23.1 | 36.4 | 54.5 | 69.3 | 77.5 | 82.8 | 81.9 | 70.1 | 56.6 | 35.8 | 21.9 | 52.2 |
| Minimum Temp °F | -4.4 | 2.4 | 15.8 | 29.8 | 42.5 | 51.7 | 56.1 | 54.1 | 43.5 | 32.0 | 17.1 | 2.3 | 28.6 |
| Mean Temp °F | 6.0 | 12.8 | 26.1 | 42.2 | 55.9 | 64.6 | 69.5 | 68.0 | 56.8 | 44.4 | 26.5 | 12.1 | 40.4 |
| Days Max Temp ≥ 90 °F | 0 | 0 | 0 | 0 | 1 | 3 | 6 | 7 | 1 | 0 | 0 | 0 | 18 |
| Days Max Temp ≤ 32 °F | 25 | 19 | 11 | 1 | 0 | 0 | 0 | 0 | 0 | 1 | 11 | 23 | 91 |
| Days Min Temp ≤ 32 °F | 31 | 28 | 29 | 19 | 5 | 0 | 0 | 0 | 3 | 16 | 28 | 31 | 190 |
| Days Min Temp ≤ 0 °F | 19 | 13 | 5 | 0 | 0 | 0 | 0 | 0 | 0 | 0 | 3 | 14 | 54 |
| Heating Degree Days | 1828 | 1471 | 1199 | 681 | 304 | 87 | 23 | 49 | 268 | 633 | 1150 | 1636 | 9329 |
| Cooling Degree Days | 0 | 0 | 0 | 4 | 38 | 91 | 183 | 157 | 26 | 0 | 0 | 0 | 499 |
| Total Precipitation (") | 0.32 | 0.25 | 0.53 | 1.21 | 1.96 | 2.87 | 2.64 | 2.01 | 1.57 | 1.13 | 0.49 | 0.34 | 15.32 |
| Days ≥ 0.1" Precip | 1 | 1 | 2 | 3 | 4 | 6 | 5 | 4 | 4 | 2 | 2 | 1 | 35 |
| Total Snowfall (") | 5.8 | 3.5 | 4.9 | 2.3 | 0.2 | 0.0 | 0.0 | 0.0 | 0.1 | 1.7 | 5.1 | 5.4 | 29.0 |
| Days ≥ 1" Snow Depth | 28 | 24 | 19 | 5 | 0 | 0 | 0 | 0 | 0 | 1 | 11 | 23 | 111 |

## DUNN CENTER 2 SW *Dunn County* ELEVATION 2192 ft LAT/LONG 47° 21 ' N / 102° 37 ' W

| | JAN | FEB | MAR | APR | MAY | JUN | JUL | AUG | SEP | OCT | NOV | DEC | YEAR |
|---|---|---|---|---|---|---|---|---|---|---|---|---|---|
| Maximum Temp °F | 21.0 | 27.5 | 39.3 | 54.4 | 67.8 | 75.9 | 83.4 | 82.7 | 71.1 | 57.5 | 38.5 | 26.6 | 53.8 |
| Minimum Temp °F | -0.4 | 5.7 | 16.9 | 29.9 | 41.9 | 51.1 | 55.6 | 53.6 | 43.3 | 32.1 | 18.1 | 5.6 | 29.5 |
| Mean Temp °F | 10.3 | 16.6 | 28.1 | 42.2 | 54.9 | 63.6 | 69.5 | 68.2 | 57.2 | 44.8 | 28.3 | 16.1 | 41.7 |
| Days Max Temp ≥ 90 °F | 0 | 0 | 0 | 0 | 1 | 3 | 7 | 8 | 2 | 0 | 0 | 0 | 21 |
| Days Max Temp ≤ 32 °F | 21 | 16 | 9 | 1 | 0 | 0 | 0 | 0 | 0 | 1 | 9 | 18 | 75 |
| Days Min Temp ≤ 32 °F | 31 | 28 | 29 | 19 | 5 | 0 | 0 | 0 | 3 | 16 | 27 | 30 | 188 |
| Days Min Temp ≤ 0 °F | 16 | 11 | 4 | 0 | 0 | 0 | 0 | 0 | 0 | 0 | 3 | 10 | 44 |
| Heating Degree Days | 1693 | 1362 | 1137 | 679 | 328 | 103 | 26 | 48 | 258 | 620 | 1095 | 1509 | 8858 |
| Cooling Degree Days | 0 | 0 | 0 | 2 | 29 | 91 | 187 | 164 | 34 | 0 | 0 | 0 | 507 |
| Total Precipitation (") | 0.43 | 0.35 | 0.61 | 1.72 | 2.29 | 3.64 | 2.13 | 1.73 | 1.61 | 1.21 | 0.55 | 0.42 | 16.69 |
| Days ≥ 0.1" Precip | 1 | 1 | 2 | 4 | 5 | 7 | 5 | 4 | 4 | 3 | 2 | 1 | 39 |
| Total Snowfall (") | 6.8 | 5.5 | 6.4 | 5.7 | 0.8 | 0.0 | 0.0 | 0.0 | 0.4 | 1.8 | 5.9 | 6.2 | 39.5 |
| Days ≥ 1" Snow Depth | 25 | 21 | 14 | 4 | 0 | 0 | 0 | 0 | 0 | 1 | 9 | 19 | 93 |

## ELLENDALE *Dickey County* ELEVATION 1460 ft LAT/LONG 46° 1 ' N / 98° 32 ' W

| | JAN | FEB | MAR | APR | MAY | JUN | JUL | AUG | SEP | OCT | NOV | DEC | YEAR |
|---|---|---|---|---|---|---|---|---|---|---|---|---|---|
| Maximum Temp °F | 19.9 | 26.3 | 39.3 | 57.0 | 70.9 | 78.8 | 85.3 | 84.1 | 73.4 | 59.6 | 38.2 | 24.9 | 54.8 |
| Minimum Temp °F | -0.7 | 5.6 | 19.0 | 31.7 | 42.8 | 52.3 | 57.7 | 55.4 | 45.0 | 33.6 | 18.9 | 5.3 | 30.6 |
| Mean Temp °F | 9.6 | 16.0 | 29.2 | 44.4 | 56.9 | 65.6 | 71.5 | 69.8 | 59.2 | 46.6 | 28.6 | 15.1 | 42.7 |
| Days Max Temp ≥ 90 °F | 0 | 0 | 0 | 0 | 1 | 3 | 9 | 8 | 2 | 0 | 0 | 0 | 23 |
| Days Max Temp ≤ 32 °F | 23 | 17 | 9 | 1 | 0 | 0 | 0 | 0 | 0 | 0 | 10 | 21 | 81 |
| Days Min Temp ≤ 32 °F | 31 | 28 | 28 | 17 | 4 | 0 | 0 | 0 | 2 | 14 | 28 | 31 | 183 |
| Days Min Temp ≤ 0 °F | 17 | 11 | 3 | 0 | 0 | 0 | 0 | 0 | 0 | 0 | 2 | 11 | 44 |
| Heating Degree Days | 1715 | 1379 | 1104 | 615 | 271 | 68 | 12 | 31 | 211 | 565 | 1086 | 1542 | 8599 |
| Cooling Degree Days | 0 | 0 | 0 | 5 | 35 | 97 | 224 | 186 | 41 | 1 | 0 | 0 | 589 |
| Total Precipitation (") | 0.43 | 0.41 | 0.94 | 2.31 | 2.97 | 3.70 | 2.64 | 2.54 | 2.03 | 1.57 | 0.64 | 0.34 | 20.52 |
| Days ≥ 0.1" Precip | 2 | 1 | 2 | 5 | 6 | 6 | 5 | 5 | 4 | 3 | 2 | 1 | 42 |
| Total Snowfall (") | 6.8 | 5.6 | 6.3 | 4.4 | 0.0 | 0.0 | 0.0 | 0.0 | 0.0 | 0.7 | 6.7 | 5.6 | 36.1 |
| Days ≥ 1" Snow Depth | *24* | *19* | *10* | 1 | 0 | 0 | 0 | 0 | 0 | 0 | 7 | *15* | 76 |

## FAIRFIELD *Billings County* ELEVATION 2753 ft LAT/LONG 47° 11 ' N / 103° 13 ' W

| | JAN | FEB | MAR | APR | MAY | JUN | JUL | AUG | SEP | OCT | NOV | DEC | YEAR |
|---|---|---|---|---|---|---|---|---|---|---|---|---|---|
| Maximum Temp °F | 21.1 | 27.2 | 38.5 | 53.6 | 66.3 | 74.8 | 81.8 | 81.0 | 68.7 | 56.1 | 37.3 | 25.6 | 52.7 |
| Minimum Temp °F | 1.9 | 7.5 | 18.1 | 30.0 | 41.2 | 50.4 | 55.4 | 53.7 | 43.4 | 32.5 | 18.5 | 6.6 | 29.9 |
| Mean Temp °F | 11.5 | 17.4 | 28.3 | 41.8 | 53.8 | 62.6 | 68.6 | 67.4 | 56.1 | 44.3 | 27.9 | 16.1 | 41.3 |
| Days Max Temp ≥ 90 °F | 0 | 0 | 0 | 0 | 0 | 2 | 6 | 6 | 2 | 0 | 0 | 0 | 16 |
| Days Max Temp ≤ 32 °F | 22 | 16 | 10 | 2 | 0 | 0 | 0 | 0 | 0 | 1 | 10 | 20 | 81 |
| Days Min Temp ≤ 32 °F | 31 | 28 | 29 | 19 | 5 | 0 | 0 | 0 | 3 | 15 | 28 | 30 | 188 |
| Days Min Temp ≤ 0 °F | 14 | 10 | 3 | 0 | 0 | 0 | 0 | 0 | 0 | 0 | 2 | 10 | 39 |
| Heating Degree Days | 1655 | 1340 | 1131 | 690 | 355 | 121 | 33 | 58 | 287 | 635 | 1108 | 1510 | 8923 |
| Cooling Degree Days | 0 | 0 | 0 | 1 | 19 | 68 | 156 | 144 | 24 | 0 | 0 | 0 | 412 |
| Total Precipitation (") | 0.34 | 0.24 | 0.50 | 1.54 | 2.29 | 3.30 | 2.01 | 1.54 | 1.59 | 1.00 | 0.43 | 0.35 | 15.13 |
| Days ≥ 0.1" Precip | 1 | 1 | 2 | 4 | 5 | 7 | 5 | 4 | 3 | 3 | 2 | 1 | 38 |
| Total Snowfall (") | 5.5 | 4.2 | 4.8 | 5.1 | 1.2 | 0.0 | 0.0 | 0.0 | 0.4 | 1.7 | 4.6 | 5.3 | 32.8 |
| Days ≥ 1" Snow Depth | 23 | 19 | 14 | 5 | 0 | 0 | 0 | 0 | 0 | 1 | 9 | 18 | 89 |

## FARGO HECTOR FIELD *Cass County* ELEVATION 896 ft LAT/LONG 46° 54 ' N / 96° 48 ' W

| | JAN | FEB | MAR | APR | MAY | JUN | JUL | AUG | SEP | OCT | NOV | DEC | YEAR |
|---|---|---|---|---|---|---|---|---|---|---|---|---|---|
| Maximum Temp °F | 15.1 | 21.2 | 34.9 | 54.1 | 68.9 | 76.9 | 82.4 | 80.7 | 69.4 | 56.0 | 35.8 | 20.9 | 51.4 |
| Minimum Temp °F | -3.5 | 3.1 | 17.9 | 32.2 | 44.2 | 53.8 | 58.8 | 56.6 | 46.0 | 34.3 | 19.2 | 4.3 | 30.6 |
| Mean Temp °F | 5.8 | 12.2 | 26.4 | 43.2 | 56.6 | 65.3 | 70.6 | 68.7 | 57.7 | 45.1 | 27.5 | 12.6 | 41.0 |
| Days Max Temp ≥ 90 °F | 0 | 0 | 0 | 0 | 1 | 2 | 5 | 5 | 1 | 0 | 0 | 0 | 14 |
| Days Max Temp ≤ 32 °F | 27 | 21 | 12 | 1 | 0 | 0 | 0 | 0 | 0 | 1 | 12 | 25 | 99 |
| Days Min Temp ≤ 32 °F | 31 | 28 | 27 | 16 | 4 | 0 | 0 | 0 | 2 | 13 | 27 | 31 | 179 |
| Days Min Temp ≤ 0 °F | 19 | 13 | 4 | 0 | 0 | 0 | 0 | 0 | 0 | 0 | 2 | 13 | 51 |
| Heating Degree Days | 1833 | 1488 | 1190 | 652 | 288 | 78 | 18 | 41 | 249 | 610 | 1118 | 1620 | 9185 |
| Cooling Degree Days | 0 | 0 | 0 | 6 | 44 | 103 | 204 | 169 | 33 | 2 | 0 | 0 | 561 |
| Total Precipitation (") | 0.69 | 0.48 | 1.09 | 1.73 | 2.32 | 3.12 | 2.90 | 2.32 | 1.88 | 1.73 | 0.85 | 0.65 | 19.76 |
| Days ≥ 0.1" Precip | 2 | 1 | 3 | 4 | 5 | 6 | 5 | 5 | 4 | 4 | 2 | 2 | 43 |
| Total Snowfall (") | 10.9 | 6.2 | 7.6 | 3.0 | 0.0 | 0.0 | 0.0 | 0.0 | 0.0 | 0.7 | 6.6 | 8.0 | 43.0 |
| Days ≥ 1" Snow Depth | 28 | 24 | 18 | 3 | 0 | 0 | 0 | 0 | 0 | 0 | 8 | 22 | 103 |

## FESSENDEN *Wells County* ELEVATION 1611 ft LAT/LONG 47° 39 ' N / 99° 37 ' W

| | JAN | FEB | MAR | APR | MAY | JUN | JUL | AUG | SEP | OCT | NOV | DEC | YEAR |
|---|---|---|---|---|---|---|---|---|---|---|---|---|---|
| Maximum Temp °F | 17.1 | 23.3 | 37.1 | 55.5 | 70.6 | 77.8 | 83.0 | *82.3* | *71.8* | 57.9 | 36.4 | *23.4* | 53.0 |
| Minimum Temp °F | -3.1 | 2.9 | 17.1 | 30.6 | 42.6 | 52.0 | 56.2 | *53.7* | *44.1* | 32.5 | 17.4 | *3.7* | 29.1 |
| Mean Temp °F | 7.0 | 13.1 | 27.1 | 43.1 | 56.6 | 64.9 | 69.6 | *68.0* | *58.0* | 45.2 | 26.9 | *13.6* | 41.1 |
| Days Max Temp ≥ 90 °F | 0 | 0 | 0 | 0 | 1 | *3* | 6 | *7* | 2 | 0 | 0 | 0 | 19 |
| Days Max Temp ≤ 32 °F | 25 | 19 | 10 | 1 | 0 | 0 | 0 | 0 | 0 | 1 | 11 | 21 | 88 |
| Days Min Temp ≤ 32 °F | 31 | 28 | 29 | 18 | 5 | 0 | 0 | 0 | 2 | 15 | 27 | 29 | 184 |
| Days Min Temp ≤ 0 °F | 19 | 13 | 4 | 0 | 0 | 0 | 0 | 0 | 0 | 0 | 3 | 13 | 52 |
| Heating Degree Days | 1795 | 1460 | 1167 | 653 | 283 | 80 | 22 | *47* | *242* | 607 | 1136 | *1588* | 9080 |
| Cooling Degree Days | *0* | *0* | *0* | *6* | *39* | *97* | *168* | *129* | na | *1* | *0* | na | na |
| Total Precipitation (") | 0.59 | 0.55 | 0.71 | 1.38 | 2.15 | 3.40 | 2.81 | 2.07 | *1.77* | 1.14 | 0.52 | *0.52* | 17.61 |
| Days ≥ 0.1" Precip | 2 | 2 | 2 | 4 | 5 | 7 | 5 | *5* | 4 | 3 | 2 | 2 | 43 |
| Total Snowfall (") | na | na | na | *1.6* | 0.3 | 0.0 | 0.0 | 0.0 | 0.1 | *0.5* | *3.2* | na | na |
| Days ≥ 1" Snow Depth | na | na | na | 2 | 0 | 0 | 0 | 0 | 0 | 0 | *5* | na | na |

## FORMAN 5 SSE *Sargent County* ELEVATION 1250 ft LAT/LONG 46° 2 ' N / 97° 36 ' W

| | JAN | FEB | MAR | APR | MAY | JUN | JUL | AUG | SEP | OCT | NOV | DEC | YEAR |
|---|---|---|---|---|---|---|---|---|---|---|---|---|---|
| Maximum Temp °F | 17.2 | 23.8 | 37.2 | 54.9 | 69.2 | 77.6 | 84.0 | 82.6 | 71.1 | 57.6 | 37.9 | 23.7 | 53.1 |
| Minimum Temp °F | -3.4 | 3.1 | 17.6 | 32.1 | 44.2 | 53.5 | 58.4 | 55.8 | 44.7 | 32.9 | 18.8 | 4.5 | 30.2 |
| Mean Temp °F | 6.9 | 13.5 | 27.4 | 43.5 | 56.7 | 65.6 | 71.2 | 69.2 | 57.9 | 45.3 | 28.4 | 14.1 | 41.6 |
| Days Max Temp ≥ 90 °F | 0 | 0 | 0 | 0 | 1 | 3 | 8 | 7 | 2 | 0 | 0 | 0 | 21 |
| Days Max Temp ≤ 32 °F | 25 | 19 | 10 | 1 | 0 | 0 | 0 | 0 | 0 | 0 | 10 | 22 | 87 |
| Days Min Temp ≤ 32 °F | 31 | 28 | 28 | 16 | 4 | 0 | 0 | 0 | 2 | 16 | 28 | 31 | 184 |
| Days Min Temp ≤ 0 °F | 18 | 13 | 4 | 0 | 0 | 0 | 0 | 0 | 0 | 0 | 2 | 12 | 49 |
| Heating Degree Days | 1781 | 1451 | 1159 | 641 | 284 | 77 | 17 | 41 | 244 | 606 | 1092 | 1572 | 8965 |
| Cooling Degree Days | 0 | 0 | 0 | 5 | 43 | 106 | 218 | 178 | 37 | 2 | 0 | 0 | 589 |
| Total Precipitation (") | 0.57 | 0.49 | 1.17 | 1.90 | 2.39 | 3.65 | 2.72 | 2.24 | 1.85 | 1.32 | 0.87 | 0.49 | 19.66 |
| Days ≥ 0.1" Precip | 2 | 2 | 3 | 5 | 5 | 7 | 5 | 5 | 4 | 3 | 2 | 2 | 45 |
| Total Snowfall (") | 7.5 | 5.1 | 5.1 | 2.3 | 0.0 | 0.0 | 0.0 | 0.0 | 0.0 | *0.4* | 6.2 | 5.0 | 31.6 |
| Days ≥ 1" Snow Depth | *22* | *19* | 11 | 2 | 0 | 0 | 0 | 0 | 0 | 0 | 5 | *14* | 73 |

## FORT YATES 4 SW *Sioux County* ELEVATION 1670 ft LAT/LONG 46° 5 ' N / 100° 38 ' W

| | JAN | FEB | MAR | APR | MAY | JUN | JUL | AUG | SEP | OCT | NOV | DEC | YEAR |
|---|---|---|---|---|---|---|---|---|---|---|---|---|---|
| Maximum Temp °F | 23.5 | 29.2 | 41.3 | 57.1 | 70.0 | 78.8 | 85.6 | 84.2 | 73.7 | 61.0 | 41.3 | 28.4 | 56.2 |
| Minimum Temp °F | 2.9 | 8.6 | 20.0 | 33.6 | 45.9 | 55.5 | 61.0 | 58.7 | 47.8 | 36.0 | 22.1 | 9.2 | 33.4 |
| Mean Temp °F | 13.2 | 18.9 | 30.7 | 45.4 | 58.0 | 67.2 | 73.3 | 71.5 | 60.8 | 48.5 | 31.7 | 18.8 | 44.8 |
| Days Max Temp ≥ 90 °F | 0 | 0 | 0 | 0 | 1 | 3 | 9 | 9 | 2 | 0 | 0 | 0 | 24 |
| Days Max Temp ≤ 32 °F | 20 | 16 | 8 | 1 | 0 | 0 | 0 | 0 | 0 | 0 | 7 | 18 | 70 |
| Days Min Temp ≤ 32 °F | 31 | 27 | 27 | 14 | 2 | 0 | 0 | 0 | 1 | 11 | 26 | 31 | 170 |
| Days Min Temp ≤ 0 °F | 14 | 9 | 3 | 0 | 0 | 0 | 0 | 0 | 0 | 0 | 1 | 8 | 35 |
| Heating Degree Days | 1602 | 1294 | 1057 | 588 | 247 | 52 | 9 | 20 | 176 | 506 | 990 | 1427 | 7968 |
| Cooling Degree Days | 0 | 0 | 0 | 10 | 53 | 148 | 290 | 245 | 60 | 3 | 0 | 0 | 809 |
| Total Precipitation (") | 0.27 | 0.30 | 0.71 | 1.47 | 1.90 | 2.56 | 1.84 | 1.50 | 1.31 | 0.96 | 0.33 | 0.25 | 13.40 |
| Days ≥ 0.1" Precip | 1 | 1 | 2 | 3 | 5 | 6 | 4 | 3 | 3 | 2 | 1 | 1 | 32 |
| Total Snowfall (") | 4.4 | 5.1 | 5.2 | 1.5 | 0.4 | 0.0 | 0.0 | 0.0 | 0.1 | 0.3 | 3.9 | 5.1 | 26.0 |
| Days ≥ 1" Snow Depth | 20 | 16 | 9 | 1 | 0 | 0 | 0 | 0 | 0 | 0 | 6 | 15 | 67 |

## FORTUNA 1 W *Divide County* ELEVATION 2349 ft LAT/LONG 48° 55' N / 103° 49' W

| | JAN | FEB | MAR | APR | MAY | JUN | JUL | AUG | SEP | OCT | NOV | DEC | YEAR |
|---|---|---|---|---|---|---|---|---|---|---|---|---|---|
| Maximum Temp °F | 15.0 | 21.4 | 35.3 | 52.3 | 65.8 | 74.1 | 80.6 | 79.5 | 67.9 | 54.3 | 33.8 | 20.8 | 50.1 |
| Minimum Temp °F | -4.8 | 2.0 | 14.8 | 28.3 | 40.0 | 49.0 | 53.4 | 51.1 | 40.7 | 29.0 | 14.9 | 1.5 | 26.7 |
| Mean Temp °F | 5.1 | 11.4 | 25.1 | 40.3 | 53.0 | 61.6 | 66.9 | 65.4 | 54.3 | 41.7 | 24.4 | 11.2 | 38.4 |
| Days Max Temp ≥ 90 °F | 0 | 0 | 0 | 0 | 0 | 2 | 5 | 5 | 1 | 0 | 0 | 0 | 13 |
| Days Max Temp ≤ 32 °F | 26 | 20 | 13 | 2 | 0 | 0 | 0 | 0 | 0 | 2 | 13 | 23 | 99 |
| Days Min Temp ≤ 32 °F | 31 | 28 | 30 | 21 | 6 | 0 | 0 | 0 | 5 | 21 | 29 | 31 | 202 |
| Days Min Temp ≤ 0 °F | 19 | 13 | 5 | 0 | 0 | 0 | 0 | 0 | 0 | 0 | 4 | 13 | 54 |
| Heating Degree Days | 1856 | 1510 | 1233 | 735 | 378 | 144 | 49 | 87 | 331 | 716 | 1214 | 1665 | 9918 |
| Cooling Degree Days | 0 | 0 | 0 | 1 | 15 | 58 | 112 | 108 | 14 | 0 | 0 | 0 | 308 |
| Total Precipitation (") | 0.43 | 0.38 | 0.65 | 1.06 | 2.08 | 2.64 | 2.48 | 1.82 | 1.33 | 0.75 | 0.33 | 0.39 | 14.34 |
| Days ≥ 0.1" Precip | 2 | *2* | *2* | 3 | 5 | 6 | 5 | 4 | 3 | 2 | 1 | 1 | 36 |
| Total Snowfall (") | *5.9* | na | na | *2.6* | 0.8 | 0.0 | 0.0 | 0.0 | 0.1 | *1.3* | na | na | na |
| Days ≥ 1" Snow Depth | na | na | na | na | 0 | 0 | 0 | 0 | 0 | *0* | na | na | na |

## FOXHOLM 7 N *Ward County* ELEVATION 1611 ft LAT/LONG 48° 27' N / 101° 35' W

| | JAN | FEB | MAR | APR | MAY | JUN | JUL | AUG | SEP | OCT | NOV | DEC | YEAR |
|---|---|---|---|---|---|---|---|---|---|---|---|---|---|
| Maximum Temp °F | 17.6 | 24.3 | 37.8 | 55.0 | 69.0 | 77.7 | 83.7 | 82.7 | 70.5 | 57.3 | 37.2 | *24.2* | 53.1 |
| Minimum Temp °F | -3.9 | 3.6 | 15.5 | 28.9 | 41.3 | 50.5 | 54.4 | 52.1 | 42.0 | 31.8 | 16.7 | *3.4* | 28.0 |
| Mean Temp °F | 6.9 | 13.8 | 26.7 | 42.0 | 55.3 | 64.1 | 69.0 | *67.4* | 56.3 | 44.5 | 27.0 | *13.8* | 40.6 |
| Days Max Temp ≥ 90 °F | 0 | 0 | 0 | 0 | 1 | 3 | 7 | 8 | 2 | 0 | 0 | 0 | 21 |
| Days Max Temp ≤ 32 °F | 22 | 18 | 10 | 1 | 0 | 0 | 0 | 0 | 0 | 1 | 10 | 19 | 81 |
| Days Min Temp ≤ 32 °F | 29 | 27 | 30 | 20 | 5 | 0 | 0 | 0 | 3 | 15 | 26 | 28 | 183 |
| Days Min Temp ≤ 0 °F | 18 | 12 | 5 | 0 | 0 | 0 | 0 | 0 | 0 | 0 | 3 | 12 | 50 |
| Heating Degree Days | 1801 | 1441 | 1182 | 684 | 315 | 94 | 25 | *58* | 278 | 628 | 1136 | *1576* | 9218 |
| Cooling Degree Days | *0* | 0 | 0 | 3 | 32 | *94* | *164* | *143* | *15* | *0* | *0* | na | na |
| Total Precipitation (") | 0.56 | 0.38 | 0.73 | 1.57 | 1.89 | 3.08 | 2.37 | 1.92 | 1.78 | 1.25 | 0.55 | 0.53 | 16.61 |
| Days ≥ 0.1" Precip | 2 | 1 | 2 | 4 | 4 | 6 | 5 | 4 | 4 | 3 | 1 | 2 | 38 |
| Total Snowfall (") | 8.5 | 5.7 | 5.6 | 3.8 | 0.2 | 0.0 | 0.0 | 0.0 | 0.1 | 1.8 | 5.7 | 7.5 | 38.9 |
| Days ≥ 1" Snow Depth | 25 | 20 | 16 | 4 | 0 | 0 | 0 | 0 | 0 | 1 | 9 | 21 | 96 |

## FULLERTON 1 ESE *Dickey County* ELEVATION 1440 ft LAT/LONG 46° 10' N / 98° 25' W

| | JAN | FEB | MAR | APR | MAY | JUN | JUL | AUG | SEP | OCT | NOV | DEC | YEAR |
|---|---|---|---|---|---|---|---|---|---|---|---|---|---|
| Maximum Temp °F | 18.9 | 25.8 | 39.1 | 57.3 | 71.3 | 79.0 | 85.4 | 83.8 | 73.2 | 59.3 | 38.0 | 24.4 | 54.6 |
| Minimum Temp °F | -1.7 | 4.9 | 18.5 | 31.5 | 43.0 | 52.4 | 57.2 | 54.7 | 44.5 | 33.0 | 18.8 | 4.9 | 30.1 |
| Mean Temp °F | 8.6 | 15.4 | 28.8 | 44.5 | 57.2 | 65.7 | 71.3 | 69.3 | 58.9 | 46.2 | 28.5 | 14.6 | 42.4 |
| Days Max Temp ≥ 90 °F | 0 | 0 | 0 | 0 | 1 | 3 | 9 | 8 | 2 | 0 | 0 | 0 | 23 |
| Days Max Temp ≤ 32 °F | 24 | 18 | 9 | 1 | 0 | 0 | 0 | 0 | 0 | 0 | 10 | 22 | 84 |
| Days Min Temp ≤ 32 °F | 31 | 28 | 29 | 17 | 4 | 0 | 0 | 0 | 3 | 15 | 28 | 31 | 186 |
| Days Min Temp ≤ 0 °F | 17 | 12 | 3 | 0 | 0 | 0 | 0 | 0 | 0 | 0 | 2 | 11 | 45 |
| Heating Degree Days | 1745 | 1397 | 1114 | 613 | 265 | 68 | 14 | 38 | 219 | 578 | 1090 | 1557 | 8698 |
| Cooling Degree Days | 0 | 0 | 0 | 4 | 36 | 101 | 211 | 164 | 36 | 1 | 0 | 0 | 553 |
| Total Precipitation (") | 0.66 | 0.59 | 1.33 | 2.33 | 2.79 | 3.34 | 2.57 | 2.30 | 1.84 | 1.51 | 0.94 | 0.44 | 20.64 |
| Days ≥ 0.1" Precip | 2 | 2 | 3 | 5 | 5 | 7 | 5 | 5 | 4 | 3 | 2 | 2 | 45 |
| Total Snowfall (") | 7.6 | *6.8* | *7.7* | 3.6 | 0.1 | 0.0 | 0.0 | 0.0 | 0.0 | 0.7 | 6.9 | *4.7* | 38.1 |
| Days ≥ 1" Snow Depth | *23* | *21* | *13* | 2 | 0 | 0 | 0 | 0 | 0 | 0 | 8 | na | na |

## GACKLE *Logan County* ELEVATION 1952 ft LAT/LONG 46° 38' N / 99° 8' W

| | JAN | FEB | MAR | APR | MAY | JUN | JUL | AUG | SEP | OCT | NOV | DEC | YEAR |
|---|---|---|---|---|---|---|---|---|---|---|---|---|---|
| Maximum Temp °F | 18.0 | 24.5 | 37.3 | 55.3 | 69.5 | 77.9 | 84.1 | 82.7 | 71.5 | 57.8 | 36.8 | 23.2 | 53.2 |
| Minimum Temp °F | -1.5 | 4.8 | 17.6 | 31.3 | 43.4 | 52.9 | 58.2 | 55.8 | 45.3 | 33.7 | 18.6 | 4.9 | 30.4 |
| Mean Temp °F | 8.3 | 14.7 | 27.5 | 43.3 | 56.4 | 65.4 | 71.2 | 69.3 | 58.4 | 45.8 | 27.7 | 14.1 | 41.8 |
| Days Max Temp ≥ 90 °F | 0 | 0 | 0 | 0 | 1 | 2 | 8 | 7 | 2 | 0 | 0 | 0 | 20 |
| Days Max Temp ≤ 32 °F | 25 | 19 | 10 | 1 | 0 | 0 | 0 | 0 | 0 | 1 | 12 | 23 | 91 |
| Days Min Temp ≤ 32 °F | 31 | 28 | 29 | 17 | 4 | 0 | 0 | 0 | 2 | 14 | 28 | 31 | 184 |
| Days Min Temp ≤ 0 °F | 17 | 11 | 4 | 0 | 0 | 0 | 0 | 0 | 0 | 0 | 2 | 12 | 46 |
| Heating Degree Days | 1757 | 1416 | 1157 | 646 | 287 | 74 | 16 | 37 | 232 | 590 | 1112 | 1575 | 8899 |
| Cooling Degree Days | 0 | 0 | 0 | 5 | 34 | 96 | 211 | 175 | 37 | 1 | 0 | 0 | 559 |
| Total Precipitation (") | 0.40 | 0.32 | 0.97 | 1.73 | 2.49 | 3.58 | 2.87 | 1.94 | 1.90 | 1.28 | 0.63 | 0.36 | 18.47 |
| Days ≥ 0.1" Precip | 1 | 1 | 2 | 5 | 6 | 7 | 5 | 5 | 4 | 3 | 2 | 1 | 42 |
| Total Snowfall (") | 7.0 | 4.8 | 7.5 | 3.5 | 0.4 | 0.0 | 0.0 | 0.0 | 0.1 | 1.2 | 5.9 | 5.5 | 35.9 |
| Days ≥ 1" Snow Depth | 25 | 23 | 16 | 3 | 0 | 0 | 0 | 0 | 0 | 0 | 8 | 19 | 94 |

## GARRISON 1 NNW *McLean County* ELEVATION 1926 ft LAT/LONG 47° 38' N / 101° 25' W

| | JAN | FEB | MAR | APR | MAY | JUN | JUL | AUG | SEP | OCT | NOV | DEC | YEAR |
|---|---|---|---|---|---|---|---|---|---|---|---|---|---|
| Maximum Temp °F | 17.4 | 23.5 | 37.3 | 53.0 | 66.8 | 75.3 | 81.6 | 81.2 | 68.8 | 56.3 | *36.0* | *23.7* | 51.7 |
| Minimum Temp °F | -3.5 | 2.5 | 16.2 | 29.1 | 41.2 | 50.5 | 54.9 | 52.7 | 42.1 | 30.8 | *16.4* | *4.2* | 28.1 |
| Mean Temp °F | 6.9 | 13.0 | 26.8 | 41.0 | 54.0 | 62.9 | 68.2 | 67.0 | 55.5 | 43.6 | *26.2* | *13.9* | 39.9 |
| Days Max Temp ≥ 90 °F | 0 | 0 | 0 | 0 | 0 | 2 | 5 | 6 | 1 | 0 | 0 | *0* | 14 |
| Days Max Temp ≤ 32 °F | 25 | 19 | 11 | 2 | 0 | 0 | 0 | 0 | 0 | 1 | *11* | *21* | 90 |
| Days Min Temp ≤ 32 °F | 31 | 28 | 30 | 19 | 5 | 0 | 0 | 0 | 4 | 17 | *28* | *30* | 192 |
| Days Min Temp ≤ 0 °F | 18 | 13 | 4 | 0 | 0 | 0 | 0 | 0 | 0 | 0 | 3 | *11* | 49 |
| Heating Degree Days | 1799 | 1464 | 1179 | 714 | 349 | 116 | 35 | 61 | 299 | 657 | *1157* | *1578* | 9408 |
| Cooling Degree Days | 0 | 0 | 0 | 2 | 20 | 70 | 137 | 138 | 20 | 0 | 0 | *0* | 387 |
| Total Precipitation (") | 0.42 | 0.33 | 0.60 | 1.35 | 1.82 | 3.11 | 2.49 | 1.75 | 1.50 | 1.06 | 0.43 | 0.43 | 15.29 |
| Days ≥ 0.1" Precip | 1 | 1 | *2* | 3 | 4 | 6 | 5 | 4 | 3 | 3 | 2 | *2* | 36 |
| Total Snowfall (") | na | na | na | na | *0.2* | 0.0 | 0.0 | 0.0 | 0.1 | na | na | na | na |
| Days ≥ 1" Snow Depth | na | na | na | *2* | 0 | 0 | 0 | 0 | 0 | 1 | na | na | na |

## GRAFTON *Walsh County* ELEVATION 830 ft LAT/LONG 48° 25' N / 97° 25' W

| | JAN | FEB | MAR | APR | MAY | JUN | JUL | AUG | SEP | OCT | NOV | DEC | YEAR |
|---|---|---|---|---|---|---|---|---|---|---|---|---|---|
| Maximum Temp °F | 14.0 | 21.0 | 34.3 | 54.6 | 70.7 | 78.3 | 82.9 | 81.8 | 70.5 | 56.1 | 34.7 | 19.6 | 51.5 |
| Minimum Temp °F | -5.1 | 1.2 | 15.6 | 31.3 | 43.8 | 52.9 | 57.2 | 54.6 | 45.3 | 33.9 | 18.3 | 2.3 | 29.3 |
| Mean Temp °F | 4.6 | 11.1 | 25.0 | 43.0 | 57.3 | 65.6 | 70.1 | 68.2 | 57.9 | 45.0 | 26.5 | 11.0 | 40.4 |
| Days Max Temp ≥ 90 °F | 0 | 0 | 0 | 0 | 2 | 3 | 5 | 6 | 1 | 0 | 0 | 0 | 17 |
| Days Max Temp ≤ 32 °F | 27 | 21 | 13 | 1 | 0 | 0 | 0 | 0 | 0 | 1 | 12 | 25 | 100 |
| Days Min Temp ≤ 32 °F | 31 | 28 | 28 | 17 | 4 | 0 | 0 | 0 | 2 | 13 | 27 | 31 | 181 |
| Days Min Temp ≤ 0 °F | 20 | 14 | 6 | 0 | 0 | 0 | 0 | 0 | 0 | 0 | 3 | 14 | 57 |
| Heating Degree Days | 1871 | 1516 | 1235 | 657 | 277 | 74 | 18 | 46 | 242 | 612 | 1148 | 1672 | 9368 |
| Cooling Degree Days | 0 | 0 | 0 | 6 | 58 | 112 | 192 | 166 | 33 | 1 | 0 | 0 | 568 |
| Total Precipitation (") | 0.56 | 0.42 | 0.88 | 1.42 | 2.24 | 3.02 | 2.75 | 2.34 | 1.87 | 1.33 | 0.69 | 0.47 | 17.99 |
| Days ≥ 0.1" Precip | 2 | 1 | 3 | 3 | 5 | 6 | 6 | 5 | 4 | 3 | 2 | 2 | 42 |
| Total Snowfall (") | na | na | na | *1.0* | 0.0 | 0.0 | 0.0 | 0.0 | 0.0 | 0.0 | na | na | na |
| Days ≥ 1" Snow Depth | na | na | na | *1* | 0 | 0 | 0 | 0 | 0 | 0 | na | na | na |

## GRAND FORKS INTL AP *Grand Forks County* ELEVATION 852 ft LAT/LONG 47° 57' N / 97° 11' W

| | JAN | FEB | MAR | APR | MAY | JUN | JUL | AUG | SEP | OCT | NOV | DEC | YEAR |
|---|---|---|---|---|---|---|---|---|---|---|---|---|---|
| Maximum Temp °F | 13.8 | 19.7 | 33.0 | 52.2 | 68.2 | 75.9 | 80.9 | 79.9 | 68.3 | 54.5 | 33.9 | 19.7 | 50.0 |
| Minimum Temp °F | -5.7 | 0.9 | 16.0 | 31.3 | 42.9 | 52.0 | 56.2 | 53.7 | 44.1 | 32.9 | 17.9 | 2.4 | 28.7 |
| Mean Temp °F | 3.9 | 10.3 | 24.5 | 41.7 | 55.6 | 64.0 | 68.6 | 66.8 | 56.2 | 43.7 | 25.9 | 11.1 | 39.4 |
| Days Max Temp ≥ 90 °F | 0 | 0 | 0 | 0 | 1 | 2 | 4 | 5 | 1 | 0 | 0 | 0 | 13 |
| Days Max Temp ≤ 32 °F | 27 | 22 | 13 | 2 | 0 | 0 | 0 | 0 | 0 | 1 | 13 | 25 | 103 |
| Days Min Temp ≤ 32 °F | 30 | 28 | 28 | 17 | 4 | 0 | 0 | 0 | 2 | 15 | 28 | 31 | 183 |
| Days Min Temp ≤ 0 °F | 20 | 14 | 6 | 0 | 0 | 0 | 0 | 0 | 0 | 0 | 3 | 15 | 58 |
| Heating Degree Days | 1892 | 1541 | 1249 | 693 | 315 | 98 | 29 | 61 | 284 | 654 | 1166 | 1667 | 9649 |
| Cooling Degree Days | 0 | 0 | 0 | 3 | 41 | 79 | 159 | 139 | 25 | 1 | 0 | 0 | 447 |
| Total Precipitation (") | 0.73 | 0.47 | 0.91 | 1.38 | 2.12 | 3.01 | 2.91 | 2.62 | 2.15 | 1.40 | 0.76 | 0.61 | 19.07 |
| Days ≥ 0.1" Precip | 2 | 1 | 3 | 3 | 5 | 6 | 6 | 6 | 4 | 3 | 2 | 2 | 43 |
| Total Snowfall (") | 9.6 | 5.3 | 6.9 | 2.8 | 0.1 | 0.0 | 0.0 | 0.0 | 0.0 | 1.1 | 6.3 | 7.0 | 39.1 |
| Days ≥ 1" Snow Depth | 29 | 24 | 19 | 3 | 0 | 0 | 0 | 0 | 0 | 1 | 10 | 23 | 109 |

## GRAND FORKS UNIV *Grand Forks County* ELEVATION 830 ft LAT/LONG 47° 55' N / 97° 5' W

| | JAN | FEB | MAR | APR | MAY | JUN | JUL | AUG | SEP | OCT | NOV | DEC | YEAR |
|---|---|---|---|---|---|---|---|---|---|---|---|---|---|
| Maximum Temp °F | 13.9 | 20.7 | 34.3 | 54.3 | 70.2 | 77.7 | 82.1 | 81.2 | 69.8 | 55.8 | 34.8 | 20.2 | 51.3 |
| Minimum Temp °F | -3.8 | 2.4 | 17.1 | 32.2 | 44.2 | 53.7 | 58.0 | 55.8 | 45.8 | 34.7 | 19.4 | 3.8 | 30.3 |
| Mean Temp °F | 5.0 | 11.6 | 25.7 | 43.3 | 57.3 | 65.7 | 70.0 | 68.5 | 57.8 | 45.3 | 27.1 | 12.0 | 40.8 |
| Days Max Temp ≥ 90 °F | 0 | 0 | 0 | 0 | 1 | 2 | 4 | 5 | 1 | 0 | 0 | 0 | 13 |
| Days Max Temp ≤ 32 °F | 27 | 22 | 12 | 1 | 0 | 0 | 0 | 0 | 0 | 1 | 13 | 25 | 101 |
| Days Min Temp ≤ 32 °F | 31 | 28 | 28 | 16 | 4 | 0 | 0 | 0 | 2 | 12 | 27 | 31 | 179 |
| Days Min Temp ≤ 0 °F | 20 | 14 | 5 | 0 | 0 | 0 | 0 | 0 | 0 | 0 | 2 | 13 | 54 |
| Heating Degree Days | 1857 | 1506 | 1211 | 649 | 272 | 69 | 19 | 43 | 243 | 606 | 1130 | 1639 | 9244 |
| Cooling Degree Days | 0 | 0 | 0 | 6 | 54 | 106 | 200 | 171 | 33 | 1 | 0 | 0 | 571 |
| Total Precipitation (") | 0.76 | 0.49 | 0.92 | 1.37 | 2.02 | 3.24 | 2.90 | 2.75 | 2.12 | 1.33 | 0.72 | 0.57 | 19.19 |
| Days ≥ 0.1" Precip | 2 | 2 | 3 | 3 | 5 | 7 | 6 | 5 | 4 | 3 | 2 | 2 | 44 |
| Total Snowfall (") | 9.3 | 4.9 | 6.3 | 1.8 | 0.0 | 0.0 | 0.0 | 0.0 | 0.0 | 0.8 | 4.4 | 6.9 | 34.4 |
| Days ≥ 1" Snow Depth | *28* | *24* | 16 | 2 | 0 | 0 | 0 | 0 | 0 | 0 | 8 | 23 | 101 |

## GRANVILLE *McHenry County* ELEVATION 1503 ft LAT/LONG 48° 16' N / 100° 51' W

| | JAN | FEB | MAR | APR | MAY | JUN | JUL | AUG | SEP | OCT | NOV | DEC | YEAR |
|---|---|---|---|---|---|---|---|---|---|---|---|---|---|
| Maximum Temp °F | 16.2 | 22.7 | 36.5 | 54.4 | 68.9 | 77.4 | 83.1 | 82.6 | 70.5 | 57.0 | 36.7 | 22.5 | 52.4 |
| Minimum Temp °F | -4.5 | 2.1 | 16.1 | 29.5 | 41.5 | 50.6 | 55.1 | 52.6 | 42.7 | 31.5 | 16.5 | 2.3 | 28.0 |
| Mean Temp °F | 5.9 | 12.4 | 26.3 | 42.0 | 55.2 | 64.1 | 69.1 | 67.6 | 56.6 | 44.2 | 26.6 | 12.4 | 40.2 |
| Days Max Temp ≥ 90 °F | 0 | 0 | 0 | 0 | 1 | 3 | 7 | 8 | 2 | 0 | 0 | 0 | 21 |
| Days Max Temp ≤ 32 °F | 24 | 19 | 10 | 1 | 0 | 0 | 0 | 0 | 0 | 1 | 11 | 22 | 88 |
| Days Min Temp ≤ 32 °F | 31 | 27 | 29 | 19 | 5 | 0 | 0 | 0 | 3 | 16 | 28 | 31 | 189 |
| Days Min Temp ≤ 0 °F | 19 | 13 | 5 | 0 | 0 | 0 | 0 | 0 | 0 | 0 | 3 | 13 | 53 |
| Heating Degree Days | 1831 | 1479 | 1192 | 686 | 322 | 99 | 28 | 58 | 279 | 638 | 1144 | 1626 | 9382 |
| Cooling Degree Days | 0 | 0 | 0 | 3 | 32 | 89 | 156 | 153 | 26 | 1 | 0 | 0 | 460 |
| Total Precipitation (") | 0.42 | 0.35 | 0.85 | 1.67 | 2.11 | 3.11 | 2.62 | 2.02 | 1.71 | 1.24 | 0.54 | 0.46 | 17.10 |
| Days ≥ 0.1" Precip | 2 | 1 | 2 | 4 | 5 | 6 | 5 | 4 | 4 | 2 | 2 | *1* | 38 |
| Total Snowfall (") | 7.4 | 4.8 | 5.5 | 4.1 | 0.1 | 0.0 | 0.0 | 0.0 | 0.2 | 2.4 | 4.7 | 5.6 | 34.8 |
| Days ≥ 1" Snow Depth | 24 | 18 | 14 | 4 | 0 | 0 | 0 | 0 | 0 | 1 | 8 | *17* | 86 |

## HANSBORO 4 NNE *Towner County* ELEVATION 1542 ft LAT/LONG 49° 0' N / 99° 21' W

| | JAN | FEB | MAR | APR | MAY | JUN | JUL | AUG | SEP | OCT | NOV | DEC | YEAR |
|---|---|---|---|---|---|---|---|---|---|---|---|---|---|
| Maximum Temp °F | 13.7 | 20.0 | 32.9 | 51.8 | 67.5 | 75.5 | 79.9 | 79.2 | 68.4 | 54.4 | 33.1 | 19.5 | 49.7 |
| Minimum Temp °F | -6.9 | -0.6 | 13.3 | 28.5 | 40.1 | 48.8 | 53.3 | 50.9 | 41.4 | 30.2 | 14.5 | 0.5 | 26.2 |
| Mean Temp °F | 3.4 | 9.7 | 23.2 | 40.1 | 53.8 | 62.2 | 66.6 | 65.1 | 54.9 | 42.3 | 23.8 | 10.0 | 37.9 |
| Days Max Temp ≥ 90 °F | 0 | 0 | 0 | 0 | 1 | 2 | 3 | 4 | 1 | 0 | 0 | 0 | 11 |
| Days Max Temp ≤ 32 °F | 26 | 22 | 13 | 2 | 0 | 0 | 0 | 0 | 0 | 1 | 14 | 25 | 103 |
| Days Min Temp ≤ 32 °F | 31 | 28 | 30 | 21 | 7 | 0 | 0 | 0 | 5 | 18 | 29 | 31 | 200 |
| Days Min Temp ≤ 0 °F | 21 | 15 | 7 | 0 | 0 | 0 | 0 | 0 | 0 | 0 | 4 | 15 | 62 |
| Heating Degree Days | 1908 | 1559 | 1291 | 742 | 359 | 128 | 45 | 86 | 314 | 696 | 1230 | 1704 | 10062 |
| Cooling Degree Days | 0 | 0 | 0 | 2 | 25 | 55 | 102 | 104 | 16 | 0 | 0 | 0 | 304 |
| Total Precipitation (") | 0.63 | 0.42 | 0.77 | 1.16 | 2.28 | 2.92 | 2.94 | 2.75 | 1.80 | 1.19 | 0.59 | 0.49 | 17.94 |
| Days ≥ 0.1" Precip | 2 | 1 | 2 | 3 | 6 | 6 | 6 | 5 | 4 | 3 | 2 | 1 | 41 |
| Total Snowfall (") | 7.3 | 5.5 | 6.9 | 2.2 | 0.4 | 0.0 | 0.0 | 0.0 | 0.2 | 2.6 | *5.8* | 6.6 | 37.5 |
| Days ≥ 1" Snow Depth | 26 | 23 | 19 | 5 | 0 | 0 | 0 | 0 | 0 | *1* | *13* | 23 | 110 |

## HEBRON *Morton County* ELEVATION 2162 ft LAT/LONG 46° 54' N / 102° 3' W

| | JAN | FEB | MAR | APR | MAY | JUN | JUL | AUG | SEP | OCT | NOV | DEC | YEAR |
|---|---|---|---|---|---|---|---|---|---|---|---|---|---|
| Maximum Temp °F | *19.9* | *26.1* | *39.2* | *54.0* | *67.5* | *76.1* | *82.8* | *82.3* | *71.3* | *57.8* | *39.2* | *26.8* | 53.6 |
| Minimum Temp °F | *-1.7* | *4.5* | *16.0* | *29.3* | *40.6* | *50.5* | *54.0* | *52.1* | *41.6* | *30.1* | *17.6* | *5.5* | 28.3 |
| Mean Temp °F | *9.1* | *15.3* | *27.6* | *41.6* | *54.1* | *63.3* | *68.4* | *67.2* | *56.5* | *44.0* | *28.5* | *16.1* | 41.0 |
| Days Max Temp ≥ 90 °F | *0* | *0* | *0* | *0* | *1* | *3* | *7* | *8* | *2* | *0* | *0* | *0* | 21 |
| Days Max Temp ≤ 32 °F | *22* | *17* | *10* | *1* | *0* | *0* | *0* | *0* | *0* | *1* | *9* | *19* | 79 |
| Days Min Temp ≤ 32 °F | *31* | *28* | *29* | *20* | *6* | *0* | *0* | *0* | *4* | *19* | *28* | *31* | 196 |
| Days Min Temp ≤ 0 °F | *17* | *12* | *4* | *0* | *0* | *0* | *0* | *0* | *0* | *0* | *3* | *11* | 47 |
| Heating Degree Days | *1731* | *1399* | *1151* | *696* | *350* | *116* | *40* | *62* | *279* | *646* | *1090* | *1508* | 9068 |
| Cooling Degree Days | na | na | na | na | na | na | na | na | na | na | na | na | na |
| Total Precipitation (") | *0.31* | *0.22* | *0.51* | *1.86* | *2.57* | *3.48* | *2.34* | *1.57* | *1.65* | *1.04* | *0.43* | *0.32* | 16.30 |
| Days ≥ 0.1" Precip | *1* | *1* | *1* | *5* | *6* | *7* | *5* | *4* | *3* | *2* | *2* | *1* | 38 |
| Total Snowfall (") | *5.1* | *4.1* | *5.7* | *3.0* | *0.8* | *0.0* | *0.0* | *0.0* | *0.1* | *1.9* | *5.0* | *5.6* | 31.3 |
| Days ≥ 1" Snow Depth | *25* | *22* | *15* | *5* | *0* | *0* | *0* | *0* | *0* | *1* | *9* | *19* | 96 |

## HETTINGER *Adams County* ELEVATION 2680 ft LAT/LONG 46° 0' N / 102° 38' W

| | JAN | FEB | MAR | APR | MAY | JUN | JUL | AUG | SEP | OCT | NOV | DEC | YEAR |
|---|---|---|---|---|---|---|---|---|---|---|---|---|---|
| Maximum Temp °F | 24.0 | 29.8 | 41.0 | 55.4 | 67.1 | 76.3 | 84.0 | 83.3 | 71.5 | 58.7 | 40.4 | 28.7 | 55.0 |
| Minimum Temp °F | 2.7 | 8.7 | 18.9 | 30.5 | 40.8 | 50.4 | 55.1 | 53.0 | 41.8 | 30.9 | 18.2 | 7.2 | 29.9 |
| Mean Temp °F | 13.4 | 19.3 | 30.0 | 43.0 | 54.0 | 63.3 | 69.6 | 68.2 | 56.7 | 44.8 | 29.3 | 18.0 | 42.5 |
| Days Max Temp ≥ 90 °F | 0 | 0 | 0 | 0 | 0 | 2 | 8 | 8 | 2 | 0 | 0 | 0 | 20 |
| Days Max Temp ≤ 32 °F | 19 | 15 | 8 | 1 | 0 | 0 | 0 | 0 | 0 | 1 | 8 | *18* | 70 |
| Days Min Temp ≤ 32 °F | 30 | 28 | 29 | 17 | 5 | 0 | 0 | 0 | 4 | *18* | 28 | 31 | 190 |
| Days Min Temp ≤ 0 °F | 14 | 9 | 3 | 0 | 0 | 0 | 0 | 0 | 0 | 0 | 3 | 9 | 38 |
| Heating Degree Days | 1598 | 1286 | 1079 | 654 | 348 | 112 | 30 | 48 | 272 | 619 | 1064 | 1453 | 8563 |
| Cooling Degree Days | 0 | 0 | 0 | 1 | 17 | 73 | 175 | 150 | 29 | 1 | 0 | 0 | 446 |
| Total Precipitation (") | 0.30 | 0.26 | 0.58 | 1.74 | 2.55 | 3.32 | 2.18 | 1.46 | 1.39 | 1.17 | 0.48 | 0.33 | 15.76 |
| Days ≥ 0.1" Precip | 1 | 1 | 2 | 4 | 6 | 6 | 4 | 3 | 3 | 2 | 2 | 1 | 35 |
| Total Snowfall (") | 4.9 | 5.0 | 6.5 | 3.9 | 1.0 | 0.0 | 0.0 | 0.0 | 0.6 | 1.7 | 4.7 | 5.3 | 33.6 |
| Days ≥ 1" Snow Depth | 25 | 19 | 13 | 4 | 0 | 0 | 0 | 0 | 0 | 1 | 8 | 18 | 88 |

## HILLSBORO 3 N *Traill County* ELEVATION 909 ft LAT/LONG 47° 27 ' N / 97° 4 ' W

| | JAN | FEB | MAR | APR | MAY | JUN | JUL | AUG | SEP | OCT | NOV | DEC | YEAR |
|---|---|---|---|---|---|---|---|---|---|---|---|---|---|
| Maximum Temp °F | 14.7 | 21.5 | 34.7 | 53.8 | 69.8 | 77.8 | 83.0 | 81.9 | 70.3 | 56.3 | 35.5 | 20.9 | 51.7 |
| Minimum Temp °F | -3.9 | 2.3 | 16.7 | 31.8 | 43.8 | 53.2 | 57.8 | 55.5 | 45.2 | 34.0 | 18.8 | 4.0 | 29.9 |
| Mean Temp °F | 5.4 | 11.9 | 25.7 | 42.9 | 56.8 | 65.5 | 70.5 | 68.7 | 57.8 | 45.2 | 27.2 | 12.5 | 40.8 |
| Days Max Temp ≥ 90 °F | 0 | 0 | 0 | 0 | 1 | 3 | 6 | 6 | 1 | 0 | 0 | 0 | 17 |
| Days Max Temp ≤ 32 °F | 27 | 21 | 12 | 1 | 0 | 0 | 0 | 0 | 0 | 1 | 13 | 25 | 100 |
| Days Min Temp ≤ 32 °F | 31 | 28 | 28 | 16 | 3 | 0 | 0 | 0 | 2 | 14 | 28 | 31 | 181 |
| Days Min Temp ≤ 0 °F | 19 | 13 | 4 | 0 | 0 | 0 | 0 | 0 | 0 | 0 | 2 | 13 | 51 |
| Heating Degree Days | 1846 | 1496 | 1210 | 661 | 280 | 71 | 16 | 35 | 243 | 609 | 1131 | 1625 | 9223 |
| Cooling Degree Days | 0 | 0 | 0 | ***5*** | ***44*** | 86 | 170 | 152 | 25 | 1 | 0 | 0 | 483 |
| Total Precipitation (") | 0.43 | 0.41 | 0.84 | 1.77 | 2.33 | 3.65 | 3.00 | 2.65 | 1.93 | 1.54 | 0.61 | 0.50 | 19.66 |
| Days ≥ 0.1" Precip | 1 | 2 | 3 | 4 | 6 | 7 | 5 | 5 | 5 | 3 | 2 | 2 | 45 |
| Total Snowfall (") | na | na | na | na | 0.0 | 0.0 | 0.0 | 0.0 | 0.0 | 0.8 | ***2.5*** | na | na |
| Days ≥ 1" Snow Depth | ***25*** | ***21*** | ***14*** | 2 | 0 | 0 | 0 | 0 | 0 | 0 | 5 | ***18*** | 85 |

## HURDSFIELD 8 SW *Wells County* ELEVATION 1939 ft LAT/LONG 47° 21 ' N / 100° 1 ' W

| | JAN | FEB | MAR | APR | MAY | JUN | JUL | AUG | SEP | OCT | NOV | DEC | YEAR |
|---|---|---|---|---|---|---|---|---|---|---|---|---|---|
| Maximum Temp °F | 15.9 | 21.8 | 34.4 | 51.8 | 66.2 | 74.6 | 81.3 | 80.9 | 68.9 | 55.2 | 35.4 | 21.5 | 50.7 |
| Minimum Temp °F | -4.7 | 1.1 | 14.0 | 27.9 | ***40.4*** | 49.5 | 54.4 | 51.8 | 41.1 | 29.5 | 16.2 | 1.8 | 26.9 |
| Mean Temp °F | 5.3 | 11.4 | 24.3 | 39.9 | ***53.3*** | 62.1 | 67.9 | 66.4 | 55.0 | 42.4 | 25.8 | 11.6 | 38.8 |
| Days Max Temp ≥ 90 °F | 0 | 0 | 0 | 0 | 0 | 1 | ***4*** | 5 | 1 | 0 | 0 | 0 | 11 |
| Days Max Temp ≤ 32 °F | 26 | 21 | ***12*** | 2 | 0 | 0 | 0 | 0 | 0 | 1 | ***12*** | ***22*** | 96 |
| Days Min Temp ≤ 32 °F | 31 | 28 | ***30*** | ***22*** | 6 | 0 | 0 | 0 | 5 | ***19*** | 29 | 31 | 201 |
| Days Min Temp ≤ 0 °F | ***18*** | 13 | ***5*** | 0 | 0 | 0 | 0 | 0 | 0 | 0 | 3 | 14 | 53 |
| Heating Degree Days | 1850 | 1511 | 1254 | 749 | ***374*** | 135 | 38 | 71 | 316 | 694 | 1167 | 1655 | 9814 |
| Cooling Degree Days | 0 | 0 | 0 | 2 | 19 | 55 | 122 | 102 | 16 | 0 | 0 | 0 | 316 |
| Total Precipitation (") | 0.49 | 0.40 | 0.67 | 1.49 | 2.25 | 3.32 | 2.42 | 1.92 | 1.56 | 1.23 | 0.55 | 0.45 | 16.75 |
| Days ≥ 0.1" Precip | 2 | 1 | ***2*** | 4 | ***6*** | 6 | ***6*** | 4 | 3 | 3 | 2 | 1 | 40 |
| Total Snowfall (") | 6.3 | 5.0 | 6.0 | 3.0 | 0.5 | 0.0 | 0.0 | 0.0 | 0.2 | 2.2 | 5.5 | ***6.5*** | 35.2 |
| Days ≥ 1" Snow Depth | ***28*** | ***24*** | ***18*** | 3 | 0 | 0 | 0 | 0 | 0 | 1 | ***11*** | ***22*** | 107 |

## JAMESTOWN MUNI AP *Stutsman County* ELEVATION 1496 ft LAT/LONG 46° 55 ' N / 98° 41 ' W

| | JAN | FEB | MAR | APR | MAY | JUN | JUL | AUG | SEP | OCT | NOV | DEC | YEAR |
|---|---|---|---|---|---|---|---|---|---|---|---|---|---|
| Maximum Temp °F | 17.1 | 23.6 | 36.4 | 54.0 | 68.7 | 77.0 | 83.5 | 82.0 | 70.0 | 56.4 | 36.2 | 22.5 | 52.3 |
| Minimum Temp °F | -2.0 | 4.7 | 18.0 | 31.3 | 43.3 | 52.9 | 58.0 | 55.4 | 44.8 | 33.5 | 18.6 | 4.7 | 30.3 |
| Mean Temp °F | 7.4 | 14.2 | 27.2 | 42.7 | 56.0 | 65.0 | 70.8 | 68.7 | 57.4 | 44.9 | 27.4 | 13.6 | 41.3 |
| Days Max Temp ≥ 90 °F | 0 | 0 | 0 | 0 | 1 | 2 | 7 | 7 | 2 | 0 | 0 | 0 | 19 |
| Days Max Temp ≤ 32 °F | 25 | 19 | 11 | 1 | 0 | 0 | 0 | 0 | 0 | 1 | 12 | 23 | 92 |
| Days Min Temp ≤ 32 °F | 30 | 28 | 29 | 18 | 4 | 0 | 0 | 0 | 2 | 14 | 28 | 31 | 184 |
| Days Min Temp ≤ 0 °F | 17 | 12 | 4 | 0 | 0 | 0 | 0 | 0 | 0 | 0 | 2 | 12 | 47 |
| Heating Degree Days | 1783 | 1430 | 1164 | 666 | 299 | 83 | 16 | 42 | 255 | 615 | 1121 | 1589 | 9063 |
| Cooling Degree Days | 0 | 0 | 0 | 5 | 39 | 99 | 214 | 174 | 31 | 1 | 0 | 0 | 563 |
| Total Precipitation (") | 0.65 | 0.48 | 0.89 | 1.49 | 1.93 | 3.21 | 2.89 | 1.96 | 1.78 | 1.11 | 0.61 | 0.47 | 17.47 |
| Days ≥ 0.1" Precip | 2 | 1 | 3 | 4 | 5 | 7 | 5 | 5 | 4 | 3 | 2 | 2 | 43 |
| Total Snowfall (") | 8.6 | 5.6 | 6.7 | 3.3 | 0.5 | 0.0 | 0.0 | 0.0 | 0.0 | 0.7 | 5.5 | 5.8 | 36.7 |
| Days ≥ 1" Snow Depth | 27 | 22 | 15 | 3 | 0 | 0 | 0 | 0 | 0 | 1 | 10 | 22 | 100 |

## JAMESTOWN STATE HOSP *Stutsman County* ELEVATION 1467 ft LAT/LONG 46° 53 ' N / 98° 41 ' W

| | JAN | FEB | MAR | APR | MAY | JUN | JUL | AUG | SEP | OCT | NOV | DEC | YEAR |
|---|---|---|---|---|---|---|---|---|---|---|---|---|---|
| Maximum Temp °F | 16.7 | 23.9 | 36.9 | 54.9 | 70.0 | 78.3 | 84.4 | 82.9 | 71.1 | 57.2 | 36.5 | 22.7 | 53.0 |
| Minimum Temp °F | -1.5 | 5.6 | 18.7 | 31.7 | 43.5 | 52.8 | 57.9 | 55.4 | 44.9 | 33.9 | 18.9 | 5.3 | 30.6 |
| Mean Temp °F | 7.6 | 14.8 | 27.8 | 43.3 | 56.8 | 65.6 | 71.2 | 69.2 | 58.1 | 45.6 | 27.7 | 14.0 | 41.8 |
| Days Max Temp ≥ 90 °F | 0 | 0 | 0 | 0 | 1 | 3 | 8 | 8 | 2 | 0 | 0 | 0 | 22 |
| Days Max Temp ≤ 32 °F | 26 | 19 | 10 | 1 | 0 | 0 | 0 | 0 | 0 | 1 | 11 | 23 | 91 |
| Days Min Temp ≤ 32 °F | 31 | 28 | 29 | 17 | 4 | 0 | 0 | 0 | 2 | 14 | 27 | 31 | 183 |
| Days Min Temp ≤ 0 °F | 18 | 11 | 3 | 0 | 0 | 0 | 0 | 0 | 0 | 0 | 2 | 12 | 46 |
| Heating Degree Days | 1777 | 1414 | 1146 | 646 | 278 | 74 | 14 | 38 | 240 | 595 | 1112 | 1575 | 8909 |
| Cooling Degree Days | 0 | 0 | 0 | 4 | 37 | 103 | 212 | 175 | 33 | 1 | 0 | 0 | 565 |
| Total Precipitation (") | 0.49 | 0.27 | 0.74 | 1.46 | 2.03 | 3.40 | 2.95 | 2.17 | 1.90 | 1.17 | 0.54 | 0.38 | 17.50 |
| Days ≥ 0.1" Precip | 2 | 1 | 2 | 4 | 5 | 7 | 5 | 5 | 4 | 3 | 2 | 2 | 42 |
| Total Snowfall (") | na | ***5.1*** | ***7.6*** | ***1.3*** | 0.3 | 0.0 | 0.0 | 0.0 | 0.0 | ***0.3*** | ***6.5*** | na | na |
| Days ≥ 1" Snow Depth | na | na | na | ***1*** | 0 | 0 | 0 | 0 | 0 | ***0*** | na | na | na |

## KEENE 3 S *McKenzie County* ELEVATION 2470 ft LAT/LONG 47° 50 ' N / 102° 55 ' W

| | JAN | FEB | MAR | APR | MAY | JUN | JUL | AUG | SEP | OCT | NOV | DEC | YEAR |
|---|---|---|---|---|---|---|---|---|---|---|---|---|---|
| Maximum Temp °F | 20.2 | 27.0 | 40.9 | 56.7 | 69.4 | 77.2 | *83.6* | *82.6* | *71.9* | 57.9 | 37.5 | 25.0 | 54.2 |
| Minimum Temp °F | -1.1 | 5.8 | 17.9 | 29.4 | 40.6 | 49.3 | 53.9 | 52.5 | 42.7 | 31.2 | 17.0 | 4.6 | 28.7 |
| Mean Temp °F | 9.6 | 16.4 | 29.4 | 43.1 | 55.0 | 63.3 | *68.8* | *67.5* | *57.4* | 44.5 | 27.3 | 14.8 | 41.4 |
| Days Max Temp ≥ 90 °F | 0 | 0 | 0 | 0 | 1 | 3 | 7 | *7* | 2 | 0 | 0 | 0 | 20 |
| Days Max Temp ≤ 32 °F | 22 | 16 | 8 | 1 | 0 | 0 | 0 | *0* | 0 | 1 | 11 | 19 | 78 |
| Days Min Temp ≤ 32 °F | 31 | 28 | 29 | 19 | 6 | 1 | 0 | 0 | 3 | 17 | 28 | 31 | 193 |
| Days Min Temp ≤ 0 °F | 16 | 10 | 3 | 0 | 0 | 0 | 0 | 0 | 0 | 0 | 3 | 12 | 44 |
| Heating Degree Days | 1715 | 1368 | 1096 | 653 | 325 | 116 | *37* | *60* | *257* | 628 | 1126 | 1552 | 8933 |
| Cooling Degree Days | 0 | 0 | 0 | 3 | 28 | 90 | 169 | 157 | 35 | 1 | 0 | 0 | 483 |
| Total Precipitation (") | 0.36 | 0.29 | 0.50 | 1.40 | 2.15 | 3.21 | 2.36 | 1.59 | 1.61 | 0.99 | 0.46 | 0.39 | 15.31 |
| Days ≥ 0.1" Precip | 1 | 1 | 2 | 3 | 5 | 7 | 5 | 4 | 4 | 3 | 1 | *1* | 37 |
| Total Snowfall (") | na | na | na | na | 0.2 | 0.0 | 0.0 | 0.0 | 0.2 | *0.9* | na | na | na |
| Days ≥ 1" Snow Depth | *21* | na | *10* | *2* | 0 | 0 | 0 | 0 | 0 | *1* | *6* | *17* | na |

## KENMARE 1 WSW *Ward County* ELEVATION 1801 ft LAT/LONG 48° 40 ' N / 102° 5 ' W

| | JAN | FEB | MAR | APR | MAY | JUN | JUL | AUG | SEP | OCT | NOV | DEC | YEAR |
|---|---|---|---|---|---|---|---|---|---|---|---|---|---|
| Maximum Temp °F | 16.4 | 23.0 | 36.0 | 52.4 | 66.5 | 75.1 | 81.1 | 80.2 | 68.3 | 55.3 | 35.8 | 22.4 | 51.0 |
| Minimum Temp °F | -4.5 | 1.9 | 14.6 | 27.9 | 40.3 | 49.5 | 53.5 | 50.5 | 40.6 | 29.9 | 15.8 | 2.5 | 26.9 |
| Mean Temp °F | 6.0 | 12.2 | 25.3 | 40.2 | 53.4 | 62.3 | 67.4 | 65.4 | 54.5 | 42.7 | 25.9 | 12.5 | 39.0 |
| Days Max Temp ≥ 90 °F | 0 | 0 | 0 | 0 | 1 | 2 | 5 | 5 | 1 | 0 | 0 | 0 | 14 |
| Days Max Temp ≤ 32 °F | 24 | 19 | 11 | 2 | 0 | 0 | 0 | 0 | 0 | 1 | 11 | 21 | 89 |
| Days Min Temp ≤ 32 °F | 31 | 28 | 30 | 21 | 6 | 0 | 0 | 0 | 5 | 19 | 28 | 30 | 198 |
| Days Min Temp ≤ 0 °F | 19 | 13 | 5 | 0 | 0 | 0 | 0 | 0 | 0 | 0 | 4 | 13 | 54 |
| Heating Degree Days | 1826 | 1484 | 1220 | 737 | 368 | 130 | 45 | 86 | 328 | 686 | 1169 | 1627 | 9706 |
| Cooling Degree Days | 0 | 0 | 0 | 1 | 21 | 64 | 117 | 110 | 19 | 0 | 0 | 0 | 332 |
| Total Precipitation (") | 0.64 | 0.48 | 0.69 | 1.38 | 1.87 | 2.71 | 2.44 | 1.80 | 1.95 | 1.05 | 0.47 | 0.49 | 15.97 |
| Days ≥ 0.1" Precip | 2 | 2 | 2 | 3 | 5 | 6 | 6 | 4 | 4 | 3 | 2 | 2 | 41 |
| Total Snowfall (") | *9.0* | *6.1* | 5.8 | 4.3 | 0.6 | 0.0 | 0.0 | 0.0 | 0.1 | 2.1 | *4.2* | *5.8* | 38.0 |
| Days ≥ 1" Snow Depth | na | na | na | na | 0 | 0 | 0 | 0 | 0 | 1 | na | na | na |

## LA MOURE *LaMoure County* ELEVATION 1381 ft LAT/LONG 46° 21 ' N / 98° 18 ' W

| | JAN | FEB | MAR | APR | MAY | JUN | JUL | AUG | SEP | OCT | NOV | DEC | YEAR |
|---|---|---|---|---|---|---|---|---|---|---|---|---|---|
| Maximum Temp °F | 18.5 | 24.9 | 38.8 | 55.3 | 69.7 | 77.7 | 84.0 | 82.6 | 71.1 | 58.4 | 38.4 | 24.4 | 53.7 |
| Minimum Temp °F | -4.0 | 2.4 | 17.3 | 30.6 | 41.9 | 51.9 | 56.4 | 53.0 | 42.3 | 30.8 | 17.0 | 3.2 | 28.6 |
| Mean Temp °F | 7.5 | 13.7 | 28.1 | 43.0 | 55.9 | 64.9 | 70.2 | 67.8 | 56.7 | 44.6 | 27.7 | 13.8 | 41.2 |
| Days Max Temp ≥ 90 °F | 0 | 0 | 0 | 0 | 1 | 2 | 7 | 7 | 1 | 0 | 0 | 0 | 18 |
| Days Max Temp ≤ 32 °F | 24 | 18 | 9 | 1 | 0 | 0 | 0 | 0 | 0 | 0 | 10 | 22 | 84 |
| Days Min Temp ≤ 32 °F | 31 | 28 | 29 | 18 | 5 | 0 | 0 | 0 | 4 | 18 | 28 | 31 | 192 |
| Days Min Temp ≤ 0 °F | 19 | 12 | 4 | 0 | 0 | 0 | 0 | 0 | 0 | 0 | 3 | 13 | 51 |
| Heating Degree Days | 1791 | 1442 | 1136 | 657 | 299 | 87 | 20 | 53 | 270 | 628 | 1122 | 1583 | 9088 |
| Cooling Degree Days | 0 | 0 | 0 | 4 | 31 | 91 | 182 | 135 | 22 | 1 | 0 | 0 | 466 |
| Total Precipitation (") | 0.59 | 0.53 | 1.18 | 2.17 | 2.56 | 3.90 | 3.00 | 2.36 | 2.00 | 1.40 | 0.75 | 0.45 | 20.89 |
| Days ≥ 0.1" Precip | 2 | 2 | *3* | 5 | 5 | 8 | 6 | 5 | 4 | 3 | 2 | 2 | 47 |
| Total Snowfall (") | *9.3* | *6.7* | *6.7* | 3.3 | 0.1 | 0.0 | 0.0 | 0.0 | 0.0 | *0.6* | 7.1 | 5.8 | 39.6 |
| Days ≥ 1" Snow Depth | *27* | *21* | na | *2* | 0 | 0 | 0 | 0 | 0 | 0 | *8* | *18* | na |

## LANGDON EXPERIMENT F *Cavalier County* ELEVATION 1621 ft LAT/LONG 48° 45 ' N / 98° 20 ' W

| | JAN | FEB | MAR | APR | MAY | JUN | JUL | AUG | SEP | OCT | NOV | DEC | YEAR |
|---|---|---|---|---|---|---|---|---|---|---|---|---|---|
| Maximum Temp °F | 10.4 | 17.1 | 30.3 | 49.5 | 65.8 | 73.5 | 77.9 | 77.3 | 66.3 | 52.1 | 31.4 | 16.4 | 47.3 |
| Minimum Temp °F | -9.1 | -3.2 | 11.0 | 27.8 | 39.5 | 49.2 | 53.5 | 51.1 | 41.0 | 29.5 | 13.9 | -1.7 | 25.2 |
| Mean Temp °F | 0.7 | 7.0 | 20.7 | 38.7 | 52.7 | 61.4 | 65.7 | 64.2 | 53.7 | 40.9 | 22.7 | 7.4 | 36.3 |
| Days Max Temp ≥ 90 °F | 0 | 0 | 0 | 0 | 1 | 1 | 2 | 3 | 1 | 0 | 0 | 0 | 8 |
| Days Max Temp ≤ 32 °F | 29 | 24 | 16 | 3 | 0 | 0 | 0 | 0 | 0 | 2 | 16 | 28 | 118 |
| Days Min Temp ≤ 32 °F | 31 | 28 | 30 | 22 | 8 | 0 | 0 | 0 | 5 | 20 | 29 | 31 | 204 |
| Days Min Temp ≤ 0 °F | 22 | 17 | 8 | 0 | 0 | 0 | 0 | 0 | 0 | 0 | 5 | 17 | 69 |
| Heating Degree Days | 1995 | 1635 | 1369 | 785 | 392 | 149 | 61 | 98 | 350 | 742 | 1264 | 1784 | 10624 |
| Cooling Degree Days | 0 | 0 | 0 | 2 | 22 | 46 | 86 | 82 | 13 | 0 | 0 | 0 | 251 |
| Total Precipitation (") | 0.47 | 0.35 | 0.65 | 1.19 | 2.23 | 3.06 | 3.14 | 2.60 | 1.81 | 1.33 | 0.54 | 0.44 | 17.81 |
| Days ≥ 0.1" Precip | 2 | 1 | 2 | 3 | 5 | 7 | 7 | 6 | 5 | 3 | 2 | 1 | 44 |
| Total Snowfall (") | 7.0 | 5.1 | 6.2 | 3.7 | 0.9 | 0.0 | 0.0 | 0.0 | 0.0 | 2.1 | 6.2 | 6.0 | 37.2 |
| Days ≥ 1" Snow Depth | 30 | 27 | 24 | 6 | 0 | 0 | 0 | 0 | 0 | 1 | 15 | 27 | 130 |

## LEEDS *Benson County* ELEVATION 1522 ft LAT/LONG 48° 17' N / 99° 26' W

| | JAN | FEB | MAR | APR | MAY | JUN | JUL | AUG | SEP | OCT | NOV | DEC | YEAR |
|---|---|---|---|---|---|---|---|---|---|---|---|---|---|
| Maximum Temp °F | 13.5 | 20.2 | 33.0 | 51.4 | 67.2 | 75.5 | 80.5 | 80.0 | 68.1 | 54.5 | 33.8 | 19.5 | 49.8 |
| Minimum Temp °F | -6.7 | -0.5 | 13.0 | 28.4 | 40.4 | 50.5 | 54.6 | 51.6 | 41.8 | 30.4 | 15.4 | 0.4 | 26.6 |
| Mean Temp °F | 3.4 | 9.9 | 23.0 | 39.9 | 53.8 | 63.0 | 67.6 | 65.8 | 54.9 | 42.5 | 24.7 | 9.9 | 38.2 |
| Days Max Temp ≥ 90 °F | 0 | 0 | 0 | 0 | 1 | 2 | 3 | 5 | 1 | 0 | 0 | 0 | 12 |
| Days Max Temp ≤ 32 °F | 27 | 22 | 13 | 2 | 0 | 0 | 0 | 0 | 0 | 1 | 14 | 25 | 104 |
| Days Min Temp ≤ 32 °F | 31 | 28 | 30 | 21 | 7 | 0 | 0 | 0 | 4 | 18 | 29 | 31 | 199 |
| Days Min Temp ≤ 0 °F | 21 | 15 | 7 | 0 | 0 | 0 | 0 | 0 | 0 | 0 | 4 | 16 | 63 |
| Heating Degree Days | 1909 | 1554 | 1296 | 746 | 358 | 116 | 39 | 77 | 316 | 693 | 1204 | 1703 | 10011 |
| Cooling Degree Days | 0 | 0 | 0 | 2 | 22 | 63 | 123 | 108 | 17 | 1 | 0 | 0 | 336 |
| Total Precipitation (") | 0.68 | 0.46 | 0.84 | 1.40 | 2.04 | 2.91 | 3.03 | 2.07 | 1.72 | 1.37 | 0.65 | 0.54 | 17.71 |
| Days ≥ 0.1" Precip | 2 | 2 | 2 | 4 | 5 | 6 | 6 | 5 | 4 | 3 | 2 | 2 | 43 |
| Total Snowfall (") | 8.1 | 4.7 | *5.8* | 3.2 | 0.7 | 0.0 | 0.0 | 0.0 | 0.2 | 1.9 | 5.8 | 6.5 | 36.9 |
| Days ≥ 1" Snow Depth | na | na | na | 2 | 0 | 0 | 0 | 0 | 0 | *0* | na | na | na |

## LINTON *Emmons County* ELEVATION 1713 ft LAT/LONG 46° 16' N / 100° 14' W

| | JAN | FEB | MAR | APR | MAY | JUN | JUL | AUG | SEP | OCT | NOV | DEC | YEAR |
|---|---|---|---|---|---|---|---|---|---|---|---|---|---|
| Maximum Temp °F | 19.7 | 26.8 | 40.8 | 57.6 | 71.7 | 79.8 | *87.0* | *84.7* | 73.7 | 60.8 | 40.0 | 26.8 | 55.8 |
| Minimum Temp °F | -3.8 | 3.1 | 17.2 | 30.9 | 42.8 | 51.5 | *57.0* | *54.2* | 43.5 | 30.8 | 16.4 | 4.6 | 29.0 |
| Mean Temp °F | 8.0 | 14.9 | 29.0 | 44.2 | 57.2 | 65.7 | *72.0* | *69.6* | 58.6 | 45.8 | 28.2 | 15.7 | 42.4 |
| Days Max Temp ≥ 90 °F | 0 | 0 | 0 | 0 | 2 | 5 | 10 | 10 | 3 | 0 | 0 | 0 | 30 |
| Days Max Temp ≤ 32 °F | 23 | 17 | 8 | 0 | 0 | 0 | 0 | 0 | 0 | 0 | 8 | 19 | 75 |
| Days Min Temp ≤ 32 °F | 31 | 27 | 28 | 18 | 5 | 0 | 0 | 0 | 4 | 17 | 28 | 30 | 188 |
| Days Min Temp ≤ 0 °F | 18 | 12 | 4 | 0 | 0 | 0 | 0 | 0 | 0 | 0 | 2 | 11 | 47 |
| Heating Degree Days | 1765 | 1407 | 1107 | 620 | 267 | 78 | *15* | *38* | 230 | 590 | 1097 | 1523 | 8737 |
| Cooling Degree Days | *0* | *0* | *0* | *7* | *40* | *102* | na | na | *27* | *1* | *0* | *0* | na |
| Total Precipitation (") | 0.41 | 0.39 | *0.86* | 1.75 | 2.24 | 2.96 | 2.21 | 1.65 | 1.50 | 1.01 | 0.50 | *0.48* | 15.96 |
| Days ≥ 0.1" Precip | *2* | *2* | *2* | 5 | 5 | 6 | 5 | 3 | 3 | 2 | 2 | *1* | 38 |
| Total Snowfall (") | 6.7 | *5.1* | na | *3.2* | 0.4 | 0.0 | 0.0 | 0.0 | 0.0 | *0.5* | na | *5.9* | na |
| Days ≥ 1" Snow Depth | na | na | *11* | *1* | 0 | 0 | 0 | 0 | 0 | 0 | *5* | na | na |

## LISBON *Ransom County* ELEVATION 1152 ft LAT/LONG 46° 26' N / 97° 41' W

| | JAN | FEB | MAR | APR | MAY | JUN | JUL | AUG | SEP | OCT | NOV | DEC | YEAR |
|---|---|---|---|---|---|---|---|---|---|---|---|---|---|
| Maximum Temp °F | 18.3 | 24.6 | 38.3 | 56.6 | 70.5 | 78.5 | 84.3 | 82.7 | 71.9 | 58.9 | 38.2 | 24.6 | 54.0 |
| Minimum Temp °F | -3.2 | 2.7 | 17.5 | 31.2 | 42.7 | 51.4 | 56.6 | 53.8 | 43.6 | 32.5 | 18.4 | 4.3 | 29.3 |
| Mean Temp °F | 7.5 | 13.7 | 27.9 | 44.0 | 56.5 | 64.9 | 70.4 | 68.3 | 57.8 | 45.7 | 28.3 | 14.5 | 41.6 |
| Days Max Temp ≥ 90 °F | 0 | 0 | 0 | 0 | 1 | 3 | 7 | 7 | 2 | 0 | 0 | 0 | 20 |
| Days Max Temp ≤ 32 °F | 24 | 19 | 9 | 1 | 0 | 0 | 0 | 0 | 0 | 0 | 9 | 22 | 84 |
| Days Min Temp ≤ 32 °F | 30 | 28 | 28 | 18 | 5 | 0 | 0 | 0 | 3 | 16 | 28 | 31 | 187 |
| Days Min Temp ≤ 0 °F | 18 | 13 | 4 | 0 | 0 | 0 | 0 | 0 | 0 | 0 | 2 | 12 | 49 |
| Heating Degree Days | 1780 | 1447 | 1143 | 626 | 285 | 83 | 19 | 46 | 247 | 592 | 1096 | 1560 | 8924 |
| Cooling Degree Days | 0 | 0 | 0 | 3 | 35 | 93 | 191 | 159 | 28 | 1 | 0 | 0 | 510 |
| Total Precipitation (") | 0.56 | 0.42 | 0.94 | 1.81 | 2.52 | 3.28 | 2.73 | 2.30 | 1.96 | 1.47 | 0.81 | 0.53 | 19.33 |
| Days ≥ 0.1" Precip | *2* | 2 | *3* | 4 | 5 | 6 | 5 | 5 | 4 | 3 | 2 | 2 | 43 |
| Total Snowfall (") | *8.8* | *4.8* | na | *1.6* | 0.1 | 0.0 | 0.0 | 0.0 | 0.0 | *0.3* | *7.0* | na | na |
| Days ≥ 1" Snow Depth | *27* | na | na | *2* | 0 | 0 | 0 | 0 | 0 | *0* | na | na | na |

## MANDAN EXP STN *Morton County* ELEVATION 1752 ft LAT/LONG 46° 48' N / 100° 54' W

| | JAN | FEB | MAR | APR | MAY | JUN | JUL | AUG | SEP | OCT | NOV | DEC | YEAR |
|---|---|---|---|---|---|---|---|---|---|---|---|---|---|
| Maximum Temp °F | 19.3 | 25.3 | 38.2 | 54.1 | 67.8 | 76.2 | 83.0 | 81.7 | 70.0 | 57.4 | 38.2 | 25.0 | 53.0 |
| Minimum Temp °F | -1.7 | 4.8 | 17.6 | 30.8 | 42.6 | 52.2 | 57.3 | 54.7 | 43.5 | 31.8 | 18.1 | 4.7 | 29.7 |
| Mean Temp °F | 8.8 | 15.1 | 27.9 | 42.5 | 55.2 | 64.2 | 70.2 | 68.2 | 56.8 | 44.6 | 28.2 | 14.9 | 41.4 |
| Days Max Temp ≥ 90 °F | 0 | 0 | 0 | 0 | 1 | 2 | 7 | 7 | 2 | 0 | 0 | 0 | 19 |
| Days Max Temp ≤ 32 °F | 23 | 18 | 10 | 1 | 0 | 0 | 0 | 0 | 0 | 1 | 10 | 20 | 83 |
| Days Min Temp ≤ 32 °F | 31 | 28 | 29 | 18 | 4 | 0 | 0 | 0 | 3 | 16 | 28 | 31 | 188 |
| Days Min Temp ≤ 0 °F | 17 | 12 | 4 | 0 | 0 | 0 | 0 | 0 | 0 | 0 | 2 | 11 | 46 |
| Heating Degree Days | 1741 | 1406 | 1142 | 671 | 319 | 97 | 23 | 51 | 271 | 627 | 1098 | 1550 | 8996 |
| Cooling Degree Days | 0 | 0 | 0 | 3 | 30 | 91 | 194 | 160 | 29 | 1 | 0 | 0 | 508 |
| Total Precipitation (") | 0.33 | 0.30 | 0.55 | 1.65 | 2.17 | 2.96 | 2.65 | 1.93 | 1.63 | 1.19 | 0.51 | 0.36 | 16.23 |
| Days ≥ 0.1" Precip | 1 | 1 | 2 | 4 | 5 | 7 | 5 | 4 | 4 | 3 | 1 | 1 | 38 |
| Total Snowfall (") | 7.2 | 5.8 | 6.4 | 3.5 | 0.8 | 0.0 | 0.0 | 0.0 | 0.3 | 1.2 | *6.1* | *6.3* | 37.6 |
| Days ≥ 1" Snow Depth | *21* | na | *12* | 2 | 0 | 0 | 0 | 0 | 0 | 1 | *5* | na | na |

## MAX *McLean County* ELEVATION 2090 ft LAT/LONG 47° 49 ' N / 101° 17 ' W

| | JAN | FEB | MAR | APR | MAY | JUN | JUL | AUG | SEP | OCT | NOV | DEC | YEAR |
|---|---|---|---|---|---|---|---|---|---|---|---|---|---|
| Maximum Temp °F | 16.0 | 22.4 | 35.4 | 52.3 | 66.8 | 75.3 | 81.9 | 80.6 | 68.3 | 55.0 | 35.1 | 21.9 | 50.9 |
| Minimum Temp °F | -3.9 | 2.1 | 15.2 | 29.0 | 41.1 | 50.4 | 55.0 | 52.2 | 41.7 | 30.2 | 15.9 | 2.5 | 27.6 |
| Mean Temp °F | 6.1 | 12.3 | 25.3 | 40.7 | 54.0 | 62.9 | 68.6 | 66.4 | 55.0 | 42.6 | 25.4 | 12.2 | 39.3 |
| Days Max Temp ≥ 90 °F | 0 | 0 | 0 | 0 | 1 | 2 | 5 | 5 | 1 | 0 | 0 | 0 | 14 |
| Days Max Temp ≤ 32 °F | 26 | 20 | 12 | 2 | 0 | 0 | 0 | 0 | 0 | 1 | 13 | 23 | 97 |
| Days Min Temp ≤ 32 °F | 31 | 28 | 30 | 20 | 5 | 0 | 0 | 0 | 4 | 18 | 29 | 31 | 196 |
| Days Min Temp ≤ 0 °F | 19 | 13 | 5 | 0 | 0 | 0 | 0 | 0 | 0 | 0 | 3 | 13 | 53 |
| Heating Degree Days | 1826 | 1484 | 1225 | 724 | 354 | 118 | 32 | 67 | 315 | 686 | 1181 | 1632 | 9644 |
| Cooling Degree Days | 0 | 0 | 0 | 3 | 25 | 72 | 151 | 130 | 20 | 0 | 0 | 0 | 401 |
| Total Precipitation (") | 0.49 | 0.34 | 0.66 | 1.60 | 2.05 | 3.26 | 2.75 | 1.71 | 1.74 | 1.27 | 0.52 | 0.44 | 16.83 |
| Days ≥ 0.1" Precip | 2 | 1 | 2 | 4 | 5 | 7 | 5 | 4 | 4 | 3 | 2 | 2 | 41 |
| Total Snowfall (") | 7.6 | 5.4 | 6.7 | 3.3 | 0.4 | 0.0 | 0.0 | 0.0 | 0.3 | 3.4 | 7.0 | 6.0 | 40.1 |
| Days ≥ 1" Snow Depth | 29 | 24 | 19 | 5 | 0 | 0 | 0 | 0 | 0 | 2 | 11 | 26 | 116 |

## MC CLUSKY *Sheridan County* ELEVATION 1942 ft LAT/LONG 47° 29 ' N / 100° 28 ' W

| | JAN | FEB | MAR | APR | MAY | JUN | JUL | AUG | SEP | OCT | NOV | DEC | YEAR |
|---|---|---|---|---|---|---|---|---|---|---|---|---|---|
| Maximum Temp °F | 17.3 | 23.8 | 37.2 | 54.2 | 69.6 | 78.2 | 84.0 | 82.8 | 71.3 | 57.3 | 36.1 | 23.4 | 52.9 |
| Minimum Temp °F | -1.8 | 4.4 | 17.3 | 30.2 | 42.8 | 52.3 | 56.9 | 54.7 | 44.8 | 32.6 | 17.9 | 5.2 | 29.8 |
| Mean Temp °F | 7.8 | 14.1 | 27.3 | 42.2 | 56.2 | 65.3 | 70.5 | 68.8 | 58.1 | 45.0 | 27.1 | 14.3 | 41.4 |
| Days Max Temp ≥ 90 °F | 0 | 0 | 0 | 0 | 1 | 3 | 7 | 8 | 2 | 0 | 0 | 0 | 21 |
| Days Max Temp ≤ 32 °F | 25 | 19 | 11 | 1 | 0 | 0 | 0 | 0 | 0 | 1 | 11 | 22 | 90 |
| Days Min Temp ≤ 32 °F | 31 | 28 | 29 | 19 | 5 | 0 | 0 | 0 | 2 | 15 | 28 | 31 | 188 |
| Days Min Temp ≤ 0 °F | 17 | 12 | 4 | 0 | 0 | 0 | 0 | 0 | 0 | 0 | 3 | 11 | 47 |
| Heating Degree Days | 1772 | 1432 | 1162 | 678 | 293 | 80 | 17 | 43 | 240 | 613 | 1132 | 1567 | 9029 |
| Cooling Degree Days | 0 | 0 | 0 | *4* | 35 | 106 | 194 | 170 | 35 | 0 | 0 | 0 | 544 |
| Total Precipitation (") | 0.57 | 0.44 | 0.77 | 1.61 | 1.94 | 3.46 | 2.61 | 1.95 | 1.74 | 1.18 | 0.60 | 0.54 | 17.41 |
| Days ≥ 0.1" Precip | 2 | 1 | 2 | 4 | 5 | 7 | 6 | 5 | 4 | 2 | 2 | 2 | 42 |
| Total Snowfall (") | 8.3 | 6.0 | 6.4 | 3.5 | 0.6 | 0.0 | 0.0 | 0.0 | 0.1 | 2.6 | 6.3 | 6.9 | 40.7 |
| Days ≥ 1" Snow Depth | 27 | 22 | 16 | 3 | 0 | 0 | 0 | 0 | 0 | 1 | 10 | 22 | 101 |

## MC HENRY 3 W *Foster County* ELEVATION 1512 ft LAT/LONG 47° 38 ' N / 98° 35 ' W

| | JAN | FEB | MAR | APR | MAY | JUN | JUL | AUG | SEP | OCT | NOV | DEC | YEAR |
|---|---|---|---|---|---|---|---|---|---|---|---|---|---|
| Maximum Temp °F | 13.6 | 20.0 | 33.3 | 51.4 | 67.2 | 75.1 | 80.8 | 79.5 | 67.9 | 54.8 | 34.3 | 19.8 | 49.8 |
| Minimum Temp °F | -7.3 | -1.2 | 13.7 | 28.7 | 41.0 | 50.5 | 55.1 | 52.4 | 42.0 | 30.3 | 15.3 | 0.4 | 26.7 |
| Mean Temp °F | 3.1 | 9.4 | 23.5 | 40.1 | 54.1 | 62.8 | 68.0 | 66.0 | 55.0 | 42.6 | 24.8 | 10.1 | 38.3 |
| Days Max Temp ≥ 90 °F | 0 | 0 | 0 | 0 | 1 | 2 | 4 | 4 | 1 | 0 | 0 | 0 | 12 |
| Days Max Temp ≤ 32 °F | 27 | 22 | 13 | 2 | 0 | 0 | 0 | 0 | 0 | 1 | 14 | 26 | 105 |
| Days Min Temp ≤ 32 °F | 31 | 28 | 30 | 21 | 6 | 0 | 0 | 0 | 4 | 19 | 29 | 31 | 199 |
| Days Min Temp ≤ 0 °F | 21 | 15 | 7 | 0 | 0 | 0 | 0 | 0 | 0 | 0 | 4 | 15 | 62 |
| Heating Degree Days | 1918 | 1568 | 1279 | 743 | 349 | 118 | 33 | 71 | 314 | 689 | 1199 | 1698 | 9979 |
| Cooling Degree Days | 0 | 0 | 0 | 3 | 24 | 63 | 130 | 109 | 17 | 0 | 0 | 0 | 346 |
| Total Precipitation (") | 0.56 | 0.42 | 0.85 | 1.49 | 2.14 | 3.51 | 2.84 | 2.67 | 2.05 | 1.27 | 0.81 | 0.57 | 19.18 |
| Days ≥ 0.1" Precip | 2 | 1 | 2 | 4 | 5 | 7 | 6 | 6 | 4 | 3 | 3 | 2 | 45 |
| Total Snowfall (") | 7.8 | 5.6 | 6.6 | 3.9 | 0.5 | 0.0 | 0.0 | 0.0 | 0.2 | 2.3 | 8.4 | 7.0 | 42.3 |
| Days ≥ 1" Snow Depth | 29 | 26 | 20 | 4 | 0 | 0 | 0 | 0 | 0 | 1 | 13 | 25 | 118 |

## MC LEOD 3 E *Richland County* ELEVATION 1079 ft LAT/LONG 46° 24 ' N / 97° 14 ' W

| | JAN | FEB | MAR | APR | MAY | JUN | JUL | AUG | SEP | OCT | NOV | DEC | YEAR |
|---|---|---|---|---|---|---|---|---|---|---|---|---|---|
| Maximum Temp °F | 17.9 | 24.5 | 38.1 | 57.0 | 71.6 | 79.2 | 84.5 | 82.9 | 72.3 | 59.0 | 38.1 | 23.7 | 54.1 |
| Minimum Temp °F | -3.4 | 3.2 | 17.6 | 31.6 | 44.1 | 53.4 | 58.4 | 56.0 | 45.9 | 34.1 | 19.1 | 4.3 | 30.4 |
| Mean Temp °F | 7.2 | 13.9 | 27.8 | 44.3 | 57.9 | 66.3 | 71.5 | 69.5 | 59.1 | 46.6 | 28.6 | 14.0 | 42.2 |
| Days Max Temp ≥ 90 °F | 0 | 0 | 0 | 0 | 1 | 3 | 7 | 6 | 2 | 0 | 0 | 0 | 19 |
| Days Max Temp ≤ 32 °F | 25 | 19 | 9 | 1 | 0 | 0 | 0 | 0 | 0 | 0 | 10 | 22 | 86 |
| Days Min Temp ≤ 32 °F | 31 | 28 | 28 | 17 | 4 | 0 | 0 | 0 | 2 | 14 | 28 | 31 | 183 |
| Days Min Temp ≤ 0 °F | 19 | 12 | 4 | 0 | 0 | 0 | 0 | 0 | 0 | 0 | 2 | 12 | 49 |
| Heating Degree Days | 1789 | 1439 | 1146 | 618 | 252 | 62 | 12 | 33 | 216 | 566 | 1086 | 1575 | 8794 |
| Cooling Degree Days | 0 | 0 | 0 | 5 | 47 | 116 | 220 | 179 | 42 | 3 | 0 | 0 | 612 |
| Total Precipitation (") | 0.54 | 0.42 | 0.89 | 1.47 | 2.39 | 3.33 | 3.06 | 2.45 | 1.90 | 1.53 | 0.78 | 0.49 | 19.25 |
| Days ≥ 0.1" Precip | 2 | 1 | 3 | 4 | 5 | 6 | 5 | 5 | 4 | 3 | 2 | 2 | 42 |
| Total Snowfall (") | *6.6* | *4.5* | *4.6* | 2.6 | 0.0 | 0.0 | 0.0 | 0.0 | 0.0 | 0.5 | 4.6 | *4.7* | 28.1 |
| Days ≥ 1" Snow Depth | 28 | 23 | 14 | 3 | 0 | 0 | 0 | 0 | 0 | 1 | 9 | 20 | 98 |

## MC VILLE *Nelson County* ELEVATION 1470 ft LAT/LONG 47° 46 ' N / 98° 10 ' W

| | JAN | FEB | MAR | APR | MAY | JUN | JUL | AUG | SEP | OCT | NOV | DEC | YEAR |
|---|---|---|---|---|---|---|---|---|---|---|---|---|---|
| Maximum Temp °F | 13.9 | 21.0 | 34.4 | 53.4 | 69.2 | 77.1 | 82.9 | 81.3 | 69.7 | 55.6 | 33.9 | 19.6 | 51.0 |
| Minimum Temp °F | -5.4 | 1.1 | 15.3 | 30.0 | 42.5 | 52.0 | 56.3 | 53.7 | 43.9 | 31.9 | 16.8 | 1.8 | 28.3 |
| Mean Temp °F | 4.3 | 11.1 | 24.9 | 41.8 | 55.9 | 64.6 | 69.5 | 67.4 | 56.8 | 43.8 | 25.4 | 10.7 | 39.7 |
| Days Max Temp ≥ 90 °F | 0 | 0 | 0 | 0 | 1 | 3 | 6 | 6 | 1 | 0 | 0 | 0 | 17 |
| Days Max Temp ≤ 32 °F | 27 | 22 | 13 | 2 | 0 | 0 | 0 | 0 | 0 | 1 | 14 | 26 | 105 |
| Days Min Temp ≤ 32 °F | 31 | 28 | 29 | 19 | 5 | 0 | 0 | 0 | 4 | 16 | 29 | 31 | 192 |
| Days Min Temp ≤ 0 °F | 20 | 14 | 6 | 0 | 0 | 0 | 0 | 0 | 0 | 0 | 3 | 15 | 58 |
| Heating Degree Days | 1881 | 1518 | 1237 | 693 | 303 | 88 | 24 | 57 | 271 | 651 | 1183 | 1678 | 9584 |
| Cooling Degree Days | 0 | 0 | 0 | 4 | 34 | 82 | 174 | 148 | 26 | 0 | 0 | 0 | 468 |
| Total Precipitation (") | 0.57 | 0.33 | 0.92 | 1.31 | 2.27 | 3.33 | 3.13 | 2.36 | 2.24 | 1.27 | 0.76 | 0.48 | 18.97 |
| Days ≥ 0.1" Precip | na | na | *2* | 3 | 6 | 6 | 6 | 5 | 4 | 3 | *2* | na | na |
| Total Snowfall (") | na | na | na | *2.5* | 0.3 | 0.0 | 0.0 | 0.0 | 0.0 | 1.2 | na | na | na |
| Days ≥ 1" Snow Depth | na | na | *15* | *1* | 0 | 0 | 0 | 0 | 0 | 0 | na | *21* | na |

## MEDORA *Billings County* ELEVATION 2251 ft LAT/LONG 46° 58 ' N / 103° 30 ' W

| | JAN | FEB | MAR | APR | MAY | JUN | JUL | AUG | SEP | OCT | NOV | DEC | YEAR |
|---|---|---|---|---|---|---|---|---|---|---|---|---|---|
| Maximum Temp °F | 27.4 | 34.1 | 45.3 | 59.2 | 71.4 | 80.2 | 87.4 | 87.0 | 75.1 | 61.8 | 42.8 | 31.5 | 58.6 |
| Minimum Temp °F | 2.8 | 9.2 | 19.4 | 30.4 | 41.2 | 50.6 | 55.2 | 53.1 | 42.0 | 31.1 | 18.5 | 7.6 | 30.1 |
| Mean Temp °F | 15.2 | 21.7 | 32.4 | 44.8 | 56.3 | 65.4 | 71.3 | 70.1 | 58.6 | 46.5 | 30.7 | 19.6 | 44.4 |
| Days Max Temp ≥ 90 °F | 0 | 0 | 0 | 0 | 1 | 4 | 12 | 13 | 4 | 0 | 0 | 0 | 34 |
| Days Max Temp ≤ 32 °F | 17 | 12 | 6 | 0 | 0 | 0 | 0 | 0 | 0 | 0 | 6 | 14 | 55 |
| Days Min Temp ≤ 32 °F | 30 | 27 | 28 | 18 | 5 | 0 | 0 | 0 | 4 | 17 | 28 | 31 | 188 |
| Days Min Temp ≤ 0 °F | 14 | 8 | 2 | 0 | 0 | 0 | 0 | 0 | 0 | 0 | 2 | 9 | 35 |
| Heating Degree Days | 1541 | 1216 | 1004 | 602 | 288 | 77 | 17 | 31 | 226 | 571 | 1022 | 1403 | 7998 |
| Cooling Degree Days | 0 | 0 | 0 | 3 | 34 | 116 | 237 | 211 | 39 | 1 | 0 | 0 | 641 |
| Total Precipitation (") | 0.42 | 0.32 | 0.60 | 1.52 | 2.34 | 3.22 | 2.06 | 1.38 | 1.47 | 0.99 | 0.46 | 0.42 | 15.20 |
| Days ≥ 0.1" Precip | 1 | 1 | 2 | 4 | 5 | 7 | 5 | 3 | 3 | 2 | 2 | 1 | 36 |
| Total Snowfall (") | 6.3 | 4.8 | 5.4 | 3.7 | 1.0 | 0.0 | 0.0 | 0.0 | 0.3 | 1.3 | 4.2 | 6.0 | 33.0 |
| Days ≥ 1" Snow Depth | 23 | 17 | 9 | 2 | 0 | 0 | 0 | 0 | 0 | 1 | 8 | 17 | 77 |

## MINOT EXPERIMENT STN *Ward County* ELEVATION 1772 ft LAT/LONG 48° 11 ' N / 101° 18 ' W

| | JAN | FEB | MAR | APR | MAY | JUN | JUL | AUG | SEP | OCT | NOV | DEC | YEAR |
|---|---|---|---|---|---|---|---|---|---|---|---|---|---|
| Maximum Temp °F | 16.3 | 22.4 | 35.5 | 52.2 | 66.8 | 75.4 | 81.0 | 80.1 | 68.0 | 54.8 | 35.1 | 22.2 | 50.8 |
| Minimum Temp °F | -3.3 | 3.0 | 15.7 | 29.4 | 41.6 | 51.2 | 55.5 | 52.9 | 42.9 | 31.5 | 17.0 | 3.2 | 28.4 |
| Mean Temp °F | 6.5 | 12.7 | 25.6 | 40.8 | 54.3 | 63.3 | 68.3 | 66.6 | 55.4 | 43.2 | 26.1 | 12.7 | 39.6 |
| Days Max Temp ≥ 90 °F | 0 | 0 | 0 | 0 | 1 | 2 | 4 | 5 | 1 | 0 | 0 | 0 | 13 |
| Days Max Temp ≤ 32 °F | 25 | 20 | 12 | 2 | 0 | 0 | 0 | 0 | 0 | 1 | 12 | 23 | 95 |
| Days Min Temp ≤ 32 °F | 31 | 28 | 30 | 19 | 5 | 0 | 0 | 0 | 3 | 17 | 28 | 31 | 192 |
| Days Min Temp ≤ 0 °F | 19 | 13 | 5 | 0 | 0 | 0 | 0 | 0 | 0 | 0 | 3 | 13 | 53 |
| Heating Degree Days | 1812 | 1472 | 1215 | 721 | 347 | 112 | 34 | 68 | 303 | 670 | 1161 | 1617 | 9532 |
| Cooling Degree Days | 0 | 0 | 0 | 3 | 26 | 74 | 137 | 128 | 21 | 1 | 0 | 0 | 390 |
| Total Precipitation (") | 0.74 | 0.51 | 0.90 | 1.85 | 2.21 | 2.87 | 2.61 | 1.94 | 1.85 | 1.27 | 0.82 | 0.67 | 18.24 |
| Days ≥ 0.1" Precip | 2 | 2 | 3 | 4 | 4 | 6 | 5 | 4 | 4 | 3 | 2 | 2 | 41 |
| Total Snowfall (") | 9.9 | 6.7 | 8.1 | 5.0 | 0.5 | 0.0 | 0.0 | 0.0 | 0.3 | 2.8 | 7.6 | 8.5 | 49.4 |
| Days ≥ 1" Snow Depth | 29 | 23 | 17 | 5 | 1 | 0 | 0 | 0 | 0 | 2 | 12 | 24 | 113 |

## MINOT FAA AP *Ward County* ELEVATION 1726 ft LAT/LONG 48° 15 ' N / 101° 17 ' W

| | JAN | FEB | MAR | APR | MAY | JUN | JUL | AUG | SEP | OCT | NOV | DEC | YEAR |
|---|---|---|---|---|---|---|---|---|---|---|---|---|---|
| Maximum Temp °F | 16.8 | 22.9 | 36.0 | 52.9 | 66.6 | 74.9 | 81.3 | 79.9 | 67.6 | 54.8 | 35.1 | 23.0 | 51.0 |
| Minimum Temp °F | -0.4 | 6.3 | 18.6 | 31.8 | 43.3 | 52.5 | 57.5 | 55.0 | 45.0 | 34.1 | 19.4 | 6.8 | 30.8 |
| Mean Temp °F | 8.0 | 14.6 | 27.3 | 42.4 | 55.0 | 63.7 | 69.4 | 67.5 | 56.3 | 44.4 | 27.3 | 14.9 | 40.9 |
| Days Max Temp ≥ 90 °F | 0 | 0 | 0 | 0 | 1 | 2 | 5 | 5 | 1 | 0 | 0 | 0 | 14 |
| Days Max Temp ≤ 32 °F | 25 | 20 | 12 | 1 | 0 | 0 | 0 | 0 | 0 | 1 | 12 | 22 | 93 |
| Days Min Temp ≤ 32 °F | 30 | 28 | 28 | 16 | 3 | 0 | 0 | 0 | 2 | 13 | 27 | 31 | 178 |
| Days Min Temp ≤ 0 °F | 16 | 11 | 3 | 0 | 0 | 0 | 0 | 0 | 0 | 0 | 2 | 10 | 42 |
| Heating Degree Days | 1763 | 1418 | 1160 | 674 | 325 | 102 | 22 | 58 | 281 | 631 | 1126 | 1548 | 9108 |
| Cooling Degree Days | *0* | 0 | 0 | 2 | 30 | 78 | 166 | 145 | 25 | 1 | 0 | 0 | 447 |
| Total Precipitation (") | 0.79 | 0.53 | 1.00 | 1.95 | 2.24 | 2.99 | 2.60 | 1.86 | 1.84 | 1.24 | 0.78 | 0.74 | 18.56 |
| Days ≥ 0.1" Precip | 3 | 2 | 3 | 4 | 5 | 6 | 5 | 4 | 4 | 3 | 2 | 2 | 43 |
| Total Snowfall (") | 8.6 | 5.5 | 7.3 | 6.6 | 0.8 | 0.0 | 0.0 | 0.0 | 0.2 | 2.8 | 6.6 | 7.9 | 46.3 |
| Days ≥ 1" Snow Depth | 27 | 22 | 16 | 4 | 0 | 0 | 0 | 0 | 0 | 1 | 10 | 22 | 102 |

## MOFFIT 3 SE *Burleigh County* ELEVATION 1762 ft LAT/LONG 46° 40 ' N / 100° 14 ' W

| | JAN | FEB | MAR | APR | MAY | JUN | JUL | AUG | SEP | OCT | NOV | DEC | YEAR |
|---|---|---|---|---|---|---|---|---|---|---|---|---|---|
| Maximum Temp °F | 20.6 | 27.4 | 39.6 | 56.5 | 70.6 | 79.1 | 85.7 | 84.9 | 73.5 | 59.9 | 39.6 | *26.3* | 55.3 |
| Minimum Temp °F | -1.9 | 5.2 | 16.8 | 30.8 | 42.3 | 51.9 | 56.5 | 54.0 | 43.5 | 31.6 | 17.7 | *5.0* | 29.5 |
| Mean Temp °F | 9.4 | 16.4 | 28.3 | 43.7 | 56.6 | 65.6 | 71.1 | 69.5 | 58.5 | 45.8 | 28.7 | *15.7* | 42.4 |
| Days Max Temp ≥ 90 °F | 0 | 0 | 0 | 0 | 1 | 3 | 9 | 10 | 2 | 0 | 0 | 0 | 25 |
| Days Max Temp ≤ 32 °F | 22 | 16 | 8 | 1 | 0 | 0 | 0 | 0 | 0 | 0 | 8 | na | na |
| Days Min Temp ≤ 32 °F | 31 | 28 | 29 | 18 | 4 | 0 | 0 | 0 | 3 | 15 | 28 | 31 | 187 |
| Days Min Temp ≤ 0 °F | 16 | 10 | 4 | 0 | 0 | 0 | 0 | 0 | 0 | 0 | 2 | na | na |
| Heating Degree Days | 1722 | 1368 | 1132 | 635 | 282 | 76 | 16 | 37 | 232 | 589 | 1082 | *1524* | 8695 |
| Cooling Degree Days | 0 | 0 | 0 | 5 | 35 | 111 | 229 | 205 | 41 | 1 | 0 | 0 | 627 |
| Total Precipitation (") | 0.30 | 0.31 | 0.64 | 1.51 | 2.05 | 3.03 | 2.56 | 1.88 | 1.88 | 1.12 | 0.43 | 0.34 | 16.05 |
| Days ≥ 0.1" Precip | 1 | 1 | 1 | 4 | 5 | 6 | 4 | 4 | 3 | 2 | 1 | 1 | 33 |
| Total Snowfall (") | 4.6 | 4.2 | 5.7 | 3.0 | 0.7 | 0.0 | 0.0 | 0.0 | 0.2 | 0.7 | 4.6 | *4.7* | 28.4 |
| Days ≥ 1" Snow Depth | 22 | *18* | *13* | 2 | 0 | 0 | 0 | 0 | 0 | 0 | 6 | *16* | 77 |

## MOHALL *Renville County* ELEVATION 1650 ft LAT/LONG 48° 46 ' N / 101° 31 ' W

| | JAN | FEB | MAR | APR | MAY | JUN | JUL | AUG | SEP | OCT | NOV | DEC | YEAR |
|---|---|---|---|---|---|---|---|---|---|---|---|---|---|
| Maximum Temp °F | 15.0 | 20.7 | *35.2* | 52.0 | 66.8 | 75.3 | 80.5 | 80.2 | 68.3 | 54.6 | *34.8* | 21.2 | 50.4 |
| Minimum Temp °F | -5.4 | -0.1 | *14.5* | 28.4 | 40.5 | 50.1 | 54.2 | 51.2 | 40.7 | 29.4 | 15.9 | *1.2* | 26.7 |
| Mean Temp °F | 4.8 | 10.3 | *24.9* | 40.5 | 53.7 | 62.7 | 67.4 | 65.7 | 54.5 | 42.0 | *25.3* | *11.1* | 38.6 |
| Days Max Temp ≥ 90 °F | 0 | 0 | *0* | 0 | 1 | 2 | 4 | 5 | 1 | 0 | 0 | 0 | 13 |
| Days Max Temp ≤ 32 °F | 26 | 21 | *12* | 2 | 0 | 0 | 0 | 0 | 0 | 1 | 12 | *23* | 97 |
| Days Min Temp ≤ 32 °F | 31 | 28 | *30* | 21 | 6 | 0 | 0 | 0 | 5 | 20 | 28 | 31 | 200 |
| Days Min Temp ≤ 0 °F | 19 | 14 | *6* | 0 | 0 | 0 | 0 | 0 | 0 | 0 | 3 | *15* | 57 |
| Heating Degree Days | 1858 | 1539 | *1238* | 729 | 361 | 120 | 40 | 78 | 324 | 706 | *1186* | *1669* | 9848 |
| Cooling Degree Days | 0 | 0 | 0 | 1 | 22 | 66 | 112 | 113 | 17 | 0 | 0 | 0 | 331 |
| Total Precipitation (") | 0.57 | 0.37 | 0.65 | 1.46 | 2.08 | 2.91 | 2.77 | 2.10 | 1.88 | 1.36 | 0.53 | 0.44 | 17.12 |
| Days ≥ 0.1" Precip | 2 | *1* | *2* | 4 | 5 | 7 | 6 | 4 | 4 | 3 | 2 | 2 | 42 |
| Total Snowfall (") | 7.7 | *4.1* | *4.8* | *3.8* | 0.3 | 0.0 | 0.0 | 0.0 | 0.0 | *1.2* | na | *5.4* | na |
| Days ≥ 1" Snow Depth | 27 | 22 | *16* | 4 | 0 | 0 | 0 | 0 | 0 | 2 | *11* | *23* | 105 |

## MOTT *Hettinger County* ELEVATION 2441 ft LAT/LONG 46° 22 ' N / 102° 20 ' W

| | JAN | FEB | MAR | APR | MAY | JUN | JUL | AUG | SEP | OCT | NOV | DEC | YEAR |
|---|---|---|---|---|---|---|---|---|---|---|---|---|---|
| Maximum Temp °F | 23.7 | 29.9 | 41.2 | 55.3 | 67.8 | 76.4 | 83.4 | 82.6 | 71.3 | 58.6 | 40.1 | 28.5 | 54.9 |
| Minimum Temp °F | 0.8 | 6.9 | 18.1 | 29.7 | 41.0 | 50.6 | 55.4 | 53.0 | 41.9 | 30.3 | 16.9 | 5.3 | 29.2 |
| Mean Temp °F | 12.2 | 18.4 | 29.7 | 42.5 | 54.5 | 63.5 | 69.4 | 67.8 | 56.6 | 44.5 | 28.5 | 17.0 | 42.1 |
| Days Max Temp ≥ 90 °F | 0 | 0 | 0 | 0 | 0 | 3 | 8 | 8 | 2 | 0 | 0 | 0 | 21 |
| Days Max Temp ≤ 32 °F | 20 | 15 | 8 | 1 | 0 | 0 | 0 | 0 | 0 | 1 | 9 | 17 | 71 |
| Days Min Temp ≤ 32 °F | 31 | 28 | 30 | 19 | 5 | 0 | 0 | 0 | 4 | 19 | 28 | 31 | 195 |
| Days Min Temp ≤ 0 °F | 15 | 9 | 3 | 0 | 0 | 0 | 0 | 0 | 0 | 0 | 3 | 11 | 41 |
| Heating Degree Days | 1631 | 1309 | 1088 | 669 | 335 | 104 | 28 | 50 | 273 | 629 | 1088 | 1485 | 8689 |
| Cooling Degree Days | 0 | 0 | 0 | 2 | 20 | 76 | 183 | 156 | 29 | 0 | 0 | 0 | 466 |
| Total Precipitation (") | 0.38 | 0.35 | 0.78 | 1.98 | 2.60 | 3.45 | 2.11 | 1.63 | 1.31 | 1.03 | 0.49 | 0.43 | 16.54 |
| Days ≥ 0.1" Precip | 1 | 1 | 2 | 4 | 5 | 7 | 5 | 4 | 3 | 2 | 2 | 1 | 37 |
| Total Snowfall (") | 5.8 | 4.7 | 5.5 | 3.0 | 1.1 | 0.0 | 0.0 | 0.0 | 0.4 | 1.8 | 5.5 | 6.1 | 33.9 |
| Days ≥ 1" Snow Depth | 20 | 16 | 9 | 3 | 0 | 0 | 0 | 0 | 0 | 1 | 7 | *17* | 73 |

## NAPOLEON *Logan County* ELEVATION 1962 ft LAT/LONG 46° 29 ' N / 99° 45 ' W

| | JAN | FEB | MAR | APR | MAY | JUN | JUL | AUG | SEP | OCT | NOV | DEC | YEAR |
|---|---|---|---|---|---|---|---|---|---|---|---|---|---|
| Maximum Temp °F | 17.7 | 23.9 | 36.7 | 53.4 | 67.6 | 76.0 | 83.1 | 81.6 | 69.9 | 56.9 | 37.5 | 23.1 | 52.3 |
| Minimum Temp °F | -3.0 | 3.0 | 16.1 | 30.1 | 42.0 | 51.8 | 56.7 | 54.0 | 43.1 | 31.5 | 17.7 | 3.6 | 28.9 |
| Mean Temp °F | 7.3 | 13.5 | 26.4 | 41.8 | 54.8 | 63.9 | 69.8 | 67.8 | 56.5 | 44.2 | 27.6 | 13.4 | 40.6 |
| Days Max Temp ≥ 90 °F | 0 | 0 | 0 | 0 | 1 | 2 | 7 | 6 | 1 | 0 | 0 | 0 | 17 |
| Days Max Temp ≤ 32 °F | 25 | 19 | 11 | 1 | 0 | 0 | 0 | 0 | 0 | 1 | 11 | 23 | 91 |
| Days Min Temp ≤ 32 °F | 31 | 28 | 29 | 19 | 5 | 0 | 0 | 0 | 3 | 17 | 29 | 31 | 192 |
| Days Min Temp ≤ 0 °F | 18 | 12 | 5 | 0 | 0 | 0 | 0 | 0 | 0 | 0 | 2 | 12 | 49 |
| Heating Degree Days | 1785 | 1450 | 1190 | 692 | 331 | 100 | 25 | 54 | 278 | 638 | 1114 | 1596 | 9253 |
| Cooling Degree Days | 0 | 0 | 0 | 3 | 26 | 79 | 180 | 150 | 28 | 1 | 0 | 0 | 467 |
| Total Precipitation (") | 0.51 | 0.43 | 0.97 | 1.76 | 2.39 | 3.33 | 2.73 | 2.00 | 1.77 | 1.35 | 0.69 | 0.45 | 18.38 |
| Days ≥ 0.1" Precip | 2 | 1 | 3 | 4 | 5 | 6 | 5 | 4 | 3 | 3 | 2 | 1 | 39 |
| Total Snowfall (") | *7.6* | 6.4 | *7.3* | *3.1* | 0.8 | 0.0 | 0.0 | 0.0 | 0.2 | 1.2 | 7.2 | 7.0 | 40.8 |
| Days ≥ 1" Snow Depth | *27* | *24* | *17* | *2* | 0 | 0 | 0 | 0 | 0 | 1 | *10* | *22* | 103 |

## NEW ENGLAND *Hettinger County* ELEVATION 2621 ft LAT/LONG 46° 33 ' N / 102° 52 ' W

| | JAN | FEB | MAR | APR | MAY | JUN | JUL | AUG | SEP | OCT | NOV | DEC | YEAR |
|---|---|---|---|---|---|---|---|---|---|---|---|---|---|
| Maximum Temp °F | 24.1 | 29.7 | 41.7 | 55.4 | 68.4 | 76.9 | 84.0 | 82.9 | 71.1 | 58.2 | 39.8 | 29.4 | 55.1 |
| Minimum Temp °F | 2.5 | 8.2 | 18.6 | 29.9 | 41.2 | 50.4 | 55.3 | 53.7 | 43.0 | 31.7 | 18.4 | 8.0 | 30.1 |
| Mean Temp °F | 13.3 | 19.0 | 30.1 | 42.7 | 54.8 | 63.7 | 69.7 | 68.3 | 57.1 | 45.0 | 29.1 | 18.7 | 42.6 |
| Days Max Temp ≥ 90 °F | 0 | 0 | 0 | 0 | 1 | 2 | 8 | 8 | 2 | 0 | 0 | 0 | 21 |
| Days Max Temp ≤ 32 °F | 20 | 15 | 8 | 1 | 0 | 0 | 0 | 0 | 0 | 1 | 9 | 17 | 71 |
| Days Min Temp ≤ 32 °F | 31 | 28 | 29 | 19 | 5 | 0 | 0 | 0 | 3 | 16 | 28 | 31 | 190 |
| Days Min Temp ≤ 0 °F | 14 | 9 | 3 | 0 | 0 | 0 | 0 | 0 | 0 | 0 | 2 | 9 | 37 |
| Heating Degree Days | 1597 | 1294 | 1074 | 664 | 324 | 103 | 23 | 41 | 260 | 614 | 1069 | 1429 | 8492 |
| Cooling Degree Days | 0 | 0 | 0 | 2 | 20 | 82 | 182 | 161 | 30 | 1 | 0 | 0 | 478 |
| Total Precipitation (") | 0.45 | 0.37 | 0.74 | 1.83 | 2.43 | 3.59 | 1.97 | 1.68 | 1.46 | 1.22 | 0.49 | 0.45 | 16.68 |
| Days ≥ 0.1" Precip | 2 | 1 | 2 | 5 | 5 | 7 | 5 | 4 | 4 | 3 | 2 | 2 | 42 |
| Total Snowfall (") | *8.5* | 6.1 | 7.6 | 5.3 | 0.9 | 0.0 | 0.0 | 0.0 | 0.0 | 2.0 | 5.3 | 7.9 | 43.6 |
| Days ≥ 1" Snow Depth | na | na | na | 4 | 0 | 0 | 0 | 0 | 0 | 1 | na | na | na |

## NEW SALEM *Morton County* ELEVATION 2129 ft LAT/LONG 46° 51 ' N / 101° 28 ' W

| | JAN | FEB | MAR | APR | MAY | JUN | JUL | AUG | SEP | OCT | NOV | DEC | YEAR |
|---|---|---|---|---|---|---|---|---|---|---|---|---|---|
| Maximum Temp °F | 20.2 | 26.6 | 39.3 | 55.4 | 68.9 | 77.1 | 84.0 | 82.8 | 71.5 | 57.8 | 37.7 | 25.1 | 53.9 |
| Minimum Temp °F | -0.2 | 6.0 | 17.6 | 29.8 | 41.5 | 50.6 | 55.5 | 53.6 | 43.3 | 31.8 | 17.7 | 5.2 | 29.4 |
| Mean Temp °F | 10.0 | 16.4 | 28.4 | 42.6 | 55.2 | 63.9 | 69.8 | 68.2 | 57.4 | 44.8 | 27.7 | 15.2 | 41.6 |
| Days Max Temp ≥ 90 °F | 0 | 0 | 0 | 0 | 1 | 3 | 8 | 8 | 2 | 0 | 0 | 0 | 22 |
| Days Max Temp ≤ 32 °F | 22 | 17 | 9 | 1 | 0 | 0 | 0 | 0 | 0 | 1 | 10 | 20 | 80 |
| Days Min Temp ≤ 32 °F | 31 | 28 | 29 | 19 | 5 | 0 | 0 | 0 | 3 | 16 | 28 | 31 | 190 |
| Days Min Temp ≤ 0 °F | 16 | 11 | 4 | 0 | 0 | 0 | 0 | 0 | 0 | 0 | 2 | 11 | 44 |
| Heating Degree Days | 1703 | 1368 | 1128 | 668 | 317 | 99 | 24 | 48 | 254 | 619 | 1111 | 1539 | 8878 |
| Cooling Degree Days | 0 | 0 | 0 | 4 | 27 | 85 | 190 | 165 | 33 | 1 | 0 | 0 | 505 |
| Total Precipitation (") | 0.49 | 0.44 | 0.84 | 1.98 | 2.37 | 3.24 | 2.54 | 2.04 | 1.58 | 1.20 | 0.64 | 0.52 | 17.88 |
| Days ≥ 0.1" Precip | 2 | 2 | 2 | 5 | 5 | 7 | 6 | 4 | 4 | 3 | 2 | 2 | 44 |
| Total Snowfall (") | 6.9 | 6.3 | 7.3 | 4.5 | 0.8 | 0.0 | 0.0 | 0.0 | 0.3 | 2.5 | 7.2 | 6.4 | 42.2 |
| Days ≥ 1" Snow Depth | 20 | 17 | 10 | 4 | 0 | 0 | 0 | 0 | 0 | 1 | 8 | 15 | 75 |

## OAKES 2 S *Dickey County* ELEVATION 1601 ft LAT/LONG 46° 8 ' N / 98° 5 ' W

| | JAN | FEB | MAR | APR | MAY | JUN | JUL | AUG | SEP | OCT | NOV | DEC | YEAR |
|---|---|---|---|---|---|---|---|---|---|---|---|---|---|
| Maximum Temp °F | 17.5 | 23.8 | 37.9 | 55.7 | 70.0 | 78.3 | 84.3 | 83.2 | 72.1 | 58.2 | 37.9 | 24.7 | 53.6 |
| Minimum Temp °F | -5.1 | 1.5 | 16.8 | 30.7 | 42.7 | 52.6 | 57.4 | 54.7 | 43.9 | 31.4 | 17.1 | 3.2 | 28.9 |
| Mean Temp °F | 6.3 | 12.7 | 27.4 | 43.3 | 56.4 | 65.5 | 70.9 | 69.0 | 58.0 | 44.8 | 27.5 | 14.0 | 41.3 |
| Days Max Temp ≥ 90 °F | 0 | 0 | 0 | 0 | 1 | 3 | 8 | 8 | 2 | 0 | 0 | 0 | 22 |
| Days Max Temp ≤ 32 °F | 25 | 19 | 10 | 1 | 0 | 0 | 0 | 0 | 0 | 0 | 10 | 22 | 87 |
| Days Min Temp ≤ 32 °F | 31 | 28 | 29 | 18 | 4 | 0 | 0 | 0 | 2 | 17 | 28 | 31 | 188 |
| Days Min Temp ≤ 0 °F | 19 | 14 | 4 | 0 | 0 | 0 | 0 | 0 | 0 | 0 | 3 | 13 | 53 |
| Heating Degree Days | 1818 | 1473 | 1162 | 647 | 291 | 77 | 17 | 40 | 244 | 619 | 1118 | 1578 | 9084 |
| Cooling Degree Days | 0 | 0 | 0 | 4 | 36 | 104 | 209 | 169 | 34 | 1 | 0 | 0 | 557 |
| Total Precipitation (") | *0.51* | *0.51* | 0.98 | 2.08 | 2.36 | 3.47 | 2.36 | 1.88 | 2.03 | 1.39 | *0.74* | *0.46* | 18.77 |
| Days ≥ 0.1" Precip | na | na | *2* | 4 | 5 | 7 | 5 | 4 | 4 | 3 | *2* | na | na |
| Total Snowfall (") | na | na | na | *3.6* | 0.0 | 0.0 | 0.0 | 0.0 | 0.0 | 0.6 | *7.7* | na | na |
| Days ≥ 1" Snow Depth | na | na | na | *1* | 0 | 0 | 0 | 0 | 0 | 0 | na | na | na |

## PARK RIVER *Walsh County* ELEVATION 991 ft LAT/LONG 48° 23 ' N / 97° 45 ' W

| | JAN | FEB | MAR | APR | MAY | JUN | JUL | AUG | SEP | OCT | NOV | DEC | YEAR |
|---|---|---|---|---|---|---|---|---|---|---|---|---|---|
| Maximum Temp °F | 14.9 | 22.2 | 35.1 | 54.6 | 70.7 | 78.4 | 82.8 | 81.7 | 71.1 | 57.0 | 35.4 | 20.9 | 52.1 |
| Minimum Temp °F | -4.0 | 2.2 | 15.9 | 30.6 | 43.0 | 52.4 | 57.3 | 54.4 | 45.0 | 33.8 | 18.4 | 3.3 | 29.4 |
| Mean Temp °F | 5.5 | 12.2 | 25.5 | 42.6 | 56.9 | 65.4 | 70.1 | 68.1 | 58.1 | 45.5 | 27.0 | 12.1 | 40.8 |
| Days Max Temp ≥ 90 °F | 0 | 0 | 0 | 0 | 2 | 3 | 5 | 6 | 1 | 0 | 0 | 0 | 17 |
| Days Max Temp ≤ 32 °F | 26 | 21 | 12 | 1 | 0 | 0 | 0 | 0 | 0 | 1 | 12 | 23 | 96 |
| Days Min Temp ≤ 32 °F | 31 | 28 | 29 | 18 | 4 | 0 | 0 | 0 | 2 | 13 | 28 | 30 | 183 |
| Days Min Temp ≤ 0 °F | 20 | 13 | 5 | 0 | 0 | 0 | 0 | 0 | 0 | 0 | 2 | 13 | 53 |
| Heating Degree Days | 1845 | 1487 | 1215 | 668 | 282 | 72 | 17 | 44 | 236 | 600 | 1135 | 1635 | 9236 |
| Cooling Degree Days | 0 | 0 | 0 | 4 | 49 | 98 | 182 | 149 | 30 | 2 | 0 | 0 | 514 |
| Total Precipitation (") | 0.61 | 0.46 | 0.90 | 1.70 | 2.36 | 3.22 | 3.15 | 2.51 | 1.90 | 1.44 | 0.72 | 0.56 | 19.53 |
| Days ≥ 0.1" Precip | 2 | 2 | 3 | 4 | 5 | 7 | 6 | 5 | 4 | 3 | 2 | 2 | 45 |
| Total Snowfall (") | na | na | na | *1.1* | 0.0 | 0.0 | 0.0 | 0.0 | 0.0 | 0.9 | na | na | na |
| Days ≥ 1" Snow Depth | na | *21* | *16* | 3 | 0 | 0 | 0 | 0 | 0 | 0 | *9* | na | na |

## PEMBINA *Pembina County* ELEVATION 791 ft LAT/LONG 48° 58 ' N / 97° 14 ' W

| | JAN | FEB | MAR | APR | MAY | JUN | JUL | AUG | SEP | OCT | NOV | DEC | YEAR |
|---|---|---|---|---|---|---|---|---|---|---|---|---|---|
| Maximum Temp °F | 10.5 | 17.1 | 30.9 | 51.3 | 67.5 | 75.4 | 79.9 | 78.9 | 67.8 | 53.3 | 32.8 | 17.5 | 48.6 |
| Minimum Temp °F | -10.5 | -4.7 | 11.1 | 28.0 | 40.4 | 49.9 | 53.9 | 50.9 | 41.1 | 29.8 | 14.4 | -1.8 | 25.2 |
| Mean Temp °F | 0.0 | 6.2 | 21.1 | 39.7 | 53.9 | 62.7 | 67.0 | 64.9 | 54.5 | 41.6 | 23.6 | 7.8 | 36.9 |
| Days Max Temp ≥ 90 °F | 0 | 0 | 0 | 0 | 1 | 2 | 3 | 4 | 1 | 0 | 0 | 0 | 11 |
| Days Max Temp ≤ 32 °F | 28 | 24 | 16 | 2 | 0 | 0 | 0 | 0 | 0 | 1 | 15 | 27 | 113 |
| Days Min Temp ≤ 32 °F | 31 | 28 | 30 | 21 | 7 | 0 | 0 | 0 | 5 | 19 | 29 | 31 | 201 |
| Days Min Temp ≤ 0 °F | 24 | 18 | 8 | 1 | 0 | 0 | 0 | 0 | 0 | 0 | 5 | 17 | 73 |
| Heating Degree Days | 2019 | 1658 | 1355 | 753 | 364 | 126 | 44 | 89 | 327 | 719 | 1235 | 1769 | 10458 |
| Cooling Degree Days | 0 | 0 | 0 | 2 | 31 | 59 | 102 | 93 | 12 | 0 | 0 | 0 | 299 |
| Total Precipitation (") | 0.46 | 0.32 | 0.71 | 1.18 | 2.19 | 3.28 | 2.78 | 2.54 | 2.13 | 1.29 | 0.66 | 0.46 | 18.00 |
| Days ≥ 0.1" Precip | 2 | 1 | *2* | 3 | 5 | 7 | 6 | 5 | 5 | 4 | 2 | 2 | 44 |
| Total Snowfall (") | *6.0* | *3.7* | na | *1.1* | 0.1 | 0.0 | 0.0 | 0.0 | 0.0 | 1.2 | *5.9* | *5.7* | na |
| Days ≥ 1" Snow Depth | na | *27* | na | *2* | 0 | 0 | 0 | 0 | 0 | 1 | na | *25* | na |

## PETERSBURG 2 N *Nelson County* ELEVATION 1522 ft LAT/LONG 48° 1 ' N / 98° 0 ' W

| | JAN | FEB | MAR | APR | MAY | JUN | JUL | AUG | SEP | OCT | NOV | DEC | YEAR |
|---|---|---|---|---|---|---|---|---|---|---|---|---|---|
| Maximum Temp °F | 12.5 | 18.5 | 31.4 | 49.8 | 65.6 | 74.2 | 79.1 | 78.3 | 67.2 | 53.4 | 32.8 | 18.6 | 48.4 |
| Minimum Temp °F | -8.6 | -2.7 | 12.0 | 28.3 | 40.4 | 50.0 | 53.8 | 51.0 | 40.8 | 29.4 | 14.6 | -1.2 | 25.7 |
| Mean Temp °F | 1.9 | 7.9 | 21.8 | 39.1 | 53.0 | 62.1 | 66.5 | 64.7 | 54.0 | 41.4 | 23.7 | 8.7 | 37.1 |
| Days Max Temp ≥ 90 °F | 0 | 0 | 0 | 0 | 0 | 1 | 2 | 3 | 1 | 0 | 0 | 0 | 7 |
| Days Max Temp ≤ 32 °F | 28 | 24 | 15 | 3 | 0 | 0 | 0 | 0 | 0 | 1 | 16 | 26 | 113 |
| Days Min Temp ≤ 32 °F | 31 | 28 | 30 | 21 | 7 | 0 | 0 | 0 | 4 | 20 | 29 | 31 | 201 |
| Days Min Temp ≤ 0 °F | 22 | 16 | 7 | 0 | 0 | 0 | 0 | 0 | 0 | 0 | 4 | 17 | 66 |
| Heating Degree Days | 1955 | 1610 | 1334 | 773 | 383 | 133 | 50 | 93 | 340 | 723 | 1234 | 1743 | 10371 |
| Cooling Degree Days | 0 | 0 | 0 | 3 | 25 | 56 | 104 | 98 | 13 | 0 | 0 | 0 | 299 |
| Total Precipitation (") | 0.71 | 0.41 | 0.98 | 1.40 | 2.23 | 3.33 | 3.06 | 2.62 | 2.13 | 1.47 | 0.81 | 0.53 | 19.68 |
| Days ≥ 0.1" Precip | 3 | 1 | 3 | 3 | 6 | 7 | 6 | 6 | 5 | 4 | 3 | 2 | 49 |
| Total Snowfall (") | *9.0* | na | *7.6* | *2.4* | 0.4 | 0.0 | 0.0 | 0.0 | 0.2 | 2.1 | *6.8* | *6.7* | na |
| Days ≥ 1" Snow Depth | na | na | na | *1* | 0 | 0 | 0 | 0 | 0 | 0 | na | na | na |

## PETTIBONE *Kidder County* ELEVATION 1860 ft LAT/LONG 47° 7 ' N / 99° 31 ' W

| | JAN | FEB | MAR | APR | MAY | JUN | JUL | AUG | SEP | OCT | NOV | DEC | YEAR |
|---|---|---|---|---|---|---|---|---|---|---|---|---|---|
| Maximum Temp °F | 15.6 | 22.5 | 36.0 | 54.2 | 68.3 | 76.6 | 82.9 | 82.0 | 69.9 | 56.2 | 35.9 | 21.0 | 51.8 |
| Minimum Temp °F | -4.4 | 2.7 | 15.9 | 30.4 | 41.7 | 51.5 | 56.5 | 54.2 | 43.7 | 31.8 | 17.2 | 2.4 | 28.6 |
| Mean Temp °F | 5.6 | 12.6 | 26.0 | 42.3 | 55.0 | 64.1 | 69.7 | 68.1 | 56.8 | 44.0 | 26.5 | 11.7 | 40.2 |
| Days Max Temp ≥ 90 °F | 0 | 0 | 0 | 0 | 1 | 2 | 6 | 7 | 1 | 0 | 0 | 0 | 17 |
| Days Max Temp ≤ 32 °F | 26 | *20* | 11 | 1 | 0 | 0 | 0 | 0 | 0 | 1 | *12* | *23* | 94 |
| Days Min Temp ≤ 32 °F | 31 | 28 | 29 | *20* | 5 | 0 | 0 | 0 | 3 | 17 | *29* | 31 | 193 |
| Days Min Temp ≤ 0 °F | *18* | *13* | 4 | 0 | 0 | 0 | 0 | 0 | 0 | 0 | 2 | na | na |
| Heating Degree Days | 1839 | 1475 | 1205 | 677 | 324 | 99 | 22 | 50 | 269 | 644 | 1145 | 1648 | 9397 |
| Cooling Degree Days | 0 | 0 | 0 | 4 | 28 | 87 | 184 | 157 | 26 | 0 | 0 | 0 | 486 |
| Total Precipitation (") | 0.50 | 0.38 | 0.71 | 1.54 | 2.12 | 3.57 | 2.46 | 1.84 | 1.94 | 1.28 | 0.57 | 0.42 | 17.33 |
| Days ≥ 0.1" Precip | 2 | 2 | 2 | 4 | 5 | 7 | 5 | 4 | 4 | 3 | 2 | 2 | 42 |
| Total Snowfall (") | *7.2* | 4.8 | *5.2* | *3.3* | 0.3 | 0.0 | 0.0 | 0.0 | 0.0 | 1.3 | *5.8* | *5.0* | 32.9 |
| Days ≥ 1" Snow Depth | na | na | na | 2 | 0 | 0 | 0 | 0 | 0 | 1 | *6* | na | na |

## POWERS LAKE 1 N *Burke County* ELEVATION 2211 ft LAT/LONG 48° 34 ' N / 102° 39 ' W

| | JAN | FEB | MAR | APR | MAY | JUN | JUL | AUG | SEP | OCT | NOV | DEC | YEAR |
|---|---|---|---|---|---|---|---|---|---|---|---|---|---|
| Maximum Temp °F | 15.4 | 21.9 | 34.8 | 51.8 | 66.3 | 74.1 | 80.2 | 79.3 | 67.4 | 54.7 | 34.7 | 21.6 | 50.2 |
| Minimum Temp °F | -6.5 | 0.6 | 13.5 | 27.3 | 38.9 | 48.3 | 53.0 | 50.5 | 39.9 | 27.9 | 13.8 | 0.2 | 25.6 |
| Mean Temp °F | 4.5 | 11.3 | 24.2 | 39.6 | 52.6 | 61.2 | 66.7 | 64.9 | 53.7 | 41.3 | 24.3 | 10.9 | 37.9 |
| Days Max Temp ≥ 90 °F | 0 | 0 | 0 | 0 | 1 | 2 | 4 | 5 | 1 | 0 | 0 | 0 | 13 |
| Days Max Temp ≤ 32 °F | 26 | 20 | 12 | 2 | 0 | 0 | 0 | 0 | 0 | 2 | 13 | 23 | 98 |
| Days Min Temp ≤ 32 °F | 31 | 28 | 30 | 22 | 8 | 0 | 0 | 0 | 5 | 22 | 29 | 31 | 206 |
| Days Min Temp ≤ 0 °F | 20 | 14 | 7 | 0 | 0 | 0 | 0 | 0 | 0 | 0 | 4 | 15 | 60 |
| Heating Degree Days | 1874 | 1511 | 1258 | 756 | 390 | 156 | 53 | 93 | 349 | 728 | 1214 | 1670 | 10052 |
| Cooling Degree Days | 0 | 0 | 0 | 1 | 20 | 59 | 113 | 106 | 18 | 0 | 0 | 0 | 317 |
| Total Precipitation (") | 0.40 | 0.32 | 0.60 | 1.34 | 2.08 | 2.79 | 2.77 | 2.01 | 1.83 | 0.96 | 0.44 | 0.33 | 15.87 |
| Days ≥ 0.1" Precip | 2 | 1 | 2 | 3 | 5 | 6 | 6 | 4 | 4 | 3 | 2 | 1 | 39 |
| Total Snowfall (") | 6.1 | 4.7 | 6.3 | 5.2 | 0.8 | 0.0 | 0.0 | 0.0 | 0.5 | 2.8 | 5.4 | 5.2 | 37.0 |
| Days ≥ 1" Snow Depth | 28 | 21 | 15 | 5 | 1 | 0 | 0 | 0 | 0 | 2 | 11 | 21 | 104 |

## PRETTY ROCK *Grant County* ELEVATION 2480 ft LAT/LONG 46° 10 ' N / 101° 51 ' W

| | JAN | FEB | MAR | APR | MAY | JUN | JUL | AUG | SEP | OCT | NOV | DEC | YEAR |
|---|---|---|---|---|---|---|---|---|---|---|---|---|---|
| Maximum Temp °F | 23.2 | 29.0 | 41.5 | 55.7 | 69.0 | 77.4 | 85.0 | 83.7 | 72.4 | 58.7 | 39.7 | 27.8 | 55.3 |
| Minimum Temp °F | 2.1 | 7.9 | 19.4 | 30.2 | 41.4 | 51.1 | 55.8 | 53.9 | 43.3 | 32.2 | 18.5 | 7.1 | 30.2 |
| Mean Temp °F | 12.7 | 18.5 | 30.4 | 43.0 | 55.2 | 64.3 | 70.4 | 68.9 | 57.9 | 45.5 | 29.1 | 17.4 | 42.8 |
| Days Max Temp ≥ 90 °F | 0 | 0 | 0 | 0 | 1 | 3 | 10 | 9 | 2 | 0 | 0 | 0 | 25 |
| Days Max Temp ≤ 32 °F | 20 | 16 | 8 | 1 | 0 | 0 | 0 | 0 | 0 | 1 | 9 | 18 | 73 |
| Days Min Temp ≤ 32 °F | 31 | 28 | 29 | 19 | 5 | 0 | 0 | 0 | 3 | 15 | 28 | 31 | 189 |
| Days Min Temp ≤ 0 °F | 14 | 9 | 2 | 0 | 0 | 0 | 0 | 0 | 0 | 0 | 2 | 9 | 36 |
| Heating Degree Days | 1618 | 1310 | 1066 | 657 | 314 | 94 | 21 | 41 | 244 | 599 | 1070 | 1469 | 8503 |
| Cooling Degree Days | 0 | 0 | 0 | 3 | 21 | 92 | 204 | 171 | 35 | 1 | 0 | 0 | 527 |
| Total Precipitation (") | 0.30 | 0.31 | 0.87 | 1.93 | 2.76 | 3.25 | 2.35 | 1.72 | 1.42 | 1.17 | 0.47 | 0.35 | 16.90 |
| Days ≥ 0.1" Precip | 1 | 1 | 2 | 5 | 5 | 7 | 4 | 3 | 3 | 3 | 1 | 1 | 36 |
| Total Snowfall (") | 6.6 | 7.6 | 10.4 | 6.7 | 1.4 | 0.0 | 0.0 | 0.0 | 0.5 | 2.6 | 6.9 | 6.6 | 49.3 |
| Days ≥ 1" Snow Depth | 24 | 20 | 14 | 4 | 0 | 0 | 0 | 0 | 0 | 1 | 9 | 18 | 90 |

## RICHARDTON ABBEY *Stark County* ELEVATION 2470 ft LAT/LONG 46° 53 ' N / 102° 19 ' W

| | JAN | FEB | MAR | APR | MAY | JUN | JUL | AUG | SEP | OCT | NOV | DEC | YEAR |
|---|---|---|---|---|---|---|---|---|---|---|---|---|---|
| Maximum Temp °F | 21.8 | 28.0 | 40.2 | 55.0 | 68.1 | 76.3 | 83.0 | 81.7 | 70.2 | 56.9 | 38.2 | 26.6 | 53.8 |
| Minimum Temp °F | 3.2 | 9.0 | 19.6 | 31.4 | 42.9 | 52.0 | 57.1 | 55.6 | 45.3 | 34.1 | 19.9 | 8.4 | 31.5 |
| Mean Temp °F | 12.5 | 18.5 | 29.9 | 43.2 | 55.5 | 64.2 | 70.1 | 68.7 | 57.8 | 45.5 | 29.1 | 17.5 | 42.7 |
| Days Max Temp ≥ 90 °F | 0 | 0 | 0 | 0 | 0 | 2 | 7 | 6 | 1 | 0 | 0 | 0 | 16 |
| Days Max Temp ≤ 32 °F | 21 | 16 | 9 | 1 | 0 | 0 | 0 | 0 | 0 | 1 | 10 | 19 | 77 |
| Days Min Temp ≤ 32 °F | 30 | 27 | 28 | 17 | 4 | 0 | 0 | 0 | 2 | 13 | 27 | 31 | 179 |
| Days Min Temp ≤ 0 °F | 13 | 9 | 2 | 0 | 0 | 0 | 0 | 0 | 0 | 0 | 2 | 9 | 35 |
| Heating Degree Days | 1623 | 1307 | 1081 | 648 | 307 | 96 | 22 | 41 | 245 | 598 | 1071 | 1467 | 8506 |
| Cooling Degree Days | 0 | 0 | 0 | 3 | 25 | 92 | 197 | 167 | 35 | 1 | 0 | 0 | 520 |
| Total Precipitation (") | 0.46 | 0.38 | 0.83 | 1.99 | 2.58 | 3.77 | 2.21 | 1.76 | 1.55 | 1.23 | 0.63 | 0.51 | 17.90 |
| Days ≥ 0.1" Precip | 1 | 1 | 2 | 5 | 5 | 7 | 5 | 4 | 4 | 3 | 2 | 2 | 41 |
| Total Snowfall (") | 6.3 | 5.2 | 8.5 | 7.0 | 1.5 | 0.0 | 0.0 | 0.0 | 0.6 | 2.9 | 6.6 | 6.7 | 45.3 |
| Days ≥ 1" Snow Depth | 25 | 20 | 14 | 5 | 1 | 0 | 0 | 0 | 0 | 2 | 11 | 20 | 98 |

## ROLLA 3 NW *Rolette County* ELEVATION 1923 ft LAT/LONG 48° 54 ' N / 99° 40 ' W

| | JAN | FEB | MAR | APR | MAY | JUN | JUL | AUG | SEP | OCT | NOV | DEC | YEAR |
|---|---|---|---|---|---|---|---|---|---|---|---|---|---|
| Maximum Temp °F | 13.2 | 19.1 | 30.9 | 47.9 | 64.0 | 71.8 | 76.7 | 75.8 | 64.1 | 51.3 | 31.6 | 18.4 | 47.1 |
| Minimum Temp °F | -5.8 | -0.3 | 12.8 | 27.1 | 40.3 | 50.1 | 55.2 | 52.8 | 42.4 | 30.9 | 15.3 | 0.7 | 26.8 |
| Mean Temp °F | 3.7 | 9.4 | 21.9 | 37.5 | 52.2 | 61.0 | 66.0 | 64.3 | 53.3 | 41.1 | 23.5 | 9.6 | 37.0 |
| Days Max Temp ≥ 90 °F | 0 | 0 | 0 | 0 | 0 | 1 | 1 | 2 | 0 | 0 | 0 | 0 | 4 |
| Days Max Temp ≤ 32 °F | 27 | 22 | 16 | 4 | 0 | 0 | 0 | 0 | 0 | 2 | 16 | 26 | 113 |
| Days Min Temp ≤ 32 °F | 31 | 28 | 30 | 21 | 6 | 0 | 0 | 0 | 3 | 18 | 28 | 31 | 196 |
| Days Min Temp ≤ 0 °F | 20 | 15 | 6 | 1 | 0 | 0 | 0 | 0 | 0 | 0 | 4 | 15 | 61 |
| Heating Degree Days | 1903 | 1567 | 1331 | 819 | 404 | 158 | 56 | 99 | 359 | 733 | 1240 | 1715 | 10384 |
| Cooling Degree Days | 0 | 0 | 0 | 1 | 17 | 44 | 85 | 81 | 11 | 0 | 0 | 0 | 239 |
| Total Precipitation (") | 0.50 | 0.46 | 0.66 | 1.25 | 2.23 | 3.22 | 2.96 | 2.53 | 2.10 | 1.19 | 0.60 | 0.50 | 18.20 |
| Days ≥ 0.1" Precip | 2 | 1 | 2 | 3 | 5 | 6 | 6 | 6 | 5 | 3 | 2 | 2 | 43 |
| Total Snowfall (") | na | na | *7.7* | ***4.8*** | 0.8 | 0.0 | 0.0 | 0.0 | 0.1 | 1.3 | na | na | na |
| Days ≥ 1" Snow Depth | na | na | na | ***6*** | 0 | 0 | 0 | 0 | 0 | ***0*** | na | na | na |

## RUGBY *Pierce County* ELEVATION 1542 ft LAT/LONG 48° 23 ' N / 100° 0 ' W

| | JAN | FEB | MAR | APR | MAY | JUN | JUL | AUG | SEP | OCT | NOV | DEC | YEAR |
|---|---|---|---|---|---|---|---|---|---|---|---|---|---|
| Maximum Temp °F | 14.9 | 21.9 | 35.1 | 54.0 | 69.2 | 77.6 | 82.5 | 81.3 | 69.1 | 55.9 | 34.6 | 20.8 | 51.4 |
| Minimum Temp °F | -5.0 | 2.0 | 15.9 | 30.6 | 42.9 | 52.1 | 56.3 | 54.3 | 43.8 | 32.3 | 17.1 | 1.8 | 28.7 |
| Mean Temp °F | 5.0 | 12.0 | 25.5 | 42.3 | 56.0 | 64.9 | 69.4 | 67.8 | 56.5 | 44.1 | 25.9 | 11.3 | 40.1 |
| Days Max Temp ≥ 90 °F | 0 | 0 | 0 | 0 | 1 | 3 | 5 | 6 | 1 | 0 | 0 | 0 | 16 |
| Days Max Temp ≤ 32 °F | 27 | 21 | 11 | 1 | 0 | 0 | 0 | 0 | 0 | 1 | 12 | 24 | 97 |
| Days Min Temp ≤ 32 °F | 31 | 28 | 29 | 17 | 4 | 0 | 0 | 0 | 2 | 15 | 28 | 31 | 185 |
| Days Min Temp ≤ 0 °F | 19 | 13 | 5 | 0 | 0 | 0 | 0 | 0 | 0 | 0 | 3 | 14 | 54 |
| Heating Degree Days | 1859 | 1493 | 1218 | 676 | 299 | 89 | 24 | 52 | 276 | 641 | 1166 | 1661 | 9454 |
| Cooling Degree Days | 0 | 0 | 0 | 4 | 33 | 90 | 151 | 139 | 20 | 0 | 0 | 0 | 437 |
| Total Precipitation (") | 0.49 | 0.34 | 0.68 | 1.37 | 2.04 | 2.90 | 2.95 | 2.37 | 1.91 | 1.20 | 0.54 | 0.52 | 17.31 |
| Days ≥ 0.1" Precip | 2 | 1 | 2 | 3 | 6 | 6 | 6 | 5 | 4 | 3 | 2 | 2 | 42 |
| Total Snowfall (") | na | na | na | ***3.6*** | ***0.6*** | 0.0 | 0.0 | 0.0 | 0.0 | 1.3 | na | na | na |
| Days ≥ 1" Snow Depth | na | na | na | ***2*** | ***0*** | 0 | 0 | 0 | 0 | 0 | na | na | na |

## SHARON *Steele County* ELEVATION 1522 ft LAT/LONG 47° 36 ' N / 97° 54 ' W

| | JAN | FEB | MAR | APR | MAY | JUN | JUL | AUG | SEP | OCT | NOV | DEC | YEAR |
|---|---|---|---|---|---|---|---|---|---|---|---|---|---|
| Maximum Temp °F | 13.1 | 20.2 | 33.5 | 52.6 | 67.8 | 74.9 | 79.7 | 79.3 | 68.6 | 55.1 | 33.7 | 19.0 | 49.8 |
| Minimum Temp °F | -5.2 | 1.4 | 15.4 | 30.2 | 42.6 | 51.7 | 56.1 | 54.0 | 44.0 | 32.8 | 16.8 | 1.9 | 28.5 |
| Mean Temp °F | 4.0 | 10.9 | 24.5 | 41.4 | 55.2 | 63.3 | 67.9 | 66.7 | 56.3 | 44.0 | 25.3 | 10.5 | 39.2 |
| Days Max Temp ≥ 90 °F | 0 | 0 | 0 | 0 | 0 | 1 | 2 | 3 | 1 | 0 | 0 | 0 | 7 |
| Days Max Temp ≤ 32 °F | 28 | 22 | 13 | 2 | 0 | 0 | 0 | 0 | 0 | 1 | 14 | 26 | 106 |
| Days Min Temp ≤ 32 °F | 31 | 28 | 30 | 19 | 4 | 0 | 0 | 0 | 2 | 15 | 28 | 31 | 188 |
| Days Min Temp ≤ 0 °F | 20 | 14 | 5 | 0 | 0 | 0 | 0 | 0 | 0 | 0 | 3 | 15 | 57 |
| Heating Degree Days | 1891 | 1526 | 1251 | 703 | 320 | 106 | 31 | 62 | 280 | 645 | 1185 | 1687 | 9687 |
| Cooling Degree Days | 0 | 0 | 0 | 3 | 29 | 64 | 132 | 117 | 20 | 0 | 0 | 0 | 365 |
| Total Precipitation (") | 0.67 | 0.47 | 1.08 | 1.64 | 2.60 | 3.66 | 3.47 | 2.60 | 2.25 | 1.47 | 0.81 | 0.58 | 21.30 |
| Days ≥ 0.1" Precip | 2 | 2 | 3 | 4 | 6 | 7 | 7 | 6 | 5 | 4 | 3 | 2 | 51 |
| Total Snowfall (") | 8.0 | 5.6 | 6.3 | 3.3 | 0.3 | 0.0 | 0.0 | 0.0 | 0.0 | 1.4 | 6.4 | 6.5 | 37.8 |
| Days ≥ 1" Snow Depth | 30 | 25 | 19 | 3 | 0 | 0 | 0 | 0 | 0 | 1 | 12 | 25 | 115 |

## STANLEY 3 NNW *Mountrail County* ELEVATION 2260 ft LAT/LONG 48° 19 ' N / 102° 24 ' W

| | JAN | FEB | MAR | APR | MAY | JUN | JUL | AUG | SEP | OCT | NOV | DEC | YEAR |
|---|---|---|---|---|---|---|---|---|---|---|---|---|---|
| Maximum Temp °F | 15.2 | 22.3 | 35.3 | 51.7 | 66.0 | 74.1 | 80.3 | 79.8 | 67.0 | 54.2 | 34.4 | 21.0 | 50.1 |
| Minimum Temp °F | -4.7 | 2.3 | 13.8 | 27.3 | 38.5 | 48.3 | 52.6 | 49.5 | 39.0 | 28.1 | 14.9 | 1.3 | 25.9 |
| Mean Temp °F | 5.3 | 12.3 | 24.6 | 39.5 | 52.2 | 61.3 | 66.5 | 64.7 | 53.0 | 41.2 | 24.7 | 11.2 | 38.0 |
| Days Max Temp ≥ 90 °F | 0 | 0 | 0 | 0 | 0 | 1 | 4 | 5 | 1 | 0 | 0 | 0 | 11 |
| Days Max Temp ≤ 32 °F | 26 | 20 | 12 | 2 | 0 | 0 | 0 | 0 | 0 | 1 | 13 | 24 | 98 |
| Days Min Temp ≤ 32 °F | 31 | 28 | 30 | 22 | 7 | 0 | 0 | 0 | 6 | 22 | 29 | 31 | 206 |
| Days Min Temp ≤ 0 °F | 19 | 13 | 6 | 0 | 0 | 0 | 0 | 0 | 0 | 0 | 4 | 14 | 56 |
| Heating Degree Days | 1852 | 1483 | 1247 | 759 | 399 | 147 | 52 | 93 | 366 | 731 | 1204 | 1666 | 9999 |
| Cooling Degree Days | 0 | 0 | 0 | 0 | 14 | 53 | 101 | 95 | 13 | 0 | 0 | 0 | 276 |
| Total Precipitation (") | 0.55 | 0.41 | 0.82 | 1.70 | 2.56 | 3.80 | 2.87 | 2.16 | 2.19 | 1.12 | 0.63 | 0.54 | 19.35 |
| Days ≥ 0.1" Precip | 2 | 1 | 2 | 4 | 6 | 8 | 7 | 5 | 5 | 3 | 2 | 2 | 47 |
| Total Snowfall (") | *7.3* | *4.5* | 8.3 | 6.4 | 1.1 | 0.0 | 0.0 | 0.0 | 0.4 | 2.9 | *6.6* | 7.0 | 44.5 |
| Days ≥ 1" Snow Depth | 29 | *23* | *18* | 6 | 1 | 0 | 0 | 0 | 0 | *2* | na | *26* | na |

## STEELE *Kidder County* ELEVATION 1860 ft LAT/LONG 46° 51 ' N / 99° 55 ' W

| | JAN | FEB | MAR | APR | MAY | JUN | JUL | AUG | SEP | OCT | NOV | DEC | YEAR |
|---|---|---|---|---|---|---|---|---|---|---|---|---|---|
| Maximum Temp °F | 17.4 | *25.8* | *37.7* | *54.8* | *69.0* | *76.4* | *83.1* | *83.4* | *71.8* | *58.4* | 37.9 | 23.5 | 53.3 |
| Minimum Temp °F | -3.2 | *4.6* | *16.9* | *30.6* | *42.1* | *51.5* | *56.3* | *54.3* | *44.0* | *32.4* | 17.7 | 4.0 | 29.3 |
| Mean Temp °F | 7.1 | *15.2* | *27.3* | *42.5* | *55.6* | *64.0* | *69.7* | *68.9* | *58.0* | *45.4* | 27.8 | 13.8 | 41.3 |
| Days Max Temp ≥ 90 °F | 0 | *0* | *0* | *0* | *1* | *2* | *6* | *8* | *2* | *0* | 0 | 0 | 19 |
| Days Max Temp ≤ 32 °F | 25 | *18* | *10* | *1* | *0* | *0* | *0* | *0* | *0* | *1* | 10 | 23 | 88 |
| Days Min Temp ≤ 32 °F | 31 | *28* | *29* | *19* | *5* | *0* | *0* | *0* | *3* | *15* | 28 | 31 | 189 |
| Days Min Temp ≤ 0 °F | 18 | *12* | *4* | *0* | *0* | *0* | *0* | *0* | *0* | *0* | 3 | 12 | 49 |
| Heating Degree Days | 1792 | *1401* | *1162* | *670* | *309* | *97* | *22* | *42* | *247* | *603* | 1109 | 1582 | 9036 |
| Cooling Degree Days | *0* | na | na | na | na | na | na | na | na | *1* | *0* | *0* | na |
| Total Precipitation (") | 0.47 | *0.46* | *0.95* | *1.76* | *2.51* | *3.51* | *2.68* | *1.50* | *1.96* | *1.32* | 0.63 | 0.48 | 18.23 |
| Days ≥ 0.1" Precip | 2 | *2* | *3* | *4* | *6* | *7* | *6* | *4* | *3* | *3* | 2 | 2 | 44 |
| Total Snowfall (") | na | na | na | na | *1.2* | *0.0* | *0.0* | *0.0* | *0.0* | *1.6* | na | na | na |
| Days ≥ 1" Snow Depth | na | na | na | na | *0* | *0* | *0* | *0* | *0* | *1* | na | na | na |

## TIOGA 1 E *Williams County* ELEVATION 2280 ft LAT/LONG 48° 24 ' N / 102° 56 ' W

| | JAN | FEB | MAR | APR | MAY | JUN | JUL | AUG | SEP | OCT | NOV | DEC | YEAR |
|---|---|---|---|---|---|---|---|---|---|---|---|---|---|
| Maximum Temp °F | 15.4 | 22.0 | 35.4 | 52.3 | 66.2 | 74.7 | 81.0 | 80.3 | 67.7 | 54.6 | 34.3 | 21.4 | 50.4 |
| Minimum Temp °F | -5.4 | 1.3 | 13.9 | 27.8 | 39.9 | 49.0 | 53.2 | 50.6 | 39.8 | 28.1 | 14.2 | 0.9 | 26.1 |
| Mean Temp °F | 5.0 | 11.7 | 24.7 | 40.1 | 53.1 | 61.9 | 67.1 | 65.5 | 53.7 | 41.4 | 24.3 | 11.2 | 38.3 |
| Days Max Temp ≥ 90 °F | 0 | 0 | 0 | 0 | 1 | 2 | 5 | 6 | 1 | 0 | 0 | 0 | 15 |
| Days Max Temp ≤ 32 °F | 25 | 20 | 13 | 2 | 0 | 0 | 0 | 0 | 0 | 1 | 14 | 23 | 98 |
| Days Min Temp ≤ 32 °F | 31 | 28 | 30 | 21 | 7 | 0 | 0 | 0 | 6 | 21 | 29 | 31 | 204 |
| Days Min Temp ≤ 0 °F | 19 | 14 | 6 | 0 | 0 | 0 | 0 | 0 | 0 | 0 | 5 | 14 | 58 |
| Heating Degree Days | 1858 | 1502 | 1243 | 741 | 378 | 141 | 51 | 88 | 350 | 726 | 1216 | 1665 | 9959 |
| Cooling Degree Days | 0 | 0 | 0 | 1 | 20 | 61 | 121 | 116 | 18 | 0 | 0 | 0 | 337 |
| Total Precipitation (") | 0.46 | 0.33 | 0.53 | 1.27 | 2.03 | 2.53 | 2.10 | 1.87 | 1.71 | 0.84 | 0.46 | 0.37 | 14.50 |
| Days ≥ 0.1" Precip | 2 | 1 | 2 | 3 | 5 | 6 | 5 | 4 | 4 | 2 | 2 | 1 | 37 |
| Total Snowfall (") | 6.2 | 4.5 | 5.3 | 4.6 | 1.2 | 0.0 | 0.0 | 0.0 | 0.6 | 2.0 | 5.4 | 5.1 | 34.9 |
| Days ≥ 1" Snow Depth | 29 | 24 | 16 | 4 | 0 | 0 | 0 | 0 | 0 | 1 | 12 | 24 | 110 |

## TOWNER 2 NE *McHenry County* ELEVATION 1480 ft LAT/LONG 48° 21 ' N / 100° 24 ' W

| | JAN | FEB | MAR | APR | MAY | JUN | JUL | AUG | SEP | OCT | NOV | DEC | YEAR |
|---|---|---|---|---|---|---|---|---|---|---|---|---|---|
| Maximum Temp °F | 14.7 | 21.8 | 35.1 | 53.2 | 68.2 | 77.1 | 82.7 | 81.9 | 69.6 | 55.7 | 34.8 | 20.7 | 51.3 |
| Minimum Temp °F | -7.2 | -0.6 | 13.3 | 28.1 | 40.5 | 50.0 | 53.8 | 50.9 | 40.7 | 29.3 | 15.1 | 0.1 | 26.2 |
| Mean Temp °F | 3.7 | 10.6 | 24.4 | 40.7 | 54.3 | 63.6 | 68.3 | 66.7 | 55.3 | ***42.6*** | 25.0 | 10.4 | 38.8 |
| Days Max Temp ≥ 90 °F | 0 | 0 | 0 | 0 | 1 | 3 | ***5*** | 7 | 1 | 0 | 0 | 0 | 17 |
| Days Max Temp ≤ 32 °F | 26 | 21 | 12 | 2 | 0 | 0 | 0 | 0 | 0 | 1 | 12 | 24 | 98 |
| Days Min Temp ≤ 32 °F | 31 | 28 | 30 | 22 | 6 | 0 | 0 | 0 | 5 | 19 | 29 | 31 | 201 |
| Days Min Temp ≤ 0 °F | 20 | 14 | 6 | 0 | 0 | 0 | 0 | 0 | 0 | 0 | 4 | 15 | 59 |
| Heating Degree Days | 1901 | 1533 | 1252 | 724 | 343 | 109 | 35 | 68 | 307 | ***688*** | 1195 | 1689 | 9844 |
| Cooling Degree Days | 0 | 0 | 0 | 2 | 23 | 71 | 132 | 124 | 16 | 0 | 0 | 0 | 368 |
| Total Precipitation (") | 0.55 | 0.44 | 0.63 | 1.40 | 1.77 | 2.53 | 2.47 | 2.17 | 1.87 | 1.37 | 0.53 | 0.54 | 16.27 |
| Days ≥ 0.1" Precip | 2 | 1 | 2 | 3 | 4 | 6 | 5 | 4 | 4 | 3 | 2 | 2 | 38 |
| Total Snowfall (") | 7.6 | 5.6 | ***4.5*** | ***2.7*** | 0.4 | 0.0 | 0.0 | 0.0 | 0.0 | ***1.8*** | ***6.1*** | ***6.3*** | 35.0 |
| Days ≥ 1" Snow Depth | 29 | 25 | 17 | 4 | 0 | 0 | 0 | 0 | 0 | 1 | 10 | 23 | 109 |

## TROTTERS 3 SSE *Golden Valley County* ELEVATION 2421 ft LAT/LONG 47° 20 ' N / 103° 54 ' W

| | JAN | FEB | MAR | APR | MAY | JUN | JUL | AUG | SEP | OCT | NOV | DEC | YEAR |
|---|---|---|---|---|---|---|---|---|---|---|---|---|---|
| Maximum Temp °F | 22.5 | 28.6 | 40.8 | 55.5 | 67.6 | 76.2 | 83.7 | 82.8 | 70.6 | 57.2 | 38.3 | 26.9 | 54.2 |
| Minimum Temp °F | 2.8 | 9.0 | 19.5 | 30.6 | 41.7 | 50.6 | 55.6 | 54.0 | 43.6 | 32.8 | 18.9 | 7.9 | 30.6 |
| Mean Temp °F | 12.7 | 18.8 | 30.2 | 43.1 | 54.7 | 63.4 | 69.7 | 68.4 | 57.1 | 45.0 | 28.7 | 17.4 | 42.4 |
| Days Max Temp ≥ 90 °F | 0 | 0 | 0 | 0 | 1 | 2 | 8 | 8 | 2 | 0 | 0 | 0 | 21 |
| Days Max Temp ≤ 32 °F | 21 | 15 | 8 | 1 | 0 | 0 | 0 | 0 | 0 | 1 | 9 | 19 | 74 |
| Days Min Temp ≤ 32 °F | 30 | 27 | 28 | 18 | 4 | 0 | 0 | 0 | 3 | 14 | 27 | 30 | 181 |
| Days Min Temp ≤ 0 °F | 14 | 9 | 3 | 0 | 0 | 0 | 0 | 0 | 0 | 0 | 3 | 9 | 38 |
| Heating Degree Days | 1618 | 1299 | 1073 | 652 | 330 | 107 | 25 | 47 | 262 | 614 | 1084 | 1471 | 8582 |
| Cooling Degree Days | 0 | 0 | 0 | 2 | 24 | 80 | 194 | 174 | 33 | 1 | 0 | 0 | 508 |
| Total Precipitation (") | 0.35 | 0.30 | 0.50 | 1.31 | 2.12 | 3.23 | 1.82 | 1.35 | 1.69 | 1.00 | 0.50 | 0.41 | 14.58 |
| Days ≥ 0.1" Precip | 1 | 1 | 2 | 4 | 5 | 6 | 4 | 3 | 4 | 2 | 2 | 1 | 35 |
| Total Snowfall (") | 5.3 | 4.0 | 4.8 | 4.6 | 1.2 | 0.0 | 0.0 | 0.0 | 0.4 | 1.9 | 4.9 | 5.9 | 33.0 |
| Days ≥ 1" Snow Depth | 27 | 21 | 12 | 3 | 0 | 0 | 0 | 0 | 0 | 1 | 10 | 23 | 97 |

## TURTLE LAKE *McLean County* ELEVATION 1903 ft LAT/LONG 47° 31 ' N / 100° 53 ' W

| | JAN | FEB | MAR | APR | MAY | JUN | JUL | AUG | SEP | OCT | NOV | DEC | YEAR |
|---|---|---|---|---|---|---|---|---|---|---|---|---|---|
| Maximum Temp °F | 18.1 | 24.6 | 37.9 | 54.6 | 68.7 | 76.6 | 82.9 | 81.9 | 70.1 | 56.7 | 36.5 | 23.3 | 52.7 |
| Minimum Temp °F | -2.4 | 4.1 | 16.8 | 29.6 | 41.7 | 51.3 | 55.9 | 53.6 | 43.4 | 31.7 | 17.1 | 4.0 | 28.9 |
| Mean Temp °F | 8.1 | 14.4 | 27.4 | 42.2 | 55.2 | 64.0 | 69.4 | 67.8 | 56.7 | 44.3 | 26.8 | 13.6 | 40.8 |
| Days Max Temp ≥ 90 °F | 0 | 0 | 0 | 0 | 1 | 2 | 7 | 7 | 2 | 0 | 0 | 0 | 19 |
| Days Max Temp ≤ 32 °F | 25 | 19 | 10 | 1 | 0 | 0 | 0 | 0 | 0 | 1 | 11 | 22 | 89 |
| Days Min Temp ≤ 32 °F | 31 | 28 | 29 | 20 | 5 | 0 | 0 | 0 | 3 | 16 | 28 | 31 | 191 |
| Days Min Temp ≤ 0 °F | 18 | 12 | 4 | 0 | 0 | 0 | 0 | 0 | 0 | 0 | 3 | 12 | 49 |
| Heating Degree Days | 1760 | 1424 | 1160 | 680 | 318 | 98 | 25 | 52 | 272 | 636 | 1138 | 1589 | 9152 |
| Cooling Degree Days | 0 | 0 | 0 | 2 | 23 | 70 | 147 | 126 | 22 | 1 | 0 | 0 | 391 |
| Total Precipitation (") | 0.62 | 0.47 | 0.80 | 1.62 | 2.07 | 3.47 | 2.73 | 2.00 | 1.66 | 1.17 | 0.62 | 0.58 | 17.81 |
| Days ≥ 0.1" Precip | 2 | 2 | 2 | 4 | 5 | 7 | 6 | 4 | 4 | 2 | 2 | 2 | 42 |
| Total Snowfall (") | ***6.5*** | na | ***4.1*** | ***0.9*** | 0.3 | 0.0 | 0.0 | 0.0 | 0.1 | 1.4 | ***6.4*** | na | na |
| Days ≥ 1" Snow Depth | ***24*** | ***22*** | ***13*** | 2 | 0 | 0 | 0 | 0 | 0 | 1 | ***7*** | ***18*** | 87 |

## TUTTLE *Kidder County* ELEVATION 1880 ft LAT/LONG 47° 8 ' N / 100° 0 ' W

| | JAN | FEB | MAR | APR | MAY | JUN | JUL | AUG | SEP | OCT | NOV | DEC | YEAR |
|---|---|---|---|---|---|---|---|---|---|---|---|---|---|
| Maximum Temp °F | 16.7 | 22.7 | 35.9 | 53.0 | 67.8 | 76.0 | 82.6 | 81.7 | 69.7 | 56.5 | 36.7 | 22.7 | 51.8 |
| Minimum Temp °F | -3.9 | 2.0 | 15.5 | 29.7 | 41.5 | 50.8 | 55.4 | 52.5 | 41.9 | 30.2 | 16.0 | 2.3 | 27.8 |
| Mean Temp °F | 6.4 | 12.4 | 25.7 | 41.4 | 54.7 | 63.5 | 69.0 | 67.1 | 55.8 | 43.4 | 26.4 | 12.5 | 39.9 |
| Days Max Temp ≥ 90 °F | 0 | 0 | 0 | 0 | 0 | 2 | 6 | 7 | 1 | 0 | 0 | 0 | 16 |
| Days Max Temp ≤ 32 °F | 25 | 20 | 11 | 2 | 0 | 0 | 0 | 0 | 0 | 1 | 11 | 23 | 93 |
| Days Min Temp ≤ 32 °F | 31 | 28 | 30 | 20 | 5 | 0 | 0 | 0 | 4 | 19 | 29 | 31 | 197 |
| Days Min Temp ≤ 0 °F | 18 | 13 | 5 | 0 | 0 | 0 | 0 | 0 | 0 | 0 | 3 | 14 | 53 |
| Heating Degree Days | 1815 | 1481 | 1211 | 704 | 334 | 108 | 28 | 62 | 294 | 665 | 1152 | 1623 | 9477 |
| Cooling Degree Days | 0 | 0 | 0 | 2 | 24 | 75 | 160 | 137 | 25 | 1 | 0 | 0 | 424 |
| Total Precipitation (") | 0.45 | 0.37 | 0.68 | 1.51 | 2.14 | 3.29 | 2.47 | 1.75 | 1.86 | 1.11 | 0.47 | 0.37 | 16.47 |
| Days ≥ 0.1" Precip | 2 | 1 | 2 | 4 | 5 | 7 | 5 | 5 | 4 | 2 | 2 | 1 | 40 |
| Total Snowfall (") | 6.9 | 5.7 | 6.9 | 4.2 | 1.1 | 0.0 | 0.0 | 0.0 | 0.2 | 2.2 | 5.8 | 5.8 | 38.8 |
| Days ≥ 1" Snow Depth | 28 | 23 | 17 | 4 | 0 | 0 | 0 | 0 | 0 | 1 | 8 | 21 | 102 |

## UNDERWOOD *McLean County* ELEVATION 2031 ft LAT/LONG 47° 27 ' N / 101° 8 ' W

| | JAN | FEB | MAR | APR | MAY | JUN | JUL | AUG | SEP | OCT | NOV | DEC | YEAR |
|---|---|---|---|---|---|---|---|---|---|---|---|---|---|
| Maximum Temp °F | 18.5 | 25.0 | 38.6 | 55.3 | 69.3 | 77.3 | 83.4 | 82.4 | 70.7 | 57.3 | 36.4 | 23.5 | 53.1 |
| Minimum Temp °F | -0.4 | 5.6 | 17.6 | 30.7 | 42.6 | 51.9 | 56.5 | 54.4 | 44.5 | 33.0 | 18.6 | 5.5 | 30.0 |
| Mean Temp °F | 9.1 | 15.3 | 28.1 | 43.0 | 56.0 | 64.6 | 70.0 | 68.4 | 57.6 | 45.2 | 27.5 | 14.5 | 41.6 |
| Days Max Temp ≥ 90 °F | 0 | 0 | 0 | 0 | 1 | 2 | 7 | 7 | 1 | 0 | 0 | 0 | 18 |
| Days Max Temp ≤ 32 °F | 24 | 18 | 9 | 1 | 0 | 0 | 0 | 0 | 0 | 1 | 11 | 22 | 86 |
| Days Min Temp ≤ 32 °F | 31 | 28 | 29 | 19 | 4 | 0 | 0 | 0 | 2 | 15 | 28 | 31 | 187 |
| Days Min Temp ≤ 0 °F | 17 | 11 | 4 | 0 | 0 | 0 | 0 | 0 | 0 | 0 | 3 | 11 | 46 |
| Heating Degree Days | 1731 | 1398 | 1137 | 654 | 299 | 88 | 20 | 44 | 247 | 609 | 1117 | 1562 | 8906 |
| Cooling Degree Days | 0 | 0 | 0 | 5 | 33 | 100 | 190 | 171 | 33 | 2 | ***0*** | 0 | 534 |
| Total Precipitation (") | 0.50 | 0.45 | 0.70 | 1.82 | 2.09 | 3.48 | 2.49 | 1.81 | 1.78 | 1.14 | 0.59 | 0.56 | 17.41 |
| Days ≥ 0.1" Precip | 1 | 2 | 2 | 4 | 5 | 7 | 5 | 4 | 4 | 2 | 2 | 2 | 40 |
| Total Snowfall (") | na | na | na | 1.7 | 0.8 | 0.0 | 0.0 | 0.0 | 0.1 | 1.9 | ***3.7*** | na | na |
| Days ≥ 1" Snow Depth | na | na | na | 4 | 0 | 0 | 0 | 0 | 0 | 1 | ***6*** | na | na |

## UPHAM 3 N *McHenry County* ELEVATION 1430 ft LAT/LONG 48° 37 ' N / 100° 44 ' W

| | JAN | FEB | MAR | APR | MAY | JUN | JUL | AUG | SEP | OCT | NOV | DEC | YEAR |
|---|---|---|---|---|---|---|---|---|---|---|---|---|---|
| Maximum Temp °F | 14.0 | 21.5 | 35.0 | 53.6 | 68.3 | 76.3 | 81.8 | 81.0 | 69.3 | 55.7 | 34.9 | 20.4 | 51.0 |
| Minimum Temp °F | -9.5 | -3.2 | 11.7 | 27.6 | 40.3 | 49.5 | 53.2 | 50.6 | 39.9 | 27.4 | 13.4 | -2.1 | 24.9 |
| Mean Temp °F | 2.2 | 9.2 | 23.3 | 40.6 | 54.3 | 62.9 | 67.5 | 65.9 | 54.6 | 41.6 | 24.2 | 9.2 | 38.0 |
| Days Max Temp ≥ 90 °F | 0 | 0 | 0 | 0 | 1 | 2 | 5 | 6 | 1 | 0 | 0 | 0 | 15 |
| Days Max Temp ≤ 32 °F | 27 | 21 | 12 | 1 | 0 | 0 | 0 | 0 | 0 | 1 | 12 | 24 | 98 |
| Days Min Temp ≤ 32 °F | 31 | 28 | 30 | 22 | 6 | 0 | 0 | 0 | 6 | 23 | 29 | 31 | 206 |
| Days Min Temp ≤ 0 °F | 22 | 16 | 7 | 0 | 0 | 0 | 0 | 0 | 0 | 0 | 5 | 17 | 67 |
| Heating Degree Days | 1945 | 1573 | 1285 | 726 | 344 | 118 | 41 | 76 | 324 | 720 | 1217 | 1729 | 10098 |
| Cooling Degree Days | 0 | 0 | 0 | 3 | 26 | 69 | 122 | 120 | 14 | 0 | 0 | 0 | 354 |
| Total Precipitation (") | 0.55 | 0.37 | 0.64 | 1.44 | 1.90 | 2.95 | 2.52 | 2.19 | 1.78 | 1.19 | 0.65 | 0.58 | 16.76 |
| Days ≥ 0.1" Precip | 2 | 1 | 2 | 4 | 4 | 7 | 5 | 5 | 4 | 3 | 2 | 2 | 41 |
| Total Snowfall (") | 9.5 | 5.9 | 6.4 | 4.5 | 0.3 | 0.0 | 0.0 | 0.0 | 0.2 | 2.6 | 6.8 | 9.2 | 45.4 |
| Days ≥ 1" Snow Depth | 30 | 26 | ***19*** | 4 | 0 | 0 | 0 | 0 | 0 | 1 | 12 | 25 | 117 |

## VALLEY CITY 3 NNW *Barnes County* ELEVATION 1230 ft LAT/LONG 46° 55 ' N / 98° 0 ' W

| | JAN | FEB | MAR | APR | MAY | JUN | JUL | AUG | SEP | OCT | NOV | DEC | YEAR |
|---|---|---|---|---|---|---|---|---|---|---|---|---|---|
| Maximum Temp °F | 16.3 | 22.4 | 35.6 | 53.4 | 68.2 | 75.7 | 81.5 | 80.2 | 69.3 | 56.6 | 36.4 | 21.9 | 51.5 |
| Minimum Temp °F | -5.2 | 0.7 | 14.9 | 28.8 | 41.1 | 50.7 | 55.6 | 52.4 | 41.8 | 30.3 | 16.2 | 1.7 | 27.4 |
| Mean Temp °F | 5.5 | 11.6 | 25.3 | 41.1 | 54.7 | 63.3 | 68.6 | 66.3 | 55.6 | 43.5 | 26.3 | 11.9 | 39.5 |
| Days Max Temp ≥ 90 °F | 0 | 0 | 0 | 0 | 1 | 1 | 4 | 5 | 1 | 0 | 0 | 0 | 12 |
| Days Max Temp ≤ 32 °F | 26 | 20 | 11 | 1 | 0 | 0 | 0 | 0 | 0 | 1 | 11 | 23 | 93 |
| Days Min Temp ≤ 32 °F | 31 | 28 | 29 | 20 | 6 | 0 | 0 | 0 | 4 | 19 | 29 | 31 | 197 |
| Days Min Temp ≤ 0 °F | 20 | 14 | 5 | 0 | 0 | 0 | 0 | 0 | 0 | 0 | 3 | 14 | 56 |
| Heating Degree Days | 1842 | 1506 | 1225 | 711 | 334 | 109 | 27 | 67 | 301 | 661 | 1155 | 1645 | 9583 |
| Cooling Degree Days | 0 | 0 | 0 | 3 | 24 | 68 | 136 | 116 | 18 | 1 | 0 | 0 | 366 |
| Total Precipitation (") | 0.55 | 0.38 | 0.82 | 1.52 | 2.50 | 3.50 | 2.69 | 2.39 | 2.06 | 1.39 | 0.66 | 0.47 | 18.93 |
| Days ≥ 0.1" Precip | 2 | 1 | 2 | 3 | 6 | 7 | 5 | 5 | 4 | 3 | 2 | 2 | 42 |
| Total Snowfall (") | 8.2 | 5.4 | 5.6 | 3.1 | 0.1 | 0.0 | 0.0 | 0.0 | 0.0 | 0.6 | 5.2 | 6.2 | 34.4 |
| Days ≥ 1" Snow Depth | ***25*** | ***24*** | 16 | 3 | 0 | 0 | 0 | 0 | 0 | 0 | ***7*** | ***21*** | 96 |

## VELVA *McHenry County* ELEVATION 1512 ft LAT/LONG 48° 3 ' N / 100° 55 ' W

| | JAN | FEB | MAR | APR | MAY | JUN | JUL | AUG | SEP | OCT | NOV | DEC | YEAR |
|---|---|---|---|---|---|---|---|---|---|---|---|---|---|
| Maximum Temp °F | 18.2 | 24.8 | 38.8 | 55.2 | 69.8 | 77.9 | 83.4 | 82.5 | 70.6 | 57.8 | 37.1 | 23.4 | 53.3 |
| Minimum Temp °F | -3.2 | 3.4 | 16.7 | 29.8 | 42.1 | 51.6 | 56.0 | 53.5 | 42.9 | 32.0 | 17.7 | 3.1 | 28.8 |
| Mean Temp °F | 7.5 | 14.1 | 27.8 | 42.5 | 55.9 | 64.7 | 69.8 | 68.0 | 56.7 | 44.9 | 27.4 | 13.2 | 41.0 |
| Days Max Temp ≥ 90 °F | 0 | 0 | 0 | 0 | 1 | 3 | 7 | 7 | 2 | 0 | 0 | 0 | 20 |
| Days Max Temp ≤ 32 °F | 23 | 18 | 10 | 1 | 0 | 0 | 0 | 0 | 0 | 1 | 11 | 22 | 86 |
| Days Min Temp ≤ 32 °F | 31 | 28 | 29 | 18 | 5 | 0 | 0 | 0 | 3 | 16 | 28 | 31 | 189 |
| Days Min Temp ≤ 0 °F | 18 | 13 | 4 | 0 | 0 | 0 | 0 | 0 | 0 | 0 | 3 | 14 | 52 |
| Heating Degree Days | 1779 | 1430 | 1147 | 673 | 302 | 86 | 24 | 48 | 271 | 617 | 1122 | 1600 | 9099 |
| Cooling Degree Days | 0 | 0 | ***0*** | ***4*** | 34 | 94 | 176 | 155 | 28 | 1 | 0 | 0 | 492 |
| Total Precipitation (") | 0.68 | 0.43 | 0.71 | 1.75 | 2.09 | 3.22 | 2.85 | 1.84 | 1.69 | 1.53 | 0.66 | 0.57 | 18.02 |
| Days ≥ 0.1" Precip | 2 | ***2*** | 2 | 3 | 4 | 6 | 5 | 4 | 4 | 3 | 2 | 2 | 39 |
| Total Snowfall (") | 8.6 | ***5.7*** | ***5.6*** | ***3.9*** | 0.1 | 0.0 | 0.0 | 0.0 | 0.3 | 2.2 | 5.2 | 7.2 | 38.8 |
| Days ≥ 1" Snow Depth | na | na | na | ***2*** | 0 | 0 | 0 | 0 | 0 | 1 | na | na | na |

## WAHPETON 3 N *Richland County* ELEVATION 961 ft LAT/LONG 46° 16' N / 96° 36' W

| | JAN | FEB | MAR | APR | MAY | JUN | JUL | AUG | SEP | OCT | NOV | DEC | YEAR |
|---|---|---|---|---|---|---|---|---|---|---|---|---|---|
| Maximum Temp °F | 17.3 | 23.5 | 37.2 | 56.4 | 71.4 | 79.4 | 84.4 | 82.2 | 71.4 | 57.7 | 37.4 | 23.4 | 53.5 |
| Minimum Temp °F | -1.5 | 5.0 | 19.3 | 33.6 | 45.8 | 55.0 | 59.7 | 57.4 | 47.6 | 36.2 | 21.1 | 6.3 | 32.1 |
| Mean Temp °F | 7.9 | 14.3 | 28.3 | 44.9 | 58.6 | 67.3 | 72.1 | 69.9 | 59.5 | 47.0 | 29.3 | 14.9 | 42.8 |
| Days Max Temp ≥ 90 °F | 0 | 0 | 0 | 0 | 1 | 3 | 7 | 6 | 1 | 0 | 0 | 0 | 18 |
| Days Max Temp ≤ 32 °F | 26 | 20 | 10 | 1 | 0 | 0 | 0 | 0 | 0 | 0 | 11 | 23 | 91 |
| Days Min Temp ≤ 32 °F | 31 | 28 | 27 | 14 | 2 | 0 | 0 | 0 | 1 | 11 | 26 | 31 | 171 |
| Days Min Temp ≤ 0 °F | 18 | 11 | 3 | 0 | 0 | 0 | 0 | 0 | 0 | 0 | 2 | 11 | 45 |
| Heating Degree Days | 1768 | 1428 | 1131 | 598 | 233 | 48 | 9 | 28 | 204 | 555 | 1066 | 1548 | 8616 |
| Cooling Degree Days | 0 | 0 | 0 | 6 | 55 | 128 | 236 | 190 | 43 | 1 | 0 | 0 | 659 |
| Total Precipitation (") | 0.53 | 0.36 | 0.92 | 2.11 | 2.64 | 3.67 | 3.38 | 2.77 | 2.44 | 1.69 | 0.71 | 0.47 | 21.69 |
| Days ≥ 0.1" Precip | 2 | 1 | 3 | 5 | 6 | 7 | 6 | 5 | 5 | 4 | 2 | 2 | 48 |
| Total Snowfall (") | 6.8 | 4.0 | 5.1 | 2.5 | 0.0 | 0.0 | 0.0 | 0.0 | 0.0 | 0.5 | 4.3 | 4.5 | 27.7 |
| Days ≥ 1" Snow Depth | 27 | 22 | 17 | 3 | 0 | 0 | 0 | 0 | 0 | 0 | 8 | 19 | 96 |

## WATFORD CITY 14 S *McKenzie County* ELEVATION 1972 ft LAT/LONG 47° 36' N / 103° 17' W

| | JAN | FEB | MAR | APR | MAY | JUN | JUL | AUG | SEP | OCT | NOV | DEC | YEAR |
|---|---|---|---|---|---|---|---|---|---|---|---|---|---|
| Maximum Temp °F | 24.3 | 31.7 | 43.9 | 58.8 | 71.6 | 79.7 | 87.2 | 86.6 | 74.5 | 61.4 | 41.4 | 30.1 | 57.6 |
| Minimum Temp °F | 0.9 | 8.2 | 19.5 | 31.0 | 42.4 | 51.2 | 56.0 | 54.0 | 43.4 | 32.5 | 18.9 | 7.4 | 30.4 |
| Mean Temp °F | 12.6 | 20.0 | 31.8 | 44.9 | 57.0 | 65.5 | 71.6 | 70.3 | 59.0 | 47.0 | 30.2 | 18.8 | 44.1 |
| Days Max Temp ≥ 90 °F | 0 | 0 | 0 | 0 | 1 | 4 | 13 | 13 | 4 | 0 | 0 | 0 | 35 |
| Days Max Temp ≤ 32 °F | 19 | 13 | 7 | 1 | 0 | 0 | 0 | 0 | 0 | 0 | 7 | 16 | 63 |
| Days Min Temp ≤ 32 °F | 30 | 27 | 28 | 17 | 5 | 0 | 0 | 0 | 3 | 15 | 27 | 30 | 182 |
| Days Min Temp ≤ 0 °F | 15 | 9 | 2 | 0 | 0 | 0 | 0 | 0 | 0 | 0 | 2 | 9 | 37 |
| Heating Degree Days | 1623 | 1265 | 1024 | 599 | 270 | 75 | 17 | 33 | 220 | 552 | 1039 | 1428 | 8145 |
| Cooling Degree Days | 0 | 0 | 0 | 4 | 36 | 106 | 229 | 211 | 43 | 2 | 0 | 0 | 631 |
| Total Precipitation (") | 0.39 | 0.31 | 0.56 | 1.44 | 2.21 | 3.04 | 2.07 | 1.51 | 1.76 | 1.16 | 0.43 | 0.40 | 15.28 |
| Days ≥ 0.1" Precip | 1 | 1 | 2 | 4 | 5 | 7 | 5 | 4 | 4 | 2 | 1 | 1 | 37 |
| Total Snowfall (") | 6.1 | 5.3 | 4.5 | 3.7 | 0.6 | 0.0 | 0.0 | 0.0 | 0.1 | 1.6 | 4.1 | 7.0 | 33.0 |
| Days ≥ 1" Snow Depth | 23 | 18 | 8 | 2 | 0 | 0 | 0 | 0 | 0 | 1 | 6 | 16 | 74 |

## WESTHOPE *Bottineau County* ELEVATION 1512 ft LAT/LONG 48° 55' N / 101° 1' W

| | JAN | FEB | MAR | APR | MAY | JUN | JUL | AUG | SEP | OCT | NOV | DEC | YEAR |
|---|---|---|---|---|---|---|---|---|---|---|---|---|---|
| Maximum Temp °F | 13.2 | 21.0 | 35.0 | 54.2 | 69.2 | 76.6 | 81.1 | 80.8 | 70.0 | 55.9 | 34.1 | 19.3 | 50.9 |
| Minimum Temp °F | -7.1 | 0.9 | 15.2 | 29.3 | 41.5 | 50.4 | 54.1 | 51.9 | 42.0 | 30.6 | 15.6 | 0.3 | 27.1 |
| Mean Temp °F | 3.0 | 10.9 | 25.2 | 41.8 | 55.4 | 63.5 | 67.6 | 66.4 | 56.0 | 43.3 | 24.9 | 9.8 | 39.0 |
| Days Max Temp ≥ 90 °F | 0 | 0 | 0 | 0 | 1 | 2 | 4 | 5 | 1 | 0 | 0 | 0 | 13 |
| Days Max Temp ≤ 32 °F | 26 | 21 | 11 | 2 | 0 | 0 | 0 | 0 | 0 | 1 | 13 | 25 | 99 |
| Days Min Temp ≤ 32 °F | 31 | 28 | 30 | 19 | 5 | 0 | 0 | 0 | 4 | 17 | 29 | 31 | 194 |
| Days Min Temp ≤ 0 °F | 20 | 14 | 5 | 0 | 0 | 0 | 0 | 0 | 0 | 0 | 4 | 15 | 58 |
| Heating Degree Days | 1920 | 1523 | 1229 | 691 | 316 | 105 | 36 | 67 | 286 | 667 | 1198 | 1707 | 9745 |
| Cooling Degree Days | 0 | 0 | 0 | 2 | 32 | 78 | 129 | 133 | 21 | 0 | 0 | 0 | 395 |
| Total Precipitation (") | 0.41 | 0.29 | 0.58 | 1.29 | 1.83 | 2.83 | 2.60 | 2.11 | 1.86 | 1.08 | 0.45 | 0.47 | 15.80 |
| Days ≥ 0.1" Precip | 1 | 1 | 2 | 3 | 5 | 6 | 5 | 4 | 4 | 2 | 1 | 1 | 35 |
| Total Snowfall (") | 7.1 | 4.6 | 4.9 | 3.5 | 0.4 | 0.0 | 0.0 | 0.0 | 0.0 | 2.3 | 5.0 | 6.9 | 34.7 |
| Days ≥ 1" Snow Depth | 28 | 26 | 21 | 5 | 0 | 0 | 0 | 0 | 0 | 2 | 12 | 26 | 120 |

## WILDROSE *Williams County* ELEVATION 2260 ft LAT/LONG 48° 38' N / 103° 10' W

| | JAN | FEB | MAR | APR | MAY | JUN | JUL | AUG | SEP | OCT | NOV | DEC | YEAR |
|---|---|---|---|---|---|---|---|---|---|---|---|---|---|
| Maximum Temp °F | 15.5 | 22.3 | 35.8 | 52.7 | 66.3 | 74.8 | 80.7 | 79.7 | 67.1 | 54.4 | 34.5 | 21.8 | 50.5 |
| Minimum Temp °F | -4.1 | 2.6 | 15.2 | 28.5 | 40.4 | 49.6 | 53.9 | 50.9 | 40.7 | 29.7 | 16.0 | 2.9 | 27.2 |
| Mean Temp °F | 5.7 | 12.5 | 25.6 | 40.6 | 53.4 | 62.3 | 67.3 | 65.3 | 53.9 | 42.1 | 25.2 | 12.4 | 38.9 |
| Days Max Temp ≥ 90 °F | 0 | 0 | 0 | 0 | 0 | 2 | 5 | 5 | 1 | 0 | 0 | 0 | 13 |
| Days Max Temp ≤ 32 °F | 25 | 20 | 12 | 2 | 0 | 0 | 0 | 0 | 0 | 1 | 13 | 23 | 96 |
| Days Min Temp ≤ 32 °F | 31 | 28 | 30 | 21 | 6 | 0 | 0 | 0 | 4 | 20 | 29 | 31 | 200 |
| Days Min Temp ≤ 0 °F | 18 | 12 | 5 | 0 | 0 | 0 | 0 | 0 | 0 | 0 | 4 | 13 | 52 |
| Heating Degree Days | 1837 | 1479 | 1217 | 725 | 368 | 133 | 43 | 88 | 343 | 704 | 1187 | 1628 | 9752 |
| Cooling Degree Days | 0 | 0 | 0 | 0 | 19 | 64 | 122 | 119 | 18 | 0 | 0 | 0 | 342 |
| Total Precipitation (") | 0.44 | 0.32 | 0.52 | 1.17 | 2.24 | 2.38 | 2.39 | 1.84 | 1.62 | 0.80 | 0.40 | 0.45 | 14.57 |
| Days ≥ 0.1" Precip | 1 | 1 | 2 | 3 | 5 | 6 | 5 | 4 | 4 | 2 | 1 | 1 | 35 |
| Total Snowfall (") | 7.0 | 4.6 | 5.4 | 3.5 | 0.4 | 0.0 | 0.0 | 0.0 | 0.1 | 1.7 | 4.4 | 5.9 | 33.0 |
| Days ≥ 1" Snow Depth | 24 | 19 | 12 | 3 | 0 | 0 | 0 | 0 | 0 | 2 | 9 | 20 | 89 |

## WILLISTON EXP FARM *Williams County* ELEVATION 2113 ft LAT/LONG 48° 8 ' N / 103° 45 ' W

| | JAN | FEB | MAR | APR | MAY | JUN | JUL | AUG | SEP | OCT | NOV | DEC | YEAR |
|---|---|---|---|---|---|---|---|---|---|---|---|---|---|
| Maximum Temp °F | 19.8 | 27.5 | 41.2 | 57.2 | 70.1 | 78.3 | 84.9 | 84.2 | 72.2 | 59.2 | 38.0 | 25.1 | 54.8 |
| Minimum Temp °F | 0.2 | 6.9 | 18.8 | 30.9 | 42.7 | 51.5 | 56.4 | 54.7 | 44.2 | 33.0 | 18.6 | 6.0 | 30.3 |
| Mean Temp °F | 10.0 | 17.3 | 30.0 | 44.1 | 56.4 | 64.9 | 70.7 | 69.5 | 58.2 | 46.2 | 28.3 | 15.6 | 42.6 |
| Days Max Temp ≥ 90 °F | 0 | 0 | 0 | 0 | 1 | 4 | 9 | 10 | 3 | 0 | 0 | 0 | 27 |
| Days Max Temp ≤ 32 °F | 22 | 16 | 8 | 1 | 0 | 0 | 0 | 0 | 0 | 1 | 9 | 20 | 77 |
| Days Min Temp ≤ 32 °F | 31 | 28 | 28 | 18 | 4 | 0 | 0 | 0 | 2 | 14 | 28 | 31 | 184 |
| Days Min Temp ≤ 0 °F | 16 | 10 | 3 | 0 | 0 | 0 | 0 | 0 | 0 | 0 | 3 | 11 | 43 |
| Heating Degree Days | 1703 | 1343 | 1077 | 623 | 285 | 83 | 18 | 39 | 236 | 578 | 1094 | 1528 | 8607 |
| Cooling Degree Days | 0 | 0 | 0 | 2 | 34 | 104 | 207 | 198 | 40 | 1 | 0 | 0 | 586 |
| Total Precipitation (") | 0.44 | 0.25 | 0.55 | 1.25 | 2.28 | 2.76 | 2.04 | 1.64 | 1.60 | 0.87 | 0.42 | 0.45 | 14.55 |
| Days ≥ 0.1" Precip | 1 | 1 | 2 | 3 | 5 | 6 | 5 | 4 | 4 | 2 | 1 | 1 | 35 |
| Total Snowfall (") | 5.3 | 3.7 | 4.8 | 3.3 | 0.7 | 0.0 | 0.0 | 0.0 | 0.2 | 1.6 | 3.9 | 5.6 | 29.1 |
| Days ≥ 1" Snow Depth | 24 | 17 | 11 | 3 | 0 | 0 | 0 | 0 | 0 | 1 | 7 | 18 | 81 |

## WILLISTON SLOULIN AP *Williams County* ELEVATION 1900 ft LAT/LONG 48° 11 ' N / 103° 38 ' W

| | JAN | FEB | MAR | APR | MAY | JUN | JUL | AUG | SEP | OCT | NOV | DEC | YEAR |
|---|---|---|---|---|---|---|---|---|---|---|---|---|---|
| Maximum Temp °F | 19.3 | 26.6 | 40.2 | 56.0 | 68.7 | 77.5 | 84.2 | 82.8 | 70.2 | 57.3 | 37.3 | 24.5 | 53.7 |
| Minimum Temp °F | -2.2 | 5.2 | 18.2 | 30.7 | 42.2 | 51.3 | 56.3 | 54.3 | 43.0 | 31.3 | 16.7 | 3.7 | 29.2 |
| Mean Temp °F | 8.6 | 15.9 | 29.2 | 43.4 | 55.5 | 64.4 | 70.3 | 68.6 | 56.6 | 44.3 | 27.0 | 14.1 | 41.5 |
| Days Max Temp ≥ 90 °F | 0 | 0 | 0 | 0 | 1 | 3 | 8 | 9 | 2 | 0 | 0 | 0 | 23 |
| Days Max Temp ≤ 32 °F | 23 | 17 | 9 | 1 | 0 | 0 | 0 | 0 | 0 | 1 | 10 | 20 | 81 |
| Days Min Temp ≤ 32 °F | 31 | 28 | 29 | 18 | 4 | 0 | 0 | 0 | 3 | 16 | 29 | 31 | 189 |
| Days Min Temp ≤ 0 °F | 17 | 11 | 3 | 0 | 0 | 0 | 0 | 0 | 0 | 0 | 3 | 12 | 46 |
| Heating Degree Days | 1747 | 1381 | 1104 | 643 | 309 | 90 | 19 | 47 | 272 | 635 | 1133 | 1574 | 8954 |
| Cooling Degree Days | 0 | 0 | 0 | 2 | 29 | 91 | 198 | 182 | 25 | 0 | 0 | 0 | 527 |
| Total Precipitation (") | 0.55 | 0.37 | 0.68 | 1.26 | 2.00 | 2.39 | 1.91 | 1.42 | 1.37 | 0.78 | 0.48 | 0.57 | 13.78 |
| Days ≥ 0.1" Precip | 2 | 1 | 2 | 3 | 5 | 5 | 4 | 3 | 3 | 2 | 2 | 2 | 34 |
| Total Snowfall (") | 7.9 | 5.3 | 6.4 | 4.6 | 0.8 | 0.0 | 0.0 | 0.0 | 0.4 | 1.6 | 5.1 | 7.6 | 39.7 |
| Days ≥ 1" Snow Depth | 26 | 19 | 12 | 3 | 0 | 0 | 0 | 0 | 0 | 1 | 9 | 21 | 91 |

## WILLOW CITY *Bottineau County* ELEVATION 1470 ft LAT/LONG 48° 36 ' N / 100° 17 ' W

| | JAN | FEB | MAR | APR | MAY | JUN | JUL | AUG | SEP | OCT | NOV | DEC | YEAR |
|---|---|---|---|---|---|---|---|---|---|---|---|---|---|
| Maximum Temp °F | 12.4 | 19.4 | 33.2 | 51.9 | 67.0 | 75.2 | 80.4 | 79.7 | 67.8 | 54.7 | 34.0 | 18.9 | 49.6 |
| Minimum Temp °F | -9.7 | -3.5 | 11.3 | 27.8 | 40.1 | 49.6 | 53.4 | 50.3 | 39.6 | 28.0 | 13.6 | -2.2 | 24.9 |
| Mean Temp °F | 1.4 | 8.0 | 22.3 | 39.9 | 53.6 | 62.4 | 67.0 | 65.0 | 53.7 | 41.4 | 23.8 | 8.3 | 37.2 |
| Days Max Temp ≥ 90 °F | 0 | 0 | 0 | 0 | 1 | 2 | 4 | 5 | 1 | 0 | 0 | 0 | 13 |
| Days Max Temp ≤ 32 °F | 27 | 22 | 13 | 2 | 0 | 0 | 0 | 0 | 0 | 1 | 13 | 25 | 103 |
| Days Min Temp ≤ 32 °F | 31 | 28 | 30 | 21 | 7 | 0 | 0 | 0 | 6 | 22 | 30 | 31 | 206 |
| Days Min Temp ≤ 0 °F | 22 | 16 | 8 | 0 | 0 | 0 | 0 | 0 | 0 | 0 | 4 | 18 | 68 |
| Heating Degree Days | 1972 | 1607 | 1319 | 748 | 364 | 128 | 48 | 89 | 345 | 726 | 1231 | 1755 | 10332 |
| Cooling Degree Days | 0 | 0 | 0 | 1 | 24 | 65 | 115 | 109 | 13 | 0 | 0 | 0 | 327 |
| Total Precipitation (") | 0.55 | 0.32 | 0.70 | 1.26 | 1.70 | 2.79 | 2.68 | 2.45 | 1.79 | 1.11 | 0.50 | 0.46 | 16.31 |
| Days ≥ 0.1" Precip | 2 | 1 | 2 | 4 | 4 | 6 | 6 | 5 | 4 | 3 | 2 | 2 | 41 |
| Total Snowfall (") | 9.0 | 4.4 | 5.7 | 2.4 | 0.1 | 0.0 | 0.0 | 0.0 | 0.1 | 1.0 | 5.1 | 6.9 | 34.7 |
| Days ≥ 1" Snow Depth | ***24*** | na | ***16*** | 3 | 0 | 0 | 0 | 0 | 0 | 0 | ***6*** | ***19*** | na |

## WILTON *McLean County* ELEVATION 2162 ft LAT/LONG 47° 9 ' N / 100° 48 ' W

| | JAN | FEB | MAR | APR | MAY | JUN | JUL | AUG | SEP | OCT | NOV | DEC | YEAR |
|---|---|---|---|---|---|---|---|---|---|---|---|---|---|
| Maximum Temp °F | ***17.3*** | 22.7 | ***35.9*** | 52.3 | 66.6 | 74.3 | 80.9 | 80.2 | ***69.4*** | 54.6 | 35.4 | 21.6 | 50.9 |
| Minimum Temp °F | ***-1.2*** | 4.2 | ***16.2*** | 30.5 | 42.3 | 51.2 | 55.6 | 53.3 | ***43.6*** | 31.2 | 17.1 | 3.4 | 29.0 |
| Mean Temp °F | ***8.0*** | 13.5 | ***26.3*** | 41.4 | 54.4 | 62.8 | 68.2 | 66.8 | ***56.5*** | 42.9 | 26.3 | 12.5 | 40.0 |
| Days Max Temp ≥ 90 °F | 0 | 0 | 0 | 0 | 0 | 2 | 4 | 5 | 1 | 0 | 0 | 0 | 12 |
| Days Max Temp ≤ 32 °F | ***24*** | ***20*** | ***11*** | 2 | 0 | 0 | 0 | 0 | 0 | 1 | 12 | ***23*** | 93 |
| Days Min Temp ≤ 32 °F | 30 | 28 | ***29*** | ***18*** | 4 | 0 | 0 | 0 | 2 | 17 | 28 | 31 | 187 |
| Days Min Temp ≤ 0 °F | ***17*** | ***12*** | ***4*** | 0 | 0 | 0 | 0 | 0 | 0 | 0 | 3 | 13 | 49 |
| Heating Degree Days | ***1763*** | 1451 | ***1191*** | 704 | 340 | 121 | 33 | 66 | ***279*** | 678 | 1157 | 1623 | 9406 |
| Cooling Degree Days | 0 | 0 | 0 | 2 | 23 | 67 | 132 | 121 | na | 1 | 0 | 0 | na |
| Total Precipitation (") | 0.52 | 0.40 | 0.68 | 1.62 | 2.20 | 3.79 | 2.86 | 2.02 | 1.80 | 1.30 | 0.51 | 0.42 | 18.12 |
| Days ≥ 0.1" Precip | 2 | 1 | 2 | 4 | 4 | 7 | 5 | 4 | 4 | 3 | 2 | 1 | 39 |
| Total Snowfall (") | na | na | na | ***3.5*** | 0.8 | 0.0 | 0.0 | 0.0 | 0.2 | 2.2 | 4.6 | ***5.2*** | na |
| Days ≥ 1" Snow Depth | na | na | na | ***2*** | 0 | 0 | 0 | 0 | 0 | 1 | ***8*** | na | na |

**WISHEK** ***McIntosh County*** ELEVATION 2011 ft LAT/LONG 46° 16 ' N / 99° 34 ' W

| | JAN | FEB | MAR | APR | MAY | JUN | JUL | AUG | SEP | OCT | NOV | DEC | YEAR |
|---|---|---|---|---|---|---|---|---|---|---|---|---|---|
| Maximum Temp °F | 17.7 | 24.0 | 36.6 | 52.9 | 67.1 | 75.4 | 82.3 | 81.0 | 69.8 | 56.6 | 37.2 | 23.5 | 52.0 |
| Minimum Temp °F | -3.5 | 2.9 | 15.2 | 28.5 | 40.2 | 49.7 | 54.4 | 51.6 | 41.1 | 29.7 | 15.8 | 3.0 | 27.4 |
| Mean Temp °F | 7.1 | 13.5 | 25.9 | 40.7 | 53.6 | 62.6 | 68.4 | 66.3 | 55.5 | 43.2 | 26.5 | 13.3 | 39.7 |
| Days Max Temp ≥ 90 °F | 0 | 0 | 0 | 0 | 0 | 2 | 6 | 5 | 1 | 0 | 0 | 0 | 14 |
| Days Max Temp ≤ 32 °F | 25 | 18 | 11 | 1 | 0 | 0 | 0 | 0 | 0 | 1 | 11 | 22 | 89 |
| Days Min Temp ≤ 32 °F | 31 | 28 | 29 | 21 | 7 | 0 | 0 | 0 | 5 | 19 | 29 | 31 | 200 |
| Days Min Temp ≤ 0 °F | 18 | *12* | 5 | 0 | 0 | 0 | 0 | 0 | 0 | 0 | 3 | 13 | 51 |
| Heating Degree Days | 1797 | 1445 | 1205 | 722 | 361 | 122 | 36 | 72 | 305 | 670 | 1149 | 1599 | 9483 |
| Cooling Degree Days | 0 | 0 | 0 | 2 | 19 | 60 | 143 | 119 | 21 | 1 | 0 | 0 | 365 |
| Total Precipitation (") | 0.39 | 0.39 | 0.92 | 1.90 | 2.32 | 3.64 | 2.69 | 2.14 | 1.48 | 1.14 | 0.52 | 0.42 | 17.95 |
| Days ≥ 0.1" Precip | 2 | *1* | *3* | 5 | 6 | 7 | 5 | 4 | 4 | 3 | 2 | 1 | 43 |
| Total Snowfall (") | *6.2* | *7.1* | 6.2 | 4.0 | 0.6 | 0.0 | 0.0 | 0.0 | 0.0 | 1.2 | 7.3 | 5.2 | 37.8 |
| Days ≥ 1" Snow Depth | na | na | na | 2 | 0 | 0 | 0 | 0 | 0 | 0 | 7 | na | na |

## JANUARY MINIMUM TEMPERATURE °F

### LOWEST

| Rank | Place | Value |
|---|---|---|
| 1 | Pembina | -10.5 |
| 2 | Willow City | -9.7 |
| 3 | Belcourt | -9.6 |
| 4 | Upham | -9.5 |
| 5 | Langdon | -9.1 |
| 6 | Petersburg | -8.6 |
| 7 | Bottineau | -8.1 |
| 8 | McHenry | -7.3 |
| 9 | Cavalier | -7.2 |
|  | Towner | -7.2 |
| 11 | Westhope | -7.1 |
| 12 | Hansboro | -6.9 |
| 13 | Leeds | -6.7 |
| 14 | Powers Lake | -6.5 |
| 15 | Rolla | -5.8 |
| 16 | Grand Forks-Intl | -5.7 |
| 17 | McVille | -5.4 |
|  | Mohall | -5.4 |
|  | Tioga | -5.4 |
| 20 | Sharon | -5.2 |
|  | Valley City | -5.2 |
| 22 | Bowbells | -5.1 |
|  | Grafton | -5.1 |
|  | Oakes | -5.1 |
| 25 | Rugby | -5.0 |

### HIGHEST

| Rank | Place | Value |
|---|---|---|
| 1 | Bowman | 3.7 |
| 2 | Amidon | 3.6 |
|  | Dickinson-Muni | 3.6 |
| 4 | Richardton | 3.2 |
| 5 | Fort Yates | 2.9 |
| 6 | Medora | 2.8 |
|  | Trotters | 2.8 |
| 8 | Hettinger | 2.7 |
| 9 | New England | 2.5 |
| 10 | Pretty Rock | 2.1 |
| 11 | Fairfield | 1.9 |
| 12 | Watford City | 0.9 |
| 13 | Mott | 0.8 |
| 14 | Williston-Exp | 0.2 |
| 15 | New Salem | -0.2 |
| 16 | Dunn Center | -0.4 |
|  | Minot-FAA | -0.4 |
|  | Underwood | -0.4 |
| 19 | Center | -0.5 |
| 20 | Carson | -0.6 |
| 21 | Butte | -0.7 |
|  | Ellendale | -0.7 |
| 23 | Dickinson-Exp | -0.9 |
| 24 | Keene | -1.1 |
| 25 | Wilton | -1.2 |

## JULY MAXIMUM TEMPERATURE °F

### HIGHEST

| Rank | Place | Value |
|---|---|---|
| 1 | Medora | 87.4 |
| 2 | Watford City | 87.2 |
| 3 | Linton | 87.0 |
| 4 | Breien | 86.3 |
| 5 | Moffit | 85.7 |
| 6 | Fort Yates | 85.6 |
| 7 | Fullerton | 85.4 |
| 8 | Ellendale | 85.3 |
| 9 | Pretty Rock | 85.0 |
| 10 | Williston-Exp | 84.9 |
| 11 | McLeod | 84.5 |
| 12 | Butte | 84.4 |
|  | Jamestown-Hosp | 84.4 |
|  | Wahpeton | 84.4 |
| 15 | Lisbon | 84.3 |
|  | Oakes | 84.3 |
| 17 | Beulah | 84.2 |
|  | Bismarck | 84.2 |
|  | Williston-Sloulin | 84.2 |
| 20 | Gackle | 84.1 |
| 21 | Forman | 84.0 |
|  | Hettinger | 84.0 |
|  | La Moure | 84.0 |
|  | McClusky | 84.0 |
|  | New England | 84.0 |

### LOWEST

| Rank | Place | Value |
|---|---|---|
| 1 | Rolla | 76.7 |
| 2 | Langdon | 77.9 |
| 3 | Belcourt | 78.7 |
| 4 | Petersburg | 79.1 |
| 5 | Cavalier | 79.5 |
| 6 | Bottineau | 79.7 |
|  | Sharon | 79.7 |
| 8 | Hansboro | 79.9 |
|  | Pembina | 79.9 |
| 10 | Powers Lake | 80.2 |
| 11 | Stanley | 80.3 |
| 12 | Carrington | 80.4 |
|  | Willow City | 80.4 |
| 14 | Devils Lake | 80.5 |
|  | Leeds | 80.5 |
|  | Mohall | 80.5 |
| 17 | Fortuna | 80.6 |
| 18 | Wildrose | 80.7 |
| 19 | McHenry | 80.8 |
| 20 | Bowbells | 80.9 |
|  | Grand Forks-Intl | 80.9 |
|  | Wilton | 80.9 |
| 23 | Minot-Exp | 81.0 |
|  | Tioga | 81.0 |
| 25 | Kenmare | 81.1 |

## ANNUAL PRECIPITATION (")

### HIGHEST

| Rank | Place | Value |
|---|---|---|
| 1 | Wahpeton | 21.69 |
| 2 | Sharon | 21.30 |
| 3 | La Moure | 20.89 |
| 4 | Fullerton | 20.64 |
| 5 | Ellendale | 20.52 |
| 6 | Cooperstown | 20.46 |
| 7 | Fargo | 19.76 |
| 8 | Petersburg | 19.68 |
| 9 | Forman | 19.66 |
|  | Hillsboro | 19.66 |
| 11 | Park River | 19.53 |
| 12 | Stanley | 19.35 |
| 13 | Lisbon | 19.33 |
| 14 | McLeod | 19.25 |
| 15 | Grand Forks-Univ | 19.19 |
| 16 | McHenry | 19.18 |
| 17 | Grand Forks-Intl | 19.07 |
| 18 | McVille | 18.97 |
| 19 | Valley City | 18.93 |
| 20 | Oakes | 18.77 |
| 21 | Minot-FAA | 18.56 |
| 22 | Gackle | 18.47 |
| 23 | Belcourt | 18.46 |
| 24 | Devils Lake | 18.44 |
| 25 | Colgate | 18.43 |

### LOWEST

| Rank | Place | Value |
|---|---|---|
| 1 | Fort Yates | 13.40 |
| 2 | Williston-Sloulin | 13.78 |
| 3 | Fortuna | 14.34 |
| 4 | Tioga | 14.50 |
| 5 | Williston-Exp | 14.55 |
| 6 | Wildrose | 14.57 |
| 7 | Trotters | 14.58 |
| 8 | Crosby | 14.81 |
| 9 | Fairfield | 15.13 |
| 10 | Medora | 15.20 |
| 11 | Watford City | 15.28 |
| 12 | Garrison | 15.29 |
| 13 | Keene | 15.31 |
| 14 | Drake | 15.32 |
| 15 | Amidon | 15.40 |
| 16 | Bowman | 15.47 |
| 17 | Hettinger | 15.76 |
| 18 | Westhope | 15.80 |
| 19 | Dickinson-Muni | 15.85 |
| 20 | Powers Lake | 15.87 |
| 21 | Linton | 15.96 |
| 22 | Kenmare | 15.97 |
| 23 | Bismarck | 16.01 |
| 24 | Moffit | 16.05 |
| 25 | Mandan | 16.23 |

## ANNUAL SNOWFALL (")

### HIGHEST

| Rank | Place | Value |
|---|---|---|
| 1 | Bowman | 51.0 |
| 2 | Minot-Exp | 49.4 |
| 3 | Pretty Rock | 49.3 |
| 4 | Bismarck | 47.8 |
| 5 | Minot-FAA | 46.3 |
| 6 | Upham | 45.4 |
| 7 | Richardton | 45.3 |
| 8 | Belcourt | 45.0 |
| 9 | Stanley | 44.5 |
| 10 | Carrington | 43.9 |
| 11 | New England | 43.6 |
| 12 | Fargo | 43.0 |
| 13 | McHenry | 42.3 |
| 14 | New Salem | 42.2 |
| 15 | Napoleon | 40.8 |
| 16 | McClusky | 40.7 |
| 17 | Max | 40.1 |
| 18 | Williston-Sloulin | 39.7 |
| 19 | La Moure | 39.6 |
| 20 | Dunn Center | 39.5 |
| 21 | Bottineau | 39.1 |
|  | Grand Forks-Intl | 39.1 |
| 23 | Crosby | 39.0 |
| 24 | Foxholm | 38.9 |
| 25 | Tuttle | 38.8 |

### LOWEST

| Rank | Place | Value |
|---|---|---|
| 1 | Fort Yates | 26.0 |
| 2 | Wahpeton | 27.7 |
| 3 | Ashley | 28.0 |
| 4 | McLeod | 28.1 |
| 5 | Moffit | 28.4 |
| 6 | Drake | 29.0 |
| 7 | Williston-Exp | 29.1 |
| 8 | Beulah | 29.5 |
| 9 | Colgate | 30.0 |
| 10 | Center | 30.9 |
| 11 | Hebron | 31.3 |
| 12 | Forman | 31.6 |
| 13 | Butte | 31.7 |
| 14 | Fairfield | 32.8 |
| 15 | Pettibone | 32.9 |
| 16 | Medora | 33.0 |
|  | Trotters | 33.0 |
|  | Watford City | 33.0 |
|  | Wildrose | 33.0 |
| 20 | Carson | 33.1 |
| 21 | Hettinger | 33.6 |
| 22 | Mott | 33.9 |
| 23 | Grand Forks-Univ | 34.4 |
|  | Valley City | 34.4 |
| 25 | Westhope | 34.7 |

# OHIO

PHYSICAL FEATURES AND GENERAL CLIMATE. The climate of Ohio is remarkably varied. Less than one-half of its area is occupied by typical plains, while most of eastern and much of southern Ohio is decidedly hilly. Topography ranges in elevation from 430 feet above sea level at the junction of the Great Miami and Ohio Rivers up to 1,550 feet on a summit near Bellefontaine. In addition to this high point there are innumerable other hills which rise above 1,400 feet (mean sea level). These are located mainly along the dividing line between the Ohio River and Lake Erie drainage basins. Large areas in the State have elevations above 1,000 feet. An extensive area in northwestern Ohio is occupied by a flat lake plain -- once the bottom of glacial Lake Maumee which was much larger than the present Lake Erie. The greater part of eastern Ohio is within the Allegheny Plateau, an unglaciated area consisting of picturesque hills many of which rise above 1,300 feet and between which there are many winding rivers and streams.

The Ohio River, which forms the southern and southeastern boundaries of Ohio, and its tributaries drain the greater portion of the State. A number of streams drain northward into Lake Erie. Although this area comprises nearly a third of the State, the divide between the two drainages is only 20 to 40 miles from the lake shore for a distance of more than 100 miles until it dips south of the arrowhead-shaped Maumee Basin. The largest streams in this region are the Maumee, Sandusky, and Cuyahoga Rivers. Principal tributaries flowing southward into the Ohio River include the Muskingum in the east, the Scioto in the central section, and the Great Miami in the west. A small portion in the west-central region drains westward into the Wabash River basin of Indiana.

Located west of the Appalachian Mountains, Ohio has a climate essentially continental in nature, characterized by moderate extremes of heat and cold, and wetness and dryness. Summers are moderately warm and humid, with occasional days when temperatures exceed 100°F.; winters are reasonably cold, with an average of about 2 days of subzero weather; and autumns are predominately cool, dry, and invigorating. Spring is the wettest season and vegetation grows luxuriantly.

PRECIPITATION. Annual precipitation is slightly in excess of the national average and is well distributed, though with peaks in early spring and summer. In spite of the relatively small range in latitude and the compact shape of Ohio, rainfall varies considerably in amount and seasonal distribution. This is accounted for not only by the presence of Lake Erie on the north, but also by its topography and proximity to rain producing storm paths. Annual precipitation averages about 38 inches, being most generous in spring (about 4 inches in April) and least in the fall (about 2.5 inches in October). Greatest amounts are measured in the southwest where Wilmington has an average of 44.36 inches; and the lake shore is driest, Gilbralter Island having a normal of only 29.06 inches.

The southern half of the State is visited more frequently by productive rainstorms which, together with the general roughness of terrain, accounts for the larger total precipitation. The lifting of moist air masses over the hills tends to increase the yield of rainfall, especially in winter and spring. There is a marked tendency during the cold season for northeastern counties to receive snowfall amounts substantially in excess of those measured elsewhere. Northerly winds have a long fetch across Lake Huron and the widest part of Lake Erie, thus picking up moisture and heat from the lakes. This moisture is then forced to condense as the air is lifted abruptly over the divide a short distance from the lake. Average snowfall ranges from 60 inches in parts of Lake and adjoining counties down to 16 inches or less along the Ohio River.

TEMPERATURE. The normal annual temperature for the State ranges from 49.6° F. at Hiram in Portage County up to 56.9° F. at Portsmouth on the Ohio River. Variations over the State are due mainly to differences in latitude and topography, but the immediate lake shore area experiences a moderating effect due to its proximity to a large body of water. Widest temperature ranges are found generally among the eastern hills. In an average year, 90° heat may be expected about 20 times in summer with 100°F. or more once or twice. Readings of zero or lower are generally to be expected on 2 to 4 days each winter, and these are just as likely to occur in the south as the north. However, 1 winter out of 6 or 8 will pass without experiencing zero readings anywhere in the State.

OTHER CLIMATIC ELEMENTS. The growing season, as defined by the period 32° F. or higher, ranges widely because of latitude and proximity to Lake Erie. The longest is about 200 days on the lake shore and the shortest is in the northeastern valleys within the Ohio River drainage. Dates of the average last freezing temperature in spring range from April 15 to May 18 and the mean first freeze date in fall varies from September 30 to November 6, the latter being on the western lake shore.

Damaging windstorms are mostly associated with heavy thunderstorms or line squalls. Three or four tornadoes may be expected to strike in Ohio each year. Most tornadoes, however, are of limited effect having paths that are short and narrow.

Most floods in Ohio are caused by unusual precipitation. The storms causing floods may bring rainfall of unusual intensity or of unusual duration and extent. Some floods may be caused by a series of ordinary storms which follow one another in rapid succession. Others may result from rain falling at relatively high temperatures on snow-covered areas. At times, though infrequent, flood conditions are caused or aggravated by ice gorges, especially in the tributary streams. Severe thunderstorms frequently cause local flash flooding. General flooding occurs most frequently during January to March and rarely occurs during August to October.

# COUNTY INDEX

***Morgan County***
MCCONNELSVILLE LOCK7

***Muskingum County***
PHILO 3 SW
ZANESVILLE MUNI AP

***Ottawa County***
PUT-IN-BAY

***Paulding County***
PAULDING

***Perry County***
NEW LEXINGTON 2 NW

***Pickaway County***
CIRCLEVILLE

***Pike County***
WAVERLY

***Portage County***
HIRAM

***Preble County***
EATON

***Putnam County***
PANDORA

***Richland County***
MANSFIELD 5 W
MANSFIELD LAHM AP

***Sandusky County***
FREMONT

***Scioto County***
PORTSMOUTH SCIOTOVIL

***Seneca County***
TIFFIN

***Summit County***
AKRON-CANTON REG AP

***Trumbull County***
MINERAL RIDGE W WKS
WARREN 3 S
YOUNGSTOWN MUNI AP

***Tuscarawas County***
NEW PHILADELPHIA

***Union County***
IRWIN
MARYSVILLE

***Van Wert County***
VAN WERT

***Warren County***
FRANKLIN

***Washington County***
MARIETTA WWTP

***Wayne County***
WOOSTER EXP STN

***Williams County***
MONTPELIER

***Wood County***
BOWLING GREEN WWTP
HOYTVILLE 2 NE

***Wyandot County***
UPPER SANDUSKY

## ELEVATION INDEX

| FEET | STATION NAME |
|---|---|
| 502 | CHILO MELDAHL L&D |
| 502 | CINCINNATI LUNKEN AP |
| 540 | PORTSMOUTH SCIOTOVIL |
| 581 | GALLIPOLIS |
| 581 | MARIETTA WWTP |
| 581 | PUT-IN-BAY |
| 600 | PAINESVILLE 4 NW |
| 600 | WAVERLY |
| 630 | SANDUSKY |
| 640 | FREMONT |
| 669 | MCCONNELSVILLE LOCK7 |
| 669 | TOLEDO EXPRESS AP |
| 679 | BOWLING GREEN WWTP |
| 679 | CIRCLEVILLE |
| 689 | ASHTABULA |
| 689 | FRANKLIN |
| 702 | HOYTVILLE 2 NE |
| 712 | DEFIANCE |
| 722 | NORWALK WWTP |
| 732 | ELYRIA 3 E |
| 732 | PAULDING |
| 751 | COSHOCTON WPC PLANT |
| 751 | DAYTON MCD |
| 751 | WAUSEON WATER PLANT |
| 761 | COLUMBUS VLY CROSSNG |
| 761 | PANDORA |
| 761 | TIFFIN |
| 761 | TOM JENKINS DAM-BURR |
| 771 | FINDLAY WPCC |
| 781 | VAN WERT |
| 804 | CLEVELAND HOPKINS AP |
| 804 | FINDLAY AIRPORT |
| 820 | OBERLIN |
| 833 | COLUMBUS INTL AP |
| 840 | LANCASTER 2 NW |
| 840 | NEWARK WATER WORKS |
| 850 | DELAWARE |
| 850 | UPPER SANDUSKY |
| 860 | CELINA 3 NE |
| 860 | LIMA WWTP |
| 860 | MONTPELIER |
| 860 | WESTERVILLE |
| 879 | CAMBRIDGE |
| 879 | CARPENTER 2 S |
| 879 | RIPLEY EXP FARM |
| 881 | ZANESVILLE MUNI AP |
| 889 | MINERAL RIDGE W WKS |
| 889 | NEW LEXINGTON 2 NW |
| 902 | NEW PHILADELPHIA |
| 902 | WARREN 3 S |
| 922 | MARION 2 N |
| 922 | XENIA 6 SSE |
| 961 | KENTON |
| 961 | WASHINGTON COURT HSE |
| 981 | MILLERSBURG |
| 981 | SPRINGFIELD NEW WW |
| 991 | STEUBENVILLE |
| 1001 | BUCYRUS |
| 1001 | MARYSVILLE |
| 1010 | DANVILLE 2 W |
| 1010 | IRWIN |
| 1017 | DAYTON INTL ARPT |
| 1020 | LONDON |
| 1020 | PHILO 3 SW |
| 1030 | EATON |
| 1030 | WILMINGTON 3 N |

| FEET | STATION NAME |
|---|---|
| 1030 | WOOSTER EXP STN |
| 1040 | DORSET |
| 1040 | GREENVILLE WATER PLT |
| 1050 | FREDERICKTOWN 4 S |
| 1050 | URBANA WWTP |
| 1102 | HILLSBORO |
| 1132 | BARNESVILLE |
| 1152 | MILLPORT 2 NW |
| 1161 | CANFIELD 1 S |
| 1178 | YOUNGSTOWN MUNI AP |
| 1180 | CHIPPEWA LAKE |
| 1191 | BELLEFONTAINE |
| 1191 | COSHOCTON ARG RESEAR |
| 1204 | CENTERBURG 2 SE |
| 1214 | AKRON-CANTON REG AP |
| 1240 | CADIZ |
| 1260 | CHARDON |
| 1260 | HIRAM |
| 1270 | ASHLAND 2 SW |
| 1289 | MANSFIELD 5 W |
| 1302 | MANSFIELD LAHM AP |

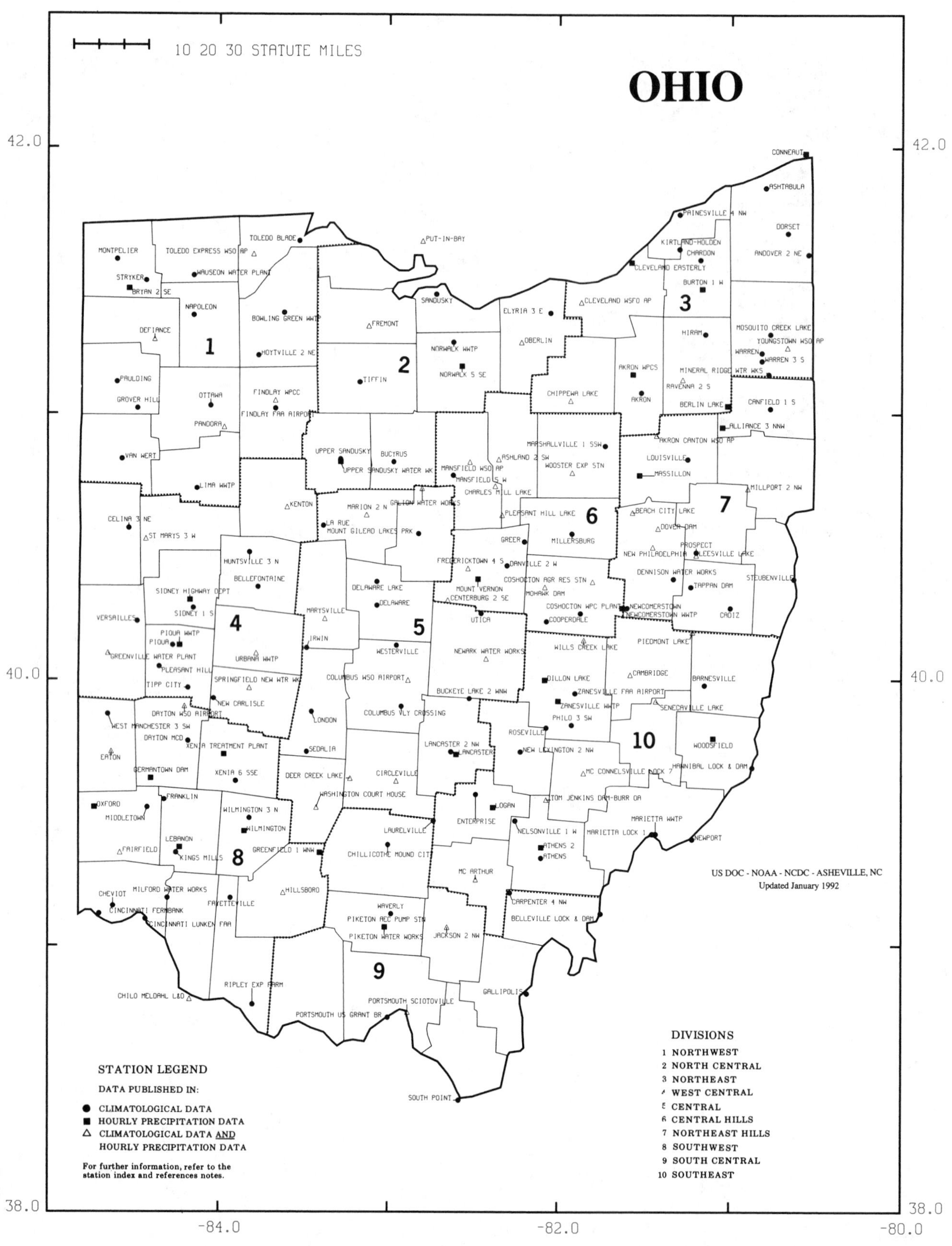

10 20 30 STATUTE MILES
OHIO
42.0
40.0
38.0
-84.0
-82.0
-80.0
CONNEAUT
ASHTABULA
PAINESVILLE 4 NW
DORSET
KIRTLAND-HOLDEN
CHARDON
ANDOVER 2 NE
CLEVELAND EASTERLY
BURTON 1 W
TOLEDO BLADE
PUT-IN-BAY
MONTPELIER
TOLEDO EXPRESS WSO AP
WAUSEON WATER PLANT
STRYKER
BRYAN 2 SE
NAPOLEON
SANDUSKY
ELYRIA 3 E
CLEVELAND WSFO AP
BOWLING GREEN WWTP
FREMONT
DEFIANCE
OBERLIN
HIRAM
MOSQUITO CREEK LAKE
YOUNGSTOWN WSO AP
WARREN
WARREN 3 S
HOYTVILLE 2 NE
NORWALK WWTP
NORWALK 5 SE
AKRON WPCS
MINERAL RIDGE WTR WKS
RAVENNA 2 S
PAULDING
TIFFIN
GROVER HILL
OTTAWA
FINDLAY WPCC
FINDLAY FAA AIRPORT
CHIPPEWA LAKE
AKRON
BERLIN LAKE
CANFIELD 1 S
PANDORA
ALLIANCE 3 NNW
MARSHALLVILLE 1 SSW
AKRON CANTON WSO AP
UPPER SANDUSKY
BUCYRUS
UPPER SANDUSKY WATER WK
ASHLAND 2 SW
LOUISVILLE
VAN WERT
MANSFIELD WSO AP
MANSFIELD 5 W
WOOSTER EXP STN
MASSILLON
LIMA WWTP
CHARLES MILL LAKE
MILLPORT 2 NW
KENTON
MARION 2 N
GALION WATER WORKS
PLEASANT HILL LAKE
BEACH CITY LAKE
CELINA 3 NE
LA RUE
MOUNT GILEAD LAKES PRK
DOVER DAM
ST MARYS 3 W
GREER
MILLERSBURG
PROSPECT
NEW PHILADELPHIA
LEESVILLE LAKE
HUNTSVILLE 3 N
FREDERICKTOWN 4 S
DANVILLE 2 W
BELLEFONTAINE
COSHOCTON AGR RES STN
DENNISON WATER WORKS
STEUBENVILLE
SIDNEY HIGHWAY DEPT
DELAWARE LAKE
MOUNT VERNON
MOHAWK DAM
TAPPAN DAM
CENTERBURG 2 SE
SIDNEY 1 S
DELAWARE
COSHOCTON WPC PLANT
NEWCOMERSTOWN
VERSAILLES
MARYSVILLE
UTICA
NEWCOMERSTOWN WWTP
CADIZ
COOPERDALE
PIQUA WWTP
IRWIN
PIQUA
URBANA WWTP
WESTERVILLE
NEWARK WATER WORKS
WILLS CREEK LAKE
PIEDMONT LAKE
GREENVILLE WATER PLANT
PLEASANT HILL
CAMBRIDGE
SPRINGFIELD NEW WTR WK
COLUMBUS WSO AIRPORT
DILLON LAKE
BARNESVILLE
TIPP CITY
BUCKEYE LAKE 2 WNW
ZANESVILLE FAA AIRPORT
NEW CARLISLE
ZANESVILLE WWTP
DAYTON WSO AIRPORT
COLUMBUS VLY CROSSING
SENECAVILLE LAKE
LONDON
PHILO 3 SW
WEST MANCHESTER 3 SW
ROSEVILLE
DAYTON MCD
XENIA TREATMENT PLANT
LANCASTER 2 NW
LANCASTER
NEW LEXINGTON 2 NW
WOODSFIELD
EATON
SEDALIA
GERMANTOWN DAM
XENIA 6 SSE
DEER CREEK LAKE
CIRCLEVILLE
MC CONNELSVILLE LOCK 7
HANNIBAL LOCK & DAM
WASHINGTON COURT HOUSE
FRANKLIN
TOM JENKINS DAM-BURR OA
OXFORD
LOGAN
WILMINGTON 3 N
MIDDLETOWN
ENTERPRISE
MARIETTA WWTP
WILMINGTON
LAURELVILLE
NELSONVILLE 1 W
MARIETTA LOCK 1
LEBANON
NEWPORT
FAIRFIELD
GREENFIELD 1 WNW
ATHENS 2
KINGS MILLS
CHILLICOTHE MOUND CITY
ATHENS
MC ARTHUR
CHEVIOT
MILFORD WATER WORKS
HILLSBORO
CINCINNATI FERNBANK
FAYETTEVILLE
CARPENTER 4 NW
WAVERLY
CINCINNATI LUNKEN FAA
PIKETON REC PUMP STN
BELLEVILLE LOCK & DAM
PIKETON WATER WORKS
JACKSON 2 NW
RIPLEY EXP FARM
GALLIPOLIS
CHILO MELDAHL L&D
PORTSMOUTH SCIOTOVILLE
PORTSMOUTH US GRANT BR
SOUTH POINT
1
2
3
4
5
6
7
8
9
10
US DOC - NOAA - NCDC - ASHEVILLE, NC
Updated January 1992
DIVISIONS
1 NORTHWEST
2 NORTH CENTRAL
3 NORTHEAST
4 WEST CENTRAL
5 CENTRAL
6 CENTRAL HILLS
7 NORTHEAST HILLS
8 SOUTHWEST
9 SOUTH CENTRAL
10 SOUTHEAST
STATION LEGEND
DATA PUBLISHED IN:
● CLIMATOLOGICAL DATA
■ HOURLY PRECIPITATION DATA
△ CLIMATOLOGICAL DATA AND HOURLY PRECIPITATION DATA
For further information, refer to the station index and references notes.

## AKRON-CANTON REG AP *Summit County* ELEVATION 1214 ft LAT/LONG 40° 55 ' N / 81° 26 ' W

| | JAN | FEB | MAR | APR | MAY | JUN | JUL | AUG | SEP | OCT | NOV | DEC | YEAR |
|---|---|---|---|---|---|---|---|---|---|---|---|---|---|
| Maximum Temp °F | 33.3 | 36.2 | 46.9 | 59.3 | 69.6 | 78.6 | 82.4 | 80.7 | 73.4 | 61.7 | 49.5 | 38.5 | 59.2 |
| Minimum Temp °F | 18.0 | 19.6 | 28.4 | 38.2 | 48.2 | 57.3 | 61.8 | 60.3 | 53.6 | 42.5 | 34.1 | 24.5 | 40.5 |
| Mean Temp °F | 25.7 | 27.9 | 37.7 | 48.8 | 58.9 | 68.0 | 72.1 | 70.5 | 63.5 | 52.1 | 41.8 | 31.5 | 49.9 |
| Days Max Temp ≥ 90 °F | 0 | 0 | 0 | 0 | 0 | 2 | 3 | 2 | 0 | 0 | 0 | 0 | 7 |
| Days Max Temp ≤ 32 °F | 14 | 11 | 4 | 0 | 0 | 0 | 0 | 0 | 0 | 0 | 1 | 9 | 39 |
| Days Min Temp ≤ 32 °F | 28 | 24 | 21 | 9 | 1 | 0 | 0 | 0 | 0 | 4 | 14 | 25 | 126 |
| Days Min Temp ≤ 0 °F | 3 | 2 | 0 | 0 | 0 | 0 | 0 | 0 | 0 | 0 | 0 | 1 | 6 |
| Heating Degree Days | 1212 | 1041 | 840 | 487 | 221 | 45 | 7 | 15 | 115 | 398 | 688 | 1031 | 6100 |
| Cooling Degree Days | 0 | 0 | 1 | 9 | 47 | 134 | 256 | 199 | 77 | 5 | 0 | 0 | 728 |
| Total Precipitation (") | 2.38 | 2.25 | 3.12 | 3.15 | 3.81 | 3.25 | 3.96 | 3.40 | 3.39 | 2.43 | 3.23 | 2.99 | 37.36 |
| Days ≥ 0.1" Precip | 6 | 6 | 8 | 8 | 8 | 7 | 7 | 6 | 6 | 6 | 7 | 8 | 83 |
| Total Snowfall (") | 12.4 | 10.1 | 8.6 | 2.6 | 0.2 | 0.0 | 0.0 | 0.0 | 0.0 | 0.6 | 3.7 | 9.1 | 47.3 |
| Days ≥ 1" Snow Depth | 16 | 13 | 6 | 1 | 0 | 0 | 0 | 0 | 0 | 0 | 2 | 9 | 47 |

## ASHLAND 2 SW *Ashland County* ELEVATION 1270 ft LAT/LONG 40° 50 ' N / 82° 21 ' W

| | JAN | FEB | MAR | APR | MAY | JUN | JUL | AUG | SEP | OCT | NOV | DEC | YEAR |
|---|---|---|---|---|---|---|---|---|---|---|---|---|---|
| Maximum Temp °F | 31.8 | 35.6 | 46.4 | 59.3 | 70.3 | 78.7 | 83.0 | 81.0 | 74.4 | 62.1 | 49.0 | 36.1 | 59.0 |
| Minimum Temp °F | 15.1 | 17.5 | 26.3 | 36.1 | 46.6 | 55.4 | 59.8 | 57.4 | 51.3 | 40.1 | 31.6 | 21.7 | 38.2 |
| Mean Temp °F | 23.4 | 26.6 | 36.4 | 47.7 | 58.5 | 67.1 | 71.4 | 69.2 | 62.8 | 51.1 | 40.4 | 29.5 | 48.7 |
| Days Max Temp ≥ 90 °F | 0 | 0 | 0 | 0 | 0 | 2 | 4 | 2 | 1 | 0 | 0 | 0 | 9 |
| Days Max Temp ≤ 32 °F | 16 | 11 | 4 | 0 | 0 | 0 | 0 | 0 | 0 | 0 | 2 | 10 | 43 |
| Days Min Temp ≤ 32 °F | 29 | 25 | 23 | 12 | 2 | 0 | 0 | 0 | 1 | 7 | 17 | 27 | 143 |
| Days Min Temp ≤ 0 °F | 5 | 3 | 1 | 0 | 0 | 0 | 0 | 0 | 0 | 0 | 0 | 1 | 10 |
| Heating Degree Days | 1282 | 1078 | 880 | 518 | 235 | 59 | 11 | 25 | 130 | 430 | 733 | 1093 | 6474 |
| Cooling Degree Days | 0 | 0 | 0 | 6 | 35 | 119 | 216 | 151 | 63 | 4 | 0 | 0 | 594 |
| Total Precipitation (") | 2.32 | 2.10 | 2.76 | 3.39 | 4.25 | 3.72 | 4.52 | 3.94 | 3.39 | 2.38 | 3.45 | 2.69 | 38.91 |
| Days ≥ 0.1" Precip | 6 | 5 | 7 | 8 | 9 | 7 | 7 | 7 | 6 | 6 | 8 | 7 | 83 |
| Total Snowfall (") | 10.1 | 8.4 | 5.7 | 1.7 | 0.1 | 0.0 | 0.0 | 0.0 | 0.0 | 0.1 | 2.0 | *7.4* | 35.5 |
| Days ≥ 1" Snow Depth | 17 | 13 | 5 | 1 | 0 | 0 | 0 | 0 | 0 | 0 | 1 | 10 | 47 |

## ASHTABULA *Ashtabula County* ELEVATION 689 ft LAT/LONG 41° 51 ' N / 80° 48 ' W

| | JAN | FEB | MAR | APR | MAY | JUN | JUL | AUG | SEP | OCT | NOV | DEC | YEAR |
|---|---|---|---|---|---|---|---|---|---|---|---|---|---|
| Maximum Temp °F | 32.1 | 33.7 | 43.0 | 55.5 | 66.7 | 76.3 | 80.6 | 79.1 | 72.7 | 61.0 | 49.5 | 37.9 | 57.3 |
| Minimum Temp °F | 17.7 | 17.7 | 26.8 | 36.9 | 46.9 | 56.3 | 61.3 | 60.0 | 53.3 | 42.8 | 35.0 | 25.3 | 40.0 |
| Mean Temp °F | 25.0 | 25.7 | 34.9 | 46.2 | 56.8 | 66.4 | 71.0 | 69.6 | 63.0 | 51.9 | 42.2 | 31.7 | 48.7 |
| Days Max Temp ≥ 90 °F | 0 | 0 | 0 | 0 | 0 | 1 | 2 | 1 | 0 | 0 | 0 | 0 | 4 |
| Days Max Temp ≤ 32 °F | 16 | 14 | 7 | 1 | 0 | 0 | 0 | 0 | 0 | 0 | 1 | 9 | 48 |
| Days Min Temp ≤ 32 °F | 28 | 26 | 23 | 11 | 1 | 0 | 0 | 0 | 0 | 3 | 12 | 25 | 129 |
| Days Min Temp ≤ 0 °F | 3 | 2 | 0 | 0 | 0 | 0 | 0 | 0 | 0 | 0 | 0 | 0 | 5 |
| Heating Degree Days | 1234 | 1102 | 926 | 567 | 279 | 71 | 14 | 23 | 123 | 405 | 676 | 1027 | 6447 |
| Cooling Degree Days | 0 | 0 | 1 | 8 | *34* | 111 | 209 | 171 | 67 | 5 | 0 | 0 | 606 |
| Total Precipitation (") | 2.10 | 1.83 | 2.28 | 3.22 | 3.12 | 4.02 | 3.93 | 4.06 | 4.03 | 3.66 | 3.97 | 3.12 | 39.34 |
| Days ≥ 0.1" Precip | 6 | 5 | 7 | 7 | 8 | 7 | 6 | 7 | 8 | 9 | 10 | 9 | 89 |
| Total Snowfall (") | na | *11.6* | na | *0.8* | *0.0* | *0.0* | *0.0* | *0.0* | *0.0* | *0.1* | *3.2* | *11.5* | na |
| Days ≥ 1" Snow Depth | na | *15* | na | *1* | *0* | *0* | *0* | *0* | *0* | *0* | *1* | *9* | na |

## BARNESVILLE *Belmont County* ELEVATION 1132 ft LAT/LONG 39° 58 ' N / 81° 10 ' W

| | JAN | FEB | MAR | APR | MAY | JUN | JUL | AUG | SEP | OCT | NOV | DEC | YEAR |
|---|---|---|---|---|---|---|---|---|---|---|---|---|---|
| Maximum Temp °F | 34.3 | 37.7 | 48.9 | 60.5 | 70.3 | 78.5 | 82.1 | 80.7 | 74.4 | 62.6 | 51.1 | 39.6 | 60.1 |
| Minimum Temp °F | 15.6 | 17.7 | 27.1 | 36.5 | 46.1 | 54.9 | 59.6 | 57.8 | 50.8 | 38.6 | 31.6 | 22.7 | 38.3 |
| Mean Temp °F | 25.0 | 27.7 | 38.0 | 48.5 | 58.2 | 66.7 | 70.9 | 69.2 | 62.7 | 50.6 | 41.4 | 31.2 | 49.2 |
| Days Max Temp ≥ 90 °F | 0 | 0 | 0 | 0 | 0 | 1 | 2 | 1 | 0 | 0 | 0 | 0 | 4 |
| Days Max Temp ≤ 32 °F | 14 | 10 | 3 | 0 | 0 | 0 | 0 | 0 | 0 | 0 | 1 | 9 | 37 |
| Days Min Temp ≤ 32 °F | 29 | 25 | 22 | 11 | 2 | 0 | 0 | 0 | 0 | 8 | 17 | 25 | 139 |
| Days Min Temp ≤ 0 °F | 4 | 3 | 0 | 0 | 0 | 0 | 0 | 0 | 0 | 0 | 0 | 1 | 8 |
| Heating Degree Days | 1234 | 1045 | 829 | 495 | 235 | 57 | 12 | 24 | 128 | 443 | 702 | 1041 | 6245 |
| Cooling Degree Days | 0 | 0 | 1 | 9 | 37 | 125 | 237 | 180 | 68 | 5 | 1 | 0 | 663 |
| Total Precipitation (") | 2.74 | 2.60 | 3.52 | 3.82 | 4.37 | 4.40 | 4.90 | 3.84 | 3.44 | 2.95 | 3.66 | 3.20 | 43.44 |
| Days ≥ 0.1" Precip | 7 | 6 | 9 | 9 | 9 | 8 | 8 | 7 | 6 | 6 | 8 | 8 | 91 |
| Total Snowfall (") | 11.4 | 7.9 | 5.1 | 1.2 | 0.0 | 0.0 | 0.0 | 0.0 | 0.0 | 0.0 | 1.5 | 5.5 | 32.6 |
| Days ≥ 1" Snow Depth | 15 | 12 | 5 | 0 | 0 | 0 | 0 | 0 | 0 | 0 | 2 | 7 | 41 |

## BELLEFONTAINE *Logan County* ELEVATION 1191 ft LAT/LONG 40° 22 ' N / 83° 46 ' W

| | JAN | FEB | MAR | APR | MAY | JUN | JUL | AUG | SEP | OCT | NOV | DEC | YEAR |
|---|---|---|---|---|---|---|---|---|---|---|---|---|---|
| Maximum Temp °F | 32.2 | 36.6 | 48.2 | 60.9 | 71.6 | 79.9 | 83.3 | 81.5 | 75.6 | 63.4 | 49.7 | 38.1 | 60.1 |
| Minimum Temp °F | 16.8 | 19.4 | 29.0 | 39.5 | 49.7 | 58.4 | 62.4 | 60.5 | 53.9 | 42.7 | 33.4 | 23.4 | 40.8 |
| Mean Temp °F | 24.5 | 28.0 | 38.6 | 50.2 | 60.7 | 69.2 | 72.9 | 71.0 | 64.7 | 53.1 | 41.6 | 30.7 | 50.4 |
| Days Max Temp ≥ 90 °F | 0 | 0 | 0 | 0 | 0 | 2 | 4 | 2 | 1 | 0 | 0 | 0 | 9 |
| Days Max Temp ≤ 32 °F | 15 | 11 | 3 | 0 | 0 | 0 | 0 | 0 | 0 | 0 | 2 | 9 | 40 |
| Days Min Temp ≤ 32 °F | 28 | 25 | 20 | 8 | 1 | 0 | 0 | 0 | 0 | 5 | 15 | 25 | 127 |
| Days Min Temp ≤ 0 °F | 4 | 2 | 0 | 0 | 0 | 0 | 0 | 0 | 0 | 0 | 0 | 1 | 7 |
| Heating Degree Days | 1248 | 1038 | 810 | 445 | 182 | 34 | 5 | 13 | 96 | 374 | 696 | 1054 | 5995 |
| Cooling Degree Days | 0 | 0 | 1 | 8 | 58 | 164 | 271 | 218 | 96 | 11 | 0 | 0 | 827 |
| Total Precipitation (") | 2.11 | 1.91 | 2.73 | 3.37 | 3.93 | 3.79 | 3.76 | 3.29 | 2.87 | 2.40 | 3.21 | 2.95 | 36.32 |
| Days ≥ 0.1" Precip | 6 | 5 | 7 | 8 | 8 | 7 | 6 | 6 | 5 | 6 | 7 | 7 | 78 |
| Total Snowfall (") | *7.4* | 4.5 | 2.6 | 0.4 | 0.0 | 0.0 | 0.0 | 0.0 | 0.0 | 0.2 | 1.4 | *4.3* | 20.8 |
| Days ≥ 1" Snow Depth | 12 | 7 | 2 | 0 | 0 | 0 | 0 | 0 | 0 | 0 | 1 | 4 | 26 |

## BOWLING GREEN WWTP *Wood County* ELEVATION 679 ft LAT/LONG 41° 23 ' N / 83° 38 ' W

| | JAN | FEB | MAR | APR | MAY | JUN | JUL | AUG | SEP | OCT | NOV | DEC | YEAR |
|---|---|---|---|---|---|---|---|---|---|---|---|---|---|
| Maximum Temp °F | 31.5 | 35.0 | 46.7 | 60.2 | 71.9 | 81.4 | 84.6 | 82.3 | 75.9 | 63.6 | 49.6 | 37.3 | 60.0 |
| Minimum Temp °F | 16.3 | 18.8 | 27.8 | 37.7 | 48.2 | 58.1 | 61.9 | 59.4 | 52.7 | 41.6 | 32.7 | 22.8 | 39.8 |
| Mean Temp °F | 23.9 | 26.9 | 37.3 | 49.0 | 60.1 | 69.8 | 73.3 | 70.9 | 64.3 | 52.6 | 41.2 | 30.1 | 50.0 |
| Days Max Temp ≥ 90 °F | 0 | 0 | 0 | 0 | 1 | 4 | 6 | 3 | 1 | 0 | 0 | 0 | 15 |
| Days Max Temp ≤ 32 °F | 16 | 12 | 4 | 0 | 0 | 0 | 0 | 0 | 0 | 0 | 1 | 9 | 42 |
| Days Min Temp ≤ 32 °F | 29 | 26 | 23 | 10 | 1 | 0 | 0 | 0 | 0 | 5 | 16 | 26 | 136 |
| Days Min Temp ≤ 0 °F | 4 | 2 | 0 | 0 | 0 | 0 | 0 | 0 | 0 | 0 | 0 | 1 | 7 |
| Heating Degree Days | 1267 | 1069 | 853 | 482 | 199 | 32 | 4 | 15 | 104 | 387 | 709 | 1076 | 6197 |
| Cooling Degree Days | 0 | 0 | 0 | 9 | 51 | 169 | 268 | 194 | 81 | 8 | 0 | 0 | 780 |
| Total Precipitation (") | 1.66 | 1.48 | 2.34 | 3.11 | 3.46 | 3.57 | 3.93 | 3.00 | 2.75 | 2.49 | 2.95 | 2.63 | 33.37 |
| Days ≥ 0.1" Precip | 4 | 4 | 6 | 7 | 7 | 7 | 7 | 6 | 6 | 6 | 7 | 7 | 74 |
| Total Snowfall (") | *6.8* | *6.2* | *3.1* | *0.8* | *0.0* | *0.0* | *0.0* | *0.0* | *0.0* | *0.0* | *0.6* | *5.0* | 22.5 |
| Days ≥ 1" Snow Depth | *14* | *13* | *4* | *1* | *0* | *0* | *0* | *0* | *0* | *0* | *0* | *6* | 38 |

## BUCYRUS *Crawford County* ELEVATION 1001 ft LAT/LONG 40° 48 ' N / 82° 58 ' W

| | JAN | FEB | MAR | APR | MAY | JUN | JUL | AUG | SEP | OCT | NOV | DEC | YEAR |
|---|---|---|---|---|---|---|---|---|---|---|---|---|---|
| Maximum Temp °F | 31.6 | 34.5 | 45.7 | 58.8 | 70.1 | 79.2 | 82.9 | 80.9 | 74.2 | 61.8 | 48.8 | 36.9 | 58.8 |
| Minimum Temp °F | 15.6 | 17.3 | 26.6 | 36.8 | 46.9 | 56.4 | 60.8 | 58.4 | 51.5 | 40.2 | 32.0 | 22.4 | 38.7 |
| Mean Temp °F | 23.6 | 25.9 | 36.2 | 47.8 | 58.5 | 67.8 | 71.9 | 69.7 | 62.9 | 51.0 | 40.5 | 29.7 | 48.8 |
| Days Max Temp ≥ 90 °F | 0 | 0 | 0 | 0 | 0 | 2 | 4 | 2 | 0 | 0 | 0 | 0 | 8 |
| Days Max Temp ≤ 32 °F | 16 | 12 | 4 | 0 | 0 | 0 | 0 | 0 | 0 | 0 | 2 | 10 | 44 |
| Days Min Temp ≤ 32 °F | 29 | 25 | 23 | 11 | 1 | 0 | 0 | 0 | 0 | 6 | 17 | 26 | 138 |
| Days Min Temp ≤ 0 °F | 4 | 3 | 0 | 0 | 0 | 0 | 0 | 0 | 0 | 0 | 0 | 1 | 8 |
| Heating Degree Days | 1275 | 1097 | 887 | 516 | 237 | 51 | 9 | 23 | 130 | 433 | 729 | 1087 | 6474 |
| Cooling Degree Days | 0 | 0 | 1 | 7 | 43 | 140 | 246 | 183 | 73 | 6 | 0 | 0 | 699 |
| Total Precipitation (") | 2.13 | 1.84 | 2.66 | 3.37 | 4.01 | 4.06 | 4.60 | 3.79 | 3.31 | 2.31 | 3.27 | 2.74 | 38.09 |
| Days ≥ 0.1" Precip | 6 | 5 | 7 | 8 | 7 | 7 | 8 | 6 | 6 | 6 | 8 | 6 | 80 |
| Total Snowfall (") | 7.9 | 7.5 | 4.8 | 0.9 | 0.1 | 0.0 | 0.0 | 0.0 | 0.0 | 0.1 | 1.6 | 4.8 | 27.7 |
| Days ≥ 1" Snow Depth | 16 | 12 | 5 | 0 | 0 | 0 | 0 | 0 | 0 | 0 | 1 | 7 | 41 |

## CADIZ *Harrison County* ELEVATION 1240 ft LAT/LONG 40° 16 ' N / 81° 0 ' W

| | JAN | FEB | MAR | APR | MAY | JUN | JUL | AUG | SEP | OCT | NOV | DEC | YEAR |
|---|---|---|---|---|---|---|---|---|---|---|---|---|---|
| Maximum Temp °F | 34.9 | 39.0 | 50.4 | 62.3 | 71.6 | 79.4 | 82.9 | 81.6 | 75.6 | 64.1 | 51.4 | 40.4 | 61.1 |
| Minimum Temp °F | 18.8 | 20.9 | 29.7 | 39.6 | 49.5 | 57.9 | 62.5 | 60.9 | 54.4 | 42.7 | 34.0 | 24.8 | 41.3 |
| Mean Temp °F | 26.9 | 30.0 | 40.1 | 50.9 | 60.6 | 68.7 | 72.7 | 71.3 | 65.0 | 53.4 | 42.8 | 32.6 | 51.3 |
| Days Max Temp ≥ 90 °F | 0 | 0 | 0 | 0 | 0 | 1 | 3 | 2 | 1 | 0 | 0 | 0 | 7 |
| Days Max Temp ≤ 32 °F | 13 | 9 | 3 | 0 | 0 | 0 | 0 | 0 | 0 | 0 | 1 | 8 | 34 |
| Days Min Temp ≤ 32 °F | 27 | 24 | 20 | 8 | 1 | 0 | 0 | 0 | 0 | 4 | 14 | 25 | 123 |
| Days Min Temp ≤ 0 °F | 3 | 1 | 0 | 0 | 0 | 0 | 0 | 0 | 0 | 0 | 0 | 1 | 5 |
| Heating Degree Days | 1175 | 983 | 767 | 427 | 181 | 35 | 5 | 11 | 88 | 361 | 661 | 998 | 5692 |
| Cooling Degree Days | 0 | 0 | 1 | 14 | 51 | 145 | 264 | 209 | 94 | 10 | 1 | 0 | 789 |
| Total Precipitation (") | 2.55 | 2.29 | 3.18 | 3.34 | 4.12 | 3.87 | 4.50 | 4.10 | 3.09 | 2.65 | 3.24 | 2.95 | 39.88 |
| Days ≥ 0.1" Precip | 7 | 6 | 8 | 8 | 9 | 7 | 8 | 7 | 6 | 6 | 8 | 7 | 87 |
| Total Snowfall (") | *11.7* | na | *4.7* | *1.4* | *0.0* | *0.0* | *0.0* | *0.0* | *0.0* | *0.1* | na | na | na |
| Days ≥ 1" Snow Depth | na | na | na | *0* | *0* | *0* | *0* | *0* | *0* | *0* | na | na | na |

## CAMBRIDGE *Guernsey County* ELEVATION 879 ft LAT/LONG 40° 2 ' N / 81° 35 ' W

| | JAN | FEB | MAR | APR | MAY | JUN | JUL | AUG | SEP | OCT | NOV | DEC | YEAR |
|---|---|---|---|---|---|---|---|---|---|---|---|---|---|
| Maximum Temp °F | 36.9 | 41.1 | 52.9 | 64.9 | 74.2 | 81.9 | 85.0 | 83.5 | 77.3 | 65.7 | 53.3 | 42.1 | 63.2 |
| Minimum Temp °F | 19.3 | 21.3 | 30.1 | 39.1 | 48.1 | 56.8 | 61.6 | 60.1 | 53.5 | 41.3 | 33.9 | 25.3 | 40.9 |
| Mean Temp °F | 28.1 | 31.2 | 41.5 | 52.1 | 61.2 | 69.3 | 73.3 | 71.8 | 65.4 | 53.5 | 43.6 | 33.7 | 52.1 |
| Days Max Temp ≥ 90 °F | 0 | 0 | 0 | 0 | 0 | 3 | 6 | 3 | 1 | 0 | 0 | 0 | 13 |
| Days Max Temp ≤ 32 °F | 11 | 7 | 1 | 0 | 0 | 0 | 0 | 0 | 0 | 0 | 1 | 6 | 26 |
| Days Min Temp ≤ 32 °F | 27 | 23 | 19 | 8 | 2 | 0 | 0 | 0 | 0 | 6 | 14 | 23 | 122 |
| Days Min Temp ≤ 0 °F | 3 | 2 | 0 | 0 | 0 | 0 | 0 | 0 | 0 | 0 | 0 | 1 | 6 |
| Heating Degree Days | 1137 | 947 | 722 | 392 | 163 | 28 | 4 | 9 | 83 | 358 | 636 | 962 | 5441 |
| Cooling Degree Days | 0 | 0 | 1 | 13 | 56 | 172 | 300 | 242 | 113 | 11 | 1 | 0 | 909 |
| Total Precipitation (") | 2.45 | 2.34 | 3.17 | 3.39 | 3.89 | 3.51 | 4.51 | 3.73 | 2.98 | 2.61 | 3.34 | 2.86 | 38.78 |
| Days ≥ 0.1" Precip | 7 | 6 | 8 | 7 | 8 | 7 | 8 | 7 | 6 | 6 | 7 | 7 | 84 |
| Total Snowfall (") | 8.1 | 4.9 | 3.1 | 0.7 | 0.0 | 0.0 | 0.0 | 0.0 | 0.0 | 0.0 | 0.9 | 3.1 | 20.8 |
| Days ≥ 1" Snow Depth | 11 | 7 | 3 | 0 | 0 | 0 | 0 | 0 | 0 | 0 | 1 | 4 | 26 |

## CANFIELD 1 S *Mahoning County* ELEVATION 1161 ft LAT/LONG 41° 0 ' N / 80° 45 ' W

| | JAN | FEB | MAR | APR | MAY | JUN | JUL | AUG | SEP | OCT | NOV | DEC | YEAR |
|---|---|---|---|---|---|---|---|---|---|---|---|---|---|
| Maximum Temp °F | 34.1 | 36.7 | 48.5 | 60.7 | 71.2 | 79.4 | 83.1 | 81.6 | 74.7 | 62.9 | 50.2 | 38.8 | 60.2 |
| Minimum Temp °F | 15.9 | 16.6 | 26.1 | 35.0 | 44.6 | 53.4 | 57.6 | 55.8 | 49.5 | 38.4 | 31.5 | 22.0 | 37.2 |
| Mean Temp °F | 25.0 | 26.7 | 37.3 | 47.9 | 57.9 | 66.4 | 70.4 | 68.8 | 62.1 | 50.7 | 40.9 | 30.4 | 48.7 |
| Days Max Temp ≥ 90 °F | 0 | 0 | 0 | 0 | 0 | 1 | 4 | 2 | 0 | 0 | 0 | 0 | 7 |
| Days Max Temp ≤ 32 °F | 14 | 10 | 3 | 0 | 0 | 0 | 0 | 0 | 0 | 0 | 1 | 9 | 37 |
| Days Min Temp ≤ 32 °F | 28 | 25 | 23 | 13 | 4 | 0 | 0 | 0 | 1 | 9 | 17 | 26 | 146 |
| Days Min Temp ≤ 0 °F | 5 | 3 | 1 | 0 | 0 | 0 | 0 | 0 | 0 | 0 | 0 | 2 | 11 |
| Heating Degree Days | 1233 | 1072 | 851 | 511 | 243 | 65 | 16 | 30 | 141 | 441 | 718 | 1065 | 6386 |
| Cooling Degree Days | 0 | 0 | 1 | 5 | 30 | 107 | 204 | 152 | 57 | 3 | 0 | 0 | 559 |
| Total Precipitation (") | 1.89 | 1.72 | 2.82 | 2.91 | 3.69 | 3.79 | 4.31 | 3.50 | 3.63 | 2.68 | 3.12 | 2.58 | 36.64 |
| Days ≥ 0.1" Precip | 5 | 5 | 7 | 7 | 9 | 8 | 7 | 7 | 7 | 6 | 8 | 6 | 82 |
| Total Snowfall (") | na | na | ***4.0*** | 0.7 | 0.0 | 0.0 | 0.0 | 0.0 | 0.0 | 0.3 | ***1.4*** | na | na |
| Days ≥ 1" Snow Depth | na | na | na | 1 | 0 | 0 | 0 | 0 | 0 | 0 | 1 | na | na |

## CARPENTER 2 S *Meigs County* ELEVATION 879 ft LAT/LONG 39° 12 ' N / 82° 17 ' W

| | JAN | FEB | MAR | APR | MAY | JUN | JUL | AUG | SEP | OCT | NOV | DEC | YEAR |
|---|---|---|---|---|---|---|---|---|---|---|---|---|---|
| Maximum Temp °F | 38.3 | 42.8 | 55.0 | 65.9 | 74.6 | 82.3 | 85.1 | 83.7 | 77.6 | 66.8 | 55.0 | 44.2 | 64.3 |
| Minimum Temp °F | 19.0 | 21.3 | 30.8 | 39.9 | 48.8 | 56.9 | 61.1 | ***59.6*** | 52.7 | 40.2 | 33.2 | 25.7 | 40.8 |
| Mean Temp °F | 28.7 | 31.8 | 42.9 | 52.9 | 61.7 | 69.6 | 73.1 | ***71.7*** | 65.0 | 53.6 | 44.1 | 35.0 | 52.5 |
| Days Max Temp ≥ 90 °F | 0 | 0 | 0 | 0 | 0 | 3 | 6 | 4 | 1 | 0 | 0 | 0 | 14 |
| Days Max Temp ≤ 32 °F | 10 | 6 | 1 | 0 | 0 | 0 | 0 | 0 | 0 | 0 | 0 | 5 | 22 |
| Days Min Temp ≤ 32 °F | 27 | 23 | 18 | 8 | 2 | 0 | 0 | 0 | 0 | 8 | 15 | 23 | 124 |
| Days Min Temp ≤ 0 °F | 3 | 2 | 0 | 0 | 0 | 0 | 0 | 0 | 0 | 0 | 0 | 0 | 5 |
| Heating Degree Days | 1119 | 930 | 681 | 368 | 153 | 24 | 5 | ***8*** | 90 | 360 | 623 | 922 | 5283 |
| Cooling Degree Days | ***0*** | 0 | 3 | 15 | 59 | ***172*** | ***289*** | 233 | 102 | 13 | 3 | 0 | 889 |
| Total Precipitation (") | 2.49 | 2.48 | 3.30 | 3.61 | 4.31 | 3.59 | 4.54 | 3.96 | 3.45 | 2.69 | 3.02 | 3.01 | 40.45 |
| Days ≥ 0.1" Precip | 6 | 6 | 7 | 8 | 9 | ***6*** | 8 | 7 | 6 | 6 | 7 | 7 | 83 |
| Total Snowfall (") | ***7.1*** | 5.1 | 2.8 | 0.8 | 0.0 | 0.0 | 0.0 | 0.0 | 0.0 | 0.2 | 0.8 | 2.6 | 19.4 |
| Days ≥ 1" Snow Depth | na | ***5*** | 1 | 0 | 0 | 0 | 0 | 0 | 0 | 0 | 0 | ***2*** | na |

## CELINA 3 NE *Mercer County* ELEVATION 860 ft LAT/LONG 40° 34 ' N / 84° 32 ' W

| | JAN | FEB | MAR | APR | MAY | JUN | JUL | AUG | SEP | OCT | NOV | DEC | YEAR |
|---|---|---|---|---|---|---|---|---|---|---|---|---|---|
| Maximum Temp °F | 32.4 | 36.5 | 48.8 | 62.0 | 73.0 | 81.6 | 84.7 | 82.5 | 76.7 | 64.4 | 50.0 | 38.1 | 60.9 |
| Minimum Temp °F | 17.3 | 19.7 | 29.8 | 40.0 | 50.2 | 59.5 | 63.4 | 60.8 | 54.5 | 43.4 | 34.3 | 24.1 | 41.4 |
| Mean Temp °F | 24.9 | 28.2 | 39.3 | 51.0 | 61.6 | 70.6 | 74.0 | 71.7 | 65.6 | 53.9 | 42.2 | 31.1 | 51.2 |
| Days Max Temp ≥ 90 °F | 0 | 0 | 0 | 0 | 0 | 4 | 6 | 3 | 1 | 0 | 0 | 0 | 14 |
| Days Max Temp ≤ 32 °F | 15 | 11 | 3 | 0 | 0 | 0 | 0 | 0 | 0 | 0 | 1 | 8 | 38 |
| Days Min Temp ≤ 32 °F | 28 | 24 | 20 | 7 | 1 | 0 | 0 | 0 | 0 | 4 | 14 | 25 | 123 |
| Days Min Temp ≤ 0 °F | 4 | 3 | 0 | 0 | 0 | 0 | 0 | 0 | 0 | 0 | 0 | 1 | 8 |
| Heating Degree Days | 1237 | 1034 | 790 | 421 | 164 | 24 | 3 | 11 | 86 | 350 | 677 | 1043 | 5840 |
| Cooling Degree Days | 0 | 0 | 1 | 11 | 71 | 201 | 304 | 232 | 110 | 14 | 0 | 0 | 944 |
| Total Precipitation (") | 2.08 | 1.97 | 2.85 | 3.48 | 3.60 | 3.57 | 4.10 | 3.48 | 2.98 | 2.35 | 3.19 | 2.83 | 36.48 |
| Days ≥ 0.1" Precip | 6 | 5 | 7 | 8 | 8 | 7 | 7 | 6 | 6 | 6 | 7 | 7 | 80 |
| Total Snowfall (") | 10.0 | 8.5 | 5.5 | 1.1 | 0.0 | 0.0 | 0.0 | 0.0 | 0.0 | 0.3 | 2.4 | 6.5 | 34.3 |
| Days ≥ 1" Snow Depth | 13 | 10 | 3 | 0 | 0 | 0 | 0 | 0 | 0 | 0 | 1 | 6 | 33 |

## CENTERBURG 2 SE *Knox County* ELEVATION 1204 ft LAT/LONG 40° 18 ' N / 82° 39 ' W

| | JAN | FEB | MAR | APR | MAY | JUN | JUL | AUG | SEP | OCT | NOV | DEC | YEAR |
|---|---|---|---|---|---|---|---|---|---|---|---|---|---|
| Maximum Temp °F | 31.8 | 36.1 | 47.2 | 59.6 | 70.0 | 78.5 | 82.1 | 80.7 | 74.2 | 62.2 | 49.3 | 37.5 | 59.1 |
| Minimum Temp °F | 15.1 | 18.0 | 27.5 | 37.5 | 48.0 | 56.7 | 60.8 | 59.0 | 52.2 | 40.3 | 32.1 | 22.4 | 39.1 |
| Mean Temp °F | 23.6 | 27.1 | 37.3 | 48.6 | 59.1 | 67.6 | 71.5 | 69.9 | 63.3 | 51.2 | 40.7 | 29.8 | 49.1 |
| Days Max Temp ≥ 90 °F | 0 | 0 | 0 | 0 | 0 | 1 | 3 | 2 | 0 | 0 | 0 | 0 | 6 |
| Days Max Temp ≤ 32 °F | 15 | 11 | 4 | 0 | 0 | 0 | 0 | 0 | 0 | 0 | 2 | 10 | 42 |
| Days Min Temp ≤ 32 °F | 29 | 25 | 22 | 10 | 1 | 0 | 0 | 0 | 0 | 7 | 17 | 26 | 137 |
| Days Min Temp ≤ 0 °F | 4 | 3 | 0 | 0 | 0 | 0 | 0 | 0 | 0 | 0 | 0 | 1 | 8 |
| Heating Degree Days | 1275 | 1063 | 853 | 493 | 218 | 48 | 9 | 20 | 120 | 426 | 724 | 1084 | 6333 |
| Cooling Degree Days | 0 | 0 | 1 | 9 | 43 | 136 | 242 | 191 | 71 | *6* | 0 | 0 | 699 |
| Total Precipitation (") | 2.36 | 2.06 | 2.88 | 3.50 | 3.93 | 4.20 | 4.59 | 3.74 | 3.35 | 2.73 | 3.80 | 2.99 | 40.13 |
| Days ≥ 0.1" Precip | 6 | 5 | 8 | 8 | 7 | 7 | 8 | 7 | 6 | 6 | 8 | 7 | 83 |
| Total Snowfall (") | na | na | na | *0.3* | 0.0 | 0.0 | 0.0 | 0.0 | 0.0 | 0.1 | na | na | na |
| Days ≥ 1" Snow Depth | na | na | *3* | 1 | 0 | 0 | 0 | 0 | 0 | 0 | *1* | na | na |

## CHARDON *Geauga County* ELEVATION 1260 ft LAT/LONG 41° 35 ' N / 81° 12 ' W

| | JAN | FEB | MAR | APR | MAY | JUN | JUL | AUG | SEP | OCT | NOV | DEC | YEAR |
|---|---|---|---|---|---|---|---|---|---|---|---|---|---|
| Maximum Temp °F | 31.2 | 34.0 | 44.1 | 56.6 | 67.6 | 76.4 | 80.1 | 78.7 | 71.9 | 60.3 | 48.1 | 36.6 | 57.1 |
| Minimum Temp °F | 14.1 | 14.8 | 24.3 | 34.4 | 44.2 | 53.3 | 58.0 | 56.5 | 49.9 | 39.4 | 32.0 | 21.8 | 36.9 |
| Mean Temp °F | 22.7 | 24.4 | 34.2 | 45.5 | 55.9 | 64.8 | 69.1 | 67.6 | 60.9 | 49.9 | 40.1 | 29.2 | 47.0 |
| Days Max Temp ≥ 90 °F | 0 | 0 | 0 | 0 | 0 | 1 | 2 | 1 | 0 | 0 | 0 | 0 | 4 |
| Days Max Temp ≤ 32 °F | 17 | 13 | 6 | 1 | 0 | 0 | 0 | 0 | 0 | 0 | 2 | 11 | 50 |
| Days Min Temp ≤ 32 °F | 29 | 26 | 24 | 14 | 3 | 0 | 0 | 0 | 1 | 6 | 17 | 27 | 147 |
| Days Min Temp ≤ 0 °F | 5 | 4 | 1 | 0 | 0 | 0 | 0 | 0 | 0 | 0 | 0 | 1 | 11 |
| Heating Degree Days | 1306 | 1141 | 947 | 582 | 297 | 91 | 28 | 39 | 168 | 466 | 741 | 1103 | 6909 |
| Cooling Degree Days | 0 | 0 | 0 | 5 | 23 | 87 | 170 | 124 | 50 | 3 | 0 | 0 | 462 |
| Total Precipitation (") | 3.09 | 2.57 | 3.38 | 3.62 | 4.13 | 4.43 | 4.03 | 4.47 | 4.26 | 3.85 | 4.30 | 4.14 | 46.27 |
| Days ≥ 0.1" Precip | 9 | 8 | 9 | 9 | 9 | 8 | 8 | 8 | 9 | 9 | 11 | 11 | 108 |
| Total Snowfall (") | 27.8 | 20.6 | 15.3 | 3.3 | 0.0 | 0.0 | 0.0 | 0.0 | 0.0 | 0.9 | 8.9 | 24.5 | 101.3 |
| Days ≥ 1" Snow Depth | 24 | 21 | 13 | 1 | 0 | 0 | 0 | 0 | 0 | 0 | 5 | 18 | 82 |

## CHILO MELDAHL L&D *Clermont County* ELEVATION 502 ft LAT/LONG 38° 47 ' N / 84° 8 ' W

| | JAN | FEB | MAR | APR | MAY | JUN | JUL | AUG | SEP | OCT | NOV | DEC | YEAR |
|---|---|---|---|---|---|---|---|---|---|---|---|---|---|
| Maximum Temp °F | 38.3 | 42.5 | 52.9 | 65.3 | 73.9 | 82.5 | 86.3 | 85.1 | 78.9 | 67.2 | 54.8 | 43.4 | 64.3 |
| Minimum Temp °F | 19.7 | 22.1 | 31.1 | 40.6 | 49.9 | 59.2 | 64.2 | 62.9 | 56.2 | 43.8 | 35.2 | 26.1 | 42.6 |
| Mean Temp °F | 29.1 | 32.5 | 42.0 | 53.0 | 62.0 | 70.9 | 75.4 | 74.0 | 67.6 | 55.6 | 45.0 | 34.8 | 53.5 |
| Days Max Temp ≥ 90 °F | 0 | 0 | 0 | 0 | 0 | 4 | 9 | 7 | 2 | 0 | 0 | 0 | 22 |
| Days Max Temp ≤ 32 °F | 10 | 6 | 1 | 0 | 0 | 0 | 0 | 0 | 0 | 0 | 0 | 5 | 22 |
| Days Min Temp ≤ 32 °F | 27 | 23 | 18 | 6 | 0 | 0 | 0 | 0 | 0 | 2 | 12 | 23 | 111 |
| Days Min Temp ≤ 0 °F | 2 | 1 | 0 | 0 | 0 | 0 | 0 | 0 | 0 | 0 | 0 | 1 | 4 |
| Heating Degree Days | 1106 | 913 | 707 | 363 | 146 | 19 | 1 | 3 | 55 | 300 | 594 | 930 | 5137 |
| Cooling Degree Days | 0 | 0 | 0 | *12* | 56 | 209 | 349 | 306 | 144 | 18 | 0 | 0 | 1094 |
| Total Precipitation (") | 2.82 | 2.88 | 3.99 | 3.92 | 4.23 | 3.67 | 4.19 | 4.13 | 3.49 | 2.91 | 3.49 | 3.31 | 43.03 |
| Days ≥ 0.1" Precip | 6 | 5 | 8 | 8 | 7 | 6 | 7 | 6 | 5 | 5 | 7 | 7 | 77 |
| Total Snowfall (") | na | na | *1.3* | 0.0 | 0.0 | 0.0 | 0.0 | 0.0 | 0.0 | 0.0 | *0.2* | *1.7* | na |
| Days ≥ 1" Snow Depth | *6* | *6* | *1* | 0 | 0 | 0 | 0 | 0 | 0 | 0 | 0 | 1 | 14 |

## CHIPPEWA LAKE *Medina County* ELEVATION 1180 ft LAT/LONG 41° 3 ' N / 81° 56 ' W

| | JAN | FEB | MAR | APR | MAY | JUN | JUL | AUG | SEP | OCT | NOV | DEC | YEAR |
|---|---|---|---|---|---|---|---|---|---|---|---|---|---|
| Maximum Temp °F | 32.7 | 36.1 | 47.4 | 60.3 | 70.9 | 79.4 | 83.1 | 81.5 | 74.7 | 62.8 | 50.0 | 37.9 | 59.7 |
| Minimum Temp °F | 16.4 | 17.6 | 26.9 | 36.3 | 46.1 | 55.1 | 59.4 | 57.7 | 51.3 | 40.7 | 32.7 | 22.8 | 38.6 |
| Mean Temp °F | 24.6 | 26.9 | 37.2 | 48.3 | 58.5 | 67.3 | 71.3 | 69.6 | 63.0 | 51.8 | 41.4 | 30.4 | 49.2 |
| Days Max Temp ≥ 90 °F | 0 | 0 | 0 | 0 | 0 | 2 | 4 | 2 | 1 | 0 | 0 | 0 | 9 |
| Days Max Temp ≤ 32 °F | 15 | 11 | 4 | 0 | 0 | 0 | 0 | 0 | 0 | 0 | 1 | 10 | 41 |
| Days Min Temp ≤ 32 °F | 28 | 25 | 22 | 12 | 2 | 0 | 0 | 0 | 0 | 6 | 16 | 26 | 137 |
| Days Min Temp ≤ 0 °F | 4 | 3 | 0 | 0 | 0 | 0 | 0 | 0 | 0 | 0 | 0 | 1 | 8 |
| Heating Degree Days | 1246 | 1070 | 856 | 500 | 230 | 52 | 12 | 21 | 123 | 409 | 702 | 1067 | 6288 |
| Cooling Degree Days | 0 | 0 | 1 | 7 | 38 | 129 | 237 | 185 | 73 | 6 | 0 | 0 | 676 |
| Total Precipitation (") | 2.12 | 2.06 | 2.97 | 3.23 | 3.79 | 3.61 | 3.96 | 3.32 | 3.51 | 2.37 | 3.55 | 3.01 | 37.50 |
| Days ≥ 0.1" Precip | 7 | 6 | 8 | 8 | 8 | 7 | 7 | 6 | 6 | 7 | 8 | 8 | 86 |
| Total Snowfall (") | 10.6 | 9.2 | 7.5 | 2.2 | 0.1 | 0.0 | 0.0 | 0.0 | 0.0 | 0.2 | 3.3 | 8.1 | 41.2 |
| Days ≥ 1" Snow Depth | 16 | 14 | 6 | 1 | 0 | 0 | 0 | 0 | 0 | 0 | 2 | 10 | 49 |

## CINCINNATI LUNKEN AP *Hamilton County* ELEVATION 502 ft LAT/LONG 39° 6' N / 84° 26' W

| | JAN | FEB | MAR | APR | MAY | JUN | JUL | AUG | SEP | OCT | NOV | DEC | YEAR |
|---|---|---|---|---|---|---|---|---|---|---|---|---|---|
| Maximum Temp °F | 38.5 | 43.1 | 54.2 | 65.7 | 74.9 | 83.2 | 86.7 | 84.9 | 78.4 | 67.0 | 54.5 | 43.7 | 64.6 |
| Minimum Temp °F | 22.0 | 24.8 | 34.2 | 43.5 | 53.0 | 61.9 | 66.6 | 64.5 | 57.5 | 44.7 | 36.4 | 27.8 | 44.7 |
| Mean Temp °F | 30.3 | 34.0 | 44.2 | 54.6 | 64.0 | 72.5 | 76.7 | 74.7 | 68.0 | 55.9 | 45.5 | 35.8 | 54.7 |
| Days Max Temp ≥ 90 °F | 0 | 0 | 0 | 0 | 1 | 5 | 10 | 7 | 2 | 0 | 0 | 0 | 25 |
| Days Max Temp ≤ 32 °F | 10 | 6 | 1 | 0 | 0 | 0 | 0 | 0 | 0 | 0 | 0 | 5 | 22 |
| Days Min Temp ≤ 32 °F | 25 | 21 | 15 | 3 | 0 | 0 | 0 | 0 | 0 | 3 | 12 | 21 | 100 |
| Days Min Temp ≤ 0 °F | 1 | 0 | 0 | 0 | 0 | 0 | 0 | 0 | 0 | 0 | 0 | 0 | 1 |
| Heating Degree Days | 1068 | 870 | 640 | 320 | 113 | 11 | 0 | 2 | 52 | 295 | 580 | 899 | 4850 |
| Cooling Degree Days | 0 | 0 | 2 | 16 | 90 | 244 | 386 | 321 | 149 | 19 | 2 | 0 | 1229 |
| Total Precipitation (") | 2.63 | 2.46 | 3.69 | 3.84 | 4.20 | 3.50 | 4.10 | 3.84 | 3.22 | 2.75 | 3.48 | 3.06 | 40.77 |
| Days ≥ 0.1" Precip | 6 | 5 | 8 | 8 | 8 | 7 | 7 | 7 | 6 | 5 | 7 | 7 | 81 |
| Total Snowfall (") | 4.2 | 3.7 | 2.9 | 0.2 | 0.0 | 0.0 | 0.0 | 0.0 | 0.0 | 0.1 | 0.6 | 1.6 | 13.3 |
| Days ≥ 1" Snow Depth | 7 | 6 | 2 | 0 | 0 | 0 | 0 | 0 | 0 | 0 | 0 | 2 | 17 |

## CIRCLEVILLE *Pickaway County* ELEVATION 679 ft LAT/LONG 39° 36' N / 82° 57' W

| | JAN | FEB | MAR | APR | MAY | JUN | JUL | AUG | SEP | OCT | NOV | DEC | YEAR |
|---|---|---|---|---|---|---|---|---|---|---|---|---|---|
| Maximum Temp °F | 36.9 | 40.9 | 52.8 | 64.6 | 74.2 | 82.3 | 85.6 | 84.1 | 78.2 | 66.6 | 53.7 | 42.1 | 63.5 |
| Minimum Temp °F | 20.6 | 22.3 | 31.5 | 40.8 | 50.5 | 59.3 | 63.5 | 61.2 | 54.5 | 42.9 | 35.0 | 26.5 | 42.4 |
| Mean Temp °F | 28.8 | 31.6 | 42.2 | 52.7 | 62.4 | 70.9 | 74.6 | 72.7 | 66.4 | 54.7 | 44.3 | 34.3 | 53.0 |
| Days Max Temp ≥ 90 °F | 0 | 0 | 0 | 0 | 1 | 4 | 7 | 5 | 1 | 0 | 0 | 0 | 18 |
| Days Max Temp ≤ 32 °F | 11 | 7 | 1 | 0 | 0 | 0 | 0 | 0 | 0 | 0 | 1 | 6 | 26 |
| Days Min Temp ≤ 32 °F | 26 | 22 | 18 | 7 | 1 | 0 | 0 | 0 | 0 | 5 | 13 | 22 | 114 |
| Days Min Temp ≤ 0 °F | 2 | 2 | 0 | 0 | 0 | 0 | 0 | 0 | 0 | 0 | 0 | 1 | 5 |
| Heating Degree Days | 1115 | 935 | 703 | 374 | 145 | 20 | 2 | 6 | 70 | 328 | 614 | 944 | 5256 |
| Cooling Degree Days | 0 | 0 | 2 | 13 | 65 | 193 | 312 | 251 | 113 | 18 | 2 | 0 | 969 |
| Total Precipitation (") | 2.20 | 2.06 | 2.85 | 3.58 | 4.65 | 3.46 | 4.00 | 3.98 | 3.34 | 2.55 | 3.29 | 2.64 | 38.60 |
| Days ≥ 0.1" Precip | 6 | 5 | 7 | 8 | 8 | 7 | 7 | 6 | 5 | 6 | 7 | 7 | 79 |
| Total Snowfall (") | *5.3* | 4.4 | 1.8 | 0.4 | 0.0 | 0.0 | 0.0 | 0.0 | 0.0 | 0.1 | 0.8 | 1.9 | 14.7 |
| Days ≥ 1" Snow Depth | 9 | 6 | 2 | 0 | 0 | 0 | 0 | 0 | 0 | 0 | 0 | *4* | 21 |

## CLEVELAND HOPKINS AP *Cuyahoga County* ELEVATION 804 ft LAT/LONG 41° 24' N / 81° 51' W

| | JAN | FEB | MAR | APR | MAY | JUN | JUL | AUG | SEP | OCT | NOV | DEC | YEAR |
|---|---|---|---|---|---|---|---|---|---|---|---|---|---|
| Maximum Temp °F | 32.7 | 35.4 | 46.0 | 58.1 | 68.9 | 77.9 | 82.2 | 80.3 | 73.5 | 61.8 | 50.0 | 38.4 | 58.8 |
| Minimum Temp °F | 18.2 | 19.9 | 28.7 | 38.2 | 48.0 | 57.4 | 62.1 | 60.9 | 54.3 | 43.3 | 35.0 | 25.2 | 40.9 |
| Mean Temp °F | 25.5 | 27.7 | 37.4 | 48.2 | 58.5 | 67.7 | 72.2 | 70.6 | 63.9 | 52.6 | 42.5 | 31.9 | 49.9 |
| Days Max Temp ≥ 90 °F | 0 | 0 | 0 | 0 | 0 | 2 | 4 | 2 | 1 | 0 | 0 | 0 | 9 |
| Days Max Temp ≤ 32 °F | 15 | 12 | 5 | 0 | 0 | 0 | 0 | 0 | 0 | 0 | 1 | 9 | 42 |
| Days Min Temp ≤ 32 °F | 28 | 24 | 21 | 9 | 1 | 0 | 0 | 0 | 0 | 3 | 13 | 24 | 123 |
| Days Min Temp ≤ 0 °F | 3 | 2 | 0 | 0 | 0 | 0 | 0 | 0 | 0 | 0 | 0 | 1 | 6 |
| Heating Degree Days | 1219 | 1048 | 850 | 506 | 237 | 54 | 8 | 15 | 110 | 386 | 668 | 1021 | 6122 |
| Cooling Degree Days | 0 | 0 | 1 | 9 | 47 | 136 | 259 | 199 | 84 | 7 | 1 | 0 | 743 |
| Total Precipitation (") | 2.27 | 2.22 | 2.87 | 3.11 | 3.57 | 3.79 | 3.62 | 3.51 | 3.60 | 2.58 | 3.44 | 3.19 | 37.77 |
| Days ≥ 0.1" Precip | 7 | 6 | 8 | 8 | 8 | 7 | 6 | 7 | 7 | 7 | 8 | 8 | 87 |
| Total Snowfall (") | 15.0 | 13.8 | 10.6 | 2.3 | 0.1 | 0.0 | 0.0 | 0.0 | 0.0 | 0.5 | 4.5 | 11.9 | 58.7 |
| Days ≥ 1" Snow Depth | 17 | 15 | 7 | 1 | 0 | 0 | 0 | 0 | 0 | 0 | 2 | 11 | 53 |

## COLUMBUS INTL AP *Franklin County* ELEVATION 833 ft LAT/LONG 39° 59' N / 82° 52' W

| | JAN | FEB | MAR | APR | MAY | JUN | JUL | AUG | SEP | OCT | NOV | DEC | YEAR |
|---|---|---|---|---|---|---|---|---|---|---|---|---|---|
| Maximum Temp °F | 34.5 | 38.5 | 50.5 | 62.4 | 72.5 | 81.0 | 84.4 | 82.7 | 76.0 | 64.1 | 51.4 | 40.2 | 61.5 |
| Minimum Temp °F | 19.3 | 21.7 | 31.1 | 40.4 | 50.3 | 59.1 | 63.8 | 61.9 | 55.0 | 43.0 | 34.7 | 25.8 | 42.2 |
| Mean Temp °F | 27.0 | 30.1 | 40.8 | 51.4 | 61.4 | 70.1 | 74.1 | 72.3 | 65.5 | 53.6 | 43.1 | 33.0 | 51.9 |
| Days Max Temp ≥ 90 °F | 0 | 0 | 0 | 0 | 0 | 4 | 6 | 4 | 1 | 0 | 0 | 0 | 15 |
| Days Max Temp ≤ 32 °F | 13 | 9 | 2 | 0 | 0 | 0 | 0 | 0 | 0 | 0 | 1 | 8 | 33 |
| Days Min Temp ≤ 32 °F | 27 | 23 | 19 | 6 | 1 | 0 | 0 | 0 | 0 | 4 | 13 | 23 | 116 |
| Days Min Temp ≤ 0 °F | 3 | 1 | 0 | 0 | 0 | 0 | 0 | 0 | 0 | 0 | 0 | 1 | 5 |
| Heating Degree Days | 1173 | 979 | 745 | 411 | 168 | 27 | 3 | 9 | 84 | 358 | 652 | 986 | 5595 |
| Cooling Degree Days | 0 | 0 | 1 | 11 | 68 | 187 | 314 | 256 | 112 | 12 | 1 | 0 | 962 |
| Total Precipitation (") | 2.34 | 2.16 | 2.90 | 3.20 | 3.97 | 3.96 | 4.72 | 3.65 | 3.06 | 2.26 | 3.35 | 2.87 | 38.44 |
| Days ≥ 0.1" Precip | 6 | 6 | 7 | 7 | 8 | 7 | 8 | 7 | 6 | 6 | 7 | 7 | 82 |
| Total Snowfall (") | 9.3 | 6.4 | 4.4 | 1.1 | 0.0 | 0.0 | 0.0 | 0.0 | 0.0 | 0.2 | 1.9 | 4.6 | 27.9 |
| Days ≥ 1" Snow Depth | 12 | 8 | 3 | 0 | 0 | 0 | 0 | 0 | 0 | 0 | 1 | 4 | 28 |

## COLUMBUS VLY CROSSNG *Franklin County* ELEVATION 761 ft LAT/LONG 39° 56 ' N / 82° 57 ' W

| | JAN | FEB | MAR | APR | MAY | JUN | JUL | AUG | SEP | OCT | NOV | DEC | YEAR |
|---|---|---|---|---|---|---|---|---|---|---|---|---|---|
| Maximum Temp °F | 35.9 | 40.3 | 52.2 | 64.3 | 74.0 | 82.2 | 85.3 | 83.4 | 77.5 | 66.1 | 53.1 | 41.4 | 63.0 |
| Minimum Temp °F | 19.6 | 22.2 | 31.5 | 40.8 | 50.5 | 59.5 | 63.7 | 61.5 | 54.8 | 43.0 | 34.6 | 25.6 | 42.3 |
| Mean Temp °F | 27.8 | 31.3 | 41.9 | 52.6 | 62.3 | 70.9 | 74.5 | 72.5 | 66.2 | 54.6 | 43.9 | 33.5 | 52.7 |
| Days Max Temp ≥ 90 °F | 0 | 0 | 0 | 0 | 1 | 4 | 7 | 4 | 1 | 0 | 0 | 0 | 17 |
| Days Max Temp ≤ 32 °F | 12 | 8 | 2 | 0 | 0 | 0 | 0 | 0 | 0 | 0 | 0 | 6 | 28 |
| Days Min Temp ≤ 32 °F | 27 | 23 | 18 | 6 | 0 | 0 | 0 | 0 | 0 | 5 | 14 | 24 | 117 |
| Days Min Temp ≤ 0 °F | 3 | 1 | 0 | 0 | 0 | 0 | 0 | 0 | 0 | 0 | 0 | 1 | 5 |
| Heating Degree Days | 1148 | 946 | 711 | 376 | 143 | 18 | 1 | 5 | 72 | 330 | 627 | 969 | 5346 |
| Cooling Degree Days | 0 | 0 | 1 | 10 | 65 | 198 | 314 | 246 | 114 | 13 | 1 | 0 | 962 |
| Total Precipitation (") | 2.24 | 1.88 | 2.92 | 3.62 | 4.36 | 3.89 | 4.31 | 4.20 | 3.09 | 2.43 | 3.50 | 2.79 | 39.23 |
| Days ≥ 0.1" Precip | 6 | 5 | 7 | 8 | 8 | 7 | 7 | 7 | 5 | 6 | 7 | 7 | 80 |
| Total Snowfall (") | 7.7 | 5.4 | 2.1 | 0.6 | 0.0 | 0.0 | 0.0 | 0.0 | 0.0 | 0.0 | 1.1 | 3.3 | 20.2 |
| Days ≥ 1" Snow Depth | na | na | *1* | 0 | 0 | 0 | 0 | 0 | 0 | 0 | *0* | *3* | na |

## COSHOCTON ARG RESEAR *Coshocton County* ELEVATION 1191 ft LAT/LONG 40° 22 ' N / 81° 48 ' W

| | JAN | FEB | MAR | APR | MAY | JUN | JUL | AUG | SEP | OCT | NOV | DEC | YEAR |
|---|---|---|---|---|---|---|---|---|---|---|---|---|---|
| Maximum Temp °F | 32.8 | 35.9 | 47.1 | 59.0 | 69.2 | 77.8 | 81.7 | 80.1 | 73.6 | 61.8 | 49.8 | 38.2 | 58.9 |
| Minimum Temp °F | 17.1 | 19.3 | 28.8 | 39.7 | 50.2 | 58.7 | 63.1 | 61.7 | 55.0 | 42.9 | 33.8 | 23.8 | 41.2 |
| Mean Temp °F | 24.9 | 27.6 | 38.0 | 49.4 | 59.8 | 68.3 | 72.4 | 70.9 | 64.3 | 52.4 | 41.8 | 31.0 | 50.1 |
| Days Max Temp ≥ 90 °F | 0 | 0 | 0 | 0 | 0 | 1 | 2 | 2 | 0 | 0 | 0 | 0 | 5 |
| Days Max Temp ≤ 32 °F | 15 | 12 | 4 | 0 | 0 | 0 | 0 | 0 | 0 | 0 | 2 | 10 | 43 |
| Days Min Temp ≤ 32 °F | 29 | 24 | 21 | 7 | 0 | 0 | 0 | 0 | 0 | 4 | 15 | 25 | 125 |
| Days Min Temp ≤ 0 °F | 3 | 2 | 0 | 0 | 0 | 0 | 0 | 0 | 0 | 0 | 0 | 1 | 6 |
| Heating Degree Days | 1234 | 1050 | 832 | 472 | 205 | 41 | 6 | 13 | 101 | 394 | 691 | 1046 | 6085 |
| Cooling Degree Days | 0 | 0 | 1 | 11 | 50 | 154 | 270 | 220 | 92 | 10 | 1 | 0 | 809 |
| Total Precipitation (") | 2.11 | 2.00 | 2.95 | 3.23 | 3.73 | 3.63 | 4.42 | 3.30 | 3.09 | 2.38 | 3.15 | 2.62 | 36.61 |
| Days ≥ 0.1" Precip | 5 | 5 | 7 | 7 | 7 | 7 | 7 | 6 | 6 | 6 | 7 | 7 | 77 |
| Total Snowfall (") | na | na | na | 0.0 | 0.0 | 0.0 | 0.0 | 0.0 | 0.0 | 0.0 | *0.5* | na | na |
| Days ≥ 1" Snow Depth | na | na | na | 0 | 0 | 0 | 0 | 0 | 0 | 0 | *0* | na | na |

## COSHOCTON WPC PLANT *Coshocton County* ELEVATION 751 ft LAT/LONG 40° 17 ' N / 81° 52 ' W

| | JAN | FEB | MAR | APR | MAY | JUN | JUL | AUG | SEP | OCT | NOV | DEC | YEAR |
|---|---|---|---|---|---|---|---|---|---|---|---|---|---|
| Maximum Temp °F | 35.6 | 39.7 | 51.0 | 62.7 | 72.6 | 80.7 | 84.1 | 82.5 | 76.0 | 64.6 | 52.3 | 41.0 | 61.9 |
| Minimum Temp °F | 17.6 | 19.7 | 28.8 | 37.9 | 47.6 | 56.4 | 60.8 | 59.0 | 52.1 | 39.8 | 32.2 | 23.7 | 39.6 |
| Mean Temp °F | 26.6 | 29.7 | 39.9 | 50.3 | 60.1 | 68.6 | 72.4 | 70.8 | 64.1 | 52.2 | 42.3 | 32.4 | 50.8 |
| Days Max Temp ≥ 90 °F | 0 | 0 | 0 | 0 | 0 | 3 | 5 | 3 | 1 | 0 | 0 | 0 | 12 |
| Days Max Temp ≤ 32 °F | 12 | 8 | 2 | 0 | 0 | 0 | 0 | 0 | 0 | 0 | 1 | 6 | 29 |
| Days Min Temp ≤ 32 °F | 28 | 24 | 21 | 10 | 1 | 0 | 0 | 0 | 0 | 7 | 17 | 25 | 133 |
| Days Min Temp ≤ 0 °F | 3 | 2 | 0 | 0 | 0 | 0 | 0 | 0 | 0 | 0 | 0 | 1 | 6 |
| Heating Degree Days | 1182 | 990 | 771 | 441 | 190 | 35 | 5 | 13 | 103 | 395 | 676 | 1004 | 5805 |
| Cooling Degree Days | 0 | 0 | 1 | 8 | 50 | 151 | 268 | 211 | 84 | 6 | 1 | 0 | 780 |
| Total Precipitation (") | 2.33 | 2.34 | 3.09 | 3.63 | 3.92 | 3.78 | 4.90 | 3.93 | 3.31 | 2.60 | 3.52 | 3.00 | 40.35 |
| Days ≥ 0.1" Precip | 6 | 6 | 8 | 8 | 8 | 8 | 8 | 7 | 6 | 6 | 8 | 7 | 86 |
| Total Snowfall (") | 8.8 | 6.0 | 3.4 | 0.8 | 0.0 | 0.0 | 0.0 | 0.0 | 0.0 | 0.0 | 1.2 | 4.1 | 24.3 |
| Days ≥ 1" Snow Depth | 12 | 8 | 3 | 0 | 0 | 0 | 0 | 0 | 0 | 0 | 1 | 5 | 29 |

## DANVILLE 2 W *Knox County* ELEVATION 1010 ft LAT/LONG 40° 27 ' N / 82° 16 ' W

| | JAN | FEB | MAR | APR | MAY | JUN | JUL | AUG | SEP | OCT | NOV | DEC | YEAR |
|---|---|---|---|---|---|---|---|---|---|---|---|---|---|
| Maximum Temp °F | 33.8 | 37.9 | 49.8 | 62.0 | 72.2 | 80.6 | 84.4 | 82.7 | 76.1 | 64.2 | 51.1 | 39.4 | 61.2 |
| Minimum Temp °F | 15.6 | 17.7 | 26.7 | 35.2 | 44.6 | 53.7 | 58.2 | 56.4 | 49.5 | 37.6 | 30.8 | 22.3 | 37.4 |
| Mean Temp °F | 24.7 | 27.8 | 38.3 | 48.6 | 58.4 | 67.2 | 71.4 | 69.6 | 62.8 | 50.9 | 41.0 | 30.9 | 49.3 |
| Days Max Temp ≥ 90 °F | 0 | 0 | 0 | 0 | 0 | 2 | 5 | 3 | 1 | 0 | 0 | 0 | 11 |
| Days Max Temp ≤ 32 °F | 14 | 10 | 3 | 0 | 0 | 0 | 0 | 0 | 0 | 0 | 1 | 8 | 36 |
| Days Min Temp ≤ 32 °F | 28 | 25 | 22 | 13 | 3 | 0 | 0 | 0 | 1 | 11 | 18 | 26 | 147 |
| Days Min Temp ≤ 0 °F | 5 | 3 | 0 | 0 | 0 | 0 | 0 | 0 | 0 | 0 | 0 | 2 | 10 |
| Heating Degree Days | 1241 | 1043 | 823 | 489 | 230 | 52 | 11 | 24 | 130 | 434 | 715 | 1052 | 6244 |
| Cooling Degree Days | 0 | 0 | 0 | 3 | 34 | 118 | 226 | 171 | 67 | 5 | 0 | 0 | 624 |
| Total Precipitation (") | 2.45 | 2.31 | 3.18 | 3.53 | 3.97 | 4.14 | 4.46 | 3.67 | 3.41 | 2.58 | 3.59 | 2.99 | 40.28 |
| Days ≥ 0.1" Precip | 6 | 6 | 8 | 8 | 8 | 8 | 8 | 6 | 6 | 6 | 8 | 7 | 85 |
| Total Snowfall (") | 11.0 | 8.4 | 4.5 | 1.3 | 0.0 | 0.0 | 0.0 | 0.0 | 0.0 | 0.0 | 1.7 | 6.6 | 33.5 |
| Days ≥ 1" Snow Depth | 16 | 11 | 4 | 1 | 0 | 0 | 0 | 0 | 0 | 0 | 1 | 8 | 41 |

## DAYTON INTL ARPT *Montgomery County* ELEVATION 1017 ft LAT/LONG 39° 54' N / 84° 12' W

| | JAN | FEB | MAR | APR | MAY | JUN | JUL | AUG | SEP | OCT | NOV | DEC | YEAR |
|---|---|---|---|---|---|---|---|---|---|---|---|---|---|
| Maximum Temp °F | 34.1 | 38.2 | 49.9 | 62.1 | 72.4 | 81.5 | 85.1 | 83.0 | 76.1 | 63.9 | 51.0 | 39.7 | 61.4 |
| Minimum Temp °F | 18.5 | 21.1 | 30.9 | 40.8 | 51.0 | 60.1 | 64.5 | 62.2 | 55.1 | 43.4 | 34.6 | 24.9 | 42.3 |
| Mean Temp °F | 26.3 | 29.7 | 40.4 | 51.4 | 61.7 | 70.8 | 74.8 | 72.6 | 65.6 | 53.7 | 42.8 | 32.3 | 51.8 |
| Days Max Temp ≥ 90 °F | 0 | 0 | 0 | 0 | 0 | 4 | 7 | 4 | 1 | 0 | 0 | 0 | 16 |
| Days Max Temp ≤ 32 °F | 13 | 9 | 3 | 0 | 0 | 0 | 0 | 0 | 0 | 0 | 1 | 8 | 34 |
| Days Min Temp ≤ 32 °F | 27 | 23 | 18 | 6 | 0 | 0 | 0 | 0 | 0 | 4 | 14 | 24 | 116 |
| Days Min Temp ≤ 0 °F | 3 | 2 | 0 | 0 | 0 | 0 | 0 | 0 | 0 | 0 | 0 | 1 | 6 |
| Heating Degree Days | 1193 | 991 | 756 | 411 | 164 | 21 | 2 | 7 | 86 | 357 | 660 | 1008 | 5656 |
| Cooling Degree Days | 0 | 0 | 2 | 11 | 66 | 195 | 323 | 255 | 109 | 12 | 1 | 0 | 974 |
| Total Precipitation (") | 2.36 | 2.14 | 3.15 | 3.74 | 3.97 | 4.00 | 3.63 | 3.34 | 2.69 | 2.56 | 3.37 | 3.06 | 38.01 |
| Days ≥ 0.1" Precip | 6 | 5 | 8 | 8 | 8 | 7 | 7 | 6 | 5 | 6 | 7 | 7 | 80 |
| Total Snowfall (") | 8.7 | 6.8 | 5.2 | 0.8 | 0.0 | 0.0 | 0.0 | 0.0 | 0.0 | 0.3 | 2.0 | 4.7 | 28.5 |
| Days ≥ 1" Snow Depth | 11 | 9 | 3 | 0 | 0 | 0 | 0 | 0 | 0 | 0 | 1 | 5 | 29 |

## DAYTON MCD *Montgomery County* ELEVATION 751 ft LAT/LONG 39° 46' N / 84° 11' W

| | JAN | FEB | MAR | APR | MAY | JUN | JUL | AUG | SEP | OCT | NOV | DEC | YEAR |
|---|---|---|---|---|---|---|---|---|---|---|---|---|---|
| Maximum Temp °F | 34.9 | 39.1 | 50.4 | 63.0 | 73.8 | 83.0 | 86.8 | 84.9 | 78.2 | 65.4 | 52.2 | 40.6 | 62.7 |
| Minimum Temp °F | 20.1 | 22.5 | 32.1 | 42.7 | 53.1 | 62.3 | 66.9 | 64.4 | 57.3 | 45.2 | 36.3 | 26.5 | 44.1 |
| Mean Temp °F | 27.5 | 30.8 | 41.3 | 52.9 | 63.5 | 72.7 | 76.9 | 74.7 | 67.8 | 55.3 | 44.3 | 33.6 | 53.4 |
| Days Max Temp ≥ 90 °F | 0 | 0 | 0 | 0 | 1 | 6 | 11 | 7 | 2 | 0 | 0 | 0 | 27 |
| Days Max Temp ≤ 32 °F | 13 | 9 | 3 | 0 | 0 | 0 | 0 | 0 | 0 | 0 | 1 | 7 | 33 |
| Days Min Temp ≤ 32 °F | 26 | 22 | 17 | 4 | 0 | 0 | 0 | 0 | 0 | 2 | 11 | 22 | 104 |
| Days Min Temp ≤ 0 °F | 2 | 1 | 0 | 0 | 0 | 0 | 0 | 0 | 0 | 0 | 0 | 1 | 4 |
| Heating Degree Days | 1155 | 959 | 732 | 375 | 134 | 14 | 1 | 3 | 60 | 315 | 616 | 966 | 5330 |
| Cooling Degree Days | 0 | 0 | 3 | 19 | 92 | 254 | 397 | 320 | 151 | 23 | 1 | 0 | 1260 |
| Total Precipitation (") | 2.42 | 2.32 | 3.04 | 3.76 | 4.31 | 3.97 | 3.97 | 3.07 | 2.87 | 2.61 | 3.40 | 2.91 | 38.65 |
| Days ≥ 0.1" Precip | 6 | 6 | 7 | 8 | 7 | 7 | 7 | 6 | 5 | 6 | 7 | 7 | 79 |
| Total Snowfall (") | 5.8 | 4.0 | 2.5 | 0.2 | 0.0 | 0.0 | 0.0 | 0.0 | 0.0 | 0.0 | 0.8 | 3.0 | 16.3 |
| Days ≥ 1" Snow Depth | *10* | 7 | 2 | 0 | 0 | 0 | 0 | 0 | 0 | 0 | 0 | 4 | 23 |

## DEFIANCE *Defiance County* ELEVATION 712 ft LAT/LONG 41° 17' N / 84° 23' W

| | JAN | FEB | MAR | APR | MAY | JUN | JUL | AUG | SEP | OCT | NOV | DEC | YEAR |
|---|---|---|---|---|---|---|---|---|---|---|---|---|---|
| Maximum Temp °F | 31.0 | 34.1 | 45.5 | 59.1 | 71.2 | 80.8 | 84.4 | 82.5 | 75.3 | 62.8 | 48.6 | 36.7 | 59.3 |
| Minimum Temp °F | 14.0 | 15.8 | 25.5 | 36.3 | 46.6 | 56.5 | 60.9 | 58.5 | 51.5 | 39.9 | 31.3 | 21.2 | 38.2 |
| Mean Temp °F | 22.5 | 25.0 | 35.5 | 47.7 | 59.0 | 68.7 | 72.7 | 70.5 | 63.5 | 51.4 | 40.0 | 29.0 | 48.8 |
| Days Max Temp ≥ 90 °F | 0 | 0 | 0 | 0 | 1 | 4 | 7 | 4 | 1 | 0 | 0 | 0 | 17 |
| Days Max Temp ≤ 32 °F | 16 | 13 | 4 | 0 | 0 | 0 | 0 | 0 | 0 | 0 | 1 | 10 | 44 |
| Days Min Temp ≤ 32 °F | 29 | 26 | 24 | 11 | 1 | 0 | 0 | 0 | 0 | 7 | 18 | 27 | 143 |
| Days Min Temp ≤ 0 °F | 5 | 4 | 0 | 0 | 0 | 0 | 0 | 0 | 0 | 0 | 0 | 2 | 11 |
| Heating Degree Days | 1310 | 1124 | 907 | 519 | 226 | 43 | 7 | 17 | 120 | 424 | 744 | 1108 | 6549 |
| Cooling Degree Days | 0 | 0 | 1 | 9 | 50 | 164 | 277 | 214 | 84 | 8 | 0 | 0 | 807 |
| Total Precipitation (") | 1.83 | 1.64 | 2.52 | 3.32 | 3.70 | 3.69 | 3.75 | 2.87 | 3.36 | 2.68 | 3.25 | 2.90 | 35.51 |
| Days ≥ 0.1" Precip | 5 | 4 | 6 | 7 | 7 | 7 | 7 | 6 | 7 | 6 | 7 | 7 | 76 |
| Total Snowfall (") | 6.3 | 5.9 | 3.3 | 0.6 | 0.0 | 0.0 | 0.0 | 0.0 | 0.0 | 0.1 | 1.6 | 5.0 | 22.8 |
| Days ≥ 1" Snow Depth | 16 | 13 | 4 | 0 | 0 | 0 | 0 | 0 | 0 | 0 | 1 | 8 | 42 |

## DELAWARE *Delaware County* ELEVATION 850 ft LAT/LONG 40° 18' N / 83° 4' W

| | JAN | FEB | MAR | APR | MAY | JUN | JUL | AUG | SEP | OCT | NOV | DEC | YEAR |
|---|---|---|---|---|---|---|---|---|---|---|---|---|---|
| Maximum Temp °F | 33.3 | 36.5 | 47.9 | 60.6 | 71.2 | 79.9 | 83.8 | 82.2 | 75.7 | 63.5 | 50.3 | 38.6 | 60.3 |
| Minimum Temp °F | 15.4 | 17.6 | 27.2 | 37.2 | 47.1 | 56.5 | 60.9 | 58.4 | 51.2 | 39.2 | 31.3 | 22.5 | 38.7 |
| Mean Temp °F | 24.4 | 27.1 | 37.6 | 48.9 | 59.2 | 68.2 | 72.4 | 70.3 | 63.5 | 51.4 | 40.8 | 30.6 | 49.5 |
| Days Max Temp ≥ 90 °F | 0 | 0 | 0 | 0 | 0 | 2 | 5 | 3 | 1 | 0 | 0 | 0 | 11 |
| Days Max Temp ≤ 32 °F | 14 | 11 | 3 | 0 | 0 | 0 | 0 | 0 | 0 | 0 | 1 | 9 | 38 |
| Days Min Temp ≤ 32 °F | 29 | 25 | 23 | 10 | 1 | 0 | 0 | 0 | 0 | 8 | 18 | 26 | 140 |
| Days Min Temp ≤ 0 °F | 5 | 3 | 0 | 0 | 0 | 0 | 0 | 0 | 0 | 0 | 0 | 1 | 9 |
| Heating Degree Days | 1253 | 1064 | 844 | 482 | 217 | 43 | 7 | 18 | 118 | 422 | 719 | 1061 | 6248 |
| Cooling Degree Days | 0 | 0 | 1 | 8 | 48 | 148 | 263 | 198 | 81 | 9 | 0 | 0 | 756 |
| Total Precipitation (") | 2.20 | 1.89 | 2.60 | 3.44 | 3.89 | 3.72 | 4.37 | 3.41 | 2.89 | 2.41 | 3.67 | 2.82 | 37.31 |
| Days ≥ 0.1" Precip | 6 | 5 | 7 | 8 | 8 | 7 | 7 | 6 | 5 | 6 | 7 | 7 | 79 |
| Total Snowfall (") | 7.9 | 5.0 | 2.9 | 0.8 | 0.0 | 0.0 | 0.0 | 0.0 | 0.0 | 0.0 | 0.9 | 4.0 | 21.5 |
| Days ≥ 1" Snow Depth | 13 | 9 | 3 | 0 | 0 | 0 | 0 | 0 | 0 | 0 | 1 | 5 | 31 |

## DORSET *Ashtabula County* ELEVATION 1040 ft LAT/LONG 41° 41' N / 80° 38' W

| | JAN | FEB | MAR | APR | MAY | JUN | JUL | AUG | SEP | OCT | NOV | DEC | YEAR |
|---|---|---|---|---|---|---|---|---|---|---|---|---|---|
| Maximum Temp °F | 31.1 | 33.8 | 44.0 | 56.7 | 67.9 | 76.8 | 81.1 | 79.6 | 72.8 | 60.9 | 48.6 | 36.5 | 57.5 |
| Minimum Temp °F | 14.0 | 14.7 | 24.5 | 34.4 | 43.6 | 52.8 | 57.2 | 55.9 | 49.6 | 39.2 | 32.0 | 21.8 | 36.6 |
| Mean Temp °F | 22.6 | 24.3 | 34.3 | 45.6 | 55.8 | 64.8 | 69.2 | 67.8 | 61.3 | 50.1 | 40.3 | 29.2 | 47.1 |
| Days Max Temp ≥ 90 °F | 0 | 0 | 0 | 0 | 0 | 1 | 3 | 1 | 0 | 0 | 0 | 0 | 5 |
| Days Max Temp ≤ 32 °F | 17 | 13 | 6 | 1 | 0 | 0 | 0 | 0 | 0 | 0 | 2 | 11 | 50 |
| Days Min Temp ≤ 32 °F | 29 | 26 | 24 | 14 | 4 | 0 | 0 | 0 | 1 | 7 | 17 | 26 | 148 |
| Days Min Temp ≤ 0 °F | 6 | 4 | 1 | 0 | 0 | 0 | 0 | 0 | 0 | 0 | 0 | 2 | 13 |
| Heating Degree Days | 1309 | 1144 | 945 | 580 | 303 | 92 | 29 | 40 | 164 | 459 | 734 | 1103 | 6902 |
| Cooling Degree Days | 0 | 0 | 0 | 7 | 30 | 94 | 183 | 141 | 66 | 3 | 1 | 0 | 525 |
| Total Precipitation (") | 2.63 | 2.27 | 3.13 | 3.42 | 3.76 | 4.37 | 4.29 | 4.24 | 4.32 | 3.71 | 4.12 | 3.28 | 43.54 |
| Days ≥ 0.1" Precip | 8 | 7 | 9 | 9 | 9 | 8 | 7 | 8 | 8 | 9 | 11 | 9 | 102 |
| Total Snowfall (") | 17.9 | 13.8 | 12.6 | 2.8 | 0.0 | 0.0 | 0.0 | 0.0 | 0.0 | 0.6 | 7.9 | 18.3 | 73.9 |
| Days ≥ 1" Snow Depth | 22 | 19 | 11 | 2 | 0 | 0 | 0 | 0 | 0 | 0 | 5 | 16 | 75 |

## EATON *Preble County* ELEVATION 1030 ft LAT/LONG 39° 45' N / 84° 38' W

| | JAN | FEB | MAR | APR | MAY | JUN | JUL | AUG | SEP | OCT | NOV | DEC | YEAR |
|---|---|---|---|---|---|---|---|---|---|---|---|---|---|
| Maximum Temp °F | 33.5 | 37.9 | 49.5 | 61.8 | 72.6 | 81.2 | 85.0 | 83.3 | 77.0 | 64.5 | 50.9 | 39.0 | 61.4 |
| Minimum Temp °F | 14.7 | 17.3 | 27.2 | 36.8 | 47.5 | 56.4 | 60.5 | 58.4 | 51.3 | 38.9 | 31.1 | 21.9 | 38.5 |
| Mean Temp °F | 24.2 | 27.6 | 38.4 | 49.1 | 60.2 | 68.8 | 72.7 | 70.9 | 64.2 | 51.8 | 41.1 | 30.6 | 50.0 |
| Days Max Temp ≥ 90 °F | 0 | 0 | 0 | 0 | 0 | 4 | 7 | 4 | 1 | 0 | 0 | 0 | 16 |
| Days Max Temp ≤ 32 °F | 14 | 10 | 3 | 0 | 0 | 0 | 0 | 0 | 0 | 0 | 1 | 8 | 36 |
| Days Min Temp ≤ 32 °F | 29 | 25 | 23 | 10 | 1 | 0 | 0 | 0 | 0 | 8 | 17 | 26 | 139 |
| Days Min Temp ≤ 0 °F | 6 | 3 | 0 | 0 | 0 | 0 | 0 | 0 | 0 | 0 | 0 | 2 | 11 |
| Heating Degree Days | 1259 | 1051 | 819 | 476 | 192 | 35 | 6 | 11 | 105 | 410 | 714 | 1061 | 6139 |
| Cooling Degree Days | 0 | 0 | 1 | *5* | 51 | 163 | 264 | 210 | 88 | 9 | 0 | 0 | 791 |
| Total Precipitation (") | 2.41 | 2.26 | 3.24 | 3.91 | 4.59 | 3.47 | 3.93 | 3.46 | 2.98 | 2.76 | 3.60 | 3.17 | 39.78 |
| Days ≥ 0.1" Precip | 6 | 6 | 8 | 8 | 8 | 6 | 7 | 6 | 5 | 6 | 8 | 7 | 81 |
| Total Snowfall (") | na | na | na | *0.6* | *0.0* | *0.0* | *0.0* | *0.0* | *0.0* | *0.0* | na | na | na |
| Days ≥ 1" Snow Depth | *12* | *9* | *4* | *0* | *0* | *0* | *0* | *0* | *0* | *0* | *1* | *6* | 32 |

## ELYRIA 3 E *Lorain County* ELEVATION 732 ft LAT/LONG 41° 23' N / 82° 3' W

| | JAN | FEB | MAR | APR | MAY | JUN | JUL | AUG | SEP | OCT | NOV | DEC | YEAR |
|---|---|---|---|---|---|---|---|---|---|---|---|---|---|
| Maximum Temp °F | 33.6 | 36.7 | 47.4 | 60.4 | 71.3 | 80.3 | 84.3 | 82.3 | 75.6 | 63.8 | 50.9 | 39.1 | 60.5 |
| Minimum Temp °F | 18.4 | 19.7 | 28.7 | 38.4 | 48.0 | 57.4 | 62.0 | 60.5 | 54.3 | 43.3 | 34.9 | 25.1 | 40.9 |
| Mean Temp °F | 26.0 | 28.2 | 38.1 | 49.4 | 59.7 | 68.9 | 73.2 | 71.4 | 65.0 | 53.6 | 43.0 | 32.1 | 50.7 |
| Days Max Temp ≥ 90 °F | 0 | 0 | 0 | 0 | 1 | 3 | 7 | 4 | 1 | 0 | 0 | 0 | 16 |
| Days Max Temp ≤ 32 °F | 14 | 11 | 4 | 0 | 0 | 0 | 0 | 0 | 0 | 0 | 1 | 8 | 38 |
| Days Min Temp ≤ 32 °F | 28 | 25 | 21 | 9 | 1 | 0 | 0 | 0 | 0 | 3 | 13 | 24 | 124 |
| Days Min Temp ≤ 0 °F | 3 | 2 | 0 | 0 | 0 | 0 | 0 | 0 | 0 | 0 | 0 | 1 | 6 |
| Heating Degree Days | 1202 | 1033 | 829 | 470 | 208 | 39 | 6 | 12 | 90 | 357 | 655 | 1013 | 5914 |
| Cooling Degree Days | 0 | 0 | 1 | 11 | 55 | 161 | 291 | 225 | 99 | 9 | 0 | 0 | 852 |
| Total Precipitation (") | 2.23 | 2.06 | 2.59 | 2.90 | 3.65 | 3.98 | 3.50 | 3.55 | 3.27 | 2.52 | 3.32 | 3.10 | 36.67 |
| Days ≥ 0.1" Precip | 7 | 6 | 7 | 8 | 8 | 7 | 7 | 7 | 7 | 7 | 8 | 8 | 87 |
| Total Snowfall (") | 11.9 | 11.0 | 7.2 | 1.5 | 0.0 | 0.0 | 0.0 | 0.0 | 0.0 | 0.1 | 2.9 | 9.3 | 43.9 |
| Days ≥ 1" Snow Depth | 16 | 13 | 5 | 0 | 0 | 0 | 0 | 0 | 0 | 0 | 1 | 9 | 44 |

## FINDLAY AIRPORT *Hancock County* ELEVATION 804 ft LAT/LONG 41° 1' N / 83° 40' W

| | JAN | FEB | MAR | APR | MAY | JUN | JUL | AUG | SEP | OCT | NOV | DEC | YEAR |
|---|---|---|---|---|---|---|---|---|---|---|---|---|---|
| Maximum Temp °F | 30.7 | 34.0 | 45.7 | 58.7 | 69.9 | 78.9 | 82.6 | 80.4 | 73.9 | 61.6 | 48.3 | 36.3 | 58.4 |
| Minimum Temp °F | 17.1 | 19.4 | 28.8 | 39.0 | 49.4 | 58.8 | 62.9 | 60.6 | 53.8 | 42.5 | 33.7 | 23.7 | 40.8 |
| Mean Temp °F | 24.0 | 26.7 | 37.3 | 48.9 | 59.7 | 68.9 | 72.8 | 70.6 | 63.9 | 52.1 | 41.0 | 30.0 | 49.7 |
| Days Max Temp ≥ 90 °F | 0 | 0 | 0 | 0 | 0 | 2 | 4 | 2 | 1 | 0 | 0 | 0 | 9 |
| Days Max Temp ≤ 32 °F | 17 | 13 | 4 | 0 | 0 | 0 | 0 | 0 | 0 | 0 | 2 | 11 | 47 |
| Days Min Temp ≤ 32 °F | 28 | 24 | 20 | 8 | 0 | 0 | 0 | 0 | 0 | 4 | 15 | 25 | 124 |
| Days Min Temp ≤ 0 °F | 4 | 2 | 0 | 0 | 0 | 0 | 0 | 0 | 0 | 0 | 0 | 1 | 7 |
| Heating Degree Days | 1263 | 1074 | 853 | 485 | 210 | 37 | 5 | 16 | 111 | 402 | 712 | 1076 | 6244 |
| Cooling Degree Days | 0 | 0 | 1 | 8 | 55 | 157 | 267 | 202 | 85 | 8 | 0 | 0 | 783 |
| Total Precipitation (") | 1.72 | 1.49 | 2.45 | 3.09 | 3.63 | 3.79 | 4.09 | 3.62 | 3.03 | 2.16 | 2.97 | 2.55 | 34.59 |
| Days ≥ 0.1" Precip | 4 | 4 | 7 | 8 | 7 | 7 | 7 | 6 | 7 | 5 | 7 | 6 | 75 |
| Total Snowfall (") | 6.9 | 5.7 | 3.5 | 0.8 | 0.0 | 0.0 | 0.0 | 0.0 | 0.0 | 0.1 | 1.5 | 5.4 | 23.9 |
| Days ≥ 1" Snow Depth | 14 | 11 | 4 | 1 | 0 | 0 | 0 | 0 | 0 | 0 | 1 | 7 | 38 |

## FINDLAY WPCC *Hancock County* ELEVATION 771 ft LAT/LONG 41° 3 ' N / 83° 40 ' W

| | JAN | FEB | MAR | APR | MAY | JUN | JUL | AUG | SEP | OCT | NOV | DEC | YEAR |
|---|---|---|---|---|---|---|---|---|---|---|---|---|---|
| Maximum Temp °F | 30.9 | 34.4 | 46.1 | 59.2 | 70.7 | 79.7 | 83.3 | 80.9 | 74.1 | 61.7 | 48.0 | 36.2 | 58.8 |
| Minimum Temp °F | 16.6 | 19.0 | 28.4 | 38.8 | 49.4 | 58.8 | 63.0 | 60.5 | 53.7 | 42.3 | 33.0 | 23.1 | 40.6 |
| Mean Temp °F | 23.8 | 26.7 | 37.3 | 49.0 | 60.1 | 69.3 | 73.2 | 70.8 | 63.9 | 52.0 | 40.5 | 29.6 | 49.7 |
| Days Max Temp ≥ 90 °F | 0 | 0 | 0 | 0 | 0 | 3 | 4 | 2 | 1 | 0 | 0 | 0 | 10 |
| Days Max Temp ≤ 32 °F | 17 | 13 | 4 | 0 | 0 | 0 | 0 | 0 | 0 | 0 | 2 | 11 | 47 |
| Days Min Temp ≤ 32 °F | 29 | 25 | 21 | 8 | 1 | 0 | 0 | 0 | 0 | 4 | 16 | 26 | 130 |
| Days Min Temp ≤ 0 °F | 4 | 2 | 0 | 0 | 0 | 0 | 0 | 0 | 0 | 0 | 0 | 1 | 7 |
| Heating Degree Days | 1271 | 1074 | 853 | 481 | 200 | 34 | 5 | 14 | 110 | 403 | 728 | 1089 | 6262 |
| Cooling Degree Days | 0 | 0 | 1 | 9 | 61 | 175 | 293 | 223 | 95 | 8 | 0 | 0 | 865 |
| Total Precipitation (") | 2.02 | 1.84 | 2.71 | 3.12 | 3.85 | 3.79 | 4.15 | 3.99 | 3.13 | 2.28 | 3.03 | 2.88 | 36.79 |
| Days ≥ 0.1" Precip | 5 | 5 | 7 | 7 | 7 | 7 | 7 | 6 | 6 | 6 | 7 | 6 | 76 |
| Total Snowfall (") | 9.2 | 6.9 | 4.1 | 1.3 | 0.1 | 0.0 | 0.0 | 0.0 | 0.0 | 0.2 | 1.6 | *6.3* | 29.7 |
| Days ≥ 1" Snow Depth | 16 | 13 | 4 | 0 | 0 | 0 | 0 | 0 | 0 | 0 | 1 | 8 | 42 |

## FRANKLIN *Warren County* ELEVATION 689 ft LAT/LONG 39° 33 ' N / 84° 18 ' W

| | JAN | FEB | MAR | APR | MAY | JUN | JUL | AUG | SEP | OCT | NOV | DEC | YEAR |
|---|---|---|---|---|---|---|---|---|---|---|---|---|---|
| Maximum Temp °F | 35.9 | 39.7 | 51.0 | 62.8 | 72.9 | 81.8 | 85.5 | 83.9 | 77.5 | 65.3 | 52.8 | 41.4 | 62.5 |
| Minimum Temp °F | 18.3 | 20.4 | 30.1 | 40.2 | 49.5 | 59.0 | 63.2 | 60.4 | 52.9 | 40.3 | 33.4 | 24.6 | 41.0 |
| Mean Temp °F | 27.1 | 30.0 | 40.6 | 51.6 | 61.3 | 70.4 | 74.4 | 72.2 | 65.2 | 52.8 | 43.1 | 33.0 | 51.8 |
| Days Max Temp ≥ 90 °F | 0 | 0 | 0 | 0 | 0 | 4 | 7 | 5 | 2 | 0 | 0 | 0 | 18 |
| Days Max Temp ≤ 32 °F | 12 | 9 | 2 | 0 | 0 | 0 | 0 | 0 | 0 | 0 | 1 | 6 | 30 |
| Days Min Temp ≤ 32 °F | 27 | 24 | 20 | 7 | 1 | 0 | 0 | 0 | 0 | 7 | 15 | 24 | 125 |
| Days Min Temp ≤ 0 °F | 3 | 2 | 0 | 0 | 0 | 0 | 0 | 0 | 0 | 0 | 0 | 1 | 6 |
| Heating Degree Days | 1168 | 980 | 752 | 407 | 168 | 25 | 3 | 8 | 90 | 382 | 651 | 984 | 5618 |
| Cooling Degree Days | 0 | 0 | 1 | 8 | 51 | 179 | 300 | 232 | 97 | 10 | 1 | 0 | 879 |
| Total Precipitation (") | 2.43 | 2.35 | 3.15 | 3.74 | 4.37 | 3.27 | 3.96 | 3.41 | 2.98 | 2.93 | 3.49 | 3.00 | 39.08 |
| Days ≥ 0.1" Precip | 5 | 6 | 7 | 8 | 8 | 7 | 7 | 6 | 5 | 6 | 7 | 7 | 79 |
| Total Snowfall (") | *2.6* | *3.0* | *2.4* | 0.0 | 0.0 | 0.0 | 0.0 | 0.0 | 0.0 | 0.0 | 0.8 | *1.8* | 10.6 |
| Days ≥ 1" Snow Depth | *9* | *7* | 2 | 0 | 0 | 0 | 0 | 0 | 0 | 0 | 1 | *2* | 21 |

## FREDERICKTOWN 4 S *Knox County* ELEVATION 1050 ft LAT/LONG 40° 25 ' N / 82° 32 ' W

| | JAN | FEB | MAR | APR | MAY | JUN | JUL | AUG | SEP | OCT | NOV | DEC | YEAR |
|---|---|---|---|---|---|---|---|---|---|---|---|---|---|
| Maximum Temp °F | 32.3 | 36.0 | 47.5 | 59.8 | 70.6 | 79.4 | 83.3 | 81.7 | 75.1 | 62.9 | 50.0 | 38.5 | 59.8 |
| Minimum Temp °F | 13.0 | 15.5 | 25.7 | 35.7 | 45.3 | 54.5 | 58.1 | 55.6 | 48.6 | 37.3 | 30.0 | 21.4 | 36.7 |
| Mean Temp °F | 22.7 | 25.8 | 36.6 | 47.8 | 58.0 | 67.0 | 70.7 | 68.7 | 61.9 | 50.2 | 40.0 | 30.0 | 48.3 |
| Days Max Temp ≥ 90 °F | 0 | 0 | 0 | 0 | 0 | 2 | 4 | 2 | 1 | 0 | 0 | 0 | 9 |
| Days Max Temp ≤ 32 °F | 15 | 11 | 4 | 0 | 0 | 0 | 0 | 0 | 0 | 0 | 1 | 9 | 40 |
| Days Min Temp ≤ 32 °F | 29 | 26 | 24 | 12 | 2 | 0 | 0 | 0 | 1 | 11 | 19 | 26 | 150 |
| Days Min Temp ≤ 0 °F | 7 | 4 | 1 | 0 | 0 | 0 | 0 | 0 | 0 | 0 | 0 | 2 | 14 |
| Heating Degree Days | 1306 | 1102 | 874 | 515 | 247 | 57 | 13 | 30 | 149 | 457 | 742 | 1079 | 6571 |
| Cooling Degree Days | 0 | 0 | 0 | 5 | 35 | 115 | 205 | 144 | 54 | *3* | 0 | 0 | 561 |
| Total Precipitation (") | 2.57 | 2.16 | 3.11 | 3.52 | 4.20 | 3.92 | 4.24 | 3.61 | 3.45 | 2.52 | 3.43 | 3.10 | 39.83 |
| Days ≥ 0.1" Precip | 6 | 5 | 7 | 8 | 8 | 8 | 7 | 6 | 6 | 6 | 8 | 7 | 82 |
| Total Snowfall (") | *7.2* | 6.0 | *3.4* | 0.7 | 0.1 | 0.0 | 0.0 | 0.0 | 0.0 | 0.0 | 1.4 | na | na |
| Days ≥ 1" Snow Depth | *13* | *9* | *3* | 0 | 0 | 0 | 0 | 0 | 0 | 0 | 1 | na | na |

## FREMONT *Sandusky County* ELEVATION 640 ft LAT/LONG 41° 20 ' N / 83° 7 ' W

| | JAN | FEB | MAR | APR | MAY | JUN | JUL | AUG | SEP | OCT | NOV | DEC | YEAR |
|---|---|---|---|---|---|---|---|---|---|---|---|---|---|
| Maximum Temp °F | 31.2 | 33.9 | 44.7 | 58.0 | 69.7 | 79.1 | 83.2 | 81.2 | 74.3 | 62.0 | 48.8 | 36.4 | 58.5 |
| Minimum Temp °F | 15.6 | 17.6 | 27.1 | 37.8 | 48.3 | 58.1 | 62.2 | 59.7 | 52.5 | 40.4 | 32.5 | 22.1 | 39.5 |
| Mean Temp °F | 23.4 | 25.8 | 35.9 | 47.9 | 59.0 | 68.6 | 72.7 | 70.5 | 63.4 | 51.2 | 40.7 | 29.3 | 49.0 |
| Days Max Temp ≥ 90 °F | 0 | 0 | 0 | 0 | 0 | 3 | 5 | 2 | 1 | 0 | 0 | 0 | 11 |
| Days Max Temp ≤ 32 °F | 16 | 13 | 5 | 0 | 0 | 0 | 0 | 0 | 0 | 0 | 1 | 10 | 45 |
| Days Min Temp ≤ 32 °F | 29 | 25 | 23 | 9 | 1 | 0 | 0 | 0 | 0 | 6 | 16 | 26 | 135 |
| Days Min Temp ≤ 0 °F | 5 | 3 | 0 | 0 | 0 | 0 | 0 | 0 | 0 | 0 | 0 | 1 | 9 |
| Heating Degree Days | 1281 | 1101 | 895 | 513 | 225 | 44 | 7 | 18 | 121 | 427 | 724 | 1102 | 6458 |
| Cooling Degree Days | 0 | 0 | 1 | 8 | 51 | 155 | 268 | 202 | 78 | 7 | 0 | 0 | 770 |
| Total Precipitation (") | 1.97 | 1.73 | 2.62 | 3.04 | 3.71 | 4.10 | 3.79 | 3.31 | 3.23 | 2.47 | 3.01 | 2.83 | 35.81 |
| Days ≥ 0.1" Precip | 5 | 5 | 7 | 8 | 8 | 7 | 6 | 6 | 6 | 6 | 8 | 7 | 79 |
| Total Snowfall (") | *5.3* | *5.5* | *3.4* | 0.3 | 0.0 | 0.0 | 0.0 | 0.0 | 0.0 | 0.0 | *0.5* | na | na |
| Days ≥ 1" Snow Depth | *8* | *6* | *3* | 0 | 0 | 0 | 0 | 0 | 0 | 0 | *0* | na | na |

## GALLIPOLIS *Gallia County* ELEVATION 581 ft LAT/LONG 38° 49 ' N / 82° 11 ' W

| | JAN | FEB | MAR | APR | MAY | JUN | JUL | AUG | SEP | OCT | NOV | DEC | YEAR |
|---|---|---|---|---|---|---|---|---|---|---|---|---|---|
| Maximum Temp °F | 41.9 | 46.6 | 58.2 | 68.6 | 77.0 | 84.3 | 87.3 | 85.9 | 80.0 | 69.3 | 57.8 | 46.8 | 67.0 |
| Minimum Temp °F | 21.9 | 23.8 | 32.4 | 40.9 | 50.0 | 58.9 | 63.8 | 62.1 | 55.6 | 43.1 | 35.0 | 27.1 | 42.9 |
| Mean Temp °F | 31.9 | 35.3 | 45.3 | 54.7 | 63.5 | 71.6 | 75.5 | 74.1 | 67.9 | 56.2 | 46.4 | 37.0 | 55.0 |
| Days Max Temp ≥ 90 °F | 0 | 0 | 0 | 0 | 2 | 6 | 11 | 8 | 3 | 0 | 0 | 0 | 30 |
| Days Max Temp ≤ 32 °F | 7 | 4 | 0 | 0 | 0 | 0 | 0 | 0 | 0 | 0 | 0 | 3 | 14 |
| Days Min Temp ≤ 32 °F | 25 | 22 | 16 | 7 | 1 | 0 | 0 | 0 | 0 | 5 | 13 | 22 | 111 |
| Days Min Temp ≤ 0 °F | 2 | 1 | 0 | 0 | 0 | 0 | 0 | 0 | 0 | 0 | 0 | 0 | 3 |
| Heating Degree Days | 1020 | 834 | 608 | 318 | 118 | 14 | 1 | 3 | 50 | 284 | 553 | 862 | 4665 |
| Cooling Degree Days | 0 | 0 | 4 | 18 | 87 | 227 | 364 | 307 | 149 | 22 | 2 | 0 | 1180 |
| Total Precipitation (") | 2.70 | 2.78 | 3.56 | 3.49 | 3.81 | 3.51 | 4.65 | 3.72 | 3.11 | 2.79 | 3.11 | 3.21 | 40.44 |
| Days ≥ 0.1" Precip | 7 | 7 | 8 | 8 | 8 | 7 | 8 | 7 | 6 | 6 | 7 | 7 | 86 |
| Total Snowfall (") | 6.8 | *5.3* | 2.7 | 0.0 | 0.0 | 0.0 | 0.0 | 0.0 | 0.0 | 0.0 | 0.6 | *1.8* | 17.2 |
| Days ≥ 1" Snow Depth | *8* | *6* | 1 | 0 | 0 | 0 | 0 | 0 | 0 | 0 | 0 | *1* | 16 |

## GREENVILLE WATER PLT *Darke County* ELEVATION 1040 ft LAT/LONG 40° 6 ' N / 84° 38 ' W

| | JAN | FEB | MAR | APR | MAY | JUN | JUL | AUG | SEP | OCT | NOV | DEC | YEAR |
|---|---|---|---|---|---|---|---|---|---|---|---|---|---|
| Maximum Temp °F | 32.2 | 35.9 | 47.8 | 60.5 | 71.6 | 80.5 | 84.3 | 82.4 | 76.3 | 63.8 | 50.1 | 38.1 | 60.3 |
| Minimum Temp °F | 14.4 | 16.5 | 27.3 | 37.6 | 47.6 | 56.9 | 60.8 | 57.6 | 50.3 | 38.5 | 31.0 | 21.3 | 38.3 |
| Mean Temp °F | 23.3 | 26.2 | 37.5 | 49.1 | 59.6 | 68.7 | 72.6 | 70.0 | 63.3 | 51.2 | 40.6 | 29.7 | 49.3 |
| Days Max Temp ≥ 90 °F | 0 | 0 | 0 | 0 | 0 | 4 | 6 | 4 | 1 | 0 | 0 | 0 | 15 |
| Days Max Temp ≤ 32 °F | 15 | 11 | 3 | 0 | 0 | 0 | 0 | 0 | 0 | 0 | 2 | 9 | 40 |
| Days Min Temp ≤ 32 °F | 29 | 26 | 23 | 10 | 1 | 0 | 0 | 0 | 0 | 9 | 18 | 26 | 142 |
| Days Min Temp ≤ 0 °F | 6 | 4 | 1 | 0 | 0 | 0 | 0 | 0 | 0 | 0 | 0 | 2 | 13 |
| Heating Degree Days | 1286 | 1087 | 845 | 477 | 208 | 40 | 7 | 20 | 124 | 429 | 727 | 1086 | 6336 |
| Cooling Degree Days | 0 | 0 | 1 | 7 | 49 | 159 | 260 | 186 | 80 | 8 | 0 | 0 | 750 |
| Total Precipitation (") | 2.03 | 1.99 | 2.80 | 3.53 | 4.06 | 3.69 | 4.21 | 3.31 | 2.70 | 2.73 | 3.25 | 2.79 | 37.09 |
| Days ≥ 0.1" Precip | 5 | 5 | 7 | 8 | 8 | 6 | 7 | 6 | 5 | 6 | 7 | 7 | 77 |
| Total Snowfall (") | 7.3 | *6.1* | 3.2 | 0.5 | 0.0 | 0.0 | 0.0 | 0.0 | 0.0 | 0.2 | 1.2 | *3.6* | 22.1 |
| Days ≥ 1" Snow Depth | na | na | *2* | 0 | 0 | 0 | 0 | 0 | 0 | 0 | na | na | na |

## HILLSBORO *Highland County* ELEVATION 1102 ft LAT/LONG 39° 12 ' N / 83° 37 ' W

| | JAN | FEB | MAR | APR | MAY | JUN | JUL | AUG | SEP | OCT | NOV | DEC | YEAR |
|---|---|---|---|---|---|---|---|---|---|---|---|---|---|
| Maximum Temp °F | 35.5 | 39.7 | 51.2 | 62.9 | 71.9 | 79.7 | 83.3 | 81.8 | 76.1 | 64.7 | 52.5 | 41.0 | 61.7 |
| Minimum Temp °F | 19.3 | 21.8 | 31.5 | 41.7 | 51.2 | 59.7 | 64.2 | 62.0 | 55.4 | 43.7 | 34.9 | 25.4 | 42.6 |
| Mean Temp °F | 27.4 | 30.8 | 41.4 | 52.3 | 61.6 | 69.7 | 73.8 | 71.9 | 65.8 | 54.2 | 43.7 | 33.2 | 52.2 |
| Days Max Temp ≥ 90 °F | 0 | 0 | 0 | 0 | 0 | 1 | 3 | 2 | 1 | 0 | 0 | 0 | 7 |
| Days Max Temp ≤ 32 °F | 12 | 9 | 2 | 0 | 0 | 0 | 0 | 0 | 0 | 0 | 1 | 7 | 31 |
| Days Min Temp ≤ 32 °F | 27 | 23 | 18 | 5 | 0 | 0 | 0 | 0 | 0 | 4 | 14 | 23 | 114 |
| Days Min Temp ≤ 0 °F | 3 | 1 | 0 | 0 | 0 | 0 | 0 | 0 | 0 | 0 | 0 | 1 | 5 |
| Heating Degree Days | 1160 | 959 | 727 | 389 | 159 | 26 | 2 | 9 | 79 | 341 | 634 | 979 | 5464 |
| Cooling Degree Days | 0 | 0 | 2 | 17 | 57 | 178 | 296 | 239 | 109 | 18 | 1 | 0 | 917 |
| Total Precipitation (") | 2.70 | 2.50 | 3.65 | 4.00 | 4.51 | 3.60 | 4.51 | 4.11 | 3.56 | 2.88 | 3.37 | 3.05 | 42.44 |
| Days ≥ 0.1" Precip | 6 | 5 | 8 | 8 | 8 | 7 | 7 | 6 | 6 | 6 | 7 | 7 | 81 |
| Total Snowfall (") | 6.5 | 4.6 | 3.3 | 0.4 | 0.0 | 0.0 | 0.0 | 0.0 | 0.0 | 0.2 | 1.2 | 3.0 | 19.2 |
| Days ≥ 1" Snow Depth | 10 | 8 | 3 | 0 | 0 | 0 | 0 | 0 | 0 | 0 | 1 | 4 | 26 |

## HIRAM *Portage County* ELEVATION 1260 ft LAT/LONG 41° 19 ' N / 81° 9 ' W

| | JAN | FEB | MAR | APR | MAY | JUN | JUL | AUG | SEP | OCT | NOV | DEC | YEAR |
|---|---|---|---|---|---|---|---|---|---|---|---|---|---|
| Maximum Temp °F | 31.8 | 35.5 | 46.1 | 59.2 | 69.8 | 77.6 | 81.8 | 80.0 | 73.0 | 61.2 | 49.0 | 36.9 | 58.5 |
| Minimum Temp °F | 16.4 | 18.2 | 27.2 | 37.6 | 47.8 | 56.1 | 61.0 | 59.6 | 53.0 | 41.8 | 33.1 | 22.5 | 39.5 |
| Mean Temp °F | 24.2 | 26.9 | 36.5 | 48.4 | 58.8 | 66.9 | 71.4 | 69.9 | 63.0 | 51.5 | 41.1 | 29.7 | 49.0 |
| Days Max Temp ≥ 90 °F | 0 | 0 | 0 | 0 | 0 | 1 | 2 | 1 | 0 | 0 | 0 | 0 | 4 |
| Days Max Temp ≤ 32 °F | 17 | 12 | 4 | 0 | 0 | 0 | 0 | 0 | 0 | 0 | 1 | 11 | 45 |
| Days Min Temp ≤ 32 °F | 29 | 25 | 22 | 10 | 1 | 0 | 0 | 0 | 0 | 4 | 15 | 27 | 133 |
| Days Min Temp ≤ 0 °F | 3 | 2 | 0 | 0 | 0 | 0 | 0 | 0 | 0 | 0 | 0 | 1 | 6 |
| Heating Degree Days | 1259 | 1070 | 876 | 499 | 225 | 59 | 8 | 17 | 121 | 416 | 711 | 1087 | 6348 |
| Cooling Degree Days | 0 | 0 | 1 | 10 | 39 | 117 | 232 | 182 | 67 | 4 | 0 | 0 | 652 |
| Total Precipitation (") | 2.63 | 2.29 | 3.37 | 3.40 | 3.86 | 3.96 | 3.99 | 3.74 | 4.22 | 3.15 | 3.82 | 3.65 | 42.08 |
| Days ≥ 0.1" Precip | 8 | 7 | 9 | 9 | 8 | 8 | 7 | 7 | 8 | 8 | 9 | 9 | 97 |
| Total Snowfall (") | 16.0 | 12.4 | 10.1 | 1.5 | 0.0 | 0.0 | 0.0 | 0.0 | 0.0 | 0.2 | 5.6 | 15.3 | 61.1 |
| Days ≥ 1" Snow Depth | 21 | 17 | 8 | 1 | 0 | 0 | 0 | 0 | 0 | 0 | 3 | 15 | 65 |

## HOYTVILLE 2 NE *Wood County* ELEVATION 702 ft LAT/LONG 41° 13 ' N / 83° 46 ' W

| | JAN | FEB | MAR | APR | MAY | JUN | JUL | AUG | SEP | OCT | NOV | DEC | YEAR |
|---|---|---|---|---|---|---|---|---|---|---|---|---|---|
| Maximum Temp °F | 31.1 | 34.2 | 46.1 | 59.7 | 71.4 | 80.4 | 83.7 | 81.5 | 75.2 | 63.0 | 48.9 | 36.8 | 59.3 |
| Minimum Temp °F | 15.0 | 17.2 | 26.8 | 36.9 | 47.5 | 57.4 | 61.2 | 58.3 | 51.5 | 40.2 | 32.4 | 22.1 | 38.9 |
| Mean Temp °F | 23.2 | 25.7 | 36.5 | 48.4 | 59.5 | 68.9 | 72.5 | 69.9 | 63.4 | 51.6 | 40.7 | 29.4 | 49.1 |
| Days Max Temp ≥ 90 °F | 0 | 0 | 0 | 0 | 1 | 4 | 5 | 3 | 1 | 0 | 0 | 0 | 14 |
| Days Max Temp ≤ 32 °F | 16 | 13 | 4 | 0 | 0 | 0 | 0 | 0 | 0 | 0 | 1 | 10 | 44 |
| Days Min Temp ≤ 32 °F | 29 | 25 | 23 | 11 | 1 | 0 | 0 | 0 | 0 | 7 | 17 | 27 | 140 |
| Days Min Temp ≤ 0 °F | 5 | 3 | 0 | 0 | 0 | 0 | 0 | 0 | 0 | 0 | 0 | 2 | 10 |
| Heating Degree Days | 1291 | 1102 | 878 | 501 | 216 | 40 | 7 | 23 | 123 | 418 | 722 | 1096 | 6417 |
| Cooling Degree Days | 0 | 0 | 1 | 9 | 52 | 153 | 259 | 185 | 77 | 7 | 0 | 0 | 743 |
| Total Precipitation (") | 1.72 | 1.55 | 2.45 | 3.09 | 3.49 | 3.60 | 3.65 | 3.26 | 2.92 | 2.39 | 3.02 | 2.61 | 33.75 |
| Days ≥ 0.1" Precip | 5 | 4 | 6 | 7 | 7 | 6 | 7 | 6 | 6 | 5 | 7 | 6 | 72 |
| Total Snowfall (") | 6.3 | 5.7 | 3.9 | 1.0 | 0.1 | 0.0 | 0.0 | 0.0 | 0.0 | 0.0 | 1.6 | 5.4 | 24.0 |
| Days ≥ 1" Snow Depth | 15 | 11 | 4 | 1 | 0 | 0 | 0 | 0 | 0 | 0 | 1 | 8 | 40 |

## IRWIN *Union County* ELEVATION 1010 ft LAT/LONG 40° 7 ' N / 83° 29 ' W

| | JAN | FEB | MAR | APR | MAY | JUN | JUL | AUG | SEP | OCT | NOV | DEC | YEAR |
|---|---|---|---|---|---|---|---|---|---|---|---|---|---|
| Maximum Temp °F | 34.5 | 39.2 | 51.1 | 63.9 | 74.3 | 82.5 | 85.5 | 83.9 | 78.0 | 66.3 | 51.9 | 39.6 | 62.6 |
| Minimum Temp °F | 17.0 | 19.7 | 28.9 | 38.5 | 48.5 | 57.1 | 60.8 | 58.3 | 51.5 | 40.5 | 32.7 | 23.3 | 39.7 |
| Mean Temp °F | 25.8 | 29.4 | 40.0 | 51.3 | 61.4 | 69.9 | 73.2 | 71.1 | 64.8 | 53.4 | 42.3 | 31.4 | 51.2 |
| Days Max Temp ≥ 90 °F | 0 | 0 | 0 | 0 | 1 | 4 | 7 | 4 | 1 | 0 | 0 | 0 | 17 |
| Days Max Temp ≤ 32 °F | 13 | 9 | 2 | 0 | 0 | 0 | 0 | 0 | 0 | 0 | 1 | 7 | 32 |
| Days Min Temp ≤ 32 °F | 28 | 24 | 21 | 9 | 1 | 0 | 0 | 0 | 1 | 7 | 16 | 25 | 132 |
| Days Min Temp ≤ 0 °F | 5 | 2 | 0 | 0 | 0 | 0 | 0 | 0 | 0 | 0 | 0 | 1 | 8 |
| Heating Degree Days | 1209 | 998 | 771 | 414 | 164 | 25 | 5 | 13 | 93 | 363 | 674 | 1035 | 5764 |
| Cooling Degree Days | 0 | 0 | 1 | 9 | 61 | 176 | 283 | 223 | 96 | 10 | 0 | 0 | 859 |
| Total Precipitation (") | 2.13 | 1.91 | 2.59 | 3.50 | 3.97 | 4.08 | 4.63 | 3.58 | 3.12 | 2.50 | 3.24 | 2.63 | 37.88 |
| Days ≥ 0.1" Precip | 5 | *5* | 6 | 7 | 7 | 7 | 7 | 6 | 5 | 6 | 7 | 6 | 74 |
| Total Snowfall (") | 6.0 | 4.5 | 2.9 | 0.5 | 0.0 | 0.0 | 0.0 | 0.0 | 0.0 | 0.1 | 1.2 | na | na |
| Days ≥ 1" Snow Depth | 10 | 8 | 2 | 0 | 0 | 0 | 0 | 0 | 0 | 0 | 1 | *4* | 25 |

## KENTON *Hardin County* ELEVATION 961 ft LAT/LONG 40° 39 ' N / 83° 37 ' W

| | JAN | FEB | MAR | APR | MAY | JUN | JUL | AUG | SEP | OCT | NOV | DEC | YEAR |
|---|---|---|---|---|---|---|---|---|---|---|---|---|---|
| Maximum Temp °F | 32.6 | 36.5 | 47.5 | 60.6 | 72.0 | 80.9 | 84.8 | 82.7 | 76.0 | 63.7 | 49.8 | 37.6 | 60.4 |
| Minimum Temp °F | 15.7 | 18.2 | 27.9 | 38.1 | 48.5 | 57.6 | 61.9 | 59.2 | 52.2 | 40.8 | 32.3 | 22.1 | 39.5 |
| Mean Temp °F | 24.3 | 27.6 | 37.7 | 49.4 | 60.3 | 69.3 | 73.5 | 71.0 | 64.2 | 52.3 | 41.0 | 29.9 | 50.0 |
| Days Max Temp ≥ 90 °F | 0 | 0 | 0 | 0 | 1 | 4 | 7 | 4 | 1 | 0 | 0 | 0 | 17 |
| Days Max Temp ≤ 32 °F | 15 | 10 | 4 | 0 | 0 | 0 | 0 | 0 | 0 | 0 | 2 | 9 | 40 |
| Days Min Temp ≤ 32 °F | 29 | 25 | 22 | 9 | 1 | 0 | 0 | 0 | 0 | 6 | 16 | 26 | 134 |
| Days Min Temp ≤ 0 °F | 4 | 3 | 0 | 0 | 0 | 0 | 0 | 0 | 0 | 0 | 0 | 1 | 8 |
| Heating Degree Days | 1256 | 1051 | 839 | 471 | 195 | 37 | 5 | 16 | 108 | 396 | 710 | 1082 | 6166 |
| Cooling Degree Days | 0 | 0 | 1 | 13 | 63 | 182 | 301 | 230 | 95 | 9 | 0 | 0 | 894 |
| Total Precipitation (") | 2.06 | 1.95 | 2.83 | 3.41 | 4.00 | 3.61 | 4.30 | 3.24 | 2.98 | 2.11 | 3.15 | 2.84 | 36.48 |
| Days ≥ 0.1" Precip | 5 | 5 | 7 | 8 | 7 | 7 | 7 | 6 | 6 | 6 | 7 | 7 | 78 |
| Total Snowfall (") | na | na | *4.0* | 0.5 | 0.0 | 0.0 | 0.0 | 0.0 | 0.0 | 0.0 | *1.9* | *5.2* | na |
| Days ≥ 1" Snow Depth | na | na | na | 0 | 0 | 0 | 0 | 0 | 0 | 0 | *1* | na | na |

## LANCASTER 2 NW *Fairfield County* ELEVATION 840 ft LAT/LONG 39° 43 ' N / 82° 36 ' W

| | JAN | FEB | MAR | APR | MAY | JUN | JUL | AUG | SEP | OCT | NOV | DEC | YEAR |
|---|---|---|---|---|---|---|---|---|---|---|---|---|---|
| Maximum Temp °F | 34.9 | 38.3 | 49.7 | 61.6 | 71.8 | 80.7 | 84.6 | 82.8 | 76.9 | 65.1 | 52.1 | 40.4 | 61.6 |
| Minimum Temp °F | 16.6 | 18.9 | 28.1 | 37.6 | 47.2 | 56.8 | 61.2 | 59.0 | 52.0 | 39.8 | 32.3 | 23.0 | 39.4 |
| Mean Temp °F | 25.8 | 28.6 | 38.9 | 49.6 | 59.5 | 68.8 | 72.9 | 70.9 | 64.5 | 52.5 | 42.3 | 31.7 | 50.5 |
| Days Max Temp ≥ 90 °F | 0 | 0 | 0 | 0 | 0 | 3 | 6 | 3 | 1 | 0 | 0 | 0 | 13 |
| Days Max Temp ≤ 32 °F | 13 | 10 | 3 | 0 | 0 | 0 | 0 | 0 | 0 | 0 | 1 | 7 | 34 |
| Days Min Temp ≤ 32 °F | 28 | 24 | 22 | 10 | 1 | 0 | 0 | 0 | 0 | 8 | 16 | 25 | 134 |
| Days Min Temp ≤ 0 °F | 4 | 2 | 0 | 0 | 0 | 0 | 0 | 0 | 0 | 0 | 0 | 1 | 7 |
| Heating Degree Days | 1209 | 1019 | 804 | 465 | 208 | 38 | 6 | 16 | 105 | 390 | 677 | 1026 | 5963 |
| Cooling Degree Days | 0 | 0 | 2 | 11 | 45 | 164 | 278 | 224 | 103 | 12 | 1 | 0 | 840 |
| Total Precipitation (") | 2.17 | 2.10 | 2.66 | 3.19 | 3.95 | 3.66 | 4.30 | 3.58 | 2.88 | 2.30 | 3.21 | 2.72 | 36.72 |
| Days ≥ 0.1" Precip | 5 | 5 | 6 | 6 | 7 | 6 | 6 | 6 | 5 | 5 | 7 | 6 | 70 |
| Total Snowfall (") | na | na | na | 0.7 | 0.0 | 0.0 | 0.0 | 0.0 | 0.0 | 0.0 | *0.4* | na | na |
| Days ≥ 1" Snow Depth | na | na | na | 0 | 0 | 0 | 0 | 0 | 0 | 0 | *0* | na | na |

## LIMA WWTP *Allen County* ELEVATION 860 ft LAT/LONG 40° 43 ' N / 84° 7 ' W

| | JAN | FEB | MAR | APR | MAY | JUN | JUL | AUG | SEP | OCT | NOV | DEC | YEAR |
|---|---|---|---|---|---|---|---|---|---|---|---|---|---|
| Maximum Temp °F | 32.9 | 37.0 | 48.2 | 61.2 | 72.5 | 81.1 | 84.8 | 82.5 | 76.3 | 64.2 | 50.5 | 38.7 | 60.8 |
| Minimum Temp °F | 17.4 | 20.0 | 29.5 | 39.7 | 50.1 | 59.4 | 63.6 | 61.5 | 54.9 | 43.4 | 34.3 | 24.2 | 41.5 |
| Mean Temp °F | 25.2 | 28.5 | 38.9 | 50.5 | 61.4 | 70.3 | 74.2 | 72.0 | 65.6 | 53.8 | 42.5 | 31.5 | 51.2 |
| Days Max Temp ≥ 90 °F | 0 | 0 | 0 | 0 | 1 | 3 | 6 | 4 | 1 | 0 | 0 | 0 | 15 |
| Days Max Temp ≤ 32 °F | 14 | 11 | 3 | 0 | 0 | 0 | 0 | 0 | 0 | 0 | 1 | 8 | 37 |
| Days Min Temp ≤ 32 °F | 28 | 24 | 20 | 7 | 1 | 0 | 0 | 0 | 0 | 4 | 14 | 24 | 122 |
| Days Min Temp ≤ 0 °F | 4 | 2 | 0 | 0 | 0 | 0 | 0 | 0 | 0 | 0 | 0 | 1 | 7 |
| Heating Degree Days | 1228 | 1022 | 804 | 439 | 173 | 26 | 3 | 10 | 85 | 354 | 669 | 1031 | 5844 |
| Cooling Degree Days | 0 | 0 | 2 | 13 | 72 | 196 | 318 | 255 | 117 | 14 | 0 | 0 | 987 |
| Total Precipitation (") | 2.07 | 1.79 | 2.63 | 3.35 | 3.94 | 3.93 | 4.05 | 3.27 | 3.44 | 2.33 | 3.30 | 2.72 | 36.82 |
| Days ≥ 0.1" Precip | 5 | 4 | 6 | 8 | 8 | 7 | 7 | 6 | 6 | 6 | 7 | 7 | 77 |
| Total Snowfall (") | na | na | na | 0.1 | 0.0 | 0.0 | 0.0 | 0.0 | 0.0 | 0.0 | ***0.9*** | na | na |
| Days ≥ 1" Snow Depth | na | na | na | 0 | 0 | 0 | 0 | 0 | 0 | 0 | ***1*** | na | na |

## LONDON *Madison County* ELEVATION 1020 ft LAT/LONG 39° 53 ' N / 83° 27 ' W

| | JAN | FEB | MAR | APR | MAY | JUN | JUL | AUG | SEP | OCT | NOV | DEC | YEAR |
|---|---|---|---|---|---|---|---|---|---|---|---|---|---|
| Maximum Temp °F | 34.4 | 38.8 | 50.8 | 62.9 | 73.3 | 81.7 | 85.2 | 82.9 | 76.9 | 65.1 | 51.6 | 40.0 | 62.0 |
| Minimum Temp °F | 18.1 | 20.4 | 29.8 | 39.0 | 48.9 | 57.8 | 61.7 | 59.5 | 52.7 | 41.2 | 33.4 | 24.5 | 40.6 |
| Mean Temp °F | 26.3 | 29.6 | 40.3 | 51.0 | 61.1 | 69.8 | 73.5 | 71.2 | 64.8 | 53.1 | 42.5 | 32.2 | 51.3 |
| Days Max Temp ≥ 90 °F | 0 | 0 | 0 | 0 | 0 | 4 | 6 | 3 | 1 | 0 | 0 | 0 | 14 |
| Days Max Temp ≤ 32 °F | 13 | 9 | 2 | 0 | 0 | 0 | 0 | 0 | 0 | 0 | 1 | 7 | 32 |
| Days Min Temp ≤ 32 °F | 27 | 24 | 20 | 9 | 1 | 0 | 0 | 0 | 0 | 6 | 15 | 24 | 126 |
| Days Min Temp ≤ 0 °F | 4 | 2 | 0 | 0 | 0 | 0 | 0 | 0 | 0 | 0 | 0 | 1 | 7 |
| Heating Degree Days | 1194 | 992 | 759 | 422 | 170 | 28 | 4 | 14 | 96 | 371 | 668 | 1009 | 5727 |
| Cooling Degree Days | 0 | 0 | 1 | 9 | 61 | 183 | 296 | 225 | 99 | 12 | 0 | 0 | 886 |
| Total Precipitation (") | 2.13 | 2.14 | 2.76 | 3.47 | 4.01 | 3.95 | 3.79 | 3.52 | 3.09 | 2.48 | 3.43 | 2.88 | 37.65 |
| Days ≥ 0.1" Precip | 5 | 6 | 7 | 7 | 8 | 7 | 6 | 6 | 5 | 6 | 7 | 7 | 77 |
| Total Snowfall (") | na | na | na | na | ***0.0*** | ***0.0*** | ***0.0*** | ***0.0*** | ***0.0*** | ***0.2*** | na | na | na |
| Days ≥ 1" Snow Depth | na | na | na | na | ***0*** | ***0*** | ***0*** | ***0*** | ***0*** | ***0*** | na | na | na |

## MANSFIELD 5 W *Richland County* ELEVATION 1289 ft LAT/LONG 40° 45 ' N / 82° 38 ' W

| | JAN | FEB | MAR | APR | MAY | JUN | JUL | AUG | SEP | OCT | NOV | DEC | YEAR |
|---|---|---|---|---|---|---|---|---|---|---|---|---|---|
| Maximum Temp °F | 31.6 | 35.2 | 46.6 | 59.4 | 69.7 | 78.1 | 81.8 | 80.0 | 73.3 | 61.7 | 48.9 | 37.2 | 58.6 |
| Minimum Temp °F | 14.9 | 17.3 | 26.6 | 36.5 | 46.5 | 54.9 | 59.2 | 57.4 | 51.2 | 40.3 | 31.7 | 21.7 | 38.2 |
| Mean Temp °F | 23.3 | 26.3 | 36.6 | 48.0 | 58.1 | 66.5 | 70.5 | 68.7 | 62.3 | 51.0 | 40.3 | 29.5 | 48.4 |
| Days Max Temp ≥ 90 °F | 0 | 0 | 0 | 0 | 0 | 1 | 2 | 1 | 0 | 0 | 0 | 0 | 4 |
| Days Max Temp ≤ 32 °F | 16 | 12 | 5 | 0 | 0 | 0 | 0 | 0 | 0 | 0 | 2 | 11 | 46 |
| Days Min Temp ≤ 32 °F | 29 | 25 | 23 | 12 | 2 | 0 | 0 | 0 | 1 | 7 | 17 | 26 | 142 |
| Days Min Temp ≤ 0 °F | 5 | 3 | 1 | 0 | 0 | 0 | 0 | 0 | 0 | 0 | 0 | 2 | 11 |
| Heating Degree Days | 1287 | 1087 | 874 | 511 | 241 | 62 | 17 | 27 | 140 | 433 | 733 | 1095 | 6507 |
| Cooling Degree Days | 0 | 0 | 1 | 8 | 36 | 114 | 213 | 158 | 67 | 6 | 0 | 0 | 603 |
| Total Precipitation (") | 1.91 | 1.66 | 2.70 | 3.45 | 4.15 | 4.08 | 4.26 | 3.60 | 3.26 | 2.41 | 3.16 | 2.50 | 37.14 |
| Days ≥ 0.1" Precip | 5 | 5 | 7 | 8 | 9 | 7 | 7 | 7 | 6 | 6 | 7 | 6 | 80 |
| Total Snowfall (") | na | na | na | 0.4 | 0.0 | 0.0 | 0.0 | 0.0 | 0.0 | 0.0 | ***1.7*** | ***5.0*** | na |
| Days ≥ 1" Snow Depth | na | na | ***4*** | 0 | 0 | 0 | 0 | 0 | 0 | 0 | ***2*** | na | na |

## MANSFIELD LAHM AP *Richland County* ELEVATION 1302 ft LAT/LONG 40° 49 ' N / 82° 31 ' W

| | JAN | FEB | MAR | APR | MAY | JUN | JUL | AUG | SEP | OCT | NOV | DEC | YEAR |
|---|---|---|---|---|---|---|---|---|---|---|---|---|---|
| Maximum Temp °F | 32.0 | 35.1 | 46.3 | 58.8 | 69.2 | 78.0 | 82.0 | 80.1 | 73.3 | 61.6 | 49.0 | 37.4 | 58.6 |
| Minimum Temp °F | 17.3 | 19.3 | 28.4 | 38.5 | 48.4 | 57.4 | 62.2 | 60.5 | 53.8 | 42.7 | 33.9 | 23.7 | 40.5 |
| Mean Temp °F | 24.7 | 27.2 | 37.4 | 48.6 | 58.8 | 67.7 | 72.2 | 70.3 | 63.6 | 52.2 | 41.4 | 30.6 | 49.6 |
| Days Max Temp ≥ 90 °F | 0 | 0 | 0 | 0 | 0 | 1 | 3 | 1 | 0 | 0 | 0 | 0 | 5 |
| Days Max Temp ≤ 32 °F | 16 | 12 | 4 | 0 | 0 | 0 | 0 | 0 | 0 | 0 | 2 | 11 | 45 |
| Days Min Temp ≤ 32 °F | 28 | 24 | 21 | 9 | 1 | 0 | 0 | 0 | 0 | 4 | 15 | 25 | 127 |
| Days Min Temp ≤ 0 °F | 4 | 2 | 0 | 0 | 0 | 0 | 0 | 0 | 0 | 0 | 0 | 1 | 7 |
| Heating Degree Days | 1245 | 1060 | 850 | 492 | 226 | 50 | 8 | 19 | 117 | 399 | 701 | 1060 | 6227 |
| Cooling Degree Days | 0 | 0 | 1 | 9 | 45 | 135 | 246 | 186 | 77 | 7 | 0 | 0 | 706 |
| Total Precipitation (") | 2.26 | 1.98 | 3.19 | 3.86 | 4.46 | 4.08 | 4.31 | 4.07 | 3.53 | 2.47 | 3.93 | 3.21 | 41.35 |
| Days ≥ 0.1" Precip | 6 | 5 | 8 | 9 | 9 | 7 | 7 | 7 | 6 | 6 | 8 | 7 | 85 |
| Total Snowfall (") | 11.7 | 9.7 | 6.7 | 1.9 | 0.1 | 0.0 | 0.0 | 0.0 | 0.0 | 0.5 | 2.5 | 8.2 | 41.3 |
| Days ≥ 1" Snow Depth | 16 | 14 | 6 | 1 | 0 | 0 | 0 | 0 | 0 | 0 | 2 | 10 | 49 |

## MARIETTA WWTP *Washington County* ELEVATION 581 ft LAT/LONG 39° 25' N / 81° 26' W

| | JAN | FEB | MAR | APR | MAY | JUN | JUL | AUG | SEP | OCT | NOV | DEC | YEAR |
|---|---|---|---|---|---|---|---|---|---|---|---|---|---|
| Maximum Temp °F | 39.5 | 43.4 | 54.8 | 66.0 | 75.2 | 82.8 | 86.1 | 84.7 | 78.4 | 67.2 | 55.4 | 44.7 | 64.9 |
| Minimum Temp °F | 21.7 | 23.4 | 32.2 | 41.1 | 50.5 | 59.2 | 64.0 | 62.4 | 55.7 | 43.4 | 35.5 | 27.3 | 43.0 |
| Mean Temp °F | 30.6 | 33.4 | 43.5 | 53.6 | 62.8 | 71.0 | 75.1 | 73.6 | 67.1 | 55.3 | 45.4 | 36.0 | 54.0 |
| Days Max Temp ≥ 90 °F | 0 | 0 | 0 | 0 | 1 | 5 | 8 | 6 | 2 | 0 | 0 | 0 | 22 |
| Days Max Temp ≤ 32 °F | 9 | 6 | 1 | 0 | 0 | 0 | 0 | 0 | 0 | 0 | 0 | 4 | 20 |
| Days Min Temp ≤ 32 °F | 26 | 22 | 17 | 6 | 1 | 0 | 0 | 0 | 0 | 4 | 13 | 22 | 111 |
| Days Min Temp ≤ 0 °F | 2 | 1 | 0 | 0 | 0 | 0 | 0 | 0 | 0 | 0 | 0 | 0 | 3 |
| Heating Degree Days | 1060 | 885 | 660 | 348 | 130 | 18 | 1 | 4 | 58 | 307 | 582 | 891 | 4944 |
| Cooling Degree Days | 0 | 0 | 2 | 13 | 70 | 205 | 341 | 278 | 122 | 15 | 1 | 0 | 1047 |
| Total Precipitation (") | 2.85 | 2.63 | 3.46 | 3.25 | 3.97 | 3.71 | 4.14 | 4.03 | 3.21 | 2.77 | 3.25 | 3.31 | 40.58 |
| Days ≥ 0.1" Precip | 7 | 6 | 8 | 8 | 9 | 7 | 8 | 7 | 6 | 6 | 7 | 7 | 86 |
| Total Snowfall (") | *6.3* | na | *2.7* | 0.5 | 0.0 | 0.0 | 0.0 | 0.0 | 0.0 | 0.0 | *0.5* | *2.5* | na |
| Days ≥ 1" Snow Depth | na | na | *1* | 0 | 0 | 0 | 0 | 0 | 0 | 0 | *0* | *2* | na |

## MARION 2 N *Marion County* ELEVATION 922 ft LAT/LONG 40° 36' N / 83° 10' W

| | JAN | FEB | MAR | APR | MAY | JUN | JUL | AUG | SEP | OCT | NOV | DEC | YEAR |
|---|---|---|---|---|---|---|---|---|---|---|---|---|---|
| Maximum Temp °F | 32.3 | 36.2 | 47.5 | 60.6 | 71.5 | 80.8 | 84.3 | 82.5 | 75.7 | 63.4 | 50.0 | 37.9 | 60.2 |
| Minimum Temp °F | 15.5 | 17.9 | 27.6 | 37.8 | 48.1 | 57.6 | 61.5 | 58.7 | 51.8 | 40.4 | 32.3 | 22.6 | 39.3 |
| Mean Temp °F | 23.9 | 27.1 | 37.6 | 49.2 | 59.8 | 69.2 | 72.9 | 70.6 | 63.7 | 51.9 | 41.1 | 30.3 | 49.8 |
| Days Max Temp ≥ 90 °F | 0 | 0 | 0 | 0 | 1 | 4 | 6 | 4 | 1 | 0 | 0 | 0 | 16 |
| Days Max Temp ≤ 32 °F | 15 | 11 | 4 | 0 | 0 | 0 | 0 | 0 | 0 | 0 | 1 | 9 | 40 |
| Days Min Temp ≤ 32 °F | 29 | 25 | 22 | 10 | 1 | 0 | 0 | 0 | 0 | 6 | 17 | 26 | 136 |
| Days Min Temp ≤ 0 °F | 4 | 3 | 0 | 0 | 0 | 0 | 0 | 0 | 0 | 0 | 0 | 1 | 8 |
| Heating Degree Days | 1268 | 1063 | 843 | 475 | 209 | 38 | 6 | 19 | 118 | 407 | 709 | 1068 | 6223 |
| Cooling Degree Days | 0 | 0 | 2 | 10 | 54 | 166 | 272 | 203 | 85 | 9 | 0 | 0 | 801 |
| Total Precipitation (") | 2.19 | 1.66 | 2.34 | 3.54 | 4.10 | 3.95 | 4.57 | 3.53 | 3.17 | 2.61 | 3.23 | 2.85 | 37.74 |
| Days ≥ 0.1" Precip | 5 | 5 | 6 | 8 | 8 | 7 | 7 | 6 | 6 | 6 | 8 | 7 | 79 |
| Total Snowfall (") | 8.7 | 6.1 | *3.7* | 0.6 | 0.0 | 0.0 | 0.0 | 0.0 | 0.0 | 0.0 | 1.3 | *4.2* | 24.6 |
| Days ≥ 1" Snow Depth | *14* | *11* | *3* | 0 | 0 | 0 | 0 | 0 | 0 | 0 | 1 | *5* | 34 |

## MARYSVILLE *Union County* ELEVATION 1001 ft LAT/LONG 40° 14' N / 83° 22' W

| | JAN | FEB | MAR | APR | MAY | JUN | JUL | AUG | SEP | OCT | NOV | DEC | YEAR |
|---|---|---|---|---|---|---|---|---|---|---|---|---|---|
| Maximum Temp °F | 33.4 | 37.1 | 48.9 | 61.3 | 72.0 | 80.5 | 84.0 | 82.0 | 75.1 | 63.0 | 50.0 | 38.5 | 60.5 |
| Minimum Temp °F | 17.1 | 19.4 | 28.9 | 38.9 | 49.3 | 58.0 | 62.3 | 60.1 | 53.3 | 41.8 | 33.0 | 23.6 | 40.5 |
| Mean Temp °F | 25.2 | 28.3 | 38.9 | 50.1 | 60.7 | 69.2 | 73.2 | 71.1 | 64.3 | 52.4 | 41.5 | 31.1 | 50.5 |
| Days Max Temp ≥ 90 °F | 0 | 0 | 0 | 0 | 0 | 3 | 5 | 3 | 1 | 0 | 0 | 0 | 12 |
| Days Max Temp ≤ 32 °F | 14 | 10 | 3 | 0 | 0 | 0 | 0 | 0 | 0 | 0 | 1 | 9 | 37 |
| Days Min Temp ≤ 32 °F | 28 | 24 | 21 | 8 | 1 | 0 | 0 | 0 | 0 | 5 | 15 | 25 | 127 |
| Days Min Temp ≤ 0 °F | 4 | 3 | 0 | 0 | 0 | 0 | 0 | 0 | 0 | 0 | 0 | 1 | 8 |
| Heating Degree Days | 1225 | 1032 | 803 | 448 | 183 | 32 | 5 | 13 | 104 | 390 | 698 | 1045 | 5978 |
| Cooling Degree Days | 0 | 0 | 1 | 9 | 61 | 169 | 291 | 225 | 93 | 8 | 1 | 0 | 858 |
| Total Precipitation (") | 2.13 | 1.90 | 2.67 | 3.23 | 3.81 | 3.93 | 4.13 | 3.16 | 2.87 | 2.45 | 3.13 | 2.69 | 36.10 |
| Days ≥ 0.1" Precip | 5 | 5 | 7 | 7 | 7 | 7 | 7 | 7 | 5 | 6 | 7 | 7 | 77 |
| Total Snowfall (") | 6.3 | 5.4 | 3.4 | 0.8 | 0.0 | 0.0 | 0.0 | 0.0 | 0.0 | 0.1 | 1.2 | 3.6 | 20.8 |
| Days ≥ 1" Snow Depth | 12 | 8 | 3 | 0 | 0 | 0 | 0 | 0 | 0 | 0 | 1 | 5 | 29 |

## MCCONNELSVILLE LOCK7 *Morgan County* ELEVATION 669 ft LAT/LONG 39° 39' N / 81° 51' W

| | JAN | FEB | MAR | APR | MAY | JUN | JUL | AUG | SEP | OCT | NOV | DEC | YEAR |
|---|---|---|---|---|---|---|---|---|---|---|---|---|---|
| Maximum Temp °F | 38.1 | 41.5 | 53.0 | 64.9 | 74.6 | 82.2 | 85.6 | 84.2 | 78.0 | 66.5 | 54.6 | 43.2 | 63.9 |
| Minimum Temp °F | 16.6 | 18.4 | 27.5 | 36.9 | 46.7 | 55.9 | 61.0 | 59.4 | 52.1 | 39.0 | 31.5 | 23.0 | 39.0 |
| Mean Temp °F | 27.4 | 30.0 | 40.3 | 50.9 | 60.6 | 69.1 | 73.3 | 71.8 | 65.1 | 52.8 | 43.1 | 33.1 | 51.5 |
| Days Max Temp ≥ 90 °F | 0 | 0 | 0 | 0 | 1 | 4 | 8 | 6 | 2 | 0 | 0 | 0 | 21 |
| Days Max Temp ≤ 32 °F | 10 | 6 | 1 | 0 | 0 | 0 | 0 | 0 | 0 | 0 | 0 | 5 | 22 |
| Days Min Temp ≤ 32 °F | 28 | 25 | 22 | 11 | 2 | 0 | 0 | 0 | 0 | 8 | 17 | 25 | 138 |
| Days Min Temp ≤ 0 °F | 4 | 3 | 0 | 0 | 0 | 0 | 0 | 0 | 0 | 0 | 0 | 1 | 8 |
| Heating Degree Days | 1159 | 982 | 760 | 424 | 177 | 31 | 4 | 9 | 89 | 380 | 650 | 981 | 5646 |
| Cooling Degree Days | 0 | 0 | 1 | 8 | 42 | 148 | 271 | 218 | 89 | 7 | 0 | 0 | 784 |
| Total Precipitation (") | 2.71 | 2.52 | 3.40 | 3.69 | 4.18 | 3.66 | 4.84 | 4.16 | 3.13 | 2.74 | 3.39 | 3.22 | 41.64 |
| Days ≥ 0.1" Precip | 7 | 7 | 8 | 8 | 9 | 6 | 8 | 7 | 6 | 6 | 7 | 8 | 87 |
| Total Snowfall (") | 8.6 | 6.6 | 3.7 | 0.9 | 0.0 | 0.0 | 0.0 | 0.0 | 0.0 | 0.0 | 0.8 | 3.2 | 23.8 |
| Days ≥ 1" Snow Depth | na | *7* | *2* | 0 | 0 | 0 | 0 | 0 | 0 | 0 | 1 | *3* | na |

## MILLERSBURG *Holmes County* ELEVATION 981 ft LAT/LONG 40° 33 ' N / 81° 56 ' W

| | JAN | FEB | MAR | APR | MAY | JUN | JUL | AUG | SEP | OCT | NOV | DEC | YEAR |
|---|---|---|---|---|---|---|---|---|---|---|---|---|---|
| Maximum Temp °F | 34.8 | 37.8 | 49.7 | 61.4 | 71.0 | 79.8 | 83.2 | 81.8 | 75.2 | 63.5 | 50.9 | 39.4 | 60.7 |
| Minimum Temp °F | 17.4 | 18.9 | 28.5 | 37.3 | 46.6 | 55.9 | 60.0 | 58.2 | 51.9 | 40.1 | 32.5 | 23.4 | 39.2 |
| Mean Temp °F | 26.1 | 28.4 | 39.1 | 49.3 | 58.8 | 67.9 | 71.7 | 70.0 | 63.5 | 51.8 | 41.8 | 31.4 | 50.0 |
| Days Max Temp ≥ 90 °F | 0 | 0 | 0 | 0 | 0 | 2 | 4 | 2 | 1 | 0 | 0 | 0 | 9 |
| Days Max Temp ≤ 32 °F | 13 | 10 | 2 | 0 | 0 | 0 | 0 | 0 | 0 | 0 | 1 | 8 | 34 |
| Days Min Temp ≤ 32 °F | 28 | 25 | 21 | 11 | 2 | 0 | 0 | 0 | 0 | 7 | 17 | 25 | 136 |
| Days Min Temp ≤ 0 °F | 3 | 2 | 0 | 0 | 0 | 0 | 0 | 0 | 0 | 0 | 0 | 1 | 6 |
| Heating Degree Days | 1199 | 1028 | 795 | 470 | 220 | 42 | 8 | 17 | 113 | 408 | 691 | 1033 | 6024 |
| Cooling Degree Days | na | na | na | na | na | na | na | 184 | na | na | na | na | na |
| Total Precipitation (") | 2.28 | 1.94 | 2.89 | 3.31 | 3.92 | 4.35 | 4.16 | 3.66 | 3.47 | 2.52 | 3.12 | 2.50 | 38.12 |
| Days ≥ 0.1" Precip | 5 | 5 | 8 | 7 | 8 | 7 | 8 | 7 | 6 | 5 | 7 | 6 | 79 |
| Total Snowfall (") | 9.1 | 6.8 | 4.0 | 0.5 | 0.0 | 0.0 | 0.0 | 0.0 | 0.0 | 0.0 | 1.0 | 5.3 | 26.7 |
| Days ≥ 1" Snow Depth | 14 | 11 | 4 | 0 | 0 | 0 | 0 | 0 | 0 | 0 | 1 | 6 | 36 |

## MILLPORT 2 NW *Columbiana County* ELEVATION 1152 ft LAT/LONG 40° 43 ' N / 80° 54 ' W

| | JAN | FEB | MAR | APR | MAY | JUN | JUL | AUG | SEP | OCT | NOV | DEC | YEAR |
|---|---|---|---|---|---|---|---|---|---|---|---|---|---|
| Maximum Temp °F | 34.1 | 38.1 | 49.2 | 61.4 | 71.0 | 79.5 | 83.2 | 81.6 | 74.8 | 63.0 | 50.6 | 39.3 | 60.5 |
| Minimum Temp °F | 16.1 | 17.8 | 26.9 | 35.7 | 45.0 | 53.5 | 58.2 | 56.3 | 49.8 | 38.6 | 31.4 | 22.7 | 37.7 |
| Mean Temp °F | 25.2 | 28.0 | 38.1 | 48.6 | 58.0 | 66.6 | 70.7 | 69.0 | 62.3 | 50.8 | 41.1 | 31.1 | 49.1 |
| Days Max Temp ≥ 90 °F | 0 | 0 | 0 | 0 | 0 | 2 | 4 | 2 | 1 | 0 | 0 | 0 | 9 |
| Days Max Temp ≤ 32 °F | 13 | 9 | 3 | 0 | 0 | 0 | 0 | 0 | 0 | 0 | 1 | 8 | 34 |
| Days Min Temp ≤ 32 °F | 28 | 25 | 22 | 12 | 3 | 0 | 0 | 0 | 1 | 10 | 18 | 25 | 144 |
| Days Min Temp ≤ 0 °F | 5 | 3 | 1 | 0 | 0 | 0 | 0 | 0 | 0 | 0 | 0 | 1 | 10 |
| Heating Degree Days | 1229 | 1038 | 828 | 490 | 238 | 58 | 13 | 26 | 136 | 436 | 712 | 1045 | 6249 |
| Cooling Degree Days | 0 | 0 | 0 | 5 | 30 | 113 | 220 | 165 | 62 | 4 | 0 | 0 | 599 |
| Total Precipitation (") | 2.40 | 2.30 | 3.19 | 3.09 | 3.95 | 3.52 | 4.24 | 3.10 | 3.31 | 2.50 | 3.39 | 3.11 | 38.10 |
| Days ≥ 0.1" Precip | 7 | 7 | 8 | 8 | 9 | 7 | 8 | 7 | 7 | 6 | 8 | 8 | 90 |
| Total Snowfall (") | 7.9 | 7.1 | 6.4 | 1.3 | 0.0 | 0.0 | 0.0 | 0.0 | 0.0 | 0.1 | 2.7 | 6.3 | 31.8 |
| Days ≥ 1" Snow Depth | na | na | na | 0 | 0 | 0 | 0 | 0 | 0 | 0 | 1 | na | na |

## MINERAL RIDGE W WKS *Trumbull County* ELEVATION 889 ft LAT/LONG 41° 9 ' N / 80° 47 ' W

| | JAN | FEB | MAR | APR | MAY | JUN | JUL | AUG | SEP | OCT | NOV | DEC | YEAR |
|---|---|---|---|---|---|---|---|---|---|---|---|---|---|
| Maximum Temp °F | 34.8 | 38.7 | 49.9 | 62.7 | 73.4 | 81.6 | 85.5 | 83.6 | 76.9 | 64.9 | 51.7 | 40.3 | 62.0 |
| Minimum Temp °F | 17.8 | 19.2 | 27.6 | 37.0 | 46.4 | 55.2 | 60.0 | 58.6 | 52.6 | 41.2 | 33.6 | 25.0 | 39.5 |
| Mean Temp °F | 26.3 | 29.0 | 38.8 | 49.9 | 60.0 | 68.5 | 72.8 | 71.2 | 64.7 | 53.1 | 42.7 | 32.7 | 50.8 |
| Days Max Temp ≥ 90 °F | 0 | 0 | 0 | 0 | 0 | 4 | 7 | 4 | 1 | 0 | 0 | 0 | 16 |
| Days Max Temp ≤ 32 °F | 13 | 9 | 3 | 0 | 0 | 0 | 0 | 0 | 0 | 0 | 1 | 7 | 33 |
| Days Min Temp ≤ 32 °F | 28 | 24 | 22 | 10 | 2 | 0 | 0 | 0 | 0 | 6 | 15 | 25 | 132 |
| Days Min Temp ≤ 0 °F | 3 | 2 | 0 | 0 | 0 | 0 | 0 | 0 | 0 | 0 | 0 | 1 | 6 |
| Heating Degree Days | 1192 | 1011 | 807 | 455 | 198 | 42 | 6 | 14 | 96 | 371 | 663 | 995 | 5850 |
| Cooling Degree Days | 0 | 0 | 1 | 7 | 52 | 147 | 276 | 218 | 98 | 8 | 0 | 0 | 807 |
| Total Precipitation (") | 2.11 | 1.81 | 2.68 | 2.86 | 3.35 | 4.03 | 4.08 | 3.37 | 3.82 | 2.48 | 3.10 | 2.59 | 36.28 |
| Days ≥ 0.1" Precip | 5 | 5 | 7 | 7 | 8 | 8 | 7 | 6 | 7 | 6 | 7 | 7 | 80 |
| Total Snowfall (") | 9.9 | 7.9 | 8.1 | 1.2 | 0.1 | 0.0 | 0.0 | 0.0 | 0.0 | 0.1 | 2.5 | 8.2 | 38.0 |
| Days ≥ 1" Snow Depth | 16 | 14 | 5 | 0 | 0 | 0 | 0 | 0 | 0 | 0 | 1 | 8 | 44 |

## MONTPELIER *Williams County* ELEVATION 860 ft LAT/LONG 41° 35 ' N / 84° 36 ' W

| | JAN | FEB | MAR | APR | MAY | JUN | JUL | AUG | SEP | OCT | NOV | DEC | YEAR |
|---|---|---|---|---|---|---|---|---|---|---|---|---|---|
| Maximum Temp °F | 30.3 | 33.8 | 45.0 | 58.7 | 70.8 | 80.1 | 83.8 | 81.9 | 74.7 | 62.0 | 48.1 | 35.9 | 58.8 |
| Minimum Temp °F | 13.3 | 15.3 | 25.3 | 35.8 | 45.6 | 55.2 | 59.5 | 57.0 | 49.6 | 38.3 | 30.4 | 20.6 | 37.2 |
| Mean Temp °F | 21.8 | 24.6 | 35.2 | 47.2 | 58.2 | 67.6 | 71.6 | 69.5 | 62.2 | 50.2 | 39.3 | 28.3 | 48.0 |
| Days Max Temp ≥ 90 °F | 0 | 0 | 0 | 0 | 1 | 3 | 5 | 3 | 1 | 0 | 0 | 0 | 13 |
| Days Max Temp ≤ 32 °F | 17 | 13 | 4 | 0 | 0 | 0 | 0 | 0 | 0 | 0 | 2 | 11 | 47 |
| Days Min Temp ≤ 32 °F | 30 | 27 | 25 | 12 | 2 | 0 | 0 | 0 | 1 | 9 | 19 | 28 | 153 |
| Days Min Temp ≤ 0 °F | 6 | 4 | 0 | 0 | 0 | 0 | 0 | 0 | 0 | 0 | 0 | 2 | 12 |
| Heating Degree Days | 1332 | 1135 | 917 | 531 | 241 | 51 | 10 | 23 | 141 | 458 | 764 | 1132 | 6735 |
| Cooling Degree Days | 0 | 0 | 0 | 7 | 40 | 139 | 240 | 178 | 60 | 3 | 0 | 0 | 667 |
| Total Precipitation (") | 1.84 | 1.76 | 2.78 | 3.46 | 3.48 | 3.34 | 3.60 | 3.19 | 3.30 | 2.53 | 3.31 | 2.95 | 35.54 |
| Days ≥ 0.1" Precip | 5 | 5 | 7 | 8 | 7 | 7 | 7 | 6 | 6 | 6 | 7 | 7 | 78 |
| Total Snowfall (") | 8.6 | 9.6 | 5.1 | 0.9 | 0.0 | 0.0 | 0.0 | 0.0 | 0.1 | 0.2 | 2.4 | na | na |
| Days ≥ 1" Snow Depth | 17 | 14 | 5 | 1 | 0 | 0 | 0 | 0 | 0 | 0 | 2 | 8 | 47 |

## NEW LEXINGTON 2 NW *Perry County* ELEVATION 889 ft LAT/LONG 39° 44' N / 82° 13' W

| | JAN | FEB | MAR | APR | MAY | JUN | JUL | AUG | SEP | OCT | NOV | DEC | YEAR |
|---|---|---|---|---|---|---|---|---|---|---|---|---|---|
| Maximum Temp °F | 36.7 | 41.3 | 53.0 | 64.7 | 74.2 | 81.8 | 84.9 | 83.2 | 77.2 | 65.9 | 53.3 | 41.9 | 63.2 |
| Minimum Temp °F | 17.0 | 19.1 | 28.2 | 37.2 | 46.8 | 55.5 | 60.2 | 58.5 | 51.5 | 39.1 | 31.5 | 23.3 | 39.0 |
| Mean Temp °F | 26.9 | 30.2 | 40.6 | 51.0 | 60.5 | 68.6 | 72.6 | 70.9 | 64.3 | 52.5 | 42.4 | 32.6 | 51.1 |
| Days Max Temp ≥ 90 °F | 0 | 0 | 0 | 0 | 1 | 3 | 7 | 4 | 1 | 0 | 0 | 0 | 16 |
| Days Max Temp ≤ 32 °F | 11 | 7 | 2 | 0 | 0 | 0 | 0 | 0 | 0 | 0 | 1 | 6 | 27 |
| Days Min Temp ≤ 32 °F | 28 | 24 | 21 | 11 | 2 | 0 | 0 | 0 | 1 | 9 | 17 | 25 | 138 |
| Days Min Temp ≤ 0 °F | 4 | 2 | 0 | 0 | 0 | 0 | 0 | 0 | 0 | 0 | 0 | 1 | 7 |
| Heating Degree Days | 1175 | 976 | 750 | 421 | 180 | 33 | 5 | 13 | 99 | 386 | 673 | 997 | 5708 |
| Cooling Degree Days | 0 | 0 | 0 | 7 | 42 | 140 | 256 | 201 | 86 | 7 | 0 | 0 | 739 |
| Total Precipitation (") | 2.67 | 2.58 | 3.40 | 3.82 | 4.25 | 3.85 | 4.83 | 3.80 | 2.95 | 2.53 | 3.46 | 3.09 | 41.23 |
| Days ≥ 0.1" Precip | 7 | 6 | 8 | 8 | 9 | 7 | 8 | 7 | 5 | 6 | 8 | 8 | 87 |
| Total Snowfall (") | 8.9 | 6.2 | 3.4 | 0.5 | 0.0 | 0.0 | 0.0 | 0.0 | 0.0 | 0.0 | 0.9 | 3.2 | 23.1 |
| Days ≥ 1" Snow Depth | 14 | 9 | 3 | 0 | 0 | 0 | 0 | 0 | 0 | 0 | 1 | 5 | 32 |

## NEW PHILADELPHIA *Tuscarawas County* ELEVATION 902 ft LAT/LONG 40° 30' N / 81° 27' W

| | JAN | FEB | MAR | APR | MAY | JUN | JUL | AUG | SEP | OCT | NOV | DEC | YEAR |
|---|---|---|---|---|---|---|---|---|---|---|---|---|---|
| Maximum Temp °F | 34.4 | 37.9 | 49.1 | 61.1 | 71.9 | 80.6 | 84.4 | 82.9 | 76.1 | 63.7 | 51.3 | 39.7 | 61.1 |
| Minimum Temp °F | 16.8 | 18.3 | 27.0 | 36.5 | 46.0 | 55.4 | 60.0 | 58.3 | 51.4 | 39.3 | 32.3 | 23.6 | 38.7 |
| Mean Temp °F | 25.6 | 28.2 | 38.1 | 48.8 | 59.0 | 68.1 | 72.2 | 70.6 | 63.8 | 51.6 | 41.8 | 31.7 | 50.0 |
| Days Max Temp ≥ 90 °F | 0 | 0 | 0 | 0 | 0 | 3 | 6 | 4 | 1 | 0 | 0 | 0 | 14 |
| Days Max Temp ≤ 32 °F | 14 | 10 | 3 | 0 | 0 | 0 | 0 | 0 | 0 | 0 | 1 | 8 | 36 |
| Days Min Temp ≤ 32 °F | 28 | 24 | 22 | 12 | 2 | 0 | 0 | 0 | 0 | 8 | 16 | 25 | 137 |
| Days Min Temp ≤ 0 °F | 4 | 3 | 0 | 0 | 0 | 0 | 0 | 0 | 0 | 0 | 0 | 1 | 8 |
| Heating Degree Days | 1215 | 1034 | 829 | 486 | 223 | 47 | 9 | 19 | 115 | 417 | 689 | 1027 | 6110 |
| Cooling Degree Days | 0 | 0 | 1 | 9 | 43 | 142 | 263 | 207 | 86 | 8 | 1 | 0 | 760 |
| Total Precipitation (") | 2.47 | 2.39 | 3.31 | 3.38 | 3.91 | 3.85 | 4.49 | 4.04 | 3.19 | 2.44 | 3.36 | 2.93 | 39.76 |
| Days ≥ 0.1" Precip | 6 | 6 | 8 | 8 | 8 | 7 | 7 | 7 | 6 | 6 | 8 | 7 | 84 |
| Total Snowfall (") | 10.2 | 7.3 | 5.4 | 1.5 | 0.0 | 0.0 | 0.0 | 0.0 | 0.0 | 0.0 | 1.4 | 5.3 | 31.1 |
| Days ≥ 1" Snow Depth | 15 | 12 | 5 | 0 | 0 | 0 | 0 | 0 | 0 | 0 | 1 | 6 | 39 |

## NEWARK WATER WORKS *Licking County* ELEVATION 840 ft LAT/LONG 40° 5' N / 82° 25' W

| | JAN | FEB | MAR | APR | MAY | JUN | JUL | AUG | SEP | OCT | NOV | DEC | YEAR |
|---|---|---|---|---|---|---|---|---|---|---|---|---|---|
| Maximum Temp °F | 35.3 | 39.5 | 51.4 | 63.5 | 73.6 | 81.7 | 85.0 | 83.1 | 76.5 | 64.6 | 52.0 | 40.6 | 62.2 |
| Minimum Temp °F | 18.9 | 21.1 | 30.5 | 39.7 | 48.9 | 57.4 | 61.9 | 60.1 | 53.4 | 41.6 | 33.6 | 24.9 | 41.0 |
| Mean Temp °F | 27.2 | 30.3 | 40.9 | 51.6 | 61.2 | 69.6 | 73.4 | 71.6 | 65.0 | 53.1 | 42.8 | 32.7 | 51.6 |
| Days Max Temp ≥ 90 °F | 0 | 0 | 0 | 0 | 0 | 3 | 6 | 3 | 1 | 0 | 0 | 0 | 13 |
| Days Max Temp ≤ 32 °F | 12 | 8 | 2 | 0 | 0 | 0 | 0 | 0 | 0 | 0 | 1 | 7 | 30 |
| Days Min Temp ≤ 32 °F | 27 | 23 | 19 | 8 | 1 | 0 | 0 | 0 | 0 | 6 | 15 | 24 | 123 |
| Days Min Temp ≤ 0 °F | 3 | 2 | 0 | 0 | 0 | 0 | 0 | 0 | 0 | 0 | 0 | 1 | 6 |
| Heating Degree Days | 1167 | 972 | 740 | 404 | 165 | 27 | 4 | 10 | 92 | 370 | 659 | 993 | 5603 |
| Cooling Degree Days | 0 | 0 | 1 | 9 | 54 | 165 | 286 | 226 | 98 | 8 | 1 | 0 | 848 |
| Total Precipitation (") | 2.70 | 2.39 | 3.24 | 3.79 | 4.23 | 4.09 | 4.70 | 3.99 | 3.18 | 2.66 | 3.52 | 3.11 | 41.60 |
| Days ≥ 0.1" Precip | 7 | 6 | 8 | 8 | 8 | 7 | 8 | 7 | 6 | 6 | 8 | 7 | 86 |
| Total Snowfall (") | 8.2 | 6.0 | 3.0 | 0.9 | 0.0 | 0.0 | 0.0 | 0.0 | 0.0 | 0.0 | 0.6 | *3.3* | 22.0 |
| Days ≥ 1" Snow Depth | 12 | 9 | 3 | 0 | 0 | 0 | 0 | 0 | 0 | 0 | 0 | 5 | 29 |

## NORWALK WWTP *Huron County* ELEVATION 722 ft LAT/LONG 41° 15' N / 82° 37' W

| | JAN | FEB | MAR | APR | MAY | JUN | JUL | AUG | SEP | OCT | NOV | DEC | YEAR |
|---|---|---|---|---|---|---|---|---|---|---|---|---|---|
| Maximum Temp °F | 32.3 | 34.9 | 45.4 | 58.3 | 69.6 | 78.8 | 82.7 | 80.8 | 74.4 | 62.4 | 49.6 | 37.7 | 58.9 |
| Minimum Temp °F | 16.1 | 17.8 | 27.1 | 37.0 | 47.2 | 56.7 | 61.1 | 59.1 | 52.2 | 40.9 | 33.2 | 23.1 | 39.3 |
| Mean Temp °F | 24.3 | 26.4 | 36.3 | 47.7 | 58.4 | 67.8 | 71.9 | 70.0 | 63.3 | 51.7 | 41.5 | 30.4 | 49.1 |
| Days Max Temp ≥ 90 °F | 0 | 0 | 0 | 0 | 0 | 3 | 5 | 2 | 1 | 0 | 0 | 0 | 11 |
| Days Max Temp ≤ 32 °F | 15 | 12 | 4 | 0 | 0 | 0 | 0 | 0 | 0 | 0 | 1 | 9 | 41 |
| Days Min Temp ≤ 32 °F | 29 | 25 | 23 | 11 | 1 | 0 | 0 | 0 | 0 | 5 | 16 | 26 | 136 |
| Days Min Temp ≤ 0 °F | 4 | 3 | 0 | 0 | 0 | 0 | 0 | 0 | 0 | 0 | 0 | 1 | 8 |
| Heating Degree Days | 1257 | 1084 | 885 | 522 | 242 | 58 | 13 | 24 | 125 | 415 | 701 | 1065 | 6391 |
| Cooling Degree Days | 0 | 0 | 1 | 10 | 46 | 144 | 248 | 189 | 84 | 8 | 0 | 0 | 730 |
| Total Precipitation (") | 2.01 | 1.71 | 2.58 | 3.06 | 3.59 | 4.03 | 4.24 | 3.60 | 3.12 | 2.34 | 3.11 | 2.78 | 36.17 |
| Days ≥ 0.1" Precip | 6 | 5 | 7 | 8 | 8 | 7 | 7 | 7 | 7 | 6 | 7 | 7 | 82 |
| Total Snowfall (") | 8.6 | 6.9 | 4.8 | 0.7 | 0.0 | 0.0 | 0.0 | 0.0 | 0.0 | 0.0 | 0.9 | 5.4 | 27.3 |
| Days ≥ 1" Snow Depth | 15 | 13 | 5 | 1 | 0 | 0 | 0 | 0 | 0 | 0 | 1 | 9 | 44 |

## OBERLIN *Lorain County* ELEVATION 820 ft LAT/LONG 41° 17 ' N / 82° 13 ' W

| | JAN | FEB | MAR | APR | MAY | JUN | JUL | AUG | SEP | OCT | NOV | DEC | YEAR |
|---|---|---|---|---|---|---|---|---|---|---|---|---|---|
| Maximum Temp °F | 33.1 | 36.4 | 47.3 | 60.4 | 71.5 | 80.3 | 84.2 | 82.3 | 75.5 | 63.5 | 50.4 | 38.3 | 60.3 |
| Minimum Temp °F | 15.3 | 17.0 | 26.0 | 35.9 | 45.6 | 54.8 | 59.5 | 57.3 | 50.6 | 39.7 | 31.8 | 21.9 | 37.9 |
| Mean Temp °F | 24.2 | 26.7 | 36.7 | 48.1 | 58.5 | 67.5 | 71.9 | 69.9 | 63.1 | 51.6 | 41.1 | 30.1 | 49.1 |
| Days Max Temp ≥ 90 °F | 0 | 0 | 0 | 0 | 0 | 4 | 7 | 4 | 1 | 0 | 0 | 0 | 16 |
| Days Max Temp ≤ 32 °F | 15 | 11 | 3 | 0 | 0 | 0 | 0 | 0 | 0 | 0 | 1 | 9 | 39 |
| Days Min Temp ≤ 32 °F | 29 | 26 | 24 | 12 | 2 | 0 | 0 | 0 | 0 | 7 | 17 | 27 | 144 |
| Days Min Temp ≤ 0 °F | 4 | 3 | 0 | 0 | 0 | 0 | 0 | 0 | 0 | 0 | 0 | 1 | 8 |
| Heating Degree Days | 1258 | 1074 | 872 | 506 | 235 | 56 | 11 | 21 | 125 | 416 | 710 | 1075 | 6359 |
| Cooling Degree Days | 0 | 0 | 1 | 8 | 44 | 136 | 246 | 183 | 74 | 6 | 0 | 0 | 698 |
| Total Precipitation (") | 2.05 | 1.89 | 2.53 | 2.95 | 3.69 | 3.78 | 3.82 | 3.22 | 3.16 | 2.38 | 3.32 | 2.80 | 35.59 |
| Days ≥ 0.1" Precip | 6 | 5 | 7 | 8 | 8 | 8 | 7 | 6 | 6 | 6 | 8 | 7 | 82 |
| Total Snowfall (") | 11.1 | ***10.4*** | ***6.9*** | 1.4 | 0.0 | 0.0 | 0.0 | 0.0 | 0.0 | 0.0 | 2.1 | 8.5 | 40.4 |
| Days ≥ 1" Snow Depth | na | na | na | ***1*** | 0 | 0 | ***0*** | ***0*** | ***0*** | ***0*** | na | na | na |

## PAINESVILLE 4 NW *Lake County* ELEVATION 600 ft LAT/LONG 41° 45 ' N / 81° 18 ' W

| | JAN | FEB | MAR | APR | MAY | JUN | JUL | AUG | SEP | OCT | NOV | DEC | YEAR |
|---|---|---|---|---|---|---|---|---|---|---|---|---|---|
| Maximum Temp °F | 33.9 | 35.8 | 45.4 | 56.6 | 67.5 | 76.8 | 81.2 | 79.8 | 74.2 | 63.1 | 51.4 | 40.1 | 58.8 |
| Minimum Temp °F | 19.6 | 20.4 | 28.2 | 38.6 | 49.0 | 58.4 | 63.1 | 62.2 | 56.2 | 45.6 | 36.7 | 26.9 | 42.1 |
| Mean Temp °F | 26.8 | 28.1 | 36.8 | 47.6 | 58.3 | 67.7 | 72.2 | 71.0 | 65.2 | 54.4 | 44.1 | 33.5 | 50.5 |
| Days Max Temp ≥ 90 °F | 0 | 0 | 0 | 0 | 0 | 1 | 2 | 1 | 0 | 0 | 0 | 0 | 4 |
| Days Max Temp ≤ 32 °F | 14 | 12 | 4 | 0 | 0 | 0 | 0 | 0 | 0 | 0 | 0 | 6 | 36 |
| Days Min Temp ≤ 32 °F | 27 | 25 | 22 | 7 | 0 | 0 | 0 | 0 | 0 | 1 | 10 | 23 | 115 |
| Days Min Temp ≤ 0 °F | 2 | 1 | 0 | 0 | 0 | 0 | 0 | 0 | 0 | 0 | 0 | 0 | 3 |
| Heating Degree Days | 1178 | 1035 | 867 | 521 | 238 | 48 | 5 | 9 | 79 | 334 | 624 | 969 | 5907 |
| Cooling Degree Days | 0 | 0 | 1 | 6 | 38 | 128 | 251 | 203 | 94 | 9 | 0 | 0 | 730 |
| Total Precipitation (") | 2.03 | 1.68 | 2.55 | 2.86 | 3.19 | 3.72 | 3.28 | 3.70 | 3.87 | 3.30 | 3.73 | 3.00 | 36.91 |
| Days ≥ 0.1" Precip | 7 | 5 | 7 | 7 | 8 | 7 | 6 | 7 | 7 | 8 | 9 | 8 | 86 |
| Total Snowfall (") | na | na | na | ***1.3*** | 0.0 | ***0.0*** | 0.0 | 0.0 | 0.0 | ***0.0*** | na | na | na |
| Days ≥ 1" Snow Depth | 16 | 13 | 5 | 0 | 0 | 0 | 0 | 0 | 0 | 0 | ***1*** | 9 | 44 |

## PANDORA *Putnam County* ELEVATION 761 ft LAT/LONG 40° 59 ' N / 83° 57 ' W

| | JAN | FEB | MAR | APR | MAY | JUN | JUL | AUG | SEP | OCT | NOV | DEC | YEAR |
|---|---|---|---|---|---|---|---|---|---|---|---|---|---|
| Maximum Temp °F | 31.5 | 34.9 | 46.8 | 60.0 | 71.5 | 80.7 | 83.8 | 81.6 | 75.1 | 62.8 | 48.9 | 37.0 | 59.5 |
| Minimum Temp °F | 16.2 | 18.5 | 28.2 | 38.3 | 48.9 | 58.3 | 62.0 | 59.3 | 52.7 | 41.6 | 33.1 | 23.1 | 40.0 |
| Mean Temp °F | 23.9 | 26.8 | 37.5 | 49.2 | 60.3 | 69.5 | 72.9 | 70.5 | 63.9 | 52.3 | 41.0 | 30.1 | 49.8 |
| Days Max Temp ≥ 90 °F | 0 | 0 | 0 | 0 | 1 | 4 | 5 | 3 | 1 | 0 | 0 | 0 | 14 |
| Days Max Temp ≤ 32 °F | 16 | 12 | 4 | 0 | 0 | 0 | 0 | 0 | 0 | 0 | 2 | 10 | 44 |
| Days Min Temp ≤ 32 °F | 29 | 25 | 22 | 9 | 1 | 0 | 0 | 0 | 0 | 5 | 15 | 26 | 132 |
| Days Min Temp ≤ 0 °F | 5 | 3 | 0 | 0 | 0 | 0 | 0 | 0 | 0 | 0 | 0 | 1 | 9 |
| Heating Degree Days | 1269 | 1073 | 846 | 477 | 199 | 33 | 6 | 18 | 112 | 397 | 713 | 1076 | 6219 |
| Cooling Degree Days | 0 | 0 | 1 | 10 | 63 | 171 | 282 | 211 | 91 | 9 | 0 | 0 | 838 |
| Total Precipitation (") | 1.96 | 1.75 | 2.70 | 3.20 | 3.54 | 3.73 | 3.69 | 3.17 | 3.22 | 2.26 | 3.06 | 2.86 | 35.14 |
| Days ≥ 0.1" Precip | 5 | 5 | 7 | 8 | 7 | 7 | 7 | 6 | 6 | 6 | 7 | 7 | 78 |
| Total Snowfall (") | 8.5 | 7.8 | 5.0 | 1.3 | 0.0 | 0.0 | 0.0 | 0.0 | 0.0 | 0.1 | 2.5 | 6.8 | 32.0 |
| Days ≥ 1" Snow Depth | 16 | 13 | 5 | 1 | 0 | 0 | 0 | 0 | 0 | 0 | 2 | 8 | 45 |

## PAULDING *Paulding County* ELEVATION 732 ft LAT/LONG 41° 7 ' N / 84° 35 ' W

| | JAN | FEB | MAR | APR | MAY | JUN | JUL | AUG | SEP | OCT | NOV | DEC | YEAR |
|---|---|---|---|---|---|---|---|---|---|---|---|---|---|
| Maximum Temp °F | 30.5 | 33.9 | 45.8 | 59.2 | 70.9 | 80.5 | 84.2 | 82.0 | 75.1 | 62.4 | 48.8 | 36.6 | 59.2 |
| Minimum Temp °F | 13.1 | 15.5 | 25.7 | 36.3 | 46.6 | 56.3 | 60.1 | 57.1 | 50.0 | 38.1 | 30.5 | 20.6 | 37.5 |
| Mean Temp °F | 21.8 | 24.7 | 35.7 | 47.8 | 58.8 | 68.4 | 72.2 | 69.5 | 62.7 | 50.3 | 39.7 | 28.7 | 48.4 |
| Days Max Temp ≥ 90 °F | 0 | 0 | 0 | 0 | 0 | 4 | 6 | 3 | 1 | 0 | 0 | 0 | 14 |
| Days Max Temp ≤ 32 °F | 16 | 13 | 4 | 0 | 0 | 0 | 0 | 0 | 0 | 0 | 1 | 10 | 44 |
| Days Min Temp ≤ 32 °F | 30 | 27 | 24 | 11 | 2 | 0 | 0 | 0 | 1 | 8 | 19 | 27 | 149 |
| Days Min Temp ≤ 0 °F | 6 | 3 | 1 | 0 | 0 | 0 | 0 | 0 | 0 | 0 | 0 | 2 | 12 |
| Heating Degree Days | 1332 | 1131 | 900 | 515 | 229 | 44 | 8 | 23 | 132 | 455 | 753 | 1119 | 6641 |
| Cooling Degree Days | 0 | 0 | 0 | 6 | 48 | 152 | 252 | 175 | 75 | 5 | 0 | 0 | 713 |
| Total Precipitation (") | 1.82 | 1.59 | 2.52 | 3.32 | 3.70 | 3.31 | 3.45 | 2.73 | 3.14 | 2.60 | 3.17 | 2.83 | 34.18 |
| Days ≥ 0.1" Precip | 4 | 4 | 6 | 7 | 7 | 6 | 6 | 5 | 6 | 6 | 6 | 6 | 69 |
| Total Snowfall (") | ***4.9*** | ***5.5*** | ***3.4*** | 0.4 | 0.0 | 0.0 | 0.0 | 0.0 | 0.0 | 0.1 | 1.4 | ***4.3*** | 20.0 |
| Days ≥ 1" Snow Depth | na | na | na | 0 | 0 | 0 | 0 | 0 | 0 | 0 | ***1*** | na | na |

## PHILO 3 SW *Muskingum County* ELEVATION 1020 ft LAT/LONG 39° 50' N / 81° 55' W

| | JAN | FEB | MAR | APR | MAY | JUN | JUL | AUG | SEP | OCT | NOV | DEC | YEAR |
|---|---|---|---|---|---|---|---|---|---|---|---|---|---|
| Maximum Temp °F | 35.3 | 39.7 | 51.4 | 62.9 | 72.0 | 79.4 | 82.6 | 81.4 | 75.1 | 64.2 | 51.8 | 40.5 | 61.4 |
| Minimum Temp °F | 19.9 | 22.1 | 31.2 | 40.9 | 49.9 | 57.7 | 61.7 | 60.0 | 53.7 | 42.7 | 34.7 | 25.4 | 41.7 |
| Mean Temp °F | 27.6 | 30.9 | 41.3 | 51.9 | 60.9 | 68.6 | 72.3 | 70.8 | 64.4 | 53.4 | 43.3 | 33.0 | 51.5 |
| Days Max Temp ≥ 90 °F | 0 | 0 | 0 | 0 | 0 | 1 | 4 | 2 | 1 | 0 | 0 | 0 | 8 |
| Days Max Temp ≤ 32 °F | 13 | 9 | 2 | 0 | 0 | 0 | 0 | 0 | 0 | 0 | 1 | 8 | 33 |
| Days Min Temp ≤ 32 °F | 27 | 23 | 19 | 7 | 1 | 0 | 0 | 0 | 0 | 5 | 14 | 23 | 119 |
| Days Min Temp ≤ 0 °F | 2 | 1 | 0 | 0 | 0 | 0 | 0 | 0 | 0 | 0 | 0 | 1 | 4 |
| Heating Degree Days | 1153 | 957 | 732 | 400 | 173 | 35 | 5 | 14 | 100 | 362 | 646 | 986 | 5563 |
| Cooling Degree Days | 0 | 0 | 3 | 15 | 51 | 142 | 250 | 209 | 89 | 11 | 2 | 0 | 772 |
| Total Precipitation (") | 1.92 | 2.03 | 2.69 | 3.08 | 3.93 | 3.83 | 4.39 | 3.66 | 2.96 | 2.46 | 3.07 | 2.56 | 36.58 |
| Days ≥ 0.1" Precip | 6 | 6 | 7 | 7 | 8 | 7 | 7 | 7 | 6 | 6 | 7 | 7 | 81 |
| Total Snowfall (") | 8.9 | 5.5 | 4.4 | 1.1 | 0.0 | 0.0 | 0.0 | 0.0 | 0.0 | 0.0 | 1.1 | 3.8 | 24.8 |
| Days ≥ 1" Snow Depth | 13 | 9 | 3 | 0 | 0 | 0 | 0 | 0 | 0 | 0 | 1 | 5 | 31 |

## PORTSMOUTH SCIOTOVIL *Scioto County* ELEVATION 540 ft LAT/LONG 38° 45' N / 82° 53' W

| | JAN | FEB | MAR | APR | MAY | JUN | JUL | AUG | SEP | OCT | NOV | DEC | YEAR |
|---|---|---|---|---|---|---|---|---|---|---|---|---|---|
| Maximum Temp °F | 39.6 | 43.7 | 55.2 | 66.6 | 75.5 | 83.1 | 86.5 | 85.0 | 78.8 | 67.5 | 56.1 | 45.0 | 65.2 |
| Minimum Temp °F | 21.9 | 24.4 | 33.7 | 42.7 | 51.7 | 59.9 | 64.3 | 62.3 | 55.5 | 43.4 | 35.5 | 27.3 | 43.6 |
| Mean Temp °F | 30.8 | 34.0 | 44.5 | 54.6 | 63.6 | 71.5 | 75.4 | 73.7 | 67.2 | 55.5 | 45.8 | 36.2 | 54.4 |
| Days Max Temp ≥ 90 °F | 0 | 0 | 0 | 0 | 1 | 5 | 9 | 7 | 2 | 0 | 0 | 0 | 24 |
| Days Max Temp ≤ 32 °F | 9 | 5 | 1 | 0 | 0 | 0 | 0 | 0 | 0 | 0 | 0 | 4 | 19 |
| Days Min Temp ≤ 32 °F | 26 | 22 | 15 | 4 | 0 | 0 | 0 | 0 | 0 | 4 | 12 | 21 | 104 |
| Days Min Temp ≤ 0 °F | 2 | 0 | 0 | 0 | 0 | 0 | 0 | 0 | 0 | 0 | 0 | 1 | 3 |
| Heating Degree Days | 1055 | 867 | 633 | 325 | 119 | 17 | 1 | 5 | 63 | 308 | 571 | 888 | 4852 |
| Cooling Degree Days | 0 | 0 | 2 | 15 | 78 | 203 | 320 | 254 | 112 | 14 | 1 | 0 | 999 |
| Total Precipitation (") | 2.78 | 2.70 | 3.56 | 3.61 | 4.13 | 3.57 | 3.96 | 4.10 | 3.36 | 2.55 | 3.09 | 3.21 | 40.62 |
| Days ≥ 0.1" Precip | 6 | 6 | 8 | 8 | 8 | 7 | 8 | 6 | 5 | 6 | 7 | 7 | 82 |
| Total Snowfall (") | 5.9 | 4.6 | 2.7 | 0.3 | 0.0 | 0.0 | 0.0 | 0.0 | 0.0 | 0.0 | 0.6 | 1.7 | 15.8 |
| Days ≥ 1" Snow Depth | 7 | 6 | 2 | 0 | 0 | 0 | 0 | 0 | 0 | 0 | 0 | 2 | 17 |

## PUT-IN-BAY *Ottawa County* ELEVATION 581 ft LAT/LONG 41° 39' N / 82° 50' W

| | JAN | FEB | MAR | APR | MAY | JUN | JUL | AUG | SEP | OCT | NOV | DEC | YEAR |
|---|---|---|---|---|---|---|---|---|---|---|---|---|---|
| Maximum Temp °F | 31.5 | 33.9 | 43.4 | 56.2 | 67.6 | 77.8 | 82.3 | 80.6 | 74.0 | 61.4 | 49.0 | 37.2 | 57.9 |
| Minimum Temp °F | 18.4 | 19.6 | 28.5 | 39.2 | 50.9 | 61.0 | 66.3 | 65.6 | 59.2 | 47.3 | 36.8 | 25.8 | 43.2 |
| Mean Temp °F | 24.9 | 26.8 | 35.9 | 47.7 | 59.3 | 69.4 | 74.3 | 73.1 | 66.6 | 54.4 | 42.9 | 31.5 | 50.6 |
| Days Max Temp ≥ 90 °F | 0 | 0 | 0 | 0 | 0 | 1 | 4 | 2 | 1 | 0 | 0 | 0 | 8 |
| Days Max Temp ≤ 32 °F | 15 | 13 | 4 | 0 | 0 | 0 | 0 | 0 | 0 | 0 | 1 | 8 | 41 |
| Days Min Temp ≤ 32 °F | 29 | 25 | 21 | 5 | 0 | 0 | 0 | 0 | 0 | 0 | 9 | 24 | 113 |
| Days Min Temp ≤ 0 °F | 2 | 1 | 0 | 0 | 0 | 0 | 0 | 0 | 0 | 0 | 0 | 1 | 4 |
| Heating Degree Days | 1235 | 1071 | 895 | 517 | 208 | 29 | 2 | 3 | 61 | 332 | 657 | 1030 | 6040 |
| Cooling Degree Days | 0 | 0 | 0 | 7 | 39 | 162 | 309 | 259 | 107 | 8 | 0 | 0 | 891 |
| Total Precipitation (") | 1.71 | 1.44 | 4.86 | 2.93 | 3.41 | 3.49 | 3.11 | 3.39 | 3.05 | 2.43 | 2.86 | 2.39 | 35.07 |
| Days ≥ 0.1" Precip | 5 | 4 | 6 | 7 | 7 | 6 | 6 | 6 | 6 | 6 | 7 | 6 | 72 |
| Total Snowfall (") | 7.5 | 5.8 | *3.2* | 0.4 | 0.0 | 0.0 | 0.0 | 0.0 | 0.0 | 0.0 | 0.4 | 3.7 | 21.0 |
| Days ≥ 1" Snow Depth | 16 | 12 | 5 | 0 | 0 | 0 | 0 | 0 | 0 | 0 | 0 | 7 | 40 |

## RIPLEY EXP FARM *Brown County* ELEVATION 879 ft LAT/LONG 38° 47' N / 83° 51' W

| | JAN | FEB | MAR | APR | MAY | JUN | JUL | AUG | SEP | OCT | NOV | DEC | YEAR |
|---|---|---|---|---|---|---|---|---|---|---|---|---|---|
| Maximum Temp °F | 37.8 | 42.2 | 53.4 | 64.6 | 73.4 | 81.8 | 84.9 | 83.4 | 77.6 | 66.2 | 53.8 | 43.2 | 63.5 |
| Minimum Temp °F | 20.0 | 22.3 | 31.8 | 41.7 | 50.8 | 59.3 | 63.5 | 61.5 | 54.7 | 42.6 | 34.7 | 25.7 | 42.4 |
| Mean Temp °F | 28.9 | 32.3 | 42.6 | 53.2 | 62.1 | 70.6 | 74.2 | 72.5 | 66.2 | 54.4 | 44.3 | 34.5 | 53.0 |
| Days Max Temp ≥ 90 °F | 0 | 0 | 0 | 0 | 0 | 3 | 6 | 4 | 1 | 0 | 0 | 0 | 14 |
| Days Max Temp ≤ 32 °F | 10 | 7 | 1 | 0 | 0 | 0 | 0 | 0 | 0 | 0 | 1 | 5 | 24 |
| Days Min Temp ≤ 32 °F | 27 | 22 | 18 | 5 | 0 | 0 | 0 | 0 | 0 | 5 | 14 | 23 | 114 |
| Days Min Temp ≤ 0 °F | 3 | 2 | 0 | 0 | 0 | 0 | 0 | 0 | 0 | 0 | 0 | 1 | 6 |
| Heating Degree Days | 1112 | 918 | 689 | 361 | 147 | 21 | 2 | 7 | 74 | 336 | 616 | 940 | 5223 |
| Cooling Degree Days | 0 | 0 | 2 | 14 | 67 | 194 | 312 | 256 | 113 | 15 | 1 | 0 | 974 |
| Total Precipitation (") | 2.56 | 2.58 | 3.79 | 4.27 | 4.73 | 3.81 | 4.81 | 4.28 | 3.51 | 2.93 | 3.54 | 3.43 | 44.24 |
| Days ≥ 0.1" Precip | 5 | 5 | 8 | 9 | 9 | 7 | 8 | 7 | 6 | 6 | 7 | 7 | 84 |
| Total Snowfall (") | 6.8 | 5.3 | 3.8 | 0.5 | 0.0 | 0.0 | 0.0 | 0.0 | 0.0 | 0.1 | 1.0 | 3.1 | 20.6 |
| Days ≥ 1" Snow Depth | 10 | 8 | 2 | 0 | 0 | 0 | 0 | 0 | 0 | 0 | 1 | 4 | 25 |

## SANDUSKY *Erie County* ELEVATION 630 ft LAT/LONG 41° 27 ' N / 82° 43 ' W

| | JAN | FEB | MAR | APR | MAY | JUN | JUL | AUG | SEP | OCT | NOV | DEC | YEAR |
|---|---|---|---|---|---|---|---|---|---|---|---|---|---|
| Maximum Temp °F | 32.3 | 34.3 | 43.8 | 56.7 | 67.6 | 77.6 | 82.2 | 80.4 | 73.7 | 61.7 | 49.5 | 37.5 | 58.1 |
| Minimum Temp °F | 17.8 | 19.4 | 28.6 | 39.4 | 50.4 | 60.3 | 65.0 | 63.1 | 55.9 | 44.3 | 35.1 | 24.4 | 42.0 |
| Mean Temp °F | 25.0 | 26.9 | 36.2 | 48.1 | 59.0 | 69.0 | 73.6 | 71.8 | 64.9 | 53.0 | 42.3 | 31.0 | 50.1 |
| Days Max Temp ≥ 90 °F | 0 | 0 | 0 | 0 | 0 | 2 | 4 | 2 | 1 | 0 | 0 | 0 | 9 |
| Days Max Temp ≤ 32 °F | 15 | 13 | 5 | 0 | 0 | 0 | 0 | 0 | 0 | 0 | 1 | 9 | 43 |
| Days Min Temp ≤ 32 °F | 28 | 25 | 21 | 6 | 0 | 0 | 0 | 0 | 0 | 2 | 12 | 25 | 119 |
| Days Min Temp ≤ 0 °F | 3 | 2 | 0 | 0 | 0 | 0 | 0 | 0 | 0 | 0 | 0 | 1 | 6 |
| Heating Degree Days | 1233 | 1069 | 886 | 510 | 222 | 37 | 3 | 7 | 92 | 373 | 674 | 1048 | 6154 |
| Cooling Degree Days | 0 | 0 | 1 | 10 | 48 | 158 | 286 | 229 | 93 | 8 | 0 | 0 | 833 |
| Total Precipitation (") | 1.83 | 1.61 | 2.39 | 2.87 | 3.46 | 4.07 | 3.68 | 3.52 | 3.07 | 2.25 | 2.98 | 2.64 | 34.37 |
| Days ≥ 0.1" Precip | 5 | 4 | 6 | 7 | 7 | 7 | 6 | 6 | 6 | 6 | 7 | 6 | 73 |
| Total Snowfall (") | 8.3 | 6.7 | 3.3 | 0.6 | 0.0 | 0.0 | 0.0 | 0.0 | 0.0 | 0.0 | 0.3 | 5.1 | 24.3 |
| Days ≥ 1" Snow Depth | 14 | 9 | 4 | 0 | 0 | 0 | 0 | 0 | 0 | 0 | 0 | 6 | 33 |

## SPRINGFIELD NEW WW *Clark County* ELEVATION 981 ft LAT/LONG 39° 57 ' N / 83° 45 ' W

| | JAN | FEB | MAR | APR | MAY | JUN | JUL | AUG | SEP | OCT | NOV | DEC | YEAR |
|---|---|---|---|---|---|---|---|---|---|---|---|---|---|
| Maximum Temp °F | 33.2 | 37.2 | 48.6 | 60.4 | 71.5 | 80.0 | 83.8 | 82.3 | 76.0 | 63.6 | 51.0 | 39.2 | 60.6 |
| Minimum Temp °F | 15.9 | 18.2 | 28.7 | 38.1 | 48.5 | 58.0 | 61.9 | 59.2 | 51.9 | 39.9 | 32.6 | 23.0 | 39.7 |
| Mean Temp °F | 24.6 | 27.7 | 38.7 | 49.3 | 60.1 | 69.0 | 72.9 | 70.7 | 64.0 | 51.8 | 41.8 | 31.1 | 50.1 |
| Days Max Temp ≥ 90 °F | 0 | 0 | 0 | 0 | 0 | 2 | 5 | 3 | 1 | 0 | 0 | 0 | 11 |
| Days Max Temp ≤ 32 °F | 14 | 10 | 3 | 0 | 0 | 0 | 0 | 0 | 0 | 0 | 1 | 8 | 36 |
| Days Min Temp ≤ 32 °F | 29 | 24 | 21 | 10 | 1 | 0 | 0 | 0 | 0 | 7 | 16 | 25 | 133 |
| Days Min Temp ≤ 0 °F | 4 | 4 | 0 | 0 | 0 | 0 | 0 | 0 | 0 | 0 | 0 | 1 | 9 |
| Heating Degree Days | 1246 | 1047 | 810 | 474 | 197 | 38 | 6 | 16 | 112 | 412 | 690 | 1044 | 6092 |
| Cooling Degree Days | 0 | 0 | 1 | 7 | 52 | 172 | 278 | 211 | 88 | 9 | 0 | 0 | 818 |
| Total Precipitation (") | 2.12 | 1.86 | 2.67 | 3.23 | 4.12 | 4.26 | 4.16 | 3.84 | 3.01 | 2.65 | 3.10 | 2.76 | 37.78 |
| Days ≥ 0.1" Precip | 5 | 5 | 7 | 7 | 8 | 7 | 7 | 6 | 5 | 6 | 6 | 7 | 76 |
| Total Snowfall (") | na | na | na | 0.1 | 0.0 | 0.0 | 0.0 | 0.0 | 0.0 | 0.1 | 0.2 | na | na |
| Days ≥ 1" Snow Depth | na | na | na | 0 | 0 | 0 | 0 | 0 | 0 | 0 | na | na | na |

## STEUBENVILLE *Jefferson County* ELEVATION 991 ft LAT/LONG 40° 23 ' N / 80° 38 ' W

| | JAN | FEB | MAR | APR | MAY | JUN | JUL | AUG | SEP | OCT | NOV | DEC | YEAR |
|---|---|---|---|---|---|---|---|---|---|---|---|---|---|
| Maximum Temp °F | 35.9 | 39.8 | 50.8 | 62.5 | 72.0 | 80.2 | 83.5 | 82.1 | 75.6 | 64.1 | 52.3 | 41.4 | 61.7 |
| Minimum Temp °F | 19.5 | 21.6 | 30.1 | 39.3 | 49.1 | 58.0 | 62.9 | 61.7 | 55.1 | 43.2 | 34.8 | 25.7 | 41.8 |
| Mean Temp °F | 27.8 | 30.7 | 40.5 | 50.9 | 60.6 | 69.1 | 73.2 | 71.9 | 65.4 | 53.6 | 43.6 | 33.5 | 51.7 |
| Days Max Temp ≥ 90 °F | 0 | 0 | 0 | 0 | 0 | 2 | 4 | 2 | 1 | 0 | 0 | 0 | 9 |
| Days Max Temp ≤ 32 °F | 12 | 8 | 2 | 0 | 0 | 0 | 0 | 0 | 0 | 0 | 1 | 7 | 30 |
| Days Min Temp ≤ 32 °F | 27 | 23 | 20 | 8 | 1 | 0 | 0 | 0 | 0 | 3 | 13 | 24 | 119 |
| Days Min Temp ≤ 0 °F | 2 | 1 | 0 | 0 | 0 | 0 | 0 | 0 | 0 | 0 | 0 | 0 | 3 |
| Heating Degree Days | 1148 | 961 | 754 | 424 | 178 | 31 | 4 | 8 | 81 | 354 | 637 | 968 | 5548 |
| Cooling Degree Days | 0 | 0 | 2 | 10 | 49 | 158 | 283 | 230 | 98 | 9 | 1 | 0 | 840 |
| Total Precipitation (") | 2.64 | 2.33 | 3.35 | 3.10 | 4.03 | 3.97 | 4.29 | 3.71 | 3.19 | 2.60 | 3.30 | 3.01 | 39.52 |
| Days ≥ 0.1" Precip | 7 | 6 | 9 | 8 | 8 | 8 | 8 | 7 | 6 | 6 | 8 | 8 | 89 |
| Total Snowfall (") | na | na | na | 0.1 | 0.0 | 0.0 | 0.0 | 0.0 | 0.0 | 0.2 | 1.4 | na | na |
| Days ≥ 1" Snow Depth | na | na | na | 0 | 0 | 0 | 0 | 0 | 0 | 0 | 0 | na | na |

## TIFFIN *Seneca County* ELEVATION 761 ft LAT/LONG 41° 7 ' N / 83° 10 ' W

| | JAN | FEB | MAR | APR | MAY | JUN | JUL | AUG | SEP | OCT | NOV | DEC | YEAR |
|---|---|---|---|---|---|---|---|---|---|---|---|---|---|
| Maximum Temp °F | 32.1 | 35.5 | 47.6 | 60.6 | 71.8 | 80.4 | 84.2 | 82.0 | 75.6 | 63.1 | 49.6 | 37.7 | 60.0 |
| Minimum Temp °F | 17.4 | 19.6 | 29.3 | 39.1 | 49.2 | 58.3 | 62.8 | 60.3 | 53.8 | 42.1 | 34.1 | 23.9 | 40.8 |
| Mean Temp °F | 24.7 | 27.6 | 38.4 | 49.8 | 60.6 | 69.4 | 73.5 | 71.2 | 64.7 | 52.6 | 41.8 | 30.8 | 50.4 |
| Days Max Temp ≥ 90 °F | 0 | 0 | 0 | 0 | 1 | 3 | 6 | 3 | 1 | 0 | 0 | 0 | 14 |
| Days Max Temp ≤ 32 °F | 15 | 12 | 3 | 0 | 0 | 0 | 0 | 0 | 0 | 0 | 1 | 9 | 40 |
| Days Min Temp ≤ 32 °F | 28 | 24 | 20 | 8 | 1 | 0 | 0 | 0 | 0 | 5 | 14 | 25 | 125 |
| Days Min Temp ≤ 0 °F | 3 | 2 | 0 | 0 | 0 | 0 | 0 | 0 | 0 | 0 | 0 | 1 | 6 |
| Heating Degree Days | 1241 | 1050 | 817 | 459 | 189 | 37 | 4 | 14 | 97 | 388 | 688 | 1053 | 6037 |
| Cooling Degree Days | 0 | 0 | 1 | 12 | 61 | 168 | 294 | 215 | 94 | 8 | 0 | 0 | 853 |
| Total Precipitation (") | 2.18 | 1.84 | 2.66 | 3.20 | 3.82 | 3.82 | 3.58 | 3.60 | 3.33 | 2.40 | 3.21 | 3.01 | 36.65 |
| Days ≥ 0.1" Precip | 5 | 5 | 7 | 8 | 8 | 7 | 7 | 6 | 6 | 6 | 8 | 7 | 80 |
| Total Snowfall (") | 8.2 | 6.5 | 4.1 | 1.2 | 0.0 | 0.0 | 0.0 | 0.0 | 0.0 | 0.0 | 1.6 | 6.8 | 28.4 |
| Days ≥ 1" Snow Depth | 15 | 13 | 4 | 1 | 0 | 0 | 0 | 0 | 0 | 0 | 1 | 8 | 42 |

## TOLEDO EXPRESS AP *Lucas County* ELEVATION 669 ft LAT/LONG 41° 36 ' N / 83° 48 ' W

| | JAN | FEB | MAR | APR | MAY | JUN | JUL | AUG | SEP | OCT | NOV | DEC | YEAR |
|---|---|---|---|---|---|---|---|---|---|---|---|---|---|
| Maximum Temp °F | 30.5 | 33.7 | 45.5 | 59.1 | 70.6 | 79.8 | 83.5 | 81.5 | 74.2 | 62.0 | 48.3 | 36.1 | 58.7 |
| Minimum Temp °F | 15.5 | 17.4 | 26.8 | 37.0 | 46.8 | 56.3 | 61.0 | 58.7 | 51.5 | 40.0 | 31.8 | 21.8 | 38.7 |
| Mean Temp °F | 23.0 | 25.6 | 36.2 | 48.0 | 58.8 | 68.1 | 72.3 | 70.1 | 62.9 | 51.0 | 40.1 | 29.0 | 48.8 |
| Days Max Temp ≥ 90 °F | 0 | 0 | 0 | 0 | 1 | 3 | 5 | 3 | 1 | 0 | 0 | 0 | 13 |
| Days Max Temp ≤ 32 °F | 17 | 13 | 4 | 0 | 0 | 0 | 0 | 0 | 0 | 0 | 2 | 11 | 47 |
| Days Min Temp ≤ 32 °F | 29 | 25 | 22 | 10 | 1 | 0 | 0 | 0 | 0 | 7 | 17 | 26 | 137 |
| Days Min Temp ≤ 0 °F | 5 | 3 | 0 | 0 | 0 | 0 | 0 | 0 | 0 | 0 | 0 | 1 | 9 |
| Heating Degree Days | 1295 | 1106 | 886 | 510 | 229 | 46 | 7 | 20 | 132 | 434 | 740 | 1109 | 6514 |
| Cooling Degree Days | 0 | 0 | 1 | 8 | 48 | 146 | 266 | 197 | 75 | 5 | 0 | 0 | 746 |
| Total Precipitation (") | 1.87 | 1.74 | 2.56 | 3.10 | 2.97 | 3.75 | 3.28 | 3.22 | 2.91 | 2.30 | 3.00 | 3.06 | 33.76 |
| Days ≥ 0.1" Precip | 5 | 5 | 7 | 8 | 7 | 7 | 6 | 6 | 6 | 6 | 6 | 7 | 76 |
| Total Snowfall (") | 9.9 | 8.6 | 6.2 | 1.5 | 0.0 | 0.0 | 0.0 | 0.0 | 0.0 | 0.2 | 3.0 | 8.7 | 38.1 |
| Days ≥ 1" Snow Depth | 17 | 14 | 5 | 1 | 0 | 0 | 0 | 0 | 0 | 0 | 2 | 10 | 49 |

## TOM JENKINS DAM-BURR *Athens County* ELEVATION 761 ft LAT/LONG 39° 33 ' N / 82° 4 ' W

| | JAN | FEB | MAR | APR | MAY | JUN | JUL | AUG | SEP | OCT | NOV | DEC | YEAR |
|---|---|---|---|---|---|---|---|---|---|---|---|---|---|
| Maximum Temp °F | na | *39.8* | *51.7* | *63.7* | na | *80.7* | *83.6* | na | na | *65.0* | *52.7* | *42.0* | na |
| Minimum Temp °F | na | *15.1* | *24.8* | *33.8* | na | *52.4* | *57.1* | na | na | *37.1* | *29.4* | *21.3* | na |
| Mean Temp °F | na | *27.3* | *38.3* | *48.8* | na | *66.6* | *70.4* | na | na | *51.1* | *41.0* | *31.7* | na |
| Days Max Temp ≥ 90 °F | *0* | *0* | *0* | *0* | *0* | *2* | *5* | *2* | *1* | *0* | *0* | *0* | 10 |
| Days Max Temp ≤ 32 °F | *13* | *8* | *2* | *0* | *0* | *0* | *0* | *0* | *0* | *0* | *1* | *6* | 30 |
| Days Min Temp ≤ 32 °F | *28* | *25* | *23* | *15* | *5* | *0* | *0* | *0* | *1* | *11* | *19* | *25* | 152 |
| Days Min Temp ≤ 0 °F | *6* | *5* | *0* | *0* | *0* | *0* | *0* | *0* | *0* | *0* | *0* | *1* | 12 |
| Heating Degree Days | na | *1057* | *822* | *485* | na | *59* | *15* | na | na | *429* | *713* | *1025* | na |
| Cooling Degree Days | na | na | na | na | na | na | na | na | na | na | na | na | na |
| Total Precipitation (") | *2.53* | *2.28* | *3.08* | *3.22* | *4.25* | *3.79* | *4.48* | *3.88* | *3.35* | *2.82* | *3.26* | *2.87* | 39.81 |
| Days ≥ 0.1" Precip | *7* | *5* | *7* | *8* | *8* | *7* | *8* | *7* | *6* | *6* | *8* | *7* | 84 |
| Total Snowfall (") | *10.2* | *7.0* | *3.4* | *1.1* | *0.0* | *0.0* | *0.0* | *0.0* | *0.0* | *0.0* | *1.2* | *2.7* | 25.6 |
| Days ≥ 1" Snow Depth | *12* | *9* | *3* | *0* | *0* | *0* | *0* | *0* | *0* | *0* | *1* | *3* | 28 |

## UPPER SANDUSKY *Wyandot County* ELEVATION 850 ft LAT/LONG 40° 50 ' N / 83° 17 ' W

| | JAN | FEB | MAR | APR | MAY | JUN | JUL | AUG | SEP | OCT | NOV | DEC | YEAR |
|---|---|---|---|---|---|---|---|---|---|---|---|---|---|
| Maximum Temp °F | 32.4 | 36.2 | 47.7 | 60.8 | 71.8 | 80.8 | 84.5 | 82.7 | 76.2 | 63.8 | 49.9 | 38.1 | 60.4 |
| Minimum Temp °F | 17.1 | 19.2 | 28.8 | 38.6 | 48.6 | 57.9 | 61.9 | 59.7 | 53.1 | 41.8 | 33.5 | 23.7 | 40.3 |
| Mean Temp °F | 24.8 | 27.6 | 38.2 | 49.7 | 60.2 | 69.4 | 73.2 | 71.3 | 64.7 | 52.8 | 41.7 | 30.9 | 50.4 |
| Days Max Temp ≥ 90 °F | 0 | 0 | 0 | 0 | 1 | 4 | 6 | 4 | 1 | 0 | 0 | 0 | 16 |
| Days Max Temp ≤ 32 °F | 15 | 11 | 3 | 0 | 0 | 0 | 0 | 0 | 0 | 0 | 1 | 9 | 39 |
| Days Min Temp ≤ 32 °F | 28 | 24 | 20 | 9 | 1 | 0 | 0 | 0 | 0 | 5 | 15 | 25 | 127 |
| Days Min Temp ≤ 0 °F | 4 | 2 | 0 | 0 | 0 | 0 | 0 | 0 | 0 | 0 | 0 | 1 | 7 |
| Heating Degree Days | 1240 | 1049 | 823 | 462 | 198 | 35 | 5 | 14 | 98 | 382 | 692 | 1051 | 6049 |
| Cooling Degree Days | 0 | 0 | 1 | 9 | 54 | 163 | 276 | 219 | 90 | 9 | 0 | 0 | 821 |
| Total Precipitation (") | 2.03 | 1.72 | 2.64 | 3.24 | 4.06 | 3.32 | 4.25 | 3.16 | 3.27 | 2.15 | 3.27 | 2.71 | 35.82 |
| Days ≥ 0.1" Precip | 5 | 5 | 7 | 8 | 8 | 6 | 7 | 6 | 6 | 6 | 8 | 7 | 79 |
| Total Snowfall (") | 8.4 | 6.8 | 4.6 | 1.1 | 0.0 | 0.0 | 0.0 | 0.0 | 0.0 | 0.0 | 1.7 | 6.2 | 28.8 |
| Days ≥ 1" Snow Depth | 14 | 12 | 5 | 0 | 0 | 0 | 0 | 0 | 0 | 0 | 1 | 7 | 39 |

## URBANA WWTP *Champaign County* ELEVATION 1050 ft LAT/LONG 40° 7 ' N / 83° 46 ' W

| | JAN | FEB | MAR | APR | MAY | JUN | JUL | AUG | SEP | OCT | NOV | DEC | YEAR |
|---|---|---|---|---|---|---|---|---|---|---|---|---|---|
| Maximum Temp °F | 33.0 | 36.6 | 47.9 | 60.4 | 71.3 | 80.2 | 84.2 | 82.4 | 76.0 | 63.4 | 50.3 | 38.7 | 60.4 |
| Minimum Temp °F | 15.7 | 18.0 | 28.1 | 38.2 | 48.3 | 57.4 | 61.2 | 58.2 | 51.1 | 39.9 | 32.3 | 22.8 | 39.3 |
| Mean Temp °F | 24.4 | 27.3 | 38.0 | 49.3 | 59.8 | 68.9 | 72.7 | 70.3 | 63.6 | 51.7 | 41.3 | 30.8 | 49.8 |
| Days Max Temp ≥ 90 °F | 0 | 0 | 0 | 0 | 0 | 3 | 5 | 3 | 1 | 0 | 0 | 0 | 12 |
| Days Max Temp ≤ 32 °F | 14 | 11 | 3 | 0 | 0 | 0 | 0 | 0 | 0 | 0 | 1 | 9 | 38 |
| Days Min Temp ≤ 32 °F | 29 | 25 | 22 | 9 | 1 | 0 | 0 | 0 | 0 | 7 | 16 | 25 | 134 |
| Days Min Temp ≤ 0 °F | 5 | 3 | 0 | 0 | 0 | 0 | 0 | 0 | 0 | 0 | 0 | 1 | 9 |
| Heating Degree Days | 1253 | 1057 | 828 | 471 | 202 | 36 | 6 | 17 | 115 | 414 | 704 | 1053 | 6156 |
| Cooling Degree Days | 0 | 0 | 0 | 6 | 46 | 154 | 263 | 194 | 80 | 7 | 0 | 0 | 750 |
| Total Precipitation (") | 2.18 | 2.00 | 2.83 | 3.38 | 4.26 | 4.13 | 4.57 | 3.39 | 3.06 | 2.70 | 3.24 | 2.84 | 38.58 |
| Days ≥ 0.1" Precip | 5 | 5 | 7 | 8 | 8 | 7 | 7 | 6 | 6 | 6 | 7 | 7 | 79 |
| Total Snowfall (") | na | na | na | 0.6 | 0.0 | 0.0 | 0.0 | 0.0 | 0.0 | 0.2 | na | na | na |
| Days ≥ 1" Snow Depth | na | na | na | 0 | 0 | 0 | 0 | 0 | 0 | 0 | *0* | na | na |

## VAN WERT *Van Wert County* ELEVATION 781 ft LAT/LONG 40° 52' N / 84° 35' W

| | JAN | FEB | MAR | APR | MAY | JUN | JUL | AUG | SEP | OCT | NOV | DEC | YEAR |
|---|---|---|---|---|---|---|---|---|---|---|---|---|---|
| Maximum Temp °F | 32.4 | 36.0 | 47.9 | 61.4 | 73.0 | 81.9 | 85.3 | 83.2 | 76.5 | 64.1 | 49.8 | 37.8 | 60.8 |
| Minimum Temp °F | 16.4 | 18.2 | 28.2 | 38.7 | 49.1 | 58.6 | 62.5 | 59.9 | 52.8 | 41.3 | 32.9 | 23.0 | 40.1 |
| Mean Temp °F | 24.4 | 27.1 | 38.0 | 50.0 | 61.1 | 70.3 | 73.9 | 71.6 | 64.7 | 52.7 | 41.4 | 30.4 | 50.5 |
| Days Max Temp ≥ 90 °F | 0 | 0 | 0 | 0 | 1 | 5 | 7 | 5 | 1 | 0 | 0 | 0 | 19 |
| Days Max Temp ≤ 32 °F | 15 | 10 | 3 | 0 | 0 | 0 | 0 | 0 | 0 | 0 | 1 | 8 | 37 |
| Days Min Temp ≤ 32 °F | 29 | 25 | 21 | 8 | 1 | 0 | 0 | 0 | 0 | 5 | 16 | 26 | 131 |
| Days Min Temp ≤ 0 °F | 4 | 3 | 0 | 0 | 0 | 0 | 0 | 0 | 0 | 0 | 0 | 1 | 8 |
| Heating Degree Days | 1252 | 1063 | 829 | 452 | 178 | 27 | 4 | 12 | 100 | 384 | 702 | 1065 | 6068 |
| Cooling Degree Days | 0 | 0 | 1 | 10 | 65 | 189 | 299 | 224 | 93 | 9 | 0 | 0 | 890 |
| Total Precipitation (") | 1.99 | 1.84 | 2.73 | 3.49 | 3.79 | 3.96 | 3.64 | 3.16 | 3.24 | 2.61 | 3.18 | 2.96 | 36.59 |
| Days ≥ 0.1" Precip | 5 | 5 | 7 | 8 | 7 | 7 | 7 | 6 | 6 | 6 | 7 | 6 | 77 |
| Total Snowfall (") | 10.0 | 8.3 | 4.8 | 1.2 | 0.0 | 0.0 | 0.0 | 0.0 | 0.0 | 0.2 | 3.0 | 8.5 | 36.0 |
| Days ≥ 1" Snow Depth | 14 | 13 | 5 | 0 | 0 | 0 | 0 | 0 | 0 | 0 | 1 | 7 | 40 |

## WARREN 3 S *Trumbull County* ELEVATION 902 ft LAT/LONG 41° 15' N / 80° 51' W

| | JAN | FEB | MAR | APR | MAY | JUN | JUL | AUG | SEP | OCT | NOV | DEC | YEAR |
|---|---|---|---|---|---|---|---|---|---|---|---|---|---|
| Maximum Temp °F | 33.8 | 37.2 | 48.2 | 60.7 | 71.1 | 79.4 | 83.3 | 81.5 | 74.3 | 62.6 | 50.5 | 39.2 | 60.2 |
| Minimum Temp °F | 15.9 | 17.0 | 25.9 | 35.0 | 44.5 | 53.4 | 57.9 | 56.5 | 50.0 | 38.9 | 31.6 | 22.7 | 37.4 |
| Mean Temp °F | 24.9 | 27.1 | 37.1 | 47.9 | 57.8 | 66.4 | 70.6 | 69.0 | 62.2 | 50.8 | 41.1 | 30.9 | 48.8 |
| Days Max Temp ≥ 90 °F | 0 | 0 | 0 | 0 | 0 | 2 | 5 | 2 | 1 | 0 | 0 | 0 | 10 |
| Days Max Temp ≤ 32 °F | 14 | 10 | 3 | 0 | 0 | 0 | 0 | 0 | 0 | 0 | 1 | 8 | 36 |
| Days Min Temp ≤ 32 °F | 29 | 25 | 24 | 13 | 4 | 0 | 0 | 0 | 0 | 8 | 18 | 26 | 147 |
| Days Min Temp ≤ 0 °F | 4 | 3 | 0 | 0 | 0 | 0 | 0 | 0 | 0 | 0 | 0 | 1 | 8 |
| Heating Degree Days | 1237 | 1063 | 861 | 512 | 245 | 62 | 14 | 25 | 141 | 437 | 712 | 1049 | 6358 |
| Cooling Degree Days | 0 | 0 | 1 | 6 | 32 | 108 | 215 | 164 | 63 | 3 | 0 | 0 | 592 |
| Total Precipitation (") | 2.15 | 1.67 | 2.85 | 2.94 | 3.56 | 3.89 | 4.13 | 3.32 | 3.79 | 2.65 | 3.20 | 2.77 | 36.92 |
| Days ≥ 0.1" Precip | 6 | 5 | 8 | 8 | 8 | 7 | 7 | 7 | 7 | 7 | 8 | 7 | 85 |
| Total Snowfall (") | 10.6 | 9.5 | 6.4 | 0.5 | 0.1 | 0.0 | 0.0 | 0.0 | 0.0 | 0.0 | 1.7 | 8.0 | 36.8 |
| Days ≥ 1" Snow Depth | 16 | 12 | 4 | 0 | 0 | 0 | 0 | 0 | 0 | 0 | 1 | 9 | 42 |

## WASHINGTON COURT HSE *Fayette County* ELEVATION 961 ft LAT/LONG 39° 31' N / 83° 25' W

| | JAN | FEB | MAR | APR | MAY | JUN | JUL | AUG | SEP | OCT | NOV | DEC | YEAR |
|---|---|---|---|---|---|---|---|---|---|---|---|---|---|
| Maximum Temp °F | 35.4 | 39.7 | 51.4 | 63.4 | 72.5 | 79.9 | 82.7 | 81.7 | 76.1 | 65.5 | 52.5 | 40.8 | 61.8 |
| Minimum Temp °F | 20.0 | 22.4 | 31.7 | 41.2 | 51.6 | 60.4 | 64.4 | 62.4 | 55.8 | 43.9 | 34.8 | 25.8 | 42.9 |
| Mean Temp °F | 27.7 | 31.1 | 41.6 | 52.4 | 62.0 | 70.2 | 73.6 | 72.1 | 66.0 | 54.7 | 43.7 | 33.3 | 52.4 |
| Days Max Temp ≥ 90 °F | 0 | 0 | 0 | 0 | 0 | 2 | 3 | 3 | 1 | 0 | 0 | 0 | 9 |
| Days Max Temp ≤ 32 °F | 12 | 8 | 2 | 0 | 0 | 0 | 0 | 0 | 0 | 0 | 1 | 7 | 30 |
| Days Min Temp ≤ 32 °F | 27 | 23 | 18 | 6 | 0 | 0 | 0 | 0 | 0 | 3 | 14 | 23 | 114 |
| Days Min Temp ≤ 0 °F | 2 | 1 | 0 | 0 | 0 | 0 | 0 | 0 | 0 | 0 | 0 | 1 | 4 |
| Heating Degree Days | 1149 | 950 | 721 | 383 | 149 | 21 | 2 | 7 | 74 | 326 | 634 | 976 | 5392 |
| Cooling Degree Days | 0 | 0 | 2 | 12 | 68 | 188 | 288 | 243 | 118 | 14 | 1 | 0 | 934 |
| Total Precipitation (") | 2.36 | 2.30 | 3.35 | 3.50 | 4.63 | 3.58 | 3.90 | 3.68 | 2.93 | 2.62 | 3.25 | 2.82 | 38.92 |
| Days ≥ 0.1" Precip | 6 | 5 | 8 | 8 | 8 | 7 | 7 | 6 | 6 | 6 | 7 | 7 | 81 |
| Total Snowfall (") | 8.5 | 6.5 | 4.4 | 0.6 | 0.0 | 0.0 | 0.0 | 0.0 | 0.0 | 0.2 | 1.2 | 3.7 | 25.1 |
| Days ≥ 1" Snow Depth | 11 | 8 | 3 | 0 | 0 | 0 | 0 | 0 | 0 | 0 | 1 | 4 | 27 |

## WAUSEON WATER PLANT *Fulton County* ELEVATION 751 ft LAT/LONG 41° 31' N / 84° 9' W

| | JAN | FEB | MAR | APR | MAY | JUN | JUL | AUG | SEP | OCT | NOV | DEC | YEAR |
|---|---|---|---|---|---|---|---|---|---|---|---|---|---|
| Maximum Temp °F | 30.7 | 34.4 | 46.3 | 60.1 | 71.7 | 81.1 | 84.1 | 81.8 | 75.4 | 63.1 | 48.8 | 36.4 | 59.5 |
| Minimum Temp °F | 15.1 | 17.2 | 27.2 | 37.4 | 47.5 | 57.1 | 60.5 | 58.1 | 51.4 | 40.4 | 31.9 | 21.9 | 38.8 |
| Mean Temp °F | 22.8 | 25.8 | 36.8 | 48.8 | 59.6 | 69.1 | 72.4 | 70.0 | 63.4 | 51.8 | 40.4 | 29.2 | 49.2 |
| Days Max Temp ≥ 90 °F | 0 | 0 | 0 | 0 | 0 | 4 | 5 | 3 | 1 | 0 | 0 | 0 | 13 |
| Days Max Temp ≤ 32 °F | 16 | 12 | 3 | 0 | 0 | 0 | 0 | 0 | 0 | 0 | 1 | 10 | 42 |
| Days Min Temp ≤ 32 °F | 29 | 26 | 23 | 10 | 1 | 0 | 0 | 0 | 0 | 6 | 17 | 27 | 139 |
| Days Min Temp ≤ 0 °F | 5 | 3 | 0 | 0 | 0 | 0 | 0 | 0 | 0 | 0 | 0 | 1 | 9 |
| Heating Degree Days | 1300 | 1100 | 870 | 485 | 207 | 34 | 6 | 18 | 115 | 410 | 732 | 1104 | 6381 |
| Cooling Degree Days | 0 | 0 | 0 | 8 | 58 | 177 | 274 | 203 | 87 | 6 | 0 | 0 | 813 |
| Total Precipitation (") | 1.84 | 1.53 | 2.48 | 3.32 | 3.40 | 3.59 | 3.61 | 3.36 | 3.22 | 2.72 | 3.21 | 2.87 | 35.15 |
| Days ≥ 0.1" Precip | 4 | 4 | 6 | 8 | 7 | 7 | 7 | 6 | 6 | 6 | 7 | 7 | 75 |
| Total Snowfall (") | 7.8 | 7.8 | 5.0 | 1.1 | 0.1 | 0.0 | 0.0 | 0.0 | 0.0 | 0.1 | 2.8 | 6.6 | 31.3 |
| Days ≥ 1" Snow Depth | 14 | 13 | 5 | 0 | 0 | 0 | 0 | 0 | 0 | 0 | 2 | 7 | 41 |

## WAVERLY *Pike County* ELEVATION 600 ft LAT/LONG 39° 8 ' N / 82° 59 ' W

| | JAN | FEB | MAR | APR | MAY | JUN | JUL | AUG | SEP | OCT | NOV | DEC | YEAR |
|---|---|---|---|---|---|---|---|---|---|---|---|---|---|
| Maximum Temp °F | 38.7 | 43.1 | 55.1 | 66.8 | 75.3 | 82.8 | 86.1 | 84.7 | 78.6 | 67.5 | 55.7 | 44.1 | 64.9 |
| Minimum Temp °F | 18.6 | 20.8 | 30.1 | 38.9 | 48.7 | 57.6 | 62.5 | 60.3 | 53.3 | 40.1 | 32.5 | 24.2 | 40.6 |
| Mean Temp °F | 28.6 | 32.0 | 42.6 | 52.9 | 62.0 | 70.2 | 74.3 | 72.5 | 65.9 | 53.8 | 44.1 | 34.1 | 52.8 |
| Days Max Temp ≥ 90 °F | 0 | 0 | 0 | 0 | 1 | 4 | 8 | 6 | 1 | 0 | 0 | 0 | 20 |
| Days Max Temp ≤ 32 °F | 8 | 5 | 1 | 0 | 0 | 0 | 0 | 0 | 0 | 0 | 0 | 5 | 19 |
| Days Min Temp ≤ 32 °F | 27 | 24 | 19 | 9 | 1 | 0 | 0 | 0 | 0 | 8 | 16 | 24 | 128 |
| Days Min Temp ≤ 0 °F | 3 | 2 | 0 | 0 | 0 | 0 | 0 | 0 | 0 | 0 | 0 | 1 | 6 |
| Heating Degree Days | 1119 | 925 | 693 | 367 | 150 | 23 | 3 | 9 | 78 | 350 | 620 | 950 | 5287 |
| Cooling Degree Days | 0 | 0 | 1 | 11 | 62 | 186 | 313 | 249 | 108 | 14 | 1 | 0 | 945 |
| Total Precipitation (") | 2.45 | 2.19 | 3.62 | 3.64 | 4.19 | 3.28 | 4.08 | 4.31 | 2.97 | 2.45 | 3.22 | 2.92 | 39.32 |
| Days ≥ 0.1" Precip | 5 | 5 | 8 | 8 | 8 | 6 | 7 | 7 | 6 | 5 | 7 | 7 | 79 |
| Total Snowfall (") | *5.0* | na | 2.9 | 0.1 | 0.0 | 0.0 | 0.0 | 0.0 | 0.0 | 0.0 | 0.6 | *1.9* | na |
| Days ≥ 1" Snow Depth | na | na | *1* | 0 | 0 | 0 | 0 | 0 | 0 | 0 | 0 | na | na |

## WESTERVILLE *Franklin County* ELEVATION 860 ft LAT/LONG 40° 7 ' N / 82° 56 ' W

| | JAN | FEB | MAR | APR | MAY | JUN | JUL | AUG | SEP | OCT | NOV | DEC | YEAR |
|---|---|---|---|---|---|---|---|---|---|---|---|---|---|
| Maximum Temp °F | 35.3 | 39.9 | 52.0 | 64.2 | 74.3 | 82.3 | 85.4 | 83.8 | 77.6 | 66.1 | 52.6 | 40.8 | 62.9 |
| Minimum Temp °F | 17.5 | 19.6 | 29.6 | 38.8 | 48.5 | 57.2 | 61.7 | 59.6 | 52.8 | 41.0 | 33.5 | 24.4 | 40.4 |
| Mean Temp °F | 26.4 | 29.8 | 40.8 | 51.5 | 61.4 | 69.8 | 73.6 | 71.7 | 65.2 | 53.6 | 43.1 | 32.6 | 51.6 |
| Days Max Temp ≥ 90 °F | 0 | 0 | 0 | 0 | 0 | 4 | 7 | 4 | 1 | 0 | 0 | 0 | 16 |
| Days Max Temp ≤ 32 °F | 12 | 8 | 2 | 0 | 0 | 0 | 0 | 0 | 0 | 0 | 1 | 7 | 30 |
| Days Min Temp ≤ 32 °F | 28 | 24 | 20 | 9 | 1 | 0 | 0 | 0 | 0 | 6 | 15 | 24 | 127 |
| Days Min Temp ≤ 0 °F | 4 | 3 | 0 | 0 | 0 | 0 | 0 | 0 | 0 | 0 | 0 | 1 | 8 |
| Heating Degree Days | 1191 | 989 | 745 | 407 | 163 | 25 | 4 | 10 | 89 | 358 | 652 | 996 | 5629 |
| Cooling Degree Days | 0 | 0 | 2 | 12 | 65 | 188 | 312 | 250 | 115 | 13 | 1 | 0 | 958 |
| Total Precipitation (") | 2.34 | 2.12 | 2.77 | 3.55 | 4.09 | 4.14 | 4.36 | 3.54 | 3.16 | 2.59 | 3.50 | 2.91 | 39.07 |
| Days ≥ 0.1" Precip | 6 | 5 | 7 | 8 | 8 | 7 | 7 | 7 | 6 | 6 | 8 | 7 | 82 |
| Total Snowfall (") | 7.3 | 5.4 | 2.3 | 0.7 | 0.0 | 0.0 | 0.0 | 0.0 | 0.0 | 0.0 | 0.6 | 3.4 | 19.7 |
| Days ≥ 1" Snow Depth | 12 | 7 | 2 | 0 | 0 | 0 | 0 | 0 | 0 | 0 | 0 | 4 | 25 |

## WILMINGTON 3 N *Clinton County* ELEVATION 1030 ft LAT/LONG 39° 27 ' N / 83° 50 ' W

| | JAN | FEB | MAR | APR | MAY | JUN | JUL | AUG | SEP | OCT | NOV | DEC | YEAR |
|---|---|---|---|---|---|---|---|---|---|---|---|---|---|
| Maximum Temp °F | 35.0 | 38.8 | 50.4 | 62.2 | 72.3 | 80.8 | 84.2 | 82.8 | 76.8 | 64.8 | 51.8 | 40.4 | 61.7 |
| Minimum Temp °F | 18.1 | 20.2 | 29.9 | 39.5 | 49.2 | 57.8 | 61.9 | 59.4 | 52.9 | 41.4 | 33.4 | 24.4 | 40.7 |
| Mean Temp °F | 26.6 | 29.5 | 40.2 | 50.9 | 60.8 | 69.3 | 73.1 | 71.1 | 64.9 | 53.1 | 42.6 | 32.4 | 51.2 |
| Days Max Temp ≥ 90 °F | 0 | 0 | 0 | 0 | 0 | 3 | 6 | 4 | 1 | 0 | 0 | 0 | 14 |
| Days Max Temp ≤ 32 °F | 13 | 9 | 2 | 0 | 0 | 0 | 0 | 0 | 0 | 0 | 1 | 7 | 32 |
| Days Min Temp ≤ 32 °F | 27 | 24 | 20 | 8 | 1 | 0 | 0 | 0 | 0 | 6 | 15 | 24 | 125 |
| Days Min Temp ≤ 0 °F | 4 | 3 | 0 | 0 | 0 | 0 | 0 | 0 | 0 | 0 | 0 | 1 | 8 |
| Heating Degree Days | 1185 | 995 | 763 | 427 | 178 | 31 | 5 | 15 | 97 | 373 | 666 | 1004 | 5739 |
| Cooling Degree Days | 0 | 0 | 1 | 10 | 50 | 166 | 276 | 215 | 99 | 12 | 1 | 0 | 830 |
| Total Precipitation (") | 2.63 | 2.47 | 3.43 | 3.90 | 4.83 | 3.70 | 4.27 | 3.50 | 3.07 | 2.81 | 3.63 | 2.89 | 41.13 |
| Days ≥ 0.1" Precip | 6 | 6 | 8 | 9 | 8 | 7 | 8 | 6 | 6 | 6 | 7 | 7 | 84 |
| Total Snowfall (") | 8.0 | 6.8 | 4.4 | 0.7 | 0.0 | 0.0 | 0.0 | 0.0 | 0.0 | 0.2 | 2.0 | 3.5 | 25.6 |
| Days ≥ 1" Snow Depth | 11 | 9 | 3 | 0 | 0 | 0 | 0 | 0 | 0 | 0 | 1 | 4 | 28 |

## WOOSTER EXP STN *Wayne County* ELEVATION 1030 ft LAT/LONG 40° 47 ' N / 81° 56 ' W

| | JAN | FEB | MAR | APR | MAY | JUN | JUL | AUG | SEP | OCT | NOV | DEC | YEAR |
|---|---|---|---|---|---|---|---|---|---|---|---|---|---|
| Maximum Temp °F | 32.2 | 35.6 | 46.9 | 59.0 | 69.3 | 78.0 | 81.8 | 79.9 | 72.9 | 61.2 | 49.0 | 37.5 | 58.6 |
| Minimum Temp °F | 17.1 | 19.0 | 28.2 | 37.4 | 47.3 | 56.1 | 60.3 | 58.5 | 51.8 | 40.7 | 33.1 | 23.6 | 39.4 |
| Mean Temp °F | 24.6 | 27.3 | 37.5 | 48.2 | 58.4 | 67.1 | 71.1 | 69.2 | 62.4 | 50.9 | 41.1 | 30.5 | 49.0 |
| Days Max Temp ≥ 90 °F | 0 | 0 | 0 | 0 | 0 | 1 | 2 | 1 | 0 | 0 | 0 | 0 | 4 |
| Days Max Temp ≤ 32 °F | 16 | 11 | 4 | 0 | 0 | 0 | 0 | 0 | 0 | 0 | 2 | 10 | 43 |
| Days Min Temp ≤ 32 °F | 28 | 24 | 22 | 10 | 1 | 0 | 0 | 0 | 0 | 6 | 16 | 26 | 133 |
| Days Min Temp ≤ 0 °F | 4 | 2 | 0 | 0 | 0 | 0 | 0 | 0 | 0 | 0 | 0 | 1 | 7 |
| Heating Degree Days | 1245 | 1057 | 845 | 502 | 234 | 55 | 11 | 24 | 136 | 434 | 712 | 1062 | 6317 |
| Cooling Degree Days | 0 | 0 | 1 | 6 | 39 | 124 | 230 | 171 | 65 | 5 | 0 | 0 | 641 |
| Total Precipitation (") | 2.07 | 1.90 | 2.68 | 3.17 | 3.97 | 3.62 | 4.04 | 3.85 | 3.45 | 2.34 | 3.16 | 2.65 | 36.90 |
| Days ≥ 0.1" Precip | 5 | 5 | 7 | 8 | 8 | 7 | 7 | 7 | 6 | 6 | 7 | 7 | 80 |
| Total Snowfall (") | 10.0 | 7.8 | 5.4 | 1.2 | 0.0 | 0.0 | 0.0 | 0.0 | 0.0 | 0.1 | 2.1 | 6.5 | 33.1 |
| Days ≥ 1" Snow Depth | 16 | 13 | 6 | 1 | 0 | 0 | 0 | 0 | 0 | 0 | 2 | 10 | 48 |

## XENIA 6 SSE *Greene County* ELEVATION 922 ft LAT/LONG 39° 38' N / 83° 54' W

| | JAN | FEB | MAR | APR | MAY | JUN | JUL | AUG | SEP | OCT | NOV | DEC | YEAR |
|---|---|---|---|---|---|---|---|---|---|---|---|---|---|
| Maximum Temp °F | 35.5 | 40.0 | 52.1 | 63.8 | 73.2 | 80.6 | 83.7 | 82.0 | 76.2 | 64.9 | 52.3 | 41.0 | 62.1 |
| Minimum Temp °F | 18.9 | 21.5 | 31.6 | 40.7 | 50.2 | 58.7 | 62.4 | 59.7 | 53.2 | 41.8 | 33.8 | 24.7 | 41.4 |
| Mean Temp °F | 27.2 | 30.8 | 41.9 | 52.2 | 61.7 | 69.7 | 73.1 | 70.9 | 64.7 | 53.4 | 43.1 | 32.9 | 51.8 |
| Days Max Temp ≥ 90 °F | 0 | 0 | 0 | 0 | 0 | 2 | 4 | 2 | 0 | 0 | 0 | 0 | 8 |
| Days Max Temp ≤ 32 °F | 12 | 8 | 1 | 0 | 0 | 0 | 0 | 0 | 0 | 0 | 1 | 6 | 28 |
| Days Min Temp ≤ 32 °F | 27 | 23 | 18 | 7 | 1 | 0 | 0 | 0 | 0 | 5 | 15 | 24 | 120 |
| Days Min Temp ≤ 0 °F | 3 | 2 | 0 | 0 | 0 | 0 | 0 | 0 | 0 | 0 | 0 | 1 | 6 |
| Heating Degree Days | 1164 | 959 | 712 | 386 | 154 | 24 | 4 | 11 | 93 | 363 | 651 | 988 | 5509 |
| Cooling Degree Days | 0 | 0 | 2 | 12 | 65 | 181 | 282 | 211 | 98 | 12 | 1 | 0 | 864 |
| Total Precipitation (") | 2.37 | 2.26 | 3.08 | 3.77 | 4.34 | 3.58 | 4.06 | 3.73 | 3.04 | 2.80 | 3.39 | 2.98 | 39.40 |
| Days ≥ 0.1" Precip | 6 | 6 | 7 | 8 | 8 | 7 | 8 | 7 | 6 | 6 | 7 | 7 | 83 |
| Total Snowfall (") | 7.8 | 5.6 | 3.8 | 0.4 | 0.0 | 0.0 | 0.0 | 0.0 | 0.0 | 0.2 | 1.4 | *3.3* | 22.5 |
| Days ≥ 1" Snow Depth | na | na | *1* | 0 | 0 | 0 | 0 | 0 | 0 | 0 | *1* | na | na |

## YOUNGSTOWN MUNI AP *Trumbull County* ELEVATION 1178 ft LAT/LONG 41° 15' N / 80° 40' W

| | JAN | FEB | MAR | APR | MAY | JUN | JUL | AUG | SEP | OCT | NOV | DEC | YEAR |
|---|---|---|---|---|---|---|---|---|---|---|---|---|---|
| Maximum Temp °F | 31.2 | 34.4 | 45.3 | 58.2 | 68.9 | 77.6 | 81.7 | 79.9 | 72.4 | 60.6 | 48.3 | 36.9 | 58.0 |
| Minimum Temp °F | 16.9 | 18.4 | 27.1 | 37.0 | 46.2 | 54.8 | 59.4 | 57.9 | 51.4 | 41.3 | 33.6 | 23.7 | 39.0 |
| Mean Temp °F | 24.1 | 26.4 | 36.2 | 47.7 | 57.6 | 66.2 | 70.6 | 68.9 | 61.9 | 50.9 | 40.9 | 30.3 | 48.5 |
| Days Max Temp ≥ 90 °F | 0 | 0 | 0 | 0 | 0 | 1 | 3 | 2 | 0 | 0 | 0 | 0 | 6 |
| Days Max Temp ≤ 32 °F | 17 | 13 | 6 | 0 | 0 | 0 | 0 | 0 | 0 | 0 | 2 | 11 | 49 |
| Days Min Temp ≤ 32 °F | 28 | 24 | 22 | 11 | 2 | 0 | 0 | 0 | 0 | 5 | 15 | 25 | 132 |
| Days Min Temp ≤ 0 °F | 3 | 2 | 0 | 0 | 0 | 0 | 0 | 0 | 0 | 0 | 0 | 1 | 6 |
| Heating Degree Days | 1262 | 1082 | 886 | 522 | 258 | 70 | 14 | 29 | 147 | 434 | 716 | 1068 | 6488 |
| Cooling Degree Days | 0 | 0 | 1 | 10 | 38 | 111 | 212 | 163 | 63 | 5 | 1 | 0 | 604 |
| Total Precipitation (") | 2.28 | 2.02 | 3.09 | 3.04 | 3.42 | 3.77 | 4.29 | 3.45 | 3.74 | 2.52 | 3.27 | 3.00 | 37.89 |
| Days ≥ 0.1" Precip | 6 | 6 | 8 | 8 | 8 | 7 | 7 | 6 | 7 | 6 | 8 | 8 | 85 |
| Total Snowfall (") | 13.8 | 11.4 | 10.9 | 2.3 | 0.2 | 0.0 | 0.0 | 0.0 | 0.0 | 0.6 | 4.9 | 12.3 | 56.4 |
| Days ≥ 1" Snow Depth | 17 | 14 | 7 | 1 | 0 | 0 | 0 | 0 | 0 | 0 | 3 | 12 | 54 |

## ZANESVILLE MUNI AP *Muskingum County* ELEVATION 881 ft LAT/LONG 39° 57' N / 81° 54' W

| | JAN | FEB | MAR | APR | MAY | JUN | JUL | AUG | SEP | OCT | NOV | DEC | YEAR |
|---|---|---|---|---|---|---|---|---|---|---|---|---|---|
| Maximum Temp °F | 35.5 | 39.7 | 51.3 | 62.5 | 72.2 | 80.4 | 83.7 | 82.1 | 75.4 | 63.8 | 52.2 | 40.7 | 61.6 |
| Minimum Temp °F | 19.2 | 21.2 | 30.8 | 39.9 | 49.4 | 57.9 | 62.6 | 60.9 | 53.6 | 41.7 | 34.1 | 25.0 | 41.4 |
| Mean Temp °F | 27.4 | 30.5 | 41.0 | 51.2 | 60.8 | 69.2 | 73.2 | 71.5 | 64.5 | 52.8 | 43.2 | 32.9 | 51.5 |
| Days Max Temp ≥ 90 °F | 0 | 0 | 0 | 0 | 0 | 2 | 5 | 3 | 1 | 0 | 0 | 0 | 11 |
| Days Max Temp ≤ 32 °F | 13 | 8 | 2 | 0 | 0 | 0 | 0 | 0 | 0 | 0 | 1 | 7 | 31 |
| Days Min Temp ≤ 32 °F | 27 | 23 | 19 | 7 | 1 | 0 | 0 | 0 | 0 | 5 | 14 | 24 | 120 |
| Days Min Temp ≤ 0 °F | 3 | 2 | 0 | 0 | 0 | 0 | 0 | 0 | 0 | 0 | 0 | 1 | 6 |
| Heating Degree Days | 1159 | 968 | 737 | 416 | 177 | 31 | 4 | 12 | 98 | 381 | 650 | 989 | 5622 |
| Cooling Degree Days | 0 | 0 | 2 | 10 | 54 | 162 | 284 | 225 | 96 | 10 | 1 | 0 | 844 |
| Total Precipitation (") | 2.28 | 2.26 | 3.18 | 3.48 | 3.92 | 3.83 | 4.62 | 3.90 | 3.12 | 2.40 | 3.27 | 2.81 | 39.07 |
| Days ≥ 0.1" Precip | 6 | 6 | 8 | 8 | 8 | 7 | 7 | 7 | 6 | 6 | 7 | 7 | 83 |
| Total Snowfall (") | 8.1 | 4.7 | 3.6 | 1.4 | 0.0 | 0.0 | 0.0 | 0.0 | 0.0 | 0.1 | 1.3 | 3.8 | 23.0 |
| Days ≥ 1" Snow Depth | 12 | 8 | 3 | 0 | 0 | 0 | 0 | 0 | 0 | 0 | 1 | 5 | 29 |

## JANUARY MINIMUM TEMPERATURE °F

### LOWEST

| | | |
|---|---|---|
| 1 | Fredericktown | 13.0 |
| 2 | Paulding | 13.1 |
| 3 | Montpelier | 13.3 |
| 4 | Defiance | 14.0 |
| | Dorset | 14.0 |
| 6 | Chardon | 14.1 |
| 7 | Greenville | 14.4 |
| 8 | Eaton | 14.7 |
| 9 | Mansfield-5 W | 14.9 |
| 10 | Hoytville | 15.0 |
| 11 | Ashland | 15.1 |
| | Centerburg | 15.1 |
| | Wauseon | 15.1 |
| 14 | Oberlin | 15.3 |
| 15 | Delaware | 15.4 |
| 16 | Marion | 15.5 |
| | Toledo | 15.5 |
| 18 | Barnesville | 15.6 |
| | Bucyrus | 15.6 |
| | Danville | 15.6 |
| | Fremont | 15.6 |
| 22 | Kenton | 15.7 |
| | Urbana | 15.7 |
| 24 | Canfield | 15.9 |
| | Springfield | 15.9 |

### HIGHEST

| | | |
|---|---|---|
| 1 | Cincinnati | 22.0 |
| 2 | Gallipolis | 21.9 |
| | Portsmouth | 21.9 |
| 4 | Marietta | 21.7 |
| 5 | Circleville | 20.6 |
| 6 | Dayton-Mcd | 20.1 |
| 7 | Ripley | 20.0 |
| | Wshingtn Crt Hse | 20.0 |
| 9 | Philo | 19.9 |
| 10 | Chilo Mldhl L&D | 19.7 |
| 11 | Columbus-Vly | 19.6 |
| | Painesville | 19.6 |
| 13 | Steubenville | 19.5 |
| 14 | Cambridge | 19.3 |
| | Columbus-Intl | 19.3 |
| | Hillsboro | 19.3 |
| 17 | Zanesville | 19.2 |
| 18 | Carpenter | 19.0 |
| 19 | Newark | 18.9 |
| | Xenia | 18.9 |
| 21 | Cadiz | 18.8 |
| 22 | Waverly | 18.6 |
| 23 | Dayton-Intl | 18.5 |
| 24 | Elyria | 18.4 |
| | Put-in-Bay | 18.4 |

## JULY MAXIMUM TEMPERATURE °F

### HIGHEST

| | | |
|---|---|---|
| 1 | Gallipolis | 87.3 |
| 2 | Dayton-Mcd | 86.8 |
| 3 | Cincinnati | 86.7 |
| 4 | Portsmouth | 86.5 |
| 5 | Chilo Mldhl L&D | 86.3 |
| 6 | Marietta | 86.1 |
| | Waverly | 86.1 |
| 8 | Circleville | 85.6 |
| | McConnlsville Lk | 85.6 |
| 10 | Franklin | 85.5 |
| | Irwin | 85.5 |
| | Mineral Ridge | 85.5 |
| 13 | Westerville | 85.4 |
| 14 | Columbus-Vly | 85.3 |
| | Van Wert | 85.3 |
| 16 | London | 85.2 |
| 17 | Carpenter | 85.1 |
| | Dayton-Intl | 85.1 |
| 19 | Cambridge | 85.0 |
| | Eaton | 85.0 |
| | Newark | 85.0 |
| 22 | New Lexington | 84.9 |
| | Ripley | 84.9 |
| 24 | Kenton | 84.8 |
| | Lima | 84.8 |

### LOWEST

| | | |
|---|---|---|
| 1 | Chardon | 80.1 |
| 2 | Ashtabula | 80.6 |
| 3 | Dorset | 81.1 |
| 4 | Painesville | 81.2 |
| 5 | Coshocton-Arg | 81.7 |
| | Youngstown | 81.7 |
| 7 | Hiram | 81.8 |
| | Mansfield-5 W | 81.8 |
| | Wooster | 81.8 |
| 10 | Mansfield-Lahm | 82.0 |
| 11 | Barnesville | 82.1 |
| | Centerburg | 82.1 |
| 13 | Cleveland | 82.2 |
| | Sandusky | 82.2 |
| 15 | Put-in-Bay | 82.3 |
| 16 | Akron | 82.4 |
| 17 | Findlay-Airpot | 82.6 |
| | Philo | 82.6 |
| 19 | Norwalk | 82.7 |
| | Wshingtn Crt Hse | 82.7 |
| 21 | Bucyrus | 82.9 |
| | Cadiz | 82.9 |
| 23 | Ashland | 83.0 |
| 24 | Canfield | 83.1 |
| | Chippewa Lake | 83.1 |

## ANNUAL PRECIPITATION (")

### HIGHEST

| | | |
|---|---|---|
| 1 | Chardon | 46.27 |
| 2 | Ripley | 44.24 |
| 3 | Dorset | 43.54 |
| 4 | Barnesville | 43.44 |
| 5 | Chilo Mldhl L&D | 43.03 |
| 6 | Hillsboro | 42.44 |
| 7 | Hiram | 42.08 |
| 8 | McConnelsvlle Lk | 41.64 |
| 9 | Newark | 41.60 |
| 10 | Mansfield-Lahm | 41.35 |
| 11 | New Lexington | 41.23 |
| 12 | Wilmington | 41.13 |
| 13 | Cincinnati | 40.77 |
| 14 | Portsmouth | 40.62 |
| 15 | Marietta | 40.58 |
| 16 | Carpenter | 40.45 |
| 17 | Gallipolis | 40.44 |
| 18 | Coshocton-Wpc | 40.35 |
| 19 | Danville | 40.28 |
| 20 | Centerburg | 40.13 |
| 21 | Cadiz | 39.88 |
| 22 | Fredericktown | 39.83 |
| 23 | Tom Jenkins Dam | 39.81 |
| 24 | Eaton | 39.78 |
| 25 | New Philadelphia | 39.76 |

### LOWEST

| | | |
|---|---|---|
| 1 | Bowling Green | 33.37 |
| 2 | Hoytville | 33.75 |
| 3 | Toledo | 33.76 |
| 4 | Paulding | 34.18 |
| 5 | Sandusky | 34.37 |
| 6 | Findlay-Airpot | 34.59 |
| 7 | Put-in-Bay | 35.07 |
| 8 | Pandora | 35.14 |
| 9 | Wauseon | 35.15 |
| 10 | Defiance | 35.51 |
| 11 | Montpelier | 35.54 |
| 12 | Oberlin | 35.59 |
| 13 | Fremont | 35.81 |
| 14 | Upper Sandusky | 35.82 |
| 15 | Marysville | 36.10 |
| 16 | Norwalk | 36.17 |
| 17 | Mineral Ridge | 36.28 |
| 18 | Bellefontaine | 36.32 |
| 19 | Celina | 36.48 |
| | Kenton | 36.48 |
| 21 | Philo | 36.58 |
| 22 | Van Wert | 36.59 |
| 23 | Coshocton-Arg | 36.61 |
| 24 | Canfield | 36.64 |
| 25 | Tiffin | 36.65 |

## ANNUAL SNOWFALL (")

### HIGHEST

| | | |
|---|---|---|
| 1 | Chardon | 101.3 |
| 2 | Dorset | 73.9 |
| 3 | Hiram | 61.1 |
| 4 | Cleveland | 58.7 |
| 5 | Youngstown | 56.4 |
| 6 | Akron | 47.3 |
| 7 | Elyria | 43.9 |
| 8 | Mansfield-Lahm | 41.3 |
| 9 | Chippewa Lake | 41.2 |
| 10 | Oberlin | 40.4 |
| 11 | Toledo | 38.1 |
| 12 | Mineral Ridge | 38.0 |
| 13 | Warren | 36.8 |
| 14 | Van Wert | 36.0 |
| 15 | Ashland | 35.5 |
| 16 | Celina | 34.3 |
| 17 | Danville | 33.5 |
| 18 | Wooster | 33.1 |
| 19 | Barnesville | 32.6 |
| 20 | Pandora | 32.0 |
| 21 | Millport | 31.8 |
| 22 | Wauseon | 31.3 |
| 23 | New Philadelphia | 31.1 |
| 24 | Findlay-WPCC | 29.7 |
| 25 | Upper Sandusky | 28.8 |

### LOWEST

| | | |
|---|---|---|
| 1 | Franklin | 10.6 |
| 2 | Cincinnati | 13.3 |
| 3 | Circleville | 14.7 |
| 4 | Portsmouth | 15.8 |
| 5 | Dayton-Mcd | 16.3 |
| 6 | Gallipolis | 17.2 |
| 7 | Hillsboro | 19.2 |
| 8 | Carpenter | 19.4 |
| 9 | Westerville | 19.7 |
| 10 | Paulding | 20.0 |
| 11 | Columbus-Vly | 20.2 |
| 12 | Ripley | 20.6 |
| 13 | Bellefontaine | 20.8 |
| | Cambridge | 20.8 |
| | Marysville | 20.8 |
| 16 | Put-in-Bay | 21.0 |
| 17 | Delaware | 21.5 |
| 18 | Newark | 22.0 |
| 19 | Greenville | 22.1 |
| 20 | Bowling Green | 22.5 |
| | Xenia | 22.5 |
| 22 | Defiance | 22.8 |
| 23 | Zanesville | 23.0 |
| 24 | New Lexington | 23.1 |
| 25 | McConnelsvlle Lk | 23.8 |

# OKLAHOMA

PHYSICAL FEATURES. Oklahoma is located in the southern Great Plains. Of the 50 states, it ranks 18th in size with an area of approximately 70,000 square miles, only 935 of which are covered by lakes and ponds. Its northern boundary is about 465 miles in length and its southern boundary 315 miles in length. Greatest depth is 222 miles.

The terrain is mostly rolling plains, sloping downward from west to east. The plains are broken by scattered hilly areas where most points are 600 feet or less above the adjacent countryside, and by a mountainous area in the southeast where some peaks rise more than 2,000 feet above their base. The hilly areas consist of the Wichita Mountains, with some isolated peaks, in the southwest; the Arbuckle Mountains in the south-central; and an extension of the Ozarks in the northeast. The Ouachita Mountains occupy much of the southeast. Elevations in the State range from 4,976 feet above sea level on Black Mesa in the northwestern corner of the Panhandle, to about 305 feet above sea level in the bed of the Red River where it leaves Oklahoma at the southeastern corner of the State.

Oklahoma lies entirely within the drainage basin of the Mississippi River. The two main rivers in the State are the Arkansas which drains the northern two-thirds of Oklahoma and the Red River which drains the southern third and forms the State's southern boundary. Principal tributaries of the Arkansas are the Verdigris, Grand (Neosho), Illinois, Cimarron, North Canadian, and Canadian Rivers. The Red draws largely from the North Fork of the Red, Washita, Boggy, and Little Rivers.

In western Oklahoma, rivers tend to be broad, shallow, sand choked and dry or nearly dry much of the time. Basins are mostly long and narrow. In the east, rivers are fairly swift and clear and basins more oval in form. Most lakes are manmade and were built for flood control, irrigation, municipal water storage, recreational, and hydroelectric power purposes.

GENERAL CLIMATE. The climate of Oklahoma is mostly continental in type, as in all of the central Great Plains. Warm, moist air moving northward from the Gulf of Mexico exerts much influence at times, particularly over the southern and more eastern sections of the State where, as a result, humidities and cloudiness are generally greater and precipitation considerably heavier than in the western and northern sections. Summers are long and occasionally very hot. Winters are shorter and less rigorous than those of the more northern Plains States. Periods of extreme cold are infrequent.

The mean annual temperature over the State ranges from 64° along the southern border to about 60° along the northern border. It then decreases westward across the Panhandle to about 57° in Cimarron County. Temperatures of 90° or higher occur, on an average, about 85 days per year in the western Panhandle and in the northeast corner of the State. In the southwest, the average is about 120 days, and in the southeast from 95 to 100 days. Temperatures of 100° or higher are common over the State from May well into September. In the southwest part of the State the average number of 100° days is 20 to 25 per year. Other sections of the State will average somewhat less, but very seldom will any location in the State not reach a 100° temperature sometime during the summer months.

Low humidities and good southerly breezes usually accompany the high summer temperatures and somewhat lessen their discomforting effect. Occasionally strong, hot winds accompany the high daytime temperatures; this combination produces rapid evaporation and often injures crops. When these conditions persist for long periods of time, droughts develop and occasionally become severe. Nights are generally comfortable because the clear skies and dry air allows for rapid cooling after sunset.

Temperatures of 32° or less occur on an average of 55 to 65 days per year along the southern tier of counties and from 90 to 100 days per year along the Kansas border in the north-central and northeastern sections of the State. In the Panhandle, days with 32° or less occur, on an average, 125 to 140 days per year.

The average length of the growing season, or freeze-free period, ranges from 168 days in the northwestern corner of the Panhandle, to about 225 days along the Red River in the south-central and southeastern sections of Oklahoma.

Freezing temperatures have occurred as late as April 20 along the southern border and as late as May 15 in the extreme northwest and in the Panhandle. Fall freezes have occurred as early as September 20 in the Panhandle and as early as October 9 along the southern border. Frozen soil is not a major problem, nor much of a deterrent to seasonal activities. The average maximum depth that frost penetrates the soil ranges from less than 3 inches in the southeastern corner of the State to more than 10 inches in the extreme northwestern portion.

PRECIPITATION. The geographical distribution of rainfall decreases sharply from east to west. Average annual precipitation ranges from about 56 inches in the southeastern corner of the State, to 15 inches in the extreme western Panhandle. Frequency of rainfall, as determined from the average number of days with 0.01 inch or more, varies from 95 to 100 days a year in the extreme east to from 70 to 80 days a year over the western third of the State.

Excessively heavy rains occur at times. Amounts of 10 inches or more within a 24-hour period have been recorded. Floods may occur during any season. They occur with greater frequency, however, from May to July and in September and October, representing periods when storms are of greater magnitude and rains of greatest intensities. In general, floods in other seasons are the result of more abnormal and persistent buildup of soil moisture plus a concurrent increase in stream-flow due to prolonged rains.

SNOWFALL. The geographical distribution of annual snowfall is usually almost the reverse of the annual precipitation pattern and ranges, on an average, from approximately 2 inches in the southeastern corner of the State to approximately 20 inches in the western sections of the Panhandle. Snow rarely remains on the ground more than a few days. At times, strong winds with heavy snowfalls cause bad drifting and occasionally produce blizzard conditions.

OTHER CLIMATIC ELEMENTS. Relative humidity averages about 10 percent higher in the eastern portion of the State because of lower elevations and more frequent inflow of Gulf moisture. Summer afternoon and early evening relative humidities are considerably lower than those of winter.

Oklahoma, along with other states in the southern Great Plains, has at times been subject to droughts of varying degree and duration, although drought years have been far less frequent than dry summers and falls. Average annual lake evaporation varies from about 48 inches in the extreme eastern sections of the State to as high as 65 inches in the southwestern corner. In the western Panhandle approximately 58 inches of water is evaporated each year.

Prevailing winds are southerly although northerly winds predominate during the winter months. Average yearly wind speeds vary from 9 m.p.h. in the east to approximately 14 m.p.h. in the west. March and April are the windiest months, and July and August the calmest.

Thunderstorms occur, on an average, on 50 to 60 days per year in the eastern half of the State and from 40 to 50 days per year in the western half. Some of the more severe thunderstorms are accompanied by tornadoes and damaging hail, and approximately 75 percent of these occur during the spring season.

Skies are preponderantly clear in western and central sections and about equally clear and cloudy in eastern sections. Sunshine records show an annual average of 68 percent of the possible amount at Oklahoma City and 63 percent at Tulsa. Summer is the period of greatest possible sunshine and winter the least.

## COUNTY INDEX

*Alfalfa County*
CHEROKEE
GRT SALT PLAINS DAM
HELENA 1 SSE

*Atoka County*
ATOKA DAM

*Beaver County*
BEAVER
GATE

*Beckham County*
ELK CITY
ERICK 4 E

*Blaine County*
GEARY
OKEENE
WATONGA

*Bryan County*
DURANT USDA

*Caddo County*
ANADARKO
CARNEGIE 2 ENE

*Canadian County*
EL RENO 1 N

*Carter County*
ARDMORE
HEALDTON

*Cherokee County*
TAHLEQUAH

*Choctaw County*
BOSWELL 4 NNW
HUGO

*Cimarron County*
BOISE CITY 2 E
KENTON

*Comanche County*
CHATTANOOGA 3 NE
LAWTON
WICHITA MTN WL REF

*Cotton County*
WALTERS

*Craig County*
VINITA 2 N

*Creek County*
BRISTOW

*Custer County*
CLINTON
WEATHERFORD

*Delaware County*
KANSAS 1 ESE

*Dewey County*
TALOGA

*Ellis County*
ARNETT
GAGE AIRPORT

*Garfield County*
ENID

*Garvin County*
LINDSAY 2 W
PAULS VALLEY 4 WSW

*Grady County*
CHICKASHA EXP STN

*Grant County*
JEFFERSON

*Greer County*
MANGUM RESEARCH STN

*Harmon County*
HOLLIS

*Harper County*
BUFFALO

*Haskell County*
MC CURTAIN 1 SE

*Hughes County*
HOLDENVILLE

*Jackson County*
ALTUS IRIG RES STN

*Jefferson County*
WAURIKA

*Kay County*
NEWKIRK
PONCA CITY MUNI AP

*Kingfisher County*
HENNESSEY 4 ESE
KINGFISHER 2 SE

*Kiowa County*
ALTUS DAM
HOBART MUNICIPAL AP

*Latimer County*
WILBURTON 9 ENE

*Lincoln County*
CHANDLER 1
MEEKER 4 W

*Logan County*
GUTHRIE

*Love County*
MARIETTA

*McClain County*
BLANCHARD 2 SSW
PURCELL 5 SW

*McCurtain County*
BEAR MOUNTAIN TOWER
BROKEN BOW DAM
IDABEL
SMITHVILLE

*McIntosh County*
HANNA

*Marshall County*
MADILL

### Mayes County
PRYOR 6 N
SPAVINAW

### Muskogee County
MUSKOGEE
WEBBERS FALLS 5 WSW

### Noble County
BILLINGS
PERRY

### Nowata County
NOWATA

### Okfuskee County
OKEMAH

### Oklahoma County
OKLAHOMA CITY ROGERS

### Okmulgee County
OKMULGEE WATER WORKS

### Osage County
BARNSDALL
BARTLESVILLE PHILLIP
PAWHUSKA

### Ottawa County
MIAMI

### Pawnee County
MANNFORD 6 NW
RALSTON

### Payne County
CUSHING
STILLWATER 2 W

### Pittsburg County
MCALESTER MUNI AP

### Pontotoc County
ADA

### Pushmataha County
ANTLERS
TUSKAHOMA

### Roger Mills County
HAMMON 3 SSW
REYDON

### Rogers County
CLAREMORE 2 ENE

### Seminole County
SEMINOLE

### Sequoyah County
SALLISAW 2 NE

### Stephens County
DUNCAN

### Texas County
GOODWELL RESEARCH
HOOKER

### Tillman County
FREDERICK

### Tulsa County
BIXBY
TULSA INTL AP

### Wagoner County
WAGONER

### Woods County
ALVA
FREEDOM
WAYNOKA

### Woodward County
FORT SUPPLY DAM
MUTUAL

## ELEVATION INDEX

| FEET | STATION NAME |
|---|---|
| 502 | IDABEL |
| 512 | ANTLERS |
| 522 | MUSKOGEE |
| 531 | BOSWELL 4 NNW |
| 531 | SALLISAW 2 NE |
| 550 | WEBBERS FALLS 5 WSW |
| 571 | MC CURTAIN 1 SE |
| 571 | TUSKAHOMA |
| 591 | WAGONER |
| 600 | CLAREMORE 2 ENE |
| 600 | HUGO |
| 605 | BIXBY |
| 630 | ATOKA DAM |
| 630 | RALSTON |
| 640 | DURANT USDA |
| 640 | PRYOR 6 N |
| 640 | WILBURTON 9 ENE |
| 659 | BROKEN BOW DAM |
| 676 | TULSA INTL AP |
| 679 | HANNA |
| 679 | SPAVINAW |
| 712 | VINITA 2 N |
| 715 | BARTLESVILLE PHILLIP |
| 715 | OKMULGEE WATER WORKS |
| 722 | NOWATA |
| 741 | BARNSDALL |
| 741 | MANNFORD 6 NW |
| 771 | MADILL |
| 778 | MCALESTER MUNI AP |
| 791 | TAHLEQUAH |
| 801 | BEAR MOUNTAIN TOWER |
| 801 | MIAMI |
| 820 | BRISTOW |
| 840 | PAWHUSKA |
| 840 | SMITHVILLE |
| 850 | MARIETTA |
| 860 | CHANDLER 1 |
| 879 | ARDMORE |
| 879 | PAULS VALLEY 4 WSW |
| 879 | STILLWATER 2 W |
| 902 | HOLDENVILLE |
| 912 | HEALDTON |
| 912 | WAURIKA |
| 925 | MEEKER 4 W |
| 935 | OKEMAH |
| 971 | CUSHING |
| 981 | LINDSAY 2 W |
| 997 | PONCA CITY MUNI AP |
| 1000 | BILLINGS |
| 1001 | WALTERS |
| 1010 | SEMINOLE |
| 1020 | ADA |
| 1030 | GUTHRIE |
| 1030 | PERRY |
| 1043 | PURCELL 5 SW |
| 1060 | JEFFERSON |
| 1060 | KINGFISHER 2 SE |

| FEET | STATION NAME |
|---|---|
| 1089 | CHICKASHA EXP STN |
| 1122 | KANSAS 1 ESE |
| 1132 | DUNCAN |
| | |
| 1142 | LAWTON |
| 1152 | CHATTANOOGA 3 NE |
| 1152 | NEWKIRK |
| 1161 | HENNESSEY 4 ESE |
| 1195 | ANADARKO |
| | |
| 1201 | CHEROKEE |
| 1201 | GRT SALT PLAINS DAM |
| 1201 | OKEENE |
| 1250 | BLANCHARD 2 SSW |
| 1270 | ENID |
| | |
| 1280 | OKLAHOMA CITY ROGERS |
| 1289 | FREDERICK |
| 1322 | EL RENO 1 N |
| 1362 | ALVA |
| 1362 | HELENA 1 SSE |
| | |
| 1381 | ALTUS IRIG RES STN |
| 1503 | CARNEGIE 2 ENE |
| 1522 | CLINTON |
| 1522 | MANGUM RESEARCH STN |
| 1522 | WATONGA |
| | |
| 1529 | WAYNOKA |
| 1532 | ALTUS DAM |
| 1542 | FREEDOM |
| 1542 | HOBART MUNICIPAL AP |
| 1552 | GEARY |
| | |
| 1601 | WICHITA MTN WL REF |
| 1611 | HOLLIS |
| 1640 | WEATHERFORD |
| 1752 | TALOGA |
| 1780 | HAMMON 3 SSW |
| | |
| 1801 | BUFFALO |
| 1841 | MUTUAL |
| 1962 | ELK CITY |
| 1991 | ERICK 4 E |
| 2080 | FORT SUPPLY DAM |
| | |
| 2195 | GAGE AIRPORT |
| 2200 | REYDON |
| 2221 | GATE |
| 2461 | ARNETT |
| 2461 | BEAVER |
| | |
| 2992 | HOOKER |
| 3304 | GOODWELL RESEARCH |
| 4173 | BOISE CITY 2 E |
| 4354 | KENTON |

OKLAHOMA
10 20 30 STATUTE MILES
STATION LEGEND
DATA PUBLISHED IN:
● CLIMATOLOGICAL DATA
■ HOURLY PRECIPITATION DATA
△ CLIMATOLOGICAL DATA AND HOURLY PRECIPITATION DATA
For further information, refer to the station index and references notes.
DIVISIONS
1 PANHANDLE
2 NORTH CENTRAL
3 NORTHEAST
4 WEST CENTRAL
5 CENTRAL
6 EAST CENTRAL
7 SOUTHWEST
8 SOUTH CENTRAL
9 SOUTHEAST
US DOC - NOAA - NCDC - ASHEVILLE, NC Updated January 1992
36.0
34.0
-102.0
-100.0
-98.0
-96.0
-94.0

## ADA *Pontotoc County* ELEVATION 1020 ft LAT/LONG 34° 47 ' N / 96° 41 ' W

| | JAN | FEB | MAR | APR | MAY | JUN | JUL | AUG | SEP | OCT | NOV | DEC | YEAR |
|---|---|---|---|---|---|---|---|---|---|---|---|---|---|
| Maximum Temp °F | 50.8 | 56.0 | 65.1 | 74.2 | 80.3 | 87.6 | 92.9 | 92.9 | 85.0 | 75.3 | 63.0 | 53.8 | 73.1 |
| Minimum Temp °F | 28.3 | 31.8 | 40.4 | 49.4 | 57.4 | 65.8 | 70.4 | 68.7 | 61.0 | 50.4 | 39.5 | 31.2 | 49.5 |
| Mean Temp °F | 39.6 | 43.9 | 52.8 | 61.8 | 68.9 | 76.8 | 81.7 | 80.8 | 73.0 | 62.8 | 51.3 | 42.5 | 61.3 |
| Days Max Temp ≥ 90 °F | 0 | 0 | 0 | 0 | 2 | 11 | 24 | 24 | 9 | 1 | 0 | 0 | 71 |
| Days Max Temp ≤ 32 °F | 3 | 1 | 0 | 0 | 0 | 0 | 0 | 0 | 0 | 0 | 0 | 1 | 5 |
| Days Min Temp ≤ 32 °F | 20 | 15 | 7 | 1 | 0 | 0 | 0 | 0 | 0 | 1 | 7 | 16 | 67 |
| Days Min Temp ≤ 0 °F | 0 | 0 | 0 | 0 | 0 | 0 | 0 | 0 | 0 | 0 | 0 | 0 | 0 |
| Heating Degree Days | 783 | 590 | 385 | 138 | 28 | 1 | 0 | 0 | 19 | 131 | 411 | 690 | 3176 |
| Cooling Degree Days | 0 | 0 | 8 | 45 | 160 | 370 | 536 | 523 | 285 | 62 | 6 | 0 | 1995 |
| Total Precipitation (") | 1.59 | 2.26 | 3.16 | 3.74 | 5.80 | 4.22 | 2.63 | 3.02 | 4.95 | 4.41 | 2.86 | 2.32 | 40.96 |
| Days ≥ 0.1" Precip | 3 | 4 | 5 | 6 | 7 | 6 | 4 | 4 | 6 | 5 | 4 | 4 | 58 |
| Total Snowfall (") | na | *1.8* | 1.1 | 0.0 | 0.0 | 0.0 | 0.0 | 0.0 | 0.0 | 0.0 | 0.2 | 0.7 | na |
| Days ≥ 1" Snow Depth | *1* | na | 0 | 0 | 0 | 0 | 0 | 0 | 0 | 0 | 0 | 0 | na |

## ALTUS DAM *Kiowa County* ELEVATION 1532 ft LAT/LONG 34° 53 ' N / 99° 18 ' W

| | JAN | FEB | MAR | APR | MAY | JUN | JUL | AUG | SEP | OCT | NOV | DEC | YEAR |
|---|---|---|---|---|---|---|---|---|---|---|---|---|---|
| Maximum Temp °F | 48.6 | 54.2 | 64.6 | 74.3 | 81.6 | 90.8 | 96.1 | 94.0 | 85.1 | 75.0 | 61.0 | 51.8 | 73.1 |
| Minimum Temp °F | 25.4 | 29.6 | 38.8 | 48.6 | 58.1 | 67.6 | 72.3 | 70.4 | 62.0 | 49.6 | 37.1 | 28.3 | 49.0 |
| Mean Temp °F | 37.0 | 41.9 | 51.7 | 61.5 | 69.8 | 79.2 | 84.2 | 82.2 | 73.6 | 62.3 | 49.1 | 40.1 | 61.1 |
| Days Max Temp ≥ 90 °F | 0 | 0 | 1 | 1 | 6 | 19 | 27 | 24 | 11 | 2 | 0 | 0 | 91 |
| Days Max Temp ≤ 32 °F | 4 | 2 | 0 | 0 | 0 | 0 | 0 | 0 | 0 | 0 | 0 | 2 | 8 |
| Days Min Temp ≤ 32 °F | 23 | 17 | 8 | 1 | 0 | 0 | 0 | 0 | 0 | 1 | 10 | 22 | 82 |
| Days Min Temp ≤ 0 °F | 0 | 0 | 0 | 0 | 0 | 0 | 0 | 0 | 0 | 0 | 0 | 0 | 0 |
| Heating Degree Days | 860 | 644 | 416 | 160 | 35 | 1 | 0 | 1 | 24 | 150 | 477 | 766 | 3534 |
| Cooling Degree Days | 0 | 1 | 10 | 59 | 184 | 425 | 605 | 561 | 299 | 70 | 6 | 0 | 2220 |
| Total Precipitation (") | 0.92 | 1.11 | 1.82 | 2.20 | 4.72 | 3.81 | 1.86 | 2.48 | 3.23 | 2.62 | 1.53 | 1.01 | 27.31 |
| Days ≥ 0.1" Precip | 2 | 3 | 3 | 4 | 6 | 5 | 3 | 4 | 4 | 4 | 3 | 2 | 43 |
| Total Snowfall (") | na | na | 0.1 | 0.0 | 0.0 | 0.0 | 0.0 | 0.0 | 0.0 | 0.0 | 0.1 | *1.1* | na |
| Days ≥ 1" Snow Depth | na | na | 0 | 0 | 0 | 0 | 0 | 0 | 0 | 0 | 0 | *0* | na |

## ALTUS IRIG RES STN *Jackson County* ELEVATION 1381 ft LAT/LONG 34° 35 ' N / 99° 20 ' W

| | JAN | FEB | MAR | APR | MAY | JUN | JUL | AUG | SEP | OCT | NOV | DEC | YEAR |
|---|---|---|---|---|---|---|---|---|---|---|---|---|---|
| Maximum Temp °F | 52.4 | 57.9 | 68.3 | 77.6 | 84.3 | 93.2 | 97.6 | 95.5 | 87.5 | 78.0 | 64.5 | 54.6 | 76.0 |
| Minimum Temp °F | 26.2 | 30.0 | 38.7 | 47.9 | 57.4 | 66.6 | 70.7 | 68.9 | 61.7 | 49.8 | 38.2 | 29.2 | 48.8 |
| Mean Temp °F | 39.4 | 44.0 | 53.5 | 62.8 | 70.9 | 79.9 | 84.2 | 82.3 | 74.7 | 63.9 | 51.4 | 41.9 | 62.4 |
| Days Max Temp ≥ 90 °F | 0 | 0 | 1 | 3 | 8 | 22 | 28 | 26 | 14 | 4 | 0 | 0 | 106 |
| Days Max Temp ≤ 32 °F | 3 | 1 | 0 | 0 | 0 | 0 | 0 | 0 | 0 | 0 | 0 | 1 | 5 |
| Days Min Temp ≤ 32 °F | 24 | 17 | 8 | 2 | 0 | 0 | 0 | 0 | 0 | 1 | 9 | 21 | 82 |
| Days Min Temp ≤ 0 °F | 0 | 0 | 0 | 0 | 0 | 0 | 0 | 0 | 0 | 0 | 0 | 0 | 0 |
| Heating Degree Days | 788 | 588 | 365 | 128 | 24 | 0 | 0 | 0 | 17 | 114 | 410 | 709 | 3143 |
| Cooling Degree Days | 0 | 0 | 14 | 66 | 213 | 449 | 601 | 569 | 328 | 86 | 8 | 0 | 2334 |
| Total Precipitation (") | 0.96 | 1.05 | 1.57 | 2.23 | 4.70 | 3.68 | 1.79 | 2.69 | 3.39 | 2.39 | 1.38 | 1.02 | 26.85 |
| Days ≥ 0.1" Precip | 2 | 3 | 3 | 4 | 6 | 5 | 3 | 4 | 4 | 3 | 3 | 2 | 42 |
| Total Snowfall (") | *1.6* | na | 0.1 | 0.0 | 0.0 | 0.0 | 0.0 | 0.0 | 0.0 | 0.0 | 0.2 | 0.4 | na |
| Days ≥ 1" Snow Depth | *1* | *1* | 0 | 0 | 0 | 0 | 0 | 0 | 0 | 0 | 0 | 0 | 2 |

## ALVA *Woods County* ELEVATION 1362 ft LAT/LONG 36° 48 ' N / 98° 40 ' W

| | JAN | FEB | MAR | APR | MAY | JUN | JUL | AUG | SEP | OCT | NOV | DEC | YEAR |
|---|---|---|---|---|---|---|---|---|---|---|---|---|---|
| Maximum Temp °F | *46.0* | *51.8* | *62.4* | *73.1* | *80.2* | *91.0* | *96.6* | *94.1* | *85.2* | *75.5* | *59.0* | *49.6* | 72.0 |
| Minimum Temp °F | *22.8* | *26.4* | *35.9* | *46.5* | *55.4* | *65.4* | *70.3* | *68.1* | *60.0* | *47.9* | *35.2* | *26.5* | 46.7 |
| Mean Temp °F | *34.4* | *39.2* | *49.2* | *59.8* | *67.8* | *78.3* | *83.4* | *81.1* | *72.7* | *61.7* | *47.1* | *38.1* | 59.4 |
| Days Max Temp ≥ 90 °F | *0* | *0* | *0* | *1* | *5* | na | na | na | *11* | *3* | *0* | *0* | na |
| Days Max Temp ≤ 32 °F | *6* | *3* | *0* | *0* | *0* | *0* | *0* | *0* | *0* | *0* | *0* | *3* | 12 |
| Days Min Temp ≤ 32 °F | *26* | *21* | na | *2* | *0* | *0* | *0* | *0* | *0* | *1* | *11* | *24* | na |
| Days Min Temp ≤ 0 °F | *1* | *0* | *0* | *0* | *0* | *0* | *0* | *0* | *0* | *0* | *0* | *0* | 1 |
| Heating Degree Days | *940* | *722* | *491* | *195* | *57* | *2* | *0* | *0* | *25* | *169* | *533* | *828* | 3962 |
| Cooling Degree Days | na | na | na | na | na | na | na | na | na | na | na | na | na |
| Total Precipitation (") | *0.68* | *0.89* | *1.83* | *2.51* | *4.14* | *3.18* | *2.66* | *3.51* | *2.41* | *1.30* | *1.41* | *0.78* | 25.30 |
| Days ≥ 0.1" Precip | *2* | *2* | na | *4* | *6* | *5* | *5* | *5* | *4* | *2* | *2* | *2* | na |
| Total Snowfall (") | na | na | na | *0.7* | *0.0* | *0.0* | *0.0* | *0.0* | *0.0* | *0.0* | *1.0* | na | na |
| Days ≥ 1" Snow Depth | na | na | na | *0* | *0* | *0* | *0* | *0* | *0* | *0* | *0* | na | na |

## ANADARKO *Caddo County* ELEVATION 1195 ft LAT/LONG 35° 6 ' N / 98° 14 ' W

| | JAN | FEB | MAR | APR | MAY | JUN | JUL | AUG | SEP | OCT | NOV | DEC | YEAR |
|---|---|---|---|---|---|---|---|---|---|---|---|---|---|
| Maximum Temp °F | 49.1 | 54.9 | 65.0 | 74.6 | 81.5 | 89.9 | 95.4 | 94.0 | 86.1 | 75.5 | 62.6 | 52.5 | 73.4 |
| Minimum Temp °F | 24.3 | 28.7 | 37.8 | 47.7 | 56.6 | 65.3 | 69.8 | 68.1 | 61.0 | 48.5 | 36.8 | 27.5 | 47.7 |
| Mean Temp °F | 36.7 | 41.8 | 51.5 | 61.2 | 69.1 | 77.6 | 82.7 | 81.1 | 73.6 | 62.0 | 49.6 | 40.0 | 60.6 |
| Days Max Temp ≥ 90 °F | 0 | 0 | 0 | 1 | 4 | 16 | 26 | 24 | 11 | 2 | 0 | 0 | 84 |
| Days Max Temp ≤ 32 °F | 3 | 2 | 0 | 0 | 0 | 0 | 0 | 0 | 0 | 0 | 0 | 2 | 7 |
| Days Min Temp ≤ 32 °F | *24* | 18 | 10 | 2 | 0 | 0 | 0 | 0 | 0 | 2 | 10 | 22 | 88 |
| Days Min Temp ≤ 0 °F | 0 | 0 | 0 | 0 | 0 | 0 | 0 | 0 | 0 | 0 | 0 | 0 | 0 |
| Heating Degree Days | 869 | 647 | 425 | 162 | 35 | 1 | 0 | 0 | 20 | 154 | 456 | 767 | 3536 |
| Cooling Degree Days | 0 | 0 | 9 | 43 | 159 | *381* | 546 | 514 | 288 | 61 | *5* | 0 | 2006 |
| Total Precipitation (") | 1.07 | 1.48 | 2.20 | 2.60 | 4.79 | 3.63 | 2.33 | 2.61 | 3.56 | 2.70 | 1.65 | 1.42 | 30.04 |
| Days ≥ 0.1" Precip | 2 | 3 | 4 | 4 | 6 | 4 | 4 | 4 | 5 | 4 | 3 | 3 | 46 |
| Total Snowfall (") | na | *2.6* | 0.8 | 0.0 | 0.0 | 0.0 | 0.0 | 0.0 | 0.0 | 0.0 | 0.0 | 0.8 | na |
| Days ≥ 1" Snow Depth | *1* | *1* | 0 | 0 | 0 | 0 | 0 | 0 | 0 | 0 | 0 | 0 | 2 |

## ANTLERS *Pushmataha County* ELEVATION 512 ft LAT/LONG 34° 14 ' N / 95° 37 ' W

| | JAN | FEB | MAR | APR | MAY | JUN | JUL | AUG | SEP | OCT | NOV | DEC | YEAR |
|---|---|---|---|---|---|---|---|---|---|---|---|---|---|
| Maximum Temp °F | 52.3 | 57.2 | 66.4 | 75.6 | 81.3 | 88.7 | 93.5 | 93.8 | 86.5 | 76.5 | 64.2 | 54.7 | 74.2 |
| Minimum Temp °F | 28.4 | 32.3 | 40.7 | 49.7 | 57.5 | 65.7 | 68.9 | 67.6 | 61.1 | 49.2 | 40.5 | 31.9 | 49.5 |
| Mean Temp °F | 40.6 | 44.8 | 53.6 | 62.7 | 69.4 | 77.2 | 81.2 | 80.7 | 73.8 | 62.9 | 52.4 | 43.3 | 61.9 |
| Days Max Temp ≥ 90 °F | 0 | 0 | 0 | 1 | 1 | 15 | 26 | 24 | *11* | 1 | 0 | 0 | 79 |
| Days Max Temp ≤ 32 °F | 2 | 1 | 0 | 0 | 0 | 0 | 0 | 0 | 0 | 0 | 0 | 1 | 4 |
| Days Min Temp ≤ 32 °F | 21 | 14 | *6* | 1 | 0 | 0 | 0 | 0 | 0 | 1 | 7 | *16* | 66 |
| Days Min Temp ≤ 0 °F | 0 | 0 | 0 | 0 | 0 | 0 | 0 | 0 | 0 | 0 | 0 | 0 | 0 |
| Heating Degree Days | 748 | 564 | 359 | 125 | 27 | 1 | 0 | 0 | 18 | 129 | 383 | *665* | 3019 |
| Cooling Degree Days | 0 | 1 | 11 | 62 | 187 | 390 | 508 | *516* | 295 | 73 | 14 | 2 | 2059 |
| Total Precipitation (") | 2.17 | 2.80 | 3.76 | 4.49 | 6.32 | 4.64 | 3.16 | 2.64 | 4.49 | 4.80 | 3.59 | 3.29 | 46.15 |
| Days ≥ 0.1" Precip | 4 | 4 | 5 | 6 | 7 | 6 | 5 | 4 | 5 | 5 | 4 | *5* | 60 |
| Total Snowfall (") | *0.7* | 1.1 | 0.0 | 0.0 | 0.0 | 0.0 | 0.0 | 0.0 | 0.0 | 0.0 | 0.0 | 0.1 | 1.9 |
| Days ≥ 1" Snow Depth | *1* | 0 | 0 | 0 | 0 | 0 | 0 | 0 | 0 | 0 | 0 | 0 | 1 |

## ARDMORE *Carter County* ELEVATION 879 ft LAT/LONG 34° 10 ' N / 97° 8 ' W

| | JAN | FEB | MAR | APR | MAY | JUN | JUL | AUG | SEP | OCT | NOV | DEC | YEAR |
|---|---|---|---|---|---|---|---|---|---|---|---|---|---|
| Maximum Temp °F | 52.8 | 58.0 | 67.6 | 76.0 | 82.2 | 89.8 | 94.5 | 94.3 | 87.1 | 77.3 | 64.8 | 56.0 | 75.0 |
| Minimum Temp °F | 30.9 | 34.5 | 43.3 | 52.5 | 60.7 | 68.7 | 72.6 | 71.5 | 64.8 | 53.7 | 42.9 | 34.0 | 52.5 |
| Mean Temp °F | 41.9 | 46.3 | 55.5 | 64.3 | 71.5 | 79.3 | 83.6 | 82.9 | 76.0 | 65.5 | 53.9 | 45.0 | 63.8 |
| Days Max Temp ≥ 90 °F | 0 | 0 | 0 | 0 | 3 | 17 | 26 | 26 | 12 | 2 | 0 | 0 | 86 |
| Days Max Temp ≤ 32 °F | 2 | 1 | 0 | 0 | 0 | 0 | 0 | 0 | 0 | 0 | 0 | 1 | 4 |
| Days Min Temp ≤ 32 °F | 18 | 12 | 5 | 0 | 0 | 0 | 0 | 0 | 0 | 0 | 5 | 13 | 53 |
| Days Min Temp ≤ 0 °F | 0 | 0 | 0 | 0 | 0 | 0 | 0 | 0 | 0 | 0 | 0 | 0 | 0 |
| Heating Degree Days | 711 | 524 | 311 | 97 | 14 | 1 | 0 | 0 | 11 | 91 | 342 | 614 | 2716 |
| Cooling Degree Days | 0 | 2 | 17 | 69 | 207 | 422 | 560 | 571 | 335 | 99 | 14 | 1 | 2297 |
| Total Precipitation (") | 1.65 | 2.18 | 2.95 | 3.59 | 5.47 | 3.91 | 2.32 | 2.59 | 4.42 | 4.16 | 2.42 | 2.04 | 37.70 |
| Days ≥ 0.1" Precip | 3 | 4 | 5 | 6 | 7 | 5 | 3 | 4 | 5 | 5 | 4 | 3 | 54 |
| Total Snowfall (") | 1.9 | *1.0* | *0.5* | 0.0 | 0.0 | 0.0 | 0.0 | 0.0 | 0.0 | 0.0 | 0.2 | 0.3 | 3.9 |
| Days ≥ 1" Snow Depth | *1* | *0* | *0* | 0 | 0 | 0 | 0 | 0 | 0 | 0 | 0 | *1* | 2 |

## ARNETT *Ellis County* ELEVATION 2461 ft LAT/LONG 36° 8 ' N / 99° 46 ' W

| | JAN | FEB | MAR | APR | MAY | JUN | JUL | AUG | SEP | OCT | NOV | DEC | YEAR |
|---|---|---|---|---|---|---|---|---|---|---|---|---|---|
| Maximum Temp °F | 45.2 | 50.2 | 59.3 | 70.1 | 77.2 | 86.7 | 92.6 | 90.9 | 82.2 | 71.7 | 57.5 | 47.8 | 69.3 |
| Minimum Temp °F | 20.1 | 24.1 | 32.9 | 43.2 | 52.8 | 62.6 | 67.3 | 65.5 | 57.0 | 44.7 | 32.4 | 23.6 | 43.8 |
| Mean Temp °F | 32.6 | 37.2 | 46.2 | 56.7 | 65.0 | 74.7 | 80.0 | 78.2 | 69.7 | 58.2 | 45.0 | 35.8 | 56.6 |
| Days Max Temp ≥ 90 °F | 0 | 0 | 0 | 1 | 2 | 12 | 22 | 19 | 8 | 1 | 0 | 0 | 65 |
| Days Max Temp ≤ 32 °F | 7 | 4 | 1 | 0 | 0 | 0 | 0 | 0 | 0 | 0 | 1 | 4 | 17 |
| Days Min Temp ≤ 32 °F | 29 | 22 | 15 | 3 | 0 | 0 | 0 | 0 | 0 | 3 | 15 | 27 | 114 |
| Days Min Temp ≤ 0 °F | 1 | 0 | 0 | 0 | 0 | 0 | 0 | 0 | 0 | 0 | 0 | 1 | 2 |
| Heating Degree Days | 997 | 779 | 580 | 268 | 92 | 8 | 1 | 2 | 50 | 237 | 595 | 900 | 4509 |
| Cooling Degree Days | 0 | 0 | 2 | 27 | 97 | 303 | 463 | 433 | 204 | 29 | 2 | 0 | 1560 |
| Total Precipitation (") | 0.50 | 0.95 | 1.65 | 1.92 | 4.44 | 3.45 | 1.88 | 2.44 | 2.55 | 1.80 | 1.35 | 0.83 | 23.76 |
| Days ≥ 0.1" Precip | 2 | 2 | 3 | 4 | 6 | 6 | 4 | 4 | 4 | 3 | 3 | 2 | 43 |
| Total Snowfall (") | *1.3* | 3.2 | 1.8 | 0.5 | 0.0 | 0.0 | 0.0 | 0.0 | 0.0 | 0.0 | 2.1 | *0.8* | 9.7 |
| Days ≥ 1" Snow Depth | 4 | 4 | 1 | 0 | 0 | 0 | 0 | 0 | 0 | 0 | 1 | 3 | 13 |

## ATOKA DAM *Atoka County* ELEVATION 630 ft LAT/LONG 34° 26' N / 96° 5' W

| | JAN | FEB | MAR | APR | MAY | JUN | JUL | AUG | SEP | OCT | NOV | DEC | YEAR |
|---|---|---|---|---|---|---|---|---|---|---|---|---|---|
| Maximum Temp °F | na | 54.8 | 63.6 | 73.4 | na | 87.7 | na | 93.9 | 86.0 | 75.7 | na | na | na |
| Minimum Temp °F | 28.3 | 31.7 | 40.7 | 49.8 | na | 66.8 | na | 70.1 | 62.8 | 51.2 | na | na | na |
| Mean Temp °F | na | 43.5 | 52.2 | 61.6 | na | 77.2 | na | 82.0 | 74.5 | 63.5 | na | na | na |
| Days Max Temp ≥ 90 °F | 0 | 0 | 0 | 0 | 1 | 13 | 23 | 24 | 10 | 1 | 0 | 0 | 72 |
| Days Max Temp ≤ 32 °F | 3 | 1 | 0 | 0 | 0 | 0 | 0 | 0 | 0 | 0 | 0 | 1 | 5 |
| Days Min Temp ≤ 32 °F | 20 | 13 | 6 | 1 | 0 | 0 | 0 | 0 | 0 | 0 | na | 15 | na |
| Days Min Temp ≤ 0 °F | 0 | 0 | 0 | 0 | 0 | 0 | 0 | 0 | 0 | 0 | 0 | 0 | 0 |
| Heating Degree Days | na | 589 | 402 | 158 | na | 2 | na | 0 | na | 119 | na | na | na |
| Cooling Degree Days | na | 0 | 12 | 54 | na | 392 | na | 576 | 327 | 84 | na | na | na |
| Total Precipitation (") | 2.14 | 2.60 | 3.63 | 4.55 | 5.84 | 4.34 | 2.54 | 2.44 | 4.97 | 3.93 | 3.25 | 2.70 | 42.93 |
| Days ≥ 0.1" Precip | 4 | 4 | 5 | 6 | 6 | 5 | 4 | 3 | 5 | 4 | 4 | 4 | 54 |
| Total Snowfall (") | 1.0 | 1.2 | 0.2 | 0.0 | 0.0 | 0.0 | 0.0 | 0.0 | 0.0 | 0.0 | 0.1 | 0.1 | 2.6 |
| Days ≥ 1" Snow Depth | 1 | 0 | 0 | 0 | 0 | 0 | 0 | 0 | 0 | 0 | 0 | 0 | 1 |

## BARNSDALL *Osage County* ELEVATION 741 ft LAT/LONG 36° 33' N / 96° 10' W

| | JAN | FEB | MAR | APR | MAY | JUN | JUL | AUG | SEP | OCT | NOV | DEC | YEAR |
|---|---|---|---|---|---|---|---|---|---|---|---|---|---|
| Maximum Temp °F | 46.4 | 52.6 | 63.2 | 73.9 | 80.0 | 88.1 | 94.2 | 93.5 | 84.9 | 74.1 | 60.5 | 50.2 | 71.8 |
| Minimum Temp °F | 23.3 | 27.6 | 37.4 | 48.0 | 56.3 | 65.4 | 69.9 | 67.6 | 60.1 | 47.6 | 36.8 | 27.3 | 47.3 |
| Mean Temp °F | 34.9 | 40.1 | 50.3 | 61.0 | 68.2 | 76.7 | 82.1 | 80.6 | 72.5 | 60.9 | 48.7 | 38.8 | 59.6 |
| Days Max Temp ≥ 90 °F | 0 | 0 | 0 | 1 | 2 | 12 | 24 | 23 | 9 | 1 | 0 | 0 | 72 |
| Days Max Temp ≤ 32 °F | 5 | 2 | 0 | 0 | 0 | 0 | 0 | 0 | 0 | 0 | 0 | 2 | 9 |
| Days Min Temp ≤ 32 °F | 25 | 19 | 12 | 2 | 0 | 0 | 0 | 0 | 0 | 2 | 11 | 22 | 93 |
| Days Min Temp ≤ 0 °F | 1 | 0 | 0 | 0 | 0 | 0 | 0 | 0 | 0 | 0 | 0 | 1 | 2 |
| Heating Degree Days | 928 | 698 | 460 | 171 | 44 | 2 | 0 | 0 | 28 | 179 | 489 | 806 | 3805 |
| Cooling Degree Days | 0 | 1 | 11 | 53 | 144 | 357 | 531 | 504 | 266 | 51 | 6 | 0 | 1924 |
| Total Precipitation (") | 1.47 | 2.00 | 3.62 | 3.77 | 5.09 | 4.85 | 3.18 | 3.21 | 5.23 | 3.31 | 3.17 | 2.17 | 41.07 |
| Days ≥ 0.1" Precip | 3 | 4 | 5 | 6 | 8 | 7 | 4 | 5 | 6 | 5 | 4 | 4 | 61 |
| Total Snowfall (") | 3.2 | 3.1 | 1.1 | 0.0 | 0.0 | 0.0 | 0.0 | 0.0 | 0.0 | 0.0 | 0.3 | 2.0 | 9.7 |
| Days ≥ 1" Snow Depth | na | na | 0 | 0 | 0 | 0 | 0 | 0 | 0 | 0 | 0 | 1 | na |

## BARTLESVILLE PHILLIP *Osage County* ELEVATION 715 ft LAT/LONG 36° 45' N / 96° 0' W

| | JAN | FEB | MAR | APR | MAY | JUN | JUL | AUG | SEP | OCT | NOV | DEC | YEAR |
|---|---|---|---|---|---|---|---|---|---|---|---|---|---|
| Maximum Temp °F | 46.4 | 52.5 | 63.3 | 74.1 | 80.3 | 88.4 | 94.3 | 93.4 | 84.9 | 74.4 | 60.5 | 49.9 | 71.9 |
| Minimum Temp °F | 23.2 | 27.4 | 37.3 | 47.9 | 56.4 | 65.3 | 69.7 | 67.2 | 59.9 | 47.5 | 36.9 | 27.6 | 47.2 |
| Mean Temp °F | 34.8 | 40.0 | 50.3 | 61.0 | 68.4 | 76.9 | 82.0 | 80.3 | 72.4 | 61.0 | 48.7 | 38.8 | 59.6 |
| Days Max Temp ≥ 90 °F | 0 | 0 | 0 | 1 | 2 | 14 | 25 | 23 | 9 | 1 | 0 | 0 | 75 |
| Days Max Temp ≤ 32 °F | 5 | 3 | 0 | 0 | 0 | 0 | 0 | 0 | 0 | 0 | 0 | 2 | 10 |
| Days Min Temp ≤ 32 °F | 25 | 20 | 12 | 2 | 0 | 0 | 0 | 0 | 0 | 2 | 11 | 21 | 93 |
| Days Min Temp ≤ 0 °F | 1 | 0 | 0 | 0 | 0 | 0 | 0 | 0 | 0 | 0 | 0 | 0 | 1 |
| Heating Degree Days | 929 | 700 | 459 | 171 | 42 | 2 | 0 | 0 | 27 | 175 | 487 | 807 | 3799 |
| Cooling Degree Days | 0 | 1 | 11 | 55 | 157 | 379 | 542 | 498 | 268 | 54 | 7 | 1 | 1973 |
| Total Precipitation (") | 1.37 | 1.73 | 3.19 | 3.64 | 4.54 | 4.05 | 2.87 | 2.72 | 4.44 | 3.45 | 2.97 | 1.96 | 36.93 |
| Days ≥ 0.1" Precip | 3 | 3 | 5 | 6 | 7 | 6 | 4 | 4 | 5 | 5 | 4 | 4 | 56 |
| Total Snowfall (") | 3.2 | 2.8 | 1.9 | 0.0 | 0.0 | 0.0 | 0.0 | 0.0 | 0.0 | 0.0 | 0.6 | 1.4 | 9.9 |
| Days ≥ 1" Snow Depth | 4 | 3 | 1 | 0 | 0 | 0 | 0 | 0 | 0 | 0 | 0 | 2 | 10 |

## BEAR MOUNTAIN TOWER *McCurtain County* ELEVATION 801 ft LAT/LONG 34° 8' N / 94° 57' W

| | JAN | FEB | MAR | APR | MAY | JUN | JUL | AUG | SEP | OCT | NOV | DEC | YEAR |
|---|---|---|---|---|---|---|---|---|---|---|---|---|---|
| Maximum Temp °F | 52.7 | 57.9 | 67.0 | 75.5 | 80.9 | na | 94.3 | 93.6 | 86.5 | 76.9 | 65.5 | 55.8 | na |
| Minimum Temp °F | 30.6 | 34.4 | 42.3 | 50.5 | 57.5 | na | 69.3 | 68.0 | 61.7 | 51.3 | 41.5 | 33.3 | na |
| Mean Temp °F | 41.7 | 46.2 | 54.7 | 63.0 | 69.2 | na | 81.9 | 80.8 | 74.1 | 64.1 | 53.5 | 44.5 | na |
| Days Max Temp ≥ 90 °F | 0 | 0 | 0 | 0 | 1 | 10 | 25 | 24 | 11 | 2 | 0 | 0 | 73 |
| Days Max Temp ≤ 32 °F | 1 | 0 | 0 | 0 | 0 | 0 | 0 | 0 | 0 | 0 | 0 | 1 | 2 |
| Days Min Temp ≤ 32 °F | 18 | 12 | 6 | 0 | 0 | 0 | 0 | 0 | 0 | 1 | 6 | 14 | 57 |
| Days Min Temp ≤ 0 °F | 0 | 0 | 0 | 0 | 0 | 0 | 0 | 0 | 0 | 0 | 0 | 0 | 0 |
| Heating Degree Days | 717 | 528 | 328 | 114 | 24 | na | 0 | 0 | 13 | 108 | 349 | 630 | na |
| Cooling Degree Days | 0 | 1 | 12 | 60 | na | na | na | na | na | 88 | na | na | na |
| Total Precipitation (") | 2.78 | 3.73 | 4.69 | 4.94 | 6.81 | 4.13 | 3.44 | 3.48 | 5.02 | 5.03 | 4.39 | 4.71 | 53.15 |
| Days ≥ 0.1" Precip | 5 | 5 | 6 | 6 | 8 | 6 | 5 | 5 | 6 | 5 | 5 | 6 | 68 |
| Total Snowfall (") | 1.3 | 1.7 | 0.2 | 0.0 | 0.0 | 0.0 | 0.0 | 0.0 | 0.0 | 0.0 | 0.2 | 0.5 | 3.9 |
| Days ≥ 1" Snow Depth | 1 | 1 | 0 | 0 | 0 | 0 | 0 | 0 | 0 | 0 | 0 | 0 | 2 |

## BEAVER *Beaver County* ELEVATION 2461 ft LAT/LONG 36° 49 ' N / 100° 31 ' W

| | JAN | FEB | MAR | APR | MAY | JUN | JUL | AUG | SEP | OCT | NOV | DEC | YEAR |
|---|---|---|---|---|---|---|---|---|---|---|---|---|---|
| Maximum Temp °F | 45.7 | 51.4 | 60.4 | 71.3 | 78.9 | 89.1 | 95.0 | 92.7 | 84.4 | 73.7 | 58.7 | 48.8 | 70.8 |
| Minimum Temp °F | 17.8 | 21.6 | 30.3 | 40.8 | 51.0 | 61.1 | 66.5 | 64.5 | 55.4 | 41.7 | 28.9 | 20.5 | 41.7 |
| Mean Temp °F | 31.8 | 36.5 | 45.4 | 56.1 | 65.0 | 75.2 | 80.8 | 78.6 | 69.9 | 57.7 | 43.8 | 34.7 | 56.3 |
| Days Max Temp ≥ 90 °F | 0 | 0 | 0 | 1 | 4 | 16 | 25 | 21 | 11 | 3 | 0 | 0 | 81 |
| Days Max Temp ≤ 32 °F | 6 | 4 | 1 | 0 | 0 | 0 | 0 | 0 | 0 | 0 | 1 | 4 | 16 |
| Days Min Temp ≤ 32 °F | 30 | 25 | 18 | 5 | 0 | 0 | 0 | 0 | 0 | 4 | 21 | 29 | 132 |
| Days Min Temp ≤ 0 °F | 2 | 1 | 0 | 0 | 0 | 0 | 0 | 0 | 0 | 0 | 0 | 1 | 4 |
| Heating Degree Days | 1024 | 797 | 603 | 286 | 99 | 11 | 1 | 1 | 51 | 253 | 629 | 933 | 4688 |
| Cooling Degree Days | 0 | 0 | 1 | 26 | 100 | 322 | 497 | 450 | 221 | 31 | 1 | 0 | 1649 |
| Total Precipitation (") | 0.50 | 0.80 | 1.42 | 1.65 | 3.12 | 3.36 | 2.83 | 2.89 | 1.73 | 1.15 | 1.04 | 0.65 | 21.14 |
| Days ≥ 0.1" Precip | 1 | 2 | 3 | 3 | 5 | 5 | 4 | 5 | 4 | 2 | 2 | 2 | 38 |
| Total Snowfall (") | *1.7* | *3.4* | *0.9* | 0.2 | 0.0 | 0.0 | 0.0 | 0.0 | 0.0 | 0.2 | *1.1* | *1.4* | 8.9 |
| Days ≥ 1" Snow Depth | *1* | na | *1* | 0 | 0 | 0 | 0 | 0 | 0 | 0 | *1* | *1* | na |

## BILLINGS *Noble County* ELEVATION 1000 ft LAT/LONG 36° 32 ' N / 97° 27 ' W

| | JAN | FEB | MAR | APR | MAY | JUN | JUL | AUG | SEP | OCT | NOV | DEC | YEAR |
|---|---|---|---|---|---|---|---|---|---|---|---|---|---|
| Maximum Temp °F | 44.4 | 49.5 | 60.3 | 70.6 | 78.8 | 88.9 | 95.0 | 93.2 | 84.0 | 73.5 | 58.3 | 46.6 | 70.3 |
| Minimum Temp °F | 23.3 | 26.8 | 35.9 | 45.2 | 55.5 | 65.7 | 70.9 | 68.7 | 60.5 | 47.7 | 35.6 | 26.2 | 46.8 |
| Mean Temp °F | 33.9 | 38.2 | *48.2* | 57.9 | 67.1 | 77.2 | 83.0 | 81.0 | 72.3 | 60.6 | 46.9 | 36.5 | 58.6 |
| Days Max Temp ≥ 90 °F | 0 | 0 | 0 | 0 | 1 | 15 | 26 | 23 | 9 | 1 | 0 | 0 | 75 |
| Days Max Temp ≤ 32 °F | 6 | *4* | 0 | 0 | 0 | 0 | 0 | 0 | 0 | 0 | 0 | 3 | 13 |
| Days Min Temp ≤ 32 °F | 26 | 19 | *11* | 2 | 0 | 0 | 0 | 0 | 0 | 2 | 12 | 23 | 95 |
| Days Min Temp ≤ 0 °F | 1 | 0 | 0 | 0 | 0 | 0 | 0 | 0 | 0 | 0 | 0 | 1 | 2 |
| Heating Degree Days | 959 | 751 | *517* | 234 | 58 | 3 | 0 | 0 | 29 | *175* | 537 | *878* | 4141 |
| Cooling Degree Days | 0 | 0 | 3 | 27 | 122 | 365 | 552 | 506 | 265 | 41 | 4 | 0 | 1885 |
| Total Precipitation (") | 1.05 | 1.40 | 2.70 | 3.41 | 4.68 | 4.23 | 3.05 | 3.24 | 3.85 | 2.57 | 2.42 | 1.48 | 34.08 |
| Days ≥ 0.1" Precip | 2 | 3 | 5 | 5 | 6 | 6 | 4 | 4 | 4 | 3 | 3 | 3 | 48 |
| Total Snowfall (") | 2.3 | 3.1 | 1.5 | 0.0 | 0.0 | 0.0 | 0.0 | 0.0 | 0.0 | 0.0 | 0.5 | 1.3 | 8.7 |
| Days ≥ 1" Snow Depth | *1* | *1* | 1 | 0 | 0 | 0 | 0 | 0 | 0 | 0 | 0 | *1* | 4 |

## BIXBY *Tulsa County* ELEVATION 605 ft LAT/LONG 35° 59 ' N / 95° 53 ' W

| | JAN | FEB | MAR | APR | MAY | JUN | JUL | AUG | SEP | OCT | NOV | DEC | YEAR |
|---|---|---|---|---|---|---|---|---|---|---|---|---|---|
| Maximum Temp °F | 45.8 | 51.8 | 61.5 | 72.0 | 78.9 | 87.2 | 93.0 | 92.1 | 83.3 | 74.4 | 60.3 | *50.1* | 70.9 |
| Minimum Temp °F | *23.1* | 27.3 | 36.6 | 46.9 | 56.0 | 65.0 | 69.0 | 66.8 | 59.2 | 46.6 | 36.0 | *27.7* | 46.7 |
| Mean Temp °F | *34.4* | 39.6 | 49.1 | 59.5 | 67.5 | 76.1 | 80.9 | 79.5 | 71.3 | 60.5 | 48.1 | *38.9* | 58.8 |
| Days Max Temp ≥ 90 °F | 0 | 0 | 0 | 0 | 1 | 12 | *23* | *21* | 7 | 1 | 0 | 0 | 65 |
| Days Max Temp ≤ 32 °F | 5 | 2 | 0 | 0 | 0 | 0 | 0 | 0 | 0 | 0 | 0 | *2* | 9 |
| Days Min Temp ≤ 32 °F | *26* | *20* | 11 | 2 | 0 | 0 | 0 | 0 | 0 | 2 | 12 | *22* | 95 |
| Days Min Temp ≤ 0 °F | 1 | 0 | 0 | 0 | 0 | 0 | 0 | 0 | 0 | 0 | 0 | 0 | 1 |
| Heating Degree Days | *942* | 714 | 495 | 200 | 52 | 2 | 0 | 0 | 35 | 182 | 498 | *804* | 3924 |
| Cooling Degree Days | *0* | *0* | 6 | 37 | 138 | 359 | 504 | 485 | 237 | 43 | 6 | *0* | 1815 |
| Total Precipitation (") | 1.47 | 1.85 | 3.21 | 3.65 | 5.02 | 4.25 | 2.37 | 2.70 | 4.86 | 3.79 | 3.35 | 2.31 | 38.83 |
| Days ≥ 0.1" Precip | *2* | *4* | 4 | 6 | 7 | 5 | 3 | 3 | 6 | 5 | 4 | 4 | 53 |
| Total Snowfall (") | *1.4* | na | 0.9 | 0.0 | 0.0 | 0.0 | 0.0 | 0.0 | 0.0 | 0.0 | 0.2 | 0.9 | na |
| Days ≥ 1" Snow Depth | *1* | na | 0 | 0 | 0 | 0 | 0 | 0 | 0 | 0 | 0 | *0* | na |

## BLANCHARD 2 SSW *McClain County* ELEVATION 1250 ft LAT/LONG 35° 8 ' N / 97° 40 ' W

| | JAN | FEB | MAR | APR | MAY | JUN | JUL | AUG | SEP | OCT | NOV | DEC | YEAR |
|---|---|---|---|---|---|---|---|---|---|---|---|---|---|
| Maximum Temp °F | 48.8 | 54.5 | 64.2 | 74.0 | 80.3 | 88.1 | 94.1 | 93.2 | 85.2 | 74.9 | 61.5 | 52.1 | 72.6 |
| Minimum Temp °F | 27.0 | 30.6 | 39.3 | 49.4 | 58.1 | 65.8 | 70.2 | 69.1 | 61.8 | 49.9 | 38.8 | 30.1 | 49.2 |
| Mean Temp °F | 38.0 | 42.6 | 51.8 | 61.7 | 69.2 | 76.9 | 82.2 | 81.2 | 73.5 | 62.4 | 50.2 | 41.1 | 60.9 |
| Days Max Temp ≥ 90 °F | 0 | 0 | 0 | 1 | 2 | 13 | 24 | 23 | 9 | 1 | 0 | 0 | 73 |
| Days Max Temp ≤ 32 °F | 4 | 2 | 0 | 0 | 0 | 0 | 0 | 0 | 0 | 0 | 0 | 2 | 8 |
| Days Min Temp ≤ 32 °F | 21 | 16 | 8 | 1 | 0 | 0 | 0 | 0 | 0 | 1 | 9 | 19 | 75 |
| Days Min Temp ≤ 0 °F | 0 | 0 | 0 | 0 | 0 | 0 | 0 | 0 | 0 | 0 | 0 | 0 | 0 |
| Heating Degree Days | 831 | 628 | 416 | 153 | 31 | 2 | 0 | 0 | 21 | 142 | 445 | 733 | 3402 |
| Cooling Degree Days | 0 | 0 | 12 | 52 | 169 | 361 | 542 | 523 | *289* | 64 | *8* | 0 | 2020 |
| Total Precipitation (") | 1.16 | 1.65 | 2.45 | 3.15 | 5.25 | 3.46 | 2.57 | 2.70 | 4.17 | 3.25 | 1.90 | 1.69 | 33.40 |
| Days ≥ 0.1" Precip | 2 | 3 | 4 | 5 | 7 | 5 | 4 | 4 | 5 | 4 | 3 | 3 | 49 |
| Total Snowfall (") | *2.5* | *2.0* | *0.9* | 0.1 | 0.0 | 0.0 | 0.0 | 0.0 | 0.0 | 0.0 | 0.7 | 0.7 | 6.9 |
| Days ≥ 1" Snow Depth | *2* | *1* | *0* | 0 | 0 | 0 | 0 | 0 | 0 | 0 | *0* | *0* | 3 |

## BOISE CITY 2 E *Cimarron County* ELEVATION 4173 ft LAT/LONG 36° 44 ' N / 102° 30 ' W

| | JAN | FEB | MAR | APR | MAY | JUN | JUL | AUG | SEP | OCT | NOV | DEC | YEAR |
|---|---|---|---|---|---|---|---|---|---|---|---|---|---|
| Maximum Temp °F | 49.7 | 53.9 | 61.9 | 71.5 | 78.4 | 88.6 | 92.8 | 90.0 | 83.0 | 73.3 | 59.5 | 50.8 | 71.1 |
| Minimum Temp °F | 18.9 | 22.2 | 29.2 | 38.6 | 47.3 | 57.4 | 62.4 | 60.5 | 52.4 | 40.4 | 28.7 | 21.0 | 39.9 |
| Mean Temp °F | 34.3 | 38.1 | 45.6 | 55.0 | 62.9 | 73.0 | 77.6 | 75.3 | 67.7 | 56.9 | 44.2 | 35.9 | 55.5 |
| Days Max Temp ≥ 90 °F | 0 | 0 | 0 | 1 | 3 | 15 | 24 | 19 | 7 | 1 | 0 | 0 | 70 |
| Days Max Temp ≤ 32 °F | 4 | 2 | 1 | 0 | 0 | 0 | 0 | 0 | 0 | 0 | 1 | 3 | 11 |
| Days Min Temp ≤ 32 °F | 30 | 25 | 19 | 7 | 1 | 0 | 0 | 0 | 0 | 4 | 20 | 28 | 134 |
| Days Min Temp ≤ 0 °F | 2 | 1 | 0 | 0 | 0 | 0 | 0 | 0 | 0 | 0 | 0 | 1 | 4 |
| Heating Degree Days | 944 | 755 | 596 | 302 | 119 | 13 | 1 | 2 | 54 | 260 | 618 | 894 | 4558 |
| Cooling Degree Days | 0 | 0 | 0 | 12 | 59 | 262 | 402 | 344 | 156 | 13 | 0 | 0 | 1248 |
| Total Precipitation (") | 0.35 | 0.42 | 0.95 | 1.23 | 2.73 | 2.79 | 2.65 | 2.58 | 1.77 | 0.90 | 0.70 | 0.45 | 17.52 |
| Days ≥ 0.1" Precip | 1 | 1 | 3 | 3 | 5 | 5 | 5 | 5 | 4 | 2 | 2 | 2 | 38 |
| Total Snowfall (") | 6.2 | 4.1 | 6.1 | 2.3 | 0.4 | 0.0 | 0.0 | 0.0 | 0.1 | 1.1 | 3.0 | 5.2 | 28.5 |
| Days ≥ 1" Snow Depth | 5 | *2* | 2 | 0 | 0 | 0 | 0 | 0 | 0 | 0 | 1 | 3 | 13 |

## BOSWELL 4 NNW *Choctaw County* ELEVATION 531 ft LAT/LONG 34° 5 ' N / 95° 54 ' W

| | JAN | FEB | MAR | APR | MAY | JUN | JUL | AUG | SEP | OCT | NOV | DEC | YEAR |
|---|---|---|---|---|---|---|---|---|---|---|---|---|---|
| Maximum Temp °F | 52.5 | 57.4 | 66.7 | 74.4 | 80.6 | 88.4 | 93.3 | 93.3 | 86.1 | 76.2 | 64.3 | 55.5 | 74.1 |
| Minimum Temp °F | 29.7 | 34.0 | 42.7 | 51.6 | 58.9 | 66.7 | 70.5 | 68.9 | 62.7 | 51.0 | 41.8 | 33.3 | 51.0 |
| Mean Temp °F | 41.1 | 45.7 | 54.9 | 63.0 | 69.8 | 77.4 | 82.0 | 81.2 | 74.4 | 63.5 | 53.1 | 44.4 | 62.5 |
| Days Max Temp ≥ 90 °F | 0 | 0 | 0 | 0 | 1 | 13 | 25 | 24 | 10 | 1 | 0 | 0 | 74 |
| Days Max Temp ≤ 32 °F | *1* | 1 | 0 | 0 | 0 | 0 | 0 | 0 | 0 | 0 | 0 | 1 | 3 |
| Days Min Temp ≤ 32 °F | *19* | *13* | 5 | 1 | 0 | 0 | 0 | 0 | 0 | 1 | 6 | 15 | 60 |
| Days Min Temp ≤ 0 °F | 0 | 0 | 0 | 0 | 0 | 0 | 0 | 0 | 0 | 0 | 0 | 0 | 0 |
| Heating Degree Days | 734 | 538 | 322 | 117 | 22 | 0 | 0 | 0 | 14 | 118 | 363 | 631 | 2859 |
| Cooling Degree Days | 0 | 1 | 17 | 68 | 185 | 389 | 532 | 539 | 316 | 85 | 14 | 2 | 2148 |
| Total Precipitation (") | 2.17 | 3.21 | 3.73 | 4.14 | 6.06 | 4.32 | 2.78 | 2.72 | 4.22 | 4.56 | 3.85 | 2.90 | 44.66 |
| Days ≥ 0.1" Precip | *4* | 5 | 5 | 6 | *7* | 6 | 4 | 3 | 5 | 5 | 5 | 5 | 60 |
| Total Snowfall (") | na | *0.3* | 0.2 | 0.0 | 0.0 | 0.0 | 0.0 | 0.0 | 0.0 | 0.0 | 0.0 | 0.5 | na |
| Days ≥ 1" Snow Depth | na | na | 0 | 0 | 0 | 0 | 0 | 0 | 0 | 0 | 0 | *0* | na |

## BRISTOW *Creek County* ELEVATION 820 ft LAT/LONG 35° 50 ' N / 96° 24 ' W

| | JAN | FEB | MAR | APR | MAY | JUN | JUL | AUG | SEP | OCT | NOV | DEC | YEAR |
|---|---|---|---|---|---|---|---|---|---|---|---|---|---|
| Maximum Temp °F | 49.0 | 55.5 | 65.2 | 75.1 | 80.9 | 88.4 | 94.5 | 94.0 | 85.7 | 76.0 | 62.1 | 52.6 | 73.3 |
| Minimum Temp °F | 25.1 | 29.0 | 38.5 | 48.9 | 56.7 | 65.4 | 69.3 | 67.4 | 59.8 | 48.8 | 38.1 | 28.6 | 48.0 |
| Mean Temp °F | 37.0 | 42.3 | 51.9 | 62.1 | 68.8 | 77.0 | 81.9 | 80.8 | 72.8 | 62.4 | 50.1 | 40.6 | 60.6 |
| Days Max Temp ≥ 90 °F | 0 | 0 | 0 | 1 | 2 | 14 | 26 | 24 | 10 | 1 | 0 | 0 | 78 |
| Days Max Temp ≤ 32 °F | 4 | 1 | 0 | 0 | 0 | 0 | 0 | 0 | 0 | 0 | 0 | 2 | 7 |
| Days Min Temp ≤ 32 °F | 24 | 18 | 10 | 1 | 0 | 0 | 0 | 0 | 0 | 2 | *10* | 21 | 86 |
| Days Min Temp ≤ 0 °F | 1 | 0 | 0 | 0 | 0 | 0 | 0 | 0 | 0 | 0 | 0 | 0 | 1 |
| Heating Degree Days | 861 | 637 | 414 | 145 | 38 | 1 | 0 | 0 | 24 | 148 | 449 | 751 | 3468 |
| Cooling Degree Days | 0 | 0 | 12 | 63 | 168 | 378 | 540 | 519 | 275 | 62 | 9 | 1 | 2027 |
| Total Precipitation (") | 1.43 | 2.03 | 3.09 | 3.52 | 5.64 | 4.02 | 2.33 | 2.76 | 4.82 | 3.40 | 3.00 | 2.46 | 38.50 |
| Days ≥ 0.1" Precip | 3 | 4 | 5 | 6 | 7 | 5 | 4 | 5 | 5 | 4 | 4 | 4 | 56 |
| Total Snowfall (") | 3.0 | 2.7 | 1.3 | 0.0 | 0.0 | 0.0 | 0.0 | 0.0 | 0.0 | 0.0 | 0.5 | 1.9 | 9.4 |
| Days ≥ 1" Snow Depth | 4 | *3* | 1 | 0 | 0 | 0 | 0 | 0 | 0 | 0 | 0 | 2 | 10 |

## BROKEN BOW DAM *McCurtain County* ELEVATION 659 ft LAT/LONG 34° 9 ' N / 94° 41 ' W

| | JAN | FEB | MAR | APR | MAY | JUN | JUL | AUG | SEP | OCT | NOV | DEC | YEAR |
|---|---|---|---|---|---|---|---|---|---|---|---|---|---|
| Maximum Temp °F | 51.8 | 56.7 | 65.9 | 74.7 | 80.9 | 88.6 | 93.3 | 93.2 | 86.2 | 76.2 | 64.1 | 55.4 | 73.9 |
| Minimum Temp °F | 27.9 | 31.1 | 39.3 | 48.1 | 56.9 | 64.6 | 67.9 | 66.9 | 61.1 | 49.0 | 39.5 | 31.5 | 48.7 |
| Mean Temp °F | 39.8 | 43.9 | 52.6 | 61.4 | 68.9 | 76.7 | 80.6 | 80.1 | 73.7 | 62.6 | 51.9 | 43.5 | 61.3 |
| Days Max Temp ≥ 90 °F | 0 | 0 | 0 | 0 | 2 | 14 | 24 | *23* | 11 | 2 | 0 | 0 | 76 |
| Days Max Temp ≤ 32 °F | 1 | 0 | 0 | 0 | 0 | 0 | 0 | 0 | 0 | 0 | 0 | 1 | 2 |
| Days Min Temp ≤ 32 °F | 23 | 16 | 7 | 1 | 0 | 0 | 0 | 0 | 0 | 1 | 8 | 18 | 74 |
| Days Min Temp ≤ 0 °F | 0 | 0 | 0 | 0 | 0 | 0 | 0 | 0 | 0 | 0 | 0 | 0 | 0 |
| Heating Degree Days | 773 | 589 | 384 | 145 | 29 | 1 | 0 | 0 | 13 | 131 | 395 | 662 | 3122 |
| Cooling Degree Days | 0 | 0 | 7 | 42 | 162 | 365 | 494 | 488 | 286 | 64 | 7 | 1 | 1916 |
| Total Precipitation (") | 3.22 | 3.62 | 5.13 | 4.80 | 6.98 | 4.34 | 4.06 | 2.63 | 4.69 | 5.16 | 4.78 | 4.84 | 54.25 |
| Days ≥ 0.1" Precip | 5 | 6 | 6 | 6 | 8 | 6 | 5 | 4 | 6 | 6 | 6 | 6 | 70 |
| Total Snowfall (") | *0.4* | *0.4* | 0.0 | 0.0 | 0.0 | 0.0 | 0.0 | 0.0 | 0.0 | 0.0 | 0.0 | 0.1 | 0.9 |
| Days ≥ 1" Snow Depth | *0* | na | 0 | 0 | 0 | 0 | 0 | 0 | 0 | 0 | 0 | 0 | na |

## BUFFALO *Harper County* ELEVATION 1801 ft LAT/LONG 36° 50 ' N / 99° 38 ' W

| | JAN | FEB | MAR | APR | MAY | JUN | JUL | AUG | SEP | OCT | NOV | DEC | YEAR |
|---|---|---|---|---|---|---|---|---|---|---|---|---|---|
| Maximum Temp °F | 48.8 | 55.5 | 65.0 | 75.4 | 82.3 | 92.2 | 97.8 | 96.1 | 87.6 | 77.0 | 60.6 | 51.2 | 74.1 |
| Minimum Temp °F | 20.6 | 24.7 | 33.6 | 43.7 | 53.1 | 63.3 | 68.4 | 66.4 | 58.2 | 45.3 | 32.3 | 23.5 | 44.4 |
| Mean Temp °F | 34.7 | 40.1 | 49.3 | 59.6 | 67.7 | 77.8 | 83.1 | 81.2 | 72.9 | 61.1 | 46.5 | 37.4 | 59.3 |
| Days Max Temp ≥ 90 °F | 0 | 0 | 1 | 3 | 7 | 20 | 27 | 25 | 14 | 4 | 0 | 0 | 101 |
| Days Max Temp ≤ 32 °F | 5 | 2 | 0 | 0 | 0 | 0 | 0 | 0 | 0 | 0 | 0 | 2 | 9 |
| Days Min Temp ≤ 32 °F | 28 | 22 | 15 | 4 | 0 | 0 | 0 | 0 | 0 | 3 | 15 | 26 | 113 |
| Days Min Temp ≤ 0 °F | 1 | 1 | 0 | 0 | 0 | 0 | 0 | 0 | 0 | 0 | 0 | 1 | 3 |
| Heating Degree Days | 932 | 697 | 488 | 204 | 58 | 3 | 0 | 0 | 28 | 176 | 550 | 849 | 3985 |
| Cooling Degree Days | 0 | 0 | 7 | 50 | 148 | 392 | 571 | 532 | 298 | 58 | 2 | 0 | 2058 |
| Total Precipitation (") | 0.49 | 1.00 | 1.89 | 2.43 | 4.31 | 3.54 | 2.66 | 3.38 | 2.76 | 1.85 | 1.54 | 0.86 | 26.71 |
| Days ≥ 0.1" Precip | 2 | 3 | 5 | 5 | 7 | 6 | 5 | 5 | 4 | 3 | 3 | 3 | 51 |
| Total Snowfall (") | *2.0* | 4.4 | *1.5* | 0.2 | 0.0 | 0.0 | 0.0 | 0.0 | 0.0 | 0.0 | 1.3 | *1.8* | 11.2 |
| Days ≥ 1" Snow Depth | na | *0* | *0* | 0 | 0 | 0 | 0 | 0 | 0 | 0 | 0 | na | na |

## CARNEGIE 2 ENE *Caddo County* ELEVATION 1503 ft LAT/LONG 35° 6 ' N / 98° 36 ' W

| | JAN | FEB | MAR | APR | MAY | JUN | JUL | AUG | SEP | OCT | NOV | DEC | YEAR |
|---|---|---|---|---|---|---|---|---|---|---|---|---|---|
| Maximum Temp °F | 49.6 | 55.1 | 65.0 | 74.7 | 81.8 | 91.0 | 96.0 | 94.5 | 86.3 | 76.0 | 62.0 | 52.5 | 73.7 |
| Minimum Temp °F | 24.8 | 29.3 | 38.2 | 48.1 | 57.2 | 66.4 | 70.5 | 68.6 | 61.2 | 49.3 | 37.6 | 28.4 | 48.3 |
| Mean Temp °F | 37.2 | 42.2 | 51.7 | 61.4 | 69.5 | 78.7 | 83.3 | 81.6 | 73.7 | 62.6 | 49.8 | 40.5 | 61.0 |
| Days Max Temp ≥ 90 °F | 0 | 0 | 0 | 1 | 4 | 18 | 26 | 24 | 12 | 2 | 0 | 0 | 87 |
| Days Max Temp ≤ 32 °F | 4 | 2 | 0 | 0 | 0 | 0 | 0 | 0 | 0 | 0 | 0 | 2 | 8 |
| Days Min Temp ≤ 32 °F | 25 | 18 | 10 | 2 | 0 | 0 | 0 | 0 | 0 | 1 | 10 | 21 | 87 |
| Days Min Temp ≤ 0 °F | 0 | 0 | 0 | 0 | 0 | 0 | 0 | 0 | 0 | 0 | 0 | 0 | 0 |
| Heating Degree Days | 855 | 637 | 419 | 157 | 34 | 1 | 0 | 0 | 22 | 143 | 454 | 753 | 3475 |
| Cooling Degree Days | 0 | 0 | 10 | 55 | 184 | 423 | 577 | 546 | 305 | 74 | 6 | 0 | 2180 |
| Total Precipitation (") | 1.02 | 1.32 | 2.08 | 2.70 | 5.54 | 4.05 | 2.12 | 2.49 | 3.58 | 2.37 | 1.66 | 1.22 | 30.15 |
| Days ≥ 0.1" Precip | 2 | 3 | 4 | 4 | 6 | 5 | 3 | 4 | 4 | 4 | 3 | 2 | 44 |
| Total Snowfall (") | 2.9 | 2.6 | 0.5 | 0.0 | 0.0 | 0.0 | 0.0 | 0.0 | 0.0 | 0.0 | 0.4 | 1.3 | 7.7 |
| Days ≥ 1" Snow Depth | 3 | 2 | 0 | 0 | 0 | 0 | 0 | 0 | 0 | 0 | 0 | 1 | 6 |

## CHANDLER 1 *Lincoln County* ELEVATION 860 ft LAT/LONG 35° 42 ' N / 96° 53 ' W

| | JAN | FEB | MAR | APR | MAY | JUN | JUL | AUG | SEP | OCT | NOV | DEC | YEAR |
|---|---|---|---|---|---|---|---|---|---|---|---|---|---|
| Maximum Temp °F | 48.3 | 53.6 | 63.7 | 73.6 | 79.8 | 87.9 | 94.0 | 93.0 | 84.5 | 74.4 | 61.4 | *52.3* | 72.2 |
| Minimum Temp °F | 26.6 | 30.5 | 40.0 | 49.8 | 57.8 | 66.7 | 70.9 | 69.1 | 61.9 | 50.4 | 39.3 | *30.3* | 49.4 |
| Mean Temp °F | 37.4 | 42.0 | 51.9 | 61.7 | 68.8 | 77.3 | 82.5 | 81.1 | 73.2 | 62.4 | 50.4 | *41.4* | 60.8 |
| Days Max Temp ≥ 90 °F | 0 | 0 | 0 | 0 | 2 | 13 | 23 | 23 | 8 | 1 | 0 | 0 | 70 |
| Days Max Temp ≤ 32 °F | 4 | 2 | 0 | 0 | 0 | 0 | 0 | 0 | 0 | 0 | 0 | *2* | 8 |
| Days Min Temp ≤ 32 °F | *22* | 16 | 8 | 1 | 0 | 0 | 0 | 0 | 0 | 1 | 8 | na | na |
| Days Min Temp ≤ 0 °F | 1 | 0 | 0 | 0 | 0 | 0 | 0 | 0 | 0 | 0 | 0 | 0 | 1 |
| Heating Degree Days | 847 | 642 | 415 | 151 | 35 | 1 | 0 | 0 | 21 | 144 | 438 | *727* | 3421 |
| Cooling Degree Days | 0 | *0* | 13 | 53 | *168* | *389* | 549 | 514 | 276 | 59 | 8 | *1* | 2030 |
| Total Precipitation (") | 1.39 | 1.96 | 2.90 | 3.37 | 5.17 | 3.73 | 2.46 | 2.64 | 4.42 | 3.17 | 2.68 | 1.70 | 35.59 |
| Days ≥ 0.1" Precip | *2* | 3 | 4 | 5 | 7 | 5 | 3 | 4 | 5 | 4 | 4 | *3* | 49 |
| Total Snowfall (") | 3.3 | *2.9* | 0.9 | 0.0 | 0.0 | 0.0 | 0.0 | 0.0 | 0.0 | 0.0 | 0.4 | *1.3* | 8.8 |
| Days ≥ 1" Snow Depth | 3 | 2 | 0 | 0 | 0 | 0 | 0 | 0 | 0 | 0 | 0 | *1* | 6 |

## CHATTANOOGA 3 NE *Comanche County* ELEVATION 1152 ft LAT/LONG 34° 25 ' N / 98° 39 ' W

| | JAN | FEB | MAR | APR | MAY | JUN | JUL | AUG | SEP | OCT | NOV | DEC | YEAR |
|---|---|---|---|---|---|---|---|---|---|---|---|---|---|
| Maximum Temp °F | 51.7 | 57.5 | 67.2 | 76.7 | 83.9 | 93.0 | 98.1 | 96.7 | 88.3 | 78.0 | 64.0 | 54.7 | 75.8 |
| Minimum Temp °F | 26.5 | 30.3 | 38.9 | 48.5 | 57.4 | 66.3 | 70.4 | 69.0 | 62.1 | 50.3 | 38.5 | 29.5 | 49.0 |
| Mean Temp °F | 39.1 | 43.9 | 53.0 | 62.7 | 70.7 | 79.7 | 84.3 | 82.9 | 75.2 | 64.2 | 51.3 | 42.1 | 62.4 |
| Days Max Temp ≥ 90 °F | 0 | 0 | 1 | 2 | 7 | 22 | 28 | 27 | 15 | 4 | 0 | 0 | 106 |
| Days Max Temp ≤ 32 °F | 3 | 1 | 0 | 0 | 0 | 0 | 0 | 0 | 0 | 0 | 0 | 1 | 5 |
| Days Min Temp ≤ 32 °F | 23 | 17 | 8 | 1 | 0 | 0 | 0 | 0 | 0 | 1 | 9 | 20 | 79 |
| Days Min Temp ≤ 0 °F | 0 | 0 | 0 | 0 | 0 | 0 | 0 | 0 | 0 | 0 | 0 | 0 | 0 |
| Heating Degree Days | 795 | 588 | 377 | 129 | 23 | 0 | 0 | 0 | 15 | 112 | 411 | 702 | 3152 |
| Cooling Degree Days | 0 | 0 | 9 | 57 | 211 | 451 | 609 | 590 | 346 | 95 | 7 | 0 | 2375 |
| Total Precipitation (") | 1.04 | 1.42 | 2.22 | 2.58 | 4.88 | 3.56 | 2.27 | 2.62 | 3.61 | 2.77 | 1.52 | 1.45 | 29.94 |
| Days ≥ 0.1" Precip | 2 | 3 | 4 | 4 | 6 | 5 | 3 | 4 | 5 | 3 | 3 | 3 | 45 |
| Total Snowfall (") | *1.2* | *1.1* | 0.4 | 0.0 | 0.0 | 0.0 | 0.0 | 0.0 | 0.0 | 0.0 | 0.0 | *0.3* | 3.0 |
| Days ≥ 1" Snow Depth | *1* | na | 0 | 0 | 0 | 0 | 0 | 0 | 0 | 0 | 0 | 0 | na |

## CHEROKEE *Alfalfa County* ELEVATION 1201 ft LAT/LONG 36° 45 ' N / 98° 21 ' W

| | JAN | FEB | MAR | APR | MAY | JUN | JUL | AUG | SEP | OCT | NOV | DEC | YEAR |
|---|---|---|---|---|---|---|---|---|---|---|---|---|---|
| Maximum Temp °F | 46.6 | 53.0 | 63.3 | 73.7 | 81.7 | 91.9 | 97.0 | 95.3 | 86.5 | 75.6 | 59.7 | 49.8 | 72.8 |
| Minimum Temp °F | 22.3 | 26.4 | 35.7 | 45.8 | 55.2 | 65.6 | 70.8 | 69.1 | 60.5 | 47.9 | 35.1 | 25.9 | 46.7 |
| Mean Temp °F | 34.4 | 39.8 | 49.5 | 59.8 | 68.4 | 78.8 | 83.9 | 82.2 | 73.5 | 61.6 | 47.5 | 37.9 | 59.8 |
| Days Max Temp ≥ 90 °F | 0 | 0 | 0 | 1 | 5 | 20 | 27 | 24 | 12 | 2 | 0 | 0 | 91 |
| Days Max Temp ≤ 32 °F | 5 | 2 | 0 | 0 | 0 | 0 | 0 | 0 | 0 | 0 | 0 | 2 | 9 |
| Days Min Temp ≤ 32 °F | 27 | 21 | 12 | 2 | 0 | 0 | 0 | 0 | 0 | 1 | 12 | 25 | 100 |
| Days Min Temp ≤ 0 °F | 1 | 1 | 0 | 0 | 0 | 0 | 0 | 0 | 0 | 0 | 0 | 1 | 3 |
| Heating Degree Days | 940 | 714 | 481 | 195 | 49 | 1 | 0 | 0 | 24 | 165 | 522 | 834 | 3925 |
| Cooling Degree Days | 0 | 0 | 5 | 44 | 166 | 436 | 600 | 573 | 310 | 63 | 3 | 0 | 2200 |
| Total Precipitation (") | 0.86 | 1.18 | 2.53 | 2.41 | 4.32 | 3.64 | 2.91 | 3.09 | 2.80 | 1.92 | 1.64 | 1.21 | 28.51 |
| Days ≥ 0.1" Precip | 1 | 3 | 4 | 4 | 6 | 5 | 4 | 4 | 4 | 3 | 3 | 3 | 44 |
| Total Snowfall (") | *2.9* | *4.4* | *2.5* | 0.2 | 0.0 | 0.0 | 0.0 | 0.0 | 0.0 | 0.0 | 0.6 | 2.3 | 12.9 |
| Days ≥ 1" Snow Depth | *2* | *1* | *1* | 0 | 0 | 0 | 0 | 0 | 0 | 0 | 0 | *1* | 5 |

## CHICKASHA EXP STN *Grady County* ELEVATION 1089 ft LAT/LONG 35° 3 ' N / 97° 55 ' W

| | JAN | FEB | MAR | APR | MAY | JUN | JUL | AUG | SEP | OCT | NOV | DEC | YEAR |
|---|---|---|---|---|---|---|---|---|---|---|---|---|---|
| Maximum Temp °F | 49.6 | 55.9 | 65.6 | 75.7 | 82.2 | 90.3 | 95.3 | 93.5 | 86.1 | 76.3 | 62.9 | 53.2 | 73.9 |
| Minimum Temp °F | 25.6 | 29.7 | 39.0 | 49.0 | 58.0 | 67.0 | 70.8 | 68.7 | 61.0 | 49.1 | 37.8 | 29.1 | 48.7 |
| Mean Temp °F | 37.6 | 42.8 | 52.4 | 62.4 | 70.1 | 78.7 | 83.1 | 81.1 | 73.6 | 62.7 | 50.4 | 41.2 | 61.3 |
| Days Max Temp ≥ 90 °F | 0 | 0 | 0 | 1 | 5 | 18 | 26 | 25 | 11 | 2 | 0 | 0 | 88 |
| Days Max Temp ≤ 32 °F | 4 | 2 | 0 | 0 | 0 | 0 | 0 | 0 | 0 | 0 | 0 | 2 | 8 |
| Days Min Temp ≤ 32 °F | 24 | 18 | 9 | 1 | 0 | 0 | 0 | 0 | 0 | 1 | 10 | 20 | 83 |
| Days Min Temp ≤ 0 °F | 0 | 0 | 0 | 0 | 0 | 0 | 0 | 0 | 0 | 0 | 0 | 0 | 0 |
| Heating Degree Days | 842 | 621 | 399 | 139 | 29 | 0 | 0 | 0 | 20 | 139 | 440 | 732 | 3361 |
| Cooling Degree Days | 0 | 1 | 10 | 58 | 192 | 416 | 564 | 532 | 293 | 69 | 6 | 0 | 2141 |
| Total Precipitation (") | 1.17 | 1.67 | 2.55 | 3.34 | 5.05 | 3.66 | 2.07 | 2.88 | 3.79 | 3.37 | 1.91 | 1.61 | 33.07 |
| Days ≥ 0.1" Precip | 2 | 4 | 5 | 5 | 7 | 5 | 3 | 5 | 6 | 4 | 3 | 3 | 52 |
| Total Snowfall (") | 1.2 | na | 0.2 | 0.0 | 0.0 | 0.0 | 0.0 | 0.0 | 0.0 | 0.0 | 0.1 | 0.8 | na |
| Days ≥ 1" Snow Depth | *0* | *0* | 0 | 0 | 0 | 0 | 0 | 0 | 0 | 0 | 0 | 0 | 0 |

## CLAREMORE 2 ENE *Rogers County* ELEVATION 600 ft LAT/LONG 36° 19 ' N / 95° 37 ' W

| | JAN | FEB | MAR | APR | MAY | JUN | JUL | AUG | SEP | OCT | NOV | DEC | YEAR |
|---|---|---|---|---|---|---|---|---|---|---|---|---|---|
| Maximum Temp °F | 44.7 | 50.5 | 60.4 | 70.9 | 77.8 | 86.2 | 92.5 | 91.7 | 83.3 | 72.9 | 59.7 | 49.3 | 70.0 |
| Minimum Temp °F | 22.1 | 26.4 | 36.3 | 46.6 | 55.4 | 65.0 | 69.5 | 67.2 | 60.1 | 46.8 | 36.4 | 26.9 | 46.6 |
| Mean Temp °F | 33.4 | 38.5 | 48.4 | 58.8 | 66.6 | 75.6 | 81.0 | 79.5 | 71.7 | 59.9 | 48.0 | 38.1 | 58.3 |
| Days Max Temp ≥ 90 °F | 0 | 0 | 0 | 0 | 1 | 10 | 22 | 20 | 8 | 1 | 0 | 0 | 62 |
| Days Max Temp ≤ 32 °F | 6 | 3 | 1 | 0 | 0 | 0 | 0 | 0 | 0 | 0 | 0 | 3 | 13 |
| Days Min Temp ≤ 32 °F | 27 | 20 | 12 | 2 | 0 | 0 | 0 | 0 | 0 | 2 | 12 | 22 | 97 |
| Days Min Temp ≤ 0 °F | 1 | 0 | 0 | 0 | 0 | 0 | 0 | 0 | 0 | 0 | 0 | 0 | 1 |
| Heating Degree Days | 972 | 742 | 516 | 219 | 65 | 3 | 0 | 0 | 36 | 201 | 508 | 826 | 4088 |
| Cooling Degree Days | 0 | 0 | 7 | 36 | 126 | 347 | 514 | 482 | 262 | 45 | 7 | 0 | 1826 |
| Total Precipitation (") | 1.72 | 2.09 | 3.49 | 3.99 | 5.02 | 4.62 | 3.11 | 3.08 | 4.64 | 3.79 | 3.63 | 2.56 | 41.74 |
| Days ≥ 0.1" Precip | 4 | 4 | 6 | 6 | 8 | 6 | 4 | 4 | 6 | 5 | 5 | 5 | 63 |
| Total Snowfall (") | 2.9 | 3.1 | 1.8 | 0.0 | 0.0 | 0.0 | 0.0 | 0.0 | 0.0 | 0.0 | 0.3 | 1.6 | 9.7 |
| Days ≥ 1" Snow Depth | 4 | 2 | 1 | 0 | 0 | 0 | 0 | 0 | 0 | 0 | 0 | 2 | 9 |

## CLINTON *Custer County* ELEVATION 1522 ft LAT/LONG 35° 31 ' N / 98° 59 ' W

| | JAN | FEB | MAR | APR | MAY | JUN | JUL | AUG | SEP | OCT | NOV | DEC | YEAR |
|---|---|---|---|---|---|---|---|---|---|---|---|---|---|
| Maximum Temp °F | 48.9 | 54.6 | 64.7 | 74.5 | 81.7 | 91.2 | 96.9 | 95.3 | 86.4 | 75.3 | 61.4 | 51.7 | 73.6 |
| Minimum Temp °F | 24.8 | 28.9 | 37.6 | 47.2 | 56.5 | 66.1 | 70.4 | 68.8 | 61.0 | 48.8 | 36.9 | 28.0 | 47.9 |
| Mean Temp °F | 36.9 | 41.8 | 51.2 | 60.9 | 69.1 | 78.7 | 83.7 | 82.1 | 73.7 | 62.1 | 49.2 | 39.8 | 60.8 |
| Days Max Temp ≥ 90 °F | 0 | 0 | 0 | 1 | 5 | 19 | 27 | 25 | 12 | 1 | 0 | 0 | 90 |
| Days Max Temp ≤ 32 °F | 4 | 2 | 0 | 0 | 0 | 0 | 0 | 0 | 0 | 0 | 0 | 2 | 8 |
| Days Min Temp ≤ 32 °F | 24 | 18 | 9 | 1 | 0 | 0 | 0 | 0 | 0 | 1 | 10 | 22 | 85 |
| Days Min Temp ≤ 0 °F | 0 | 0 | 0 | 0 | 0 | 0 | 0 | 0 | 0 | 0 | 0 | 0 | 0 |
| Heating Degree Days | 866 | 649 | 431 | 166 | 38 | 1 | 0 | 0 | 21 | 149 | 471 | 773 | 3565 |
| Cooling Degree Days | 0 | 0 | 7 | 48 | 182 | 432 | 598 | 568 | 315 | 63 | 4 | 0 | 2217 |
| Total Precipitation (") | 1.03 | 1.12 | 2.25 | 2.40 | 5.01 | 3.72 | 2.21 | 3.11 | 3.49 | 2.62 | 1.71 | 1.14 | 29.81 |
| Days ≥ 0.1" Precip | 2 | 3 | 4 | 4 | 6 | 5 | 4 | 5 | 5 | 4 | 3 | 3 | 48 |
| Total Snowfall (") | 2.8 | 2.5 | 0.8 | 0.0 | 0.0 | 0.0 | 0.0 | 0.0 | 0.0 | 0.0 | 0.7 | *1.7* | 8.5 |
| Days ≥ 1" Snow Depth | *0* | na | 0 | 0 | 0 | 0 | 0 | 0 | 0 | 0 | 0 | na | na |

## CUSHING *Payne County* ELEVATION 971 ft LAT/LONG 35° 59 ' N / 96° 46 ' W

| | JAN | FEB | MAR | APR | MAY | JUN | JUL | AUG | SEP | OCT | NOV | DEC | YEAR |
|---|---|---|---|---|---|---|---|---|---|---|---|---|---|
| Maximum Temp °F | 45.4 | 51.0 | 61.0 | 71.8 | 78.2 | 86.8 | 92.9 | 92.2 | 83.5 | 73.2 | 59.7 | 49.6 | 70.4 |
| Minimum Temp °F | 23.6 | 27.8 | 37.4 | 48.0 | 56.9 | 65.8 | 70.6 | 68.8 | 60.9 | 49.0 | 37.7 | 28.1 | 47.9 |
| Mean Temp °F | 34.5 | 39.4 | 49.2 | 59.9 | 67.5 | 76.3 | 81.8 | 80.6 | 72.3 | 61.1 | 48.8 | 38.9 | 59.2 |
| Days Max Temp ≥ 90 °F | 0 | 0 | 0 | 0 | 1 | 11 | 23 | 22 | 8 | 1 | 0 | 0 | 66 |
| Days Max Temp ≤ 32 °F | 5 | 3 | 0 | 0 | 0 | 0 | 0 | 0 | 0 | 0 | 1 | 3 | 12 |
| Days Min Temp ≤ 32 °F | 25 | 18 | 10 | 1 | 0 | 0 | 0 | 0 | 0 | 1 | 9 | 21 | 85 |
| Days Min Temp ≤ 0 °F | 1 | 0 | 0 | 0 | 0 | 0 | 0 | 0 | 0 | 0 | 0 | 0 | 1 |
| Heating Degree Days | 937 | 715 | 491 | 190 | 52 | 2 | 0 | 0 | 29 | 168 | 486 | 803 | 3873 |
| Cooling Degree Days | 0 | 0 | 7 | 42 | 141 | 361 | 531 | 508 | 265 | 50 | 6 | 0 | 1911 |
| Total Precipitation (") | 1.15 | 1.92 | 2.92 | 3.57 | 5.55 | 4.17 | 2.75 | 2.80 | 4.09 | 2.99 | 2.73 | 1.76 | 36.40 |
| Days ≥ 0.1" Precip | 2 | 4 | 5 | 5 | 7 | 6 | 4 | 4 | 6 | 4 | 4 | 3 | 54 |
| Total Snowfall (") | *1.6* | na | 0.8 | 0.0 | 0.0 | 0.0 | 0.0 | 0.0 | 0.0 | 0.0 | 0.3 | 1.2 | na |
| Days ≥ 1" Snow Depth | *2* | na | 0 | 0 | 0 | 0 | 0 | 0 | 0 | 0 | 0 | *0* | na |

## DUNCAN *Stephens County* ELEVATION 1132 ft LAT/LONG 34° 30 ' N / 97° 57 ' W

| | JAN | FEB | MAR | APR | MAY | JUN | JUL | AUG | SEP | OCT | NOV | DEC | YEAR |
|---|---|---|---|---|---|---|---|---|---|---|---|---|---|
| Maximum Temp °F | 50.3 | 55.5 | 65.4 | 74.5 | 81.1 | 89.2 | 94.9 | 93.8 | 85.8 | 75.8 | 63.3 | 53.9 | 73.6 |
| Minimum Temp °F | 27.6 | 31.5 | 40.6 | 50.3 | 58.3 | 67.1 | 71.3 | 69.9 | 62.6 | 51.1 | 40.0 | 31.0 | 50.1 |
| Mean Temp °F | 39.0 | 43.5 | 53.0 | 62.4 | 69.7 | 78.1 | 83.1 | 81.9 | 74.2 | 63.5 | 51.7 | 42.5 | 61.9 |
| Days Max Temp ≥ 90 °F | 0 | 0 | 0 | 1 | 3 | 15 | 26 | *25* | *11* | 2 | 0 | 0 | 83 |
| Days Max Temp ≤ 32 °F | 3 | 2 | 0 | 0 | 0 | 0 | 0 | 0 | 0 | 0 | 0 | *2* | 7 |
| Days Min Temp ≤ 32 °F | 21 | 14 | *6* | 1 | 0 | 0 | 0 | 0 | 0 | 0 | *7* | *17* | 66 |
| Days Min Temp ≤ 0 °F | 0 | 0 | 0 | 0 | 0 | 0 | 0 | 0 | 0 | 0 | 0 | 0 | 0 |
| Heating Degree Days | 800 | 601 | 379 | 134 | 29 | 1 | 0 | 0 | 20 | 125 | 401 | 691 | 3181 |
| Cooling Degree Days | 0 | 1 | 11 | 56 | 174 | 399 | *564* | 539 | 303 | 79 | 8 | 0 | 2134 |
| Total Precipitation (") | 1.31 | 1.75 | 2.39 | 3.15 | 5.23 | 3.93 | 2.47 | 2.52 | 4.49 | 3.45 | 1.92 | 1.80 | 34.41 |
| Days ≥ 0.1" Precip | 2 | 3 | 4 | 5 | 6 | 5 | 3 | 4 | 6 | 5 | 3 | 3 | 49 |
| Total Snowfall (") | *1.7* | *1.5* | 0.5 | 0.0 | 0.0 | 0.0 | 0.0 | 0.0 | 0.0 | 0.0 | 0.2 | *0.7* | 4.6 |
| Days ≥ 1" Snow Depth | *1* | *0* | 0 | 0 | 0 | 0 | 0 | 0 | 0 | 0 | 0 | *0* | 1 |

## DURANT USDA *Bryan County* ELEVATION 640 ft LAT/LONG 33° 59 ' N / 96° 23 ' W

| | JAN | FEB | MAR | APR | MAY | JUN | JUL | AUG | SEP | OCT | NOV | DEC | YEAR |
|---|---|---|---|---|---|---|---|---|---|---|---|---|---|
| Maximum Temp °F | 50.0 | 55.9 | 65.6 | 74.4 | 80.8 | 88.8 | 93.9 | 93.6 | 85.8 | 76.3 | 63.9 | 54.3 | 73.6 |
| Minimum Temp °F | 27.5 | 32.4 | 41.4 | 50.3 | 58.6 | 66.8 | 70.6 | 69.2 | 61.9 | 50.6 | 40.4 | 31.6 | 50.1 |
| Mean Temp °F | 38.8 | 44.2 | 53.5 | 62.4 | 69.6 | 77.8 | 82.3 | 81.4 | 73.9 | 63.5 | 52.3 | 43.0 | 61.9 |
| Days Max Temp ≥ 90 °F | 0 | 0 | 0 | 0 | 2 | 15 | 26 | *25* | 11 | 2 | 0 | 0 | 81 |
| Days Max Temp ≤ 32 °F | *3* | 2 | 0 | 0 | 0 | 0 | 0 | 0 | 0 | 0 | 0 | 1 | 6 |
| Days Min Temp ≤ 32 °F | *22* | 14 | 7 | 1 | 0 | 0 | 0 | 0 | 0 | 1 | 7 | *16* | 68 |
| Days Min Temp ≤ 0 °F | *0* | 0 | 0 | 0 | 0 | 0 | 0 | 0 | 0 | 0 | 0 | 0 | 0 |
| Heating Degree Days | 807 | 582 | 366 | 134 | 31 | 1 | 0 | 0 | 18 | 123 | 385 | 676 | 3123 |
| Cooling Degree Days | 0 | 1 | 11 | 56 | 172 | 389 | 539 | 528 | 299 | 78 | 10 | 1 | 2084 |
| Total Precipitation (") | 2.06 | 2.86 | 3.80 | 4.54 | 6.17 | 4.94 | 2.70 | 2.79 | 5.32 | 4.48 | 3.21 | 2.65 | 45.52 |
| Days ≥ 0.1" Precip | 3 | 5 | 5 | 6 | 7 | 6 | 4 | 4 | 5 | 5 | 5 | 5 | 60 |
| Total Snowfall (") | na | *1.8* | 0.2 | 0.0 | 0.0 | 0.0 | 0.0 | 0.0 | 0.0 | 0.0 | 0.1 | *0.3* | na |
| Days ≥ 1" Snow Depth | na | *0* | 0 | 0 | 0 | 0 | 0 | 0 | 0 | 0 | 0 | *0* | na |

## EL RENO 1 N *Canadian County* ELEVATION 1322 ft LAT/LONG 35° 33 ' N / 97° 58 ' W

| | JAN | FEB | MAR | APR | MAY | JUN | JUL | AUG | SEP | OCT | NOV | DEC | YEAR |
|---|---|---|---|---|---|---|---|---|---|---|---|---|---|
| Maximum Temp °F | 47.3 | 53.1 | 63.3 | 73.2 | 80.3 | 88.7 | 94.2 | 93.0 | 84.9 | 74.2 | 60.1 | 50.6 | 71.9 |
| Minimum Temp °F | 24.8 | 28.9 | 37.8 | 47.8 | 56.5 | 65.4 | 70.0 | 68.4 | 60.8 | 49.2 | 37.1 | 28.2 | 47.9 |
| Mean Temp °F | 36.1 | 41.1 | 50.5 | 60.5 | 68.4 | 77.0 | 82.1 | 80.7 | 72.9 | 61.8 | 48.6 | 39.4 | 59.9 |
| Days Max Temp ≥ 90 °F | 0 | 0 | 0 | 1 | 3 | 14 | 25 | 23 | 9 | 1 | 0 | 0 | 76 |
| Days Max Temp ≤ 32 °F | 4 | 2 | 0 | 0 | 0 | 0 | 0 | 0 | 0 | 0 | 0 | 2 | 8 |
| Days Min Temp ≤ 32 °F | 24 | 18 | 10 | 2 | 0 | 0 | 0 | 0 | 0 | 1 | 10 | 21 | 86 |
| Days Min Temp ≤ 0 °F | 0 | 0 | 0 | 0 | 0 | 0 | 0 | 0 | 0 | 0 | 0 | 0 | 0 |
| Heating Degree Days | 889 | 670 | 453 | 178 | 42 | 1 | 0 | 0 | 23 | 159 | 490 | 787 | 3692 |
| Cooling Degree Days | 0 | 0 | 8 | 47 | 156 | 371 | 536 | 513 | 265 | 58 | 3 | 0 | 1957 |
| Total Precipitation (") | 1.03 | 1.30 | 2.48 | 2.93 | 5.66 | 4.33 | 2.37 | 2.91 | 3.90 | 2.61 | 2.03 | 1.13 | 32.68 |
| Days ≥ 0.1" Precip | 2 | 3 | 4 | 4 | 7 | 6 | 4 | 4 | 5 | 4 | 3 | 3 | 49 |
| Total Snowfall (") | 2.2 | 2.3 | 0.9 | 0.1 | 0.0 | 0.0 | 0.0 | 0.0 | 0.0 | 0.0 | 0.5 | 0.8 | 6.8 |
| Days ≥ 1" Snow Depth | *0* | *1* | 0 | 0 | 0 | 0 | 0 | 0 | 0 | 0 | 0 | *0* | 1 |

## ELK CITY *Beckham County* ELEVATION 1962 ft LAT/LONG 35° 25' N / 99° 23' W

| | JAN | FEB | MAR | APR | MAY | JUN | JUL | AUG | SEP | OCT | NOV | DEC | YEAR |
|---|---|---|---|---|---|---|---|---|---|---|---|---|---|
| Maximum Temp °F | 48.4 | 54.0 | 64.1 | 73.8 | 81.0 | 89.6 | 94.9 | ***92.6*** | 85.3 | 74.4 | 60.2 | 50.8 | 72.4 |
| Minimum Temp °F | 24.3 | 28.2 | 36.9 | 46.5 | 55.6 | 64.8 | 69.1 | ***67.2*** | 59.8 | 47.8 | 36.0 | 27.5 | 47.0 |
| Mean Temp °F | 36.4 | 41.1 | 50.5 | 60.2 | 68.3 | 77.3 | 82.0 | ***79.9*** | 72.6 | 61.1 | 48.1 | 39.2 | 59.7 |
| Days Max Temp ≥ 90 °F | 0 | 0 | 0 | 1 | 4 | 16 | 26 | ***22*** | 10 | 1 | 0 | 0 | 80 |
| Days Max Temp ≤ 32 °F | 4 | 2 | 0 | 0 | 0 | 0 | 0 | 0 | 0 | 0 | 0 | 2 | 8 |
| Days Min Temp ≤ 32 °F | 25 | 19 | 10 | 2 | 0 | 0 | 0 | 0 | 0 | 1 | 12 | 22 | 91 |
| Days Min Temp ≤ 0 °F | 0 | 0 | 0 | 0 | 0 | 0 | 0 | 0 | 0 | 0 | 0 | 0 | 0 |
| Heating Degree Days | 880 | 667 | 450 | 178 | 43 | 2 | 0 | ***1*** | 22 | 168 | 501 | 794 | 3706 |
| Cooling Degree Days | 0 | 0 | 6 | 35 | 142 | 366 | 518 | 486 | 256 | 51 | 1 | 0 | 1861 |
| Total Precipitation (") | 0.79 | 1.16 | 2.18 | 2.16 | 4.69 | 3.55 | 2.03 | 2.89 | 3.05 | 2.03 | 1.48 | 0.86 | 26.87 |
| Days ≥ 0.1" Precip | 2 | 2 | 4 | 4 | 6 | 5 | 3 | 4 | 4 | 3 | 3 | 2 | 42 |
| Total Snowfall (") | na | na | 0.4 | 0.1 | 0.0 | 0.0 | 0.0 | 0.0 | 0.0 | 0.1 | 0.2 | ***1.9*** | na |
| Days ≥ 1" Snow Depth | na | na | 0 | 0 | 0 | 0 | 0 | 0 | 0 | 0 | 0 | ***1*** | na |

## ENID *Garfield County* ELEVATION 1270 ft LAT/LONG 36° 24' N / 97° 53' W

| | JAN | FEB | MAR | APR | MAY | JUN | JUL | AUG | SEP | OCT | NOV | DEC | YEAR |
|---|---|---|---|---|---|---|---|---|---|---|---|---|---|
| Maximum Temp °F | 45.2 | 51.6 | 61.7 | 72.0 | 79.9 | 89.7 | 95.0 | 93.2 | 84.6 | 73.7 | 58.6 | 49.1 | 71.2 |
| Minimum Temp °F | 25.2 | 29.3 | 38.2 | 48.4 | 57.4 | 66.6 | 71.4 | 69.7 | 61.9 | 50.1 | 37.9 | 29.1 | 48.8 |
| Mean Temp °F | 35.2 | 40.5 | 50.0 | 60.2 | 68.7 | 78.2 | 83.2 | 81.4 | 73.3 | 61.9 | 48.3 | 39.1 | 60.0 |
| Days Max Temp ≥ 90 °F | 0 | 0 | 0 | 0 | 3 | 17 | 25 | 23 | 10 | 1 | 0 | 0 | 79 |
| Days Max Temp ≤ 32 °F | 6 | 3 | 0 | 0 | 0 | 0 | 0 | 0 | 0 | 0 | 0 | 3 | 12 |
| Days Min Temp ≤ 32 °F | 23 | 17 | 9 | 1 | 0 | 0 | 0 | 0 | 0 | 1 | 9 | 20 | 80 |
| Days Min Temp ≤ 0 °F | 0 | 0 | 0 | 0 | 0 | 0 | 0 | 0 | 0 | 0 | 0 | 0 | 0 |
| Heating Degree Days | 915 | 686 | 468 | 185 | 42 | 1 | 0 | 0 | 25 | 156 | 499 | 795 | 3772 |
| Cooling Degree Days | 0 | 0 | 6 | 46 | 162 | 405 | 572 | 533 | 293 | 57 | 4 | 0 | 2078 |
| Total Precipitation (") | 1.02 | 1.49 | 2.25 | 3.08 | 4.87 | 4.11 | 2.65 | 3.44 | 3.17 | 2.98 | 2.04 | 1.27 | 32.37 |
| Days ≥ 0.1" Precip | 2 | 3 | 4 | 5 | 7 | 6 | 4 | 5 | 5 | 4 | 3 | 3 | 51 |
| Total Snowfall (") | ***2.3*** | na | 1.8 | 0.0 | 0.0 | 0.0 | 0.0 | 0.0 | 0.0 | 0.0 | ***0.3*** | 1.2 | na |
| Days ≥ 1" Snow Depth | ***1*** | na | 1 | 0 | 0 | 0 | 0 | 0 | 0 | 0 | ***0*** | ***1*** | na |

## ERICK 4 E *Beckham County* ELEVATION 1991 ft LAT/LONG 35° 12' N / 99° 48' W

| | JAN | FEB | MAR | APR | MAY | JUN | JUL | AUG | SEP | OCT | NOV | DEC | YEAR |
|---|---|---|---|---|---|---|---|---|---|---|---|---|---|
| Maximum Temp °F | 50.4 | 55.7 | 65.1 | 74.7 | 81.4 | 90.0 | 95.4 | 93.6 | 85.9 | 75.5 | 62.0 | 52.6 | 73.5 |
| Minimum Temp °F | 23.2 | 27.3 | 35.7 | 45.2 | 54.3 | 63.6 | 67.8 | 66.0 | 58.6 | 46.5 | 34.9 | 26.2 | 45.8 |
| Mean Temp °F | 36.8 | 41.5 | 50.4 | 60.0 | 67.9 | 76.9 | 81.6 | 79.8 | 72.3 | 61.0 | 48.5 | 39.4 | 59.7 |
| Days Max Temp ≥ 90 °F | 0 | 0 | 0 | 1 | 5 | 17 | 26 | 23 | 11 | 2 | 0 | 0 | 85 |
| Days Max Temp ≤ 32 °F | 4 | 2 | 0 | 0 | 0 | 0 | 0 | 0 | 0 | 0 | 0 | 2 | 8 |
| Days Min Temp ≤ 32 °F | 27 | 19 | 11 | 2 | 0 | 0 | 0 | 0 | 0 | 2 | 12 | 24 | 97 |
| Days Min Temp ≤ 0 °F | 1 | 0 | 0 | 0 | 0 | 0 | 0 | 0 | 0 | 0 | 0 | 0 | 1 |
| Heating Degree Days | 868 | 657 | 451 | 182 | 47 | 2 | 0 | 1 | 25 | 166 | 490 | 787 | 3676 |
| Cooling Degree Days | 0 | 0 | 4 | 38 | 145 | 362 | 527 | 486 | 259 | 47 | 3 | 0 | 1871 |
| Total Precipitation (") | 0.61 | 0.89 | 1.81 | 2.13 | 4.29 | 3.28 | 1.90 | 2.51 | 3.06 | 2.20 | 1.25 | 0.79 | 24.72 |
| Days ≥ 0.1" Precip | 2 | 2 | 4 | 4 | 6 | 5 | 4 | 4 | 4 | 3 | 3 | 2 | 43 |
| Total Snowfall (") | 3.0 | 3.1 | 1.5 | 0.2 | 0.0 | 0.0 | 0.0 | 0.0 | 0.0 | 0.1 | 1.3 | 2.1 | 11.3 |
| Days ≥ 1" Snow Depth | 3 | 3 | 1 | 0 | 0 | 0 | 0 | 0 | 0 | 0 | 1 | 2 | 10 |

## FORT SUPPLY DAM *Woodward County* ELEVATION 2080 ft LAT/LONG 36° 33' N / 99° 35' W

| | JAN | FEB | MAR | APR | MAY | JUN | JUL | AUG | SEP | OCT | NOV | DEC | YEAR |
|---|---|---|---|---|---|---|---|---|---|---|---|---|---|
| Maximum Temp °F | ***46.5*** | 51.5 | 61.4 | 71.5 | 78.2 | 87.3 | 93.2 | 91.6 | 83.5 | 72.9 | 58.8 | 48.9 | 70.4 |
| Minimum Temp °F | 20.0 | 23.9 | 33.8 | 44.3 | 53.9 | 63.7 | 68.5 | 66.6 | 57.9 | 45.2 | 32.4 | 23.3 | 44.5 |
| Mean Temp °F | ***33.4*** | 37.7 | 47.6 | 57.9 | 66.1 | 75.5 | 80.9 | 79.1 | 70.7 | 59.1 | 45.5 | 36.1 | 57.5 |
| Days Max Temp ≥ 90 °F | 0 | 0 | 0 | 1 | 3 | 13 | 23 | 20 | 9 | 2 | 0 | 0 | 71 |
| Days Max Temp ≤ 32 °F | 6 | 3 | 1 | 0 | 0 | 0 | 0 | 0 | 0 | 0 | 1 | 3 | 14 |
| Days Min Temp ≤ 32 °F | ***27*** | 22 | 14 | 3 | 0 | 0 | 0 | 0 | 0 | 3 | 15 | 27 | 111 |
| Days Min Temp ≤ 0 °F | 1 | 1 | 0 | 0 | 0 | 0 | 0 | 0 | 0 | 0 | 0 | 1 | 3 |
| Heating Degree Days | ***973*** | 759 | 544 | 242 | 78 | 7 | 0 | 2 | 40 | 224 | 580 | 889 | 4338 |
| Cooling Degree Days | 0 | 0 | 3 | 32 | 101 | 312 | 484 | 454 | 220 | 32 | 2 | 0 | 1640 |
| Total Precipitation (") | 0.51 | 0.92 | 1.63 | 1.84 | 3.79 | 2.99 | 2.12 | 2.83 | 2.20 | 1.54 | 1.34 | 0.78 | 22.49 |
| Days ≥ 0.1" Precip | 2 | 2 | 3 | 4 | 6 | 5 | 4 | 5 | 4 | 2 | 3 | 2 | 42 |
| Total Snowfall (") | 3.0 | 4.0 | 2.0 | 0.9 | 0.0 | 0.0 | 0.0 | 0.0 | 0.0 | 0.1 | 1.0 | 1.8 | 12.8 |
| Days ≥ 1" Snow Depth | ***4*** | 3 | 2 | 0 | 0 | 0 | 0 | 0 | 0 | 0 | 1 | 2 | 12 |

## FREDERICK *Tillman County* ELEVATION 1289 ft LAT/LONG 34° 22 ' N / 99° 0 ' W

| | JAN | FEB | MAR | APR | MAY | JUN | JUL | AUG | SEP | OCT | NOV | DEC | YEAR |
|---|---|---|---|---|---|---|---|---|---|---|---|---|---|
| Maximum Temp °F | 51.4 | 56.9 | 66.8 | 76.4 | 83.5 | 92.3 | 97.7 | 95.8 | 87.3 | 77.0 | 63.5 | 54.7 | 75.3 |
| Minimum Temp °F | 26.8 | 30.6 | 39.3 | 49.1 | 57.5 | 66.7 | 71.4 | 69.6 | 62.1 | 50.4 | 38.6 | 29.7 | 49.3 |
| Mean Temp °F | 39.1 | 43.8 | 53.1 | 62.7 | 70.6 | 79.5 | 84.6 | 82.7 | 74.7 | 63.7 | 51.1 | 42.4 | 62.3 |
| Days Max Temp ≥ 90 °F | 0 | 0 | 1 | 2 | 8 | 21 | 27 | 26 | 14 | 3 | 0 | 0 | 102 |
| Days Max Temp ≤ 32 °F | 3 | 2 | 0 | 0 | 0 | 0 | 0 | 0 | 0 | 0 | 0 | 1 | 6 |
| Days Min Temp ≤ 32 °F | 23 | 16 | 7 | 1 | 0 | 0 | 0 | 0 | 0 | 0 | 8 | 20 | 75 |
| Days Min Temp ≤ 0 °F | 0 | 0 | 0 | 0 | 0 | 0 | 0 | 0 | 0 | 0 | 0 | 0 | 0 |
| Heating Degree Days | 795 | 594 | 379 | 131 | 27 | 1 | 0 | 0 | 17 | 120 | 417 | 695 | 3176 |
| Cooling Degree Days | 0 | 0 | 9 | 58 | 186 | 422 | 601 | 561 | 310 | 72 | 7 | 0 | 2226 |
| Total Precipitation (") | 1.07 | 1.35 | 2.07 | 2.36 | 4.53 | 3.50 | 2.29 | 2.85 | 3.47 | 2.75 | 1.64 | 1.22 | 29.10 |
| Days ≥ 0.1" Precip | 2 | 3 | 4 | 4 | 6 | 4 | 3 | 4 | 4 | 4 | 3 | 3 | 44 |
| Total Snowfall (") | 1.3 | 1.9 | 0.2 | 0.0 | 0.0 | 0.0 | 0.0 | 0.0 | 0.0 | 0.0 | 0.2 | 0.8 | 4.4 |
| Days ≥ 1" Snow Depth | *1* | 1 | 0 | 0 | 0 | 0 | 0 | 0 | 0 | 0 | 0 | *0* | 2 |

## FREEDOM *Woods County* ELEVATION 1542 ft LAT/LONG 36° 46 ' N / 99° 7 ' W

| | JAN | FEB | MAR | APR | MAY | JUN | JUL | AUG | SEP | OCT | NOV | DEC | YEAR |
|---|---|---|---|---|---|---|---|---|---|---|---|---|---|
| Maximum Temp °F | 47.5 | 53.5 | 63.4 | 73.9 | 81.6 | 91.2 | 96.5 | 94.6 | 86.2 | 75.4 | 60.1 | 50.2 | 72.8 |
| Minimum Temp °F | 20.2 | 24.6 | 34.0 | 44.7 | 54.4 | 64.3 | 68.9 | 66.9 | 58.3 | 45.0 | 32.1 | 22.9 | 44.7 |
| Mean Temp °F | 33.9 | 39.1 | 48.5 | 59.3 | 68.1 | 77.7 | 82.7 | 80.8 | 72.3 | 60.3 | 46.1 | 36.6 | 58.8 |
| Days Max Temp ≥ 90 °F | 0 | 0 | 0 | 2 | 5 | 18 | 27 | 24 | 12 | 3 | 0 | 0 | 91 |
| Days Max Temp ≤ 32 °F | 5 | 3 | 0 | 0 | 0 | 0 | 0 | 0 | 0 | 0 | 0 | 2 | 10 |
| Days Min Temp ≤ 32 °F | 28 | 22 | 14 | 3 | 0 | 0 | 0 | 0 | 0 | 4 | 16 | 27 | 114 |
| Days Min Temp ≤ 0 °F | 1 | 1 | 0 | 0 | 0 | 0 | 0 | 0 | 0 | 0 | 0 | 1 | 3 |
| Heating Degree Days | 957 | 724 | 512 | 210 | 53 | 3 | 0 | 1 | 30 | 196 | 562 | 875 | 4123 |
| Cooling Degree Days | 0 | 0 | 4 | 43 | 146 | 386 | 545 | 503 | 258 | 45 | 3 | 0 | 1933 |
| Total Precipitation (") | 0.59 | 0.90 | 1.93 | 2.29 | 3.77 | 2.87 | 2.44 | 2.84 | 2.30 | 1.79 | 1.41 | 0.88 | 24.01 |
| Days ≥ 0.1" Precip | 1 | 2 | 4 | 4 | 5 | 5 | 4 | 4 | 4 | 3 | 2 | 2 | 40 |
| Total Snowfall (") | *1.0* | na | 0.7 | 0.1 | 0.0 | 0.0 | 0.0 | 0.0 | 0.0 | 0.1 | 0.8 | na | na |
| Days ≥ 1" Snow Depth | *1* | na | 0 | 0 | 0 | 0 | 0 | 0 | 0 | 0 | 0 | *1* | na |

## GAGE AIRPORT *Ellis County* ELEVATION 2195 ft LAT/LONG 36° 18 ' N / 99° 46 ' W

| | JAN | FEB | MAR | APR | MAY | JUN | JUL | AUG | SEP | OCT | NOV | DEC | YEAR |
|---|---|---|---|---|---|---|---|---|---|---|---|---|---|
| Maximum Temp °F | 46.8 | 52.3 | 61.2 | 71.5 | 78.5 | 88.3 | 94.1 | 92.1 | 83.6 | 73.1 | 58.6 | 49.2 | 70.8 |
| Minimum Temp °F | 20.1 | 24.5 | 33.1 | 43.3 | 52.9 | 63.2 | 67.9 | 66.2 | 57.6 | 44.7 | 31.9 | 22.8 | 44.0 |
| Mean Temp °F | 33.5 | 38.4 | 47.2 | 57.4 | 65.7 | 75.8 | 81.0 | 79.2 | 70.6 | 58.9 | 45.3 | 36.0 | 57.4 |
| Days Max Temp ≥ 90 °F | 0 | 0 | 0 | 1 | 3 | 14 | 24 | 21 | 10 | 2 | 0 | 0 | 75 |
| Days Max Temp ≤ 32 °F | 6 | 3 | 1 | 0 | 0 | 0 | 0 | 0 | 0 | 0 | 1 | 4 | 15 |
| Days Min Temp ≤ 32 °F | 28 | 22 | 15 | 4 | 0 | 0 | 0 | 0 | 0 | 3 | 17 | 27 | 116 |
| Days Min Temp ≤ 0 °F | 2 | 0 | 0 | 0 | 0 | 0 | 0 | 0 | 0 | 0 | 0 | 1 | 3 |
| Heating Degree Days | 970 | 745 | 551 | 253 | 82 | 6 | 0 | 1 | 45 | 227 | 586 | 893 | 4359 |
| Cooling Degree Days | 0 | 0 | 4 | 33 | 112 | 333 | 502 | 464 | 237 | 40 | 2 | 0 | 1727 |
| Total Precipitation (") | 0.45 | 0.76 | 1.56 | 1.93 | 3.61 | 2.85 | 1.69 | 2.60 | 1.93 | 1.51 | 1.08 | 0.75 | 20.72 |
| Days ≥ 0.1" Precip | 1 | 2 | 3 | 4 | 6 | 5 | 4 | 4 | 3 | 2 | 2 | 2 | 38 |
| Total Snowfall (") | 3.3 | 4.4 | 2.4 | 0.7 | 0.0 | 0.0 | 0.0 | 0.0 | 0.0 | 0.2 | 1.6 | 2.8 | 15.4 |
| Days ≥ 1" Snow Depth | 5 | 3 | 1 | 0 | 0 | 0 | 0 | 0 | 0 | 0 | 1 | 3 | 13 |

## GATE *Beaver County* ELEVATION 2221 ft LAT/LONG 36° 51 ' N / 100° 3 ' W

| | JAN | FEB | MAR | APR | MAY | JUN | JUL | AUG | SEP | OCT | NOV | DEC | YEAR |
|---|---|---|---|---|---|---|---|---|---|---|---|---|---|
| Maximum Temp °F | 47.1 | 52.6 | 62.0 | 72.6 | 80.1 | 90.2 | 95.9 | 93.8 | 85.4 | 74.6 | 59.4 | 49.7 | 72.0 |
| Minimum Temp °F | 21.0 | 24.9 | 33.6 | 43.8 | 53.5 | 63.2 | 68.3 | 66.2 | 57.9 | 45.4 | 32.5 | 24.1 | 44.5 |
| Mean Temp °F | 34.0 | 38.8 | 47.8 | 58.2 | 66.8 | 76.7 | 82.1 | 80.0 | 71.7 | 60.0 | 46.1 | 37.0 | 58.3 |
| Days Max Temp ≥ 90 °F | 0 | 0 | 0 | 1 | 5 | 17 | 25 | 23 | 11 | 3 | 0 | 0 | 85 |
| Days Max Temp ≤ 32 °F | 6 | 3 | 1 | 0 | 0 | 0 | 0 | 0 | 0 | 0 | 1 | 3 | 14 |
| Days Min Temp ≤ 32 °F | 28 | 22 | 14 | 3 | 0 | 0 | 0 | 0 | 0 | 2 | 15 | 26 | 110 |
| Days Min Temp ≤ 0 °F | 1 | 1 | 0 | 0 | 0 | 0 | 0 | 0 | 0 | 0 | 0 | 1 | 3 |
| Heating Degree Days | 952 | 735 | 534 | 234 | 71 | 6 | 0 | 1 | 35 | 204 | 563 | 863 | 4198 |
| Cooling Degree Days | 0 | 0 | 4 | 38 | 127 | 351 | 525 | 481 | 250 | 41 | 3 | 0 | 1820 |
| Total Precipitation (") | 0.62 | 0.81 | 1.69 | 1.84 | 3.11 | 2.85 | 2.36 | 2.90 | 2.11 | 1.48 | 1.17 | 0.73 | 21.67 |
| Days ≥ 0.1" Precip | 2 | 2 | 3 | 3 | 5 | 5 | 4 | 4 | 4 | 3 | 2 | 2 | 39 |
| Total Snowfall (") | *3.8* | 3.7 | 4.8 | 0.9 | 0.1 | 0.0 | 0.0 | 0.0 | 0.0 | 0.4 | 2.9 | 3.4 | 20.0 |
| Days ≥ 1" Snow Depth | 5 | *3* | *2* | 0 | 0 | 0 | 0 | 0 | 0 | 0 | 1 | *3* | 14 |

## GEARY *Blaine County* ELEVATION 1552 ft LAT/LONG 35° 38' N / 98° 19' W

| | JAN | FEB | MAR | APR | MAY | JUN | JUL | AUG | SEP | OCT | NOV | DEC | YEAR |
|---|---|---|---|---|---|---|---|---|---|---|---|---|---|
| Maximum Temp °F | 46.7 | 52.1 | 62.4 | 72.3 | 79.6 | 88.4 | 94.1 | 93.0 | 84.3 | 73.2 | 59.4 | 49.8 | 71.3 |
| Minimum Temp °F | 24.9 | 28.9 | 37.7 | 47.5 | 56.3 | 65.5 | 70.1 | 68.5 | 60.9 | 49.3 | 37.1 | 28.4 | 47.9 |
| Mean Temp °F | 35.8 | 40.5 | 50.1 | 59.9 | 68.0 | 76.9 | 82.1 | 80.8 | 72.6 | 61.3 | 48.2 | 39.1 | 59.6 |
| Days Max Temp ≥ 90 °F | 0 | 0 | 0 | 0 | 3 | 14 | 24 | 23 | 9 | 1 | 0 | 0 | 74 |
| Days Max Temp ≤ 32 °F | 5 | 3 | 0 | 0 | 0 | 0 | 0 | 0 | 0 | 0 | 0 | 2 | 10 |
| Days Min Temp ≤ 32 °F | 24 | 17 | 9 | 1 | 0 | 0 | 0 | 0 | 0 | 1 | 10 | 21 | 83 |
| Days Min Temp ≤ 0 °F | 0 | 0 | 0 | 0 | 0 | 0 | 0 | 0 | 0 | 0 | 0 | 0 | 0 |
| Heating Degree Days | 896 | 685 | 464 | 189 | 48 | 2 | 0 | 0 | 26 | 168 | 499 | 795 | 3772 |
| Cooling Degree Days | 0 | 0 | 6 | 37 | 142 | 364 | 529 | 509 | 267 | 46 | 2 | 0 | 1902 |
| Total Precipitation (") | 0.77 | 1.17 | 2.08 | 2.61 | 4.93 | 3.74 | 1.99 | 2.43 | 3.45 | 2.11 | 1.83 | 1.11 | 28.22 |
| Days ≥ 0.1" Precip | 1 | 3 | 4 | 4 | 6 | 5 | 3 | 4 | 4 | 3 | 3 | 2 | 42 |
| Total Snowfall (") | ***1.8*** | ***1.7*** | 0.7 | 0.0 | 0.0 | 0.0 | 0.0 | 0.0 | 0.0 | 0.0 | 0.6 | ***0.8*** | 5.6 |
| Days ≥ 1" Snow Depth | ***1*** | ***1*** | 0 | 0 | 0 | 0 | 0 | 0 | 0 | 0 | 0 | ***1*** | 3 |

## GOODWELL RESEARCH *Texas County* ELEVATION 3304 ft LAT/LONG 36° 36' N / 101° 37' W

| | JAN | FEB | MAR | APR | MAY | JUN | JUL | AUG | SEP | OCT | NOV | DEC | YEAR |
|---|---|---|---|---|---|---|---|---|---|---|---|---|---|
| Maximum Temp °F | 47.1 | 52.2 | 60.6 | 70.5 | 77.8 | 87.8 | 93.0 | 90.6 | 83.0 | 72.4 | 58.3 | 49.0 | 70.2 |
| Minimum Temp °F | 18.3 | 21.9 | 29.4 | 39.3 | 48.8 | 58.7 | 64.1 | 61.9 | 53.5 | 41.0 | 29.0 | 20.5 | 40.5 |
| Mean Temp °F | 32.8 | 37.1 | 45.0 | 54.9 | 63.3 | 73.3 | 78.6 | 76.3 | 68.4 | 56.8 | 43.7 | 34.7 | 55.4 |
| Days Max Temp ≥ 90 °F | 0 | 0 | 0 | 1 | 3 | 14 | 23 | 19 | 9 | 1 | 0 | 0 | 70 |
| Days Max Temp ≤ 32 °F | 6 | 3 | 1 | 0 | 0 | 0 | 0 | 0 | 0 | 0 | 1 | 4 | 15 |
| Days Min Temp ≤ 32 °F | 30 | 25 | 20 | 6 | 1 | 0 | 0 | 0 | 0 | 4 | 20 | 29 | 135 |
| Days Min Temp ≤ 0 °F | 2 | 1 | 0 | 0 | 0 | 0 | 0 | 0 | 0 | 0 | 0 | 1 | 4 |
| Heating Degree Days | 993 | 782 | 612 | 309 | 121 | 14 | 1 | 3 | 57 | 272 | 632 | 934 | 4730 |
| Cooling Degree Days | 0 | 0 | 1 | 14 | 64 | 267 | 427 | 381 | 172 | 19 | 0 | 0 | 1345 |
| Total Precipitation (") | 0.31 | 0.38 | 0.89 | 1.29 | 3.14 | 2.70 | 2.42 | 2.18 | 1.62 | 0.94 | 0.66 | 0.34 | 16.87 |
| Days ≥ 0.1" Precip | 1 | 1 | 3 | 3 | 5 | 5 | 5 | 4 | 3 | 2 | 2 | 1 | 35 |
| Total Snowfall (") | 4.1 | 2.7 | 2.6 | 1.3 | 0.1 | 0.0 | 0.0 | 0.0 | 0.1 | 0.3 | 1.5 | 2.9 | 15.6 |
| Days ≥ 1" Snow Depth | 5 | 3 | 2 | 0 | 0 | 0 | 0 | 0 | 0 | 0 | 1 | 3 | 14 |

## GRT SALT PLAINS DAM *Alfalfa County* ELEVATION 1201 ft LAT/LONG 36° 45' N / 98° 7' W

| | JAN | FEB | MAR | APR | MAY | JUN | JUL | AUG | SEP | OCT | NOV | DEC | YEAR |
|---|---|---|---|---|---|---|---|---|---|---|---|---|---|
| Maximum Temp °F | ***45.1*** | 50.4 | 61.6 | 71.3 | 79.9 | 90.2 | 96.1 | 94.1 | 85.3 | ***73.6*** | 59.3 | ***48.2*** | 71.3 |
| Minimum Temp °F | 22.2 | 26.1 | 35.9 | 46.1 | 55.9 | 65.2 | 70.1 | 67.9 | 59.5 | ***47.4*** | 35.4 | ***25.8*** | 46.5 |
| Mean Temp °F | ***33.7*** | 38.2 | 48.8 | 58.7 | 67.9 | 77.7 | ***83.2*** | 81.0 | ***72.6*** | ***60.5*** | 47.3 | ***37.0*** | 58.9 |
| Days Max Temp ≥ 90 °F | 0 | 0 | 0 | 1 | 4 | ***17*** | ***26*** | 23 | 10 | 2 | 0 | 0 | 83 |
| Days Max Temp ≤ 32 °F | 6 | 3 | 0 | 0 | 0 | 0 | 0 | 0 | 0 | 0 | 0 | 3 | 12 |
| Days Min Temp ≤ 32 °F | 27 | 20 | ***12*** | 2 | 0 | 0 | 0 | 0 | 0 | 1 | ***11*** | ***24*** | 97 |
| Days Min Temp ≤ 0 °F | 1 | 1 | 0 | 0 | 0 | 0 | 0 | 0 | 0 | 0 | 0 | 1 | 3 |
| Heating Degree Days | ***963*** | ***748*** | ***522*** | ***226*** | ***56*** | ***4*** | ***0*** | 0 | ***28*** | ***190*** | ***528*** | ***861*** | 4126 |
| Cooling Degree Days | 0 | 0 | 5 | 38 | 143 | 392 | 564 | 518 | 272 | 46 | 4 | 0 | 1982 |
| Total Precipitation (") | 0.72 | 1.07 | 2.32 | 2.75 | 4.27 | 3.59 | 2.75 | 3.37 | 3.21 | 2.12 | 1.73 | 0.88 | 28.78 |
| Days ≥ 0.1" Precip | 1 | 2 | 4 | 4 | 6 | 5 | 4 | 5 | 4 | 3 | ***2*** | 2 | 42 |
| Total Snowfall (") | na | ***2.6*** | 0.2 | 0.1 | 0.0 | 0.0 | 0.0 | 0.0 | 0.0 | 0.0 | ***0.4*** | ***1.4*** | na |
| Days ≥ 1" Snow Depth | ***2*** | na | ***0*** | 0 | 0 | 0 | 0 | 0 | 0 | 0 | 0 | 1 | na |

## GUTHRIE *Logan County* ELEVATION 1030 ft LAT/LONG 35° 53' N / 97° 27' W

| | JAN | FEB | MAR | APR | MAY | JUN | JUL | AUG | SEP | OCT | NOV | DEC | YEAR |
|---|---|---|---|---|---|---|---|---|---|---|---|---|---|
| Maximum Temp °F | 48.0 | 54.2 | 64.4 | 74.6 | 81.4 | 89.5 | 95.5 | 94.9 | 86.0 | 75.3 | 61.4 | 51.8 | 73.1 |
| Minimum Temp °F | 25.5 | 29.8 | 39.0 | 49.2 | 57.6 | 66.7 | 71.2 | 69.6 | 62.1 | 50.2 | 38.6 | 29.6 | 49.1 |
| Mean Temp °F | 36.8 | 42.0 | 51.7 | 61.9 | 69.5 | 78.1 | 83.4 | 82.3 | 74.1 | 62.8 | 50.0 | 40.7 | 61.1 |
| Days Max Temp ≥ 90 °F | 0 | 0 | 0 | 1 | 4 | 16 | 27 | 25 | 11 | 2 | 0 | 0 | 86 |
| Days Max Temp ≤ 32 °F | 4 | 2 | 0 | 0 | 0 | 0 | 0 | 0 | 0 | 0 | 0 | 2 | 8 |
| Days Min Temp ≤ 32 °F | 24 | 17 | 9 | 1 | 0 | 0 | 0 | 0 | 0 | 1 | 9 | 19 | 80 |
| Days Min Temp ≤ 0 °F | 1 | 0 | 0 | 0 | 0 | 0 | 0 | 0 | 0 | 0 | 0 | 0 | 1 |
| Heating Degree Days | 867 | 642 | 421 | 154 | 35 | 1 | 0 | 0 | 22 | 141 | 449 | 744 | 3476 |
| Cooling Degree Days | 0 | 1 | 16 | 71 | 197 | 420 | 590 | 573 | 320 | 77 | 8 | 0 | 2273 |
| Total Precipitation (") | 1.35 | 1.77 | 2.99 | 3.08 | 5.19 | 4.08 | 2.16 | 2.32 | 4.11 | 2.72 | 2.62 | 1.85 | 34.24 |
| Days ≥ 0.1" Precip | 3 | 4 | 5 | 5 | 6 | 6 | 4 | 4 | 5 | 4 | 4 | 3 | 53 |
| Total Snowfall (") | ***2.1*** | ***1.8*** | 0.8 | 0.0 | 0.0 | 0.0 | 0.0 | 0.0 | 0.0 | 0.0 | 0.3 | ***1.4*** | 6.4 |
| Days ≥ 1" Snow Depth | ***1*** | na | 0 | 0 | 0 | 0 | 0 | 0 | 0 | 0 | 0 | ***1*** | na |

## HAMMON 3 SSW *Roger Mills County* ELEVATION 1780 ft LAT/LONG 35° 38 ' N / 99° 23 ' W

| | JAN | FEB | MAR | APR | MAY | JUN | JUL | AUG | SEP | OCT | NOV | DEC | YEAR |
|---|---|---|---|---|---|---|---|---|---|---|---|---|---|
| Maximum Temp °F | 47.9 | 53.4 | 62.8 | 73.0 | 80.0 | 89.5 | 95.5 | 93.0 | 84.6 | 74.3 | 59.3 | 50.3 | 72.0 |
| Minimum Temp °F | 20.8 | 25.1 | 33.6 | 44.1 | 53.1 | 63.6 | 68.4 | 66.6 | 57.7 | 44.7 | 32.7 | 23.6 | 44.5 |
| Mean Temp °F | 34.4 | 39.3 | 48.2 | 58.6 | 66.6 | 76.6 | 82.0 | 79.8 | 71.2 | 59.6 | 46.0 | 37.0 | 58.3 |
| Days Max Temp ≥ 90 °F | 0 | 0 | 0 | 1 | 4 | 16 | 26 | 23 | 10 | 1 | 0 | 0 | 81 |
| Days Max Temp ≤ 32 °F | 5 | 3 | 0 | 0 | 0 | 0 | 0 | 0 | 0 | 0 | 0 | 3 | 11 |
| Days Min Temp ≤ 32 °F | 28 | 22 | 15 | 3 | 0 | 0 | 0 | 0 | 0 | 3 | 15 | *27* | 113 |
| Days Min Temp ≤ 0 °F | 1 | 0 | 0 | 0 | 0 | 0 | 0 | 0 | 0 | 0 | 0 | 0 | 1 |
| Heating Degree Days | 943 | 718 | 521 | 221 | 70 | 5 | 0 | 1 | 38 | 204 | 564 | 863 | 4148 |
| Cooling Degree Days | 0 | 0 | 2 | 25 | 110 | 338 | 516 | 475 | 233 | 31 | 2 | 0 | 1732 |
| Total Precipitation (") | 0.74 | 0.94 | 1.96 | 2.14 | 4.41 | 3.44 | 2.05 | 2.95 | 3.00 | 1.95 | 1.49 | 0.85 | 25.92 |
| Days ≥ 0.1" Precip | 2 | 2 | 4 | 4 | 6 | 5 | 3 | 4 | 5 | 3 | 3 | 2 | 43 |
| Total Snowfall (") | 1.6 | 2.4 | 1.0 | 0.3 | 0.0 | 0.0 | 0.0 | 0.0 | 0.0 | 0.0 | 0.3 | 1.8 | 7.4 |
| Days ≥ 1" Snow Depth | *2* | *1* | 0 | 0 | 0 | 0 | 0 | 0 | 0 | 0 | *0* | *1* | 4 |

## HANNA *McIntosh County* ELEVATION 679 ft LAT/LONG 35° 12 ' N / 95° 53 ' W

| | JAN | FEB | MAR | APR | MAY | JUN | JUL | AUG | SEP | OCT | NOV | DEC | YEAR |
|---|---|---|---|---|---|---|---|---|---|---|---|---|---|
| Maximum Temp °F | 49.2 | 54.9 | 64.8 | 74.4 | 80.3 | 87.9 | 93.4 | 93.1 | 85.2 | 75.4 | 62.5 | 52.8 | 72.8 |
| Minimum Temp °F | 27.4 | 31.4 | 40.8 | 50.1 | 58.1 | 66.5 | 70.0 | 68.2 | 61.3 | 49.8 | 39.9 | 31.2 | 49.6 |
| Mean Temp °F | 38.3 | 43.2 | 52.8 | 62.3 | 69.2 | 77.2 | 81.7 | 80.7 | 73.3 | 62.6 | 51.2 | 42.0 | 61.2 |
| Days Max Temp ≥ 90 °F | 0 | 0 | 0 | 0 | 1 | 12 | 24 | 24 | 9 | 1 | 0 | 0 | 71 |
| Days Max Temp ≤ 32 °F | 3 | 1 | 0 | 0 | 0 | 0 | 0 | 0 | 0 | 0 | 0 | 1 | 5 |
| Days Min Temp ≤ 32 °F | 22 | 16 | 8 | 1 | 0 | 0 | 0 | 0 | 0 | 1 | 8 | 18 | 74 |
| Days Min Temp ≤ 0 °F | 0 | 0 | 0 | 0 | 0 | 0 | 0 | 0 | 0 | 0 | 0 | 0 | 0 |
| Heating Degree Days | 821 | 611 | 385 | 136 | 30 | 1 | 0 | 0 | 20 | 138 | 417 | 706 | 3265 |
| Cooling Degree Days | 0 | 1 | 12 | 60 | 170 | 385 | 526 | 516 | 284 | 69 | 10 | 1 | 2034 |
| Total Precipitation (") | 1.86 | 2.47 | 4.06 | 4.31 | 6.22 | 3.83 | 2.62 | 2.75 | 5.11 | 4.40 | 3.82 | 2.81 | 44.26 |
| Days ≥ 0.1" Precip | 4 | 4 | 6 | 6 | 7 | 6 | 4 | 4 | 6 | 5 | 5 | 5 | 62 |
| Total Snowfall (") | *1.6* | 2.4 | 1.2 | 0.0 | 0.0 | 0.0 | 0.0 | 0.0 | 0.0 | 0.0 | 0.4 | 1.1 | 6.7 |
| Days ≥ 1" Snow Depth | *1* | *1* | 0 | 0 | 0 | 0 | 0 | 0 | 0 | 0 | 0 | 1 | 3 |

## HEALDTON *Carter County* ELEVATION 912 ft LAT/LONG 34° 14 ' N / 97° 30 ' W

| | JAN | FEB | MAR | APR | MAY | JUN | JUL | AUG | SEP | OCT | NOV | DEC | YEAR |
|---|---|---|---|---|---|---|---|---|---|---|---|---|---|
| Maximum Temp °F | 52.2 | 57.8 | 67.3 | 76.1 | 82.2 | 90.0 | 96.0 | *96.0* | 87.0 | 77.1 | 64.5 | 55.5 | 75.1 |
| Minimum Temp °F | 27.2 | 31.2 | 40.3 | 49.6 | 57.5 | 66.0 | 69.6 | *68.2* | 61.5 | 50.0 | 39.2 | 30.7 | 49.3 |
| Mean Temp °F | 39.7 | 44.5 | 53.8 | 62.9 | 69.8 | 78.1 | 82.7 | *82.1* | 74.2 | 63.6 | 51.9 | 43.2 | 62.2 |
| Days Max Temp ≥ 90 °F | 0 | 0 | 0 | 1 | 4 | 17 | 27 | 24 | 12 | 2 | 0 | 0 | 87 |
| Days Max Temp ≤ 32 °F | 2 | 1 | 0 | 0 | 0 | 0 | 0 | 0 | 0 | 0 | 0 | 1 | 4 |
| Days Min Temp ≤ 32 °F | 22 | 16 | 8 | 2 | 0 | 0 | 0 | 0 | 0 | 1 | 9 | 18 | 76 |
| Days Min Temp ≤ 0 °F | 0 | 0 | 0 | 0 | 0 | 0 | 0 | 0 | 0 | 0 | 0 | 0 | 0 |
| Heating Degree Days | 777 | 575 | 358 | 127 | 27 | 1 | 0 | *0* | 17 | 123 | 397 | 671 | 3073 |
| Cooling Degree Days | 0 | 1 | 15 | 63 | 185 | 403 | *560* | na | 302 | 82 | 12 | 1 | na |
| Total Precipitation (") | 1.57 | 2.03 | 2.92 | 3.64 | 5.54 | 4.18 | 2.23 | 2.42 | 4.66 | 3.64 | 2.20 | 1.98 | 37.01 |
| Days ≥ 0.1" Precip | 3 | 4 | 5 | 5 | 6 | 5 | 3 | 4 | 5 | 5 | 4 | 3 | 52 |
| Total Snowfall (") | *1.2* | 1.5 | 0.6 | 0.0 | 0.0 | 0.0 | 0.0 | 0.0 | 0.0 | 0.0 | 0.2 | *0.7* | 4.2 |
| Days ≥ 1" Snow Depth | *2* | *1* | 0 | 0 | 0 | 0 | 0 | 0 | 0 | 0 | 0 | *0* | 3 |

## HELENA 1 SSE *Alfalfa County* ELEVATION 1362 ft LAT/LONG 36° 33 ' N / 98° 16 ' W

| | JAN | FEB | MAR | APR | MAY | JUN | JUL | AUG | SEP | OCT | NOV | DEC | YEAR |
|---|---|---|---|---|---|---|---|---|---|---|---|---|---|
| Maximum Temp °F | 43.5 | 49.0 | 59.3 | 69.9 | 78.5 | 88.8 | 94.7 | 93.1 | 83.9 | 72.6 | 57.5 | 47.2 | 69.8 |
| Minimum Temp °F | 20.7 | 24.5 | 33.8 | 43.8 | 53.7 | 63.9 | 68.8 | 67.1 | 58.8 | 46.1 | 33.6 | 24.3 | 44.9 |
| Mean Temp °F | 32.2 | 36.8 | 46.6 | 56.9 | 66.1 | 76.4 | 81.8 | 80.1 | 71.3 | 59.4 | 45.5 | 35.8 | 57.4 |
| Days Max Temp ≥ 90 °F | 0 | 0 | 0 | 0 | 3 | 15 | 25 | 22 | 9 | 1 | 0 | 0 | 75 |
| Days Max Temp ≤ 32 °F | 7 | 4 | 1 | 0 | 0 | 0 | 0 | 0 | 0 | 0 | 1 | 4 | 17 |
| Days Min Temp ≤ 32 °F | 28 | 22 | 14 | 3 | 0 | 0 | 0 | 0 | 0 | 2 | 14 | 26 | 109 |
| Days Min Temp ≤ 0 °F | 1 | 1 | 0 | 0 | 0 | 0 | 0 | 0 | 0 | 0 | 0 | 1 | 3 |
| Heating Degree Days | 1011 | 790 | 568 | 263 | 79 | 4 | 0 | 1 | 39 | 212 | 579 | 899 | 4445 |
| Cooling Degree Days | 0 | 0 | 2 | 31 | 124 | 365 | 538 | 503 | 256 | 39 | 2 | 0 | 1860 |
| Total Precipitation (") | 0.80 | 1.19 | 2.39 | 2.71 | 4.35 | 3.61 | 2.79 | 2.84 | 3.11 | 2.18 | 1.83 | 1.08 | 28.88 |
| Days ≥ 0.1" Precip | 2 | 3 | 4 | 5 | 6 | 6 | 5 | 5 | 5 | 4 | 3 | 3 | 51 |
| Total Snowfall (") | 3.8 | 4.8 | 2.9 | 0.7 | 0.0 | 0.0 | 0.0 | 0.0 | 0.0 | 0.1 | 1.4 | 3.8 | 17.5 |
| Days ≥ 1" Snow Depth | 7 | 5 | 2 | 0 | 0 | 0 | 0 | 0 | 0 | 0 | 1 | 4 | 19 |

## HENNESSEY 4 ESE *Kingfisher County* ELEVATION 1161 ft LAT/LONG 36° 6 ' N / 97° 54 ' W

| | JAN | FEB | MAR | APR | MAY | JUN | JUL | AUG | SEP | OCT | NOV | DEC | YEAR |
|---|---|---|---|---|---|---|---|---|---|---|---|---|---|
| Maximum Temp °F | 45.6 | 51.8 | 61.9 | 71.9 | 80.3 | 89.8 | 95.1 | 94.0 | 85.2 | 73.9 | 59.2 | 49.2 | 71.5 |
| Minimum Temp °F | 24.5 | 28.9 | 37.5 | 47.2 | 56.5 | 65.8 | 70.2 | 68.8 | 60.8 | 49.3 | 37.0 | 28.2 | 47.9 |
| Mean Temp °F | 35.1 | 40.4 | 49.7 | 59.6 | 68.4 | 77.8 | 82.7 | 81.4 | 73.1 | 61.6 | 48.2 | 38.7 | 59.7 |
| Days Max Temp ≥ 90 °F | 0 | 0 | 0 | 1 | 3 | 16 | 25 | 24 | *10* | 1 | 0 | 0 | 80 |
| Days Max Temp ≤ 32 °F | 5 | 3 | 0 | 0 | 0 | 0 | 0 | 0 | 0 | 0 | 0 | 2 | 10 |
| Days Min Temp ≤ 32 °F | 25 | 18 | 10 | 2 | 0 | 0 | 0 | 0 | 0 | 1 | 10 | 21 | 87 |
| Days Min Temp ≤ 0 °F | 1 | 0 | 0 | 0 | 0 | 0 | 0 | 0 | 0 | 0 | 0 | 1 | 2 |
| Heating Degree Days | 922 | 689 | 477 | 200 | 47 | 2 | 0 | 0 | 26 | 162 | 501 | 807 | 3833 |
| Cooling Degree Days | *0* | 0 | 5 | 31 | 148 | 390 | 541 | 525 | 276 | 54 | 3 | 0 | 1973 |
| Total Precipitation (") | 0.85 | 1.26 | 2.39 | 3.09 | 5.06 | 3.95 | 2.45 | 2.96 | 3.75 | 2.20 | 2.09 | 1.20 | 31.25 |
| Days ≥ 0.1" Precip | 2 | 3 | 4 | 5 | 6 | 6 | 4 | 4 | 5 | 4 | 3 | 3 | 49 |
| Total Snowfall (") | *1.9* | *2.7* | 1.1 | 0.0 | 0.0 | 0.0 | 0.0 | 0.0 | 0.0 | 0.0 | 0.2 | *2.1* | 8.0 |
| Days ≥ 1" Snow Depth | *1* | *2* | 0 | 0 | 0 | 0 | 0 | 0 | 0 | 0 | 0 | na | na |

## HOBART MUNICIPAL AP *Kiowa County* ELEVATION 1542 ft LAT/LONG 35° 1 ' N / 99° 5 ' W

| | JAN | FEB | MAR | APR | MAY | JUN | JUL | AUG | SEP | OCT | NOV | DEC | YEAR |
|---|---|---|---|---|---|---|---|---|---|---|---|---|---|
| Maximum Temp °F | 47.7 | 53.2 | 63.2 | 73.2 | 80.5 | 90.3 | 95.2 | 93.1 | 84.4 | 73.8 | 60.1 | 50.6 | 72.1 |
| Minimum Temp °F | 25.0 | 29.2 | 37.8 | 47.4 | 56.9 | 66.4 | 71.0 | 69.4 | 61.9 | 49.6 | 37.5 | 28.4 | 48.4 |
| Mean Temp °F | 36.4 | 41.2 | 50.5 | 60.3 | 68.7 | 78.3 | 83.1 | 81.3 | 73.2 | 61.8 | 48.8 | 39.5 | 60.3 |
| Days Max Temp ≥ 90 °F | 0 | 0 | 0 | 1 | 4 | 18 | 25 | 22 | 10 | 1 | 0 | 0 | 81 |
| Days Max Temp ≤ 32 °F | 5 | 2 | 0 | 0 | 0 | 0 | 0 | 0 | 0 | 0 | 0 | 2 | 9 |
| Days Min Temp ≤ 32 °F | 25 | 18 | 9 | 2 | 0 | 0 | 0 | 0 | 0 | 1 | 9 | 22 | 86 |
| Days Min Temp ≤ 0 °F | 0 | 0 | 0 | 0 | 0 | 0 | 0 | 0 | 0 | 0 | 0 | 0 | 0 |
| Heating Degree Days | 881 | 665 | 451 | 181 | 45 | 2 | 0 | 1 | 26 | 159 | 482 | 783 | 3676 |
| Cooling Degree Days | 0 | 0 | 7 | 47 | 175 | 412 | 578 | 538 | 298 | 65 | 4 | 0 | 2124 |
| Total Precipitation (") | 0.87 | 1.03 | 1.77 | 2.23 | 4.70 | 3.01 | 2.33 | 2.59 | 3.41 | 2.59 | 1.49 | 1.06 | 27.08 |
| Days ≥ 0.1" Precip | 2 | 3 | 3 | 4 | 7 | 5 | 4 | 4 | 4 | 4 | 3 | 2 | 45 |
| Total Snowfall (") | 2.1 | 2.5 | 0.4 | 0.0 | 0.0 | 0.0 | 0.0 | 0.0 | 0.0 | 0.0 | 0.3 | 1.4 | 6.7 |
| Days ≥ 1" Snow Depth | 3 | 2 | 0 | 0 | 0 | 0 | 0 | 0 | 0 | 0 | 0 | 1 | 6 |

## HOLDENVILLE *Hughes County* ELEVATION 902 ft LAT/LONG 35° 5 ' N / 96° 24 ' W

| | JAN | FEB | MAR | APR | MAY | JUN | JUL | AUG | SEP | OCT | NOV | DEC | YEAR |
|---|---|---|---|---|---|---|---|---|---|---|---|---|---|
| Maximum Temp °F | 49.6 | 55.2 | 64.7 | 74.0 | 80.1 | 87.7 | 93.6 | 93.6 | 85.3 | 75.5 | 62.5 | 53.0 | 72.9 |
| Minimum Temp °F | 27.6 | 31.8 | 40.5 | 49.9 | 58.0 | 66.2 | 70.1 | 68.7 | 61.7 | 50.7 | 40.3 | 31.6 | 49.8 |
| Mean Temp °F | 38.5 | 43.5 | 52.6 | 62.0 | 69.1 | 77.0 | 81.9 | 81.1 | 73.5 | 63.1 | 51.4 | 42.3 | 61.3 |
| Days Max Temp ≥ 90 °F | 0 | 0 | 0 | 0 | 1 | 11 | 24 | 24 | 9 | 1 | 0 | 0 | 70 |
| Days Max Temp ≤ 32 °F | 3 | 1 | 0 | 0 | 0 | 0 | 0 | 0 | 0 | 0 | 0 | 1 | 5 |
| Days Min Temp ≤ 32 °F | 21 | 15 | 8 | 1 | 0 | 0 | 0 | 0 | 0 | 1 | 7 | 17 | 70 |
| Days Min Temp ≤ 0 °F | 0 | 0 | 0 | 0 | 0 | 0 | 0 | 0 | 0 | 0 | 0 | 0 | 0 |
| Heating Degree Days | 815 | 601 | 390 | 140 | 30 | 1 | 0 | 0 | 19 | 127 | 410 | 698 | 3231 |
| Cooling Degree Days | 0 | 1 | 11 | 55 | 164 | 377 | 525 | 523 | 289 | 67 | 8 | 1 | 2021 |
| Total Precipitation (") | 1.54 | 2.03 | 3.29 | 4.26 | 5.64 | 4.08 | 2.72 | 2.84 | 4.56 | 4.31 | 3.16 | 2.35 | 40.78 |
| Days ≥ 0.1" Precip | 3 | 4 | 5 | 6 | 7 | 6 | 4 | 4 | 6 | 5 | 4 | 4 | 58 |
| Total Snowfall (") | 1.9 | 2.3 | 0.8 | 0.0 | 0.0 | 0.0 | 0.0 | 0.0 | 0.0 | 0.0 | 0.4 | 1.0 | 6.4 |
| Days ≥ 1" Snow Depth | *1* | *0* | 0 | 0 | 0 | 0 | 0 | 0 | 0 | 0 | 0 | 0 | 1 |

## HOLLIS *Harmon County* ELEVATION 1611 ft LAT/LONG 34° 41 ' N / 99° 55 ' W

| | JAN | FEB | MAR | APR | MAY | JUN | JUL | AUG | SEP | OCT | NOV | DEC | YEAR |
|---|---|---|---|---|---|---|---|---|---|---|---|---|---|
| Maximum Temp °F | 52.7 | 58.4 | 68.6 | 78.0 | 84.7 | 93.6 | 97.8 | 95.8 | 87.6 | 77.8 | 64.0 | 54.7 | 76.1 |
| Minimum Temp °F | 24.4 | 28.8 | 37.3 | 47.0 | 56.3 | 65.9 | 69.6 | 67.8 | 60.3 | 48.0 | 36.2 | 27.5 | 47.4 |
| Mean Temp °F | 38.6 | 43.6 | 53.0 | 62.5 | 70.5 | 79.8 | 83.6 | 81.7 | 73.9 | 62.9 | 50.1 | 41.1 | 61.8 |
| Days Max Temp ≥ 90 °F | 0 | 0 | 1 | 3 | 10 | 22 | 28 | 26 | 14 | 3 | 0 | 0 | 107 |
| Days Max Temp ≤ 32 °F | 3 | 1 | 0 | 0 | 0 | 0 | 0 | 0 | 0 | 0 | 0 | 1 | 5 |
| Days Min Temp ≤ 32 °F | 26 | 19 | 9 | 2 | 0 | 0 | 0 | 0 | 0 | 1 | 11 | 23 | 91 |
| Days Min Temp ≤ 0 °F | 0 | 0 | 0 | 0 | 0 | 0 | 0 | 0 | 0 | 0 | 0 | 0 | 0 |
| Heating Degree Days | 812 | 598 | 379 | 132 | 26 | 1 | 0 | 0 | 17 | 130 | 443 | 734 | 3272 |
| Cooling Degree Days | 0 | 1 | 8 | 55 | 193 | 439 | *548* | 524 | *284* | 59 | 4 | 0 | 2115 |
| Total Precipitation (") | 0.69 | 1.03 | 1.39 | 2.22 | 3.72 | 3.68 | 1.65 | 2.55 | 3.09 | 2.30 | 1.15 | 0.78 | 24.25 |
| Days ≥ 0.1" Precip | 2 | 3 | 3 | 4 | 5 | 5 | 3 | 4 | 4 | 3 | 3 | 2 | 41 |
| Total Snowfall (") | 1.7 | 2.5 | 0.7 | 0.1 | 0.0 | 0.0 | 0.0 | 0.0 | 0.0 | 0.0 | 0.4 | 1.1 | 6.5 |
| Days ≥ 1" Snow Depth | *1* | *1* | 0 | 0 | 0 | 0 | 0 | 0 | 0 | 0 | 0 | 1 | 3 |

## HOOKER *Texas County* ELEVATION 2992 ft LAT/LONG 36° 52 ' N / 101° 12 ' W

| | JAN | FEB | MAR | APR | MAY | JUN | JUL | AUG | SEP | OCT | NOV | DEC | YEAR |
|---|---|---|---|---|---|---|---|---|---|---|---|---|---|
| Maximum Temp °F | 47.0 | 52.4 | 61.6 | 71.6 | 79.1 | 89.4 | 94.3 | 91.7 | 83.9 | 73.5 | 58.3 | 49.0 | 71.0 |
| Minimum Temp °F | 18.8 | 22.8 | 30.7 | 40.6 | 50.1 | 60.2 | 64.9 | 62.9 | 54.7 | 42.0 | 29.6 | 21.3 | 41.6 |
| Mean Temp °F | 32.9 | 37.6 | 46.1 | 56.1 | 64.6 | 74.8 | 79.6 | 77.3 | 69.3 | 57.8 | 44.0 | 35.2 | 56.3 |
| Days Max Temp ≥ 90 °F | 0 | 0 | 0 | 1 | 4 | 15 | 24 | 21 | 9 | 2 | 0 | 0 | 76 |
| Days Max Temp ≤ 32 °F | 6 | 3 | 1 | 0 | 0 | 0 | 0 | 0 | 0 | 0 | 1 | 4 | 15 |
| Days Min Temp ≤ 32 °F | 29 | 25 | 18 | 5 | 0 | 0 | 0 | 0 | 0 | 4 | 19 | 29 | 129 |
| Days Min Temp ≤ 0 °F | 2 | 1 | 0 | 0 | 0 | 0 | 0 | 0 | 0 | 0 | 0 | 1 | 4 |
| Heating Degree Days | 987 | 767 | 580 | 279 | 97 | 8 | 0 | 2 | 46 | 248 | 625 | 918 | 4557 |
| Cooling Degree Days | 0 | 0 | 1 | 21 | 85 | 304 | 460 | 416 | 198 | 26 | 0 | 0 | 1511 |
| Total Precipitation (") | 0.45 | 0.53 | 1.15 | 1.38 | 2.90 | 2.77 | 2.42 | 2.45 | 1.88 | 0.90 | 0.76 | 0.49 | 18.08 |
| Days ≥ 0.1" Precip | 1 | 2 | 3 | 3 | 5 | 5 | 5 | 4 | 3 | 2 | 2 | 2 | 37 |
| Total Snowfall (") | 4.0 | 3.9 | 3.6 | 0.9 | 0.1 | 0.0 | 0.0 | 0.0 | 0.0 | 0.3 | 1.7 | *3.3* | 17.8 |
| Days ≥ 1" Snow Depth | 5 | 3 | 2 | 0 | 0 | 0 | 0 | 0 | 0 | 0 | 1 | *3* | 14 |

## HUGO *Choctaw County* ELEVATION 600 ft LAT/LONG 34° 1 ' N / 95° 30 ' W

| | JAN | FEB | MAR | APR | MAY | JUN | JUL | AUG | SEP | OCT | NOV | DEC | YEAR |
|---|---|---|---|---|---|---|---|---|---|---|---|---|---|
| Maximum Temp °F | 52.8 | 58.2 | 67.5 | 75.9 | 81.8 | 89.2 | 94.1 | 94.2 | 87.5 | 77.2 | 65.0 | 55.9 | 74.9 |
| Minimum Temp °F | 31.4 | 35.2 | 43.5 | 52.1 | 59.6 | 67.2 | 70.4 | 69.1 | 62.8 | 51.7 | 42.4 | 34.4 | 51.7 |
| Mean Temp °F | 42.2 | 46.8 | 55.5 | 64.0 | 70.7 | 78.2 | 82.3 | 81.6 | 75.2 | 64.5 | 53.7 | 45.2 | 63.3 |
| Days Max Temp ≥ 90 °F | 0 | 0 | 0 | 0 | 2 | 14 | 26 | 26 | 13 | 1 | 0 | 0 | 82 |
| Days Max Temp ≤ 32 °F | 1 | 0 | 0 | 0 | 0 | 0 | 0 | 0 | 0 | 0 | 0 | 1 | 2 |
| Days Min Temp ≤ 32 °F | 18 | 12 | 5 | 0 | 0 | 0 | 0 | 0 | 0 | 0 | 5 | 14 | 54 |
| Days Min Temp ≤ 0 °F | 0 | 0 | 0 | 0 | 0 | 0 | 0 | 0 | 0 | 0 | 0 | 0 | 0 |
| Heating Degree Days | 702 | 510 | 307 | 96 | 16 | 0 | 0 | 0 | 12 | 101 | 346 | 610 | 2700 |
| Cooling Degree Days | 0 | 3 | 18 | 72 | 198 | 410 | 538 | 538 | 321 | 86 | 13 | 1 | 2198 |
| Total Precipitation (") | 2.40 | 3.49 | 4.17 | 4.22 | 6.29 | 4.62 | 2.87 | 2.46 | 4.38 | 4.73 | 4.12 | 3.77 | 47.52 |
| Days ≥ 0.1" Precip | 5 | 5 | 6 | 6 | 7 | 6 | 4 | 4 | 5 | 6 | 5 | 6 | 65 |
| Total Snowfall (") | *0.9* | 1.2 | 0.3 | 0.0 | 0.0 | 0.0 | 0.0 | 0.0 | 0.0 | 0.0 | 0.1 | 0.6 | 3.1 |
| Days ≥ 1" Snow Depth | na | 1 | 0 | 0 | 0 | 0 | 0 | 0 | 0 | 0 | 0 | 1 | na |

## IDABEL *McCurtain County* ELEVATION 502 ft LAT/LONG 33° 54 ' N / 94° 49 ' W

| | JAN | FEB | MAR | APR | MAY | JUN | JUL | AUG | SEP | OCT | NOV | DEC | YEAR |
|---|---|---|---|---|---|---|---|---|---|---|---|---|---|
| Maximum Temp °F | 52.6 | 57.8 | 66.7 | 75.2 | 81.4 | 88.7 | 92.8 | 92.8 | 86.3 | 76.7 | 64.8 | 55.9 | 74.3 |
| Minimum Temp °F | 29.5 | 33.4 | 41.4 | 49.9 | 57.9 | 66.1 | 69.6 | 68.4 | 61.9 | 50.1 | 40.2 | 32.7 | 50.1 |
| Mean Temp °F | 41.1 | 45.7 | 54.1 | 62.6 | 69.6 | 77.4 | 81.2 | 80.6 | 74.1 | 63.4 | 52.5 | 44.3 | 62.2 |
| Days Max Temp ≥ 90 °F | 0 | 0 | 0 | 0 | 1 | 15 | 24 | 24 | 11 | 1 | 0 | 0 | 76 |
| Days Max Temp ≤ 32 °F | 1 | 0 | 0 | 0 | 0 | 0 | 0 | 0 | 0 | 0 | 0 | 1 | 2 |
| Days Min Temp ≤ 32 °F | 20 | 13 | 6 | 1 | 0 | 0 | 0 | 0 | 0 | 1 | 7 | 16 | 64 |
| Days Min Temp ≤ 0 °F | 0 | 0 | 0 | 0 | 0 | 0 | 0 | 0 | 0 | 0 | 0 | 0 | 0 |
| Heating Degree Days | 735 | 541 | 345 | 123 | 24 | 1 | 0 | 0 | 14 | 119 | 378 | 636 | 2916 |
| Cooling Degree Days | 0 | *1* | 13 | 52 | 166 | 383 | 510 | 503 | 298 | 77 | 10 | 1 | 2014 |
| Total Precipitation (") | 3.00 | 3.58 | 4.50 | 4.61 | 6.21 | 4.22 | 3.46 | 2.42 | 4.01 | 4.99 | 4.31 | 4.36 | 49.67 |
| Days ≥ 0.1" Precip | 5 | 5 | 6 | 6 | 7 | 6 | 5 | 4 | 5 | 5 | 5 | 6 | 65 |
| Total Snowfall (") | *0.6* | *1.0* | 0.1 | 0.0 | 0.0 | 0.0 | 0.0 | 0.0 | 0.0 | 0.0 | 0.0 | *0.3* | 2.0 |
| Days ≥ 1" Snow Depth | na | na | 0 | 0 | 0 | 0 | 0 | 0 | 0 | 0 | 0 | 0 | na |

## JEFFERSON *Grant County* ELEVATION 1060 ft LAT/LONG 36° 43 ' N / 97° 48 ' W

| | JAN | FEB | MAR | APR | MAY | JUN | JUL | AUG | SEP | OCT | NOV | DEC | YEAR |
|---|---|---|---|---|---|---|---|---|---|---|---|---|---|
| Maximum Temp °F | 45.8 | 52.0 | 62.5 | 72.8 | 80.9 | 91.2 | 96.4 | 94.7 | 85.9 | 74.9 | 59.4 | 49.2 | 72.1 |
| Minimum Temp °F | 22.8 | 26.7 | 36.1 | 46.4 | 55.8 | 65.6 | 70.3 | 68.3 | 60.2 | 48.2 | 35.7 | 26.3 | 46.9 |
| Mean Temp °F | 34.3 | 39.4 | 49.4 | 59.6 | 68.4 | 78.4 | 83.3 | 81.5 | 73.0 | 61.5 | 47.6 | 37.8 | 59.5 |
| Days Max Temp ≥ 90 °F | 0 | 0 | 0 | 1 | 4 | 20 | 26 | 24 | 11 | 2 | 0 | 0 | 88 |
| Days Max Temp ≤ 32 °F | 5 | 3 | 0 | 0 | 0 | 0 | 0 | 0 | 0 | 0 | 0 | 3 | 11 |
| Days Min Temp ≤ 32 °F | 26 | 20 | 13 | 2 | 0 | 0 | 0 | 0 | 0 | 2 | 12 | 24 | 99 |
| Days Min Temp ≤ 0 °F | 1 | 1 | 0 | 0 | 0 | 0 | 0 | 0 | 0 | 0 | 0 | 1 | 3 |
| Heating Degree Days | 944 | 717 | 487 | 200 | 47 | 1 | 0 | 0 | 25 | 165 | 519 | 836 | 3941 |
| Cooling Degree Days | 0 | 0 | 7 | 44 | 159 | 414 | 575 | 537 | 283 | 55 | 4 | 0 | 2078 |
| Total Precipitation (") | 0.90 | 1.27 | 2.53 | 3.10 | 4.75 | 4.10 | 3.45 | 3.36 | 3.44 | 2.57 | 2.15 | 1.23 | 32.85 |
| Days ≥ 0.1" Precip | 2 | 3 | 4 | 5 | 7 | 6 | 5 | 5 | 5 | 4 | 3 | 3 | 52 |
| Total Snowfall (") | *1.7* | 2.7 | 1.1 | 0.2 | 0.0 | 0.0 | 0.0 | 0.0 | 0.0 | 0.0 | 0.5 | *1.3* | 7.5 |
| Days ≥ 1" Snow Depth | *3* | 3 | 0 | 0 | 0 | 0 | 0 | 0 | 0 | 0 | 0 | *1* | 7 |

## KANSAS 1 ESE *Delaware County* ELEVATION 1122 ft LAT/LONG 36° 12' N / 94° 48' W

| | JAN | FEB | MAR | APR | MAY | JUN | JUL | AUG | SEP | OCT | NOV | DEC | YEAR |
|---|---|---|---|---|---|---|---|---|---|---|---|---|---|
| Maximum Temp °F | 46.5 | 52.2 | 61.9 | 72.0 | 77.7 | 85.0 | 90.9 | 90.3 | 82.5 | 72.5 | 59.7 | 50.2 | 70.1 |
| Minimum Temp °F | 25.9 | 29.9 | 39.1 | 48.4 | 56.0 | 64.4 | 68.6 | 67.1 | 60.2 | 49.2 | 38.9 | 29.9 | 48.1 |
| Mean Temp °F | 36.2 | 41.1 | 50.5 | 60.2 | 66.9 | 74.7 | 79.8 | 78.7 | 71.4 | 60.9 | 49.3 | 40.1 | 59.2 |
| Days Max Temp ≥ 90 °F | 0 | 0 | 0 | 0 | 0 | 5 | 18 | 18 | 6 | 0 | 0 | 0 | 47 |
| Days Max Temp ≤ 32 °F | 4 | 2 | 0 | 0 | 0 | 0 | 0 | 0 | 0 | 0 | 0 | 2 | 8 |
| Days Min Temp ≤ 32 °F | 22 | 17 | 9 | 1 | 0 | 0 | 0 | 0 | 0 | 1 | 8 | 18 | 76 |
| Days Min Temp ≤ 0 °F | 1 | 0 | 0 | 0 | 0 | 0 | 0 | 0 | 0 | 0 | 0 | 1 | 2 |
| Heating Degree Days | 885 | 668 | 452 | 180 | 52 | 3 | 0 | 0 | 29 | 171 | 466 | 765 | 3671 |
| Cooling Degree Days | 0 | 0 | 8 | 40 | 114 | 312 | 475 | 452 | 233 | 45 | 4 | 0 | 1683 |
| Total Precipitation (") | 2.30 | 2.46 | 4.12 | 4.58 | 5.30 | 5.00 | 2.63 | 3.48 | 5.57 | 4.34 | 4.45 | 3.54 | 47.77 |
| Days ≥ 0.1" Precip | 4 | 5 | 7 | 7 | 8 | 7 | 5 | 5 | 7 | 5 | 5 | 5 | 70 |
| Total Snowfall (") | 4.0 | 4.0 | 3.2 | 0.0 | 0.0 | 0.0 | 0.0 | 0.0 | 0.0 | 0.1 | 1.0 | 2.5 | 14.8 |
| Days ≥ 1" Snow Depth | 5 | 4 | 1 | 0 | 0 | 0 | 0 | 0 | 0 | 0 | 1 | 1 | 12 |

## KENTON *Cimarron County* ELEVATION 4354 ft LAT/LONG 36° 55' N / 102° 58' W

| | JAN | FEB | MAR | APR | MAY | JUN | JUL | AUG | SEP | OCT | NOV | DEC | YEAR |
|---|---|---|---|---|---|---|---|---|---|---|---|---|---|
| Maximum Temp °F | 49.7 | 53.4 | 61.1 | 69.9 | 78.0 | 87.9 | 92.1 | 89.4 | 82.4 | 72.7 | 59.6 | 50.7 | 70.6 |
| Minimum Temp °F | 18.0 | 21.6 | 28.6 | 38.6 | 47.9 | 57.3 | 62.9 | 61.1 | 52.4 | 39.3 | 27.8 | 19.9 | 39.6 |
| Mean Temp °F | 33.9 | 37.5 | 44.9 | 54.3 | 63.0 | 72.6 | 77.5 | 75.2 | 67.4 | 56.0 | 43.7 | 35.3 | 55.1 |
| Days Max Temp ≥ 90 °F | 0 | 0 | 0 | 1 | 3 | 13 | 22 | 17 | 7 | 1 | 0 | 0 | 64 |
| Days Max Temp ≤ 32 °F | *3* | 2 | 1 | 0 | 0 | 0 | 0 | 0 | 0 | 0 | 1 | *3* | 10 |
| Days Min Temp ≤ 32 °F | *30* | 25 | 21 | 7 | 1 | 0 | 0 | 0 | 0 | 6 | 22 | *29* | 141 |
| Days Min Temp ≤ 0 °F | 1 | 1 | 0 | 0 | 0 | 0 | 0 | 0 | 0 | 0 | 0 | 1 | 3 |
| Heating Degree Days | 957 | 769 | 617 | 327 | 117 | 14 | 1 | 3 | 61 | 285 | 632 | 912 | 4695 |
| Cooling Degree Days | 0 | 0 | 1 | 11 | 55 | 241 | 387 | 329 | 144 | 9 | 0 | 0 | 1177 |
| Total Precipitation (") | 0.29 | 0.32 | 0.80 | 1.23 | 2.41 | 2.16 | 3.29 | 2.91 | 1.78 | 0.99 | 0.59 | 0.30 | 17.07 |
| Days ≥ 0.1" Precip | 1 | 1 | 2 | 3 | 5 | 5 | 6 | 6 | 4 | 2 | 2 | *1* | 38 |
| Total Snowfall (") | *4.1* | *3.4* | *4.5* | *1.0* | 0.2 | 0.0 | 0.0 | 0.0 | 0.2 | 0.2 | *2.3* | na | na |
| Days ≥ 1" Snow Depth | na | na | na | 1 | 0 | 0 | 0 | 0 | 0 | 0 | *2* | na | na |

## KINGFISHER 2 SE *Kingfisher County* ELEVATION 1060 ft LAT/LONG 35° 52' N / 97° 56' W

| | JAN | FEB | MAR | APR | MAY | JUN | JUL | AUG | SEP | OCT | NOV | DEC | YEAR |
|---|---|---|---|---|---|---|---|---|---|---|---|---|---|
| Maximum Temp °F | 47.3 | 53.2 | 63.6 | 73.4 | 80.9 | 90.0 | 95.7 | 94.4 | 85.6 | 74.8 | 60.3 | 50.6 | 72.5 |
| Minimum Temp °F | 24.4 | 28.8 | 37.8 | 47.8 | 56.6 | 66.1 | 70.5 | 68.9 | 61.2 | 49.3 | 37.3 | 28.1 | 48.1 |
| Mean Temp °F | 35.9 | 41.0 | 50.7 | 60.6 | 68.8 | 78.1 | 83.1 | 81.7 | 73.5 | 62.1 | 48.8 | 39.4 | 60.3 |
| Days Max Temp ≥ 90 °F | 0 | 0 | 0 | 1 | 4 | 17 | 26 | 24 | 11 | 1 | 0 | 0 | 84 |
| Days Max Temp ≤ 32 °F | 5 | 2 | 0 | 0 | 0 | 0 | 0 | 0 | 0 | 0 | 0 | 2 | 9 |
| Days Min Temp ≤ 32 °F | 25 | 18 | 10 | 2 | 0 | 0 | 0 | 0 | 0 | 1 | 10 | 22 | 88 |
| Days Min Temp ≤ 0 °F | 1 | 0 | 0 | 0 | 0 | 0 | 0 | 0 | 0 | 0 | 0 | 0 | 1 |
| Heating Degree Days | 895 | 670 | 448 | 178 | 43 | 1 | 0 | 0 | 24 | 154 | 483 | 787 | 3683 |
| Cooling Degree Days | 0 | 0 | 9 | 47 | 156 | 393 | 556 | 525 | 284 | 58 | 4 | 0 | 2032 |
| Total Precipitation (") | 1.06 | 1.42 | 2.30 | 3.14 | 4.90 | 4.04 | 2.03 | 2.70 | 3.85 | 2.23 | 2.15 | 1.39 | 31.21 |
| Days ≥ 0.1" Precip | 2 | 3 | 4 | 5 | 6 | 5 | 4 | 4 | 5 | 3 | 4 | 3 | 48 |
| Total Snowfall (") | *1.5* | *1.0* | 0.9 | 0.0 | 0.0 | 0.0 | 0.0 | 0.0 | 0.0 | 0.0 | 0.3 | 1.2 | 4.9 |
| Days ≥ 1" Snow Depth | *1* | *1* | 0 | 0 | 0 | 0 | 0 | 0 | 0 | 0 | 0 | 1 | 3 |

## LAWTON *Comanche County* ELEVATION 1142 ft LAT/LONG 34° 36' N / 98° 24' W

| | JAN | FEB | MAR | APR | MAY | JUN | JUL | AUG | SEP | OCT | NOV | DEC | YEAR |
|---|---|---|---|---|---|---|---|---|---|---|---|---|---|
| Maximum Temp °F | 51.1 | 55.3 | 65.2 | 75.0 | 81.9 | 90.2 | 95.7 | 94.6 | 86.2 | *76.0* | 62.7 | 53.1 | 73.9 |
| Minimum Temp °F | 26.3 | 30.0 | 39.4 | 49.6 | 58.3 | 66.9 | 71.3 | 69.7 | 61.8 | *50.1* | 37.8 | 29.4 | 49.2 |
| Mean Temp °F | 38.6 | 42.7 | 52.2 | 62.3 | 70.1 | 78.6 | 83.5 | 82.2 | 74.0 | *63.1* | 50.3 | 41.3 | 61.6 |
| Days Max Temp ≥ 90 °F | 0 | 0 | 0 | 1 | 5 | *18* | 27 | 25 | 12 | *2* | 0 | 0 | 90 |
| Days Max Temp ≤ 32 °F | 3 | 2 | 0 | 0 | 0 | 0 | 0 | 0 | 0 | 0 | 0 | 2 | 7 |
| Days Min Temp ≤ 32 °F | *24* | 16 | *7* | 1 | 0 | 0 | 0 | 0 | 0 | *0* | *9* | *21* | 78 |
| Days Min Temp ≤ 0 °F | 0 | 0 | 0 | 0 | 0 | 0 | 0 | 0 | 0 | 0 | 0 | 0 | 0 |
| Heating Degree Days | 811 | 624 | 400 | 135 | 26 | 1 | 0 | 0 | 21 | *127* | 439 | 728 | 3312 |
| Cooling Degree Days | 0 | 0 | 6 | 53 | 183 | 402 | 576 | 555 | 296 | *57* | 5 | 0 | 2133 |
| Total Precipitation (") | 1.14 | 1.27 | 2.17 | 2.66 | 5.14 | 3.51 | 1.99 | 2.19 | 3.56 | 2.76 | 1.63 | 1.48 | 29.50 |
| Days ≥ 0.1" Precip | 2 | 3 | 4 | 4 | 6 | 4 | 3 | 4 | 5 | *3* | 3 | 3 | 44 |
| Total Snowfall (") | *1.1* | na | 0.3 | 0.0 | 0.0 | 0.0 | 0.0 | 0.0 | 0.0 | 0.0 | 0.0 | na | na |
| Days ≥ 1" Snow Depth | *1* | na | 0 | 0 | 0 | 0 | 0 | 0 | 0 | 0 | 0 | na | na |

## LINDSAY 2 W *Garvin County* ELEVATION 981 ft LAT/LONG 34° 50 ' N / 97° 37 ' W

| | JAN | FEB | MAR | APR | MAY | JUN | JUL | AUG | SEP | OCT | NOV | DEC | YEAR |
|---|---|---|---|---|---|---|---|---|---|---|---|---|---|
| Maximum Temp °F | *50.6* | *56.7* | *66.5* | 76.1 | *81.8* | *89.9* | *95.4* | *94.3* | *85.8* | *76.7* | *63.7* | *54.6* | 74.3 |
| Minimum Temp °F | *26.4* | *29.8* | *38.9* | 49.4 | *57.2* | *66.0* | *70.1* | *67.7* | *60.6* | *49.4* | *38.2* | *29.7* | 48.6 |
| Mean Temp °F | *38.5* | *43.3* | *52.7* | 62.8 | *69.6* | *78.1* | *82.7* | *81.0* | *73.3* | *63.1* | *51.0* | *42.2* | 61.5 |
| Days Max Temp ≥ 90 °F | *0* | *0* | 0 | 1 | *3* | *16* | *26* | *23* | *10* | *1* | *0* | *0* | 80 |
| Days Max Temp ≤ 32 °F | *3* | *1* | 0 | 0 | 0 | 0 | 0 | 0 | 0 | *0* | *0* | *1* | 5 |
| Days Min Temp ≤ 32 °F | *22* | *18* | *9* | 1 | 0 | 0 | 0 | 0 | 0 | *1* | *10* | *20* | 81 |
| Days Min Temp ≤ 0 °F | *0* | *0* | 0 | 0 | 0 | 0 | 0 | 0 | 0 | *0* | *0* | *0* | 0 |
| Heating Degree Days | *814* | *608* | *391* | 128 | *31* | *0* | *0* | *0* | *19* | *130* | *422* | *702* | 3245 |
| Cooling Degree Days | na | na | na | *63* | *182* | na | na | na | *277* | na | na | na | na |
| Total Precipitation (") | *1.51* | *1.69* | 2.53 | 3.55 | 5.82 | 3.56 | 2.22 | 2.43 | 4.51 | *3.22* | *2.10* | *1.74* | 34.88 |
| Days ≥ 0.1" Precip | *3* | *3* | 5 | 5 | 7 | 5 | *4* | 4 | *5* | *4* | *3* | *3* | 51 |
| Total Snowfall (") | na | na | *1.0* | 0.0 | 0.0 | 0.0 | 0.0 | 0.0 | 0.0 | *0.0* | *0.3* | na | na |
| Days ≥ 1" Snow Depth | na | na | na | 0 | 0 | 0 | 0 | 0 | 0 | *0* | *0* | na | na |

## MADILL *Marshall County* ELEVATION 771 ft LAT/LONG 34° 5 ' N / 96° 46 ' W

| | JAN | FEB | MAR | APR | MAY | JUN | JUL | AUG | SEP | OCT | NOV | DEC | YEAR |
|---|---|---|---|---|---|---|---|---|---|---|---|---|---|
| Maximum Temp °F | 51.4 | 56.7 | 66.1 | 75.1 | 81.3 | 89.0 | 94.6 | 94.3 | 86.5 | 76.5 | 64.2 | 54.8 | 74.2 |
| Minimum Temp °F | 30.1 | 34.3 | 42.9 | 52.1 | 59.8 | 67.8 | 71.7 | 70.3 | 63.7 | 52.6 | 42.4 | 33.7 | 51.8 |
| Mean Temp °F | 40.8 | 45.5 | 54.5 | 63.6 | 70.5 | 78.4 | 83.2 | 82.3 | 75.1 | 64.6 | 53.3 | 44.3 | 63.0 |
| Days Max Temp ≥ 90 °F | 0 | 0 | 0 | 1 | 2 | 15 | 26 | 25 | 12 | 2 | 0 | 0 | 83 |
| Days Max Temp ≤ 32 °F | 2 | 1 | 0 | 0 | 0 | 0 | 0 | 0 | 0 | 0 | 0 | 1 | 4 |
| Days Min Temp ≤ 32 °F | 18 | 12 | 5 | 1 | 0 | 0 | 0 | 0 | 0 | 0 | 5 | 14 | 55 |
| Days Min Temp ≤ 0 °F | 0 | 0 | 0 | 0 | 0 | 0 | 0 | 0 | 0 | 0 | 0 | 0 | 0 |
| Heating Degree Days | 744 | 544 | 336 | 108 | 20 | 1 | 0 | 0 | 15 | 105 | 358 | 637 | 2868 |
| Cooling Degree Days | 0 | 1 | 17 | 71 | 200 | 414 | 561 | 554 | 332 | 95 | 15 | 1 | 2261 |
| Total Precipitation (") | 2.06 | 2.72 | 3.40 | 3.95 | 5.77 | 4.78 | 2.27 | 2.69 | 4.84 | 4.61 | 2.71 | 2.40 | 42.20 |
| Days ≥ 0.1" Precip | 4 | 5 | 5 | 6 | 7 | 6 | 3 | 4 | 5 | 5 | 4 | 4 | 58 |
| Total Snowfall (") | *1.4* | 1.6 | 0.0 | 0.0 | 0.0 | 0.0 | 0.0 | 0.0 | 0.0 | 0.0 | 0.1 | 0.8 | 3.9 |
| Days ≥ 1" Snow Depth | *1* | *1* | 0 | 0 | 0 | 0 | 0 | 0 | 0 | 0 | 0 | 0 | 2 |

## MANGUM RESEARCH STN *Greer County* ELEVATION 1522 ft LAT/LONG 34° 50 ' N / 99° 26 ' W

| | JAN | FEB | MAR | APR | MAY | JUN | JUL | AUG | SEP | OCT | NOV | DEC | YEAR |
|---|---|---|---|---|---|---|---|---|---|---|---|---|---|
| Maximum Temp °F | 51.5 | 57.7 | 68.1 | 77.9 | 84.6 | 93.6 | 98.7 | 96.6 | 88.2 | 77.8 | 63.6 | 53.8 | 76.0 |
| Minimum Temp °F | 25.3 | 29.3 | 37.9 | 47.1 | 56.8 | 65.6 | 69.1 | 67.6 | 60.4 | 48.7 | 37.1 | 28.4 | 47.8 |
| Mean Temp °F | 38.4 | 43.5 | 53.0 | 62.5 | 70.7 | 79.6 | 83.9 | 82.1 | 74.3 | 63.2 | 50.4 | 41.1 | 61.9 |
| Days Max Temp ≥ 90 °F | 0 | 0 | 1 | 3 | 9 | 22 | 29 | 26 | 15 | 4 | 0 | 0 | 109 |
| Days Max Temp ≤ 32 °F | 3 | 1 | 0 | 0 | 0 | 0 | 0 | 0 | 0 | 0 | 0 | 1 | 5 |
| Days Min Temp ≤ 32 °F | 25 | 18 | 9 | 1 | 0 | 0 | 0 | 0 | 0 | 1 | 10 | 22 | 86 |
| Days Min Temp ≤ 0 °F | 0 | 0 | 0 | 0 | 0 | 0 | 0 | 0 | 0 | 0 | 0 | 0 | 0 |
| Heating Degree Days | 818 | 600 | 377 | 133 | 24 | 0 | 0 | 0 | 17 | 125 | 434 | 734 | 3262 |
| Cooling Degree Days | 0 | 0 | 11 | 60 | 204 | 434 | 583 | 548 | 307 | 69 | 3 | 0 | 2219 |
| Total Precipitation (") | 0.87 | 1.07 | 1.61 | 2.01 | 4.48 | 3.92 | 2.02 | 2.35 | 3.35 | 2.64 | 1.32 | 0.96 | 26.60 |
| Days ≥ 0.1" Precip | 2 | 3 | 3 | 3 | 6 | 5 | 3 | 4 | 4 | 4 | 3 | 2 | 42 |
| Total Snowfall (") | 1.3 | *1.7* | 0.3 | 0.1 | 0.0 | 0.0 | 0.0 | 0.0 | 0.0 | 0.0 | 0.0 | 1.2 | 4.6 |
| Days ≥ 1" Snow Depth | *1* | 1 | 0 | 0 | 0 | 0 | 0 | 0 | 0 | 0 | 0 | 0 | 2 |

## MANNFORD 6 NW *Pawnee County* ELEVATION 741 ft LAT/LONG 36° 10 ' N / 96° 23 ' W

| | JAN | FEB | MAR | APR | MAY | JUN | JUL | AUG | SEP | OCT | NOV | DEC | YEAR |
|---|---|---|---|---|---|---|---|---|---|---|---|---|---|
| Maximum Temp °F | 47.8 | 53.9 | 64.2 | 74.6 | 79.7 | 88.0 | 94.8 | 94.2 | 85.1 | 74.8 | 61.4 | 51.1 | 72.5 |
| Minimum Temp °F | 23.9 | 28.6 | 38.9 | 49.0 | 56.3 | 65.3 | 69.1 | 67.0 | 59.7 | 48.6 | 37.1 | 27.9 | 47.6 |
| Mean Temp °F | 35.9 | 41.3 | 51.6 | 61.8 | 68.0 | 76.6 | 82.0 | 80.5 | 72.4 | 61.8 | 49.1 | 39.5 | 60.0 |
| Days Max Temp ≥ 90 °F | 0 | 0 | 0 | 1 | 2 | 13 | *25* | 24 | 10 | 2 | 0 | 0 | 77 |
| Days Max Temp ≤ 32 °F | 4 | 2 | 0 | 0 | 0 | 0 | 0 | 0 | 0 | 0 | 0 | 2 | 8 |
| Days Min Temp ≤ 32 °F | 25 | 18 | 9 | *1* | 0 | 0 | 0 | 0 | 0 | 2 | 11 | *20* | 86 |
| Days Min Temp ≤ 0 °F | 1 | 0 | 0 | 0 | 0 | 0 | 0 | 0 | 0 | 0 | 0 | 1 | 2 |
| Heating Degree Days | 896 | 665 | 426 | 157 | 44 | 2 | 0 | 0 | 30 | 161 | 476 | 785 | 3642 |
| Cooling Degree Days | 0 | 1 | 18 | 70 | 153 | 370 | 544 | 502 | 273 | 65 | 9 | 1 | 2006 |
| Total Precipitation (") | 1.45 | 2.05 | 3.22 | 3.42 | 5.04 | 3.93 | 2.47 | 3.34 | 4.38 | 3.23 | 3.05 | 2.06 | 37.64 |
| Days ≥ 0.1" Precip | 3 | 4 | 5 | 6 | 7 | 6 | 4 | 4 | 6 | 5 | 4 | 4 | 58 |
| Total Snowfall (") | *2.9* | *2.5* | 1.1 | 0.0 | 0.0 | 0.0 | 0.0 | 0.0 | 0.0 | 0.0 | 0.2 | *1.5* | 8.2 |
| Days ≥ 1" Snow Depth | na | *2* | 0 | 0 | 0 | 0 | 0 | 0 | 0 | 0 | *0* | *1* | na |

## MARIETTA *Love County* ELEVATION 850 ft LAT/LONG 33° 56' N / 97° 7' W

| | JAN | FEB | MAR | APR | MAY | JUN | JUL | AUG | SEP | OCT | NOV | DEC | YEAR |
|---|---|---|---|---|---|---|---|---|---|---|---|---|---|
| Maximum Temp °F | 52.0 | 57.6 | 66.9 | 75.3 | 81.1 | 88.9 | 94.5 | 94.4 | 86.6 | 76.8 | 64.5 | 55.5 | 74.5 |
| Minimum Temp °F | 30.0 | 33.9 | 42.6 | 51.7 | 59.5 | 67.5 | 71.5 | 70.2 | 63.8 | 53.0 | 42.1 | 33.2 | 51.6 |
| Mean Temp °F | 41.0 | 45.8 | 54.8 | 63.5 | 70.3 | 78.2 | 83.0 | 82.3 | 75.2 | 64.9 | 53.3 | 44.4 | 63.1 |
| Days Max Temp ≥ 90 °F | 0 | 0 | 0 | 0 | 2 | 14 | 26 | 26 | 12 | 2 | 0 | 0 | 82 |
| Days Max Temp ≤ 32 °F | 2 | 1 | 0 | 0 | 0 | 0 | 0 | 0 | 0 | 0 | 0 | 1 | 4 |
| Days Min Temp ≤ 32 °F | 18 | 12 | 5 | 1 | 0 | 0 | 0 | 0 | 0 | 0 | 6 | 15 | 57 |
| Days Min Temp ≤ 0 °F | 0 | 0 | 0 | 0 | 0 | 0 | 0 | 0 | 0 | 0 | 0 | 0 | 0 |
| Heating Degree Days | 737 | 537 | 330 | 109 | 20 | 0 | 0 | 0 | 13 | 101 | 359 | 634 | 2840 |
| Cooling Degree Days | 0 | 2 | 19 | 77 | 206 | 422 | 577 | 582 | 349 | 109 | 17 | 2 | 2362 |
| Total Precipitation (") | 1.61 | 2.37 | 3.22 | 3.62 | 5.35 | 4.02 | 2.22 | 2.59 | 4.36 | 4.17 | 2.40 | 2.25 | 38.18 |
| Days ≥ 0.1" Precip | 4 | 4 | 5 | 5 | 7 | 5 | 3 | 4 | 5 | 5 | 4 | 4 | 55 |
| Total Snowfall (") | *1.7* | *1.6* | 0.6 | 0.0 | 0.0 | 0.0 | 0.0 | 0.0 | 0.0 | 0.0 | 0.3 | 0.9 | 5.1 |
| Days ≥ 1" Snow Depth | *1* | *1* | 0 | 0 | 0 | 0 | 0 | 0 | 0 | 0 | 0 | 1 | 3 |

## MC CURTAIN 1 SE *Haskell County* ELEVATION 571 ft LAT/LONG 35° 9' N / 94° 57' W

| | JAN | FEB | MAR | APR | MAY | JUN | JUL | AUG | SEP | OCT | NOV | DEC | YEAR |
|---|---|---|---|---|---|---|---|---|---|---|---|---|---|
| Maximum Temp °F | 50.5 | 55.8 | 65.4 | 74.9 | 80.4 | 88.1 | 93.8 | 93.2 | 85.4 | 75.8 | 63.5 | 54.0 | 73.4 |
| Minimum Temp °F | 28.7 | 32.6 | 41.9 | 51.0 | 58.5 | 66.9 | 71.1 | 69.2 | 62.4 | 51.0 | 41.1 | 32.3 | 50.6 |
| Mean Temp °F | 39.6 | 44.2 | 53.7 | 63.0 | 69.5 | 77.5 | 82.5 | 81.2 | 73.9 | 63.4 | 52.4 | 43.2 | 62.0 |
| Days Max Temp ≥ 90 °F | 0 | 0 | 0 | 0 | 1 | 13 | 25 | 24 | 10 | 1 | 0 | 0 | 74 |
| Days Max Temp ≤ 32 °F | 3 | 1 | 0 | 0 | 0 | 0 | 0 | 0 | 0 | 0 | 0 | 1 | 5 |
| Days Min Temp ≤ 32 °F | 20 | 14 | 7 | 1 | 0 | 0 | 0 | 0 | 0 | 1 | 7 | 16 | 66 |
| Days Min Temp ≤ 0 °F | 0 | 0 | 0 | 0 | 0 | 0 | 0 | 0 | 0 | 0 | 0 | 0 | 0 |
| Heating Degree Days | 781 | 583 | 365 | 127 | 30 | 1 | 0 | 0 | 19 | 128 | 384 | 671 | 3089 |
| Cooling Degree Days | 0 | 3 | 23 | 75 | 185 | 402 | 564 | 536 | 308 | 83 | 12 | 2 | 2193 |
| Total Precipitation (") | 2.26 | 2.97 | 4.10 | 4.83 | 6.16 | 4.49 | 3.11 | 2.87 | 4.92 | 4.09 | 4.85 | 3.28 | 47.93 |
| Days ≥ 0.1" Precip | 5 | 5 | 6 | 7 | 8 | 6 | 4 | 5 | 6 | 5 | 5 | 5 | 67 |
| Total Snowfall (") | 3.0 | 2.7 | 1.0 | 0.0 | 0.0 | 0.0 | 0.0 | 0.0 | 0.0 | 0.0 | 0.2 | 0.7 | 7.6 |
| Days ≥ 1" Snow Depth | 3 | 2 | 1 | 0 | 0 | 0 | 0 | 0 | 0 | 0 | 0 | 1 | 7 |

## MCALESTER MUNI AP *Pittsburg County* ELEVATION 778 ft LAT/LONG 34° 53' N / 95° 47' W

| | JAN | FEB | MAR | APR | MAY | JUN | JUL | AUG | SEP | OCT | NOV | DEC | YEAR |
|---|---|---|---|---|---|---|---|---|---|---|---|---|---|
| Maximum Temp °F | 48.5 | 53.9 | 63.7 | 73.2 | 79.1 | 87.1 | 92.5 | 92.3 | 84.4 | 74.6 | 62.2 | 52.3 | 72.0 |
| Minimum Temp °F | 27.6 | 31.7 | 41.1 | 50.4 | 58.3 | 67.4 | 71.1 | 69.9 | 62.6 | 50.7 | 40.4 | 31.5 | 50.2 |
| Mean Temp °F | 38.1 | 42.8 | 52.4 | 61.8 | 68.7 | 77.3 | 81.9 | 81.1 | 73.5 | 62.7 | 51.3 | 41.9 | 61.1 |
| Days Max Temp ≥ 90 °F | 0 | 0 | 0 | 0 | 1 | 11 | 22 | 22 | 9 | 1 | 0 | 0 | 66 |
| Days Max Temp ≤ 32 °F | 4 | 2 | 0 | 0 | 0 | 0 | 0 | 0 | 0 | 0 | 0 | 2 | 8 |
| Days Min Temp ≤ 32 °F | 22 | 15 | 7 | 1 | 0 | 0 | 0 | 0 | 0 | 1 | 8 | 17 | 71 |
| Days Min Temp ≤ 0 °F | 0 | 0 | 0 | 0 | 0 | 0 | 0 | 0 | 0 | 0 | 0 | 0 | 0 |
| Heating Degree Days | 827 | 621 | 399 | 149 | 38 | 1 | 0 | 0 | 22 | 144 | 414 | 709 | 3324 |
| Cooling Degree Days | 0 | 1 | 15 | 60 | 163 | 390 | 539 | 535 | 303 | 82 | 12 | 1 | 2101 |
| Total Precipitation (") | 2.15 | 2.76 | 3.91 | 4.36 | 6.10 | 4.33 | 2.68 | 3.13 | 4.91 | 4.88 | 3.76 | 3.03 | 46.00 |
| Days ≥ 0.1" Precip | 4 | 4 | 6 | 6 | 8 | 6 | 4 | 5 | 6 | 5 | 5 | 5 | 64 |
| Total Snowfall (") | 2.5 | 2.1 | 0.8 | 0.0 | 0.0 | 0.0 | 0.0 | 0.0 | 0.0 | 0.0 | 0.2 | 0.8 | 6.4 |
| Days ≥ 1" Snow Depth | 3 | 2 | 0 | 0 | 0 | 0 | 0 | 0 | 0 | 0 | 0 | 1 | 6 |

## MEEKER 4 W *Lincoln County* ELEVATION 925 ft LAT/LONG 35° 30' N / 96° 59' W

| | JAN | FEB | MAR | APR | MAY | JUN | JUL | AUG | SEP | OCT | NOV | DEC | YEAR |
|---|---|---|---|---|---|---|---|---|---|---|---|---|---|
| Maximum Temp °F | 47.2 | 53.6 | 63.0 | 73.4 | 79.2 | 87.2 | 93.3 | 92.9 | 84.2 | 74.4 | 61.0 | 51.7 | 71.8 |
| Minimum Temp °F | 25.3 | 29.5 | 38.4 | 48.8 | 57.3 | 65.8 | 70.2 | 68.3 | 60.8 | 49.4 | 37.7 | 29.4 | 48.4 |
| Mean Temp °F | 36.3 | 41.6 | 50.7 | 61.1 | 68.2 | 76.6 | 81.7 | 80.6 | 72.5 | 61.9 | 49.4 | 40.6 | 60.1 |
| Days Max Temp ≥ 90 °F | 0 | 0 | 0 | 0 | 1 | 11 | 23 | 22 | 8 | 1 | 0 | 0 | 66 |
| Days Max Temp ≤ 32 °F | 4 | 2 | 0 | 0 | 0 | 0 | 0 | 0 | 0 | 0 | 0 | 2 | 8 |
| Days Min Temp ≤ 32 °F | 23 | 17 | 9 | 2 | 0 | 0 | 0 | 0 | 0 | 1 | 10 | 19 | 81 |
| Days Min Temp ≤ 0 °F | 1 | 0 | 0 | 0 | 0 | 0 | 0 | 0 | 0 | 0 | 0 | 0 | 1 |
| Heating Degree Days | 884 | 657 | 444 | 167 | 44 | 2 | *0* | 0 | 27 | 156 | 468 | 751 | 3600 |
| Cooling Degree Days | 0 | 1 | *13* | 57 | 156 | 368 | 531 | *511* | 267 | 62 | 8 | 1 | 1975 |
| Total Precipitation (") | 1.18 | 2.01 | 2.92 | 3.32 | 5.52 | 4.28 | 2.15 | 2.39 | 4.43 | 3.74 | 2.62 | 1.70 | 36.26 |
| Days ≥ 0.1" Precip | 2 | *3* | 5 | 5 | *7* | 5 | 3 | 4 | 5 | 4 | 4 | 3 | 50 |
| Total Snowfall (") | *1.0* | *1.4* | 0.4 | 0.0 | 0.0 | 0.0 | 0.0 | 0.0 | 0.0 | 0.0 | 0.2 | *0.7* | 3.7 |
| Days ≥ 1" Snow Depth | *1* | na | 0 | 0 | 0 | 0 | 0 | 0 | 0 | 0 | 0 | *0* | na |

## MIAMI *Ottawa County* ELEVATION 801 ft LAT/LONG 36° 53 ' N / 94° 53 ' W

| | JAN | FEB | MAR | APR | MAY | JUN | JUL | AUG | SEP | OCT | NOV | DEC | YEAR |
|---|---|---|---|---|---|---|---|---|---|---|---|---|---|
| Maximum Temp °F | 44.9 | 50.1 | 60.8 | 71.5 | 78.3 | 85.9 | 91.9 | 91.3 | 83.2 | 72.3 | 59.0 | 48.4 | 69.8 |
| Minimum Temp °F | 22.2 | 26.4 | 36.2 | 46.1 | 54.6 | 63.8 | 68.0 | 65.7 | 58.3 | 46.2 | 36.2 | 26.6 | 45.9 |
| Mean Temp °F | 33.6 | 38.3 | 48.5 | 58.8 | 66.5 | 74.9 | 80.0 | 78.5 | 70.8 | 59.2 | 47.6 | 37.5 | 57.9 |
| Days Max Temp ≥ 90 °F | 0 | 0 | 0 | 0 | 1 | 9 | 21 | 20 | 7 | 0 | 0 | 0 | 58 |
| Days Max Temp ≤ 32 °F | 6 | 3 | 0 | 0 | 0 | 0 | 0 | 0 | 0 | 0 | 0 | 3 | 12 |
| Days Min Temp ≤ 32 °F | 26 | 20 | 12 | 2 | 0 | 0 | 0 | 0 | 0 | 2 | 12 | 22 | 96 |
| Days Min Temp ≤ 0 °F | 1 | 1 | 0 | 0 | 0 | 0 | 0 | 0 | 0 | 0 | 0 | 1 | 3 |
| Heating Degree Days | 966 | 747 | 510 | 215 | 62 | 3 | 0 | 1 | 39 | 212 | 520 | 838 | 4113 |
| Cooling Degree Days | 0 | 0 | 6 | 34 | 110 | 311 | 470 | 441 | 221 | 33 | 3 | 0 | 1629 |
| Total Precipitation (") | 1.76 | 2.15 | 3.89 | 4.28 | 5.11 | 4.25 | 3.74 | 3.76 | 5.16 | 3.90 | 4.06 | 2.78 | 44.84 |
| Days ≥ 0.1" Precip | 4 | 4 | 6 | 6 | 8 | 6 | 5 | 5 | 6 | 5 | 5 | 4 | 64 |
| Total Snowfall (") | na | na | *1.3* | 0.0 | 0.0 | 0.0 | 0.0 | 0.0 | 0.0 | 0.0 | *0.3* | *1.6* | na |
| Days ≥ 1" Snow Depth | na | na | *0* | 0 | 0 | 0 | 0 | 0 | 0 | 0 | *0* | *2* | na |

## MUSKOGEE *Muskogee County* ELEVATION 522 ft LAT/LONG 35° 46 ' N / 95° 18 ' W

| | JAN | FEB | MAR | APR | MAY | JUN | JUL | AUG | SEP | OCT | NOV | DEC | YEAR |
|---|---|---|---|---|---|---|---|---|---|---|---|---|---|
| Maximum Temp °F | 47.6 | 53.4 | 63.9 | 73.7 | 79.8 | 87.4 | 93.4 | 92.7 | 84.6 | 74.1 | 61.2 | 50.9 | 71.9 |
| Minimum Temp °F | 26.9 | 31.3 | 40.7 | 50.2 | 58.0 | 66.4 | 70.6 | 68.9 | 61.8 | 50.2 | 39.9 | 31.0 | 49.7 |
| Mean Temp °F | 37.3 | 42.4 | 52.3 | 62.0 | 68.9 | 76.9 | 82.0 | 80.8 | 73.2 | 62.2 | 50.6 | 40.9 | 60.8 |
| Days Max Temp ≥ 90 °F | 0 | 0 | 0 | 0 | 1 | 11 | 24 | 22 | 9 | 1 | 0 | 0 | 68 |
| Days Max Temp ≤ 32 °F | 4 | 2 | 0 | 0 | 0 | 0 | 0 | 0 | 0 | 0 | 0 | 2 | 8 |
| Days Min Temp ≤ 32 °F | 22 | 16 | 7 | 1 | 0 | 0 | 0 | 0 | 0 | 1 | 7 | 17 | 71 |
| Days Min Temp ≤ 0 °F | 0 | 0 | 0 | 0 | 0 | 0 | 0 | 0 | 0 | 0 | 0 | 0 | 0 |
| Heating Degree Days | 852 | 632 | 401 | 143 | 33 | 1 | 0 | 0 | 22 | 144 | 432 | 740 | 3400 |
| Cooling Degree Days | 0 | 0 | 15 | 65 | 171 | 382 | 539 | *524* | 290 | 66 | 8 | 1 | 2061 |
| Total Precipitation (") | 1.97 | 2.41 | 3.64 | 4.27 | 5.22 | 4.52 | 2.52 | 2.93 | 4.76 | 4.44 | 3.85 | 3.08 | 43.61 |
| Days ≥ 0.1" Precip | 4 | 5 | 6 | 6 | 7 | 6 | 3 | 4 | 6 | 5 | 5 | 4 | 61 |
| Total Snowfall (") | *2.1* | *2.8* | 0.9 | 0.0 | 0.0 | 0.0 | 0.0 | 0.0 | 0.0 | 0.0 | 0.2 | *1.3* | 7.3 |
| Days ≥ 1" Snow Depth | na | *1* | 0 | 0 | 0 | 0 | 0 | 0 | 0 | 0 | 0 | *0* | na |

## MUTUAL *Woodward County* ELEVATION 1841 ft LAT/LONG 36° 14 ' N / 99° 10 ' W

| | JAN | FEB | MAR | APR | MAY | JUN | JUL | AUG | SEP | OCT | NOV | DEC | YEAR |
|---|---|---|---|---|---|---|---|---|---|---|---|---|---|
| Maximum Temp °F | 45.5 | 50.5 | 60.4 | 70.7 | 78.3 | 88.7 | 95.3 | 93.4 | 84.1 | 72.9 | 58.0 | 48.4 | 70.5 |
| Minimum Temp °F | 21.4 | 25.2 | 34.0 | 44.0 | 53.4 | 63.5 | 68.3 | 66.5 | 58.1 | 45.7 | 33.7 | 24.8 | 44.9 |
| Mean Temp °F | 33.5 | 37.9 | 47.2 | 57.4 | 65.9 | 76.2 | 81.8 | 80.0 | 71.1 | 59.3 | 45.9 | 36.6 | 57.7 |
| Days Max Temp ≥ 90 °F | 0 | 0 | 0 | 1 | 3 | 15 | 25 | 22 | 10 | 2 | 0 | 0 | 78 |
| Days Max Temp ≤ 32 °F | 6 | 4 | 1 | 0 | 0 | 0 | 0 | 0 | 0 | 0 | 1 | 4 | 16 |
| Days Min Temp ≤ 32 °F | 28 | 22 | 14 | 3 | 0 | 0 | 0 | 0 | 0 | 2 | 14 | 26 | 109 |
| Days Min Temp ≤ 0 °F | 1 | 1 | 0 | 0 | 0 | 0 | 0 | 0 | 0 | 0 | 0 | 1 | 3 |
| Heating Degree Days | 970 | 759 | 549 | 254 | 81 | 6 | 0 | 1 | 42 | 215 | 568 | 873 | 4318 |
| Cooling Degree Days | 0 | 0 | 3 | 27 | 99 | 332 | 514 | 481 | 233 | 36 | 2 | 0 | 1727 |
| Total Precipitation (") | 0.66 | 1.00 | 2.03 | 2.51 | 4.41 | 3.12 | 2.32 | 2.32 | 2.51 | 1.66 | 1.44 | 0.83 | 24.81 |
| Days ≥ 0.1" Precip | 2 | 2 | 4 | 4 | 6 | 5 | 4 | 4 | 4 | 3 | 3 | 2 | 43 |
| Total Snowfall (") | 3.2 | 3.3 | 2.3 | 0.6 | 0.0 | 0.0 | 0.0 | 0.0 | 0.0 | 0.1 | 1.1 | 2.8 | 13.4 |
| Days ≥ 1" Snow Depth | 5 | 3 | 1 | 0 | 0 | 0 | 0 | 0 | 0 | 0 | 1 | 3 | 13 |

## NEWKIRK *Kay County* ELEVATION 1152 ft LAT/LONG 36° 53 ' N / 97° 3 ' W

| | JAN | FEB | MAR | APR | MAY | JUN | JUL | AUG | SEP | OCT | NOV | DEC | YEAR |
|---|---|---|---|---|---|---|---|---|---|---|---|---|---|
| Maximum Temp °F | 43.0 | *50.2* | *60.3* | 71.2 | 78.9 | *87.8* | *93.3* | *91.7* | *84.0* | 72.8 | 57.9 | 47.0 | 69.8 |
| Minimum Temp °F | 23.4 | *27.5* | *37.0* | 47.3 | 56.8 | *66.2* | *70.6* | *68.6* | *61.3* | 49.8 | 36.9 | 27.5 | 47.7 |
| Mean Temp °F | 33.2 | *38.9* | *48.9* | 59.3 | 67.9 | *77.0* | *82.0* | *80.2* | *72.7* | 61.3 | 47.4 | 37.3 | 58.8 |
| Days Max Temp ≥ 90 °F | 0 | 0 | 0 | 0 | 2 | *12* | *23* | *20* | *7* | 1 | 0 | 0 | 65 |
| Days Max Temp ≤ 32 °F | 6 | *3* | 1 | 0 | 0 | 0 | 0 | 0 | 0 | 0 | 1 | *3* | 14 |
| Days Min Temp ≤ 32 °F | 25 | *19* | *10* | 2 | 0 | 0 | 0 | 0 | 0 | *1* | 11 | *22* | 90 |
| Days Min Temp ≤ 0 °F | 1 | 0 | 0 | 0 | 0 | 0 | 0 | 0 | 0 | 0 | 0 | 1 | 2 |
| Heating Degree Days | 964 | *722* | *501* | 209 | 52 | *2* | *0* | *0* | *26* | 171 | 525 | 851 | 4023 |
| Cooling Degree Days | 0 | *0* | *8* | 38 | *147* | *375* | *514* | *477* | *266* | *53* | *4* | 0 | 1882 |
| Total Precipitation (") | 0.91 | 1.19 | 2.33 | 3.78 | 4.90 | 4.81 | 3.63 | 3.42 | 3.97 | 3.01 | 2.48 | 1.47 | 35.90 |
| Days ≥ 0.1" Precip | 2 | *3* | 4 | 6 | *6* | *6* | 4 | *4* | *5* | 4 | *3* | 3 | 50 |
| Total Snowfall (") | na | na | *1.2* | 0.0 | 0.0 | 0.0 | 0.0 | 0.0 | 0.0 | 0.0 | 0.4 | na | na |
| Days ≥ 1" Snow Depth | na | na | *0* | 0 | 0 | 0 | 0 | 0 | 0 | 0 | *0* | na | na |

## NOWATA *Nowata County* ELEVATION 722 ft LAT/LONG 36° 42' N / 95° 38' W

| | JAN | FEB | MAR | APR | MAY | JUN | JUL | AUG | SEP | OCT | NOV | DEC | YEAR |
|---|---|---|---|---|---|---|---|---|---|---|---|---|---|
| Maximum Temp °F | 45.2 | 51.3 | 62.0 | 72.6 | 79.1 | 87.4 | 93.6 | 92.7 | 84.1 | 73.5 | 60.2 | 49.3 | 70.9 |
| Minimum Temp °F | 23.9 | 28.0 | 37.8 | 47.6 | 56.3 | 65.2 | 69.7 | 67.9 | 60.2 | 48.1 | 37.5 | 27.8 | 47.5 |
| Mean Temp °F | 34.6 | 39.7 | 49.9 | 60.1 | 67.7 | 76.3 | 81.7 | 80.3 | 72.2 | 60.7 | 48.9 | 38.6 | 59.2 |
| Days Max Temp ≥ 90 °F | 0 | 0 | 0 | 0 | 1 | 12 | 23 | 22 | 8 | 1 | 0 | 0 | 67 |
| Days Max Temp ≤ 32 °F | 5 | 2 | 0 | 0 | 0 | 0 | 0 | 0 | 0 | 0 | 0 | 2 | 9 |
| Days Min Temp ≤ 32 °F | 25 | 18 | 10 | 2 | 0 | 0 | 0 | 0 | 0 | 1 | 10 | 21 | 87 |
| Days Min Temp ≤ 0 °F | 1 | 0 | 0 | 0 | 0 | 0 | 0 | 0 | 0 | 0 | 0 | 0 | 1 |
| Heating Degree Days | 936 | 710 | 470 | 185 | 46 | 2 | 0 | 0 | 28 | 175 | 482 | 814 | 3848 |
| Cooling Degree Days | 0 | 1 | 8 | 44 | 134 | 361 | 524 | 493 | 257 | 47 | 5 | 1 | 1875 |
| Total Precipitation (") | 1.65 | 1.87 | 3.60 | 3.99 | 4.54 | 4.58 | 2.82 | 3.06 | 5.08 | 3.54 | 3.45 | 2.34 | 40.52 |
| Days ≥ 0.1" Precip | 4 | 4 | 6 | 6 | 7 | 6 | 4 | 4 | 5 | 4 | 4 | 4 | 58 |
| Total Snowfall (") | 2.5 | *3.3* | 1.7 | 0.0 | 0.0 | 0.0 | 0.0 | 0.0 | 0.0 | 0.0 | 0.5 | 1.9 | 9.9 |
| Days ≥ 1" Snow Depth | na | na | 1 | 0 | 0 | 0 | 0 | 0 | 0 | 0 | 0 | 1 | na |

## OKEENE *Blaine County* ELEVATION 1201 ft LAT/LONG 36° 7' N / 98° 19' W

| | JAN | FEB | MAR | APR | MAY | JUN | JUL | AUG | SEP | OCT | NOV | DEC | YEAR |
|---|---|---|---|---|---|---|---|---|---|---|---|---|---|
| Maximum Temp °F | 48.7 | 54.9 | 64.3 | 74.3 | 81.7 | 91.2 | 96.4 | 95.1 | 86.5 | 75.9 | 61.0 | 51.3 | 73.4 |
| Minimum Temp °F | 23.8 | 28.0 | 36.8 | 46.7 | 55.8 | 65.4 | 70.2 | 68.1 | 60.4 | 48.7 | 36.5 | 27.4 | 47.3 |
| Mean Temp °F | 36.3 | 41.5 | 50.6 | 60.5 | 68.7 | 78.3 | 83.3 | 81.6 | 73.5 | 62.3 | 48.7 | 39.4 | 60.4 |
| Days Max Temp ≥ 90 °F | 0 | 0 | 0 | 1 | 5 | 19 | 27 | 25 | 12 | 2 | 0 | 0 | 91 |
| Days Max Temp ≤ 32 °F | 4 | 2 | 0 | 0 | 0 | 0 | 0 | 0 | 0 | 0 | 0 | 2 | 8 |
| Days Min Temp ≤ 32 °F | 26 | 19 | 11 | 2 | 0 | 0 | 0 | 0 | 0 | 1 | 10 | 23 | 92 |
| Days Min Temp ≤ 0 °F | 0 | 0 | 0 | 0 | 0 | 0 | 0 | 0 | 0 | 0 | 0 | 1 | 1 |
| Heating Degree Days | 885 | 657 | 449 | 179 | 42 | 1 | 0 | 0 | 23 | 148 | 483 | 788 | 3655 |
| Cooling Degree Days | 0 | 0 | 6 | 47 | 161 | 407 | 572 | 539 | 293 | 58 | 4 | 0 | 2087 |
| Total Precipitation (") | 0.81 | 1.25 | 2.24 | 2.63 | 4.86 | 3.92 | 2.36 | 2.80 | 3.43 | 2.47 | 1.88 | 1.16 | 29.81 |
| Days ≥ 0.1" Precip | 2 | 3 | 4 | 5 | 6 | 5 | 4 | 4 | 5 | 3 | 3 | 3 | 47 |
| Total Snowfall (") | *2.9* | na | 1.2 | 0.3 | 0.0 | 0.0 | 0.0 | 0.0 | 0.0 | 0.2 | 0.5 | *2.8* | na |
| Days ≥ 1" Snow Depth | na | na | 0 | 0 | 0 | 0 | 0 | 0 | 0 | 0 | 0 | *1* | na |

## OKEMAH *Okfuskee County* ELEVATION 935 ft LAT/LONG 35° 26' N / 96° 18' W

| | JAN | FEB | MAR | APR | MAY | JUN | JUL | AUG | SEP | OCT | NOV | DEC | YEAR |
|---|---|---|---|---|---|---|---|---|---|---|---|---|---|
| Maximum Temp °F | 48.0 | 53.7 | 63.7 | 73.6 | 79.7 | 87.5 | 93.4 | 93.0 | 84.7 | 74.5 | 61.3 | 51.6 | 72.1 |
| Minimum Temp °F | 27.6 | 31.6 | 40.6 | 50.2 | 58.0 | 66.2 | 70.1 | 68.8 | 62.1 | 51.3 | 40.5 | 31.5 | 49.9 |
| Mean Temp °F | 37.8 | 42.7 | 52.2 | 61.9 | 68.9 | 76.9 | 81.8 | 80.9 | 73.4 | 62.9 | 50.9 | 41.6 | 61.0 |
| Days Max Temp ≥ 90 °F | 0 | 0 | 0 | 0 | 1 | 12 | 24 | 22 | 9 | 1 | 0 | 0 | 69 |
| Days Max Temp ≤ 32 °F | 4 | 2 | 0 | 0 | 0 | 0 | 0 | 0 | 0 | 0 | 0 | 2 | 8 |
| Days Min Temp ≤ 32 °F | 21 | 15 | 7 | 1 | 0 | 0 | 0 | 0 | 0 | 0 | 7 | 16 | 67 |
| Days Min Temp ≤ 0 °F | 0 | 0 | 0 | 0 | 0 | 0 | 0 | 0 | 0 | 0 | 0 | 0 | 0 |
| Heating Degree Days | 837 | 623 | 403 | 141 | 32 | 1 | 0 | 0 | 20 | 130 | 422 | 719 | 3328 |
| Cooling Degree Days | 0 | 1 | 11 | 58 | 165 | 379 | 534 | 524 | 290 | 67 | 8 | 1 | 2038 |
| Total Precipitation (") | 1.59 | 2.01 | 3.15 | 4.07 | 5.38 | 4.39 | 2.93 | 2.81 | 4.89 | 3.97 | 3.27 | 2.32 | 40.78 |
| Days ≥ 0.1" Precip | 3 | 4 | 5 | 6 | 7 | 6 | 4 | 4 | 5 | 4 | 4 | 4 | 56 |
| Total Snowfall (") | *1.6* | *1.5* | 1.1 | 0.0 | 0.0 | 0.0 | 0.0 | 0.0 | 0.0 | 0.0 | 0.4 | *0.9* | 5.5 |
| Days ≥ 1" Snow Depth | 3 | 2 | 1 | 0 | 0 | 0 | 0 | 0 | 0 | 0 | 0 | 0 | 6 |

## OKLAHOMA CITY ROGERS *Oklahoma County* ELEVATION 1280 ft LAT/LONG 35° 24' N / 97° 36' W

| | JAN | FEB | MAR | APR | MAY | JUN | JUL | AUG | SEP | OCT | NOV | DEC | YEAR |
|---|---|---|---|---|---|---|---|---|---|---|---|---|---|
| Maximum Temp °F | 46.6 | 52.1 | 62.2 | 71.9 | 78.6 | 87.4 | 93.2 | 92.1 | 83.6 | 73.0 | 59.9 | 50.4 | 70.9 |
| Minimum Temp °F | 25.9 | 29.9 | 38.9 | 48.8 | 57.6 | 66.4 | 70.8 | 69.5 | 62.0 | 50.1 | 38.2 | 29.3 | 48.9 |
| Mean Temp °F | 36.2 | 41.1 | 50.6 | 60.4 | 68.1 | 76.9 | 82.0 | 80.8 | 72.8 | 61.6 | 49.1 | 39.9 | 60.0 |
| Days Max Temp ≥ 90 °F | 0 | 0 | 0 | 0 | 2 | 12 | 22 | 22 | 8 | 1 | 0 | 0 | 67 |
| Days Max Temp ≤ 32 °F | 5 | 2 | 0 | 0 | 0 | 0 | 0 | 0 | 0 | 0 | 0 | 2 | 9 |
| Days Min Temp ≤ 32 °F | 23 | 17 | 8 | 1 | 0 | 0 | 0 | 0 | 0 | 1 | 8 | 19 | 77 |
| Days Min Temp ≤ 0 °F | 0 | 0 | 0 | 0 | 0 | 0 | 0 | 0 | 0 | 0 | 0 | 0 | 0 |
| Heating Degree Days | 885 | 669 | 449 | 177 | 43 | 1 | 0 | 0 | 25 | 161 | 476 | 772 | 3658 |
| Cooling Degree Days | 0 | 0 | 8 | 45 | 154 | 378 | 539 | 517 | 273 | 54 | 5 | 0 | 1973 |
| Total Precipitation (") | 1.18 | 1.62 | 2.72 | 2.90 | 5.66 | 4.35 | 2.43 | 2.63 | 4.13 | 3.28 | 1.99 | 1.63 | 34.52 |
| Days ≥ 0.1" Precip | 2 | 4 | 5 | 5 | 7 | 6 | 4 | 4 | 5 | 4 | 3 | 3 | 52 |
| Total Snowfall (") | 2.7 | 2.4 | 1.3 | 0.0 | 0.0 | 0.0 | 0.0 | 0.0 | 0.0 | 0.0 | 0.7 | 1.8 | 8.9 |
| Days ≥ 1" Snow Depth | 3 | 2 | 0 | 0 | 0 | 0 | 0 | 0 | 0 | 0 | 0 | 2 | 7 |

## OKMULGEE WATER WORKS *Okmulgee County* ELEVATION 715 ft LAT/LONG 35° 40 ' N / 95° 57 ' W

| | JAN | FEB | MAR | APR | MAY | JUN | JUL | AUG | SEP | OCT | NOV | DEC | YEAR |
|---|---|---|---|---|---|---|---|---|---|---|---|---|---|
| Maximum Temp °F | 48.7 | 54.7 | 65.1 | 74.4 | 80.1 | 87.7 | 93.6 | 93.4 | 85.3 | 75.8 | 62.4 | 53.2 | 72.9 |
| Minimum Temp °F | 24.5 | 29.0 | 39.1 | 48.8 | 56.8 | 65.4 | 68.3 | 66.5 | 59.5 | 47.8 | 37.7 | 29.1 | 47.7 |
| Mean Temp °F | 36.6 | 41.8 | 52.1 | 61.6 | 68.5 | 76.6 | 80.8 | 79.9 | 72.4 | 61.8 | 50.1 | 41.2 | 60.3 |
| Days Max Temp ≥ 90 °F | 0 | 0 | 0 | 0 | 2 | 13 | 25 | 24 | 10 | 2 | 0 | 0 | 76 |
| Days Max Temp ≤ 32 °F | 4 | 2 | 0 | 0 | 0 | 0 | 0 | 0 | 0 | 0 | 0 | 2 | 8 |
| Days Min Temp ≤ 32 °F | 24 | 18 | 9 | 1 | 0 | 0 | 0 | 0 | 0 | 2 | 10 | 20 | 84 |
| Days Min Temp ≤ 0 °F | 0 | 0 | 0 | 0 | 0 | 0 | 0 | 0 | 0 | 0 | 0 | 0 | 0 |
| Heating Degree Days | 873 | 652 | 406 | 155 | 39 | 1 | 0 | 0 | 27 | 157 | 448 | 731 | 3489 |
| Cooling Degree Days | 0 | 1 | 9 | 46 | 143 | 356 | 488 | 477 | 250 | 54 | 7 | 1 | 1832 |
| Total Precipitation (") | 1.69 | 2.21 | 3.54 | 4.36 | 5.52 | 4.36 | 2.51 | 2.96 | 4.31 | 4.12 | 3.58 | 2.54 | 41.70 |
| Days ≥ 0.1" Precip | 3 | 4 | 5 | 6 | 7 | 6 | 4 | 4 | 5 | 4 | 5 | 4 | 57 |
| Total Snowfall (") | na | na | *1.4* | 0.0 | 0.0 | 0.0 | 0.0 | 0.0 | 0.0 | 0.0 | *0.1* | 0.4 | na |
| Days ≥ 1" Snow Depth | na | na | *0* | 0 | 0 | 0 | 0 | 0 | 0 | 0 | 0 | *0* | na |

## PAULS VALLEY 4 WSW *Garvin County* ELEVATION 879 ft LAT/LONG 34° 45 ' N / 97° 13 ' W

| | JAN | FEB | MAR | APR | MAY | JUN | JUL | AUG | SEP | OCT | NOV | DEC | YEAR |
|---|---|---|---|---|---|---|---|---|---|---|---|---|---|
| Maximum Temp °F | 51.5 | 56.9 | 66.3 | 75.9 | 81.9 | 89.5 | 94.9 | 94.5 | 86.3 | 76.7 | 63.8 | 54.5 | 74.4 |
| Minimum Temp °F | 28.1 | 31.7 | 40.6 | 50.4 | 58.3 | 66.9 | 70.9 | 69.3 | 62.3 | 50.3 | 39.7 | 31.2 | 50.0 |
| Mean Temp °F | 39.8 | 44.3 | 53.4 | 63.2 | 70.1 | 78.2 | 82.9 | 81.9 | 74.3 | 63.5 | 51.8 | 42.9 | 62.2 |
| Days Max Temp ≥ 90 °F | 0 | 0 | 0 | 1 | 3 | 17 | 26 | 25 | 11 | 2 | 0 | 0 | 85 |
| Days Max Temp ≤ 32 °F | 2 | 1 | 0 | 0 | 0 | 0 | 0 | 0 | 0 | 0 | 0 | 1 | 4 |
| Days Min Temp ≤ 32 °F | 21 | 15 | 7 | 1 | 0 | 0 | 0 | 0 | 0 | 1 | 8 | 18 | 71 |
| Days Min Temp ≤ 0 °F | 0 | 0 | 0 | 0 | 0 | 0 | 0 | 0 | 0 | 0 | 0 | 0 | 0 |
| Heating Degree Days | 772 | 578 | 371 | 123 | 25 | 1 | 0 | 0 | 19 | 121 | 400 | 681 | 3091 |
| Cooling Degree Days | 0 | 1 | 16 | 67 | 177 | 390 | 545 | 536 | 302 | 74 | 12 | 1 | 2121 |
| Total Precipitation (") | 1.64 | 1.96 | 2.74 | 3.48 | 6.05 | 4.07 | 2.51 | 2.14 | 4.43 | 3.92 | 2.58 | 1.92 | 37.44 |
| Days ≥ 0.1" Precip | 3 | 4 | 4 | 5 | 7 | 6 | 4 | 4 | 5 | 4 | 4 | 4 | 54 |
| Total Snowfall (") | 2.4 | *2.0* | 1.2 | 0.0 | 0.0 | 0.0 | 0.0 | 0.0 | 0.0 | 0.0 | 0.3 | 1.4 | 7.3 |
| Days ≥ 1" Snow Depth | *2* | *1* | 0 | 0 | 0 | 0 | 0 | 0 | 0 | 0 | 0 | *0* | 3 |

## PAWHUSKA *Osage County* ELEVATION 840 ft LAT/LONG 36° 40 ' N / 96° 21 ' W

| | JAN | FEB | MAR | APR | MAY | JUN | JUL | AUG | SEP | OCT | NOV | DEC | YEAR |
|---|---|---|---|---|---|---|---|---|---|---|---|---|---|
| Maximum Temp °F | 46.0 | 52.3 | 63.0 | 73.4 | 79.3 | 87.1 | 93.0 | 92.4 | 84.2 | 73.9 | 59.9 | 49.7 | 71.2 |
| Minimum Temp °F | 22.7 | 27.1 | 37.0 | 47.5 | 55.9 | 65.1 | 69.7 | 67.4 | 59.9 | 47.6 | 36.5 | 26.9 | 46.9 |
| Mean Temp °F | 34.4 | 39.7 | 50.0 | 60.5 | 67.6 | 76.1 | 81.4 | 80.0 | 72.1 | 60.8 | 48.3 | 38.3 | 59.1 |
| Days Max Temp ≥ 90 °F | 0 | 0 | 0 | 1 | 1 | 10 | 23 | 22 | 8 | 1 | 0 | 0 | 66 |
| Days Max Temp ≤ 32 °F | 5 | 2 | 0 | 0 | 0 | 0 | 0 | 0 | 0 | 0 | 0 | 2 | 9 |
| Days Min Temp ≤ 32 °F | 25 | 20 | 12 | 2 | 0 | 0 | 0 | 0 | 0 | 2 | 12 | 22 | 95 |
| Days Min Temp ≤ 0 °F | 1 | 1 | 0 | 0 | 0 | 0 | 0 | 0 | 0 | 0 | 0 | 1 | 3 |
| Heating Degree Days | 941 | 709 | 468 | 184 | 50 | 2 | 0 | 0 | 29 | 182 | 501 | 820 | 3886 |
| Cooling Degree Days | 0 | 0 | 11 | 52 | 141 | 352 | 521 | 489 | 263 | 52 | 6 | 1 | 1888 |
| Total Precipitation (") | 1.36 | 1.99 | 3.06 | 4.01 | 5.21 | 4.78 | 3.44 | 3.33 | 4.70 | 3.31 | 3.06 | 1.90 | 40.15 |
| Days ≥ 0.1" Precip | 3 | 4 | 5 | 6 | 8 | 6 | 5 | 5 | 5 | 5 | 4 | 4 | 60 |
| Total Snowfall (") | 3.4 | 3.0 | 1.2 | 0.0 | 0.0 | 0.0 | 0.0 | 0.0 | 0.0 | 0.0 | 0.5 | 2.1 | 10.2 |
| Days ≥ 1" Snow Depth | 4 | 3 | 0 | 0 | 0 | 0 | 0 | 0 | 0 | 0 | 0 | 2 | 9 |

## PERRY *Noble County* ELEVATION 1030 ft LAT/LONG 36° 17 ' N / 97° 18 ' W

| | JAN | FEB | MAR | APR | MAY | JUN | JUL | AUG | SEP | OCT | NOV | DEC | YEAR |
|---|---|---|---|---|---|---|---|---|---|---|---|---|---|
| Maximum Temp °F | 47.2 | 53.3 | 63.6 | 74.3 | 81.0 | 89.3 | 94.9 | 94.0 | 85.7 | 75.3 | 60.8 | 50.7 | 72.5 |
| Minimum Temp °F | 24.6 | 28.8 | 38.0 | 48.3 | 57.1 | 66.3 | 70.7 | 69.2 | 61.4 | 49.6 | 37.7 | 28.3 | 48.3 |
| Mean Temp °F | 36.0 | 41.1 | 50.8 | 61.3 | 69.1 | 77.8 | 82.8 | 81.6 | 73.6 | 62.5 | 49.4 | 39.5 | 60.5 |
| Days Max Temp ≥ 90 °F | 0 | 0 | 0 | 1 | 3 | 16 | 26 | 25 | 11 | 1 | 0 | 0 | 83 |
| Days Max Temp ≤ 32 °F | 5 | 2 | 0 | 0 | 0 | 0 | 0 | 0 | 0 | 0 | 0 | 2 | 9 |
| Days Min Temp ≤ 32 °F | 24 | 18 | 10 | 2 | 0 | 0 | 0 | 0 | 0 | 1 | 10 | 20 | 85 |
| Days Min Temp ≤ 0 °F | 0 | 0 | 0 | 0 | 0 | 0 | 0 | 0 | 0 | 0 | 0 | 1 | 1 |
| Heating Degree Days | 894 | 668 | 446 | 165 | 39 | 2 | 0 | 0 | 22 | 144 | 468 | 782 | 3630 |
| Cooling Degree Days | 0 | 0 | 12 | 61 | 178 | 405 | 558 | 539 | 299 | 63 | 7 | 0 | 2122 |
| Total Precipitation (") | 1.01 | 1.55 | 2.70 | 3.28 | 5.28 | 3.93 | 2.74 | 3.24 | 4.27 | 2.66 | 2.25 | 1.40 | 34.31 |
| Days ≥ 0.1" Precip | 2 | 3 | 5 | 5 | 7 | 6 | 4 | 5 | 5 | 4 | 4 | 3 | 53 |
| Total Snowfall (") | *1.5* | 1.9 | 1.3 | 0.0 | 0.0 | 0.0 | 0.0 | 0.0 | 0.0 | 0.0 | 0.4 | na | na |
| Days ≥ 1" Snow Depth | *1* | na | 0 | 0 | 0 | 0 | 0 | 0 | 0 | 0 | 0 | *1* | na |

## PONCA CITY MUNI AP *Kay County* ELEVATION 997 ft LAT/LONG 36° 44 ' N / 97° 6 ' W

| | JAN | FEB | MAR | APR | MAY | JUN | JUL | AUG | SEP | OCT | NOV | DEC | YEAR |
|---|---|---|---|---|---|---|---|---|---|---|---|---|---|
| Maximum Temp °F | 42.9 | 49.1 | 60.0 | 71.0 | 78.5 | 87.8 | 93.8 | 92.3 | 83.1 | 72.1 | 57.6 | 46.9 | 69.6 |
| Minimum Temp °F | 23.1 | 27.1 | 36.7 | 47.1 | 56.6 | 66.4 | 71.2 | 69.5 | 61.2 | 48.7 | 36.6 | 27.0 | 47.6 |
| Mean Temp °F | 33.0 | 38.2 | 48.4 | 59.1 | 67.5 | 77.1 | 82.5 | 80.9 | 72.2 | 60.4 | 47.1 | 36.9 | 58.6 |
| Days Max Temp ≥ 90 °F | 0 | 0 | 0 | 1 | 2 | 13 | 23 | 21 | 8 | 1 | 0 | 0 | 69 |
| Days Max Temp ≤ 32 °F | 7 | 4 | 1 | 0 | 0 | 0 | 0 | 0 | 0 | 0 | 1 | 4 | 17 |
| Days Min Temp ≤ 32 °F | 26 | 19 | 12 | 2 | 0 | 0 | 0 | 0 | 0 | 1 | 12 | 23 | 95 |
| Days Min Temp ≤ 0 °F | 1 | 1 | 0 | 0 | 0 | 0 | 0 | 0 | 0 | 0 | 0 | 1 | 3 |
| Heating Degree Days | 984 | 751 | 516 | 217 | 60 | 3 | 0 | 0 | 34 | 193 | 535 | 863 | 4156 |
| Cooling Degree Days | 0 | 0 | 9 | 50 | 159 | 396 | 574 | 539 | 288 | 55 | 5 | 0 | 2075 |
| Total Precipitation (") | 1.09 | 1.37 | 2.57 | 3.32 | 4.77 | 4.10 | 3.48 | 3.28 | 3.97 | 2.79 | 2.33 | 1.50 | 34.57 |
| Days ≥ 0.1" Precip | 2 | 3 | 5 | 5 | 7 | 6 | 5 | 4 | 5 | 4 | 3 | 3 | 52 |
| Total Snowfall (") | 2.2 | 2.9 | 1.1 | 0.0 | 0.0 | 0.0 | 0.0 | 0.0 | 0.0 | 0.0 | 0.3 | 1.7 | 8.2 |
| Days ≥ 1" Snow Depth | 4 | 3 | 0 | 0 | 0 | 0 | 0 | 0 | 0 | 0 | 0 | 2 | 9 |

## PRYOR 6 N *Mayes County* ELEVATION 640 ft LAT/LONG 36° 24 ' N / 95° 18 ' W

| | JAN | FEB | MAR | APR | MAY | JUN | JUL | AUG | SEP | OCT | NOV | DEC | YEAR |
|---|---|---|---|---|---|---|---|---|---|---|---|---|---|
| Maximum Temp °F | 45.9 | 50.9 | 61.4 | 72.3 | 78.5 | 86.4 | 92.6 | ***91.9*** | 83.4 | ***73.0*** | 60.2 | 49.4 | 70.5 |
| Minimum Temp °F | 22.5 | 26.8 | 36.3 | 46.7 | 55.3 | 64.2 | 68.8 | ***67.0*** | 59.0 | ***46.1*** | 36.0 | 27.0 | 46.3 |
| Mean Temp °F | 34.4 | 39.1 | 48.9 | 59.4 | 66.9 | 75.3 | 80.6 | ***79.3*** | 71.3 | ***59.6*** | 48.2 | 38.2 | 58.4 |
| Days Max Temp ≥ 90 °F | 0 | 0 | 0 | 0 | 0 | ***9*** | ***22*** | ***19*** | ***7*** | 1 | 0 | 0 | 58 |
| Days Max Temp ≤ 32 °F | 5 | 3 | 0 | 0 | 0 | 0 | 0 | 0 | 0 | 0 | 0 | 3 | 11 |
| Days Min Temp ≤ 32 °F | 26 | ***20*** | ***13*** | 2 | 0 | 0 | 0 | 0 | 0 | 2 | ***11*** | 22 | 96 |
| Days Min Temp ≤ 0 °F | 1 | 0 | 0 | 0 | 0 | 0 | 0 | 0 | 0 | 0 | 0 | 1 | 2 |
| Heating Degree Days | 944 | 727 | 502 | 200 | 57 | 2 | 0 | ***1*** | 38 | ***206*** | 502 | 825 | 4004 |
| Cooling Degree Days | 0 | 0 | 6 | 27 | 114 | 325 | 474 | ***444*** | 220 | ***29*** | 4 | 0 | 1643 |
| Total Precipitation (") | 1.90 | 2.07 | 3.46 | 4.08 | 4.63 | 4.63 | 2.87 | 3.41 | 4.97 | 3.97 | 3.77 | 2.54 | 42.30 |
| Days ≥ 0.1" Precip | 3 | 4 | 5 | 6 | 7 | 6 | 4 | 5 | 6 | 5 | ***5*** | 4 | 60 |
| Total Snowfall (") | ***1.9*** | na | 1.6 | 0.0 | 0.0 | 0.0 | 0.0 | 0.0 | 0.0 | 0.0 | ***0.4*** | ***1.8*** | na |
| Days ≥ 1" Snow Depth | ***4*** | na | 1 | 0 | 0 | 0 | 0 | 0 | 0 | 0 | ***0*** | ***2*** | na |

## PURCELL 5 SW *McClain County* ELEVATION 1043 ft LAT/LONG 34° 58 ' N / 97° 26 ' W

| | JAN | FEB | MAR | APR | MAY | JUN | JUL | AUG | SEP | OCT | NOV | DEC | YEAR |
|---|---|---|---|---|---|---|---|---|---|---|---|---|---|
| Maximum Temp °F | 49.4 | 55.3 | 65.5 | 74.9 | 81.2 | 89.0 | 94.8 | 94.3 | 86.0 | 75.8 | 62.5 | 52.9 | 73.5 |
| Minimum Temp °F | 25.8 | 30.1 | 39.5 | 49.3 | 57.8 | 66.4 | 70.5 | 68.8 | 61.7 | 49.6 | 38.2 | 29.4 | 48.9 |
| Mean Temp °F | 37.6 | 42.7 | 52.5 | 62.1 | 69.5 | 77.7 | 82.6 | 81.6 | 73.9 | 62.7 | 50.4 | 41.2 | 61.2 |
| Days Max Temp ≥ 90 °F | 0 | 0 | 0 | 1 | 3 | 15 | 25 | 25 | 11 | 2 | 0 | 0 | 82 |
| Days Max Temp ≤ 32 °F | 3 | 2 | 0 | 0 | 0 | 0 | 0 | 0 | 0 | 0 | 0 | 2 | 7 |
| Days Min Temp ≤ 32 °F | 23 | 17 | 9 | 1 | 0 | 0 | 0 | 0 | 0 | 1 | 9 | 20 | 80 |
| Days Min Temp ≤ 0 °F | 0 | 0 | 0 | 0 | 0 | 0 | 0 | 0 | 0 | 0 | 0 | 0 | 0 |
| Heating Degree Days | 842 | 624 | 395 | 142 | 31 | 1 | 0 | 0 | 20 | 140 | 439 | 732 | 3366 |
| Cooling Degree Days | 0 | 1 | 12 | 57 | 173 | 381 | 542 | 529 | 295 | 67 | 7 | 0 | 2064 |
| Total Precipitation (") | 1.49 | 2.13 | 3.15 | 3.70 | 5.98 | 4.14 | 2.68 | 2.81 | 4.53 | 4.10 | 2.47 | 2.24 | 39.42 |
| Days ≥ 0.1" Precip | 3 | 4 | 5 | 5 | 8 | 6 | 4 | 4 | 6 | 5 | 4 | 4 | 58 |
| Total Snowfall (") | 2.9 | 2.1 | 1.0 | 0.0 | 0.0 | 0.0 | 0.0 | 0.0 | 0.0 | 0.0 | 0.5 | 1.3 | 7.8 |
| Days ≥ 1" Snow Depth | 4 | 2 | 1 | 0 | 0 | 0 | 0 | 0 | 0 | 0 | 0 | 1 | 8 |

## RALSTON *Pawnee County* ELEVATION 630 ft LAT/LONG 36° 30 ' N / 96° 44 ' W

| | JAN | FEB | MAR | APR | MAY | JUN | JUL | AUG | SEP | OCT | NOV | DEC | YEAR |
|---|---|---|---|---|---|---|---|---|---|---|---|---|---|
| Maximum Temp °F | 46.7 | 52.9 | 63.3 | 73.8 | 80.5 | 88.3 | 94.3 | 93.5 | 85.2 | 74.9 | 60.8 | 50.4 | 72.1 |
| Minimum Temp °F | 23.5 | 27.7 | 37.5 | 47.7 | 56.3 | 65.4 | 69.2 | 67.4 | 59.9 | 47.5 | 36.6 | 27.0 | 47.1 |
| Mean Temp °F | 35.1 | 40.3 | 50.4 | 60.8 | 68.4 | 76.9 | 81.8 | 80.4 | 72.6 | 61.2 | 48.7 | 38.8 | 59.6 |
| Days Max Temp ≥ 90 °F | 0 | 0 | 0 | 1 | 3 | 14 | 26 | 23 | 10 | 1 | 0 | 0 | 78 |
| Days Max Temp ≤ 32 °F | 5 | 2 | 0 | 0 | 0 | 0 | 0 | 0 | 0 | 0 | 0 | 2 | 9 |
| Days Min Temp ≤ 32 °F | 25 | 19 | 11 | 2 | 0 | 0 | 0 | 0 | 0 | 2 | 11 | 22 | 92 |
| Days Min Temp ≤ 0 °F | 1 | 0 | 0 | 0 | 0 | 0 | 0 | 0 | 0 | 0 | 0 | 1 | 2 |
| Heating Degree Days | 918 | 691 | 458 | 176 | 42 | 1 | 0 | 0 | 27 | 171 | 486 | 807 | 3777 |
| Cooling Degree Days | 0 | 0 | 10 | 54 | 163 | 378 | 539 | 508 | 278 | 60 | 6 | 0 | 1996 |
| Total Precipitation (") | 1.25 | 1.72 | 3.02 | 3.52 | 5.14 | 4.28 | 2.78 | 3.39 | 4.41 | 2.89 | 2.54 | 1.71 | 36.65 |
| Days ≥ 0.1" Precip | 3 | 4 | 5 | 6 | 7 | 7 | 5 | 5 | 5 | 4 | 4 | 3 | 58 |
| Total Snowfall (") | 2.9 | 3.4 | 1.7 | 0.1 | 0.0 | 0.0 | 0.0 | 0.0 | 0.0 | 0.0 | 0.7 | 2.1 | 10.9 |
| Days ≥ 1" Snow Depth | 4 | 4 | 1 | 0 | 0 | 0 | 0 | 0 | 0 | 0 | 0 | 2 | 11 |

## REYDON *Roger Mills County* ELEVATION 2200 ft LAT/LONG 35° 39 ' N / 99° 55 ' W

| | JAN | FEB | MAR | APR | MAY | JUN | JUL | AUG | SEP | OCT | NOV | DEC | YEAR |
|---|---|---|---|---|---|---|---|---|---|---|---|---|---|
| Maximum Temp °F | 48.1 | 53.7 | 62.6 | 73.3 | 79.7 | 88.0 | 93.6 | 91.5 | 83.8 | 74.1 | 60.1 | 50.5 | 71.6 |
| Minimum Temp °F | 22.9 | 27.1 | 35.5 | 45.7 | 54.3 | 63.2 | 67.7 | 65.8 | 58.3 | 46.7 | 34.8 | 25.9 | 45.7 |
| Mean Temp °F | 35.5 | 40.5 | 49.0 | 59.5 | 67.0 | 75.6 | 80.7 | 78.7 | 71.1 | 60.4 | 47.5 | 38.2 | 58.6 |
| Days Max Temp ≥ 90 °F | 0 | 0 | 0 | 1 | 4 | *14* | 23 | *20* | 9 | 1 | 0 | 0 | 72 |
| Days Max Temp ≤ 32 °F | 5 | 2 | 1 | 0 | 0 | 0 | 0 | 0 | 0 | 0 | 0 | 3 | 11 |
| Days Min Temp ≤ 32 °F | 26 | 19 | 12 | 2 | 0 | 0 | 0 | 0 | 0 | 2 | 11 | 23 | 95 |
| Days Min Temp ≤ 0 °F | 1 | 0 | 0 | 0 | 0 | 0 | 0 | 0 | 0 | 0 | 0 | 1 | 2 |
| Heating Degree Days | 907 | 686 | 495 | 198 | 63 | 4 | 0 | 1 | 34 | 180 | 521 | 823 | 3912 |
| Cooling Degree Days | 0 | 0 | 5 | 52 | 142 | *337* | 506 | 473 | 251 | 48 | *3* | 0 | 1817 |
| Total Precipitation (") | 0.55 | 0.90 | 1.64 | 2.21 | 4.04 | 3.47 | 1.74 | 2.47 | 2.84 | 1.65 | 1.03 | 0.67 | 23.21 |
| Days ≥ 0.1" Precip | 2 | 2 | 3 | 4 | 6 | 5 | 3 | 4 | 4 | 3 | 2 | 2 | 40 |
| Total Snowfall (") | *3.3* | *3.4* | 1.9 | 0.3 | 0.0 | 0.0 | 0.0 | 0.0 | 0.0 | 0.0 | 1.2 | 2.0 | 12.1 |
| Days ≥ 1" Snow Depth | 2 | *2* | 1 | 0 | 0 | 0 | 0 | 0 | 0 | 0 | 0 | *1* | 6 |

## SALLISAW 2 NE *Sequoyah County* ELEVATION 531 ft LAT/LONG 35° 28 ' N / 94° 47 ' W

| | JAN | FEB | MAR | APR | MAY | JUN | JUL | AUG | SEP | OCT | NOV | DEC | YEAR |
|---|---|---|---|---|---|---|---|---|---|---|---|---|---|
| Maximum Temp °F | 49.4 | 55.0 | 64.6 | 74.6 | 80.6 | 88.2 | 93.7 | 92.7 | 85.5 | 75.4 | 62.4 | 52.5 | 72.9 |
| Minimum Temp °F | 26.9 | 30.9 | 39.9 | 49.4 | 57.6 | 65.9 | 69.7 | 68.5 | 61.7 | 49.9 | 39.4 | 30.6 | 49.2 |
| Mean Temp °F | 38.3 | 43.0 | 52.3 | 62.1 | 69.1 | 77.1 | 81.7 | 80.6 | 73.6 | 62.7 | 50.9 | 41.5 | 61.1 |
| Days Max Temp ≥ 90 °F | 0 | 0 | 0 | 0 | 1 | 13 | 24 | 23 | 9 | 1 | 0 | 0 | 71 |
| Days Max Temp ≤ 32 °F | 3 | 1 | 0 | 0 | 0 | 0 | 0 | 0 | 0 | 0 | 0 | 1 | 5 |
| Days Min Temp ≤ 32 °F | 23 | 16 | 7 | 1 | 0 | 0 | 0 | 0 | 0 | 1 | 8 | 18 | 74 |
| Days Min Temp ≤ 0 °F | 0 | 0 | 0 | 0 | 0 | 0 | 0 | 0 | 0 | 0 | 0 | 0 | 0 |
| Heating Degree Days | 821 | 616 | 398 | 141 | 29 | 1 | 0 | 0 | 18 | 133 | 423 | 722 | 3302 |
| Cooling Degree Days | 0 | 0 | 9 | 50 | 157 | 371 | 519 | 501 | 284 | 58 | 6 | 1 | 1956 |
| Total Precipitation (") | 2.05 | 2.79 | 4.15 | 4.61 | 5.90 | 3.92 | 2.59 | 3.36 | 4.52 | 4.72 | 4.36 | 2.97 | 45.94 |
| Days ≥ 0.1" Precip | 3 | 4 | 6 | 6 | 7 | 6 | 4 | 4 | 6 | 5 | 5 | 5 | 61 |
| Total Snowfall (") | 2.3 | 2.0 | 0.8 | 0.0 | 0.0 | 0.0 | 0.0 | 0.0 | 0.0 | 0.0 | 0.4 | 1.1 | 6.6 |
| Days ≥ 1" Snow Depth | *2* | 1 | 0 | 0 | 0 | 0 | 0 | 0 | 0 | 0 | 0 | 1 | 4 |

## SEMINOLE *Seminole County* ELEVATION 1010 ft LAT/LONG 35° 16 ' N / 96° 40 ' W

| | JAN | FEB | MAR | APR | MAY | JUN | JUL | AUG | SEP | OCT | NOV | DEC | YEAR |
|---|---|---|---|---|---|---|---|---|---|---|---|---|---|
| Maximum Temp °F | 50.2 | 56.5 | 66.2 | 76.0 | 81.8 | 89.5 | 95.2 | 94.8 | 86.6 | 76.7 | 63.4 | 53.7 | 74.2 |
| Minimum Temp °F | 27.8 | 32.0 | 41.1 | 50.8 | 58.7 | 67.1 | 71.4 | 69.7 | 62.4 | 51.2 | 40.5 | 31.6 | 50.4 |
| Mean Temp °F | 39.1 | 44.2 | 53.6 | 63.4 | 70.3 | 78.4 | 83.3 | 82.3 | 74.4 | 64.0 | 52.0 | 42.6 | 62.3 |
| Days Max Temp ≥ 90 °F | 0 | 0 | 0 | 1 | 3 | 16 | 27 | 26 | 12 | 2 | 0 | 0 | 87 |
| Days Max Temp ≤ 32 °F | 3 | 1 | 0 | 0 | 0 | 0 | 0 | 0 | 0 | 0 | 0 | 2 | 6 |
| Days Min Temp ≤ 32 °F | 21 | 15 | 7 | 1 | 0 | 0 | 0 | 0 | 0 | 0 | 7 | 17 | 68 |
| Days Min Temp ≤ 0 °F | 0 | 0 | 0 | 0 | 0 | 0 | 0 | 0 | 0 | 0 | 0 | 0 | 0 |
| Heating Degree Days | 797 | 580 | 363 | 116 | 24 | 1 | 0 | 0 | 15 | 114 | 394 | 686 | 3090 |
| Cooling Degree Days | 0 | 2 | 17 | 73 | 200 | 415 | 577 | 564 | 310 | 79 | 11 | 1 | 2249 |
| Total Precipitation (") | 1.60 | 2.16 | 3.22 | 3.98 | 5.41 | 4.42 | 2.43 | 2.80 | 4.71 | 3.97 | 3.09 | 2.15 | 39.94 |
| Days ≥ 0.1" Precip | 3 | 4 | 5 | 6 | 7 | 6 | 4 | 4 | 5 | 5 | 4 | 4 | 57 |
| Total Snowfall (") | *1.7* | *1.4* | 0.6 | 0.0 | 0.0 | 0.0 | 0.0 | 0.0 | 0.0 | 0.0 | 0.2 | *0.4* | 4.3 |
| Days ≥ 1" Snow Depth | na | *0* | 0 | 0 | 0 | 0 | 0 | 0 | 0 | 0 | 0 | *0* | na |

## SMITHVILLE *McCurtain County* ELEVATION 840 ft LAT/LONG 34° 28 ' N / 94° 40 ' W

| | JAN | FEB | MAR | APR | MAY | JUN | JUL | AUG | SEP | OCT | NOV | DEC | YEAR |
|---|---|---|---|---|---|---|---|---|---|---|---|---|---|
| Maximum Temp °F | 51.2 | 55.7 | 64.6 | 73.4 | 79.0 | 86.8 | 92.0 | 91.4 | 84.5 | 74.8 | 62.8 | 53.8 | 72.5 |
| Minimum Temp °F | 27.1 | 30.2 | 38.9 | 47.2 | 54.9 | 63.2 | 66.6 | 64.8 | 58.2 | 46.6 | 37.8 | 30.4 | 47.2 |
| Mean Temp °F | 39.3 | 43.0 | 51.8 | 60.3 | 67.0 | 75.1 | 79.3 | 78.1 | 71.4 | 60.7 | 50.4 | 42.3 | 59.9 |
| Days Max Temp ≥ 90 °F | 0 | 0 | 0 | 0 | 0 | *9* | *22* | 20 | 8 | 1 | 0 | 0 | 60 |
| Days Max Temp ≤ 32 °F | 1 | 1 | 0 | 0 | 0 | 0 | 0 | 0 | 0 | 0 | 0 | 1 | 3 |
| Days Min Temp ≤ 32 °F | 22 | 18 | 9 | 2 | 0 | 0 | 0 | 0 | 0 | 3 | 11 | 19 | 84 |
| Days Min Temp ≤ 0 °F | 0 | 0 | 0 | 0 | 0 | 0 | 0 | 0 | 0 | 0 | 0 | 0 | 0 |
| Heating Degree Days | 791 | 616 | 412 | 170 | 47 | 2 | 0 | 0 | 25 | 169 | 439 | 699 | 3370 |
| Cooling Degree Days | 0 | 0 | 7 | 34 | 114 | 311 | 440 | 425 | *228* | 46 | 6 | 1 | 1612 |
| Total Precipitation (") | 3.37 | 3.72 | 5.22 | 5.12 | 7.38 | 4.35 | 4.30 | 3.19 | 4.76 | 6.06 | 4.62 | 4.94 | 57.03 |
| Days ≥ 0.1" Precip | 5 | 5 | 6 | 8 | 8 | 6 | 5 | 5 | 5 | 5 | 6 | 6 | 70 |
| Total Snowfall (") | *1.4* | *1.8* | 0.0 | 0.0 | 0.0 | 0.0 | 0.0 | 0.0 | 0.0 | 0.0 | 0.1 | 0.3 | 3.6 |
| Days ≥ 1" Snow Depth | na | *0* | 0 | 0 | 0 | 0 | 0 | 0 | 0 | 0 | 0 | 0 | na |

## SPAVINAW *Mayes County* ELEVATION 679 ft LAT/LONG 36° 23' N / 95° 3' W

| | JAN | FEB | MAR | APR | MAY | JUN | JUL | AUG | SEP | OCT | NOV | DEC | YEAR |
|---|---|---|---|---|---|---|---|---|---|---|---|---|---|
| Maximum Temp °F | 48.4 | 53.8 | 63.3 | 73.2 | 79.2 | 86.6 | 92.3 | 91.8 | 83.9 | 73.7 | 61.3 | 51.5 | 71.6 |
| Minimum Temp °F | 26.8 | 30.6 | 39.9 | 49.6 | 57.8 | 66.7 | 71.4 | 69.5 | 62.4 | 51.5 | 40.5 | 30.9 | 49.8 |
| Mean Temp °F | 37.7 | 42.2 | 51.6 | 61.4 | 68.6 | 76.7 | 81.9 | 80.7 | 73.2 | 62.6 | 50.9 | 41.2 | 60.7 |
| Days Max Temp ≥ 90 °F | 0 | 0 | 0 | 0 | 0 | 9 | 22 | 21 | 7 | 0 | 0 | 0 | 59 |
| Days Max Temp ≤ 32 °F | 3 | 2 | 0 | 0 | 0 | 0 | 0 | 0 | 0 | 0 | 0 | 2 | 7 |
| Days Min Temp ≤ 32 °F | 22 | 17 | 9 | 1 | 0 | 0 | 0 | 0 | 0 | 1 | 7 | 18 | 75 |
| Days Min Temp ≤ 0 °F | 1 | 0 | 0 | 0 | 0 | 0 | 0 | 0 | 0 | 0 | 0 | 0 | 1 |
| Heating Degree Days | 841 | 639 | 420 | 159 | 37 | 1 | 0 | 0 | 21 | 139 | 425 | 733 | 3415 |
| Cooling Degree Days | 0 | 0 | 12 | 56 | 155 | 375 | 545 | 515 | 291 | 70 | 9 | 1 | 2029 |
| Total Precipitation (") | 1.80 | 2.05 | 3.55 | 4.32 | 4.68 | 4.50 | 2.89 | 3.46 | 4.81 | 3.72 | 4.19 | 2.91 | 42.88 |
| Days ≥ 0.1" Precip | 4 | 4 | 6 | 7 | 7 | 6 | 4 | 5 | 6 | 5 | 5 | 5 | 64 |
| Total Snowfall (") | 2.0 | 2.0 | 2.1 | 0.0 | 0.0 | 0.0 | 0.0 | 0.0 | 0.0 | 0.0 | 0.1 | 1.7 | 7.9 |
| Days ≥ 1" Snow Depth | 4 | 2 | 1 | 0 | 0 | 0 | 0 | 0 | 0 | 0 | 0 | 2 | 9 |

## STILLWATER 2 W *Payne County* ELEVATION 879 ft LAT/LONG 36° 7' N / 97° 5' W

| | JAN | FEB | MAR | APR | MAY | JUN | JUL | AUG | SEP | OCT | NOV | DEC | YEAR |
|---|---|---|---|---|---|---|---|---|---|---|---|---|---|
| Maximum Temp °F | 46.0 | 51.4 | 61.4 | 71.8 | 78.7 | 87.1 | 93.2 | 92.4 | 83.9 | 73.6 | 60.2 | 50.0 | 70.8 |
| Minimum Temp °F | 22.8 | 27.1 | 36.8 | 47.1 | 56.3 | 65.7 | 70.3 | 68.3 | 60.3 | 47.4 | 36.8 | 27.1 | 47.2 |
| Mean Temp °F | 34.4 | 39.3 | 49.1 | 59.5 | 67.5 | 76.5 | 81.7 | 80.4 | 72.1 | 60.5 | 48.5 | 38.6 | 59.0 |
| Days Max Temp ≥ 90 °F | 0 | 0 | 0 | 1 | 2 | 12 | 23 | 22 | 9 | 1 | 0 | 0 | 70 |
| Days Max Temp ≤ 32 °F | 5 | 3 | 0 | 0 | 0 | 0 | 0 | 0 | 0 | 0 | 0 | 3 | 11 |
| Days Min Temp ≤ 32 °F | 27 | 20 | 12 | 2 | 0 | 0 | 0 | 0 | 0 | 2 | 11 | 23 | 97 |
| Days Min Temp ≤ 0 °F | 1 | 0 | 0 | 0 | 0 | 0 | 0 | 0 | 0 | 0 | 0 | 1 | 2 |
| Heating Degree Days | 940 | 721 | 495 | 202 | 56 | 2 | 0 | 0 | 32 | 185 | 494 | 812 | 3939 |
| Cooling Degree Days | 0 | 1 | 6 | 44 | 141 | 359 | 525 | 503 | 264 | 46 | 6 | 1 | 1896 |
| Total Precipitation (") | 1.23 | 1.58 | 2.82 | 3.35 | 5.39 | 3.98 | 2.72 | 2.88 | 4.08 | 2.82 | 2.40 | 1.54 | 34.79 |
| Days ≥ 0.1" Precip | 2 | 3 | 5 | 5 | 7 | 6 | 4 | 4 | 5 | 4 | 4 | 3 | 52 |
| Total Snowfall (") | 2.6 | 2.3 | 1.5 | 0.1 | 0.0 | 0.0 | 0.0 | 0.0 | 0.0 | 0.0 | 0.3 | 2.2 | 9.0 |
| Days ≥ 1" Snow Depth | 3 | 3 | 0 | 0 | 0 | 0 | 0 | 0 | 0 | 0 | 0 | 2 | 8 |

## TAHLEQUAH *Cherokee County* ELEVATION 791 ft LAT/LONG 35° 54' N / 94° 58' W

| | JAN | FEB | MAR | APR | MAY | JUN | JUL | AUG | SEP | OCT | NOV | DEC | YEAR |
|---|---|---|---|---|---|---|---|---|---|---|---|---|---|
| Maximum Temp °F | 48.0 | 53.2 | 63.4 | 73.2 | 79.2 | 86.6 | 92.3 | 92.0 | 84.2 | 73.6 | 61.1 | 51.2 | 71.5 |
| Minimum Temp °F | 26.0 | 29.5 | 39.1 | 48.6 | 56.4 | 64.5 | 68.2 | 66.9 | 60.4 | 49.2 | 39.0 | 29.7 | 48.1 |
| Mean Temp °F | 37.0 | 41.1 | 51.2 | 60.9 | 67.9 | 75.5 | 80.2 | 79.5 | 72.2 | 61.6 | 50.1 | 40.4 | 59.8 |
| Days Max Temp ≥ 90 °F | 0 | 0 | 0 | 0 | 0 | 9 | 21 | 22 | 8 | 0 | 0 | 0 | 60 |
| Days Max Temp ≤ 32 °F | 3 | 2 | 0 | 0 | 0 | 0 | 0 | 0 | 0 | 0 | 0 | 2 | 7 |
| Days Min Temp ≤ 32 °F | 23 | 17 | 9 | 2 | 0 | 0 | 0 | 0 | 0 | 2 | 9 | 19 | 81 |
| Days Min Temp ≤ 0 °F | 1 | 0 | 0 | 0 | 0 | 0 | 0 | 0 | 0 | 0 | 0 | 0 | 1 |
| Heating Degree Days | 862 | 668 | 430 | 169 | 46 | 2 | 0 | 0 | 28 | 162 | 448 | 755 | 3570 |
| Cooling Degree Days | 0 | 0 | 9 | 49 | 135 | 334 | 494 | 473 | 252 | 51 | 6 | 1 | 1804 |
| Total Precipitation (") | 2.13 | 2.48 | 3.98 | 4.36 | 5.45 | 4.64 | 2.82 | 3.60 | 5.33 | 4.63 | 4.21 | 3.33 | 46.96 |
| Days ≥ 0.1" Precip | 4 | 4 | 6 | 7 | 8 | 6 | 4 | 5 | 7 | 5 | 5 | 5 | 66 |
| Total Snowfall (") | 2.2 | 2.2 | 1.1 | 0.0 | 0.0 | 0.0 | 0.0 | 0.0 | 0.0 | 0.0 | 0.5 | 1.5 | 7.5 |
| Days ≥ 1" Snow Depth | *2* | *1* | 1 | 0 | 0 | 0 | 0 | 0 | 0 | 0 | 0 | 1 | 5 |

## TALOGA *Dewey County* ELEVATION 1752 ft LAT/LONG 36° 3' N / 98° 58' W

| | JAN | FEB | MAR | APR | MAY | JUN | JUL | AUG | SEP | OCT | NOV | DEC | YEAR |
|---|---|---|---|---|---|---|---|---|---|---|---|---|---|
| Maximum Temp °F | 48.1 | 53.6 | 63.2 | 73.0 | 80.4 | 90.0 | 95.8 | 93.7 | 85.1 | 74.5 | 60.0 | 50.5 | 72.3 |
| Minimum Temp °F | 22.0 | 26.2 | 35.1 | 44.8 | 53.8 | 64.0 | 68.0 | 66.3 | 58.6 | 46.0 | 33.9 | 25.0 | 45.3 |
| Mean Temp °F | 35.1 | 39.9 | 49.2 | 58.9 | 67.1 | 77.0 | 81.9 | 80.1 | 71.9 | 60.3 | 46.9 | 37.8 | 58.8 |
| Days Max Temp ≥ 90 °F | 0 | 0 | 0 | 1 | 4 | 17 | 27 | 23 | 11 | 1 | 0 | 0 | 84 |
| Days Max Temp ≤ 32 °F | 5 | 2 | 0 | 0 | 0 | 0 | 0 | 0 | 0 | 0 | 0 | 2 | 9 |
| Days Min Temp ≤ 32 °F | 27 | 21 | 13 | 3 | 0 | 0 | 0 | 0 | 0 | 3 | 14 | 25 | 106 |
| Days Min Temp ≤ 0 °F | 1 | 0 | 0 | 0 | 0 | 0 | 0 | 0 | 0 | 0 | 0 | 1 | 2 |
| Heating Degree Days | 920 | 701 | 489 | 209 | 61 | 2 | 0 | 1 | 30 | 189 | 536 | 837 | 3975 |
| Cooling Degree Days | 0 | 0 | 4 | 31 | 134 | 372 | 530 | 497 | 254 | 43 | 1 | 0 | 1866 |
| Total Precipitation (") | 0.75 | 1.03 | 1.96 | 2.62 | 5.13 | 3.46 | 2.20 | 2.55 | 2.90 | 2.13 | 1.86 | 0.79 | 27.38 |
| Days ≥ 0.1" Precip | 2 | 3 | 4 | 4 | 6 | 5 | 3 | 4 | 4 | 3 | 3 | 2 | 43 |
| Total Snowfall (") | 3.7 | 3.9 | 1.7 | 0.6 | 0.0 | 0.0 | 0.0 | 0.0 | 0.0 | 0.1 | 1.7 | 3.5 | 15.2 |
| Days ≥ 1" Snow Depth | 4 | 4 | 1 | 0 | 0 | 0 | 0 | 0 | 0 | 0 | 1 | 3 | 13 |

## TULSA INTL AP *Tulsa County* ELEVATION 676 ft LAT/LONG 36° 12 ' N / 95° 54 ' W

| | JAN | FEB | MAR | APR | MAY | JUN | JUL | AUG | SEP | OCT | NOV | DEC | YEAR |
|---|---|---|---|---|---|---|---|---|---|---|---|---|---|
| Maximum Temp °F | 45.7 | 51.6 | 62.2 | 72.7 | 79.2 | 87.8 | 93.7 | 92.6 | 83.8 | 73.4 | 59.9 | 49.7 | 71.0 |
| Minimum Temp °F | 25.7 | 29.8 | 39.6 | 49.9 | 58.6 | 67.8 | 72.8 | 70.5 | 62.6 | 50.3 | 39.1 | 29.7 | 49.7 |
| Mean Temp °F | 35.7 | 40.7 | 50.9 | 61.3 | 68.9 | 77.8 | 83.3 | 81.6 | 73.2 | 61.9 | 49.5 | 39.7 | 60.4 |
| Days Max Temp ≥ 90 °F | 0 | 0 | 0 | 1 | 2 | 13 | 24 | 22 | 9 | 1 | 0 | 0 | 72 |
| Days Max Temp ≤ 32 °F | 6 | 3 | 0 | 0 | 0 | 0 | 0 | 0 | 0 | 0 | 0 | 3 | 12 |
| Days Min Temp ≤ 32 °F | 23 | 17 | 8 | 1 | 0 | 0 | 0 | 0 | 0 | 0 | 8 | 19 | 76 |
| Days Min Temp ≤ 0 °F | 0 | 0 | 0 | 0 | 0 | 0 | 0 | 0 | 0 | 0 | 0 | 0 | 0 |
| Heating Degree Days | 901 | 679 | 441 | 163 | 39 | 1 | 0 | 0 | 24 | 158 | 464 | 778 | 3648 |
| Cooling Degree Days | 0 | 0 | 10 | 58 | 176 | 411 | 588 | 547 | 288 | 66 | 7 | 1 | 2152 |
| Total Precipitation (") | 1.56 | 1.97 | 3.30 | 3.90 | 5.68 | 4.45 | 2.70 | 2.95 | 4.57 | 3.92 | 3.17 | 2.32 | 40.49 |
| Days ≥ 0.1" Precip | 3 | 4 | 5 | 6 | 7 | 6 | 4 | 4 | 5 | 4 | 5 | 4 | 57 |
| Total Snowfall (") | 2.8 | 2.5 | 1.7 | 0.0 | 0.0 | 0.0 | 0.0 | 0.0 | 0.0 | 0.0 | 0.5 | 1.7 | 9.2 |
| Days ≥ 1" Snow Depth | 4 | 2 | 1 | 0 | 0 | 0 | 0 | 0 | 0 | 0 | 0 | 2 | 9 |

## TUSKAHOMA *Pushmataha County* ELEVATION 571 ft LAT/LONG 34° 38 ' N / 95° 15 ' W

| | JAN | FEB | MAR | APR | MAY | JUN | JUL | AUG | SEP | OCT | NOV | DEC | YEAR |
|---|---|---|---|---|---|---|---|---|---|---|---|---|---|
| Maximum Temp °F | *52.0* | *57.9* | *67.6* | *75.6* | *81.3* | *88.5* | *94.1* | *93.8* | *86.3* | *76.7* | 63.8 | 55.4 | 74.4 |
| Minimum Temp °F | *28.2* | *32.5* | *41.4* | *49.5* | *57.8* | *66.0* | *69.4* | *68.3* | *62.0* | *50.4* | 40.2 | 31.9 | 49.8 |
| Mean Temp °F | *40.2* | *45.2* | *54.5* | *62.6* | *69.6* | *77.2* | *81.8* | *81.1* | *74.2* | *63.6* | 52.1 | 43.7 | 62.2 |
| Days Max Temp ≥ 90 °F | *0* | *0* | *0* | *0* | *1* | *13* | *24* | *24* | *11* | *2* | 0 | 0 | 75 |
| Days Max Temp ≤ 32 °F | *2* | *1* | *0* | *0* | *0* | *0* | *0* | *0* | *0* | *0* | 0 | *1* | 4 |
| Days Min Temp ≤ 32 °F | *20* | *15* | *8* | *2* | *0* | *0* | *0* | *0* | *0* | *2* | 9 | 17 | 73 |
| Days Min Temp ≤ 0 °F | *0* | *0* | *0* | *0* | *0* | *0* | *0* | *0* | *0* | *0* | 0 | 0 | 0 |
| Heating Degree Days | *763* | *554* | *338* | *137* | *30* | *1* | *0* | *0* | *22* | *125* | 395 | 656 | 3021 |
| Cooling Degree Days | 0 | 2 | 20 | 67 | 177 | 380 | 528 | 516 | 302 | 87 | 14 | 2 | 2095 |
| Total Precipitation (") | *2.12* | *2.90* | *4.13* | *4.82* | *6.98* | *5.19* | *3.91* | *2.96* | *5.19* | *5.00* | 4.71 | 3.18 | 51.09 |
| Days ≥ 0.1" Precip | *4* | *4* | *6* | *7* | *8* | *7* | *5* | *4* | *6* | *5* | 6 | 5 | 67 |
| Total Snowfall (") | *1.9* | *1.8* | *0.2* | *0.0* | *0.0* | *0.0* | *0.0* | *0.0* | *0.0* | *0.0* | 0.2 | 0.3 | 4.4 |
| Days ≥ 1" Snow Depth | *2* | *1* | *0* | *0* | *0* | *0* | *0* | *0* | *0* | *0* | 0 | 0 | 3 |

## VINITA 2 N *Craig County* ELEVATION 712 ft LAT/LONG 36° 39 ' N / 95° 9 ' W

| | JAN | FEB | MAR | APR | MAY | JUN | JUL | AUG | SEP | OCT | NOV | DEC | YEAR |
|---|---|---|---|---|---|---|---|---|---|---|---|---|---|
| Maximum Temp °F | 44.5 | 50.5 | 61.4 | 71.7 | 77.9 | 86.3 | 92.0 | 91.4 | 83.2 | 72.6 | 59.3 | 49.0 | 70.0 |
| Minimum Temp °F | 23.3 | 27.5 | 36.8 | 46.9 | 55.3 | 64.2 | 68.5 | 66.2 | 59.3 | 47.4 | 37.0 | 27.5 | 46.7 |
| Mean Temp °F | 33.9 | 39.0 | 49.1 | 59.3 | 66.6 | 75.3 | 80.3 | 78.8 | 71.3 | 60.0 | 48.2 | 38.3 | 58.3 |
| Days Max Temp ≥ 90 °F | 0 | 0 | 0 | 0 | 0 | 9 | 21 | 19 | 7 | 0 | 0 | 0 | 56 |
| Days Max Temp ≤ 32 °F | 5 | 3 | 0 | 0 | 0 | 0 | 0 | 0 | 0 | 0 | 0 | 3 | 11 |
| Days Min Temp ≤ 32 °F | 25 | 19 | 12 | 2 | 0 | 0 | 0 | 0 | 0 | 2 | 11 | 22 | 93 |
| Days Min Temp ≤ 0 °F | 1 | 0 | 0 | 0 | 0 | 0 | 0 | 0 | 0 | 0 | 0 | 1 | 2 |
| Heating Degree Days | 956 | 726 | 493 | 202 | 58 | 3 | 0 | 0 | 33 | 193 | 501 | 821 | 3986 |
| Cooling Degree Days | 0 | 0 | 7 | 38 | 120 | 341 | 502 | 467 | 249 | 45 | 5 | 0 | 1774 |
| Total Precipitation (") | 1.85 | 2.05 | 3.77 | 3.98 | 5.03 | 4.45 | 3.08 | 3.42 | 5.30 | 3.95 | 4.15 | 2.88 | 43.91 |
| Days ≥ 0.1" Precip | 4 | 4 | 6 | 7 | 7 | 6 | 4 | 5 | 6 | 5 | 5 | 4 | 63 |
| Total Snowfall (") | 4.0 | 3.1 | 2.2 | 0.0 | 0.0 | 0.0 | 0.0 | 0.0 | 0.0 | 0.0 | 0.5 | 2.2 | 12.0 |
| Days ≥ 1" Snow Depth | 6 | 4 | 1 | 0 | 0 | 0 | 0 | 0 | 0 | 0 | 0 | 2 | 13 |

## WAGONER *Wagoner County* ELEVATION 591 ft LAT/LONG 35° 58 ' N / 95° 22 ' W

| | JAN | FEB | MAR | APR | MAY | JUN | JUL | AUG | SEP | OCT | NOV | DEC | YEAR |
|---|---|---|---|---|---|---|---|---|---|---|---|---|---|
| Maximum Temp °F | 47.6 | 53.2 | 63.6 | 73.5 | 79.3 | 87.0 | 93.0 | 92.1 | 84.0 | 73.9 | 61.3 | 51.7 | 71.7 |
| Minimum Temp °F | 26.6 | 30.7 | 40.0 | 49.8 | 57.9 | 66.5 | 70.6 | 69.0 | 62.0 | 50.6 | 40.0 | 30.9 | 49.6 |
| Mean Temp °F | 37.1 | 42.0 | 51.9 | 61.7 | 68.6 | 76.8 | 81.8 | 80.5 | 73.0 | 62.3 | 50.7 | 41.5 | 60.7 |
| Days Max Temp ≥ 90 °F | 0 | 0 | 0 | 0 | 0 | 10 | 23 | 21 | 7 | 1 | 0 | 0 | 62 |
| Days Max Temp ≤ 32 °F | 4 | 2 | 0 | 0 | 0 | 0 | 0 | 0 | 0 | 0 | 0 | 2 | 8 |
| Days Min Temp ≤ 32 °F | 22 | 16 | 8 | 1 | 0 | 0 | 0 | 0 | 0 | 1 | 8 | 18 | 74 |
| Days Min Temp ≤ 0 °F | 1 | 0 | 0 | 0 | 0 | 0 | 0 | 0 | 0 | 0 | 0 | 0 | 1 |
| Heating Degree Days | 858 | 645 | 413 | 151 | 36 | 1 | 0 | 0 | 22 | 146 | 431 | 721 | 3424 |
| Cooling Degree Days | 0 | 1 | 12 | 58 | 159 | 377 | 538 | 514 | 280 | 65 | 9 | 1 | 2014 |
| Total Precipitation (") | 2.01 | 2.19 | 3.54 | 4.48 | 5.14 | 5.29 | 2.62 | 3.12 | 4.75 | 4.24 | 3.97 | 2.83 | 44.18 |
| Days ≥ 0.1" Precip | 4 | 5 | 6 | 7 | 7 | 6 | 3 | 4 | 6 | 5 | 5 | 4 | 62 |
| Total Snowfall (") | 2.8 | 2.4 | 1.5 | 0.0 | 0.0 | 0.0 | 0.0 | 0.0 | 0.0 | 0.0 | 0.3 | 1.1 | 8.1 |
| Days ≥ 1" Snow Depth | 4 | 2 | 1 | 0 | 0 | 0 | 0 | 0 | 0 | 0 | 0 | 1 | 8 |

## WALTERS *Cotton County* ELEVATION 1001 ft LAT/LONG 34° 21' N / 98° 19' W

| | JAN | FEB | MAR | APR | MAY | JUN | JUL | AUG | SEP | OCT | NOV | DEC | YEAR |
|---|---|---|---|---|---|---|---|---|---|---|---|---|---|
| Maximum Temp °F | 52.5 | 58.0 | 67.7 | 76.9 | 83.7 | 91.9 | 97.3 | 96.4 | 88.0 | 77.5 | 64.9 | *55.2* | 75.8 |
| Minimum Temp °F | 27.2 | 31.4 | 39.8 | 49.4 | 58.0 | 66.5 | 70.3 | 69.0 | 62.0 | 50.4 | 38.9 | *30.1* | 49.4 |
| Mean Temp °F | 39.9 | 44.7 | 53.8 | 63.1 | 70.8 | 79.2 | 83.8 | 82.7 | 75.0 | 63.9 | 51.9 | *42.6* | 62.6 |
| Days Max Temp ≥ 90 °F | 0 | 0 | 1 | 1 | 6 | 19 | 28 | 27 | 14 | 2 | 0 | 0 | 98 |
| Days Max Temp ≤ 32 °F | 2 | 1 | 0 | 0 | 0 | 0 | 0 | 0 | 0 | 0 | 0 | 1 | 4 |
| Days Min Temp ≤ 32 °F | 22 | 15 | *7* | 1 | 0 | 0 | 0 | 0 | 0 | 1 | 8 | *17* | 71 |
| Days Min Temp ≤ 0 °F | 0 | 0 | 0 | 0 | 0 | 0 | 0 | 0 | 0 | 0 | 0 | 0 | 0 |
| Heating Degree Days | 774 | 565 | 355 | 118 | 21 | 0 | 0 | 0 | 16 | 117 | 391 | *687* | 3044 |
| Cooling Degree Days | 0 | 1 | 14 | 66 | 210 | 433 | 588 | 584 | 333 | 92 | 9 | *0* | 2330 |
| Total Precipitation (") | 1.56 | 1.84 | 2.57 | 2.99 | 5.15 | 4.02 | 2.47 | 2.65 | 4.30 | 3.34 | 1.95 | 1.36 | 34.20 |
| Days ≥ 0.1" Precip | 3 | 4 | 4 | 5 | 6 | 5 | 4 | 4 | 5 | 4 | 3 | 3 | 50 |
| Total Snowfall (") | *1.4* | na | 0.8 | 0.0 | 0.0 | 0.0 | 0.0 | 0.0 | 0.0 | 0.0 | 0.1 | *0.5* | na |
| Days ≥ 1" Snow Depth | *0* | na | 0 | 0 | 0 | 0 | 0 | 0 | 0 | 0 | 0 | *0* | na |

## WATONGA *Blaine County* ELEVATION 1522 ft LAT/LONG 35° 50' N / 98° 24' W

| | JAN | FEB | MAR | APR | MAY | JUN | JUL | AUG | SEP | OCT | NOV | DEC | YEAR |
|---|---|---|---|---|---|---|---|---|---|---|---|---|---|
| Maximum Temp °F | 46.9 | 52.4 | 62.3 | 72.4 | 79.7 | 88.9 | 94.4 | 92.8 | 84.3 | 73.7 | 59.4 | 49.8 | 71.4 |
| Minimum Temp °F | 24.5 | 28.5 | 37.4 | 47.3 | 56.1 | 65.7 | 70.4 | 68.6 | 60.8 | 49.1 | 36.6 | 27.7 | 47.7 |
| Mean Temp °F | 35.7 | 40.5 | 49.9 | 59.8 | 67.9 | 77.3 | 82.4 | 80.7 | 72.6 | 61.4 | 48.0 | 38.7 | 59.6 |
| Days Max Temp ≥ 90 °F | 0 | 0 | 0 | 1 | 3 | 15 | 25 | 23 | 9 | 1 | 0 | 0 | 77 |
| Days Max Temp ≤ 32 °F | 5 | 2 | 0 | 0 | 0 | 0 | 0 | 0 | 0 | 0 | 0 | 3 | 10 |
| Days Min Temp ≤ 32 °F | 25 | 18 | 10 | 2 | 0 | 0 | 0 | 0 | 0 | 1 | 11 | 22 | 89 |
| Days Min Temp ≤ 0 °F | 1 | 0 | 0 | 0 | 0 | 0 | 0 | 0 | 0 | 0 | 0 | 0 | 1 |
| Heating Degree Days | 900 | 686 | 471 | 194 | 51 | 2 | 0 | 0 | 27 | 166 | 506 | 807 | 3810 |
| Cooling Degree Days | 0 | 0 | 8 | 46 | 152 | 388 | 552 | 521 | 279 | 57 | 4 | 0 | 2007 |
| Total Precipitation (") | 1.01 | 1.27 | 2.34 | 2.68 | 4.88 | 3.70 | 2.27 | 2.46 | 3.22 | 2.29 | 1.92 | 1.20 | 29.24 |
| Days ≥ 0.1" Precip | 2 | 3 | 4 | 4 | 7 | 5 | 4 | 4 | 5 | 3 | 3 | 3 | 47 |
| Total Snowfall (") | 2.9 | 3.2 | 1.4 | 0.5 | 0.0 | 0.0 | 0.0 | 0.0 | 0.0 | 0.0 | 0.9 | 2.8 | 11.7 |
| Days ≥ 1" Snow Depth | 3 | 3 | 1 | 0 | 0 | 0 | 0 | 0 | 0 | 0 | 0 | 2 | 9 |

## WAURIKA *Jefferson County* ELEVATION 912 ft LAT/LONG 34° 10' N / 98° 0' W

| | JAN | FEB | MAR | APR | MAY | JUN | JUL | AUG | SEP | OCT | NOV | DEC | YEAR |
|---|---|---|---|---|---|---|---|---|---|---|---|---|---|
| Maximum Temp °F | 53.5 | 59.2 | 68.7 | 77.2 | 83.6 | 91.5 | 96.9 | 95.8 | 87.8 | 78.2 | 65.6 | 57.1 | 76.3 |
| Minimum Temp °F | 28.2 | 32.3 | 41.0 | 50.6 | 58.8 | 67.1 | 71.1 | 69.7 | 62.6 | 51.5 | 40.0 | 31.8 | 50.4 |
| Mean Temp °F | 40.9 | 45.8 | 54.9 | 63.9 | 71.3 | 79.3 | 84.0 | 82.8 | 75.2 | 64.8 | 52.8 | 44.5 | 63.4 |
| Days Max Temp ≥ 90 °F | 0 | 0 | 1 | 2 | 6 | 20 | 28 | 27 | 14 | 3 | 0 | 0 | 101 |
| Days Max Temp ≤ 32 °F | 2 | 1 | 0 | 0 | 0 | 0 | 0 | 0 | 0 | 0 | 0 | 1 | 4 |
| Days Min Temp ≤ 32 °F | 20 | 14 | 7 | 1 | 0 | 0 | 0 | 0 | 0 | 1 | 8 | 17 | 68 |
| Days Min Temp ≤ 0 °F | 0 | 0 | 0 | 0 | 0 | 0 | 0 | 0 | 0 | 0 | 0 | 0 | 0 |
| Heating Degree Days | 741 | 537 | 329 | 108 | 19 | 0 | 0 | 0 | 14 | 104 | 370 | 631 | 2853 |
| Cooling Degree Days | 0 | 2 | 20 | 78 | 224 | 435 | 594 | 578 | 343 | 104 | 13 | 1 | 2392 |
| Total Precipitation (") | 1.23 | 1.73 | 2.38 | 3.08 | 4.93 | 3.59 | 1.81 | 2.54 | 3.70 | 3.00 | 1.67 | 1.73 | 31.39 |
| Days ≥ 0.1" Precip | 3 | 4 | 4 | 4 | 6 | 4 | 3 | 4 | 5 | 4 | 3 | 3 | 47 |
| Total Snowfall (") | *1.6* | *1.4* | 0.9 | 0.0 | 0.0 | 0.0 | 0.0 | 0.0 | 0.0 | 0.0 | 0.2 | 0.5 | 4.6 |
| Days ≥ 1" Snow Depth | *1* | na | 0 | 0 | 0 | 0 | 0 | 0 | 0 | 0 | 0 | *0* | na |

## WAYNOKA *Woods County* ELEVATION 1529 ft LAT/LONG 36° 35' N / 98° 51' W

| | JAN | FEB | MAR | APR | MAY | JUN | JUL | AUG | SEP | OCT | NOV | DEC | YEAR |
|---|---|---|---|---|---|---|---|---|---|---|---|---|---|
| Maximum Temp °F | 47.4 | 53.3 | 63.3 | 73.7 | 81.0 | 90.2 | 95.7 | 93.8 | 85.5 | 75.0 | 60.2 | 49.8 | 72.4 |
| Minimum Temp °F | 22.4 | 26.4 | 35.7 | 45.6 | 55.3 | 65.1 | 70.1 | 68.2 | 59.7 | 47.0 | 34.3 | 25.2 | 46.3 |
| Mean Temp °F | 34.9 | 39.8 | 49.5 | 59.7 | 68.2 | 77.7 | 82.9 | 80.9 | 72.6 | 61.0 | 47.3 | 37.5 | 59.3 |
| Days Max Temp ≥ 90 °F | 0 | 0 | 0 | 1 | 4 | 17 | 26 | 22 | 11 | 2 | 0 | 0 | 83 |
| Days Max Temp ≤ 32 °F | 5 | 3 | 0 | 0 | 0 | 0 | 0 | 0 | 0 | 0 | 0 | 3 | 11 |
| Days Min Temp ≤ 32 °F | 26 | 21 | 12 | 3 | 0 | 0 | 0 | 0 | 0 | 2 | 13 | 25 | 102 |
| Days Min Temp ≤ 0 °F | 1 | 0 | 0 | 0 | 0 | 0 | 0 | 0 | 0 | 0 | 0 | 1 | 2 |
| Heating Degree Days | 926 | 703 | 485 | 200 | 51 | 2 | 0 | 0 | 28 | 179 | 529 | 845 | 3948 |
| Cooling Degree Days | 0 | 0 | 6 | 43 | 148 | 377 | 552 | 503 | 271 | 50 | 3 | 0 | 1953 |
| Total Precipitation (") | 0.64 | 1.10 | 1.93 | 2.19 | 4.67 | 3.01 | 2.40 | 2.99 | 2.52 | 1.76 | 1.55 | 0.91 | 25.67 |
| Days ≥ 0.1" Precip | 2 | 3 | 4 | 4 | 6 | 5 | 4 | 4 | 4 | 3 | 3 | 2 | 44 |
| Total Snowfall (") | 2.8 | *3.7* | 2.4 | 0.2 | 0.0 | 0.0 | 0.0 | 0.0 | 0.0 | 0.0 | 0.8 | *2.0* | 11.9 |
| Days ≥ 1" Snow Depth | *2* | *2* | 1 | 0 | 0 | 0 | 0 | 0 | 0 | 0 | 0 | *1* | 6 |

## WEATHERFORD *Custer County* ELEVATION 1640 ft LAT/LONG 35° 32 ' N / 98° 42 ' W

| | JAN | FEB | MAR | APR | MAY | JUN | JUL | AUG | SEP | OCT | NOV | DEC | YEAR |
|---|---|---|---|---|---|---|---|---|---|---|---|---|---|
| Maximum Temp °F | 47.3 | 53.2 | 63.1 | 73.2 | 80.9 | 90.3 | 95.6 | 93.8 | 85.1 | 74.5 | 60.1 | 50.2 | 72.3 |
| Minimum Temp °F | 24.6 | 28.7 | 37.4 | 47.2 | 56.2 | 65.5 | 69.9 | 68.2 | 60.3 | 48.7 | 37.1 | 27.6 | 47.6 |
| Mean Temp °F | 36.0 | 40.9 | 50.3 | 60.2 | 68.6 | 77.9 | 82.8 | 81.1 | 72.7 | 61.6 | 48.6 | 38.9 | 60.0 |
| Days Max Temp ≥ 90 °F | 0 | 0 | 0 | 1 | 5 | 18 | 27 | 23 | 10 | 1 | 0 | 0 | 85 |
| Days Max Temp ≤ 32 °F | 4 | 2 | 0 | 0 | 0 | 0 | 0 | 0 | 0 | 0 | 0 | 3 | 9 |
| Days Min Temp ≤ 32 °F | 25 | 18 | 9 | 1 | 0 | 0 | 0 | 0 | 0 | 1 | 10 | 22 | 86 |
| Days Min Temp ≤ 0 °F | 0 | 0 | 0 | 0 | 0 | 0 | 0 | 0 | 0 | 0 | 0 | 1 | 1 |
| Heating Degree Days | 894 | 674 | 456 | 182 | 45 | 2 | 0 | 1 | 28 | 161 | 486 | 801 | 3730 |
| Cooling Degree Days | 0 | 0 | 5 | 40 | 153 | 388 | 552 | 522 | 270 | 53 | 3 | 0 | 1986 |
| Total Precipitation (") | 0.89 | 1.14 | 2.03 | 2.31 | 4.85 | 3.58 | 2.19 | 2.96 | 3.42 | 2.36 | 1.72 | 1.08 | 28.53 |
| Days ≥ 0.1" Precip | 2 | 3 | 4 | 4 | 6 | 5 | 3 | 5 | 5 | 3 | 3 | 2 | 45 |
| Total Snowfall (") | 2.4 | 2.6 | 0.9 | 0.1 | 0.0 | 0.0 | 0.0 | 0.0 | 0.0 | 0.1 | 0.5 | 1.9 | 8.5 |
| Days ≥ 1" Snow Depth | 3 | 2 | 0 | 0 | 0 | 0 | 0 | 0 | 0 | 0 | 0 | 2 | 7 |

## WEBBERS FALLS 5 WSW *Muskogee County* ELEVATION 550 ft LAT/LONG 35° 29 ' N / 95° 12 ' W

| | JAN | FEB | MAR | APR | MAY | JUN | JUL | AUG | SEP | OCT | NOV | DEC | YEAR |
|---|---|---|---|---|---|---|---|---|---|---|---|---|---|
| Maximum Temp °F | 46.9 | 52.4 | 62.6 | 72.6 | 79.3 | 87.7 | 93.3 | 92.0 | 84.5 | 74.3 | 61.8 | 52.0 | 71.6 |
| Minimum Temp °F | 24.3 | 28.8 | 38.4 | 47.9 | 56.8 | 65.7 | 69.6 | 67.5 | 60.6 | 47.8 | 38.0 | 29.2 | 47.9 |
| Mean Temp °F | 35.6 | 40.6 | 50.6 | 60.3 | 68.1 | 76.7 | 81.5 | 79.8 | 72.5 | 61.0 | 49.9 | 40.6 | 59.8 |
| Days Max Temp ≥ 90 °F | 0 | 0 | 0 | 0 | 1 | 13 | 24 | 21 | 9 | 1 | 0 | 0 | 69 |
| Days Max Temp ≤ 32 °F | 4 | 2 | 0 | 0 | 0 | 0 | 0 | 0 | 0 | 0 | 0 | 1 | 7 |
| Days Min Temp ≤ 32 °F | 25 | 18 | 9 | 1 | 0 | 0 | 0 | 0 | 0 | 1 | 9 | 21 | 84 |
| Days Min Temp ≤ 0 °F | 1 | 0 | 0 | 0 | 0 | 0 | 0 | 0 | 0 | 0 | 0 | 0 | 1 |
| Heating Degree Days | 901 | 681 | 450 | 179 | 45 | 1 | 0 | 0 | 26 | 171 | 452 | 749 | 3655 |
| Cooling Degree Days | 0 | 0 | 6 | 40 | 145 | 366 | 517 | 478 | 268 | 53 | 6 | 0 | 1879 |
| Total Precipitation (") | 2.00 | 2.62 | 4.06 | 4.51 | 5.79 | 3.86 | 2.46 | 2.97 | 5.03 | 4.79 | 3.86 | 3.13 | 45.08 |
| Days ≥ 0.1" Precip | 4 | 4 | 6 | 7 | 8 | 5 | 4 | 4 | 6 | 5 | 5 | 5 | 63 |
| Total Snowfall (") | 2.6 | 1.7 | 0.9 | 0.0 | 0.0 | 0.0 | 0.0 | 0.0 | 0.0 | 0.0 | 0.1 | 0.5 | 5.8 |
| Days ≥ 1" Snow Depth | na | na | 0 | 0 | 0 | 0 | 0 | 0 | 0 | 0 | 0 | 0 | na |

## WICHITA MTN WL REF *Comanche County* ELEVATION 1601 ft LAT/LONG 34° 43 ' N / 98° 41 ' W

| | JAN | FEB | MAR | APR | MAY | JUN | JUL | AUG | SEP | OCT | NOV | DEC | YEAR |
|---|---|---|---|---|---|---|---|---|---|---|---|---|---|
| Maximum Temp °F | 49.2 | 54.9 | 64.5 | 74.0 | 80.7 | 89.3 | 95.1 | 93.7 | 85.4 | 75.4 | 61.7 | 52.4 | 73.0 |
| Minimum Temp °F | 24.1 | 28.8 | 37.6 | 47.1 | 55.7 | 64.8 | 69.3 | 67.4 | 59.6 | 48.0 | 36.2 | 27.5 | 47.2 |
| Mean Temp °F | 36.7 | 41.9 | 51.1 | 60.6 | 68.3 | 77.1 | 82.2 | 80.6 | 72.5 | 61.7 | 49.0 | 40.0 | 60.1 |
| Days Max Temp ≥ 90 °F | 0 | 0 | 0 | 1 | 4 | 15 | 26 | 24 | 10 | 1 | 0 | 0 | 81 |
| Days Max Temp ≤ 32 °F | 4 | 2 | 0 | 0 | 0 | 0 | 0 | 0 | 0 | 0 | 0 | 2 | 8 |
| Days Min Temp ≤ 32 °F | 25 | 19 | 10 | 1 | 0 | 0 | 0 | 0 | 0 | 1 | 11 | 22 | 89 |
| Days Min Temp ≤ 0 °F | 0 | 0 | 0 | 0 | 0 | 0 | 0 | 0 | 0 | 0 | 0 | 0 | 0 |
| Heating Degree Days | 872 | 645 | 434 | 174 | 42 | 2 | 0 | 1 | 28 | 158 | 481 | 771 | 3608 |
| Cooling Degree Days | 0 | 0 | 7 | 42 | 142 | 356 | 534 | 502 | 257 | 51 | 3 | 0 | 1894 |
| Total Precipitation (") | 1.30 | 1.47 | 2.43 | 2.84 | 5.30 | 3.65 | 2.47 | 2.28 | 3.88 | 2.94 | 2.10 | 1.47 | 32.13 |
| Days ≥ 0.1" Precip | 2 | 3 | 4 | 4 | 6 | 5 | 4 | 4 | 4 | 3 | 3 | 3 | 45 |
| Total Snowfall (") | 1.3 | 1.6 | 0.6 | 0.0 | 0.0 | 0.0 | 0.0 | 0.0 | 0.0 | 0.0 | 0.1 | 0.7 | 4.3 |
| Days ≥ 1" Snow Depth | 1 | 1 | 0 | 0 | 0 | 0 | 0 | 0 | 0 | 0 | 0 | 0 | 2 |

## WILBURTON 9 ENE *Latimer County* ELEVATION 640 ft LAT/LONG 34° 57 ' N / 95° 10 ' W

| | JAN | FEB | MAR | APR | MAY | JUN | JUL | AUG | SEP | OCT | NOV | DEC | YEAR |
|---|---|---|---|---|---|---|---|---|---|---|---|---|---|
| Maximum Temp °F | 50.7 | 57.0 | 66.0 | 75.1 | 80.6 | 88.1 | 94.2 | 92.8 | 85.2 | 75.9 | 62.9 | 54.4 | 73.6 |
| Minimum Temp °F | 27.3 | 30.7 | 40.1 | 49.1 | 56.1 | 64.4 | 68.1 | 66.0 | 59.7 | 48.0 | 38.1 | 30.1 | 48.1 |
| Mean Temp °F | 39.0 | 43.9 | 53.1 | 62.1 | 68.4 | 76.3 | 81.2 | 79.5 | 72.5 | 61.9 | 50.6 | 42.3 | 60.9 |
| Days Max Temp ≥ 90 °F | 0 | 0 | 0 | 0 | 1 | na | 25 | 22 | na | 1 | 0 | 0 | na |
| Days Max Temp ≤ 32 °F | 3 | 1 | 0 | 0 | 0 | 0 | 0 | 0 | 0 | 0 | 0 | 1 | 5 |
| Days Min Temp ≤ 32 °F | 21 | na | 8 | 2 | 0 | 0 | 0 | 0 | 0 | 2 | 9 | na | na |
| Days Min Temp ≤ 0 °F | 1 | 0 | 0 | 0 | 0 | 0 | 0 | 0 | 0 | 0 | 0 | 0 | 1 |
| Heating Degree Days | 799 | 580 | 378 | 136 | 37 | 2 | 0 | 0 | 23 | 151 | 432 | 700 | 3238 |
| Cooling Degree Days | na | na | na | na | na | 361 | na | na | na | na | na | na | na |
| Total Precipitation (") | 2.52 | 2.87 | 4.11 | 5.40 | 6.30 | 4.70 | 3.61 | 2.90 | 5.29 | 4.33 | 4.74 | 3.23 | 50.00 |
| Days ≥ 0.1" Precip | 5 | 4 | 6 | 8 | 8 | 6 | 4 | 5 | 6 | 5 | 5 | 5 | 67 |
| Total Snowfall (") | na | na | 0.5 | 0.0 | 0.0 | 0.0 | 0.0 | 0.0 | 0.0 | 0.0 | 0.3 | 0.7 | na |
| Days ≥ 1" Snow Depth | na | na | na | 0 | 0 | 0 | 0 | 0 | 0 | 0 | 0 | na | na |

## JANUARY MINIMUM TEMPERATURE °F

### LOWEST

| | | |
|---|---|---|
| 1 | Beaver | 17.8 |
| 2 | Kenton | 18.0 |
| 3 | Goodwell | 18.3 |
| 4 | Hooker | 18.8 |
| 5 | Boise City | 18.9 |
| 6 | Fort Supply Dam | 20.0 |
| 7 | Arnett | 20.1 |
| | Gage | 20.1 |
| 9 | Freedom | 20.2 |
| 10 | Buffalo | 20.6 |
| 11 | Helena | 20.7 |
| 12 | Hammon | 20.8 |
| 13 | Gate | 21.0 |
| 14 | Mutual | 21.4 |
| 15 | Taloga | 22.0 |
| 16 | Claremore | 22.1 |
| 17 | Great Slt Plns Dm | 22.2 |
| | Miami | 22.2 |
| 19 | Cherokee | 22.3 |
| 20 | Waynoka | 22.4 |
| 21 | Pryor | 22.5 |
| 22 | Pawhuska | 22.7 |
| 23 | Alva | 22.8 |
| | Jefferson | 22.8 |
| | Stillwater | 22.8 |

### HIGHEST

| | | |
|---|---|---|
| 1 | Hugo | 31.4 |
| 2 | Ardmore | 30.9 |
| 3 | Bear Mountain | 30.6 |
| 4 | Madill | 30.1 |
| 5 | Marietta | 30.0 |
| 6 | Boswell | 29.7 |
| 7 | Idabel | 29.5 |
| 8 | McCurtain | 28.7 |
| 9 | Antlers | 28.4 |
| 10 | Ada | 28.3 |
| | Atoka Dam | 28.3 |
| 12 | Tuskahoma | 28.2 |
| | Waurika | 28.2 |
| 14 | Pauls Valley | 28.1 |
| 15 | Broken Bow Dam | 27.9 |
| 16 | Seminole | 27.8 |
| 17 | Duncan | 27.6 |
| | Holdenville | 27.6 |
| | McAlester | 27.6 |
| | Okemah | 27.6 |
| 21 | Durant | 27.5 |
| 22 | Hanna | 27.4 |
| 23 | Wilburton | 27.3 |
| 24 | Healdton | 27.2 |
| | Walters | 27.2 |

## JULY MAXIMUM TEMPERATURE °F

### HIGHEST

| | | |
|---|---|---|
| 1 | Mangum | 98.7 |
| 2 | Chattanooga | 98.1 |
| 3 | Buffalo | 97.8 |
| | Hollis | 97.8 |
| 5 | Frederick | 97.7 |
| 6 | Altus | 97.6 |
| 7 | Walters | 97.3 |
| 8 | Cherokee | 97.0 |
| 9 | Clinton | 96.9 |
| | Waurika | 96.9 |
| 11 | Alva | 96.6 |
| 12 | Freedom | 96.5 |
| 13 | Jefferson | 96.4 |
| | Okeene | 96.4 |
| 15 | Altus Dam | 96.1 |
| | Great Slt Plns Dm | 96.1 |
| 17 | Carnegie | 96.0 |
| | Healdton | 96.0 |
| 19 | Gate | 95.9 |
| 20 | Taloga | 95.8 |
| 21 | Kingfisher | 95.7 |
| | Lawton | 95.7 |
| | Waynoka | 95.7 |
| 24 | Weatherford | 95.6 |
| 25 | Guthrie | 95.5 |

### LOWEST

| | | |
|---|---|---|
| 1 | Kansas | 90.9 |
| 2 | Miami | 91.9 |
| 3 | Smithville | 92.0 |
| | Vinita | 92.0 |
| 5 | Kenton | 92.1 |
| 6 | Spavinaw | 92.3 |
| | Tahlequah | 92.3 |
| 8 | Claremore | 92.5 |
| | McAlester | 92.5 |
| 10 | Arnett | 92.6 |
| | Pryor | 92.6 |
| 12 | Boise City | 92.8 |
| | Idabel | 92.8 |
| 14 | Ada | 92.9 |
| | Cushing | 92.9 |
| 16 | Bixby | 93.0 |
| | Goodwell | 93.0 |
| | Pawhuska | 93.0 |
| | Wagoner | 93.0 |
| 20 | Fort Supply Dam | 93.2 |
| | Oklahoma City | 93.2 |
| | Stillwater | 93.2 |
| 23 | Boswell | 93.3 |
| | Broken Bow Dam | 93.3 |
| | Meeker | 93.3 |

## ANNUAL PRECIPITATION (")

### HIGHEST

| | | |
|---|---|---|
| 1 | Smithville | 57.03 |
| 2 | Broken Bow Dam | 54.25 |
| 3 | Bear Mountain | 53.15 |
| 4 | Tuskahoma | 51.09 |
| 5 | Wilburton | 50.00 |
| 6 | Idabel | 49.67 |
| 7 | McCurtain | 47.93 |
| 8 | Kansas | 47.77 |
| 9 | Hugo | 47.52 |
| 10 | Tahlequah | 46.96 |
| 11 | Antlers | 46.15 |
| 12 | McAlester | 46.00 |
| 13 | Sallisaw | 45.94 |
| 14 | Durant | 45.52 |
| 15 | Webbers Falls | 45.08 |
| 16 | Miami | 44.84 |
| 17 | Boswell | 44.66 |
| 18 | Hanna | 44.26 |
| 19 | Wagoner | 44.18 |
| 20 | Vinita | 43.91 |
| 21 | Muskogee | 43.61 |
| 22 | Atoka Dam | 42.93 |
| 23 | Spavinaw | 42.88 |
| 24 | Pryor | 42.30 |
| 25 | Madill | 42.20 |

### LOWEST

| | | |
|---|---|---|
| 1 | Goodwell | 16.87 |
| 2 | Kenton | 17.07 |
| 3 | Boise City | 17.52 |
| 4 | Hooker | 18.08 |
| 5 | Gage | 20.72 |
| 6 | Beaver | 21.14 |
| 7 | Gate | 21.67 |
| 8 | Fort Supply Dam | 22.49 |
| 9 | Reydon | 23.21 |
| 10 | Arnett | 23.76 |
| 11 | Freedom | 24.01 |
| 12 | Hollis | 24.25 |
| 13 | Erick | 24.72 |
| 14 | Mutual | 24.81 |
| 15 | Alva | 25.30 |
| 16 | Waynoka | 25.67 |
| 17 | Hammon | 25.92 |
| 18 | Mangum | 26.60 |
| 19 | Buffalo | 26.71 |
| 20 | Altus | 26.85 |
| 21 | Elk City | 26.87 |
| 22 | Hobart | 27.08 |
| 23 | Altus Dam | 27.31 |
| 24 | Taloga | 27.38 |
| 25 | Geary | 28.22 |

## ANNUAL SNOWFALL (")

### HIGHEST

| | | |
|---|---|---|
| 1 | Boise City | 28.5 |
| 2 | Gate | 20.0 |
| 3 | Hooker | 17.8 |
| 4 | Helena | 17.5 |
| 5 | Goodwell | 15.6 |
| 6 | Gage | 15.4 |
| 7 | Taloga | 15.2 |
| 8 | Kansas | 14.8 |
| 9 | Mutual | 13.4 |
| 10 | Cherokee | 12.9 |
| 11 | Fort Supply Dam | 12.8 |
| 12 | Reydon | 12.1 |
| 13 | Vinita | 12.0 |
| 14 | Waynoka | 11.9 |
| 15 | Watonga | 11.7 |
| 16 | Erick | 11.3 |
| 17 | Buffalo | 11.2 |
| 18 | Ralston | 10.9 |
| 19 | Pawhuska | 10.2 |
| 20 | Bartlesville | 9.9 |
| | Nowata | 9.9 |
| 22 | Arnett | 9.7 |
| | Barnsdall | 9.7 |
| | Claremore | 9.7 |
| 25 | Bristow | 9.4 |

### LOWEST

| | | |
|---|---|---|
| 1 | Broken Bow Dam | 0.9 |
| 2 | Antlers | 1.9 |
| 3 | Idabel | 2.0 |
| 4 | Atoka Dam | 2.6 |
| 5 | Chattanooga | 3.0 |
| 6 | Hugo | 3.1 |
| 7 | Smithville | 3.6 |
| 8 | Meeker | 3.7 |
| 9 | Ardmore | 3.9 |
| | Bear Mountain | 3.9 |
| | Madill | 3.9 |
| 12 | Healdton | 4.2 |
| 13 | Seminole | 4.3 |
| | Wichita Mountain | 4.3 |
| 15 | Frederick | 4.4 |
| | Tuskahoma | 4.4 |
| 17 | Duncan | 4.6 |
| | Mangum | 4.6 |
| | Waurika | 4.6 |
| 20 | Kingfisher | 4.9 |
| 21 | Marietta | 5.1 |
| 22 | Okemah | 5.5 |
| 23 | Geary | 5.6 |
| 24 | Webbers Falls | 5.8 |
| 25 | Guthrie | 6.4 |

# OREGON

PHYSICAL FEATURES AND GENERAL CLIMATE. Oregon enjoys a mild though varied climate with only a rare occurrence of devastating weather elements. The single most important geographic feature of the climate of Oregon is the Pacific Ocean whose coastline makes up the western border. Because of the normal movement of air masses from west to east, most of the systems moving across Oregon have been modified extensively in traveling over the Pacific. As a result, winter minimum and summer maximum temperatures in the west, and to a lesser extent in the eastern portion, are greatly moderated. The occurrence of extreme low or high temperatures is generally associated with the occasional invasion of the continental air masses. The unlimited supply of moisture available to those air masses that move across the Pacific is largely responsible for the abundant rainfall over western Oregon and the higher elevations of the eastern portion.

Beginning near and following the coast the full length of the State, the Coast Range is the farthest west of the three mountain ranges that exert an important influence on Oregon's climate. This range, athwart the path of the moisture laden marine air moving in from the Pacific, forces it to rise as it moves eastward. The resultant cooling and condensation produces some of the heaviest annual rainfalls in the United States along the higher western slopes, and materially reduces the available moisture in the air.

The Cascade Mountains parallel the Coast Range about 75 miles to the east. The Cascades rise from the broad valley of the Willamette eastward to an average height of about 5,000 feet, with a few peaks over 10,000 feet. One of these, Mount Hood, at an elevation of 11,245 feet, is the highest point in the State. Once again, the air masses from the west are forced to ascend causing them to give up additional moisture. The rain potential of the marine air, however, was greatly reduced by passage over the Coast Range; therefore, the rainfall on the west slopes of the Cascades at a corresponding elevation is only about one-half to two-thirds as great as on the Coast Range. Precipitation amounts decrease rapidly once the crest is crossed and descent down the eastward side begins.

Cutting through both the Cascade and the Coast Ranges, the Columbia River Gorge offers ready passage of marine air from the Pacific. Temperatures are moderated to the east in both summer and winter. Continental air occasionally passes in reverse and produces the more extreme low temperatures in the western valleys. Winding through the rugged terrain that makes up much of Oregon are the Columbia and Snake River Basins, the valleys of the many streams that head in the mountains and several very wide plateau regions. The Columbia Plateau covers about two-thirds of the State's total area and extends from the eastern border westward to the eastern slopes of the Cascade Mountains and from the southern border north to the Columbia River. Its elevation ranges from 4,000 to 6,000 feet and because of its arid nature and scant vegetation, summer heating and winter cooling often become extreme.

TEMPERATURE. Few states have greater temperature extremes than Oregon. The most extreme temperatures generally occur east of the Cascades. In the coastal sections they never drop as low as zero and on very few occasions pass the 100°F. mark. Here the mean of the coldest month, January, is 45° F., only 15° less than that of July, the warmest month. In the Willamette Valley, mean temperatures average 38° F. in January and 66° F. in July. In the inland valleys of the southwest the average summer temperatures are about 5° higher than in the northwest and maximums of 90° F. or more occur 40 to 50 days a year. In south-central Oregon the median annual maximum temperatures over a period of years have been between 95° and 100° F.; in most other areas east of the Cascades this variance is between 100° and 105° F. Median annual minimum temperatures for eastern Oregon vary from near zero in the more protected areas of the Columbia Basin to -26° F. in the high mountain and plateau regions.

PRECIPITATION. The average annual rainfall in Oregon varies from less than 8 inches in drier Plateau Regions to as much as 200 inches at points along the upper west slopes of the Coast Range. The State as a whole has a very definite winter rainfall climate. West of the Cascades about one-half of the annual total precipitation falls from December through February; about one-fourth in the spring and fall, and very little during the summer months. East of the Cascades the differences are not as pronounced, with slightly more precipitation in winter than in spring and fall, while only about 10 percent falls during the summer. Along the coast the normal annual total is from 75 to 90 inches, and increases up the west slopes of the Coast Range to almost 200 inches near the crest. Amounts decrease on the eastern slopes and in the Willamette Valley. On the western slopes of the Cascades there is again a marked increase in precipitation with elevation as annual averages range up to 75 inches. Amounts decrease rapidly on the east side. The annual average precipitation for the great plateau of the State is often less than 8 inches. In the Columbia River Basin and the Blue Mountains, totals are about 15 to 20 inches; however, some of the mountain regions receive as much as 35 inches.

SNOWFALL. In the high Cascades, where the State's heaviest snowfalls occur, it appears that annual average totals can range from 300 to 550 inches. Winter precipitation along the Coast Range, due to its lower elevations, occurs largely in the form of rain, although it too is occasionally subject to very heavy snows. In the Blue Mountains, seasonal totals range between 150 to 300 inches and depths on the ground may occasionally exceed 120 inches. The periods of continuous snow cover vary with elevation. On the peaks of the Cascades higher than 7,000 feet above sea level, it persists in glacial form the year around. In most mountain areas above 4,500 feet, snow cover lasts from early December until the latter part of April. Along the coast the average annual snowfall is only 1 to 3 inches, with many years in which there is no measurable amount.

STORMS. Hailstorms occur each year, but are generally light and cover very small areas. Practically all of these storms occur east of the Cascades. In the western part of the State thunderstorms occur in the valleys an average of 4 or 5 days a year and are not usually severe. In the eastern part, they occur on 12 to 15 days with heavier precipitation and greater wind damage. It is in the mountain areas that these storms occur most frequently and each year many forest fires are started by the accompanying lightning.

Several times each year winds of hurricane force (74 m.p.h. and over) strike the Oregon coast. They sometimes move inland to the western valleys and up the Columbia Gorge. The few tornadoes reported have been short lived. The prevailing wind direction is influenced by the surrounding terrain. In the Columbia Gorge the prevailing direction of the wind follows the orientation of the gorge at that point. Similarly, in the Willamette Valley prevailing directions are aligned north-south with the valley. The very strong winds, of course, are determined by the direction of the major storm movements.

FLOODS. Most of the State is drained into the Pacific Ocean through the Columbia River. The Snake River makes up more than half of Oregon's eastern border and drains practically all of the State east of the Blue Mountains. The west slope of the Coast Range and all areas south of the Willamette Basin and west of the summit of the Cascades are drained directly into the Pacific Ocean by three large river systems -- the Umpqua, Rogue and Smith Rivers. The only major river draining south central Oregon is the Klamath. The remainder of the area lying south of the Deschutes and John Day Basins and between the Cascades and the Blue Mountains has only internal drainage into brackish lakes. Many of these lakes become dry during the summer months.

Major flooding in the Willamette Basin and the coastal streams may result from several days of moderate to heavy rain extending over the entire Basin. When combined with sharply rising air temperatures and a warm southerly wind, the melting of a heavy snowpack on the middle and upper slopes of the Coast Range and/or the Cascades greatly increases the flood potential. Flooding in the main channel of the Columbia River usually occurs during late spring and early summer when snow melt in the mountains is most rapid. Simultaneous occurrences of heavy, warm rain over large parts of the Columbia Basin have, on occasion, produced some very damaging floods.

During the early morning hours the relative humidity is greatest and there is little variation at this time between winter and summer readings in eastern and western Oregon. In contrast, the afternoon averages, when the relative humidities are least, show a very marked difference between summer and winter and also between the areas east and west of the Cascades.

SUNSHINE. The north coastal area has the least sunshine, while the southeast corner of the State has the most. The sun will shine about 20 percent of the time possible in the coastal area and 45 percent of the time possible in the southeast. These values increase in April to values of 50 to 70 percent; in July to 55 percent in the northwest and 90 percent in the southeast. By October they have declined to 40 and 65 percent respectively.

# COUNTY INDEX

### *Baker County*
BAKER MUNICIPAL AP
HALFWAY
HUNTINGTON
RICHLAND

### *Benton County*
CORVALLIS STATE UNIV
CORVALLIS WATER BURE

### *Clackamas County*
ESTACADA 2 SE
GOVERNMENT CAMP
N WILLAMETTE EXP STN
OREGON CITY
PORTLAND HEADWORKS W
SCOTTS MILLS 9 SE
THREE LYNX

### *Clatsop County*
ASTORIA CLATSOP ARPT
SEASIDE

### *Columbia County*
CLATSKANIE
VERNONIA 2

### *Coos County*
BANDON 2 NNE
NORTH BEND MUNI AP
POWERS

### *Crook County*
BARNES STN
OCHOCO RS
PAULINA
PRINEVILLE 4 NW

### *Curry County*
BROOKINGS 2 SE
GOLD BEACH RS
ILLAHE

### *Deschutes County*
BEND
BROTHERS
REDMOND ROBERTS FLD
SISTERS
WICKIUP DAM

### *Douglas County*
DRAIN
ELKTON 3 SW
RIDDLE
ROSEBURG KQEN
TOKETEE FALLS

### *Gilliam County*
ARLINGTON
CONDON

### *Grant County*
AUSTIN 3 S
JOHN DAY
LONG CREEK
MONUMENT 2
SENECA

### *Harney County*
MALHEUR REFUGE HDQ
P RANCH REFUGE
SQUAW BUTTE
WHITEHORSE RANCH

### *Hood River County*
HOOD RIVER EXP STN

### *Jackson County*
ASHLAND
HOWARD PRAIRIE DAM
MEDFORD EXP STN
MEDFORD JACKSON CO
PROSPECT 2 SW
RUCH

### *Jefferson County*
GRIZZLY
MADRAS
PELTON DAM

### *Josephine County*
CAVE JUNCTION 1 WNW
GRANTS PASS
SEXTON SUMMIT

### *Klamath County*
CHEMULT
CRATER LAKE N P HQ
KLAMATH FALLS 2 SSW

### *Lake County*
ADEL
ALKALI LAKE
FREMONT 5 NW
HART MOUNTAIN REFUGE
LAKEVIEW 2 NNW
PAISLEY
SILVER LAKE R S
SUMMER LAKE 1 S

### *Lane County*
COTTAGE GROVE 1 S
COTTAGE GROVE DAM
DORENA DAM
EUGENE MAHLON SWEET
FERN RIDGE DAM
LEABURG 1 SW
LOOKOUT POINT DAM
OAKRIDGE FISH HATCHE

### *Lincoln County*
NEWPORT
OTIS 2 NE
TIDEWATER

### *Linn County*
BELKNAP SPRINGS 8 N
CASCADIA
MARION FORKS FISH HA

### *Malheur County*
BEULAH
IRONSIDE 2 W
JUNTURA 9 ENE
MALHEUR BRCH EXP STN
MC DERMITT 26 N
NYSSA
ONTARIO KSRV
OWYHEE DAM
RIVERSIDE 7 SSW
ROCKVILLE 5 N
ROME 2 NW
SHEAVILLE 1 SE
VALE

### *Marion County*
DETROIT DAM
SALEM MCNARY FIELD
SILVER CREEK FALLS
SILVERTON
STAYTON

### *Morrow County*
HEPPNER

### *Multnomah County*

BONNEVILLE DAM
PORTLAND INTL ARPT
TROUTDALE SUBSTATION

### *Polk County*

DALLAS 2 NE
FALLS CITY 2

### *Sherman County*

KENT
MORO

### *Tillamook County*

CLOVERDALE
TILLAMOOK 1 W

### *Umatilla County*

HERMISTON 1 SE
MILTON FREEWATER
PENDLETON BR EXP STN
PENDLETON MUNICPL AP
PILOT ROCK 1 SE
UKIAH

### *Union County*

ELGIN
LA GRANDE
UNION EXP STN

### *Wallowa County*

WALLOWA

### *Wasco County*

ANTELOPE 1 NW
DUFUR
THE DALLES

### *Washington County*

FOREST GROVE
HILLSBORO

### *Wheeler County*

FOSSIL
MITCHELL 2 NW

### *Yamhill County*

MC MINNVILLE

## ELEVATION INDEX

| FEET | STATION NAME |
|---|---|
| 10 | ASTORIA CLATSOP ARPT |
| 10 | SEASIDE |
| 22 | CLATSKANIE |
| 29 | TROUTDALE SUBSTATION |
| 30 | NORTH BEND MUNI AP |
| 39 | PORTLAND INTL ARPT |
| 39 | TIDEWATER |
| 39 | TILLAMOOK 1 W |
| 49 | GOLD BEACH RS |
| 59 | CLOVERDALE |
| 89 | BONNEVILLE DAM |
| 102 | THE DALLES |
| 122 | NEWPORT |
| 131 | ELKTON 3 SW |
| 151 | MC MINNVILLE |
| 151 | N WILLAMETTE EXP STN |
| 151 | OTIS 2 NE |
| 164 | BROOKINGS 2 SE |
| 171 | OREGON CITY |
| 180 | FOREST GROVE |
| 200 | HILLSBORO |
| 210 | SALEM MCNARY FIELD |
| 249 | BANDON 2 NNE |
| 259 | CORVALLIS STATE UNIV |
| 302 | DRAIN |
| 302 | POWERS |
| 331 | DALLAS 2 NE |
| 351 | ARLINGTON |
| 367 | EUGENE MAHLON SWEET |
| 371 | FALLS CITY 2 |
| 371 | ILLAHE |
| 381 | FERN RIDGE DAM |
| 408 | SILVERTON |
| 410 | ESTACADA 2 SE |
| 430 | STAYTON |
| 469 | ROSEBURG KQEN |
| 502 | HOOD RIVER EXP STN |
| 512 | CORVALLIS WATER BURE |
| 620 | HERMISTON 1 SE |
| 630 | VERNONIA 2 |
| 650 | COTTAGE GROVE 1 S |
| 659 | RIDDLE |
| 679 | LEABURG 1 SW |
| 712 | LOOKOUT POINT DAM |
| 748 | PORTLAND HEADWORKS W |
| 758 | DORENA DAM |

| FEET | STATION NAME |
|---|---|
| 801 | CASCADIA |
| 820 | COTTAGE GROVE DAM |
| 932 | GRANTS PASS |
| 1060 | MILTON FREEWATER |
| 1122 | THREE LYNX |
| 1299 | DETROIT DAM |
| 1312 | OAKRIDGE FISH HATCHE |
| 1329 | MEDFORD JACKSON CO |
| 1332 | CAVE JUNCTION 1 WNW |
| 1332 | DUFUR |
| 1342 | SILVER CREEK FALLS |
| 1411 | PELTON DAM |
| 1460 | MEDFORD EXP STN |
| 1489 | PENDLETON BR EXP STN |
| 1493 | PENDLETON MUNICPL AP |
| 1552 | RUCH |
| 1640 | PILOT ROCK 1 SE |
| 1781 | ASHLAND |
| 1870 | MORO |
| 1952 | HEPPNER |
| 2001 | MONUMENT 2 |
| 2060 | TOKETEE FALLS |
| 2152 | BELKNAP SPRINGS 8 N |
| 2152 | HUNTINGTON |
| 2152 | ONTARIO KSRV |
| 2192 | NYSSA |
| 2221 | RICHLAND |
| 2221 | SCOTTS MILLS 9 SE |
| 2241 | VALE |
| 2251 | MALHEUR BRCH EXP STN |
| 2303 | MADRAS |
| 2402 | OWYHEE DAM |
| 2451 | MARION FORKS FISH HA |
| 2480 | PROSPECT 2 SW |
| 2661 | FOSSIL |
| 2680 | ELGIN |
| 2680 | HALFWAY |
| 2690 | ANTELOPE 1 NW |
| 2713 | KENT |
| 2772 | UNION EXP STN |
| 2782 | LA GRANDE |
| 2831 | JUNTURA 9 ENE |
| 2871 | PRINEVILLE 4 NW |
| 2904 | CONDON |
| 2904 | MITCHELL 2 NW |
| 2953 | WALLOWA |
| 3071 | JOHN DAY |
| 3084 | REDMOND ROBERTS FLD |
| 3182 | SISTERS |
| 3232 | RIVERSIDE 7 SSW |

| FEET | STATION NAME |
|---|---|
| 3271 | BEULAH |
| 3343 | UKIAH |
| 3373 | BAKER MUNICIPAL AP |
| 3402 | ROME 2 NW |
| 3596 | BEND |
| 3642 | GRIZZLY |
| 3671 | ROCKVILLE 5 N |
| 3681 | PAULINA |
| 3724 | LONG CREEK |
| 3836 | SEXTON SUMMIT |
| 3924 | IRONSIDE 2 W |
| 3973 | BARNES STN |
| 3983 | GOVERNMENT CAMP |
| 3983 | OCHOCO RS |
| 4091 | KLAMATH FALLS 2 SSW |
| 4114 | MALHEUR REFUGE HDQ |
| 4193 | SUMMER LAKE 1 S |
| 4213 | AUSTIN 3 S |
| 4213 | P RANCH REFUGE |
| 4334 | ALKALI LAKE |
| 4334 | WICKIUP DAM |
| 4373 | PAISLEY |
| 4380 | WHITEHORSE RANCH |
| 4382 | SILVER LAKE R S |
| 4462 | MC DERMITT 26 N |
| 4573 | ADEL |
| 4573 | HOWARD PRAIRIE DAM |
| 4583 | SHEAVILLE 1 SE |
| 4609 | FREMONT 5 NW |
| 4642 | BROTHERS |
| 4682 | SQUAW BUTTE |
| 4705 | SENECA |
| 4754 | CHEMULT |
| 4774 | LAKEVIEW 2 NNW |
| 5623 | HART MOUNTAIN REFUGE |
| 6483 | CRATER LAKE N P HQ |

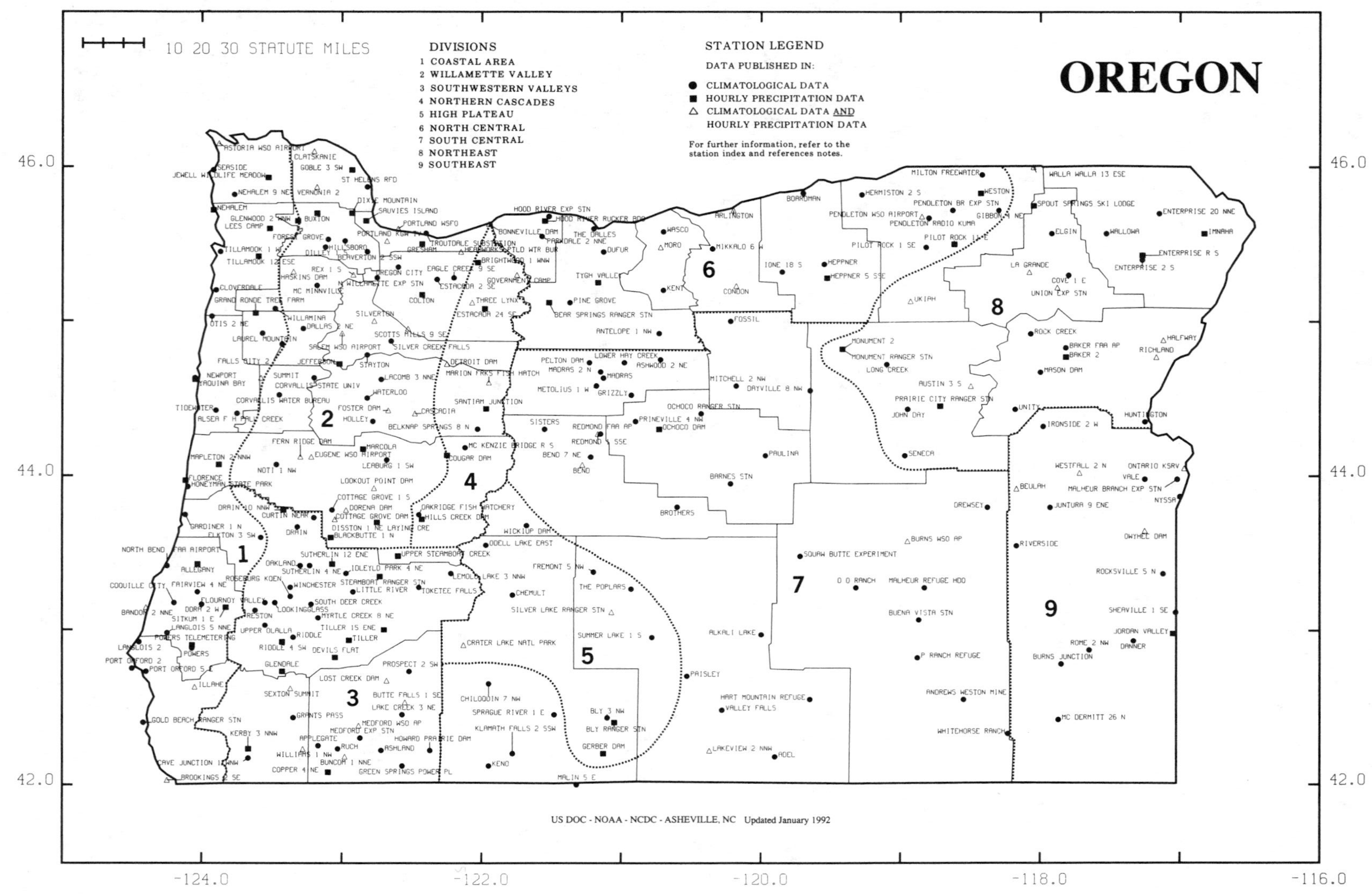
OREGON
10 20 30 STATUTE MILES
DIVISIONS
1 COASTAL AREA
2 WILLAMETTE VALLEY
3 SOUTHWESTERN VALLEYS
4 NORTHERN CASCADES
5 HIGH PLATEAU
6 NORTH CENTRAL
7 SOUTH CENTRAL
8 NORTHEAST
9 SOUTHEAST
STATION LEGEND
DATA PUBLISHED IN:
● CLIMATOLOGICAL DATA
■ HOURLY PRECIPITATION DATA
△ CLIMATOLOGICAL DATA AND HOURLY PRECIPITATION DATA
For further information, refer to the station index and references notes.
US DOC - NOAA - NCDC - ASHEVILLE, NC Updated January 1992

## ADEL *Lake County* ELEVATION 4573 ft LAT/LONG 42° 10 ' N / 119° 54 ' W

| | JAN | FEB | MAR | APR | MAY | JUN | JUL | AUG | SEP | OCT | NOV | DEC | YEAR |
|---|---|---|---|---|---|---|---|---|---|---|---|---|---|
| Maximum Temp °F | 42.5 | 46.2 | 52.2 | 59.1 | 69.8 | 78.5 | 88.2 | 86.5 | 77.9 | 66.5 | 50.3 | 42.3 | 63.3 |
| Minimum Temp °F | 22.3 | 25.0 | 28.7 | 32.1 | 39.3 | 46.3 | 52.5 | 50.6 | 42.4 | 34.4 | 28.2 | 22.0 | 35.3 |
| Mean Temp °F | 32.4 | 35.8 | 40.5 | 45.6 | 54.6 | 62.4 | 70.4 | 68.6 | 60.2 | 50.5 | 39.3 | 32.2 | 49.4 |
| Days Max Temp ≥ 90 °F | 0 | 0 | 0 | 0 | 1 | 4 | 13 | 11 | 2 | 0 | 0 | 0 | 31 |
| Days Max Temp ≤ 32 °F | 4 | 1 | 0 | 0 | 0 | 0 | 0 | 0 | 0 | 0 | 0 | 5 | 10 |
| Days Min Temp ≤ 32 °F | 26 | 22 | 21 | 16 | 5 | 0 | 0 | 0 | 3 | 11 | 20 | 25 | 149 |
| Days Min Temp ≤ 0 °F | 1 | 1 | 0 | 0 | 0 | 0 | 0 | 0 | 0 | 0 | 0 | 1 | 3 |
| Heating Degree Days | 1003 | 818 | 752 | 574 | 327 | 134 | 21 | 37 | 173 | 446 | 765 | 1010 | 6060 |
| Cooling Degree Days | 0 | na | na | 1 | na | 66 | 183 | na | na | 3 | na | na | na |
| Total Precipitation (") | 1.05 | 1.02 | 0.89 | 0.95 | 0.83 | 0.83 | 0.38 | 0.54 | 0.46 | 0.59 | 1.22 | 1.03 | 9.79 |
| Days ≥ 0.1" Precip | 3 | 3 | 3 | 3 | 2 | 2 | 1 | 2 | 2 | 2 | 5 | 3 | 31 |
| Total Snowfall (") | na | 4.6 | 3.6 | 1.3 | 0.8 | 0.0 | 0.0 | 0.0 | 0.1 | 0.6 | 2.9 | 6.2 | na |
| Days ≥ 1" Snow Depth | na | 3 | 2 | 0 | 0 | 0 | 0 | 0 | 0 | 0 | 2 | na | na |

## ALKALI LAKE *Lake County* ELEVATION 4334 ft LAT/LONG 42° 58 ' N / 120° 0 ' W

| | JAN | FEB | MAR | APR | MAY | JUN | JUL | AUG | SEP | OCT | NOV | DEC | YEAR |
|---|---|---|---|---|---|---|---|---|---|---|---|---|---|
| Maximum Temp °F | 41.9 | 46.5 | 51.9 | 59.4 | 69.6 | 78.0 | 87.6 | 86.5 | 77.9 | 65.6 | 49.7 | 41.7 | 63.0 |
| Minimum Temp °F | 19.1 | 22.8 | 25.5 | 28.8 | 34.9 | 42.3 | 47.0 | 45.7 | 36.9 | 29.3 | 24.5 | 18.4 | 31.3 |
| Mean Temp °F | 30.5 | 34.7 | 38.7 | 44.1 | 52.3 | 60.2 | 67.3 | 66.1 | 57.4 | 47.5 | 37.1 | 30.0 | 47.2 |
| Days Max Temp ≥ 90 °F | 0 | 0 | 0 | 0 | 1 | 4 | 15 | 13 | 3 | 0 | 0 | 0 | 36 |
| Days Max Temp ≤ 32 °F | 5 | 2 | 0 | 0 | 0 | 0 | 0 | 0 | 0 | 0 | 1 | 5 | 13 |
| Days Min Temp ≤ 32 °F | 26 | 23 | 24 | 21 | 12 | 2 | 0 | 1 | 7 | 20 | 23 | 27 | 186 |
| Days Min Temp ≤ 0 °F | 3 | 1 | 0 | 0 | 0 | 0 | 0 | 0 | 0 | 0 | 0 | 2 | 6 |
| Heating Degree Days | 1061 | 849 | 807 | 620 | 389 | 178 | 50 | 62 | 234 | 538 | 827 | 1077 | 6692 |
| Cooling Degree Days | 0 | 0 | 0 | 0 | 8 | 42 | 120 | 103 | 15 | 0 | 0 | 0 | 288 |
| Total Precipitation (") | 0.57 | 0.47 | 0.73 | 0.84 | 1.10 | 1.11 | 0.59 | 0.74 | 0.52 | 0.65 | 0.76 | 0.64 | 8.72 |
| Days ≥ 0.1" Precip | 2 | 2 | 3 | 3 | 3 | 3 | 1 | 2 | 1 | 2 | 2 | 2 | 26 |
| Total Snowfall (") | 3.6 | 2.5 | 2.4 | 1.7 | 0.7 | 0.1 | 0.0 | 0.0 | 0.0 | 0.3 | 2.5 | 5.2 | 19.0 |
| Days ≥ 1" Snow Depth | 7 | 4 | 2 | 0 | 0 | 0 | 0 | 0 | 0 | 0 | 2 | 6 | 21 |

## ANTELOPE 1 NW *Wasco County* ELEVATION 2690 ft LAT/LONG 44° 53 ' N / 120° 45 ' W

| | JAN | FEB | MAR | APR | MAY | JUN | JUL | AUG | SEP | OCT | NOV | DEC | YEAR |
|---|---|---|---|---|---|---|---|---|---|---|---|---|---|
| Maximum Temp °F | 40.0 | 45.5 | 52.3 | 58.6 | 67.5 | 75.7 | 84.2 | 84.1 | 75.2 | 63.6 | 48.0 | 39.7 | 61.2 |
| Minimum Temp °F | 24.6 | 27.5 | 30.4 | 33.2 | 39.1 | 45.7 | 50.5 | 50.5 | 44.0 | 36.6 | 30.1 | 24.7 | 36.4 |
| Mean Temp °F | 32.3 | 36.5 | 41.4 | 45.9 | 53.3 | 60.7 | 67.4 | 67.4 | 59.6 | 50.1 | 39.1 | 32.2 | 48.8 |
| Days Max Temp ≥ 90 °F | 0 | 0 | 0 | 0 | 0 | 3 | 10 | 9 | 1 | 0 | 0 | 0 | 23 |
| Days Max Temp ≤ 32 °F | 6 | 3 | 0 | 0 | 0 | 0 | 0 | 0 | 0 | 0 | 1 | 6 | 16 |
| Days Min Temp ≤ 32 °F | 25 | 21 | 20 | 14 | 6 | 1 | 0 | 0 | 2 | 9 | 18 | 25 | 141 |
| Days Min Temp ≤ 0 °F | 1 | 0 | 0 | 0 | 0 | 0 | 0 | 0 | 0 | 0 | 0 | 1 | 2 |
| Heating Degree Days | 1006 | 798 | 726 | 566 | 366 | 174 | 59 | 57 | 191 | 456 | 768 | 1009 | 6176 |
| Cooling Degree Days | 0 | 0 | 0 | 0 | 10 | 37 | 101 | 108 | 30 | 2 | 0 | 0 | 288 |
| Total Precipitation (") | 1.60 | 1.02 | 1.18 | 1.03 | 1.14 | 1.09 | 0.45 | 0.66 | 0.74 | 0.86 | 1.82 | 1.51 | 13.10 |
| Days ≥ 0.1" Precip | 5 | 3 | 4 | 3 | 3 | 3 | 2 | 2 | 2 | 3 | 6 | 5 | 41 |
| Total Snowfall (") | 6.8 | 3.2 | 2.0 | 1.3 | 0.3 | 0.0 | 0.0 | 0.0 | 0.0 | 0.3 | 2.9 | 6.8 | 23.6 |
| Days ≥ 1" Snow Depth | 5 | 2 | 0 | 0 | 0 | 0 | 0 | 0 | 0 | 0 | 1 | 4 | 12 |

## ARLINGTON *Gilliam County* ELEVATION 351 ft LAT/LONG 45° 43 ' N / 120° 11 ' W

| | JAN | FEB | MAR | APR | MAY | JUN | JUL | AUG | SEP | OCT | NOV | DEC | YEAR |
|---|---|---|---|---|---|---|---|---|---|---|---|---|---|
| Maximum Temp °F | 40.7 | 47.3 | 57.0 | 65.7 | 75.8 | 83.6 | 91.5 | 90.0 | 80.2 | 65.6 | 49.8 | 41.0 | 65.7 |
| Minimum Temp °F | 29.2 | 32.0 | 36.4 | 41.9 | 48.9 | 56.1 | 61.2 | 60.6 | 52.0 | 42.3 | 35.7 | 29.7 | 43.8 |
| Mean Temp °F | 35.0 | 39.7 | 46.8 | 53.9 | 62.4 | 69.9 | 76.4 | 75.3 | 66.1 | 53.9 | 42.8 | 35.4 | 54.8 |
| Days Max Temp ≥ 90 °F | 0 | 0 | 0 | 0 | 3 | 9 | 18 | 17 | 4 | 0 | 0 | 0 | 51 |
| Days Max Temp ≤ 32 °F | 6 | 2 | 0 | 0 | 0 | 0 | 0 | 0 | 0 | 0 | 1 | 5 | 14 |
| Days Min Temp ≤ 32 °F | 17 | 14 | 9 | 3 | 0 | 0 | 0 | 0 | 0 | 3 | 9 | 18 | 73 |
| Days Min Temp ≤ 0 °F | 1 | 0 | 0 | 0 | 0 | 0 | 0 | 0 | 0 | 0 | 0 | 1 | 2 |
| Heating Degree Days | 923 | 708 | 559 | 332 | 131 | 26 | 2 | 3 | 65 | 341 | 660 | 912 | 4662 |
| Cooling Degree Days | 0 | 0 | 0 | 7 | 63 | 167 | 343 | 316 | 104 | 6 | 0 | 0 | 1006 |
| Total Precipitation (") | 1.32 | 0.80 | 0.71 | 0.62 | 0.57 | 0.39 | 0.22 | 0.31 | 0.40 | 0.60 | 1.21 | 1.36 | 8.51 |
| Days ≥ 0.1" Precip | 4 | 3 | 3 | 2 | 2 | 1 | 1 | 1 | 1 | 2 | 4 | 4 | 28 |
| Total Snowfall (") | 3.1 | 1.6 | 0.2 | 0.0 | 0.0 | 0.0 | 0.0 | 0.0 | 0.0 | 0.1 | 0.8 | 2.8 | 8.6 |
| Days ≥ 1" Snow Depth | 3 | 1 | 0 | 0 | 0 | 0 | 0 | 0 | 0 | 0 | 1 | 3 | 8 |

## ASHLAND *Jackson County* ELEVATION 1781 ft LAT/LONG 42° 13 ' N / 122° 43 ' W

| | JAN | FEB | MAR | APR | MAY | JUN | JUL | AUG | SEP | OCT | NOV | DEC | YEAR |
|---|---|---|---|---|---|---|---|---|---|---|---|---|---|
| Maximum Temp °F | 46.7 | 52.7 | 57.1 | 62.3 | 70.8 | 78.8 | 87.1 | 86.1 | 79.4 | 67.4 | 52.4 | 45.7 | 65.5 |
| Minimum Temp °F | 29.9 | 32.0 | 34.0 | 36.6 | 41.9 | 48.0 | 51.8 | 51.4 | 45.6 | 38.8 | 33.9 | 29.9 | 39.5 |
| Mean Temp °F | 38.4 | 42.4 | 45.6 | 49.5 | 56.4 | 63.4 | 69.5 | 68.8 | 62.5 | 53.2 | 43.2 | 37.8 | 52.6 |
| Days Max Temp ≥ 90 °F | 0 | 0 | 0 | 0 | 1 | 5 | 13 | 12 | 4 | 0 | 0 | 0 | 35 |
| Days Max Temp ≤ 32 °F | 1 | 0 | 0 | 0 | 0 | 0 | 0 | 0 | 0 | 0 | 0 | 1 | 2 |
| Days Min Temp ≤ 32 °F | 20 | 16 | 13 | 8 | 2 | 0 | 0 | 0 | 0 | 5 | 14 | 20 | 98 |
| Days Min Temp ≤ 0 °F | 0 | 0 | 0 | 0 | 0 | 0 | 0 | 0 | 0 | 0 | 0 | 0 | 0 |
| Heating Degree Days | 819 | 633 | 595 | 459 | 274 | 105 | 23 | 25 | 114 | 363 | 648 | 835 | 4893 |
| Cooling Degree Days | 0 | 0 | 0 | 1 | 14 | 59 | 144 | 133 | 43 | 4 | 0 | 0 | 398 |
| Total Precipitation (") | 2.36 | 1.73 | 1.93 | 1.63 | 1.37 | 0.97 | 0.39 | 0.59 | 0.89 | 1.35 | 2.76 | 2.74 | 18.71 |
| Days ≥ 0.1" Precip | 6 | 5 | 6 | 5 | 4 | 3 | 1 | 1 | 3 | 4 | 8 | 7 | 53 |
| Total Snowfall (") | 2.5 | 1.3 | 0.5 | 0.3 | 0.0 | 0.0 | 0.0 | 0.0 | 0.0 | 0.0 | 0.1 | 2.0 | 6.7 |
| Days ≥ 1" Snow Depth | 2 | 1 | 0 | 0 | 0 | 0 | 0 | 0 | 0 | 0 | 0 | 2 | 5 |

## ASTORIA CLATSOP ARPT *Clatsop County* ELEVATION 10 ft LAT/LONG 46° 9 ' N / 123° 53 ' W

| | JAN | FEB | MAR | APR | MAY | JUN | JUL | AUG | SEP | OCT | NOV | DEC | YEAR |
|---|---|---|---|---|---|---|---|---|---|---|---|---|---|
| Maximum Temp °F | 47.9 | 51.1 | 53.6 | 56.1 | 60.5 | 64.4 | 67.7 | 69.0 | 67.8 | 61.1 | 53.2 | 48.2 | 58.4 |
| Minimum Temp °F | 36.1 | 37.2 | 38.2 | 40.3 | 44.8 | 49.7 | 52.5 | 52.8 | 49.1 | 43.9 | 40.0 | 36.6 | 43.4 |
| Mean Temp °F | 42.0 | 44.2 | 45.9 | 48.3 | 52.7 | 57.0 | 60.2 | 60.9 | 58.4 | 52.5 | 46.6 | 42.4 | 50.9 |
| Days Max Temp ≥ 90 °F | 0 | 0 | 0 | 0 | 0 | 0 | 0 | 0 | 0 | 0 | 0 | 0 | 0 |
| Days Max Temp ≤ 32 °F | 1 | 0 | 0 | 0 | 0 | 0 | 0 | 0 | 0 | 0 | 0 | 1 | 2 |
| Days Min Temp ≤ 32 °F | 10 | 7 | 5 | 2 | 0 | 0 | 0 | 0 | 0 | 1 | 5 | 9 | 39 |
| Days Min Temp ≤ 0 °F | 0 | 0 | 0 | 0 | 0 | 0 | 0 | 0 | 0 | 0 | 0 | 0 | 0 |
| Heating Degree Days | 705 | 580 | 585 | 496 | 375 | 235 | 147 | 127 | 196 | 380 | 544 | 694 | 5064 |
| Cooling Degree Days | 0 | 0 | 0 | 0 | 1 | 3 | 3 | 7 | 6 | 1 | 0 | 0 | 21 |
| Total Precipitation (") | 9.68 | 7.22 | 6.76 | 4.87 | 3.04 | 2.43 | 1.10 | 1.27 | 2.77 | 5.45 | 9.82 | 10.45 | 64.86 |
| Days ≥ 0.1" Precip | 16 | 14 | 14 | 12 | 8 | 6 | 3 | 3 | 6 | 10 | 16 | 17 | 125 |
| Total Snowfall (") | 2.4 | 0.4 | 0.4 | 0.1 | 0.0 | 0.0 | 0.0 | 0.0 | 0.0 | 0.0 | 0.2 | 0.8 | 4.3 |
| Days ≥ 1" Snow Depth | 1 | 0 | 0 | 0 | 0 | 0 | 0 | 0 | 0 | 0 | 0 | 1 | 2 |

## AUSTIN 3 S *Grant County* ELEVATION 4213 ft LAT/LONG 44° 35 ' N / 118° 30 ' W

| | JAN | FEB | MAR | APR | MAY | JUN | JUL | AUG | SEP | OCT | NOV | DEC | YEAR |
|---|---|---|---|---|---|---|---|---|---|---|---|---|---|
| Maximum Temp °F | 34.6 | 40.5 | 46.7 | 54.1 | 63.2 | 71.7 | 81.3 | 81.7 | 72.1 | 60.3 | 42.7 | 34.6 | 57.0 |
| Minimum Temp °F | 10.7 | 14.5 | 19.7 | 25.6 | 31.2 | 37.0 | 39.5 | 38.2 | 30.9 | 25.0 | 20.5 | 12.1 | 25.4 |
| Mean Temp °F | 22.7 | 27.5 | 33.3 | 39.9 | 47.2 | 54.4 | 60.4 | 60.0 | 51.5 | 42.6 | 31.5 | 23.3 | 41.2 |
| Days Max Temp ≥ 90 °F | 0 | 0 | 0 | 0 | 0 | 1 | 6 | 6 | 1 | 0 | 0 | 0 | 14 |
| Days Max Temp ≤ 32 °F | 11 | 4 | 1 | 0 | 0 | 0 | 0 | 0 | 0 | 0 | 3 | 11 | 30 |
| Days Min Temp ≤ 32 °F | 31 | 28 | 30 | 26 | *18* | 8 | 3 | 5 | 19 | 27 | 28 | 31 | 254 |
| Days Min Temp ≤ 0 °F | 8 | 4 | 1 | 0 | 0 | 0 | 0 | 0 | 0 | 0 | 1 | 5 | 19 |
| Heating Degree Days | 1304 | 1052 | 977 | 747 | 546 | 319 | 161 | 172 | 401 | 686 | 998 | 1284 | 8647 |
| Cooling Degree Days | 0 | 0 | 0 | 0 | 1 | 6 | 18 | 22 | 1 | 0 | 0 | 0 | 48 |
| Total Precipitation (") | 2.81 | 1.87 | 1.92 | 1.47 | 1.51 | 1.64 | 0.81 | 1.08 | 0.98 | 1.23 | 2.60 | 2.85 | 20.77 |
| Days ≥ 0.1" Precip | 9 | 7 | 7 | 5 | 5 | 5 | 2 | 3 | 3 | 4 | 8 | 9 | 67 |
| Total Snowfall (") | 22.6 | 14.4 | 10.8 | 5.1 | 0.4 | 0.0 | 0.0 | 0.0 | 0.1 | 1.3 | *14.2* | 23.4 | 92.3 |
| Days ≥ 1" Snow Depth | 30 | 27 | 23 | 5 | 0 | 0 | 0 | 0 | 0 | 1 | 12 | 29 | 127 |

## BAKER MUNICIPAL AP *Baker County* ELEVATION 3373 ft LAT/LONG 44° 50 ' N / 117° 50 ' W

| | JAN | FEB | MAR | APR | MAY | JUN | JUL | AUG | SEP | OCT | NOV | DEC | YEAR |
|---|---|---|---|---|---|---|---|---|---|---|---|---|---|
| Maximum Temp °F | 34.1 | 41.3 | 50.7 | 58.9 | 67.6 | 75.5 | 84.4 | 84.4 | 75.0 | 62.6 | 44.9 | 35.3 | 59.6 |
| Minimum Temp °F | 17.4 | 22.2 | 26.9 | 31.3 | 38.2 | 44.8 | 48.3 | 47.3 | 38.9 | 30.4 | 24.7 | 17.6 | 32.3 |
| Mean Temp °F | 25.7 | 31.8 | 38.8 | 45.2 | 52.9 | 60.1 | 66.3 | 65.9 | 57.0 | 46.5 | 34.8 | 26.4 | 45.9 |
| Days Max Temp ≥ 90 °F | 0 | 0 | 0 | 0 | 0 | 2 | 9 | 10 | 2 | 0 | 0 | 0 | 23 |
| Days Max Temp ≤ 32 °F | 12 | 4 | 0 | 0 | 0 | 0 | 0 | 0 | 0 | 0 | 2 | 10 | 28 |
| Days Min Temp ≤ 32 °F | 28 | 25 | 25 | 18 | 6 | 1 | 0 | 0 | 5 | 19 | 25 | 28 | 180 |
| Days Min Temp ≤ 0 °F | 4 | 1 | 0 | 0 | 0 | 0 | 0 | 0 | 0 | 0 | 0 | 3 | 8 |
| Heating Degree Days | 1210 | 931 | 803 | 588 | 371 | 172 | 54 | 60 | 244 | 567 | 900 | 1189 | 7089 |
| Cooling Degree Days | 0 | 0 | 0 | 0 | 6 | 34 | 92 | 93 | 11 | 0 | 0 | 0 | 236 |
| Total Precipitation (") | 0.94 | 0.61 | 0.84 | 0.86 | 1.28 | 1.35 | 0.64 | 0.96 | 0.68 | 0.59 | 0.93 | 0.92 | 10.60 |
| Days ≥ 0.1" Precip | 3 | 2 | 3 | 3 | 4 | 4 | 2 | 3 | 2 | 2 | 3 | 3 | 34 |
| Total Snowfall (") | 5.7 | 3.6 | 2.7 | 1.3 | 0.5 | 0.0 | 0.0 | 0.0 | 0.0 | 0.4 | 3.6 | 6.9 | 24.7 |
| Days ≥ 1" Snow Depth | 18 | 9 | 3 | 0 | 0 | 0 | 0 | 0 | 0 | 0 | 4 | 13 | 47 |

## BANDON 2 NNE *Coos County* ELEVATION 249 ft LAT/LONG 43° 7' N / 124° 22' W

| | JAN | FEB | MAR | APR | MAY | JUN | JUL | AUG | SEP | OCT | NOV | DEC | YEAR |
|---|---|---|---|---|---|---|---|---|---|---|---|---|---|
| Maximum Temp °F | 53.8 | 55.4 | 56.2 | 57.6 | 60.9 | 64.1 | 66.6 | 67.5 | 67.6 | 63.6 | 57.4 | 53.6 | 60.4 |
| Minimum Temp °F | 38.2 | 39.4 | 39.9 | 41.0 | 44.1 | 48.2 | 50.6 | 50.7 | 47.8 | 44.1 | 41.6 | 38.3 | 43.7 |
| Mean Temp °F | 46.0 | 47.4 | 48.1 | 49.4 | 52.5 | 56.2 | 58.6 | 59.1 | 57.7 | 53.9 | 49.5 | 46.0 | 52.0 |
| Days Max Temp ≥ 90 °F | 0 | 0 | 0 | 0 | 0 | 0 | 0 | 0 | 0 | 0 | 0 | 0 | 0 |
| Days Max Temp ≤ 32 °F | 0 | 0 | 0 | 0 | 0 | 0 | 0 | 0 | 0 | 0 | 0 | 0 | 0 |
| Days Min Temp ≤ 32 °F | 7 | 5 | 4 | 2 | 0 | 0 | 0 | 0 | 0 | 1 | 3 | 7 | 29 |
| Days Min Temp ≤ 0 °F | 0 | 0 | 0 | 0 | 0 | 0 | 0 | 0 | 0 | 0 | 0 | 0 | 0 |
| Heating Degree Days | 582 | 490 | 518 | 463 | 379 | 258 | 191 | 176 | 213 | 339 | 457 | 583 | 4649 |
| Cooling Degree Days | 0 | 0 | 0 | 0 | 0 | 0 | 1 | 1 | 1 | 2 | 0 | 0 | 5 |
| Total Precipitation (") | 9.10 | 6.81 | 7.01 | 4.49 | 2.60 | 1.57 | 0.43 | 0.95 | 1.63 | 3.75 | 8.65 | 10.14 | 57.13 |
| Days ≥ 0.1" Precip | 14 | 13 | 14 | 10 | 6 | 4 | 1 | 2 | 4 | 7 | 14 | 15 | 104 |
| Total Snowfall (") | 0.8 | 0.2 | 0.0 | 0.0 | 0.0 | 0.0 | 0.0 | 0.0 | 0.0 | 0.0 | 0.0 | 0.1 | 1.1 |
| Days ≥ 1" Snow Depth | 0 | 0 | 0 | 0 | 0 | 0 | 0 | 0 | 0 | 0 | 0 | 0 | 0 |

## BARNES STN *Crook County* ELEVATION 3973 ft LAT/LONG 43° 57' N / 120° 13' W

| | JAN | FEB | MAR | APR | MAY | JUN | JUL | AUG | SEP | OCT | NOV | DEC | YEAR |
|---|---|---|---|---|---|---|---|---|---|---|---|---|---|
| Maximum Temp °F | 38.7 | 44.3 | 50.5 | 57.5 | 66.6 | 75.1 | 84.1 | 83.4 | 75.3 | 63.7 | 46.6 | 38.9 | 60.4 |
| Minimum Temp °F | 18.3 | 22.7 | 25.9 | 28.6 | 34.7 | 41.3 | 45.7 | 44.6 | 37.0 | 30.1 | 25.0 | 18.7 | 31.1 |
| Mean Temp °F | 28.5 | 33.5 | 38.2 | 43.1 | 50.7 | 58.2 | 64.9 | 64.0 | 56.1 | 47.0 | 35.8 | 28.8 | 45.7 |
| Days Max Temp ≥ 90 °F | 0 | 0 | 0 | 0 | 0 | 2 | 9 | 8 | 2 | 0 | 0 | 0 | 21 |
| Days Max Temp ≤ 32 °F | 7 | 2 | 0 | 0 | 0 | 0 | 0 | 0 | 0 | 0 | 1 | 6 | 16 |
| Days Min Temp ≤ 32 °F | 28 | 25 | 25 | 21 | 13 | 3 | 0 | 1 | 8 | 20 | 24 | 28 | 196 |
| Days Min Temp ≤ 0 °F | 3 | 1 | 0 | 0 | 0 | 0 | 0 | 0 | 0 | 0 | 1 | 2 | 7 |
| Heating Degree Days | 1125 | 882 | 824 | 650 | 440 | 221 | 78 | 92 | 267 | 553 | 869 | 1114 | 7115 |
| Cooling Degree Days | 0 | 0 | 0 | 0 | 5 | 25 | 76 | 67 | 11 | 0 | 0 | 0 | 184 |
| Total Precipitation (") | 1.27 | 0.91 | 1.08 | 0.92 | 1.22 | 1.14 | 0.78 | 0.94 | 0.61 | 0.83 | 1.64 | 1.41 | 12.75 |
| Days ≥ 0.1" Precip | 4 | 4 | 4 | 3 | 4 | 3 | 2 | 3 | 2 | 3 | 6 | 5 | 43 |
| Total Snowfall (") | 10.1 | 7.1 | 4.2 | 1.5 | 0.6 | 0.1 | 0.0 | 0.0 | 0.1 | 1.4 | 7.6 | 12.6 | 45.3 |
| Days ≥ 1" Snow Depth | 20 | 10 | 2 | 0 | 0 | 0 | 0 | 0 | 0 | 0 | 5 | 15 | 52 |

## BELKNAP SPRINGS 8 N *Linn County* ELEVATION 2152 ft LAT/LONG 44° 18' N / 122° 2' W

| | JAN | FEB | MAR | APR | MAY | JUN | JUL | AUG | SEP | OCT | NOV | DEC | YEAR |
|---|---|---|---|---|---|---|---|---|---|---|---|---|---|
| Maximum Temp °F | 39.0 | 44.5 | 50.6 | 56.5 | 65.5 | 72.6 | 80.9 | 81.2 | 75.0 | 62.8 | 46.7 | 38.6 | 59.5 |
| Minimum Temp °F | 27.3 | 29.0 | 30.8 | 33.7 | 38.7 | 45.0 | 48.5 | 48.3 | 43.5 | 37.6 | 32.5 | 28.3 | 36.9 |
| Mean Temp °F | 33.2 | 36.8 | 40.7 | 45.2 | 52.2 | 58.8 | 64.7 | 64.8 | 59.3 | 50.2 | 39.7 | 33.5 | 48.3 |
| Days Max Temp ≥ 90 °F | 0 | 0 | 0 | 0 | 1 | 2 | 7 | 7 | 3 | 0 | 0 | 0 | 20 |
| Days Max Temp ≤ 32 °F | 3 | 1 | 0 | 0 | 0 | 0 | 0 | 0 | 0 | 0 | 0 | 2 | 6 |
| Days Min Temp ≤ 32 °F | 26 | 23 | 22 | 14 | 4 | 0 | 0 | 0 | 1 | 5 | 16 | 26 | 137 |
| Days Min Temp ≤ 0 °F | 0 | 0 | 0 | 0 | 0 | 0 | 0 | 0 | 0 | 0 | 0 | 0 | 0 |
| Heating Degree Days | 979 | 791 | 746 | 589 | 398 | 206 | 81 | 76 | 189 | 453 | 754 | 970 | 6232 |
| Cooling Degree Days | 0 | 0 | 0 | 1 | 11 | 31 | 78 | 78 | 31 | 2 | 0 | 0 | 232 |
| Total Precipitation (") | 11.37 | 8.57 | 7.56 | 5.40 | 3.51 | 2.46 | 0.96 | 1.15 | 2.32 | 5.52 | 11.37 | 12.06 | 72.25 |
| Days ≥ 0.1" Precip | 15 | 13 | 14 | 12 | 8 | 6 | 3 | 3 | 5 | 9 | 16 | 16 | 120 |
| Total Snowfall (") | 23.8 | 15.7 | 10.3 | 2.0 | 0.1 | 0.0 | 0.0 | 0.0 | 0.0 | 0.3 | 5.8 | 20.2 | 78.2 |
| Days ≥ 1" Snow Depth | 20 | 16 | 12 | 2 | 0 | 0 | 0 | 0 | 0 | 0 | 4 | 17 | 71 |

## BEND *Deschutes County* ELEVATION 3596 ft LAT/LONG 44° 4' N / 121° 19' W

| | JAN | FEB | MAR | APR | MAY | JUN | JUL | AUG | SEP | OCT | NOV | DEC | YEAR |
|---|---|---|---|---|---|---|---|---|---|---|---|---|---|
| Maximum Temp °F | 41.3 | 45.8 | 51.6 | 57.4 | 65.6 | 73.4 | 81.2 | 81.1 | 73.3 | 63.2 | 48.1 | 41.2 | 60.3 |
| Minimum Temp °F | 22.2 | 24.2 | 26.5 | 29.4 | 35.1 | 41.3 | 45.4 | 44.9 | 37.7 | 31.7 | 26.9 | 22.3 | 32.3 |
| Mean Temp °F | 31.8 | 35.1 | 39.1 | 43.5 | 50.4 | 57.4 | 63.3 | 63.0 | 55.5 | 47.5 | 37.5 | 31.8 | 46.3 |
| Days Max Temp ≥ 90 °F | 0 | 0 | 0 | 0 | 0 | 1 | 5 | 6 | 1 | 0 | 0 | 0 | 13 |
| Days Max Temp ≤ 32 °F | 5 | 2 | 0 | 0 | 0 | 0 | 0 | 0 | 0 | 0 | 1 | 4 | 12 |
| Days Min Temp ≤ 32 °F | 26 | 24 | 25 | 20 | 12 | 4 | 1 | 1 | 8 | 17 | 22 | 26 | 186 |
| Days Min Temp ≤ 0 °F | 1 | 1 | 0 | 0 | 0 | 0 | 0 | 0 | 0 | 0 | 0 | 1 | 3 |
| Heating Degree Days | 1022 | 839 | 798 | 640 | 450 | 246 | 109 | 113 | 286 | 536 | 818 | 1022 | 6879 |
| Cooling Degree Days | 0 | 0 | 0 | 0 | 6 | 25 | 64 | 62 | 14 | 1 | 0 | 0 | 172 |
| Total Precipitation (") | 1.86 | 0.89 | 0.93 | 0.64 | 0.75 | 0.90 | 0.59 | 0.56 | 0.47 | 0.63 | 1.41 | 1.70 | 11.33 |
| Days ≥ 0.1" Precip | 5 | 3 | 3 | 2 | 2 | 3 | 2 | 2 | 2 | 2 | 4 | 5 | 35 |
| Total Snowfall (") | 9.5 | 4.6 | 3.4 | 1.8 | 0.2 | 0.0 | 0.0 | 0.0 | 0.0 | 0.3 | 5.1 | 9.4 | 34.3 |
| Days ≥ 1" Snow Depth | 10 | 5 | 2 | 1 | 0 | 0 | 0 | 0 | 0 | 0 | 3 | 10 | 31 |

## BEULAH *Malheur County* ELEVATION 3271 ft LAT/LONG 43° 55 ' N / 118° 10 ' W

| | JAN | FEB | MAR | APR | MAY | JUN | JUL | AUG | SEP | OCT | NOV | DEC | YEAR |
|---|---|---|---|---|---|---|---|---|---|---|---|---|---|
| Maximum Temp °F | 36.5 | 44.4 | 53.5 | 61.9 | 71.8 | 80.3 | 89.8 | 88.9 | 79.9 | *67.4* | *49.0* | 37.8 | 63.4 |
| Minimum Temp °F | 16.4 | 22.1 | 27.2 | 32.4 | 40.5 | 47.3 | 53.0 | 51.2 | 41.4 | *31.2* | na | 16.7 | na |
| Mean Temp °F | 26.4 | 33.2 | 40.4 | 47.1 | 56.2 | 63.8 | 71.5 | 70.1 | 60.6 | *49.3* | na | 27.3 | na |
| Days Max Temp ≥ 90 °F | 0 | 0 | 0 | 0 | 1 | 6 | 18 | 17 | 5 | 0 | 0 | 0 | 47 |
| Days Max Temp ≤ 32 °F | 9 | 2 | 0 | 0 | 0 | 0 | 0 | 0 | 0 | 0 | 1 | 6 | 18 |
| Days Min Temp ≤ 32 °F | 29 | 26 | 24 | 16 | 4 | 0 | 0 | 0 | 4 | 18 | *25* | 28 | 174 |
| Days Min Temp ≤ 0 °F | 4 | 1 | 0 | 0 | 0 | 0 | 0 | 0 | 0 | 0 | 0 | 3 | 8 |
| Heating Degree Days | 1188 | 891 | 757 | 530 | 281 | 109 | 20 | 26 | 160 | *479* | na | 1163 | na |
| Cooling Degree Days | 0 | 0 | 0 | 0 | 18 | 68 | 194 | 168 | 34 | 0 | 0 | 0 | 482 |
| Total Precipitation (") | 1.36 | 0.96 | 1.10 | 0.80 | 0.97 | 0.91 | 0.38 | 0.59 | 0.52 | 0.72 | 1.53 | 1.57 | 11.41 |
| Days ≥ 0.1" Precip | 4 | 3 | 4 | 3 | 3 | 3 | 1 | 2 | 1 | 2 | *5* | 4 | 35 |
| Total Snowfall (") | 9.3 | 4.0 | 1.5 | 0.1 | 0.1 | 0.0 | 0.0 | 0.0 | 0.0 | 0.2 | *4.3* | 10.9 | 30.4 |
| Days ≥ 1" Snow Depth | 17 | 6 | 1 | 0 | 0 | 0 | 0 | 0 | 0 | 0 | 2 | 14 | 40 |

## BONNEVILLE DAM *Multnomah County* ELEVATION 89 ft LAT/LONG 45° 38 ' N / 121° 57 ' W

| | JAN | FEB | MAR | APR | MAY | JUN | JUL | AUG | SEP | OCT | NOV | DEC | YEAR |
|---|---|---|---|---|---|---|---|---|---|---|---|---|---|
| Maximum Temp °F | 42.9 | 47.4 | 54.3 | 59.5 | 66.8 | 72.4 | 78.4 | 79.3 | 73.8 | 63.4 | 51.0 | 43.6 | 61.1 |
| Minimum Temp °F | 33.2 | 35.7 | 38.3 | 42.0 | 47.1 | 52.5 | 56.6 | 56.6 | 52.8 | 46.7 | 40.2 | 34.7 | 44.7 |
| Mean Temp °F | 38.0 | 41.6 | 46.3 | 50.8 | 57.0 | 62.5 | 67.5 | 68.0 | 63.3 | 55.1 | 45.6 | 39.2 | 52.9 |
| Days Max Temp ≥ 90 °F | 0 | 0 | 0 | 0 | 0 | 1 | 3 | 3 | 1 | 0 | 0 | 0 | 8 |
| Days Max Temp ≤ 32 °F | 3 | 1 | 0 | 0 | 0 | 0 | 0 | 0 | 0 | 0 | 1 | 2 | 7 |
| Days Min Temp ≤ 32 °F | 12 | 7 | 3 | 1 | 0 | 0 | 0 | 0 | 0 | 0 | 2 | 9 | 34 |
| Days Min Temp ≤ 0 °F | 0 | 0 | 0 | 0 | 0 | 0 | 0 | 0 | 0 | 0 | 0 | 0 | 0 |
| Heating Degree Days | 829 | 655 | 572 | 420 | 257 | 115 | 31 | 25 | 97 | 305 | 574 | 794 | 4674 |
| Cooling Degree Days | 0 | 0 | 0 | 1 | 17 | 48 | 120 | 138 | 59 | 7 | 0 | 0 | 390 |
| Total Precipitation (") | 11.30 | 8.80 | 7.49 | 5.78 | 3.45 | 2.70 | 1.00 | 1.50 | 3.04 | 5.55 | 11.14 | 12.25 | 74.00 |
| Days ≥ 0.1" Precip | 16 | 13 | 14 | 12 | 9 | 6 | 3 | 3 | 6 | 10 | 16 | 16 | 124 |
| Total Snowfall (") | 9.0 | 3.6 | 0.9 | 0.0 | 0.0 | 0.0 | 0.0 | 0.0 | 0.0 | 0.0 | 1.3 | 3.2 | 18.0 |
| Days ≥ 1" Snow Depth | 5 | 2 | 0 | 0 | 0 | 0 | 0 | 0 | 0 | 0 | 1 | 1 | 9 |

## BROOKINGS 2 SE *Curry County* ELEVATION 164 ft LAT/LONG 42° 3 ' N / 124° 17 ' W

| | JAN | FEB | MAR | APR | MAY | JUN | JUL | AUG | SEP | OCT | NOV | DEC | YEAR |
|---|---|---|---|---|---|---|---|---|---|---|---|---|---|
| Maximum Temp °F | 54.8 | 56.3 | 57.7 | 59.7 | 63.6 | 66.5 | 67.9 | 67.6 | 68.5 | 64.7 | 58.5 | 54.6 | 61.7 |
| Minimum Temp °F | 41.1 | 42.1 | 42.2 | 43.1 | 46.4 | 49.7 | 51.4 | 52.3 | 51.2 | 48.0 | 44.5 | 41.0 | 46.1 |
| Mean Temp °F | 48.0 | 49.2 | 50.0 | 51.4 | 55.0 | 58.1 | 59.7 | 60.0 | 59.8 | 56.4 | 51.5 | 47.8 | 53.9 |
| Days Max Temp ≥ 90 °F | 0 | 0 | 0 | 0 | 0 | 0 | 0 | 0 | 1 | 0 | 0 | 0 | 1 |
| Days Max Temp ≤ 32 °F | 0 | 0 | 0 | 0 | 0 | 0 | 0 | 0 | 0 | 0 | 0 | 0 | 0 |
| Days Min Temp ≤ 32 °F | 2 | 1 | 0 | 0 | 0 | 0 | 0 | 0 | 0 | 0 | 0 | 2 | 5 |
| Days Min Temp ≤ 0 °F | 0 | 0 | 0 | 0 | 0 | 0 | 0 | 0 | 0 | 0 | 0 | 0 | 0 |
| Heating Degree Days | 519 | 439 | 458 | 402 | 306 | 208 | 166 | 156 | 166 | 267 | 398 | 525 | 4010 |
| Cooling Degree Days | 0 | 0 | 0 | 2 | 3 | 11 | 12 | 11 | 15 | 9 | 1 | 0 | 64 |
| Total Precipitation (") | 11.08 | 8.73 | 9.03 | 5.60 | 3.35 | 1.65 | 0.51 | 1.23 | 2.08 | 5.23 | 10.52 | 12.37 | 71.38 |
| Days ≥ 0.1" Precip | 14 | 13 | 14 | 10 | 6 | 3 | 1 | 2 | 3 | 7 | 14 | 15 | 102 |
| Total Snowfall (") | 0.2 | 0.3 | 0.1 | 0.0 | 0.0 | 0.0 | 0.0 | 0.0 | 0.0 | 0.0 | 0.0 | 0.1 | 0.7 |
| Days ≥ 1" Snow Depth | 0 | 0 | 0 | 0 | 0 | 0 | 0 | 0 | 0 | 0 | 0 | 0 | 0 |

## BROTHERS *Deschutes County* ELEVATION 4642 ft LAT/LONG 43° 48 ' N / 120° 36 ' W

| | JAN | FEB | MAR | APR | MAY | JUN | JUL | AUG | SEP | OCT | NOV | DEC | YEAR |
|---|---|---|---|---|---|---|---|---|---|---|---|---|---|
| Maximum Temp °F | 37.5 | 42.6 | 48.6 | 56.0 | 65.1 | 72.9 | 81.8 | 81.0 | 73.1 | 62.4 | 45.4 | 38.0 | 58.7 |
| Minimum Temp °F | 17.2 | 20.8 | 23.1 | 25.5 | 31.5 | 38.0 | 42.6 | 41.5 | 34.3 | 28.3 | 22.6 | 17.2 | 28.6 |
| Mean Temp °F | 27.4 | 31.7 | 35.9 | 40.8 | 48.3 | 55.5 | 62.2 | 61.3 | 53.7 | 45.4 | 34.0 | 27.6 | 43.7 |
| Days Max Temp ≥ 90 °F | 0 | 0 | 0 | 0 | 0 | 1 | 6 | 5 | 1 | 0 | 0 | 0 | 13 |
| Days Max Temp ≤ 32 °F | 8 | 3 | 1 | 0 | 0 | 0 | 0 | 0 | 0 | 0 | 2 | 7 | 21 |
| Days Min Temp ≤ 32 °F | 28 | 26 | 28 | 24 | 17 | 8 | 3 | 4 | 12 | 22 | 25 | 29 | 226 |
| Days Min Temp ≤ 0 °F | 3 | 1 | 0 | 0 | 0 | 0 | 0 | 0 | 0 | 0 | 1 | 2 | 7 |
| Heating Degree Days | 1159 | 932 | 896 | 720 | 512 | 292 | 133 | 151 | 336 | 601 | 923 | 1152 | 7807 |
| Cooling Degree Days | 0 | 0 | 0 | 0 | 4 | 16 | 54 | 43 | 7 | 0 | 0 | 0 | 124 |
| Total Precipitation (") | 1.00 | 0.47 | 0.70 | 0.65 | 1.05 | 0.98 | 0.61 | 0.79 | 0.48 | 0.70 | 1.23 | 1.01 | 9.67 |
| Days ≥ 0.1" Precip | 3 | 2 | 3 | 2 | 3 | 3 | 2 | 2 | 1 | 3 | 4 | 3 | 31 |
| Total Snowfall (") | *6.2* | 2.7 | 2.0 | *1.1* | 0.7 | 0.0 | 0.0 | 0.0 | 0.0 | 0.7 | *3.8* | 8.1 | 25.3 |
| Days ≥ 1" Snow Depth | *16* | 5 | 1 | 0 | 0 | 0 | 0 | 0 | 0 | 0 | 4 | 11 | 37 |

## CASCADIA *Linn County* ELEVATION 801 ft LAT/LONG 44° 23 ' N / 122° 30 ' W

| | JAN | FEB | MAR | APR | MAY | JUN | JUL | AUG | SEP | OCT | NOV | DEC | YEAR |
|---|---|---|---|---|---|---|---|---|---|---|---|---|---|
| Maximum Temp °F | 45.2 | 50.6 | 54.9 | 59.0 | 65.0 | 71.0 | 77.8 | 79.5 | 74.3 | 64.2 | 51.2 | 44.4 | 61.4 |
| Minimum Temp °F | 31.3 | 32.8 | 34.6 | 37.2 | 41.3 | 46.6 | 48.7 | 47.9 | 43.5 | 38.5 | 35.7 | 32.2 | 39.2 |
| Mean Temp °F | 38.3 | 41.7 | 44.8 | 48.1 | 53.2 | 58.9 | 63.3 | 63.7 | 58.8 | 51.4 | 43.5 | 38.3 | 50.3 |
| Days Max Temp ≥ 90 °F | 0 | 0 | 0 | 0 | 0 | 1 | 2 | 3 | 2 | 0 | 0 | 0 | 8 |
| Days Max Temp ≤ 32 °F | 1 | 1 | 0 | 0 | 0 | 0 | 0 | 0 | 0 | 0 | 0 | 1 | 3 |
| Days Min Temp ≤ 32 °F | 18 | 15 | 13 | 6 | 2 | 0 | 0 | 0 | 1 | 6 | 10 | 16 | 87 |
| Days Min Temp ≤ 0 °F | 0 | 0 | 0 | 0 | 0 | 0 | 0 | 0 | 0 | 0 | 0 | 0 | 0 |
| Heating Degree Days | 821 | 652 | 620 | 501 | 362 | 193 | 89 | 79 | 191 | 416 | 640 | 820 | 5384 |
| Cooling Degree Days | 0 | 0 | 0 | 0 | 6 | 18 | 41 | 44 | 13 | 1 | 0 | 0 | 123 |
| Total Precipitation (") | 8.35 | 6.30 | 6.63 | 5.55 | 4.06 | 2.87 | 0.94 | 1.37 | 2.49 | 4.78 | 9.53 | 8.93 | 61.80 |
| Days ≥ 0.1" Precip | 15 | 13 | 15 | 12 | 9 | 6 | 3 | 3 | 5 | 10 | 15 | 16 | 122 |
| Total Snowfall (") | 3.5 | 2.3 | 0.6 | 0.0 | 0.0 | 0.0 | 0.0 | 0.0 | 0.0 | 0.0 | 0.0 | 1.0 | 7.4 |
| Days ≥ 1" Snow Depth | *2* | 1 | 0 | 0 | 0 | 0 | 0 | 0 | 0 | 0 | 1 | *1* | 5 |

## CAVE JUNCTION 1 WNW *Josephine County* ELEVATION 1332 ft LAT/LONG 42° 10 ' N / 123° 39 ' W

| | JAN | FEB | MAR | APR | MAY | JUN | JUL | AUG | SEP | OCT | NOV | DEC | YEAR |
|---|---|---|---|---|---|---|---|---|---|---|---|---|---|
| Maximum Temp °F | 47.1 | 53.3 | 58.5 | 64.0 | 73.2 | 80.8 | 88.9 | 88.6 | 82.2 | 69.8 | 52.9 | 45.9 | 67.1 |
| Minimum Temp °F | 32.0 | 33.2 | 34.7 | 36.4 | 40.9 | 46.7 | 50.1 | 49.1 | 44.2 | 39.0 | 36.2 | 32.5 | 39.6 |
| Mean Temp °F | 39.6 | 43.3 | 46.6 | 50.2 | 57.1 | 63.8 | 69.5 | 68.8 | 63.3 | 54.4 | 44.6 | 39.2 | 53.4 |
| Days Max Temp ≥ 90 °F | 0 | 0 | 0 | 0 | 2 | 6 | 16 | 15 | 8 | 1 | 0 | 0 | 48 |
| Days Max Temp ≤ 32 °F | 0 | 0 | 0 | 0 | 0 | 0 | 0 | 0 | 0 | 0 | 0 | 0 | 0 |
| Days Min Temp ≤ 32 °F | 17 | 14 | 13 | 9 | 2 | 0 | 0 | 0 | 1 | 5 | 9 | 16 | 86 |
| Days Min Temp ≤ 0 °F | 0 | 0 | 0 | 0 | 0 | 0 | 0 | 0 | 0 | 0 | 0 | 0 | 0 |
| Heating Degree Days | 782 | 607 | 563 | 437 | 252 | 94 | 22 | 21 | 92 | 325 | 606 | 792 | 4593 |
| Cooling Degree Days | 0 | 0 | 0 | 1 | 15 | 62 | 169 | 154 | 55 | 4 | 0 | 0 | 460 |
| Total Precipitation (") | 10.41 | 8.01 | 7.38 | 3.77 | 1.94 | 0.66 | 0.22 | 0.65 | 1.30 | 3.58 | 9.75 | 10.93 | 58.60 |
| Days ≥ 0.1" Precip | 11 | 10 | 11 | 8 | 4 | 2 | 1 | 1 | 3 | 6 | 12 | 12 | 81 |
| Total Snowfall (") | 4.6 | 4.3 | 2.1 | 1.0 | 0.0 | 0.0 | 0.0 | 0.0 | 0.0 | 0.0 | 1.0 | 4.0 | 17.0 |
| Days ≥ 1" Snow Depth | 2 | 1 | 0 | 0 | 0 | 0 | 0 | 0 | 0 | 0 | 0 | 2 | 5 |

## CHEMULT *Klamath County* ELEVATION 4754 ft LAT/LONG 43° 15 ' N / 121° 47 ' W

| | JAN | FEB | MAR | APR | MAY | JUN | JUL | AUG | SEP | OCT | NOV | DEC | YEAR |
|---|---|---|---|---|---|---|---|---|---|---|---|---|---|
| Maximum Temp °F | na | na | na | na | na | 73.6 | 81.8 | 80.9 | 73.9 | 62.6 | na | na | na |
| Minimum Temp °F | na | *17.5* | na | na | na | 35.5 | 38.7 | 37.6 | 32.0 | 26.3 | na | na | na |
| Mean Temp °F | na | na | na | na | na | 54.6 | 60.5 | 59.3 | 52.9 | 44.5 | na | na | na |
| Days Max Temp ≥ 90 °F | 0 | 0 | 0 | 0 | 0 | 1 | 5 | 5 | 1 | 0 | 0 | 0 | 12 |
| Days Max Temp ≤ 32 °F | 6 | 2 | 1 | 0 | 0 | 0 | 0 | 0 | 0 | 0 | 2 | 7 | 18 |
| Days Min Temp ≤ 32 °F | 25 | 23 | 25 | 22 | 19 | 11 | 5 | 7 | 17 | 24 | 23 | 25 | 226 |
| Days Min Temp ≤ 0 °F | 4 | 2 | 0 | 0 | 0 | 0 | 0 | 0 | 0 | 0 | 1 | 3 | 10 |
| Heating Degree Days | na | na | na | na | na | 313 | 159 | 188 | 357 | 630 | na | na | na |
| Cooling Degree Days | na | na | na | na | na | *7* | *24* | *17* | *4* | *0* | na | na | na |
| Total Precipitation (") | 4.10 | 2.70 | 2.48 | 1.13 | 0.83 | 0.82 | 0.54 | 0.67 | 0.79 | 1.35 | 3.96 | 4.18 | 23.55 |
| Days ≥ 0.1" Precip | 7 | 6 | 6 | 3 | 2 | 2 | 1 | 2 | 2 | 3 | 7 | *7* | 48 |
| Total Snowfall (") | *30.4* | *18.6* | *13.9* | 4.0 | 0.4 | 0.4 | 0.0 | 0.0 | 0.0 | 1.9 | *18.0* | *29.6* | 117.2 |
| Days ≥ 1" Snow Depth | 25 | 23 | 21 | 8 | 1 | 0 | 0 | 0 | 0 | 1 | 9 | 22 | 110 |

## CLATSKANIE *Columbia County* ELEVATION 22 ft LAT/LONG 46° 6 ' N / 123° 12 ' W

| | JAN | FEB | MAR | APR | MAY | JUN | JUL | AUG | SEP | OCT | NOV | DEC | YEAR |
|---|---|---|---|---|---|---|---|---|---|---|---|---|---|
| Maximum Temp °F | 44.6 | 49.7 | 54.7 | 58.4 | 64.3 | 68.5 | 73.4 | 74.5 | 71.4 | 61.7 | 50.5 | 44.3 | 59.7 |
| Minimum Temp °F | 33.7 | 35.1 | 37.3 | 39.6 | 44.6 | 49.5 | 52.8 | 53.4 | 49.5 | 43.1 | 38.1 | 34.3 | 42.6 |
| Mean Temp °F | 39.2 | 42.4 | 46.0 | 49.0 | 54.5 | 59.0 | 63.1 | 64.0 | 60.4 | 52.4 | 44.3 | 39.3 | 51.1 |
| Days Max Temp ≥ 90 °F | 0 | 0 | 0 | 0 | 0 | 0 | 1 | 1 | 0 | 0 | 0 | 0 | 2 |
| Days Max Temp ≤ 32 °F | 1 | 0 | 0 | 0 | 0 | 0 | 0 | 0 | 0 | 0 | 0 | 1 | 2 |
| Days Min Temp ≤ 32 °F | 12 | 10 | 6 | 3 | 0 | 0 | 0 | 0 | 0 | 1 | 6 | 12 | 50 |
| Days Min Temp ≤ 0 °F | 0 | 0 | 0 | 0 | 0 | 0 | 0 | 0 | 0 | 0 | 0 | 0 | 0 |
| Heating Degree Days | 795 | 631 | 582 | 472 | 324 | 184 | 87 | 67 | 145 | 383 | 613 | 789 | 5072 |
| Cooling Degree Days | 0 | 0 | 0 | 0 | 5 | 13 | 36 | 48 | 15 | 1 | 0 | 0 | 118 |
| Total Precipitation (") | 8.76 | 6.14 | 5.76 | 3.90 | 2.48 | 1.74 | 0.76 | 1.04 | 2.16 | 4.14 | 8.26 | 9.37 | 54.51 |
| Days ≥ 0.1" Precip | 15 | 13 | 13 | 10 | 7 | 5 | 2 | 3 | 5 | 9 | 15 | 16 | 113 |
| Total Snowfall (") | 3.3 | 1.3 | 0.5 | 0.0 | 0.0 | 0.0 | 0.0 | 0.0 | 0.0 | 0.0 | 0.4 | 2.3 | 7.8 |
| Days ≥ 1" Snow Depth | 3 | 1 | 0 | 0 | 0 | 0 | 0 | 0 | 0 | 0 | 1 | 2 | 7 |

## CLOVERDALE *Tillamook County* ELEVATION 59 ft LAT/LONG 45° 13 ' N / 123° 54 ' W

| | JAN | FEB | MAR | APR | MAY | JUN | JUL | AUG | SEP | OCT | NOV | DEC | YEAR |
|---|---|---|---|---|---|---|---|---|---|---|---|---|---|
| Maximum Temp °F | 50.3 | 53.6 | 55.5 | 57.8 | 62.7 | 66.2 | 70.2 | 71.3 | 70.3 | 64.2 | 55.2 | 50.3 | 60.6 |
| Minimum Temp °F | 37.2 | 38.5 | 38.8 | 40.1 | 43.6 | 47.5 | 49.4 | 50.1 | 48.4 | 44.2 | 40.8 | 37.3 | 43.0 |
| Mean Temp °F | 43.8 | 46.0 | 47.2 | 49.0 | 53.2 | 56.9 | 59.8 | 60.7 | 59.4 | 54.2 | 48.0 | 43.9 | 51.8 |
| Days Max Temp ≥ 90 °F | 0 | 0 | 0 | 0 | 0 | 0 | 0 | 0 | 1 | 0 | 0 | 0 | 1 |
| Days Max Temp ≤ 32 °F | 0 | 0 | 0 | 0 | 0 | 0 | 0 | 0 | 0 | 0 | 0 | 1 | 1 |
| Days Min Temp ≤ 32 °F | 9 | 5 | 5 | 2 | 0 | 0 | 0 | 0 | 0 | 0 | 3 | 7 | 31 |
| Days Min Temp ≤ 0 °F | 0 | 0 | 0 | 0 | 0 | 0 | 0 | 0 | 0 | 0 | 0 | 0 | 0 |
| Heating Degree Days | 650 | 528 | 546 | 474 | 360 | 240 | 157 | 131 | 171 | 330 | 503 | 649 | 4739 |
| Cooling Degree Days | 0 | 0 | 0 | 0 | 1 | 3 | 3 | 7 | 7 | 3 | 0 | 0 | 24 |
| Total Precipitation (") | 12.42 | 9.02 | 9.13 | 6.05 | 4.23 | 3.23 | 1.50 | 1.54 | 3.61 | 6.37 | 11.85 | 13.15 | 82.10 |
| Days ≥ 0.1" Precip | 17 | 15 | 16 | 13 | 10 | 7 | 3 | 3 | 7 | 11 | 17 | 17 | 136 |
| Total Snowfall (") | 1.2 | 0.0 | 0.3 | 0.1 | 0.0 | 0.0 | 0.0 | 0.0 | 0.0 | 0.0 | 0.1 | 0.4 | 2.1 |
| Days ≥ 1" Snow Depth | 1 | 0 | 0 | 0 | 0 | 0 | 0 | 0 | 0 | 0 | 0 | 0 | 1 |

## CONDON *Gilliam County* ELEVATION 2904 ft LAT/LONG 45° 14 ' N / 120° 11 ' W

| | JAN | FEB | MAR | APR | MAY | JUN | JUL | AUG | SEP | OCT | NOV | DEC | YEAR |
|---|---|---|---|---|---|---|---|---|---|---|---|---|---|
| Maximum Temp °F | 38.4 | 43.4 | 50.6 | 56.8 | 65.9 | 73.8 | 82.0 | 81.3 | 72.4 | 61.3 | 46.0 | 38.9 | 59.2 |
| Minimum Temp °F | 23.7 | 27.1 | 30.2 | 33.4 | 39.1 | 45.1 | 50.0 | 50.1 | 43.4 | 36.0 | 29.5 | 24.0 | 36.0 |
| Mean Temp °F | 31.1 | 35.3 | 40.4 | 45.1 | 52.5 | 59.5 | 66.0 | 65.8 | 57.9 | 48.7 | 37.8 | 31.5 | 47.6 |
| Days Max Temp ≥ 90 °F | 0 | 0 | 0 | 0 | 0 | 1 | 6 | 5 | 0 | 0 | 0 | 0 | 12 |
| Days Max Temp ≤ 32 °F | 8 | 4 | 0 | 0 | 0 | 0 | 0 | 0 | 0 | 0 | 2 | 8 | 22 |
| Days Min Temp ≤ 32 °F | 25 | 21 | 20 | 14 | 6 | 0 | 0 | 0 | 1 | 10 | 19 | 25 | 141 |
| Days Min Temp ≤ 0 °F | 1 | 0 | 0 | 0 | 0 | 0 | 0 | 0 | 0 | 0 | 0 | 1 | 2 |
| Heating Degree Days | 1046 | 833 | 756 | 589 | 385 | 192 | 69 | 69 | 223 | 499 | 811 | 1032 | 6504 |
| Cooling Degree Days | 0 | 0 | 0 | 0 | 9 | 29 | 90 | 90 | 17 | 1 | 0 | 0 | 236 |
| Total Precipitation (") | 1.53 | 1.18 | 1.26 | 1.29 | 1.24 | 1.02 | 0.55 | 0.69 | 0.67 | 0.99 | 1.89 | 1.60 | 13.91 |
| Days ≥ 0.1" Precip | 5 | 4 | 4 | 4 | 4 | 3 | 1 | 2 | 2 | 3 | 7 | 5 | 44 |
| Total Snowfall (") | 5.6 | 4.6 | 2.5 | 1.1 | 0.1 | 0.0 | 0.0 | 0.0 | 0.0 | 0.5 | 4.8 | 6.1 | 25.3 |
| Days ≥ 1" Snow Depth | 9 | 4 | 1 | 0 | 0 | 0 | 0 | 0 | 0 | 0 | 3 | 7 | 24 |

## CORVALLIS STATE UNIV *Benton County* ELEVATION 259 ft LAT/LONG 44° 34 ' N / 123° 17 ' W

| | JAN | FEB | MAR | APR | MAY | JUN | JUL | AUG | SEP | OCT | NOV | DEC | YEAR |
|---|---|---|---|---|---|---|---|---|---|---|---|---|---|
| Maximum Temp °F | 45.7 | 50.4 | 55.6 | 59.8 | 66.8 | 73.2 | 80.5 | 81.5 | 76.0 | 64.7 | 52.2 | 45.6 | 62.7 |
| Minimum Temp °F | 33.2 | 35.1 | 37.4 | 39.5 | 43.5 | 48.8 | 51.3 | 51.3 | 47.9 | 41.7 | 37.9 | 33.9 | 41.8 |
| Mean Temp °F | 39.5 | 42.8 | 46.5 | 49.6 | 55.2 | 61.0 | 65.9 | 66.4 | 62.0 | 53.2 | 45.0 | 39.7 | 52.2 |
| Days Max Temp ≥ 90 °F | 0 | 0 | 0 | 0 | 0 | 1 | 5 | 5 | 2 | 0 | 0 | 0 | 13 |
| Days Max Temp ≤ 32 °F | 1 | 0 | 0 | 0 | 0 | 0 | 0 | 0 | 0 | 0 | 0 | 1 | 2 |
| Days Min Temp ≤ 32 °F | 14 | 10 | 6 | 3 | 0 | 0 | 0 | 0 | 0 | 2 | 6 | 12 | 53 |
| Days Min Temp ≤ 0 °F | 0 | 0 | 0 | 0 | 0 | 0 | 0 | 0 | 0 | 0 | 0 | 0 | 0 |
| Heating Degree Days | 784 | 621 | 566 | 455 | 305 | 146 | 49 | 40 | 117 | 360 | 592 | 776 | 4811 |
| Cooling Degree Days | 0 | 0 | 0 | 0 | 10 | 31 | 85 | 92 | 39 | 2 | 0 | 0 | 259 |
| Total Precipitation (") | 6.69 | 4.90 | 4.24 | 2.72 | 1.97 | 1.37 | 0.55 | 0.86 | 1.44 | 3.06 | 6.49 | 7.74 | 42.03 |
| Days ≥ 0.1" Precip | 13 | 11 | 11 | 8 | 6 | 4 | 2 | 2 | 4 | 7 | 13 | 13 | 94 |
| Total Snowfall (") | 2.5 | 1.9 | 0.1 | 0.0 | 0.0 | 0.0 | 0.0 | 0.0 | 0.0 | 0.0 | 0.2 | 1.7 | 6.4 |
| Days ≥ 1" Snow Depth | 1 | 1 | 0 | 0 | 0 | 0 | 0 | 0 | 0 | 0 | 0 | 1 | 3 |

## CORVALLIS WATER BURE *Benton County* ELEVATION 512 ft LAT/LONG 44° 30 ' N / 123° 27 ' W

| | JAN | FEB | MAR | APR | MAY | JUN | JUL | AUG | SEP | OCT | NOV | DEC | YEAR |
|---|---|---|---|---|---|---|---|---|---|---|---|---|---|
| Maximum Temp °F | 44.9 | 49.4 | 54.5 | 59.0 | 65.9 | 71.8 | 78.5 | 79.2 | 73.8 | 63.3 | 50.9 | 44.4 | 61.3 |
| Minimum Temp °F | 31.8 | 33.6 | 35.7 | 37.9 | 42.3 | 47.4 | 50.2 | 50.4 | 47.4 | 41.0 | 36.3 | 32.4 | 40.5 |
| Mean Temp °F | 38.4 | 41.5 | 45.1 | 48.5 | 54.1 | 59.6 | 64.3 | 64.8 | 60.6 | 52.2 | 43.6 | 38.5 | 50.9 |
| Days Max Temp ≥ 90 °F | 0 | 0 | 0 | 0 | 0 | 1 | 4 | 3 | 1 | 0 | 0 | 0 | 9 |
| Days Max Temp ≤ 32 °F | 1 | 0 | 0 | 0 | 0 | 0 | 0 | 0 | 0 | 0 | 0 | 1 | 2 |
| Days Min Temp ≤ 32 °F | 17 | 13 | 10 | 5 | 1 | 0 | 0 | 0 | 0 | 1 | 9 | 16 | 72 |
| Days Min Temp ≤ 0 °F | 0 | 0 | 0 | 0 | 0 | 0 | 0 | 0 | 0 | 0 | 0 | 0 | 0 |
| Heating Degree Days | 819 | 656 | 610 | 489 | 337 | 178 | 75 | 64 | 149 | 393 | 635 | 817 | 5222 |
| Cooling Degree Days | 0 | 0 | 0 | 0 | 6 | 20 | 51 | 53 | 21 | 1 | 0 | 0 | 152 |
| Total Precipitation (") | 11.51 | 8.14 | 7.34 | 4.27 | 2.49 | 1.47 | 0.44 | 0.76 | 1.69 | 4.14 | 10.06 | 12.23 | 64.54 |
| Days ≥ 0.1" Precip | 15 | 13 | 13 | 10 | 7 | 4 | 1 | 2 | 4 | 7 | 15 | 16 | 107 |
| Total Snowfall (") | 2.1 | 3.0 | 0.3 | 0.0 | 0.0 | 0.0 | 0.0 | 0.0 | 0.0 | 0.0 | 0.3 | 1.5 | 7.2 |
| Days ≥ 1" Snow Depth | *1* | 2 | 0 | 0 | 0 | 0 | 0 | 0 | 0 | 0 | 0 | *2* | 5 |

## COTTAGE GROVE 1 S *Lane County* ELEVATION 650 ft LAT/LONG 43° 47' N / 123° 4' W

| | JAN | FEB | MAR | APR | MAY | JUN | JUL | AUG | SEP | OCT | NOV | DEC | YEAR |
|---|---|---|---|---|---|---|---|---|---|---|---|---|---|
| Maximum Temp °F | 48.2 | 53.2 | 57.7 | 61.8 | 68.3 | 74.4 | 81.6 | 82.1 | 76.9 | 66.0 | 53.6 | 47.6 | 64.3 |
| Minimum Temp °F | 32.9 | 34.1 | 35.6 | 37.3 | 41.0 | 45.7 | 47.7 | 47.6 | 43.7 | 40.0 | 37.3 | 33.3 | 39.7 |
| Mean Temp °F | 40.6 | 43.7 | 46.6 | 49.6 | 54.7 | 60.1 | 64.7 | 64.9 | 60.3 | 53.0 | 45.5 | 40.5 | 52.0 |
| Days Max Temp ≥ 90 °F | 0 | 0 | 0 | 0 | 0 | 1 | 5 | 5 | 2 | 0 | 0 | 0 | 13 |
| Days Max Temp ≤ 32 °F | 1 | 0 | 0 | 0 | 0 | 0 | 0 | 0 | 0 | 0 | 0 | 1 | 2 |
| Days Min Temp ≤ 32 °F | 15 | 11 | 11 | 7 | 3 | 0 | 0 | 0 | 2 | 4 | 7 | 14 | 74 |
| Days Min Temp ≤ 0 °F | 0 | 0 | 0 | 0 | 0 | 0 | 0 | 0 | 0 | 0 | 0 | 0 | 0 |
| Heating Degree Days | 750 | 595 | 563 | 457 | 316 | 157 | 61 | 57 | 149 | 364 | 578 | 753 | 4800 |
| Cooling Degree Days | 0 | 0 | 0 | 0 | 6 | 17 | 54 | 60 | 17 | 1 | 0 | 0 | 155 |
| Total Precipitation (") | 6.36 | 4.69 | 4.97 | 3.85 | 2.48 | 1.43 | 0.56 | 0.98 | 1.54 | 3.33 | 7.03 | 6.89 | 44.11 |
| Days ≥ 0.1" Precip | 12 | 11 | 12 | 10 | 7 | 4 | 2 | 2 | 4 | 7 | 13 | 13 | 97 |
| Total Snowfall (") | 2.7 | 1.5 | 0.3 | 0.0 | 0.0 | 0.0 | 0.0 | 0.0 | 0.0 | 0.0 | 0.2 | 1.4 | 6.1 |
| Days ≥ 1" Snow Depth | 1 | 1 | 0 | 0 | 0 | 0 | 0 | 0 | 0 | 0 | 0 | 1 | 3 |

## COTTAGE GROVE DAM *Lane County* ELEVATION 820 ft LAT/LONG 43° 43' N / 123° 3' W

| | JAN | FEB | MAR | APR | MAY | JUN | JUL | AUG | SEP | OCT | NOV | DEC | YEAR |
|---|---|---|---|---|---|---|---|---|---|---|---|---|---|
| Maximum Temp °F | 46.3 | 51.2 | 55.4 | 59.2 | 65.7 | 71.7 | 78.9 | 79.8 | 74.8 | 64.5 | 51.9 | 46.0 | 62.1 |
| Minimum Temp °F | 32.2 | 33.6 | 35.5 | 38.0 | 42.5 | 47.8 | 50.7 | 50.7 | 46.6 | 41.1 | 36.5 | 32.6 | 40.7 |
| Mean Temp °F | 39.3 | 42.4 | 45.5 | 48.6 | 54.1 | 59.8 | 64.8 | 65.3 | 60.7 | 52.8 | 44.2 | 39.3 | 51.4 |
| Days Max Temp ≥ 90 °F | 0 | 0 | 0 | 0 | 0 | 1 | 3 | 4 | 2 | 0 | 0 | 0 | 10 |
| Days Max Temp ≤ 32 °F | 1 | 0 | 0 | 0 | 0 | 0 | 0 | 0 | 0 | 0 | 0 | 1 | 2 |
| Days Min Temp ≤ 32 °F | 17 | 13 | 10 | 4 | 0 | 0 | 0 | 0 | 0 | 2 | 8 | 16 | 70 |
| Days Min Temp ≤ 0 °F | 0 | 0 | 0 | 0 | 0 | 0 | 0 | 0 | 0 | 0 | 0 | 0 | 0 |
| Heating Degree Days | 791 | 631 | 598 | 484 | 335 | 173 | 67 | 55 | 145 | 372 | 617 | 788 | 5056 |
| Cooling Degree Days | 0 | 0 | 0 | 0 | 7 | 18 | 56 | 63 | 25 | 3 | 0 | 0 | 172 |
| Total Precipitation (") | 6.60 | 5.00 | 5.20 | 4.10 | 2.74 | 1.61 | 0.66 | 1.09 | 1.71 | 3.57 | 7.44 | 7.26 | 46.98 |
| Days ≥ 0.1" Precip | 12 | 11 | 12 | 11 | 7 | 5 | 2 | 3 | 4 | 8 | 14 | 13 | 102 |
| Total Snowfall (") | 1.1 | 1.7 | 0.4 | 0.0 | 0.0 | 0.0 | 0.0 | 0.0 | 0.0 | 0.0 | 0.1 | 0.6 | 3.9 |
| Days ≥ 1" Snow Depth | 1 | 1 | 0 | 0 | 0 | 0 | 0 | 0 | 0 | 0 | 0 | 1 | 3 |

## CRATER LAKE N P HQ *Klamath County* ELEVATION 6483 ft LAT/LONG 42° 54' N / 122° 8' W

| | JAN | FEB | MAR | APR | MAY | JUN | JUL | AUG | SEP | OCT | NOV | DEC | YEAR |
|---|---|---|---|---|---|---|---|---|---|---|---|---|---|
| Maximum Temp °F | 34.3 | 35.1 | 37.2 | 41.8 | 49.8 | 58.0 | 67.7 | 68.7 | 62.1 | 52.3 | 38.3 | 33.9 | 48.3 |
| Minimum Temp °F | 17.7 | 18.0 | 18.8 | 21.5 | 27.5 | 33.8 | 39.7 | 40.3 | 35.8 | 30.2 | 22.2 | 18.0 | 27.0 |
| Mean Temp °F | 26.0 | 26.6 | 28.1 | 31.7 | 38.7 | 45.9 | 53.7 | 54.5 | 49.0 | 41.3 | 30.3 | 26.0 | 37.7 |
| Days Max Temp ≥ 90 °F | 0 | 0 | 0 | 0 | 0 | 0 | 0 | 0 | 0 | 0 | 0 | 0 | 0 |
| Days Max Temp ≤ 32 °F | 14 | 12 | 10 | 6 | 1 | 0 | 0 | 0 | 0 | 2 | 9 | 15 | 69 |
| Days Min Temp ≤ 32 °F | 31 | 28 | 31 | 27 | 23 | 13 | 4 | 3 | 9 | 19 | 27 | 31 | 246 |
| Days Min Temp ≤ 0 °F | 1 | 1 | 0 | 0 | 0 | 0 | 0 | 0 | 0 | 0 | 0 | 1 | 3 |
| Heating Degree Days | 1203 | 1078 | 1139 | 993 | 809 | 567 | 345 | 324 | 474 | 727 | 1035 | 1203 | 9897 |
| Cooling Degree Days | 0 | 0 | 0 | 0 | 0 | 0 | 2 | 4 | 1 | 0 | 0 | 0 | 7 |
| Total Precipitation (") | 9.82 | 7.53 | 7.56 | 4.92 | 3.06 | 2.01 | 0.78 | 1.20 | 2.17 | 4.44 | 10.40 | 10.05 | 63.94 |
| Days ≥ 0.1" Precip | 14 | 13 | 15 | 11 | 7 | 5 | 2 | 3 | 4 | 8 | 14 | 16 | 112 |
| Total Snowfall (") | 84.3 | 73.2 | 75.9 | 43.0 | 17.3 | 4.1 | 0.6 | 0.3 | 4.1 | 20.1 | 68.8 | 83.6 | 475.3 |
| Days ≥ 1" Snow Depth | 31 | 28 | 31 | 30 | 30 | 18 | 3 | 0 | 1 | 9 | 25 | 30 | 236 |

## DALLAS 2 NE *Polk County* ELEVATION 331 ft LAT/LONG 44° 56' N / 123° 19' W

| | JAN | FEB | MAR | APR | MAY | JUN | JUL | AUG | SEP | OCT | NOV | DEC | YEAR |
|---|---|---|---|---|---|---|---|---|---|---|---|---|---|
| Maximum Temp °F | 46.0 | 51.0 | 56.5 | 61.2 | 68.8 | 75.1 | 82.3 | 82.8 | 77.5 | 65.8 | 51.7 | 45.0 | 63.6 |
| Minimum Temp °F | 33.3 | 34.9 | 36.7 | 38.5 | 42.6 | 47.2 | 49.4 | 49.2 | 46.7 | 41.5 | 37.3 | 33.5 | 40.9 |
| Mean Temp °F | 39.7 | 43.0 | 46.6 | 49.9 | 55.9 | 61.2 | 65.8 | 66.0 | 62.1 | 53.6 | 44.6 | 39.3 | 52.3 |
| Days Max Temp ≥ 90 °F | 0 | 0 | 0 | 0 | 0 | 2 | 6 | 6 | 3 | 0 | 0 | 0 | 17 |
| Days Max Temp ≤ 32 °F | 1 | 0 | 0 | 0 | 0 | 0 | 0 | 0 | 0 | 0 | 0 | 1 | 2 |
| Days Min Temp ≤ 32 °F | 14 | 10 | 8 | 5 | 1 | 0 | 0 | 0 | 0 | 2 | 7 | 13 | 60 |
| Days Min Temp ≤ 0 °F | 0 | 0 | 0 | 0 | 0 | 0 | 0 | 0 | 0 | 0 | 0 | 0 | 0 |
| Heating Degree Days | 778 | 616 | 564 | 448 | 286 | 141 | 48 | 44 | 118 | 348 | 607 | 792 | 4790 |
| Cooling Degree Days | 0 | 0 | 0 | 0 | 12 | 32 | 79 | 82 | 45 | 4 | 0 | 0 | 254 |
| Total Precipitation (") | 7.85 | 5.72 | 5.12 | 2.91 | 1.91 | 1.34 | 0.49 | 0.70 | 1.43 | 3.37 | 7.34 | 9.04 | 47.22 |
| Days ≥ 0.1" Precip | 13 | 10 | 10 | 8 | 5 | 4 | 1 | 2 | 4 | 6 | 13 | 13 | 89 |
| Total Snowfall (") | 1.9 | 1.8 | 0.3 | 0.0 | 0.0 | 0.0 | 0.0 | 0.0 | 0.0 | 0.0 | 0.3 | 2.6 | 6.9 |
| Days ≥ 1" Snow Depth | 1 | 1 | 0 | 0 | 0 | 0 | 0 | 0 | 0 | 0 | 0 | 1 | 3 |

## DETROIT DAM *Marion County* ELEVATION 1299 ft LAT/LONG 44° 43 ' N / 122° 15 ' W

| | JAN | FEB | MAR | APR | MAY | JUN | JUL | AUG | SEP | OCT | NOV | DEC | YEAR |
|---|---|---|---|---|---|---|---|---|---|---|---|---|---|
| Maximum Temp °F | 43.2 | 47.4 | 52.5 | 56.9 | 64.2 | 70.1 | 77.5 | 78.4 | 72.9 | 62.1 | 49.3 | 43.4 | 59.8 |
| Minimum Temp °F | 33.1 | 34.5 | 36.2 | 38.8 | 43.8 | 49.3 | 53.1 | 53.9 | 50.3 | 44.6 | 38.7 | 34.2 | 42.5 |
| Mean Temp °F | 38.2 | 41.0 | 44.4 | 47.9 | 54.0 | 59.7 | 65.3 | 66.1 | 61.6 | 53.4 | 44.0 | 38.9 | 51.2 |
| Days Max Temp ≥ 90 °F | 0 | 0 | 0 | 0 | 0 | 1 | 3 | 3 | 1 | 0 | 0 | 0 | 8 |
| Days Max Temp ≤ 32 °F | 1 | 0 | 0 | 0 | 0 | 0 | 0 | 0 | 0 | 0 | 0 | 1 | 2 |
| Days Min Temp ≤ 32 °F | 12 | 8 | 4 | 1 | 0 | 0 | 0 | 0 | 0 | 0 | 3 | 9 | 37 |
| Days Min Temp ≤ 0 °F | 0 | 0 | 0 | 0 | 0 | 0 | 0 | 0 | 0 | 0 | 0 | 0 | 0 |
| Heating Degree Days | 825 | 672 | 632 | 507 | 343 | 184 | 71 | 55 | 136 | 356 | 623 | 804 | 5208 |
| Cooling Degree Days | 0 | 0 | 0 | 0 | 12 | 29 | 80 | 93 | 44 | 4 | 0 | 0 | 262 |
| Total Precipitation (") | 12.66 | 9.67 | 8.98 | 7.00 | 4.89 | 3.38 | 0.95 | 1.48 | 3.23 | 6.35 | 13.38 | 13.54 | 85.51 |
| Days ≥ 0.1" Precip | 16 | 14 | 16 | 14 | 10 | 7 | 3 | 3 | 6 | 10 | 16 | 17 | 132 |
| Total Snowfall (") | 4.6 | 4.1 | 1.0 | 0.2 | 0.0 | 0.0 | 0.0 | 0.0 | 0.0 | 0.0 | 0.8 | 3.1 | 13.8 |
| Days ≥ 1" Snow Depth | 5 | 3 | 1 | 0 | 0 | 0 | 0 | 0 | 0 | 0 | 1 | 3 | 13 |

## DORENA DAM *Lane County* ELEVATION 758 ft LAT/LONG 43° 47 ' N / 122° 58 ' W

| | JAN | FEB | MAR | APR | MAY | JUN | JUL | AUG | SEP | OCT | NOV | DEC | YEAR |
|---|---|---|---|---|---|---|---|---|---|---|---|---|---|
| Maximum Temp °F | 46.8 | 51.7 | 55.9 | 59.6 | 66.1 | 72.1 | 79.4 | 80.3 | 75.3 | 64.9 | 52.3 | 46.4 | 62.6 |
| Minimum Temp °F | 31.9 | 33.1 | 35.0 | 37.8 | 42.0 | 47.1 | 49.8 | 49.4 | 45.5 | 40.4 | 36.4 | 32.7 | 40.1 |
| Mean Temp °F | 39.4 | 42.4 | 45.4 | 48.7 | 54.1 | 59.7 | 64.6 | 64.9 | 60.4 | 52.7 | 44.4 | 39.6 | 51.4 |
| Days Max Temp ≥ 90 °F | 0 | 0 | 0 | 0 | 0 | 1 | 3 | 4 | 2 | 0 | 0 | 0 | 10 |
| Days Max Temp ≤ 32 °F | 1 | 0 | 0 | 0 | 0 | 0 | 0 | 0 | 0 | 0 | 0 | 1 | 2 |
| Days Min Temp ≤ 32 °F | 17 | 13 | 11 | 5 | 1 | 0 | 0 | 0 | 0 | 3 | 9 | 14 | 73 |
| Days Min Temp ≤ 0 °F | 0 | 0 | 0 | 0 | 0 | 0 | 0 | 0 | 0 | 0 | 0 | 0 | 0 |
| Heating Degree Days | 787 | 631 | 600 | 483 | 337 | 175 | 69 | 62 | 152 | 376 | 611 | 781 | 5064 |
| Cooling Degree Days | 0 | 0 | 0 | 0 | 9 | 22 | 61 | 67 | 24 | 2 | 0 | 0 | 185 |
| Total Precipitation (") | 6.10 | 4.75 | 5.02 | 4.03 | 2.77 | 1.75 | 0.75 | 1.11 | 1.84 | 3.40 | 7.05 | 6.68 | 45.25 |
| Days ≥ 0.1" Precip | 13 | 11 | 12 | 11 | 7 | 5 | 2 | 3 | 4 | 8 | 13 | 13 | 102 |
| Total Snowfall (") | 3.1 | 1.7 | 0.3 | 0.1 | 0.0 | 0.0 | 0.0 | 0.0 | 0.0 | 0.0 | 0.1 | 1.0 | 6.3 |
| Days ≥ 1" Snow Depth | 1 | 1 | 0 | 0 | 0 | 0 | 0 | 0 | 0 | 0 | 0 | 1 | 3 |

## DRAIN *Douglas County* ELEVATION 302 ft LAT/LONG 43° 40 ' N / 123° 19 ' W

| | JAN | FEB | MAR | APR | MAY | JUN | JUL | AUG | SEP | OCT | NOV | DEC | YEAR |
|---|---|---|---|---|---|---|---|---|---|---|---|---|---|
| Maximum Temp °F | 48.1 | 53.4 | 58.4 | 62.6 | 69.7 | 75.6 | 82.9 | 83.2 | 78.5 | 67.2 | 54.0 | 47.5 | 65.1 |
| Minimum Temp °F | 34.2 | 35.5 | 37.1 | 39.3 | 43.3 | 48.3 | 50.6 | 50.6 | 46.2 | 42.1 | 39.1 | 34.9 | 41.8 |
| Mean Temp °F | 41.2 | 44.5 | 47.8 | 50.9 | 56.5 | 62.0 | 66.8 | 66.9 | 62.4 | 54.7 | 46.6 | 41.2 | 53.5 |
| Days Max Temp ≥ 90 °F | 0 | 0 | 0 | 0 | 1 | 2 | 7 | 7 | 4 | 0 | 0 | 0 | 21 |
| Days Max Temp ≤ 32 °F | 0 | 0 | 0 | 0 | 0 | 0 | 0 | 0 | 0 | 0 | 0 | 1 | 1 |
| Days Min Temp ≤ 32 °F | 12 | 8 | 7 | 3 | 1 | 0 | 0 | 0 | 1 | 2 | 4 | 9 | 47 |
| Days Min Temp ≤ 0 °F | 0 | 0 | 0 | 0 | 0 | 0 | 0 | 0 | 0 | 0 | 0 | 0 | 0 |
| Heating Degree Days | 732 | 573 | 527 | 416 | 267 | 123 | 37 | 34 | 108 | 315 | 547 | 730 | 4409 |
| Cooling Degree Days | 0 | 0 | 0 | 1 | 16 | 38 | 99 | 103 | 42 | 4 | 0 | 0 | 303 |
| Total Precipitation (") | 6.97 | 5.22 | 4.98 | 3.56 | 2.13 | 1.23 | 0.48 | 0.88 | 1.27 | 3.30 | 7.39 | 7.70 | 45.11 |
| Days ≥ 0.1" Precip | 12 | 11 | 11 | 10 | 6 | 4 | 1 | 2 | 3 | 7 | 14 | 14 | 95 |
| Total Snowfall (") | *0.0* | 1.3 | 0.0 | 0.0 | 0.0 | 0.0 | 0.0 | 0.0 | 0.0 | 0.0 | 0.2 | 0.8 | 2.3 |
| Days ≥ 1" Snow Depth | *0* | 0 | 0 | 0 | 0 | 0 | 0 | 0 | 0 | 0 | 0 | *0* | 0 |

## DUFUR *Wasco County* ELEVATION 1332 ft LAT/LONG 45° 27 ' N / 121° 8 ' W

| | JAN | FEB | MAR | APR | MAY | JUN | JUL | AUG | SEP | OCT | NOV | DEC | YEAR |
|---|---|---|---|---|---|---|---|---|---|---|---|---|---|
| Maximum Temp °F | 40.3 | 46.9 | 55.3 | 62.2 | 71.1 | 77.9 | 85.3 | 84.9 | 77.0 | 64.4 | 48.5 | 40.4 | 62.9 |
| Minimum Temp °F | 24.5 | 27.7 | 30.6 | 33.7 | 38.8 | 44.7 | 48.4 | 48.3 | 42.8 | 35.5 | 30.2 | 25.1 | 35.9 |
| Mean Temp °F | 32.5 | 37.4 | 43.0 | 48.0 | 55.0 | 61.3 | 66.8 | 66.6 | 59.9 | 50.0 | 39.4 | 32.8 | 49.4 |
| Days Max Temp ≥ 90 °F | 0 | 0 | 0 | 0 | 1 | 4 | 11 | 10 | 2 | 0 | 0 | 0 | 28 |
| Days Max Temp ≤ 32 °F | 6 | 2 | 0 | 0 | 0 | 0 | 0 | 0 | 0 | 0 | 1 | 6 | 15 |
| Days Min Temp ≤ 32 °F | 26 | 21 | 19 | 13 | 6 | 1 | 0 | 0 | 2 | 11 | 18 | 26 | 143 |
| Days Min Temp ≤ 0 °F | 1 | 0 | 0 | 0 | 0 | 0 | 0 | 0 | 0 | 0 | 0 | 1 | 2 |
| Heating Degree Days | 1001 | 775 | 676 | 505 | 314 | 147 | 51 | 49 | 168 | 459 | 762 | 992 | 5899 |
| Cooling Degree Days | 0 | 0 | 0 | 1 | 17 | 47 | 123 | 120 | 30 | 1 | 0 | 0 | 339 |
| Total Precipitation (") | 1.95 | 1.23 | 1.17 | 0.81 | 0.74 | 0.66 | 0.28 | 0.48 | 0.52 | 0.87 | 1.73 | 1.99 | 12.43 |
| Days ≥ 0.1" Precip | 6 | 4 | 4 | 3 | 2 | 2 | 1 | 2 | 2 | 3 | 5 | 6 | 40 |
| Total Snowfall (") | 7.1 | 3.4 | 0.9 | 0.0 | 0.0 | 0.0 | 0.0 | 0.0 | 0.0 | 0.2 | 2.9 | 7.8 | 22.3 |
| Days ≥ 1" Snow Depth | 9 | 3 | 0 | 0 | 0 | 0 | 0 | 0 | 0 | 0 | 2 | 7 | 21 |

## ELGIN *Union County* ELEVATION 2680 ft LAT/LONG 45° 33 ' N / 117° 55 ' W

| | JAN | FEB | MAR | APR | MAY | JUN | JUL | AUG | SEP | OCT | NOV | DEC | YEAR |
|---|---|---|---|---|---|---|---|---|---|---|---|---|---|
| Maximum Temp °F | 37.6 | 44.6 | 52.8 | 60.8 | 69.8 | 77.7 | 87.2 | 87.7 | 78.3 | 65.2 | 47.2 | 38.7 | 62.3 |
| Minimum Temp °F | 21.3 | 24.8 | 28.3 | 32.2 | 37.7 | 43.6 | 46.0 | 45.0 | 37.5 | 30.9 | 28.0 | 22.4 | 33.1 |
| Mean Temp °F | 29.4 | 34.8 | 40.6 | 46.5 | 53.7 | 60.7 | 66.6 | 66.4 | 57.9 | 48.1 | 37.6 | 30.6 | 47.7 |
| Days Max Temp ≥ 90 °F | 0 | 0 | 0 | 0 | 1 | 4 | *14* | 14 | 5 | 0 | 0 | 0 | 38 |
| Days Max Temp ≤ 32 °F | *8* | 2 | 0 | 0 | 0 | 0 | 0 | 0 | 0 | 0 | 1 | 6 | 17 |
| Days Min Temp ≤ 32 °F | 26 | 22 | 22 | 16 | 6 | 1 | 0 | 1 | 8 | 19 | 21 | 25 | 167 |
| Days Min Temp ≤ 0 °F | 3 | 1 | 0 | 0 | 0 | 0 | 0 | 0 | 0 | 0 | 0 | 2 | 6 |
| Heating Degree Days | 1095 | 846 | 751 | 547 | 348 | 158 | 49 | 52 | 219 | 518 | 815 | 1061 | 6459 |
| Cooling Degree Days | 0 | 0 | 0 | 0 | 9 | 35 | 88 | 94 | 12 | 0 | 0 | 0 | 238 |
| Total Precipitation (") | 3.24 | 2.37 | 2.09 | 1.70 | 1.73 | 1.56 | 0.78 | 0.85 | 0.96 | 1.63 | 3.07 | 3.20 | 23.18 |
| Days ≥ 0.1" Precip | 8 | 6 | 7 | 6 | 5 | 5 | 2 | 2 | 3 | 5 | 9 | *8* | 66 |
| Total Snowfall (") | *15.5* | 7.3 | 3.3 | 0.9 | 0.0 | 0.0 | 0.0 | 0.0 | 0.0 | 0.1 | 6.8 | *15.2* | 49.1 |
| Days ≥ 1" Snow Depth | *17* | 7 | 1 | 0 | 0 | 0 | 0 | 0 | 0 | 0 | 3 | *13* | 41 |

## ELKTON 3 SW *Douglas County* ELEVATION 131 ft LAT/LONG 43° 39 ' N / 123° 36 ' W

| | JAN | FEB | MAR | APR | MAY | JUN | JUL | AUG | SEP | OCT | NOV | DEC | YEAR |
|---|---|---|---|---|---|---|---|---|---|---|---|---|---|
| Maximum Temp °F | 49.2 | 54.1 | 59.5 | 63.9 | 70.9 | 76.9 | 83.5 | 84.3 | 80.1 | *67.9* | *54.5* | 47.9 | 66.1 |
| Minimum Temp °F | 36.5 | 37.8 | 39.4 | 41.2 | 45.0 | 49.3 | 51.8 | 52.0 | 48.6 | *45.0* | 41.4 | 36.9 | 43.7 |
| Mean Temp °F | 42.9 | 46.0 | 49.5 | 52.6 | 58.0 | 63.1 | 67.7 | 68.2 | 64.4 | *56.5* | *48.0* | 42.4 | 54.9 |
| Days Max Temp ≥ 90 °F | 0 | 0 | 0 | 0 | 1 | 3 | 8 | 7 | 4 | 0 | 0 | 0 | 23 |
| Days Max Temp ≤ 32 °F | 0 | 0 | 0 | 0 | 0 | 0 | 0 | 0 | 0 | 0 | 0 | 0 | 0 |
| Days Min Temp ≤ 32 °F | 8 | 5 | 3 | 1 | 0 | 0 | 0 | 0 | 0 | 1 | 2 | 7 | 27 |
| Days Min Temp ≤ 0 °F | 0 | 0 | 0 | 0 | 0 | 0 | 0 | 0 | 0 | 0 | 0 | 0 | 0 |
| Heating Degree Days | 680 | 530 | 475 | 367 | 226 | 96 | 24 | 15 | 68 | *262* | *504* | 694 | 3941 |
| Cooling Degree Days | 0 | 0 | 0 | 2 | 20 | 44 | 109 | 114 | 60 | 5 | 0 | 0 | 354 |
| Total Precipitation (") | 8.41 | 6.01 | 5.84 | 3.73 | 1.98 | 1.06 | 0.35 | 0.72 | 1.39 | 3.52 | 8.42 | 9.52 | 50.95 |
| Days ≥ 0.1" Precip | 13 | *11* | 12 | *8* | 6 | 3 | 1 | 2 | 3 | na | *13* | *14* | na |
| Total Snowfall (") | 1.9 | 0.2 | 0.1 | 0.1 | 0.0 | 0.0 | 0.0 | 0.0 | 0.0 | 0.0 | 0.1 | 0.4 | 2.8 |
| Days ≥ 1" Snow Depth | 0 | 0 | 0 | 0 | 0 | 0 | 0 | 0 | 0 | 0 | 0 | 0 | 0 |

## ESTACADA 2 SE *Clackamas County* ELEVATION 410 ft LAT/LONG 45° 16 ' N / 122° 19 ' W

| | JAN | FEB | MAR | APR | MAY | JUN | JUL | AUG | SEP | OCT | NOV | DEC | YEAR |
|---|---|---|---|---|---|---|---|---|---|---|---|---|---|
| Maximum Temp °F | 45.4 | 50.0 | 55.4 | 60.2 | 67.0 | 72.6 | 78.8 | 78.8 | 73.0 | 61.1 | 50.8 | 45.1 | 61.5 |
| Minimum Temp °F | 33.9 | 36.0 | 37.7 | 40.1 | 44.6 | 49.4 | 52.5 | 52.4 | 48.7 | 43.3 | 38.6 | 34.3 | 42.6 |
| Mean Temp °F | 39.7 | 43.0 | 46.6 | 50.2 | 55.9 | 61.0 | 65.7 | 65.6 | 60.9 | 52.2 | 44.7 | 39.8 | 52.1 |
| Days Max Temp ≥ 90 °F | 0 | 0 | 0 | 0 | 0 | 1 | 4 | 4 | 2 | 0 | 0 | 0 | 11 |
| Days Max Temp ≤ 32 °F | 1 | 0 | 0 | 0 | 0 | 0 | 0 | 0 | 0 | 0 | 0 | 1 | 2 |
| Days Min Temp ≤ 32 °F | 12 | 8 | 5 | 2 | 0 | 0 | 0 | 0 | 0 | 1 | 5 | 11 | 44 |
| Days Min Temp ≤ 0 °F | 0 | 0 | 0 | 0 | 0 | 0 | 0 | 0 | 0 | 0 | 0 | 0 | 0 |
| Heating Degree Days | 778 | 615 | 564 | 439 | 286 | 141 | 47 | 46 | 142 | 392 | 602 | 777 | 4829 |
| Cooling Degree Days | 0 | 0 | 0 | 1 | 11 | 26 | 69 | 77 | 26 | 2 | 0 | 0 | 212 |
| Total Precipitation (") | 8.31 | 6.24 | 5.80 | 4.96 | 3.60 | 2.67 | 1.03 | 1.40 | 2.45 | 4.79 | 8.11 | 8.66 | 58.02 |
| Days ≥ 0.1" Precip | 14 | 12 | 13 | 12 | 9 | 6 | 3 | 3 | 5 | 9 | 14 | 16 | 116 |
| Total Snowfall (") | 1.1 | 0.9 | 0.1 | 0.0 | 0.0 | 0.0 | 0.0 | 0.0 | 0.0 | 0.0 | 0.3 | 1.3 | 3.7 |
| Days ≥ 1" Snow Depth | 1 | 0 | 0 | 0 | 0 | 0 | 0 | 0 | 0 | 0 | 0 | 1 | 2 |

## EUGENE MAHLON SWEET *Lane County* ELEVATION 367 ft LAT/LONG 44° 7 ' N / 123° 13 ' W

| | JAN | FEB | MAR | APR | MAY | JUN | JUL | AUG | SEP | OCT | NOV | DEC | YEAR |
|---|---|---|---|---|---|---|---|---|---|---|---|---|---|
| Maximum Temp °F | 46.4 | 51.2 | 56.5 | 60.7 | 67.6 | 74.3 | 82.1 | 82.2 | 76.6 | 64.9 | 52.3 | 46.0 | 63.4 |
| Minimum Temp °F | 33.6 | 35.3 | 37.4 | 39.4 | 43.2 | 48.4 | 51.7 | 52.0 | 47.9 | 41.9 | 38.0 | 34.2 | 41.9 |
| Mean Temp °F | 40.0 | 43.3 | 46.9 | 50.1 | 55.5 | 61.4 | 66.9 | 67.1 | 62.3 | 53.4 | 45.2 | 40.1 | 52.7 |
| Days Max Temp ≥ 90 °F | 0 | 0 | 0 | 0 | 0 | 1 | 6 | 6 | 2 | 0 | 0 | 0 | 15 |
| Days Max Temp ≤ 32 °F | 2 | 0 | 0 | 0 | 0 | 0 | 0 | 0 | 0 | 0 | 0 | 1 | 3 |
| Days Min Temp ≤ 32 °F | 13 | 10 | 6 | 3 | 0 | 0 | 0 | 0 | 0 | 2 | 7 | 12 | 53 |
| Days Min Temp ≤ 0 °F | 0 | 0 | 0 | 0 | 0 | 0 | 0 | 0 | 0 | 0 | 0 | 0 | 0 |
| Heating Degree Days | 768 | 607 | 553 | 441 | 294 | 132 | 34 | 28 | 109 | 354 | 587 | 765 | 4672 |
| Cooling Degree Days | 0 | 0 | 0 | 0 | 8 | 26 | 86 | 95 | 35 | 2 | 0 | 0 | 252 |
| Total Precipitation (") | 7.79 | 5.47 | 5.48 | 3.50 | 2.33 | 1.54 | 0.58 | 1.10 | 1.57 | 3.46 | 8.13 | 8.62 | 49.57 |
| Days ≥ 0.1" Precip | 12 | 10 | 11 | 8 | 6 | 4 | 1 | 2 | 4 | 7 | 13 | 13 | 91 |
| Total Snowfall (") | 3.1 | 1.2 | 0.2 | 0.0 | 0.0 | 0.0 | 0.0 | 0.0 | 0.0 | 0.0 | 0.2 | 1.5 | 6.2 |
| Days ≥ 1" Snow Depth | 1 | 1 | 0 | 0 | 0 | 0 | 0 | 0 | 0 | 0 | 0 | 1 | 3 |

## FALLS CITY 2 *Polk County* ELEVATION 371 ft LAT/LONG 44° 52 ' N / 123° 26 ' W

| | JAN | FEB | MAR | APR | MAY | JUN | JUL | AUG | SEP | OCT | NOV | DEC | YEAR |
|---|---|---|---|---|---|---|---|---|---|---|---|---|---|
| Maximum Temp °F | 45.9 | 50.5 | 55.5 | 59.8 | 66.7 | 72.7 | 80.0 | 81.2 | 75.8 | 64.8 | 52.2 | 45.5 | 62.5 |
| Minimum Temp °F | 31.5 | 33.4 | 35.0 | 37.3 | 41.2 | 46.2 | 48.8 | 48.9 | 46.0 | 40.3 | 36.0 | 32.2 | 39.7 |
| Mean Temp °F | 38.7 | 42.0 | 45.3 | 48.6 | 54.0 | 59.5 | 64.4 | 65.1 | 60.9 | 52.5 | 44.1 | 38.9 | 51.2 |
| Days Max Temp ≥ 90 °F | 0 | 0 | 0 | 0 | 0 | 1 | 5 | 5 | 3 | 0 | 0 | 0 | 14 |
| Days Max Temp ≤ 32 °F | 1 | 0 | 0 | 0 | 0 | 0 | 0 | 0 | 0 | 0 | 0 | 1 | 2 |
| Days Min Temp ≤ 32 °F | 19 | 14 | 11 | 6 | 1 | 0 | 0 | 0 | 0 | 2 | 10 | 16 | 79 |
| Days Min Temp ≤ 0 °F | 0 | 0 | 0 | 0 | 0 | 0 | 0 | 0 | 0 | 0 | 0 | 0 | 0 |
| Heating Degree Days | 807 | 642 | 604 | 487 | 341 | 182 | 73 | 60 | 144 | 381 | 620 | 802 | 5143 |
| Cooling Degree Days | 0 | 0 | 0 | 0 | 9 | 24 | 62 | 73 | 34 | 2 | 0 | 0 | 204 |
| Total Precipitation (") | 11.30 | 8.64 | 7.39 | 4.06 | 2.31 | 1.39 | 0.48 | 0.80 | 1.59 | 4.42 | 10.34 | 13.07 | 65.79 |
| Days ≥ 0.1" Precip | 14 | 12 | 13 | 9 | 6 | 4 | 1 | 2 | 4 | 8 | 15 | 15 | 103 |
| Total Snowfall (") | 4.2 | 2.5 | 0.8 | 0.2 | 0.0 | 0.0 | 0.0 | 0.0 | 0.0 | 0.0 | 0.5 | 3.8 | 12.0 |
| Days ≥ 1" Snow Depth | 4 | 2 | 1 | 0 | 0 | 0 | 0 | 0 | 0 | 0 | 0 | 3 | 10 |

## FERN RIDGE DAM *Lane County* ELEVATION 381 ft LAT/LONG 44° 7 ' N / 123° 18 ' W

| | JAN | FEB | MAR | APR | MAY | JUN | JUL | AUG | SEP | OCT | NOV | DEC | YEAR |
|---|---|---|---|---|---|---|---|---|---|---|---|---|---|
| Maximum Temp °F | ***46.2*** | 50.7 | 56.4 | 60.0 | 66.6 | 73.7 | 80.9 | 81.5 | 76.1 | ***64.9*** | ***52.8*** | ***46.4*** | 63.0 |
| Minimum Temp °F | ***32.8*** | 34.4 | 37.0 | 39.0 | 43.1 | 48.8 | 51.9 | 52.4 | 48.8 | ***42.9*** | ***38.1*** | ***34.0*** | 41.9 |
| Mean Temp °F | ***39.5*** | 42.6 | 46.8 | 49.5 | 54.9 | 61.3 | 66.4 | 66.9 | 62.4 | ***53.9*** | ***45.4*** | ***40.2*** | 52.5 |
| Days Max Temp ≥ 90 °F | 0 | 0 | 0 | 0 | 0 | 1 | 5 | 5 | 2 | 0 | 0 | 0 | 13 |
| Days Max Temp ≤ 32 °F | 1 | 0 | 0 | 0 | 0 | 0 | 0 | 0 | 0 | 0 | 0 | 1 | 2 |
| Days Min Temp ≤ 32 °F | 14 | 10 | 6 | 2 | 0 | 0 | 0 | 0 | 0 | 1 | 6 | 11 | 50 |
| Days Min Temp ≤ 0 °F | 0 | 0 | 0 | 0 | 0 | 0 | 0 | 0 | 0 | 0 | 0 | 0 | 0 |
| Heating Degree Days | ***783*** | 627 | 559 | 458 | 312 | 139 | 45 | 34 | 108 | ***340*** | ***583*** | ***762*** | 4750 |
| Cooling Degree Days | na | ***0*** | 0 | 0 | ***9*** | 36 | ***100*** | 110 | ***49*** | na | na | na | na |
| Total Precipitation (") | 6.33 | 4.79 | 4.29 | 2.87 | 1.84 | 1.26 | 0.44 | 0.77 | 1.25 | 2.80 | 6.73 | 7.52 | 40.89 |
| Days ≥ 0.1" Precip | 11 | 10 | 10 | 8 | 5 | 3 | 1 | 2 | 4 | 6 | 12 | 12 | 84 |
| Total Snowfall (") | 1.5 | 0.5 | 0.1 | 0.0 | 0.0 | 0.0 | 0.0 | 0.0 | 0.0 | 0.0 | 0.1 | 1.0 | 3.2 |
| Days ≥ 1" Snow Depth | 1 | 1 | 0 | 0 | 0 | 0 | 0 | 0 | 0 | 0 | 0 | 1 | 3 |

## FOREST GROVE *Washington County* ELEVATION 180 ft LAT/LONG 45° 32 ' N / 123° 6 ' W

| | JAN | FEB | MAR | APR | MAY | JUN | JUL | AUG | SEP | OCT | NOV | DEC | YEAR |
|---|---|---|---|---|---|---|---|---|---|---|---|---|---|
| Maximum Temp °F | 45.8 | 51.0 | 56.8 | 61.7 | 69.2 | 75.3 | 82.1 | 83.1 | 77.1 | 65.3 | 52.3 | 45.4 | 63.8 |
| Minimum Temp °F | 32.6 | 34.4 | 36.9 | 39.5 | 44.3 | 49.6 | 53.0 | 52.7 | 48.3 | 41.3 | 37.2 | 33.4 | 41.9 |
| Mean Temp °F | 39.2 | 42.7 | 46.9 | 50.6 | 56.8 | 62.5 | 67.5 | 67.9 | 62.7 | 53.3 | 44.8 | 39.4 | 52.9 |
| Days Max Temp ≥ 90 °F | 0 | 0 | 0 | 0 | 1 | 2 | 7 | 8 | 3 | 0 | 0 | 0 | 21 |
| Days Max Temp ≤ 32 °F | 1 | 0 | 0 | 0 | 0 | 0 | 0 | 0 | 0 | 0 | 0 | 1 | 2 |
| Days Min Temp ≤ 32 °F | 15 | 11 | 7 | 4 | 0 | 0 | 0 | 0 | 0 | 2 | 8 | 13 | 60 |
| Days Min Temp ≤ 0 °F | 0 | 0 | 0 | 0 | 0 | 0 | 0 | 0 | 0 | 0 | 0 | 0 | 0 |
| Heating Degree Days | 793 | 622 | 555 | 426 | 264 | 121 | 39 | 33 | 112 | 358 | 600 | 786 | 4709 |
| Cooling Degree Days | 0 | 0 | 0 | 1 | 21 | 50 | 120 | 134 | 53 | 3 | 0 | 0 | 382 |
| Total Precipitation (") | 6.92 | 5.06 | 4.50 | 2.71 | 1.73 | 1.35 | 0.48 | 0.82 | 1.51 | 3.21 | 6.50 | 8.07 | 42.86 |
| Days ≥ 0.1" Precip | 13 | 11 | 11 | 7 | 5 | 4 | 2 | 2 | 4 | 7 | 13 | 14 | 93 |
| Total Snowfall (") | 2.1 | ***1.4*** | 0.3 | 0.0 | 0.0 | 0.0 | 0.0 | 0.0 | 0.0 | 0.0 | 0.8 | 2.2 | 6.8 |
| Days ≥ 1" Snow Depth | 2 | 0 | 0 | 0 | 0 | 0 | 0 | 0 | 0 | 0 | 0 | 1 | 3 |

## FOSSIL *Wheeler County* ELEVATION 2661 ft LAT/LONG 45° 0 ' N / 120° 12 ' W

| | JAN | FEB | MAR | APR | MAY | JUN | JUL | AUG | SEP | OCT | NOV | DEC | YEAR |
|---|---|---|---|---|---|---|---|---|---|---|---|---|---|
| Maximum Temp °F | 41.2 | 46.4 | 52.4 | 58.7 | 67.4 | 75.0 | 83.5 | 83.7 | 75.7 | 64.8 | 48.5 | 41.5 | 61.6 |
| Minimum Temp °F | 24.2 | 26.5 | 28.7 | 31.5 | 36.1 | 42.2 | 45.3 | 45.6 | 39.8 | 33.7 | 29.3 | 25.1 | 34.0 |
| Mean Temp °F | 32.7 | 36.5 | 40.6 | 45.0 | 51.8 | 58.6 | 64.4 | 64.7 | 57.7 | 49.3 | 38.9 | 33.3 | 47.8 |
| Days Max Temp ≥ 90 °F | 0 | 0 | 0 | 0 | 0 | 2 | ***8*** | 8 | 2 | 0 | 0 | 0 | 20 |
| Days Max Temp ≤ 32 °F | 6 | 2 | 0 | 0 | 0 | 0 | 0 | 0 | 0 | 0 | 2 | 5 | 15 |
| Days Min Temp ≤ 32 °F | 24 | 21 | 21 | 18 | 10 | 2 | 0 | 0 | 5 | 13 | 18 | 23 | 155 |
| Days Min Temp ≤ 0 °F | 1 | 1 | 0 | 0 | 0 | 0 | 0 | 0 | 0 | 0 | 0 | 1 | 3 |
| Heating Degree Days | 994 | 799 | 752 | 592 | 407 | 210 | 87 | 77 | 223 | 482 | 775 | 975 | 6373 |
| Cooling Degree Days | 0 | 0 | 0 | 0 | 6 | 22 | 74 | 73 | 13 | 1 | 0 | 0 | 189 |
| Total Precipitation (") | 1.62 | 1.14 | 1.40 | 1.39 | 1.35 | 1.16 | 0.57 | 0.80 | 0.77 | 1.20 | 1.85 | 1.56 | 14.81 |
| Days ≥ 0.1" Precip | 5 | 4 | 5 | 4 | 4 | 4 | 2 | 2 | 3 | 4 | 6 | 5 | 48 |
| Total Snowfall (") | na | 2.0 | ***0.8*** | 0.3 | 0.2 | 0.0 | 0.0 | 0.0 | 0.0 | 0.5 | 2.3 | na | na |
| Days ≥ 1" Snow Depth | na | 2 | 0 | 0 | 0 | 0 | 0 | 0 | 0 | 0 | 1 | ***5*** | na |

## FREMONT 5 NW *Lake County* ELEVATION 4609 ft LAT/LONG 43° 23 ' N / 121° 12 ' W

| | JAN | FEB | MAR | APR | MAY | JUN | JUL | AUG | SEP | OCT | NOV | DEC | YEAR |
|---|---|---|---|---|---|---|---|---|---|---|---|---|---|
| Maximum Temp °F | 38.6 | 43.8 | 49.5 | 57.3 | 67.6 | 75.3 | ***84.0*** | 83.3 | 75.8 | 64.1 | 46.8 | 38.7 | 60.4 |
| Minimum Temp °F | 15.1 | 18.6 | 20.9 | 22.9 | 28.0 | 34.3 | ***36.3*** | ***35.6*** | 28.2 | 22.5 | 19.1 | 13.6 | 24.6 |
| Mean Temp °F | 27.1 | 31.5 | 35.2 | 40.1 | 47.8 | 54.8 | ***60.2*** | ***59.5*** | 51.9 | 43.3 | 33.0 | 26.0 | 42.5 |
| Days Max Temp ≥ 90 °F | 0 | 0 | 0 | 0 | 0 | 2 | ***8*** | ***7*** | 1 | 0 | 0 | 0 | 18 |
| Days Max Temp ≤ 32 °F | 8 | 2 | 1 | 0 | 0 | 0 | 0 | 0 | 0 | 0 | 1 | ***7*** | 19 |
| Days Min Temp ≤ 32 °F | 28 | 26 | 29 | 26 | ***22*** | ***12*** | ***9*** | ***9*** | ***20*** | 27 | 27 | 29 | 264 |
| Days Min Temp ≤ 0 °F | ***5*** | 2 | 0 | 0 | 0 | 0 | 0 | 0 | 0 | 0 | 2 | 5 | 14 |
| Heating Degree Days | 1166 | 941 | 914 | 740 | 526 | 306 | ***169*** | ***183*** | 387 | 666 | 953 | 1202 | 8153 |
| Cooling Degree Days | ***0*** | 0 | 0 | 0 | 1 | 7 | ***18*** | na | ***1*** | 0 | 0 | 0 | na |
| Total Precipitation (") | 1.66 | 0.91 | 1.38 | 0.73 | 0.84 | 0.93 | 0.53 | 0.71 | 0.38 | 0.77 | 1.66 | 1.52 | 12.02 |
| Days ≥ 0.1" Precip | 4 | 3 | 3 | 2 | 3 | 3 | 1 | 1 | 1 | 2 | ***4*** | 4 | 31 |
| Total Snowfall (") | ***9.0*** | 6.0 | 4.8 | 2.7 | 0.8 | 0.0 | 0.0 | 0.0 | 0.0 | 0.7 | 5.4 | ***10.1*** | 39.5 |
| Days ≥ 1" Snow Depth | ***14*** | 7 | 3 | 1 | 0 | 0 | 0 | 0 | 0 | 0 | ***4*** | ***12*** | 41 |

## GOLD BEACH RS *Curry County* ELEVATION 49 ft LAT/LONG 42° 24 ' N / 124° 25 ' W

| | JAN | FEB | MAR | APR | MAY | JUN | JUL | AUG | SEP | OCT | NOV | DEC | YEAR |
|---|---|---|---|---|---|---|---|---|---|---|---|---|---|
| Maximum Temp °F | ***54.6*** | 55.8 | 56.6 | 58.3 | 61.4 | 65.2 | 68.1 | 68.6 | 68.0 | 64.4 | ***57.7*** | ***54.6*** | 61.1 |
| Minimum Temp °F | ***40.3*** | 41.1 | 41.5 | 42.7 | 45.4 | 49.4 | 51.0 | 51.9 | 50.5 | 47.3 | 43.4 | ***40.4*** | 45.4 |
| Mean Temp °F | ***47.5*** | 48.5 | 49.1 | 50.5 | 53.4 | 57.3 | 59.6 | 60.3 | 59.3 | 55.8 | ***50.6*** | ***47.5*** | 53.3 |
| Days Max Temp ≥ 90 °F | 0 | 0 | 0 | 0 | 0 | 0 | 0 | 0 | 0 | 0 | 0 | 0 | 0 |
| Days Max Temp ≤ 32 °F | 0 | 0 | 0 | 0 | 0 | 0 | 0 | 0 | 0 | 0 | 0 | 0 | 0 |
| Days Min Temp ≤ 32 °F | 3 | 2 | 1 | 0 | 0 | 0 | 0 | 0 | 0 | 0 | 1 | 3 | 10 |
| Days Min Temp ≤ 0 °F | 0 | 0 | 0 | 0 | 0 | 0 | 0 | 0 | 0 | 0 | 0 | 0 | 0 |
| Heating Degree Days | ***535*** | 459 | 486 | 429 | 352 | 224 | 163 | 142 | 171 | 281 | ***426*** | ***535*** | 4203 |
| Cooling Degree Days | ***0*** | ***0*** | 0 | 0 | ***0*** | 1 | 1 | 0 | 3 | 4 | 0 | ***0*** | 9 |
| Total Precipitation (") | 12.07 | 9.85 | 10.16 | 6.33 | 3.52 | 1.72 | 0.44 | 1.17 | 2.30 | 5.01 | 11.55 | 13.82 | 77.94 |
| Days ≥ 0.1" Precip | na | na | na | na | na | ***3*** | 1 | 2 | 3 | na | na | na | na |
| Total Snowfall (") | 0.2 | 0.0 | 0.0 | 0.0 | 0.0 | 0.0 | 0.0 | 0.0 | 0.0 | 0.0 | 0.0 | 0.1 | 0.3 |
| Days ≥ 1" Snow Depth | 0 | 0 | 0 | 0 | 0 | 0 | 0 | 0 | 0 | 0 | 0 | 0 | 0 |

## GOVERNMENT CAMP *Clackamas County* ELEVATION 3983 ft LAT/LONG 45° 18 ' N / 121° 45 ' W

| | JAN | FEB | MAR | APR | MAY | JUN | JUL | AUG | SEP | OCT | NOV | DEC | YEAR |
|---|---|---|---|---|---|---|---|---|---|---|---|---|---|
| Maximum Temp °F | 35.3 | 38.2 | 41.3 | 45.2 | 53.0 | 59.7 | 67.5 | 68.2 | 62.3 | 53.5 | 40.3 | 35.8 | 50.0 |
| Minimum Temp °F | 23.8 | 25.6 | 27.7 | 30.3 | 35.2 | 41.0 | 46.1 | 46.7 | 42.4 | 36.4 | 29.3 | 24.8 | 34.1 |
| Mean Temp °F | 29.6 | 31.9 | 34.5 | 37.8 | 44.1 | 50.4 | 56.8 | 57.5 | 52.4 | 45.0 | 34.8 | 30.3 | 42.1 |
| Days Max Temp ≥ 90 °F | 0 | 0 | 0 | 0 | 0 | 0 | 0 | 0 | 0 | 0 | 0 | 0 | 0 |
| Days Max Temp ≤ 32 °F | 11 | 7 | 5 | 2 | 0 | 0 | 0 | 0 | 0 | 0 | 5 | 10 | 40 |
| Days Min Temp ≤ 32 °F | 27 | 24 | 25 | 21 | 12 | 2 | 0 | 0 | 2 | 9 | 21 | 27 | 170 |
| Days Min Temp ≤ 0 °F | 1 | 0 | 0 | 0 | 0 | 0 | 0 | 0 | 0 | 0 | 0 | 1 | 2 |
| Heating Degree Days | 1091 | 927 | 937 | 810 | 642 | 436 | 268 | 251 | 378 | 615 | 899 | 1068 | 8322 |
| Cooling Degree Days | 0 | 0 | 0 | 0 | 3 | 5 | 21 | 24 | 9 | 1 | 0 | 0 | 63 |
| Total Precipitation (") | 13.30 | 9.57 | 8.20 | 7.25 | 4.68 | 3.55 | 1.29 | 1.66 | 3.56 | 6.17 | 12.32 | 13.51 | 85.06 |
| Days ≥ 0.1" Precip | 17 | 14 | 16 | 13 | 11 | 8 | 3 | 4 | 6 | 10 | 16 | 17 | 135 |
| Total Snowfall (") | 57.7 | 42.7 | 38.3 | 27.6 | 7.0 | 0.2 | 0.0 | 0.0 | 0.3 | 6.4 | 36.8 | 49.5 | 266.5 |
| Days ≥ 1" Snow Depth | 30 | 26 | 28 | 22 | 7 | 0 | 0 | 0 | 0 | 3 | 17 | 29 | 162 |

## GRANTS PASS *Josephine County* ELEVATION 932 ft LAT/LONG 42° 26 ' N / 123° 19 ' W

| | JAN | FEB | MAR | APR | MAY | JUN | JUL | AUG | SEP | OCT | NOV | DEC | YEAR |
|---|---|---|---|---|---|---|---|---|---|---|---|---|---|
| Maximum Temp °F | 47.9 | 54.8 | 61.0 | 66.7 | 75.2 | 82.3 | 89.8 | 89.7 | 83.7 | 70.6 | 53.9 | 46.1 | 68.5 |
| Minimum Temp °F | 33.0 | 34.3 | 36.2 | 38.7 | 43.7 | 49.6 | 53.0 | 52.2 | 46.3 | 40.6 | 37.4 | 33.4 | 41.5 |
| Mean Temp °F | 40.5 | 44.6 | 48.6 | 52.7 | 59.5 | 65.9 | 71.4 | 70.9 | 65.0 | 55.6 | 45.6 | 39.8 | 55.0 |
| Days Max Temp ≥ 90 °F | 0 | 0 | 0 | 0 | 3 | 8 | 17 | 17 | 9 | 1 | 0 | 0 | 55 |
| Days Max Temp ≤ 32 °F | 0 | 0 | 0 | 0 | 0 | 0 | 0 | 0 | 0 | 0 | 0 | 1 | 1 |
| Days Min Temp ≤ 32 °F | 15 | 11 | 9 | 5 | 1 | 0 | 0 | 0 | 0 | 4 | 7 | 13 | 65 |
| Days Min Temp ≤ 0 °F | 0 | 0 | 0 | 0 | 0 | 0 | 0 | 0 | 0 | 0 | 0 | 0 | 0 |
| Heating Degree Days | 753 | 570 | 501 | 364 | 195 | 66 | 11 | 11 | 67 | 289 | 574 | 775 | 4176 |
| Cooling Degree Days | 0 | 0 | 0 | 4 | 32 | 94 | 203 | 196 | 76 | 7 | 0 | 0 | 612 |
| Total Precipitation (") | 4.97 | 3.67 | 3.37 | 1.85 | 1.08 | 0.56 | 0.34 | 0.46 | 0.85 | 2.16 | 5.12 | 5.53 | 29.96 |
| Days ≥ 0.1" Precip | 9 | 8 | 8 | 5 | 3 | 2 | 1 | 2 | 2 | 5 | 9 | 10 | 64 |
| Total Snowfall (") | 1.8 | 0.4 | 0.2 | 0.0 | 0.0 | 0.0 | 0.0 | 0.0 | 0.0 | 0.0 | 0.0 | 1.1 | 3.5 |
| Days ≥ 1" Snow Depth | 1 | 0 | 0 | 0 | 0 | 0 | 0 | 0 | 0 | 0 | 0 | 1 | 2 |

## GRIZZLY *Jefferson County* ELEVATION 3642 ft LAT/LONG 44° 31 ' N / 120° 56 ' W

| | JAN | FEB | MAR | APR | MAY | JUN | JUL | AUG | SEP | OCT | NOV | DEC | YEAR |
|---|---|---|---|---|---|---|---|---|---|---|---|---|---|
| Maximum Temp °F | 40.2 | 44.4 | 50.3 | 55.7 | 64.7 | 73.1 | 82.0 | 82.1 | 72.8 | 63.1 | 47.1 | 40.8 | 59.7 |
| Minimum Temp °F | 21.8 | 24.2 | 26.2 | 28.2 | 33.2 | 38.8 | 42.0 | 42.1 | 36.7 | 31.2 | 26.7 | 21.9 | 31.1 |
| Mean Temp °F | 31.1 | 34.4 | 38.3 | 42.1 | 49.0 | 56.0 | 62.1 | 62.1 | 54.8 | 47.1 | 37.0 | 31.4 | 45.4 |
| Days Max Temp ≥ 90 °F | 0 | 0 | 0 | 0 | 0 | 1 | 7 | 7 | 1 | 0 | 0 | 0 | 16 |
| Days Max Temp ≤ 32 °F | 6 | 2 | 0 | 0 | 0 | 0 | 0 | 0 | 0 | 0 | 1 | 5 | 14 |
| Days Min Temp ≤ 32 °F | 25 | 24 | 25 | *21* | 14 | 6 | 2 | 1 | 8 | *17* | *22* | *26* | 191 |
| Days Min Temp ≤ 0 °F | 2 | 1 | 0 | 0 | 0 | 0 | 0 | 0 | 0 | 0 | 1 | 1 | 5 |
| Heating Degree Days | 1044 | 859 | 822 | 681 | 492 | 277 | 131 | 128 | 300 | 547 | 835 | 1036 | 7152 |
| Cooling Degree Days | *0* | 0 | 0 | *0* | 5 | 13 | 36 | 38 | *3* | *0* | 0 | 0 | 95 |
| Total Precipitation (") | 1.53 | 1.04 | 1.10 | 1.05 | 1.20 | 0.99 | 0.42 | 0.73 | 0.61 | 0.87 | 1.74 | 1.17 | 12.45 |
| Days ≥ 0.1" Precip | 5 | 4 | *4* | 4 | 4 | 3 | 1 | 2 | 2 | 3 | *5* | *4* | 41 |
| Total Snowfall (") | 7.7 | 4.4 | 3.5 | 1.9 | 0.3 | 0.0 | 0.0 | 0.0 | 0.0 | 0.6 | 4.3 | 7.5 | 30.2 |
| Days ≥ 1" Snow Depth | *7* | *4* | 2 | 1 | 0 | 0 | 0 | 0 | 0 | 0 | *2* | *7* | 23 |

## HALFWAY *Baker County* ELEVATION 2680 ft LAT/LONG 44° 53 ' N / 117° 7 ' W

| | JAN | FEB | MAR | APR | MAY | JUN | JUL | AUG | SEP | OCT | NOV | DEC | YEAR |
|---|---|---|---|---|---|---|---|---|---|---|---|---|---|
| Maximum Temp °F | 32.9 | 40.1 | 51.7 | 62.1 | 71.3 | 79.0 | 87.8 | 87.1 | 77.6 | 64.6 | 45.8 | 34.3 | 61.2 |
| Minimum Temp °F | 15.2 | 19.5 | 26.3 | 31.4 | 37.3 | 43.8 | 47.6 | 46.4 | 38.6 | 30.6 | 25.0 | 16.5 | 31.5 |
| Mean Temp °F | 24.1 | 29.8 | 39.0 | 46.8 | 54.4 | 61.4 | 67.7 | 66.7 | 58.1 | 47.7 | 35.4 | 25.4 | 46.4 |
| Days Max Temp ≥ 90 °F | 0 | 0 | 0 | 0 | 1 | 4 | 15 | 13 | 3 | 0 | 0 | 0 | 36 |
| Days Max Temp ≤ 32 °F | 13 | 4 | 0 | 0 | 0 | 0 | 0 | 0 | 0 | 0 | 2 | 11 | 30 |
| Days Min Temp ≤ 32 °F | 30 | 26 | 24 | 18 | 8 | 1 | 0 | 0 | 6 | 19 | 24 | 30 | 186 |
| Days Min Temp ≤ 0 °F | 5 | 2 | 0 | 0 | 0 | 0 | 0 | 0 | 0 | 0 | 0 | 4 | 11 |
| Heating Degree Days | 1261 | 986 | 798 | 540 | 329 | 140 | 36 | 46 | 212 | 531 | 881 | 1220 | 6980 |
| Cooling Degree Days | 0 | 0 | 0 | 0 | 8 | 40 | 112 | 104 | 11 | 0 | 0 | 0 | 275 |
| Total Precipitation (") | 3.38 | 2.28 | 1.82 | 1.50 | 1.41 | 1.33 | 0.50 | 0.71 | 0.80 | 1.28 | 2.99 | 3.41 | 21.41 |
| Days ≥ 0.1" Precip | 8 | 6 | 6 | 4 | 4 | 4 | 2 | 2 | 2 | 3 | 8 | 8 | 57 |
| Total Snowfall (") | 24.9 | 10.5 | 3.7 | 0.5 | 0.0 | 0.0 | 0.0 | 0.0 | 0.0 | 0.7 | 11.9 | 20.9 | 73.1 |
| Days ≥ 1" Snow Depth | 28 | 23 | 10 | 0 | 0 | 0 | 0 | 0 | 0 | 0 | 6 | 22 | 89 |

## HART MOUNTAIN REFUGE *Lake County* ELEVATION 5623 ft LAT/LONG 42° 33 ' N / 119° 39 ' W

| | JAN | FEB | MAR | APR | MAY | JUN | JUL | AUG | SEP | OCT | NOV | DEC | YEAR |
|---|---|---|---|---|---|---|---|---|---|---|---|---|---|
| Maximum Temp °F | 39.4 | 42.1 | 46.1 | 53.3 | 63.0 | 71.5 | 81.1 | 80.3 | 71.7 | 60.9 | 45.9 | 39.1 | 57.9 |
| Minimum Temp °F | 18.8 | 20.9 | 23.0 | 26.5 | 32.5 | 38.8 | 43.6 | 43.4 | 37.0 | 30.5 | 23.6 | 17.9 | 29.7 |
| Mean Temp °F | 29.1 | 31.5 | 34.5 | 39.9 | 47.8 | 55.2 | 62.4 | 61.9 | 54.4 | 45.7 | 34.7 | 28.5 | 43.8 |
| Days Max Temp ≥ 90 °F | 0 | 0 | 0 | 0 | 0 | 0 | 4 | 3 | 0 | 0 | 0 | 0 | 7 |
| Days Max Temp ≤ 32 °F | 6 | 4 | 2 | 0 | 0 | 0 | 0 | 0 | 0 | 0 | 3 | 7 | 22 |
| Days Min Temp ≤ 32 °F | 28 | 26 | 27 | 23 | 16 | 6 | 2 | 2 | 8 | 18 | 24 | 29 | 209 |
| Days Min Temp ≤ 0 °F | 2 | 1 | 0 | 0 | 0 | 0 | 0 | 0 | 0 | 0 | 1 | 2 | 6 |
| Heating Degree Days | 1104 | 940 | 939 | 746 | 528 | 298 | 117 | 130 | 315 | 589 | 902 | 1127 | 7735 |
| Cooling Degree Days | 0 | 0 | 0 | 0 | 1 | 10 | 43 | 39 | 5 | 0 | 0 | 0 | 98 |
| Total Precipitation (") | 0.89 | 0.75 | 1.25 | 1.39 | 1.49 | 1.36 | 0.47 | 0.60 | 0.80 | 0.96 | 1.22 | 1.06 | 12.24 |
| Days ≥ 0.1" Precip | 3 | 3 | 4 | 4 | 4 | 4 | 1 | 2 | 2 | 3 | 4 | 3 | 37 |
| Total Snowfall (") | *6.3* | 6.2 | 7.7 | 4.0 | 2.3 | 0.6 | 0.0 | 0.0 | 0.6 | 2.1 | 5.4 | *8.1* | 43.3 |
| Days ≥ 1" Snow Depth | 11 | 8 | 6 | 2 | 1 | 0 | 0 | 0 | 0 | 1 | 5 | *10* | 44 |

## HEPPNER *Morrow County* ELEVATION 1952 ft LAT/LONG 45° 21 ' N / 119° 33 ' W

| | JAN | FEB | MAR | APR | MAY | JUN | JUL | AUG | SEP | OCT | NOV | DEC | YEAR |
|---|---|---|---|---|---|---|---|---|---|---|---|---|---|
| Maximum Temp °F | 41.7 | 47.1 | 54.1 | 60.6 | 69.6 | 77.6 | 85.5 | 84.9 | 75.7 | 64.3 | 49.8 | 42.2 | 62.8 |
| Minimum Temp °F | 25.9 | 29.5 | 33.3 | 36.5 | 42.6 | 48.8 | 52.7 | 52.9 | 46.1 | 38.6 | 32.2 | 26.7 | 38.8 |
| Mean Temp °F | 33.9 | 38.3 | 43.7 | 48.6 | 56.1 | 63.2 | 69.2 | 68.9 | 61.0 | 51.5 | 41.0 | 34.4 | 50.8 |
| Days Max Temp ≥ 90 °F | 0 | 0 | 0 | 0 | 1 | 4 | 11 | 10 | 2 | 0 | 0 | 0 | 28 |
| Days Max Temp ≤ 32 °F | 7 | 2 | 0 | 0 | 0 | 0 | 0 | 0 | 0 | 0 | 2 | 6 | 17 |
| Days Min Temp ≤ 32 °F | 23 | 18 | 15 | 8 | 1 | 0 | 0 | 0 | 1 | 5 | 15 | 23 | 109 |
| Days Min Temp ≤ 0 °F | 1 | 0 | 0 | 0 | 0 | 0 | 0 | 0 | 0 | 0 | 0 | 1 | 2 |
| Heating Degree Days | 959 | 747 | 653 | 487 | 285 | 117 | 33 | 31 | 153 | 415 | 713 | 942 | 5535 |
| Cooling Degree Days | 0 | 0 | 0 | 1 | 24 | 66 | 156 | 153 | 45 | 3 | 0 | 0 | 448 |
| Total Precipitation (") | 1.51 | 1.06 | 1.50 | 1.31 | 1.46 | 1.03 | 0.41 | 0.63 | 0.75 | 1.04 | 1.78 | 1.37 | 13.85 |
| Days ≥ 0.1" Precip | 5 | 4 | 5 | 4 | 4 | 3 | 1 | 2 | 2 | 3 | 6 | 5 | 44 |
| Total Snowfall (") | 6.2 | 2.5 | 0.8 | 0.1 | 0.0 | 0.0 | 0.0 | 0.0 | 0.0 | 0.3 | 2.4 | 4.2 | 16.5 |
| Days ≥ 1" Snow Depth | 8 | 4 | 1 | 0 | 0 | 0 | 0 | 0 | 0 | 0 | 2 | 7 | 22 |

## HERMISTON 1 SE *Umatilla County* ELEVATION 620 ft LAT/LONG 45° 49 ' N / 119° 17 ' W

| | JAN | FEB | MAR | APR | MAY | JUN | JUL | AUG | SEP | OCT | NOV | DEC | YEAR |
|---|---|---|---|---|---|---|---|---|---|---|---|---|---|
| Maximum Temp °F | 40.4 | 47.6 | 57.3 | 64.8 | 73.8 | 80.9 | 88.2 | 87.4 | 78.4 | 66.1 | 50.4 | 41.3 | 64.7 |
| Minimum Temp °F | 25.9 | 29.1 | 33.8 | 39.1 | 46.3 | 53.2 | 57.7 | 56.6 | 47.8 | 37.6 | 32.3 | 26.5 | 40.5 |
| Mean Temp °F | 33.1 | 38.4 | 45.6 | 52.0 | 60.0 | 67.1 | 73.0 | 72.0 | 63.1 | 51.9 | 41.4 | 33.9 | 52.6 |
| Days Max Temp ≥ 90 °F | 0 | 0 | 0 | 0 | 2 | 6 | 14 | 12 | 3 | 0 | 0 | 0 | 37 |
| Days Max Temp ≤ 32 °F | 7 | 2 | 0 | 0 | 0 | 0 | 0 | 0 | 0 | 0 | 1 | 6 | 16 |
| Days Min Temp ≤ 32 °F | 23 | 19 | 13 | 5 | 0 | 0 | 0 | 0 | 1 | 8 | 14 | 23 | 106 |
| Days Min Temp ≤ 0 °F | 1 | 0 | 0 | 0 | 0 | 0 | 0 | 0 | 0 | 0 | 0 | 1 | 2 |
| Heating Degree Days | 981 | 744 | 595 | 386 | 185 | 56 | 11 | 12 | 111 | 402 | 701 | 956 | 5140 |
| Cooling Degree Days | 0 | 0 | 0 | 4 | 45 | 115 | 240 | 222 | 61 | 2 | 0 | 0 | 689 |
| Total Precipitation (") | 1.32 | 0.82 | 0.77 | 0.71 | 0.64 | 0.55 | 0.23 | 0.39 | 0.43 | 0.59 | 1.31 | 1.27 | 9.03 |
| Days ≥ 0.1" Precip | 5 | 3 | 3 | 2 | 2 | 2 | 1 | 1 | 2 | 2 | 5 | 5 | 33 |
| Total Snowfall (") | 4.0 | 2.0 | 0.3 | 0.0 | 0.0 | 0.0 | 0.0 | 0.0 | 0.0 | 0.0 | 1.3 | 2.6 | 10.2 |
| Days ≥ 1" Snow Depth | 8 | 3 | 0 | 0 | 0 | 0 | 0 | 0 | 0 | 0 | 1 | *4* | 16 |

## HILLSBORO *Washington County* ELEVATION 200 ft LAT/LONG 45° 31 ' N / 123° 0 ' W

| | JAN | FEB | MAR | APR | MAY | JUN | JUL | AUG | SEP | OCT | NOV | DEC | YEAR |
|---|---|---|---|---|---|---|---|---|---|---|---|---|---|
| Maximum Temp °F | 45.7 | 50.9 | 56.4 | 61.1 | 68.1 | 73.6 | 80.0 | 80.7 | 75.6 | 64.7 | 52.2 | 45.5 | 62.9 |
| Minimum Temp °F | 33.2 | 34.9 | 37.0 | 39.4 | 43.8 | 49.3 | 52.2 | 51.7 | 47.2 | 41.0 | 37.3 | 33.7 | 41.7 |
| Mean Temp °F | 39.5 | 42.9 | 46.8 | 50.2 | 56.0 | 61.5 | 66.2 | 66.2 | 61.4 | 52.9 | 44.8 | 39.6 | 52.3 |
| Days Max Temp ≥ 90 °F | 0 | 0 | 0 | 0 | 0 | 2 | 5 | 5 | 2 | 0 | 0 | 0 | 14 |
| Days Max Temp ≤ 32 °F | 1 | 0 | 0 | 0 | 0 | 0 | 0 | 0 | 0 | 0 | 0 | 1 | 2 |
| Days Min Temp ≤ 32 °F | 15 | 11 | 7 | 3 | 0 | 0 | 0 | 0 | 0 | 2 | 7 | 13 | 58 |
| Days Min Temp ≤ 0 °F | 0 | 0 | 0 | 0 | 0 | 0 | 0 | 0 | 0 | 0 | 0 | 0 | 0 |
| Heating Degree Days | 784 | 617 | 559 | 436 | 283 | 135 | 47 | 46 | 128 | 370 | 599 | 779 | 4783 |
| Cooling Degree Days | 0 | 0 | 0 | 1 | 14 | 38 | 85 | 91 | 34 | 1 | 0 | 0 | 264 |
| Total Precipitation (") | 5.71 | 4.14 | 3.68 | 2.29 | 1.61 | 1.42 | 0.56 | 0.99 | 1.47 | 2.69 | 5.43 | 6.67 | 36.66 |
| Days ≥ 0.1" Precip | 12 | 10 | 10 | 7 | 5 | 4 | 2 | 2 | 4 | 7 | 13 | 13 | 89 |
| Total Snowfall (") | 1.7 | 0.8 | 0.1 | 0.0 | 0.0 | 0.0 | 0.0 | 0.0 | 0.0 | 0.0 | 0.3 | 1.4 | 4.3 |
| Days ≥ 1" Snow Depth | 1 | 0 | 0 | 0 | 0 | 0 | 0 | 0 | 0 | 0 | 0 | 1 | 2 |

## HOOD RIVER EXP STN *Hood River County* ELEVATION 502 ft LAT/LONG 45° 41 ' N / 121° 31 ' W

| | JAN | FEB | MAR | APR | MAY | JUN | JUL | AUG | SEP | OCT | NOV | DEC | YEAR |
|---|---|---|---|---|---|---|---|---|---|---|---|---|---|
| Maximum Temp °F | 40.5 | 46.4 | 54.2 | 60.2 | 68.2 | 74.1 | 80.4 | 80.9 | 74.6 | 63.7 | 49.2 | 41.2 | 61.1 |
| Minimum Temp °F | 28.3 | 31.1 | 34.5 | 38.6 | 44.3 | 50.2 | 53.8 | 53.1 | 45.9 | 38.1 | 34.3 | 29.6 | 40.2 |
| Mean Temp °F | 34.4 | 38.8 | 44.4 | 49.4 | 56.3 | 62.1 | 67.1 | 67.0 | 60.3 | 50.9 | 41.8 | 35.4 | 50.7 |
| Days Max Temp ≥ 90 °F | 0 | 0 | 0 | 0 | 1 | 2 | 6 | 6 | 2 | 0 | 0 | 0 | 17 |
| Days Max Temp ≤ 32 °F | 5 | 1 | 0 | 0 | 0 | 0 | 0 | 0 | 0 | 0 | 1 | 4 | 11 |
| Days Min Temp ≤ 32 °F | 20 | 16 | 12 | 5 | 1 | 0 | 0 | 0 | 1 | 6 | 12 | 20 | 93 |
| Days Min Temp ≤ 0 °F | 0 | 0 | 0 | 0 | 0 | 0 | 0 | 0 | 0 | 0 | 0 | 0 | 0 |
| Heating Degree Days | 940 | 734 | 633 | 461 | 280 | 131 | 47 | 45 | 162 | 430 | 690 | 909 | 5462 |
| Cooling Degree Days | 0 | 0 | 0 | 2 | 21 | 50 | 118 | 118 | 31 | 1 | 0 | 0 | 341 |
| Total Precipitation (") | 5.48 | 3.78 | 2.63 | 1.66 | 0.93 | 0.75 | 0.28 | 0.56 | 1.05 | 2.24 | 4.98 | 5.71 | 30.05 |
| Days ≥ 0.1" Precip | 11 | 8 | 7 | 5 | 3 | 2 | 1 | 2 | 3 | 5 | 12 | 11 | 70 |
| Total Snowfall (") | 14.1 | 6.8 | 1.0 | 0.0 | 0.0 | 0.0 | 0.0 | 0.0 | 0.0 | 0.1 | 3.7 | 9.5 | 35.2 |
| Days ≥ 1" Snow Depth | 12 | 6 | 1 | 0 | 0 | 0 | 0 | 0 | 0 | 0 | 2 | 7 | 28 |

## HOWARD PRAIRIE DAM *Jackson County* ELEVATION 4573 ft LAT/LONG 42° 13 ' N / 122° 22 ' W

| | JAN | FEB | MAR | APR | MAY | JUN | JUL | AUG | SEP | OCT | NOV | DEC | YEAR |
|---|---|---|---|---|---|---|---|---|---|---|---|---|---|
| Maximum Temp °F | 37.1 | 42.2 | 46.6 | 52.3 | 61.9 | 70.0 | 78.8 | 78.8 | 72.4 | 60.9 | 43.3 | 36.1 | 56.7 |
| Minimum Temp °F | 19.3 | 21.4 | 24.5 | 28.0 | 33.4 | 39.9 | 43.9 | 43.3 | 37.8 | 32.2 | 26.2 | 20.8 | 30.9 |
| Mean Temp °F | 28.3 | 31.8 | 35.5 | 40.2 | 47.7 | 55.0 | 61.4 | 61.1 | 55.1 | 46.6 | 34.8 | 28.4 | 43.8 |
| Days Max Temp ≥ 90 °F | 0 | 0 | 0 | 0 | 0 | 0 | 3 | 3 | 0 | 0 | 0 | 0 | 6 |
| Days Max Temp ≤ 32 °F | 7 | 2 | 1 | 0 | 0 | 0 | 0 | 0 | 0 | 0 | 2 | 9 | 21 |
| Days Min Temp ≤ 32 °F | 29 | 27 | 29 | 25 | 15 | 4 | 0 | 0 | 5 | 17 | 26 | 29 | 206 |
| Days Min Temp ≤ 0 °F | 1 | 1 | 0 | 0 | 0 | 0 | 0 | 0 | 0 | 0 | 0 | 1 | 3 |
| Heating Degree Days | 1132 | 930 | 906 | 737 | 532 | 304 | 141 | 147 | 294 | 564 | 901 | 1126 | 7714 |
| Cooling Degree Days | 0 | 0 | 0 | 0 | 2 | 11 | 37 | 33 | 6 | 0 | 0 | 0 | 89 |
| Total Precipitation (") | 4.75 | 3.35 | 3.53 | 2.26 | 1.82 | 1.28 | 0.47 | 0.70 | 1.00 | 2.03 | 4.91 | 5.31 | 31.41 |
| Days ≥ 0.1" Precip | 9 | 9 | 10 | 7 | 5 | 4 | 1 | 2 | 3 | 5 | 11 | 11 | 77 |
| Total Snowfall (") | 26.6 | 20.7 | 20.6 | 10.2 | 3.5 | 0.2 | 0.0 | 0.0 | 0.2 | 1.8 | 20.1 | 31.9 | 135.8 |
| Days ≥ 1" Snow Depth | 30 | 27 | 23 | 10 | 2 | 0 | 0 | 0 | 0 | 1 | 13 | 27 | 133 |

## HUNTINGTON *Baker County* ELEVATION 2152 ft LAT/LONG 44° 21' N / 117° 16' W

| | JAN | FEB | MAR | APR | MAY | JUN | JUL | AUG | SEP | OCT | NOV | DEC | YEAR |
|---|---|---|---|---|---|---|---|---|---|---|---|---|---|
| Maximum Temp °F | 35.8 | 44.2 | 55.3 | 64.5 | 74.5 | 83.4 | 92.9 | 91.4 | 80.5 | 66.6 | 48.1 | 37.6 | 64.6 |
| Minimum Temp °F | 20.1 | 25.2 | 31.8 | 38.5 | 47.2 | 55.5 | 63.4 | 60.8 | 49.8 | 37.7 | 28.5 | 21.1 | 40.0 |
| Mean Temp °F | 28.0 | 34.7 | 43.6 | 51.5 | 60.8 | 69.5 | 78.2 | 76.0 | 65.1 | 52.2 | 38.3 | 29.3 | 52.3 |
| Days Max Temp ≥ 90 °F | 0 | 0 | 0 | 0 | 2 | 9 | 22 | 20 | 6 | 0 | 0 | 0 | 59 |
| Days Max Temp ≤ 32 °F | 10 | 3 | 0 | 0 | 0 | 0 | 0 | 0 | 0 | 0 | 1 | 6 | 20 |
| Days Min Temp ≤ 32 °F | 28 | 23 | 17 | 6 | 1 | 0 | 0 | 0 | 0 | 8 | 21 | 28 | 132 |
| Days Min Temp ≤ 0 °F | 2 | 1 | 0 | 0 | 0 | 0 | 0 | 0 | 0 | 0 | 0 | 1 | 4 |
| Heating Degree Days | 1140 | 848 | 658 | 403 | 176 | 47 | 4 | 9 | 92 | 395 | 795 | 1099 | 5666 |
| Cooling Degree Days | 0 | 0 | 0 | 6 | 54 | 171 | 384 | 337 | 100 | 5 | 0 | 0 | 1057 |
| Total Precipitation (") | 1.80 | 1.37 | 1.29 | 0.85 | 1.01 | 0.92 | 0.37 | 0.65 | 0.57 | 0.76 | 1.84 | 1.98 | 13.41 |
| Days ≥ 0.1" Precip | 5 | 4 | 5 | 3 | 3 | 3 | 1 | 2 | 2 | 3 | 6 | 5 | 42 |
| Total Snowfall (") | 9.0 | 3.2 | 0.4 | 0.0 | 0.0 | 0.0 | 0.0 | 0.0 | 0.0 | 0.0 | 3.0 | 8.2 | 23.8 |
| Days ≥ 1" Snow Depth | 15 | 6 | 0 | 0 | 0 | 0 | 0 | 0 | 0 | 0 | 2 | 12 | 35 |

## ILLAHE *Curry County* ELEVATION 371 ft LAT/LONG 42° 39' N / 124° 4' W

| | JAN | FEB | MAR | APR | MAY | JUN | JUL | AUG | SEP | OCT | NOV | DEC | YEAR |
|---|---|---|---|---|---|---|---|---|---|---|---|---|---|
| Maximum Temp °F | 49.6 | 54.8 | 59.3 | 64.6 | 72.2 | 79.3 | 87.1 | 87.3 | 83.5 | 69.6 | 55.5 | 49.0 | 67.6 |
| Minimum Temp °F | 35.7 | 37.7 | 38.5 | 39.5 | 43.7 | 48.6 | 51.5 | 51.3 | 48.6 | 43.9 | 40.3 | 35.8 | 42.9 |
| Mean Temp °F | 42.7 | 46.3 | 48.9 | 52.0 | 58.0 | 64.0 | 69.4 | 69.4 | 66.0 | 56.7 | 48.1 | 42.5 | 55.3 |
| Days Max Temp ≥ 90 °F | 0 | 0 | 0 | 0 | 2 | 4 | 12 | 13 | 9 | 1 | 0 | 0 | 41 |
| Days Max Temp ≤ 32 °F | 0 | 0 | 0 | 0 | 0 | 0 | 0 | 0 | 0 | 0 | 0 | 0 | 0 |
| Days Min Temp ≤ 32 °F | 10 | 6 | 4 | 2 | 0 | 0 | 0 | 0 | 0 | 1 | 3 | 10 | 36 |
| Days Min Temp ≤ 0 °F | 0 | 0 | 0 | 0 | 0 | 0 | 0 | 0 | 0 | 0 | 0 | 0 | 0 |
| Heating Degree Days | 686 | 522 | 492 | 383 | 228 | 84 | 16 | 14 | 50 | 254 | 501 | 691 | 3921 |
| Cooling Degree Days | 0 | 0 | 0 | 1 | 18 | 53 | 166 | 168 | 97 | 8 | 0 | 0 | 511 |
| Total Precipitation (") | 13.06 | 11.20 | 10.27 | 5.33 | 2.96 | 1.31 | 0.32 | 0.90 | 2.21 | 5.37 | 12.81 | 14.66 | 80.40 |
| Days ≥ 0.1" Precip | 13 | 13 | 13 | 8 | 5 | 2 | 1 | 2 | 3 | 7 | 13 | 14 | 94 |
| Total Snowfall (") | 6.8 | 2.0 | 0.5 | 0.3 | 0.0 | 0.0 | 0.0 | 0.0 | 0.0 | 0.0 | 0.1 | 2.4 | 12.1 |
| Days ≥ 1" Snow Depth | 1 | 1 | 0 | 0 | 0 | 0 | 0 | 0 | 0 | 0 | 0 | 1 | 3 |

## IRONSIDE 2 W *Malheur County* ELEVATION 3924 ft LAT/LONG 44° 19' N / 117° 59' W

| | JAN | FEB | MAR | APR | MAY | JUN | JUL | AUG | SEP | OCT | NOV | DEC | YEAR |
|---|---|---|---|---|---|---|---|---|---|---|---|---|---|
| Maximum Temp °F | 32.7 | 39.2 | 49.5 | 58.5 | 66.9 | 75.6 | 86.3 | 86.5 | 75.3 | 63.1 | 44.5 | 34.3 | 59.4 |
| Minimum Temp °F | 13.6 | 19.1 | 26.2 | 31.9 | 38.7 | 45.9 | 52.9 | 52.4 | 42.6 | 32.5 | 23.3 | 15.4 | 32.9 |
| Mean Temp °F | 23.3 | 29.2 | 38.0 | 45.2 | 52.8 | 60.8 | 69.6 | 69.4 | 59.0 | 47.8 | 33.9 | 24.9 | 46.2 |
| Days Max Temp ≥ 90 °F | 0 | 0 | 0 | 0 | 0 | 3 | 13 | 13 | 2 | 0 | 0 | 0 | 31 |
| Days Max Temp ≤ 32 °F | na | 5 | 1 | 0 | 0 | 0 | 0 | 0 | 0 | 0 | 2 | 11 | na |
| Days Min Temp ≤ 32 °F | na | 26 | 25 | na | 6 | 1 | 0 | 0 | 3 | na | na | 29 | na |
| Days Min Temp ≤ 0 °F | 3 | 1 | 0 | 0 | 0 | 0 | 0 | 0 | 0 | 0 | 0 | 2 | 6 |
| Heating Degree Days | na | 1007 | 829 | 590 | 380 | 175 | 38 | 41 | 211 | 528 | 927 | 1242 | na |
| Cooling Degree Days | 0 | 0 | 0 | 0 | 13 | 55 | 174 | 172 | 41 | 2 | 0 | 0 | 457 |
| Total Precipitation (") | 1.44 | 0.94 | 0.92 | 0.76 | 1.04 | 1.04 | 0.53 | 0.78 | 0.55 | 0.61 | 1.46 | 1.65 | 11.72 |
| Days ≥ 0.1" Precip | na | 2 | 3 | 2 | 4 | 3 | 2 | 2 | 2 | 2 | 4 | 4 | na |
| Total Snowfall (") | na | 5.9 | 2.5 | 0.5 | 0.2 | 0.0 | 0.0 | 0.0 | 0.0 | 0.5 | 6.9 | 12.6 | na |
| Days ≥ 1" Snow Depth | 19 | 13 | 5 | 1 | 0 | 0 | 0 | 0 | 0 | 0 | 6 | 16 | 60 |

## JOHN DAY *Grant County* ELEVATION 3071 ft LAT/LONG 44° 25' N / 118° 57' W

| | JAN | FEB | MAR | APR | MAY | JUN | JUL | AUG | SEP | OCT | NOV | DEC | YEAR |
|---|---|---|---|---|---|---|---|---|---|---|---|---|---|
| Maximum Temp °F | 40.5 | 47.1 | 53.6 | 60.2 | 69.3 | 78.0 | 87.6 | 87.4 | 78.0 | 66.1 | 49.5 | 41.8 | 63.3 |
| Minimum Temp °F | 21.3 | 24.8 | 28.5 | 32.5 | 38.9 | 45.1 | 48.6 | 48.0 | 40.4 | 33.0 | 27.9 | 22.0 | 34.2 |
| Mean Temp °F | 31.0 | 36.0 | 41.1 | 46.4 | 54.1 | 61.6 | 68.1 | 67.7 | 59.2 | 49.6 | 38.7 | 31.9 | 48.8 |
| Days Max Temp ≥ 90 °F | 0 | 0 | 0 | 0 | 1 | 5 | 15 | 15 | 5 | 1 | 0 | 0 | 42 |
| Days Max Temp ≤ 32 °F | 6 | 2 | 0 | 0 | 0 | 0 | 0 | 0 | 0 | 0 | 1 | 4 | 13 |
| Days Min Temp ≤ 32 °F | 27 | 24 | 23 | 16 | 5 | 0 | 0 | 0 | 4 | 14 | 22 | 27 | 162 |
| Days Min Temp ≤ 0 °F | 1 | 1 | 0 | 0 | 0 | 0 | 0 | 0 | 0 | 0 | 0 | 1 | 3 |
| Heating Degree Days | 1048 | 813 | 734 | 553 | 340 | 151 | 42 | 45 | 196 | 473 | 783 | 1018 | 6196 |
| Cooling Degree Days | 0 | 0 | 0 | 1 | 15 | 55 | 132 | 134 | 27 | 2 | 0 | 0 | 366 |
| Total Precipitation (") | 1.16 | 0.77 | 1.21 | 1.28 | 1.67 | 1.43 | 0.62 | 1.04 | 0.77 | 0.90 | 1.46 | 1.32 | 13.63 |
| Days ≥ 0.1" Precip | 4 | 3 | 4 | 4 | 5 | 5 | 2 | 3 | 3 | 3 | 5 | 5 | 46 |
| Total Snowfall (") | 5.8 | 3.5 | 2.8 | 0.8 | 0.0 | 0.0 | 0.0 | 0.0 | 0.0 | 0.4 | 2.7 | 7.1 | 23.1 |
| Days ≥ 1" Snow Depth | 12 | 5 | 2 | 0 | 0 | 0 | 0 | 0 | 0 | 0 | 3 | 9 | 31 |

## JUNTURA 9 ENE *Malheur County* ELEVATION 2831 ft LAT/LONG 43° 48' N / 117° 56' W

| | JAN | FEB | MAR | APR | MAY | JUN | JUL | AUG | SEP | OCT | NOV | DEC | YEAR |
|---|---|---|---|---|---|---|---|---|---|---|---|---|---|
| Maximum Temp °F | *37.9* | *46.9* | *55.5* | 64.6 | 74.7 | 83.2 | *93.5* | 91.6 | *80.8* | *67.4* | 49.7 | *38.7* | 65.4 |
| Minimum Temp °F | *21.4* | *26.1* | *29.4* | 34.5 | 42.4 | 49.8 | *55.9* | 54.4 | *44.9* | *34.5* | 26.8 | *20.8* | 36.7 |
| Mean Temp °F | *29.7* | *36.5* | *42.5* | 49.6 | 58.6 | 66.6 | *74.6* | 73.0 | *62.9* | *51.0* | 38.2 | *29.8* | 51.1 |
| Days Max Temp ≥ 90 °F | 0 | *0* | *0* | 0 | 2 | 9 | *23* | *20* | *6* | 0 | 0 | 0 | 60 |
| Days Max Temp ≤ 32 °F | 7 | *2* | *0* | 0 | 0 | 0 | *0* | 0 | *0* | 0 | 1 | 6 | 16 |
| Days Min Temp ≤ 32 °F | 26 | *23* | *20* | 12 | 2 | 0 | *0* | 0 | *2* | *13* | 23 | 27 | 148 |
| Days Min Temp ≤ 0 °F | 2 | *0* | *0* | 0 | 0 | 0 | *0* | 0 | *0* | 0 | 0 | 1 | 3 |
| Heating Degree Days | *1088* | *798* | *691* | 458 | 219 | 69 | *8* | 12 | *119* | *429* | 796 | *1085* | 5772 |
| Cooling Degree Days | na | na | na | *2* | *29* | *108* | na | *254* | na | na | *0* | na | na |
| Total Precipitation (") | *1.16* | *0.88* | *0.91* | 0.83 | 0.96 | 1.06 | *0.44* | 0.53 | *0.51* | *0.69* | *1.28* | *1.40* | 10.65 |
| Days ≥ 0.1" Precip | *4* | *3* | *3* | 2 | 3 | 3 | *1* | 2 | *2* | 2 | *5* | *4* | 34 |
| Total Snowfall (") | *6.1* | na | na | *0.1* | 0.0 | *0.0* | *0.0* | *0.0* | *0.0* | *0.2* | *2.0* | *7.6* | na |
| Days ≥ 1" Snow Depth | *10* | *3* | *0* | 0 | 0 | 0 | *0* | 0 | *0* | 0 | *1* | *9* | 23 |

## KENT *Sherman County* ELEVATION 2713 ft LAT/LONG 45° 12' N / 120° 41' W

| | JAN | FEB | MAR | APR | MAY | JUN | JUL | AUG | SEP | OCT | NOV | DEC | YEAR |
|---|---|---|---|---|---|---|---|---|---|---|---|---|---|
| Maximum Temp °F | 38.2 | 43.5 | 51.0 | 57.4 | 66.7 | *75.0* | 83.7 | 83.7 | 74.6 | 63.1 | 47.1 | 39.3 | 60.3 |
| Minimum Temp °F | 23.2 | 26.6 | 30.3 | 33.3 | 39.0 | *45.9* | 51.3 | 51.7 | 45.0 | 37.0 | 29.3 | 23.8 | 36.4 |
| Mean Temp °F | 30.8 | 35.1 | 40.7 | 45.4 | 53.0 | *60.4* | 67.5 | 67.7 | 59.8 | 50.1 | 38.2 | 31.6 | 48.4 |
| Days Max Temp ≥ 90 °F | 0 | 0 | 0 | 0 | 0 | 3 | 9 | 9 | 2 | 0 | 0 | 0 | 23 |
| Days Max Temp ≤ 32 °F | 9 | 4 | 0 | 0 | 0 | 0 | 0 | 0 | 0 | 0 | 2 | 7 | 22 |
| Days Min Temp ≤ 32 °F | 26 | 22 | 20 | 14 | 6 | 0 | 0 | 0 | 1 | 8 | 20 | 26 | 143 |
| Days Min Temp ≤ 0 °F | 2 | 0 | 0 | 0 | 0 | 0 | 0 | 0 | 0 | 0 | 0 | 1 | 3 |
| Heating Degree Days | 1056 | 837 | 748 | 585 | 376 | *182* | 62 | 54 | 188 | 457 | 798 | 1035 | 6378 |
| Cooling Degree Days | 0 | 0 | 0 | 0 | 15 | *55* | 137 | 147 | 51 | 5 | 0 | 0 | 410 |
| Total Precipitation (") | 1.41 | 1.00 | 1.05 | 0.97 | 0.93 | 0.78 | 0.52 | 0.53 | 0.55 | 0.78 | 1.63 | 1.51 | 11.66 |
| Days ≥ 0.1" Precip | 4 | 3 | 4 | 3 | 3 | 2 | 1 | 2 | 2 | 2 | 5 | 4 | 35 |
| Total Snowfall (") | na | 3.3 | 1.4 | 0.7 | 0.0 | 0.0 | 0.0 | 0.0 | 0.0 | 0.3 | 3.0 | *4.9* | na |
| Days ≥ 1" Snow Depth | 10 | 5 | 2 | 0 | 0 | 0 | 0 | 0 | 0 | 0 | 3 | 7 | 27 |

## KLAMATH FALLS 2 SSW *Klamath County* ELEVATION 4091 ft LAT/LONG 42° 13' N / 121° 47' W

| | JAN | FEB | MAR | APR | MAY | JUN | JUL | AUG | SEP | OCT | NOV | DEC | YEAR |
|---|---|---|---|---|---|---|---|---|---|---|---|---|---|
| Maximum Temp °F | 39.3 | 45.6 | 51.6 | 58.7 | 68.6 | 76.7 | 85.4 | 84.5 | 76.7 | 64.6 | 47.2 | 38.8 | 61.5 |
| Minimum Temp °F | 21.0 | 25.0 | 28.4 | 31.7 | 38.9 | 45.8 | 51.4 | 50.0 | 43.1 | 34.7 | 27.4 | 21.4 | 34.9 |
| Mean Temp °F | 30.2 | 35.3 | 40.0 | 45.2 | 53.8 | 61.3 | 68.4 | 67.2 | 59.9 | 49.7 | 37.3 | 30.1 | 48.2 |
| Days Max Temp ≥ 90 °F | 0 | 0 | 0 | 0 | 1 | 3 | 10 | 9 | 2 | 0 | 0 | 0 | 25 |
| Days Max Temp ≤ 32 °F | 5 | 1 | 0 | 0 | 0 | 0 | 0 | 0 | 0 | 0 | 1 | 5 | 12 |
| Days Min Temp ≤ 32 °F | 27 | 23 | 22 | 17 | 7 | 1 | 0 | 0 | 2 | 12 | 22 | 28 | 161 |
| Days Min Temp ≤ 0 °F | 1 | 0 | 0 | 0 | 0 | 0 | 0 | 0 | 0 | 0 | 0 | 1 | 2 |
| Heating Degree Days | 1072 | 831 | 768 | 588 | 354 | 157 | 41 | 52 | 176 | 470 | 824 | 1075 | 6408 |
| Cooling Degree Days | 0 | 0 | 0 | 0 | 17 | 57 | 162 | 136 | 33 | 1 | 0 | 0 | 406 |
| Total Precipitation (") | 1.76 | 1.23 | 1.39 | 0.80 | 0.94 | 0.68 | 0.35 | 0.59 | 0.54 | 0.90 | 1.96 | 2.01 | 13.15 |
| Days ≥ 0.1" Precip | 5 | 4 | 4 | 3 | 3 | 2 | 1 | 2 | 2 | 3 | 6 | 6 | 41 |
| Total Snowfall (") | 8.2 | 4.6 | 2.9 | 0.7 | 0.1 | 0.0 | 0.0 | 0.0 | 0.0 | 0.4 | 4.7 | 10.9 | 32.5 |
| Days ≥ 1" Snow Depth | 13 | 4 | 2 | 0 | 0 | 0 | 0 | 0 | 0 | 0 | 3 | 10 | 32 |

## LA GRANDE *Union County* ELEVATION 2782 ft LAT/LONG 45° 20' N / 118° 5' W

| | JAN | FEB | MAR | APR | MAY | JUN | JUL | AUG | SEP | OCT | NOV | DEC | YEAR |
|---|---|---|---|---|---|---|---|---|---|---|---|---|---|
| Maximum Temp °F | 37.7 | 43.2 | 51.4 | 58.7 | 67.9 | 76.4 | 85.6 | 86.1 | 76.3 | 63.4 | 46.4 | 38.5 | 61.0 |
| Minimum Temp °F | 23.5 | 26.7 | 30.6 | 34.9 | 41.6 | 48.4 | 52.3 | 51.5 | 43.3 | 35.4 | 30.1 | 24.4 | 36.9 |
| Mean Temp °F | 30.6 | 35.0 | 41.0 | 46.8 | 54.8 | 62.4 | 68.9 | 68.8 | 59.8 | 49.4 | 38.3 | 31.5 | 48.9 |
| Days Max Temp ≥ 90 °F | 0 | 0 | 0 | 0 | 1 | 3 | 11 | 12 | 3 | 0 | 0 | 0 | 30 |
| Days Max Temp ≤ 32 °F | 8 | 2 | 0 | 0 | 0 | 0 | 0 | 0 | 0 | 0 | 1 | 6 | 17 |
| Days Min Temp ≤ 32 °F | 26 | 21 | 19 | 11 | 2 | 0 | 0 | 0 | 2 | 11 | 18 | 25 | 135 |
| Days Min Temp ≤ 0 °F | 1 | 0 | 0 | 0 | 0 | 0 | 0 | 0 | 0 | 0 | 0 | 1 | 2 |
| Heating Degree Days | 1060 | 842 | 737 | 539 | 321 | 134 | 35 | 38 | 183 | 475 | 795 | 1030 | 6189 |
| Cooling Degree Days | 0 | 0 | 0 | 1 | 14 | 60 | 152 | 158 | 30 | 0 | 0 | 0 | 415 |
| Total Precipitation (") | 1.93 | 1.44 | 1.46 | 1.50 | 1.74 | 1.54 | 0.66 | 0.91 | 0.87 | 1.24 | 2.04 | 1.84 | 17.17 |
| Days ≥ 0.1" Precip | 6 | 4 | 5 | 5 | 5 | 5 | 2 | 3 | 3 | 4 | 6 | 6 | 54 |
| Total Snowfall (") | 8.3 | 4.2 | 1.6 | 0.6 | 0.0 | 0.0 | 0.0 | 0.0 | 0.0 | 0.2 | 3.1 | 6.9 | 24.9 |
| Days ≥ 1" Snow Depth | 12 | 4 | 1 | 0 | 0 | 0 | 0 | 0 | 0 | 0 | 2 | 10 | 29 |

## LAKEVIEW 2 NNW *Lake County* ELEVATION 4774 ft LAT/LONG 42° 11 ' N / 120° 21 ' W

| | JAN | FEB | MAR | APR | MAY | JUN | JUL | AUG | SEP | OCT | NOV | DEC | YEAR |
|---|---|---|---|---|---|---|---|---|---|---|---|---|---|
| Maximum Temp °F | 37.8 | 42.5 | 48.5 | 55.9 | 66.0 | 74.0 | 83.9 | 82.6 | 75.0 | 62.8 | 46.0 | 38.4 | 59.5 |
| Minimum Temp °F | 19.5 | 23.1 | 27.0 | 30.9 | 37.7 | 44.2 | 50.4 | 48.2 | 41.4 | 33.1 | 26.0 | 20.0 | 33.5 |
| Mean Temp °F | 28.7 | 32.8 | 37.7 | 43.4 | 51.9 | 59.2 | 67.2 | 65.5 | 58.2 | 48.0 | 36.0 | 29.2 | 46.5 |
| Days Max Temp ≥ 90 °F | 0 | 0 | 0 | 0 | 0 | 1 | 8 | 6 | 1 | 0 | 0 | 0 | 16 |
| Days Max Temp ≤ 32 °F | 7 | 3 | 1 | 0 | 0 | 0 | 0 | 0 | 0 | 0 | 1 | 7 | 19 |
| Days Min Temp ≤ 32 °F | 28 | 25 | 24 | 18 | 8 | 2 | 0 | 0 | 3 | 15 | 23 | 28 | 174 |
| Days Min Temp ≤ 0 °F | 2 | 1 | 0 | 0 | 0 | 0 | 0 | 0 | 0 | 0 | 0 | 2 | 5 |
| Heating Degree Days | 1120 | 901 | 838 | 641 | 405 | 199 | 50 | 72 | 216 | 520 | 863 | 1102 | 6927 |
| Cooling Degree Days | 0 | 0 | 0 | 0 | 7 | 28 | 127 | 91 | 26 | 1 | 0 | 0 | 280 |
| Total Precipitation (") | 1.90 | 1.44 | 1.55 | 1.20 | 1.32 | 1.11 | 0.39 | 0.51 | 0.64 | 1.01 | 2.02 | 1.87 | 14.96 |
| Days ≥ 0.1" Precip | 5 | 5 | 5 | 4 | 4 | 3 | 1 | 1 | 2 | 3 | 6 | 6 | 45 |
| Total Snowfall (") | 13.0 | 8.6 | 6.9 | 4.3 | 1.4 | 0.0 | 0.0 | 0.0 | 0.2 | 1.4 | 7.8 | 13.5 | 57.1 |
| Days ≥ 1" Snow Depth | *16* | 10 | 4 | 1 | 0 | 0 | 0 | 0 | 0 | 1 | *4* | *13* | 49 |

## LEABURG 1 SW *Lane County* ELEVATION 679 ft LAT/LONG 44° 6 ' N / 122° 41 ' W

| | JAN | FEB | MAR | APR | MAY | JUN | JUL | AUG | SEP | OCT | NOV | DEC | YEAR |
|---|---|---|---|---|---|---|---|---|---|---|---|---|---|
| Maximum Temp °F | 47.0 | 52.0 | 56.9 | 60.9 | 67.7 | 74.0 | 81.6 | 82.4 | 76.4 | 65.6 | 52.7 | 46.2 | 63.6 |
| Minimum Temp °F | 33.2 | 34.7 | 36.6 | 39.2 | 43.3 | 48.2 | 50.5 | 50.3 | 46.9 | 42.2 | 37.9 | 33.7 | 41.4 |
| Mean Temp °F | 40.1 | 43.4 | 46.7 | 50.1 | 55.5 | 61.1 | 66.1 | 66.4 | 61.7 | 53.9 | 45.3 | 40.0 | 52.5 |
| Days Max Temp ≥ 90 °F | 0 | 0 | 0 | 0 | 0 | 2 | 7 | 6 | 3 | 0 | 0 | 0 | 18 |
| Days Max Temp ≤ 32 °F | 0 | 0 | 0 | 0 | 0 | 0 | 0 | 0 | 0 | 0 | 0 | 1 | 1 |
| Days Min Temp ≤ 32 °F | 14 | 10 | 7 | 3 | 0 | 0 | 0 | 0 | 0 | 1 | 5 | 12 | 52 |
| Days Min Temp ≤ 0 °F | 0 | 0 | 0 | 0 | 0 | 0 | 0 | 0 | 0 | 0 | 0 | 0 | 0 |
| Heating Degree Days | 764 | 603 | 559 | 443 | 296 | 144 | 49 | 41 | 123 | 339 | 583 | 769 | 4713 |
| Cooling Degree Days | 0 | 0 | 0 | 1 | 11 | 31 | 82 | 86 | 33 | 4 | 0 | 0 | 248 |
| Total Precipitation (") | 8.75 | 6.84 | 6.93 | 5.47 | 3.73 | 2.58 | 0.81 | 1.23 | 2.53 | 4.87 | 10.03 | 9.45 | 63.22 |
| Days ≥ 0.1" Precip | 15 | 13 | 14 | 12 | 8 | 6 | 2 | 3 | 5 | 9 | 15 | 16 | 118 |
| Total Snowfall (") | 3.4 | 1.8 | 0.5 | 0.0 | 0.0 | 0.0 | 0.0 | 0.0 | 0.0 | 0.0 | 0.3 | 1.7 | 7.7 |
| Days ≥ 1" Snow Depth | 2 | 2 | 0 | 0 | 0 | 0 | 0 | 0 | 0 | 0 | 0 | 2 | 6 |

## LONG CREEK *Grant County* ELEVATION 3724 ft LAT/LONG 44° 43 ' N / 119° 6 ' W

| | JAN | FEB | MAR | APR | MAY | JUN | JUL | AUG | SEP | OCT | NOV | DEC | YEAR |
|---|---|---|---|---|---|---|---|---|---|---|---|---|---|
| Maximum Temp °F | 38.8 | 44.9 | 50.7 | 56.4 | 65.3 | 73.6 | 82.3 | 82.3 | 73.0 | 62.2 | 46.9 | 39.6 | 59.7 |
| Minimum Temp °F | 21.3 | 24.5 | 26.8 | 30.0 | 35.4 | 41.1 | 44.7 | 44.7 | 38.1 | 33.0 | 27.0 | 21.5 | 32.3 |
| Mean Temp °F | 30.1 | 34.8 | 38.8 | 43.3 | 50.4 | 57.4 | 63.5 | 63.5 | 55.7 | 47.6 | 37.0 | 30.6 | 46.1 |
| Days Max Temp ≥ 90 °F | 0 | 0 | 0 | 0 | 0 | 1 | 6 | 6 | 1 | 0 | 0 | 0 | 14 |
| Days Max Temp ≤ 32 °F | 7 | 2 | 0 | 0 | 0 | 0 | 0 | 0 | 0 | 0 | 2 | 6 | 17 |
| Days Min Temp ≤ 32 °F | 26 | 23 | 23 | 20 | 10 | 2 | 0 | 1 | 6 | 15 | 22 | 27 | 175 |
| Days Min Temp ≤ 0 °F | 2 | 1 | 0 | 0 | 0 | 0 | 0 | 0 | 0 | 0 | 0 | 1 | 4 |
| Heating Degree Days | 1076 | 846 | 807 | 645 | *447* | 237 | 97 | 96 | 278 | 532 | 833 | 1061 | 6955 |
| Cooling Degree Days | 0 | 0 | 0 | 0 | *5* | 16 | 53 | 56 | 8 | 1 | 0 | 0 | 139 |
| Total Precipitation (") | 1.52 | 1.12 | 1.51 | 1.45 | 1.70 | 1.41 | 0.80 | 0.92 | 0.73 | 1.18 | 1.76 | 1.55 | 15.65 |
| Days ≥ 0.1" Precip | 6 | 4 | 5 | 5 | 5 | 4 | 2 | 2 | 2 | 4 | 6 | 5 | 50 |
| Total Snowfall (") | 7.7 | 5.3 | 5.0 | 2.9 | 0.5 | 0.0 | 0.0 | 0.0 | 0.0 | 0.7 | *5.0* | 8.9 | 36.0 |
| Days ≥ 1" Snow Depth | *10* | *3* | 1 | 0 | 0 | 0 | 0 | 0 | 0 | 0 | 2 | *8* | 24 |

## LOOKOUT POINT DAM *Lane County* ELEVATION 712 ft LAT/LONG 43° 55 ' N / 122° 46 ' W

| | JAN | FEB | MAR | APR | MAY | JUN | JUL | AUG | SEP | OCT | NOV | DEC | YEAR |
|---|---|---|---|---|---|---|---|---|---|---|---|---|---|
| Maximum Temp °F | 47.1 | 51.5 | 55.7 | 59.5 | 66.1 | 72.1 | 79.6 | 80.3 | 74.8 | 64.5 | 52.9 | 47.1 | 62.6 |
| Minimum Temp °F | 34.6 | 36.5 | 38.8 | 41.1 | 45.1 | 49.8 | 53.2 | 53.3 | 50.1 | 45.4 | 40.1 | 35.5 | 43.6 |
| Mean Temp °F | 40.8 | 44.0 | 47.3 | 50.3 | 55.6 | 61.0 | 66.5 | 66.8 | 62.5 | 55.0 | 46.5 | 41.3 | 53.1 |
| Days Max Temp ≥ 90 °F | 0 | 0 | 0 | 0 | 0 | 1 | 4 | 4 | 1 | 0 | 0 | 0 | 10 |
| Days Max Temp ≤ 32 °F | 1 | 0 | 0 | 0 | 0 | 0 | 0 | 0 | 0 | 0 | 0 | 1 | 2 |
| Days Min Temp ≤ 32 °F | 11 | 6 | 2 | 1 | 0 | 0 | 0 | 0 | 0 | 0 | 3 | 9 | 32 |
| Days Min Temp ≤ 0 °F | 0 | 0 | 0 | 0 | 0 | 0 | 0 | 0 | 0 | 0 | 0 | 0 | 0 |
| Heating Degree Days | 742 | 585 | 543 | 433 | 293 | 145 | 46 | 36 | 107 | 308 | 547 | 727 | 4512 |
| Cooling Degree Days | 0 | 0 | 0 | 1 | 11 | 29 | 88 | 95 | 40 | 5 | 0 | 0 | 269 |
| Total Precipitation (") | 5.91 | 4.50 | 4.78 | 3.86 | 3.05 | 2.08 | 0.79 | 1.17 | 1.83 | 3.19 | 6.74 | 6.41 | 44.31 |
| Days ≥ 0.1" Precip | 12 | 11 | 12 | 11 | 8 | 5 | 2 | 3 | 5 | 8 | 13 | 13 | 103 |
| Total Snowfall (") | 0.7 | *0.6* | 0.0 | 0.0 | 0.0 | 0.0 | 0.0 | 0.0 | 0.0 | 0.0 | 0.0 | 0.7 | 2.0 |
| Days ≥ 1" Snow Depth | 1 | 1 | 0 | 0 | 0 | 0 | 0 | 0 | 0 | 0 | 0 | 1 | 3 |

## MADRAS *Jefferson County* ELEVATION 2303 ft LAT/LONG 44° 38 ' N / 121° 8 ' W

| | JAN | FEB | MAR | APR | MAY | JUN | JUL | AUG | SEP | OCT | NOV | DEC | YEAR |
|---|---|---|---|---|---|---|---|---|---|---|---|---|---|
| Maximum Temp °F | 42.8 | 49.0 | 56.5 | 62.6 | 71.8 | 79.3 | 87.3 | 86.8 | 78.5 | 66.3 | 50.7 | 42.7 | 64.5 |
| Minimum Temp °F | 23.3 | 26.1 | 28.3 | 31.7 | 37.3 | 43.6 | 47.0 | 46.3 | 39.5 | 32.5 | 28.5 | 23.7 | 34.0 |
| Mean Temp °F | 33.1 | 37.6 | 42.4 | 47.3 | 54.6 | 61.6 | 67.1 | 66.6 | 59.0 | 49.4 | 39.6 | 33.2 | 49.3 |
| Days Max Temp ≥ 90 °F | 0 | 0 | 0 | 0 | 1 | 5 | 14 | 12 | 4 | 0 | 0 | 0 | 36 |
| Days Max Temp ≤ 32 °F | 5 | 1 | 0 | 0 | 0 | 0 | 0 | 0 | 0 | 0 | 1 | 5 | 12 |
| Days Min Temp ≤ 32 °F | 24 | 21 | 21 | 16 | 9 | 1 | 0 | 0 | 5 | 16 | 19 | 25 | 157 |
| Days Min Temp ≤ 0 °F | 1 | 1 | 0 | 0 | 0 | 0 | 0 | 0 | 0 | 0 | 0 | 1 | 3 |
| Heating Degree Days | 983 | 768 | 696 | 527 | 326 | 144 | 46 | 48 | 190 | 478 | 756 | 979 | 5941 |
| Cooling Degree Days | 0 | 0 | 0 | 1 | 16 | 49 | 117 | 101 | 24 | 1 | 0 | 0 | 309 |
| Total Precipitation (") | 1.28 | 0.79 | 0.85 | 0.76 | 0.78 | 0.79 | 0.48 | 0.51 | 0.49 | 0.72 | 1.42 | 1.20 | 10.07 |
| Days ≥ 0.1" Precip | 4 | 3 | 3 | 2 | 3 | 2 | 1 | 2 | 1 | 2 | 4 | 4 | 31 |
| Total Snowfall (") | 3.8 | 1.7 | 0.2 | 0.0 | 0.0 | 0.0 | 0.0 | 0.0 | 0.0 | 0.0 | 2.1 | ***2.8*** | 10.6 |
| Days ≥ 1" Snow Depth | 6 | 2 | 0 | 0 | 0 | 0 | 0 | 0 | 0 | 0 | 2 | ***4*** | 14 |

## MALHEUR BRCH EXP STN *Malheur County* ELEVATION 2251 ft LAT/LONG 43° 58 ' N / 117° 1 ' W

| | JAN | FEB | MAR | APR | MAY | JUN | JUL | AUG | SEP | OCT | NOV | DEC | YEAR |
|---|---|---|---|---|---|---|---|---|---|---|---|---|---|
| Maximum Temp °F | 33.9 | 42.8 | 54.8 | 64.0 | 73.7 | 82.3 | 90.9 | 89.9 | 79.0 | 65.2 | 47.2 | 35.7 | 63.3 |
| Minimum Temp °F | 19.0 | 24.6 | 31.2 | 37.3 | 45.3 | 52.6 | 57.9 | 55.7 | 45.9 | 35.7 | 28.1 | 20.3 | 37.8 |
| Mean Temp °F | 26.4 | 33.7 | 43.0 | 50.7 | 59.5 | 67.5 | 74.4 | 72.8 | 62.5 | 50.5 | 37.6 | 28.0 | 50.6 |
| Days Max Temp ≥ 90 °F | 0 | 0 | 0 | 0 | 2 | 7 | 20 | 18 | 4 | 0 | 0 | 0 | 51 |
| Days Max Temp ≤ 32 °F | 12 | 3 | 0 | 0 | 0 | 0 | 0 | 0 | 0 | 0 | 1 | 9 | 25 |
| Days Min Temp ≤ 32 °F | 28 | 24 | 19 | 8 | 1 | 0 | 0 | 0 | 1 | 10 | 22 | 28 | 141 |
| Days Min Temp ≤ 0 °F | 3 | 1 | 0 | 0 | 0 | 0 | 0 | 0 | 0 | 0 | 0 | 2 | 6 |
| Heating Degree Days | 1188 | 877 | 675 | 426 | 199 | 59 | 8 | 14 | 127 | 444 | 813 | 1140 | 5970 |
| Cooling Degree Days | 0 | 0 | 0 | 4 | 43 | 131 | 277 | 259 | 58 | 1 | 0 | 0 | 773 |
| Total Precipitation (") | 1.29 | 0.94 | 1.00 | 0.78 | 0.81 | 0.81 | 0.23 | 0.49 | 0.49 | 0.65 | 1.31 | 1.32 | 10.12 |
| Days ≥ 0.1" Precip | 4 | 3 | 3 | 3 | 3 | 3 | 1 | 1 | 2 | 2 | 5 | 4 | 34 |
| Total Snowfall (") | 7.3 | 3.0 | 0.7 | 0.0 | 0.0 | 0.0 | 0.0 | 0.0 | 0.0 | 0.1 | 2.5 | 6.5 | 20.1 |
| Days ≥ 1" Snow Depth | 18 | 10 | 1 | 0 | 0 | 0 | 0 | 0 | 0 | 0 | 2 | 12 | 43 |

## MALHEUR REFUGE HDQ *Harney County* ELEVATION 4114 ft LAT/LONG 43° 17 ' N / 118° 50 ' W

| | JAN | FEB | MAR | APR | MAY | JUN | JUL | AUG | SEP | OCT | NOV | DEC | YEAR |
|---|---|---|---|---|---|---|---|---|---|---|---|---|---|
| Maximum Temp °F | 37.3 | 44.3 | 50.5 | 58.2 | 67.9 | 75.5 | 84.2 | 82.8 | 75.0 | ***62.9*** | ***46.7*** | 37.7 | 60.3 |
| Minimum Temp °F | 17.5 | 22.3 | 25.7 | 30.1 | 37.8 | 44.4 | 49.3 | 47.6 | 38.5 | ***30.2*** | ***23.8*** | 17.4 | 32.1 |
| Mean Temp °F | 27.4 | 33.4 | 38.1 | 44.2 | 52.9 | 60.0 | 66.8 | 65.2 | 56.8 | ***46.6*** | ***35.3*** | 27.6 | 46.2 |
| Days Max Temp ≥ 90 °F | 0 | 0 | 0 | 0 | 0 | 2 | 7 | 7 | 1 | 0 | 0 | 0 | 17 |
| Days Max Temp ≤ 32 °F | 8 | 2 | 0 | 0 | 0 | 0 | 0 | 0 | 0 | 0 | 1 | 8 | 19 |
| Days Min Temp ≤ 32 °F | 27 | 24 | 24 | 19 | 7 | 1 | 0 | 0 | 6 | 17 | 22 | 27 | 174 |
| Days Min Temp ≤ 0 °F | 2 | 1 | 0 | 0 | 0 | 0 | 0 | 0 | 0 | 0 | 0 | 3 | 6 |
| Heating Degree Days | 1158 | 884 | 826 | 618 | 374 | 177 | 51 | 67 | 252 | ***564*** | ***887*** | 1153 | 7011 |
| Cooling Degree Days | ***0*** | ***0*** | ***0*** | ***0*** | 5 | 36 | 98 | 74 | 9 | ***0*** | ***0*** | ***0*** | 222 |
| Total Precipitation (") | 0.91 | 0.60 | 0.99 | 0.80 | 1.12 | 0.96 | 0.41 | 0.68 | 0.49 | 0.70 | 1.18 | 0.93 | 9.77 |
| Days ≥ 0.1" Precip | 3 | 2 | 3 | 2 | 4 | 3 | 1 | 2 | 2 | 2 | 4 | ***3*** | 31 |
| Total Snowfall (") | ***6.2*** | ***3.9*** | 2.6 | 1.6 | 0.3 | 0.0 | 0.0 | 0.0 | 0.0 | 0.2 | 3.0 | ***8.4*** | 26.2 |
| Days ≥ 1" Snow Depth | ***10*** | ***7*** | 2 | 0 | 0 | 0 | 0 | 0 | 0 | 0 | 2 | ***12*** | 33 |

## MARION FORKS FISH HA *Linn County* ELEVATION 2451 ft LAT/LONG 44° 37 ' N / 121° 57 ' W

| | JAN | FEB | MAR | APR | MAY | JUN | JUL | AUG | SEP | OCT | NOV | DEC | YEAR |
|---|---|---|---|---|---|---|---|---|---|---|---|---|---|
| Maximum Temp °F | 38.5 | 43.8 | 49.4 | 54.8 | 63.9 | 71.7 | 79.8 | 79.7 | 72.4 | 61.1 | 45.4 | 38.0 | 58.2 |
| Minimum Temp °F | 25.9 | 27.5 | 29.5 | 32.7 | 37.3 | 43.4 | 46.2 | 45.1 | 39.6 | 34.4 | 30.8 | 26.8 | 34.9 |
| Mean Temp °F | 32.2 | 35.7 | 39.5 | 43.8 | 50.7 | 57.6 | 63.1 | 62.4 | 56.0 | 47.7 | 38.1 | 32.5 | 46.6 |
| Days Max Temp ≥ 90 °F | 0 | 0 | 0 | 0 | 1 | 2 | 6 | 5 | 2 | 0 | 0 | 0 | 16 |
| Days Max Temp ≤ 32 °F | 3 | 1 | 0 | 0 | 0 | 0 | 0 | 0 | 0 | 0 | 0 | 2 | 6 |
| Days Min Temp ≤ 32 °F | 28 | 24 | 24 | 17 | 6 | 1 | 0 | 0 | 3 | 12 | 19 | 28 | 162 |
| Days Min Temp ≤ 0 °F | 0 | 0 | 0 | 0 | 0 | 0 | 0 | 0 | 0 | 0 | 0 | 0 | 0 |
| Heating Degree Days | 1010 | 821 | 785 | 630 | 442 | 238 | 110 | 118 | 270 | 528 | 800 | 1002 | 6754 |
| Cooling Degree Days | 0 | 0 | 0 | 0 | 7 | 21 | 50 | 44 | 8 | 0 | 0 | 0 | 130 |
| Total Precipitation (") | 10.63 | 7.67 | 6.93 | 4.93 | 3.36 | 2.38 | 1.01 | 1.21 | 2.22 | 4.89 | 10.38 | 10.68 | 66.29 |
| Days ≥ 0.1" Precip | 14 | 12 | 13 | 12 | 9 | 6 | 3 | 3 | 5 | 9 | 15 | 15 | 116 |
| Total Snowfall (") | 25.5 | 19.5 | 13.7 | 7.2 | 0.4 | 0.0 | 0.0 | 0.0 | 0.0 | 0.6 | 9.7 | 24.0 | 100.6 |
| Days ≥ 1" Snow Depth | 24 | 19 | 15 | 5 | 0 | 0 | 0 | 0 | 0 | 0 | 7 | 20 | 90 |

## MC DERMITT 26 N *Malheur County* ELEVATION 4462 ft LAT/LONG 42° 25' N / 117° 52' W

| | JAN | FEB | MAR | APR | MAY | JUN | JUL | AUG | SEP | OCT | NOV | DEC | YEAR |
|---|---|---|---|---|---|---|---|---|---|---|---|---|---|
| Maximum Temp °F | 40.6 | 46.8 | 53.6 | 61.4 | 71.3 | 80.9 | 91.1 | 89.1 | 79.5 | 67.6 | 50.0 | 41.2 | 64.4 |
| Minimum Temp °F | 18.6 | 22.5 | 25.7 | 29.6 | 36.8 | 43.8 | 49.2 | 47.6 | 39.6 | 31.8 | 24.9 | 18.6 | 32.4 |
| Mean Temp °F | 29.6 | 34.7 | 39.7 | 45.5 | 54.2 | 62.4 | 70.1 | 68.4 | 59.6 | 49.7 | 37.4 | 29.9 | 48.4 |
| Days Max Temp ≥ 90 °F | 0 | 0 | 0 | 0 | 1 | 7 | 20 | 17 | 4 | 0 | 0 | 0 | 49 |
| Days Max Temp ≤ 32 °F | 6 | 2 | 0 | 0 | 0 | 0 | 0 | 0 | 0 | 0 | 1 | 6 | 15 |
| Days Min Temp ≤ 32 °F | 27 | 24 | 24 | 19 | 10 | 2 | 0 | 1 | 6 | 17 | 23 | 27 | 180 |
| Days Min Temp ≤ 0 °F | 2 | 1 | 0 | 0 | 0 | 0 | 0 | 0 | 0 | 0 | 0 | 2 | 5 |
| Heating Degree Days | 1089 | 849 | 779 | 578 | 341 | 138 | 23 | 41 | 189 | 468 | 822 | 1080 | 6397 |
| Cooling Degree Days | 0 | 0 | 0 | 1 | 15 | 68 | 179 | 156 | 30 | 0 | 0 | 0 | 449 |
| Total Precipitation (") | 0.72 | 0.58 | 0.93 | 0.97 | 1.28 | 0.98 | 0.41 | 0.57 | 0.48 | 0.62 | 0.95 | 0.83 | 9.32 |
| Days ≥ 0.1" Precip | 2 | 2 | 3 | 4 | 4 | 3 | 1 | 2 | 2 | 2 | 3 | 2 | 30 |
| Total Snowfall (") | 4.2 | 3.2 | 3.9 | 1.5 | 0.4 | 0.0 | 0.0 | 0.0 | 0.0 | 0.1 | 1.6 | *3.9* | 18.8 |
| Days ≥ 1" Snow Depth | *4* | *1* | 1 | 0 | 0 | 0 | 0 | 0 | 0 | 0 | 1 | na | na |

## MC MINNVILLE *Yamhill County* ELEVATION 151 ft LAT/LONG 45° 12' N / 123° 12' W

| | JAN | FEB | MAR | APR | MAY | JUN | JUL | AUG | SEP | OCT | NOV | DEC | YEAR |
|---|---|---|---|---|---|---|---|---|---|---|---|---|---|
| Maximum Temp °F | 46.1 | 51.2 | 56.9 | 61.8 | 69.3 | 75.4 | 82.1 | 82.9 | 77.1 | 65.3 | 52.8 | 45.9 | 63.9 |
| Minimum Temp °F | 33.5 | 35.8 | 37.1 | 39.2 | 43.1 | 47.6 | 50.0 | 50.2 | 46.9 | 41.7 | 38.1 | 33.9 | 41.4 |
| Mean Temp °F | 39.8 | 43.5 | 47.1 | 50.5 | 56.1 | 61.5 | 66.1 | 66.6 | 62.0 | 53.6 | 45.5 | 39.9 | 52.7 |
| Days Max Temp ≥ 90 °F | 0 | 0 | 0 | 0 | 0 | 2 | 6 | 7 | 3 | 0 | 0 | 0 | 18 |
| Days Max Temp ≤ 32 °F | 1 | 0 | 0 | 0 | 0 | 0 | 0 | 0 | 0 | 0 | 0 | 1 | 2 |
| Days Min Temp ≤ 32 °F | 13 | 9 | 7 | 5 | 1 | 0 | 0 | 0 | 0 | 2 | 6 | 12 | 55 |
| Days Min Temp ≤ 0 °F | 0 | 0 | 0 | 0 | 0 | 0 | 0 | 0 | 0 | 0 | 0 | 0 | 0 |
| Heating Degree Days | 773 | 600 | 550 | 428 | 277 | 133 | 46 | 42 | 123 | 351 | 579 | 745 | 4647 |
| Cooling Degree Days | 0 | 0 | 0 | 0 | 16 | 42 | 89 | 108 | 47 | 4 | 0 | 0 | 306 |
| Total Precipitation (") | 6.79 | 5.11 | 4.55 | 2.78 | 1.81 | 1.05 | 0.44 | 0.64 | 1.56 | 3.05 | 5.68 | 7.71 | 41.17 |
| Days ≥ 0.1" Precip | 11 | 10 | 11 | 7 | 5 | 3 | 1 | 1 | 4 | *6* | 11 | *13* | 83 |
| Total Snowfall (") | *1.5* | 0.7 | 0.1 | 0.0 | 0.0 | 0.0 | 0.0 | 0.0 | 0.0 | 0.0 | 0.0 | 1.9 | 4.2 |
| Days ≥ 1" Snow Depth | *1* | 0 | 0 | 0 | 0 | 0 | 0 | 0 | 0 | 0 | 0 | 0 | 1 |

## MEDFORD EXP STN *Jackson County* ELEVATION 1460 ft LAT/LONG 42° 18' N / 122° 52' W

| | JAN | FEB | MAR | APR | MAY | JUN | JUL | AUG | SEP | OCT | NOV | DEC | YEAR |
|---|---|---|---|---|---|---|---|---|---|---|---|---|---|
| Maximum Temp °F | 46.3 | 53.4 | 58.9 | 64.6 | 73.2 | 80.8 | 88.4 | 88.0 | 81.3 | 68.2 | 52.3 | 45.0 | 66.7 |
| Minimum Temp °F | 30.3 | 31.9 | 34.5 | 36.9 | 41.6 | 47.8 | 50.9 | 50.7 | 44.2 | 37.4 | 34.2 | 30.6 | 39.2 |
| Mean Temp °F | 38.3 | 42.7 | 46.7 | 50.8 | 57.5 | 64.3 | 69.7 | 69.4 | 62.8 | 52.9 | 43.3 | 37.8 | 53.0 |
| Days Max Temp ≥ 90 °F | 0 | 0 | 0 | 0 | 2 | 6 | 15 | 14 | 6 | 0 | 0 | 0 | 43 |
| Days Max Temp ≤ 32 °F | 0 | 0 | 0 | 0 | 0 | 0 | 0 | 0 | 0 | 0 | 0 | 1 | 1 |
| Days Min Temp ≤ 32 °F | 19 | 16 | 13 | 9 | 3 | 0 | 0 | 0 | 1 | 8 | 13 | 18 | 100 |
| Days Min Temp ≤ 0 °F | 0 | 0 | 0 | 0 | 0 | 0 | 0 | 0 | 0 | 0 | 0 | 0 | 0 |
| Heating Degree Days | 819 | 623 | 561 | 421 | 243 | 90 | 22 | 17 | 107 | 371 | 645 | 835 | 4754 |
| Cooling Degree Days | 0 | 0 | 0 | 2 | 18 | 65 | 147 | 141 | 41 | 2 | 0 | 0 | 416 |
| Total Precipitation (") | 2.83 | 2.00 | 2.06 | 1.46 | 1.18 | 0.80 | 0.39 | 0.59 | 0.98 | 1.47 | 3.24 | 3.24 | 20.24 |
| Days ≥ 0.1" Precip | 6 | 5 | 5 | 4 | 4 | 2 | 1 | 2 | 2 | 4 | 7 | 7 | 49 |
| Total Snowfall (") | 2.0 | *0.5* | 0.2 | 0.2 | 0.0 | 0.0 | 0.0 | 0.0 | 0.0 | 0.0 | 0.0 | 1.1 | 4.0 |
| Days ≥ 1" Snow Depth | 1 | 0 | 0 | 0 | 0 | 0 | 0 | 0 | 0 | 0 | 0 | 1 | 2 |

## MEDFORD JACKSON CO *Jackson County* ELEVATION 1329 ft LAT/LONG 42° 22' N / 122° 52' W

| | JAN | FEB | MAR | APR | MAY | JUN | JUL | AUG | SEP | OCT | NOV | DEC | YEAR |
|---|---|---|---|---|---|---|---|---|---|---|---|---|---|
| Maximum Temp °F | 46.4 | 53.6 | 58.9 | 64.3 | 73.3 | 82.0 | 90.7 | 90.1 | 83.2 | 69.8 | 52.4 | 44.5 | 67.4 |
| Minimum Temp °F | 30.8 | 32.4 | 35.3 | 38.2 | 43.4 | 50.2 | 54.7 | 54.5 | 47.9 | 39.9 | 34.8 | 31.0 | 41.1 |
| Mean Temp °F | 38.6 | 43.0 | 47.1 | 51.3 | 58.4 | 66.1 | 72.7 | 72.4 | 65.6 | 54.9 | 43.6 | 37.8 | 54.3 |
| Days Max Temp ≥ 90 °F | 0 | 0 | 0 | 0 | 2 | 8 | 18 | 18 | 9 | 1 | 0 | 0 | 56 |
| Days Max Temp ≤ 32 °F | 1 | 0 | 0 | 0 | 0 | 0 | 0 | 0 | 0 | 0 | 0 | 2 | 3 |
| Days Min Temp ≤ 32 °F | 19 | 15 | 10 | 6 | 1 | 0 | 0 | 0 | 0 | 4 | 12 | 17 | 84 |
| Days Min Temp ≤ 0 °F | 0 | 0 | 0 | 0 | 0 | 0 | 0 | 0 | 0 | 0 | 0 | 0 | 0 |
| Heating Degree Days | 812 | 614 | 548 | 407 | 223 | 65 | 9 | 7 | 69 | 314 | 635 | 837 | 4540 |
| Cooling Degree Days | 0 | 0 | 0 | 3 | 31 | 106 | 247 | 253 | 103 | 11 | 0 | 0 | 754 |
| Total Precipitation (") | 2.56 | 1.88 | 1.73 | 1.17 | 1.02 | 0.66 | 0.30 | 0.51 | 0.82 | 1.25 | 2.98 | 2.88 | 17.76 |
| Days ≥ 0.1" Precip | 6 | 5 | 5 | 4 | 3 | 2 | 1 | 2 | 2 | 4 | 7 | 7 | 48 |
| Total Snowfall (") | 2.7 | 0.9 | 0.6 | 0.1 | 0.0 | 0.0 | 0.0 | 0.0 | 0.0 | 0.0 | 0.4 | 1.9 | 6.6 |
| Days ≥ 1" Snow Depth | 1 | 0 | 0 | 0 | 0 | 0 | 0 | 0 | 0 | 0 | 0 | 1 | 2 |

## MILTON FREEWATER *Umatilla County* ELEVATION 1060 ft LAT/LONG 45° 56 ' N / 118° 24 ' W

| | JAN | FEB | MAR | APR | MAY | JUN | JUL | AUG | SEP | OCT | NOV | DEC | YEAR |
|---|---|---|---|---|---|---|---|---|---|---|---|---|---|
| Maximum Temp °F | 41.6 | 47.7 | 56.5 | 63.9 | 72.4 | 80.4 | 88.3 | 87.4 | 77.6 | 65.3 | 50.6 | 42.4 | 64.5 |
| Minimum Temp °F | 27.8 | 32.1 | 37.3 | 42.2 | 48.3 | 54.6 | 59.4 | 58.3 | 50.3 | 41.5 | 34.3 | 28.4 | 42.9 |
| Mean Temp °F | 34.7 | 39.9 | 46.9 | 53.0 | 60.4 | 67.5 | 73.9 | 72.9 | 64.0 | 53.4 | 42.5 | 35.4 | 53.7 |
| Days Max Temp ≥ 90 °F | 0 | 0 | 0 | 0 | 1 | 5 | 15 | 13 | 2 | 0 | 0 | 0 | 36 |
| Days Max Temp ≤ 32 °F | 8 | 3 | 0 | 0 | 0 | 0 | 0 | 0 | 0 | 0 | 2 | 7 | 20 |
| Days Min Temp ≤ 32 °F | 20 | 14 | 7 | 1 | 0 | 0 | 0 | 0 | 1 | 4 | 11 | 20 | 78 |
| Days Min Temp ≤ 0 °F | 1 | 0 | 0 | 0 | 0 | 0 | 0 | 0 | 0 | 0 | 0 | 1 | 2 |
| Heating Degree Days | 933 | 701 | 553 | 357 | 176 | 51 | 8 | 10 | 100 | 356 | 670 | 910 | 4825 |
| Cooling Degree Days | 0 | 0 | 0 | 6 | 47 | 129 | 273 | 261 | 85 | 6 | 0 | 0 | 807 |
| Total Precipitation (") | 1.77 | 1.16 | 1.54 | 1.31 | 1.45 | 1.05 | 0.59 | 0.77 | 0.74 | 1.05 | 1.98 | 1.61 | 15.02 |
| Days ≥ 0.1" Precip | 6 | 4 | 5 | 4 | 4 | 3 | 2 | 2 | 2 | 3 | 6 | 5 | 46 |
| Total Snowfall (") | 4.8 | 2.6 | 0.5 | 0.0 | 0.0 | 0.0 | 0.0 | 0.0 | 0.0 | 0.1 | 0.9 | 4.4 | 13.3 |
| Days ≥ 1" Snow Depth | 6 | 2 | 0 | 0 | 0 | 0 | 0 | 0 | 0 | 0 | 1 | 4 | 13 |

## MITCHELL 2 NW *Wheeler County* ELEVATION 2904 ft LAT/LONG 44° 34 ' N / 120° 9 ' W

| | JAN | FEB | MAR | APR | MAY | JUN | JUL | AUG | SEP | OCT | NOV | DEC | YEAR |
|---|---|---|---|---|---|---|---|---|---|---|---|---|---|
| Maximum Temp °F | 41.4 | 47.3 | 53.7 | 59.7 | 68.9 | 77.3 | 85.7 | 85.5 | 76.5 | 65.0 | 49.0 | 41.8 | 62.7 |
| Minimum Temp °F | 23.8 | 27.0 | 30.5 | 33.5 | 39.6 | 46.7 | 50.8 | 50.7 | 43.0 | 35.6 | 29.3 | 24.3 | 36.2 |
| Mean Temp °F | 32.6 | 37.1 | 42.1 | 46.6 | 54.2 | 62.0 | 68.3 | 68.1 | 59.8 | 50.3 | 39.2 | 33.0 | 49.4 |
| Days Max Temp ≥ 90 °F | 0 | 0 | 0 | 0 | 1 | 5 | 12 | 12 | 3 | 0 | 0 | 0 | 33 |
| Days Max Temp ≤ 32 °F | 6 | 2 | 0 | 0 | 0 | 0 | 0 | 0 | 0 | 0 | 1 | 5 | 14 |
| Days Min Temp ≤ 32 °F | 25 | ***21*** | ***20*** | 15 | 4 | 0 | 0 | 0 | 1 | 10 | 19 | ***25*** | 140 |
| Days Min Temp ≤ 0 °F | 1 | 1 | 0 | 0 | 0 | 0 | 0 | 0 | 0 | 0 | 0 | 1 | 3 |
| Heating Degree Days | 997 | 781 | 702 | 546 | 343 | 150 | 45 | 48 | 183 | 451 | 769 | 986 | 6001 |
| Cooling Degree Days | 0 | 0 | 0 | 1 | 23 | 71 | 157 | 168 | 47 | 4 | 0 | 0 | 471 |
| Total Precipitation (") | 0.95 | 0.59 | 0.99 | 1.16 | 1.48 | 1.36 | 0.64 | 0.77 | 0.68 | 0.76 | 1.28 | 1.03 | 11.69 |
| Days ≥ 0.1" Precip | 3 | 2 | 4 | 4 | 4 | 4 | 2 | 2 | 2 | 3 | 4 | 3 | 37 |
| Total Snowfall (") | ***3.6*** | ***2.1*** | 1.6 | 0.5 | 0.0 | 0.0 | 0.0 | 0.0 | 0.0 | 0.7 | 2.6 | na | na |
| Days ≥ 1" Snow Depth | ***8*** | 3 | 1 | 0 | 0 | 0 | 0 | 0 | 0 | 0 | 2 | ***6*** | 20 |

## MONUMENT 2 *Grant County* ELEVATION 2001 ft LAT/LONG 44° 49 ' N / 119° 25 ' W

| | JAN | FEB | MAR | APR | MAY | JUN | JUL | AUG | SEP | OCT | NOV | DEC | YEAR |
|---|---|---|---|---|---|---|---|---|---|---|---|---|---|
| Maximum Temp °F | 42.0 | ***49.2*** | 56.9 | 64.0 | 71.9 | 80.9 | 90.3 | 90.1 | ***80.1*** | 68.4 | ***51.9*** | ***42.6*** | 65.7 |
| Minimum Temp °F | 21.9 | 25.5 | 29.6 | 33.7 | 39.6 | 46.5 | ***49.8*** | 48.4 | 40.7 | 32.1 | ***28.5*** | ***22.1*** | 34.9 |
| Mean Temp °F | ***32.1*** | ***37.6*** | 43.3 | 48.9 | ***55.9*** | 63.9 | ***69.9*** | 69.2 | ***60.4*** | 50.3 | ***40.2*** | ***32.4*** | 50.3 |
| Days Max Temp ≥ 90 °F | 0 | 0 | 0 | 0 | 2 | 7 | 17 | 18 | 7 | 0 | 0 | 0 | 51 |
| Days Max Temp ≤ 32 °F | 4 | 1 | 0 | 0 | 0 | 0 | 0 | 0 | 0 | 0 | 1 | 3 | 9 |
| Days Min Temp ≤ 32 °F | 25 | 22 | 20 | 12 | 4 | 0 | 0 | 0 | 2 | 15 | 20 | 25 | 145 |
| Days Min Temp ≤ 0 °F | 2 | 1 | 0 | 0 | 0 | 0 | 0 | 0 | 0 | 0 | 0 | 1 | 4 |
| Heating Degree Days | ***1011*** | ***767*** | 671 | 484 | ***290*** | 106 | ***24*** | 27 | ***168*** | 451 | ***736*** | ***1003*** | 5738 |
| Cooling Degree Days | na | na | ***0*** | ***2*** | na | ***77*** | na | 163 | na | ***1*** | na | na | na |
| Total Precipitation (") | 1.40 | 1.02 | 1.40 | 1.34 | 1.44 | 1.27 | 0.61 | 0.76 | 0.64 | 0.89 | 1.57 | 1.38 | 13.72 |
| Days ≥ 0.1" Precip | 5 | 4 | 5 | 4 | 4 | 4 | 2 | 2 | 2 | 3 | 5 | 4 | 44 |
| Total Snowfall (") | 4.8 | 2.8 | 1.9 | 0.1 | 0.0 | 0.0 | 0.0 | 0.0 | 0.0 | 0.1 | 2.0 | 6.1 | 17.8 |
| Days ≥ 1" Snow Depth | 8 | 3 | 1 | 0 | 0 | 0 | 0 | 0 | 0 | 0 | 1 | 7 | 20 |

## MORO *Sherman County* ELEVATION 1870 ft LAT/LONG 45° 29 ' N / 120° 43 ' W

| | JAN | FEB | MAR | APR | MAY | JUN | JUL | AUG | SEP | OCT | NOV | DEC | YEAR |
|---|---|---|---|---|---|---|---|---|---|---|---|---|---|
| Maximum Temp °F | 37.8 | 43.3 | 51.2 | 57.6 | 66.0 | 73.9 | 81.5 | 81.6 | 73.3 | 61.9 | 46.6 | 38.4 | 59.4 |
| Minimum Temp °F | 24.3 | 27.9 | 32.0 | 36.0 | 42.0 | 48.7 | 53.9 | 53.3 | 45.5 | 36.9 | 30.9 | 25.2 | 38.0 |
| Mean Temp °F | 31.1 | 35.6 | 41.6 | 46.8 | 54.0 | 61.3 | 67.7 | 67.5 | 59.4 | 49.4 | 38.8 | 31.8 | 48.7 |
| Days Max Temp ≥ 90 °F | 0 | 0 | 0 | 0 | 0 | 2 | 8 | 7 | 1 | 0 | 0 | 0 | 18 |
| Days Max Temp ≤ 32 °F | 8 | 4 | 0 | 0 | 0 | 0 | 0 | 0 | 0 | 0 | 2 | 8 | 22 |
| Days Min Temp ≤ 32 °F | 25 | 21 | 17 | 9 | 2 | 0 | 0 | 0 | 1 | 8 | 17 | 25 | 125 |
| Days Min Temp ≤ 0 °F | 1 | 1 | 0 | 0 | 0 | 0 | 0 | 0 | 0 | 0 | 0 | 1 | 3 |
| Heating Degree Days | 1045 | 823 | 718 | 540 | 346 | 164 | 61 | 59 | 194 | 478 | 781 | 1021 | 6230 |
| Cooling Degree Days | 0 | 0 | 0 | 1 | 17 | 56 | 137 | 138 | 37 | 2 | 0 | 0 | 388 |
| Total Precipitation (") | 1.55 | 0.95 | 1.03 | 0.80 | 0.73 | 0.61 | 0.32 | 0.43 | 0.49 | 0.79 | 1.65 | 1.51 | 10.86 |
| Days ≥ 0.1" Precip | 4 | 3 | 3 | 2 | 3 | 2 | 1 | 1 | 2 | 2 | 5 | 5 | 33 |
| Total Snowfall (") | 6.3 | 3.3 | 1.1 | 0.1 | 0.0 | 0.0 | 0.0 | 0.0 | 0.0 | 0.3 | 2.3 | 6.3 | 19.7 |
| Days ≥ 1" Snow Depth | 11 | 5 | 1 | 0 | 0 | 0 | 0 | 0 | 0 | 0 | 2 | 9 | 28 |

## N WILLAMETTE EXP STN *Clackamas County* ELEVATION 151 ft LAT/LONG 45° 17 ' N / 122° 45 ' W

| | JAN | FEB | MAR | APR | MAY | JUN | JUL | AUG | SEP | OCT | NOV | DEC | YEAR |
|---|---|---|---|---|---|---|---|---|---|---|---|---|---|
| Maximum Temp °F | 46.3 | 51.1 | 56.1 | 60.1 | 67.1 | 73.6 | 80.2 | 80.7 | 75.2 | 64.3 | 52.6 | 46.1 | 62.8 |
| Minimum Temp °F | 32.7 | 34.4 | 36.9 | 39.8 | 44.5 | 49.8 | 52.6 | 52.5 | 48.5 | 41.4 | 37.3 | 33.3 | 42.0 |
| Mean Temp °F | 39.5 | 42.8 | 46.5 | 50.0 | 55.8 | 61.7 | 66.4 | 66.6 | 61.9 | 52.9 | 45.0 | 39.7 | 52.4 |
| Days Max Temp ≥ 90 °F | 0 | 0 | 0 | 0 | 1 | 2 | 5 | 5 | 2 | 0 | 0 | 0 | 15 |
| Days Max Temp ≤ 32 °F | 1 | 0 | 0 | 0 | 0 | 0 | 0 | 0 | 0 | 0 | 0 | 1 | 2 |
| Days Min Temp ≤ 32 °F | 15 | 11 | 7 | 3 | 0 | 0 | 0 | 0 | 0 | 2 | 8 | 14 | 60 |
| Days Min Temp ≤ 0 °F | 0 | 0 | 0 | 0 | 0 | 0 | 0 | 0 | 0 | 0 | 0 | 0 | 0 |
| Heating Degree Days | 783 | 621 | 567 | 445 | 289 | 132 | 44 | 41 | 124 | 370 | 594 | 777 | 4787 |
| Cooling Degree Days | 0 | 0 | 0 | 1 | 15 | 43 | 93 | 106 | 41 | 3 | 0 | 0 | 302 |
| Total Precipitation (") | 5.87 | 4.47 | 3.90 | 2.90 | 2.20 | 1.71 | 0.73 | 0.90 | 1.74 | 3.31 | 5.93 | 6.91 | 40.57 |
| Days ≥ 0.1" Precip | 13 | 11 | 12 | 9 | 6 | 5 | 2 | 2 | 4 | 7 | 13 | 13 | 97 |
| Total Snowfall (") | 1.3 | 0.3 | 0.1 | 0.0 | 0.0 | 0.0 | 0.0 | 0.0 | 0.0 | 0.0 | 0.1 | 0.6 | 2.4 |
| Days ≥ 1" Snow Depth | 0 | 0 | 0 | 0 | 0 | 0 | 0 | 0 | 0 | 0 | 0 | 1 | 1 |

## NEWPORT *Lincoln County* ELEVATION 122 ft LAT/LONG 44° 38 ' N / 124° 3 ' W

| | JAN | FEB | MAR | APR | MAY | JUN | JUL | AUG | SEP | OCT | NOV | DEC | YEAR |
|---|---|---|---|---|---|---|---|---|---|---|---|---|---|
| Maximum Temp °F | 50.4 | 52.9 | 54.1 | 55.6 | 59.0 | 62.2 | 64.6 | 65.6 | 65.6 | 61.2 | 54.7 | 50.4 | 58.0 |
| Minimum Temp °F | 37.9 | 38.7 | 39.2 | 40.3 | 44.0 | 48.2 | 50.0 | 50.2 | 48.5 | 44.8 | 41.5 | 38.1 | 43.4 |
| Mean Temp °F | 44.2 | 45.8 | 46.6 | 48.0 | 51.5 | 55.2 | 57.4 | 58.0 | 57.1 | 53.0 | 48.1 | 44.3 | 50.8 |
| Days Max Temp ≥ 90 °F | 0 | 0 | 0 | 0 | 0 | 0 | 0 | 0 | 0 | 0 | 0 | 0 | 0 |
| Days Max Temp ≤ 32 °F | 0 | 0 | 0 | 0 | 0 | 0 | 0 | 0 | 0 | 0 | 0 | 1 | 1 |
| Days Min Temp ≤ 32 °F | 7 | 4 | 4 | 2 | 0 | 0 | 0 | 0 | 0 | 0 | 2 | 6 | 25 |
| Days Min Temp ≤ 0 °F | 0 | 0 | 0 | 0 | 0 | 0 | 0 | 0 | 0 | 0 | 0 | 0 | 0 |
| Heating Degree Days | 639 | 534 | 562 | 504 | 413 | 288 | 230 | 213 | 235 | 367 | 499 | 635 | 5119 |
| Cooling Degree Days | 0 | 0 | 0 | 0 | 0 | 1 | 1 | 2 | 5 | 2 | 0 | 0 | 11 |
| Total Precipitation (") | 10.78 | 7.76 | 7.59 | 4.99 | 3.40 | 2.72 | 0.99 | 1.16 | 2.44 | 5.22 | 10.19 | 11.99 | 69.23 |
| Days ≥ 0.1" Precip | 17 | 13 | 15 | 12 | 9 | 6 | 2 | 3 | 5 | 9 | 16 | 17 | 124 |
| Total Snowfall (") | 0.6 | 0.2 | 0.0 | 0.0 | 0.0 | 0.0 | 0.0 | 0.0 | 0.0 | 0.0 | 0.0 | 0.5 | 1.3 |
| Days ≥ 1" Snow Depth | 0 | 0 | 0 | 0 | 0 | 0 | 0 | 0 | 0 | 0 | 0 | 0 | 0 |

## NORTH BEND MUNI AP *Coos County* ELEVATION 30 ft LAT/LONG 43° 25 ' N / 124° 15 ' W

| | JAN | FEB | MAR | APR | MAY | JUN | JUL | AUG | SEP | OCT | NOV | DEC | YEAR |
|---|---|---|---|---|---|---|---|---|---|---|---|---|---|
| Maximum Temp °F | 52.0 | 54.2 | 55.3 | 56.7 | 60.6 | 63.8 | 66.3 | 67.4 | 67.1 | 63.0 | 56.8 | 52.3 | 59.6 |
| Minimum Temp °F | 39.3 | 40.6 | 41.4 | 42.8 | 46.8 | 50.7 | 52.7 | 53.2 | 50.5 | 46.4 | 42.9 | 39.5 | 45.6 |
| Mean Temp °F | 45.7 | 47.5 | 48.4 | 49.8 | 53.7 | 57.2 | 59.5 | 60.3 | 58.8 | 54.8 | 49.8 | 45.9 | 52.6 |
| Days Max Temp ≥ 90 °F | 0 | 0 | 0 | 0 | 0 | 0 | 0 | 0 | 0 | 0 | 0 | 0 | 0 |
| Days Max Temp ≤ 32 °F | 0 | 0 | 0 | 0 | 0 | 0 | 0 | 0 | 0 | 0 | 0 | 0 | 0 |
| Days Min Temp ≤ 32 °F | 5 | 3 | 1 | 0 | 0 | 0 | 0 | 0 | 0 | 0 | 2 | 4 | 15 |
| Days Min Temp ≤ 0 °F | 0 | 0 | 0 | 0 | 0 | 0 | 0 | 0 | 0 | 0 | 0 | 0 | 0 |
| Heating Degree Days | 592 | 489 | 508 | 450 | 344 | 226 | 163 | 140 | 182 | 312 | 448 | 584 | 4438 |
| Cooling Degree Days | 0 | 0 | 0 | 0 | 0 | 0 | 0 | 1 | 2 | 1 | 0 | 0 | 4 |
| Total Precipitation (") | 10.04 | 7.47 | 7.56 | 5.00 | 2.86 | 1.67 | 0.49 | 0.98 | 1.73 | 4.51 | 9.93 | 11.43 | 63.67 |
| Days ≥ 0.1" Precip | 15 | 13 | 14 | 11 | 7 | 4 | 1 | 2 | 4 | 8 | 16 | 15 | 110 |
| Total Snowfall (") | 1.3 | 0.2 | 0.1 | 0.0 | 0.0 | 0.0 | 0.0 | 0.0 | 0.0 | 0.0 | 0.1 | 0.2 | 1.9 |
| Days ≥ 1" Snow Depth | 0 | 0 | 0 | 0 | 0 | 0 | 0 | 0 | 0 | 0 | 0 | 0 | 0 |

## NYSSA *Malheur County* ELEVATION 2192 ft LAT/LONG 43° 52 ' N / 117° 0 ' W

| | JAN | FEB | MAR | APR | MAY | JUN | JUL | AUG | SEP | OCT | NOV | DEC | YEAR |
|---|---|---|---|---|---|---|---|---|---|---|---|---|---|
| Maximum Temp °F | 34.3 | 43.6 | 55.4 | 64.2 | 73.5 | 81.8 | 90.6 | 89.4 | 78.6 | 65.5 | 47.9 | 36.0 | 63.4 |
| Minimum Temp °F | 20.6 | 25.5 | 31.8 | 38.4 | 46.2 | 53.9 | 59.0 | 57.0 | 46.8 | 36.7 | 29.1 | 21.8 | 38.9 |
| Mean Temp °F | 27.5 | 34.6 | 43.6 | 51.3 | 59.8 | 67.9 | 74.8 | 73.2 | 62.7 | 51.1 | 38.5 | 28.9 | 51.2 |
| Days Max Temp ≥ 90 °F | 0 | 0 | 0 | 0 | 1 | 7 | 19 | 18 | 3 | 0 | 0 | 0 | 48 |
| Days Max Temp ≤ 32 °F | 12 | 3 | 0 | 0 | 0 | 0 | 0 | 0 | 0 | 0 | 1 | 10 | 26 |
| Days Min Temp ≤ 32 °F | 27 | 23 | 17 | 5 | 0 | 0 | 0 | 0 | 1 | 8 | 20 | 27 | 128 |
| Days Min Temp ≤ 0 °F | 2 | 1 | 0 | 0 | 0 | 0 | 0 | 0 | 0 | 0 | 0 | 1 | 4 |
| Heating Degree Days | 1156 | 853 | 656 | 406 | 193 | 55 | 7 | 12 | 119 | 424 | 788 | 1111 | 5780 |
| Cooling Degree Days | 0 | 0 | 0 | 6 | 47 | 152 | 310 | 286 | 61 | 1 | 0 | 0 | 863 |
| Total Precipitation (") | 1.26 | 1.00 | 0.93 | 0.79 | 0.84 | 0.87 | 0.26 | 0.45 | 0.55 | 0.65 | 1.31 | 1.33 | 10.24 |
| Days ≥ 0.1" Precip | 4 | 3 | 3 | 3 | 3 | 3 | 1 | 1 | 2 | 2 | 5 | 4 | 34 |
| Total Snowfall (") | 6.1 | 2.2 | 0.3 | 0.0 | 0.0 | 0.0 | 0.0 | 0.0 | 0.0 | 0.1 | 1.6 | 6.8 | 17.1 |
| Days ≥ 1" Snow Depth | *15* | 4 | 0 | 0 | 0 | 0 | 0 | 0 | 0 | 0 | 2 | 9 | 30 |

## OAKRIDGE FISH HATCHE *Lane County* ELEVATION 1312 ft LAT/LONG 43° 45 ' N / 122° 26 ' W

| | JAN | FEB | MAR | APR | MAY | JUN | JUL | AUG | SEP | OCT | NOV | DEC | YEAR |
|---|---|---|---|---|---|---|---|---|---|---|---|---|---|
| Maximum Temp °F | 46.0 | 51.9 | 56.7 | 61.2 | 68.3 | 74.6 | 81.9 | 82.5 | 77.3 | 66.9 | 51.7 | 44.9 | 63.7 |
| Minimum Temp °F | 29.9 | 31.3 | 33.8 | 37.2 | 41.8 | 47.2 | 49.7 | 49.5 | 44.5 | 38.6 | 34.6 | 30.6 | 39.1 |
| Mean Temp °F | 38.0 | 41.6 | 45.2 | 49.2 | 55.1 | 60.9 | 65.8 | 66.0 | 60.9 | 52.8 | 43.2 | 37.8 | 51.4 |
| Days Max Temp ≥ 90 °F | 0 | 0 | 0 | 0 | 1 | 2 | 6 | 6 | 4 | 0 | 0 | 0 | 19 |
| Days Max Temp ≤ 32 °F | 0 | 0 | 0 | 0 | 0 | 0 | 0 | 0 | 0 | 0 | 0 | 1 | 1 |
| Days Min Temp ≤ 32 °F | 21 | 17 | 14 | 5 | 1 | 0 | 0 | 0 | 1 | 4 | 11 | 19 | 93 |
| Days Min Temp ≤ 0 °F | 0 | 0 | 0 | 0 | 0 | 0 | 0 | 0 | 0 | 0 | 0 | 0 | 0 |
| Heating Degree Days | 831 | 653 | 606 | 468 | 311 | 150 | 52 | 48 | 145 | 374 | 648 | 838 | 5124 |
| Cooling Degree Days | 0 | 0 | 0 | 1 | 12 | 29 | 70 | 76 | 28 | 2 | 0 | 0 | 218 |
| Total Precipitation (") | 6.26 | 4.55 | 4.61 | 3.80 | 2.58 | 1.69 | 0.60 | 1.16 | 1.63 | 3.17 | 6.90 | 6.65 | 43.60 |
| Days ≥ 0.1" Precip | 13 | 11 | 12 | 11 | 8 | 5 | 2 | 3 | 4 | 7 | 13 | 13 | 102 |
| Total Snowfall (") | 3.9 | 2.1 | 1.1 | 0.1 | 0.0 | 0.0 | 0.0 | 0.0 | 0.0 | 0.0 | 0.6 | 2.0 | 9.8 |
| Days ≥ 1" Snow Depth | 2 | 1 | 1 | 0 | 0 | 0 | 0 | 0 | 0 | 0 | 0 | 1 | 5 |

## OCHOCO RS *Crook County* ELEVATION 3983 ft LAT/LONG 44° 24 ' N / 120° 26 ' W

| | JAN | FEB | MAR | APR | MAY | JUN | JUL | AUG | SEP | OCT | NOV | DEC | YEAR |
|---|---|---|---|---|---|---|---|---|---|---|---|---|---|
| Maximum Temp °F | 35.7 | 41.2 | 48.3 | 55.2 | 64.1 | 72.7 | 81.3 | 82.2 | 73.8 | 61.9 | 43.4 | 35.4 | 57.9 |
| Minimum Temp °F | 16.4 | 19.6 | 23.4 | 26.8 | 31.8 | 37.6 | 40.9 | 40.6 | 33.7 | 28.0 | 23.6 | 17.8 | 28.4 |
| Mean Temp °F | 26.0 | 30.4 | 35.9 | 41.0 | 47.9 | 55.1 | 61.2 | 61.4 | 53.8 | 45.0 | 33.5 | 26.6 | 43.2 |
| Days Max Temp ≥ 90 °F | 0 | 0 | 0 | 0 | 0 | 1 | 6 | 7 | 1 | 0 | 0 | 0 | 15 |
| Days Max Temp ≤ 32 °F | ***10*** | 4 | 1 | 0 | 0 | 0 | 0 | 0 | 0 | 0 | 3 | 10 | 28 |
| Days Min Temp ≤ 32 °F | 29 | 27 | 29 | 25 | 18 | 7 | 2 | 2 | 13 | ***24*** | 27 | 30 | 233 |
| Days Min Temp ≤ 0 °F | 3 | 1 | 0 | 0 | 0 | 0 | 0 | 0 | 0 | 0 | 0 | 2 | 6 |
| Heating Degree Days | 1204 | 970 | 895 | 708 | 524 | 299 | 145 | 142 | 334 | 614 | 938 | 1187 | 7960 |
| Cooling Degree Days | 0 | 0 | 0 | 0 | 3 | 10 | 31 | 37 | 6 | 0 | 0 | 0 | 87 |
| Total Precipitation (") | 2.16 | 1.44 | 1.39 | 1.13 | 1.28 | 1.27 | 0.74 | 0.90 | 0.88 | 1.20 | 2.31 | 2.08 | 16.78 |
| Days ≥ 0.1" Precip | 6 | 5 | 5 | 4 | 4 | 3 | 2 | 3 | 3 | 4 | ***7*** | na | na |
| Total Snowfall (") | ***9.9*** | 7.1 | 4.5 | 2.2 | 0.5 | 0.0 | 0.0 | 0.0 | 0.0 | 0.9 | 7.3 | ***11.9*** | 44.3 |
| Days ≥ 1" Snow Depth | na | ***13*** | 8 | 1 | 0 | 0 | 0 | 0 | 0 | 0 | 5 | ***14*** | na |

## ONTARIO KSRV *Malheur County* ELEVATION 2152 ft LAT/LONG 44° 3 ' N / 116° 58 ' W

| | JAN | FEB | MAR | APR | MAY | JUN | JUL | AUG | SEP | OCT | NOV | DEC | YEAR |
|---|---|---|---|---|---|---|---|---|---|---|---|---|---|
| Maximum Temp °F | 35.3 | 44.5 | 56.8 | 66.2 | 76.5 | 85.5 | 95.0 | 93.2 | 81.9 | 67.3 | 48.3 | 36.8 | 65.6 |
| Minimum Temp °F | 19.5 | 24.6 | 30.6 | 36.5 | 44.4 | 51.5 | 57.5 | 54.5 | 44.3 | 34.2 | 27.5 | 20.0 | 37.1 |
| Mean Temp °F | 27.4 | 34.6 | 43.7 | 51.4 | 60.5 | 68.6 | 76.3 | 73.9 | 63.1 | 50.8 | 37.9 | 28.4 | 51.4 |
| Days Max Temp ≥ 90 °F | 0 | 0 | 0 | 0 | 3 | 11 | 25 | 22 | 7 | 0 | 0 | 0 | 68 |
| Days Max Temp ≤ 32 °F | 11 | 3 | 0 | 0 | 0 | 0 | 0 | 0 | 0 | 0 | 1 | 7 | 22 |
| Days Min Temp ≤ 32 °F | 28 | 24 | 19 | 9 | 2 | 0 | 0 | 0 | 1 | 13 | 22 | 28 | 146 |
| Days Min Temp ≤ 0 °F | 2 | 1 | 0 | 0 | 0 | 0 | 0 | 0 | 0 | 0 | 0 | 2 | 5 |
| Heating Degree Days | 1157 | 852 | 652 | 405 | 177 | 48 | 4 | 10 | 113 | 435 | 805 | 1126 | 5784 |
| Cooling Degree Days | 0 | 0 | 0 | 3 | 46 | 152 | 333 | 284 | 57 | 1 | 0 | 0 | 876 |
| Total Precipitation (") | 1.35 | 0.84 | 0.82 | 0.68 | 0.76 | 0.73 | 0.21 | 0.41 | 0.49 | 0.59 | 1.32 | 1.42 | 9.62 |
| Days ≥ 0.1" Precip | 4 | 3 | 3 | 2 | 3 | 2 | 1 | 1 | 2 | 2 | 5 | 5 | 33 |
| Total Snowfall (") | 7.2 | 2.3 | 0.4 | 0.0 | 0.0 | 0.0 | 0.0 | 0.0 | 0.0 | 0.0 | 2.6 | 6.2 | 18.7 |
| Days ≥ 1" Snow Depth | 14 | 7 | 0 | 0 | 0 | 0 | 0 | 0 | 0 | 0 | 1 | 10 | 32 |

## OREGON CITY *Clackamas County* ELEVATION 171 ft LAT/LONG 45° 21 ' N / 122° 36 ' W

| | JAN | FEB | MAR | APR | MAY | JUN | JUL | AUG | SEP | OCT | NOV | DEC | YEAR |
|---|---|---|---|---|---|---|---|---|---|---|---|---|---|
| Maximum Temp °F | 47.0 | 52.4 | 57.6 | 62.4 | 70.1 | 76.0 | 82.1 | 82.3 | 77.2 | 65.4 | 53.1 | 46.5 | 64.3 |
| Minimum Temp °F | 35.1 | 37.1 | 39.3 | 41.9 | 46.8 | 51.9 | 55.1 | 55.3 | 51.5 | 45.0 | 39.7 | 35.4 | 44.5 |
| Mean Temp °F | 41.1 | 44.7 | 48.5 | 52.2 | 58.5 | 64.0 | 68.6 | 68.8 | 64.4 | 55.2 | 46.4 | 41.0 | 54.5 |
| Days Max Temp ≥ 90 °F | 0 | 0 | 0 | 0 | 1 | 2 | 6 | 6 | 3 | 0 | 0 | 0 | 18 |
| Days Max Temp ≤ 32 °F | 1 | 0 | 0 | 0 | 0 | 0 | 0 | 0 | 0 | 0 | 0 | 1 | 2 |
| Days Min Temp ≤ 32 °F | 11 | 7 | 3 | 1 | 0 | 0 | 0 | 0 | 0 | 0 | 4 | 10 | 36 |
| Days Min Temp ≤ 0 °F | 0 | 0 | 0 | 0 | 0 | 0 | 0 | 0 | 0 | 0 | 0 | 0 | 0 |
| Heating Degree Days | 736 | 566 | 506 | 379 | 218 | 88 | 19 | 17 | 77 | 301 | 550 | 737 | 4194 |
| Cooling Degree Days | 0 | 0 | 0 | 2 | 28 | 62 | 133 | 144 | 72 | 6 | 0 | 0 | 447 |
| Total Precipitation (") | 6.96 | 5.01 | 4.46 | 3.53 | 2.45 | 1.90 | 0.78 | 1.12 | 1.93 | 3.52 | 6.45 | 7.69 | 45.80 |
| Days ≥ 0.1" Precip | 13 | 11 | 11 | 9 | 7 | 4 | 2 | 3 | 5 | 8 | 13 | 14 | 100 |
| Total Snowfall (") | 1.3 | 0.8 | 0.1 | 0.0 | 0.0 | 0.0 | 0.0 | 0.0 | 0.0 | 0.0 | 0.1 | 1.1 | 3.4 |
| Days ≥ 1" Snow Depth | 1 | 0 | 0 | 0 | 0 | 0 | 0 | 0 | 0 | 0 | 0 | 1 | 2 |

## OTIS 2 NE *Lincoln County* ELEVATION 151 ft LAT/LONG 45° 2' N / 123° 56' W

| | JAN | FEB | MAR | APR | MAY | JUN | JUL | AUG | SEP | OCT | NOV | DEC | YEAR |
|---|---|---|---|---|---|---|---|---|---|---|---|---|---|
| Maximum Temp °F | 47.0 | 51.1 | 54.6 | 57.1 | 61.9 | 65.7 | 69.6 | 70.9 | 69.6 | 61.5 | 52.1 | 46.7 | 59.0 |
| Minimum Temp °F | 36.2 | 37.5 | 38.5 | 39.8 | 43.5 | 47.7 | 49.6 | 50.4 | 48.5 | 44.5 | 40.1 | 36.4 | 42.7 |
| Mean Temp °F | 41.6 | 44.4 | 46.5 | 48.4 | 52.7 | 56.7 | 59.6 | 60.7 | 59.0 | 53.0 | 46.1 | 41.6 | 50.9 |
| Days Max Temp ≥ 90 °F | 0 | 0 | 0 | 0 | 0 | 0 | 0 | 0 | 0 | 0 | 0 | 0 | 0 |
| Days Max Temp ≤ 32 °F | 0 | 0 | 0 | 0 | 0 | 0 | 0 | 0 | 0 | 0 | 0 | 1 | 1 |
| Days Min Temp ≤ 32 °F | 10 | 6 | 5 | 3 | 0 | 0 | 0 | 0 | 0 | 1 | 4 | 9 | 38 |
| Days Min Temp ≤ 0 °F | 0 | 0 | 0 | 0 | 0 | 0 | 0 | 0 | 0 | 0 | 0 | 0 | 0 |
| Heating Degree Days | 718 | 576 | 564 | 490 | 374 | 247 | 165 | 135 | 179 | 365 | 560 | 719 | 5092 |
| Cooling Degree Days | 0 | 0 | 0 | 0 | 1 | 4 | 4 | 9 | 6 | 1 | 0 | 0 | 25 |
| Total Precipitation (") | 14.62 | 10.87 | 10.73 | 7.28 | 4.82 | 3.65 | 1.64 | 1.71 | 3.73 | 7.45 | 13.87 | 15.76 | 96.13 |
| Days ≥ 0.1" Precip | ***16*** | ***15*** | ***15*** | ***13*** | 10 | ***6*** | ***4*** | 3 | 6 | ***11*** | ***18*** | ***18*** | 135 |
| Total Snowfall (") | 1.8 | 0.1 | 0.4 | 0.0 | 0.0 | 0.0 | 0.0 | 0.0 | 0.0 | 0.0 | 0.1 | 0.8 | 3.2 |
| Days ≥ 1" Snow Depth | 1 | 0 | 0 | 0 | 0 | 0 | 0 | 0 | 0 | 0 | 0 | 1 | 2 |

## OWYHEE DAM *Malheur County* ELEVATION 2402 ft LAT/LONG 43° 39' N / 117° 15' W

| | JAN | FEB | MAR | APR | MAY | JUN | JUL | AUG | SEP | OCT | NOV | DEC | YEAR |
|---|---|---|---|---|---|---|---|---|---|---|---|---|---|
| Maximum Temp °F | 38.5 | 45.7 | 55.8 | 64.3 | 74.2 | 83.2 | 92.2 | 91.1 | 80.2 | 67.2 | 50.0 | 39.4 | 65.2 |
| Minimum Temp °F | 22.1 | 26.8 | 32.4 | 37.4 | 43.7 | 50.0 | 53.8 | 53.1 | 46.1 | 37.7 | 29.7 | 22.3 | 37.9 |
| Mean Temp °F | 30.3 | 36.3 | 44.1 | 50.9 | 59.0 | 66.6 | 73.0 | 72.1 | 63.2 | 52.5 | 39.9 | 30.9 | 51.6 |
| Days Max Temp ≥ 90 °F | 0 | 0 | 0 | 0 | 2 | 9 | 21 | 20 | 6 | 0 | 0 | 0 | 58 |
| Days Max Temp ≤ 32 °F | 8 | 2 | 0 | 0 | 0 | 0 | 0 | 0 | 0 | 0 | 1 | 6 | 17 |
| Days Min Temp ≤ 32 °F | 27 | 21 | 15 | 7 | 1 | 0 | 0 | 0 | 1 | 7 | 18 | 28 | 125 |
| Days Min Temp ≤ 0 °F | 1 | 0 | 0 | 0 | 0 | 0 | 0 | 0 | 0 | 0 | 0 | 1 | 2 |
| Heating Degree Days | 1069 | 804 | 639 | 420 | 211 | 65 | 10 | 15 | 114 | 384 | 747 | 1053 | 5531 |
| Cooling Degree Days | 0 | 0 | 0 | 4 | 42 | 136 | 272 | 261 | 74 | 6 | 0 | 0 | 795 |
| Total Precipitation (") | 0.97 | 0.75 | 0.86 | 0.87 | 0.88 | 1.11 | 0.36 | 0.57 | 0.53 | 0.58 | 0.97 | 1.04 | 9.49 |
| Days ≥ 0.1" Precip | 3 | 3 | 3 | 3 | 3 | 3 | 1 | 2 | 2 | 2 | 4 | 4 | 33 |
| Total Snowfall (") | 3.5 | ***0.9*** | 0.3 | 0.0 | 0.0 | 0.0 | 0.0 | 0.0 | 0.0 | 0.0 | 0.7 | ***3.8*** | 9.2 |
| Days ≥ 1" Snow Depth | ***7*** | 2 | 0 | 0 | 0 | 0 | 0 | 0 | 0 | 0 | 1 | ***5*** | 15 |

## P RANCH REFUGE *Harney County* ELEVATION 4213 ft LAT/LONG 42° 49' N / 118° 53' W

| | JAN | FEB | MAR | APR | MAY | JUN | JUL | AUG | SEP | OCT | NOV | DEC | YEAR |
|---|---|---|---|---|---|---|---|---|---|---|---|---|---|
| Maximum Temp °F | 41.4 | 47.4 | 53.8 | 60.7 | 68.7 | 76.8 | ***85.4*** | 85.3 | 78.1 | 66.6 | ***50.6*** | 41.6 | 63.0 |
| Minimum Temp °F | 19.9 | 23.4 | 26.5 | 30.6 | 37.5 | 43.3 | ***46.4*** | 44.5 | 35.9 | 29.9 | 25.7 | 20.3 | 32.0 |
| Mean Temp °F | 30.7 | 35.4 | 40.2 | 45.7 | 53.1 | 60.0 | ***66.0*** | 64.9 | 57.0 | 48.3 | ***38.4*** | 31.0 | 47.6 |
| Days Max Temp ≥ 90 °F | 0 | 0 | 0 | 0 | 0 | 2 | ***8*** | ***10*** | 2 | 0 | 0 | 0 | 22 |
| Days Max Temp ≤ 32 °F | 6 | 1 | 0 | 0 | 0 | 0 | 0 | 0 | 0 | 0 | 1 | 5 | 13 |
| Days Min Temp ≤ 32 °F | 26 | 22 | 23 | 19 | 7 | 1 | 0 | 1 | na | 19 | 22 | 25 | na |
| Days Min Temp ≤ 0 °F | 2 | 1 | 0 | 0 | 0 | 0 | 0 | 0 | 0 | 0 | 0 | 2 | 5 |
| Heating Degree Days | 1057 | 827 | 763 | 573 | 367 | 172 | ***51*** | ***68*** | 241 | 512 | ***791*** | 1047 | 6469 |
| Cooling Degree Days | ***0*** | 0 | 0 | 0 | 6 | ***25*** | na | ***63*** | 7 | ***1*** | ***0*** | 0 | na |
| Total Precipitation (") | 1.03 | 0.90 | 1.15 | 1.13 | 1.36 | 1.18 | 0.39 | 0.74 | 0.71 | 0.89 | 1.29 | 1.23 | 12.00 |
| Days ≥ 0.1" Precip | ***3*** | 3 | 4 | 4 | 4 | 3 | 1 | 2 | 2 | 3 | 4 | 4 | 37 |
| Total Snowfall (") | 4.0 | ***2.4*** | 1.9 | 1.1 | 0.3 | 0.0 | 0.0 | 0.0 | 0.0 | 0.4 | 1.7 | ***4.1*** | 15.9 |
| Days ≥ 1" Snow Depth | ***2*** | ***1*** | 1 | 0 | 0 | 0 | 0 | 0 | 0 | 0 | 1 | ***5*** | 10 |

## PAISLEY *Lake County* ELEVATION 4373 ft LAT/LONG 42° 42' N / 120° 33' W

| | JAN | FEB | MAR | APR | MAY | JUN | JUL | AUG | SEP | OCT | NOV | DEC | YEAR |
|---|---|---|---|---|---|---|---|---|---|---|---|---|---|
| Maximum Temp °F | 41.9 | 46.8 | 52.1 | 59.1 | 68.5 | 76.1 | 84.7 | 83.9 | 76.3 | 65.3 | 48.9 | 41.2 | 62.1 |
| Minimum Temp °F | 22.1 | 24.9 | 27.8 | 31.8 | 38.7 | 45.3 | 49.1 | 48.1 | 40.5 | 33.3 | 26.0 | 21.5 | 34.1 |
| Mean Temp °F | 32.0 | 35.9 | 40.1 | 45.5 | 53.6 | 60.7 | 66.9 | 66.0 | 58.4 | 49.3 | 37.5 | 31.4 | 48.1 |
| Days Max Temp ≥ 90 °F | 0 | 0 | 0 | 0 | 0 | 2 | 8 | 8 | 1 | 0 | 0 | 0 | 19 |
| Days Max Temp ≤ 32 °F | 5 | 2 | 0 | 0 | 0 | 0 | 0 | 0 | 0 | 0 | 1 | 5 | 13 |
| Days Min Temp ≤ 32 °F | 25 | 22 | 21 | 17 | 7 | 1 | 0 | 0 | 5 | ***14*** | 22 | 25 | 159 |
| Days Min Temp ≤ 0 °F | 1 | 0 | 0 | 0 | 0 | 0 | 0 | 0 | 0 | 0 | 0 | 1 | 2 |
| Heating Degree Days | 1016 | 816 | 766 | 579 | 353 | 163 | 49 | 64 | 210 | 480 | 819 | 1032 | 6347 |
| Cooling Degree Days | 0 | 0 | 0 | 1 | 11 | 40 | 105 | 96 | 21 | 1 | 0 | 0 | 275 |
| Total Precipitation (") | 1.34 | 0.79 | 1.02 | 0.77 | 0.92 | 1.08 | 0.46 | 0.68 | 0.53 | 0.64 | 1.11 | 1.31 | 10.65 |
| Days ≥ 0.1" Precip | 3 | 3 | 3 | 3 | 3 | 3 | 1 | 2 | 1 | 2 | 3 | 4 | 31 |
| Total Snowfall (") | ***2.9*** | 2.8 | 3.0 | 1.6 | 0.6 | 0.0 | 0.0 | 0.0 | 0.0 | 0.3 | 2.4 | 4.6 | 18.2 |
| Days ≥ 1" Snow Depth | 5 | 3 | 2 | 0 | 0 | 0 | 0 | 0 | 0 | 0 | 2 | ***5*** | 17 |

## PAULINA *Crook County* ELEVATION 3681 ft LAT/LONG 44° 8 ' N / 119° 58 ' W

| | JAN | FEB | MAR | APR | MAY | JUN | JUL | AUG | SEP | OCT | NOV | DEC | YEAR |
|---|---|---|---|---|---|---|---|---|---|---|---|---|---|
| Maximum Temp °F | 38.2 | 45.0 | 52.1 | 59.7 | 68.9 | 77.3 | 86.7 | 86.1 | 78.2 | 66.0 | 47.9 | 39.0 | 62.1 |
| Minimum Temp °F | 16.6 | 21.7 | 25.2 | 27.7 | 33.7 | 40.3 | 43.5 | 41.8 | 33.3 | 26.7 | 23.6 | 17.1 | 29.3 |
| Mean Temp °F | 27.4 | 33.4 | 38.7 | 43.7 | 51.3 | 58.8 | 65.1 | 64.0 | 55.8 | 46.4 | 35.8 | 28.1 | 45.7 |
| Days Max Temp ≥ 90 °F | 0 | 0 | 0 | 0 | 0 | 4 | 14 | 12 | 4 | 0 | 0 | 0 | 34 |
| Days Max Temp ≤ 32 °F | 8 | 2 | 0 | 0 | 0 | 0 | 0 | 0 | 0 | 0 | 1 | 6 | 17 |
| Days Min Temp ≤ 32 °F | 28 | 25 | 26 | 22 | 14 | 4 | 1 | 2 | 14 | 23 | 24 | 27 | 210 |
| Days Min Temp ≤ 0 °F | 4 | 1 | 0 | 0 | 0 | 0 | 0 | 0 | 0 | 0 | 1 | 3 | 9 |
| Heating Degree Days | 1159 | 888 | 810 | 631 | 420 | 203 | 67 | 86 | 274 | 572 | 870 | 1138 | 7118 |
| Cooling Degree Days | 0 | 0 | 0 | 0 | 3 | 22 | 67 | 56 | 5 | 0 | 0 | 0 | 153 |
| Total Precipitation (") | 1.27 | 0.84 | 1.03 | 0.82 | 1.14 | 1.08 | 0.65 | 0.69 | 0.50 | 0.80 | 1.37 | 1.17 | 11.36 |
| Days ≥ 0.1" Precip | 4 | 3 | 4 | 3 | 3 | 3 | 2 | 2 | 2 | 3 | 4 | 4 | 37 |
| Total Snowfall (") | *6.8* | 3.0 | 2.5 | 0.7 | 0.1 | 0.0 | 0.0 | 0.0 | 0.0 | 0.3 | 3.2 | 8.5 | 25.1 |
| Days ≥ 1" Snow Depth | *15* | 5 | 1 | 0 | 0 | 0 | 0 | 0 | 0 | 0 | 2 | 9 | 32 |

## PELTON DAM *Jefferson County* ELEVATION 1411 ft LAT/LONG 44° 44 ' N / 121° 14 ' W

| | JAN | FEB | MAR | APR | MAY | JUN | JUL | AUG | SEP | OCT | NOV | DEC | YEAR |
|---|---|---|---|---|---|---|---|---|---|---|---|---|---|
| Maximum Temp °F | 45.1 | 52.1 | 60.3 | 67.2 | 76.9 | 84.6 | 93.1 | 92.4 | 83.6 | 70.4 | 53.1 | 44.9 | 68.6 |
| Minimum Temp °F | 25.8 | 28.5 | 30.9 | 34.7 | 40.5 | 47.0 | 50.7 | 50.2 | 43.7 | 36.2 | 31.6 | 26.5 | 37.2 |
| Mean Temp °F | 35.4 | 40.4 | 45.7 | 51.0 | 58.7 | 65.7 | 71.9 | 71.3 | 63.8 | 53.2 | 42.3 | 35.6 | 52.9 |
| Days Max Temp ≥ 90 °F | 0 | 0 | 0 | 0 | 4 | 10 | 21 | 20 | 9 | 1 | 0 | 0 | 65 |
| Days Max Temp ≤ 32 °F | 4 | 1 | 0 | 0 | 0 | 0 | 0 | 0 | 0 | 0 | 1 | 3 | 9 |
| Days Min Temp ≤ 32 °F | 24 | 20 | 19 | 12 | 4 | 0 | 0 | 0 | 1 | 10 | 16 | 24 | 130 |
| Days Min Temp ≤ 0 °F | 1 | 0 | 0 | 0 | 0 | 0 | 0 | 0 | 0 | 0 | 0 | 1 | 2 |
| Heating Degree Days | 909 | 688 | 593 | 416 | 216 | 72 | 12 | 12 | 96 | 362 | 675 | 903 | 4954 |
| Cooling Degree Days | 0 | 0 | 0 | 4 | 37 | 100 | 231 | 227 | 76 | 5 | 0 | 0 | 680 |
| Total Precipitation (") | 1.53 | 0.85 | 0.83 | 0.73 | 0.71 | 0.60 | 0.35 | 0.46 | 0.43 | 0.62 | 1.42 | 1.39 | 9.92 |
| Days ≥ 0.1" Precip | 5 | 3 | 3 | 2 | 3 | 2 | 1 | 2 | 1 | 2 | 4 | 4 | 32 |
| Total Snowfall (") | *1.2* | na | 0.0 | 0.0 | 0.0 | 0.0 | 0.0 | 0.0 | 0.0 | 0.0 | *0.8* | na | na |
| Days ≥ 1" Snow Depth | *1* | *0* | 0 | 0 | 0 | 0 | 0 | 0 | 0 | 0 | *0* | na | na |

## PENDLETON BR EXP STN *Umatilla County* ELEVATION 1489 ft LAT/LONG 45° 42 ' N / 118° 41 ' W

| | JAN | FEB | MAR | APR | MAY | JUN | JUL | AUG | SEP | OCT | NOV | DEC | YEAR |
|---|---|---|---|---|---|---|---|---|---|---|---|---|---|
| Maximum Temp °F | 40.1 | 46.3 | 54.4 | 61.6 | 70.0 | 78.5 | 88.0 | 87.3 | 77.5 | 65.4 | 49.5 | 41.2 | 63.3 |
| Minimum Temp °F | 24.9 | 28.9 | 32.3 | 36.1 | 41.8 | 47.7 | 51.6 | 51.0 | 42.9 | 34.4 | 31.2 | 25.7 | 37.4 |
| Mean Temp °F | 32.5 | 37.7 | 43.4 | 48.9 | 56.0 | 63.2 | 69.8 | 69.2 | 60.2 | 49.9 | 40.4 | 33.5 | 50.4 |
| Days Max Temp ≥ 90 °F | 0 | 0 | 0 | 0 | 1 | 5 | 15 | 13 | 3 | 0 | 0 | 0 | 37 |
| Days Max Temp ≤ 32 °F | 8 | 3 | 0 | 0 | 0 | 0 | 0 | 0 | 0 | 0 | 2 | 7 | 20 |
| Days Min Temp ≤ 32 °F | 24 | 19 | 16 | 9 | 2 | 0 | 0 | 0 | 2 | 13 | 16 | 23 | 124 |
| Days Min Temp ≤ 0 °F | 2 | 1 | 0 | 0 | 0 | 0 | 0 | 0 | 0 | 0 | 0 | 1 | 4 |
| Heating Degree Days | 1001 | 766 | 663 | 478 | 287 | 117 | 28 | 31 | 171 | 462 | 732 | 971 | 5707 |
| Cooling Degree Days | 0 | 0 | 0 | 1 | 17 | 68 | 175 | 164 | 37 | 2 | 0 | 0 | 464 |
| Total Precipitation (") | 2.04 | 1.45 | 1.70 | 1.53 | 1.48 | 1.06 | 0.44 | 0.73 | 0.76 | 1.20 | 2.15 | 1.88 | 16.42 |
| Days ≥ 0.1" Precip | 6 | 5 | 6 | 5 | 4 | 3 | 1 | 2 | 2 | 3 | 7 | 6 | 50 |
| Total Snowfall (") | 6.5 | 3.1 | 0.8 | 0.1 | 0.0 | 0.0 | 0.0 | 0.0 | 0.0 | 0.2 | 2.4 | 5.6 | 18.7 |
| Days ≥ 1" Snow Depth | 11 | 4 | 1 | 0 | 0 | 0 | 0 | 0 | 0 | 0 | 2 | 7 | 25 |

## PENDLETON MUNICPL AP *Umatilla County* ELEVATION 1493 ft LAT/LONG 45° 41 ' N / 118° 51 ' W

| | JAN | FEB | MAR | APR | MAY | JUN | JUL | AUG | SEP | OCT | NOV | DEC | YEAR |
|---|---|---|---|---|---|---|---|---|---|---|---|---|---|
| Maximum Temp °F | 39.7 | 46.1 | 54.4 | 61.5 | 70.4 | 79.0 | 87.4 | 86.1 | 76.5 | 63.9 | 48.5 | 40.5 | 62.8 |
| Minimum Temp °F | 27.3 | 31.1 | 35.3 | 39.6 | 46.1 | 52.8 | 58.0 | 57.7 | 49.9 | 41.0 | 33.8 | 27.9 | 41.7 |
| Mean Temp °F | 33.5 | 38.6 | 44.8 | 50.6 | 58.3 | 65.9 | 72.7 | 72.0 | 63.2 | 52.5 | 41.2 | 34.2 | 52.3 |
| Days Max Temp ≥ 90 °F | 0 | 0 | 0 | 0 | 1 | 5 | 14 | 12 | 3 | 0 | 0 | 0 | 35 |
| Days Max Temp ≤ 32 °F | 8 | 3 | 0 | 0 | 0 | 0 | 0 | 0 | 0 | 0 | 2 | 8 | 21 |
| Days Min Temp ≤ 32 °F | 20 | 15 | 9 | 3 | 0 | 0 | 0 | 0 | 0 | 3 | 11 | 20 | 81 |
| Days Min Temp ≤ 0 °F | 1 | 0 | 0 | 0 | 0 | 0 | 0 | 0 | 0 | 0 | 0 | 1 | 2 |
| Heating Degree Days | 970 | 739 | 618 | 427 | 226 | 72 | 12 | 13 | 113 | 385 | 708 | 948 | 5231 |
| Cooling Degree Days | 0 | 0 | 0 | 2 | 27 | 87 | 218 | 213 | 62 | 4 | 0 | 0 | 613 |
| Total Precipitation (") | 1.55 | 1.12 | 1.12 | 1.04 | 1.08 | 0.74 | 0.41 | 0.59 | 0.56 | 0.87 | 1.57 | 1.49 | 12.14 |
| Days ≥ 0.1" Precip | 5 | 4 | 4 | 3 | 3 | 2 | 1 | 1 | 2 | 3 | 5 | 5 | 38 |
| Total Snowfall (") | 6.3 | 3.1 | 1.1 | 0.1 | 0.0 | 0.0 | 0.0 | 0.0 | 0.0 | 0.3 | 2.0 | 5.2 | 18.1 |
| Days ≥ 1" Snow Depth | 9 | 3 | 1 | 0 | 0 | 0 | 0 | 0 | 0 | 0 | 2 | 6 | 21 |

## PILOT ROCK 1 SE *Umatilla County* ELEVATION 1640 ft LAT/LONG 45° 29' N / 118° 50' W

| | JAN | FEB | MAR | APR | MAY | JUN | JUL | AUG | SEP | OCT | NOV | DEC | YEAR |
|---|---|---|---|---|---|---|---|---|---|---|---|---|---|
| Maximum Temp °F | 42.1 | 47.8 | 55.2 | 61.9 | 70.6 | 79.4 | 88.5 | 87.8 | 78.2 | 65.9 | 50.9 | 42.8 | 64.3 |
| Minimum Temp °F | 25.4 | 28.7 | 32.4 | 36.0 | 42.0 | 48.2 | 51.7 | 51.7 | 44.5 | 36.7 | 31.2 | 25.9 | 37.9 |
| Mean Temp °F | 33.8 | 38.2 | 43.8 | 49.0 | 56.3 | 63.8 | 70.1 | 69.8 | 61.4 | 51.4 | 41.1 | 34.4 | 51.1 |
| Days Max Temp ≥ 90 °F | 0 | 0 | 0 | 0 | 1 | 5 | 15 | 14 | 4 | 0 | 0 | 0 | 39 |
| Days Max Temp ≤ 32 °F | 7 | 3 | 0 | 0 | 0 | 0 | 0 | 0 | 0 | 0 | 2 | 7 | 19 |
| Days Min Temp ≤ 32 °F | 23 | 19 | 15 | 9 | 2 | 0 | 0 | 0 | 1 | 8 | 16 | 23 | 116 |
| Days Min Temp ≤ 0 °F | 2 | 1 | 0 | 0 | 0 | 0 | 0 | 0 | 0 | 0 | 0 | 1 | 4 |
| Heating Degree Days | 962 | 748 | 649 | 475 | 278 | 108 | 27 | 27 | 146 | 419 | 711 | 943 | 5493 |
| Cooling Degree Days | 0 | 0 | 0 | 2 | 19 | 72 | 175 | 178 | 47 | 3 | 0 | 0 | 496 |
| Total Precipitation (") | 1.49 | 1.03 | 1.41 | 1.37 | 1.42 | 1.25 | 0.40 | 0.79 | 0.69 | 0.90 | 1.67 | 1.38 | 13.80 |
| Days ≥ 0.1" Precip | 5 | 4 | 5 | 4 | 4 | 4 | 1 | 2 | 2 | 3 | 6 | 5 | 45 |
| Total Snowfall (") | 6.1 | 3.4 | 1.8 | 0.3 | 0.0 | 0.0 | 0.0 | 0.0 | 0.0 | 0.2 | 3.0 | 5.6 | 20.4 |
| Days ≥ 1" Snow Depth | 7 | 3 | 1 | 0 | 0 | 0 | 0 | 0 | 0 | 0 | 2 | 5 | 18 |

## PORTLAND HEADWORKS W *Clackamas County* ELEVATION 748 ft LAT/LONG 45° 27' N / 122° 9' W

| | JAN | FEB | MAR | APR | MAY | JUN | JUL | AUG | SEP | OCT | NOV | DEC | YEAR |
|---|---|---|---|---|---|---|---|---|---|---|---|---|---|
| Maximum Temp °F | 45.2 | 49.7 | 54.5 | 59.5 | 66.9 | 72.2 | 78.4 | 78.9 | 73.7 | 63.7 | 51.3 | 45.0 | 61.6 |
| Minimum Temp °F | 34.0 | 35.9 | 37.3 | 39.9 | 44.3 | 49.0 | 52.3 | 53.2 | 50.1 | 44.8 | 39.1 | 34.6 | 42.9 |
| Mean Temp °F | 39.6 | 42.8 | 46.0 | 49.7 | 55.6 | 60.6 | 65.4 | 66.1 | 61.9 | 54.3 | 45.2 | 39.8 | 52.3 |
| Days Max Temp ≥ 90 °F | 0 | 0 | 0 | 0 | 1 | 1 | 3 | 4 | 2 | 0 | 0 | 0 | 11 |
| Days Max Temp ≤ 32 °F | 1 | 0 | 0 | 0 | 0 | 0 | 0 | 0 | 0 | 0 | 0 | 1 | 2 |
| Days Min Temp ≤ 32 °F | 12 | 8 | 5 | 1 | 0 | 0 | 0 | 0 | 0 | 0 | 4 | 10 | 40 |
| Days Min Temp ≤ 0 °F | 0 | 0 | 0 | 0 | 0 | 0 | 0 | 0 | 0 | 0 | 0 | 0 | 0 |
| Heating Degree Days | 780 | 621 | 584 | 453 | 293 | 150 | 55 | 45 | 122 | 331 | 588 | 773 | 4795 |
| Cooling Degree Days | 0 | 0 | 0 | 1 | 15 | 29 | 80 | 98 | 43 | 8 | 0 | 0 | 274 |
| Total Precipitation (") | 10.91 | 8.34 | 7.89 | 6.93 | 4.83 | 3.90 | 1.60 | 2.04 | 3.88 | 6.19 | 10.47 | 11.41 | 78.39 |
| Days ≥ 0.1" Precip | 16 | 14 | 16 | 14 | 10 | 8 | 4 | 4 | 7 | 11 | 16 | 17 | 137 |
| Total Snowfall (") | 4.6 | 2.2 | 1.1 | 0.2 | 0.1 | 0.0 | 0.0 | 0.0 | 0.0 | 0.0 | 1.5 | 2.5 | 12.2 |
| Days ≥ 1" Snow Depth | 3 | 2 | 1 | 0 | 0 | 0 | 0 | 0 | 0 | 0 | 1 | 2 | 9 |

## PORTLAND INTL ARPT *Multnomah County* ELEVATION 39 ft LAT/LONG 45° 36' N / 122° 36' W

| | JAN | FEB | MAR | APR | MAY | JUN | JUL | AUG | SEP | OCT | NOV | DEC | YEAR |
|---|---|---|---|---|---|---|---|---|---|---|---|---|---|
| Maximum Temp °F | 45.6 | 51.0 | 56.8 | 61.0 | 68.0 | 74.0 | 80.2 | 80.7 | 75.3 | 64.4 | 52.6 | 45.8 | 62.9 |
| Minimum Temp °F | 34.1 | 36.1 | 38.8 | 41.9 | 47.5 | 53.1 | 56.8 | 57.1 | 52.2 | 45.1 | 39.6 | 35.0 | 44.8 |
| Mean Temp °F | 39.8 | 43.6 | 47.8 | 51.5 | 57.7 | 63.6 | 68.5 | 68.9 | 63.8 | 54.8 | 46.1 | 40.4 | 53.9 |
| Days Max Temp ≥ 90 °F | 0 | 0 | 0 | 0 | 0 | 1 | 4 | 5 | 2 | 0 | 0 | 0 | 12 |
| Days Max Temp ≤ 32 °F | 2 | 0 | 0 | 0 | 0 | 0 | 0 | 0 | 0 | 0 | 0 | 1 | 3 |
| Days Min Temp ≤ 32 °F | 12 | 8 | 4 | 1 | 0 | 0 | 0 | 0 | 0 | 1 | 4 | 10 | 40 |
| Days Min Temp ≤ 0 °F | 0 | 0 | 0 | 0 | 0 | 0 | 0 | 0 | 0 | 0 | 0 | 0 | 0 |
| Heating Degree Days | 774 | 599 | 526 | 400 | 232 | 87 | 20 | 15 | 81 | 312 | 559 | 755 | 4360 |
| Cooling Degree Days | 0 | 0 | 0 | 1 | 18 | 50 | 132 | 152 | 59 | 3 | 0 | 0 | 415 |
| Total Precipitation (") | 5.21 | 3.75 | 3.41 | 2.49 | 2.03 | 1.52 | 0.66 | 1.01 | 1.67 | 2.82 | 5.07 | 6.03 | 35.67 |
| Days ≥ 0.1" Precip | 11 | 9 | 10 | 8 | 6 | 4 | 2 | 2 | 4 | 7 | 12 | 12 | 87 |
| Total Snowfall (") | 1.8 | 1.1 | 0.1 | 0.0 | 0.0 | 0.0 | 0.0 | 0.0 | 0.0 | 0.0 | 0.5 | 1.8 | 5.3 |
| Days ≥ 1" Snow Depth | 1 | 0 | 0 | 0 | 0 | 0 | 0 | 0 | 0 | 0 | 0 | 1 | 2 |

## POWERS *Coos County* ELEVATION 302 ft LAT/LONG 42° 53' N / 124° 4' W

| | JAN | FEB | MAR | APR | MAY | JUN | JUL | AUG | SEP | OCT | NOV | DEC | YEAR |
|---|---|---|---|---|---|---|---|---|---|---|---|---|---|
| Maximum Temp °F | 52.8 | 56.4 | 58.7 | 61.9 | 67.6 | 72.5 | 78.0 | 79.2 | 77.1 | 69.1 | 57.8 | 52.2 | 65.3 |
| Minimum Temp °F | 34.3 | 35.8 | 37.6 | 39.9 | 43.9 | 48.4 | 51.0 | 50.6 | 46.7 | 42.2 | 38.6 | 34.8 | 42.0 |
| Mean Temp °F | 43.6 | 46.1 | 48.2 | 50.9 | 55.8 | 60.5 | 64.5 | 64.9 | 61.9 | 55.7 | 48.2 | 43.5 | 53.7 |
| Days Max Temp ≥ 90 °F | 0 | 0 | 0 | 0 | 0 | 1 | 1 | 2 | 2 | 1 | 0 | 0 | 7 |
| Days Max Temp ≤ 32 °F | 0 | 0 | 0 | 0 | 0 | 0 | 0 | 0 | 0 | 0 | 0 | 0 | 0 |
| Days Min Temp ≤ 32 °F | 14 | 9 | 6 | 2 | 0 | 0 | 0 | 0 | 0 | 1 | 6 | 11 | 49 |
| Days Min Temp ≤ 0 °F | 0 | 0 | 0 | 0 | 0 | 0 | 0 | 0 | 0 | 0 | 0 | 0 | 0 |
| Heating Degree Days | 657 | 527 | 515 | 416 | 285 | 145 | 54 | 47 | 110 | 285 | 496 | 659 | 4196 |
| Cooling Degree Days | 0 | 0 | 0 | 1 | 8 | 15 | 46 | 53 | 29 | 6 | 0 | 0 | 158 |
| Total Precipitation (") | 9.45 | 7.19 | 7.38 | 4.96 | 2.46 | 1.04 | 0.35 | 0.74 | 1.68 | 3.41 | 8.78 | 10.47 | 57.91 |
| Days ≥ 0.1" Precip | 13 | 12 | 13 | 11 | 6 | 3 | 1 | 2 | 3 | 7 | 14 | 14 | 99 |
| Total Snowfall (") | 1.1 | 0.7 | 0.1 | 0.0 | 0.0 | 0.0 | 0.0 | 0.0 | 0.0 | 0.0 | 0.0 | 0.5 | 2.4 |
| Days ≥ 1" Snow Depth | 1 | 0 | 0 | 0 | 0 | 0 | 0 | 0 | 0 | 0 | 0 | 0 | 1 |

## PRINEVILLE 4 NW *Crook County* ELEVATION 2871 ft LAT/LONG 44° 19 ' N / 120° 53 ' W

| | JAN | FEB | MAR | APR | MAY | JUN | JUL | AUG | SEP | OCT | NOV | DEC | YEAR |
|---|---|---|---|---|---|---|---|---|---|---|---|---|---|
| Maximum Temp °F | 42.8 | 48.9 | 55.3 | 61.5 | 70.2 | 78.2 | 86.7 | 86.1 | 78.0 | 66.4 | 50.3 | 42.6 | 63.9 |
| Minimum Temp °F | 21.8 | 24.4 | 25.8 | 28.5 | 34.8 | 40.8 | 43.3 | 42.3 | 35.5 | 29.6 | 26.2 | 21.4 | 31.2 |
| Mean Temp °F | 32.3 | 36.7 | 40.6 | 45.0 | 52.5 | 59.5 | 65.0 | 64.2 | 56.8 | 48.0 | 38.3 | 32.0 | 47.6 |
| Days Max Temp ≥ 90 °F | 0 | 0 | 0 | 0 | 1 | 5 | 13 | 12 | 4 | 0 | 0 | 0 | 35 |
| Days Max Temp ≤ 32 °F | 4 | 1 | 0 | 0 | 0 | 0 | 0 | 0 | 0 | 0 | 1 | 4 | 10 |
| Days Min Temp ≤ 32 °F | 27 | 23 | 25 | 21 | 12 | 3 | 1 | 1 | 10 | 21 | 22 | 27 | 193 |
| Days Min Temp ≤ 0 °F | 1 | 1 | 0 | 0 | 0 | 0 | 0 | 0 | 0 | 0 | 0 | 1 | 3 |
| Heating Degree Days | 1006 | 793 | 750 | 593 | 384 | 182 | 69 | 81 | 246 | 518 | 795 | 1015 | 6432 |
| Cooling Degree Days | 0 | 0 | 0 | 0 | 6 | 26 | 67 | 62 | 8 | 0 | 0 | 0 | 169 |
| Total Precipitation (") | 1.16 | 0.84 | 0.89 | 0.76 | 0.88 | 0.97 | 0.51 | 0.53 | 0.43 | 0.76 | 1.33 | 1.17 | 10.23 |
| Days ≥ 0.1" Precip | 4 | 3 | 3 | 2 | 3 | 2 | 1 | 1 | 2 | 2 | 4 | 4 | 31 |
| Total Snowfall (") | *3.2* | 1.9 | 0.5 | 0.4 | 0.0 | 0.0 | 0.0 | 0.0 | 0.0 | 0.1 | *2.1* | 3.5 | 11.7 |
| Days ≥ 1" Snow Depth | *4* | 1 | 0 | 0 | 0 | 0 | 0 | 0 | 0 | 0 | na | *3* | na |

## PROSPECT 2 SW *Jackson County* ELEVATION 2480 ft LAT/LONG 42° 44 ' N / 122° 31 ' W

| | JAN | FEB | MAR | APR | MAY | JUN | JUL | AUG | SEP | OCT | NOV | DEC | YEAR |
|---|---|---|---|---|---|---|---|---|---|---|---|---|---|
| Maximum Temp °F | 46.8 | 52.1 | 56.3 | 61.6 | 70.6 | 78.2 | 86.7 | 86.9 | 80.9 | 69.4 | 52.0 | 45.4 | 65.6 |
| Minimum Temp °F | 28.2 | 29.8 | 31.6 | 33.8 | 38.6 | 44.4 | 48.0 | 47.4 | 42.2 | 36.6 | 32.6 | 28.6 | 36.8 |
| Mean Temp °F | 37.5 | 41.0 | 44.0 | 47.7 | 54.6 | 61.3 | 67.4 | 67.2 | 61.6 | 53.0 | 42.3 | 37.0 | 51.2 |
| Days Max Temp ≥ 90 °F | 0 | 0 | 0 | 0 | 1 | 4 | 13 | 13 | 7 | 1 | 0 | 0 | 39 |
| Days Max Temp ≤ 32 °F | 0 | 0 | 0 | 0 | 0 | 0 | 0 | 0 | 0 | 0 | 0 | 1 | 1 |
| Days Min Temp ≤ 32 °F | 23 | 20 | 19 | 14 | 6 | 1 | 0 | 0 | 2 | 8 | 17 | 23 | 133 |
| Days Min Temp ≤ 0 °F | 0 | 0 | 0 | 0 | 0 | 0 | 0 | 0 | 0 | 0 | 0 | 0 | 0 |
| Heating Degree Days | 845 | 672 | 645 | 512 | 324 | 148 | 44 | 42 | 132 | 369 | 675 | 860 | 5268 |
| Cooling Degree Days | 0 | 0 | 0 | 1 | 11 | 45 | 114 | 110 | 45 | 6 | 0 | 0 | 332 |
| Total Precipitation (") | 6.00 | 4.53 | 4.53 | 2.87 | 2.23 | 1.08 | 0.56 | 0.95 | 1.35 | 3.04 | 6.48 | 6.30 | 39.92 |
| Days ≥ 0.1" Precip | 12 | 10 | 10 | 8 | 6 | 3 | 1 | 2 | 3 | 6 | 12 | 12 | 85 |
| Total Snowfall (") | 14.2 | 7.5 | 6.3 | 2.5 | 0.1 | 0.0 | 0.0 | 0.0 | 0.1 | 0.3 | 3.1 | 12.4 | 46.5 |
| Days ≥ 1" Snow Depth | 12 | 5 | 2 | 0 | 0 | 0 | 0 | 0 | 0 | 0 | 2 | 8 | 29 |

## REDMOND ROBERTS FLD *Deschutes County* ELEVATION 3084 ft LAT/LONG 44° 16 ' N / 121° 9 ' W

| | JAN | FEB | MAR | APR | MAY | JUN | JUL | AUG | SEP | OCT | NOV | DEC | YEAR |
|---|---|---|---|---|---|---|---|---|---|---|---|---|---|
| Maximum Temp °F | 41.1 | 46.6 | 53.1 | 59.3 | 67.8 | 76.3 | 84.9 | 83.9 | 75.2 | 64.4 | 48.6 | 41.3 | 61.9 |
| Minimum Temp °F | 22.0 | 24.9 | 26.6 | 29.7 | 35.6 | 42.4 | 46.8 | 46.5 | 39.1 | 32.5 | 27.2 | 21.9 | 32.9 |
| Mean Temp °F | 31.6 | 35.8 | 39.9 | 44.5 | 51.7 | 59.4 | 65.9 | 65.3 | 57.2 | 48.5 | 37.9 | 31.6 | 47.4 |
| Days Max Temp ≥ 90 °F | 0 | 0 | 0 | 0 | 1 | 4 | 11 | 10 | 2 | 0 | 0 | 0 | 28 |
| Days Max Temp ≤ 32 °F | 6 | 3 | 0 | 0 | 0 | 0 | 0 | 0 | 0 | 0 | 2 | 5 | 16 |
| Days Min Temp ≤ 32 °F | 26 | 23 | 25 | 20 | 11 | 2 | 0 | 0 | 6 | 15 | 22 | 27 | 177 |
| Days Min Temp ≤ 0 °F | 2 | 1 | 0 | 0 | 0 | 0 | 0 | 0 | 0 | 0 | 0 | 1 | 4 |
| Heating Degree Days | 1030 | 819 | 771 | 608 | 410 | 197 | 73 | 77 | 244 | 506 | 806 | 1027 | 6568 |
| Cooling Degree Days | 0 | 0 | 0 | 0 | 9 | 33 | 90 | 86 | 17 | 1 | 0 | 0 | 236 |
| Total Precipitation (") | 1.02 | 0.59 | 0.70 | 0.61 | 0.76 | 0.73 | 0.49 | 0.50 | 0.41 | 0.52 | 1.04 | 0.90 | 8.27 |
| Days ≥ 0.1" Precip | 3 | 2 | 3 | 2 | 2 | 2 | 2 | 2 | 1 | 2 | 3 | 3 | 27 |
| Total Snowfall (") | 5.2 | 3.2 | 1.8 | 1.0 | 0.0 | 0.1 | 0.0 | 0.0 | 0.0 | 0.2 | 3.0 | 5.3 | 19.8 |
| Days ≥ 1" Snow Depth | 8 | 4 | 1 | 0 | 0 | 0 | 0 | 0 | 0 | 0 | 2 | 7 | 22 |

## RICHLAND *Baker County* ELEVATION 2221 ft LAT/LONG 44° 46 ' N / 117° 10 ' W

| | JAN | FEB | MAR | APR | MAY | JUN | JUL | AUG | SEP | OCT | NOV | DEC | YEAR |
|---|---|---|---|---|---|---|---|---|---|---|---|---|---|
| Maximum Temp °F | *38.9* | 46.0 | 56.1 | 66.6 | 75.3 | 83.2 | 91.0 | 91.0 | 81.7 | 68.6 | 50.8 | 39.8 | 65.7 |
| Minimum Temp °F | *21.0* | 24.6 | 29.7 | 35.0 | 41.3 | 48.4 | 52.6 | *51.9* | 43.2 | 33.5 | 26.8 | 21.0 | 35.8 |
| Mean Temp °F | *30.0* | 35.3 | 42.9 | 50.8 | 58.3 | 65.8 | 71.7 | *71.6* | 62.5 | 51.1 | 38.8 | 30.4 | 50.8 |
| Days Max Temp ≥ 90 °F | 0 | 0 | 0 | 0 | 2 | *8* | 20 | 19 | 7 | 0 | 0 | 0 | 56 |
| Days Max Temp ≤ 32 °F | *6* | 2 | 0 | 0 | 0 | 0 | 0 | 0 | 0 | 0 | 1 | 5 | 14 |
| Days Min Temp ≤ 32 °F | *28* | 24 | 21 | 12 | 3 | *0* | 0 | *0* | 3 | 14 | 24 | *29* | 158 |
| Days Min Temp ≤ 0 °F | *2* | 1 | 0 | 0 | 0 | 0 | 0 | *0* | 0 | 0 | 0 | 1 | 4 |
| Heating Degree Days | *1080* | 832 | 679 | 424 | 223 | 75 | 27 | *14* | 122 | 427 | 778 | 1065 | 5746 |
| Cooling Degree Days | *0* | *0* | 0 | 8 | *30* | *89* | *197* | na | *58* | *2* | *0* | *0* | na |
| Total Precipitation (") | *1.60* | 0.90 | 1.04 | 1.05 | 1.19 | 1.14 | 0.64 | 0.87 | 0.53 | 0.75 | 1.56 | 1.31 | 12.58 |
| Days ≥ 0.1" Precip | *5* | 3 | 4 | 3 | 4 | *3* | 2 | 2 | 2 | 3 | 5 | *4* | 40 |
| Total Snowfall (") | *6.8* | *3.1* | *0.4* | 0.1 | 0.0 | 0.0 | 0.0 | 0.0 | 0.0 | 0.1 | 2.8 | na | na |
| Days ≥ 1" Snow Depth | *7* | *3* | *0* | 0 | 0 | 0 | 0 | 0 | 0 | 0 | 1 | *6* | 17 |

## RIDDLE *Douglas County* ELEVATION 659 ft LAT/LONG 42° 58 ' N / 123° 21 ' W

| | JAN | FEB | MAR | APR | MAY | JUN | JUL | AUG | SEP | OCT | NOV | DEC | YEAR |
|---|---|---|---|---|---|---|---|---|---|---|---|---|---|
| Maximum Temp °F | 49.4 | 54.9 | 59.3 | 63.4 | 70.1 | 76.7 | 83.6 | 84.0 | 79.3 | 68.9 | 54.8 | 48.3 | 66.1 |
| Minimum Temp °F | 34.2 | 35.6 | 37.4 | 39.4 | 44.0 | 49.4 | 52.9 | 52.4 | 46.8 | 42.0 | 39.1 | 34.8 | 42.3 |
| Mean Temp °F | 41.8 | 45.3 | 48.4 | 51.4 | 57.1 | 63.1 | 68.3 | 68.2 | 63.1 | 55.5 | 46.9 | 41.6 | 54.2 |
| Days Max Temp ≥ 90 °F | 0 | 0 | 0 | 0 | 1 | 3 | 7 | 8 | 5 | 1 | 0 | 0 | 25 |
| Days Max Temp ≤ 32 °F | 0 | 0 | 0 | 0 | 0 | 0 | 0 | 0 | 0 | 0 | 0 | 0 | 0 |
| Days Min Temp ≤ 32 °F | 12 | 8 | 6 | 3 | 0 | 0 | 0 | 0 | 0 | 2 | 4 | 10 | 45 |
| Days Min Temp ≤ 0 °F | 0 | 0 | 0 | 0 | 0 | 0 | 0 | 0 | 0 | 0 | 0 | 0 | 0 |
| Heating Degree Days | 713 | 550 | 508 | 401 | 252 | 101 | 22 | 22 | 99 | 293 | 535 | 720 | 4216 |
| Cooling Degree Days | 0 | 0 | 0 | 2 | 20 | 54 | 131 | 128 | 51 | 6 | 0 | 0 | 392 |
| Total Precipitation (") | 4.72 | 3.29 | 3.20 | 2.12 | 1.19 | 0.72 | 0.34 | 0.67 | 1.05 | 2.05 | 5.13 | 5.35 | 29.83 |
| Days ≥ 0.1" Precip | 10 | 7 | 9 | 6 | 4 | 2 | 1 | 2 | 3 | 5 | 11 | 11 | 71 |
| Total Snowfall (") | 2.2 | 1.1 | 0.3 | 0.0 | 0.0 | 0.0 | 0.0 | 0.0 | 0.0 | 0.0 | 0.3 | 1.1 | 5.0 |
| Days ≥ 1" Snow Depth | 1 | 0 | 0 | 0 | 0 | 0 | 0 | 0 | 0 | 0 | 0 | 1 | 2 |

## RIVERSIDE 7 SSW *Malheur County* ELEVATION 3232 ft LAT/LONG 43° 34 ' N / 118° 8 ' W

| | JAN | FEB | MAR | APR | MAY | JUN | JUL | AUG | SEP | OCT | NOV | DEC | YEAR |
|---|---|---|---|---|---|---|---|---|---|---|---|---|---|
| Maximum Temp °F | 37.1 | 45.0 | 54.0 | 62.3 | 72.0 | 81.0 | 89.9 | 88.0 | 78.6 | 65.9 | 48.6 | 37.6 | 63.3 |
| Minimum Temp °F | 19.0 | 24.0 | 28.1 | 32.0 | 39.3 | 46.8 | 52.2 | 50.5 | 41.0 | 31.7 | 25.6 | 19.0 | 34.1 |
| Mean Temp °F | 28.1 | 34.5 | 41.1 | 47.2 | 55.6 | 63.9 | 71.0 | 69.3 | 59.8 | 48.8 | 37.1 | 28.3 | 48.7 |
| Days Max Temp ≥ 90 °F | 0 | 0 | 0 | 0 | 1 | 7 | 18 | 15 | 4 | 0 | 0 | 0 | 45 |
| Days Max Temp ≤ 32 °F | 10 | 2 | 0 | 0 | 0 | 0 | 0 | 0 | 0 | 0 | 1 | 8 | 21 |
| Days Min Temp ≤ 32 °F | 27 | 23 | 22 | *16* | 6 | 1 | 0 | 0 | 4 | 16 | *23* | 28 | 166 |
| Days Min Temp ≤ 0 °F | 3 | 1 | 0 | 0 | 0 | 0 | 0 | 0 | 0 | 0 | 0 | 3 | 7 |
| Heating Degree Days | 1138 | 853 | 734 | 529 | 293 | 106 | 20 | 31 | 179 | 496 | 829 | 1131 | 6339 |
| Cooling Degree Days | 0 | 0 | 0 | 1 | 12 | 79 | 201 | 160 | 26 | 0 | 0 | 0 | 479 |
| Total Precipitation (") | 1.05 | 0.82 | 1.04 | 0.77 | 0.93 | 0.91 | 0.48 | 0.54 | 0.46 | 0.63 | 1.04 | 1.14 | 9.81 |
| Days ≥ 0.1" Precip | 3 | 3 | 3 | 3 | 3 | 3 | 1 | 2 | 1 | 2 | 4 | *4* | 32 |
| Total Snowfall (") | 4.8 | 2.6 | 1.2 | 0.3 | 0.0 | 0.0 | 0.0 | 0.0 | 0.0 | 0.0 | 2.1 | 6.2 | 17.2 |
| Days ≥ 1" Snow Depth | 12 | 3 | 1 | 0 | 0 | 0 | 0 | 0 | 0 | 0 | 2 | 10 | 28 |

## ROCKVILLE 5 N *Malheur County* ELEVATION 3671 ft LAT/LONG 43° 22 ' N / 117° 7 ' W

| | JAN | FEB | MAR | APR | MAY | JUN | JUL | AUG | SEP | OCT | NOV | DEC | YEAR |
|---|---|---|---|---|---|---|---|---|---|---|---|---|---|
| Maximum Temp °F | 37.4 | 44.1 | 53.2 | 61.9 | 70.9 | 79.3 | 88.2 | 87.0 | 77.0 | 64.1 | 47.7 | 38.3 | 62.4 |
| Minimum Temp °F | 18.2 | 22.2 | 25.4 | 29.4 | 35.9 | 42.3 | 46.3 | 45.0 | 36.4 | 28.5 | 23.5 | 18.0 | 30.9 |
| Mean Temp °F | 27.8 | 33.2 | 39.3 | 45.7 | 53.5 | 60.8 | 67.3 | 66.0 | 56.7 | 46.3 | 35.6 | 28.2 | 46.7 |
| Days Max Temp ≥ 90 °F | 0 | 0 | 0 | 0 | 1 | 4 | 16 | 13 | 2 | 0 | 0 | 0 | 36 |
| Days Max Temp ≤ 32 °F | 9 | 3 | 0 | 0 | 0 | 0 | 0 | 0 | 0 | 0 | 1 | 8 | 21 |
| Days Min Temp ≤ 32 °F | 28 | 25 | 25 | 20 | 11 | 3 | 0 | 1 | 9 | 21 | 24 | 28 | 195 |
| Days Min Temp ≤ 0 °F | 3 | 1 | 0 | 0 | 0 | 0 | 0 | 0 | 0 | 0 | 1 | 3 | 8 |
| Heating Degree Days | 1146 | 892 | 789 | 574 | 355 | 156 | 40 | 52 | 252 | 574 | 875 | 1134 | 6839 |
| Cooling Degree Days | 0 | 0 | 0 | 0 | 6 | 43 | 110 | 100 | 10 | 1 | 0 | 0 | 270 |
| Total Precipitation (") | 0.98 | 0.84 | 1.26 | 1.30 | 1.41 | 1.27 | 0.49 | 0.60 | 0.64 | 0.84 | 1.22 | 1.04 | 11.89 |
| Days ≥ 0.1" Precip | 5 | 4 | 5 | 5 | 5 | 4 | 2 | 2 | 2 | 3 | 5 | 5 | 47 |
| Total Snowfall (") | 4.9 | 3.0 | 2.0 | 0.7 | 0.2 | 0.0 | 0.0 | 0.0 | 0.0 | 0.2 | 2.1 | 5.5 | 18.6 |
| Days ≥ 1" Snow Depth | 8 | 3 | 1 | 0 | 0 | 0 | 0 | 0 | 0 | 0 | 2 | 7 | 21 |

## ROME 2 NW *Malheur County* ELEVATION 3402 ft LAT/LONG 42° 50 ' N / 117° 38 ' W

| | JAN | FEB | MAR | APR | MAY | JUN | JUL | AUG | SEP | OCT | NOV | DEC | YEAR |
|---|---|---|---|---|---|---|---|---|---|---|---|---|---|
| Maximum Temp °F | 39.8 | 47.6 | 55.4 | 63.4 | 73.4 | 82.6 | 92.0 | 90.6 | 80.3 | 67.9 | 50.7 | 40.4 | 65.3 |
| Minimum Temp °F | 17.7 | 23.0 | 26.0 | 30.5 | 38.7 | 46.2 | 51.4 | 48.6 | 38.8 | 30.2 | 23.7 | 18.0 | 32.7 |
| Mean Temp °F | 28.8 | 35.3 | 40.8 | 47.0 | 56.1 | 64.4 | 71.7 | 69.6 | 59.5 | 49.1 | 37.3 | 29.2 | 49.1 |
| Days Max Temp ≥ 90 °F | 0 | 0 | 0 | 0 | 2 | 8 | 21 | 19 | *5* | 0 | 0 | 0 | 55 |
| Days Max Temp ≤ 32 °F | 6 | 2 | 0 | 0 | 0 | 0 | 0 | 0 | 0 | 0 | 1 | 6 | 15 |
| Days Min Temp ≤ 32 °F | 29 | 25 | 24 | 18 | 6 | 1 | 0 | 0 | *5* | 19 | 24 | 29 | 180 |
| Days Min Temp ≤ 0 °F | 3 | 1 | 0 | 0 | 0 | 0 | 0 | 0 | 0 | 0 | 0 | 3 | 7 |
| Heating Degree Days | 1116 | 832 | 744 | 534 | 282 | 96 | 14 | 23 | 181 | 487 | 825 | 1100 | 6234 |
| Cooling Degree Days | 0 | 0 | *0* | 1 | 17 | 86 | 206 | 172 | 25 | 0 | 0 | 0 | 507 |
| Total Precipitation (") | 0.66 | 0.47 | 0.79 | 0.75 | 0.99 | 0.96 | 0.36 | 0.42 | 0.54 | 0.49 | 0.83 | 0.68 | 7.94 |
| Days ≥ 0.1" Precip | 2 | 2 | 3 | 3 | 3 | 3 | 1 | 1 | 1 | 2 | 3 | *2* | 26 |
| Total Snowfall (") | 4.3 | 0.7 | *1.3* | 0.3 | 0.2 | 0.0 | 0.0 | 0.0 | 0.0 | 0.2 | *1.7* | 3.8 | 12.5 |
| Days ≥ 1" Snow Depth | 9 | 2 | 1 | 0 | 0 | 0 | 0 | 0 | 0 | 0 | 1 | *6* | 19 |

## ROSEBURG KQEN *Douglas County* ELEVATION 469 ft LAT/LONG 43° 12 ' N / 123° 21 ' W

| | JAN | FEB | MAR | APR | MAY | JUN | JUL | AUG | SEP | OCT | NOV | DEC | YEAR |
|---|---|---|---|---|---|---|---|---|---|---|---|---|---|
| Maximum Temp °F | 49.0 | 53.5 | 58.5 | 63.1 | 70.0 | 76.6 | 83.8 | 84.4 | 78.8 | 67.7 | 54.5 | 48.1 | 65.7 |
| Minimum Temp °F | 34.6 | 36.0 | 37.9 | 39.7 | 44.7 | 50.4 | 53.8 | 54.1 | 49.3 | 43.4 | 39.3 | 35.0 | 43.2 |
| Mean Temp °F | 41.8 | 44.8 | 48.2 | 51.4 | 57.4 | 63.5 | 68.8 | 69.3 | 64.1 | 55.6 | 46.9 | 41.6 | 54.5 |
| Days Max Temp ≥ 90 °F | 0 | 0 | 0 | 0 | 1 | 3 | 8 | 8 | 4 | 1 | 0 | 0 | 25 |
| Days Max Temp ≤ 32 °F | 0 | 0 | 0 | 0 | 0 | 0 | 0 | 0 | 0 | 0 | 0 | 1 | 1 |
| Days Min Temp ≤ 32 °F | 12 | 8 | 5 | 3 | 0 | 0 | 0 | 0 | 0 | 1 | 4 | 9 | 42 |
| Days Min Temp ≤ 0 °F | 0 | 0 | 0 | 0 | 0 | 0 | 0 | 0 | 0 | 0 | 0 | 0 | 0 |
| Heating Degree Days | 713 | 563 | 514 | 402 | 245 | 95 | 22 | 14 | 83 | 292 | 536 | 719 | 4198 |
| Cooling Degree Days | 0 | 0 | 0 | 2 | 21 | 59 | 152 | 160 | 70 | 10 | 0 | 0 | 474 |
| Total Precipitation (") | 4.88 | 3.58 | 3.57 | 2.38 | 1.50 | 0.90 | 0.46 | 0.69 | 1.13 | 2.19 | 5.29 | 5.48 | 32.05 |
| Days ≥ 0.1" Precip | 10 | 9 | 10 | 7 | 5 | 3 | 1 | 2 | 3 | 6 | 12 | 11 | 79 |
| Total Snowfall (") | 2.0 | 0.6 | 0.1 | 0.0 | 0.0 | 0.0 | 0.0 | 0.0 | 0.0 | 0.0 | 0.0 | 0.6 | 3.3 |
| Days ≥ 1" Snow Depth | 0 | 1 | 0 | 0 | 0 | 0 | 0 | 0 | 0 | 0 | 0 | *0* | 1 |

## RUCH *Jackson County* ELEVATION 1552 ft LAT/LONG 42° 14 ' N / 123° 2 ' W

| | JAN | FEB | MAR | APR | MAY | JUN | JUL | AUG | SEP | OCT | NOV | DEC | YEAR |
|---|---|---|---|---|---|---|---|---|---|---|---|---|---|
| Maximum Temp °F | 48.9 | 55.2 | 60.1 | 65.8 | 74.3 | 82.1 | 89.7 | 89.0 | 82.9 | 70.5 | 54.4 | 47.0 | 68.3 |
| Minimum Temp °F | 29.6 | 31.0 | 33.2 | 35.8 | 40.8 | 46.7 | 49.8 | 49.7 | 44.6 | 38.5 | 33.9 | 30.2 | 38.7 |
| Mean Temp °F | 39.3 | 43.2 | 46.7 | 50.8 | 57.5 | 64.5 | 69.7 | 69.4 | 63.8 | 54.5 | 44.2 | 38.6 | 53.5 |
| Days Max Temp ≥ 90 °F | 0 | 0 | 0 | 0 | 2 | 8 | 17 | 16 | 9 | 1 | 0 | 0 | 53 |
| Days Max Temp ≤ 32 °F | 0 | 0 | 0 | 0 | 0 | 0 | 0 | 0 | 0 | 0 | 0 | 0 | 0 |
| Days Min Temp ≤ 32 °F | 21 | 17 | 15 | 10 | 3 | 0 | 0 | 0 | 1 | 5 | 12 | 19 | 103 |
| Days Min Temp ≤ 0 °F | 0 | 0 | 0 | 0 | 0 | 0 | 0 | 0 | 0 | 0 | 0 | 0 | 0 |
| Heating Degree Days | 790 | 610 | 561 | 420 | 241 | 85 | 18 | 17 | 91 | 323 | 618 | 811 | 4585 |
| Cooling Degree Days | 0 | 0 | 0 | 2 | 19 | 75 | 162 | 160 | 64 | 7 | 0 | 0 | 489 |
| Total Precipitation (") | 3.92 | 2.79 | 2.78 | 1.65 | 1.13 | 0.75 | 0.42 | 0.63 | 1.00 | 1.70 | 4.08 | 4.42 | 25.27 |
| Days ≥ 0.1" Precip | 8 | 7 | 7 | 5 | 4 | 3 | 1 | 2 | 2 | 5 | 9 | 9 | 62 |
| Total Snowfall (") | 4.1 | 4.3 | 2.2 | 0.2 | 0.0 | 0.0 | 0.0 | 0.0 | 0.0 | 0.0 | 1.0 | 4.5 | 16.3 |
| Days ≥ 1" Snow Depth | 1 | 1 | 0 | 0 | 0 | 0 | 0 | 0 | 0 | 0 | 0 | 1 | 3 |

## SALEM MCNARY FIELD *Marion County* ELEVATION 210 ft LAT/LONG 44° 55 ' N / 123° 0 ' W

| | JAN | FEB | MAR | APR | MAY | JUN | JUL | AUG | SEP | OCT | NOV | DEC | YEAR |
|---|---|---|---|---|---|---|---|---|---|---|---|---|---|
| Maximum Temp °F | 46.6 | 51.5 | 56.6 | 60.8 | 67.8 | 74.6 | 81.8 | 82.2 | 76.5 | 64.7 | 52.4 | 46.2 | 63.5 |
| Minimum Temp °F | 33.2 | 34.2 | 36.0 | 38.1 | 42.7 | 48.2 | 51.1 | 51.4 | 47.1 | 41.0 | 37.3 | 33.7 | 41.2 |
| Mean Temp °F | 39.9 | 42.9 | 46.3 | 49.5 | 55.3 | 61.4 | 66.5 | 66.8 | 61.8 | 52.9 | 44.9 | 40.0 | 52.4 |
| Days Max Temp ≥ 90 °F | 0 | 0 | 0 | 0 | 0 | 2 | 6 | 6 | 3 | 0 | 0 | 0 | 17 |
| Days Max Temp ≤ 32 °F | 1 | 0 | 0 | 0 | 0 | 0 | 0 | 0 | 0 | 0 | 0 | 1 | 2 |
| Days Min Temp ≤ 32 °F | 14 | 11 | 9 | 6 | 1 | 0 | 0 | 0 | 0 | 3 | 8 | 13 | 65 |
| Days Min Temp ≤ 0 °F | 0 | 0 | 0 | 0 | 0 | 0 | 0 | 0 | 0 | 0 | 0 | 0 | 0 |
| Heating Degree Days | 770 | 618 | 573 | 458 | 300 | 132 | 39 | 34 | 118 | 370 | 596 | 768 | 4776 |
| Cooling Degree Days | 0 | 0 | 0 | 0 | 10 | 31 | 87 | 99 | 34 | 2 | 0 | 0 | 263 |
| Total Precipitation (") | 5.98 | 4.34 | 3.83 | 2.57 | 1.91 | 1.46 | 0.54 | 0.76 | 1.48 | 3.05 | 5.99 | 6.72 | 38.63 |
| Days ≥ 0.1" Precip | 12 | 10 | 11 | 8 | 6 | 4 | 2 | 2 | 4 | 7 | 13 | 13 | 92 |
| Total Snowfall (") | 2.3 | 1.7 | 0.1 | 0.0 | 0.0 | 0.0 | 0.0 | 0.0 | 0.0 | 0.0 | 0.4 | 2.4 | 6.9 |
| Days ≥ 1" Snow Depth | 1 | 1 | 0 | 0 | 0 | 0 | 0 | 0 | 0 | 0 | 0 | 1 | 3 |

## SCOTTS MILLS 9 SE *Clackamas County* ELEVATION 2221 ft LAT/LONG 44° 58 ' N / 122° 32 ' W

| | JAN | FEB | MAR | APR | MAY | JUN | JUL | AUG | SEP | OCT | NOV | DEC | YEAR |
|---|---|---|---|---|---|---|---|---|---|---|---|---|---|
| Maximum Temp °F | 43.8 | 46.7 | 49.0 | 52.6 | 59.3 | 65.0 | 71.5 | 72.2 | 67.7 | 58.9 | 47.8 | 43.1 | 56.5 |
| Minimum Temp °F | 31.7 | 33.0 | 33.7 | 35.4 | 40.2 | 45.0 | 48.6 | 49.3 | 46.8 | 41.4 | 35.2 | 31.8 | 39.3 |
| Mean Temp °F | 37.8 | 39.9 | 41.4 | 44.1 | 49.7 | 55.0 | 60.0 | 60.8 | 57.3 | 50.2 | 41.6 | 37.5 | 47.9 |
| Days Max Temp ≥ 90 °F | 0 | 0 | 0 | 0 | 0 | 0 | 1 | 1 | 0 | 0 | 0 | 0 | 2 |
| Days Max Temp ≤ 32 °F | 2 | 1 | 0 | 0 | 0 | 0 | 0 | 0 | 0 | 0 | 1 | 2 | 6 |
| Days Min Temp ≤ 32 °F | 16 | 14 | 14 | 11 | 3 | 0 | 0 | 0 | 0 | 2 | 11 | 17 | 88 |
| Days Min Temp ≤ 0 °F | 0 | 0 | 0 | 0 | 0 | 0 | 0 | 0 | 0 | 0 | 0 | 0 | 0 |
| Heating Degree Days | 837 | 703 | 725 | 622 | 470 | 302 | 175 | 160 | 248 | 455 | 697 | 847 | 6241 |
| Cooling Degree Days | 0 | 0 | 0 | 0 | 6 | 11 | 29 | 39 | 24 | 3 | 0 | 0 | 112 |
| Total Precipitation (") | 11.83 | 8.76 | 8.66 | 6.65 | 4.78 | 3.38 | 1.26 | 1.62 | 3.27 | 6.13 | 11.50 | 12.58 | 80.42 |
| Days ≥ 0.1" Precip | 16 | 14 | 15 | 13 | 10 | 7 | 3 | 3 | 6 | 10 | 16 | 17 | 130 |
| Total Snowfall (") | 18.4 | 13.9 | 12.6 | 6.4 | 0.7 | 0.0 | 0.0 | 0.0 | 0.0 | 0.3 | 7.1 | 16.3 | 75.7 |
| Days ≥ 1" Snow Depth | 11 | 7 | 7 | 2 | 0 | 0 | 0 | 0 | 0 | 0 | 4 | 10 | 41 |

## SEASIDE *Clatsop County* ELEVATION 10 ft LAT/LONG 45° 59 ' N / 123° 55 ' W

| | JAN | FEB | MAR | APR | MAY | JUN | JUL | AUG | SEP | OCT | NOV | DEC | YEAR |
|---|---|---|---|---|---|---|---|---|---|---|---|---|---|
| Maximum Temp °F | 51.1 | 53.9 | 55.7 | 57.8 | 61.8 | 64.9 | 67.8 | 68.9 | 69.6 | 63.6 | 55.6 | 51.2 | 60.2 |
| Minimum Temp °F | 37.0 | 38.1 | 38.8 | 41.0 | 44.9 | 49.1 | 51.7 | 52.1 | 49.0 | 44.6 | 40.7 | 37.3 | 43.7 |
| Mean Temp °F | 44.1 | 46.0 | 47.3 | 49.4 | 53.4 | 57.0 | 59.8 | 60.5 | 59.3 | 54.1 | 48.2 | 44.3 | 52.0 |
| Days Max Temp ≥ 90 °F | 0 | 0 | 0 | 0 | 0 | 0 | 0 | 0 | 1 | 0 | 0 | 0 | 1 |
| Days Max Temp ≤ 32 °F | 0 | 0 | 0 | 0 | 0 | 0 | 0 | 0 | 0 | 0 | 0 | 1 | 1 |
| Days Min Temp ≤ 32 °F | 9 | 6 | 5 | 2 | 0 | 0 | 0 | 0 | 0 | 1 | 4 | 8 | 35 |
| Days Min Temp ≤ 0 °F | 0 | 0 | 0 | 0 | 0 | 0 | 0 | 0 | 0 | 0 | 0 | 0 | 0 |
| Heating Degree Days | 642 | 529 | 543 | 461 | 353 | 234 | 158 | 137 | 177 | 332 | 497 | 635 | 4698 |
| Cooling Degree Days | 0 | 0 | 0 | 0 | 1 | 2 | 3 | 7 | 9 | 2 | 0 | 0 | 24 |
| Total Precipitation (") | 10.83 | 8.78 | 7.85 | 5.65 | 3.55 | 2.83 | 1.49 | 1.41 | 2.98 | 5.94 | 10.67 | 11.52 | 73.50 |
| Days ≥ 0.1" Precip | 17 | 14 | 15 | 12 | 9 | 6 | 3 | 3 | 6 | 10 | 17 | 18 | 130 |
| Total Snowfall (") | 1.2 | 0.1 | 0.0 | 0.0 | 0.0 | 0.0 | 0.0 | 0.0 | 0.0 | 0.0 | 0.0 | 0.2 | 1.5 |
| Days ≥ 1" Snow Depth | 0 | 0 | 0 | 0 | 0 | 0 | 0 | 0 | 0 | 0 | 0 | 0 | 0 |

## SENECA *Grant County* ELEVATION 4705 ft LAT/LONG 44° 9 ' N / 118° 58 ' W

| | JAN | FEB | MAR | APR | MAY | JUN | JUL | AUG | SEP | OCT | NOV | DEC | YEAR |
|---|---|---|---|---|---|---|---|---|---|---|---|---|---|
| Maximum Temp °F | 33.3 | 38.8 | 45.1 | 52.5 | 61.4 | 69.9 | 79.9 | 80.0 | 70.7 | 59.4 | 42.9 | 34.2 | 55.7 |
| Minimum Temp °F | 8.8 | 13.2 | 20.2 | 25.4 | 31.4 | 36.6 | 38.1 | 35.8 | 27.6 | 21.1 | 19.1 | 11.1 | 24.0 |
| Mean Temp °F | 21.1 | 26.0 | 32.7 | 39.0 | 46.4 | 53.3 | 59.0 | 57.9 | 49.3 | 40.2 | 31.0 | 22.7 | 39.9 |
| Days Max Temp ≥ 90 °F | 0 | 0 | 0 | 0 | 0 | 0 | 4 | 4 | 0 | 0 | 0 | 0 | 8 |
| Days Max Temp ≤ 32 °F | 13 | 6 | 1 | 0 | 0 | 0 | 0 | 0 | 0 | 0 | 3 | 12 | 35 |
| Days Min Temp ≤ 32 °F | 30 | 27 | 29 | 25 | 18 | ***8*** | 6 | 9 | 22 | 28 | 28 | 30 | 260 |
| Days Min Temp ≤ 0 °F | 9 | 5 | 1 | 0 | 0 | 0 | 0 | 0 | 0 | 0 | 2 | 7 | 24 |
| Heating Degree Days | 1355 | 1080 | 996 | 776 | 568 | 351 | 194 | 223 | 466 | 761 | 1012 | 1307 | 9089 |
| Cooling Degree Days | 0 | 0 | 0 | 0 | 0 | 6 | 12 | 10 | 0 | 0 | 0 | 0 | 28 |
| Total Precipitation (") | 1.34 | 0.97 | 1.19 | 0.96 | 1.37 | 1.19 | 0.62 | 0.89 | 0.63 | 0.84 | 1.48 | 1.58 | 13.06 |
| Days ≥ 0.1" Precip | 4 | 4 | 4 | 3 | 5 | 4 | 2 | 2 | 2 | 3 | 6 | 5 | 44 |
| Total Snowfall (") | na | ***8.9*** | ***6.0*** | 2.3 | 0.8 | 0.1 | 0.0 | 0.0 | 0.0 | 1.0 | ***6.8*** | na | na |
| Days ≥ 1" Snow Depth | ***25*** | ***18*** | ***8*** | 2 | 0 | 0 | 0 | 0 | 0 | 0 | ***6*** | ***19*** | 78 |

## SEXTON SUMMIT *Josephine County* ELEVATION 3836 ft LAT/LONG 42° 37 ' N / 123° 22 ' W

| | JAN | FEB | MAR | APR | MAY | JUN | JUL | AUG | SEP | OCT | NOV | DEC | YEAR |
|---|---|---|---|---|---|---|---|---|---|---|---|---|---|
| Maximum Temp °F | 41.4 | 43.9 | 45.8 | 50.9 | 60.1 | 67.5 | 74.8 | 75.0 | 69.0 | 59.0 | 45.7 | 40.9 | 56.2 |
| Minimum Temp °F | 31.5 | 32.7 | 32.2 | 33.9 | 39.5 | 45.7 | 51.6 | 52.7 | 49.3 | 43.5 | 35.5 | 31.5 | 40.0 |
| Mean Temp °F | 36.4 | 38.3 | 39.0 | 42.4 | 49.8 | 56.6 | 63.2 | 63.9 | 59.2 | 51.3 | 40.6 | 36.3 | 48.1 |
| Days Max Temp ≥ 90 °F | 0 | 0 | 0 | 0 | 0 | 0 | 1 | 1 | 0 | 0 | 0 | 0 | 2 |
| Days Max Temp ≤ 32 °F | 5 | 3 | 2 | 1 | 0 | 0 | 0 | 0 | 0 | 0 | 2 | 6 | 19 |
| Days Min Temp ≤ 32 °F | 17 | 14 | 17 | 15 | 6 | 0 | 0 | 0 | 0 | 3 | 12 | 18 | 102 |
| Days Min Temp ≤ 0 °F | 0 | 0 | 0 | 0 | 0 | 0 | 0 | 0 | 0 | 0 | 0 | 0 | 0 |
| Heating Degree Days | 878 | 747 | 799 | 672 | 476 | 273 | 124 | 116 | 219 | 432 | 726 | 884 | 6346 |
| Cooling Degree Days | ***0*** | 0 | ***0*** | ***1*** | ***14*** | ***26*** | ***59*** | ***85*** | ***52*** | 16 | 0 | 0 | 253 |
| Total Precipitation (") | 5.57 | 3.73 | 3.76 | 2.17 | 1.22 | 0.89 | 0.36 | 0.73 | 1.23 | 2.73 | 5.58 | 5.67 | 33.64 |
| Days ≥ 0.1" Precip | 9 | 8 | 10 | 6 | 4 | 3 | 1 | 2 | 3 | 5 | 11 | 11 | 73 |
| Total Snowfall (") | 18.4 | 15.5 | 15.5 | 9.6 | 1.7 | 0.1 | 0.0 | 0.0 | 0.1 | 1.2 | 9.2 | 21.8 | 93.1 |
| Days ≥ 1" Snow Depth | 13 | 11 | 9 | 4 | 0 | 0 | 0 | 0 | 0 | 1 | 5 | 12 | 55 |

## SHEAVILLE 1 SE *Malheur County* ELEVATION 4583 ft LAT/LONG 43° 7 ' N / 117° 3 ' W

| | JAN | FEB | MAR | APR | MAY | JUN | JUL | AUG | SEP | OCT | NOV | DEC | YEAR |
|---|---|---|---|---|---|---|---|---|---|---|---|---|---|
| Maximum Temp °F | 36.6 | 42.8 | 50.0 | 57.2 | 67.1 | 76.5 | 87.0 | 85.6 | ***75.8*** | 64.0 | 46.4 | ***37.4*** | 60.5 |
| Minimum Temp °F | 16.8 | 21.4 | 25.8 | 29.8 | 36.5 | 43.3 | 49.4 | 47.9 | 38.8 | 31.1 | 23.2 | ***16.7*** | 31.7 |
| Mean Temp °F | 26.7 | 32.1 | 38.0 | 43.5 | 51.8 | 59.9 | 68.2 | 66.8 | ***57.4*** | 47.6 | 34.8 | ***27.1*** | 46.2 |
| Days Max Temp ≥ 90 °F | 0 | 0 | 0 | 0 | 0 | 3 | ***13*** | ***11*** | 2 | 0 | 0 | 0 | 29 |
| Days Max Temp ≤ 32 °F | 9 | 3 | 1 | 0 | 0 | 0 | 0 | 0 | 0 | 0 | 2 | 8 | 23 |
| Days Min Temp ≤ 32 °F | 28 | 26 | 25 | 18 | 9 | 2 | 0 | 0 | 5 | 18 | 25 | 28 | 184 |
| Days Min Temp ≤ 0 °F | 3 | 1 | 0 | 0 | 0 | 0 | 0 | 0 | 0 | 0 | 1 | 3 | 8 |
| Heating Degree Days | 1181 | 920 | 831 | 638 | 408 | 186 | 37 | 62 | ***242*** | 534 | 899 | ***1169*** | 7107 |
| Cooling Degree Days | 0 | 0 | 0 | 0 | 7 | 47 | ***149*** | 139 | ***26*** | 0 | ***0*** | ***0*** | 368 |
| Total Precipitation (") | 1.60 | 1.43 | 1.50 | 1.39 | 1.26 | 1.32 | 0.47 | 0.67 | 0.74 | 0.97 | 1.78 | 1.61 | 14.74 |
| Days ≥ 0.1" Precip | 5 | 5 | 4 | 4 | 4 | 3 | 1 | 2 | 2 | 3 | 5 | 5 | 43 |
| Total Snowfall (") | na | 6.5 | 4.4 | ***1.8*** | 0.4 | 0.0 | 0.0 | 0.0 | 0.0 | 0.5 | ***6.1*** | na | na |
| Days ≥ 1" Snow Depth | ***21*** | 12 | 4 | 1 | 0 | 0 | 0 | 0 | 0 | 0 | ***5*** | ***13*** | 56 |

## SILVER CREEK FALLS *Marion County* ELEVATION 1342 ft LAT/LONG 44° 53 ' N / 122° 39 ' W

| | JAN | FEB | MAR | APR | MAY | JUN | JUL | AUG | SEP | OCT | NOV | DEC | YEAR |
|---|---|---|---|---|---|---|---|---|---|---|---|---|---|
| Maximum Temp °F | 44.1 | 48.6 | 52.9 | 57.7 | 64.8 | 70.3 | 76.9 | 77.1 | 72.6 | 61.7 | 49.4 | 43.6 | 60.0 |
| Minimum Temp °F | 30.3 | 31.8 | 33.3 | 35.5 | 39.6 | 44.8 | 47.4 | 47.0 | 43.4 | 38.9 | 34.6 | 31.4 | 38.2 |
| Mean Temp °F | 37.2 | 40.2 | 43.1 | 46.6 | 52.2 | 57.6 | 62.2 | 62.1 | 58.1 | 50.3 | 41.9 | 37.5 | 49.1 |
| Days Max Temp ≥ 90 °F | 0 | 0 | 0 | 0 | 0 | 1 | 2 | 3 | 1 | 0 | 0 | 0 | 7 |
| Days Max Temp ≤ 32 °F | 1 | 0 | 0 | 0 | 0 | 0 | 0 | 0 | 0 | 0 | 0 | 2 | 3 |
| Days Min Temp ≤ 32 °F | 19 | 16 | 13 | 9 | 3 | 0 | 0 | 0 | 1 | 5 | 11 | 16 | 93 |
| Days Min Temp ≤ 0 °F | 0 | 0 | 0 | 0 | 0 | 0 | 0 | 0 | 0 | 0 | 0 | 0 | 0 |
| Heating Degree Days | 849 | 694 | 672 | 543 | 392 | ***229*** | 115 | 117 | 210 | 451 | ***688*** | 846 | 5806 |
| Cooling Degree Days | 0 | 0 | ***0*** | 0 | ***6*** | 9 | 31 | ***27*** | 8 | 4 | 0 | 0 | 85 |
| Total Precipitation (") | 11.23 | 8.57 | 8.25 | 6.61 | 4.62 | 3.41 | 1.12 | 1.58 | 2.84 | 5.99 | 11.08 | 11.73 | 77.03 |
| Days ≥ 0.1" Precip | ***14*** | na | 13 | ***11*** | ***8*** | ***6*** | 2 | 3 | 4 | 8 | ***14*** | ***15*** | na |
| Total Snowfall (") | ***4.5*** | 4.7 | 2.3 | 0.4 | 0.0 | 0.0 | 0.0 | 0.0 | 0.0 | 0.0 | 1.4 | ***2.5*** | 15.8 |
| Days ≥ 1" Snow Depth | na | 2 | 1 | 0 | 0 | 0 | 0 | 0 | 0 | 0 | 0 | ***2*** | na |

## SILVER LAKE R S *Lake County* ELEVATION 4382 ft LAT/LONG 43° 8 ' N / 121° 4 ' W

| | JAN | FEB | MAR | APR | MAY | JUN | JUL | AUG | SEP | OCT | NOV | DEC | YEAR |
|---|---|---|---|---|---|---|---|---|---|---|---|---|---|
| Maximum Temp °F | ***39.7*** | ***43.4*** | ***49.9*** | ***58.1*** | ***66.8*** | ***75.2*** | ***83.2*** | 83.4 | ***75.5*** | 64.0 | 46.4 | ***38.4*** | 60.3 |
| Minimum Temp °F | ***20.7*** | 22.4 | ***25.1*** | ***27.8*** | ***33.3*** | ***40.0*** | ***43.6*** | 43.5 | ***36.3*** | 29.8 | 25.2 | ***19.2*** | 30.6 |
| Mean Temp °F | ***30.2*** | ***33.0*** | ***37.7*** | ***43.0*** | ***50.1*** | ***57.6*** | ***63.4*** | 63.5 | ***55.9*** | 46.9 | 35.9 | ***28.8*** | 45.5 |
| Days Max Temp ≥ 90 °F | 0 | 0 | 0 | 0 | 0 | 2 | ***8*** | 7 | ***1*** | 0 | 0 | 0 | 18 |
| Days Max Temp ≤ 32 °F | ***6*** | ***3*** | 0 | 0 | 0 | 0 | ***0*** | 0 | ***0*** | 0 | 2 | ***7*** | 18 |
| Days Min Temp ≤ 32 °F | ***26*** | ***23*** | 25 | ***21*** | 14 | ***5*** | ***2*** | 2 | ***9*** | 19 | ***23*** | ***26*** | 195 |
| Days Min Temp ≤ 0 °F | ***2*** | ***1*** | 0 | 0 | 0 | 0 | ***0*** | 0 | ***0*** | 0 | 1 | 2 | 6 |
| Heating Degree Days | ***1071*** | ***895*** | ***841*** | ***654*** | ***458*** | ***231*** | ***104*** | 103 | ***271*** | 554 | 868 | ***1114*** | 7164 |
| Cooling Degree Days | 0 | 0 | 0 | 0 | 3 | 18 | 50 | 57 | 6 | 0 | 0 | 0 | 134 |
| Total Precipitation (") | ***0.92*** | ***0.66*** | 0.75 | ***0.62*** | 0.95 | 0.80 | ***0.53*** | 0.57 | ***0.61*** | ***0.61*** | 1.11 | ***1.15*** | 9.28 |
| Days ≥ 0.1" Precip | na | ***2*** | ***2*** | ***2*** | ***2*** | 2 | ***2*** | 2 | ***2*** | ***2*** | ***3*** | ***3*** | na |
| Total Snowfall (") | na | ***2.6*** | ***2.8*** | ***1.5*** | 0.6 | 0.0 | ***0.0*** | 0.0 | ***0.0*** | ***0.2*** | ***2.1*** | na | na |
| Days ≥ 1" Snow Depth | na | ***3*** | 2 | 0 | 0 | 0 | ***0*** | 0 | ***0*** | 0 | ***1*** | na | na |

## SILVERTON *Marion County* ELEVATION 408 ft LAT/LONG 45° 0 ' N / 122° 46 ' W

| | JAN | FEB | MAR | APR | MAY | JUN | JUL | AUG | SEP | OCT | NOV | DEC | YEAR |
|---|---|---|---|---|---|---|---|---|---|---|---|---|---|
| Maximum Temp °F | 45.7 | 50.7 | 55.4 | 59.4 | 66.2 | 72.0 | 78.7 | 79.4 | 74.2 | 63.5 | 52.0 | 45.6 | 61.9 |
| Minimum Temp °F | 33.0 | 35.3 | 37.9 | 40.5 | 45.4 | 50.5 | 53.7 | 54.0 | 50.1 | 43.5 | 38.1 | 33.5 | 43.0 |
| Mean Temp °F | 39.4 | 43.0 | 46.7 | 50.0 | 55.8 | 61.3 | 66.3 | 66.8 | 62.2 | 53.6 | 45.1 | 39.6 | 52.5 |
| Days Max Temp ≥ 90 °F | 0 | 0 | 0 | 0 | 0 | 1 | 3 | 3 | 2 | 0 | 0 | 0 | 9 |
| Days Max Temp ≤ 32 °F | 1 | 0 | 0 | 0 | 0 | 0 | 0 | 0 | 0 | 0 | 0 | 1 | 2 |
| Days Min Temp ≤ 32 °F | 15 | 9 | 5 | 1 | 0 | 0 | 0 | 0 | 0 | 1 | 7 | 14 | 52 |
| Days Min Temp ≤ 0 °F | 0 | 0 | 0 | 0 | 0 | 0 | 0 | 0 | 0 | 0 | 0 | 0 | 0 |
| Heating Degree Days | 787 | 615 | 561 | 444 | 289 | 142 | 48 | 39 | 118 | 350 | 591 | 781 | 4765 |
| Cooling Degree Days | 0 | 0 | 0 | 1 | 16 | 39 | 91 | 105 | 46 | 3 | 0 | 0 | 301 |
| Total Precipitation (") | 6.38 | 4.92 | 4.65 | 3.59 | 2.70 | 1.95 | 0.81 | 1.07 | 1.97 | 3.46 | 6.64 | 7.32 | 45.46 |
| Days ≥ 0.1" Precip | 14 | 11 | 12 | 11 | 7 | 5 | 2 | 3 | 5 | 9 | 14 | 14 | 107 |
| Total Snowfall (") | 1.4 | 1.2 | 0.0 | 0.0 | 0.0 | 0.0 | 0.0 | 0.0 | 0.0 | 0.0 | 0.3 | 1.6 | 4.5 |
| Days ≥ 1" Snow Depth | 1 | 1 | 0 | 0 | 0 | 0 | 0 | 0 | 0 | 0 | 0 | 1 | 3 |

## SISTERS *Deschutes County* ELEVATION 3182 ft LAT/LONG 44° 17 ' N / 121° 32 ' W

| | JAN | FEB | MAR | APR | MAY | JUN | JUL | AUG | SEP | OCT | NOV | DEC | YEAR |
|---|---|---|---|---|---|---|---|---|---|---|---|---|---|
| Maximum Temp °F | ***40.8*** | 45.2 | 51.9 | 57.9 | ***66.7*** | 75.2 | 83.8 | 83.7 | 75.3 | 63.7 | 47.1 | 40.6 | 61.0 |
| Minimum Temp °F | ***20.6*** | 22.8 | 25.8 | 28.0 | ***33.0*** | 38.8 | 41.4 | 40.9 | 34.5 | 28.9 | 25.0 | 20.2 | 30.0 |
| Mean Temp °F | ***30.7*** | ***34.2*** | 38.8 | 42.7 | ***49.9*** | 57.0 | 62.7 | 62.3 | 55.0 | 46.3 | 36.1 | 30.4 | 45.5 |
| Days Max Temp ≥ 90 °F | 0 | 0 | 0 | 0 | 0 | 2 | 9 | 9 | 2 | 0 | 0 | 0 | 22 |
| Days Max Temp ≤ 32 °F | 5 | 2 | 0 | 0 | 0 | 0 | 0 | 0 | 0 | 0 | 1 | 5 | 13 |
| Days Min Temp ≤ 32 °F | 27 | 23 | 25 | 22 | 15 | 7 | 3 | 3 | 12 | 21 | 23 | 27 | 208 |
| Days Min Temp ≤ 0 °F | 2 | 1 | 0 | 0 | 0 | 0 | 0 | 0 | 0 | 0 | 1 | 2 | 6 |
| Heating Degree Days | ***1057*** | ***865*** | 805 | 664 | ***463*** | 250 | 116 | 124 | 300 | 573 | 862 | 1069 | 7148 |
| Cooling Degree Days | ***0*** | ***0*** | 0 | 0 | na | 18 | 44 | 49 | 6 | ***0*** | 0 | ***0*** | na |
| Total Precipitation (") | 2.45 | 1.52 | 1.16 | 0.81 | 0.66 | 0.62 | 0.46 | 0.46 | 0.47 | 0.93 | 2.03 | 2.14 | 13.71 |
| Days ≥ 0.1" Precip | 5 | ***4*** | 4 | 2 | 2 | 2 | 1 | 2 | 1 | 3 | 5 | 5 | 36 |
| Total Snowfall (") | na | na | ***3.7*** | 0.4 | 0.0 | 0.0 | 0.0 | 0.0 | 0.0 | 0.3 | ***5.2*** | ***7.5*** | na |
| Days ≥ 1" Snow Depth | 12 | 7 | 2 | 0 | 0 | 0 | 0 | 0 | 0 | 0 | ***3*** | ***7*** | 31 |

## SQUAW BUTTE *Harney County* ELEVATION 4682 ft LAT/LONG 43° 29 ' N / 119° 41 ' W

| | JAN | FEB | MAR | APR | MAY | JUN | JUL | AUG | SEP | OCT | NOV | DEC | YEAR |
|---|---|---|---|---|---|---|---|---|---|---|---|---|---|
| Maximum Temp °F | *35.1* | 41.1 | *47.8* | 55.3 | 64.6 | 73.7 | 82.7 | 82.4 | 73.5 | 61.6 | 45.1 | *36.9* | 58.3 |
| Minimum Temp °F | *18.2* | 22.4 | *26.1* | 30.2 | 36.8 | 44.2 | 50.3 | 49.9 | 43.0 | 34.9 | 26.2 | *20.1* | 33.5 |
| Mean Temp °F | *26.7* | 31.8 | *37.0* | 42.8 | 50.7 | 59.0 | 66.6 | 66.2 | 58.3 | 48.3 | 35.7 | *28.6* | 46.0 |
| Days Max Temp ≥ 90 °F | 0 | 0 | 0 | 0 | 0 | 2 | 7 | 7 | 1 | 0 | 0 | 0 | 17 |
| Days Max Temp ≤ 32 °F | 10 | 3 | 1 | 0 | 0 | 0 | 0 | 0 | 0 | 0 | 3 | *8* | 25 |
| Days Min Temp ≤ 32 °F | 27 | 25 | 24 | 18 | 10 | 2 | 0 | 0 | 3 | 11 | 22 | 27 | 169 |
| Days Min Temp ≤ 0 °F | *1* | 0 | 0 | 0 | 0 | 0 | 0 | 0 | 0 | 0 | 0 | 1 | 2 |
| Heating Degree Days | *1179* | 932 | *860* | 659 | 443 | 212 | 72 | 76 | 226 | 514 | 872 | *1123* | 7168 |
| Cooling Degree Days | *0* | *0* | *0* | *0* | 10 | 41 | 122 | 116 | 42 | *3* | *0* | na | na |
| Total Precipitation (") | 1.17 | 0.79 | 1.08 | 0.86 | 1.12 | 0.92 | 0.43 | 0.75 | 0.61 | 0.78 | 1.31 | 1.23 | 11.05 |
| Days ≥ 0.1" Precip | *4* | 3 | 3 | 3 | 4 | 3 | 1 | 2 | 2 | 2 | 4 | na | na |
| Total Snowfall (") | na | *5.7* | *4.7* | *1.3* | 0.5 | 0.0 | 0.0 | 0.0 | 0.0 | 0.9 | *5.7* | na | na |
| Days ≥ 1" Snow Depth | *21* | *11* | *5* | 1 | 0 | 0 | 0 | 0 | 0 | 0 | *5* | na | na |

## STAYTON *Marion County* ELEVATION 430 ft LAT/LONG 44° 47 ' N / 122° 49 ' W

| | JAN | FEB | MAR | APR | MAY | JUN | JUL | AUG | SEP | OCT | NOV | DEC | YEAR |
|---|---|---|---|---|---|---|---|---|---|---|---|---|---|
| Maximum Temp °F | 46.6 | 51.3 | 56.2 | 60.2 | 67.2 | 73.5 | 79.9 | 80.7 | 75.2 | 64.5 | 52.5 | 46.3 | 62.8 |
| Minimum Temp °F | 32.8 | 35.1 | 37.7 | 40.1 | 44.1 | 49.1 | 51.5 | 51.2 | 47.8 | 42.5 | 38.0 | 33.8 | 42.0 |
| Mean Temp °F | 39.8 | 43.3 | 47.0 | 50.2 | 55.7 | 61.3 | 65.7 | 66.0 | 61.5 | 53.6 | 45.3 | 40.1 | 52.5 |
| Days Max Temp ≥ 90 °F | 0 | 0 | 0 | 0 | 0 | 1 | 4 | 5 | 2 | 0 | 0 | 0 | 12 |
| Days Max Temp ≤ 32 °F | 1 | 0 | 0 | 0 | 0 | 0 | 0 | 0 | 0 | 0 | 0 | 1 | 2 |
| Days Min Temp ≤ 32 °F | 15 | 10 | 6 | 2 | 0 | 0 | 0 | 0 | 0 | 1 | 7 | 13 | 54 |
| Days Min Temp ≤ 0 °F | 0 | 0 | 0 | 0 | 0 | 0 | 0 | 0 | 0 | 0 | 0 | 0 | 0 |
| Heating Degree Days | 775 | 608 | 551 | 438 | 289 | 136 | 49 | 45 | 126 | 350 | 584 | 766 | 4717 |
| Cooling Degree Days | *0* | 0 | 0 | 0 | 11 | 31 | 71 | 78 | 31 | 2 | 0 | 0 | 224 |
| Total Precipitation (") | 7.20 | 5.53 | 4.91 | 3.96 | 2.90 | 2.36 | 0.78 | 1.23 | 2.17 | 3.94 | 7.88 | 8.21 | 51.07 |
| Days ≥ 0.1" Precip | 14 | 12 | 13 | 11 | 8 | 6 | 2 | 3 | 5 | 9 | 14 | 14 | 111 |
| Total Snowfall (") | na | 1.0 | 0.0 | 0.0 | 0.0 | 0.0 | 0.0 | 0.0 | 0.0 | 0.0 | 0.0 | 0.7 | na |
| Days ≥ 1" Snow Depth | na | 0 | 0 | 0 | 0 | 0 | 0 | 0 | 0 | 0 | 0 | 0 | na |

## SUMMER LAKE 1 S *Lake County* ELEVATION 4193 ft LAT/LONG 42° 57 ' N / 120° 47 ' W

| | JAN | FEB | MAR | APR | MAY | JUN | JUL | AUG | SEP | OCT | NOV | DEC | YEAR |
|---|---|---|---|---|---|---|---|---|---|---|---|---|---|
| Maximum Temp °F | 41.2 | 46.4 | 51.5 | 58.3 | 68.2 | 76.4 | 85.6 | 84.7 | 76.6 | 64.9 | 48.6 | 41.1 | 62.0 |
| Minimum Temp °F | 23.3 | 26.5 | 29.6 | 33.1 | 39.8 | 46.7 | 51.5 | 50.1 | 42.6 | 34.6 | 28.4 | 23.3 | 35.8 |
| Mean Temp °F | 32.3 | 36.5 | 40.6 | 45.7 | 54.0 | 61.5 | 68.5 | 67.4 | 59.6 | 49.8 | 38.5 | 32.2 | 48.9 |
| Days Max Temp ≥ 90 °F | 0 | 0 | 0 | 0 | 0 | 2 | 11 | 10 | 2 | 0 | 0 | 0 | 25 |
| Days Max Temp ≤ 32 °F | 5 | 2 | 0 | 0 | 0 | 0 | 0 | 0 | 0 | 0 | 1 | 4 | 12 |
| Days Min Temp ≤ 32 °F | 25 | 22 | 21 | 15 | 5 | 0 | 0 | 0 | 2 | 13 | 21 | 26 | 150 |
| Days Min Temp ≤ 0 °F | 1 | 0 | 0 | 0 | 0 | 0 | 0 | 0 | 0 | 0 | 0 | 1 | 2 |
| Heating Degree Days | 1007 | 799 | 751 | 571 | 343 | 149 | 36 | 44 | 183 | 465 | 787 | 1008 | 6143 |
| Cooling Degree Days | 0 | 0 | 0 | 1 | 14 | 61 | 148 | 133 | 34 | 1 | 0 | 0 | 392 |
| Total Precipitation (") | 1.36 | 0.98 | 1.09 | 0.78 | 1.00 | 1.02 | 0.50 | 0.60 | 0.54 | 0.76 | 1.58 | 1.52 | 11.73 |
| Days ≥ 0.1" Precip | 4 | 3 | 3 | 3 | 3 | 3 | 1 | 2 | 2 | 3 | 5 | 4 | 36 |
| Total Snowfall (") | 5.6 | 4.0 | 3.1 | 1.2 | 0.2 | 0.0 | 0.0 | 0.0 | 0.0 | 0.1 | 2.8 | 6.0 | 23.0 |
| Days ≥ 1" Snow Depth | 6 | 2 | 1 | 0 | 0 | 0 | 0 | 0 | 0 | 0 | 2 | 6 | 17 |

## THE DALLES *Wasco County* ELEVATION 102 ft LAT/LONG 45° 36 ' N / 121° 12 ' W

| | JAN | FEB | MAR | APR | MAY | JUN | JUL | AUG | SEP | OCT | NOV | DEC | YEAR |
|---|---|---|---|---|---|---|---|---|---|---|---|---|---|
| Maximum Temp °F | *42.8* | *49.0* | *58.5* | *66.1* | *74.1* | *80.8* | *87.9* | *88.1* | *81.0* | *68.6* | *51.9* | *43.0* | 66.0 |
| Minimum Temp °F | *30.1* | *32.4* | *36.9* | *42.4* | *48.8* | *55.4* | *60.2* | *59.3* | *51.2* | *42.2* | *35.7* | *30.7* | 43.8 |
| Mean Temp °F | *36.5* | *40.7* | *47.8* | *54.3* | *61.4* | *68.1* | *74.0* | *72.8* | *66.1* | *55.4* | *43.8* | *36.9* | 54.8 |
| Days Max Temp ≥ 90 °F | *0* | *0* | *0* | *0* | *2* | *7* | *14* | *14* | *6* | *0* | *0* | *0* | 43 |
| Days Max Temp ≤ 32 °F | *4* | *1* | *0* | *0* | *0* | *0* | *0* | *0* | *0* | *0* | *1* | *3* | 9 |
| Days Min Temp ≤ 32 °F | *17* | *13* | *6* | *1* | *0* | *0* | *0* | *0* | *0* | *1* | *9* | *17* | 64 |
| Days Min Temp ≤ 0 °F | *0* | *0* | *0* | *0* | *0* | *0* | *0* | *0* | *0* | *0* | *0* | *0* | 0 |
| Heating Degree Days | *878* | *679* | *529* | *320* | *151* | *42* | *6* | *6* | *58* | *298* | *628* | *866* | 4461 |
| Cooling Degree Days | 0 | 0 | 0 | 5 | 54 | 142 | 286 | 291 | 106 | 9 | 0 | 0 | 893 |
| Total Precipitation (") | *2.26* | *1.81* | *1.11* | *0.79* | *0.49* | *0.42* | *0.19* | *0.51* | *0.45* | *0.96* | *2.03* | *2.62* | 13.64 |
| Days ≥ 0.1" Precip | na | *5* | *4* | *3* | *2* | *1* | *1* | *2* | *1* | *3* | *6* | *6* | na |
| Total Snowfall (") | na | *1.3* | *0.4* | *0.0* | *0.0* | *0.0* | *0.0* | *0.0* | *0.0* | *0.0* | *1.6* | na | na |
| Days ≥ 1" Snow Depth | na | na | *0* | *0* | *0* | *0* | *0* | *0* | *0* | *0* | na | na | na |

## THREE LYNX *Clackamas County* ELEVATION 1122 ft LAT/LONG 45° 7 ' N / 122° 4 ' W

| | JAN | FEB | MAR | APR | MAY | JUN | JUL | AUG | SEP | OCT | NOV | DEC | YEAR |
|---|---|---|---|---|---|---|---|---|---|---|---|---|---|
| Maximum Temp °F | 41.5 | 46.5 | 52.6 | 57.7 | 64.8 | 70.5 | 77.4 | 78.5 | 73.3 | 62.1 | 48.1 | 41.6 | 59.6 |
| Minimum Temp °F | 30.5 | 32.2 | 34.4 | 37.4 | 42.1 | 47.3 | 50.5 | 50.3 | 46.2 | 40.4 | 35.8 | 31.5 | 39.9 |
| Mean Temp °F | 36.0 | 39.4 | 43.5 | 47.6 | 53.5 | 58.9 | 64.0 | 64.4 | 59.7 | 51.3 | 42.0 | 36.6 | 49.7 |
| Days Max Temp ≥ 90 °F | 0 | 0 | 0 | 0 | 1 | 1 | 3 | 4 | 2 | 0 | 0 | 0 | 11 |
| Days Max Temp ≤ 32 °F | 2 | 1 | 0 | 0 | 0 | 0 | 0 | 0 | 0 | 0 | 0 | 1 | 4 |
| Days Min Temp ≤ 32 °F | 18 | 14 | 11 | 4 | 0 | 0 | 0 | 0 | 0 | 2 | 8 | 16 | 73 |
| Days Min Temp ≤ 0 °F | 0 | 0 | 0 | 0 | 0 | 0 | 0 | 0 | 0 | 0 | 0 | 0 | 0 |
| Heating Degree Days | 892 | 717 | 658 | 517 | 358 | 198 | 86 | 76 | 175 | 419 | 684 | 874 | 5654 |
| Cooling Degree Days | 0 | 0 | 0 | 0 | 10 | 22 | 55 | 63 | 22 | 1 | 0 | 0 | 173 |
| Total Precipitation (") | 10.92 | 8.01 | 7.30 | 5.67 | 3.92 | 2.80 | 0.92 | 1.17 | 2.76 | 5.23 | 10.56 | 11.28 | 70.54 |
| Days ≥ 0.1" Precip | 16 | 13 | 15 | 13 | 10 | 7 | 3 | 3 | 6 | 10 | 16 | 16 | 128 |
| Total Snowfall (") | 8.8 | *2.8* | *1.3* | 0.5 | 0.0 | 0.0 | 0.0 | 0.0 | 0.0 | 0.0 | 2.4 | 5.7 | 21.5 |
| Days ≥ 1" Snow Depth | 7 | 3 | 1 | 0 | 0 | 0 | 0 | 0 | 0 | 0 | 1 | 3 | 15 |

## TIDEWATER *Lincoln County* ELEVATION 39 ft LAT/LONG 44° 24 ' N / 123° 53 ' W

| | JAN | FEB | MAR | APR | MAY | JUN | JUL | AUG | SEP | OCT | NOV | DEC | YEAR |
|---|---|---|---|---|---|---|---|---|---|---|---|---|---|
| Maximum Temp °F | 50.5 | 55.3 | 58.4 | 61.1 | 66.3 | 70.5 | 74.7 | 76.3 | 75.4 | 67.4 | 55.9 | 49.8 | 63.5 |
| Minimum Temp °F | 36.5 | 37.8 | 38.9 | 40.2 | 44.3 | 48.8 | 52.1 | 52.4 | 49.8 | 45.6 | 40.9 | 37.0 | 43.7 |
| Mean Temp °F | 43.5 | 46.5 | 48.7 | 50.7 | 55.3 | 59.7 | 63.4 | 64.4 | 62.6 | 56.5 | 48.4 | 43.4 | 53.6 |
| Days Max Temp ≥ 90 °F | 0 | 0 | 0 | 0 | 0 | 1 | 1 | 1 | 2 | 0 | 0 | 0 | 5 |
| Days Max Temp ≤ 32 °F | 0 | 0 | 0 | 0 | 0 | 0 | 0 | 0 | 0 | 0 | 0 | 0 | 0 |
| Days Min Temp ≤ 32 °F | 10 | 6 | 4 | 2 | 0 | 0 | 0 | 0 | 0 | 0 | 3 | 9 | 34 |
| Days Min Temp ≤ 0 °F | 0 | 0 | 0 | 0 | 0 | 0 | 0 | 0 | 0 | 0 | 0 | 0 | 0 |
| Heating Degree Days | 658 | 515 | 500 | 423 | 297 | 164 | 69 | 48 | 90 | 260 | 490 | 662 | 4176 |
| Cooling Degree Days | 0 | 0 | 0 | 0 | 6 | 11 | 33 | 38 | 29 | 2 | 0 | 0 | 119 |
| Total Precipitation (") | 14.42 | 10.75 | 10.43 | 6.82 | 4.12 | 2.60 | 0.84 | 1.21 | 2.69 | 5.98 | 13.43 | 15.33 | 88.62 |
| Days ≥ 0.1" Precip | 17 | 15 | 16 | 13 | 9 | 6 | 2 | 3 | 5 | 10 | 17 | 18 | 131 |
| Total Snowfall (") | 0.3 | 0.6 | 0.0 | 0.0 | 0.0 | 0.0 | 0.0 | 0.0 | 0.0 | 0.0 | 0.0 | 0.5 | 1.4 |
| Days ≥ 1" Snow Depth | 0 | 0 | 0 | 0 | 0 | 0 | 0 | 0 | 0 | 0 | 0 | 0 | 0 |

## TILLAMOOK 1 W *Tillamook County* ELEVATION 39 ft LAT/LONG 45° 29 ' N / 123° 51 ' W

| | JAN | FEB | MAR | APR | MAY | JUN | JUL | AUG | SEP | OCT | NOV | DEC | YEAR |
|---|---|---|---|---|---|---|---|---|---|---|---|---|---|
| Maximum Temp °F | 50.0 | 53.0 | 54.9 | 57.1 | 61.2 | 64.7 | 67.3 | 68.9 | 69.0 | 62.8 | 54.5 | 49.6 | 59.4 |
| Minimum Temp °F | 35.8 | 36.8 | 37.1 | 38.9 | 42.7 | 46.9 | 49.5 | 49.6 | 46.5 | 41.9 | 39.0 | 36.0 | 41.7 |
| Mean Temp °F | 42.9 | 44.9 | 46.0 | 48.0 | 52.0 | 55.8 | 58.4 | 59.3 | 57.7 | 52.4 | 46.8 | 42.9 | 50.6 |
| Days Max Temp ≥ 90 °F | 0 | 0 | 0 | 0 | 0 | 0 | 0 | 0 | 0 | 0 | 0 | 0 | 0 |
| Days Max Temp ≤ 32 °F | 0 | 0 | 0 | 0 | 0 | 0 | 0 | 0 | 0 | 0 | 0 | 1 | 1 |
| Days Min Temp ≤ 32 °F | 11 | 8 | 9 | 5 | 1 | 0 | 0 | 0 | 1 | 3 | 7 | 10 | 55 |
| Days Min Temp ≤ 0 °F | 0 | 0 | 0 | 0 | 0 | 0 | 0 | 0 | 0 | 0 | 0 | 0 | 0 |
| Heating Degree Days | 679 | 561 | 582 | 504 | 397 | 271 | 200 | 175 | 217 | 387 | 541 | 680 | 5194 |
| Cooling Degree Days | 0 | 0 | 0 | 0 | 1 | 1 | 2 | 6 | 4 | 2 | 0 | 0 | 16 |
| Total Precipitation (") | 13.27 | 9.61 | 9.56 | 6.70 | 4.42 | 3.24 | 1.53 | 1.44 | 3.56 | 6.99 | 12.68 | 14.19 | 87.19 |
| Days ≥ 0.1" Precip | 17 | 15 | 16 | 13 | 10 | 7 | 4 | 4 | 7 | 10 | 18 | 18 | 139 |
| Total Snowfall (") | 0.7 | 0.4 | 0.3 | 0.0 | 0.0 | 0.0 | 0.0 | 0.0 | 0.0 | 0.0 | 0.2 | 0.2 | 1.8 |
| Days ≥ 1" Snow Depth | 0 | 0 | 0 | 0 | 0 | 0 | 0 | 0 | 0 | 0 | 0 | 0 | 0 |

## TOKETEE FALLS *Douglas County* ELEVATION 2060 ft LAT/LONG 43° 17 ' N / 122° 27 ' W

| | JAN | FEB | MAR | APR | MAY | JUN | JUL | AUG | SEP | OCT | NOV | DEC | YEAR |
|---|---|---|---|---|---|---|---|---|---|---|---|---|---|
| Maximum Temp °F | 42.2 | 48.2 | 54.3 | 60.8 | 70.4 | 77.9 | 85.7 | 85.5 | 77.5 | 63.0 | 47.6 | 41.3 | 62.9 |
| Minimum Temp °F | 29.3 | 30.8 | 32.9 | 36.0 | 41.0 | 47.0 | 50.2 | 49.4 | 44.2 | 38.1 | 33.6 | 29.7 | 38.5 |
| Mean Temp °F | 35.8 | 39.5 | 43.7 | 48.4 | 55.7 | 62.5 | 68.0 | 67.5 | 60.9 | 50.6 | 40.6 | 35.5 | 50.7 |
| Days Max Temp ≥ 90 °F | 0 | 0 | 0 | 0 | 2 | 4 | 12 | 11 | 4 | 0 | 0 | 0 | 33 |
| Days Max Temp ≤ 32 °F | 1 | 0 | 0 | 0 | 0 | 0 | 0 | 0 | 0 | 0 | 0 | 1 | 2 |
| Days Min Temp ≤ 32 °F | 23 | 18 | 16 | 9 | 2 | 0 | 0 | 0 | 1 | 5 | 14 | 22 | 110 |
| Days Min Temp ≤ 0 °F | 0 | 0 | 0 | 0 | 0 | 0 | 0 | 0 | 0 | 0 | 0 | 0 | 0 |
| Heating Degree Days | 898 | 713 | 655 | 491 | 296 | 124 | 34 | 35 | 148 | 441 | 725 | 907 | 5467 |
| Cooling Degree Days | 0 | 0 | 0 | 1 | 19 | 48 | 117 | 110 | 27 | 0 | 0 | 0 | 322 |
| Total Precipitation (") | 6.57 | 5.04 | 5.18 | 3.91 | 2.69 | 1.57 | 0.70 | 1.11 | 1.58 | 3.59 | 7.47 | 6.93 | 46.34 |
| Days ≥ 0.1" Precip | 13 | 11 | 12 | 10 | 7 | 4 | 2 | 3 | 4 | 7 | 14 | 13 | 100 |
| Total Snowfall (") | 12.0 | 9.8 | 5.2 | 0.2 | 0.0 | 0.0 | 0.0 | 0.0 | 0.0 | 0.1 | 4.1 | *10.6* | 42.0 |
| Days ≥ 1" Snow Depth | 13 | 8 | 3 | 0 | 0 | 0 | 0 | 0 | 0 | 0 | 3 | 10 | 37 |

## TROUTDALE SUBSTATION *Multnomah County* ELEVATION 29 ft LAT/LONG 45° 33 ' N / 122° 24 ' W

| | JAN | FEB | MAR | APR | MAY | JUN | JUL | AUG | SEP | OCT | NOV | DEC | YEAR |
|---|---|---|---|---|---|---|---|---|---|---|---|---|---|
| Maximum Temp °F | 45.6 | 50.3 | 57.1 | 62.0 | 69.0 | 75.1 | 81.4 | 82.1 | 76.3 | 65.1 | 53.3 | 46.2 | 63.6 |
| Minimum Temp °F | 33.9 | 36.4 | 38.7 | 41.7 | 46.4 | 51.7 | 54.6 | 54.7 | 50.1 | 44.4 | 39.8 | 34.8 | 43.9 |
| Mean Temp °F | 39.8 | 43.4 | 47.9 | 51.9 | 57.7 | 63.4 | 68.0 | 68.4 | 63.2 | 54.8 | 46.6 | 40.5 | 53.8 |
| Days Max Temp ≥ 90 °F | 0 | 0 | 0 | 0 | 0 | 2 | 6 | 6 | 2 | 0 | 0 | 0 | 16 |
| Days Max Temp ≤ 32 °F | 2 | 0 | 0 | 0 | 0 | 0 | 0 | 0 | 0 | 0 | 0 | 1 | 3 |
| Days Min Temp ≤ 32 °F | 12 | 6 | 4 | 1 | 0 | 0 | 0 | 0 | 0 | 0 | 4 | 10 | 37 |
| Days Min Temp ≤ 0 °F | 0 | 0 | 0 | 0 | 0 | 0 | 0 | 0 | 0 | 0 | 0 | 0 | 0 |
| Heating Degree Days | 774 | 605 | 523 | 388 | 232 | 97 | 28 | 23 | 95 | 312 | 546 | 752 | 4375 |
| Cooling Degree Days | 0 | 0 | 0 | 2 | 19 | 55 | 125 | *132* | 52 | 3 | 0 | 0 | 388 |
| Total Precipitation (") | 6.36 | 4.90 | 4.05 | 3.51 | 2.56 | 2.02 | 0.87 | 1.24 | 2.04 | 3.36 | 6.08 | 7.04 | 44.03 |
| Days ≥ 0.1" Precip | 13 | 11 | 11 | 10 | 7 | 5 | 2 | 3 | 4 | 8 | 13 | 13 | 100 |
| Total Snowfall (") | 0.4 | *0.6* | 0.5 | 0.0 | 0.0 | 0.0 | 0.0 | 0.0 | 0.0 | 0.0 | 0.5 | 1.0 | 3.0 |
| Days ≥ 1" Snow Depth | 1 | 1 | 0 | 0 | 0 | 0 | 0 | 0 | 0 | 0 | 0 | 1 | 3 |

## UKIAH *Umatilla County* ELEVATION 3343 ft LAT/LONG 45° 7 ' N / 118° 56 ' W

| | JAN | FEB | MAR | APR | MAY | JUN | JUL | AUG | SEP | OCT | NOV | DEC | YEAR |
|---|---|---|---|---|---|---|---|---|---|---|---|---|---|
| Maximum Temp °F | 36.4 | 42.8 | 49.6 | 56.2 | 64.7 | 73.3 | 81.9 | 82.4 | 73.6 | 62.9 | 46.0 | 37.2 | 58.9 |
| Minimum Temp °F | 14.8 | 19.0 | 23.3 | 27.8 | 32.6 | 38.8 | 40.1 | 39.1 | 31.6 | 25.6 | 23.0 | 16.3 | 27.7 |
| Mean Temp °F | 25.7 | 31.0 | 36.5 | 42.0 | 48.7 | 56.1 | 61.0 | 60.8 | 52.6 | 44.2 | 34.5 | 26.8 | 43.3 |
| Days Max Temp ≥ 90 °F | 0 | 0 | 0 | 0 | 0 | 1 | 7 | 7 | 2 | 0 | 0 | 0 | 17 |
| Days Max Temp ≤ 32 °F | 9 | 3 | 1 | 0 | 0 | 0 | 0 | 0 | 0 | 0 | 2 | 8 | 23 |
| Days Min Temp ≤ 32 °F | 29 | 27 | 28 | 23 | 15 | 6 | 3 | 5 | 17 | 26 | 26 | 29 | 234 |
| Days Min Temp ≤ 0 °F | 5 | 2 | 0 | 0 | 0 | 0 | 0 | 0 | 0 | 0 | 1 | 4 | 12 |
| Heating Degree Days | 1213 | 956 | 877 | 683 | 501 | 273 | 148 | 153 | 368 | 637 | 909 | 1179 | 7897 |
| Cooling Degree Days | 0 | 0 | 0 | 0 | 1 | 9 | 24 | 22 | 2 | 0 | 0 | 0 | 58 |
| Total Precipitation (") | 1.93 | 1.29 | 1.33 | 1.32 | 1.64 | 1.34 | 0.69 | 0.92 | 0.82 | 1.22 | 1.91 | 1.87 | 16.28 |
| Days ≥ 0.1" Precip | 7 | 5 | 5 | 5 | 4 | 4 | 2 | 3 | 3 | 3 | *7* | 6 | 54 |
| Total Snowfall (") | 9.7 | 5.6 | 3.9 | 1.2 | 0.2 | 0.0 | 0.0 | 0.0 | 0.0 | 0.3 | 5.7 | 9.6 | 36.2 |
| Days ≥ 1" Snow Depth | 18 | 8 | 4 | 1 | 0 | 0 | 0 | 0 | 0 | 0 | 4 | 13 | 48 |

## UNION EXP STN *Union County* ELEVATION 2772 ft LAT/LONG 45° 13 ' N / 117° 53 ' W

| | JAN | FEB | MAR | APR | MAY | JUN | JUL | AUG | SEP | OCT | NOV | DEC | YEAR |
|---|---|---|---|---|---|---|---|---|---|---|---|---|---|
| Maximum Temp °F | 36.8 | 43.0 | 51.0 | 58.1 | 66.3 | 74.0 | 83.2 | 84.1 | 74.3 | 62.6 | 46.5 | 38.0 | 59.8 |
| Minimum Temp °F | 24.0 | 27.3 | 30.0 | 34.1 | 39.8 | 46.0 | 49.6 | 48.8 | 41.0 | 33.9 | 30.3 | 24.9 | 35.8 |
| Mean Temp °F | 30.4 | 35.2 | 40.5 | 46.1 | 53.1 | 60.0 | 66.4 | 66.5 | 57.7 | 48.2 | 38.4 | 31.5 | 47.8 |
| Days Max Temp ≥ 90 °F | 0 | 0 | 0 | 0 | 0 | 2 | 9 | 9 | 2 | 0 | 0 | 0 | 22 |
| Days Max Temp ≤ 32 °F | 9 | 3 | 1 | 0 | 0 | 0 | 0 | 0 | 0 | 0 | 1 | 6 | 20 |
| Days Min Temp ≤ 32 °F | 25 | 20 | 20 | 13 | 4 | 0 | 0 | 0 | 3 | 13 | 18 | 24 | 140 |
| Days Min Temp ≤ 0 °F | 1 | 0 | 0 | 0 | 0 | 0 | 0 | 0 | 0 | 0 | 0 | 1 | 2 |
| Heating Degree Days | 1065 | 835 | 753 | 560 | 371 | 181 | 60 | 62 | 231 | 512 | 791 | 1032 | 6453 |
| Cooling Degree Days | 0 | 0 | 0 | 1 | 10 | 41 | 103 | 115 | 18 | 0 | 0 | 0 | 288 |
| Total Precipitation (") | 1.19 | 0.91 | 1.18 | 1.32 | 1.76 | 1.63 | 0.68 | 0.94 | 0.96 | 0.97 | 1.39 | 1.14 | 14.07 |
| Days ≥ 0.1" Precip | 4 | 3 | 4 | 4 | 5 | 5 | 2 | 2 | 3 | 3 | 5 | 3 | 43 |
| Total Snowfall (") | 6.4 | 3.4 | 2.4 | 0.6 | 0.1 | 0.0 | 0.0 | 0.0 | 0.0 | 0.2 | 3.0 | 5.5 | 21.6 |
| Days ≥ 1" Snow Depth | 13 | 6 | 2 | 0 | 0 | 0 | 0 | 0 | 0 | 0 | 4 | 10 | 35 |

## VALE *Malheur County* ELEVATION 2241 ft LAT/LONG 43° 59 ' N / 117° 15 ' W

| | JAN | FEB | MAR | APR | MAY | JUN | JUL | AUG | SEP | OCT | NOV | DEC | YEAR |
|---|---|---|---|---|---|---|---|---|---|---|---|---|---|
| Maximum Temp °F | 35.7 | 44.9 | 57.0 | 66.2 | 76.1 | 84.5 | 93.9 | 91.8 | 80.5 | 66.6 | 48.6 | 37.0 | 65.2 |
| Minimum Temp °F | 18.8 | 24.4 | 30.3 | 35.7 | 44.1 | 51.5 | 57.0 | 54.6 | 44.6 | 34.2 | 27.5 | 19.6 | 36.9 |
| Mean Temp °F | 27.3 | 34.7 | 43.7 | 50.9 | 60.1 | 68.2 | 75.5 | 73.2 | 62.6 | 50.4 | 38.1 | 28.2 | 51.1 |
| Days Max Temp ≥ 90 °F | 0 | 0 | 0 | 0 | 3 | 11 | 24 | 20 | 5 | 0 | 0 | 0 | 63 |
| Days Max Temp ≤ 32 °F | 11 | 3 | 0 | 0 | 0 | 0 | 0 | 0 | 0 | 0 | 1 | 8 | 23 |
| Days Min Temp ≤ 32 °F | 28 | 24 | 20 | 11 | 2 | 0 | 0 | 0 | 2 | 13 | 22 | 28 | 150 |
| Days Min Temp ≤ 0 °F | 3 | 1 | 0 | 0 | 0 | 0 | 0 | 0 | 0 | 0 | 0 | 2 | 6 |
| Heating Degree Days | 1164 | 851 | 654 | 419 | 190 | 56 | 6 | 13 | 126 | 446 | 800 | 1133 | 5858 |
| Cooling Degree Days | 0 | 0 | 0 | 4 | 52 | 157 | 319 | 273 | 62 | 1 | 0 | 0 | 868 |
| Total Precipitation (") | 1.18 | 0.86 | 0.90 | 0.77 | 0.81 | 0.84 | 0.35 | 0.49 | 0.50 | 0.63 | 1.18 | 1.28 | 9.79 |
| Days ≥ 0.1" Precip | 4 | 3 | 3 | 3 | 3 | 3 | 1 | 1 | 2 | 2 | 4 | 4 | 33 |
| Total Snowfall (") | *4.9* | *1.0* | 0.2 | 0.0 | 0.0 | 0.0 | 0.0 | 0.0 | 0.0 | 0.1 | 1.8 | *4.9* | 12.9 |
| Days ≥ 1" Snow Depth | *9* | 3 | 0 | 0 | 0 | 0 | 0 | 0 | 0 | 0 | 1 | na | na |

## VERNONIA 2 *Columbia County* ELEVATION 630 ft LAT/LONG 45° 52' N / 123° 11' W

| | JAN | FEB | MAR | APR | MAY | JUN | JUL | AUG | SEP | OCT | NOV | DEC | YEAR |
|---|---|---|---|---|---|---|---|---|---|---|---|---|---|
| Maximum Temp °F | 44.3 | 49.4 | 54.1 | 58.0 | 64.3 | 69.5 | 75.3 | 76.7 | 72.5 | 62.6 | 50.4 | 44.0 | 60.1 |
| Minimum Temp °F | 29.2 | 30.7 | 33.3 | 35.7 | 39.9 | 44.6 | 47.1 | 46.7 | 42.4 | 36.9 | 33.5 | 30.4 | 37.5 |
| Mean Temp °F | 36.8 | 40.1 | 43.8 | 46.9 | 52.1 | 57.1 | 61.2 | 61.7 | 57.5 | 49.8 | 42.0 | 37.2 | 48.9 |
| Days Max Temp ≥ 90 °F | 0 | 0 | 0 | 0 | 0 | 1 | 2 | 3 | 1 | 0 | 0 | 0 | 7 |
| Days Max Temp ≤ 32 °F | 1 | 0 | 0 | 0 | 0 | 0 | 0 | 0 | 0 | 0 | 0 | 1 | 2 |
| Days Min Temp ≤ 32 °F | 20 | 17 | 15 | 9 | 2 | 0 | 0 | 0 | 2 | 7 | 13 | 18 | 103 |
| Days Min Temp ≤ 0 °F | 0 | 0 | 0 | 0 | 0 | 0 | 0 | 0 | 0 | 0 | 0 | 0 | 0 |
| Heating Degree Days | 869 | 696 | 652 | 536 | 394 | 240 | 140 | 127 | 226 | 464 | 684 | 855 | 5883 |
| Cooling Degree Days | 0 | 0 | 0 | 0 | 4 | 9 | 26 | 31 | 9 | 0 | 0 | 0 | 79 |
| Total Precipitation (") | 7.23 | 5.45 | 5.11 | 3.64 | 2.22 | 1.67 | 0.63 | 0.88 | 2.08 | 3.76 | 6.93 | 8.03 | 47.63 |
| Days ≥ 0.1" Precip | 14 | 12 | 13 | 10 | 7 | 5 | 2 | 3 | 6 | 9 | 15 | 15 | 111 |
| Total Snowfall (") | 4.4 | 3.5 | 0.4 | 0.0 | 0.0 | 0.0 | 0.0 | 0.0 | 0.0 | 0.0 | 1.2 | 2.8 | 12.3 |
| Days ≥ 1" Snow Depth | *3* | *2* | 0 | 0 | 0 | 0 | 0 | 0 | 0 | 0 | 1 | 3 | 9 |

## WALLOWA *Wallowa County* ELEVATION 2953 ft LAT/LONG 45° 34' N / 117° 32' W

| | JAN | FEB | MAR | APR | MAY | JUN | JUL | AUG | SEP | OCT | NOV | DEC | YEAR |
|---|---|---|---|---|---|---|---|---|---|---|---|---|---|
| Maximum Temp °F | 34.7 | 41.9 | 51.2 | 59.6 | 68.1 | 76.0 | 84.4 | 84.6 | 75.6 | 62.5 | 44.6 | 35.4 | 59.9 |
| Minimum Temp °F | 18.5 | 22.1 | 26.4 | 31.2 | 37.0 | 43.0 | 45.4 | 44.2 | 37.1 | 30.0 | 25.8 | 19.0 | 31.6 |
| Mean Temp °F | 26.6 | 32.0 | 38.8 | 45.4 | 52.6 | 59.5 | 64.9 | 64.5 | 56.4 | 46.3 | 35.2 | 27.3 | 45.8 |
| Days Max Temp ≥ 90 °F | 0 | 0 | 0 | 0 | 0 | 3 | 10 | 10 | 3 | 0 | 0 | 0 | 26 |
| Days Max Temp ≤ 32 °F | 11 | 3 | 0 | 0 | 0 | 0 | 0 | 0 | 0 | 0 | 2 | 10 | 26 |
| Days Min Temp ≤ 32 °F | 28 | 25 | 24 | 17 | 8 | 2 | 0 | 1 | 9 | 20 | 24 | 28 | 186 |
| Days Min Temp ≤ 0 °F | 3 | 1 | 0 | 0 | 0 | 0 | 0 | 0 | 0 | 0 | 0 | 3 | 7 |
| Heating Degree Days | 1183 | 925 | 805 | 581 | 380 | 181 | 65 | 75 | 257 | 573 | 887 | 1164 | 7076 |
| Cooling Degree Days | 0 | 0 | 0 | 0 | 4 | 24 | 70 | 73 | 7 | 0 | 0 | 0 | 178 |
| Total Precipitation (") | 1.95 | 1.29 | 1.29 | 1.27 | 1.67 | 1.53 | 0.94 | 0.93 | 1.15 | 1.33 | 1.88 | 1.79 | 17.02 |
| Days ≥ 0.1" Precip | 6 | 4 | 5 | 4 | 5 | 5 | 3 | 3 | 3 | 5 | 6 | 6 | 55 |
| Total Snowfall (") | 12.3 | 6.2 | 3.8 | 0.9 | 0.1 | 0.0 | 0.0 | 0.0 | 0.0 | 0.3 | 6.3 | 10.4 | 40.3 |
| Days ≥ 1" Snow Depth | 19 | 12 | 4 | 0 | 0 | 0 | 0 | 0 | 0 | 0 | 5 | 14 | 54 |

## WHITEHORSE RANCH *Harney County* ELEVATION 4380 ft LAT/LONG 42° 20' N / 118° 14' W

| | JAN | FEB | MAR | APR | MAY | JUN | JUL | AUG | SEP | OCT | NOV | DEC | YEAR |
|---|---|---|---|---|---|---|---|---|---|---|---|---|---|
| Maximum Temp °F | 40.9 | 47.1 | 53.3 | 60.3 | 69.2 | 77.6 | 86.4 | 84.8 | 76.7 | *65.5* | 50.2 | 41.7 | 62.8 |
| Minimum Temp °F | 18.7 | 21.8 | 26.5 | 30.6 | 37.5 | 44.2 | 50.4 | 49.8 | 41.5 | *33.4* | 25.5 | 18.8 | 33.2 |
| Mean Temp °F | 29.8 | 34.6 | 40.0 | 45.5 | 53.4 | 60.9 | 68.4 | 67.3 | 59.1 | *49.5* | 37.9 | 30.3 | 48.1 |
| Days Max Temp ≥ 90 °F | 0 | 0 | 0 | 0 | 0 | 3 | 11 | 9 | 1 | *0* | 0 | 0 | 24 |
| Days Max Temp ≤ 32 °F | 6 | 2 | 0 | 0 | 0 | 0 | 0 | 0 | 0 | *0* | 1 | 5 | 14 |
| Days Min Temp ≤ 32 °F | 27 | 24 | 23 | 17 | 8 | 1 | 0 | 0 | 4 | *13* | 23 | 27 | 167 |
| Days Min Temp ≤ 0 °F | 2 | 1 | 0 | 0 | 0 | 0 | 0 | 0 | 0 | *0* | 0 | 2 | 5 |
| Heating Degree Days | 1084 | 853 | 769 | 578 | 360 | 162 | 35 | 50 | 194 | *475* | *798* | 1068 | 6426 |
| Cooling Degree Days | 0 | 0 | 0 | 0 | 10 | 43 | 153 | 145 | 30 | *1* | *0* | 0 | 382 |
| Total Precipitation (") | *0.57* | 0.62 | 0.81 | 0.99 | 0.78 | 0.59 | 0.22 | 0.80 | 0.53 | *0.61* | 0.83 | 0.81 | 8.16 |
| Days ≥ 0.1" Precip | *2* | 2 | 2 | 3 | 2 | 2 | 1 | 2 | 2 | *2* | 3 | 3 | 26 |
| Total Snowfall (") | na | *2.1* | 2.3 | 0.9 | 0.1 | 0.0 | 0.0 | 0.0 | 0.2 | 0.1 | *1.8* | *2.6* | na |
| Days ≥ 1" Snow Depth | na | 1 | *1* | 0 | 0 | 0 | 0 | 0 | 0 | 0 | *1* | na | na |

## WICKIUP DAM *Deschutes County* ELEVATION 4334 ft LAT/LONG 43° 41' N / 121° 42' W

| | JAN | FEB | MAR | APR | MAY | JUN | JUL | AUG | SEP | OCT | NOV | DEC | YEAR |
|---|---|---|---|---|---|---|---|---|---|---|---|---|---|
| Maximum Temp °F | 37.7 | 42.2 | 47.0 | 53.3 | 62.9 | 71.0 | 79.8 | 80.1 | 72.4 | 61.2 | 44.7 | 37.8 | 57.5 |
| Minimum Temp °F | 17.2 | 19.7 | 23.8 | 28.1 | 34.0 | 40.3 | 43.9 | 42.4 | 35.3 | 29.1 | 24.9 | 18.6 | 29.8 |
| Mean Temp °F | 27.5 | 31.0 | 35.4 | 40.7 | 48.4 | 55.7 | 61.9 | 61.3 | 53.9 | 45.2 | 34.8 | 28.2 | 43.7 |
| Days Max Temp ≥ 90 °F | 0 | 0 | 0 | 0 | 0 | 1 | 5 | 4 | 1 | 0 | 0 | 0 | 11 |
| Days Max Temp ≤ 32 °F | 8 | 3 | 1 | 0 | 0 | 0 | 0 | 0 | 0 | 0 | 1 | 7 | 20 |
| Days Min Temp ≤ 32 °F | 29 | 27 | 29 | 24 | 14 | 3 | 0 | 1 | 10 | 24 | 27 | 30 | 218 |
| Days Min Temp ≤ 0 °F | 3 | 1 | 0 | 0 | 0 | 0 | 0 | 0 | 0 | 0 | 0 | 2 | 6 |
| Heating Degree Days | 1157 | 954 | 910 | 722 | 508 | 284 | 133 | 142 | 330 | 606 | 899 | 1134 | 7779 |
| Cooling Degree Days | 0 | 0 | 0 | 0 | 3 | 14 | 45 | 34 | 5 | 0 | 0 | 0 | 101 |
| Total Precipitation (") | 3.44 | 2.26 | 1.93 | 1.21 | 0.93 | 1.06 | 0.69 | 0.84 | 0.76 | 1.35 | 3.06 | 3.43 | 20.96 |
| Days ≥ 0.1" Precip | 8 | 7 | 6 | 4 | 3 | 3 | 2 | 2 | 2 | 4 | 8 | 9 | 58 |
| Total Snowfall (") | 20.9 | 14.9 | 10.7 | 4.2 | 0.5 | 0.0 | 0.0 | 0.0 | 0.0 | 1.9 | 12.4 | 20.2 | 85.7 |
| Days ≥ 1" Snow Depth | 27 | 22 | 15 | 4 | 0 | 0 | 0 | 0 | 0 | 1 | 10 | 21 | 100 |

## JANUARY MINIMUM TEMPERATURE °F

### LOWEST

| | | |
|---|---|---|
| 1 | Seneca | 8.8 |
| 2 | Austin | 10.7 |
| 3 | Ironside | 13.6 |
| 4 | Ukiah | 14.8 |
| 5 | Fremont | 15.1 |
| 6 | Halfway | 15.2 |
| 7 | Beulah | 16.4 |
| | Ochoco | 16.4 |
| 9 | Paulina | 16.6 |
| 10 | Sheaville | 16.8 |
| 11 | Brothers | 17.2 |
| | Wickiup Dam | 17.2 |
| 13 | Baker | 17.4 |
| 14 | Malheur-Refuge | 17.5 |
| 15 | Crater Lake | 17.7 |
| | Rome | 17.7 |
| 17 | Rockville | 18.2 |
| | Squaw Butte | 18.2 |
| 19 | Barnes | 18.3 |
| 20 | Wallowa | 18.5 |
| 21 | McDermitt | 18.6 |
| 22 | Whitehorse | 18.7 |
| 23 | Hart Mountain | 18.8 |
| | Vale | 18.8 |
| 25 | Malheur-Brch | 19.0 |

### HIGHEST

| | | |
|---|---|---|
| 1 | Brookings | 41.1 |
| 2 | Gold Beach | 40.3 |
| 3 | North Bend | 39.3 |
| 4 | Bandon | 38.2 |
| 5 | Newport | 37.9 |
| 6 | Cloverdale | 37.2 |
| 7 | Seaside | 37.0 |
| 8 | Elkton | 36.5 |
| | Tidewater | 36.5 |
| 10 | Otis | 36.2 |
| 11 | Astoria | 36.1 |
| 12 | Tillamook | 35.8 |
| 13 | Illahe | 35.7 |
| 14 | Oregon City | 35.1 |
| 15 | Lookout Point Dm | 34.6 |
| | Roseburg | 34.6 |
| 17 | Powers | 34.3 |
| 18 | Drain | 34.2 |
| | Riddle | 34.2 |
| 20 | Portland-Intl | 34.1 |
| 21 | Portland-Hdwrks | 34.0 |
| 22 | Estacada | 33.9 |
| | Troutdale | 33.9 |
| 24 | Clatskanie | 33.7 |
| 25 | Eugene | 33.6 |

## JULY MAXIMUM TEMPERATURE °F

### HIGHEST

| | | |
|---|---|---|
| 1 | Ontario | 95.0 |
| 2 | Vale | 93.9 |
| 3 | Juntura | 93.5 |
| 4 | Pelton Dam | 93.1 |
| 5 | Huntington | 92.9 |
| 6 | Owyhee Dam | 92.2 |
| 7 | Rome | 92.0 |
| 8 | Arlington | 91.5 |
| 9 | McDermitt | 91.1 |
| 10 | Richland | 91.0 |
| 11 | Malheur-Brch | 90.9 |
| 12 | Medford-Jackson | 90.7 |
| 13 | Nyssa | 90.6 |
| 14 | Monument | 90.3 |
| 15 | Riverside | 89.9 |
| 16 | Beulah | 89.8 |
| | Grants Pass | 89.8 |
| 18 | Ruch | 89.7 |
| 19 | Cave Junction | 88.9 |
| 20 | Pilot Rock | 88.5 |
| 21 | Medford-Exp | 88.4 |
| 22 | Milton Freewater | 88.3 |
| 23 | Adel | 88.2 |
| | Hermiston | 88.2 |
| | Rockville | 88.2 |

### LOWEST

| | | |
|---|---|---|
| 1 | Newport | 64.6 |
| 2 | North Bend | 66.3 |
| 3 | Bandon | 66.6 |
| 4 | Tillamook | 67.3 |
| 5 | Government Cmp | 67.5 |
| 6 | Astoria | 67.7 |
| | Crater Lake | 67.7 |
| 8 | Seaside | 67.8 |
| 9 | Brookings | 67.9 |
| 10 | Gold Beach | 68.1 |
| 11 | Otis | 69.6 |
| 12 | Cloverdale | 70.2 |
| 13 | Scotts Mills | 71.5 |
| 14 | Clatskanie | 73.4 |
| 15 | Tidewater | 74.7 |
| 16 | Sexton Summit | 74.8 |
| 17 | Vernonia | 75.3 |
| 18 | Silver Creek Falls | 76.9 |
| 19 | Three Lynx | 77.4 |
| 20 | Detroit Dam | 77.5 |
| 21 | Cascadia | 77.8 |
| 22 | Powers | 78.0 |
| 23 | Bonneville Dam | 78.4 |
| | Portland-Hdwrks | 78.4 |
| 25 | Corvallis-Water | 78.5 |

## ANNUAL PRECIPITATION (")

### HIGHEST

| | | |
|---|---|---|
| 1 | Otis | 96.13 |
| 2 | Tidewater | 88.62 |
| 3 | Tillamook | 87.19 |
| 4 | Detroit Dam | 85.51 |
| 5 | Government Cmp | 85.06 |
| 6 | Cloverdale | 82.10 |
| 7 | Scotts Mills | 80.42 |
| 8 | Illahe | 80.40 |
| 9 | Portland-Hdwrks | 78.39 |
| 10 | Gold Beach | 77.94 |
| 11 | Silver Creek Falls | 77.03 |
| 12 | Bonneville Dam | 74.00 |
| 13 | Seaside | 73.50 |
| 14 | Belknap Springs | 72.25 |
| 15 | Brookings | 71.38 |
| 16 | Three Lynx | 70.54 |
| 17 | Newport | 69.23 |
| 18 | Marion | 66.29 |
| 19 | Falls City | 65.79 |
| 20 | Astoria | 64.86 |
| 21 | Corvallis-Water | 64.54 |
| 22 | Crater Lake | 63.94 |
| 23 | North Bend | 63.67 |
| 24 | Leaburg | 63.22 |
| 25 | Cascadia | 61.80 |

### LOWEST

| | | |
|---|---|---|
| 1 | Rome | 7.94 |
| 2 | Whitehorse | 8.16 |
| 3 | Redmond | 8.27 |
| 4 | Arlington | 8.51 |
| 5 | Alkali Lake | 8.72 |
| 6 | Hermiston | 9.03 |
| 7 | Silver Lake | 9.28 |
| 8 | McDermitt | 9.32 |
| 9 | Owyhee Dam | 9.49 |
| 10 | Ontario | 9.62 |
| 11 | Brothers | 9.67 |
| 12 | Malheur-Refuge | 9.77 |
| 13 | Adel | 9.79 |
| | Vale | 9.79 |
| 15 | Riverside | 9.81 |
| 16 | Pelton Dam | 9.92 |
| 17 | Madras | 10.07 |
| 18 | Malheur-Brch | 10.12 |
| 19 | Prineville | 10.23 |
| 20 | Nyssa | 10.24 |
| 21 | Baker | 10.60 |
| 22 | Juntura | 10.65 |
| | Paisley | 10.65 |
| 24 | Moro | 10.86 |
| 25 | Squaw Butte | 11.05 |

## ANNUAL SNOWFALL (")

### HIGHEST

| | | |
|---|---|---|
| 1 | Crater Lake | 475.3 |
| 2 | Government Cmp | 266.5 |
| 3 | Hward Prairie Dm | 135.8 |
| 4 | Chemult | 117.2 |
| 5 | Marion | 100.6 |
| 6 | Sexton Summit | 93.1 |
| 7 | Austin | 92.3 |
| 8 | Wickiup Dam | 85.7 |
| 9 | Belknap Springs | 78.2 |
| 10 | Scotts Mills | 75.7 |
| 11 | Halfway | 73.1 |
| 12 | Lakeview | 57.1 |
| 13 | Elgin | 49.1 |
| 14 | Prospect | 46.5 |
| 15 | Barnes | 45.3 |
| 16 | Ochoco | 44.3 |
| 17 | Hart Mountain | 43.3 |
| 18 | Toketee Falls | 42.0 |
| 19 | Wallowa | 40.3 |
| 20 | Fremont | 39.5 |
| 21 | Ukiah | 36.2 |
| 22 | Long Creek | 36.0 |
| 23 | Hood River | 35.2 |
| 24 | Bend | 34.3 |
| 25 | Klamath Falls | 32.5 |

### LOWEST

| | | |
|---|---|---|
| 1 | Gold Beach | 0.3 |
| 2 | Brookings | 0.7 |
| 3 | Bandon | 1.1 |
| 4 | Newport | 1.3 |
| 5 | Tidewater | 1.4 |
| 6 | Seaside | 1.5 |
| 7 | Tillamook | 1.8 |
| 8 | North Bend | 1.9 |
| 9 | Lookout Point Dm | 2.0 |
| 10 | Cloverdale | 2.1 |
| 11 | Drain | 2.3 |
| 12 | North Willamette | 2.4 |
| | Powers | 2.4 |
| 14 | Elkton | 2.8 |
| 15 | Troutdale | 3.0 |
| 16 | Fern Ridge Dam | 3.2 |
| | Otis | 3.2 |
| 18 | Roseburg | 3.3 |
| 19 | Oregon City | 3.4 |
| 20 | Grants Pass | 3.5 |
| 21 | Estacada | 3.7 |
| 22 | Cottage Grve Dm | 3.9 |
| 23 | Medford-Exp | 4.0 |
| 24 | McMinnville | 4.2 |
| 25 | Astoria | 4.3 |

# PENNSYLVANIA

PHYSICAL FEATURES. The erratic course of the Delaware River is the only natural boundary of Pennsylvania. All others are arbitrary boundaries that do not conform to physical features. Notable contrasts in topography, climate, and soils exist. Within this 45,126-square-mile area lies a great variety of physical land forms of which the most notable is the Appalachian Mountain system composed of two ranges, the Blue Ridge and the Allegheny. These mountains divide the Commonwealth into three major topographical sections. In addition, two plain areas of relatively small size also exist, one in the southeast and the other in the northwest.

In the extreme southeast is the Coastal Plain situated along the Delaware River and covering an area 50 miles long and 10 miles wide. The land is low, flat, and poorly drained. Bordering the Coastal Plain and extending 60 to 80 miles northwest to the Blue Ridge is the Piedmont Plateau, with elevations ranging from 100 to 500 feet and including rolling or undulating uplands, low hills, fertile valleys, and well-drained soils. Just northwest of the Piedmont and between the Blue Ridge and Allegheny Mountains is the Ridge and Valley Region, 80 to 100 miles wide and characterized by parallel ridges and valleys oriented northeast-southwest. The mountain ridges vary from 1,300 to 1,600 feet above sea level. North and west of the Ridge and Valley Region and extending to the New York and Ohio borders is the area known as the Allegheny Plateau. This is the largest natural division of the State and occupies more than half the area. It is crossed by many deep narrow valleys and drained by the Delaware, Susquehanna, Allegheny, and Monongahela River systems. Elevations are generally 1,000 to 2,000 feet above sea level. Bordering Lake Erie is a narrow 40-mile strip of flat, rich land 3 to 4 miles wide called the Lake Erie Plain.

GENERAL CLIMATE. Pennsylvania is generally considered to have a humid continental type of climate, but the varied physiographic features have a marked effect on the weather and climate of the various sections within the State. The prevailing westerly winds carry most of the weather disturbances that affect Pennsylvania from the interior of the continent, so that the Atlantic Ocean has only limited influence upon the climate of the State.

TEMPERATURE. Throughout the State temperatures generally remain between 0° and 100° and average from near 47° annually in the north-central mountains to 57° annually in the extreme southeast. Summers are generally warm, averaging about 68° along Lake Erie to 74° in southeastern counties. High temperatures, 90° or above, occur on the average of 10 to 20 days per year in most sections. During the coldest months temperatures average near the freezing point with daily minimum readings sometimes near 0° or below. Freezing temperatures occur on the average of 100 or more days annually with the greatest number of occurrences in mountainous regions.

PRECIPITATION. Precipitation is fairly evenly distributed throughout the year. Annual amounts generally range between 34 to 52 inches, while the majority of places receive 38 to 46 inches. Greatest amounts usually occur in spring and summer months, while February is the driest month. Precipitation tends to be somewhat greater in eastern sections due primarily to coastal storms which occasionally frequent the area. During the warm season these storms bring heavy rain, while in winter heavy snow or a mixture of rain and snow may be produced. Thunderstorms, which average between 30 and 35 per year, are concentrated in the warm months and are responsible for most of the summertime rainfall. Winter precipitation is usually 3 to 4 inches less than summer rainfall and is produced most frequently from northeastward-moving storms. When temperatures are low enough these storms sometimes cause heavy snow which may accumulate to 20 inches or more. Annual snowfall ranges between wide limits from year to year and place to place. Some years are quite lean as snowfall may total less than 10 inches while other years may produce upwards to 100 inches mostly in northern and mountainous areas. Measurable snow generally occurs between November 20 and March 15 although snow has been observed as early as the beginning of October and as late as May, especially in northern counties. Greatest monthly amounts usually fall in December and January, however, greatest amounts from individual storms generally occur in March as the moisture supply increases with the annual march of temperature.

STORMS. Hurricanes or low pressure systems with a tropical origin seldom affect the State. However, the tornado does occur in Pennsylvania. At least one tornado has been noted in almost all counties. On the average, 5 or 6 tornadoes are observed annually in Pennsylvania, and the State ranks 27th nationally. June is the month of highest frequency, followed closely by July and August. Principal areas of tornado concentration are in the extreme northwest, the Southwest Plateau, and the Southeastern Piedmont.

CLIMATIC AREAS. The topographic features of Pennsylvania divide the State into four rather distinct climatic areas: (1) the Southeastern Coastal Plain and Piedmont Plateau, (2) the Ridge and Valley Province, (3) the Allegheny Plateau, and (4) the Lake Erie Plain.

In the Southeastern Coastal Plain and Piedmont Plateau summers are long and at times uncomfortably hot. Daily temperatures reach 90° or above on the average of 25 days during the summer season. From about July 1 to the middle of September this area occasionally experiences uncomfortably warm periods, 4 to 5 days to a week in length, during which light wind movement and high relative humidity make conditions oppressive. In general, the winters are comparatively mild, with an average of less than 100 days with minimum temperatures below the freezing point. Average annual precipitation in the area ranges from about 30 inches in the lower Susquehanna Valley to about 46 in Chester County. Under the influence of an occasional severe coastal storm, a normal month's rainfall, or more, may occur within a period of 48 hours. The average seasonal snowfall is about 30 inches, and fields are ordinarily snow covered about one-third of the time during the winter season.

The Ridge and Valley Province is not rugged enough for a true mountain type of climate, but it does have many of the characteristics of such a climate. The mountain-and-valley influence on the air movements causes somewhat greater temperature extremes than are experienced in the southeastern part of the State where the modifying coastal and Chesapeake Bay influence hold them relatively constant, and the daily range of temperature increases somewhat under the valley influences. The effects of nocturnal radiation in the valleys and the tendency for cool airmasses to flow down them at night result in a shortening of the growing season by causing freezes later in spring and earlier in fall than would otherwise occur. The annual precipitation in this area has a mean value of 3 or 4 inches more than in the southeastern part of the State, but its geographic distribution is less uniform. The mountain ridges are high enough to have some deflecting influence on general storm winds, while summer showers and thunderstorms are often shunted up the valleys. Seasonal snowfall of the Ridge and Valley Province varies considerably within short distances.

The Allegheny Plateau is fairly typical of a continental type of climate, with changeable temperatures and more frequent precipitation than other parts of the State. In the more northerly sections the influence of latitude, together with higher elevation and radiation conditions, serve to make this the coldest area in the State. Occasionally, winter minimum temperatures are severe. The daily temperature range is fairly large. Annual precipitation has a mean of about 41 inches, ranging from less than 35 inches to more than 45 inches. The seasonal snowfall averages 54 inches in northern areas, while southern sections receive several inches less. Fields are normally snow covered three-fourths of the time during the winter season. Although average annual precipitation is about equal to that for the State as a whole, it usually occurs in smaller amounts at more frequent intervals.

Although the Lake Erie Plain is of relatively small size, it has a unique and agriculturally advantageous climate typical of the coastal areas surrounding much of the Great Lakes. Both in spring and autumn the lake water exerts a retarding influence on the temperature regime and the freeze-free season is extended about 45 days. In the autumn this prevents early freezing temperatures. Annual precipitation totals about 34.5 inches, which is fairly evenly distributed throughout the year. Snowfall exceeds 54 inches per year, with heavy snows sometimes experienced late in April.

## COUNTY INDEX

*Allegheny County*
PITTSBURGH GR P'BURG

*Armstrong County*
FORD CITY 4 S DAM
PUTNEYVILLE 2 SE DAM

*Beaver County*
MONTGOMERY L&D

*Bradford County*
TOWANDA 1 ESE

*Butler County*
SLIPPERY ROCK 1 SSW

*Cambria County*
EBENSBURG SEWAGE PLA
JOHNSTOWN

*Carbon County*
PALMERTON

*Centre County*
PHILIPSBURG 8 E
STATE COLLEGE

*Chester County*
PHOENIXVILLE 1 E

*Clarion County*
CLARION 3 SW

*Crawford County*
JAMESTOWN 2 NW
MEADVILLE 1 S
TITUSVILLE WTR WORKS

*Cumberland County*
BLOSERVILLE 1 N
SHIPPENSBURG

*Delaware County*
MARCUS HOOK

*Elk County*
RIDGWAY

*Erie County*
CORRY
ERIE INTL ARPT

*Fayette County*
UNIONTOWN 1 NE

*Forest County*
TIONESTA 2 SE LAKE

*Franklin County*
CHAMBERSBURG 1 ESE

*Greene County*
WAYNESBURG 1 E

*Indiana County*
INDIANA 3 SE
MARION CENTER 2 SE

*Jefferson County*
BROOKVILLE SWG PLT
DUBOIS FAA AP

*Lancaster County*
HOLTWOOD
LANDISVILLE 2 NW

*Lawrence County*
NEW CASTLE 1 N

*Lebanon County*
LEBANON 2 W

*Lehigh County*
ALLENTOWN A-B-E INTL

*Luzerne County*
FRANCIS E WALTER DAM
WILKES-BARRE SCRANTN

*Lycoming County*
WILLIAMSPRT-LYCOMING

*McKean County*
BRADFORD 4 SW RES 5
BRADFORD REGIONAL AP
KANE 1 NNE

*Mercer County*
GREENVILLE 2 NE
MERCER

*Mifflin County*
LEWISTOWN

*Monroe County*
STROUDSBURG
TOBYHANNA

*Montgomery County*
GRATERFORD 1 E

*Perry County*
NEWPORT RIVER

*Philadelphia County*
PHILADELPHIA INTL AP

*Somerset County*
CONFLUENCE 1 SW DAM

*Susquehanna County*
MONTROSE

*Tioga County*
WELLSBORO 4 SSE

*Union County*
LAURELTON ST VILLAGE

*Venango County*
FRANKLIN

*Warren County*
WARREN

*Washington County*
DONORA 1 SW

*Wayne County*
HAWLEY 1 E
PLEASANT MOUNT 1 W

*Westmoreland County*
DERRY 4 SW
DONEGAL 2 NW
SALINA 3 W

*York County*
HANOVER
YORK 3 SSW PUMP STN

## ELEVATION INDEX

| FEET | STATION NAME |
|---|---|
| 10 | MARCUS HOOK |
| 26 | PHILADELPHIA INTL AP |
| 102 | PHOENIXVILLE 1 E |
| 190 | HOLTWOOD |
| 239 | GRATERFORD 1 E |
| 361 | STROUDSBURG |
| 381 | NEWPORT RIVER |
| 384 | ALLENTOWN A-B-E INTL |
| 390 | YORK 3 SSW PUMP STN |
| 410 | LANDISVILLE 2 NW |
| 410 | PALMERTON |
| 479 | LEWISTOWN |
| 489 | LEBANON 2 W |
| 522 | WILLIAMSPRT-LYCOMING |
| 600 | HANOVER |
| 640 | CHAMBERSBURG 1 ESE |
| 650 | BLOSERVILLE 1 N |
| 689 | MONTGOMERY L&D |
| 712 | SHIPPENSBURG |
| 738 | ERIE INTL ARPT |
| 761 | TOWANDA 1 ESE |
| 810 | DONORA 1 SW |
| 825 | NEW CASTLE 1 N |
| 879 | LAURELTON ST VILLAGE |
| 930 | WILKES-BARRE SCRANTN |
| 942 | WAYNESBURG 1 E |
| 951 | FORD CITY 4 S DAM |
| 981 | HAWLEY 1 E |
| 991 | FRANKLIN |
| 1030 | GREENVILLE 2 NE |
| 1040 | UNIONTOWN 1 NE |
| 1050 | JAMESTOWN 2 NW |
| 1070 | MEADVILLE 1 S |
| 1102 | INDIANA 3 SE |
| 1109 | SALINA 3 W |
| 1112 | CLARION 3 SW |
| 1152 | DERRY 4 SW |
| 1181 | STATE COLLEGE |
| 1191 | WARREN |
| 1201 | TIONESTA 2 SE LAKE |
| 1211 | BROOKVILLE SWG PLT |
| 1211 | JOHNSTOWN |
| 1220 | TITUSVILLE WTR WORKS |
| 1224 | PITTSBURGH GR P'BURG |
| 1230 | MERCER |
| 1270 | PUTNEYVILLE 2 SE DAM |

| FEET | STATION NAME |
|---|---|
| 1352 | SLIPPERY ROCK 1 SSW |
| 1371 | RIDGWAY |
| 1430 | CORRY |
| 1489 | CONFLUENCE 1 SW DAM |
| 1509 | FRANCIS E WALTER DAM |
| 1611 | MARION CENTER 2 SE |
| 1680 | BRADFORD 4 SW RES 5 |
| 1680 | MONTROSE |
| 1752 | DONEGAL 2 NW |
| 1752 | KANE 1 NNE |
| 1800 | PLEASANT MOUNT 1 W |
| 1814 | DUBOIS FAA AP |
| 1923 | WELLSBORO 4 SSE |
| 1942 | EBENSBURG SEWAGE PLA |
| 1950 | TOBYHANNA |
| 2000 | PHILIPSBURG 8 E |
| 2118 | BRADFORD REGIONAL AP |

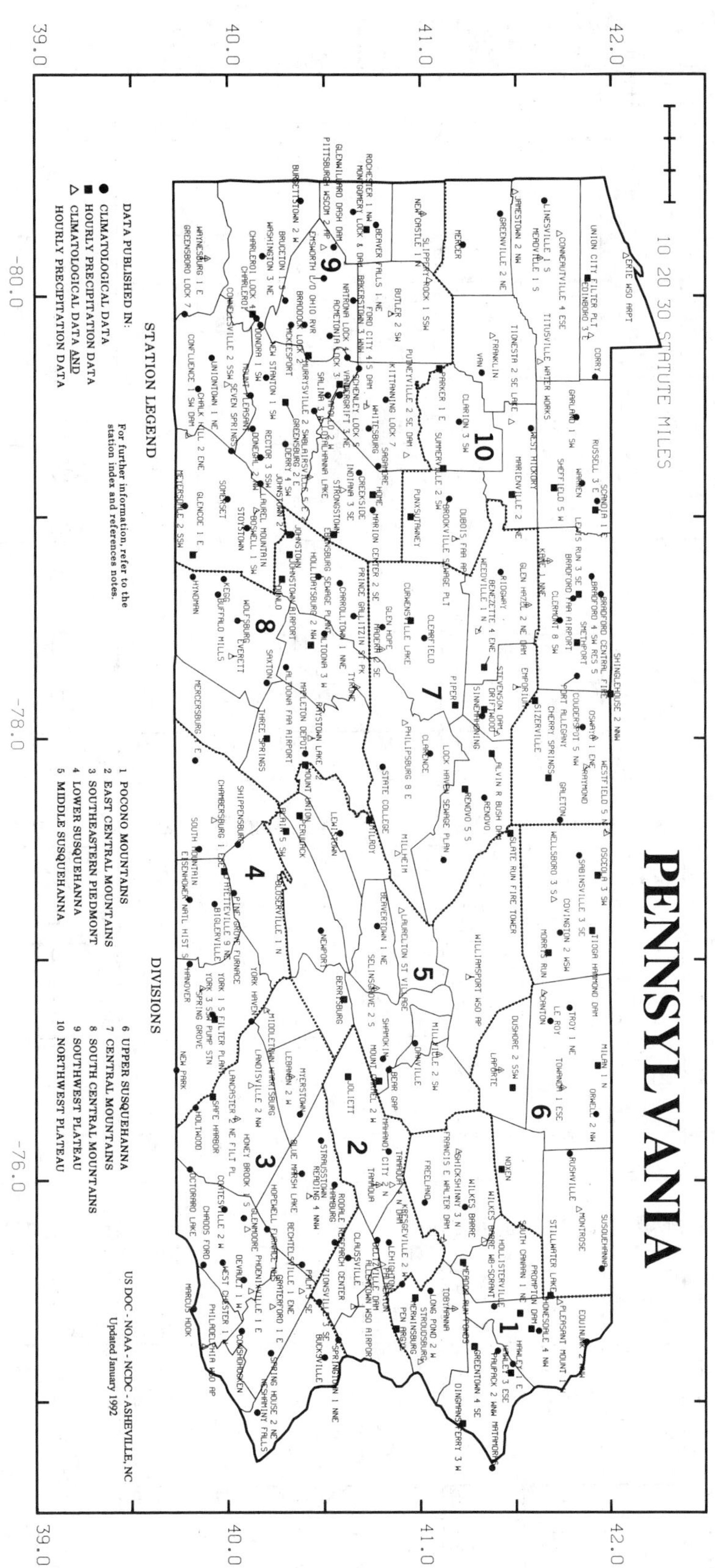
PENNSYLVANIA
10 20 30 STATUTE MILES
STATION LEGEND
DATA PUBLISHED IN:
● CLIMATOLOGICAL DATA
■ HOURLY PRECIPITATION DATA
△ CLIMATOLOGICAL DATA AND HOURLY PRECIPITATION DATA
For further information, refer to the station index and references notes.
DIVISIONS
1 POCONO MOUNTAINS
2 EAST CENTRAL MOUNTAINS
3 SOUTHEASTERN PIEDMONT
4 LOWER SUSQUEHANNA
5 MIDDLE SUSQUEHANNA
6 UPPER SUSQUEHANNA
7 CENTRAL MOUNTAINS
8 SOUTH CENTRAL MOUNTAINS
9 SOUTHWEST PLATEAU
10 NORTHWEST PLATEAU
US DOC - NOAA - NCDC - ASHEVILLE, NC
Updated January 1992

## ALLENTOWN A-B-E INTL *Lehigh County* ELEVATION 384 ft LAT/LONG 40° 39 ' N / 75° 26 ' W

| | JAN | FEB | MAR | APR | MAY | JUN | JUL | AUG | SEP | OCT | NOV | DEC | YEAR |
|---|---|---|---|---|---|---|---|---|---|---|---|---|---|
| Maximum Temp °F | 34.7 | 38.3 | 48.6 | 60.9 | 71.5 | 80.2 | 84.6 | 82.5 | 74.9 | 63.5 | 52.0 | 40.2 | 61.0 |
| Minimum Temp °F | 19.3 | 21.3 | 29.9 | 39.3 | 49.7 | 59.0 | 64.0 | 62.2 | 54.3 | 42.6 | 34.5 | 25.2 | 41.8 |
| Mean Temp °F | 27.0 | 29.8 | 39.2 | 50.1 | 60.6 | 69.6 | 74.3 | 72.4 | 64.6 | 53.0 | 43.3 | 32.7 | 51.4 |
| Days Max Temp ≥ 90 °F | 0 | 0 | 0 | 0 | 1 | 3 | 7 | 4 | 1 | 0 | 0 | 0 | 16 |
| Days Max Temp ≤ 32 °F | 12 | 8 | 1 | 0 | 0 | 0 | 0 | 0 | 0 | 0 | 0 | 6 | 27 |
| Days Min Temp ≤ 32 °F | 28 | 24 | 19 | 6 | 0 | 0 | 0 | 0 | 0 | 4 | 13 | 25 | 119 |
| Days Min Temp ≤ 0 °F | 1 | 0 | 0 | 0 | 0 | 0 | 0 | 0 | 0 | 0 | 0 | 0 | 1 |
| Heating Degree Days | 1171 | 987 | 792 | 445 | 175 | 25 | 2 | 8 | 92 | 370 | 646 | 995 | 5708 |
| Cooling Degree Days | 0 | 0 | 0 | 6 | 55 | 178 | 317 | 245 | 92 | 9 | 0 | 0 | 902 |
| Total Precipitation (") | 3.11 | 2.78 | 3.48 | 3.43 | 4.41 | 3.86 | 4.34 | 4.43 | 4.01 | 3.13 | 3.82 | 3.55 | 44.35 |
| Days ≥ 0.1" Precip | 6 | 6 | 7 | 7 | 8 | 7 | 7 | 7 | 6 | 5 | 7 | 7 | 80 |
| Total Snowfall (") | 9.5 | 9.3 | 5.6 | 0.8 | 0.0 | 0.0 | 0.0 | 0.0 | 0.0 | 0.1 | 1.4 | 5.0 | 31.7 |
| Days ≥ 1" Snow Depth | 14 | 10 | 4 | 0 | 0 | 0 | 0 | 0 | 0 | 0 | 1 | 5 | 34 |

## BLOSERVILLE 1 N *Cumberland County* ELEVATION 650 ft LAT/LONG 40° 16 ' N / 77° 22 ' W

| | JAN | FEB | MAR | APR | MAY | JUN | JUL | AUG | SEP | OCT | NOV | DEC | YEAR |
|---|---|---|---|---|---|---|---|---|---|---|---|---|---|
| Maximum Temp °F | 35.8 | 39.1 | 48.6 | 60.5 | 71.1 | 79.7 | 84.3 | 82.8 | 75.4 | 63.6 | 52.0 | 40.8 | 61.1 |
| Minimum Temp °F | 19.3 | 21.0 | 29.5 | 39.4 | 48.9 | 57.5 | 62.6 | 61.0 | 53.5 | 41.9 | 34.1 | 24.8 | 41.1 |
| Mean Temp °F | 27.7 | 30.1 | 39.1 | 50.0 | 60.0 | 68.6 | 73.5 | 71.9 | 64.5 | 52.8 | 43.1 | 32.8 | 51.2 |
| Days Max Temp ≥ 90 °F | 0 | 0 | 0 | 0 | 0 | 3 | 7 | 4 | 1 | 0 | 0 | 0 | 15 |
| Days Max Temp ≤ 32 °F | 11 | 7 | 2 | 0 | 0 | 0 | 0 | 0 | 0 | 0 | 0 | 6 | 26 |
| Days Min Temp ≤ 32 °F | 28 | 25 | 20 | 6 | 1 | 0 | 0 | 0 | 0 | 3 | 14 | 26 | 123 |
| Days Min Temp ≤ 0 °F | 1 | 0 | 0 | 0 | 0 | 0 | 0 | 0 | 0 | 0 | 0 | 0 | 1 |
| Heating Degree Days | 1149 | 986 | 798 | 452 | 192 | 35 | 5 | 9 | 94 | 378 | 652 | 991 | 5741 |
| Cooling Degree Days | 0 | 0 | 1 | 6 | 47 | 158 | 296 | 227 | 77 | 7 | 1 | 0 | 820 |
| Total Precipitation (") | 2.62 | 2.48 | 3.24 | 3.23 | 3.99 | 3.68 | 3.85 | 3.43 | 3.77 | 3.19 | 3.34 | 2.96 | 39.78 |
| Days ≥ 0.1" Precip | 6 | 5 | 7 | 7 | 8 | 7 | 7 | 6 | 6 | 5 | 6 | 6 | 76 |
| Total Snowfall (") | *7.7* | 7.4 | 5.9 | 0.3 | 0.0 | 0.0 | 0.0 | 0.0 | 0.0 | 0.0 | 1.6 | 4.7 | 27.6 |
| Days ≥ 1" Snow Depth | *5* | *3* | *1* | 0 | 0 | 0 | 0 | 0 | 0 | 0 | 0 | *2* | 11 |

## BRADFORD 4 SW RES 5 *McKean County* ELEVATION 1680 ft LAT/LONG 41° 57 ' N / 78° 44 ' W

| | JAN | FEB | MAR | APR | MAY | JUN | JUL | AUG | SEP | OCT | NOV | DEC | YEAR |
|---|---|---|---|---|---|---|---|---|---|---|---|---|---|
| Maximum Temp °F | 29.7 | 32.6 | 42.4 | 55.7 | 67.0 | 75.0 | 78.4 | 76.5 | 69.4 | 58.5 | 46.0 | 34.6 | 55.5 |
| Minimum Temp °F | 10.8 | 11.7 | 20.8 | 31.6 | 40.7 | 49.1 | 53.6 | 52.3 | 45.9 | 35.5 | 28.7 | 18.5 | 33.3 |
| Mean Temp °F | 20.3 | 22.2 | 31.7 | 43.7 | 53.9 | 62.1 | 66.1 | 64.4 | 57.7 | 47.0 | 37.4 | 26.5 | 44.4 |
| Days Max Temp ≥ 90 °F | 0 | 0 | 0 | 0 | 0 | 0 | 1 | 0 | 0 | 0 | 0 | 0 | 1 |
| Days Max Temp ≤ 32 °F | 18 | 15 | 7 | 1 | 0 | 0 | 0 | 0 | 0 | 0 | 3 | 13 | 57 |
| Days Min Temp ≤ 32 °F | 30 | 27 | 27 | 18 | 7 | 1 | 0 | 0 | 2 | 13 | 21 | 29 | 175 |
| Days Min Temp ≤ 0 °F | 7 | 5 | 2 | 0 | 0 | 0 | 0 | 0 | 0 | 0 | 0 | 3 | 17 |
| Heating Degree Days | 1382 | 1204 | 1026 | 635 | 350 | 128 | 52 | 73 | 233 | 551 | 822 | 1186 | 7642 |
| Cooling Degree Days | 0 | 0 | 0 | 2 | 11 | 43 | 97 | 61 | 19 | 0 | 0 | 0 | 233 |
| Total Precipitation (") | 2.55 | 2.34 | 3.27 | 3.42 | 3.89 | 5.28 | 4.60 | 4.15 | 4.73 | 3.79 | 4.10 | 3.88 | 46.00 |
| Days ≥ 0.1" Precip | 8 | 7 | 8 | 9 | 9 | 9 | 9 | 8 | 9 | 9 | 11 | 11 | 107 |
| Total Snowfall (") | 21.5 | 18.0 | 12.8 | 2.9 | 0.2 | 0.0 | 0.0 | 0.0 | 0.0 | 0.3 | 9.0 | 23.6 | 88.3 |
| Days ≥ 1" Snow Depth | 26 | 25 | 17 | 2 | 0 | 0 | 0 | 0 | 0 | 0 | 6 | 20 | 96 |

## BRADFORD REGIONAL AP *McKean County* ELEVATION 2118 ft LAT/LONG 41° 48 ' N / 78° 38 ' W

| | JAN | FEB | MAR | APR | MAY | JUN | JUL | AUG | SEP | OCT | NOV | DEC | YEAR |
|---|---|---|---|---|---|---|---|---|---|---|---|---|---|
| Maximum Temp °F | 26.9 | 30.0 | 40.2 | 53.2 | 64.4 | 72.6 | 76.6 | 74.9 | 67.3 | 55.8 | 43.1 | 32.0 | 53.1 |
| Minimum Temp °F | 12.3 | 12.9 | 21.9 | 32.4 | 41.1 | 49.4 | 54.2 | 52.8 | 46.4 | 36.5 | 28.6 | 18.4 | 33.9 |
| Mean Temp °F | 19.6 | 21.5 | 31.0 | 42.8 | 52.8 | 61.0 | 65.4 | 63.9 | 56.9 | 46.2 | 35.9 | 25.2 | 43.5 |
| Days Max Temp ≥ 90 °F | 0 | 0 | 0 | 0 | 0 | 0 | 1 | 0 | 0 | 0 | 0 | 0 | 1 |
| Days Max Temp ≤ 32 °F | 21 | 17 | 9 | 1 | 0 | 0 | 0 | 0 | 0 | 0 | 6 | 17 | 71 |
| Days Min Temp ≤ 32 °F | 30 | 26 | 26 | 16 | 7 | 1 | 0 | 0 | 2 | 11 | 21 | 28 | 168 |
| Days Min Temp ≤ 0 °F | 6 | 5 | 1 | 0 | 0 | 0 | 0 | 0 | 0 | 0 | 0 | 3 | 15 |
| Heating Degree Days | 1398 | 1223 | 1046 | 660 | 381 | 154 | 65 | 89 | 258 | 577 | 867 | 1226 | 7944 |
| Cooling Degree Days | 0 | 0 | 0 | 3 | 11 | 39 | 95 | 63 | 21 | 0 | 0 | 0 | 232 |
| Total Precipitation (") | 3.15 | 2.83 | 3.73 | 3.52 | 4.04 | 5.21 | 4.42 | 4.19 | 4.31 | 3.34 | 4.04 | 3.99 | 46.77 |
| Days ≥ 0.1" Precip | 8 | 8 | 9 | 8 | 9 | 9 | 8 | 8 | 8 | 8 | 10 | 10 | 103 |
| Total Snowfall (") | 18.4 | 17.1 | 14.9 | 3.6 | 0.6 | 0.0 | 0.0 | 0.0 | 0.0 | 1.4 | 8.0 | 20.8 | 84.8 |
| Days ≥ 1" Snow Depth | 26 | 24 | 17 | 3 | 0 | 0 | 0 | 0 | 0 | 0 | 8 | 21 | 99 |

## BROOKVILLE SWG PLT *Jefferson County* ELEVATION 1211 ft LAT/LONG 41° 9 ' N / 79° 5 ' W

| | JAN | FEB | MAR | APR | MAY | JUN | JUL | AUG | SEP | OCT | NOV | DEC | YEAR |
|---|---|---|---|---|---|---|---|---|---|---|---|---|---|
| Maximum Temp °F | 33.5 | 37.2 | 47.5 | 60.6 | 70.6 | 78.3 | 81.7 | 80.0 | 73.2 | 62.3 | 50.2 | 38.4 | 59.5 |
| Minimum Temp °F | 15.4 | 16.3 | 24.6 | 33.9 | 43.1 | 51.5 | 56.2 | 55.0 | 48.7 | 37.5 | 31.1 | 21.8 | 36.3 |
| Mean Temp °F | 24.5 | 26.8 | 36.1 | 47.3 | 56.9 | 64.9 | 69.0 | 67.5 | 61.0 | 49.9 | 40.7 | 30.1 | 47.9 |
| Days Max Temp ≥ 90 °F | 0 | 0 | 0 | 0 | 0 | 1 | 2 | 1 | 0 | 0 | 0 | 0 | 4 |
| Days Max Temp ≤ 32 °F | 13 | 10 | 3 | 0 | 0 | 0 | 0 | 0 | 0 | 0 | 1 | 8 | 35 |
| Days Min Temp ≤ 32 °F | 29 | 26 | 24 | 14 | 5 | 0 | 0 | 0 | 1 | 10 | 18 | 26 | 153 |
| Days Min Temp ≤ 0 °F | 5 | 4 | 1 | 0 | 0 | 0 | 0 | 0 | 0 | 0 | 0 | 2 | 12 |
| Heating Degree Days | 1249 | 1073 | 889 | 528 | 265 | 76 | 21 | 34 | 158 | 463 | 723 | 1074 | 6553 |
| Cooling Degree Days | 0 | 0 | 0 | 3 | 19 | 73 | 158 | 114 | 39 | 2 | 0 | 0 | 408 |
| Total Precipitation (") | 2.78 | 2.50 | 3.51 | 3.43 | 4.15 | 4.85 | 4.93 | 4.15 | 4.25 | 3.14 | 3.65 | 3.47 | 44.81 |
| Days ≥ 0.1" Precip | 7 | 7 | 8 | 8 | 9 | 8 | 9 | 7 | 8 | 7 | 8 | 8 | 94 |
| Total Snowfall (") | 14.3 | 10.9 | 8.1 | 1.3 | 0.0 | 0.0 | 0.0 | 0.0 | 0.0 | 0.0 | 2.3 | 10.1 | 47.0 |
| Days ≥ 1" Snow Depth | 20 | 18 | 8 | 1 | 0 | 0 | 0 | 0 | 0 | 0 | 2 | 13 | 62 |

## CHAMBERSBURG 1 ESE *Franklin County* ELEVATION 640 ft LAT/LONG 39° 56 ' N / 77° 39 ' W

| | JAN | FEB | MAR | APR | MAY | JUN | JUL | AUG | SEP | OCT | NOV | DEC | YEAR |
|---|---|---|---|---|---|---|---|---|---|---|---|---|---|
| Maximum Temp °F | 35.8 | 40.0 | 50.4 | 62.7 | 72.6 | 81.1 | 84.9 | 83.3 | 76.1 | 64.0 | 52.5 | 41.3 | 62.1 |
| Minimum Temp °F | 19.6 | 22.0 | 30.2 | 39.6 | 49.5 | 58.0 | 62.5 | 60.8 | 53.3 | 41.3 | 33.7 | 25.3 | 41.3 |
| Mean Temp °F | 27.7 | 31.0 | 40.3 | 51.2 | 61.1 | 69.6 | 73.7 | 72.1 | 64.7 | 52.6 | 43.1 | 33.3 | 51.7 |
| Days Max Temp ≥ 90 °F | 0 | 0 | 0 | 0 | 1 | 3 | 7 | 5 | 1 | 0 | 0 | 0 | 17 |
| Days Max Temp ≤ 32 °F | 11 | 7 | 1 | 0 | 0 | 0 | 0 | 0 | 0 | 0 | 0 | 5 | 24 |
| Days Min Temp ≤ 32 °F | 28 | 24 | 19 | 7 | 1 | 0 | 0 | 0 | 0 | 5 | 14 | 25 | 123 |
| Days Min Temp ≤ 0 °F | 1 | 0 | 0 | 0 | 0 | 0 | 0 | 0 | 0 | 0 | 0 | 0 | 1 |
| Heating Degree Days | 1149 | 952 | 760 | 417 | 167 | 26 | 3 | 9 | 90 | 382 | 650 | 976 | 5581 |
| Cooling Degree Days | 0 | 0 | 2 | 10 | 56 | 177 | 305 | 237 | 87 | 9 | 0 | 0 | 883 |
| Total Precipitation (") | 2.77 | 2.61 | 3.51 | 3.36 | 4.01 | 3.59 | 3.57 | 3.47 | 3.28 | 3.12 | 3.34 | 3.32 | 39.95 |
| Days ≥ 0.1" Precip | 6 | 6 | 7 | 7 | 8 | 6 | 7 | 7 | 6 | 5 | 6 | 6 | 77 |
| Total Snowfall (") | 10.2 | 8.5 | 5.9 | 0.5 | 0.0 | 0.0 | 0.0 | 0.0 | 0.0 | 0.0 | 2.0 | 5.3 | 32.4 |
| Days ≥ 1" Snow Depth | 14 | 11 | 4 | 0 | 0 | 0 | 0 | 0 | 0 | 0 | 1 | 5 | 35 |

## CLARION 3 SW *Clarion County* ELEVATION 1112 ft LAT/LONG 41° 12 ' N / 79° 26 ' W

| | JAN | FEB | MAR | APR | MAY | JUN | JUL | AUG | SEP | OCT | NOV | DEC | YEAR |
|---|---|---|---|---|---|---|---|---|---|---|---|---|---|
| Maximum Temp °F | 33.0 | 36.6 | 47.6 | 61.1 | 72.8 | 80.1 | 83.5 | 81.3 | 74.0 | 62.1 | 48.9 | 37.5 | 59.9 |
| Minimum Temp °F | 15.3 | 16.0 | 24.4 | 34.0 | 43.6 | 52.0 | 56.9 | 56.1 | 49.5 | 38.1 | 30.9 | 21.7 | 36.5 |
| Mean Temp °F | 24.2 | 26.3 | 35.9 | 47.5 | 58.3 | 66.0 | 70.2 | 68.7 | 61.8 | 50.1 | 39.9 | 29.6 | 48.2 |
| Days Max Temp ≥ 90 °F | 0 | 0 | 0 | 0 | 1 | 3 | 5 | 3 | 1 | 0 | 0 | 0 | 13 |
| Days Max Temp ≤ 32 °F | 15 | 10 | 3 | 0 | 0 | 0 | 0 | 0 | 0 | 0 | 2 | 9 | 39 |
| Days Min Temp ≤ 32 °F | 29 | 27 | 25 | 14 | 4 | 0 | 0 | 0 | 1 | 9 | 19 | 27 | 155 |
| Days Min Temp ≤ 0 °F | 4 | 4 | 1 | 0 | 0 | 0 | 0 | 0 | 0 | 0 | 0 | 1 | 10 |
| Heating Degree Days | 1258 | 1087 | 898 | 521 | 233 | 66 | 16 | 26 | 142 | 456 | 746 | 1090 | 6539 |
| Cooling Degree Days | 0 | 0 | 0 | 2 | 34 | 111 | 213 | 160 | 59 | 2 | 0 | 0 | 581 |
| Total Precipitation (") | 2.75 | 2.41 | 3.29 | 3.54 | 4.11 | 5.11 | 4.79 | 4.36 | 4.57 | 3.38 | 4.07 | 3.43 | 45.81 |
| Days ≥ 0.1" Precip | 8 | 7 | 8 | 8 | 9 | 8 | 8 | 8 | 8 | 8 | 8 | 8 | 96 |
| Total Snowfall (") | ***11.2*** | 8.9 | ***4.9*** | 1.1 | 0.0 | 0.0 | 0.0 | 0.0 | 0.0 | 0.0 | 1.7 | ***9.1*** | 36.9 |
| Days ≥ 1" Snow Depth | 20 | 18 | 8 | 1 | 0 | 0 | 0 | 0 | 0 | 0 | 2 | ***12*** | 61 |

## CONFLUENCE 1 SW DAM *Somerset County* ELEVATION 1489 ft LAT/LONG 39° 48 ' N / 79° 22 ' W

| | JAN | FEB | MAR | APR | MAY | JUN | JUL | AUG | SEP | OCT | NOV | DEC | YEAR |
|---|---|---|---|---|---|---|---|---|---|---|---|---|---|
| Maximum Temp °F | 35.4 | 38.9 | 49.1 | 61.2 | 71.4 | 80.0 | 83.5 | 82.2 | 75.5 | 63.7 | 51.8 | 40.2 | 61.1 |
| Minimum Temp °F | 15.4 | 16.7 | 24.8 | 34.2 | 43.5 | 51.9 | 57.0 | 56.2 | 49.5 | 37.3 | 29.8 | 21.3 | 36.5 |
| Mean Temp °F | 25.4 | 27.8 | 37.0 | 47.7 | 57.4 | 66.0 | 70.3 | 69.2 | 62.6 | 50.5 | 40.8 | 30.8 | 48.8 |
| Days Max Temp ≥ 90 °F | 0 | 0 | 0 | 0 | 0 | 2 | 5 | 3 | 1 | 0 | 0 | 0 | 11 |
| Days Max Temp ≤ 32 °F | 12 | 8 | 3 | 0 | 0 | 0 | 0 | 0 | 0 | 0 | 1 | 8 | 32 |
| Days Min Temp ≤ 32 °F | 29 | 26 | 25 | 14 | 3 | 0 | 0 | 0 | 1 | 10 | 20 | 27 | 155 |
| Days Min Temp ≤ 0 °F | 4 | 3 | 0 | 0 | 0 | 0 | 0 | 0 | 0 | 0 | 0 | 1 | 8 |
| Heating Degree Days | 1222 | 1043 | 862 | 514 | 251 | 62 | 13 | 23 | 126 | 444 | 719 | 1055 | 6334 |
| Cooling Degree Days | 0 | 0 | 0 | 3 | 26 | 99 | 203 | 166 | 60 | 2 | 0 | 0 | 559 |
| Total Precipitation (") | 3.25 | 2.70 | 3.86 | 3.86 | 4.39 | 3.69 | 4.89 | 3.88 | 3.93 | 3.06 | 3.51 | 3.64 | 44.66 |
| Days ≥ 0.1" Precip | 8 | 7 | 9 | 9 | 9 | 8 | 9 | 8 | 8 | 7 | 8 | 9 | 99 |
| Total Snowfall (") | 19.1 | 13.8 | 10.5 | 1.8 | 0.0 | 0.0 | 0.0 | 0.0 | 0.0 | 0.0 | 3.7 | 11.7 | 60.6 |
| Days ≥ 1" Snow Depth | 20 | 18 | 9 | 1 | 0 | 0 | 0 | 0 | 0 | 0 | 3 | 13 | 64 |

## CORRY *Erie County* ELEVATION 1430 ft LAT/LONG 41° 55' N / 79° 38' W

| | JAN | FEB | MAR | APR | MAY | JUN | JUL | AUG | SEP | OCT | NOV | DEC | YEAR |
|---|---|---|---|---|---|---|---|---|---|---|---|---|---|
| Maximum Temp °F | 30.9 | 34.2 | 44.5 | 57.5 | 68.6 | 76.4 | 80.0 | 78.0 | 71.0 | 59.9 | 47.3 | 36.0 | 57.0 |
| Minimum Temp °F | 15.2 | 15.6 | 24.0 | 34.1 | 43.6 | 52.4 | 56.9 | 55.8 | 49.8 | 39.6 | 31.8 | 22.0 | 36.7 |
| Mean Temp °F | 23.1 | 24.9 | 34.3 | 45.9 | 56.1 | 64.4 | 68.5 | 66.9 | 60.4 | 49.8 | 39.6 | 29.0 | 46.9 |
| Days Max Temp ≥ 90 °F | 0 | 0 | 0 | 0 | 0 | 1 | 1 | 0 | 0 | 0 | 0 | 0 | 2 |
| Days Max Temp ≤ 32 °F | 17 | 13 | 6 | 0 | 0 | 0 | 0 | 0 | 0 | 0 | 2 | 11 | 49 |
| Days Min Temp ≤ 32 °F | 29 | 26 | 24 | 14 | 4 | 0 | 0 | 0 | 1 | 8 | 18 | 27 | 151 |
| Days Min Temp ≤ 0 °F | 4 | 4 | 1 | 0 | 0 | 0 | 0 | 0 | 0 | 0 | 0 | 1 | 10 |
| Heating Degree Days | 1294 | 1125 | 944 | 571 | 291 | 90 | 28 | 43 | 174 | 467 | 755 | 1109 | 6891 |
| Cooling Degree Days | 0 | 0 | 0 | 3 | 23 | 75 | 153 | 113 | 43 | 2 | 0 | 0 | 412 |
| Total Precipitation (") | 3.15 | 2.67 | 3.53 | 3.55 | 3.70 | 4.65 | 4.85 | 4.67 | 4.49 | 3.90 | 4.44 | 4.35 | 47.95 |
| Days ≥ 0.1" Precip | 9 | 8 | 10 | 9 | 9 | 9 | 8 | 8 | 8 | 10 | 11 | 12 | 111 |
| Total Snowfall (") | 32.8 | 22.9 | 18.1 | 5.4 | 0.3 | 0.0 | 0.0 | 0.0 | 0.0 | 1.1 | 15.1 | 35.6 | 131.3 |
| Days ≥ 1" Snow Depth | 25 | 22 | 13 | 1 | 0 | 0 | 0 | 0 | 0 | 0 | 6 | 19 | 86 |

## DERRY 4 SW *Westmoreland County* ELEVATION 1152 ft LAT/LONG 40° 20' N / 79° 18' W

| | JAN | FEB | MAR | APR | MAY | JUN | JUL | AUG | SEP | OCT | NOV | DEC | YEAR |
|---|---|---|---|---|---|---|---|---|---|---|---|---|---|
| Maximum Temp °F | ***35.9*** | 39.2 | 49.8 | 61.5 | ***72.0*** | ***79.8*** | ***83.9*** | ***81.8*** | ***75.6*** | ***63.6*** | ***52.1*** | ***41.2*** | 61.4 |
| Minimum Temp °F | ***17.7*** | 19.6 | 27.5 | 36.7 | ***46.7*** | ***55.1*** | ***60.2*** | ***58.5*** | ***51.9*** | 40.7 | 33.1 | ***24.3*** | 39.3 |
| Mean Temp °F | ***26.9*** | 29.5 | 38.7 | 49.1 | ***59.3*** | ***67.5*** | ***72.1*** | ***70.2*** | ***63.7*** | ***52.2*** | ***42.6*** | ***32.8*** | 50.4 |
| Days Max Temp ≥ 90 °F | ***0*** | 0 | 0 | 0 | 0 | ***2*** | ***5*** | ***3*** | ***1*** | ***0*** | 0 | 0 | 11 |
| Days Max Temp ≤ 32 °F | ***12*** | 8 | 3 | 0 | 0 | ***0*** | ***0*** | ***0*** | ***0*** | ***0*** | 1 | ***7*** | 31 |
| Days Min Temp ≤ 32 °F | ***27*** | ***24*** | ***21*** | ***11*** | ***3*** | ***0*** | ***0*** | ***0*** | ***1*** | 7 | ***15*** | ***23*** | 132 |
| Days Min Temp ≤ 0 °F | ***4*** | ***2*** | ***0*** | 0 | 0 | ***0*** | ***0*** | ***0*** | ***0*** | 0 | 0 | ***1*** | 7 |
| Heating Degree Days | ***1174*** | 998 | 811 | 476 | ***213*** | ***52*** | ***7*** | ***20*** | ***115*** | ***398*** | ***667*** | ***995*** | 5926 |
| Cooling Degree Days | 0 | 0 | 2 | 7 | 40 | 128 | ***251*** | 191 | 73 | 6 | 0 | 0 | 698 |
| Total Precipitation (") | ***3.06*** | 2.99 | 4.01 | 4.31 | ***4.74*** | ***4.79*** | ***5.22*** | ***4.14*** | ***4.40*** | 3.08 | 4.08 | 3.61 | 48.43 |
| Days ≥ 0.1" Precip | ***9*** | 9 | 10 | ***11*** | ***11*** | ***9*** | ***9*** | ***8*** | ***8*** | 7 | ***11*** | ***10*** | 112 |
| Total Snowfall (") | ***13.4*** | ***11.3*** | ***8.6*** | 1.5 | 0.0 | ***0.0*** | ***0.0*** | ***0.0*** | ***0.0*** | 0.2 | ***2.8*** | ***8.4*** | 46.2 |
| Days ≥ 1" Snow Depth | ***14*** | ***12*** | ***5*** | 1 | 0 | ***0*** | ***0*** | ***0*** | ***0*** | 0 | ***1*** | 7 | 40 |

## DONEGAL 2 NW *Westmoreland County* ELEVATION 1752 ft LAT/LONG 40° 7' N / 79° 23' W

| | JAN | FEB | MAR | APR | MAY | JUN | JUL | AUG | SEP | OCT | NOV | DEC | YEAR |
|---|---|---|---|---|---|---|---|---|---|---|---|---|---|
| Maximum Temp °F | 33.3 | 36.8 | 46.6 | 58.0 | 68.5 | 76.4 | 80.3 | 78.9 | 72.6 | 60.7 | 49.3 | 38.9 | 58.4 |
| Minimum Temp °F | 15.6 | 17.8 | 25.4 | 34.7 | 44.3 | 52.8 | 57.6 | 56.1 | 49.9 | 38.9 | 31.2 | 21.9 | 37.2 |
| Mean Temp °F | 24.5 | 27.3 | 36.0 | 46.4 | 56.4 | 64.6 | 69.0 | 67.6 | 61.3 | 49.7 | 40.3 | 30.4 | 47.8 |
| Days Max Temp ≥ 90 °F | 0 | 0 | 0 | 0 | 0 | 1 | 2 | 1 | 0 | 0 | 0 | 0 | 4 |
| Days Max Temp ≤ 32 °F | 16 | 11 | 5 | 0 | 0 | 0 | 0 | 0 | 0 | 0 | 3 | 10 | 45 |
| Days Min Temp ≤ 32 °F | 29 | 25 | 23 | 14 | 3 | 0 | 0 | 0 | 1 | 9 | 19 | 26 | 149 |
| Days Min Temp ≤ 0 °F | 5 | 3 | 0 | 0 | 0 | 0 | 0 | 0 | 0 | 0 | 0 | 2 | 10 |
| Heating Degree Days | 1250 | 1058 | 892 | 555 | 279 | 86 | 23 | 39 | 154 | 471 | 736 | 1067 | 6610 |
| Cooling Degree Days | 0 | 0 | 0 | 3 | 18 | 79 | 170 | 134 | 57 | 4 | 0 | 0 | 465 |
| Total Precipitation (") | 3.00 | 2.65 | 3.52 | 3.86 | 4.24 | 4.51 | 4.57 | 4.58 | 3.94 | 3.08 | 3.62 | 3.57 | 45.14 |
| Days ≥ 0.1" Precip | 8 | 7 | 8 | 9 | 9 | 8 | 7 | 7 | 6 | 7 | 9 | 8 | 93 |
| Total Snowfall (") | na | na | na | na | 0.0 | 0.0 | 0.0 | 0.0 | 0.0 | 0.0 | na | na | na |
| Days ≥ 1" Snow Depth | na | na | na | 1 | 0 | 0 | 0 | 0 | 0 | 0 | na | na | na |

## DONORA 1 SW *Washington County* ELEVATION 810 ft LAT/LONG 40° 11' N / 79° 51' W

| | JAN | FEB | MAR | APR | MAY | JUN | JUL | AUG | SEP | OCT | NOV | DEC | YEAR |
|---|---|---|---|---|---|---|---|---|---|---|---|---|---|
| Maximum Temp °F | 39.1 | 42.8 | 54.2 | 65.5 | 74.6 | 82.2 | 85.6 | 84.0 | 78.3 | 67.1 | 55.2 | 43.9 | 64.4 |
| Minimum Temp °F | 20.5 | 22.0 | 30.5 | 38.9 | 48.4 | 57.3 | 61.9 | 61.1 | 54.9 | 42.7 | 34.6 | 25.8 | 41.6 |
| Mean Temp °F | 29.8 | 32.4 | 42.4 | 52.3 | 61.6 | 69.8 | 73.8 | 72.6 | 66.6 | 54.9 | 44.9 | 34.9 | 53.0 |
| Days Max Temp ≥ 90 °F | 0 | 0 | 0 | 0 | 1 | 4 | 6 | 5 | 1 | 0 | 0 | 0 | 17 |
| Days Max Temp ≤ 32 °F | 9 | 6 | 1 | 0 | 0 | 0 | 0 | 0 | 0 | 0 | 0 | 5 | 21 |
| Days Min Temp ≤ 32 °F | 27 | 23 | 18 | 8 | 1 | 0 | 0 | 0 | 0 | 4 | 13 | 23 | 117 |
| Days Min Temp ≤ 0 °F | 2 | 1 | 0 | 0 | 0 | 0 | 0 | 0 | 0 | 0 | 0 | 1 | 4 |
| Heating Degree Days | 1083 | 915 | 695 | 383 | 160 | 26 | 1 | 6 | 63 | 318 | 596 | 928 | 5174 |
| Cooling Degree Days | 0 | 0 | 1 | 10 | 61 | 174 | 306 | 251 | 121 | 12 | 1 | 0 | 937 |
| Total Precipitation (") | 2.31 | 2.04 | 3.31 | 2.98 | 3.77 | 3.45 | 3.67 | 3.84 | 2.96 | 2.36 | 2.88 | 2.67 | 36.24 |
| Days ≥ 0.1" Precip | 6 | 5 | 8 | 7 | 8 | 7 | 7 | 6 | 6 | 6 | 7 | 7 | 80 |
| Total Snowfall (") | na | na | ***4.8*** | 0.2 | 0.0 | 0.0 | 0.0 | 0.0 | 0.0 | 0.0 | ***0.7*** | na | na |
| Days ≥ 1" Snow Depth | na | na | ***2*** | 0 | 0 | 0 | 0 | 0 | 0 | 0 | ***0*** | na | na |

## DUBOIS FAA AP *Jefferson County* ELEVATION 1814 ft LAT/LONG 41° 11 ' N / 78° 54 ' W

| | JAN | FEB | MAR | APR | MAY | JUN | JUL | AUG | SEP | OCT | NOV | DEC | YEAR |
|---|---|---|---|---|---|---|---|---|---|---|---|---|---|
| Maximum Temp °F | 29.7 | 33.2 | 43.4 | 56.5 | 67.3 | 75.4 | 79.2 | 77.3 | 69.7 | 58.0 | 46.0 | 35.0 | 55.9 |
| Minimum Temp °F | 16.0 | 17.5 | 25.9 | 36.4 | 46.3 | 54.6 | 59.3 | 57.9 | 51.0 | 40.1 | 32.1 | 22.0 | 38.3 |
| Mean Temp °F | 22.8 | 25.4 | 34.7 | 46.4 | 56.8 | 65.0 | 69.2 | 67.7 | 60.4 | 49.1 | 39.1 | 28.5 | 47.1 |
| Days Max Temp ≥ 90 °F | 0 | 0 | 0 | 0 | 0 | 0 | 1 | 0 | 0 | 0 | 0 | 0 | 1 |
| Days Max Temp ≤ 32 °F | 19 | 14 | 6 | 0 | 0 | 0 | 0 | 0 | 0 | 0 | 3 | 13 | 55 |
| Days Min Temp ≤ 32 °F | 29 | 25 | 23 | 11 | 2 | 0 | 0 | 0 | 0 | 6 | 16 | 27 | 139 |
| Days Min Temp ≤ 0 °F | 4 | 2 | 0 | 0 | 0 | 0 | 0 | 0 | 0 | 0 | 0 | 1 | 7 |
| Heating Degree Days | 1298 | 1112 | 933 | 555 | 273 | 79 | 19 | 34 | 172 | 488 | 771 | 1125 | 6859 |
| Cooling Degree Days | 0 | 0 | 1 | 6 | 26 | 79 | 168 | 123 | 39 | 1 | 0 | 0 | 443 |
| Total Precipitation (") | 2.81 | 2.64 | 3.67 | 3.29 | 3.94 | 4.68 | 3.95 | 4.08 | 3.83 | 2.88 | 3.40 | 3.52 | 42.69 |
| Days ≥ 0.1" Precip | 8 | 7 | 8 | 8 | 9 | 8 | 8 | 7 | 7 | 7 | 8 | 9 | 94 |
| Total Snowfall (") | 14.2 | 13.2 | 10.8 | 2.5 | 0.1 | 0.0 | 0.0 | 0.0 | 0.0 | 0.6 | 5.0 | 12.4 | 58.8 |
| Days ≥ 1" Snow Depth | 22 | 20 | 11 | 2 | 0 | 0 | 0 | 0 | 0 | 0 | 4 | 16 | 75 |

## EBENSBURG SEWAGE PLA *Cambria County* ELEVATION 1942 ft LAT/LONG 40° 28 ' N / 78° 44 ' W

| | JAN | FEB | MAR | APR | MAY | JUN | JUL | AUG | SEP | OCT | NOV | DEC | YEAR |
|---|---|---|---|---|---|---|---|---|---|---|---|---|---|
| Maximum Temp °F | 34.0 | 37.5 | 47.4 | 59.9 | 69.8 | 77.7 | 81.2 | 79.6 | 72.9 | 61.8 | 49.9 | 38.5 | 59.2 |
| Minimum Temp °F | 15.0 | 15.7 | 23.9 | 33.4 | 42.3 | 50.4 | 55.4 | 54.1 | 48.0 | 36.8 | 30.1 | 20.8 | 35.5 |
| Mean Temp °F | 24.5 | 26.6 | 35.7 | 46.7 | 56.1 | 64.1 | 68.3 | 66.9 | 60.5 | 49.3 | 40.0 | 29.7 | 47.4 |
| Days Max Temp ≥ 90 °F | 0 | 0 | 0 | 0 | 0 | 1 | 1 | 1 | 0 | 0 | 0 | 0 | 3 |
| Days Max Temp ≤ 32 °F | 14 | 10 | 3 | 0 | 0 | 0 | 0 | 0 | 0 | 0 | 2 | 9 | 38 |
| Days Min Temp ≤ 32 °F | 29 | 26 | 25 | 15 | 6 | 1 | 0 | 0 | 2 | 11 | 19 | 27 | 161 |
| Days Min Temp ≤ 0 °F | 5 | 4 | 1 | 0 | 0 | 0 | 0 | 0 | 0 | 0 | 0 | 2 | 12 |
| Heating Degree Days | 1249 | 1077 | 901 | 544 | 283 | 89 | 26 | 42 | 169 | 479 | 742 | 1088 | 6689 |
| Cooling Degree Days | 0 | 0 | 0 | 2 | 17 | 74 | 158 | 116 | 44 | 2 | 0 | 0 | 413 |
| Total Precipitation (") | 3.82 | 3.28 | 4.51 | 4.17 | 4.72 | 4.36 | 4.99 | 4.01 | 4.18 | 3.25 | 4.07 | 3.91 | 49.27 |
| Days ≥ 0.1" Precip | 10 | 9 | 10 | 10 | 10 | 9 | 9 | 8 | 8 | 8 | 10 | 10 | 111 |
| Total Snowfall (") | 28.1 | 24.9 | 20.8 | 5.9 | 0.0 | 0.0 | 0.0 | 0.0 | 0.0 | 0.5 | 7.3 | 20.2 | 107.7 |
| Days ≥ 1" Snow Depth | 24 | 22 | 16 | 2 | 0 | 0 | 0 | 0 | 0 | 0 | 4 | 17 | 85 |

## ERIE INTL ARPT *Erie County* ELEVATION 738 ft LAT/LONG 42° 5 ' N / 80° 11 ' W

| | JAN | FEB | MAR | APR | MAY | JUN | JUL | AUG | SEP | OCT | NOV | DEC | YEAR |
|---|---|---|---|---|---|---|---|---|---|---|---|---|---|
| Maximum Temp °F | 31.8 | 33.2 | 42.7 | 54.5 | 65.4 | 74.5 | 79.0 | 77.7 | 71.1 | 59.8 | 48.5 | 37.7 | 56.3 |
| Minimum Temp °F | 18.8 | 18.6 | 26.7 | 36.9 | 46.9 | 56.6 | 62.0 | 61.1 | 54.6 | 44.3 | 35.6 | 26.0 | 40.7 |
| Mean Temp °F | 25.4 | 25.9 | 34.7 | 45.7 | 56.2 | 65.5 | 70.5 | 69.4 | 62.9 | 52.1 | 42.1 | 31.8 | 48.5 |
| Days Max Temp ≥ 90 °F | 0 | 0 | 0 | 0 | 0 | 1 | 1 | 1 | 0 | 0 | 0 | 0 | 3 |
| Days Max Temp ≤ 32 °F | 16 | 14 | 7 | 0 | 0 | 0 | 0 | 0 | 0 | 0 | 1 | 9 | 47 |
| Days Min Temp ≤ 32 °F | 28 | 25 | 23 | 11 | 1 | 0 | 0 | 0 | 0 | 2 | 11 | 24 | 125 |
| Days Min Temp ≤ 0 °F | 2 | 2 | 0 | 0 | 0 | 0 | 0 | 0 | 0 | 0 | 0 | 0 | 4 |
| Heating Degree Days | 1222 | 1097 | 932 | 577 | 296 | 79 | 14 | 19 | 123 | 400 | 681 | 1021 | 6461 |
| Cooling Degree Days | 0 | 0 | 1 | 8 | 37 | 110 | 230 | 191 | 77 | 7 | 1 | 0 | 662 |
| Total Precipitation (") | 2.30 | 2.23 | 3.07 | 3.22 | 3.34 | 4.11 | 3.49 | 4.15 | 4.64 | 3.88 | 4.10 | 3.56 | 42.09 |
| Days ≥ 0.1" Precip | 7 | 6 | 8 | 8 | 8 | 7 | 6 | 7 | 8 | 9 | 11 | 9 | 94 |
| Total Snowfall (") | 25.0 | 17.6 | 11.0 | 2.4 | 0.0 | 0.0 | 0.0 | 0.0 | 0.0 | 0.3 | 9.7 | 22.4 | 88.4 |
| Days ≥ 1" Snow Depth | 21 | 17 | 8 | 1 | 0 | 0 | 0 | 0 | 0 | 0 | 3 | 13 | 63 |

## FORD CITY 4 S DAM *Armstrong County* ELEVATION 951 ft LAT/LONG 40° 43 ' N / 79° 30 ' W

| | JAN | FEB | MAR | APR | MAY | JUN | JUL | AUG | SEP | OCT | NOV | DEC | YEAR |
|---|---|---|---|---|---|---|---|---|---|---|---|---|---|
| Maximum Temp °F | 34.4 | 37.7 | 48.1 | 60.7 | 71.2 | 79.0 | 83.2 | 81.6 | 74.9 | 63.1 | 51.3 | 40.0 | 60.4 |
| Minimum Temp °F | 15.9 | 17.1 | 25.9 | 35.5 | 44.7 | 53.5 | 58.8 | 57.2 | 50.4 | 38.2 | 31.7 | 22.6 | 37.6 |
| Mean Temp °F | 25.2 | 27.5 | 37.0 | 48.1 | 58.0 | 66.3 | 71.1 | 69.4 | 62.7 | 50.6 | 41.5 | 31.3 | 49.1 |
| Days Max Temp ≥ 90 °F | 0 | 0 | 0 | 0 | 0 | 1 | 4 | 3 | 1 | 0 | 0 | 0 | 9 |
| Days Max Temp ≤ 32 °F | 13 | 9 | 3 | 0 | 0 | 0 | 0 | 0 | 0 | 0 | 1 | 7 | 33 |
| Days Min Temp ≤ 32 °F | 28 | 25 | 24 | 12 | 3 | 0 | 0 | 0 | 0 | 8 | 17 | 26 | 143 |
| Days Min Temp ≤ 0 °F | 5 | 3 | 0 | 0 | 0 | 0 | 0 | 0 | 0 | 0 | 0 | 1 | 9 |
| Heating Degree Days | 1228 | 1054 | 861 | 507 | 239 | 63 | 11 | 22 | 128 | 443 | 699 | 1037 | 6292 |
| Cooling Degree Days | 0 | 0 | 1 | 7 | 33 | 115 | 235 | 179 | 69 | 4 | 1 | 0 | 644 |
| Total Precipitation (") | 2.61 | 2.27 | 3.32 | 3.02 | 4.09 | 4.20 | 4.16 | 4.19 | 3.68 | 2.78 | 3.37 | 2.96 | 40.65 |
| Days ≥ 0.1" Precip | 7 | 6 | 8 | 8 | 9 | 8 | 8 | 7 | 7 | 7 | 8 | 8 | 91 |
| Total Snowfall (") | 11.2 | 8.6 | 6.5 | 0.8 | 0.1 | 0.0 | 0.0 | 0.0 | 0.0 | 0.0 | 1.9 | 6.5 | 35.6 |
| Days ≥ 1" Snow Depth | 16 | 13 | 6 | 1 | 0 | 0 | 0 | 0 | 0 | 0 | 1 | 8 | 45 |

## FRANCIS E WALTER DAM *Luzerne County* ELEVATION 1509 ft LAT/LONG 41° 7 ' N / 75° 44 ' W

| | JAN | FEB | MAR | APR | MAY | JUN | JUL | AUG | SEP | OCT | NOV | DEC | YEAR |
|---|---|---|---|---|---|---|---|---|---|---|---|---|---|
| Maximum Temp °F | ***30.4*** | 33.3 | 42.9 | 55.3 | 66.9 | 74.5 | 78.9 | 77.4 | 69.3 | 58.7 | ***47.0*** | ***35.5*** | 55.8 |
| Minimum Temp °F | ***11.2*** | 12.3 | 21.9 | 32.1 | 42.2 | 50.1 | 54.9 | 53.7 | 46.0 | 35.1 | ***27.9*** | ***17.4*** | 33.7 |
| Mean Temp °F | ***20.8*** | 22.9 | 32.4 | 43.7 | 54.6 | 62.4 | 66.9 | 65.6 | 57.7 | 46.9 | ***37.5*** | ***26.4*** | 44.8 |
| Days Max Temp ≥ 90 °F | 0 | 0 | 0 | 0 | 0 | 0 | 1 | 0 | 0 | 0 | 0 | 0 | 1 |
| Days Max Temp ≤ 32 °F | ***18*** | ***14*** | ***5*** | 0 | 0 | 0 | 0 | 0 | 0 | 0 | ***2*** | ***11*** | 50 |
| Days Min Temp ≤ 32 °F | ***30*** | ***27*** | ***26*** | ***16*** | 5 | 0 | 0 | 0 | 2 | 14 | ***21*** | ***28*** | 169 |
| Days Min Temp ≤ 0 °F | ***5*** | ***4*** | 1 | 0 | 0 | 0 | 0 | 0 | 0 | 0 | 0 | ***2*** | 12 |
| Heating Degree Days | ***1359*** | ***1179*** | ***1007*** | 636 | 327 | 123 | 43 | 63 | 234 | 556 | ***815*** | ***1189*** | 7531 |
| Cooling Degree Days | 0 | 0 | 0 | 1 | 12 | 43 | 110 | 81 | 19 | 1 | 0 | 0 | 267 |
| Total Precipitation (") | 2.66 | 2.40 | 3.16 | 3.67 | 4.50 | 4.57 | 4.56 | 4.33 | 4.35 | 3.45 | 3.78 | 3.10 | 44.53 |
| Days ≥ 0.1" Precip | ***5*** | 5 | 7 | 7 | 8 | 8 | 7 | 7 | 7 | 6 | ***6*** | 6 | 79 |
| Total Snowfall (") | 13.3 | 11.4 | 9.8 | 2.8 | 0.2 | 0.0 | 0.0 | 0.0 | 0.0 | 0.1 | 3.6 | 9.2 | 50.4 |
| Days ≥ 1" Snow Depth | 23 | 21 | 14 | 2 | 0 | 0 | 0 | 0 | 0 | 0 | 2 | 12 | 74 |

## FRANKLIN *Venango County* ELEVATION 991 ft LAT/LONG 41° 23 ' N / 79° 49 ' W

| | JAN | FEB | MAR | APR | MAY | JUN | JUL | AUG | SEP | OCT | NOV | DEC | YEAR |
|---|---|---|---|---|---|---|---|---|---|---|---|---|---|
| Maximum Temp °F | 33.0 | 35.3 | 45.6 | 58.8 | 69.9 | 78.6 | 82.6 | 80.7 | 73.1 | 61.2 | 49.2 | 37.9 | 58.8 |
| Minimum Temp °F | 15.7 | 16.2 | 24.4 | 34.5 | 43.7 | 53.2 | 57.9 | 56.8 | 50.5 | 39.2 | 32.1 | 22.7 | 37.2 |
| Mean Temp °F | 24.4 | 25.8 | 35.0 | 46.7 | 56.8 | 65.9 | 70.2 | 68.8 | 61.8 | 50.2 | 40.7 | 30.3 | 48.1 |
| Days Max Temp ≥ 90 °F | 0 | 0 | 0 | 0 | 0 | 2 | 4 | 2 | 0 | 0 | 0 | 0 | 8 |
| Days Max Temp ≤ 32 °F | 14 | 11 | 4 | 0 | 0 | 0 | 0 | 0 | 0 | 0 | 1 | 9 | 39 |
| Days Min Temp ≤ 32 °F | 29 | 26 | 25 | 14 | 3 | 0 | 0 | 0 | 0 | 6 | 16 | 26 | 145 |
| Days Min Temp ≤ 0 °F | 4 | 3 | 1 | 0 | 0 | 0 | 0 | 0 | 0 | 0 | 0 | 1 | 9 |
| Heating Degree Days | 1253 | 1101 | 922 | 547 | 273 | 66 | 15 | 26 | 143 | 452 | 723 | 1068 | 6589 |
| Cooling Degree Days | 0 | 0 | 0 | 4 | 26 | 98 | 206 | 160 | 57 | 2 | 0 | 0 | 553 |
| Total Precipitation (") | 2.51 | 2.31 | 3.33 | 3.55 | 3.74 | 4.72 | 4.97 | 4.12 | 4.22 | 3.21 | 3.91 | 3.29 | 43.88 |
| Days ≥ 0.1" Precip | 7 | 6 | 8 | 9 | 9 | 8 | 8 | 8 | 8 | 8 | 9 | 9 | 97 |
| Total Snowfall (") | 15.4 | 12.1 | ***9.7*** | 1.4 | 0.0 | 0.0 | 0.0 | 0.0 | 0.0 | 0.1 | 2.7 | ***11.9*** | 53.3 |
| Days ≥ 1" Snow Depth | 22 | 17 | 8 | 1 | 0 | 0 | 0 | 0 | 0 | 0 | 2 | 12 | 62 |

## GRATERFORD 1 E *Montgomery County* ELEVATION 239 ft LAT/LONG 40° 14 ' N / 75° 26 ' W

| | JAN | FEB | MAR | APR | MAY | JUN | JUL | AUG | SEP | OCT | NOV | DEC | YEAR |
|---|---|---|---|---|---|---|---|---|---|---|---|---|---|
| Maximum Temp °F | 36.4 | 39.1 | 48.6 | 60.6 | 70.7 | 79.8 | 84.6 | 82.8 | 76.1 | 64.3 | 53.6 | 42.0 | 61.5 |
| Minimum Temp °F | 18.6 | 20.2 | 28.7 | 38.3 | 48.0 | 56.6 | 61.7 | 59.6 | 51.4 | 38.8 | 33.1 | 24.2 | 39.9 |
| Mean Temp °F | 27.5 | 29.6 | 38.7 | 49.5 | 59.4 | 68.2 | 73.2 | 71.2 | 63.8 | 51.5 | 43.4 | 33.1 | 50.8 |
| Days Max Temp ≥ 90 °F | 0 | 0 | 0 | 0 | 0 | 3 | 7 | 4 | 1 | 0 | 0 | 0 | 15 |
| Days Max Temp ≤ 32 °F | 11 | 7 | 2 | 0 | 0 | 0 | 0 | 0 | 0 | 0 | 0 | 5 | 25 |
| Days Min Temp ≤ 32 °F | 29 | 25 | 21 | 8 | 1 | 0 | 0 | 0 | 1 | 9 | 15 | 25 | 134 |
| Days Min Temp ≤ 0 °F | 2 | 1 | 0 | 0 | 0 | 0 | 0 | 0 | 0 | 0 | 0 | 0 | 3 |
| Heating Degree Days | 1160 | 991 | 810 | 465 | 207 | 45 | 6 | 17 | 112 | 417 | 641 | 982 | 5853 |
| Cooling Degree Days | 0 | 0 | 1 | 5 | 37 | 146 | 262 | 179 | 68 | 8 | 1 | 0 | 707 |
| Total Precipitation (") | 3.38 | 2.57 | 3.42 | 3.57 | 4.20 | 3.32 | 4.56 | 4.20 | 3.90 | 3.04 | 3.48 | 3.26 | 42.90 |
| Days ≥ 0.1" Precip | 6 | ***5*** | 6 | ***6*** | ***7*** | ***6*** | 7 | 6 | ***6*** | 5 | ***6*** | 6 | 72 |
| Total Snowfall (") | na | na | na | 0.3 | 0.0 | 0.0 | 0.0 | 0.0 | 0.0 | 0.0 | ***0.4*** | ***3.3*** | na |
| Days ≥ 1" Snow Depth | na | na | na | 0 | 0 | 0 | 0 | 0 | 0 | 0 | 0 | na | na |

## GREENVILLE 2 NE *Mercer County* ELEVATION 1030 ft LAT/LONG 41° 24 ' N / 80° 23 ' W

| | JAN | FEB | MAR | APR | MAY | JUN | JUL | AUG | SEP | OCT | NOV | DEC | YEAR |
|---|---|---|---|---|---|---|---|---|---|---|---|---|---|
| Maximum Temp °F | 33.3 | 36.6 | 47.1 | 60.1 | 70.8 | 79.2 | 83.0 | 81.5 | 74.8 | 63.0 | 50.1 | 38.4 | 59.8 |
| Minimum Temp °F | 15.7 | 16.7 | 25.6 | 35.4 | 45.1 | 54.0 | 58.2 | 56.4 | 49.9 | 39.0 | 32.1 | 22.4 | 37.5 |
| Mean Temp °F | 24.5 | 26.6 | 36.4 | 47.7 | 58.0 | 66.6 | 70.6 | 69.0 | 62.3 | 51.0 | 41.1 | 30.4 | 48.7 |
| Days Max Temp ≥ 90 °F | 0 | 0 | 0 | 0 | 0 | 2 | 4 | 2 | 1 | 0 | 0 | 0 | 9 |
| Days Max Temp ≤ 32 °F | 15 | 11 | 4 | 0 | 0 | 0 | 0 | 0 | 0 | 0 | 1 | 9 | 40 |
| Days Min Temp ≤ 32 °F | 29 | 26 | 23 | 13 | 3 | 0 | 0 | 0 | 1 | 8 | 17 | 26 | 146 |
| Days Min Temp ≤ 0 °F | 4 | 3 | 1 | 0 | 0 | 0 | 0 | 0 | 0 | 0 | 0 | 1 | 9 |
| Heating Degree Days | 1249 | 1077 | 880 | 517 | 245 | 62 | 15 | 26 | 137 | 432 | 710 | 1067 | 6417 |
| Cooling Degree Days | 0 | 0 | 0 | 6 | 34 | 110 | 201 | 145 | 51 | 2 | 0 | 0 | 549 |
| Total Precipitation (") | 2.04 | 1.79 | 2.83 | 3.26 | 3.76 | 4.30 | 4.03 | 3.69 | 4.09 | 2.97 | 3.60 | 2.80 | 39.16 |
| Days ≥ 0.1" Precip | 6 | 5 | 8 | 8 | 9 | 8 | 8 | 7 | 8 | 7 | 8 | 8 | 90 |
| Total Snowfall (") | 13.8 | 11.5 | 10.1 | 2.2 | 0.1 | 0.0 | 0.0 | 0.0 | 0.0 | 0.2 | 4.4 | 12.6 | 54.9 |
| Days ≥ 1" Snow Depth | 16 | 13 | 6 | 1 | 0 | 0 | 0 | 0 | 0 | 0 | 2 | 12 | 50 |

## HANOVER *York County* ELEVATION 600 ft LAT/LONG 39° 48 ' N / 76° 59 ' W

| | JAN | FEB | MAR | APR | MAY | JUN | JUL | AUG | SEP | OCT | NOV | DEC | YEAR |
|---|---|---|---|---|---|---|---|---|---|---|---|---|---|
| Maximum Temp °F | 37.8 | 41.1 | 51.1 | 62.8 | 73.4 | 82.1 | 86.8 | 85.0 | 77.9 | 65.6 | 54.5 | 43.3 | 63.5 |
| Minimum Temp °F | 20.5 | 21.9 | 30.2 | 39.7 | 49.4 | 58.9 | 63.3 | 61.4 | 53.9 | 42.0 | 34.5 | 25.6 | 41.8 |
| Mean Temp °F | 29.1 | 31.5 | 40.7 | 51.3 | 61.4 | 70.6 | 75.1 | 73.3 | 65.9 | 53.8 | 44.5 | 34.5 | 52.6 |
| Days Max Temp ≥ 90 °F | 0 | 0 | 0 | 0 | 1 | 5 | 11 | 7 | 2 | 0 | 0 | 0 | 26 |
| Days Max Temp ≤ 32 °F | 9 | 6 | 1 | 0 | 0 | 0 | 0 | 0 | 0 | 0 | 0 | 4 | 20 |
| Days Min Temp ≤ 32 °F | 27 | 24 | 19 | 5 | 0 | 0 | 0 | 0 | 0 | 5 | 13 | 25 | 118 |
| Days Min Temp ≤ 0 °F | 1 | 0 | 0 | 0 | 0 | 0 | 0 | 0 | 0 | 0 | 0 | 0 | 1 |
| Heating Degree Days | 1105 | 939 | 749 | 414 | 158 | 23 | 3 | 7 | 75 | 350 | 611 | 940 | 5374 |
| Cooling Degree Days | *0* | *0* | *3* | 9 | 55 | *196* | *339* | 256 | 99 | *12* | *1* | *0* | 970 |
| Total Precipitation (") | 2.92 | 2.57 | 3.32 | 3.35 | 3.93 | 3.56 | 2.86 | 3.33 | 3.88 | 3.00 | 3.14 | 3.20 | 39.06 |
| Days ≥ 0.1" Precip | 6 | 6 | 7 | 6 | 7 | 7 | 6 | 6 | 5 | 5 | 6 | 6 | 73 |
| Total Snowfall (") | 9.4 | *9.3* | *3.6* | 0.3 | 0.0 | 0.0 | 0.0 | 0.0 | 0.0 | 0.1 | 1.2 | 4.1 | 28.0 |
| Days ≥ 1" Snow Depth | na | *7* | 2 | 0 | 0 | 0 | 0 | 0 | 0 | 0 | *0* | 3 | na |

## HAWLEY 1 E *Wayne County* ELEVATION 981 ft LAT/LONG 41° 29 ' N / 75° 10 ' W

| | JAN | FEB | MAR | APR | MAY | JUN | JUL | AUG | SEP | OCT | NOV | DEC | YEAR |
|---|---|---|---|---|---|---|---|---|---|---|---|---|---|
| Maximum Temp °F | 33.0 | 35.7 | 44.7 | 56.9 | 68.1 | 75.2 | 79.6 | 78.1 | 70.7 | 61.1 | 49.8 | 37.6 | 57.5 |
| Minimum Temp °F | 11.7 | 13.0 | 22.3 | 32.6 | 42.5 | 51.6 | 56.6 | 55.5 | 47.8 | 35.9 | 29.2 | 19.0 | 34.8 |
| Mean Temp °F | 22.4 | 24.4 | 33.5 | 44.8 | 55.3 | 63.4 | 68.1 | 66.8 | 59.3 | 48.5 | 39.5 | 28.3 | 46.2 |
| Days Max Temp ≥ 90 °F | 0 | 0 | 0 | 0 | 0 | 1 | 2 | 1 | 0 | 0 | 0 | 0 | 4 |
| Days Max Temp ≤ 32 °F | 14 | 11 | 4 | 0 | 0 | 0 | 0 | 0 | 0 | 0 | 1 | 9 | 39 |
| Days Min Temp ≤ 32 °F | 30 | 27 | 27 | 15 | 4 | 0 | 0 | 0 | 1 | 12 | 21 | 29 | 166 |
| Days Min Temp ≤ 0 °F | 6 | 5 | 1 | 0 | 0 | 0 | 0 | 0 | 0 | 0 | 0 | 2 | 14 |
| Heating Degree Days | 1314 | 1141 | 969 | 602 | 308 | 101 | 29 | 44 | 196 | 505 | 759 | 1130 | 7098 |
| Cooling Degree Days | 0 | 0 | 0 | 2 | 16 | 64 | 147 | 112 | 32 | 2 | 0 | 0 | 375 |
| Total Precipitation (") | 2.64 | 2.52 | 3.03 | 3.48 | 4.01 | 3.88 | 3.50 | 3.69 | 3.67 | 3.03 | 3.55 | 3.13 | 40.13 |
| Days ≥ 0.1" Precip | 6 | 6 | 6 | 7 | 8 | 8 | 7 | 7 | 7 | 6 | 7 | 6 | 81 |
| Total Snowfall (") | 10.4 | 10.1 | 9.3 | 2.4 | 0.1 | 0.0 | 0.0 | 0.0 | 0.0 | 0.1 | 3.3 | 10.2 | 45.9 |
| Days ≥ 1" Snow Depth | 23 | 19 | 11 | 1 | 0 | 0 | 0 | 0 | 0 | 0 | 2 | 13 | 69 |

## HOLTWOOD *Lancaster County* ELEVATION 190 ft LAT/LONG 39° 50 ' N / 76° 20 ' W

| | JAN | FEB | MAR | APR | MAY | JUN | JUL | AUG | SEP | OCT | NOV | DEC | YEAR |
|---|---|---|---|---|---|---|---|---|---|---|---|---|---|
| Maximum Temp °F | 36.7 | 39.7 | 49.1 | 60.8 | 71.5 | 80.3 | 84.7 | 83.0 | 75.9 | 64.1 | 53.1 | 42.1 | 61.8 |
| Minimum Temp °F | 22.3 | 24.2 | 32.5 | 41.9 | 52.2 | 61.7 | 66.7 | 65.5 | 57.9 | 46.0 | 36.9 | 27.9 | 44.6 |
| Mean Temp °F | 29.6 | 32.0 | 40.8 | 51.4 | 61.9 | 71.0 | 75.7 | 74.3 | 66.9 | 55.1 | 45.0 | 35.0 | 53.2 |
| Days Max Temp ≥ 90 °F | 0 | 0 | 0 | 0 | 0 | 2 | 6 | 4 | 1 | 0 | 0 | 0 | 13 |
| Days Max Temp ≤ 32 °F | 10 | 6 | 1 | 0 | 0 | 0 | 0 | 0 | 0 | 0 | 0 | 4 | 21 |
| Days Min Temp ≤ 32 °F | 27 | 23 | 15 | 3 | 0 | 0 | 0 | 0 | 0 | 1 | 9 | 22 | 100 |
| Days Min Temp ≤ 0 °F | 1 | 0 | 0 | 0 | 0 | 0 | 0 | 0 | 0 | 0 | 0 | 0 | 1 |
| Heating Degree Days | 1092 | 926 | 743 | 408 | 143 | 14 | 1 | 2 | 55 | 310 | 593 | 923 | 5210 |
| Cooling Degree Days | 0 | 0 | 0 | 5 | 57 | 202 | 352 | 283 | 112 | 12 | 0 | 0 | 1023 |
| Total Precipitation (") | 2.72 | 2.16 | 2.88 | 3.01 | 3.66 | 3.18 | 3.63 | 3.31 | 3.31 | 2.77 | 3.08 | 3.05 | 36.76 |
| Days ≥ 0.1" Precip | 6 | 5 | 6 | 6 | 7 | 6 | 6 | 6 | 5 | 5 | 6 | 6 | 70 |
| Total Snowfall (") | na | na | na | 0.1 | 0.0 | 0.0 | 0.0 | 0.0 | 0.0 | 0.0 | 0.2 | na | na |
| Days ≥ 1" Snow Depth | na | na | na | 0 | 0 | 0 | 0 | 0 | 0 | 0 | 0 | na | na |

## INDIANA 3 SE *Indiana County* ELEVATION 1102 ft LAT/LONG 40° 36 ' N / 79° 7 ' W

| | JAN | FEB | MAR | APR | MAY | JUN | JUL | AUG | SEP | OCT | NOV | DEC | YEAR |
|---|---|---|---|---|---|---|---|---|---|---|---|---|---|
| Maximum Temp °F | 35.1 | 38.5 | 49.3 | 61.7 | 71.3 | 79.1 | 82.4 | 80.7 | 73.9 | 62.8 | 51.3 | 40.4 | 60.5 |
| Minimum Temp °F | 17.4 | 18.6 | 27.3 | 36.3 | 45.6 | 54.1 | 59.0 | 58.0 | 51.7 | 40.2 | 32.6 | 23.8 | 38.7 |
| Mean Temp °F | 26.2 | 28.6 | 38.3 | 49.0 | 58.5 | 66.6 | 70.7 | 69.4 | 62.8 | 51.5 | 42.0 | 32.1 | 49.6 |
| Days Max Temp ≥ 90 °F | 0 | 0 | 0 | 0 | 0 | 1 | 3 | 1 | 0 | 0 | 0 | 0 | 5 |
| Days Max Temp ≤ 32 °F | 12 | 9 | 3 | 0 | 0 | 0 | 0 | 0 | 0 | 0 | 1 | 7 | 32 |
| Days Min Temp ≤ 32 °F | 28 | 25 | 22 | 12 | 3 | 0 | 0 | 0 | 0 | 7 | 16 | 25 | 138 |
| Days Min Temp ≤ 0 °F | 3 | 2 | 0 | 0 | 0 | 0 | 0 | 0 | 0 | 0 | 0 | 1 | 6 |
| Heating Degree Days | 1195 | 1023 | 820 | 477 | 222 | 53 | 9 | 18 | 121 | 415 | 684 | 1013 | 6050 |
| Cooling Degree Days | 0 | 0 | 0 | 4 | 25 | 104 | 202 | 159 | 63 | 4 | 0 | 0 | 561 |
| Total Precipitation (") | 3.27 | 2.89 | 3.96 | 3.54 | 4.46 | 4.52 | 5.08 | 3.99 | 4.04 | 3.00 | 3.82 | 3.40 | 45.97 |
| Days ≥ 0.1" Precip | 9 | 7 | 9 | 9 | 9 | 9 | 8 | 8 | 8 | 7 | 9 | 8 | 100 |
| Total Snowfall (") | 17.0 | 13.6 | 9.8 | 1.8 | 0.0 | 0.0 | 0.0 | 0.0 | 0.0 | 0.0 | 2.8 | 9.1 | 54.1 |
| Days ≥ 1" Snow Depth | 18 | 15 | 7 | 1 | 0 | 0 | 0 | 0 | 0 | 0 | 2 | 10 | 53 |

## JAMESTOWN 2 NW *Crawford County* ELEVATION 1050 ft LAT/LONG 41° 30 ' N / 80° 28 ' W

| | JAN | FEB | MAR | APR | MAY | JUN | JUL | AUG | SEP | OCT | NOV | DEC | YEAR |
|---|---|---|---|---|---|---|---|---|---|---|---|---|---|
| Maximum Temp °F | 32.0 | 34.4 | 44.8 | 57.3 | 68.5 | 77.4 | 81.5 | 79.7 | 72.9 | 61.1 | 48.8 | 37.2 | 58.0 |
| Minimum Temp °F | 13.8 | 14.9 | 24.4 | 34.7 | 44.3 | 53.4 | 57.7 | 56.1 | 49.5 | 38.7 | 31.6 | 21.6 | 36.7 |
| Mean Temp °F | 22.9 | 24.6 | 34.6 | 46.1 | 56.4 | 65.4 | 69.6 | 67.9 | 61.2 | 49.9 | 40.2 | 29.4 | 47.3 |
| Days Max Temp ≥ 90 °F | 0 | 0 | 0 | 0 | 0 | 1 | 3 | 1 | 0 | 0 | 0 | 0 | 5 |
| Days Max Temp ≤ 32 °F | 16 | 13 | 6 | 0 | 0 | 0 | 0 | 0 | 0 | 0 | 2 | 10 | 47 |
| Days Min Temp ≤ 32 °F | 30 | 26 | 24 | 14 | 3 | 0 | 0 | 0 | 1 | 8 | 18 | 27 | 151 |
| Days Min Temp ≤ 0 °F | 6 | 4 | 1 | 0 | 0 | 0 | 0 | 0 | 0 | 0 | 0 | 1 | 12 |
| Heating Degree Days | 1299 | 1134 | 935 | 567 | 285 | 80 | 23 | 37 | 159 | 462 | 737 | 1096 | 6814 |
| Cooling Degree Days | 0 | 0 | 1 | 6 | 28 | 100 | 192 | 141 | 53 | 3 | 0 | 0 | 524 |
| Total Precipitation (") | 2.35 | 2.24 | 2.98 | 3.23 | 3.81 | 3.95 | 4.28 | 4.06 | 3.97 | 3.10 | 3.69 | 3.27 | 40.93 |
| Days ≥ 0.1" Precip | 7 | 6 | 9 | 8 | 9 | 8 | 8 | 7 | 8 | 8 | 9 | 9 | 96 |
| Total Snowfall (") | 17.1 | 13.1 | 11.4 | 2.5 | 0.0 | 0.0 | 0.0 | 0.0 | 0.0 | 0.5 | 5.3 | 16.4 | 66.3 |
| Days ≥ 1" Snow Depth | 21 | 18 | 10 | 1 | 0 | 0 | 0 | 0 | 0 | 0 | 4 | 15 | 69 |

## JOHNSTOWN *Cambria County* ELEVATION 1211 ft LAT/LONG 40° 20 ' N / 78° 55 ' W

| | JAN | FEB | MAR | APR | MAY | JUN | JUL | AUG | SEP | OCT | NOV | DEC | YEAR |
|---|---|---|---|---|---|---|---|---|---|---|---|---|---|
| Maximum Temp °F | 36.6 | 39.8 | 50.2 | 63.1 | 73.5 | 82.0 | 86.2 | 84.2 | 76.8 | 64.6 | 52.6 | 41.1 | 62.6 |
| Minimum Temp °F | 19.4 | 21.3 | 29.2 | 39.1 | 48.4 | 56.2 | 61.0 | 59.3 | 53.0 | 41.2 | 34.0 | 24.8 | 40.6 |
| Mean Temp °F | 28.1 | 30.5 | 39.8 | 51.1 | 61.0 | 69.2 | 73.7 | 71.7 | 64.9 | 52.9 | 43.3 | 32.8 | 51.6 |
| Days Max Temp ≥ 90 °F | 0 | 0 | 0 | 0 | 1 | 5 | 10 | 6 | 2 | 0 | 0 | 0 | 24 |
| Days Max Temp ≤ 32 °F | 11 | 7 | 2 | 0 | 0 | 0 | 0 | 0 | 0 | 0 | 1 | 6 | 27 |
| Days Min Temp ≤ 32 °F | 27 | 24 | 20 | 8 | 1 | 0 | 0 | 0 | 0 | 5 | 15 | 24 | 124 |
| Days Min Temp ≤ 0 °F | 2 | 1 | 0 | 0 | 0 | 0 | 0 | 0 | 0 | 0 | 0 | 0 | 3 |
| Heating Degree Days | 1139 | 958 | 777 | 421 | 172 | 33 | 4 | 11 | 92 | 376 | 644 | 991 | 5618 |
| Cooling Degree Days | 0 | 0 | 1 | 11 | 63 | 172 | 296 | 231 | 99 | 10 | 0 | 0 | 883 |
| Total Precipitation (") | 3.68 | 3.42 | 4.04 | 3.66 | 4.45 | 4.41 | 5.02 | 4.16 | 3.78 | 3.34 | 3.76 | 3.64 | 47.36 |
| Days ≥ 0.1" Precip | 9 | *9* | 9 | *9* | 9 | 8 | 8 | 8 | 7 | 7 | 9 | *9* | 101 |
| Total Snowfall (") | *11.8* | *9.1* | *8.1* | 0.5 | 0.0 | 0.0 | 0.0 | 0.0 | 0.0 | 0.0 | 2.3 | *7.3* | 39.1 |
| Days ≥ 1" Snow Depth | na | *9* | *4* | 0 | 0 | 0 | 0 | 0 | 0 | 0 | 1 | *5* | na |

## KANE 1 NNE *McKean County* ELEVATION 1752 ft LAT/LONG 41° 41 ' N / 78° 48 ' W

| | JAN | FEB | MAR | APR | MAY | JUN | JUL | AUG | SEP | OCT | NOV | DEC | YEAR |
|---|---|---|---|---|---|---|---|---|---|---|---|---|---|
| Maximum Temp °F | 29.2 | 32.2 | 42.1 | 54.8 | 66.6 | 74.6 | 78.6 | 76.7 | 69.5 | 57.9 | 45.5 | 34.0 | 55.1 |
| Minimum Temp °F | 10.9 | 10.8 | 19.8 | 30.5 | 38.8 | 47.5 | 52.1 | 50.5 | 44.2 | 34.3 | 28.1 | 18.3 | 32.2 |
| Mean Temp °F | 20.1 | 21.5 | 30.9 | 42.7 | 52.7 | 61.1 | 65.3 | 63.7 | 56.8 | 46.1 | 36.8 | 26.2 | 43.7 |
| Days Max Temp ≥ 90 °F | 0 | 0 | 0 | 0 | 0 | 0 | 1 | 0 | 0 | 0 | 0 | 0 | 1 |
| Days Max Temp ≤ 32 °F | 19 | 15 | 7 | 1 | 0 | 0 | 0 | 0 | 0 | 0 | 3 | 13 | 58 |
| Days Min Temp ≤ 32 °F | 30 | 27 | 27 | 19 | 10 | 1 | 0 | 0 | 4 | 15 | 22 | 28 | 183 |
| Days Min Temp ≤ 0 °F | 7 | 7 | 2 | 0 | 0 | 0 | 0 | 0 | 0 | 0 | 0 | 3 | 19 |
| Heating Degree Days | 1386 | 1223 | 1049 | 664 | 381 | 151 | 64 | 91 | 257 | 580 | 839 | 1197 | 7882 |
| Cooling Degree Days | 0 | 0 | 0 | 1 | 8 | 38 | 90 | 57 | 17 | 1 | 0 | 0 | 212 |
| Total Precipitation (") | 2.99 | 2.59 | 3.60 | 3.61 | 4.11 | 4.92 | 4.40 | 4.24 | 4.34 | 3.46 | 4.08 | 3.87 | 46.21 |
| Days ≥ 0.1" Precip | 8 | 8 | 9 | 9 | 9 | 9 | 8 | 8 | 8 | 9 | 10 | 10 | 105 |
| Total Snowfall (") | 22.0 | 18.6 | 14.7 | 3.3 | 0.2 | 0.0 | 0.0 | 0.0 | 0.0 | 0.5 | 8.1 | 21.7 | 89.1 |
| Days ≥ 1" Snow Depth | 27 | 25 | 18 | 2 | 0 | 0 | 0 | 0 | 0 | 0 | 7 | 21 | 100 |

## LANDISVILLE 2 NW *Lancaster County* ELEVATION 410 ft LAT/LONG 40° 6 ' N / 76° 25 ' W

| | JAN | FEB | MAR | APR | MAY | JUN | JUL | AUG | SEP | OCT | NOV | DEC | YEAR |
|---|---|---|---|---|---|---|---|---|---|---|---|---|---|
| Maximum Temp °F | 36.9 | 40.6 | 51.1 | 63.2 | 73.5 | 82.1 | 85.7 | 84.3 | 77.4 | 65.9 | 53.8 | 42.2 | 63.1 |
| Minimum Temp °F | 19.3 | 21.1 | 30.1 | 38.7 | 48.7 | 57.9 | 62.0 | 60.2 | 52.6 | 41.1 | 33.7 | 25.3 | 40.9 |
| Mean Temp °F | 28.1 | 30.9 | 40.6 | 51.0 | 61.2 | 70.0 | 73.9 | 72.3 | 65.0 | 53.5 | 43.8 | 33.8 | 52.0 |
| Days Max Temp ≥ 90 °F | 0 | 0 | 0 | 0 | 1 | 4 | 7 | 5 | 2 | 0 | 0 | 0 | 19 |
| Days Max Temp ≤ 32 °F | 10 | 6 | 1 | 0 | 0 | 0 | 0 | 0 | 0 | 0 | 0 | 5 | 22 |
| Days Min Temp ≤ 32 °F | 27 | 24 | 19 | 8 | 1 | 0 | 0 | 0 | 0 | 7 | 15 | 24 | 125 |
| Days Min Temp ≤ 0 °F | 2 | 1 | 0 | 0 | 0 | 0 | 0 | 0 | 0 | 0 | 0 | 0 | 3 |
| Heating Degree Days | 1137 | 958 | 749 | 421 | 162 | 22 | 3 | 10 | 88 | 358 | 631 | 960 | 5499 |
| Cooling Degree Days | 0 | 0 | 1 | 7 | 61 | 195 | 315 | 254 | 98 | 13 | 0 | 0 | 944 |
| Total Precipitation (") | 2.77 | 2.27 | 3.15 | 3.41 | 4.12 | 4.22 | 5.07 | 3.38 | 3.70 | 3.06 | 3.54 | 3.20 | 41.89 |
| Days ≥ 0.1" Precip | 6 | 5 | 7 | 7 | 8 | 7 | 7 | 6 | 6 | 6 | 7 | 6 | 78 |
| Total Snowfall (") | 8.7 | 7.9 | 3.9 | 0.4 | 0.0 | 0.0 | 0.0 | 0.0 | 0.0 | 0.1 | 0.8 | 4.1 | 25.9 |
| Days ≥ 1" Snow Depth | 13 | 9 | 4 | 0 | 0 | 0 | 0 | 0 | 0 | 0 | 0 | 4 | 30 |

## LAURELTON ST VILLAGE *Union County* ELEVATION 879 ft LAT/LONG 40° 54 ' N / 77° 13 ' W

| | JAN | FEB | MAR | APR | MAY | JUN | JUL | AUG | SEP | OCT | NOV | DEC | YEAR |
|---|---|---|---|---|---|---|---|---|---|---|---|---|---|
| Maximum Temp °F | 36.3 | 40.4 | 50.9 | 64.6 | 75.1 | 82.7 | 86.5 | 84.6 | 76.8 | 65.4 | 52.4 | 40.4 | 63.0 |
| Minimum Temp °F | 16.6 | 18.5 | 26.6 | 36.5 | 46.1 | 54.6 | 59.4 | 57.9 | 50.8 | 39.5 | 31.4 | 22.2 | 38.3 |
| Mean Temp °F | 26.5 | 29.5 | 38.8 | 50.6 | 60.6 | 68.7 | 72.9 | 71.3 | 63.8 | 52.4 | 41.9 | 31.3 | 50.7 |
| Days Max Temp ≥ 90 °F | 0 | 0 | 0 | 0 | 2 | 4 | 9 | 6 | 2 | 0 | 0 | 0 | 23 |
| Days Max Temp ≤ 32 °F | 11 | 6 | 1 | 0 | 0 | 0 | 0 | 0 | 0 | 0 | 0 | 5 | 23 |
| Days Min Temp ≤ 32 °F | 30 | 26 | 23 | 11 | 2 | 0 | 0 | 0 | 0 | 7 | 17 | 27 | 143 |
| Days Min Temp ≤ 0 °F | 3 | 2 | 0 | 0 | 0 | 0 | 0 | 0 | 0 | 0 | 0 | 1 | 6 |
| Heating Degree Days | 1188 | 997 | 806 | 434 | 178 | 31 | 4 | 11 | 104 | 388 | 686 | 1036 | 5863 |
| Cooling Degree Days | 0 | 0 | 0 | 9 | 53 | 151 | 280 | 216 | 77 | 6 | 0 | 0 | 792 |
| Total Precipitation (") | 2.63 | 2.77 | 3.59 | 3.40 | 4.19 | 4.35 | 4.30 | 3.57 | 4.28 | 3.49 | 3.93 | 3.14 | 43.64 |
| Days ≥ 0.1" Precip | 5 | 6 | 7 | 7 | 8 | 8 | 8 | 6 | 7 | 6 | 7 | 6 | 81 |
| Total Snowfall (") | na | na | na | 0.4 | 0.0 | 0.0 | 0.0 | 0.0 | 0.0 | 0.0 | *0.8* | na | na |
| Days ≥ 1" Snow Depth | *13* | *12* | *6* | 0 | 0 | 0 | 0 | 0 | 0 | 0 | 1 | *6* | 38 |

## LEBANON 2 W *Lebanon County* ELEVATION 489 ft LAT/LONG 40° 19 ' N / 76° 28 ' W

| | JAN | FEB | MAR | APR | MAY | JUN | JUL | AUG | SEP | OCT | NOV | DEC | YEAR |
|---|---|---|---|---|---|---|---|---|---|---|---|---|---|
| Maximum Temp °F | 36.4 | 39.6 | 49.8 | 62.1 | 72.1 | 81.0 | 85.3 | 83.4 | 75.8 | 64.8 | 53.2 | 41.7 | 62.1 |
| Minimum Temp °F | 19.1 | 20.9 | 29.5 | 39.0 | 48.3 | 57.4 | 62.4 | 60.4 | *53.2* | 41.2 | 33.3 | 25.0 | 40.8 |
| Mean Temp °F | 27.7 | 30.3 | 39.6 | 50.5 | 60.3 | 69.2 | 73.9 | 71.8 | *64.5* | 52.9 | 43.3 | 33.3 | 51.4 |
| Days Max Temp ≥ 90 °F | 0 | 0 | 0 | 0 | 0 | 3 | 8 | 5 | 2 | 0 | 0 | 0 | 18 |
| Days Max Temp ≤ 32 °F | 10 | 7 | 1 | 0 | 0 | 0 | 0 | 0 | 0 | 0 | 0 | 4 | 22 |
| Days Min Temp ≤ 32 °F | 28 | 24 | 20 | *7* | 1 | 0 | 0 | 0 | 0 | 5 | 15 | 25 | 125 |
| Days Min Temp ≤ 0 °F | 2 | 1 | 0 | 0 | 0 | 0 | 0 | 0 | 0 | 0 | 0 | 0 | 3 |
| Heating Degree Days | 1149 | 975 | 780 | 434 | 183 | 29 | 2 | 11 | *98* | 374 | 647 | 978 | 5660 |
| Cooling Degree Days | 0 | 0 | 0 | 5 | 45 | 157 | 290 | 218 | 81 | 9 | 0 | 0 | 805 |
| Total Precipitation (") | 2.88 | 2.42 | 3.18 | 3.63 | 4.64 | 3.98 | 4.80 | 3.63 | 3.95 | 3.14 | 3.74 | 3.25 | 43.24 |
| Days ≥ 0.1" Precip | 6 | 6 | 7 | 7 | 9 | *7* | 7 | 6 | 6 | 5 | 7 | *6* | 79 |
| Total Snowfall (") | na | na | *3.8* | 0.1 | 0.0 | 0.0 | 0.0 | 0.0 | 0.0 | 0.0 | 0.8 | na | na |
| Days ≥ 1" Snow Depth | na | na | *2* | 0 | 0 | 0 | 0 | 0 | 0 | 0 | 0 | na | na |

## LEWISTOWN *Mifflin County* ELEVATION 479 ft LAT/LONG 40° 35 ' N / 77° 35 ' W

| | JAN | FEB | MAR | APR | MAY | JUN | JUL | AUG | SEP | OCT | NOV | DEC | YEAR |
|---|---|---|---|---|---|---|---|---|---|---|---|---|---|
| Maximum Temp °F | 35.8 | 39.3 | 49.6 | 62.3 | 72.8 | 80.9 | 85.1 | 83.5 | 75.5 | 64.1 | 52.0 | 40.1 | 61.8 |
| Minimum Temp °F | 18.8 | 20.4 | 28.5 | 38.3 | 47.3 | 55.9 | 61.2 | 59.8 | 52.5 | 40.6 | 33.3 | 24.8 | 40.1 |
| Mean Temp °F | 27.3 | 29.8 | 39.1 | 50.3 | 60.1 | 68.5 | 73.2 | 71.7 | 64.2 | 52.4 | 42.7 | 32.5 | 51.0 |
| Days Max Temp ≥ 90 °F | 0 | 0 | 0 | 0 | 1 | 3 | 7 | 5 | 1 | 0 | 0 | 0 | 17 |
| Days Max Temp ≤ 32 °F | 11 | 7 | 1 | 0 | 0 | 0 | 0 | 0 | 0 | 0 | 0 | 5 | 24 |
| Days Min Temp ≤ 32 °F | 28 | 25 | 21 | 7 | 1 | 0 | 0 | 0 | 0 | 6 | 15 | 25 | 128 |
| Days Min Temp ≤ 0 °F | 2 | 1 | 0 | 0 | 0 | 0 | 0 | 0 | 0 | 0 | 0 | 0 | 3 |
| Heating Degree Days | 1162 | 986 | 797 | 442 | 187 | 33 | 4 | 10 | 98 | 390 | 664 | 1001 | 5774 |
| Cooling Degree Days | 0 | 0 | 0 | 6 | 41 | 132 | 266 | 208 | *67* | 5 | 0 | 0 | 725 |
| Total Precipitation (") | 2.23 | 2.32 | 3.31 | 3.01 | 4.07 | 4.25 | 4.27 | 3.29 | 3.32 | 2.92 | 3.41 | 2.94 | 39.34 |
| Days ≥ 0.1" Precip | 6 | 5 | 7 | 7 | 8 | 7 | 7 | 6 | 7 | 5 | 7 | 6 | 78 |
| Total Snowfall (") | na | na | na | 0.2 | 0.0 | 0.0 | 0.0 | 0.0 | 0.0 | 0.0 | *0.9* | na | na |
| Days ≥ 1" Snow Depth | na | na | *3* | 0 | 0 | 0 | 0 | 0 | 0 | 0 | *0* | na | na |

## MARCUS HOOK *Delaware County* ELEVATION 10 ft LAT/LONG 39° 49 ' N / 75° 25 ' W

| | JAN | FEB | MAR | APR | MAY | JUN | JUL | AUG | SEP | OCT | NOV | DEC | YEAR |
|---|---|---|---|---|---|---|---|---|---|---|---|---|---|
| Maximum Temp °F | 38.5 | 42.0 | 51.1 | 63.0 | 73.7 | 82.9 | 87.2 | 85.1 | 77.2 | 65.0 | 54.2 | 43.8 | 63.6 |
| Minimum Temp °F | 26.6 | 28.6 | 36.0 | 45.3 | 55.2 | 64.3 | 69.4 | 68.2 | 61.2 | 49.8 | 41.1 | 32.2 | 48.2 |
| Mean Temp °F | 32.6 | 35.3 | 43.6 | 54.2 | 64.5 | 73.6 | 78.4 | 76.7 | 69.2 | 57.4 | 47.7 | 38.1 | 55.9 |
| Days Max Temp ≥ 90 °F | 0 | 0 | 0 | 0 | 1 | 6 | 12 | 8 | 2 | 0 | 0 | 0 | 29 |
| Days Max Temp ≤ 32 °F | 8 | 5 | 1 | 0 | 0 | 0 | 0 | 0 | 0 | 0 | 0 | 3 | 17 |
| Days Min Temp ≤ 32 °F | 22 | 19 | 10 | 1 | 0 | 0 | 0 | 0 | 0 | 0 | 4 | 15 | 71 |
| Days Min Temp ≤ 0 °F | 0 | 0 | 0 | 0 | 0 | 0 | 0 | 0 | 0 | 0 | 0 | 0 | 0 |
| Heating Degree Days | 999 | 831 | 659 | 331 | 99 | 6 | 0 | 1 | 33 | 247 | 514 | 829 | 4549 |
| Cooling Degree Days | 0 | 0 | 1 | 12 | 102 | 284 | 444 | 359 | 162 | 20 | 1 | 0 | 1385 |
| Total Precipitation (") | 2.71 | 2.78 | 3.46 | 3.51 | 4.20 | 3.49 | 4.22 | 3.46 | 3.85 | 2.85 | 3.45 | 3.38 | 41.36 |
| Days ≥ 0.1" Precip | 5 | 5 | 6 | 6 | 7 | 6 | 6 | 5 | 5 | 4 | 6 | 6 | 67 |
| Total Snowfall (") | na | 5.7 | *1.9* | 0.0 | 0.0 | 0.0 | 0.0 | 0.0 | 0.0 | 0.0 | 0.3 | *1.6* | na |
| Days ≥ 1" Snow Depth | na | *3* | *1* | 0 | 0 | 0 | 0 | 0 | 0 | 0 | 0 | *1* | na |

## MARION CENTER 2 SE *Indiana County* ELEVATION 1611 ft LAT/LONG 40° 45 ' N / 79° 2 ' W

| | JAN | FEB | MAR | APR | MAY | JUN | JUL | AUG | SEP | OCT | NOV | DEC | YEAR |
|---|---|---|---|---|---|---|---|---|---|---|---|---|---|
| Maximum Temp °F | 31.2 | 34.9 | 45.5 | 58.6 | 68.6 | 76.0 | 79.5 | 78.0 | 70.6 | 59.3 | 47.8 | 36.6 | 57.2 |
| Minimum Temp °F | 15.3 | 17.3 | 25.6 | 35.8 | 46.0 | 54.2 | 58.8 | 58.0 | 51.6 | 40.1 | 31.9 | 22.0 | 38.1 |
| Mean Temp °F | 23.3 | 26.1 | 35.5 | 47.3 | 57.3 | 65.1 | 69.2 | 68.0 | 61.1 | 49.8 | 39.9 | 29.3 | 47.7 |
| Days Max Temp ≥ 90 °F | 0 | 0 | 0 | 0 | 0 | 0 | 1 | 1 | 0 | 0 | 0 | 0 | 2 |
| Days Max Temp ≤ 32 °F | 17 | 12 | 5 | 0 | 0 | 0 | 0 | 0 | 0 | 0 | 3 | 11 | 48 |
| Days Min Temp ≤ 32 °F | 29 | 25 | 23 | 12 | 2 | 0 | 0 | 0 | 0 | 7 | 17 | 26 | 141 |
| Days Min Temp ≤ 0 °F | 4 | 3 | 1 | 0 | 0 | 0 | 0 | 0 | 0 | 0 | 0 | 1 | 9 |
| Heating Degree Days | 1286 | 1092 | 906 | 531 | 256 | 75 | 19 | 28 | 153 | 466 | 747 | 1100 | 6659 |
| Cooling Degree Days | 0 | 0 | 0 | 5 | 23 | 75 | 163 | 130 | 40 | 1 | 0 | 0 | 437 |
| Total Precipitation (") | 3.51 | 3.14 | 4.29 | 3.96 | 4.48 | 4.65 | 5.14 | 4.39 | 4.34 | 3.20 | 4.09 | 3.98 | 49.17 |
| Days ≥ 0.1" Precip | 9 | 8 | 10 | 9 | 9 | 9 | 9 | 8 | 8 | 8 | 9 | 9 | 105 |
| Total Snowfall (") | 19.2 | 16.6 | 15.6 | 3.8 | 0.1 | 0.0 | 0.0 | 0.0 | 0.0 | 0.4 | 6.1 | 14.1 | 75.9 |
| Days ≥ 1" Snow Depth | 25 | 24 | 16 | 3 | 0 | 0 | 0 | 0 | 0 | 0 | 5 | 19 | 92 |

## MEADVILLE 1 S *Crawford County* ELEVATION 1070 ft LAT/LONG 41° 38 ' N / 80° 10 ' W

| | JAN | FEB | MAR | APR | MAY | JUN | JUL | AUG | SEP | OCT | NOV | DEC | YEAR |
|---|---|---|---|---|---|---|---|---|---|---|---|---|---|
| Maximum Temp °F | 32.0 | 34.5 | 44.7 | 57.3 | 68.8 | 77.1 | 81.0 | 79.2 | 72.5 | 60.7 | 48.6 | 37.3 | 57.8 |
| Minimum Temp °F | 14.6 | 15.1 | 23.6 | 33.7 | 43.0 | 52.1 | 56.8 | 55.6 | 49.3 | 38.8 | 31.7 | 22.0 | 36.4 |
| Mean Temp °F | 23.3 | 24.8 | 34.2 | 45.5 | 55.9 | 64.6 | 68.9 | 67.5 | 60.9 | 49.8 | 40.2 | 29.7 | 47.1 |
| Days Max Temp ≥ 90 °F | 0 | 0 | 0 | 0 | 0 | 1 | 2 | 1 | 0 | 0 | 0 | 0 | 4 |
| Days Max Temp ≤ 32 °F | 16 | 12 | 6 | 0 | 0 | 0 | 0 | 0 | 0 | 0 | 2 | 10 | 46 |
| Days Min Temp ≤ 32 °F | 29 | 26 | 25 | 15 | 4 | 0 | 0 | 0 | 0 | 7 | 18 | 27 | 151 |
| Days Min Temp ≤ 0 °F | 5 | 4 | 1 | 0 | 0 | 0 | 0 | 0 | 0 | 0 | 0 | 1 | 11 |
| Heating Degree Days | 1285 | 1128 | 949 | 581 | 296 | 89 | 26 | 39 | 164 | 467 | 739 | 1089 | 6852 |
| Cooling Degree Days | 0 | 0 | 0 | 4 | 21 | 81 | 170 | 127 | 47 | 2 | 0 | 0 | 452 |
| Total Precipitation (") | 2.73 | 2.50 | 3.21 | 3.20 | 3.72 | 4.47 | 4.52 | 4.62 | 4.25 | 3.73 | 4.15 | 3.72 | 44.82 |
| Days ≥ 0.1" Precip | 8 | 6 | 9 | 8 | 9 | 8 | 8 | 8 | 8 | 8 | 10 | 10 | 100 |
| Total Snowfall (") | 29.3 | 21.4 | 18.4 | 5.3 | 0.1 | 0.0 | 0.0 | 0.0 | 0.0 | 1.1 | 11.6 | 25.1 | 112.3 |
| Days ≥ 1" Snow Depth | 24 | 21 | 13 | 2 | 0 | 0 | 0 | 0 | 0 | 0 | 5 | 18 | 83 |

## MERCER *Mercer County* ELEVATION 1230 ft LAT/LONG 41° 15 ' N / 80° 15 ' W

| | JAN | FEB | MAR | APR | MAY | JUN | JUL | AUG | SEP | OCT | NOV | DEC | YEAR |
|---|---|---|---|---|---|---|---|---|---|---|---|---|---|
| Maximum Temp °F | 32.9 | 36.6 | 47.6 | 59.7 | 69.7 | 77.7 | 81.5 | 79.9 | 73.3 | 61.8 | 49.4 | 38.1 | 59.0 |
| Minimum Temp °F | 16.6 | 18.4 | 26.9 | 36.3 | 45.6 | 53.8 | 58.3 | 56.9 | 50.7 | 40.0 | 32.5 | 22.9 | 38.2 |
| Mean Temp °F | 24.8 | 27.5 | 37.3 | 48.0 | 57.7 | 65.8 | 69.9 | 68.4 | 62.1 | 50.9 | 41.0 | 30.5 | 48.7 |
| Days Max Temp ≥ 90 °F | 0 | 0 | 0 | 0 | 0 | 1 | 3 | 1 | 0 | 0 | 0 | 0 | 5 |
| Days Max Temp ≤ 32 °F | 14 | 11 | 3 | 0 | 0 | 0 | 0 | 0 | 0 | 0 | 1 | 9 | 38 |
| Days Min Temp ≤ 32 °F | 28 | 25 | 22 | 12 | 3 | 0 | 0 | 0 | 1 | 7 | 16 | 25 | 139 |
| Days Min Temp ≤ 0 °F | 4 | 2 | 0 | 0 | 0 | 0 | 0 | 0 | 0 | 0 | 0 | 1 | 7 |
| Heating Degree Days | 1241 | 1052 | 854 | 508 | 249 | 69 | 17 | 29 | 140 | 433 | 715 | 1064 | 6371 |
| Cooling Degree Days | 0 | 0 | 1 | 7 | 29 | 91 | 181 | 141 | 55 | 4 | 0 | 0 | 509 |
| Total Precipitation (") | 2.60 | 2.51 | 3.32 | 3.27 | 3.57 | 4.24 | 4.36 | 3.90 | 4.20 | 2.88 | 3.80 | 3.33 | 41.98 |
| Days ≥ 0.1" Precip | 7 | 7 | 8 | 8 | 9 | 8 | 7 | 7 | 8 | 7 | 9 | 9 | 94 |
| Total Snowfall (") | 13.2 | 11.2 | 9.2 | 2.1 | 0.1 | 0.0 | 0.0 | 0.0 | 0.0 | 0.1 | 4.6 | 11.4 | 51.9 |
| Days ≥ 1" Snow Depth | 20 | 17 | 9 | 1 | 0 | 0 | 0 | 0 | 0 | 0 | 3 | 14 | 64 |

## MONTGOMERY L&D *Beaver County* ELEVATION 689 ft LAT/LONG 40° 39 ' N / 80° 23 ' W

| | JAN | FEB | MAR | APR | MAY | JUN | JUL | AUG | SEP | OCT | NOV | DEC | YEAR |
|---|---|---|---|---|---|---|---|---|---|---|---|---|---|
| Maximum Temp °F | 36.2 | 40.0 | 51.0 | 63.7 | 73.6 | 81.3 | 84.5 | 82.2 | 75.5 | 63.9 | 52.3 | 41.2 | 62.1 |
| Minimum Temp °F | 20.0 | 21.6 | 29.4 | 38.5 | 47.8 | 56.7 | 61.5 | 60.4 | 54.3 | 42.8 | 35.0 | 26.0 | 41.2 |
| Mean Temp °F | 28.2 | 30.8 | 40.2 | 51.1 | 60.7 | 69.0 | 73.0 | 71.3 | 64.9 | 53.4 | 43.7 | 33.6 | 51.7 |
| Days Max Temp ≥ 90 °F | 0 | 0 | 0 | 0 | 0 | 3 | 5 | 2 | 0 | 0 | 0 | 0 | 10 |
| Days Max Temp ≤ 32 °F | 11 | 7 | 2 | 0 | 0 | 0 | 0 | 0 | 0 | 0 | 1 | 5 | 26 |
| Days Min Temp ≤ 32 °F | 26 | 23 | 19 | 9 | 1 | 0 | 0 | 0 | 0 | 4 | 12 | 23 | 117 |
| Days Min Temp ≤ 0 °F | 2 | 1 | 0 | 0 | 0 | 0 | 0 | 0 | 0 | 0 | 0 | 1 | 4 |
| Heating Degree Days | 1135 | 958 | 761 | 417 | 173 | 30 | 3 | 9 | 87 | 360 | 634 | 966 | 5533 |
| Cooling Degree Days | 0 | 0 | 0 | 9 | 46 | 156 | 285 | 221 | 98 | 7 | 1 | 0 | 823 |
| Total Precipitation (") | 2.40 | 2.10 | 3.16 | 3.01 | 3.89 | 3.38 | 4.12 | 3.37 | 3.47 | 2.37 | 3.07 | 2.93 | 37.27 |
| Days ≥ 0.1" Precip | 6 | 6 | 8 | 8 | 9 | 7 | 8 | 6 | 7 | 6 | 8 | 7 | 86 |
| Total Snowfall (") | ***7.0*** | ***4.1*** | ***4.2*** | 0.2 | 0.0 | 0.0 | 0.0 | 0.0 | 0.0 | 0.0 | 0.7 | ***3.9*** | 20.1 |
| Days ≥ 1" Snow Depth | 13 | 10 | 5 | 0 | 0 | 0 | 0 | 0 | 0 | 0 | 1 | 6 | 35 |

## MONTROSE *Susquehanna County* ELEVATION 1680 ft LAT/LONG 41° 50 ' N / 75° 49 ' W

| | JAN | FEB | MAR | APR | MAY | JUN | JUL | AUG | SEP | OCT | NOV | DEC | YEAR |
|---|---|---|---|---|---|---|---|---|---|---|---|---|---|
| Maximum Temp °F | 28.7 | 31.5 | 40.9 | 53.6 | 65.6 | 74.0 | 78.6 | 76.7 | 68.8 | 57.5 | 45.2 | 33.7 | 54.6 |
| Minimum Temp °F | 11.9 | 13.1 | 22.1 | 33.2 | 43.3 | 52.2 | 56.7 | 55.2 | 47.7 | 36.9 | 29.0 | 18.6 | 35.0 |
| Mean Temp °F | 20.3 | 22.3 | 31.6 | 43.4 | 54.5 | 63.1 | 67.6 | 65.9 | 58.3 | 47.2 | 37.2 | 26.2 | 44.8 |
| Days Max Temp ≥ 90 °F | 0 | 0 | 0 | 0 | 0 | 0 | 1 | 0 | 0 | 0 | 0 | 0 | 1 |
| Days Max Temp ≤ 32 °F | 19 | 15 | 7 | 1 | 0 | 0 | 0 | 0 | 0 | 0 | 3 | 14 | 59 |
| Days Min Temp ≤ 32 °F | 30 | 27 | 26 | 15 | 3 | 0 | 0 | 0 | 1 | 10 | 20 | 29 | 161 |
| Days Min Temp ≤ 0 °F | 6 | 5 | 1 | 0 | 0 | 0 | 0 | 0 | 0 | 0 | 0 | 2 | 14 |
| Heating Degree Days | 1382 | 1202 | 1033 | 646 | 337 | 112 | 35 | 57 | 223 | 547 | 831 | 1200 | 7605 |
| Cooling Degree Days | 0 | 0 | 0 | 2 | 18 | 64 | 141 | 97 | 28 | 1 | 0 | 0 | 351 |
| Total Precipitation (") | 2.79 | 2.66 | 3.35 | 3.70 | 4.03 | 4.31 | 4.12 | 3.93 | 3.88 | 3.41 | 3.96 | 3.36 | 43.50 |
| Days ≥ 0.1" Precip | 7 | 7 | 8 | 8 | 8 | 9 | 8 | 8 | 7 | 7 | 8 | 8 | 93 |
| Total Snowfall (") | 20.3 | 17.7 | 17.5 | 6.4 | 0.5 | 0.0 | 0.0 | 0.0 | 0.0 | 1.0 | 9.7 | 17.0 | 90.1 |
| Days ≥ 1" Snow Depth | 24 | 23 | 15 | 3 | 0 | 0 | 0 | 0 | 0 | 0 | 5 | 15 | 85 |

## NEW CASTLE 1 N *Lawrence County* ELEVATION 825 ft LAT/LONG 41° 1 ' N / 80° 22 ' W

| | JAN | FEB | MAR | APR | MAY | JUN | JUL | AUG | SEP | OCT | NOV | DEC | YEAR |
|---|---|---|---|---|---|---|---|---|---|---|---|---|---|
| Maximum Temp °F | 34.8 | 38.5 | 49.4 | 62.0 | 72.3 | 80.5 | 84.0 | 82.2 | 75.8 | 64.2 | 51.5 | 40.1 | 61.3 |
| Minimum Temp °F | 17.4 | 18.4 | 26.2 | 35.4 | 44.9 | 53.9 | 58.6 | 57.5 | 51.2 | 39.6 | 32.6 | 23.8 | 38.3 |
| Mean Temp °F | 26.1 | 28.5 | 37.8 | 48.7 | 58.6 | 67.2 | 71.3 | 69.9 | 63.5 | 51.9 | 42.1 | 32.0 | 49.8 |
| Days Max Temp ≥ 90 °F | 0 | 0 | 0 | 0 | 0 | 2 | 5 | 2 | 1 | 0 | 0 | 0 | 10 |
| Days Max Temp ≤ 32 °F | 13 | 9 | 3 | 0 | 0 | 0 | 0 | 0 | 0 | 0 | 1 | 7 | 33 |
| Days Min Temp ≤ 32 °F | 28 | 25 | 23 | 13 | 3 | 0 | 0 | 0 | 0 | 7 | 16 | 25 | 140 |
| Days Min Temp ≤ 0 °F | 4 | 2 | 0 | 0 | 0 | 0 | 0 | 0 | 0 | 0 | 0 | 1 | 7 |
| Heating Degree Days | 1198 | 1025 | 836 | 488 | 227 | 54 | 10 | 20 | 115 | 405 | 681 | 1017 | 6076 |
| Cooling Degree Days | 0 | 0 | 0 | 5 | 30 | 110 | 216 | 171 | 69 | 4 | 0 | 0 | 605 |
| Total Precipitation (") | 2.13 | 2.10 | 2.97 | 2.98 | 3.47 | 4.01 | 4.21 | 3.61 | 3.67 | 2.65 | 3.18 | 2.82 | 37.80 |
| Days ≥ 0.1" Precip | 6 | 6 | 7 | 7 | 8 | 8 | 8 | 7 | 7 | 6 | 8 | 7 | 85 |
| Total Snowfall (") | ***9.5*** | 7.3 | 5.7 | 0.7 | 0.0 | 0.0 | 0.0 | 0.0 | 0.0 | 0.0 | 2.0 | 4.4 | 29.6 |
| Days ≥ 1" Snow Depth | 14 | 11 | 4 | 0 | 0 | 0 | 0 | 0 | 0 | 0 | 1 | 6 | 36 |

## NEWPORT RIVER *Perry County* ELEVATION 381 ft LAT/LONG 40° 29 ' N / 77° 8 ' W

| | JAN | FEB | MAR | APR | MAY | JUN | JUL | AUG | SEP | OCT | NOV | DEC | YEAR |
|---|---|---|---|---|---|---|---|---|---|---|---|---|---|
| Maximum Temp °F | 36.2 | 39.9 | 50.5 | 62.8 | 72.0 | 80.6 | 84.9 | 83.0 | 75.4 | 63.5 | 52.5 | 40.6 | 61.8 |
| Minimum Temp °F | 18.4 | 20.1 | 28.5 | 37.9 | 47.6 | 57.0 | 62.0 | 60.8 | 53.1 | 41.4 | 32.8 | 24.3 | 40.3 |
| Mean Temp °F | 27.3 | 30.0 | 39.6 | 50.3 | 59.8 | 68.8 | 73.5 | 71.9 | 64.3 | 52.5 | 42.7 | 32.5 | 51.1 |
| Days Max Temp ≥ 90 °F | 0 | 0 | 0 | 0 | 0 | 3 | 8 | 4 | 1 | 0 | 0 | 0 | 16 |
| Days Max Temp ≤ 32 °F | 11 | 7 | 1 | 0 | 0 | 0 | 0 | 0 | 0 | 0 | 0 | 6 | 25 |
| Days Min Temp ≤ 32 °F | 28 | 25 | 21 | 8 | 1 | 0 | 0 | 0 | 0 | 5 | 15 | 25 | 128 |
| Days Min Temp ≤ 0 °F | 2 | 1 | 0 | 0 | 0 | 0 | 0 | 0 | 0 | 0 | 0 | 0 | 3 |
| Heating Degree Days | 1161 | 982 | 783 | 442 | 194 | 32 | 5 | 10 | 97 | 386 | 662 | 1000 | 5754 |
| Cooling Degree Days | 0 | 0 | 1 | 7 | 43 | ***147*** | ***288*** | ***228*** | ***81*** | ***7*** | ***0*** | ***0*** | 802 |
| Total Precipitation (") | 2.45 | 2.59 | 3.26 | 2.99 | 4.06 | 3.82 | 3.64 | 3.53 | 3.73 | 3.41 | 3.49 | 3.22 | 40.19 |
| Days ≥ 0.1" Precip | 6 | 6 | 7 | 7 | 8 | 7 | 6 | 7 | 6 | 6 | 7 | 6 | 79 |
| Total Snowfall (") | ***9.1*** | ***6.0*** | ***3.8*** | 0.3 | 0.0 | 0.0 | 0.0 | 0.0 | 0.0 | 0.0 | 1.5 | ***3.6*** | 24.3 |
| Days ≥ 1" Snow Depth | 16 | 12 | 3 | 0 | 0 | 0 | 0 | 0 | 0 | 0 | 1 | ***6*** | 38 |

## PALMERTON *Carbon County* ELEVATION 410 ft LAT/LONG 40° 48 ' N / 75° 37 ' W

| | JAN | FEB | MAR | APR | MAY | JUN | JUL | AUG | SEP | OCT | NOV | DEC | YEAR |
|---|---|---|---|---|---|---|---|---|---|---|---|---|---|
| Maximum Temp °F | 36.5 | 39.4 | 48.6 | 61.1 | 72.2 | 81.1 | 85.7 | ***84.2*** | ***75.3*** | 64.5 | 52.3 | 41.1 | 61.8 |
| Minimum Temp °F | 19.2 | 20.8 | 29.1 | 38.1 | 48.1 | 57.0 | 62.3 | ***60.8*** | ***52.6*** | 41.3 | 33.8 | 25.1 | 40.7 |
| Mean Temp °F | 27.9 | 30.1 | 38.9 | 49.6 | 60.2 | 69.1 | 74.0 | ***72.5*** | ***64.0*** | 52.9 | 43.1 | 33.1 | 51.3 |
| Days Max Temp ≥ 90 °F | 0 | 0 | 0 | 0 | 1 | 5 | 9 | 7 | 1 | 0 | 0 | 0 | 23 |
| Days Max Temp ≤ 32 °F | 10 | 7 | 1 | 0 | 0 | 0 | 0 | 0 | 0 | 0 | 0 | 5 | 23 |
| Days Min Temp ≤ 32 °F | 27 | 23 | 20 | 8 | 1 | 0 | 0 | 0 | 0 | 6 | 14 | 25 | 124 |
| Days Min Temp ≤ 0 °F | 2 | 1 | 0 | 0 | 0 | 0 | 0 | 0 | 0 | 0 | 0 | 0 | 3 |
| Heating Degree Days | 1145 | 980 | 803 | 459 | 187 | 33 | 4 | ***9*** | ***107*** | 376 | 652 | 981 | 5736 |
| Cooling Degree Days | ***0*** | ***0*** | 1 | 7 | ***46*** | ***169*** | ***310*** | ***252*** | ***83*** | ***11*** | ***0*** | ***0*** | 879 |
| Total Precipitation (") | 2.57 | 2.34 | 3.21 | 3.27 | 4.54 | 3.98 | 4.05 | 4.23 | 4.09 | 3.32 | 3.69 | 3.03 | 42.32 |
| Days ≥ 0.1" Precip | 4 | 4 | 6 | 5 | 8 | 6 | 7 | 6 | 5 | 5 | 6 | 5 | 67 |
| Total Snowfall (") | 5.4 | 5.8 | 3.8 | 0.6 | 0.0 | 0.0 | 0.0 | 0.0 | 0.0 | 0.1 | 1.2 | 3.2 | 20.1 |
| Days ≥ 1" Snow Depth | ***12*** | ***9*** | 4 | 0 | 0 | 0 | 0 | 0 | 0 | 0 | 1 | ***4*** | 30 |

## PHILADELPHIA INTL AP *Philadelphia County* ELEVATION 26 ft LAT/LONG 39° 53 ' N / 75° 14 ' W

| | JAN | FEB | MAR | APR | MAY | JUN | JUL | AUG | SEP | OCT | NOV | DEC | YEAR |
|---|---|---|---|---|---|---|---|---|---|---|---|---|---|
| Maximum Temp °F | 38.5 | 41.8 | 51.5 | 63.2 | 73.5 | 82.1 | 86.6 | 85.1 | 77.8 | 66.2 | 55.4 | 44.4 | 63.8 |
| Minimum Temp °F | 23.6 | 25.4 | 33.4 | 42.6 | 53.2 | 62.4 | 68.1 | 67.0 | 59.3 | 47.0 | 38.3 | 29.2 | 45.8 |
| Mean Temp °F | 31.1 | 33.6 | 42.5 | 52.9 | 63.4 | 72.3 | 77.3 | 76.1 | 68.6 | 56.6 | 46.9 | 36.8 | 54.8 |
| Days Max Temp ≥ 90 °F | 0 | 0 | 0 | 0 | 1 | 5 | 10 | 7 | 2 | 0 | 0 | 0 | 25 |
| Days Max Temp ≤ 32 °F | 9 | 6 | 1 | 0 | 0 | 0 | 0 | 0 | 0 | 0 | 0 | 3 | 19 |
| Days Min Temp ≤ 32 °F | 26 | 22 | 14 | 2 | 0 | 0 | 0 | 0 | 0 | 1 | 8 | 20 | 93 |
| Days Min Temp ≤ 0 °F | 0 | 0 | 0 | 0 | 0 | 0 | 0 | 0 | 0 | 0 | 0 | 0 | 0 |
| Heating Degree Days | 1045 | 879 | 693 | 365 | 115 | 10 | 0 | 2 | 40 | 272 | 540 | 867 | 4828 |
| Cooling Degree Days | 0 | 0 | 1 | 10 | 82 | 250 | 414 | 350 | 153 | 22 | 1 | 0 | 1283 |
| Total Precipitation (") | 3.17 | 2.68 | 3.64 | 3.50 | 3.88 | 3.57 | 4.51 | 3.86 | 3.45 | 2.68 | 3.18 | 3.47 | 41.59 |
| Days ≥ 0.1" Precip | 6 | 5 | 7 | 6 | 7 | 6 | 6 | 6 | 5 | 4 | 6 | 6 | 70 |
| Total Snowfall (") | 7.0 | 7.0 | 3.6 | 0.5 | 0.0 | 0.0 | 0.0 | 0.0 | 0.0 | 0.1 | 0.5 | 2.6 | 21.3 |
| Days ≥ 1" Snow Depth | 7 | 5 | 1 | 0 | 0 | 0 | 0 | 0 | 0 | 0 | 0 | 2 | 15 |

## PHILIPSBURG 8 E *Centre County* ELEVATION 2000 ft LAT/LONG 40° 55 ' N / 78° 4 ' W

| | JAN | FEB | MAR | APR | MAY | JUN | JUL | AUG | SEP | OCT | NOV | DEC | YEAR |
|---|---|---|---|---|---|---|---|---|---|---|---|---|---|
| Maximum Temp °F | 29.6 | 33.3 | 42.9 | 56.2 | 66.6 | 74.5 | 78.2 | 76.3 | 68.7 | 57.4 | 45.7 | 34.6 | 55.3 |
| Minimum Temp °F | 14.6 | 16.2 | 24.7 | 34.6 | 43.8 | 51.8 | 56.3 | 55.0 | 48.0 | 38.1 | 30.8 | 20.6 | 36.2 |
| Mean Temp °F | 22.1 | 24.8 | 33.8 | 45.4 | 55.2 | 63.2 | 67.3 | 65.7 | 58.4 | 47.8 | 38.3 | 27.6 | 45.8 |
| Days Max Temp ≥ 90 °F | 0 | 0 | 0 | 0 | 0 | 0 | 1 | 0 | 0 | 0 | 0 | 0 | 1 |
| Days Max Temp ≤ 32 °F | 19 | 14 | 6 | 0 | 0 | 0 | 0 | 0 | 0 | 0 | 3 | 13 | 55 |
| Days Min Temp ≤ 32 °F | 29 | 25 | 24 | 14 | 4 | 0 | 0 | 0 | 2 | 9 | 18 | 27 | 152 |
| Days Min Temp ≤ 0 °F | 5 | 3 | 1 | 0 | 0 | 0 | 0 | 0 | 0 | 0 | 0 | 2 | 11 |
| Heating Degree Days | 1320 | 1130 | 959 | 584 | 310 | 108 | 42 | 64 | 219 | 527 | 795 | 1151 | 7209 |
| Cooling Degree Days | 0 | 0 | 0 | 4 | 16 | 56 | 130 | 92 | 27 | 1 | 0 | 0 | 326 |
| Total Precipitation (") | 2.29 | 2.48 | 3.23 | 3.15 | 3.93 | 4.47 | 4.39 | 3.71 | 3.91 | 3.07 | 3.56 | 2.72 | 40.91 |
| Days ≥ 0.1" Precip | 6 | 6 | 7 | 7 | 9 | 8 | 8 | 7 | 7 | 6 | 8 | 7 | 86 |
| Total Snowfall (") | 11.4 | 13.8 | 11.3 | 1.6 | 0.0 | 0.0 | 0.0 | 0.0 | 0.0 | 0.3 | 4.1 | 9.0 | 51.5 |
| Days ≥ 1" Snow Depth | 22 | 19 | 12 | 1 | 0 | 0 | 0 | 0 | 0 | 0 | 3 | 13 | 70 |

## PHOENIXVILLE 1 E *Chester County* ELEVATION 102 ft LAT/LONG 40° 7 ' N / 75° 30 ' W

| | JAN | FEB | MAR | APR | MAY | JUN | JUL | AUG | SEP | OCT | NOV | DEC | YEAR |
|---|---|---|---|---|---|---|---|---|---|---|---|---|---|
| Maximum Temp °F | 39.4 | 42.8 | 52.6 | 64.2 | 74.8 | 82.6 | 86.6 | 84.9 | 77.9 | 66.6 | 55.8 | 44.0 | 64.4 |
| Minimum Temp °F | 19.7 | 21.8 | 30.4 | 39.1 | 49.2 | 57.9 | 62.7 | 60.4 | 53.4 | 41.7 | 33.7 | 25.0 | 41.2 |
| Mean Temp °F | 29.6 | 32.3 | 41.5 | 51.7 | 62.0 | 70.3 | 74.7 | 72.7 | 65.7 | 54.2 | 44.7 | 34.5 | 52.8 |
| Days Max Temp ≥ 90 °F | 0 | 0 | 0 | 0 | 1 | 4 | 9 | 7 | 2 | 0 | 0 | 0 | 23 |
| Days Max Temp ≤ 32 °F | 7 | 4 | 0 | 0 | 0 | 0 | 0 | 0 | 0 | 0 | 0 | 3 | 14 |
| Days Min Temp ≤ 32 °F | 28 | 24 | 19 | 7 | 1 | 0 | 0 | 0 | 0 | 6 | 14 | 25 | 124 |
| Days Min Temp ≤ 0 °F | 1 | 1 | 0 | 0 | 0 | 0 | 0 | 0 | 0 | 0 | 0 | 0 | 2 |
| Heating Degree Days | 1091 | 918 | 722 | 401 | 142 | 19 | 2 | 7 | 76 | 340 | 602 | 939 | 5259 |
| Cooling Degree Days | *0* | 0 | 1 | 8 | *59* | 175 | 316 | 232 | 89 | 11 | 0 | 0 | 891 |
| Total Precipitation (") | 3.13 | 2.69 | 3.58 | 3.42 | 4.20 | 3.80 | 4.25 | 3.58 | 3.84 | 2.96 | 3.65 | 3.73 | 42.83 |
| Days ≥ 0.1" Precip | 6 | 6 | 7 | 6 | 8 | 6 | 6 | 6 | 6 | 5 | 6 | 6 | 74 |
| Total Snowfall (") | na | na | *1.1* | 0.1 | 0.0 | 0.0 | 0.0 | 0.0 | 0.0 | 0.0 | *0.3* | *2.0* | na |
| Days ≥ 1" Snow Depth | na | na | na | 0 | 0 | 0 | 0 | 0 | 0 | 0 | *0* | *1* | na |

## PITTSBURGH GR P'BURG *Allegheny County* ELEVATION 1224 ft LAT/LONG 40° 30 ' N / 80° 13 ' W

| | JAN | FEB | MAR | APR | MAY | JUN | JUL | AUG | SEP | OCT | NOV | DEC | YEAR |
|---|---|---|---|---|---|---|---|---|---|---|---|---|---|
| Maximum Temp °F | 34.1 | 37.6 | 48.9 | 60.9 | 70.7 | 79.2 | 82.9 | 81.0 | 74.2 | 62.1 | 50.4 | 39.6 | 60.1 |
| Minimum Temp °F | 19.0 | 21.0 | 29.8 | 39.3 | 48.8 | 57.3 | 62.3 | 60.7 | 53.8 | 42.2 | 34.2 | 25.3 | 41.1 |
| Mean Temp °F | 26.6 | 29.3 | 39.4 | 50.1 | 59.8 | 68.3 | 72.6 | 70.9 | 64.0 | 52.2 | 42.4 | 32.5 | 50.7 |
| Days Max Temp ≥ 90 °F | 0 | 0 | 0 | 0 | 0 | 2 | 4 | 2 | 0 | 0 | 0 | 0 | 8 |
| Days Max Temp ≤ 32 °F | 14 | 10 | 3 | 0 | 0 | 0 | 0 | 0 | 0 | 0 | 1 | 9 | 37 |
| Days Min Temp ≤ 32 °F | 27 | 23 | 20 | 8 | 1 | 0 | 0 | 0 | 0 | 4 | 14 | 24 | 121 |
| Days Min Temp ≤ 0 °F | 2 | 1 | 0 | 0 | 0 | 0 | 0 | 0 | 0 | 0 | 0 | 1 | 4 |
| Heating Degree Days | 1184 | 1001 | 788 | 449 | 200 | 41 | 5 | 14 | 105 | 397 | 674 | 1002 | 5860 |
| Cooling Degree Days | 0 | 0 | 1 | 12 | 49 | 148 | 271 | 216 | 89 | 7 | 0 | 0 | 793 |
| Total Precipitation (") | 2.64 | 2.33 | 3.35 | 2.95 | 3.69 | 3.66 | 3.83 | 3.47 | 3.14 | 2.34 | 2.95 | 2.98 | 37.33 |
| Days ≥ 0.1" Precip | 7 | 6 | 8 | 8 | 8 | 7 | 8 | 6 | 6 | 6 | 7 | 7 | 84 |
| Total Snowfall (") | 12.3 | 9.1 | 8.7 | 1.5 | 0.1 | 0.0 | 0.0 | 0.0 | 0.0 | 0.4 | 3.1 | 7.7 | 42.9 |
| Days ≥ 1" Snow Depth | 14 | 12 | 6 | 1 | 0 | 0 | 0 | 0 | 0 | 0 | 2 | 7 | 42 |

## PLEASANT MOUNT 1 W *Wayne County* ELEVATION 1800 ft LAT/LONG 41° 44' N / 75° 27' W

| | JAN | FEB | MAR | APR | MAY | JUN | JUL | AUG | SEP | OCT | NOV | DEC | YEAR |
|---|---|---|---|---|---|---|---|---|---|---|---|---|---|
| Maximum Temp °F | 27.9 | 30.8 | 39.5 | 52.5 | 64.3 | 72.2 | 76.9 | 75.3 | 67.5 | 56.8 | 44.5 | 32.8 | 53.4 |
| Minimum Temp °F | 9.5 | 10.6 | 19.3 | 31.1 | 41.0 | 49.8 | 54.2 | 52.9 | 45.3 | 34.7 | 27.0 | 16.2 | 32.6 |
| Mean Temp °F | 18.7 | 20.7 | 29.4 | 41.8 | 52.7 | 61.0 | 65.6 | 64.1 | 56.4 | 45.8 | 35.8 | 24.6 | 43.1 |
| Days Max Temp ≥ 90 °F | 0 | 0 | 0 | 0 | 0 | 0 | 0 | 0 | 0 | 0 | 0 | 0 | 0 |
| Days Max Temp ≤ 32 °F | 21 | 16 | 9 | 1 | 0 | 0 | 0 | 0 | 0 | 0 | 3 | 15 | 65 |
| Days Min Temp ≤ 32 °F | 31 | 27 | 28 | 18 | 5 | 0 | 0 | 0 | 2 | 14 | 23 | 30 | 178 |
| Days Min Temp ≤ 0 °F | 7 | 6 | 1 | 0 | 0 | 0 | 0 | 0 | 0 | 0 | 0 | 3 | 17 |
| Heating Degree Days | 1428 | 1244 | 1096 | 690 | 383 | 150 | 58 | 82 | 268 | 591 | 870 | 1247 | 8107 |
| Cooling Degree Days | 0 | 0 | 0 | 0 | 8 | 36 | 92 | 66 | 16 | 1 | 0 | 0 | 219 |
| Total Precipitation (") | 3.05 | 2.79 | 3.43 | 4.03 | 4.79 | 4.75 | 4.18 | 4.43 | 4.42 | 3.98 | 4.42 | 3.69 | 47.96 |
| Days ≥ 0.1" Precip | 7 | 6 | 7 | 8 | 9 | 9 | 8 | 8 | 7 | 7 | 8 | 8 | 92 |
| Total Snowfall (") | 17.9 | 15.5 | 14.6 | 4.1 | 0.4 | 0.0 | 0.0 | 0.0 | 0.0 | 0.4 | 7.7 | 14.9 | 75.5 |
| Days ≥ 1" Snow Depth | 27 | 26 | 22 | 5 | 0 | 0 | 0 | 0 | 0 | 0 | 6 | 19 | 105 |

## PUTNEYVILLE 2 SE DAM *Armstrong County* ELEVATION 1270 ft LAT/LONG 40° 55' N / 79° 17' W

| | JAN | FEB | MAR | APR | MAY | JUN | JUL | AUG | SEP | OCT | NOV | DEC | YEAR |
|---|---|---|---|---|---|---|---|---|---|---|---|---|---|
| Maximum Temp °F | 33.1 | 35.9 | 46.3 | 59.3 | 70.0 | 78.2 | 82.0 | 80.2 | 73.0 | 61.2 | 49.3 | 38.1 | 58.9 |
| Minimum Temp °F | 14.4 | 15.6 | 24.0 | 34.0 | 43.4 | 52.1 | 56.9 | 55.5 | 48.8 | 37.5 | 30.6 | 21.3 | 36.2 |
| Mean Temp °F | 23.8 | 25.8 | 35.2 | 46.7 | 56.7 | 65.1 | 69.5 | 67.9 | 60.9 | 49.4 | 40.0 | 29.7 | 47.6 |
| Days Max Temp ≥ 90 °F | 0 | 0 | 0 | 0 | 0 | 1 | 4 | 2 | 0 | 0 | 0 | 0 | 7 |
| Days Max Temp ≤ 32 °F | 15 | 11 | 4 | 0 | 0 | 0 | 0 | 0 | 0 | 0 | 2 | 9 | 41 |
| Days Min Temp ≤ 32 °F | 29 | 26 | 25 | 15 | 4 | 0 | 0 | 0 | 1 | 10 | 19 | 27 | 156 |
| Days Min Temp ≤ 0 °F | 5 | 4 | 1 | 0 | 0 | 0 | 0 | 0 | 0 | 0 | 0 | 2 | 12 |
| Heating Degree Days | 1270 | 1101 | 919 | 548 | 274 | 78 | 22 | 35 | 164 | 480 | 744 | 1088 | 6723 |
| Cooling Degree Days | 0 | 0 | 0 | 5 | 26 | 87 | 186 | 139 | 49 | 2 | 0 | 0 | 494 |
| Total Precipitation (") | 2.94 | 2.65 | 3.75 | 3.41 | 4.04 | 4.27 | 4.64 | 4.14 | 4.05 | 3.17 | 3.62 | 3.43 | 44.11 |
| Days ≥ 0.1" Precip | 8 | 7 | 9 | 8 | 9 | 8 | 8 | 8 | 8 | 7 | 8 | 8 | 96 |
| Total Snowfall (") | 11.8 | 10.1 | 8.3 | 0.7 | 0.0 | 0.0 | 0.0 | 0.0 | 0.0 | 0.0 | 2.3 | 7.4 | 40.6 |
| Days ≥ 1" Snow Depth | 19 | 17 | 9 | 1 | 0 | 0 | 0 | 0 | 0 | 0 | 2 | 11 | 59 |

## RIDGWAY *Elk County* ELEVATION 1371 ft LAT/LONG 41° 26' N / 78° 44' W

| | JAN | FEB | MAR | APR | MAY | JUN | JUL | AUG | SEP | OCT | NOV | DEC | YEAR |
|---|---|---|---|---|---|---|---|---|---|---|---|---|---|
| Maximum Temp °F | 31.5 | 34.5 | 44.5 | 57.2 | 68.3 | 76.1 | 79.9 | 78.3 | 71.4 | 60.0 | 47.9 | 36.6 | 57.2 |
| Minimum Temp °F | 12.3 | 12.9 | 21.3 | 31.1 | 39.9 | 48.7 | 53.7 | 52.6 | 46.2 | 35.1 | 28.9 | 19.5 | 33.5 |
| Mean Temp °F | 22.0 | 23.7 | 32.9 | 44.2 | 54.1 | 62.4 | 66.8 | 65.5 | 58.8 | 47.6 | 38.4 | 28.0 | 45.4 |
| Days Max Temp ≥ 90 °F | 0 | 0 | 0 | 0 | 0 | 0 | 1 | 1 | 0 | 0 | 0 | 0 | 2 |
| Days Max Temp ≤ 32 °F | 16 | 12 | 5 | 0 | 0 | 0 | 0 | 0 | 0 | 0 | 2 | 10 | 45 |
| Days Min Temp ≤ 32 °F | 30 | 26 | 26 | 18 | 8 | 1 | 0 | 0 | 2 | 14 | 20 | 27 | 172 |
| Days Min Temp ≤ 0 °F | 7 | 6 | 1 | 0 | 0 | 0 | 0 | 0 | 0 | 0 | 0 | 3 | 17 |
| Heating Degree Days | 1327 | 1159 | 989 | 619 | 341 | 122 | 43 | 59 | 206 | 533 | 791 | 1139 | 7328 |
| Cooling Degree Days | 0 | 0 | 0 | 1 | 10 | 43 | 108 | 76 | 23 | 1 | 0 | 0 | 262 |
| Total Precipitation (") | 2.53 | 2.32 | 3.32 | 3.44 | 4.34 | 4.77 | 4.79 | 4.11 | 3.94 | 3.11 | 3.66 | 3.17 | 43.50 |
| Days ≥ 0.1" Precip | 7 | 6 | 8 | 9 | 9 | 8 | 9 | 8 | 8 | 8 | 9 | 8 | 97 |
| Total Snowfall (") | 14.9 | 13.7 | 8.8 | 1.7 | 0.0 | 0.0 | 0.0 | 0.0 | 0.0 | 0.1 | 3.5 | 12.4 | 55.1 |
| Days ≥ 1" Snow Depth | 23 | 21 | 11 | 1 | 0 | 0 | 0 | 0 | 0 | 0 | 4 | 16 | 76 |

## SALINA 3 W *Westmoreland County* ELEVATION 1109 ft LAT/LONG 40° 31' N / 79° 33' W

| | JAN | FEB | MAR | APR | MAY | JUN | JUL | AUG | SEP | OCT | NOV | DEC | YEAR |
|---|---|---|---|---|---|---|---|---|---|---|---|---|---|
| Maximum Temp °F | 35.8 | 39.2 | 49.9 | 62.3 | 72.1 | 80.1 | 83.6 | 81.8 | 75.3 | 63.7 | 52.0 | 41.1 | 61.4 |
| Minimum Temp °F | 16.9 | 18.3 | 26.5 | 35.7 | 45.0 | 53.5 | 58.4 | 56.8 | 50.5 | 39.2 | 32.3 | 23.2 | 38.0 |
| Mean Temp °F | 26.4 | 28.8 | 38.2 | 49.0 | 58.5 | 66.8 | 71.0 | 69.2 | 62.9 | 51.4 | 42.2 | 32.2 | 49.7 |
| Days Max Temp ≥ 90 °F | 0 | 0 | 0 | 0 | 0 | 2 | 5 | 2 | 1 | 0 | 0 | 0 | 10 |
| Days Max Temp ≤ 32 °F | 12 | 9 | 2 | 0 | 0 | 0 | 0 | 0 | 0 | 0 | 1 | 7 | 31 |
| Days Min Temp ≤ 32 °F | 28 | 25 | 22 | 13 | 3 | 0 | 0 | 0 | 1 | 9 | 17 | 25 | 143 |
| Days Min Temp ≤ 0 °F | 4 | 3 | 0 | 0 | 0 | 0 | 0 | 0 | 0 | 0 | 0 | 1 | 8 |
| Heating Degree Days | 1191 | 1017 | 824 | 477 | 226 | 55 | 11 | 24 | 124 | 419 | 679 | 1011 | 6058 |
| Cooling Degree Days | 0 | 0 | 0 | 5 | 28 | 105 | 204 | 142 | 60 | 4 | 0 | 0 | 548 |
| Total Precipitation (") | 2.42 | 2.27 | 3.25 | 3.16 | 4.04 | 3.77 | 4.54 | 3.76 | 3.55 | 2.64 | 3.42 | 2.89 | 39.71 |
| Days ≥ 0.1" Precip | 7 | 6 | 7 | 8 | 9 | 8 | 9 | 7 | 7 | 7 | 8 | 7 | 90 |
| Total Snowfall (") | ***10.0*** | 7.9 | 4.4 | 0.4 | 0.0 | 0.0 | 0.0 | 0.0 | 0.0 | 0.0 | 1.9 | 4.9 | 29.5 |
| Days ≥ 1" Snow Depth | 15 | 13 | 4 | 0 | 0 | 0 | 0 | 0 | 0 | 0 | 1 | 7 | 40 |

## SHIPPENSBURG *Cumberland County* ELEVATION 712 ft LAT/LONG 40° 3 ' N / 77° 32 ' W

| | JAN | FEB | MAR | APR | MAY | JUN | JUL | AUG | SEP | OCT | NOV | DEC | YEAR |
|---|---|---|---|---|---|---|---|---|---|---|---|---|---|
| Maximum Temp °F | 35.8 | 39.6 | 50.2 | 62.5 | 72.7 | 81.1 | 85.2 | 83.5 | 76.1 | 64.5 | 52.0 | 40.7 | 62.0 |
| Minimum Temp °F | 20.5 | 22.6 | 30.5 | 40.3 | 50.1 | 58.9 | 63.5 | 61.8 | 54.7 | 43.1 | 34.8 | 26.1 | 42.2 |
| Mean Temp °F | 28.2 | 31.1 | 40.4 | 51.4 | 61.4 | 70.0 | 74.4 | 72.7 | 65.4 | 53.8 | 43.4 | 33.4 | 52.1 |
| Days Max Temp ≥ 90 °F | 0 | 0 | 0 | 0 | 1 | 3 | 8 | 5 | 1 | 0 | 0 | 0 | 18 |
| Days Max Temp ≤ 32 °F | 12 | 7 | 1 | 0 | 0 | 0 | 0 | 0 | 0 | 0 | 1 | 6 | 27 |
| Days Min Temp ≤ 32 °F | 27 | 24 | 19 | 5 | 0 | 0 | 0 | 0 | 0 | 4 | 13 | 24 | 116 |
| Days Min Temp ≤ 0 °F | 1 | 0 | 0 | 0 | 0 | 0 | 0 | 0 | 0 | 0 | 0 | 0 | 1 |
| Heating Degree Days | 1135 | 950 | 758 | 411 | 159 | 25 | 3 | 8 | 82 | 350 | 643 | 973 | 5497 |
| Cooling Degree Days | 0 | 0 | 1 | 10 | 63 | 189 | 328 | 254 | 96 | 11 | 0 | 0 | 952 |
| Total Precipitation (") | 2.69 | 2.53 | 3.55 | 3.14 | 3.75 | 3.46 | 3.42 | 3.02 | 3.40 | 2.88 | 3.29 | 3.12 | 38.25 |
| Days ≥ 0.1" Precip | 6 | 6 | 7 | 7 | 7 | 7 | 7 | 6 | 6 | 5 | 6 | 6 | 76 |
| Total Snowfall (") | 10.8 | 9.6 | 7.3 | 0.7 | 0.0 | 0.0 | 0.0 | 0.0 | 0.0 | 0.0 | 2.3 | 5.8 | 36.5 |
| Days ≥ 1" Snow Depth | 14 | 11 | 5 | 0 | 0 | 0 | 0 | 0 | 0 | 0 | 1 | 6 | 37 |

## SLIPPERY ROCK 1 SSW *Butler County* ELEVATION 1352 ft LAT/LONG 41° 4 ' N / 80° 3 ' W

| | JAN | FEB | MAR | APR | MAY | JUN | JUL | AUG | SEP | OCT | NOV | DEC | YEAR |
|---|---|---|---|---|---|---|---|---|---|---|---|---|---|
| Maximum Temp °F | 33.0 | 36.4 | 46.5 | 58.9 | 69.9 | 78.1 | 81.9 | 80.3 | 73.7 | 62.0 | 49.8 | 38.3 | 59.1 |
| Minimum Temp °F | 13.9 | 15.4 | 24.0 | 33.0 | 42.6 | 51.0 | 55.4 | 53.7 | 47.6 | 36.5 | 29.6 | 20.5 | 35.3 |
| Mean Temp °F | 23.5 | 25.9 | 35.3 | 46.0 | 56.3 | 64.6 | 68.7 | 67.1 | 60.7 | 49.3 | 39.7 | 29.4 | 47.2 |
| Days Max Temp ≥ 90 °F | 0 | 0 | 0 | 0 | 0 | 1 | 3 | 1 | 0 | 0 | 0 | 0 | 5 |
| Days Max Temp ≤ 32 °F | 15 | 10 | 4 | 0 | 0 | 0 | 0 | 0 | 0 | 0 | 2 | 9 | 40 |
| Days Min Temp ≤ 32 °F | 29 | 26 | 25 | 16 | 5 | 0 | 0 | 0 | 1 | 12 | 20 | 27 | 161 |
| Days Min Temp ≤ 0 °F | 6 | 4 | 1 | 0 | 0 | 0 | 0 | 0 | 0 | 0 | 0 | 2 | 13 |
| Heating Degree Days | 1281 | 1098 | 915 | 566 | 285 | 87 | 26 | 42 | 168 | 483 | 752 | 1098 | 6801 |
| Cooling Degree Days | 0 | 0 | 0 | 2 | 18 | 69 | 151 | 113 | 43 | 2 | 0 | 0 | 398 |
| Total Precipitation (") | 2.69 | 2.25 | 3.34 | 3.14 | 3.75 | 4.40 | 4.44 | 4.07 | 3.84 | 3.01 | 3.53 | 3.18 | 41.64 |
| Days ≥ 0.1" Precip | 7 | 7 | 9 | 8 | 9 | 8 | 8 | 7 | 7 | 7 | 8 | 9 | 94 |
| Total Snowfall (") | 13.1 | 9.5 | 7.4 | 1.0 | 0.0 | 0.0 | 0.0 | 0.0 | 0.0 | 0.0 | 2.7 | 8.7 | 42.4 |
| Days ≥ 1" Snow Depth | 19 | 16 | 7 | 1 | 0 | 0 | 0 | 0 | 0 | 0 | 2 | 12 | 57 |

## STATE COLLEGE *Centre County* ELEVATION 1181 ft LAT/LONG 40° 48 ' N / 77° 52 ' W

| | JAN | FEB | MAR | APR | MAY | JUN | JUL | AUG | SEP | OCT | NOV | DEC | YEAR |
|---|---|---|---|---|---|---|---|---|---|---|---|---|---|
| Maximum Temp °F | 32.2 | 35.9 | 45.8 | 58.7 | 69.7 | 78.0 | 81.9 | 80.3 | 72.7 | 60.7 | 48.9 | 37.6 | 58.5 |
| Minimum Temp °F | 16.7 | 18.5 | 26.6 | 37.4 | 47.6 | 56.2 | 60.8 | 59.0 | 51.8 | 40.1 | 32.4 | 23.2 | 39.2 |
| Mean Temp °F | 24.5 | 27.2 | 36.2 | 48.1 | 58.7 | 67.1 | 71.4 | 69.7 | 62.3 | 50.4 | 40.7 | 30.4 | 48.9 |
| Days Max Temp ≥ 90 °F | 0 | 0 | 0 | 0 | 0 | 1 | 3 | 2 | 0 | 0 | 0 | 0 | 6 |
| Days Max Temp ≤ 32 °F | 15 | 11 | 4 | 0 | 0 | 0 | 0 | 0 | 0 | 0 | 1 | 9 | 40 |
| Days Min Temp ≤ 32 °F | 29 | 25 | 23 | 9 | 1 | 0 | 0 | 0 | 0 | 6 | *16* | 26 | 135 |
| Days Min Temp ≤ 0 °F | 3 | 1 | 0 | 0 | 0 | 0 | 0 | 0 | 0 | 0 | 0 | 1 | 5 |
| Heating Degree Days | 1250 | 1060 | 886 | 511 | 231 | 54 | 10 | 22 | 135 | 450 | 725 | 1064 | 6398 |
| Cooling Degree Days | 0 | 0 | 1 | 7 | 36 | 116 | 221 | 165 | 55 | 4 | 0 | 0 | 605 |
| Total Precipitation (") | 2.48 | 2.50 | 3.42 | 2.99 | 3.68 | 3.94 | 3.75 | 3.23 | 3.28 | 2.86 | 3.44 | 2.83 | 38.40 |
| Days ≥ 0.1" Precip | 6 | 6 | 7 | 7 | 8 | 7 | 8 | 6 | 7 | 6 | 7 | 6 | 81 |
| Total Snowfall (") | 12.9 | 11.4 | 11.8 | 1.5 | 0.0 | 0.0 | 0.0 | 0.0 | 0.0 | 0.1 | 3.3 | 7.9 | 48.9 |
| Days ≥ 1" Snow Depth | 18 | 14 | 8 | 1 | 0 | 0 | 0 | 0 | 0 | 0 | *2* | 8 | 51 |

## STROUDSBURG *Monroe County* ELEVATION 361 ft LAT/LONG 40° 59 ' N / 75° 10 ' W

| | JAN | FEB | MAR | APR | MAY | JUN | JUL | AUG | SEP | OCT | NOV | DEC | YEAR |
|---|---|---|---|---|---|---|---|---|---|---|---|---|---|
| Maximum Temp °F | 35.7 | 39.4 | 49.8 | 62.8 | 74.3 | 81.9 | 86.5 | 84.1 | 76.1 | 64.3 | 51.5 | 40.2 | 62.2 |
| Minimum Temp °F | 17.7 | 19.4 | 27.7 | 37.0 | 46.7 | 54.7 | 60.0 | 58.7 | 51.5 | 39.9 | 32.7 | 23.5 | 39.1 |
| Mean Temp °F | 26.7 | 29.5 | 38.8 | 49.9 | 60.5 | 68.3 | 73.3 | 71.4 | 63.9 | 52.1 | 42.1 | 31.9 | 50.7 |
| Days Max Temp ≥ 90 °F | 0 | 0 | 0 | 0 | 1 | 4 | 9 | 5 | 1 | 0 | 0 | 0 | 20 |
| Days Max Temp ≤ 32 °F | 11 | 6 | 1 | 0 | 0 | 0 | 0 | 0 | 0 | 0 | 0 | 6 | 24 |
| Days Min Temp ≤ 32 °F | 29 | 25 | 22 | 10 | 2 | 0 | 0 | 0 | 0 | 8 | 16 | 26 | 138 |
| Days Min Temp ≤ 0 °F | 2 | 2 | 0 | 0 | 0 | 0 | 0 | 0 | 0 | 0 | 0 | 1 | 5 |
| Heating Degree Days | 1180 | 998 | 806 | 453 | 176 | 35 | 5 | 12 | 107 | 398 | 679 | 1021 | 5870 |
| Cooling Degree Days | 0 | 0 | 1 | 7 | 50 | 153 | 295 | 225 | 87 | 8 | 0 | 0 | 826 |
| Total Precipitation (") | 3.28 | 2.95 | 3.74 | 3.87 | 4.87 | 4.29 | 4.32 | 4.53 | 4.53 | 3.68 | 4.29 | 3.94 | 48.29 |
| Days ≥ 0.1" Precip | 6 | 6 | 7 | 7 | 8 | 8 | 7 | 7 | 6 | 6 | 7 | 6 | 81 |
| Total Snowfall (") | 10.2 | 10.1 | 7.2 | 1.3 | 0.0 | 0.0 | 0.0 | 0.0 | 0.0 | 0.1 | 2.0 | 7.0 | 37.9 |
| Days ≥ 1" Snow Depth | 20 | 14 | 6 | 0 | 0 | 0 | 0 | 0 | 0 | 0 | 1 | 8 | 49 |

## TIONESTA 2 SE LAKE *Forest County* ELEVATION 1201 ft LAT/LONG 41° 29 ' N / 79° 26 ' W

| | JAN | FEB | MAR | APR | MAY | JUN | JUL | AUG | SEP | OCT | NOV | DEC | YEAR |
|---|---|---|---|---|---|---|---|---|---|---|---|---|---|
| Maximum Temp °F | 31.7 | 34.0 | 44.4 | 57.8 | 69.1 | 77.1 | 80.9 | 79.5 | 72.4 | 60.3 | 48.4 | 36.7 | 57.7 |
| Minimum Temp °F | 12.3 | 12.4 | 21.4 | 32.0 | 41.9 | 51.1 | *56.2* | 55.3 | 48.5 | 36.5 | 29.5 | 19.5 | 34.7 |
| Mean Temp °F | 22.1 | 23.2 | 32.9 | 44.9 | 55.5 | 64.1 | *68.5* | 67.4 | 60.5 | 48.4 | 39.0 | 28.2 | 46.2 |
| Days Max Temp ≥ 90 °F | 0 | 0 | 0 | 0 | 0 | 1 | 2 | 1 | 0 | 0 | 0 | 0 | 4 |
| Days Max Temp ≤ 32 °F | 16 | 13 | 5 | 0 | 0 | 0 | 0 | 0 | 0 | 0 | 2 | 10 | 46 |
| Days Min Temp ≤ 32 °F | 30 | 27 | 27 | 16 | 5 | 0 | *0* | 0 | 0 | 10 | 20 | 28 | 163 |
| Days Min Temp ≤ 0 °F | 6 | 5 | 1 | 0 | 0 | 0 | *0* | 0 | 0 | 0 | 0 | 2 | 14 |
| Heating Degree Days | 1324 | 1174 | 987 | 599 | 308 | 94 | *25* | 36 | 171 | 507 | 774 | 1136 | 7135 |
| Cooling Degree Days | *0* | *0* | *0* | 2 | *21* | *69* | na | *122* | *38* | 1 | 0 | 0 | na |
| Total Precipitation (") | 2.59 | 2.20 | 3.27 | 3.45 | 3.76 | 4.73 | 4.81 | 4.20 | 4.16 | 3.36 | 3.72 | 3.28 | 43.53 |
| Days ≥ 0.1" Precip | 7 | 6 | 8 | 8 | 9 | 8 | 9 | 8 | 8 | 8 | 9 | 8 | 96 |
| Total Snowfall (") | 14.5 | 12.4 | 11.1 | 2.3 | 0.0 | 0.0 | 0.0 | 0.0 | 0.0 | 0.3 | 3.9 | 13.1 | 57.6 |
| Days ≥ 1" Snow Depth | 22 | 21 | 12 | 1 | 0 | 0 | 0 | 0 | 0 | 0 | 3 | 15 | 74 |

## TITUSVILLE WTR WORKS *Crawford County* ELEVATION 1220 ft LAT/LONG 41° 38 ' N / 79° 42 ' W

| | JAN | FEB | MAR | APR | MAY | JUN | JUL | AUG | SEP | OCT | NOV | DEC | YEAR |
|---|---|---|---|---|---|---|---|---|---|---|---|---|---|
| Maximum Temp °F | 31.5 | 34.2 | 44.1 | 56.9 | 68.4 | 77.0 | 81.0 | 79.0 | 71.9 | 60.6 | 47.9 | 36.2 | 57.4 |
| Minimum Temp °F | 12.2 | 12.7 | 21.6 | 31.8 | 41.1 | 50.2 | 54.7 | 52.9 | 46.7 | 35.6 | 29.0 | 19.4 | 34.0 |
| Mean Temp °F | 21.9 | 23.5 | 32.9 | 44.4 | 54.8 | 63.6 | 67.9 | 65.9 | 59.3 | 48.1 | 38.5 | 27.8 | 45.7 |
| Days Max Temp ≥ 90 °F | 0 | 0 | 0 | 0 | 0 | 1 | 2 | 1 | 0 | 0 | 0 | 0 | 4 |
| Days Max Temp ≤ 32 °F | 17 | 13 | 6 | 1 | 0 | 0 | 0 | 0 | 0 | 0 | 2 | 11 | 50 |
| Days Min Temp ≤ 32 °F | 30 | 27 | 26 | 17 | 6 | 0 | 0 | 0 | 1 | 12 | 21 | 27 | 167 |
| Days Min Temp ≤ 0 °F | 6 | 6 | 1 | 0 | 0 | 0 | 0 | 0 | 0 | 0 | 0 | 2 | 15 |
| Heating Degree Days | 1331 | 1168 | 989 | 615 | 325 | 105 | 37 | 56 | 200 | 518 | 789 | 1146 | 7279 |
| Cooling Degree Days | 0 | 0 | 0 | 2 | 21 | 73 | 152 | 101 | 37 | 1 | 0 | 0 | 387 |
| Total Precipitation (") | 2.40 | 2.22 | 3.12 | 3.61 | 3.99 | 4.59 | 4.42 | 4.33 | 4.57 | 3.62 | 4.04 | 3.29 | 44.20 |
| Days ≥ 0.1" Precip | 7 | 6 | 8 | 9 | 9 | 8 | 8 | 8 | 8 | 8 | 10 | 9 | 98 |
| Total Snowfall (") | 18.8 | 14.7 | 12.5 | 3.2 | 0.1 | 0.0 | 0.0 | 0.0 | 0.0 | 0.5 | 7.1 | 17.4 | 74.3 |
| Days ≥ 1" Snow Depth | 24 | 22 | 12 | 2 | 0 | 0 | 0 | 0 | 0 | 0 | 5 | 18 | 83 |

## TOBYHANNA *Monroe County* ELEVATION 1950 ft LAT/LONG 41° 11 ' N / 75° 25 ' W

| | JAN | FEB | MAR | APR | MAY | JUN | JUL | AUG | SEP | OCT | NOV | DEC | YEAR |
|---|---|---|---|---|---|---|---|---|---|---|---|---|---|
| Maximum Temp °F | 30.3 | 33.1 | 42.4 | 54.9 | 66.4 | 74.0 | 78.2 | 76.6 | 68.8 | 58.2 | 46.3 | 35.3 | 55.4 |
| Minimum Temp °F | 12.9 | 13.9 | 22.4 | 32.6 | 42.3 | 50.4 | 55.5 | 54.6 | 47.3 | 36.9 | 29.0 | 19.3 | 34.8 |
| Mean Temp °F | 21.6 | 23.6 | 32.4 | 43.8 | 54.4 | 62.2 | 66.9 | 65.6 | 58.1 | 47.5 | 37.7 | 27.3 | 45.1 |
| Days Max Temp ≥ 90 °F | 0 | 0 | 0 | 0 | 0 | 0 | 1 | 0 | 0 | 0 | 0 | 0 | 1 |
| Days Max Temp ≤ 32 °F | 17 | 14 | 6 | 0 | 0 | 0 | 0 | 0 | 0 | 0 | 2 | 13 | 52 |
| Days Min Temp ≤ 32 °F | 28 | 26 | 26 | 16 | 5 | 0 | 0 | 0 | 2 | 12 | 20 | 28 | 163 |
| Days Min Temp ≤ 0 °F | 5 | 4 | 1 | 0 | 0 | 0 | 0 | 0 | 0 | 0 | 0 | 2 | 12 |
| Heating Degree Days | 1339 | 1166 | 1004 | 630 | 332 | 121 | 40 | 59 | 224 | 535 | 813 | 1157 | 7420 |
| Cooling Degree Days | *0* | *0* | 0 | 1 | 8 | *39* | *106* | 81 | *20* | 1 | *0* | *0* | 256 |
| Total Precipitation (") | 3.36 | 3.24 | 3.96 | 4.34 | 4.62 | 4.28 | 4.17 | 4.06 | 4.35 | 3.78 | 4.48 | 3.70 | 48.34 |
| Days ≥ 0.1" Precip | 6 | 6 | 7 | 8 | 8 | 8 | 7 | 7 | 7 | 6 | 7 | 7 | 84 |
| Total Snowfall (") | 17.0 | 16.3 | 14.2 | 4.6 | 0.4 | 0.0 | 0.0 | 0.0 | 0.0 | 0.4 | 5.9 | 14.2 | 73.0 |
| Days ≥ 1" Snow Depth | 22 | 21 | 13 | 2 | 0 | 0 | 0 | 0 | 0 | 0 | 3 | 14 | 75 |

## TOWANDA 1 ESE *Bradford County* ELEVATION 761 ft LAT/LONG 41° 46 ' N / 76° 26 ' W

| | JAN | FEB | MAR | APR | MAY | JUN | JUL | AUG | SEP | OCT | NOV | DEC | YEAR |
|---|---|---|---|---|---|---|---|---|---|---|---|---|---|
| Maximum Temp °F | 33.4 | 36.3 | 46.2 | 59.4 | 70.4 | 78.4 | 82.6 | 80.7 | 73.0 | 61.7 | 49.6 | 38.3 | 59.2 |
| Minimum Temp °F | 15.3 | 16.7 | 25.5 | 35.3 | 45.0 | 54.0 | 58.7 | 57.5 | 50.3 | 39.2 | 31.8 | 22.4 | 37.6 |
| Mean Temp °F | 24.4 | 26.5 | 35.9 | 47.4 | 57.7 | 66.2 | 70.7 | 69.2 | 61.7 | 50.5 | 40.7 | 30.4 | 48.4 |
| Days Max Temp ≥ 90 °F | 0 | 0 | 0 | 0 | 0 | 1 | 4 | 2 | 0 | 0 | 0 | 0 | 7 |
| Days Max Temp ≤ 32 °F | 14 | 11 | 3 | 0 | 0 | 0 | 0 | 0 | 0 | 0 | 1 | 8 | 37 |
| Days Min Temp ≤ 32 °F | 29 | 25 | 23 | 13 | 2 | 0 | 0 | 0 | 0 | 8 | 17 | 26 | 143 |
| Days Min Temp ≤ 0 °F | 4 | 3 | 0 | 0 | 0 | 0 | 0 | 0 | 0 | 0 | 0 | 1 | 8 |
| Heating Degree Days | 1254 | 1079 | 897 | 525 | 246 | 61 | 13 | 24 | 146 | 446 | 723 | 1067 | 6481 |
| Cooling Degree Days | 0 | 0 | 0 | 4 | 28 | 103 | 211 | 156 | 52 | 3 | 0 | 0 | 557 |
| Total Precipitation (") | 1.92 | 2.04 | 2.59 | 2.94 | 3.32 | 3.63 | 3.15 | 3.11 | 3.30 | 2.59 | 3.12 | 2.42 | 34.13 |
| Days ≥ 0.1" Precip | 5 | 5 | 6 | 7 | 8 | 8 | 7 | 7 | 6 | 5 | 6 | 6 | 76 |
| Total Snowfall (") | 10.9 | 9.9 | 8.7 | 1.9 | 0.1 | 0.0 | 0.0 | 0.0 | 0.0 | 0.3 | 3.0 | 8.1 | 42.9 |
| Days ≥ 1" Snow Depth | 19 | 17 | 9 | 1 | 0 | 0 | 0 | 0 | 0 | 0 | 2 | 10 | 58 |

## UNIONTOWN 1 NE *Fayette County* ELEVATION 1040 ft LAT/LONG 39° 54 ' N / 79° 44 ' W

| | JAN | FEB | MAR | APR | MAY | JUN | JUL | AUG | SEP | OCT | NOV | DEC | YEAR |
|---|---|---|---|---|---|---|---|---|---|---|---|---|---|
| Maximum Temp °F | 38.2 | 41.0 | 52.1 | 62.9 | 72.7 | 80.6 | 84.1 | 82.5 | 76.3 | 64.8 | 53.8 | 43.2 | 62.7 |
| Minimum Temp °F | 19.6 | 20.7 | 28.5 | 37.5 | 46.9 | 55.8 | 60.7 | 58.8 | 52.0 | 40.0 | 32.7 | 25.1 | 39.9 |
| Mean Temp °F | 29.0 | 30.9 | 40.4 | 50.2 | 59.8 | 68.2 | 72.4 | 70.7 | 64.2 | 52.4 | 43.3 | 34.2 | 51.3 |
| Days Max Temp ≥ 90 °F | 0 | 0 | 0 | 0 | 0 | 2 | 5 | 3 | 1 | 0 | 0 | 0 | 11 |
| Days Max Temp ≤ 32 °F | 10 | 7 | 2 | 0 | 0 | 0 | 0 | 0 | 0 | 0 | 1 | 5 | 25 |
| Days Min Temp ≤ 32 °F | 26 | 23 | 20 | 11 | 2 | 0 | 0 | 0 | 0 | 8 | 16 | 24 | 130 |
| Days Min Temp ≤ 0 °F | 3 | 2 | 0 | 0 | 0 | 0 | 0 | 0 | 0 | 0 | 0 | 1 | 6 |
| Heating Degree Days | 1109 | 957 | 758 | 446 | 198 | 45 | 7 | 16 | 104 | 391 | 645 | 949 | 5625 |
| Cooling Degree Days | 0 | 0 | 1 | 7 | 39 | *123* | 246 | 186 | 77 | 6 | 0 | 0 | 685 |
| Total Precipitation (") | 2.80 | 2.55 | 3.68 | 3.69 | 4.31 | 3.97 | 4.60 | 3.88 | 3.50 | 2.84 | 3.38 | 3.18 | 42.38 |
| Days ≥ 0.1" Precip | 8 | 7 | 8 | 9 | 9 | 8 | 8 | 7 | 7 | 7 | 8 | 8 | 94 |
| Total Snowfall (") | na | na | na | 0.4 | 0.0 | 0.0 | 0.0 | 0.0 | 0.0 | 0.0 | *1.3* | na | na |
| Days ≥ 1" Snow Depth | na | na | na | 0 | 0 | 0 | 0 | 0 | 0 | 0 | *1* | na | na |

## WARREN *Warren County* ELEVATION 1191 ft LAT/LONG 41° 50 ' N / 79° 9 ' W

| | JAN | FEB | MAR | APR | MAY | JUN | JUL | AUG | SEP | OCT | NOV | DEC | YEAR |
|---|---|---|---|---|---|---|---|---|---|---|---|---|---|
| Maximum Temp °F | 32.1 | 35.2 | 45.1 | 58.6 | 69.9 | 78.4 | 82.2 | 80.1 | 72.5 | 60.9 | 48.4 | 36.9 | 58.4 |
| Minimum Temp °F | 15.7 | 16.0 | 23.8 | 33.8 | 43.6 | 52.6 | 57.6 | 56.7 | 50.2 | 39.2 | 31.8 | 22.2 | 36.9 |
| Mean Temp °F | 23.9 | 25.6 | 34.5 | 46.2 | 56.8 | 65.5 | 69.9 | 68.4 | 61.4 | 50.1 | 40.1 | 29.6 | 47.7 |
| Days Max Temp ≥ 90 °F | 0 | 0 | 0 | 0 | 0 | 2 | 4 | 2 | 0 | 0 | 0 | 0 | 8 |
| Days Max Temp ≤ 32 °F | 16 | 12 | 5 | 0 | 0 | 0 | 0 | 0 | 0 | 0 | 2 | 10 | 45 |
| Days Min Temp ≤ 32 °F | 29 | 26 | 25 | 15 | 4 | 0 | 0 | 0 | 0 | 7 | 17 | 26 | 149 |
| Days Min Temp ≤ 0 °F | 4 | 3 | 1 | 0 | 0 | 0 | 0 | 0 | 0 | 0 | 0 | 1 | 9 |
| Heating Degree Days | 1266 | 1106 | 939 | 561 | 276 | 73 | 17 | 29 | 151 | 458 | 741 | 1092 | 6709 |
| Cooling Degree Days | 0 | 0 | 0 | 5 | 28 | 92 | 192 | 148 | 51 | 2 | 0 | 0 | 518 |
| Total Precipitation (") | 2.74 | 2.36 | 3.34 | 3.58 | 3.89 | 4.94 | 4.34 | 4.38 | 4.40 | 3.47 | 4.11 | 3.75 | 45.30 |
| Days ≥ 0.1" Precip | 8 | 7 | 9 | 9 | 9 | 9 | 8 | 8 | 8 | 9 | 11 | 10 | 105 |
| Total Snowfall (") | 18.1 | 13.3 | 10.1 | 2.2 | 0.0 | 0.0 | 0.0 | 0.0 | 0.0 | 0.4 | 6.3 | 17.0 | 67.4 |
| Days ≥ 1" Snow Depth | 23 | 21 | 10 | 1 | 0 | 0 | 0 | 0 | 0 | 0 | 4 | 16 | 75 |

## WAYNESBURG 1 E *Greene County* ELEVATION 942 ft LAT/LONG 39° 54 ' N / 80° 10 ' W

| | JAN | FEB | MAR | APR | MAY | JUN | JUL | AUG | SEP | OCT | NOV | DEC | YEAR |
|---|---|---|---|---|---|---|---|---|---|---|---|---|---|
| Maximum Temp °F | 37.3 | 40.7 | 51.4 | 62.5 | 72.2 | 80.2 | 83.8 | 82.2 | 76.2 | 64.8 | 53.8 | 42.4 | 62.3 |
| Minimum Temp °F | 16.8 | 18.3 | 26.6 | 35.5 | 44.8 | 53.8 | 58.6 | 57.1 | 50.0 | 37.3 | 30.5 | 23.1 | 37.7 |
| Mean Temp °F | 27.1 | 29.5 | 39.0 | 49.0 | 58.5 | 67.0 | 71.2 | 69.7 | 63.1 | 51.0 | 42.1 | 32.7 | 50.0 |
| Days Max Temp ≥ 90 °F | 0 | 0 | 0 | 0 | 0 | 2 | 5 | 3 | 1 | 0 | 0 | 0 | 11 |
| Days Max Temp ≤ 32 °F | 11 | 8 | 2 | 0 | 0 | 0 | 0 | 0 | 0 | 0 | 1 | 6 | 28 |
| Days Min Temp ≤ 32 °F | 28 | 24 | 23 | 13 | 3 | 0 | 0 | 0 | 0 | 11 | 19 | 25 | 146 |
| Days Min Temp ≤ 0 °F | 4 | 3 | 0 | 0 | 0 | 0 | 0 | 0 | 0 | 0 | 0 | 1 | 8 |
| Heating Degree Days | 1168 | 996 | 800 | 478 | 226 | 54 | 10 | 21 | 120 | 430 | 680 | 993 | 5976 |
| Cooling Degree Days | 0 | 0 | 0 | 6 | 33 | 119 | 228 | 173 | 70 | 5 | 0 | 0 | 634 |
| Total Precipitation (") | 2.64 | 2.29 | 3.49 | 3.31 | 3.97 | 3.53 | 4.14 | 3.97 | 3.14 | 2.46 | 3.21 | 2.84 | 38.99 |
| Days ≥ 0.1" Precip | 7 | 6 | 8 | 8 | 9 | 7 | 8 | 7 | 7 | 6 | 8 | 7 | 88 |
| Total Snowfall (") | 10.9 | 6.8 | 5.6 | 1.0 | 0.0 | 0.0 | 0.0 | 0.0 | 0.0 | 0.0 | 1.8 | 4.3 | 30.4 |
| Days ≥ 1" Snow Depth | 12 | 9 | 4 | 0 | 0 | 0 | 0 | 0 | 0 | 0 | 1 | 5 | 31 |

## WELLSBORO 4 SSE *Tioga County* ELEVATION 1923 ft LAT/LONG 41° 43 ' N / 77° 16 ' W

| | JAN | FEB | MAR | APR | MAY | JUN | JUL | AUG | SEP | OCT | NOV | DEC | YEAR |
|---|---|---|---|---|---|---|---|---|---|---|---|---|---|
| Maximum Temp °F | 28.5 | 31.2 | 40.4 | 53.4 | 64.4 | 72.3 | 76.7 | 75.3 | 67.9 | 57.0 | 45.1 | 33.4 | 53.8 |
| Minimum Temp °F | 12.1 | 13.2 | 21.3 | 32.6 | 42.3 | 50.9 | 55.7 | 54.1 | 47.0 | 37.0 | 29.0 | 18.7 | 34.5 |
| Mean Temp °F | 20.4 | 22.2 | 30.9 | 43.0 | 53.4 | 61.6 | 66.2 | 64.7 | 57.5 | 47.0 | 37.1 | 26.1 | 44.2 |
| Days Max Temp ≥ 90 °F | 0 | 0 | 0 | 0 | 0 | 0 | 1 | 0 | 0 | 0 | 0 | 0 | 1 |
| Days Max Temp ≤ 32 °F | 19 | 15 | 8 | 1 | 0 | 0 | 0 | 0 | 0 | 0 | 3 | 14 | 60 |
| Days Min Temp ≤ 32 °F | 29 | 27 | 26 | 16 | 4 | 0 | 0 | 0 | 1 | 10 | 20 | 29 | 162 |
| Days Min Temp ≤ 0 °F | 6 | 4 | 1 | 0 | 0 | 0 | 0 | 0 | 0 | 0 | 0 | 2 | 13 |
| Heating Degree Days | 1378 | 1202 | 1051 | 656 | 366 | 140 | 48 | 71 | 239 | 553 | 831 | 1200 | 7735 |
| Cooling Degree Days | *0* | 0 | 0 | 3 | 13 | 46 | 105 | 70 | 18 | 1 | 0 | *0* | 256 |
| Total Precipitation (") | 1.64 | 1.69 | 2.35 | 2.44 | 3.26 | 4.24 | 3.48 | 3.07 | 3.20 | 2.72 | 2.85 | 2.21 | 33.15 |
| Days ≥ 0.1" Precip | 4 | 5 | 5 | 6 | 8 | 8 | 8 | 6 | 6 | 6 | 6 | 5 | 73 |
| Total Snowfall (") | 11.2 | 13.2 | *13.6* | 3.7 | 0.3 | 0.0 | 0.0 | 0.0 | 0.0 | 0.6 | 4.7 | na | na |
| Days ≥ 1" Snow Depth | 22 | 20 | 13 | 2 | 0 | 0 | 0 | 0 | 0 | 0 | 3 | 15 | 75 |

## WILKES-BARRE SCRANTN *Luzerne County* ELEVATION 930 ft LAT/LONG 41° 20 ' N / 75° 44 ' W

| | JAN | FEB | MAR | APR | MAY | JUN | JUL | AUG | SEP | OCT | NOV | DEC | YEAR |
|---|---|---|---|---|---|---|---|---|---|---|---|---|---|
| Maximum Temp °F | 32.1 | 35.1 | 45.4 | 58.3 | 69.5 | 77.6 | 82.0 | 79.9 | 71.9 | 60.6 | 48.9 | 37.4 | 58.2 |
| Minimum Temp °F | 17.8 | 19.4 | 28.2 | 38.4 | 48.4 | 56.7 | 61.8 | 60.2 | 52.8 | 41.8 | 34.0 | 24.0 | 40.3 |
| Mean Temp °F | 25.0 | 27.3 | 36.8 | 48.4 | 59.0 | 67.2 | 71.9 | 70.1 | 62.4 | 51.2 | 41.5 | 30.7 | 49.3 |
| Days Max Temp ≥ 90 °F | 0 | 0 | 0 | 0 | 0 | 2 | 3 | 2 | 0 | 0 | 0 | 0 | 7 |
| Days Max Temp ≤ 32 °F | 15 | 11 | 4 | 0 | 0 | 0 | 0 | 0 | 0 | 0 | 1 | 10 | 41 |
| Days Min Temp ≤ 32 °F | 28 | 24 | 21 | 8 | 1 | 0 | 0 | 0 | 0 | 4 | 13 | 25 | 124 |
| Days Min Temp ≤ 0 °F | 2 | 1 | 0 | 0 | 0 | 0 | 0 | 0 | 0 | 0 | 0 | 1 | 4 |
| Heating Degree Days | 1234 | 1058 | 868 | 498 | 216 | 50 | 8 | 18 | 132 | 425 | 699 | 1055 | 6261 |
| Cooling Degree Days | 0 | 0 | 1 | 6 | 41 | 121 | 243 | 184 | 63 | 5 | 0 | 0 | 664 |
| Total Precipitation (") | 2.04 | 2.02 | 2.64 | 3.19 | 3.77 | 3.85 | 3.72 | 3.42 | 3.65 | 2.82 | 3.14 | 2.53 | 36.79 |
| Days ≥ 0.1" Precip | 5 | 5 | 6 | 7 | 8 | 8 | 7 | 7 | 6 | 5 | 7 | 6 | 77 |
| Total Snowfall (") | 12.2 | 10.3 | 9.1 | 2.6 | 0.1 | 0.0 | 0.0 | 0.0 | 0.0 | 0.1 | 3.8 | 8.3 | 46.5 |
| Days ≥ 1" Snow Depth | 17 | 15 | 6 | 1 | 0 | 0 | 0 | 0 | 0 | 0 | 2 | 9 | 50 |

## WILLIAMSPRT-LYCOMING *Lycoming County* ELEVATION 522 ft LAT/LONG 41° 15 ' N / 76° 55 ' W

| | JAN | FEB | MAR | APR | MAY | JUN | JUL | AUG | SEP | OCT | NOV | DEC | YEAR |
|---|---|---|---|---|---|---|---|---|---|---|---|---|---|
| Maximum Temp °F | 33.5 | 37.1 | 47.7 | 60.6 | 71.4 | 79.4 | 83.5 | 81.6 | 73.6 | 61.9 | 50.0 | 38.7 | 59.9 |
| Minimum Temp °F | 17.7 | 19.5 | 28.4 | 38.5 | 48.1 | 56.7 | 61.8 | 60.7 | 53.2 | 41.3 | 33.7 | 24.4 | 40.3 |
| Mean Temp °F | 25.6 | 28.3 | 38.1 | 49.6 | 59.8 | 68.0 | 72.7 | 71.1 | 63.4 | 51.7 | 41.8 | 31.6 | 50.1 |
| Days Max Temp ≥ 90 °F | 0 | 0 | 0 | 0 | 1 | 2 | 5 | 3 | 1 | 0 | 0 | 0 | 12 |
| Days Max Temp ≤ 32 °F | 14 | 9 | 2 | 0 | 0 | 0 | 0 | 0 | 0 | 0 | 0 | 7 | 32 |
| Days Min Temp ≤ 32 °F | 28 | 24 | 21 | 8 | 1 | 0 | 0 | 0 | 0 | 5 | 14 | 24 | 125 |
| Days Min Temp ≤ 0 °F | 2 | 1 | 0 | 0 | 0 | 0 | 0 | 0 | 0 | 0 | 0 | 1 | 4 |
| Heating Degree Days | 1214 | 1029 | 828 | 463 | 194 | 37 | 5 | 12 | 112 | 410 | 688 | 1030 | 6022 |
| Cooling Degree Days | 0 | 0 | 0 | 7 | 45 | 140 | 274 | 212 | 71 | 5 | 0 | 0 | 754 |
| Total Precipitation (") | 2.44 | 2.57 | 3.26 | 3.26 | 3.82 | 4.30 | 4.20 | 3.48 | 3.71 | 3.27 | 3.84 | 3.05 | 41.20 |
| Days ≥ 0.1" Precip | 6 | 5 | 7 | 7 | 8 | 8 | 8 | 7 | 7 | 6 | 7 | 6 | 82 |
| Total Snowfall (") | 11.6 | 9.8 | 8.2 | 1.2 | 0.0 | 0.0 | 0.0 | 0.0 | 0.0 | 0.1 | 2.9 | 7.2 | 41.0 |
| Days ≥ 1" Snow Depth | 16 | 14 | 6 | 0 | 0 | 0 | 0 | 0 | 0 | 0 | 1 | 7 | 44 |

## YORK 3 SSW PUMP STN *York County* ELEVATION 390 ft LAT/LONG 39° 55 ' N / 76° 45 ' W

| | JAN | FEB | MAR | APR | MAY | JUN | JUL | AUG | SEP | OCT | NOV | DEC | YEAR |
|---|---|---|---|---|---|---|---|---|---|---|---|---|---|
| Maximum Temp °F | 38.7 | 42.8 | 53.2 | 64.9 | 75.1 | 83.1 | 86.8 | 85.1 | 78.2 | 67.0 | 54.8 | 43.5 | 64.4 |
| Minimum Temp °F | 19.8 | 21.9 | 30.2 | 38.7 | 48.6 | 57.5 | 62.5 | 60.9 | 53.8 | 41.7 | 33.9 | 25.7 | 41.3 |
| Mean Temp °F | 29.3 | 32.4 | 41.7 | 51.9 | 61.9 | 70.3 | 74.7 | 73.0 | 66.0 | 54.4 | 44.3 | 34.6 | 52.9 |
| Days Max Temp ≥ 90 °F | 0 | 0 | 0 | 0 | 1 | 5 | 10 | 7 | 2 | 0 | 0 | 0 | 25 |
| Days Max Temp ≤ 32 °F | 8 | 5 | 1 | 0 | 0 | 0 | 0 | 0 | 0 | 0 | 0 | 4 | 18 |
| Days Min Temp ≤ 32 °F | 27 | 24 | 19 | 8 | 1 | 0 | 0 | 0 | 0 | 6 | 15 | 24 | 124 |
| Days Min Temp ≤ 0 °F | 2 | 1 | 0 | 0 | 0 | 0 | 0 | 0 | 0 | 0 | 0 | 0 | 3 |
| Heating Degree Days | 1101 | 915 | 716 | 395 | 148 | 23 | 2 | 7 | 75 | 333 | 614 | 935 | 5264 |
| Cooling Degree Days | 0 | 0 | 1 | 7 | 62 | 194 | 328 | 255 | 107 | 14 | 1 | 0 | 969 |
| Total Precipitation (") | 2.89 | 2.63 | 3.46 | 3.50 | 3.91 | 4.07 | 3.69 | 3.57 | 3.71 | 3.01 | 3.42 | 3.35 | 41.21 |
| Days ≥ 0.1" Precip | 6 | 6 | 7 | 7 | 8 | 7 | 7 | 6 | 6 | 5 | 7 | 6 | 78 |
| Total Snowfall (") | 9.2 | 8.8 | 4.3 | 0.4 | 0.0 | 0.0 | 0.0 | 0.0 | 0.0 | 0.1 | 1.2 | 4.1 | 28.1 |
| Days ≥ 1" Snow Depth | 11 | 8 | 2 | 0 | 0 | 0 | 0 | 0 | 0 | 0 | 1 | 3 | 25 |

## JANUARY MINIMUM TEMPERATURE °F

### LOWEST

| | | |
|---|---|---|
| 1 | Pleasant Mount | 9.5 |
| 2 | Bradford-4 SW | 10.8 |
| 3 | Kane | 10.9 |
| 4 | Francis Wltr Dm | 11.2 |
| 5 | Hawley | 11.7 |
| 6 | Montrose | 11.9 |
| 7 | Wellsboro | 12.1 |
| 8 | Titusville | 12.2 |
| 9 | Bradford-Regional | 12.3 |
| | Ridgway | 12.3 |
| | Tionesta | 12.3 |
| 12 | Tobyhanna | 12.9 |
| 13 | Jamestown | 13.8 |
| 14 | Slippery Rock | 13.9 |
| 15 | Putneyville | 14.4 |
| 16 | Meadville | 14.6 |
| | Philipsburg | 14.6 |
| 18 | Ebensburg | 15.0 |
| 19 | Corry | 15.2 |
| 20 | Clarion | 15.3 |
| | Marion Center | 15.3 |
| | Towanda | 15.3 |
| 23 | Brookville | 15.4 |
| | Confluence | 15.4 |
| 25 | Donegal | 15.6 |

### HIGHEST

| | | |
|---|---|---|
| 1 | Marcus Hook | 26.6 |
| 2 | Philadelphia | 23.6 |
| 3 | Holtwood | 22.3 |
| 4 | Donora | 20.5 |
| | Hanover | 20.5 |
| | Shippensburg | 20.5 |
| 7 | Montgomry L&D | 20.0 |
| 8 | York | 19.8 |
| 9 | Phoenixville | 19.7 |
| 10 | Chambersburg | 19.6 |
| | Uniontown | 19.6 |
| 12 | Johnstown | 19.4 |
| 13 | Allentown | 19.3 |
| | Bloserville | 19.3 |
| | Landisville | 19.3 |
| 16 | Palmerton | 19.2 |
| 17 | Lebanon | 19.1 |
| 18 | Pittsburgh | 19.0 |
| 19 | Erie | 18.8 |
| | Lewistown | 18.8 |
| 21 | Graterford | 18.6 |
| 22 | Newport River | 18.4 |
| 23 | Wilkes-Barre | 17.8 |
| 24 | Derry | 17.7 |
| | Stroudsburg | 17.7 |

## JULY MAXIMUM TEMPERATURE °F

### HIGHEST

| | | |
|---|---|---|
| 1 | Marcus Hook | 87.2 |
| 2 | Hanover | 86.8 |
| | York | 86.8 |
| 4 | Philadelphia | 86.6 |
| | Phoenixville | 86.6 |
| 6 | Laurelton | 86.5 |
| | Stroudsburg | 86.5 |
| 8 | Johnstown | 86.2 |
| 9 | Landisville | 85.7 |
| | Palmerton | 85.7 |
| 11 | Donora | 85.6 |
| 12 | Lebanon | 85.3 |
| 13 | Shippensburg | 85.2 |
| 14 | Lewistown | 85.1 |
| 15 | Chambersburg | 84.9 |
| | Newport River | 84.9 |
| 17 | Holtwood | 84.7 |
| 18 | Allentown | 84.6 |
| | Graterford | 84.6 |
| 20 | Montgomry L& D | 84.5 |
| 21 | Bloserville | 84.3 |
| 22 | Uniontown | 84.1 |
| 23 | New Castle | 84.0 |
| 24 | Derry | 83.9 |
| 25 | Waynesburg | 83.8 |

### LOWEST

| | | |
|---|---|---|
| 1 | Bradford-Regional | 76.6 |
| 2 | Wellsboro | 76.7 |
| 3 | Pleasant Mount | 76.9 |
| 4 | Philipsburg | 78.2 |
| | Tobyhanna | 78.2 |
| 6 | Bradford-4 SW | 78.4 |
| 7 | Kane | 78.6 |
| | Montrose | 78.6 |
| 9 | Francis Wltr Dm | 78.9 |
| 10 | Erie | 79.0 |
| 11 | Dubois | 79.2 |
| 12 | Marion Center | 79.5 |
| 13 | Hawley | 79.6 |
| 14 | Ridgway | 79.9 |
| 15 | Corry | 80.0 |
| 16 | Donegal | 80.3 |
| 17 | Tionesta | 80.9 |
| 18 | Meadville | 81.0 |
| | Titusville | 81.0 |
| 20 | Ebensburg | 81.2 |
| 21 | Jamestown | 81.5 |
| | Mercer | 81.5 |
| 23 | Brookville | 81.7 |
| 24 | Slippery Rock | 81.9 |
| | State College | 81.9 |

## ANNUAL PRECIPITATION (")

### HIGHEST

| | | |
|---|---|---|
| 1 | Ebensburg | 49.27 |
| 2 | Marion Center | 49.17 |
| 3 | Derry | 48.43 |
| 4 | Tobyhanna | 48.34 |
| 5 | Stroudsburg | 48.29 |
| 6 | Pleasant Mount | 47.96 |
| 7 | Corry | 47.95 |
| 8 | Johnstown | 47.36 |
| 9 | Bradford-Regional | 46.77 |
| 10 | Kane | 46.21 |
| 11 | Bradford-4 SW | 46.00 |
| 12 | Indiana | 45.97 |
| 13 | Clarion | 45.81 |
| 14 | Warren | 45.30 |
| 15 | Donegal | 45.14 |
| 16 | Meadville | 44.82 |
| 17 | Brookville | 44.81 |
| 18 | Confluence | 44.66 |
| 19 | Francis Wltr Dm | 44.53 |
| 20 | Allentown | 44.35 |
| 21 | Titusville | 44.20 |
| 22 | Putneyville | 44.11 |
| 23 | Franklin | 43.88 |
| 24 | Laurelton | 43.64 |
| 25 | Tionesta | 43.53 |

### LOWEST

| | | |
|---|---|---|
| 1 | Wellsboro | 33.15 |
| 2 | Towanda | 34.13 |
| 3 | Donora | 36.24 |
| 4 | Holtwood | 36.76 |
| 5 | Wilkes-Barre | 36.79 |
| 6 | Montgomry L&D | 37.27 |
| 7 | Pittsburgh | 37.33 |
| 8 | New Castle | 37.80 |
| 9 | Shippensburg | 38.25 |
| 10 | State College | 38.40 |
| 11 | Waynesburg | 38.99 |
| 12 | Hanover | 39.06 |
| 13 | Greenville | 39.16 |
| 14 | Lewistown | 39.34 |
| 15 | Salina | 39.71 |
| 16 | Bloserville | 39.78 |
| 17 | Chambersburg | 39.95 |
| 18 | Hawley | 40.13 |
| 19 | Newport River | 40.19 |
| 20 | Ford City | 40.65 |
| 21 | Philipsburg | 40.91 |
| 22 | Jamestown | 40.93 |
| 23 | Williamsport | 41.20 |
| 24 | York | 41.21 |
| 25 | Marcus Hook | 41.36 |

## ANNUAL SNOWFALL (")

### HIGHEST

| | | |
|---|---|---|
| 1 | Corry | 131.3 |
| 2 | Meadville | 112.3 |
| 3 | Ebensburg | 107.7 |
| 4 | Montrose | 90.1 |
| 5 | Kane | 89.1 |
| 6 | Erie | 88.4 |
| 7 | Bradford-4 SW | 88.3 |
| 8 | Bradford-Regional | 84.8 |
| 9 | Marion Center | 75.9 |
| 10 | Pleasant Mount | 75.5 |
| 11 | Titusville | 74.3 |
| 12 | Tobyhanna | 73.0 |
| 13 | Warren | 67.4 |
| 14 | Jamestown | 66.3 |
| 15 | Confluence | 60.6 |
| 16 | Dubois | 58.8 |
| 17 | Tionesta | 57.6 |
| 18 | Ridgway | 55.1 |
| 19 | Greenville | 54.9 |
| 20 | Indiana | 54.1 |
| 21 | Franklin | 53.3 |
| 22 | Mercer | 51.9 |
| 23 | Philipsburg | 51.5 |
| 24 | Francis Wltr Da | 50.4 |
| 25 | State College | 48.9 |

### LOWEST

| | | |
|---|---|---|
| 1 | Montgomry L&D | 20.1 |
| | Palmerton | 20.1 |
| 3 | Philadelphia | 21.3 |
| 4 | Newport River | 24.3 |
| 5 | Landisville | 25.9 |
| 6 | Bloserville | 27.6 |
| 7 | Hanover | 28.0 |
| 8 | York | 28.1 |
| 9 | Salina | 29.5 |
| 10 | New Castle | 29.6 |
| 11 | Waynesburg | 30.4 |
| 12 | Allentown | 31.7 |
| 13 | Chambersburg | 32.4 |
| 14 | Ford City | 35.6 |
| 15 | Shippensburg | 36.5 |
| 16 | Clarion | 36.9 |
| 17 | Stroudsburg | 37.9 |
| 18 | Johnstown | 39.1 |
| 19 | Putneyville | 40.6 |
| 20 | Williamsport | 41.0 |
| 21 | Slippery Rock | 42.4 |
| 22 | Pittsburgh | 42.9 |
| | Towanda | 42.9 |
| 24 | Hawley | 45.9 |
| 25 | Derry | 46.2 |

# RHODE ISLAND

PHYSICAL FEATURES. Rhode Island, the smallest of the states, shares the southeastern corner of New England with a portion of Massachusetts. The State extends for 50 miles in a north-south direction and has an average width of about 30 miles. The total area, including Block Island some 10 miles offshore, is 1,497 square miles of which Narragansett Bay occupies about 25 percent. There are three topographical divisions of the State. A narrow coastal plain with an elevation of less than 100 feet occurs along the south shore and around Narragansett Bay. A second division with gently rolling uplands of up to 200 feet elevation lies to the north and east of the Bay. The western two-thirds of Rhode Island consists of predominantly hilly uplands of mostly 200 to 600 feet elevation, rising to a maximum of 800 feet above sea level in the northwest corner of the State.

Narragansett Bay has a very irregular shoreline, indented by numerous small bays or coves and the mouths of the Taunton and Blackstone Rivers. The Bay contains several islands of which the one known as Aquidneck, or Rhode Island, is the largest. The shore line facing Long Island Sound is about 20 miles long. No point in the State is more than 25 miles from the ocean. The Blackstone River in northeastern Rhode Island is the principal river. A number of smaller rivers or brooks originating in the western uplands of the State or in southeastern Massachusetts empty into Narragansett Bay or Long Island Sound.

GENERAL CLIMATE. The chief characteristics of Rhode Island's climate may be summarized as follows: (1) equable distribution of precipitation among the four seasons; (2) large ranges of temperature both daily and annual; (3) great differences in the same season of different years; and (4) considerable diversity of the weather over short periods of time. These characteristics are modified by nearness to the Bay or ocean, elevation, and nature of the terrain.

Rhode Island lies in the "prevailing westerlies", the belt of generally eastward air movement which encircles the globe in middle latitudes. Embedded in this circulation are extensive masses of air originating in higher and lower latitudes and interacting to produce storm systems. A large number of these systems and air-mass fronts pass near or over Rhode Island in a year.

Air masses affecting the State belong to three types: (1) cold, dry air pouring down from subarctic North America; (2) warm, moist air streaming up on a long overland journey from the Gulf of Mexico and adjacent waters; and (3) cool, damp air moving from the North Atlantic. Because the atmospheric flow is usually from continental areas, Rhode Island is more influenced by the first two types than it is by the third. The ocean constitutes an important modifying factor, particularly in southeast sections of the State, but does not dominate the climate as it would if the prevailing circulation was onshore.

The procession of contrasting air masses and the relatively frequent passage of "Lows" bring about a roughly twice-weekly alternation from fair to cloudy or stormy weather, usually attended by abrupt changes in temperature, moisture, sunshine, wind direction, and speed. There is no regular or persistent rhythm to this sequence, and it is sometimes interrupted by periods of several days, or infrequently of a few weeks, with the same weather pattern.

TEMPERATURE. The mean annual temperature ranges from 48° to 49° F., except near the south shore, Narragansett Bay, and in the area around Providence, where it is 50° to 51° F. Southwestern Rhode Island, from 4 to 10 miles inland, exhibits a coolness not suggested by the nearness to the ocean or the general elevation of 50 to 150 feet. Here the annual mean temperature is not more than 48° F., making the section as cool as the cooler areas of the northwest interior.

The average daily minimum temperature in January and February is 19° to 20° F. over about two-thirds of the State, increasing to near 25° F. in immediate coastal sections. The number of days with minimum temperatures of zero or below averages 1 or less per year in the Bay and coastal areas, increasing to about 5 per year in most of the interior. A maximum temperature of 32° F. or lower occurs on an average of 20 to 25 days per year along the shoreline and 30 to 40 days in the remainder of the State. Summer temperatures are considerably influenced by proximity to the coastal waters and the frequent onshore flow of air during the warmer months. The average July maximum temperature is about 80° F., except in the northwestern interior where it is a few degrees higher. The greatest number of hot days occurs in the metropolitan areas and in parts of the northern interior. Here, about 8 to 10 days of temperatures 90° F. or higher may be expected per year. Near the immediate coast the occurrence of 90° F. temperatures is limited to 1 day in the average summer, if it occurs at all. The length of the freeze-free season, as limited by the occurrence of temperatures of 32° F. or lower, averages from 155 to 180 days in most of the State. Climatic differences of temperature in this small State are very striking in the fall season. Autumnal coloration of foliage will be past its peak of brilliance in the northwestern interior before leaves have begun to noticeably turn color in the Newport area of the southeast.

PRECIPITATION. The climate of Rhode Island is characterized by the rather even distribution of precipitation throughout the year. Storm centers and their accompanying fronts are the principal year-round producers of precipitation. Storms moving up the Atlantic coast generally yield the heaviest amounts of rain and snow. Bands and patches of thunderstorms or convective showers contribute considerable precipitation in the summer and make up the difference resulting from decreased activity of the storm centers. In comparison with the general storms, these are of brief duration, but they yield the heaviest local rainfall.

Annual precipitation averages 42 to 46 inches over most of the State, with a tendency for decreasing amounts from west to east. It varies from about 40 inches in the immediate southeastern Bay area and on Block Island to 48 inches in the western uplands. Total precipitation in the freeze-free season of April through October shows similar differences over the State with an average of 22 to 24 inches near the Bay and 26 to 29 inches in the western interior. While there are no pronounced wet and dry months as in other climates, the months of May through July are relatively dry in proximity of the Bay. Measurable precipitation falls on an average of 1 day in 3 or on approximately 120 days per year. Periods of 5 days or more of successive daily precipitation occur a few times during most years. Extended periods of little or no precipitation are observed nearly every summer or early fall. Such a period may last from 10 to 20 days.

SNOWFALL. The average annual snowfall in Rhode Island increases from about 20 inches on Block Island and along the southeast shores of Narragansett Bay to from 40 to 55 inches in the western third of the State. Most of the snow falls in January and February; however, there are occasional winters when in coastal sections, particularly, heavier monthly amounts will occur in December or March. In the western and northern portions of the State the first snowfall of 1 inch or more usually occurs in mid or late November. The southeastern Bay area does not observe measurable snow before December in the great majority of years. The last measurable snowfall usually occurs by late March in the populous areas of the State, although an April snowstorm is by no means rare. The average number of days with 1 inch or more of snow on the ground also increases from the shore areas to the western interior. In the latter, a snowcover prevails most of the time from mid or late December to about mid-March. Near the Bay a snow cover does not last more than a few days unless a heavy snowstorm is followed by prolonged cold temperatures.

WINDS AND STORMS. The prevailing wind in Rhode Island is northwesterly from December through March, and southwesterly in the remaining months. An important feature of the climate is the sea breeze which affects a considerable portion of the State's area. From approximately late spring to mid autumn this cool onshore wind blows during the afternoon hours and penetrates from 5 to 10 miles inland. Since much of Rhode Island is within 10 miles of the Sound or Bay, the relatively cool summer maximum temperatures can be accounted for. Aside from hurricanes, coastal storms or "northeasters" are the most serious weather hazard in Rhode Island. They generate very strong winds and heavy rains, and produce the greatest snowfalls in the winter. Hurricanes or storms of tropical origin occasionally affect the State during the summer or fall months. Localized thunderstorms with heavy and intense rainfall on occasions cause damaging flash floods in the small as well as the larger streams of the State.

OTHER CLIMATIC ELEMENTS. The percentage of possible sunshine averages 55 to 60 percent, ranging from about 50 percent in the winter months to a little over 60 percent during the summer. The average number of clear and cloudy days per year are about equal. The highest number of clear days per month usually occurs in September or October, while the maximum number of cloudy days are noted in December and January. Heavy fog is observed on an average of about 50 days per year in the southeastern areas of the Bay. This number decreases to 30 or 35 along the western and northern shores of the Bay, and to about 25 days in the western interior.

## COUNTY INDEX

***Kent County***
PROVIDENCE GREEN ST

***Newport County***
NEWPORT ROSE

***Washington County***
BLOCK IS STATE AP
KINGSTON

## ELEVATION INDEX

| FEET | STATION NAME |
|---|---|
| 20 | NEWPORT ROSE |
| 66 | PROVIDENCE GREEN ST |
| 102 | KINGSTON |
| 108 | BLOCK IS STATE AP |

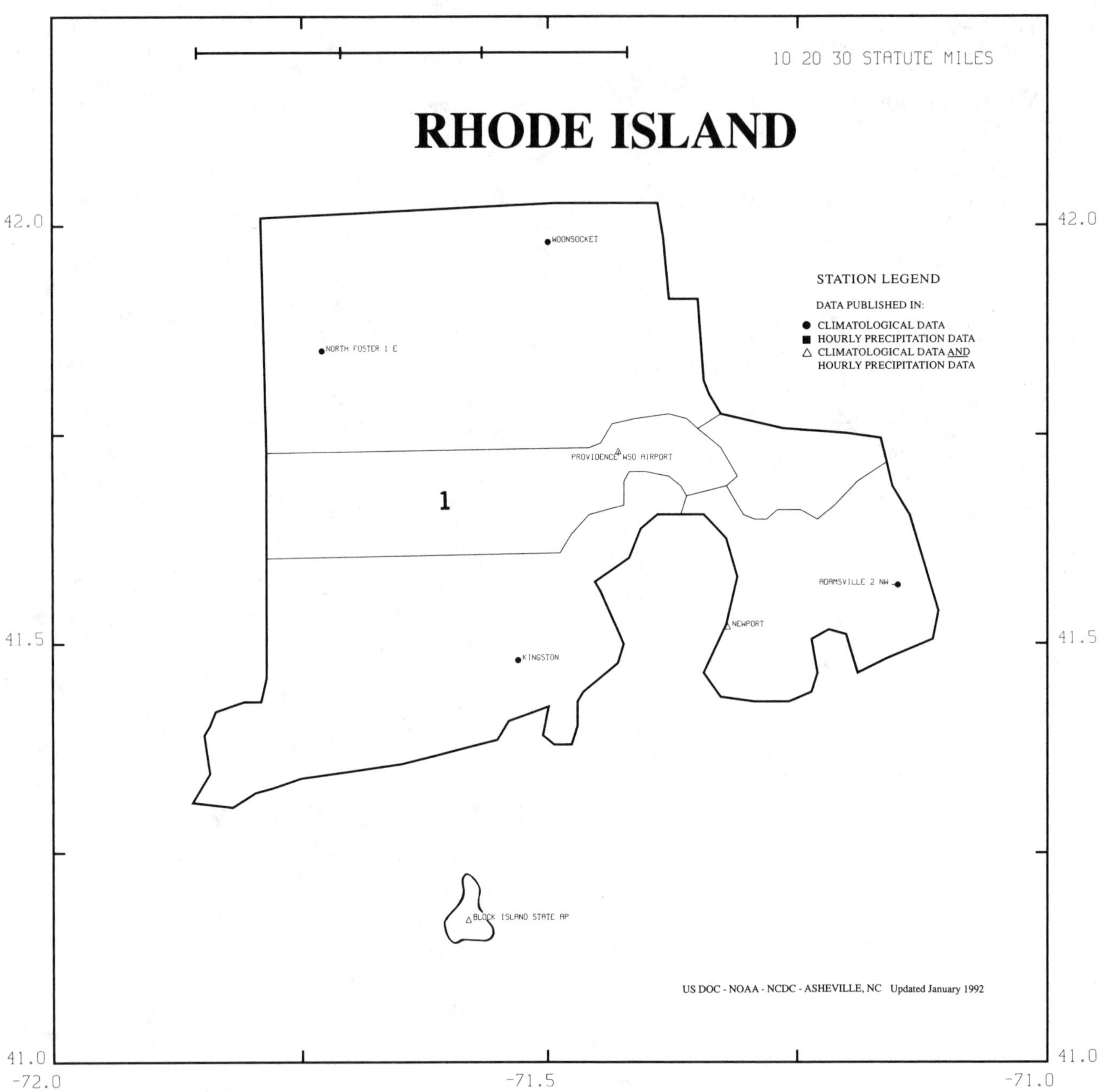
10 20 30 STATUTE MILES
RHODE ISLAND
WOONSOCKET
NORTH FOSTER 1 E
PROVIDENCE WSO AIRPORT
1
ADAMSVILLE 2 NW
NEWPORT
KINGSTON
BLOCK ISLAND STATE AP
STATION LEGEND
DATA PUBLISHED IN:
● CLIMATOLOGICAL DATA
■ HOURLY PRECIPITATION DATA
△ CLIMATOLOGICAL DATA AND HOURLY PRECIPITATION DATA
US DOC - NOAA - NCDC - ASHEVILLE, NC Updated January 1992
42.0
41.5
41.0
-72.0
-71.5
-71.0

## BLOCK IS STATE AP *Washington County* ELEVATION 108 ft LAT/LONG 41° 10 ' N / 71° 35 ' W

| | JAN | FEB | MAR | APR | MAY | JUN | JUL | AUG | SEP | OCT | NOV | DEC | YEAR |
|---|---|---|---|---|---|---|---|---|---|---|---|---|---|
| Maximum Temp °F | 37.7 | 37.8 | 43.6 | 52.0 | 61.2 | 70.4 | 76.7 | 76.4 | 70.0 | 60.8 | 52.1 | 43.1 | 56.8 |
| Minimum Temp °F | 24.9 | 25.4 | 31.4 | 39.2 | 48.1 | 57.2 | 64.0 | 64.1 | 57.6 | 48.5 | 40.4 | 30.5 | 44.3 |
| Mean Temp °F | 31.3 | 31.6 | 37.6 | 45.7 | 54.7 | 63.8 | 70.4 | 70.3 | 63.8 | 54.7 | 46.2 | 36.9 | 50.6 |
| Days Max Temp ≥ 90 °F | 0 | 0 | 0 | 0 | 0 | 0 | 0 | 0 | 0 | 0 | 0 | 0 | 0 |
| Days Max Temp ≤ 32 °F | 8 | 6 | 1 | 0 | 0 | 0 | 0 | 0 | 0 | 0 | 0 | 4 | 19 |
| Days Min Temp ≤ 32 °F | 23 | 21 | 16 | 3 | 0 | 0 | 0 | 0 | 0 | 0 | 5 | 17 | 85 |
| Days Min Temp ≤ 0 °F | 0 | 0 | 0 | 0 | 0 | 0 | 0 | 0 | 0 | 0 | 0 | 0 | 0 |
| Heating Degree Days | 1038 | 936 | 844 | 574 | 317 | 71 | 5 | 7 | 79 | 317 | 556 | 865 | 5609 |
| Cooling Degree Days | 0 | 0 | 0 | 0 | 4 | 47 | 190 | 174 | 50 | 4 | 0 | 0 | 469 |
| Total Precipitation (") | 3.11 | 2.97 | 3.94 | 3.53 | 3.26 | 2.74 | 2.69 | 3.08 | 2.85 | 2.77 | 4.06 | 3.81 | 38.81 |
| Days ≥ 0.1" Precip | 6 | 5 | 7 | 6 | 6 | 5 | 5 | 4 | 5 | 5 | 7 | 7 | 68 |
| Total Snowfall (") | na | na | na | na | na | na | na | na | na | na | na | na | na |
| Days ≥ 1" Snow Depth | na | na | na | na | na | na | na | na | na | na | na | na | na |

## KINGSTON *Washington County* ELEVATION 102 ft LAT/LONG 41° 29 ' N / 71° 32 ' W

| | JAN | FEB | MAR | APR | MAY | JUN | JUL | AUG | SEP | OCT | NOV | DEC | YEAR |
|---|---|---|---|---|---|---|---|---|---|---|---|---|---|
| Maximum Temp °F | 37.9 | 39.5 | 47.3 | 57.7 | 67.8 | 76.3 | 81.5 | 80.3 | 73.5 | 63.7 | 53.1 | 42.8 | 60.1 |
| Minimum Temp °F | 17.8 | 19.5 | 27.4 | 35.6 | 44.9 | 54.0 | 60.0 | 59.1 | 51.0 | 40.1 | 33.0 | 23.7 | 38.8 |
| Mean Temp °F | 27.9 | 29.5 | 37.4 | 46.7 | 56.4 | 65.2 | 70.8 | 69.7 | 62.3 | 51.9 | 43.1 | 33.3 | 49.5 |
| Days Max Temp ≥ 90 °F | 0 | 0 | 0 | 0 | 0 | 1 | 2 | 1 | 0 | 0 | 0 | 0 | 4 |
| Days Max Temp ≤ 32 °F | 9 | 6 | 1 | 0 | 0 | 0 | 0 | 0 | 0 | 0 | 0 | 5 | 21 |
| Days Min Temp ≤ 32 °F | 28 | 25 | 22 | 11 | 2 | 0 | 0 | 0 | 1 | 8 | 16 | 25 | 138 |
| Days Min Temp ≤ 0 °F | 2 | 2 | 0 | 0 | 0 | 0 | 0 | 0 | 0 | 0 | 0 | 1 | 5 |
| Heating Degree Days | 1145 | 995 | 848 | 544 | 274 | 67 | 10 | 20 | 130 | 403 | 650 | 977 | 6063 |
| Cooling Degree Days | 0 | 0 | 0 | 0 | 14 | 84 | 208 | 177 | 60 | 5 | 0 | 0 | 548 |
| Total Precipitation (") | 4.19 | 3.65 | 4.90 | 4.32 | 4.17 | 3.67 | 3.23 | 4.06 | 3.92 | 3.74 | 5.34 | 4.91 | 50.10 |
| Days ≥ 0.1" Precip | 7 | 6 | 7 | 7 | 7 | 6 | 5 | 6 | 6 | 6 | 7 | 7 | 77 |
| Total Snowfall (") | 9.9 | 9.4 | 5.0 | 0.5 | 0.0 | 0.0 | 0.0 | 0.0 | 0.0 | 0.1 | 0.7 | 5.2 | 30.8 |
| Days ≥ 1" Snow Depth | 12 | 10 | 4 | 0 | 0 | 0 | 0 | 0 | 0 | 0 | 0 | 4 | 30 |

## NEWPORT ROSE *Newport County* ELEVATION 20 ft LAT/LONG 41° 31 ' N / 71° 19 ' W

| | JAN | FEB | MAR | APR | MAY | JUN | JUL | AUG | SEP | OCT | NOV | DEC | YEAR |
|---|---|---|---|---|---|---|---|---|---|---|---|---|---|
| Maximum Temp °F | 38.0 | 39.0 | 45.8 | 55.0 | 64.2 | 72.6 | 78.6 | 78.0 | 71.9 | 62.6 | 53.0 | 43.2 | 58.5 |
| Minimum Temp °F | 22.6 | 23.7 | 30.3 | 38.2 | 47.4 | 56.8 | 63.4 | 63.5 | 56.7 | 47.0 | 38.3 | 28.1 | 43.0 |
| Mean Temp °F | 30.4 | 31.4 | 38.1 | 46.6 | 55.8 | 64.7 | 71.0 | 70.7 | 64.3 | 54.8 | 45.6 | 35.7 | 50.8 |
| Days Max Temp ≥ 90 °F | 0 | 0 | 0 | 0 | 0 | 0 | 1 | 0 | 0 | 0 | 0 | 0 | 1 |
| Days Max Temp ≤ 32 °F | 8 | 6 | 1 | 0 | 0 | 0 | 0 | 0 | 0 | 0 | 0 | 4 | 19 |
| Days Min Temp ≤ 32 °F | 26 | 22 | 19 | 5 | 0 | 0 | 0 | 0 | 0 | 1 | 7 | 21 | 101 |
| Days Min Temp ≤ 0 °F | 1 | 0 | 0 | 0 | 0 | 0 | 0 | 0 | 0 | 0 | 0 | 0 | 1 |
| Heating Degree Days | 1066 | 943 | 828 | 544 | 286 | 59 | 4 | 7 | 74 | 313 | 575 | 901 | 5600 |
| Cooling Degree Days | 0 | 0 | 0 | 0 | 7 | 67 | 207 | 197 | 63 | 5 | 0 | 0 | 546 |
| Total Precipitation (") | 3.83 | 3.50 | 4.56 | 3.99 | 3.69 | 3.00 | 2.88 | 3.56 | 3.45 | 3.29 | 4.73 | 4.53 | 45.01 |
| Days ≥ 0.1" Precip | 7 | 6 | 7 | 6 | 7 | 6 | 5 | 6 | 6 | 6 | 7 | 7 | 76 |
| Total Snowfall (") | *7.7* | na | na | 0.2 | 0.0 | 0.0 | 0.0 | 0.0 | 0.0 | 0.0 | 0.5 | *2.6* | na |
| Days ≥ 1" Snow Depth | na | na | na | 0 | 0 | 0 | 0 | 0 | 0 | 0 | 0 | na | na |

## PROVIDENCE GREEN ST *Kent County* ELEVATION 66 ft LAT/LONG 41° 44 ' N / 71° 26 ' W

| | JAN | FEB | MAR | APR | MAY | JUN | JUL | AUG | SEP | OCT | NOV | DEC | YEAR |
|---|---|---|---|---|---|---|---|---|---|---|---|---|---|
| Maximum Temp °F | 36.2 | 38.2 | 46.7 | 57.9 | 68.2 | 77.2 | 82.4 | 81.0 | 73.5 | 63.1 | 52.5 | 41.6 | 59.9 |
| Minimum Temp °F | 19.4 | 21.3 | 29.5 | 38.6 | 48.3 | 57.5 | 63.9 | 62.7 | 54.1 | 42.9 | 35.1 | 25.1 | 41.5 |
| Mean Temp °F | 27.8 | 29.8 | 38.1 | 48.3 | 58.3 | 67.4 | 73.2 | 71.9 | 63.8 | 53.0 | 43.8 | 33.4 | 50.7 |
| Days Max Temp ≥ 90 °F | 0 | 0 | 0 | 0 | 1 | 2 | 4 | 3 | 1 | 0 | 0 | 0 | 11 |
| Days Max Temp ≤ 32 °F | 11 | 8 | 1 | 0 | 0 | 0 | 0 | 0 | 0 | 0 | 0 | 6 | 26 |
| Days Min Temp ≤ 32 °F | 28 | 24 | 19 | 5 | 0 | 0 | 0 | 0 | 0 | 4 | 13 | 24 | 117 |
| Days Min Temp ≤ 0 °F | 1 | 1 | 0 | 0 | 0 | 0 | 0 | 0 | 0 | 0 | 0 | 0 | 2 |
| Heating Degree Days | 1146 | 988 | 827 | 497 | 226 | 44 | 3 | 10 | 102 | 371 | 630 | 973 | 5817 |
| Cooling Degree Days | 0 | 0 | 0 | 2 | 28 | 128 | 275 | 231 | 78 | 8 | 0 | 0 | 750 |
| Total Precipitation (") | 3.84 | 3.46 | 4.47 | 3.98 | 3.63 | 3.23 | 3.17 | 3.93 | 3.44 | 3.41 | 4.50 | 4.61 | 45.67 |
| Days ≥ 0.1" Precip | 7 | 6 | 8 | 6 | 7 | 6 | 6 | 6 | 5 | 6 | 7 | 8 | 78 |
| Total Snowfall (") | 9.7 | 10.2 | 6.6 | 0.7 | 0.2 | 0.0 | 0.0 | 0.0 | 0.0 | 0.1 | 1.1 | 6.4 | 35.0 |
| Days ≥ 1" Snow Depth | 13 | 11 | 4 | 0 | 0 | 0 | 0 | 0 | 0 | 0 | 1 | 6 | 35 |

## JANUARY MINIMUM TEMPERATURE °F

### LOWEST

| | | |
|---|---|---|
| 1 | Kingston | 17.8 |
| 2 | Providence | 19.4 |
| 3 | Newport | 22.6 |
| 4 | Block | 24.9 |

### HIGHEST

| | | |
|---|---|---|
| 1 | Block | 24.9 |
| 2 | Newport | 22.6 |
| 3 | Providence | 19.4 |
| 4 | Kingston | 17.8 |

## JULY MAXIMUM TEMPERATURE °F

### HIGHEST

| | | |
|---|---|---|
| 1 | Providence | 82.4 |
| 2 | Kingston | 81.5 |
| 3 | Newport | 78.6 |
| 4 | Block | 76.7 |

### LOWEST

| | | |
|---|---|---|
| 1 | Block | 76.7 |
| 2 | Newport | 78.6 |
| 3 | Kingston | 81.5 |
| 4 | Providence | 82.4 |

## ANNUAL PRECIPITATION (")

### HIGHEST

| | | |
|---|---|---|
| 1 | Kingston | 50.10 |
| 2 | Providence | 45.67 |
| 3 | Newport | 45.01 |
| 4 | Block | 38.81 |

### LOWEST

| | | |
|---|---|---|
| 1 | Block | 38.81 |
| 2 | Newport | 45.01 |
| 3 | Providence | 45.67 |
| 4 | Kingston | 50.10 |

## ANNUAL SNOWFALL (")

### HIGHEST

| | | |
|---|---|---|
| 1 | Providence | 35.0 |
| 2 | Kingston | 30.8 |

### LOWEST

| | | |
|---|---|---|
| 1 | Kingston | 30.8 |
| 2 | Providence | 35.0 |

# SOUTH CAROLINA

PHYSICAL FEATURES. South Carolina is located on the southeastern coast of the United States between the southern part of the Appalachian Mountains and the Atlantic Ocean. Its north-south extent is 220 miles, from 32° to 35.2° N. latitude. The mountains in the extreme northwestern part of the State are 240 miles from the coastline. The coastline is 185 miles long and oriented southwest to northeast.

South Carolina shares some common topographic features with several eastern seaboard states. All of these features have a southwest to northeast orientation and extend across the whole State. The Blue Ridge Range of the Appalachian Mountains lies in the extreme northwestern part of the State. Elevations range from 1,000 to 2,000 feet with several peaks going over 3,000 feet. Sassafras Mountain, at 3,554 feet elevation, is the highest point in the State. The Mountain Region covers less than 10 percent of the State's area and to its southeast lies the Piedmont Plateau. The Plateau extends nearly to the center of the State with elevations decreasing northwest to southeast from 1,000 to 500 feet. There is a narrow hilly region where the Plateau descends to the Coastal Plain. In South Carolina this "fall line" region is known as the "Sand Hills"; elevations range from 500 to 200 feet. The width of the Sand Hills area is about 30 to 40 miles. Between the Sand Hills and the Atlantic Ocean lies the Coastal Plain. The Plain is broad and nearly level with elevations mostly between 50 and 200 feet. About 40 percent of the area of the State lies in the Coastal Plain.

All of the State's rivers drain southeast from the Mountain Region or Piedmont Plateau toward the ocean. There are three major and one minor river-basin systems. The Santee is the largest and drains the entire center portion of the State. The Savannah drains the western part of the State. The third major system is the Pee Dee, located in the northeastern section. The Edisto is a lesser river system lying between the Santee and Savannah.

There are many low sea islands separated from the mainland by shallow straits, sounds, and coastal streams. The Intracoastal Waterway can be found along much of the coastline.

GENERAL CLIMATE. Several major factors combine to give South Carolina a pleasant, mild, and humid climate. It is located at a relatively low latitude (32° to 35° N.) and most of the State is under 1,000 feet in elevation. It has a long coastline along which moves the warm Gulf Stream current. The mountains to the north and west block or delay many cold air masses approaching from those directions. Even the deep cold air masses which cross the mountains rapidly are warmed somewhat as the air is heated by compression when it descends on the southeastern side. This effect can be seen on the maps of minimum temperature in January and to a lesser degree in July, where a fairly large area of relatively higher temperature appears just southeast of the mountains.

It is convenient for climatic discussion to divide the State into areas coinciding closely with the topographic features already discussed. Six areas can be defined: (1) the Outer Coastal Plain, (2) the Inner Coastal Plain, (3) the Sand Hills, (4) the Lower Piedmont Plateau, (5) the Upper Piedmont Plateau, and (6) the Mountain Region.

TEMPERATURE. Lower temperatures can be expected in the Upper Piedmont and Mountain Region, where latitude, elevation and distance inland all have large values. Higher temperatures will result from smaller values of the three factors, as are found along the southern coast. There is a gradual decrease in annual average temperature northwestward from 68° at the coast to 58° at the edge of the mountains. Within the Mountain Region, variations in temperature are due almost entirely to elevation differences. The ocean waters have very small daily and annual changes in temperature when compared with the land surface. The air over the coastal water is cooler than the air over the land in summer and warmer than the air over land in winter, and this has a controlling effect on the temperatures of locations on and very near the coast. The highest temperatures are found in the central part of the State with the coast being 4° to 5° cooler. Clouds and rainfall have a minor effect on temperature. Maximum temperatures in summer are reduced slightly in areas where afternoon cloudiness and rain are persistent. Such an area is found along the Outer Coastal Plain where sea breezes produce clouds and rain nearly every summer day and dissipate at night.

Summers are rather hot and air conditioning is desirable at elevations below 500 feet. Fall and spring are mild and winters are rather cool at elevations above 500 feet.

PRECIPITATION. Rainfall is adequate in all parts of the State. Annual rainfall averages up to 80 inches in the highest part of the Mountain Region and less than 42 inches in parts of the Inner Coastal Plain and the Sand Hills. The Mountain Region is wet with amounts of 56 inches or more, the Upper Piedmont is relatively wet with amounts of 48 to 55 inches, the Lower Piedmont is relatively dry with amounts of 43 to 47 inches, the Outer Coastal Plain is relatively wet with amounts of 48 to 53 inches, and the Inner Coastal Plain is relatively dry with amounts of 38 to 47 inches. The Sand Hills area is less clear cut but is in general a relatively wet strip with a small dry area imbedded in it a few miles south of Columbia. The immediate south coast is also on the dry side. The driest period is in October and November when there is little cyclonic storm activity. Rainfall increases gradually and reaches a peak in March when cyclone and cold front activity are at a maximum. There is a general decrease again to a dry period from late April through early June. From the latter part of June through early September is a wet period primarily due to thunderstorm and shower activity which reaches its peak in July, the wettest summer month. The summer maximum stretches a little into the fall along the coast due to occasional tropical storm activity.

Solid forms of precipitation include snow, sleet, and hail. Hail is not too frequent but does occur with spring thunderstorms from March through early May. Snow and sleet may occur separately or together or mixed with rain during the winter months of December through February. Snow may occur from one to three times in winter. Seldom do accumulations remain very long on the ground except in the mountains. Statewide snows of notable amounts can occur when a cyclonic storm moves northeastward along or just off the coast. Freezing rain also occurs from one to three times per winter in the northern half of the State. Severe drought occurs about once in 15 years with less severe and less widespread droughts about once in 7 or 8 years.

OTHER CLIMATIC ELEMENTS. The percent of possible sunshine received varies over the State in a way similar to the variation in cloudiness and precipitation. Values in winter range from 50 to 60 percent, in summer from 60 to 70, with the dry periods in spring and fall receiving 70 to 75 percent. The variation in relative humidity with time of day is considerably greater than day to day and month to month variations. Highest values of 80 to 90 percent or more are reached at about sunrise and the lowest values of 45 to 50 percent occur an hour or two after local noon. There is about a 10 percent difference between winter and summer, with summer being the higher of the two seasons. The prevailing surface winds tend to be either from northeast or southwest due to the presence and orientation of the Appalachian Mountains. Winds of all directions occur throughout the State during the year, but the prevailing directions by seasons are: spring--southwest; summer--south and southwest; autumn--northeast; and winter--northeast and southwest.

STORMS. Severe weather comes to South Carolina occasionally in the form of violent thunderstorms, tornadoes and hurricanes. Although thunderstorms are common in the summer months, the really violent ones generally accompany the squall lines and active cold fronts of spring. Generally, they bring high winds, hail and considerable lightning, and sometimes spawn a tornado (average of 7 or 8 a year). Sixty percent of the tornadoes occur from March through June with April being the peak month with 25 percent. Tropical storms or hurricanes affect the State about one year out of two. Most of the occurrences are tropical storms which do little damage, frequently bringing rains at a time when they are needed. Most of the hurricanes affect only the Outer Coastal Plain. If they do come far inland, they decrease in intensity quite rapidly. Considerable flooding accompanies hurricanes which come very far inland and high tides occur along the coast to the north and east of the storm centers.

There is minor flooding somewhere in the State every year. It can occur on any of the many streams and rivers. There is a major flood about once every 7 or 8 years.

There have been many earth tremors in South Carolina over the years. The southern part of the Coastal Plain is indicated as earthquake prone.

## COUNTY INDEX

***Abbeville County***
CALHOUN FALLS

***Aiken County***
AIKEN 4 NE

***Anderson County***
ANDERSON
ANDERSON COUNTY AP
WEST PELZER

***Bamberg County***
BAMBERG

***Barnwell County***
BLACKVILLE 3 W

***Beaufort County***
BEAUFORT 7 SW
HILTON HEAD
YEMASSEE

***Berkeley County***
PINOPOLIS DAM

***Charleston County***
CHARLESTON CITY
CHARLESTON INTL ARPT
SULLIVANS ISLAND

***Cherokee County***
NINETY NINE ISLANDS

***Chester County***
CHESTER 1 NW

***Chesterfield County***
CHERAW

***Colleton County***
EDISTO ISLAND 3 SW
WALTERBORO 2 S

***Darlington County***
DARLINGTON
FLORENCE 8 NE

***Dillon County***
DILLON

***Dorchester County***
SUMMERVILLE

***Edgefield County***
JOHNSTON 4 SW

***Fairfield County***
PARR
WINNSBORO

***Florence County***
FLORENCE CTY CNTY AP
LAKE CITY 1 SE

***Georgetown County***
ANDREWS
BROOKGREEN GARDENS
GEORGETOWN 2 E

***Greenwood County***
GREENWOOD 3 SW

***Hampton County***
HAMPTON

***Horry County***
CONWAY

***Jasper County***
RIDGELAND 5 NE

***Kershaw County***
CAMDEN 3 W

***Laurens County***
LAURENS

***Lee County***
BISHOPVILLE 8 NNW

***Lexington County***
COLUMBIA METRO AP
PELION 4 NW

***McCormick County***
CLARK HILL 1 W

***Marion County***
MARION

***Marlboro County***
MCCOLL 3 NNW

***Newberry County***
LITTLE MOUNTAIN
NEWBERRY

***Oconee County***
WALHALLA

***Orangeburg County***
ORANGEBURG 2

***Pickens County***
CLEMSON UNIV
PICKENS

***Richland County***
COLUMBIA UNIV OF SC
SANDHILL EXP STN

***Saluda County***
SALUDA

***Spartanburg County***
GREER GREENV'L-SPART

***Sumter County***
SUMTER

***Union County***
SANTUCK
UNION 8 SW

***Williamsburg County***
KINGSTREE 1 SE

***York County***
WINTHROP UNIVERSITY

## ELEVATION INDEX

| FEET | STATION NAME |
|---|---|
| 10 | EDISTO ISLAND 3 SW |
| 10 | GEORGETOWN 2 E |
| 10 | SULLIVANS ISLAND |
| 20 | BEAUFORT 7 SW |
| 20 | BROOKGREEN GARDENS |

| FEET | STATION NAME |
|---|---|
| 20 | HILTON HEAD |
| 20 | RIDGELAND 5 NE |
| 23 | YEMASSEE |
| 30 | ANDREWS |
| 30 | CONWAY |
| 30 | PINOPOLIS DAM |
| 35 | SUMMERVILLE |
| 46 | CHARLESTON INTL ARPT |
| 49 | CHARLESTON CITY |
| 49 | KINGSTREE 1 SE |
| 69 | MARION |
| 69 | WALTERBORO 2 S |
| 79 | LAKE CITY 1 SE |
| 89 | HAMPTON |
| 112 | DILLON |
| 120 | FLORENCE 8 NE |
| 141 | CHERAW |
| 148 | FLORENCE CTY CNTY AP |
| 171 | BAMBERG |
| 171 | SUMTER |
| 180 | DARLINGTON |
| 190 | MCCOLL 3 NNW |
| 220 | CAMDEN 3 W |
| 220 | COLUMBIA METRO AP |
| 239 | ORANGEBURG 2 |
| 249 | BISHOPVILLE 8 NNW |
| 280 | PARR |
| 302 | BLACKVILLE 3 W |
| 322 | COLUMBIA UNIV OF SC |
| 381 | CLARK HILL 1 W |
| 400 | AIKEN 4 NE |
| 440 | SANDHILL EXP STN |
| 449 | PELION 4 NW |
| 479 | NEWBERRY |
| 502 | NINETY NINE ISLANDS |
| 512 | SANTUCK |
| 522 | CALHOUN FALLS |
| 541 | SALUDA |
| 551 | UNION 8 SW |
| 551 | WINNSBORO |
| 561 | CHESTER 1 NW |
| 589 | LAURENS |
| 591 | JOHNSTON 4 SW |
| 615 | GREENWOOD 3 SW |
| 689 | WINTHROP UNIVERSITY |
| 712 | LITTLE MOUNTAIN |
| 761 | ANDERSON |
| 764 | ANDERSON COUNTY AP |
| 830 | CLEMSON UNIV |
| 850 | WEST PELZER |
| 957 | GREER GREENV'L-SPART |
| 981 | WALHALLA |
| 1040 | PICKENS |

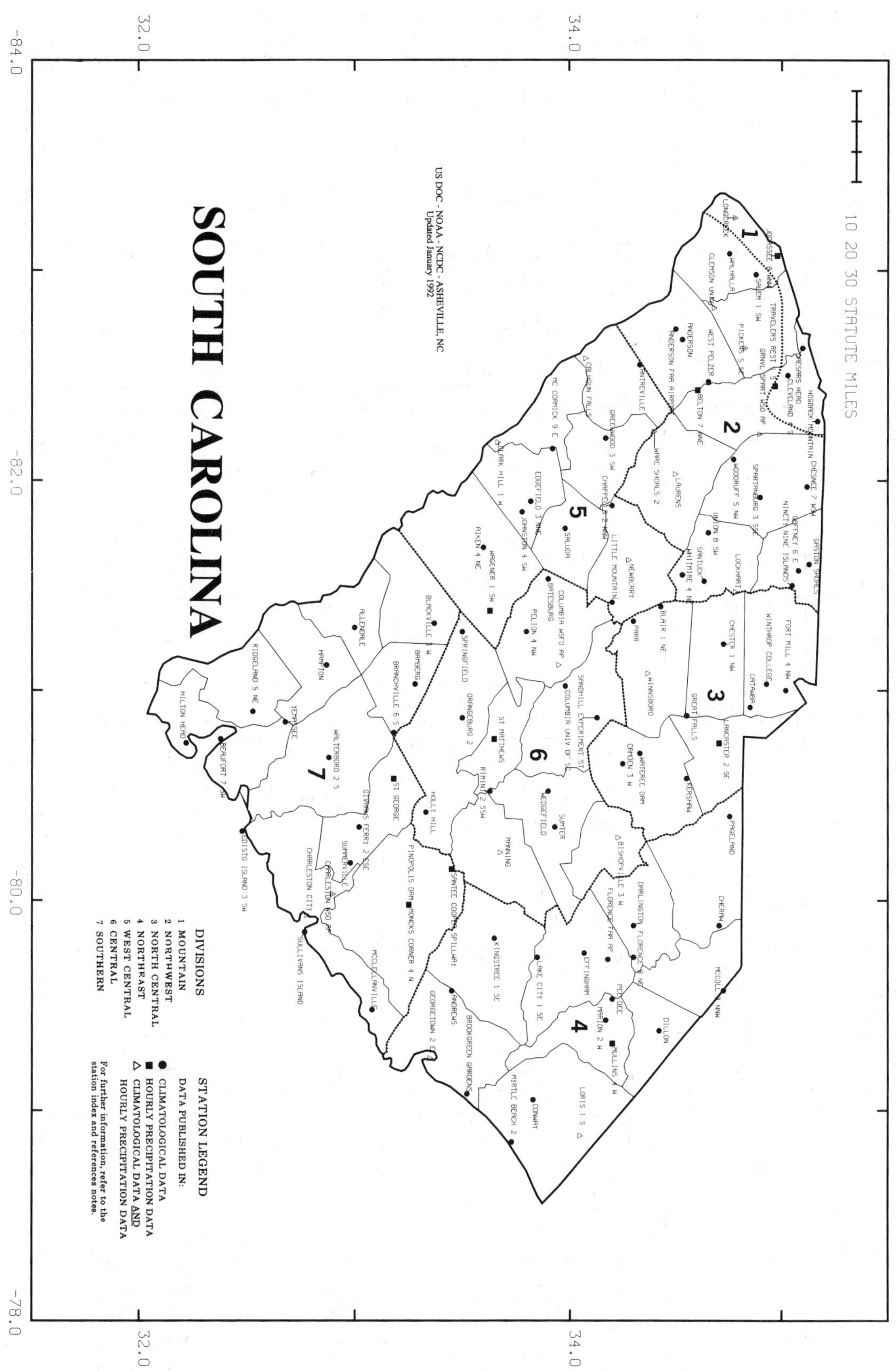
SOUTH CAROLINA
US DOC - NOAA - NCDC - ASHEVILLE, NC
Updated January 1992
10 20 30 STATUTE MILES
DIVISIONS
1 MOUNTAIN
2 NORTHWEST
3 NORTH CENTRAL
4 NORTHEAST
5 WEST CENTRAL
6 CENTRAL
7 SOUTHERN
STATION LEGEND
DATA PUBLISHED IN:
● CLIMATOLOGICAL DATA
■ HOURLY PRECIPITATION DATA
△ CLIMATOLOGICAL DATA AND HOURLY PRECIPITATION DATA
For further information, refer to the station index and references notes.

## AIKEN 4 NE *Aiken County* ELEVATION 400 ft LAT/LONG 33° 36 ' N / 81° 41 ' W

| | JAN | FEB | MAR | APR | MAY | JUN | JUL | AUG | SEP | OCT | NOV | DEC | YEAR |
|---|---|---|---|---|---|---|---|---|---|---|---|---|---|
| Maximum Temp °F | 57.5 | 61.8 | 70.0 | 78.2 | 84.0 | 89.9 | 92.6 | 90.8 | 86.3 | 77.5 | 68.6 | 60.5 | 76.5 |
| Minimum Temp °F | 33.4 | 35.7 | 42.2 | 48.9 | 57.2 | 65.2 | 69.1 | 68.2 | 62.4 | 50.9 | 41.2 | 35.2 | 50.8 |
| Mean Temp °F | 45.5 | 48.8 | 56.2 | 63.6 | 70.7 | 77.5 | 80.9 | 79.5 | 74.4 | 64.2 | 55.0 | 47.9 | 63.7 |
| Days Max Temp ≥ 90 °F | 0 | 0 | 0 | 2 | 6 | 16 | 23 | 19 | 9 | 1 | 0 | 0 | 76 |
| Days Max Temp ≤ 32 °F | 0 | 0 | 0 | 0 | 0 | 0 | 0 | 0 | 0 | 0 | 0 | 0 | 0 |
| Days Min Temp ≤ 32 °F | 15 | 12 | 6 | 2 | 0 | 0 | 0 | 0 | 0 | 1 | 7 | 14 | 57 |
| Days Min Temp ≤ 0 °F | 0 | 0 | 0 | 0 | 0 | 0 | 0 | 0 | 0 | 0 | 0 | 0 | 0 |
| Heating Degree Days | 598 | 455 | 285 | 108 | 18 | 0 | 0 | 0 | 5 | 101 | 307 | 521 | 2398 |
| Cooling Degree Days | 0 | 4 | 21 | 79 | 215 | 424 | 542 | 479 | 298 | 86 | 15 | 2 | 2165 |
| Total Precipitation (") | 4.65 | 4.45 | 5.35 | 3.54 | 4.24 | 5.41 | 5.45 | 5.28 | 3.62 | 3.33 | 3.21 | 3.86 | 52.39 |
| Days ≥ 0.1" Precip | 8 | 7 | 7 | 5 | 7 | 7 | 9 | 8 | 5 | 5 | 5 | 7 | 80 |
| Total Snowfall (") | ***0.3*** | ***0.8*** | 0.0 | 0.0 | 0.0 | 0.0 | 0.0 | 0.0 | 0.0 | 0.0 | 0.0 | 0.0 | 1.1 |
| Days ≥ 1" Snow Depth | ***0*** | ***0*** | 0 | 0 | 0 | 0 | 0 | 0 | 0 | 0 | 0 | 0 | 0 |

## ANDERSON *Anderson County* ELEVATION 761 ft LAT/LONG 34° 31 ' N / 82° 39 ' W

| | JAN | FEB | MAR | APR | MAY | JUN | JUL | AUG | SEP | OCT | NOV | DEC | YEAR |
|---|---|---|---|---|---|---|---|---|---|---|---|---|---|
| Maximum Temp °F | 53.9 | 58.4 | 66.7 | 75.4 | 81.4 | 87.9 | 90.6 | 89.0 | 83.8 | 74.7 | 65.2 | 56.5 | 73.6 |
| Minimum Temp °F | 30.6 | 32.8 | 39.0 | 46.9 | 55.5 | 63.2 | 67.4 | 66.6 | 60.5 | 48.6 | 40.0 | 33.1 | 48.7 |
| Mean Temp °F | 42.3 | 45.6 | 52.8 | 61.2 | 68.5 | 75.6 | 79.0 | 77.8 | 72.2 | 61.6 | 52.7 | 44.8 | 61.2 |
| Days Max Temp ≥ 90 °F | 0 | 0 | 0 | 0 | 2 | 12 | 19 | 15 | 5 | 0 | 0 | 0 | 53 |
| Days Max Temp ≤ 32 °F | 1 | 0 | 0 | 0 | 0 | 0 | 0 | 0 | 0 | 0 | 0 | 0 | 1 |
| Days Min Temp ≤ 32 °F | 18 | 14 | 9 | 2 | 0 | 0 | 0 | 0 | 0 | 1 | 7 | 16 | 67 |
| Days Min Temp ≤ 0 °F | 0 | 0 | 0 | 0 | 0 | 0 | 0 | 0 | 0 | 0 | 0 | 0 | 0 |
| Heating Degree Days | 698 | 542 | 376 | 152 | 30 | 2 | 0 | 0 | 12 | 142 | 367 | 619 | 2940 |
| Cooling Degree Days | 0 | 0 | 4 | 43 | 148 | 353 | 465 | 417 | 236 | 48 | 5 | 1 | 1720 |
| Total Precipitation (") | 5.05 | 4.15 | 5.27 | 3.32 | 4.11 | 3.76 | 4.50 | 4.01 | 3.96 | 3.97 | 3.98 | 4.54 | 50.62 |
| Days ≥ 0.1" Precip | 8 | 7 | 8 | 6 | 7 | 6 | 7 | 7 | 6 | 5 | 6 | 7 | 80 |
| Total Snowfall (") | ***1.0*** | 0.8 | 0.3 | 0.0 | 0.0 | 0.0 | 0.0 | 0.0 | 0.0 | 0.0 | 0.0 | 0.0 | 2.1 |
| Days ≥ 1" Snow Depth | ***0*** | ***0*** | 0 | 0 | 0 | 0 | 0 | 0 | 0 | 0 | 0 | 0 | 0 |

## ANDERSON COUNTY AP *Anderson County* ELEVATION 764 ft LAT/LONG 34° 30 ' N / 82° 43 ' W

| | JAN | FEB | MAR | APR | MAY | JUN | JUL | AUG | SEP | OCT | NOV | DEC | YEAR |
|---|---|---|---|---|---|---|---|---|---|---|---|---|---|
| Maximum Temp °F | 51.5 | 56.0 | 64.8 | 73.6 | 80.3 | 87.3 | 90.0 | 88.0 | 82.6 | 73.1 | 63.9 | 55.1 | 72.2 |
| Minimum Temp °F | 32.0 | 34.5 | 41.8 | 49.5 | 57.8 | 66.0 | 69.8 | 68.8 | 62.8 | 50.5 | 41.7 | 35.4 | 50.9 |
| Mean Temp °F | 41.8 | 45.3 | 53.3 | 61.6 | 69.1 | 76.7 | 79.9 | 78.4 | 72.7 | 61.9 | 52.8 | 45.3 | 61.6 |
| Days Max Temp ≥ 90 °F | 0 | 0 | 0 | 0 | 2 | 11 | 18 | 13 | 4 | 0 | 0 | 0 | 48 |
| Days Max Temp ≤ 32 °F | 1 | 0 | 0 | 0 | 0 | 0 | 0 | 0 | 0 | 0 | 0 | 0 | 1 |
| Days Min Temp ≤ 32 °F | 16 | 13 | 6 | 1 | 0 | 0 | 0 | 0 | 0 | 1 | 6 | 13 | 56 |
| Days Min Temp ≤ 0 °F | 0 | 0 | 0 | 0 | 0 | 0 | 0 | 0 | 0 | 0 | 0 | 0 | 0 |
| Heating Degree Days | 713 | 551 | 363 | 145 | 30 | 1 | 0 | 0 | 10 | 144 | 365 | 607 | 2929 |
| Cooling Degree Days | 0 | 0 | 8 | 45 | 161 | 374 | 487 | 428 | 242 | 54 | 5 | 2 | 1806 |
| Total Precipitation (") | 4.19 | 3.87 | 4.92 | 3.04 | 3.98 | 3.44 | 4.08 | 3.80 | 3.90 | 3.19 | 3.47 | 3.81 | 45.69 |
| Days ≥ 0.1" Precip | 7 | 7 | 8 | 6 | 6 | 5 | 7 | 6 | 5 | 5 | 6 | 6 | 74 |
| Total Snowfall (") | 1.3 | 1.2 | 0.6 | 0.0 | 0.0 | 0.0 | 0.0 | 0.0 | 0.0 | 0.0 | 0.0 | 0.3 | 3.4 |
| Days ≥ 1" Snow Depth | 1 | 1 | 0 | 0 | 0 | 0 | 0 | 0 | 0 | 0 | 0 | 0 | 2 |

## ANDREWS *Georgetown County* ELEVATION 30 ft LAT/LONG 33° 27 ' N / 79° 34 ' W

| | JAN | FEB | MAR | APR | MAY | JUN | JUL | AUG | SEP | OCT | NOV | DEC | YEAR |
|---|---|---|---|---|---|---|---|---|---|---|---|---|---|
| Maximum Temp °F | 56.9 | 61.2 | ***69.8*** | 77.5 | ***83.6*** | 88.3 | 91.5 | ***89.8*** | 84.8 | ***76.7*** | ***69.2*** | ***61.0*** | 75.9 |
| Minimum Temp °F | 34.4 | 37.0 | 44.6 | 51.0 | ***59.2*** | 66.3 | 70.6 | ***70.1*** | 64.8 | ***53.7*** | ***45.1*** | ***37.5*** | 52.9 |
| Mean Temp °F | 45.7 | 49.2 | ***57.2*** | 64.3 | ***71.4*** | 77.4 | 81.1 | ***80.0*** | 74.8 | ***65.3*** | ***57.2*** | ***49.3*** | 64.4 |
| Days Max Temp ≥ 90 °F | 0 | 0 | 0 | ***2*** | ***5*** | 14 | 21 | ***17*** | ***6*** | ***0*** | ***0*** | 0 | 65 |
| Days Max Temp ≤ 32 °F | 0 | 0 | 0 | 0 | 0 | 0 | 0 | ***0*** | 0 | ***0*** | ***0*** | 0 | 0 |
| Days Min Temp ≤ 32 °F | 14 | ***10*** | ***3*** | 0 | 0 | 0 | 0 | ***0*** | 0 | ***0*** | ***4*** | ***11*** | 42 |
| Days Min Temp ≤ 0 °F | 0 | 0 | 0 | 0 | 0 | 0 | 0 | ***0*** | 0 | ***0*** | ***0*** | 0 | 0 |
| Heating Degree Days | 593 | 445 | ***258*** | 95 | ***17*** | 1 | 0 | ***0*** | 3 | ***83*** | ***252*** | ***487*** | 2234 |
| Cooling Degree Days | 1 | 7 | 23 | 75 | 218 | 401 | 529 | 465 | 305 | 102 | 26 | 5 | 2157 |
| Total Precipitation (") | 4.18 | 3.34 | 3.72 | ***2.67*** | ***3.92*** | 4.75 | 5.96 | ***5.51*** | 4.39 | ***3.45*** | 2.67 | ***3.39*** | 47.95 |
| Days ≥ 0.1" Precip | ***7*** | 6 | ***6*** | ***5*** | ***7*** | 7 | 8 | ***8*** | ***5*** | ***4*** | ***4*** | ***6*** | 73 |
| Total Snowfall (") | 0.3 | 0.1 | 0.1 | 0.0 | 0.0 | 0.0 | 0.0 | 0.0 | 0.0 | 0.0 | 0.0 | 0.4 | 0.9 |
| Days ≥ 1" Snow Depth | ***0*** | 0 | 0 | 0 | 0 | 0 | 0 | 0 | 0 | 0 | ***0*** | 0 | 0 |

## BAMBERG *Bamberg County* ELEVATION 171 ft LAT/LONG 33° 17 ' N / 81° 3 ' W

| | JAN | FEB | MAR | APR | MAY | JUN | JUL | AUG | SEP | OCT | NOV | DEC | YEAR |
|---|---|---|---|---|---|---|---|---|---|---|---|---|---|
| Maximum Temp °F | 57.0 | 61.5 | 70.4 | 78.2 | 83.9 | 88.9 | 91.5 | 89.6 | 84.9 | 76.1 | 67.7 | 60.0 | 75.8 |
| Minimum Temp °F | 35.3 | 37.3 | 44.0 | 50.7 | 58.8 | 66.1 | 70.0 | 69.3 | 64.4 | 52.9 | 44.3 | 38.1 | 52.6 |
| Mean Temp °F | 46.2 | 49.4 | 57.2 | 64.5 | 71.4 | 77.5 | 80.7 | 79.4 | 74.7 | 64.6 | 56.0 | 49.1 | 64.2 |
| Days Max Temp ≥ 90 °F | 0 | 0 | 0 | 1 | 6 | 14 | 21 | 16 | 6 | 0 | 0 | 0 | 64 |
| Days Max Temp ≤ 32 °F | 0 | 0 | 0 | 0 | 0 | 0 | 0 | 0 | 0 | 0 | 0 | 0 | 0 |
| Days Min Temp ≤ 32 °F | 13 | 10 | 4 | 0 | 0 | 0 | 0 | 0 | 0 | 0 | 5 | 10 | 42 |
| Days Min Temp ≤ 0 °F | 0 | 0 | 0 | 0 | 0 | 0 | 0 | 0 | 0 | 0 | 0 | 0 | 0 |
| Heating Degree Days | 578 | 439 | 258 | 90 | 12 | 0 | 0 | 0 | 4 | 96 | 282 | 492 | 2251 |
| Cooling Degree Days | 1 | 6 | 27 | 78 | 225 | 413 | 522 | 468 | 303 | 91 | 20 | 5 | 2159 |
| Total Precipitation (") | 4.21 | 3.78 | 4.44 | 2.76 | 3.86 | 5.53 | 5.09 | 5.60 | 3.36 | 2.89 | 2.81 | 3.51 | 47.84 |
| Days ≥ 0.1" Precip | 8 | 6 | 7 | 5 | 7 | 7 | 8 | 8 | 5 | 5 | 5 | 6 | 77 |
| Total Snowfall (") | 0.3 | 1.0 | 0.0 | 0.0 | 0.0 | 0.0 | 0.0 | 0.0 | 0.0 | 0.0 | 0.0 | 0.1 | 1.4 |
| Days ≥ 1" Snow Depth | 0 | 0 | 0 | 0 | 0 | 0 | 0 | 0 | 0 | 0 | 0 | 0 | 0 |

## BEAUFORT 7 SW *Beaufort County* ELEVATION 20 ft LAT/LONG 32° 23 ' N / 80° 46 ' W

| | JAN | FEB | MAR | APR | MAY | JUN | JUL | AUG | SEP | OCT | NOV | DEC | YEAR |
|---|---|---|---|---|---|---|---|---|---|---|---|---|---|
| Maximum Temp °F | 58.9 | 62.3 | 69.6 | 76.9 | 82.9 | 87.6 | 90.2 | 88.8 | 84.7 | 77.3 | 69.4 | 61.9 | 75.9 |
| Minimum Temp °F | 39.5 | 41.3 | 48.1 | 55.2 | 63.2 | 69.7 | 73.0 | 72.4 | 68.2 | 58.4 | 49.6 | 42.7 | 56.8 |
| Mean Temp °F | 49.2 | 51.9 | 58.9 | 66.1 | 73.1 | 78.7 | 81.7 | 80.6 | 76.5 | 67.9 | 59.5 | 52.4 | 66.4 |
| Days Max Temp ≥ 90 °F | 0 | 0 | 0 | 1 | 3 | 10 | 19 | 14 | 4 | 1 | 0 | 0 | 52 |
| Days Max Temp ≤ 32 °F | 0 | 0 | 0 | 0 | 0 | 0 | 0 | 0 | 0 | 0 | 0 | 0 | 0 |
| Days Min Temp ≤ 32 °F | 8 | 6 | 1 | 0 | 0 | 0 | 0 | 0 | 0 | 0 | 1 | 5 | 21 |
| Days Min Temp ≤ 0 °F | 0 | 0 | 0 | 0 | 0 | 0 | 0 | 0 | 0 | 0 | 0 | 0 | 0 |
| Heating Degree Days | 486 | 369 | 209 | 60 | 4 | 0 | 0 | 0 | 1 | 48 | 193 | 391 | 1761 |
| Cooling Degree Days | 2 | 6 | 26 | 96 | 264 | 445 | 544 | 500 | 356 | 144 | 40 | 7 | 2430 |
| Total Precipitation (") | 4.13 | 2.97 | 4.12 | 2.74 | 3.84 | 5.71 | 5.91 | 7.68 | 4.88 | 3.20 | 2.62 | 3.20 | 51.00 |
| Days ≥ 0.1" Precip | 7 | 6 | 6 | 5 | 6 | 8 | 8 | 9 | 7 | 4 | 4 | 6 | 76 |
| Total Snowfall (") | 0.1 | 0.2 | 0.0 | 0.0 | 0.0 | 0.0 | 0.0 | 0.0 | 0.0 | 0.0 | 0.0 | 0.2 | 0.5 |
| Days ≥ 1" Snow Depth | 0 | 0 | 0 | 0 | 0 | 0 | 0 | 0 | 0 | 0 | 0 | 0 | 0 |

## BISHOPVILLE 8 NNW *Lee County* ELEVATION 249 ft LAT/LONG 34° 14 ' N / 80° 18 ' W

| | JAN | FEB | MAR | APR | MAY | JUN | JUL | AUG | SEP | OCT | NOV | DEC | YEAR |
|---|---|---|---|---|---|---|---|---|---|---|---|---|---|
| Maximum Temp °F | 52.3 | 57.1 | 65.7 | 74.8 | 81.5 | 87.8 | 90.5 | 88.6 | 83.9 | 74.9 | 66.1 | 57.5 | 73.4 |
| Minimum Temp °F | 30.7 | 33.0 | 40.1 | 48.2 | 57.1 | 65.1 | 68.7 | 67.5 | 61.7 | 49.7 | 40.5 | 34.0 | 49.7 |
| Mean Temp °F | 41.6 | 45.1 | 52.9 | 61.5 | 69.3 | 76.5 | 79.7 | 78.1 | 72.8 | 62.4 | 53.3 | 45.8 | 61.6 |
| Days Max Temp ≥ 90 °F | 0 | 0 | 0 | 1 | 4 | 12 | 18 | 15 | 6 | 0 | 0 | 0 | 56 |
| Days Max Temp ≤ 32 °F | 1 | 0 | 0 | 0 | 0 | 0 | 0 | 0 | 0 | 0 | 0 | 0 | 1 |
| Days Min Temp ≤ 32 °F | 18 | 15 | 8 | 1 | 0 | 0 | 0 | 0 | 0 | 1 | 8 | 16 | 67 |
| Days Min Temp ≤ 0 °F | 0 | 0 | 0 | 0 | 0 | 0 | 0 | 0 | 0 | 0 | 0 | 0 | 0 |
| Heating Degree Days | 720 | 557 | 380 | 155 | 33 | 2 | 0 | 0 | 12 | 137 | 355 | 591 | 2942 |
| Cooling Degree Days | 0 | 3 | 13 | 53 | 174 | 388 | 486 | 427 | 256 | 71 | 10 | 3 | 1884 |
| Total Precipitation (") | 3.93 | 3.32 | 4.19 | 2.70 | 3.88 | 4.35 | 4.99 | 4.74 | 3.27 | 3.15 | 2.86 | 3.24 | 44.62 |
| Days ≥ 0.1" Precip | 7 | 6 | 7 | 5 | 6 | 7 | 8 | 7 | 5 | 4 | 5 | 6 | 73 |
| Total Snowfall (") | 0.6 | 1.1 | 0.4 | 0.0 | 0.0 | 0.0 | 0.0 | 0.0 | 0.0 | 0.0 | 0.0 | 0.2 | 2.3 |
| Days ≥ 1" Snow Depth | 1 | 1 | 0 | 0 | 0 | 0 | 0 | 0 | 0 | 0 | 0 | 0 | 2 |

## BLACKVILLE 3 W *Barnwell County* ELEVATION 302 ft LAT/LONG 33° 21 ' N / 81° 16 ' W

| | JAN | FEB | MAR | APR | MAY | JUN | JUL | AUG | SEP | OCT | NOV | DEC | YEAR |
|---|---|---|---|---|---|---|---|---|---|---|---|---|---|
| Maximum Temp °F | 57.8 | 62.0 | 70.3 | 78.2 | 84.0 | 89.4 | 91.8 | 90.1 | 86.2 | 77.8 | 69.5 | 61.4 | 76.5 |
| Minimum Temp °F | 34.7 | 36.8 | 43.7 | 50.5 | 58.4 | 65.3 | 68.9 | 68.0 | 62.9 | 51.9 | 43.4 | 37.4 | 51.8 |
| Mean Temp °F | 46.3 | 49.4 | 57.0 | 64.4 | 71.2 | 77.4 | 80.4 | 79.1 | 74.6 | 64.9 | 56.5 | 49.4 | 64.2 |
| Days Max Temp ≥ 90 °F | 0 | 0 | 0 | 1 | 5 | 15 | 22 | 18 | 9 | 1 | 0 | 0 | 71 |
| Days Max Temp ≤ 32 °F | 0 | 0 | 0 | 0 | 0 | 0 | 0 | 0 | 0 | 0 | 0 | 0 | 0 |
| Days Min Temp ≤ 32 °F | 14 | 11 | 5 | 1 | 0 | 0 | 0 | 0 | 0 | 1 | 6 | 11 | 49 |
| Days Min Temp ≤ 0 °F | 0 | 0 | 0 | 0 | 0 | 0 | 0 | 0 | 0 | 0 | 0 | 0 | 0 |
| Heating Degree Days | 576 | 438 | 264 | 94 | 14 | 0 | 0 | 0 | 5 | 92 | 267 | 481 | 2231 |
| Cooling Degree Days | 1 | 6 | 26 | 79 | 219 | 407 | 510 | 454 | 304 | 100 | 21 | 5 | 2132 |
| Total Precipitation (") | 4.19 | 3.87 | 4.58 | 2.80 | 3.86 | 5.32 | 4.81 | 5.01 | 3.13 | 3.03 | 2.58 | 3.42 | 46.60 |
| Days ≥ 0.1" Precip | 7 | 6 | 7 | 5 | 7 | 8 | 8 | 8 | 5 | 4 | 5 | 6 | 76 |
| Total Snowfall (") | 0.1 | 1.1 | 0.0 | 0.0 | 0.0 | 0.0 | 0.0 | 0.0 | 0.0 | 0.0 | 0.0 | 0.1 | 1.3 |
| Days ≥ 1" Snow Depth | 0 | 0 | 0 | 0 | 0 | 0 | 0 | 0 | 0 | 0 | 0 | 0 | 0 |

## BROOKGREEN GARDENS *Georgetown County* ELEVATION 20 ft LAT/LONG 33° 31' N / 79° 5' W

| | JAN | FEB | MAR | APR | MAY | JUN | JUL | AUG | SEP | OCT | NOV | DEC | YEAR |
|---|---|---|---|---|---|---|---|---|---|---|---|---|---|
| Maximum Temp °F | 57.3 | 60.6 | 67.8 | 75.6 | 81.8 | 87.0 | 90.3 | 88.6 | 84.6 | 76.6 | 68.7 | 61.0 | 75.0 |
| Minimum Temp °F | 35.2 | 37.0 | 43.9 | 50.7 | 59.1 | 66.3 | 70.6 | 69.7 | 64.9 | 53.6 | 44.9 | 38.2 | 52.8 |
| Mean Temp °F | 46.2 | 48.8 | 55.9 | 63.2 | 70.5 | 76.7 | 80.5 | 79.2 | 74.8 | 65.2 | 56.8 | 49.6 | 64.0 |
| Days Max Temp ≥ 90 °F | 0 | 0 | 0 | 1 | 2 | 8 | 18 | 12 | 4 | 0 | 0 | 0 | 45 |
| Days Max Temp ≤ 32 °F | 0 | 0 | 0 | 0 | 0 | 0 | 0 | 0 | 0 | 0 | 0 | 0 | 0 |
| Days Min Temp ≤ 32 °F | 13 | 10 | 5 | 1 | 0 | 0 | 0 | 0 | 0 | 0 | 5 | 11 | 45 |
| Days Min Temp ≤ 0 °F | 0 | 0 | 0 | 0 | 0 | 0 | 0 | 0 | 0 | 0 | 0 | 0 | 0 |
| Heating Degree Days | 577 | 454 | 295 | 116 | 18 | 1 | 0 | 0 | 3 | 88 | 263 | 475 | 2290 |
| Cooling Degree Days | 1 | 5 | 18 | 62 | 196 | 384 | 512 | 454 | 299 | 101 | 27 | 6 | 2065 |
| Total Precipitation (") | 4.35 | 3.52 | 4.45 | 2.66 | 4.23 | 4.70 | 5.88 | 6.95 | 4.92 | 3.84 | 3.30 | 4.09 | 52.89 |
| Days ≥ 0.1" Precip | 8 | 6 | 7 | 5 | 7 | 7 | 9 | 9 | 6 | 4 | 5 | 7 | 80 |
| Total Snowfall (") | 0.2 | 0.4 | 0.2 | 0.0 | 0.0 | 0.0 | 0.0 | 0.0 | 0.0 | 0.0 | 0.0 | 0.4 | 1.2 |
| Days ≥ 1" Snow Depth | 0 | 0 | 0 | 0 | 0 | 0 | 0 | 0 | 0 | 0 | 0 | 0 | 0 |

## CALHOUN FALLS *Abbeville County* ELEVATION 522 ft LAT/LONG 34° 5' N / 82° 35' W

| | JAN | FEB | MAR | APR | MAY | JUN | JUL | AUG | SEP | OCT | NOV | DEC | YEAR |
|---|---|---|---|---|---|---|---|---|---|---|---|---|---|
| Maximum Temp °F | 52.5 | 57.1 | 65.9 | 74.7 | 81.4 | 87.8 | 91.3 | 89.9 | 84.4 | 74.7 | 65.3 | 55.8 | 73.4 |
| Minimum Temp °F | 30.9 | 32.7 | 39.6 | 47.0 | 56.1 | 64.3 | 68.7 | 67.8 | 61.5 | 48.8 | 39.7 | 33.5 | 49.2 |
| Mean Temp °F | 41.7 | 45.0 | 52.7 | 60.9 | 68.8 | 76.1 | 80.0 | 78.9 | 73.0 | 61.8 | 52.5 | 44.6 | 61.3 |
| Days Max Temp ≥ 90 °F | 0 | 0 | 0 | 0 | 3 | 12 | 20 | 17 | 7 | 0 | 0 | 0 | 59 |
| Days Max Temp ≤ 32 °F | 1 | 0 | 0 | 0 | 0 | 0 | 0 | 0 | 0 | 0 | 0 | 0 | 1 |
| Days Min Temp ≤ 32 °F | 18 | 15 | 8 | 1 | 0 | 0 | 0 | 0 | 0 | 1 | 8 | 16 | 67 |
| Days Min Temp ≤ 0 °F | 0 | 0 | 0 | 0 | 0 | 0 | 0 | 0 | 0 | 0 | 0 | 0 | 0 |
| Heating Degree Days | 716 | 560 | 381 | 160 | 33 | 2 | 0 | 0 | 11 | 146 | 374 | 628 | 3011 |
| Cooling Degree Days | 0 | 1 | 8 | 42 | 165 | 372 | 513 | 463 | 266 | 60 | 8 | 2 | 1900 |
| Total Precipitation (") | 4.80 | 4.16 | 5.10 | 3.36 | 4.03 | 4.09 | 4.62 | 3.46 | 3.26 | 3.13 | 3.49 | 3.77 | 47.27 |
| Days ≥ 0.1" Precip | 8 | 7 | 8 | 6 | 6 | 6 | 7 | 6 | 5 | 4 | 6 | 7 | 76 |
| Total Snowfall (") | 0.4 | 0.6 | 0.5 | 0.0 | 0.0 | 0.0 | 0.0 | 0.0 | 0.0 | 0.0 | 0.1 | 0.1 | 1.7 |
| Days ≥ 1" Snow Depth | 0 | 0 | 0 | 0 | 0 | 0 | 0 | 0 | 0 | 0 | 0 | 0 | 0 |

## CAMDEN 3 W *Kershaw County* ELEVATION 220 ft LAT/LONG 34° 15' N / 80° 37' W

| | JAN | FEB | MAR | APR | MAY | JUN | JUL | AUG | SEP | OCT | NOV | DEC | YEAR |
|---|---|---|---|---|---|---|---|---|---|---|---|---|---|
| Maximum Temp °F | 53.0 | 57.4 | 65.9 | 74.3 | 80.3 | 86.4 | 89.6 | 87.5 | 82.7 | 73.5 | 65.0 | 56.7 | 72.7 |
| Minimum Temp °F | 28.4 | 30.1 | 37.1 | 45.3 | 54.7 | 63.4 | 68.1 | 67.0 | 60.6 | 47.5 | 37.9 | 31.0 | 47.6 |
| Mean Temp °F | 40.7 | 43.8 | 51.5 | 59.8 | 67.6 | 74.9 | 78.9 | 77.3 | 71.7 | 60.5 | 51.5 | 43.9 | 60.2 |
| Days Max Temp ≥ 90 °F | 0 | 0 | 0 | 1 | 2 | 8 | 16 | 12 | 4 | 0 | 0 | 0 | 43 |
| Days Max Temp ≤ 32 °F | 1 | 0 | 0 | 0 | 0 | 0 | 0 | 0 | 0 | 0 | 0 | 0 | 1 |
| Days Min Temp ≤ 32 °F | 21 | 18 | 11 | 3 | 0 | 0 | 0 | 0 | 0 | 2 | 11 | 19 | 85 |
| Days Min Temp ≤ 0 °F | 0 | 0 | 0 | 0 | 0 | 0 | 0 | 0 | 0 | 0 | 0 | 0 | 0 |
| Heating Degree Days | 746 | 594 | 417 | 186 | 46 | 3 | 0 | 0 | 15 | 174 | 405 | 650 | 3236 |
| Cooling Degree Days | 0 | 2 | 9 | 33 | 131 | 329 | 462 | 398 | 228 | 46 | 5 | 2 | 1645 |
| Total Precipitation (") | 4.13 | 3.39 | 4.38 | 2.85 | 3.44 | 4.33 | 5.15 | 4.74 | 3.62 | 3.31 | 3.04 | 3.37 | 45.75 |
| Days ≥ 0.1" Precip | 8 | 6 | 7 | 5 | 6 | 7 | 8 | 7 | 5 | 4 | 5 | 6 | 74 |
| Total Snowfall (") | *0.3* | 0.8 | 0.2 | 0.0 | 0.0 | 0.0 | 0.0 | 0.0 | 0.0 | 0.0 | 0.0 | 0.1 | 1.4 |
| Days ≥ 1" Snow Depth | *0* | 0 | 0 | 0 | 0 | 0 | 0 | 0 | 0 | 0 | 0 | 0 | 0 |

## CHARLESTON CITY *Charleston County* ELEVATION 49 ft LAT/LONG 32° 47' N / 79° 56' W

| | JAN | FEB | MAR | APR | MAY | JUN | JUL | AUG | SEP | OCT | NOV | DEC | YEAR |
|---|---|---|---|---|---|---|---|---|---|---|---|---|---|
| Maximum Temp °F | 56.2 | 59.0 | 65.4 | 73.0 | 79.4 | 84.9 | 88.4 | 87.0 | 82.8 | 75.0 | 67.3 | 60.0 | 73.2 |
| Minimum Temp °F | 41.4 | 43.7 | 50.5 | 58.2 | 66.2 | 72.6 | 75.9 | 75.0 | 71.0 | 60.9 | 52.3 | 45.1 | 59.4 |
| Mean Temp °F | 48.8 | 51.4 | 58.0 | 65.6 | 72.8 | 78.8 | 82.1 | 81.0 | 76.9 | 68.0 | 59.8 | 52.6 | 66.3 |
| Days Max Temp ≥ 90 °F | 0 | 0 | 0 | 0 | 1 | 5 | 12 | 10 | 3 | 0 | 0 | 0 | 31 |
| Days Max Temp ≤ 32 °F | 0 | 0 | 0 | 0 | 0 | 0 | 0 | 0 | 0 | 0 | 0 | 0 | 0 |
| Days Min Temp ≤ 32 °F | 5 | 3 | 0 | 0 | 0 | 0 | 0 | 0 | 0 | 0 | 0 | 2 | 10 |
| Days Min Temp ≤ 0 °F | 0 | 0 | 0 | 0 | 0 | 0 | 0 | 0 | 0 | 0 | 0 | 0 | 0 |
| Heating Degree Days | 496 | 382 | 232 | 68 | 6 | 0 | 0 | 0 | 1 | 44 | 183 | 384 | 1796 |
| Cooling Degree Days | 1 | 5 | 20 | 85 | 255 | 436 | 553 | 506 | 364 | 142 | 39 | 7 | 2413 |
| Total Precipitation (") | 3.57 | 2.71 | 4.20 | 2.31 | 3.38 | 5.26 | 5.54 | 7.23 | 5.19 | 2.98 | 2.35 | 2.90 | 47.62 |
| Days ≥ 0.1" Precip | 6 | 5 | 6 | 4 | 5 | 7 | 7 | 9 | 6 | 4 | 4 | 6 | 69 |
| Total Snowfall (") | na | na | na | na | na | na | na | na | na | na | na | na | na |
| Days ≥ 1" Snow Depth | na | na | na | na | na | na | na | na | na | na | na | na | na |

## CHARLESTON INTL ARPT *Charleston County* ELEVATION 46 ft LAT/LONG 32° 54 ' N / 80° 2 ' W

| | JAN | FEB | MAR | APR | MAY | JUN | JUL | AUG | SEP | OCT | NOV | DEC | YEAR |
|---|---|---|---|---|---|---|---|---|---|---|---|---|---|
| Maximum Temp °F | 58.0 | 61.2 | 68.8 | 76.2 | 82.7 | 87.6 | 90.7 | 89.1 | 85.0 | 77.1 | 69.5 | 61.8 | 75.6 |
| Minimum Temp °F | 37.3 | 39.4 | 46.5 | 53.0 | 61.8 | 69.0 | 73.0 | 72.3 | 67.6 | 56.1 | 47.0 | 40.3 | 55.3 |
| Mean Temp °F | 47.7 | 50.3 | 57.6 | 64.6 | 72.3 | 78.4 | 81.9 | 80.7 | 76.3 | 66.6 | 58.3 | 51.1 | 65.5 |
| Days Max Temp ≥ 90 °F | 0 | 0 | 0 | 1 | 4 | 11 | 19 | 16 | 6 | 0 | 0 | 0 | 57 |
| Days Max Temp ≤ 32 °F | 0 | 0 | 0 | 0 | 0 | 0 | 0 | 0 | 0 | 0 | 0 | 0 | 0 |
| Days Min Temp ≤ 32 °F | 11 | 8 | 2 | 0 | 0 | 0 | 0 | 0 | 0 | 0 | 2 | 8 | 31 |
| Days Min Temp ≤ 0 °F | 0 | 0 | 0 | 0 | 0 | 0 | 0 | 0 | 0 | 0 | 0 | 0 | 0 |
| Heating Degree Days | 533 | 415 | 250 | 92 | 10 | 0 | 0 | 0 | 2 | 67 | 229 | 433 | 2031 |
| Cooling Degree Days | 2 | 9 | 31 | 90 | 248 | 438 | 562 | 514 | 357 | 130 | 41 | 9 | 2431 |
| Total Precipitation (") | 3.92 | 2.99 | 4.32 | 2.56 | 4.08 | 5.84 | 6.34 | 7.19 | 4.89 | 3.10 | 2.66 | 3.24 | 51.13 |
| Days ≥ 0.1" Precip | 7 | 6 | 7 | 5 | 6 | 7 | 8 | 9 | 6 | 4 | 4 | 6 | 75 |
| Total Snowfall (") | 0.1 | 0.4 | 0.1 | 0.0 | 0.0 | 0.0 | 0.0 | 0.0 | 0.0 | 0.0 | 0.0 | 0.4 | 1.0 |
| Days ≥ 1" Snow Depth | 0 | 0 | 0 | 0 | 0 | 0 | 0 | 0 | 0 | 0 | 0 | 0 | 0 |

## CHERAW *Chesterfield County* ELEVATION 141 ft LAT/LONG 34° 42 ' N / 79° 54 ' W

| | JAN | FEB | MAR | APR | MAY | JUN | JUL | AUG | SEP | OCT | NOV | DEC | YEAR |
|---|---|---|---|---|---|---|---|---|---|---|---|---|---|
| Maximum Temp °F | 51.8 | 56.0 | 64.6 | 74.4 | 80.9 | 87.4 | 90.6 | 88.2 | 83.4 | 73.7 | 64.8 | 55.8 | 72.6 |
| Minimum Temp °F | 28.8 | 31.0 | 38.1 | 46.4 | 55.3 | 63.8 | 68.5 | 67.5 | 61.4 | 48.3 | 39.0 | 32.0 | 48.3 |
| Mean Temp °F | 40.1 | 43.6 | 51.3 | 60.4 | 68.1 | 75.7 | 79.6 | 77.9 | 72.4 | 61.0 | 51.8 | 44.0 | 60.5 |
| Days Max Temp ≥ 90 °F | 0 | 0 | 0 | 1 | 3 | 12 | 18 | 14 | 5 | 0 | 0 | 0 | 53 |
| Days Max Temp ≤ 32 °F | 1 | 0 | 0 | 0 | 0 | 0 | 0 | 0 | 0 | 0 | 0 | 0 | 1 |
| Days Min Temp ≤ 32 °F | *20* | *17* | 9 | 1 | 0 | 0 | 0 | 0 | 0 | 1 | 9 | *17* | 74 |
| Days Min Temp ≤ 0 °F | 0 | 0 | 0 | 0 | 0 | 0 | 0 | 0 | 0 | 0 | 0 | 0 | 0 |
| Heating Degree Days | 766 | 600 | 427 | 175 | 41 | 3 | 0 | 0 | 13 | 163 | 397 | 646 | 3231 |
| Cooling Degree Days | 0 | 2 | 11 | 45 | 156 | 363 | 497 | 427 | 254 | 57 | 8 | 2 | 1822 |
| Total Precipitation (") | 4.09 | 3.67 | 4.36 | 2.90 | 3.70 | 4.77 | 5.42 | 5.43 | 3.70 | 3.60 | 2.75 | 3.34 | 47.73 |
| Days ≥ 0.1" Precip | 7 | 6 | 7 | 5 | 7 | 7 | 8 | 7 | 5 | 4 | 5 | 6 | 74 |
| Total Snowfall (") | 1.0 | 1.4 | 0.5 | 0.0 | 0.0 | 0.0 | 0.0 | 0.0 | 0.0 | 0.0 | 0.0 | 0.3 | 3.2 |
| Days ≥ 1" Snow Depth | 1 | 1 | 0 | 0 | 0 | 0 | 0 | 0 | 0 | 0 | 0 | 0 | 2 |

## CHESTER 1 NW *Chester County* ELEVATION 561 ft LAT/LONG 34° 42 ' N / 81° 12 ' W

| | JAN | FEB | MAR | APR | MAY | JUN | JUL | AUG | SEP | OCT | NOV | DEC | YEAR |
|---|---|---|---|---|---|---|---|---|---|---|---|---|---|
| Maximum Temp °F | 52.1 | 57.0 | 65.7 | 74.8 | 80.4 | 87.0 | 90.3 | 88.4 | 83.5 | 74.2 | 65.1 | 55.8 | 72.9 |
| Minimum Temp °F | 30.3 | 32.8 | 40.0 | 47.5 | 56.0 | 64.2 | 68.5 | 67.7 | 61.5 | 48.8 | 40.3 | 33.5 | 49.3 |
| Mean Temp °F | 41.3 | 44.9 | 52.9 | 61.2 | 68.3 | 75.6 | 79.4 | 78.1 | 72.5 | 61.5 | 52.7 | 44.7 | 61.1 |
| Days Max Temp ≥ 90 °F | 0 | 0 | 0 | 0 | 2 | 10 | 18 | 14 | 6 | 0 | 0 | 0 | 50 |
| Days Max Temp ≤ 32 °F | 1 | 0 | 0 | 0 | 0 | 0 | 0 | 0 | 0 | 0 | 0 | 0 | 1 |
| Days Min Temp ≤ 32 °F | 18 | 15 | 8 | 2 | 0 | 0 | 0 | 0 | 0 | 2 | 8 | 16 | 69 |
| Days Min Temp ≤ 0 °F | 0 | 0 | 0 | 0 | 0 | 0 | 0 | 0 | 0 | 0 | 0 | 0 | 0 |
| Heating Degree Days | 731 | 562 | 379 | 158 | 40 | 3 | 0 | 0 | 13 | 156 | 369 | 625 | 3036 |
| Cooling Degree Days | 0 | 1 | 9 | 42 | 135 | 343 | 469 | 405 | 231 | 52 | 7 | 2 | 1696 |
| Total Precipitation (") | 4.35 | 3.72 | 4.98 | 3.03 | 3.66 | 4.38 | 4.14 | 5.23 | 3.91 | 3.34 | 3.44 | 3.55 | 47.73 |
| Days ≥ 0.1" Precip | 7 | 6 | 8 | 5 | 6 | 7 | 7 | 6 | 5 | 4 | 5 | 6 | 72 |
| Total Snowfall (") | 1.3 | 1.7 | 0.6 | 0.0 | 0.0 | 0.0 | 0.0 | 0.0 | 0.0 | 0.0 | 0.0 | 0.3 | 3.9 |
| Days ≥ 1" Snow Depth | 1 | 1 | 0 | 0 | 0 | 0 | 0 | 0 | 0 | 0 | 0 | 0 | 2 |

## CLARK HILL 1 W *McCormick County* ELEVATION 381 ft LAT/LONG 33° 40 ' N / 82° 11 ' W

| | JAN | FEB | MAR | APR | MAY | JUN | JUL | AUG | SEP | OCT | NOV | DEC | YEAR |
|---|---|---|---|---|---|---|---|---|---|---|---|---|---|
| Maximum Temp °F | 53.8 | 58.7 | 67.1 | 75.9 | 82.8 | 89.0 | 92.2 | 90.8 | 86.0 | 76.5 | 67.1 | 57.7 | 74.8 |
| Minimum Temp °F | 30.8 | 33.3 | 40.6 | 48.7 | 56.7 | 64.3 | 67.9 | 66.9 | 61.1 | 48.8 | 40.6 | 33.5 | 49.4 |
| Mean Temp °F | 42.3 | 46.0 | 53.9 | 62.3 | 69.8 | 76.7 | 80.1 | 78.8 | 73.6 | 62.7 | 53.9 | 45.6 | 62.1 |
| Days Max Temp ≥ 90 °F | 0 | 0 | 0 | 1 | 5 | 14 | 22 | 20 | 10 | 1 | 0 | 0 | 73 |
| Days Max Temp ≤ 32 °F | 1 | 0 | 0 | 0 | 0 | 0 | 0 | 0 | 0 | 0 | 0 | 0 | 1 |
| Days Min Temp ≤ 32 °F | 18 | 14 | 7 | 1 | 0 | 0 | 0 | 0 | 0 | 1 | 7 | 16 | 64 |
| Days Min Temp ≤ 0 °F | 0 | 0 | 0 | 0 | 0 | 0 | 0 | 0 | 0 | 0 | 0 | 0 | 0 |
| Heating Degree Days | 698 | 531 | 348 | 134 | 24 | 1 | 0 | 0 | 9 | 130 | 337 | 598 | 2810 |
| Cooling Degree Days | 0 | 2 | 13 | 55 | 186 | 387 | 509 | 450 | 276 | 69 | 11 | 2 | 1960 |
| Total Precipitation (") | 4.57 | 3.90 | 5.02 | 3.24 | 3.60 | 3.63 | 4.67 | 4.18 | 2.78 | 3.61 | 3.04 | 3.83 | 46.07 |
| Days ≥ 0.1" Precip | 7 | 6 | 7 | 5 | 6 | 6 | 7 | 7 | 5 | 4 | 5 | 6 | 71 |
| Total Snowfall (") | na | na | na | na | na | na | na | na | na | na | na | na | na |
| Days ≥ 1" Snow Depth | na | na | na | na | *0* | na | *0* | *0* | na | *0* | *0* | *0* | na |

## CLEMSON UNIV *Pickens County* ELEVATION 830 ft LAT/LONG 34° 40 ' N / 82° 50 ' W

| | JAN | FEB | MAR | APR | MAY | JUN | JUL | AUG | SEP | OCT | NOV | DEC | YEAR |
|---|---|---|---|---|---|---|---|---|---|---|---|---|---|
| Maximum Temp °F | 51.6 | 56.1 | 64.2 | 73.1 | 79.4 | 86.0 | 89.5 | 87.7 | 82.5 | 73.2 | 64.1 | 55.1 | 71.9 |
| Minimum Temp °F | 29.3 | 31.8 | 39.0 | 47.0 | 55.5 | 63.4 | 67.6 | 66.8 | 60.7 | 48.3 | 39.6 | 32.7 | 48.5 |
| Mean Temp °F | 40.5 | 43.9 | 51.6 | 60.1 | 67.5 | 74.7 | 78.6 | 77.3 | 71.6 | 60.8 | 51.9 | 43.9 | 60.2 |
| Days Max Temp ≥ 90 °F | 0 | 0 | 0 | 0 | 1 | 9 | 16 | 12 | 4 | 0 | 0 | 0 | 42 |
| Days Max Temp ≤ 32 °F | 1 | 0 | 0 | 0 | 0 | 0 | 0 | 0 | 0 | 0 | 0 | 0 | 1 |
| Days Min Temp ≤ 32 °F | 19 | 16 | 9 | 1 | 0 | 0 | 0 | 0 | 0 | 1 | 8 | 17 | 71 |
| Days Min Temp ≤ 0 °F | 0 | 0 | 0 | 0 | 0 | 0 | 0 | 0 | 0 | 0 | 0 | 0 | 0 |
| Heating Degree Days | 752 | 589 | 413 | 176 | 43 | 2 | 0 | 0 | 15 | 164 | 391 | 648 | 3193 |
| Cooling Degree Days | 0 | 0 | 5 | 33 | 125 | 320 | 445 | 395 | 216 | 41 | 3 | 1 | 1584 |
| Total Precipitation (") | 5.20 | 4.75 | 5.78 | 3.64 | 4.42 | 3.88 | 4.52 | 4.77 | 3.82 | 4.17 | 4.16 | 4.55 | 53.66 |
| Days ≥ 0.1" Precip | 8 | 7 | 8 | 6 | 8 | 6 | 8 | 7 | 6 | 5 | 7 | 8 | 84 |
| Total Snowfall (") | 1.6 | 1.5 | 0.8 | 0.0 | 0.0 | 0.0 | 0.0 | 0.0 | 0.0 | 0.0 | 0.1 | 0.3 | 4.3 |
| Days ≥ 1" Snow Depth | 1 | 1 | 0 | 0 | 0 | 0 | 0 | 0 | 0 | 0 | 0 | 0 | 2 |

## COLUMBIA METRO AP *Lexington County* ELEVATION 220 ft LAT/LONG 33° 57 ' N / 81° 7 ' W

| | JAN | FEB | MAR | APR | MAY | JUN | JUL | AUG | SEP | OCT | NOV | DEC | YEAR |
|---|---|---|---|---|---|---|---|---|---|---|---|---|---|
| Maximum Temp °F | 55.4 | 59.6 | 68.2 | 76.9 | 83.2 | 89.1 | 92.2 | 90.2 | 85.3 | 76.3 | 67.5 | 59.1 | 75.3 |
| Minimum Temp °F | 32.7 | 34.7 | 42.0 | 49.6 | 58.3 | 66.2 | 70.5 | 69.4 | 63.5 | 50.5 | 41.3 | 35.4 | 51.2 |
| Mean Temp °F | 44.1 | 47.2 | 55.1 | 63.3 | 70.8 | 77.7 | 81.4 | 79.8 | 74.4 | 63.4 | 54.5 | 47.3 | 63.2 |
| Days Max Temp ≥ 90 °F | 0 | 0 | 0 | 2 | 6 | 15 | 22 | 18 | 9 | 1 | 0 | 0 | 73 |
| Days Max Temp ≤ 32 °F | 1 | 0 | 0 | 0 | 0 | 0 | 0 | 0 | 0 | 0 | 0 | 0 | 1 |
| Days Min Temp ≤ 32 °F | 16 | 13 | 6 | 1 | 0 | 0 | 0 | 0 | 0 | 1 | 8 | 14 | 59 |
| Days Min Temp ≤ 0 °F | 0 | 0 | 0 | 0 | 0 | 0 | 0 | 0 | 0 | 0 | 0 | 0 | 0 |
| Heating Degree Days | 643 | 500 | 319 | 122 | 22 | 1 | 0 | 0 | 7 | 119 | 324 | 548 | 2605 |
| Cooling Degree Days | 1 | 5 | 19 | 70 | 207 | 418 | 545 | 474 | 295 | 79 | 15 | 4 | 2132 |
| Total Precipitation (") | 4.39 | 3.79 | 4.88 | 3.06 | 3.78 | 5.04 | 5.64 | 5.93 | 3.63 | 3.10 | 2.88 | 3.55 | 49.67 |
| Days ≥ 0.1" Precip | 7 | 6 | 7 | 5 | 6 | 7 | 8 | 8 | 5 | 4 | 4 | 6 | 73 |
| Total Snowfall (") | 0.6 | 1.2 | 0.3 | 0.0 | 0.0 | 0.0 | 0.0 | 0.0 | 0.0 | 0.0 | 0.0 | 0.1 | 2.2 |
| Days ≥ 1" Snow Depth | 1 | 0 | 0 | 0 | 0 | 0 | 0 | 0 | 0 | 0 | 0 | 0 | 1 |

## COLUMBIA UNIV OF SC *Richland County* ELEVATION 322 ft LAT/LONG 34° 0 ' N / 81° 1 ' W

| | JAN | FEB | MAR | APR | MAY | JUN | JUL | AUG | SEP | OCT | NOV | DEC | YEAR |
|---|---|---|---|---|---|---|---|---|---|---|---|---|---|
| Maximum Temp °F | 56.4 | 61.2 | 69.6 | 78.5 | 84.6 | 90.4 | 93.2 | 90.9 | 86.4 | 77.1 | 68.1 | 60.2 | 76.4 |
| Minimum Temp °F | 35.7 | 38.4 | 45.5 | 52.7 | 60.7 | 67.6 | 71.3 | 70.4 | 65.2 | 53.8 | 45.3 | 39.2 | 53.8 |
| Mean Temp °F | 46.1 | 49.8 | 57.6 | 65.6 | 72.7 | 79.0 | 82.3 | 80.7 | 75.8 | 65.5 | 56.7 | 49.7 | 65.1 |
| Days Max Temp ≥ 90 °F | 0 | 0 | 0 | 2 | 7 | 17 | 24 | 21 | 10 | 1 | 0 | 0 | 82 |
| Days Max Temp ≤ 32 °F | 0 | 0 | 0 | 0 | 0 | 0 | 0 | 0 | 0 | 0 | 0 | 0 | 0 |
| Days Min Temp ≤ 32 °F | 13 | 9 | 3 | 0 | 0 | 0 | 0 | 0 | 0 | 0 | 3 | 9 | 37 |
| Days Min Temp ≤ 0 °F | 0 | 0 | 0 | 0 | 0 | 0 | 0 | 0 | 0 | 0 | 0 | 0 | 0 |
| Heating Degree Days | 581 | 429 | 252 | 79 | 10 | 0 | 0 | 0 | 3 | 78 | 263 | 473 | 2168 |
| Cooling Degree Days | 1 | 7 | 31 | 106 | 265 | 464 | 577 | 508 | 348 | 114 | 23 | 4 | 2448 |
| Total Precipitation (") | 4.15 | 3.68 | 4.67 | 2.83 | 3.56 | 5.11 | 5.53 | 4.69 | 3.41 | 3.05 | 3.02 | 3.41 | 47.11 |
| Days ≥ 0.1" Precip | 7 | 6 | 7 | 5 | 7 | 7 | 8 | 7 | 5 | 4 | 5 | 6 | 74 |
| Total Snowfall (") | 0.2 | 1.1 | 0.1 | 0.0 | 0.0 | 0.0 | 0.0 | 0.0 | 0.0 | 0.0 | 0.0 | 0.0 | 1.4 |
| Days ≥ 1" Snow Depth | 0 | 0 | 0 | 0 | 0 | 0 | 0 | 0 | 0 | 0 | 0 | 0 | 0 |

## CONWAY *Horry County* ELEVATION 30 ft LAT/LONG 33° 50 ' N / 79° 3 ' W

| | JAN | FEB | MAR | APR | MAY | JUN | JUL | AUG | SEP | OCT | NOV | DEC | YEAR |
|---|---|---|---|---|---|---|---|---|---|---|---|---|---|
| Maximum Temp °F | 56.4 | 60.0 | 67.7 | 75.7 | 82.2 | 87.6 | 90.8 | 89.0 | 84.8 | 76.5 | 68.8 | 60.1 | 75.0 |
| Minimum Temp °F | 33.7 | 35.9 | 43.2 | 50.8 | 59.3 | 66.9 | 71.1 | 70.2 | 64.8 | 52.7 | 43.8 | 36.7 | 52.4 |
| Mean Temp °F | 45.1 | 47.9 | 55.4 | 63.3 | 70.7 | 77.3 | 81.0 | 79.7 | 74.8 | 64.6 | 56.4 | 48.4 | 63.7 |
| Days Max Temp ≥ 90 °F | 0 | 0 | 0 | 1 | 4 | 12 | 20 | 16 | 7 | 0 | 0 | 0 | 60 |
| Days Max Temp ≤ 32 °F | 0 | 0 | 0 | 0 | 0 | 0 | 0 | 0 | 0 | 0 | 0 | 0 | 0 |
| Days Min Temp ≤ 32 °F | 15 | 12 | 4 | 0 | 0 | 0 | 0 | 0 | 0 | 0 | 4 | 12 | 47 |
| Days Min Temp ≤ 0 °F | 0 | 0 | 0 | 0 | 0 | 0 | 0 | 0 | 0 | 0 | 0 | 0 | 0 |
| Heating Degree Days | 612 | 479 | 307 | 117 | 20 | 2 | 0 | 0 | 3 | 96 | 274 | 511 | 2421 |
| Cooling Degree Days | 0 | 5 | 17 | 66 | 214 | 409 | 533 | 478 | 314 | 102 | 24 | 6 | 2168 |
| Total Precipitation (") | 4.34 | 3.37 | 4.24 | 2.91 | 4.57 | 4.64 | 6.52 | 6.42 | 4.88 | 3.00 | 2.59 | 3.59 | 51.07 |
| Days ≥ 0.1" Precip | 8 | 6 | 7 | 5 | 7 | 7 | 10 | 9 | 6 | 4 | 4 | 6 | 79 |
| Total Snowfall (") | 0.2 | 0.8 | 0.0 | 0.0 | 0.0 | 0.0 | 0.0 | 0.0 | 0.0 | 0.0 | 0.0 | 0.5 | 1.5 |
| Days ≥ 1" Snow Depth | 0 | 0 | 0 | 0 | 0 | 0 | 0 | 0 | 0 | 0 | 0 | 0 | 0 |

## DARLINGTON *Darlington County* ELEVATION 180 ft LAT/LONG 34° 17' N / 79° 52' W

| | JAN | FEB | MAR | APR | MAY | JUN | JUL | AUG | SEP | OCT | NOV | DEC | YEAR |
|---|---|---|---|---|---|---|---|---|---|---|---|---|---|
| Maximum Temp °F | 55.6 | 60.2 | 68.6 | 77.3 | 83.5 | 89.1 | 91.8 | 89.7 | 85.4 | 76.6 | 67.8 | 59.3 | 75.4 |
| Minimum Temp °F | 33.2 | 35.3 | 42.5 | 49.8 | 58.1 | 65.8 | 69.9 | 68.7 | 62.9 | 50.8 | 41.9 | 36.2 | 51.3 |
| Mean Temp °F | 44.4 | 47.8 | 55.6 | 63.6 | 70.8 | 77.5 | 80.9 | 79.1 | 74.1 | 63.7 | 54.9 | 47.8 | 63.4 |
| Days Max Temp ≥ 90 °F | 0 | 0 | 0 | 1 | 5 | 15 | 22 | 18 | 8 | 1 | 0 | 0 | 70 |
| Days Max Temp ≤ 32 °F | 0 | 0 | 0 | 0 | 0 | 0 | 0 | 0 | 0 | 0 | 0 | 0 | 0 |
| Days Min Temp ≤ 32 °F | 16 | 12 | 6 | 1 | 0 | 0 | 0 | 0 | 0 | 1 | 7 | 12 | 55 |
| Days Min Temp ≤ 0 °F | 0 | 0 | 0 | 0 | 0 | 0 | 0 | 0 | 0 | 0 | 0 | 0 | 0 |
| Heating Degree Days | 632 | 483 | 304 | 111 | 19 | 1 | 0 | 0 | 8 | 110 | 311 | 530 | 2509 |
| Cooling Degree Days | 1 | 5 | 22 | 76 | 215 | 415 | 529 | 451 | 291 | 76 | 15 | 4 | 2100 |
| Total Precipitation (") | 4.12 | 3.49 | 4.53 | 2.66 | 3.65 | 4.51 | 5.01 | 5.65 | 3.39 | 3.09 | 2.68 | 3.67 | 46.45 |
| Days ≥ 0.1" Precip | 8 | 6 | 8 | 5 | 7 | 7 | 8 | 8 | 5 | 4 | 5 | 7 | 78 |
| Total Snowfall (") | 0.3 | 0.9 | 0.2 | 0.0 | 0.0 | 0.0 | 0.0 | 0.0 | 0.0 | 0.0 | 0.0 | 0.0 | 1.4 |
| Days ≥ 1" Snow Depth | 0 | 0 | 0 | 0 | 0 | 0 | 0 | 0 | 0 | 0 | 0 | 0 | 0 |

## DILLON *Dillon County* ELEVATION 112 ft LAT/LONG 34° 25' N / 79° 22' W

| | JAN | FEB | MAR | APR | MAY | JUN | JUL | AUG | SEP | OCT | NOV | DEC | YEAR |
|---|---|---|---|---|---|---|---|---|---|---|---|---|---|
| Maximum Temp °F | 53.5 | 57.5 | 66.0 | 75.0 | 82.0 | 87.9 | 91.1 | 89.5 | 84.7 | 75.3 | 66.7 | 57.7 | 73.9 |
| Minimum Temp °F | 30.6 | 32.6 | 39.6 | 47.5 | 56.2 | 64.3 | 69.1 | 67.9 | 61.5 | 48.5 | 39.8 | 33.3 | 49.2 |
| Mean Temp °F | 42.1 | 45.2 | 52.8 | 61.2 | 69.1 | 76.1 | 80.1 | 78.7 | 73.1 | 62.0 | 53.3 | 45.5 | 61.6 |
| Days Max Temp ≥ 90 °F | 0 | 0 | 0 | 1 | 4 | 13 | 20 | 17 | 8 | 1 | 0 | 0 | 64 |
| Days Max Temp ≤ 32 °F | 1 | 0 | 0 | 0 | 0 | 0 | 0 | 0 | 0 | 0 | 0 | 0 | 1 |
| Days Min Temp ≤ 32 °F | 18 | 15 | 8 | 1 | 0 | 0 | 0 | 0 | 0 | 2 | 9 | 16 | 69 |
| Days Min Temp ≤ 0 °F | 0 | 0 | 0 | 0 | 0 | 0 | 0 | 0 | 0 | 0 | 0 | 0 | 0 |
| Heating Degree Days | 704 | 558 | 383 | 164 | 37 | 3 | 0 | 0 | 12 | 149 | 359 | 601 | 2970 |
| Cooling Degree Days | 0 | 3 | 12 | 52 | 184 | 376 | 516 | 441 | 268 | 67 | 14 | 4 | 1937 |
| Total Precipitation (") | 3.71 | 3.19 | 4.48 | 2.93 | 3.34 | 4.45 | 5.23 | 5.58 | 3.23 | 2.89 | 2.80 | 3.45 | 45.28 |
| Days ≥ 0.1" Precip | 8 | 6 | 7 | 5 | 7 | 6 | 8 | 8 | 5 | 4 | 5 | 6 | 75 |
| Total Snowfall (") | 0.4 | *0.5* | 0.5 | 0.0 | 0.0 | 0.0 | 0.0 | 0.0 | 0.0 | 0.0 | 0.0 | 0.4 | 1.8 |
| Days ≥ 1" Snow Depth | 0 | *0* | 0 | 0 | 0 | 0 | 0 | 0 | 0 | 0 | 0 | 0 | 0 |

## EDISTO ISLAND 3 SW *Colleton County* ELEVATION 10 ft LAT/LONG 32° 29' N / 80° 20' W

| | JAN | FEB | MAR | APR | MAY | JUN | JUL | AUG | SEP | OCT | NOV | DEC | YEAR |
|---|---|---|---|---|---|---|---|---|---|---|---|---|---|
| Maximum Temp °F | *55.8* | *58.8* | *65.2* | *73.4* | *79.9* | *86.2* | *89.4* | *88.1* | *84.5* | *76.7* | *68.2* | *59.9* | 73.8 |
| Minimum Temp °F | *36.9* | *39.2* | *46.7* | *54.6* | *63.1* | *70.6* | *74.1* | *72.9* | *68.5* | *57.4* | *49.0* | *40.8* | 56.2 |
| Mean Temp °F | *46.4* | *49.0* | *56.0* | *64.0* | *71.6* | *78.4* | *81.8* | *80.6* | *76.5* | *67.1* | *58.6* | *50.6* | 65.1 |
| Days Max Temp ≥ 90 °F | *0* | *0* | *0* | *0* | *1* | *6* | *14* | *10* | *4* | *0* | *0* | *0* | 35 |
| Days Max Temp ≤ 32 °F | *0* | *0* | *0* | *0* | *0* | *0* | *0* | *0* | *0* | *0* | *0* | *0* | 0 |
| Days Min Temp ≤ 32 °F | *10* | *7* | *1* | *0* | *0* | *0* | *0* | *0* | *0* | *0* | *1* | *6* | 25 |
| Days Min Temp ≤ 0 °F | *0* | *0* | *0* | *0* | *0* | *0* | *0* | *0* | *0* | *0* | *0* | *0* | 0 |
| Heating Degree Days | *570* | *447* | *281* | *91* | *10* | *0* | *0* | *0* | *1* | *58* | *210* | *444* | 2112 |
| Cooling Degree Days | *0* | 1 | 9 | *65* | *223* | 423 | 538 | *498* | 356 | 138 | 31 | 4 | 2286 |
| Total Precipitation (") | *3.70* | *3.14* | *4.41* | *2.69* | *3.48* | *3.95* | *4.64* | *7.44* | *5.01* | *3.89* | *3.36* | *3.64* | 49.35 |
| Days ≥ 0.1" Precip | *7* | *6* | *6* | *4* | *5* | *6* | *6* | *8* | *6* | *5* | *5* | *6* | 70 |
| Total Snowfall (") | *0.1* | *0.0* | *0.0* | *0.0* | *0.0* | *0.0* | *0.0* | *0.0* | *0.0* | *0.0* | *0.0* | *0.0* | 0.1 |
| Days ≥ 1" Snow Depth | *0* | *0* | *0* | *0* | *0* | *0* | *0* | *0* | *0* | *0* | *0* | *0* | 0 |

## FLORENCE 8 NE *Darlington County* ELEVATION 120 ft LAT/LONG 34° 18' N / 79° 44' W

| | JAN | FEB | MAR | APR | MAY | JUN | JUL | AUG | SEP | OCT | NOV | DEC | YEAR |
|---|---|---|---|---|---|---|---|---|---|---|---|---|---|
| Maximum Temp °F | 53.6 | 57.5 | 65.7 | 74.9 | 81.6 | 87.6 | 90.7 | 88.7 | 84.3 | 75.2 | 66.9 | 57.8 | 73.7 |
| Minimum Temp °F | 32.3 | 34.6 | 42.4 | 50.3 | 58.9 | 66.4 | 70.5 | 69.0 | 62.9 | 50.4 | 42.1 | 35.2 | 51.3 |
| Mean Temp °F | 43.0 | 46.1 | 54.1 | 62.7 | 70.3 | 77.0 | 80.6 | 78.9 | 73.6 | 62.8 | 54.5 | 46.5 | 62.5 |
| Days Max Temp ≥ 90 °F | 0 | 0 | 0 | 1 | 4 | 12 | 19 | 15 | 7 | 1 | 0 | 0 | 59 |
| Days Max Temp ≤ 32 °F | 1 | 0 | 0 | 0 | 0 | 0 | 0 | 0 | 0 | 0 | 0 | 0 | 1 |
| Days Min Temp ≤ 32 °F | 17 | 13 | 5 | 1 | 0 | 0 | 0 | 0 | 0 | 1 | 6 | 14 | 57 |
| Days Min Temp ≤ 0 °F | 0 | 0 | 0 | 0 | 0 | 0 | 0 | 0 | 0 | 0 | 0 | 0 | 0 |
| Heating Degree Days | 677 | 532 | 347 | 133 | 26 | 2 | 0 | 0 | 9 | 130 | 323 | 570 | 2749 |
| Cooling Degree Days | 0 | 3 | 15 | 66 | 204 | 405 | 526 | 452 | 280 | 78 | 16 | 5 | 2050 |
| Total Precipitation (") | 4.06 | 3.22 | 4.43 | 2.54 | 3.60 | 4.28 | 5.54 | 5.68 | 3.34 | 2.94 | 2.75 | 3.59 | 45.97 |
| Days ≥ 0.1" Precip | 8 | 6 | 7 | 5 | 6 | 7 | 8 | 8 | 5 | 4 | 4 | 6 | 74 |
| Total Snowfall (") | 0.1 | 0.5 | 0.4 | 0.0 | 0.0 | 0.0 | 0.0 | 0.0 | 0.0 | 0.0 | 0.0 | 0.2 | 1.2 |
| Days ≥ 1" Snow Depth | *0* | 0 | 0 | 0 | 0 | 0 | 0 | 0 | 0 | 0 | 0 | 0 | 0 |

## FLORENCE CTY CNTY AP *Florence County* ELEVATION 148 ft LAT/LONG 34° 11' N / 79° 43' W

| | JAN | FEB | MAR | APR | MAY | JUN | JUL | AUG | SEP | OCT | NOV | DEC | YEAR |
|---|---|---|---|---|---|---|---|---|---|---|---|---|---|
| Maximum Temp °F | 54.8 | 58.6 | 67.2 | 76.0 | 82.4 | 88.1 | 90.9 | 88.9 | 84.5 | 75.6 | 67.3 | 58.6 | 74.4 |
| Minimum Temp °F | 34.4 | 36.6 | 43.7 | 50.8 | 59.6 | 67.2 | 71.4 | 70.2 | 64.7 | 52.5 | 43.8 | 37.2 | 52.7 |
| Mean Temp °F | 44.6 | 47.6 | 55.5 | 63.4 | 71.0 | 77.7 | 81.2 | 79.6 | 74.7 | 64.1 | 55.6 | 48.0 | 63.6 |
| Days Max Temp ≥ 90 °F | 0 | 0 | 0 | 1 | 4 | 13 | 19 | 15 | 7 | 0 | 0 | 0 | 59 |
| Days Max Temp ≤ 32 °F | 1 | 0 | 0 | 0 | 0 | 0 | 0 | 0 | 0 | 0 | 0 | 0 | 1 |
| Days Min Temp ≤ 32 °F | 14 | 11 | 4 | 0 | 0 | 0 | 0 | 0 | 0 | 0 | 5 | 11 | 45 |
| Days Min Temp ≤ 0 °F | 0 | 0 | 0 | 0 | 0 | 0 | 0 | 0 | 0 | 0 | 0 | 0 | 0 |
| Heating Degree Days | 628 | 489 | 311 | 117 | 20 | 1 | 0 | 0 | 6 | 107 | 297 | 527 | 2503 |
| Cooling Degree Days | 1 | 6 | 22 | 77 | 218 | 416 | 538 | 469 | 301 | 87 | 20 | 6 | 2161 |
| Total Precipitation (") | 3.71 | 3.07 | 4.09 | 2.60 | 3.49 | 4.27 | 5.50 | 5.51 | 3.27 | 2.73 | 2.65 | 3.46 | 44.35 |
| Days ≥ 0.1" Precip | 7 | 6 | 7 | 5 | 7 | 6 | 8 | 7 | 5 | 4 | 5 | 6 | 73 |
| Total Snowfall (") | 0.8 | 1.4 | 0.4 | 0.0 | 0.0 | 0.0 | 0.0 | 0.0 | 0.0 | 0.0 | 0.0 | 0.3 | 2.9 |
| Days ≥ 1" Snow Depth | 1 | 1 | 0 | 0 | 0 | 0 | 0 | 0 | 0 | 0 | 0 | 0 | 2 |

## GEORGETOWN 2 E *Georgetown County* ELEVATION 10 ft LAT/LONG 33° 23' N / 79° 17' W

| | JAN | FEB | MAR | APR | MAY | JUN | JUL | AUG | SEP | OCT | NOV | DEC | YEAR |
|---|---|---|---|---|---|---|---|---|---|---|---|---|---|
| Maximum Temp °F | 58.7 | 61.8 | 69.2 | 76.0 | 82.4 | 87.3 | 90.4 | 88.9 | 85.1 | 77.2 | 70.0 | 63.1 | 75.8 |
| Minimum Temp °F | 36.6 | 37.9 | 45.0 | 51.6 | 60.0 | 67.2 | 71.2 | 70.5 | 66.1 | 55.1 | 46.6 | 39.5 | 53.9 |
| Mean Temp °F | 47.7 | 49.9 | 57.1 | 63.8 | 71.3 | 77.2 | 80.8 | 79.8 | 75.7 | 66.2 | 58.3 | 51.7 | 65.0 |
| Days Max Temp ≥ 90 °F | 0 | 0 | 0 | 1 | 2 | 10 | 18 | 14 | 5 | 0 | 0 | 0 | 50 |
| Days Max Temp ≤ 32 °F | 0 | 0 | 0 | 0 | 0 | 0 | 0 | 0 | 0 | 0 | 0 | 0 | 0 |
| Days Min Temp ≤ 32 °F | 12 | 9 | 3 | 0 | 0 | 0 | 0 | 0 | 0 | 0 | 2 | 8 | 34 |
| Days Min Temp ≤ 0 °F | 0 | 0 | 0 | 0 | 0 | 0 | 0 | 0 | 0 | 0 | 0 | 0 | 0 |
| Heating Degree Days | 532 | 426 | 259 | 98 | 13 | 1 | 0 | 0 | 2 | 67 | 222 | 413 | 2033 |
| Cooling Degree Days | 2 | 6 | 22 | 69 | 214 | 402 | 526 | 470 | 334 | 115 | 29 | 7 | 2196 |
| Total Precipitation (") | 4.43 | 3.38 | 4.22 | 2.45 | 4.35 | 5.66 | 6.18 | 7.09 | 5.28 | 4.10 | 2.97 | 3.73 | 53.84 |
| Days ≥ 0.1" Precip | 7 | 5 | 6 | 5 | 6 | 7 | 8 | 9 | 6 | 4 | 5 | 6 | 74 |
| Total Snowfall (") | 0.2 | 0.1 | 0.1 | 0.0 | 0.0 | 0.0 | 0.0 | 0.0 | 0.0 | 0.0 | 0.0 | 0.0 | 0.4 |
| Days ≥ 1" Snow Depth | 0 | 0 | 0 | 0 | 0 | 0 | 0 | 0 | 0 | 0 | 0 | 0 | 0 |

## GREENWOOD 3 SW *Greenwood County* ELEVATION 615 ft LAT/LONG 34° 10' N / 82° 12' W

| | JAN | FEB | MAR | APR | MAY | JUN | JUL | AUG | SEP | OCT | NOV | DEC | YEAR |
|---|---|---|---|---|---|---|---|---|---|---|---|---|---|
| Maximum Temp °F | 51.7 | 56.4 | 65.1 | 74.1 | 80.7 | 87.3 | 90.6 | 88.8 | 83.3 | 73.4 | 64.4 | 55.2 | 72.6 |
| Minimum Temp °F | 29.1 | 30.9 | 37.7 | 45.7 | 54.3 | 62.8 | 67.1 | 66.2 | 59.8 | 46.9 | 38.2 | 31.9 | 47.6 |
| Mean Temp °F | 40.4 | 43.7 | 51.4 | 59.9 | 67.5 | 75.0 | 78.9 | 77.5 | 71.6 | 60.2 | 51.3 | 43.6 | 60.1 |
| Days Max Temp ≥ 90 °F | 0 | 0 | 0 | 0 | 2 | 11 | 19 | 15 | 5 | 0 | 0 | 0 | 52 |
| Days Max Temp ≤ 32 °F | 1 | 0 | 0 | 0 | 0 | 0 | 0 | 0 | 0 | 0 | 0 | 0 | 1 |
| Days Min Temp ≤ 32 °F | 20 | 18 | 11 | 2 | 0 | 0 | 0 | 0 | 0 | 2 | 10 | 18 | 81 |
| Days Min Temp ≤ 0 °F | 0 | 0 | 0 | 0 | 0 | 0 | 0 | 0 | 0 | 0 | 0 | 0 | 0 |
| Heating Degree Days | 755 | 596 | 420 | 185 | 46 | 3 | 0 | 0 | 17 | 181 | 409 | 658 | 3270 |
| Cooling Degree Days | 0 | 0 | 5 | 33 | 127 | 329 | 462 | 406 | 227 | 44 | 5 | 1 | 1639 |
| Total Precipitation (") | 4.72 | 4.07 | 4.99 | 3.24 | 3.82 | 3.63 | 4.30 | 3.43 | 3.20 | 3.48 | 3.48 | 3.82 | 46.18 |
| Days ≥ 0.1" Precip | 8 | 7 | 8 | 6 | 6 | 6 | 7 | 6 | 5 | 4 | 5 | 6 | 74 |
| Total Snowfall (") | *0.9* | 1.0 | 0.5 | 0.0 | 0.0 | 0.0 | 0.0 | 0.0 | 0.0 | 0.0 | 0.0 | 0.1 | 2.5 |
| Days ≥ 1" Snow Depth | *1* | 1 | 0 | 0 | 0 | 0 | 0 | 0 | 0 | 0 | 0 | 0 | 2 |

## GREER GREENV'L-SPART *Spartanburg County* ELEVATION 957 ft LAT/LONG 34° 54' N / 82° 13' W

| | JAN | FEB | MAR | APR | MAY | JUN | JUL | AUG | SEP | OCT | NOV | DEC | YEAR |
|---|---|---|---|---|---|---|---|---|---|---|---|---|---|
| Maximum Temp °F | 50.2 | 54.4 | 63.3 | 72.2 | 78.9 | 85.6 | 88.6 | 86.7 | 81.1 | 71.7 | 62.2 | 53.7 | 70.7 |
| Minimum Temp °F | 30.2 | 32.6 | 39.7 | 47.4 | 56.1 | 64.2 | 68.4 | 67.2 | 61.1 | 48.8 | 40.0 | 33.8 | 49.1 |
| Mean Temp °F | 40.3 | 43.5 | 51.5 | 59.9 | 67.5 | 75.0 | 78.5 | 77.0 | 71.1 | 60.3 | 51.1 | 43.7 | 60.0 |
| Days Max Temp ≥ 90 °F | 0 | 0 | 0 | 0 | 1 | 8 | 14 | 9 | 3 | 0 | 0 | 0 | 35 |
| Days Max Temp ≤ 32 °F | 1 | 0 | 0 | 0 | 0 | 0 | 0 | 0 | 0 | 0 | 0 | 0 | 1 |
| Days Min Temp ≤ 32 °F | 19 | 15 | 8 | 1 | 0 | 0 | 0 | 0 | 0 | 1 | 7 | 15 | 66 |
| Days Min Temp ≤ 0 °F | 0 | 0 | 0 | 0 | 0 | 0 | 0 | 0 | 0 | 0 | 0 | 0 | 0 |
| Heating Degree Days | 761 | 600 | 417 | 182 | 46 | 2 | 0 | 0 | 18 | 177 | 413 | 653 | 3269 |
| Cooling Degree Days | 0 | 0 | 5 | 32 | 130 | 335 | 452 | 385 | 202 | 40 | 3 | 1 | 1585 |
| Total Precipitation (") | 4.07 | 4.19 | 5.34 | 3.39 | 4.59 | 4.31 | 4.61 | 4.03 | 3.93 | 3.93 | 3.63 | 3.94 | 49.96 |
| Days ≥ 0.1" Precip | 7 | 7 | 8 | 6 | 7 | 6 | 8 | 7 | 6 | 5 | 6 | 6 | 79 |
| Total Snowfall (") | 2.6 | 1.9 | 1.3 | 0.0 | 0.0 | 0.0 | 0.0 | 0.0 | 0.0 | 0.0 | 0.1 | 0.5 | 6.4 |
| Days ≥ 1" Snow Depth | 2 | 1 | 0 | 0 | 0 | 0 | 0 | 0 | 0 | 0 | 0 | 0 | 3 |

## HAMPTON *Hampton County* ELEVATION 89 ft LAT/LONG 32° 52' N / 81° 7' W

| | JAN | FEB | MAR | APR | MAY | JUN | JUL | AUG | SEP | OCT | NOV | DEC | YEAR |
|---|---|---|---|---|---|---|---|---|---|---|---|---|---|
| Maximum Temp °F | 59.4 | 63.5 | 71.6 | 79.0 | 84.6 | 89.8 | 92.3 | 90.5 | 86.3 | 78.2 | 70.1 | 62.7 | 77.3 |
| Minimum Temp °F | 36.2 | 38.5 | 45.4 | 51.7 | 59.7 | 66.7 | 70.4 | 69.5 | 64.7 | 53.4 | 44.7 | 38.8 | 53.3 |
| Mean Temp °F | 47.8 | 51.0 | 58.5 | 65.4 | 72.2 | 78.3 | 81.4 | 80.1 | 75.5 | 65.8 | 57.5 | 50.8 | 65.4 |
| Days Max Temp ≥ 90 °F | 0 | 0 | 0 | 1 | 6 | 17 | 24 | 20 | 10 | 1 | 0 | 0 | 79 |
| Days Max Temp ≤ 32 °F | 0 | 0 | 0 | 0 | 0 | 0 | 0 | 0 | 0 | 0 | 0 | 0 | 0 |
| Days Min Temp ≤ 32 °F | 12 | 9 | 3 | 0 | 0 | 0 | 0 | 0 | 0 | 0 | 5 | 9 | 38 |
| Days Min Temp ≤ 0 °F | 0 | 0 | 0 | 0 | 0 | 0 | 0 | 0 | 0 | 0 | 0 | 0 | 0 |
| Heating Degree Days | 529 | 395 | 225 | 75 | 8 | 0 | 0 | 0 | 2 | 75 | 246 | 441 | 1996 |
| Cooling Degree Days | 2 | 10 | 32 | 90 | 246 | 442 | 546 | 495 | 339 | 120 | 35 | 8 | 2365 |
| Total Precipitation (") | 4.16 | 3.38 | 4.43 | 3.05 | 3.93 | 6.02 | 5.27 | 6.07 | 3.68 | 2.84 | 2.60 | 3.22 | 48.65 |
| Days ≥ 0.1" Precip | 7 | 6 | 7 | 5 | 6 | 8 | 8 | 9 | 5 | 4 | 4 | 6 | 75 |
| Total Snowfall (") | 0.3 | 0.8 | 0.0 | 0.0 | 0.0 | 0.0 | 0.0 | 0.0 | 0.0 | 0.0 | 0.0 | 0.1 | 1.2 |
| Days ≥ 1" Snow Depth | 0 | 0 | 0 | 0 | 0 | 0 | 0 | 0 | 0 | 0 | 0 | 0 | 0 |

## HILTON HEAD *Beaufort County* ELEVATION 20 ft LAT/LONG 32° 13' N / 80° 45' W

| | JAN | FEB | MAR | APR | MAY | JUN | JUL | AUG | SEP | OCT | NOV | DEC | YEAR |
|---|---|---|---|---|---|---|---|---|---|---|---|---|---|
| Maximum Temp °F | 59.5 | 62.4 | 69.1 | 76.1 | 81.9 | 86.8 | 89.4 | 88.3 | 84.4 | 77.2 | 69.7 | 62.5 | 75.6 |
| Minimum Temp °F | 38.6 | 40.4 | 47.5 | 54.0 | 62.1 | 68.8 | 72.3 | 71.7 | 68.0 | 57.6 | 48.6 | 41.4 | 55.9 |
| Mean Temp °F | 49.0 | 51.4 | 58.4 | 65.1 | 72.0 | 77.8 | 80.9 | 80.0 | 76.2 | 67.4 | 59.2 | 52.0 | 65.8 |
| Days Max Temp ≥ 90 °F | 0 | 0 | 0 | 1 | 2 | 7 | 15 | 11 | 3 | 0 | 0 | 0 | 39 |
| Days Max Temp ≤ 32 °F | 0 | 0 | 0 | 0 | 0 | 0 | 0 | 0 | 0 | 0 | 0 | 0 | 0 |
| Days Min Temp ≤ 32 °F | 9 | 7 | 2 | 0 | 0 | 0 | 0 | 0 | 0 | 0 | 1 | 7 | 26 |
| Days Min Temp ≤ 0 °F | 0 | 0 | 0 | 0 | 0 | 0 | 0 | 0 | 0 | 0 | 0 | 0 | 0 |
| Heating Degree Days | 490 | 383 | 223 | 74 | 6 | 0 | 0 | 0 | 1 | 53 | 200 | 403 | 1833 |
| Cooling Degree Days | 2 | 7 | 27 | 81 | 236 | 425 | 528 | 492 | 354 | 145 | 40 | 8 | 2345 |
| Total Precipitation (") | 4.18 | 3.17 | 4.15 | 2.86 | 3.75 | 4.87 | 5.75 | 8.27 | 5.45 | 3.66 | 2.64 | 3.25 | 52.00 |
| Days ≥ 0.1" Precip | 7 | 6 | 6 | 5 | 5 | 6 | 8 | 9 | 7 | 4 | 4 | 6 | 73 |
| Total Snowfall (") | 0.1 | 0.0 | 0.0 | 0.0 | 0.0 | 0.0 | 0.0 | 0.0 | 0.0 | 0.0 | 0.0 | 0.0 | 0.1 |
| Days ≥ 1" Snow Depth | 0 | 0 | 0 | 0 | 0 | 0 | 0 | 0 | 0 | 0 | 0 | 0 | 0 |

## JOHNSTON 4 SW *Edgefield County* ELEVATION 591 ft LAT/LONG 33° 53' N / 81° 49' W

| | JAN | FEB | MAR | APR | MAY | JUN | JUL | AUG | SEP | OCT | NOV | DEC | YEAR |
|---|---|---|---|---|---|---|---|---|---|---|---|---|---|
| Maximum Temp °F | 52.7 | 57.1 | 65.5 | 74.4 | 81.0 | 87.5 | 90.9 | 88.9 | 83.8 | 74.7 | 65.9 | 56.6 | 73.3 |
| Minimum Temp °F | 31.4 | 33.5 | 40.8 | 48.5 | 56.8 | 64.5 | 68.2 | 67.0 | 61.3 | 49.4 | 41.1 | 34.2 | 49.7 |
| Mean Temp °F | 42.1 | 45.3 | 53.2 | 61.5 | 68.9 | 76.0 | 79.5 | 78.0 | 72.6 | 62.1 | 53.5 | 45.4 | 61.5 |
| Days Max Temp ≥ 90 °F | 0 | 0 | 0 | 1 | 3 | 12 | 19 | 15 | 7 | 0 | 0 | 0 | 57 |
| Days Max Temp ≤ 32 °F | 1 | 0 | 0 | 0 | 0 | 0 | 0 | 0 | 0 | 0 | 0 | 0 | 1 |
| Days Min Temp ≤ 32 °F | 18 | 14 | 7 | 1 | 0 | 0 | 0 | 0 | 0 | 1 | 6 | 16 | 63 |
| Days Min Temp ≤ 0 °F | 0 | 0 | 0 | 0 | 0 | 0 | 0 | 0 | 0 | 0 | 0 | 0 | 0 |
| Heating Degree Days | 704 | 552 | 369 | 153 | 33 | 2 | 0 | 0 | 12 | 139 | 348 | 601 | 2913 |
| Cooling Degree Days | 0 | 2 | 10 | 49 | 170 | 372 | 495 | 426 | 258 | 61 | 11 | 2 | 1856 |
| Total Precipitation (") | 4.75 | 4.02 | 4.99 | 3.46 | 3.66 | 4.79 | 4.52 | 4.79 | 3.39 | 3.19 | 3.00 | 3.55 | 48.11 |
| Days ≥ 0.1" Precip | 7 | 7 | 7 | 5 | 6 | 7 | 7 | 7 | 5 | 4 | 5 | 6 | 73 |
| Total Snowfall (") | 0.6 | 1.4 | 0.2 | 0.0 | 0.0 | 0.0 | 0.0 | 0.0 | 0.0 | 0.0 | 0.0 | 0.1 | 2.3 |
| Days ≥ 1" Snow Depth | 0 | 0 | 0 | 0 | 0 | 0 | 0 | 0 | 0 | 0 | 0 | 0 | 0 |

## KINGSTREE 1 SE *Williamsburg County* ELEVATION 49 ft LAT/LONG 33° 40' N / 79° 49' W

| | JAN | FEB | MAR | APR | MAY | JUN | JUL | AUG | SEP | OCT | NOV | DEC | YEAR |
|---|---|---|---|---|---|---|---|---|---|---|---|---|---|
| Maximum Temp °F | 55.9 | 59.6 | 67.9 | 76.1 | 82.9 | 88.5 | 91.7 | 89.8 | 85.2 | 76.5 | 68.4 | 59.7 | 75.2 |
| Minimum Temp °F | 31.0 | 32.9 | 40.4 | 47.6 | 56.0 | 63.8 | 68.0 | 67.1 | 61.1 | 48.5 | 39.6 | 33.3 | 49.1 |
| Mean Temp °F | 43.5 | 46.3 | 54.1 | 61.9 | 69.5 | 76.2 | 79.9 | 78.5 | 73.2 | 62.5 | 54.0 | 46.5 | 62.2 |
| Days Max Temp ≥ 90 °F | 0 | 0 | 0 | 1 | 5 | 14 | *21* | 18 | 8 | 1 | 0 | 0 | 68 |
| Days Max Temp ≤ 32 °F | 1 | 0 | 0 | 0 | 0 | 0 | 0 | 0 | 0 | 0 | 0 | 0 | 1 |
| Days Min Temp ≤ 32 °F | 18 | 15 | 7 | 1 | 0 | 0 | 0 | 0 | 0 | 1 | 9 | 17 | 68 |
| Days Min Temp ≤ 0 °F | 0 | 0 | 0 | 0 | 0 | 0 | 0 | 0 | 0 | 0 | 0 | 0 | 0 |
| Heating Degree Days | 661 | 526 | 345 | 144 | 30 | 2 | 0 | 0 | 9 | 134 | 338 | 570 | 2759 |
| Cooling Degree Days | 0 | 4 | 14 | 53 | 178 | 373 | 496 | 435 | 269 | 71 | 16 | 5 | 1914 |
| Total Precipitation (") | 4.39 | 3.44 | 4.48 | 2.87 | 4.04 | 5.14 | 5.18 | 5.98 | 3.84 | 3.58 | 2.49 | 3.80 | 49.23 |
| Days ≥ 0.1" Precip | 9 | 6 | 8 | 5 | 7 | 8 | 8 | 8 | 5 | 5 | 5 | 7 | 81 |
| Total Snowfall (") | 0.3 | 1.1 | 0.2 | 0.0 | 0.0 | 0.0 | 0.0 | 0.0 | 0.0 | 0.0 | 0.0 | 0.3 | 1.9 |
| Days ≥ 1" Snow Depth | 0 | 1 | 0 | 0 | 0 | 0 | 0 | 0 | 0 | 0 | 0 | 0 | 1 |

## LAKE CITY 1 SE *Florence County* ELEVATION 79 ft LAT/LONG 33° 51 ' N / 79° 44 ' W

| | JAN | FEB | MAR | APR | MAY | JUN | JUL | AUG | SEP | OCT | NOV | DEC | YEAR |
|---|---|---|---|---|---|---|---|---|---|---|---|---|---|
| Maximum Temp °F | 55.4 | 59.1 | 67.2 | 75.9 | 82.4 | 87.9 | 91.3 | 89.6 | 85.1 | 76.2 | 68.3 | 59.0 | 74.8 |
| Minimum Temp °F | 32.5 | 34.6 | 41.5 | 48.5 | 57.3 | 65.3 | 69.6 | 68.8 | 63.2 | 50.6 | 41.9 | 34.8 | 50.7 |
| Mean Temp °F | 44.0 | 46.9 | 54.4 | 62.2 | 69.9 | 76.6 | 80.5 | 79.2 | 74.2 | 63.4 | 55.1 | 46.9 | 62.8 |
| Days Max Temp ≥ 90 °F | 0 | 0 | 0 | 1 | 3 | *13* | *21* | 18 | 7 | 1 | 0 | 0 | 64 |
| Days Max Temp ≤ 32 °F | 1 | 0 | 0 | 0 | 0 | 0 | 0 | 0 | 0 | 0 | 0 | 0 | 1 |
| Days Min Temp ≤ 32 °F | 17 | 14 | 6 | 1 | 0 | 0 | 0 | 0 | 0 | 1 | 7 | 14 | 60 |
| Days Min Temp ≤ 0 °F | 0 | 0 | 0 | 0 | 0 | 0 | 0 | 0 | 0 | 0 | 0 | 0 | 0 |
| Heating Degree Days | 649 | 508 | 337 | 137 | 27 | 2 | 0 | 0 | 6 | 120 | 307 | 557 | 2650 |
| Cooling Degree Days | 1 | 5 | 16 | 49 | 173 | 375 | 504 | 447 | 281 | 77 | 22 | 3 | 1953 |
| Total Precipitation (") | 3.94 | 3.41 | 4.84 | 2.85 | 3.68 | 4.47 | 5.59 | 5.97 | 4.01 | 3.05 | 2.55 | 3.75 | 48.11 |
| Days ≥ 0.1" Precip | 7 | 6 | 7 | 5 | 6 | 6 | 8 | 8 | 5 | 4 | 4 | 6 | 72 |
| Total Snowfall (") | 0.5 | *1.2* | 0.3 | 0.0 | 0.0 | 0.0 | 0.0 | 0.0 | 0.0 | 0.0 | 0.0 | 0.3 | 2.3 |
| Days ≥ 1" Snow Depth | *0* | 0 | 0 | 0 | 0 | 0 | 0 | 0 | 0 | 0 | 0 | 0 | 0 |

## LAURENS *Laurens County* ELEVATION 589 ft LAT/LONG 34° 30 ' N / 82° 2 ' W

| | JAN | FEB | MAR | APR | MAY | JUN | JUL | AUG | SEP | OCT | NOV | DEC | YEAR |
|---|---|---|---|---|---|---|---|---|---|---|---|---|---|
| Maximum Temp °F | 52.0 | 56.4 | 65.0 | 74.1 | 80.7 | 87.7 | 91.0 | 89.2 | 83.6 | 73.7 | 64.7 | 55.5 | 72.8 |
| Minimum Temp °F | 28.3 | 29.9 | 37.6 | 45.9 | 54.4 | 63.1 | 67.3 | 65.8 | 59.1 | 46.0 | 36.7 | 30.6 | 47.1 |
| Mean Temp °F | 40.1 | 43.0 | 51.3 | 60.0 | 67.6 | 75.4 | 79.2 | 77.5 | 71.4 | 59.8 | 50.7 | 43.1 | 59.9 |
| Days Max Temp ≥ 90 °F | 0 | 0 | 0 | 0 | 2 | 12 | 19 | 16 | 6 | 0 | 0 | 0 | 55 |
| Days Max Temp ≤ 32 °F | 1 | 0 | 0 | 0 | 0 | 0 | 0 | 0 | 0 | 0 | 0 | 0 | 1 |
| Days Min Temp ≤ 32 °F | 21 | 18 | 10 | 2 | 0 | 0 | 0 | 0 | 0 | 3 | 11 | 19 | 84 |
| Days Min Temp ≤ 0 °F | 0 | 0 | 0 | 0 | 0 | 0 | 0 | 0 | 0 | 0 | 0 | 0 | 0 |
| Heating Degree Days | 765 | 617 | 422 | 180 | 46 | 2 | 0 | 0 | 18 | 186 | 427 | 675 | 3338 |
| Cooling Degree Days | 0 | 0 | 6 | 36 | 136 | 344 | 464 | 403 | 217 | 35 | 4 | 1 | 1646 |
| Total Precipitation (") | 4.67 | 4.13 | 5.06 | 3.19 | 4.13 | 3.75 | 3.35 | 3.87 | 3.70 | 3.69 | 3.76 | 3.96 | 47.26 |
| Days ≥ 0.1" Precip | 7 | 7 | *8* | 6 | 6 | 6 | 7 | 6 | 6 | 5 | 6 | 7 | 77 |
| Total Snowfall (") | 1.0 | 0.6 | 0.5 | 0.0 | 0.0 | 0.0 | 0.0 | 0.0 | 0.0 | 0.0 | 0.2 | 0.4 | 2.7 |
| Days ≥ 1" Snow Depth | *0* | 0 | 0 | 0 | 0 | 0 | 0 | 0 | 0 | 0 | 0 | 0 | 0 |

## LITTLE MOUNTAIN *Newberry County* ELEVATION 712 ft LAT/LONG 34° 12 ' N / 81° 25 ' W

| | JAN | FEB | MAR | APR | MAY | JUN | JUL | AUG | SEP | OCT | NOV | DEC | YEAR |
|---|---|---|---|---|---|---|---|---|---|---|---|---|---|
| Maximum Temp °F | 53.4 | 57.8 | 66.4 | 75.0 | 81.1 | 87.5 | 90.5 | 88.5 | 83.5 | 74.2 | 65.5 | 56.9 | 73.4 |
| Minimum Temp °F | 33.8 | 36.0 | 43.3 | 50.8 | 58.7 | 65.9 | 69.6 | 68.5 | 63.1 | 51.6 | 43.5 | 36.8 | 51.8 |
| Mean Temp °F | 43.6 | 46.9 | 54.9 | 62.9 | 69.9 | 76.7 | 80.1 | 78.5 | 73.3 | 62.9 | 54.5 | 46.9 | 62.6 |
| Days Max Temp ≥ 90 °F | 0 | 0 | 0 | 0 | 2 | 10 | 18 | 13 | 5 | 0 | 0 | 0 | 48 |
| Days Max Temp ≤ 32 °F | 1 | 0 | 0 | 0 | 0 | 0 | 0 | 0 | 0 | 0 | 0 | 0 | 1 |
| Days Min Temp ≤ 32 °F | 14 | 11 | 5 | 1 | 0 | 0 | 0 | 0 | 0 | 0 | 5 | 11 | 47 |
| Days Min Temp ≤ 0 °F | 0 | 0 | 0 | 0 | 0 | 0 | 0 | 0 | 0 | 0 | 0 | 0 | 0 |
| Heating Degree Days | 657 | 506 | 322 | 124 | 26 | 1 | 0 | 0 | 11 | 125 | 319 | 558 | 2649 |
| Cooling Degree Days | 0 | 2 | 15 | 61 | 184 | 376 | 490 | 422 | 258 | 70 | 12 | 3 | 1893 |
| Total Precipitation (") | 4.50 | 3.76 | 5.08 | 3.07 | 3.67 | 3.96 | 4.89 | 4.98 | 4.04 | 3.37 | 3.07 | 3.71 | 48.10 |
| Days ≥ 0.1" Precip | 7 | 7 | 8 | 5 | 6 | 6 | 7 | 7 | 5 | 4 | 5 | 6 | 73 |
| Total Snowfall (") | 0.8 | 1.5 | 0.7 | 0.0 | 0.0 | 0.0 | 0.0 | 0.0 | 0.0 | 0.0 | 0.0 | 0.2 | 3.2 |
| Days ≥ 1" Snow Depth | 1 | 1 | 0 | 0 | 0 | 0 | 0 | 0 | 0 | 0 | 0 | 0 | 2 |

## MARION *Marion County* ELEVATION 69 ft LAT/LONG 34° 11 ' N / 79° 23 ' W

| | JAN | FEB | MAR | APR | MAY | JUN | JUL | AUG | SEP | OCT | NOV | DEC | YEAR |
|---|---|---|---|---|---|---|---|---|---|---|---|---|---|
| Maximum Temp °F | 56.0 | 60.7 | 69.2 | 77.5 | 83.7 | 89.6 | 92.4 | *90.2* | 85.8 | 76.8 | 67.8 | *59.4* | 75.8 |
| Minimum Temp °F | 33.5 | 34.9 | 42.6 | 49.5 | 57.9 | 65.2 | 69.4 | *68.0* | 62.6 | 50.5 | *42.1* | 35.6 | 51.0 |
| Mean Temp °F | 44.8 | 47.8 | 55.9 | 63.5 | 70.8 | 77.5 | 81.0 | *79.1* | 74.2 | 63.7 | *54.9* | *47.6* | 63.4 |
| Days Max Temp ≥ 90 °F | 0 | 0 | 0 | 2 | 5 | 17 | 24 | *19* | *8* | 1 | 0 | 0 | 76 |
| Days Max Temp ≤ 32 °F | 0 | 0 | 0 | 0 | 0 | 0 | 0 | 0 | 0 | 0 | 0 | 0 | 0 |
| Days Min Temp ≤ 32 °F | 15 | 12 | 6 | 1 | 0 | 0 | 0 | 0 | 0 | 1 | *7* | *13* | 55 |
| Days Min Temp ≤ 0 °F | 0 | 0 | 0 | 0 | 0 | 0 | 0 | 0 | 0 | 0 | 0 | 0 | 0 |
| Heating Degree Days | 623 | 484 | 297 | 115 | 19 | 1 | 0 | 0 | 6 | 115 | *313* | *537* | 2510 |
| Cooling Degree Days | 1 | *5* | 22 | 74 | 211 | 413 | 540 | *459* | *307* | *94* | 19 | 3 | 2148 |
| Total Precipitation (") | 3.84 | 3.75 | 4.50 | 2.98 | 4.26 | 4.28 | 6.14 | 6.08 | 3.60 | 3.16 | 2.90 | 3.46 | 48.95 |
| Days ≥ 0.1" Precip | 7 | 6 | 6 | 5 | 7 | 6 | 8 | *8* | *5* | *4* | *5* | *6* | 73 |
| Total Snowfall (") | *0.3* | *1.4* | 0.0 | 0.0 | 0.0 | 0.0 | 0.0 | 0.0 | 0.0 | 0.0 | 0.0 | 0.4 | 2.1 |
| Days ≥ 1" Snow Depth | 0 | *0* | 0 | 0 | 0 | 0 | 0 | 0 | 0 | 0 | *0* | 0 | 0 |

## MCCOLL 3 NNW *Marlboro County* ELEVATION 190 ft LAT/LONG 34° 40 ' N / 79° 33 ' W

| | JAN | FEB | MAR | APR | MAY | JUN | JUL | AUG | SEP | OCT | NOV | DEC | YEAR |
|---|---|---|---|---|---|---|---|---|---|---|---|---|---|
| Maximum Temp °F | 54.3 | 58.5 | 66.9 | 76.3 | 82.5 | 88.9 | 91.4 | 89.2 | 84.6 | 75.2 | 66.4 | 57.7 | 74.3 |
| Minimum Temp °F | 32.3 | 34.9 | 41.8 | 49.6 | 58.3 | 65.9 | 69.5 | 68.3 | 62.9 | 50.6 | 42.0 | 35.1 | 50.9 |
| Mean Temp °F | 43.5 | 46.8 | 54.4 | 62.9 | 70.5 | 77.5 | 80.4 | 78.8 | 73.8 | 62.9 | 54.2 | 46.4 | 62.7 |
| Days Max Temp ≥ 90 °F | 0 | 0 | 0 | 2 | 5 | 16 | 22 | 17 | 7 | 0 | 0 | 0 | 69 |
| Days Max Temp ≤ 32 °F | 1 | 0 | 0 | 0 | 0 | 0 | 0 | 0 | 0 | 0 | 0 | 0 | 1 |
| Days Min Temp ≤ 32 °F | 16 | 13 | 6 | 1 | 0 | 0 | 0 | 0 | 0 | 1 | 7 | 13 | 57 |
| Days Min Temp ≤ 0 °F | 0 | 0 | 0 | 0 | 0 | 0 | 0 | 0 | 0 | 0 | 0 | 0 | 0 |
| Heating Degree Days | 661 | 509 | 338 | 125 | 22 | 1 | 0 | 0 | 8 | 129 | 331 | 572 | 2696 |
| Cooling Degree Days | 0 | 1 | 13 | 69 | 195 | 398 | 505 | 436 | 266 | 69 | 12 | 3 | 1967 |
| Total Precipitation (") | 3.54 | 3.37 | 3.86 | 2.34 | 3.25 | 3.99 | 4.54 | 4.16 | 3.05 | 2.70 | 2.60 | 2.88 | 40.28 |
| Days ≥ 0.1" Precip | 6 | 6 | 7 | 4 | 6 | 6 | 8 | 7 | 5 | 4 | 5 | 5 | 69 |
| Total Snowfall (") | 0.7 | 1.0 | 0.4 | 0.0 | 0.0 | 0.0 | 0.0 | 0.0 | 0.0 | 0.0 | 0.0 | 0.3 | 2.4 |
| Days ≥ 1" Snow Depth | 1 | 0 | 0 | 0 | 0 | 0 | 0 | 0 | 0 | 0 | 0 | 0 | 1 |

## NEWBERRY *Newberry County* ELEVATION 479 ft LAT/LONG 34° 17 ' N / 81° 37 ' W

| | JAN | FEB | MAR | APR | MAY | JUN | JUL | AUG | SEP | OCT | NOV | DEC | YEAR |
|---|---|---|---|---|---|---|---|---|---|---|---|---|---|
| Maximum Temp °F | 54.6 | 59.5 | 68.3 | 76.9 | 83.0 | 89.1 | 92.0 | 90.1 | 84.9 | 75.6 | 66.3 | 57.9 | 74.9 |
| Minimum Temp °F | 31.9 | 34.0 | 40.8 | 48.4 | 56.9 | 64.7 | 68.9 | 67.8 | 61.9 | 49.9 | 41.0 | 34.6 | 50.1 |
| Mean Temp °F | 43.3 | 46.8 | 54.6 | 62.7 | 69.9 | 76.9 | 80.5 | 78.9 | 73.4 | 62.8 | 53.7 | 46.3 | 62.5 |
| Days Max Temp ≥ 90 °F | 0 | 0 | 0 | 1 | 4 | 14 | 22 | 18 | 7 | 0 | 0 | 0 | 66 |
| Days Max Temp ≤ 32 °F | 1 | 0 | 0 | 0 | 0 | 0 | 0 | 0 | 0 | 0 | 0 | 0 | 1 |
| Days Min Temp ≤ 32 °F | 17 | 14 | 7 | 1 | 0 | 0 | 0 | 0 | 0 | 1 | 7 | 15 | 62 |
| Days Min Temp ≤ 0 °F | 0 | 0 | 0 | 0 | 0 | 0 | 0 | 0 | 0 | 0 | 0 | 0 | 0 |
| Heating Degree Days | 667 | 509 | 327 | 124 | 24 | 1 | 0 | 0 | 9 | 127 | 343 | 576 | 2707 |
| Cooling Degree Days | 0 | 1 | 12 | 61 | 192 | 397 | 522 | 457 | 278 | 73 | 10 | 3 | 2006 |
| Total Precipitation (") | 4.39 | 3.93 | 4.86 | 3.06 | 3.76 | 4.63 | 4.32 | 5.08 | 4.16 | 3.37 | 3.32 | 3.66 | 48.54 |
| Days ≥ 0.1" Precip | 7 | 7 | 7 | 5 | 7 | 6 | 7 | 6 | 5 | 4 | 5 | 6 | 72 |
| Total Snowfall (") | 0.8 | 0.9 | 0.5 | 0.0 | 0.0 | 0.0 | 0.0 | 0.0 | 0.0 | 0.0 | 0.0 | 0.2 | 2.4 |
| Days ≥ 1" Snow Depth | 0 | 0 | 0 | 0 | 0 | 0 | 0 | 0 | 0 | 0 | 0 | 0 | 0 |

## NINETY NINE ISLANDS *Cherokee County* ELEVATION 502 ft LAT/LONG 35° 3 ' N / 81° 30 ' W

| | JAN | FEB | MAR | APR | MAY | JUN | JUL | AUG | SEP | OCT | NOV | DEC | YEAR |
|---|---|---|---|---|---|---|---|---|---|---|---|---|---|
| Maximum Temp °F | 51.3 | 55.6 | 64.1 | 73.1 | 79.4 | 86.0 | 89.5 | 87.9 | 82.3 | 73.1 | 63.8 | 54.8 | 71.7 |
| Minimum Temp °F | 26.5 | 28.6 | 35.3 | 43.1 | 52.4 | 61.0 | 65.9 | 65.1 | 58.3 | 45.0 | 35.4 | 29.3 | 45.5 |
| Mean Temp °F | 39.0 | 42.1 | 49.7 | 58.1 | 65.9 | 73.5 | 77.7 | 76.5 | 70.3 | 59.1 | 49.6 | 42.1 | 58.6 |
| Days Max Temp ≥ 90 °F | 0 | 0 | 0 | 0 | 2 | 8 | 17 | 13 | 5 | 0 | 0 | 0 | 45 |
| Days Max Temp ≤ 32 °F | 1 | 0 | 0 | 0 | 0 | 0 | 0 | 0 | 0 | 0 | 0 | 0 | 1 |
| Days Min Temp ≤ 32 °F | 22 | 19 | 13 | 5 | 0 | 0 | 0 | 0 | 0 | 4 | 14 | 20 | 97 |
| Days Min Temp ≤ 0 °F | 0 | 0 | 0 | 0 | 0 | 0 | 0 | 0 | 0 | 0 | 0 | 0 | 0 |
| Heating Degree Days | 800 | 640 | 469 | 221 | 64 | 4 | 0 | 0 | 23 | 208 | 459 | 705 | 3593 |
| Cooling Degree Days | 0 | 0 | 2 | 19 | 94 | 279 | 418 | 363 | 183 | 32 | 2 | 1 | 1393 |
| Total Precipitation (") | 4.25 | 3.98 | 5.10 | 2.94 | 4.39 | 3.82 | 4.29 | 5.04 | 3.71 | 3.72 | 3.51 | 3.81 | 48.56 |
| Days ≥ 0.1" Precip | 7 | 7 | 8 | 6 | 7 | 6 | 7 | 7 | 5 | 5 | 6 | 7 | 78 |
| Total Snowfall (") | 1.8 | 1.5 | 0.8 | 0.0 | 0.0 | 0.0 | 0.0 | 0.0 | 0.0 | 0.0 | 0.0 | 0.6 | 4.7 |
| Days ≥ 1" Snow Depth | 1 | 0 | 0 | 0 | 0 | 0 | 0 | 0 | 0 | 0 | 0 | 0 | 1 |

## ORANGEBURG 2 *Orangeburg County* ELEVATION 239 ft LAT/LONG 33° 29 ' N / 80° 52 ' W

| | JAN | FEB | MAR | APR | MAY | JUN | JUL | AUG | SEP | OCT | NOV | DEC | YEAR |
|---|---|---|---|---|---|---|---|---|---|---|---|---|---|
| Maximum Temp °F | 55.4 | 59.5 | 68.0 | 76.6 | 82.6 | 88.1 | 91.0 | 89.4 | 84.8 | 76.0 | 67.8 | 59.3 | 74.9 |
| Minimum Temp °F | 33.2 | 35.4 | 43.0 | 50.3 | 58.4 | 65.9 | 69.8 | 68.9 | 63.4 | 51.2 | 42.5 | 35.9 | 51.5 |
| Mean Temp °F | 44.3 | 47.5 | 55.5 | 63.5 | 70.6 | 77.1 | 80.4 | 79.2 | 74.1 | 63.6 | 55.2 | 47.6 | 63.2 |
| Days Max Temp ≥ 90 °F | 0 | 0 | 0 | 1 | 4 | 13 | 20 | 17 | 7 | 0 | 0 | 0 | 62 |
| Days Max Temp ≤ 32 °F | 1 | 0 | 0 | 0 | 0 | 0 | 0 | 0 | 0 | 0 | 0 | 0 | 1 |
| Days Min Temp ≤ 32 °F | 15 | 12 | 5 | 0 | 0 | 0 | 0 | 0 | 0 | 0 | 6 | 13 | 51 |
| Days Min Temp ≤ 0 °F | 0 | 0 | 0 | 0 | 0 | 0 | 0 | 0 | 0 | 0 | 0 | 0 | 0 |
| Heating Degree Days | 636 | 492 | 305 | 112 | 20 | 1 | 0 | 0 | 8 | 112 | 303 | 536 | 2525 |
| Cooling Degree Days | 1 | 5 | 20 | 68 | 204 | 396 | 508 | 454 | 293 | 80 | 19 | 5 | 2053 |
| Total Precipitation (") | 4.24 | 3.57 | 4.38 | 2.63 | 3.86 | 4.77 | 5.63 | 5.31 | 3.68 | 3.05 | 2.70 | 3.31 | 47.13 |
| Days ≥ 0.1" Precip | 8 | 6 | 7 | 5 | 6 | 7 | 8 | 8 | 5 | 5 | 4 | 6 | 75 |
| Total Snowfall (") | 0.1 | 0.5 | 0.1 | 0.0 | 0.0 | 0.0 | 0.0 | 0.0 | 0.0 | 0.0 | 0.0 | 0.1 | 0.8 |
| Days ≥ 1" Snow Depth | 0 | 0 | 0 | 0 | 0 | 0 | 0 | 0 | 0 | 0 | 0 | 0 | 0 |

## PARR *Fairfield County* ELEVATION 280 ft LAT/LONG 34° 18 ' N / 81° 20 ' W

| | JAN | FEB | MAR | APR | MAY | JUN | JUL | AUG | SEP | OCT | NOV | DEC | YEAR |
|---|---|---|---|---|---|---|---|---|---|---|---|---|---|
| Maximum Temp °F | 54.7 | 59.5 | 67.8 | 76.6 | 82.9 | 88.9 | 92.1 | 90.6 | 85.7 | 76.3 | 67.3 | 58.0 | 75.0 |
| Minimum Temp °F | 30.1 | 31.9 | 38.8 | 46.4 | 55.1 | 63.5 | 68.3 | 67.2 | 61.0 | 47.9 | 39.0 | 32.3 | 48.5 |
| Mean Temp °F | 42.4 | 45.7 | 53.3 | 61.5 | 69.0 | 76.3 | 80.2 | 78.9 | 73.4 | 62.1 | 53.2 | 45.2 | 61.8 |
| Days Max Temp ≥ 90 °F | 0 | 0 | 0 | 1 | 4 | 15 | 23 | 19 | 9 | 0 | 0 | 0 | 71 |
| Days Max Temp ≤ 32 °F | 0 | 0 | 0 | 0 | 0 | 0 | 0 | 0 | 0 | 0 | 0 | 0 | 0 |
| Days Min Temp ≤ 32 °F | 19 | 17 | 10 | 3 | 0 | 0 | 0 | 0 | 0 | 2 | 10 | 17 | 78 |
| Days Min Temp ≤ 0 °F | 0 | 0 | 0 | 0 | 0 | 0 | 0 | 0 | 0 | 0 | 0 | 0 | 0 |
| Heating Degree Days | 693 | 540 | 365 | 149 | 32 | 2 | 0 | 0 | 10 | 140 | 357 | 610 | 2898 |
| Cooling Degree Days | 0 | 1 | 8 | 45 | 158 | 373 | 502 | 440 | 265 | 55 | 8 | 1 | 1856 |
| Total Precipitation (") | 4.38 | 3.67 | 4.83 | 2.93 | 3.54 | 4.30 | 4.22 | 4.20 | 3.53 | 3.18 | 2.89 | 3.44 | 45.11 |
| Days ≥ 0.1" Precip | 7 | 6 | 7 | 5 | 6 | 6 | 7 | 6 | 4 | 4 | 5 | 6 | 69 |
| Total Snowfall (") | 0.5 | 0.5 | 0.2 | 0.0 | 0.0 | 0.0 | 0.0 | 0.0 | 0.0 | 0.0 | 0.0 | 0.1 | 1.3 |
| Days ≥ 1" Snow Depth | 0 | 0 | 0 | 0 | 0 | 0 | 0 | 0 | 0 | 0 | 0 | 0 | 0 |

## PELION 4 NW *Lexington County* ELEVATION 449 ft LAT/LONG 33° 48 ' N / 81° 16 ' W

| | JAN | FEB | MAR | APR | MAY | JUN | JUL | AUG | SEP | OCT | NOV | DEC | YEAR |
|---|---|---|---|---|---|---|---|---|---|---|---|---|---|
| Maximum Temp °F | 55.1 | 59.5 | 68.3 | 76.8 | 83.2 | 89.1 | 91.4 | 89.6 | 85.0 | 75.9 | 66.9 | 58.8 | 75.0 |
| Minimum Temp °F | 32.5 | 34.4 | 41.6 | 48.6 | 56.9 | 64.7 | 69.0 | 67.8 | 62.1 | 49.9 | 41.1 | 35.1 | 50.3 |
| Mean Temp °F | 43.8 | 47.0 | 55.0 | 62.7 | 70.1 | 76.9 | 80.2 | 78.7 | 73.6 | 62.9 | 54.0 | 47.0 | 62.7 |
| Days Max Temp ≥ 90 °F | 0 | 0 | 0 | 1 | 4 | 15 | 20 | 17 | 8 | 0 | 0 | 0 | 65 |
| Days Max Temp ≤ 32 °F | 1 | 0 | 0 | 0 | 0 | 0 | 0 | 0 | 0 | 0 | 0 | 0 | 1 |
| Days Min Temp ≤ 32 °F | 16 | 13 | 7 | 2 | 0 | 0 | 0 | 0 | 0 | 2 | 8 | 14 | 62 |
| Days Min Temp ≤ 0 °F | 0 | 0 | 0 | 0 | 0 | 0 | 0 | 0 | 0 | 0 | 0 | 0 | 0 |
| Heating Degree Days | 650 | 505 | 320 | 125 | 23 | 1 | 0 | 0 | 7 | 125 | 335 | 556 | 2647 |
| Cooling Degree Days | 1 | 4 | 17 | 61 | 188 | 391 | 505 | 440 | 276 | 73 | 13 | 4 | 1973 |
| Total Precipitation (") | 4.46 | 3.87 | 5.03 | 3.21 | 3.48 | 5.08 | 5.80 | 5.50 | 3.91 | 3.08 | 3.00 | 3.65 | 50.07 |
| Days ≥ 0.1" Precip | 8 | 6 | 8 | 5 | 6 | 7 | 9 | 8 | 6 | 4 | 5 | 6 | 78 |
| Total Snowfall (") | 0.3 | 0.9 | 0.0 | 0.0 | 0.0 | 0.0 | 0.0 | 0.0 | 0.0 | 0.0 | 0.0 | 0.0 | 1.2 |
| Days ≥ 1" Snow Depth | 0 | 0 | 0 | 0 | 0 | 0 | 0 | 0 | 0 | 0 | 0 | 0 | 0 |

## PICKENS *Pickens County* ELEVATION 1040 ft LAT/LONG 34° 51 ' N / 82° 39 ' W

| | JAN | FEB | MAR | APR | MAY | JUN | JUL | AUG | SEP | OCT | NOV | DEC | YEAR |
|---|---|---|---|---|---|---|---|---|---|---|---|---|---|
| Maximum Temp °F | 51.2 | 55.9 | 65.0 | 73.6 | 79.3 | 85.5 | 88.5 | 87.0 | 81.8 | 72.5 | 62.8 | 54.1 | 71.4 |
| Minimum Temp °F | 31.3 | 34.0 | 41.2 | 49.0 | 56.3 | 63.5 | 67.1 | 66.3 | 61.3 | 50.4 | 41.6 | 34.7 | 49.7 |
| Mean Temp °F | 41.2 | 45.0 | 53.2 | 61.3 | 67.8 | 74.5 | 77.8 | 76.7 | 71.5 | 61.5 | 52.2 | 44.4 | 60.6 |
| Days Max Temp ≥ 90 °F | 0 | 0 | 0 | 0 | 1 | 7 | 13 | 10 | 3 | 0 | 0 | 0 | 34 |
| Days Max Temp ≤ 32 °F | 1 | 1 | 0 | 0 | 0 | 0 | 0 | 0 | 0 | 0 | 0 | 0 | 2 |
| Days Min Temp ≤ 32 °F | 17 | 13 | 6 | 1 | 0 | 0 | 0 | 0 | 0 | 1 | 6 | 13 | 57 |
| Days Min Temp ≤ 0 °F | 0 | 0 | 0 | 0 | 0 | 0 | 0 | 0 | 0 | 0 | 0 | 0 | 0 |
| Heating Degree Days | 730 | 560 | 366 | 147 | 36 | 2 | 0 | 0 | 13 | 148 | 380 | 631 | 3013 |
| Cooling Degree Days | 0 | 0 | 7 | 45 | 133 | 319 | 425 | 374 | 215 | 51 | 3 | 0 | 1572 |
| Total Precipitation (") | 5.25 | 4.84 | 5.75 | 3.79 | 4.99 | 4.60 | 4.82 | 4.79 | 3.69 | 4.59 | 4.39 | 4.72 | 56.22 |
| Days ≥ 0.1" Precip | 8 | 7 | 8 | 6 | 8 | 7 | 7 | 7 | 6 | 5 | 7 | 8 | 84 |
| Total Snowfall (") | *1.7* | *1.3* | 0.4 | 0.0 | 0.0 | 0.0 | 0.0 | 0.0 | 0.0 | 0.0 | 0.1 | 0.5 | 4.0 |
| Days ≥ 1" Snow Depth | na | *0* | *0* | 0 | 0 | 0 | 0 | 0 | 0 | 0 | 0 | 0 | na |

## PINOPOLIS DAM *Berkeley County* ELEVATION 30 ft LAT/LONG 33° 15 ' N / 79° 59 ' W

| | JAN | FEB | MAR | APR | MAY | JUN | JUL | AUG | SEP | OCT | NOV | DEC | YEAR |
|---|---|---|---|---|---|---|---|---|---|---|---|---|---|
| Maximum Temp °F | 56.2 | 59.4 | 67.6 | 75.7 | 82.4 | 88.1 | 91.5 | 89.7 | 85.3 | 76.6 | 68.5 | 60.1 | 75.1 |
| Minimum Temp °F | 34.5 | 36.3 | 43.2 | 50.4 | 58.8 | 66.1 | 70.5 | 69.6 | 64.4 | 52.9 | 44.2 | 37.3 | 52.4 |
| Mean Temp °F | 45.4 | 47.9 | 55.4 | 63.1 | 70.7 | 77.1 | 81.0 | 79.7 | 74.9 | 64.8 | 56.4 | 48.7 | 63.8 |
| Days Max Temp ≥ 90 °F | 0 | 0 | 0 | 1 | 3 | 13 | 22 | 18 | 8 | 0 | 0 | 0 | 65 |
| Days Max Temp ≤ 32 °F | 0 | 0 | 0 | 0 | 0 | 0 | 0 | 0 | 0 | 0 | 0 | 0 | 0 |
| Days Min Temp ≤ 32 °F | 14 | 10 | 4 | 0 | 0 | 0 | 0 | 0 | 0 | 0 | 4 | 12 | 44 |
| Days Min Temp ≤ 0 °F | 0 | 0 | 0 | 0 | 0 | 0 | 0 | 0 | 0 | 0 | 0 | 0 | 0 |
| Heating Degree Days | 602 | 479 | 307 | 117 | 19 | 1 | 0 | 0 | 5 | 95 | 272 | 502 | 2399 |
| Cooling Degree Days | 0 | 5 | 17 | 63 | 205 | 402 | 532 | 478 | 306 | 95 | 20 | 7 | 2130 |
| Total Precipitation (") | 4.23 | 3.23 | 4.37 | 2.78 | 4.25 | 5.80 | 6.09 | 6.46 | 4.28 | 3.11 | 2.43 | 3.33 | 50.36 |
| Days ≥ 0.1" Precip | 8 | 6 | 7 | 5 | 7 | 7 | 9 | 9 | 6 | 4 | 4 | 6 | 78 |
| Total Snowfall (") | 0.1 | 0.0 | 0.0 | 0.0 | 0.0 | 0.0 | 0.0 | 0.0 | 0.0 | 0.0 | 0.0 | 0.0 | 0.1 |
| Days ≥ 1" Snow Depth | 0 | 0 | 0 | 0 | 0 | 0 | 0 | 0 | 0 | 0 | 0 | 0 | 0 |

## RIDGELAND 5 NE *Jasper County* ELEVATION 20 ft LAT/LONG 32° 32 ' N / 80° 54 ' W

| | JAN | FEB | MAR | APR | MAY | JUN | JUL | AUG | SEP | OCT | NOV | DEC | YEAR |
|---|---|---|---|---|---|---|---|---|---|---|---|---|---|
| Maximum Temp °F | 59.2 | 63.3 | 71.4 | 78.7 | 84.1 | 88.7 | 91.8 | 89.9 | 85.4 | 77.4 | 69.6 | 62.4 | 76.8 |
| Minimum Temp °F | 36.6 | 38.7 | 45.5 | 52.3 | 60.2 | 66.7 | 70.2 | 69.5 | 65.1 | 54.6 | 45.7 | 39.4 | 53.7 |
| Mean Temp °F | 48.0 | 51.0 | 58.5 | 65.6 | 72.1 | 77.7 | 81.0 | 79.8 | 75.3 | 66.0 | 57.7 | 51.0 | 65.3 |
| Days Max Temp ≥ 90 °F | 0 | 0 | 0 | 2 | 5 | 13 | 22 | 18 | 7 | 1 | 0 | 0 | 68 |
| Days Max Temp ≤ 32 °F | 0 | 0 | 0 | 0 | 0 | 0 | 0 | 0 | 0 | 0 | 0 | 0 | 0 |
| Days Min Temp ≤ 32 °F | 12 | 9 | 3 | 0 | 0 | 0 | 0 | 0 | 0 | 0 | 4 | 9 | 37 |
| Days Min Temp ≤ 0 °F | 0 | 0 | 0 | 0 | 0 | 0 | 0 | 0 | 0 | 0 | 0 | 0 | 0 |
| Heating Degree Days | 523 | 394 | 225 | 73 | 6 | 0 | 0 | 0 | 2 | 75 | 239 | 432 | 1969 |
| Cooling Degree Days | 2 | 9 | 32 | 87 | 241 | 420 | 524 | 476 | 325 | 121 | 30 | 8 | 2275 |
| Total Precipitation (") | 4.25 | 3.38 | 4.28 | 3.00 | 4.37 | 5.32 | 5.69 | 6.81 | 4.80 | 3.14 | 2.72 | 3.50 | 51.26 |
| Days ≥ 0.1" Precip | 8 | 6 | 7 | 5 | 7 | 8 | 9 | 10 | 7 | 4 | 5 | 6 | 82 |
| Total Snowfall (") | 0.1 | 0.4 | 0.0 | 0.0 | 0.0 | 0.0 | 0.0 | 0.0 | 0.0 | 0.0 | 0.0 | 0.2 | 0.7 |
| Days ≥ 1" Snow Depth | 0 | 0 | 0 | 0 | 0 | 0 | 0 | 0 | 0 | 0 | 0 | 0 | 0 |

## SALUDA *Saluda County* ELEVATION 541 ft LAT/LONG 34° 0 ' N / 81° 47 ' W

| | JAN | FEB | MAR | APR | MAY | JUN | JUL | AUG | SEP | OCT | NOV | DEC | YEAR |
|---|---|---|---|---|---|---|---|---|---|---|---|---|---|
| Maximum Temp °F | 53.2 | 57.9 | 66.3 | 75.6 | 82.2 | 88.7 | 92.1 | 90.2 | 85.0 | 75.2 | 65.8 | 56.7 | 74.1 |
| Minimum Temp °F | 29.4 | 31.6 | 39.1 | 47.1 | 55.4 | 63.7 | 67.6 | 66.3 | 60.2 | 47.1 | 38.2 | 32.4 | 48.2 |
| Mean Temp °F | 41.4 | 44.8 | 52.7 | 61.4 | 68.8 | 76.2 | 79.9 | 78.3 | 72.6 | 61.2 | 52.1 | 44.5 | 61.2 |
| Days Max Temp ≥ 90 °F | 0 | 0 | 0 | 1 | 4 | 13 | 22 | *19* | 9 | 0 | 0 | 0 | 68 |
| Days Max Temp ≤ 32 °F | 1 | 0 | 0 | 0 | 0 | 0 | 0 | 0 | 0 | 0 | 0 | 0 | 1 |
| Days Min Temp ≤ 32 °F | 19 | 16 | *8* | 2 | 0 | 0 | 0 | 0 | 0 | 2 | 10 | *18* | 75 |
| Days Min Temp ≤ 0 °F | 0 | 0 | 0 | 0 | 0 | 0 | 0 | 0 | 0 | 0 | 0 | 0 | 0 |
| Heating Degree Days | 725 | 568 | 383 | 154 | 36 | 2 | 0 | 0 | 13 | 159 | 389 | 628 | 3057 |
| Cooling Degree Days | 0 | 2 | 11 | 55 | 174 | 387 | 513 | 442 | 260 | 57 | 8 | 2 | 1911 |
| Total Precipitation (") | 4.68 | 3.95 | 5.02 | 3.30 | 3.79 | 4.44 | 4.77 | 4.73 | 3.15 | 3.25 | 3.15 | 3.80 | 48.03 |
| Days ≥ 0.1" Precip | 8 | 6 | 7 | 5 | 6 | 6 | 7 | 6 | 5 | 4 | 5 | *6* | 71 |
| Total Snowfall (") | 0.5 | 0.8 | 0.3 | 0.0 | 0.0 | 0.0 | 0.0 | 0.0 | 0.0 | 0.0 | 0.0 | 0.2 | 1.8 |
| Days ≥ 1" Snow Depth | 0 | *0* | 0 | 0 | 0 | 0 | 0 | 0 | 0 | 0 | 0 | 0 | 0 |

## SANDHILL EXP STN *Richland County* ELEVATION 440 ft LAT/LONG 34° 8 ' N / 80° 52 ' W

| | JAN | FEB | MAR | APR | MAY | JUN | JUL | AUG | SEP | OCT | NOV | DEC | YEAR |
|---|---|---|---|---|---|---|---|---|---|---|---|---|---|
| Maximum Temp °F | 53.0 | 57.3 | 65.6 | 74.7 | 81.0 | 87.2 | 90.7 | 88.5 | 83.8 | 74.4 | 65.9 | 57.2 | 73.3 |
| Minimum Temp °F | 31.9 | 34.2 | 41.7 | 50.3 | 58.3 | 65.6 | 69.8 | 68.5 | 62.7 | 50.9 | 42.6 | 35.2 | 51.0 |
| Mean Temp °F | 42.5 | 45.8 | 53.7 | 62.5 | 69.6 | 76.4 | 80.2 | 78.5 | 73.3 | 62.7 | 54.3 | 46.2 | 62.1 |
| Days Max Temp ≥ 90 °F | 0 | 0 | 0 | 1 | 3 | 11 | 19 | 15 | 6 | 0 | 0 | 0 | 55 |
| Days Max Temp ≤ 32 °F | 1 | 0 | 0 | 0 | 0 | 0 | 0 | 0 | 0 | 0 | 0 | 0 | 1 |
| Days Min Temp ≤ 32 °F | 17 | 13 | 6 | 0 | 0 | 0 | 0 | 0 | 0 | 1 | 5 | 13 | 55 |
| Days Min Temp ≤ 0 °F | 0 | 0 | 0 | 0 | 0 | 0 | 0 | 0 | 0 | 0 | 0 | 0 | 0 |
| Heating Degree Days | 691 | 538 | 358 | 136 | 28 | 2 | 0 | 0 | 10 | 128 | 328 | 579 | 2798 |
| Cooling Degree Days | 0 | 4 | 13 | 64 | 189 | 385 | 507 | 432 | 265 | 68 | 10 | 2 | 1939 |
| Total Precipitation (") | 4.38 | 3.40 | 4.64 | 2.99 | 3.71 | 4.58 | 5.24 | 4.89 | 3.56 | 3.08 | 3.02 | 3.39 | 46.88 |
| Days ≥ 0.1" Precip | 7 | 6 | 7 | 5 | 6 | 7 | 8 | 7 | 5 | 4 | 5 | 7 | 74 |
| Total Snowfall (") | 0.4 | 0.0 | 0.2 | 0.0 | 0.0 | 0.0 | 0.0 | 0.0 | 0.0 | 0.0 | 0.0 | 0.0 | 0.6 |
| Days ≥ 1" Snow Depth | 0 | 0 | 0 | 0 | 0 | 0 | 0 | 0 | 0 | 0 | 0 | 0 | 0 |

## SANTUCK *Union County* ELEVATION 512 ft LAT/LONG 34° 38 ' N / 81° 30 ' W

| | JAN | FEB | MAR | APR | MAY | JUN | JUL | AUG | SEP | OCT | NOV | DEC | YEAR |
|---|---|---|---|---|---|---|---|---|---|---|---|---|---|
| Maximum Temp °F | 52.5 | 57.4 | 66.3 | 75.3 | 81.1 | 87.4 | 90.5 | 88.3 | 82.5 | 72.7 | 64.0 | 55.6 | 72.8 |
| Minimum Temp °F | 31.5 | 33.8 | 40.9 | 48.7 | 56.9 | 64.5 | 68.8 | 67.6 | 61.7 | 49.8 | 41.2 | 34.5 | 50.0 |
| Mean Temp °F | 42.0 | 45.6 | 53.6 | 62.0 | 69.0 | 76.0 | 79.6 | 78.0 | 72.1 | 61.3 | 52.6 | 45.1 | 61.4 |
| Days Max Temp ≥ 90 °F | 0 | 0 | 0 | 1 | 3 | 11 | 19 | 13 | 4 | 0 | 0 | 0 | 51 |
| Days Max Temp ≤ 32 °F | 1 | 0 | 0 | 0 | 0 | 0 | 0 | 0 | 0 | 0 | 0 | 0 | 1 |
| Days Min Temp ≤ 32 °F | 17 | 14 | 7 | 1 | 0 | 0 | 0 | 0 | 0 | 1 | 7 | 14 | 61 |
| Days Min Temp ≤ 0 °F | 0 | 0 | 0 | 0 | 0 | 0 | 0 | 0 | 0 | 0 | 0 | 0 | 0 |
| Heating Degree Days | 706 | 543 | 356 | 139 | 30 | 2 | 0 | 0 | 14 | 158 | 370 | 613 | 2931 |
| Cooling Degree Days | 0 | 2 | 11 | 53 | 162 | 362 | 485 | 412 | 229 | 52 | 6 | 2 | 1776 |
| Total Precipitation (") | 4.27 | 4.05 | 5.14 | 2.90 | 3.70 | 4.09 | 4.19 | 4.39 | 3.83 | 3.41 | 3.34 | 3.62 | 46.93 |
| Days ≥ 0.1" Precip | 7 | 7 | 8 | 6 | 6 | 6 | 7 | 6 | 5 | 5 | 6 | 7 | 76 |
| Total Snowfall (") | 1.5 | 1.6 | 0.6 | 0.0 | 0.0 | 0.0 | 0.0 | 0.0 | 0.0 | 0.0 | 0.1 | 0.5 | 4.3 |
| Days ≥ 1" Snow Depth | 1 | 1 | 0 | 0 | 0 | 0 | 0 | 0 | 0 | 0 | 0 | 0 | 2 |

## SULLIVANS ISLAND *Charleston County* ELEVATION 10 ft LAT/LONG 32° 46 ' N / 79° 52 ' W

| | JAN | FEB | MAR | APR | MAY | JUN | JUL | AUG | SEP | OCT | NOV | DEC | YEAR |
|---|---|---|---|---|---|---|---|---|---|---|---|---|---|
| Maximum Temp °F | 57.3 | 60.4 | 66.7 | 74.0 | 80.8 | ***85.9*** | 88.9 | 88.0 | ***84.1*** | 77.1 | 69.3 | 61.4 | 74.5 |
| Minimum Temp °F | 39.1 | 40.6 | 47.4 | 55.4 | 63.0 | ***70.5*** | 73.7 | ***73.0*** | ***68.6*** | 58.7 | 49.9 | 42.5 | 56.9 |
| Mean Temp °F | 48.2 | 50.6 | 57.1 | ***64.8*** | 71.9 | ***78.3*** | ***81.3*** | ***80.6*** | ***76.3*** | 91.3 | 59.6 | 52.1 | 67.7 |
| Days Max Temp ≥ 90 °F | 0 | 0 | 0 | 0 | 1 | ***6*** | 13 | ***10*** | 3 | 0 | 0 | 0 | 33 |
| Days Max Temp ≤ 32 °F | 0 | 0 | 0 | 0 | 0 | 0 | 0 | 0 | 0 | 0 | 0 | 0 | 0 |
| Days Min Temp ≤ 32 °F | 8 | 5 | 1 | 0 | 0 | 0 | 0 | 0 | 0 | 0 | 1 | ***5*** | 20 |
| Days Min Temp ≤ 0 °F | 0 | 0 | 0 | 0 | 0 | 0 | 0 | 0 | 0 | 0 | 0 | 0 | 0 |
| Heating Degree Days | 515 | 400 | 249 | ***75*** | 9 | ***0*** | ***0*** | ***0*** | ***1*** | 41 | 178 | 397 | 1865 |
| Cooling Degree Days | 0 | 2 | 11 | ***70*** | 238 | 436 | 537 | 491 | ***343*** | 148 | 37 | 5 | 2318 |
| Total Precipitation (") | 4.22 | 3.20 | 4.37 | 2.72 | 3.45 | 5.06 | 4.81 | 7.10 | 4.28 | 4.00 | 3.15 | 3.66 | 50.02 |
| Days ≥ 0.1" Precip | 7 | 6 | 7 | ***5*** | 5 | ***6*** | 7 | 8 | ***5*** | 4 | 5 | 6 | 71 |
| Total Snowfall (") | 0.0 | 0.0 | 0.0 | 0.0 | 0.0 | 0.0 | 0.0 | 0.0 | 0.0 | 0.0 | 0.0 | 0.1 | 0.1 |
| Days ≥ 1" Snow Depth | 0 | 0 | 0 | 0 | 0 | 0 | 0 | 0 | 0 | 0 | 0 | 0 | 0 |

## SUMMERVILLE *Dorchester County* ELEVATION 35 ft LAT/LONG 32° 59 ' N / 80° 11 ' W

| | JAN | FEB | MAR | APR | MAY | JUN | JUL | AUG | SEP | OCT | NOV | DEC | YEAR |
|---|---|---|---|---|---|---|---|---|---|---|---|---|---|
| Maximum Temp °F | 57.4 | 60.9 | 68.9 | 76.5 | 82.5 | 87.6 | 90.6 | 89.2 | 85.0 | 76.8 | 69.4 | 61.2 | 75.5 |
| Minimum Temp °F | 34.0 | 36.1 | 43.2 | 49.8 | 58.6 | 66.3 | 70.3 | 69.7 | 64.6 | 52.6 | 43.5 | 36.8 | 52.1 |
| Mean Temp °F | 45.8 | 48.5 | 56.1 | 63.2 | 70.6 | 77.0 | 80.5 | 79.5 | 74.8 | 64.7 | 56.5 | 49.0 | 63.9 |
| Days Max Temp ≥ 90 °F | 0 | 0 | 0 | 1 | 4 | 11 | 19 | 16 | 7 | 1 | 0 | 0 | 59 |
| Days Max Temp ≤ 32 °F | 0 | 0 | 0 | 0 | 0 | 0 | 0 | 0 | 0 | 0 | 0 | 0 | 0 |
| Days Min Temp ≤ 32 °F | 15 | 11 | 5 | 1 | 0 | 0 | 0 | 0 | 0 | 1 | 5 | 12 | 50 |
| Days Min Temp ≤ 0 °F | 0 | 0 | 0 | 0 | 0 | 0 | 0 | 0 | 0 | 0 | 0 | 0 | 0 |
| Heating Degree Days | 592 | 464 | 291 | 118 | 20 | 1 | 0 | 0 | 4 | 96 | 272 | 493 | 2351 |
| Cooling Degree Days | 0 | 5 | 19 | 64 | 204 | 398 | 513 | 471 | 309 | 97 | 28 | 6 | 2114 |
| Total Precipitation (") | 4.36 | 3.33 | 4.62 | 2.90 | 4.36 | 5.88 | 6.17 | 6.81 | 4.98 | 3.25 | 2.68 | 3.38 | 52.72 |
| Days ≥ 0.1" Precip | 8 | 6 | 6 | 5 | 7 | 8 | 9 | 9 | 6 | 4 | 5 | 6 | 79 |
| Total Snowfall (") | 0.1 | 0.6 | 0.0 | 0.0 | 0.0 | 0.0 | 0.0 | 0.0 | 0.0 | 0.0 | 0.0 | 0.2 | 0.9 |
| Days ≥ 1" Snow Depth | 0 | 0 | 0 | 0 | 0 | 0 | 0 | 0 | 0 | 0 | 0 | 0 | 0 |

## SUMTER *Sumter County* ELEVATION 171 ft LAT/LONG 33° 56 ' N / 80° 19 ' W

| | JAN | FEB | MAR | APR | MAY | JUN | JUL | AUG | SEP | OCT | NOV | DEC | YEAR |
|---|---|---|---|---|---|---|---|---|---|---|---|---|---|
| Maximum Temp °F | 55.7 | 59.8 | 68.2 | 76.8 | 83.1 | 88.6 | 91.6 | 89.6 | 85.2 | 76.2 | 67.9 | 59.4 | 75.2 |
| Minimum Temp °F | 32.9 | 35.3 | 42.1 | 49.7 | 57.9 | 65.3 | 69.4 | 68.0 | 62.4 | 50.4 | 41.9 | 35.7 | 50.9 |
| Mean Temp °F | 44.3 | 47.6 | 55.2 | 63.3 | 70.5 | 77.0 | 80.5 | 78.8 | 73.8 | 63.3 | 54.9 | 47.6 | 63.1 |
| Days Max Temp ≥ 90 °F | 0 | 0 | 0 | 1 | 4 | 14 | 22 | 18 | 8 | 0 | 0 | 0 | 67 |
| Days Max Temp ≤ 32 °F | 0 | 0 | 0 | 0 | 0 | 0 | 0 | 0 | 0 | 0 | 0 | 0 | 0 |
| Days Min Temp ≤ 32 °F | 16 | 13 | 6 | 1 | 0 | 0 | 0 | 0 | 0 | 1 | 7 | 13 | 57 |
| Days Min Temp ≤ 0 °F | 0 | 0 | 0 | 0 | 0 | 0 | 0 | 0 | 0 | 0 | 0 | 0 | 0 |
| Heating Degree Days | 636 | 489 | 317 | 118 | 21 | 1 | 0 | 0 | 8 | 118 | 311 | 539 | 2558 |
| Cooling Degree Days | 0 | 4 | 15 | 59 | 185 | 384 | 508 | 438 | 272 | 69 | 12 | 3 | 1949 |
| Total Precipitation (") | 4.21 | 3.50 | 4.61 | 2.77 | 3.68 | 5.57 | 5.72 | 4.98 | 3.65 | 2.99 | 2.74 | 3.46 | 47.88 |
| Days ≥ 0.1" Precip | 8 | 6 | 7 | 5 | 6 | 7 | 8 | 7 | 5 | 4 | 5 | 6 | 74 |
| Total Snowfall (") | ***0.0*** | ***0.2*** | 0.0 | 0.0 | 0.0 | 0.0 | 0.0 | 0.0 | 0.0 | 0.0 | 0.0 | 0.0 | 0.2 |
| Days ≥ 1" Snow Depth | 0 | ***0*** | 0 | 0 | 0 | 0 | 0 | 0 | 0 | 0 | 0 | 0 | 0 |

## UNION 8 SW *Union County* ELEVATION 551 ft LAT/LONG 34° 39 ' N / 81° 45 ' W

| | JAN | FEB | MAR | APR | MAY | JUN | JUL | AUG | SEP | OCT | NOV | DEC | YEAR |
|---|---|---|---|---|---|---|---|---|---|---|---|---|---|
| Maximum Temp °F | 51.6 | 56.2 | 64.8 | 74.3 | 80.7 | 87.2 | 90.6 | 88.8 | 83.4 | 73.8 | 64.4 | 55.4 | 72.6 |
| Minimum Temp °F | 26.7 | 28.5 | 35.3 | 43.4 | 52.2 | 60.9 | 65.5 | 64.4 | 57.6 | 44.4 | 36.1 | 29.7 | 45.4 |
| Mean Temp °F | 39.2 | 42.4 | 50.1 | 58.9 | 66.5 | 74.1 | 78.1 | 76.7 | 70.5 | 59.1 | 50.3 | 42.5 | 59.0 |
| Days Max Temp ≥ 90 °F | 0 | 0 | 0 | 1 | 3 | 11 | 19 | 15 | 7 | 0 | 0 | 0 | 56 |
| Days Max Temp ≤ 32 °F | 1 | 0 | 0 | 0 | 0 | 0 | 0 | 0 | 0 | 0 | 0 | 0 | 1 |
| Days Min Temp ≤ 32 °F | 22 | 19 | 13 | 4 | 0 | 0 | 0 | 0 | 0 | 3 | 12 | 20 | 93 |
| Days Min Temp ≤ 0 °F | 0 | 0 | 0 | 0 | 0 | 0 | 0 | 0 | 0 | 0 | 0 | 0 | 0 |
| Heating Degree Days | 793 | 630 | 462 | 209 | 62 | 5 | 0 | 0 | 24 | 205 | 438 | 690 | 3518 |
| Cooling Degree Days | 0 | 0 | 6 | 30 | 117 | 311 | 441 | 382 | 204 | 35 | 3 | 1 | 1530 |
| Total Precipitation (") | 4.92 | 4.20 | 5.45 | 3.27 | 3.84 | 4.11 | 3.96 | 4.48 | 3.84 | 3.95 | 3.72 | 4.02 | 49.76 |
| Days ≥ 0.1" Precip | 7 | 7 | 8 | 5 | 7 | 6 | 7 | 6 | 5 | 5 | 6 | 7 | 76 |
| Total Snowfall (") | ***0.9*** | 0.9 | 0.1 | 0.0 | 0.0 | 0.0 | 0.0 | 0.0 | 0.0 | 0.0 | 0.0 | 0.3 | 2.2 |
| Days ≥ 1" Snow Depth | ***0*** | ***0*** | 0 | 0 | 0 | 0 | 0 | 0 | 0 | 0 | 0 | 0 | 0 |

## WALHALLA *Oconee County* ELEVATION 981 ft LAT/LONG 34° 45' N / 83° 5' W

| | JAN | FEB | MAR | APR | MAY | JUN | JUL | AUG | SEP | OCT | NOV | DEC | YEAR |
|---|---|---|---|---|---|---|---|---|---|---|---|---|---|
| Maximum Temp °F | 51.5 | 56.1 | 64.4 | 72.9 | 78.8 | 85.0 | 88.2 | 86.5 | 81.3 | 72.4 | 62.9 | 54.5 | 71.2 |
| Minimum Temp °F | 29.2 | 30.9 | 37.6 | 44.5 | 52.9 | 60.9 | 65.0 | 64.4 | 58.8 | 46.9 | 38.2 | 32.0 | 46.8 |
| Mean Temp °F | 40.4 | 43.5 | 51.0 | 58.7 | 65.9 | 73.0 | 76.6 | 75.5 | 70.1 | 59.7 | 50.6 | 43.2 | 59.0 |
| Days Max Temp ≥ 90 °F | 0 | 0 | 0 | 0 | 1 | 6 | 13 | 8 | 2 | 0 | 0 | 0 | 30 |
| Days Max Temp ≤ 32 °F | 1 | 0 | 0 | 0 | 0 | 0 | 0 | 0 | 0 | 0 | 0 | 0 | 1 |
| Days Min Temp ≤ 32 °F | 19 | 17 | 11 | 4 | 0 | 0 | 0 | 0 | 0 | 3 | 10 | 18 | 82 |
| Days Min Temp ≤ 0 °F | 0 | 0 | 0 | 0 | 0 | 0 | 0 | 0 | 0 | 0 | 0 | 0 | 0 |
| Heating Degree Days | 757 | 600 | 428 | 203 | 57 | 4 | 0 | 0 | 20 | 189 | 429 | 668 | 3355 |
| Cooling Degree Days | 0 | 0 | 3 | 19 | 89 | 264 | 380 | 337 | 178 | 32 | 1 | 1 | 1304 |
| Total Precipitation (") | 5.54 | 5.14 | 6.39 | 4.23 | 5.68 | 4.76 | 5.06 | 5.54 | 4.38 | 4.46 | 4.75 | 4.96 | 60.89 |
| Days ≥ 0.1" Precip | 8 | 8 | 9 | 7 | 8 | 7 | 8 | 8 | 7 | 5 | 7 | 8 | 90 |
| Total Snowfall (") | 2.2 | 1.5 | 0.8 | 0.0 | 0.0 | 0.0 | 0.0 | 0.0 | 0.0 | 0.0 | 0.1 | 0.5 | 5.1 |
| Days ≥ 1" Snow Depth | 1 | 1 | 0 | 0 | 0 | 0 | 0 | 0 | 0 | 0 | 0 | 0 | 2 |

## WALTERBORO 2 S *Colleton County* ELEVATION 69 ft LAT/LONG 32° 54' N / 80° 40' W

| | JAN | FEB | MAR | APR | MAY | JUN | JUL | AUG | SEP | OCT | NOV | DEC | YEAR |
|---|---|---|---|---|---|---|---|---|---|---|---|---|---|
| Maximum Temp °F | 58.2 | 62.3 | 70.7 | 78.5 | 84.2 | 89.0 | 91.8 | 90.0 | 85.5 | 77.3 | 69.5 | 62.0 | 76.6 |
| Minimum Temp °F | 35.0 | 37.5 | 44.7 | 50.3 | 58.4 | 65.6 | 69.8 | 69.2 | 64.6 | 52.8 | 44.0 | 37.8 | 52.5 |
| Mean Temp °F | 46.6 | 49.9 | 57.8 | 64.4 | 71.3 | 77.4 | 80.8 | 79.6 | 75.0 | 65.1 | 56.7 | 49.9 | 64.5 |
| Days Max Temp ≥ 90 °F | 0 | 0 | 0 | 2 | 6 | 15 | 22 | 18 | 8 | 1 | 0 | 0 | 72 |
| Days Max Temp ≤ 32 °F | 0 | 0 | 0 | 0 | 0 | 0 | 0 | 0 | 0 | 0 | 0 | 0 | 0 |
| Days Min Temp ≤ 32 °F | 13 | 10 | 4 | 1 | 0 | 0 | 0 | 0 | 0 | 0 | 6 | 11 | 45 |
| Days Min Temp ≤ 0 °F | 0 | 0 | 0 | 0 | 0 | 0 | 0 | 0 | 0 | 0 | 0 | 0 | 0 |
| Heating Degree Days | 565 | 426 | 245 | 88 | 13 | 1 | 0 | 0 | 4 | 91 | 264 | 467 | 2164 |
| Cooling Degree Days | 1 | 7 | 29 | 75 | ***227*** | ***419*** | 534 | ***487*** | 326 | ***111*** | 27 | ***6*** | 2249 |
| Total Precipitation (") | 3.74 | 3.56 | 4.77 | 2.76 | 4.45 | 5.27 | 5.88 | 5.81 | 4.11 | 2.84 | 2.46 | 3.74 | 49.39 |
| Days ≥ 0.1" Precip | ***6*** | 6 | 6 | 5 | 7 | 7 | 9 | 8 | 6 | 4 | 4 | 6 | 74 |
| Total Snowfall (") | 0.0 | 0.3 | 0.0 | 0.0 | 0.0 | 0.0 | 0.0 | 0.0 | 0.0 | 0.0 | 0.0 | 0.2 | 0.5 |
| Days ≥ 1" Snow Depth | 0 | 0 | 0 | 0 | 0 | 0 | 0 | 0 | 0 | 0 | 0 | 0 | 0 |

## WEST PELZER *Anderson County* ELEVATION 850 ft LAT/LONG 34° 39' N / 82° 28' W

| | JAN | FEB | MAR | APR | MAY | JUN | JUL | AUG | SEP | OCT | NOV | DEC | YEAR |
|---|---|---|---|---|---|---|---|---|---|---|---|---|---|
| Maximum Temp °F | 51.7 | 56.6 | 65.4 | 73.9 | 80.1 | 87.0 | 90.1 | 88.4 | 82.7 | 73.2 | 63.7 | 54.9 | 72.3 |
| Minimum Temp °F | 30.3 | 32.6 | 40.0 | 47.4 | 55.9 | 63.8 | 67.7 | 66.8 | 60.9 | 48.7 | 40.0 | 33.3 | 48.9 |
| Mean Temp °F | 41.0 | 44.6 | 52.7 | 60.7 | 68.0 | 75.4 | 78.9 | 77.6 | 71.8 | 61.0 | 51.9 | 44.1 | 60.6 |
| Days Max Temp ≥ 90 °F | 0 | 0 | 0 | 0 | 1 | 10 | 18 | 13 | 4 | 0 | 0 | 0 | 46 |
| Days Max Temp ≤ 32 °F | 1 | 0 | 0 | 0 | 0 | 0 | 0 | 0 | 0 | 0 | 0 | 0 | 1 |
| Days Min Temp ≤ 32 °F | 18 | 15 | 8 | 2 | 0 | 0 | 0 | 0 | 0 | 1 | 8 | 16 | 68 |
| Days Min Temp ≤ 0 °F | 0 | 0 | 0 | 0 | 0 | 0 | 0 | 0 | 0 | 0 | 0 | 0 | 0 |
| Heating Degree Days | 737 | 569 | 381 | 162 | 38 | 2 | 0 | 0 | 15 | 161 | 389 | 642 | 3096 |
| Cooling Degree Days | 0 | 0 | 7 | 40 | 146 | 349 | 471 | 418 | 230 | 49 | 4 | 1 | 1715 |
| Total Precipitation (") | 4.91 | 4.25 | 5.41 | 3.41 | 4.56 | 3.88 | 4.04 | 3.63 | 4.16 | 3.89 | 3.80 | 4.25 | 50.19 |
| Days ≥ 0.1" Precip | 8 | 7 | 8 | 6 | 7 | 6 | 7 | 6 | 6 | 5 | 6 | 7 | 79 |
| Total Snowfall (") | 1.3 | 1.6 | 0.9 | 0.0 | 0.0 | 0.0 | 0.0 | 0.0 | 0.0 | 0.0 | 0.0 | 0.3 | 4.1 |
| Days ≥ 1" Snow Depth | 1 | 1 | 0 | 0 | 0 | 0 | 0 | 0 | 0 | 0 | 0 | 0 | 2 |

## WINNSBORO *Fairfield County* ELEVATION 551 ft LAT/LONG 34° 23' N / 81° 5' W

| | JAN | FEB | MAR | APR | MAY | JUN | JUL | AUG | SEP | OCT | NOV | DEC | YEAR |
|---|---|---|---|---|---|---|---|---|---|---|---|---|---|
| Maximum Temp °F | 52.4 | 57.0 | 65.7 | 74.8 | 80.9 | 87.2 | 90.6 | 88.3 | 83.1 | 74.2 | 65.4 | 56.3 | 73.0 |
| Minimum Temp °F | 30.3 | 32.3 | 40.1 | 48.4 | 57.0 | 64.7 | 69.3 | 68.0 | 61.7 | 49.3 | 40.8 | 33.6 | 49.6 |
| Mean Temp °F | 41.4 | 44.7 | 52.9 | 61.5 | 69.0 | 76.0 | 80.0 | 78.1 | 72.4 | 61.7 | 53.1 | 44.9 | 61.3 |
| Days Max Temp ≥ 90 °F | 0 | 0 | 0 | 1 | 2 | 11 | 19 | 14 | 6 | 0 | 0 | 0 | 53 |
| Days Max Temp ≤ 32 °F | 1 | 0 | 0 | 0 | 0 | 0 | 0 | 0 | 0 | 0 | 0 | 0 | 1 |
| Days Min Temp ≤ 32 °F | 18 | 15 | 8 | 1 | 0 | 0 | 0 | 0 | 0 | 1 | 7 | 15 | 65 |
| Days Min Temp ≤ 0 °F | 0 | 0 | 0 | 0 | 0 | 0 | 0 | 0 | 0 | 0 | 0 | 0 | 0 |
| Heating Degree Days | 725 | 568 | 378 | 153 | 34 | 3 | 0 | 0 | 14 | 147 | 360 | 617 | 2999 |
| Cooling Degree Days | 0 | 2 | 13 | 53 | 175 | 377 | 510 | 430 | 245 | 60 | 7 | 2 | 1874 |
| Total Precipitation (") | 4.49 | 3.61 | 4.80 | 3.03 | 3.58 | 4.31 | 4.34 | 4.20 | 3.34 | 3.30 | 2.99 | 3.41 | 45.40 |
| Days ≥ 0.1" Precip | 8 | 5 | 7 | 5 | 6 | 6 | 7 | 6 | 5 | 4 | 5 | 6 | 70 |
| Total Snowfall (") | 0.8 | 0.9 | 0.1 | 0.0 | 0.0 | 0.0 | 0.0 | 0.0 | 0.0 | 0.0 | 0.0 | 0.2 | 2.0 |
| Days ≥ 1" Snow Depth | ***0*** | ***0*** | 0 | 0 | 0 | 0 | 0 | 0 | 0 | 0 | 0 | 0 | 0 |

## WINTHROP UNIVERSITY *York County* ELEVATION 689 ft LAT/LONG 34° 56' N / 81° 2' W

| | JAN | FEB | MAR | APR | MAY | JUN | JUL | AUG | SEP | OCT | NOV | DEC | YEAR |
|---|---|---|---|---|---|---|---|---|---|---|---|---|---|
| Maximum Temp °F | 50.9 | 55.4 | 64.3 | 73.4 | 79.8 | 86.4 | 89.7 | 87.7 | 82.2 | 72.5 | 63.3 | 54.5 | 71.7 |
| Minimum Temp °F | 31.6 | 34.0 | 41.4 | 49.5 | 57.9 | 65.6 | 69.8 | 68.5 | 62.5 | 50.8 | 42.1 | 35.2 | 50.7 |
| Mean Temp °F | 41.3 | 44.7 | 52.9 | 61.5 | 68.9 | 76.0 | 79.8 | 78.1 | 72.4 | 61.7 | 52.7 | 44.8 | 61.2 |
| Days Max Temp ≥ 90 °F | 0 | 0 | 0 | 0 | 2 | 8 | 16 | 12 | 4 | 0 | 0 | 0 | 42 |
| Days Max Temp ≤ 32 °F | 1 | 0 | 0 | 0 | 0 | 0 | 0 | 0 | 0 | 0 | 0 | 0 | 1 |
| Days Min Temp ≤ 32 °F | 17 | 14 | 6 | 1 | 0 | 0 | 0 | 0 | 0 | 0 | 6 | 13 | 57 |
| Days Min Temp ≤ 0 °F | 0 | 0 | 0 | 0 | 0 | 0 | 0 | 0 | 0 | 0 | 0 | 0 | 0 |
| Heating Degree Days | 729 | 567 | 379 | 149 | 32 | 2 | 0 | 0 | 12 | 144 | 367 | 619 | 3000 |
| Cooling Degree Days | 0 | 1 | 8 | 49 | 161 | 370 | 495 | 425 | 244 | 56 | 4 | 2 | 1815 |
| Total Precipitation (") | 4.34 | 3.95 | 5.17 | 2.84 | 3.89 | 4.29 | 4.06 | 4.54 | 4.10 | 3.73 | 3.49 | 3.63 | 48.03 |
| Days ≥ 0.1" Precip | 7 | 7 | 8 | 5 | 7 | 7 | 7 | 6 | 5 | 5 | 6 | 7 | 77 |
| Total Snowfall (") | 1.9 | 1.2 | 0.7 | 0.0 | 0.0 | 0.0 | 0.0 | 0.0 | 0.0 | 0.0 | 0.1 | 0.5 | 4.4 |
| Days ≥ 1" Snow Depth | 2 | 1 | 1 | 0 | 0 | 0 | 0 | 0 | 0 | 0 | 0 | 0 | 4 |

## YEMASSEE *Beaufort County* ELEVATION 23 ft LAT/LONG 32° 41' N / 80° 51' W

| | JAN | FEB | MAR | APR | MAY | JUN | JUL | AUG | SEP | OCT | NOV | DEC | YEAR |
|---|---|---|---|---|---|---|---|---|---|---|---|---|---|
| Maximum Temp °F | 59.6 | 63.1 | 71.2 | 78.8 | 85.0 | 90.0 | 92.5 | 90.7 | 86.4 | 78.2 | 70.2 | 63.0 | 77.4 |
| Minimum Temp °F | 35.1 | 37.0 | 43.8 | 50.2 | 57.7 | 65.3 | 69.3 | 68.5 | 63.5 | 52.3 | 43.4 | 37.2 | 51.9 |
| Mean Temp °F | 47.4 | 50.1 | 57.5 | 64.5 | 71.4 | 77.7 | 80.9 | 79.6 | 75.0 | 65.3 | 56.8 | 50.1 | 64.7 |
| Days Max Temp ≥ 90 °F | 0 | 0 | 0 | 2 | 7 | 18 | 24 | 20 | 9 | 1 | 0 | 0 | 81 |
| Days Max Temp ≤ 32 °F | 0 | 0 | 0 | 0 | 0 | 0 | 0 | 0 | 0 | 0 | 0 | 0 | 0 |
| Days Min Temp ≤ 32 °F | 13 | 10 | 4 | 1 | 0 | 0 | 0 | 0 | 0 | 1 | 5 | 11 | 45 |
| Days Min Temp ≤ 0 °F | 0 | 0 | 0 | 0 | 0 | 0 | 0 | 0 | 0 | 0 | 0 | 0 | 0 |
| Heating Degree Days | 542 | 421 | 252 | 88 | 11 | 0 | 0 | 0 | 3 | 84 | 261 | 460 | 2122 |
| Cooling Degree Days | 2 | 7 | 26 | 76 | 222 | 418 | 524 | 470 | 320 | 108 | 28 | 7 | 2208 |
| Total Precipitation (") | 4.06 | 3.59 | 4.53 | 3.29 | 4.26 | 5.77 | 5.60 | 7.03 | 4.52 | 3.27 | 2.42 | 3.45 | 51.79 |
| Days ≥ 0.1" Precip | 6 | 6 | 6 | 5 | 6 | 7 | 9 | 9 | 6 | 4 | 4 | 5 | 73 |
| Total Snowfall (") | 0.1 | 0.2 | 0.0 | 0.0 | 0.0 | 0.0 | 0.0 | 0.0 | 0.0 | 0.0 | 0.0 | 0.1 | 0.4 |
| Days ≥ 1" Snow Depth | 0 | 0 | 0 | 0 | 0 | 0 | 0 | 0 | 0 | 0 | 0 | 0 | 0 |

## JANUARY MINIMUM TEMPERATURE °F

### LOWEST

| | | |
|---|---|---|
| 1 | Ninety Nine Islnds | 26.5 |
| 2 | Union | 26.7 |
| 3 | Laurens | 28.3 |
| 4 | Camden | 28.4 |
| 5 | Cheraw | 28.8 |
| 6 | Greenwood | 29.1 |
| 7 | Walhalla | 29.2 |
| 8 | Clemson | 29.3 |
| 9 | Saluda | 29.4 |
| 10 | Parr | 30.1 |
| 11 | Greer | 30.2 |
| 12 | Chester | 30.3 |
| | West Pelzer | 30.3 |
| | Winnsboro | 30.3 |
| 15 | Anderson | 30.6 |
| | Dillon | 30.6 |
| 17 | Bishopville | 30.7 |
| 18 | Clark Hill | 30.8 |
| 19 | Calhoun Falls | 30.9 |
| 20 | Kingstree | 31.0 |
| 21 | Pickens | 31.3 |
| 22 | Johnston | 31.4 |
| 23 | Santuck | 31.5 |
| 24 | Winthrop | 31.6 |
| 25 | Newberry | 31.9 |

### HIGHEST

| | | |
|---|---|---|
| 1 | Charleston | 41.4 |
| 2 | Beaufort | 39.5 |
| 3 | Sullivans Island | 39.1 |
| 4 | Hilton Head | 38.6 |
| 5 | Charleston-Intl | 37.3 |
| 6 | Edisto Island | 36.9 |
| 7 | Georgetown | 36.6 |
| | Ridgeland | 36.6 |
| 9 | Hampton | 36.2 |
| 10 | Columbia-U of SC | 35.7 |
| 11 | Bamberg | 35.3 |
| 12 | Brookgreen Grdns | 35.2 |
| 13 | Yemassee | 35.1 |
| 14 | Walterboro | 35.0 |
| 15 | Blackville | 34.7 |
| 16 | Pinopolis Dam | 34.5 |
| 17 | Andrews | 34.4 |
| | Florence-Cty | 34.4 |
| 19 | Summerville | 34.0 |
| 20 | Little Mountain | 33.8 |
| 21 | Conway | 33.7 |
| 22 | Marion | 33.5 |
| 23 | Aiken | 33.4 |
| 24 | Darlington | 33.2 |
| | Orangeburg | 33.2 |

## JULY MAXIMUM TEMPERATURE °F

### HIGHEST

| | | |
|---|---|---|
| 1 | Columbia-U of SC | 93.2 |
| 2 | Aiken | 92.6 |
| 3 | Yemassee | 92.5 |
| 4 | Marion | 92.4 |
| 5 | Hampton | 92.3 |
| 6 | Clark Hill | 92.2 |
| | Columbia-Metro | 92.2 |
| 8 | Parr | 92.1 |
| | Saluda | 92.1 |
| 10 | Newberry | 92.0 |
| 11 | Blackville | 91.8 |
| | Darlington | 91.8 |
| | Ridgeland | 91.8 |
| | Walterboro | 91.8 |
| 15 | Kingstree | 91.7 |
| 16 | Sumter | 91.6 |
| 17 | Andrews | 91.5 |
| | Bamberg | 91.5 |
| | Pinopolis Dam | 91.5 |
| 20 | McColl | 91.4 |
| | Pelion | 91.4 |
| 22 | Calhoun Falls | 91.3 |
| | Lake City | 91.3 |
| 24 | Dillon | 91.1 |
| 25 | Laurens | 91.0 |

### LOWEST

| | | |
|---|---|---|
| 1 | Walhalla | 88.2 |
| 2 | Charleston | 88.4 |
| 3 | Pickens | 88.5 |
| 4 | Greer | 88.6 |
| 5 | Sullivans Island | 88.9 |
| 6 | Edisto Island | 89.4 |
| | Hilton Head | 89.4 |
| 8 | Clemson | 89.5 |
| | Ninety Nine Islnds | 89.5 |
| 10 | Camden | 89.6 |
| 11 | Winthrop | 89.7 |
| 12 | Anderson-County | 90.0 |
| 13 | West Pelzer | 90.1 |
| 14 | Beaufort | 90.2 |
| 15 | Brookgreen Grdns | 90.3 |
| | Chester | 90.3 |
| 17 | Georgetown | 90.4 |
| 18 | Bishopville | 90.5 |
| | Little Mountain | 90.5 |
| | Santuck | 90.5 |
| 21 | Anderson | 90.6 |
| | Cheraw | 90.6 |
| | Greenwood | 90.6 |
| | Summerville | 90.6 |
| | Union | 90.6 |

## ANNUAL PRECIPITATION (")

### HIGHEST

| | | |
|---|---|---|
| 1 | Walhalla | 60.89 |
| 2 | Pickens | 56.22 |
| 3 | Georgetown | 53.84 |
| 4 | Clemson | 53.66 |
| 5 | Brookgreen Grdns | 52.89 |
| 6 | Summerville | 52.72 |
| 7 | Aiken | 52.39 |
| 8 | Hilton Head | 52.00 |
| 9 | Yemassee | 51.79 |
| 10 | Ridgeland | 51.26 |
| 11 | Charleston-Intl | 51.13 |
| 12 | Conway | 51.07 |
| 13 | Beaufort | 51.00 |
| 14 | Anderson | 50.62 |
| 15 | Pinopolis Dam | 50.36 |
| 16 | West Pelzer | 50.19 |
| 17 | Pelion | 50.07 |
| 18 | Sullivans Island | 50.02 |
| 19 | Greer | 49.96 |
| 20 | Union | 49.76 |
| 21 | Columbia-Metro | 49.67 |
| 22 | Walterboro | 49.39 |
| 23 | Edisto Island | 49.35 |
| 24 | Kingstree | 49.23 |
| 25 | Marion | 48.95 |

### LOWEST

| | | |
|---|---|---|
| 1 | McColl | 40.28 |
| 2 | Florence-Cty | 44.35 |
| 3 | Bishopville | 44.62 |
| 4 | Parr | 45.11 |
| 5 | Dillon | 45.28 |
| 6 | Winnsboro | 45.40 |
| 7 | Anderson-County | 45.69 |
| 8 | Camden | 45.75 |
| 9 | Florence-8 NE | 45.97 |
| 10 | Clark Hill | 46.07 |
| 11 | Greenwood | 46.18 |
| 12 | Darlington | 46.45 |
| 13 | Blackville | 46.60 |
| 14 | Sandhill | 46.88 |
| 15 | Santuck | 46.93 |
| 16 | Columbia-U of SC | 47.11 |
| 17 | Orangeburg | 47.13 |
| 18 | Laurens | 47.26 |
| 19 | Calhoun Falls | 47.27 |
| 20 | Charleston | 47.62 |
| 21 | Cheraw | 47.73 |
| | Chester | 47.73 |
| 23 | Bamberg | 47.84 |
| 24 | Sumter | 47.88 |
| 25 | Andrews | 47.95 |

## ANNUAL SNOWFALL (")

### HIGHEST

| | | |
|---|---|---|
| 1 | Greer | 6.4 |
| 2 | Walhalla | 5.1 |
| 3 | Ninety Nine Islnds | 4.7 |
| 4 | Winthrop | 4.4 |
| 5 | Clemson | 4.3 |
| | Santuck | 4.3 |
| 7 | West Pelzer | 4.1 |
| 8 | Pickens | 4.0 |
| 9 | Chester | 3.9 |
| 10 | Anderson-County | 3.4 |
| 11 | Cheraw | 3.2 |
| | Little Mountain | 3.2 |
| 13 | Florence-Cty | 2.9 |
| 14 | Laurens | 2.7 |
| 15 | Greenwood | 2.5 |
| 16 | McColl | 2.4 |
| | Newberry | 2.4 |
| 18 | Bishopville | 2.3 |
| | Johnston | 2.3 |
| | Lake City | 2.3 |
| 21 | Columbia-Metro | 2.2 |
| | Union | 2.2 |
| 23 | Anderson | 2.1 |
| | Marion | 2.1 |
| 25 | Winnsboro | 2.0 |

### LOWEST

| | | |
|---|---|---|
| 1 | Edisto Island | 0.1 |
| | Hilton Head | 0.1 |
| | Pinopolis Dam | 0.1 |
| | Sullivans Island | 0.1 |
| 5 | Sumter | 0.2 |
| 6 | Georgetown | 0.4 |
| | Yemassee | 0.4 |
| 8 | Beaufort | 0.5 |
| | Walterboro | 0.5 |
| 10 | Sandhill | 0.6 |
| 11 | Ridgeland | 0.7 |
| 12 | Orangeburg | 0.8 |
| 13 | Andrews | 0.9 |
| | Summerville | 0.9 |
| 15 | Charleston-Intl | 1.0 |
| 16 | Aiken | 1.1 |
| 17 | Brookgreen Grdns | 1.2 |
| | Florence-8 NE | 1.2 |
| | Hampton | 1.2 |
| | Pelion | 1.2 |
| 21 | Blackville | 1.3 |
| | Parr | 1.3 |
| 23 | Bamberg | 1.4 |
| | Camden | 1.4 |
| | Columbia-U of SC | 1.4 |

# SOUTH DAKOTA

PHYSICAL FEATURES. Rolling plains are the main feature of South Dakota, varying from nearly level land to hilly ridges, and increasing in elevation from the eastern border to the western edge of the State. The general elevation above sea level in the extreme eastern portion is about 1,500 feet, and in the extreme west is about 3,000 feet, except in the Black Hills area. The Black Hills, an isolated group of forest-covered mountains, have a climate of their own.

The soil covering the State was laid down in past ages by glaciers, water, and wind. There are occasional outcroppings of bedrock. The Missouri River and its tributaries drain all of South Dakota except for a small portion of the northeastern part of the State. Some of this small drainage area is in the headwaters of the Red River of the North in the Hudson Bay Drainage, and the remainder is in the headwater area of the Minnesota River which forms a part of the upper Mississippi River Basin.

South Dakota is bisected by the Missouri River which flows in a southerly direction to Pierre and then turns to the south-southeast where it forms the South Dakota-Nebraska State line. West of the Missouri lies a country of canyons, broad upland flats, and buttes. The principal tributaries which drain this region are the Grand, and the Moreau and Cheyenne, which drain the Black Hills, and the White. To the east of the Missouri is mostly prairie land with numerous small ponds and lakes, some of which dry up in periods of droughts. The principal rivers in this area are the James and the Big Sioux. The larger of the two, the James River, has an extremely low slope and consequently is sluggish and meanders. Water falling on much of the eastern area does not reach the stream valleys at all, but lies in depressions until it evaporates or soaks into the ground.

GENERAL CLIMATE. Since South Dakota is situated in the heart of the North American Continent, it is near the paths of many cyclones and anticyclones, and has the extremes of summer heat and winter cold that are characteristic of continental climates. Rapid fluctuations in temperature are common. Partly because of the great distance from any large body of water, the ranges of daily, monthly, and annual temperatures are very large. Temperatures of 100° F. or higher are experienced in some part of the State each summer, and on rare occasions such readings have been noted as early as April and as late as October. These high temperatures are usually attended by low humidity, which greatly reduces the oppressiveness of the heat. Below-zero temperatures occur frequently on midwinter mornings, but it is not often that the temperature stays below zero during the entire day. In the north, subzero temperatures can occur in October and April.

Warm "chinook" winds and frequent sunny skies make the Black Hills area the warmest part of the State in winter. Also, because of the tendency for very cold air masses to stay at low elevations, some of the Arctic air outbreaks that blanket the eastern counties do not reach the higher counties in the west. During summer, the higher elevation of the Black Hills results in that section having cooler temperatures than the rest of the State. At this season, the central and southeastern counties are warmest. The freeze-free season is shortest high in the Black Hills where brief freezing has been known to occur at any time of summer. Elsewhere, the first autumn freeze generally occurs in mid-September in the northwest, in late September in the central and east, and in the first week of October in the extreme southeastern corner. The average time of the last freeze in spring ranges from early May in the southeast to late May in the northwest.

PRECIPITATION. The annual precipitation decreases northwestward from about 25 inches in the extreme southeast to less than 13 inches in part of the northwest. The Black Hills are again an exception, varying from 16 inches in their southern portion to almost 25 inches in the northern, where rain and snow are often formed when the prevailing winds are abruptly forced up the mountainsides. Most of the State's precipitation occurs from April through September. On the average, it reaches a maximum during June, and decreases sharply in early July. The least precipitation is received during winter.

SNOWFALL. Occasionally there is heavy snowfall in winter and the amount of snow on the ground accumulates to a considerable depth, but as a rule the snow cover is not great. Wind usually accompanies the snow, causing a large proportion of it to collect in gullies and behind windbreaks. In the worst storms, isolated drifts many feet deep may block roads, while windswept fields nearby are nearly bare of snow. Snow that falls early in the season seldom stays on the ground very long. After the ground has frozen deeply and the days become very short, it remains longer. Once a snow cover is present, there is a tendency for it to continue, since the temperature falls to much lower levels over snow than over bare ground. Snowfall reaches a maximum in February and early March, and decreases markedly near the end of March. Violent, cold winds carrying snow picked up from the ground, commonly called "blizzards", are not very frequent.

STORMS. Rainstorms occur most frequently in early summer, hailstorms are most frequent in midsummer, and lightning does its worst damage in late summer. In dry seasons, and particularly in the west in late summer, thunderstorm bases may be as high as 2 miles above the ground; consequently, the rain showers may evaporate before reaching ground. When the thundershowers do reach the ground during summer, there is a high incidence of hail. Tornadoes are not as frequent as in states farther south and east. Much more damage is caused by straight-line thunderstorm winds. Such winds are not impeded by trees or other obstacles on the open prairie, and speeds near the ground become very high.

The most serious flooding has been caused by rapid melting of a heavy snow pack and aggravated by ice jams. Heavy rainfall alone causes severe floods on tributary streams, especially in the eastern part of the State. Intense local storms result in flash flooding along minor tributaries.

OTHER CLIMATIC ELEMENTS. South Dakota has considerable fair weather. The air is generally clear with excellent visibility, since much of it arrives by way of the Rocky Mountains and Canada. The wind is most frequently from the south and southeast during the summer, and from the north and northwest during the winter. Wind speeds are often moderate to fresh at midday and almost calm at night, averaging 11 or 12 m.p.h. on a year-round basis.

## COUNTY INDEX

***Marshall County***
BRITTON

***Meade County***
FAITH
FORT MEADE

***Mellette County***
CEDAR BUTTE
WOOD

***Miner County***
HOWARD

***Minnehaha County***
SIOUX FALLS FOSS FLD

***Moody County***
FLANDREAU

***Pennington County***
MT RUSHMORE NATL MEM
PACTOLA DAM
RAPID CITY
RAPID CITY REGINL AP
WASTA

***Perkins County***
BISON
LEMMON
USTA 8 WNW KELLY RCH

***Potter County***
GETTYSBURG

***Roberts County***
SISSETON 2 E
SUMMIT 1 W

***Sanborn County***
FORESTBURG 3 NE

***Shannon County***
PORCUPINE 11 N

***Spink County***
MELLETTE
REDFIELD 2 NE

***Stanley County***
OAHE DAM

***Sully County***
ONIDA 4 NW

***Todd County***
MISSION 14 S

***Tripp County***
WINNER

***Turner County***
MARION

***Walworth County***
MOBRIDGE 2 NNW
SELBY

***Yankton County***
YANKTON 2 E

***Ziebach County***
DUPREE
DUPREE 15 SSE
GLAD VALLEY 2 W

## ELEVATION INDEX

| FEET | STATION NAME |
|---|---|
| 1142 | MILBANK 2 SSW |
| 1201 | SISSETON 2 E |
| 1220 | VERMILLION 2 SE |
| 1230 | FORESTBURG 3 NE |
| 1260 | CENTERVILLE 6 SE |
| 1283 | YANKTON 2 E |
| 1289 | MITCHELL |
| 1290 | REDFIELD 2 NE |
| 1296 | ABERDEEN REGIONAL AP |
| 1296 | HURON REGIONAL AP |
| 1302 | COLUMBIA 8 N |
| 1302 | MELLETTE |
| 1322 | MENNO |
| 1352 | ALEXANDRIA |
| 1352 | BRITTON |
| 1352 | CANTON 4 WNW |
| 1421 | BRIDGEWATER |
| 1421 | TYNDALL |
| 1424 | SIOUX FALLS FOSS FLD |
| 1440 | MARION |
| 1440 | WAGNER |
| 1470 | OAHE DAM |
| 1490 | PICKSTOWN |
| 1512 | ARMOUR |
| 1532 | IPSWICH |
| 1562 | FLANDREAU |
| 1562 | HOWARD |
| 1581 | FAULKTON |
| 1591 | LEOLA |
| 1591 | MILLER |
| 1611 | POLLOCK |
| 1621 | BROOKINGS 2 NE |
| 1631 | WHITE LAKE |
| 1640 | WESSINGTON SPRINGS |
| 1680 | ACADEMY 2 NE |
| 1690 | CASTLEWOOD |
| 1690 | KENNEBEC |
| 1696 | MOBRIDGE 2 NNW |
| 1703 | WENTWORTH 2 WNW |
| 1710 | MADISON 2 E |
| 1722 | PIERRE MUNICIPAL AP |
| 1732 | DE SMET |
| 1746 | WATERTOWN MUNI AP |
| 1752 | GANN VALLEY 4 NW |
| 1781 | CLARK |
| 1801 | CLEAR LAKE |
| 1801 | HARROLD 12 SSW |
| 1831 | STEPHAN 1 ENE |
| 1831 | WAUBAY NWR |
| 1850 | ONIDA 4 NW |
| 1870 | HIGHMORE 23 N |
| 1870 | WEBSTER |
| 1880 | EUREKA |
| 1880 | MIDLAND |
| 1880 | SELBY |
| 1890 | HIGHMORE 1 W |
| 1972 | WINNER |
| 2001 | BONESTEEL |
| 2011 | SUMMIT 1 W |
| 2060 | GETTYSBURG |
| 2100 | DUPREE 15 SSE |
| 2160 | GREGORY |
| 2162 | TIMBER LAKE |
| 2182 | WOOD |
| 2241 | PHILIP 2 N |
| 2270 | MC INTOSH 6 SE |
| 2290 | CEDAR BUTTE |
| 2313 | MURDO |
| 2323 | WASTA |
| 2342 | MILESVILLE 8 NE |
| 2352 | DUPREE |

| FEET | STATION NAME |
|---|---|
| 2382 | USTA 8 WNW KELLY RCH |
| 2414 | COTTONWOOD 2 E |
| 2441 | INTERIOR 3 NE |
| 2470 | LONG VALLEY |
| 2592 | FAITH |
| 2592 | LEMMON |
| 2731 | ZEONA 10 SSW |
| 2780 | BISON |
| 2792 | PORCUPINE 11 N |
| 2802 | MISSION 14 S |
| 2831 | RALPH 1 N |
| 2851 | LUDLOW |
| 2881 | NEWELL |
| 2904 | GLAD VALLEY 2 W |
| 2982 | HARRINGTON |
| 2992 | REDIG 11 NE |
| 3012 | BELLE FOURCHE |
| 3123 | CAMP CROOK |
| 3162 | RAPID CITY REGINL AP |
| 3304 | FORT MEADE |
| 3343 | OELRICHS |
| 3373 | RAPID CITY |
| 3553 | ARDMORE 2 N |
| 3553 | HOT SPRINGS |
| 3753 | SPEARFISH |
| 4534 | PACTOLA DAM |
| 4554 | DEADWOOD |
| 5154 | MT RUSHMORE NATL MEM |
| 5253 | LEAD |
| 5325 | CUSTER |

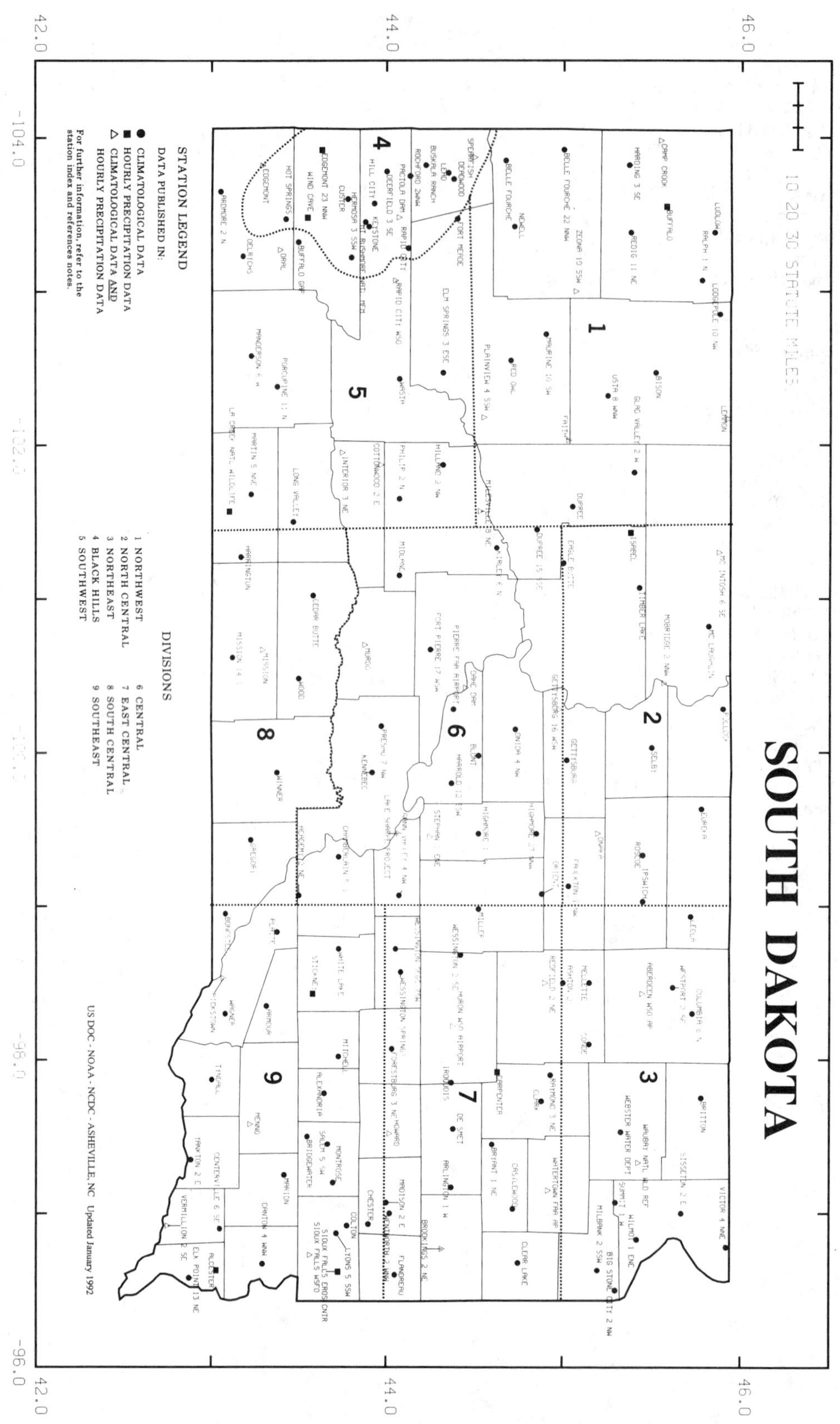
SOUTH DAKOTA
10 20 30 STATUTE MILES
42.0
44.0
46.0
-104.0
-102.0
-96.0
STATION LEGEND
DATA PUBLISHED IN:
● CLIMATOLOGICAL DATA
■ HOURLY PRECIPITATION DATA
△ CLIMATOLOGICAL DATA AND HOURLY PRECIPITATION DATA
For further information, refer to the station index and references notes.
DIVISIONS
1 NORTHWEST
2 NORTH CENTRAL
3 NORTHEAST
4 BLACK HILLS
5 SOUTHWEST
6 CENTRAL
7 EAST CENTRAL
8 SOUTH CENTRAL
9 SOUTHEAST
US DOC - NOAA - NCDC - ASHEVILLE, NC Updated January 1992

## ABERDEEN REGIONAL AP *Brown County* ELEVATION 1296 ft LAT/LONG 45° 27 ' N / 98° 26 ' W

| | JAN | FEB | MAR | APR | MAY | JUN | JUL | AUG | SEP | OCT | NOV | DEC | YEAR |
|---|---|---|---|---|---|---|---|---|---|---|---|---|---|
| Maximum Temp °F | 21.2 | 26.8 | 40.1 | 57.3 | 70.0 | 78.5 | 85.1 | 84.0 | 73.0 | 59.6 | 39.8 | 26.3 | 55.1 |
| Minimum Temp °F | -0.1 | 6.7 | 20.4 | 33.1 | 44.6 | 54.3 | 59.4 | 57.0 | 46.3 | 34.1 | 20.2 | 6.6 | 31.9 |
| Mean Temp °F | 10.5 | 16.8 | 30.3 | 45.2 | 57.4 | 66.5 | 72.3 | 70.5 | 59.7 | 46.8 | 30.0 | 16.5 | 43.5 |
| Days Max Temp ≥ 90 °F | 0 | 0 | 0 | 0 | 0 | 3 | 9 | 8 | 2 | 0 | 0 | 0 | 22 |
| Days Max Temp ≤ 32 °F | 23 | 17 | 9 | 0 | 0 | 0 | 0 | 0 | 0 | 0 | 8 | 20 | 77 |
| Days Min Temp ≤ 32 °F | 31 | 28 | 27 | 15 | 3 | 0 | 0 | 0 | 2 | 13 | 27 | 31 | 177 |
| Days Min Temp ≤ 0 °F | 16 | 10 | 2 | 0 | 0 | 0 | 0 | 0 | 0 | 0 | 2 | 10 | 40 |
| Heating Degree Days | 1686 | 1357 | 1068 | 590 | 260 | 61 | 10 | 29 | 205 | 558 | 1043 | 1500 | 8367 |
| Cooling Degree Days | 0 | 0 | 0 | 5 | 39 | 112 | 243 | 196 | 47 | 2 | 0 | 0 | 644 |
| Total Precipitation (") | 0.40 | 0.43 | 1.24 | 2.02 | 2.29 | 3.33 | 2.92 | 2.20 | 1.74 | 1.17 | 0.63 | 0.38 | 18.75 |
| Days ≥ 0.1" Precip | 1 | 1 | 3 | 4 | 5 | 6 | 6 | 4 | 3 | 3 | 2 | 1 | 39 |
| Total Snowfall (") | 6.3 | 6.5 | 7.5 | 3.9 | 0.1 | 0.0 | 0.0 | 0.0 | 0.0 | 1.0 | 5.7 | 5.7 | 36.7 |
| Days ≥ 1" Snow Depth | 21 | 17 | 11 | 2 | 0 | 0 | 0 | 0 | 0 | 0 | 6 | 15 | 72 |

## ACADEMY 2 NE *Charles Mix County* ELEVATION 1680 ft LAT/LONG 43° 27 ' N / 99° 5 ' W

| | JAN | FEB | MAR | APR | MAY | JUN | JUL | AUG | SEP | OCT | NOV | DEC | YEAR |
|---|---|---|---|---|---|---|---|---|---|---|---|---|---|
| Maximum Temp °F | 30.2 | 35.8 | 47.6 | 61.9 | 73.3 | 82.5 | 89.1 | 87.8 | 77.9 | 64.4 | 45.5 | 34.0 | 60.8 |
| Minimum Temp °F | 7.1 | 12.6 | 23.1 | 34.7 | 45.7 | 55.4 | 60.7 | 58.3 | 48.3 | 36.3 | 23.2 | 11.9 | 34.8 |
| Mean Temp °F | 18.7 | 24.3 | 35.4 | 48.3 | 59.5 | 69.0 | 74.9 | 73.1 | 63.1 | 50.4 | 34.4 | 23.0 | 47.8 |
| Days Max Temp ≥ 90 °F | 0 | 0 | 0 | 0 | 2 | 6 | 15 | 13 | 5 | 1 | 0 | 0 | 42 |
| Days Max Temp ≤ 32 °F | 16 | 11 | 5 | 0 | 0 | 0 | 0 | 0 | 0 | 0 | 5 | 13 | 50 |
| Days Min Temp ≤ 32 °F | 31 | 27 | 26 | 13 | 2 | 0 | 0 | 0 | 1 | 11 | 25 | 31 | 167 |
| Days Min Temp ≤ 0 °F | 11 | 6 | 1 | 0 | 0 | 0 | 0 | 0 | 0 | 0 | 1 | 6 | 25 |
| Heating Degree Days | 1431 | 1144 | 913 | 501 | 205 | 37 | 8 | 16 | 142 | 453 | 912 | 1297 | 7059 |
| Cooling Degree Days | 0 | 0 | 0 | 11 | 41 | 154 | 292 | 244 | 85 | 7 | 0 | 0 | 834 |
| Total Precipitation (") | 0.46 | 0.59 | 1.43 | 2.59 | 3.33 | 3.50 | 2.84 | 2.11 | 2.02 | 1.58 | 0.89 | 0.49 | 21.83 |
| Days ≥ 0.1" Precip | 2 | 1 | 4 | 6 | 6 | 6 | 5 | 4 | 4 | 3 | 2 | 2 | 45 |
| Total Snowfall (") | 6.3 | 6.6 | 8.0 | 3.3 | 0.0 | 0.0 | 0.0 | 0.0 | 0.3 | 1.6 | 6.7 | 7.0 | 39.8 |
| Days ≥ 1" Snow Depth | 20 | 13 | 9 | 1 | 0 | 0 | 0 | 0 | 0 | 1 | 6 | 14 | 64 |

## ALEXANDRIA *Hanson County* ELEVATION 1352 ft LAT/LONG 43° 39 ' N / 97° 46 ' W

| | JAN | FEB | MAR | APR | MAY | JUN | JUL | AUG | SEP | OCT | NOV | DEC | YEAR |
|---|---|---|---|---|---|---|---|---|---|---|---|---|---|
| Maximum Temp °F | 26.2 | 32.1 | 45.6 | 61.4 | 73.0 | 82.1 | 87.1 | 85.2 | 75.9 | 63.0 | 44.0 | 30.4 | 58.8 |
| Minimum Temp °F | 6.7 | 12.8 | 24.7 | 36.9 | 48.3 | 57.9 | 63.0 | 60.8 | 50.9 | 38.8 | 25.1 | 12.4 | 36.5 |
| Mean Temp °F | 16.5 | 22.4 | 35.2 | 49.2 | 60.7 | 70.0 | 75.1 | 73.0 | 63.4 | 50.9 | 34.5 | 21.4 | 47.7 |
| Days Max Temp ≥ 90 °F | 0 | 0 | 0 | 0 | 1 | 5 | 12 | 9 | 3 | 0 | 0 | 0 | 30 |
| Days Max Temp ≤ 32 °F | 19 | 13 | 5 | 0 | 0 | 0 | 0 | 0 | 0 | 0 | 5 | 16 | 58 |
| Days Min Temp ≤ 32 °F | 31 | 27 | 23 | 10 | 1 | 0 | 0 | 0 | 1 | 8 | 23 | 30 | 154 |
| Days Min Temp ≤ 0 °F | 11 | 7 | 1 | 0 | 0 | 0 | 0 | 0 | 0 | 0 | 1 | 6 | 26 |
| Heating Degree Days | 1501 | 1196 | 918 | 477 | 178 | 26 | 4 | 11 | 129 | 436 | 907 | 1344 | 7127 |
| Cooling Degree Days | 0 | 0 | 0 | 13 | 57 | 192 | 316 | 265 | 92 | 6 | 0 | 0 | 941 |
| Total Precipitation (") | 0.37 | 0.49 | 1.33 | 2.32 | 2.94 | 3.62 | 2.79 | 2.66 | 2.27 | 1.58 | 0.95 | 0.46 | 21.78 |
| Days ≥ 0.1" Precip | 1 | 2 | 3 | 5 | 6 | 6 | 5 | 4 | 4 | 3 | 3 | 2 | 44 |
| Total Snowfall (") | 4.9 | 5.1 | 5.2 | 2.0 | 0.0 | 0.0 | 0.0 | 0.0 | 0.0 | 0.7 | 4.1 | 6.2 | 28.2 |
| Days ≥ 1" Snow Depth | 21 | 16 | 7 | 1 | 0 | 0 | 0 | 0 | 0 | 0 | 5 | 16 | 66 |

## ARDMORE 2 N *Fall River County* ELEVATION 3553 ft LAT/LONG 43° 4 ' N / 103° 39 ' W

| | JAN | FEB | MAR | APR | MAY | JUN | JUL | AUG | SEP | OCT | NOV | DEC | YEAR |
|---|---|---|---|---|---|---|---|---|---|---|---|---|---|
| Maximum Temp °F | 34.3 | 39.8 | 50.0 | 61.0 | 70.9 | 81.4 | 89.0 | 87.9 | 77.3 | 64.2 | 47.2 | 35.9 | 61.6 |
| Minimum Temp °F | 5.9 | 10.6 | 20.4 | 30.4 | 40.4 | 49.7 | 55.8 | 53.1 | 41.6 | 29.7 | 17.9 | 7.1 | 30.2 |
| Mean Temp °F | 20.2 | 25.1 | 35.2 | 45.7 | 55.7 | 65.6 | 72.4 | 70.5 | 59.5 | 47.0 | 32.5 | 21.5 | 45.9 |
| Days Max Temp ≥ 90 °F | 0 | 0 | 0 | 0 | 1 | 7 | 15 | 15 | 4 | 0 | 0 | 0 | 42 |
| Days Max Temp ≤ 32 °F | 12 | 8 | 3 | 0 | 0 | 0 | 0 | 0 | 0 | 0 | 4 | 11 | 38 |
| Days Min Temp ≤ 32 °F | 31 | 28 | 28 | 18 | 5 | 0 | 0 | 0 | 5 | 19 | 27 | 31 | 192 |
| Days Min Temp ≤ 0 °F | 10 | 6 | 1 | 0 | 0 | 0 | 0 | 0 | 0 | 0 | 2 | 8 | 27 |
| Heating Degree Days | 1383 | 1120 | 917 | 572 | 294 | 75 | 11 | 19 | 198 | 555 | 969 | 1342 | 7455 |
| Cooling Degree Days | 0 | 0 | 0 | 2 | 15 | 105 | 246 | 209 | 43 | 1 | 0 | 0 | 621 |
| Total Precipitation (") | 0.39 | 0.51 | 0.94 | 1.80 | 2.73 | 2.74 | 2.22 | 1.57 | 1.25 | 1.16 | 0.55 | 0.47 | 16.33 |
| Days ≥ 0.1" Precip | 1 | 2 | 3 | 4 | 6 | 6 | 5 | 4 | 4 | 3 | 2 | 1 | 41 |
| Total Snowfall (") | 6.6 | 7.3 | 8.1 | 5.2 | 0.9 | 0.1 | 0.0 | 0.0 | 0.8 | 2.9 | 5.8 | 8.2 | 45.9 |
| Days ≥ 1" Snow Depth | 20 | 14 | 8 | 2 | 0 | 0 | 0 | 0 | 0 | 1 | 6 | 16 | 67 |

## ARMOUR *Douglas County* ELEVATION 1512 ft LAT/LONG 43° 19 ' N / 98° 21 ' W

| | JAN | FEB | MAR | APR | MAY | JUN | JUL | AUG | SEP | OCT | NOV | DEC | YEAR |
|---|---|---|---|---|---|---|---|---|---|---|---|---|---|
| Maximum Temp °F | 28.7 | 35.1 | 47.9 | 62.0 | 73.7 | 83.1 | 88.6 | 87.0 | 77.2 | 63.9 | 45.1 | 32.9 | 60.4 |
| Minimum Temp °F | 7.6 | 13.0 | 24.4 | 36.0 | 47.2 | 57.1 | 62.2 | 60.0 | 49.7 | 37.2 | 24.1 | 12.3 | 35.9 |
| Mean Temp °F | 18.3 | 24.0 | 36.3 | 49.1 | 60.5 | 70.1 | 75.5 | 73.5 | 63.4 | 50.6 | 34.6 | 22.6 | 48.2 |
| Days Max Temp ≥ 90 °F | 0 | 0 | 0 | 0 | 1 | 7 | 14 | 12 | 4 | 0 | 0 | 0 | 38 |
| Days Max Temp ≤ 32 °F | 17 | 12 | 4 | 0 | 0 | 0 | 0 | 0 | 0 | 0 | 5 | 14 | 52 |
| Days Min Temp ≤ 32 °F | 31 | 27 | 24 | 11 | 1 | 0 | 0 | 0 | 1 | 10 | 24 | 31 | 160 |
| Days Min Temp ≤ 0 °F | 10 | 6 | 1 | 0 | 0 | 0 | 0 | 0 | 0 | 0 | 1 | 6 | 24 |
| Heating Degree Days | 1442 | 1150 | 883 | 479 | 183 | 27 | 4 | 11 | 131 | 444 | 905 | 1307 | 6966 |
| Cooling Degree Days | 0 | 0 | 0 | 14 | 57 | 197 | 330 | 284 | 99 | 7 | 0 | 0 | 988 |
| Total Precipitation (") | 0.54 | 0.73 | 1.60 | 2.50 | 3.25 | 3.58 | 2.99 | 2.16 | 2.27 | 1.61 | 0.98 | 0.72 | 22.93 |
| Days ≥ 0.1" Precip | 2 | 2 | 3 | 5 | 6 | 6 | 6 | 4 | 4 | 3 | 3 | 2 | 46 |
| Total Snowfall (") | 6.2 | 6.0 | 5.2 | 1.7 | 0.0 | 0.0 | 0.0 | 0.0 | 0.0 | 0.6 | 5.0 | 7.7 | 32.4 |
| Days ≥ 1" Snow Depth | *18* | *13* | 5 | 1 | 0 | 0 | 0 | 0 | 0 | 0 | 5 | 14 | 56 |

## BELLE FOURCHE *Butte County* ELEVATION 3012 ft LAT/LONG 44° 42 ' N / 103° 49 ' W

| | JAN | FEB | MAR | APR | MAY | JUN | JUL | AUG | SEP | OCT | NOV | DEC | YEAR |
|---|---|---|---|---|---|---|---|---|---|---|---|---|---|
| Maximum Temp °F | 34.6 | 40.1 | 48.4 | 60.9 | 71.3 | 80.8 | 88.1 | 87.3 | 76.8 | 64.6 | 47.5 | 37.4 | 61.5 |
| Minimum Temp °F | 9.4 | 13.5 | 21.4 | 32.3 | 42.5 | 51.9 | 57.0 | 54.4 | 43.4 | 32.5 | 21.5 | 12.1 | 32.7 |
| Mean Temp °F | 22.0 | 26.8 | 34.9 | 46.6 | 56.9 | 66.4 | 72.6 | 70.9 | 60.1 | 48.6 | 34.5 | 24.7 | 47.1 |
| Days Max Temp ≥ 90 °F | 0 | 0 | 0 | 0 | 1 | 5 | 14 | 13 | 4 | 0 | 0 | 0 | 37 |
| Days Max Temp ≤ 32 °F | 13 | 8 | 4 | 0 | 0 | 0 | 0 | 0 | 0 | 0 | 4 | 10 | 39 |
| Days Min Temp ≤ 32 °F | 29 | 27 | 27 | 15 | 3 | 0 | 0 | 0 | 3 | 16 | 26 | 30 | 176 |
| Days Min Temp ≤ 0 °F | 9 | 5 | 2 | 0 | 0 | 0 | 0 | 0 | 0 | 0 | 1 | 6 | 23 |
| Heating Degree Days | 1328 | 1072 | 926 | 546 | 261 | 60 | 10 | 17 | 184 | 504 | 908 | 1242 | 7058 |
| Cooling Degree Days | 0 | 0 | 0 | 4 | 21 | 112 | 236 | 206 | 38 | 1 | 0 | 0 | 618 |
| Total Precipitation (") | 0.41 | 0.50 | 1.10 | 1.86 | 2.91 | 2.91 | 1.93 | 1.32 | 1.33 | 1.39 | 0.64 | 0.61 | 16.91 |
| Days ≥ 0.1" Precip | 2 | 2 | 3 | 4 | 6 | 6 | 5 | 3 | 3 | 3 | 2 | 2 | 41 |
| Total Snowfall (") | 6.8 | *5.2* | 8.9 | 3.7 | 0.9 | 0.1 | 0.0 | 0.0 | 0.3 | 1.8 | *5.1* | *7.9* | 40.7 |
| Days ≥ 1" Snow Depth | na | *9* | *5* | 1 | 0 | 0 | 0 | 0 | 0 | 0 | 4 | na | na |

## BISON *Perkins County* ELEVATION 2780 ft LAT/LONG 45° 31 ' N / 102° 28 ' W

| | JAN | FEB | MAR | APR | MAY | JUN | JUL | AUG | SEP | OCT | NOV | DEC | YEAR |
|---|---|---|---|---|---|---|---|---|---|---|---|---|---|
| Maximum Temp °F | 26.1 | 31.6 | 42.5 | 56.8 | 68.8 | 78.0 | 85.4 | 85.0 | 73.5 | 60.1 | 41.4 | 30.0 | 56.6 |
| Minimum Temp °F | 5.5 | 10.6 | 20.4 | 31.6 | 42.4 | 51.8 | 57.2 | 55.2 | 44.7 | 33.9 | 20.6 | 9.9 | 32.0 |
| Mean Temp °F | 15.8 | 21.1 | 31.5 | 44.2 | 55.7 | 64.9 | 71.3 | 70.1 | 59.1 | 47.0 | 31.0 | 20.0 | 44.3 |
| Days Max Temp ≥ 90 °F | 0 | 0 | 0 | 0 | 1 | 3 | 11 | 10 | 2 | 0 | 0 | 0 | 27 |
| Days Max Temp ≤ 32 °F | 18 | 14 | 7 | 1 | 0 | 0 | 0 | 0 | 0 | 1 | 8 | 17 | 66 |
| Days Min Temp ≤ 32 °F | 30 | 27 | 28 | 17 | 3 | 0 | 0 | 0 | 3 | 13 | 26 | 30 | 177 |
| Days Min Temp ≤ 0 °F | 12 | 7 | 2 | 0 | 0 | 0 | 0 | 0 | 0 | 0 | 2 | 8 | 31 |
| Heating Degree Days | 1522 | 1233 | 1033 | 619 | 302 | 84 | 16 | 28 | 214 | 553 | 1014 | 1390 | 8008 |
| Cooling Degree Days | 0 | 0 | 0 | 3 | 23 | 99 | 226 | 191 | 50 | 3 | 0 | 0 | 595 |
| Total Precipitation (") | 0.38 | 0.46 | 1.22 | 2.26 | 2.50 | 2.98 | 2.41 | 1.44 | 1.24 | 1.16 | 0.57 | 0.51 | 17.13 |
| Days ≥ 0.1" Precip | 2 | 2 | 3 | 5 | 6 | 7 | 5 | 4 | 3 | 2 | 2 | 2 | 43 |
| Total Snowfall (") | 5.1 | *5.9* | *7.0* | *4.3* | 0.3 | 0.0 | 0.0 | 0.0 | 0.3 | 0.9 | *3.9* | na | na |
| Days ≥ 1" Snow Depth | na | na | na | na | 0 | 0 | 0 | 0 | 0 | 0 | *2* | na | na |

## BONESTEEL *Gregory County* ELEVATION 2001 ft LAT/LONG 43° 5 ' N / 98° 57 ' W

| | JAN | FEB | MAR | APR | MAY | JUN | JUL | AUG | SEP | OCT | NOV | DEC | YEAR |
|---|---|---|---|---|---|---|---|---|---|---|---|---|---|
| Maximum Temp °F | 29.6 | 34.9 | 47.1 | 60.5 | 71.9 | 81.3 | 87.3 | 86.1 | 76.3 | 63.4 | 45.0 | 34.0 | 59.8 |
| Minimum Temp °F | 7.4 | 12.9 | 23.6 | 35.2 | 46.2 | 55.9 | 62.2 | 59.6 | 49.1 | 36.6 | 23.4 | 13.1 | 35.4 |
| Mean Temp °F | 18.5 | 23.9 | 35.4 | 47.9 | 59.0 | 68.6 | 74.8 | 72.9 | 62.7 | 49.9 | 34.3 | 23.6 | 47.6 |
| Days Max Temp ≥ 90 °F | 0 | 0 | 0 | 0 | 1 | 5 | 13 | 11 | 4 | 0 | 0 | 0 | 34 |
| Days Max Temp ≤ 32 °F | 16 | 12 | 5 | 0 | 0 | 0 | 0 | 0 | 0 | 0 | 5 | 13 | 51 |
| Days Min Temp ≤ 32 °F | 30 | 27 | 24 | 12 | 2 | 0 | 0 | 0 | 1 | 10 | 25 | 30 | 161 |
| Days Min Temp ≤ 0 °F | 11 | 6 | 1 | 0 | 0 | 0 | 0 | 0 | 0 | 0 | 1 | 5 | 24 |
| Heating Degree Days | 1437 | 1152 | 916 | 512 | 217 | 40 | 8 | 16 | 146 | 466 | 916 | 1279 | 7105 |
| Cooling Degree Days | 0 | 0 | 0 | 6 | 35 | 140 | 291 | 247 | 84 | 4 | 0 | 0 | 807 |
| Total Precipitation (") | 0.36 | 0.67 | 1.72 | 2.85 | 3.89 | 4.06 | 3.34 | 2.81 | 2.78 | 1.80 | 0.93 | 0.51 | 25.72 |
| Days ≥ 0.1" Precip | 1 | 2 | 4 | 6 | 7 | 6 | 6 | 4 | 4 | 3 | 3 | 2 | 48 |
| Total Snowfall (") | 6.5 | 6.4 | 7.4 | 4.1 | 0.0 | 0.0 | 0.0 | 0.0 | 0.1 | 1.7 | 6.9 | 7.3 | 40.4 |
| Days ≥ 1" Snow Depth | 18 | 13 | 8 | 2 | 0 | 0 | 0 | 0 | 0 | 1 | 6 | 14 | 62 |

## BRIDGEWATER *McCook County* ELEVATION 1421 ft LAT/LONG 43° 33 ' N / 97° 30 ' W

| | JAN | FEB | MAR | APR | MAY | JUN | JUL | AUG | SEP | OCT | NOV | DEC | YEAR |
|---|---|---|---|---|---|---|---|---|---|---|---|---|---|
| Maximum Temp °F | 25.7 | 32.3 | 45.7 | 61.1 | 73.5 | 83.1 | 87.5 | 85.3 | 75.3 | 63.1 | 43.1 | ***30.9*** | 58.9 |
| Minimum Temp °F | 4.8 | 11.2 | 23.5 | 35.5 | 47.3 | 57.3 | 61.9 | 59.3 | 49.5 | 37.3 | 23.3 | ***11.8*** | 35.2 |
| Mean Temp °F | 15.2 | 21.8 | 34.6 | 48.3 | 60.5 | 70.2 | 74.7 | 72.3 | 62.4 | 50.2 | 33.3 | ***21.3*** | 47.1 |
| Days Max Temp ≥ 90 °F | 0 | 0 | 0 | 0 | 1 | 7 | 12 | 9 | 3 | 0 | 0 | 0 | 32 |
| Days Max Temp ≤ 32 °F | 20 | 14 | 5 | 0 | 0 | 0 | 0 | 0 | 0 | 0 | 6 | ***16*** | 61 |
| Days Min Temp ≤ 32 °F | 31 | 27 | 25 | 12 | 2 | 0 | 0 | 0 | 1 | 10 | 25 | ***31*** | 164 |
| Days Min Temp ≤ 0 °F | 12 | 7 | 1 | 0 | 0 | 0 | 0 | 0 | 0 | 0 | 1 | ***6*** | 27 |
| Heating Degree Days | 1539 | 1213 | 933 | 501 | 184 | 26 | 6 | 14 | 146 | 457 | 946 | ***1347*** | 7312 |
| Cooling Degree Days | ***0*** | 0 | 0 | 15 | ***51*** | ***195*** | ***282*** | 220 | 75 | 5 | ***0*** | ***0*** | 843 |
| Total Precipitation (") | 0.52 | 0.50 | 1.49 | 2.23 | 3.30 | 3.56 | 3.26 | 3.16 | 2.71 | 1.80 | 1.05 | 0.56 | 24.14 |
| Days ≥ 0.1" Precip | ***2*** | 2 | 4 | 6 | 6 | 7 | 6 | 5 | 5 | 3 | 3 | ***2*** | 51 |
| Total Snowfall (") | na | na | ***3.9*** | ***0.9*** | 0.0 | 0.0 | 0.0 | 0.0 | 0.0 | 0.0 | na | na | na |
| Days ≥ 1" Snow Depth | na | na | na | 0 | 0 | 0 | 0 | 0 | 0 | 0 | na | na | na |

## BRITTON *Marshall County* ELEVATION 1352 ft LAT/LONG 45° 47 ' N / 97° 44 ' W

| | JAN | FEB | MAR | APR | MAY | JUN | JUL | AUG | SEP | OCT | NOV | DEC | YEAR |
|---|---|---|---|---|---|---|---|---|---|---|---|---|---|
| Maximum Temp °F | 19.8 | 26.2 | 40.2 | 58.0 | 71.6 | 79.6 | 85.8 | 84.2 | 73.8 | 60.2 | 39.1 | 25.5 | 55.3 |
| Minimum Temp °F | -1.4 | 5.6 | 19.7 | 33.0 | 44.9 | 53.9 | 59.1 | 57.1 | 46.7 | 34.8 | 19.9 | 6.0 | 31.6 |
| Mean Temp °F | 9.2 | 16.0 | 30.0 | 45.5 | 58.3 | 66.8 | 72.5 | 70.7 | 60.3 | 47.5 | 29.5 | 15.8 | 43.5 |
| Days Max Temp ≥ 90 °F | 0 | 0 | 0 | 0 | 1 | 3 | 10 | 8 | 2 | 0 | 0 | 0 | 24 |
| Days Max Temp ≤ 32 °F | 24 | 17 | 8 | 0 | 0 | 0 | 0 | 0 | 0 | 0 | 9 | 20 | 78 |
| Days Min Temp ≤ 32 °F | 31 | 28 | 27 | 15 | 3 | 0 | 0 | 0 | 2 | 13 | 27 | 31 | 177 |
| Days Min Temp ≤ 0 °F | 17 | 11 | 3 | 0 | 0 | 0 | 0 | 0 | 0 | 0 | 2 | 10 | 43 |
| Heating Degree Days | 1727 | 1380 | 1080 | 583 | 244 | 61 | 13 | 30 | 196 | 538 | 1057 | 1520 | 8429 |
| Cooling Degree Days | 0 | 0 | 0 | 8 | 49 | 129 | 259 | 214 | 57 | 2 | 0 | 0 | 718 |
| Total Precipitation (") | 0.52 | 0.47 | 0.87 | 1.91 | 2.63 | 3.69 | 3.07 | 2.07 | 1.91 | 1.35 | 0.74 | 0.41 | 19.64 |
| Days ≥ 0.1" Precip | 2 | 2 | 2 | 5 | 5 | 7 | 5 | 4 | 4 | 3 | 2 | 2 | 43 |
| Total Snowfall (") | na | na | na | ***2.2*** | 0.0 | 0.0 | 0.0 | 0.0 | 0.0 | 0.6 | 6.5 | ***6.7*** | na |
| Days ≥ 1" Snow Depth | 26 | ***24*** | ***14*** | ***1*** | 0 | 0 | 0 | 0 | 0 | 0 | 8 | ***18*** | 91 |

## BROOKINGS 2 NE *Brookings County* ELEVATION 1621 ft LAT/LONG 44° 18 ' N / 96° 45 ' W

| | JAN | FEB | MAR | APR | MAY | JUN | JUL | AUG | SEP | OCT | NOV | DEC | YEAR |
|---|---|---|---|---|---|---|---|---|---|---|---|---|---|
| Maximum Temp °F | 20.4 | 25.5 | 38.6 | 54.6 | 68.0 | 76.9 | 82.0 | 79.8 | 70.3 | 57.7 | 39.4 | 25.6 | 53.2 |
| Minimum Temp °F | -0.8 | 5.0 | 19.2 | 32.2 | 43.5 | 53.6 | 58.2 | 55.4 | 45.2 | 32.9 | 20.0 | 6.6 | 30.9 |
| Mean Temp °F | 9.8 | 15.3 | 28.9 | 43.4 | 55.8 | 65.3 | 70.1 | 67.6 | 57.8 | 45.3 | 29.7 | 16.1 | 42.1 |
| Days Max Temp ≥ 90 °F | 0 | 0 | 0 | 0 | 0 | 2 | 5 | 4 | 1 | 0 | 0 | 0 | 12 |
| Days Max Temp ≤ 32 °F | 24 | 18 | 10 | 1 | 0 | 0 | 0 | 0 | 0 | 0 | 9 | 21 | 83 |
| Days Min Temp ≤ 32 °F | 31 | 28 | 28 | 17 | 4 | 0 | 0 | 0 | 3 | 15 | 28 | 31 | 185 |
| Days Min Temp ≤ 0 °F | 16 | 11 | 3 | 0 | 0 | 0 | 0 | 0 | 0 | 0 | 2 | 10 | 42 |
| Heating Degree Days | 1709 | 1400 | 1111 | 642 | 302 | 81 | 22 | 51 | 246 | 603 | 1052 | 1510 | 8729 |
| Cooling Degree Days | 0 | 0 | 0 | 3 | 25 | 98 | 178 | 136 | 35 | 1 | 0 | 0 | 476 |
| Total Precipitation (") | 0.32 | 0.45 | 1.21 | 2.15 | 2.78 | 4.83 | 3.01 | 2.84 | 2.50 | 1.66 | 0.92 | 0.34 | 23.01 |
| Days ≥ 0.1" Precip | 1 | 1 | 3 | 6 | 6 | 7 | 6 | 5 | 5 | 3 | 2 | 1 | 46 |
| Total Snowfall (") | 5.9 | 6.1 | 7.2 | 2.0 | 0.0 | 0.0 | 0.0 | 0.0 | 0.0 | 0.6 | 5.1 | 6.8 | 33.7 |
| Days ≥ 1" Snow Depth | 20 | 17 | 10 | 3 | 0 | 0 | 0 | 0 | 0 | 0 | 6 | 16 | 72 |

## CAMP CROOK *Harding County* ELEVATION 3123 ft LAT/LONG 45° 32 ' N / 103° 59 ' W

| | JAN | FEB | MAR | APR | MAY | JUN | JUL | AUG | SEP | OCT | NOV | DEC | YEAR |
|---|---|---|---|---|---|---|---|---|---|---|---|---|---|
| Maximum Temp °F | 29.1 | 34.9 | 45.3 | 58.3 | 69.2 | 78.7 | 87.0 | 86.8 | 75.5 | 61.9 | 43.5 | 32.7 | 58.6 |
| Minimum Temp °F | 5.3 | 10.2 | 19.5 | 30.4 | 40.2 | 49.6 | 54.3 | 52.0 | 41.3 | 30.5 | 18.8 | 8.7 | 30.1 |
| Mean Temp °F | 17.2 | 22.5 | 32.5 | 44.4 | 54.7 | 64.2 | 70.6 | 69.4 | 58.5 | 46.2 | 31.2 | 20.7 | 44.3 |
| Days Max Temp ≥ 90 °F | 0 | 0 | 0 | 0 | 1 | 4 | 12 | 13 | 3 | 0 | 0 | 0 | 33 |
| Days Max Temp ≤ 32 °F | 16 | 11 | 5 | 1 | 0 | 0 | 0 | 0 | 0 | 0 | 6 | 14 | 53 |
| Days Min Temp ≤ 32 °F | 30 | 27 | 28 | 18 | 5 | 0 | 0 | 0 | 5 | 18 | 28 | 30 | 189 |
| Days Min Temp ≤ 0 °F | 12 | 7 | 2 | 0 | 0 | 0 | 0 | 0 | 0 | 0 | 2 | 8 | 31 |
| Heating Degree Days | 1478 | 1194 | 1003 | 609 | 325 | 95 | 22 | 33 | 227 | 577 | 1008 | 1367 | 7938 |
| Cooling Degree Days | 0 | 0 | 0 | 1 | 19 | 92 | 207 | 188 | 36 | 1 | 0 | 0 | 544 |
| Total Precipitation (") | 0.28 | 0.29 | 0.59 | 1.49 | 2.49 | 2.65 | 2.04 | 1.18 | 1.15 | 1.02 | 0.40 | 0.36 | 13.94 |
| Days ≥ 0.1" Precip | 1 | 1 | 2 | 4 | 5 | 6 | 4 | 3 | 3 | 2 | 1 | 1 | 33 |
| Total Snowfall (") | ***5.2*** | 5.4 | ***5.4*** | ***3.5*** | 0.6 | 0.0 | 0.0 | 0.0 | 0.5 | 1.2 | ***5.1*** | ***5.4*** | 32.3 |
| Days ≥ 1" Snow Depth | ***19*** | 15 | ***9*** | 2 | 1 | 0 | 0 | 0 | 0 | 1 | 6 | 16 | 69 |

## CANTON 4 WNW *Lincoln County* ELEVATION 1352 ft LAT/LONG 43° 18 ' N / 96° 40 ' W

| | JAN | FEB | MAR | APR | MAY | JUN | JUL | AUG | SEP | OCT | NOV | DEC | YEAR |
|---|---|---|---|---|---|---|---|---|---|---|---|---|---|
| Maximum Temp °F | 25.6 | 31.7 | 45.4 | 61.5 | 73.8 | 82.6 | 86.3 | 83.6 | 75.4 | 63.0 | 43.6 | 29.9 | 58.5 |
| Minimum Temp °F | 4.2 | 10.7 | 23.3 | 35.6 | 47.1 | 56.7 | 61.2 | 58.5 | 48.9 | 36.6 | 23.2 | 10.4 | 34.7 |
| Mean Temp °F | 15.0 | 21.2 | 34.4 | 48.6 | 60.5 | 69.7 | 73.8 | 71.1 | 62.2 | 49.8 | 33.4 | 20.2 | 46.7 |
| Days Max Temp ≥ 90 °F | 0 | 0 | 0 | 1 | 1 | 6 | 10 | 6 | 2 | 0 | 0 | 0 | 26 |
| Days Max Temp ≤ 32 °F | 20 | 14 | 5 | 0 | 0 | 0 | 0 | 0 | 0 | 0 | 5 | 17 | 61 |
| Days Min Temp ≤ 32 °F | 31 | 27 | 25 | 12 | 2 | 0 | 0 | 0 | 1 | 11 | 25 | 31 | 165 |
| Days Min Temp ≤ 0 °F | 13 | 7 | 2 | 0 | 0 | 0 | 0 | 0 | 0 | 0 | 1 | 7 | 30 |
| Heating Degree Days | 1547 | 1230 | 942 | 496 | 187 | 28 | 6 | 18 | 146 | 469 | 940 | 1385 | 7394 |
| Cooling Degree Days | 0 | 0 | 0 | 14 | 60 | 177 | 266 | 200 | 76 | 4 | 0 | 0 | 797 |
| Total Precipitation (") | 0.38 | 0.46 | 1.40 | 2.41 | 2.88 | 3.91 | 3.12 | 3.09 | 2.52 | 1.89 | 1.04 | 0.63 | 23.73 |
| Days ≥ 0.1" Precip | 1 | 1 | 3 | 5 | 6 | 6 | 5 | 5 | 5 | 3 | 3 | 2 | 45 |
| Total Snowfall (") | 6.3 | 4.8 | 5.2 | 1.5 | 0.0 | 0.0 | 0.0 | 0.0 | 0.0 | 0.6 | 4.2 | 7.5 | 30.1 |
| Days ≥ 1" Snow Depth | 22 | 18 | 7 | 0 | 0 | 0 | 0 | 0 | 0 | 0 | 5 | 17 | 69 |

## CASTLEWOOD *Hamlin County* ELEVATION 1690 ft LAT/LONG 44° 43 ' N / 97° 2 ' W

| | JAN | FEB | MAR | APR | MAY | JUN | JUL | AUG | SEP | OCT | NOV | DEC | YEAR |
|---|---|---|---|---|---|---|---|---|---|---|---|---|---|
| Maximum Temp °F | 21.9 | 28.1 | 41.1 | 57.8 | 71.1 | 79.4 | 84.9 | 83.0 | 73.0 | 60.0 | 40.4 | 26.9 | 55.6 |
| Minimum Temp °F | -0.7 | 5.8 | 19.2 | 31.9 | 43.5 | 53.4 | 58.2 | 55.9 | 45.5 | 33.2 | 19.9 | 6.3 | 31.0 |
| Mean Temp °F | 10.6 | 17.0 | 30.2 | 44.9 | 57.3 | 66.4 | 71.6 | 69.4 | 59.3 | 46.6 | 30.2 | 16.6 | 43.3 |
| Days Max Temp ≥ 90 °F | 0 | 0 | 0 | 0 | 0 | 3 | 9 | 7 | 2 | 0 | 0 | 0 | 21 |
| Days Max Temp ≤ 32 °F | 23 | 16 | 7 | 0 | 0 | 0 | 0 | 0 | 0 | 0 | 8 | 20 | 74 |
| Days Min Temp ≤ 32 °F | 31 | 28 | 28 | 17 | 4 | 0 | 0 | 0 | 2 | 16 | 27 | 31 | 184 |
| Days Min Temp ≤ 0 °F | 16 | 10 | 3 | 0 | 0 | 0 | 0 | 0 | 0 | 0 | 2 | 10 | 41 |
| Heating Degree Days | 1683 | 1351 | 1072 | 601 | 259 | 61 | 17 | 34 | 211 | 565 | 1038 | 1494 | 8386 |
| Cooling Degree Days | 0 | 0 | 0 | 5 | 29 | 106 | 211 | 166 | 42 | 1 | 0 | 0 | 560 |
| Total Precipitation (") | 0.65 | 0.61 | 1.35 | 2.06 | 2.84 | 4.26 | 3.14 | 2.69 | 2.37 | 1.81 | 0.98 | 0.56 | 23.32 |
| Days ≥ 0.1" Precip | 2 | 2 | 4 | 5 | 6 | 7 | 6 | 5 | 4 | 4 | 3 | 2 | 50 |
| Total Snowfall (") | 6.4 | 5.0 | 5.5 | 1.1 | 0.0 | 0.0 | 0.0 | 0.0 | 0.0 | 0.5 | 4.6 | 4.7 | 27.8 |
| Days ≥ 1" Snow Depth | 25 | 20 | 11 | 1 | 0 | 0 | 0 | 0 | 0 | 0 | 5 | 17 | 79 |

## CEDAR BUTTE *Mellette County* ELEVATION 2290 ft LAT/LONG 43° 35 ' N / 101° 1 ' W

| | JAN | FEB | MAR | APR | MAY | JUN | JUL | AUG | SEP | OCT | NOV | DEC | YEAR |
|---|---|---|---|---|---|---|---|---|---|---|---|---|---|
| Maximum Temp °F | 33.6 | 38.6 | 49.3 | 61.9 | 73.2 | 83.1 | 91.0 | 89.9 | 79.0 | 65.6 | 47.7 | 36.7 | 62.5 |
| Minimum Temp °F | 9.4 | 14.8 | 23.9 | 35.1 | 45.8 | 55.1 | 60.7 | 59.3 | 48.8 | 37.3 | 24.3 | 14.1 | 35.7 |
| Mean Temp °F | 21.4 | 26.7 | 36.6 | 48.4 | 59.6 | 69.0 | 75.9 | 74.6 | 63.9 | 51.5 | 36.1 | 25.4 | 49.1 |
| Days Max Temp ≥ 90 °F | 0 | 0 | 0 | 0 | 2 | 7 | 16 | 16 | 6 | 1 | 0 | 0 | 48 |
| Days Max Temp ≤ 32 °F | 13 | 10 | 4 | 0 | 0 | 0 | 0 | 0 | 0 | 0 | 4 | 11 | 42 |
| Days Min Temp ≤ 32 °F | 29 | 26 | 25 | 12 | 2 | 0 | 0 | 0 | 1 | 9 | 23 | 29 | 156 |
| Days Min Temp ≤ 0 °F | 9 | 5 | 1 | 0 | 0 | 0 | 0 | 0 | 0 | 0 | 1 | 5 | 21 |
| Heating Degree Days | 1352 | 1077 | 869 | 501 | 210 | 37 | 5 | 9 | 128 | 422 | 859 | 1225 | 6694 |
| Cooling Degree Days | 0 | 0 | 0 | 12 | 54 | 164 | 334 | 304 | 110 | 10 | 0 | 0 | 988 |
| Total Precipitation (") | 0.30 | 0.39 | 1.19 | 2.03 | 2.94 | 3.54 | 2.64 | 1.95 | 1.37 | 1.21 | 0.55 | 0.40 | 18.51 |
| Days ≥ 0.1" Precip | 1 | 1 | 3 | 4 | 6 | 6 | 5 | 4 | 3 | 3 | 2 | 1 | 39 |
| Total Snowfall (") | 4.6 | 6.3 | 7.1 | 4.2 | 0.2 | 0.0 | 0.0 | 0.0 | 0.2 | 1.6 | 5.2 | 5.8 | 35.2 |
| Days ≥ 1" Snow Depth | 18 | 13 | 8 | 2 | 0 | 0 | 0 | 0 | 0 | 1 | 6 | 15 | 63 |

## CENTERVILLE 6 SE *Clay County* ELEVATION 1260 ft LAT/LONG 43° 3 ' N / 96° 54 ' W

| | JAN | FEB | MAR | APR | MAY | JUN | JUL | AUG | SEP | OCT | NOV | DEC | YEAR |
|---|---|---|---|---|---|---|---|---|---|---|---|---|---|
| Maximum Temp °F | 25.3 | 31.0 | 44.2 | 60.1 | 72.1 | 81.5 | 85.5 | 83.4 | 74.8 | 62.4 | 43.8 | 29.4 | 57.8 |
| Minimum Temp °F | 3.6 | 9.9 | 22.6 | 35.1 | 46.7 | 56.6 | 60.8 | 58.0 | 47.5 | 35.3 | 22.7 | 9.8 | 34.1 |
| Mean Temp °F | 14.5 | 20.5 | 33.4 | 47.6 | 59.4 | 69.1 | 73.2 | 70.7 | 61.2 | 48.9 | 33.3 | 19.6 | 45.9 |
| Days Max Temp ≥ 90 °F | 0 | 0 | 0 | 0 | 1 | 5 | 9 | 7 | 2 | 0 | 0 | 0 | 24 |
| Days Max Temp ≤ 32 °F | 20 | 15 | 6 | 0 | 0 | 0 | 0 | 0 | 0 | 0 | 5 | 17 | 63 |
| Days Min Temp ≤ 32 °F | 31 | 28 | 26 | 12 | 2 | 0 | 0 | 0 | 2 | 12 | 25 | 31 | 169 |
| Days Min Temp ≤ 0 °F | 13 | 8 | 2 | 0 | 0 | 0 | 0 | 0 | 0 | 0 | 1 | 7 | 31 |
| Heating Degree Days | 1562 | 1251 | 973 | 523 | 214 | 38 | 8 | 22 | 172 | 499 | 944 | 1402 | 7608 |
| Cooling Degree Days | 0 | 0 | 0 | 13 | 52 | 168 | 257 | 196 | 65 | 3 | 0 | 0 | 754 |
| Total Precipitation (") | 0.46 | 0.69 | 1.43 | 2.29 | 3.47 | 4.07 | 3.56 | 2.92 | 2.55 | 1.84 | 1.10 | 0.65 | 25.03 |
| Days ≥ 0.1" Precip | 2 | 2 | 3 | 5 | 7 | 7 | 6 | 6 | 5 | 4 | 3 | 2 | 52 |
| Total Snowfall (") | 5.8 | 5.4 | 5.3 | 1.6 | 0.0 | 0.0 | 0.0 | 0.0 | 0.0 | 0.9 | 4.8 | 6.7 | 30.5 |
| Days ≥ 1" Snow Depth | na | na | 7 | 1 | 0 | 0 | 0 | 0 | 0 | 0 | 2 | na | na |

## CLARK *Clark County* ELEVATION 1781 ft LAT/LONG 44° 52 ' N / 97° 44 ' W

| | JAN | FEB | MAR | APR | MAY | JUN | JUL | AUG | SEP | OCT | NOV | DEC | YEAR |
|---|---|---|---|---|---|---|---|---|---|---|---|---|---|
| Maximum Temp °F | 21.2 | 27.1 | 40.0 | 56.6 | 69.6 | 77.5 | 83.6 | 81.9 | 72.0 | 59.1 | 39.2 | 26.0 | 54.5 |
| Minimum Temp °F | 0.7 | 7.0 | 19.7 | 32.9 | 45.1 | 54.9 | 60.1 | 57.8 | 47.5 | 35.2 | 20.6 | 7.3 | 32.4 |
| Mean Temp °F | 11.0 | 17.1 | 29.9 | 44.8 | 57.4 | 66.2 | 71.9 | 69.9 | 59.8 | 47.2 | 29.9 | 16.7 | 43.5 |
| Days Max Temp ≥ 90 °F | 0 | 0 | 0 | 0 | 0 | 2 | 7 | 5 | 2 | 0 | 0 | 0 | 16 |
| Days Max Temp ≤ 32 °F | 23 | 17 | 8 | 1 | 0 | 0 | 0 | 0 | 0 | 0 | 9 | 20 | 78 |
| Days Min Temp ≤ 32 °F | 31 | 28 | 28 | 16 | 3 | 0 | 0 | 0 | 1 | 13 | 27 | 31 | 178 |
| Days Min Temp ≤ 0 °F | 16 | 9 | 3 | 0 | 0 | 0 | 0 | 0 | 0 | 0 | 2 | 9 | 39 |
| Heating Degree Days | 1671 | 1348 | 1083 | 603 | 258 | 62 | 14 | 30 | 203 | 550 | 1046 | 1493 | 8361 |
| Cooling Degree Days | 0 | 0 | 0 | 5 | 32 | 104 | 224 | 175 | 46 | 2 | 0 | 0 | 588 |
| Total Precipitation (") | 0.56 | 0.63 | 1.29 | 2.16 | 2.65 | 3.84 | 3.09 | 2.77 | 1.94 | 1.53 | 0.97 | 0.55 | 21.98 |
| Days ≥ 0.1" Precip | 2 | 2 | 4 | 5 | 6 | 7 | 6 | 5 | 4 | 3 | 3 | 2 | 49 |
| Total Snowfall (") | 7.0 | 6.7 | 5.8 | 1.5 | 0.0 | 0.0 | 0.0 | 0.0 | 0.1 | 0.5 | 5.4 | 6.3 | 33.3 |
| Days ≥ 1" Snow Depth | 28 | 24 | 15 | 2 | 0 | 0 | 0 | 0 | 0 | 0 | 6 | 18 | 93 |

## CLEAR LAKE *Deuel County* ELEVATION 1801 ft LAT/LONG 44° 45 ' N / 96° 41 ' W

| | JAN | FEB | MAR | APR | MAY | JUN | JUL | AUG | SEP | OCT | NOV | DEC | YEAR |
|---|---|---|---|---|---|---|---|---|---|---|---|---|---|
| Maximum Temp °F | 21.1 | 26.7 | 39.7 | 56.6 | 70.0 | 78.2 | 83.2 | 80.5 | 71.2 | 58.7 | 39.2 | 25.8 | 54.2 |
| Minimum Temp °F | 1.7 | 7.4 | 20.6 | 33.3 | 45.4 | 55.2 | 60.2 | 58.1 | 48.3 | 36.3 | 21.8 | 8.4 | 33.1 |
| Mean Temp °F | 11.4 | 17.1 | 30.3 | 45.0 | 57.7 | 66.7 | 71.7 | 69.3 | 59.8 | 47.5 | 30.5 | 17.1 | 43.7 |
| Days Max Temp ≥ 90 °F | 0 | 0 | 0 | 0 | 0 | 2 | 6 | 4 | 1 | 0 | 0 | 0 | 13 |
| Days Max Temp ≤ 32 °F | 24 | 18 | 8 | 1 | 0 | 0 | 0 | 0 | 0 | 0 | 9 | 21 | 81 |
| Days Min Temp ≤ 32 °F | 31 | 28 | 27 | 15 | 3 | 0 | 0 | 0 | 1 | 11 | 26 | 31 | 173 |
| Days Min Temp ≤ 0 °F | 15 | 10 | 2 | 0 | 0 | 0 | 0 | 0 | 0 | 0 | 1 | 9 | 37 |
| Heating Degree Days | 1658 | 1349 | 1068 | 597 | 250 | 58 | 13 | 32 | 198 | 537 | 1028 | 1480 | 8268 |
| Cooling Degree Days | 0 | 0 | 0 | 6 | 35 | 115 | 215 | 169 | 48 | 2 | 0 | 0 | 590 |
| Total Precipitation (") | 0.73 | 0.66 | 1.69 | 2.40 | 2.98 | 4.39 | 3.34 | 3.12 | 2.33 | 1.94 | 1.21 | 0.58 | 25.37 |
| Days ≥ 0.1" Precip | 2 | 2 | 4 | 6 | 6 | 7 | 6 | 6 | 5 | 4 | 3 | 2 | 53 |
| Total Snowfall (") | 10.3 | 8.3 | 10.6 | 3.8 | 0.7 | 0.0 | 0.0 | 0.0 | 0.1 | 1.4 | 8.2 | 8.3 | 51.7 |
| Days ≥ 1" Snow Depth | 25 | 21 | *13* | 2 | 0 | 0 | 0 | 0 | 0 | 0 | 9 | 18 | 88 |

## COLUMBIA 8 N *Brown County* ELEVATION 1302 ft LAT/LONG 45° 44 ' N / 98° 18 ' W

| | JAN | FEB | MAR | APR | MAY | JUN | JUL | AUG | SEP | OCT | NOV | DEC | YEAR |
|---|---|---|---|---|---|---|---|---|---|---|---|---|---|
| Maximum Temp °F | 19.7 | 26.1 | 39.1 | 56.6 | 69.7 | 78.3 | 84.9 | 83.1 | 72.6 | 59.2 | 38.6 | 24.9 | 54.4 |
| Minimum Temp °F | -2.2 | 4.8 | 18.6 | 32.4 | 44.8 | 54.1 | 58.9 | 56.3 | 45.5 | 33.3 | 19.4 | 4.9 | 30.9 |
| Mean Temp °F | 8.8 | 15.5 | 28.9 | 44.5 | 57.3 | 66.2 | 71.9 | 69.7 | 59.1 | 46.2 | 29.0 | 14.9 | 42.7 |
| Days Max Temp ≥ 90 °F | 0 | 0 | 0 | 0 | 0 | 2 | 8 | 6 | 2 | 0 | 0 | 0 | 18 |
| Days Max Temp ≤ 32 °F | 24 | 18 | 9 | 1 | 0 | 0 | 0 | 0 | 0 | 0 | 9 | 21 | 82 |
| Days Min Temp ≤ 32 °F | 31 | 28 | 28 | 16 | 3 | 0 | 0 | 0 | 2 | 15 | 28 | 31 | 182 |
| Days Min Temp ≤ 0 °F | 18 | 11 | 3 | 0 | 0 | 0 | 0 | 0 | 0 | 0 | 2 | 11 | 45 |
| Heating Degree Days | 1741 | 1395 | 1114 | 610 | 262 | 62 | 12 | 31 | 215 | 575 | 1073 | 1547 | 8637 |
| Cooling Degree Days | 0 | 0 | 0 | 4 | 33 | 101 | 214 | 163 | 31 | 1 | 0 | 0 | 547 |
| Total Precipitation (") | 0.55 | 0.50 | 1.40 | 2.13 | 2.50 | 3.30 | 2.81 | 2.20 | 1.92 | 1.33 | 0.73 | 0.49 | 19.86 |
| Days ≥ 0.1" Precip | 2 | 2 | 3 | 4 | 5 | 6 | 5 | 4 | 3 | 3 | 2 | 2 | 41 |
| Total Snowfall (") | 7.0 | 6.7 | 8.0 | 3.5 | 0.0 | 0.0 | 0.0 | 0.0 | 0.0 | 0.9 | 7.3 | 6.6 | 40.0 |
| Days ≥ 1" Snow Depth | 26 | 22 | 15 | 2 | 0 | 0 | 0 | 0 | 0 | 0 | 9 | 20 | 94 |

## COTTONWOOD 2 E *Jackson County* ELEVATION 2414 ft LAT/LONG 43° 58 ' N / 101° 53 ' W

| | JAN | FEB | MAR | APR | MAY | JUN | JUL | AUG | SEP | OCT | NOV | DEC | YEAR |
|---|---|---|---|---|---|---|---|---|---|---|---|---|---|
| Maximum Temp °F | 32.1 | 37.4 | 47.3 | 60.7 | 71.8 | 81.4 | 89.4 | 89.0 | 78.5 | 64.0 | 46.3 | 35.4 | 61.1 |
| Minimum Temp °F | 6.2 | 10.5 | 20.2 | 31.7 | 42.7 | 52.5 | 57.9 | 55.6 | 44.2 | 31.4 | 18.9 | 8.8 | 31.7 |
| Mean Temp °F | 19.2 | 23.9 | 33.8 | 46.2 | 57.3 | 66.9 | 73.7 | 72.3 | 61.4 | 47.7 | 32.6 | 22.1 | 46.4 |
| Days Max Temp ≥ 90 °F | 0 | 0 | 0 | 0 | 2 | 6 | 16 | 16 | 6 | 1 | 0 | 0 | 47 |
| Days Max Temp ≤ 32 °F | 14 | 10 | 5 | 0 | 0 | 0 | 0 | 0 | 0 | 0 | 5 | 12 | 46 |
| Days Min Temp ≤ 32 °F | 31 | 28 | 28 | 16 | 4 | 0 | 0 | 0 | 3 | 16 | *28* | 31 | 185 |
| Days Min Temp ≤ 0 °F | 10 | 7 | 2 | 0 | 0 | 0 | 0 | 0 | 0 | 0 | 2 | 8 | 29 |
| Heating Degree Days | 1414 | 1155 | 960 | 560 | 263 | 60 | 11 | 20 | 174 | 532 | 964 | 1323 | 7436 |
| Cooling Degree Days | 0 | 0 | 0 | 5 | 33 | 121 | 278 | 244 | 61 | 3 | 0 | 0 | 745 |
| Total Precipitation (") | 0.32 | 0.51 | 1.06 | 1.81 | 2.84 | 3.31 | 2.23 | 1.60 | 1.16 | 1.15 | 0.63 | 0.40 | 17.02 |
| Days ≥ 0.1" Precip | 1 | 2 | 3 | 4 | 5 | 7 | 4 | 3 | 3 | 3 | 2 | 2 | 39 |
| Total Snowfall (") | 5.1 | 7.5 | 8.7 | 4.4 | 0.1 | 0.0 | 0.0 | 0.0 | 0.0 | 1.2 | 5.5 | 6.5 | 39.0 |
| Days ≥ 1" Snow Depth | 19 | 14 | 10 | 2 | 0 | 0 | 0 | 0 | 0 | 1 | *7* | 16 | 69 |

## CUSTER *Custer County* ELEVATION 5325 ft LAT/LONG 43° 46' N / 103° 36' W

| | JAN | FEB | MAR | APR | MAY | JUN | JUL | AUG | SEP | OCT | NOV | DEC | YEAR |
|---|---|---|---|---|---|---|---|---|---|---|---|---|---|
| Maximum Temp °F | 34.3 | 38.1 | 44.0 | 52.4 | 62.6 | 72.1 | 79.5 | 78.5 | 69.3 | 58.0 | 43.6 | 36.6 | 55.8 |
| Minimum Temp °F | 9.4 | 12.4 | 18.5 | 27.4 | 36.6 | 45.3 | 51.0 | 48.8 | 39.3 | 29.3 | 19.0 | 11.4 | 29.0 |
| Mean Temp °F | 21.9 | 25.3 | 31.3 | 39.9 | 49.6 | 58.7 | 65.3 | 63.6 | 54.3 | 43.7 | 31.3 | 24.0 | 42.4 |
| Days Max Temp ≥ 90 °F | 0 | 0 | 0 | 0 | 0 | 1 | 4 | 2 | 0 | 0 | 0 | 0 | 7 |
| Days Max Temp ≤ 32 °F | 13 | 8 | 5 | 1 | 0 | 0 | 0 | 0 | 0 | 1 | 5 | 10 | 43 |
| Days Min Temp ≤ 32 °F | 30 | 27 | 29 | 23 | 10 | 2 | 0 | 0 | 7 | 19 | 27 | 29 | 203 |
| Days Min Temp ≤ 0 °F | 8 | 5 | 2 | 0 | 0 | 0 | 0 | 0 | 0 | 0 | 2 | 6 | 23 |
| Heating Degree Days | 1332 | 1115 | 1039 | 746 | 472 | 209 | 70 | 95 | 327 | 654 | 1003 | 1264 | 8326 |
| Cooling Degree Days | 0 | 0 | 0 | 0 | 2 | 37 | 101 | 74 | 15 | 0 | 0 | 0 | 229 |
| Total Precipitation (") | 0.37 | 0.61 | 0.99 | 1.93 | 3.11 | 3.27 | 3.24 | 2.12 | 1.41 | 1.12 | 0.63 | 0.56 | 19.36 |
| Days ≥ 0.1" Precip | 1 | 2 | 3 | 5 | 6 | 7 | 7 | 5 | 4 | 3 | 2 | 2 | 47 |
| Total Snowfall (") | *5.1* | na | na | na | 0.5 | 0.0 | 0.0 | 0.0 | 0.4 | 2.2 | na | na | na |
| Days ≥ 1" Snow Depth | na | na | na | na | 0 | 0 | 0 | 0 | 0 | 1 | na | na | na |

## DE SMET *Kingsbury County* ELEVATION 1732 ft LAT/LONG 44° 23' N / 97° 33' W

| | JAN | FEB | MAR | APR | MAY | JUN | JUL | AUG | SEP | OCT | NOV | DEC | YEAR |
|---|---|---|---|---|---|---|---|---|---|---|---|---|---|
| Maximum Temp °F | 22.9 | 29.0 | 41.5 | 58.5 | 71.0 | 79.7 | 85.2 | 83.3 | 73.3 | 60.0 | 40.6 | 27.3 | 56.0 |
| Minimum Temp °F | 2.5 | 9.0 | 21.4 | 34.5 | 46.0 | 55.5 | 60.8 | 58.5 | 48.7 | 36.3 | 22.2 | 8.7 | 33.7 |
| Mean Temp °F | 12.9 | 19.2 | 31.5 | 46.5 | 58.6 | 67.7 | 73.0 | 70.9 | 61.0 | 48.2 | 31.4 | 18.0 | 44.9 |
| Days Max Temp ≥ 90 °F | 0 | 0 | 0 | 0 | 0 | 3 | 9 | 7 | 2 | 0 | 0 | 0 | 21 |
| Days Max Temp ≤ 32 °F | 22 | 16 | 7 | 0 | 0 | 0 | 0 | 0 | 0 | 0 | 8 | 19 | 72 |
| Days Min Temp ≤ 32 °F | 31 | 28 | 27 | 13 | 2 | 0 | 0 | 0 | 1 | 11 | 26 | 31 | 170 |
| Days Min Temp ≤ 0 °F | 14 | 9 | 2 | 0 | 0 | 0 | 0 | 0 | 0 | 0 | 1 | 8 | 34 |
| Heating Degree Days | 1611 | 1287 | 1030 | 554 | 226 | 45 | 9 | 21 | 174 | 517 | 1002 | 1452 | 7928 |
| Cooling Degree Days | 0 | 0 | 0 | 8 | 38 | 128 | 246 | 200 | 59 | 3 | 0 | 0 | 682 |
| Total Precipitation (") | 0.63 | 0.66 | 1.52 | 2.33 | 2.95 | 4.20 | 3.12 | 2.61 | 2.24 | 1.67 | 0.97 | 0.53 | 23.43 |
| Days ≥ 0.1" Precip | 2 | 2 | 4 | 5 | 6 | 7 | 5 | 5 | 4 | 3 | 3 | 2 | 48 |
| Total Snowfall (") | 6.9 | 7.1 | 7.4 | 2.1 | 0.0 | 0.0 | 0.0 | 0.0 | 0.0 | 0.7 | 6.5 | 6.4 | 37.1 |
| Days ≥ 1" Snow Depth | *24* | *20* | 14 | 1 | 0 | 0 | 0 | 0 | 0 | 0 | *5* | *18* | 82 |

## DEADWOOD *Lawrence County* ELEVATION 4554 ft LAT/LONG 44° 23' N / 103° 43' W

| | JAN | FEB | MAR | APR | MAY | JUN | JUL | AUG | SEP | OCT | NOV | DEC | YEAR |
|---|---|---|---|---|---|---|---|---|---|---|---|---|---|
| Maximum Temp °F | 33.7 | 37.0 | 43.9 | 53.5 | 64.2 | 73.8 | 81.0 | 80.0 | 69.3 | 56.9 | 42.6 | 35.8 | 56.0 |
| Minimum Temp °F | 11.4 | 14.4 | 20.5 | 29.3 | 38.8 | 47.7 | 53.5 | 51.6 | 41.9 | 31.9 | 21.0 | 14.2 | 31.3 |
| Mean Temp °F | 22.6 | 25.7 | 32.2 | 41.4 | 51.6 | 60.8 | 67.3 | 65.8 | 55.7 | 44.5 | 31.8 | 25.0 | 43.7 |
| Days Max Temp ≥ 90 °F | 0 | 0 | 0 | 0 | 0 | 1 | 5 | 3 | 0 | 0 | 0 | 0 | 9 |
| Days Max Temp ≤ 32 °F | 13 | 10 | 5 | 2 | 0 | 0 | 0 | 0 | 0 | 1 | 6 | 12 | 49 |
| Days Min Temp ≤ 32 °F | 30 | 27 | 28 | 20 | 7 | 1 | 0 | 0 | 4 | 16 | 27 | 30 | 190 |
| Days Min Temp ≤ 0 °F | 7 | 4 | 2 | 0 | 0 | 0 | 0 | 0 | 0 | 0 | 1 | 4 | 18 |
| Heating Degree Days | 1309 | 1100 | 1009 | 701 | 416 | 164 | 45 | 65 | 291 | 629 | 989 | 1234 | 7952 |
| Cooling Degree Days | 0 | 0 | 0 | 1 | 6 | 54 | 133 | 103 | 18 | 0 | 0 | 0 | 315 |
| Total Precipitation (") | 1.29 | 1.22 | 2.34 | 3.55 | 4.64 | 3.76 | 2.40 | 1.93 | 2.04 | 1.99 | 1.37 | 1.45 | 27.98 |
| Days ≥ 0.1" Precip | 4 | 4 | 5 | 7 | 8 | 8 | 5 | 4 | 4 | 4 | 4 | 5 | 62 |
| Total Snowfall (") | 14.6 | 15.7 | 22.3 | 16.6 | 2.7 | 0.1 | 0.0 | 0.0 | 0.6 | 7.2 | 14.8 | 19.3 | 113.9 |
| Days ≥ 1" Snow Depth | na | na | na | na | na | 0 | 0 | 0 | 0 | na | na | na | na |

## DUPREE *Ziebach County* ELEVATION 2352 ft LAT/LONG 45° 3' N / 101° 36' W

| | JAN | FEB | MAR | APR | MAY | JUN | JUL | AUG | SEP | OCT | NOV | DEC | YEAR |
|---|---|---|---|---|---|---|---|---|---|---|---|---|---|
| Maximum Temp °F | 27.2 | 32.8 | 43.6 | 58.9 | 71.3 | 80.2 | 88.2 | 87.4 | 76.1 | 62.6 | 42.9 | 30.3 | 58.5 |
| Minimum Temp °F | 5.5 | 10.8 | 20.9 | 32.6 | 43.7 | 52.9 | 58.3 | 56.2 | 45.7 | 34.2 | 20.8 | 9.4 | 32.6 |
| Mean Temp °F | 16.4 | 21.8 | 32.3 | 45.7 | 57.5 | 66.6 | 73.3 | 71.8 | 61.0 | 48.4 | 31.9 | 19.9 | 45.5 |
| Days Max Temp ≥ 90 °F | 0 | 0 | 0 | 0 | 1 | 5 | 14 | 13 | 4 | 0 | 0 | 0 | 37 |
| Days Max Temp ≤ 32 °F | 17 | 13 | 7 | 0 | 0 | 0 | 0 | 0 | 0 | 0 | 7 | 16 | 60 |
| Days Min Temp ≤ 32 °F | 30 | 28 | 27 | 15 | 3 | 0 | 0 | 0 | 2 | 13 | 27 | 30 | 175 |
| Days Min Temp ≤ 0 °F | 12 | 7 | 2 | 0 | 0 | 0 | 0 | 0 | 0 | 0 | 2 | 8 | 31 |
| Heating Degree Days | 1504 | 1214 | 1008 | 576 | 255 | 62 | 10 | 20 | 181 | 509 | 987 | 1393 | 7719 |
| Cooling Degree Days | 0 | 0 | 0 | 6 | 37 | 115 | 269 | 234 | 62 | 3 | 0 | 0 | 726 |
| Total Precipitation (") | 0.26 | 0.47 | 1.11 | 1.83 | 2.87 | 3.36 | 2.28 | 1.53 | 1.14 | 1.20 | 0.47 | 0.38 | 16.90 |
| Days ≥ 0.1" Precip | 1 | 2 | 3 | 4 | 6 | 6 | 5 | 4 | 3 | 3 | 2 | 1 | 40 |
| Total Snowfall (") | 4.6 | 7.0 | 8.1 | 4.7 | 0.6 | 0.0 | 0.0 | 0.0 | 0.0 | 1.0 | 4.4 | 6.2 | 36.6 |
| Days ≥ 1" Snow Depth | 21 | 16 | 11 | 2 | 0 | 0 | 0 | 0 | 0 | 0 | 6 | 17 | 73 |

## DUPREE 15 SSE *Ziebach County* ELEVATION 2100 ft LAT/LONG 44° 51' N / 101° 27' W

| | JAN | FEB | MAR | APR | MAY | JUN | JUL | AUG | SEP | OCT | NOV | DEC | YEAR |
|---|---|---|---|---|---|---|---|---|---|---|---|---|---|
| Maximum Temp °F | 26.9 | 32.5 | 43.7 | 58.5 | 70.4 | 80.4 | 87.8 | 87.1 | 75.7 | 62.1 | 43.4 | 30.8 | 58.3 |
| Minimum Temp °F | 4.7 | 10.2 | 21.0 | 33.0 | 43.7 | 53.6 | 58.8 | 56.7 | 45.3 | 33.3 | 19.7 | 9.2 | 32.4 |
| Mean Temp °F | 15.8 | 21.3 | 32.4 | 45.8 | 57.2 | 67.0 | 73.3 | 72.0 | 60.5 | 47.7 | 31.6 | 20.0 | 45.4 |
| Days Max Temp ≥ 90 °F | 0 | 0 | 0 | 0 | 1 | 5 | 14 | 13 | 4 | 0 | 0 | 0 | 37 |
| Days Max Temp ≤ 32 °F | 18 | 14 | 7 | 1 | 0 | 0 | 0 | 0 | 0 | 0 | 6 | 16 | 62 |
| Days Min Temp ≤ 32 °F | 31 | 28 | 27 | 14 | 3 | 0 | 0 | 0 | 3 | 14 | 27 | 31 | 178 |
| Days Min Temp ≤ 0 °F | 12 | 7 | 2 | 0 | 0 | 0 | 0 | 0 | 0 | 0 | 2 | 7 | 30 |
| Heating Degree Days | 1521 | 1227 | 1003 | 573 | 266 | 57 | 11 | 19 | 188 | 530 | 998 | 1389 | 7782 |
| Cooling Degree Days | 0 | 0 | 0 | 5 | 33 | 132 | 273 | 245 | 60 | 3 | 0 | 0 | 751 |
| Total Precipitation (") | 0.25 | 0.46 | 0.98 | 1.62 | 2.66 | 3.24 | 2.55 | 1.51 | 1.16 | 1.11 | 0.51 | 0.35 | 16.40 |
| Days ≥ 0.1" Precip | 1 | 2 | 3 | 4 | 5 | 7 | 5 | 4 | 3 | 2 | 1 | 1 | 38 |
| Total Snowfall (") | 4.2 | 6.5 | 6.8 | 5.6 | 0.5 | 0.0 | 0.0 | 0.0 | 0.0 | 1.1 | 4.6 | 5.7 | 35.0 |
| Days ≥ 1" Snow Depth | 18 | 15 | 8 | 2 | 0 | 0 | 0 | 0 | 0 | 0 | 7 | 15 | 65 |

## EUREKA *McPherson County* ELEVATION 1880 ft LAT/LONG 45° 46' N / 99° 37' W

| | JAN | FEB | MAR | APR | MAY | JUN | JUL | AUG | SEP | OCT | NOV | DEC | YEAR |
|---|---|---|---|---|---|---|---|---|---|---|---|---|---|
| Maximum Temp °F | 19.8 | 26.3 | 39.7 | 56.8 | 69.9 | 78.1 | 85.1 | 83.8 | 72.9 | 59.2 | 38.4 | 24.9 | 54.6 |
| Minimum Temp °F | -0.8 | 6.0 | 18.1 | 31.2 | 43.0 | 52.2 | 57.1 | 55.0 | 44.4 | 32.8 | 18.4 | 5.7 | 30.3 |
| Mean Temp °F | 9.5 | 16.2 | 28.9 | 44.0 | 56.5 | 65.2 | 71.1 | 69.4 | 58.7 | 46.1 | 28.5 | 15.3 | 42.4 |
| Days Max Temp ≥ 90 °F | 0 | 0 | 0 | 0 | 1 | 3 | 9 | 8 | 2 | 0 | 0 | 0 | 23 |
| Days Max Temp ≤ 32 °F | 23 | 18 | 9 | 1 | 0 | 0 | 0 | 0 | 0 | 0 | 10 | 22 | 83 |
| Days Min Temp ≤ 32 °F | 31 | 28 | 29 | 18 | 4 | 0 | 0 | 0 | 3 | 15 | 28 | 31 | 187 |
| Days Min Temp ≤ 0 °F | 16 | 10 | 3 | 0 | 0 | 0 | 0 | 0 | 0 | 0 | 2 | 11 | 42 |
| Heating Degree Days | 1717 | 1374 | 1112 | 625 | 282 | 75 | 17 | 34 | 226 | 582 | 1090 | 1537 | 8671 |
| Cooling Degree Days | 0 | 0 | 0 | 4 | 28 | 93 | 211 | 174 | 40 | 1 | 0 | 0 | 551 |
| Total Precipitation (") | 0.28 | 0.35 | 0.94 | 1.96 | 2.44 | 3.12 | 2.73 | 2.23 | 1.29 | 1.33 | 0.63 | 0.35 | 17.65 |
| Days ≥ 0.1" Precip | 1 | 1 | 2 | 4 | 5 | 6 | 5 | 4 | 3 | 3 | 2 | 1 | 37 |
| Total Snowfall (") | *6.3* | 6.9 | *7.9* | 5.1 | 0.3 | 0.0 | 0.0 | 0.0 | 0.1 | 1.1 | 7.2 | 7.2 | 42.1 |
| Days ≥ 1" Snow Depth | 26 | 23 | 17 | 4 | 0 | 0 | 0 | 0 | 0 | 0 | *9* | 17 | 96 |

## FAITH *Meade County* ELEVATION 2592 ft LAT/LONG 45° 2' N / 102° 2' W

| | JAN | FEB | MAR | APR | MAY | JUN | JUL | AUG | SEP | OCT | NOV | DEC | YEAR |
|---|---|---|---|---|---|---|---|---|---|---|---|---|---|
| Maximum Temp °F | 26.6 | 31.8 | 43.2 | 58.1 | 70.3 | 79.7 | 87.4 | 86.2 | 75.4 | 61.4 | 42.6 | 30.5 | 57.8 |
| Minimum Temp °F | 6.2 | 11.2 | 21.3 | 33.1 | 43.8 | 53.1 | 58.9 | 57.1 | 46.7 | 34.7 | 21.2 | 10.6 | 33.2 |
| Mean Temp °F | 16.4 | 21.6 | 32.3 | 45.6 | 57.1 | 66.4 | 73.2 | 71.7 | 61.1 | 48.1 | 31.9 | 20.6 | 45.5 |
| Days Max Temp ≥ 90 °F | 0 | 0 | 0 | 0 | 1 | 4 | 12 | 11 | 3 | 0 | 0 | 0 | 31 |
| Days Max Temp ≤ 32 °F | 18 | 14 | 7 | 1 | 0 | 0 | 0 | 0 | 0 | 0 | 7 | 16 | 63 |
| Days Min Temp ≤ 32 °F | 30 | 27 | 27 | 14 | 3 | 0 | 0 | 0 | 2 | 12 | 26 | 30 | 171 |
| Days Min Temp ≤ 0 °F | 11 | 7 | 2 | 0 | 0 | 0 | 0 | 0 | 0 | 0 | 2 | 7 | 29 |
| Heating Degree Days | 1502 | 1222 | 1008 | 578 | 265 | 65 | 10 | 19 | 176 | 519 | 988 | 1371 | 7723 |
| Cooling Degree Days | 0 | 0 | 0 | 4 | 31 | 120 | 271 | 237 | 62 | 4 | 0 | 0 | 729 |
| Total Precipitation (") | 0.33 | 0.55 | 1.08 | 1.87 | 2.75 | 2.96 | 2.61 | 1.34 | 1.15 | 1.12 | 0.52 | 0.41 | 16.69 |
| Days ≥ 0.1" Precip | 1 | 2 | 3 | 4 | 5 | 6 | 5 | 3 | 3 | 3 | 2 | 1 | 38 |
| Total Snowfall (") | 5.2 | 7.1 | 9.1 | 6.1 | 0.4 | 0.1 | 0.0 | 0.0 | 0.1 | 1.0 | 5.5 | 6.2 | 40.8 |
| Days ≥ 1" Snow Depth | 20 | 16 | 10 | 2 | 0 | 0 | 0 | 0 | 0 | 0 | 7 | 17 | 72 |

## FAULKTON *Faulk County* ELEVATION 1581 ft LAT/LONG 45° 2' N / 99° 6' W

| | JAN | FEB | MAR | APR | MAY | JUN | JUL | AUG | SEP | OCT | NOV | DEC | YEAR |
|---|---|---|---|---|---|---|---|---|---|---|---|---|---|
| Maximum Temp °F | 23.5 | 29.4 | 42.5 | 58.8 | 71.6 | 80.1 | 86.9 | 85.8 | 75.3 | 61.9 | 41.3 | 27.8 | 57.1 |
| Minimum Temp °F | 2.0 | 8.3 | 20.6 | 32.8 | 44.0 | 53.6 | 58.6 | 56.3 | 46.0 | 34.5 | 20.4 | 7.6 | 32.1 |
| Mean Temp °F | 12.8 | 18.9 | 31.6 | 45.8 | 57.8 | 66.9 | 72.8 | 71.1 | 60.7 | 48.2 | 30.9 | 17.7 | 44.6 |
| Days Max Temp ≥ 90 °F | 0 | 0 | 0 | 0 | 1 | 4 | 12 | 11 | 4 | 0 | 0 | 0 | 32 |
| Days Max Temp ≤ 32 °F | 21 | 16 | 6 | 1 | 0 | 0 | 0 | 0 | 0 | 0 | 8 | 18 | 70 |
| Days Min Temp ≤ 32 °F | 31 | 28 | 27 | 15 | 4 | 0 | 0 | 0 | 2 | 12 | 27 | 31 | 177 |
| Days Min Temp ≤ 0 °F | 15 | 9 | 2 | 0 | 0 | 0 | 0 | 0 | 0 | 0 | 1 | 9 | 36 |
| Heating Degree Days | 1615 | 1297 | 1032 | 572 | 250 | 55 | 12 | 25 | 187 | 518 | 1019 | 1460 | 8042 |
| Cooling Degree Days | 0 | 0 | 0 | 6 | 38 | 116 | 249 | 210 | 60 | 4 | 0 | 0 | 683 |
| Total Precipitation (") | 0.33 | 0.56 | 1.34 | 2.12 | 2.61 | 3.10 | 2.70 | 2.43 | 1.62 | 1.29 | 0.70 | 0.39 | 19.19 |
| Days ≥ 0.1" Precip | 1 | 2 | 3 | 4 | 5 | 6 | 5 | 4 | 3 | 3 | 2 | 1 | 39 |
| Total Snowfall (") | 4.7 | 7.5 | *6.1* | *2.7* | 0.0 | 0.0 | 0.0 | 0.0 | 0.0 | 0.9 | *5.8* | 5.1 | 32.8 |
| Days ≥ 1" Snow Depth | na | na | na | *1* | 0 | 0 | 0 | 0 | 0 | 0 | *5* | na | na |

## FLANDREAU *Moody County* ELEVATION 1562 ft LAT/LONG 44° 3 ' N / 96° 36 ' W

| | JAN | FEB | MAR | APR | MAY | JUN | JUL | AUG | SEP | OCT | NOV | DEC | YEAR |
|---|---|---|---|---|---|---|---|---|---|---|---|---|---|
| Maximum Temp °F | 21.9 | 27.1 | 39.8 | 55.8 | 69.3 | 78.2 | 83.1 | 81.0 | 71.6 | 59.2 | 40.7 | 26.9 | 54.6 |
| Minimum Temp °F | -0.5 | 5.3 | 19.8 | 33.2 | 44.8 | 54.9 | 59.5 | 56.3 | 46.1 | 33.4 | 20.7 | 7.0 | 31.7 |
| Mean Temp °F | 10.7 | 16.2 | 29.8 | 44.5 | 57.0 | 66.6 | 71.3 | 68.7 | 58.9 | 46.3 | 30.7 | 17.0 | 43.1 |
| Days Max Temp ≥ 90 °F | 0 | 0 | 0 | 0 | 0 | 3 | 7 | 5 | 1 | 0 | 0 | 0 | 16 |
| Days Max Temp ≤ 32 °F | 23 | 17 | 9 | 1 | 0 | 0 | 0 | 0 | 0 | 0 | 8 | 20 | 78 |
| Days Min Temp ≤ 32 °F | 31 | 28 | 27 | 14 | 3 | 0 | 0 | 0 | 2 | 14 | 27 | 31 | 177 |
| Days Min Temp ≤ 0 °F | 16 | 11 | 3 | 0 | 0 | 0 | 0 | 0 | 0 | 0 | 1 | 10 | 41 |
| Heating Degree Days | 1679 | 1371 | 1084 | 611 | 269 | 64 | 16 | 40 | 219 | 575 | 1022 | 1483 | 8433 |
| Cooling Degree Days | 0 | 0 | 0 | 5 | 29 | 118 | 206 | 152 | 40 | 1 | 0 | 0 | 551 |
| Total Precipitation (") | 0.40 | 0.47 | 1.19 | 2.19 | 2.83 | 4.16 | 2.97 | 2.85 | 2.73 | 1.92 | 0.96 | 0.44 | 23.11 |
| Days ≥ 0.1" Precip | 1 | 2 | 3 | 5 | 6 | 7 | 5 | 5 | 5 | 3 | 2 | 1 | 45 |
| Total Snowfall (") | 5.7 | 5.6 | 6.4 | 1.2 | 0.0 | 0.0 | 0.0 | 0.0 | 0.0 | 1.0 | 4.8 | 6.2 | 30.9 |
| Days ≥ 1" Snow Depth | *24* | 21 | 13 | 1 | 0 | 0 | 0 | 0 | 0 | 0 | *2* | na | na |

## FORESTBURG 3 NE *Sanborn County* ELEVATION 1230 ft LAT/LONG 44° 2 ' N / 98° 4 ' W

| | JAN | FEB | MAR | APR | MAY | JUN | JUL | AUG | SEP | OCT | NOV | DEC | YEAR |
|---|---|---|---|---|---|---|---|---|---|---|---|---|---|
| Maximum Temp °F | 25.1 | 31.2 | 44.4 | 60.4 | 72.0 | 81.3 | 87.1 | 85.5 | 75.9 | 62.3 | 43.2 | 29.8 | 58.2 |
| Minimum Temp °F | 3.4 | 9.7 | 22.5 | 35.0 | 46.2 | 56.1 | 61.2 | 58.7 | 48.6 | 36.6 | 22.9 | 9.7 | 34.2 |
| Mean Temp °F | 14.3 | 20.5 | 33.5 | 47.7 | 59.1 | 68.7 | 74.1 | 72.1 | 62.3 | 49.5 | 33.1 | 19.8 | 46.2 |
| Days Max Temp ≥ 90 °F | 0 | 0 | 0 | 0 | 1 | 5 | 11 | 10 | 3 | 0 | 0 | 0 | 30 |
| Days Max Temp ≤ 32 °F | 20 | 14 | 5 | 0 | 0 | 0 | 0 | 0 | 0 | 0 | 6 | 17 | 62 |
| Days Min Temp ≤ 32 °F | 31 | 28 | 25 | 12 | 2 | 0 | 0 | 0 | 1 | 11 | 26 | 31 | 167 |
| Days Min Temp ≤ 0 °F | 13 | 8 | 2 | 0 | 0 | 0 | 0 | 0 | 0 | 0 | 1 | 8 | 32 |
| Heating Degree Days | 1569 | 1252 | 970 | 517 | 212 | 37 | 6 | 17 | 151 | 479 | 951 | 1396 | 7557 |
| Cooling Degree Days | 0 | 0 | 0 | 9 | 42 | 160 | 289 | 242 | 79 | 4 | 0 | 0 | 825 |
| Total Precipitation (") | 0.47 | 0.61 | 1.51 | 2.45 | 2.89 | 3.45 | 2.96 | 2.08 | 1.99 | 1.60 | 1.03 | 0.54 | 21.58 |
| Days ≥ 0.1" Precip | 2 | 2 | 4 | 6 | 5 | 6 | 6 | 4 | 4 | 3 | 3 | 2 | 47 |
| Total Snowfall (") | 5.2 | 5.7 | 6.4 | 2.3 | 0.0 | 0.0 | 0.0 | 0.0 | 0.0 | 1.4 | 4.7 | 5.9 | 31.6 |
| Days ≥ 1" Snow Depth | 22 | 19 | 9 | 1 | 0 | 0 | 0 | 0 | 0 | 1 | 5 | 16 | 73 |

## FORT MEADE *Meade County* ELEVATION 3304 ft LAT/LONG 44° 24 ' N / 103° 28 ' W

| | JAN | FEB | MAR | APR | MAY | JUN | JUL | AUG | SEP | OCT | NOV | DEC | YEAR |
|---|---|---|---|---|---|---|---|---|---|---|---|---|---|
| Maximum Temp °F | 35.6 | 39.7 | 47.6 | 59.2 | 69.7 | 79.1 | 86.8 | 85.9 | 75.5 | 63.0 | 46.3 | 38.1 | 60.5 |
| Minimum Temp °F | 12.5 | 16.2 | 23.7 | 33.8 | 43.9 | 53.0 | 58.9 | 57.0 | 47.1 | 36.2 | 23.9 | 15.3 | 35.1 |
| Mean Temp °F | 24.1 | 28.0 | 35.7 | 46.5 | 56.8 | 66.0 | 72.9 | 71.5 | 61.3 | 49.6 | 35.1 | 26.7 | 47.9 |
| Days Max Temp ≥ 90 °F | 0 | 0 | 0 | 0 | 0 | 4 | 12 | 11 | 3 | 0 | 0 | 0 | 30 |
| Days Max Temp ≤ 32 °F | 12 | 8 | 4 | 0 | 0 | 0 | 0 | 0 | 0 | 0 | 5 | 10 | 39 |
| Days Min Temp ≤ 32 °F | 29 | 26 | 25 | 13 | 2 | 0 | 0 | 0 | 2 | 10 | 24 | 29 | 160 |
| Days Min Temp ≤ 0 °F | 8 | 4 | 1 | 0 | 0 | 0 | 0 | 0 | 0 | 0 | 1 | 4 | 18 |
| Heating Degree Days | 1263 | 1039 | 903 | 552 | 269 | 70 | 10 | 18 | 168 | 474 | 891 | 1181 | 6838 |
| Cooling Degree Days | 0 | 0 | 0 | 7 | 28 | 116 | 267 | 231 | 62 | 4 | 0 | 0 | 715 |
| Total Precipitation (") | 0.52 | 0.68 | 1.43 | 2.79 | 3.43 | 3.53 | 2.16 | 1.53 | 1.34 | 1.49 | 0.89 | 0.65 | 20.44 |
| Days ≥ 0.1" Precip | 2 | 2 | 4 | 5 | 6 | 6 | 5 | 4 | 3 | 3 | 3 | 2 | 45 |
| Total Snowfall (") | *5.8* | *7.0* | *9.4* | 5.4 | 1.4 | 0.0 | 0.0 | 0.0 | 0.1 | 1.1 | *4.7* | na | na |
| Days ≥ 1" Snow Depth | 15 | 14 | 10 | 3 | 1 | 0 | 0 | 0 | 0 | 1 | 6 | 14 | 64 |

## GANN VALLEY 4 NW *Buffalo County* ELEVATION 1752 ft LAT/LONG 44° 2 ' N / 98° 59 ' W

| | JAN | FEB | MAR | APR | MAY | JUN | JUL | AUG | SEP | OCT | NOV | DEC | YEAR |
|---|---|---|---|---|---|---|---|---|---|---|---|---|---|
| Maximum Temp °F | 25.7 | 31.8 | 44.9 | 60.4 | 72.2 | 81.5 | 88.4 | 87.0 | 77.0 | 62.9 | 43.1 | 30.3 | 58.8 |
| Minimum Temp °F | 3.2 | 9.4 | 21.1 | 33.3 | 44.6 | 54.7 | 60.1 | 57.8 | 47.1 | 35.0 | 21.1 | 8.9 | 33.0 |
| Mean Temp °F | 14.4 | 20.6 | 33.0 | 46.9 | 58.4 | 68.1 | 74.3 | 72.4 | 62.1 | 49.0 | 32.1 | 19.6 | 45.9 |
| Days Max Temp ≥ 90 °F | 0 | 0 | 0 | 0 | 1 | 5 | 15 | 13 | 5 | 0 | 0 | 0 | 39 |
| Days Max Temp ≤ 32 °F | 19 | 14 | 6 | 0 | 0 | 0 | 0 | 0 | 0 | 0 | 6 | 16 | 61 |
| Days Min Temp ≤ 32 °F | 31 | 28 | 27 | 14 | 3 | 0 | 0 | 0 | 2 | 12 | 27 | 31 | 175 |
| Days Min Temp ≤ 0 °F | 13 | 8 | 2 | 0 | 0 | 0 | 0 | 0 | 0 | 0 | 1 | 8 | 32 |
| Heating Degree Days | 1563 | 1247 | 985 | 541 | 231 | 45 | 10 | 19 | 161 | 494 | 980 | 1400 | 7676 |
| Cooling Degree Days | 0 | 0 | 0 | 8 | 35 | 140 | 280 | 242 | 79 | 3 | 0 | 0 | 787 |
| Total Precipitation (") | 0.22 | 0.45 | 1.12 | 1.99 | 2.68 | 3.16 | 2.55 | 2.10 | 1.67 | 1.40 | 0.64 | 0.35 | 18.33 |
| Days ≥ 0.1" Precip | 1 | 1 | 3 | 5 | 5 | 6 | 6 | 4 | 4 | 3 | 2 | 1 | 41 |
| Total Snowfall (") | 3.6 | 5.6 | 5.3 | 2.7 | 0.0 | 0.0 | 0.0 | 0.0 | 0.0 | 0.6 | 4.4 | 4.8 | 27.0 |
| Days ≥ 1" Snow Depth | 21 | 17 | 8 | 1 | 0 | 0 | 0 | 0 | 0 | 0 | 5 | 13 | 65 |

## GETTYSBURG *Potter County* ELEVATION 2060 ft LAT/LONG 45° 1' N / 99° 57' W

| | JAN | FEB | MAR | APR | MAY | JUN | JUL | AUG | SEP | OCT | NOV | DEC | YEAR |
|---|---|---|---|---|---|---|---|---|---|---|---|---|---|
| Maximum Temp °F | 22.5 | 28.2 | 40.4 | 56.4 | 69.0 | 78.0 | 85.5 | 84.3 | 73.3 | 59.7 | 40.1 | 27.4 | 55.4 |
| Minimum Temp °F | 1.9 | 7.6 | 19.1 | 32.2 | 43.5 | 53.3 | 58.4 | 56.0 | 45.4 | 33.6 | 20.2 | 7.8 | 31.6 |
| Mean Temp °F | 12.2 | 17.9 | 29.8 | 44.3 | 56.3 | 65.7 | 72.0 | 70.2 | 59.4 | 46.6 | 30.2 | 17.6 | 43.5 |
| Days Max Temp ≥ 90 °F | 0 | 0 | 0 | 0 | 1 | 3 | 11 | 9 | 2 | 0 | 0 | 0 | 26 |
| Days Max Temp ≤ 32 °F | 21 | 16 | 9 | 1 | 0 | 0 | 0 | 0 | 0 | 0 | 9 | 18 | 74 |
| Days Min Temp ≤ 32 °F | 31 | 28 | 29 | 16 | 3 | 0 | 0 | 0 | 2 | 14 | 27 | 31 | 181 |
| Days Min Temp ≤ 0 °F | 15 | 10 | 3 | 0 | 0 | 0 | 0 | 0 | 0 | 0 | 1 | 9 | 38 |
| Heating Degree Days | 1634 | 1325 | 1086 | 618 | 288 | 73 | 16 | 30 | 212 | 564 | 1039 | 1464 | 8349 |
| Cooling Degree Days | 0 | 0 | 0 | 5 | 31 | 104 | 229 | 192 | 49 | 2 | 0 | 0 | 612 |
| Total Precipitation (") | 0.37 | 0.55 | 1.20 | 2.10 | 2.52 | 3.02 | 2.67 | 2.15 | 1.29 | 1.27 | 0.69 | 0.51 | 18.34 |
| Days ≥ 0.1" Precip | 1 | 2 | 3 | 5 | 5 | 7 | 5 | 4 | 3 | 3 | 2 | 2 | 42 |
| Total Snowfall (") | na | na | na | *3.2* | 0.1 | 0.0 | 0.0 | 0.0 | 0.0 | 1.0 | na | na | na |
| Days ≥ 1" Snow Depth | na | na | na | *1* | 0 | 0 | 0 | 0 | 0 | 0 | na | na | na |

## GLAD VALLEY 2 W *Ziebach County* ELEVATION 2904 ft LAT/LONG 45° 24' N / 101° 47' W

| | JAN | FEB | MAR | APR | MAY | JUN | JUL | AUG | SEP | OCT | NOV | DEC | YEAR |
|---|---|---|---|---|---|---|---|---|---|---|---|---|---|
| Maximum Temp °F | 25.3 | 30.4 | 41.8 | 56.5 | 68.6 | 78.0 | *85.2* | *84.1* | *73.0* | *60.0* | 41.4 | 29.5 | 56.2 |
| Minimum Temp °F | 3.5 | 8.9 | 19.7 | 31.7 | 42.3 | 52.1 | *58.0* | *55.2* | *44.9* | *33.0* | 19.8 | 8.3 | 31.5 |
| Mean Temp °F | 14.4 | 19.7 | 30.8 | 44.1 | 55.5 | 65.1 | *71.6* | *69.7* | *59.0* | *46.5* | 30.6 | 18.9 | 43.8 |
| Days Max Temp ≥ 90 °F | 0 | 0 | 0 | 0 | 0 | 3 | 10 | *9* | 2 | 0 | 0 | 0 | 24 |
| Days Max Temp ≤ 32 °F | 19 | 15 | 8 | 1 | 0 | 0 | 0 | 0 | 0 | 0 | 8 | 16 | 67 |
| Days Min Temp ≤ 32 °F | 31 | 27 | 28 | *16* | 4 | 0 | 0 | 0 | 2 | *13* | 27 | 29 | 177 |
| Days Min Temp ≤ 0 °F | 13 | 9 | 2 | 0 | 0 | 0 | 0 | 0 | 0 | 0 | 2 | 8 | 34 |
| Heating Degree Days | 1564 | 1273 | 1054 | 631 | 310 | 85 | *17* | *34* | *221* | *567* | 1025 | 1424 | 8205 |
| Cooling Degree Days | 0 | 0 | 0 | *3* | *18* | *100* | *232* | na | na | na | 0 | *0* | na |
| Total Precipitation (") | 0.34 | 0.53 | 1.01 | 2.05 | 2.96 | 2.98 | 2.49 | 1.41 | 1.22 | 1.13 | 0.53 | 0.43 | 17.08 |
| Days ≥ 0.1" Precip | 1 | 2 | 3 | 5 | 6 | 6 | 5 | 3 | 3 | 2 | 2 | 1 | 39 |
| Total Snowfall (") | 5.8 | 7.8 | 10.4 | 8.0 | 1.1 | 0.0 | 0.0 | 0.0 | 0.4 | *1.7* | 6.3 | 7.3 | 48.8 |
| Days ≥ 1" Snow Depth | *21* | *16* | *14* | *3* | 0 | 0 | 0 | 0 | 0 | 0 | 7 | *16* | 77 |

## GREGORY *Gregory County* ELEVATION 2160 ft LAT/LONG 43° 14' N / 99° 26' W

| | JAN | FEB | MAR | APR | MAY | JUN | JUL | AUG | SEP | OCT | NOV | DEC | YEAR |
|---|---|---|---|---|---|---|---|---|---|---|---|---|---|
| Maximum Temp °F | 32.0 | 37.7 | 49.0 | 62.6 | 73.7 | 82.9 | 88.9 | 87.8 | 78.3 | 66.1 | 47.4 | 35.8 | 61.8 |
| Minimum Temp °F | 7.7 | 12.4 | 22.8 | 34.3 | 44.7 | 54.8 | 60.2 | 58.1 | 48.4 | 36.1 | 23.0 | 12.4 | 34.6 |
| Mean Temp °F | 19.9 | 25.1 | 35.9 | 48.5 | 59.2 | 68.9 | 74.6 | 73.0 | 63.4 | 51.1 | 35.2 | 24.1 | 48.2 |
| Days Max Temp ≥ 90 °F | 0 | 0 | 0 | 0 | 1 | 6 | 15 | 14 | 5 | 1 | 0 | 0 | 42 |
| Days Max Temp ≤ 32 °F | 15 | 11 | 4 | 0 | 0 | 0 | 0 | 0 | 0 | 0 | 4 | 12 | 46 |
| Days Min Temp ≤ 32 °F | 30 | 27 | 26 | 13 | 3 | 0 | 0 | 0 | 1 | 11 | 25 | 30 | 166 |
| Days Min Temp ≤ 0 °F | 10 | 6 | 2 | 0 | 0 | 0 | 0 | 0 | 0 | 0 | 1 | 7 | 26 |
| Heating Degree Days | 1393 | 1121 | 895 | 496 | 212 | 37 | 7 | 13 | 133 | 432 | 887 | 1261 | 6887 |
| Cooling Degree Days | 0 | 0 | 0 | 13 | 43 | 167 | 300 | 259 | 91 | 7 | 0 | 0 | 880 |
| Total Precipitation (") | 0.59 | 0.64 | 1.80 | 2.78 | 3.31 | 3.62 | 3.29 | 2.52 | 2.53 | 1.66 | 1.13 | 0.78 | 24.65 |
| Days ≥ 0.1" Precip | 2 | 2 | 4 | 6 | 7 | 7 | 6 | 5 | 5 | 4 | 3 | 3 | 54 |
| Total Snowfall (") | *8.4* | *6.6* | *8.8* | 3.7 | 0.1 | 0.0 | 0.0 | 0.0 | 0.4 | 1.6 | 7.1 | 9.0 | 45.7 |
| Days ≥ 1" Snow Depth | na | na | na | *0* | 0 | 0 | 0 | 0 | 0 | *0* | na | na | na |

## HARRINGTON *Bennett County* ELEVATION 2982 ft LAT/LONG 43° 10' N / 101° 16' W

| | JAN | FEB | MAR | APR | MAY | JUN | JUL | AUG | SEP | OCT | NOV | DEC | YEAR |
|---|---|---|---|---|---|---|---|---|---|---|---|---|---|
| Maximum Temp °F | 32.3 | 37.4 | 47.1 | 59.8 | 70.9 | 80.4 | 87.1 | 85.8 | 76.4 | 63.5 | 46.0 | 35.3 | 60.2 |
| Minimum Temp °F | 7.6 | 12.4 | 21.5 | 32.0 | 42.6 | 52.4 | 58.1 | 55.9 | 45.4 | 33.3 | 20.6 | 10.9 | 32.7 |
| Mean Temp °F | 20.0 | 24.9 | 34.3 | 45.9 | 56.8 | 66.4 | 72.7 | 70.9 | 61.0 | 48.4 | 33.3 | 23.1 | 46.5 |
| Days Max Temp ≥ 90 °F | 0 | 0 | 0 | 0 | 1 | 5 | 13 | 12 | 4 | 0 | 0 | 0 | 35 |
| Days Max Temp ≤ 32 °F | 14 | 10 | 5 | 0 | 0 | 0 | 0 | 0 | 0 | 0 | 5 | 12 | 46 |
| Days Min Temp ≤ 32 °F | 31 | 28 | 27 | 16 | 4 | 0 | 0 | 0 | 3 | 14 | 27 | 31 | 181 |
| Days Min Temp ≤ 0 °F | 10 | 6 | 2 | 0 | 0 | 0 | 0 | 0 | 0 | 0 | 1 | 6 | 25 |
| Heating Degree Days | 1392 | 1126 | 945 | 569 | 272 | 65 | 13 | 23 | 179 | 510 | 944 | 1293 | 7331 |
| Cooling Degree Days | 0 | 0 | 0 | 5 | 28 | 118 | 248 | 210 | 64 | 3 | 0 | 0 | 676 |
| Total Precipitation (") | 0.39 | 0.51 | 1.26 | 1.96 | 3.05 | 3.22 | 2.78 | 2.04 | 1.43 | 1.17 | 0.68 | 0.47 | 18.96 |
| Days ≥ 0.1" Precip | 1 | 2 | 3 | 4 | 6 | 6 | 6 | 4 | 3 | 3 | 2 | 2 | 42 |
| Total Snowfall (") | 5.9 | 6.7 | 10.0 | 6.0 | 0.7 | 0.0 | 0.0 | 0.0 | 0.2 | 3.0 | 7.3 | 6.9 | 46.7 |
| Days ≥ 1" Snow Depth | 18 | 14 | 8 | 2 | 0 | 0 | 0 | 0 | 0 | 1 | 7 | 16 | 66 |

## HARROLD 12 SSW *Hughes County* ELEVATION 1801 ft LAT/LONG 44° 22 ' N / 99° 48 ' W

| | JAN | FEB | MAR | APR | MAY | JUN | JUL | AUG | SEP | OCT | NOV | DEC | YEAR |
|---|---|---|---|---|---|---|---|---|---|---|---|---|---|
| Maximum Temp °F | 25.6 | 32.4 | 45.1 | 60.2 | 72.2 | 81.6 | 89.0 | 88.0 | 77.8 | 63.9 | 43.9 | 30.6 | 59.2 |
| Minimum Temp °F | 0.5 | 8.1 | 19.7 | 31.6 | 43.0 | 53.4 | 58.7 | 56.3 | 44.8 | 31.9 | 18.9 | 6.4 | 31.1 |
| Mean Temp °F | 13.1 | 20.1 | 32.4 | 45.9 | 57.6 | 67.6 | 73.9 | 72.2 | 61.3 | 47.9 | 31.5 | 18.6 | 45.2 |
| Days Max Temp ≥ 90 °F | 0 | 0 | 0 | 0 | 2 | 6 | 14 | 14 | 5 | 1 | 0 | 0 | 42 |
| Days Max Temp ≤ 32 °F | 19 | 13 | 6 | 0 | 0 | 0 | 0 | 0 | 0 | 0 | 6 | 16 | 60 |
| Days Min Temp ≤ 32 °F | 31 | 28 | 28 | 16 | 5 | 0 | 0 | 0 | 3 | 16 | 28 | 31 | 186 |
| Days Min Temp ≤ 0 °F | 15 | 9 | 3 | 0 | 0 | 0 | 0 | 0 | 0 | 0 | 1 | 9 | 37 |
| Heating Degree Days | 1606 | 1263 | 1003 | 571 | 260 | 54 | 12 | 23 | 179 | 526 | 1003 | 1433 | 7933 |
| Cooling Degree Days | 0 | 0 | 0 | 6 | 44 | 136 | 281 | 254 | 74 | 4 | 0 | 0 | 799 |
| Total Precipitation (") | 0.34 | 0.54 | 1.18 | 2.02 | 2.20 | 3.09 | 2.60 | 2.07 | 1.16 | 1.31 | 0.69 | 0.42 | 17.62 |
| Days ≥ 0.1" Precip | 1 | 1 | 3 | 5 | 5 | 6 | 5 | 4 | 3 | 3 | 2 | 1 | 39 |
| Total Snowfall (") | 6.8 | 7.9 | 9.3 | 4.5 | 0.3 | 0.0 | 0.0 | 0.0 | 0.0 | 1.7 | 7.0 | 8.1 | 45.6 |
| Days ≥ 1" Snow Depth | 21 | 15 | 9 | 1 | 0 | 0 | 0 | 0 | 0 | 1 | 8 | 17 | 72 |

## HIGHMORE 1 W *Hyde County* ELEVATION 1890 ft LAT/LONG 44° 31 ' N / 99° 28 ' W

| | JAN | FEB | MAR | APR | MAY | JUN | JUL | AUG | SEP | OCT | NOV | DEC | YEAR |
|---|---|---|---|---|---|---|---|---|---|---|---|---|---|
| Maximum Temp °F | 24.9 | 31.3 | 43.5 | 60.0 | 72.2 | 81.2 | 88.4 | 87.1 | 76.6 | 62.5 | 42.7 | 29.8 | 58.4 |
| Minimum Temp °F | 3.0 | 9.5 | 20.6 | 33.0 | 43.9 | 53.6 | 59.3 | 57.3 | 47.0 | 34.9 | 21.6 | 9.1 | 32.7 |
| Mean Temp °F | 14.0 | 20.4 | 32.1 | 46.6 | 58.1 | 67.4 | 73.8 | 72.2 | 61.8 | 48.7 | 32.2 | 19.5 | 45.6 |
| Days Max Temp ≥ 90 °F | 0 | 0 | 0 | 0 | 1 | 6 | 14 | 12 | 4 | 0 | 0 | 0 | 37 |
| Days Max Temp ≤ 32 °F | 19 | 14 | 7 | 0 | 0 | 0 | 0 | 0 | 0 | 0 | 6 | 17 | 63 |
| Days Min Temp ≤ 32 °F | 31 | 28 | 27 | 15 | 4 | 0 | 0 | 0 | 2 | 12 | 26 | 31 | 176 |
| Days Min Temp ≤ 0 °F | 14 | 8 | 2 | 0 | 0 | 0 | 0 | 0 | 0 | 0 | 1 | 8 | 33 |
| Heating Degree Days | 1578 | 1254 | 1015 | 551 | 242 | 52 | 11 | 21 | 167 | 502 | 979 | 1405 | 7777 |
| Cooling Degree Days | 0 | 0 | 0 | 7 | 38 | 126 | 270 | 237 | 71 | 4 | 0 | 0 | 753 |
| Total Precipitation (") | 0.34 | 0.53 | 1.27 | 2.38 | 2.64 | 3.37 | 3.26 | 2.23 | 1.75 | 1.46 | 0.70 | 0.42 | 20.35 |
| Days ≥ 0.1" Precip | 1 | 2 | 3 | 6 | 6 | 7 | 6 | 5 | 4 | 3 | 3 | 1 | 47 |
| Total Snowfall (") | 6.1 | 8.3 | 8.9 | 3.6 | 0.1 | 0.0 | 0.0 | 0.0 | 0.0 | *1.3* | 5.4 | 6.4 | 40.1 |
| Days ≥ 1" Snow Depth | *19* | 14 | 7 | 1 | 0 | 0 | 0 | 0 | 0 | 0 | 5 | 13 | 59 |

## HIGHMORE 23 N *Hyde County* ELEVATION 1870 ft LAT/LONG 44° 52 ' N / 99° 27 ' W

| | JAN | FEB | MAR | APR | MAY | JUN | JUL | AUG | SEP | OCT | NOV | DEC | YEAR |
|---|---|---|---|---|---|---|---|---|---|---|---|---|---|
| Maximum Temp °F | 24.1 | 30.4 | 43.1 | 59.4 | 71.8 | 80.3 | 87.4 | 86.2 | 75.9 | 62.4 | 42.0 | 28.5 | 57.6 |
| Minimum Temp °F | 1.7 | 8.2 | 19.8 | 31.7 | 42.6 | 52.4 | 57.4 | 55.3 | 45.0 | 33.3 | 19.7 | 7.4 | 31.2 |
| Mean Temp °F | 12.9 | 19.3 | 31.5 | 45.5 | 57.2 | 66.4 | 72.4 | 70.8 | 60.5 | 47.9 | 30.9 | 18.0 | 44.4 |
| Days Max Temp ≥ 90 °F | 0 | 0 | 0 | 0 | 1 | 4 | 12 | 11 | 4 | 0 | 0 | 0 | 32 |
| Days Max Temp ≤ 32 °F | 20 | 15 | 7 | 1 | 0 | 0 | 0 | 0 | 0 | 0 | 7 | 18 | 68 |
| Days Min Temp ≤ 32 °F | 31 | 28 | 28 | 17 | 5 | 0 | 0 | 0 | 3 | 14 | 28 | 31 | 185 |
| Days Min Temp ≤ 0 °F | 15 | 9 | 2 | 0 | 0 | 0 | 0 | 0 | 0 | 0 | 2 | 9 | 37 |
| Heating Degree Days | 1611 | 1284 | 1032 | 580 | 260 | 62 | 13 | 27 | 190 | 526 | 1018 | 1452 | 8055 |
| Cooling Degree Days | 0 | 0 | 0 | 5 | 31 | 117 | 242 | 211 | 62 | 3 | 0 | 0 | 671 |
| Total Precipitation (") | 0.39 | 0.59 | 1.13 | 2.10 | 2.27 | 2.98 | 2.72 | 2.02 | 1.54 | 1.25 | 0.69 | 0.52 | 18.20 |
| Days ≥ 0.1" Precip | 2 | 2 | 3 | 4 | 5 | 6 | 5 | 4 | 3 | 2 | 2 | 2 | 40 |
| Total Snowfall (") | 4.7 | 6.9 | 7.8 | 3.9 | 0.1 | 0.0 | 0.0 | 0.0 | 0.0 | 1.2 | 6.2 | 5.9 | 36.7 |
| Days ≥ 1" Snow Depth | 23 | *17* | 11 | 2 | 0 | 0 | 0 | 0 | 0 | 0 | 6 | 14 | 73 |

## HOT SPRINGS *Fall River County* ELEVATION 3553 ft LAT/LONG 43° 26 ' N / 103° 28 ' W

| | JAN | FEB | MAR | APR | MAY | JUN | JUL | AUG | SEP | OCT | NOV | DEC | YEAR |
|---|---|---|---|---|---|---|---|---|---|---|---|---|---|
| Maximum Temp °F | 38.0 | 42.9 | 51.6 | 61.5 | 71.5 | 81.6 | 88.9 | 87.8 | 78.4 | 66.0 | 48.7 | 39.6 | 63.0 |
| Minimum Temp °F | 11.4 | 15.1 | 22.6 | 32.1 | 41.7 | 50.6 | 56.5 | 54.2 | 43.7 | 33.1 | 22.3 | 13.8 | 33.1 |
| Mean Temp °F | 24.7 | 29.0 | 37.1 | 46.9 | 56.7 | 66.1 | 72.7 | 71.0 | 61.1 | 49.6 | 35.5 | 26.7 | 48.1 |
| Days Max Temp ≥ 90 °F | 0 | 0 | 0 | 0 | 1 | 6 | 15 | 14 | 5 | 0 | 0 | 0 | 41 |
| Days Max Temp ≤ 32 °F | 9 | 6 | 3 | 0 | 0 | 0 | 0 | 0 | 0 | 0 | 3 | 8 | 29 |
| Days Min Temp ≤ 32 °F | 30 | 27 | 27 | 15 | 4 | 0 | 0 | 0 | 3 | 15 | 26 | 30 | 177 |
| Days Min Temp ≤ 0 °F | 8 | 4 | 1 | 0 | 0 | 0 | 0 | 0 | 0 | 0 | 1 | 4 | 18 |
| Heating Degree Days | 1243 | 1009 | 858 | 539 | 268 | 66 | 10 | 18 | 166 | 474 | 879 | 1180 | 6710 |
| Cooling Degree Days | 0 | 0 | 0 | 3 | 20 | 112 | 256 | 232 | 58 | 2 | 0 | 0 | 683 |
| Total Precipitation (") | 0.32 | 0.44 | 0.83 | 1.65 | 2.76 | 2.83 | 2.58 | 1.52 | 1.29 | 1.06 | 0.49 | 0.38 | 16.15 |
| Days ≥ 0.1" Precip | 1 | 2 | 2 | 4 | 6 | 6 | 5 | 3 | 3 | 3 | 2 | 1 | 38 |
| Total Snowfall (") | 5.9 | 6.1 | 6.4 | 3.6 | 0.1 | 0.0 | 0.0 | 0.0 | 0.1 | 1.7 | 4.1 | 6.9 | 34.9 |
| Days ≥ 1" Snow Depth | 15 | 11 | 5 | 2 | 0 | 0 | 0 | 0 | 0 | 1 | 5 | 13 | 52 |

## HOWARD *Miner County* ELEVATION 1562 ft LAT/LONG 44° 1' N / 97° 31' W

| | JAN | FEB | MAR | APR | MAY | JUN | JUL | AUG | SEP | OCT | NOV | DEC | YEAR |
|---|---|---|---|---|---|---|---|---|---|---|---|---|---|
| Maximum Temp °F | 23.2 | 29.6 | 43.4 | 59.5 | 72.2 | 80.8 | 86.7 | 84.3 | 74.2 | 60.7 | 41.5 | 27.9 | 57.0 |
| Minimum Temp °F | 2.9 | 9.1 | 22.0 | 34.8 | 46.1 | 55.7 | 60.8 | 58.4 | 48.2 | 36.0 | 22.4 | 9.0 | 33.8 |
| Mean Temp °F | 13.1 | 19.4 | 32.7 | 47.2 | 59.2 | 68.3 | 73.8 | 71.4 | 61.2 | 48.4 | 32.0 | 18.4 | 45.4 |
| Days Max Temp ≥ 90 °F | 0 | 0 | 0 | 0 | 1 | 4 | 11 | 9 | 2 | 0 | 0 | 0 | 27 |
| Days Max Temp ≤ 32 °F | 22 | 15 | 6 | 0 | 0 | 0 | 0 | 0 | 0 | 0 | 7 | 19 | 69 |
| Days Min Temp ≤ 32 °F | 31 | 28 | 26 | 13 | 2 | 0 | 0 | 0 | 1 | 11 | 25 | 31 | 168 |
| Days Min Temp ≤ 0 °F | 13 | 8 | 2 | 0 | 0 | 0 | 0 | 0 | 0 | 0 | 1 | 8 | 32 |
| Heating Degree Days | 1606 | 1283 | 995 | 534 | 214 | 41 | 9 | 20 | 173 | 511 | 985 | 1439 | 7810 |
| Cooling Degree Days | 0 | 0 | 0 | 9 | 43 | 145 | 266 | 201 | 60 | 3 | 0 | 0 | 727 |
| Total Precipitation (") | 0.51 | 0.67 | 1.45 | 2.31 | 2.75 | 4.04 | 2.94 | 2.77 | 2.18 | 1.65 | 1.07 | 0.59 | 22.93 |
| Days ≥ 0.1" Precip | 1 | 2 | 4 | 6 | 6 | 7 | 6 | 5 | 5 | 3 | 3 | 2 | 50 |
| Total Snowfall (") | *5.8* | *6.2* | *7.1* | *2.0* | 0.0 | 0.0 | 0.0 | 0.0 | 0.1 | 0.7 | *4.7* | *6.4* | 33.0 |
| Days ≥ 1" Snow Depth | na | na | na | *0* | 0 | 0 | 0 | 0 | 0 | 0 | *2* | na | na |

## HURON REGIONAL AP *Beadle County* ELEVATION 1296 ft LAT/LONG 44° 23' N / 98° 13' W

| | JAN | FEB | MAR | APR | MAY | JUN | JUL | AUG | SEP | OCT | NOV | DEC | YEAR |
|---|---|---|---|---|---|---|---|---|---|---|---|---|---|
| Maximum Temp °F | 24.3 | 29.7 | 42.7 | 58.5 | 70.6 | 80.2 | 86.4 | 84.4 | 74.0 | 60.6 | 42.0 | 28.9 | 56.9 |
| Minimum Temp °F | 2.9 | 9.5 | 22.2 | 34.2 | 45.0 | 54.9 | 60.7 | 58.2 | 47.4 | 35.0 | 22.0 | 9.0 | 33.4 |
| Mean Temp °F | 13.6 | 19.6 | 32.5 | 46.4 | 57.9 | 67.6 | 73.6 | 71.3 | 60.7 | 47.8 | 32.0 | 19.0 | 45.2 |
| Days Max Temp ≥ 90 °F | 0 | 0 | 0 | 0 | 1 | 4 | 11 | 9 | 3 | 0 | 0 | 0 | 28 |
| Days Max Temp ≤ 32 °F | 20 | 15 | 6 | 0 | 0 | 0 | 0 | 0 | 0 | 0 | 7 | 17 | 65 |
| Days Min Temp ≤ 32 °F | 31 | 27 | 26 | 13 | 2 | 0 | 0 | 0 | 1 | 12 | 26 | 31 | 169 |
| Days Min Temp ≤ 0 °F | 14 | 9 | 2 | 0 | 0 | 0 | 0 | 0 | 0 | 0 | 1 | 8 | 34 |
| Heating Degree Days | 1590 | 1275 | 1001 | 556 | 245 | 50 | 8 | 24 | 185 | 528 | 982 | 1420 | 7864 |
| Cooling Degree Days | 0 | 0 | 0 | 7 | 38 | 142 | 273 | 227 | 67 | 3 | 0 | 0 | 757 |
| Total Precipitation (") | 0.44 | 0.63 | 1.60 | 2.32 | 2.62 | 3.65 | 2.81 | 2.06 | 1.79 | 1.38 | 0.82 | 0.47 | 20.59 |
| Days ≥ 0.1" Precip | 1 | 2 | 4 | 5 | 5 | 6 | 5 | 4 | 4 | 3 | 2 | 1 | 42 |
| Total Snowfall (") | 7.4 | 8.5 | 9.1 | 2.4 | 0.1 | 0.0 | 0.0 | 0.0 | 0.0 | 0.9 | 5.7 | 7.2 | 41.3 |
| Days ≥ 1" Snow Depth | 21 | 18 | 10 | 1 | 0 | 0 | 0 | 0 | 0 | 0 | 6 | 14 | 70 |

## INTERIOR 3 NE *Jackson County* ELEVATION 2441 ft LAT/LONG 43° 45' N / 101° 57' W

| | JAN | FEB | MAR | APR | MAY | JUN | JUL | AUG | SEP | OCT | NOV | DEC | YEAR |
|---|---|---|---|---|---|---|---|---|---|---|---|---|---|
| Maximum Temp °F | 34.8 | 40.0 | 50.3 | 62.8 | 73.5 | 83.3 | 91.1 | 90.4 | 79.9 | 66.9 | 47.1 | 37.7 | 63.2 |
| Minimum Temp °F | 11.8 | 16.0 | 25.3 | 36.1 | 46.6 | 55.9 | 62.0 | 60.1 | 50.0 | 37.7 | 24.7 | 15.4 | 36.8 |
| Mean Temp °F | 23.2 | 28.0 | 37.8 | 49.4 | 60.1 | 69.6 | 76.5 | 75.2 | 65.0 | 52.4 | 35.9 | 26.6 | 50.0 |
| Days Max Temp ≥ 90 °F | 0 | 0 | 0 | 0 | 2 | 8 | 17 | 18 | 7 | 1 | 0 | 0 | 53 |
| Days Max Temp ≤ 32 °F | 12 | 9 | 4 | 0 | 0 | 0 | 0 | 0 | 0 | 0 | 4 | 10 | 39 |
| Days Min Temp ≤ 32 °F | 29 | 26 | 23 | 10 | 1 | 0 | 0 | 0 | 1 | 8 | *23* | 28 | 149 |
| Days Min Temp ≤ 0 °F | 8 | 4 | 1 | 0 | 0 | 0 | 0 | 0 | 0 | 0 | 1 | 5 | 19 |
| Heating Degree Days | 1289 | 1038 | 836 | 466 | 197 | 34 | 5 | 9 | 112 | 395 | 864 | 1184 | 6429 |
| Cooling Degree Days | 0 | 0 | 0 | 14 | 62 | 191 | 364 | 343 | 121 | 11 | 0 | 0 | 1106 |
| Total Precipitation (") | 0.32 | 0.46 | 0.92 | 2.00 | 2.83 | 2.85 | 2.13 | 1.68 | 1.22 | 1.15 | 0.51 | 0.34 | 16.41 |
| Days ≥ 0.1" Precip | 1 | 1 | 2 | 5 | 5 | 6 | 5 | 4 | 3 | 3 | 2 | 1 | 38 |
| Total Snowfall (") | 2.8 | 4.9 | *4.1* | *2.3* | 0.1 | 0.0 | 0.0 | 0.0 | 0.0 | 0.4 | *3.2* | 3.8 | 21.6 |
| Days ≥ 1" Snow Depth | na | na | *5* | 1 | 0 | 0 | 0 | 0 | 0 | *0* | na | *10* | na |

## IPSWICH *Edmunds County* ELEVATION 1532 ft LAT/LONG 45° 27' N / 99° 1' W

| | JAN | FEB | MAR | APR | MAY | JUN | JUL | AUG | SEP | OCT | NOV | DEC | YEAR |
|---|---|---|---|---|---|---|---|---|---|---|---|---|---|
| Maximum Temp °F | 22.0 | 28.0 | 41.6 | 58.5 | 71.3 | 79.6 | 86.2 | 85.2 | 74.3 | 61.0 | 40.3 | 26.9 | 56.2 |
| Minimum Temp °F | -0.5 | 5.5 | 18.6 | 31.5 | 42.8 | 52.6 | 57.4 | 54.9 | 44.3 | 32.5 | 18.4 | 5.4 | 30.3 |
| Mean Temp °F | 10.7 | 16.8 | 30.1 | 45.0 | 57.1 | 66.1 | 71.8 | 70.0 | 59.3 | 46.7 | 29.4 | 16.2 | 43.3 |
| Days Max Temp ≥ 90 °F | 0 | 0 | 0 | 0 | 1 | 3 | 11 | 10 | 3 | 0 | 0 | 0 | 28 |
| Days Max Temp ≤ 32 °F | 21 | 17 | 7 | 0 | 0 | 0 | 0 | 0 | 0 | 0 | 8 | 20 | 73 |
| Days Min Temp ≤ 32 °F | 31 | 28 | 29 | 17 | 4 | 0 | 0 | 0 | 3 | 16 | 28 | 31 | 187 |
| Days Min Temp ≤ 0 °F | 16 | 11 | 3 | 0 | 0 | 0 | 0 | 0 | 0 | 0 | 2 | 11 | 43 |
| Heating Degree Days | 1680 | 1356 | 1074 | 597 | 263 | 63 | 12 | 29 | 210 | 561 | 1062 | 1508 | 8415 |
| Cooling Degree Days | 0 | 0 | 0 | 4 | 28 | 103 | 221 | 178 | 41 | 1 | 0 | 0 | 576 |
| Total Precipitation (") | 0.40 | 0.42 | 1.11 | 2.21 | 2.44 | 3.49 | 2.89 | 2.13 | 1.50 | 1.18 | 0.66 | 0.38 | 18.81 |
| Days ≥ 0.1" Precip | 1 | 1 | 2 | 5 | 5 | 6 | 6 | 4 | 3 | 2 | 2 | 1 | 38 |
| Total Snowfall (") | 5.1 | 5.9 | 6.5 | 3.1 | 0.0 | 0.0 | 0.0 | 0.0 | 0.0 | 0.5 | 5.4 | 5.4 | 31.9 |
| Days ≥ 1" Snow Depth | 23 | 18 | 11 | 2 | 0 | 0 | 0 | 0 | 0 | 0 | 6 | 15 | 75 |

## KENNEBEC *Lyman County* ELEVATION 1690 ft LAT/LONG 43° 54 ' N / 99° 51 ' W

| | JAN | FEB | MAR | APR | MAY | JUN | JUL | AUG | SEP | OCT | NOV | DEC | YEAR |
|---|---|---|---|---|---|---|---|---|---|---|---|---|---|
| Maximum Temp °F | 29.4 | 35.2 | 47.4 | 62.3 | 74.0 | 83.5 | 90.9 | 89.3 | 79.3 | 65.1 | 45.2 | 33.3 | 61.2 |
| Minimum Temp °F | 6.0 | 11.4 | 22.6 | 34.1 | 45.4 | 55.7 | 61.2 | 59.1 | 48.4 | 35.8 | 22.0 | 10.5 | 34.4 |
| Mean Temp °F | 17.7 | 23.3 | 35.0 | 48.2 | 59.7 | 69.6 | 76.0 | 74.2 | 63.9 | 50.5 | 33.6 | 21.9 | 47.8 |
| Days Max Temp ≥ 90 °F | 0 | 0 | 0 | 0 | 2 | 7 | 17 | 16 | 6 | 1 | 0 | 0 | 49 |
| Days Max Temp ≤ 32 °F | 16 | 12 | 5 | 0 | 0 | 0 | 0 | 0 | 0 | 0 | 5 | 14 | 52 |
| Days Min Temp ≤ 32 °F | 30 | 28 | 26 | 13 | 3 | 0 | 0 | 0 | 2 | 12 | 26 | 31 | 171 |
| Days Min Temp ≤ 0 °F | 11 | 7 | 2 | 0 | 0 | 0 | 0 | 0 | 0 | 0 | 1 | 7 | 28 |
| Heating Degree Days | 1462 | 1171 | 922 | 503 | 205 | 35 | 6 | 13 | 135 | 451 | 935 | 1330 | 7168 |
| Cooling Degree Days | 0 | 0 | 0 | 12 | 56 | 180 | 345 | 303 | 106 | 6 | 0 | 0 | 1008 |
| Total Precipitation (") | 0.24 | 0.44 | 1.19 | 2.03 | 2.78 | 3.07 | 2.49 | 2.09 | 1.34 | 1.24 | 0.55 | 0.38 | 17.84 |
| Days ≥ 0.1" Precip | 1 | 2 | 3 | 5 | 6 | 6 | 5 | 4 | 4 | 3 | 2 | 1 | 42 |
| Total Snowfall (") | 3.9 | 6.3 | 8.0 | 2.0 | 0.1 | 0.0 | 0.0 | 0.0 | 0.1 | 1.0 | 4.1 | 5.4 | 30.9 |
| Days ≥ 1" Snow Depth | 14 | 11 | 6 | 1 | 0 | 0 | 0 | 0 | 0 | 0 | 4 | 11 | 47 |

## LEAD *Lawrence County* ELEVATION 5253 ft LAT/LONG 44° 21 ' N / 103° 46 ' W

| | JAN | FEB | MAR | APR | MAY | JUN | JUL | AUG | SEP | OCT | NOV | DEC | YEAR |
|---|---|---|---|---|---|---|---|---|---|---|---|---|---|
| Maximum Temp °F | 34.0 | 36.9 | 42.3 | 51.5 | 62.2 | 71.9 | 79.3 | 78.2 | 68.3 | 56.3 | 42.2 | 35.6 | 54.9 |
| Minimum Temp °F | 14.3 | 17.0 | 22.3 | 30.5 | 39.9 | 48.9 | 55.1 | 53.9 | 44.6 | 34.6 | 23.6 | 16.6 | 33.4 |
| Mean Temp °F | 24.2 | 27.0 | 32.3 | 41.0 | 51.1 | 60.4 | 67.2 | 66.1 | 56.5 | 45.5 | 32.9 | 26.1 | 44.2 |
| Days Max Temp ≥ 90 °F | 0 | 0 | 0 | 0 | 0 | 1 | 2 | 1 | 0 | 0 | 0 | 0 | 4 |
| Days Max Temp ≤ 32 °F | 13 | 9 | 7 | 2 | 0 | 0 | 0 | 0 | 0 | 1 | 7 | 12 | 51 |
| Days Min Temp ≤ 32 °F | 28 | 26 | 26 | 19 | 6 | 0 | 0 | 0 | 3 | 13 | 23 | 28 | 172 |
| Days Min Temp ≤ 0 °F | 6 | 3 | 1 | 0 | 0 | 0 | 0 | 0 | 0 | 0 | 1 | 4 | 15 |
| Heating Degree Days | 1260 | 1068 | 1006 | 713 | 431 | 177 | 49 | 68 | 279 | 599 | 956 | 1199 | 7805 |
| Cooling Degree Days | 0 | 0 | 0 | 1 | 7 | 50 | 126 | 105 | 27 | 0 | 0 | 0 | 316 |
| Total Precipitation (") | 1.30 | 1.33 | 2.52 | 3.96 | 4.22 | 3.67 | 2.42 | 1.98 | 1.87 | 2.16 | 1.69 | 1.51 | 28.63 |
| Days ≥ 0.1" Precip | 4 | 4 | 6 | 8 | 8 | 8 | 6 | 5 | 4 | 5 | 5 | 5 | 68 |
| Total Snowfall (") | 21.3 | 21.1 | 34.7 | 33.3 | 7.1 | 0.4 | 0.0 | 0.0 | 2.4 | 12.5 | 23.0 | 25.0 | 180.8 |
| Days ≥ 1" Snow Depth | 26 | 22 | 17 | 10 | 2 | 0 | 0 | 0 | 0 | 4 | 14 | 21 | 116 |

## LEMMON *Perkins County* ELEVATION 2592 ft LAT/LONG 45° 56 ' N / 102° 10 ' W

| | JAN | FEB | MAR | APR | MAY | JUN | JUL | AUG | SEP | OCT | NOV | DEC | YEAR |
|---|---|---|---|---|---|---|---|---|---|---|---|---|---|
| Maximum Temp °F | 24.1 | 30.0 | 41.3 | 55.7 | 68.1 | 76.8 | 83.9 | 82.8 | 71.6 | 58.6 | 40.3 | 28.3 | 55.1 |
| Minimum Temp °F | 3.6 | 9.2 | 19.3 | 31.1 | 42.2 | 51.6 | 57.1 | 55.2 | 44.5 | 33.2 | 19.9 | 8.4 | 31.3 |
| Mean Temp °F | 13.9 | 19.6 | 30.3 | 43.4 | 55.2 | 64.2 | 70.5 | 69.0 | 58.1 | 45.9 | 30.1 | 18.4 | 43.2 |
| Days Max Temp ≥ 90 °F | 0 | 0 | 0 | 0 | 0 | 3 | 8 | 8 | 2 | 0 | 0 | 0 | 21 |
| Days Max Temp ≤ 32 °F | 20 | 15 | 8 | 1 | 0 | 0 | 0 | 0 | 0 | 1 | 8 | 18 | 71 |
| Days Min Temp ≤ 32 °F | 31 | 28 | 29 | 18 | 4 | 0 | 0 | 0 | 3 | 14 | 27 | 31 | 185 |
| Days Min Temp ≤ 0 °F | 13 | 9 | 2 | 0 | 0 | 0 | 0 | 0 | 0 | 0 | 2 | 8 | 34 |
| Heating Degree Days | 1579 | 1276 | 1068 | 642 | 315 | 95 | 19 | 38 | 240 | 585 | 1039 | 1438 | 8334 |
| Cooling Degree Days | 0 | 0 | 0 | 3 | 23 | 94 | 212 | 183 | 42 | 2 | 0 | 0 | 559 |
| Total Precipitation (") | 0.50 | 0.52 | 1.08 | 2.08 | 2.67 | 3.43 | 2.82 | 1.73 | 1.35 | 1.14 | 0.63 | 0.64 | 18.59 |
| Days ≥ 0.1" Precip | 2 | 2 | 3 | 6 | 6 | 7 | 5 | 4 | 3 | 2 | 2 | 2 | 44 |
| Total Snowfall (") | ***4.8*** | ***5.8*** | ***7.1*** | ***5.3*** | 0.2 | ***0.0*** | 0.0 | 0.0 | 0.3 | ***1.5*** | ***6.3*** | ***6.2*** | 37.5 |
| Days ≥ 1" Snow Depth | 24 | 19 | 13 | 3 | 0 | 0 | 0 | 0 | 0 | 1 | 9 | 19 | 88 |

## LEOLA *McPherson County* ELEVATION 1591 ft LAT/LONG 45° 43 ' N / 98° 56 ' W

| | JAN | FEB | MAR | APR | MAY | JUN | JUL | AUG | SEP | OCT | NOV | DEC | YEAR |
|---|---|---|---|---|---|---|---|---|---|---|---|---|---|
| Maximum Temp °F | 20.6 | 27.1 | 39.6 | 56.7 | 70.4 | 78.7 | 85.3 | 83.9 | 73.0 | 59.4 | 39.0 | 25.6 | 54.9 |
| Minimum Temp °F | 0.0 | 6.6 | 19.0 | 31.6 | 43.3 | 52.7 | 57.8 | 55.5 | 44.9 | 33.5 | 19.1 | 6.1 | 30.8 |
| Mean Temp °F | 10.3 | 16.8 | 29.3 | 44.2 | 56.9 | 65.7 | 71.6 | 69.7 | 59.0 | 46.5 | 29.1 | 15.9 | 42.9 |
| Days Max Temp ≥ 90 °F | 0 | 0 | 0 | 0 | 1 | 3 | 9 | 8 | 2 | 0 | 0 | 0 | 23 |
| Days Max Temp ≤ 32 °F | 23 | 17 | 9 | 1 | 0 | 0 | 0 | 0 | 0 | 0 | 10 | 21 | 81 |
| Days Min Temp ≤ 32 °F | 31 | 28 | 28 | 17 | 3 | 0 | 0 | 0 | 2 | 15 | 28 | 31 | 183 |
| Days Min Temp ≤ 0 °F | 16 | 10 | 3 | 0 | 0 | 0 | 0 | 0 | 0 | 0 | 2 | 10 | 41 |
| Heating Degree Days | 1693 | 1355 | 1100 | 619 | 271 | 66 | 13 | 30 | 215 | 570 | 1072 | 1518 | 8522 |
| Cooling Degree Days | 0 | 0 | 0 | 4 | 29 | 100 | 218 | 177 | 40 | 2 | 0 | 0 | 570 |
| Total Precipitation (") | 0.47 | 0.51 | 1.25 | 2.20 | 2.57 | 3.26 | 2.67 | 2.05 | 1.60 | 1.18 | 0.78 | 0.43 | 18.97 |
| Days ≥ 0.1" Precip | 2 | 2 | 3 | 5 | 6 | 6 | 6 | 4 | 4 | 3 | 2 | 2 | 45 |
| Total Snowfall (") | ***5.5*** | ***7.8*** | ***7.9*** | 4.4 | 0.1 | 0.0 | 0.0 | 0.0 | 0.1 | 0.4 | ***8.3*** | ***5.6*** | 40.1 |
| Days ≥ 1" Snow Depth | na | na | ***12*** | ***1*** | 0 | 0 | 0 | 0 | 0 | 0 | na | na | na |

## LONG VALLEY *Jackson County* ELEVATION 2470 ft LAT/LONG 43° 28' N / 101° 30' W

| | JAN | FEB | MAR | APR | MAY | JUN | JUL | AUG | SEP | OCT | NOV | DEC | YEAR |
|---|---|---|---|---|---|---|---|---|---|---|---|---|---|
| Maximum Temp °F | 34.9 | 39.7 | 48.4 | 60.5 | 71.7 | 81.4 | 88.9 | 88.0 | 77.5 | 64.7 | 47.4 | 37.8 | 61.7 |
| Minimum Temp °F | 10.5 | 14.5 | 22.9 | 33.6 | 43.9 | 53.6 | 59.4 | 57.3 | 47.2 | 35.7 | 22.9 | 13.7 | 34.6 |
| Mean Temp °F | 22.7 | 27.1 | 35.7 | 47.1 | 57.8 | 67.5 | 74.2 | 72.7 | 62.4 | 50.2 | 35.2 | 25.8 | 48.2 |
| Days Max Temp ≥ 90 °F | 0 | 0 | 0 | 0 | 1 | 6 | 15 | 14 | 5 | 0 | 0 | 0 | 41 |
| Days Max Temp ≤ 32 °F | 12 | 8 | 4 | 0 | 0 | 0 | 0 | 0 | 0 | 0 | 5 | 10 | 39 |
| Days Min Temp ≤ 32 °F | 29 | 26 | 25 | 14 | 3 | 0 | 0 | 0 | 2 | 11 | 24 | 29 | 163 |
| Days Min Temp ≤ 0 °F | 9 | 5 | 1 | 0 | 0 | 0 | 0 | 0 | 0 | 0 | 1 | 5 | 21 |
| Heating Degree Days | 1306 | 1063 | 904 | 536 | 248 | 54 | 9 | 16 | 157 | 459 | 887 | 1209 | 6848 |
| Cooling Degree Days | 0 | 0 | 0 | 8 | 33 | 134 | 275 | 246 | 79 | 6 | 0 | 0 | 781 |
| Total Precipitation (") | 0.28 | 0.42 | 1.23 | 2.14 | 2.71 | 3.16 | 2.67 | 1.82 | 1.31 | 1.04 | 0.54 | 0.36 | 17.68 |
| Days ≥ 0.1" Precip | 1 | 1 | 3 | 5 | 6 | 6 | 5 | 4 | 3 | 3 | 2 | 1 | 40 |
| Total Snowfall (") | 4.1 | 5.5 | 10.2 | 5.9 | 0.2 | 0.0 | 0.0 | 0.0 | 0.4 | 1.7 | 5.9 | 5.6 | 39.5 |
| Days ≥ 1" Snow Depth | 18 | 13 | 9 | 3 | 0 | 0 | 0 | 0 | 0 | 1 | 8 | 16 | 68 |

## LUDLOW *Harding County* ELEVATION 2851 ft LAT/LONG 45° 52' N / 103° 25' W

| | JAN | FEB | MAR | APR | MAY | JUN | JUL | AUG | SEP | OCT | NOV | DEC | YEAR |
|---|---|---|---|---|---|---|---|---|---|---|---|---|---|
| Maximum Temp °F | 26.8 | 32.2 | 42.5 | 56.5 | 68.4 | 77.2 | 85.1 | 84.7 | 73.1 | 59.6 | 41.3 | 30.6 | 56.5 |
| Minimum Temp °F | 5.4 | 10.7 | 19.9 | 31.1 | 41.2 | 50.3 | 55.7 | 53.5 | 43.4 | 32.5 | 19.6 | 9.2 | 31.0 |
| Mean Temp °F | 16.1 | 21.5 | 31.2 | 43.9 | 54.8 | 63.8 | 70.4 | 69.1 | 58.3 | 46.1 | 30.5 | 19.9 | 43.8 |
| Days Max Temp ≥ 90 °F | 0 | 0 | 0 | 0 | 0 | 3 | 10 | 11 | 3 | 0 | 0 | 0 | 27 |
| Days Max Temp ≤ 32 °F | 17 | 13 | 7 | 1 | 0 | 0 | 0 | 0 | 0 | 1 | 8 | 15 | 62 |
| Days Min Temp ≤ 32 °F | 30 | 27 | 28 | 18 | 5 | 0 | 0 | 0 | 3 | 15 | 27 | 30 | 183 |
| Days Min Temp ≤ 0 °F | 12 | 7 | 2 | 0 | 0 | 0 | 0 | 0 | 0 | 0 | 2 | 8 | 31 |
| Heating Degree Days | 1511 | 1225 | 1040 | 628 | 323 | 102 | 20 | 39 | 235 | 582 | 1030 | 1392 | 8127 |
| Cooling Degree Days | 0 | 0 | 0 | 2 | 17 | 82 | 205 | 176 | 40 | 3 | 0 | 0 | 525 |
| Total Precipitation (") | 0.36 | 0.33 | 0.75 | 1.95 | 2.75 | 3.34 | 2.14 | 1.23 | 1.31 | 1.19 | 0.52 | 0.46 | 16.33 |
| Days ≥ 0.1" Precip | 1 | 1 | 2 | 4 | 6 | 7 | 5 | 3 | 3 | 3 | 2 | 2 | 39 |
| Total Snowfall (") | na | na | na | na | 0.6 | 0.0 | 0.0 | 0.0 | 0.2 | *1.7* | na | na | na |
| Days ≥ 1" Snow Depth | na | na | na | na | 0 | 0 | 0 | 0 | 0 | 1 | *4* | na | na |

## MADISON 2 E *Lake County* ELEVATION 1710 ft LAT/LONG 44° 0' N / 97° 4' W

| | JAN | FEB | MAR | APR | MAY | JUN | JUL | AUG | SEP | OCT | NOV | DEC | YEAR |
|---|---|---|---|---|---|---|---|---|---|---|---|---|---|
| Maximum Temp °F | 21.9 | 27.3 | 40.8 | 57.4 | 70.0 | 78.9 | 83.8 | 82.1 | 72.6 | 59.8 | 40.6 | 27.0 | 55.2 |
| Minimum Temp °F | 2.2 | 8.0 | 21.3 | 34.4 | 46.0 | 55.5 | 60.4 | 57.9 | 48.2 | 35.9 | 22.1 | 8.7 | 33.4 |
| Mean Temp °F | 12.1 | 17.7 | 31.1 | 45.9 | 58.0 | 67.2 | 72.1 | 70.0 | 60.4 | 47.9 | 31.4 | 17.9 | 44.3 |
| Days Max Temp ≥ 90 °F | 0 | 0 | 0 | 0 | 0 | 3 | 7 | 5 | 2 | 0 | 0 | 0 | 17 |
| Days Max Temp ≤ 32 °F | 23 | 17 | 8 | 1 | 0 | 0 | 0 | 0 | 0 | 0 | 8 | 20 | 77 |
| Days Min Temp ≤ 32 °F | 31 | 28 | 27 | 13 | 2 | 0 | 0 | 0 | 1 | 11 | 26 | 31 | 170 |
| Days Min Temp ≤ 0 °F | 14 | 9 | 2 | 0 | 0 | 0 | 0 | 0 | 0 | 0 | 1 | 9 | 35 |
| Heating Degree Days | 1638 | 1331 | 1044 | 569 | 241 | 52 | 12 | 28 | 185 | 527 | 1003 | 1456 | 8086 |
| Cooling Degree Days | 0 | 0 | 0 | 7 | 31 | 119 | 214 | 171 | 49 | 2 | 0 | 0 | 593 |
| Total Precipitation (") | 0.53 | 0.78 | 1.73 | 2.42 | 3.02 | 3.95 | 2.84 | 2.98 | 2.42 | 1.84 | 1.20 | 0.77 | 24.48 |
| Days ≥ 0.1" Precip | 2 | 2 | 4 | 6 | 6 | 7 | 5 | 5 | 5 | 3 | 3 | 2 | 50 |
| Total Snowfall (") | 6.5 | 7.0 | 7.7 | 2.6 | 0.1 | 0.0 | 0.0 | 0.0 | 0.1 | 0.8 | 5.9 | 7.4 | 38.1 |
| Days ≥ 1" Snow Depth | *25* | *21* | 11 | 2 | 0 | 0 | 0 | 0 | 0 | 0 | 5 | *17* | 81 |

## MARION *Turner County* ELEVATION 1440 ft LAT/LONG 43° 25' N / 97° 15' W

| | JAN | FEB | MAR | APR | MAY | JUN | JUL | AUG | SEP | OCT | NOV | DEC | YEAR |
|---|---|---|---|---|---|---|---|---|---|---|---|---|---|
| Maximum Temp °F | 24.2 | 29.6 | 43.5 | 59.1 | 71.7 | 80.7 | 85.8 | 83.5 | 74.1 | 61.2 | 42.5 | 28.5 | 57.0 |
| Minimum Temp °F | 3.5 | 9.0 | 22.2 | 34.9 | 46.6 | 56.6 | 61.4 | 58.7 | 48.4 | 35.9 | 22.5 | 9.4 | 34.1 |
| Mean Temp °F | 13.9 | 19.3 | 32.8 | 47.0 | 59.2 | 68.7 | 73.6 | 71.1 | 61.3 | 48.6 | 32.5 | 19.0 | 45.6 |
| Days Max Temp ≥ 90 °F | 0 | 0 | 0 | 0 | 1 | 4 | 10 | 7 | 2 | 0 | 0 | 0 | 24 |
| Days Max Temp ≤ 32 °F | 21 | 15 | 6 | 1 | 0 | 0 | 0 | 0 | 0 | 0 | 6 | 18 | 67 |
| Days Min Temp ≤ 32 °F | 30 | 27 | 26 | 13 | 2 | 0 | 0 | 0 | 1 | 11 | 26 | 31 | 167 |
| Days Min Temp ≤ 0 °F | 13 | 8 | 2 | 0 | 0 | 0 | 0 | 0 | 0 | 0 | 1 | 8 | 32 |
| Heating Degree Days | 1582 | 1283 | 990 | 538 | 215 | 38 | 7 | 18 | 168 | 505 | 968 | 1421 | 7733 |
| Cooling Degree Days | 0 | 0 | 0 | 9 | 43 | 157 | 270 | 205 | 63 | 2 | 0 | 0 | 749 |
| Total Precipitation (") | 0.58 | 0.62 | 1.74 | 2.72 | 3.25 | 3.87 | 2.87 | 2.85 | 2.89 | 1.84 | 1.30 | 0.68 | 25.21 |
| Days ≥ 0.1" Precip | 2 | 2 | 4 | 6 | 6 | 7 | 6 | 5 | 6 | 4 | 3 | 2 | 53 |
| Total Snowfall (") | 7.0 | 6.6 | 6.8 | 2.7 | 0.0 | 0.0 | 0.0 | 0.0 | 0.0 | 0.5 | 6.4 | 7.4 | 37.4 |
| Days ≥ 1" Snow Depth | na | na | *8* | *1* | 0 | 0 | 0 | 0 | 0 | 0 | *3* | *14* | na |

## MC INTOSH 6 SE *Corson County* ELEVATION 2270 ft LAT/LONG 45° 55 ' N / 101° 21 ' W

| | JAN | FEB | MAR | APR | MAY | JUN | JUL | AUG | SEP | OCT | NOV | DEC | YEAR |
|---|---|---|---|---|---|---|---|---|---|---|---|---|---|
| Maximum Temp °F | 23.7 | 29.6 | 41.3 | 56.8 | 69.8 | 78.7 | 86.1 | 84.5 | 72.8 | 59.6 | 39.8 | 27.0 | 55.8 |
| Minimum Temp °F | 3.1 | 9.0 | 20.0 | 32.2 | 43.6 | 53.0 | 58.4 | 56.3 | 45.4 | 34.0 | 19.7 | 7.7 | 31.9 |
| Mean Temp °F | 13.5 | 19.4 | 30.6 | 44.5 | 56.8 | 65.9 | 72.3 | 70.4 | 59.2 | 46.8 | 29.8 | 17.4 | 43.9 |
| Days Max Temp ≥ 90 °F | 0 | 0 | 0 | 0 | 1 | 3 | 11 | 10 | 2 | 0 | 0 | 0 | 27 |
| Days Max Temp ≤ 32 °F | 20 | 15 | 8 | 1 | 0 | 0 | 0 | 0 | 0 | 0 | 9 | 19 | 72 |
| Days Min Temp ≤ 32 °F | 31 | 28 | 28 | 16 | 3 | 0 | 0 | 0 | 2 | 13 | 27 | 31 | 179 |
| Days Min Temp ≤ 0 °F | 14 | 8 | 2 | 0 | 0 | 0 | 0 | 0 | 0 | 0 | 2 | 9 | 35 |
| Heating Degree Days | 1593 | 1283 | 1059 | 610 | 275 | 71 | 11 | 25 | 212 | 559 | 1048 | 1473 | 8219 |
| Cooling Degree Days | 0 | 0 | 0 | 4 | 31 | 109 | 234 | 193 | 46 | 2 | 0 | 0 | 619 |
| Total Precipitation (") | 0.31 | 0.38 | 0.83 | 1.80 | 2.36 | 3.05 | 2.36 | 1.79 | 1.35 | 1.19 | 0.48 | 0.43 | 16.33 |
| Days ≥ 0.1" Precip | 1 | 1 | 2 | 4 | 5 | 6 | 5 | 4 | 3 | 3 | 2 | 1 | 37 |
| Total Snowfall (") | 4.7 | 4.8 | 7.6 | 4.7 | 0.4 | 0.0 | 0.0 | 0.0 | 0.2 | 0.9 | 5.6 | 5.4 | 34.3 |
| Days ≥ 1" Snow Depth | 24 | 18 | 14 | 4 | 0 | 0 | 0 | 0 | 0 | 1 | 10 | 20 | 91 |

## MELLETTE *Spink County* ELEVATION 1302 ft LAT/LONG 45° 9 ' N / 98° 30 ' W

| | JAN | FEB | MAR | APR | MAY | JUN | JUL | AUG | SEP | OCT | NOV | DEC | YEAR |
|---|---|---|---|---|---|---|---|---|---|---|---|---|---|
| Maximum Temp °F | 21.3 | 27.7 | 41.1 | 58.0 | 71.1 | 79.5 | 86.2 | 84.6 | 73.6 | 60.4 | 40.7 | 26.6 | 55.9 |
| Minimum Temp °F | -1.7 | 4.7 | 19.1 | 31.7 | 43.5 | 53.4 | 58.3 | 55.3 | 44.0 | 31.9 | 18.4 | 4.8 | 30.3 |
| Mean Temp °F | 9.8 | 16.2 | 30.1 | 44.9 | 57.3 | 66.4 | 72.3 | 70.0 | 58.8 | 46.2 | 29.6 | 15.5 | 43.1 |
| Days Max Temp ≥ 90 °F | 0 | 0 | 0 | 0 | 1 | 4 | 10 | 9 | 2 | 0 | 0 | 0 | 26 |
| Days Max Temp ≤ 32 °F | 23 | 17 | 7 | 1 | 0 | 0 | 0 | 0 | 0 | 0 | 8 | 19 | 75 |
| Days Min Temp ≤ 32 °F | 31 | 28 | 28 | 17 | 4 | 0 | 0 | 0 | 3 | 17 | 28 | 31 | 187 |
| Days Min Temp ≤ 0 °F | 17 | 11 | 3 | 0 | 0 | 0 | 0 | 0 | 0 | 0 | 2 | 11 | 44 |
| Heating Degree Days | 1708 | 1372 | 1075 | 600 | 262 | 66 | 13 | 33 | 226 | 579 | 1056 | 1528 | 8518 |
| Cooling Degree Days | 0 | 0 | 0 | 4 | 38 | 118 | 229 | 188 | 42 | 1 | 0 | 0 | 620 |
| Total Precipitation (") | 0.42 | 0.53 | 1.36 | 2.49 | 2.58 | 3.70 | 2.77 | 2.92 | 2.08 | 1.22 | 0.79 | 0.45 | 21.31 |
| Days ≥ 0.1" Precip | 2 | 2 | 3 | 5 | 5 | 7 | 5 | 4 | 4 | 3 | 2 | 1 | 43 |
| Total Snowfall (") | *5.9* | 6.3 | 5.9 | *3.7* | 0.0 | 0.0 | 0.0 | 0.0 | 0.0 | 0.5 | *6.0* | 6.0 | 34.3 |
| Days ≥ 1" Snow Depth | na | *16* | *11* | *1* | 0 | 0 | 0 | 0 | 0 | 0 | *7* | *16* | na |

## MENNO *Hutchinson County* ELEVATION 1322 ft LAT/LONG 43° 14 ' N / 97° 35 ' W

| | JAN | FEB | MAR | APR | MAY | JUN | JUL | AUG | SEP | OCT | NOV | DEC | YEAR |
|---|---|---|---|---|---|---|---|---|---|---|---|---|---|
| Maximum Temp °F | 27.6 | 33.8 | 47.0 | 62.6 | 74.4 | 83.8 | 88.1 | 86.1 | 77.3 | 64.2 | 44.7 | 31.2 | 60.1 |
| Minimum Temp °F | 5.8 | 12.2 | 24.3 | 36.2 | 47.6 | 57.1 | 62.0 | 59.6 | 49.6 | 37.5 | 23.7 | 11.5 | 35.6 |
| Mean Temp °F | 16.7 | 23.1 | 35.8 | 49.5 | 61.1 | 70.5 | 75.1 | 72.9 | 63.4 | 50.9 | 34.2 | 21.3 | 47.9 |
| Days Max Temp ≥ 90 °F | 0 | 0 | 0 | 0 | 1 | 7 | 13 | 11 | 4 | 0 | 0 | 0 | 36 |
| Days Max Temp ≤ 32 °F | 18 | 13 | 4 | 0 | 0 | 0 | 0 | 0 | 0 | 0 | 5 | 16 | 56 |
| Days Min Temp ≤ 32 °F | 31 | 27 | 24 | 11 | 2 | 0 | 0 | 0 | 1 | 10 | 25 | 31 | 162 |
| Days Min Temp ≤ 0 °F | 12 | 7 | 1 | 0 | 0 | 0 | 0 | 0 | 0 | 0 | 1 | 6 | 27 |
| Heating Degree Days | 1494 | 1177 | 899 | 468 | 171 | 23 | 5 | 12 | 129 | 437 | 916 | 1349 | 7080 |
| Cooling Degree Days | 0 | 0 | 0 | 15 | 61 | 198 | 304 | 252 | 93 | 6 | 0 | 0 | 929 |
| Total Precipitation (") | 0.46 | 0.56 | 1.53 | 2.38 | 3.31 | 3.67 | 3.48 | 2.37 | 2.51 | 1.67 | 1.12 | 0.61 | 23.67 |
| Days ≥ 0.1" Precip | 1 | 2 | 3 | 5 | 6 | 6 | 6 | 5 | 5 | 4 | 3 | 2 | 48 |
| Total Snowfall (") | 6.3 | 6.1 | 6.3 | 2.4 | 0.0 | 0.0 | 0.0 | 0.0 | 0.0 | 1.0 | 5.3 | 8.0 | 35.4 |
| Days ≥ 1" Snow Depth | 22 | 16 | 7 | 1 | 0 | 0 | 0 | 0 | 0 | 0 | 5 | 17 | 68 |

## MIDLAND *Haakon County* ELEVATION 1880 ft LAT/LONG 44° 4 ' N / 101° 9 ' W

| | JAN | FEB | MAR | APR | MAY | JUN | JUL | AUG | SEP | OCT | NOV | DEC | YEAR |
|---|---|---|---|---|---|---|---|---|---|---|---|---|---|
| Maximum Temp °F | 32.1 | 37.4 | 48.6 | 62.7 | 74.0 | 83.0 | 90.8 | 89.6 | 78.8 | 65.3 | 46.8 | 35.6 | 62.1 |
| Minimum Temp °F | 6.1 | 10.5 | 21.3 | 33.3 | 44.4 | 54.4 | 59.9 | 57.6 | 45.9 | 33.4 | 20.1 | 10.0 | 33.1 |
| Mean Temp °F | 19.1 | 24.0 | 35.0 | 47.9 | 59.1 | 68.7 | 75.3 | 73.7 | 62.3 | 49.4 | 33.5 | 22.8 | 47.6 |
| Days Max Temp ≥ 90 °F | 0 | 0 | 0 | 0 | 2 | 8 | 18 | 17 | 6 | 1 | 0 | 0 | 52 |
| Days Max Temp ≤ 32 °F | 14 | 11 | 4 | 0 | 0 | 0 | 0 | 0 | 0 | 0 | 5 | 12 | 46 |
| Days Min Temp ≤ 32 °F | 31 | 28 | 26 | 14 | 3 | 0 | 0 | 0 | 3 | 14 | 27 | 30 | 176 |
| Days Min Temp ≤ 0 °F | 11 | 7 | 1 | 0 | 0 | 0 | 0 | 0 | 0 | 0 | 1 | 7 | 27 |
| Heating Degree Days | 1417 | 1152 | 925 | 511 | 217 | 42 | 9 | 15 | 158 | 481 | 939 | 1300 | 7166 |
| Cooling Degree Days | 0 | 0 | 0 | 7 | 46 | 151 | 308 | 277 | 76 | 2 | 0 | 0 | 867 |
| Total Precipitation (") | 0.28 | 0.44 | 1.24 | 1.79 | 2.53 | 2.98 | 2.25 | 1.81 | 1.27 | 1.03 | 0.54 | 0.36 | 16.52 |
| Days ≥ 0.1" Precip | 1 | 1 | 3 | 4 | 5 | 6 | 5 | 4 | 3 | 2 | 1 | 1 | 36 |
| Total Snowfall (") | na | *6.5* | *6.3* | *1.6* | 0.3 | 0.0 | 0.0 | 0.0 | 0.0 | 0.7 | na | na | na |
| Days ≥ 1" Snow Depth | na | na | na | 1 | 0 | 0 | 0 | 0 | 0 | 0 | na | na | na |

## MILBANK 2 SSW *Grant County* ELEVATION 1142 ft LAT/LONG 45° 13' N / 96° 37' W

| | JAN | FEB | MAR | APR | MAY | JUN | JUL | AUG | SEP | OCT | NOV | DEC | YEAR |
|---|---|---|---|---|---|---|---|---|---|---|---|---|---|
| Maximum Temp °F | 20.8 | 25.9 | 39.6 | 56.0 | 70.5 | 79.9 | 84.8 | 82.2 | 72.9 | 60.1 | 41.0 | 27.0 | 55.1 |
| Minimum Temp °F | -0.6 | 5.0 | 19.7 | 32.3 | 44.3 | 54.8 | 59.1 | 56.7 | 46.4 | 35.0 | 21.4 | 7.7 | 31.8 |
| Mean Temp °F | 10.1 | 15.5 | 29.7 | 44.2 | 57.5 | 67.4 | 71.9 | 69.5 | 59.7 | 47.6 | 31.2 | 17.4 | 43.5 |
| Days Max Temp ≥ 90 °F | 0 | 0 | 0 | 0 | 1 | 4 | 9 | 6 | 2 | 0 | 0 | 0 | 22 |
| Days Max Temp ≤ 32 °F | 23 | 18 | 9 | 1 | 0 | 0 | 0 | 0 | 0 | 0 | 7 | 20 | 78 |
| Days Min Temp ≤ 32 °F | 31 | 28 | 27 | 16 | 3 | 0 | 0 | 0 | 1 | 13 | 26 | 31 | 176 |
| Days Min Temp ≤ 0 °F | 18 | 12 | 3 | 0 | 0 | 0 | 0 | 0 | 0 | 0 | 1 | 10 | 44 |
| Heating Degree Days | 1698 | 1393 | 1088 | 620 | 259 | 50 | 12 | 32 | 198 | 538 | 1006 | 1471 | 8365 |
| Cooling Degree Days | ***0*** | 0 | 0 | 5 | 36 | ***122*** | ***201*** | 150 | 34 | 3 | ***0*** | ***0*** | 551 |
| Total Precipitation (") | 0.50 | 0.43 | 1.25 | 2.29 | 2.39 | 3.65 | 3.19 | 2.64 | 1.90 | 1.97 | 0.98 | 0.43 | 21.62 |
| Days ≥ 0.1" Precip | 2 | 2 | 3 | 5 | 6 | 6 | 6 | 5 | 4 | 4 | 2 | 1 | 46 |
| Total Snowfall (") | 8.3 | 6.9 | 7.3 | 2.0 | 0.1 | 0.0 | 0.0 | 0.0 | 0.0 | 0.2 | 4.1 | 5.5 | 34.4 |
| Days ≥ 1" Snow Depth | na | na | ***8*** | ***2*** | 0 | ***0*** | 0 | 0 | 0 | 0 | ***4*** | ***12*** | na |

## MILESVILLE 8 NE *Haakon County* ELEVATION 2342 ft LAT/LONG 44° 31' N / 101° 37' W

| | JAN | FEB | MAR | APR | MAY | JUN | JUL | AUG | SEP | OCT | NOV | DEC | YEAR |
|---|---|---|---|---|---|---|---|---|---|---|---|---|---|
| Maximum Temp °F | 28.5 | 34.3 | 45.6 | 59.8 | 71.4 | 80.8 | 88.6 | 87.5 | 76.7 | 63.1 | 44.2 | 32.1 | 59.4 |
| Minimum Temp °F | 6.5 | 11.7 | 22.1 | 34.1 | 44.7 | 54.4 | 60.0 | 58.1 | 47.3 | 35.4 | 21.7 | 10.5 | 33.9 |
| Mean Temp °F | 17.5 | 23.0 | 33.9 | 47.0 | 58.2 | 67.6 | 74.4 | 72.8 | 62.0 | 49.3 | 33.0 | 21.3 | 46.7 |
| Days Max Temp ≥ 90 °F | 0 | 0 | 0 | 0 | 1 | 5 | 15 | 13 | 5 | 0 | 0 | 0 | 39 |
| Days Max Temp ≤ 32 °F | 16 | 13 | 6 | 0 | 0 | 0 | 0 | 0 | 0 | 0 | 6 | 15 | 56 |
| Days Min Temp ≤ 32 °F | 31 | 27 | 27 | 13 | 2 | 0 | 0 | 0 | 1 | 11 | 26 | 30 | 168 |
| Days Min Temp ≤ 0 °F | 11 | 6 | 2 | 0 | 0 | 0 | 0 | 0 | 0 | 0 | 1 | 7 | 27 |
| Heating Degree Days | 1466 | 1174 | 957 | 538 | 236 | 50 | 8 | 15 | 161 | 485 | 954 | 1348 | 7392 |
| Cooling Degree Days | 0 | 0 | 0 | 6 | 36 | 133 | 295 | 257 | 73 | 4 | 0 | 0 | 804 |
| Total Precipitation (") | 0.35 | 0.54 | 1.17 | 1.82 | 3.36 | 3.12 | 2.62 | 2.04 | 1.36 | 1.31 | 0.60 | 0.52 | 18.81 |
| Days ≥ 0.1" Precip | 1 | 1 | 3 | 5 | 6 | 6 | 5 | 4 | 3 | 3 | 2 | 2 | 41 |
| Total Snowfall (") | 6.6 | 9.5 | 10.7 | 6.5 | 0.7 | 0.1 | 0.0 | 0.0 | 0.1 | 1.8 | 6.5 | 8.6 | 51.1 |
| Days ≥ 1" Snow Depth | 21 | 15 | 10 | 2 | 0 | 0 | 0 | 0 | 0 | 1 | 7 | 17 | 73 |

## MILLER *Hand County* ELEVATION 1591 ft LAT/LONG 44° 31' N / 98° 59' W

| | JAN | FEB | MAR | APR | MAY | JUN | JUL | AUG | SEP | OCT | NOV | DEC | YEAR |
|---|---|---|---|---|---|---|---|---|---|---|---|---|---|
| Maximum Temp °F | 25.6 | 31.6 | 43.9 | 59.7 | 72.1 | 80.6 | 87.4 | 86.0 | 75.6 | 62.3 | 42.4 | 29.7 | 58.1 |
| Minimum Temp °F | 4.9 | 10.8 | 22.2 | 34.2 | 45.9 | 55.7 | 61.0 | 58.8 | 48.6 | 36.8 | 22.7 | 10.2 | 34.3 |
| Mean Temp °F | 15.3 | 21.2 | 33.1 | 46.9 | 59.0 | 68.2 | 74.2 | 72.5 | 62.1 | 49.6 | 32.6 | 20.0 | 46.2 |
| Days Max Temp ≥ 90 °F | 0 | 0 | 0 | 0 | 1 | 4 | 13 | 11 | 3 | 0 | 0 | 0 | 32 |
| Days Max Temp ≤ 32 °F | 19 | 14 | 6 | 0 | 0 | 0 | 0 | 0 | 0 | 0 | 7 | 17 | 63 |
| Days Min Temp ≤ 32 °F | 31 | 27 | 26 | 13 | 2 | 0 | 0 | 0 | 1 | 10 | 25 | 31 | 166 |
| Days Min Temp ≤ 0 °F | 13 | 8 | 2 | 0 | 0 | 0 | 0 | 0 | 0 | 0 | 1 | 7 | 31 |
| Heating Degree Days | 1536 | 1232 | 983 | 541 | 218 | 42 | 9 | 17 | 160 | 476 | 967 | 1390 | 7571 |
| Cooling Degree Days | 0 | 0 | 0 | 8 | 44 | 139 | 285 | 240 | 77 | 4 | 0 | 0 | 797 |
| Total Precipitation (") | 0.38 | 0.59 | 1.21 | 2.17 | 2.50 | 3.11 | 2.60 | 2.06 | 1.77 | 1.33 | 0.73 | 0.47 | 18.92 |
| Days ≥ 0.1" Precip | 1 | 2 | 3 | 5 | 5 | 6 | 5 | 5 | 3 | 3 | 2 | 1 | 41 |
| Total Snowfall (") | 5.6 | ***8.2*** | 6.2 | 2.0 | 0.1 | 0.0 | 0.0 | 0.0 | 0.0 | 1.2 | 4.8 | 5.4 | 33.5 |
| Days ≥ 1" Snow Depth | na | na | 7 | 1 | 0 | 0 | 0 | 0 | 0 | 0 | 4 | na | na |

## MISSION 14 S *Todd County* ELEVATION 2802 ft LAT/LONG 43° 7' N / 100° 36' W

| | JAN | FEB | MAR | APR | MAY | JUN | JUL | AUG | SEP | OCT | NOV | DEC | YEAR |
|---|---|---|---|---|---|---|---|---|---|---|---|---|---|
| Maximum Temp °F | 31.6 | 36.8 | 47.2 | 60.0 | 71.1 | 80.7 | 87.1 | 85.7 | 76.0 | 63.4 | 45.3 | 35.4 | 60.0 |
| Minimum Temp °F | 8.2 | 12.8 | 21.8 | 32.8 | 43.6 | 53.6 | 59.2 | 56.6 | 46.7 | 35.1 | 22.2 | 12.9 | 33.8 |
| Mean Temp °F | 19.9 | 24.8 | 34.5 | 46.5 | 57.4 | 67.2 | 73.2 | 71.2 | 61.3 | 49.2 | 33.8 | 24.1 | 46.9 |
| Days Max Temp ≥ 90 °F | 0 | 0 | 0 | 0 | 1 | 5 | 13 | 10 | 4 | 0 | 0 | 0 | 33 |
| Days Max Temp ≤ 32 °F | 15 | 11 | 6 | 1 | 0 | 0 | 0 | 0 | 0 | 0 | 6 | 12 | 51 |
| Days Min Temp ≤ 32 °F | 30 | 27 | 27 | 15 | 3 | 0 | 0 | 0 | 1 | 12 | 26 | 30 | 171 |
| Days Min Temp ≤ 0 °F | 10 | 6 | 2 | 0 | 0 | 0 | 0 | 0 | 0 | 0 | 1 | 5 | 24 |
| Heating Degree Days | 1393 | 1128 | 938 | 554 | 256 | 55 | 11 | 21 | 171 | 486 | 930 | 1261 | 7204 |
| Cooling Degree Days | 0 | 0 | 0 | 7 | 30 | 135 | 259 | 209 | 70 | 5 | 0 | 0 | 715 |
| Total Precipitation (") | 0.36 | 0.53 | 1.28 | 1.96 | 3.07 | 3.12 | 2.89 | 2.25 | 1.56 | 1.15 | 0.67 | 0.47 | 19.31 |
| Days ≥ 0.1" Precip | 1 | 2 | 3 | 4 | 6 | 6 | 6 | 5 | 4 | 3 | 2 | 2 | 44 |
| Total Snowfall (") | 5.4 | 6.5 | 9.1 | 4.3 | 0.2 | 0.0 | 0.0 | 0.0 | 0.2 | 1.5 | 5.7 | 5.8 | 38.7 |
| Days ≥ 1" Snow Depth | 18 | ***15*** | ***9*** | 3 | 0 | 0 | 0 | 0 | 0 | 1 | 6 | ***13*** | 65 |

## MITCHELL *Davison County* ELEVATION 1289 ft LAT/LONG 43° 43 ' N / 98° 1 ' W

| | JAN | FEB | MAR | APR | MAY | JUN | JUL | AUG | SEP | OCT | NOV | DEC | YEAR |
|---|---|---|---|---|---|---|---|---|---|---|---|---|---|
| Maximum Temp °F | 25.4 | 31.4 | 44.7 | 59.8 | 72.2 | 81.4 | 87.2 | 85.5 | 75.7 | 62.2 | 43.8 | 30.4 | 58.3 |
| Minimum Temp °F | 4.7 | 10.9 | 23.0 | 35.8 | 47.3 | 57.4 | 62.5 | 59.9 | 49.4 | 36.8 | 23.5 | 10.7 | 35.2 |
| Mean Temp °F | 15.1 | 21.2 | 33.9 | 47.8 | 59.8 | 69.4 | 74.9 | 72.7 | 62.6 | 49.5 | 33.7 | 20.6 | 46.8 |
| Days Max Temp ≥ 90 °F | 0 | 0 | 0 | 0 | 1 | 5 | 12 | 10 | 3 | 0 | 0 | 0 | 31 |
| Days Max Temp ≤ 32 °F | 20 | 14 | 6 | 0 | 0 | 0 | 0 | 0 | 0 | 0 | 6 | 17 | 63 |
| Days Min Temp ≤ 32 °F | 31 | 28 | 25 | 11 | 1 | 0 | 0 | 0 | 1 | 10 | 25 | 31 | 163 |
| Days Min Temp ≤ 0 °F | 13 | 8 | 1 | 0 | 0 | 0 | 0 | 0 | 0 | 0 | 1 | 7 | 30 |
| Heating Degree Days | 1544 | 1233 | 958 | 515 | 202 | 35 | 6 | 15 | 147 | 477 | 934 | 1372 | 7438 |
| Cooling Degree Days | 0 | 0 | 0 | 10 | 49 | 171 | 301 | 248 | 76 | 3 | 0 | 0 | 858 |
| Total Precipitation (") | 0.43 | 0.63 | 1.52 | 2.52 | 3.01 | 3.73 | 2.61 | 2.15 | 2.32 | 1.40 | 1.08 | 0.61 | 22.01 |
| Days ≥ 0.1" Precip | 1 | 2 | 3 | 6 | 6 | 6 | 5 | 4 | 4 | 3 | 2 | 2 | 44 |
| Total Snowfall (") | 5.8 | 6.7 | *5.3* | 1.5 | 0.0 | 0.0 | 0.0 | 0.0 | 0.0 | 0.2 | 3.3 | na | na |
| Days ≥ 1" Snow Depth | na | na | *8* | 1 | 0 | 0 | 0 | 0 | 0 | 0 | 3 | na | na |

## MOBRIDGE 2 NNW *Walworth County* ELEVATION 1696 ft LAT/LONG 45° 34 ' N / 100° 27 ' W

| | JAN | FEB | MAR | APR | MAY | JUN | JUL | AUG | SEP | OCT | NOV | DEC | YEAR |
|---|---|---|---|---|---|---|---|---|---|---|---|---|---|
| Maximum Temp °F | 22.9 | 29.5 | 41.5 | 56.8 | 69.3 | 77.9 | 84.7 | 83.0 | 72.0 | 59.1 | 40.5 | 27.4 | 55.4 |
| Minimum Temp °F | 2.4 | 8.6 | 20.3 | 33.2 | 45.2 | 55.3 | 61.0 | 58.9 | 48.3 | 36.1 | 22.0 | 8.8 | 33.3 |
| Mean Temp °F | 12.7 | 19.1 | 30.9 | 45.0 | 57.3 | 66.6 | 72.9 | 71.0 | 60.2 | 47.6 | 31.2 | 18.1 | 44.4 |
| Days Max Temp ≥ 90 °F | 0 | 0 | 0 | 0 | 1 | 3 | 9 | 7 | 2 | 0 | 0 | 0 | 22 |
| Days Max Temp ≤ 32 °F | 21 | 16 | 8 | 1 | 0 | 0 | 0 | 0 | 0 | 0 | 8 | 19 | 73 |
| Days Min Temp ≤ 32 °F | 31 | 28 | 28 | 14 | 2 | 0 | 0 | 0 | 1 | 10 | 27 | 31 | 172 |
| Days Min Temp ≤ 0 °F | 14 | 9 | 2 | 0 | 0 | 0 | 0 | 0 | 0 | 0 | 1 | 8 | 34 |
| Heating Degree Days | 1619 | 1291 | 1050 | 596 | 260 | 55 | 9 | 21 | 190 | 533 | 1006 | 1447 | 8077 |
| Cooling Degree Days | 0 | 0 | 0 | 5 | 33 | 119 | 256 | 216 | 50 | 2 | 0 | 0 | 681 |
| Total Precipitation (") | 0.34 | 0.43 | 1.12 | 1.96 | 2.45 | 3.10 | 2.17 | 1.75 | 1.21 | 1.15 | 0.57 | 0.46 | 16.71 |
| Days ≥ 0.1" Precip | 1 | 2 | 3 | 5 | 6 | 6 | 5 | 4 | 3 | 2 | 2 | 1 | 40 |
| Total Snowfall (") | 4.8 | 6.0 | 8.6 | 3.9 | 0.4 | 0.0 | 0.0 | 0.0 | 0.0 | 0.6 | 5.4 | 6.3 | 36.0 |
| Days ≥ 1" Snow Depth | 24 | 18 | 13 | 2 | 0 | 0 | 0 | 0 | 0 | 0 | 7 | 18 | 82 |

## MT RUSHMORE NATL MEM *Pennington County* ELEVATION 5154 ft LAT/LONG 43° 53 ' N / 103° 27 ' W

| | JAN | FEB | MAR | APR | MAY | JUN | JUL | AUG | SEP | OCT | NOV | DEC | YEAR |
|---|---|---|---|---|---|---|---|---|---|---|---|---|---|
| Maximum Temp °F | 35.5 | 38.4 | 44.1 | 53.0 | 63.9 | 73.8 | 81.3 | 79.8 | 70.0 | 57.8 | 43.5 | 36.8 | 56.5 |
| Minimum Temp °F | 16.1 | 18.9 | 24.1 | 32.5 | 42.1 | 51.6 | 58.4 | 57.3 | 48.0 | 37.2 | 25.6 | 18.6 | 35.9 |
| Mean Temp °F | 25.8 | 28.6 | 34.1 | 42.8 | 53.0 | 62.7 | 69.9 | 68.6 | 59.0 | 47.5 | 34.6 | 27.7 | 46.2 |
| Days Max Temp ≥ 90 °F | 0 | 0 | 0 | 0 | 0 | 2 | 6 | 4 | 1 | 0 | 0 | 0 | 13 |
| Days Max Temp ≤ 32 °F | 11 | 8 | 6 | 1 | 0 | 0 | 0 | 0 | 0 | 1 | 6 | 11 | 44 |
| Days Min Temp ≤ 32 °F | 27 | 24 | 24 | 16 | 4 | 0 | 0 | 0 | 2 | 10 | 22 | 27 | 156 |
| Days Min Temp ≤ 0 °F | 5 | 3 | 1 | 0 | 0 | 0 | 0 | 0 | 0 | 0 | 1 | 3 | 13 |
| Heating Degree Days | 1209 | 1020 | 950 | 661 | 378 | 136 | 34 | 45 | 226 | 538 | 907 | 1149 | 7253 |
| Cooling Degree Days | 0 | 0 | 0 | 2 | 14 | 80 | 188 | 159 | 50 | 3 | 0 | 0 | 496 |
| Total Precipitation (") | 0.34 | 0.53 | 1.03 | 2.12 | 3.70 | 3.84 | 3.12 | 2.01 | 1.51 | 1.17 | 0.57 | 0.48 | 20.42 |
| Days ≥ 0.1" Precip | 1 | 2 | 3 | 5 | 7 | 8 | 7 | 5 | 4 | 3 | 2 | 2 | 49 |
| Total Snowfall (") | 5.6 | 7.8 | 10.3 | 12.2 | 1.9 | 0.2 | 0.0 | 0.0 | 0.6 | 2.8 | 7.5 | 7.0 | 55.9 |
| Days ≥ 1" Snow Depth | 12 | 12 | 10 | 5 | 1 | 0 | 0 | 0 | 0 | 1 | *9* | 14 | 64 |

## MURDO *Jones County* ELEVATION 2313 ft LAT/LONG 43° 53 ' N / 100° 43 ' W

| | JAN | FEB | MAR | APR | MAY | JUN | JUL | AUG | SEP | OCT | NOV | DEC | YEAR |
|---|---|---|---|---|---|---|---|---|---|---|---|---|---|
| Maximum Temp °F | 30.3 | 36.0 | 46.8 | 60.5 | 71.9 | 81.6 | 89.1 | 87.8 | 77.3 | 63.7 | 45.2 | 33.8 | 60.3 |
| Minimum Temp °F | 7.9 | 12.6 | 22.4 | 33.9 | 44.8 | 54.7 | 60.4 | 58.4 | 47.9 | 36.4 | 22.8 | 11.8 | 34.5 |
| Mean Temp °F | 19.1 | 24.3 | 34.6 | 47.2 | 58.4 | 68.2 | 74.8 | 73.1 | 62.6 | 50.1 | 34.0 | 22.8 | 47.4 |
| Days Max Temp ≥ 90 °F | 0 | 0 | 0 | 0 | 2 | 6 | 15 | 14 | 5 | 1 | 0 | 0 | 43 |
| Days Max Temp ≤ 32 °F | 15 | 11 | 5 | 0 | 0 | 0 | 0 | 0 | 0 | 0 | 6 | 13 | 50 |
| Days Min Temp ≤ 32 °F | 30 | 27 | 26 | 13 | 2 | 0 | 0 | 0 | 1 | 10 | 25 | 30 | 164 |
| Days Min Temp ≤ 0 °F | 11 | 6 | 1 | 0 | 0 | 0 | 0 | 0 | 0 | 0 | 1 | 6 | 25 |
| Heating Degree Days | 1417 | 1142 | 935 | 531 | 234 | 44 | 7 | 13 | 153 | 464 | 923 | 1301 | 7164 |
| Cooling Degree Days | 0 | 0 | 0 | 7 | 42 | 148 | 306 | 275 | 91 | 5 | 0 | 0 | 874 |
| Total Precipitation (") | 0.40 | 0.49 | 1.55 | 2.22 | 2.61 | 3.22 | 2.59 | 1.68 | 1.21 | 1.34 | 0.71 | 0.59 | 18.61 |
| Days ≥ 0.1" Precip | 1 | 2 | 4 | 5 | 6 | 6 | 5 | 4 | 3 | 3 | 2 | 2 | 43 |
| Total Snowfall (") | 4.7 | 5.4 | 8.5 | 3.4 | 0.4 | 0.0 | 0.0 | 0.0 | 0.2 | 1.4 | 5.8 | 6.7 | 36.5 |
| Days ≥ 1" Snow Depth | na | *8* | *6* | 2 | 0 | 0 | 0 | 0 | 0 | 1 | *3* | na | na |

## NEWELL *Butte County* ELEVATION 2881 ft LAT/LONG 44° 44 ' N / 103° 27 ' W

| | JAN | FEB | MAR | APR | MAY | JUN | JUL | AUG | SEP | OCT | NOV | DEC | YEAR |
|---|---|---|---|---|---|---|---|---|---|---|---|---|---|
| Maximum Temp °F | 29.3 | 34.4 | 44.5 | 57.3 | 67.8 | 77.7 | 86.1 | 85.1 | 74.1 | 61.3 | 44.0 | 33.2 | 57.9 |
| Minimum Temp °F | 6.5 | 10.9 | 20.9 | 32.1 | 42.7 | 52.4 | 58.0 | 55.7 | 44.7 | 32.2 | 20.1 | 9.4 | 32.1 |
| Mean Temp °F | 17.9 | 22.7 | 32.8 | 44.7 | 55.3 | 65.1 | 72.1 | 70.4 | 59.4 | 46.8 | 32.1 | 21.3 | 45.0 |
| Days Max Temp ≥ 90 °F | 0 | 0 | 0 | 0 | 1 | 4 | 11 | 10 | 3 | 0 | 0 | 0 | 29 |
| Days Max Temp ≤ 32 °F | 16 | 12 | 6 | 1 | 0 | 0 | 0 | 0 | 0 | 0 | 6 | 14 | 55 |
| Days Min Temp ≤ 32 °F | 31 | 28 | 28 | 15 | 4 | 0 | 0 | 0 | 2 | 15 | 27 | 31 | 181 |
| Days Min Temp ≤ 0 °F | 10 | 6 | 2 | 0 | 0 | 0 | 0 | 0 | 0 | 0 | 1 | 7 | 26 |
| Heating Degree Days | 1453 | 1189 | 993 | 603 | 311 | 85 | 16 | 26 | 209 | 560 | 982 | 1349 | 7776 |
| Cooling Degree Days | 0 | 0 | 0 | 3 | 22 | 107 | 243 | 210 | 46 | 2 | 0 | 0 | 633 |
| Total Precipitation (") | 0.34 | 0.41 | 0.86 | 1.67 | 2.61 | 2.66 | 1.84 | 1.32 | 1.14 | 1.10 | 0.52 | 0.37 | 14.84 |
| Days ≥ 0.1" Precip | 1 | 1 | 2 | 5 | 6 | 6 | 5 | 3 | 3 | 2 | 2 | 2 | 38 |
| Total Snowfall (") | *4.8* | *4.9* | na | *1.9* | 0.8 | 0.1 | 0.0 | 0.0 | 0.1 | 1.0 | 3.7 | *5.4* | na |
| Days ≥ 1" Snow Depth | na | *12* | *8* | *1* | 0 | 0 | 0 | 0 | 0 | 1 | *4* | *11* | na |

## OAHE DAM *Stanley County* ELEVATION 1470 ft LAT/LONG 44° 26 ' N / 100° 25 ' W

| | JAN | FEB | MAR | APR | MAY | JUN | JUL | AUG | SEP | OCT | NOV | DEC | YEAR |
|---|---|---|---|---|---|---|---|---|---|---|---|---|---|
| Maximum Temp °F | 26.8 | 31.5 | 43.3 | 57.8 | 69.8 | 80.3 | 88.6 | 86.9 | 75.4 | 62.0 | 43.8 | 31.9 | 58.2 |
| Minimum Temp °F | 6.7 | 11.3 | 21.8 | 34.4 | 45.6 | 55.9 | 61.9 | 59.8 | 49.1 | 37.5 | 24.0 | 12.6 | 35.1 |
| Mean Temp °F | 16.8 | 21.6 | 32.6 | 46.1 | 57.7 | 68.2 | 75.3 | 73.5 | 62.3 | 49.8 | 33.9 | 22.3 | 46.7 |
| Days Max Temp ≥ 90 °F | 0 | 0 | 0 | 0 | 1 | 5 | 14 | 13 | 4 | 0 | 0 | 0 | 37 |
| Days Max Temp ≤ 32 °F | 18 | 14 | 7 | 1 | 0 | 0 | 0 | 0 | 0 | 0 | 6 | 15 | 61 |
| Days Min Temp ≤ 32 °F | 31 | 27 | 27 | 12 | 1 | 0 | 0 | 0 | 1 | 8 | 24 | 30 | 161 |
| Days Min Temp ≤ 0 °F | 11 | 7 | 1 | 0 | 0 | 0 | 0 | 0 | 0 | 0 | 1 | 5 | 25 |
| Heating Degree Days | 1491 | 1219 | 998 | 563 | 251 | 46 | 6 | 12 | 157 | 471 | 926 | 1318 | 7458 |
| Cooling Degree Days | 0 | 0 | 0 | 6 | 42 | 155 | 333 | 292 | 79 | 6 | 0 | 0 | 913 |
| Total Precipitation (") | 0.20 | 0.35 | 0.76 | 1.59 | 2.19 | 2.68 | 2.23 | 1.43 | 1.15 | 0.86 | 0.37 | 0.27 | 14.08 |
| Days ≥ 0.1" Precip | 1 | 1 | 2 | 4 | 5 | 5 | 4 | 3 | 3 | 2 | 1 | 1 | 32 |
| Total Snowfall (") | na | na | na | *0.2* | 0.0 | 0.0 | 0.0 | 0.0 | 0.0 | 0.0 | na | na | na |
| Days ≥ 1" Snow Depth | na | na | na | *0* | 0 | 0 | 0 | 0 | 0 | 0 | na | na | na |

## OELRICHS *Fall River County* ELEVATION 3343 ft LAT/LONG 43° 11 ' N / 103° 14 ' W

| | JAN | FEB | MAR | APR | MAY | JUN | JUL | AUG | SEP | OCT | NOV | DEC | YEAR |
|---|---|---|---|---|---|---|---|---|---|---|---|---|---|
| Maximum Temp °F | 34.1 | 40.0 | 50.1 | 61.4 | 72.0 | 82.1 | 90.3 | 89.1 | 78.9 | 65.1 | 47.0 | 36.4 | 62.2 |
| Minimum Temp °F | 10.7 | 14.9 | 22.8 | 32.4 | 42.1 | 51.3 | 57.7 | 55.7 | 45.1 | 33.6 | 21.9 | 13.0 | 33.4 |
| Mean Temp °F | 22.4 | 27.5 | 36.5 | 47.0 | 57.1 | 66.7 | 74.0 | 72.4 | 62.0 | 49.4 | 34.5 | 24.7 | 47.9 |
| Days Max Temp ≥ 90 °F | 0 | 0 | 0 | 0 | 1 | 7 | 17 | 16 | 6 | 0 | 0 | 0 | 47 |
| Days Max Temp ≤ 32 °F | 12 | 8 | 3 | 0 | 0 | 0 | 0 | 0 | 0 | 0 | 4 | 11 | 38 |
| Days Min Temp ≤ 32 °F | 30 | 27 | 26 | 15 | 3 | 0 | 0 | 0 | 2 | 13 | 26 | 30 | 172 |
| Days Min Temp ≤ 0 °F | 8 | 5 | 1 | 0 | 0 | 0 | 0 | 0 | 0 | 0 | 1 | 5 | 20 |
| Heating Degree Days | 1314 | 1053 | 878 | 536 | 257 | 60 | 8 | 13 | 153 | 480 | 909 | 1242 | 6903 |
| Cooling Degree Days | 0 | 0 | 0 | 2 | 20 | 119 | 285 | 260 | 68 | 2 | 0 | 0 | 756 |
| Total Precipitation (") | 0.38 | 0.49 | 0.97 | 1.90 | 3.03 | 2.91 | 2.24 | 1.71 | 1.22 | 1.14 | 0.60 | 0.47 | 17.06 |
| Days ≥ 0.1" Precip | 1 | 2 | 3 | 4 | 6 | 6 | 5 | 4 | 3 | 3 | 2 | 2 | 41 |
| Total Snowfall (") | 6.4 | 6.6 | 8.7 | 5.8 | 1.0 | 0.1 | 0.0 | 0.0 | 0.3 | 2.9 | 6.3 | 7.8 | 45.9 |
| Days ≥ 1" Snow Depth | *18* | *10* | *6* | 1 | 0 | 0 | 0 | 0 | 0 | 1 | 4 | *13* | 53 |

## ONIDA 4 NW *Sully County* ELEVATION 1850 ft LAT/LONG 44° 44 ' N / 100° 9 ' W

| | JAN | FEB | MAR | APR | MAY | JUN | JUL | AUG | SEP | OCT | NOV | DEC | YEAR |
|---|---|---|---|---|---|---|---|---|---|---|---|---|---|
| Maximum Temp °F | 25.2 | 31.7 | 43.5 | 60.5 | 72.7 | 81.6 | 88.9 | 87.4 | 76.7 | 62.9 | 41.8 | 29.2 | 58.5 |
| Minimum Temp °F | 4.6 | 10.3 | 21.8 | 33.8 | 45.4 | 54.5 | 60.2 | 58.2 | 47.6 | 36.0 | 21.8 | 9.7 | 33.7 |
| Mean Temp °F | 14.9 | 21.0 | 32.7 | 47.2 | 59.1 | 68.1 | 74.6 | 72.8 | 62.2 | 49.4 | 31.8 | 19.5 | 46.1 |
| Days Max Temp ≥ 90 °F | 0 | 0 | 0 | 0 | 1 | 6 | 14 | 13 | 4 | 0 | 0 | 0 | 38 |
| Days Max Temp ≤ 32 °F | 19 | 14 | 6 | 1 | 0 | 0 | 0 | 0 | 0 | 0 | 7 | 17 | 64 |
| Days Min Temp ≤ 32 °F | 31 | 28 | 27 | *14* | 2 | 0 | 0 | 0 | 2 | 11 | 26 | 31 | 172 |
| Days Min Temp ≤ 0 °F | 12 | 8 | 2 | 0 | 0 | 0 | 0 | 0 | 0 | 0 | 1 | 7 | 30 |
| Heating Degree Days | 1549 | 1235 | 994 | 534 | 216 | 48 | 8 | 16 | 159 | 480 | 990 | 1406 | 7635 |
| Cooling Degree Days | 0 | 0 | 0 | 8 | 42 | 143 | 292 | 256 | 80 | 4 | 0 | 0 | 825 |
| Total Precipitation (") | 0.51 | 0.63 | 1.38 | 1.82 | 2.49 | 2.85 | 2.76 | 2.01 | 1.29 | 1.34 | 0.79 | 0.57 | 18.44 |
| Days ≥ 0.1" Precip | 2 | 2 | 4 | 5 | 5 | 6 | 5 | 4 | 3 | 3 | 3 | 2 | 44 |
| Total Snowfall (") | 5.2 | *7.0* | *9.2* | *3.3* | 0.0 | 0.0 | 0.0 | 0.0 | 0.0 | 1.3 | 6.5 | *6.0* | 38.5 |
| Days ≥ 1" Snow Depth | 21 | *17* | *9* | na | 0 | 0 | 0 | 0 | 0 | 0 | 8 | *19* | na |

## PACTOLA DAM *Pennington County* ELEVATION 4534 ft LAT/LONG 44° 4' N / 103° 30' W

| | JAN | FEB | MAR | APR | MAY | JUN | JUL | AUG | SEP | OCT | NOV | DEC | YEAR |
|---|---|---|---|---|---|---|---|---|---|---|---|---|---|
| Maximum Temp °F | 35.3 | 38.4 | 43.4 | 51.6 | 61.7 | 71.4 | 78.9 | 78.3 | ***68.0*** | 57.2 | 43.5 | 36.7 | 55.4 |
| Minimum Temp °F | 8.5 | 10.4 | 16.8 | 25.5 | 34.6 | 43.5 | 48.7 | 46.5 | ***36.5*** | 27.3 | 18.2 | 10.2 | 27.2 |
| Mean Temp °F | 21.9 | 24.4 | 30.1 | 38.6 | 48.2 | 57.5 | 63.8 | 62.4 | ***52.3*** | 42.3 | 30.9 | 23.5 | 41.3 |
| Days Max Temp ≥ 90 °F | 0 | 0 | 0 | 0 | 0 | 1 | 3 | 2 | 1 | 0 | 0 | 0 | 7 |
| Days Max Temp ≤ 32 °F | 11 | 8 | 6 | 2 | 0 | 0 | 0 | 0 | 0 | 1 | 6 | ***11*** | 45 |
| Days Min Temp ≤ 32 °F | 30 | 28 | ***30*** | ***24*** | ***12*** | 2 | 0 | 0 | ***9*** | ***23*** | ***29*** | 30 | 217 |
| Days Min Temp ≤ 0 °F | 8 | 6 | 3 | 0 | 0 | 0 | 0 | 0 | 0 | 0 | 1 | 6 | 24 |
| Heating Degree Days | 1328 | 1138 | 1076 | 791 | ***513*** | 238 | 92 | 117 | ***381*** | 697 | 1010 | 1281 | 8662 |
| Cooling Degree Days | 0 | 0 | 0 | 0 | 1 | 21 | 67 | 45 | 4 | 0 | 0 | 0 | 138 |
| Total Precipitation (") | 0.28 | 0.45 | 0.94 | 2.33 | 3.45 | 3.93 | 3.17 | 2.06 | 1.33 | 1.25 | 0.58 | 0.41 | 20.18 |
| Days ≥ 0.1" Precip | 1 | 2 | 3 | 5 | ***7*** | 8 | 7 | 5 | 3 | 3 | 2 | 1 | 47 |
| Total Snowfall (") | 3.5 | 5.5 | 9.9 | 11.1 | 1.9 | 0.1 | 0.0 | 0.0 | 0.2 | 2.4 | 5.6 | ***5.2*** | 45.4 |
| Days ≥ 1" Snow Depth | 15 | 13 | 12 | 7 | 1 | 0 | 0 | 0 | 0 | 2 | 8 | ***16*** | 74 |

## PHILIP 2 N *Haakon County* ELEVATION 2241 ft LAT/LONG 44° 4' N / 101° 39' W

| | JAN | FEB | MAR | APR | MAY | JUN | JUL | AUG | SEP | OCT | NOV | DEC | YEAR |
|---|---|---|---|---|---|---|---|---|---|---|---|---|---|
| Maximum Temp °F | 30.2 | 36.6 | 46.5 | 59.8 | 71.1 | 81.6 | 89.3 | ***88.6*** | ***76.9*** | ***63.6*** | ***46.3*** | ***35.2*** | 60.5 |
| Minimum Temp °F | 6.7 | 12.0 | 21.9 | 33.2 | 44.1 | ***53.8*** | ***59.2*** | ***57.2*** | ***46.0*** | 34.3 | ***21.6*** | ***11.0*** | 33.4 |
| Mean Temp °F | 18.5 | 24.3 | 34.2 | 46.6 | 57.6 | ***67.4*** | ***74.2*** | ***72.9*** | ***61.5*** | ***48.9*** | ***34.0*** | ***23.1*** | 46.9 |
| Days Max Temp ≥ 90 °F | 0 | 0 | 0 | 0 | 2 | 6 | 15 | 15 | 5 | 0 | 0 | ***0*** | 43 |
| Days Max Temp ≤ 32 °F | 15 | 11 | 6 | 0 | 0 | 0 | 0 | 0 | 0 | 0 | ***5*** | ***12*** | 49 |
| Days Min Temp ≤ 32 °F | 31 | 27 | 27 | 14 | 3 | 0 | 0 | 0 | 2 | 13 | ***26*** | ***30*** | 173 |
| Days Min Temp ≤ 0 °F | 10 | 6 | 2 | 0 | 0 | 0 | 0 | 0 | 0 | 0 | ***1*** | ***6*** | 25 |
| Heating Degree Days | 1437 | 1144 | 946 | 550 | 252 | ***52*** | ***7*** | ***17*** | ***174*** | ***496*** | ***923*** | ***1292*** | 7290 |
| Cooling Degree Days | ***0*** | ***0*** | ***0*** | ***6*** | ***38*** | ***121*** | na | na | na | ***5*** | ***0*** | ***0*** | na |
| Total Precipitation (") | 0.32 | 0.49 | 1.01 | 1.89 | 2.81 | 3.08 | 2.20 | 1.60 | 1.15 | 1.03 | 0.55 | 0.37 | 16.50 |
| Days ≥ 0.1" Precip | 1 | 2 | 2 | 4 | 5 | 6 | 5 | 3 | 3 | 3 | ***2*** | ***1*** | 37 |
| Total Snowfall (") | 3.9 | 7.3 | 6.0 | 3.4 | 0.3 | 0.0 | 0.0 | 0.0 | 0.0 | 0.6 | ***3.4*** | ***4.7*** | 29.6 |
| Days ≥ 1" Snow Depth | ***19*** | ***13*** | ***9*** | 2 | 0 | 0 | 0 | 0 | ***0*** | ***0*** | 5 | ***14*** | 62 |

## PICKSTOWN *Charles Mix County* ELEVATION 1490 ft LAT/LONG 43° 4' N / 98° 32' W

| | JAN | FEB | MAR | APR | MAY | JUN | JUL | AUG | SEP | OCT | NOV | DEC | YEAR |
|---|---|---|---|---|---|---|---|---|---|---|---|---|---|
| Maximum Temp °F | 30.0 | 35.4 | 47.2 | 61.0 | 72.4 | 82.1 | 87.8 | 86.3 | 76.1 | 63.4 | 45.5 | 33.8 | 60.1 |
| Minimum Temp °F | 8.7 | 14.4 | 24.8 | 36.7 | 47.8 | 57.7 | 63.5 | 61.6 | 51.4 | 39.3 | 25.9 | 14.2 | 37.2 |
| Mean Temp °F | 19.4 | 24.9 | 36.0 | 48.9 | 60.1 | 69.9 | 75.6 | 74.0 | 63.8 | 51.4 | 35.8 | 24.0 | 48.7 |
| Days Max Temp ≥ 90 °F | 0 | 0 | 0 | 0 | 1 | 6 | 13 | 12 | 4 | 0 | 0 | 0 | 36 |
| Days Max Temp ≤ 32 °F | 16 | 12 | 5 | 0 | 0 | 0 | 0 | 0 | 0 | 0 | 5 | 13 | 51 |
| Days Min Temp ≤ 32 °F | 30 | 27 | 24 | 10 | 1 | 0 | 0 | 0 | 1 | 6 | 22 | 30 | 151 |
| Days Min Temp ≤ 0 °F | 10 | 5 | 1 | 0 | 0 | 0 | 0 | 0 | 0 | 0 | 1 | 5 | 22 |
| Heating Degree Days | 1409 | 1126 | 891 | 487 | 193 | 30 | 4 | 9 | 128 | 422 | 871 | 1265 | 6835 |
| Cooling Degree Days | 0 | 0 | 0 | 15 | 51 | 186 | 326 | 288 | 101 | 6 | 0 | 0 | 973 |
| Total Precipitation (") | 0.43 | 0.52 | 1.52 | 2.51 | 3.08 | 3.84 | 2.72 | 2.26 | 2.30 | 1.64 | 0.90 | 0.61 | 22.33 |
| Days ≥ 0.1" Precip | 1 | 1 | 3 | 5 | 6 | 7 | 5 | 4 | 4 | 3 | 2 | 2 | 43 |
| Total Snowfall (") | 4.0 | 3.9 | 4.1 | 1.0 | 0.0 | 0.0 | 0.0 | 0.0 | 0.0 | 0.6 | 3.4 | 5.2 | 22.2 |
| Days ≥ 1" Snow Depth | 21 | 13 | 8 | 1 | 0 | 0 | 0 | 0 | 0 | 0 | 4 | 15 | 62 |

## PIERRE MUNICIPAL AP *Hughes County* ELEVATION 1722 ft LAT/LONG 44° 23' N / 100° 17' W

| | JAN | FEB | MAR | APR | MAY | JUN | JUL | AUG | SEP | OCT | NOV | DEC | YEAR |
|---|---|---|---|---|---|---|---|---|---|---|---|---|---|
| Maximum Temp °F | 26.9 | 32.5 | 44.7 | 59.5 | 71.3 | 81.1 | 89.1 | 87.3 | 76.3 | 62.1 | 43.5 | 31.3 | 58.8 |
| Minimum Temp °F | 6.8 | 12.1 | 23.3 | 34.9 | 45.7 | 55.7 | 61.6 | 59.6 | 48.5 | 36.6 | 23.6 | 12.0 | 35.0 |
| Mean Temp °F | 16.9 | 22.3 | 34.1 | 47.2 | 58.5 | 68.4 | 75.4 | 73.5 | 62.4 | 49.4 | 33.6 | 21.7 | 46.9 |
| Days Max Temp ≥ 90 °F | 0 | 0 | 0 | 0 | 1 | 6 | 15 | 13 | 5 | 0 | 0 | 0 | 40 |
| Days Max Temp ≤ 32 °F | 18 | 13 | 6 | 1 | 0 | 0 | 0 | 0 | 0 | 0 | 6 | 16 | 60 |
| Days Min Temp ≤ 32 °F | 30 | 27 | 25 | 11 | 2 | 0 | 0 | 0 | 1 | 9 | 25 | 30 | 160 |
| Days Min Temp ≤ 0 °F | 11 | 6 | 1 | 0 | 0 | 0 | 0 | 0 | 0 | 0 | 1 | 6 | 25 |
| Heating Degree Days | 1484 | 1199 | 953 | 532 | 230 | 43 | 6 | 14 | 156 | 482 | 937 | 1338 | 7374 |
| Cooling Degree Days | 0 | 0 | 0 | 7 | 44 | 157 | 339 | 292 | 87 | 5 | 0 | 0 | 931 |
| Total Precipitation (") | 0.45 | 0.54 | 1.14 | 2.00 | 2.68 | 3.58 | 2.77 | 1.77 | 1.54 | 1.35 | 0.64 | 0.55 | 19.01 |
| Days ≥ 0.1" Precip | 1 | 2 | 3 | 5 | 6 | 6 | 5 | 4 | 3 | 3 | 2 | 2 | 42 |
| Total Snowfall (") | 4.3 | 5.4 | 6.8 | 2.7 | 0.1 | 0.0 | 0.0 | 0.0 | 0.0 | 1.0 | 4.5 | 5.3 | 30.1 |
| Days ≥ 1" Snow Depth | 19 | 14 | 8 | 1 | 0 | 0 | 0 | 0 | 0 | 0 | 5 | 13 | 60 |

## POLLOCK *Campbell County* ELEVATION 1611 ft LAT/LONG 45° 55' N / 100° 17' W

| | JAN | FEB | MAR | APR | MAY | JUN | JUL | AUG | SEP | OCT | NOV | DEC | YEAR |
|---|---|---|---|---|---|---|---|---|---|---|---|---|---|
| Maximum Temp °F | 22.4 | 29.0 | 41.9 | 58.3 | 71.6 | 80.0 | 86.9 | 85.2 | 74.2 | 60.4 | 40.2 | 26.3 | 56.4 |
| Minimum Temp °F | -0.9 | 5.8 | 18.9 | 31.9 | 43.7 | 53.5 | 58.8 | 56.2 | 45.6 | 33.2 | 19.3 | 4.7 | 30.9 |
| Mean Temp °F | 10.8 | 17.4 | 30.4 | 45.1 | 57.7 | 66.8 | 72.9 | 70.8 | 59.9 | 46.9 | 29.8 | 15.6 | 43.7 |
| Days Max Temp ≥ 90 °F | 0 | 0 | 0 | 0 | 1 | 4 | 12 | 10 | 2 | 0 | 0 | 0 | 29 |
| Days Max Temp ≤ 32 °F | 21 | 16 | 7 | 0 | 0 | 0 | 0 | 0 | 0 | 0 | 8 | 20 | 72 |
| Days Min Temp ≤ 32 °F | 31 | 28 | 28 | 16 | 3 | 0 | 0 | 0 | 2 | 15 | 28 | 31 | 182 |
| Days Min Temp ≤ 0 °F | 17 | 11 | 3 | 0 | 0 | 0 | 0 | 0 | 0 | 0 | 2 | 11 | 44 |
| Heating Degree Days | 1679 | 1340 | 1065 | 593 | 250 | 52 | 9 | 22 | 193 | 556 | 1050 | 1527 | 8336 |
| Cooling Degree Days | 0 | 0 | 0 | 4 | 38 | 122 | 260 | 215 | 46 | 1 | 0 | 0 | 686 |
| Total Precipitation (") | 0.37 | 0.45 | 1.06 | 1.95 | 2.34 | 2.95 | 2.39 | 1.94 | 1.26 | 1.21 | 0.61 | 0.42 | 16.95 |
| Days ≥ 0.1" Precip | 1 | 2 | 3 | 5 | 5 | 6 | 5 | 4 | 3 | 3 | 2 | 1 | 40 |
| Total Snowfall (") | ***6.7*** | ***6.9*** | na | ***2.6*** | 0.3 | 0.0 | 0.0 | 0.0 | 0.0 | 0.5 | 5.2 | ***7.0*** | na |
| Days ≥ 1" Snow Depth | na | na | na | na | 0 | 0 | 0 | 0 | 0 | 0 | na | na | na |

## PORCUPINE 11 N *Shannon County* ELEVATION 2792 ft LAT/LONG 43° 27' N / 102° 27' W

| | JAN | FEB | MAR | APR | MAY | JUN | JUL | AUG | SEP | OCT | NOV | DEC | YEAR |
|---|---|---|---|---|---|---|---|---|---|---|---|---|---|
| Maximum Temp °F | 33.8 | 40.2 | 49.8 | 61.2 | 71.7 | 81.9 | 89.5 | 88.8 | 78.4 | 65.0 | 47.2 | 36.9 | 62.0 |
| Minimum Temp °F | 8.7 | 13.7 | 22.7 | 32.7 | 43.0 | 52.6 | 58.5 | 56.3 | 44.7 | 32.3 | 20.4 | 10.9 | 33.0 |
| Mean Temp °F | 21.3 | 27.0 | 36.3 | 47.0 | 57.4 | 67.2 | 74.1 | 72.6 | 61.6 | 48.7 | 33.8 | 23.9 | 47.6 |
| Days Max Temp ≥ 90 °F | 0 | 0 | 0 | 0 | 1 | 7 | 16 | 16 | 6 | 0 | 0 | 0 | 46 |
| Days Max Temp ≤ 32 °F | 13 | 8 | 4 | 0 | 0 | 0 | 0 | 0 | 0 | 0 | 5 | 11 | 41 |
| Days Min Temp ≤ 32 °F | 30 | 27 | 26 | 15 | 4 | 0 | 0 | 0 | 4 | 15 | 26 | 30 | 177 |
| Days Min Temp ≤ 0 °F | 9 | 5 | 1 | 0 | 0 | 0 | 0 | 0 | 0 | 0 | 1 | 6 | 22 |
| Heating Degree Days | 1350 | 1068 | 884 | 536 | 257 | 61 | 12 | 15 | 171 | 502 | 928 | 1268 | 7052 |
| Cooling Degree Days | 0 | 0 | 0 | 5 | 34 | 137 | 290 | 262 | 69 | 2 | 0 | 0 | 799 |
| Total Precipitation (") | 0.27 | 0.39 | 0.93 | 1.78 | 2.63 | 3.04 | 2.67 | 1.52 | 1.16 | 1.14 | 0.55 | 0.40 | 16.48 |
| Days ≥ 0.1" Precip | 1 | 1 | 2 | 4 | 6 | 6 | 5 | 4 | 3 | 3 | 2 | 1 | 38 |
| Total Snowfall (") | 3.5 | 5.7 | ***4.8*** | 3.1 | 0.1 | 0.0 | 0.0 | 0.0 | 0.0 | 1.8 | ***5.6*** | ***5.5*** | 30.1 |
| Days ≥ 1" Snow Depth | 18 | 12 | 7 | 2 | 0 | 0 | 0 | 0 | 0 | 1 | 7 | 15 | 62 |

## RALPH 1 N *Harding County* ELEVATION 2831 ft LAT/LONG 45° 48' N / 103° 6' W

| | JAN | FEB | MAR | APR | MAY | JUN | JUL | AUG | SEP | OCT | NOV | DEC | YEAR |
|---|---|---|---|---|---|---|---|---|---|---|---|---|---|
| Maximum Temp °F | 27.5 | 33.2 | 44.2 | 57.8 | 69.5 | 78.0 | 85.6 | 85.3 | 74.1 | 61.1 | 42.7 | 31.4 | 57.5 |
| Minimum Temp °F | 3.2 | 8.4 | 18.7 | 29.8 | 40.2 | 49.3 | 54.0 | 51.8 | 40.5 | 29.6 | 17.0 | 7.1 | 29.1 |
| Mean Temp °F | 15.4 | 20.8 | 31.5 | 43.8 | 54.9 | 63.6 | 69.9 | 68.6 | 57.3 | 45.4 | 29.8 | 19.2 | 43.4 |
| Days Max Temp ≥ 90 °F | 0 | 0 | 0 | 0 | 1 | 3 | 10 | 11 | 3 | 0 | 0 | 0 | 28 |
| Days Max Temp ≤ 32 °F | 17 | 12 | 6 | 1 | 0 | 0 | 0 | 0 | 0 | 0 | 7 | 15 | 58 |
| Days Min Temp ≤ 32 °F | 31 | 28 | 29 | 19 | 6 | 0 | 0 | 0 | 5 | 18 | 28 | 31 | 195 |
| Days Min Temp ≤ 0 °F | 13 | 8 | 3 | 0 | 0 | 0 | 0 | 0 | 0 | 0 | 2 | 9 | 35 |
| Heating Degree Days | 1536 | 1233 | 1034 | 630 | 322 | 99 | 23 | 39 | 253 | 604 | 1047 | 1412 | 8232 |
| Cooling Degree Days | 0 | 0 | 0 | 1 | 17 | 79 | 199 | 171 | 33 | 1 | 0 | 0 | 501 |
| Total Precipitation (") | 0.38 | 0.32 | 0.68 | 1.85 | 2.69 | 3.09 | 2.14 | 1.28 | 1.23 | 1.12 | 0.51 | 0.40 | 15.69 |
| Days ≥ 0.1" Precip | 1 | 1 | 2 | 4 | 6 | 6 | 5 | 3 | 3 | 2 | 2 | 2 | 37 |
| Total Snowfall (") | na | na | na | ***4.0*** | 0.7 | 0.0 | 0.0 | 0.0 | 0.6 | ***1.4*** | na | na | na |
| Days ≥ 1" Snow Depth | na | ***12*** | na | ***2*** | 1 | 0 | 0 | 0 | 0 | 1 | ***7*** | na | na |

## RAPID CITY *Pennington County* ELEVATION 3373 ft LAT/LONG 44° 4' N / 103° 16' W

| | JAN | FEB | MAR | APR | MAY | JUN | JUL | AUG | SEP | OCT | NOV | DEC | YEAR |
|---|---|---|---|---|---|---|---|---|---|---|---|---|---|
| Maximum Temp °F | 36.1 | 40.5 | 48.2 | 58.7 | 68.6 | 78.3 | 85.8 | 85.0 | 75.1 | 62.9 | 47.3 | 39.7 | 60.5 |
| Minimum Temp °F | 11.9 | 15.9 | 23.7 | 33.3 | 43.2 | 52.7 | 58.5 | 56.5 | 46.4 | 35.7 | 23.9 | 15.6 | 34.8 |
| Mean Temp °F | 24.0 | 28.2 | 36.0 | 46.0 | 55.9 | 65.5 | 72.2 | 70.8 | 60.8 | 49.3 | 35.6 | 27.7 | 47.7 |
| Days Max Temp ≥ 90 °F | 0 | 0 | 0 | 0 | 0 | 3 | 10 | 10 | 3 | 0 | 0 | 0 | 26 |
| Days Max Temp ≤ 32 °F | 11 | 8 | 5 | 1 | 0 | 0 | 0 | 0 | 0 | 0 | 4 | 9 | 38 |
| Days Min Temp ≤ 32 °F | 29 | 26 | 26 | 14 | 3 | 0 | 0 | 0 | 2 | 10 | 24 | 29 | 163 |
| Days Min Temp ≤ 0 °F | 8 | 4 | 1 | 0 | 0 | 0 | 0 | 0 | 0 | 0 | 1 | 4 | 18 |
| Heating Degree Days | 1264 | 1032 | 894 | 564 | 294 | 78 | 14 | 20 | 176 | 482 | 875 | 1152 | 6845 |
| Cooling Degree Days | 0 | 0 | 0 | 5 | 21 | 102 | 233 | 197 | 47 | 3 | 0 | 0 | 608 |
| Total Precipitation (") | 0.34 | 0.44 | 0.89 | 2.02 | 3.22 | 3.03 | 2.62 | 2.06 | 1.16 | 1.19 | 0.50 | 0.38 | 17.85 |
| Days ≥ 0.1" Precip | 1 | 2 | 3 | 4 | 7 | 6 | 6 | 4 | 3 | 3 | 2 | 1 | 42 |
| Total Snowfall (") | 4.4 | 5.8 | 7.7 | 6.2 | 0.9 | 0.1 | 0.0 | 0.0 | 0.1 | 1.4 | 4.8 | 5.1 | 36.5 |
| Days ≥ 1" Snow Depth | ***10*** | 7 | 5 | 2 | 0 | 0 | 0 | 0 | 0 | 1 | 5 | ***9*** | 39 |

## RAPID CITY REGINL AP *Pennington County* ELEVATION 3162 ft LAT/LONG 44° 3 ' N / 103° 4 ' W

| | JAN | FEB | MAR | APR | MAY | JUN | JUL | AUG | SEP | OCT | NOV | DEC | YEAR |
|---|---|---|---|---|---|---|---|---|---|---|---|---|---|
| Maximum Temp °F | 33.7 | 37.9 | 46.8 | 58.2 | 68.4 | 77.8 | 85.9 | 85.2 | 74.9 | 62.0 | 45.8 | 36.4 | 59.4 |
| Minimum Temp °F | 10.9 | 14.8 | 23.0 | 32.6 | 42.6 | 51.9 | 58.0 | 56.3 | 45.8 | 34.7 | 22.6 | 13.5 | 33.9 |
| Mean Temp °F | 22.3 | 26.4 | 34.9 | 45.4 | 55.5 | 64.9 | 71.9 | 70.8 | 60.4 | 48.4 | 34.2 | 25.0 | 46.7 |
| Days Max Temp ≥ 90 °F | 0 | 0 | 0 | 0 | 0 | 3 | 11 | 11 | 3 | 0 | 0 | 0 | 28 |
| Days Max Temp ≤ 32 °F | 13 | 10 | 5 | 1 | 0 | 0 | 0 | 0 | 0 | 0 | 5 | 11 | 45 |
| Days Min Temp ≤ 32 °F | 30 | 27 | 26 | 15 | 3 | 0 | 0 | 0 | 2 | 12 | 25 | 30 | 170 |
| Days Min Temp ≤ 0 °F | 8 | 4 | 1 | 0 | 0 | 0 | 0 | 0 | 0 | 0 | 1 | 4 | 18 |
| Heating Degree Days | 1318 | 1083 | 925 | 583 | 303 | 87 | 14 | 23 | 190 | 512 | 917 | 1235 | 7190 |
| Cooling Degree Days | 0 | 0 | 0 | 4 | 20 | 98 | 242 | 222 | 60 | 4 | 0 | 0 | 650 |
| Total Precipitation (") | 0.38 | 0.50 | 1.01 | 1.91 | 2.63 | 2.88 | 2.06 | 1.62 | 1.20 | 1.15 | 0.62 | 0.46 | 16.42 |
| Days ≥ 0.1" Precip | 1 | 2 | 3 | 5 | 6 | 6 | 4 | 4 | 3 | 3 | 2 | 2 | 41 |
| Total Snowfall (") | 4.7 | 6.7 | 8.8 | 7.0 | 0.8 | 0.0 | 0.0 | 0.0 | 0.2 | 1.7 | 6.0 | 6.0 | 41.9 |
| Days ≥ 1" Snow Depth | 15 | 13 | 9 | 3 | 0 | 0 | 0 | 0 | 0 | 1 | 7 | 13 | 61 |

## REDFIELD 2 NE *Spink County* ELEVATION 1290 ft LAT/LONG 44° 54 ' N / 98° 30 ' W

| | JAN | FEB | MAR | APR | MAY | JUN | JUL | AUG | SEP | OCT | NOV | DEC | YEAR |
|---|---|---|---|---|---|---|---|---|---|---|---|---|---|
| Maximum Temp °F | 21.0 | 27.9 | 41.1 | 58.2 | 71.0 | 79.8 | 86.6 | 85.1 | 74.4 | 60.6 | 41.0 | 27.5 | 56.2 |
| Minimum Temp °F | -0.9 | 6.5 | 19.9 | 32.5 | 43.8 | 54.0 | 59.2 | 56.6 | 45.7 | 33.2 | 19.8 | 6.5 | 31.4 |
| Mean Temp °F | 10.1 | 17.2 | 30.6 | 45.4 | 57.4 | 66.9 | 72.9 | 70.9 | 60.1 | 46.9 | 30.4 | 17.0 | 43.8 |
| Days Max Temp ≥ 90 °F | 0 | 0 | 0 | 0 | 1 | 4 | 11 | 10 | 3 | 0 | 0 | 0 | 29 |
| Days Max Temp ≤ 32 °F | 23 | 17 | 8 | 1 | 0 | 0 | 0 | 0 | 0 | 0 | 8 | 19 | 76 |
| Days Min Temp ≤ 32 °F | 31 | 28 | 27 | 16 | 4 | 0 | 0 | 0 | 2 | 15 | 28 | 31 | 182 |
| Days Min Temp ≤ 0 °F | 16 | 10 | 3 | 0 | 0 | 0 | 0 | 0 | 0 | 0 | 1 | 10 | 40 |
| Heating Degree Days | 1701 | 1345 | 1064 | 584 | 261 | 59 | 11 | 27 | 200 | 558 | 1032 | 1484 | 8326 |
| Cooling Degree Days | 0 | 0 | 0 | 4 | 37 | 119 | 243 | 204 | 51 | 2 | 0 | 0 | 660 |
| Total Precipitation (") | 0.34 | 0.58 | 1.16 | 2.22 | 2.69 | 3.40 | 2.88 | 2.38 | 1.81 | 1.36 | 0.62 | 0.38 | 19.82 |
| Days ≥ 0.1" Precip | 1 | 2 | 3 | 5 | 5 | 6 | 5 | 4 | 4 | 3 | 2 | 1 | 41 |
| Total Snowfall (") | ***4.6*** | 6.6 | ***6.6*** | 2.7 | 0.0 | 0.0 | 0.0 | 0.0 | 0.0 | 0.5 | ***5.1*** | ***5.8*** | 31.9 |
| Days ≥ 1" Snow Depth | ***23*** | ***18*** | 11 | 2 | 0 | 0 | 0 | 0 | 0 | 0 | ***6*** | ***14*** | 74 |

## REDIG 11 NE *Harding County* ELEVATION 2992 ft LAT/LONG 45° 22 ' N / 103° 24 ' W

| | JAN | FEB | MAR | APR | MAY | JUN | JUL | AUG | SEP | OCT | NOV | DEC | YEAR |
|---|---|---|---|---|---|---|---|---|---|---|---|---|---|
| Maximum Temp °F | 27.7 | 33.0 | 43.2 | 56.4 | 67.8 | 76.8 | 84.9 | 84.3 | 73.2 | 60.2 | 42.6 | 31.5 | 56.8 |
| Minimum Temp °F | 5.3 | 10.2 | 19.4 | 30.5 | 40.8 | 50.0 | 55.5 | 53.4 | 42.9 | 32.0 | 19.2 | 9.1 | 30.7 |
| Mean Temp °F | 16.5 | 21.6 | 31.4 | 43.5 | 54.3 | 63.4 | 70.2 | 68.9 | 58.1 | 46.1 | 30.9 | 20.3 | 43.8 |
| Days Max Temp ≥ 90 °F | 0 | 0 | 0 | 0 | 0 | 3 | 9 | 10 | 2 | 0 | 0 | 0 | 24 |
| Days Max Temp ≤ 32 °F | 17 | 13 | 7 | 1 | 0 | 0 | 0 | 0 | 0 | 1 | 7 | 15 | 61 |
| Days Min Temp ≤ 32 °F | 31 | 28 | 29 | 19 | 5 | 0 | 0 | 0 | 3 | 16 | 28 | 31 | 190 |
| Days Min Temp ≤ 0 °F | 12 | 7 | 3 | 0 | 0 | 0 | 0 | 0 | 0 | 0 | 2 | 8 | 32 |
| Heating Degree Days | 1499 | 1219 | 1037 | 640 | 335 | 107 | 20 | 36 | 236 | 580 | 1015 | 1379 | 8103 |
| Cooling Degree Days | 0 | 0 | 0 | 1 | 14 | 75 | 193 | 169 | 35 | 2 | 0 | 0 | 489 |
| Total Precipitation (") | 0.27 | 0.37 | 0.76 | 1.68 | 2.71 | 3.06 | 2.25 | 1.29 | 1.07 | 1.02 | 0.42 | 0.32 | 15.22 |
| Days ≥ 0.1" Precip | 1 | 1 | 2 | 4 | 6 | 7 | 5 | 3 | 3 | 2 | 2 | 1 | 37 |
| Total Snowfall (") | 6.9 | 8.0 | 9.0 | 7.5 | 1.5 | 0.0 | 0.0 | 0.0 | 0.4 | 1.9 | 7.0 | 7.3 | 49.5 |
| Days ≥ 1" Snow Depth | 21 | 16 | 11 | 3 | 1 | 0 | 0 | 0 | 0 | 1 | 8 | 17 | 78 |

## SELBY *Walworth County* ELEVATION 1880 ft LAT/LONG 45° 30 ' N / 100° 2 ' W

| | JAN | FEB | MAR | APR | MAY | JUN | JUL | AUG | SEP | OCT | NOV | DEC | YEAR |
|---|---|---|---|---|---|---|---|---|---|---|---|---|---|
| Maximum Temp °F | 21.9 | 27.8 | 40.2 | 56.2 | 69.5 | 78.3 | 85.3 | 83.8 | 72.7 | 59.4 | 39.6 | 26.6 | 55.1 |
| Minimum Temp °F | 0.7 | 6.7 | 18.7 | 31.9 | 43.7 | 53.3 | 58.5 | 56.2 | 44.8 | 33.0 | 19.2 | 6.7 | 31.1 |
| Mean Temp °F | 11.3 | 17.3 | 29.5 | 44.0 | 56.6 | 65.8 | 71.9 | 70.0 | 58.8 | 46.2 | 29.5 | 16.7 | 43.1 |
| Days Max Temp ≥ 90 °F | 0 | 0 | 0 | 0 | 1 | 3 | 9 | 9 | 2 | 0 | 0 | 0 | 24 |
| Days Max Temp ≤ 32 °F | 22 | 17 | 9 | 1 | 0 | 0 | 0 | 0 | 0 | 0 | 9 | 20 | 78 |
| Days Min Temp ≤ 32 °F | 31 | 28 | 28 | 17 | 3 | 0 | 0 | 0 | 2 | 15 | 28 | 31 | 183 |
| Days Min Temp ≤ 0 °F | 15 | 10 | 3 | 0 | 0 | 0 | 0 | 0 | 0 | 0 | 2 | 10 | 40 |
| Heating Degree Days | 1661 | 1342 | 1094 | 625 | 282 | 71 | 16 | 32 | 225 | 575 | 1059 | 1493 | 8475 |
| Cooling Degree Days | 0 | 0 | 0 | 5 | 33 | 102 | 219 | 180 | 39 | 2 | 0 | 0 | 580 |
| Total Precipitation (") | 0.31 | 0.43 | 1.05 | 2.05 | 2.46 | 3.18 | 2.50 | 1.96 | 1.15 | 1.18 | 0.68 | 0.42 | 17.37 |
| Days ≥ 0.1" Precip | 1 | 1 | 3 | 5 | 5 | 6 | 5 | 4 | 3 | 3 | 2 | 1 | 39 |
| Total Snowfall (") | 3.4 | 5.5 | 6.5 | 3.7 | 0.0 | 0.0 | 0.0 | 0.0 | 0.0 | 0.6 | 5.1 | 5.4 | 30.2 |
| Days ≥ 1" Snow Depth | 26 | 22 | 18 | 4 | 0 | 0 | 0 | 0 | 0 | 1 | 11 | 22 | 104 |

## SIOUX FALLS FOSS FLD *Minnehaha County* ELEVATION 1424 ft LAT/LONG 43° 34 ' N / 96° 44 ' W

| | JAN | FEB | MAR | APR | MAY | JUN | JUL | AUG | SEP | OCT | NOV | DEC | YEAR |
|---|---|---|---|---|---|---|---|---|---|---|---|---|---|
| Maximum Temp °F | 24.2 | 29.6 | 43.1 | 58.9 | 71.0 | 80.3 | 85.6 | 83.0 | 73.1 | 60.5 | 42.2 | 28.5 | 56.7 |
| Minimum Temp °F | 3.6 | 10.0 | 22.9 | 35.0 | 46.4 | 56.3 | 62.0 | 59.5 | 48.8 | 35.8 | 22.4 | 9.6 | 34.4 |
| Mean Temp °F | 13.9 | 19.8 | 33.0 | 47.0 | 58.7 | 68.3 | 73.8 | 71.3 | 61.0 | 48.2 | 32.3 | 19.1 | 45.5 |
| Days Max Temp ≥ 90 °F | 0 | 0 | 0 | 0 | 1 | 4 | 10 | 7 | 2 | 0 | 0 | 0 | 24 |
| Days Max Temp ≤ 32 °F | 21 | 16 | 7 | 0 | 0 | 0 | 0 | 0 | 0 | 0 | 7 | 18 | 69 |
| Days Min Temp ≤ 32 °F | 31 | 28 | 25 | 12 | 2 | 0 | 0 | 0 | 1 | 12 | 26 | 31 | 168 |
| Days Min Temp ≤ 0 °F | 13 | 8 | 1 | 0 | 0 | 0 | 0 | 0 | 0 | 0 | 1 | 8 | 31 |
| Heating Degree Days | 1579 | 1271 | 985 | 541 | 229 | 44 | 7 | 21 | 177 | 518 | 973 | 1419 | 7764 |
| Cooling Degree Days | 0 | 0 | 0 | 10 | 47 | 161 | 281 | 220 | 68 | 3 | 0 | 0 | 790 |
| Total Precipitation (") | 0.54 | 0.57 | 1.62 | 2.59 | 3.15 | 3.71 | 2.75 | 2.96 | 2.98 | 1.86 | 1.20 | 0.67 | 24.60 |
| Days ≥ 0.1" Precip | 2 | 2 | 4 | 5 | 6 | 6 | 5 | 5 | 5 | 3 | 3 | 2 | 48 |
| Total Snowfall (") | 7.2 | 6.9 | 7.7 | 2.6 | 0.0 | 0.0 | 0.0 | 0.0 | 0.0 | 0.9 | 6.1 | 7.9 | 39.3 |
| Days ≥ 1" Snow Depth | 23 | 19 | 10 | 1 | 0 | 0 | 0 | 0 | 0 | 0 | 6 | 17 | 76 |

## SISSETON 2 E *Roberts County* ELEVATION 1201 ft LAT/LONG 45° 40 ' N / 97° 3 ' W

| | JAN | FEB | MAR | APR | MAY | JUN | JUL | AUG | SEP | OCT | NOV | DEC | YEAR |
|---|---|---|---|---|---|---|---|---|---|---|---|---|---|
| Maximum Temp °F | 21.2 | 27.1 | 40.4 | 57.5 | 71.2 | 79.2 | 85.0 | 83.1 | 72.9 | 60.0 | 40.0 | 26.8 | 55.4 |
| Minimum Temp °F | 1.1 | 7.6 | 21.4 | 33.8 | 45.7 | 54.8 | 60.3 | 58.1 | 48.1 | 36.5 | 22.0 | 8.0 | 33.1 |
| Mean Temp °F | 11.2 | 17.4 | 30.9 | 45.7 | 58.5 | 67.0 | 72.7 | 70.6 | 60.5 | 48.3 | 31.0 | 17.4 | 44.3 |
| Days Max Temp ≥ 90 °F | 0 | 0 | 0 | 0 | 1 | 3 | 9 | 7 | 2 | 0 | 0 | 0 | 22 |
| Days Max Temp ≤ 32 °F | 23 | 17 | 8 | 1 | 0 | 0 | 0 | 0 | 0 | 0 | 8 | 19 | 76 |
| Days Min Temp ≤ 32 °F | 31 | 28 | 26 | 14 | 2 | 0 | 0 | 0 | 1 | 10 | 26 | 31 | 169 |
| Days Min Temp ≤ 0 °F | 16 | 10 | 2 | 0 | 0 | 0 | 0 | 0 | 0 | 0 | 1 | 10 | 39 |
| Heating Degree Days | 1666 | 1339 | 1050 | 577 | 234 | 52 | 9 | 24 | 186 | 515 | 1013 | 1469 | 8134 |
| Cooling Degree Days | 0 | 0 | 0 | 7 | 44 | 113 | 243 | 193 | 52 | 3 | 0 | 0 | 655 |
| Total Precipitation (") | 0.58 | 0.56 | 1.34 | 2.28 | 2.78 | 3.32 | 2.90 | 2.60 | 1.93 | 1.56 | 1.05 | 0.50 | 21.40 |
| Days ≥ 0.1" Precip | 2 | 2 | 4 | 5 | 6 | 6 | 5 | 5 | 4 | 4 | 3 | 2 | 48 |
| Total Snowfall (") | 7.7 | 6.6 | 7.6 | 3.4 | 0.0 | 0.0 | 0.0 | 0.0 | 0.0 | 0.6 | 7.1 | 5.7 | 38.7 |
| Days ≥ 1" Snow Depth | 24 | 18 | 9 | 1 | 0 | 0 | 0 | 0 | 0 | 0 | 7 | 15 | 74 |

## SPEARFISH *Lawrence County* ELEVATION 3753 ft LAT/LONG 44° 29 ' N / 103° 51 ' W

| | JAN | FEB | MAR | APR | MAY | JUN | JUL | AUG | SEP | OCT | NOV | DEC | YEAR |
|---|---|---|---|---|---|---|---|---|---|---|---|---|---|
| Maximum Temp °F | 35.5 | 38.6 | 45.7 | 56.4 | 67.1 | 76.3 | 84.6 | 83.4 | 73.1 | 60.6 | 45.4 | 37.6 | 58.7 |
| Minimum Temp °F | 12.7 | 16.1 | 23.3 | 32.9 | 42.7 | 51.3 | 57.7 | 55.5 | 45.9 | 35.6 | 23.9 | 16.1 | 34.5 |
| Mean Temp °F | 24.2 | 27.4 | 34.5 | 44.7 | 54.9 | 63.8 | 71.2 | 69.5 | 59.5 | 48.1 | 34.6 | 26.9 | 46.6 |
| Days Max Temp ≥ 90 °F | 0 | 0 | 0 | 0 | 0 | 2 | 9 | 8 | 2 | 0 | 0 | 0 | 21 |
| Days Max Temp ≤ 32 °F | 11 | 9 | 5 | 1 | 0 | 0 | 0 | 0 | 0 | 1 | 5 | 10 | 42 |
| Days Min Temp ≤ 32 °F | 28 | 25 | 25 | 14 | 3 | 0 | 0 | 0 | 1 | 10 | 23 | 28 | 157 |
| Days Min Temp ≤ 0 °F | 8 | 4 | 1 | 0 | 0 | 0 | 0 | 0 | 0 | 0 | 1 | ***5*** | 19 |
| Heating Degree Days | 1260 | 1056 | 938 | 605 | 323 | 104 | 17 | 31 | 206 | 518 | 903 | 1174 | 7135 |
| Cooling Degree Days | 0 | 0 | 0 | 3 | 19 | 81 | 230 | 182 | 48 | 2 | ***0*** | 0 | 565 |
| Total Precipitation (") | 0.55 | 0.72 | 1.49 | 2.67 | 3.33 | 3.81 | 2.02 | 1.78 | 1.65 | 1.82 | 0.87 | 0.82 | 21.53 |
| Days ≥ 0.1" Precip | 2 | 2 | 4 | 5 | 7 | 7 | 5 | 4 | 4 | 4 | 3 | 3 | 50 |
| Total Snowfall (") | ***9.5*** | ***10.8*** | ***15.8*** | 8.9 | 0.9 | 0.0 | 0.0 | 0.0 | 0.3 | 4.0 | ***7.2*** | ***10.5*** | 67.9 |
| Days ≥ 1" Snow Depth | ***19*** | ***13*** | 9 | 4 | 0 | 0 | 0 | 0 | 0 | 2 | ***7*** | ***15*** | 69 |

## STEPHAN 1 ENE *Hyde County* ELEVATION 1831 ft LAT/LONG 44° 15 ' N / 99° 27 ' W

| | JAN | FEB | MAR | APR | MAY | JUN | JUL | AUG | SEP | OCT | NOV | DEC | YEAR |
|---|---|---|---|---|---|---|---|---|---|---|---|---|---|
| Maximum Temp °F | 25.9 | 31.6 | 44.6 | 60.3 | 72.4 | 81.5 | 88.4 | 87.1 | 76.5 | 62.7 | 43.2 | 30.5 | 58.7 |
| Minimum Temp °F | 2.0 | 7.9 | 19.5 | 31.5 | 42.5 | 52.6 | 58.1 | 55.8 | 45.4 | 33.3 | 19.2 | 7.6 | 31.3 |
| Mean Temp °F | 14.0 | 19.8 | 32.1 | 45.9 | 57.5 | 67.0 | 73.3 | 71.5 | 61.0 | 48.0 | 31.3 | 19.1 | 45.0 |
| Days Max Temp ≥ 90 °F | 0 | 0 | 0 | 0 | 1 | 5 | 14 | 13 | 4 | 0 | 0 | 0 | 37 |
| Days Max Temp ≤ 32 °F | 19 | 14 | 6 | 0 | 0 | 0 | 0 | 0 | 0 | 0 | 7 | 16 | 62 |
| Days Min Temp ≤ 32 °F | 31 | 28 | 28 | 17 | 4 | 0 | 0 | 0 | 3 | 15 | 28 | 31 | 185 |
| Days Min Temp ≤ 0 °F | 14 | 9 | 3 | 0 | 0 | 0 | 0 | 0 | 0 | 0 | 2 | 9 | 37 |
| Heating Degree Days | 1579 | 1271 | 1013 | 569 | 253 | 54 | 9 | 23 | 177 | 523 | 1004 | 1418 | 7893 |
| Cooling Degree Days | 0 | 0 | 0 | 4 | 32 | 124 | 262 | 229 | 65 | 2 | 0 | 0 | 718 |
| Total Precipitation (") | 0.32 | 0.51 | 1.16 | 1.90 | 2.37 | 3.11 | 2.63 | 1.96 | 1.63 | 1.38 | 0.60 | 0.36 | 17.93 |
| Days ≥ 0.1" Precip | 1 | 2 | 2 | 4 | 5 | 6 | 5 | 4 | 3 | 3 | 2 | 1 | 38 |
| Total Snowfall (") | ***4.6*** | ***8.6*** | ***7.7*** | 2.8 | 0.1 | 0.0 | 0.0 | 0.0 | 0.1 | 0.9 | 4.1 | ***6.3*** | 35.2 |
| Days ≥ 1" Snow Depth | na | na | na | na | 0 | 0 | 0 | 0 | 0 | 0 | na | na | na |

## SUMMIT 1 W *Roberts County* ELEVATION 2011 ft LAT/LONG 45° 19' N / 97° 2' W

| | JAN | FEB | MAR | APR | MAY | JUN | JUL | AUG | SEP | OCT | NOV | DEC | YEAR |
|---|---|---|---|---|---|---|---|---|---|---|---|---|---|
| Maximum Temp °F | 18.7 | 24.1 | 37.1 | 54.6 | 68.2 | 76.7 | 82.1 | 80.2 | 70.4 | 57.5 | 36.9 | 23.7 | 52.5 |
| Minimum Temp °F | -1.2 | 4.5 | 17.8 | 31.1 | 43.0 | 52.6 | 57.2 | 55.2 | 45.3 | 33.4 | 18.4 | 5.1 | 30.2 |
| Mean Temp °F | 8.8 | 14.3 | 27.5 | 42.9 | 55.6 | 64.7 | 69.6 | 67.7 | 57.9 | 45.5 | 27.7 | 14.4 | 41.4 |
| Days Max Temp ≥ 90 °F | 0 | 0 | 0 | 0 | 0 | 1 | 5 | 4 | 1 | 0 | 0 | 0 | 11 |
| Days Max Temp ≤ 32 °F | 25 | 20 | 11 | 1 | 0 | 0 | 0 | 0 | 0 | 0 | 11 | 23 | 91 |
| Days Min Temp ≤ 32 °F | 31 | 28 | 28 | 18 | 5 | 0 | 0 | 0 | 3 | 15 | 27 | 31 | 186 |
| Days Min Temp ≤ 0 °F | 17 | 11 | 3 | 0 | 0 | 0 | 0 | 0 | 0 | 0 | 2 | 12 | 45 |
| Heating Degree Days | 1741 | 1426 | 1154 | 658 | 306 | 87 | 26 | 50 | 243 | 601 | 1112 | 1564 | 8968 |
| Cooling Degree Days | 0 | 0 | 0 | 4 | 26 | 87 | 166 | 134 | 33 | 1 | 0 | 0 | 451 |
| Total Precipitation (") | 0.47 | 0.47 | 1.31 | 2.21 | 2.70 | 3.80 | 3.37 | 2.86 | 2.11 | 1.53 | 0.79 | 0.36 | 21.98 |
| Days ≥ 0.1" Precip | 2 | 2 | 3 | 5 | 6 | 7 | 5 | 5 | 4 | 4 | 2 | 1 | 46 |
| Total Snowfall (") | 8.1 | 8.1 | 8.9 | 4.4 | 0.1 | 0.1 | 0.0 | 0.0 | 0.0 | 1.2 | *7.0* | 6.0 | 43.9 |
| Days ≥ 1" Snow Depth | 25 | 23 | 15 | 3 | 0 | 0 | 0 | 0 | 0 | 1 | 9 | 18 | 94 |

## TIMBER LAKE *Dewey County* ELEVATION 2162 ft LAT/LONG 45° 26' N / 101° 4' W

| | JAN | FEB | MAR | APR | MAY | JUN | JUL | AUG | SEP | OCT | NOV | DEC | YEAR |
|---|---|---|---|---|---|---|---|---|---|---|---|---|---|
| Maximum Temp °F | 24.4 | 30.2 | 41.7 | 57.3 | 69.9 | 78.5 | 85.3 | 83.8 | 73.0 | 59.7 | 40.7 | 28.2 | 56.1 |
| Minimum Temp °F | 4.0 | 9.7 | 20.7 | 33.1 | 44.3 | 53.8 | 59.3 | 57.3 | 46.6 | 34.9 | 20.8 | 8.9 | 32.8 |
| Mean Temp °F | 14.2 | 20.0 | 31.2 | 45.2 | 57.1 | 66.2 | 72.3 | 70.6 | 59.8 | 47.3 | 30.8 | 18.6 | 44.4 |
| Days Max Temp ≥ 90 °F | 0 | 0 | 0 | 0 | 1 | 3 | 10 | 9 | 2 | 0 | 0 | 0 | 25 |
| Days Max Temp ≤ 32 °F | 20 | 15 | 7 | 1 | 0 | 0 | 0 | 0 | 0 | 0 | 8 | 18 | 69 |
| Days Min Temp ≤ 32 °F | 31 | 28 | 28 | 15 | 3 | 0 | 0 | 0 | 2 | 12 | 27 | 31 | 177 |
| Days Min Temp ≤ 0 °F | 13 | 8 | 2 | 0 | 0 | 0 | 0 | 0 | 0 | 0 | 2 | 8 | 33 |
| Heating Degree Days | 1571 | 1266 | 1041 | 590 | 265 | 66 | 14 | 24 | 201 | 541 | 1020 | 1434 | 8033 |
| Cooling Degree Days | 0 | 0 | 0 | 5 | 33 | 110 | 245 | 206 | 49 | 2 | 0 | 0 | 650 |
| Total Precipitation (") | 0.35 | 0.52 | 1.19 | 2.01 | 2.67 | 3.08 | 2.27 | 1.89 | 1.21 | 1.30 | 0.63 | 0.53 | 17.65 |
| Days ≥ 0.1" Precip | 1 | 2 | 3 | 5 | 6 | 6 | 5 | 4 | 3 | 3 | 2 | 2 | 42 |
| Total Snowfall (") | 4.9 | 6.8 | 8.4 | 5.5 | 0.5 | 0.0 | 0.0 | 0.0 | 0.0 | 0.9 | 5.8 | 6.6 | 39.4 |
| Days ≥ 1" Snow Depth | 24 | 17 | 14 | 3 | 0 | 0 | 0 | 0 | 0 | 1 | 8 | 19 | 86 |

## TYNDALL *Bon Homme County* ELEVATION 1421 ft LAT/LONG 43° 0' N / 97° 52' W

| | JAN | FEB | MAR | APR | MAY | JUN | JUL | AUG | SEP | OCT | NOV | DEC | YEAR |
|---|---|---|---|---|---|---|---|---|---|---|---|---|---|
| Maximum Temp °F | 28.6 | 34.5 | 47.1 | 62.2 | 73.5 | 83.2 | 87.9 | 85.7 | 76.8 | 63.7 | 45.0 | 32.5 | 60.1 |
| Minimum Temp °F | 7.5 | 13.3 | 25.2 | 36.9 | 48.6 | 58.0 | 62.9 | 60.9 | 50.8 | 38.4 | 24.8 | 12.7 | 36.7 |
| Mean Temp °F | 18.1 | 23.9 | 36.2 | 49.6 | 61.1 | 70.6 | 75.3 | 73.3 | 63.8 | 51.1 | 34.9 | 22.6 | 48.4 |
| Days Max Temp ≥ 90 °F | 0 | 0 | 0 | 0 | 1 | 7 | 13 | 11 | 4 | 0 | 0 | 0 | 36 |
| Days Max Temp ≤ 32 °F | 17 | 12 | 5 | 0 | 0 | 0 | 0 | 0 | 0 | 0 | 5 | 15 | 54 |
| Days Min Temp ≤ 32 °F | 31 | 27 | 24 | 10 | 1 | 0 | 0 | 0 | 1 | 8 | 23 | 30 | 155 |
| Days Min Temp ≤ 0 °F | 10 | 6 | 1 | 0 | 0 | 0 | 0 | 0 | 0 | 0 | 1 | 6 | 24 |
| Heating Degree Days | 1450 | 1156 | 887 | 467 | 173 | 26 | 4 | 9 | 125 | 431 | 897 | 1308 | 6933 |
| Cooling Degree Days | 0 | 0 | 0 | 16 | 59 | 204 | 323 | 272 | 103 | 6 | 0 | 0 | 983 |
| Total Precipitation (") | 0.47 | 0.68 | 1.51 | 2.35 | 3.32 | 3.41 | 3.66 | 2.29 | 2.44 | 1.61 | 1.19 | 0.69 | 23.62 |
| Days ≥ 0.1" Precip | 2 | 2 | 3 | 6 | 6 | 6 | 6 | 5 | 5 | 4 | 3 | 2 | 50 |
| Total Snowfall (") | 6.1 | 6.2 | 5.0 | 2.4 | 0.0 | 0.0 | 0.0 | 0.0 | 0.0 | 0.6 | 5.9 | 7.7 | 33.9 |
| Days ≥ 1" Snow Depth | 23 | 14 | 7 | 0 | 0 | 0 | 0 | 0 | 0 | 0 | 4 | 15 | 63 |

## USTA 8 WNW KELLY RCH *Perkins County* ELEVATION 2382 ft LAT/LONG 45° 15' N / 102° 19' W

| | JAN | FEB | MAR | APR | MAY | JUN | JUL | AUG | SEP | OCT | NOV | DEC | YEAR |
|---|---|---|---|---|---|---|---|---|---|---|---|---|---|
| Maximum Temp °F | 26.8 | 32.9 | *43.8* | *59.1* | 70.8 | 79.8 | 87.1 | 86.7 | 75.0 | 61.3 | 42.4 | 30.4 | 58.0 |
| Minimum Temp °F | 1.1 | 7.4 | 18.2 | *30.7* | 41.5 | 51.1 | 56.6 | 53.9 | 42.1 | 29.8 | 16.4 | 4.5 | 29.4 |
| Mean Temp °F | 13.9 | 20.3 | *31.0* | *44.9* | 56.2 | 65.5 | 71.8 | 70.3 | 58.5 | 45.5 | 29.5 | 17.5 | 43.7 |
| Days Max Temp ≥ 90 °F | 0 | 0 | 0 | 0 | 1 | 4 | 12 | 12 | 3 | 0 | 0 | 0 | 32 |
| Days Max Temp ≤ 32 °F | 17 | 13 | *6* | 0 | 0 | 0 | 0 | 0 | 0 | 0 | 7 | 15 | 58 |
| Days Min Temp ≤ 32 °F | 30 | 28 | 29 | 16 | 5 | 0 | 0 | 0 | 5 | 18 | 29 | 30 | 190 |
| Days Min Temp ≤ 0 °F | 15 | 9 | 3 | 0 | 0 | 0 | 0 | 0 | 0 | 0 | 3 | 10 | 40 |
| Heating Degree Days | 1579 | 1255 | *1047* | *599* | 287 | 74 | 15 | 28 | 228 | 596 | 1061 | 1468 | 8237 |
| Cooling Degree Days | 0 | 0 | *0* | na | 23 | *100* | *240* | *220* | 37 | *0* | 0 | 0 | na |
| Total Precipitation (") | 0.18 | 0.27 | 0.78 | *1.59* | 2.56 | 2.82 | 2.20 | 1.60 | 1.08 | 0.97 | 0.40 | 0.27 | 14.72 |
| Days ≥ 0.1" Precip | 1 | 1 | *2* | 3 | 5 | 6 | 4 | 3 | 2 | 2 | 1 | 1 | 31 |
| Total Snowfall (") | 3.1 | 5.2 | 6.3 | *3.6* | 0.5 | 0.0 | 0.0 | 0.0 | 0.1 | 0.8 | 4.2 | 4.4 | 28.2 |
| Days ≥ 1" Snow Depth | *17* | *12* | *9* | 1 | 0 | 0 | 0 | 0 | 0 | 0 | 4 | *13* | 56 |

## VERMILLION 2 SE *Clay County* ELEVATION 1220 ft LAT/LONG 42° 47 ' N / 96° 53 ' W

| | JAN | FEB | MAR | APR | MAY | JUN | JUL | AUG | SEP | OCT | NOV | DEC | YEAR |
|---|---|---|---|---|---|---|---|---|---|---|---|---|---|
| Maximum Temp °F | 29.7 | 35.8 | 49.1 | 64.5 | 75.5 | 84.7 | 88.6 | 86.2 | 78.1 | 65.8 | 46.7 | 33.4 | 61.5 |
| Minimum Temp °F | 6.9 | 12.6 | 24.7 | 36.6 | 48.0 | 58.0 | 63.0 | 60.6 | 50.3 | 37.7 | 24.6 | 12.7 | 36.3 |
| Mean Temp °F | 18.3 | 24.2 | 36.9 | 50.6 | 61.7 | 71.4 | 75.8 | 73.4 | 64.2 | 51.7 | 35.7 | 23.0 | 48.9 |
| Days Max Temp ≥ 90 °F | 0 | 0 | 0 | 1 | 2 | 8 | 14 | 10 | 4 | 0 | 0 | 0 | 39 |
| Days Max Temp ≤ 32 °F | 17 | 12 | 3 | 0 | 0 | 0 | 0 | 0 | 0 | 0 | 4 | 13 | 49 |
| Days Min Temp ≤ 32 °F | 31 | 27 | 24 | 11 | 2 | 0 | 0 | 0 | 1 | 10 | 23 | 30 | 159 |
| Days Min Temp ≤ 0 °F | 11 | 6 | 1 | 0 | 0 | 0 | 0 | 0 | 0 | 0 | 1 | 5 | 24 |
| Heating Degree Days | 1441 | 1145 | 865 | 441 | 164 | 20 | 4 | 10 | 118 | 413 | 873 | 1296 | 6790 |
| Cooling Degree Days | 0 | 0 | 0 | 21 | 74 | 244 | 350 | 293 | 117 | 9 | 0 | 0 | 1108 |
| Total Precipitation (") | 0.42 | 0.50 | 1.72 | 2.50 | 3.64 | 3.74 | 3.54 | 2.96 | 2.45 | 1.96 | 1.25 | 0.68 | 25.36 |
| Days ≥ 0.1" Precip | 1 | 1 | 4 | 5 | 6 | 6 | 6 | 5 | 4 | 4 | 3 | 2 | 47 |
| Total Snowfall (") | 5.7 | 5.2 | 5.2 | 2.4 | 0.0 | 0.0 | 0.0 | 0.0 | 0.0 | 0.9 | 5.1 | 7.3 | 31.8 |
| Days ≥ 1" Snow Depth | *19* | 14 | 7 | 1 | 0 | 0 | 0 | 0 | 0 | 0 | 4 | 17 | 62 |

## WAGNER *Charles Mix County* ELEVATION 1440 ft LAT/LONG 43° 5 ' N / 98° 17 ' W

| | JAN | FEB | MAR | APR | MAY | JUN | JUL | AUG | SEP | OCT | NOV | DEC | YEAR |
|---|---|---|---|---|---|---|---|---|---|---|---|---|---|
| Maximum Temp °F | 30.0 | 35.5 | 48.3 | 63.5 | 75.1 | 84.9 | 90.3 | 88.3 | 78.1 | 64.7 | 45.8 | 33.4 | 61.5 |
| Minimum Temp °F | 8.9 | 14.4 | 25.3 | 37.2 | 48.8 | 58.4 | 64.0 | 61.8 | 51.4 | 39.2 | 25.3 | 14.0 | 37.4 |
| Mean Temp °F | 19.4 | 25.0 | 36.8 | 50.4 | 61.9 | 71.7 | 77.1 | 75.0 | 64.8 | 52.0 | 35.6 | 23.7 | 49.4 |
| Days Max Temp ≥ 90 °F | 0 | 0 | 0 | 1 | 2 | 9 | 18 | 14 | 5 | 0 | 0 | 0 | 49 |
| Days Max Temp ≤ 32 °F | 16 | 12 | 4 | 0 | 0 | 0 | 0 | 0 | 0 | 0 | 4 | 14 | 50 |
| Days Min Temp ≤ 32 °F | 30 | 27 | 24 | 9 | 1 | 0 | 0 | 0 | 1 | 7 | 23 | 30 | 152 |
| Days Min Temp ≤ 0 °F | 9 | 6 | 1 | 0 | 0 | 0 | 0 | 0 | 0 | 0 | 1 | 5 | 22 |
| Heating Degree Days | 1407 | 1124 | 866 | 446 | 154 | 18 | 3 | 7 | 112 | 404 | 876 | 1274 | 6691 |
| Cooling Degree Days | 0 | 0 | 0 | 18 | 68 | 224 | 366 | 313 | 115 | 8 | 0 | 0 | 1112 |
| Total Precipitation (") | 0.58 | 0.81 | 1.63 | 2.62 | 3.53 | 3.57 | 3.00 | 2.61 | 2.79 | 1.82 | 1.20 | 0.79 | 24.95 |
| Days ≥ 0.1" Precip | 2 | 2 | 4 | 5 | 7 | 6 | 6 | 4 | 5 | 3 | 3 | 3 | 50 |
| Total Snowfall (") | 6.3 | 7.0 | 5.8 | 2.4 | 0.0 | 0.0 | 0.0 | 0.0 | 0.0 | 0.8 | 6.0 | 9.2 | 37.5 |
| Days ≥ 1" Snow Depth | 18 | 11 | 5 | 0 | 0 | 0 | 0 | 0 | 0 | 0 | 4 | 13 | 51 |

## WASTA *Pennington County* ELEVATION 2323 ft LAT/LONG 44° 4 ' N / 102° 26 ' W

| | JAN | FEB | MAR | APR | MAY | JUN | JUL | AUG | SEP | OCT | NOV | DEC | YEAR |
|---|---|---|---|---|---|---|---|---|---|---|---|---|---|
| Maximum Temp °F | 33.4 | 39.1 | 49.3 | 61.9 | 72.6 | 81.8 | 89.5 | 88.0 | 77.1 | 64.0 | 47.0 | 36.3 | 61.7 |
| Minimum Temp °F | 8.2 | 13.0 | 22.8 | 33.6 | 44.3 | 54.0 | 59.7 | 57.5 | 46.4 | 34.2 | 21.5 | 11.1 | 33.9 |
| Mean Temp °F | 20.8 | 26.0 | 36.1 | 47.8 | 58.5 | 67.9 | 74.6 | 72.8 | 61.8 | 49.1 | 34.3 | 23.7 | 47.8 |
| Days Max Temp ≥ 90 °F | 0 | 0 | 0 | 0 | 2 | 6 | 16 | 15 | 4 | 0 | 0 | 0 | 43 |
| Days Max Temp ≤ 32 °F | 13 | 9 | 4 | 0 | 0 | 0 | 0 | 0 | 0 | 0 | 4 | 10 | 40 |
| Days Min Temp ≤ 32 °F | 31 | 28 | 26 | 13 | 3 | 0 | 0 | 0 | 2 | 12 | 27 | 31 | 173 |
| Days Min Temp ≤ 0 °F | 9 | 5 | 1 | 0 | 0 | 0 | 0 | 0 | 0 | 0 | 1 | 6 | 22 |
| Heating Degree Days | 1363 | 1094 | 890 | 513 | 225 | 46 | 7 | 12 | 157 | 487 | 915 | 1274 | 6983 |
| Cooling Degree Days | 0 | 0 | 0 | 5 | 36 | 146 | 306 | 267 | 63 | 2 | 0 | 0 | 825 |
| Total Precipitation (") | 0.33 | 0.42 | 0.97 | 2.00 | 2.58 | 2.75 | 2.04 | 1.66 | 1.20 | 1.23 | 0.58 | 0.38 | 16.14 |
| Days ≥ 0.1" Precip | 1 | 1 | 3 | 4 | 5 | 5 | 4 | 4 | 3 | 3 | 2 | 2 | 37 |
| Total Snowfall (") | 5.0 | 6.0 | 7.9 | 4.1 | 0.1 | 0.0 | 0.0 | 0.0 | 0.0 | 1.2 | 4.9 | 5.4 | 34.6 |
| Days ≥ 1" Snow Depth | 17 | 12 | 7 | 1 | 0 | 0 | 0 | 0 | 0 | 0 | 5 | 14 | 56 |

## WATERTOWN MUNI AP *Codington County* ELEVATION 1746 ft LAT/LONG 44° 55 ' N / 97° 9 ' W

| | JAN | FEB | MAR | APR | MAY | JUN | JUL | AUG | SEP | OCT | NOV | DEC | YEAR |
|---|---|---|---|---|---|---|---|---|---|---|---|---|---|
| Maximum Temp °F | 20.3 | 25.9 | 39.0 | 55.5 | 68.7 | 77.7 | 83.5 | 81.3 | 70.7 | 57.5 | 38.5 | 25.2 | 53.7 |
| Minimum Temp °F | -0.3 | 6.2 | 19.2 | 32.5 | 44.4 | 54.3 | 59.4 | 56.9 | 46.7 | 34.6 | 20.3 | 6.8 | 31.8 |
| Mean Temp °F | 10.0 | 16.1 | 29.1 | 44.0 | 56.6 | 66.0 | 71.5 | 69.1 | 58.7 | 46.1 | 29.4 | 16.0 | 42.7 |
| Days Max Temp ≥ 90 °F | 0 | 0 | 0 | 0 | 0 | 2 | 7 | 5 | 2 | 0 | 0 | 0 | 16 |
| Days Max Temp ≤ 32 °F | 24 | 18 | 9 | 1 | 0 | 0 | 0 | 0 | 0 | 0 | 10 | 22 | 84 |
| Days Min Temp ≤ 32 °F | 31 | 28 | 28 | 16 | 3 | 0 | 0 | 0 | 2 | 13 | 27 | 31 | 179 |
| Days Min Temp ≤ 0 °F | 16 | 11 | 3 | 0 | 0 | 0 | 0 | 0 | 0 | 0 | 1 | 10 | 41 |
| Heating Degree Days | 1703 | 1376 | 1106 | 625 | 281 | 68 | 15 | 37 | 224 | 580 | 1060 | 1514 | 8589 |
| Cooling Degree Days | 0 | 0 | 0 | 4 | 34 | 112 | 225 | 175 | 43 | 1 | 0 | 0 | 594 |
| Total Precipitation (") | 0.55 | 0.54 | 1.32 | 2.23 | 2.71 | 4.09 | 2.94 | 2.77 | 2.07 | 1.84 | 0.91 | 0.49 | 22.46 |
| Days ≥ 0.1" Precip | 2 | 2 | 4 | 5 | 5 | 7 | 6 | 5 | 4 | 3 | 3 | 2 | 48 |
| Total Snowfall (") | 5.9 | 5.7 | 6.5 | 1.6 | 0.1 | 0.0 | 0.0 | 0.0 | 0.0 | 0.9 | 4.9 | 4.8 | 30.4 |
| Days ≥ 1" Snow Depth | 25 | 21 | 13 | 2 | 0 | 0 | 0 | 0 | 0 | 0 | 6 | 19 | 86 |

## WAUBAY NWR *Day County* ELEVATION 1831 ft LAT/LONG 45° 26' N / 97° 20' W

| | JAN | FEB | MAR | APR | MAY | JUN | JUL | AUG | SEP | OCT | NOV | DEC | YEAR |
|---|---|---|---|---|---|---|---|---|---|---|---|---|---|
| Maximum Temp °F | 20.6 | 25.9 | 38.8 | 56.1 | 69.4 | 77.7 | 83.7 | 81.9 | 72.1 | 58.9 | 38.6 | 25.5 | 54.1 |
| Minimum Temp °F | -0.2 | 5.1 | 17.9 | 32.1 | 44.7 | 54.4 | 59.3 | 57.1 | 46.9 | 35.4 | 20.2 | 6.5 | 31.6 |
| Mean Temp °F | 10.2 | 15.6 | 28.3 | 44.2 | 57.1 | 66.1 | 71.5 | 69.6 | 59.6 | 47.2 | 29.4 | 16.0 | 42.9 |
| Days Max Temp ≥ 90 °F | 0 | 0 | 0 | 0 | 0 | 1 | 6 | 6 | 1 | 0 | 0 | 0 | 14 |
| Days Max Temp ≤ 32 °F | 24 | 18 | 9 | 1 | 0 | 0 | 0 | 0 | 0 | 0 | 10 | 21 | 83 |
| Days Min Temp ≤ 32 °F | 31 | 28 | 28 | 17 | 3 | 0 | 0 | 0 | 1 | 12 | 27 | 31 | 178 |
| Days Min Temp ≤ 0 °F | 16 | 11 | 4 | 0 | 0 | 0 | 0 | 0 | 0 | 0 | 2 | 10 | 43 |
| Heating Degree Days | 1695 | 1390 | 1129 | 621 | 264 | 61 | 13 | 31 | 203 | 548 | 1067 | 1511 | 8533 |
| Cooling Degree Days | 0 | 0 | 0 | 5 | 30 | 101 | 217 | 173 | 44 | 1 | 0 | 0 | 571 |
| Total Precipitation (") | 0.48 | 0.53 | 1.00 | 1.83 | 2.61 | 3.51 | 3.13 | 2.72 | 1.88 | 1.43 | 0.76 | 0.42 | 20.30 |
| Days ≥ 0.1" Precip | 2 | 2 | 3 | 5 | 6 | 7 | 5 | 5 | 4 | 3 | 2 | 1 | 45 |
| Total Snowfall (") | 7.2 | 6.7 | 5.7 | 3.2 | 0.1 | 0.0 | 0.0 | 0.0 | 0.1 | 0.7 | 5.8 | 5.6 | 35.1 |
| Days ≥ 1" Snow Depth | 27 | 22 | 16 | 2 | 0 | 0 | 0 | 0 | 0 | 0 | 9 | 19 | 95 |

## WEBSTER *Day County* ELEVATION 1870 ft LAT/LONG 45° 20' N / 97° 32' W

| | JAN | FEB | MAR | APR | MAY | JUN | JUL | AUG | SEP | OCT | NOV | DEC | YEAR |
|---|---|---|---|---|---|---|---|---|---|---|---|---|---|
| Maximum Temp °F | 19.7 | 25.7 | 39.1 | 56.1 | 69.9 | 78.1 | 84.2 | 82.1 | 71.3 | 57.7 | 38.1 | 24.6 | 53.9 |
| Minimum Temp °F | -0.6 | 6.1 | 19.1 | 32.5 | 44.7 | 54.1 | 59.0 | 56.9 | 46.1 | 34.6 | 19.8 | 6.3 | 31.6 |
| Mean Temp °F | 9.6 | 15.9 | 29.1 | 44.3 | 57.3 | 66.2 | 71.6 | 69.5 | 58.7 | 46.2 | 29.0 | 15.5 | 42.7 |
| Days Max Temp ≥ 90 °F | 0 | 0 | 0 | 0 | 0 | 2 | 8 | 6 | 1 | 0 | 0 | 0 | 17 |
| Days Max Temp ≤ 32 °F | 24 | 18 | 9 | 1 | 0 | 0 | 0 | 0 | 0 | 0 | 10 | 22 | 84 |
| Days Min Temp ≤ 32 °F | 31 | 28 | 28 | 16 | 3 | 0 | 0 | 0 | 2 | 14 | 27 | 31 | 180 |
| Days Min Temp ≤ 0 °F | 17 | 10 | 3 | 0 | 0 | 0 | 0 | 0 | 0 | 0 | 2 | 11 | 43 |
| Heating Degree Days | 1716 | 1381 | 1106 | 616 | 261 | 66 | 14 | 33 | 225 | 579 | 1073 | 1531 | 8601 |
| Cooling Degree Days | 0 | 0 | 0 | 6 | 37 | 114 | 231 | 181 | 44 | 1 | 0 | 0 | 614 |
| Total Precipitation (") | 0.67 | 0.60 | 1.11 | 1.93 | 2.58 | 3.53 | 3.37 | 2.86 | 1.94 | 1.43 | 0.80 | 0.58 | 21.40 |
| Days ≥ 0.1" Precip | 2 | 2 | 3 | 5 | 5 | 7 | 6 | 5 | 4 | 3 | 3 | 2 | 47 |
| Total Snowfall (") | na | na | na | na | 0.1 | 0.0 | 0.0 | 0.0 | 0.1 | 0.3 | na | na | na |
| Days ≥ 1" Snow Depth | na | na | na | *1* | 0 | 0 | 0 | 0 | 0 | 0 | na | na | na |

## WENTWORTH 2 WNW *Lake County* ELEVATION 1703 ft LAT/LONG 44° 0' N / 96° 58' W

| | JAN | FEB | MAR | APR | MAY | JUN | JUL | AUG | SEP | OCT | NOV | DEC | YEAR |
|---|---|---|---|---|---|---|---|---|---|---|---|---|---|
| Maximum Temp °F | 23.9 | 29.5 | 42.7 | 59.3 | 71.6 | 80.0 | 84.9 | 82.7 | 73.4 | 60.6 | 41.6 | 28.2 | 56.5 |
| Minimum Temp °F | 3.3 | 9.2 | 22.0 | 34.5 | 46.3 | 55.9 | 60.6 | 58.3 | 48.5 | 36.7 | 22.8 | 9.6 | 34.0 |
| Mean Temp °F | 13.6 | 19.4 | 32.4 | 46.9 | 59.0 | 68.0 | 72.8 | 70.5 | 61.0 | 48.7 | 32.2 | 18.9 | 45.3 |
| Days Max Temp ≥ 90 °F | 0 | 0 | 0 | 0 | 0 | 3 | 8 | 6 | 2 | 0 | 0 | 0 | 19 |
| Days Max Temp ≤ 32 °F | 21 | 15 | 6 | 0 | 0 | 0 | 0 | 0 | 0 | 0 | 7 | 19 | 68 |
| Days Min Temp ≤ 32 °F | 31 | 28 | 26 | 14 | 2 | 0 | 0 | 0 | 1 | 11 | 26 | 31 | 170 |
| Days Min Temp ≤ 0 °F | 13 | 8 | 2 | 0 | 0 | 0 | 0 | 0 | 0 | 0 | 1 | 8 | 32 |
| Heating Degree Days | 1589 | 1282 | 1004 | 541 | 217 | 43 | 9 | 24 | 171 | 502 | 977 | 1423 | 7782 |
| Cooling Degree Days | 0 | 0 | 0 | 8 | 41 | 142 | 245 | 195 | 59 | 2 | 0 | 0 | 692 |
| Total Precipitation (") | 0.50 | 0.74 | 1.52 | 2.30 | 2.91 | 4.23 | 2.89 | 2.95 | 2.50 | 1.74 | 1.14 | 0.62 | 24.04 |
| Days ≥ 0.1" Precip | 2 | 2 | 4 | 6 | 6 | 7 | 5 | 5 | 5 | 4 | 3 | 2 | 51 |
| Total Snowfall (") | 6.4 | 7.1 | 7.5 | 2.4 | 0.0 | 0.0 | 0.0 | 0.0 | 0.0 | 1.1 | 6.4 | 6.9 | 37.8 |
| Days ≥ 1" Snow Depth | 25 | 20 | 11 | 1 | 0 | 0 | 0 | 0 | 0 | 0 | 8 | 19 | 84 |

## WESSINGTON SPRINGS *Jerauld County* ELEVATION 1640 ft LAT/LONG 44° 5' N / 98° 34' W

| | JAN | FEB | MAR | APR | MAY | JUN | JUL | AUG | SEP | OCT | NOV | DEC | YEAR |
|---|---|---|---|---|---|---|---|---|---|---|---|---|---|
| Maximum Temp °F | 25.9 | 31.2 | 43.9 | 59.4 | 71.6 | 80.8 | 86.8 | 85.4 | 75.4 | 61.4 | 42.3 | 30.0 | 57.8 |
| Minimum Temp °F | 7.0 | 12.5 | 23.9 | 35.9 | 47.6 | 57.5 | 62.8 | 60.8 | 50.9 | 39.2 | 24.8 | 12.6 | 36.3 |
| Mean Temp °F | 16.5 | 21.9 | 33.9 | 47.7 | 59.6 | 69.2 | 74.8 | 73.1 | 63.2 | 50.4 | 33.6 | 21.3 | 47.1 |
| Days Max Temp ≥ 90 °F | 0 | 0 | 0 | 0 | 1 | 4 | 11 | 10 | 3 | 0 | 0 | 0 | 29 |
| Days Max Temp ≤ 32 °F | 19 | 14 | 6 | 0 | 0 | 0 | 0 | 0 | 0 | 0 | 7 | 17 | 63 |
| Days Min Temp ≤ 32 °F | 30 | 27 | 24 | 11 | 1 | 0 | 0 | 0 | 1 | 7 | 23 | 30 | 154 |
| Days Min Temp ≤ 0 °F | 11 | 7 | 1 | 0 | 0 | 0 | 0 | 0 | 0 | 0 | 1 | 6 | 26 |
| Heating Degree Days | 1500 | 1212 | 957 | 522 | 204 | 30 | 5 | 12 | 133 | 455 | 936 | 1349 | 7315 |
| Cooling Degree Days | 0 | 0 | 0 | 10 | 53 | 171 | 310 | 276 | 92 | 9 | 0 | 0 | 921 |
| Total Precipitation (") | 0.36 | 0.58 | 1.78 | 2.56 | 3.36 | 3.50 | 2.83 | 2.32 | 2.03 | 1.58 | 0.96 | 0.46 | 22.32 |
| Days ≥ 0.1" Precip | 1 | 2 | 4 | 5 | 6 | 6 | 6 | 4 | 4 | 3 | 2 | 2 | 45 |
| Total Snowfall (") | 4.3 | 6.3 | 7.4 | 3.2 | 0.0 | 0.0 | 0.0 | 0.0 | 0.0 | 1.1 | 4.7 | 5.4 | 32.4 |
| Days ≥ 1" Snow Depth | 21 | 18 | 9 | 1 | 0 | 0 | 0 | 0 | 0 | 1 | 6 | 14 | 70 |

## WHITE LAKE *Aurora County* ELEVATION 1631 ft LAT/LONG 43° 44 ' N / 98° 43 ' W

| | JAN | FEB | MAR | APR | MAY | JUN | JUL | AUG | SEP | OCT | NOV | DEC | YEAR |
|---|---|---|---|---|---|---|---|---|---|---|---|---|---|
| Maximum Temp °F | 27.7 | 33.3 | 45.9 | 61.5 | 73.1 | 82.3 | 88.6 | 87.1 | 77.6 | 63.5 | 44.1 | 31.7 | 59.7 |
| Minimum Temp °F | 5.7 | 11.6 | 23.0 | 34.9 | 46.2 | 55.8 | 61.3 | 59.0 | 48.9 | 36.7 | 23.0 | 10.9 | 34.7 |
| Mean Temp °F | 16.7 | 22.5 | 34.5 | 48.2 | 59.7 | 69.1 | 74.9 | 73.1 | 63.3 | 50.1 | 33.6 | 21.3 | 47.3 |
| Days Max Temp ≥ 90 °F | 0 | 0 | 0 | 0 | 1 | 6 | 14 | 13 | 4 | 0 | 0 | 0 | 38 |
| Days Max Temp ≤ 32 °F | 18 | 13 | 5 | 0 | 0 | 0 | 0 | 0 | 0 | 0 | 5 | 15 | 56 |
| Days Min Temp ≤ 32 °F | 31 | 28 | 25 | 13 | 2 | 0 | 0 | 0 | 1 | 10 | 25 | 31 | 166 |
| Days Min Temp ≤ 0 °F | 12 | 7 | 1 | 0 | 0 | 0 | 0 | 0 | 0 | 0 | 1 | 7 | 28 |
| Heating Degree Days | 1492 | 1195 | 939 | 503 | 202 | 36 | 5 | 13 | 131 | 460 | 937 | 1347 | 7260 |
| Cooling Degree Days | 0 | 0 | 0 | 10 | 48 | 162 | 301 | 253 | 88 | 5 | 0 | 0 | 867 |
| Total Precipitation (") | 0.34 | 0.58 | 1.44 | 2.48 | 3.04 | 3.32 | 2.69 | 2.15 | 1.89 | 1.41 | 0.89 | 0.46 | 20.69 |
| Days ≥ 0.1" Precip | 1 | 2 | 3 | 5 | 6 | 6 | 5 | 4 | 3 | 3 | 2 | 2 | 42 |
| Total Snowfall (") | 4.4 | 6.7 | 5.6 | 1.8 | 0.0 | 0.0 | 0.0 | 0.0 | 0.0 | 0.9 | 4.5 | 5.3 | 29.2 |
| Days ≥ 1" Snow Depth | *18* | *15* | 8 | 1 | 0 | 0 | 0 | 0 | 0 | 0 | *4* | *12* | 58 |

## WINNER *Tripp County* ELEVATION 1972 ft LAT/LONG 43° 23 ' N / 99° 52 ' W

| | JAN | FEB | MAR | APR | MAY | JUN | JUL | AUG | SEP | OCT | NOV | DEC | YEAR |
|---|---|---|---|---|---|---|---|---|---|---|---|---|---|
| Maximum Temp °F | 32.4 | 38.2 | 49.3 | 62.8 | 74.3 | 83.6 | 90.4 | 88.7 | 79.1 | 65.5 | 46.4 | 35.8 | 62.2 |
| Minimum Temp °F | 10.6 | 15.6 | 24.9 | 36.3 | 47.3 | 57.2 | 62.6 | 60.6 | 50.5 | 39.0 | 25.6 | 15.1 | 37.1 |
| Mean Temp °F | 21.5 | 27.0 | 37.2 | 49.6 | 60.8 | 70.5 | 76.5 | 74.7 | 64.8 | 52.3 | 36.0 | 25.5 | 49.7 |
| Days Max Temp ≥ 90 °F | 0 | 0 | 0 | 1 | 2 | 8 | 17 | 15 | 6 | 1 | 0 | 0 | 50 |
| Days Max Temp ≤ 32 °F | 14 | 10 | 4 | 0 | 0 | 0 | 0 | 0 | 0 | 0 | 5 | 12 | 45 |
| Days Min Temp ≤ 32 °F | 29 | 26 | 23 | 11 | 2 | 0 | 0 | 0 | 1 | 8 | 23 | 29 | 152 |
| Days Min Temp ≤ 0 °F | 9 | 5 | 1 | 0 | 0 | 0 | 0 | 0 | 0 | 0 | 1 | 5 | 21 |
| Heating Degree Days | 1344 | 1068 | 857 | 469 | 183 | 29 | 5 | 9 | 115 | 399 | 863 | 1219 | 6560 |
| Cooling Degree Days | 0 | 0 | 0 | 18 | 66 | 201 | 357 | 311 | 123 | 10 | 0 | 0 | 1086 |
| Total Precipitation (") | 0.53 | 0.64 | 1.64 | 2.70 | 3.56 | 3.55 | 3.29 | 2.24 | 1.92 | 1.56 | 0.89 | 0.57 | 23.09 |
| Days ≥ 0.1" Precip | 2 | 2 | 4 | 6 | 7 | 6 | 7 | 5 | 4 | 3 | 3 | 2 | 51 |
| Total Snowfall (") | 7.6 | 7.6 | 8.6 | 3.7 | 0.2 | 0.0 | 0.0 | 0.0 | 0.2 | 1.5 | 6.9 | 7.6 | 43.9 |
| Days ≥ 1" Snow Depth | *19* | 11 | 7 | 1 | 0 | 0 | 0 | 0 | 0 | 0 | 4 | 12 | 54 |

## WOOD *Mellette County* ELEVATION 2182 ft LAT/LONG 43° 30 ' N / 100° 29 ' W

| | JAN | FEB | MAR | APR | MAY | JUN | JUL | AUG | SEP | OCT | NOV | DEC | YEAR |
|---|---|---|---|---|---|---|---|---|---|---|---|---|---|
| Maximum Temp °F | 32.9 | 38.3 | 48.9 | 61.5 | 72.8 | 82.6 | 90.0 | 89.0 | 79.1 | 65.4 | 47.6 | 36.5 | 62.1 |
| Minimum Temp °F | 9.4 | 14.1 | 23.4 | 35.0 | 45.3 | 55.7 | 61.3 | 59.5 | 49.1 | 37.3 | 24.7 | 13.8 | 35.7 |
| Mean Temp °F | 21.2 | 26.2 | 36.2 | 48.3 | 59.1 | 69.2 | 75.7 | 74.3 | 64.1 | 51.4 | 36.2 | 25.2 | 48.9 |
| Days Max Temp ≥ 90 °F | 0 | 0 | 0 | 0 | 1 | 7 | 17 | 16 | 6 | 1 | 0 | 0 | 48 |
| Days Max Temp ≤ 32 °F | 14 | 10 | 5 | 0 | 0 | 0 | 0 | 0 | 0 | 0 | 4 | 11 | 44 |
| Days Min Temp ≤ 32 °F | 30 | 26 | 25 | 13 | 3 | 0 | 0 | 0 | 1 | 9 | 24 | 30 | 161 |
| Days Min Temp ≤ 0 °F | 10 | 5 | 1 | 0 | 0 | 0 | 0 | 0 | 0 | 0 | 1 | 5 | 22 |
| Heating Degree Days | 1354 | 1090 | 887 | 503 | 220 | 39 | 7 | 11 | 131 | 425 | 857 | 1228 | 6752 |
| Cooling Degree Days | *0* | *0* | *0* | *12* | *43* | *159* | *312* | *296* | *112* | *8* | *1* | *0* | 943 |
| Total Precipitation (") | 0.37 | 0.48 | 1.20 | 2.10 | 2.99 | 2.93 | 2.96 | 1.86 | 1.46 | 1.36 | 0.65 | 0.53 | 18.89 |
| Days ≥ 0.1" Precip | 1 | 1 | *3* | *5* | *6* | 6 | 6 | 4 | 3 | 3 | 2 | 2 | 42 |
| Total Snowfall (") | na | *6.2* | *7.5* | *3.6* | *0.2* | 0.0 | 0.0 | 0.0 | 0.0 | *1.5* | *4.8* | na | na |
| Days ≥ 1" Snow Depth | na | *11* | *7* | *1* | *0* | 0 | 0 | 0 | 0 | *1* | *5* | na | na |

## YANKTON 2 E *Yankton County* ELEVATION 1283 ft LAT/LONG 42° 55 ' N / 97° 23 ' W

| | JAN | FEB | MAR | APR | MAY | JUN | JUL | AUG | SEP | OCT | NOV | DEC | YEAR |
|---|---|---|---|---|---|---|---|---|---|---|---|---|---|
| Maximum Temp °F | 27.6 | 32.9 | 45.6 | 60.7 | 72.8 | 82.1 | 87.3 | 85.1 | 75.8 | 63.4 | 45.1 | 32.7 | 59.3 |
| Minimum Temp °F | 6.1 | 11.5 | 22.9 | 35.3 | 47.1 | 57.2 | 62.1 | 59.5 | 49.4 | 37.2 | 24.2 | 12.5 | 35.4 |
| Mean Temp °F | 16.9 | 22.2 | 34.3 | 48.0 | 60.0 | 69.6 | 74.7 | 72.3 | 62.6 | 50.3 | 34.6 | 22.7 | 47.4 |
| Days Max Temp ≥ 90 °F | 0 | 0 | 0 | 1 | 1 | 6 | 12 | 9 | 4 | 0 | 0 | 0 | 33 |
| Days Max Temp ≤ 32 °F | 18 | 14 | 6 | 0 | 0 | 0 | 0 | 0 | 0 | 0 | 5 | 14 | 57 |
| Days Min Temp ≤ 32 °F | 31 | 28 | 26 | 12 | 1 | 0 | 0 | 0 | 1 | 9 | 25 | 31 | 164 |
| Days Min Temp ≤ 0 °F | 11 | 7 | 1 | 0 | 0 | 0 | 0 | 0 | 0 | 0 | 1 | 5 | 25 |
| Heating Degree Days | 1488 | 1204 | 946 | 512 | 198 | 34 | 6 | 14 | 147 | 454 | 905 | 1306 | 7214 |
| Cooling Degree Days | 0 | 0 | 0 | 15 | 54 | 196 | 314 | 259 | 95 | 6 | 0 | 0 | 939 |
| Total Precipitation (") | 0.41 | 0.51 | 1.59 | 2.17 | 3.64 | 3.99 | 3.20 | 2.74 | 2.37 | 1.78 | 1.10 | 0.56 | 24.06 |
| Days ≥ 0.1" Precip | 1 | 2 | 3 | 5 | 6 | 6 | 5 | 5 | 5 | 4 | 2 | 1 | 45 |
| Total Snowfall (") | 5.5 | 5.4 | 4.7 | 1.9 | 0.0 | 0.0 | 0.0 | 0.0 | 0.0 | 0.4 | 5.3 | 5.7 | 28.9 |
| Days ≥ 1" Snow Depth | 19 | 14 | 8 | 1 | 0 | 0 | 0 | 0 | 0 | 0 | 4 | 14 | 60 |

## ZEONA 10 SSW *Butte County* ELEVATION 2731 ft LAT/LONG 45° 4' N / 103° 0' W

| | JAN | FEB | MAR | APR | MAY | JUN | JUL | AUG | SEP | OCT | NOV | DEC | YEAR |
|---|---|---|---|---|---|---|---|---|---|---|---|---|---|
| Maximum Temp °F | *27.3* | *32.6* | *44.2* | *57.4* | *68.7* | 78.7 | 86.8 | 86.4 | 75.5 | 61.0 | 43.4 | 31.8 | 57.8 |
| Minimum Temp °F | *4.2* | *9.4* | *20.0* | *31.0* | *41.9* | 51.5 | 56.9 | 54.2 | 43.0 | 30.2 | 18.1 | 7.3 | 30.6 |
| Mean Temp °F | *15.8* | *21.0* | *32.1* | *44.2* | *55.3* | 65.1 | 71.9 | 70.3 | 59.3 | 45.6 | 30.8 | 19.6 | 44.3 |
| Days Max Temp ≥ 90 °F | 0 | *0* | *0* | *0* | *1* | 4 | *12* | 12 | *4* | 0 | 0 | 0 | 33 |
| Days Max Temp ≤ 32 °F | *17* | *13* | *6* | *1* | *0* | 0 | 0 | 0 | 0 | 1 | *7* | 15 | 60 |
| Days Min Temp ≤ 32 °F | *31* | *28* | *28* | *17* | *5* | 0 | 0 | 0 | 3 | 18 | *28* | 31 | 189 |
| Days Min Temp ≤ 0 °F | *12* | *8* | *2* | *0* | *0* | 0 | 0 | 0 | 0 | 0 | 2 | *8* | 32 |
| Heating Degree Days | *1521* | *1238* | *1013* | *618* | *314* | 86 | 17 | 27 | 210 | 595 | 1026 | 1402 | 8067 |
| Cooling Degree Days | 0 | 0 | 0 | 2 | 25 | 108 | 251 | 209 | 47 | 1 | 0 | 0 | 643 |
| Total Precipitation (") | 0.31 | 0.49 | 0.82 | 1.57 | 2.54 | 3.03 | 2.15 | 1.45 | 0.97 | 0.98 | 0.51 | 0.46 | 15.28 |
| Days ≥ 0.1" Precip | 1 | 2 | *2* | *4* | 5 | 6 | 4 | 3 | 3 | 2 | 2 | 2 | 36 |
| Total Snowfall (") | *4.8* | *6.3* | *7.0* | *4.9* | 1.2 | 0.0 | 0.0 | 0.0 | 0.2 | *0.8* | 4.6 | 6.4 | 36.2 |
| Days ≥ 1" Snow Depth | *14* | *11* | *8* | *2* | 0 | 0 | 0 | 0 | 0 | 0 | 3 | *11* | 49 |

## JANUARY MINIMUM TEMPERATURE °F

### LOWEST

| | | |
|---|---|---|
| 1 | Columbia | -2.2 |
| 2 | Mellette | -1.7 |
| 3 | Britton | -1.4 |
| 4 | Summit | -1.2 |
| 5 | Pollock | -0.9 |
| | Redfield | -0.9 |
| 7 | Brookings | -0.8 |
| | Eureka | -0.8 |
| 9 | Castlewood | -0.7 |
| 10 | Milbank | -0.6 |
| | Webster | -0.6 |
| 12 | Flandreau | -0.5 |
| | Ipswich | -0.5 |
| 14 | Watertown | -0.3 |
| 15 | Waubay | -0.2 |
| 16 | Aberdeen | -0.1 |
| 17 | Leola | 0.0 |
| 18 | Harrold | 0.5 |
| 19 | Clark | 0.7 |
| | Selby | 0.7 |
| 21 | Sisseton | 1.1 |
| | Usta | 1.1 |
| 23 | Clear Lake | 1.7 |
| | Highmore-23 N | 1.7 |
| 25 | Gettysburg | 1.9 |

### HIGHEST

| | | |
|---|---|---|
| 1 | Mt. Rushmore | 16.1 |
| 2 | Lead | 14.3 |
| 3 | Spearfish | 12.7 |
| 4 | Fort Meade | 12.5 |
| 5 | Rapid City | 11.9 |
| 6 | Interior | 11.8 |
| 7 | Deadwood | 11.4 |
| | Hot Springs | 11.4 |
| 9 | Rapid City-Reginl | 10.9 |
| 10 | Oelrichs | 10.7 |
| 11 | Winner | 10.6 |
| 12 | Long Valley | 10.5 |
| 13 | Belle Fourche | 9.4 |
| | Cedar Butte | 9.4 |
| | Custer | 9.4 |
| | Wood | 9.4 |
| 17 | Wagner | 8.9 |
| 18 | Pickstown | 8.7 |
| | Porcupine | 8.7 |
| 20 | Pactola Dam | 8.5 |
| 21 | Mission | 8.2 |
| | Wasta | 8.2 |
| 23 | Murdo | 7.9 |
| 24 | Gregory | 7.7 |
| 25 | Armour | 7.6 |

## JULY MAXIMUM TEMPERATURE °F

### HIGHEST

| | | |
|---|---|---|
| 1 | Interior | 91.1 |
| 2 | Cedar Butte | 91.0 |
| 3 | Kennebec | 90.9 |
| 4 | Midland | 90.8 |
| 5 | Winner | 90.4 |
| 6 | Oelrichs | 90.3 |
| | Wagner | 90.3 |
| 8 | Wood | 90.0 |
| 9 | Porcupine | 89.5 |
| | Wasta | 89.5 |
| 11 | Cottonwood | 89.4 |
| 12 | Philip | 89.3 |
| 13 | Academy | 89.1 |
| | Murdo | 89.1 |
| | Pierre | 89.1 |
| 16 | Ardmore | 89.0 |
| | Harrold | 89.0 |
| 18 | Gregory | 88.9 |
| | Hot Springs | 88.9 |
| | Long Valley | 88.9 |
| | Onida | 88.9 |
| 22 | Armour | 88.6 |
| | Milesville | 88.6 |
| | Oahe Dam | 88.6 |
| | Vermillion | 88.6 |

### LOWEST

| | | |
|---|---|---|
| 1 | Pactola Dam | 78.9 |
| 2 | Lead | 79.3 |
| 3 | Custer | 79.5 |
| 4 | Deadwood | 81.0 |
| 5 | Mt. Rushmore | 81.3 |
| 6 | Brookings | 82.0 |
| 7 | Summit | 82.1 |
| 8 | Flandreau | 83.1 |
| 9 | Clear Lake | 83.2 |
| 10 | Watertown | 83.5 |
| 11 | Clark | 83.6 |
| 12 | Waubay | 83.7 |
| 13 | Madison | 83.8 |
| 14 | Lemmon | 83.9 |
| 15 | Webster | 84.2 |
| 16 | Spearfish | 84.6 |
| 17 | Mobridge | 84.7 |
| 18 | Milbank | 84.8 |
| 19 | Castlewood | 84.9 |
| | Columbia | 84.9 |
| | Redig | 84.9 |
| | Wentworth | 84.9 |
| 23 | Sisseton | 85.0 |
| 24 | Aberdeen | 85.1 |
| | Eureka | 85.1 |

## ANNUAL PRECIPITATION (")

### HIGHEST

| | | |
|---|---|---|
| 1 | Lead | 28.63 |
| 2 | Deadwood | 27.98 |
| 3 | Bonesteel | 25.72 |
| 4 | Clear Lake | 25.37 |
| 5 | Vermillion | 25.36 |
| 6 | Marion | 25.21 |
| 7 | Centerville | 25.03 |
| 8 | Wagner | 24.95 |
| 9 | Gregory | 24.65 |
| 10 | Sioux Falls | 24.60 |
| 11 | Madison | 24.48 |
| 12 | Bridgewater | 24.14 |
| 13 | Yankton | 24.06 |
| 14 | Wentworth | 24.04 |
| 15 | Canton | 23.73 |
| 16 | Menno | 23.67 |
| 17 | Tyndall | 23.62 |
| 18 | De Smet | 23.43 |
| 19 | Castlewood | 23.32 |
| 20 | Flandreau | 23.11 |
| 21 | Winner | 23.09 |
| 22 | Brookings | 23.01 |
| 23 | Armour | 22.93 |
| | Howard | 22.93 |
| 25 | Watertown | 22.46 |

### LOWEST

| | | |
|---|---|---|
| 1 | Camp Crook | 13.94 |
| 2 | Oahe Dam | 14.08 |
| 3 | Usta | 14.72 |
| 4 | Newell | 14.84 |
| 5 | Redig | 15.22 |
| 6 | Zeona | 15.28 |
| 7 | Ralph | 15.69 |
| 8 | Wasta | 16.14 |
| 9 | Hot Springs | 16.15 |
| 10 | Ardmore | 16.33 |
| | Ludlow | 16.33 |
| | McIntosh | 16.33 |
| 13 | Dupree-15 SSE | 16.40 |
| 14 | Interior | 16.41 |
| 15 | Rapid City-Reginl | 16.42 |
| 16 | Porcupine | 16.48 |
| 17 | Philip | 16.50 |
| 18 | Midland | 16.52 |
| 19 | Faith | 16.69 |
| 20 | Mobridge | 16.71 |
| 21 | Dupree | 16.90 |
| 22 | Belle Fourche | 16.91 |
| 23 | Pollock | 16.95 |
| 24 | Cottonwood | 17.02 |
| 25 | Oelrichs | 17.06 |

## ANNUAL SNOWFALL (")

### HIGHEST

| | | |
|---|---|---|
| 1 | Lead | 180.8 |
| 2 | Deadwood | 113.9 |
| 3 | Spearfish | 67.9 |
| 4 | Mt. Rushmore | 55.9 |
| 5 | Clear Lake | 51.7 |
| 6 | Milesville | 51.1 |
| 7 | Redig | 49.5 |
| 8 | Glad Valley | 48.8 |
| 9 | Harrington | 46.7 |
| 10 | Ardmore | 45.9 |
| | Oelrichs | 45.9 |
| 12 | Gregory | 45.7 |
| 13 | Harrold | 45.6 |
| 14 | Pactola Dam | 45.4 |
| 15 | Summit | 43.9 |
| | Winner | 43.9 |
| 17 | Eureka | 42.1 |
| 18 | Rapid City-Reginl | 41.9 |
| 19 | Huron | 41.3 |
| 20 | Faith | 40.8 |
| 21 | Belle Fourche | 40.7 |
| 22 | Bonesteel | 40.4 |
| 23 | Highmore-1 W | 40.1 |
| | Leola | 40.1 |
| 25 | Columbia | 40.0 |

### LOWEST

| | | |
|---|---|---|
| 1 | Interior | 21.6 |
| 2 | Pickstown | 22.2 |
| 3 | Gann Valley | 27.0 |
| 4 | Castlewood | 27.8 |
| 5 | Alexandria | 28.2 |
| | Usta | 28.2 |
| 7 | Yankton | 28.9 |
| 8 | White Lake | 29.2 |
| 9 | Philip | 29.6 |
| 10 | Canton | 30.1 |
| | Pierre | 30.1 |
| | Porcupine | 30.1 |
| 13 | Selby | 30.2 |
| 14 | Watertown | 30.4 |
| 15 | Centerville | 30.5 |
| 16 | Flandreau | 30.9 |
| | Kennebec | 30.9 |
| 18 | Forestburg | 31.6 |
| 19 | Vermillion | 31.8 |
| 20 | Ipswich | 31.9 |
| | Redfield | 31.9 |
| 22 | Camp Crook | 32.3 |
| 23 | Armour | 32.4 |
| | Wessington Sprgs | 32.4 |
| 25 | Faulkton | 32.8 |

# TENNESSEE

PHYSICAL FEATURES. The topography of Tennessee is quite varied, stretching from the lowlands of the Mississippi Valley to the mountain peaks in the east. The westernmost part of the State, between the bluffs overlooking the Mississippi River and the western valley of the Tennessee River, is a region of gently rolling plains sloping gradually from 200 to 250 feet in the west to about 600 feet above sea level in the hills overlooking the Tennessee River. The hilly Highland Rim, in a wide circle touching the Tennessee River Valley on the west and the Cumberland Plateau on the east, together with the enclosed Central Basin makes up the whole of Middle Tennessee. The Highland Rim ranges from about 600 feet in elevation along the Tennessee River to 1,000 feet in the east and rises 300 to 400 feet above the Central Basin. The Cumberland Plateau, with an average elevation of 2,000 feet above sea level, extends roughly northeast-southwest across the State in a belt 30 to 50 miles wide, being bounded on the west by the Highland Rim and overlooking the Great Valley of East Tennessee on the east. The Great Valley, paralleling the Plateau to the west and the Great Smoky Mountains to the east, is a funnel-shaped valley varying in width from about 30 miles in the south to about 90 miles in the north. Within the valley, which slopes from 1,500 feet in the north to 700 feet above sea level in the south, are a series of northeast-southwest ridges. Along the Tennessee-North Carolina border lie the Great Smoky Mountains, the most rugged and elevated portion of Tennessee, with numerous peaks from 4,000 to 6,000 feet above sea level.

Tennessee, except for a small area east of Chattanooga, lies entirely within the drainage of the Mississippi River system. The extreme western section of the State is drained through several relatively small rivers directly into the Mississippi River. Otherwise drainage is into either the Cumberland or Tennessee Rivers, both of which flow northward near the end of their courses to join the Ohio River along the Kentucky-Illinois border.

TEMPERATURE. Most aspects of the State's climate are related to the widely varying topography within its borders. The decrease of temperature with elevation is quite apparent, amounting to, on the average, 3° F. per 1,000 feet increase in elevation. Thus higher portions of the State, such as the Cumberland Plateau and the mountains of the east, have lower average temperatures than the Great Valley of East Tennessee, which they flank, and other lower parts of the State. In the Great Valley temperature increases from north to south. Across the State, the average annual temperature varies from over 62° F. in the extreme southwest to near 45° F. atop the highest peaks of the east. While most of the State can be described as having a warm, humid summer and a mild winter, this must be qualified to include variations with elevation. Thus with increasing elevations, summers become cooler and more pleasant while winters become colder and more blustery.

Length of growing season (freeze-free period) is linked to topography in a way similar to temperature, varying from 237 days at low-lying Memphis to near 130 days on the highest mountains in the east. Most of the State is included in the range 180 to 220 days.

PRECIPITATION. Since the principal source of moist air for this area is the Gulf region, there exists a gradual decrease of average precipitation from south to north. This effect is largely obscured, however, by the overruling influence of topography. Air forced to ascend cools and condenses out a portion of its moisture charge; thus average precipitation is generally greater at higher elevations. This is apparent in all parts of the State. In West Tennessee average annual precipitation ranges from 46 to 54 inches, increasing from Mississippi bottomlands to the slight hills farther east. In Middle Tennessee the variation is from a minimum of 45 inches in the Central Basin to 50 to 55 inches in the surrounding hilly Highland Rim. Over the elevated Cumberland Plateau average annual precipitation is generally from 50 to 55 inches. In contrast, average annual precipitation in the Great Valley of East Tennessee increases from near 40 inches in northern portions to over 50 inches in the south. The northern minimum, lowest for the entire State, results from the shielding influence of the Great Smoky Mountains to the southeast and the Cumberland Plateau to the northwest. The mountainous eastern border of the State is its wettest part, having average annual precipitation ranging up to about 80 inches on the higher, well-exposed peaks of the Smokies.

Over most of the State greatest precipitation occurs during the winter and early spring due to the more frequent passage of large scale storms over and near the State during those months. A secondary maximum of precipitation occurs in midsummer in response to shower and thundershower activity. This is especially pronounced in the mountains of the east where July rainfall exceeds the precipitation of any other month. Lightest precipitation, observed in the fall, is brought about by the maximum occurrence of slow moving, rain suppressing high pressure areas during that season. Although all parts of Tennessee are generally well supplied with precipitation, there occurs on the average one or more prolonged dry spells each year during summer and fall.

The most important flood season is during the winter and early spring (December through March) when the frequent migratory storms bring general rains of high intensity. During this period both widespread flooding and local flash floods can occur. During summer, heavy thunderstorm rains frequently result in local flash flooding. In the fall, while flood producing rains are rare, decadent hurricanes on occasion cause serious floods in the east.

Average snowfall varies from 4 to 6 inches in southern and western parts of the State and in most of the Great Valley of East Tennessee, to more than 10 inches over the northern Cumberland Plateau and the mountains of the east. Over most of the State, due to relatively mild winter temperatures, a snow cover rarely persists for more than a few days.

STORMS. Severe storms are relatively infrequent in Tennessee, being east of the center of tornado activity, south of most blizzard conditions, and too far inland to be often affected by hurricanes. On the average 4 or 5 tornadoes are observed in the State each year, with greatest frequency in March, when 1 or 2 usually occur. Tornado occurrence is not evenly distributed throughout the State, being largely confined to areas west of the Cumberland Plateau. Annual expectancy is less than in bordering locations to the south and west. Damage from tropical storms is rare, occurring only about once every 18 years. Hailstorms at a given locality are observed 2 or 3 times a year and damaging glaze storms occur in the State every 5 or 6 years. Thunderstorms are frequent and severe thunderstorms with damaging winds are experienced at scattered locations throughout the State each year during the warm season.

## COUNTY INDEX

***Anderson County***
NORRIS
OAK RIDGE

***Bedford County***
SHELBYVILLE 3

***Blount County***
KNOXVILLE MCG TYSON

***Bradley County***
CLEVELAND FILTER PLT

***Cannon County***
WOODBURY 1 WNW

***Cheatham County***
KINGSTON SPRINGS

***Cocke County***
NEWPORT 1 NW

***Coffee County***
TULLAHOMA

***Cumberland County***
CROSSVILLE EXP STN
CROSSVILLE MEMORIAL

***Davidson County***
NASHVILLE METRO AP

***Dickson County***
DICKSON

***Dyer County***
DYERSBURG MUNI AP

***Fayette County***
MOSCOW

***Fentress County***
ALLARDT

***Giles County***
PULASKI WATER PLANT

***Greene County***
GREENEVILLE EXP STN

***Grundy County***
MONTEAGLE

***Hamilton County***
CHATTANOOGA LOVELL

***Hardeman County***
BOLIVAR WATERWORKS

***Hardin County***
SAVANNAH 6 SW

***Hawkins County***
ROGERSVILLE 1 NE

***Haywood County***
BROWNSVILLE

***Henderson County***
LEXINGTON

***Henry County***
PARIS 2 NW

***Hickman County***
CENTERVILLE WATER PL

***Lauderdale County***
RIPLEY

***Lawrence County***
LAWRENCEBURG FLT PLT

***Loudon County***
LENOIR CITY

***McMinn County***
ATHENS

***McNairy County***
SELMER

***Macon County***
LAFAYETTE

***Madison County***
JACKSON EXP STN
JACKSON MCKELLAR-SIP

***Marshall County***
LEWISBURG EXP STN

***Maury County***
COLUMBIA 3 WNW

***Montgomery County***
CLARKSVILLE SWG PLT

***Obion County***
SAMBURG WILDLIFE REF
UNION CITY

***Overton County***
LIVINGSTON RADIO WLI

***Perry County***
LINDEN 2

***Polk County***
COPPERHILL

***Putnam County***
COOKEVILLE

***Rhea County***
DAYTON

***Robertson County***
SPRINGFIELD EXP STN

***Rutherford County***
MURFREESBORO 5 N

***Scott County***
ONEIDA

***Sevier County***
GATLINBURG 2 SW
SEVIERVILLE 1 SE

***Shelby County***
MEMPHIS INTL ARPT

***Stewart County***
DOVER 1 W

***Sullivan County***
BRISTOL TRI CITY AP
KINGSPORT

***Sumner County***
PORTLAND SEWAGE PLAN

***Tipton County***
COVINGTON 1 W

***Warren County***
MC MINNVILLE

***Wayne County***
WAYNESBORO

***Weakley County***
DRESDEN
MARTIN UNIV OF TENN

***White County***
SPARTA

***Williamson County***
FRANKLIN SEWAGE PLT

***Wilson County***
LEBANON 3 W

## ELEVATION INDEX

| FEET | STATION NAME |
|---|---|
| 258 | MEMPHIS INTL ARPT |
| 308 | SAMBURG WILDLIFE REF |
| 312 | COVINGTON 1 W |
| 331 | RIPLEY |
| 338 | DYERSBURG MUNI AP |
| 341 | BOLIVAR WATERWORKS |
| 341 | UNION CITY |
| 351 | MOSCOW |
| 361 | BROWNSVILLE |
| 390 | MARTIN UNIV OF TENN |
| 400 | JACKSON EXP STN |
| 420 | DRESDEN |
| 420 | JACKSON MCKELLAR-SIP |
| 420 | SAVANNAH 6 SW |
| 449 | KINGSTON SPRINGS |
| 470 | SELMER |
| 489 | DOVER 1 W |
| 502 | CLARKSVILLE SWG PLT |
| 502 | LINDEN 2 |
| 531 | LEXINGTON |
| 535 | LEBANON 3 W |
| 551 | MURFREESBORO 5 N |
| 580 | PARIS 2 NW |
| 581 | NASHVILLE METRO AP |
| 630 | CENTERVILLE WATER PL |
| 630 | FRANKLIN SEWAGE PLT |
| 640 | PORTLAND SEWAGE PLAN |
| 650 | PULASKI WATER PLANT |
| 669 | COLUMBIA 3 WNW |
| 689 | CHATTANOOGA LOVELL |
| 722 | SHELBYVILLE 3 |
| 751 | SPRINGFIELD EXP STN |
| 751 | WAYNESBORO |
| 751 | WOODBURY 1 WNW |
| 781 | LENOIR CITY |
| 791 | LEWISBURG EXP STN |
| 800 | CLEVELAND FILTER PLT |
| 801 | DICKSON |
| 830 | DAYTON |
| 902 | MC MINNVILLE |
| 902 | SPARTA |
| 905 | KNOXVILLE MCG TYSON |
| 905 | OAK RIDGE |
| 919 | ATHENS |
| 932 | SEVIERVILLE 1 SE |
| 951 | LAWRENCEBURG FLT PLT |
| 991 | LAFAYETTE |
| 1020 | LIVINGSTON RADIO WLI |
| 1070 | TULLAHOMA |
| 1089 | ROGERSVILLE 1 NE |
| 1102 | NEWPORT 1 NW |
| 1110 | NORRIS |
| 1152 | COOKEVILLE |
| 1280 | KINGSPORT |
| 1319 | GREENEVILLE EXP STN |
| 1450 | GATLINBURG 2 SW |
| 1450 | ONEIDA |
| 1565 | BRISTOL TRI CITY AP |
| 1621 | COPPERHILL |
| 1670 | ALLARDT |
| 1811 | CROSSVILLE EXP STN |
| 1870 | CROSSVILLE MEMORIAL |
| 1926 | MONTEAGLE |

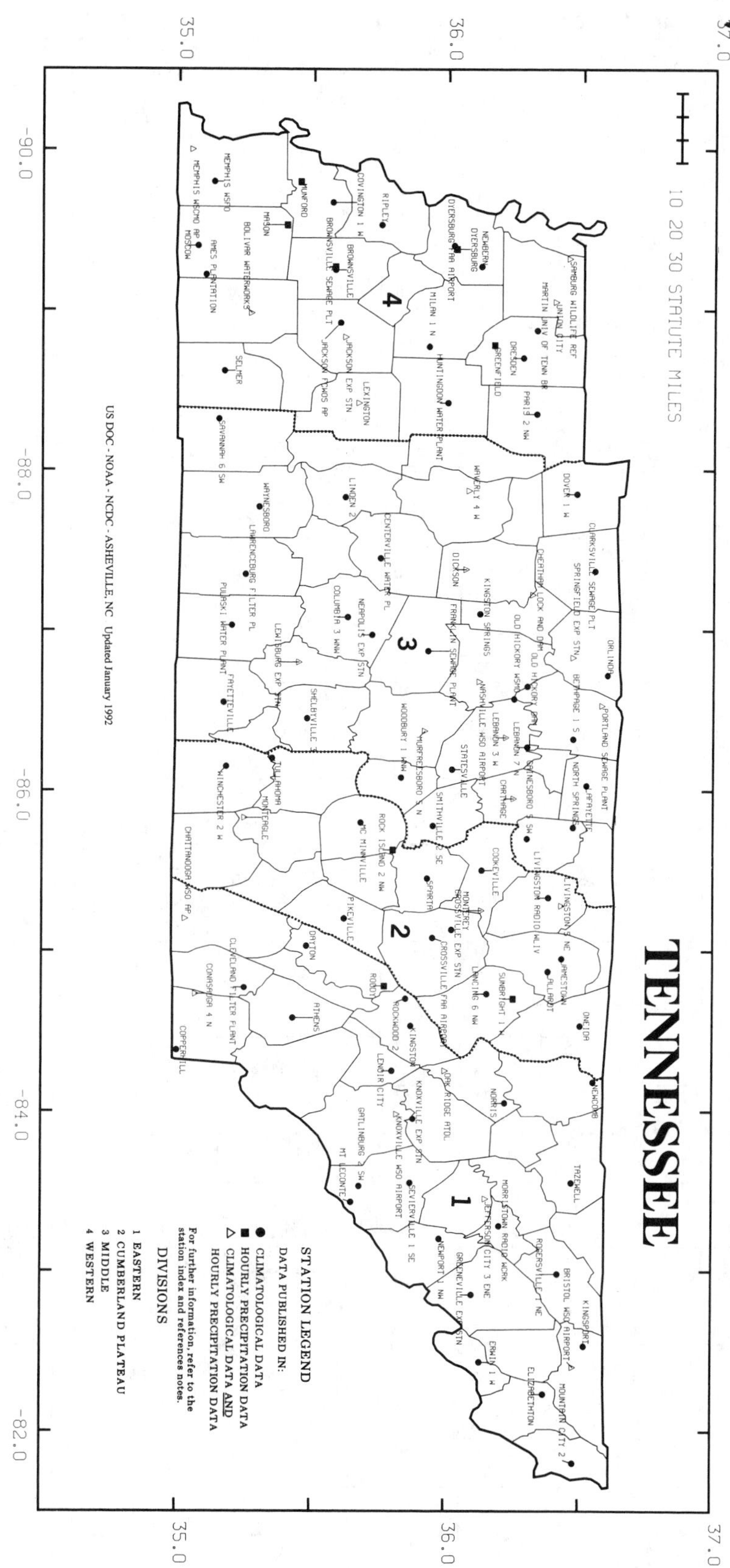
TENNESSEE
10 20 30 STATUTE MILES
US DOC - NOAA - NCDC - ASHEVILLE, NC Updated January 1992
STATION LEGEND
DATA PUBLISHED IN:
● CLIMATOLOGICAL DATA
■ HOURLY PRECIPITATION DATA
△ CLIMATOLOGICAL DATA AND HOURLY PRECIPITATION DATA
For further information, refer to the station index and reference notes.
DIVISIONS
1 EASTERN
2 CUMBERLAND PLATEAU
3 MIDDLE
4 WESTERN
35.0
36.0
37.0
-90.0
-88.0
-86.0
-84.0
-82.0

## ALLARDT *Fentress County* ELEVATION 1670 ft LAT/LONG 36° 23 ' N / 84° 53 ' W

| | JAN | FEB | MAR | APR | MAY | JUN | JUL | AUG | SEP | OCT | NOV | DEC | YEAR |
|---|---|---|---|---|---|---|---|---|---|---|---|---|---|
| Maximum Temp °F | 43.6 | 47.8 | 57.5 | 67.9 | 74.0 | 81.2 | 84.2 | 82.9 | 77.0 | 67.3 | 56.8 | 47.6 | 65.6 |
| Minimum Temp °F | 25.4 | 27.7 | 35.8 | 44.5 | 51.7 | 59.7 | 64.0 | 62.5 | 56.7 | 45.1 | 36.9 | 29.5 | 45.0 |
| Mean Temp °F | 34.5 | 37.8 | 46.7 | 56.2 | 62.9 | 70.5 | 74.1 | 72.8 | 66.9 | 56.2 | 46.9 | 38.6 | 55.3 |
| Days Max Temp ≥ 90 °F | 0 | 0 | 0 | 0 | 0 | 1 | 4 | 2 | 1 | 0 | 0 | 0 | 8 |
| Days Max Temp ≤ 32 °F | 5 | 3 | 1 | 0 | 0 | 0 | 0 | 0 | 0 | 0 | 0 | 2 | 11 |
| Days Min Temp ≤ 32 °F | 22 | 19 | 13 | 4 | 0 | 0 | 0 | 0 | 0 | 3 | 11 | 20 | 92 |
| Days Min Temp ≤ 0 °F | 1 | 0 | 0 | 0 | 0 | 0 | 0 | 0 | 0 | 0 | 0 | 0 | 1 |
| Heating Degree Days | 938 | 762 | 562 | 273 | 113 | 11 | 1 | 2 | 54 | 277 | 537 | 811 | 4341 |
| Cooling Degree Days | 0 | 0 | 2 | 16 | 56 | 196 | 310 | 257 | 125 | 15 | 1 | 0 | 978 |
| Total Precipitation (") | 4.44 | 4.25 | 5.28 | 4.05 | 5.24 | 4.70 | 5.25 | 4.54 | 3.91 | 3.25 | 4.36 | 5.05 | 54.32 |
| Days ≥ 0.1" Precip | 8 | 8 | 9 | 8 | 9 | 7 | 8 | 7 | 6 | 6 | 8 | 9 | 93 |
| Total Snowfall (") | 5.8 | 5.6 | 2.6 | 0.3 | 0.0 | 0.0 | 0.0 | 0.0 | 0.0 | 0.1 | 1.0 | 2.4 | 17.8 |
| Days ≥ 1" Snow Depth | 6 | 5 | 1 | 0 | 0 | 0 | 0 | 0 | 0 | 0 | 0 | 2 | 14 |

## ATHENS *McMinn County* ELEVATION 919 ft LAT/LONG 35° 26 ' N / 84° 34 ' W

| | JAN | FEB | MAR | APR | MAY | JUN | JUL | AUG | SEP | OCT | NOV | DEC | YEAR |
|---|---|---|---|---|---|---|---|---|---|---|---|---|---|
| Maximum Temp °F | 46.4 | 51.3 | 60.9 | 70.9 | 77.8 | 85.4 | 88.4 | 87.2 | 81.4 | 71.1 | 60.4 | 50.7 | 69.3 |
| Minimum Temp °F | 25.2 | 27.5 | 35.7 | 44.0 | 52.3 | 60.8 | 65.3 | 64.2 | 57.8 | 44.2 | 36.0 | 28.9 | 45.2 |
| Mean Temp °F | 35.9 | 39.4 | 48.3 | 57.5 | 65.1 | 73.2 | 76.9 | 75.7 | 69.6 | 57.7 | 48.3 | 39.9 | 57.3 |
| Days Max Temp ≥ 90 °F | 0 | 0 | 0 | 0 | 1 | 6 | 14 | 10 | 3 | 0 | 0 | 0 | 34 |
| Days Max Temp ≤ 32 °F | 3 | 2 | 0 | 0 | 0 | 0 | 0 | 0 | 0 | 0 | 0 | 1 | 6 |
| Days Min Temp ≤ 32 °F | 23 | 20 | 13 | 4 | 0 | 0 | 0 | 0 | 0 | 3 | 12 | 21 | 96 |
| Days Min Temp ≤ 0 °F | 1 | 0 | 0 | 0 | 0 | 0 | 0 | 0 | 0 | 0 | 0 | 0 | 1 |
| Heating Degree Days | 897 | 716 | 515 | 243 | 81 | 6 | 0 | 0 | 29 | 239 | 497 | 773 | 3996 |
| Cooling Degree Days | 0 | 0 | 4 | 22 | 95 | 273 | 398 | 352 | 180 | 24 | 2 | 0 | 1350 |
| Total Precipitation (") | 5.56 | 4.91 | 6.21 | 4.65 | 4.77 | 3.76 | 4.99 | 4.20 | 4.68 | 3.64 | 4.71 | 5.64 | 57.72 |
| Days ≥ 0.1" Precip | 9 | 8 | 9 | 8 | 8 | 7 | 9 | 7 | 7 | 6 | 7 | 8 | 93 |
| Total Snowfall (") | *3.1* | *2.1* | 1.0 | 0.4 | 0.0 | 0.0 | 0.0 | 0.0 | 0.0 | 0.0 | 0.1 | 0.2 | 6.9 |
| Days ≥ 1" Snow Depth | na | *1* | 0 | 0 | 0 | 0 | 0 | 0 | 0 | 0 | 0 | 0 | na |

## BOLIVAR WATERWORKS *Hardeman County* ELEVATION 341 ft LAT/LONG 35° 16 ' N / 88° 57 ' W

| | JAN | FEB | MAR | APR | MAY | JUN | JUL | AUG | SEP | OCT | NOV | DEC | YEAR |
|---|---|---|---|---|---|---|---|---|---|---|---|---|---|
| Maximum Temp °F | 47.0 | 51.9 | 61.7 | 72.0 | 78.8 | 86.7 | 90.0 | 88.7 | 82.8 | 72.7 | 61.5 | 51.6 | 70.5 |
| Minimum Temp °F | 27.2 | 30.2 | 38.9 | 48.2 | 56.1 | 64.5 | 68.5 | 66.2 | 59.5 | 46.5 | 38.4 | 31.3 | 48.0 |
| Mean Temp °F | 37.1 | 41.0 | 50.3 | 60.1 | 67.5 | 75.6 | 79.3 | 77.5 | 71.2 | 59.6 | 50.0 | 41.5 | 59.2 |
| Days Max Temp ≥ 90 °F | 0 | 0 | 0 | 0 | 1 | 10 | 18 | 14 | 6 | 0 | 0 | 0 | 49 |
| Days Max Temp ≤ 32 °F | 4 | 2 | 0 | 0 | 0 | 0 | 0 | 0 | 0 | 0 | 0 | 2 | 8 |
| Days Min Temp ≤ 32 °F | 22 | 18 | 9 | 2 | 0 | 0 | 0 | 0 | 0 | 2 | 10 | 18 | 81 |
| Days Min Temp ≤ 0 °F | 0 | 0 | 0 | 0 | 0 | 0 | 0 | 0 | 0 | 0 | 0 | 0 | 0 |
| Heating Degree Days | 860 | 672 | 457 | 188 | 51 | 2 | 0 | 0 | 28 | 203 | 448 | 722 | 3631 |
| Cooling Degree Days | 0 | 1 | 9 | 44 | 133 | 339 | 468 | 417 | 226 | 44 | 7 | 1 | 1689 |
| Total Precipitation (") | 3.99 | 4.40 | 5.47 | 5.41 | 5.61 | 3.97 | 3.88 | 3.38 | 4.08 | 3.34 | 4.91 | 5.62 | 54.06 |
| Days ≥ 0.1" Precip | 7 | 7 | 8 | 7 | 8 | 5 | 6 | 5 | 6 | 5 | 6 | 8 | 78 |
| Total Snowfall (") | *1.4* | 1.0 | 0.5 | 0.0 | 0.0 | 0.0 | 0.0 | 0.0 | 0.0 | 0.0 | 0.1 | 0.1 | 3.1 |
| Days ≥ 1" Snow Depth | *1* | 0 | 0 | 0 | 0 | 0 | 0 | 0 | 0 | 0 | 0 | 0 | 1 |

## BRISTOL TRI CITY AP *Sullivan County* ELEVATION 1565 ft LAT/LONG 36° 29 ' N / 82° 24 ' W

| | JAN | FEB | MAR | APR | MAY | JUN | JUL | AUG | SEP | OCT | NOV | DEC | YEAR |
|---|---|---|---|---|---|---|---|---|---|---|---|---|---|
| Maximum Temp °F | 44.0 | 48.4 | 58.6 | 68.0 | 75.1 | 82.5 | 85.3 | 84.2 | 79.0 | 68.8 | 58.3 | 48.7 | 66.7 |
| Minimum Temp °F | 25.1 | 27.2 | 35.3 | 43.3 | 51.8 | 60.2 | 64.6 | 63.2 | 56.6 | 44.3 | 35.9 | 28.9 | 44.7 |
| Mean Temp °F | 34.6 | 37.8 | 47.0 | 55.7 | 63.5 | 71.4 | 74.9 | 73.7 | 67.9 | 56.6 | 47.1 | 38.8 | 55.8 |
| Days Max Temp ≥ 90 °F | 0 | 0 | 0 | 0 | 0 | 2 | 6 | 3 | 2 | 0 | 0 | 0 | 13 |
| Days Max Temp ≤ 32 °F | 5 | 3 | 0 | 0 | 0 | 0 | 0 | 0 | 0 | 0 | 0 | 2 | 10 |
| Days Min Temp ≤ 32 °F | 23 | 20 | 13 | 4 | 0 | 0 | 0 | 0 | 0 | 3 | 13 | 20 | 96 |
| Days Min Temp ≤ 0 °F | 1 | 0 | 0 | 0 | 0 | 0 | 0 | 0 | 0 | 0 | 0 | 0 | 1 |
| Heating Degree Days | 937 | 762 | 553 | 284 | 105 | 10 | 1 | 1 | 42 | 268 | 531 | 805 | 4299 |
| Cooling Degree Days | 0 | 0 | 1 | 12 | 69 | 223 | 334 | 280 | 128 | 13 | 0 | 0 | 1060 |
| Total Precipitation (") | 3.18 | 3.48 | 3.78 | 3.28 | 3.98 | 3.55 | 4.49 | 3.23 | 3.13 | 2.50 | 2.92 | 3.50 | 41.02 |
| Days ≥ 0.1" Precip | 8 | 7 | 8 | 7 | 9 | 8 | 8 | 7 | 6 | 5 | 7 | 7 | 87 |
| Total Snowfall (") | 6.1 | 4.8 | 1.8 | 0.7 | 0.0 | 0.0 | 0.0 | 0.0 | 0.0 | 0.0 | 0.5 | 2.5 | 16.4 |
| Days ≥ 1" Snow Depth | 5 | 4 | 1 | 0 | 0 | 0 | 0 | 0 | 0 | 0 | 0 | 1 | 11 |

## BROWNSVILLE *Haywood County* ELEVATION 361 ft LAT/LONG 35° 36 ' N / 89° 16 ' W

| | JAN | FEB | MAR | APR | MAY | JUN | JUL | AUG | SEP | OCT | NOV | DEC | YEAR |
|---|---|---|---|---|---|---|---|---|---|---|---|---|---|
| Maximum Temp °F | 46.2 | 51.6 | 61.6 | 72.4 | 79.6 | 87.7 | 90.8 | 89.3 | 83.4 | 73.4 | 61.4 | 51.2 | 70.7 |
| Minimum Temp °F | 28.0 | 31.5 | 40.6 | 50.0 | 58.3 | 66.3 | 70.1 | 67.8 | 61.1 | 48.4 | 40.5 | 32.4 | 49.6 |
| Mean Temp °F | 37.1 | 41.5 | 51.1 | 61.2 | 69.0 | 77.0 | 80.5 | 78.5 | 72.3 | 60.9 | 50.9 | 41.8 | 60.1 |
| Days Max Temp ≥ 90 °F | 0 | 0 | 0 | 0 | 2 | 13 | 20 | 16 | 7 | 1 | 0 | 0 | 59 |
| Days Max Temp ≤ 32 °F | 5 | 2 | 0 | 0 | 0 | 0 | 0 | 0 | 0 | 0 | 0 | 2 | 9 |
| Days Min Temp ≤ 32 °F | 21 | 16 | 8 | 1 | 0 | 0 | 0 | 0 | 0 | 1 | 8 | 16 | 71 |
| Days Min Temp ≤ 0 °F | 0 | 0 | 0 | 0 | 0 | 0 | 0 | 0 | 0 | 0 | 0 | 0 | 0 |
| Heating Degree Days | 857 | 657 | 434 | 168 | 37 | 1 | 0 | 0 | 22 | 176 | 423 | 712 | 3487 |
| Cooling Degree Days | 0 | 1 | 10 | 53 | 161 | 374 | 499 | 438 | 245 | 51 | 8 | 2 | 1842 |
| Total Precipitation (") | 4.01 | 4.44 | 5.14 | 5.27 | 5.66 | 3.74 | 3.89 | 2.99 | 3.74 | 3.13 | 4.67 | 5.75 | 52.43 |
| Days ≥ 0.1" Precip | 7 | 7 | 8 | 7 | 8 | 6 | 6 | 4 | 5 | 5 | 6 | 7 | 76 |
| Total Snowfall (") | 3.2 | 2.3 | 1.1 | 0.0 | 0.0 | 0.0 | 0.0 | 0.0 | 0.0 | 0.0 | 0.1 | 0.3 | 7.0 |
| Days ≥ 1" Snow Depth | 4 | 2 | 0 | 0 | 0 | 0 | 0 | 0 | 0 | 0 | 0 | 0 | 6 |

## CENTERVILLE WATER PL *Hickman County* ELEVATION 630 ft LAT/LONG 35° 47 ' N / 87° 28 ' W

| | JAN | FEB | MAR | APR | MAY | JUN | JUL | AUG | SEP | OCT | NOV | DEC | YEAR |
|---|---|---|---|---|---|---|---|---|---|---|---|---|---|
| Maximum Temp °F | 48.3 | 54.3 | 64.5 | 75.1 | 80.5 | 86.8 | 90.1 | 89.0 | 83.3 | 73.7 | 62.0 | 52.5 | 71.7 |
| Minimum Temp °F | 24.3 | 26.6 | 34.5 | 42.9 | 51.4 | 59.5 | 63.9 | 62.3 | 56.0 | 43.0 | 34.8 | 28.7 | 44.0 |
| Mean Temp °F | 36.3 | 40.5 | 49.5 | 59.0 | 66.0 | 73.2 | 77.0 | 75.7 | 69.7 | 58.4 | 48.4 | 40.6 | 57.9 |
| Days Max Temp ≥ 90 °F | 0 | 0 | 0 | 1 | 1 | 9 | 18 | 15 | 5 | 0 | 0 | 0 | 49 |
| Days Max Temp ≤ 32 °F | 3 | 1 | 0 | 0 | 0 | 0 | 0 | 0 | 0 | 0 | 0 | 1 | 5 |
| Days Min Temp ≤ 32 °F | 23 | 20 | 14 | 6 | 1 | 0 | 0 | 0 | 0 | 6 | 14 | 19 | 103 |
| Days Min Temp ≤ 0 °F | 1 | 0 | 0 | 0 | 0 | 0 | 0 | 0 | 0 | 0 | 0 | 0 | 1 |
| Heating Degree Days | 882 | 685 | 479 | 206 | 66 | 5 | 0 | 1 | 35 | 227 | 493 | 750 | 3829 |
| Cooling Degree Days | 0 | 0 | 7 | 30 | 108 | 270 | 399 | 354 | 186 | 30 | 3 | 1 | 1388 |
| Total Precipitation (") | 3.81 | 4.32 | 5.38 | 4.92 | 5.72 | 3.94 | 4.27 | 3.22 | 3.58 | 3.36 | 5.08 | 5.63 | 53.23 |
| Days ≥ 0.1" Precip | 7 | 7 | 8 | 8 | 8 | 6 | 7 | 5 | 6 | 5 | 7 | 8 | 82 |
| Total Snowfall (") | 1.5 | 1.4 | 0.6 | 0.0 | 0.0 | 0.0 | 0.0 | 0.0 | 0.0 | 0.0 | 0.0 | 0.3 | 3.8 |
| Days ≥ 1" Snow Depth | *1* | 0 | 0 | 0 | 0 | 0 | 0 | 0 | 0 | 0 | 0 | 0 | 1 |

## CHATTANOOGA LOVELL *Hamilton County* ELEVATION 689 ft LAT/LONG 35° 2 ' N / 85° 12 ' W

| | JAN | FEB | MAR | APR | MAY | JUN | JUL | AUG | SEP | OCT | NOV | DEC | YEAR |
|---|---|---|---|---|---|---|---|---|---|---|---|---|---|
| Maximum Temp °F | 47.8 | 52.8 | 62.2 | 72.4 | 78.7 | 86.1 | 89.3 | 88.1 | 82.0 | 71.7 | 61.2 | 51.8 | 70.3 |
| Minimum Temp °F | 28.9 | 31.3 | 39.0 | 47.1 | 55.5 | 64.0 | 68.7 | 67.8 | 61.5 | 48.1 | 38.9 | 32.3 | 48.6 |
| Mean Temp °F | 38.4 | 42.1 | 50.6 | 59.8 | 67.1 | 75.1 | 79.0 | 78.0 | 71.8 | 59.9 | 50.1 | 42.1 | 59.5 |
| Days Max Temp ≥ 90 °F | 0 | 0 | 0 | 0 | 1 | 8 | 16 | 12 | 4 | 0 | 0 | 0 | 41 |
| Days Max Temp ≤ 32 °F | 3 | 1 | 0 | 0 | 0 | 0 | 0 | 0 | 0 | 0 | 0 | 1 | 5 |
| Days Min Temp ≤ 32 °F | 20 | 16 | 8 | 2 | 0 | 0 | 0 | 0 | 0 | 1 | 9 | 17 | 73 |
| Days Min Temp ≤ 0 °F | 0 | 0 | 0 | 0 | 0 | 0 | 0 | 0 | 0 | 0 | 0 | 0 | 0 |
| Heating Degree Days | 819 | 641 | 443 | 186 | 49 | 2 | 0 | 0 | 16 | 185 | 443 | 704 | 3488 |
| Cooling Degree Days | 0 | 0 | 6 | 35 | 134 | 339 | 477 | 434 | 228 | 37 | 3 | 1 | 1694 |
| Total Precipitation (") | 4.85 | 4.76 | 5.89 | 4.05 | 4.32 | 3.54 | 4.77 | 3.67 | 4.29 | 3.32 | 4.57 | 5.07 | 53.10 |
| Days ≥ 0.1" Precip | 8 | 7 | 9 | 7 | 8 | 6 | 8 | 7 | 6 | 5 | 7 | 8 | 86 |
| Total Snowfall (") | 2.3 | 1.2 | 1.0 | 0.1 | 0.0 | 0.0 | 0.0 | 0.0 | 0.0 | 0.0 | 0.0 | 0.3 | 4.9 |
| Days ≥ 1" Snow Depth | 2 | 1 | 0 | 0 | 0 | 0 | 0 | 0 | 0 | 0 | 0 | 0 | 3 |

## CLARKSVILLE SWG PLT *Montgomery County* ELEVATION 502 ft LAT/LONG 36° 31 ' N / 87° 22 ' W

| | JAN | FEB | MAR | APR | MAY | JUN | JUL | AUG | SEP | OCT | NOV | DEC | YEAR |
|---|---|---|---|---|---|---|---|---|---|---|---|---|---|
| Maximum Temp °F | 44.6 | 49.5 | 60.4 | 71.2 | 78.2 | 86.3 | 90.5 | 88.8 | 82.7 | 71.9 | 60.5 | 49.7 | 69.5 |
| Minimum Temp °F | 23.5 | 26.4 | 35.6 | 44.8 | 53.2 | 62.2 | 66.8 | 64.7 | 57.8 | 44.4 | 36.4 | 28.1 | 45.3 |
| Mean Temp °F | 34.1 | 38.0 | 48.0 | 58.0 | 65.7 | 74.3 | 78.7 | 76.8 | 70.3 | 58.1 | 48.5 | 38.8 | 57.4 |
| Days Max Temp ≥ 90 °F | 0 | 0 | 0 | 0 | 2 | 10 | 19 | 15 | 6 | 0 | 0 | 0 | 52 |
| Days Max Temp ≤ 32 °F | 6 | 2 | 0 | 0 | 0 | 0 | 0 | 0 | 0 | 0 | 0 | 2 | 10 |
| Days Min Temp ≤ 32 °F | 24 | 20 | 14 | 3 | 0 | 0 | 0 | 0 | 0 | 3 | 12 | 21 | 97 |
| Days Min Temp ≤ 0 °F | 1 | 1 | 0 | 0 | 0 | 0 | 0 | 0 | 0 | 0 | 0 | 0 | 2 |
| Heating Degree Days | 953 | 756 | 525 | 237 | 77 | 5 | 0 | 0 | 35 | 238 | 493 | 805 | 4124 |
| Cooling Degree Days | 0 | 0 | 4 | 34 | 104 | 305 | 457 | 393 | 200 | 33 | 4 | 0 | 1534 |
| Total Precipitation (") | 3.81 | 4.30 | 5.36 | 4.41 | 4.70 | 3.87 | 4.11 | 3.50 | 3.58 | 3.33 | 4.54 | 5.33 | 50.84 |
| Days ≥ 0.1" Precip | 7 | 7 | 8 | 7 | 8 | 7 | 7 | 6 | 6 | 6 | 7 | 7 | 83 |
| Total Snowfall (") | ***2.6*** | 3.1 | 0.7 | 0.0 | 0.0 | 0.0 | 0.0 | 0.0 | 0.0 | 0.0 | 0.1 | ***0.8*** | 7.3 |
| Days ≥ 1" Snow Depth | ***2*** | *1* | 0 | 0 | 0 | 0 | 0 | 0 | 0 | 0 | 0 | ***0*** | 3 |

## CLEVELAND FILTER PLT *Bradley County* ELEVATION 800 ft LAT/LONG 35° 15 ' N / 84° 46 ' W

| | JAN | FEB | MAR | APR | MAY | JUN | JUL | AUG | SEP | OCT | NOV | DEC | YEAR |
|---|---|---|---|---|---|---|---|---|---|---|---|---|---|
| Maximum Temp °F | 47.8 | 52.9 | 62.0 | 71.8 | 78.0 | 85.1 | 88.4 | 87.3 | 81.4 | 71.4 | 60.8 | 51.4 | 69.9 |
| Minimum Temp °F | 27.0 | 29.5 | 36.8 | 44.5 | 52.9 | 61.5 | 65.9 | 64.8 | 58.5 | 45.7 | 36.3 | 30.2 | 46.1 |
| Mean Temp °F | 37.4 | 41.2 | 49.4 | 58.2 | 65.5 | 73.3 | 77.2 | 76.1 | 70.0 | 58.6 | 48.6 | 40.7 | 58.0 |
| Days Max Temp ≥ 90 °F | 0 | 0 | 0 | 0 | 0 | 5 | 14 | 9 | 3 | 0 | 0 | 0 | 31 |
| Days Max Temp ≤ 32 °F | 3 | 1 | 0 | 0 | 0 | 0 | 0 | 0 | 0 | 0 | 0 | 1 | 5 |
| Days Min Temp ≤ 32 °F | 22 | 18 | 12 | 3 | 0 | 0 | 0 | 0 | 0 | 3 | 12 | 19 | 89 |
| Days Min Temp ≤ 0 °F | 0 | 0 | 0 | 0 | 0 | 0 | 0 | 0 | 0 | 0 | 0 | 0 | 0 |
| Heating Degree Days | 848 | 667 | 480 | 223 | 72 | 4 | 0 | 0 | 26 | 216 | 486 | 745 | 3767 |
| Cooling Degree Days | 0 | 0 | 5 | 22 | 98 | 282 | 416 | 367 | 185 | 28 | 2 | 0 | 1405 |
| Total Precipitation (") | 4.86 | 4.57 | 6.09 | 4.09 | 5.03 | 4.25 | 4.84 | 3.56 | 4.45 | 3.53 | 4.50 | 5.26 | 55.03 |
| Days ≥ 0.1" Precip | 8 | 8 | 9 | 7 | 8 | 7 | 8 | 6 | 6 | 6 | 7 | 8 | 88 |
| Total Snowfall (") | *1.5* | *0.9* | 0.1 | 0.2 | 0.0 | 0.0 | 0.0 | 0.0 | 0.0 | 0.0 | 0.0 | 0.3 | 3.0 |
| Days ≥ 1" Snow Depth | *1* | *0* | 0 | 0 | 0 | 0 | 0 | 0 | 0 | 0 | 0 | 0 | 1 |

## COLUMBIA 3 WNW *Maury County* ELEVATION 669 ft LAT/LONG 35° 37 ' N / 87° 3 ' W

| | JAN | FEB | MAR | APR | MAY | JUN | JUL | AUG | SEP | OCT | NOV | DEC | YEAR |
|---|---|---|---|---|---|---|---|---|---|---|---|---|---|
| Maximum Temp °F | 46.5 | 51.3 | 61.1 | 71.3 | 77.9 | 85.7 | 89.2 | 88.0 | 82.0 | 71.7 | 61.0 | 51.4 | 69.8 |
| Minimum Temp °F | 25.2 | 28.2 | 35.9 | 44.8 | 53.3 | 61.5 | 66.2 | 64.6 | 57.8 | 44.3 | 36.5 | 29.6 | 45.7 |
| Mean Temp °F | 35.9 | 39.8 | 48.5 | 57.9 | 65.6 | 73.6 | 77.8 | 76.3 | 69.9 | 58.0 | 48.8 | 40.5 | 57.7 |
| Days Max Temp ≥ 90 °F | 0 | 0 | 0 | 0 | 0 | 8 | 16 | 12 | 5 | 0 | 0 | 0 | 41 |
| Days Max Temp ≤ 32 °F | 4 | 2 | 0 | 0 | 0 | 0 | 0 | 0 | 0 | 0 | 0 | 1 | 7 |
| Days Min Temp ≤ 32 °F | 23 | 19 | 13 | 4 | 0 | 0 | 0 | 0 | 0 | 4 | 12 | 19 | 94 |
| Days Min Temp ≤ 0 °F | 1 | 0 | 0 | 0 | 0 | 0 | 0 | 0 | 0 | 0 | 0 | 0 | 1 |
| Heating Degree Days | 894 | 705 | 510 | 237 | 74 | 6 | 0 | 1 | 36 | 237 | 483 | 752 | 3935 |
| Cooling Degree Days | 0 | 0 | 4 | 19 | 95 | 267 | 407 | 356 | 174 | 22 | 2 | 1 | 1347 |
| Total Precipitation (") | 4.17 | 4.31 | 5.91 | 5.00 | 5.83 | 3.97 | 4.66 | 3.69 | 3.74 | 3.64 | 4.53 | 5.43 | 54.88 |
| Days ≥ 0.1" Precip | 7 | 7 | 8 | 8 | 8 | 6 | 7 | 5 | 6 | 5 | 7 | 8 | 82 |
| Total Snowfall (") | *2.2* | 1.3 | 0.2 | 0.0 | 0.0 | 0.0 | 0.0 | 0.0 | 0.0 | 0.0 | 0.3 | 0.4 | 4.4 |
| Days ≥ 1" Snow Depth | *1* | *0* | 0 | 0 | 0 | 0 | 0 | 0 | 0 | 0 | 0 | 0 | 1 |

## COOKEVILLE *Putnam County* ELEVATION 1152 ft LAT/LONG 36° 10 ' N / 85° 30 ' W

| | JAN | FEB | MAR | APR | MAY | JUN | JUL | AUG | SEP | OCT | NOV | DEC | YEAR |
|---|---|---|---|---|---|---|---|---|---|---|---|---|---|
| Maximum Temp °F | 44.6 | 48.7 | 58.8 | 69.1 | 76.3 | 83.9 | 87.3 | 86.5 | 80.8 | 70.5 | 59.4 | 49.4 | 67.9 |
| Minimum Temp °F | 24.3 | 26.3 | 35.1 | 44.1 | 52.0 | 60.7 | 64.9 | 63.2 | 56.5 | 43.9 | 36.1 | 28.8 | 44.7 |
| Mean Temp °F | 34.5 | 37.6 | 47.0 | 56.6 | 64.2 | 72.3 | 76.1 | 74.9 | 68.7 | 57.2 | 47.8 | 39.1 | 56.3 |
| Days Max Temp ≥ 90 °F | 0 | 0 | 0 | 0 | 0 | 5 | 11 | 8 | 3 | 0 | 0 | 0 | 27 |
| Days Max Temp ≤ 32 °F | 6 | 3 | 0 | 0 | 0 | 0 | 0 | 0 | 0 | 0 | 0 | 2 | 11 |
| Days Min Temp ≤ 32 °F | 23 | 20 | 14 | 4 | 0 | 0 | 0 | 0 | 0 | 4 | 13 | 21 | 99 |
| Days Min Temp ≤ 0 °F | 2 | 1 | 0 | 0 | 0 | 0 | 0 | 0 | 0 | 0 | 0 | 0 | 3 |
| Heating Degree Days | 940 | 768 | 558 | 270 | 101 | 11 | 0 | 1 | 45 | 261 | 514 | 795 | 4264 |
| Cooling Degree Days | 0 | 0 | 4 | 17 | 82 | 239 | 368 | 315 | 150 | 23 | 2 | 0 | 1200 |
| Total Precipitation (") | 4.77 | 4.46 | 5.73 | 4.56 | 5.44 | 3.90 | 5.35 | 4.36 | 4.16 | 3.48 | 4.75 | 5.71 | 56.67 |
| Days ≥ 0.1" Precip | 9 | 8 | 9 | 8 | 9 | 7 | 8 | 7 | 6 | 6 | 7 | 8 | 92 |
| Total Snowfall (") | 3.6 | *2.9* | 1.0 | 0.1 | 0.0 | 0.0 | 0.0 | 0.0 | 0.0 | 0.0 | 0.9 | 0.9 | 9.4 |
| Days ≥ 1" Snow Depth | 5 | 3 | 1 | 0 | 0 | 0 | 0 | 0 | 0 | 0 | 0 | 1 | 10 |

## COPPERHILL *Polk County* ELEVATION 1621 ft LAT/LONG 35° 0 ' N / 84° 23 ' W

| | JAN | FEB | MAR | APR | MAY | JUN | JUL | AUG | SEP | OCT | NOV | DEC | YEAR |
|---|---|---|---|---|---|---|---|---|---|---|---|---|---|
| Maximum Temp °F | 48.0 | 52.2 | 60.9 | 71.0 | 77.9 | 84.5 | 87.9 | 86.7 | 81.1 | 71.6 | 61.1 | 52.1 | 69.6 |
| Minimum Temp °F | 26.3 | 27.7 | 34.7 | 42.1 | 50.4 | 58.4 | 63.5 | 62.5 | 56.2 | 43.4 | 35.3 | 29.0 | 44.1 |
| Mean Temp °F | 37.2 | 40.0 | 47.8 | 56.5 | 64.2 | 71.5 | 75.7 | 74.6 | 68.7 | 57.5 | 48.2 | 40.6 | 56.9 |
| Days Max Temp ≥ 90 °F | 0 | 0 | 0 | 0 | 1 | 5 | 12 | 9 | 2 | 0 | 0 | 0 | 29 |
| Days Max Temp ≤ 32 °F | 3 | 1 | 0 | 0 | 0 | 0 | 0 | 0 | 0 | 0 | 0 | 1 | 5 |
| Days Min Temp ≤ 32 °F | 22 | 20 | 14 | 5 | 0 | 0 | 0 | 0 | 0 | 5 | 14 | 20 | 100 |
| Days Min Temp ≤ 0 °F | 1 | 0 | 0 | 0 | 0 | 0 | 0 | 0 | 0 | 0 | 0 | 0 | 1 |
| Heating Degree Days | 855 | 701 | 527 | 263 | 90 | 9 | 1 | 1 | 33 | 242 | 498 | 749 | 3969 |
| Cooling Degree Days | *0* | 0 | 2 | 13 | 75 | *218* | 360 | 300 | 144 | 18 | 1 | 0 | 1131 |
| Total Precipitation (") | 5.35 | 5.36 | 6.29 | 4.66 | 5.15 | 4.17 | 5.61 | 5.00 | 4.36 | 3.70 | 4.78 | 5.08 | 59.51 |
| Days ≥ 0.1" Precip | 9 | 7 | 9 | 7 | 8 | 8 | 9 | 8 | 7 | 5 | 7 | 8 | 92 |
| Total Snowfall (") | *1.1* | 1.4 | 1.6 | 0.0 | 0.0 | 0.0 | 0.0 | 0.0 | 0.0 | 0.0 | 0.1 | 0.4 | 4.6 |
| Days ≥ 1" Snow Depth | 2 | 1 | 0 | 0 | 0 | 0 | 0 | 0 | 0 | 0 | 0 | 0 | 3 |

## COVINGTON 1 W *Tipton County* ELEVATION 312 ft LAT/LONG 35° 33 ' N / 89° 39 ' W

| | JAN | FEB | MAR | APR | MAY | JUN | JUL | AUG | SEP | OCT | NOV | DEC | YEAR |
|---|---|---|---|---|---|---|---|---|---|---|---|---|---|
| Maximum Temp °F | 44.8 | 49.8 | 60.1 | 71.3 | 79.4 | 87.9 | 91.1 | 89.2 | 83.1 | 73.0 | 60.5 | 50.1 | 70.0 |
| Minimum Temp °F | 26.7 | 30.0 | 39.3 | 49.0 | 57.1 | 65.4 | 69.2 | 66.6 | 59.8 | 47.6 | 39.5 | 31.4 | 48.5 |
| Mean Temp °F | 35.8 | 39.9 | 49.7 | 60.2 | 68.3 | 76.7 | 80.2 | 77.9 | 71.5 | 60.4 | 50.0 | 40.8 | 59.3 |
| Days Max Temp ≥ 90 °F | 0 | 0 | 0 | 0 | 2 | 14 | 21 | 16 | 7 | 0 | 0 | 0 | 60 |
| Days Max Temp ≤ 32 °F | 5 | 3 | 0 | 0 | 0 | 0 | 0 | 0 | 0 | 0 | 0 | 2 | 10 |
| Days Min Temp ≤ 32 °F | 22 | 18 | 9 | 1 | 0 | 0 | 0 | 0 | 0 | 1 | 8 | 18 | 77 |
| Days Min Temp ≤ 0 °F | 1 | 0 | 0 | 0 | 0 | 0 | 0 | 0 | 0 | 0 | 0 | 0 | 1 |
| Heating Degree Days | 899 | 702 | 474 | 191 | 46 | 2 | 0 | 0 | 27 | 188 | 449 | 745 | 3723 |
| Cooling Degree Days | 0 | 1 | 9 | 53 | 157 | 374 | 499 | 432 | 238 | 53 | 8 | 1 | 1825 |
| Total Precipitation (") | 3.98 | 4.38 | 5.18 | 5.60 | 5.35 | 3.78 | 3.84 | 2.82 | 3.98 | 3.16 | 5.16 | 5.83 | 53.06 |
| Days ≥ 0.1" Precip | 7 | 7 | 8 | 8 | 8 | 6 | 6 | 5 | 6 | 5 | 7 | 7 | 80 |
| Total Snowfall (") | 3.6 | 2.8 | 1.2 | 0.0 | 0.0 | 0.0 | 0.0 | 0.0 | 0.0 | 0.0 | 0.1 | 0.4 | 8.1 |
| Days ≥ 1" Snow Depth | 4 | 2 | 1 | 0 | 0 | 0 | 0 | 0 | 0 | 0 | 0 | 0 | 7 |

## CROSSVILLE EXP STN *Cumberland County* ELEVATION 1811 ft LAT/LONG 36° 1 ' N / 85° 8 ' W

| | JAN | FEB | MAR | APR | MAY | JUN | JUL | AUG | SEP | OCT | NOV | DEC | YEAR |
|---|---|---|---|---|---|---|---|---|---|---|---|---|---|
| Maximum Temp °F | 41.9 | 46.1 | 55.6 | 65.6 | 72.9 | 80.3 | 83.7 | 82.6 | 76.8 | 66.9 | 56.3 | 46.7 | 64.6 |
| Minimum Temp °F | 21.7 | 24.3 | 33.5 | 42.7 | 50.3 | 57.9 | 61.8 | 60.3 | 54.3 | 42.2 | 34.8 | 26.5 | 42.5 |
| Mean Temp °F | 31.8 | 35.3 | 44.6 | 54.2 | 61.6 | 69.1 | 72.8 | 71.5 | 65.5 | 54.6 | 45.6 | 36.6 | 53.6 |
| Days Max Temp ≥ 90 °F | 0 | 0 | 0 | 0 | 0 | 1 | 4 | 3 | 1 | 0 | 0 | 0 | 9 |
| Days Max Temp ≤ 32 °F | 7 | 5 | 1 | 0 | 0 | 0 | 0 | 0 | 0 | 0 | 1 | 3 | 17 |
| Days Min Temp ≤ 32 °F | 25 | 21 | 16 | 5 | 0 | 0 | 0 | 0 | 0 | 5 | 14 | 21 | 107 |
| Days Min Temp ≤ 0 °F | 2 | 1 | 0 | 0 | 0 | 0 | 0 | 0 | 0 | 0 | 0 | 0 | 3 |
| Heating Degree Days | 1023 | 833 | 627 | 327 | 141 | 22 | 3 | 5 | 71 | 324 | 576 | 873 | 4825 |
| Cooling Degree Days | 0 | 0 | 1 | 9 | 50 | 169 | 277 | 226 | 97 | 10 | 0 | 0 | 839 |
| Total Precipitation (") | 5.19 | 4.75 | 6.23 | 4.84 | 5.64 | 4.12 | 5.42 | 4.02 | 4.00 | 3.76 | 5.12 | 6.18 | 59.27 |
| Days ≥ 0.1" Precip | 9 | 8 | 9 | 9 | 9 | 7 | 8 | 7 | 7 | 6 | 8 | 8 | 95 |
| Total Snowfall (") | 5.4 | 5.1 | 2.7 | 0.3 | 0.0 | 0.0 | 0.0 | 0.0 | 0.0 | 0.0 | 1.0 | 1.7 | 16.2 |
| Days ≥ 1" Snow Depth | 6 | 5 | 1 | 0 | 0 | 0 | 0 | 0 | 0 | 0 | 1 | 2 | 15 |

## CROSSVILLE MEMORIAL *Cumberland County* ELEVATION 1870 ft LAT/LONG 35° 57 ' N / 85° 5 ' W

| | JAN | FEB | MAR | APR | MAY | JUN | JUL | AUG | SEP | OCT | NOV | DEC | YEAR |
|---|---|---|---|---|---|---|---|---|---|---|---|---|---|
| Maximum Temp °F | 43.0 | 47.5 | 57.0 | 66.9 | 73.6 | 80.6 | 83.7 | 82.7 | 76.7 | 66.9 | 56.6 | 47.6 | 65.2 |
| Minimum Temp °F | 25.2 | 28.0 | 36.2 | 44.5 | 51.8 | 59.6 | 64.2 | 63.0 | 56.7 | 44.4 | 36.9 | 30.1 | 45.0 |
| Mean Temp °F | 34.1 | 37.8 | 46.6 | 55.7 | 62.7 | 70.1 | 74.0 | 72.9 | 66.7 | 55.7 | 46.8 | 38.9 | 55.2 |
| Days Max Temp ≥ 90 °F | 0 | 0 | 0 | 0 | 0 | 1 | 4 | 3 | 1 | 0 | 0 | 0 | 9 |
| Days Max Temp ≤ 32 °F | 6 | 4 | 1 | 0 | 0 | 0 | 0 | 0 | 0 | 0 | 0 | 3 | 14 |
| Days Min Temp ≤ 32 °F | 23 | 19 | 13 | 4 | 0 | 0 | 0 | 0 | 0 | 4 | 11 | 19 | 93 |
| Days Min Temp ≤ 0 °F | 1 | 0 | 0 | 0 | 0 | 0 | 0 | 0 | 0 | 0 | 0 | 0 | 1 |
| Heating Degree Days | 949 | 763 | 564 | 286 | 121 | 15 | 2 | 3 | 59 | 293 | 542 | 803 | 4400 |
| Cooling Degree Days | 0 | 0 | 2 | 14 | 61 | 190 | 313 | 266 | 120 | 13 | 1 | 0 | 980 |
| Total Precipitation (") | 4.60 | 4.30 | 5.93 | 4.68 | 5.27 | 4.04 | 5.13 | 4.12 | 3.91 | 3.35 | 4.92 | 5.42 | 55.67 |
| Days ≥ 0.1" Precip | 8 | 8 | 10 | 8 | 9 | 7 | 9 | 6 | 6 | 6 | 8 | 8 | 93 |
| Total Snowfall (") | 4.5 | 4.0 | 2.2 | 0.3 | 0.0 | 0.0 | 0.0 | 0.0 | 0.0 | 0.0 | 0.8 | 1.5 | 13.3 |
| Days ≥ 1" Snow Depth | 6 | 3 | 1 | 0 | 0 | 0 | 0 | 0 | 0 | 0 | 0 | 1 | 11 |

## DAYTON *Rhea County* ELEVATION 830 ft LAT/LONG 35° 29 ' N / 85° 2 ' W

| | JAN | FEB | MAR | APR | MAY | JUN | JUL | AUG | SEP | OCT | NOV | DEC | YEAR |
|---|---|---|---|---|---|---|---|---|---|---|---|---|---|
| Maximum Temp °F | 46.9 | 52.5 | 62.6 | 72.9 | 79.1 | 86.1 | 88.8 | 87.7 | 81.8 | 71.5 | 60.0 | 50.5 | 70.0 |
| Minimum Temp °F | 27.7 | 30.1 | 37.6 | 45.4 | 53.4 | 61.2 | 65.6 | 64.6 | 58.9 | 46.5 | 38.1 | 31.3 | 46.7 |
| Mean Temp °F | 37.4 | 41.3 | 50.1 | 59.2 | 66.3 | 73.7 | 77.2 | 76.2 | 70.4 | 59.0 | 49.1 | 40.9 | 58.4 |
| Days Max Temp ≥ 90 °F | 0 | 0 | 0 | 0 | 1 | 7 | 15 | 11 | 4 | 0 | 0 | 0 | 38 |
| Days Max Temp ≤ 32 °F | 2 | 1 | 0 | 0 | 0 | 0 | 0 | 0 | 0 | 0 | 0 | 1 | 4 |
| Days Min Temp ≤ 32 °F | 20 | 17 | 10 | 3 | 0 | 0 | 0 | 0 | 0 | 2 | 10 | 18 | 80 |
| Days Min Temp ≤ 0 °F | 0 | 0 | 0 | 0 | 0 | 0 | 0 | 0 | 0 | 0 | 0 | 0 | 0 |
| Heating Degree Days | 850 | 663 | 457 | 196 | 55 | 3 | 0 | 0 | 21 | 204 | 472 | 739 | 3660 |
| Cooling Degree Days | 0 | 0 | 3 | 26 | 102 | 281 | 402 | 360 | 185 | 28 | 1 | 0 | 1388 |
| Total Precipitation (") | 4.91 | 4.79 | 6.21 | 4.39 | 5.21 | 3.78 | 4.67 | 4.19 | 4.44 | 3.53 | 4.74 | 5.92 | 56.78 |
| Days ≥ 0.1" Precip | 8 | 8 | 9 | 7 | 8 | 7 | 8 | 7 | 6 | 6 | 7 | 8 | 89 |
| Total Snowfall (") | 2.4 | 1.7 | 0.7 | 0.1 | 0.0 | 0.0 | 0.0 | 0.0 | 0.0 | 0.0 | 0.0 | 0.5 | 5.4 |
| Days ≥ 1" Snow Depth | 2 | 1 | 0 | 0 | 0 | 0 | 0 | 0 | 0 | 0 | 0 | 0 | 3 |

## DICKSON *Dickson County* ELEVATION 801 ft LAT/LONG 36° 4 ' N / 87° 24 ' W

| | JAN | FEB | MAR | APR | MAY | JUN | JUL | AUG | SEP | OCT | NOV | DEC | YEAR |
|---|---|---|---|---|---|---|---|---|---|---|---|---|---|
| Maximum Temp °F | 45.6 | 51.0 | 61.5 | 72.3 | 78.7 | 85.9 | 89.3 | 88.1 | 81.7 | 71.1 | 59.8 | 50.4 | 69.6 |
| Minimum Temp °F | 25.9 | 28.8 | 37.2 | 46.1 | 54.5 | 62.5 | 66.9 | 65.3 | 59.3 | 46.6 | 38.1 | 30.3 | 46.8 |
| Mean Temp °F | 35.7 | 39.9 | 49.4 | 59.3 | 66.6 | 74.2 | 78.1 | 76.7 | 70.5 | 58.9 | 48.9 | 40.4 | 58.2 |
| Days Max Temp ≥ 90 °F | 0 | 0 | 0 | 0 | 1 | 8 | 16 | 12 | 4 | 0 | 0 | 0 | 41 |
| Days Max Temp ≤ 32 °F | 5 | 2 | 0 | 0 | 0 | 0 | 0 | 0 | 0 | 0 | 0 | 2 | 9 |
| Days Min Temp ≤ 32 °F | 22 | 19 | 12 | 3 | 0 | 0 | 0 | 0 | 0 | 3 | 10 | 19 | 88 |
| Days Min Temp ≤ 0 °F | 1 | 0 | 0 | 0 | 0 | 0 | 0 | 0 | 0 | 0 | 0 | 0 | 1 |
| Heating Degree Days | 900 | 702 | 485 | 204 | 58 | 2 | 0 | 0 | 29 | 217 | 478 | 756 | 3831 |
| Cooling Degree Days | 0 | 0 | 6 | 30 | 111 | 288 | 430 | 382 | 201 | 32 | 3 | 1 | 1484 |
| Total Precipitation (") | 3.82 | 4.50 | 5.39 | 4.77 | 5.63 | 4.30 | 4.42 | 3.50 | 4.01 | 3.43 | 5.16 | 5.43 | 54.36 |
| Days ≥ 0.1" Precip | 7 | 7 | 8 | 8 | 8 | 7 | 7 | 6 | 6 | 5 | 7 | 7 | 83 |
| Total Snowfall (") | *2.4* | 2.5 | 0.9 | 0.1 | 0.0 | 0.0 | 0.0 | 0.0 | 0.0 | 0.0 | 0.4 | 1.2 | 7.5 |
| Days ≥ 1" Snow Depth | 3 | 2 | 0 | 0 | 0 | 0 | 0 | 0 | 0 | 0 | 0 | 1 | 6 |

## DOVER 1 W *Stewart County* ELEVATION 489 ft LAT/LONG 36° 30 ' N / 87° 51 ' W

| | JAN | FEB | MAR | APR | MAY | JUN | JUL | AUG | SEP | OCT | NOV | DEC | YEAR |
|---|---|---|---|---|---|---|---|---|---|---|---|---|---|
| Maximum Temp °F | 43.5 | 48.7 | 59.7 | 70.7 | 77.4 | 85.0 | 88.7 | 87.4 | 81.2 | 70.9 | 59.1 | 48.7 | 68.4 |
| Minimum Temp °F | 23.2 | 26.4 | 36.1 | 45.9 | 53.5 | 62.1 | 66.3 | 63.8 | 57.2 | 44.2 | 36.8 | 28.1 | 45.3 |
| Mean Temp °F | 33.4 | 37.6 | 47.9 | 58.3 | 65.4 | 73.6 | 77.5 | 75.5 | 69.2 | 57.6 | 48.0 | 38.4 | 56.9 |
| Days Max Temp ≥ 90 °F | 0 | 0 | 0 | 0 | 1 | 7 | 15 | 12 | 4 | 0 | 0 | 0 | 39 |
| Days Max Temp ≤ 32 °F | 6 | 3 | 0 | 0 | 0 | 0 | 0 | 0 | 0 | 0 | 0 | 3 | 12 |
| Days Min Temp ≤ 32 °F | 25 | 20 | 13 | 3 | 0 | 0 | 0 | 0 | 0 | 4 | 11 | 20 | 96 |
| Days Min Temp ≤ 0 °F | 1 | 0 | 0 | 0 | 0 | 0 | 0 | 0 | 0 | 0 | 0 | 0 | 1 |
| Heating Degree Days | 973 | 769 | 530 | 235 | 82 | 6 | 1 | 1 | 46 | 254 | 507 | 817 | 4221 |
| Cooling Degree Days | 0 | 0 | 7 | 43 | 102 | 276 | 419 | *339* | 179 | 30 | 4 | 1 | 1400 |
| Total Precipitation (") | 4.07 | 4.48 | 5.19 | 4.90 | 4.75 | 3.92 | 4.41 | 3.62 | 3.87 | 3.53 | 4.83 | 5.30 | 52.87 |
| Days ≥ 0.1" Precip | 7 | 7 | 8 | 8 | 7 | 6 | 6 | 5 | 6 | 6 | 7 | 7 | 80 |
| Total Snowfall (") | 3.9 | 3.6 | 1.0 | 0.0 | 0.0 | 0.0 | 0.0 | 0.0 | 0.0 | 0.0 | 0.4 | 0.9 | 9.8 |
| Days ≥ 1" Snow Depth | 5 | 3 | 0 | 0 | 0 | 0 | 0 | 0 | 0 | 0 | 0 | 1 | 9 |

## DRESDEN *Weakley County* ELEVATION 420 ft LAT/LONG 36° 17 ' N / 88° 43 ' W

| | JAN | FEB | MAR | APR | MAY | JUN | JUL | AUG | SEP | OCT | NOV | DEC | YEAR |
|---|---|---|---|---|---|---|---|---|---|---|---|---|---|
| Maximum Temp °F | 43.1 | 48.2 | 59.2 | 70.3 | 78.2 | 86.4 | 89.6 | 88.2 | 82.5 | 71.8 | 59.5 | 48.9 | 68.8 |
| Minimum Temp °F | 23.9 | 27.4 | 37.2 | 46.9 | 55.4 | 63.6 | 67.3 | 64.8 | 58.5 | 46.0 | 37.7 | 29.2 | 46.5 |
| Mean Temp °F | 33.5 | 37.7 | 48.2 | 58.6 | 66.8 | 75.1 | 78.5 | 76.5 | 70.5 | 58.9 | 48.6 | 39.1 | 57.7 |
| Days Max Temp ≥ 90 °F | 0 | 0 | 0 | 0 | 1 | 10 | 18 | 13 | 6 | 0 | 0 | 0 | 48 |
| Days Max Temp ≤ 32 °F | 6 | 4 | 0 | 0 | 0 | 0 | 0 | 0 | 0 | 0 | 0 | 2 | 12 |
| Days Min Temp ≤ 32 °F | 24 | 19 | 12 | 2 | 0 | 0 | 0 | 0 | 0 | 3 | 10 | 20 | 90 |
| Days Min Temp ≤ 0 °F | 1 | 0 | 0 | 0 | 0 | 0 | 0 | 0 | 0 | 0 | 0 | 0 | 1 |
| Heating Degree Days | 970 | 766 | 521 | 225 | 62 | 4 | 1 | 1 | 34 | 222 | 489 | 798 | 4093 |
| Cooling Degree Days | 0 | 0 | 8 | 39 | 124 | 328 | 453 | 391 | 215 | 41 | 5 | 1 | 1605 |
| Total Precipitation (") | 4.07 | 4.40 | 4.92 | 5.12 | 5.12 | 4.36 | 5.01 | 3.83 | 3.83 | 3.40 | 4.73 | 5.37 | 54.16 |
| Days ≥ 0.1" Precip | 7 | 7 | 8 | 8 | 8 | 6 | 6 | 5 | 5 | 5 | 7 | 7 | 79 |
| Total Snowfall (") | 3.5 | *2.6* | 1.0 | 0.0 | 0.0 | 0.0 | 0.0 | 0.0 | 0.0 | 0.0 | 0.2 | 0.5 | 7.8 |
| Days ≥ 1" Snow Depth | *4* | 2 | 1 | 0 | 0 | 0 | 0 | 0 | 0 | 0 | 0 | 0 | 7 |

## DYERSBURG MUNI AP *Dyer County* ELEVATION 338 ft LAT/LONG 36° 1 ' N / 89° 24 ' W

| | JAN | FEB | MAR | APR | MAY | JUN | JUL | AUG | SEP | OCT | NOV | DEC | YEAR |
|---|---|---|---|---|---|---|---|---|---|---|---|---|---|
| Maximum Temp °F | 45.1 | 50.4 | 60.8 | 71.7 | 79.6 | 87.7 | 90.8 | 88.6 | 82.4 | 72.4 | 60.0 | 49.8 | 69.9 |
| Minimum Temp °F | 29.3 | 33.1 | 42.1 | 51.7 | 59.7 | 67.7 | 71.4 | 68.9 | 62.4 | 50.7 | 42.2 | 33.9 | 51.1 |
| Mean Temp °F | 37.2 | 41.8 | 51.5 | 61.7 | 69.7 | 77.8 | 81.1 | 78.8 | 72.4 | 61.6 | 51.1 | 41.9 | 60.6 |
| Days Max Temp ≥ 90 °F | 0 | 0 | 0 | 0 | 2 | 12 | 20 | 14 | 5 | 0 | 0 | 0 | 53 |
| Days Max Temp ≤ 32 °F | 5 | 2 | 0 | 0 | 0 | 0 | 0 | 0 | 0 | 0 | 0 | 2 | 9 |
| Days Min Temp ≤ 32 °F | 19 | 14 | 6 | 1 | 0 | 0 | 0 | 0 | 0 | 0 | 6 | 14 | 60 |
| Days Min Temp ≤ 0 °F | 0 | 0 | 0 | 0 | 0 | 0 | 0 | 0 | 0 | 0 | 0 | 0 | 0 |
| Heating Degree Days | 854 | 650 | 423 | 154 | 29 | 0 | 0 | 0 | 19 | 157 | 417 | 710 | 3413 |
| Cooling Degree Days | 0 | 0 | 12 | 61 | 181 | 399 | 530 | 455 | 254 | 59 | 8 | 1 | 1960 |
| Total Precipitation (") | 3.69 | 4.18 | 4.56 | 4.73 | 4.83 | 4.13 | 4.08 | 3.20 | 3.44 | 3.23 | 4.69 | 5.42 | 50.18 |
| Days ≥ 0.1" Precip | 6 | 7 | 7 | 7 | 7 | 6 | 6 | 4 | 5 | 5 | 7 | 7 | 74 |
| Total Snowfall (") | 3.1 | 3.1 | 1.4 | 0.0 | 0.0 | 0.0 | 0.0 | 0.0 | 0.0 | 0.0 | 0.0 | 0.6 | 8.2 |
| Days ≥ 1" Snow Depth | 5 | 3 | 1 | 0 | 0 | 0 | 0 | 0 | 0 | 0 | 0 | 1 | 10 |

## FRANKLIN SEWAGE PLT *Williamson County* ELEVATION 630 ft LAT/LONG 35° 56 ' N / 86° 51 ' W

| | JAN | FEB | MAR | APR | MAY | JUN | JUL | AUG | SEP | OCT | NOV | DEC | YEAR |
|---|---|---|---|---|---|---|---|---|---|---|---|---|---|
| Maximum Temp °F | 47.0 | 52.4 | 62.3 | 72.3 | 79.0 | 86.7 | 90.1 | 88.5 | 83.0 | 72.7 | 61.1 | 51.6 | 70.6 |
| Minimum Temp °F | 26.0 | 28.4 | 36.9 | 45.4 | 53.5 | 61.7 | 65.9 | 63.8 | 57.4 | 44.6 | 37.3 | 30.1 | 45.9 |
| Mean Temp °F | 36.5 | 40.4 | 49.5 | 58.9 | 66.3 | 74.2 | 78.0 | 76.2 | 70.2 | 58.7 | 49.3 | 40.9 | 58.3 |
| Days Max Temp ≥ 90 °F | 0 | 0 | 0 | 0 | 1 | 10 | 19 | 14 | 6 | 0 | 0 | 0 | 50 |
| Days Max Temp ≤ 32 °F | 4 | 1 | 0 | 0 | 0 | 0 | 0 | 0 | 0 | 0 | 0 | 1 | 6 |
| Days Min Temp ≤ 32 °F | 22 | 19 | 12 | 4 | 0 | 0 | 0 | 0 | 0 | 4 | 11 | 19 | 91 |
| Days Min Temp ≤ 0 °F | 1 | 0 | 0 | 0 | 0 | 0 | 0 | 0 | 0 | 0 | 0 | 0 | 1 |
| Heating Degree Days | 876 | 688 | 481 | 214 | 67 | 5 | 0 | 1 | 31 | 222 | 469 | 742 | 3796 |
| Cooling Degree Days | 0 | 0 | 7 | 26 | 105 | 286 | 420 | 353 | 185 | 32 | 2 | 1 | 1417 |
| Total Precipitation (") | 3.97 | 4.17 | 5.61 | 4.35 | 5.33 | 3.75 | 4.62 | 3.80 | 3.86 | 3.26 | 4.76 | 5.51 | 52.99 |
| Days ≥ 0.1" Precip | 7 | 7 | 8 | 7 | 8 | 6 | 8 | 6 | 6 | 5 | 7 | 7 | 82 |
| Total Snowfall (") | *2.3* | *1.4* | 0.6 | 0.0 | 0.0 | 0.0 | 0.0 | 0.0 | 0.0 | 0.0 | 0.3 | 0.4 | 5.0 |
| Days ≥ 1" Snow Depth | *1* | *1* | 0 | 0 | 0 | 0 | 0 | 0 | 0 | 0 | 0 | 0 | 2 |

## GATLINBURG 2 SW *Sevier County* ELEVATION 1450 ft LAT/LONG 35° 41 ' N / 83° 32 ' W

| | JAN | FEB | MAR | APR | MAY | JUN | JUL | AUG | SEP | OCT | NOV | DEC | YEAR |
|---|---|---|---|---|---|---|---|---|---|---|---|---|---|
| Maximum Temp °F | 47.3 | 51.3 | 61.0 | 70.4 | 76.0 | 82.4 | 84.8 | 83.4 | 78.4 | 69.2 | 60.6 | 51.6 | 68.0 |
| Minimum Temp °F | 24.4 | 26.1 | 33.4 | 41.0 | 48.8 | 57.0 | 61.3 | 60.3 | 54.6 | 41.4 | 33.6 | 27.4 | 42.4 |
| Mean Temp °F | 35.9 | 38.7 | 47.2 | 55.7 | 62.4 | 69.7 | 73.0 | 71.8 | 66.5 | 55.3 | 47.1 | 39.5 | 55.2 |
| Days Max Temp ≥ 90 °F | 0 | 0 | 0 | 0 | 0 | 2 | 5 | 3 | 1 | 0 | 0 | 0 | 11 |
| Days Max Temp ≤ 32 °F | 4 | 2 | 0 | 0 | 0 | 0 | 0 | 0 | 0 | 0 | 0 | 1 | 7 |
| Days Min Temp ≤ 32 °F | 23 | 21 | 17 | 7 | 1 | 0 | 0 | 0 | 0 | 6 | 15 | 22 | 112 |
| Days Min Temp ≤ 0 °F | 1 | 0 | 0 | 0 | 0 | 0 | 0 | 0 | 0 | 0 | 0 | 0 | 1 |
| Heating Degree Days | 896 | 737 | 547 | 289 | 123 | 16 | 1 | 1 | 50 | 302 | 530 | 782 | 4274 |
| Cooling Degree Days | 0 | 0 | 2 | 13 | 51 | 180 | 275 | 223 | 101 | 10 | 1 | 0 | 856 |
| Total Precipitation (") | 4.36 | 4.22 | 5.40 | 4.31 | 5.29 | 5.22 | 6.25 | 5.24 | 4.39 | 2.96 | 3.95 | 4.50 | 56.09 |
| Days ≥ 0.1" Precip | 8 | 8 | 9 | 8 | 10 | 9 | 10 | 9 | 7 | 5 | 7 | 8 | 98 |
| Total Snowfall (") | *3.7* | na | *2.1* | *0.9* | *0.0* | *0.0* | *0.0* | *0.0* | *0.0* | *0.0* | *0.1* | *1.0* | na |
| Days ≥ 1" Snow Depth | na | na | *0* | *0* | *0* | *0* | *0* | 0 | 0 | 0 | 0 | *0* | na |

## GREENEVILLE EXP STN *Greene County* ELEVATION 1319 ft LAT/LONG 36° 6 ' N / 82° 51 ' W

| | JAN | FEB | MAR | APR | MAY | JUN | JUL | AUG | SEP | OCT | NOV | DEC | YEAR |
|---|---|---|---|---|---|---|---|---|---|---|---|---|---|
| Maximum Temp °F | 46.1 | 50.5 | 60.1 | 69.7 | 77.0 | 84.1 | 87.1 | 85.9 | 80.7 | 70.3 | 60.1 | 50.7 | 68.5 |
| Minimum Temp °F | 24.1 | 26.0 | 33.9 | 41.6 | 50.7 | 59.4 | 64.0 | 62.5 | 56.1 | 42.7 | 34.2 | 27.4 | 43.6 |
| Mean Temp °F | 35.2 | 38.3 | 47.0 | 55.7 | 63.8 | 71.8 | 75.6 | 74.3 | 68.5 | 56.6 | 47.2 | 39.1 | 56.1 |
| Days Max Temp ≥ 90 °F | 0 | 0 | 0 | 0 | 0 | 4 | 10 | 6 | 2 | 0 | 0 | 0 | 22 |
| Days Max Temp ≤ 32 °F | 4 | 2 | 0 | 0 | 0 | 0 | 0 | 0 | 0 | 0 | 0 | 2 | 8 |
| Days Min Temp ≤ 32 °F | 23 | 21 | 15 | 6 | 0 | 0 | 0 | 0 | 0 | 5 | 15 | 22 | 107 |
| Days Min Temp ≤ 0 °F | 1 | 1 | 0 | 0 | 0 | 0 | 0 | 0 | 0 | 0 | 0 | 0 | 2 |
| Heating Degree Days | 919 | 747 | 552 | 287 | 103 | 10 | 0 | 1 | 39 | 272 | 529 | 798 | 4257 |
| Cooling Degree Days | 0 | 0 | 1 | 11 | 69 | 219 | 343 | 285 | 138 | 16 | 1 | 1 | 1084 |
| Total Precipitation (") | 3.25 | 3.40 | 4.04 | 3.49 | 4.21 | 3.92 | 4.74 | 3.77 | 3.18 | 2.57 | 2.94 | 3.33 | 42.84 |
| Days ≥ 0.1" Precip | 8 | 7 | 8 | 7 | 8 | 8 | 8 | 7 | 6 | 6 | 6 | 7 | 86 |
| Total Snowfall (") | *4.8* | 3.1 | 1.4 | 0.8 | 0.0 | 0.0 | 0.0 | 0.0 | 0.0 | 0.0 | 0.1 | 1.1 | 11.3 |
| Days ≥ 1" Snow Depth | *2* | *1* | 0 | 0 | 0 | 0 | 0 | 0 | 0 | 0 | 0 | *1* | 4 |

## JACKSON EXP STN *Madison County* ELEVATION 400 ft LAT/LONG 35° 37 ' N / 88° 51 ' W

| | JAN | FEB | MAR | APR | MAY | JUN | JUL | AUG | SEP | OCT | NOV | DEC | YEAR |
|---|---|---|---|---|---|---|---|---|---|---|---|---|---|
| Maximum Temp °F | 45.0 | 50.0 | 60.1 | 71.0 | 78.0 | 86.0 | 89.2 | 88.2 | 82.3 | 72.1 | 60.4 | 50.3 | 69.4 |
| Minimum Temp °F | 27.2 | 30.2 | 39.4 | 48.7 | 57.3 | 65.3 | 69.3 | 67.2 | 60.6 | 47.7 | 39.6 | 31.7 | 48.7 |
| Mean Temp °F | 36.1 | 40.1 | 49.8 | 59.9 | 67.7 | 75.7 | 79.3 | 77.7 | 71.5 | 59.9 | 50.0 | 41.0 | 59.1 |
| Days Max Temp ≥ 90 °F | 0 | 0 | 0 | 0 | 1 | 9 | 17 | 13 | 6 | 0 | 0 | 0 | 46 |
| Days Max Temp ≤ 32 °F | 5 | 3 | 0 | 0 | 0 | 0 | 0 | 0 | 0 | 0 | 0 | 2 | 10 |
| Days Min Temp ≤ 32 °F | 22 | 17 | 9 | 2 | 0 | 0 | 0 | 0 | 0 | 2 | 9 | 17 | 78 |
| Days Min Temp ≤ 0 °F | 0 | 0 | 0 | 0 | 0 | 0 | 0 | 0 | 0 | 0 | 0 | 0 | 0 |
| Heating Degree Days | 889 | 697 | 474 | 200 | 52 | 3 | 0 | 0 | 29 | 198 | 450 | 737 | 3729 |
| Cooling Degree Days | 0 | 1 | 9 | 49 | 142 | 339 | 467 | 421 | 233 | 49 | 7 | 1 | 1718 |
| Total Precipitation (") | 3.91 | 4.30 | 4.96 | 5.04 | 5.83 | 4.45 | 4.27 | 2.78 | 3.96 | 3.44 | 4.99 | 5.50 | 53.43 |
| Days ≥ 0.1" Precip | 7 | 7 | 8 | 7 | 8 | 7 | 6 | 5 | 6 | 5 | 7 | 7 | 80 |
| Total Snowfall (") | *1.9* | *2.1* | 0.8 | 0.0 | 0.0 | 0.0 | 0.0 | 0.0 | 0.0 | 0.0 | 0.1 | 0.2 | 5.1 |
| Days ≥ 1" Snow Depth | 3 | 2 | 0 | 0 | 0 | 0 | 0 | 0 | 0 | 0 | 0 | 0 | 5 |

## JACKSON MCKELLAR-SIP *Madison County* ELEVATION 420 ft LAT/LONG 35° 36 ' N / 88° 55 ' W

| | JAN | FEB | MAR | APR | MAY | JUN | JUL | AUG | SEP | OCT | NOV | DEC | YEAR |
|---|---|---|---|---|---|---|---|---|---|---|---|---|---|
| Maximum Temp °F | 46.6 | 51.8 | 62.0 | 72.6 | 79.7 | 87.7 | 91.0 | 89.6 | 83.4 | 73.1 | 61.0 | 51.3 | 70.8 |
| Minimum Temp °F | 28.0 | 31.4 | 40.3 | 49.5 | 57.6 | 65.7 | 69.5 | 67.3 | 60.8 | 48.4 | 40.1 | 32.3 | 49.2 |
| Mean Temp °F | 37.3 | 41.7 | 51.1 | 61.1 | 68.7 | 76.8 | 80.3 | 78.5 | 72.1 | 60.8 | 50.5 | 41.8 | 60.1 |
| Days Max Temp ≥ 90 °F | 0 | 0 | 0 | 0 | 2 | 13 | 20 | 17 | 7 | 0 | 0 | 0 | 59 |
| Days Max Temp ≤ 32 °F | 4 | 2 | 0 | 0 | 0 | 0 | 0 | 0 | 0 | 0 | 0 | 2 | 8 |
| Days Min Temp ≤ 32 °F | 21 | 16 | 8 | 1 | 0 | 0 | 0 | 0 | 0 | 1 | 8 | 17 | 72 |
| Days Min Temp ≤ 0 °F | 0 | 0 | 0 | 0 | 0 | 0 | 0 | 0 | 0 | 0 | 0 | 0 | 0 |
| Heating Degree Days | 852 | 653 | 434 | 169 | 38 | 1 | 0 | 0 | 22 | 177 | 434 | 713 | 3493 |
| Cooling Degree Days | 0 | 1 | 11 | 55 | 158 | 366 | 498 | 443 | 245 | 54 | 8 | 2 | 1841 |
| Total Precipitation (") | 3.98 | 4.31 | 4.92 | 5.35 | 5.82 | 4.51 | 4.32 | 3.00 | 3.84 | 3.36 | 4.87 | 5.45 | 53.73 |
| Days ≥ 0.1" Precip | 6 | 7 | 8 | 8 | 8 | 6 | 6 | 5 | 6 | 5 | 7 | 7 | 79 |
| Total Snowfall (") | 2.8 | 2.1 | 1.1 | 0.0 | 0.0 | 0.0 | 0.0 | 0.0 | 0.0 | 0.0 | 0.1 | 0.2 | 6.3 |
| Days ≥ 1" Snow Depth | 3 | 2 | 0 | 0 | 0 | 0 | 0 | 0 | 0 | 0 | 0 | 0 | 5 |

## KINGSPORT *Sullivan County* ELEVATION 1280 ft LAT/LONG 36° 31 ' N / 82° 30 ' W

| | JAN | FEB | MAR | APR | MAY | JUN | JUL | AUG | SEP | OCT | NOV | DEC | YEAR |
|---|---|---|---|---|---|---|---|---|---|---|---|---|---|
| Maximum Temp °F | 46.0 | 51.0 | 61.5 | 71.6 | 78.2 | 84.6 | 87.5 | 86.2 | 80.8 | 70.9 | 59.9 | 50.2 | 69.0 |
| Minimum Temp °F | 27.2 | 29.0 | 36.9 | 44.7 | 53.0 | 60.8 | 65.2 | 64.1 | 58.0 | 46.1 | 37.9 | 31.0 | 46.2 |
| Mean Temp °F | 36.6 | 40.0 | 49.2 | 58.2 | 65.6 | 72.7 | 76.3 | 75.2 | 69.4 | 58.5 | 48.9 | 40.6 | 57.6 |
| Days Max Temp ≥ 90 °F | 0 | 0 | 0 | 0 | 0 | 3 | 10 | 6 | 2 | 0 | 0 | 0 | 21 |
| Days Max Temp ≤ 32 °F | 4 | 2 | 0 | 0 | 0 | 0 | 0 | 0 | 0 | 0 | 0 | 1 | 7 |
| Days Min Temp ≤ 32 °F | 21 | 18 | 11 | 3 | 0 | 0 | 0 | 0 | 0 | 2 | 10 | 18 | 83 |
| Days Min Temp ≤ 0 °F | 1 | 0 | 0 | 0 | 0 | 0 | 0 | 0 | 0 | 0 | 0 | 0 | 1 |
| Heating Degree Days | 873 | 699 | 484 | 222 | 67 | 6 | 0 | 1 | 27 | 215 | 477 | 749 | 3820 |
| Cooling Degree Days | 0 | 0 | 3 | 25 | 96 | 259 | 384 | 333 | 168 | 24 | 1 | 0 | 1293 |
| Total Precipitation (") | 3.59 | 3.67 | 4.00 | 3.38 | 4.11 | 3.73 | 4.68 | 3.97 | 3.21 | 2.88 | 3.12 | 3.69 | 44.03 |
| Days ≥ 0.1" Precip | 8 | 7 | 8 | 7 | 8 | 8 | 9 | 7 | 6 | 6 | 7 | 7 | 88 |
| Total Snowfall (") | 5.6 | 4.4 | 1.6 | 0.4 | 0.0 | 0.0 | 0.0 | 0.0 | 0.0 | 0.0 | 0.3 | 2.2 | 14.5 |
| Days ≥ 1" Snow Depth | 5 | 3 | 1 | 0 | 0 | 0 | 0 | 0 | 0 | 0 | 0 | 1 | 10 |

## KINGSTON SPRINGS *Cheatham County* ELEVATION 449 ft LAT/LONG 36° 7 ' N / 87° 6 ' W

| | JAN | FEB | MAR | APR | MAY | JUN | JUL | AUG | SEP | OCT | NOV | DEC | YEAR |
|---|---|---|---|---|---|---|---|---|---|---|---|---|---|
| Maximum Temp °F | 45.0 | 49.7 | 60.4 | 71.0 | 77.9 | 85.7 | 89.3 | 88.7 | 82.3 | 72.4 | 60.8 | 50.2 | 69.5 |
| Minimum Temp °F | 22.6 | 24.6 | 33.8 | 42.4 | 51.3 | 60.2 | 65.0 | 63.1 | 55.9 | 42.0 | 34.3 | 26.9 | 43.5 |
| Mean Temp °F | 33.9 | 37.3 | 47.1 | 56.8 | 64.7 | 73.0 | 77.2 | 75.9 | 69.3 | 57.3 | 47.7 | 38.7 | 56.6 |
| Days Max Temp ≥ 90 °F | 0 | 0 | 0 | 0 | 1 | 8 | 17 | 14 | 6 | 0 | 0 | 0 | 46 |
| Days Max Temp ≤ 32 °F | 6 | 3 | 0 | 0 | 0 | 0 | 0 | 0 | 0 | 0 | 0 | 2 | 11 |
| Days Min Temp ≤ 32 °F | 24 | 22 | 16 | 5 | 0 | 0 | 0 | 0 | 0 | 6 | 14 | 21 | 108 |
| Days Min Temp ≤ 0 °F | 2 | 0 | 0 | 0 | 0 | 0 | 0 | 0 | 0 | 0 | 0 | 0 | 2 |
| Heating Degree Days | 958 | 776 | 552 | 265 | 94 | 9 | 0 | 1 | 42 | 257 | 515 | 808 | 4277 |
| Cooling Degree Days | 0 | 0 | 7 | 24 | 96 | 269 | 416 | *376* | 179 | 29 | 3 | 0 | 1399 |
| Total Precipitation (") | 3.69 | 4.54 | 5.30 | 4.41 | 5.12 | 3.94 | 4.00 | 3.43 | 3.79 | 3.64 | 4.63 | 5.23 | 51.72 |
| Days ≥ 0.1" Precip | 7 | 7 | 8 | 7 | 8 | 6 | 7 | 6 | 6 | 5 | 7 | 7 | 81 |
| Total Snowfall (") | 2.3 | *1.4* | 0.5 | 0.0 | 0.0 | 0.0 | 0.0 | 0.0 | 0.0 | 0.0 | 0.1 | 0.7 | 5.0 |
| Days ≥ 1" Snow Depth | *2* | *1* | 0 | 0 | 0 | 0 | 0 | 0 | 0 | 0 | 0 | 0 | 3 |

## KNOXVILLE MCG TYSON *Blount County* ELEVATION 905 ft LAT/LONG 35° 49 ' N / 83° 59 ' W

| | JAN | FEB | MAR | APR | MAY | JUN | JUL | AUG | SEP | OCT | NOV | DEC | YEAR |
|---|---|---|---|---|---|---|---|---|---|---|---|---|---|
| Maximum Temp °F | 46.2 | 51.2 | 61.1 | 70.9 | 77.5 | 84.7 | 87.6 | 86.7 | 81.0 | 70.4 | 60.0 | 50.7 | 69.0 |
| Minimum Temp °F | 28.3 | 30.8 | 38.8 | 47.2 | 55.5 | 63.7 | 68.1 | 67.2 | 60.7 | 47.8 | 39.0 | 32.0 | 48.3 |
| Mean Temp °F | 37.3 | 41.0 | 50.0 | 59.1 | 66.5 | 74.2 | 77.9 | 76.9 | 70.9 | 59.1 | 49.5 | 41.4 | 58.7 |
| Days Max Temp ≥ 90 °F | 0 | 0 | 0 | 0 | 0 | 5 | 12 | 9 | 3 | 0 | 0 | 0 | 29 |
| Days Max Temp ≤ 32 °F | 3 | 1 | 0 | 0 | 0 | 0 | 0 | 0 | 0 | 0 | 0 | 1 | 5 |
| Days Min Temp ≤ 32 °F | 20 | 16 | 9 | 2 | 0 | 0 | 0 | 0 | 0 | 1 | 8 | 17 | 73 |
| Days Min Temp ≤ 0 °F | 1 | 0 | 0 | 0 | 0 | 0 | 0 | 0 | 0 | 0 | 0 | 0 | 1 |
| Heating Degree Days | 853 | 670 | 463 | 203 | 62 | 3 | 0 | 0 | 21 | 204 | 460 | 726 | 3665 |
| Cooling Degree Days | 0 | 0 | 4 | 27 | 114 | 300 | 432 | 392 | 202 | 33 | 2 | 1 | 1507 |
| Total Precipitation (") | 4.16 | 4.00 | 5.11 | 3.70 | 4.08 | 4.04 | 4.54 | 3.33 | 3.13 | 2.88 | 3.76 | 4.66 | 47.39 |
| Days ≥ 0.1" Precip | 8 | 7 | 9 | 7 | 8 | 7 | 8 | 6 | 5 | 5 | 7 | 8 | 85 |
| Total Snowfall (") | 4.4 | 3.6 | 1.8 | 0.7 | 0.0 | 0.0 | 0.0 | 0.0 | 0.0 | 0.0 | 0.1 | 1.1 | 11.7 |
| Days ≥ 1" Snow Depth | 3 | 2 | 1 | 0 | 0 | 0 | 0 | 0 | 0 | 0 | 0 | 0 | 6 |

## LAFAYETTE *Macon County* ELEVATION 991 ft LAT/LONG 36° 32 ' N / 86° 2 ' W

| | JAN | FEB | MAR | APR | MAY | JUN | JUL | AUG | SEP | OCT | NOV | DEC | YEAR |
|---|---|---|---|---|---|---|---|---|---|---|---|---|---|
| Maximum Temp °F | 44.9 | 50.3 | 60.9 | 71.3 | 78.0 | 85.2 | 88.4 | 87.2 | 81.3 | 71.3 | 59.9 | 50.0 | 69.1 |
| Minimum Temp °F | 26.0 | 29.1 | 37.7 | 46.7 | 54.3 | 62.4 | 66.2 | 64.3 | 58.1 | 46.2 | 38.6 | 30.6 | 46.7 |
| Mean Temp °F | 35.5 | 39.7 | 49.3 | 59.0 | 66.2 | 73.8 | 77.3 | 75.8 | 69.8 | 58.7 | 49.3 | 40.3 | 57.9 |
| Days Max Temp ≥ 90 °F | 0 | 0 | 0 | 0 | 1 | 6 | 13 | 10 | 3 | 0 | 0 | 0 | 33 |
| Days Max Temp ≤ 32 °F | 5 | 2 | 0 | 0 | 0 | 0 | 0 | 0 | 0 | 0 | 0 | 2 | 9 |
| Days Min Temp ≤ 32 °F | 22 | 18 | 11 | 2 | 0 | 0 | 0 | 0 | 0 | 2 | 9 | 18 | 82 |
| Days Min Temp ≤ 0 °F | 1 | 0 | 0 | 0 | 0 | 0 | 0 | 0 | 0 | 0 | 0 | 0 | 1 |
| Heating Degree Days | 909 | 707 | 489 | 211 | 67 | 5 | 0 | 1 | 32 | 217 | 467 | 760 | 3865 |
| Cooling Degree Days | 0 | 0 | 7 | 40 | 122 | 301 | 428 | 377 | 194 | 35 | 3 | 0 | 1507 |
| Total Precipitation (") | 4.28 | 4.39 | 5.47 | 4.27 | 5.40 | 4.42 | 4.84 | 4.15 | 4.35 | 3.67 | 4.86 | 5.82 | 55.92 |
| Days ≥ 0.1" Precip | 8 | 7 | 9 | 8 | 8 | 7 | 7 | 6 | 6 | 6 | 7 | 7 | 86 |
| Total Snowfall (") | 4.7 | *3.8* | 1.2 | 0.1 | 0.0 | 0.0 | 0.0 | 0.0 | 0.0 | 0.0 | 0.2 | 1.1 | 11.1 |
| Days ≥ 1" Snow Depth | na | *1* | 0 | 0 | 0 | 0 | 0 | 0 | 0 | 0 | 0 | 1 | na |

## LAWRENCEBURG FLT PLT *Lawrence County* ELEVATION 951 ft LAT/LONG 35° 16 ' N / 87° 19 ' W

| | JAN | FEB | MAR | APR | MAY | JUN | JUL | AUG | SEP | OCT | NOV | DEC | YEAR |
|---|---|---|---|---|---|---|---|---|---|---|---|---|---|
| Maximum Temp °F | 48.1 | 53.2 | 62.8 | 72.6 | 78.3 | 85.3 | 88.4 | 87.4 | 81.8 | 72.3 | 61.6 | 52.2 | 70.3 |
| Minimum Temp °F | 27.1 | 29.6 | 37.8 | 45.6 | 53.6 | 61.2 | 65.4 | 63.9 | 58.0 | 45.5 | 37.8 | 31.0 | 46.4 |
| Mean Temp °F | 37.6 | 41.4 | 50.3 | 59.1 | 66.0 | 73.3 | 76.9 | 75.7 | 69.9 | 59.0 | 49.7 | 41.6 | 58.4 |
| Days Max Temp ≥ 90 °F | 0 | 0 | 0 | 0 | 1 | 6 | 13 | 11 | 3 | 0 | 0 | 0 | 34 |
| Days Max Temp ≤ 32 °F | 3 | 1 | 0 | 0 | 0 | 0 | 0 | 0 | 0 | 0 | 0 | 1 | 5 |
| Days Min Temp ≤ 32 °F | 22 | 17 | 12 | 3 | 0 | 0 | 0 | 0 | 0 | 4 | 11 | 17 | 86 |
| Days Min Temp ≤ 0 °F | 1 | 0 | 0 | 0 | 0 | 0 | 0 | 0 | 0 | 0 | 0 | 0 | 1 |
| Heating Degree Days | 843 | 660 | 455 | 202 | 64 | 4 | 0 | 1 | 32 | 211 | 455 | 720 | 3647 |
| Cooling Degree Days | 0 | 0 | 6 | 26 | 106 | 267 | 388 | 350 | 185 | 32 | 3 | 2 | 1365 |
| Total Precipitation (") | 4.77 | 4.60 | 6.33 | 4.75 | 5.99 | 3.54 | 4.43 | 3.83 | 4.42 | 3.63 | 5.13 | 5.77 | 57.19 |
| Days ≥ 0.1" Precip | 7 | 7 | 8 | 7 | 8 | 6 | 8 | 6 | 6 | 5 | 7 | 8 | 83 |
| Total Snowfall (") | 2.2 | 1.9 | 0.3 | 0.0 | 0.0 | 0.0 | 0.0 | 0.0 | 0.0 | 0.0 | 0.2 | 0.6 | 5.2 |
| Days ≥ 1" Snow Depth | *2* | *1* | 0 | 0 | 0 | 0 | 0 | 0 | 0 | 0 | 0 | 0 | 3 |

## LEBANON 3 W *Wilson County* ELEVATION 535 ft LAT/LONG 36° 13 ' N / 86° 20 ' W

| | JAN | FEB | MAR | APR | MAY | JUN | JUL | AUG | SEP | OCT | NOV | DEC | YEAR |
|---|---|---|---|---|---|---|---|---|---|---|---|---|---|
| Maximum Temp °F | 45.4 | 50.3 | 60.4 | 71.0 | 78.4 | 86.5 | 90.0 | 88.8 | 82.6 | 71.9 | 60.1 | 50.3 | 69.6 |
| Minimum Temp °F | 24.2 | 26.8 | 35.9 | 44.9 | 53.0 | 61.9 | 66.3 | 64.3 | 57.2 | 43.8 | 36.4 | 28.7 | 45.3 |
| Mean Temp °F | 34.8 | 38.6 | 48.2 | 58.0 | 65.7 | 74.2 | 78.2 | 76.6 | 69.9 | 57.9 | 48.3 | 39.5 | 57.5 |
| Days Max Temp ≥ 90 °F | 0 | 0 | 0 | 0 | 1 | 10 | 18 | 14 | 6 | 0 | 0 | 0 | 49 |
| Days Max Temp ≤ 32 °F | 5 | 2 | 0 | 0 | 0 | 0 | 0 | 0 | 0 | 0 | 0 | 2 | 9 |
| Days Min Temp ≤ 32 °F | 23 | 20 | 13 | 4 | 0 | 0 | 0 | 0 | 0 | 5 | 12 | 20 | 97 |
| Days Min Temp ≤ 0 °F | 1 | 0 | 0 | 0 | 0 | 0 | 0 | 0 | 0 | 0 | 0 | 0 | 1 |
| Heating Degree Days | 929 | 740 | 521 | 241 | 81 | 6 | 0 | 1 | 38 | 247 | 498 | 783 | 4085 |
| Cooling Degree Days | 0 | 0 | 7 | 33 | 115 | 295 | 437 | 379 | 189 | 34 | 4 | 1 | 1494 |
| Total Precipitation (") | 4.05 | 4.18 | 5.30 | 4.31 | 4.95 | 3.96 | 4.44 | 4.24 | 3.90 | 3.46 | 4.55 | 5.33 | 52.67 |
| Days ≥ 0.1" Precip | 7 | 7 | 8 | 7 | 8 | 6 | 7 | 6 | 6 | 5 | 7 | 8 | 82 |
| Total Snowfall (") | 2.6 | 2.9 | 0.3 | 0.0 | 0.0 | 0.0 | 0.0 | 0.0 | 0.0 | 0.0 | 0.3 | 0.5 | 6.6 |
| Days ≥ 1" Snow Depth | 4 | 2 | 0 | 0 | 0 | 0 | 0 | 0 | 0 | 0 | 0 | 1 | 7 |

## LENOIR CITY *Loudon County* ELEVATION 781 ft LAT/LONG 35° 48 ' N / 84° 15 ' W

| | JAN | FEB | MAR | APR | MAY | JUN | JUL | AUG | SEP | OCT | NOV | DEC | YEAR |
|---|---|---|---|---|---|---|---|---|---|---|---|---|---|
| Maximum Temp °F | 46.1 | 50.7 | 60.6 | 70.7 | 77.5 | 85.0 | 88.1 | 87.3 | 81.6 | 71.2 | 60.5 | 50.6 | 69.2 |
| Minimum Temp °F | 25.9 | 27.8 | 36.0 | 44.6 | 53.3 | 62.2 | 66.6 | 65.3 | 58.8 | 45.2 | 36.9 | 29.6 | 46.0 |
| Mean Temp °F | 36.1 | 39.3 | 48.3 | 57.7 | 65.4 | 73.6 | 77.4 | 76.3 | 70.2 | 58.2 | 48.7 | 40.1 | 57.6 |
| Days Max Temp ≥ 90 °F | 0 | 0 | 0 | 0 | 0 | 5 | 12 | 10 | 3 | 0 | 0 | 0 | 30 |
| Days Max Temp ≤ 32 °F | 3 | 2 | 0 | 0 | 0 | 0 | 0 | 0 | 0 | 0 | 0 | 1 | 6 |
| Days Min Temp ≤ 32 °F | 23 | 19 | 12 | 3 | 0 | 0 | 0 | 0 | 0 | 2 | 11 | 20 | 90 |
| Days Min Temp ≤ 0 °F | 1 | 0 | 0 | 0 | 0 | 0 | 0 | 0 | 0 | 0 | 0 | 0 | 1 |
| Heating Degree Days | 890 | 720 | 513 | 238 | 76 | 5 | 0 | 0 | 25 | 227 | 483 | 765 | 3942 |
| Cooling Degree Days | 0 | 0 | 3 | 26 | 111 | 305 | 431 | 386 | 204 | 29 | 2 | 0 | 1497 |
| Total Precipitation (") | 4.73 | 4.58 | 5.82 | 4.24 | 4.71 | 4.09 | 4.83 | 4.14 | 3.48 | 3.27 | 4.16 | 5.32 | 53.37 |
| Days ≥ 0.1" Precip | 9 | 8 | 9 | 8 | 8 | 7 | 8 | 7 | 6 | 5 | 7 | 8 | 90 |
| Total Snowfall (") | 3.9 | 3.9 | 1.1 | 0.4 | 0.0 | 0.0 | 0.0 | 0.0 | 0.0 | 0.0 | 0.0 | 0.9 | 10.2 |
| Days ≥ 1" Snow Depth | 3 | 2 | 0 | 0 | 0 | 0 | 0 | 0 | 0 | 0 | 0 | 0 | 5 |

## LEWISBURG EXP STN *Marshall County* ELEVATION 791 ft LAT/LONG 35° 27 ' N / 86° 48 ' W

| | JAN | FEB | MAR | APR | MAY | JUN | JUL | AUG | SEP | OCT | NOV | DEC | YEAR |
|---|---|---|---|---|---|---|---|---|---|---|---|---|---|
| Maximum Temp °F | 45.9 | 50.4 | 59.9 | 70.3 | 77.5 | 85.3 | 88.6 | 88.1 | 81.9 | 71.4 | 60.2 | 50.6 | 69.2 |
| Minimum Temp °F | 24.5 | 26.9 | 35.3 | 44.0 | 52.4 | 61.2 | 65.7 | 63.6 | 57.0 | 43.9 | 36.1 | 28.7 | 44.9 |
| Mean Temp °F | 35.2 | 38.7 | 47.6 | 57.1 | 65.0 | 73.3 | 77.2 | 75.8 | 69.5 | 57.7 | 48.2 | 39.7 | 57.1 |
| Days Max Temp ≥ 90 °F | 0 | 0 | 0 | 0 | 1 | 7 | 14 | 13 | 5 | 0 | 0 | 0 | 40 |
| Days Max Temp ≤ 32 °F | 5 | 2 | 0 | 0 | 0 | 0 | 0 | 0 | 0 | 0 | 0 | 2 | 9 |
| Days Min Temp ≤ 32 °F | 23 | 20 | 14 | 5 | 0 | 0 | 0 | 0 | 0 | 5 | 13 | 19 | 99 |
| Days Min Temp ≤ 0 °F | 1 | 0 | 0 | 0 | 0 | 0 | 0 | 0 | 0 | 0 | 0 | 0 | 1 |
| Heating Degree Days | 916 | 738 | 537 | 257 | 90 | 8 | 0 | 1 | 42 | 251 | 500 | 779 | 4119 |
| Cooling Degree Days | 0 | 0 | 5 | 24 | 99 | 274 | 403 | 360 | 185 | 33 | 3 | 1 | 1387 |
| Total Precipitation (") | 4.41 | 4.10 | 5.97 | 4.60 | 5.68 | 4.05 | 4.86 | 3.27 | 4.17 | 3.90 | 4.84 | 5.35 | 55.20 |
| Days ≥ 0.1" Precip | 7 | 7 | 8 | 7 | 8 | 7 | 8 | 5 | 6 | 5 | 7 | 8 | 83 |
| Total Snowfall (") | *2.5* | *0.5* | *0.2* | 0.0 | 0.0 | 0.0 | 0.0 | 0.0 | 0.0 | 0.0 | 0.6 | 0.6 | 4.4 |
| Days ≥ 1" Snow Depth | *3* | 1 | 0 | 0 | 0 | 0 | 0 | 0 | 0 | 0 | 0 | 1 | 5 |

## LEXINGTON *Henderson County* ELEVATION 531 ft LAT/LONG 35° 41 ' N / 88° 23 ' W

| | JAN | FEB | MAR | APR | MAY | JUN | JUL | AUG | SEP | OCT | NOV | DEC | YEAR |
|---|---|---|---|---|---|---|---|---|---|---|---|---|---|
| Maximum Temp °F | 46.4 | 51.3 | *60.9* | *71.8* | *78.3* | 86.5 | 90.1 | 88.5 | 83.1 | 73.0 | 61.0 | 51.2 | 70.2 |
| Minimum Temp °F | 26.1 | 29.0 | *37.6* | *46.6* | *54.9* | 64.2 | 68.2 | 66.4 | 60.2 | 46.7 | 38.2 | 30.6 | 47.4 |
| Mean Temp °F | 36.3 | 40.2 | *49.2* | *59.2* | *66.6* | 75.3 | 79.1 | 77.5 | 71.7 | 59.9 | 49.6 | 41.0 | 58.8 |
| Days Max Temp ≥ 90 °F | 0 | 0 | *0* | *0* | *1* | 9 | 19 | 13 | 5 | 0 | 0 | 0 | 47 |
| Days Max Temp ≤ 32 °F | 4 | 2 | *0* | *0* | *0* | 0 | 0 | 0 | 0 | 0 | 0 | 2 | 8 |
| Days Min Temp ≤ 32 °F | 22 | 18 | *10* | *2* | *0* | 0 | 0 | 0 | 0 | 2 | 10 | 18 | 82 |
| Days Min Temp ≤ 0 °F | 1 | 0 | *0* | *0* | *0* | 0 | 0 | 0 | 0 | 0 | 0 | 0 | 1 |
| Heating Degree Days | 883 | 694 | *486* | *204* | *62* | 2 | 0 | 0 | 23 | 195 | 460 | 740 | 3749 |
| Cooling Degree Days | 0 | 0 | na | na | na | *356* | 482 | 427 | *252* | 46 | 3 | *3* | na |
| Total Precipitation (") | *3.83* | 4.00 | *5.09* | *4.74* | *5.84* | 3.96 | *4.41* | 2.91 | 3.76 | 3.42 | 4.41 | *4.49* | 50.86 |
| Days ≥ 0.1" Precip | na | 6 | na | *8* | *8* | 5 | *6* | 4 | 5 | 4 | 6 | 6 | na |
| Total Snowfall (") | *2.4* | *1.8* | *0.8* | *0.0* | *0.0* | 0.0 | 0.0 | 0.0 | 0.0 | 0.0 | 0.1 | 0.4 | 5.5 |
| Days ≥ 1" Snow Depth | *3* | 1 | *0* | *0* | *0* | 0 | 0 | 0 | 0 | 0 | 0 | 0 | 4 |

## LINDEN 2 *Perry County* ELEVATION 502 ft LAT/LONG 35° 37 ' N / 87° 50 ' W

| | JAN | FEB | MAR | APR | MAY | JUN | JUL | AUG | SEP | OCT | NOV | DEC | YEAR |
|---|---|---|---|---|---|---|---|---|---|---|---|---|---|
| Maximum Temp °F | 46.6 | 51.8 | 61.8 | 72.2 | 78.3 | 86.0 | 89.4 | 88.1 | 82.5 | 72.3 | 61.2 | 51.2 | 70.1 |
| Minimum Temp °F | 24.3 | 27.1 | 35.4 | 43.8 | 52.5 | 61.0 | 65.7 | 63.4 | 57.0 | 43.2 | 35.8 | 28.5 | 44.8 |
| Mean Temp °F | 35.4 | 39.4 | 48.6 | 58.0 | 65.4 | 73.5 | 77.5 | 75.8 | 69.8 | 57.8 | 48.5 | 39.9 | 57.5 |
| Days Max Temp ≥ 90 °F | 0 | 0 | 0 | 0 | 1 | 9 | 17 | 13 | 5 | 0 | 0 | 0 | 45 |
| Days Max Temp ≤ 32 °F | 4 | 2 | 0 | 0 | 0 | 0 | 0 | 0 | 0 | 0 | 0 | 2 | 8 |
| Days Min Temp ≤ 32 °F | 24 | 20 | 14 | 5 | 0 | 0 | 0 | 0 | 0 | 5 | 13 | 21 | 102 |
| Days Min Temp ≤ 0 °F | 1 | 0 | 0 | 0 | 0 | 0 | 0 | 0 | 0 | 0 | 0 | 0 | 1 |
| Heating Degree Days | 903 | 705 | 514 | 235 | 81 | 7 | 0 | 1 | 36 | 245 | 491 | 773 | 3991 |
| Cooling Degree Days | *0* | 1 | 7 | 22 | 89 | 268 | 400 | 340 | 174 | 27 | 3 | 1 | 1332 |
| Total Precipitation (") | 4.68 | 4.61 | 5.58 | 4.70 | 6.16 | 4.52 | 4.47 | 3.62 | 3.85 | 3.54 | 5.06 | 6.05 | 56.84 |
| Days ≥ 0.1" Precip | *7* | *7* | 8 | 7 | 8 | 6 | 6 | 5 | 6 | 5 | 7 | 8 | 80 |
| Total Snowfall (") | na | na | *0.5* | 0.0 | 0.0 | 0.0 | 0.0 | 0.0 | 0.0 | 0.0 | 0.1 | *0.4* | na |
| Days ≥ 1" Snow Depth | na | na | *0* | 0 | 0 | 0 | 0 | 0 | 0 | 0 | 0 | *0* | na |

## LIVINGSTON RADIO WLI *Overton County* ELEVATION 1020 ft LAT/LONG 36° 23 ' N / 85° 18 ' W

| | JAN | FEB | MAR | APR | MAY | JUN | JUL | AUG | SEP | OCT | NOV | DEC | YEAR |
|---|---|---|---|---|---|---|---|---|---|---|---|---|---|
| Maximum Temp °F | 45.7 | 50.8 | 61.0 | 71.1 | 77.1 | 84.7 | 87.7 | 86.6 | 80.6 | 70.5 | 59.8 | 50.7 | 68.9 |
| Minimum Temp °F | 26.1 | 28.3 | 36.7 | 44.9 | 52.1 | 60.3 | 64.6 | 62.7 | 56.6 | 44.4 | 37.0 | 30.6 | 45.4 |
| Mean Temp °F | 36.0 | 39.6 | 48.9 | 58.0 | 64.6 | 72.5 | 76.2 | 74.7 | 68.6 | 57.5 | 48.4 | 40.7 | 57.1 |
| Days Max Temp ≥ 90 °F | 0 | 0 | 0 | 0 | 1 | 4 | 12 | 7 | 3 | 0 | 0 | 0 | 27 |
| Days Max Temp ≤ 32 °F | 4 | 2 | 0 | 0 | 0 | 0 | 0 | 0 | 0 | 0 | 0 | 2 | 8 |
| Days Min Temp ≤ 32 °F | 21 | 18 | 12 | 4 | 0 | 0 | 0 | 0 | 0 | 4 | 11 | 18 | 88 |
| Days Min Temp ≤ 0 °F | 1 | 0 | 0 | 0 | 0 | 0 | 0 | 0 | 0 | 0 | 0 | 0 | 1 |
| Heating Degree Days | 894 | 711 | 500 | 230 | 89 | 7 | 1 | 1 | 39 | 248 | 493 | 748 | 3961 |
| Cooling Degree Days | 0 | 0 | 8 | 30 | 101 | 269 | 401 | 351 | 173 | 27 | 3 | 1 | 1364 |
| Total Precipitation (") | 4.15 | 3.89 | 5.34 | 4.17 | 5.10 | 3.92 | 5.36 | 4.20 | 3.69 | 3.01 | 4.34 | 5.26 | 52.43 |
| Days ≥ 0.1" Precip | 8 | 7 | 8 | *7* | 8 | 7 | 7 | 6 | 6 | 5 | 7 | 7 | 83 |
| Total Snowfall (") | *4.3* | 3.3 | 1.0 | 0.2 | 0.0 | 0.0 | 0.0 | 0.0 | 0.0 | 0.0 | 0.9 | 1.2 | 10.9 |
| Days ≥ 1" Snow Depth | *3* | *1* | 0 | 0 | 0 | 0 | 0 | 0 | 0 | 0 | 0 | *0* | 4 |

## MARTIN UNIV OF TENN *Weakley County* ELEVATION 390 ft LAT/LONG 36° 19 ' N / 88° 51 ' W

| | JAN | FEB | MAR | APR | MAY | JUN | JUL | AUG | SEP | OCT | NOV | DEC | YEAR |
|---|---|---|---|---|---|---|---|---|---|---|---|---|---|
| Maximum Temp °F | 43.8 | 49.0 | 59.8 | 70.6 | 78.5 | 86.9 | 90.1 | 88.8 | 82.8 | 72.2 | 59.9 | 49.1 | 69.3 |
| Minimum Temp °F | 24.5 | 27.8 | 37.5 | 47.2 | 55.5 | 64.0 | 67.8 | 65.1 | 58.2 | 45.9 | 37.8 | 29.5 | 46.7 |
| Mean Temp °F | 34.1 | 38.4 | 48.7 | 58.9 | 67.1 | 75.5 | 79.0 | 77.0 | 70.5 | 59.1 | 48.9 | 39.3 | 58.0 |
| Days Max Temp ≥ 90 °F | 0 | 0 | 0 | 0 | 2 | 11 | 19 | 15 | 6 | 0 | 0 | 0 | 53 |
| Days Max Temp ≤ 32 °F | 6 | 3 | 0 | 0 | 0 | 0 | 0 | 0 | 0 | 0 | 0 | 2 | 11 |
| Days Min Temp ≤ 32 °F | 24 | 19 | 12 | 2 | 0 | 0 | 0 | 0 | 0 | 3 | 10 | 19 | 89 |
| Days Min Temp ≤ 0 °F | 1 | 0 | 0 | 0 | 0 | 0 | 0 | 0 | 0 | 0 | 0 | 0 | 1 |
| Heating Degree Days | 951 | 744 | 507 | 217 | 60 | 3 | 0 | 1 | 35 | 218 | 480 | 791 | 4007 |
| Cooling Degree Days | 0 | 0 | 6 | 33 | 114 | 316 | 449 | 381 | 192 | 34 | 3 | 1 | 1529 |
| Total Precipitation (") | 3.83 | 4.19 | 4.82 | 5.22 | 4.87 | 4.44 | 4.97 | 3.25 | 3.95 | 3.76 | 4.91 | 5.41 | 53.62 |
| Days ≥ 0.1" Precip | 6 | 7 | 7 | 8 | 8 | 6 | 6 | 5 | 5 | 6 | 7 | 7 | 78 |
| Total Snowfall (") | 3.0 | *2.0* | 0.9 | 0.0 | 0.0 | 0.0 | 0.0 | 0.0 | 0.0 | 0.0 | 0.0 | 0.5 | 6.4 |
| Days ≥ 1" Snow Depth | *2* | 1 | 0 | 0 | 0 | 0 | 0 | 0 | 0 | 0 | 0 | 0 | 3 |

## MC MINNVILLE *Warren County* ELEVATION 902 ft LAT/LONG 35° 41 ' N / 85° 48 ' W

| | JAN | FEB | MAR | APR | MAY | JUN | JUL | AUG | SEP | OCT | NOV | DEC | YEAR |
|---|---|---|---|---|---|---|---|---|---|---|---|---|---|
| Maximum Temp °F | 47.4 | 52.0 | 61.7 | 71.8 | 78.0 | 85.4 | 88.1 | 87.0 | 81.1 | 71.5 | 60.9 | 52.1 | 69.8 |
| Minimum Temp °F | 28.3 | 30.7 | 38.7 | 46.6 | 54.1 | 62.0 | 66.2 | 65.1 | 59.2 | 46.7 | 38.7 | 32.4 | 47.4 |
| Mean Temp °F | 37.9 | 41.4 | 50.2 | 59.2 | 66.0 | 73.8 | 77.2 | 76.1 | 70.1 | 59.1 | 49.8 | 42.3 | 58.6 |
| Days Max Temp ≥ 90 °F | 0 | 0 | 0 | 0 | 0 | 6 | 13 | 9 | 3 | 0 | 0 | 0 | 31 |
| Days Max Temp ≤ 32 °F | 4 | 2 | 0 | 0 | 0 | 0 | 0 | 0 | 0 | 0 | 0 | 1 | 7 |
| Days Min Temp ≤ 32 °F | 20 | 17 | 10 | 2 | 0 | 0 | 0 | 0 | 0 | 3 | 10 | 17 | 79 |
| Days Min Temp ≤ 0 °F | 1 | 0 | 0 | 0 | 0 | 0 | 0 | 0 | 0 | 0 | 0 | 0 | 1 |
| Heating Degree Days | 835 | 662 | 460 | 203 | 67 | 4 | 0 | 0 | 28 | 210 | 452 | 698 | 3619 |
| Cooling Degree Days | 0 | 1 | 8 | 34 | 115 | 289 | 412 | 366 | 196 | 37 | 5 | 2 | 1465 |
| Total Precipitation (") | 4.26 | 4.24 | 5.75 | 4.30 | 5.23 | 4.20 | 4.74 | 3.55 | 3.83 | 3.34 | 4.61 | 5.40 | 53.45 |
| Days ≥ 0.1" Precip | 8 | 8 | 9 | 7 | 8 | 7 | 8 | 7 | 6 | 5 | 7 | 8 | 88 |
| Total Snowfall (") | 2.9 | 2.4 | 0.6 | 0.0 | 0.0 | 0.0 | 0.0 | 0.0 | 0.0 | 0.0 | 0.2 | 0.8 | 6.9 |
| Days ≥ 1" Snow Depth | 2 | 1 | 0 | 0 | 0 | 0 | 0 | 0 | 0 | 0 | 0 | 0 | 3 |

## MEMPHIS INTL ARPT *Shelby County* ELEVATION 258 ft LAT/LONG 35° 3 ' N / 89° 59 ' W

| | JAN | FEB | MAR | APR | MAY | JUN | JUL | AUG | SEP | OCT | NOV | DEC | YEAR |
|---|---|---|---|---|---|---|---|---|---|---|---|---|---|
| Maximum Temp °F | 48.1 | 53.2 | 62.9 | 73.3 | 80.5 | 88.5 | 91.6 | 89.9 | 84.0 | 74.0 | 62.1 | 52.6 | 71.7 |
| Minimum Temp °F | 31.1 | 34.5 | 43.2 | 52.7 | 61.3 | 69.7 | 73.8 | 71.7 | 64.8 | 52.5 | 43.1 | 35.1 | 52.8 |
| Mean Temp °F | 39.6 | 43.9 | 53.1 | 63.0 | 70.9 | 79.1 | 82.7 | 80.8 | 74.4 | 63.3 | 52.6 | 43.9 | 62.3 |
| Days Max Temp ≥ 90 °F | 0 | 0 | 0 | 0 | 2 | 14 | 22 | 18 | 8 | 0 | 0 | 0 | 64 |
| Days Max Temp ≤ 32 °F | 3 | 1 | 0 | 0 | 0 | 0 | 0 | 0 | 0 | 0 | 0 | 1 | 5 |
| Days Min Temp ≤ 32 °F | 18 | 13 | 5 | 0 | 0 | 0 | 0 | 0 | 0 | 0 | 4 | 13 | 53 |
| Days Min Temp ≤ 0 °F | 0 | 0 | 0 | 0 | 0 | 0 | 0 | 0 | 0 | 0 | 0 | 0 | 0 |
| Heating Degree Days | 781 | 591 | 378 | 132 | 21 | 0 | 0 | 0 | 13 | 128 | 377 | 650 | 3071 |
| Cooling Degree Days | 0 | 2 | 16 | 79 | 220 | 450 | 579 | 528 | 319 | 85 | 14 | 3 | 2295 |
| Total Precipitation (") | 3.86 | 4.29 | 5.24 | 5.59 | 5.12 | 3.74 | 3.58 | 3.31 | 3.51 | 3.24 | 5.20 | 5.77 | 52.45 |
| Days ≥ 0.1" Precip | 7 | 7 | 7 | 7 | 7 | 6 | 6 | 5 | 5 | 5 | 7 | 7 | 76 |
| Total Snowfall (") | 2.4 | 1.5 | 1.0 | 0.0 | 0.0 | 0.0 | 0.0 | 0.0 | 0.0 | 0.0 | 0.1 | 0.1 | 5.1 |
| Days ≥ 1" Snow Depth | 2 | 1 | 0 | 0 | 0 | 0 | 0 | 0 | 0 | 0 | 0 | 0 | 3 |

## MONTEAGLE *Grundy County* ELEVATION 1926 ft LAT/LONG 35° 15 ' N / 85° 51 ' W

| | JAN | FEB | MAR | APR | MAY | JUN | JUL | AUG | SEP | OCT | NOV | DEC | YEAR |
|---|---|---|---|---|---|---|---|---|---|---|---|---|---|
| Maximum Temp °F | 44.0 | 48.8 | 57.4 | 67.8 | 74.0 | 81.1 | 83.9 | 83.2 | 77.5 | 67.9 | 57.4 | 48.2 | 65.9 |
| Minimum Temp °F | 26.7 | 29.8 | 37.8 | 46.7 | 54.2 | 61.8 | 65.4 | 64.3 | 58.9 | 47.8 | 39.0 | 31.1 | 47.0 |
| Mean Temp °F | 35.4 | 39.3 | 47.6 | 57.2 | 64.1 | 71.5 | 74.7 | 73.8 | 68.2 | 57.9 | 48.2 | 39.7 | 56.5 |
| Days Max Temp ≥ 90 °F | 0 | 0 | 0 | 0 | 0 | 1 | 4 | 3 | 1 | 0 | 0 | 0 | 9 |
| Days Max Temp ≤ 32 °F | 6 | 3 | 0 | 0 | 0 | 0 | 0 | 0 | 0 | 0 | 0 | 2 | 11 |
| Days Min Temp ≤ 32 °F | 20 | 17 | 11 | 2 | 0 | 0 | 0 | 0 | 0 | 1 | 9 | 18 | 78 |
| Days Min Temp ≤ 0 °F | 1 | 0 | 0 | 0 | 0 | 0 | 0 | 0 | 0 | 0 | 0 | 0 | 1 |
| Heating Degree Days | 911 | 719 | 535 | 249 | 90 | 8 | 1 | 1 | 39 | 233 | 497 | 778 | 4061 |
| Cooling Degree Days | 0 | 0 | 3 | 19 | 71 | 214 | 324 | 288 | 144 | 21 | 2 | 0 | 1086 |
| Total Precipitation (") | 5.51 | 5.33 | 6.39 | 5.06 | 5.63 | 4.28 | 5.60 | 4.09 | 4.91 | 4.30 | 5.60 | 6.16 | 62.86 |
| Days ≥ 0.1" Precip | 8 | 8 | 9 | *8* | 8 | 7 | 9 | 7 | 7 | 6 | 8 | 9 | 94 |
| Total Snowfall (") | *3.3* | 2.1 | 1.7 | 0.0 | 0.0 | 0.0 | 0.0 | 0.0 | 0.0 | 0.0 | 0.7 | 1.1 | 8.9 |
| Days ≥ 1" Snow Depth | *4* | *1* | 1 | 0 | 0 | 0 | 0 | 0 | 0 | 0 | 0 | 1 | 7 |

## MOSCOW *Fayette County* ELEVATION 351 ft LAT/LONG 35° 4 ' N / 89° 24 ' W

| | JAN | FEB | MAR | APR | MAY | JUN | JUL | AUG | SEP | OCT | NOV | DEC | YEAR |
|---|---|---|---|---|---|---|---|---|---|---|---|---|---|
| Maximum Temp °F | 48.6 | 53.7 | 64.2 | 73.5 | 79.4 | ***86.7*** | 89.7 | 88.8 | 83.1 | 73.6 | 63.0 | 53.4 | 71.5 |
| Minimum Temp °F | 28.3 | 31.4 | 40.7 | 49.0 | 56.9 | ***64.8*** | 68.4 | 66.1 | 59.2 | 46.5 | 39.6 | 32.6 | 48.6 |
| Mean Temp °F | 38.5 | 42.6 | 52.5 | 61.3 | 68.2 | ***75.8*** | 79.1 | 77.5 | 71.2 | 60.0 | 51.3 | 43.0 | 60.1 |
| Days Max Temp ≥ 90 °F | 0 | 0 | 0 | 0 | 1 | ***10*** | 17 | 16 | 6 | 0 | 0 | 0 | 50 |
| Days Max Temp ≤ 32 °F | ***3*** | 1 | 0 | 0 | 0 | 0 | 0 | 0 | 0 | 0 | 0 | 1 | 5 |
| Days Min Temp ≤ 32 °F | ***20*** | ***16*** | 8 | 1 | 0 | 0 | 0 | 0 | 0 | 3 | 8 | ***16*** | 72 |
| Days Min Temp ≤ 0 °F | 0 | 0 | 0 | 0 | 0 | 0 | 0 | 0 | 0 | 0 | 0 | 0 | 0 |
| Heating Degree Days | 813 | 628 | 392 | 160 | 39 | ***1*** | 0 | 0 | 26 | 191 | 413 | 672 | 3335 |
| Cooling Degree Days | 0 | 1 | 12 | 54 | 163 | 364 | 482 | 431 | 250 | 52 | 10 | 2 | 1821 |
| Total Precipitation (") | 3.95 | 4.45 | 5.18 | 5.63 | 5.22 | 3.83 | 4.32 | 3.15 | 4.02 | 3.25 | 5.17 | 5.43 | 53.60 |
| Days ≥ 0.1" Precip | ***6*** | ***6*** | 7 | 7 | 7 | 5 | 5 | 5 | 5 | 4 | 6 | ***6*** | 69 |
| Total Snowfall (") | na | na | 0.6 | 0.0 | 0.0 | 0.0 | 0.0 | 0.0 | 0.0 | 0.0 | 0.1 | 0.1 | na |
| Days ≥ 1" Snow Depth | na | na | 0 | 0 | 0 | 0 | 0 | 0 | 0 | 0 | 0 | 0 | na |

## MURFREESBORO 5 N *Rutherford County* ELEVATION 551 ft LAT/LONG 35° 55 ' N / 86° 22 ' W

| | JAN | FEB | MAR | APR | MAY | JUN | JUL | AUG | SEP | OCT | NOV | DEC | YEAR |
|---|---|---|---|---|---|---|---|---|---|---|---|---|---|
| Maximum Temp °F | 46.0 | 50.6 | 60.7 | 71.0 | 78.0 | 86.0 | 89.5 | 88.5 | 82.3 | 71.9 | 60.8 | 50.9 | 69.7 |
| Minimum Temp °F | 25.4 | 27.6 | 36.5 | 45.4 | 53.9 | 62.2 | 66.9 | 65.0 | 58.1 | 44.8 | 37.4 | 29.7 | 46.1 |
| Mean Temp °F | 35.7 | 39.2 | 48.6 | 58.2 | 66.0 | 74.1 | 78.2 | 76.7 | 70.2 | 58.4 | 49.1 | 40.4 | 57.9 |
| Days Max Temp ≥ 90 °F | 0 | 0 | 0 | 0 | 1 | 9 | 17 | 13 | 5 | 0 | 0 | 0 | 45 |
| Days Max Temp ≤ 32 °F | 5 | 2 | 0 | 0 | 0 | 0 | 0 | 0 | 0 | 0 | 0 | 2 | 9 |
| Days Min Temp ≤ 32 °F | 23 | 19 | 12 | 3 | 0 | 0 | 0 | 0 | 0 | 3 | 11 | 19 | 90 |
| Days Min Temp ≤ 0 °F | 1 | 0 | 0 | 0 | 0 | 0 | 0 | 0 | 0 | 0 | 0 | 0 | 1 |
| Heating Degree Days | 901 | 724 | 506 | 231 | 73 | 5 | 0 | 1 | 34 | 231 | 474 | 757 | 3937 |
| Cooling Degree Days | 0 | 0 | 5 | 29 | 111 | 290 | 431 | 380 | 192 | 34 | 4 | 0 | 1476 |
| Total Precipitation (") | 4.21 | 3.97 | 5.59 | 4.29 | 5.54 | 4.06 | 4.84 | 3.84 | 4.43 | 3.41 | 4.55 | 5.36 | 54.09 |
| Days ≥ 0.1" Precip | 7 | 7 | 8 | 7 | 8 | 7 | 7 | 6 | 6 | 5 | 7 | 8 | 83 |
| Total Snowfall (") | ***2.5*** | 1.1 | 0.9 | 0.0 | 0.0 | 0.0 | 0.0 | 0.0 | 0.0 | 0.0 | 0.3 | 0.6 | 5.4 |
| Days ≥ 1" Snow Depth | ***2*** | 1 | 0 | 0 | 0 | 0 | 0 | 0 | 0 | 0 | 0 | 0 | 3 |

## NASHVILLE METRO AP *Davidson County* ELEVATION 581 ft LAT/LONG 36° 7 ' N / 86° 41 ' W

| | JAN | FEB | MAR | APR | MAY | JUN | JUL | AUG | SEP | OCT | NOV | DEC | YEAR |
|---|---|---|---|---|---|---|---|---|---|---|---|---|---|
| Maximum Temp °F | 46.0 | 50.8 | 61.1 | 71.3 | 78.5 | 86.5 | 89.7 | 88.3 | 82.2 | 71.8 | 60.1 | 50.7 | 69.8 |
| Minimum Temp °F | 27.3 | 30.1 | 39.0 | 47.7 | 56.4 | 64.8 | 69.3 | 67.8 | 61.2 | 48.4 | 39.7 | 31.8 | 48.6 |
| Mean Temp °F | 36.7 | 40.5 | 50.1 | 59.5 | 67.5 | 75.7 | 79.5 | 78.1 | 71.7 | 60.1 | 49.9 | 41.3 | 59.2 |
| Days Max Temp ≥ 90 °F | 0 | 0 | 0 | 0 | 1 | 9 | 17 | 12 | 5 | 0 | 0 | 0 | 44 |
| Days Max Temp ≤ 32 °F | 5 | 2 | 0 | 0 | 0 | 0 | 0 | 0 | 0 | 0 | 0 | 2 | 9 |
| Days Min Temp ≤ 32 °F | 22 | 17 | 9 | 2 | 0 | 0 | 0 | 0 | 0 | 1 | 8 | 17 | 76 |
| Days Min Temp ≤ 0 °F | 1 | 0 | 0 | 0 | 0 | 0 | 0 | 0 | 0 | 0 | 0 | 0 | 1 |
| Heating Degree Days | 870 | 686 | 466 | 201 | 55 | 1 | 0 | 0 | 23 | 190 | 451 | 729 | 3672 |
| Cooling Degree Days | 0 | 1 | 10 | 41 | 141 | 345 | 485 | 430 | 234 | 48 | 6 | 1 | 1742 |
| Total Precipitation (") | 3.57 | 3.66 | 4.64 | 4.12 | 4.99 | 3.72 | 4.05 | 3.28 | 3.51 | 2.81 | 4.14 | 4.74 | 47.23 |
| Days ≥ 0.1" Precip | 6 | 7 | 8 | 7 | 7 | 6 | 7 | 6 | 5 | 5 | 7 | 8 | 79 |
| Total Snowfall (") | 3.9 | 3.3 | 1.2 | 0.0 | 0.0 | 0.0 | 0.0 | 0.0 | 0.0 | 0.0 | 0.3 | 1.0 | 9.7 |
| Days ≥ 1" Snow Depth | 4 | 2 | 0 | 0 | 0 | 0 | 0 | 0 | 0 | 0 | 0 | 1 | 7 |

## NEWPORT 1 NW *Cocke County* ELEVATION 1102 ft LAT/LONG 35° 58 ' N / 83° 10 ' W

| | JAN | FEB | MAR | APR | MAY | JUN | JUL | AUG | SEP | OCT | NOV | DEC | YEAR |
|---|---|---|---|---|---|---|---|---|---|---|---|---|---|
| Maximum Temp °F | 46.0 | 50.7 | 60.7 | 70.4 | 77.0 | 84.3 | 87.7 | 86.4 | 81.3 | 70.7 | 60.6 | 50.6 | 68.9 |
| Minimum Temp °F | 24.8 | 26.6 | 34.9 | 43.3 | 51.9 | 60.7 | 65.1 | 63.8 | 56.9 | 43.2 | 34.8 | 27.8 | 44.5 |
| Mean Temp °F | 35.4 | 38.7 | 47.8 | 56.9 | 64.5 | 72.5 | 76.4 | 75.1 | 69.1 | 57.0 | 47.7 | 39.2 | 56.7 |
| Days Max Temp ≥ 90 °F | 0 | 0 | 0 | 0 | 0 | 5 | 12 | 9 | 3 | 0 | 0 | 0 | 29 |
| Days Max Temp ≤ 32 °F | 4 | 2 | 0 | 0 | 0 | 0 | 0 | 0 | 0 | 0 | 0 | 1 | 7 |
| Days Min Temp ≤ 32 °F | 23 | 20 | 13 | 4 | 0 | 0 | 0 | 0 | 0 | 4 | 14 | 22 | 100 |
| Days Min Temp ≤ 0 °F | 1 | 0 | 0 | 0 | 0 | 0 | 0 | 0 | 0 | 0 | 0 | 0 | 1 |
| Heating Degree Days | 910 | 736 | 527 | 256 | 91 | 8 | 0 | 1 | 33 | 258 | 513 | 793 | 4126 |
| Cooling Degree Days | 0 | 0 | 2 | 17 | 87 | 259 | 384 | 329 | 160 | 18 | 1 | 0 | 1257 |
| Total Precipitation (") | 3.45 | 3.55 | 4.40 | 3.67 | 4.65 | 3.71 | 4.52 | 3.99 | 3.15 | 2.49 | 3.14 | 3.67 | 44.39 |
| Days ≥ 0.1" Precip | 8 | 7 | 9 | 7 | 9 | 8 | 8 | 7 | 6 | 5 | 7 | 7 | 88 |
| Total Snowfall (") | 5.6 | 3.2 | 1.5 | 0.1 | 0.0 | 0.0 | 0.0 | 0.0 | 0.0 | 0.0 | 0.2 | 1.3 | 11.9 |
| Days ≥ 1" Snow Depth | 4 | 2 | 1 | 0 | 0 | 0 | 0 | 0 | 0 | 0 | 0 | 1 | 8 |

## NORRIS *Anderson County* ELEVATION 1110 ft LAT/LONG 36° 13 ' N / 84° 3 ' W

| | JAN | FEB | MAR | APR | MAY | JUN | JUL | AUG | SEP | OCT | NOV | DEC | YEAR |
|---|---|---|---|---|---|---|---|---|---|---|---|---|---|
| Maximum Temp °F | 45.6 | ***51.6*** | 61.9 | 72.2 | 78.1 | 84.3 | 87.5 | 86.2 | 81.1 | 71.1 | 60.2 | 49.8 | 69.1 |
| Minimum Temp °F | 25.0 | ***26.9*** | 34.4 | 41.9 | 50.7 | 59.3 | 63.7 | 63.1 | 57.1 | 44.2 | 35.4 | 28.3 | 44.2 |
| Mean Temp °F | 35.2 | ***39.2*** | 48.1 | 57.1 | 64.4 | 71.8 | 75.7 | 74.8 | 69.2 | 57.7 | 47.8 | 39.0 | 56.7 |
| Days Max Temp ≥ 90 °F | 0 | 0 | 0 | 0 | 1 | 4 | 12 | 7 | 2 | 0 | 0 | 0 | 26 |
| Days Max Temp ≤ 32 °F | 3 | ***1*** | 0 | 0 | 0 | 0 | 0 | 0 | 0 | 0 | 0 | 1 | 5 |
| Days Min Temp ≤ 32 °F | 23 | ***20*** | 14 | 5 | 0 | 0 | 0 | 0 | 0 | 4 | 12 | 21 | 99 |
| Days Min Temp ≤ 0 °F | 1 | ***0*** | 0 | 0 | 0 | 0 | 0 | 0 | 0 | 0 | 0 | 0 | 1 |
| Heating Degree Days | 917 | ***722*** | 518 | 248 | 85 | 8 | 0 | 1 | 29 | 238 | 511 | 799 | 4076 |
| Cooling Degree Days | 0 | ***0*** | 2 | 12 | 77 | 222 | 348 | 309 | 152 | 18 | 1 | 0 | 1141 |
| Total Precipitation (") | 4.48 | 4.20 | 5.55 | 4.33 | 4.78 | 4.81 | 4.70 | 4.32 | 3.66 | 3.22 | 4.63 | 5.17 | 53.85 |
| Days ≥ 0.1" Precip | 8 | ***7*** | 9 | 7 | 8 | 7 | 8 | 7 | 6 | 6 | 7 | 8 | 88 |
| Total Snowfall (") | ***3.0*** | ***2.6*** | 0.8 | 0.4 | 0.0 | 0.0 | 0.0 | 0.0 | 0.0 | 0.0 | 0.0 | 0.9 | 7.7 |
| Days ≥ 1" Snow Depth | ***2*** | na | 0 | 0 | 0 | 0 | 0 | 0 | 0 | 0 | 0 | 0 | na |

## OAK RIDGE *Anderson County* ELEVATION 905 ft LAT/LONG 36° 2 ' N / 84° 14 ' W

| | JAN | FEB | MAR | APR | MAY | JUN | JUL | AUG | SEP | OCT | NOV | DEC | YEAR |
|---|---|---|---|---|---|---|---|---|---|---|---|---|---|
| Maximum Temp °F | 45.1 | 50.2 | 60.3 | 70.7 | 77.4 | 84.5 | 87.5 | 86.4 | 80.4 | 70.2 | 58.8 | 48.9 | 68.4 |
| Minimum Temp °F | 26.5 | 28.5 | 36.2 | 44.4 | 53.0 | 61.3 | 66.0 | 65.0 | 58.7 | 45.6 | 36.5 | 30.0 | 46.0 |
| Mean Temp °F | 35.8 | 39.4 | 48.3 | 57.6 | 65.3 | 72.9 | 76.8 | 75.7 | 69.5 | 57.9 | 47.6 | 39.5 | 57.2 |
| Days Max Temp ≥ 90 °F | 0 | 0 | 0 | 0 | 0 | 5 | 12 | 8 | 2 | 0 | 0 | 0 | 27 |
| Days Max Temp ≤ 32 °F | 4 | 2 | 0 | 0 | 0 | 0 | 0 | 0 | 0 | 0 | 0 | 2 | 8 |
| Days Min Temp ≤ 32 °F | 22 | 19 | 12 | 3 | 0 | 0 | 0 | 0 | 0 | 2 | 11 | 20 | 89 |
| Days Min Temp ≤ 0 °F | 1 | 0 | 0 | 0 | 0 | 0 | 0 | 0 | 0 | 0 | 0 | 0 | 1 |
| Heating Degree Days | 898 | 718 | 514 | 239 | 79 | 6 | 0 | 0 | 30 | 234 | 515 | 785 | 4018 |
| Cooling Degree Days | 0 | 0 | 2 | 23 | 102 | 270 | 404 | 354 | 180 | 25 | 1 | 0 | 1361 |
| Total Precipitation (") | 4.61 | 4.38 | 5.63 | 4.10 | 4.65 | 4.35 | 5.26 | 3.70 | 3.88 | 3.19 | 4.56 | 5.50 | 53.81 |
| Days ≥ 0.1" Precip | 8 | 8 | 9 | 7 | 8 | 8 | 8 | 7 | 6 | 6 | 7 | 8 | 90 |
| Total Snowfall (") | 3.6 | 3.3 | 1.2 | 0.2 | 0.0 | 0.0 | 0.0 | 0.0 | 0.0 | 0.0 | 0.1 | 1.4 | 9.8 |
| Days ≥ 1" Snow Depth | 3 | 2 | 0 | 0 | 0 | 0 | 0 | 0 | 0 | 0 | 0 | 1 | 6 |

## ONEIDA *Scott County* ELEVATION 1450 ft LAT/LONG 36° 30 ' N / 84° 31 ' W

| | JAN | FEB | MAR | APR | MAY | JUN | JUL | AUG | SEP | OCT | NOV | DEC | YEAR |
|---|---|---|---|---|---|---|---|---|---|---|---|---|---|
| Maximum Temp °F | 43.4 | 47.9 | 58.1 | 68.4 | 74.7 | 81.9 | 85.4 | 84.4 | 78.4 | 68.7 | 58.3 | 47.9 | 66.5 |
| Minimum Temp °F | 21.4 | 23.6 | 31.6 | 39.4 | 47.7 | 57.0 | 61.7 | 60.1 | 53.9 | 40.6 | 32.7 | 25.3 | 41.3 |
| Mean Temp °F | 32.5 | 35.8 | 44.8 | 53.9 | 61.2 | 69.5 | 73.6 | 72.3 | 66.2 | 54.7 | 45.5 | 36.7 | 53.9 |
| Days Max Temp ≥ 90 °F | 0 | 0 | 0 | 0 | 0 | 2 | 7 | 5 | 1 | 0 | 0 | 0 | 15 |
| Days Max Temp ≤ 32 °F | 6 | 4 | 1 | 0 | 0 | 0 | 0 | 0 | 0 | 0 | 0 | 3 | 14 |
| Days Min Temp ≤ 32 °F | 25 | 22 | 18 | 9 | 1 | 0 | 0 | 0 | 0 | 7 | ***16*** | ***23*** | 121 |
| Days Min Temp ≤ 0 °F | 2 | 1 | 0 | 0 | 0 | 0 | 0 | 0 | 0 | 0 | 0 | 0 | 3 |
| Heating Degree Days | 1001 | 818 | 623 | 336 | 155 | 21 | 3 | 5 | 65 | 324 | 579 | 873 | 4803 |
| Cooling Degree Days | 0 | 0 | 2 | ***7*** | 47 | 175 | ***306*** | ***253*** | 107 | ***9*** | ***1*** | 1 | 908 |
| Total Precipitation (") | 4.23 | 4.26 | 5.56 | 4.22 | 5.29 | 4.51 | 5.21 | 4.58 | 3.97 | 4.07 | 4.55 | 4.56 | 55.01 |
| Days ≥ 0.1" Precip | 9 | 8 | 10 | 8 | 9 | 8 | 9 | 8 | 7 | 7 | 8 | 9 | 100 |
| Total Snowfall (") | na | na | na | 0.1 | 0.0 | 0.0 | 0.0 | 0.0 | 0.0 | 0.0 | na | na | na |
| Days ≥ 1" Snow Depth | na | na | ***0*** | 0 | 0 | 0 | 0 | 0 | 0 | 0 | ***0*** | ***0*** | na |

## PARIS 2 NW *Henry County* ELEVATION 580 ft LAT/LONG 36° 20 ' N / 88° 21 ' W

| | JAN | FEB | MAR | APR | MAY | JUN | JUL | AUG | SEP | OCT | NOV | DEC | YEAR |
|---|---|---|---|---|---|---|---|---|---|---|---|---|---|
| Maximum Temp °F | 42.9 | 48.2 | 59.1 | 70.1 | 77.3 | 85.2 | 88.8 | 87.6 | 81.0 | 69.8 | 58.5 | 48.3 | 68.1 |
| Minimum Temp °F | 24.2 | 27.1 | 36.2 | 45.5 | 53.7 | 62.3 | 66.7 | 64.6 | 58.1 | 45.3 | 37.4 | 28.9 | 45.8 |
| Mean Temp °F | 33.5 | 37.7 | 47.7 | 57.8 | 65.6 | 73.8 | 77.7 | 76.1 | 69.6 | 57.6 | 48.0 | 38.6 | 57.0 |
| Days Max Temp ≥ 90 °F | 0 | 0 | 0 | 0 | 1 | 7 | 15 | 11 | 4 | 0 | 0 | 0 | 38 |
| Days Max Temp ≤ 32 °F | 7 | 4 | 0 | 0 | 0 | 0 | 0 | 0 | 0 | 0 | 0 | 3 | 14 |
| Days Min Temp ≤ 32 °F | 24 | 20 | 13 | 3 | 0 | 0 | 0 | 0 | 0 | 2 | 11 | 20 | 93 |
| Days Min Temp ≤ 0 °F | 1 | 0 | 0 | 0 | 0 | 0 | 0 | 0 | 0 | 0 | 0 | 0 | 1 |
| Heating Degree Days | 970 | 766 | 535 | 243 | 77 | 6 | 0 | 1 | 41 | 252 | 508 | 812 | 4211 |
| Cooling Degree Days | 0 | 0 | 6 | 32 | 99 | 284 | 415 | 362 | 181 | 28 | 4 | 1 | 1412 |
| Total Precipitation (") | 4.10 | 4.42 | 5.06 | 4.89 | 4.66 | 4.17 | 4.25 | 4.02 | 3.93 | 3.58 | 4.85 | 5.36 | 53.29 |
| Days ≥ 0.1" Precip | 7 | 7 | 8 | 7 | 8 | 6 | 7 | 6 | 6 | 5 | 7 | 7 | 81 |
| Total Snowfall (") | 4.2 | 3.4 | 1.5 | 0.0 | 0.0 | 0.0 | 0.0 | 0.0 | 0.0 | 0.0 | 0.2 | 0.8 | 10.1 |
| Days ≥ 1" Snow Depth | 5 | 3 | 1 | 0 | 0 | 0 | 0 | 0 | 0 | 0 | 0 | 0 | 9 |

## PORTLAND SEWAGE PLAN *Sumner County* ELEVATION 640 ft LAT/LONG 36° 37 ' N / 86° 34 ' W

| | JAN | FEB | MAR | APR | MAY | JUN | JUL | AUG | SEP | OCT | NOV | DEC | YEAR |
|---|---|---|---|---|---|---|---|---|---|---|---|---|---|
| Maximum Temp °F | 43.2 | 48.0 | 58.8 | 69.7 | 76.8 | 84.4 | 88.2 | 86.6 | 80.9 | 70.0 | 58.6 | 48.3 | 67.8 |
| Minimum Temp °F | 24.4 | 27.7 | 36.7 | 46.1 | 54.6 | 62.9 | 66.9 | 64.8 | 58.3 | 45.7 | 37.8 | 29.0 | 46.2 |
| Mean Temp °F | 33.8 | 37.9 | 47.8 | 58.0 | 65.7 | 73.7 | 77.6 | 75.7 | 69.6 | 57.9 | 48.2 | 38.7 | 57.1 |
| Days Max Temp ≥ 90 °F | 0 | 0 | 0 | 0 | 0 | 6 | 14 | 10 | 4 | 0 | 0 | 0 | 34 |
| Days Max Temp ≤ 32 °F | 6 | 3 | 0 | 0 | 0 | 0 | 0 | 0 | 0 | 0 | 0 | 2 | 11 |
| Days Min Temp ≤ 32 °F | 23 | 19 | 12 | 3 | 0 | 0 | 0 | 0 | 0 | 3 | 10 | 20 | 90 |
| Days Min Temp ≤ 0 °F | 1 | 0 | 0 | 0 | 0 | 0 | 0 | 0 | 0 | 0 | 0 | 0 | 1 |
| Heating Degree Days | 959 | 755 | 531 | 239 | 80 | 7 | 0 | 1 | 38 | 243 | 500 | 811 | 4164 |
| Cooling Degree Days | 0 | 0 | 7 | 37 | 123 | 300 | 443 | 371 | 193 | 30 | 3 | 0 | 1507 |
| Total Precipitation (") | 3.74 | 4.09 | 5.20 | 4.29 | 5.10 | 4.50 | 4.44 | 3.89 | 3.62 | 3.33 | 4.53 | 4.99 | 51.72 |
| Days ≥ 0.1" Precip | 6 | 6 | 8 | 8 | 8 | 7 | 7 | 6 | 6 | 5 | 7 | 7 | 81 |
| Total Snowfall (") | 3.2 | 2.4 | 0.8 | 0.0 | 0.0 | 0.0 | 0.0 | 0.0 | 0.0 | 0.0 | 0.5 | 0.9 | 7.8 |
| Days ≥ 1" Snow Depth | *4* | *2* | 0 | 0 | 0 | 0 | 0 | 0 | 0 | 0 | 0 | 1 | 7 |

## PULASKI WATER PLANT *Giles County* ELEVATION 650 ft LAT/LONG 35° 12 ' N / 87° 2 ' W

| | JAN | FEB | MAR | APR | MAY | JUN | JUL | AUG | SEP | OCT | NOV | DEC | YEAR |
|---|---|---|---|---|---|---|---|---|---|---|---|---|---|
| Maximum Temp °F | 49.4 | 54.7 | 64.3 | 74.2 | 80.3 | 87.3 | 90.3 | 89.5 | 84.0 | 74.3 | 63.0 | 53.8 | 72.1 |
| Minimum Temp °F | 28.4 | 31.2 | 39.3 | 47.8 | 55.1 | 63.0 | 67.4 | 65.7 | 59.4 | 46.8 | 39.1 | 32.4 | 48.0 |
| Mean Temp °F | 38.9 | 43.0 | 51.8 | 61.0 | 67.7 | 75.2 | 78.8 | 77.6 | 71.7 | 60.6 | 51.1 | 43.1 | 60.0 |
| Days Max Temp ≥ 90 °F | 0 | 0 | 0 | 0 | 2 | 10 | 19 | 16 | 7 | 0 | 0 | 0 | 54 |
| Days Max Temp ≤ 32 °F | 2 | 1 | 0 | 0 | 0 | 0 | 0 | 0 | 0 | 0 | 0 | 1 | 4 |
| Days Min Temp ≤ 32 °F | 20 | 16 | 9 | 2 | 0 | 0 | 0 | 0 | 0 | 3 | 9 | 16 | 75 |
| Days Min Temp ≤ 0 °F | 0 | 0 | 0 | 0 | 0 | 0 | 0 | 0 | 0 | 0 | 0 | 0 | 0 |
| Heating Degree Days | 802 | 616 | 412 | 162 | 44 | 2 | 0 | 0 | 20 | 175 | 415 | 673 | 3321 |
| Cooling Degree Days | 0 | 1 | 11 | 42 | 139 | 326 | 453 | 413 | 234 | 49 | 5 | 2 | 1675 |
| Total Precipitation (") | 4.53 | 4.59 | 5.97 | 4.27 | 5.30 | 3.90 | 4.55 | 3.53 | 4.25 | 3.61 | 4.88 | 5.75 | 55.13 |
| Days ≥ 0.1" Precip | 7 | 7 | 8 | 7 | 7 | 7 | 7 | 5 | 6 | 5 | 7 | 8 | 81 |
| Total Snowfall (") | *1.8* | 1.1 | 0.4 | 0.0 | 0.0 | 0.0 | 0.0 | 0.0 | 0.0 | 0.0 | 0.2 | 0.4 | 3.9 |
| Days ≥ 1" Snow Depth | 2 | 1 | 0 | 0 | 0 | 0 | 0 | 0 | 0 | 0 | 0 | 0 | 3 |

## RIPLEY *Lauderdale County* ELEVATION 331 ft LAT/LONG 35° 45 ' N / 89° 32 ' W

| | JAN | FEB | MAR | APR | MAY | JUN | JUL | AUG | SEP | OCT | NOV | DEC | YEAR |
|---|---|---|---|---|---|---|---|---|---|---|---|---|---|
| Maximum Temp °F | 45.1 | 50.2 | 60.5 | 71.5 | 79.3 | 87.3 | 90.9 | 89.2 | 83.1 | 73.4 | 60.6 | 50.3 | 70.1 |
| Minimum Temp °F | 26.8 | 30.0 | 39.5 | 49.2 | 57.4 | 65.7 | 69.8 | 67.5 | 60.8 | 48.6 | 39.8 | 31.0 | 48.8 |
| Mean Temp °F | 36.0 | 40.1 | 50.0 | 60.4 | 68.3 | 76.5 | 80.4 | 78.4 | 72.0 | 61.0 | 50.2 | 40.7 | 59.5 |
| Days Max Temp ≥ 90 °F | 0 | 0 | 0 | 0 | 2 | 12 | 20 | 16 | 7 | 0 | 0 | 0 | 57 |
| Days Max Temp ≤ 32 °F | 5 | 3 | 0 | 0 | 0 | 0 | 0 | 0 | 0 | 0 | 0 | 2 | 10 |
| Days Min Temp ≤ 32 °F | 22 | 17 | 9 | 1 | 0 | 0 | 0 | 0 | 0 | 1 | 7 | 18 | 75 |
| Days Min Temp ≤ 0 °F | 1 | 0 | 0 | 0 | 0 | 0 | 0 | 0 | 0 | 0 | 0 | 0 | 1 |
| Heating Degree Days | 893 | 696 | 466 | 182 | 42 | 2 | 0 | 0 | 24 | 172 | 443 | 748 | 3668 |
| Cooling Degree Days | 0 | 1 | 9 | 54 | 155 | 370 | 509 | 454 | 257 | 58 | 6 | 1 | 1874 |
| Total Precipitation (") | 3.76 | 3.91 | 5.03 | 5.13 | 5.32 | 4.08 | 3.86 | 2.54 | 4.23 | 3.40 | 5.35 | 5.70 | 52.31 |
| Days ≥ 0.1" Precip | 7 | 6 | 7 | 8 | 8 | 6 | 6 | 4 | 6 | 5 | 7 | 7 | 77 |
| Total Snowfall (") | 2.3 | 2.2 | 1.2 | 0.0 | 0.0 | 0.0 | 0.0 | 0.0 | 0.0 | 0.0 | 0.1 | 0.3 | 6.1 |
| Days ≥ 1" Snow Depth | *2* | 1 | 0 | 0 | 0 | 0 | 0 | 0 | 0 | 0 | 0 | 0 | 3 |

## ROGERSVILLE 1 NE *Hawkins County* ELEVATION 1089 ft LAT/LONG 36° 22 ' N / 83° 0 ' W

| | JAN | FEB | MAR | APR | MAY | JUN | JUL | AUG | SEP | OCT | NOV | DEC | YEAR |
|---|---|---|---|---|---|---|---|---|---|---|---|---|---|
| Maximum Temp °F | 45.2 | 50.2 | 60.7 | 70.5 | 77.2 | 83.8 | 87.0 | 85.7 | 80.3 | 69.9 | 59.0 | 49.3 | 68.2 |
| Minimum Temp °F | 25.3 | 27.4 | 35.4 | 43.2 | 51.1 | 58.7 | 63.0 | 61.9 | 55.9 | 43.5 | 35.8 | 28.8 | 44.2 |
| Mean Temp °F | 35.3 | 38.8 | 48.1 | 56.9 | 64.2 | 71.3 | 75.0 | 73.8 | 68.1 | 56.7 | 47.5 | 39.1 | 56.2 |
| Days Max Temp ≥ 90 °F | 0 | 0 | 0 | 0 | 0 | 3 | 9 | 6 | 1 | 0 | 0 | 0 | 19 |
| Days Max Temp ≤ 32 °F | 4 | 1 | 0 | 0 | 0 | 0 | 0 | 0 | 0 | 0 | 0 | 1 | 6 |
| Days Min Temp ≤ 32 °F | 22 | 19 | 13 | 4 | 0 | 0 | 0 | 0 | 0 | 4 | 12 | 20 | 94 |
| Days Min Temp ≤ 0 °F | 1 | 0 | 0 | 0 | 0 | 0 | 0 | 0 | 0 | 0 | 0 | 0 | 1 |
| Heating Degree Days | 914 | 732 | 518 | 253 | 89 | 10 | 1 | 1 | 38 | 263 | 520 | 796 | 4135 |
| Cooling Degree Days | 0 | 0 | 1 | 16 | 74 | 222 | 343 | 289 | 137 | 16 | 0 | 0 | 1098 |
| Total Precipitation (") | 3.71 | 3.63 | 4.06 | 3.78 | 4.55 | 3.47 | 4.19 | 3.77 | 3.41 | 2.85 | 3.41 | 4.33 | 45.16 |
| Days ≥ 0.1" Precip | 7 | 7 | 7 | 6 | 7 | 6 | 7 | 6 | 5 | 5 | 7 | 6 | 76 |
| Total Snowfall (") | *3.5* | *1.8* | 0.8 | 0.5 | 0.0 | 0.0 | 0.0 | 0.0 | 0.0 | 0.0 | 0.0 | 0.9 | 7.5 |
| Days ≥ 1" Snow Depth | *3* | 2 | 0 | 0 | 0 | 0 | 0 | 0 | 0 | 0 | 0 | 0 | 5 |

## SAMBURG WILDLIFE REF *Obion County* ELEVATION 308 ft LAT/LONG 36° 27' N / 89° 19' W

| | JAN | FEB | MAR | APR | MAY | JUN | JUL | AUG | SEP | OCT | NOV | DEC | YEAR |
|---|---|---|---|---|---|---|---|---|---|---|---|---|---|
| Maximum Temp °F | 42.8 | 48.0 | 58.6 | 70.3 | 78.7 | 87.0 | 90.6 | 88.6 | 82.2 | 72.1 | 59.7 | *48.3* | 68.9 |
| Minimum Temp °F | 25.0 | 28.7 | 38.8 | 48.0 | 56.3 | 64.3 | 68.3 | 65.7 | 59.1 | 46.5 | 38.9 | *29.9* | 47.5 |
| Mean Temp °F | 33.9 | 38.4 | 48.7 | 59.1 | 67.5 | 75.7 | 79.5 | 77.2 | 70.7 | 59.4 | 49.3 | *39.1* | 58.2 |
| Days Max Temp ≥ 90 °F | 0 | 0 | 0 | 0 | 1 | 11 | 19 | 13 | 5 | 0 | 0 | 0 | 49 |
| Days Max Temp ≤ 32 °F | 7 | 4 | 0 | 0 | 0 | 0 | 0 | 0 | 0 | 0 | 0 | 2 | 13 |
| Days Min Temp ≤ 32 °F | 23 | 19 | 10 | 2 | 0 | 0 | 0 | 0 | 0 | 2 | 8 | 18 | 82 |
| Days Min Temp ≤ 0 °F | 1 | 0 | 0 | 0 | 0 | 0 | 0 | 0 | 0 | 0 | 0 | 0 | 1 |
| Heating Degree Days | 957 | 744 | 505 | 214 | 53 | 3 | 0 | 1 | 30 | 211 | 468 | *796* | 3982 |
| Cooling Degree Days | 0 | 0 | 6 | 38 | *126* | 326 | *472* | 399 | *202* | *38* | *4* | na | na |
| Total Precipitation (") | 3.68 | 3.99 | 4.65 | 5.08 | 5.25 | 4.14 | 3.60 | 3.35 | 3.55 | 3.56 | 4.60 | 5.33 | 50.78 |
| Days ≥ 0.1" Precip | 6 | 6 | 7 | 7 | 6 | 6 | 5 | 4 | 5 | 5 | 6 | 6 | 69 |
| Total Snowfall (") | *3.6* | 2.8 | 1.4 | 0.0 | 0.0 | 0.0 | 0.0 | 0.0 | 0.0 | 0.0 | 0.3 | 0.5 | 8.6 |
| Days ≥ 1" Snow Depth | *3* | 2 | 0 | 0 | 0 | 0 | 0 | 0 | 0 | 0 | 0 | 0 | 5 |

## SAVANNAH 6 SW *Hardin County* ELEVATION 420 ft LAT/LONG 35° 9' N / 88° 19' W

| | JAN | FEB | MAR | APR | MAY | JUN | JUL | AUG | SEP | OCT | NOV | DEC | YEAR |
|---|---|---|---|---|---|---|---|---|---|---|---|---|---|
| Maximum Temp °F | 49.3 | 55.2 | 65.4 | 75.5 | 80.9 | 88.1 | 91.3 | 90.3 | 84.4 | 74.7 | 63.0 | 53.6 | 72.6 |
| Minimum Temp °F | 28.6 | 31.4 | 40.2 | 48.6 | 56.2 | 64.1 | 67.9 | 66.3 | 60.2 | 48.2 | 40.1 | 32.9 | 48.7 |
| Mean Temp °F | 39.0 | 43.1 | 52.8 | 62.1 | 68.6 | 76.1 | 79.6 | 78.3 | 72.3 | 61.5 | 51.6 | 43.3 | 60.7 |
| Days Max Temp ≥ 90 °F | 0 | 0 | 0 | 0 | 1 | 13 | 20 | 18 | 7 | 0 | 0 | 0 | 59 |
| Days Max Temp ≤ 32 °F | 3 | 1 | 0 | 0 | 0 | 0 | 0 | 0 | 0 | 0 | 0 | 1 | 5 |
| Days Min Temp ≤ 32 °F | 20 | 15 | 9 | 2 | 0 | 0 | 0 | 0 | 0 | 2 | 9 | 16 | 73 |
| Days Min Temp ≤ 0 °F | 0 | 0 | 0 | 0 | 0 | 0 | 0 | 0 | 0 | 0 | 0 | 0 | 0 |
| Heating Degree Days | 800 | 611 | 385 | 145 | 34 | 1 | 0 | 0 | 18 | 159 | 402 | 667 | 3222 |
| Cooling Degree Days | 0 | 1 | 14 | 59 | 157 | 359 | 484 | 447 | 257 | 66 | 8 | 2 | 1854 |
| Total Precipitation (") | 4.58 | 4.63 | 5.86 | 5.28 | 6.55 | 4.42 | 4.16 | 3.30 | 4.24 | 3.72 | 5.55 | 6.01 | 58.30 |
| Days ≥ 0.1" Precip | 7 | 7 | 8 | 7 | 8 | 6 | 6 | 5 | 6 | 5 | 7 | 8 | 80 |
| Total Snowfall (") | *2.4* | *1.3* | 0.6 | 0.0 | 0.0 | 0.0 | 0.0 | 0.0 | 0.0 | 0.0 | 0.0 | 0.3 | 4.6 |
| Days ≥ 1" Snow Depth | na | *1* | 0 | 0 | 0 | 0 | 0 | 0 | 0 | 0 | 0 | 0 | na |

## SELMER *McNairy County* ELEVATION 470 ft LAT/LONG 35° 10' N / 88° 37' W

| | JAN | FEB | MAR | APR | MAY | JUN | JUL | AUG | SEP | OCT | NOV | DEC | YEAR |
|---|---|---|---|---|---|---|---|---|---|---|---|---|---|
| Maximum Temp °F | 46.9 | 52.0 | 61.8 | 72.6 | 79.2 | 86.7 | 90.3 | 89.2 | 83.2 | 73.2 | 62.0 | 52.0 | 70.8 |
| Minimum Temp °F | 25.8 | 28.5 | 37.4 | 45.9 | 54.3 | 62.3 | 66.3 | 64.3 | 57.9 | 45.0 | 37.4 | 30.3 | 46.3 |
| Mean Temp °F | 36.4 | 40.3 | 49.6 | 59.3 | 66.8 | 74.5 | 78.2 | 76.8 | 70.6 | 59.1 | 49.7 | 41.2 | 58.5 |
| Days Max Temp ≥ 90 °F | 0 | 0 | 0 | 0 | 1 | 10 | 18 | 16 | 6 | 0 | 0 | 0 | 51 |
| Days Max Temp ≤ 32 °F | 4 | 2 | 0 | 0 | 0 | 0 | 0 | 0 | 0 | 0 | 0 | 2 | 8 |
| Days Min Temp ≤ 32 °F | 23 | 19 | 12 | 3 | 0 | 0 | 0 | 0 | 0 | 4 | 12 | 19 | 92 |
| Days Min Temp ≤ 0 °F | 0 | 0 | 0 | 0 | 0 | 0 | 0 | 0 | 0 | 0 | 0 | 0 | 0 |
| Heating Degree Days | 881 | 691 | 478 | 207 | 62 | 4 | 1 | 1 | 33 | 213 | 459 | 733 | 3763 |
| Cooling Degree Days | 0 | 1 | 10 | 44 | 136 | 325 | 445 | 406 | 224 | 42 | 6 | 1 | 1640 |
| Total Precipitation (") | 4.63 | 4.66 | 5.77 | 5.29 | 6.11 | 3.98 | 4.35 | 2.72 | 4.14 | 3.56 | 5.44 | 5.97 | 56.62 |
| Days ≥ 0.1" Precip | 8 | 7 | 8 | 7 | 8 | 6 | 6 | 5 | 6 | 5 | 7 | 8 | 81 |
| Total Snowfall (") | *1.8* | *1.3* | 0.6 | 0.0 | 0.0 | 0.0 | 0.0 | 0.0 | 0.0 | 0.0 | 0.0 | 0.3 | 4.0 |
| Days ≥ 1" Snow Depth | *1* | *0* | 0 | 0 | 0 | 0 | 0 | 0 | 0 | 0 | 0 | 0 | 1 |

## SEVIERVILLE 1 SE *Sevier County* ELEVATION 932 ft LAT/LONG 35° 52' N / 83° 33' W

| | JAN | FEB | MAR | APR | MAY | JUN | JUL | AUG | SEP | OCT | NOV | DEC | YEAR |
|---|---|---|---|---|---|---|---|---|---|---|---|---|---|
| Maximum Temp °F | 46.1 | 51.0 | 61.3 | 70.9 | 77.7 | 84.5 | 87.3 | 86.3 | 80.7 | 70.3 | 59.7 | 50.6 | 68.9 |
| Minimum Temp °F | 24.2 | 26.2 | 33.9 | 41.7 | 50.7 | 59.1 | 63.5 | 62.3 | 55.7 | 42.4 | 33.6 | 27.4 | 43.4 |
| Mean Temp °F | 35.2 | 38.6 | 47.6 | 56.3 | 64.2 | 71.8 | 75.4 | 74.3 | 68.2 | 56.4 | 46.7 | 39.0 | 56.1 |
| Days Max Temp ≥ 90 °F | 0 | 0 | 0 | 0 | 0 | 5 | 12 | 7 | 3 | 0 | 0 | 0 | 27 |
| Days Max Temp ≤ 32 °F | 3 | 2 | 0 | 0 | 0 | 0 | 0 | 0 | 0 | 0 | 0 | 1 | 6 |
| Days Min Temp ≤ 32 °F | 23 | 20 | 15 | 6 | 0 | 0 | 0 | 0 | 0 | 5 | 14 | 21 | 104 |
| Days Min Temp ≤ 0 °F | 1 | 0 | 0 | 0 | 0 | 0 | 0 | 0 | 0 | 0 | 0 | 0 | 1 |
| Heating Degree Days | 916 | 737 | 535 | 265 | 88 | 8 | 0 | 1 | 36 | 275 | 545 | 800 | 4206 |
| Cooling Degree Days | 0 | 0 | 3 | 16 | 81 | 258 | 375 | 337 | 162 | 23 | 2 | 0 | 1257 |
| Total Precipitation (") | 3.54 | 3.55 | 4.24 | 3.52 | 4.35 | 3.73 | 4.01 | 3.39 | 3.30 | 2.63 | 3.17 | 3.59 | 43.02 |
| Days ≥ 0.1" Precip | 7 | 6 | 7 | 7 | 8 | 6 | 8 | 6 | 6 | 5 | 6 | 7 | 79 |
| Total Snowfall (") | na | *1.9* | 0.9 | 0.6 | 0.0 | 0.0 | 0.0 | 0.0 | 0.0 | 0.0 | 0.1 | 0.6 | na |
| Days ≥ 1" Snow Depth | *2* | *1* | 0 | 0 | 0 | 0 | 0 | 0 | 0 | 0 | 0 | 0 | 3 |

## SHELBYVILLE 3 *Bedford County* ELEVATION 722 ft LAT/LONG 35° 29 ' N / 86° 29 ' W

| | JAN | FEB | MAR | APR | MAY | JUN | JUL | AUG | SEP | OCT | NOV | DEC | YEAR |
|---|---|---|---|---|---|---|---|---|---|---|---|---|---|
| Maximum Temp °F | 48.1 | 53.1 | 62.8 | 72.9 | 79.5 | 86.9 | 89.7 | 88.9 | 83.1 | 72.9 | 61.6 | 52.4 | 71.0 |
| Minimum Temp °F | 27.9 | 30.4 | 38.8 | 46.6 | 54.8 | 62.5 | 66.8 | 65.5 | 59.2 | 46.8 | 38.9 | 32.1 | 47.5 |
| Mean Temp °F | 38.0 | 41.8 | 50.9 | 59.8 | 67.2 | 74.7 | 78.3 | 77.2 | 71.2 | 59.9 | 50.3 | 42.3 | 59.3 |
| Days Max Temp ≥ 90 °F | 0 | 0 | 0 | 0 | 2 | 10 | 17 | 14 | 6 | 0 | 0 | 0 | 49 |
| Days Max Temp ≤ 32 °F | 3 | 1 | 0 | 0 | 0 | 0 | 0 | 0 | 0 | 0 | 0 | 1 | 5 |
| Days Min Temp ≤ 32 °F | 20 | 17 | 10 | 3 | 0 | 0 | 0 | 0 | 0 | 3 | 10 | 17 | 80 |
| Days Min Temp ≤ 0 °F | 1 | 0 | 0 | 0 | 0 | 0 | 0 | 0 | 0 | 0 | 0 | 0 | 1 |
| Heating Degree Days | 830 | 650 | 438 | 191 | 53 | 3 | 0 | 0 | 25 | 193 | 438 | 698 | 3519 |
| Cooling Degree Days | 0 | 1 | 8 | 35 | 138 | 314 | 445 | 402 | 222 | 47 | 4 | 1 | 1617 |
| Total Precipitation (") | 4.52 | 4.29 | 5.83 | 4.27 | 5.48 | 4.13 | 5.12 | 3.33 | 4.16 | 3.87 | 5.07 | 5.59 | 55.66 |
| Days ≥ 0.1" Precip | 8 | 8 | 9 | 7 | 8 | 7 | 7 | 6 | 6 | 5 | 7 | 8 | 86 |
| Total Snowfall (") | 2.8 | 1.9 | 0.8 | 0.0 | 0.0 | 0.0 | 0.0 | 0.0 | 0.0 | 0.0 | 0.4 | 0.6 | 6.5 |
| Days ≥ 1" Snow Depth | 2 | 2 | 0 | 0 | 0 | 0 | 0 | 0 | 0 | 0 | 0 | 0 | 4 |

## SPARTA *White County* ELEVATION 902 ft LAT/LONG 35° 57 ' N / 85° 30 ' W

| | JAN | FEB | MAR | APR | MAY | JUN | JUL | AUG | SEP | OCT | NOV | DEC | YEAR |
|---|---|---|---|---|---|---|---|---|---|---|---|---|---|
| Maximum Temp °F | 46.8 | 50.9 | 61.1 | 71.8 | 77.6 | 85.5 | 88.4 | 87.8 | 81.8 | 71.4 | 61.2 | 50.7 | 69.6 |
| Minimum Temp °F | 26.0 | 28.0 | 36.0 | 43.9 | 51.2 | 59.7 | 64.2 | 62.6 | 56.7 | 43.1 | 36.3 | 29.2 | 44.7 |
| Mean Temp °F | 36.4 | 39.5 | 48.6 | 58.0 | 64.5 | 72.6 | 76.3 | 75.2 | 69.3 | 57.3 | 48.8 | 40.0 | 57.2 |
| Days Max Temp ≥ 90 °F | 0 | 0 | 0 | 0 | 0 | 7 | 14 | 11 | 4 | 0 | 0 | 0 | 36 |
| Days Max Temp ≤ 32 °F | 4 | 2 | 0 | 0 | 0 | 0 | 0 | 0 | 0 | 0 | 0 | 1 | 7 |
| Days Min Temp ≤ 32 °F | 21 | 18 | 12 | 4 | 0 | 0 | 0 | 0 | 0 | 5 | *11* | 19 | 90 |
| Days Min Temp ≤ 0 °F | 1 | 0 | 0 | 0 | 0 | 0 | 0 | 0 | 0 | 0 | 0 | 0 | 1 |
| Heating Degree Days | 878 | 713 | 506 | 228 | 92 | 10 | 1 | 1 | 32 | 256 | 479 | 766 | 3962 |
| Cooling Degree Days | 0 | 1 | *5* | 28 | 95 | 275 | 396 | *366* | *189* | *26* | 3 | 0 | 1384 |
| Total Precipitation (") | 4.63 | 4.39 | 5.58 | 4.54 | 5.53 | 4.00 | 4.90 | 4.01 | 4.09 | 3.32 | 4.64 | 5.73 | 55.36 |
| Days ≥ 0.1" Precip | na | na | *8* | *7* | *8* | 6 | 8 | 6 | 6 | *4* | *6* | na | na |
| Total Snowfall (") | na | na | 0.2 | 0.0 | 0.0 | 0.0 | 0.0 | 0.0 | 0.0 | 0.0 | 0.2 | *0.1* | na |
| Days ≥ 1" Snow Depth | na | na | 0 | 0 | 0 | 0 | 0 | 0 | 0 | 0 | 0 | 0 | na |

## SPRINGFIELD EXP STN *Robertson County* ELEVATION 751 ft LAT/LONG 36° 28 ' N / 86° 50 ' W

| | JAN | FEB | MAR | APR | MAY | JUN | JUL | AUG | SEP | OCT | NOV | DEC | YEAR |
|---|---|---|---|---|---|---|---|---|---|---|---|---|---|
| Maximum Temp °F | 42.4 | 47.3 | 57.8 | 68.6 | 76.2 | 84.5 | 88.4 | 87.1 | 80.8 | 69.6 | 57.9 | 47.9 | 67.4 |
| Minimum Temp °F | 23.3 | 26.0 | 35.3 | 44.8 | 53.2 | 61.8 | 65.8 | 63.6 | 57.2 | 44.9 | 37.0 | 28.4 | 45.1 |
| Mean Temp °F | 32.9 | 36.7 | 46.5 | 56.7 | 64.8 | 73.2 | 77.2 | 75.4 | 69.1 | 57.3 | 47.5 | 38.1 | 56.3 |
| Days Max Temp ≥ 90 °F | 0 | 0 | 0 | 0 | 0 | 6 | 15 | 11 | 4 | 0 | 0 | 0 | 36 |
| Days Max Temp ≤ 32 °F | 7 | 4 | 1 | 0 | 0 | 0 | 0 | 0 | 0 | 0 | 0 | 3 | 15 |
| Days Min Temp ≤ 32 °F | 24 | 20 | 14 | 4 | 0 | 0 | 0 | 0 | 0 | 4 | 11 | 20 | 97 |
| Days Min Temp ≤ 0 °F | 2 | 1 | 0 | 0 | 0 | 0 | 0 | 0 | 0 | 0 | 0 | 0 | 3 |
| Heating Degree Days | 990 | 793 | 571 | 273 | 98 | 9 | 1 | 2 | 48 | 263 | 522 | 826 | 4396 |
| Cooling Degree Days | 0 | 0 | 5 | 29 | 97 | 267 | 400 | 342 | 178 | 31 | 3 | 1 | 1353 |
| Total Precipitation (") | 3.57 | 4.04 | 4.94 | 4.25 | 5.26 | 4.21 | 4.13 | 3.26 | 3.53 | 3.28 | 4.44 | 4.96 | 49.87 |
| Days ≥ 0.1" Precip | 7 | 7 | 7 | 8 | 8 | 7 | 6 | 6 | 6 | 5 | 7 | 7 | 81 |
| Total Snowfall (") | 4.8 | 4.4 | 1.1 | 0.1 | 0.0 | 0.0 | 0.0 | 0.0 | 0.0 | 0.0 | 0.2 | 1.1 | 11.7 |
| Days ≥ 1" Snow Depth | 6 | 4 | 1 | 0 | 0 | 0 | 0 | 0 | 0 | 0 | 0 | 1 | 12 |

## TULLAHOMA *Coffee County* ELEVATION 1070 ft LAT/LONG 35° 22 ' N / 86° 12 ' W

| | JAN | FEB | MAR | APR | MAY | JUN | JUL | AUG | SEP | OCT | NOV | DEC | YEAR |
|---|---|---|---|---|---|---|---|---|---|---|---|---|---|
| Maximum Temp °F | 47.1 | 51.7 | 61.7 | 71.8 | 77.8 | 84.9 | 87.7 | 87.0 | 81.4 | 71.6 | 60.7 | 51.4 | 69.6 |
| Minimum Temp °F | 28.1 | 30.4 | 38.6 | 46.9 | 54.6 | 62.3 | 66.6 | 65.3 | 59.3 | 46.9 | 38.8 | 31.9 | 47.5 |
| Mean Temp °F | 37.6 | 41.1 | 50.1 | 59.4 | 66.2 | 73.7 | 77.2 | 76.2 | 70.4 | 59.3 | 49.8 | 41.7 | 58.6 |
| Days Max Temp ≥ 90 °F | 0 | 0 | 0 | 0 | 0 | 5 | 12 | 9 | 3 | 0 | 0 | 0 | 29 |
| Days Max Temp ≤ 32 °F | 3 | 2 | 0 | 0 | 0 | 0 | 0 | 0 | 0 | 0 | 0 | 1 | 6 |
| Days Min Temp ≤ 32 °F | 20 | 17 | 10 | 2 | 0 | 0 | 0 | 0 | 0 | 2 | 9 | 17 | 77 |
| Days Min Temp ≤ 0 °F | 1 | 0 | 0 | 0 | 0 | 0 | 0 | 0 | 0 | 0 | 0 | 0 | 1 |
| Heating Degree Days | 842 | 668 | 458 | 196 | 60 | 4 | 0 | 0 | 27 | 202 | 452 | 716 | 3625 |
| Cooling Degree Days | 0 | 0 | 5 | 25 | 102 | 272 | 395 | 355 | 190 | 31 | 3 | 1 | 1379 |
| Total Precipitation (") | 5.05 | 4.90 | 6.23 | 4.86 | 5.25 | 3.94 | 4.85 | 3.51 | 4.07 | 3.90 | 5.05 | 5.80 | 57.41 |
| Days ≥ 0.1" Precip | 8 | 8 | 9 | 7 | 8 | 6 | 8 | 6 | 7 | 6 | 8 | 8 | 89 |
| Total Snowfall (") | *2.6* | *2.1* | 1.1 | 0.0 | 0.0 | 0.0 | 0.0 | 0.0 | 0.0 | 0.0 | 0.2 | *0.7* | 6.7 |
| Days ≥ 1" Snow Depth | *2* | 1 | 0 | 0 | 0 | 0 | 0 | 0 | 0 | 0 | 0 | 0 | 3 |

## UNION CITY *Obion County* ELEVATION 341 ft LAT/LONG 36° 25' N / 89° 3' W

| | JAN | FEB | MAR | APR | MAY | JUN | JUL | AUG | SEP | OCT | NOV | DEC | YEAR |
|---|---|---|---|---|---|---|---|---|---|---|---|---|---|
| Maximum Temp °F | 42.3 | 47.4 | 58.1 | 69.2 | 77.4 | 85.8 | 89.1 | 87.5 | 81.5 | 70.9 | 58.8 | 48.1 | 68.0 |
| Minimum Temp °F | 24.1 | 27.5 | 36.9 | 46.8 | 55.7 | 64.1 | 68.0 | 64.7 | 57.6 | 44.8 | 37.3 | 29.2 | 46.4 |
| Mean Temp °F | 33.2 | 37.5 | 47.5 | 58.0 | 66.5 | 75.0 | 78.6 | 76.1 | 69.6 | 57.9 | 48.1 | 38.7 | 57.2 |
| Days Max Temp ≥ 90 °F | 0 | 0 | 0 | 0 | 1 | 9 | 16 | 12 | 4 | 0 | 0 | 0 | 42 |
| Days Max Temp ≤ 32 °F | 7 | 4 | 0 | 0 | 0 | 0 | 0 | 0 | 0 | 0 | 0 | 3 | 14 |
| Days Min Temp ≤ 32 °F | 24 | 20 | 12 | 2 | 0 | 0 | 0 | 0 | 0 | 3 | 11 | 20 | 92 |
| Days Min Temp ≤ 0 °F | 1 | 0 | 0 | 0 | 0 | 0 | 0 | 0 | 0 | 0 | 0 | 0 | 1 |
| Heating Degree Days | 979 | 771 | 539 | 237 | 66 | 4 | 0 | 1 | 40 | 245 | 505 | 810 | 4197 |
| Cooling Degree Days | 0 | 0 | 4 | 32 | 120 | 324 | 449 | 375 | 186 | 32 | 3 | 1 | 1526 |
| Total Precipitation (") | 3.92 | 3.96 | 4.83 | 4.94 | 4.83 | 4.53 | 4.16 | 3.37 | 3.54 | 3.92 | 4.92 | 5.26 | 52.18 |
| Days ≥ 0.1" Precip | 7 | 6 | 8 | 8 | 8 | 6 | 6 | 5 | 5 | 5 | 7 | 7 | 78 |
| Total Snowfall (") | 4.1 | 4.2 | 2.0 | 0.0 | 0.0 | 0.0 | 0.0 | 0.0 | 0.0 | 0.1 | 0.3 | 1.2 | 11.9 |
| Days ≥ 1" Snow Depth | 5 | *3* | 1 | 0 | 0 | 0 | 0 | 0 | 0 | 0 | 0 | 1 | 10 |

## WAYNESBORO *Wayne County* ELEVATION 751 ft LAT/LONG 35° 19' N / 87° 46' W

| | JAN | FEB | MAR | APR | MAY | JUN | JUL | AUG | SEP | OCT | NOV | DEC | YEAR |
|---|---|---|---|---|---|---|---|---|---|---|---|---|---|
| Maximum Temp °F | 46.6 | 51.6 | 61.3 | 72.0 | 78.0 | 85.6 | 89.1 | 88.1 | 82.5 | 72.2 | 61.2 | 51.3 | 70.0 |
| Minimum Temp °F | 23.7 | 25.8 | 34.2 | 42.8 | 51.1 | 59.9 | 64.6 | 62.2 | 55.9 | 42.3 | 34.5 | 28.1 | 43.8 |
| Mean Temp °F | 35.2 | 38.7 | 47.8 | 57.4 | 64.6 | 72.8 | 76.9 | 75.2 | 69.2 | 57.2 | 47.9 | 39.7 | 56.9 |
| Days Max Temp ≥ 90 °F | 0 | 0 | 0 | 0 | 1 | 8 | 17 | 13 | 5 | 0 | 0 | 0 | 44 |
| Days Max Temp ≤ 32 °F | 4 | 2 | 0 | 0 | 0 | 0 | 0 | 0 | 0 | 0 | 0 | 2 | 8 |
| Days Min Temp ≤ 32 °F | 24 | 21 | 16 | 6 | 0 | 0 | 0 | 0 | 0 | 7 | 15 | 20 | 109 |
| Days Min Temp ≤ 0 °F | 1 | 0 | 0 | 0 | 0 | 0 | 0 | 0 | 0 | 0 | 0 | 0 | 1 |
| Heating Degree Days | 917 | 735 | 531 | 248 | 93 | 9 | 1 | 2 | 42 | 258 | 509 | 777 | 4122 |
| Cooling Degree Days | 0 | 0 | 5 | 23 | 83 | 262 | 390 | 338 | 180 | 26 | 2 | 1 | 1310 |
| Total Precipitation (") | 4.47 | 4.71 | 6.19 | 5.10 | 6.39 | 4.16 | 5.13 | 3.74 | 3.95 | 3.83 | 5.47 | 5.95 | 59.09 |
| Days ≥ 0.1" Precip | 7 | 7 | 8 | 7 | 8 | 6 | 7 | 6 | 6 | 6 | 7 | 8 | 83 |
| Total Snowfall (") | 2.1 | 0.7 | 0.7 | 0.0 | 0.0 | 0.0 | 0.0 | 0.0 | 0.0 | 0.0 | 0.2 | 0.5 | 4.2 |
| Days ≥ 1" Snow Depth | 2 | 1 | 0 | 0 | 0 | 0 | 0 | 0 | 0 | 0 | 0 | 0 | 3 |

## WOODBURY 1 WNW *Cannon County* ELEVATION 751 ft LAT/LONG 35° 50' N / 86° 5' W

| | JAN | FEB | MAR | APR | MAY | JUN | JUL | AUG | SEP | OCT | NOV | DEC | YEAR |
|---|---|---|---|---|---|---|---|---|---|---|---|---|---|
| Maximum Temp °F | 48.1 | 53.0 | 62.7 | 72.8 | 79.2 | 86.6 | 89.4 | 88.7 | 83.0 | 73.3 | 62.1 | 52.9 | 71.0 |
| Minimum Temp °F | 25.6 | 27.9 | 36.0 | 43.8 | 52.0 | 60.1 | 64.8 | 63.2 | 56.5 | 43.6 | 35.7 | 29.4 | 44.9 |
| Mean Temp °F | 36.9 | 40.5 | 49.4 | 58.3 | 65.6 | 73.4 | 77.1 | 76.0 | 69.8 | 58.5 | 48.9 | 41.2 | 58.0 |
| Days Max Temp ≥ 90 °F | 0 | 0 | 0 | 0 | 1 | 9 | 16 | 14 | 5 | 0 | 0 | 0 | 45 |
| Days Max Temp ≤ 32 °F | 3 | 1 | 0 | 0 | 0 | 0 | 0 | 0 | 0 | 0 | 0 | 1 | 5 |
| Days Min Temp ≤ 32 °F | 22 | 19 | 13 | 5 | 0 | 0 | 0 | 0 | 0 | 5 | 13 | 19 | 96 |
| Days Min Temp ≤ 0 °F | 1 | 0 | 0 | 0 | 0 | 0 | 0 | 0 | 0 | 0 | 0 | 0 | 1 |
| Heating Degree Days | 866 | 685 | 482 | 221 | 72 | 5 | 0 | 0 | 32 | 225 | 479 | 733 | 3800 |
| Cooling Degree Days | 0 | 0 | 6 | 23 | 109 | 287 | 417 | 370 | 181 | 31 | 2 | 1 | 1427 |
| Total Precipitation (") | 4.52 | 4.36 | 5.98 | 4.49 | 5.63 | 3.81 | 4.94 | 4.44 | 4.30 | 3.80 | 4.56 | 5.72 | 56.55 |
| Days ≥ 0.1" Precip | 9 | 8 | 9 | 8 | 8 | 7 | 8 | 7 | 6 | 6 | 7 | 8 | 91 |
| Total Snowfall (") | na | *0.9* | *0.8* | 0.0 | 0.0 | 0.0 | 0.0 | 0.0 | 0.0 | 0.0 | 0.5 | 0.7 | na |
| Days ≥ 1" Snow Depth | na | 1 | 0 | 0 | 0 | 0 | 0 | 0 | 0 | 0 | 0 | 0 | na |

## JANUARY MINIMUM TEMPERATURE °F

| LOWEST | | | HIGHEST | | |
|---|---|---|---|---|---|
| 1 | Oneida | 21.4 | 1 | Memphis | 31.1 |
| 2 | Crossville-Exp | 21.7 | 2 | Dyersburg | 29.3 |
| 3 | Kingston Springs | 22.6 | 3 | Chattanooga | 28.9 |
| 4 | Dover | 23.2 | 4 | Savannah | 28.6 |
| 5 | Springfield | 23.3 | 5 | Pulaski | 28.4 |
| 6 | Clarksville | 23.5 | 6 | Knoxville | 28.3 |
| 7 | Waynesboro | 23.7 | | McMinnville | 28.3 |
| 8 | Dresden | 23.9 | | Moscow | 28.3 |
| 9 | Greeneville | 24.1 | 9 | Tullahoma | 28.1 |
| | Union City | 24.1 | 10 | Brownsville | 28.0 |
| 11 | Lebanon | 24.2 | | Jackson-McKellar | 28.0 |
| | Paris | 24.2 | 12 | Shelbyville | 27.9 |
| | Sevierville | 24.2 | 13 | Dayton | 27.7 |
| 14 | Centerville | 24.3 | 14 | Nashville | 27.3 |
| | Cookeville | 24.3 | 15 | Bolivar | 27.2 |
| | Linden | 24.3 | | Jackson-Exp | 27.2 |
| 17 | Gatlinburg | 24.4 | | Kingsport | 27.2 |
| | Portland | 24.4 | 18 | Lawrenceburg | 27.1 |
| 19 | Lewisburg | 24.5 | 19 | Cleveland | 27.0 |
| | Martin | 24.5 | 20 | Ripley | 26.8 |
| 21 | Newport | 24.8 | 21 | Covington | 26.7 |
| 22 | Norris | 25.0 | | Monteagle | 26.7 |
| | Samburg | 25.0 | 23 | Oak Ridge | 26.5 |
| 24 | Bristol | 25.1 | 24 | Copperhill | 26.3 |
| 25 | Athens | 25.2 | 25 | Lexington | 26.1 |

## JULY MAXIMUM TEMPERATURE °F

| HIGHEST | | | LOWEST | | |
|---|---|---|---|---|---|
| 1 | Memphis | 91.6 | 1 | Crossville-Exp | 83.7 |
| 2 | Savannah | 91.3 | | Crossville-Mem | 83.7 |
| 3 | Covington | 91.1 | 3 | Monteagle | 83.9 |
| 4 | Jackson-McKellar | 91.0 | 4 | Allardt | 84.2 |
| 5 | Ripley | 90.9 | 5 | Gatlinburg | 84.8 |
| 6 | Brownsville | 90.8 | 6 | Bristol | 85.3 |
| | Dyersburg | 90.8 | 7 | Oneida | 85.4 |
| 8 | Samburg | 90.6 | 8 | Rogersville | 87.0 |
| 9 | Clarksville | 90.5 | 9 | Greeneville | 87.1 |
| 10 | Pulaski | 90.3 | 10 | Cookeville | 87.3 |
| | Selmer | 90.3 | | Sevierville | 87.3 |
| 12 | Centerville | 90.1 | 12 | Kingsport | 87.5 |
| | Franklin | 90.1 | | Norris | 87.5 |
| | Lexington | 90.1 | | Oak Ridge | 87.5 |
| | Martin | 90.1 | 15 | Knoxville | 87.6 |
| 16 | Bolivar | 90.0 | 16 | Livingston | 87.7 |
| | Lebanon | 90.0 | | Newport | 87.7 |
| 18 | Moscow | 89.7 | | Tullahoma | 87.7 |
| | Nashville | 89.7 | 19 | Copperhill | 87.9 |
| | Shelbyville | 89.7 | 20 | Lenoir City | 88.1 |
| 21 | Dresden | 89.6 | | McMinnville | 88.1 |
| 22 | Murfreesboro | 89.5 | 22 | Portland | 88.2 |
| 23 | Linden | 89.4 | 23 | Athens | 88.4 |
| | Woodbury | 89.4 | | Cleveland | 88.4 |
| 25 | Chattanooga | 89.3 | | Lafayette | 88.4 |

## ANNUAL PRECIPITATION (")

| HIGHEST | | | LOWEST | | |
|---|---|---|---|---|---|
| 1 | Monteagle | 62.86 | 1 | Bristol | 41.02 |
| 2 | Copperhill | 59.51 | 2 | Greeneville | 42.84 |
| 3 | Crossville-Exp | 59.27 | 3 | Sevierville | 43.02 |
| 4 | Waynesboro | 59.09 | 4 | Kingsport | 44.03 |
| 5 | Savannah | 58.30 | 5 | Newport | 44.39 |
| 6 | Athens | 57.72 | 6 | Rogersville | 45.16 |
| 7 | Tullahoma | 57.41 | 7 | Nashville | 47.23 |
| 8 | Lawrenceburg | 57.19 | 8 | Knoxville | 47.39 |
| 9 | Linden | 56.84 | 9 | Springfield | 49.87 |
| 10 | Dayton | 56.78 | 10 | Dyersburg | 50.18 |
| 11 | Cookeville | 56.67 | 11 | Samburg | 50.78 |
| 12 | Selmer | 56.62 | 12 | Clarksville | 50.84 |
| 13 | Woodbury | 56.55 | 13 | Lexington | 50.86 |
| 14 | Gatlinburg | 56.09 | 14 | Kingston Springs | 51.72 |
| 15 | Lafayette | 55.92 | | Portland | 51.72 |
| 16 | Crossville-Mem | 55.67 | 16 | Union City | 52.18 |
| 17 | Shelbyville | 55.66 | 17 | Ripley | 52.31 |
| 18 | Sparta | 55.36 | 18 | Brownsville | 52.43 |
| 19 | Lewisburg | 55.20 | | Livingston | 52.43 |
| 20 | Pulaski | 55.13 | 20 | Memphis | 52.45 |
| 21 | Cleveland | 55.03 | 21 | Lebanon | 52.67 |
| 22 | Oneida | 55.01 | 22 | Dover | 52.87 |
| 23 | Columbia | 54.88 | 23 | Franklin | 52.99 |
| 24 | Dickson | 54.36 | 24 | Covington | 53.06 |
| 25 | Allardt | 54.32 | 25 | Chattanooga | 53.10 |

## ANNUAL SNOWFALL (")

| HIGHEST | | | LOWEST | | |
|---|---|---|---|---|---|
| 1 | Allardt | 17.8 | 1 | Cleveland | 3.0 |
| 2 | Bristol | 16.4 | 2 | Bolivar | 3.1 |
| 3 | Crossville-Exp | 16.2 | 3 | Centerville | 3.8 |
| 4 | Kingsport | 14.5 | 4 | Pulaski | 3.9 |
| 5 | Crossville-Mem | 13.3 | 5 | Selmer | 4.0 |
| 6 | Newport | 11.9 | 6 | Waynesboro | 4.2 |
| | Union City | 11.9 | 7 | Columbia | 4.4 |
| 8 | Knoxville | 11.7 | | Lewisburg | 4.4 |
| | Springfield | 11.7 | 9 | Copperhill | 4.6 |
| 10 | Greeneville | 11.3 | | Savannah | 4.6 |
| 11 | Lafayette | 11.1 | 11 | Chattanooga | 4.9 |
| 12 | Livingston | 10.9 | 12 | Franklin | 5.0 |
| 13 | Lenoir City | 10.2 | | Kingston Springs | 5.0 |
| 14 | Paris | 10.1 | 14 | Jackson-Exp | 5.1 |
| 15 | Dover | 9.8 | | Memphis | 5.1 |
| | Oak Ridge | 9.8 | 16 | Lawrenceburg | 5.2 |
| 17 | Nashville | 9.7 | 17 | Dayton | 5.4 |
| 18 | Cookeville | 9.4 | | Murfreesboro | 5.4 |
| 19 | Monteagle | 8.9 | 19 | Lexington | 5.5 |
| 20 | Samburg | 8.6 | 20 | Ripley | 6.1 |
| 21 | Dyersburg | 8.2 | 21 | Jackson-McKellar | 6.3 |
| 22 | Covington | 8.1 | 22 | Martin | 6.4 |
| 23 | Dresden | 7.8 | 23 | Shelbyville | 6.5 |
| | Portland | 7.8 | 24 | Lebanon | 6.6 |
| 25 | Norris | 7.7 | 25 | Tullahoma | 6.7 |

# TEXAS

PHYSICAL FEATURES. Texas has been called "the crossroads of North American geology." Within the State's boundaries four great physiographic subdivisions of the North American Continent come together. These are: the Gulf Coastal Forested Plain, Great Western Lower Plains, Great Western High Plains, and the Rocky Mountain Region. Texas may be described as a vast amphitheater, sloping upward from sea level along the coast of the Gulf of Mexico to more than 4,000 feet general elevation along the Texas-New Mexico line. While much of the State is relatively flat, there are 90 mountains a mile or more high, all of them in the Trans-Pecos region. Guadalupe Peak, at 8,751 feet, is the State's highest.

Texas contains 267,339 square miles or 7.4 percent of the Nation's total area. In straight-line distance, Texas extends 801 miles from north to south and 773 miles from east to west. The boundary of Texas extends 4,137 miles. The Rio Grande forms the longest segment of the boundary, 1,569 miles. The second longest segment, 726 miles, is formed by the Red River. The tidewater coastline extends 624 miles. Texas ranks second only to Alaska among the 50 states in the volume of its inland water with nearly 6,000 square miles of lakes and streams. Most Texas rivers parallel each other and flow directly into the Gulf, but the Canadian, Red, and Sulphur Rivers are part of the Mississippi River system. The Brazos is the largest river between the Rio Grande and the Red and third in size of all rivers flowing either partly or wholly in Texas. Other principal rivers are the Colorado, Trinity, Sabine, Nueces, Neches, and Guadalupe.

GENERAL CLIMATE. Wedged between the warm waters of the Gulf of Mexico and the high plateaus and mountain ranges of the North American Continent, Texas has diverse meteorological and climatological conditions. Continental, marine, and mountain types of climates are all found in Texas, the marine climate modified by surges of continental air. The High Plains, separated from the Lower Plains by the Cap Rock Escarpment, lies in a cool-temperature climatic zone. Most of the remainder of the State lies in a warm-temperature subtropical zone.

The proximity to the Gulf of Mexico, the persistent southerly and southeasterly flow of warm tropical maritime air into Texas from around the westward extension of the Azores High, and adequate rainfall combine to produce a humid subtropical climate with hot summers across the eastern third of the State. The Gulf moisture supply gradually decreases westward and is cut off more frequently during the colder months by intrusions of drier polar air from the north and west; as a result, most of Central Texas, as far north as the High Plains, has a subtropical climate with dry winters and humid summers. This region is semi-arid. As the distance from the Gulf increases westward, the summer moisture supply continues to decrease gradually, producing a subtropical steppe climate across a broad section along the middle Rio Grande Valley that extends as far west as the Pecos Valley. The area west of the Pecos is mostly arid subtropical. The mountain climates in the Trans-Pecos are cooler throughout the year than those of the adjacent lowlands.

Stretching over the largest level plain of its kind in the United States, the High Plains rise gradually from about 2,700 feet on the east to more than 4,000 feet in spots along the New Mexico border. The combination of high elevation, remoteness from moisture source regions, and frequent intrusions of dry polar airmasses, results in a dry steppe climate with relatively mild winters.

While the changes in climate across Texas are considerable, they are nevertheless gradual; no natural boundary separates the moist East from the dry West or the cool North from the warm South.

PRECIPITATION. Rainfall in Texas is not evenly distributed over the State and varies greatly from year to year. Average annual rainfall along the Louisiana border exceeds 56 inches, and in the western extremity of the State, is less than 8 inches. Except along the upper Texas coast, it is possible for one or two thunderstorms to account for the entire month's rainfall. Torrential rains of 10 to 20 inches or more may accompany a tropical storm as it moves inland across the Texas coast. Rains occur most frequently in late spring as a result of squall-line thunderstorms; consequently, most areas of the State show a peak in May. Rainfall in the Pecos Valley, most of southern Texas, the lower Rio Grande Valley, and in the coastal section, shows a peak in September, with a secondary peak in May. On the High Plains a significant percentage of the total annual precipitation occurs during the summer months (following the May peak).

Throughout the central part of the State, July and August are relatively dry months. In the mountainous Trans-Pecos area of West Texas, afternoon thundershowers during July, August, and September account for most of the annual rainfall. Throughout most of East Texas, rainfall is fairly evenly distributed throughout the year. East of about 96° W. longitude, annual rainfall exceeds average potential evapotranspiration. West of this meridian, average potential evapotranspiration exceeds annual average rainfall.

FLOOD AND DROUGHT. In most of Texas a large portion of the annual rainfall occurs within short periods of time, resulting in excessive run-off and frequently producing damaging floods. Flood stage is reached on some Texas streams nearly every year. From the early days of Texas history recorded by Spaniards exploring the Southwest, drought has been a re-occurring problem. A drought in Central Texas dried up the San Gabriel River in 1756, forcing the abandonment of a settlement of missionaries and Indians. Stephen F. Austin's first colonists also were hurt by drought. Their initial corn crop was snuffed out in 1822, forcing the once ambitious farmers into desperate hunters. In most years, some sections of the State receive less than normal rainfall, while other sections receive a greater than normal supply. Severe drought or excessively wet conditions rarely exist over the entire State at the same time.

TEMPERATURE. The vast land area of Texas experiences a wide range of temperatures. The High Plains experiences rather low temperatures in winter, while there are several separate areas within the State that experience very high temperatures in summer. Extended periods of subfreezing temperatures are rare, even on the High Plains. In South Texas, subfreezing temperatures associated with Arctic airmasses ordinarily are confined to several hours prior to sunrise, and seasons may pass with no subfreezing temperatures at all. In summer, the temperature contrast is much less pronounced from north to south with daily highs generally in the 90's. August is the hottest month.

OTHER CLIMATIC ELEMENTS. Relative humidity is highest in the coastal region, and decreases gradually inland, as the distance from the Gulf of Mexico increases. Mean annual relative humidity at noon varies from slightly more than 60 percent near the coast to around 35 percent in the El Paso area. As temperatures increase, relative humidities generally decrease.

Sunshine is abundant in the extreme southwestern section of the State, decreasing gradually eastward. On an average, the western Trans-Pecos receives 80 percent of the total possible sunshine annually, while the Upper Coast receives only 60 percent.

Significant amounts of snowfall are confined almost entirely to the mountainous Trans-Pecos region and the High Plains. Measurable snow falls south of the High Plains but usually melts almost as fast as it falls. Blizzards may occur in extreme West Texas or Northwest Texas during the winter or early spring months, but are rare. Blizzards are characterized by subfreezing temperatures, very strong winds, and considerable blowing or drifting snow.

STORMS. Tropical cyclones are a threat to all sections of the Texas coast during the summer and fall. Those tropical cyclones with sustained wind speeds of 64 knots (74 m.p.h.) or greater are known as hurricanes. Virtually all tropical cyclones which have affected the Texas coast originated in the Gulf of Mexico, the Caribbean Sea, or the southern part of the North Atlantic Ocean. The season extends from June to October; storms are most frequent in August and September, and rarely affect the Texas coast after the first days of October. The average frequency for the entire Texas coast is approximately one per year.

Tornadoes have occurred in Texas during all seasons; however, they have occurred with greatest frequency during April, May, and June. Approximately one-fourth of the total annual number of tornadoes occur in the month of May alone. Hailstorms occur in all parts of the State. The most frequent and most damaging of these occur in spring and early summer. Thunderstorms, from which most damaging local weather develops (tornadoes, hail, windstorms, and high intensity showers) occur on about 60 days each year in the extreme eastern section of the State. The mean annual number of thunderstorm days decreases to about 40 in extreme West Texas, and to 30 in the lower Rio Grande Valley.

Blowing dust and sand may occur occasionally in West Texas where strong winds are more frequent and vegetation is sparse.

## COUNTY INDEX

***Anderson County***
PALESTINE 2 NE

***Andrews County***
ANDREWS

***Angelina County***
LUFKIN ANGELINA CO

***Aransas County***
ROCKPORT

***Archer County***
ARCHER CITY

***Armstrong County***
CLAUDE

***Atascosa County***
CHARLOTTE 5 NNW
POTEET

***Austin County***
SEALY

***Bailey County***
MULESHOE 1

***Bandera County***
MEDINA 2 W

***Bastrop County***
ELGIN
RED ROCK
SMITHVILLE

***Baylor County***
SEYMOUR

***Bee County***
BEEVILLE 5 NE

***Bell County***
STILLHOUSE HOLLOW DA
TEMPLE

***Bexar County***
SAN ANTONIO INTL AP

***Blanco County***
BLANCO
JOHNSON CITY

***Borden County***
GAIL

***Bosque County***
WHITNEY DAM

***Brazoria County***
ANGLETON 2 W
FREEPORT 2 NW

***Brazos County***
COLLEGE STN EASTERWD

***Brewster County***
ALPINE
CHISOS BASIN
MARATHON
PANTHER JUNCTION

***Briscoe County***
SILVERTON

***Brooks County***
FALFURRIAS

***Brown County***
BROWNWOOD

***Burleson County***
SOMERVILLE DAM

***Burnet County***
BURNET

***Caldwell County***
LULING

***Calhoun County***
POINT COMFORT
PORT O CONNOR

***Callahan County***
PUTNAM

***Cameron County***
BROWNSVILLE INTL AP
HARLINGEN

***Castro County***
DIMMITT 2 N

***Chambers County***
ANAHUAC

***Cherokee County***
JACKSONVILLE
RUSK

***Childress County***
CHILDRESS MUNI AP

***Clay County***
HENRIETTA

***Cochran County***
MORTON

***Coke County***
ROBERT LEE

***Coleman County***
COLEMAN
HORDS CREEK DAM

***Collin County***
LAVON DAM
MC KINNEY 3 S

***Collingsworth County***
WELLINGTON

***Colorado County***
COLUMBUS

***Comal County***
CANYON DAM
NEW BRAUNFELS

***Comanche County***
PROCTOR RESERVOIR

***Concho County***
PAINT ROCK

***Coryell County***
EVANT 1 SSW
GATESVILLE 4 SSE

***Cottle County***
PADUCAH

***Crane County***
CRANE

***Crockett County***
OZONA 2 SSW

***Crosby County***
CROSBYTON

***Culberson County***
VAN HORN

***Dallam County***
DALHART MUNI AP

***Dallas County***
DALLAS LOVE FIELD

***Dawson County***
LAMESA 1 SSE

***Deaf Smith County***
HEREFORD

***Denton County***
DENTON 2 SE
PILOT POINT

***DeWitt County***
CUERO

***Dimmit County***
CARRIZO SPRINGS
CATARINA

***Donley County***
CLARENDON

***Duval County***
BENAVIDES 2
FREER

***Eastland County***
EASTLAND
RISING STAR

***Ector County***
PENWELL

***Edwards County***
CARTA VALLEY
ROCKSPRINGS

***Ellis County***
BARDWELL DAM
FERRIS
WAXAHACHIE

***El Paso County***
EL PASO INTL ARPT
LA TUNA 1 S
YSLETA

***Erath County***
DUBLIN

***Falls County***
MARLIN 3 NE

***Fannin County***
BONHAM 3 NNE

***Fayette County***
FLATONIA
LA GRANGE

***Fisher County***
ROTAN

***Floyd County***
FLOYDADA

***Fort Bend County***
SUGAR LAND
THOMPSONS 3 WSW

***Freestone County***
FAIRFIELD 3 W

***Frio County***
DILLEY
PEARSALL

***Gaines County***
SEMINOLE

***Galveston County***
GALVESTON

***Garza County***
POST 3 ENE

***Gillespie County***
FREDERICKSBURG

***Glasscock County***
GARDEN CITY

***Goliad County***
GOLIAD

***Gonzales County***
GONZALES 2 N
NIXON

***Gray County***
MC LEAN
PAMPA 2

***Grayson County***
DENISON DAM
SHERMAN

***Hale County***
PLAINVIEW

***Hall County***
MEMPHIS
TURKEY

***Hamilton County***
HICO

***Hansford County***
GRUVER
SPEARMAN

***Hardeman County***
QUANAH 5 SE

***Harris County***
HOUSTON WM HOBBY AP

***Harrison County***
MARSHALL

***Hartley County***
BRAVO

***Haskell County***
HASKELL

***Hays County***
SAN MARCOS

***Hemphill County***
CANADIAN

***Henderson County***
ATHENS 3 SSE

***Hidalgo County***
MC COOK
MISSION 4 W
WESLACO 2 E

***Hill County***
HILLSBORO

***Hockley County***
LEVELLAND

***Hopkins County***
SULPHUR SPRINGS

***Houston County***
CROCKETT

***Howard County***
BIG SPRING

***Hudspeth County***
CORNUDAS SERVICE STN
SIERRA BLANCA 2 E

***Hunt County***
GREENVILLE

***Hutchinson County***
BORGER

***Jack County***
JACKSBORO

***Jeff Davis County***
MOUNT LOCKE

***Jefferson County***
PORT ARTHUR JEFFERSN

***Jim Hogg County***
HEBBRONVILLE

***Jim Wells County***
ALICE
MATHIS 4 SSW

***Johnson County***
CLEBURNE

***Jones County***
ANSON

***Kaufman County***
KAUFMAN 3 SE

***Kendall County***
BOERNE

***Kent County***
JAYTON

***King County***
GUTHRIE

***Kinney County***
BRACKETTVILLE

***Kleberg County***
KINGSVILLE

***Knox County***
MUNDAY
TRUSCOTT 2 NW

***Lamar County***
PARIS

***Lamb County***
LITTLEFIELD 2 NW
OLTON

***Lampasas County***
LAMPASAS

***La Salle County***
FOWLERTON

***Lavaca County***
HALLETTSVILLE 2 N
YOAKUM

***Lee County***
LEXINGTON

***Leon County***
CENTERVILLE

***Liberty County***
CLEVELAND
LIBERTY

***Limestone County***
MEXIA

***Lipscomb County***
FOLLETT
LIPSCOMB

***Llano County***
LLANO

***Lubbock County***
LUBBOCK REGIONAL AP

***Lynn County***
TAHOKA

***McCulloch County***
BRADY

***McLennan County***
MC GREGOR
WACO DAM
WACO MADISN COOPR AP

***McMullen County***
TILDEN 4 SSE

***Madison County***
MADISONVILLE

***Mason County***
MASON

***Matagorda County***
BAY CITY WATERWORKS
MATAGORDA 2
PALACIOS MUNI ARPT

***Maverick County***
EAGLE PASS

***Menard County***
MENARD

***Midland County***
MIDLAND 4 ENE
MIDLAND REGIONAL TER

***Milam County***
CAMERON

***Mills County***
GOLDTHWAITE 1 WSW

***Montague County***
BOWIE

***Montgomery County***
CONROE

***Moore County***
DUMAS

***Morris County***
DAINGERFIELD 9 S

***Motley County***
MATADOR

***Navarro County***
CORSICANA
NAVARRO MILLS DAM

***Nolan County***
ROSCOE

***Nueces County***
CHAPMAN RANCH
CORPUS CHRISTI INTL
ROBSTOWN

***Ochiltree County***
PERRYTON

***Palo Pinto County***
MINERAL WELLS MUN AP

***Panola County***
CARTHAGE

***Parker County***
WEATHERFORD

***Parmer County***
FRIONA

***Pecos County***
BAKERSFIELD 2 NW
FORT STOCKTON
SHEFFIELD

***Polk County***
LIVINGSTON 2 NNE

***Potter County***
AMARILLO INTL ARPT

***Presidio County***
CANDELARIA
MARFA # 2
PRESIDIO

***Rains County***
EMORY

***Randall County***
CANYON

***Reagan County***
BIG LAKE 2
COPE RANCH

***Real County***
CAMP WOOD

***Red River County***
CLARKSVILLE 2 NE

***Reeves County***
BALMORHEA
PECOS
RED BLUFF DAM

***Roberts County***
MIAMI

***Robertson County***
FRANKLIN

***Runnels County***
BALLINGER 2 NW

***Rusk County***
HENDERSON

***San Jacinto County***
COLDSPRING 5 SSW

***San Patricio County***
SINTON

***San Saba County***
SAN SABA

***Scurry County***
SNYDER

***Shackelford County***
ALBANY

***Shelby County***
CENTER

***Sherman County***
STRATFORD

***Somervell County***
GLEN ROSE 2 W

***Starr County***
FALCON DAM
RIO GRANDE CITY 3 W

***Stephens County***
BRECKENRIDGE

***Sterling County***
STERLING CITY

***Stonewall County***
ASPERMONT

***Sutton County***
SONORA

***Swisher County***
TULIA

***Tarrant County***
BENBROOK DAM
GRAPEVINE DAM

***Taylor County***
ABILENE MUNI AP

***Terrell County***
SANDERSON

***Terry County***
BROWNFIELD 2

***Throckmorton County***
THROCKMORTON

***Titus County***
MOUNT PLEASANT

### *Tom Green County*
SAN ANGELO MATHIS FD
WATER VALLEY

### *Travis County*
AUSTIN MUNICIPAL AP

### *Trinity County*
GROVETON

### *Upshur County*
GILMER 2 W

### *Upton County*
MC CAMEY

### *Val Verde County*
AMISTAD DAM
DEL RIO INTL AP
LANGTRY
PANDALE 2 NE

### *Van Zandt County*
WILLS POINT

### *Victoria County*
VICTORIA REGIONAL AP

### *Walker County*
HUNTSVILLE

### *Ward County*
GRANDFALLS 3 SSE
MONAHANS

### *Washington County*
BRENHAM
WASHINGTON STATE PAR

### *Webb County*
ENCINAL
LAREDO 2

### *Wharton County*
DANEVANG 1 W
NEW GULF
PIERCE 1 E

### *Wheeler County*
SHAMROCK 2

### *Wichita County*
WICHITA FALLS MUN AP

### *Wilbarger County*
VERNON 4 S

### *Willacy County*
PORT MANSFIELD
RAYMONDVILLE

### *Williamson County*
TAYLOR

### *Wilson County*
FLORESVILLE

### *Winkler County*
WINK WINKLER CO AP

### *Wise County*
BRIDGEPORT

### *Yoakum County*
PLAINS

### *Young County*
GRAHAM
OLNEY

### *Zapata County*
ZAPATA 3 SW

### *Zavala County*
CRYSTAL CITY

## ELEVATION INDEX

| FEET | STATION NAME |
|---|---|
| 7 | ROCKPORT |
| 10 | FREEPORT 2 NW |
| 10 | MATAGORDA 2 |
| 10 | PORT MANSFIELD |
| 10 | PORT O CONNOR |
| 16 | PALACIOS MUNI ARPT |
| 20 | ANAHUAC |
| 20 | POINT COMFORT |
| 30 | ANGLETON 2 W |
| 30 | CHAPMAN RANCH |
| 30 | PORT ARTHUR JEFFERSN |
| 30 | RAYMONDVILLE |
| 33 | BROWNSVILLE INTL AP |
| 35 | LIBERTY |
| 39 | HARLINGEN |
| 41 | CORPUS CHRISTI INTL |
| 49 | BAY CITY WATERWORKS |
| 49 | SINTON |
| 59 | GALVESTON |
| 59 | KINGSVILLE |
| 69 | DANEVANG 1 W |
| 69 | NEW GULF |
| 72 | THOMPSONS 3 WSW |
| 79 | ROBSTOWN |
| 79 | SUGAR LAND |
| 79 | WESLACO 2 E |
| 85 | HOUSTON WM HOBBY AP |
| 102 | PIERCE 1 E |
| 104 | VICTORIA REGIONAL AP |
| 112 | FALFURRIAS |
| 112 | MATHIS 4 SSW |
| 131 | MISSION 4 W |
| 161 | CLEVELAND |
| 161 | GOLIAD |
| 171 | ALICE |
| 180 | CUERO |
| 180 | RIO GRANDE CITY 3 W |
| 190 | LIVINGSTON 2 NNE |
| 200 | SEALY |
| 210 | COLUMBUS |
| 210 | WASHINGTON STATE PAR |
| 220 | BEEVILLE 5 NE |
| 220 | MC COOK |
| 239 | HALLETTSVILLE 2 N |
| 239 | MADISONVILLE |
| 245 | CONROE |
| 259 | TILDEN 4 SSE |
| 263 | SOMERVILLE DAM |
| 279 | LA GRANGE |
| 289 | ZAPATA 3 SW |
| 292 | LUFKIN ANGELINA CO |
| 302 | CARTHAGE |
| 302 | DAINGERFIELD 9 S |
| 302 | GONZALES 2 N |
| 318 | COLLEGE STN EASTERWD |
| 322 | FALCON DAM |
| 322 | GROVETON |
| 331 | SMITHVILLE |
| 347 | FOWLERTON |
| 351 | BRENHAM |

| FEET | STATION NAME |
|---|---|
| 351 | CENTER |
| 351 | CENTERVILLE |
| 351 | COLDSPRING 5 SSW |
| 351 | CROCKETT |
| 371 | GILMER 2 W |
| 381 | BENAVIDES 2 |
| 381 | MARSHALL |
| 381 | YOAKUM |
| 390 | CAMERON |
| 390 | FLORESVILLE |
| 400 | HUNTSVILLE |
| 400 | LULING |
| 400 | MARLIN 3 NE |
| 400 | NIXON |
| 420 | KAUFMAN 3 SE |
| 420 | MOUNT PLEASANT |
| 430 | LAREDO 2 |
| 435 | FAIRFIELD 3 W |
| 440 | CHARLOTTE 5 NNW |
| 440 | CLARKSVILLE 2 NE |
| 440 | CORSICANA |
| 440 | FERRIS |
| 454 | NAVARRO MILLS DAM |
| 459 | BARDWELL DAM |
| 459 | FRANKLIN |
| 459 | LEXINGTON |
| 460 | PALESTINE 2 NE |
| 463 | EMORY |
| 469 | FLATONIA |
| 489 | ATHENS 3 SSE |
| 489 | RUSK |
| 502 | DALLAS LOVE FIELD |
| 502 | HENDERSON |
| 502 | LAVON DAM |
| 502 | RED ROCK |
| 502 | SULPHUR SPRINGS |
| 502 | WACO DAM |
| 509 | WACO MADISN COOPR AP |
| 522 | WILLS POINT |
| 531 | FREER |
| 531 | MEXIA |
| 531 | WHITNEY DAM |
| 551 | HEBBRONVILLE |
| 561 | WAXAHACHIE |
| 571 | BONHAM 3 NNE |
| 571 | DILLEY |
| 571 | JACKSONVILLE |
| 571 | TAYLOR |
| 581 | CATARINA |
| 581 | CRYSTAL CITY |

| FEET | STATION NAME |
|---|---|
| 581 | ELGIN |
| 581 | GRAPEVINE DAM |
| 590 | ENCINAL |
| 591 | PARIS |
| 610 | DENISON DAM |
| 610 | GREENVILLE |
| 610 | MC KINNEY 3 S |
| 612 | SAN MARCOS |
| 620 | AUSTIN MUNICIPAL AP |
| 620 | GLEN ROSE 2 W |
| 620 | POTEET |
| 630 | DENTON 2 SE |
| 630 | HILLSBORO |
| 630 | PEARSALL |
| 650 | CARRIZO SPRINGS |
| 669 | PILOT POINT |
| 702 | TEMPLE |
| 712 | STILLHOUSE HOLLOW DA |
| 722 | NEW BRAUNFELS |
| 732 | MC GREGOR |
| 741 | EAGLE PASS |
| 751 | SHERMAN |
| 761 | CLEBURNE |
| 771 | BENBROOK DAM |
| 791 | GATESVILLE 4 SSE |
| 791 | SAN ANTONIO INTL AP |
| 820 | BRIDGEPORT |
| 922 | HENRIETTA |
| 938 | MINERAL WELLS MUN AP |
| 1001 | CANYON DAM |
| 1010 | HICO |
| 1020 | LAMPASAS |
| 1026 | DEL RIO INTL AP |
| 1027 | WICHITA FALLS MUN AP |
| 1040 | ARCHER CITY |
| 1040 | GRAHAM |
| 1040 | LLANO |
| 1050 | BRACKETTVILLE |
| 1070 | JACKSBORO |
| 1102 | WEATHERFORD |
| 1112 | BOWIE |
| 1161 | AMISTAD DAM |
| 1181 | OLNEY |
| 1191 | JOHNSON CITY |
| 1201 | SAN SABA |
| 1202 | VERNON 4 S |
| 1211 | BRECKENRIDGE |
| 1220 | PROCTOR RESERVOIR |
| 1240 | EVANT 1 SSW |
| 1280 | SEYMOUR |

| FEET | STATION NAME |
|---|---|
| 1302 | BURNET |
| 1312 | LANGTRY |
| 1322 | THROCKMORTON |
| 1352 | BLANCO |
| 1385 | BROWNWOOD |
| 1421 | BOERNE |
| 1421 | EASTLAND |
| 1430 | ALBANY |
| 1430 | MASON |
| 1450 | CAMP WOOD |
| 1470 | DUBLIN |
| 1480 | MUNDAY |
| 1503 | QUANAH 5 SE |
| 1532 | HASKELL |
| 1552 | TRUSCOTT 2 NW |
| 1581 | GOLDTHWAITE 1 WSW |
| 1591 | PANDALE 2 NE |
| 1591 | PUTNAM |
| 1631 | MEDINA 2 W |
| 1631 | PAINT ROCK |
| 1631 | RISING STAR |
| 1670 | ASPERMONT |
| 1682 | BALLINGER 2 NW |
| 1713 | COLEMAN |
| 1722 | BRADY |
| 1722 | FREDERICKSBURG |
| 1752 | ANSON |
| 1752 | GUTHRIE |
| 1755 | ABILENE MUNI AP |
| 1781 | ROBERT LEE |
| 1870 | PADUCAH |
| 1903 | CARTA VALLEY |
| 1903 | SAN ANGELO MATHIS FD |
| 1923 | ROTAN |
| 1942 | HORDS CREEK DAM |
| 1955 | CHILDRESS MUNI AP |
| 1962 | MENARD |
| 1982 | WELLINGTON |
| 2011 | JAYTON |
| 2070 | MEMPHIS |
| 2113 | WATER VALLEY |
| 2201 | SHEFFIELD |
| 2241 | SONORA |
| 2290 | STERLING CITY |
| 2342 | CANADIAN |
| 2346 | OZONA 2 SSW |
| 2352 | MATADOR |
| 2352 | SHAMROCK 2 |
| 2352 | TURKEY |
| 2382 | ROSCOE |

| FEET | STATION NAME |
|---|---|
| 2411 | ROCKSPRINGS |
| 2431 | LIPSCOMB |
| 2441 | GRANDFALLS 3 SSE |
| 2451 | MC CAMEY |
| 2451 | SNYDER |
| 2533 | BAKERSFIELD 2 NW |
| 2533 | BIG SPRING |
| 2533 | GAIL |
| 2546 | POST 3 ENE |
| 2552 | MC LEAN |
| 2570 | COPE RANCH |
| 2582 | PECOS |
| 2589 | PRESIDIO |
| 2602 | CRANE |
| 2621 | MONAHANS |
| 2631 | GARDEN CITY |
| 2680 | BIG LAKE 2 |
| 2743 | MIAMI |
| 2743 | MIDLAND 4 ENE |
| 2800 | RED BLUFF DAM |
| 2802 | FOLLETT |
| 2802 | SANDERSON |
| 2815 | WINK WINKLER CO AP |
| 2858 | MIDLAND REGIONAL TER |
| 2871 | CLARENDON |
| 2881 | CANDELARIA |
| 2933 | FORT STOCKTON |
| 2933 | PERRYTON |
| 2943 | PENWELL |
| 2972 | LAMESA 1 SSE |
| 3091 | TAHOKA |
| 3104 | SPEARMAN |
| 3113 | CROSBYTON |
| 3143 | BORGER |
| 3173 | GRUVER |
| 3182 | FLOYDADA |
| 3212 | PAMPA 2 |
| 3215 | LUBBOCK REGIONAL AP |
| 3222 | BALMORHEA |
| 3281 | SEMINOLE |
| 3304 | BROWNFIELD 2 |
| 3373 | PLAINVIEW |
| 3402 | CLAUDE |
| 3412 | ANDREWS |
| 3481 | SILVERTON |
| 3491 | LITTLEFIELD 2 NW |
| 3504 | PLAINS |
| 3504 | TULIA |
| 3553 | LEVELLAND |
| 3583 | CANYON |

| FEET | STATION NAME |
|---|---|
| 3609 | AMARILLO INTL ARPT |
| 3612 | OLTON |
| 3654 | DUMAS |
| 3671 | YSLETA |
| 3704 | MULESHOE 1 |
| 3704 | STRATFORD |
| 3763 | MORTON |
| 3802 | LA TUNA 1 S |
| 3802 | PANTHER JUNCTION |
| 3842 | HEREFORD |
| 3862 | DIMMITT 2 N |
| 3940 | EL PASO INTL ARPT |
| 3955 | VAN HORN |
| 3996 | DALHART MUNI AP |
| 4012 | FRIONA |
| 4052 | MARATHON |
| 4153 | BRAVO |
| 4482 | CORNUDAS SERVICE STN |
| 4535 | SIERRA BLANCA 2 E |
| 4564 | ALPINE |
| 4692 | MARFA # 2 |
| 5282 | CHISOS BASIN |
| 6795 | MOUNT LOCKE |

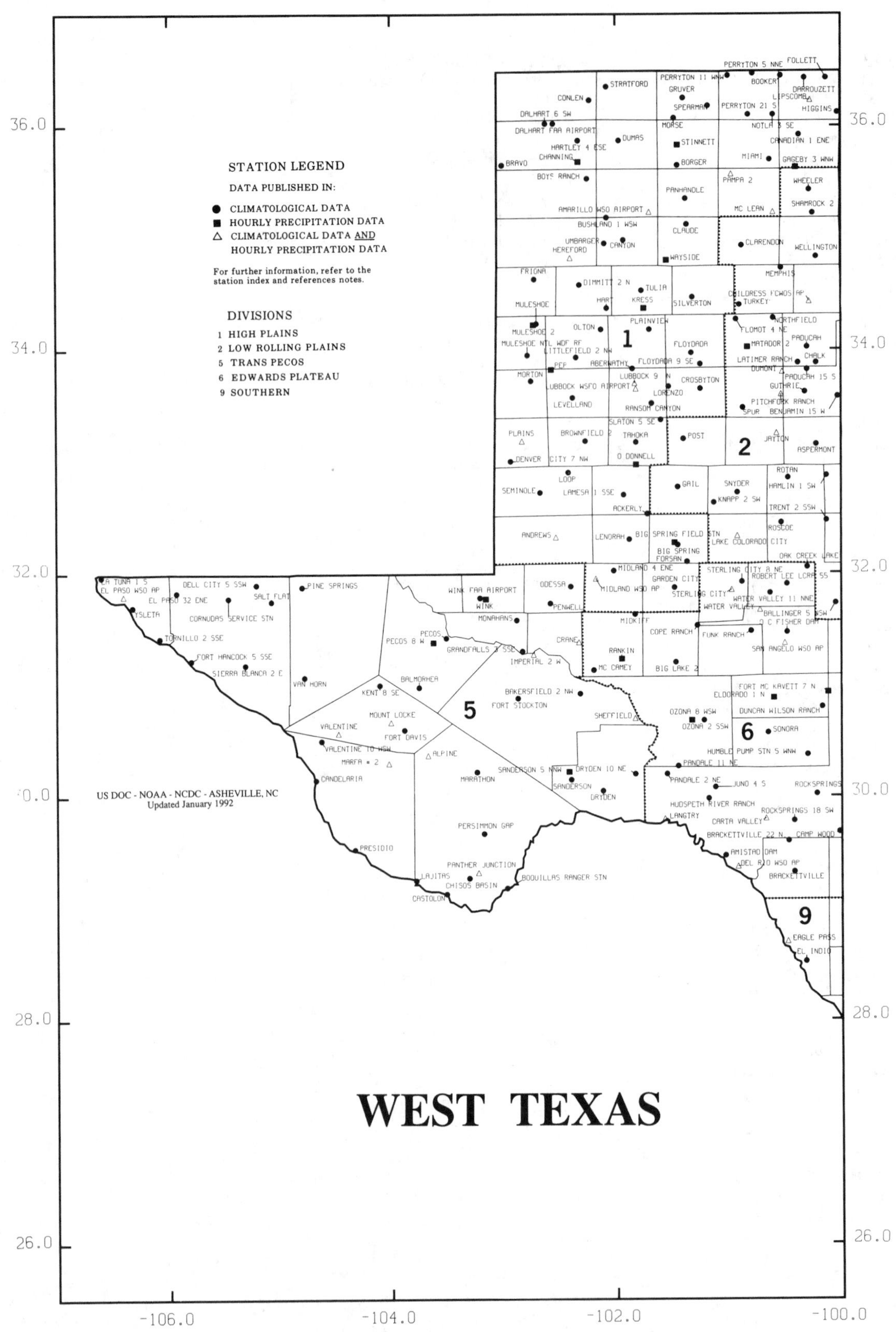
STATION LEGEND
DATA PUBLISHED IN:
● CLIMATOLOGICAL DATA
■ HOURLY PRECIPITATION DATA
△ CLIMATOLOGICAL DATA AND HOURLY PRECIPITATION DATA
For further information, refer to the station index and references notes.
DIVISIONS
1 HIGH PLAINS
2 LOW ROLLING PLAINS
5 TRANS PECOS
6 EDWARDS PLATEAU
9 SOUTHERN
US DOC - NOAA - NCDC - ASHEVILLE, NC
Updated January 1992
WEST TEXAS

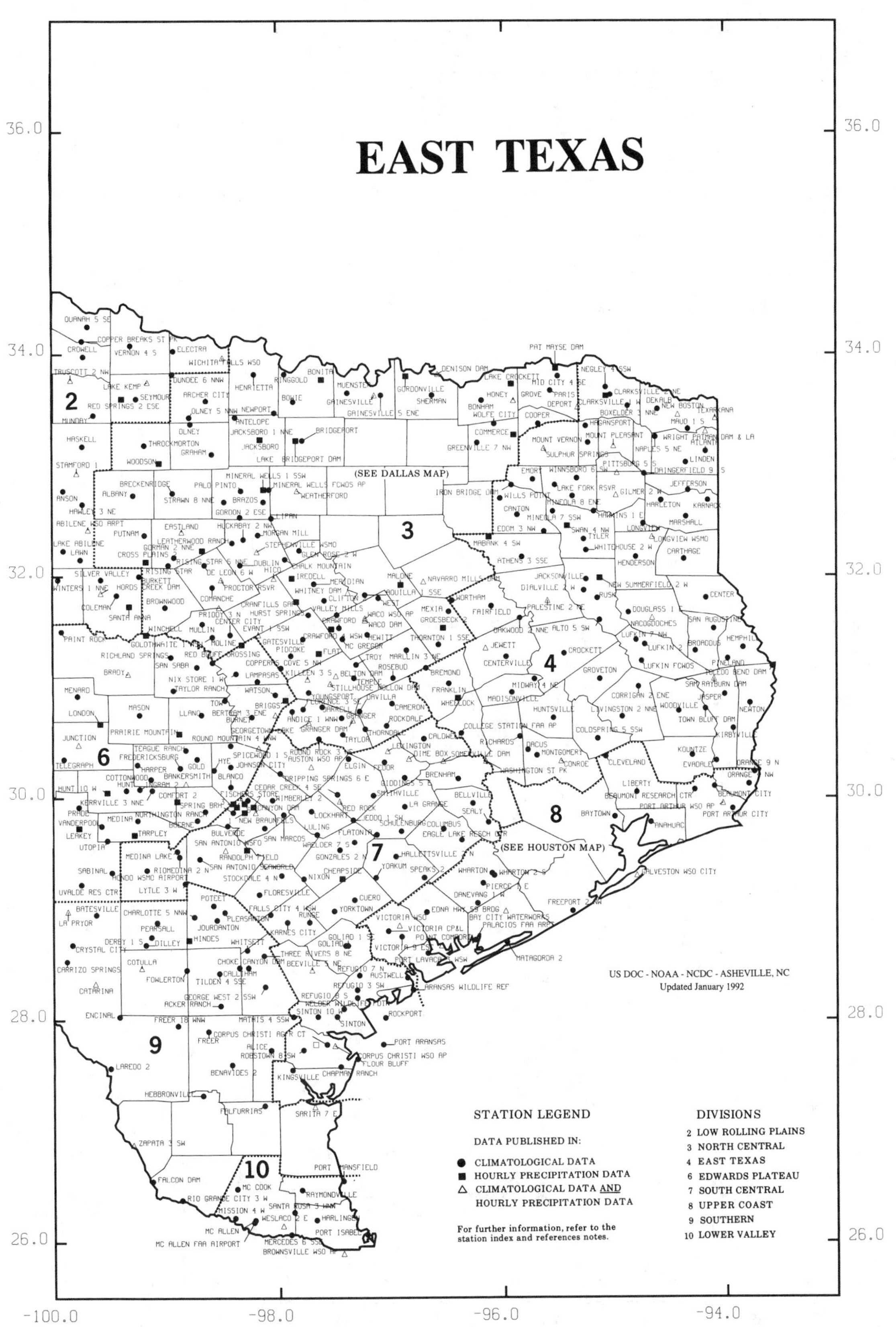

EAST TEXAS
36.0
34.0
32.0
30.0
28.0
26.0
-100.0
-98.0
-96.0
-94.0
(SEE DALLAS MAP)
(SEE HOUSTON MAP)
2
3
4
6
7
8
9
10
US DOC - NOAA - NCDC - ASHEVILLE, NC
Updated January 1992
STATION LEGEND
DATA PUBLISHED IN:
● CLIMATOLOGICAL DATA
■ HOURLY PRECIPITATION DATA
△ CLIMATOLOGICAL DATA AND HOURLY PRECIPITATION DATA
For further information, refer to the station index and references notes.
DIVISIONS
2 LOW ROLLING PLAINS
3 NORTH CENTRAL
4 EAST TEXAS
6 EDWARDS PLATEAU
7 SOUTH CENTRAL
8 UPPER COAST
9 SOUTHERN
10 LOWER VALLEY

# DALLAS

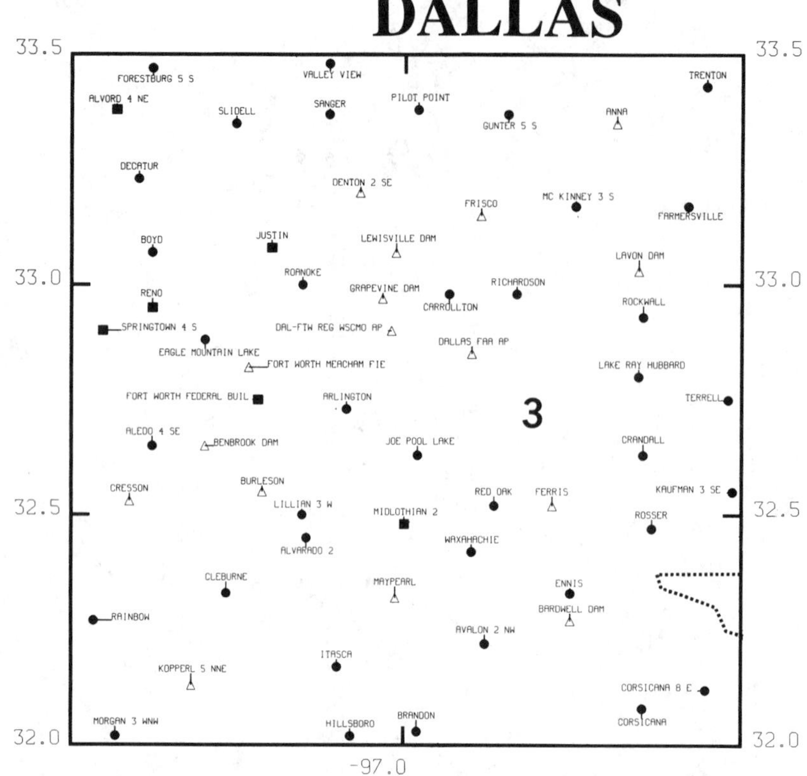


# HOUSTON

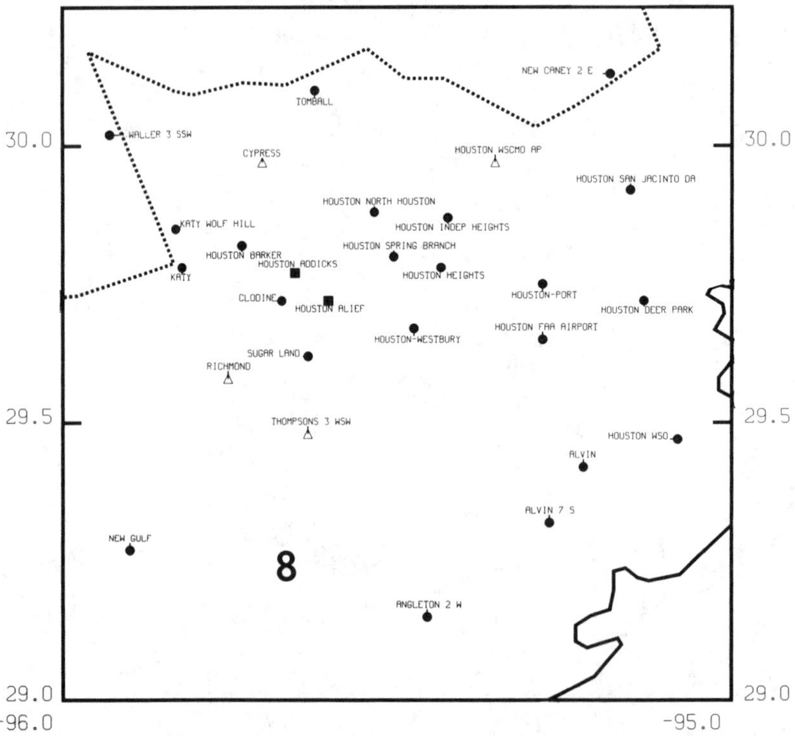

## ABILENE MUNI AP *Taylor County* ELEVATION 1755 ft LAT/LONG 32° 26 ' N / 99° 41 ' W

| | JAN | FEB | MAR | APR | MAY | JUN | JUL | AUG | SEP | OCT | NOV | DEC | YEAR |
|---|---|---|---|---|---|---|---|---|---|---|---|---|---|
| Maximum Temp °F | 54.6 | 59.4 | 68.6 | 77.3 | 83.5 | 90.9 | 94.5 | 93.3 | 85.9 | 76.9 | 65.2 | 57.1 | 75.6 |
| Minimum Temp °F | 31.4 | 35.5 | 43.4 | 52.6 | 60.6 | 68.7 | 72.4 | 71.3 | 64.5 | 54.1 | 42.5 | 34.2 | 52.6 |
| Mean Temp °F | 43.0 | 47.5 | 56.0 | 65.0 | 72.1 | 79.8 | 83.4 | 82.3 | 75.2 | 65.5 | 53.9 | 45.7 | 64.1 |
| Days Max Temp ≥ 90 °F | 0 | 0 | 1 | 3 | 7 | 19 | 26 | 24 | 11 | 2 | 0 | 0 | 93 |
| Days Max Temp ≤ 32 °F | 2 | 1 | 0 | 0 | 0 | 0 | 0 | 0 | 0 | 0 | 0 | 1 | 4 |
| Days Min Temp ≤ 32 °F | 17 | 11 | 4 | 0 | 0 | 0 | 0 | 0 | 0 | 0 | 5 | 14 | 51 |
| Days Min Temp ≤ 0 °F | 0 | 0 | 0 | 0 | 0 | 0 | 0 | 0 | 0 | 0 | 0 | 0 | 0 |
| Heating Degree Days | 675 | 492 | 301 | 98 | 20 | 0 | 0 | 0 | 15 | 96 | 345 | 593 | 2635 |
| Cooling Degree Days | 0 | 3 | 28 | 104 | 254 | 451 | 576 | 565 | 335 | 124 | 20 | 1 | 2461 |
| Total Precipitation (") | 1.03 | 1.20 | 1.34 | 1.98 | 3.14 | 2.73 | 1.89 | 2.65 | 3.15 | 2.68 | 1.40 | 1.28 | 24.47 |
| Days ≥ 0.1" Precip | 3 | 3 | 3 | 3 | 5 | 4 | 3 | 4 | 4 | 4 | 3 | 3 | 42 |
| Total Snowfall (") | 2.1 | 1.0 | 0.6 | 0.0 | 0.0 | 0.0 | 0.0 | 0.0 | 0.0 | 0.0 | 0.7 | 1.0 | 5.4 |
| Days ≥ 1" Snow Depth | 1 | 1 | 0 | 0 | 0 | 0 | 0 | 0 | 0 | 0 | 0 | 1 | 3 |

## ALBANY *Shackelford County* ELEVATION 1430 ft LAT/LONG 32° 44 ' N / 99° 17 ' W

| | JAN | FEB | MAR | APR | MAY | JUN | JUL | AUG | SEP | OCT | NOV | DEC | YEAR |
|---|---|---|---|---|---|---|---|---|---|---|---|---|---|
| Maximum Temp °F | 56.1 | 60.1 | 69.7 | 77.9 | 83.9 | 91.6 | 95.7 | 95.0 | 87.2 | 78.8 | 67.2 | ***59.0*** | 76.9 |
| Minimum Temp °F | 30.3 | 33.6 | 41.4 | 50.6 | 58.1 | 66.5 | 70.3 | 68.6 | 61.7 | 51.0 | 40.3 | ***33.3*** | 50.5 |
| Mean Temp °F | 43.2 | 46.9 | 55.6 | 64.3 | 71.1 | 79.1 | 83.0 | 81.8 | 74.5 | 64.9 | 53.7 | ***46.2*** | 63.7 |
| Days Max Temp ≥ 90 °F | 0 | 0 | 1 | 3 | 8 | 20 | 27 | 26 | 14 | 3 | 0 | 0 | 102 |
| Days Max Temp ≤ 32 °F | 2 | 1 | 0 | 0 | 0 | 0 | 0 | 0 | 0 | 0 | 0 | ***1*** | 4 |
| Days Min Temp ≤ 32 °F | 19 | 13 | 6 | 1 | 0 | 0 | 0 | 0 | 0 | 1 | 7 | ***15*** | 62 |
| Days Min Temp ≤ 0 °F | 0 | 0 | 0 | 0 | 0 | 0 | 0 | 0 | 0 | 0 | 0 | 0 | 0 |
| Heating Degree Days | 669 | 507 | 311 | 105 | 25 | 1 | 0 | 0 | 17 | 104 | 352 | ***579*** | 2670 |
| Cooling Degree Days | 0 | 5 | 28 | 95 | 234 | 439 | 562 | 551 | 327 | 120 | 24 | ***1*** | 2386 |
| Total Precipitation (") | 1.28 | 1.67 | 1.76 | 2.74 | 4.15 | 2.98 | 1.83 | 3.12 | 3.52 | 2.98 | 1.46 | 1.72 | 29.21 |
| Days ≥ 0.1" Precip | 3 | 3 | 3 | 4 | 5 | 4 | 3 | 4 | 4 | 4 | 3 | ***3*** | 43 |
| Total Snowfall (") | ***1.0*** | ***0.8*** | 0.8 | 0.0 | 0.0 | 0.0 | 0.0 | 0.0 | 0.0 | 0.0 | 0.3 | 0.2 | 3.1 |
| Days ≥ 1" Snow Depth | ***0*** | ***0*** | 0 | 0 | 0 | 0 | 0 | 0 | 0 | 0 | 0 | ***0*** | 0 |

## ALICE *Jim Wells County* ELEVATION 171 ft LAT/LONG 27° 45 ' N / 98° 4 ' W

| | JAN | FEB | MAR | APR | MAY | JUN | JUL | AUG | SEP | OCT | NOV | DEC | YEAR |
|---|---|---|---|---|---|---|---|---|---|---|---|---|---|
| Maximum Temp °F | 66.4 | 70.0 | 78.1 | 84.1 | 87.7 | 93.1 | 95.8 | 96.1 | 91.8 | 85.4 | 77.2 | 69.9 | 83.0 |
| Minimum Temp °F | 43.8 | 46.4 | 54.1 | 61.6 | 67.6 | 72.1 | 73.4 | 73.4 | 70.0 | 61.7 | 53.5 | 46.7 | 60.4 |
| Mean Temp °F | 55.1 | 58.2 | 66.1 | 72.9 | 77.7 | 82.6 | 84.6 | 84.8 | 80.9 | 73.6 | 65.4 | 58.3 | 71.7 |
| Days Max Temp ≥ 90 °F | 0 | 1 | 3 | 7 | 12 | 24 | 29 | 29 | 21 | 10 | 2 | 0 | 138 |
| Days Max Temp ≤ 32 °F | 0 | 0 | 0 | 0 | 0 | 0 | 0 | 0 | 0 | 0 | 0 | 0 | 0 |
| Days Min Temp ≤ 32 °F | 4 | 2 | 0 | 0 | 0 | 0 | 0 | 0 | 0 | 0 | 1 | 2 | 9 |
| Days Min Temp ≤ 0 °F | 0 | 0 | 0 | 0 | 0 | 0 | 0 | 0 | 0 | 0 | 0 | 0 | 0 |
| Heating Degree Days | 327 | 223 | 85 | 16 | 1 | 0 | 0 | 0 | 1 | 14 | 107 | 246 | 1020 |
| Cooling Degree Days | 20 | 45 | 129 | 252 | 411 | 552 | 624 | 639 | 496 | 309 | 134 | 50 | 3661 |
| Total Precipitation (") | 1.45 | 1.63 | 1.27 | 1.76 | 3.55 | 3.65 | 1.93 | 2.85 | 5.24 | 3.05 | 1.34 | 1.28 | 29.00 |
| Days ≥ 0.1" Precip | 4 | 3 | 2 | 3 | 5 | 5 | 3 | 4 | 6 | 3 | 2 | 3 | 43 |
| Total Snowfall (") | 0.0 | 0.0 | 0.0 | 0.0 | 0.0 | 0.0 | 0.0 | 0.0 | 0.0 | 0.0 | 0.0 | 0.0 | 0.0 |
| Days ≥ 1" Snow Depth | 0 | 0 | 0 | 0 | 0 | 0 | 0 | 0 | 0 | 0 | 0 | 0 | 0 |

## ALPINE *Brewster County* ELEVATION 4564 ft LAT/LONG 30° 21 ' N / 103° 40 ' W

| | JAN | FEB | MAR | APR | MAY | JUN | JUL | AUG | SEP | OCT | NOV | DEC | YEAR |
|---|---|---|---|---|---|---|---|---|---|---|---|---|---|
| Maximum Temp °F | 60.6 | 64.6 | 71.5 | 78.9 | ***84.6*** | 90.4 | 89.0 | 87.2 | 82.5 | 77.6 | 68.5 | 62.2 | 76.5 |
| Minimum Temp °F | 30.6 | 33.0 | 38.7 | 45.8 | ***53.2*** | 61.2 | 63.0 | 61.3 | 56.5 | 47.0 | 37.8 | 31.9 | 46.7 |
| Mean Temp °F | 45.6 | 48.8 | 55.1 | 62.4 | ***68.9*** | 75.9 | 76.0 | 74.3 | 69.5 | 62.3 | 53.1 | 47.1 | 61.6 |
| Days Max Temp ≥ 90 °F | 0 | 0 | 0 | 1 | 7 | 18 | 15 | 11 | 4 | 1 | 0 | 0 | 57 |
| Days Max Temp ≤ 32 °F | 0 | 0 | 0 | 0 | 0 | 0 | 0 | 0 | 0 | 0 | 0 | 0 | 0 |
| Days Min Temp ≤ 32 °F | 18 | 13 | 8 | 2 | 0 | 0 | 0 | 0 | 0 | 1 | 9 | 16 | 67 |
| Days Min Temp ≤ 0 °F | 0 | 0 | 0 | 0 | 0 | 0 | 0 | 0 | 0 | 0 | 0 | 0 | 0 |
| Heating Degree Days | 595 | 450 | 306 | 122 | ***28*** | 1 | 0 | 1 | 26 | 123 | 351 | 548 | 2551 |
| Cooling Degree Days | 0 | 1 | 5 | 41 | na | 341 | 354 | 308 | 175 | ***45*** | 4 | 0 | na |
| Total Precipitation (") | 0.50 | 0.53 | 0.42 | 0.59 | 1.25 | 2.07 | 3.09 | 3.06 | 3.33 | 1.48 | 0.56 | 0.63 | 17.51 |
| Days ≥ 0.1" Precip | 1 | 1 | 1 | 1 | 3 | 4 | 6 | 6 | 6 | 3 | 1 | 1 | 34 |
| Total Snowfall (") | ***0.5*** | ***0.2*** | 0.0 | 0.0 | 0.0 | 0.0 | 0.0 | 0.0 | 0.0 | 0.0 | ***0.3*** | 0.0 | 1.0 |
| Days ≥ 1" Snow Depth | ***0*** | ***0*** | 0 | 0 | 0 | 0 | 0 | 0 | 0 | 0 | 0 | 0 | 0 |

## AMARILLO INTL ARPT *Potter County* ELEVATION 3609 ft LAT/LONG 35° 14 ' N / 101° 42 ' W

| | JAN | FEB | MAR | APR | MAY | JUN | JUL | AUG | SEP | OCT | NOV | DEC | YEAR |
|---|---|---|---|---|---|---|---|---|---|---|---|---|---|
| Maximum Temp °F | 48.7 | 53.0 | 61.6 | 71.3 | 78.5 | 87.4 | 91.0 | 88.4 | 81.4 | 71.7 | 58.8 | 50.5 | 70.2 |
| Minimum Temp °F | 21.8 | 25.7 | 32.9 | 42.0 | 51.4 | 61.1 | 65.7 | 63.9 | 56.3 | 44.5 | 32.1 | 23.8 | 43.4 |
| Mean Temp °F | 35.3 | 39.3 | 47.3 | 56.6 | 64.9 | 74.3 | 78.4 | 76.2 | 68.9 | 58.1 | 45.4 | 37.2 | 56.8 |
| Days Max Temp ≥ 90 °F | 0 | 0 | 0 | 1 | 4 | 13 | 20 | 16 | 6 | 1 | 0 | 0 | 61 |
| Days Max Temp ≤ 32 °F | 5 | 3 | 1 | 0 | 0 | 0 | 0 | 0 | 0 | 0 | 1 | 3 | 13 |
| Days Min Temp ≤ 32 °F | 27 | 22 | 14 | 4 | 0 | 0 | 0 | 0 | 0 | 2 | 16 | 27 | 112 |
| Days Min Temp ≤ 0 °F | 1 | 0 | 0 | 0 | 0 | 0 | 0 | 0 | 0 | 0 | 0 | 1 | 2 |
| Heating Degree Days | 915 | 717 | 543 | 265 | 90 | 7 | 0 | 2 | 48 | 239 | 581 | 856 | 4263 |
| Cooling Degree Days | 0 | 0 | 1 | 21 | 89 | 298 | 419 | 363 | 175 | 24 | 1 | 0 | 1391 |
| Total Precipitation (") | 0.59 | 0.54 | 0.99 | 1.02 | 2.45 | 3.53 | 2.55 | 3.05 | 1.87 | 1.32 | 0.62 | 0.50 | 19.03 |
| Days ≥ 0.1" Precip | 2 | 1 | 3 | 3 | 5 | 6 | 4 | 6 | 4 | 3 | 2 | 2 | 41 |
| Total Snowfall (") | 4.2 | 3.8 | 2.6 | 0.6 | 0.0 | 0.0 | 0.0 | 0.0 | 0.0 | 0.4 | 1.8 | 2.8 | 16.2 |
| Days ≥ 1" Snow Depth | 5 | 3 | 1 | 0 | 0 | 0 | 0 | 0 | 0 | 0 | 1 | 2 | 12 |

## AMISTAD DAM *Val Verde County* ELEVATION 1161 ft LAT/LONG 29° 28 ' N / 101° 2 ' W

| | JAN | FEB | MAR | APR | MAY | JUN | JUL | AUG | SEP | OCT | NOV | DEC | YEAR |
|---|---|---|---|---|---|---|---|---|---|---|---|---|---|
| Maximum Temp °F | 61.4 | 66.0 | 75.3 | 83.2 | 88.1 | 93.9 | 96.3 | 96.3 | 90.5 | 81.6 | 71.3 | 63.6 | 80.6 |
| Minimum Temp °F | 38.2 | 41.8 | 49.8 | 58.2 | 65.1 | 71.0 | 73.1 | 72.7 | 68.0 | 58.5 | 48.2 | 40.7 | 57.1 |
| Mean Temp °F | 49.8 | 53.9 | 62.6 | 70.7 | 76.6 | 82.5 | 84.8 | 84.5 | 79.2 | 70.1 | 59.8 | 52.1 | 68.9 |
| Days Max Temp ≥ 90 °F | 0 | 0 | 2 | 7 | 13 | 24 | 28 | 28 | 19 | 5 | 0 | 0 | 126 |
| Days Max Temp ≤ 32 °F | 0 | 0 | 0 | 0 | 0 | 0 | 0 | 0 | 0 | 0 | 0 | 0 | 0 |
| Days Min Temp ≤ 32 °F | 7 | 4 | 1 | 0 | 0 | 0 | 0 | 0 | 0 | 0 | 1 | 5 | 18 |
| Days Min Temp ≤ 0 °F | 0 | 0 | 0 | 0 | 0 | 0 | 0 | 0 | 0 | 0 | 0 | 0 | 0 |
| Heating Degree Days | 466 | 314 | 135 | 26 | 2 | 0 | 0 | 0 | 3 | 30 | 189 | 393 | 1558 |
| Cooling Degree Days | 0 | 10 | 73 | 207 | 392 | 554 | 634 | 644 | 455 | 219 | 47 | 3 | 3238 |
| Total Precipitation (") | 0.65 | 0.91 | 0.83 | 1.65 | 2.42 | 2.26 | 2.03 | 1.92 | 3.34 | 1.85 | 0.96 | 0.80 | 19.62 |
| Days ≥ 0.1" Precip | 2 | 2 | 2 | 3 | 4 | 3 | 3 | 2 | 4 | 3 | 2 | 2 | 32 |
| Total Snowfall (") | 0.1 | 0.0 | 0.0 | 0.0 | 0.0 | 0.0 | 0.0 | 0.0 | 0.0 | 0.0 | 0.0 | 0.0 | 0.1 |
| Days ≥ 1" Snow Depth | 0 | 0 | 0 | 0 | 0 | 0 | 0 | 0 | 0 | 0 | 0 | 0 | 0 |

## ANAHUAC *Chambers County* ELEVATION 20 ft LAT/LONG 29° 46 ' N / 94° 41 ' W

| | JAN | FEB | MAR | APR | MAY | JUN | JUL | AUG | SEP | OCT | NOV | DEC | YEAR |
|---|---|---|---|---|---|---|---|---|---|---|---|---|---|
| Maximum Temp °F | 60.5 | 64.2 | 71.1 | 78.1 | 84.0 | 89.9 | 92.0 | 92.1 | 88.7 | 81.5 | 72.0 | 64.2 | 78.2 |
| Minimum Temp °F | 41.0 | 43.8 | 51.0 | 59.1 | 66.0 | 72.2 | 74.0 | 73.2 | 68.5 | 58.6 | 50.2 | 44.0 | 58.5 |
| Mean Temp °F | 50.9 | 54.0 | 61.1 | 68.7 | 75.0 | 81.1 | 83.1 | 82.7 | 78.6 | 70.1 | 61.1 | 54.1 | 68.4 |
| Days Max Temp ≥ 90 °F | 0 | 0 | 0 | 0 | 3 | 17 | 24 | 24 | 15 | 2 | 0 | 0 | 85 |
| Days Max Temp ≤ 32 °F | 0 | 0 | 0 | 0 | 0 | 0 | 0 | 0 | 0 | 0 | 0 | 0 | 0 |
| Days Min Temp ≤ 32 °F | 6 | 4 | 1 | 0 | 0 | 0 | 0 | 0 | 0 | 0 | 1 | 3 | 15 |
| Days Min Temp ≤ 0 °F | 0 | 0 | 0 | 0 | 0 | 0 | 0 | 0 | 0 | 0 | 0 | 0 | 0 |
| Heating Degree Days | 441 | 314 | 158 | 34 | 1 | 0 | 0 | 0 | 0 | 31 | 172 | 348 | 1499 |
| Cooling Degree Days | 5 | 11 | 45 | 140 | 321 | 497 | 571 | 564 | 421 | 201 | 68 | 18 | 2862 |
| Total Precipitation (") | 4.34 | 2.93 | 3.27 | 3.98 | 5.51 | 5.80 | 4.59 | 4.83 | 5.81 | 4.35 | 4.25 | 4.22 | 53.88 |
| Days ≥ 0.1" Precip | 7 | 5 | 5 | 4 | 6 | 6 | 7 | 7 | 7 | 5 | 5 | 6 | 70 |
| Total Snowfall (") | 0.0 | 0.0 | 0.0 | 0.0 | 0.0 | 0.0 | 0.0 | 0.0 | 0.0 | 0.0 | 0.0 | 0.0 | 0.0 |
| Days ≥ 1" Snow Depth | 0 | 0 | 0 | 0 | 0 | 0 | 0 | 0 | 0 | 0 | 0 | 0 | 0 |

## ANDREWS *Andrews County* ELEVATION 3412 ft LAT/LONG 32° 19 ' N / 102° 33 ' W

| | JAN | FEB | MAR | APR | MAY | JUN | JUL | AUG | SEP | OCT | NOV | DEC | YEAR |
|---|---|---|---|---|---|---|---|---|---|---|---|---|---|
| Maximum Temp °F | 57.3 | 62.9 | 71.2 | 80.0 | 86.6 | 93.2 | 93.8 | 92.3 | 86.1 | 78.4 | 66.4 | 59.3 | 77.3 |
| Minimum Temp °F | 29.8 | 33.6 | 40.5 | 49.0 | 57.5 | 65.1 | 67.8 | 66.5 | 60.7 | 50.6 | 39.4 | 31.9 | 49.4 |
| Mean Temp °F | 43.6 | 48.3 | 55.9 | 64.5 | 72.1 | 79.2 | 80.8 | 79.4 | 73.4 | 64.5 | 53.0 | 45.6 | 63.4 |
| Days Max Temp ≥ 90 °F | 0 | 0 | 0 | 3 | 11 | 23 | 26 | 23 | 11 | 2 | 0 | 0 | 99 |
| Days Max Temp ≤ 32 °F | 1 | 0 | 0 | 0 | 0 | 0 | 0 | 0 | 0 | 0 | 0 | 1 | 2 |
| Days Min Temp ≤ 32 °F | 20 | 13 | 6 | 1 | 0 | 0 | 0 | 0 | 0 | 0 | 7 | 16 | 63 |
| Days Min Temp ≤ 0 °F | 0 | 0 | 0 | 0 | 0 | 0 | 0 | 0 | 0 | 0 | 0 | 0 | 0 |
| Heating Degree Days | 656 | 467 | 293 | 94 | 17 | 1 | 0 | 0 | 16 | 97 | 359 | 594 | 2594 |
| Cooling Degree Days | 0 | 2 | 16 | 89 | 253 | 444 | 509 | 480 | 293 | 96 | 7 | 0 | 2189 |
| Total Precipitation (") | 0.47 | 0.55 | 0.62 | 0.92 | 1.73 | 1.89 | 2.52 | 1.84 | 2.32 | 1.47 | 0.63 | 0.53 | 15.49 |
| Days ≥ 0.1" Precip | 2 | 2 | 2 | 2 | 3 | 3 | 4 | 4 | 4 | 3 | 1 | 2 | 32 |
| Total Snowfall (") | 1.9 | 0.8 | 0.0 | 0.1 | 0.0 | 0.0 | 0.0 | 0.0 | 0.0 | 0.1 | 0.8 | 1.0 | 4.7 |
| Days ≥ 1" Snow Depth | *1* | 0 | 0 | 0 | 0 | 0 | 0 | 0 | 0 | 0 | 0 | *0* | 1 |

## ANGLETON 2 W *Brazoria County* ELEVATION 30 ft LAT/LONG 29° 12' N / 95° 23' W

| | JAN | FEB | MAR | APR | MAY | JUN | JUL | AUG | SEP | OCT | NOV | DEC | YEAR |
|---|---|---|---|---|---|---|---|---|---|---|---|---|---|
| Maximum Temp °F | 62.0 | 64.7 | 71.5 | 77.9 | 83.2 | 88.9 | 91.5 | 91.4 | 87.7 | 81.2 | 72.6 | 65.5 | 78.2 |
| Minimum Temp °F | 41.6 | 44.2 | 51.4 | 59.1 | 65.5 | 70.8 | 72.4 | 71.9 | 68.4 | 58.8 | 51.0 | 44.4 | 58.3 |
| Mean Temp °F | 51.8 | 54.5 | 61.5 | 68.5 | 74.4 | 79.9 | 82.0 | 81.7 | 78.1 | 70.1 | 61.8 | 55.0 | 68.3 |
| Days Max Temp ≥ 90 °F | 0 | 0 | 0 | 0 | 1 | 14 | 24 | 24 | 12 | 2 | 0 | 0 | 77 |
| Days Max Temp ≤ 32 °F | 0 | 0 | 0 | 0 | 0 | 0 | 0 | 0 | 0 | 0 | 0 | 0 | 0 |
| Days Min Temp ≤ 32 °F | 6 | 4 | 1 | 0 | 0 | 0 | 0 | 0 | 0 | 0 | 1 | 4 | 16 |
| Days Min Temp ≤ 0 °F | 0 | 0 | 0 | 0 | 0 | 0 | 0 | 0 | 0 | 0 | 0 | 0 | 0 |
| Heating Degree Days | 413 | 303 | 154 | 37 | 2 | 0 | 0 | 0 | 1 | 30 | 162 | 325 | 1427 |
| Cooling Degree Days | 6 | 15 | 55 | 141 | 308 | 471 | 545 | 544 | 411 | 212 | 79 | 23 | 2810 |
| Total Precipitation (") | 5.21 | 3.80 | 3.66 | 3.82 | 5.70 | 6.54 | 4.85 | 4.97 | 6.93 | 4.47 | 4.57 | 4.26 | 58.78 |
| Days ≥ 0.1" Precip | 8 | 6 | 4 | 4 | 6 | 6 | 6 | 7 | 7 | 5 | 6 | 7 | 72 |
| Total Snowfall (") | 0.1 | 0.1 | 0.0 | 0.0 | 0.0 | 0.0 | 0.0 | 0.0 | 0.0 | 0.0 | 0.0 | 0.0 | 0.2 |
| Days ≥ 1" Snow Depth | 0 | 0 | 0 | 0 | 0 | 0 | 0 | 0 | 0 | 0 | 0 | 0 | 0 |

## ANSON *Jones County* ELEVATION 1752 ft LAT/LONG 32° 45' N / 99° 54' W

| | JAN | FEB | MAR | APR | MAY | JUN | JUL | AUG | SEP | OCT | NOV | DEC | YEAR |
|---|---|---|---|---|---|---|---|---|---|---|---|---|---|
| Maximum Temp °F | 56.8 | 61.9 | 71.4 | 80.2 | 85.9 | 92.9 | 96.2 | 94.7 | 87.4 | 79.0 | *67.2* | *59.5* | 77.8 |
| Minimum Temp °F | 31.5 | 35.6 | 43.3 | 52.6 | 60.5 | 68.6 | 72.1 | 70.7 | 63.9 | 53.5 | *42.4* | *34.6* | 52.4 |
| Mean Temp °F | 44.2 | 48.8 | 57.4 | 66.4 | 73.3 | 80.8 | 84.2 | 82.7 | 75.7 | 66.3 | *54.8* | *47.0* | 65.1 |
| Days Max Temp ≥ 90 °F | 0 | 0 | 1 | *4* | 11 | 22 | 28 | 25 | 14 | 3 | 0 | 0 | 108 |
| Days Max Temp ≤ 32 °F | 2 | 1 | 0 | 0 | 0 | 0 | 0 | 0 | 0 | 0 | 0 | 1 | 4 |
| Days Min Temp ≤ 32 °F | 16 | 11 | 5 | 1 | 0 | 0 | 0 | 0 | 0 | 0 | *5* | *13* | 51 |
| Days Min Temp ≤ 0 °F | 0 | 0 | 0 | 0 | 0 | 0 | 0 | 0 | 0 | 0 | 0 | 0 | 0 |
| Heating Degree Days | 640 | 457 | 267 | 75 | 15 | 0 | 0 | 0 | 13 | 82 | *317* | *553* | 2419 |
| Cooling Degree Days | *0* | *4* | *28* | *107* | 268 | 461 | 576 | 555 | *317* | *111* | *16* | *1* | 2444 |
| Total Precipitation (") | 1.20 | 1.63 | 1.21 | 2.29 | 3.45 | 2.74 | 2.20 | 2.89 | 4.48 | 2.50 | 1.19 | 1.38 | 27.16 |
| Days ≥ 0.1" Precip | 3 | 3 | 3 | 4 | 5 | 4 | 3 | 4 | 5 | 4 | *2* | *3* | 43 |
| Total Snowfall (") | *1.4* | na | 0.6 | 0.0 | 0.0 | 0.0 | 0.0 | 0.0 | 0.0 | 0.0 | *0.1* | *0.5* | na |
| Days ≥ 1" Snow Depth | *1* | *1* | 0 | 0 | 0 | 0 | 0 | 0 | 0 | 0 | *0* | 0 | 2 |

## ARCHER CITY *Archer County* ELEVATION 1040 ft LAT/LONG 33° 36' N / 98° 38' W

| | JAN | FEB | MAR | APR | MAY | JUN | JUL | AUG | SEP | OCT | NOV | DEC | YEAR |
|---|---|---|---|---|---|---|---|---|---|---|---|---|---|
| Maximum Temp °F | 54.7 | 59.4 | 69.2 | 78.1 | 84.3 | 91.9 | 97.2 | 96.4 | 87.8 | 78.5 | 66.4 | 57.4 | 76.8 |
| Minimum Temp °F | 29.0 | 33.2 | 41.6 | 51.3 | 59.3 | 67.7 | 71.9 | 70.5 | 63.1 | 52.3 | 41.4 | 32.4 | 51.1 |
| Mean Temp °F | 41.8 | 46.3 | 55.3 | 64.7 | 71.8 | 79.8 | 84.5 | 83.5 | 75.5 | 65.4 | 53.9 | 44.9 | 64.0 |
| Days Max Temp ≥ 90 °F | 0 | 0 | 1 | 2 | 7 | 21 | 29 | 27 | 15 | 4 | 0 | 0 | 106 |
| Days Max Temp ≤ 32 °F | 2 | 1 | 0 | 0 | 0 | 0 | 0 | 0 | 0 | 0 | 0 | 1 | 4 |
| Days Min Temp ≤ 32 °F | 20 | 13 | 5 | 1 | 0 | 0 | 0 | 0 | 0 | 0 | 6 | 16 | 61 |
| Days Min Temp ≤ 0 °F | 0 | 0 | 0 | 0 | 0 | 0 | 0 | 0 | 0 | 0 | 0 | 0 | 0 |
| Heating Degree Days | 713 | 522 | 322 | 98 | 18 | 0 | 0 | 0 | 14 | 95 | 342 | 618 | 2742 |
| Cooling Degree Days | 0 | 2 | 26 | 94 | 241 | 453 | 616 | *624* | 360 | 116 | 17 | 1 | 2550 |
| Total Precipitation (") | 1.12 | 1.81 | 1.98 | 2.65 | 4.50 | 3.13 | 1.83 | 2.50 | 3.90 | 3.36 | 1.62 | 1.57 | 29.97 |
| Days ≥ 0.1" Precip | 2 | 4 | 4 | 4 | 6 | 5 | 3 | 4 | 5 | 5 | 3 | 3 | 48 |
| Total Snowfall (") | 2.0 | 1.1 | 1.0 | 0.0 | 0.0 | 0.0 | 0.0 | 0.0 | 0.0 | 0.0 | 0.5 | 1.1 | 5.7 |
| Days ≥ 1" Snow Depth | 1 | 1 | 0 | 0 | 0 | 0 | 0 | 0 | 0 | 0 | 0 | 0 | 2 |

## ASPERMONT *Stonewall County* ELEVATION 1670 ft LAT/LONG 33° 9' N / 100° 13' W

| | JAN | FEB | MAR | APR | MAY | JUN | JUL | AUG | SEP | OCT | NOV | DEC | YEAR |
|---|---|---|---|---|---|---|---|---|---|---|---|---|---|
| Maximum Temp °F | 53.9 | 59.4 | 69.1 | 78.5 | 85.7 | 93.7 | 97.4 | 95.6 | 86.9 | 77.5 | 65.0 | 57.0 | 76.6 |
| Minimum Temp °F | 27.9 | 31.6 | 39.1 | 48.5 | 57.2 | 66.2 | 69.7 | 68.4 | 61.2 | 50.2 | 38.0 | 30.2 | 49.0 |
| Mean Temp °F | 40.9 | 45.5 | 54.1 | 63.5 | 71.5 | 80.0 | 83.6 | 82.0 | 74.1 | 63.9 | 51.5 | 43.5 | 62.8 |
| Days Max Temp ≥ 90 °F | 0 | 0 | 1 | 4 | 11 | 23 | 28 | 25 | 13 | 3 | 0 | 0 | 108 |
| Days Max Temp ≤ 32 °F | 3 | 1 | 0 | 0 | 0 | 0 | 0 | 0 | 0 | 0 | 0 | 1 | 5 |
| Days Min Temp ≤ 32 °F | 22 | 15 | 7 | 1 | 0 | 0 | 0 | 0 | 0 | 1 | 8 | 19 | 73 |
| Days Min Temp ≤ 0 °F | 0 | 0 | 0 | 0 | 0 | 0 | 0 | 0 | 0 | 0 | 0 | 0 | 0 |
| Heating Degree Days | 739 | 544 | 348 | 123 | 27 | 1 | 0 | 0 | 19 | 119 | 405 | 661 | 2986 |
| Cooling Degree Days | 0 | 2 | 16 | 85 | 243 | 466 | 593 | 569 | 314 | 87 | 7 | 0 | 2382 |
| Total Precipitation (") | 0.90 | 1.30 | 1.30 | 1.83 | 3.39 | 2.69 | 1.45 | 2.56 | 3.37 | 2.54 | 1.16 | 1.01 | 23.50 |
| Days ≥ 0.1" Precip | 2 | 3 | 3 | 3 | 5 | 4 | 3 | 4 | 4 | 4 | 2 | 2 | 39 |
| Total Snowfall (") | 2.3 | *0.8* | 0.6 | 0.0 | 0.0 | 0.0 | 0.0 | 0.0 | 0.0 | 0.0 | 0.4 | 0.4 | 4.5 |
| Days ≥ 1" Snow Depth | *0* | *0* | 0 | 0 | 0 | 0 | 0 | 0 | 0 | 0 | 0 | 0 | 0 |

## ATHENS 3 SSE *Henderson County* ELEVATION 489 ft LAT/LONG 32° 13' N / 95° 51' W

| | JAN | FEB | MAR | APR | MAY | JUN | JUL | AUG | SEP | OCT | NOV | DEC | YEAR |
|---|---|---|---|---|---|---|---|---|---|---|---|---|---|
| Maximum Temp °F | 57.4 | 62.5 | 70.5 | 77.8 | 83.3 | 90.3 | 94.7 | 95.3 | 88.4 | 79.2 | 68.8 | 60.4 | 77.4 |
| Minimum Temp °F | 35.1 | 38.1 | 45.4 | 53.9 | 61.0 | 68.4 | 70.9 | 70.1 | 64.9 | 54.1 | 44.8 | 37.4 | 53.7 |
| Mean Temp °F | 46.3 | 50.3 | 58.0 | 65.9 | 72.2 | 79.3 | 82.8 | 82.7 | 76.7 | 66.7 | 56.8 | 48.9 | 65.6 |
| Days Max Temp ≥ 90 °F | 0 | 0 | 0 | 0 | 3 | 18 | 27 | 27 | 15 | 2 | 0 | 0 | 92 |
| Days Max Temp ≤ 32 °F | 1 | 0 | 0 | 0 | 0 | 0 | 0 | 0 | 0 | 0 | 0 | 0 | 1 |
| Days Min Temp ≤ 32 °F | 14 | 9 | 3 | 0 | 0 | 0 | 0 | 0 | 0 | 0 | 4 | 10 | 40 |
| Days Min Temp ≤ 0 °F | 0 | 0 | 0 | 0 | 0 | 0 | 0 | 0 | 0 | 0 | 0 | 0 | 0 |
| Heating Degree Days | 578 | 415 | 241 | 68 | 8 | 0 | 0 | 0 | 5 | 68 | 271 | 499 | 2153 |
| Cooling Degree Days | 1 | 5 | 26 | 90 | 226 | 425 | 536 | 557 | 361 | 131 | 28 | 6 | 2392 |
| Total Precipitation (") | 2.54 | 3.46 | 3.63 | 3.75 | 5.27 | 3.41 | 1.67 | 2.17 | 3.56 | 4.85 | 3.47 | 3.59 | 41.37 |
| Days ≥ 0.1" Precip | 5 | 5 | 5 | 5 | 6 | 5 | 3 | 3 | 4 | 5 | 4 | 5 | 55 |
| Total Snowfall (") | 0.6 | 0.4 | 0.0 | 0.0 | 0.0 | 0.0 | 0.0 | 0.0 | 0.0 | 0.0 | 0.0 | 0.1 | 1.1 |
| Days ≥ 1" Snow Depth | 0 | 0 | 0 | 0 | 0 | 0 | 0 | 0 | 0 | 0 | 0 | 0 | 0 |

## AUSTIN MUNICIPAL AP *Travis County* ELEVATION 620 ft LAT/LONG 30° 18' N / 97° 42' W

| | JAN | FEB | MAR | APR | MAY | JUN | JUL | AUG | SEP | OCT | NOV | DEC | YEAR |
|---|---|---|---|---|---|---|---|---|---|---|---|---|---|
| Maximum Temp °F | 59.1 | 63.6 | 71.9 | 79.2 | 84.3 | 90.9 | 94.8 | 95.1 | 89.2 | 80.9 | 70.3 | 62.5 | 78.5 |
| Minimum Temp °F | 39.4 | 42.7 | 50.3 | 58.7 | 65.4 | 71.6 | 74.0 | 73.8 | 69.1 | 59.5 | 49.2 | 42.0 | 58.0 |
| Mean Temp °F | 49.3 | 53.1 | 61.1 | 69.0 | 74.8 | 81.3 | 84.4 | 84.5 | 79.1 | 70.2 | 59.8 | 52.3 | 68.2 |
| Days Max Temp ≥ 90 °F | 0 | 0 | 1 | 2 | 6 | 21 | 28 | 28 | 17 | 4 | 0 | 0 | 107 |
| Days Max Temp ≤ 32 °F | 0 | 0 | 0 | 0 | 0 | 0 | 0 | 0 | 0 | 0 | 0 | 0 | 0 |
| Days Min Temp ≤ 32 °F | 7 | 4 | 1 | 0 | 0 | 0 | 0 | 0 | 0 | 0 | 1 | 5 | 18 |
| Days Min Temp ≤ 0 °F | 0 | 0 | 0 | 0 | 0 | 0 | 0 | 0 | 0 | 0 | 0 | 0 | 0 |
| Heating Degree Days | 489 | 342 | 173 | 37 | 3 | 0 | 0 | 0 | 2 | 34 | 204 | 399 | 1683 |
| Cooling Degree Days | 6 | 16 | 61 | 166 | 333 | 511 | 624 | 642 | 454 | 224 | 62 | 13 | 3112 |
| Total Precipitation (") | 2.17 | 2.33 | 2.07 | 2.64 | 5.34 | 3.21 | 1.80 | 2.30 | 3.11 | 3.61 | 2.36 | 2.47 | 33.41 |
| Days ≥ 0.1" Precip | 4 | 4 | 4 | 4 | 6 | 5 | 3 | 3 | 5 | 5 | 4 | 4 | 51 |
| Total Snowfall (") | 0.4 | 0.4 | 0.1 | 0.0 | 0.0 | 0.0 | 0.0 | 0.0 | 0.0 | 0.0 | 0.1 | 0.0 | 1.0 |
| Days ≥ 1" Snow Depth | 0 | 0 | 0 | 0 | 0 | 0 | 0 | 0 | 0 | 0 | 0 | 0 | 0 |

## BAKERSFIELD 2 NW *Pecos County* ELEVATION 2533 ft LAT/LONG 30° 53' N / 102° 18' W

| | JAN | FEB | MAR | APR | MAY | JUN | JUL | AUG | SEP | OCT | NOV | DEC | YEAR |
|---|---|---|---|---|---|---|---|---|---|---|---|---|---|
| Maximum Temp °F | 59.4 | 64.3 | 73.0 | 82.1 | 87.9 | 93.5 | 94.9 | 94.1 | 87.6 | 79.2 | 68.8 | 61.7 | 78.9 |
| Minimum Temp °F | 32.8 | 36.7 | 44.6 | 53.5 | 61.3 | 68.7 | 71.2 | 70.1 | 64.2 | 54.0 | 42.8 | 34.9 | 52.9 |
| Mean Temp °F | 46.1 | 50.6 | 58.8 | 67.8 | 74.6 | 81.1 | 83.1 | 82.2 | 75.9 | 66.6 | 55.8 | 48.3 | 65.9 |
| Days Max Temp ≥ 90 °F | 0 | 0 | 1 | 6 | 14 | 23 | 27 | 26 | 14 | 3 | 0 | 0 | 114 |
| Days Max Temp ≤ 32 °F | 1 | 0 | 0 | 0 | 0 | 0 | 0 | 0 | 0 | 0 | 0 | 1 | 2 |
| Days Min Temp ≤ 32 °F | 15 | 9 | 4 | 0 | 0 | 0 | 0 | 0 | 0 | 0 | 4 | 12 | 44 |
| Days Min Temp ≤ 0 °F | 0 | 0 | 0 | 0 | 0 | 0 | 0 | 0 | 0 | 0 | 0 | 0 | 0 |
| Heating Degree Days | 579 | 405 | 222 | 58 | 11 | 1 | 0 | 0 | 13 | 73 | 285 | 511 | 2158 |
| Cooling Degree Days | 0 | 5 | 39 | 155 | 327 | 503 | 579 | 559 | 366 | 143 | 20 | 0 | 2696 |
| Total Precipitation (") | 0.52 | 0.68 | 0.45 | 0.86 | 1.77 | 1.81 | 1.25 | 1.80 | 3.40 | 2.23 | 0.70 | 0.57 | 16.04 |
| Days ≥ 0.1" Precip | 2 | 2 | 1 | 2 | 3 | 3 | 3 | 3 | 4 | 3 | 2 | 2 | 30 |
| Total Snowfall (") | 0.6 | 0.3 | 0.1 | 0.0 | 0.0 | 0.0 | 0.0 | 0.0 | 0.0 | 0.0 | 0.3 | 0.1 | 1.4 |
| Days ≥ 1" Snow Depth | 0 | 0 | 0 | 0 | 0 | 0 | 0 | 0 | 0 | 0 | 0 | 0 | 0 |

## BALLINGER 2 NW *Runnels County* ELEVATION 1682 ft LAT/LONG 31° 44' N / 100° 3' W

| | JAN | FEB | MAR | APR | MAY | JUN | JUL | AUG | SEP | OCT | NOV | DEC | YEAR |
|---|---|---|---|---|---|---|---|---|---|---|---|---|---|
| Maximum Temp °F | 58.6 | 64.6 | 73.4 | 81.9 | 86.4 | 92.3 | 95.1 | 94.0 | 87.3 | 79.8 | 68.4 | 61.0 | 78.6 |
| Minimum Temp °F | 31.0 | 35.0 | 43.2 | 52.3 | 60.4 | 67.6 | 70.0 | 69.0 | 62.7 | 51.9 | 41.3 | 33.2 | 51.5 |
| Mean Temp °F | 44.8 | 49.8 | 58.3 | 67.1 | 73.4 | 80.0 | 82.5 | 81.5 | 75.0 | 65.9 | 54.9 | 47.0 | 65.0 |
| Days Max Temp ≥ 90 °F | 0 | 0 | 1 | 6 | 11 | 22 | 27 | 25 | 13 | 4 | 0 | 0 | 109 |
| Days Max Temp ≤ 32 °F | 1 | 0 | 0 | 0 | 0 | 0 | 0 | 0 | 0 | 0 | 0 | 1 | 2 |
| Days Min Temp ≤ 32 °F | 18 | 11 | 5 | 1 | 0 | 0 | 0 | 0 | 0 | 1 | 6 | 15 | 57 |
| Days Min Temp ≤ 0 °F | 0 | 0 | 0 | 0 | 0 | 0 | 0 | 0 | 0 | 0 | 0 | 0 | 0 |
| Heating Degree Days | 619 | 425 | 241 | 63 | 11 | 0 | 0 | 0 | 11 | 81 | 317 | 554 | 2322 |
| Cooling Degree Days | 0 | 3 | 37 | 135 | 290 | 466 | 558 | 548 | 325 | 128 | 23 | 2 | 2515 |
| Total Precipitation (") | 1.06 | 1.40 | 1.32 | 1.86 | 3.63 | 2.61 | 1.54 | 2.44 | 3.56 | 2.41 | 1.19 | 1.34 | 24.36 |
| Days ≥ 0.1" Precip | 3 | 3 | 3 | 3 | 5 | 4 | 3 | 3 | 5 | 4 | 2 | 2 | 40 |
| Total Snowfall (") | na | *0.6* | 0.1 | 0.0 | 0.0 | 0.0 | 0.0 | 0.0 | 0.0 | 0.0 | 0.3 | 0.1 | na |
| Days ≥ 1" Snow Depth | *0* | *0* | 0 | 0 | 0 | 0 | 0 | 0 | 0 | 0 | 0 | 0 | 0 |

## BALMORHEA *Reeves County* ELEVATION 3222 ft LAT/LONG 30° 59 ' N / 103° 45 ' W

| | JAN | FEB | MAR | APR | MAY | JUN | JUL | AUG | SEP | OCT | NOV | DEC | YEAR |
|---|---|---|---|---|---|---|---|---|---|---|---|---|---|
| Maximum Temp °F | 61.2 | 66.0 | 73.7 | 82.3 | 88.8 | 95.4 | 95.7 | 93.3 | 87.3 | 79.9 | 70.6 | 62.2 | 79.7 |
| Minimum Temp °F | 30.1 | 33.2 | 39.8 | 48.1 | 55.5 | 63.7 | 66.3 | 64.8 | 58.8 | 48.3 | 37.7 | 30.8 | 48.1 |
| Mean Temp °F | 45.7 | 49.7 | 56.8 | 65.2 | 72.2 | 79.6 | 81.0 | 79.1 | 73.1 | 64.1 | 54.2 | 46.6 | 63.9 |
| Days Max Temp ≥ 90 °F | 0 | 0 | 1 | 6 | 16 | 25 | 26 | 24 | 14 | 5 | 0 | 0 | 117 |
| Days Max Temp ≤ 32 °F | 1 | 0 | 0 | 0 | 0 | 0 | 0 | 0 | 0 | 0 | 0 | 1 | 2 |
| Days Min Temp ≤ 32 °F | 19 | 13 | 6 | 1 | 0 | 0 | 0 | 0 | 0 | 1 | 8 | 18 | 66 |
| Days Min Temp ≤ 0 °F | 0 | 0 | 0 | 0 | 0 | 0 | 0 | 0 | 0 | 0 | 0 | 0 | 0 |
| Heating Degree Days | 593 | 429 | 268 | 83 | 17 | 1 | 0 | 0 | 18 | 101 | 324 | 565 | 2399 |
| Cooling Degree Days | 0 | 3 | 16 | 84 | 235 | 427 | 492 | 446 | 263 | 74 | 7 | 0 | 2047 |
| Total Precipitation (") | 0.60 | 0.60 | 0.36 | 0.60 | 1.56 | 1.13 | 1.91 | 2.47 | 3.14 | 1.22 | 0.65 | 0.57 | 14.81 |
| Days ≥ 0.1" Precip | 2 | 1 | 1 | 1 | 3 | 3 | 4 | 4 | 5 | 3 | 1 | 1 | 29 |
| Total Snowfall (") | 1.7 | 0.7 | 0.1 | 0.0 | 0.0 | 0.0 | 0.0 | 0.0 | 0.0 | 0.0 | 0.6 | 0.3 | 3.4 |
| Days ≥ 1" Snow Depth | *1* | 0 | 0 | 0 | 0 | 0 | 0 | 0 | 0 | 0 | 0 | 0 | 1 |

## BARDWELL DAM *Ellis County* ELEVATION 459 ft LAT/LONG 32° 16 ' N / 96° 38 ' W

| | JAN | FEB | MAR | APR | MAY | JUN | JUL | AUG | SEP | OCT | NOV | DEC | YEAR |
|---|---|---|---|---|---|---|---|---|---|---|---|---|---|
| Maximum Temp °F | 54.0 | 59.0 | 67.8 | 75.7 | 82.1 | 89.9 | 94.6 | 94.8 | 87.9 | 78.4 | 66.7 | 58.0 | 75.7 |
| Minimum Temp °F | 31.6 | 35.8 | 44.8 | 53.5 | 61.3 | 68.6 | 71.9 | 70.8 | 64.2 | 53.2 | 43.4 | 35.0 | 52.8 |
| Mean Temp °F | 42.8 | 47.4 | 56.3 | 64.6 | 71.7 | 79.3 | 83.3 | 82.9 | 76.0 | 65.8 | 55.0 | 46.5 | 64.3 |
| Days Max Temp ≥ 90 °F | 0 | 0 | 0 | 0 | 3 | 18 | 27 | 27 | 15 | 2 | 0 | 0 | 92 |
| Days Max Temp ≤ 32 °F | 2 | 1 | 0 | 0 | 0 | 0 | 0 | 0 | 0 | 0 | 0 | 1 | 4 |
| Days Min Temp ≤ 32 °F | 18 | 10 | 3 | 0 | 0 | 0 | 0 | 0 | 0 | 0 | 4 | 12 | 47 |
| Days Min Temp ≤ 0 °F | 0 | 0 | 0 | 0 | 0 | 0 | 0 | 0 | 0 | 0 | 0 | 0 | 0 |
| Heating Degree Days | 681 | 492 | 284 | 91 | 13 | 0 | 0 | 0 | 10 | 84 | 311 | 568 | 2534 |
| Cooling Degree Days | 0 | 1 | 20 | 71 | 223 | 424 | 564 | 571 | 348 | 112 | 21 | 2 | 2357 |
| Total Precipitation (") | 2.16 | 2.86 | 2.98 | 3.15 | 5.42 | 3.37 | 2.12 | 1.68 | 3.66 | 4.33 | 2.94 | 2.82 | 37.49 |
| Days ≥ 0.1" Precip | 4 | 5 | 5 | 4 | 7 | 5 | 3 | 3 | 4 | 4 | 4 | 4 | 52 |
| Total Snowfall (") | *0.2* | 0.0 | 0.0 | 0.0 | 0.0 | 0.0 | 0.0 | 0.0 | 0.0 | 0.0 | 0.0 | 0.0 | 0.2 |
| Days ≥ 1" Snow Depth | *0* | *0* | 0 | 0 | 0 | 0 | 0 | 0 | 0 | 0 | 0 | 0 | 0 |

## BAY CITY WATERWORKS *Matagorda County* ELEVATION 49 ft LAT/LONG 28° 59 ' N / 95° 58 ' W

| | JAN | FEB | MAR | APR | MAY | JUN | JUL | AUG | SEP | OCT | NOV | DEC | YEAR |
|---|---|---|---|---|---|---|---|---|---|---|---|---|---|
| Maximum Temp °F | 62.1 | 65.1 | 72.2 | 78.5 | 83.8 | 89.3 | 92.0 | 92.1 | 88.4 | 82.6 | 72.9 | 66.1 | 78.8 |
| Minimum Temp °F | 42.9 | 45.1 | 52.7 | 60.9 | 67.2 | 72.3 | 74.1 | 73.2 | 68.9 | 60.4 | 52.5 | 45.6 | 59.7 |
| Mean Temp °F | 52.5 | 55.1 | 62.5 | 69.7 | 75.5 | 80.8 | 83.1 | 82.7 | 78.8 | 71.7 | 62.7 | 55.9 | 69.3 |
| Days Max Temp ≥ 90 °F | 0 | 0 | 0 | 0 | 3 | 16 | 26 | 26 | 16 | 3 | 0 | 0 | 90 |
| Days Max Temp ≤ 32 °F | 0 | 0 | 0 | 0 | 0 | 0 | 0 | 0 | 0 | 0 | 0 | 0 | 0 |
| Days Min Temp ≤ 32 °F | 5 | 3 | 1 | 0 | 0 | 0 | 0 | 0 | 0 | 0 | 1 | 3 | 13 |
| Days Min Temp ≤ 0 °F | 0 | 0 | 0 | 0 | 0 | 0 | 0 | 0 | 0 | 0 | 0 | 0 | 0 |
| Heating Degree Days | 395 | 285 | 131 | 26 | 2 | 0 | 0 | 0 | 1 | 20 | 146 | 304 | 1310 |
| Cooling Degree Days | 10 | 17 | 65 | 169 | *337* | 497 | 567 | 570 | 435 | 246 | 93 | 32 | 3038 |
| Total Precipitation (") | 4.06 | 3.09 | 2.61 | 3.28 | 5.43 | 4.64 | 4.18 | 3.76 | 5.36 | 5.69 | 3.51 | 3.26 | 48.87 |
| Days ≥ 0.1" Precip | 6 | 5 | 4 | 3 | 5 | 5 | 5 | 6 | 6 | 5 | 4 | 5 | 59 |
| Total Snowfall (") | 0.0 | 0.0 | 0.0 | 0.0 | 0.0 | 0.0 | 0.0 | 0.0 | 0.0 | 0.0 | 0.0 | 0.0 | 0.0 |
| Days ≥ 1" Snow Depth | 0 | 0 | 0 | 0 | 0 | 0 | 0 | 0 | 0 | 0 | 0 | 0 | 0 |

## BEEVILLE 5 NE *Bee County* ELEVATION 220 ft LAT/LONG 28° 27 ' N / 97° 42 ' W

| | JAN | FEB | MAR | APR | MAY | JUN | JUL | AUG | SEP | OCT | NOV | DEC | YEAR |
|---|---|---|---|---|---|---|---|---|---|---|---|---|---|
| Maximum Temp °F | 63.4 | 66.8 | 74.6 | 81.1 | 85.5 | 91.0 | 94.3 | 94.4 | 90.2 | 83.3 | 74.2 | 67.1 | 80.5 |
| Minimum Temp °F | 41.6 | 44.1 | 51.9 | 59.8 | 66.1 | 70.9 | 72.4 | 72.0 | 68.4 | 59.7 | 51.5 | 44.6 | 58.6 |
| Mean Temp °F | 52.5 | 55.5 | 63.3 | 70.4 | 75.8 | 81.0 | 83.3 | 83.2 | 79.3 | 71.5 | 62.9 | 55.9 | 69.6 |
| Days Max Temp ≥ 90 °F | 0 | 0 | 1 | 2 | 7 | 21 | 27 | 27 | 18 | 6 | 0 | 0 | 109 |
| Days Max Temp ≤ 32 °F | 0 | 0 | 0 | 0 | 0 | 0 | 0 | 0 | 0 | 0 | 0 | 0 | 0 |
| Days Min Temp ≤ 32 °F | 6 | 3 | 1 | 0 | 0 | 0 | 0 | 0 | 0 | 0 | 1 | 4 | 15 |
| Days Min Temp ≤ 0 °F | 0 | 0 | 0 | 0 | 0 | 0 | 0 | 0 | 0 | 0 | 0 | 0 | 0 |
| Heating Degree Days | 397 | 280 | 128 | 29 | 2 | 0 | 0 | 0 | 1 | 24 | 146 | 305 | 1312 |
| Cooling Degree Days | 10 | 23 | 77 | 190 | 356 | 509 | 590 | 591 | 456 | 249 | 95 | 34 | 3180 |
| Total Precipitation (") | 2.19 | 2.05 | 1.56 | 2.57 | 4.11 | 3.86 | 2.63 | 3.03 | 4.94 | 3.40 | 1.78 | 1.96 | 34.08 |
| Days ≥ 0.1" Precip | 4 | 4 | 3 | 4 | 6 | 5 | 4 | 4 | 6 | 4 | 3 | 3 | 50 |
| Total Snowfall (") | 0.0 | 0.0 | 0.0 | 0.0 | 0.0 | 0.0 | 0.0 | 0.0 | 0.0 | 0.0 | 0.0 | 0.0 | 0.0 |
| Days ≥ 1" Snow Depth | 0 | 0 | 0 | 0 | 0 | 0 | 0 | 0 | 0 | 0 | 0 | 0 | 0 |

## BENAVIDES 2 *Duval County* ELEVATION 381 ft LAT/LONG 27° 36' N / 98° 25' W

| | JAN | FEB | MAR | APR | MAY | JUN | JUL | AUG | SEP | OCT | NOV | DEC | YEAR |
|---|---|---|---|---|---|---|---|---|---|---|---|---|---|
| Maximum Temp °F | ***66.6*** | 70.4 | 78.9 | 85.7 | ***89.2*** | 94.3 | ***96.5*** | 96.9 | 91.7 | 85.8 | 77.4 | ***69.8*** | 83.6 |
| Minimum Temp °F | ***43.1*** | ***45.6*** | 53.2 | 61.4 | ***67.6*** | 71.7 | 73.0 | 72.3 | 69.0 | 61.0 | ***52.1*** | 45.3 | 59.6 |
| Mean Temp °F | ***54.9*** | ***57.9*** | 66.1 | 73.6 | ***78.4*** | ***83.0*** | ***84.8*** | 84.7 | 80.3 | 73.3 | ***64.8*** | ***57.8*** | 71.6 |
| Days Max Temp ≥ 90 °F | 0 | 1 | 3 | 9 | 16 | 24 | 28 | 29 | 20 | 9 | 1 | 0 | 140 |
| Days Max Temp ≤ 32 °F | 0 | 0 | 0 | 0 | 0 | 0 | 0 | 0 | 0 | 0 | 0 | 0 | 0 |
| Days Min Temp ≤ 32 °F | ***5*** | 3 | 1 | 0 | 0 | 0 | 0 | 0 | 0 | 0 | 1 | ***4*** | 14 |
| Days Min Temp ≤ 0 °F | 0 | 0 | 0 | 0 | 0 | 0 | 0 | 0 | 0 | 0 | 0 | 0 | 0 |
| Heating Degree Days | ***331*** | ***229*** | 91 | 15 | 1 | 0 | 0 | 0 | 1 | 16 | ***114*** | ***261*** | 1059 |
| Cooling Degree Days | 26 | ***52*** | 143 | 283 | ***454*** | ***579*** | ***644*** | ***651*** | ***488*** | 306 | ***134*** | 49 | 3809 |
| Total Precipitation (") | 1.47 | 1.66 | 1.08 | 1.76 | 3.63 | 3.10 | 1.46 | 2.24 | 4.64 | 1.84 | 0.94 | 1.11 | 24.93 |
| Days ≥ 0.1" Precip | ***3*** | 3 | 2 | 2 | ***4*** | 3 | 2 | 3 | 4 | 2 | ***2*** | 2 | 32 |
| Total Snowfall (") | 0.0 | 0.0 | 0.0 | 0.0 | 0.0 | 0.0 | 0.0 | 0.0 | 0.0 | 0.0 | 0.0 | 0.0 | 0.0 |
| Days ≥ 1" Snow Depth | 0 | 0 | 0 | 0 | 0 | 0 | 0 | 0 | 0 | 0 | 0 | 0 | 0 |

## BENBROOK DAM *Tarrant County* ELEVATION 771 ft LAT/LONG 32° 39' N / 97° 27' W

| | JAN | FEB | MAR | APR | MAY | JUN | JUL | AUG | SEP | OCT | NOV | DEC | YEAR |
|---|---|---|---|---|---|---|---|---|---|---|---|---|---|
| Maximum Temp °F | 53.7 | 58.5 | 67.1 | 75.7 | 81.8 | 90.1 | 95.2 | 95.0 | 87.4 | 77.8 | 66.1 | 57.4 | 75.5 |
| Minimum Temp °F | 31.1 | 35.2 | 43.6 | 53.2 | 60.5 | 68.4 | 72.0 | 71.2 | 64.8 | 54.0 | 43.5 | 34.9 | 52.7 |
| Mean Temp °F | 42.4 | 46.9 | 55.4 | 64.5 | 71.2 | 79.3 | 83.6 | 83.1 | 76.1 | 65.9 | 54.8 | 46.2 | 64.1 |
| Days Max Temp ≥ 90 °F | 0 | 0 | 0 | 1 | 4 | 18 | 27 | 26 | 14 | 2 | 0 | 0 | 92 |
| Days Max Temp ≤ 32 °F | 2 | 1 | 0 | 0 | 0 | 0 | 0 | 0 | 0 | 0 | 0 | 1 | 4 |
| Days Min Temp ≤ 32 °F | 18 | 11 | 4 | 0 | 0 | 0 | 0 | 0 | 0 | 0 | 4 | 12 | 49 |
| Days Min Temp ≤ 0 °F | 0 | 0 | 0 | 0 | 0 | 0 | 0 | 0 | 0 | 0 | 0 | 0 | 0 |
| Heating Degree Days | 693 | 508 | 311 | 96 | 17 | 1 | 0 | 0 | 10 | 86 | 319 | 578 | 2619 |
| Cooling Degree Days | 0 | 3 | 19 | 83 | 214 | 439 | 585 | 592 | 360 | 118 | 23 | 1 | 2437 |
| Total Precipitation (") | 1.73 | 2.29 | 2.61 | 3.34 | 4.76 | 3.36 | 2.35 | 2.14 | 3.33 | 3.90 | 1.95 | 2.25 | 34.01 |
| Days ≥ 0.1" Precip | 4 | 4 | 4 | 5 | 6 | 5 | 3 | 3 | 4 | 5 | 4 | 4 | 51 |
| Total Snowfall (") | ***0.0*** | 0.1 | 0.0 | 0.0 | 0.0 | 0.0 | 0.0 | 0.0 | 0.0 | 0.0 | 0.1 | 0.1 | 0.3 |
| Days ≥ 1" Snow Depth | 0 | 0 | 0 | 0 | 0 | 0 | 0 | 0 | 0 | 0 | 0 | 0 | 0 |

## BIG LAKE 2 *Reagan County* ELEVATION 2680 ft LAT/LONG 31° 12' N / 101° 28' W

| | JAN | FEB | MAR | APR | MAY | JUN | JUL | AUG | SEP | OCT | NOV | DEC | YEAR |
|---|---|---|---|---|---|---|---|---|---|---|---|---|---|
| Maximum Temp °F | 56.8 | 61.5 | 70.0 | 79.0 | 85.4 | 91.6 | 93.9 | 93.1 | 86.3 | 77.5 | 66.8 | 59.0 | 76.7 |
| Minimum Temp °F | 29.0 | 32.5 | 41.1 | 50.3 | 58.0 | 65.4 | 67.5 | 65.9 | 59.8 | 49.8 | 39.0 | 30.8 | 49.1 |
| Mean Temp °F | 43.0 | 47.0 | 55.6 | 64.7 | 71.7 | 78.5 | 80.7 | 79.5 | 73.1 | 63.7 | 52.9 | 44.9 | 62.9 |
| Days Max Temp ≥ 90 °F | 0 | 0 | 1 | 4 | 10 | 20 | 25 | 25 | 12 | 2 | 0 | 0 | 99 |
| Days Max Temp ≤ 32 °F | 1 | 1 | 0 | 0 | 0 | 0 | 0 | 0 | 0 | 0 | 0 | 1 | 3 |
| Days Min Temp ≤ 32 °F | 21 | 14 | 6 | 1 | 0 | 0 | 0 | 0 | 0 | 1 | 8 | 18 | 69 |
| Days Min Temp ≤ 0 °F | 0 | 0 | 0 | 0 | 0 | 0 | 0 | 0 | 0 | 0 | 0 | 0 | 0 |
| Heating Degree Days | 676 | 502 | 302 | 97 | 21 | 2 | 0 | 0 | 20 | 114 | 362 | 617 | 2713 |
| Cooling Degree Days | 0 | 1 | 18 | 91 | 238 | 422 | 509 | 491 | 293 | 90 | 9 | 0 | 2162 |
| Total Precipitation (") | 0.71 | 0.96 | 0.87 | 1.43 | 2.57 | 1.76 | 1.95 | 2.04 | 3.21 | 2.17 | 0.83 | 0.88 | 19.38 |
| Days ≥ 0.1" Precip | 2 | 2 | 2 | 3 | 5 | 3 | 3 | 3 | 4 | 3 | 2 | 2 | 34 |
| Total Snowfall (") | 0.9 | 0.0 | 0.0 | 0.0 | 0.0 | 0.0 | 0.0 | 0.0 | 0.0 | 0.0 | 0.4 | 0.0 | 1.3 |
| Days ≥ 1" Snow Depth | ***0*** | 0 | 0 | 0 | 0 | 0 | 0 | 0 | 0 | 0 | 0 | 0 | 0 |

## BIG SPRING *Howard County* ELEVATION 2533 ft LAT/LONG 32° 15' N / 101° 27' W

| | JAN | FEB | MAR | APR | MAY | JUN | JUL | AUG | SEP | OCT | NOV | DEC | YEAR |
|---|---|---|---|---|---|---|---|---|---|---|---|---|---|
| Maximum Temp °F | 55.9 | 60.6 | 69.8 | 78.8 | 85.1 | 91.5 | 94.0 | 92.5 | 85.8 | 77.8 | 66.7 | 58.5 | 76.4 |
| Minimum Temp °F | 29.6 | 33.5 | 41.2 | 50.4 | 58.8 | 66.8 | 70.7 | 69.2 | 62.6 | 52.0 | 40.4 | 32.4 | 50.6 |
| Mean Temp °F | 42.9 | 47.0 | 55.5 | 64.6 | 72.0 | 79.2 | 82.4 | 80.9 | 74.2 | 64.9 | 53.5 | 45.5 | 63.6 |
| Days Max Temp ≥ 90 °F | 0 | 0 | 1 | 4 | 10 | 19 | 25 | 23 | 11 | 2 | 0 | 0 | 95 |
| Days Max Temp ≤ 32 °F | 2 | 1 | 0 | 0 | 0 | 0 | 0 | 0 | 0 | 0 | 0 | 1 | 4 |
| Days Min Temp ≤ 32 °F | 20 | 13 | 5 | 1 | 0 | 0 | 0 | 0 | 0 | 0 | 6 | 16 | 61 |
| Days Min Temp ≤ 0 °F | 0 | 0 | 0 | 0 | 0 | 0 | 0 | 0 | 0 | 0 | 0 | 0 | 0 |
| Heating Degree Days | 679 | 503 | 307 | 105 | 21 | 1 | 0 | 0 | 18 | 100 | 347 | 599 | 2680 |
| Cooling Degree Days | ***0*** | 0 | 17 | 97 | 249 | 432 | 558 | 527 | 320 | 102 | 10 | 0 | 2312 |
| Total Precipitation (") | 0.79 | 0.92 | 0.74 | 1.44 | 3.06 | 2.44 | 1.87 | 2.32 | 3.75 | 1.75 | 0.71 | 0.70 | 20.49 |
| Days ≥ 0.1" Precip | 2 | 2 | 2 | 2 | 5 | 4 | 3 | 4 | 5 | 3 | 2 | 2 | 36 |
| Total Snowfall (") | ***0.9*** | ***0.5*** | 0.2 | 0.0 | 0.0 | 0.0 | 0.0 | 0.0 | 0.0 | 0.0 | 0.4 | 0.4 | 2.4 |
| Days ≥ 1" Snow Depth | ***1*** | ***0*** | 0 | 0 | 0 | 0 | 0 | 0 | 0 | 0 | 0 | 0 | 1 |

## BLANCO *Blanco County* ELEVATION 1352 ft LAT/LONG 30° 6' N / 98° 25' W

| | JAN | FEB | MAR | APR | MAY | JUN | JUL | AUG | SEP | OCT | NOV | DEC | YEAR |
|---|---|---|---|---|---|---|---|---|---|---|---|---|---|
| Maximum Temp °F | 58.4 | 62.4 | 70.6 | 78.1 | 83.0 | 89.6 | 93.7 | 94.2 | 88.0 | 79.8 | 69.1 | 61.4 | 77.4 |
| Minimum Temp °F | 33.6 | 36.6 | 44.5 | 52.8 | 60.5 | 67.5 | 69.8 | 69.0 | 64.1 | 53.8 | 43.7 | 36.2 | 52.7 |
| Mean Temp °F | 46.0 | 49.6 | 57.6 | 65.5 | 71.8 | 78.6 | 81.8 | 81.6 | 76.0 | 66.8 | 56.4 | 48.8 | 65.0 |
| Days Max Temp ≥ 90 °F | 0 | 0 | 0 | 1 | 5 | 16 | 26 | 27 | 14 | 2 | 0 | 0 | 91 |
| Days Max Temp ≤ 32 °F | 1 | 0 | 0 | 0 | 0 | 0 | 0 | 0 | 0 | 0 | 0 | 0 | 1 |
| Days Min Temp ≤ 32 °F | 15 | 10 | 4 | 1 | 0 | 0 | 0 | 0 | 0 | 0 | 6 | 13 | 49 |
| Days Min Temp ≤ 0 °F | 0 | 0 | 0 | 0 | 0 | 0 | 0 | 0 | 0 | 0 | 0 | 0 | 0 |
| Heating Degree Days | 585 | 434 | 254 | 82 | 13 | 0 | 0 | 0 | 7 | 74 | 281 | 497 | 2227 |
| Cooling Degree Days | 1 | 4 | 30 | 92 | 231 | 413 | 521 | 533 | 350 | 144 | 33 | 4 | 2356 |
| Total Precipitation (") | 2.21 | 2.31 | 2.46 | 2.83 | 4.99 | 3.67 | 2.16 | 2.22 | 3.78 | 3.91 | 2.24 | 2.42 | 35.20 |
| Days ≥ 0.1" Precip | 5 | 4 | 5 | 4 | 7 | 5 | 3 | 3 | 6 | 4 | 4 | 4 | 54 |
| Total Snowfall (") | 0.3 | 0.5 | 0.1 | 0.0 | 0.0 | 0.0 | 0.0 | 0.0 | 0.0 | 0.0 | 0.0 | 0.0 | 0.9 |
| Days ≥ 1" Snow Depth | 0 | 0 | 0 | 0 | 0 | 0 | 0 | 0 | 0 | 0 | 0 | 0 | 0 |

## BOERNE *Kendall County* ELEVATION 1421 ft LAT/LONG 29° 47' N / 98° 44' W

| | JAN | FEB | MAR | APR | MAY | JUN | JUL | AUG | SEP | OCT | NOV | DEC | YEAR |
|---|---|---|---|---|---|---|---|---|---|---|---|---|---|
| Maximum Temp °F | 59.3 | 63.6 | 71.7 | 78.6 | 83.1 | 89.2 | 92.5 | 93.2 | 87.9 | 80.0 | 69.8 | 62.2 | 77.6 |
| Minimum Temp °F | 34.4 | 37.2 | 44.8 | 53.2 | 60.6 | 67.1 | 69.1 | 68.0 | 63.6 | 53.9 | 43.7 | 36.9 | 52.7 |
| Mean Temp °F | 46.8 | 50.4 | 58.3 | 65.9 | 71.9 | 78.2 | 80.8 | 80.6 | 75.8 | 67.0 | 56.8 | 49.5 | 65.2 |
| Days Max Temp ≥ 90 °F | 0 | 0 | 1 | 1 | 4 | 15 | 25 | 26 | 13 | 2 | 0 | 0 | 87 |
| Days Max Temp ≤ 32 °F | 0 | 0 | 0 | 0 | 0 | 0 | 0 | 0 | 0 | 0 | 0 | 0 | 0 |
| Days Min Temp ≤ 32 °F | 14 | 10 | 4 | 1 | 0 | 0 | 0 | 0 | 0 | 0 | 6 | 12 | 47 |
| Days Min Temp ≤ 0 °F | 0 | 0 | 0 | 0 | 0 | 0 | 0 | 0 | 0 | 0 | 0 | 0 | 0 |
| Heating Degree Days | 557 | 409 | 234 | 72 | 11 | 0 | 0 | 0 | 5 | 68 | 268 | 476 | 2100 |
| Cooling Degree Days | 1 | 5 | 32 | 100 | 239 | 415 | 506 | 518 | 351 | 153 | 33 | 5 | 2358 |
| Total Precipitation (") | 2.01 | 2.36 | 2.33 | 3.12 | 4.86 | 3.91 | 2.19 | 2.96 | 4.22 | 3.54 | 2.71 | 2.43 | 36.64 |
| Days ≥ 0.1" Precip | 5 | 4 | 4 | 4 | 6 | 5 | 3 | 4 | 6 | 5 | 4 | 4 | 54 |
| Total Snowfall (") | 0.3 | 0.2 | 0.0 | 0.0 | 0.0 | 0.0 | 0.0 | 0.0 | 0.0 | 0.0 | 0.0 | 0.0 | 0.5 |
| Days ≥ 1" Snow Depth | 0 | 0 | 0 | 0 | 0 | 0 | 0 | 0 | 0 | 0 | 0 | 0 | 0 |

## BONHAM 3 NNE *Fannin County* ELEVATION 571 ft LAT/LONG 33° 36' N / 96° 11' W

| | JAN | FEB | MAR | APR | MAY | JUN | JUL | AUG | SEP | OCT | NOV | DEC | YEAR |
|---|---|---|---|---|---|---|---|---|---|---|---|---|---|
| Maximum Temp °F | 52.9 | 57.9 | 67.2 | 75.6 | 81.8 | 89.3 | 93.9 | 94.1 | 86.2 | 76.4 | 64.6 | 55.9 | 74.7 |
| Minimum Temp °F | 31.0 | 34.9 | 43.2 | 51.8 | 59.7 | 67.6 | 71.2 | 70.1 | 63.3 | 52.1 | 41.9 | 33.9 | 51.7 |
| Mean Temp °F | 41.9 | 46.4 | 55.2 | 63.7 | 70.8 | 78.5 | 82.6 | 82.1 | 74.8 | 64.3 | 53.3 | 44.9 | 63.2 |
| Days Max Temp ≥ 90 °F | 0 | 0 | 0 | 0 | 2 | 16 | 25 | 25 | 11 | 1 | 0 | 0 | 80 |
| Days Max Temp ≤ 32 °F | 2 | 1 | 0 | 0 | 0 | 0 | 0 | 0 | 0 | 0 | 0 | 1 | 4 |
| Days Min Temp ≤ 32 °F | 18 | 12 | 5 | 1 | 0 | 0 | 0 | 0 | 0 | 1 | 6 | 14 | 57 |
| Days Min Temp ≤ 0 °F | 0 | 0 | 0 | 0 | 0 | 0 | 0 | 0 | 0 | 0 | 0 | 0 | 0 |
| Heating Degree Days | 709 | 519 | 318 | 105 | 18 | 0 | 0 | 0 | 14 | 108 | 358 | 618 | 2767 |
| Cooling Degree Days | 0 | 1 | 16 | 65 | 193 | 410 | 541 | 546 | 321 | 90 | 14 | 1 | 2198 |
| Total Precipitation (") | 2.17 | 3.30 | 3.78 | 3.78 | 6.12 | 4.48 | 3.46 | 2.27 | 4.25 | 5.05 | 3.42 | 3.12 | 45.20 |
| Days ≥ 0.1" Precip | 4 | 5 | 6 | 6 | 7 | 5 | 4 | 4 | 5 | 5 | 4 | 5 | 60 |
| Total Snowfall (") | *0.5* | 1.1 | 0.3 | 0.0 | 0.0 | 0.0 | 0.0 | 0.0 | 0.0 | 0.0 | 0.0 | 0.4 | 2.3 |
| Days ≥ 1" Snow Depth | 0 | *0* | 0 | 0 | 0 | 0 | 0 | 0 | 0 | 0 | 0 | 0 | 0 |

## BORGER *Hutchinson County* ELEVATION 3143 ft LAT/LONG 35° 39' N / 101° 27' W

| | JAN | FEB | MAR | APR | MAY | JUN | JUL | AUG | SEP | OCT | NOV | DEC | YEAR |
|---|---|---|---|---|---|---|---|---|---|---|---|---|---|
| Maximum Temp °F | 49.9 | 55.0 | 63.9 | 73.1 | 80.0 | 88.9 | 92.8 | 90.4 | 83.6 | 74.2 | 60.2 | 51.4 | 72.0 |
| Minimum Temp °F | 23.5 | 27.3 | 34.7 | 43.7 | 52.5 | 61.8 | 66.5 | 64.8 | 57.3 | 46.1 | 34.0 | 26.0 | 44.9 |
| Mean Temp °F | 36.7 | 41.2 | 49.3 | 58.4 | 66.3 | 75.4 | 79.7 | 77.6 | 70.5 | 60.1 | 47.1 | 38.7 | 58.4 |
| Days Max Temp ≥ 90 °F | 0 | 0 | 0 | 1 | 4 | 15 | 23 | 19 | 8 | 1 | 0 | 0 | 71 |
| Days Max Temp ≤ 32 °F | 4 | 2 | 1 | 0 | 0 | 0 | 0 | 0 | 0 | 0 | 1 | 3 | 11 |
| Days Min Temp ≤ 32 °F | 26 | 20 | 12 | 3 | 0 | 0 | 0 | 0 | 0 | 2 | 13 | 24 | 100 |
| Days Min Temp ≤ 0 °F | 0 | 0 | 0 | 0 | 0 | 0 | 0 | 0 | 0 | 0 | 0 | 1 | 1 |
| Heating Degree Days | 870 | 666 | 484 | 217 | 68 | 6 | 0 | 1 | 34 | 189 | 531 | 808 | 3874 |
| Cooling Degree Days | 0 | 0 | 3 | 26 | 101 | 315 | 451 | 401 | 206 | 35 | 2 | 0 | 1540 |
| Total Precipitation (") | 0.59 | 0.75 | 1.30 | 1.43 | 3.08 | 3.39 | 2.66 | 3.09 | 1.93 | 1.36 | 0.82 | 0.58 | 20.98 |
| Days ≥ 0.1" Precip | 2 | 2 | 3 | 3 | 5 | 5 | 4 | 6 | 3 | 3 | 2 | 2 | 40 |
| Total Snowfall (") | 5.3 | 6.0 | 4.2 | 1.0 | 0.1 | 0.0 | 0.0 | 0.0 | 0.0 | 0.6 | 2.5 | 3.6 | 23.3 |
| Days ≥ 1" Snow Depth | 4 | 3 | 1 | 0 | 0 | 0 | 0 | 0 | 0 | 0 | 1 | 2 | 11 |

## BOWIE *Montague County* ELEVATION 1112 ft LAT/LONG 33° 34 ' N / 97° 51 ' W

| | JAN | FEB | MAR | APR | MAY | JUN | JUL | AUG | SEP | OCT | NOV | DEC | YEAR |
|---|---|---|---|---|---|---|---|---|---|---|---|---|---|
| Maximum Temp °F | 53.9 | 58.9 | 68.3 | 76.8 | 82.6 | 90.3 | 95.7 | 95.0 | 87.1 | 77.4 | 65.2 | 56.9 | 75.7 |
| Minimum Temp °F | 31.5 | 35.2 | 43.7 | 53.0 | 60.9 | 68.9 | 72.9 | 71.9 | 65.0 | 54.3 | 43.1 | 34.6 | 52.9 |
| Mean Temp °F | 42.8 | 47.0 | 56.0 | 65.0 | 71.8 | 79.6 | 84.3 | 83.5 | 76.1 | 65.9 | 54.2 | 45.7 | 64.3 |
| Days Max Temp ≥ 90 °F | 0 | 0 | 1 | 1 | 5 | 18 | 27 | 26 | 13 | 2 | 0 | 0 | 93 |
| Days Max Temp ≤ 32 °F | 2 | 1 | 0 | 0 | 0 | 0 | 0 | 0 | 0 | 0 | 0 | 1 | 4 |
| Days Min Temp ≤ 32 °F | 16 | 11 | 4 | 0 | 0 | 0 | 0 | 0 | 0 | 0 | 5 | 12 | 48 |
| Days Min Temp ≤ 0 °F | 0 | 0 | 0 | 0 | 0 | 0 | 0 | 0 | 0 | 0 | 0 | 0 | 0 |
| Heating Degree Days | 684 | 505 | 298 | 90 | 16 | 1 | 0 | 0 | 12 | 85 | 337 | 592 | 2620 |
| Cooling Degree Days | 0 | 3 | 22 | 88 | 218 | 429 | 582 | 582 | 344 | 101 | 18 | 1 | 2388 |
| Total Precipitation (") | 1.50 | 2.15 | 2.55 | 2.88 | 5.17 | 3.48 | 1.88 | 2.27 | 3.94 | 3.96 | 2.00 | 1.88 | 33.66 |
| Days ≥ 0.1" Precip | 3 | 5 | 4 | 5 | 6 | 5 | 3 | 4 | 5 | 5 | 4 | 3 | 52 |
| Total Snowfall (") | ***0.5*** | ***0.8*** | 0.3 | 0.0 | 0.0 | 0.0 | 0.0 | 0.0 | 0.0 | 0.0 | 0.3 | 0.4 | 2.3 |
| Days ≥ 1" Snow Depth | ***0*** | 0 | 0 | 0 | 0 | 0 | 0 | 0 | 0 | 0 | 0 | 0 | 0 |

## BRACKETTVILLE *Kinney County* ELEVATION 1050 ft LAT/LONG 29° 19 ' N / 100° 25 ' W

| | JAN | FEB | MAR | APR | MAY | JUN | JUL | AUG | SEP | OCT | NOV | DEC | YEAR |
|---|---|---|---|---|---|---|---|---|---|---|---|---|---|
| Maximum Temp °F | ***62.6*** | 67.3 | 75.4 | 82.9 | 87.2 | 92.6 | 95.3 | 95.5 | 90.5 | 82.6 | 73.2 | 65.2 | 80.9 |
| Minimum Temp °F | 37.2 | 40.4 | 48.5 | 56.6 | 64.1 | 69.6 | 72.0 | 71.4 | 66.6 | 57.4 | 46.1 | 39.0 | 55.7 |
| Mean Temp °F | ***49.8*** | 53.7 | 62.0 | 69.8 | 75.7 | 81.1 | 83.7 | 83.5 | 78.5 | 70.0 | 59.6 | 52.2 | 68.3 |
| Days Max Temp ≥ 90 °F | 0 | 0 | 1 | 6 | 11 | 23 | 28 | 28 | 20 | 5 | 0 | 0 | 122 |
| Days Max Temp ≤ 32 °F | 0 | 0 | 0 | 0 | 0 | 0 | 0 | 0 | 0 | 0 | 0 | 0 | 0 |
| Days Min Temp ≤ 32 °F | ***10*** | 5 | 2 | 0 | 0 | 0 | 0 | 0 | 0 | 0 | 3 | 9 | 29 |
| Days Min Temp ≤ 0 °F | 0 | 0 | 0 | 0 | 0 | 0 | 0 | 0 | 0 | 0 | 0 | 0 | 0 |
| Heating Degree Days | ***466*** | 319 | 143 | 29 | 3 | 0 | 0 | 0 | 3 | 31 | 193 | 395 | 1582 |
| Cooling Degree Days | ***2*** | 8 | 60 | 174 | 358 | 503 | 600 | 608 | 441 | 221 | ***43*** | 6 | 3024 |
| Total Precipitation (") | 0.93 | 1.26 | 0.97 | 2.38 | 2.83 | 2.80 | 1.70 | 2.04 | 3.04 | 2.22 | 1.23 | 0.94 | 22.34 |
| Days ≥ 0.1" Precip | 2 | 2 | 2 | 3 | 4 | 4 | 3 | 3 | 4 | 3 | 2 | ***2*** | 34 |
| Total Snowfall (") | 0.5 | 0.1 | 0.0 | 0.0 | 0.0 | 0.0 | 0.0 | 0.0 | 0.0 | 0.0 | 0.0 | 0.0 | 0.6 |
| Days ≥ 1" Snow Depth | 0 | 0 | 0 | 0 | 0 | 0 | 0 | 0 | 0 | 0 | 0 | 0 | 0 |

## BRADY *McCulloch County* ELEVATION 1722 ft LAT/LONG 31° 8 ' N / 99° 21 ' W

| | JAN | FEB | MAR | APR | MAY | JUN | JUL | AUG | SEP | OCT | NOV | DEC | YEAR |
|---|---|---|---|---|---|---|---|---|---|---|---|---|---|
| Maximum Temp °F | 57.4 | 61.7 | 70.7 | 79.2 | 84.0 | 91.1 | 95.2 | 94.2 | 87.6 | 79.3 | 68.0 | 60.2 | 77.4 |
| Minimum Temp °F | 31.2 | 34.9 | 43.0 | 51.4 | 59.2 | 66.2 | 69.1 | 68.1 | 62.4 | 52.0 | 41.6 | 34.0 | 51.1 |
| Mean Temp °F | 44.3 | 48.3 | 56.9 | 65.3 | 71.6 | 78.7 | 82.2 | 81.2 | 75.0 | 65.7 | 54.8 | 47.1 | 64.3 |
| Days Max Temp ≥ 90 °F | 0 | 0 | 1 | 3 | 7 | 20 | 28 | 26 | 14 | 3 | 0 | 0 | 102 |
| Days Max Temp ≤ 32 °F | 1 | 1 | 0 | 0 | 0 | 0 | 0 | 0 | 0 | 0 | 0 | 1 | 3 |
| Days Min Temp ≤ 32 °F | 18 | 12 | 5 | 1 | 0 | 0 | 0 | 0 | 0 | 0 | 7 | 14 | 57 |
| Days Min Temp ≤ 0 °F | 0 | 0 | 0 | 0 | 0 | 0 | 0 | 0 | 0 | 0 | 0 | 0 | 0 |
| Heating Degree Days | 636 | 468 | 273 | 86 | 18 | 1 | 0 | 0 | 12 | 87 | 317 | 550 | 2448 |
| Cooling Degree Days | 0 | 5 | 31 | 104 | 247 | 430 | 553 | 551 | 341 | 131 | 25 | 2 | 2420 |
| Total Precipitation (") | 1.19 | 1.70 | 1.40 | 2.11 | 3.82 | 2.93 | 2.33 | 2.53 | 3.52 | 2.51 | 1.44 | 1.58 | 27.06 |
| Days ≥ 0.1" Precip | 3 | 4 | 3 | 4 | 6 | 5 | 3 | 3 | 5 | 4 | 3 | 3 | 46 |
| Total Snowfall (") | 0.3 | 0.6 | 0.0 | 0.0 | 0.0 | 0.0 | 0.0 | 0.0 | 0.0 | 0.0 | 0.1 | 0.1 | 1.1 |
| Days ≥ 1" Snow Depth | 0 | 0 | 0 | 0 | 0 | 0 | 0 | 0 | 0 | 0 | 0 | 0 | 0 |

## BRAVO *Hartley County* ELEVATION 4153 ft LAT/LONG 35° 39 ' N / 103° 0 ' W

| | JAN | FEB | MAR | APR | MAY | JUN | JUL | AUG | SEP | OCT | NOV | DEC | YEAR |
|---|---|---|---|---|---|---|---|---|---|---|---|---|---|
| Maximum Temp °F | 51.6 | 56.0 | 64.6 | 73.0 | 79.7 | 88.0 | 91.0 | 88.9 | 82.5 | 73.8 | 60.8 | 53.0 | 71.9 |
| Minimum Temp °F | 20.5 | 24.0 | 30.2 | 38.8 | 47.9 | 57.3 | 62.0 | 60.6 | 53.2 | 41.1 | 29.4 | 22.1 | 40.6 |
| Mean Temp °F | 36.1 | 40.0 | 47.4 | 55.9 | 63.8 | 72.7 | 76.5 | 74.8 | 67.9 | 57.5 | 45.1 | 37.6 | 56.3 |
| Days Max Temp ≥ 90 °F | 0 | 0 | 0 | 1 | 3 | 13 | 20 | 16 | 5 | 1 | 0 | 0 | 59 |
| Days Max Temp ≤ 32 °F | 3 | 2 | 0 | 0 | 0 | 0 | 0 | 0 | 0 | 0 | 0 | 2 | 7 |
| Days Min Temp ≤ 32 °F | 28 | 24 | 19 | 7 | 1 | 0 | 0 | 0 | 0 | 4 | 18 | 28 | 129 |
| Days Min Temp ≤ 0 °F | 1 | 0 | 0 | 0 | 0 | 0 | 0 | 0 | 0 | 0 | 0 | 1 | 2 |
| Heating Degree Days | 888 | 697 | 539 | 276 | 96 | 9 | 1 | 2 | 46 | 245 | 590 | 843 | 4232 |
| Cooling Degree Days | 0 | 0 | 0 | 10 | 58 | 248 | 375 | 324 | 141 | 12 | 0 | 0 | 1168 |
| Total Precipitation (") | 0.40 | 0.47 | 0.64 | 0.92 | 2.43 | 2.79 | 2.61 | 3.27 | 2.09 | 1.03 | 0.65 | 0.39 | 17.69 |
| Days ≥ 0.1" Precip | 1 | 2 | 2 | 2 | 4 | 5 | 5 | 5 | 4 | 2 | 1 | 1 | 34 |
| Total Snowfall (") | ***0.6*** | ***0.5*** | ***0.0*** | 0.4 | 0.0 | 0.0 | 0.0 | 0.0 | 0.0 | 0.1 | 0.4 | 0.5 | 2.5 |
| Days ≥ 1" Snow Depth | ***0*** | ***0*** | 0 | 0 | 0 | 0 | 0 | 0 | 0 | 0 | 0 | 0 | 0 |

## BRECKENRIDGE *Stephens County* ELEVATION 1211 ft LAT/LONG 32° 45 ' N / 98° 56 ' W

| | JAN | FEB | MAR | APR | MAY | JUN | JUL | AUG | SEP | OCT | NOV | DEC | YEAR |
|---|---|---|---|---|---|---|---|---|---|---|---|---|---|
| Maximum Temp °F | 55.0 | 59.1 | 68.8 | 78.6 | 84.4 | 92.3 | 96.8 | 95.9 | 88.8 | 78.3 | 67.2 | 58.0 | 76.9 |
| Minimum Temp °F | 29.4 | 33.3 | 41.7 | 52.1 | 59.7 | 68.5 | 72.0 | 70.7 | 63.5 | 51.7 | 40.4 | 31.8 | 51.2 |
| Mean Temp °F | 42.3 | 46.2 | 55.3 | 65.4 | 72.1 | 80.4 | 84.4 | 83.3 | 76.2 | 65.0 | 53.8 | 44.9 | 64.1 |
| Days Max Temp ≥ 90 °F | 0 | 0 | 1 | 3 | 8 | 22 | 28 | 27 | 16 | 4 | 0 | 0 | 109 |
| Days Max Temp ≤ 32 °F | 2 | 1 | 0 | 0 | 0 | 0 | 0 | 0 | 0 | 0 | 0 | 1 | 4 |
| Days Min Temp ≤ 32 °F | 20 | 14 | 6 | 1 | 0 | 0 | 0 | 0 | 0 | 0 | 7 | 17 | 65 |
| Days Min Temp ≤ 0 °F | 0 | 0 | 0 | 0 | 0 | 0 | 0 | 0 | 0 | 0 | 0 | 0 | 0 |
| Heating Degree Days | 699 | 527 | 320 | 87 | 19 | 0 | 0 | 0 | 13 | 108 | 343 | 615 | 2731 |
| Cooling Degree Days | 0 | 4 | 24 | 97 | 258 | 463 | 600 | 597 | 368 | na | 17 | 1 | na |
| Total Precipitation (") | 1.35 | 1.50 | 1.86 | 2.65 | 4.00 | 2.76 | 1.71 | 1.97 | 3.01 | 3.46 | 1.32 | 1.51 | 27.10 |
| Days ≥ 0.1" Precip | 2 | 3 | 3 | 4 | 5 | 4 | 3 | 3 | 4 | 5 | 3 | 3 | 42 |
| Total Snowfall (") | 0.8 | 0.6 | 0.6 | 0.0 | 0.0 | 0.0 | 0.0 | 0.0 | 0.0 | 0.0 | 0.1 | 0.2 | 2.3 |
| Days ≥ 1" Snow Depth | na | na | 0 | 0 | 0 | 0 | 0 | 0 | 0 | 0 | na | 0 | na |

## BRENHAM *Washington County* ELEVATION 351 ft LAT/LONG 30° 9 ' N / 96° 24 ' W

| | JAN | FEB | MAR | APR | MAY | JUN | JUL | AUG | SEP | OCT | NOV | DEC | YEAR |
|---|---|---|---|---|---|---|---|---|---|---|---|---|---|
| Maximum Temp °F | 59.9 | 64.2 | 72.5 | 79.9 | 85.3 | 91.7 | 95.6 | 96.2 | 90.4 | 82.3 | 71.6 | 63.5 | 79.4 |
| Minimum Temp °F | 39.0 | 41.8 | 49.3 | 57.8 | 64.4 | 70.6 | 72.9 | 72.3 | 67.6 | 58.0 | 48.9 | 41.9 | 57.0 |
| Mean Temp °F | 49.5 | 53.0 | 60.9 | 68.9 | 74.9 | 81.2 | 84.3 | 84.3 | 79.0 | 70.1 | 60.3 | 52.8 | 68.3 |
| Days Max Temp ≥ 90 °F | 0 | 0 | 0 | 1 | 7 | 22 | 29 | 28 | 19 | 6 | 0 | 0 | 112 |
| Days Max Temp ≤ 32 °F | 1 | 0 | 0 | 0 | 0 | 0 | 0 | 0 | 0 | 0 | 0 | 0 | 1 |
| Days Min Temp ≤ 32 °F | 8 | 5 | 1 | 0 | 0 | 0 | 0 | 0 | 0 | 0 | 1 | 6 | 21 |
| Days Min Temp ≤ 0 °F | 0 | 0 | 0 | 0 | 0 | 0 | 0 | 0 | 0 | 0 | 0 | 0 | 0 |
| Heating Degree Days | 485 | 345 | 176 | 39 | 3 | 0 | 0 | 0 | 2 | 37 | 195 | 389 | 1671 |
| Cooling Degree Days | 6 | 15 | 61 | 163 | 329 | 507 | 608 | 628 | 449 | 219 | 67 | 18 | 3070 |
| Total Precipitation (") | 3.41 | 3.00 | 2.91 | 3.49 | 5.78 | 4.26 | 1.92 | 2.92 | 4.78 | 4.27 | 3.68 | 3.37 | 43.79 |
| Days ≥ 0.1" Precip | 6 | 5 | 5 | 4 | 7 | 5 | 4 | 4 | 6 | 5 | 5 | 6 | 62 |
| Total Snowfall (") | 0.2 | 0.0 | 0.0 | 0.0 | 0.0 | 0.0 | 0.0 | 0.0 | 0.0 | 0.0 | 0.1 | 0.0 | 0.3 |
| Days ≥ 1" Snow Depth | 0 | 0 | 0 | 0 | 0 | 0 | 0 | 0 | 0 | 0 | 0 | 0 | 0 |

## BRIDGEPORT *Wise County* ELEVATION 820 ft LAT/LONG 33° 13 ' N / 97° 45 ' W

| | JAN | FEB | MAR | APR | MAY | JUN | JUL | AUG | SEP | OCT | NOV | DEC | YEAR |
|---|---|---|---|---|---|---|---|---|---|---|---|---|---|
| Maximum Temp °F | 54.9 | 59.1 | 68.3 | 77.4 | 83.5 | 91.6 | 97.6 | 96.9 | 88.8 | 78.8 | 66.6 | 58.6 | 76.8 |
| Minimum Temp °F | 29.3 | 32.9 | 41.3 | 50.7 | 58.6 | 66.5 | 70.0 | 68.3 | 61.6 | 50.2 | 40.4 | 32.2 | 50.2 |
| Mean Temp °F | 42.1 | 46.1 | 54.8 | 64.1 | 71.1 | 79.0 | 83.8 | 82.7 | 75.2 | 64.5 | 53.5 | 45.4 | 63.5 |
| Days Max Temp ≥ 90 °F | 0 | 0 | 1 | 2 | 6 | 20 | 28 | 27 | 16 | 4 | 0 | 0 | 104 |
| Days Max Temp ≤ 32 °F | 2 | 1 | 0 | 0 | 0 | 0 | 0 | 0 | 0 | 0 | 0 | 0 | 3 |
| Days Min Temp ≤ 32 °F | 20 | 14 | 6 | 1 | 0 | 0 | 0 | 0 | 0 | 0 | 8 | 16 | 65 |
| Days Min Temp ≤ 0 °F | 0 | 0 | 0 | 0 | 0 | 0 | 0 | 0 | 0 | 0 | 0 | 0 | 0 |
| Heating Degree Days | 703 | 531 | 333 | 107 | 22 | 1 | 0 | 0 | 16 | 110 | 355 | 603 | 2781 |
| Cooling Degree Days | 0 | 3 | 22 | 84 | 218 | 422 | 570 | 556 | 332 | 98 | 20 | 2 | 2327 |
| Total Precipitation (") | 1.62 | 2.12 | 2.56 | 2.89 | 5.63 | 3.38 | 2.15 | 1.97 | 3.51 | 4.08 | 1.95 | 1.90 | 33.76 |
| Days ≥ 0.1" Precip | 3 | 4 | 4 | 5 | 6 | 5 | 3 | 3 | 5 | 5 | 3 | 4 | 50 |
| Total Snowfall (") | 1.0 | 1.2 | 0.1 | 0.0 | 0.0 | 0.0 | 0.0 | 0.0 | 0.0 | 0.0 | 0.4 | 0.3 | 3.0 |
| Days ≥ 1" Snow Depth | 1 | 0 | 0 | 0 | 0 | 0 | 0 | 0 | 0 | 0 | 0 | 0 | 1 |

## BROWNFIELD 2 *Terry County* ELEVATION 3304 ft LAT/LONG 33° 11 ' N / 102° 15 ' W

| | JAN | FEB | MAR | APR | MAY | JUN | JUL | AUG | SEP | OCT | NOV | DEC | YEAR |
|---|---|---|---|---|---|---|---|---|---|---|---|---|---|
| Maximum Temp °F | 53.4 | 58.7 | 67.3 | 76.3 | 83.4 | 90.8 | 92.5 | 90.5 | 83.9 | 76.0 | 64.1 | 55.9 | 74.4 |
| Minimum Temp °F | 25.3 | 28.3 | 34.8 | 44.0 | 53.0 | 61.7 | 65.0 | 63.7 | 57.2 | 46.0 | 34.9 | 27.3 | 45.1 |
| Mean Temp °F | 39.3 | 43.5 | 51.1 | 60.1 | 68.2 | 76.3 | 78.8 | 77.1 | 70.6 | 61.0 | 49.5 | 41.7 | 59.8 |
| Days Max Temp ≥ 90 °F | 0 | 0 | 0 | 2 | 8 | 19 | 23 | 20 | 9 | 2 | 0 | 0 | 83 |
| Days Max Temp ≤ 32 °F | 3 | 1 | 0 | 0 | 0 | 0 | 0 | 0 | 0 | 0 | 0 | 2 | 6 |
| Days Min Temp ≤ 32 °F | 26 | 20 | 12 | 2 | 0 | 0 | 0 | 0 | 0 | 2 | 12 | 24 | 98 |
| Days Min Temp ≤ 0 °F | 0 | 0 | 0 | 0 | 0 | 0 | 0 | 0 | 0 | 0 | 0 | 0 | 0 |
| Heating Degree Days | 788 | 600 | 430 | 178 | 48 | 4 | 0 | 1 | 33 | 164 | 459 | 717 | 3422 |
| Cooling Degree Days | 0 | 0 | 3 | 39 | 160 | 356 | 447 | 405 | 215 | 47 | 2 | 0 | 1674 |
| Total Precipitation (") | 0.56 | 0.68 | 0.81 | 1.02 | 2.84 | 2.91 | 2.33 | 2.40 | 2.66 | 1.59 | 0.77 | 0.61 | 19.18 |
| Days ≥ 0.1" Precip | 2 | 2 | 2 | 2 | 4 | 4 | 4 | 4 | 4 | 3 | 2 | 2 | 35 |
| Total Snowfall (") | 2.2 | 1.8 | 1.6 | 0.2 | 0.0 | 0.0 | 0.0 | 0.0 | 0.0 | 0.1 | 0.8 | 1.7 | 8.4 |
| Days ≥ 1" Snow Depth | 2 | 1 | 1 | 0 | 0 | 0 | 0 | 0 | 0 | 0 | 0 | 1 | 5 |

## BROWNSVILLE INTL AP *Cameron County* ELEVATION 33 ft LAT/LONG 25° 54 ' N / 97° 26 ' W

| | JAN | FEB | MAR | APR | MAY | JUN | JUL | AUG | SEP | OCT | NOV | DEC | YEAR |
|---|---|---|---|---|---|---|---|---|---|---|---|---|---|
| Maximum Temp °F | 69.1 | 72.1 | 78.4 | 83.7 | 87.4 | 91.1 | 93.1 | 93.3 | 90.0 | 85.0 | 78.1 | 72.2 | 82.8 |
| Minimum Temp °F | 50.5 | 52.8 | 59.2 | 66.3 | 71.6 | 75.0 | 75.7 | 75.4 | 72.8 | 66.1 | 58.9 | 53.1 | 64.8 |
| Mean Temp °F | 59.8 | 62.5 | 68.8 | 75.0 | 79.5 | 83.1 | 84.4 | 84.4 | 81.5 | 75.6 | 68.5 | 62.6 | 73.8 |
| Days Max Temp ≥ 90 °F | 0 | 0 | 1 | 3 | 10 | 23 | 27 | 28 | 19 | 6 | 0 | 0 | 117 |
| Days Max Temp ≤ 32 °F | 0 | 0 | 0 | 0 | 0 | 0 | 0 | 0 | 0 | 0 | 0 | 0 | 0 |
| Days Min Temp ≤ 32 °F | 1 | 0 | 0 | 0 | 0 | 0 | 0 | 0 | 0 | 0 | 0 | 1 | 2 |
| Days Min Temp ≤ 0 °F | 0 | 0 | 0 | 0 | 0 | 0 | 0 | 0 | 0 | 0 | 0 | 0 | 0 |
| Heating Degree Days | 212 | 131 | 50 | 6 | 0 | 0 | 0 | 0 | 0 | 6 | 63 | 154 | 622 |
| Cooling Degree Days | 52 | 76 | 177 | 301 | 469 | 573 | 626 | 629 | 512 | 358 | 187 | 91 | 4051 |
| Total Precipitation (") | 1.71 | 1.24 | 0.60 | 1.91 | 3.01 | 2.91 | 1.94 | 2.92 | 5.69 | 3.13 | 1.65 | 1.18 | 27.89 |
| Days ≥ 0.1" Precip | 4 | 2 | 1 | 2 | 3 | 4 | 3 | 5 | 6 | 4 | 3 | 3 | 40 |
| Total Snowfall (") | 0.0 | 0.0 | 0.0 | 0.0 | 0.0 | 0.0 | 0.0 | 0.0 | 0.0 | 0.0 | 0.0 | 0.0 | 0.0 |
| Days ≥ 1" Snow Depth | 0 | 0 | 0 | 0 | 0 | 0 | 0 | 0 | 0 | 0 | 0 | 0 | 0 |

## BROWNWOOD *Brown County* ELEVATION 1385 ft LAT/LONG 31° 43 ' N / 99° 0 ' W

| | JAN | FEB | MAR | APR | MAY | JUN | JUL | AUG | SEP | OCT | NOV | DEC | YEAR |
|---|---|---|---|---|---|---|---|---|---|---|---|---|---|
| Maximum Temp °F | 57.0 | 62.0 | 70.7 | 79.6 | 85.3 | 92.2 | 96.3 | 95.7 | 88.5 | 80.3 | 68.1 | 60.4 | 78.0 |
| Minimum Temp °F | 31.2 | 34.8 | 42.8 | 52.2 | 60.2 | 67.8 | 70.8 | 69.7 | 63.3 | 52.6 | 41.9 | 33.4 | 51.7 |
| Mean Temp °F | 44.1 | 48.4 | 56.7 | 65.9 | 72.8 | 80.0 | 83.6 | 82.7 | 75.9 | 66.4 | 55.1 | 46.9 | 64.9 |
| Days Max Temp ≥ 90 °F | 0 | 0 | 1 | 4 | 9 | 22 | 28 | 27 | 16 | 4 | 0 | 0 | 111 |
| Days Max Temp ≤ 32 °F | 1 | 1 | 0 | 0 | 0 | 0 | 0 | 0 | 0 | 0 | 0 | 1 | 3 |
| Days Min Temp ≤ 32 °F | 18 | 11 | 5 | 1 | 0 | 0 | 0 | 0 | 0 | 0 | 6 | 15 | 56 |
| Days Min Temp ≤ 0 °F | 0 | 0 | 0 | 0 | 0 | 0 | 0 | 0 | 0 | 0 | 0 | 0 | 0 |
| Heating Degree Days | 641 | 465 | 279 | 78 | 13 | 1 | 0 | 0 | 8 | 76 | 310 | 556 | 2427 |
| Cooling Degree Days | 1 | 3 | 23 | 103 | 255 | 440 | 563 | 568 | 349 | 126 | 23 | 1 | 2455 |
| Total Precipitation (") | 1.48 | 1.90 | 1.85 | 2.54 | 3.95 | 3.23 | 1.70 | 2.28 | 3.30 | 2.72 | 1.55 | 1.78 | 28.28 |
| Days ≥ 0.1" Precip | 3 | 3 | 3 | 4 | 5 | 4 | 3 | 4 | 4 | 3 | 3 | 3 | 42 |
| Total Snowfall (") | 0.9 | 0.4 | 0.0 | 0.0 | 0.0 | 0.0 | 0.0 | 0.0 | 0.0 | 0.0 | 0.2 | 0.2 | 1.7 |
| Days ≥ 1" Snow Depth | *0* | 0 | 0 | 0 | 0 | 0 | 0 | 0 | 0 | 0 | 0 | 0 | 0 |

## BURNET *Burnet County* ELEVATION 1302 ft LAT/LONG 30° 46 ' N / 98° 13 ' W

| | JAN | FEB | MAR | APR | MAY | JUN | JUL | AUG | SEP | OCT | NOV | DEC | YEAR |
|---|---|---|---|---|---|---|---|---|---|---|---|---|---|
| Maximum Temp °F | 58.3 | 62.5 | 70.5 | 77.8 | 82.7 | 89.3 | 93.3 | 93.3 | 87.5 | 79.2 | 68.6 | 61.1 | 77.0 |
| Minimum Temp °F | 33.5 | 37.0 | 45.4 | 53.7 | 61.0 | 67.8 | 70.1 | 69.5 | 64.0 | 53.6 | 43.9 | 36.2 | 53.0 |
| Mean Temp °F | 45.9 | 49.8 | 58.0 | 65.8 | 71.9 | 78.6 | 81.7 | 81.5 | 75.8 | 66.4 | 56.2 | 48.7 | 65.0 |
| Days Max Temp ≥ 90 °F | 0 | 0 | 1 | 1 | 4 | 16 | 26 | 26 | 13 | 2 | 0 | 0 | 89 |
| Days Max Temp ≤ 32 °F | 1 | 0 | 0 | 0 | 0 | 0 | 0 | 0 | 0 | 0 | 0 | 1 | 2 |
| Days Min Temp ≤ 32 °F | 16 | 10 | 4 | 1 | 0 | 0 | 0 | 0 | 0 | 0 | 5 | 12 | 48 |
| Days Min Temp ≤ 0 °F | 0 | 0 | 0 | 0 | 0 | 0 | 0 | 0 | 0 | 0 | 0 | 0 | 0 |
| Heating Degree Days | 588 | 428 | 244 | 76 | 12 | 0 | 0 | 0 | 8 | 77 | 281 | 502 | 2216 |
| Cooling Degree Days | 1 | 3 | 28 | 95 | 232 | 414 | 533 | 536 | 344 | 133 | 25 | 3 | 2347 |
| Total Precipitation (") | 1.91 | 2.12 | 2.23 | 2.68 | 4.99 | 3.66 | 1.95 | 2.12 | 3.48 | 3.42 | 2.03 | 1.88 | 32.47 |
| Days ≥ 0.1" Precip | 4 | 4 | 4 | 4 | 6 | 5 | 3 | 3 | 5 | 5 | 4 | 3 | 50 |
| Total Snowfall (") | 0.3 | 0.6 | 0.0 | 0.0 | 0.0 | 0.0 | 0.0 | 0.0 | 0.0 | 0.0 | 0.0 | 0.0 | 0.9 |
| Days ≥ 1" Snow Depth | 0 | 0 | 0 | 0 | 0 | 0 | 0 | 0 | 0 | 0 | 0 | 0 | 0 |

## CAMERON *Milam County* ELEVATION 390 ft LAT/LONG 30° 51 ' N / 96° 59 ' W

| | JAN | FEB | MAR | APR | MAY | JUN | JUL | AUG | SEP | OCT | NOV | DEC | YEAR |
|---|---|---|---|---|---|---|---|---|---|---|---|---|---|
| Maximum Temp °F | 59.9 | 64.4 | 72.7 | 79.7 | 84.6 | 90.9 | 95.2 | 95.6 | 89.4 | 81.0 | 70.9 | 63.1 | 79.0 |
| Minimum Temp °F | 38.4 | 41.5 | 49.2 | 57.3 | 64.3 | 70.4 | 72.7 | 72.3 | 67.2 | 57.7 | 48.3 | 40.9 | 56.7 |
| Mean Temp °F | 49.2 | 53.0 | 61.0 | 68.6 | 74.4 | 80.7 | 84.0 | 84.0 | 78.4 | 69.4 | 59.6 | 52.0 | 67.9 |
| Days Max Temp ≥ 90 °F | 0 | 0 | 1 | 1 | 5 | 20 | 28 | 28 | 17 | 4 | 0 | 0 | 104 |
| Days Max Temp ≤ 32 °F | 1 | 0 | 0 | 0 | 0 | 0 | 0 | 0 | 0 | 0 | 0 | 0 | 1 |
| Days Min Temp ≤ 32 °F | 10 | 6 | 2 | 0 | 0 | 0 | 0 | 0 | 0 | 0 | 2 | 7 | 27 |
| Days Min Temp ≤ 0 °F | 0 | 0 | 0 | 0 | 0 | 0 | 0 | 0 | 0 | 0 | 0 | 0 | 0 |
| Heating Degree Days | 494 | 346 | 178 | 42 | 3 | 0 | 0 | 0 | 2 | 45 | 209 | 410 | 1729 |
| Cooling Degree Days | 7 | 15 | 62 | 154 | 312 | 488 | 603 | 615 | 426 | 200 | 63 | 16 | 2961 |
| Total Precipitation (") | 2.51 | 2.80 | 2.47 | 3.28 | 5.38 | 3.03 | 1.87 | 1.76 | 3.67 | 3.65 | 2.89 | 2.82 | 36.13 |
| Days ≥ 0.1" Precip | 5 | 5 | 4 | 4 | 7 | 5 | 3 | 3 | 5 | 4 | 4 | 5 | 54 |
| Total Snowfall (") | 0.1 | *0.0* | 0.0 | 0.0 | 0.0 | 0.0 | 0.0 | 0.0 | 0.0 | 0.0 | 0.0 | 0.0 | 0.1 |
| Days ≥ 1" Snow Depth | 0 | 0 | 0 | 0 | 0 | 0 | 0 | 0 | 0 | 0 | 0 | 0 | 0 |

## CAMP WOOD *Real County* ELEVATION 1450 ft LAT/LONG 29° 40 ' N / 100° 1 ' W

| | JAN | FEB | MAR | APR | MAY | JUN | JUL | AUG | SEP | OCT | NOV | DEC | YEAR |
|---|---|---|---|---|---|---|---|---|---|---|---|---|---|
| Maximum Temp °F | 61.4 | 65.6 | 73.6 | 80.5 | 85.7 | 91.4 | 94.6 | 94.3 | 89.1 | *81.2* | 71.1 | 63.9 | 79.4 |
| Minimum Temp °F | 34.0 | 37.4 | 45.4 | 53.9 | 61.7 | 68.0 | 70.0 | 69.1 | 64.4 | *54.5* | 43.7 | 36.0 | 53.2 |
| Mean Temp °F | 47.7 | 51.5 | 59.5 | 67.2 | 73.8 | 79.7 | 82.3 | 81.7 | 76.8 | *67.8* | 57.5 | 50.0 | 66.3 |
| Days Max Temp ≥ 90 °F | 0 | 0 | 1 | 3 | 9 | 21 | 27 | 26 | 17 | 3 | 0 | 0 | 107 |
| Days Max Temp ≤ 32 °F | 0 | 0 | 0 | 0 | 0 | 0 | 0 | 0 | 0 | 0 | 0 | 0 | 0 |
| Days Min Temp ≤ 32 °F | 15 | 9 | 3 | 0 | 0 | 0 | 0 | 0 | 0 | 0 | 5 | 13 | 45 |
| Days Min Temp ≤ 0 °F | 0 | 0 | 0 | 0 | 0 | 0 | 0 | 0 | 0 | 0 | 0 | 0 | 0 |
| Heating Degree Days | 531 | 378 | 203 | 57 | 5 | 0 | 0 | 0 | 5 | *46* | 247 | 461 | 1933 |
| Cooling Degree Days | 0 | 2 | 33 | 114 | 288 | 450 | 545 | 535 | 372 | 144 | 28 | 2 | 2513 |
| Total Precipitation (") | 1.28 | 1.52 | 1.39 | 2.59 | 3.24 | 3.27 | 2.35 | 3.05 | 3.16 | 3.39 | 1.73 | 1.43 | 28.40 |
| Days ≥ 0.1" Precip | 3 | 3 | 3 | 4 | 5 | 4 | 3 | 3 | 5 | 4 | 3 | 3 | 43 |
| Total Snowfall (") | 0.0 | 0.3 | 0.1 | 0.0 | 0.0 | 0.0 | 0.0 | 0.0 | 0.0 | 0.0 | 0.0 | 0.0 | 0.4 |
| Days ≥ 1" Snow Depth | 0 | 0 | 0 | 0 | 0 | 0 | 0 | 0 | 0 | 0 | 0 | 0 | 0 |

## CANADIAN *Hemphill County* ELEVATION 2342 ft LAT/LONG 35° 55 ' N / 100° 22 ' W

| | JAN | FEB | MAR | APR | MAY | JUN | JUL | AUG | SEP | OCT | NOV | DEC | YEAR |
|---|---|---|---|---|---|---|---|---|---|---|---|---|---|
| Maximum Temp °F | 51.2 | 56.3 | 65.3 | 75.3 | 81.7 | 90.3 | 95.7 | 94.0 | 86.3 | 76.2 | 62.1 | 52.8 | 73.9 |
| Minimum Temp °F | 22.2 | 26.2 | 34.6 | 44.1 | 53.6 | 63.0 | 67.4 | 65.4 | 57.6 | 45.1 | 32.9 | 24.2 | 44.7 |
| Mean Temp °F | 36.7 | 41.3 | 49.9 | 59.8 | 67.7 | 76.7 | 81.6 | 79.7 | 72.0 | 60.7 | 47.6 | 38.5 | 59.4 |
| Days Max Temp ≥ 90 °F | 0 | 0 | 0 | 2 | 6 | 17 | 26 | 24 | 12 | 3 | 0 | 0 | 90 |
| Days Max Temp ≤ 32 °F | 4 | 2 | 0 | 0 | 0 | 0 | 0 | 0 | 0 | 0 | 0 | 2 | 8 |
| Days Min Temp ≤ 32 °F | 27 | 21 | 13 | 3 | 0 | 0 | 0 | 0 | 0 | 3 | 14 | 26 | 107 |
| Days Min Temp ≤ 0 °F | 1 | 0 | 0 | 0 | 0 | 0 | 0 | 0 | 0 | 0 | 0 | 1 | 2 |
| Heating Degree Days | 870 | 663 | 467 | 193 | 52 | 3 | 0 | 1 | 29 | 179 | 518 | 813 | 3788 |
| Cooling Degree Days | 0 | 0 | 4 | 44 | 143 | 358 | 518 | 479 | 256 | 45 | 2 | 0 | 1849 |
| Total Precipitation (") | 0.38 | 0.72 | 1.39 | 1.51 | 3.52 | 3.11 | 1.87 | 2.60 | 2.34 | 1.44 | 0.84 | 0.57 | 20.29 |
| Days ≥ 0.1" Precip | 1 | 2 | 3 | 3 | 6 | 5 | 4 | 4 | 4 | 3 | 2 | 2 | 39 |
| Total Snowfall (") | 2.5 | na | 1.9 | 0.5 | 0.0 | 0.0 | 0.0 | 0.0 | 0.0 | 0.1 | 1.1 | *3.5* | na |
| Days ≥ 1" Snow Depth | 1 | na | 0 | 0 | 0 | 0 | 0 | 0 | 0 | 0 | 0 | *1* | na |

## CANDELARIA *Presidio County* ELEVATION 2881 ft LAT/LONG 30° 9 ' N / 104° 41 ' W

| | JAN | FEB | MAR | APR | MAY | JUN | JUL | AUG | SEP | OCT | NOV | DEC | YEAR |
|---|---|---|---|---|---|---|---|---|---|---|---|---|---|
| Maximum Temp °F | 65.8 | 71.8 | 80.2 | 88.8 | 95.6 | 101.6 | 99.8 | 97.2 | 92.1 | 84.9 | 73.8 | 66.5 | 84.8 |
| Minimum Temp °F | 31.6 | 34.5 | 40.2 | 47.5 | 55.5 | 64.3 | 67.1 | 65.2 | 60.4 | 49.0 | 37.2 | 31.9 | 48.7 |
| Mean Temp °F | 48.7 | 53.2 | 60.2 | 68.2 | 75.6 | 83.0 | 83.5 | 81.2 | 76.3 | 67.0 | 55.5 | 49.3 | 66.8 |
| Days Max Temp ≥ 90 °F | 0 | 0 | 3 | 16 | 26 | 29 | 29 | 28 | 21 | 8 | 0 | 0 | 160 |
| Days Max Temp ≤ 32 °F | 0 | 0 | 0 | 0 | 0 | 0 | 0 | 0 | 0 | 0 | 0 | 0 | 0 |
| Days Min Temp ≤ 32 °F | 17 | 12 | 5 | 1 | 0 | 0 | 0 | 0 | 0 | 0 | 9 | 17 | 61 |
| Days Min Temp ≤ 0 °F | 0 | 0 | 0 | 0 | 0 | 0 | 0 | 0 | 0 | 0 | 0 | 0 | 0 |
| Heating Degree Days | 498 | 329 | 167 | 36 | 1 | 0 | 0 | 0 | 5 | 45 | 283 | 482 | 1846 |
| Cooling Degree Days | 0 | 2 | 19 | 122 | 326 | 544 | 579 | 524 | 359 | 120 | 5 | 0 | 2600 |
| Total Precipitation (") | 0.35 | 0.33 | 0.25 | 0.34 | 0.69 | 1.70 | 2.28 | 2.60 | 2.56 | 1.16 | 0.42 | 0.51 | 13.19 |
| Days ≥ 0.1" Precip | 1 | 1 | 1 | 1 | 2 | 3 | 5 | 5 | 5 | 2 | 1 | 1 | 28 |
| Total Snowfall (") | 0.1 | 0.0 | 0.0 | 0.0 | 0.0 | 0.0 | 0.0 | 0.0 | 0.0 | 0.0 | 0.0 | 0.0 | 0.1 |
| Days ≥ 1" Snow Depth | 0 | 0 | 0 | 0 | 0 | 0 | 0 | 0 | 0 | 0 | 0 | 0 | 0 |

## CANYON *Randall County* ELEVATION 3583 ft LAT/LONG 34° 59 ' N / 101° 56 ' W

| | JAN | FEB | MAR | APR | MAY | JUN | JUL | AUG | SEP | OCT | NOV | DEC | YEAR |
|---|---|---|---|---|---|---|---|---|---|---|---|---|---|
| Maximum Temp °F | 52.3 | 57.0 | 65.3 | 74.3 | 81.2 | 89.4 | 92.0 | 89.4 | 83.7 | 74.9 | 61.9 | 54.1 | 73.0 |
| Minimum Temp °F | 23.3 | 27.0 | 34.1 | 43.0 | 51.9 | 61.1 | 65.7 | 63.8 | 56.6 | 45.0 | 33.3 | 25.7 | 44.2 |
| Mean Temp °F | 37.8 | 42.0 | 49.7 | 58.6 | 66.6 | 75.3 | 78.9 | 76.6 | 70.2 | 60.0 | 47.6 | 39.9 | 58.6 |
| Days Max Temp ≥ 90 °F | 0 | 0 | 0 | 1 | 5 | 15 | 22 | 17 | 8 | 1 | 0 | 0 | 69 |
| Days Max Temp ≤ 32 °F | 3 | 2 | 1 | 0 | 0 | 0 | 0 | 0 | 0 | 0 | 0 | 2 | 8 |
| Days Min Temp ≤ 32 °F | 26 | 20 | 13 | 3 | 0 | 0 | 0 | 0 | 0 | 2 | 15 | 25 | 104 |
| Days Min Temp ≤ 0 °F | 0 | 0 | 0 | 0 | 0 | 0 | 0 | 0 | 0 | 0 | 0 | 0 | 0 |
| Heating Degree Days | 837 | 642 | 470 | 210 | 61 | 5 | 0 | 1 | 34 | 186 | 515 | 771 | 3732 |
| Cooling Degree Days | 0 | 0 | 3 | 25 | 116 | 327 | 442 | 384 | 203 | 32 | 1 | 0 | 1533 |
| Total Precipitation (") | 0.49 | 0.53 | 0.86 | 0.98 | 2.71 | 3.42 | 2.24 | 3.24 | 1.99 | 1.56 | 0.67 | 0.52 | 19.21 |
| Days ≥ 0.1" Precip | 1 | 1 | 2 | 2 | 5 | 5 | 4 | 5 | 3 | 3 | 2 | 1 | 34 |
| Total Snowfall (") | 2.7 | 2.7 | 1.3 | 0.1 | 0.0 | 0.0 | 0.0 | 0.0 | 0.0 | 0.0 | 1.1 | 2.4 | 10.3 |
| Days ≥ 1" Snow Depth | 2 | 1 | 0 | 0 | 0 | 0 | 0 | 0 | 0 | 0 | 0 | 1 | 4 |

## CANYON DAM *Comal County* ELEVATION 1001 ft LAT/LONG 29° 52' N / 98° 12' W

| | JAN | FEB | MAR | APR | MAY | JUN | JUL | AUG | SEP | OCT | NOV | DEC | YEAR |
|---|---|---|---|---|---|---|---|---|---|---|---|---|---|
| Maximum Temp °F | 59.2 | 63.3 | 71.3 | 78.5 | 83.4 | 90.2 | 94.0 | 94.3 | 88.6 | 80.6 | 70.5 | 62.8 | 78.1 |
| Minimum Temp °F | 37.5 | 40.3 | 47.9 | 56.4 | 63.1 | 69.3 | 71.6 | 71.2 | 67.0 | 57.9 | 48.3 | 40.7 | 55.9 |
| Mean Temp °F | 48.3 | 51.8 | 59.6 | 67.5 | 73.3 | 79.8 | 82.8 | 82.8 | 77.8 | 69.3 | 59.4 | 51.8 | 67.0 |
| Days Max Temp ≥ 90 °F | 0 | 0 | 0 | 1 | 5 | 18 | 27 | 27 | 16 | 3 | 0 | 0 | 97 |
| Days Max Temp ≤ 32 °F | 0 | 0 | 0 | 0 | 0 | 0 | 0 | 0 | 0 | 0 | 0 | 0 | 0 |
| Days Min Temp ≤ 32 °F | 9 | 5 | 1 | 0 | 0 | 0 | 0 | 0 | 0 | 0 | 2 | 6 | 23 |
| Days Min Temp ≤ 0 °F | 0 | 0 | 0 | 0 | 0 | 0 | 0 | 0 | 0 | 0 | 0 | 0 | 0 |
| Heating Degree Days | 514 | 372 | 199 | 50 | 6 | 0 | 0 | 0 | 3 | 42 | 208 | 410 | 1804 |
| Cooling Degree Days | 2 | *6* | 33 | 121 | 264 | 451 | 556 | 575 | 406 | 191 | 52 | 8 | 2665 |
| Total Precipitation (") | 2.49 | 2.18 | 2.22 | 3.08 | 5.00 | 3.30 | 2.24 | 2.60 | 4.39 | 3.63 | 2.77 | 2.41 | 36.31 |
| Days ≥ 0.1" Precip | 5 | 4 | 4 | 4 | 7 | 4 | 3 | 4 | 6 | 5 | 4 | 4 | 54 |
| Total Snowfall (") | 0.4 | 0.2 | 0.0 | 0.0 | 0.0 | 0.0 | 0.0 | 0.0 | 0.0 | 0.0 | 0.0 | 0.0 | 0.6 |
| Days ≥ 1" Snow Depth | *0* | 0 | 0 | 0 | 0 | 0 | 0 | 0 | 0 | 0 | 0 | 0 | 0 |

## CARRIZO SPRINGS *Dimmit County* ELEVATION 650 ft LAT/LONG 28° 32' N / 99° 52' W

| | JAN | FEB | MAR | APR | MAY | JUN | JUL | AUG | SEP | OCT | NOV | DEC | YEAR |
|---|---|---|---|---|---|---|---|---|---|---|---|---|---|
| Maximum Temp °F | 64.7 | 69.2 | 78.0 | 85.7 | *89.5* | 95.5 | *98.9* | *98.5* | *93.2* | *84.7* | *75.3* | *66.9* | 83.3 |
| Minimum Temp °F | 41.0 | 44.2 | 51.3 | 60.5 | *66.6* | 71.8 | *73.6* | *73.3* | *69.5* | *60.0* | *49.8* | *42.7* | 58.7 |
| Mean Temp °F | 52.8 | 56.7 | 64.7 | 73.1 | *78.1* | 83.6 | *86.3* | *85.9* | *81.4* | *72.4* | *62.6* | *54.8* | 71.0 |
| Days Max Temp ≥ 90 °F | 0 | 1 | 3 | *10* | *15* | 26 | *29* | *29* | *22* | *9* | *1* | *0* | 145 |
| Days Max Temp ≤ 32 °F | 0 | 0 | 0 | 0 | *0* | 0 | *0* | *0* | *0* | 0 | *0* | *0* | 0 |
| Days Min Temp ≤ 32 °F | 5 | 3 | 1 | 0 | *0* | 0 | *0* | *0* | *0* | 0 | *1* | *5* | 15 |
| Days Min Temp ≤ 0 °F | 0 | 0 | 0 | 0 | *0* | 0 | *0* | *0* | *0* | 0 | *0* | *0* | 0 |
| Heating Degree Days | 381 | 252 | 108 | 16 | *1* | 0 | *0* | *0* | *1* | *24* | *143* | *324* | 1250 |
| Cooling Degree Days | 10 | 26 | 106 | 259 | 437 | 581 | 678 | 678 | 514 | 280 | *81* | *14* | 3664 |
| Total Precipitation (") | 1.14 | 1.17 | 0.86 | 1.79 | *3.56* | 2.53 | *1.27* | *2.33* | *2.58* | *2.54* | *1.06* | *1.09* | 21.92 |
| Days ≥ 0.1" Precip | 3 | 2 | 2 | 3 | *4* | 4 | *2* | *3* | *4* | *3* | *2* | *3* | 35 |
| Total Snowfall (") | 0.0 | 0.0 | 0.0 | 0.0 | *0.0* | 0.0 | *0.0* | *0.0* | *0.0* | 0.0 | *0.0* | *0.0* | 0.0 |
| Days ≥ 1" Snow Depth | 0 | 0 | 0 | 0 | *0* | 0 | *0* | *0* | *0* | 0 | *0* | *0* | 0 |

## CARTA VALLEY *Edwards County* ELEVATION 1903 ft LAT/LONG 29° 48' N / 100° 40' W

| | JAN | FEB | MAR | APR | MAY | JUN | JUL | AUG | SEP | OCT | NOV | DEC | YEAR |
|---|---|---|---|---|---|---|---|---|---|---|---|---|---|
| Maximum Temp °F | 62.7 | 67.0 | 75.1 | 82.3 | 86.7 | 91.6 | 94.5 | 94.3 | 89.5 | 81.2 | 71.6 | 64.9 | 80.1 |
| Minimum Temp °F | 35.0 | 38.5 | 46.5 | 55.1 | 63.1 | 69.0 | 71.1 | 70.7 | 65.9 | 55.7 | 44.7 | 36.9 | 54.4 |
| Mean Temp °F | 48.9 | 52.8 | 60.9 | 68.7 | 74.9 | 80.3 | 82.9 | 82.5 | 77.7 | 68.4 | 58.2 | 50.9 | 67.3 |
| Days Max Temp ≥ 90 °F | 0 | 0 | 1 | 5 | 10 | 20 | 27 | 26 | 17 | 3 | 0 | 0 | 109 |
| Days Max Temp ≤ 32 °F | 0 | 0 | 0 | 0 | 0 | 0 | 0 | 0 | 0 | 0 | 0 | 0 | 0 |
| Days Min Temp ≤ 32 °F | 13 | 8 | 3 | 0 | 0 | 0 | 0 | 0 | 0 | 0 | 5 | 11 | 40 |
| Days Min Temp ≤ 0 °F | 0 | 0 | 0 | 0 | 0 | 0 | 0 | 0 | 0 | 0 | 0 | 0 | 0 |
| Heating Degree Days | 495 | 344 | 171 | 40 | 5 | 0 | 0 | 0 | 4 | 46 | 226 | 432 | 1763 |
| Cooling Degree Days | *1* | *4* | *47* | 157 | *357* | 499 | *584* | 590 | 423 | *179* | *33* | *3* | 2877 |
| Total Precipitation (") | 0.81 | 1.16 | 1.18 | 1.98 | 3.01 | 2.44 | 2.70 | 2.26 | 2.74 | 2.43 | 1.13 | 0.81 | 22.65 |
| Days ≥ 0.1" Precip | 2 | 2 | 2 | 3 | 4 | 3 | 3 | 3 | 4 | 3 | 2 | 2 | 33 |
| Total Snowfall (") | 0.5 | 0.2 | 0.0 | 0.0 | 0.0 | 0.0 | 0.0 | 0.0 | 0.0 | 0.0 | 0.0 | 0.0 | 0.7 |
| Days ≥ 1" Snow Depth | 0 | 0 | 0 | 0 | 0 | 0 | 0 | 0 | 0 | 0 | 0 | 0 | 0 |

## CARTHAGE *Panola County* ELEVATION 302 ft LAT/LONG 32° 9' N / 94° 20' W

| | JAN | FEB | MAR | APR | MAY | JUN | JUL | AUG | SEP | OCT | NOV | DEC | YEAR |
|---|---|---|---|---|---|---|---|---|---|---|---|---|---|
| Maximum Temp °F | 56.5 | 61.6 | 70.0 | 77.7 | 83.3 | 90.0 | 93.9 | 93.8 | 87.9 | 79.1 | 68.1 | 60.2 | 76.8 |
| Minimum Temp °F | 34.1 | 37.1 | 44.9 | 52.9 | 60.5 | 67.4 | 70.4 | 69.1 | 63.4 | 52.4 | 43.3 | 36.8 | 52.7 |
| Mean Temp °F | 45.3 | 49.4 | 57.5 | 65.3 | 71.9 | 78.8 | 82.2 | 81.5 | 75.7 | 65.8 | 55.7 | 48.5 | 64.8 |
| Days Max Temp ≥ 90 °F | 0 | 0 | 0 | 0 | 4 | 18 | 27 | 26 | 14 | 2 | 0 | 0 | 91 |
| Days Max Temp ≤ 32 °F | 1 | 0 | 0 | 0 | 0 | 0 | 0 | 0 | 0 | 0 | 0 | 1 | 2 |
| Days Min Temp ≤ 32 °F | 15 | 10 | 4 | 0 | 0 | 0 | 0 | 0 | 0 | 0 | 4 | 12 | 45 |
| Days Min Temp ≤ 0 °F | 0 | 0 | 0 | 0 | 0 | 0 | 0 | 0 | 0 | 0 | 0 | 0 | 0 |
| Heating Degree Days | 607 | 439 | 259 | 83 | 11 | 0 | 0 | 0 | 8 | 87 | 295 | 509 | 2298 |
| Cooling Degree Days | 1 | 4 | 26 | 82 | 212 | 414 | 532 | 519 | 327 | 105 | 23 | 6 | 2251 |
| Total Precipitation (") | 4.39 | 3.92 | 3.78 | 4.48 | 5.32 | 4.57 | 3.14 | 2.69 | 3.98 | 4.56 | 4.78 | 4.96 | 50.57 |
| Days ≥ 0.1" Precip | 7 | 6 | 6 | 5 | 7 | 5 | 5 | 4 | 5 | 5 | 6 | 7 | 68 |
| Total Snowfall (") | 0.7 | 0.5 | 0.2 | 0.0 | 0.0 | 0.0 | 0.0 | 0.0 | 0.0 | 0.0 | 0.0 | 0.2 | 1.6 |
| Days ≥ 1" Snow Depth | 0 | 0 | 0 | 0 | 0 | 0 | 0 | 0 | 0 | 0 | 0 | 0 | 0 |

## CATARINA *Dimmit County* ELEVATION 581 ft LAT/LONG 28° 22 ' N / 99° 40 ' W

| | JAN | FEB | MAR | APR | MAY | JUN | JUL | AUG | SEP | OCT | NOV | DEC | YEAR |
|---|---|---|---|---|---|---|---|---|---|---|---|---|---|
| Maximum Temp °F | 66.5 | 71.7 | 81.2 | 88.1 | ***92.4*** | 96.8 | 99.5 | 99.0 | 93.7 | 86.6 | ***77.1*** | ***68.8*** | 85.1 |
| Minimum Temp °F | 41.6 | 44.9 | 52.4 | 60.5 | 67.3 | 71.9 | 73.2 | 72.7 | 68.8 | 60.8 | ***51.1*** | ***43.2*** | 59.0 |
| Mean Temp °F | 54.1 | 58.3 | 66.8 | 74.3 | ***79.9*** | 84.4 | 86.3 | 85.9 | 81.3 | 73.7 | ***64.1*** | ***56.0*** | 72.1 |
| Days Max Temp ≥ 90 °F | 0 | 1 | 5 | ***14*** | 20 | 27 | 30 | 30 | 23 | 11 | ***2*** | ***0*** | 163 |
| Days Max Temp ≤ 32 °F | 0 | 0 | 0 | 0 | 0 | 0 | 0 | 0 | 0 | 0 | ***0*** | ***0*** | 0 |
| Days Min Temp ≤ 32 °F | 5 | 2 | 1 | 0 | 0 | 0 | 0 | 0 | 0 | 0 | ***1*** | ***4*** | 13 |
| Days Min Temp ≤ 0 °F | 0 | 0 | 0 | 0 | 0 | 0 | 0 | 0 | 0 | 0 | 0 | 0 | 0 |
| Heating Degree Days | 348 | 213 | 79 | 14 | 0 | 0 | 0 | 0 | 1 | 12 | ***115*** | ***294*** | 1076 |
| Cooling Degree Days | 9 | 32 | 127 | 286 | ***478*** | 611 | 679 | 675 | 508 | 303 | ***103*** | ***21*** | 3832 |
| Total Precipitation (") | 0.98 | 1.06 | 0.90 | 1.80 | ***2.59*** | 2.93 | 1.05 | 2.39 | 2.79 | 2.64 | 1.07 | 1.11 | 21.31 |
| Days ≥ 0.1" Precip | 3 | 2 | 1 | 2 | ***4*** | 3 | 2 | 3 | 3 | 3 | 2 | 2 | 30 |
| Total Snowfall (") | 0.2 | 0.0 | 0.0 | 0.0 | 0.0 | 0.0 | 0.0 | 0.0 | 0.0 | 0.0 | 0.0 | 0.0 | 0.2 |
| Days ≥ 1" Snow Depth | 0 | 0 | 0 | 0 | 0 | 0 | 0 | 0 | 0 | 0 | ***0*** | 0 | 0 |

## CENTER *Shelby County* ELEVATION 351 ft LAT/LONG 31° 48 ' N / 94° 11 ' W

| | JAN | FEB | MAR | APR | MAY | JUN | JUL | AUG | SEP | OCT | NOV | DEC | YEAR |
|---|---|---|---|---|---|---|---|---|---|---|---|---|---|
| Maximum Temp °F | 56.4 | 61.4 | 69.2 | 77.1 | 83.0 | 89.6 | 93.6 | 93.4 | 87.8 | 78.9 | 68.4 | 60.4 | 76.6 |
| Minimum Temp °F | 33.5 | 36.0 | 43.0 | 51.6 | 59.5 | 66.7 | 70.0 | 68.5 | 62.8 | 50.9 | 42.1 | 35.7 | 51.7 |
| Mean Temp °F | 45.0 | 48.7 | 56.1 | 64.4 | 71.3 | 78.2 | 81.8 | 81.0 | 75.2 | 65.0 | 55.3 | 48.1 | 64.2 |
| Days Max Temp ≥ 90 °F | 0 | 0 | 0 | 0 | 4 | 17 | 26 | 25 | 14 | 2 | 0 | 0 | 88 |
| Days Max Temp ≤ 32 °F | 1 | 0 | 0 | 0 | 0 | 0 | 0 | 0 | 0 | 0 | 0 | 0 | 1 |
| Days Min Temp ≤ 32 °F | 16 | 12 | 5 | 1 | 0 | 0 | 0 | 0 | 0 | 1 | 6 | 14 | 55 |
| Days Min Temp ≤ 0 °F | 0 | 0 | 0 | 0 | 0 | 0 | 0 | 0 | 0 | 0 | 0 | 0 | 0 |
| Heating Degree Days | 619 | 459 | 292 | 98 | 14 | 0 | 0 | 0 | 10 | 99 | 310 | 525 | 2426 |
| Cooling Degree Days | 2 | 6 | 28 | 83 | 223 | 419 | 544 | 529 | 344 | 113 | 29 | 7 | 2327 |
| Total Precipitation (") | 4.59 | 4.46 | 3.94 | 4.19 | 5.64 | 4.69 | 3.15 | 3.74 | 4.28 | 4.51 | 4.21 | 4.84 | 52.24 |
| Days ≥ 0.1" Precip | 7 | 7 | 6 | 5 | 7 | 6 | 5 | 5 | 6 | 5 | 6 | 7 | 72 |
| Total Snowfall (") | 0.8 | 0.2 | 0.1 | 0.0 | 0.0 | 0.0 | 0.0 | 0.0 | 0.0 | 0.0 | 0.1 | 0.0 | 1.2 |
| Days ≥ 1" Snow Depth | 1 | 0 | 0 | 0 | 0 | 0 | 0 | 0 | 0 | 0 | 0 | 0 | 1 |

## CENTERVILLE *Leon County* ELEVATION 351 ft LAT/LONG 31° 15 ' N / 95° 58 ' W

| | JAN | FEB | MAR | APR | MAY | JUN | JUL | AUG | SEP | OCT | NOV | DEC | YEAR |
|---|---|---|---|---|---|---|---|---|---|---|---|---|---|
| Maximum Temp °F | 57.0 | 61.5 | 69.9 | 77.6 | 83.1 | 90.0 | 94.2 | 94.7 | 88.5 | 79.8 | 68.9 | 61.0 | 77.2 |
| Minimum Temp °F | 34.6 | 37.5 | 45.4 | 53.9 | 61.1 | 68.2 | 70.8 | 69.6 | 64.3 | 53.0 | 44.0 | 37.2 | 53.3 |
| Mean Temp °F | 45.8 | 49.5 | 57.7 | 65.8 | 72.1 | 79.1 | 82.5 | 82.2 | 76.4 | 66.4 | 56.5 | 49.1 | 65.3 |
| Days Max Temp ≥ 90 °F | 0 | 0 | 0 | 0 | 4 | 18 | 27 | 27 | 16 | 3 | 0 | 0 | 95 |
| Days Max Temp ≤ 32 °F | 1 | 0 | 0 | 0 | 0 | 0 | 0 | 0 | 0 | 0 | 0 | 0 | 1 |
| Days Min Temp ≤ 32 °F | 14 | 9 | 3 | 1 | 0 | 0 | 0 | 0 | 0 | 0 | 5 | 11 | 43 |
| Days Min Temp ≤ 0 °F | 0 | 0 | 0 | 0 | 0 | 0 | 0 | 0 | 0 | 0 | 0 | 0 | 0 |
| Heating Degree Days | 593 | 436 | 253 | 78 | 12 | 0 | 0 | 0 | 7 | 78 | 281 | 493 | 2231 |
| Cooling Degree Days | 2 | 5 | 32 | 100 | 237 | 433 | 548 | 553 | 364 | 131 | 36 | 9 | 2450 |
| Total Precipitation (") | 3.27 | 3.26 | 3.32 | 3.92 | 5.05 | 3.88 | 2.37 | 2.43 | 3.75 | 4.50 | 3.27 | 3.56 | 42.58 |
| Days ≥ 0.1" Precip | 6 | 5 | 5 | 5 | 6 | 6 | 4 | 4 | 5 | 5 | 5 | 6 | 62 |
| Total Snowfall (") | 0.4 | 0.3 | 0.0 | 0.0 | 0.0 | 0.0 | 0.0 | 0.0 | 0.0 | 0.0 | 0.0 | 0.0 | 0.7 |
| Days ≥ 1" Snow Depth | 0 | 0 | 0 | 0 | 0 | 0 | 0 | 0 | 0 | 0 | 0 | 0 | 0 |

## CHAPMAN RANCH *Nueces County* ELEVATION 30 ft LAT/LONG 27° 35 ' N / 97° 27 ' W

| | JAN | FEB | MAR | APR | MAY | JUN | JUL | AUG | SEP | OCT | NOV | DEC | YEAR |
|---|---|---|---|---|---|---|---|---|---|---|---|---|---|
| Maximum Temp °F | 66.1 | 69.1 | 75.9 | 81.0 | 84.7 | 89.3 | 91.9 | 92.1 | 89.4 | 83.5 | 75.6 | 69.2 | 80.7 |
| Minimum Temp °F | 46.7 | 49.6 | 56.1 | 63.8 | 69.4 | 73.5 | 74.7 | 74.3 | 71.7 | 64.2 | 56.1 | 49.3 | 62.5 |
| Mean Temp °F | 56.4 | 59.3 | 66.0 | 72.4 | 77.1 | 81.5 | 83.3 | 83.3 | 80.6 | 73.9 | 65.9 | 59.3 | 71.6 |
| Days Max Temp ≥ 90 °F | 0 | 0 | 1 | 1 | 2 | 15 | 27 | 26 | 16 | 3 | 0 | 0 | 91 |
| Days Max Temp ≤ 32 °F | 0 | 0 | 0 | 0 | 0 | 0 | 0 | 0 | 0 | 0 | 0 | 0 | 0 |
| Days Min Temp ≤ 32 °F | 3 | 1 | 1 | 0 | 0 | 0 | 0 | 0 | 0 | 0 | 0 | 2 | 7 |
| Days Min Temp ≤ 0 °F | 0 | 0 | 0 | 0 | 0 | 0 | 0 | 0 | 0 | 0 | 0 | 0 | 0 |
| Heating Degree Days | 287 | 190 | 80 | 13 | 0 | 0 | 0 | 0 | 0 | 10 | 92 | 219 | 891 |
| Cooling Degree Days | 21 | 42 | 111 | 233 | 389 | 513 | 580 | 580 | 483 | 301 | 132 | 47 | 3432 |
| Total Precipitation (") | 1.78 | 2.05 | 1.26 | 2.19 | 3.91 | 3.52 | 1.77 | 3.26 | 5.50 | 3.55 | 1.55 | 1.41 | 31.75 |
| Days ≥ 0.1" Precip | 4 | 3 | 2 | 3 | 4 | 4 | 2 | 4 | 6 | 4 | 3 | 3 | 42 |
| Total Snowfall (") | 0.0 | 0.0 | 0.0 | 0.0 | 0.0 | 0.0 | 0.0 | 0.0 | 0.0 | 0.0 | 0.0 | 0.0 | 0.0 |
| Days ≥ 1" Snow Depth | 0 | 0 | 0 | 0 | 0 | 0 | 0 | 0 | 0 | 0 | 0 | 0 | 0 |

## CHARLOTTE 5 NNW *Atascosa County* ELEVATION 440 ft LAT/LONG 28° 56 ' N / 98° 45 ' W

| | JAN | FEB | MAR | APR | MAY | JUN | JUL | AUG | SEP | OCT | NOV | DEC | YEAR |
|---|---|---|---|---|---|---|---|---|---|---|---|---|---|
| Maximum Temp °F | 65.8 | 70.3 | 78.6 | 84.9 | 88.6 | 93.8 | 96.7 | 96.9 | 92.2 | 85.1 | 75.9 | 68.6 | 83.1 |
| Minimum Temp °F | 41.0 | 43.9 | 51.3 | 59.3 | 66.0 | 70.9 | 72.3 | 71.9 | 68.3 | 59.6 | 50.6 | 43.5 | 58.2 |
| Mean Temp °F | 53.4 | 57.1 | 65.0 | 72.1 | 77.3 | 82.4 | 84.5 | 84.4 | 80.3 | 72.4 | 63.3 | 56.1 | 70.7 |
| Days Max Temp ≥ 90 °F | 0 | 0 | 2 | 8 | 13 | 25 | 29 | 29 | 22 | 9 | 1 | 0 | 138 |
| Days Max Temp ≤ 32 °F | 0 | 0 | 0 | 0 | 0 | 0 | 0 | 0 | 0 | 0 | 0 | 0 | 0 |
| Days Min Temp ≤ 32 °F | 7 | 4 | 1 | 0 | 0 | 0 | 0 | 0 | 0 | 0 | 2 | 5 | 19 |
| Days Min Temp ≤ 0 °F | 0 | 0 | 0 | 0 | 0 | 0 | 0 | 0 | 0 | 0 | 0 | 0 | 0 |
| Heating Degree Days | 368 | 243 | 103 | 20 | 1 | 0 | 0 | 0 | 1 | 21 | 137 | 295 | 1189 |
| Cooling Degree Days | 12 | 30 | 102 | 233 | 406 | 546 | 625 | 632 | 476 | 273 | 100 | 30 | 3465 |
| Total Precipitation (") | 1.53 | 1.73 | 1.15 | 2.47 | 4.05 | 3.01 | 1.46 | 2.56 | 3.62 | 3.04 | 1.64 | 1.53 | 27.79 |
| Days ≥ 0.1" Precip | 3 | 3 | 3 | 4 | 5 | 4 | 3 | 3 | 5 | 4 | 3 | 3 | 43 |
| Total Snowfall (") | 0.1 | 0.1 | 0.0 | 0.0 | 0.0 | 0.0 | 0.0 | 0.0 | 0.0 | 0.0 | 0.0 | 0.0 | 0.2 |
| Days ≥ 1" Snow Depth | 0 | 0 | 0 | 0 | 0 | 0 | 0 | 0 | 0 | 0 | 0 | 0 | 0 |

## CHILDRESS MUNI AP *Childress County* ELEVATION 1955 ft LAT/LONG 34° 26 ' N / 100° 17 ' W

| | JAN | FEB | MAR | APR | MAY | JUN | JUL | AUG | SEP | OCT | NOV | DEC | YEAR |
|---|---|---|---|---|---|---|---|---|---|---|---|---|---|
| Maximum Temp °F | 51.5 | 56.2 | 65.8 | 75.6 | 82.4 | 91.0 | 95.2 | 93.3 | 85.3 | 75.5 | 62.5 | 53.7 | 74.0 |
| Minimum Temp °F | 25.8 | 29.9 | 38.2 | 47.9 | 56.9 | 65.9 | 70.5 | 68.7 | 61.1 | 49.4 | 37.6 | 28.9 | 48.4 |
| Mean Temp °F | 38.7 | 43.1 | 52.0 | 61.8 | 69.7 | 78.5 | 82.9 | 81.0 | 73.3 | 62.5 | 50.1 | 41.3 | 61.2 |
| Days Max Temp ≥ 90 °F | 0 | 0 | 1 | 3 | 7 | 19 | 26 | 23 | 12 | 2 | 0 | 0 | 93 |
| Days Max Temp ≤ 32 °F | 4 | 2 | 0 | 0 | 0 | 0 | 0 | 0 | 0 | 0 | 0 | 2 | 8 |
| Days Min Temp ≤ 32 °F | 24 | 17 | 8 | 1 | 0 | 0 | 0 | 0 | 0 | 1 | 9 | 21 | 81 |
| Days Min Temp ≤ 0 °F | 0 | 0 | 0 | 0 | 0 | 0 | 0 | 0 | 0 | 0 | 0 | 0 | 0 |
| Heating Degree Days | 809 | 612 | 409 | 152 | 35 | 2 | 0 | 1 | 24 | 145 | 446 | 728 | 3363 |
| Cooling Degree Days | 0 | 1 | 13 | 63 | 189 | 402 | 553 | 531 | 295 | 71 | 7 | 0 | 2125 |
| Total Precipitation (") | 0.62 | 0.92 | 1.24 | 1.77 | 3.19 | 2.97 | 1.82 | 2.18 | 2.47 | 2.01 | 1.08 | 0.81 | 21.08 |
| Days ≥ 0.1" Precip | 2 | 3 | 3 | 4 | 5 | 5 | 3 | 4 | 4 | 3 | 2 | 2 | 40 |
| Total Snowfall (") | 2.3 | 2.2 | 1.1 | 0.2 | 0.0 | 0.0 | 0.0 | 0.0 | 0.0 | 0.1 | 1.1 | 2.1 | 9.1 |
| Days ≥ 1" Snow Depth | 2 | 2 | 0 | 0 | 0 | 0 | 0 | 0 | 0 | 0 | 0 | 2 | 6 |

## CHISOS BASIN *Brewster County* ELEVATION 5282 ft LAT/LONG 29° 15 ' N / 103° 15 ' W

| | JAN | FEB | MAR | APR | MAY | JUN | JUL | AUG | SEP | OCT | NOV | DEC | YEAR |
|---|---|---|---|---|---|---|---|---|---|---|---|---|---|
| Maximum Temp °F | 57.4 | 60.6 | 67.9 | 75.5 | 81.5 | 85.9 | 84.3 | 82.7 | 78.3 | 73.0 | 65.0 | 59.5 | 72.6 |
| Minimum Temp °F | 36.0 | 38.2 | 43.8 | 51.3 | 57.5 | 62.8 | 63.6 | 62.4 | 58.3 | 51.4 | 43.3 | 38.3 | 50.6 |
| Mean Temp °F | 46.8 | 49.4 | 55.9 | 63.4 | 69.5 | 74.4 | 74.0 | 72.5 | 68.3 | 62.2 | 54.2 | 48.9 | 61.6 |
| Days Max Temp ≥ 90 °F | 0 | 0 | 0 | 0 | 3 | 8 | 4 | 2 | 1 | 0 | 0 | 0 | 18 |
| Days Max Temp ≤ 32 °F | 1 | 0 | 0 | 0 | 0 | 0 | 0 | 0 | 0 | 0 | 0 | 0 | 1 |
| Days Min Temp ≤ 32 °F | 10 | 6 | 3 | 1 | 0 | 0 | 0 | 0 | 0 | 0 | 4 | 8 | 32 |
| Days Min Temp ≤ 0 °F | 0 | 0 | 0 | 0 | 0 | 0 | 0 | 0 | 0 | 0 | 0 | 0 | 0 |
| Heating Degree Days | 559 | 434 | 287 | 108 | 24 | 3 | 0 | 1 | 30 | 121 | 324 | 490 | 2381 |
| Cooling Degree Days | 0 | 1 | 10 | 54 | 162 | 290 | 289 | 250 | 138 | 36 | 5 | 0 | 1235 |
| Total Precipitation (") | 0.65 | 0.73 | 0.42 | 0.61 | 1.57 | 2.23 | 3.49 | 3.80 | 3.32 | 1.84 | 0.53 | 0.62 | 19.81 |
| Days ≥ 0.1" Precip | 2 | 2 | 1 | 1 | 3 | 4 | 6 | 6 | 5 | 3 | 1 | 1 | 35 |
| Total Snowfall (") | 1.0 | 0.5 | 0.1 | 0.1 | 0.0 | 0.0 | 0.0 | 0.0 | 0.0 | 0.0 | 0.4 | 0.3 | 2.4 |
| Days ≥ 1" Snow Depth | 0 | 0 | 0 | 0 | 0 | 0 | 0 | 0 | 0 | 0 | 0 | 0 | 0 |

## CLARENDON *Donley County* ELEVATION 2871 ft LAT/LONG 34° 57 ' N / 100° 56 ' W

| | JAN | FEB | MAR | APR | MAY | JUN | JUL | AUG | SEP | OCT | NOV | DEC | YEAR |
|---|---|---|---|---|---|---|---|---|---|---|---|---|---|
| Maximum Temp °F | 50.3 | 54.8 | 63.6 | 73.5 | 80.4 | 88.8 | 94.1 | 91.9 | 84.0 | 74.2 | 61.4 | 52.7 | 72.5 |
| Minimum Temp °F | 22.0 | 25.7 | 33.6 | 43.1 | 52.4 | 61.7 | 66.1 | 64.2 | 56.6 | 44.3 | 33.0 | 24.7 | 44.0 |
| Mean Temp °F | 36.2 | 40.3 | 48.6 | 58.3 | 66.4 | 75.3 | 80.1 | 78.1 | 70.3 | 59.3 | 47.2 | 38.7 | 58.2 |
| Days Max Temp ≥ 90 °F | 0 | 0 | 0 | 1 | 5 | 15 | 24 | 21 | 9 | 2 | 0 | 0 | 77 |
| Days Max Temp ≤ 32 °F | 4 | 3 | 1 | 0 | 0 | 0 | 0 | 0 | 0 | 0 | 1 | 3 | 12 |
| Days Min Temp ≤ 32 °F | 28 | 22 | 14 | 3 | 0 | 0 | 0 | 0 | 0 | 2 | 15 | 26 | 110 |
| Days Min Temp ≤ 0 °F | 0 | 0 | 0 | 0 | 0 | 0 | 0 | 0 | 0 | 0 | 0 | 0 | 0 |
| Heating Degree Days | 886 | 692 | 505 | 227 | 70 | 5 | 0 | 1 | 40 | 211 | 529 | 808 | 3974 |
| Cooling Degree Days | 0 | 0 | 4 | 31 | 120 | 320 | 474 | 433 | 221 | 38 | 3 | 0 | 1644 |
| Total Precipitation (") | 0.64 | 0.77 | 1.25 | 1.59 | 3.49 | 3.57 | 2.14 | 2.91 | 2.66 | 1.62 | 0.86 | 0.72 | 22.22 |
| Days ≥ 0.1" Precip | 2 | 2 | 3 | 3 | 5 | 5 | 4 | 5 | 4 | 3 | 2 | 2 | 40 |
| Total Snowfall (") | 2.6 | *2.2* | 1.1 | 0.3 | 0.0 | 0.0 | 0.0 | 0.0 | 0.0 | 0.1 | 1.1 | 2.4 | 9.8 |
| Days ≥ 1" Snow Depth | 1 | 3 | 1 | 0 | 0 | 0 | 0 | 0 | 0 | 0 | 0 | 1 | 6 |

## CLARKSVILLE 2 NE *Red River County* ELEVATION 440 ft LAT/LONG 33° 36 ' N / 95° 2 ' W

| | JAN | FEB | MAR | APR | MAY | JUN | JUL | AUG | SEP | OCT | NOV | DEC | YEAR |
|---|---|---|---|---|---|---|---|---|---|---|---|---|---|
| Maximum Temp °F | 52.0 | 56.8 | 65.7 | 74.5 | 80.2 | 87.8 | 92.3 | 92.2 | 85.6 | 76.2 | 64.9 | 55.6 | 73.7 |
| Minimum Temp °F | 29.3 | 33.1 | 41.2 | 50.4 | 58.7 | 66.7 | 70.0 | 68.9 | 62.3 | 50.3 | 40.5 | 32.6 | 50.3 |
| Mean Temp °F | 40.7 | 44.9 | 53.5 | 62.4 | 69.5 | 77.3 | 81.2 | 80.6 | 74.0 | 63.3 | 52.7 | 44.1 | 62.0 |
| Days Max Temp ≥ 90 °F | 0 | 0 | 0 | 0 | 1 | 12 | 23 | 23 | 10 | 1 | 0 | 0 | 70 |
| Days Max Temp ≤ 32 °F | 2 | 1 | 0 | 0 | 0 | 0 | 0 | 0 | 0 | 0 | 0 | 1 | 4 |
| Days Min Temp ≤ 32 °F | 21 | 15 | 7 | 1 | 0 | 0 | 0 | 0 | 0 | 1 | 7 | 15 | 67 |
| Days Min Temp ≤ 0 °F | 0 | 0 | 0 | 0 | 0 | 0 | 0 | 0 | 0 | 0 | 0 | 0 | 0 |
| Heating Degree Days | 748 | 562 | 366 | 127 | 25 | 1 | 0 | 0 | 17 | 124 | 371 | 643 | 2984 |
| Cooling Degree Days | 0 | 1 | 13 | 50 | 168 | 384 | 506 | 509 | 300 | 72 | 10 | 1 | 2014 |
| Total Precipitation (") | 2.52 | 3.32 | 4.47 | 4.39 | 5.76 | 3.64 | 3.22 | 2.14 | 3.76 | 5.00 | 4.49 | 4.26 | 46.97 |
| Days ≥ 0.1" Precip | 5 | 5 | 6 | 6 | 7 | 5 | 4 | 3 | 5 | 5 | 5 | 6 | 62 |
| Total Snowfall (") | 0.0 | 0.1 | 0.1 | 0.0 | 0.0 | 0.0 | 0.0 | 0.0 | 0.0 | 0.0 | 0.0 | 0.0 | 0.2 |
| Days ≥ 1" Snow Depth | 0 | 0 | 0 | 0 | 0 | 0 | 0 | 0 | 0 | 0 | 0 | 0 | 0 |

## CLAUDE *Armstrong County* ELEVATION 3402 ft LAT/LONG 35° 7 ' N / 101° 22 ' W

| | JAN | FEB | MAR | APR | MAY | JUN | JUL | AUG | SEP | OCT | NOV | DEC | YEAR |
|---|---|---|---|---|---|---|---|---|---|---|---|---|---|
| Maximum Temp °F | 48.7 | 52.5 | 61.7 | 71.2 | 78.4 | 86.9 | 91.2 | 89.2 | 81.9 | 72.3 | 59.2 | 50.7 | 70.3 |
| Minimum Temp °F | 20.6 | 24.2 | 31.7 | 40.8 | 49.9 | 59.9 | 64.5 | 62.9 | 55.4 | 43.2 | 31.3 | 23.2 | 42.3 |
| Mean Temp °F | 34.7 | 38.2 | 46.9 | 56.0 | 64.2 | 73.4 | 77.9 | 76.1 | 68.8 | 57.8 | 45.3 | 37.0 | 56.4 |
| Days Max Temp ≥ 90 °F | 0 | 0 | 0 | 1 | 3 | 12 | 20 | 16 | 7 | 1 | 0 | 0 | 60 |
| Days Max Temp ≤ 32 °F | 5 | 3 | 1 | 0 | 0 | 0 | 0 | 0 | 0 | 0 | 1 | 3 | 13 |
| Days Min Temp ≤ 32 °F | 28 | 24 | 17 | 5 | 0 | 0 | 0 | 0 | 0 | 3 | 16 | 27 | 120 |
| Days Min Temp ≤ 0 °F | 1 | 0 | 0 | 0 | 0 | 0 | 0 | 0 | 0 | 0 | 0 | 1 | 2 |
| Heating Degree Days | 932 | 750 | 556 | 282 | 100 | 10 | 1 | 2 | 48 | 245 | 586 | 860 | 4372 |
| Cooling Degree Days | 0 | 0 | 1 | 19 | 80 | 273 | 405 | 363 | 176 | *21* | 1 | 0 | 1339 |
| Total Precipitation (") | 0.48 | 0.54 | 1.17 | 1.20 | 3.22 | 3.45 | 2.89 | 3.01 | 2.41 | 1.57 | 0.77 | 0.49 | 21.20 |
| Days ≥ 0.1" Precip | 1 | 2 | 2 | 3 | 5 | 5 | 4 | 5 | 4 | 3 | 2 | 2 | 38 |
| Total Snowfall (") | *2.9* | 3.3 | 1.6 | 0.4 | 0.0 | 0.0 | 0.0 | 0.0 | 0.0 | 0.1 | 1.0 | 2.4 | 11.7 |
| Days ≥ 1" Snow Depth | 3 | 2 | 0 | 0 | 0 | 0 | 0 | 0 | 0 | 0 | 1 | 2 | 8 |

## CLEBURNE *Johnson County* ELEVATION 761 ft LAT/LONG 32° 21 ' N / 97° 23 ' W

| | JAN | FEB | MAR | APR | MAY | JUN | JUL | AUG | SEP | OCT | NOV | DEC | YEAR |
|---|---|---|---|---|---|---|---|---|---|---|---|---|---|
| Maximum Temp °F | 56.9 | 62.0 | 70.7 | 78.7 | 84.4 | 92.1 | 96.8 | 96.8 | 89.2 | 79.9 | 68.0 | 59.9 | 78.0 |
| Minimum Temp °F | 33.7 | 37.3 | 45.4 | 54.0 | 61.6 | 68.7 | 71.7 | 71.0 | 65.2 | 54.6 | 44.5 | 36.5 | 53.7 |
| Mean Temp °F | 45.3 | 49.7 | 58.0 | 66.4 | 73.0 | 80.4 | 84.3 | 83.9 | 77.2 | 67.2 | 56.2 | 48.2 | 65.8 |
| Days Max Temp ≥ 90 °F | 0 | 0 | 0 | 1 | 7 | 21 | 28 | 27 | 17 | 4 | 0 | 0 | 105 |
| Days Max Temp ≤ 32 °F | 1 | 0 | 0 | 0 | 0 | 0 | 0 | 0 | 0 | 0 | 0 | 1 | 2 |
| Days Min Temp ≤ 32 °F | 15 | 9 | 3 | 0 | 0 | 0 | 0 | 0 | 0 | 0 | 4 | 11 | 42 |
| Days Min Temp ≤ 0 °F | 0 | 0 | 0 | 0 | 0 | 0 | 0 | 0 | 0 | 0 | 0 | 0 | 0 |
| Heating Degree Days | 606 | 431 | 244 | 66 | 9 | 0 | 0 | 0 | 7 | 69 | 285 | 519 | 2236 |
| Cooling Degree Days | 1 | 5 | 35 | 116 | 272 | 471 | 608 | 614 | 389 | 150 | 31 | 4 | 2696 |
| Total Precipitation (") | 1.90 | 2.40 | 2.93 | 3.69 | 5.73 | 3.34 | 2.05 | 2.26 | 3.19 | 3.64 | 2.17 | 2.30 | 35.60 |
| Days ≥ 0.1" Precip | 4 | 4 | 5 | 5 | 6 | 5 | 3 | 3 | 4 | 5 | 4 | 4 | 52 |
| Total Snowfall (") | 0.7 | 0.7 | 0.2 | 0.0 | 0.0 | 0.0 | 0.0 | 0.0 | 0.0 | 0.0 | 0.2 | 0.2 | 2.0 |
| Days ≥ 1" Snow Depth | 0 | 0 | 0 | 0 | 0 | 0 | 0 | 0 | 0 | 0 | 0 | 0 | 0 |

## CLEVELAND *Liberty County* ELEVATION 161 ft LAT/LONG 30° 20 ' N / 95° 5 ' W

| | JAN | FEB | MAR | APR | MAY | JUN | JUL | AUG | SEP | OCT | NOV | DEC | YEAR |
|---|---|---|---|---|---|---|---|---|---|---|---|---|---|
| Maximum Temp °F | 59.0 | 63.6 | 71.3 | 78.0 | 83.3 | 89.1 | 92.3 | 92.6 | 88.0 | 79.8 | 69.5 | 62.3 | 77.4 |
| Minimum Temp °F | 37.5 | 39.6 | 47.2 | 55.6 | 62.4 | 68.6 | 70.9 | 70.2 | 65.6 | 55.1 | 46.4 | 39.8 | 54.9 |
| Mean Temp °F | 48.3 | 51.6 | 59.3 | 66.8 | 72.9 | 78.9 | 81.7 | 81.4 | 76.9 | 67.5 | 58.0 | 51.1 | 66.2 |
| Days Max Temp ≥ 90 °F | 0 | 0 | 0 | 0 | 3 | 16 | 25 | 25 | 14 | 2 | 0 | 0 | 85 |
| Days Max Temp ≤ 32 °F | 0 | 0 | 0 | 0 | 0 | 0 | 0 | 0 | 0 | 0 | 0 | 0 | 0 |
| Days Min Temp ≤ 32 °F | 11 | 8 | 3 | 0 | 0 | 0 | 0 | 0 | 0 | 0 | 4 | 9 | 35 |
| Days Min Temp ≤ 0 °F | 0 | 0 | 0 | 0 | 0 | 0 | 0 | 0 | 0 | 0 | 0 | 0 | 0 |
| Heating Degree Days | 520 | 382 | 212 | 60 | 6 | 0 | 0 | 0 | 3 | 64 | 246 | 437 | 1930 |
| Cooling Degree Days | 4 | 11 | 42 | 107 | 260 | 434 | 525 | 531 | 373 | 158 | 45 | 13 | 2503 |
| Total Precipitation (") | 4.57 | 3.87 | 4.04 | 3.70 | 6.36 | 5.13 | 3.70 | 3.20 | 4.32 | 5.17 | 4.29 | 4.65 | 53.00 |
| Days ≥ 0.1" Precip | 7 | 6 | 6 | 5 | 6 | 7 | 6 | 5 | 7 | 5 | 6 | 7 | 73 |
| Total Snowfall (") | 0.4 | 0.2 | 0.0 | 0.0 | 0.0 | 0.0 | 0.0 | 0.0 | 0.0 | 0.0 | 0.0 | 0.0 | 0.6 |
| Days ≥ 1" Snow Depth | 0 | 0 | 0 | 0 | 0 | 0 | 0 | 0 | 0 | 0 | 0 | 0 | 0 |

## COLDSPRING 5 SSW *San Jacinto County* ELEVATION 351 ft LAT/LONG 30° 32 ' N / 95° 9 ' W

| | JAN | FEB | MAR | APR | MAY | JUN | JUL | AUG | SEP | OCT | NOV | DEC | YEAR |
|---|---|---|---|---|---|---|---|---|---|---|---|---|---|
| Maximum Temp °F | 58.6 | 62.8 | 70.6 | 77.8 | 83.4 | 89.9 | 93.1 | 93.3 | 88.3 | 80.6 | 70.2 | 62.5 | 77.6 |
| Minimum Temp °F | 36.7 | 38.8 | 46.2 | 54.5 | 61.6 | 67.9 | 70.4 | 69.3 | 65.0 | 54.0 | 45.9 | 39.2 | 54.1 |
| Mean Temp °F | 47.7 | 50.8 | 58.4 | 66.2 | 72.5 | 78.9 | 81.8 | 81.3 | 76.7 | 67.3 | 58.0 | 50.8 | 65.9 |
| Days Max Temp ≥ 90 °F | 0 | 0 | 0 | 0 | 3 | 17 | 26 | 25 | 14 | 3 | 0 | 0 | 88 |
| Days Max Temp ≤ 32 °F | 0 | 0 | 0 | 0 | 0 | 0 | 0 | 0 | 0 | 0 | 0 | 0 | 0 |
| Days Min Temp ≤ 32 °F | 12 | 9 | 4 | 0 | 0 | 0 | 0 | 0 | 0 | 0 | 4 | 10 | 39 |
| Days Min Temp ≤ 0 °F | 0 | 0 | 0 | 0 | 0 | 0 | 0 | 0 | 0 | 0 | 0 | 0 | 0 |
| Heating Degree Days | 539 | 403 | 235 | 71 | 9 | 0 | 0 | 0 | 4 | 64 | 247 | 444 | 2016 |
| Cooling Degree Days | 4 | 10 | 39 | 108 | 255 | *443* | 533 | 536 | 387 | 156 | 50 | 13 | 2534 |
| Total Precipitation (") | 4.12 | 3.46 | 3.71 | 3.61 | 5.92 | 5.62 | 2.77 | 3.33 | 4.16 | 4.48 | 4.00 | 4.82 | 50.00 |
| Days ≥ 0.1" Precip | 7 | 5 | 6 | 4 | 7 | 7 | 5 | 6 | 6 | 5 | 5 | 7 | 70 |
| Total Snowfall (") | 0.1 | 0.1 | 0.0 | 0.0 | 0.0 | 0.0 | 0.0 | 0.0 | 0.0 | 0.0 | 0.0 | 0.0 | 0.2 |
| Days ≥ 1" Snow Depth | 0 | 0 | 0 | 0 | 0 | 0 | 0 | 0 | 0 | 0 | 0 | 0 | 0 |

## COLEMAN *Coleman County* ELEVATION 1713 ft LAT/LONG 31° 50 ' N / 99° 26 ' W

| | JAN | FEB | MAR | APR | MAY | JUN | JUL | AUG | SEP | OCT | NOV | DEC | YEAR |
|---|---|---|---|---|---|---|---|---|---|---|---|---|---|
| Maximum Temp °F | 58.0 | 62.8 | 71.8 | 79.8 | 85.0 | 91.5 | 95.3 | 94.6 | 87.9 | 79.6 | 68.0 | 60.4 | 77.9 |
| Minimum Temp °F | 32.7 | 36.2 | 43.9 | 52.5 | 59.9 | 66.6 | 69.6 | 68.7 | 62.7 | 52.9 | 42.7 | 35.3 | 52.0 |
| Mean Temp °F | 45.3 | 49.5 | 57.9 | 66.2 | 72.5 | 79.1 | 82.5 | 81.6 | 75.3 | 66.3 | 55.4 | 47.9 | 65.0 |
| Days Max Temp ≥ 90 °F | 0 | 0 | 1 | 3 | 8 | 20 | 27 | 26 | 14 | 3 | 0 | 0 | 102 |
| Days Max Temp ≤ 32 °F | 1 | 1 | 0 | 0 | 0 | 0 | 0 | 0 | 0 | 0 | 0 | 1 | 3 |
| Days Min Temp ≤ 32 °F | 16 | 10 | 4 | 0 | 0 | 0 | 0 | 0 | 0 | 0 | 5 | 12 | 47 |
| Days Min Temp ≤ 0 °F | 0 | 0 | 0 | 0 | 0 | 0 | 0 | 0 | 0 | 0 | 0 | 0 | 0 |
| Heating Degree Days | 606 | 435 | 249 | 71 | 12 | 0 | 0 | 0 | 9 | 73 | 300 | 525 | 2280 |
| Cooling Degree Days | 1 | 4 | 32 | 112 | 255 | 421 | 543 | 543 | 330 | 127 | 20 | 1 | 2389 |
| Total Precipitation (") | 1.28 | 1.77 | 1.70 | 2.21 | 4.29 | 3.25 | 1.81 | 2.56 | 3.89 | 2.94 | 1.47 | 1.59 | 28.76 |
| Days ≥ 0.1" Precip | 3 | 3 | 3 | 4 | 6 | 4 | 3 | 4 | 5 | 4 | 3 | 3 | 45 |
| Total Snowfall (") | 1.0 | 0.6 | 0.3 | 0.0 | 0.0 | 0.0 | 0.0 | 0.0 | 0.0 | 0.0 | 0.5 | 0.3 | 2.7 |
| Days ≥ 1" Snow Depth | 1 | *0* | 0 | 0 | 0 | 0 | 0 | 0 | 0 | 0 | 0 | 0 | 1 |

## COLLEGE STN EASTERWD *Brazos County* ELEVATION 318 ft LAT/LONG 30° 35 ' N / 96° 22 ' W

| | JAN | FEB | MAR | APR | MAY | JUN | JUL | AUG | SEP | OCT | NOV | DEC | YEAR |
|---|---|---|---|---|---|---|---|---|---|---|---|---|---|
| Maximum Temp °F | 58.6 | 63.0 | 71.0 | 78.1 | 83.7 | 90.4 | 94.0 | 94.5 | 88.7 | 80.2 | 70.0 | 62.0 | 77.9 |
| Minimum Temp °F | 39.4 | 42.1 | 49.8 | 57.8 | 64.8 | 71.1 | 73.4 | 73.0 | 68.3 | 58.5 | 49.0 | 41.8 | 57.4 |
| Mean Temp °F | 49.0 | 52.6 | 60.4 | 68.0 | 74.3 | 80.8 | 83.7 | 83.8 | 78.5 | 69.4 | 59.5 | 51.9 | 67.7 |
| Days Max Temp ≥ 90 °F | 0 | 0 | 0 | 0 | 4 | 20 | 27 | 27 | 16 | 3 | 0 | 0 | 97 |
| Days Max Temp ≤ 32 °F | 0 | 0 | 0 | 0 | 0 | 0 | 0 | 0 | 0 | 0 | 0 | 0 | 0 |
| Days Min Temp ≤ 32 °F | 8 | 4 | 1 | 0 | 0 | 0 | 0 | 0 | 0 | 0 | 2 | 6 | 21 |
| Days Min Temp ≤ 0 °F | 0 | 0 | 0 | 0 | 0 | 0 | 0 | 0 | 0 | 0 | 0 | 0 | 0 |
| Heating Degree Days | 498 | 356 | 188 | 48 | 4 | 0 | 0 | 0 | 2 | 43 | 214 | 413 | 1766 |
| Cooling Degree Days | 6 | 14 | 54 | 142 | 311 | 500 | 599 | 615 | 440 | 205 | 62 | 17 | 2965 |
| Total Precipitation (") | 3.25 | 2.84 | 2.83 | 3.38 | 5.38 | 3.84 | 2.07 | 2.47 | 4.38 | 4.06 | 3.02 | 3.22 | 40.74 |
| Days ≥ 0.1" Precip | 5 | 5 | 5 | 4 | 6 | 5 | 3 | 4 | 5 | 5 | 5 | 5 | 57 |
| Total Snowfall (") | 0.3 | 0.1 | 0.0 | 0.0 | 0.0 | 0.0 | 0.0 | 0.0 | 0.0 | 0.0 | 0.0 | 0.0 | 0.4 |
| Days ≥ 1" Snow Depth | 0 | 0 | 0 | 0 | 0 | 0 | 0 | 0 | 0 | 0 | 0 | 0 | 0 |

## COLUMBUS *Colorado County* ELEVATION 210 ft LAT/LONG 29° 42 ' N / 96° 32 ' W

| | JAN | FEB | MAR | APR | MAY | JUN | JUL | AUG | SEP | OCT | NOV | DEC | YEAR |
|---|---|---|---|---|---|---|---|---|---|---|---|---|---|
| Maximum Temp °F | 61.7 | 65.9 | 74.0 | 80.3 | 85.1 | 91.5 | 95.0 | 96.2 | 91.4 | 84.0 | 73.5 | 65.5 | 80.3 |
| Minimum Temp °F | 37.0 | 39.9 | 47.8 | 56.4 | 63.0 | 68.8 | 70.3 | 69.5 | 64.5 | 54.4 | 46.2 | 39.4 | 54.8 |
| Mean Temp °F | 49.4 | 52.9 | 60.9 | 68.4 | 74.1 | 80.2 | 82.7 | 82.9 | 77.9 | 69.3 | 59.9 | 52.4 | 67.6 |
| Days Max Temp ≥ 90 °F | 0 | 0 | 0 | 1 | 6 | 22 | 29 | 29 | 21 | 8 | 1 | 0 | 117 |
| Days Max Temp ≤ 32 °F | 0 | 0 | 0 | 0 | 0 | 0 | 0 | 0 | 0 | 0 | 0 | 0 | 0 |
| Days Min Temp ≤ 32 °F | 11 | 7 | 2 | 0 | 0 | 0 | 0 | 0 | 0 | 0 | 4 | 9 | 33 |
| Days Min Temp ≤ 0 °F | 0 | 0 | 0 | 0 | 0 | 0 | 0 | 0 | 0 | 0 | 0 | 0 | 0 |
| Heating Degree Days | 487 | 347 | 178 | 46 | 5 | 0 | 0 | 0 | 3 | 44 | 209 | 398 | 1717 |
| Cooling Degree Days | 7 | 14 | 57 | 147 | 296 | 469 | 547 | 565 | 408 | 191 | 66 | 19 | 2786 |
| Total Precipitation (") | 3.69 | 2.83 | 2.78 | 3.56 | 6.17 | 4.40 | 2.61 | 2.93 | 4.47 | 3.67 | 3.50 | 3.25 | 43.86 |
| Days ≥ 0.1" Precip | 6 | 5 | 5 | 4 | 7 | 5 | 5 | 5 | 6 | 5 | 5 | 6 | 64 |
| Total Snowfall (") | 0.0 | 0.0 | 0.0 | 0.0 | 0.0 | 0.0 | 0.0 | 0.0 | 0.0 | 0.0 | 0.0 | 0.0 | 0.0 |
| Days ≥ 1" Snow Depth | 0 | 0 | 0 | 0 | 0 | 0 | 0 | 0 | 0 | 0 | 0 | 0 | 0 |

## CONROE *Montgomery County* ELEVATION 245 ft LAT/LONG 30° 20 ' N / 95° 29 ' W

| | JAN | FEB | MAR | APR | MAY | JUN | JUL | AUG | SEP | OCT | NOV | DEC | YEAR |
|---|---|---|---|---|---|---|---|---|---|---|---|---|---|
| Maximum Temp °F | 59.8 | 63.7 | 72.0 | 79.0 | 84.3 | 90.5 | 94.0 | 94.1 | 89.0 | 81.0 | 71.3 | 63.0 | 78.5 |
| Minimum Temp °F | 38.2 | 40.6 | 48.4 | 56.6 | 63.3 | 69.8 | 72.1 | 71.5 | 66.7 | 56.2 | 47.6 | 40.5 | 56.0 |
| Mean Temp °F | 49.0 | 52.2 | 60.3 | 67.8 | 73.8 | 80.2 | 83.1 | 82.8 | 77.9 | 68.6 | 59.4 | 51.8 | 67.2 |
| Days Max Temp ≥ 90 °F | 0 | 0 | 0 | 1 | 6 | 20 | 27 | 27 | 17 | 4 | 0 | 0 | 102 |
| Days Max Temp ≤ 32 °F | 0 | 0 | 0 | 0 | 0 | 0 | 0 | 0 | 0 | 0 | 0 | 0 | 0 |
| Days Min Temp ≤ 32 °F | 10 | 6 | 2 | 0 | 0 | 0 | 0 | 0 | 0 | 0 | 2 | 7 | 27 |
| Days Min Temp ≤ 0 °F | 0 | 0 | 0 | 0 | 0 | 0 | 0 | 0 | 0 | 0 | 0 | 0 | 0 |
| Heating Degree Days | 498 | 366 | 188 | 49 | 4 | 0 | 0 | 0 | 2 | 50 | 212 | 415 | 1784 |
| Cooling Degree Days | 3 | 9 | 52 | 131 | 289 | 478 | 581 | 576 | 404 | 173 | 50 | 11 | 2757 |
| Total Precipitation (") | 4.10 | 3.18 | 3.08 | 3.96 | 5.96 | 4.50 | 3.16 | 3.65 | 4.47 | 4.78 | 3.99 | 4.13 | 48.96 |
| Days ≥ 0.1" Precip | 7 | 5 | 5 | 4 | 7 | 6 | 5 | 6 | 6 | 5 | 5 | 6 | 67 |
| Total Snowfall (") | 0.1 | 0.0 | 0.0 | 0.0 | 0.0 | 0.0 | 0.0 | 0.0 | 0.0 | 0.0 | 0.0 | 0.0 | 0.1 |
| Days ≥ 1" Snow Depth | 0 | 0 | 0 | 0 | 0 | 0 | 0 | 0 | 0 | 0 | 0 | 0 | 0 |

## COPE RANCH *Reagan County* ELEVATION 2570 ft LAT/LONG 31° 32 ' N / 101° 17 ' W

| | JAN | FEB | MAR | APR | MAY | JUN | JUL | AUG | SEP | OCT | NOV | DEC | YEAR |
|---|---|---|---|---|---|---|---|---|---|---|---|---|---|
| Maximum Temp °F | 56.5 | 61.2 | 69.9 | 78.9 | 85.4 | 91.5 | 94.3 | 93.4 | 86.2 | 77.6 | 66.5 | 58.8 | 76.7 |
| Minimum Temp °F | 27.0 | 30.5 | 38.1 | 47.7 | 56.4 | 64.6 | 66.9 | 65.3 | 59.2 | 48.0 | 36.6 | 28.8 | 47.4 |
| Mean Temp °F | 41.8 | 45.9 | 54.0 | 63.3 | 70.9 | 78.1 | 80.6 | 79.4 | 72.7 | 62.8 | 51.5 | 43.8 | 62.1 |
| Days Max Temp ≥ 90 °F | 0 | 0 | 1 | 4 | 10 | 20 | 26 | 24 | 12 | 2 | 0 | 0 | 99 |
| Days Max Temp ≤ 32 °F | 2 | 1 | 0 | 0 | 0 | 0 | 0 | 0 | 0 | 0 | 0 | 1 | 4 |
| Days Min Temp ≤ 32 °F | 23 | 17 | 9 | 2 | 0 | 0 | 0 | 0 | 0 | 1 | 11 | 21 | 84 |
| Days Min Temp ≤ 0 °F | 0 | 0 | 0 | 0 | 0 | 0 | 0 | 0 | 0 | 0 | 0 | 0 | 0 |
| Heating Degree Days | 714 | 536 | 347 | 124 | 28 | 1 | 0 | 0 | 23 | 135 | 405 | 650 | 2963 |
| Cooling Degree Days | 0 | 1 | 12 | 80 | 227 | 404 | 496 | 477 | 268 | 74 | 7 | 0 | 2046 |
| Total Precipitation (") | 0.67 | 0.85 | 0.81 | 1.23 | 2.75 | 2.52 | 1.94 | 1.84 | 3.38 | 2.05 | 0.84 | 0.91 | 19.79 |
| Days ≥ 0.1" Precip | 2 | 2 | 2 | 3 | 4 | 3 | 3 | 3 | 4 | 3 | 2 | 2 | 33 |
| Total Snowfall (") | 0.5 | 0.2 | 0.3 | 0.0 | 0.0 | 0.0 | 0.0 | 0.0 | 0.0 | 0.0 | 0.0 | 0.1 | 1.1 |
| Days ≥ 1" Snow Depth | 0 | 0 | 0 | 0 | 0 | 0 | 0 | 0 | 0 | 0 | 0 | 0 | 0 |

## CORNUDAS SERVICE STN *Hudspeth County* ELEVATION 4482 ft LAT/LONG 31° 47 ' N / 105° 28 ' W

| | JAN | FEB | MAR | APR | MAY | JUN | JUL | AUG | SEP | OCT | NOV | DEC | YEAR |
|---|---|---|---|---|---|---|---|---|---|---|---|---|---|
| Maximum Temp °F | 58.1 | 63.3 | 70.9 | 79.4 | 87.1 | 95.6 | 94.4 | 91.8 | 86.9 | 78.7 | 66.7 | 58.9 | 77.7 |
| Minimum Temp °F | 25.6 | 28.2 | 33.1 | 40.5 | 49.1 | 58.6 | 63.3 | 61.4 | 54.6 | 43.2 | 31.9 | 26.0 | 43.0 |
| Mean Temp °F | 41.9 | 45.8 | 52.0 | 59.9 | 68.1 | 77.1 | 78.9 | 76.6 | 70.8 | 61.0 | 49.3 | 42.5 | 60.3 |
| Days Max Temp ≥ 90 °F | 0 | 0 | 0 | 2 | 12 | 25 | 25 | *21* | 12 | 2 | 0 | 0 | 99 |
| Days Max Temp ≤ 32 °F | 0 | 0 | 0 | 0 | 0 | 0 | 0 | 0 | 0 | 0 | 0 | 0 | 0 |
| Days Min Temp ≤ 32 °F | 25 | 20 | 14 | 5 | 0 | 0 | 0 | 0 | 0 | 2 | 16 | 25 | 107 |
| Days Min Temp ≤ 0 °F | 0 | 0 | 0 | 0 | 0 | 0 | 0 | 0 | 0 | 0 | 0 | 0 | 0 |
| Heating Degree Days | 710 | 535 | 397 | 166 | 32 | 0 | 0 | 0 | 17 | 148 | 463 | 691 | 3159 |
| Cooling Degree Days | *0* | 0 | 1 | *27* | 134 | *389* | 433 | *374* | *197* | 24 | 0 | 0 | 1579 |
| Total Precipitation (") | 0.60 | 0.39 | 0.19 | 0.28 | 0.68 | 1.08 | 1.71 | 2.20 | 1.94 | 0.80 | 0.38 | 0.54 | 10.79 |
| Days ≥ 0.1" Precip | 2 | 1 | 1 | 1 | 2 | 2 | 4 | 5 | 4 | 2 | 1 | 2 | 27 |
| Total Snowfall (") | 0.9 | 0.5 | 0.1 | 0.0 | 0.0 | 0.0 | 0.0 | 0.0 | 0.0 | 0.2 | 0.7 | 0.4 | 2.8 |
| Days ≥ 1" Snow Depth | 0 | 0 | 0 | 0 | 0 | 0 | 0 | 0 | 0 | 0 | 0 | 0 | 0 |

## CORPUS CHRISTI INTL *Nueces County* ELEVATION 41 ft LAT/LONG 27° 46 ' N / 97° 30 ' W

| | JAN | FEB | MAR | APR | MAY | JUN | JUL | AUG | SEP | OCT | NOV | DEC | YEAR |
|---|---|---|---|---|---|---|---|---|---|---|---|---|---|
| Maximum Temp °F | 65.3 | 68.9 | 75.8 | 81.4 | 85.7 | 90.3 | 93.1 | 93.2 | 89.6 | 83.8 | 75.5 | 68.9 | 81.0 |
| Minimum Temp °F | 46.2 | 48.6 | 55.7 | 63.3 | 69.4 | 73.7 | 74.9 | 75.0 | 72.2 | 64.1 | 55.6 | 49.2 | 62.3 |
| Mean Temp °F | 55.8 | 58.8 | 65.7 | 72.4 | 77.6 | 82.0 | 84.0 | 84.1 | 80.9 | 74.0 | 65.6 | 59.1 | 71.7 |
| Days Max Temp ≥ 90 °F | 0 | 0 | 1 | 2 | 5 | 19 | 28 | 27 | 17 | 5 | 0 | 0 | 104 |
| Days Max Temp ≤ 32 °F | 0 | 0 | 0 | 0 | 0 | 0 | 0 | 0 | 0 | 0 | 0 | 0 | 0 |
| Days Min Temp ≤ 32 °F | 3 | 1 | 0 | 0 | 0 | 0 | 0 | 0 | 0 | 0 | 0 | 2 | 6 |
| Days Min Temp ≤ 0 °F | 0 | 0 | 0 | 0 | 0 | 0 | 0 | 0 | 0 | 0 | 0 | 0 | 0 |
| Heating Degree Days | 309 | 205 | 87 | 15 | 0 | 0 | 0 | 0 | 0 | 11 | 102 | 232 | 961 |
| Cooling Degree Days | 23 | 41 | 112 | 228 | 394 | 523 | 591 | 604 | 486 | 308 | 132 | 55 | 3497 |
| Total Precipitation (") | 1.83 | 2.10 | 1.28 | 2.05 | 3.94 | 3.80 | 2.22 | 3.39 | 5.67 | 3.30 | 1.66 | 1.83 | 33.07 |
| Days ≥ 0.1" Precip | 4 | 3 | 2 | 3 | 5 | 5 | 3 | 4 | 7 | 4 | 3 | 3 | 46 |
| Total Snowfall (") | 0.0 | 0.0 | 0.0 | 0.0 | 0.0 | 0.0 | 0.0 | 0.0 | 0.0 | 0.0 | 0.0 | 0.0 | 0.0 |
| Days ≥ 1" Snow Depth | 0 | 0 | 0 | 0 | 0 | 0 | 0 | 0 | 0 | 0 | 0 | 0 | 0 |

## CORSICANA *Navarro County* ELEVATION 440 ft LAT/LONG 32° 5 ' N / 96° 28 ' W

| | JAN | FEB | MAR | APR | MAY | JUN | JUL | AUG | SEP | OCT | NOV | DEC | YEAR |
|---|---|---|---|---|---|---|---|---|---|---|---|---|---|
| Maximum Temp °F | 54.9 | 59.5 | 68.0 | 76.1 | 81.9 | 89.3 | 94.1 | 94.4 | 87.9 | 78.6 | 67.3 | 58.7 | 75.9 |
| Minimum Temp °F | 33.6 | 36.9 | 45.0 | 53.6 | 61.3 | 68.9 | 72.3 | 71.7 | 65.7 | 54.5 | 44.1 | 36.5 | 53.7 |
| Mean Temp °F | 44.3 | 48.2 | 56.5 | 64.9 | 71.6 | 79.1 | 83.2 | 83.1 | 76.8 | 66.6 | 55.7 | 47.6 | 64.8 |
| Days Max Temp ≥ 90 °F | 0 | 0 | 0 | 0 | 2 | 16 | 27 | 27 | 15 | 3 | 0 | 0 | 90 |
| Days Max Temp ≤ 32 °F | 1 | 1 | 0 | 0 | 0 | 0 | 0 | 0 | 0 | 0 | 0 | 1 | 3 |
| Days Min Temp ≤ 32 °F | 15 | 9 | 3 | 0 | 0 | 0 | 0 | 0 | 0 | 0 | 4 | 10 | 41 |
| Days Min Temp ≤ 0 °F | 0 | 0 | 0 | 0 | 0 | 0 | 0 | 0 | 0 | 0 | 0 | 0 | 0 |
| Heating Degree Days | 638 | 471 | 281 | 86 | 13 | 0 | 0 | 0 | 7 | 76 | 296 | 537 | 2405 |
| Cooling Degree Days | 1 | 3 | 23 | 80 | 223 | 425 | 566 | 583 | 384 | 131 | 25 | 3 | 2447 |
| Total Precipitation (") | 2.21 | 3.16 | 3.23 | 3.64 | 5.80 | 3.01 | 2.13 | 2.06 | 3.52 | 4.47 | 2.96 | 3.51 | 39.70 |
| Days ≥ 0.1" Precip | 5 | 5 | 5 | 5 | 7 | 5 | 3 | 3 | 5 | 5 | 5 | 5 | 58 |
| Total Snowfall (") | 0.3 | *0.4* | 0.0 | 0.0 | 0.0 | 0.0 | 0.0 | 0.0 | 0.0 | 0.0 | 0.0 | 0.1 | 0.8 |
| Days ≥ 1" Snow Depth | 0 | 0 | 0 | 0 | 0 | 0 | 0 | 0 | 0 | 0 | 0 | 0 | 0 |

## CRANE *Crane County* ELEVATION 2602 ft LAT/LONG 31° 24 ' N / 102° 20 ' W

| | JAN | FEB | MAR | APR | MAY | JUN | JUL | AUG | SEP | OCT | NOV | DEC | YEAR |
|---|---|---|---|---|---|---|---|---|---|---|---|---|---|
| Maximum Temp °F | 60.4 | 65.5 | 74.4 | 82.6 | 89.2 | 94.8 | 95.8 | 94.5 | 88.0 | 80.0 | 69.5 | 62.3 | 79.7 |
| Minimum Temp °F | 31.1 | 35.0 | 43.0 | 51.5 | 59.7 | 67.8 | 70.4 | 69.0 | 62.8 | 52.5 | 40.9 | 32.9 | 51.4 |
| Mean Temp °F | 45.8 | 50.3 | 58.8 | 67.1 | 74.5 | 81.4 | 83.1 | 81.8 | 75.5 | 66.3 | 55.3 | 47.6 | 65.6 |
| Days Max Temp ≥ 90 °F | 0 | 0 | 1 | 6 | 15 | 25 | 27 | 26 | 15 | 3 | 0 | 0 | 118 |
| Days Max Temp ≤ 32 °F | 0 | 0 | 0 | 0 | 0 | 0 | 0 | 0 | 0 | 0 | 0 | 0 | 0 |
| Days Min Temp ≤ 32 °F | 18 | 11 | 4 | 1 | 0 | 0 | 0 | 0 | 0 | 0 | 6 | 15 | 55 |
| Days Min Temp ≤ 0 °F | 0 | 0 | 0 | 0 | 0 | 0 | 0 | 0 | 0 | 0 | 0 | 0 | 0 |
| Heating Degree Days | 589 | 412 | 221 | 65 | 12 | 0 | 0 | 0 | 12 | 76 | 299 | 532 | 2218 |
| Cooling Degree Days | 0 | 5 | 36 | 144 | 331 | 512 | 580 | 545 | 356 | 138 | 19 | 0 | 2666 |
| Total Precipitation (") | 0.56 | 0.67 | 0.34 | 0.90 | 1.77 | 1.58 | 1.53 | 2.20 | 3.04 | 1.62 | 0.62 | 0.65 | 15.48 |
| Days ≥ 0.1" Precip | 2 | 2 | 1 | 2 | 3 | 3 | 3 | 3 | 4 | 3 | 2 | 2 | 30 |
| Total Snowfall (") | *0.8* | *0.0* | 0.1 | 0.0 | 0.0 | 0.0 | 0.0 | 0.0 | 0.0 | 0.0 | 0.6 | 0.4 | 1.9 |
| Days ≥ 1" Snow Depth | *0* | 0 | 0 | 0 | 0 | 0 | 0 | 0 | 0 | 0 | 0 | 0 | 0 |

## CROCKETT *Houston County* ELEVATION 351 ft LAT/LONG 31° 19 ' N / 95° 28 ' W

| | JAN | FEB | MAR | APR | MAY | JUN | JUL | AUG | SEP | OCT | NOV | DEC | YEAR |
|---|---|---|---|---|---|---|---|---|---|---|---|---|---|
| Maximum Temp °F | 56.9 | 61.5 | 69.6 | 77.4 | 83.0 | 89.5 | 93.2 | 93.8 | 88.1 | 79.7 | 68.8 | 60.7 | 76.9 |
| Minimum Temp °F | 35.2 | 37.9 | 45.3 | 53.8 | 61.3 | 68.3 | 71.0 | 70.1 | 65.3 | 54.2 | 44.6 | 37.6 | 53.7 |
| Mean Temp °F | 46.1 | 49.7 | 57.5 | 65.6 | 72.2 | 78.9 | 82.1 | 82.0 | 76.7 | 67.0 | 56.7 | 49.2 | 65.3 |
| Days Max Temp ≥ 90 °F | 0 | 0 | 0 | 0 | 3 | 17 | 26 | 26 | 14 | 3 | 0 | 0 | 89 |
| Days Max Temp ≤ 32 °F | 1 | 0 | 0 | 0 | 0 | 0 | 0 | 0 | 0 | 0 | 0 | 0 | 1 |
| Days Min Temp ≤ 32 °F | 14 | 9 | 3 | 0 | 0 | 0 | 0 | 0 | 0 | 0 | 4 | 10 | 40 |
| Days Min Temp ≤ 0 °F | 0 | 0 | 0 | 0 | 0 | 0 | 0 | 0 | 0 | 0 | 0 | 0 | 0 |
| Heating Degree Days | 585 | 430 | 258 | 76 | 9 | 0 | 0 | 0 | 4 | 68 | 276 | 491 | 2197 |
| Cooling Degree Days | 3 | 6 | 30 | 97 | 241 | 436 | 535 | 546 | 375 | 140 | 37 | 7 | 2453 |
| Total Precipitation (") | 3.75 | 3.25 | 3.43 | 4.29 | 4.91 | 3.99 | 2.93 | 2.70 | 4.45 | 4.24 | 3.67 | 3.97 | 45.58 |
| Days ≥ 0.1" Precip | 7 | 5 | 5 | 5 | 6 | 5 | 5 | 4 | 6 | 5 | 5 | 6 | 64 |
| Total Snowfall (") | 0.5 | 0.0 | 0.0 | 0.0 | 0.0 | 0.0 | 0.0 | 0.0 | 0.0 | 0.0 | 0.0 | 0.0 | 0.5 |
| Days ≥ 1" Snow Depth | 0 | 0 | 0 | 0 | 0 | 0 | 0 | 0 | 0 | 0 | 0 | 0 | 0 |

## CROSBYTON *Crosby County* ELEVATION 3113 ft LAT/LONG 33° 39 ' N / 101° 13 ' W

| | JAN | FEB | MAR | APR | MAY | JUN | JUL | AUG | SEP | OCT | NOV | DEC | YEAR |
|---|---|---|---|---|---|---|---|---|---|---|---|---|---|
| Maximum Temp °F | 51.4 | 56.4 | 65.3 | 74.8 | 81.3 | 89.0 | 92.6 | 90.3 | 82.9 | 74.3 | 62.2 | 53.9 | 72.9 |
| Minimum Temp °F | 24.3 | 27.4 | 34.8 | 44.6 | 53.6 | 62.7 | 66.2 | 64.6 | 57.7 | 46.7 | 35.0 | 27.0 | 45.4 |
| Mean Temp °F | 37.8 | 41.9 | 50.1 | 59.7 | 67.5 | 75.9 | 79.4 | 77.5 | 70.3 | 60.6 | 48.7 | 40.5 | 59.2 |
| Days Max Temp ≥ 90 °F | 0 | 0 | 0 | 2 | 6 | 16 | 23 | 20 | 8 | 1 | 0 | 0 | 76 |
| Days Max Temp ≤ 32 °F | 3 | 2 | 1 | 0 | 0 | 0 | 0 | 0 | 0 | 0 | 0 | 2 | 8 |
| Days Min Temp ≤ 32 °F | 26 | 21 | 12 | 2 | 0 | 0 | 0 | 0 | 0 | 1 | 11 | 24 | 97 |
| Days Min Temp ≤ 0 °F | 0 | 0 | 0 | 0 | 0 | 0 | 0 | 0 | 0 | 0 | 0 | 0 | 0 |
| Heating Degree Days | 835 | 645 | 461 | 191 | 58 | 4 | 0 | 1 | 37 | 178 | 485 | 754 | 3649 |
| Cooling Degree Days | 0 | 0 | 4 | 40 | 140 | 332 | 448 | 409 | 207 | 44 | 2 | 0 | 1626 |
| Total Precipitation (") | 0.63 | 0.92 | 1.11 | 1.48 | 3.08 | 2.86 | 2.13 | 3.36 | 3.50 | 2.01 | 0.89 | 0.80 | 22.77 |
| Days ≥ 0.1" Precip | 2 | 2 | 3 | 3 | 5 | 5 | 4 | 5 | 5 | 3 | 2 | 2 | 41 |
| Total Snowfall (") | 2.7 | 2.4 | 0.8 | 0.0 | 0.0 | 0.0 | 0.0 | 0.0 | 0.0 | 0.2 | 1.0 | 1.7 | 8.8 |
| Days ≥ 1" Snow Depth | *0* | *0* | 0 | 0 | 0 | 0 | 0 | 0 | 0 | 0 | 0 | *0* | 0 |

## CRYSTAL CITY *Zavala County* ELEVATION 581 ft LAT/LONG 28° 41 ' N / 99° 50 ' W

| | JAN | FEB | MAR | APR | MAY | JUN | JUL | AUG | SEP | OCT | NOV | DEC | YEAR |
|---|---|---|---|---|---|---|---|---|---|---|---|---|---|
| Maximum Temp °F | 64.9 | 69.7 | 78.6 | 84.6 | 88.7 | 94.0 | 96.7 | 96.4 | 91.2 | 83.8 | 74.0 | 67.1 | 82.5 |
| Minimum Temp °F | 42.4 | 46.0 | 53.4 | 61.7 | 68.2 | 73.1 | 74.9 | 74.6 | 70.7 | 62.2 | 51.9 | 44.7 | 60.3 |
| Mean Temp °F | 53.7 | 57.9 | 66.0 | 73.2 | 78.5 | 83.6 | 85.8 | 85.5 | 81.0 | 73.0 | 63.0 | 56.0 | 71.4 |
| Days Max Temp ≥ 90 °F | 0 | 0 | 3 | 7 | 14 | 25 | 29 | 29 | 20 | 7 | 0 | 0 | 134 |
| Days Max Temp ≤ 32 °F | 0 | 0 | 0 | 0 | 0 | 0 | 0 | 0 | 0 | 0 | 0 | 0 | 0 |
| Days Min Temp ≤ 32 °F | 5 | 2 | 1 | 0 | 0 | 0 | 0 | 0 | 0 | 0 | 1 | 3 | 12 |
| Days Min Temp ≤ 0 °F | 0 | 0 | 0 | 0 | 0 | 0 | 0 | 0 | 0 | 0 | 0 | 0 | 0 |
| Heating Degree Days | 357 | 223 | 87 | 14 | 1 | 0 | 0 | 0 | 1 | 17 | 136 | 292 | 1128 |
| Cooling Degree Days | 11 | 36 | 126 | 265 | 441 | 585 | 667 | 670 | 510 | 303 | 95 | 24 | 3733 |
| Total Precipitation (") | 1.08 | 1.19 | 0.93 | 1.89 | 2.97 | 2.75 | 1.53 | 1.81 | 2.52 | 2.49 | 1.09 | 0.94 | 21.19 |
| Days ≥ 0.1" Precip | 2 | 2 | 2 | 3 | 4 | 4 | 2 | 3 | 4 | 3 | 2 | 2 | 33 |
| Total Snowfall (") | 0.2 | 0.1 | 0.0 | 0.0 | 0.0 | 0.0 | 0.0 | 0.0 | 0.0 | 0.0 | 0.0 | 0.0 | 0.3 |
| Days ≥ 1" Snow Depth | 0 | 0 | 0 | 0 | 0 | 0 | 0 | 0 | 0 | 0 | 0 | 0 | 0 |

## CUERO *DeWitt County* ELEVATION 180 ft LAT/LONG 29° 8 ' N / 97° 19 ' W

| | JAN | FEB | MAR | APR | MAY | JUN | JUL | AUG | SEP | OCT | NOV | DEC | YEAR |
|---|---|---|---|---|---|---|---|---|---|---|---|---|---|
| Maximum Temp °F | 64.1 | 68.2 | 75.5 | *81.6* | *86.1* | *92.1* | 95.0 | 95.9 | 91.2 | 84.1 | *74.5* | 67.2 | 81.3 |
| Minimum Temp °F | 41.3 | 44.4 | 51.2 | *58.8* | *65.5* | *71.3* | 72.3 | 71.9 | 68.0 | 58.8 | *49.9* | 44.3 | 58.1 |
| Mean Temp °F | 52.7 | 56.3 | 63.4 | *70.2* | *75.8* | *81.8* | 83.7 | 83.9 | 79.6 | 71.4 | *62.2* | 55.8 | 69.7 |
| Days Max Temp ≥ 90 °F | 0 | 0 | 1 | *2* | *8* | 24 | 29 | 30 | 19 | 7 | 0 | 0 | 120 |
| Days Max Temp ≤ 32 °F | 0 | 0 | 0 | 0 | 0 | 0 | 0 | 0 | 0 | 0 | 0 | 0 | 0 |
| Days Min Temp ≤ 32 °F | 7 | 4 | 2 | 0 | 0 | 0 | 0 | 0 | 0 | 0 | *2* | 5 | 20 |
| Days Min Temp ≤ 0 °F | 0 | 0 | 0 | 0 | 0 | 0 | 0 | 0 | 0 | 0 | 0 | 0 | 0 |
| Heating Degree Days | 393 | 265 | 124 | *27* | *1* | 0 | 0 | 0 | 0 | 28 | *160* | 312 | 1310 |
| Cooling Degree Days | 13 | 29 | 80 | 179 | 354 | 513 | 588 | 604 | 452 | 242 | 95 | 34 | 3183 |
| Total Precipitation (") | 2.36 | 2.22 | 1.94 | 3.21 | 5.21 | 4.13 | 2.33 | 1.85 | 4.95 | 3.37 | 2.36 | 2.34 | 36.27 |
| Days ≥ 0.1" Precip | 5 | 4 | 3 | *4* | *6* | *5* | 3 | 3 | 5 | 4 | *4* | 4 | 50 |
| Total Snowfall (") | 0.1 | 0.0 | 0.0 | 0.0 | 0.0 | 0.0 | 0.0 | 0.0 | 0.0 | 0.0 | 0.0 | 0.0 | 0.1 |
| Days ≥ 1" Snow Depth | 0 | 0 | 0 | 0 | 0 | 0 | 0 | 0 | 0 | 0 | 0 | 0 | 0 |

## DAINGERFIELD 9 S *Morris County* ELEVATION 302 ft LAT/LONG 32° 55 ' N / 94° 42 ' W

| | JAN | FEB | MAR | APR | MAY | JUN | JUL | AUG | SEP | OCT | NOV | DEC | YEAR |
|---|---|---|---|---|---|---|---|---|---|---|---|---|---|
| Maximum Temp °F | 56.5 | 61.8 | 70.4 | 78.3 | 84.0 | 90.8 | 94.5 | 94.6 | 88.3 | 79.1 | 68.3 | 60.1 | 77.2 |
| Minimum Temp °F | 35.9 | 39.5 | 47.5 | 55.5 | 63.2 | 70.5 | 73.8 | 72.9 | 67.1 | 56.0 | 46.5 | 39.0 | 55.6 |
| Mean Temp °F | 46.2 | 50.7 | 59.1 | 66.8 | 73.5 | 80.7 | 84.2 | 83.8 | 77.7 | 67.6 | 57.4 | 49.6 | 66.4 |
| Days Max Temp ≥ 90 °F | 0 | 0 | 0 | 1 | 4 | 20 | 27 | 27 | 15 | 2 | 0 | 0 | 96 |
| Days Max Temp ≤ 32 °F | 0 | 0 | 0 | 0 | 0 | 0 | 0 | 0 | 0 | 0 | 0 | 0 | 0 |
| Days Min Temp ≤ 32 °F | 12 | 7 | 2 | 0 | 0 | 0 | 0 | 0 | 0 | 0 | 2 | 8 | 31 |
| Days Min Temp ≤ 0 °F | 0 | 0 | 0 | 0 | 0 | 0 | 0 | 0 | 0 | 0 | 0 | 0 | 0 |
| Heating Degree Days | 579 | 404 | 217 | 60 | 5 | 0 | 0 | 0 | 5 | 61 | 253 | 476 | 2060 |
| Cooling Degree Days | 1 | 8 | 37 | 118 | 272 | 474 | 598 | 601 | 400 | 143 | 35 | 5 | 2692 |
| Total Precipitation (") | 3.18 | 3.62 | 4.39 | 4.76 | 5.10 | 3.77 | 3.04 | 2.41 | 3.25 | 4.33 | 4.58 | 4.41 | 46.84 |
| Days ≥ 0.1" Precip | 6 | 6 | 6 | 6 | 6 | 5 | 4 | 4 | 4 | 5 | 6 | 6 | 64 |
| Total Snowfall (") | 0.1 | 0.1 | 0.0 | 0.0 | 0.0 | 0.0 | 0.0 | 0.0 | 0.0 | 0.0 | 0.0 | 0.3 | 0.5 |
| Days ≥ 1" Snow Depth | 0 | 0 | 0 | 0 | 0 | 0 | 0 | 0 | 0 | 0 | 0 | 0 | 0 |

## DALHART MUNI AP *Dallam County* ELEVATION 3996 ft LAT/LONG 36° 1 ' N / 102° 33 ' W

| | JAN | FEB | MAR | APR | MAY | JUN | JUL | AUG | SEP | OCT | NOV | DEC | YEAR |
|---|---|---|---|---|---|---|---|---|---|---|---|---|---|
| Maximum Temp °F | 49.0 | 53.1 | 61.2 | 70.1 | 77.4 | 87.0 | 90.7 | 88.0 | 81.0 | 71.2 | 58.7 | 50.2 | 69.8 |
| Minimum Temp °F | 19.3 | 22.9 | 29.9 | 39.1 | 48.6 | 58.3 | 63.2 | 61.4 | 53.2 | 40.9 | 28.8 | 21.1 | 40.6 |
| Mean Temp °F | 34.2 | 38.0 | 45.5 | 54.6 | 63.0 | 72.7 | 77.0 | 74.7 | 67.1 | 56.1 | 43.8 | 35.6 | 55.2 |
| Days Max Temp ≥ 90 °F | 0 | 0 | 0 | 1 | 3 | 12 | 20 | 14 | 5 | 1 | 0 | 0 | 56 |
| Days Max Temp ≤ 32 °F | 4 | 2 | 1 | 0 | 0 | 0 | 0 | 0 | 0 | 0 | 1 | 3 | 11 |
| Days Min Temp ≤ 32 °F | 29 | 25 | 19 | 6 | 1 | 0 | 0 | 0 | 0 | 4 | 20 | 29 | 133 |
| Days Min Temp ≤ 0 °F | 1 | 1 | 0 | 0 | 0 | 0 | 0 | 0 | 0 | 0 | 0 | 1 | 3 |
| Heating Degree Days | 948 | 756 | 597 | 313 | 116 | 13 | 1 | 2 | 61 | 282 | 629 | 904 | 4622 |
| Cooling Degree Days | 0 | 0 | 0 | 9 | 60 | 253 | 375 | 321 | 139 | 11 | 0 | 0 | 1168 |
| Total Precipitation (") | 0.47 | 0.44 | 0.89 | 1.21 | 2.82 | 2.51 | 3.12 | 3.23 | 1.77 | 1.03 | 0.71 | 0.49 | 18.69 |
| Days ≥ 0.1" Precip | 1 | 1 | 2 | 3 | 5 | 5 | 5 | 6 | 3 | 2 | 2 | 2 | 37 |
| Total Snowfall (") | 3.9 | 3.7 | 3.1 | 0.8 | 0.3 | 0.0 | 0.0 | 0.0 | 0.0 | 0.5 | 1.8 | 2.9 | 17.0 |
| Days ≥ 1" Snow Depth | 5 | 3 | 2 | 0 | 0 | 0 | 0 | 0 | 0 | 0 | 1 | 3 | 14 |

## DALLAS LOVE FIELD *Dallas County* ELEVATION 502 ft LAT/LONG 32° 51 ' N / 96° 51 ' W

| | JAN | FEB | MAR | APR | MAY | JUN | JUL | AUG | SEP | OCT | NOV | DEC | YEAR |
|---|---|---|---|---|---|---|---|---|---|---|---|---|---|
| Maximum Temp °F | 54.8 | 59.7 | 68.6 | 77.0 | 83.3 | 91.4 | 95.6 | 95.3 | 87.8 | 78.2 | 66.6 | 58.2 | 76.4 |
| Minimum Temp °F | 35.5 | 39.2 | 47.4 | 56.2 | 63.9 | 72.0 | 76.0 | 75.4 | 68.2 | 57.0 | 46.4 | 38.4 | 56.3 |
| Mean Temp °F | 45.2 | 49.5 | 58.0 | 66.6 | 73.6 | 81.7 | 85.9 | 85.4 | 78.0 | 67.6 | 56.5 | 48.3 | 66.4 |
| Days Max Temp ≥ 90 °F | 0 | 0 | 0 | 1 | 6 | 20 | 27 | 27 | 14 | 3 | 0 | 0 | 98 |
| Days Max Temp ≤ 32 °F | 1 | 1 | 0 | 0 | 0 | 0 | 0 | 0 | 0 | 0 | 0 | 1 | 3 |
| Days Min Temp ≤ 32 °F | 12 | 7 | 2 | 0 | 0 | 0 | 0 | 0 | 0 | 0 | 2 | 8 | 31 |
| Days Min Temp ≤ 0 °F | 0 | 0 | 0 | 0 | 0 | 0 | 0 | 0 | 0 | 0 | 0 | 0 | 0 |
| Heating Degree Days | 609 | 438 | 250 | 67 | 10 | 0 | 0 | 0 | 6 | 67 | 278 | 516 | 2241 |
| Cooling Degree Days | 2 | 7 | 40 | 121 | 287 | 515 | 659 | 664 | 421 | 158 | 34 | 4 | 2912 |
| Total Precipitation (") | 1.78 | 2.48 | 3.09 | 3.72 | 5.47 | 3.59 | 2.27 | 2.42 | 3.20 | 4.50 | 2.24 | 2.28 | 37.04 |
| Days ≥ 0.1" Precip | 4 | 4 | 5 | 5 | 7 | 5 | 3 | 3 | 5 | 5 | 4 | 4 | 54 |
| Total Snowfall (") | 0.9 | 0.7 | 0.1 | 0.0 | 0.0 | 0.0 | 0.0 | 0.0 | 0.0 | 0.0 | 0.2 | 0.2 | 2.1 |
| Days ≥ 1" Snow Depth | 1 | 0 | 0 | 0 | 0 | 0 | 0 | 0 | 0 | 0 | 0 | 0 | 1 |

## DANEVANG 1 W *Wharton County* ELEVATION 69 ft LAT/LONG 29° 3 ' N / 96° 11 ' W

| | JAN | FEB | MAR | APR | MAY | JUN | JUL | AUG | SEP | OCT | NOV | DEC | YEAR |
|---|---|---|---|---|---|---|---|---|---|---|---|---|---|
| Maximum Temp °F | 63.0 | 66.2 | 73.9 | 80.3 | 84.7 | 89.8 | 92.7 | 93.3 | 89.6 | 83.2 | 73.8 | 66.4 | 79.7 |
| Minimum Temp °F | 43.6 | 45.5 | 52.7 | 60.4 | 66.4 | 71.4 | 72.7 | 72.6 | 69.3 | 60.6 | 52.3 | 46.0 | 59.5 |
| Mean Temp °F | 53.3 | 55.9 | 63.3 | 70.4 | 75.5 | 80.6 | 82.8 | 83.0 | 79.5 | 71.9 | 63.1 | 56.3 | 69.6 |
| Days Max Temp ≥ 90 °F | 0 | 0 | 0 | 1 | 3 | 18 | 27 | 26 | 16 | 5 | 0 | 0 | 96 |
| Days Max Temp ≤ 32 °F | 0 | 0 | 0 | 0 | 0 | 0 | 0 | 0 | 0 | 0 | 0 | 0 | 0 |
| Days Min Temp ≤ 32 °F | 4 | 2 | 1 | 0 | 0 | 0 | 0 | 0 | 0 | 0 | 1 | 3 | 11 |
| Days Min Temp ≤ 0 °F | 0 | 0 | 0 | 0 | 0 | 0 | 0 | 0 | 0 | 0 | 0 | 0 | 0 |
| Heating Degree Days | 373 | 268 | 123 | 25 | 1 | 0 | 0 | 0 | 0 | 21 | 139 | 293 | 1243 |
| Cooling Degree Days | 9 | 17 | 69 | 175 | 334 | 488 | 562 | 571 | 452 | 248 | 89 | 30 | 3044 |
| Total Precipitation (") | 3.19 | 2.86 | 2.35 | 2.76 | 5.54 | 4.65 | 3.66 | 3.75 | 5.37 | 4.45 | 3.26 | 2.91 | 44.75 |
| Days ≥ 0.1" Precip | 6 | 5 | 4 | 3 | 5 | 5 | 5 | 6 | 6 | 5 | 4 | 5 | 59 |
| Total Snowfall (") | 0.1 | 0.1 | 0.0 | 0.0 | 0.0 | 0.0 | 0.0 | 0.0 | 0.0 | 0.0 | 0.0 | 0.0 | 0.2 |
| Days ≥ 1" Snow Depth | 0 | 0 | 0 | 0 | 0 | 0 | 0 | 0 | 0 | 0 | 0 | 0 | 0 |

## DEL RIO INTL AP *Val Verde County* ELEVATION 1026 ft LAT/LONG 29° 22 ' N / 100° 55 ' W

| | JAN | FEB | MAR | APR | MAY | JUN | JUL | AUG | SEP | OCT | NOV | DEC | YEAR |
|---|---|---|---|---|---|---|---|---|---|---|---|---|---|
| Maximum Temp °F | 62.2 | 67.1 | 75.9 | 83.2 | 88.1 | 93.7 | 96.0 | 95.8 | 90.2 | 81.8 | 71.5 | 64.2 | 80.8 |
| Minimum Temp °F | 39.6 | 43.4 | 51.2 | 59.3 | 66.2 | 71.8 | 74.0 | 73.6 | 68.9 | 59.8 | 48.8 | 41.5 | 58.2 |
| Mean Temp °F | 51.0 | 55.3 | 63.5 | 71.3 | 77.2 | 82.8 | 85.0 | 84.7 | 79.6 | 70.8 | 60.2 | 52.9 | 69.5 |
| Days Max Temp ≥ 90 °F | 0 | 0 | 2 | 6 | 13 | 25 | 28 | 28 | 19 | 4 | 0 | 0 | 125 |
| Days Max Temp ≤ 32 °F | 0 | 0 | 0 | 0 | 0 | 0 | 0 | 0 | 0 | 0 | 0 | 0 | 0 |
| Days Min Temp ≤ 32 °F | 6 | 3 | 1 | 0 | 0 | 0 | 0 | 0 | 0 | 0 | 1 | 4 | 15 |
| Days Min Temp ≤ 0 °F | 0 | 0 | 0 | 0 | 0 | 0 | 0 | 0 | 0 | 0 | 0 | 0 | 0 |
| Heating Degree Days | 431 | 281 | 122 | 22 | 1 | 0 | 0 | 0 | 3 | 25 | 182 | 373 | 1440 |
| Cooling Degree Days | 2 | 14 | 83 | 219 | 401 | 559 | 639 | 646 | 461 | 236 | 49 | 5 | 3314 |
| Total Precipitation (") | 0.67 | 1.07 | 0.84 | 2.00 | 2.17 | 2.13 | 2.09 | 1.53 | 2.51 | 2.13 | 0.95 | 0.79 | 18.88 |
| Days ≥ 0.1" Precip | 2 | 2 | 2 | 3 | 4 | 3 | 3 | 2 | 4 | 3 | 2 | 2 | 32 |
| Total Snowfall (") | 0.6 | 0.2 | 0.1 | 0.0 | 0.0 | 0.0 | 0.0 | 0.0 | 0.0 | 0.0 | 0.0 | 0.0 | 0.9 |
| Days ≥ 1" Snow Depth | 0 | 0 | 0 | 0 | 0 | 0 | 0 | 0 | 0 | 0 | 0 | 0 | 0 |

## DENISON DAM *Grayson County* ELEVATION 610 ft LAT/LONG 33° 49 ' N / 96° 34 ' W

| | JAN | FEB | MAR | APR | MAY | JUN | JUL | AUG | SEP | OCT | NOV | DEC | YEAR |
|---|---|---|---|---|---|---|---|---|---|---|---|---|---|
| Maximum Temp °F | 51.4 | 56.3 | 65.5 | 74.3 | 80.7 | 88.8 | 93.9 | 93.4 | 85.8 | 76.2 | 64.4 | 55.1 | 73.8 |
| Minimum Temp °F | 30.3 | 34.1 | 42.8 | 52.0 | 59.9 | 68.0 | 71.9 | 70.2 | 63.7 | 52.5 | 42.9 | 34.0 | 51.9 |
| Mean Temp °F | 40.8 | 45.3 | 54.2 | 63.1 | 70.3 | 78.4 | 82.9 | 81.8 | 74.8 | 64.4 | 53.6 | 44.6 | 62.9 |
| Days Max Temp ≥ 90 °F | 0 | 0 | 0 | 0 | 2 | 15 | 25 | 24 | 11 | 1 | 0 | 0 | 78 |
| Days Max Temp ≤ 32 °F | 3 | 1 | 0 | 0 | 0 | 0 | 0 | 0 | 0 | 0 | 0 | 1 | 5 |
| Days Min Temp ≤ 32 °F | 19 | 12 | 4 | 0 | 0 | 0 | 0 | 0 | 0 | 0 | 5 | 14 | 54 |
| Days Min Temp ≤ 0 °F | 0 | 0 | 0 | 0 | 0 | 0 | 0 | 0 | 0 | 0 | 0 | 0 | 0 |
| Heating Degree Days | 743 | 552 | 346 | 118 | 21 | 0 | 0 | 0 | 14 | 108 | 349 | 627 | 2878 |
| Cooling Degree Days | 0 | 2 | 14 | 61 | 187 | 411 | 565 | 545 | 322 | 90 | 15 | 1 | 2213 |
| Total Precipitation (") | 1.91 | 2.64 | 3.19 | 3.91 | 5.41 | 4.28 | 2.63 | 2.44 | 4.74 | 4.27 | 3.01 | 2.41 | 40.84 |
| Days ≥ 0.1" Precip | 4 | 4 | 5 | 5 | 7 | 6 | 3 | 4 | 5 | 5 | 5 | 4 | 57 |
| Total Snowfall (") | 0.2 | 0.5 | 0.0 | 0.0 | 0.0 | 0.0 | 0.0 | 0.0 | 0.0 | 0.0 | 0.0 | 0.1 | 0.8 |
| Days ≥ 1" Snow Depth | 0 | 0 | 0 | 0 | 0 | 0 | 0 | 0 | 0 | 0 | 0 | 0 | 0 |

## DENTON 2 SE *Denton County* ELEVATION 630 ft LAT/LONG 33° 12 ' N / 97° 6 ' W

| | JAN | FEB | MAR | APR | MAY | JUN | JUL | AUG | SEP | OCT | NOV | DEC | YEAR |
|---|---|---|---|---|---|---|---|---|---|---|---|---|---|
| Maximum Temp °F | 55.1 | 60.0 | 69.2 | 76.8 | 82.6 | 90.0 | 94.5 | 94.0 | 86.8 | 77.7 | 66.4 | 58.3 | 76.0 |
| Minimum Temp °F | 32.7 | 36.4 | 44.9 | 53.8 | 61.2 | 69.1 | 72.8 | 71.5 | 65.0 | 54.1 | 43.9 | 36.1 | 53.5 |
| Mean Temp °F | 44.0 | 48.1 | 57.1 | 65.3 | 71.9 | 79.5 | 83.6 | 82.8 | 76.0 | 66.0 | 55.2 | 47.5 | 64.8 |
| Days Max Temp ≥ 90 °F | 0 | 0 | 0 | 1 | 3 | 17 | 27 | 26 | 12 | 2 | 0 | 0 | 88 |
| Days Max Temp ≤ 32 °F | 2 | 1 | 0 | 0 | 0 | 0 | 0 | 0 | 0 | 0 | 0 | 1 | 4 |
| Days Min Temp ≤ 32 °F | 16 | 10 | 3 | 0 | 0 | 0 | 0 | 0 | 0 | 0 | 4 | 11 | 44 |
| Days Min Temp ≤ 0 °F | 0 | 0 | 0 | 0 | 0 | 0 | 0 | 0 | 0 | 0 | 0 | 0 | 0 |
| Heating Degree Days | 647 | 473 | 268 | 80 | 12 | 0 | 0 | 0 | 10 | 81 | 309 | 541 | 2421 |
| Cooling Degree Days | 0 | 4 | 27 | 88 | 231 | 443 | 581 | 575 | 349 | 112 | 19 | 3 | 2432 |
| Total Precipitation (") | 1.87 | 2.67 | 2.86 | 3.64 | 5.60 | 3.33 | 2.52 | 2.33 | 4.39 | 4.44 | 2.33 | 2.54 | 38.52 |
| Days ≥ 0.1" Precip | 4 | 4 | 5 | 5 | 6 | 5 | 3 | 4 | 5 | 5 | 4 | 4 | 54 |
| Total Snowfall (") | *0.2* | 0.3 | 0.1 | 0.0 | 0.0 | 0.0 | 0.0 | 0.0 | 0.0 | 0.0 | 0.0 | 0.1 | 0.7 |
| Days ≥ 1" Snow Depth | 0 | 0 | 0 | 0 | 0 | 0 | 0 | 0 | 0 | 0 | 0 | 0 | 0 |

## DILLEY *Frio County* ELEVATION 571 ft LAT/LONG 28° 40 ' N / 99° 10 ' W

| | JAN | FEB | MAR | APR | MAY | JUN | JUL | AUG | SEP | OCT | NOV | DEC | YEAR |
|---|---|---|---|---|---|---|---|---|---|---|---|---|---|
| Maximum Temp °F | 63.2 | 67.6 | 76.5 | 84.1 | 87.9 | 93.6 | 97.7 | 97.3 | 91.8 | 83.8 | 73.8 | 66.7 | 82.0 |
| Minimum Temp °F | 39.2 | 42.6 | 50.7 | 58.8 | 65.5 | 70.4 | 72.2 | 72.1 | 68.6 | 59.5 | 49.2 | 42.7 | 57.6 |
| Mean Temp °F | 51.2 | 55.1 | 63.6 | 71.5 | 76.7 | 82.0 | 84.9 | 84.7 | 80.2 | 71.6 | 61.6 | 54.7 | 69.8 |
| Days Max Temp ≥ 90 °F | 0 | 0 | 3 | *8* | 13 | 24 | 29 | 29 | 21 | 8 | 0 | 0 | 135 |
| Days Max Temp ≤ 32 °F | 0 | 0 | 0 | 0 | 0 | 0 | 0 | 0 | 0 | 0 | 0 | 0 | 0 |
| Days Min Temp ≤ 32 °F | 7 | 4 | 1 | 0 | 0 | 0 | 0 | 0 | 0 | 0 | 2 | 4 | 18 |
| Days Min Temp ≤ 0 °F | 0 | 0 | 0 | 0 | 0 | 0 | 0 | 0 | 0 | 0 | 0 | 0 | 0 |
| Heating Degree Days | 428 | 289 | 125 | 24 | 2 | 0 | 0 | 0 | 2 | 26 | 165 | 328 | 1389 |
| Cooling Degree Days | *5* | *22* | *86* | *223* | *384* | *528* | *648* | *650* | *494* | *275* | *81* | *20* | 3416 |
| Total Precipitation (") | 1.46 | 1.48 | 1.08 | 1.97 | 4.10 | 3.15 | 1.34 | 2.39 | 3.54 | 3.52 | 1.32 | 1.39 | 26.74 |
| Days ≥ 0.1" Precip | 3 | *3* | 2 | 3 | 5 | 4 | 3 | 4 | 5 | *4* | 3 | 3 | 42 |
| Total Snowfall (") | *0.0* | *0.0* | 0.0 | 0.0 | 0.0 | 0.0 | 0.0 | 0.0 | 0.0 | 0.0 | 0.0 | 0.0 | 0.0 |
| Days ≥ 1" Snow Depth | *0* | *0* | 0 | 0 | 0 | 0 | 0 | *0* | 0 | 0 | 0 | 0 | 0 |

## DIMMITT 2 N *Castro County* ELEVATION 3862 ft LAT/LONG 34° 33 ' N / 102° 19 ' W

| | JAN | FEB | MAR | APR | MAY | JUN | JUL | AUG | SEP | OCT | NOV | DEC | YEAR |
|---|---|---|---|---|---|---|---|---|---|---|---|---|---|
| Maximum Temp °F | 50.2 | 54.6 | 63.2 | 72.3 | 79.8 | 87.7 | 90.4 | 87.9 | 81.8 | 72.5 | 60.1 | 51.8 | 71.0 |
| Minimum Temp °F | 20.1 | 23.0 | 29.6 | 38.7 | 48.1 | 57.7 | 61.7 | 60.0 | 52.7 | 40.7 | 29.1 | 21.7 | 40.3 |
| Mean Temp °F | 35.1 | 38.8 | 46.4 | 55.5 | 64.0 | 72.7 | 76.0 | 74.0 | 67.3 | 56.6 | 44.6 | 36.8 | 55.7 |
| Days Max Temp ≥ 90 °F | 0 | 0 | 0 | 1 | 5 | 13 | 19 | 14 | 6 | 1 | 0 | 0 | 59 |
| Days Max Temp ≤ 32 °F | 4 | 2 | 1 | 0 | 0 | 0 | 0 | 0 | 0 | 0 | 1 | 3 | 11 |
| Days Min Temp ≤ 32 °F | 29 | 25 | 20 | 7 | 1 | 0 | 0 | 0 | 0 | 4 | 20 | 29 | 135 |
| Days Min Temp ≤ 0 °F | 0 | 0 | 0 | 0 | 0 | 0 | 0 | 0 | 0 | 0 | 0 | 1 | 1 |
| Heating Degree Days | 919 | 733 | 568 | 291 | 104 | 12 | 1 | 3 | 58 | 269 | 604 | 867 | 4429 |
| Cooling Degree Days | 0 | 0 | 0 | 12 | 77 | 251 | 351 | 301 | 139 | 11 | 0 | 0 | 1142 |
| Total Precipitation (") | 0.48 | 0.49 | 0.78 | 0.87 | 2.48 | 3.02 | 2.55 | 3.00 | 2.42 | 1.45 | 0.67 | 0.57 | 18.78 |
| Days ≥ 0.1" Precip | 1 | 2 | 2 | 2 | 5 | 5 | 4 | 5 | 5 | 3 | 2 | 2 | 38 |
| Total Snowfall (") | *2.9* | *2.3* | 1.5 | 0.4 | 0.0 | 0.0 | 0.0 | 0.0 | 0.0 | 0.1 | 1.0 | 2.7 | 10.9 |
| Days ≥ 1" Snow Depth | *2* | *1* | 1 | 0 | 0 | 0 | 0 | 0 | 0 | 0 | 1 | 1 | 6 |

## DUBLIN *Erath County* ELEVATION 1470 ft LAT/LONG 32° 5 ' N / 98° 20 ' W

| | JAN | FEB | MAR | APR | MAY | JUN | JUL | AUG | SEP | OCT | NOV | DEC | YEAR |
|---|---|---|---|---|---|---|---|---|---|---|---|---|---|
| Maximum Temp °F | 54.1 | 58.6 | 67.2 | 76.2 | 81.6 | 89.0 | 93.6 | 93.7 | 86.5 | 77.3 | 65.5 | 57.2 | 75.0 |
| Minimum Temp °F | 31.3 | 34.9 | 42.8 | 52.0 | 59.5 | 66.8 | 70.1 | 69.4 | 63.4 | 53.2 | 42.7 | 34.5 | 51.7 |
| Mean Temp °F | 42.7 | 46.8 | 55.0 | 64.1 | 70.6 | 77.9 | 81.9 | 81.6 | 74.9 | 65.3 | 54.1 | 45.9 | 63.4 |
| Days Max Temp ≥ 90 °F | 0 | 0 | 0 | 1 | 4 | 15 | 25 | 25 | 12 | 2 | 0 | 0 | 84 |
| Days Max Temp ≤ 32 °F | 2 | 1 | 0 | 0 | 0 | 0 | 0 | 0 | 0 | 0 | 0 | 1 | 4 |
| Days Min Temp ≤ 32 °F | 17 | 11 | 5 | 0 | 0 | 0 | 0 | 0 | 0 | 0 | 5 | 12 | 50 |
| Days Min Temp ≤ 0 °F | 0 | 0 | 0 | 0 | 0 | 0 | 0 | 0 | 0 | 0 | 0 | 0 | 0 |
| Heating Degree Days | 684 | 510 | 325 | 103 | 22 | 1 | 0 | 0 | 14 | 94 | 338 | 586 | 2677 |
| Cooling Degree Days | 0 | 3 | 22 | 83 | 209 | 398 | 530 | 553 | 335 | 110 | 20 | 1 | 2264 |
| Total Precipitation (") | 1.73 | 2.27 | 2.31 | 3.13 | 5.16 | 3.60 | 2.27 | 3.03 | 3.56 | 3.36 | 1.99 | 2.13 | 34.54 |
| Days ≥ 0.1" Precip | 4 | 5 | 4 | 4 | 7 | 5 | 3 | 4 | 5 | 5 | 4 | 4 | 54 |
| Total Snowfall (") | 1.3 | 1.0 | 0.3 | 0.0 | 0.0 | 0.0 | 0.0 | 0.0 | 0.0 | 0.0 | 0.9 | 0.4 | 3.9 |
| Days ≥ 1" Snow Depth | 1 | 1 | 0 | 0 | 0 | 0 | 0 | 0 | 0 | 0 | 0 | 0 | 2 |

## DUMAS *Moore County* ELEVATION 3654 ft LAT/LONG 35° 51 ' N / 101° 58 ' W

| | JAN | FEB | MAR | APR | MAY | JUN | JUL | AUG | SEP | OCT | NOV | DEC | YEAR |
|---|---|---|---|---|---|---|---|---|---|---|---|---|---|
| Maximum Temp °F | 47.5 | 51.6 | 60.2 | 70.1 | 77.7 | 87.3 | 91.8 | 89.3 | 81.6 | 71.6 | 58.1 | 49.2 | 69.7 |
| Minimum Temp °F | 20.3 | 23.5 | 30.6 | 40.0 | 49.6 | 59.6 | 65.2 | 63.4 | 55.3 | 42.9 | 30.6 | 22.7 | 42.0 |
| Mean Temp °F | 34.0 | 37.6 | 45.4 | 55.1 | 63.7 | 73.5 | 78.5 | 76.4 | 68.5 | 57.2 | 44.4 | 36.0 | 55.9 |
| Days Max Temp ≥ 90 °F | 0 | 0 | 0 | 1 | 4 | 13 | 21 | 18 | 7 | 1 | 0 | 0 | 65 |
| Days Max Temp ≤ 32 °F | 5 | 3 | 1 | 0 | 0 | 0 | 0 | 0 | 0 | 0 | 1 | 4 | 14 |
| Days Min Temp ≤ 32 °F | 28 | 24 | 18 | 6 | 1 | 0 | 0 | 0 | 0 | 3 | 18 | 28 | 126 |
| Days Min Temp ≤ 0 °F | 1 | 1 | 0 | 0 | 0 | 0 | 0 | 0 | 0 | 0 | 0 | 1 | 3 |
| Heating Degree Days | 955 | 767 | 602 | 306 | 116 | 14 | 1 | 3 | 57 | 261 | 612 | 893 | 4587 |
| Cooling Degree Days | 0 | 0 | 1 | 16 | 75 | 278 | 429 | 376 | 174 | 21 | 0 | 0 | 1370 |
| Total Precipitation (") | 0.46 | 0.63 | 0.97 | 1.12 | 2.84 | 2.91 | 2.14 | 2.38 | 1.75 | 1.10 | 0.67 | 0.50 | 17.47 |
| Days ≥ 0.1" Precip | 2 | 2 | 2 | 3 | 5 | 5 | 4 | 5 | 3 | 2 | 2 | 2 | 37 |
| Total Snowfall (") | 3.3 | 4.1 | 2.8 | 1.1 | 0.1 | 0.0 | 0.0 | 0.0 | 0.0 | 0.3 | 1.5 | 1.8 | 15.0 |
| Days ≥ 1" Snow Depth | 5 | 4 | 1 | 1 | 0 | 0 | 0 | 0 | 0 | 0 | 1 | 3 | 15 |

## EAGLE PASS *Maverick County* ELEVATION 741 ft LAT/LONG 28° 43 ' N / 100° 30 ' W

| | JAN | FEB | MAR | APR | MAY | JUN | JUL | AUG | SEP | OCT | NOV | DEC | YEAR |
|---|---|---|---|---|---|---|---|---|---|---|---|---|---|
| Maximum Temp °F | 62.6 | 67.6 | 76.9 | 84.9 | 89.0 | 94.9 | 97.7 | 97.6 | 91.7 | 83.5 | 73.4 | 65.5 | 82.1 |
| Minimum Temp °F | 39.6 | 43.4 | 51.3 | 59.5 | 66.5 | 72.0 | 74.3 | 73.7 | 69.3 | 59.9 | 49.1 | 41.8 | 58.4 |
| Mean Temp °F | 51.1 | 55.5 | 64.2 | 72.2 | 77.8 | 83.5 | 86.0 | 85.7 | 80.6 | 71.7 | 61.3 | 53.7 | 70.3 |
| Days Max Temp ≥ 90 °F | 0 | 0 | 3 | 9 | 15 | 25 | 29 | 29 | 21 | 7 | 1 | 0 | 139 |
| Days Max Temp ≤ 32 °F | 0 | 0 | 0 | 0 | 0 | 0 | 0 | 0 | 0 | 0 | 0 | 0 | 0 |
| Days Min Temp ≤ 32 °F | 6 | 3 | 1 | 0 | 0 | 0 | 0 | 0 | 0 | 0 | 2 | 4 | 16 |
| Days Min Temp ≤ 0 °F | 0 | 0 | 0 | 0 | 0 | 0 | 0 | 0 | 0 | 0 | 0 | 0 | 0 |
| Heating Degree Days | 428 | 278 | 116 | 21 | 2 | 0 | 0 | 0 | 2 | 24 | 167 | 350 | 1388 |
| Cooling Degree Days | 4 | 20 | 102 | 240 | 422 | 582 | 672 | 677 | 500 | 269 | 72 | 8 | 3568 |
| Total Precipitation (") | 0.92 | 1.07 | 0.76 | 1.97 | 3.48 | 3.24 | 2.10 | 2.18 | 2.78 | 2.19 | 0.95 | 0.95 | 22.59 |
| Days ≥ 0.1" Precip | 2 | 2 | 2 | 3 | 5 | 4 | 3 | 3 | 4 | 3 | 2 | 2 | 35 |
| Total Snowfall (") | 0.7 | 0.0 | 0.0 | 0.0 | 0.0 | 0.0 | 0.0 | 0.0 | 0.0 | 0.0 | 0.0 | 0.0 | 0.7 |
| Days ≥ 1" Snow Depth | 0 | 0 | 0 | 0 | 0 | 0 | 0 | 0 | 0 | 0 | 0 | 0 | 0 |

## EASTLAND *Eastland County* ELEVATION 1421 ft LAT/LONG 32° 24 ' N / 98° 48 ' W

| | JAN | FEB | MAR | APR | MAY | JUN | JUL | AUG | SEP | OCT | NOV | DEC | YEAR |
|---|---|---|---|---|---|---|---|---|---|---|---|---|---|
| Maximum Temp °F | 54.0 | 58.8 | 68.4 | 77.2 | 83.2 | 90.5 | 94.9 | 94.6 | 86.9 | 78.1 | 65.8 | 57.5 | 75.8 |
| Minimum Temp °F | 28.8 | 33.0 | 40.8 | 50.0 | 57.7 | 66.2 | 69.4 | 68.5 | 62.2 | 50.6 | 39.8 | 32.0 | 49.9 |
| Mean Temp °F | 41.4 | 45.9 | 54.7 | 63.6 | *70.5* | 78.4 | 82.1 | 81.6 | 74.6 | 64.2 | 52.7 | 44.8 | 62.9 |
| Days Max Temp ≥ 90 °F | 0 | 0 | 1 | 2 | 7 | 18 | 26 | 26 | 13 | 3 | 0 | 0 | 96 |
| Days Max Temp ≤ 32 °F | 2 | 1 | 0 | 0 | 0 | 0 | 0 | 0 | 0 | 0 | 0 | 1 | 4 |
| Days Min Temp ≤ 32 °F | 21 | 13 | 7 | 1 | 0 | 0 | 0 | 0 | 0 | 1 | 8 | 18 | 69 |
| Days Min Temp ≤ 0 °F | 0 | 0 | 0 | 0 | 0 | 0 | 0 | 0 | 0 | 0 | 0 | 0 | 0 |
| Heating Degree Days | 715 | 533 | *330* | 116 | *28* | 1 | 0 | 0 | 15 | 111 | 375 | 629 | 2853 |
| Cooling Degree Days | 0 | *1* | *16* | 71 | *196* | 385 | 510 | 516 | 303 | 89 | 18 | *1* | 2106 |
| Total Precipitation (") | 1.43 | 1.79 | 1.82 | 2.53 | 3.83 | 3.16 | 1.71 | 2.27 | 3.05 | 2.99 | 1.65 | 1.64 | 27.87 |
| Days ≥ 0.1" Precip | 3 | 3 | 3 | 4 | *5* | 4 | 3 | 4 | 5 | 4 | 3 | 3 | 44 |
| Total Snowfall (") | 1.3 | 0.7 | 0.8 | 0.0 | 0.0 | 0.0 | 0.0 | 0.0 | 0.0 | 0.0 | 0.5 | 0.4 | 3.7 |
| Days ≥ 1" Snow Depth | *1* | 1 | 0 | 0 | 0 | 0 | 0 | 0 | 0 | 0 | 0 | *0* | 2 |

## EL PASO INTL ARPT *El Paso County* ELEVATION 3940 ft LAT/LONG 31° 48 ' N / 106° 24 ' W

| | JAN | FEB | MAR | APR | MAY | JUN | JUL | AUG | SEP | OCT | NOV | DEC | YEAR |
|---|---|---|---|---|---|---|---|---|---|---|---|---|---|
| Maximum Temp °F | 57.5 | 63.3 | 70.4 | 79.1 | 87.0 | 96.1 | 95.5 | 92.8 | 87.4 | 78.6 | 66.3 | 58.3 | 77.7 |
| Minimum Temp °F | 30.5 | 34.6 | 40.7 | 48.3 | 56.6 | 65.4 | 69.2 | 67.4 | 60.8 | 49.1 | 37.7 | 31.2 | 49.3 |
| Mean Temp °F | 44.0 | 48.9 | 55.6 | 63.7 | 71.8 | 80.7 | 82.4 | 80.2 | 74.1 | 63.9 | 52.0 | 44.8 | 63.5 |
| Days Max Temp ≥ 90 °F | 0 | 0 | 0 | 2 | 12 | 26 | 27 | 23 | 13 | 2 | 0 | 0 | 105 |
| Days Max Temp ≤ 32 °F | 0 | 0 | 0 | 0 | 0 | 0 | 0 | 0 | 0 | 0 | 0 | 0 | 0 |
| Days Min Temp ≤ 32 °F | 19 | 12 | 5 | 1 | 0 | 0 | 0 | 0 | 0 | 0 | 8 | 18 | 63 |
| Days Min Temp ≤ 0 °F | 0 | 0 | 0 | 0 | 0 | 0 | 0 | 0 | 0 | 0 | 0 | 0 | 0 |
| Heating Degree Days | 644 | 447 | 292 | 98 | 13 | 0 | 0 | 0 | 9 | 95 | 385 | 619 | 2602 |
| Cooling Degree Days | 0 | 1 | 6 | 74 | 235 | 498 | 553 | 494 | 304 | 73 | 2 | 0 | 2240 |
| Total Precipitation (") | 0.46 | 0.42 | 0.27 | 0.23 | 0.40 | 0.74 | 1.54 | 1.70 | 1.57 | 0.74 | 0.41 | 0.75 | 9.23 |
| Days ≥ 0.1" Precip | 2 | 2 | 1 | 1 | 1 | 2 | 4 | 4 | 3 | 2 | 1 | 2 | 25 |
| Total Snowfall (") | 1.3 | 0.8 | 0.3 | 0.6 | 0.0 | 0.0 | 0.0 | 0.0 | 0.0 | 0.1 | 1.1 | 2.3 | 6.5 |
| Days ≥ 1" Snow Depth | 1 | 0 | 0 | 0 | 0 | 0 | 0 | 0 | 0 | 0 | 0 | 1 | 2 |

## ELGIN *Bastrop County* ELEVATION 581 ft LAT/LONG 30° 21 ' N / 97° 22 ' W

| | JAN | FEB | MAR | APR | MAY | JUN | JUL | AUG | SEP | OCT | NOV | DEC | YEAR |
|---|---|---|---|---|---|---|---|---|---|---|---|---|---|
| Maximum Temp °F | 60.6 | 65.1 | 73.5 | 80.4 | 85.1 | 91.5 | 95.2 | 95.6 | 89.9 | 81.9 | 71.7 | 63.8 | 79.5 |
| Minimum Temp °F | 39.6 | 42.5 | 50.1 | 58.0 | 64.6 | 70.4 | 72.6 | 72.3 | 67.9 | 58.7 | 49.3 | 42.2 | 57.4 |
| Mean Temp °F | 50.1 | 53.8 | 61.8 | 69.3 | 74.8 | 81.0 | 83.9 | 84.0 | 78.9 | 70.3 | 60.5 | 53.0 | 68.5 |
| Days Max Temp ≥ 90 °F | 0 | 0 | 1 | 1 | 6 | 22 | 28 | 29 | 18 | 4 | 0 | 0 | 109 |
| Days Max Temp ≤ 32 °F | 0 | 0 | 0 | 0 | 0 | 0 | 0 | 0 | 0 | 0 | 0 | 0 | 0 |
| Days Min Temp ≤ 32 °F | 8 | 5 | 1 | 0 | 0 | 0 | 0 | 0 | 0 | 0 | 2 | 5 | 21 |
| Days Min Temp ≤ 0 °F | 0 | 0 | 0 | 0 | 0 | 0 | 0 | 0 | 0 | 0 | 0 | 0 | 0 |
| Heating Degree Days | 465 | 326 | 159 | 35 | 3 | 0 | 0 | 0 | 2 | 35 | 188 | 381 | 1594 |
| Cooling Degree Days | 6 | 19 | 65 | 171 | 323 | 497 | 603 | 616 | 434 | 218 | 65 | 16 | 3033 |
| Total Precipitation (") | 2.65 | 2.40 | 2.29 | 2.69 | 5.18 | 3.50 | 1.77 | 1.93 | 3.46 | 3.76 | 2.96 | 2.52 | 35.11 |
| Days ≥ 0.1" Precip | 4 | 5 | 5 | 4 | 6 | 5 | 3 | 3 | 5 | 5 | 4 | 4 | 53 |
| Total Snowfall (") | 0.1 | 0.5 | 0.0 | 0.0 | 0.0 | 0.0 | 0.0 | 0.0 | 0.0 | 0.0 | 0.0 | 0.0 | 0.6 |
| Days ≥ 1" Snow Depth | 0 | 0 | 0 | 0 | 0 | 0 | 0 | 0 | 0 | 0 | 0 | 0 | 0 |

## EMORY *Rains County* ELEVATION 463 ft LAT/LONG 32° 52 ' N / 95° 44 ' W

| | JAN | FEB | MAR | APR | MAY | JUN | JUL | AUG | SEP | OCT | NOV | DEC | YEAR |
|---|---|---|---|---|---|---|---|---|---|---|---|---|---|
| Maximum Temp °F | 54.0 | 58.7 | 67.2 | 75.6 | 81.0 | 88.7 | ***93.4*** | 93.8 | 87.1 | 77.9 | 66.3 | 57.0 | 75.1 |
| Minimum Temp °F | 31.5 | 35.0 | 43.4 | 51.8 | 59.6 | 67.4 | ***70.7*** | 69.3 | 63.1 | 51.8 | 42.2 | 34.3 | 51.7 |
| Mean Temp °F | 42.8 | 46.9 | 55.3 | 63.8 | 70.3 | 78.1 | ***82.1*** | 81.6 | 75.1 | 64.9 | 54.3 | 45.7 | 63.4 |
| Days Max Temp ≥ 90 °F | 0 | 0 | 0 | 0 | 2 | 14 | 24 | 24 | 13 | 2 | 0 | 0 | 79 |
| Days Max Temp ≤ 32 °F | 1 | 1 | 0 | 0 | 0 | 0 | ***0*** | 0 | 0 | 0 | 0 | 1 | 3 |
| Days Min Temp ≤ 32 °F | 18 | 12 | 4 | 0 | 0 | 0 | ***0*** | 0 | 0 | 0 | 6 | 14 | 54 |
| Days Min Temp ≤ 0 °F | 0 | 0 | 0 | 0 | 0 | 0 | ***0*** | 0 | 0 | 0 | 0 | 0 | 0 |
| Heating Degree Days | 684 | 508 | 310 | 103 | 21 | 1 | ***0*** | 0 | 12 | 98 | 335 | 593 | 2665 |
| Cooling Degree Days | 1 | 2 | 19 | 74 | 194 | 404 | ***522*** | ***527*** | 328 | 100 | 19 | 2 | 2192 |
| Total Precipitation (") | 2.85 | 3.65 | 3.78 | 4.03 | 6.08 | 3.69 | 2.40 | 2.18 | 3.50 | 5.08 | 3.50 | 3.79 | 44.53 |
| Days ≥ 0.1" Precip | 5 | 5 | 5 | 5 | 7 | 5 | 3 | 3 | 5 | 5 | 5 | 5 | 58 |
| Total Snowfall (") | ***0.0*** | 0.3 | 0.0 | 0.0 | 0.0 | 0.0 | 0.0 | 0.0 | 0.0 | 0.0 | 0.0 | 0.3 | 0.6 |
| Days ≥ 1" Snow Depth | ***0*** | 0 | 0 | 0 | 0 | 0 | 0 | 0 | 0 | 0 | 0 | 0 | 0 |

## ENCINAL *Webb County* ELEVATION 590 ft LAT/LONG 28° 2 ' N / 99° 25 ' W

| | JAN | FEB | MAR | APR | MAY | JUN | JUL | AUG | SEP | OCT | NOV | DEC | YEAR |
|---|---|---|---|---|---|---|---|---|---|---|---|---|---|
| Maximum Temp °F | 65.1 | 69.1 | 79.6 | 86.5 | 90.7 | 96.2 | 98.4 | 98.5 | 92.9 | 85.2 | 76.0 | 68.4 | 83.9 |
| Minimum Temp °F | 39.3 | 42.4 | 50.2 | 59.0 | 65.4 | 70.5 | 71.7 | 71.5 | 68.1 | 58.7 | 48.7 | 42.3 | 57.3 |
| Mean Temp °F | 52.3 | 55.8 | 64.9 | 72.8 | 78.1 | 83.4 | 85.1 | 85.0 | 80.6 | 72.0 | 62.4 | 55.4 | 70.6 |
| Days Max Temp ≥ 90 °F | 0 | 1 | 5 | 12 | 19 | 26 | 29 | 30 | 23 | 10 | 1 | 0 | 156 |
| Days Max Temp ≤ 32 °F | 0 | 0 | 0 | 0 | 0 | 0 | 0 | 0 | 0 | 0 | 0 | 0 | 0 |
| Days Min Temp ≤ 32 °F | 7 | 4 | 1 | 0 | 0 | 0 | 0 | 0 | 0 | 0 | 2 | 5 | 19 |
| Days Min Temp ≤ 0 °F | 0 | 0 | 0 | 0 | 0 | 0 | 0 | 0 | 0 | 0 | 0 | 0 | 0 |
| Heating Degree Days | 400 | 274 | 106 | 20 | 2 | 0 | 0 | 0 | 2 | 23 | 151 | 313 | 1291 |
| Cooling Degree Days | 6 | 24 | ***109*** | 250 | 430 | 593 | 643 | ***666*** | 496 | 270 | ***87*** | ***27*** | 3601 |
| Total Precipitation (") | 1.20 | 1.09 | 0.95 | 1.77 | 3.11 | 2.39 | 1.55 | 2.22 | 3.60 | 2.56 | 0.92 | 1.09 | 22.45 |
| Days ≥ 0.1" Precip | 3 | 2 | 2 | 2 | 4 | 3 | 2 | 3 | 4 | 2 | 1 | 2 | 30 |
| Total Snowfall (") | 0.0 | 0.0 | 0.0 | 0.0 | 0.0 | 0.0 | 0.0 | 0.0 | 0.0 | 0.0 | 0.0 | 0.0 | 0.0 |
| Days ≥ 1" Snow Depth | 0 | 0 | 0 | 0 | 0 | 0 | 0 | 0 | 0 | 0 | 0 | 0 | 0 |

## EVANT 1 SSW *Coryell County* ELEVATION 1240 ft LAT/LONG 31° 27 ' N / 98° 13 ' W

| | JAN | FEB | MAR | APR | MAY | JUN | JUL | AUG | SEP | OCT | NOV | DEC | YEAR |
|---|---|---|---|---|---|---|---|---|---|---|---|---|---|
| Maximum Temp °F | 58.0 | 62.2 | 70.7 | 78.1 | 83.2 | ***90.5*** | 94.7 | 94.5 | 88.2 | 79.6 | 68.2 | 60.5 | 77.4 |
| Minimum Temp °F | 34.1 | 37.5 | 45.3 | 53.6 | 60.5 | ***67.9*** | 70.9 | 70.0 | 64.2 | 54.5 | 44.1 | 36.8 | 53.3 |
| Mean Temp °F | 46.1 | 49.9 | 58.0 | 65.9 | 71.9 | ***79.2*** | 82.8 | 82.3 | 76.2 | 67.1 | 56.2 | 48.6 | 65.4 |
| Days Max Temp ≥ 90 °F | 0 | 0 | 0 | 1 | ***5*** | ***18*** | 27 | 26 | 14 | 3 | 0 | 0 | 94 |
| Days Max Temp ≤ 32 °F | 1 | 1 | 0 | 0 | 0 | 0 | 0 | 0 | 0 | 0 | 0 | 1 | 3 |
| Days Min Temp ≤ 32 °F | 14 | 9 | 4 | 0 | 0 | 0 | 0 | 0 | 0 | 0 | 4 | 10 | 41 |
| Days Min Temp ≤ 0 °F | 0 | 0 | 0 | 0 | 0 | 0 | 0 | 0 | 0 | 0 | 0 | 0 | 0 |
| Heating Degree Days | 583 | 426 | 244 | 73 | 11 | ***0*** | 0 | 0 | 8 | 67 | 281 | 504 | 2197 |
| Cooling Degree Days | 1 | 5 | 28 | ***95*** | 226 | 429 | 552 | 552 | 352 | 138 | 25 | 2 | 2405 |
| Total Precipitation (") | 1.61 | 2.05 | 2.25 | 2.68 | 4.47 | 3.38 | 2.08 | 2.23 | 2.85 | 2.95 | 2.12 | 1.69 | 30.36 |
| Days ≥ 0.1" Precip | 3 | 4 | 4 | 4 | 6 | ***4*** | 3 | 3 | 4 | 4 | 3 | 3 | 45 |
| Total Snowfall (") | ***0.5*** | 0.6 | 0.1 | 0.0 | 0.0 | 0.0 | 0.0 | 0.0 | 0.0 | 0.0 | 0.3 | 0.0 | 1.5 |
| Days ≥ 1" Snow Depth | ***0*** | 0 | 0 | 0 | 0 | 0 | 0 | 0 | 0 | 0 | 0 | 0 | 0 |

## FAIRFIELD 3 W *Freestone County* ELEVATION 435 ft LAT/LONG 31° 44' N / 96° 6' W

| | JAN | FEB | MAR | APR | MAY | JUN | JUL | AUG | SEP | OCT | NOV | DEC | YEAR |
|---|---|---|---|---|---|---|---|---|---|---|---|---|---|
| Maximum Temp °F | 57.8 | 62.8 | 71.1 | 78.6 | 84.0 | 90.9 | ***95.1*** | 95.2 | 88.7 | 79.8 | 69.6 | 60.9 | 77.9 |
| Minimum Temp °F | 36.3 | 39.5 | 46.9 | 55.1 | 62.3 | 69.0 | ***71.6*** | 70.6 | 65.3 | 55.5 | 46.9 | 38.5 | 54.8 |
| Mean Temp °F | 47.1 | 51.2 | 59.1 | 66.9 | 73.2 | ***80.0*** | ***83.4*** | 82.9 | 77.0 | 67.7 | 58.3 | 49.7 | 66.4 |
| Days Max Temp ≥ 90 °F | 0 | 0 | 0 | 1 | 5 | 19 | ***28*** | 28 | 15 | 3 | ***0*** | 0 | 99 |
| Days Max Temp ≤ 32 °F | 1 | 0 | 0 | 0 | 0 | 0 | 0 | 0 | 0 | 0 | 0 | 0 | 1 |
| Days Min Temp ≤ 32 °F | 12 | 8 | 3 | 0 | 0 | 0 | 0 | 0 | 0 | 0 | 3 | 9 | 35 |
| Days Min Temp ≤ 0 °F | 0 | 0 | 0 | 0 | 0 | 0 | 0 | 0 | 0 | 0 | 0 | 0 | 0 |
| Heating Degree Days | 555 | 392 | 221 | 59 | 5 | 0 | ***0*** | 0 | 4 | 60 | 236 | 475 | 2007 |
| Cooling Degree Days | ***3*** | 9 | ***41*** | 118 | ***277*** | ***467*** | ***580*** | 590 | ***381*** | ***155*** | ***47*** | ***8*** | 2676 |
| Total Precipitation (") | 2.56 | 3.33 | 3.37 | 3.88 | 5.37 | 3.07 | 2.09 | 2.50 | 3.93 | 4.55 | ***3.83*** | 3.68 | 42.16 |
| Days ≥ 0.1" Precip | 5 | 5 | 5 | 5 | 6 | 5 | ***3*** | 4 | 5 | 5 | ***5*** | 5 | 58 |
| Total Snowfall (") | 0.5 | 0.3 | 0.0 | 0.0 | 0.0 | 0.0 | 0.0 | 0.0 | 0.0 | 0.0 | 0.0 | 0.1 | 0.9 |
| Days ≥ 1" Snow Depth | 0 | 0 | 0 | 0 | 0 | 0 | 0 | 0 | 0 | 0 | 0 | 0 | 0 |

## FALCON DAM *Starr County* ELEVATION 322 ft LAT/LONG 26° 33' N / 99° 8' W

| | JAN | FEB | MAR | APR | MAY | JUN | JUL | AUG | SEP | OCT | NOV | DEC | YEAR |
|---|---|---|---|---|---|---|---|---|---|---|---|---|---|
| Maximum Temp °F | 66.4 | 71.3 | 80.9 | 87.6 | 91.7 | 96.5 | 99.0 | 98.9 | 93.5 | 86.1 | 77.3 | 69.5 | 84.9 |
| Minimum Temp °F | 45.1 | 48.2 | 55.5 | 63.1 | 69.1 | 73.2 | 74.0 | 73.9 | 70.9 | 63.3 | 54.5 | 47.8 | 61.5 |
| Mean Temp °F | 55.8 | 59.7 | 68.2 | 75.4 | 80.4 | 84.9 | 86.5 | 86.4 | 82.2 | 74.7 | 66.0 | 58.7 | 73.2 |
| Days Max Temp ≥ 90 °F | 0 | 1 | 7 | 14 | 21 | 27 | 30 | 30 | 23 | 12 | 3 | 0 | 168 |
| Days Max Temp ≤ 32 °F | 0 | 0 | 0 | 0 | 0 | 0 | 0 | 0 | 0 | 0 | 0 | 0 | 0 |
| Days Min Temp ≤ 32 °F | 2 | 1 | 0 | 0 | 0 | 0 | 0 | 0 | 0 | 0 | 0 | 1 | 4 |
| Days Min Temp ≤ 0 °F | 0 | 0 | 0 | 0 | 0 | 0 | 0 | 0 | 0 | 0 | 0 | 0 | 0 |
| Heating Degree Days | 306 | 190 | 68 | 12 | 1 | 0 | 0 | 0 | 1 | 12 | 100 | 234 | 924 |
| Cooling Degree Days | 20 | 60 | 178 | 316 | 499 | 626 | 693 | 690 | 542 | 358 | 150 | 51 | 4183 |
| Total Precipitation (") | 1.06 | 1.19 | 0.50 | 1.45 | 2.73 | 2.57 | 1.62 | 2.45 | 3.75 | 1.61 | 0.96 | 0.93 | 20.82 |
| Days ≥ 0.1" Precip | 3 | 3 | 1 | 3 | 4 | 4 | 3 | 4 | 5 | 3 | 2 | 3 | 38 |
| Total Snowfall (") | 0.1 | 0.0 | 0.0 | 0.0 | 0.0 | 0.0 | 0.0 | 0.0 | 0.0 | 0.0 | 0.0 | 0.0 | 0.1 |
| Days ≥ 1" Snow Depth | 0 | 0 | 0 | 0 | 0 | 0 | 0 | 0 | 0 | 0 | 0 | 0 | 0 |

## FALFURRIAS *Brooks County* ELEVATION 112 ft LAT/LONG 27° 13' N / 98° 8' W

| | JAN | FEB | MAR | APR | MAY | JUN | JUL | AUG | SEP | OCT | NOV | DEC | YEAR |
|---|---|---|---|---|---|---|---|---|---|---|---|---|---|
| Maximum Temp °F | 66.5 | 70.5 | 78.7 | 85.5 | 88.9 | 93.7 | 96.4 | 96.6 | 92.0 | 85.4 | 76.6 | 69.7 | 83.4 |
| Minimum Temp °F | 43.4 | 46.1 | 53.1 | 61.3 | 67.4 | 71.9 | 73.1 | 72.2 | 69.2 | 60.9 | 52.1 | 46.5 | 59.8 |
| Mean Temp °F | 54.9 | 58.3 | 65.9 | 73.4 | 78.2 | 82.9 | 84.8 | 84.4 | 80.6 | 73.2 | 64.4 | 58.1 | 71.6 |
| Days Max Temp ≥ 90 °F | 0 | 1 | 3 | 10 | 16 | 24 | 28 | 29 | 21 | 10 | 1 | 0 | 143 |
| Days Max Temp ≤ 32 °F | 0 | 0 | 0 | 0 | 0 | 0 | 0 | 0 | 0 | 0 | 0 | 0 | 0 |
| Days Min Temp ≤ 32 °F | 4 | 2 | 1 | 0 | 0 | 0 | 0 | 0 | 0 | 0 | 1 | 3 | 11 |
| Days Min Temp ≤ 0 °F | 0 | 0 | 0 | 0 | 0 | 0 | 0 | 0 | 0 | 0 | 0 | 0 | 0 |
| Heating Degree Days | 335 | 222 | 93 | 18 | 1 | 0 | 0 | 0 | 1 | 18 | 127 | 257 | 1072 |
| Cooling Degree Days | 23 | 47 | 131 | 260 | 432 | 561 | 625 | 632 | 493 | 305 | 132 | 57 | 3698 |
| Total Precipitation (") | 1.42 | 1.78 | 0.78 | 1.54 | 3.49 | 3.52 | 2.05 | 2.54 | 4.78 | 2.89 | 1.11 | 1.30 | 27.20 |
| Days ≥ 0.1" Precip | 3 | 3 | 1 | 3 | 4 | 5 | 3 | 3 | 6 | 4 | 2 | 3 | 40 |
| Total Snowfall (") | 0.1 | 0.0 | 0.0 | 0.0 | 0.0 | 0.0 | 0.0 | 0.0 | 0.0 | 0.0 | 0.0 | 0.0 | 0.1 |
| Days ≥ 1" Snow Depth | 0 | 0 | 0 | 0 | 0 | 0 | 0 | 0 | 0 | 0 | 0 | 0 | 0 |

## FERRIS *Ellis County* ELEVATION 440 ft LAT/LONG 32° 33' N / 96° 40' W

| | JAN | FEB | MAR | APR | MAY | JUN | JUL | AUG | SEP | OCT | NOV | DEC | YEAR |
|---|---|---|---|---|---|---|---|---|---|---|---|---|---|
| Maximum Temp °F | 55.9 | 61.0 | 69.8 | 77.9 | 83.5 | 91.3 | 96.0 | 96.1 | 88.7 | 78.9 | 67.7 | 59.0 | 77.2 |
| Minimum Temp °F | 33.7 | 37.7 | 45.7 | 54.0 | 61.7 | 68.8 | 72.0 | 71.2 | 65.4 | 54.8 | 44.9 | 36.8 | 53.9 |
| Mean Temp °F | 44.8 | 49.4 | 57.8 | 66.0 | 72.6 | 80.1 | 84.0 | 83.7 | 77.0 | 66.9 | 56.3 | 47.9 | 65.5 |
| Days Max Temp ≥ 90 °F | 0 | 0 | 0 | 1 | 4 | 20 | 28 | 27 | 16 | 3 | 0 | 0 | 99 |
| Days Max Temp ≤ 32 °F | 1 | 1 | 0 | 0 | 0 | 0 | 0 | 0 | 0 | 0 | 0 | 1 | 3 |
| Days Min Temp ≤ 32 °F | 15 | 9 | 3 | 0 | 0 | 0 | 0 | 0 | 0 | 0 | 4 | 11 | 42 |
| Days Min Temp ≤ 0 °F | 0 | 0 | 0 | 0 | 0 | 0 | 0 | 0 | 0 | 0 | 0 | 0 | 0 |
| Heating Degree Days | 621 | 439 | 250 | 70 | 9 | 0 | 0 | 0 | 6 | 70 | 281 | 527 | 2273 |
| Cooling Degree Days | 1 | 4 | 26 | 92 | 242 | 440 | 583 | 599 | 387 | 132 | 30 | 3 | 2539 |
| Total Precipitation (") | 2.28 | 3.19 | 3.23 | 3.80 | 5.19 | 3.23 | 2.11 | 1.98 | 3.22 | 4.64 | 2.84 | 2.88 | 38.59 |
| Days ≥ 0.1" Precip | 4 | 5 | 5 | 4 | 6 | 4 | 3 | 3 | 5 | 5 | 4 | 5 | 53 |
| Total Snowfall (") | 0.7 | 0.8 | 0.1 | 0.0 | 0.0 | 0.0 | 0.0 | 0.0 | 0.0 | 0.0 | 0.1 | 0.3 | 2.0 |
| Days ≥ 1" Snow Depth | 0 | 1 | 0 | 0 | 0 | 0 | 0 | 0 | 0 | 0 | 0 | 0 | 1 |

## FLATONIA *Fayette County* ELEVATION 469 ft LAT/LONG 29° 41' N / 97° 6' W

| | JAN | FEB | MAR | APR | MAY | JUN | JUL | AUG | SEP | OCT | NOV | DEC | YEAR |
|---|---|---|---|---|---|---|---|---|---|---|---|---|---|
| Maximum Temp °F | 61.3 | 65.4 | 73.4 | 80.0 | 84.9 | 91.1 | 94.8 | 95.4 | 89.8 | 82.3 | 72.1 | 64.6 | 79.6 |
| Minimum Temp °F | 40.8 | 43.8 | 51.2 | 59.0 | 64.8 | 70.4 | 72.3 | 72.0 | 67.7 | 59.4 | 50.5 | 43.5 | 58.0 |
| Mean Temp °F | 51.1 | 54.6 | 62.3 | 69.5 | 74.9 | 80.8 | 83.6 | 83.7 | 78.7 | 70.9 | 61.3 | 54.1 | 68.8 |
| Days Max Temp ≥ 90 °F | 0 | 0 | 1 | 1 | 6 | 20 | 28 | 28 | 18 | 5 | 0 | 0 | 107 |
| Days Max Temp ≤ 32 °F | 0 | 0 | 0 | 0 | 0 | 0 | 0 | 0 | 0 | 0 | 0 | 0 | 0 |
| Days Min Temp ≤ 32 °F | 7 | 4 | 1 | 0 | 0 | 0 | 0 | 0 | 0 | 0 | 2 | 5 | 19 |
| Days Min Temp ≤ 0 °F | 0 | 0 | 0 | 0 | 0 | 0 | 0 | 0 | 0 | 0 | 0 | 0 | 0 |
| Heating Degree Days | 439 | 306 | 147 | 33 | 2 | 0 | 0 | 0 | 2 | 30 | 176 | 355 | 1490 |
| Cooling Degree Days | 7 | 21 | 64 | 167 | 319 | 487 | 586 | 602 | 433 | 226 | 73 | 24 | 3009 |
| Total Precipitation (") | 2.88 | 2.63 | 2.38 | 3.20 | 5.67 | 4.13 | 1.89 | 2.49 | 4.47 | 3.70 | 2.78 | 2.65 | 38.87 |
| Days ≥ 0.1" Precip | 5 | 5 | 4 | 4 | 7 | 5 | 4 | 4 | 5 | 4 | 4 | 5 | 56 |
| Total Snowfall (") | 0.3 | 0.1 | 0.0 | 0.0 | 0.0 | 0.0 | 0.0 | 0.0 | 0.0 | 0.0 | 0.0 | 0.0 | 0.4 |
| Days ≥ 1" Snow Depth | 0 | 0 | 0 | 0 | 0 | 0 | 0 | 0 | 0 | 0 | 0 | 0 | 0 |

## FLORESVILLE *Wilson County* ELEVATION 390 ft LAT/LONG 29° 8' N / 98° 9' W

| | JAN | FEB | MAR | APR | MAY | JUN | JUL | AUG | SEP | OCT | NOV | DEC | YEAR |
|---|---|---|---|---|---|---|---|---|---|---|---|---|---|
| Maximum Temp °F | 63.4 | 67.7 | 76.2 | 82.5 | 86.9 | 92.8 | 96.2 | 96.5 | 91.6 | 84.2 | 74.8 | 67.6 | 81.7 |
| Minimum Temp °F | 37.5 | 40.9 | 49.2 | 57.7 | 64.7 | 70.6 | 72.9 | 72.2 | 67.5 | 57.4 | 47.8 | 40.1 | 56.5 |
| Mean Temp °F | 50.6 | 54.4 | 62.7 | 70.1 | 75.9 | 81.7 | 84.6 | 84.3 | 79.6 | 70.9 | 61.4 | 54.0 | 69.2 |
| Days Max Temp ≥ 90 °F | 0 | 0 | 1 | 4 | 11 | 24 | 29 | 29 | 21 | 8 | 0 | 0 | 127 |
| Days Max Temp ≤ 32 °F | 0 | 0 | 0 | 0 | 0 | 0 | 0 | 0 | 0 | 0 | 0 | 0 | 0 |
| Days Min Temp ≤ 32 °F | 11 | 6 | 2 | 0 | 0 | 0 | 0 | 0 | 0 | 0 | 3 | 9 | 31 |
| Days Min Temp ≤ 0 °F | 0 | 0 | 0 | 0 | 0 | 0 | 0 | 0 | 0 | 0 | 0 | 0 | 0 |
| Heating Degree Days | 451 | 309 | 141 | 33 | 2 | 0 | 0 | 0 | 1 | 34 | 174 | 356 | 1501 |
| Cooling Degree Days | 6 | 16 | 67 | 182 | 358 | 526 | 629 | 634 | 458 | 235 | 76 | 18 | 3205 |
| Total Precipitation (") | 1.98 | 2.02 | 1.40 | 2.55 | 4.25 | 3.10 | 1.89 | 2.36 | 3.56 | 2.64 | 2.02 | 1.64 | 29.41 |
| Days ≥ 0.1" Precip | 4 | 4 | 3 | 3 | 5 | 4 | 3 | 4 | 4 | 4 | 3 | 3 | 44 |
| Total Snowfall (") | 0.4 | 0.0 | 0.0 | 0.0 | 0.0 | 0.0 | 0.0 | 0.0 | 0.0 | 0.0 | 0.0 | 0.0 | 0.4 |
| Days ≥ 1" Snow Depth | 0 | 0 | 0 | 0 | 0 | 0 | 0 | 0 | 0 | 0 | 0 | 0 | 0 |

## FLOYDADA *Floyd County* ELEVATION 3182 ft LAT/LONG 33° 59' N / 101° 20' W

| | JAN | FEB | MAR | APR | MAY | JUN | JUL | AUG | SEP | OCT | NOV | DEC | YEAR |
|---|---|---|---|---|---|---|---|---|---|---|---|---|---|
| Maximum Temp °F | 50.1 | 54.7 | 63.6 | 73.3 | 80.3 | 88.5 | 91.8 | 89.6 | 82.3 | 73.4 | 60.8 | 52.1 | 71.7 |
| Minimum Temp °F | 22.7 | 26.0 | 33.3 | 43.2 | 52.4 | 62.1 | 65.9 | 64.0 | 56.7 | 45.3 | 33.7 | 25.5 | 44.2 |
| Mean Temp °F | 36.4 | 40.4 | 48.5 | 58.3 | 66.4 | 75.3 | 78.9 | 76.8 | 69.5 | 59.3 | 47.3 | 38.8 | 58.0 |
| Days Max Temp ≥ 90 °F | 0 | 0 | 0 | 1 | 5 | 15 | 21 | 18 | 7 | 1 | 0 | 0 | 68 |
| Days Max Temp ≤ 32 °F | 4 | 2 | 1 | 0 | 0 | 0 | 0 | 0 | 0 | 0 | 1 | 2 | 10 |
| Days Min Temp ≤ 32 °F | 28 | 22 | 14 | 3 | 0 | 0 | 0 | 0 | 0 | 2 | 13 | 26 | 108 |
| Days Min Temp ≤ 0 °F | 0 | 0 | 0 | 0 | 0 | 0 | 0 | 0 | 0 | 0 | 0 | 0 | 0 |
| Heating Degree Days | 879 | 688 | 509 | 225 | 70 | 5 | 0 | 2 | 42 | 205 | 526 | 804 | 3955 |
| Cooling Degree Days | 0 | 0 | 3 | 30 | 124 | 331 | 450 | 403 | 195 | 35 | 1 | 0 | 1572 |
| Total Precipitation (") | 0.44 | 0.68 | 0.98 | 1.38 | 2.85 | 3.43 | 2.14 | 2.51 | 2.99 | 1.65 | 0.84 | 0.56 | 20.45 |
| Days ≥ 0.1" Precip | 1 | 2 | 2 | 3 | 5 | 5 | 4 | 4 | 5 | 3 | 2 | 2 | 38 |
| Total Snowfall (") | 2.3 | 2.0 | 0.5 | 0.2 | 0.0 | 0.0 | 0.0 | 0.0 | 0.0 | 0.2 | 0.9 | 1.1 | 7.2 |
| Days ≥ 1" Snow Depth | 2 | *1* | 0 | 0 | 0 | 0 | 0 | 0 | 0 | 0 | 0 | 1 | 4 |

## FOLLETT *Lipscomb County* ELEVATION 2802 ft LAT/LONG 36° 26' N / 100° 8' W

| | JAN | FEB | MAR | APR | MAY | JUN | JUL | AUG | SEP | OCT | NOV | DEC | YEAR |
|---|---|---|---|---|---|---|---|---|---|---|---|---|---|
| Maximum Temp °F | 45.8 | 51.1 | 59.5 | 70.2 | 77.2 | 86.6 | 92.3 | 90.8 | 82.4 | 71.7 | 57.9 | 48.5 | 69.5 |
| Minimum Temp °F | 20.1 | 24.0 | 32.2 | 42.3 | 51.9 | 61.8 | 66.6 | 65.2 | 56.5 | 44.5 | 32.4 | 23.7 | 43.4 |
| Mean Temp °F | 33.0 | 37.6 | 45.9 | 56.3 | 64.6 | 74.2 | 79.5 | 78.0 | 69.5 | 58.1 | 45.2 | 36.1 | 56.5 |
| Days Max Temp ≥ 90 °F | 0 | 0 | 0 | 1 | 3 | 12 | 21 | 19 | 8 | 1 | 0 | 0 | 65 |
| Days Max Temp ≤ 32 °F | 6 | 4 | 1 | 0 | 0 | 0 | 0 | 0 | 0 | 0 | 1 | 4 | 16 |
| Days Min Temp ≤ 32 °F | 28 | 23 | 15 | 4 | 0 | 0 | 0 | 0 | 0 | 3 | 15 | 27 | 115 |
| Days Min Temp ≤ 0 °F | 1 | 1 | 0 | 0 | 0 | 0 | 0 | 0 | 0 | 0 | 0 | 1 | 3 |
| Heating Degree Days | 986 | 768 | 589 | 277 | 99 | 10 | 1 | 2 | 51 | 242 | 589 | 888 | 4502 |
| Cooling Degree Days | 0 | 0 | 1 | 23 | 87 | 290 | 445 | 424 | 197 | 28 | 2 | 0 | 1497 |
| Total Precipitation (") | 0.54 | 0.99 | 1.96 | 1.73 | 3.59 | 3.26 | 2.35 | 2.99 | 2.11 | 1.29 | 1.09 | 0.75 | 22.65 |
| Days ≥ 0.1" Precip | 2 | 3 | 4 | 3 | 6 | 5 | 4 | 5 | 4 | 2 | 3 | 2 | 43 |
| Total Snowfall (") | 4.7 | 5.1 | 4.6 | 1.2 | 0.1 | 0.0 | 0.0 | 0.0 | 0.0 | 0.3 | 2.4 | 3.8 | 22.2 |
| Days ≥ 1" Snow Depth | 3 | *3* | *2* | 0 | 0 | 0 | 0 | 0 | 0 | 0 | 1 | *2* | 11 |

## FORT STOCKTON *Pecos County* ELEVATION 2933 ft LAT/LONG 30° 54 ' N / 102° 52 ' W

| | JAN | FEB | MAR | APR | MAY | JUN | JUL | AUG | SEP | OCT | NOV | DEC | YEAR |
|---|---|---|---|---|---|---|---|---|---|---|---|---|---|
| Maximum Temp °F | 59.7 | 64.9 | 72.9 | 82.0 | 88.1 | 94.3 | 94.9 | 93.6 | 87.6 | 80.1 | 69.9 | 62.2 | 79.2 |
| Minimum Temp °F | 30.6 | 33.6 | 40.6 | 48.9 | 57.1 | 65.4 | 67.8 | 66.2 | 60.9 | 50.1 | 39.2 | 32.3 | 49.4 |
| Mean Temp °F | 45.2 | 49.3 | 56.8 | 65.5 | 72.6 | 79.9 | 81.4 | 80.0 | 74.3 | 65.1 | 54.6 | 47.3 | 64.3 |
| Days Max Temp ≥ 90 °F | 0 | 0 | 1 | 7 | 15 | 23 | 26 | 25 | 15 | 5 | 0 | 0 | 117 |
| Days Max Temp ≤ 32 °F | 1 | 0 | 0 | 0 | 0 | 0 | 0 | 0 | 0 | 0 | 0 | 1 | 2 |
| Days Min Temp ≤ 32 °F | 18 | 13 | 6 | 1 | 0 | 0 | 0 | 0 | 0 | 0 | 7 | 16 | 61 |
| Days Min Temp ≤ 0 °F | 0 | 0 | 0 | 0 | 0 | 0 | 0 | 0 | 0 | 0 | 0 | 0 | 0 |
| Heating Degree Days | 608 | 439 | 270 | 85 | 18 | 1 | 0 | 0 | 17 | 93 | 316 | 544 | 2391 |
| Cooling Degree Days | 0 | 4 | 27 | 119 | 285 | 486 | 541 | 503 | 329 | 115 | 15 | 0 | 2424 |
| Total Precipitation (") | 0.53 | 0.52 | 0.46 | 0.72 | 1.70 | 1.58 | 1.45 | 1.93 | 2.87 | 1.54 | 0.75 | 0.65 | 14.70 |
| Days ≥ 0.1" Precip | 2 | 2 | 1 | 2 | 3 | 3 | 3 | 4 | 4 | 3 | 2 | 1 | 30 |
| Total Snowfall (") | 0.7 | ***0.0*** | 0.1 | 0.0 | 0.0 | 0.0 | 0.0 | 0.0 | 0.0 | 0.0 | 0.2 | 0.3 | 1.3 |
| Days ≥ 1" Snow Depth | ***0*** | 0 | 0 | 0 | 0 | 0 | 0 | 0 | 0 | 0 | 0 | 0 | 0 |

## FOWLERTON *La Salle County* ELEVATION 347 ft LAT/LONG 28° 29 ' N / 98° 52 ' W

| | JAN | FEB | MAR | APR | MAY | JUN | JUL | AUG | SEP | OCT | NOV | DEC | YEAR |
|---|---|---|---|---|---|---|---|---|---|---|---|---|---|
| Maximum Temp °F | ***67.2*** | 71.5 | 79.8 | 86.9 | ***89.9*** | 95.8 | 99.0 | 98.6 | ***93.6*** | 86.3 | ***77.3*** | 69.8 | 84.6 |
| Minimum Temp °F | ***40.7*** | 43.3 | 51.0 | 58.6 | 65.6 | 70.9 | 72.5 | 72.1 | ***68.3*** | 59.5 | ***49.1*** | 42.4 | 57.8 |
| Mean Temp °F | ***54.0*** | 57.4 | 65.4 | 72.8 | ***77.8*** | 83.4 | 85.8 | 85.4 | ***81.0*** | 72.9 | ***63.2*** | 56.1 | 71.3 |
| Days Max Temp ≥ 90 °F | 0 | 1 | 4 | 11 | ***17*** | 27 | 30 | 29 | 23 | 11 | 1 | 0 | 154 |
| Days Max Temp ≤ 32 °F | 0 | 0 | 0 | 0 | 0 | 0 | 0 | 0 | 0 | 0 | 0 | 0 | 0 |
| Days Min Temp ≤ 32 °F | 7 | 4 | 1 | 0 | 0 | 0 | 0 | 0 | 0 | 0 | 2 | 6 | 20 |
| Days Min Temp ≤ 0 °F | 0 | 0 | 0 | 0 | 0 | 0 | 0 | 0 | 0 | 0 | 0 | 0 | 0 |
| Heating Degree Days | ***352*** | 239 | 104 | 20 | 1 | 0 | 0 | 0 | ***1*** | 17 | ***137*** | 292 | 1163 |
| Cooling Degree Days | na | ***31*** | 98 | ***219*** | na | 573 | 667 | 656 | ***494*** | ***278*** | na | ***19*** | na |
| Total Precipitation (") | ***1.15*** | 1.14 | 0.88 | 1.69 | 3.32 | 2.33 | 1.47 | 2.24 | 3.38 | 3.30 | 1.05 | 1.17 | 23.12 |
| Days ≥ 0.1" Precip | ***2*** | 2 | 1 | 2 | ***4*** | 3 | 2 | 2 | ***4*** | 3 | 2 | 2 | 29 |
| Total Snowfall (") | 0.0 | 0.0 | 0.0 | 0.0 | 0.0 | 0.0 | 0.0 | 0.0 | 0.0 | 0.0 | 0.0 | 0.0 | 0.0 |
| Days ≥ 1" Snow Depth | 0 | 0 | 0 | 0 | 0 | 0 | 0 | 0 | 0 | 0 | 0 | 0 | 0 |

## FRANKLIN *Robertson County* ELEVATION 459 ft LAT/LONG 31° 2 ' N / 96° 29 ' W

| | JAN | FEB | MAR | APR | MAY | JUN | JUL | AUG | SEP | OCT | NOV | DEC | YEAR |
|---|---|---|---|---|---|---|---|---|---|---|---|---|---|
| Maximum Temp °F | 58.4 | 63.0 | 71.0 | 78.1 | 83.4 | 90.2 | 94.9 | 95.4 | 88.9 | 80.2 | 69.7 | 62.0 | 77.9 |
| Minimum Temp °F | 37.4 | 40.6 | 47.9 | 55.8 | 62.6 | 69.1 | 71.4 | 70.9 | 65.8 | 55.9 | 46.9 | 40.1 | 55.4 |
| Mean Temp °F | 47.9 | 51.8 | 59.5 | 67.0 | 73.1 | 79.7 | 83.2 | 83.2 | 77.4 | 68.1 | 58.3 | 51.1 | 66.7 |
| Days Max Temp ≥ 90 °F | 0 | 0 | 0 | 1 | 3 | 18 | 28 | 28 | 15 | 3 | 0 | 0 | 96 |
| Days Max Temp ≤ 32 °F | 1 | 0 | 0 | 0 | 0 | 0 | 0 | 0 | 0 | 0 | 0 | 0 | 1 |
| Days Min Temp ≤ 32 °F | 11 | 6 | 2 | 0 | 0 | 0 | 0 | 0 | 0 | 0 | 3 | 7 | 29 |
| Days Min Temp ≤ 0 °F | 0 | 0 | 0 | 0 | 0 | 0 | 0 | 0 | 0 | 0 | 0 | 0 | 0 |
| Heating Degree Days | 530 | 375 | 208 | 56 | 5 | 0 | 0 | 0 | 3 | 55 | 235 | 435 | 1902 |
| Cooling Degree Days | 4 | 9 | 41 | 116 | 266 | 448 | 570 | 581 | 385 | 162 | 44 | 9 | 2635 |
| Total Precipitation (") | 2.97 | 3.06 | 2.87 | 3.46 | 4.99 | 2.74 | 1.85 | 2.40 | 3.86 | 4.55 | 2.87 | 3.50 | 39.12 |
| Days ≥ 0.1" Precip | 5 | 5 | 5 | 5 | 6 | 5 | 3 | 3 | 5 | 6 | 4 | 5 | 57 |
| Total Snowfall (") | 0.3 | 0.6 | 0.1 | 0.0 | 0.0 | 0.0 | 0.0 | 0.0 | 0.0 | 0.0 | 0.0 | 0.0 | 1.0 |
| Days ≥ 1" Snow Depth | 0 | 0 | 0 | 0 | 0 | 0 | 0 | 0 | 0 | 0 | 0 | 0 | 0 |

## FREDERICKSBURG *Gillespie County* ELEVATION 1722 ft LAT/LONG 30° 17 ' N / 98° 52 ' W

| | JAN | FEB | MAR | APR | MAY | JUN | JUL | AUG | SEP | OCT | NOV | DEC | YEAR |
|---|---|---|---|---|---|---|---|---|---|---|---|---|---|
| Maximum Temp °F | 59.7 | 63.8 | 71.7 | 78.7 | 83.2 | 89.3 | 92.7 | 92.8 | 87.0 | 79.3 | 69.0 | 62.0 | 77.4 |
| Minimum Temp °F | 35.9 | 38.9 | 46.5 | 54.4 | 61.2 | 67.1 | 69.0 | 68.2 | 63.6 | 54.9 | 45.0 | 38.2 | 53.6 |
| Mean Temp °F | 47.8 | 51.3 | 59.1 | 66.6 | 72.3 | 78.2 | 80.9 | 80.5 | 75.3 | 67.1 | 57.0 | 50.1 | 65.5 |
| Days Max Temp ≥ 90 °F | 0 | 0 | 1 | 2 | 4 | 14 | 24 | 25 | 11 | 1 | 0 | 0 | 82 |
| Days Max Temp ≤ 32 °F | 0 | 0 | 0 | 0 | 0 | 0 | 0 | 0 | 0 | 0 | 0 | 0 | 0 |
| Days Min Temp ≤ 32 °F | 13 | 8 | 3 | 0 | 0 | 0 | 0 | 0 | 0 | 0 | 4 | 10 | 38 |
| Days Min Temp ≤ 0 °F | 0 | 0 | 0 | 0 | 0 | 0 | 0 | 0 | 0 | 0 | 0 | 0 | 0 |
| Heating Degree Days | 529 | 385 | 214 | 60 | 8 | 0 | 0 | 0 | 6 | 63 | 260 | 461 | 1986 |
| Cooling Degree Days | 1 | 5 | 37 | 108 | 245 | 413 | 498 | 502 | 323 | 144 | 27 | 4 | 2307 |
| Total Precipitation (") | 1.53 | 1.97 | 1.67 | 2.52 | 4.52 | 3.70 | 2.15 | 2.82 | 3.65 | 3.65 | 2.04 | 1.91 | 32.13 |
| Days ≥ 0.1" Precip | 4 | 4 | 4 | 4 | 6 | 5 | 3 | 4 | 5 | 4 | 3 | 3 | 49 |
| Total Snowfall (") | ***0.1*** | 0.2 | 0.1 | 0.0 | 0.0 | 0.0 | 0.0 | 0.0 | 0.0 | 0.0 | 0.0 | 0.1 | 0.5 |
| Days ≥ 1" Snow Depth | ***0*** | 0 | 0 | 0 | 0 | 0 | 0 | 0 | 0 | 0 | 0 | 0 | 0 |

## FREEPORT 2 NW *Brazoria County* ELEVATION 10 ft LAT/LONG 28° 57' N / 95° 20' W

| | JAN | FEB | MAR | APR | MAY | JUN | JUL | AUG | SEP | OCT | NOV | DEC | YEAR |
|---|---|---|---|---|---|---|---|---|---|---|---|---|---|
| Maximum Temp °F | 62.1 | 64.7 | 71.4 | 77.8 | 83.5 | 89.1 | 91.6 | 91.7 | 88.1 | 81.5 | 73.3 | 66.6 | 78.5 |
| Minimum Temp °F | 44.5 | 46.6 | 53.6 | 61.9 | 68.5 | 74.5 | 76.6 | 76.0 | 71.9 | 63.0 | 54.0 | 48.0 | 61.6 |
| Mean Temp °F | 53.3 | 55.7 | 62.6 | 69.9 | 76.0 | 81.8 | 84.2 | 83.9 | 80.0 | 72.3 | 63.7 | 57.3 | 70.1 |
| Days Max Temp ≥ 90 °F | 0 | 0 | 0 | 0 | 2 | 14 | 25 | 24 | 13 | 1 | 0 | 0 | 79 |
| Days Max Temp ≤ 32 °F | 0 | 0 | 0 | 0 | 0 | 0 | 0 | 0 | 0 | 0 | 0 | 0 | 0 |
| Days Min Temp ≤ 32 °F | 4 | 2 | 0 | 0 | 0 | 0 | 0 | 0 | 0 | 0 | 0 | 2 | 8 |
| Days Min Temp ≤ 0 °F | 0 | 0 | 0 | 0 | 0 | 0 | 0 | 0 | 0 | 0 | 0 | 0 | 0 |
| Heating Degree Days | 368 | 272 | 128 | 24 | 1 | 0 | 0 | 0 | 1 | 18 | 132 | 266 | 1210 |
| Cooling Degree Days | 8 | 17 | 60 | 177 | 363 | 529 | 613 | 610 | 464 | 261 | 100 | 38 | 3240 |
| Total Precipitation (") | 4.69 | 3.20 | 2.72 | 3.17 | 4.62 | 4.97 | 5.33 | 4.47 | 7.34 | 4.81 | 3.82 | 3.77 | 52.91 |
| Days ≥ 0.1" Precip | 7 | 5 | 4 | 3 | 5 | 5 | 6 | 7 | 7 | 5 | 5 | 6 | 65 |
| Total Snowfall (") | 0.0 | 0.0 | 0.0 | 0.0 | 0.0 | 0.0 | 0.0 | 0.0 | 0.0 | 0.0 | 0.0 | 0.0 | 0.0 |
| Days ≥ 1" Snow Depth | 0 | 0 | 0 | 0 | 0 | 0 | 0 | 0 | 0 | 0 | 0 | 0 | 0 |

## FREER *Duval County* ELEVATION 531 ft LAT/LONG 27° 52' N / 98° 37' W

| | JAN | FEB | MAR | APR | MAY | JUN | JUL | AUG | SEP | OCT | NOV | DEC | YEAR |
|---|---|---|---|---|---|---|---|---|---|---|---|---|---|
| Maximum Temp °F | *66.1* | *69.4* | 77.8 | 85.0 | 88.6 | 94.0 | 96.2 | 96.7 | 91.6 | *84.9* | *76.2* | *69.3* | 83.0 |
| Minimum Temp °F | 41.8 | *44.3* | *52.1* | 60.2 | 66.0 | *70.9* | 72.1 | *72.5* | 68.6 | *60.8* | 50.6 | *45.8* | 58.8 |
| Mean Temp °F | *53.9* | *56.6* | *65.0* | 72.7 | *77.3* | *82.6* | *84.2* | *84.5* | *79.9* | *72.7* | *63.2* | *57.6* | 70.9 |
| Days Max Temp ≥ 90 °F | *0* | *0* | 2 | 8 | 15 | 24 | 28 | 29 | 20 | *8* | *1* | *0* | 135 |
| Days Max Temp ≤ 32 °F | *0* | *0* | 0 | 0 | 0 | 0 | 0 | 0 | 0 | *0* | *0* | *0* | 0 |
| Days Min Temp ≤ 32 °F | 6 | *3* | *1* | 0 | 0 | 0 | 0 | 0 | 0 | *0* | 1 | *3* | 14 |
| Days Min Temp ≤ 0 °F | 0 | *0* | *0* | 0 | 0 | 0 | 0 | 0 | 0 | 0 | 0 | *0* | 0 |
| Heating Degree Days | *353* | *252* | *100* | 18 | *2* | 0 | *0* | *0* | *1* | *16* | *132* | *258* | 1132 |
| Cooling Degree Days | na | na | na | *262* | na | *551* | *617* | na | *478* | na | na | na | na |
| Total Precipitation (") | 1.39 | 1.39 | 1.17 | 1.72 | 4.18 | 3.48 | 1.57 | 2.24 | 3.86 | 2.97 | 1.26 | 1.26 | 26.49 |
| Days ≥ 0.1" Precip | 3 | 3 | 2 | 3 | 4 | 4 | 3 | 3 | 4 | 3 | 2 | 3 | 37 |
| Total Snowfall (") | 0.4 | 0.1 | 0.0 | 0.0 | 0.0 | 0.0 | 0.0 | 0.0 | 0.0 | 0.0 | 0.0 | 0.0 | 0.5 |
| Days ≥ 1" Snow Depth | 0 | 0 | 0 | 0 | 0 | 0 | 0 | 0 | 0 | 0 | 0 | 0 | 0 |

## FRIONA *Parmer County* ELEVATION 4012 ft LAT/LONG 34° 38' N / 102° 43' W

| | JAN | FEB | MAR | APR | MAY | JUN | JUL | AUG | SEP | OCT | NOV | DEC | YEAR |
|---|---|---|---|---|---|---|---|---|---|---|---|---|---|
| Maximum Temp °F | 50.0 | 54.4 | 62.4 | 71.5 | 79.1 | 87.7 | 90.0 | 87.8 | 81.3 | 72.3 | 59.8 | 51.6 | 70.7 |
| Minimum Temp °F | 21.5 | 24.7 | 31.3 | 40.0 | 49.4 | 58.7 | 62.7 | 61.1 | 54.0 | 42.2 | 30.9 | 23.2 | 41.6 |
| Mean Temp °F | 35.8 | 39.6 | 46.9 | 55.8 | 64.3 | 73.2 | 76.4 | 74.5 | 67.7 | 57.2 | 45.4 | 37.4 | 56.2 |
| Days Max Temp ≥ 90 °F | 0 | 0 | 0 | 0 | 4 | 14 | 19 | 14 | 6 | 1 | 0 | 0 | 58 |
| Days Max Temp ≤ 32 °F | 4 | 2 | 1 | 0 | 0 | 0 | 0 | 0 | 0 | 0 | 1 | 3 | 11 |
| Days Min Temp ≤ 32 °F | 28 | 23 | 17 | 5 | 0 | 0 | 0 | 0 | 0 | 3 | 17 | 27 | 120 |
| Days Min Temp ≤ 0 °F | 0 | 0 | 0 | 0 | 0 | 0 | 0 | 0 | 0 | 0 | 0 | 1 | 1 |
| Heating Degree Days | 900 | 711 | 554 | 282 | 98 | 11 | 1 | 2 | 53 | 252 | 581 | 848 | 4293 |
| Cooling Degree Days | 0 | 0 | 0 | 11 | 73 | 258 | 345 | 309 | 139 | 13 | 0 | 0 | 1148 |
| Total Precipitation (") | 0.53 | 0.56 | 0.77 | 0.88 | 2.15 | 2.69 | 2.17 | 3.01 | 2.12 | 1.35 | 0.71 | 0.61 | 17.55 |
| Days ≥ 0.1" Precip | 2 | 2 | 2 | 3 | 4 | 5 | 4 | 5 | 4 | 3 | 2 | 2 | 38 |
| Total Snowfall (") | 3.9 | 3.8 | 1.6 | 0.7 | 0.0 | 0.0 | 0.0 | 0.0 | 0.0 | 0.3 | 2.0 | 2.9 | 15.2 |
| Days ≥ 1" Snow Depth | 3 | 2 | 1 | 0 | 0 | 0 | 0 | 0 | 0 | 0 | 1 | 2 | 9 |

## GAIL *Borden County* ELEVATION 2533 ft LAT/LONG 32° 46' N / 101° 26' W

| | JAN | FEB | MAR | APR | MAY | JUN | JUL | AUG | SEP | OCT | NOV | DEC | YEAR |
|---|---|---|---|---|---|---|---|---|---|---|---|---|---|
| Maximum Temp °F | 56.2 | 61.4 | 70.4 | 78.7 | 85.2 | 91.9 | 94.4 | 92.5 | 85.3 | 77.5 | 66.2 | 58.5 | 76.5 |
| Minimum Temp °F | 30.9 | 34.3 | 41.6 | 50.4 | 58.3 | 66.0 | 69.3 | 68.0 | 61.4 | 52.0 | 40.5 | 33.2 | 50.5 |
| Mean Temp °F | 43.6 | 47.9 | 56.0 | 64.6 | 71.7 | 79.0 | 81.9 | 80.3 | 73.4 | 64.8 | 53.4 | 45.9 | 63.5 |
| Days Max Temp ≥ 90 °F | 0 | 0 | 1 | 3 | 10 | 20 | 26 | 23 | 10 | 3 | 0 | 0 | 96 |
| Days Max Temp ≤ 32 °F | 2 | 1 | 0 | 0 | 0 | 0 | 0 | 0 | 0 | 0 | 0 | 1 | 4 |
| Days Min Temp ≤ 32 °F | 17 | 12 | 6 | 0 | 0 | 0 | 0 | 0 | 0 | 0 | 6 | 14 | 55 |
| Days Min Temp ≤ 0 °F | 0 | 0 | 0 | 0 | 0 | 0 | 0 | 0 | 0 | 0 | 0 | 0 | 0 |
| Heating Degree Days | 657 | 480 | 296 | 100 | 20 | 1 | 0 | 0 | 18 | 98 | 352 | 586 | 2608 |
| Cooling Degree Days | 0 | 2 | 21 | 93 | 238 | 422 | 535 | 510 | 285 | 96 | 7 | 0 | 2209 |
| Total Precipitation (") | 0.66 | 0.74 | 0.82 | 1.32 | 2.93 | 2.49 | 2.47 | 2.40 | 3.03 | 1.74 | 0.76 | 0.74 | 20.10 |
| Days ≥ 0.1" Precip | 2 | 2 | 2 | 3 | 4 | 4 | 3 | 4 | 4 | 3 | 2 | 2 | 35 |
| Total Snowfall (") | na | *0.7* | 0.0 | 0.0 | 0.0 | 0.0 | 0.0 | 0.0 | 0.0 | 0.0 | 0.3 | 0.0 | na |
| Days ≥ 1" Snow Depth | *0* | *0* | 0 | 0 | 0 | 0 | 0 | 0 | 0 | 0 | 0 | 0 | 0 |

## GALVESTON *Galveston County* ELEVATION 59 ft LAT/LONG 29° 18 ' N / 94° 48 ' W

| | JAN | FEB | MAR | APR | MAY | JUN | JUL | AUG | SEP | OCT | NOV | DEC | YEAR |
|---|---|---|---|---|---|---|---|---|---|---|---|---|---|
| Maximum Temp °F | 58.8 | 60.9 | 66.9 | 73.6 | 79.8 | 85.2 | 87.5 | 87.8 | 84.7 | 77.6 | 69.0 | 62.3 | 74.5 |
| Minimum Temp °F | 47.9 | 50.4 | 56.9 | 65.1 | 71.7 | 77.4 | 79.5 | 79.3 | 75.4 | 68.1 | 58.6 | 51.6 | 65.2 |
| Mean Temp °F | 53.4 | 55.7 | 61.9 | 69.4 | 75.8 | 81.3 | 83.5 | 83.6 | 80.0 | 72.9 | 63.8 | 57.0 | 69.9 |
| Days Max Temp ≥ 90 °F | 0 | 0 | 0 | 0 | 0 | 1 | 5 | 7 | 2 | 0 | 0 | 0 | 15 |
| Days Max Temp ≤ 32 °F | 0 | 0 | 0 | 0 | 0 | 0 | 0 | 0 | 0 | 0 | 0 | 0 | 0 |
| Days Min Temp ≤ 32 °F | 2 | 1 | 0 | 0 | 0 | 0 | 0 | 0 | 0 | 0 | 0 | 1 | 4 |
| Days Min Temp ≤ 0 °F | 0 | 0 | 0 | 0 | 0 | 0 | 0 | 0 | 0 | 0 | 0 | 0 | 0 |
| Heating Degree Days | 358 | 263 | 123 | 19 | 0 | 0 | 0 | 0 | 0 | 15 | 115 | 259 | 1152 |
| Cooling Degree Days | 2 | 5 | 36 | 152 | 350 | 507 | 598 | 602 | 474 | 279 | 89 | 17 | 3111 |
| Total Precipitation (") | 3.92 | 2.60 | 2.41 | 2.74 | 4.25 | 3.95 | 3.78 | 4.29 | 5.54 | 3.10 | 3.21 | 3.26 | 43.05 |
| Days ≥ 0.1" Precip | 6 | 4 | 4 | 3 | 5 | 5 | 5 | 5 | 6 | 4 | 5 | 5 | 57 |
| Total Snowfall (") | 0.1 | 0.1 | 0.0 | 0.0 | 0.0 | 0.0 | 0.0 | 0.0 | 0.0 | 0.0 | 0.0 | 0.0 | 0.2 |
| Days ≥ 1" Snow Depth | 0 | 0 | 0 | 0 | 0 | 0 | 0 | 0 | 0 | 0 | 0 | 0 | 0 |

## GARDEN CITY *Glasscock County* ELEVATION 2631 ft LAT/LONG 31° 53 ' N / 101° 27 ' W

| | JAN | FEB | MAR | APR | MAY | JUN | JUL | AUG | SEP | OCT | NOV | DEC | YEAR |
|---|---|---|---|---|---|---|---|---|---|---|---|---|---|
| Maximum Temp °F | 55.9 | 61.2 | 70.3 | 78.4 | 85.2 | 91.6 | 93.9 | 92.8 | 85.8 | 77.4 | 66.6 | 58.9 | 76.5 |
| Minimum Temp °F | 26.1 | 30.1 | ***37.5*** | 46.3 | 55.5 | 64.4 | 67.5 | 66.4 | 60.0 | 48.8 | 36.6 | 28.8 | 47.3 |
| Mean Temp °F | 41.0 | 45.7 | ***54.0*** | 62.3 | 70.4 | 78.0 | 80.7 | 79.6 | 72.9 | 63.1 | 51.6 | 43.9 | 61.9 |
| Days Max Temp ≥ 90 °F | 0 | 0 | 1 | 4 | 10 | 19 | 24 | 23 | 10 | 2 | 0 | 0 | 93 |
| Days Max Temp ≤ 32 °F | 2 | 1 | 0 | 0 | 0 | 0 | 0 | 0 | 0 | 0 | 0 | 1 | 4 |
| Days Min Temp ≤ 32 °F | 24 | 17 | 8 | 2 | 0 | 0 | 0 | 0 | 0 | 1 | 10 | 20 | 82 |
| Days Min Temp ≤ 0 °F | 0 | 0 | 0 | 0 | 0 | 0 | 0 | 0 | 0 | 0 | 0 | 0 | 0 |
| Heating Degree Days | 734 | 540 | ***347*** | 138 | 29 | 2 | 0 | 0 | 20 | 127 | 399 | 649 | 2985 |
| Cooling Degree Days | 0 | 1 | ***12*** | 74 | 220 | 410 | 514 | 491 | 287 | 83 | 4 | 0 | 2096 |
| Total Precipitation (") | 0.77 | 0.75 | 0.60 | 1.21 | 2.31 | 1.95 | 2.04 | 2.10 | 3.30 | 1.87 | 0.77 | 0.71 | 18.38 |
| Days ≥ 0.1" Precip | 2 | 2 | 1 | 2 | 4 | 3 | 3 | 4 | 4 | 3 | 2 | 2 | 32 |
| Total Snowfall (") | na | ***0.0*** | 0.0 | 0.0 | 0.0 | 0.0 | 0.0 | 0.0 | 0.0 | 0.0 | 0.0 | ***0.0*** | na |
| Days ≥ 1" Snow Depth | ***0*** | na | 0 | 0 | 0 | 0 | 0 | 0 | 0 | 0 | 0 | ***0*** | na |

## GATESVILLE 4 SSE *Coryell County* ELEVATION 791 ft LAT/LONG 31° 26 ' N / 97° 45 ' W

| | JAN | FEB | MAR | APR | MAY | JUN | JUL | AUG | SEP | OCT | NOV | DEC | YEAR |
|---|---|---|---|---|---|---|---|---|---|---|---|---|---|
| Maximum Temp °F | 57.9 | 62.2 | 70.9 | 78.3 | 83.3 | 90.5 | 95.2 | 95.3 | 88.5 | 79.9 | 68.9 | 61.2 | 77.7 |
| Minimum Temp °F | 33.4 | 36.8 | 45.4 | 53.9 | 61.0 | 68.2 | 71.1 | 70.3 | 64.9 | 54.4 | 43.6 | 36.0 | 53.3 |
| Mean Temp °F | 45.6 | 49.6 | 58.1 | 66.1 | 72.2 | 79.3 | 83.2 | 82.8 | 76.7 | 67.1 | 56.2 | 48.6 | 65.5 |
| Days Max Temp ≥ 90 °F | 0 | 0 | 0 | 1 | 4 | 18 | 28 | 27 | 15 | 3 | 0 | 0 | 96 |
| Days Max Temp ≤ 32 °F | 1 | 0 | 0 | 0 | 0 | 0 | 0 | 0 | 0 | 0 | 0 | 1 | 2 |
| Days Min Temp ≤ 32 °F | 15 | 10 | 4 | 1 | 0 | 0 | 0 | 0 | 0 | 0 | 5 | 12 | 47 |
| Days Min Temp ≤ 0 °F | 0 | 0 | 0 | 0 | 0 | 0 | 0 | 0 | 0 | 0 | 0 | 0 | 0 |
| Heating Degree Days | 598 | 435 | 243 | 72 | 11 | 0 | 0 | 0 | 7 | 70 | 286 | 505 | 2227 |
| Cooling Degree Days | 2 | 7 | 37 | 113 | 252 | 440 | 571 | 579 | 375 | 155 | 36 | 4 | 2571 |
| Total Precipitation (") | 1.86 | 2.38 | 2.44 | 3.10 | 4.54 | 3.32 | 2.17 | 2.21 | 3.40 | 3.16 | 2.31 | 2.14 | 33.03 |
| Days ≥ 0.1" Precip | 4 | 4 | 4 | 4 | 6 | 5 | 3 | 4 | 4 | 4 | 4 | 4 | 50 |
| Total Snowfall (") | 0.4 | 0.4 | 0.0 | 0.0 | 0.0 | 0.0 | 0.0 | 0.0 | 0.0 | 0.0 | 0.1 | 0.0 | 0.9 |
| Days ≥ 1" Snow Depth | ***0*** | ***0*** | 0 | 0 | 0 | 0 | 0 | 0 | 0 | 0 | 0 | 0 | 0 |

## GILMER 2 W *Upshur County* ELEVATION 371 ft LAT/LONG 32° 42 ' N / 94° 57 ' W

| | JAN | FEB | MAR | APR | MAY | JUN | JUL | AUG | SEP | OCT | NOV | DEC | YEAR |
|---|---|---|---|---|---|---|---|---|---|---|---|---|---|
| Maximum Temp °F | 54.3 | 59.3 | 67.7 | 75.8 | 81.6 | 88.7 | 92.9 | 93.3 | 86.8 | 77.4 | 66.5 | 58.4 | 75.2 |
| Minimum Temp °F | 31.0 | 34.1 | 41.9 | 50.4 | 58.4 | 66.5 | 70.0 | 68.5 | 62.0 | 50.3 | 41.2 | 34.0 | 50.7 |
| Mean Temp °F | 42.7 | 46.7 | 54.8 | 63.1 | 70.0 | 77.6 | 81.5 | 80.9 | 74.4 | 63.9 | 53.9 | 46.2 | 63.0 |
| Days Max Temp ≥ 90 °F | 0 | 0 | 0 | 0 | 2 | 14 | 24 | 25 | 12 | 2 | 0 | 0 | 79 |
| Days Max Temp ≤ 32 °F | 1 | 1 | 0 | 0 | 0 | 0 | 0 | 0 | 0 | 0 | 0 | 1 | 3 |
| Days Min Temp ≤ 32 °F | 19 | 14 | 6 | 1 | 0 | 0 | 0 | 0 | 0 | 1 | 7 | 15 | 63 |
| Days Min Temp ≤ 0 °F | 0 | 0 | 0 | 0 | 0 | 0 | 0 | 0 | 0 | 0 | 0 | 0 | 0 |
| Heating Degree Days | 690 | 514 | 330 | 120 | 23 | 1 | 0 | 0 | 14 | 121 | 347 | 581 | 2741 |
| Cooling Degree Days | 1 | 3 | 21 | 65 | 187 | 394 | 528 | 526 | 322 | 96 | 22 | 4 | 2169 |
| Total Precipitation (") | 3.22 | 3.82 | 4.34 | 4.69 | 5.09 | 3.64 | 2.91 | 2.35 | 3.93 | 4.54 | 4.49 | 4.37 | 47.39 |
| Days ≥ 0.1" Precip | 6 | 6 | 6 | 6 | 7 | 5 | 4 | 4 | 5 | 6 | 6 | 6 | 67 |
| Total Snowfall (") | 0.7 | 0.8 | 0.1 | 0.0 | 0.0 | 0.0 | 0.0 | 0.0 | 0.0 | 0.0 | 0.1 | 0.5 | 2.2 |
| Days ≥ 1" Snow Depth | 0 | 0 | 0 | 0 | 0 | 0 | 0 | 0 | 0 | 0 | 0 | 0 | 0 |

## GLEN ROSE 2 W *Somervell County* ELEVATION 620 ft LAT/LONG 32° 14 ' N / 97° 45 ' W

| | JAN | FEB | MAR | APR | MAY | JUN | JUL | AUG | SEP | OCT | NOV | DEC | YEAR |
|---|---|---|---|---|---|---|---|---|---|---|---|---|---|
| Maximum Temp °F | 58.5 | 63.6 | 72.5 | 80.4 | 85.6 | 92.9 | 97.6 | 97.5 | 90.2 | 81.0 | 69.4 | 61.4 | 79.2 |
| Minimum Temp °F | 29.5 | 33.5 | 42.4 | 51.3 | 59.1 | 66.9 | 69.5 | 68.0 | 61.9 | 51.1 | 40.6 | 32.3 | 50.5 |
| Mean Temp °F | 44.0 | 48.6 | 57.4 | 65.9 | 72.4 | 79.9 | 83.6 | 82.8 | 76.1 | 66.1 | 55.0 | 46.8 | 64.9 |
| Days Max Temp ≥ 90 °F | 0 | 0 | 1 | 3 | 9 | 23 | 29 | 28 | 18 | 5 | 0 | 0 | 116 |
| Days Max Temp ≤ 32 °F | 1 | 0 | 0 | 0 | 0 | 0 | 0 | 0 | 0 | 0 | 0 | 0 | 1 |
| Days Min Temp ≤ 32 °F | 21 | 14 | 7 | 2 | 0 | 0 | 0 | 0 | 0 | 2 | 8 | 17 | 71 |
| Days Min Temp ≤ 0 °F | 0 | 0 | 0 | 0 | 0 | 0 | 0 | 0 | 0 | 0 | 0 | 0 | 0 |
| Heating Degree Days | 646 | 462 | 264 | 82 | 13 | 0 | 0 | 0 | 12 | 90 | 320 | 562 | 2451 |
| Cooling Degree Days | 0 | 4 | 31 | 99 | 237 | 437 | 562 | 557 | 331 | 125 | 23 | 4 | 2410 |
| Total Precipitation (") | 1.77 | 2.26 | 2.70 | 3.19 | 5.81 | 3.61 | 2.07 | 2.07 | 3.46 | 3.75 | 1.95 | 2.29 | 34.93 |
| Days ≥ 0.1" Precip | 4 | 4 | 4 | 5 | 7 | 5 | 3 | 3 | 5 | 5 | 4 | 4 | 53 |
| Total Snowfall (") | 0.7 | 0.8 | 0.4 | 0.0 | 0.0 | 0.0 | 0.0 | 0.0 | 0.0 | 0.0 | 0.2 | 0.3 | 2.4 |
| Days ≥ 1" Snow Depth | 1 | 0 | 0 | 0 | 0 | 0 | 0 | 0 | 0 | 0 | 0 | 0 | 1 |

## GOLDTHWAITE 1 WSW *Mills County* ELEVATION 1581 ft LAT/LONG 31° 27 ' N / 98° 34 ' W

| | JAN | FEB | MAR | APR | MAY | JUN | JUL | AUG | SEP | OCT | NOV | DEC | YEAR |
|---|---|---|---|---|---|---|---|---|---|---|---|---|---|
| Maximum Temp °F | 58.2 | 62.4 | 70.9 | 78.5 | 82.9 | 90.1 | 93.8 | 93.4 | 87.2 | 78.8 | 68.0 | 60.9 | 77.1 |
| Minimum Temp °F | 34.7 | 38.2 | 46.0 | 54.7 | 61.3 | 68.2 | 70.9 | 70.4 | 65.3 | 55.7 | 45.3 | 37.7 | 54.0 |
| Mean Temp °F | 46.5 | 50.4 | 58.4 | 66.6 | 72.2 | 79.1 | 82.4 | 81.9 | 76.3 | 67.3 | 56.7 | 49.3 | 65.6 |
| Days Max Temp ≥ 90 °F | 0 | 0 | 0 | 2 | 4 | 16 | 26 | 24 | 12 | 2 | 0 | 0 | 86 |
| Days Max Temp ≤ 32 °F | 1 | 1 | 0 | 0 | 0 | 0 | 0 | 0 | 0 | 0 | 0 | 1 | 3 |
| Days Min Temp ≤ 32 °F | 14 | 8 | 3 | 0 | 0 | 0 | 0 | 0 | 0 | 0 | 3 | 9 | 37 |
| Days Min Temp ≤ 0 °F | 0 | 0 | 0 | 0 | 0 | 0 | 0 | 0 | 0 | 0 | 0 | 0 | 0 |
| Heating Degree Days | 570 | 412 | 233 | 63 | 11 | 0 | 0 | 0 | 7 | 64 | 269 | 483 | 2112 |
| Cooling Degree Days | 1 | 5 | 35 | 117 | 240 | 427 | 540 | 539 | 352 | 151 | 31 | 2 | 2440 |
| Total Precipitation (") | 1.43 | 2.06 | 1.99 | 2.34 | 3.88 | 3.06 | 1.55 | 2.14 | 3.25 | 3.22 | 1.87 | 1.81 | 28.60 |
| Days ≥ 0.1" Precip | 3 | 4 | 4 | 4 | 5 | 4 | 3 | 3 | 4 | 4 | 3 | 3 | 44 |
| Total Snowfall (") | *0.1* | 0.3 | 0.1 | 0.0 | 0.0 | 0.0 | 0.0 | 0.0 | 0.0 | 0.0 | 0.1 | 0.0 | 0.6 |
| Days ≥ 1" Snow Depth | *0* | 0 | 0 | 0 | 0 | 0 | 0 | 0 | 0 | 0 | 0 | 0 | 0 |

## GOLIAD *Goliad County* ELEVATION 161 ft LAT/LONG 28° 40 ' N / 97° 23 ' W

| | JAN | FEB | MAR | APR | MAY | JUN | JUL | AUG | SEP | OCT | NOV | DEC | YEAR |
|---|---|---|---|---|---|---|---|---|---|---|---|---|---|
| Maximum Temp °F | 66.6 | 70.0 | 77.2 | 82.5 | 86.9 | 92.2 | 94.8 | 95.4 | 91.4 | 85.2 | 76.2 | 69.5 | 82.3 |
| Minimum Temp °F | 43.1 | 45.4 | 52.8 | 60.2 | 66.4 | 71.2 | 72.7 | 72.4 | 68.6 | 60.1 | 52.0 | 45.6 | 59.2 |
| Mean Temp °F | 54.9 | 57.7 | 65.0 | 71.4 | 76.7 | 81.7 | 83.8 | 83.9 | 80.0 | 72.7 | 64.2 | 57.6 | 70.8 |
| Days Max Temp ≥ 90 °F | 0 | 0 | 1 | 3 | 10 | 23 | 29 | 29 | 21 | 8 | 1 | 0 | 125 |
| Days Max Temp ≤ 32 °F | 0 | 0 | 0 | 0 | 0 | 0 | 0 | 0 | 0 | 0 | 0 | 0 | 0 |
| Days Min Temp ≤ 32 °F | 6 | 3 | 1 | 0 | 0 | 0 | 0 | 0 | 0 | 0 | 2 | 4 | 16 |
| Days Min Temp ≤ 0 °F | 0 | 0 | 0 | 0 | 0 | 0 | 0 | 0 | 0 | 0 | 0 | 0 | 0 |
| Heating Degree Days | 333 | 228 | 99 | 19 | 1 | 0 | 0 | 0 | 0 | 19 | 125 | 264 | 1088 |
| Cooling Degree Days | 20 | 36 | 106 | 221 | 389 | 534 | 603 | 621 | 476 | 292 | 118 | 51 | 3467 |
| Total Precipitation (") | 2.45 | 2.24 | 1.83 | 3.23 | 4.66 | 4.47 | 3.31 | 3.48 | 4.96 | 4.05 | 2.00 | 2.18 | 38.86 |
| Days ≥ 0.1" Precip | 5 | 4 | 3 | 4 | 5 | 5 | 5 | 5 | 6 | 4 | 3 | 3 | 52 |
| Total Snowfall (") | 0.2 | 0.1 | 0.0 | 0.0 | 0.0 | 0.0 | 0.0 | 0.0 | 0.0 | 0.0 | 0.0 | 0.0 | 0.3 |
| Days ≥ 1" Snow Depth | 0 | 0 | 0 | 0 | 0 | 0 | 0 | 0 | 0 | 0 | 0 | 0 | 0 |

## GONZALES 2 N *Gonzales County* ELEVATION 302 ft LAT/LONG 29° 30 ' N / 97° 27 ' W

| | JAN | FEB | MAR | APR | MAY | JUN | JUL | AUG | SEP | OCT | NOV | DEC | YEAR |
|---|---|---|---|---|---|---|---|---|---|---|---|---|---|
| Maximum Temp °F | 60.4 | 64.9 | 73.1 | 80.2 | 85.0 | 91.2 | 95.0 | 95.5 | 90.3 | 82.4 | 72.3 | 64.3 | 79.5 |
| Minimum Temp °F | 38.1 | 41.2 | 49.1 | 57.2 | 64.3 | 70.2 | 72.3 | 71.8 | 67.3 | 57.4 | 47.9 | 41.0 | 56.5 |
| Mean Temp °F | 49.3 | 53.1 | 61.1 | 68.7 | 74.7 | 80.8 | 83.7 | 83.7 | 78.8 | 70.0 | 60.1 | 52.7 | 68.1 |
| Days Max Temp ≥ 90 °F | 0 | 0 | 1 | 2 | 7 | 21 | 28 | 29 | 19 | 5 | 0 | 0 | 112 |
| Days Max Temp ≤ 32 °F | 0 | 0 | 0 | 0 | 0 | 0 | 0 | 0 | 0 | 0 | 0 | 0 | 0 |
| Days Min Temp ≤ 32 °F | 9 | 5 | 2 | 0 | 0 | 0 | 0 | 0 | 0 | 0 | 2 | 6 | 24 |
| Days Min Temp ≤ 0 °F | 0 | 0 | 0 | 0 | 0 | 0 | 0 | 0 | 0 | 0 | 0 | 0 | 0 |
| Heating Degree Days | 489 | 343 | 173 | 43 | 4 | 0 | 0 | 0 | 1 | 38 | 198 | 391 | 1680 |
| Cooling Degree Days | 5 | 12 | 54 | 150 | 314 | 489 | 596 | 600 | 431 | 207 | 62 | 15 | 2935 |
| Total Precipitation (") | 2.53 | 2.34 | 2.12 | 3.34 | 5.31 | 4.19 | 1.62 | 2.39 | 3.78 | 3.36 | 2.50 | 2.35 | 35.83 |
| Days ≥ 0.1" Precip | 5 | 4 | 4 | 4 | 6 | 5 | 3 | 4 | 5 | 4 | 4 | 4 | 52 |
| Total Snowfall (") | 0.0 | 0.0 | 0.0 | 0.0 | 0.0 | 0.0 | 0.0 | 0.0 | 0.0 | 0.0 | 0.0 | 0.0 | 0.0 |
| Days ≥ 1" Snow Depth | 0 | 0 | 0 | 0 | 0 | 0 | 0 | 0 | 0 | 0 | 0 | 0 | 0 |

## GRAHAM *Young County* ELEVATION 1040 ft LAT/LONG 33° 5 ' N / 98° 35 ' W

| | JAN | FEB | MAR | APR | MAY | JUN | JUL | AUG | SEP | OCT | NOV | DEC | YEAR |
|---|---|---|---|---|---|---|---|---|---|---|---|---|---|
| Maximum Temp °F | 54.5 | 59.3 | 68.8 | 77.8 | 83.2 | 90.7 | 96.1 | 95.8 | 88.2 | 78.9 | 66.9 | 58.1 | 76.5 |
| Minimum Temp °F | 27.3 | 31.5 | 40.2 | 49.7 | 58.3 | 67.1 | 70.9 | 69.7 | 62.3 | 50.0 | 39.1 | 30.2 | 49.7 |
| Mean Temp °F | 40.9 | 45.4 | 54.6 | 63.7 | 70.8 | 78.9 | 83.5 | 82.8 | 75.3 | 64.4 | 53.0 | 44.2 | 63.1 |
| Days Max Temp ≥ 90 °F | 0 | 0 | 1 | 3 | 7 | 19 | 27 | 26 | 16 | 4 | 0 | 0 | 103 |
| Days Max Temp ≤ 32 °F | 3 | 1 | 0 | 0 | 0 | 0 | 0 | 0 | 0 | 0 | 0 | 1 | 5 |
| Days Min Temp ≤ 32 °F | 23 | 16 | 7 | 1 | 0 | 0 | 0 | 0 | 0 | 1 | 9 | 20 | 77 |
| Days Min Temp ≤ 0 °F | 0 | 0 | 0 | 0 | 0 | 0 | 0 | 0 | 0 | 0 | 0 | 0 | 0 |
| Heating Degree Days | 740 | 548 | 339 | 116 | 27 | 1 | 0 | 0 | 17 | 112 | 365 | 638 | 2903 |
| Cooling Degree Days | 0 | 2 | 19 | 79 | 212 | 425 | 577 | 576 | 334 | 95 | 13 | 1 | 2333 |
| Total Precipitation (") | 1.21 | 1.80 | 1.90 | 2.71 | 4.85 | 3.37 | 2.04 | 2.35 | 4.17 | 3.65 | 1.62 | 1.71 | 31.38 |
| Days ≥ 0.1" Precip | 3 | 4 | 4 | 4 | 6 | 4 | 3 | 4 | 5 | 5 | 3 | 4 | 49 |
| Total Snowfall (") | 1.1 | 1.1 | 0.6 | 0.0 | 0.0 | 0.0 | 0.0 | 0.0 | 0.0 | 0.0 | 0.2 | 0.3 | 3.3 |
| Days ≥ 1" Snow Depth | *0* | *0* | 0 | 0 | 0 | 0 | 0 | 0 | 0 | 0 | 0 | 0 | 0 |

## GRANDFALLS 3 SSE *Ward County* ELEVATION 2441 ft LAT/LONG 31° 18 ' N / 102° 50 ' W

| | JAN | FEB | MAR | APR | MAY | JUN | JUL | AUG | SEP | OCT | NOV | DEC | YEAR |
|---|---|---|---|---|---|---|---|---|---|---|---|---|---|
| Maximum Temp °F | 59.8 | 65.3 | 74.4 | 83.2 | 89.7 | 96.1 | 98.0 | 96.2 | 89.5 | 81.0 | 70.5 | 61.6 | 80.4 |
| Minimum Temp °F | 26.3 | 30.0 | 37.9 | 47.0 | 55.8 | *66.1* | 68.9 | 67.1 | 60.7 | 48.5 | 35.7 | 27.7 | 47.6 |
| Mean Temp °F | 43.1 | 47.7 | 56.2 | 65.1 | 72.8 | *81.1* | 83.5 | 81.7 | 75.1 | 64.8 | 53.1 | 44.7 | 64.1 |
| Days Max Temp ≥ 90 °F | 0 | 0 | 2 | 8 | 17 | 24 | *29* | 27 | 17 | 6 | 0 | 0 | 130 |
| Days Max Temp ≤ 32 °F | 1 | 0 | 0 | 0 | 0 | 0 | 0 | 0 | 0 | 0 | 0 | 1 | 2 |
| Days Min Temp ≤ 32 °F | 24 | 18 | 8 | 2 | 0 | 0 | 0 | 0 | 0 | 1 | *11* | 22 | 86 |
| Days Min Temp ≤ 0 °F | 0 | 0 | 0 | 0 | 0 | 0 | 0 | 0 | 0 | 0 | 0 | 0 | 0 |
| Heating Degree Days | 673 | 484 | 282 | 90 | 16 | 0 | 0 | 0 | 15 | 97 | 355 | 623 | 2635 |
| Cooling Degree Days | 0 | 1 | 16 | 99 | 277 | 518 | 582 | 541 | 339 | 104 | 8 | 0 | 2485 |
| Total Precipitation (") | 0.48 | 0.69 | 0.39 | 0.85 | 1.69 | 1.31 | 1.21 | 1.61 | 2.34 | 1.35 | 0.66 | 0.60 | 13.18 |
| Days ≥ 0.1" Precip | 1 | 2 | 1 | 1 | 3 | 2 | 2 | 3 | 4 | 2 | 1 | 1 | 23 |
| Total Snowfall (") | 1.6 | *0.2* | 0.1 | 0.0 | 0.0 | 0.0 | 0.0 | 0.0 | 0.0 | 0.0 | 0.3 | 0.3 | 2.5 |
| Days ≥ 1" Snow Depth | na | *0* | 0 | 0 | 0 | 0 | 0 | 0 | 0 | 0 | 0 | 0 | na |

## GRAPEVINE DAM *Tarrant County* ELEVATION 581 ft LAT/LONG 32° 58 ' N / 97° 3 ' W

| | JAN | FEB | MAR | APR | MAY | JUN | JUL | AUG | SEP | OCT | NOV | DEC | YEAR |
|---|---|---|---|---|---|---|---|---|---|---|---|---|---|
| Maximum Temp °F | 53.2 | 58.2 | 67.0 | 75.4 | 81.8 | 89.9 | 94.8 | 94.6 | 87.4 | 77.7 | 65.9 | 57.1 | 75.3 |
| Minimum Temp °F | 30.8 | 34.7 | 43.1 | 52.3 | 60.4 | 68.2 | 72.2 | 71.2 | 64.5 | 53.1 | 43.1 | 34.4 | 52.3 |
| Mean Temp °F | 42.0 | 46.5 | 55.1 | 63.9 | 71.1 | 79.1 | 83.5 | 82.9 | 76.0 | 65.4 | 54.5 | 45.8 | 63.8 |
| Days Max Temp ≥ 90 °F | 0 | 0 | 0 | 1 | 4 | 18 | 27 | 26 | 14 | 3 | 0 | 0 | 93 |
| Days Max Temp ≤ 32 °F | 2 | 1 | 0 | 0 | 0 | 0 | 0 | 0 | 0 | 0 | 0 | 1 | 4 |
| Days Min Temp ≤ 32 °F | 18 | 12 | 4 | 0 | 0 | 0 | 0 | 0 | 0 | 0 | 4 | 13 | 51 |
| Days Min Temp ≤ 0 °F | 0 | 0 | 0 | 0 | 0 | 0 | 0 | 0 | 0 | 0 | 0 | 0 | 0 |
| Heating Degree Days | 706 | 519 | 319 | 103 | 18 | 1 | 0 | 0 | 10 | 92 | 328 | 590 | 2686 |
| Cooling Degree Days | 0 | 3 | 16 | 69 | 214 | 430 | 578 | 583 | 359 | 111 | 20 | 2 | 2385 |
| Total Precipitation (") | 1.72 | 2.39 | 2.62 | 3.71 | 5.13 | 3.33 | 2.16 | 2.09 | 3.44 | 3.94 | 2.33 | 2.31 | 35.17 |
| Days ≥ 0.1" Precip | 4 | 4 | 4 | 5 | 7 | 4 | 3 | 4 | 4 | 5 | 4 | 5 | 53 |
| Total Snowfall (") | 0.1 | *0.0* | 0.0 | 0.0 | 0.0 | 0.0 | 0.0 | 0.0 | 0.0 | 0.0 | 0.0 | 0.1 | 0.2 |
| Days ≥ 1" Snow Depth | *0* | 0 | 0 | 0 | 0 | 0 | 0 | 0 | 0 | 0 | 0 | 0 | 0 |

## GREENVILLE *Hunt County* ELEVATION 610 ft LAT/LONG 33° 12 ' N / 96° 13 ' W

| | JAN | FEB | MAR | APR | MAY | JUN | JUL | AUG | SEP | OCT | NOV | DEC | YEAR |
|---|---|---|---|---|---|---|---|---|---|---|---|---|---|
| Maximum Temp °F | 51.7 | 56.3 | 65.2 | 74.2 | 80.6 | 89.1 | 93.6 | 93.7 | 86.4 | 76.6 | 64.8 | 55.5 | 74.0 |
| Minimum Temp °F | 29.7 | 33.3 | 41.7 | 50.9 | 59.3 | 67.4 | 71.0 | 69.8 | 63.4 | 51.5 | 41.3 | 32.9 | 51.0 |
| Mean Temp °F | 40.8 | 44.8 | 53.5 | 62.6 | 70.0 | 78.2 | 82.3 | 81.8 | 74.9 | 64.1 | 53.1 | 44.2 | 62.5 |
| Days Max Temp ≥ 90 °F | 0 | 0 | 0 | 0 | 1 | 15 | 25 | 25 | 11 | 2 | 0 | 0 | 79 |
| Days Max Temp ≤ 32 °F | 2 | 1 | 0 | 0 | 0 | 0 | 0 | 0 | 0 | 0 | 0 | 1 | 4 |
| Days Min Temp ≤ 32 °F | 20 | 14 | 6 | 1 | 0 | 0 | 0 | 0 | 0 | 0 | 6 | 16 | 63 |
| Days Min Temp ≤ 0 °F | 0 | 0 | 0 | 0 | 0 | 0 | 0 | 0 | 0 | 0 | 0 | 0 | 0 |
| Heating Degree Days | 745 | 563 | 365 | 125 | 23 | 1 | 0 | 0 | 12 | 112 | 366 | 639 | 2951 |
| Cooling Degree Days | 0 | 1 | 13 | 57 | 190 | 410 | 549 | 554 | 335 | 93 | 16 | 2 | 2220 |
| Total Precipitation (") | 2.31 | 3.18 | 3.67 | 4.16 | 5.75 | 3.69 | 2.77 | 2.33 | 4.07 | 4.77 | 3.23 | 2.86 | 42.79 |
| Days ≥ 0.1" Precip | 4 | 5 | 6 | 5 | 7 | 5 | 4 | 4 | 5 | 5 | 5 | 5 | 60 |
| Total Snowfall (") | 1.0 | 1.4 | 0.2 | 0.0 | 0.0 | 0.0 | 0.0 | 0.0 | 0.0 | 0.0 | 0.1 | 0.3 | 3.0 |
| Days ≥ 1" Snow Depth | 1 | 1 | 0 | 0 | 0 | 0 | 0 | 0 | 0 | 0 | 0 | 1 | 3 |

## GROVETON *Trinity County* ELEVATION 322 ft LAT/LONG 31° 3 ' N / 95° 8 ' W

| | JAN | FEB | MAR | APR | MAY | JUN | JUL | AUG | SEP | OCT | NOV | DEC | YEAR |
|---|---|---|---|---|---|---|---|---|---|---|---|---|---|
| Maximum Temp °F | 60.1 | 64.8 | 72.9 | 79.6 | 85.0 | 90.8 | 94.3 | 94.7 | 89.2 | 81.6 | 71.2 | 63.4 | 79.0 |
| Minimum Temp °F | 36.7 | 39.4 | 46.1 | 53.8 | 61.2 | 67.9 | 70.8 | 69.7 | 64.4 | 53.9 | 44.6 | 38.1 | 53.9 |
| Mean Temp °F | 48.4 | 52.1 | 59.5 | 66.7 | 73.1 | 79.4 | 82.6 | 82.2 | 76.9 | 67.7 | 57.8 | 50.9 | 66.4 |
| Days Max Temp ≥ 90 °F | 0 | 0 | 0 | 1 | 6 | 20 | 28 | 28 | 16 | 4 | 0 | 0 | 103 |
| Days Max Temp ≤ 32 °F | 0 | 0 | 0 | 0 | 0 | 0 | 0 | 0 | 0 | 0 | 0 | 0 | 0 |
| Days Min Temp ≤ 32 °F | 11 | 7 | 2 | 0 | 0 | 0 | 0 | 0 | 0 | 0 | 4 | 9 | 33 |
| Days Min Temp ≤ 0 °F | 0 | 0 | 0 | 0 | 0 | 0 | 0 | 0 | 0 | 0 | 0 | 0 | 0 |
| Heating Degree Days | 516 | 366 | 203 | 53 | 4 | 0 | 0 | 0 | 2 | 54 | 242 | 437 | 1877 |
| Cooling Degree Days | 4 | 7 | 37 | 113 | 273 | 461 | 563 | 565 | 382 | 153 | 31 | 6 | 2595 |
| Total Precipitation (") | 3.83 | 3.44 | 3.67 | 3.33 | 5.43 | 4.68 | 3.36 | 3.16 | 4.02 | 4.23 | 3.91 | 4.47 | 47.53 |
| Days ≥ 0.1" Precip | 7 | 5 | 5 | 4 | 6 | 6 | 5 | 5 | 5 | 5 | 5 | 6 | 64 |
| Total Snowfall (") | 0.3 | 0.0 | 0.0 | 0.0 | 0.0 | 0.0 | 0.0 | 0.0 | 0.0 | 0.0 | 0.0 | 0.0 | 0.3 |
| Days ≥ 1" Snow Depth | 0 | 0 | 0 | 0 | 0 | 0 | 0 | 0 | 0 | 0 | 0 | 0 | 0 |

## GRUVER *Hansford County* ELEVATION 3173 ft LAT/LONG 36° 15 ' N / 101° 24 ' W

| | JAN | FEB | MAR | APR | MAY | JUN | JUL | AUG | SEP | OCT | NOV | DEC | YEAR |
|---|---|---|---|---|---|---|---|---|---|---|---|---|---|
| Maximum Temp °F | 49.1 | 54.7 | 63.9 | 73.7 | 80.9 | 90.3 | 94.8 | 92.3 | 84.7 | 74.6 | 59.6 | 50.5 | 72.4 |
| Minimum Temp °F | 20.0 | 23.6 | 31.4 | 41.1 | 50.3 | 60.1 | 64.8 | 62.9 | 54.8 | 42.3 | 30.2 | 22.2 | 42.0 |
| Mean Temp °F | 34.6 | 39.2 | 47.6 | 57.4 | 65.6 | 75.2 | 79.9 | 77.6 | 69.8 | 58.5 | 44.9 | 36.4 | 57.2 |
| Days Max Temp ≥ 90 °F | 0 | 0 | 0 | 1 | 6 | 17 | 26 | 22 | 10 | 2 | 0 | 0 | 84 |
| Days Max Temp ≤ 32 °F | 4 | 2 | 1 | 0 | 0 | 0 | 0 | 0 | 0 | 0 | 1 | 3 | 11 |
| Days Min Temp ≤ 32 °F | 29 | 24 | 17 | 5 | 0 | 0 | 0 | 0 | 0 | 3 | 18 | 28 | 124 |
| Days Min Temp ≤ 0 °F | 1 | 1 | 0 | 0 | 0 | 0 | 0 | 0 | 0 | 0 | 0 | 1 | 3 |
| Heating Degree Days | 936 | 723 | 534 | 245 | 79 | 5 | 0 | 1 | 42 | 227 | 595 | 879 | 4266 |
| Cooling Degree Days | 0 | 0 | 2 | 21 | 89 | 304 | 444 | 395 | 187 | 22 | 1 | 0 | 1465 |
| Total Precipitation (") | 0.51 | 0.63 | 1.16 | 1.31 | 2.95 | 3.21 | 2.79 | 2.48 | 1.76 | 1.21 | 0.79 | 0.55 | 19.35 |
| Days ≥ 0.1" Precip | 2 | 2 | 2 | 3 | 5 | 5 | 5 | 4 | 4 | 2 | 2 | 2 | 38 |
| Total Snowfall (") | 3.0 | 3.5 | *2.7* | 1.0 | 0.2 | 0.0 | 0.0 | 0.0 | 0.0 | 0.5 | 1.8 | 2.7 | 15.4 |
| Days ≥ 1" Snow Depth | *2* | *1* | *0* | 0 | 0 | 0 | 0 | 0 | 0 | 0 | 0 | 1 | 4 |

## GUTHRIE *King County* ELEVATION 1752 ft LAT/LONG 33° 37 ' N / 100° 19 ' W

| | JAN | FEB | MAR | APR | MAY | JUN | JUL | AUG | SEP | OCT | NOV | DEC | YEAR |
|---|---|---|---|---|---|---|---|---|---|---|---|---|---|
| Maximum Temp °F | 54.0 | 59.0 | 68.5 | 78.6 | 85.6 | 93.3 | 97.4 | 95.7 | 87.8 | 78.5 | 66.0 | 56.4 | 76.7 |
| Minimum Temp °F | 25.1 | 28.8 | 36.2 | 45.7 | 55.3 | 64.4 | 68.8 | 67.1 | 59.7 | 47.7 | 35.8 | 27.5 | 46.8 |
| Mean Temp °F | 39.6 | 43.9 | 52.2 | 62.2 | 70.5 | 78.9 | 83.1 | 81.5 | 73.8 | 63.1 | 50.9 | 42.0 | 61.8 |
| Days Max Temp ≥ 90 °F | 0 | 0 | 1 | 4 | 11 | 23 | 28 | 26 | 15 | 4 | 0 | 0 | 112 |
| Days Max Temp ≤ 32 °F | 2 | 1 | 0 | 0 | 0 | 0 | 0 | 0 | 0 | 0 | 0 | 2 | 5 |
| Days Min Temp ≤ 32 °F | 25 | 19 | 10 | 2 | 0 | 0 | 0 | 0 | 0 | 1 | 11 | 23 | 91 |
| Days Min Temp ≤ 0 °F | 0 | 0 | 0 | 0 | 0 | 0 | 0 | 0 | 0 | 0 | 0 | 0 | 0 |
| Heating Degree Days | 781 | 589 | 400 | 140 | 30 | 1 | 0 | 0 | 19 | 127 | 420 | 708 | 3215 |
| Cooling Degree Days | 0 | 1 | 11 | 64 | 215 | 425 | 568 | 545 | 299 | 77 | 6 | 0 | 2211 |
| Total Precipitation (") | 1.05 | 1.31 | 1.17 | 1.72 | 3.65 | 3.01 | 1.90 | 3.00 | 3.44 | 2.41 | 0.94 | 0.93 | 24.53 |
| Days ≥ 0.1" Precip | 2 | 3 | 2 | 3 | 5 | 5 | 3 | 4 | 4 | 3 | 2 | 2 | 38 |
| Total Snowfall (") | 2.2 | 1.4 | 0.6 | 0.0 | 0.0 | 0.0 | 0.0 | 0.0 | 0.0 | 0.0 | 1.0 | 0.8 | 6.0 |
| Days ≥ 1" Snow Depth | 1 | 1 | 0 | 0 | 0 | 0 | 0 | 0 | 0 | 0 | 1 | 0 | 3 |

## HALLETTSVILLE 2 N *Lavaca County* ELEVATION 239 ft LAT/LONG 29° 27 ' N / 96° 56 ' W

| | JAN | FEB | MAR | APR | MAY | JUN | JUL | AUG | SEP | OCT | NOV | DEC | YEAR |
|---|---|---|---|---|---|---|---|---|---|---|---|---|---|
| Maximum Temp °F | 63.2 | 66.5 | 74.3 | 80.7 | 85.2 | 91.2 | 94.6 | 95.3 | 90.5 | 83.3 | 73.5 | 66.0 | 80.4 |
| Minimum Temp °F | 41.4 | 43.7 | 51.3 | 59.1 | 65.2 | 70.7 | 72.3 | 71.9 | 67.8 | 58.9 | 50.4 | 43.9 | 58.0 |
| Mean Temp °F | 52.3 | 55.1 | 62.8 | 69.9 | 75.3 | 81.0 | 83.5 | 83.6 | 79.2 | 71.1 | 61.9 | 55.0 | 69.2 |
| Days Max Temp ≥ 90 °F | 0 | 0 | 0 | 1 | 6 | 21 | 28 | 28 | 19 | 6 | 0 | 0 | 109 |
| Days Max Temp ≤ 32 °F | 0 | 0 | 0 | 0 | 0 | 0 | 0 | 0 | 0 | 0 | 0 | 0 | 0 |
| Days Min Temp ≤ 32 °F | 6 | 4 | 1 | 0 | 0 | 0 | 0 | 0 | 0 | 0 | 2 | 5 | 18 |
| Days Min Temp ≤ 0 °F | 0 | 0 | 0 | 0 | 0 | 0 | 0 | 0 | 0 | 0 | 0 | 0 | 0 |
| Heating Degree Days | 404 | 291 | 136 | 30 | 2 | 0 | 0 | 0 | 1 | 27 | 162 | 331 | 1384 |
| Cooling Degree Days | 10 | 20 | 65 | 172 | 330 | 488 | 578 | 588 | 433 | 230 | 79 | 28 | 3021 |
| Total Precipitation (") | 3.09 | 2.60 | 2.40 | 3.43 | 6.37 | 4.09 | 2.37 | 2.76 | 4.64 | 3.60 | 3.33 | 2.78 | 41.46 |
| Days ≥ 0.1" Precip | 5 | 4 | 4 | 4 | 7 | 5 | 4 | 5 | 6 | 4 | 4 | 5 | 57 |
| Total Snowfall (") | 0.1 | 0.0 | 0.0 | 0.0 | 0.0 | 0.0 | 0.0 | 0.0 | 0.0 | 0.0 | 0.0 | 0.0 | 0.1 |
| Days ≥ 1" Snow Depth | 0 | 0 | 0 | 0 | 0 | 0 | 0 | 0 | 0 | 0 | 0 | 0 | 0 |

## HARLINGEN *Cameron County* ELEVATION 39 ft LAT/LONG 26° 12 ' N / 97° 42 ' W

| | JAN | FEB | MAR | APR | MAY | JUN | JUL | AUG | SEP | OCT | NOV | DEC | YEAR |
|---|---|---|---|---|---|---|---|---|---|---|---|---|---|
| Maximum Temp °F | 67.6 | 71.4 | 79.0 | 84.5 | 88.0 | 92.2 | 94.2 | 94.8 | 91.0 | 85.2 | 77.7 | 71.2 | 83.1 |
| Minimum Temp °F | 47.4 | 49.9 | 56.6 | 64.0 | 69.1 | 72.7 | 73.6 | 73.4 | 70.9 | 63.7 | 56.4 | 50.3 | 62.3 |
| Mean Temp °F | 57.5 | 60.7 | 67.8 | 74.3 | 78.6 | 82.5 | 83.9 | 84.1 | 81.0 | 74.5 | 67.0 | 60.8 | 72.7 |
| Days Max Temp ≥ 90 °F | 0 | 0 | 2 | 6 | 13 | 24 | 29 | 28 | 20 | 7 | 1 | 0 | 130 |
| Days Max Temp ≤ 32 °F | 0 | 0 | 0 | 0 | 0 | 0 | 0 | 0 | 0 | 0 | 0 | 0 | 0 |
| Days Min Temp ≤ 32 °F | 2 | 1 | 0 | 0 | 0 | 0 | 0 | 0 | 0 | 0 | 0 | 1 | 4 |
| Days Min Temp ≤ 0 °F | 0 | 0 | 0 | 0 | 0 | 0 | 0 | 0 | 0 | 0 | 0 | 0 | 0 |
| Heating Degree Days | 267 | 169 | 66 | 10 | 1 | 0 | 0 | 0 | 1 | 10 | 85 | 196 | 805 |
| Cooling Degree Days | 41 | 64 | 166 | 286 | 440 | *562* | 617 | 623 | 508 | 332 | 173 | 76 | 3888 |
| Total Precipitation (") | 1.86 | 1.87 | 0.95 | 2.29 | 3.22 | 3.06 | 1.87 | 2.85 | 5.65 | 2.49 | 1.37 | 1.55 | 29.03 |
| Days ≥ 0.1" Precip | 4 | 3 | 2 | 2 | 4 | 4 | 3 | 4 | 7 | 4 | 3 | 3 | 43 |
| Total Snowfall (") | 0.0 | 0.0 | 0.0 | 0.0 | 0.0 | 0.0 | 0.0 | 0.0 | 0.0 | 0.0 | 0.0 | 0.0 | 0.0 |
| Days ≥ 1" Snow Depth | 0 | 0 | 0 | 0 | 0 | 0 | 0 | 0 | 0 | 0 | 0 | 0 | 0 |

## HASKELL *Haskell County* ELEVATION 1532 ft LAT/LONG 33° 9 ' N / 99° 44 ' W

| | JAN | FEB | MAR | APR | MAY | JUN | JUL | AUG | SEP | OCT | NOV | DEC | YEAR |
|---|---|---|---|---|---|---|---|---|---|---|---|---|---|
| Maximum Temp °F | 53.6 | 58.5 | 68.7 | 78.2 | 84.5 | 92.0 | 96.1 | 95.0 | 87.0 | 77.6 | 65.1 | 56.7 | 76.1 |
| Minimum Temp °F | 28.3 | 32.2 | 40.6 | 50.5 | 59.4 | 67.6 | 72.0 | 70.4 | 63.1 | 51.8 | 40.4 | 31.5 | 50.7 |
| Mean Temp °F | 41.0 | 45.4 | 54.7 | 64.4 | 72.0 | 79.8 | 84.1 | 82.7 | 75.1 | 64.7 | 52.8 | 44.1 | 63.4 |
| Days Max Temp ≥ 90 °F | 0 | 0 | 1 | 4 | 9 | 20 | 27 | 25 | 15 | 3 | 0 | 0 | 104 |
| Days Max Temp ≤ 32 °F | 3 | 1 | 0 | 0 | 0 | 0 | 0 | 0 | 0 | 0 | 0 | 1 | 5 |
| Days Min Temp ≤ 32 °F | 21 | 15 | 6 | 1 | 0 | 0 | 0 | 0 | 0 | 0 | 6 | 16 | 65 |
| Days Min Temp ≤ 0 °F | 0 | 0 | 0 | 0 | 0 | 0 | 0 | 0 | 0 | 0 | 0 | 0 | 0 |
| Heating Degree Days | 737 | 549 | 337 | 110 | 23 | 1 | 0 | 0 | 18 | 108 | 372 | 641 | 2896 |
| Cooling Degree Days | 0 | 2 | 23 | 99 | 251 | 447 | 602 | 586 | 340 | 109 | 12 | 0 | 2471 |
| Total Precipitation (") | 1.05 | 1.57 | 1.35 | 2.25 | 3.64 | 2.68 | 1.73 | 3.04 | 3.32 | 2.56 | 1.12 | 1.31 | 25.62 |
| Days ≥ 0.1" Precip | 2 | 3 | 3 | 4 | 5 | 4 | 3 | 4 | 5 | 4 | 2 | 3 | 42 |
| Total Snowfall (") | 2.0 | 1.5 | 0.5 | 0.0 | 0.0 | 0.0 | 0.0 | 0.0 | 0.0 | 0.0 | 0.8 | 0.7 | 5.5 |
| Days ≥ 1" Snow Depth | 1 | 1 | 0 | 0 | 0 | 0 | 0 | 0 | 0 | 0 | 0 | 0 | 2 |

## HEBBRONVILLE *Jim Hogg County* ELEVATION 551 ft LAT/LONG 27° 18 ' N / 98° 41 ' W

| | JAN | FEB | MAR | APR | MAY | JUN | JUL | AUG | SEP | OCT | NOV | DEC | YEAR |
|---|---|---|---|---|---|---|---|---|---|---|---|---|---|
| Maximum Temp °F | 66.1 | 70.7 | 79.2 | 85.5 | 89.9 | 94.4 | 96.5 | 97.2 | 92.0 | 85.6 | 76.5 | 69.7 | 83.6 |
| Minimum Temp °F | 42.8 | 46.0 | 53.7 | 61.5 | 67.7 | 72.0 | 72.8 | 72.2 | 69.1 | 60.9 | 52.1 | 45.7 | 59.7 |
| Mean Temp °F | 54.5 | 58.4 | 66.5 | 73.5 | 78.8 | 83.3 | 84.7 | 84.7 | 80.6 | 73.3 | 64.3 | 57.9 | 71.7 |
| Days Max Temp ≥ 90 °F | 0 | 1 | 4 | 9 | 17 | 26 | 29 | 29 | 22 | 10 | 1 | 0 | 148 |
| Days Max Temp ≤ 32 °F | 0 | 0 | 0 | 0 | 0 | 0 | 0 | 0 | 0 | 0 | 0 | 0 | 0 |
| Days Min Temp ≤ 32 °F | 4 | 2 | 1 | 0 | 0 | 0 | 0 | 0 | 0 | 0 | 1 | 3 | 11 |
| Days Min Temp ≤ 0 °F | 0 | 0 | 0 | 0 | 0 | 0 | 0 | 0 | 0 | 0 | 0 | 0 | 0 |
| Heating Degree Days | 344 | 219 | 88 | 19 | 1 | 0 | 0 | 0 | 1 | 17 | 125 | 253 | 1067 |
| Cooling Degree Days | 17 | 43 | 132 | 268 | 437 | 568 | 631 | 638 | 496 | 300 | 117 | 44 | 3691 |
| Total Precipitation (") | 1.31 | 1.38 | 0.94 | 1.72 | 3.53 | 3.27 | 1.50 | 2.23 | 4.17 | 1.87 | 1.08 | 1.26 | 24.26 |
| Days ≥ 0.1" Precip | 4 | 3 | 2 | 3 | 4 | 4 | 3 | 3 | 5 | 3 | 2 | 3 | 39 |
| Total Snowfall (") | 0.0 | 0.0 | 0.0 | 0.0 | 0.0 | 0.0 | 0.0 | 0.0 | 0.0 | 0.0 | 0.0 | 0.0 | 0.0 |
| Days ≥ 1" Snow Depth | 0 | 0 | 0 | 0 | 0 | 0 | 0 | 0 | 0 | 0 | 0 | 0 | 0 |

## HENDERSON *Rusk County* ELEVATION 502 ft LAT/LONG 32° 9 ' N / 94° 48 ' W

| | JAN | FEB | MAR | APR | MAY | JUN | JUL | AUG | SEP | OCT | NOV | DEC | YEAR |
|---|---|---|---|---|---|---|---|---|---|---|---|---|---|
| Maximum Temp °F | 54.8 | 59.5 | 67.8 | 76.0 | 81.8 | 88.9 | 92.7 | 92.7 | 86.5 | 77.7 | 67.0 | 58.8 | 75.4 |
| Minimum Temp °F | 33.4 | 36.4 | 43.6 | 52.2 | 60.0 | 67.7 | 71.1 | 69.9 | 64.1 | 52.8 | 43.3 | 36.2 | 52.6 |
| Mean Temp °F | 44.1 | 48.0 | 55.7 | 64.2 | 70.9 | 78.3 | 81.9 | 81.3 | 75.3 | 65.3 | 55.2 | 47.5 | 64.0 |
| Days Max Temp ≥ 90 °F | 0 | 0 | 0 | 0 | 2 | 15 | 24 | 24 | 12 | 1 | 0 | 0 | 78 |
| Days Max Temp ≤ 32 °F | 1 | 1 | 0 | 0 | 0 | 0 | 0 | 0 | 0 | 0 | 0 | 1 | 3 |
| Days Min Temp ≤ 32 °F | 16 | 11 | 4 | 0 | 0 | 0 | 0 | 0 | 0 | 0 | 5 | 12 | 48 |
| Days Min Temp ≤ 0 °F | 0 | 0 | 0 | 0 | 0 | 0 | 0 | 0 | 0 | 0 | 0 | 0 | 0 |
| Heating Degree Days | 644 | 479 | 304 | 103 | 16 | 1 | 0 | 0 | 9 | 94 | 312 | 538 | 2500 |
| Cooling Degree Days | 1 | 5 | 23 | 80 | 206 | 409 | 530 | 528 | 338 | 112 | 24 | 4 | 2260 |
| Total Precipitation (") | 3.79 | 3.86 | 3.89 | 4.29 | 5.49 | 4.60 | 2.81 | 2.70 | 3.62 | 4.56 | 4.37 | 4.18 | 48.16 |
| Days ≥ 0.1" Precip | 6 | 6 | 5 | 5 | 7 | 5 | 5 | 4 | 5 | 5 | 5 | 6 | 64 |
| Total Snowfall (") | 0.5 | 0.5 | 0.0 | 0.0 | 0.0 | 0.0 | 0.0 | 0.0 | 0.0 | 0.0 | 0.0 | 0.2 | 1.2 |
| Days ≥ 1" Snow Depth | 0 | 0 | 0 | 0 | 0 | 0 | 0 | 0 | 0 | 0 | 0 | 0 | 0 |

## HENRIETTA *Clay County* ELEVATION 922 ft LAT/LONG 33° 48 ' N / 98° 12 ' W

| | JAN | FEB | MAR | APR | MAY | JUN | JUL | AUG | SEP | OCT | NOV | DEC | YEAR |
|---|---|---|---|---|---|---|---|---|---|---|---|---|---|
| Maximum Temp °F | 52.7 | 57.5 | 67.3 | 76.0 | 82.3 | 90.4 | 96.4 | 95.6 | 87.0 | 77.3 | 65.1 | 56.2 | 75.3 |
| Minimum Temp °F | 26.7 | 30.9 | 40.1 | 49.5 | 58.1 | 66.7 | 70.9 | 69.4 | 61.9 | 50.0 | 38.8 | 30.2 | 49.4 |
| Mean Temp °F | 39.7 | 44.2 | 53.7 | 62.8 | 70.2 | 78.6 | 83.7 | 82.5 | 74.5 | 63.7 | 52.0 | 43.2 | 62.4 |
| Days Max Temp ≥ 90 °F | 0 | 0 | 1 | 2 | 5 | 18 | 27 | 26 | 14 | 3 | 0 | 0 | 96 |
| Days Max Temp ≤ 32 °F | 2 | 1 | 0 | 0 | 0 | 0 | 0 | 0 | 0 | 0 | 0 | 1 | 4 |
| Days Min Temp ≤ 32 °F | 23 | 16 | 7 | 1 | 0 | 0 | 0 | 0 | 0 | 1 | 8 | 20 | 76 |
| Days Min Temp ≤ 0 °F | 0 | 0 | 0 | 0 | 0 | 0 | 0 | 0 | 0 | 0 | 0 | 0 | 0 |
| Heating Degree Days | 778 | 581 | 360 | 132 | 29 | 1 | 0 | 0 | 20 | 123 | 393 | 670 | 3087 |
| Cooling Degree Days | 0 | 1 | 13 | 68 | 204 | 423 | 590 | 581 | 335 | 88 | 11 | 1 | 2315 |
| Total Precipitation (") | 1.48 | 2.05 | 2.46 | 2.92 | 4.53 | 3.76 | 1.76 | 2.63 | 3.93 | 3.28 | 1.52 | 1.84 | 32.16 |
| Days ≥ 0.1" Precip | 3 | 4 | 4 | 4 | 5 | 5 | 3 | 4 | 5 | 5 | 3 | 3 | 48 |
| Total Snowfall (") | *0.1* | *0.1* | 0.2 | 0.0 | 0.0 | 0.0 | 0.0 | 0.0 | 0.0 | 0.0 | 0.0 | 0.1 | 0.5 |
| Days ≥ 1" Snow Depth | 0 | *0* | 0 | 0 | 0 | 0 | 0 | 0 | 0 | 0 | 0 | 0 | 0 |

## HEREFORD *Deaf Smith County* ELEVATION 3842 ft LAT/LONG 34° 48 ' N / 102° 28 ' W

| | JAN | FEB | MAR | APR | MAY | JUN | JUL | AUG | SEP | OCT | NOV | DEC | YEAR |
|---|---|---|---|---|---|---|---|---|---|---|---|---|---|
| Maximum Temp °F | 49.4 | 53.8 | 62.3 | 71.6 | 79.2 | 87.8 | 90.5 | 88.1 | 81.5 | 72.2 | 59.7 | 51.2 | 70.6 |
| Minimum Temp °F | 20.9 | 24.0 | 30.8 | 39.8 | 49.4 | 58.8 | 63.3 | 61.6 | 54.0 | 42.0 | 30.3 | 22.7 | 41.5 |
| Mean Temp °F | 35.2 | 38.9 | 46.5 | 55.7 | 64.3 | 73.3 | 76.9 | 74.9 | 67.7 | 57.1 | 45.0 | 37.0 | 56.0 |
| Days Max Temp ≥ 90 °F | 0 | 0 | 0 | 1 | 4 | 14 | 19 | 15 | 6 | 1 | 0 | 0 | 60 |
| Days Max Temp ≤ 32 °F | 4 | 2 | 1 | 0 | 0 | 0 | 0 | 0 | 0 | 0 | 1 | 3 | 11 |
| Days Min Temp ≤ 32 °F | 29 | 24 | 18 | 5 | 0 | 0 | 0 | 0 | 0 | 3 | 19 | 28 | 126 |
| Days Min Temp ≤ 0 °F | 0 | 0 | 0 | 0 | 0 | 0 | 0 | 0 | 0 | 0 | 0 | 1 | 1 |
| Heating Degree Days | 918 | 729 | 566 | 286 | 99 | 11 | 1 | 2 | 55 | 256 | 590 | 862 | 4375 |
| Cooling Degree Days | 0 | 0 | 1 | 17 | 88 | 287 | 392 | 346 | 162 | 17 | 0 | 0 | 1310 |
| Total Precipitation (") | 0.45 | 0.51 | 0.83 | 0.87 | 2.06 | 2.98 | 2.16 | 3.13 | 2.09 | 1.45 | 0.68 | 0.55 | 17.76 |
| Days ≥ 0.1" Precip | 1 | 2 | 2 | 2 | 4 | 5 | 4 | 5 | 4 | 3 | 2 | 1 | 35 |
| Total Snowfall (") | 3.2 | *2.8* | 1.3 | 0.3 | 0.0 | 0.0 | 0.0 | 0.0 | 0.0 | 0.1 | 1.1 | 2.7 | 11.5 |
| Days ≥ 1" Snow Depth | 3 | 2 | 1 | 0 | 0 | 0 | 0 | 0 | 0 | 0 | 1 | 3 | 10 |

## HICO *Hamilton County* ELEVATION 1010 ft LAT/LONG 32° 0 ' N / 98° 1 ' W

| | JAN | FEB | MAR | APR | MAY | JUN | JUL | AUG | SEP | OCT | NOV | DEC | YEAR |
|---|---|---|---|---|---|---|---|---|---|---|---|---|---|
| Maximum Temp °F | 58.0 | 62.2 | 70.6 | 78.8 | 84.2 | 91.4 | 96.1 | 95.6 | 88.6 | 79.7 | 68.1 | 60.2 | 77.8 |
| Minimum Temp °F | 32.2 | 35.8 | 43.9 | 52.6 | 60.5 | 67.3 | 70.9 | 69.6 | 63.6 | 53.0 | 42.8 | 34.7 | 52.2 |
| Mean Temp °F | 45.1 | 49.0 | 57.3 | 65.7 | 72.4 | 79.4 | 83.5 | 82.6 | 76.1 | 66.4 | 55.5 | 47.5 | 65.0 |
| Days Max Temp ≥ 90 °F | 0 | 0 | 0 | 2 | 6 | 21 | 28 | 27 | 15 | 3 | 0 | 0 | 102 |
| Days Max Temp ≤ 32 °F | 1 | 1 | 0 | 0 | 0 | 0 | 0 | 0 | 0 | 0 | 0 | 1 | 3 |
| Days Min Temp ≤ 32 °F | 17 | 11 | 5 | 1 | 0 | 0 | 0 | 0 | 0 | 0 | 6 | 13 | 53 |
| Days Min Temp ≤ 0 °F | 0 | 0 | 0 | 0 | 0 | 0 | 0 | 0 | 0 | 0 | 0 | 0 | 0 |
| Heating Degree Days | 611 | 451 | 264 | 75 | 11 | 0 | 0 | 0 | 9 | 78 | 302 | 537 | 2338 |
| Cooling Degree Days | 0 | 4 | 27 | 91 | 244 | 426 | 574 | 566 | 346 | 126 | 25 | 2 | 2431 |
| Total Precipitation (") | 1.83 | 2.23 | 2.48 | 2.99 | 4.97 | 2.93 | 2.12 | 2.49 | 3.48 | 3.50 | 2.13 | 2.00 | 33.15 |
| Days ≥ 0.1" Precip | 4 | 4 | 4 | 4 | 6 | 4 | 3 | 4 | 5 | 4 | 4 | 4 | 50 |
| Total Snowfall (") | *1.0* | 0.7 | 0.2 | 0.0 | 0.5 | 0.0 | 0.0 | 0.0 | 0.0 | 0.0 | 0.4 | 0.3 | 3.1 |
| Days ≥ 1" Snow Depth | 1 | 1 | 0 | 0 | 0 | 0 | 0 | 0 | 0 | 0 | 0 | 0 | 2 |

## HILLSBORO *Hill County* ELEVATION 630 ft LAT/LONG 32° 1 ' N / 97° 8 ' W

| | JAN | FEB | MAR | APR | MAY | JUN | JUL | AUG | SEP | OCT | NOV | DEC | YEAR |
|---|---|---|---|---|---|---|---|---|---|---|---|---|---|
| Maximum Temp °F | 56.1 | 60.8 | 69.4 | 77.3 | 82.8 | 90.3 | 94.7 | 94.9 | 88.5 | 79.3 | 68.0 | 59.7 | 76.8 |
| Minimum Temp °F | 33.7 | 37.4 | 45.3 | 54.1 | 61.6 | 69.2 | 72.4 | 72.0 | 65.7 | 55.1 | 44.6 | 36.8 | 54.0 |
| Mean Temp °F | 44.9 | 49.1 | 57.4 | 65.7 | 72.2 | 79.8 | 83.6 | 83.5 | 77.1 | 67.2 | 56.3 | 48.3 | 65.4 |
| Days Max Temp ≥ 90 °F | 0 | 0 | 0 | 1 | 3 | 18 | 28 | 27 | 16 | 3 | 0 | 0 | 96 |
| Days Max Temp ≤ 32 °F | 1 | 0 | 0 | 0 | 0 | 0 | 0 | 0 | 0 | 0 | 0 | 1 | 2 |
| Days Min Temp ≤ 32 °F | 15 | 9 | 4 | 0 | 0 | 0 | 0 | 0 | 0 | 0 | 4 | 11 | 43 |
| Days Min Temp ≤ 0 °F | 0 | 0 | 0 | 0 | 0 | 0 | 0 | 0 | 0 | 0 | 0 | 0 | 0 |
| Heating Degree Days | 619 | 447 | 261 | 76 | 11 | 0 | 0 | 0 | 7 | 72 | 281 | 516 | 2290 |
| Cooling Degree Days | 2 | 6 | 35 | 112 | 260 | 458 | 586 | 601 | 396 | 160 | 34 | 5 | 2655 |
| Total Precipitation (") | 1.97 | 2.80 | 3.12 | 3.06 | 5.21 | 3.62 | 2.11 | 2.03 | 3.20 | 4.17 | 2.50 | 2.70 | 36.49 |
| Days ≥ 0.1" Precip | 4 | 5 | 5 | 5 | 7 | 5 | 3 | 3 | 4 | 5 | 4 | 5 | 55 |
| Total Snowfall (") | 0.5 | 0.3 | 0.1 | 0.0 | 0.0 | 0.0 | 0.0 | 0.0 | 0.0 | 0.0 | 0.1 | 0.0 | 1.0 |
| Days ≥ 1" Snow Depth | 0 | 0 | 0 | 0 | 0 | 0 | 0 | 0 | 0 | 0 | 0 | 0 | 0 |

## HORDS CREEK DAM *Coleman County* ELEVATION 1942 ft LAT/LONG 31° 51 ' N / 99° 34 ' W

| | JAN | FEB | MAR | APR | MAY | JUN | JUL | AUG | SEP | OCT | NOV | DEC | YEAR |
|---|---|---|---|---|---|---|---|---|---|---|---|---|---|
| Maximum Temp °F | 54.9 | 59.3 | 68.4 | 77.1 | 82.6 | 89.5 | 93.1 | 92.6 | 85.7 | 77.6 | 65.9 | 57.9 | 75.4 |
| Minimum Temp °F | 29.8 | 33.4 | 41.7 | 50.8 | 58.5 | 65.9 | 69.3 | 68.3 | 61.6 | 51.5 | 40.9 | 32.6 | 50.4 |
| Mean Temp °F | 42.4 | 46.4 | 55.1 | 64.0 | 70.6 | 77.7 | 81.2 | 80.5 | 73.7 | 64.6 | 53.4 | 45.3 | 62.9 |
| Days Max Temp ≥ 90 °F | 0 | 0 | 0 | 2 | 6 | 17 | 25 | 24 | 11 | 2 | 0 | 0 | 87 |
| Days Max Temp ≤ 32 °F | 2 | 1 | 0 | 0 | 0 | 0 | 0 | 0 | 0 | 0 | 0 | 1 | 4 |
| Days Min Temp ≤ 32 °F | 19 | 13 | 5 | 0 | 0 | 0 | 0 | 0 | 0 | 1 | 6 | 15 | 59 |
| Days Min Temp ≤ 0 °F | 0 | 0 | 0 | 0 | 0 | 0 | 0 | 0 | 0 | 0 | 0 | 0 | 0 |
| Heating Degree Days | 694 | 520 | 321 | 108 | 24 | 1 | 0 | 0 | 17 | 100 | 354 | 605 | 2744 |
| Cooling Degree Days | 0 | 1 | 21 | 84 | 214 | 393 | 513 | 511 | 294 | 96 | 16 | 1 | 2144 |
| Total Precipitation (") | 1.17 | 1.50 | 1.63 | 2.02 | 3.65 | 3.13 | 1.62 | 2.09 | 3.55 | 2.65 | 1.29 | 1.41 | 25.71 |
| Days ≥ 0.1" Precip | 2 | 3 | 3 | 3 | 5 | 4 | 3 | 3 | 5 | 3 | 3 | 3 | 40 |
| Total Snowfall (") | ***0.0*** | ***0.0*** | 0.1 | 0.0 | 0.0 | 0.0 | 0.0 | 0.0 | 0.0 | 0.0 | 0.0 | 0.0 | 0.1 |
| Days ≥ 1" Snow Depth | ***0*** | ***0*** | 0 | 0 | 0 | 0 | 0 | 0 | 0 | 0 | 0 | 0 | 0 |

## HOUSTON WM HOBBY AP *Harris County* ELEVATION 85 ft LAT/LONG 29° 39 ' N / 95° 17 ' W

| | JAN | FEB | MAR | APR | MAY | JUN | JUL | AUG | SEP | OCT | NOV | DEC | YEAR |
|---|---|---|---|---|---|---|---|---|---|---|---|---|---|
| Maximum Temp °F | 61.5 | 65.2 | 72.4 | 79.0 | 84.4 | 89.9 | 92.2 | 91.9 | 88.0 | 81.1 | 72.0 | 65.0 | 78.6 |
| Minimum Temp °F | 43.5 | 46.1 | 53.1 | 60.8 | 67.2 | 73.0 | 74.8 | 74.7 | 71.2 | 61.6 | 52.7 | 46.3 | 60.4 |
| Mean Temp °F | 52.5 | 55.7 | 62.8 | 69.9 | 75.9 | 81.5 | 83.5 | 83.4 | 79.6 | 71.4 | 62.4 | 55.7 | 69.5 |
| Days Max Temp ≥ 90 °F | 0 | 0 | 0 | 1 | 4 | 18 | 26 | 25 | 14 | 2 | 0 | 0 | 90 |
| Days Max Temp ≤ 32 °F | 0 | 0 | 0 | 0 | 0 | 0 | 0 | 0 | 0 | 0 | 0 | 0 | 0 |
| Days Min Temp ≤ 32 °F | 4 | 2 | 1 | 0 | 0 | 0 | 0 | 0 | 0 | 0 | 1 | 3 | 11 |
| Days Min Temp ≤ 0 °F | 0 | 0 | 0 | 0 | 0 | 0 | 0 | 0 | 0 | 0 | 0 | 0 | 0 |
| Heating Degree Days | 397 | 275 | 133 | 26 | 1 | 0 | 0 | 0 | 0 | 23 | 152 | 311 | 1318 |
| Cooling Degree Days | 8 | 22 | 68 | 171 | 349 | 506 | 583 | 586 | 449 | 234 | 82 | 31 | 3089 |
| Total Precipitation (") | 4.06 | 3.27 | 3.06 | 3.55 | 5.91 | 6.55 | 4.71 | 4.50 | 5.31 | 5.17 | 4.06 | 3.51 | 53.66 |
| Days ≥ 0.1" Precip | 6 | 5 | 5 | 4 | 6 | 6 | 6 | 6 | 7 | 5 | 5 | 5 | 66 |
| Total Snowfall (") | 0.0 | 0.0 | 0.0 | 0.0 | 0.0 | 0.0 | 0.0 | 0.0 | 0.0 | 0.0 | 0.0 | 0.0 | 0.0 |
| Days ≥ 1" Snow Depth | 0 | 0 | 0 | 0 | 0 | 0 | 0 | 0 | 0 | 0 | 0 | 0 | 0 |

## HUNTSVILLE *Walker County* ELEVATION 400 ft LAT/LONG 30° 44 ' N / 95° 34 ' W

| | JAN | FEB | MAR | APR | MAY | JUN | JUL | AUG | SEP | OCT | NOV | DEC | YEAR |
|---|---|---|---|---|---|---|---|---|---|---|---|---|---|
| Maximum Temp °F | 57.7 | 62.3 | 70.6 | 78.3 | 84.1 | 90.6 | 94.2 | 94.1 | 88.4 | 80.0 | 69.6 | 61.4 | 77.6 |
| Minimum Temp °F | 38.5 | 41.5 | 49.3 | 57.5 | 63.9 | 70.1 | 72.4 | 71.8 | 66.9 | 57.5 | 48.7 | 41.4 | 56.6 |
| Mean Temp °F | 48.1 | 51.9 | 60.0 | 67.9 | 74.0 | 80.4 | 83.3 | 83.0 | 77.7 | 68.8 | 59.2 | 51.4 | 67.1 |
| Days Max Temp ≥ 90 °F | 0 | 0 | 0 | 0 | 6 | 20 | 27 | 27 | 16 | 3 | 0 | 0 | 99 |
| Days Max Temp ≤ 32 °F | 1 | 0 | 0 | 0 | 0 | 0 | 0 | 0 | 0 | 0 | 0 | 0 | 1 |
| Days Min Temp ≤ 32 °F | 9 | 6 | 1 | 0 | 0 | 0 | 0 | 0 | 0 | 0 | 2 | 6 | 24 |
| Days Min Temp ≤ 0 °F | 0 | 0 | 0 | 0 | 0 | 0 | 0 | 0 | 0 | 0 | 0 | 0 | 0 |
| Heating Degree Days | 524 | 373 | 199 | 50 | 6 | 0 | 0 | 0 | 4 | 51 | 220 | 426 | 1853 |
| Cooling Degree Days | 4 | 12 | 53 | 142 | 294 | 476 | 579 | 585 | 411 | 186 | 56 | 12 | 2810 |
| Total Precipitation (") | 4.06 | 3.33 | 3.38 | 3.63 | 5.53 | 4.47 | 2.24 | 3.49 | 4.59 | 4.16 | 4.08 | 4.20 | 47.16 |
| Days ≥ 0.1" Precip | 6 | 5 | 5 | 4 | 7 | 6 | 5 | 5 | 6 | 5 | 5 | 6 | 65 |
| Total Snowfall (") | 0.2 | 0.1 | 0.0 | 0.0 | 0.0 | 0.0 | 0.0 | 0.0 | 0.0 | 0.0 | 0.0 | 0.0 | 0.3 |
| Days ≥ 1" Snow Depth | 0 | 0 | 0 | 0 | 0 | 0 | 0 | 0 | 0 | 0 | 0 | 0 | 0 |

## JACKSBORO *Jack County* ELEVATION 1070 ft LAT/LONG 33° 13 ' N / 98° 10 ' W

| | JAN | FEB | MAR | APR | MAY | JUN | JUL | AUG | SEP | OCT | NOV | DEC | YEAR |
|---|---|---|---|---|---|---|---|---|---|---|---|---|---|
| Maximum Temp °F | 55.7 | 60.6 | 69.5 | 77.8 | 83.4 | 90.9 | 95.6 | 95.1 | 87.7 | 78.7 | 67.0 | 59.1 | 76.8 |
| Minimum Temp °F | 31.5 | 35.6 | 43.9 | 53.0 | 60.6 | 68.6 | 72.5 | 71.4 | 64.5 | 54.0 | 43.0 | 34.4 | 52.8 |
| Mean Temp °F | 43.7 | 48.1 | 56.7 | 65.4 | 72.0 | 79.8 | 83.9 | 83.3 | 76.1 | 66.4 | 55.0 | 46.8 | 64.8 |
| Days Max Temp ≥ 90 °F | 0 | 0 | 1 | 2 | 6 | 19 | 27 | 26 | 14 | 3 | 0 | 0 | 98 |
| Days Max Temp ≤ 32 °F | 2 | 1 | 0 | 0 | 0 | 0 | 0 | 0 | 0 | 0 | 0 | 1 | 4 |
| Days Min Temp ≤ 32 °F | 17 | 11 | 4 | 0 | 0 | 0 | 0 | 0 | 0 | 0 | 5 | 13 | 50 |
| Days Min Temp ≤ 0 °F | 0 | 0 | 0 | 0 | 0 | 0 | 0 | 0 | 0 | 0 | 0 | 0 | 0 |
| Heating Degree Days | 656 | 477 | 286 | 86 | 17 | 0 | 0 | 0 | 11 | 81 | 315 | 562 | 2491 |
| Cooling Degree Days | 0 | 5 | 26 | 91 | 229 | 436 | 577 | 592 | 354 | 119 | 20 | 2 | 2451 |
| Total Precipitation (") | 1.37 | 1.78 | 2.17 | 2.68 | 4.91 | 3.02 | 2.29 | 2.19 | 3.75 | 3.61 | 1.79 | 1.74 | 31.30 |
| Days ≥ 0.1" Precip | 3 | 4 | 4 | 4 | 6 | 4 | 3 | 3 | 5 | 5 | 3 | 3 | 47 |
| Total Snowfall (") | 0.4 | ***0.5*** | 0.3 | 0.0 | 0.0 | 0.0 | 0.0 | 0.0 | 0.0 | 0.0 | 0.0 | 0.2 | 1.4 |
| Days ≥ 1" Snow Depth | ***0*** | 0 | 0 | 0 | 0 | 0 | 0 | 0 | 0 | 0 | 0 | 0 | 0 |

## JACKSONVILLE *Cherokee County* ELEVATION 571 ft LAT/LONG 31° 58 ' N / 95° 16 ' W

| | JAN | FEB | MAR | APR | MAY | JUN | JUL | AUG | SEP | OCT | NOV | DEC | YEAR |
|---|---|---|---|---|---|---|---|---|---|---|---|---|---|
| Maximum Temp °F | 57.5 | 62.5 | 70.8 | 78.1 | 83.2 | 90.0 | 94.0 | 93.9 | 87.7 | 79.4 | 68.7 | 60.6 | 77.2 |
| Minimum Temp °F | 36.5 | 39.5 | 47.2 | 55.1 | 62.0 | 69.1 | 71.7 | 70.8 | 65.7 | 55.7 | 46.5 | 39.6 | 55.0 |
| Mean Temp °F | 47.0 | 51.0 | 59.0 | 66.6 | 72.7 | 79.6 | 82.9 | 82.4 | 76.7 | 67.5 | 57.6 | 50.2 | 66.1 |
| Days Max Temp ≥ 90 °F | 0 | 0 | 0 | 0 | 3 | 18 | 25 | 25 | 13 | 2 | 0 | 0 | 86 |
| Days Max Temp ≤ 32 °F | 1 | 0 | 0 | 0 | 0 | 0 | 0 | 0 | 0 | 0 | 0 | 1 | 2 |
| Days Min Temp ≤ 32 °F | 12 | 7 | 2 | 0 | 0 | 0 | 0 | 0 | 0 | 0 | 3 | 8 | 32 |
| Days Min Temp ≤ 0 °F | 0 | 0 | 0 | 0 | 0 | 0 | 0 | 0 | 0 | 0 | 0 | 0 | 0 |
| Heating Degree Days | 555 | 394 | 218 | 58 | 6 | 0 | 0 | 0 | 3 | 60 | 250 | 462 | 2006 |
| Cooling Degree Days | 1 | *6* | 34 | 109 | 252 | 451 | *560* | 563 | *368* | 141 | 31 | *10* | 2526 |
| Total Precipitation (") | 3.60 | 3.55 | 3.75 | 4.21 | 5.33 | 4.20 | 2.44 | 2.12 | 4.48 | 4.59 | 4.29 | 3.73 | 46.29 |
| Days ≥ 0.1" Precip | 6 | 5 | 5 | 5 | 6 | 5 | 4 | 4 | 5 | 5 | 5 | 6 | 61 |
| Total Snowfall (") | *0.5* | 0.2 | 0.2 | 0.0 | 0.0 | 0.0 | 0.0 | 0.0 | 0.0 | 0.0 | 0.0 | 0.0 | 0.9 |
| Days ≥ 1" Snow Depth | *0* | 0 | 0 | 0 | 0 | 0 | 0 | 0 | 0 | 0 | 0 | 0 | 0 |

## JAYTON *Kent County* ELEVATION 2011 ft LAT/LONG 33° 15 ' N / 100° 34 ' W

| | JAN | FEB | MAR | APR | MAY | JUN | JUL | AUG | SEP | OCT | NOV | DEC | YEAR |
|---|---|---|---|---|---|---|---|---|---|---|---|---|---|
| Maximum Temp °F | 53.4 | 58.1 | 67.4 | 77.4 | 83.9 | 91.9 | 96.0 | 93.8 | 85.8 | 76.8 | 64.8 | 56.2 | 75.5 |
| Minimum Temp °F | 25.9 | 29.5 | 36.9 | 46.8 | 56.0 | 65.2 | 69.4 | 67.6 | 60.2 | 48.7 | 36.8 | 28.5 | 47.6 |
| Mean Temp °F | 39.7 | 43.8 | 52.2 | 62.1 | 70.0 | 78.6 | 82.8 | 80.8 | 73.0 | 62.8 | 50.8 | 42.4 | 61.6 |
| Days Max Temp ≥ 90 °F | 0 | 0 | 1 | 4 | 9 | 20 | 26 | 23 | 12 | 3 | 0 | 0 | 98 |
| Days Max Temp ≤ 32 °F | 3 | 2 | 0 | 0 | 0 | 0 | 0 | 0 | 0 | 0 | 0 | 2 | 7 |
| Days Min Temp ≤ 32 °F | 24 | 18 | 9 | 2 | 0 | 0 | 0 | 0 | 0 | 1 | 10 | 22 | 86 |
| Days Min Temp ≤ 0 °F | 0 | 0 | 0 | 0 | 0 | 0 | 0 | 0 | 0 | 0 | 0 | 0 | 0 |
| Heating Degree Days | 778 | 592 | 402 | 150 | 37 | 2 | 0 | 1 | 27 | 140 | 425 | 696 | 3250 |
| Cooling Degree Days | 0 | 1 | 9 | 63 | 196 | 406 | 554 | 515 | 288 | 77 | 5 | 0 | 2114 |
| Total Precipitation (") | 0.88 | 1.13 | 1.11 | 1.64 | 3.01 | 2.70 | 1.54 | 2.92 | 3.10 | 2.25 | 0.88 | 0.91 | 22.07 |
| Days ≥ 0.1" Precip | 2 | 2 | 2 | 3 | 5 | 4 | 3 | 5 | 4 | 4 | 2 | 2 | 38 |
| Total Snowfall (") | na | *0.4* | 0.1 | 0.0 | 0.0 | 0.0 | 0.0 | 0.0 | 0.0 | 0.0 | 0.0 | 0.1 | na |
| Days ≥ 1" Snow Depth | *0* | na | 0 | 0 | 0 | 0 | 0 | 0 | 0 | 0 | 0 | 0 | na |

## JOHNSON CITY *Blanco County* ELEVATION 1191 ft LAT/LONG 30° 17 ' N / 98° 24 ' W

| | JAN | FEB | MAR | APR | MAY | JUN | JUL | AUG | SEP | OCT | NOV | DEC | YEAR |
|---|---|---|---|---|---|---|---|---|---|---|---|---|---|
| Maximum Temp °F | *59.5* | 62.6 | 71.2 | 79.1 | 84.2 | 90.7 | 94.8 | 94.4 | 88.2 | 80.2 | 70.2 | *62.0* | 78.1 |
| Minimum Temp °F | *34.1* | 37.4 | 45.8 | 54.0 | 61.7 | 68.4 | 70.7 | 69.6 | 64.3 | 54.4 | 44.5 | *36.5* | 53.5 |
| Mean Temp °F | *46.8* | 50.0 | 58.5 | 66.6 | 73.0 | 79.6 | 82.7 | 82.1 | 76.3 | 67.3 | 57.4 | *49.3* | 65.8 |
| Days Max Temp ≥ 90 °F | 0 | 0 | 1 | 2 | 6 | 20 | 28 | 27 | 15 | 3 | 0 | 0 | 102 |
| Days Max Temp ≤ 32 °F | 0 | 0 | 0 | 0 | 0 | 0 | 0 | 0 | 0 | 0 | 0 | 0 | 0 |
| Days Min Temp ≤ 32 °F | *15* | 9 | 4 | 0 | 0 | 0 | 0 | 0 | 0 | 0 | 5 | *12* | 45 |
| Days Min Temp ≤ 0 °F | 0 | 0 | 0 | 0 | 0 | 0 | 0 | 0 | 0 | 0 | 0 | 0 | 0 |
| Heating Degree Days | *563* | 421 | 234 | 71 | 9 | 0 | 0 | 0 | 7 | 66 | 254 | *484* | 2109 |
| Cooling Degree Days | 1 | 5 | 42 | 121 | 274 | 449 | 558 | 559 | 365 | 158 | 35 | 5 | 2572 |
| Total Precipitation (") | 2.10 | 2.36 | 2.09 | 2.25 | 4.93 | 3.30 | 2.03 | 2.41 | 3.78 | 3.91 | 2.10 | 2.07 | 33.33 |
| Days ≥ 0.1" Precip | *4* | 4 | 4 | 4 | 6 | 5 | 3 | 3 | 5 | 4 | 3 | 3 | 48 |
| Total Snowfall (") | *0.4* | 0.7 | 0.3 | 0.0 | 0.0 | 0.0 | 0.0 | 0.0 | 0.0 | 0.0 | 0.0 | 0.0 | 1.4 |
| Days ≥ 1" Snow Depth | *0* | *0* | 0 | 0 | 0 | 0 | 0 | 0 | 0 | 0 | 0 | 0 | 0 |

## KAUFMAN 3 SE *Kaufman County* ELEVATION 420 ft LAT/LONG 32° 33 ' N / 96° 16 ' W

| | JAN | FEB | MAR | APR | MAY | JUN | JUL | AUG | SEP | OCT | NOV | DEC | YEAR |
|---|---|---|---|---|---|---|---|---|---|---|---|---|---|
| Maximum Temp °F | 54.0 | 58.5 | 67.4 | 75.9 | 82.2 | 90.0 | 95.0 | 95.5 | 88.4 | 78.8 | 66.9 | 57.7 | 75.9 |
| Minimum Temp °F | 32.4 | 36.0 | 44.2 | 52.9 | 60.9 | 68.6 | 72.1 | 71.3 | 65.0 | 53.7 | 43.7 | 35.6 | 53.0 |
| Mean Temp °F | 43.2 | 47.3 | 55.8 | 64.4 | 71.6 | 79.3 | 83.6 | 83.4 | 76.7 | 66.2 | 55.4 | 46.7 | 64.5 |
| Days Max Temp ≥ 90 °F | 0 | 0 | 0 | 0 | 3 | 18 | 27 | 27 | 15 | 3 | 0 | 0 | 93 |
| Days Max Temp ≤ 32 °F | 1 | 1 | 0 | 0 | 0 | 0 | 0 | 0 | 0 | 0 | 0 | 1 | 3 |
| Days Min Temp ≤ 32 °F | 16 | 11 | 4 | 0 | 0 | 0 | 0 | 0 | 0 | 0 | 4 | 12 | 47 |
| Days Min Temp ≤ 0 °F | 0 | 0 | 0 | 0 | 0 | 0 | 0 | 0 | 0 | 0 | 0 | 0 | 0 |
| Heating Degree Days | 669 | 496 | 299 | 95 | 13 | 0 | 0 | 0 | 8 | 81 | 302 | 563 | 2526 |
| Cooling Degree Days | 1 | 3 | 20 | 80 | 228 | 443 | 581 | 595 | 384 | 129 | 22 | 3 | 2489 |
| Total Precipitation (") | 2.44 | 3.28 | 3.35 | 3.41 | 5.11 | 2.87 | 2.13 | 2.09 | 3.51 | 4.94 | 3.42 | 3.34 | 39.89 |
| Days ≥ 0.1" Precip | 4 | 5 | 5 | 4 | 6 | 4 | 3 | 3 | 5 | 5 | 4 | 4 | 52 |
| Total Snowfall (") | 0.4 | *0.6* | 0.0 | 0.0 | 0.0 | 0.0 | 0.0 | 0.0 | 0.0 | 0.0 | 0.1 | 0.3 | 1.4 |
| Days ≥ 1" Snow Depth | *0* | *0* | 0 | 0 | 0 | 0 | 0 | 0 | 0 | 0 | 0 | 0 | 0 |

## KINGSVILLE *Kleberg County* ELEVATION 59 ft LAT/LONG 27° 32 ' N / 97° 53 ' W

| | JAN | FEB | MAR | APR | MAY | JUN | JUL | AUG | SEP | OCT | NOV | DEC | YEAR |
|---|---|---|---|---|---|---|---|---|---|---|---|---|---|
| Maximum Temp °F | 67.9 | 71.9 | 79.2 | 84.6 | 88.2 | 92.6 | 95.2 | 95.5 | 91.7 | 86.0 | 77.7 | 71.3 | 83.5 |
| Minimum Temp °F | 44.8 | 47.9 | 54.7 | 62.5 | 68.0 | 72.1 | 73.3 | 73.3 | 70.3 | 62.1 | 54.0 | *47.1* | 60.8 |
| Mean Temp °F | 56.4 | 59.8 | 67.0 | 73.6 | 78.2 | 82.4 | 84.3 | 84.5 | 81.0 | 74.1 | 65.9 | *59.3* | 72.2 |
| Days Max Temp ≥ 90 °F | 0 | 0 | 2 | 6 | 11 | 24 | 29 | 29 | 21 | 9 | 1 | 0 | 132 |
| Days Max Temp ≤ 32 °F | 0 | 0 | 0 | 0 | 0 | 0 | 0 | 0 | 0 | 0 | 0 | 0 | 0 |
| Days Min Temp ≤ 32 °F | 4 | 1 | 1 | 0 | 0 | 0 | 0 | 0 | 0 | 0 | 1 | 3 | 10 |
| Days Min Temp ≤ 0 °F | 0 | 0 | 0 | 0 | 0 | 0 | 0 | 0 | 0 | 0 | 0 | 0 | 0 |
| Heating Degree Days | 293 | 183 | 72 | 15 | 0 | 0 | 0 | 0 | 0 | 11 | 97 | *225* | 896 |
| Cooling Degree Days | 25 | 49 | 128 | 290 | 437 | 551 | 617 | 642 | *512* | 328 | 146 | na | na |
| Total Precipitation (") | 1.63 | 1.91 | 1.09 | 1.72 | 4.02 | 4.28 | 2.10 | 2.96 | 4.47 | 3.16 | 1.45 | 1.23 | 30.02 |
| Days ≥ 0.1" Precip | 3 | 3 | 2 | 3 | 4 | 4 | 3 | 5 | 6 | 4 | 2 | 2 | 41 |
| Total Snowfall (") | 0.0 | 0.0 | 0.0 | 0.0 | 0.0 | 0.0 | 0.0 | 0.0 | 0.0 | 0.0 | 0.0 | 0.0 | 0.0 |
| Days ≥ 1" Snow Depth | 0 | 0 | 0 | 0 | 0 | 0 | 0 | 0 | 0 | 0 | 0 | 0 | 0 |

## LA GRANGE *Fayette County* ELEVATION 279 ft LAT/LONG 29° 55 ' N / 96° 53 ' W

| | JAN | FEB | MAR | APR | MAY | JUN | JUL | AUG | SEP | OCT | NOV | DEC | YEAR |
|---|---|---|---|---|---|---|---|---|---|---|---|---|---|
| Maximum Temp °F | 61.8 | 65.9 | 74.1 | 80.7 | 85.2 | 91.3 | 95.2 | 95.7 | 89.8 | 82.5 | 72.3 | 64.9 | 80.0 |
| Minimum Temp °F | 40.6 | 43.0 | 50.7 | 58.7 | 65.1 | 71.0 | 73.0 | 72.6 | 68.1 | 59.0 | 50.0 | 42.9 | 57.9 |
| Mean Temp °F | 51.2 | 54.5 | 62.4 | 69.7 | 75.1 | 81.2 | 84.1 | 84.2 | 79.0 | 70.8 | 61.2 | 53.9 | 68.9 |
| Days Max Temp ≥ 90 °F | 0 | 0 | 1 | 1 | 6 | 21 | 28 | 29 | 17 | 5 | 0 | 0 | 108 |
| Days Max Temp ≤ 32 °F | 0 | 0 | 0 | 0 | 0 | 0 | 0 | 0 | 0 | 0 | 0 | 0 | 0 |
| Days Min Temp ≤ 32 °F | 8 | 4 | 1 | 0 | 0 | 0 | 0 | 0 | 0 | 0 | 2 | 5 | 20 |
| Days Min Temp ≤ 0 °F | 0 | 0 | 0 | 0 | 0 | 0 | 0 | 0 | 0 | 0 | 0 | 0 | 0 |
| Heating Degree Days | 435 | 306 | 145 | 31 | 1 | 0 | 0 | 0 | 1 | 33 | 177 | 356 | 1485 |
| Cooling Degree Days | 8 | 17 | 69 | 177 | 341 | 514 | 616 | 631 | 445 | 238 | 77 | 22 | 3155 |
| Total Precipitation (") | 3.10 | 3.10 | 2.46 | 3.30 | 5.54 | 4.14 | 2.22 | 2.77 | 4.25 | 4.30 | 2.97 | 2.97 | 41.12 |
| Days ≥ 0.1" Precip | 5 | 5 | 4 | 4 | 7 | 5 | 3 | 4 | 6 | 5 | 5 | 5 | 58 |
| Total Snowfall (") | 0.1 | 0.0 | 0.0 | 0.0 | 0.0 | 0.0 | 0.0 | 0.0 | 0.0 | 0.0 | 0.0 | 0.0 | 0.1 |
| Days ≥ 1" Snow Depth | 0 | 0 | 0 | 0 | 0 | 0 | 0 | 0 | 0 | 0 | 0 | 0 | 0 |

## LA TUNA 1 S *El Paso County* ELEVATION 3802 ft LAT/LONG 31° 59 ' N / 106° 36 ' W

| | JAN | FEB | MAR | APR | MAY | JUN | JUL | AUG | SEP | OCT | NOV | DEC | YEAR |
|---|---|---|---|---|---|---|---|---|---|---|---|---|---|
| Maximum Temp °F | 58.3 | 64.2 | 71.5 | 80.0 | 87.8 | 96.5 | 96.8 | 94.0 | 88.9 | 80.3 | 67.9 | 58.6 | 78.7 |
| Minimum Temp °F | 28.8 | 32.8 | 38.9 | 46.5 | 55.1 | 64.5 | 68.8 | 66.5 | 60.2 | 48.4 | 36.4 | 29.3 | 48.0 |
| Mean Temp °F | 43.6 | 48.5 | 55.2 | 63.3 | 71.5 | 80.5 | 82.7 | 80.3 | 74.5 | 64.4 | 52.2 | 44.0 | 63.4 |
| Days Max Temp ≥ 90 °F | 0 | 0 | 0 | 2 | 13 | 27 | 28 | 26 | 16 | 3 | 0 | 0 | 115 |
| Days Max Temp ≤ 32 °F | 0 | 0 | 0 | 0 | 0 | 0 | 0 | 0 | 0 | 0 | 0 | 0 | 0 |
| Days Min Temp ≤ 32 °F | 22 | 13 | 6 | 1 | 0 | 0 | 0 | 0 | 0 | 1 | 9 | 22 | 74 |
| Days Min Temp ≤ 0 °F | 0 | 0 | 0 | 0 | 0 | 0 | 0 | 0 | 0 | 0 | 0 | 0 | 0 |
| Heating Degree Days | 658 | 459 | 302 | 96 | 11 | 0 | 0 | 0 | 6 | 81 | 378 | 645 | 2636 |
| Cooling Degree Days | 0 | 1 | 5 | 56 | 219 | 478 | 541 | 473 | 298 | 63 | 1 | 0 | 2135 |
| Total Precipitation (") | 0.46 | 0.36 | 0.24 | 0.11 | 0.36 | 0.81 | 1.40 | 2.23 | 1.15 | 0.83 | 0.35 | 0.71 | 9.01 |
| Days ≥ 0.1" Precip | 2 | 1 | 1 | 0 | 1 | 2 | 3 | 4 | 3 | 2 | 1 | 2 | 22 |
| Total Snowfall (") | 0.0 | 0.0 | 0.0 | 0.0 | 0.0 | 0.0 | 0.0 | 0.0 | 0.0 | 0.0 | 0.0 | 0.1 | 0.1 |
| Days ≥ 1" Snow Depth | 0 | 0 | 0 | 0 | 0 | 0 | 0 | 0 | 0 | 0 | 0 | *0* | 0 |

## LAMESA 1 SSE *Dawson County* ELEVATION 2972 ft LAT/LONG 32° 42 ' N / 101° 56 ' W

| | JAN | FEB | MAR | APR | MAY | JUN | JUL | AUG | SEP | OCT | NOV | DEC | YEAR |
|---|---|---|---|---|---|---|---|---|---|---|---|---|---|
| Maximum Temp °F | 54.2 | 59.8 | 68.6 | 77.9 | 85.2 | 92.2 | 94.4 | 92.7 | 85.7 | 77.2 | 65.6 | 56.7 | 75.9 |
| Minimum Temp °F | 25.8 | 29.3 | 36.2 | 45.8 | 54.9 | 63.4 | 66.9 | 65.3 | 59.2 | 47.8 | 36.2 | 28.4 | 46.6 |
| Mean Temp °F | 40.0 | 44.6 | 52.4 | 62.0 | 70.1 | 77.8 | 80.7 | 79.1 | 72.5 | 62.5 | 50.9 | 42.6 | 61.3 |
| Days Max Temp ≥ 90 °F | 0 | 0 | 0 | 3 | 11 | 20 | 25 | 23 | 11 | 3 | 0 | 0 | 96 |
| Days Max Temp ≤ 32 °F | 2 | 1 | 0 | 0 | 0 | 0 | 0 | 0 | 0 | 0 | 0 | 1 | 4 |
| Days Min Temp ≤ 32 °F | 25 | 19 | 10 | 2 | 0 | 0 | 0 | 0 | 0 | 1 | 10 | 22 | 89 |
| Days Min Temp ≤ 0 °F | 0 | 0 | 0 | 0 | 0 | 0 | 0 | 0 | 0 | 0 | 0 | 0 | 0 |
| Heating Degree Days | 768 | 571 | 389 | 142 | 31 | 2 | 0 | 1 | 22 | 133 | 418 | 688 | 3165 |
| Cooling Degree Days | 0 | 1 | 5 | 53 | 189 | 377 | 476 | 442 | 245 | 55 | 2 | 0 | 1845 |
| Total Precipitation (") | 0.62 | 0.74 | 0.88 | 1.02 | 2.47 | 2.68 | 2.24 | 2.00 | 3.54 | 1.71 | 0.87 | 0.71 | 19.48 |
| Days ≥ 0.1" Precip | 2 | 2 | 2 | 2 | 4 | 4 | 4 | 4 | 4 | 3 | 2 | 2 | 35 |
| Total Snowfall (") | *1.2* | 0.7 | 0.1 | 0.0 | 0.0 | 0.0 | 0.0 | 0.0 | 0.0 | 0.0 | 0.5 | 0.7 | 3.2 |
| Days ≥ 1" Snow Depth | na | *0* | 0 | 0 | 0 | 0 | 0 | 0 | 0 | 0 | 0 | 0 | na |

## LAMPASAS *Lampasas County* ELEVATION 1020 ft LAT/LONG 31° 3 ' N / 98° 11 ' W

| | JAN | FEB | MAR | APR | MAY | JUN | JUL | AUG | SEP | OCT | NOV | DEC | YEAR |
|---|---|---|---|---|---|---|---|---|---|---|---|---|---|
| Maximum Temp °F | 57.3 | 61.4 | 69.8 | 77.8 | 82.6 | 89.7 | 94.2 | 94.4 | 87.9 | 79.3 | 68.4 | 60.3 | 76.9 |
| Minimum Temp °F | 30.2 | 33.7 | 41.9 | 50.6 | 58.7 | 66.4 | 69.0 | 68.0 | 62.4 | 51.2 | 40.4 | 32.9 | 50.4 |
| Mean Temp °F | 43.8 | 47.6 | 55.9 | 64.2 | 70.6 | 78.1 | 81.6 | 81.2 | 75.2 | 65.3 | 54.4 | 46.6 | 63.7 |
| Days Max Temp ≥ 90 °F | 0 | 0 | 1 | 1 | 4 | 17 | 27 | 27 | 14 | 3 | 0 | 0 | 94 |
| Days Max Temp ≤ 32 °F | 1 | 1 | 0 | 0 | 0 | 0 | 0 | 0 | 0 | 0 | 0 | 0 | 2 |
| Days Min Temp ≤ 32 °F | 20 | 14 | 7 | 2 | 0 | 0 | 0 | 0 | 0 | 1 | 9 | 17 | 70 |
| Days Min Temp ≤ 0 °F | 0 | 0 | 0 | 0 | 0 | 0 | 0 | 0 | 0 | 0 | 0 | 0 | 0 |
| Heating Degree Days | 652 | 489 | 302 | 105 | 21 | 1 | 0 | 0 | 11 | 95 | 331 | 565 | 2572 |
| Cooling Degree Days | 1 | 3 | 25 | 77 | 206 | 401 | 519 | 522 | 323 | 115 | 23 | 3 | 2218 |
| Total Precipitation (") | 1.71 | 2.25 | 2.25 | 2.60 | 4.63 | 3.18 | 1.73 | 2.39 | 2.97 | 3.39 | 2.10 | 2.02 | 31.22 |
| Days ≥ 0.1" Precip | 4 | 4 | 4 | 4 | 7 | 5 | 3 | 3 | 5 | 4 | 4 | 4 | 51 |
| Total Snowfall (") | 0.3 | 0.5 | 0.0 | 0.0 | 0.0 | 0.0 | 0.0 | 0.0 | 0.0 | 0.0 | 0.2 | 0.0 | 1.0 |
| Days ≥ 1" Snow Depth | 0 | 0 | 0 | 0 | 0 | 0 | 0 | 0 | 0 | 0 | 0 | 0 | 0 |

## LANGTRY *Val Verde County* ELEVATION 1312 ft LAT/LONG 29° 48 ' N / 101° 34 ' W

| | JAN | FEB | MAR | APR | MAY | JUN | JUL | AUG | SEP | OCT | NOV | DEC | YEAR |
|---|---|---|---|---|---|---|---|---|---|---|---|---|---|
| Maximum Temp °F | 61.6 | 66.7 | 76.3 | 84.3 | 90.0 | ***95.3*** | ***97.3*** | 97.2 | 91.4 | 81.9 | 70.8 | 63.6 | 81.4 |
| Minimum Temp °F | 33.7 | 38.4 | 47.7 | 56.8 | 65.2 | ***71.9*** | ***74.1*** | 73.7 | 67.4 | 57.2 | 44.0 | 35.7 | 55.5 |
| Mean Temp °F | 47.7 | 52.6 | 62.0 | 70.6 | 77.6 | ***83.6*** | ***85.7*** | 85.5 | 79.4 | 69.6 | 57.5 | 49.7 | 68.5 |
| Days Max Temp ≥ 90 °F | 0 | 0 | 2 | 8 | 18 | ***25*** | ***29*** | 29 | 21 | 5 | 0 | 0 | 137 |
| Days Max Temp ≤ 32 °F | 0 | 0 | 0 | 0 | 0 | 0 | ***0*** | 0 | 0 | 0 | 0 | 0 | 0 |
| Days Min Temp ≤ 32 °F | ***15*** | 7 | 2 | 0 | 0 | 0 | ***0*** | 0 | 0 | 0 | 4 | ***11*** | 39 |
| Days Min Temp ≤ 0 °F | 0 | 0 | 0 | 0 | 0 | 0 | 0 | 0 | 0 | 0 | 0 | 0 | 0 |
| Heating Degree Days | 531 | 347 | 144 | 31 | 2 | ***0*** | ***0*** | 0 | 5 | 41 | 245 | 468 | 1814 |
| Cooling Degree Days | 0 | 4 | 64 | 210 | 422 | 590 | 673 | 678 | 467 | 211 | 34 | 1 | 3354 |
| Total Precipitation (") | 0.55 | 0.79 | 0.58 | 1.04 | 1.97 | 1.54 | 1.56 | ***1.63*** | 2.54 | 1.57 | 0.68 | 0.61 | 15.06 |
| Days ≥ 0.1" Precip | 1 | 2 | ***1*** | 2 | 3 | ***3*** | ***2*** | ***2*** | 3 | 3 | 2 | 1 | 25 |
| Total Snowfall (") | ***0.3*** | ***0.2*** | 0.0 | 0.0 | 0.0 | 0.0 | 0.0 | ***0.0*** | 0.0 | 0.0 | 0.0 | 0.0 | 0.5 |
| Days ≥ 1" Snow Depth | ***0*** | ***0*** | 0 | 0 | 0 | 0 | 0 | ***0*** | 0 | 0 | 0 | 0 | 0 |

## LAREDO 2 *Webb County* ELEVATION 430 ft LAT/LONG 27° 34 ' N / 99° 30 ' W

| | JAN | FEB | MAR | APR | MAY | JUN | JUL | AUG | SEP | OCT | NOV | DEC | YEAR |
|---|---|---|---|---|---|---|---|---|---|---|---|---|---|
| Maximum Temp °F | 66.4 | 71.6 | 80.9 | 88.0 | 92.4 | 97.3 | 99.7 | 99.3 | 93.5 | 86.3 | 77.0 | 69.1 | 85.1 |
| Minimum Temp °F | 43.7 | 47.6 | 55.6 | 63.1 | 69.5 | 73.7 | 75.3 | 74.7 | 70.9 | 63.1 | 53.6 | 46.4 | 61.4 |
| Mean Temp °F | 55.1 | 59.7 | 68.3 | 75.6 | 81.0 | 85.5 | 87.5 | 87.0 | 82.2 | 74.7 | 65.3 | 57.8 | 73.3 |
| Days Max Temp ≥ 90 °F | 0 | 1 | 5 | 14 | 21 | 28 | 30 | 30 | 23 | 11 | 2 | 0 | 165 |
| Days Max Temp ≤ 32 °F | 0 | 0 | 0 | 0 | 0 | 0 | 0 | 0 | 0 | 0 | 0 | 0 | 0 |
| Days Min Temp ≤ 32 °F | 4 | 2 | 0 | 0 | 0 | 0 | 0 | 0 | 0 | 0 | 1 | 3 | 10 |
| Days Min Temp ≤ 0 °F | 0 | 0 | 0 | 0 | 0 | 0 | 0 | 0 | 0 | 0 | 0 | 0 | 0 |
| Heating Degree Days | 325 | 191 | 65 | 11 | 1 | 0 | 0 | 0 | 1 | 13 | 109 | 254 | 970 |
| Cooling Degree Days | 19 | 53 | 169 | 332 | 523 | 658 | 731 | 733 | 554 | 349 | 130 | 39 | 4290 |
| Total Precipitation (") | 0.94 | 1.02 | 0.72 | 1.70 | 2.90 | 3.02 | 1.48 | 2.50 | 3.20 | 2.54 | 1.05 | 1.02 | 22.09 |
| Days ≥ 0.1" Precip | 3 | 2 | 1 | 2 | 4 | 4 | 3 | 3 | 4 | 3 | 2 | 2 | 33 |
| Total Snowfall (") | 0.1 | 0.1 | 0.0 | 0.0 | 0.0 | 0.0 | 0.0 | 0.0 | 0.0 | 0.0 | 0.0 | 0.0 | 0.2 |
| Days ≥ 1" Snow Depth | 0 | 0 | 0 | 0 | 0 | 0 | 0 | 0 | 0 | 0 | 0 | 0 | 0 |

## LAVON DAM *Collin County* ELEVATION 502 ft LAT/LONG 33° 2 ' N / 96° 29 ' W

| | JAN | FEB | MAR | APR | MAY | JUN | JUL | AUG | SEP | OCT | NOV | DEC | YEAR |
|---|---|---|---|---|---|---|---|---|---|---|---|---|---|
| Maximum Temp °F | 52.5 | 57.4 | 66.3 | 74.9 | 81.3 | 89.2 | 93.8 | 93.9 | 87.1 | 77.5 | 65.4 | 56.5 | 74.6 |
| Minimum Temp °F | 31.1 | 35.2 | 43.6 | 53.2 | 61.1 | 68.8 | 72.2 | 71.2 | 64.9 | 53.5 | 43.2 | 34.9 | 52.7 |
| Mean Temp °F | 41.8 | 46.4 | 55.0 | 64.1 | 71.3 | 79.0 | 83.0 | 82.6 | 76.0 | 65.4 | 54.3 | 45.8 | 63.7 |
| Days Max Temp ≥ 90 °F | 0 | 0 | 0 | 0 | 2 | 16 | 27 | 26 | 13 | 2 | 0 | 0 | 86 |
| Days Max Temp ≤ 32 °F | 2 | 1 | 0 | 0 | 0 | 0 | 0 | 0 | 0 | 0 | 0 | 1 | 4 |
| Days Min Temp ≤ 32 °F | 18 | 11 | 4 | 0 | 0 | 0 | 0 | 0 | 0 | 0 | 4 | 12 | 49 |
| Days Min Temp ≤ 0 °F | 0 | 0 | 0 | 0 | 0 | 0 | 0 | 0 | 0 | 0 | 0 | 0 | 0 |
| Heating Degree Days | 714 | 521 | 321 | 96 | 14 | 1 | 0 | 0 | 8 | 87 | 333 | 591 | 2686 |
| Cooling Degree Days | 0 | 2 | 14 | 75 | 218 | 433 | 562 | 570 | 355 | 104 | 19 | 1 | 2353 |
| Total Precipitation (") | 2.19 | 2.88 | 3.14 | 4.03 | 5.80 | 3.74 | 2.38 | 1.84 | 4.01 | 4.16 | 2.82 | 2.90 | 39.89 |
| Days ≥ 0.1" Precip | 4 | 5 | 5 | 5 | 7 | 5 | 3 | 3 | 5 | 5 | 5 | 5 | 57 |
| Total Snowfall (") | ***0.0*** | 0.0 | 0.0 | 0.0 | 0.0 | 0.0 | 0.0 | 0.0 | 0.0 | 0.0 | 0.0 | 0.0 | 0.0 |
| Days ≥ 1" Snow Depth | ***0*** | 0 | 0 | 0 | 0 | 0 | 0 | 0 | 0 | 0 | 0 | 0 | 0 |

## LEVELLAND *Hockley County* ELEVATION 3553 ft LAT/LONG 33° 34 ' N / 102° 23 ' W

| | JAN | FEB | MAR | APR | MAY | JUN | JUL | AUG | SEP | OCT | NOV | DEC | YEAR |
|---|---|---|---|---|---|---|---|---|---|---|---|---|---|
| Maximum Temp °F | 53.0 | 57.9 | 66.4 | 75.5 | 83.2 | 90.9 | 92.1 | 89.8 | 83.0 | 75.2 | 63.3 | 55.3 | 73.8 |
| Minimum Temp °F | 23.2 | 26.1 | 32.5 | 42.3 | 51.4 | 60.4 | 63.6 | 61.9 | 55.2 | 43.5 | 32.6 | 25.2 | 43.2 |
| Mean Temp °F | 38.1 | 42.0 | 49.5 | 59.0 | 67.3 | 75.7 | 77.8 | 75.9 | 69.1 | 59.3 | 48.0 | 40.3 | 58.5 |
| Days Max Temp ≥ 90 °F | 0 | 0 | 0 | 2 | 8 | 19 | 22 | 18 | 8 | 1 | 0 | 0 | 78 |
| Days Max Temp ≤ 32 °F | 3 | 1 | 0 | 0 | 0 | 0 | 0 | 0 | 0 | 0 | 0 | 2 | 6 |
| Days Min Temp ≤ 32 °F | 28 | 23 | 14 | 3 | 0 | 0 | 0 | 0 | 0 | 2 | 15 | 26 | 111 |
| Days Min Temp ≤ 0 °F | 0 | 0 | 0 | 0 | 0 | 0 | 0 | 0 | 0 | 0 | 0 | 0 | 0 |
| Heating Degree Days | 826 | 642 | 477 | 205 | 56 | 5 | 0 | 1 | 40 | 199 | 505 | 759 | 3715 |
| Cooling Degree Days | 0 | 0 | 2 | 29 | 140 | 343 | 420 | 371 | 181 | 29 | 1 | 0 | 1516 |
| Total Precipitation (") | 0.54 | 0.61 | 0.63 | 0.99 | 2.12 | 2.54 | 2.41 | 3.14 | 3.34 | 1.59 | 0.76 | 0.72 | 19.39 |
| Days ≥ 0.1" Precip | 1 | 2 | 2 | 3 | 4 | 4 | 5 | 5 | 4 | 3 | 2 | 2 | 37 |
| Total Snowfall (") | 2.4 | 2.2 | 0.6 | 0.2 | 0.0 | 0.0 | 0.0 | 0.0 | 0.0 | 0.4 | 0.7 | 2.6 | 9.1 |
| Days ≥ 1" Snow Depth | 2 | 1 | 1 | 0 | 0 | 0 | 0 | 0 | 0 | 0 | 0 | 2 | 6 |

## LEXINGTON *Lee County* ELEVATION 459 ft LAT/LONG 30° 25 ' N / 97° 1 ' W

| | JAN | FEB | MAR | APR | MAY | JUN | JUL | AUG | SEP | OCT | NOV | DEC | YEAR |
|---|---|---|---|---|---|---|---|---|---|---|---|---|---|
| Maximum Temp °F | 58.0 | 62.6 | 70.8 | 78.0 | 83.0 | 89.7 | 93.5 | 93.9 | 88.4 | 80.1 | 69.6 | 61.8 | 77.4 |
| Minimum Temp °F | 36.6 | 40.5 | 48.6 | 56.7 | 63.7 | 70.0 | 72.5 | 72.0 | 67.2 | 56.6 | 47.7 | 40.0 | 56.0 |
| Mean Temp °F | 47.3 | 51.6 | 59.7 | 67.3 | 73.4 | 79.8 | 83.0 | 83.0 | 77.8 | 68.4 | 58.7 | 50.9 | 66.7 |
| Days Max Temp ≥ 90 °F | 0 | 0 | 0 | 1 | 3 | 18 | 27 | 28 | 15 | 3 | 0 | 0 | 95 |
| Days Max Temp ≤ 32 °F | 1 | 0 | 0 | 0 | 0 | 0 | 0 | 0 | 0 | 0 | 0 | 0 | 1 |
| Days Min Temp ≤ 32 °F | 11 | 6 | 2 | 0 | 0 | 0 | 0 | 0 | 0 | 0 | 2 | 7 | 28 |
| Days Min Temp ≤ 0 °F | 0 | 0 | 0 | 0 | 0 | 0 | 0 | 0 | 0 | 0 | 0 | 0 | 0 |
| Heating Degree Days | 546 | 380 | 203 | 55 | 7 | 0 | 0 | 0 | 4 | 52 | 229 | 440 | 1916 |
| Cooling Degree Days | 2 | 8 | 44 | 124 | 279 | 458 | 579 | 584 | 404 | 179 | 50 | 10 | 2721 |
| Total Precipitation (") | 2.67 | 2.63 | 2.46 | 2.90 | 5.30 | 3.56 | 1.73 | 2.20 | 3.61 | 4.42 | 2.92 | 2.85 | 37.25 |
| Days ≥ 0.1" Precip | 5 | 5 | 4 | 4 | 7 | 5 | 3 | 3 | 5 | 4 | 4 | 5 | 54 |
| Total Snowfall (") | 0.0 | 0.2 | 0.0 | 0.0 | 0.0 | 0.0 | 0.0 | 0.0 | 0.0 | 0.0 | 0.0 | 0.0 | 0.2 |
| Days ≥ 1" Snow Depth | 0 | 0 | 0 | 0 | 0 | 0 | 0 | 0 | 0 | 0 | 0 | 0 | 0 |

## LIBERTY *Liberty County* ELEVATION 35 ft LAT/LONG 30° 4 ' N / 94° 48 ' W

| | JAN | FEB | MAR | APR | MAY | JUN | JUL | AUG | SEP | OCT | NOV | DEC | YEAR |
|---|---|---|---|---|---|---|---|---|---|---|---|---|---|
| Maximum Temp °F | 60.8 | 64.6 | 71.9 | 78.7 | 84.6 | 90.2 | 92.9 | 93.3 | 89.0 | 81.6 | 72.0 | 64.4 | 78.7 |
| Minimum Temp °F | 40.0 | 42.5 | 49.7 | 57.7 | 64.3 | 70.3 | 72.4 | 71.9 | 67.0 | 56.7 | 48.5 | 42.4 | 57.0 |
| Mean Temp °F | 50.4 | 53.6 | 60.8 | 68.2 | 74.4 | 80.2 | 82.7 | 82.6 | 78.0 | 69.2 | 60.3 | 53.4 | 67.8 |
| Days Max Temp ≥ 90 °F | 0 | 0 | 0 | 0 | 4 | 20 | 27 | 27 | 16 | 4 | 0 | 0 | 98 |
| Days Max Temp ≤ 32 °F | 0 | 0 | 0 | 0 | 0 | 0 | 0 | 0 | 0 | 0 | 0 | 0 | 0 |
| Days Min Temp ≤ 32 °F | 8 | 5 | 1 | 0 | 0 | 0 | 0 | 0 | 0 | 0 | 2 | 6 | 22 |
| Days Min Temp ≤ 0 °F | 0 | 0 | 0 | 0 | 0 | 0 | 0 | 0 | 0 | 0 | 0 | 0 | 0 |
| Heating Degree Days | 455 | 328 | 171 | 43 | 2 | 0 | 0 | 0 | 1 | 41 | 195 | 368 | 1604 |
| Cooling Degree Days | 5 | 15 | 56 | 142 | 316 | 489 | 577 | 579 | 415 | 202 | 69 | 21 | 2886 |
| Total Precipitation (") | 4.41 | 3.92 | 3.63 | 3.91 | 6.24 | 6.43 | 4.28 | 4.12 | 5.16 | 5.64 | 5.33 | 4.98 | 58.05 |
| Days ≥ 0.1" Precip | 7 | 6 | 5 | 4 | 7 | 7 | 7 | 6 | 7 | 5 | 6 | 7 | 74 |
| Total Snowfall (") | 0.2 | 0.0 | 0.0 | 0.0 | 0.0 | 0.0 | 0.0 | 0.0 | 0.0 | 0.0 | 0.0 | 0.0 | 0.2 |
| Days ≥ 1" Snow Depth | 0 | 0 | 0 | 0 | 0 | 0 | 0 | 0 | 0 | 0 | 0 | 0 | 0 |

## LIPSCOMB *Lipscomb County* ELEVATION 2431 ft LAT/LONG 36° 14 ' N / 100° 17 ' W

| | JAN | FEB | MAR | APR | MAY | JUN | JUL | AUG | SEP | OCT | NOV | DEC | YEAR |
|---|---|---|---|---|---|---|---|---|---|---|---|---|---|
| Maximum Temp °F | 47.1 | 52.0 | 61.9 | 71.5 | 79.3 | 88.7 | 94.7 | 93.0 | 84.6 | 73.6 | 58.9 | 49.8 | 71.3 |
| Minimum Temp °F | 17.2 | 21.4 | 30.3 | 40.6 | 50.8 | 61.3 | 66.0 | 64.1 | 55.1 | 41.5 | 28.9 | 20.1 | 41.4 |
| Mean Temp °F | 32.2 | 36.7 | 46.3 | 55.9 | 65.1 | 75.0 | 80.4 | 78.6 | 69.9 | 57.6 | 44.0 | 35.0 | 56.4 |
| Days Max Temp ≥ 90 °F | 0 | 0 | 0 | 1 | 4 | 15 | 25 | 22 | 11 | 2 | 0 | 0 | 80 |
| Days Max Temp ≤ 32 °F | 6 | 4 | 1 | 0 | 0 | 0 | 0 | 0 | 0 | 0 | 1 | 4 | 16 |
| Days Min Temp ≤ 32 °F | 30 | 25 | 18 | 6 | 0 | 0 | 0 | 0 | 0 | 6 | 20 | 29 | 134 |
| Days Min Temp ≤ 0 °F | 2 | 1 | 0 | 0 | 0 | 0 | 0 | 0 | 0 | 0 | 0 | 1 | 4 |
| Heating Degree Days | 1010 | 792 | 577 | 289 | 96 | 9 | 1 | 2 | 50 | 255 | 626 | 924 | 4631 |
| Cooling Degree Days | 0 | 0 | 3 | 24 | 106 | 322 | 485 | 449 | 224 | 30 | 1 | 0 | 1644 |
| Total Precipitation (") | 0.46 | 0.78 | 1.59 | 1.64 | 3.79 | 3.04 | 2.18 | 2.65 | 1.94 | 1.26 | 0.98 | 0.65 | 20.96 |
| Days ≥ 0.1" Precip | 2 | 2 | 3 | 3 | 6 | 5 | 4 | 4 | 4 | 2 | 2 | 2 | 39 |
| Total Snowfall (") | 3.0 | 4.3 | 3.4 | 0.8 | 0.0 | 0.0 | 0.0 | 0.0 | 0.0 | 0.1 | 1.8 | 3.8 | 17.2 |
| Days ≥ 1" Snow Depth | 3 | 3 | 1 | 0 | 0 | 0 | 0 | 0 | 0 | 0 | 1 | 3 | 11 |

## LITTLEFIELD 2 NW *Lamb County* ELEVATION 3491 ft LAT/LONG 33° 55' N / 102° 20' W

| | JAN | FEB | MAR | APR | MAY | JUN | JUL | AUG | SEP | OCT | NOV | DEC | YEAR |
|---|---|---|---|---|---|---|---|---|---|---|---|---|---|
| Maximum Temp °F | 51.2 | 56.7 | 65.2 | 73.9 | 81.4 | 89.5 | 91.4 | 88.8 | 82.3 | 73.8 | 61.6 | 53.9 | 72.5 |
| Minimum Temp °F | 22.6 | 26.2 | 32.9 | 42.2 | 51.7 | 61.0 | 64.9 | 62.8 | 55.5 | 43.9 | 32.9 | 24.8 | 43.4 |
| Mean Temp °F | 36.9 | 41.5 | 49.1 | 58.1 | 66.6 | 75.3 | 78.1 | 75.8 | 68.9 | 58.9 | 47.3 | 39.4 | 58.0 |
| Days Max Temp ≥ 90 °F | 0 | 0 | 0 | 1 | 6 | 16 | 21 | 16 | 7 | 1 | 0 | 0 | 68 |
| Days Max Temp ≤ 32 °F | 4 | 2 | 0 | 0 | 0 | 0 | 0 | 0 | 0 | 0 | 0 | 2 | 8 |
| Days Min Temp ≤ 32 °F | 28 | 22 | 15 | 4 | 0 | 0 | 0 | 0 | 0 | 2 | 14 | 26 | 111 |
| Days Min Temp ≤ 0 °F | 0 | 0 | 0 | 0 | 0 | 0 | 0 | 0 | 0 | 0 | 0 | 1 | 1 |
| Heating Degree Days | 863 | 658 | 488 | 226 | 66 | 5 | 0 | 2 | 42 | 209 | 525 | 788 | 3872 |
| Cooling Degree Days | 0 | 0 | 1 | 21 | 115 | 315 | 412 | 356 | 170 | 21 | 1 | 0 | 1412 |
| Total Precipitation (") | 0.55 | 0.51 | 0.72 | 1.03 | 2.41 | 2.99 | 2.36 | 3.00 | 2.42 | 1.48 | 0.71 | 0.65 | 18.83 |
| Days ≥ 0.1" Precip | 2 | 2 | 2 | 3 | 4 | 5 | 4 | 5 | 4 | 3 | 2 | 2 | 38 |
| Total Snowfall (") | na | 1.4 | 0.7 | 0.4 | 0.0 | 0.0 | 0.0 | 0.0 | 0.0 | 0.0 | 0.8 | *1.9* | na |
| Days ≥ 1" Snow Depth | na | na | 0 | 0 | 0 | 0 | 0 | 0 | 0 | 0 | 0 | na | na |

## LIVINGSTON 2 NNE *Polk County* ELEVATION 190 ft LAT/LONG 30° 42' N / 94° 57' W

| | JAN | FEB | MAR | APR | MAY | JUN | JUL | AUG | SEP | OCT | NOV | DEC | YEAR |
|---|---|---|---|---|---|---|---|---|---|---|---|---|---|
| Maximum Temp °F | 58.5 | 62.8 | 70.8 | 78.2 | 83.8 | 90.2 | 93.7 | 93.9 | 88.4 | 80.4 | 70.4 | 62.3 | 77.8 |
| Minimum Temp °F | 36.2 | 37.9 | 45.3 | 53.8 | 61.2 | 68.0 | 70.9 | 69.7 | 64.6 | 53.4 | 45.3 | 38.1 | 53.7 |
| Mean Temp °F | 47.4 | 50.4 | 58.1 | 66.0 | 72.6 | 79.2 | 82.3 | 81.8 | 76.6 | 66.9 | 57.9 | 50.3 | 65.8 |
| Days Max Temp ≥ 90 °F | 0 | 0 | 0 | 0 | 4 | 19 | 27 | 27 | 15 | 3 | 0 | 0 | 95 |
| Days Max Temp ≤ 32 °F | 0 | 0 | 0 | 0 | 0 | 0 | 0 | 0 | 0 | 0 | 0 | 0 | 0 |
| Days Min Temp ≤ 32 °F | 14 | 11 | 5 | 1 | 0 | 0 | 0 | 0 | 0 | 0 | 5 | 12 | 48 |
| Days Min Temp ≤ 0 °F | 0 | 0 | 0 | 0 | 0 | 0 | 0 | 0 | 0 | 0 | 0 | 0 | 0 |
| Heating Degree Days | 548 | 416 | 246 | 77 | 9 | 0 | 0 | 0 | 5 | 75 | 255 | 464 | 2095 |
| Cooling Degree Days | 3 | 10 | 39 | 105 | 252 | 444 | 551 | 548 | 372 | 151 | 50 | 15 | 2540 |
| Total Precipitation (") | 4.31 | 3.68 | 3.92 | 3.92 | 5.94 | 4.72 | 3.36 | 3.11 | 4.24 | 4.07 | 4.24 | 4.92 | 50.43 |
| Days ≥ 0.1" Precip | 7 | 5 | 6 | 5 | 7 | 6 | 6 | 5 | 5 | 5 | 6 | 7 | 70 |
| Total Snowfall (") | 0.3 | 0.2 | 0.0 | 0.0 | 0.0 | 0.0 | 0.0 | 0.0 | 0.0 | 0.0 | 0.0 | 0.0 | 0.5 |
| Days ≥ 1" Snow Depth | 0 | 0 | 0 | 0 | 0 | 0 | 0 | 0 | 0 | 0 | 0 | 0 | 0 |

## LLANO *Llano County* ELEVATION 1040 ft LAT/LONG 30° 45' N / 98° 40' W

| | JAN | FEB | MAR | APR | MAY | JUN | JUL | AUG | SEP | OCT | NOV | DEC | YEAR |
|---|---|---|---|---|---|---|---|---|---|---|---|---|---|
| Maximum Temp °F | 59.2 | 63.2 | 71.7 | 79.8 | 84.7 | 91.8 | 96.1 | 95.7 | 89.3 | 80.8 | 69.9 | 62.2 | 78.7 |
| Minimum Temp °F | 32.1 | 35.6 | 44.5 | 53.9 | 61.8 | 69.2 | 71.8 | 70.6 | 64.4 | 53.6 | 42.5 | 34.3 | 52.9 |
| Mean Temp °F | 45.7 | 49.4 | 58.2 | 66.8 | 73.3 | 80.5 | 84.0 | 83.2 | 76.9 | 67.2 | 56.2 | 48.3 | 65.8 |
| Days Max Temp ≥ 90 °F | 0 | 0 | 1 | 3 | 8 | 22 | 28 | 28 | 17 | 4 | 0 | 0 | 111 |
| Days Max Temp ≤ 32 °F | 1 | 0 | 0 | 0 | 0 | 0 | 0 | 0 | 0 | 0 | 0 | 0 | 1 |
| Days Min Temp ≤ 32 °F | 17 | 11 | 4 | 0 | 0 | 0 | 0 | 0 | 0 | 0 | 6 | 15 | 53 |
| Days Min Temp ≤ 0 °F | 0 | 0 | 0 | 0 | 0 | 0 | 0 | 0 | 0 | 0 | 0 | 0 | 0 |
| Heating Degree Days | 595 | 437 | 243 | 69 | 9 | 0 | 0 | 0 | 7 | 69 | 284 | 516 | 2229 |
| Cooling Degree Days | 1 | 3 | 38 | 124 | 288 | 481 | 603 | 603 | 387 | 161 | 33 | 4 | 2726 |
| Total Precipitation (") | 1.27 | 1.87 | 1.68 | 2.43 | 4.16 | 2.99 | 1.71 | 2.40 | 2.68 | 2.86 | 1.86 | 1.69 | 27.60 |
| Days ≥ 0.1" Precip | 3 | 4 | 3 | 4 | 6 | 4 | 3 | 3 | 4 | 4 | 3 | 3 | 44 |
| Total Snowfall (") | 0.5 | 0.6 | 0.0 | 0.0 | 0.0 | 0.0 | 0.0 | 0.0 | 0.0 | 0.0 | 0.0 | 0.0 | 1.1 |
| Days ≥ 1" Snow Depth | 0 | 0 | 0 | 0 | 0 | 0 | 0 | 0 | 0 | 0 | 0 | 0 | 0 |

## LUBBOCK REGIONAL AP *Lubbock County* ELEVATION 3215 ft LAT/LONG 33° 38' N / 101° 51' W

| | JAN | FEB | MAR | APR | MAY | JUN | JUL | AUG | SEP | OCT | NOV | DEC | YEAR |
|---|---|---|---|---|---|---|---|---|---|---|---|---|---|
| Maximum Temp °F | 52.7 | 57.7 | 66.4 | 75.5 | 82.8 | 90.5 | 92.1 | 89.7 | 83.1 | 74.7 | 62.8 | 54.6 | 73.6 |
| Minimum Temp °F | 25.4 | 29.0 | 36.8 | 46.5 | 55.5 | 64.5 | 68.0 | 66.2 | 59.1 | 47.7 | 36.1 | 27.7 | 46.9 |
| Mean Temp °F | 39.1 | 43.4 | 51.6 | 61.0 | 69.2 | 77.5 | 80.1 | 78.0 | 71.1 | 61.2 | 49.5 | 41.1 | 60.2 |
| Days Max Temp ≥ 90 °F | 0 | 0 | 0 | 1 | 7 | 17 | 22 | 18 | 8 | 1 | 0 | 0 | 74 |
| Days Max Temp ≤ 32 °F | 3 | 1 | 0 | 0 | 0 | 0 | 0 | 0 | 0 | 0 | 0 | 2 | 6 |
| Days Min Temp ≤ 32 °F | 25 | 18 | 9 | 1 | 0 | 0 | 0 | 0 | 0 | 1 | 10 | 22 | 86 |
| Days Min Temp ≤ 0 °F | 0 | 0 | 0 | 0 | 0 | 0 | 0 | 0 | 0 | 0 | 0 | 0 | 0 |
| Heating Degree Days | 797 | 603 | 414 | 163 | 40 | 2 | 0 | 1 | 30 | 161 | 462 | 733 | 3406 |
| Cooling Degree Days | 0 | 1 | 6 | 53 | 183 | 389 | 491 | 446 | 235 | 52 | 2 | 0 | 1858 |
| Total Precipitation (") | 0.48 | 0.68 | 0.87 | 1.09 | 2.44 | 2.76 | 2.26 | 2.47 | 2.64 | 1.78 | 0.75 | 0.60 | 18.82 |
| Days ≥ 0.1" Precip | 1 | 2 | 2 | 2 | 4 | 4 | 4 | 4 | 4 | 3 | 2 | 2 | 34 |
| Total Snowfall (") | 2.7 | 2.4 | 1.3 | 0.2 | 0.0 | 0.0 | 0.0 | 0.0 | 0.0 | 0.3 | 1.6 | 2.0 | 10.5 |
| Days ≥ 1" Snow Depth | 2 | 2 | 1 | 0 | 0 | 0 | 0 | 0 | 0 | 0 | 1 | 2 | 8 |

## LUFKIN ANGELINA CO *Angelina County* ELEVATION 292 ft LAT/LONG 31° 14 ' N / 94° 45 ' W

| | JAN | FEB | MAR | APR | MAY | JUN | JUL | AUG | SEP | OCT | NOV | DEC | YEAR |
|---|---|---|---|---|---|---|---|---|---|---|---|---|---|
| Maximum Temp °F | 58.5 | 63.3 | 71.3 | 78.7 | 84.0 | 90.0 | 93.0 | 93.1 | 88.2 | 80.2 | 69.8 | 62.0 | 77.7 |
| Minimum Temp °F | 37.7 | 40.1 | 47.5 | 55.8 | 63.2 | 69.8 | 72.3 | 71.2 | 66.3 | 55.1 | 46.1 | 39.6 | 55.4 |
| Mean Temp °F | 48.1 | 51.8 | 59.4 | 67.2 | 73.6 | 79.9 | 82.7 | 82.2 | 77.3 | 67.7 | 57.9 | 50.8 | 66.6 |
| Days Max Temp ≥ 90 °F | 0 | 0 | 0 | 1 | 5 | 19 | 26 | 26 | 15 | 3 | 0 | 0 | 95 |
| Days Max Temp ≤ 32 °F | 0 | 0 | 0 | 0 | 0 | 0 | 0 | 0 | 0 | 0 | 0 | 0 | 0 |
| Days Min Temp ≤ 32 °F | 11 | 7 | 2 | 0 | 0 | 0 | 0 | 0 | 0 | 0 | 4 | 9 | 33 |
| Days Min Temp ≤ 0 °F | 0 | 0 | 0 | 0 | 0 | 0 | 0 | 0 | 0 | 0 | 0 | 0 | 0 |
| Heating Degree Days | 528 | 378 | 211 | 58 | 4 | 0 | 0 | 0 | 3 | 61 | 248 | 446 | 1937 |
| Cooling Degree Days | 4 | 10 | 40 | 114 | 265 | 450 | 544 | 540 | 373 | 143 | 44 | 13 | 2540 |
| Total Precipitation (") | 3.96 | 3.34 | 3.28 | 3.25 | 5.48 | 3.99 | 2.58 | 2.88 | 3.77 | 4.05 | 3.82 | 4.37 | 44.77 |
| Days ≥ 0.1" Precip | 6 | 5 | 5 | 5 | 7 | 6 | 5 | 5 | 5 | 5 | 5 | 6 | 65 |
| Total Snowfall (") | 0.4 | 0.1 | 0.0 | 0.0 | 0.0 | 0.0 | 0.0 | 0.0 | 0.0 | 0.0 | 0.0 | 0.0 | 0.5 |
| Days ≥ 1" Snow Depth | 0 | 0 | 0 | 0 | 0 | 0 | 0 | 0 | 0 | 0 | 0 | 0 | 0 |

## LULING *Caldwell County* ELEVATION 400 ft LAT/LONG 29° 40 ' N / 97° 38 ' W

| | JAN | FEB | MAR | APR | MAY | JUN | JUL | AUG | SEP | OCT | NOV | DEC | YEAR |
|---|---|---|---|---|---|---|---|---|---|---|---|---|---|
| Maximum Temp °F | 60.4 | 64.6 | 73.1 | 80.2 | 85.0 | 91.6 | 95.4 | 96.4 | 90.7 | 82.7 | 72.1 | 64.0 | 79.7 |
| Minimum Temp °F | 37.2 | 40.3 | 48.3 | 56.3 | 63.4 | 69.7 | 71.7 | 70.9 | 66.0 | 56.2 | 47.0 | 39.7 | 55.6 |
| Mean Temp °F | 48.8 | 52.5 | 60.7 | 68.3 | 74.2 | 80.7 | 83.5 | 83.7 | 78.4 | 69.5 | 59.6 | 51.9 | 67.6 |
| Days Max Temp ≥ 90 °F | 0 | 0 | 1 | 2 | 7 | 22 | 28 | 29 | 19 | 6 | 0 | 0 | 114 |
| Days Max Temp ≤ 32 °F | 0 | 0 | 0 | 0 | 0 | 0 | 0 | 0 | 0 | 0 | 0 | 0 | 0 |
| Days Min Temp ≤ 32 °F | 11 | 6 | 2 | 0 | 0 | 0 | 0 | 0 | 0 | 0 | 3 | 8 | 30 |
| Days Min Temp ≤ 0 °F | 0 | 0 | 0 | 0 | 0 | 0 | 0 | 0 | 0 | 0 | 0 | 0 | 0 |
| Heating Degree Days | 502 | 358 | 182 | 49 | 5 | 0 | 0 | 0 | 3 | 44 | 211 | 413 | 1767 |
| Cooling Degree Days | 4 | 11 | 57 | 150 | 314 | 503 | 603 | 619 | 435 | 206 | 63 | 14 | 2979 |
| Total Precipitation (") | 2.67 | 2.53 | 2.12 | 3.25 | 5.46 | 4.10 | 1.73 | 2.16 | 4.29 | 3.84 | 2.84 | 2.43 | 37.42 |
| Days ≥ 0.1" Precip | 5 | 5 | 4 | 4 | 6 | 5 | 3 | 4 | 6 | 4 | 4 | 4 | 54 |
| Total Snowfall (") | 0.4 | 0.1 | 0.0 | 0.0 | 0.0 | 0.0 | 0.0 | 0.0 | 0.0 | 0.0 | 0.0 | 0.0 | 0.5 |
| Days ≥ 1" Snow Depth | 0 | 0 | 0 | 0 | 0 | 0 | 0 | 0 | 0 | 0 | 0 | 0 | 0 |

## MADISONVILLE *Madison County* ELEVATION 239 ft LAT/LONG 30° 57 ' N / 95° 54 ' W

| | JAN | FEB | MAR | APR | MAY | JUN | JUL | AUG | SEP | OCT | NOV | DEC | YEAR |
|---|---|---|---|---|---|---|---|---|---|---|---|---|---|
| Maximum Temp °F | 60.5 | 64.7 | 72.9 | 79.9 | 85.4 | 91.7 | 95.5 | 95.9 | 90.1 | 81.8 | 71.4 | 63.6 | 79.4 |
| Minimum Temp °F | 38.9 | 41.6 | 49.0 | 56.9 | 63.7 | 70.0 | 72.4 | 71.6 | 66.8 | 56.6 | 48.1 | 41.4 | 56.4 |
| Mean Temp °F | 49.7 | 53.2 | 61.0 | 68.4 | 74.6 | 80.9 | 84.0 | 83.8 | 78.5 | 69.2 | 59.8 | 52.6 | 68.0 |
| Days Max Temp ≥ 90 °F | 0 | 0 | 0 | 1 | 6 | 23 | 29 | 29 | 18 | 4 | 0 | 0 | 110 |
| Days Max Temp ≤ 32 °F | 0 | 0 | 0 | 0 | 0 | 0 | 0 | 0 | 0 | 0 | 0 | 0 | 0 |
| Days Min Temp ≤ 32 °F | 10 | 6 | 2 | 0 | 0 | 0 | 0 | 0 | 0 | 0 | 2 | 7 | 27 |
| Days Min Temp ≤ 0 °F | 0 | 0 | 0 | 0 | 0 | 0 | 0 | 0 | 0 | 0 | 0 | 0 | 0 |
| Heating Degree Days | 478 | 338 | 174 | 44 | 3 | 0 | 0 | 0 | 2 | 44 | 208 | 395 | 1686 |
| Cooling Degree Days | 7 | 15 | 57 | 152 | 312 | 491 | 598 | 608 | 431 | 192 | 62 | 18 | 2943 |
| Total Precipitation (") | 3.56 | 3.03 | 3.37 | 3.74 | 5.40 | 4.04 | 2.43 | 2.81 | 4.25 | 4.48 | 3.61 | 3.49 | 44.21 |
| Days ≥ 0.1" Precip | 6 | 5 | 5 | 4 | 7 | 5 | 4 | 4 | 5 | 5 | 5 | 5 | 60 |
| Total Snowfall (") | 0.3 | 0.1 | 0.0 | 0.0 | 0.0 | 0.0 | 0.0 | 0.0 | 0.0 | 0.0 | 0.0 | 0.0 | 0.4 |
| Days ≥ 1" Snow Depth | 0 | 0 | 0 | 0 | 0 | 0 | 0 | 0 | 0 | 0 | 0 | 0 | 0 |

## MARATHON *Brewster County* ELEVATION 4052 ft LAT/LONG 30° 13 ' N / 103° 15 ' W

| | JAN | FEB | MAR | APR | MAY | JUN | JUL | AUG | SEP | OCT | NOV | DEC | YEAR |
|---|---|---|---|---|---|---|---|---|---|---|---|---|---|
| Maximum Temp °F | 61.2 | 65.2 | 72.5 | 80.3 | 86.1 | 91.1 | 90.2 | 88.8 | 84.1 | 78.1 | 68.7 | 62.9 | 77.4 |
| Minimum Temp °F | 28.4 | 30.5 | 36.5 | 44.4 | 52.5 | 59.5 | 61.9 | 60.8 | 55.9 | 45.4 | 35.0 | 29.2 | 45.0 |
| Mean Temp °F | 44.8 | 47.9 | 54.5 | 62.4 | 69.3 | 75.3 | 76.1 | 74.8 | 69.9 | 61.7 | 51.9 | 46.1 | 61.2 |
| Days Max Temp ≥ 90 °F | 0 | 0 | 0 | 2 | 10 | 18 | 19 | 15 | 7 | 2 | 0 | 0 | 73 |
| Days Max Temp ≤ 32 °F | 0 | 0 | 0 | 0 | 0 | 0 | 0 | 0 | 0 | 0 | 0 | 0 | 0 |
| Days Min Temp ≤ 32 °F | 21 | 17 | 9 | 2 | 0 | 0 | 0 | 0 | 0 | 2 | 12 | 20 | 83 |
| Days Min Temp ≤ 0 °F | 0 | 0 | 0 | 0 | 0 | 0 | 0 | 0 | 0 | 0 | 0 | 0 | 0 |
| Heating Degree Days | 618 | 478 | 323 | 117 | 22 | 1 | 0 | 1 | 24 | 132 | 389 | 578 | 2683 |
| Cooling Degree Days | 0 | 1 | 5 | 47 | 183 | 336 | 364 | 341 | 197 | 44 | 2 | 0 | 1520 |
| Total Precipitation (") | 0.44 | 0.46 | 0.28 | 0.62 | 1.63 | 1.82 | 2.43 | 2.30 | 3.01 | 1.46 | 0.52 | 0.57 | 15.54 |
| Days ≥ 0.1" Precip | 1 | 1 | 1 | 1 | 3 | 4 | 4 | 4 | 6 | 3 | 2 | 1 | 31 |
| Total Snowfall (") | *1.1* | 0.3 | 0.0 | 0.0 | 0.0 | 0.0 | 0.0 | 0.0 | 0.0 | 0.0 | 0.5 | 0.0 | 1.9 |
| Days ≥ 1" Snow Depth | *0* | *0* | 0 | 0 | 0 | 0 | 0 | 0 | 0 | 0 | 0 | 0 | 0 |

## MARFA # 2 *Presidio County* ELEVATION 4692 ft LAT/LONG 30° 18 ' N / 104° 1 ' W

| | JAN | FEB | MAR | APR | MAY | JUN | JUL | AUG | SEP | OCT | NOV | DEC | YEAR |
|---|---|---|---|---|---|---|---|---|---|---|---|---|---|
| Maximum Temp °F | 60.1 | 64.6 | 71.7 | 79.2 | 85.6 | 91.4 | 89.6 | 87.7 | 83.4 | 77.5 | 67.9 | 61.7 | 76.7 |
| Minimum Temp °F | 26.3 | 28.4 | 33.9 | 41.5 | 49.8 | 57.6 | 60.1 | 59.0 | 54.0 | 44.0 | 33.4 | 27.3 | 42.9 |
| Mean Temp °F | 43.2 | 46.5 | 52.8 | 60.4 | 67.8 | 74.5 | 74.8 | 73.4 | 68.7 | 60.8 | 50.7 | 44.5 | 59.8 |
| Days Max Temp ≥ 90 °F | 0 | 0 | 0 | 1 | 9 | 20 | 16 | 12 | 5 | 1 | 0 | 0 | 64 |
| Days Max Temp ≤ 32 °F | 0 | 0 | 0 | 0 | 0 | 0 | 0 | 0 | 0 | 0 | 0 | 0 | 0 |
| Days Min Temp ≤ 32 °F | 25 | 20 | 13 | 4 | 0 | 0 | 0 | 0 | 0 | 2 | 13 | 24 | 101 |
| Days Min Temp ≤ 0 °F | 0 | 0 | 0 | 0 | 0 | 0 | 0 | 0 | 0 | 0 | 0 | 0 | 0 |
| Heating Degree Days | 668 | 515 | 373 | 152 | 28 | 1 | 0 | 1 | 25 | 148 | 423 | 628 | 2962 |
| Cooling Degree Days | 0 | 0 | 0 | 19 | 126 | 309 | 322 | 290 | 153 | 23 | 0 | 0 | 1242 |
| Total Precipitation (") | 0.45 | 0.47 | 0.32 | 0.62 | 1.25 | 1.94 | 2.70 | 2.77 | 2.80 | 1.51 | 0.52 | 0.59 | 15.94 |
| Days ≥ 0.1" Precip | 1 | 1 | 1 | 2 | 3 | 4 | 5 | 6 | 5 | 3 | 1 | 2 | 34 |
| Total Snowfall (") | 1.0 | 0.8 | 0.2 | 0.0 | 0.0 | 0.0 | 0.0 | 0.0 | 0.0 | 0.0 | 0.5 | 0.5 | 3.0 |
| Days ≥ 1" Snow Depth | 0 | 0 | 0 | 0 | 0 | 0 | 0 | 0 | 0 | 0 | 0 | 0 | 0 |

## MARLIN 3 NE *Falls County* ELEVATION 400 ft LAT/LONG 31° 19 ' N / 96° 55 ' W

| | JAN | FEB | MAR | APR | MAY | JUN | JUL | AUG | SEP | OCT | NOV | DEC | YEAR |
|---|---|---|---|---|---|---|---|---|---|---|---|---|---|
| Maximum Temp °F | 58.4 | 63.3 | 71.4 | 78.5 | 83.9 | 90.8 | 95.1 | 95.3 | 89.1 | 80.6 | 69.4 | 61.3 | 78.1 |
| Minimum Temp °F | 36.4 | 39.7 | 47.6 | 55.9 | 63.2 | 69.8 | 72.3 | 71.6 | 66.4 | 56.0 | 46.5 | 39.1 | 55.4 |
| Mean Temp °F | 47.4 | 51.5 | 59.5 | 67.2 | 73.5 | 80.3 | 83.7 | 83.5 | 77.8 | 68.3 | 58.0 | 50.2 | 66.7 |
| Days Max Temp ≥ 90 °F | 0 | 0 | 0 | 1 | 4 | 20 | 28 | 28 | 16 | 3 | 0 | 0 | 100 |
| Days Max Temp ≤ 32 °F | 0 | 0 | 0 | 0 | 0 | 0 | 0 | 0 | 0 | 0 | 0 | 0 | 0 |
| Days Min Temp ≤ 32 °F | 12 | 7 | 3 | 0 | 0 | 0 | 0 | 0 | 0 | 0 | 3 | 9 | 34 |
| Days Min Temp ≤ 0 °F | 0 | 0 | 0 | 0 | 0 | 0 | 0 | 0 | 0 | 0 | 0 | 0 | 0 |
| Heating Degree Days | 543 | 381 | 208 | 55 | 5 | 0 | 0 | 0 | 3 | 55 | 244 | 460 | 1954 |
| Cooling Degree Days | 3 | 7 | 42 | 121 | 279 | 464 | 574 | 584 | 395 | 166 | 43 | 8 | 2686 |
| Total Precipitation (") | 2.23 | 2.69 | 3.20 | 3.37 | 5.72 | 3.34 | 1.93 | 2.01 | 3.40 | 4.05 | 3.04 | 3.10 | 38.08 |
| Days ≥ 0.1" Precip | 5 | 5 | 5 | 4 | 6 | 5 | 3 | 3 | 5 | 5 | 5 | 5 | 56 |
| Total Snowfall (") | 0.2 | 0.3 | 0.0 | 0.0 | 0.0 | 0.0 | 0.0 | 0.0 | 0.0 | 0.0 | 0.0 | 0.0 | 0.5 |
| Days ≥ 1" Snow Depth | 0 | 0 | 0 | 0 | 0 | 0 | 0 | 0 | 0 | 0 | 0 | 0 | 0 |

## MARSHALL *Harrison County* ELEVATION 381 ft LAT/LONG 32° 33 ' N / 94° 22 ' W

| | JAN | FEB | MAR | APR | MAY | JUN | JUL | AUG | SEP | OCT | NOV | DEC | YEAR |
|---|---|---|---|---|---|---|---|---|---|---|---|---|---|
| Maximum Temp °F | 54.3 | 59.2 | 67.4 | 75.6 | 81.5 | 88.4 | 92.2 | 91.9 | 86.2 | 77.0 | 66.6 | 58.1 | 74.9 |
| Minimum Temp °F | 32.6 | 35.9 | 43.7 | 52.1 | 59.5 | 67.2 | 70.5 | 69.5 | 63.3 | 50.8 | 42.7 | 35.5 | 51.9 |
| Mean Temp °F | 43.5 | 47.6 | 55.6 | 63.9 | 70.5 | 77.8 | 81.4 | 80.7 | 74.8 | 64.0 | 54.7 | 46.8 | 63.4 |
| Days Max Temp ≥ 90 °F | 0 | 0 | 0 | 0 | 2 | 14 | 24 | 23 | 11 | 1 | 0 | 0 | 75 |
| Days Max Temp ≤ 32 °F | 1 | 0 | 0 | 0 | 0 | 0 | 0 | 0 | 0 | 0 | 0 | 1 | 2 |
| Days Min Temp ≤ 32 °F | 17 | 11 | 4 | 1 | 0 | 0 | 0 | 0 | 0 | 1 | 5 | 12 | 51 |
| Days Min Temp ≤ 0 °F | 0 | 0 | 0 | 0 | 0 | 0 | 0 | 0 | 0 | 0 | 0 | 0 | 0 |
| Heating Degree Days | 662 | 489 | 307 | 104 | 18 | 0 | 0 | 0 | 11 | 108 | 322 | 558 | 2579 |
| Cooling Degree Days | 1 | 4 | 24 | 74 | 198 | 401 | 516 | 506 | 318 | ***85*** | 24 | 3 | 2154 |
| Total Precipitation (") | 4.00 | 4.24 | 4.20 | 4.70 | 5.33 | 4.71 | 3.09 | 2.59 | 3.75 | 4.47 | 4.29 | 4.82 | 50.19 |
| Days ≥ 0.1" Precip | 6 | 6 | 6 | 5 | 6 | 6 | 5 | 4 | 5 | 6 | 6 | 6 | 67 |
| Total Snowfall (") | 0.5 | 0.2 | 0.2 | 0.0 | 0.0 | 0.0 | 0.0 | 0.0 | 0.0 | 0.0 | 0.0 | 0.0 | 0.9 |
| Days ≥ 1" Snow Depth | 0 | 0 | 0 | 0 | 0 | 0 | 0 | 0 | 0 | 0 | 0 | 0 | 0 |

## MASON *Mason County* ELEVATION 1430 ft LAT/LONG 30° 45 ' N / 99° 14 ' W

| | JAN | FEB | MAR | APR | MAY | JUN | JUL | AUG | SEP | OCT | NOV | DEC | YEAR |
|---|---|---|---|---|---|---|---|---|---|---|---|---|---|
| Maximum Temp °F | ***58.8*** | 62.9 | 71.3 | ***79.6*** | ***84.0*** | ***90.6*** | na | ***94.0*** | ***88.1*** | ***79.2*** | ***69.3*** | na | na |
| Minimum Temp °F | ***30.9*** | 34.6 | ***42.7*** | ***52.1*** | 59.0 | ***66.1*** | na | ***67.2*** | ***62.3*** | ***51.4*** | na | na | na |
| Mean Temp °F | ***44.9*** | 48.8 | ***57.0*** | ***65.8*** | ***71.6*** | ***78.5*** | na | ***80.6*** | ***75.2*** | ***65.3*** | na | na | na |
| Days Max Temp ≥ 90 °F | 0 | 0 | 1 | 3 | 6 | 18 | 25 | 25 | 14 | 3 | 0 | 0 | 95 |
| Days Max Temp ≤ 32 °F | 1 | 0 | 0 | 0 | 0 | 0 | 0 | 0 | 0 | 0 | 0 | 0 | 1 |
| Days Min Temp ≤ 32 °F | 17 | 12 | 5 | 1 | 0 | 0 | 0 | 0 | 0 | 0 | 6 | 13 | 54 |
| Days Min Temp ≤ 0 °F | 0 | 0 | 0 | 0 | 0 | 0 | 0 | 0 | 0 | 0 | 0 | 0 | 0 |
| Heating Degree Days | ***618*** | 455 | ***272*** | ***76*** | ***17*** | ***1*** | ***0*** | ***0*** | ***10*** | ***97*** | na | na | na |
| Cooling Degree Days | ***1*** | ***5*** | ***30*** | ***104*** | 238 | ***420*** | na | ***512*** | ***331*** | na | na | na | na |
| Total Precipitation (") | 1.26 | 1.82 | 1.57 | 2.20 | 3.62 | 3.43 | 2.16 | 2.61 | 3.08 | 3.11 | 1.64 | 1.34 | 27.84 |
| Days ≥ 0.1" Precip | 2 | 3 | 3 | 3 | 5 | 4 | 3 | 4 | 4 | 3 | 3 | 2 | 39 |
| Total Snowfall (") | 0.2 | 1.0 | 0.1 | 0.0 | 0.0 | 0.0 | 0.0 | 0.0 | 0.0 | 0.0 | 0.0 | 0.0 | 1.3 |
| Days ≥ 1" Snow Depth | 0 | 0 | 0 | 0 | 0 | 0 | 0 | 0 | 0 | 0 | 0 | 0 | 0 |

## MATADOR *Motley County* ELEVATION 2352 ft LAT/LONG 34° 1 ' N / 100° 49 ' W

| | JAN | FEB | MAR | APR | MAY | JUN | JUL | AUG | SEP | OCT | NOV | DEC | YEAR |
|---|---|---|---|---|---|---|---|---|---|---|---|---|---|
| Maximum Temp °F | 53.0 | 57.7 | 66.4 | 76.2 | 82.7 | 90.6 | 95.4 | 93.0 | 85.0 | 76.2 | 64.0 | 55.5 | 74.6 |
| Minimum Temp °F | 27.1 | 30.8 | 38.2 | 47.8 | 56.2 | 65.4 | 69.9 | 68.1 | 60.6 | 49.7 | 38.3 | 29.9 | 48.5 |
| Mean Temp °F | 40.1 | 44.3 | 52.3 | 62.0 | 69.5 | 78.0 | 82.7 | 80.6 | 72.8 | 63.0 | 51.2 | 42.7 | 61.6 |
| Days Max Temp ≥ 90 °F | 0 | 0 | 1 | 3 | 7 | 18 | 26 | 23 | 11 | 3 | 0 | 0 | 92 |
| Days Max Temp ≤ 32 °F | 3 | 2 | 0 | 0 | 0 | 0 | 0 | 0 | 0 | 0 | 0 | 2 | 7 |
| Days Min Temp ≤ 32 °F | 22 | 16 | 8 | 1 | 0 | 0 | 0 | 0 | 0 | 1 | 8 | 19 | 75 |
| Days Min Temp ≤ 0 °F | 0 | 0 | 0 | 0 | 0 | 0 | 0 | 0 | 0 | 0 | 0 | 0 | 0 |
| Heating Degree Days | 766 | 581 | 402 | 153 | 39 | 2 | 0 | 0 | 27 | 139 | 416 | 684 | 3209 |
| Cooling Degree Days | 0 | 2 | 12 | 66 | 183 | 394 | 551 | 514 | 279 | 82 | 9 | 0 | 2092 |
| Total Precipitation (") | 0.68 | 0.86 | 1.12 | 1.47 | 3.05 | 3.39 | 1.98 | 2.41 | 3.14 | 1.99 | 0.98 | 0.77 | 21.84 |
| Days ≥ 0.1" Precip | 2 | 2 | 2 | 4 | 5 | 5 | 3 | 4 | 4 | 3 | 2 | 2 | 38 |
| Total Snowfall (") | 3.4 | 3.0 | 1.2 | 0.1 | 0.0 | 0.0 | 0.0 | 0.0 | 0.0 | 0.1 | 1.3 | 1.8 | 10.9 |
| Days ≥ 1" Snow Depth | 2 | 2 | 0 | 0 | 0 | 0 | 0 | 0 | 0 | 0 | 1 | 1 | 6 |

## MATAGORDA 2 *Matagorda County* ELEVATION 10 ft LAT/LONG 28° 42 ' N / 95° 58 ' W

| | JAN | FEB | MAR | APR | MAY | JUN | JUL | AUG | SEP | OCT | NOV | DEC | YEAR |
|---|---|---|---|---|---|---|---|---|---|---|---|---|---|
| Maximum Temp °F | 63.3 | 65.8 | 72.2 | 78.2 | 83.5 | 88.5 | 91.1 | 91.8 | 88.6 | 82.5 | 74.1 | 67.2 | 78.9 |
| Minimum Temp °F | 45.7 | 48.0 | 55.1 | 63.0 | 69.4 | 75.0 | 77.0 | 76.1 | 71.9 | 63.4 | 55.2 | 48.6 | 62.4 |
| Mean Temp °F | 54.5 | 56.9 | 63.7 | 70.6 | 76.5 | 81.8 | 84.1 | 84.0 | 80.2 | 73.0 | 64.7 | 57.9 | 70.7 |
| Days Max Temp ≥ 90 °F | 0 | 0 | 0 | 0 | 1 | 11 | 25 | 26 | 14 | 2 | 0 | 0 | 79 |
| Days Max Temp ≤ 32 °F | 0 | 0 | 0 | 0 | 0 | 0 | 0 | 0 | 0 | 0 | 0 | 0 | 0 |
| Days Min Temp ≤ 32 °F | 3 | 2 | 0 | 0 | 0 | 0 | 0 | 0 | 0 | 0 | 1 | 2 | 8 |
| Days Min Temp ≤ 0 °F | 0 | 0 | 0 | 0 | 0 | 0 | 0 | 0 | 0 | 0 | 0 | 0 | 0 |
| Heating Degree Days | 333 | 240 | 109 | 20 | 1 | 0 | 0 | 0 | 0 | 15 | 115 | 249 | 1082 |
| Cooling Degree Days | 12 | 20 | 75 | 194 | 374 | 526 | 609 | 615 | 481 | 289 | 121 | 42 | 3358 |
| Total Precipitation (") | 4.08 | 2.80 | 2.36 | 2.70 | 5.10 | 4.51 | 3.80 | 3.32 | 6.46 | 4.07 | 4.06 | 2.84 | 46.10 |
| Days ≥ 0.1" Precip | 6 | 4 | 4 | 3 | 5 | 5 | 4 | 5 | 6 | 5 | 5 | 5 | 57 |
| Total Snowfall (") | 0.0 | 0.1 | 0.0 | 0.0 | 0.0 | 0.0 | 0.0 | 0.0 | 0.0 | 0.0 | 0.0 | 0.0 | 0.1 |
| Days ≥ 1" Snow Depth | 0 | 0 | 0 | 0 | 0 | 0 | 0 | 0 | 0 | 0 | 0 | 0 | 0 |

## MATHIS 4 SSW *Jim Wells County* ELEVATION 112 ft LAT/LONG 28° 4 ' N / 97° 53 ' W

| | JAN | FEB | MAR | APR | MAY | JUN | JUL | AUG | SEP | OCT | NOV | DEC | YEAR |
|---|---|---|---|---|---|---|---|---|---|---|---|---|---|
| Maximum Temp °F | 64.4 | 68.4 | 76.2 | 82.5 | 86.4 | 91.6 | 94.9 | 95.4 | 90.5 | 84.1 | 74.9 | 67.9 | 81.4 |
| Minimum Temp °F | 42.9 | 45.6 | 52.7 | 60.6 | 66.6 | 71.4 | 72.6 | 72.7 | 69.9 | 61.3 | 52.6 | 45.8 | 59.6 |
| Mean Temp °F | 53.7 | 57.0 | 64.5 | 71.6 | 76.5 | 81.5 | 83.8 | 84.1 | 80.2 | 72.7 | 63.8 | 56.9 | 70.5 |
| Days Max Temp ≥ 90 °F | 0 | 0 | 2 | 4 | 9 | 22 | 28 | 29 | 19 | 8 | 1 | 0 | 122 |
| Days Max Temp ≤ 32 °F | 0 | 0 | 0 | 0 | 0 | 0 | 0 | 0 | 0 | 0 | 0 | 0 | 0 |
| Days Min Temp ≤ 32 °F | 4 | 2 | 1 | 0 | 0 | 0 | 0 | 0 | 0 | 0 | 1 | 3 | 11 |
| Days Min Temp ≤ 0 °F | 0 | 0 | 0 | 0 | 0 | 0 | 0 | 0 | 0 | 0 | 0 | 0 | 0 |
| Heating Degree Days | 365 | 246 | 107 | 21 | 1 | 0 | 0 | 0 | 1 | 16 | 130 | 279 | 1166 |
| Cooling Degree Days | 15 | 31 | 95 | 214 | 378 | 519 | 594 | 613 | 482 | 285 | 112 | 42 | 3380 |
| Total Precipitation (") | 2.28 | 2.32 | 1.69 | 2.29 | 4.59 | 4.15 | 2.74 | 2.96 | 5.25 | 3.41 | 1.70 | 1.58 | 34.96 |
| Days ≥ 0.1" Precip | 5 | 4 | 3 | 3 | 5 | 4 | 4 | 4 | 6 | 4 | 3 | 3 | 48 |
| Total Snowfall (") | 0.0 | 0.1 | 0.0 | 0.0 | 0.0 | 0.0 | 0.0 | 0.0 | 0.0 | 0.0 | 0.0 | 0.0 | 0.1 |
| Days ≥ 1" Snow Depth | 0 | 0 | 0 | 0 | 0 | 0 | 0 | 0 | 0 | 0 | 0 | 0 | 0 |

## MC CAMEY *Upton County* ELEVATION 2451 ft LAT/LONG 31° 8 ' N / 102° 12 ' W

| | JAN | FEB | MAR | APR | MAY | JUN | JUL | AUG | SEP | OCT | NOV | DEC | YEAR |
|---|---|---|---|---|---|---|---|---|---|---|---|---|---|
| Maximum Temp °F | 59.2 | 64.2 | 73.2 | 82.1 | 88.5 | 94.1 | 95.4 | 94.3 | 87.8 | 79.4 | 68.4 | 61.2 | 79.0 |
| Minimum Temp °F | 31.5 | 35.5 | 43.6 | 53.1 | 60.8 | 68.9 | 71.8 | 70.6 | 64.4 | 53.8 | 41.7 | 33.5 | 52.4 |
| Mean Temp °F | 45.4 | 49.9 | 58.4 | 67.6 | 74.7 | 81.5 | 83.6 | 82.5 | 76.1 | 66.6 | 55.0 | 47.3 | 65.7 |
| Days Max Temp ≥ 90 °F | 0 | 0 | 1 | 7 | 14 | 24 | 27 | 26 | 15 | 3 | 0 | 0 | 117 |
| Days Max Temp ≤ 32 °F | 1 | 0 | 0 | 0 | 0 | 0 | 0 | 0 | 0 | 0 | 0 | 0 | 1 |
| Days Min Temp ≤ 32 °F | 17 | 10 | 3 | 0 | 0 | 0 | 0 | 0 | 0 | 0 | 5 | 14 | 49 |
| Days Min Temp ≤ 0 °F | 0 | 0 | 0 | 0 | 0 | 0 | 0 | 0 | 0 | 0 | 0 | 0 | 0 |
| Heating Degree Days | 601 | 421 | 228 | 59 | 10 | 0 | 0 | 0 | 12 | 69 | 304 | 541 | 2245 |
| Cooling Degree Days | 0 | 3 | 35 | 150 | 330 | 519 | 596 | 572 | 372 | 139 | 16 | 0 | 2732 |
| Total Precipitation (") | 0.48 | 0.64 | 0.43 | 0.94 | 1.86 | 1.46 | 1.10 | 1.84 | 2.70 | 2.16 | 0.59 | 0.66 | 14.86 |
| Days ≥ 0.1" Precip | 1 | 2 | 1 | 2 | 3 | 2 | 2 | 3 | 4 | 3 | 1 | 2 | 26 |
| Total Snowfall (") | 0.6 | 0.1 | 0.0 | 0.0 | 0.0 | 0.0 | 0.0 | 0.0 | 0.0 | 0.0 | 0.5 | 0.0 | 1.2 |
| Days ≥ 1" Snow Depth | 0 | 0 | 0 | 0 | 0 | 0 | 0 | 0 | 0 | 0 | 0 | 0 | 0 |

## MC COOK *Hidalgo County* ELEVATION 220 ft LAT/LONG 26° 29 ' N / 98° 23 ' W

| | JAN | FEB | MAR | APR | MAY | JUN | JUL | AUG | SEP | OCT | NOV | DEC | YEAR |
|---|---|---|---|---|---|---|---|---|---|---|---|---|---|
| Maximum Temp °F | 67.2 | 71.6 | 80.1 | 86.2 | 89.0 | 93.7 | 96.1 | 97.0 | 92.5 | 85.9 | 77.5 | 70.6 | 84.0 |
| Minimum Temp °F | 45.4 | 48.3 | 55.5 | 63.2 | 68.5 | 72.0 | 73.0 | 72.5 | 69.9 | 62.7 | 54.5 | 48.4 | 61.2 |
| Mean Temp °F | 56.3 | 59.9 | 67.8 | 74.7 | 78.8 | 82.9 | 84.6 | 84.8 | 81.2 | 74.3 | 66.0 | 59.5 | 72.6 |
| Days Max Temp ≥ 90 °F | 0 | 1 | 4 | 10 | 16 | 25 | 29 | 29 | 22 | 11 | 3 | 1 | 151 |
| Days Max Temp ≤ 32 °F | 0 | 0 | 0 | 0 | 0 | 0 | 0 | 0 | 0 | 0 | 0 | 0 | 0 |
| Days Min Temp ≤ 32 °F | 3 | 2 | 0 | 0 | 0 | 0 | 0 | 0 | 0 | 0 | 0 | 2 | 7 |
| Days Min Temp ≤ 0 °F | 0 | 0 | 0 | 0 | 0 | 0 | 0 | 0 | 0 | 0 | 0 | 0 | 0 |
| Heating Degree Days | 301 | 189 | 72 | 12 | 1 | 0 | 0 | 0 | 1 | 13 | 100 | 223 | 912 |
| Cooling Degree Days | 32 | 63 | 172 | 306 | 449 | 575 | 641 | 654 | 513 | 340 | 162 | 70 | 3977 |
| Total Precipitation (") | 1.33 | 1.35 | 0.69 | 1.32 | 3.41 | 3.16 | 1.72 | 2.15 | 4.01 | 2.91 | 0.88 | 1.15 | 24.08 |
| Days ≥ 0.1" Precip | 3 | 3 | 2 | 2 | 4 | 4 | 3 | 3 | 5 | 4 | 2 | 3 | 38 |
| Total Snowfall (") | 0.0 | 0.0 | 0.0 | 0.0 | 0.0 | 0.0 | 0.0 | 0.0 | 0.0 | 0.0 | 0.0 | 0.0 | 0.0 |
| Days ≥ 1" Snow Depth | 0 | 0 | 0 | 0 | 0 | 0 | 0 | 0 | 0 | 0 | 0 | 0 | 0 |

## MC GREGOR *McLennan County* ELEVATION 732 ft LAT/LONG 31° 26 ' N / 97° 24 ' W

| | JAN | FEB | MAR | APR | MAY | JUN | JUL | AUG | SEP | OCT | NOV | DEC | YEAR |
|---|---|---|---|---|---|---|---|---|---|---|---|---|---|
| Maximum Temp °F | 55.3 | 59.9 | 68.6 | 77.0 | 83.2 | 90.8 | 95.7 | 95.5 | 88.3 | 78.9 | 67.2 | 58.8 | 76.6 |
| Minimum Temp °F | 33.4 | 36.9 | 45.2 | 54.1 | 61.7 | 69.0 | 72.3 | 71.7 | 65.5 | 55.0 | 44.8 | 36.6 | 53.9 |
| Mean Temp °F | 44.4 | 48.4 | 56.9 | 65.6 | 72.5 | 79.9 | 84.0 | 83.6 | 76.9 | 67.0 | 56.0 | 47.7 | 65.2 |
| Days Max Temp ≥ 90 °F | 0 | 0 | 0 | 1 | 5 | 19 | 28 | 27 | 15 | 3 | 0 | 0 | 98 |
| Days Max Temp ≤ 32 °F | 1 | 1 | 0 | 0 | 0 | 0 | 0 | 0 | 0 | 0 | 0 | 1 | 3 |
| Days Min Temp ≤ 32 °F | 15 | 9 | 3 | 0 | 0 | 0 | 0 | 0 | 0 | 0 | 3 | 10 | 40 |
| Days Min Temp ≤ 0 °F | 0 | 0 | 0 | 0 | 0 | 0 | 0 | 0 | 0 | 0 | 0 | 0 | 0 |
| Heating Degree Days | 633 | 466 | 274 | 80 | 12 | 0 | 0 | 0 | 7 | 72 | 289 | 532 | 2365 |
| Cooling Degree Days | 1 | 5 | 31 | 100 | 248 | 451 | 596 | 600 | 383 | 140 | 29 | 5 | 2589 |
| Total Precipitation (") | 2.00 | 2.49 | 2.62 | 2.95 | 4.70 | 3.36 | 1.99 | 2.22 | 3.27 | 3.70 | 2.49 | 2.36 | 34.15 |
| Days ≥ 0.1" Precip | 5 | 4 | 5 | 4 | 6 | 4 | 3 | 3 | 4 | 4 | 4 | 4 | 50 |
| Total Snowfall (") | 0.6 | 0.4 | 0.0 | 0.0 | 0.0 | 0.0 | 0.0 | 0.0 | 0.0 | 0.0 | 0.1 | 0.0 | 1.1 |
| Days ≥ 1" Snow Depth | 0 | 0 | 0 | 0 | 0 | 0 | 0 | 0 | 0 | 0 | 0 | 0 | 0 |

## MC KINNEY 3 S *Collin County* ELEVATION 610 ft LAT/LONG 33° 12 ' N / 96° 38 ' W

| | JAN | FEB | MAR | APR | MAY | JUN | JUL | AUG | SEP | OCT | NOV | DEC | YEAR |
|---|---|---|---|---|---|---|---|---|---|---|---|---|---|
| Maximum Temp °F | 54.7 | 59.9 | 68.8 | 76.8 | 82.5 | 90.3 | 95.1 | 95.0 | 87.7 | 78.4 | 66.4 | 57.8 | 76.1 |
| Minimum Temp °F | 33.0 | 36.6 | 44.8 | 53.4 | 61.0 | 68.8 | 72.2 | 70.7 | 64.6 | 54.0 | 44.1 | 35.9 | 53.3 |
| Mean Temp °F | 43.8 | 48.3 | 56.8 | 65.1 | 71.8 | 79.5 | 83.7 | 82.9 | 76.1 | 66.2 | 55.3 | 46.9 | 64.7 |
| Days Max Temp ≥ 90 °F | 0 | 0 | 0 | 1 | 3 | 18 | 28 | 27 | 13 | 2 | 0 | 0 | 92 |
| Days Max Temp ≤ 32 °F | 1 | 1 | 0 | 0 | 0 | 0 | 0 | 0 | 0 | 0 | 0 | 1 | 3 |
| Days Min Temp ≤ 32 °F | 15 | 10 | 4 | 0 | 0 | 0 | 0 | 0 | 0 | 0 | 5 | 12 | 46 |
| Days Min Temp ≤ 0 °F | 0 | 0 | 0 | 0 | 0 | 0 | 0 | 0 | 0 | 0 | 0 | 0 | 0 |
| Heating Degree Days | 651 | 470 | 274 | 81 | 13 | 0 | 0 | 0 | 9 | 80 | 308 | 558 | 2444 |
| Cooling Degree Days | 1 | 3 | 24 | 87 | 232 | 454 | 595 | 592 | 365 | 127 | 25 | 2 | 2507 |
| Total Precipitation (") | 2.20 | 2.98 | 3.39 | 4.08 | 5.75 | 4.03 | 2.48 | 2.35 | 3.99 | 4.13 | 3.10 | 2.74 | 41.22 |
| Days ≥ 0.1" Precip | 4 | 4 | 5 | 6 | 6 | 5 | 3 | 3 | 5 | 4 | 4 | 4 | 53 |
| Total Snowfall (") | 1.0 | 1.1 | 0.2 | 0.0 | 0.0 | 0.0 | 0.0 | 0.0 | 0.0 | 0.0 | 0.2 | 0.3 | 2.8 |
| Days ≥ 1" Snow Depth | 1 | 1 | 0 | 0 | 0 | 0 | 0 | 0 | 0 | 0 | 0 | 1 | 3 |

## MC LEAN *Gray County* ELEVATION 2552 ft LAT/LONG 35° 14 ' N / 100° 36 ' W

| | JAN | FEB | MAR | APR | MAY | JUN | JUL | AUG | SEP | OCT | NOV | DEC | YEAR |
|---|---|---|---|---|---|---|---|---|---|---|---|---|---|
| Maximum Temp °F | 49.5 | 54.4 | 63.2 | 73.2 | 79.6 | 87.9 | 92.9 | 90.8 | 83.5 | 73.7 | 60.6 | 51.4 | 71.7 |
| Minimum Temp °F | 24.5 | 28.2 | 35.8 | 45.2 | 54.1 | 63.0 | 67.2 | 65.8 | 58.4 | 47.3 | 35.5 | *27.6* | 46.1 |
| Mean Temp °F | 37.0 | 41.3 | 49.5 | 59.3 | 66.9 | 75.5 | 80.1 | 78.3 | 71.0 | 60.5 | 48.0 | *39.4* | 58.9 |
| Days Max Temp ≥ 90 °F | 0 | 0 | 0 | 1 | 4 | 14 | 22 | 20 | 8 | 1 | 0 | 0 | 70 |
| Days Max Temp ≤ 32 °F | 4 | 2 | 1 | 0 | 0 | 0 | 0 | 0 | 0 | 0 | 0 | 2 | 9 |
| Days Min Temp ≤ 32 °F | 24 | 18 | 11 | 2 | 0 | 0 | 0 | 0 | 0 | 1 | 11 | *21* | 88 |
| Days Min Temp ≤ 0 °F | 0 | 0 | 0 | 0 | 0 | 0 | 0 | 0 | 0 | 0 | 0 | 0 | 0 |
| Heating Degree Days | 860 | 662 | 479 | 202 | 61 | 5 | 0 | 1 | 33 | 183 | 504 | *791* | 3781 |
| Cooling Degree Days | 0 | 0 | 5 | 38 | 124 | 322 | 461 | 435 | 233 | 45 | 3 | 0 | 1666 |
| Total Precipitation (") | 0.60 | 0.93 | 1.58 | 1.95 | 4.08 | 3.71 | 2.46 | 2.63 | 2.80 | 2.01 | 1.01 | 0.72 | 24.48 |
| Days ≥ 0.1" Precip | 2 | 3 | 3 | 4 | 6 | 6 | 4 | 5 | 4 | 3 | 2 | 2 | 44 |
| Total Snowfall (") | *2.0* | *1.9* | 1.2 | 0.2 | 0.0 | 0.0 | 0.0 | 0.0 | 0.0 | 0.0 | 1.1 | *2.6* | 9.0 |
| Days ≥ 1" Snow Depth | na | na | 0 | 0 | 0 | 0 | 0 | 0 | 0 | 0 | *0* | *1* | na |

## MEDINA 2 W *Bandera County* ELEVATION 1631 ft LAT/LONG 29° 48 ' N / 99° 15 ' W

| | JAN | FEB | MAR | APR | MAY | JUN | JUL | AUG | SEP | OCT | NOV | DEC | YEAR |
|---|---|---|---|---|---|---|---|---|---|---|---|---|---|
| Maximum Temp °F | 60.9 | 65.4 | 73.5 | 80.5 | 84.7 | 90.8 | 93.9 | 94.1 | 88.9 | ***80.6*** | 70.7 | 63.5 | 79.0 |
| Minimum Temp °F | 32.2 | 35.7 | 44.4 | 52.5 | 60.6 | 67.0 | 68.6 | 67.6 | 62.9 | ***53.2*** | 42.6 | 35.6 | 51.9 |
| Mean Temp °F | 46.6 | 50.6 | 59.0 | 66.5 | 72.7 | 78.9 | 81.2 | 80.9 | 76.0 | ***67.0*** | 56.7 | 49.6 | 65.5 |
| Days Max Temp ≥ 90 °F | 0 | 0 | 1 | 3 | 7 | 20 | ***28*** | 27 | 17 | 3 | 0 | 0 | 106 |
| Days Max Temp ≤ 32 °F | 0 | 0 | 0 | 0 | 0 | 0 | 0 | 0 | 0 | 0 | 0 | 0 | 0 |
| Days Min Temp ≤ 32 °F | 17 | 11 | 4 | 1 | 0 | 0 | 0 | 0 | 0 | 0 | 6 | 13 | 52 |
| Days Min Temp ≤ 0 °F | 0 | 0 | 0 | 0 | 0 | 0 | 0 | 0 | 0 | 0 | 0 | 0 | 0 |
| Heating Degree Days | 566 | 405 | 218 | 62 | 8 | 0 | 0 | 0 | 7 | ***70*** | 269 | 475 | 2080 |
| Cooling Degree Days | 1 | 2 | 38 | 117 | 273 | 461 | 534 | 535 | 367 | 154 | 32 | 4 | 2518 |
| Total Precipitation (") | 1.86 | 1.97 | 2.36 | 2.72 | 4.48 | 3.78 | 2.66 | 3.07 | 4.15 | 4.12 | 2.52 | 2.13 | 35.82 |
| Days ≥ 0.1" Precip | 4 | 4 | 4 | 4 | 6 | 5 | 4 | 4 | 5 | 4 | 4 | 4 | 52 |
| Total Snowfall (") | 1.0 | 0.4 | 0.0 | 0.0 | 0.0 | 0.0 | 0.0 | 0.0 | 0.0 | 0.0 | 0.0 | 0.0 | 1.4 |
| Days ≥ 1" Snow Depth | 0 | 0 | 0 | 0 | 0 | 0 | 0 | 0 | 0 | 0 | 0 | 0 | 0 |

## MEMPHIS *Hall County* ELEVATION 2070 ft LAT/LONG 34° 43 ' N / 100° 32 ' W

| | JAN | FEB | MAR | APR | MAY | JUN | JUL | AUG | SEP | OCT | NOV | DEC | YEAR |
|---|---|---|---|---|---|---|---|---|---|---|---|---|---|
| Maximum Temp °F | 51.6 | 56.7 | 65.9 | 75.6 | 82.4 | 90.9 | 95.6 | 93.7 | 85.5 | 75.6 | 62.9 | 53.9 | 74.2 |
| Minimum Temp °F | 24.8 | 28.5 | 36.4 | 46.2 | 55.5 | 64.7 | 69.2 | 67.1 | 59.5 | 47.4 | 35.5 | 27.4 | 46.8 |
| Mean Temp °F | 38.2 | 42.6 | 51.2 | 60.9 | 69.0 | 77.7 | 82.4 | 80.4 | 72.5 | 61.6 | 49.2 | 40.6 | 60.5 |
| Days Max Temp ≥ 90 °F | 0 | 0 | 0 | 3 | 7 | 18 | 26 | 24 | 11 | 3 | 0 | 0 | 92 |
| Days Max Temp ≤ 32 °F | 3 | 2 | 0 | 0 | 0 | 0 | 0 | 0 | 0 | 0 | 0 | 2 | 7 |
| Days Min Temp ≤ 32 °F | 26 | 19 | 9 | 1 | 0 | 0 | 0 | 0 | 0 | 1 | 12 | 23 | 91 |
| Days Min Temp ≤ 0 °F | 0 | 0 | 0 | 0 | 0 | 0 | 0 | 0 | 0 | 0 | 0 | 0 | 0 |
| Heating Degree Days | 823 | 625 | 431 | 166 | 40 | 2 | 0 | 0 | 26 | 159 | 469 | 748 | 3489 |
| Cooling Degree Days | 0 | 1 | 7 | 51 | 180 | 393 | 554 | 510 | 278 | 61 | 4 | 0 | 2039 |
| Total Precipitation (") | 0.60 | 0.79 | 1.27 | 1.71 | 3.70 | 3.25 | 1.84 | 2.41 | 2.44 | 1.62 | 0.84 | 0.60 | 21.07 |
| Days ≥ 0.1" Precip | ***1*** | 2 | 3 | 3 | 5 | 5 | 3 | 4 | 4 | 3 | 2 | 2 | 37 |
| Total Snowfall (") | ***1.1*** | ***1.2*** | 0.7 | 0.0 | 0.0 | 0.0 | 0.0 | 0.0 | 0.0 | 0.0 | 0.4 | 0.7 | 4.1 |
| Days ≥ 1" Snow Depth | ***1*** | ***1*** | 0 | 0 | 0 | 0 | 0 | 0 | 0 | 0 | 0 | 1 | 3 |

## MENARD *Menard County* ELEVATION 1962 ft LAT/LONG 30° 55 ' N / 99° 48 ' W

| | JAN | FEB | MAR | APR | MAY | JUN | JUL | AUG | SEP | OCT | NOV | DEC | YEAR |
|---|---|---|---|---|---|---|---|---|---|---|---|---|---|
| Maximum Temp °F | 59.9 | 64.5 | 72.9 | 80.8 | 85.6 | 91.3 | 94.7 | 93.7 | 87.6 | 80.0 | 69.1 | 62.1 | 78.5 |
| Minimum Temp °F | 30.6 | 34.0 | 42.1 | 50.6 | 58.5 | 65.5 | 67.8 | 66.3 | 60.8 | 50.4 | 39.9 | 32.7 | 49.9 |
| Mean Temp °F | 45.3 | 49.3 | 57.5 | 65.7 | 72.1 | 78.4 | 81.2 | 80.0 | 74.3 | 65.3 | 54.5 | 47.4 | 64.2 |
| Days Max Temp ≥ 90 °F | 0 | 0 | 1 | 4 | 9 | 21 | 27 | 25 | 13 | 3 | 0 | 0 | 103 |
| Days Max Temp ≤ 32 °F | 1 | 0 | 0 | 0 | 0 | 0 | 0 | 0 | 0 | 0 | 0 | 1 | 2 |
| Days Min Temp ≤ 32 °F | 19 | 13 | 7 | 2 | 0 | 0 | 0 | 0 | 0 | 1 | 9 | 16 | 67 |
| Days Min Temp ≤ 0 °F | 0 | 0 | 0 | 0 | 0 | 0 | 0 | 0 | 0 | 0 | 0 | 0 | 0 |
| Heating Degree Days | 607 | 440 | 256 | 80 | 15 | 0 | 0 | 0 | 12 | 92 | 326 | 540 | 2368 |
| Cooling Degree Days | 1 | 3 | 32 | 108 | 251 | 423 | 520 | 505 | 306 | 119 | 20 | 1 | 2289 |
| Total Precipitation (") | 1.10 | 1.57 | 1.42 | 1.99 | 3.39 | 2.97 | 2.18 | 2.45 | 3.15 | 2.51 | 1.31 | 1.30 | 25.34 |
| Days ≥ 0.1" Precip | 3 | 3 | 3 | 4 | 5 | 4 | 3 | 4 | 4 | 4 | 2 | 2 | 41 |
| Total Snowfall (") | 1.2 | 0.7 | 0.0 | 0.0 | 0.0 | 0.0 | 0.0 | 0.0 | 0.0 | 0.0 | 0.2 | 0.2 | 2.3 |
| Days ≥ 1" Snow Depth | ***0*** | 0 | 0 | 0 | 0 | 0 | 0 | 0 | 0 | 0 | 0 | 0 | 0 |

## MEXIA *Limestone County* ELEVATION 531 ft LAT/LONG 31° 41 ' N / 96° 29 ' W

| | JAN | FEB | MAR | APR | MAY | JUN | JUL | AUG | SEP | OCT | NOV | DEC | YEAR |
|---|---|---|---|---|---|---|---|---|---|---|---|---|---|
| Maximum Temp °F | 56.5 | 61.4 | 69.9 | 77.9 | 83.3 | 91.0 | 95.4 | 96.1 | 89.6 | 80.1 | 68.6 | 60.1 | 77.5 |
| Minimum Temp °F | 33.6 | 37.1 | 45.3 | 53.8 | 60.6 | 67.9 | 71.1 | 70.0 | 64.2 | 53.7 | 44.2 | 36.5 | 53.2 |
| Mean Temp °F | 45.1 | 49.3 | 57.7 | 65.9 | 72.0 | 79.5 | 83.3 | 83.1 | 76.9 | 67.0 | 56.4 | 48.3 | 65.4 |
| Days Max Temp ≥ 90 °F | 0 | 0 | 0 | 1 | 5 | 20 | 27 | 28 | 17 | 4 | 0 | 0 | 102 |
| Days Max Temp ≤ 32 °F | 1 | 1 | 0 | 0 | 0 | 0 | 0 | 0 | 0 | 0 | 0 | 1 | 3 |
| Days Min Temp ≤ 32 °F | 15 | 9 | 3 | 0 | 0 | 0 | 0 | 0 | 0 | 0 | 4 | 10 | 41 |
| Days Min Temp ≤ 0 °F | 0 | 0 | 0 | 0 | 0 | 0 | 0 | 0 | 0 | 0 | 0 | 0 | 0 |
| Heating Degree Days | 613 | 442 | 251 | 68 | 10 | 0 | 0 | 0 | 6 | 70 | 276 | 514 | 2250 |
| Cooling Degree Days | 1 | 4 | 29 | 98 | 236 | 435 | 563 | 567 | 377 | 137 | 27 | 5 | 2479 |
| Total Precipitation (") | 2.55 | 3.25 | 3.52 | 3.46 | 5.46 | 3.47 | 1.86 | 2.57 | 4.61 | 4.33 | 3.42 | 3.69 | 42.19 |
| Days ≥ 0.1" Precip | 5 | 5 | 5 | 5 | 7 | 5 | 3 | 3 | 5 | 5 | 5 | 5 | 58 |
| Total Snowfall (") | 0.3 | 0.6 | 0.1 | 0.0 | 0.0 | 0.0 | 0.0 | 0.0 | 0.0 | 0.0 | 0.0 | 0.0 | 1.0 |
| Days ≥ 1" Snow Depth | ***0*** | 0 | 0 | 0 | 0 | 0 | 0 | 0 | 0 | 0 | 0 | 0 | 0 |

## MIAMI *Roberts County* ELEVATION 2743 ft LAT/LONG 35° 42 ' N / 100° 38 ' W

| | JAN | FEB | MAR | APR | MAY | JUN | JUL | AUG | SEP | OCT | NOV | DEC | YEAR |
|---|---|---|---|---|---|---|---|---|---|---|---|---|---|
| Maximum Temp °F | 47.4 | 51.9 | 60.7 | 71.2 | 78.4 | 87.5 | 92.8 | 90.9 | 82.9 | 72.1 | 58.7 | 50.0 | 70.4 |
| Minimum Temp °F | 20.2 | 23.9 | 32.4 | 42.1 | 51.7 | 61.6 | 66.3 | 64.5 | 56.3 | 43.3 | 31.4 | 22.8 | 43.0 |
| Mean Temp °F | 33.8 | 37.9 | 46.5 | 56.7 | 65.1 | 74.6 | 79.5 | 77.7 | 69.6 | 57.7 | 45.1 | 36.4 | 56.7 |
| Days Max Temp ≥ 90 °F | 0 | 0 | 0 | 1 | 4 | 13 | 22 | 19 | 9 | 1 | 0 | 0 | 69 |
| Days Max Temp ≤ 32 °F | 6 | 3 | 1 | 0 | 0 | 0 | 0 | 0 | 0 | 0 | 1 | 4 | 15 |
| Days Min Temp ≤ 32 °F | 29 | 23 | 15 | 4 | 0 | 0 | 0 | 0 | 0 | 3 | 17 | 28 | 119 |
| Days Min Temp ≤ 0 °F | 1 | 0 | 0 | 0 | 0 | 0 | 0 | 0 | 0 | 0 | 0 | 1 | 2 |
| Heating Degree Days | 960 | 758 | 569 | 269 | 94 | 9 | 0 | 2 | 50 | 251 | 593 | 878 | 4433 |
| Cooling Degree Days | 0 | 0 | 3 | 25 | 99 | 304 | 454 | 422 | 205 | 28 | 1 | 0 | 1541 |
| Total Precipitation (") | 0.57 | 0.84 | 1.63 | 1.69 | 3.36 | 3.24 | 2.31 | 2.40 | 2.36 | 1.65 | 1.08 | 0.61 | 21.74 |
| Days ≥ 0.1" Precip | 2 | 2 | 3 | 3 | 6 | 6 | 4 | 5 | 4 | 3 | 3 | 2 | 43 |
| Total Snowfall (") | *2.7* | 4.5 | 3.7 | 1.0 | 0.0 | 0.0 | 0.0 | 0.0 | 0.0 | 0.1 | 1.7 | *2.8* | 16.5 |
| Days ≥ 1" Snow Depth | *3* | 3 | 1 | 0 | 0 | 0 | 0 | 0 | 0 | 0 | 1 | *2* | 10 |

## MIDLAND 4 ENE *Midland County* ELEVATION 2743 ft LAT/LONG 32° 1 ' N / 102° 1 ' W

| | JAN | FEB | MAR | APR | MAY | JUN | JUL | AUG | SEP | OCT | NOV | DEC | YEAR |
|---|---|---|---|---|---|---|---|---|---|---|---|---|---|
| Maximum Temp °F | 58.4 | 64.0 | 72.2 | 81.2 | 87.4 | 93.6 | 94.9 | 93.3 | 86.9 | 79.1 | 67.5 | 60.7 | 78.3 |
| Minimum Temp °F | 29.5 | 32.9 | 40.2 | 48.8 | 57.7 | 65.5 | 68.3 | 67.1 | 60.8 | 50.4 | 38.9 | 31.3 | 49.3 |
| Mean Temp °F | 44.0 | 48.5 | 56.2 | 65.0 | 72.6 | 79.6 | 81.6 | 80.2 | 73.9 | 64.7 | 53.2 | 46.0 | 63.8 |
| Days Max Temp ≥ 90 °F | 0 | 0 | 1 | 4 | 12 | 23 | 27 | 24 | 13 | 2 | 0 | 0 | 106 |
| Days Max Temp ≤ 32 °F | 1 | 0 | 0 | 0 | 0 | 0 | 0 | 0 | 0 | 0 | 0 | 1 | 2 |
| Days Min Temp ≤ 32 °F | 20 | 14 | 6 | 1 | 0 | 0 | 0 | 0 | 0 | 1 | 7 | 17 | 66 |
| Days Min Temp ≤ 0 °F | 0 | 0 | 0 | 0 | 0 | 0 | 0 | 0 | 0 | 0 | 0 | 0 | 0 |
| Heating Degree Days | 643 | 461 | 284 | 85 | 14 | 0 | 0 | 0 | 15 | 91 | 352 | 582 | 2527 |
| Cooling Degree Days | 0 | 2 | 13 | 92 | 271 | 462 | 539 | 505 | 300 | 100 | 7 | 0 | 2291 |
| Total Precipitation (") | 0.64 | 0.60 | 0.49 | 0.81 | 2.44 | 1.28 | 1.65 | 2.14 | 3.15 | 1.33 | 0.70 | 0.55 | 15.78 |
| Days ≥ 0.1" Precip | 2 | 2 | 1 | 2 | 3 | 2 | 3 | 4 | 4 | 2 | 1 | 1 | 27 |
| Total Snowfall (") | 1.5 | 0.4 | 0.2 | 0.0 | 0.0 | 0.0 | 0.0 | 0.0 | 0.0 | 0.0 | 0.4 | 0.3 | 2.8 |
| Days ≥ 1" Snow Depth | *0* | *0* | 0 | 0 | 0 | 0 | 0 | 0 | 0 | 0 | 0 | *0* | 0 |

## MIDLAND REGIONAL TER *Midland County* ELEVATION 2858 ft LAT/LONG 31° 56 ' N / 102° 12 ' W

| | JAN | FEB | MAR | APR | MAY | JUN | JUL | AUG | SEP | OCT | NOV | DEC | YEAR |
|---|---|---|---|---|---|---|---|---|---|---|---|---|---|
| Maximum Temp °F | 56.3 | 61.7 | 70.2 | 78.7 | 85.7 | 92.3 | 93.9 | 92.3 | 85.4 | 77.1 | 66.0 | 58.8 | 76.5 |
| Minimum Temp °F | 29.3 | 33.1 | 40.4 | 49.0 | 57.8 | 65.9 | 68.6 | 67.4 | 61.2 | 50.7 | 38.8 | 31.5 | 49.5 |
| Mean Temp °F | 42.8 | 47.4 | 55.3 | 63.9 | 71.8 | 79.2 | 81.3 | 79.9 | 73.3 | 63.9 | 52.4 | 45.2 | 63.0 |
| Days Max Temp ≥ 90 °F | 0 | 0 | 0 | 3 | 11 | 21 | 26 | 22 | 10 | 2 | 0 | 0 | 95 |
| Days Max Temp ≤ 32 °F | 1 | 0 | 0 | 0 | 0 | 0 | 0 | 0 | 0 | 0 | 0 | 1 | 2 |
| Days Min Temp ≤ 32 °F | 20 | 13 | 6 | 1 | 0 | 0 | 0 | 0 | 0 | 0 | 7 | 17 | 64 |
| Days Min Temp ≤ 0 °F | 0 | 0 | 0 | 0 | 0 | 0 | 0 | 0 | 0 | 0 | 0 | 0 | 0 |
| Heating Degree Days | 680 | 490 | 308 | 107 | 21 | 1 | 0 | 0 | 17 | 107 | 374 | 608 | 2713 |
| Cooling Degree Days | 0 | 1 | 13 | 81 | 243 | 442 | 519 | 491 | 285 | 85 | 5 | 0 | 2165 |
| Total Precipitation (") | 0.51 | 0.70 | 0.55 | 0.84 | 2.10 | 1.61 | 1.81 | 1.88 | 2.51 | 1.77 | 0.68 | 0.63 | 15.59 |
| Days ≥ 0.1" Precip | 1 | 2 | 1 | 2 | 3 | 3 | 3 | 3 | 4 | 3 | 1 | 2 | 28 |
| Total Snowfall (") | 2.1 | 0.9 | 0.5 | 0.0 | 0.0 | 0.0 | 0.0 | 0.0 | 0.0 | 0.0 | 0.6 | 1.1 | 5.2 |
| Days ≥ 1" Snow Depth | 1 | 1 | 0 | 0 | 0 | 0 | 0 | 0 | 0 | 0 | 0 | 0 | 2 |

## MINERAL WELLS MUN AP *Palo Pinto County* ELEVATION 938 ft LAT/LONG 32° 47 ' N / 98° 4 ' W

| | JAN | FEB | MAR | APR | MAY | JUN | JUL | AUG | SEP | OCT | NOV | DEC | YEAR |
|---|---|---|---|---|---|---|---|---|---|---|---|---|---|
| Maximum Temp °F | *55.8* | *61.2* | *69.5* | *78.0* | *83.6* | *92.1* | *96.5* | *95.2* | *88.6* | *79.1* | *67.3* | *60.0* | 77.2 |
| Minimum Temp °F | *32.9* | *36.1* | *44.3* | *53.7* | *60.7* | *68.6* | *72.1* | *71.0* | *65.0* | *53.9* | *43.3* | *35.5* | 53.1 |
| Mean Temp °F | *44.4* | *48.7* | *56.9* | *65.9* | *72.1* | *80.4* | *84.4* | *83.1* | *76.8* | *66.6* | *55.3* | *47.8* | 65.2 |
| Days Max Temp ≥ 90 °F | *0* | *0* | *1* | *2* | *6* | *21* | *27* | *26* | *16* | *4* | *0* | *0* | 103 |
| Days Max Temp ≤ 32 °F | *2* | *0* | *0* | *0* | *0* | *0* | *0* | *0* | *0* | *0* | *0* | *1* | 3 |
| Days Min Temp ≤ 32 °F | *16* | *10* | *3* | *0* | *0* | *0* | *0* | *0* | *0* | *0* | *4* | *12* | 45 |
| Days Min Temp ≤ 0 °F | *0* | *0* | *0* | *0* | *0* | *0* | *0* | *0* | *0* | *0* | *0* | *0* | 0 |
| Heating Degree Days | *643* | *457* | *273* | *77* | *15* | *0* | *0* | *0* | *7* | *81* | *304* | *530* | 2387 |
| Cooling Degree Days | na | na | na | na | na | na | na | na | na | na | na | na | na |
| Total Precipitation (") | *1.75* | *1.96* | *2.65* | *3.44* | *5.04* | *3.12* | *2.33* | *2.43* | *3.44* | *3.73* | *1.77* | *1.69* | 33.35 |
| Days ≥ 0.1" Precip | *4* | *4* | *5* | *5* | *6* | *4* | *3* | *4* | *4* | *4* | *3* | *3* | 49 |
| Total Snowfall (") | *0.8* | *1.0* | *0.4* | *0.0* | *0.0* | *0.0* | *0.0* | *0.0* | *0.0* | *0.0* | *0.1* | *0.3* | 2.6 |
| Days ≥ 1" Snow Depth | *1* | *0* | *0* | *0* | *0* | *0* | *0* | *0* | *0* | *0* | *0* | *0* | 1 |

## MISSION 4 W *Hidalgo County* ELEVATION 131 ft LAT/LONG 26° 13' N / 98° 24' W

| | JAN | FEB | MAR | APR | MAY | JUN | JUL | AUG | SEP | OCT | NOV | DEC | YEAR |
|---|---|---|---|---|---|---|---|---|---|---|---|---|---|
| Maximum Temp °F | 68.9 | 73.0 | 81.7 | 87.5 | 90.7 | 95.2 | 97.2 | 97.8 | 93.5 | 86.9 | 79.0 | 71.9 | 85.3 |
| Minimum Temp °F | 46.7 | 49.1 | 56.4 | 64.2 | 69.7 | 73.2 | 74.5 | 73.8 | 71.3 | 63.6 | 55.6 | 48.6 | 62.2 |
| Mean Temp °F | 57.8 | 61.1 | 69.2 | 75.9 | 80.2 | 84.2 | 85.9 | 85.8 | 82.4 | 75.3 | 67.4 | 60.2 | 73.8 |
| Days Max Temp ≥ 90 °F | 1 | 1 | 6 | 13 | 20 | 27 | 29 | 30 | 24 | 13 | 4 | 0 | 168 |
| Days Max Temp ≤ 32 °F | 0 | 0 | 0 | 0 | 0 | 0 | 0 | 0 | 0 | 0 | 0 | 0 | 0 |
| Days Min Temp ≤ 32 °F | 2 | 1 | 0 | 0 | 0 | 0 | 0 | 0 | 0 | 0 | 0 | 2 | 5 |
| Days Min Temp ≤ 0 °F | 0 | 0 | 0 | 0 | 0 | 0 | 0 | 0 | 0 | 0 | 0 | 0 | 0 |
| Heating Degree Days | 265 | 169 | 61 | 10 | 1 | 0 | 0 | 0 | 1 | 9 | 86 | 206 | 808 |
| Cooling Degree Days | 44 | 77 | 201 | 343 | 491 | 616 | 681 | 682 | 550 | 367 | 181 | 70 | 4303 |
| Total Precipitation (") | 1.47 | 1.27 | 0.58 | 1.31 | 3.10 | 3.17 | 1.74 | 2.45 | 4.17 | 2.76 | 0.90 | 1.02 | 23.94 |
| Days ≥ 0.1" Precip | 3 | 3 | 1 | 2 | 4 | 4 | 3 | 3 | 5 | 4 | 2 | 2 | 36 |
| Total Snowfall (") | 0.0 | 0.0 | 0.0 | 0.0 | 0.0 | 0.0 | 0.0 | 0.0 | 0.0 | 0.0 | 0.0 | 0.0 | 0.0 |
| Days ≥ 1" Snow Depth | 0 | 0 | 0 | 0 | 0 | 0 | 0 | 0 | 0 | 0 | 0 | 0 | 0 |

## MONAHANS *Ward County* ELEVATION 2621 ft LAT/LONG 31° 35' N / 102° 53' W

| | JAN | FEB | MAR | APR | MAY | JUN | JUL | AUG | SEP | OCT | NOV | DEC | YEAR |
|---|---|---|---|---|---|---|---|---|---|---|---|---|---|
| Maximum Temp °F | 61.0 | 65.5 | 74.9 | 83.2 | 89.5 | 96.2 | 97.5 | 96.1 | 89.6 | 81.6 | 69.9 | 62.2 | 80.6 |
| Minimum Temp °F | 28.7 | 33.1 | 40.8 | 49.1 | 57.6 | 66.1 | 69.1 | 67.7 | 60.9 | 49.6 | 37.9 | 30.7 | 49.3 |
| Mean Temp °F | 44.9 | 49.4 | 57.9 | 66.2 | 73.6 | 81.2 | 83.3 | 81.9 | 75.3 | 65.5 | 53.9 | 46.5 | 65.0 |
| Days Max Temp ≥ 90 °F | 0 | 0 | 2 | 8 | 17 | 25 | 28 | 27 | 18 | 6 | 0 | 0 | 131 |
| Days Max Temp ≤ 32 °F | 0 | 0 | 0 | 0 | 0 | 0 | 0 | 0 | 0 | 0 | 0 | 0 | 0 |
| Days Min Temp ≤ 32 °F | 22 | 13 | 5 | 1 | 0 | 0 | 0 | 0 | 0 | 1 | 8 | 19 | 69 |
| Days Min Temp ≤ 0 °F | 0 | 0 | 0 | 0 | 0 | 0 | 0 | 0 | 0 | 0 | 0 | 0 | 0 |
| Heating Degree Days | 614 | 435 | 236 | 69 | 11 | 0 | 0 | 0 | 14 | 85 | 330 | 567 | 2361 |
| Cooling Degree Days | 0 | 1 | 20 | 108 | 287 | 521 | 589 | 556 | 348 | 113 | 8 | 0 | 2551 |
| Total Precipitation (") | 0.49 | 0.68 | 0.34 | 0.66 | 1.99 | 1.46 | 1.40 | 1.54 | 2.53 | 1.40 | 0.60 | 0.62 | 13.71 |
| Days ≥ 0.1" Precip | 1 | 2 | 1 | 1 | 3 | 3 | 2 | 3 | 3 | 2 | 1 | 2 | 24 |
| Total Snowfall (") | 0.6 | 0.3 | 0.2 | 0.0 | 0.0 | 0.0 | 0.0 | 0.0 | 0.0 | 0.0 | 0.3 | 0.4 | 1.8 |
| Days ≥ 1" Snow Depth | ***0*** | 0 | 0 | 0 | 0 | 0 | 0 | 0 | 0 | 0 | 0 | ***0*** | 0 |

## MORTON *Cochran County* ELEVATION 3763 ft LAT/LONG 33° 44' N / 102° 46' W

| | JAN | FEB | MAR | APR | MAY | JUN | JUL | AUG | SEP | OCT | NOV | DEC | YEAR |
|---|---|---|---|---|---|---|---|---|---|---|---|---|---|
| Maximum Temp °F | 52.5 | 57.2 | 65.3 | 74.3 | 81.8 | 89.5 | 91.0 | 88.8 | 82.2 | 74.0 | 62.5 | 54.5 | 72.8 |
| Minimum Temp °F | 22.7 | 25.7 | 31.8 | 41.0 | 50.6 | 59.7 | 63.4 | 61.6 | 54.9 | 43.4 | 32.3 | 24.6 | 42.6 |
| Mean Temp °F | 37.7 | 41.4 | 48.6 | 57.7 | 66.2 | 74.6 | 77.2 | 75.2 | 68.5 | 58.7 | 47.4 | 39.6 | 57.7 |
| Days Max Temp ≥ 90 °F | 0 | 0 | 0 | 1 | 6 | 16 | 21 | 16 | 6 | 1 | 0 | 0 | 67 |
| Days Max Temp ≤ 32 °F | 3 | 2 | 0 | 0 | 0 | 0 | 0 | 0 | 0 | 0 | 0 | 2 | 7 |
| Days Min Temp ≤ 32 °F | 28 | 23 | 16 | 5 | 0 | 0 | 0 | 0 | 0 | 2 | 15 | 26 | 115 |
| Days Min Temp ≤ 0 °F | 0 | 0 | 0 | 0 | 0 | 0 | 0 | 0 | 0 | 0 | 0 | 1 | 1 |
| Heating Degree Days | 840 | 659 | 503 | 235 | 68 | 6 | 0 | 2 | 42 | 215 | 521 | 783 | 3874 |
| Cooling Degree Days | 0 | 0 | 1 | 19 | 111 | 314 | 393 | 337 | 162 | 24 | 0 | 0 | 1361 |
| Total Precipitation (") | 0.49 | 0.56 | 0.60 | 0.92 | 2.01 | 2.75 | 2.52 | 3.15 | 2.88 | 1.58 | 0.73 | 0.59 | 18.78 |
| Days ≥ 0.1" Precip | 2 | 2 | 2 | 2 | 4 | 4 | 5 | 5 | 4 | 3 | 2 | 2 | 37 |
| Total Snowfall (") | 2.0 | 1.5 | 0.8 | 0.2 | 0.0 | 0.0 | 0.0 | 0.0 | 0.0 | 0.1 | 0.8 | 2.6 | 8.0 |
| Days ≥ 1" Snow Depth | ***0*** | ***0*** | 0 | 0 | 0 | 0 | 0 | 0 | 0 | 0 | 0 | ***0*** | 0 |

## MOUNT LOCKE *Jeff Davis County* ELEVATION 6795 ft LAT/LONG 30° 40' N / 104° 0' W

| | JAN | FEB | MAR | APR | MAY | JUN | JUL | AUG | SEP | OCT | NOV | DEC | YEAR |
|---|---|---|---|---|---|---|---|---|---|---|---|---|---|
| Maximum Temp °F | 53.2 | 56.4 | 63.4 | 71.2 | 77.7 | 84.1 | 82.4 | 80.2 | 75.8 | 70.3 | 61.2 | 55.1 | 69.3 |
| Minimum Temp °F | 31.7 | 33.3 | 37.8 | 44.8 | 51.6 | 57.7 | 58.7 | 57.8 | 54.0 | 47.4 | 38.6 | 33.6 | 45.6 |
| Mean Temp °F | 42.4 | 44.8 | 50.7 | 58.0 | 64.7 | 70.9 | 70.6 | 69.0 | 64.9 | 58.9 | 49.9 | 44.4 | 57.4 |
| Days Max Temp ≥ 90 °F | 0 | 0 | 0 | 0 | 1 | 6 | 3 | 1 | 0 | 0 | 0 | 0 | 11 |
| Days Max Temp ≤ 32 °F | 1 | 0 | 0 | 0 | 0 | 0 | 0 | 0 | 0 | 0 | 0 | 1 | 2 |
| Days Min Temp ≤ 32 °F | 16 | 12 | 9 | 3 | 0 | 0 | 0 | 0 | 0 | 2 | 8 | 13 | 63 |
| Days Min Temp ≤ 0 °F | 0 | 0 | 0 | 0 | 0 | 0 | 0 | 0 | 0 | 0 | 0 | 0 | 0 |
| Heating Degree Days | 693 | 562 | 438 | 216 | 78 | 13 | 6 | 15 | 67 | 202 | 446 | 632 | 3368 |
| Cooling Degree Days | 0 | 0 | 2 | 13 | 76 | 215 | 201 | 165 | 82 | 21 | 1 | 0 | 776 |
| Total Precipitation (") | 0.56 | 0.55 | 0.48 | 0.55 | 1.71 | 2.50 | 4.14 | 4.43 | 3.34 | 1.63 | 0.69 | 0.72 | 21.30 |
| Days ≥ 0.1" Precip | 2 | 2 | 1 | 2 | 4 | 5 | 8 | 8 | 6 | 3 | 1 | 2 | 44 |
| Total Snowfall (") | ***0.3*** | 0.1 | 0.3 | 0.0 | 0.0 | 0.0 | 0.0 | 0.0 | 0.0 | 0.0 | 0.3 | 0.7 | 1.7 |
| Days ≥ 1" Snow Depth | ***0*** | 0 | 0 | 0 | 0 | 0 | 0 | 0 | 0 | 0 | 0 | 1 | 1 |

## MOUNT PLEASANT *Titus County* ELEVATION 420 ft LAT/LONG 33° 11' N / 94° 58' W

| | JAN | FEB | MAR | APR | MAY | JUN | JUL | AUG | SEP | OCT | NOV | DEC | YEAR |
|---|---|---|---|---|---|---|---|---|---|---|---|---|---|
| Maximum Temp °F | 54.2 | 58.5 | 67.4 | 76.1 | 82.3 | 90.0 | 94.0 | 94.1 | 87.4 | 78.3 | 66.4 | 57.6 | 75.5 |
| Minimum Temp °F | 29.8 | 32.5 | 41.0 | 49.8 | 58.1 | 66.5 | 70.0 | 68.4 | 61.7 | 49.0 | 40.0 | 32.5 | 49.9 |
| Mean Temp °F | 42.0 | 45.5 | 54.2 | 63.0 | 70.2 | 78.3 | 82.0 | 81.3 | 74.6 | 63.6 | 53.3 | 45.1 | 62.8 |
| Days Max Temp ≥ 90 °F | 0 | 0 | 0 | 0 | 3 | 18 | 26 | 26 | 14 | 3 | 0 | 0 | 90 |
| Days Max Temp ≤ 32 °F | 1 | 1 | 0 | 0 | 0 | 0 | 0 | 0 | 0 | 0 | 0 | 1 | 3 |
| Days Min Temp ≤ 32 °F | 21 | 15 | 7 | 1 | 0 | 0 | 0 | 0 | 0 | 1 | 8 | 16 | 69 |
| Days Min Temp ≤ 0 °F | 0 | 0 | 0 | 0 | 0 | 0 | 0 | 0 | 0 | 0 | 0 | 0 | 0 |
| Heating Degree Days | 707 | 546 | 347 | 124 | 22 | 1 | 0 | 0 | 15 | 123 | 360 | 613 | 2858 |
| Cooling Degree Days | 0 | 2 | 18 | 65 | 185 | 413 | 542 | 530 | 318 | 92 | 16 | 3 | 2184 |
| Total Precipitation (") | 2.95 | 3.87 | 4.44 | 4.17 | 5.55 | 4.46 | 3.75 | 2.06 | 3.97 | 4.82 | 4.62 | 4.38 | 49.04 |
| Days ≥ 0.1" Precip | 5 | 5 | 6 | 6 | 7 | 5 | 4 | 4 | 5 | 5 | 5 | 6 | 63 |
| Total Snowfall (") | 0.5 | 0.8 | 0.1 | 0.0 | 0.0 | 0.0 | 0.0 | 0.0 | 0.0 | 0.0 | 0.0 | 0.4 | 1.8 |
| Days ≥ 1" Snow Depth | 1 | 1 | 0 | 0 | 0 | 0 | 0 | 0 | 0 | 0 | 0 | 0 | 2 |

## MULESHOE 1 *Bailey County* ELEVATION 3704 ft LAT/LONG 34° 13' N / 102° 43' W

| | JAN | FEB | MAR | APR | MAY | JUN | JUL | AUG | SEP | OCT | NOV | DEC | YEAR |
|---|---|---|---|---|---|---|---|---|---|---|---|---|---|
| Maximum Temp °F | 51.6 | 56.3 | 64.5 | 73.6 | 81.2 | 89.4 | 91.5 | 89.3 | 82.6 | 73.6 | 61.4 | 53.3 | 72.4 |
| Minimum Temp °F | 19.9 | 23.0 | 30.1 | 39.8 | 49.0 | 58.7 | 62.8 | 60.9 | 53.5 | 40.5 | 29.2 | 21.4 | 40.7 |
| Mean Temp °F | 35.8 | 39.7 | 47.3 | 56.7 | 65.1 | 74.0 | 77.2 | 75.1 | 68.0 | 57.1 | 45.4 | 37.4 | 56.6 |
| Days Max Temp ≥ 90 °F | 0 | 0 | 0 | 1 | 6 | 16 | 21 | 17 | 7 | 1 | 0 | 0 | 69 |
| Days Max Temp ≤ 32 °F | 3 | 1 | 1 | 0 | 0 | 0 | 0 | 0 | 0 | 0 | 1 | 3 | 9 |
| Days Min Temp ≤ 32 °F | 29 | 25 | 18 | 5 | 0 | 0 | 0 | 0 | 0 | 4 | 20 | 28 | 129 |
| Days Min Temp ≤ 0 °F | 0 | 0 | 0 | 0 | 0 | 0 | 0 | 0 | 0 | 0 | 0 | 1 | 1 |
| Heating Degree Days | 900 | 708 | 541 | 257 | 84 | 8 | 1 | 2 | 50 | 254 | 583 | 850 | 4238 |
| Cooling Degree Days | 0 | 0 | 0 | 13 | 93 | 298 | 401 | 348 | 160 | 12 | 0 | 0 | 1325 |
| Total Precipitation (") | 0.49 | 0.51 | 0.56 | 0.90 | 2.02 | 2.56 | 2.12 | 2.90 | 2.37 | 1.40 | 0.65 | 0.54 | 17.02 |
| Days ≥ 0.1" Precip | 1 | 2 | 2 | 3 | 4 | 5 | 4 | 5 | 4 | 3 | 2 | 2 | 37 |
| Total Snowfall (") | 2.6 | 2.4 | 1.3 | 0.5 | 0.0 | 0.0 | 0.0 | 0.0 | 0.0 | 0.2 | 1.0 | 2.6 | 10.6 |
| Days ≥ 1" Snow Depth | *1* | *0* | *0* | 0 | 0 | 0 | 0 | 0 | 0 | 0 | 0 | na | na |

## MUNDAY *Knox County* ELEVATION 1480 ft LAT/LONG 33° 27' N / 99° 38' W

| | JAN | FEB | MAR | APR | MAY | JUN | JUL | AUG | SEP | OCT | NOV | DEC | YEAR |
|---|---|---|---|---|---|---|---|---|---|---|---|---|---|
| Maximum Temp °F | 56.5 | 61.5 | 71.1 | 80.3 | 86.2 | 93.4 | 97.7 | 96.2 | 88.6 | 79.7 | 67.8 | 59.0 | 78.2 |
| Minimum Temp °F | 28.7 | 32.7 | 40.8 | 50.2 | 58.8 | 67.1 | 70.9 | 69.5 | 62.5 | 51.4 | 40.4 | 31.9 | 50.4 |
| Mean Temp °F | 42.7 | 47.1 | 56.0 | 65.2 | 72.5 | 80.3 | 84.3 | 82.9 | 75.6 | 65.6 | 54.1 | 45.5 | 64.3 |
| Days Max Temp ≥ 90 °F | 0 | 0 | 2 | 5 | 11 | 22 | 29 | 27 | 16 | 5 | 0 | 0 | 117 |
| Days Max Temp ≤ 32 °F | 2 | 1 | 0 | 0 | 0 | 0 | 0 | 0 | 0 | 0 | 0 | 1 | 4 |
| Days Min Temp ≤ 32 °F | 20 | 14 | 6 | 1 | 0 | 0 | 0 | 0 | 0 | 0 | 7 | 16 | 64 |
| Days Min Temp ≤ 0 °F | 0 | 0 | 0 | 0 | 0 | 0 | 0 | 0 | 0 | 0 | 0 | 0 | 0 |
| Heating Degree Days | 686 | 499 | 300 | 89 | 16 | 0 | 0 | 0 | 13 | 88 | 331 | 598 | 2620 |
| Cooling Degree Days | 0 | 1 | 25 | 100 | 259 | 461 | 602 | 590 | 355 | 119 | 15 | 1 | 2528 |
| Total Precipitation (") | 1.06 | 1.52 | 1.57 | 2.11 | 3.96 | 3.07 | 1.81 | 2.66 | 3.47 | 2.74 | 1.07 | 1.18 | 26.22 |
| Days ≥ 0.1" Precip | 2 | 3 | 3 | 3 | 5 | 4 | 3 | 4 | 4 | 4 | 2 | 3 | 40 |
| Total Snowfall (") | 1.5 | 1.5 | 0.7 | 0.0 | 0.0 | 0.0 | 0.0 | 0.0 | 0.0 | 0.0 | 0.6 | 0.5 | 4.8 |
| Days ≥ 1" Snow Depth | *1* | *1* | 0 | 0 | 0 | 0 | 0 | 0 | 0 | 0 | 0 | 0 | 2 |

## NAVARRO MILLS DAM *Navarro County* ELEVATION 454 ft LAT/LONG 31° 57' N / 96° 42' W

| | JAN | FEB | MAR | APR | MAY | JUN | JUL | AUG | SEP | OCT | NOV | DEC | YEAR |
|---|---|---|---|---|---|---|---|---|---|---|---|---|---|
| Maximum Temp °F | 54.9 | 59.3 | 68.0 | 76.2 | 82.1 | 89.7 | 94.5 | 94.8 | 88.0 | 78.7 | 67.2 | 58.6 | 76.0 |
| Minimum Temp °F | 32.8 | 36.5 | 45.1 | 53.7 | 61.6 | 69.0 | 72.2 | 71.2 | 65.0 | 53.9 | 44.1 | 35.8 | 53.4 |
| Mean Temp °F | 43.9 | 47.9 | 56.6 | 65.0 | 71.9 | 79.4 | 83.4 | 83.0 | 76.5 | 66.3 | 55.7 | 47.2 | 64.7 |
| Days Max Temp ≥ 90 °F | 0 | 0 | 0 | 1 | 3 | 17 | 27 | 27 | 15 | 3 | 0 | 0 | 93 |
| Days Max Temp ≤ 32 °F | 2 | 1 | 0 | 0 | 0 | 0 | 0 | 0 | 0 | 0 | 0 | 1 | 4 |
| Days Min Temp ≤ 32 °F | 16 | 10 | 3 | 0 | 0 | 0 | 0 | 0 | 0 | 0 | 4 | 11 | 44 |
| Days Min Temp ≤ 0 °F | 0 | 0 | 0 | 0 | 0 | 0 | 0 | 0 | 0 | 0 | 0 | 0 | 0 |
| Heating Degree Days | 650 | 478 | 280 | 85 | 13 | 0 | 0 | 0 | 8 | 80 | 298 | 548 | 2440 |
| Cooling Degree Days | 1 | 3 | 26 | 86 | 231 | 432 | 572 | 579 | 366 | 126 | 26 | 3 | 2451 |
| Total Precipitation (") | 1.99 | 2.87 | 3.23 | 3.62 | 5.61 | 3.22 | 1.63 | 2.30 | 3.51 | 4.46 | 2.93 | 3.00 | 38.37 |
| Days ≥ 0.1" Precip | 4 | 5 | 5 | 4 | 7 | 5 | 3 | 3 | 4 | 5 | 5 | 5 | 55 |
| Total Snowfall (") | *0.5* | 0.0 | 0.0 | 0.0 | 0.0 | 0.0 | 0.0 | 0.0 | 0.0 | 0.0 | 0.0 | 0.0 | 0.5 |
| Days ≥ 1" Snow Depth | *0* | 0 | 0 | 0 | 0 | 0 | 0 | 0 | 0 | 0 | 0 | 0 | 0 |

## NEW BRAUNFELS *Comal County* ELEVATION 722 ft LAT/LONG 29° 42 ' N / 98° 7 ' W

| | JAN | FEB | MAR | APR | MAY | JUN | JUL | AUG | SEP | OCT | NOV | DEC | YEAR |
|---|---|---|---|---|---|---|---|---|---|---|---|---|---|
| Maximum Temp °F | 60.0 | 64.5 | 73.2 | 80.3 | 85.2 | 91.5 | 95.2 | 95.4 | 89.9 | 81.8 | 71.7 | 63.4 | 79.3 |
| Minimum Temp °F | 37.3 | 40.1 | 47.8 | 56.3 | 63.5 | 70.0 | 72.2 | 71.7 | 66.9 | 56.9 | 47.6 | 39.9 | 55.9 |
| Mean Temp °F | 48.7 | 52.3 | 60.5 | 68.3 | 74.4 | 80.8 | 83.7 | 83.6 | 78.5 | 69.4 | 59.6 | 51.7 | 67.6 |
| Days Max Temp ≥ 90 °F | 0 | 0 | 1 | 2 | 8 | 22 | 28 | 29 | 19 | 5 | 0 | 0 | 114 |
| Days Max Temp ≤ 32 °F | 0 | 0 | 0 | 0 | 0 | 0 | 0 | 0 | 0 | 0 | 0 | 0 | 0 |
| Days Min Temp ≤ 32 °F | 11 | 6 | 2 | 0 | 0 | 0 | 0 | 0 | 0 | 0 | 3 | 7 | 29 |
| Days Min Temp ≤ 0 °F | 0 | 0 | 0 | 0 | 0 | 0 | 0 | 0 | 0 | 0 | 0 | 0 | 0 |
| Heating Degree Days | 506 | 361 | 185 | 44 | 5 | 0 | 0 | 0 | 3 | 43 | 207 | 415 | 1769 |
| Cooling Degree Days | 4 | 8 | 47 | 133 | 297 | 463 | 576 | 595 | 413 | 191 | 58 | 9 | 2794 |
| Total Precipitation (") | 2.35 | 2.32 | 1.98 | 2.77 | 5.69 | 4.20 | 1.92 | 2.59 | 3.92 | 3.67 | 2.58 | 2.69 | 36.68 |
| Days ≥ 0.1" Precip | 5 | 4 | 4 | 4 | 6 | 5 | 3 | 4 | 5 | 4 | 4 | 4 | 52 |
| Total Snowfall (") | 0.4 | 0.1 | 0.0 | 0.0 | 0.0 | 0.0 | 0.0 | 0.0 | 0.0 | 0.0 | 0.0 | 0.0 | 0.5 |
| Days ≥ 1" Snow Depth | 0 | 0 | 0 | 0 | 0 | 0 | 0 | 0 | 0 | 0 | 0 | 0 | 0 |

## NEW GULF *Wharton County* ELEVATION 69 ft LAT/LONG 29° 16 ' N / 95° 55 ' W

| | JAN | FEB | MAR | APR | MAY | JUN | JUL | AUG | SEP | OCT | NOV | DEC | YEAR |
|---|---|---|---|---|---|---|---|---|---|---|---|---|---|
| Maximum Temp °F | 61.9 | 65.2 | 72.6 | 79.3 | 84.6 | 90.2 | 92.8 | 92.9 | 89.0 | 82.5 | 73.3 | 65.6 | 79.2 |
| Minimum Temp °F | 41.7 | 44.5 | 52.2 | 60.2 | 66.0 | 71.4 | 73.1 | 72.6 | 68.4 | 59.2 | 51.5 | 44.5 | 58.8 |
| Mean Temp °F | 51.8 | 54.8 | 62.5 | 69.8 | 75.3 | 80.9 | 83.0 | 82.8 | 78.7 | 70.9 | 62.5 | 55.1 | 69.0 |
| Days Max Temp ≥ 90 °F | 0 | 0 | 0 | 0 | 4 | 20 | 27 | 27 | 17 | 4 | 0 | 0 | 99 |
| Days Max Temp ≤ 32 °F | 0 | 0 | 0 | 0 | 0 | 0 | 0 | 0 | 0 | 0 | 0 | 0 | 0 |
| Days Min Temp ≤ 32 °F | 6 | 3 | 1 | 0 | 0 | 0 | 0 | 0 | 0 | 0 | 1 | 4 | 15 |
| Days Min Temp ≤ 0 °F | 0 | 0 | 0 | 0 | 0 | 0 | 0 | 0 | 0 | 0 | 0 | 0 | 0 |
| Heating Degree Days | 417 | 296 | 140 | 30 | 2 | 0 | 0 | 0 | 1 | 29 | 152 | 327 | 1394 |
| Cooling Degree Days | 11 | 19 | 70 | 166 | 330 | 495 | 573 | 570 | 425 | 227 | 88 | 32 | 3006 |
| Total Precipitation (") | 3.31 | 2.67 | 2.61 | 3.15 | 5.28 | 4.49 | 3.99 | 3.78 | 5.35 | 4.57 | 3.66 | 3.13 | 45.99 |
| Days ≥ 0.1" Precip | 6 | 4 | 4 | 3 | 5 | 5 | 6 | 6 | 6 | 4 | 5 | 5 | 59 |
| Total Snowfall (") | 0.0 | 0.1 | 0.0 | 0.0 | 0.0 | 0.0 | 0.0 | 0.0 | 0.0 | 0.0 | 0.0 | 0.0 | 0.1 |
| Days ≥ 1" Snow Depth | 0 | 0 | 0 | 0 | 0 | 0 | 0 | 0 | 0 | 0 | 0 | 0 | 0 |

## NIXON *Gonzales County* ELEVATION 400 ft LAT/LONG 29° 16 ' N / 97° 46 ' W

| | JAN | FEB | MAR | APR | MAY | JUN | JUL | AUG | SEP | OCT | NOV | DEC | YEAR |
|---|---|---|---|---|---|---|---|---|---|---|---|---|---|
| Maximum Temp °F | 63.0 | 67.3 | 75.1 | 81.1 | 85.9 | 91.5 | 94.8 | 95.6 | 90.4 | 83.3 | 73.4 | 66.1 | 80.6 |
| Minimum Temp °F | 40.8 | 43.8 | 51.1 | 58.7 | 64.8 | 70.2 | 72.1 | 71.7 | 67.7 | 58.9 | 50.1 | 43.5 | 57.8 |
| Mean Temp °F | 51.9 | 55.6 | 63.1 | 69.9 | 75.4 | 80.9 | 83.5 | 83.6 | 79.1 | 71.1 | 61.8 | 54.8 | 69.2 |
| Days Max Temp ≥ 90 °F | 0 | 0 | 1 | 2 | 8 | 22 | 28 | 29 | 19 | 6 | 0 | 0 | 115 |
| Days Max Temp ≤ 32 °F | 0 | 0 | 0 | 0 | 0 | 0 | 0 | 0 | 0 | 0 | 0 | 0 | 0 |
| Days Min Temp ≤ 32 °F | 6 | 4 | 1 | 0 | 0 | 0 | 0 | 0 | 0 | 0 | 1 | 4 | 16 |
| Days Min Temp ≤ 0 °F | 0 | 0 | 0 | 0 | 0 | 0 | 0 | 0 | 0 | 0 | 0 | 0 | 0 |
| Heating Degree Days | 412 | 281 | 132 | 31 | 1 | 0 | 0 | 0 | 1 | 27 | 164 | 333 | 1382 |
| Cooling Degree Days | 7 | 24 | 73 | 178 | 332 | 494 | 579 | ***592*** | 439 | 233 | 77 | 24 | 3052 |
| Total Precipitation (") | 2.43 | 2.47 | 1.83 | 3.16 | 5.03 | 3.68 | 1.79 | 2.55 | 4.42 | 3.41 | 2.27 | 2.17 | 35.21 |
| Days ≥ 0.1" Precip | 5 | 4 | 3 | 4 | 6 | 5 | 3 | 4 | 5 | 4 | 3 | 4 | 50 |
| Total Snowfall (") | 0.3 | 0.1 | 0.0 | 0.0 | 0.0 | 0.0 | 0.0 | 0.0 | 0.0 | 0.0 | 0.0 | 0.0 | 0.4 |
| Days ≥ 1" Snow Depth | 0 | 0 | 0 | 0 | 0 | 0 | 0 | 0 | 0 | 0 | 0 | 0 | 0 |

## OLNEY *Young County* ELEVATION 1181 ft LAT/LONG 33° 23 ' N / 98° 45 ' W

| | JAN | FEB | MAR | APR | MAY | JUN | JUL | AUG | SEP | OCT | NOV | DEC | YEAR |
|---|---|---|---|---|---|---|---|---|---|---|---|---|---|
| Maximum Temp °F | 55.6 | 60.7 | 70.2 | 78.6 | 84.5 | 92.1 | 97.0 | 96.7 | 88.8 | 79.2 | ***66.9*** | 58.7 | 77.4 |
| Minimum Temp °F | 30.2 | 34.1 | 42.0 | 51.1 | 59.2 | 67.9 | 71.9 | 70.7 | 63.8 | 52.7 | ***41.5*** | 32.9 | 51.5 |
| Mean Temp °F | 42.8 | 47.4 | 56.1 | 64.9 | 71.9 | 80.0 | 84.6 | 83.7 | 76.3 | 66.0 | ***54.2*** | 45.8 | 64.5 |
| Days Max Temp ≥ 90 °F | 0 | 0 | 1 | 2 | 8 | 21 | 28 | 28 | 16 | 4 | 0 | 0 | 108 |
| Days Max Temp ≤ 32 °F | 2 | 1 | 0 | 0 | 0 | 0 | 0 | 0 | 0 | 0 | 0 | 1 | 4 |
| Days Min Temp ≤ 32 °F | 18 | 12 | 5 | 1 | 0 | 0 | 0 | 0 | 0 | 0 | 6 | 15 | 57 |
| Days Min Temp ≤ 0 °F | 0 | 0 | 0 | 0 | 0 | 0 | 0 | 0 | 0 | 0 | 0 | 0 | 0 |
| Heating Degree Days | 681 | 494 | 287 | 92 | 15 | 0 | 0 | 0 | 10 | 84 | 337 | 590 | 2590 |
| Cooling Degree Days | 0 | 2 | 26 | 92 | 232 | 455 | 606 | 607 | 372 | ***121*** | 18 | 1 | 2532 |
| Total Precipitation (") | 1.24 | 1.72 | 1.79 | 2.89 | 5.15 | 3.18 | 1.78 | 2.09 | 3.68 | 3.10 | 1.48 | 1.45 | 29.55 |
| Days ≥ 0.1" Precip | 3 | 3 | 4 | 4 | 6 | 5 | 3 | 4 | 4 | 4 | 3 | 3 | 46 |
| Total Snowfall (") | 0.8 | ***1.0*** | 0.5 | 0.0 | 0.0 | 0.0 | 0.0 | 0.0 | 0.0 | 0.0 | 0.2 | 0.4 | 2.9 |
| Days ≥ 1" Snow Depth | 1 | 1 | 0 | 0 | 0 | 0 | 0 | 0 | 0 | 0 | 0 | 1 | 3 |

## OLTON *Lamb County* ELEVATION 3612 ft LAT/LONG 34° 11 ' N / 102° 8 ' W

| | JAN | FEB | MAR | APR | MAY | JUN | JUL | AUG | SEP | OCT | NOV | DEC | YEAR |
|---|---|---|---|---|---|---|---|---|---|---|---|---|---|
| Maximum Temp °F | 50.4 | 55.4 | 63.9 | 72.9 | 80.2 | 87.9 | 90.0 | 87.6 | 81.3 | 72.8 | 60.5 | 52.8 | 71.3 |
| Minimum Temp °F | 21.4 | 24.4 | 31.5 | 41.4 | 50.5 | 59.4 | 62.9 | 61.1 | 53.9 | 42.2 | 31.2 | 23.6 | 42.0 |
| Mean Temp °F | 36.0 | 39.9 | 47.7 | 57.2 | 65.4 | 73.7 | 76.5 | 74.4 | 67.6 | 57.5 | 45.9 | 38.2 | 56.7 |
| Days Max Temp ≥ 90 °F | 0 | 0 | 0 | 1 | 5 | 13 | 18 | 13 | 5 | 1 | 0 | 0 | 56 |
| Days Max Temp ≤ 32 °F | 4 | 2 | 1 | 0 | 0 | 0 | 0 | 0 | 0 | 0 | 1 | 2 | 10 |
| Days Min Temp ≤ 32 °F | 28 | 24 | 16 | 4 | 0 | 0 | 0 | 0 | 0 | 4 | 16 | 27 | 119 |
| Days Min Temp ≤ 0 °F | 0 | 0 | 0 | 0 | 0 | 0 | 0 | 0 | 0 | 0 | 0 | 1 | 1 |
| Heating Degree Days | 893 | 702 | 531 | 247 | 82 | 9 | 1 | 2 | 51 | 243 | 567 | 822 | 4150 |
| Cooling Degree Days | 0 | 0 | *1* | 20 | 97 | 270 | 369 | 322 | 143 | 16 | 0 | 0 | 1238 |
| Total Precipitation (") | 0.47 | 0.47 | 0.74 | 0.93 | 2.32 | 3.12 | 2.12 | 2.81 | 2.29 | 1.43 | 0.72 | 0.57 | 17.99 |
| Days ≥ 0.1" Precip | 2 | 2 | 2 | 3 | 4 | 4 | 4 | 5 | 4 | 3 | 2 | 2 | 37 |
| Total Snowfall (") | *1.9* | 1.6 | 1.3 | 0.2 | 0.0 | 0.0 | 0.0 | 0.0 | 0.0 | 0.1 | 0.5 | 2.3 | 7.9 |
| Days ≥ 1" Snow Depth | *1* | 1 | 0 | 0 | 0 | 0 | 0 | 0 | 0 | 0 | 0 | *1* | 3 |

## OZONA 2 SSW *Crockett County* ELEVATION 2346 ft LAT/LONG 30° 43 ' N / 101° 12 ' W

| | JAN | FEB | MAR | APR | MAY | JUN | JUL | AUG | SEP | OCT | NOV | DEC | YEAR |
|---|---|---|---|---|---|---|---|---|---|---|---|---|---|
| Maximum Temp °F | 58.6 | 62.7 | 71.5 | 79.8 | 85.3 | 90.7 | 93.2 | 92.7 | 86.3 | 78.0 | 67.5 | 60.2 | 77.2 |
| Minimum Temp °F | 30.2 | 33.4 | 41.6 | 50.5 | 59.0 | 66.3 | 68.5 | 67.4 | 61.7 | 51.0 | 39.7 | 31.8 | 50.1 |
| Mean Temp °F | 44.4 | 48.1 | 56.6 | 65.1 | 72.2 | 78.5 | 80.9 | 80.1 | 74.0 | 64.5 | 53.6 | 46.0 | 63.7 |
| Days Max Temp ≥ 90 °F | 0 | 0 | 1 | 3 | 9 | 18 | 24 | 24 | 11 | 2 | 0 | 0 | 92 |
| Days Max Temp ≤ 32 °F | 1 | 0 | 0 | 0 | 0 | 0 | 0 | 0 | 0 | 0 | 0 | 1 | 2 |
| Days Min Temp ≤ 32 °F | 19 | 13 | 7 | 1 | 0 | 0 | 0 | 0 | 0 | 1 | 8 | 17 | 66 |
| Days Min Temp ≤ 0 °F | 0 | 0 | 0 | 0 | 0 | 0 | 0 | 0 | 0 | 0 | 0 | 0 | 0 |
| Heating Degree Days | 633 | 471 | 275 | 90 | 16 | 0 | 0 | 0 | 16 | 103 | 343 | 579 | 2526 |
| Cooling Degree Days | 0 | 1 | 19 | 82 | 245 | 415 | 490 | 478 | 295 | 98 | 12 | 0 | 2135 |
| Total Precipitation (") | 0.79 | 0.95 | 0.94 | 1.47 | 2.50 | 1.89 | 1.69 | 2.17 | 3.24 | 2.18 | 0.91 | 0.71 | 19.44 |
| Days ≥ 0.1" Precip | 2 | 2 | 2 | 2 | 4 | 3 | 3 | 3 | 4 | 3 | 2 | 2 | 32 |
| Total Snowfall (") | 0.1 | 0.3 | 0.0 | 0.0 | 0.0 | 0.0 | 0.0 | 0.0 | 0.0 | 0.1 | 0.1 | 0.0 | 0.6 |
| Days ≥ 1" Snow Depth | *0* | 0 | 0 | 0 | 0 | 0 | 0 | 0 | 0 | 0 | 0 | 0 | 0 |

## PADUCAH *Cottle County* ELEVATION 1870 ft LAT/LONG 34° 1 ' N / 100° 18 ' W

| | JAN | FEB | MAR | APR | MAY | JUN | JUL | AUG | SEP | OCT | NOV | DEC | YEAR |
|---|---|---|---|---|---|---|---|---|---|---|---|---|---|
| Maximum Temp °F | 52.4 | 57.1 | 66.7 | 76.7 | 83.2 | 91.3 | 96.2 | 94.2 | 85.9 | 76.5 | 63.9 | 55.5 | 75.0 |
| Minimum Temp °F | 26.3 | 30.0 | 37.8 | 47.3 | 56.4 | 65.5 | 69.7 | 68.1 | 60.6 | 48.9 | 37.4 | 29.5 | 48.1 |
| Mean Temp °F | 39.4 | 43.6 | 52.3 | 62.0 | 69.8 | 78.4 | 83.0 | 81.2 | 73.3 | 62.7 | 50.7 | 42.5 | 61.6 |
| Days Max Temp ≥ 90 °F | 0 | 0 | 1 | 3 | 7 | 19 | 27 | 24 | 12 | 3 | 0 | 0 | 96 |
| Days Max Temp ≤ 32 °F | 4 | 2 | 0 | 0 | 0 | 0 | 0 | 0 | 0 | 0 | 0 | 2 | 8 |
| Days Min Temp ≤ 32 °F | 24 | 17 | 8 | 1 | 0 | 0 | 0 | 0 | 0 | 1 | 9 | 20 | 80 |
| Days Min Temp ≤ 0 °F | 0 | 0 | 0 | 0 | 0 | 0 | 0 | 0 | 0 | 0 | 0 | 0 | 0 |
| Heating Degree Days | 788 | 599 | 401 | 147 | 34 | 2 | 0 | 0 | 24 | 138 | 428 | 691 | 3252 |
| Cooling Degree Days | 0 | 1 | 11 | 60 | 186 | 400 | 552 | 526 | 286 | 72 | 7 | 0 | 2101 |
| Total Precipitation (") | 0.84 | 1.03 | 1.12 | 1.70 | 3.51 | 3.55 | 1.69 | 2.58 | 3.14 | 2.18 | 1.06 | 1.01 | 23.41 |
| Days ≥ 0.1" Precip | 2 | 3 | 2 | 3 | 5 | 5 | 3 | 4 | 5 | 3 | 2 | 2 | 39 |
| Total Snowfall (") | 2.8 | 2.8 | 0.9 | 0.1 | 0.0 | 0.0 | 0.0 | 0.0 | 0.0 | 0.1 | 1.0 | 1.7 | 9.4 |
| Days ≥ 1" Snow Depth | 2 | 2 | 0 | 0 | 0 | 0 | 0 | 0 | 0 | 0 | 1 | 1 | 6 |

## PAINT ROCK *Concho County* ELEVATION 1631 ft LAT/LONG 31° 30 ' N / 99° 56 ' W

| | JAN | FEB | MAR | APR | MAY | JUN | JUL | AUG | SEP | OCT | NOV | DEC | YEAR |
|---|---|---|---|---|---|---|---|---|---|---|---|---|---|
| Maximum Temp °F | 60.5 | 65.6 | 74.5 | 83.0 | 88.3 | 94.2 | 97.5 | 96.6 | 89.3 | 81.6 | 70.5 | 63.1 | 80.4 |
| Minimum Temp °F | 31.7 | 35.6 | 43.5 | 52.2 | 60.2 | 67.5 | 70.3 | 69.3 | 62.9 | 52.4 | 41.6 | 33.6 | 51.7 |
| Mean Temp °F | 46.1 | 50.6 | 59.0 | 67.6 | 74.3 | 80.9 | 83.9 | 83.0 | 76.1 | 67.0 | 56.1 | 48.4 | 66.1 |
| Days Max Temp ≥ 90 °F | 0 | 0 | 2 | 7 | 14 | 25 | 29 | 27 | 16 | 5 | 0 | 0 | 125 |
| Days Max Temp ≤ 32 °F | 1 | 0 | 0 | 0 | 0 | 0 | 0 | 0 | 0 | 0 | 0 | 1 | 2 |
| Days Min Temp ≤ 32 °F | 17 | 11 | 5 | 1 | 0 | 0 | 0 | 0 | 0 | 1 | 7 | 15 | 57 |
| Days Min Temp ≤ 0 °F | 0 | 0 | 0 | 0 | 0 | 0 | 0 | 0 | 0 | 0 | 0 | 0 | 0 |
| Heating Degree Days | 579 | 404 | 223 | 58 | 9 | 0 | 0 | 0 | 8 | 65 | 284 | 511 | 2141 |
| Cooling Degree Days | 1 | 5 | 40 | 138 | 309 | 482 | 587 | 585 | 358 | 148 | 28 | 2 | 2683 |
| Total Precipitation (") | 1.22 | 1.49 | 1.44 | 1.84 | 3.50 | 3.28 | 1.79 | 1.96 | 3.99 | 2.75 | 1.24 | 1.42 | 25.92 |
| Days ≥ 0.1" Precip | 3 | 3 | 3 | 3 | 5 | 4 | 3 | 3 | 5 | 4 | 2 | 3 | 41 |
| Total Snowfall (") | 2.5 | 0.9 | 0.4 | 0.0 | 0.0 | 0.0 | 0.0 | 0.0 | 0.0 | 0.0 | 0.3 | 0.5 | 4.6 |
| Days ≥ 1" Snow Depth | *0* | 0 | 0 | 0 | 0 | 0 | 0 | 0 | 0 | 0 | 0 | 0 | 0 |

## PALACIOS MUNI ARPT *Matagorda County* ELEVATION 16 ft LAT/LONG 28° 43 ' N / 96° 15 ' W

| | JAN | FEB | MAR | APR | MAY | JUN | JUL | AUG | SEP | OCT | NOV | DEC | YEAR |
|---|---|---|---|---|---|---|---|---|---|---|---|---|---|
| Maximum Temp °F | 61.9 | 64.8 | 71.6 | 77.8 | 83.0 | 88.0 | 90.0 | 90.6 | 87.5 | 81.3 | 72.6 | 65.5 | 77.9 |
| Minimum Temp °F | 44.0 | 46.3 | 53.3 | 61.7 | 68.3 | 74.4 | 76.9 | 75.8 | 70.7 | 61.7 | 53.1 | 46.7 | 61.1 |
| Mean Temp °F | 53.0 | 55.6 | 62.5 | 69.8 | 75.7 | 81.2 | 83.5 | 83.2 | 79.2 | 71.5 | 62.9 | 56.1 | 69.5 |
| Days Max Temp ≥ 90 °F | 0 | 0 | 0 | 0 | 1 | 9 | 21 | 23 | 11 | 1 | 0 | 0 | 66 |
| Days Max Temp ≤ 32 °F | 0 | 0 | 0 | 0 | 0 | 0 | 0 | 0 | 0 | 0 | 0 | 0 | 0 |
| Days Min Temp ≤ 32 °F | 4 | 3 | 1 | 0 | 0 | 0 | 0 | 0 | 0 | 0 | 1 | 2 | 11 |
| Days Min Temp ≤ 0 °F | 0 | 0 | 0 | 0 | 0 | 0 | 0 | 0 | 0 | 0 | 0 | 0 | 0 |
| Heating Degree Days | 378 | 273 | 131 | 27 | 1 | 0 | 0 | 0 | 1 | 23 | 144 | 297 | 1275 |
| Cooling Degree Days | 10 | 15 | 61 | 171 | 351 | 502 | 584 | 587 | 444 | 245 | 93 | 32 | 3095 |
| Total Precipitation (") | 3.42 | 2.67 | 2.50 | 2.81 | 5.02 | 4.57 | 4.27 | 3.38 | 6.14 | 4.69 | 3.25 | 2.98 | 45.70 |
| Days ≥ 0.1" Precip | 6 | 4 | 3 | 3 | 5 | 5 | 5 | 5 | 6 | 4 | 4 | 5 | 55 |
| Total Snowfall (") | 0.0 | 0.1 | 0.0 | 0.0 | 0.0 | 0.0 | 0.0 | 0.0 | 0.0 | 0.0 | 0.0 | 0.0 | 0.1 |
| Days ≥ 1" Snow Depth | 0 | 0 | 0 | 0 | 0 | 0 | 0 | 0 | 0 | 0 | 0 | 0 | 0 |

## PALESTINE 2 NE *Anderson County* ELEVATION 460 ft LAT/LONG 31° 46 ' N / 95° 46 ' W

| | JAN | FEB | MAR | APR | MAY | JUN | JUL | AUG | SEP | OCT | NOV | DEC | YEAR |
|---|---|---|---|---|---|---|---|---|---|---|---|---|---|
| Maximum Temp °F | 56.2 | 60.8 | 69.2 | 76.8 | 82.4 | 89.3 | 93.4 | 93.9 | 87.6 | 79.3 | 68.1 | 59.7 | 76.4 |
| Minimum Temp °F | 35.2 | 38.3 | 46.1 | 54.9 | 61.6 | 68.2 | 71.1 | 70.1 | 64.8 | 54.6 | 45.7 | 38.0 | 54.1 |
| Mean Temp °F | 45.7 | 49.6 | 57.7 | 65.9 | 72.0 | 78.8 | 82.3 | 82.1 | 76.2 | 67.0 | 56.9 | 48.9 | 65.3 |
| Days Max Temp ≥ 90 °F | 0 | 0 | 0 | 0 | 3 | 16 | 26 | 26 | 14 | 2 | 0 | 0 | 87 |
| Days Max Temp ≤ 32 °F | 1 | 1 | 0 | 0 | 0 | 0 | 0 | 0 | 0 | 0 | 0 | 1 | 3 |
| Days Min Temp ≤ 32 °F | 13 | 9 | 3 | 0 | 0 | 0 | 0 | 0 | 0 | 0 | 3 | 9 | 37 |
| Days Min Temp ≤ 0 °F | 0 | 0 | 0 | 0 | 0 | 0 | 0 | 0 | 0 | 0 | 0 | 0 | 0 |
| Heating Degree Days | 595 | 435 | 253 | 70 | 10 | 0 | 0 | 0 | 6 | 67 | 267 | 500 | 2203 |
| Cooling Degree Days | 2 | 8 | 37 | 105 | 232 | 420 | 542 | 553 | 359 | 135 | 35 | 8 | 2436 |
| Total Precipitation (") | 3.25 | 3.45 | 3.99 | 4.39 | 5.29 | 4.29 | 2.41 | 2.77 | 3.74 | 4.63 | 3.98 | 4.08 | 46.27 |
| Days ≥ 0.1" Precip | 6 | 6 | 6 | 5 | 7 | 5 | 4 | 4 | 5 | 5 | 5 | 6 | 64 |
| Total Snowfall (") | 0.6 | 0.3 | 0.1 | 0.0 | 0.0 | 0.0 | 0.0 | 0.0 | 0.0 | 0.0 | 0.0 | 0.0 | 1.0 |
| Days ≥ 1" Snow Depth | 0 | 0 | 0 | 0 | 0 | 0 | 0 | 0 | 0 | 0 | 0 | 0 | 0 |

## PAMPA 2 *Gray County* ELEVATION 3212 ft LAT/LONG 35° 31 ' N / 100° 57 ' W

| | JAN | FEB | MAR | APR | MAY | JUN | JUL | AUG | SEP | OCT | NOV | DEC | YEAR |
|---|---|---|---|---|---|---|---|---|---|---|---|---|---|
| Maximum Temp °F | 47.4 | 51.9 | 60.7 | 70.6 | 77.8 | 86.7 | 91.6 | 89.5 | 81.8 | 71.8 | 58.4 | 49.7 | 69.8 |
| Minimum Temp °F | 21.5 | 24.9 | 32.1 | 41.7 | 51.2 | 60.9 | 65.9 | 64.3 | 56.6 | 44.8 | 32.4 | 24.3 | 43.4 |
| Mean Temp °F | 34.5 | 38.4 | 46.5 | 56.2 | 64.5 | 73.8 | 78.8 | 76.9 | 69.2 | 58.4 | 45.4 | 37.0 | 56.6 |
| Days Max Temp ≥ 90 °F | 0 | 0 | 0 | 1 | 3 | 12 | 21 | 18 | 7 | 1 | 0 | 0 | 63 |
| Days Max Temp ≤ 32 °F | 5 | 3 | 1 | 0 | 0 | 0 | 0 | 0 | 0 | 0 | 1 | 4 | 14 |
| Days Min Temp ≤ 32 °F | 27 | 22 | 16 | 4 | 0 | 0 | 0 | 0 | 0 | 2 | 15 | 26 | 112 |
| Days Min Temp ≤ 0 °F | 1 | 0 | 0 | 0 | 0 | 0 | 0 | 0 | 0 | 0 | 0 | 1 | 2 |
| Heating Degree Days | 940 | 745 | 570 | 279 | 100 | 11 | 1 | 2 | 50 | 235 | 582 | 861 | 4376 |
| Cooling Degree Days | 0 | 0 | 2 | 21 | 85 | 286 | 433 | 397 | 189 | 27 | 1 | 0 | 1441 |
| Total Precipitation (") | 0.52 | 0.84 | 1.31 | 1.46 | 2.95 | 3.54 | 2.50 | 2.51 | 2.33 | 1.44 | 1.05 | 0.54 | 20.99 |
| Days ≥ 0.1" Precip | 2 | 2 | 3 | 3 | 6 | 6 | 4 | 5 | 4 | 3 | 3 | 2 | 43 |
| Total Snowfall (") | 3.6 | 5.5 | 3.6 | 0.8 | 0.0 | 0.0 | 0.0 | 0.0 | 0.0 | 0.4 | 2.2 | 3.4 | 19.5 |
| Days ≥ 1" Snow Depth | 4 | 4 | 2 | 0 | 0 | 0 | 0 | 0 | 0 | 0 | 1 | 3 | 14 |

## PANDALE 2 NE *Val Verde County* ELEVATION 1591 ft LAT/LONG 30° 7 ' N / 101° 32 ' W

| | JAN | FEB | MAR | APR | MAY | JUN | JUL | AUG | SEP | OCT | NOV | DEC | YEAR |
|---|---|---|---|---|---|---|---|---|---|---|---|---|---|
| Maximum Temp °F | 61.6 | ***66.4*** | 75.3 | 83.2 | ***89.0*** | ***94.3*** | 95.8 | ***95.5*** | ***89.9*** | ***81.4*** | 70.1 | 63.6 | 80.5 |
| Minimum Temp °F | 32.6 | ***36.3*** | 45.3 | 54.9 | ***63.7*** | ***70.9*** | 73.4 | ***72.3*** | ***65.6*** | ***55.0*** | 42.2 | 34.6 | 53.9 |
| Mean Temp °F | 47.1 | ***51.4*** | 60.3 | 69.1 | ***76.5*** | ***82.8*** | 84.7 | ***83.9*** | ***77.8*** | ***68.2*** | 56.2 | 49.1 | 67.3 |
| Days Max Temp ≥ 90 °F | 0 | 0 | 2 | 7 | 14 | 23 | 28 | 27 | 17 | 4 | 0 | 0 | 122 |
| Days Max Temp ≤ 32 °F | 0 | 0 | 0 | 0 | 0 | 0 | 0 | 0 | 0 | 0 | 0 | 0 | 0 |
| Days Min Temp ≤ 32 °F | 17 | 10 | 3 | 0 | 0 | 0 | 0 | 0 | 0 | 0 | 6 | 14 | 50 |
| Days Min Temp ≤ 0 °F | 0 | 0 | 0 | 0 | 0 | 0 | 0 | 0 | 0 | 0 | 0 | 0 | 0 |
| Heating Degree Days | 548 | ***381*** | 186 | 42 | ***4*** | ***0*** | 0 | ***0*** | ***5*** | ***53*** | 278 | 487 | 1984 |
| Cooling Degree Days | 0 | na | 39 | 162 | ***378*** | ***554*** | 623 | ***606*** | na | ***175*** | 17 | ***0*** | na |
| Total Precipitation (") | 0.52 | 0.95 | 0.58 | 1.26 | 1.81 | 1.59 | 1.87 | 2.25 | 3.68 | 1.74 | 0.81 | 0.63 | 17.69 |
| Days ≥ 0.1" Precip | 1 | ***2*** | 1 | 2 | 3 | 2 | 3 | 3 | 4 | ***2*** | 2 | 2 | 27 |
| Total Snowfall (") | 0.1 | 0.3 | 0.0 | 0.0 | 0.0 | 0.0 | 0.0 | 0.0 | 0.0 | 0.0 | 0.0 | 0.0 | 0.4 |
| Days ≥ 1" Snow Depth | ***0*** | 0 | 0 | 0 | 0 | 0 | 0 | 0 | 0 | 0 | 0 | 0 | 0 |

## PANTHER JUNCTION *Brewster County* ELEVATION 3802 ft LAT/LONG 29° 19' N / 103° 13' W

| | JAN | FEB | MAR | APR | MAY | JUN | JUL | AUG | SEP | OCT | NOV | DEC | YEAR |
|---|---|---|---|---|---|---|---|---|---|---|---|---|---|
| Maximum Temp °F | 60.6 | 65.3 | 73.9 | 82.5 | 88.5 | 93.4 | 92.3 | 90.7 | 85.7 | 78.8 | 69.5 | 62.6 | 78.7 |
| Minimum Temp °F | 35.0 | 38.0 | 45.0 | 52.9 | 59.8 | 66.0 | 68.1 | 66.7 | 61.5 | 52.6 | 43.4 | 36.9 | 52.2 |
| Mean Temp °F | 47.8 | 51.7 | 59.5 | 67.7 | 74.2 | 79.7 | 80.2 | 78.7 | 73.6 | 65.7 | 56.5 | 49.8 | 65.4 |
| Days Max Temp ≥ 90 °F | 0 | 0 | 1 | 6 | 15 | 23 | 22 | 20 | 11 | 2 | 0 | 0 | 100 |
| Days Max Temp ≤ 32 °F | 0 | 0 | 0 | 0 | 0 | 0 | 0 | 0 | 0 | 0 | 0 | 0 | 0 |
| Days Min Temp ≤ 32 °F | 12 | 7 | 2 | 0 | 0 | 0 | 0 | 0 | 0 | 0 | 3 | 8 | 32 |
| Days Min Temp ≤ 0 °F | 0 | 0 | 0 | 0 | 0 | 0 | 0 | 0 | 0 | 0 | 0 | 0 | 0 |
| Heating Degree Days | 527 | 373 | 198 | 56 | 6 | 0 | 0 | 0 | 15 | 72 | 266 | 466 | 1979 |
| Cooling Degree Days | 0 | 5 | 34 | 131 | 297 | 447 | 482 | 437 | 287 | 98 | 18 | 0 | 2236 |
| Total Precipitation (") | 0.51 | 0.55 | 0.34 | 0.52 | 1.46 | 1.87 | 2.08 | 2.37 | 2.28 | 1.48 | 0.51 | 0.54 | 14.51 |
| Days ≥ 0.1" Precip | 2 | 2 | 1 | 1 | 3 | 4 | 4 | 5 | 4 | 2 | 1 | 1 | 30 |
| Total Snowfall (") | 0.9 | 0.1 | 0.1 | 0.1 | 0.0 | 0.0 | 0.0 | 0.0 | 0.0 | 0.0 | 0.2 | 0.1 | 1.5 |
| Days ≥ 1" Snow Depth | 0 | 0 | 0 | 0 | 0 | 0 | 0 | 0 | 0 | 0 | 0 | 0 | 0 |

## PARIS *Lamar County* ELEVATION 591 ft LAT/LONG 33° 39' N / 95° 35' W

| | JAN | FEB | MAR | APR | MAY | JUN | JUL | AUG | SEP | OCT | NOV | DEC | YEAR |
|---|---|---|---|---|---|---|---|---|---|---|---|---|---|
| Maximum Temp °F | 50.7 | 56.2 | 65.3 | 74.6 | 81.1 | 89.4 | 94.0 | 93.8 | 86.4 | 76.3 | 63.8 | 54.4 | 73.8 |
| Minimum Temp °F | 30.2 | 34.3 | 43.5 | 52.5 | 60.7 | 68.9 | 72.8 | 71.4 | 64.4 | 52.8 | 42.7 | 33.9 | 52.3 |
| Mean Temp °F | 40.5 | 45.3 | 54.4 | 63.5 | 70.9 | 79.2 | 83.4 | 82.6 | 75.4 | 64.6 | 53.3 | 44.2 | 63.1 |
| Days Max Temp ≥ 90 °F | 0 | 0 | 0 | 0 | 3 | 16 | 26 | 25 | 12 | 1 | 0 | 0 | 83 |
| Days Max Temp ≤ 32 °F | 3 | 1 | 0 | 0 | 0 | 0 | 0 | 0 | 0 | 0 | 0 | 1 | 5 |
| Days Min Temp ≤ 32 °F | 19 | 12 | 4 | 0 | 0 | 0 | 0 | 0 | 0 | 0 | 5 | 14 | 54 |
| Days Min Temp ≤ 0 °F | 0 | 0 | 0 | 0 | 0 | 0 | 0 | 0 | 0 | 0 | 0 | 0 | 0 |
| Heating Degree Days | 754 | 552 | 340 | 112 | 22 | 1 | 0 | 0 | 13 | 106 | 358 | 640 | 2898 |
| Cooling Degree Days | 0 | 2 | 19 | 73 | 209 | 440 | 583 | 578 | 347 | 94 | 13 | 1 | 2359 |
| Total Precipitation (") | 2.33 | 3.35 | 4.06 | 4.06 | 6.01 | 3.96 | 3.86 | 2.53 | 4.78 | 5.09 | 4.02 | 3.85 | 47.90 |
| Days ≥ 0.1" Precip | 5 | 5 | 6 | 6 | 7 | 6 | 5 | 4 | 5 | 5 | 5 | 6 | 65 |
| Total Snowfall (") | 0.9 | 1.3 | 0.1 | 0.0 | 0.0 | 0.0 | 0.0 | 0.0 | 0.0 | 0.0 | 0.2 | 0.5 | 3.0 |
| Days ≥ 1" Snow Depth | 1 | 1 | 0 | 0 | 0 | 0 | 0 | 0 | 0 | 0 | 0 | 0 | 2 |

## PEARSALL *Frio County* ELEVATION 630 ft LAT/LONG 28° 53' N / 99° 6' W

| | JAN | FEB | MAR | APR | MAY | JUN | JUL | AUG | SEP | OCT | NOV | DEC | YEAR |
|---|---|---|---|---|---|---|---|---|---|---|---|---|---|
| Maximum Temp °F | 64.6 | 68.8 | 78.1 | 84.0 | 88.6 | 94.1 | 97.7 | 97.5 | 92.2 | 84.6 | 74.8 | 67.0 | 82.7 |
| Minimum Temp °F | 38.9 | 41.9 | 49.9 | 57.4 | 63.9 | 68.8 | 70.2 | 69.8 | 66.4 | 57.7 | 48.2 | 40.9 | 56.2 |
| Mean Temp °F | 51.8 | 55.4 | 64.0 | 70.8 | 76.2 | 81.5 | 83.9 | 83.6 | ***79.3*** | 71.2 | 61.5 | 53.9 | 69.4 |
| Days Max Temp ≥ 90 °F | 0 | 0 | 3 | 8 | 14 | 25 | 29 | 29 | 22 | 9 | 1 | 0 | 140 |
| Days Max Temp ≤ 32 °F | 0 | 0 | 0 | 0 | 0 | 0 | 0 | 0 | 0 | 0 | 0 | 0 | 0 |
| Days Min Temp ≤ 32 °F | 8 | 4 | 1 | 0 | 0 | 0 | 0 | 0 | 0 | 0 | 2 | 6 | 21 |
| Days Min Temp ≤ 0 °F | 0 | 0 | 0 | 0 | 0 | 0 | 0 | 0 | 0 | 0 | 0 | 0 | 0 |
| Heating Degree Days | 416 | 279 | 117 | 29 | 2 | 0 | 0 | 0 | ***2*** | 29 | 169 | 348 | 1391 |
| Cooling Degree Days | 5 | 16 | 81 | 180 | 355 | ***500*** | ***588*** | ***590*** | ***434*** | 228 | 68 | ***9*** | 3054 |
| Total Precipitation (") | 1.53 | 1.59 | 1.11 | 2.29 | 3.55 | 3.34 | 1.67 | 2.47 | 3.16 | 3.28 | 1.41 | 1.36 | 26.76 |
| Days ≥ 0.1" Precip | 3 | 3 | 2 | 3 | 5 | 4 | 2 | 3 | 4 | 4 | 3 | 3 | 39 |
| Total Snowfall (") | 0.0 | 0.2 | 0.0 | 0.0 | 0.0 | 0.0 | 0.0 | 0.0 | 0.0 | 0.0 | 0.0 | 0.0 | 0.2 |
| Days ≥ 1" Snow Depth | 0 | 0 | 0 | 0 | 0 | 0 | 0 | 0 | 0 | 0 | 0 | 0 | 0 |

## PECOS *Reeves County* ELEVATION 2582 ft LAT/LONG 31° 26' N / 103° 30' W

| | JAN | FEB | MAR | APR | MAY | JUN | JUL | AUG | SEP | OCT | NOV | DEC | YEAR |
|---|---|---|---|---|---|---|---|---|---|---|---|---|---|
| Maximum Temp °F | 60.0 | 66.1 | 75.2 | 84.2 | 91.0 | 97.8 | 98.5 | 97.0 | 90.4 | 81.7 | 70.4 | 62.4 | 81.2 |
| Minimum Temp °F | 28.1 | 31.4 | 38.4 | 47.3 | 56.0 | 65.2 | 68.8 | 67.2 | 60.7 | 48.5 | 36.4 | 29.2 | 48.1 |
| Mean Temp °F | 44.1 | 48.7 | 56.8 | 65.8 | 73.5 | 81.5 | 83.7 | 82.1 | 75.5 | 65.1 | 53.4 | 45.8 | 64.7 |
| Days Max Temp ≥ 90 °F | 0 | 0 | 2 | 10 | 20 | 26 | 28 | 27 | 19 | 7 | 0 | 0 | 139 |
| Days Max Temp ≤ 32 °F | 1 | 0 | 0 | 0 | 0 | 0 | 0 | 0 | 0 | 0 | 0 | 1 | 2 |
| Days Min Temp ≤ 32 °F | 22 | 16 | 7 | 1 | 0 | 0 | 0 | 0 | 0 | 1 | 10 | 21 | 78 |
| Days Min Temp ≤ 0 °F | 0 | 0 | 0 | 0 | 0 | 0 | 0 | 0 | 0 | 0 | 0 | 0 | 0 |
| Heating Degree Days | 641 | 454 | 266 | 80 | 14 | 1 | 0 | 0 | 15 | 91 | 346 | 589 | 2497 |
| Cooling Degree Days | 0 | 2 | 16 | 100 | 284 | 501 | 584 | 546 | 337 | 101 | 9 | 0 | 2480 |
| Total Precipitation (") | 0.47 | 0.51 | 0.43 | 0.48 | 1.30 | 1.23 | 1.34 | 1.73 | 2.30 | 1.11 | 0.53 | 0.54 | 11.97 |
| Days ≥ 0.1" Precip | 1 | 1 | 1 | 1 | 3 | 2 | 3 | 3 | 4 | 2 | 1 | 2 | 24 |
| Total Snowfall (") | 1.8 | 0.8 | 0.1 | 0.1 | 0.0 | 0.0 | 0.0 | 0.0 | 0.0 | 0.1 | 0.8 | 0.6 | 4.3 |
| Days ≥ 1" Snow Depth | 1 | 0 | 0 | 0 | 0 | 0 | 0 | 0 | 0 | 0 | 0 | 0 | 1 |

## PENWELL *Ector County* ELEVATION 2943 ft LAT/LONG 31° 44 ' N / 102° 36 ' W

| | JAN | FEB | MAR | APR | MAY | JUN | JUL | AUG | SEP | OCT | NOV | DEC | YEAR |
|---|---|---|---|---|---|---|---|---|---|---|---|---|---|
| Maximum Temp °F | 57.6 | 63.0 | 72.0 | 80.1 | 87.3 | 93.4 | 95.0 | 94.0 | 87.0 | 78.6 | 68.0 | 59.5 | 78.0 |
| Minimum Temp °F | 28.6 | 32.8 | 40.5 | 49.5 | 57.9 | 66.2 | 69.2 | 67.8 | 61.6 | 50.7 | 38.9 | 30.6 | 49.5 |
| Mean Temp °F | 43.1 | 47.9 | 56.3 | 64.9 | 72.6 | 79.8 | 82.1 | 80.9 | 74.3 | 64.7 | 53.5 | 45.1 | 63.8 |
| Days Max Temp ≥ 90 °F | 0 | 0 | 1 | 4 | 13 | 22 | 27 | 25 | 13 | *3* | 0 | 0 | 108 |
| Days Max Temp ≤ 32 °F | 1 | 0 | 0 | 0 | 0 | 0 | 0 | 0 | 0 | 0 | 0 | 1 | 2 |
| Days Min Temp ≤ 32 °F | 22 | 13 | 5 | 1 | 0 | 0 | 0 | 0 | 0 | 1 | 7 | *17* | 66 |
| Days Min Temp ≤ 0 °F | 0 | 0 | 0 | 0 | 0 | 0 | 0 | 0 | 0 | 0 | 0 | 0 | 0 |
| Heating Degree Days | 669 | 477 | 279 | 94 | 17 | 1 | 0 | 0 | 17 | 98 | 345 | 609 | 2606 |
| Cooling Degree Days | 0 | 1 | 11 | 85 | 258 | 455 | 538 | 508 | 308 | 96 | 8 | 0 | 2268 |
| Total Precipitation (") | 0.41 | 0.65 | 0.48 | 0.78 | 2.08 | 1.51 | 1.38 | 1.43 | 2.49 | 1.26 | 0.66 | 0.57 | 13.70 |
| Days ≥ 0.1" Precip | 1 | 2 | 1 | 1 | 3 | 3 | 3 | 3 | 4 | 2 | 1 | 1 | 25 |
| Total Snowfall (") | 0.7 | 0.4 | 0.2 | 0.0 | 0.0 | 0.0 | 0.0 | 0.0 | 0.0 | 0.0 | 0.1 | 0.3 | 1.7 |
| Days ≥ 1" Snow Depth | 0 | 0 | 0 | 0 | 0 | 0 | 0 | 0 | 0 | 0 | 0 | 0 | 0 |

## PERRYTON *Ochiltree County* ELEVATION 2933 ft LAT/LONG 36° 28 ' N / 100° 47 ' W

| | JAN | FEB | MAR | APR | MAY | JUN | JUL | AUG | SEP | OCT | NOV | DEC | YEAR |
|---|---|---|---|---|---|---|---|---|---|---|---|---|---|
| Maximum Temp °F | 46.2 | 51.1 | 59.4 | 70.0 | 77.4 | 87.4 | 93.2 | 91.1 | 82.8 | 72.4 | 57.9 | 48.4 | 69.8 |
| Minimum Temp °F | 17.4 | 21.1 | 28.7 | 38.7 | 48.9 | 59.1 | 64.2 | 62.5 | 54.4 | 41.3 | 28.9 | 20.0 | 40.4 |
| Mean Temp °F | 31.8 | 36.1 | 44.1 | 54.4 | 63.1 | 73.2 | 78.7 | 76.8 | 68.6 | 56.9 | 43.4 | 34.1 | 55.1 |
| Days Max Temp ≥ 90 °F | 0 | 0 | 0 | 1 | 3 | 13 | 23 | 20 | 9 | 2 | 0 | 0 | 71 |
| Days Max Temp ≤ 32 °F | 6 | 4 | 1 | 0 | 0 | 0 | 0 | 0 | 0 | 0 | 1 | 4 | 16 |
| Days Min Temp ≤ 32 °F | 30 | 25 | 20 | 7 | 1 | 0 | 0 | 0 | 0 | 4 | 20 | 29 | 136 |
| Days Min Temp ≤ 0 °F | 2 | 1 | 0 | 0 | 0 | 0 | 0 | 0 | 0 | 0 | 0 | 1 | 4 |
| Heating Degree Days | 1021 | 808 | 643 | 325 | 124 | 15 | 1 | 3 | 61 | 273 | 642 | 951 | 4867 |
| Cooling Degree Days | 0 | 0 | 1 | 14 | 66 | 266 | 426 | 393 | 192 | 21 | 1 | 0 | 1380 |
| Total Precipitation (") | 0.44 | 0.65 | 1.40 | 1.45 | 3.25 | 3.08 | 2.26 | 2.60 | 1.62 | 1.23 | 1.01 | 0.56 | 19.55 |
| Days ≥ 0.1" Precip | 1 | 2 | 3 | 3 | 5 | 5 | 4 | 5 | 3 | 2 | 2 | 2 | 37 |
| Total Snowfall (") | 3.6 | 5.1 | 4.2 | 1.2 | 0.1 | 0.0 | 0.0 | 0.0 | 0.0 | 0.5 | 2.4 | 3.2 | 20.3 |
| Days ≥ 1" Snow Depth | 4 | 3 | 2 | 0 | 0 | 0 | 0 | 0 | 0 | 0 | 1 | 3 | 13 |

## PIERCE 1 E *Wharton County* ELEVATION 102 ft LAT/LONG 29° 14 ' N / 96° 11 ' W

| | JAN | FEB | MAR | APR | MAY | JUN | JUL | AUG | SEP | OCT | NOV | DEC | YEAR |
|---|---|---|---|---|---|---|---|---|---|---|---|---|---|
| Maximum Temp °F | *62.7* | 65.9 | 73.5 | 80.5 | 85.2 | 90.7 | 93.5 | 94.4 | 89.8 | 83.0 | *73.6* | *65.9* | 79.9 |
| Minimum Temp °F | *41.2* | 43.2 | 50.3 | 58.7 | 65.3 | 70.4 | 72.1 | 71.7 | 67.7 | 58.5 | *50.4* | *42.7* | 57.7 |
| Mean Temp °F | *52.0* | 54.6 | 61.9 | 69.6 | 75.3 | 80.6 | 82.8 | 83.1 | 78.8 | 70.8 | *62.0* | *54.3* | 68.8 |
| Days Max Temp ≥ 90 °F | 0 | 0 | 0 | 1 | 6 | 21 | 27 | 27 | 17 | 5 | 0 | 0 | 104 |
| Days Max Temp ≤ 32 °F | 0 | 0 | 0 | 0 | 0 | 0 | 0 | 0 | 0 | 0 | 0 | 0 | 0 |
| Days Min Temp ≤ 32 °F | 6 | 4 | 1 | 0 | 0 | 0 | 0 | 0 | 0 | 0 | 2 | 4 | 17 |
| Days Min Temp ≤ 0 °F | 0 | 0 | 0 | 0 | 0 | 0 | 0 | 0 | 0 | 0 | 0 | 0 | 0 |
| Heating Degree Days | *411* | 303 | 150 | 32 | 3 | 0 | 0 | 0 | 2 | 30 | *163* | *348* | 1442 |
| Cooling Degree Days | *10* | 17 | 66 | 179 | 363 | 500 | 590 | 606 | 446 | *243* | *94* | *27* | 3141 |
| Total Precipitation (") | 3.62 | 2.95 | 2.41 | 3.02 | 5.26 | 4.45 | 3.32 | 3.48 | 5.63 | 5.02 | 2.98 | 2.98 | 45.12 |
| Days ≥ 0.1" Precip | 6 | 5 | 4 | 4 | 5 | 5 | 5 | 6 | 7 | 5 | 4 | 5 | 61 |
| Total Snowfall (") | 0.0 | 0.0 | 0.0 | 0.0 | 0.0 | 0.0 | 0.0 | 0.0 | 0.0 | 0.0 | 0.0 | 0.0 | 0.0 |
| Days ≥ 1" Snow Depth | 0 | 0 | 0 | 0 | 0 | 0 | 0 | 0 | 0 | 0 | 0 | 0 | 0 |

## PILOT POINT *Denton County* ELEVATION 669 ft LAT/LONG 33° 24 ' N / 96° 57 ' W

| | JAN | FEB | MAR | APR | MAY | JUN | JUL | AUG | SEP | OCT | NOV | DEC | YEAR |
|---|---|---|---|---|---|---|---|---|---|---|---|---|---|
| Maximum Temp °F | 52.2 | 57.4 | 66.5 | 75.5 | 82.4 | 90.5 | 96.2 | 95.7 | 88.2 | 77.8 | 65.3 | 56.3 | 75.3 |
| Minimum Temp °F | 29.6 | 32.9 | 40.9 | 50.4 | 59.4 | 68.3 | 72.9 | 71.4 | 64.3 | 52.6 | 41.0 | 32.6 | 51.4 |
| Mean Temp °F | 40.9 | 45.1 | 53.6 | 63.0 | 70.9 | 79.4 | 84.5 | 83.6 | 76.3 | 65.2 | 53.2 | 44.5 | 63.4 |
| Days Max Temp ≥ 90 °F | 0 | 0 | 0 | 1 | 5 | 19 | 28 | 27 | 15 | 3 | 0 | 0 | 98 |
| Days Max Temp ≤ 32 °F | 2 | 1 | 0 | 0 | 0 | 0 | 0 | 0 | 0 | 0 | 0 | 1 | 4 |
| Days Min Temp ≤ 32 °F | 19 | 13 | 5 | 1 | 0 | 0 | 0 | 0 | 0 | 0 | 6 | 15 | 59 |
| Days Min Temp ≤ 0 °F | 0 | 0 | 0 | 0 | 0 | 0 | 0 | 0 | 0 | 0 | 0 | 0 | 0 |
| Heating Degree Days | 741 | 554 | 357 | 120 | 21 | 1 | 0 | 0 | 10 | 95 | 360 | 628 | 2887 |
| Cooling Degree Days | 0 | 0 | 12 | 64 | 217 | 457 | 630 | 618 | 392 | 119 | 14 | 0 | 2523 |
| Total Precipitation (") | 2.33 | 3.13 | 3.64 | 3.81 | 6.61 | 4.61 | 2.79 | 2.66 | 4.33 | 4.50 | 2.86 | 2.73 | 44.00 |
| Days ≥ 0.1" Precip | 5 | 5 | 6 | 6 | 8 | 6 | 4 | 4 | 5 | 5 | 5 | 5 | 64 |
| Total Snowfall (") | *0.6* | *0.4* | 0.2 | 0.0 | 0.0 | 0.0 | 0.0 | 0.0 | 0.0 | 0.0 | 0.1 | 0.1 | 1.4 |
| Days ≥ 1" Snow Depth | 0 | *0* | 0 | 0 | 0 | 0 | 0 | 0 | 0 | 0 | 0 | 0 | 0 |

## PLAINS *Yoakum County* ELEVATION 3504 ft LAT/LONG 33° 11 ' N / 102° 50 ' W

| | JAN | FEB | MAR | APR | MAY | JUN | JUL | AUG | SEP | OCT | NOV | DEC | YEAR |
|---|---|---|---|---|---|---|---|---|---|---|---|---|---|
| Maximum Temp °F | 52.8 | 58.6 | 66.5 | 75.4 | 82.7 | 90.1 | 91.3 | 89.6 | 83.6 | 75.4 | 63.6 | 55.8 | 73.8 |
| Minimum Temp °F | 23.2 | 26.3 | 33.2 | 42.1 | 50.6 | 59.6 | 63.1 | 61.8 | 55.4 | 43.5 | 32.1 | 25.0 | 43.0 |
| Mean Temp °F | 38.0 | 42.3 | 49.9 | 58.7 | 66.6 | 74.9 | 77.2 | 75.7 | 69.5 | 59.4 | 47.9 | 40.4 | 58.4 |
| Days Max Temp ≥ 90 °F | 0 | 0 | 0 | 1 | 7 | 17 | 20 | 17 | 8 | 1 | 0 | 0 | 71 |
| Days Max Temp ≤ 32 °F | 2 | 1 | 0 | 0 | 0 | 0 | 0 | 0 | 0 | 0 | 0 | 2 | 5 |
| Days Min Temp ≤ 32 °F | 28 | 22 | 14 | 3 | 0 | 0 | 0 | 0 | 0 | 2 | 15 | 26 | 110 |
| Days Min Temp ≤ 0 °F | 0 | 0 | 0 | 0 | 0 | 0 | 0 | 0 | 0 | 0 | 0 | 0 | 0 |
| Heating Degree Days | 831 | 639 | 462 | 207 | 59 | 4 | 0 | 1 | 36 | 190 | 508 | 756 | 3693 |
| Cooling Degree Days | *0* | *0* | 1 | *20* | 116 | 307 | 385 | 357 | 194 | 29 | 0 | 0 | 1409 |
| Total Precipitation (") | 0.48 | 0.70 | 0.63 | 1.02 | 2.34 | 2.30 | 2.57 | 2.84 | 2.51 | 1.28 | 0.73 | 0.76 | 18.16 |
| Days ≥ 0.1" Precip | 1 | 2 | 1 | 2 | 4 | 4 | 4 | 5 | 4 | 2 | 2 | 2 | 33 |
| Total Snowfall (") | na | 1.8 | 1.5 | 0.0 | 0.0 | 0.0 | 0.0 | 0.0 | 0.0 | 0.0 | 1.2 | 1.4 | na |
| Days ≥ 1" Snow Depth | na | *0* | 0 | 0 | 0 | 0 | 0 | 0 | 0 | 0 | 0 | *0* | na |

## PLAINVIEW *Hale County* ELEVATION 3373 ft LAT/LONG 34° 11 ' N / 101° 41 ' W

| | JAN | FEB | MAR | APR | MAY | JUN | JUL | AUG | SEP | OCT | NOV | DEC | YEAR |
|---|---|---|---|---|---|---|---|---|---|---|---|---|---|
| Maximum Temp °F | 50.2 | 55.4 | 64.2 | 73.2 | 80.7 | 88.7 | 91.5 | 89.0 | 82.3 | 73.6 | 60.9 | 52.7 | 71.9 |
| Minimum Temp °F | 24.7 | 28.3 | 35.7 | 44.6 | 53.8 | 63.0 | 66.7 | 64.9 | 57.9 | 46.4 | 34.9 | 26.9 | 45.7 |
| Mean Temp °F | 37.4 | 41.9 | 50.0 | 58.9 | 67.3 | 75.9 | 79.1 | 77.0 | 70.1 | 60.0 | 47.9 | 39.8 | 58.8 |
| Days Max Temp ≥ 90 °F | 0 | 0 | 0 | 1 | 5 | 15 | 21 | 17 | 7 | 1 | 0 | 0 | 67 |
| Days Max Temp ≤ 32 °F | 4 | 2 | 1 | 0 | 0 | 0 | 0 | 0 | 0 | 0 | 1 | 2 | 10 |
| Days Min Temp ≤ 32 °F | 26 | 20 | 10 | 2 | 0 | 0 | 0 | 0 | 0 | 1 | 12 | 24 | 95 |
| Days Min Temp ≤ 0 °F | 0 | 0 | 0 | 0 | 0 | 0 | 0 | 0 | 0 | 0 | 0 | 0 | 0 |
| Heating Degree Days | 847 | 655 | 462 | 208 | 59 | 5 | 0 | 1 | 36 | 186 | 506 | 773 | 3738 |
| Cooling Degree Days | 0 | 0 | 2 | 33 | 132 | 327 | 441 | 391 | 198 | 35 | 1 | 0 | 1560 |
| Total Precipitation (") | 0.52 | 0.63 | 0.85 | 1.32 | 2.96 | 3.39 | 2.48 | 2.31 | 2.33 | 1.61 | 0.75 | 0.62 | 19.77 |
| Days ≥ 0.1" Precip | 1 | 2 | 2 | 3 | 5 | 5 | 4 | 5 | 4 | 3 | 2 | 2 | 38 |
| Total Snowfall (") | 3.2 | 3.0 | 1.0 | 0.2 | 0.0 | 0.0 | 0.0 | 0.0 | 0.0 | 0.1 | 1.2 | 2.5 | 11.2 |
| Days ≥ 1" Snow Depth | 3 | 2 | 1 | 0 | 0 | 0 | 0 | 0 | 0 | 0 | 1 | 2 | 9 |

## POINT COMFORT *Calhoun County* ELEVATION 20 ft LAT/LONG 28° 40 ' N / 96° 33 ' W

| | JAN | FEB | MAR | APR | MAY | JUN | JUL | AUG | SEP | OCT | NOV | DEC | YEAR |
|---|---|---|---|---|---|---|---|---|---|---|---|---|---|
| Maximum Temp °F | 63.4 | 66.7 | 73.2 | 78.9 | 84.0 | 88.9 | 91.5 | 91.4 | 88.4 | 82.0 | 73.6 | 66.8 | 79.1 |
| Minimum Temp °F | 45.2 | 47.7 | 54.7 | 62.5 | 68.6 | 74.6 | 77.2 | 76.6 | 71.4 | 63.2 | 54.8 | 48.3 | 62.1 |
| Mean Temp °F | 54.3 | 57.3 | 64.0 | 70.7 | 76.3 | 81.8 | 84.4 | 84.1 | 80.1 | 72.8 | 64.3 | 57.6 | 70.6 |
| Days Max Temp ≥ 90 °F | 0 | 0 | 0 | 0 | 3 | 12 | 24 | 23 | 13 | 2 | 0 | 0 | 77 |
| Days Max Temp ≤ 32 °F | 0 | 0 | 0 | 0 | 0 | 0 | 0 | 0 | 0 | 0 | 0 | 0 | 0 |
| Days Min Temp ≤ 32 °F | 4 | 2 | 0 | 0 | 0 | 0 | 0 | 0 | 0 | 0 | 0 | 2 | 8 |
| Days Min Temp ≤ 0 °F | 0 | 0 | 0 | 0 | 0 | 0 | 0 | 0 | 0 | 0 | 0 | 0 | 0 |
| Heating Degree Days | 338 | 230 | 103 | 17 | 1 | 0 | 0 | 0 | 0 | 17 | 115 | 254 | 1075 |
| Cooling Degree Days | 11 | 21 | 82 | 204 | 378 | 544 | 635 | 626 | 481 | 293 | 117 | 36 | 3428 |
| Total Precipitation (") | 3.06 | 2.81 | 2.20 | 2.91 | 4.93 | 4.52 | 3.72 | 3.37 | 6.19 | 4.64 | 3.32 | 2.82 | 44.49 |
| Days ≥ 0.1" Precip | 6 | 4 | 3 | 3 | 5 | 5 | 4 | 5 | 7 | 5 | 4 | 5 | 56 |
| Total Snowfall (") | 0.0 | 0.0 | 0.0 | 0.0 | 0.0 | 0.0 | 0.0 | 0.0 | 0.0 | 0.0 | 0.0 | 0.0 | 0.0 |
| Days ≥ 1" Snow Depth | 0 | 0 | 0 | 0 | 0 | 0 | 0 | 0 | 0 | 0 | 0 | 0 | 0 |

## PORT ARTHUR JEFFERSN *Jefferson County* ELEVATION 30 ft LAT/LONG 29° 57 ' N / 94° 1 ' W

| | JAN | FEB | MAR | APR | MAY | JUN | JUL | AUG | SEP | OCT | NOV | DEC | YEAR |
|---|---|---|---|---|---|---|---|---|---|---|---|---|---|
| Maximum Temp °F | 60.8 | 64.5 | 71.5 | 78.3 | 84.0 | 89.5 | 91.7 | 91.5 | 87.9 | 80.6 | 71.4 | 64.4 | 78.0 |
| Minimum Temp °F | 42.3 | 44.7 | 51.7 | 59.7 | 66.5 | 72.4 | 74.0 | 73.5 | 69.7 | 59.6 | 51.1 | 45.2 | 59.2 |
| Mean Temp °F | 51.6 | 54.6 | 61.6 | 69.0 | 75.3 | 81.0 | 82.9 | 82.5 | 78.8 | 70.1 | 61.2 | 54.8 | 68.6 |
| Days Max Temp ≥ 90 °F | 0 | 0 | 0 | 0 | 2 | 17 | 24 | 23 | 12 | 2 | 0 | 0 | 80 |
| Days Max Temp ≤ 32 °F | 0 | 0 | 0 | 0 | 0 | 0 | 0 | 0 | 0 | 0 | 0 | 0 | 0 |
| Days Min Temp ≤ 32 °F | 5 | 3 | 1 | 0 | 0 | 0 | 0 | 0 | 0 | 0 | 1 | 4 | 14 |
| Days Min Temp ≤ 0 °F | 0 | 0 | 0 | 0 | 0 | 0 | 0 | 0 | 0 | 0 | 0 | 0 | 0 |
| Heating Degree Days | 420 | 299 | 151 | 33 | 1 | 0 | 0 | 0 | 1 | 33 | 177 | 333 | 1448 |
| Cooling Degree Days | 7 | 15 | 57 | 152 | 330 | 494 | 569 | 563 | 426 | 206 | 72 | 25 | 2916 |
| Total Precipitation (") | 5.46 | 3.53 | 3.53 | 4.00 | 6.37 | 5.98 | 5.44 | 5.32 | 5.75 | 4.90 | 4.39 | 5.15 | 59.82 |
| Days ≥ 0.1" Precip | 7 | 5 | 5 | 4 | 6 | 6 | 7 | 7 | 7 | 5 | 5 | 6 | 70 |
| Total Snowfall (") | 0.1 | 0.0 | 0.0 | 0.0 | 0.0 | 0.0 | 0.0 | 0.0 | 0.0 | 0.0 | 0.0 | 0.0 | 0.1 |
| Days ≥ 1" Snow Depth | 0 | 0 | 0 | 0 | 0 | 0 | 0 | 0 | 0 | 0 | 0 | 0 | 0 |

## PORT MANSFIELD *Willacy County* ELEVATION 10 ft LAT/LONG 26° 33' N / 97° 26' W

| | JAN | FEB | MAR | APR | MAY | JUN | JUL | AUG | SEP | OCT | NOV | DEC | YEAR |
|---|---|---|---|---|---|---|---|---|---|---|---|---|---|
| Maximum Temp °F | 63.9 | 66.9 | 72.8 | 78.2 | 82.4 | 86.7 | 88.4 | 88.4 | 86.3 | 81.2 | 74.4 | 67.5 | 78.1 |
| Minimum Temp °F | 48.1 | 51.4 | 59.6 | 66.8 | 72.0 | 76.1 | 77.0 | 76.3 | 73.0 | 66.7 | 59.1 | 51.6 | 64.8 |
| Mean Temp °F | 56.0 | 59.1 | 66.3 | 72.5 | 77.2 | 81.4 | 82.7 | 82.4 | 79.7 | 74.0 | 66.7 | 59.6 | 71.5 |
| Days Max Temp ≥ 90 °F | 0 | 0 | 1 | 1 | 2 | 3 | 9 | 11 | 5 | 1 | 0 | 0 | 33 |
| Days Max Temp ≤ 32 °F | 0 | 0 | 0 | 0 | 0 | 0 | 0 | 0 | 0 | 0 | 0 | 0 | 0 |
| Days Min Temp ≤ 32 °F | 2 | 1 | 0 | 0 | 0 | 0 | 0 | 0 | 0 | 0 | 0 | 1 | 4 |
| Days Min Temp ≤ 0 °F | 0 | 0 | 0 | 0 | 0 | 0 | 0 | 0 | 0 | 0 | 0 | 0 | 0 |
| Heating Degree Days | 298 | 194 | 71 | 11 | 0 | 0 | 0 | 0 | 0 | 11 | 85 | 215 | 885 |
| Cooling Degree Days | 19 | 40 | 121 | 226 | 384 | 509 | ***541*** | 539 | 430 | 301 | 147 | 59 | 3316 |
| Total Precipitation (") | 1.85 | 1.74 | 0.83 | 1.59 | 3.43 | 3.12 | 1.32 | 1.87 | 5.52 | 2.77 | 2.12 | 1.46 | 27.62 |
| Days ≥ 0.1" Precip | 4 | 3 | 2 | 2 | 4 | 4 | 2 | 3 | 6 | 4 | 3 | 3 | 40 |
| Total Snowfall (") | 0.0 | 0.0 | 0.0 | 0.0 | 0.0 | 0.0 | 0.0 | 0.0 | 0.0 | 0.0 | 0.0 | 0.0 | 0.0 |
| Days ≥ 1" Snow Depth | 0 | 0 | 0 | 0 | 0 | 0 | 0 | 0 | 0 | 0 | 0 | 0 | 0 |

## PORT O CONNOR *Calhoun County* ELEVATION 10 ft LAT/LONG 28° 27' N / 96° 24' W

| | JAN | FEB | MAR | APR | MAY | JUN | JUL | AUG | SEP | OCT | NOV | DEC | YEAR |
|---|---|---|---|---|---|---|---|---|---|---|---|---|---|
| Maximum Temp °F | ***61.1*** | ***63.3*** | ***70.1*** | ***76.6*** | ***82.5*** | ***87.7*** | ***90.2*** | ***90.5*** | ***87.3*** | na | ***73.0*** | ***65.8*** | na |
| Minimum Temp °F | ***45.0*** | ***47.4*** | ***55.0*** | ***63.7*** | ***70.7*** | ***76.0*** | ***78.3*** | na | ***73.8*** | na | ***56.5*** | ***48.8*** | na |
| Mean Temp °F | ***53.1*** | ***55.5*** | ***62.7*** | ***70.2*** | ***76.6*** | ***81.9*** | ***84.2*** | na | ***80.6*** | na | ***64.8*** | ***57.3*** | na |
| Days Max Temp ≥ 90 °F | ***0*** | ***0*** | ***0*** | ***0*** | ***0*** | ***7*** | ***21*** | ***22*** | ***10*** | ***1*** | ***0*** | ***0*** | 61 |
| Days Max Temp ≤ 32 °F | ***0*** | ***0*** | ***0*** | ***0*** | ***0*** | ***0*** | ***0*** | ***0*** | ***0*** | ***0*** | ***0*** | ***0*** | 0 |
| Days Min Temp ≤ 32 °F | ***3*** | ***2*** | ***0*** | ***0*** | ***0*** | ***0*** | ***0*** | ***0*** | ***0*** | ***0*** | ***0*** | ***1*** | 6 |
| Days Min Temp ≤ 0 °F | ***0*** | ***0*** | ***0*** | ***0*** | ***0*** | ***0*** | ***0*** | ***0*** | ***0*** | ***0*** | ***0*** | ***0*** | 0 |
| Heating Degree Days | ***373*** | ***271*** | ***119*** | ***20*** | ***1*** | ***0*** | ***0*** | ***0*** | ***0*** | na | ***111*** | ***260*** | na |
| Cooling Degree Days | na | na | na | na | na | na | na | na | na | na | na | na | na |
| Total Precipitation (") | ***3.35*** | ***2.68*** | ***1.82*** | ***1.88*** | ***4.48*** | ***3.42*** | ***3.65*** | ***3.75*** | 6.07 | ***4.86*** | ***2.40*** | ***2.44*** | 40.80 |
| Days ≥ 0.1" Precip | ***6*** | ***4*** | ***3*** | ***3*** | ***5*** | ***4*** | ***4*** | ***5*** | ***6*** | ***5*** | ***4*** | ***4*** | 53 |
| Total Snowfall (") | ***0.0*** | ***0.1*** | ***0.0*** | 0.0 | ***0.0*** | ***0.0*** | ***0.0*** | ***0.0*** | 0.0 | ***0.0*** | ***0.0*** | ***0.0*** | 0.1 |
| Days ≥ 1" Snow Depth | ***0*** | ***0*** | ***0*** | 0 | ***0*** | ***0*** | ***0*** | ***0*** | 0 | ***0*** | ***0*** | ***0*** | 0 |

## POST 3 ENE *Garza County* ELEVATION 2546 ft LAT/LONG 33° 12' N / 101° 20' W

| | JAN | FEB | MAR | APR | MAY | JUN | JUL | AUG | SEP | OCT | NOV | DEC | YEAR |
|---|---|---|---|---|---|---|---|---|---|---|---|---|---|
| Maximum Temp °F | 53.8 | 58.9 | 67.7 | 76.9 | 83.4 | 90.7 | 93.8 | 92.0 | 84.8 | 76.5 | 64.8 | 56.5 | 75.0 |
| Minimum Temp °F | 27.2 | 30.6 | 37.8 | 47.1 | 56.0 | 64.9 | 68.9 | 67.3 | 60.3 | 49.4 | 37.8 | 29.8 | 48.1 |
| Mean Temp °F | 40.5 | 44.7 | 52.8 | 62.0 | 69.7 | 77.8 | 81.4 | 79.7 | 72.6 | 63.0 | 51.3 | 43.2 | 61.6 |
| Days Max Temp ≥ 90 °F | 0 | 0 | 0 | 3 | 8 | 18 | 25 | 21 | 10 | 2 | 0 | 0 | 87 |
| Days Max Temp ≤ 32 °F | 2 | 1 | 0 | 0 | 0 | 0 | 0 | 0 | 0 | 0 | 0 | 2 | 5 |
| Days Min Temp ≤ 32 °F | 22 | 17 | 8 | 1 | 0 | 0 | 0 | 0 | 0 | 1 | 8 | 19 | 76 |
| Days Min Temp ≤ 0 °F | 0 | 0 | 0 | 0 | 0 | 0 | 0 | 0 | 0 | 0 | 0 | 0 | 0 |
| Heating Degree Days | 752 | 567 | 385 | 147 | 37 | 2 | 0 | 1 | 25 | 133 | 410 | 671 | 3130 |
| Cooling Degree Days | 0 | 1 | 11 | 63 | 184 | 385 | 519 | ***492*** | 278 | 77 | 5 | 0 | 2015 |
| Total Precipitation (") | 0.70 | 0.91 | 0.96 | 1.37 | 2.93 | 2.71 | 2.05 | 2.96 | 2.95 | 2.08 | 0.93 | 0.79 | 21.34 |
| Days ≥ 0.1" Precip | 2 | 2 | 2 | 3 | 5 | 4 | 4 | 4 | 4 | 3 | 2 | 2 | 37 |
| Total Snowfall (") | 1.8 | 1.8 | 1.2 | 0.1 | 0.0 | 0.0 | 0.0 | 0.0 | 0.0 | 0.1 | 1.0 | 1.0 | 7.0 |
| Days ≥ 1" Snow Depth | ***0*** | ***0*** | 0 | 0 | 0 | 0 | 0 | 0 | 0 | 0 | 0 | 0 | 0 |

## POTEET *Atascosa County* ELEVATION 620 ft LAT/LONG 29° 3' N / 98° 34' W

| | JAN | FEB | MAR | APR | MAY | JUN | JUL | AUG | SEP | OCT | NOV | DEC | YEAR |
|---|---|---|---|---|---|---|---|---|---|---|---|---|---|
| Maximum Temp °F | 62.9 | 67.0 | 75.8 | 82.4 | 86.5 | 92.8 | 95.4 | 96.2 | 91.5 | 83.5 | 74.1 | 66.3 | 81.2 |
| Minimum Temp °F | 38.9 | 41.5 | 49.9 | ***57.8*** | 65.0 | 70.5 | 72.5 | 72.1 | 68.3 | 58.7 | 49.3 | 41.3 | 57.2 |
| Mean Temp °F | 50.9 | 54.3 | 62.8 | ***70.1*** | 75.8 | 81.7 | ***84.0*** | ***84.2*** | 79.9 | 71.1 | 61.7 | 53.9 | 69.2 |
| Days Max Temp ≥ 90 °F | 0 | 0 | 1 | 5 | 9 | 23 | 28 | 29 | 20 | 7 | 0 | 0 | 122 |
| Days Max Temp ≤ 32 °F | 0 | 0 | 0 | 0 | 0 | 0 | 0 | 0 | 0 | 0 | 0 | 0 | 0 |
| Days Min Temp ≤ 32 °F | 8 | 5 | 1 | 0 | 0 | 0 | 0 | 0 | 0 | 0 | 2 | 6 | 22 |
| Days Min Temp ≤ 0 °F | 0 | 0 | 0 | 0 | 0 | 0 | 0 | 0 | 0 | 0 | 0 | 0 | 0 |
| Heating Degree Days | 438 | 308 | 135 | 32 | 2 | 0 | ***0*** | ***0*** | 1 | 30 | 164 | 356 | 1466 |
| Cooling Degree Days | 6 | 11 | 69 | ***173*** | ***353*** | 518 | ***611*** | 619 | 467 | 239 | 76 | ***18*** | 3160 |
| Total Precipitation (") | 1.61 | 2.01 | 1.42 | 2.71 | 4.54 | 3.41 | 1.80 | 2.49 | 3.81 | 3.04 | 1.69 | 1.74 | 30.27 |
| Days ≥ 0.1" Precip | 4 | 3 | 3 | 4 | 6 | 4 | 3 | 3 | 5 | 4 | 3 | 3 | 45 |
| Total Snowfall (") | 0.0 | 0.1 | 0.0 | 0.0 | 0.0 | 0.0 | 0.0 | 0.0 | 0.0 | 0.0 | 0.0 | 0.0 | 0.1 |
| Days ≥ 1" Snow Depth | 0 | 0 | 0 | 0 | 0 | 0 | 0 | 0 | 0 | 0 | 0 | 0 | 0 |

## PRESIDIO *Presidio County* ELEVATION 2589 ft LAT/LONG 29° 33 ' N / 104° 24 ' W

| | JAN | FEB | MAR | APR | MAY | JUN | JUL | AUG | SEP | OCT | NOV | DEC | YEAR |
|---|---|---|---|---|---|---|---|---|---|---|---|---|---|
| Maximum Temp °F | 68.2 | 74.0 | 82.6 | 90.6 | 96.9 | 102.3 | 100.7 | 98.8 | 94.3 | 87.3 | 77.0 | 68.9 | 86.8 |
| Minimum Temp °F | 34.2 | 38.2 | 44.9 | 53.6 | 62.0 | 70.7 | 73.0 | 71.5 | 65.9 | 54.8 | 42.3 | 35.1 | 53.9 |
| Mean Temp °F | 51.2 | 56.1 | 63.8 | 72.1 | 79.5 | 86.5 | 86.9 | 85.2 | 80.1 | 71.1 | 59.7 | 52.0 | 70.4 |
| Days Max Temp ≥ 90 °F | 0 | 0 | 6 | 20 | 28 | 29 | 30 | 29 | 24 | 12 | 1 | 0 | 179 |
| Days Max Temp ≤ 32 °F | 0 | 0 | 0 | 0 | 0 | 0 | 0 | 0 | 0 | 0 | 0 | 0 | 0 |
| Days Min Temp ≤ 32 °F | 14 | 6 | 2 | 0 | 0 | 0 | 0 | 0 | 0 | 0 | 4 | *13* | 39 |
| Days Min Temp ≤ 0 °F | 0 | 0 | 0 | 0 | 0 | 0 | 0 | 0 | 0 | 0 | 0 | 0 | 0 |
| Heating Degree Days | 422 | 252 | 100 | 16 | 1 | 0 | 0 | 0 | 3 | 20 | 180 | 397 | 1391 |
| Cooling Degree Days | 0 | 10 | 75 | 245 | 483 | 675 | 702 | 663 | 496 | 242 | 36 | 1 | 3628 |
| Total Precipitation (") | 0.34 | 0.46 | 0.21 | 0.34 | 0.68 | 1.54 | 2.10 | 1.88 | 1.98 | 0.96 | 0.38 | 0.53 | 11.40 |
| Days ≥ 0.1" Precip | 1 | 1 | 0 | 1 | 2 | 3 | 4 | 4 | 3 | 2 | 1 | 1 | 23 |
| Total Snowfall (") | 0.0 | 0.0 | 0.0 | 0.0 | 0.0 | 0.0 | 0.0 | 0.0 | 0.0 | 0.0 | 0.1 | 0.0 | 0.1 |
| Days ≥ 1" Snow Depth | 0 | 0 | 0 | 0 | 0 | 0 | 0 | 0 | 0 | 0 | 0 | 0 | 0 |

## PROCTOR RESERVOIR *Comanche County* ELEVATION 1220 ft LAT/LONG 31° 58 ' N / 98° 30 ' W

| | JAN | FEB | MAR | APR | MAY | JUN | JUL | AUG | SEP | OCT | NOV | DEC | YEAR |
|---|---|---|---|---|---|---|---|---|---|---|---|---|---|
| Maximum Temp °F | 55.7 | 60.1 | 69.0 | 77.4 | 83.0 | 90.6 | 94.9 | 94.4 | 87.2 | 78.6 | 67.2 | 59.1 | 76.4 |
| Minimum Temp °F | 30.6 | 34.6 | 43.4 | 52.9 | 60.5 | 67.8 | 70.7 | 69.6 | 63.3 | 52.4 | 42.3 | 33.9 | 51.8 |
| Mean Temp °F | 43.1 | 47.4 | 56.2 | 65.2 | 71.8 | 79.2 | 82.8 | 82.0 | 75.3 | 65.6 | 54.8 | 46.5 | 64.2 |
| Days Max Temp ≥ 90 °F | 0 | 0 | 1 | 2 | 6 | 19 | 27 | 26 | 14 | 3 | 0 | 0 | 98 |
| Days Max Temp ≤ 32 °F | 2 | 1 | 0 | 0 | 0 | 0 | 0 | 0 | 0 | 0 | 0 | 1 | 4 |
| Days Min Temp ≤ 32 °F | 19 | 12 | 4 | 0 | 0 | 0 | 0 | 0 | 0 | 0 | 5 | 14 | 54 |
| Days Min Temp ≤ 0 °F | 0 | 0 | 0 | 0 | 0 | 0 | 0 | 0 | 0 | 0 | 0 | 0 | 0 |
| Heating Degree Days | 671 | 494 | 292 | 90 | 17 | 1 | 0 | 0 | 13 | 88 | 321 | 567 | 2554 |
| Cooling Degree Days | 0 | 3 | 26 | 100 | 238 | 433 | 556 | 557 | 335 | 112 | 23 | 1 | 2384 |
| Total Precipitation (") | 1.55 | 2.05 | 2.04 | 2.91 | 4.93 | 3.30 | 1.76 | 2.07 | 3.54 | 3.14 | 1.89 | 1.74 | 30.92 |
| Days ≥ 0.1" Precip | 3 | 4 | 4 | 4 | 6 | 4 | 3 | 3 | 5 | 4 | 4 | 3 | 47 |
| Total Snowfall (") | 0.1 | 0.0 | 0.0 | 0.0 | 0.0 | 0.0 | 0.0 | 0.0 | 0.0 | 0.0 | 0.2 | 0.0 | 0.3 |
| Days ≥ 1" Snow Depth | 0 | 0 | 0 | 0 | 0 | 0 | 0 | 0 | 0 | 0 | 0 | 0 | 0 |

## PUTNAM *Callahan County* ELEVATION 1591 ft LAT/LONG 32° 22 ' N / 99° 11 ' W

| | JAN | FEB | MAR | APR | MAY | JUN | JUL | AUG | SEP | OCT | NOV | DEC | YEAR |
|---|---|---|---|---|---|---|---|---|---|---|---|---|---|
| Maximum Temp °F | 56.5 | 61.2 | 70.5 | 78.9 | 84.5 | 91.4 | 95.6 | 94.9 | 87.3 | 78.7 | 66.9 | 58.8 | 77.1 |
| Minimum Temp °F | 32.7 | 36.7 | 44.3 | 53.0 | 60.2 | 67.7 | 71.4 | 70.4 | 63.8 | 54.4 | 43.7 | 35.4 | 52.8 |
| Mean Temp °F | 44.6 | 49.0 | 57.4 | 66.0 | 72.4 | 79.6 | 83.5 | 82.7 | 75.6 | 66.6 | 55.3 | 47.2 | 65.0 |
| Days Max Temp ≥ 90 °F | 0 | 0 | 1 | 3 | 8 | 20 | 27 | 26 | 13 | 3 | 0 | 0 | 101 |
| Days Max Temp ≤ 32 °F | 2 | 1 | 0 | 0 | 0 | 0 | 0 | 0 | 0 | 0 | 0 | 1 | 4 |
| Days Min Temp ≤ 32 °F | 15 | 10 | 4 | 0 | 0 | 0 | 0 | 0 | 0 | 0 | 5 | 12 | 46 |
| Days Min Temp ≤ 0 °F | 0 | 0 | 0 | 0 | 0 | 0 | 0 | 0 | 0 | 0 | 0 | 0 | 0 |
| Heating Degree Days | 625 | 451 | 264 | 80 | 15 | 1 | 0 | 0 | 12 | 79 | 306 | 549 | 2382 |
| Cooling Degree Days | 0 | 6 | 36 | 121 | 263 | 447 | 581 | 587 | 350 | 144 | 26 | 2 | 2563 |
| Total Precipitation (") | 1.36 | 1.60 | 1.63 | 2.11 | 3.30 | 2.87 | 1.77 | 2.13 | 3.07 | 3.02 | 1.54 | 1.47 | 25.87 |
| Days ≥ 0.1" Precip | 3 | 3 | 3 | 4 | 5 | 4 | 3 | 3 | 4 | 4 | 3 | 3 | 42 |
| Total Snowfall (") | 1.9 | 1.4 | 1.1 | 0.0 | 0.0 | 0.0 | 0.0 | 0.0 | 0.0 | 0.0 | 0.9 | 0.8 | 6.1 |
| Days ≥ 1" Snow Depth | 1 | 0 | 0 | 0 | 0 | 0 | 0 | 0 | 0 | 0 | 0 | 1 | 2 |

## QUANAH 5 SE *Hardeman County* ELEVATION 1503 ft LAT/LONG 34° 15 ' N / 99° 41 ' W

| | JAN | FEB | MAR | APR | MAY | JUN | JUL | AUG | SEP | OCT | NOV | DEC | YEAR |
|---|---|---|---|---|---|---|---|---|---|---|---|---|---|
| Maximum Temp °F | 51.2 | 56.2 | 65.5 | 75.6 | 82.6 | 91.6 | 96.4 | 94.3 | 85.9 | 76.0 | 63.3 | 54.5 | 74.4 |
| Minimum Temp °F | 24.4 | 28.8 | 37.7 | 47.8 | 57.2 | 67.0 | 71.4 | 69.4 | 61.2 | 48.3 | 36.5 | 27.5 | 48.1 |
| Mean Temp °F | 37.8 | 42.6 | 51.6 | 61.7 | 69.9 | 79.3 | 83.9 | 81.9 | 73.6 | 62.1 | 49.9 | 41.0 | 61.3 |
| Days Max Temp ≥ 90 °F | 0 | 0 | 1 | 2 | 7 | 19 | 27 | 24 | 12 | 3 | 0 | 0 | 95 |
| Days Max Temp ≤ 32 °F | 3 | 2 | 0 | 0 | 0 | 0 | 0 | 0 | 0 | 0 | 0 | 1 | 6 |
| Days Min Temp ≤ 32 °F | 26 | 18 | 9 | 1 | 0 | 0 | 0 | 0 | 0 | 1 | 11 | 23 | 89 |
| Days Min Temp ≤ 0 °F | 0 | 0 | 0 | 0 | 0 | 0 | 0 | 0 | 0 | 0 | 0 | 0 | 0 |
| Heating Degree Days | 836 | 628 | 420 | 154 | 36 | 1 | 0 | 1 | 25 | 154 | 450 | 736 | 3441 |
| Cooling Degree Days | 0 | 0 | 11 | 57 | 194 | 429 | 582 | 541 | 299 | 69 | 6 | 0 | 2188 |
| Total Precipitation (") | 0.96 | 1.03 | 1.38 | 1.88 | 3.75 | 3.23 | 2.45 | 2.60 | 3.58 | 2.47 | 1.23 | 1.07 | 25.63 |
| Days ≥ 0.1" Precip | 2 | 3 | 3 | 4 | 5 | 5 | 3 | 4 | 4 | 3 | 2 | 2 | 40 |
| Total Snowfall (") | 2.1 | 2.3 | 0.7 | 0.0 | 0.0 | 0.0 | 0.0 | 0.0 | 0.0 | 0.0 | 0.6 | 1.1 | 6.8 |
| Days ≥ 1" Snow Depth | 2 | 2 | 0 | 0 | 0 | 0 | 0 | 0 | 0 | 0 | 0 | 1 | 5 |

## RAYMONDVILLE *Willacy County* ELEVATION 30 ft LAT/LONG 26° 29 ' N / 97° 47 ' W

| | JAN | FEB | MAR | APR | MAY | JUN | JUL | AUG | SEP | OCT | NOV | DEC | YEAR |
|---|---|---|---|---|---|---|---|---|---|---|---|---|---|
| Maximum Temp °F | 68.5 | 72.1 | 79.7 | 85.9 | 89.2 | 93.4 | 96.2 | 96.7 | 92.3 | 86.7 | 78.8 | 72.2 | 84.3 |
| Minimum Temp °F | 46.8 | 49.2 | 56.0 | 63.5 | 69.1 | 72.6 | 73.6 | 73.5 | 70.8 | 63.3 | 55.7 | 49.6 | 62.0 |
| Mean Temp °F | 57.6 | 60.7 | 67.9 | 74.7 | 79.1 | 83.0 | 84.9 | 85.1 | 81.6 | 75.1 | 67.2 | 60.9 | 73.2 |
| Days Max Temp ≥ 90 °F | 0 | 1 | 3 | 9 | 16 | 25 | 29 | 28 | 22 | 12 | 3 | 0 | 148 |
| Days Max Temp ≤ 32 °F | 0 | 0 | 0 | 0 | 0 | 0 | 0 | 0 | 0 | 0 | 0 | 0 | 0 |
| Days Min Temp ≤ 32 °F | 2 | 1 | 0 | 0 | 0 | 0 | 0 | 0 | 0 | 0 | 0 | 2 | 5 |
| Days Min Temp ≤ 0 °F | 0 | 0 | 0 | 0 | 0 | 0 | 0 | 0 | 0 | 0 | 0 | 0 | 0 |
| Heating Degree Days | 269 | 174 | 70 | 11 | 1 | 0 | 0 | 0 | 1 | 10 | 87 | 194 | 817 |
| Cooling Degree Days | 41 | 67 | 166 | 298 | 456 | 573 | 648 | 653 | 516 | 352 | 171 | 81 | 4022 |
| Total Precipitation (") | 1.62 | 1.67 | 1.05 | 1.56 | 3.47 | 3.43 | 1.89 | 3.15 | 5.82 | 2.45 | 1.22 | 1.15 | 28.48 |
| Days ≥ 0.1" Precip | 3 | 3 | 2 | 2 | 4 | 5 | 4 | 5 | 7 | 4 | 3 | 2 | 44 |
| Total Snowfall (") | 0.0 | 0.0 | 0.0 | 0.0 | 0.0 | 0.0 | 0.0 | 0.0 | 0.0 | 0.0 | 0.0 | 0.0 | 0.0 |
| Days ≥ 1" Snow Depth | 0 | 0 | 0 | 0 | 0 | 0 | 0 | 0 | 0 | 0 | 0 | 0 | 0 |

## RED BLUFF DAM *Reeves County* ELEVATION 2800 ft LAT/LONG 31° 54 ' N / 103° 55 ' W

| | JAN | FEB | MAR | APR | MAY | JUN | JUL | AUG | SEP | OCT | NOV | DEC | YEAR |
|---|---|---|---|---|---|---|---|---|---|---|---|---|---|
| Maximum Temp °F | 61.5 | 66.5 | 75.2 | 83.7 | 90.3 | ***97.7*** | 97.8 | 95.9 | 89.9 | 81.5 | 70.7 | ***63.1*** | 81.1 |
| Minimum Temp °F | 29.3 | 32.7 | 39.5 | 48.1 | 56.7 | ***65.6*** | 69.0 | 67.5 | 61.0 | 49.3 | 38.1 | ***30.6*** | 48.9 |
| Mean Temp °F | 45.4 | 49.6 | 57.4 | 65.9 | 73.5 | ***81.7*** | 83.5 | 81.7 | 75.5 | 65.4 | 54.4 | ***46.9*** | 65.1 |
| Days Max Temp ≥ 90 °F | 0 | 0 | 1 | 8 | 18 | ***26*** | 28 | 27 | 17 | 6 | 0 | 0 | 131 |
| Days Max Temp ≤ 32 °F | 0 | 0 | 0 | 0 | 0 | 0 | 0 | 0 | 0 | 0 | 0 | 0 | 0 |
| Days Min Temp ≤ 32 °F | 21 | 14 | 6 | 1 | 0 | 0 | 0 | 0 | 0 | 1 | 8 | ***19*** | 70 |
| Days Min Temp ≤ 0 °F | 0 | 0 | 0 | 0 | 0 | 0 | 0 | 0 | 0 | 0 | 0 | 0 | 0 |
| Heating Degree Days | 600 | 429 | 246 | 68 | 8 | ***0*** | 0 | 0 | 8 | 78 | 317 | ***556*** | 2310 |
| Cooling Degree Days | ***0*** | ***3*** | ***20*** | 116 | ***305*** | ***542*** | 594 | 552 | 343 | 108 | 11 | ***1*** | 2595 |
| Total Precipitation (") | 0.43 | 0.40 | 0.29 | 0.41 | 0.94 | ***2.05*** | 1.72 | 1.49 | 2.26 | 0.90 | 0.54 | 0.52 | 11.95 |
| Days ≥ 0.1" Precip | 1 | 1 | 1 | 1 | 2 | ***3*** | 3 | 3 | 3 | 2 | 1 | 2 | 23 |
| Total Snowfall (") | na | ***0.6*** | 0.2 | 0.0 | 0.0 | 0.0 | 0.0 | 0.0 | 0.0 | 0.0 | ***0.5*** | 0.0 | na |
| Days ≥ 1" Snow Depth | na | 0 | 0 | 0 | 0 | 0 | 0 | 0 | 0 | 0 | ***0*** | 0 | na |

## RED ROCK *Bastrop County* ELEVATION 502 ft LAT/LONG 29° 58 ' N / 97° 27 ' W

| | JAN | FEB | MAR | APR | MAY | JUN | JUL | AUG | SEP | OCT | NOV | DEC | YEAR |
|---|---|---|---|---|---|---|---|---|---|---|---|---|---|
| Maximum Temp °F | 59.1 | 63.7 | 72.0 | 78.8 | 83.7 | 90.9 | 95.3 | 96.2 | 90.0 | 81.7 | 71.0 | 62.8 | 78.8 |
| Minimum Temp °F | 35.8 | 39.3 | 47.8 | 55.4 | 62.5 | 68.8 | 71.1 | 70.1 | 64.9 | 55.3 | 46.1 | 38.7 | 54.7 |
| Mean Temp °F | 47.4 | 51.5 | 59.9 | 67.1 | 73.1 | 79.8 | 83.2 | 83.2 | 77.5 | 68.5 | 58.6 | 50.8 | 66.7 |
| Days Max Temp ≥ 90 °F | 0 | 0 | 1 | 1 | 5 | 20 | 29 | 29 | 18 | 5 | 0 | 0 | 108 |
| Days Max Temp ≤ 32 °F | 0 | 0 | 0 | 0 | 0 | 0 | 0 | 0 | 0 | 0 | 0 | 0 | 0 |
| Days Min Temp ≤ 32 °F | 12 | 8 | 3 | 0 | 0 | 0 | 0 | 0 | 0 | 0 | 4 | 9 | 36 |
| Days Min Temp ≤ 0 °F | 0 | 0 | 0 | 0 | 0 | 0 | 0 | 0 | 0 | 0 | 0 | 0 | 0 |
| Heating Degree Days | 543 | 384 | 200 | 63 | 9 | 0 | 0 | 0 | 5 | 59 | 234 | 445 | 1942 |
| Cooling Degree Days | 3 | 8 | 44 | 118 | 257 | 449 | 569 | 583 | 386 | 171 | 48 | 10 | 2646 |
| Total Precipitation (") | 2.68 | 2.60 | 2.16 | 3.50 | 5.25 | 3.63 | 1.56 | 2.22 | 3.78 | 3.91 | 3.28 | 2.74 | 37.31 |
| Days ≥ 0.1" Precip | 5 | 5 | 4 | 4 | 6 | 5 | 3 | 3 | 5 | 5 | 5 | 5 | 55 |
| Total Snowfall (") | 0.4 | 0.0 | 0.0 | 0.0 | 0.0 | 0.0 | 0.0 | 0.0 | 0.0 | 0.0 | 0.0 | 0.0 | 0.4 |
| Days ≥ 1" Snow Depth | 0 | 0 | 0 | 0 | 0 | 0 | 0 | 0 | 0 | 0 | 0 | 0 | 0 |

## RIO GRANDE CITY 3 W *Starr County* ELEVATION 180 ft LAT/LONG 26° 23 ' N / 98° 52 ' W

| | JAN | FEB | MAR | APR | MAY | JUN | JUL | AUG | SEP | OCT | NOV | DEC | YEAR |
|---|---|---|---|---|---|---|---|---|---|---|---|---|---|
| Maximum Temp °F | 67.7 | 72.5 | 81.7 | 88.1 | 90.8 | 95.8 | 98.2 | 98.7 | 93.5 | 86.4 | 78.0 | 70.7 | 85.2 |
| Minimum Temp °F | 43.9 | 46.8 | 54.2 | 62.9 | 68.7 | 72.7 | 73.8 | 73.5 | 70.2 | 61.9 | 53.0 | 46.3 | 60.7 |
| Mean Temp °F | 55.8 | 59.6 | 68.0 | 75.5 | 79.8 | 84.3 | 86.1 | 86.1 | 81.8 | 74.1 | 65.5 | 58.5 | 72.9 |
| Days Max Temp ≥ 90 °F | 1 | 1 | 7 | 14 | 20 | 27 | 30 | 30 | 23 | 12 | 3 | 1 | 169 |
| Days Max Temp ≤ 32 °F | 0 | 0 | 0 | 0 | 0 | 0 | 0 | 0 | 0 | 0 | 0 | 0 | 0 |
| Days Min Temp ≤ 32 °F | 4 | 3 | 1 | 0 | 0 | 0 | 0 | 0 | 0 | 0 | 2 | 3 | 13 |
| Days Min Temp ≤ 0 °F | 0 | 0 | 0 | 0 | 0 | 0 | 0 | 0 | 0 | 0 | 0 | 0 | 0 |
| Heating Degree Days | 310 | 196 | 76 | 12 | 1 | 0 | 0 | 0 | 1 | 16 | 111 | 244 | 967 |
| Cooling Degree Days | 24 | 60 | 176 | 321 | 473 | 602 | 681 | 683 | 538 | 335 | 153 | 57 | 4103 |
| Total Precipitation (") | 1.13 | 1.20 | 0.54 | 1.51 | 2.77 | 3.10 | 1.40 | 2.25 | 4.86 | 2.12 | 0.93 | 0.94 | 22.75 |
| Days ≥ 0.1" Precip | 3 | 3 | 1 | 2 | 4 | 4 | 2 | 3 | 5 | 3 | 2 | 2 | 34 |
| Total Snowfall (") | 0.1 | 0.0 | 0.0 | 0.0 | 0.0 | 0.0 | 0.0 | 0.0 | 0.0 | 0.0 | 0.0 | 0.0 | 0.1 |
| Days ≥ 1" Snow Depth | 0 | 0 | 0 | 0 | 0 | 0 | 0 | 0 | 0 | 0 | 0 | 0 | 0 |

## RISING STAR *Eastland County* ELEVATION 1631 ft LAT/LONG 32° 6 ' N / 98° 58 ' W

| | JAN | FEB | MAR | APR | MAY | JUN | JUL | AUG | SEP | OCT | NOV | DEC | YEAR |
|---|---|---|---|---|---|---|---|---|---|---|---|---|---|
| Maximum Temp °F | 54.3 | 58.9 | 68.1 | 77.0 | 82.4 | 89.3 | 93.9 | 93.6 | 86.1 | 77.5 | 66.2 | 57.6 | 75.4 |
| Minimum Temp °F | 29.6 | 33.4 | 41.2 | 50.5 | 58.7 | 66.5 | 69.6 | 68.6 | 62.2 | 51.4 | 40.6 | 32.5 | 50.4 |
| Mean Temp °F | 42.0 | 46.2 | 54.7 | 63.8 | 70.6 | 77.9 | 81.8 | 81.1 | 74.2 | 64.5 | 53.4 | 45.1 | 62.9 |
| Days Max Temp ≥ 90 °F | 0 | 0 | 1 | 2 | 5 | 16 | 25 | 25 | 11 | 2 | 0 | 0 | 87 |
| Days Max Temp ≤ 32 °F | 2 | 1 | 0 | 0 | 0 | 0 | 0 | 0 | 0 | 0 | 0 | 1 | 4 |
| Days Min Temp ≤ 32 °F | 19 | 13 | 6 | 1 | 0 | 0 | 0 | 0 | 0 | 1 | 7 | 16 | 63 |
| Days Min Temp ≤ 0 °F | 0 | 0 | 0 | 0 | 0 | 0 | 0 | 0 | 0 | 0 | 0 | 0 | 0 |
| Heating Degree Days | 707 | 523 | 330 | 113 | 23 | 1 | 0 | 0 | 15 | 108 | 355 | 612 | 2787 |
| Cooling Degree Days | 0 | 1 | 16 | 77 | 202 | 386 | 512 | 522 | 310 | 102 | 17 | 1 | 2146 |
| Total Precipitation (") | 1.54 | 1.87 | 2.10 | 2.68 | 4.54 | 3.95 | 1.90 | 2.14 | 3.13 | 3.39 | 1.55 | 1.65 | 30.44 |
| Days ≥ 0.1" Precip | 3 | 4 | 4 | 4 | 5 | 5 | 3 | 4 | 5 | 4 | 3 | 3 | 47 |
| Total Snowfall (") | 2.0 | 1.5 | 0.8 | 0.0 | 0.0 | 0.0 | 0.0 | 0.0 | 0.0 | 0.0 | 0.8 | 1.0 | 6.1 |
| Days ≥ 1" Snow Depth | 1 | 0 | 0 | 0 | 0 | 0 | 0 | 0 | 0 | 0 | 0 | 0 | 1 |

## ROBERT LEE *Coke County* ELEVATION 1781 ft LAT/LONG 31° 54 ' N / 100° 29 ' W

| | JAN | FEB | MAR | APR | MAY | JUN | JUL | AUG | SEP | OCT | NOV | DEC | YEAR |
|---|---|---|---|---|---|---|---|---|---|---|---|---|---|
| Maximum Temp °F | 57.0 | 61.9 | 71.1 | 80.3 | 86.1 | 92.9 | 96.3 | 95.5 | 88.0 | 79.4 | 68.1 | 59.7 | 78.0 |
| Minimum Temp °F | 29.4 | 33.0 | 41.1 | 50.1 | 59.1 | 67.4 | 70.7 | 69.2 | 62.5 | 51.3 | 39.9 | 31.8 | 50.5 |
| Mean Temp °F | 43.2 | 47.5 | 56.1 | 65.2 | 72.6 | 80.2 | 83.5 | 82.4 | 75.3 | 65.4 | 54.0 | 45.8 | 64.3 |
| Days Max Temp ≥ 90 °F | 0 | 0 | 1 | 5 | 11 | 22 | 28 | 26 | 15 | 4 | 0 | 0 | 112 |
| Days Max Temp ≤ 32 °F | 1 | 1 | 0 | 0 | 0 | 0 | 0 | 0 | 0 | 0 | 0 | 1 | 3 |
| Days Min Temp ≤ 32 °F | 21 | 14 | 6 | 1 | 0 | 0 | 0 | 0 | 0 | 1 | 7 | 17 | 67 |
| Days Min Temp ≤ 0 °F | 0 | 0 | 0 | 0 | 0 | 0 | 0 | 0 | 0 | 0 | 0 | 0 | 0 |
| Heating Degree Days | 668 | 489 | 295 | 94 | 16 | 1 | 0 | 0 | 13 | 91 | 338 | 589 | 2594 |
| Cooling Degree Days | 0 | 2 | 24 | 110 | 268 | 464 | 588 | 577 | 342 | 119 | 19 | 1 | 2514 |
| Total Precipitation (") | 0.88 | 1.30 | 1.14 | 1.83 | 3.53 | 2.89 | 1.49 | 2.07 | 3.77 | 2.65 | 1.11 | 1.08 | 23.74 |
| Days ≥ 0.1" Precip | 2 | 3 | 2 | 3 | 5 | 4 | 3 | 4 | 5 | 4 | 2 | 2 | 39 |
| Total Snowfall (") | 0.1 | 0.0 | 0.3 | 0.0 | 0.0 | 0.0 | 0.0 | 0.0 | 0.0 | 0.0 | 0.1 | 0.0 | 0.5 |
| Days ≥ 1" Snow Depth | 0 | 0 | 0 | 0 | 0 | 0 | 0 | 0 | 0 | 0 | 0 | 0 | 0 |

## ROBSTOWN *Nueces County* ELEVATION 79 ft LAT/LONG 27° 47 ' N / 97° 40 ' W

| | JAN | FEB | MAR | APR | MAY | JUN | JUL | AUG | SEP | OCT | NOV | DEC | YEAR |
|---|---|---|---|---|---|---|---|---|---|---|---|---|---|
| Maximum Temp °F | 64.8 | 68.6 | 76.4 | 82.7 | 86.4 | 91.9 | 94.8 | 95.1 | 91.1 | 84.5 | 76.4 | 68.8 | 81.8 |
| Minimum Temp °F | 44.3 | 46.9 | 54.6 | 62.7 | 68.6 | 73.1 | 74.6 | 74.7 | 71.2 | 62.9 | 54.5 | 47.6 | 61.3 |
| Mean Temp °F | 54.6 | 57.8 | 65.6 | 72.7 | 77.5 | 82.5 | 84.7 | 84.9 | 81.2 | 73.7 | 65.5 | 58.2 | 71.6 |
| Days Max Temp ≥ 90 °F | 0 | 0 | 1 | 4 | 9 | 23 | 29 | 29 | 19 | 7 | 1 | 0 | 122 |
| Days Max Temp ≤ 32 °F | 0 | 0 | 0 | 0 | 0 | 0 | 0 | 0 | 0 | 0 | 0 | 0 | 0 |
| Days Min Temp ≤ 32 °F | 4 | 2 | 0 | 0 | 0 | 0 | 0 | 0 | 0 | 0 | 0 | 2 | 8 |
| Days Min Temp ≤ 0 °F | 0 | 0 | 0 | 0 | 0 | 0 | 0 | 0 | 0 | 0 | 0 | 0 | 0 |
| Heating Degree Days | 342 | 225 | 90 | 14 | 1 | 0 | 0 | 0 | 0 | 12 | 103 | 247 | 1034 |
| Cooling Degree Days | 18 | 30 | 115 | 245 | ***404*** | 549 | 621 | 643 | 507 | 307 | 132 | 49 | 3620 |
| Total Precipitation (") | 2.11 | 2.17 | 1.49 | 2.02 | 4.16 | 3.58 | 2.53 | 3.29 | 5.51 | 3.18 | 1.75 | 1.67 | 33.46 |
| Days ≥ 0.1" Precip | 5 | 4 | 2 | 3 | 5 | 5 | 4 | 4 | 6 | 4 | 3 | 3 | 48 |
| Total Snowfall (") | 0.0 | 0.0 | 0.0 | 0.0 | 0.0 | 0.0 | 0.0 | 0.0 | 0.0 | 0.0 | 0.0 | 0.0 | 0.0 |
| Days ≥ 1" Snow Depth | 0 | 0 | 0 | 0 | 0 | 0 | 0 | 0 | 0 | 0 | 0 | 0 | 0 |

## ROCKPORT *Aransas County* ELEVATION 7 ft LAT/LONG 28° 1 ' N / 97° 3 ' W

| | JAN | FEB | MAR | APR | MAY | JUN | JUL | AUG | SEP | OCT | NOV | DEC | YEAR |
|---|---|---|---|---|---|---|---|---|---|---|---|---|---|
| Maximum Temp °F | 63.4 | 66.6 | 72.9 | 78.1 | 83.0 | 87.9 | 90.1 | 90.5 | 88.2 | 82.2 | 74.1 | 66.9 | 78.7 |
| Minimum Temp °F | 46.2 | 49.3 | 56.8 | 64.2 | 70.7 | 76.4 | 77.5 | 77.1 | 73.2 | 65.3 | 56.8 | 49.2 | 63.6 |
| Mean Temp °F | 54.8 | 58.0 | 64.9 | 71.1 | 76.8 | 82.2 | 83.8 | 83.8 | 80.7 | 73.6 | 65.5 | 58.1 | 71.1 |
| Days Max Temp ≥ 90 °F | 0 | 0 | 0 | 0 | 1 | 9 | 20 | 23 | 13 | 1 | 0 | 0 | 67 |
| Days Max Temp ≤ 32 °F | 0 | 0 | 0 | 0 | 0 | 0 | 0 | 0 | 0 | 0 | 0 | 0 | 0 |
| Days Min Temp ≤ 32 °F | 3 | 2 | 0 | 0 | 0 | 0 | 0 | 0 | 0 | 0 | 0 | 2 | 7 |
| Days Min Temp ≤ 0 °F | 0 | 0 | 0 | 0 | 0 | 0 | 0 | 0 | 0 | 0 | 0 | 0 | 0 |
| Heating Degree Days | 320 | 212 | 84 | 15 | 0 | 0 | 0 | 0 | 0 | 13 | 100 | 241 | 985 |
| Cooling Degree Days | 13 | 25 | 96 | 205 | 397 | 551 | 614 | 616 | 505 | 310 | 132 | 37 | 3501 |
| Total Precipitation (") | 2.52 | 2.20 | 1.80 | 2.08 | 4.15 | 3.82 | 2.64 | 3.40 | 6.04 | 3.86 | 2.03 | 2.17 | 36.71 |
| Days ≥ 0.1" Precip | 5 | 4 | 2 | 3 | 4 | 4 | 4 | 5 | 7 | 5 | 4 | 4 | 51 |
| Total Snowfall (") | 0.0 | 0.0 | 0.0 | 0.0 | 0.0 | 0.0 | 0.0 | 0.0 | 0.0 | 0.0 | 0.0 | 0.0 | 0.0 |
| Days ≥ 1" Snow Depth | 0 | 0 | 0 | 0 | 0 | 0 | 0 | 0 | 0 | 0 | 0 | 0 | 0 |

## ROCKSPRINGS *Edwards County* ELEVATION 2411 ft LAT/LONG 30° 1' N / 100° 13' W

| | JAN | FEB | MAR | APR | MAY | JUN | JUL | AUG | SEP | OCT | NOV | DEC | YEAR |
|---|---|---|---|---|---|---|---|---|---|---|---|---|---|
| Maximum Temp °F | 59.8 | 64.4 | 72.1 | 79.2 | 83.6 | 89.1 | 92.2 | 91.5 | 86.3 | *78.8* | *68.6* | *62.9* | 77.4 |
| Minimum Temp °F | 35.5 | 38.6 | 46.2 | 53.8 | 60.2 | 66.1 | 68.1 | 67.4 | *63.2* | *55.3* | *44.7* | *37.7* | 53.1 |
| Mean Temp °F | 47.7 | 51.5 | 59.2 | 66.5 | 72.0 | 77.6 | 80.2 | 79.5 | *74.8* | *67.1* | *56.6* | *50.3* | 65.3 |
| Days Max Temp ≥ 90 °F | 0 | 0 | 1 | 2 | 5 | 13 | 24 | *22* | 9 | 1 | *0* | 0 | 77 |
| Days Max Temp ≤ 32 °F | 0 | 0 | 0 | 0 | 0 | 0 | 0 | 0 | 0 | 0 | *0* | 0 | 0 |
| Days Min Temp ≤ 32 °F | 12 | 7 | 3 | 0 | 0 | 0 | 0 | 0 | 0 | 0 | *4* | *9* | 35 |
| Days Min Temp ≤ 0 °F | 0 | 0 | 0 | 0 | 0 | 0 | 0 | 0 | 0 | 0 | *0* | 0 | 0 |
| Heating Degree Days | 533 | 378 | 207 | 56 | 10 | 0 | 0 | 0 | *8* | *56* | *264* | *448* | 1960 |
| Cooling Degree Days | *1* | *5* | *32* | *109* | *247* | *404* | *480* | *482* | na | *140* | na | na | na |
| Total Precipitation (") | 1.01 | 1.38 | 1.13 | 2.08 | 3.16 | 2.48 | 2.17 | 2.99 | 2.63 | 2.73 | *1.26* | *0.98* | 24.00 |
| Days ≥ 0.1" Precip | 2 | 2 | 3 | 3 | 4 | 3 | 3 | *4* | 4 | 3 | *2* | 2 | 35 |
| Total Snowfall (") | *0.2* | 0.7 | 0.0 | 0.0 | 0.0 | 0.0 | 0.0 | 0.0 | 0.0 | 0.0 | *0.0* | 0.0 | 0.9 |
| Days ≥ 1" Snow Depth | *0* | 0 | 0 | 0 | 0 | 0 | 0 | 0 | 0 | 0 | *0* | 0 | 0 |

## ROSCOE *Nolan County* ELEVATION 2382 ft LAT/LONG 32° 27' N / 100° 32' W

| | JAN | FEB | MAR | APR | MAY | JUN | JUL | AUG | SEP | OCT | NOV | DEC | YEAR |
|---|---|---|---|---|---|---|---|---|---|---|---|---|---|
| Maximum Temp °F | 55.4 | 60.8 | 70.2 | 79.0 | 84.6 | 91.0 | 93.7 | 92.1 | 85.0 | 77.1 | 66.0 | 57.9 | 76.1 |
| Minimum Temp °F | 30.1 | 33.6 | 41.0 | 50.0 | 58.3 | 66.4 | 69.5 | 68.2 | 61.9 | 51.9 | 41.0 | 32.4 | 50.4 |
| Mean Temp °F | 42.8 | 47.2 | 55.6 | 64.5 | 71.5 | 78.7 | 81.7 | 80.2 | 73.5 | 64.5 | 53.5 | 45.2 | 63.2 |
| Days Max Temp ≥ 90 °F | 0 | 0 | 1 | 3 | 8 | 19 | 26 | 22 | 9 | 2 | 0 | 0 | 90 |
| Days Max Temp ≤ 32 °F | 2 | 1 | 0 | 0 | 0 | 0 | 0 | 0 | 0 | 0 | 0 | 1 | 4 |
| Days Min Temp ≤ 32 °F | 19 | 13 | 6 | 1 | 0 | 0 | 0 | 0 | 0 | 0 | 7 | 16 | 62 |
| Days Min Temp ≤ 0 °F | 0 | 0 | 0 | 0 | 0 | 0 | 0 | 0 | 0 | 0 | 0 | 0 | 0 |
| Heating Degree Days | 682 | 498 | 306 | 99 | 22 | 1 | 0 | 0 | 17 | 102 | 351 | 608 | 2686 |
| Cooling Degree Days | 0 | 2 | 20 | 87 | 231 | 412 | 524 | 496 | 285 | 97 | 15 | 0 | 2169 |
| Total Precipitation (") | 1.12 | 1.27 | 1.17 | 1.73 | 3.22 | 2.90 | 1.99 | 2.48 | 4.14 | 2.56 | 1.08 | 1.08 | 24.74 |
| Days ≥ 0.1" Precip | 3 | 3 | 3 | 3 | 5 | 4 | 3 | 4 | 5 | 4 | 2 | 2 | 41 |
| Total Snowfall (") | 2.3 | 1.3 | 0.8 | 0.0 | 0.0 | 0.0 | 0.0 | 0.0 | 0.0 | 0.0 | 1.2 | 0.6 | 6.2 |
| Days ≥ 1" Snow Depth | *1* | 0 | 0 | 0 | 0 | 0 | 0 | 0 | 0 | 0 | 0 | 0 | 1 |

## ROTAN *Fisher County* ELEVATION 1923 ft LAT/LONG 32° 52' N / 100° 28' W

| | JAN | FEB | MAR | APR | MAY | JUN | JUL | AUG | SEP | OCT | NOV | DEC | YEAR |
|---|---|---|---|---|---|---|---|---|---|---|---|---|---|
| Maximum Temp °F | 56.8 | 61.9 | 71.3 | 80.4 | 86.5 | 93.4 | 96.3 | 94.7 | 87.5 | 78.5 | 66.8 | 58.7 | 77.7 |
| Minimum Temp °F | 30.1 | 33.8 | 41.2 | 50.2 | 59.3 | 67.0 | 70.5 | 68.8 | 62.3 | 51.3 | 40.6 | 32.6 | 50.6 |
| Mean Temp °F | 43.5 | 47.9 | 56.3 | 65.3 | 72.9 | 80.2 | 83.4 | 81.8 | 74.9 | 64.9 | 53.7 | 45.6 | 64.2 |
| Days Max Temp ≥ 90 °F | 0 | 0 | 1 | 5 | 12 | 23 | 28 | 25 | 14 | 3 | 0 | 0 | 111 |
| Days Max Temp ≤ 32 °F | 2 | 1 | 0 | 0 | 0 | 0 | 0 | 0 | 0 | 0 | 0 | 1 | 4 |
| Days Min Temp ≤ 32 °F | 19 | 13 | 6 | 1 | 0 | 0 | 0 | 0 | 0 | 0 | 6 | 15 | 60 |
| Days Min Temp ≤ 0 °F | 0 | 0 | 0 | 0 | 0 | 0 | 0 | 0 | 0 | 0 | 0 | 0 | 0 |
| Heating Degree Days | 661 | 480 | 292 | 87 | 15 | 0 | 0 | 0 | 12 | 97 | 344 | 595 | 2583 |
| Cooling Degree Days | 0 | 2 | *20* | 95 | 265 | 448 | 564 | 543 | 332 | 102 | 13 | 0 | 2384 |
| Total Precipitation (") | 0.84 | 1.24 | 1.13 | 1.81 | 3.89 | 2.58 | 1.97 | 2.86 | 3.80 | 2.49 | 1.09 | 1.11 | 24.81 |
| Days ≥ 0.1" Precip | 2 | 3 | 2 | 3 | 5 | 4 | 3 | 4 | 4 | 4 | 2 | 3 | 39 |
| Total Snowfall (") | *1.5* | *1.1* | *0.1* | 0.0 | 0.0 | 0.0 | 0.0 | 0.0 | 0.0 | 0.0 | 0.4 | 0.8 | 3.9 |
| Days ≥ 1" Snow Depth | na | *0* | *0* | 0 | 0 | 0 | 0 | 0 | 0 | 0 | *0* | 0 | na |

## RUSK *Cherokee County* ELEVATION 489 ft LAT/LONG 31° 48' N / 95° 9' W

| | JAN | FEB | MAR | APR | MAY | JUN | JUL | AUG | SEP | OCT | NOV | DEC | YEAR |
|---|---|---|---|---|---|---|---|---|---|---|---|---|---|
| Maximum Temp °F | 54.8 | 59.8 | 67.7 | 75.9 | 81.4 | 88.5 | 92.6 | 93.1 | 86.9 | 78.0 | 67.1 | 58.9 | 75.4 |
| Minimum Temp °F | 35.2 | 38.5 | 46.4 | 54.5 | 61.5 | 68.0 | 70.7 | 70.0 | 64.9 | 54.8 | 45.8 | 38.3 | 54.0 |
| Mean Temp °F | 45.0 | 49.2 | 57.1 | 65.2 | 71.5 | 78.3 | 81.7 | 81.6 | 76.0 | 66.4 | 56.5 | 48.6 | 64.8 |
| Days Max Temp ≥ 90 °F | 0 | 0 | 0 | 0 | 2 | 14 | 24 | 25 | 12 | 1 | 0 | 0 | 78 |
| Days Max Temp ≤ 32 °F | 1 | 1 | 0 | 0 | 0 | 0 | 0 | 0 | 0 | 0 | 0 | 1 | 3 |
| Days Min Temp ≤ 32 °F | 12 | 8 | 2 | 0 | 0 | 0 | 0 | 0 | 0 | 0 | 2 | 8 | 32 |
| Days Min Temp ≤ 0 °F | 0 | 0 | 0 | 0 | 0 | 0 | 0 | 0 | 0 | 0 | 0 | 0 | 0 |
| Heating Degree Days | 615 | 445 | 267 | 79 | 11 | 0 | 0 | 0 | 6 | 75 | 275 | 505 | 2278 |
| Cooling Degree Days | 1 | 4 | 28 | 90 | 221 | 410 | 529 | 544 | 357 | 122 | 25 | 3 | 2334 |
| Total Precipitation (") | 4.01 | 3.65 | 4.27 | 4.29 | 5.28 | 3.95 | 2.79 | 2.10 | 4.23 | 4.87 | 4.19 | 4.48 | 48.11 |
| Days ≥ 0.1" Precip | 7 | 5 | 6 | 5 | 7 | 5 | 4 | 4 | 5 | 5 | 6 | 7 | 66 |
| Total Snowfall (") | 0.3 | 0.1 | 0.0 | 0.0 | 0.0 | 0.0 | 0.0 | 0.0 | 0.0 | 0.0 | 0.0 | 0.0 | 0.4 |
| Days ≥ 1" Snow Depth | 0 | 0 | 0 | 0 | 0 | 0 | 0 | 0 | 0 | 0 | 0 | 0 | 0 |

## SAN ANGELO MATHIS FD *Tom Green County* ELEVATION 1903 ft LAT/LONG 31° 22 ' N / 100° 30 ' W

| | JAN | FEB | MAR | APR | MAY | JUN | JUL | AUG | SEP | OCT | NOV | DEC | YEAR |
|---|---|---|---|---|---|---|---|---|---|---|---|---|---|
| Maximum Temp °F | 57.8 | 62.9 | 71.4 | 79.9 | 85.6 | 91.9 | 95.1 | 93.9 | 86.6 | 78.4 | 67.8 | 60.4 | 77.6 |
| Minimum Temp °F | 32.2 | 35.9 | 43.7 | 52.3 | 60.6 | 68.2 | 70.9 | 70.0 | 63.6 | 53.2 | 42.1 | 34.3 | 52.3 |
| Mean Temp °F | 45.0 | 49.4 | 57.5 | 66.2 | 73.1 | 80.1 | 83.0 | 82.0 | 75.1 | 65.9 | 55.0 | 47.4 | 65.0 |
| Days Max Temp ≥ 90 °F | 0 | 0 | 1 | 5 | 10 | 21 | 27 | 26 | 12 | 3 | 0 | 0 | 105 |
| Days Max Temp ≤ 32 °F | 1 | 0 | 0 | 0 | 0 | 0 | 0 | 0 | 0 | 0 | 0 | 1 | 2 |
| Days Min Temp ≤ 32 °F | 17 | 10 | 4 | 1 | 0 | 0 | 0 | 0 | 0 | 0 | 6 | 14 | 52 |
| Days Min Temp ≤ 0 °F | 0 | 0 | 0 | 0 | 0 | 0 | 0 | 0 | 0 | 0 | 0 | 0 | 0 |
| Heating Degree Days | 614 | 437 | 257 | 77 | 14 | 0 | 0 | 0 | 11 | 83 | 314 | 540 | 2347 |
| Cooling Degree Days | 0 | 3 | 31 | 110 | 267 | 445 | 552 | 542 | 321 | 119 | 21 | 1 | 2412 |
| Total Precipitation (") | 0.86 | 1.14 | 0.93 | 1.61 | 3.09 | 2.38 | 1.12 | 2.02 | 3.46 | 2.62 | 1.02 | 0.98 | 21.23 |
| Days ≥ 0.1" Precip | 2 | 3 | 2 | 3 | 5 | 4 | 2 | 3 | 4 | 4 | 2 | 2 | 36 |
| Total Snowfall (") | 1.9 | 0.6 | 0.2 | 0.0 | 0.0 | 0.0 | 0.0 | 0.0 | 0.0 | 0.0 | 0.5 | 0.3 | 3.5 |
| Days ≥ 1" Snow Depth | 1 | 0 | 0 | 0 | 0 | 0 | 0 | 0 | 0 | 0 | 0 | 0 | 1 |

## SAN ANTONIO INTL AP *Bexar County* ELEVATION 791 ft LAT/LONG 29° 32 ' N / 98° 28 ' W

| | JAN | FEB | MAR | APR | MAY | JUN | JUL | AUG | SEP | OCT | NOV | DEC | YEAR |
|---|---|---|---|---|---|---|---|---|---|---|---|---|---|
| Maximum Temp °F | 61.1 | 65.7 | 73.7 | 80.3 | 85.2 | 91.2 | 94.3 | 94.5 | 89.4 | 81.8 | 71.7 | 64.4 | 79.4 |
| Minimum Temp °F | 38.9 | 42.1 | 50.0 | 58.5 | 65.6 | 72.1 | 74.5 | 73.9 | 69.0 | 59.1 | 48.6 | 41.6 | 57.8 |
| Mean Temp °F | 50.0 | 53.9 | 61.9 | 69.5 | 75.4 | 81.7 | 84.5 | 84.2 | 79.2 | 70.5 | 60.2 | 53.0 | 68.7 |
| Days Max Temp ≥ 90 °F | 0 | 0 | 1 | 2 | 6 | 21 | 28 | 28 | 17 | 4 | 0 | 0 | 107 |
| Days Max Temp ≤ 32 °F | 0 | 0 | 0 | 0 | 0 | 0 | 0 | 0 | 0 | 0 | 0 | 0 | 0 |
| Days Min Temp ≤ 32 °F | 9 | 6 | 2 | 0 | 0 | 0 | 0 | 0 | 0 | 0 | 2 | 6 | 25 |
| Days Min Temp ≤ 0 °F | 0 | 0 | 0 | 0 | 0 | 0 | 0 | 0 | 0 | 0 | 0 | 0 | 0 |
| Heating Degree Days | 466 | 321 | 157 | 36 | 2 | 0 | 0 | 0 | 2 | 35 | 197 | 379 | 1595 |
| Cooling Degree Days | 6 | 17 | 65 | 174 | 354 | 528 | 628 | 640 | 460 | 235 | 67 | 17 | 3191 |
| Total Precipitation (") | 2.00 | 1.98 | 1.89 | 2.66 | 5.11 | 3.76 | 2.05 | 2.56 | 3.36 | 3.33 | 2.33 | 2.00 | 33.03 |
| Days ≥ 0.1" Precip | 4 | 4 | 4 | 4 | 6 | 5 | 3 | 4 | 5 | 5 | 3 | 3 | 50 |
| Total Snowfall (") | 0.6 | 0.2 | 0.0 | 0.0 | 0.0 | 0.0 | 0.0 | 0.0 | 0.0 | 0.0 | 0.0 | 0.0 | 0.8 |
| Days ≥ 1" Snow Depth | 0 | 0 | 0 | 0 | 0 | 0 | 0 | 0 | 0 | 0 | 0 | 0 | 0 |

## SAN MARCOS *Hays County* ELEVATION 612 ft LAT/LONG 29° 51 ' N / 97° 57 ' W

| | JAN | FEB | MAR | APR | MAY | JUN | JUL | AUG | SEP | OCT | NOV | DEC | YEAR |
|---|---|---|---|---|---|---|---|---|---|---|---|---|---|
| Maximum Temp °F | ***60.0*** | 64.5 | 73.0 | 79.8 | 84.8 | 91.1 | 94.7 | 95.1 | 89.7 | 82.0 | 71.4 | 63.5 | 79.1 |
| Minimum Temp °F | 36.8 | 39.3 | 47.0 | 55.6 | 63.3 | 69.8 | 71.8 | 71.3 | 66.4 | 56.2 | 46.4 | 38.9 | 55.2 |
| Mean Temp °F | ***48.5*** | 51.9 | 60.0 | 67.7 | 74.1 | 80.5 | 83.2 | 83.2 | 78.1 | 69.2 | 58.9 | 51.2 | 67.2 |
| Days Max Temp ≥ 90 °F | 0 | 0 | 1 | 2 | 7 | 21 | 28 | 28 | 19 | 4 | 0 | 0 | 110 |
| Days Max Temp ≤ 32 °F | 0 | 0 | 0 | 0 | 0 | 0 | 0 | 0 | 0 | 0 | 0 | 0 | 0 |
| Days Min Temp ≤ 32 °F | 10 | 7 | 3 | 0 | 0 | 0 | 0 | 0 | 0 | 0 | 3 | 8 | 31 |
| Days Min Temp ≤ 0 °F | 0 | 0 | 0 | 0 | 0 | 0 | 0 | 0 | 0 | 0 | 0 | 0 | 0 |
| Heating Degree Days | ***509*** | 371 | 191 | 49 | 5 | 0 | 0 | 0 | 3 | 43 | 221 | 428 | 1820 |
| Cooling Degree Days | ***3*** | 9 | 48 | 140 | 317 | 502 | 602 | 614 | 430 | 203 | 55 | 10 | 2933 |
| Total Precipitation (") | 2.46 | 2.52 | 2.05 | 2.88 | 5.78 | 4.22 | 2.05 | 2.30 | 3.53 | 3.23 | 3.04 | 2.66 | 36.72 |
| Days ≥ 0.1" Precip | ***5*** | 4 | 4 | 4 | 7 | 5 | 3 | 3 | 5 | 4 | 4 | 4 | 52 |
| Total Snowfall (") | 0.0 | 0.1 | 0.2 | 0.0 | 0.0 | 0.0 | 0.0 | 0.0 | 0.0 | 0.0 | 0.0 | 0.0 | 0.3 |
| Days ≥ 1" Snow Depth | 0 | 0 | 0 | 0 | 0 | 0 | 0 | 0 | 0 | 0 | 0 | 0 | 0 |

## SAN SABA *San Saba County* ELEVATION 1201 ft LAT/LONG 31° 11 ' N / 98° 44 ' W

| | JAN | FEB | MAR | APR | MAY | JUN | JUL | AUG | SEP | OCT | NOV | DEC | YEAR |
|---|---|---|---|---|---|---|---|---|---|---|---|---|---|
| Maximum Temp °F | 57.8 | 61.9 | 70.6 | 78.9 | 84.1 | 91.3 | 95.7 | 95.1 | 88.5 | 80.0 | 68.9 | 61.0 | 77.8 |
| Minimum Temp °F | 32.7 | 36.7 | 45.0 | 53.7 | 61.4 | 68.1 | 70.7 | 69.4 | 64.3 | 53.7 | 43.6 | 35.2 | 52.9 |
| Mean Temp °F | 45.3 | 49.3 | 57.8 | 66.3 | 72.8 | 79.7 | 83.2 | 82.3 | 76.4 | 66.8 | 56.2 | 48.1 | 65.4 |
| Days Max Temp ≥ 90 °F | 0 | 0 | 1 | 3 | 7 | 20 | 28 | 27 | 16 | 4 | 0 | 0 | 106 |
| Days Max Temp ≤ 32 °F | 1 | 1 | 0 | 0 | 0 | 0 | 0 | 0 | 0 | 0 | 0 | 1 | 3 |
| Days Min Temp ≤ 32 °F | 17 | 10 | 4 | 1 | 0 | 0 | 0 | 0 | 0 | 0 | 5 | 13 | 50 |
| Days Min Temp ≤ 0 °F | 0 | 0 | 0 | 0 | 0 | 0 | 0 | 0 | 0 | 0 | 0 | 0 | 0 |
| Heating Degree Days | 605 | 441 | 253 | 76 | 13 | 1 | 0 | 0 | 9 | 76 | 283 | 519 | 2276 |
| Cooling Degree Days | 0 | 6 | 39 | 124 | 279 | 460 | 576 | 568 | 372 | 158 | 34 | 3 | 2619 |
| Total Precipitation (") | 1.26 | 1.80 | 1.79 | 2.33 | 4.10 | 2.74 | 1.63 | 2.33 | 2.77 | 3.22 | 1.86 | 1.65 | 27.48 |
| Days ≥ 0.1" Precip | 3 | 4 | 3 | 4 | 6 | 4 | 2 | 3 | 4 | 4 | 3 | 3 | 43 |
| Total Snowfall (") | 0.2 | 0.5 | 0.0 | 0.0 | 0.0 | 0.0 | 0.0 | 0.0 | 0.0 | 0.0 | 0.0 | 0.0 | 0.7 |
| Days ≥ 1" Snow Depth | 0 | 0 | 0 | 0 | 0 | 0 | 0 | 0 | 0 | 0 | 0 | 0 | 0 |

## SANDERSON *Terrell County* ELEVATION 2802 ft LAT/LONG 30° 9' N / 102° 24' W

| | JAN | FEB | MAR | APR | MAY | JUN | JUL | AUG | SEP | OCT | NOV | DEC | YEAR |
|---|---|---|---|---|---|---|---|---|---|---|---|---|---|
| Maximum Temp °F | 59.6 | 63.5 | 72.4 | 80.8 | 86.1 | 91.1 | 91.9 | 91.5 | 85.9 | 77.8 | 68.7 | 61.7 | 77.6 |
| Minimum Temp °F | 30.6 | 33.8 | 41.9 | 51.5 | 59.4 | 66.5 | 68.7 | 67.6 | 61.7 | 50.3 | 39.6 | 32.2 | 50.3 |
| Mean Temp °F | 45.1 | 48.7 | 57.2 | 66.2 | 72.8 | 78.8 | 80.3 | 79.6 | 73.8 | 64.1 | 54.2 | 47.0 | 64.0 |
| Days Max Temp ≥ 90 °F | 0 | 0 | 1 | 5 | 11 | 19 | 22 | 21 | 11 | 2 | 0 | 0 | 92 |
| Days Max Temp ≤ 32 °F | 1 | 0 | 0 | 0 | 0 | 0 | 0 | 0 | 0 | 0 | 0 | 0 | 1 |
| Days Min Temp ≤ 32 °F | 19 | 12 | 5 | 0 | 0 | 0 | 0 | 0 | 0 | 0 | 6 | 16 | 58 |
| Days Min Temp ≤ 0 °F | 0 | 0 | 0 | 0 | 0 | 0 | 0 | 0 | 0 | 0 | 0 | 0 | 0 |
| Heating Degree Days | 609 | 455 | 255 | 70 | 13 | 1 | 0 | 0 | 16 | 101 | 325 | 552 | 2397 |
| Cooling Degree Days | 0 | 2 | 22 | 108 | 267 | 429 | 481 | 478 | 302 | 91 | 9 | 0 | 2189 |
| Total Precipitation (") | 0.43 | 0.63 | 0.37 | 1.05 | 1.68 | 1.91 | 1.48 | 1.86 | 2.65 | 1.62 | 0.73 | 0.52 | 14.93 |
| Days ≥ 0.1" Precip | 1 | 2 | 1 | 2 | 3 | 3 | 3 | 3 | 4 | 3 | 2 | 1 | 28 |
| Total Snowfall (") | 0.4 | 0.0 | 0.0 | 0.0 | 0.0 | 0.0 | 0.0 | 0.0 | 0.0 | 0.0 | 0.0 | 0.0 | 0.4 |
| Days ≥ 1" Snow Depth | 0 | 0 | 0 | 0 | 0 | 0 | 0 | 0 | 0 | 0 | 0 | 0 | 0 |

## SEALY *Austin County* ELEVATION 200 ft LAT/LONG 29° 47' N / 96° 9' W

| | JAN | FEB | MAR | APR | MAY | JUN | JUL | AUG | SEP | OCT | NOV | DEC | YEAR |
|---|---|---|---|---|---|---|---|---|---|---|---|---|---|
| Maximum Temp °F | *61.9* | *64.6* | *72.3* | *79.8* | *84.8* | *91.0* | *94.6* | *94.6* | *89.2* | *82.1* | *72.0* | *65.0* | 79.3 |
| Minimum Temp °F | *41.6* | *43.4* | *50.4* | *58.7* | *65.0* | *70.5* | *72.0* | *71.3* | *67.0* | *58.1* | *49.8* | *44.3* | 57.7 |
| Mean Temp °F | *51.8* | *54.0* | *61.4* | *69.3* | *74.9* | *80.8* | *83.4* | *83.0* | *78.1* | *70.1* | *60.9* | *54.7* | 68.5 |
| Days Max Temp ≥ 90 °F | *0* | *0* | *0* | *1* | *5* | *20* | *27* | *28* | *16* | *4* | *0* | *0* | 101 |
| Days Max Temp ≤ 32 °F | *0* | *0* | *0* | *0* | *0* | *0* | *0* | *0* | *0* | *0* | *0* | *0* | 0 |
| Days Min Temp ≤ 32 °F | *6* | *4* | *1* | *0* | *0* | *0* | *0* | *0* | *0* | *0* | *1* | *3* | 15 |
| Days Min Temp ≤ 0 °F | *0* | *0* | *0* | *0* | *0* | *0* | *0* | *0* | *0* | *0* | *0* | *0* | 0 |
| Heating Degree Days | *418* | *317* | *161* | *35* | *2* | *0* | *0* | *0* | *1* | *34* | *178* | *335* | 1481 |
| Cooling Degree Days | 8 | 16 | 55 | 146 | 314 | 481 | 572 | 571 | 405 | 202 | 60 | 20 | 2850 |
| Total Precipitation (") | *3.35* | *3.08* | *2.37* | *2.98* | *5.33* | *4.13* | *1.99* | *2.84* | *4.36* | *4.39* | *3.33* | *3.32* | 41.47 |
| Days ≥ 0.1" Precip | *5* | *5* | *4* | *3* | *6* | *5* | *4* | *4* | *5* | *5* | *4* | *5* | 55 |
| Total Snowfall (") | *0.0* | *0.0* | *0.0* | *0.0* | *0.0* | *0.0* | *0.0* | *0.0* | *0.0* | *0.0* | *0.0* | *0.0* | 0.0 |
| Days ≥ 1" Snow Depth | *0* | *0* | *0* | *0* | *0* | *0* | *0* | *0* | *0* | *0* | *0* | *0* | 0 |

## SEMINOLE *Gaines County* ELEVATION 3281 ft LAT/LONG 32° 43' N / 102° 39' W

| | JAN | FEB | MAR | APR | MAY | JUN | JUL | AUG | SEP | OCT | NOV | DEC | YEAR |
|---|---|---|---|---|---|---|---|---|---|---|---|---|---|
| Maximum Temp °F | 54.7 | 60.0 | 68.5 | 77.8 | 84.7 | 92.1 | 93.7 | 91.8 | 85.3 | 77.0 | 65.3 | 57.1 | 75.7 |
| Minimum Temp °F | 25.6 | 28.7 | 35.7 | 44.8 | 53.5 | 62.3 | 65.6 | 64.1 | 57.6 | 46.2 | 34.9 | 27.3 | 45.5 |
| Mean Temp °F | 40.2 | 44.4 | 52.1 | 61.3 | 69.1 | 77.2 | 79.7 | 78.0 | 71.5 | 61.6 | 50.1 | 42.3 | 60.6 |
| Days Max Temp ≥ 90 °F | 0 | 0 | 0 | 3 | 10 | 20 | 24 | 22 | 11 | 3 | 0 | 0 | 93 |
| Days Max Temp ≤ 32 °F | 2 | 1 | 0 | 0 | 0 | 0 | 0 | 0 | 0 | 0 | 0 | 1 | 4 |
| Days Min Temp ≤ 32 °F | 25 | 20 | 10 | 2 | 0 | 0 | 0 | 0 | 0 | 1 | 12 | 23 | 93 |
| Days Min Temp ≤ 0 °F | 0 | 0 | 0 | 0 | 0 | 0 | 0 | 0 | 0 | 0 | 0 | 0 | 0 |
| Heating Degree Days | 763 | 576 | 398 | 153 | 38 | 2 | 0 | 1 | 28 | 149 | 442 | 698 | 3248 |
| Cooling Degree Days | 0 | 0 | 5 | 48 | 181 | 385 | 470 | 426 | 232 | 50 | 2 | 0 | 1799 |
| Total Precipitation (") | 0.62 | 0.76 | 0.73 | 0.99 | 2.25 | 2.41 | 2.68 | 2.31 | 2.63 | 1.42 | 0.87 | 0.66 | 18.33 |
| Days ≥ 0.1" Precip | 2 | 2 | 2 | 2 | 4 | 4 | 4 | 4 | 4 | 3 | 2 | 2 | 35 |
| Total Snowfall (") | 2.9 | 2.1 | 1.2 | 0.3 | 0.0 | 0.0 | 0.0 | 0.0 | 0.0 | 0.1 | 1.3 | 1.8 | 9.7 |
| Days ≥ 1" Snow Depth | 2 | 1 | 1 | 0 | 0 | 0 | 0 | 0 | 0 | 0 | 1 | 2 | 7 |

## SEYMOUR *Baylor County* ELEVATION 1280 ft LAT/LONG 33° 35' N / 99° 16' W

| | JAN | FEB | MAR | APR | MAY | JUN | JUL | AUG | SEP | OCT | NOV | DEC | YEAR |
|---|---|---|---|---|---|---|---|---|---|---|---|---|---|
| Maximum Temp °F | 52.5 | 57.0 | 66.7 | 76.3 | 83.0 | 91.1 | 96.2 | 95.4 | 86.8 | 77.1 | 64.4 | 55.6 | 75.2 |
| Minimum Temp °F | 26.7 | 30.6 | 39.1 | 48.7 | 57.7 | 66.8 | 71.1 | 69.7 | 62.0 | 50.4 | 38.0 | 29.5 | 49.2 |
| Mean Temp °F | 39.6 | 43.8 | 52.9 | 62.5 | 70.3 | 79.0 | 83.7 | 82.6 | 74.4 | 63.8 | 51.2 | 42.5 | 62.2 |
| Days Max Temp ≥ 90 °F | 0 | 0 | 1 | 3 | 7 | 20 | 27 | 26 | 14 | 3 | 0 | 0 | 101 |
| Days Max Temp ≤ 32 °F | 3 | 2 | 0 | 0 | 0 | 0 | 0 | 0 | 0 | 0 | 0 | 1 | 6 |
| Days Min Temp ≤ 32 °F | 23 | 16 | 8 | 1 | 0 | 0 | 0 | 0 | 0 | 1 | 9 | 20 | 78 |
| Days Min Temp ≤ 0 °F | 0 | 0 | 0 | 0 | 0 | 0 | 0 | 0 | 0 | 0 | 0 | 0 | 0 |
| Heating Degree Days | 780 | 592 | 384 | 140 | 33 | 1 | 0 | 0 | 21 | 126 | 415 | 689 | 3181 |
| Cooling Degree Days | 0 | 0 | 14 | 67 | 207 | 427 | 588 | 581 | 325 | 94 | 9 | 0 | 2312 |
| Total Precipitation (") | 1.05 | 1.59 | 1.53 | 2.18 | 4.11 | 3.33 | 2.02 | 2.50 | 3.87 | 2.83 | 1.08 | 1.39 | 27.48 |
| Days ≥ 0.1" Precip | 3 | 3 | 3 | 4 | 6 | 4 | 3 | 4 | 4 | 4 | 3 | 3 | 44 |
| Total Snowfall (") | *0.6* | *0.3* | 0.1 | 0.0 | 0.0 | 0.0 | 0.0 | 0.0 | 0.0 | 0.0 | 0.3 | 0.4 | 1.7 |
| Days ≥ 1" Snow Depth | *0* | *0* | 0 | 0 | 0 | 0 | 0 | 0 | 0 | 0 | 0 | 0 | 0 |

## SHAMROCK 2 *Wheeler County* ELEVATION 2352 ft LAT/LONG 35° 14' N / 100° 15' W

| | JAN | FEB | MAR | APR | MAY | JUN | JUL | AUG | SEP | OCT | NOV | DEC | YEAR |
|---|---|---|---|---|---|---|---|---|---|---|---|---|---|
| Maximum Temp °F | 49.7 | 54.6 | 63.5 | 73.7 | 80.3 | 89.1 | 94.4 | 92.5 | 84.7 | 74.5 | 61.4 | 52.1 | 72.5 |
| Minimum Temp °F | 23.0 | 26.9 | 34.6 | 45.0 | 53.9 | 63.4 | 68.1 | 66.5 | 58.4 | 46.0 | 34.6 | 25.6 | 45.5 |
| Mean Temp °F | 36.4 | 40.8 | 49.1 | 59.4 | 67.1 | 76.3 | 81.3 | 79.5 | 71.6 | 60.3 | 48.0 | 38.7 | 59.0 |
| Days Max Temp ≥ 90 °F | 0 | 0 | 0 | 1 | 5 | 15 | 25 | 22 | 10 | 2 | 0 | 0 | 80 |
| Days Max Temp ≤ 32 °F | 4 | 3 | 0 | 0 | 0 | 0 | 0 | 0 | 0 | 0 | 1 | 2 | 10 |
| Days Min Temp ≤ 32 °F | 27 | 20 | 12 | 2 | 0 | 0 | 0 | 0 | 0 | 2 | 12 | 25 | 100 |
| Days Min Temp ≤ 0 °F | 1 | 0 | 0 | 0 | 0 | 0 | 0 | 0 | 0 | 0 | 0 | 0 | 1 |
| Heating Degree Days | 879 | 677 | 490 | 200 | 59 | 4 | 0 | 1 | 32 | 188 | 504 | 807 | 3841 |
| Cooling Degree Days | 0 | 0 | 3 | 34 | 121 | 336 | 504 | 471 | 243 | 41 | 2 | 0 | 1755 |
| Total Precipitation (") | 0.50 | 0.76 | 1.58 | 1.82 | 3.62 | 3.40 | 2.22 | 2.42 | 2.72 | 1.55 | 1.04 | 0.60 | 22.23 |
| Days ≥ 0.1" Precip | 1 | 2 | 3 | 3 | 6 | 5 | 3 | 4 | 4 | 3 | 2 | 1 | 37 |
| Total Snowfall (") | 1.7 | 2.5 | 0.7 | 0.1 | 0.0 | 0.0 | 0.0 | 0.0 | 0.0 | 0.0 | 0.7 | 0.8 | 6.5 |
| Days ≥ 1" Snow Depth | 1 | *1* | 0 | 0 | 0 | 0 | 0 | 0 | 0 | 0 | 0 | *0* | 2 |

## SHEFFIELD *Pecos County* ELEVATION 2201 ft LAT/LONG 30° 41' N / 101° 50' W

| | JAN | FEB | MAR | APR | MAY | JUN | JUL | AUG | SEP | OCT | NOV | DEC | YEAR |
|---|---|---|---|---|---|---|---|---|---|---|---|---|---|
| Maximum Temp °F | 61.1 | 66.0 | 75.1 | 82.9 | 88.6 | 93.9 | 95.5 | 95.1 | 88.7 | 80.2 | 69.6 | 62.4 | 79.9 |
| Minimum Temp °F | 31.0 | 34.6 | 43.4 | 51.4 | 60.5 | 68.7 | 71.4 | 70.3 | 63.6 | 51.9 | 40.0 | 32.6 | 51.6 |
| Mean Temp °F | 46.1 | 50.4 | 59.2 | 67.2 | 74.6 | 81.3 | 83.5 | 82.7 | 76.2 | 66.1 | 54.8 | 47.5 | 65.8 |
| Days Max Temp ≥ 90 °F | 0 | 0 | 1 | 6 | 14 | 23 | 28 | 27 | 15 | 3 | 0 | 0 | 117 |
| Days Max Temp ≤ 32 °F | 0 | 0 | 0 | 0 | 0 | 0 | 0 | 0 | 0 | 0 | 0 | 0 | 0 |
| Days Min Temp ≤ 32 °F | 19 | 12 | 5 | 1 | 0 | 0 | 0 | 0 | 0 | 1 | 7 | 16 | 61 |
| Days Min Temp ≤ 0 °F | 0 | 0 | 0 | 0 | 0 | 0 | 0 | 0 | 0 | 0 | 0 | 0 | 0 |
| Heating Degree Days | 576 | 407 | 204 | 60 | 7 | 0 | 0 | 0 | 10 | 75 | 309 | 534 | 2182 |
| Cooling Degree Days | 0 | 2 | 33 | 135 | 330 | 510 | 594 | 573 | 371 | 127 | 14 | 1 | 2690 |
| Total Precipitation (") | 0.51 | 0.67 | 0.52 | 1.21 | 2.03 | 1.47 | 1.30 | 1.48 | 2.44 | 1.88 | 0.76 | 0.63 | 14.90 |
| Days ≥ 0.1" Precip | 2 | 1 | 1 | 2 | 3 | 2 | 2 | 3 | 3 | 2 | 2 | 1 | 24 |
| Total Snowfall (") | 0.0 | 0.0 | 0.0 | 0.0 | 0.0 | 0.0 | 0.0 | 0.0 | 0.0 | 0.0 | 0.0 | 0.0 | 0.0 |
| Days ≥ 1" Snow Depth | 0 | 0 | 0 | 0 | 0 | 0 | 0 | 0 | 0 | 0 | 0 | 0 | 0 |

## SHERMAN *Grayson County* ELEVATION 751 ft LAT/LONG 33° 38' N / 96° 36' W

| | JAN | FEB | MAR | APR | MAY | JUN | JUL | AUG | SEP | OCT | NOV | DEC | YEAR |
|---|---|---|---|---|---|---|---|---|---|---|---|---|---|
| Maximum Temp °F | 51.4 | 56.0 | 65.3 | 74.3 | 80.8 | 89.1 | 94.2 | 93.8 | 85.9 | 75.9 | 63.9 | 54.7 | 73.8 |
| Minimum Temp °F | 30.4 | 34.2 | 42.9 | 52.1 | 59.9 | 68.2 | 72.2 | 71.0 | 63.9 | 52.5 | 42.0 | 33.7 | 51.9 |
| Mean Temp °F | 40.9 | 45.1 | 54.1 | 63.2 | 70.4 | 78.7 | 83.2 | 82.4 | 75.0 | 64.2 | 53.0 | 44.2 | 62.9 |
| Days Max Temp ≥ 90 °F | 0 | 0 | 0 | 0 | 2 | 16 | 26 | 25 | 12 | 2 | 0 | 0 | 83 |
| Days Max Temp ≤ 32 °F | 3 | 1 | 0 | 0 | 0 | 0 | 0 | 0 | 0 | 0 | 0 | 1 | 5 |
| Days Min Temp ≤ 32 °F | 19 | 12 | 5 | 0 | 0 | 0 | 0 | 0 | 0 | 0 | 5 | 14 | 55 |
| Days Min Temp ≤ 0 °F | 0 | 0 | 0 | 0 | 0 | 0 | 0 | 0 | 0 | 0 | 0 | 0 | 0 |
| Heating Degree Days | 746 | 556 | 347 | 116 | 25 | 1 | 0 | 0 | 15 | 112 | 367 | 639 | 2924 |
| Cooling Degree Days | 0 | 1 | 17 | 70 | 204 | 430 | 582 | 580 | 344 | 98 | 14 | 1 | 2341 |
| Total Precipitation (") | 1.94 | 2.89 | 3.16 | 3.86 | 6.05 | 4.29 | 2.25 | 2.21 | 4.77 | 4.76 | 3.23 | 2.56 | 41.97 |
| Days ≥ 0.1" Precip | 4 | 5 | 5 | 5 | 7 | 6 | 4 | 4 | 5 | 5 | 4 | 4 | 58 |
| Total Snowfall (") | *0.6* | *1.4* | 0.1 | 0.0 | 0.0 | 0.0 | 0.0 | 0.0 | 0.0 | 0.0 | 0.0 | 0.4 | 2.5 |
| Days ≥ 1" Snow Depth | *0* | *0* | 0 | 0 | 0 | 0 | 0 | 0 | 0 | 0 | 0 | 0 | 0 |

## SIERRA BLANCA 2 E *Hudspeth County* ELEVATION 4535 ft LAT/LONG 31° 11' N / 105° 19' W

| | JAN | FEB | MAR | APR | MAY | JUN | JUL | AUG | SEP | OCT | NOV | DEC | YEAR |
|---|---|---|---|---|---|---|---|---|---|---|---|---|---|
| Maximum Temp °F | 57.9 | 63.4 | 70.5 | 78.9 | 86.1 | 93.5 | 92.4 | 90.0 | 85.0 | 77.7 | 67.0 | 59.1 | 76.8 |
| Minimum Temp °F | 27.0 | 29.9 | 35.5 | 42.5 | 51.0 | 60.1 | 63.4 | 61.5 | 55.4 | 44.5 | 33.5 | 27.2 | 44.3 |
| Mean Temp °F | 42.5 | 46.7 | 53.0 | 60.7 | 68.6 | 76.8 | 77.9 | 75.8 | 70.2 | 61.1 | 50.3 | 43.2 | 60.6 |
| Days Max Temp ≥ 90 °F | 0 | 0 | 0 | 1 | 9 | 24 | 23 | 19 | 8 | 1 | 0 | 0 | 85 |
| Days Max Temp ≤ 32 °F | 0 | 0 | 0 | 0 | 0 | 0 | 0 | 0 | 0 | 0 | 0 | 0 | 0 |
| Days Min Temp ≤ 32 °F | 23 | 18 | 11 | 4 | 0 | 0 | 0 | 0 | 0 | 2 | 14 | 24 | 96 |
| Days Min Temp ≤ 0 °F | 0 | 0 | 0 | 0 | 0 | 0 | 0 | 0 | 0 | 0 | 0 | 0 | 0 |
| Heating Degree Days | 691 | 512 | 367 | 145 | 27 | 0 | 0 | 0 | 21 | 145 | 435 | 667 | 3010 |
| Cooling Degree Days | 0 | 0 | 2 | 26 | 155 | 373 | 409 | 348 | 190 | 27 | 0 | 0 | 1530 |
| Total Precipitation (") | 0.51 | 0.39 | 0.34 | 0.30 | 0.55 | 1.09 | 2.15 | 2.50 | 2.45 | 1.20 | 0.48 | 0.66 | 12.62 |
| Days ≥ 0.1" Precip | 2 | *1* | 1 | 1 | 2 | 3 | 4 | 5 | 5 | 3 | 2 | 2 | 31 |
| Total Snowfall (") | na | *0.0* | 0.0 | 0.0 | 0.0 | 0.0 | 0.0 | 0.0 | 0.0 | 0.0 | 0.2 | 0.0 | na |
| Days ≥ 1" Snow Depth | *0* | *0* | 0 | 0 | 0 | 0 | 0 | 0 | 0 | 0 | 0 | 0 | 0 |

## SILVERTON *Briscoe County* ELEVATION 3481 ft LAT/LONG 34° 28 ' N / 101° 18 ' W

| | JAN | FEB | MAR | APR | MAY | JUN | JUL | AUG | SEP | OCT | NOV | DEC | YEAR |
|---|---|---|---|---|---|---|---|---|---|---|---|---|---|
| Maximum Temp °F | 49.6 | 54.0 | 62.3 | 71.9 | 78.8 | 87.2 | 90.8 | 88.5 | 81.5 | 72.8 | 60.3 | 52.0 | 70.8 |
| Minimum Temp °F | 21.0 | 24.4 | 31.8 | 41.4 | 50.8 | 60.5 | 64.5 | 62.6 | 55.5 | 43.4 | 31.6 | 23.5 | 42.6 |
| Mean Temp °F | 35.3 | 39.3 | 47.1 | 56.6 | 64.8 | 73.9 | 77.7 | 75.6 | 68.5 | 58.1 | 46.0 | 37.8 | 56.7 |
| Days Max Temp ≥ 90 °F | 0 | 0 | 0 | 1 | 4 | 12 | 19 | 16 | 6 | 1 | 0 | 0 | 59 |
| Days Max Temp ≤ 32 °F | 4 | 2 | 1 | 0 | 0 | 0 | 0 | 0 | 0 | 0 | 1 | 3 | 11 |
| Days Min Temp ≤ 32 °F | 29 | 24 | 16 | 4 | 0 | 0 | 0 | 0 | 0 | 3 | 17 | 27 | 120 |
| Days Min Temp ≤ 0 °F | 0 | 0 | 0 | 0 | 0 | 0 | 0 | 0 | 0 | 0 | 0 | 0 | 0 |
| Heating Degree Days | 913 | 720 | 551 | 264 | 91 | 8 | 0 | 2 | 50 | 232 | 565 | 836 | 4232 |
| Cooling Degree Days | 0 | 0 | 2 | 19 | 96 | 288 | 412 | 365 | 179 | 25 | 1 | 0 | 1387 |
| Total Precipitation (") | 0.55 | 0.70 | 1.10 | 1.42 | 3.05 | 3.99 | 2.17 | 2.83 | 2.71 | 1.57 | 0.82 | 0.61 | 21.52 |
| Days ≥ 0.1" Precip | 2 | 2 | 3 | 4 | 6 | 6 | 4 | 6 | 4 | 3 | 2 | 2 | 44 |
| Total Snowfall (") | 2.9 | 2.8 | 1.7 | 0.3 | 0.0 | 0.0 | 0.0 | 0.0 | 0.0 | 0.2 | 1.5 | 2.7 | 12.1 |
| Days ≥ 1" Snow Depth | *1* | *0* | *0* | 0 | 0 | 0 | 0 | 0 | 0 | 0 | 0 | *1* | 2 |

## SINTON *San Patricio County* ELEVATION 49 ft LAT/LONG 28° 2 ' N / 97° 31 ' W

| | JAN | FEB | MAR | APR | MAY | JUN | JUL | AUG | SEP | OCT | NOV | DEC | YEAR |
|---|---|---|---|---|---|---|---|---|---|---|---|---|---|
| Maximum Temp °F | *65.3* | *68.8* | 76.1 | 81.4 | *85.7* | 90.9 | 93.6 | *94.0* | *90.3* | *84.0* | *75.0* | *68.7* | 81.2 |
| Minimum Temp °F | *44.7* | 47.0 | 54.9 | 62.0 | *67.8* | 73.5 | 74.8 | *74.2* | *69.9* | *61.7* | *53.4* | *46.5* | 60.9 |
| Mean Temp °F | *55.0* | *57.9* | 65.5 | 71.7 | *76.8* | 82.2 | 84.2 | *84.1* | *80.2* | *72.9* | *64.2* | *57.6* | 71.0 |
| Days Max Temp ≥ 90 °F | 0 | 0 | 1 | 2 | 7 | *21* | 27 | *28* | *19* | *6* | *0* | *0* | 111 |
| Days Max Temp ≤ 32 °F | *0* | 0 | 0 | 0 | 0 | 0 | 0 | *0* | *0* | *0* | *0* | *0* | 0 |
| Days Min Temp ≤ 32 °F | *4* | 2 | 0 | 0 | 0 | 0 | 0 | *0* | *0* | *0* | *1* | *3* | 10 |
| Days Min Temp ≤ 0 °F | 0 | 0 | 0 | 0 | 0 | 0 | 0 | *0* | *0* | *0* | *0* | *0* | 0 |
| Heating Degree Days | *332* | *223* | 90 | 17 | *1* | 0 | 0 | *0* | *0* | *17* | *121* | *269* | 1070 |
| Cooling Degree Days | na | na | *93* | *197* | *359* | *542* | *598* | *599* | *453* | *267* | *88* | na | na |
| Total Precipitation (") | 2.24 | 2.47 | 1.71 | 2.45 | 4.83 | 3.81 | 3.31 | *3.07* | *6.40* | *4.40* | *1.91* | *1.43* | 38.03 |
| Days ≥ 0.1" Precip | *5* | 4 | 3 | 3 | 5 | *5* | 4 | *4* | *6* | *5* | *3* | *3* | 50 |
| Total Snowfall (") | *0.0* | 0.1 | 0.0 | 0.0 | 0.0 | 0.0 | 0.0 | 0.0 | 0.0 | 0.0 | 0.0 | *0.0* | 0.1 |
| Days ≥ 1" Snow Depth | 0 | *0* | 0 | 0 | 0 | 0 | 0 | 0 | 0 | *0* | *0* | *0* | 0 |

## SMITHVILLE *Bastrop County* ELEVATION 331 ft LAT/LONG 30° 0 ' N / 97° 9 ' W

| | JAN | FEB | MAR | APR | MAY | JUN | JUL | AUG | SEP | OCT | NOV | DEC | YEAR |
|---|---|---|---|---|---|---|---|---|---|---|---|---|---|
| Maximum Temp °F | 59.9 | 64.0 | 72.2 | 80.0 | 84.2 | 91.3 | 95.0 | 95.9 | 90.2 | 81.7 | 71.3 | 63.9 | 79.1 |
| Minimum Temp °F | 35.8 | 39.0 | 46.6 | 55.7 | 62.6 | 69.0 | *71.4* | 69.9 | 65.4 | 55.2 | 45.9 | 39.3 | 54.7 |
| Mean Temp °F | 47.9 | 51.5 | 59.4 | 67.9 | 73.4 | 80.2 | *83.2* | 83.0 | 77.8 | 68.5 | 58.6 | 51.6 | 66.9 |
| Days Max Temp ≥ 90 °F | 0 | 0 | 0 | 2 | 6 | 21 | 28 | 29 | 18 | 5 | 0 | 0 | 109 |
| Days Max Temp ≤ 32 °F | 0 | 0 | 0 | 0 | 0 | 0 | 0 | 0 | 0 | 0 | 0 | 0 | 0 |
| Days Min Temp ≤ 32 °F | 13 | 7 | 3 | 0 | 0 | 0 | 0 | 0 | 0 | 0 | 3 | 9 | 35 |
| Days Min Temp ≤ 0 °F | 0 | 0 | 0 | 0 | 0 | 0 | 0 | 0 | 0 | 0 | 0 | 0 | 0 |
| Heating Degree Days | 530 | 385 | 211 | 52 | 7 | 0 | *0* | 0 | 3 | 56 | 230 | 421 | 1895 |
| Cooling Degree Days | 4 | 9 | 45 | 136 | 280 | 466 | 574 | 573 | 396 | 183 | 52 | 13 | 2731 |
| Total Precipitation (") | 3.03 | 2.83 | 2.51 | 3.26 | 5.74 | 3.42 | 1.93 | 2.04 | 3.85 | 4.45 | 3.14 | 3.07 | 39.27 |
| Days ≥ 0.1" Precip | 5 | 5 | 5 | 4 | 7 | 5 | 3 | 3 | 5 | 4 | 4 | 5 | 55 |
| Total Snowfall (") | 0.0 | 0.0 | 0.0 | 0.0 | 0.0 | 0.0 | 0.0 | 0.0 | 0.0 | 0.0 | 0.0 | 0.0 | 0.0 |
| Days ≥ 1" Snow Depth | 0 | 0 | 0 | 0 | 0 | 0 | 0 | 0 | 0 | 0 | 0 | 0 | 0 |

## SNYDER *Scurry County* ELEVATION 2451 ft LAT/LONG 32° 44 ' N / 100° 55 ' W

| | JAN | FEB | MAR | APR | MAY | JUN | JUL | AUG | SEP | OCT | NOV | DEC | YEAR |
|---|---|---|---|---|---|---|---|---|---|---|---|---|---|
| Maximum Temp °F | 53.6 | 58.5 | 68.3 | 77.1 | *83.8* | 90.7 | 94.0 | 92.2 | 84.9 | 76.3 | 65.3 | 56.6 | 75.1 |
| Minimum Temp °F | 26.2 | 29.9 | 37.5 | 47.3 | *56.4* | 65.0 | 68.8 | 67.1 | 60.4 | 48.8 | 37.0 | 28.5 | 47.7 |
| Mean Temp °F | 39.9 | 44.2 | 52.9 | 62.2 | *70.1* | 77.9 | 81.4 | 79.7 | 72.7 | 62.6 | 51.2 | 42.6 | 61.5 |
| Days Max Temp ≥ 90 °F | 0 | 0 | 1 | 3 | *8* | 19 | 26 | 23 | 9 | 2 | 0 | 0 | 91 |
| Days Max Temp ≤ 32 °F | 2 | 1 | 0 | 0 | 0 | 0 | 0 | 0 | 0 | 0 | 0 | 1 | 4 |
| Days Min Temp ≤ 32 °F | 24 | 18 | 9 | 2 | 0 | 0 | 0 | 0 | 0 | 1 | 9 | 21 | 84 |
| Days Min Temp ≤ 0 °F | 0 | 0 | 0 | 0 | 0 | 0 | 0 | 0 | 0 | 0 | 0 | 0 | 0 |
| Heating Degree Days | 771 | 581 | 380 | 141 | *33* | 3 | 0 | 0 | 22 | 138 | 412 | 688 | 3169 |
| Cooling Degree Days | 0 | 1 | 9 | 61 | *207* | *394* | 519 | *492* | 274 | 71 | *4* | 0 | 2032 |
| Total Precipitation (") | 0.65 | 0.96 | 0.99 | 1.72 | 3.37 | 2.80 | 1.92 | 2.61 | 3.32 | 2.35 | 0.83 | 0.84 | 22.36 |
| Days ≥ 0.1" Precip | 2 | 3 | 2 | 3 | *5* | 4 | 3 | 4 | 5 | 4 | 2 | 2 | 39 |
| Total Snowfall (") | *0.5* | 0.8 | 0.0 | 0.0 | 0.0 | 0.0 | 0.0 | 0.0 | 0.0 | 0.0 | *0.0* | *0.4* | 1.7 |
| Days ≥ 1" Snow Depth | *1* | *0* | 0 | 0 | 0 | 0 | 0 | 0 | 0 | 0 | *0* | *0* | 1 |

## SOMERVILLE DAM *Burleson County* ELEVATION 263 ft LAT/LONG 30° 20 ' N / 96° 32 ' W

| | JAN | FEB | MAR | APR | MAY | JUN | JUL | AUG | SEP | OCT | NOV | DEC | YEAR |
|---|---|---|---|---|---|---|---|---|---|---|---|---|---|
| Maximum Temp °F | 58.2 | 62.1 | 70.3 | 77.7 | 83.3 | 89.8 | 93.9 | 94.6 | 89.0 | 80.7 | 70.1 | 62.3 | 77.7 |
| Minimum Temp °F | 37.2 | 40.3 | 48.8 | 57.3 | 64.3 | 70.8 | 73.4 | 72.6 | 67.0 | 56.7 | 47.9 | 40.1 | 56.4 |
| Mean Temp °F | 47.7 | 51.2 | 59.5 | 67.6 | 73.8 | 80.4 | 83.7 | 83.6 | 78.0 | 68.7 | 59.0 | 51.3 | 67.0 |
| Days Max Temp ≥ 90 °F | 0 | 0 | 0 | 0 | 4 | 18 | 27 | 28 | 16 | 3 | 0 | 0 | 96 |
| Days Max Temp ≤ 32 °F | 1 | 0 | 0 | 0 | 0 | 0 | 0 | 0 | 0 | 0 | 0 | 0 | 1 |
| Days Min Temp ≤ 32 °F | 11 | 6 | 2 | 0 | 0 | 0 | 0 | 0 | 0 | 0 | 2 | 7 | 28 |
| Days Min Temp ≤ 0 °F | 0 | 0 | 0 | 0 | 0 | 0 | 0 | 0 | 0 | 0 | 0 | 0 | 0 |
| Heating Degree Days | 532 | 387 | 203 | 53 | 6 | 0 | 0 | 0 | 3 | 52 | 221 | 429 | 1886 |
| Cooling Degree Days | 2 | 6 | 42 | 134 | 294 | 485 | 600 | 610 | 422 | 185 | 53 | 10 | 2843 |
| Total Precipitation (") | 2.94 | 2.72 | 2.64 | 3.26 | 5.07 | 4.10 | 1.82 | 2.32 | 3.83 | 4.09 | 3.12 | 3.03 | 38.94 |
| Days ≥ 0.1" Precip | 6 | 5 | 5 | 4 | 6 | 5 | 4 | 4 | 5 | 5 | 5 | 5 | 59 |
| Total Snowfall (") | 0.0 | 0.0 | 0.0 | 0.0 | 0.0 | 0.0 | 0.0 | 0.0 | 0.0 | 0.0 | 0.0 | 0.0 | 0.0 |
| Days ≥ 1" Snow Depth | 0 | 0 | 0 | 0 | 0 | 0 | 0 | 0 | 0 | 0 | 0 | 0 | 0 |

## SONORA *Sutton County* ELEVATION 2241 ft LAT/LONG 30° 34 ' N / 100° 39 ' W

| | JAN | FEB | MAR | APR | MAY | JUN | JUL | AUG | SEP | OCT | NOV | DEC | YEAR |
|---|---|---|---|---|---|---|---|---|---|---|---|---|---|
| Maximum Temp °F | 60.7 | 65.8 | 74.3 | 82.1 | 87.7 | 92.8 | 95.6 | 94.5 | 88.5 | 80.4 | na | ***62.9*** | na |
| Minimum Temp °F | 30.2 | 33.5 | 41.7 | 51.1 | 59.0 | 65.8 | 67.6 | 66.2 | 61.2 | 50.8 | na | ***32.8*** | na |
| Mean Temp °F | 45.4 | 49.7 | 58.0 | 66.6 | 73.4 | 79.3 | 81.6 | 80.4 | 74.9 | 65.5 | na | ***47.8*** | na |
| Days Max Temp ≥ 90 °F | 0 | 0 | 1 | 5 | 12 | 22 | 27 | 26 | 15 | 3 | 0 | 0 | 111 |
| Days Max Temp ≤ 32 °F | 0 | 0 | 0 | 0 | 0 | 0 | 0 | 0 | 0 | 0 | 0 | 0 | 0 |
| Days Min Temp ≤ 32 °F | 19 | 13 | 6 | 1 | 0 | 0 | 0 | 0 | 0 | 1 | ***6*** | ***16*** | 62 |
| Days Min Temp ≤ 0 °F | 0 | 0 | 0 | 0 | 0 | 0 | 0 | 0 | 0 | 0 | 0 | 0 | 0 |
| Heating Degree Days | 600 | 431 | 233 | 65 | 8 | 0 | 0 | 0 | 8 | 82 | na | ***527*** | na |
| Cooling Degree Days | 0 | 3 | 17 | ***108*** | 270 | 436 | 522 | 496 | ***307*** | ***109*** | na | na | na |
| Total Precipitation (") | 0.93 | 1.33 | 1.05 | 1.90 | 2.51 | 2.23 | 2.09 | 2.81 | 3.28 | 2.58 | ***1.15*** | ***0.95*** | 22.81 |
| Days ≥ 0.1" Precip | 2 | 3 | 2 | 3 | 4 | 3 | 3 | 4 | 4 | 4 | ***2*** | ***2*** | 36 |
| Total Snowfall (") | 0.0 | 0.2 | 0.0 | 0.0 | 0.0 | 0.0 | 0.0 | 0.0 | 0.0 | 0.0 | ***0.0*** | ***0.0*** | 0.2 |
| Days ≥ 1" Snow Depth | 0 | 0 | 0 | 0 | 0 | 0 | 0 | 0 | 0 | 0 | ***0*** | ***0*** | 0 |

## SPEARMAN *Hansford County* ELEVATION 3104 ft LAT/LONG 36° 12 ' N / 101° 12 ' W

| | JAN | FEB | MAR | APR | MAY | JUN | JUL | AUG | SEP | OCT | NOV | DEC | YEAR |
|---|---|---|---|---|---|---|---|---|---|---|---|---|---|
| Maximum Temp °F | 49.9 | 55.2 | 64.2 | 74.2 | 80.8 | 89.9 | 94.2 | 91.7 | 84.9 | 74.7 | 60.4 | 51.6 | 72.6 |
| Minimum Temp °F | 21.5 | 25.2 | 32.8 | 42.3 | 51.6 | 61.1 | 65.7 | 64.0 | 56.4 | 44.2 | 32.1 | 24.1 | 43.4 |
| Mean Temp °F | 35.7 | 40.2 | 48.6 | 58.3 | 66.2 | 75.5 | 79.9 | 77.9 | 70.6 | 59.5 | 46.3 | 37.9 | 58.0 |
| Days Max Temp ≥ 90 °F | 0 | 0 | 0 | 1 | 5 | 17 | 25 | 21 | 10 | 2 | 0 | 0 | 81 |
| Days Max Temp ≤ 32 °F | 4 | 2 | 1 | 0 | 0 | 0 | 0 | 0 | 0 | 0 | 1 | 3 | 11 |
| Days Min Temp ≤ 32 °F | 27 | 22 | 15 | 4 | 0 | 0 | 0 | 0 | 0 | 3 | 16 | 26 | 113 |
| Days Min Temp ≤ 0 °F | 1 | 0 | 0 | 0 | 0 | 0 | 0 | 0 | 0 | 0 | 0 | 1 | 2 |
| Heating Degree Days | 901 | 693 | 507 | 224 | 72 | 6 | 0 | 1 | 35 | 204 | 557 | 834 | 4034 |
| Cooling Degree Days | 0 | 0 | 4 | 30 | 115 | 334 | 474 | 427 | 229 | 39 | 1 | 0 | 1653 |
| Total Precipitation (") | 0.51 | 0.66 | 1.33 | 1.25 | 2.83 | 3.15 | 2.64 | 2.64 | 1.98 | 1.18 | 0.94 | 0.59 | 19.70 |
| Days ≥ 0.1" Precip | 2 | 2 | 3 | 3 | 6 | 5 | 4 | 5 | 3 | 2 | 2 | 2 | 39 |
| Total Snowfall (") | 4.4 | 5.2 | 4.0 | 1.4 | 0.2 | 0.0 | 0.0 | 0.0 | 0.0 | 0.6 | 2.2 | 3.6 | 21.6 |
| Days ≥ 1" Snow Depth | 4 | 3 | 1 | 0 | 0 | 0 | 0 | 0 | 0 | 0 | 1 | 2 | 11 |

## STERLING CITY *Sterling County* ELEVATION 2290 ft LAT/LONG 31° 51 ' N / 100° 59 ' W

| | JAN | FEB | MAR | APR | MAY | JUN | JUL | AUG | SEP | OCT | NOV | DEC | YEAR |
|---|---|---|---|---|---|---|---|---|---|---|---|---|---|
| Maximum Temp °F | 57.6 | 62.4 | 71.7 | 80.7 | 86.3 | 92.7 | 95.1 | 94.1 | 87.3 | 79.2 | 68.3 | 60.9 | 78.0 |
| Minimum Temp °F | 28.5 | 31.9 | 39.7 | 48.9 | 57.2 | 65.9 | 68.6 | 67.6 | 60.6 | ***48.9*** | 38.1 | 30.5 | 48.9 |
| Mean Temp °F | 43.1 | 47.1 | 55.7 | 64.8 | 71.8 | 79.3 | 81.9 | 80.8 | 74.0 | ***64.0*** | 53.2 | 45.7 | 63.5 |
| Days Max Temp ≥ 90 °F | 0 | 0 | 1 | 5 | 11 | 22 | 26 | 25 | 13 | 3 | 0 | 0 | 106 |
| Days Max Temp ≤ 32 °F | 1 | 1 | 0 | 0 | 0 | 0 | 0 | 0 | 0 | 0 | 0 | 1 | 3 |
| Days Min Temp ≤ 32 °F | 20 | 14 | 8 | 1 | 0 | 0 | 0 | 0 | 0 | 1 | 9 | 19 | 72 |
| Days Min Temp ≤ 0 °F | 0 | 0 | 0 | 0 | 0 | 0 | 0 | 0 | 0 | 0 | 0 | 0 | 0 |
| Heating Degree Days | 672 | 498 | 305 | 98 | 20 | 1 | 0 | 0 | 19 | ***114*** | 356 | 591 | 2674 |
| Cooling Degree Days | 0 | 1 | 24 | 108 | ***248*** | 442 | 538 | 514 | 304 | ***96*** | ***15*** | 1 | 2291 |
| Total Precipitation (") | 0.89 | 0.95 | 0.93 | 1.58 | 2.88 | 2.27 | 1.38 | 2.07 | 3.72 | 1.76 | 0.92 | 1.01 | 20.36 |
| Days ≥ 0.1" Precip | 2 | 3 | 2 | 3 | 4 | 3 | 3 | 4 | 5 | 3 | 2 | 2 | 36 |
| Total Snowfall (") | 0.6 | ***0.4*** | 0.3 | 0.0 | 0.0 | 0.0 | 0.0 | 0.0 | 0.0 | 0.0 | 0.0 | 0.0 | 1.3 |
| Days ≥ 1" Snow Depth | 0 | ***0*** | 0 | 0 | 0 | 0 | 0 | 0 | 0 | 0 | 0 | 0 | 0 |

## STILLHOUSE HOLLOW DA *Bell County* ELEVATION 712 ft LAT/LONG 31° 2 ' N / 97° 32 ' W

| | JAN | FEB | MAR | APR | MAY | JUN | JUL | AUG | SEP | OCT | NOV | DEC | YEAR |
|---|---|---|---|---|---|---|---|---|---|---|---|---|---|
| Maximum Temp °F | 56.7 | 61.5 | 69.9 | 77.6 | 83.1 | 90.3 | 94.6 | 94.9 | 88.5 | 79.9 | 68.8 | 60.8 | 77.2 |
| Minimum Temp °F | 33.7 | 37.3 | 45.5 | 54.5 | 61.8 | 68.9 | 71.7 | 71.0 | 65.6 | 55.2 | 45.2 | 37.3 | 54.0 |
| Mean Temp °F | 45.2 | 49.4 | 57.8 | 66.1 | 72.5 | 79.6 | 83.2 | 82.9 | 77.1 | 67.6 | 57.0 | 49.1 | 65.6 |
| Days Max Temp ≥ 90 °F | 0 | 0 | 1 | 1 | 4 | 19 | 28 | 28 | 16 | 3 | 0 | 0 | 100 |
| Days Max Temp ≤ 32 °F | 1 | 0 | 0 | 0 | 0 | 0 | 0 | 0 | 0 | 0 | 0 | 0 | 1 |
| Days Min Temp ≤ 32 °F | 15 | 8 | 3 | 0 | 0 | 0 | 0 | 0 | 0 | 0 | 3 | 9 | 38 |
| Days Min Temp ≤ 0 °F | 0 | 0 | 0 | 0 | 0 | 0 | 0 | 0 | 0 | 0 | 0 | 0 | 0 |
| Heating Degree Days | 607 | 438 | 243 | 69 | 9 | 0 | 0 | 0 | 6 | 63 | 260 | 491 | 2186 |
| Cooling Degree Days | 1 | 5 | 29 | 102 | 256 | 445 | 568 | 577 | 383 | 156 | 33 | 6 | 2561 |
| Total Precipitation (") | 2.13 | 2.76 | 2.47 | 2.84 | 5.32 | 3.48 | 1.78 | 2.12 | 3.54 | 3.60 | 2.58 | 2.45 | 35.07 |
| Days ≥ 0.1" Precip | 4 | 4 | 5 | 4 | 6 | 5 | 3 | 4 | 5 | 5 | 5 | 4 | 54 |
| Total Snowfall (") | 0.0 | *0.3* | 0.0 | 0.0 | 0.0 | 0.0 | 0.0 | 0.0 | 0.0 | 0.0 | 0.0 | 0.0 | 0.3 |
| Days ≥ 1" Snow Depth | 0 | *0* | 0 | 0 | 0 | 0 | 0 | 0 | 0 | 0 | 0 | 0 | 0 |

## STRATFORD *Sherman County* ELEVATION 3704 ft LAT/LONG 36° 20 ' N / 102° 4 ' W

| | JAN | FEB | MAR | APR | MAY | JUN | JUL | AUG | SEP | OCT | NOV | DEC | YEAR |
|---|---|---|---|---|---|---|---|---|---|---|---|---|---|
| Maximum Temp °F | 47.2 | 51.8 | 60.2 | 70.0 | 77.4 | 86.6 | 91.3 | 89.0 | 81.5 | 71.7 | 58.5 | 48.9 | 69.5 |
| Minimum Temp °F | 18.4 | 21.8 | 28.9 | 38.1 | 47.6 | 57.7 | 63.0 | 61.2 | 52.8 | 40.4 | 28.3 | 20.5 | 39.9 |
| Mean Temp °F | 32.8 | 36.8 | 44.6 | 54.1 | 62.6 | 72.2 | 77.2 | 75.2 | 67.2 | 56.1 | 43.4 | 34.7 | 54.7 |
| Days Max Temp ≥ 90 °F | 0 | 0 | 0 | 0 | 3 | 12 | 21 | 17 | 6 | 1 | 0 | 0 | 60 |
| Days Max Temp ≤ 32 °F | 5 | 3 | 1 | 0 | 0 | 0 | 0 | 0 | 0 | 0 | 1 | 4 | 14 |
| Days Min Temp ≤ 32 °F | 30 | 26 | 20 | 7 | 1 | 0 | 0 | 0 | 0 | 4 | 21 | 29 | 138 |
| Days Min Temp ≤ 0 °F | 1 | 1 | 0 | 0 | 0 | 0 | 0 | 0 | 0 | 0 | 0 | 1 | 3 |
| Heating Degree Days | 989 | 789 | 626 | 330 | 128 | 16 | 1 | 3 | 64 | 287 | 640 | 932 | 4805 |
| Cooling Degree Days | 0 | 0 | 0 | 8 | 50 | 228 | 377 | 332 | 136 | 10 | 0 | 0 | 1141 |
| Total Precipitation (") | 0.44 | 0.49 | 0.98 | 1.25 | 2.95 | 2.49 | 2.45 | 2.79 | 1.96 | 0.91 | 0.72 | 0.47 | 17.90 |
| Days ≥ 0.1" Precip | 1 | 2 | 3 | 2 | 5 | 5 | 5 | 5 | 3 | 2 | 2 | 1 | 36 |
| Total Snowfall (") | 3.9 | 3.1 | 3.6 | 1.6 | 0.4 | 0.0 | 0.0 | 0.0 | 0.0 | 0.6 | 1.7 | 3.2 | 18.1 |
| Days ≥ 1" Snow Depth | 2 | 1 | 1 | 0 | 0 | 0 | 0 | 0 | 0 | 0 | 1 | *1* | 6 |

## SUGAR LAND *Fort Bend County* ELEVATION 79 ft LAT/LONG 29° 37 ' N / 95° 38 ' W

| | JAN | FEB | MAR | APR | MAY | JUN | JUL | AUG | SEP | OCT | NOV | DEC | YEAR |
|---|---|---|---|---|---|---|---|---|---|---|---|---|---|
| Maximum Temp °F | 61.1 | 64.6 | 72.0 | 79.3 | 84.9 | 90.4 | 93.3 | 93.1 | 88.7 | 81.7 | 72.2 | 64.7 | 78.8 |
| Minimum Temp °F | 41.3 | 43.7 | 51.4 | 59.5 | 66.3 | 72.0 | 74.0 | 73.4 | 68.8 | 59.2 | 50.7 | 43.7 | 58.7 |
| Mean Temp °F | 51.2 | 54.2 | 61.7 | 69.4 | 75.6 | 81.2 | 83.6 | 83.3 | 78.8 | 70.5 | 61.5 | 54.3 | 68.8 |
| Days Max Temp ≥ 90 °F | 0 | 0 | 0 | 0 | 5 | 20 | 27 | 26 | 16 | 3 | 0 | 0 | 97 |
| Days Max Temp ≤ 32 °F | 0 | 0 | 0 | 0 | 0 | 0 | 0 | 0 | 0 | 0 | 0 | 0 | 0 |
| Days Min Temp ≤ 32 °F | 6 | 3 | 1 | 0 | 0 | 0 | 0 | 0 | 0 | 0 | 1 | 4 | 15 |
| Days Min Temp ≤ 0 °F | 0 | 0 | 0 | 0 | 0 | 0 | 0 | 0 | 0 | 0 | 0 | 0 | 0 |
| Heating Degree Days | 436 | 314 | 155 | 35 | 2 | 0 | 0 | 0 | 1 | 33 | 175 | 348 | 1499 |
| Cooling Degree Days | 9 | 16 | 60 | 164 | 341 | 504 | 592 | 589 | 430 | 220 | 77 | 22 | 3024 |
| Total Precipitation (") | 3.85 | 3.06 | 3.09 | 3.24 | 5.27 | 5.20 | 3.49 | 4.27 | 5.43 | 4.21 | 3.88 | 3.43 | 48.42 |
| Days ≥ 0.1" Precip | 6 | 5 | 5 | 4 | 6 | 5 | 6 | 6 | 6 | 5 | 5 | 6 | 65 |
| Total Snowfall (") | 0.1 | 0.0 | 0.0 | 0.0 | 0.0 | 0.0 | 0.0 | 0.0 | 0.0 | 0.0 | 0.0 | 0.0 | 0.1 |
| Days ≥ 1" Snow Depth | 0 | 0 | 0 | 0 | 0 | 0 | 0 | 0 | 0 | 0 | 0 | 0 | 0 |

## SULPHUR SPRINGS *Hopkins County* ELEVATION 502 ft LAT/LONG 33° 9 ' N / 95° 38 ' W

| | JAN | FEB | MAR | APR | MAY | JUN | JUL | AUG | SEP | OCT | NOV | DEC | YEAR |
|---|---|---|---|---|---|---|---|---|---|---|---|---|---|
| Maximum Temp °F | 53.0 | 57.6 | 66.1 | 74.9 | 81.2 | 88.8 | 93.2 | 94.0 | 86.9 | 77.1 | 65.2 | 56.6 | 74.6 |
| Minimum Temp °F | 30.7 | 34.5 | 42.9 | 51.3 | 59.5 | 67.3 | 70.7 | 69.5 | 62.8 | 51.3 | 41.9 | 34.0 | 51.4 |
| Mean Temp °F | 41.9 | 46.1 | 54.5 | 63.1 | 70.4 | 78.1 | 82.0 | 81.8 | 74.9 | 64.1 | 53.6 | 45.4 | 63.0 |
| Days Max Temp ≥ 90 °F | 0 | 0 | 0 | 0 | 2 | 15 | 25 | 26 | 13 | 2 | 0 | 0 | 83 |
| Days Max Temp ≤ 32 °F | 2 | 1 | 0 | 0 | 0 | 0 | 0 | 0 | 0 | 0 | 0 | 1 | 4 |
| Days Min Temp ≤ 32 °F | 19 | 12 | 5 | 1 | 0 | 0 | 0 | 0 | 0 | 0 | 6 | 14 | 57 |
| Days Min Temp ≤ 0 °F | 0 | 0 | 0 | 0 | 0 | 0 | 0 | 0 | 0 | 0 | 0 | 0 | 0 |
| Heating Degree Days | 712 | 531 | 335 | 117 | 22 | 1 | 0 | 0 | 14 | 113 | 352 | 604 | 2801 |
| Cooling Degree Days | 0 | 2 | 16 | 67 | 199 | 415 | 534 | 550 | 329 | 89 | 15 | 2 | 2218 |
| Total Precipitation (") | 2.63 | 3.48 | 4.09 | 4.72 | 5.64 | 4.10 | 3.44 | 2.28 | 3.96 | 5.26 | 3.93 | 4.12 | 47.65 |
| Days ≥ 0.1" Precip | 5 | 6 | 6 | 5 | 7 | 6 | 4 | 4 | 5 | 5 | 5 | 6 | 64 |
| Total Snowfall (") | *0.0* | 0.2 | 0.3 | 0.0 | 0.0 | 0.0 | 0.0 | 0.0 | 0.0 | 0.0 | 0.0 | 0.2 | 0.7 |
| Days ≥ 1" Snow Depth | 0 | 0 | 0 | 0 | 0 | 0 | 0 | 0 | 0 | 0 | 0 | 0 | 0 |

## TAHOKA *Lynn County* ELEVATION 3091 ft LAT/LONG 33° 10 ' N / 101° 47 ' W

| | JAN | FEB | MAR | APR | MAY | JUN | JUL | AUG | SEP | OCT | NOV | DEC | YEAR |
|---|---|---|---|---|---|---|---|---|---|---|---|---|---|
| Maximum Temp °F | 53.1 | 58.5 | 67.1 | 76.2 | 82.8 | 89.9 | 91.9 | 90.2 | 83.4 | 75.4 | 63.8 | 55.9 | 74.0 |
| Minimum Temp °F | 24.7 | 27.9 | 34.8 | 44.7 | 53.9 | 62.7 | 66.2 | 64.5 | 57.9 | 46.7 | 35.0 | 27.2 | 45.5 |
| Mean Temp °F | 38.9 | 43.2 | 51.0 | 60.5 | 68.4 | 76.3 | 79.1 | 77.4 | 70.6 | 61.1 | 49.4 | 41.6 | 59.8 |
| Days Max Temp ≥ 90 °F | 0 | 0 | 0 | 2 | 7 | 17 | 22 | 19 | 8 | 2 | 0 | 0 | 77 |
| Days Max Temp ≤ 32 °F | 3 | 1 | 0 | 0 | 0 | 0 | 0 | 0 | 0 | 0 | 0 | 2 | 6 |
| Days Min Temp ≤ 32 °F | 26 | 21 | 11 | 2 | 0 | 0 | 0 | 0 | 0 | 1 | 12 | 24 | 97 |
| Days Min Temp ≤ 0 °F | 0 | 0 | 0 | 0 | 0 | 0 | 0 | 0 | 0 | 0 | 0 | 0 | 0 |
| Heating Degree Days | 803 | 608 | 432 | 172 | 48 | 2 | 0 | 1 | 34 | 163 | 462 | 719 | 3444 |
| Cooling Degree Days | 0 | 0 | 4 | 44 | 166 | 361 | 456 | 415 | 219 | 47 | 2 | 0 | 1714 |
| Total Precipitation (") | 0.63 | 0.78 | 0.86 | 1.42 | 2.77 | 2.91 | 2.49 | 2.15 | 2.75 | 1.81 | 0.88 | 0.78 | 20.23 |
| Days ≥ 0.1" Precip | 2 | 2 | 2 | 3 | 4 | 4 | 4 | 4 | 5 | 3 | 2 | 2 | 37 |
| Total Snowfall (") | 3.1 | 2.6 | 1.5 | 0.3 | 0.0 | 0.0 | 0.0 | 0.0 | 0.0 | 0.1 | 1.1 | 1.9 | 10.6 |
| Days ≥ 1" Snow Depth | 2 | 1 | 1 | 0 | 0 | 0 | 0 | 0 | 0 | 0 | 0 | *1* | 5 |

## TAYLOR *Williamson County* ELEVATION 571 ft LAT/LONG 30° 35 ' N / 97° 25 ' W

| | JAN | FEB | MAR | APR | MAY | JUN | JUL | AUG | SEP | OCT | NOV | DEC | YEAR |
|---|---|---|---|---|---|---|---|---|---|---|---|---|---|
| Maximum Temp °F | 58.0 | 62.4 | 71.5 | 79.1 | 84.2 | 90.9 | 95.2 | 95.7 | 89.7 | 80.7 | 69.7 | 61.5 | 78.2 |
| Minimum Temp °F | 35.1 | 38.1 | 46.2 | 54.7 | 62.2 | 68.9 | 71.6 | 70.9 | 65.6 | 55.3 | 45.1 | 37.8 | 54.3 |
| Mean Temp °F | 46.6 | 50.3 | 58.9 | 66.9 | 73.2 | 79.9 | 83.4 | 83.3 | 77.7 | 68.0 | 57.4 | 49.7 | 66.3 |
| Days Max Temp ≥ 90 °F | 0 | 0 | 1 | 2 | 5 | 20 | 28 | 29 | 18 | 4 | 0 | 0 | 107 |
| Days Max Temp ≤ 32 °F | 1 | 0 | 0 | 0 | 0 | 0 | 0 | 0 | 0 | 0 | 0 | 0 | 1 |
| Days Min Temp ≤ 32 °F | 13 | 8 | 3 | 0 | 0 | 0 | 0 | 0 | 0 | 0 | 4 | 9 | 37 |
| Days Min Temp ≤ 0 °F | 0 | 0 | 0 | 0 | 0 | 0 | 0 | 0 | 0 | 0 | 0 | 0 | 0 |
| Heating Degree Days | 568 | 416 | 226 | 60 | 8 | 0 | 0 | 0 | 4 | 58 | 256 | 476 | 2072 |
| Cooling Degree Days | 2 | 6 | 42 | 113 | 273 | 454 | 577 | 582 | 393 | 170 | 39 | 7 | 2658 |
| Total Precipitation (") | 2.40 | 2.68 | 2.43 | 2.94 | 5.18 | 3.33 | 1.61 | 2.08 | 3.88 | 3.80 | 2.65 | 2.61 | 35.59 |
| Days ≥ 0.1" Precip | 5 | 5 | 4 | 4 | 6 | 5 | 3 | 3 | 5 | 5 | 4 | 4 | 53 |
| Total Snowfall (") | 0.0 | 0.2 | 0.1 | 0.0 | 0.0 | 0.0 | 0.0 | 0.0 | 0.0 | 0.0 | 0.0 | 0.0 | 0.3 |
| Days ≥ 1" Snow Depth | 0 | 0 | 0 | 0 | 0 | 0 | 0 | 0 | 0 | 0 | 0 | 0 | 0 |

## TEMPLE *Bell County* ELEVATION 702 ft LAT/LONG 31° 7 ' N / 97° 20 ' W

| | JAN | FEB | MAR | APR | MAY | JUN | JUL | AUG | SEP | OCT | NOV | DEC | YEAR |
|---|---|---|---|---|---|---|---|---|---|---|---|---|---|
| Maximum Temp °F | 56.4 | 60.8 | 69.7 | 78.1 | 83.5 | 90.6 | 95.2 | 95.7 | 88.7 | 79.6 | 68.2 | 59.8 | 77.2 |
| Minimum Temp °F | 35.1 | 38.5 | 46.7 | 55.5 | 62.8 | 69.8 | 72.5 | 72.4 | 66.6 | 56.3 | 46.0 | 38.4 | 55.1 |
| Mean Temp °F | 45.8 | 49.7 | 58.2 | 66.8 | 73.2 | 80.2 | 83.9 | 84.1 | 77.6 | 68.0 | 57.1 | 49.1 | 66.1 |
| Days Max Temp ≥ 90 °F | 0 | 0 | 0 | 1 | 5 | 19 | 28 | 27 | 16 | 3 | 0 | 0 | 99 |
| Days Max Temp ≤ 32 °F | 1 | 1 | 0 | 0 | 0 | 0 | 0 | 0 | 0 | 0 | 0 | 0 | 2 |
| Days Min Temp ≤ 32 °F | 12 | 7 | 2 | 0 | 0 | 0 | 0 | 0 | 0 | 0 | 2 | 7 | 30 |
| Days Min Temp ≤ 0 °F | 0 | 0 | 0 | 0 | 0 | 0 | 0 | 0 | 0 | 0 | 0 | 0 | 0 |
| Heating Degree Days | 592 | 432 | 242 | 64 | 6 | 0 | 0 | 0 | 5 | 58 | 258 | 491 | 2148 |
| Cooling Degree Days | 1 | 8 | 37 | 116 | 265 | *457* | 581 | 606 | 387 | 155 | 31 | 5 | 2649 |
| Total Precipitation (") | 2.12 | 2.84 | 2.64 | 2.73 | 5.26 | 3.49 | 1.72 | 2.18 | 4.05 | 3.47 | 2.80 | 2.67 | 35.97 |
| Days ≥ 0.1" Precip | 5 | 5 | 4 | 4 | 6 | 5 | 3 | 3 | 5 | 4 | 4 | 4 | 52 |
| Total Snowfall (") | 0.4 | 0.6 | 0.0 | 0.0 | 0.0 | 0.0 | 0.0 | 0.0 | 0.0 | 0.0 | 0.0 | 0.0 | 1.0 |
| Days ≥ 1" Snow Depth | 0 | 0 | 0 | 0 | 0 | 0 | 0 | 0 | 0 | 0 | 0 | 0 | 0 |

## THOMPSONS 3 WSW *Fort Bend County* ELEVATION 72 ft LAT/LONG 29° 29 ' N / 95° 38 ' W

| | JAN | FEB | MAR | APR | MAY | JUN | JUL | AUG | SEP | OCT | NOV | DEC | YEAR |
|---|---|---|---|---|---|---|---|---|---|---|---|---|---|
| Maximum Temp °F | 61.5 | 64.6 | 72.1 | 79.3 | 84.8 | 90.4 | 93.2 | 93.2 | 89.1 | 82.1 | 72.9 | 65.2 | 79.0 |
| Minimum Temp °F | 41.8 | 44.0 | 51.6 | 59.2 | 65.3 | 70.9 | 72.8 | 72.5 | 68.6 | 59.0 | 51.3 | 44.4 | 58.5 |
| Mean Temp °F | 51.7 | 54.3 | 61.9 | 69.3 | 75.1 | 80.7 | 83.0 | 82.9 | 78.9 | 70.5 | 62.1 | 54.9 | 68.8 |
| Days Max Temp ≥ 90 °F | 0 | 0 | 0 | 1 | 5 | 20 | 26 | 26 | 16 | 4 | 0 | 0 | 98 |
| Days Max Temp ≤ 32 °F | 0 | 0 | 0 | 0 | 0 | 0 | 0 | 0 | 0 | 0 | 0 | 0 | 0 |
| Days Min Temp ≤ 32 °F | 6 | 3 | 1 | 0 | 0 | 0 | 0 | 0 | 0 | 0 | 1 | 3 | 14 |
| Days Min Temp ≤ 0 °F | 0 | 0 | 0 | 0 | 0 | 0 | 0 | 0 | 0 | 0 | 0 | 0 | 0 |
| Heating Degree Days | 420 | 310 | 149 | 33 | 2 | 0 | 0 | 0 | 1 | 31 | 160 | 331 | 1437 |
| Cooling Degree Days | 8 | 16 | 65 | 167 | 335 | 504 | 585 | 589 | 437 | 220 | 88 | 28 | 3042 |
| Total Precipitation (") | 3.84 | 2.82 | 2.78 | 3.34 | 5.08 | 4.76 | 3.56 | 4.18 | 5.00 | 3.73 | 3.99 | 3.35 | 46.43 |
| Days ≥ 0.1" Precip | 7 | 5 | 5 | 4 | 6 | 6 | 5 | 6 | 6 | 5 | 4 | 6 | 65 |
| Total Snowfall (") | 0.0 | 0.1 | 0.0 | 0.0 | 0.0 | 0.0 | 0.0 | 0.0 | 0.0 | 0.0 | 0.0 | 0.0 | 0.1 |
| Days ≥ 1" Snow Depth | 0 | 0 | 0 | 0 | 0 | 0 | 0 | 0 | 0 | 0 | 0 | 0 | 0 |

## THROCKMORTON *Throckmorton County* ELEVATION 1322 ft LAT/LONG 33° 11 ' N / 99° 11 ' W

| | JAN | FEB | MAR | APR | MAY | JUN | JUL | AUG | SEP | OCT | NOV | DEC | YEAR |
|---|---|---|---|---|---|---|---|---|---|---|---|---|---|
| Maximum Temp °F | 53.3 | 57.9 | 67.4 | 76.5 | 83.0 | 91.0 | 95.9 | 95.0 | 86.9 | 77.3 | 65.2 | 56.4 | 75.5 |
| Minimum Temp °F | 28.0 | 32.0 | 40.8 | 50.0 | 58.4 | 67.0 | 70.9 | 69.6 | 62.4 | 51.0 | 40.0 | 31.1 | 50.1 |
| Mean Temp °F | 40.7 | 45.0 | 54.2 | 63.3 | 70.7 | 79.0 | 83.5 | 82.3 | 74.7 | 64.2 | 52.7 | 43.8 | 62.8 |
| Days Max Temp ≥ 90 °F | 0 | 0 | 1 | 2 | 6 | 19 | 27 | 25 | 14 | 3 | 0 | 0 | 97 |
| Days Max Temp ≤ 32 °F | 3 | 2 | 0 | 0 | 0 | 0 | 0 | 0 | 0 | 0 | 0 | 1 | 6 |
| Days Min Temp ≤ 32 °F | 21 | 14 | 6 | 1 | 0 | 0 | 0 | 0 | 0 | 0 | 7 | 18 | 67 |
| Days Min Temp ≤ 0 °F | 0 | 0 | 0 | 0 | 0 | 0 | 0 | 0 | 0 | 0 | 0 | 0 | 0 |
| Heating Degree Days | 748 | 559 | 352 | 125 | 28 | 1 | 0 | 0 | 18 | 115 | 377 | 652 | 2975 |
| Cooling Degree Days | 0 | 2 | 20 | 73 | 213 | 423 | 573 | 563 | 333 | 101 | 15 | 1 | 2317 |
| Total Precipitation (") | 1.08 | 1.51 | 1.53 | 2.52 | 3.63 | 3.00 | 1.72 | 2.66 | 3.83 | 2.85 | 1.33 | 1.40 | 27.06 |
| Days ≥ 0.1" Precip | 2 | 3 | 3 | 4 | 5 | 4 | 3 | 4 | 5 | 4 | 3 | 3 | 43 |
| Total Snowfall (") | 1.5 | 1.4 | 0.7 | 0.0 | 0.0 | 0.0 | 0.0 | 0.0 | 0.0 | 0.0 | 0.5 | 0.7 | 4.8 |
| Days ≥ 1" Snow Depth | 1 | 1 | 0 | 0 | 0 | 0 | 0 | 0 | 0 | 0 | 0 | 1 | 3 |

## TILDEN 4 SSE *McMullen County* ELEVATION 259 ft LAT/LONG 28° 28 ' N / 98° 33 ' W

| | JAN | FEB | MAR | APR | MAY | JUN | JUL | AUG | SEP | OCT | NOV | DEC | YEAR |
|---|---|---|---|---|---|---|---|---|---|---|---|---|---|
| Maximum Temp °F | 64.3 | 68.8 | 77.4 | 84.5 | 88.9 | 94.6 | 97.5 | 98.1 | 92.7 | 85.0 | 75.3 | 68.1 | 82.9 |
| Minimum Temp °F | 40.3 | 43.8 | 51.6 | 59.7 | 66.3 | 71.3 | 72.9 | 72.5 | 68.4 | 59.2 | 50.2 | 43.4 | 58.3 |
| Mean Temp °F | 52.3 | 56.4 | 64.5 | 72.1 | 77.6 | 83.0 | 85.2 | 85.3 | 80.6 | 72.2 | 62.8 | 55.7 | 70.6 |
| Days Max Temp ≥ 90 °F | 0 | 1 | 3 | 8 | 15 | 26 | 30 | 30 | 23 | 10 | 1 | 0 | 147 |
| Days Max Temp ≤ 32 °F | 0 | 0 | 0 | 0 | 0 | 0 | 0 | 0 | 0 | 0 | 0 | 0 | 0 |
| Days Min Temp ≤ 32 °F | 6 | 4 | 1 | 0 | 0 | 0 | 0 | 0 | 0 | 0 | 2 | 5 | 18 |
| Days Min Temp ≤ 0 °F | 0 | 0 | 0 | 0 | 0 | 0 | 0 | 0 | 0 | 0 | 0 | 0 | 0 |
| Heating Degree Days | 400 | 263 | 112 | 23 | 2 | 0 | 0 | 0 | 1 | 26 | 152 | 307 | 1286 |
| Cooling Degree Days | 9 | 30 | 103 | 239 | 419 | 566 | 647 | 655 | 493 | 277 | 100 | 31 | 3569 |
| Total Precipitation (") | 1.49 | 1.54 | 1.08 | 2.04 | 3.48 | 3.01 | 1.70 | 2.31 | 3.65 | 2.09 | 1.13 | 1.32 | 24.84 |
| Days ≥ 0.1" Precip | 3 | 3 | 2 | 3 | 5 | 4 | 3 | 3 | 4 | 3 | 2 | 2 | 37 |
| Total Snowfall (") | 0.0 | 0.1 | 0.0 | 0.0 | 0.0 | 0.0 | 0.0 | 0.0 | 0.0 | 0.0 | 0.0 | 0.0 | 0.1 |
| Days ≥ 1" Snow Depth | 0 | 0 | 0 | 0 | 0 | 0 | 0 | 0 | 0 | 0 | 0 | 0 | 0 |

## TRUSCOTT 2 NW *Knox County* ELEVATION 1552 ft LAT/LONG 33° 45 ' N / 99° 48 ' W

| | JAN | FEB | MAR | APR | MAY | JUN | JUL | AUG | SEP | OCT | NOV | DEC | YEAR |
|---|---|---|---|---|---|---|---|---|---|---|---|---|---|
| Maximum Temp °F | 52.6 | 57.2 | 66.8 | 77.0 | 84.0 | 92.6 | 97.3 | 95.3 | 87.0 | 77.3 | 64.5 | 55.4 | 75.6 |
| Minimum Temp °F | 26.7 | 30.4 | 38.5 | 48.9 | 57.5 | 66.9 | 71.4 | 69.4 | 62.0 | 50.3 | 38.2 | 29.5 | 49.1 |
| Mean Temp °F | 39.7 | 43.8 | 52.7 | 63.0 | 70.8 | 79.8 | 84.4 | 82.4 | 74.6 | 63.8 | 51.4 | 42.4 | 62.4 |
| Days Max Temp ≥ 90 °F | 0 | 0 | 1 | 4 | 9 | 21 | 28 | 25 | 14 | 4 | 0 | 0 | 106 |
| Days Max Temp ≤ 32 °F | 4 | 2 | 0 | 0 | 0 | 0 | 0 | 0 | 0 | 0 | 0 | 2 | 8 |
| Days Min Temp ≤ 32 °F | 23 | 16 | 7 | 1 | 0 | 0 | 0 | 0 | 0 | 0 | 8 | 20 | 75 |
| Days Min Temp ≤ 0 °F | 0 | 0 | 0 | 0 | 0 | 0 | 0 | 0 | 0 | 0 | 0 | 0 | 0 |
| Heating Degree Days | 778 | 592 | 390 | 130 | 29 | 1 | 0 | 0 | 19 | 121 | 409 | 692 | 3161 |
| Cooling Degree Days | 0 | 1 | 12 | 74 | 217 | 443 | 604 | 569 | 329 | 91 | 8 | 0 | 2348 |
| Total Precipitation (") | 0.97 | 1.36 | 1.45 | 1.99 | 3.90 | 3.09 | 1.80 | 2.57 | 3.51 | 2.91 | 1.16 | 0.95 | 25.66 |
| Days ≥ 0.1" Precip | 2 | 3 | 3 | 4 | 6 | 4 | 3 | 4 | 5 | 4 | 2 | 2 | 42 |
| Total Snowfall (") | ***0.8*** | na | 0.1 | 0.0 | 0.0 | 0.0 | 0.0 | 0.0 | 0.0 | 0.0 | 0.0 | 0.3 | na |
| Days ≥ 1" Snow Depth | ***1*** | ***0*** | 0 | 0 | 0 | 0 | 0 | 0 | 0 | 0 | 0 | 0 | 1 |

## TULIA *Swisher County* ELEVATION 3504 ft LAT/LONG 34° 32 ' N / 101° 46 ' W

| | JAN | FEB | MAR | APR | MAY | JUN | JUL | AUG | SEP | OCT | NOV | DEC | YEAR |
|---|---|---|---|---|---|---|---|---|---|---|---|---|---|
| Maximum Temp °F | 50.4 | 54.9 | 63.5 | 72.9 | 79.9 | 88.2 | 91.1 | 88.8 | 82.2 | 73.2 | 60.5 | 52.2 | 71.5 |
| Minimum Temp °F | 22.1 | 25.1 | 31.9 | 40.9 | 50.4 | 60.1 | 64.4 | 62.8 | 55.6 | 43.8 | 32.0 | 24.2 | 42.8 |
| Mean Temp °F | 36.2 | 40.0 | 47.7 | 56.9 | 65.2 | 74.2 | 77.8 | 75.8 | 68.9 | 58.5 | 46.3 | 38.2 | 57.1 |
| Days Max Temp ≥ 90 °F | 0 | 0 | 0 | 1 | 5 | 15 | 21 | 17 | 7 | 1 | 0 | 0 | 67 |
| Days Max Temp ≤ 32 °F | 4 | 2 | 1 | 0 | 0 | 0 | 0 | 0 | 0 | 0 | 1 | 3 | 11 |
| Days Min Temp ≤ 32 °F | 28 | 23 | 16 | 4 | 0 | 0 | 0 | 0 | 0 | 3 | 16 | 27 | 117 |
| Days Min Temp ≤ 0 °F | 0 | 0 | 0 | 0 | 0 | 0 | 0 | 0 | 0 | 0 | 0 | 1 | 1 |
| Heating Degree Days | 885 | 699 | 531 | 256 | 86 | 8 | 0 | 2 | 47 | 223 | 555 | 823 | 4115 |
| Cooling Degree Days | 0 | 0 | 1 | 21 | 94 | 289 | 401 | 356 | 178 | 24 | 1 | 0 | 1365 |
| Total Precipitation (") | 0.54 | 0.66 | 0.97 | 1.08 | 2.63 | 3.85 | 2.21 | 2.68 | 2.44 | 1.46 | 0.76 | 0.64 | 19.92 |
| Days ≥ 0.1" Precip | 2 | 2 | 2 | 3 | 5 | 5 | 4 | 5 | 4 | 3 | 2 | 2 | 39 |
| Total Snowfall (") | 3.4 | 3.9 | 1.1 | 0.4 | 0.0 | 0.0 | 0.0 | 0.0 | 0.0 | 0.2 | 1.7 | 3.3 | 14.0 |
| Days ≥ 1" Snow Depth | 3 | 2 | 1 | 0 | 0 | 0 | 0 | 0 | 0 | 0 | 1 | 2 | 9 |

## TURKEY *Hall County* ELEVATION 2352 ft LAT/LONG 34° 24 ' N / 100° 53 ' W

| | JAN | FEB | MAR | APR | MAY | JUN | JUL | AUG | SEP | OCT | NOV | DEC | YEAR |
|---|---|---|---|---|---|---|---|---|---|---|---|---|---|
| Maximum Temp °F | 54.0 | 59.4 | 68.7 | 77.9 | 84.2 | 91.9 | 96.0 | 93.8 | 85.9 | 77.5 | 64.6 | 56.2 | 75.8 |
| Minimum Temp °F | 27.4 | 31.2 | 38.7 | 48.2 | 56.9 | 65.5 | 70.0 | 68.2 | 60.6 | 49.8 | 38.5 | 30.1 | 48.8 |
| Mean Temp °F | 40.7 | 45.3 | 53.7 | 63.0 | 70.6 | 78.8 | 83.1 | 81.0 | 73.3 | 63.7 | 51.6 | 43.2 | 62.3 |
| Days Max Temp ≥ 90 °F | 0 | 0 | 1 | 3 | 8 | 20 | 27 | 24 | 11 | 3 | 0 | 0 | 97 |
| Days Max Temp ≤ 32 °F | 3 | 1 | 0 | 0 | 0 | 0 | 0 | 0 | 0 | 0 | 0 | 1 | 5 |
| Days Min Temp ≤ 32 °F | 21 | 15 | 8 | 1 | 0 | 0 | 0 | 0 | 0 | 1 | 8 | 18 | 72 |
| Days Min Temp ≤ 0 °F | 0 | 0 | 0 | 0 | 0 | 0 | 0 | 0 | 0 | 0 | 0 | 0 | 0 |
| Heating Degree Days | 745 | 550 | 364 | 126 | 29 | 1 | 0 | 1 | 22 | 124 | 402 | 668 | 3032 |
| Cooling Degree Days | 0 | 1 | 12 | 73 | 206 | 410 | 559 | 520 | 299 | 83 | 8 | 0 | 2171 |
| Total Precipitation (") | 0.65 | 0.90 | 1.19 | 1.55 | 2.80 | 3.62 | 2.00 | 2.39 | 2.88 | 1.55 | 0.93 | 0.77 | 21.23 |
| Days ≥ 0.1" Precip | 2 | 2 | 2 | 3 | 5 | 5 | 3 | 5 | 4 | 3 | 2 | 2 | 38 |
| Total Snowfall (") | 2.8 | 2.4 | 0.9 | 0.1 | 0.0 | 0.0 | 0.0 | 0.0 | 0.0 | 0.1 | 0.9 | 1.6 | 8.8 |
| Days ≥ 1" Snow Depth | 2 | 2 | 0 | 0 | 0 | 0 | 0 | 0 | 0 | 0 | 1 | 1 | 6 |

## VAN HORN *Culberson County* ELEVATION 3955 ft LAT/LONG 31° 4 ' N / 104° 47 ' W

| | JAN | FEB | MAR | APR | MAY | JUN | JUL | AUG | SEP | OCT | NOV | DEC | YEAR |
|---|---|---|---|---|---|---|---|---|---|---|---|---|---|
| Maximum Temp °F | 57.9 | 63.2 | *71.0* | 79.9 | *87.1* | na | na | 91.4 | 85.9 | *79.0* | *67.6* | *59.8* | na |
| Minimum Temp °F | 29.0 | 32.0 | *38.8* | 46.6 | *54.9* | na | na | *64.9* | 58.6 | *48.2* | *37.4* | *29.9* | na |
| Mean Temp °F | 43.5 | 47.7 | *54.9* | 63.3 | *71.1* | na | na | *77.9* | 72.3 | *63.6* | *52.5* | *44.8* | na |
| Days Max Temp ≥ 90 °F | 0 | 0 | *0* | *2* | *14* | na | na | 20 | 11 | 2 | 0 | 0 | na |
| Days Max Temp ≤ 32 °F | 1 | 0 | *0* | 0 | *0* | *0* | *0* | 0 | 0 | 0 | 0 | 0 | 1 |
| Days Min Temp ≤ 32 °F | 21 | 15 | *7* | *1* | *0* | *0* | *0* | 0 | 0 | *0* | *8* | *19* | 71 |
| Days Min Temp ≤ 0 °F | 0 | 0 | *0* | 0 | *0* | *0* | *0* | 0 | 0 | 0 | 0 | 0 | 0 |
| Heating Degree Days | 660 | 484 | *313* | 106 | *19* | na | *0* | *1* | 21 | *100* | *375* | *621* | na |
| Cooling Degree Days | 0 | 0 | *4* | *51* | na | na | na | *425* | 243 | 63 | 4 | *0* | na |
| Total Precipitation (") | 0.54 | 0.35 | *0.22* | 0.27 | *0.78* | *1.19* | *2.04* | *2.55* | 2.52 | 1.20 | 0.69 | 0.63 | 12.98 |
| Days ≥ 0.1" Precip | 2 | 1 | *1* | 1 | *2* | *3* | *4* | 5 | 4 | *2* | 1 | 2 | 28 |
| Total Snowfall (") | 0.9 | 0.7 | *0.2* | 0.0 | *0.0* | *0.0* | *0.0* | 0.0 | 0.0 | 0.1 | 0.2 | 1.0 | 3.1 |
| Days ≥ 1" Snow Depth | *0* | *0* | *0* | 0 | *0* | *0* | *0* | 0 | 0 | 0 | *0* | 0 | 0 |

## VERNON 4 S *Wilbarger County* ELEVATION 1202 ft LAT/LONG 34° 5 ' N / 99° 18 ' W

| | JAN | FEB | MAR | APR | MAY | JUN | JUL | AUG | SEP | OCT | NOV | DEC | YEAR |
|---|---|---|---|---|---|---|---|---|---|---|---|---|---|
| Maximum Temp °F | 53.8 | 59.1 | 69.0 | 77.8 | 84.6 | 92.7 | 97.7 | 95.9 | 87.8 | 78.2 | 65.5 | 56.5 | 76.6 |
| Minimum Temp °F | 27.3 | 31.3 | 39.9 | 49.4 | 58.5 | 67.5 | 71.9 | 69.9 | 62.6 | 50.7 | 39.0 | 30.1 | 49.8 |
| Mean Temp °F | 40.6 | 45.3 | 54.4 | 63.6 | 71.5 | 80.1 | 84.8 | 83.0 | 75.2 | 64.5 | 52.3 | 43.3 | 63.2 |
| Days Max Temp ≥ 90 °F | 0 | 0 | 1 | 2 | 8 | 21 | 28 | 26 | 15 | 4 | 0 | 0 | 105 |
| Days Max Temp ≤ 32 °F | 2 | 1 | 0 | 0 | 0 | 0 | 0 | 0 | 0 | 0 | 0 | 2 | 5 |
| Days Min Temp ≤ 32 °F | 23 | 15 | 7 | 1 | 0 | 0 | 0 | 0 | 0 | 0 | 8 | 20 | 74 |
| Days Min Temp ≤ 0 °F | 0 | 0 | 0 | 0 | 0 | 0 | 0 | 0 | 0 | 0 | 0 | 0 | 0 |
| Heating Degree Days | 751 | 552 | 340 | 118 | 20 | 0 | 0 | 0 | 15 | 108 | 384 | 665 | 2953 |
| Cooling Degree Days | 0 | 1 | 13 | 67 | 219 | 447 | 606 | 575 | 326 | 88 | 9 | 0 | 2351 |
| Total Precipitation (") | 1.05 | 1.22 | 1.82 | 2.45 | 3.98 | 2.98 | 2.14 | 2.64 | 3.60 | 2.66 | 1.19 | 1.01 | 26.74 |
| Days ≥ 0.1" Precip | 2 | 3 | 3 | 4 | 6 | 4 | 3 | 4 | 5 | 4 | 2 | 2 | 42 |
| Total Snowfall (") | 1.6 | 1.0 | 0.5 | 0.0 | 0.0 | 0.0 | 0.0 | 0.0 | 0.0 | 0.0 | 0.1 | 0.2 | 3.4 |
| Days ≥ 1" Snow Depth | *1* | *0* | 0 | 0 | 0 | 0 | 0 | 0 | 0 | 0 | 0 | 0 | 1 |

## VICTORIA REGIONAL AP *Victoria County* ELEVATION 104 ft LAT/LONG 28° 51 ' N / 96° 55 ' W

| | JAN | FEB | MAR | APR | MAY | JUN | JUL | AUG | SEP | OCT | NOV | DEC | YEAR |
|---|---|---|---|---|---|---|---|---|---|---|---|---|---|
| Maximum Temp °F | 63.1 | 66.8 | 74.1 | 80.4 | 85.2 | 90.6 | 93.5 | 93.8 | 89.5 | 82.9 | 73.6 | 66.5 | 80.0 |
| Minimum Temp °F | 43.3 | 45.8 | 53.3 | 61.3 | 67.8 | 73.3 | 75.1 | 74.7 | 70.5 | 61.4 | 52.5 | 45.9 | 60.4 |
| Mean Temp °F | 53.2 | 56.3 | 63.7 | 70.9 | 76.5 | 82.0 | 84.3 | 84.2 | 80.0 | 72.2 | 63.1 | 56.2 | 70.2 |
| Days Max Temp ≥ 90 °F | 0 | 0 | 0 | 1 | 5 | 20 | 28 | 28 | 17 | 5 | 0 | 0 | 104 |
| Days Max Temp ≤ 32 °F | 0 | 0 | 0 | 0 | 0 | 0 | 0 | 0 | 0 | 0 | 0 | 0 | 0 |
| Days Min Temp ≤ 32 °F | 4 | 2 | 0 | 0 | 0 | 0 | 0 | 0 | 0 | 0 | 1 | 3 | 10 |
| Days Min Temp ≤ 0 °F | 0 | 0 | 0 | 0 | 0 | 0 | 0 | 0 | 0 | 0 | 0 | 0 | 0 |
| Heating Degree Days | 377 | 260 | 117 | 22 | 1 | 0 | 0 | 0 | 1 | 20 | 141 | 296 | 1235 |
| Cooling Degree Days | 12 | 26 | 81 | 196 | 370 | 530 | 616 | 618 | 469 | 264 | 94 | 36 | 3312 |
| Total Precipitation (") | 2.55 | 2.28 | 1.83 | 2.96 | 5.36 | 5.01 | 3.26 | 2.85 | 5.27 | 3.91 | 2.26 | 2.40 | 39.94 |
| Days ≥ 0.1" Precip | 5 | 4 | 3 | 3 | 5 | 5 | 4 | 5 | 6 | 4 | 4 | 4 | 52 |
| Total Snowfall (") | 0.1 | 0.0 | 0.0 | 0.0 | 0.0 | 0.0 | 0.0 | 0.0 | 0.0 | 0.0 | 0.0 | 0.0 | 0.1 |
| Days ≥ 1" Snow Depth | 0 | 0 | 0 | 0 | 0 | 0 | 0 | 0 | 0 | 0 | 0 | 0 | 0 |

## WACO DAM *McLennan County* ELEVATION 502 ft LAT/LONG 31° 36 ' N / 97° 13 ' W

| | JAN | FEB | MAR | APR | MAY | JUN | JUL | AUG | SEP | OCT | NOV | DEC | YEAR |
|---|---|---|---|---|---|---|---|---|---|---|---|---|---|
| Maximum Temp °F | 55.3 | 60.3 | 68.7 | 76.7 | 82.6 | 90.1 | 94.6 | 95.0 | 88.5 | 79.0 | 67.5 | 59.2 | 76.5 |
| Minimum Temp °F | 33.0 | 37.1 | 45.7 | 54.9 | 62.7 | 70.3 | 73.9 | 73.0 | 66.2 | 55.2 | 45.1 | 36.6 | 54.5 |
| Mean Temp °F | 44.2 | 48.8 | 57.2 | 65.9 | 72.7 | 80.2 | 84.3 | 84.0 | 77.4 | 67.1 | 56.3 | 47.9 | 65.5 |
| Days Max Temp ≥ 90 °F | 0 | 0 | 0 | 1 | 3 | 18 | 27 | 27 | 16 | 3 | 0 | 0 | 95 |
| Days Max Temp ≤ 32 °F | 1 | 1 | 0 | 0 | 0 | 0 | 0 | 0 | 0 | 0 | 0 | 1 | 3 |
| Days Min Temp ≤ 32 °F | 15 | 9 | 3 | 0 | 0 | 0 | 0 | 0 | 0 | 0 | 3 | 9 | 39 |
| Days Min Temp ≤ 0 °F | 0 | 0 | 0 | 0 | 0 | 0 | 0 | 0 | 0 | 0 | 0 | 0 | 0 |
| Heating Degree Days | 640 | 456 | 262 | 73 | 10 | 0 | 0 | 0 | 6 | 69 | 279 | 525 | 2320 |
| Cooling Degree Days | 0 | 4 | 29 | 100 | 258 | 463 | 610 | 620 | 399 | 145 | 30 | 3 | 2661 |
| Total Precipitation (") | 1.92 | 2.54 | 2.85 | 3.31 | 5.31 | 3.00 | 2.32 | 1.88 | 3.68 | 3.93 | 2.75 | 2.63 | 36.12 |
| Days ≥ 0.1" Precip | 5 | 4 | 5 | 4 | 7 | 5 | 3 | 3 | 5 | 5 | 4 | 4 | 54 |
| Total Snowfall (") | *0.2* | *0.2* | 0.0 | 0.0 | 0.0 | 0.0 | 0.0 | 0.0 | 0.0 | 0.0 | 0.0 | 0.0 | 0.4 |
| Days ≥ 1" Snow Depth | *0* | 0 | 0 | 0 | 0 | 0 | 0 | 0 | 0 | 0 | 0 | 0 | 0 |

## WACO MADISN COOPR AP *McLennan County* ELEVATION 509 ft LAT/LONG 31° 37 ' N / 97° 13 ' W

| | JAN | FEB | MAR | APR | MAY | JUN | JUL | AUG | SEP | OCT | NOV | DEC | YEAR |
|---|---|---|---|---|---|---|---|---|---|---|---|---|---|
| Maximum Temp °F | 56.2 | 60.9 | 69.7 | 77.9 | 84.2 | 92.0 | 96.7 | 96.9 | 89.6 | 80.0 | 68.3 | 59.6 | 77.7 |
| Minimum Temp °F | 35.4 | 39.0 | 47.0 | 55.8 | 63.6 | 71.2 | 74.7 | 74.0 | 67.6 | 57.0 | 46.6 | 38.4 | 55.9 |
| Mean Temp °F | 45.8 | 50.0 | 58.4 | 66.9 | 73.9 | 81.7 | 85.7 | 85.5 | 78.6 | 68.5 | 57.4 | 49.0 | 66.8 |
| Days Max Temp ≥ 90 °F | 0 | 0 | 0 | 1 | 7 | 22 | 29 | 28 | 18 | 4 | 0 | 0 | 109 |
| Days Max Temp ≤ 32 °F | 1 | 0 | 0 | 0 | 0 | 0 | 0 | 0 | 0 | 0 | 0 | 1 | 2 |
| Days Min Temp ≤ 32 °F | 13 | 7 | 2 | 0 | 0 | 0 | 0 | 0 | 0 | 0 | 3 | 9 | 34 |
| Days Min Temp ≤ 0 °F | 0 | 0 | 0 | 0 | 0 | 0 | 0 | 0 | 0 | 0 | 0 | 0 | 0 |
| Heating Degree Days | 589 | 423 | 238 | 64 | 7 | 0 | 0 | 0 | 4 | 57 | 258 | 496 | 2136 |
| Cooling Degree Days | 1 | 5 | 39 | 121 | 298 | 509 | 651 | 650 | 421 | 174 | 40 | 6 | 2915 |
| Total Precipitation (") | 1.70 | 2.35 | 2.49 | 3.22 | 4.92 | 2.85 | 1.93 | 1.83 | 3.47 | 3.70 | 2.43 | 2.32 | 33.21 |
| Days ≥ 0.1" Precip | 4 | 4 | 4 | 4 | 6 | 5 | 2 | 3 | 4 | 5 | 4 | 4 | 49 |
| Total Snowfall (") | 0.6 | 0.5 | 0.1 | 0.0 | 0.0 | 0.0 | 0.0 | 0.0 | 0.0 | 0.0 | 0.1 | 0.0 | 1.3 |
| Days ≥ 1" Snow Depth | 0 | 0 | 0 | 0 | 0 | 0 | 0 | 0 | 0 | 0 | 0 | 0 | 0 |

## WASHINGTON STATE PAR *Washington County* ELEVATION 210 ft LAT/LONG 30° 20 ' N / 96° 9 ' W

| | JAN | FEB | MAR | APR | MAY | JUN | JUL | AUG | SEP | OCT | NOV | DEC | YEAR |
|---|---|---|---|---|---|---|---|---|---|---|---|---|---|
| Maximum Temp °F | 59.0 | 63.0 | 71.3 | 78.5 | 84.2 | 90.7 | 94.6 | 95.0 | 88.9 | 80.9 | 70.3 | 62.5 | 78.2 |
| Minimum Temp °F | 37.9 | 41.2 | 48.9 | 57.2 | 63.9 | 70.0 | 72.1 | 71.6 | 66.7 | 56.6 | 48.2 | 40.8 | 56.3 |
| Mean Temp °F | 48.4 | 52.1 | 60.1 | 67.9 | 74.1 | 80.4 | 83.4 | 83.3 | 77.8 | 68.8 | 59.3 | 51.7 | 67.3 |
| Days Max Temp ≥ 90 °F | 0 | 0 | 0 | 1 | 5 | 20 | 27 | 28 | 15 | 4 | 0 | 0 | 100 |
| Days Max Temp ≤ 32 °F | 0 | 0 | 0 | 0 | 0 | 0 | 0 | 0 | 0 | 0 | 0 | 0 | 0 |
| Days Min Temp ≤ 32 °F | 10 | 6 | 2 | 0 | 0 | 0 | 0 | 0 | 0 | 0 | 2 | 6 | 26 |
| Days Min Temp ≤ 0 °F | 0 | 0 | 0 | 0 | 0 | 0 | 0 | 0 | 0 | 0 | 0 | 0 | 0 |
| Heating Degree Days | 515 | 365 | 195 | 53 | 6 | 0 | 0 | 0 | 3 | 50 | 218 | 420 | 1825 |
| Cooling Degree Days | 5 | 9 | 46 | 133 | 287 | 471 | 568 | 581 | 401 | 176 | 56 | 14 | 2747 |
| Total Precipitation (") | 3.57 | 2.98 | 3.17 | 3.24 | 4.88 | 4.15 | 2.18 | 2.83 | 4.56 | 4.46 | 3.09 | 3.21 | 42.32 |
| Days ≥ 0.1" Precip | 6 | 5 | 5 | 4 | 6 | 5 | 4 | 4 | 6 | 5 | 5 | 5 | 60 |
| Total Snowfall (") | 0.1 | 0.0 | 0.0 | 0.0 | 0.0 | 0.0 | 0.0 | 0.0 | 0.0 | 0.0 | 0.0 | 0.0 | 0.1 |
| Days ≥ 1" Snow Depth | 0 | 0 | 0 | 0 | 0 | 0 | 0 | 0 | 0 | 0 | 0 | 0 | 0 |

## WATER VALLEY *Tom Green County* ELEVATION 2113 ft LAT/LONG 31° 40 ' N / 100° 43 ' W

| | JAN | FEB | MAR | APR | MAY | JUN | JUL | AUG | SEP | OCT | NOV | DEC | YEAR |
|---|---|---|---|---|---|---|---|---|---|---|---|---|---|
| Maximum Temp °F | 57.7 | 62.4 | 71.4 | 79.9 | 85.4 | 91.3 | 94.6 | 93.3 | 86.5 | 78.7 | 68.4 | 60.1 | 77.5 |
| Minimum Temp °F | 28.0 | 31.6 | 39.8 | 49.0 | 57.6 | 65.7 | 69.1 | 67.5 | 61.0 | 49.7 | 38.2 | 30.0 | 48.9 |
| Mean Temp °F | 42.9 | 47.0 | 55.6 | 64.5 | 71.6 | 78.5 | 81.9 | 80.4 | 73.8 | 64.2 | 53.3 | 45.1 | 63.2 |
| Days Max Temp ≥ 90 °F | 0 | 0 | 1 | 4 | 10 | 19 | 26 | 25 | 12 | 3 | 0 | 0 | 100 |
| Days Max Temp ≤ 32 °F | 2 | 1 | 0 | 0 | 0 | 0 | 0 | 0 | 0 | 0 | 0 | 1 | 4 |
| Days Min Temp ≤ 32 °F | 22 | 15 | 7 | 1 | 0 | 0 | 0 | 0 | 0 | 1 | 9 | 19 | 74 |
| Days Min Temp ≤ 0 °F | 0 | 0 | 0 | 0 | 0 | 0 | 0 | 0 | 0 | 0 | 0 | 0 | 0 |
| Heating Degree Days | 680 | 502 | 304 | 102 | 22 | 1 | 0 | 0 | 16 | 109 | 352 | 611 | 2699 |
| Cooling Degree Days | 0 | 2 | 16 | 93 | 243 | 421 | 533 | 509 | 299 | 96 | 11 | 0 | 2223 |
| Total Precipitation (") | 0.75 | 1.12 | 0.98 | 1.53 | 3.32 | 2.73 | 1.53 | 2.98 | 3.34 | 2.58 | 1.05 | 1.09 | 23.00 |
| Days ≥ 0.1" Precip | 2 | 2 | 2 | 3 | 5 | 4 | 3 | 4 | 4 | 3 | 2 | 2 | 36 |
| Total Snowfall (") | 1.1 | 0.1 | 0.3 | 0.0 | 0.0 | 0.0 | 0.0 | 0.0 | 0.0 | 0.0 | 0.3 | 0.1 | 1.9 |
| Days ≥ 1" Snow Depth | 0 | 0 | 0 | 0 | 0 | 0 | 0 | 0 | 0 | 0 | 0 | 0 | 0 |

## WAXAHACHIE *Ellis County* ELEVATION 561 ft LAT/LONG 32° 24 ' N / 96° 51 ' W

| | JAN | FEB | MAR | APR | MAY | JUN | JUL | AUG | SEP | OCT | NOV | DEC | YEAR |
|---|---|---|---|---|---|---|---|---|---|---|---|---|---|
| Maximum Temp °F | 56.6 | 61.6 | 70.3 | 78.1 | 83.8 | 91.5 | 95.9 | 95.7 | 88.8 | 79.8 | 68.0 | 59.6 | 77.5 |
| Minimum Temp °F | 34.6 | 38.1 | 46.3 | 54.8 | 61.9 | 69.1 | 72.7 | 71.8 | 65.8 | 55.5 | 45.6 | 37.7 | 54.5 |
| Mean Temp °F | 45.6 | 49.9 | 58.3 | 66.5 | 72.9 | 80.3 | 84.3 | 83.8 | 77.3 | 67.7 | 56.8 | 48.7 | 66.0 |
| Days Max Temp ≥ 90 °F | 0 | 0 | 0 | 1 | 5 | 21 | 28 | 27 | 16 | 3 | 0 | 0 | 101 |
| Days Max Temp ≤ 32 °F | 1 | 1 | 0 | 0 | 0 | 0 | 0 | 0 | 0 | 0 | 0 | 1 | 3 |
| Days Min Temp ≤ 32 °F | 14 | 9 | 3 | 0 | 0 | 0 | 0 | 0 | 0 | 0 | 3 | 9 | 38 |
| Days Min Temp ≤ 0 °F | 0 | 0 | 0 | 0 | 0 | 0 | 0 | 0 | 0 | 0 | 0 | 0 | 0 |
| Heating Degree Days | 597 | 426 | 238 | 63 | 9 | 0 | 0 | 0 | 6 | 63 | 270 | 503 | 2175 |
| Cooling Degree Days | 1 | 6 | 32 | 105 | 251 | 466 | 603 | 605 | 389 | 150 | 32 | 3 | 2643 |
| Total Precipitation (") | 2.01 | 3.08 | 3.23 | 3.96 | 5.32 | 3.10 | 2.20 | 2.16 | 3.75 | 4.44 | 2.72 | 2.87 | 38.84 |
| Days ≥ 0.1" Precip | 4 | 5 | 5 | 5 | 7 | 5 | 3 | 3 | 5 | 5 | 5 | 5 | 57 |
| Total Snowfall (") | *0.1* | *0.3* | 0.0 | 0.0 | 0.0 | 0.0 | 0.0 | 0.0 | 0.0 | 0.0 | 0.0 | 0.0 | 0.4 |
| Days ≥ 1" Snow Depth | 0 | 0 | 0 | 0 | 0 | 0 | 0 | 0 | 0 | 0 | 0 | 0 | 0 |

## WEATHERFORD *Parker County* ELEVATION 1102 ft LAT/LONG 32° 46 ' N / 97° 49 ' W

| | JAN | FEB | MAR | APR | MAY | JUN | JUL | AUG | SEP | OCT | NOV | DEC | YEAR |
|---|---|---|---|---|---|---|---|---|---|---|---|---|---|
| Maximum Temp °F | 53.6 | 58.2 | 67.2 | 76.1 | 82.1 | 90.4 | 95.5 | 95.4 | 87.5 | 77.6 | 66.0 | 57.5 | 75.6 |
| Minimum Temp °F | 28.8 | 32.7 | 40.9 | 50.3 | 58.7 | 66.9 | 71.0 | 69.7 | 62.7 | 51.0 | 40.3 | 32.0 | 50.4 |
| Mean Temp °F | 41.2 | 45.5 | 54.1 | 63.2 | 70.4 | 78.7 | 83.3 | 82.6 | 75.1 | 64.3 | 53.2 | 44.8 | 63.0 |
| Days Max Temp ≥ 90 °F | 0 | 0 | 0 | 1 | 5 | 18 | 27 | 26 | 14 | 3 | 0 | 0 | 94 |
| Days Max Temp ≤ 32 °F | 2 | 1 | 0 | 0 | 0 | 0 | 0 | 0 | 0 | 0 | 0 | 1 | 4 |
| Days Min Temp ≤ 32 °F | 20 | 14 | 7 | 1 | 0 | 0 | 0 | 0 | 0 | 1 | 8 | 16 | 67 |
| Days Min Temp ≤ 0 °F | 0 | 0 | 0 | 0 | 0 | 0 | 0 | 0 | 0 | 0 | 0 | 0 | 0 |
| Heating Degree Days | 730 | 548 | 349 | 121 | 27 | 1 | 0 | 0 | 15 | 113 | 363 | 621 | 2888 |
| Cooling Degree Days | 0 | 1 | 15 | 71 | 208 | 421 | 567 | 573 | 335 | 96 | 16 | 0 | 2303 |
| Total Precipitation (") | 1.66 | 2.32 | 2.68 | 3.07 | 4.75 | 3.59 | 1.97 | 2.46 | 3.43 | 3.86 | 2.12 | 2.05 | 33.96 |
| Days ≥ 0.1" Precip | 3 | 4 | 4 | 4 | 6 | 5 | 3 | 4 | 4 | 5 | 4 | 4 | 50 |
| Total Snowfall (") | *0.1* | 0.6 | 0.2 | 0.0 | 0.0 | 0.0 | 0.0 | 0.0 | 0.0 | 0.0 | 0.2 | 0.3 | 1.4 |
| Days ≥ 1" Snow Depth | *0* | 0 | 0 | 0 | 0 | 0 | 0 | 0 | 0 | 0 | 0 | 0 | 0 |

## WELLINGTON *Collingsworth County* ELEVATION 1982 ft LAT/LONG 34° 52 ' N / 100° 13 ' W

| | JAN | FEB | MAR | APR | MAY | JUN | JUL | AUG | SEP | OCT | NOV | DEC | YEAR |
|---|---|---|---|---|---|---|---|---|---|---|---|---|---|
| Maximum Temp °F | 51.6 | 55.8 | 66.5 | 76.2 | 82.3 | 91.1 | 95.9 | 93.9 | 86.2 | 76.3 | 63.2 | 54.0 | 74.4 |
| Minimum Temp °F | 24.9 | 28.5 | 36.9 | 46.2 | 55.5 | 64.7 | 69.2 | 67.7 | 60.2 | 48.4 | 36.2 | 28.0 | 47.2 |
| Mean Temp °F | 38.3 | 42.1 | 51.7 | 61.2 | 68.9 | 77.9 | 82.6 | 80.8 | 73.2 | 62.4 | 49.7 | 41.0 | 60.8 |
| Days Max Temp ≥ 90 °F | 0 | 0 | 0 | 3 | 7 | 18 | 27 | 24 | 12 | 2 | 0 | 0 | 93 |
| Days Max Temp ≤ 32 °F | 4 | 2 | 0 | 0 | 0 | 0 | 0 | 0 | 0 | 0 | 0 | 2 | 8 |
| Days Min Temp ≤ 32 °F | 25 | 18 | 9 | 1 | 0 | 0 | 0 | 0 | 0 | 1 | 12 | 22 | 88 |
| Days Min Temp ≤ 0 °F | 0 | 0 | 0 | 0 | 0 | 0 | 0 | 0 | 0 | 0 | 0 | 0 | 0 |
| Heating Degree Days | 822 | 639 | 414 | 157 | 41 | 2 | 0 | 1 | 19 | 140 | 455 | 736 | 3426 |
| Cooling Degree Days | 0 | 0 | 9 | 56 | 183 | 413 | 566 | 535 | 304 | 68 | 5 | 0 | 2139 |
| Total Precipitation (") | 0.63 | 0.66 | 1.20 | 1.88 | 3.75 | 3.26 | 2.22 | 2.14 | 2.66 | 2.06 | 0.83 | 0.51 | 21.80 |
| Days ≥ 0.1" Precip | *1* | *1* | 2 | 2 | 5 | 4 | 3 | 3 | 4 | 3 | *2* | *1* | 31 |
| Total Snowfall (") | *2.0* | *2.5* | 0.7 | 0.0 | 0.0 | 0.0 | 0.0 | 0.0 | 0.0 | 0.0 | 0.5 | 1.3 | 7.0 |
| Days ≥ 1" Snow Depth | *2* | na | 0 | 0 | 0 | 0 | 0 | 0 | 0 | 0 | 0 | 1 | na |

## WESLACO 2 E *Hidalgo County* ELEVATION 79 ft LAT/LONG 26° 9 ' N / 97° 58 ' W

| | JAN | FEB | MAR | APR | MAY | JUN | JUL | AUG | SEP | OCT | NOV | DEC | YEAR |
|---|---|---|---|---|---|---|---|---|---|---|---|---|---|
| Maximum Temp °F | 69.4 | 73.2 | 80.9 | 85.8 | 88.8 | 92.9 | 95.1 | 95.7 | 91.9 | 86.6 | 78.7 | 72.4 | 84.3 |
| Minimum Temp °F | 49.3 | 52.1 | 58.9 | 65.6 | 70.4 | 73.8 | 74.5 | 74.3 | 71.7 | 65.0 | 57.9 | 52.1 | 63.8 |
| Mean Temp °F | 59.4 | 62.7 | 69.9 | 75.8 | 79.6 | 83.4 | 84.8 | 85.0 | 81.8 | 75.8 | 68.4 | 62.3 | 74.1 |
| Days Max Temp ≥ 90 °F | 0 | 0 | 3 | 8 | 15 | 25 | 29 | 29 | 22 | 11 | 2 | 0 | 144 |
| Days Max Temp ≤ 32 °F | 0 | 0 | 0 | 0 | 0 | 0 | 0 | 0 | 0 | 0 | 0 | 0 | 0 |
| Days Min Temp ≤ 32 °F | 1 | 0 | 0 | 0 | 0 | 0 | 0 | 0 | 0 | 0 | 0 | 1 | 2 |
| Days Min Temp ≤ 0 °F | 0 | 0 | 0 | 0 | 0 | 0 | 0 | 0 | 0 | 0 | 0 | 0 | 0 |
| Heating Degree Days | 222 | 131 | 44 | 6 | 1 | 0 | 0 | 0 | 0 | 6 | 67 | 165 | 642 |
| Cooling Degree Days | 44 | 77 | 198 | 320 | 465 | 580 | 638 | 647 | 520 | 361 | 194 | 88 | 4132 |
| Total Precipitation (") | 1.59 | 1.55 | 0.68 | 1.38 | 2.99 | 3.08 | 2.34 | 2.27 | 4.84 | 2.22 | 1.39 | 1.19 | 25.52 |
| Days ≥ 0.1" Precip | 3 | 3 | 2 | 2 | 4 | 4 | 3 | 4 | 7 | 3 | 3 | 3 | 41 |
| Total Snowfall (") | 0.0 | 0.0 | 0.0 | 0.0 | 0.0 | 0.0 | 0.0 | 0.0 | 0.0 | 0.0 | 0.0 | 0.0 | 0.0 |
| Days ≥ 1" Snow Depth | 0 | 0 | 0 | 0 | 0 | 0 | 0 | 0 | 0 | 0 | 0 | 0 | 0 |

## WHITNEY DAM *Bosque County* ELEVATION 531 ft LAT/LONG 31° 52 ' N / 97° 22 ' W

| | JAN | FEB | MAR | APR | MAY | JUN | JUL | AUG | SEP | OCT | NOV | DEC | YEAR |
|---|---|---|---|---|---|---|---|---|---|---|---|---|---|
| Maximum Temp °F | 55.6 | 60.1 | 69.0 | 77.2 | 83.2 | 91.0 | 95.8 | 95.8 | 88.8 | 79.3 | 67.6 | 59.2 | 76.9 |
| Minimum Temp °F | 32.3 | 36.0 | 44.8 | 53.7 | 61.5 | 68.6 | 71.3 | 70.4 | 64.9 | 53.9 | 43.8 | 35.5 | 53.1 |
| Mean Temp °F | 44.0 | 48.1 | 56.9 | 65.4 | 72.3 | 79.9 | 83.6 | 83.2 | 76.8 | 66.6 | 55.7 | 47.4 | 65.0 |
| Days Max Temp ≥ 90 °F | 0 | 0 | 0 | 1 | 5 | 20 | 28 | 27 | 16 | 4 | 0 | 0 | 101 |
| Days Max Temp ≤ 32 °F | 1 | 1 | 0 | 0 | 0 | 0 | 0 | 0 | 0 | 0 | 0 | 1 | 3 |
| Days Min Temp ≤ 32 °F | 17 | 10 | 3 | 0 | 0 | 0 | 0 | 0 | 0 | 0 | 4 | 12 | 46 |
| Days Min Temp ≤ 0 °F | 0 | 0 | 0 | 0 | 0 | 0 | 0 | 0 | 0 | 0 | 0 | 0 | 0 |
| Heating Degree Days | 646 | 474 | 272 | 81 | 12 | 0 | 0 | 0 | 7 | 75 | 295 | 543 | 2405 |
| Cooling Degree Days | 1 | 3 | 28 | 99 | 255 | 450 | 587 | 586 | 376 | 133 | 26 | 4 | 2548 |
| Total Precipitation (") | 1.80 | 2.52 | 2.75 | 3.13 | 4.90 | 3.46 | 2.10 | 2.41 | 3.22 | 4.05 | 2.40 | 2.37 | 35.11 |
| Days ≥ 0.1" Precip | 4 | 5 | 5 | 5 | 6 | 5 | 3 | 4 | 4 | 5 | 4 | 4 | 54 |
| Total Snowfall (") | *0.1* | 0.1 | 0.0 | 0.0 | 0.0 | 0.0 | 0.0 | 0.0 | 0.0 | 0.0 | 0.0 | 0.1 | 0.3 |
| Days ≥ 1" Snow Depth | 0 | 0 | 0 | 0 | 0 | 0 | 0 | 0 | 0 | 0 | 0 | 0 | 0 |

## WICHITA FALLS MUN AP *Wichita County* ELEVATION 1027 ft LAT/LONG 33° 59 ' N / 98° 31 ' W

| | JAN | FEB | MAR | APR | MAY | JUN | JUL | AUG | SEP | OCT | NOV | DEC | YEAR |
|---|---|---|---|---|---|---|---|---|---|---|---|---|---|
| Maximum Temp °F | 51.9 | 57.2 | 67.2 | 76.4 | 83.3 | 92.3 | 97.7 | 96.1 | 87.4 | 77.1 | 64.4 | 55.3 | 75.5 |
| Minimum Temp °F | 28.3 | 32.4 | 40.8 | 49.9 | 58.8 | 67.5 | 72.0 | 70.6 | 63.2 | 51.6 | 39.8 | 31.4 | 50.5 |
| Mean Temp °F | 40.1 | 44.8 | 54.0 | 63.2 | 71.1 | 79.9 | 84.9 | 83.4 | 75.3 | 64.4 | 52.1 | 43.4 | 63.0 |
| Days Max Temp ≥ 90 °F | 0 | 0 | 1 | 2 | 7 | 21 | 28 | 26 | 14 | 3 | 0 | 0 | 102 |
| Days Max Temp ≤ 32 °F | 3 | 2 | 0 | 0 | 0 | 0 | 0 | 0 | 0 | 0 | 0 | 2 | 7 |
| Days Min Temp ≤ 32 °F | 21 | 14 | 6 | 1 | 0 | 0 | 0 | 0 | 0 | 0 | 7 | 17 | 66 |
| Days Min Temp ≤ 0 °F | 0 | 0 | 0 | 0 | 0 | 0 | 0 | 0 | 0 | 0 | 0 | 0 | 0 |
| Heating Degree Days | 763 | 565 | 356 | 123 | 23 | 0 | 0 | 0 | 15 | 111 | 391 | 665 | 3012 |
| Cooling Degree Days | 0 | 1 | 18 | 73 | 216 | 450 | 617 | 596 | 344 | 96 | 11 | 1 | 2423 |
| Total Precipitation (") | 1.16 | 1.60 | 2.12 | 2.89 | 4.08 | 3.43 | 1.65 | 2.41 | 3.71 | 2.93 | 1.47 | 1.59 | 29.04 |
| Days ≥ 0.1" Precip | 2 | 3 | 4 | 4 | 6 | 4 | 3 | 4 | 5 | 4 | 3 | 3 | 45 |
| Total Snowfall (") | 2.2 | 1.5 | 0.8 | 0.0 | 0.0 | 0.0 | 0.0 | 0.0 | 0.0 | 0.0 | 0.3 | 1.0 | 5.8 |
| Days ≥ 1" Snow Depth | 2 | 1 | 0 | 0 | 0 | 0 | 0 | 0 | 0 | 0 | 0 | 0 | 3 |

## WILLS POINT *Van Zandt County* ELEVATION 522 ft LAT/LONG 32° 42 ' N / 96° 1 ' W

| | JAN | FEB | MAR | APR | MAY | JUN | JUL | AUG | SEP | OCT | NOV | DEC | YEAR |
|---|---|---|---|---|---|---|---|---|---|---|---|---|---|
| Maximum Temp °F | 53.2 | 58.0 | 66.6 | 75.2 | 81.3 | 89.2 | 94.2 | 94.6 | 87.4 | 77.6 | 66.0 | 57.0 | 75.0 |
| Minimum Temp °F | 33.0 | 36.4 | 44.5 | 52.9 | 60.9 | 68.4 | 72.0 | 71.2 | 65.1 | 54.3 | 44.0 | 36.2 | 53.2 |
| Mean Temp °F | 43.1 | 47.2 | 55.6 | 64.1 | 71.1 | 78.8 | 83.1 | 82.9 | 76.3 | 66.0 | 55.0 | 46.6 | 64.1 |
| Days Max Temp ≥ 90 °F | 0 | 0 | 0 | 0 | 1 | 15 | 26 | 26 | 13 | 2 | 0 | 0 | 83 |
| Days Max Temp ≤ 32 °F | 2 | 1 | 0 | 0 | 0 | 0 | 0 | 0 | 0 | 0 | 0 | 1 | 4 |
| Days Min Temp ≤ 32 °F | 16 | 10 | 3 | 0 | 0 | 0 | 0 | 0 | 0 | 0 | 4 | 11 | 44 |
| Days Min Temp ≤ 0 °F | 0 | 0 | 0 | 0 | 0 | 0 | 0 | 0 | 0 | 0 | 0 | 0 | 0 |
| Heating Degree Days | 674 | 499 | 307 | 97 | 15 | 1 | 0 | 0 | 9 | 83 | 314 | 566 | 2565 |
| Cooling Degree Days | 1 | 3 | 21 | 75 | 216 | 425 | 559 | 578 | 371 | 121 | 23 | 3 | 2396 |
| Total Precipitation (") | 2.73 | 3.49 | 3.73 | 4.31 | 5.63 | 4.18 | 2.13 | 2.13 | 3.84 | 5.02 | 3.66 | 3.72 | 44.57 |
| Days ≥ 0.1" Precip | 5 | 6 | 6 | 5 | 7 | 5 | 3 | 4 | 5 | 5 | 5 | 5 | 61 |
| Total Snowfall (") | 0.9 | 1.4 | 0.1 | 0.0 | 0.0 | 0.0 | 0.0 | 0.0 | 0.0 | 0.0 | 0.1 | 0.4 | 2.9 |
| Days ≥ 1" Snow Depth | 1 | 1 | 0 | 0 | 0 | 0 | 0 | 0 | 0 | 0 | 0 | 0 | 2 |

## WINK WINKLER CO AP *Winkler County* ELEVATION 2815 ft LAT/LONG 31° 47 ' N / 103° 12 ' W

| | JAN | FEB | MAR | APR | MAY | JUN | JUL | AUG | SEP | OCT | NOV | DEC | YEAR |
|---|---|---|---|---|---|---|---|---|---|---|---|---|---|
| Maximum Temp °F | 58.7 | 64.4 | 72.4 | 81.6 | 88.6 | 95.5 | 96.0 | 94.2 | 87.8 | 79.6 | 68.4 | 60.9 | 79.0 |
| Minimum Temp °F | 28.2 | 32.2 | 40.0 | 49.0 | 57.9 | 66.7 | 69.9 | 68.3 | 61.4 | 49.7 | 37.0 | 29.1 | 49.1 |
| Mean Temp °F | 43.5 | 48.3 | 56.3 | 65.3 | 73.3 | 81.1 | 83.0 | 81.3 | 74.6 | 64.7 | 52.6 | 45.1 | 64.1 |
| Days Max Temp ≥ 90 °F | 0 | 0 | 1 | 6 | 15 | 24 | 27 | 25 | 16 | 5 | 0 | 0 | 119 |
| Days Max Temp ≤ 32 °F | 1 | 0 | 0 | 0 | 0 | 0 | 0 | 0 | 0 | 0 | 0 | 0 | 1 |
| Days Min Temp ≤ 32 °F | 21 | 15 | 6 | 1 | 0 | 0 | 0 | 0 | 0 | 1 | 9 | 20 | 73 |
| Days Min Temp ≤ 0 °F | 0 | 0 | 0 | 0 | 0 | 0 | 0 | 0 | 0 | 0 | 0 | 0 | 0 |
| Heating Degree Days | 661 | 465 | 281 | 83 | 13 | 0 | 0 | 0 | 15 | 97 | 370 | 609 | 2594 |
| Cooling Degree Days | 0 | 1 | 15 | 97 | 286 | 510 | 574 | 534 | 324 | 103 | 10 | 0 | 2454 |
| Total Precipitation (") | 0.41 | 0.50 | 0.41 | 0.64 | 1.21 | 1.74 | 1.92 | 1.38 | 2.14 | 1.45 | 0.59 | 0.50 | 12.89 |
| Days ≥ 0.1" Precip | 1 | 1 | 1 | 2 | 3 | 3 | 3 | 3 | 4 | 2 | 1 | 1 | 25 |
| Total Snowfall (") | 1.9 | 0.5 | 0.4 | 0.1 | 0.0 | 0.0 | 0.0 | 0.0 | 0.0 | 0.0 | 0.5 | 0.4 | 3.8 |
| Days ≥ 1" Snow Depth | 1 | 0 | 0 | 0 | 0 | 0 | 0 | 0 | 0 | 0 | 0 | 0 | 1 |

## YOAKUM *Lavaca County* ELEVATION 381 ft LAT/LONG 29° 18 ' N / 97° 9 ' W

| | JAN | FEB | MAR | APR | MAY | JUN | JUL | AUG | SEP | OCT | NOV | DEC | YEAR |
|---|---|---|---|---|---|---|---|---|---|---|---|---|---|
| Maximum Temp °F | 62.2 | 66.4 | 74.2 | 80.8 | 85.7 | 91.6 | 94.6 | 95.5 | 90.9 | 83.5 | 73.3 | 65.8 | 80.4 |
| Minimum Temp °F | 40.4 | 43.3 | 51.2 | 58.9 | 65.2 | 70.8 | 72.2 | 71.9 | 67.8 | 58.8 | 50.2 | 43.5 | 57.9 |
| Mean Temp °F | 51.3 | 54.9 | 62.8 | 69.9 | 75.5 | 81.2 | 83.4 | 83.7 | 79.4 | 71.1 | 61.8 | 54.7 | 69.1 |
| Days Max Temp ≥ 90 °F | 0 | 0 | 1 | 2 | 7 | 21 | 28 | 29 | 20 | 7 | 0 | 0 | 115 |
| Days Max Temp ≤ 32 °F | 0 | 0 | 0 | 0 | 0 | 0 | 0 | 0 | 0 | 0 | 0 | 0 | 0 |
| Days Min Temp ≤ 32 °F | 7 | 4 | 1 | 0 | 0 | 0 | 0 | 0 | 0 | 0 | 2 | 4 | 18 |
| Days Min Temp ≤ 0 °F | 0 | 0 | 0 | 0 | 0 | 0 | 0 | 0 | 0 | 0 | 0 | 0 | 0 |
| Heating Degree Days | 433 | 296 | 139 | 31 | 2 | 0 | 0 | 0 | 1 | 28 | 168 | 338 | 1436 |
| Cooling Degree Days | 6 | 13 | 61 | 162 | 325 | 492 | 571 | 585 | 438 | 218 | 75 | 22 | 2968 |
| Total Precipitation (") | 2.73 | 2.42 | 2.19 | 3.53 | 5.10 | 4.25 | 2.71 | 3.21 | 3.89 | 3.86 | 2.93 | 2.32 | 39.14 |
| Days ≥ 0.1" Precip | 5 | 4 | 4 | 4 | 6 | 5 | 4 | 5 | 5 | 5 | 4 | 5 | 56 |
| Total Snowfall (") | 0.0 | 0.0 | 0.0 | 0.0 | 0.0 | 0.0 | 0.0 | 0.0 | 0.0 | 0.0 | 0.0 | 0.0 | 0.0 |
| Days ≥ 1" Snow Depth | 0 | 0 | 0 | 0 | 0 | 0 | 0 | 0 | 0 | 0 | 0 | 0 | 0 |

## YSLETA *El Paso County* ELEVATION 3671 ft LAT/LONG 31° 42 ' N / 106° 19 ' W

| | JAN | FEB | MAR | APR | MAY | JUN | JUL | AUG | SEP | OCT | NOV | DEC | YEAR |
|---|---|---|---|---|---|---|---|---|---|---|---|---|---|
| Maximum Temp °F | 59.7 | 65.1 | 72.6 | 81.2 | 89.0 | 98.2 | 98.1 | 94.9 | 89.5 | 81.0 | 69.3 | ***60.5*** | 79.9 |
| Minimum Temp °F | 28.4 | 32.6 | 39.0 | 46.4 | 54.7 | 63.6 | 67.8 | 65.9 | 59.2 | 47.3 | 35.7 | ***29.3*** | 47.5 |
| Mean Temp °F | 44.1 | 48.9 | 55.8 | 63.8 | 71.9 | 80.9 | 83.0 | 80.4 | 74.3 | 64.2 | 52.6 | ***44.9*** | 63.7 |
| Days Max Temp ≥ 90 °F | 0 | 0 | 0 | 3 | 16 | 27 | 29 | 27 | 17 | 4 | 0 | 0 | 123 |
| Days Max Temp ≤ 32 °F | 0 | 0 | 0 | 0 | 0 | 0 | 0 | 0 | 0 | 0 | 0 | 0 | 0 |
| Days Min Temp ≤ 32 °F | 22 | 14 | 5 | 1 | 0 | 0 | 0 | 0 | 0 | 1 | 9 | 18 | 70 |
| Days Min Temp ≤ 0 °F | 0 | 0 | 0 | 0 | 0 | 0 | 0 | 0 | 0 | 0 | 0 | 0 | 0 |
| Heating Degree Days | 642 | 449 | 285 | 92 | 12 | 0 | 0 | 0 | 9 | 88 | 366 | ***616*** | 2559 |
| Cooling Degree Days | 0 | 0 | 10 | 82 | 271 | 528 | 595 | 515 | 323 | 84 | 3 | na | na |
| Total Precipitation (") | 0.50 | 0.47 | 0.26 | 0.27 | 0.46 | 0.74 | 1.46 | 1.75 | 1.69 | 0.88 | 0.47 | ***0.70*** | 9.65 |
| Days ≥ 0.1" Precip | 2 | 2 | 1 | 1 | 1 | 2 | 3 | 4 | 3 | 2 | 1 | 2 | 24 |
| Total Snowfall (") | 0.0 | 0.0 | 0.0 | 0.0 | 0.0 | 0.0 | 0.0 | 0.0 | 0.0 | 0.0 | 0.5 | 0.1 | 0.6 |
| Days ≥ 1" Snow Depth | 0 | 0 | 0 | 0 | 0 | 0 | 0 | 0 | 0 | 0 | 0 | 0 | 0 |

## ZAPATA 3 SW *Zapata County* ELEVATION 289 ft LAT/LONG 26° 53 ' N / 99° 19 ' W

| | JAN | FEB | MAR | APR | MAY | JUN | JUL | AUG | SEP | OCT | NOV | DEC | YEAR |
|---|---|---|---|---|---|---|---|---|---|---|---|---|---|
| Maximum Temp °F | 69.3 | 73.6 | 83.1 | 89.4 | 93.2 | 97.2 | 98.9 | 99.2 | 93.9 | 87.3 | 78.4 | 70.4 | 86.2 |
| Minimum Temp °F | 46.1 | 49.2 | 56.5 | 64.5 | 70.0 | 73.7 | 74.8 | 74.7 | 71.3 | 64.0 | 55.1 | 48.2 | 62.3 |
| Mean Temp °F | 57.9 | 61.5 | 69.8 | 77.0 | 81.6 | 85.5 | 86.9 | 87.0 | 82.6 | 75.7 | 66.8 | 59.4 | 74.3 |
| Days Max Temp ≥ 90 °F | 0 | 1 | 7 | 16 | 25 | 28 | 29 | 29 | 23 | 13 | 2 | 0 | 173 |
| Days Max Temp ≤ 32 °F | 0 | 0 | 0 | 0 | 0 | 0 | 0 | 0 | 0 | 0 | 0 | 0 | 0 |
| Days Min Temp ≤ 32 °F | 2 | 1 | 0 | 0 | 0 | 0 | 0 | 0 | 0 | 0 | 0 | 1 | 4 |
| Days Min Temp ≤ 0 °F | 0 | 0 | 0 | 0 | 0 | 0 | 0 | 0 | 0 | 0 | 0 | 0 | 0 |
| Heating Degree Days | 250 | 152 | 45 | 7 | 1 | 0 | 0 | 0 | 1 | 7 | 82 | 210 | 755 |
| Cooling Degree Days | ***29*** | ***65*** | 203 | 360 | 530 | 641 | ***707*** | ***718*** | 567 | 366 | 157 | ***44*** | 4387 |
| Total Precipitation (") | 0.86 | 1.13 | 0.60 | 1.62 | 2.75 | 2.54 | 1.60 | 1.79 | 4.40 | 1.57 | 0.94 | 0.96 | 20.76 |
| Days ≥ 0.1" Precip | 3 | ***3*** | 1 | 2 | 4 | 3 | 2 | 3 | 5 | 2 | 2 | 2 | 32 |
| Total Snowfall (") | 0.3 | 0.0 | 0.0 | 0.0 | 0.0 | 0.0 | 0.0 | 0.0 | 0.0 | 0.0 | 0.0 | 0.0 | 0.3 |
| Days ≥ 1" Snow Depth | 0 | 0 | 0 | 0 | 0 | 0 | 0 | 0 | 0 | 0 | 0 | 0 | 0 |

## JANUARY MINIMUM TEMPERATURE °F

### LOWEST

| | | |
|---|---|---|
| 1 | Lipscomb | 17.2 |
| 2 | Perryton | 17.4 |
| 3 | Stratford | 18.4 |
| 4 | Dalhart | 19.3 |
| 5 | Muleshoe | 19.9 |
| 6 | Gruver | 20.0 |
| 7 | Dimmitt | 20.1 |
| | Follett | 20.1 |
| 9 | Miami | 20.2 |
| 10 | Dumas | 20.3 |
| 11 | Bravo | 20.5 |
| 12 | Claude | 20.6 |
| 13 | Hereford | 20.9 |
| 14 | Silverton | 21.0 |
| 15 | Olton | 21.4 |
| 16 | Friona | 21.5 |
| | Pampa | 21.5 |
| | Spearman | 21.5 |
| 19 | Amarillo | 21.8 |
| 20 | Clarendon | 22.0 |
| 21 | Tulia | 22.1 |
| 22 | Canadian | 22.2 |
| 23 | Littlefield | 22.6 |
| 24 | Floydada | 22.7 |
| | Morton | 22.7 |

### HIGHEST

| | | |
|---|---|---|
| 1 | Brownsville | 50.5 |
| 2 | Weslaco | 49.3 |
| 3 | Port Mansfield | 48.1 |
| 4 | Galveston | 47.9 |
| 5 | Harlingen | 47.4 |
| 6 | Raymondville | 46.8 |
| 7 | Chapman | 46.7 |
| | Mission | 46.7 |
| 9 | Corpus Christi | 46.2 |
| | Rockport | 46.2 |
| 11 | Zapata | 46.1 |
| 12 | Matagorda | 45.7 |
| 13 | McCook | 45.4 |
| 14 | Point Comfort | 45.2 |
| 15 | Falcon Dam | 45.1 |
| 16 | Port O'Connor | 45.0 |
| 17 | Kingsville | 44.8 |
| 18 | Sinton | 44.7 |
| 19 | Freeport | 44.5 |
| 20 | Robstown | 44.3 |
| 21 | Palacios | 44.0 |
| 22 | Rio Grande City | 43.9 |
| 23 | Alice | 43.8 |
| 24 | Laredo | 43.7 |
| 25 | Danevang | 43.6 |

## JULY MAXIMUM TEMPERATURE °F

### HIGHEST

| | | |
|---|---|---|
| 1 | Presidio | 100.7 |
| 2 | Candelaria | 99.8 |
| 3 | Laredo | 99.7 |
| 4 | Catarina | 99.5 |
| 5 | Falcon Dam | 99.0 |
| | Fowlerton | 99.0 |
| 7 | Carrizo Springs | 98.9 |
| | Zapata | 98.9 |
| 9 | Pecos | 98.5 |
| 10 | Encinal | 98.4 |
| 11 | Rio Grande City | 98.2 |
| 12 | Ysleta | 98.1 |
| 13 | Grandfalls | 98.0 |
| 14 | Red Bluff Dam | 97.8 |
| 15 | Dilley | 97.7 |
| | Eagle Pass | 97.7 |
| | Munday | 97.7 |
| | Pearsall | 97.7 |
| | Vernon | 97.7 |
| | Wichita Falls | 97.7 |
| 21 | Bridgeport | 97.6 |
| | Glen Rose | 97.6 |
| 23 | Monahans | 97.5 |
| | Paint Rock | 97.5 |
| | Tilden | 97.5 |

### LOWEST

| | | |
|---|---|---|
| 1 | Mount Locke | 82.4 |
| 2 | Chisos Basin | 84.3 |
| 3 | Galveston | 87.5 |
| 4 | Port Mansfield | 88.4 |
| 5 | Alpine | 89.0 |
| 6 | Marfa | 89.6 |
| 7 | Friona | 90.0 |
| | Olton | 90.0 |
| | Palacios | 90.0 |
| 10 | Rockport | 90.1 |
| 11 | Marathon | 90.2 |
| | Port O'Connor | 90.2 |
| 13 | Dimmitt | 90.4 |
| 14 | Hereford | 90.5 |
| 15 | Dalhart | 90.7 |
| 16 | Silverton | 90.8 |
| 17 | Amarillo | 91.0 |
| | Bravo | 91.0 |
| | Morton | 91.0 |
| 20 | Matagorda | 91.1 |
| | Tulia | 91.1 |
| 22 | Claude | 91.2 |
| 23 | Plains | 91.3 |
| | Stratford | 91.3 |
| 25 | Littlefield | 91.4 |

## ANNUAL PRECIPITATION (")

### HIGHEST

| | | |
|---|---|---|
| 1 | Port Arthur | 59.82 |
| 2 | Angleton | 58.78 |
| 3 | Liberty | 58.05 |
| 4 | Anahuac | 53.88 |
| 5 | Houston | 53.66 |
| 6 | Cleveland | 53.00 |
| 7 | Freeport | 52.91 |
| 8 | Center | 52.24 |
| 9 | Carthage | 50.57 |
| 10 | Livingston | 50.43 |
| 11 | Marshall | 50.19 |
| 12 | Coldspring | 50.00 |
| 13 | Mount Pleasant | 49.04 |
| 14 | Conroe | 48.96 |
| 15 | Bay City | 48.87 |
| 16 | Sugar Land | 48.42 |
| 17 | Henderson | 48.16 |
| 18 | Rusk | 48.11 |
| 19 | Paris | 47.90 |
| 20 | Sulphur Springs | 47.65 |
| 21 | Groveton | 47.53 |
| 22 | Gilmer | 47.39 |
| 23 | Huntsville | 47.16 |
| 24 | Clarksville | 46.97 |
| 25 | Daingerfield | 46.84 |

### LOWEST

| | | |
|---|---|---|
| 1 | La Tuna | 9.01 |
| 2 | El Paso | 9.23 |
| 3 | Ysleta | 9.65 |
| 4 | Cornudas | 10.79 |
| 5 | Presidio | 11.40 |
| 6 | Red Bluff Dam | 11.95 |
| 7 | Pecos | 11.97 |
| 8 | Sierra Blanca | 12.62 |
| 9 | Wink | 12.89 |
| 10 | Van Horn | 12.98 |
| 11 | Grandfalls | 13.18 |
| 12 | Candelaria | 13.19 |
| 13 | Penwell | 13.70 |
| 14 | Monahans | 13.71 |
| 15 | Panther Junction | 14.51 |
| 16 | Fort Stockton | 14.70 |
| 17 | Balmorhea | 14.81 |
| 18 | McCamey | 14.86 |
| 19 | Sheffield | 14.90 |
| 20 | Sanderson | 14.93 |
| 21 | Langtry | 15.06 |
| 22 | Crane | 15.48 |
| 23 | Andrews | 15.49 |
| 24 | Marathon | 15.54 |
| 25 | Midland-Regional | 15.59 |

## ANNUAL SNOWFALL (")

### HIGHEST

| | | |
|---|---|---|
| 1 | Borger | 23.3 |
| 2 | Follett | 22.2 |
| 3 | Spearman | 21.6 |
| 4 | Perryton | 20.3 |
| 5 | Pampa | 19.5 |
| 6 | Stratford | 18.1 |
| 7 | Lipscomb | 17.2 |
| 8 | Dalhart | 17.0 |
| 9 | Miami | 16.5 |
| 10 | Amarillo | 16.2 |
| 11 | Gruver | 15.4 |
| 12 | Friona | 15.2 |
| 13 | Dumas | 15.0 |
| 14 | Tulia | 14.0 |
| 15 | Silverton | 12.1 |
| 16 | Claude | 11.7 |
| 17 | Hereford | 11.5 |
| 18 | Plainview | 11.2 |
| 19 | Dimmitt | 10.9 |
| | Matador | 10.9 |
| 21 | Muleshoe | 10.6 |
| | Tahoka | 10.6 |
| 23 | Lubbock | 10.5 |
| 24 | Canyon | 10.3 |
| 25 | Clarendon | 9.8 |

### LOWEST*

| | | |
|---|---|---|
| 1 | Amistad Dam | 0.1 |
| | Cameron | 0.1 |
| | Candelaria | 0.1 |
| | Conroe | 0.1 |
| | Cuero | 0.1 |
| | Falcon Dam | 0.1 |
| | Falfurrias | 0.1 |
| | Hallettsville | 0.1 |
| | Hords Creek Dam | 0.1 |
| | La Grange | 0.1 |
| | La Tuna | 0.1 |
| | Matagorda | 0.1 |
| | Mathis | 0.1 |
| | New Gulf | 0.1 |
| | Palacios | 0.1 |
| | Port Arthur | 0.1 |
| | Port O'Connor | 0.1 |
| | Poteet | 0.1 |
| | Presidio | 0.1 |
| | Rio Grande City | 0.1 |
| | Sinton | 0.1 |
| | Sugar Land | 0.1 |
| | Thompsons | 0.1 |
| | Tilden | 0.1 |
| | Victoria | 0.1 |

* Does not include stations which receive no snowfall.

# UTAH

PHYSICAL FEATURES. The topography of Utah is extremely varied, with most of the State being mountainous. A series of mountains (including the Wasatch Range), which runs generally north and south through the middle of Utah, and the Uinta Mountains, which extend east and west through the northeast portion, are the principal ranges. Crest lines of these mountains are mostly above 10,000 feet. Less extensive ranges are scattered over the remainder of the State. The lowest area is the Virgin River Valley in the southwestern part with elevations between 2,500 and 3,500 feet, while the highest point is Kings Peak in the Uinta Mountains (13,498 feet).

Practically all of eastern Utah is drained by the Colorado River and its principal tributary within the State, the Green River, although neither rises within its borders. Western Utah is almost entirely within the Great Basin, with no outlet to the sea. The largest rivers in this area are the Bear, Weber, Jordan, Provo, and Sevier, the first three of which empty into Great Salt Lake. The Sevier River drains the west-central area and empties into Sevier Lake, a brackish saline basin in southwest Utah. The main streams in the eastern portion of the State flow through canyons or very narrow, confined mountain valleys and finally into desert canyons. Highest flow occurs in this region in May and June during spring runoff from melting snow.

Great Salt Lake, in northwestern Utah, lies in the Great Basin, the largest closed basin in North America. Part of this drainage area is below 4,500 feet in elevation, with the Lake being about 4,200 feet. Great Salt Lake is the largest lake at this elevation (or higher) in the world. In glacial times it was a fresh water lake occupying an area 346 miles long and 145 miles wide; but due to increased evaporation and/or reduced precipitation, it gradually shrank in size and the salinity increased. Since this large body of water now has no drainage outlet, the salt content is high, averaging about 25 percent. Thus the Lake, which never freezes over, provides a moderating effect throughout the year on temperatures in the immediate vicinity.

GENERAL CLIMATE. Essentially, Utah's climate is determined by its distance from the equator; its elevation above sea level; the location of the State with respect to the average storm paths over the Intermountain Region; and its distance from the principal moisture sources of the area, namely, the Pacific Ocean and the Gulf of Mexico. Also, the mountain ranges over the western United States, particularly the Sierra Nevada and Cascade Ranges and the Rocky Mountains, have a marked influence on the climate of the State. Pacific storms, before reaching Utah, must first cross the Sierras or Cascades. As the moist air is forced to rise over these high mountains, a large portion of the original moisture falls as precipitation. Thus, the prevailing westerly air currents reaching Utah are comparatively dry, resulting in light precipitation over most of the State.

TEMPERATURE. There are definite variations in temperature with altitude and with latitude. Naturally, the mountains and the elevated valleys have the cooler climates, with the lower areas of the State having the higher temperatures. There is about a 3° F. decrease in mean annual temperature for each 1,000-foot increase in altitude, and approximately 1.5 to 2° F. decrease in average yearly temperature for each one degree increase in latitude.

Temperatures over 100° F. occur occasionally in summer in nearly all parts of the State. However, low humidity makes these high temperatures more bearable than in more humid regions. Temperatures below zero during winter and early spring are uncommon in most areas of the State, and prolonged periods of extremely cold weather are rare. This is primarily due to the mountains east and north of the State, which act as a barrier to intensely cold continental Arctic air masses. Utah experiences relatively strong insolation during the day and rapid nocturnal cooling, resulting in wide daily ranges in temperature. Even after the hottest days, nights are usually cool over the State. On clear nights the colder air accumulates, by drainage, on the valley bottoms, while the foothills and bench areas remain relatively warm.

PRECIPITATION. Precipitation varies greatly, from an average of less than five inches annually over the Great Salt Lake Desert (west of Great Salt Lake), to more than 40 inches in some parts of the Wasatch Mountains. In the mountains, winter snows form the chief reservoirs of moisture. The areas of the State below an elevation of 4,000 feet, all in the southern part, generally receive less than 10 inches of moisture annually.

Northwestern Utah, over and along the mountains, receives appreciably more precipitation in a year than is received at similar elevations over the rest of the State, primarily due to terrain and the direction of normal storm tracks. The bulk of the moisture falling over that area can be attributed to the movement of Pacific storms through the region during the winter and spring months. In summer northwestern Utah is comparatively dry. The eastern portion receives appreciable rain from summer thunderstorms, which are usually associated with moisture-laden air masses from the Gulf of Mexico.

Snowfall is moderately heavy in the mountains, especially over the northern part. A deep snow cover seldom remains long on the ground. Runoff from melting mountain snow usually reaches a peak in April, May or early June, and sometimes causes flooding along the lower streams. However, damaging floods of this kind are infrequent. Flash floods from summer thunderstorms are more frequent, but they affect only small, local areas.

OTHER CLIMATIC ELEMENTS. Sunny skies prevail most of the year in Utah. There is an average of about 65 to 75 percent of the possible amount of sunshine at Salt Lake City during spring, summer, and fall. In winter Salt Lake City has about 50 percent of the possible sunshine.

During the late fall and winter months, anticyclones tend to settle over the Great Basin for as long as several weeks at a time. Under these conditions, smoke and haze accumulate in the lower levels of the stagnant air over the valleys of northwestern Utah, frequently becoming an obstruction to visibility. This is also true of fog which may persist for several weeks at a time.

Wind speeds are usually light to moderate, ranging below 20 miles per hour. There are only a few tornadoes in Utah as a rule, and those reported usually cause only slight damage. However, strong winds occur occasionally, sometimes attaining damaging proportions in local areas, particularly in the vicinity of the canyon mouths along the western slopes of the Wasatch Mountains. Duststorms occur occasionally, principally over western Utah. These storms are associated with the movement of low pressure disturbances through the area during the spring months.

Hailstorms may occur in limited areas during spring and summer, although the hail is usually small.

## COUNTY INDEX

### *Beaver County*
WAH WAH RANCH

### *Box Elder County*
CORINNE
GROUSE CREEK
THIOKOL PLANT 78

### *Cache County*
LOGAN RADIO KVNU
LOGAN UT STATE UNIV
RICHMOND

### *Carbon County*
SCOFIELD DAM

### *Daggett County*
FLAMING GORGE

### *Davis County*
FARMINGTON USU FIELD

### *Duchesne County*
ALTAMONT
DUCHESNE
MYTON
NEOLA

### *Emery County*
CASTLE DALE
FERRON
GREEN RIVER AVIATION

### *Garfield County*
BOULDER
BRYCE CANYON N P HQ
ESCALANTE
PANGUITCH
TROPIC

### *Grand County*
MOAB
THOMPSON

### *Iron County*
BLOWHARD MTN RADAR
CEDAR CITY MUNI AP
ENTERPRISE BERYL JCT
MODENA
PAROWAN POWER PLANT

### *Juab County*
CALLAO
FISH SPRINGS REFUGE
LEVAN
NEPHI
PARTOUN

### *Kane County*
ALTON
KANAB
ORDERVILLE

### *Millard County*
BLACK ROCK
DELTA
DESERET
FILLMORE
KANOSH
OAK CITY
SCIPIO

### *Morgan County*
MORGAN COMO SPRINGS

### *Piute County*
CIRCLEVILLE
MARYSVALE

### *Rich County*
LAKETOWN
WOODRUFF

### *Salt Lake County*
ALTA
COTTONWOOD WEIR
GARFIELD
MOUNTAIN DELL DAM
SALT LK CITY INTL AP
SILVER LAKE BRIGHTON

### *San Juan County*
ANETH PLANT
BLANDING
BLUFF
CANYONLANDS T NEEDLE
CANYONLANDS THE NECK
CEDAR POINT
HOVENWEEP NM
MEXICAN HAT
MONTICELLO
NATURAL BRIDGES N M

### *Sanpete County*
EPHRAIM SORENSENS FD
MANTI
MORONI

### *Sevier County*
KOOSHAREM
RICHFIELD RADIO KSVC
SALINA

### *Summit County*
COALVILLE
ECHO DAM
KAMAS 3 NW
WANSHIP DAM

### *Tooele County*
DUGWAY
IBAPAH
TOOELE
VERNON
WENDOVER AF AUX FLD

### *Uintah County*
DINOSAUR QUARRY AREA
FORT DUCHESNE
JENSEN
OURAY 4 NE
ROOSEVELT RADIO
VERNAL MUNICIPAL AP

### *Utah County*
FAIRFIELD
PLEASANT GROVE
SANTAQUIN CHLOR
SPANISH FORK P H
TIMPANOGOS CAVE
UTAH LAKE LEHI

### *Wasatch County*
DEER CREEK DAM
HEBER
SNAKE CREEK POWERHOU

### *Washington County*
LA VERKIN
NEW HARMONY
ST GEORGE
VEYO POWERHOUSE
ZION NATIONAL PARK

### *Wayne County*
HANKSVILLE
LOA

### *Weber County*
OGDEN PIONEER P H
OGDEN SUGAR FACTORY
PINE VIEW DAM

## ELEVATION INDEX

| FEET | STATION NAME |
|---|---|
| 2760 | ST GEORGE |
| 3202 | LA VERKIN |
| 3973 | MOAB |
| 4052 | ZION NATIONAL PARK |
| 4062 | GREEN RIVER AVIATION |
| 4232 | CORINNE |
| 4239 | WENDOVER AF AUX FLD |
| 4252 | MEXICAN HAT |
| 4252 | SALT LK CITY INTL AP |
| 4281 | OGDEN SUGAR FACTORY |
| 4308 | HANKSVILLE |
| 4314 | GARFIELD |
| 4324 | BLUFF |
| 4334 | FARMINGTON USU FIELD |
| 4338 | DUGWAY |
| 4344 | CALLAO |
| 4354 | FISH SPRINGS REFUGE |
| 4354 | OGDEN PIONEER P H |
| 4505 | LOGAN RADIO KVNU |
| 4505 | UTAH LAKE LEHI |
| 4524 | RICHMOND |
| 4541 | DESERET |
| 4600 | VEYO POWERHOUSE |
| 4603 | THIOKOL PLANT 78 |
| 4623 | ANETH PLANT |
| 4632 | DELTA |
| 4652 | OURAY 4 NE |
| 4662 | PLEASANT GROVE |
| 4715 | SPANISH FORK P H |
| 4744 | JENSEN |
| 4770 | DINOSAUR QUARRY AREA |
| 4774 | PARTOUN |
| 4783 | LOGAN UT STATE UNIV |
| 4823 | PINE VIEW DAM |
| 4823 | TOOELE |
| 4862 | BLACK ROCK |
| 4882 | FAIRFIELD |
| 4905 | OAK CITY |
| 4964 | COTTONWOOD WEIR |
| 4964 | WAH WAH RANCH |
| 4993 | FORT DUCHESNE |
| 5010 | ROOSEVELT RADIO |
| 5013 | KANAB |
| 5023 | KANOSH |
| 5043 | CANYONLANDS T NEEDLE |
| 5072 | MORGAN COMO SPRINGS |
| 5082 | MYTON |
| 5105 | FILLMORE |

| FEET | STATION NAME |
|---|---|
| 5134 | NEPHI |
| 5154 | THOMPSON |
| 5203 | SALINA |
| 5223 | ENTERPRISE BERYL JCT |
| 5253 | SANTAQUIN CHLOR |
| 5259 | VERNAL MUNICIPAL AP |
| 5282 | IBAPAH |
| 5282 | NEW HARMONY |
| 5292 | DEER CREEK DAM |
| 5299 | GROUSE CREEK |
| 5299 | RICHFIELD RADIO KSVC |
| 5305 | LEVAN |
| 5315 | SCIPIO |
| 5403 | HOVENWEEP NM |
| 5463 | ORDERVILLE |
| 5472 | ECHO DAM |
| 5476 | MODENA |
| 5492 | VERNON |
| 5505 | MOUNTAIN DELL DAM |
| 5525 | DUCHESNE |
| 5525 | TIMPANOGOS CAVE |
| 5535 | MORONI |
| 5551 | COALVILLE |
| 5584 | EPHRAIM SORENSENS FD |
| 5594 | HEBER |
| 5604 | CASTLE DALE |
| 5620 | CEDAR CITY MUNI AP |
| 5623 | CIRCLEVILLE |
| 5682 | MANTI |
| 5764 | ESCALANTE |
| 5843 | MARYSVALE |
| 5905 | CANYONLANDS THE NECK |
| 5905 | WANSHIP DAM |
| 5935 | FERRON |
| 5955 | SNAKE CREEK POWERHOU |
| 5980 | LAKETOWN |
| 6004 | NEOLA |
| 6033 | PAROWAN POWER PLANT |
| 6040 | BLANDING |
| 6270 | FLAMING GORGE |
| 6306 | TROPIC |
| 6345 | WOODRUFF |
| 6410 | KAMAS 3 NW |
| 6414 | ALTAMONT |
| 6506 | NATURAL BRIDGES N M |
| 6535 | KOOSHAREM |
| 6604 | BOULDER |
| 6653 | PANGUITCH |
| 6785 | CEDAR POINT |
| 7054 | LOA |

| FEET | STATION NAME |
|---|---|
| 7073 | MONTICELLO |
| 7155 | ALTON |
| 7634 | SCOFIELD DAM |
| 7907 | BRYCE CANYON N P HQ |
| 8707 | ALTA |
| 8707 | SILVER LAKE BRIGHTON |
| 10695 | BLOWHARD MTN RADAR |

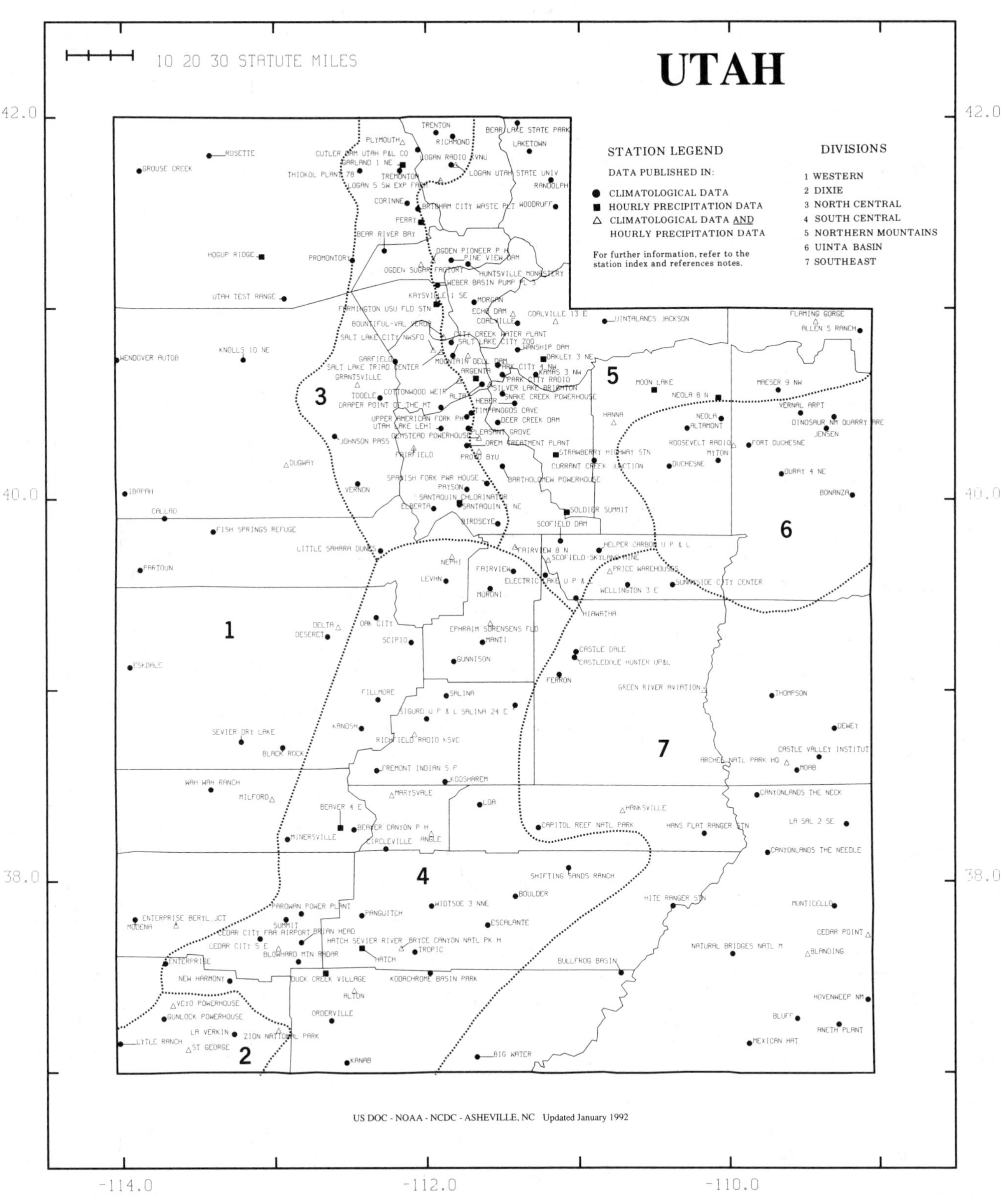
10 20 30 STATUTE MILES
UTAH
STATION LEGEND
DATA PUBLISHED IN:
CLIMATOLOGICAL DATA
HOURLY PRECIPITATION DATA
CLIMATOLOGICAL DATA AND HOURLY PRECIPITATION DATA
For further information, refer to the station index and references notes.
DIVISIONS
1 WESTERN
2 DIXIE
3 NORTH CENTRAL
4 SOUTH CENTRAL
5 NORTHERN MOUNTAINS
6 UINTA BASIN
7 SOUTHEAST
42.0
40.0
38.0
-114.0
-112.0
-110.0
US DOC - NOAA - NCDC - ASHEVILLE, NC Updated January 1992

## ALTA *Salt Lake County* ELEVATION 8707 ft LAT/LONG 40° 36' N / 111° 38' W

| | JAN | FEB | MAR | APR | MAY | JUN | JUL | AUG | SEP | OCT | NOV | DEC | YEAR |
|---|---|---|---|---|---|---|---|---|---|---|---|---|---|
| Maximum Temp °F | 30.3 | 31.3 | 34.5 | *41.3* | *52.0* | *63.1* | *71.0* | *70.1* | *60.9* | *49.2* | 36.2 | *31.0* | 47.6 |
| Minimum Temp °F | 13.0 | 14.2 | 17.3 | *23.6* | *32.2* | *41.1* | *48.4* | *47.6* | *39.7* | *30.2* | 19.2 | *13.9* | 28.4 |
| Mean Temp °F | 21.7 | 22.8 | 25.9 | *32.5* | *42.1* | *52.1* | *59.7* | *58.9* | *50.3* | *39.7* | 27.7 | *22.5* | 38.0 |
| Days Max Temp ≥ 90 °F | 0 | 0 | 0 | 0 | *0* | *0* | *0* | *0* | *0* | *0* | 0 | 0 | 0 |
| Days Max Temp ≤ 32 °F | 19 | 15 | 13 | *6* | *1* | *0* | *0* | *0* | *0* | *3* | 12 | *17* | 86 |
| Days Min Temp ≤ 32 °F | 31 | 28 | 30 | *25* | na | *4* | *0* | *0* | *6* | *17* | 27 | *31* | na |
| Days Min Temp ≤ 0 °F | 4 | 2 | 2 | *0* | *0* | *0* | *0* | *0* | *0* | *0* | 1 | *3* | 12 |
| Heating Degree Days | 1336 | 1185 | 1204 | *969* | *703* | *381* | *165* | *186* | *434* | *777* | 1112 | *1312* | 9764 |
| Cooling Degree Days | 0 | 0 | 0 | 0 | 0 | 1 | 8 | 5 | 0 | 0 | 0 | 0 | 14 |
| Total Precipitation (") | 7.00 | 6.45 | 6.92 | *6.22* | *4.13* | *1.87* | *1.87* | *1.84* | *2.77* | *4.32* | *6.39* | *6.97* | 56.75 |
| Days ≥ 0.1" Precip | *11* | 11 | *12* | *10* | na | na | *4* | *5* | na | *7* | *10* | *11* | na |
| Total Snowfall (") | 82.9 | 75.0 | 78.7 | *65.0* | *25.9* | *4.5* | *0.1* | *0.0* | *4.9* | *27.9* | 73.3 | *85.9* | 524.1 |
| Days ≥ 1" Snow Depth | *30* | 28 | 30 | na | na | *9* | *0* | *0* | *1* | *9* | *24* | *29* | na |

## ALTAMONT *Duchesne County* ELEVATION 6414 ft LAT/LONG 40° 22' N / 110° 16' W

| | JAN | FEB | MAR | APR | MAY | JUN | JUL | AUG | SEP | OCT | NOV | DEC | YEAR |
|---|---|---|---|---|---|---|---|---|---|---|---|---|---|
| Maximum Temp °F | 30.9 | 36.1 | 46.5 | 57.0 | 66.5 | 76.2 | 83.1 | 81.4 | 72.3 | 59.3 | 43.9 | 32.9 | 57.2 |
| Minimum Temp °F | 5.8 | 10.4 | 20.9 | 28.4 | 36.9 | 44.9 | 51.7 | 50.2 | 41.5 | 31.2 | 19.1 | 8.8 | 29.2 |
| Mean Temp °F | 18.4 | 23.3 | 33.8 | 42.7 | 51.7 | 60.6 | 67.4 | 65.8 | 56.9 | 45.3 | 31.6 | 20.9 | 43.2 |
| Days Max Temp ≥ 90 °F | 0 | 0 | 0 | 0 | 0 | 1 | 2 | 1 | 0 | 0 | 0 | 0 | 4 |
| Days Max Temp ≤ 32 °F | 17 | 9 | 2 | 0 | 0 | 0 | 0 | 0 | 0 | 0 | 4 | 15 | 47 |
| Days Min Temp ≤ 32 °F | 31 | 28 | 30 | 21 | 8 | 1 | 0 | 0 | 3 | 17 | 29 | 31 | 199 |
| Days Min Temp ≤ 0 °F | 9 | 4 | 1 | 0 | 0 | 0 | 0 | 0 | 0 | 0 | 1 | 6 | 21 |
| Heating Degree Days | 1439 | 1172 | 961 | 663 | 405 | 157 | 24 | 43 | 242 | 604 | 996 | 1361 | 8067 |
| Cooling Degree Days | 0 | 0 | 0 | 0 | 1 | 36 | 107 | 79 | 7 | 0 | 0 | 0 | 230 |
| Total Precipitation (") | 0.71 | 0.71 | 0.75 | 0.67 | 0.87 | 0.85 | 0.83 | 0.83 | 0.92 | 1.06 | 0.58 | 0.80 | 9.58 |
| Days ≥ 0.1" Precip | 3 | 3 | 3 | 2 | 3 | 2 | 3 | 3 | 3 | 3 | 2 | 2 | 32 |
| Total Snowfall (") | 10.1 | 8.2 | 4.8 | *1.9* | 0.1 | 0.0 | 0.0 | 0.0 | 0.0 | 1.2 | 5.0 | 10.5 | 41.8 |
| Days ≥ 1" Snow Depth | *20* | na | *7* | *1* | 0 | 0 | 0 | 0 | 0 | 1 | *3* | na | na |

## ALTON *Kane County* ELEVATION 7155 ft LAT/LONG 37° 26' N / 112° 29' W

| | JAN | FEB | MAR | APR | MAY | JUN | JUL | AUG | SEP | OCT | NOV | DEC | YEAR |
|---|---|---|---|---|---|---|---|---|---|---|---|---|---|
| Maximum Temp °F | 40.2 | 43.4 | 48.2 | 57.9 | 67.7 | 77.7 | 83.1 | 80.9 | 73.7 | 63.1 | 49.2 | 41.2 | 60.5 |
| Minimum Temp °F | 15.1 | 17.4 | 22.1 | 27.2 | 34.3 | 41.7 | 49.1 | 48.1 | 40.8 | 32.5 | 22.8 | 15.8 | 30.6 |
| Mean Temp °F | 27.7 | 30.4 | 35.2 | 42.6 | 51.0 | 59.8 | 66.1 | 64.5 | 57.3 | 47.8 | 36.0 | 28.5 | 45.6 |
| Days Max Temp ≥ 90 °F | 0 | 0 | 0 | 0 | 0 | 1 | 2 | 1 | 0 | 0 | 0 | 0 | 4 |
| Days Max Temp ≤ 32 °F | 6 | 3 | 1 | 0 | 0 | 0 | 0 | 0 | 0 | 0 | 1 | 6 | 17 |
| Days Min Temp ≤ 32 °F | 31 | 28 | 30 | 24 | 11 | 2 | 0 | 0 | 3 | 15 | 28 | 31 | 203 |
| Days Min Temp ≤ 0 °F | 2 | 1 | 0 | 0 | 0 | 0 | 0 | 0 | 0 | 0 | 0 | 1 | 4 |
| Heating Degree Days | 1149 | 969 | 917 | 666 | 427 | 170 | 31 | 55 | 229 | 525 | 863 | 1123 | 7124 |
| Cooling Degree Days | 0 | 0 | 0 | 0 | 0 | 19 | 69 | 46 | 3 | 0 | 0 | 0 | 137 |
| Total Precipitation (") | 1.86 | 1.85 | 1.80 | 1.04 | 1.00 | 0.45 | 1.58 | 1.72 | 1.39 | 1.39 | 1.66 | 1.57 | 17.31 |
| Days ≥ 0.1" Precip | 4 | 4 | 5 | 3 | 3 | 1 | 4 | 5 | 3 | 3 | 3 | 4 | 42 |
| Total Snowfall (") | 21.8 | 19.5 | 16.7 | 4.3 | 0.8 | 0.2 | 0.0 | 0.0 | 0.0 | 0.9 | 9.4 | *16.2* | 89.8 |
| Days ≥ 1" Snow Depth | *21* | *19* | 14 | 4 | 0 | 0 | 0 | 0 | 0 | 0 | *5* | *14* | 77 |

## ANETH PLANT *San Juan County* ELEVATION 4623 ft LAT/LONG 37° 15' N / 109° 20' W

| | JAN | FEB | MAR | APR | MAY | JUN | JUL | AUG | SEP | OCT | NOV | DEC | YEAR |
|---|---|---|---|---|---|---|---|---|---|---|---|---|---|
| Maximum Temp °F | *41.2* | 49.6 | 60.8 | 69.8 | *79.6* | 90.8 | *96.0* | 94.2 | 84.1 | 71.5 | 55.7 | *43.7* | 69.8 |
| Minimum Temp °F | *18.7* | 25.9 | 33.1 | 39.5 | *48.5* | 57.4 | *63.9* | *62.9* | 53.7 | 40.6 | 29.6 | *21.4* | 41.3 |
| Mean Temp °F | *30.2* | 37.8 | 47.0 | 54.7 | *64.1* | 74.2 | *79.9* | *78.6* | 69.0 | 56.1 | 42.7 | *32.5* | 55.6 |
| Days Max Temp ≥ 90 °F | 0 | 0 | 0 | 0 | 3 | 17 | 24 | 23 | 7 | 0 | 0 | 0 | 74 |
| Days Max Temp ≤ 32 °F | 5 | 1 | 0 | 0 | 0 | 0 | 0 | 0 | 0 | 0 | 0 | 3 | 9 |
| Days Min Temp ≤ 32 °F | 26 | 21 | 14 | 5 | 0 | 0 | 0 | 0 | 0 | 4 | 18 | 27 | 115 |
| Days Min Temp ≤ 0 °F | 1 | 0 | 0 | 0 | 0 | 0 | 0 | 0 | 0 | 0 | 0 | 1 | 2 |
| Heating Degree Days | *1074* | 761 | 551 | 306 | *96* | 5 | *0* | *0* | 35 | 277 | 665 | *1002* | 4772 |
| Cooling Degree Days | na | *0* | *0* | 5 | na | 281 | na | na | *139* | *4* | *0* | na | na |
| Total Precipitation (") | 0.93 | 0.74 | 0.76 | 0.59 | 0.51 | 0.22 | 0.73 | 0.74 | 0.82 | 1.09 | 0.81 | 0.90 | 8.84 |
| Days ≥ 0.1" Precip | *2* | 3 | 3 | 2 | 2 | 0 | 2 | 2 | 2 | 2 | 2 | 2 | 24 |
| Total Snowfall (") | na | na | na | 0.0 | 0.0 | 0.0 | 0.0 | 0.0 | 0.0 | 0.0 | *0.5* | na | na |
| Days ≥ 1" Snow Depth | na | na | na | 0 | 0 | 0 | 0 | 0 | 0 | 0 | *0* | na | na |

## BLACK ROCK *Millard County* ELEVATION 4862 ft LAT/LONG 38° 43 ' N / 112° 58 ' W

| | JAN | FEB | MAR | APR | MAY | JUN | JUL | AUG | SEP | OCT | NOV | DEC | YEAR |
|---|---|---|---|---|---|---|---|---|---|---|---|---|---|
| Maximum Temp °F | 40.1 | 47.8 | 57.1 | 65.9 | 75.4 | 85.7 | 92.5 | 89.8 | 80.5 | 67.9 | 52.8 | 41.3 | 66.4 |
| Minimum Temp °F | 13.6 | 19.1 | 25.3 | 30.4 | 38.0 | 45.4 | 53.4 | 51.7 | 42.1 | 31.6 | 22.7 | 14.4 | 32.3 |
| Mean Temp °F | 26.9 | 33.5 | 41.2 | 48.2 | 56.7 | 65.6 | 73.0 | 70.8 | 61.3 | 49.7 | 37.8 | 27.9 | 49.4 |
| Days Max Temp ≥ 90 °F | 0 | 0 | 0 | 0 | 1 | 11 | 24 | 17 | 4 | 0 | 0 | 0 | 57 |
| Days Max Temp ≤ 32 °F | 7 | 2 | 0 | 0 | 0 | 0 | 0 | 0 | 0 | 0 | 1 | 6 | 16 |
| Days Min Temp ≤ 32 °F | 29 | 26 | 25 | 18 | 7 | 1 | 0 | 0 | 4 | 18 | 25 | 29 | 182 |
| Days Min Temp ≤ 0 °F | 5 | 2 | 0 | 0 | 0 | 0 | 0 | 0 | 0 | 0 | 1 | 4 | 12 |
| Heating Degree Days | 1175 | 884 | 730 | 498 | 258 | 67 | 3 | 9 | 142 | 466 | 809 | 1143 | 6184 |
| Cooling Degree Days | 0 | 0 | 0 | 1 | 8 | 102 | 258 | 209 | 43 | 0 | 0 | 0 | 621 |
| Total Precipitation (") | 0.54 | 0.48 | 1.10 | 0.92 | 0.86 | 0.53 | 0.86 | 0.89 | 0.85 | 0.93 | 0.71 | 0.59 | 9.26 |
| Days ≥ 0.1" Precip | 2 | 2 | 3 | 3 | 2 | 1 | 2 | 3 | 2 | 2 | 2 | 2 | 26 |
| Total Snowfall (") | na | na | ***8.1*** | ***2.3*** | 0.5 | 0.0 | 0.0 | 0.0 | 0.1 | 0.2 | ***4.0*** | na | na |
| Days ≥ 1" Snow Depth | 13 | ***8*** | ***3*** | 0 | 0 | 0 | 0 | 0 | 0 | 0 | ***3*** | 9 | 36 |

## BLANDING *San Juan County* ELEVATION 6040 ft LAT/LONG 37° 37 ' N / 109° 28 ' W

| | JAN | FEB | MAR | APR | MAY | JUN | JUL | AUG | SEP | OCT | NOV | DEC | YEAR |
|---|---|---|---|---|---|---|---|---|---|---|---|---|---|
| Maximum Temp °F | 39.1 | 45.3 | 52.9 | 62.2 | 72.4 | 84.0 | 89.0 | 86.5 | 78.2 | 65.5 | 50.6 | 41.0 | 63.9 |
| Minimum Temp °F | 17.1 | 22.5 | 28.2 | 33.9 | 42.2 | 51.0 | 57.5 | 56.0 | 48.2 | 38.0 | 27.1 | 19.4 | 36.8 |
| Mean Temp °F | 28.1 | 33.9 | 40.6 | 48.1 | 57.4 | 67.5 | 73.3 | 71.3 | 63.2 | 51.7 | 38.9 | 30.2 | 50.4 |
| Days Max Temp ≥ 90 °F | 0 | 0 | 0 | 0 | 0 | 7 | 16 | 10 | 1 | 0 | 0 | 0 | 34 |
| Days Max Temp ≤ 32 °F | 6 | 2 | 0 | 0 | 0 | 0 | 0 | 0 | 0 | 0 | 0 | 4 | 12 |
| Days Min Temp ≤ 32 °F | 30 | 26 | 23 | 13 | 2 | 0 | 0 | 0 | 0 | 7 | 23 | 30 | 154 |
| Days Min Temp ≤ 0 °F | 2 | 0 | 0 | 0 | 0 | 0 | 0 | 0 | 0 | 0 | 0 | 1 | 3 |
| Heating Degree Days | 1137 | 871 | 750 | 503 | 239 | 47 | 0 | 5 | 98 | 406 | 778 | 1072 | 5906 |
| Cooling Degree Days | 0 | 0 | 0 | 1 | 13 | 158 | 273 | 231 | 64 | 2 | 0 | 0 | 742 |
| Total Precipitation (") | 1.45 | 1.06 | 0.97 | 0.80 | 0.79 | 0.46 | 1.34 | 1.44 | 1.18 | 1.42 | 1.10 | 1.30 | 13.31 |
| Days ≥ 0.1" Precip | 4 | 3 | 3 | 2 | 2 | 1 | 3 | 3 | 3 | 4 | 3 | 3 | 34 |
| Total Snowfall (") | 13.6 | 8.0 | 4.8 | 2.5 | 0.2 | 0.0 | 0.0 | 0.0 | 0.0 | 0.3 | 3.5 | 10.4 | 43.3 |
| Days ≥ 1" Snow Depth | ***18*** | ***13*** | 3 | 0 | 0 | 0 | 0 | 0 | 0 | 0 | 2 | ***9*** | 45 |

## BLOWHARD MTN RADAR *Iron County* ELEVATION 10695 ft LAT/LONG 37° 35 ' N / 112° 51 ' W

| | JAN | FEB | MAR | APR | MAY | JUN | JUL | AUG | SEP | OCT | NOV | DEC | YEAR |
|---|---|---|---|---|---|---|---|---|---|---|---|---|---|
| Maximum Temp °F | 27.6 | 28.0 | 30.4 | 37.2 | 45.6 | 56.2 | 62.3 | 60.3 | 53.4 | 43.7 | 33.2 | 28.1 | 42.2 |
| Minimum Temp °F | 12.5 | 12.8 | 15.1 | 21.1 | 29.6 | 40.5 | 47.2 | 45.4 | 38.6 | 29.0 | 18.5 | 13.3 | 27.0 |
| Mean Temp °F | 20.1 | 20.4 | 22.8 | 29.2 | 37.7 | 48.4 | 54.8 | 52.9 | 46.0 | 36.4 | 25.9 | 20.7 | 34.6 |
| Days Max Temp ≥ 90 °F | 0 | 0 | 0 | 0 | 0 | 0 | 0 | 0 | 0 | 0 | 0 | 0 | 0 |
| Days Max Temp ≤ 32 °F | 21 | 19 | ***18*** | ***8*** | 2 | 0 | 0 | 0 | 0 | 4 | 13 | 21 | 106 |
| Days Min Temp ≤ 32 °F | 30 | 28 | 30 | 26 | ***16*** | 6 | 0 | 0 | 5 | 19 | 27 | 30 | 217 |
| Days Min Temp ≤ 0 °F | 4 | 3 | 1 | 0 | 0 | 0 | 0 | 0 | 0 | 0 | 1 | 3 | 12 |
| Heating Degree Days | 1386 | 1252 | 1300 | 1067 | 841 | 492 | 311 | 368 | 563 | 879 | 1167 | 1368 | 10994 |
| Cooling Degree Days | 0 | 0 | 0 | 0 | 0 | 0 | 1 | 0 | 0 | 0 | 0 | 0 | 1 |
| Total Precipitation (") | 2.90 | 3.18 | 4.33 | 3.07 | 1.78 | 0.80 | 2.21 | 2.80 | 1.52 | 1.82 | 2.65 | 2.75 | 29.81 |
| Days ≥ 0.1" Precip | ***5*** | 6 | ***7*** | 5 | 4 | 2 | 5 | 6 | 3 | 4 | 5 | 5 | 57 |
| Total Snowfall (") | 34.7 | 34.5 | 46.0 | 30.3 | 13.3 | 2.0 | 0.0 | 0.1 | 2.0 | 12.5 | 29.2 | 34.4 | 239.0 |
| Days ≥ 1" Snow Depth | 31 | 28 | 31 | 29 | 26 | 7 | 0 | 0 | 1 | 8 | 23 | 30 | 214 |

## BLUFF *San Juan County* ELEVATION 4324 ft LAT/LONG 37° 17 ' N / 109° 33 ' W

| | JAN | FEB | MAR | APR | MAY | JUN | JUL | AUG | SEP | OCT | NOV | DEC | YEAR |
|---|---|---|---|---|---|---|---|---|---|---|---|---|---|
| Maximum Temp °F | 42.1 | 51.1 | 61.4 | 70.9 | 79.9 | 90.3 | 95.1 | 92.5 | 84.4 | 71.2 | 55.8 | 43.7 | 69.9 |
| Minimum Temp °F | 16.4 | 22.6 | 29.4 | 35.9 | 44.1 | 51.0 | 59.4 | 58.2 | 48.0 | 35.2 | 25.1 | 17.2 | 36.9 |
| Mean Temp °F | 29.3 | 36.9 | 45.5 | 53.4 | 62.0 | 70.7 | 77.3 | 75.4 | 66.2 | 53.2 | 40.5 | 30.5 | 53.4 |
| Days Max Temp ≥ 90 °F | 0 | 0 | 0 | 0 | 3 | 17 | 27 | 23 | 7 | 0 | 0 | 0 | 77 |
| Days Max Temp ≤ 32 °F | 5 | 1 | 0 | 0 | 0 | 0 | 0 | 0 | 0 | 0 | 0 | 3 | 9 |
| Days Min Temp ≤ 32 °F | 30 | 25 | 20 | 11 | 2 | 0 | 0 | 0 | 1 | 13 | 24 | 29 | 155 |
| Days Min Temp ≤ 0 °F | 2 | 0 | 0 | 0 | 0 | 0 | 0 | 0 | 0 | 0 | 0 | 1 | 3 |
| Heating Degree Days | 1100 | 787 | 599 | 342 | 124 | 15 | 0 | 1 | 53 | 360 | 729 | 1063 | 5173 |
| Cooling Degree Days | 0 | 0 | 0 | 4 | 43 | 211 | 389 | 349 | 104 | 2 | 0 | 0 | 1102 |
| Total Precipitation (") | 0.85 | 0.73 | 0.66 | 0.46 | 0.53 | 0.22 | 0.93 | 0.77 | 0.65 | 1.07 | 0.77 | 0.80 | 8.44 |
| Days ≥ 0.1" Precip | 3 | 3 | 2 | 1 | 2 | 1 | 2 | 3 | 2 | 3 | 3 | 2 | 27 |
| Total Snowfall (") | ***4.3*** | ***2.0*** | 0.1 | ***0.1*** | 0.0 | 0.0 | 0.0 | 0.0 | 0.0 | 0.1 | 0.5 | 4.1 | 11.2 |
| Days ≥ 1" Snow Depth | na | na | 0 | ***0*** | 0 | 0 | 0 | 0 | 0 | 0 | 0 | na | na |

## BOULDER *Garfield County* ELEVATION 6604 ft LAT/LONG 37° 55' N / 111° 25' W

| | JAN | FEB | MAR | APR | MAY | JUN | JUL | AUG | SEP | OCT | NOV | DEC | YEAR |
|---|---|---|---|---|---|---|---|---|---|---|---|---|---|
| Maximum Temp °F | 38.7 | 43.5 | 50.7 | 58.9 | 68.1 | 78.7 | 84.4 | 81.8 | 74.3 | 63.2 | 49.1 | 40.1 | 61.0 |
| Minimum Temp °F | 16.4 | 20.8 | 27.0 | 33.2 | 41.4 | 51.2 | 57.9 | 56.4 | 48.7 | 38.5 | 26.3 | 18.0 | 36.3 |
| Mean Temp °F | 27.6 | 32.2 | 38.9 | 46.1 | 54.8 | 65.0 | 71.2 | 69.1 | 61.5 | 50.9 | 37.7 | 29.1 | 48.7 |
| Days Max Temp ≥ 90 °F | 0 | 0 | 0 | 0 | 0 | 2 | 4 | 2 | 0 | 0 | 0 | 0 | 8 |
| Days Max Temp ≤ 32 °F | 7 | 2 | 0 | 0 | 0 | 0 | 0 | 0 | 0 | 0 | 1 | 5 | 15 |
| Days Min Temp ≤ 32 °F | 31 | 27 | 24 | 14 | 4 | 0 | 0 | 0 | 1 | 7 | 23 | 30 | 161 |
| Days Min Temp ≤ 0 °F | 1 | 0 | 0 | 0 | 0 | 0 | 0 | 0 | 0 | 0 | 0 | 1 | 2 |
| Heating Degree Days | 1153 | 921 | 803 | 562 | 315 | 82 | 4 | 16 | 135 | 432 | 812 | 1107 | 6342 |
| Cooling Degree Days | 0 | 0 | 0 | 1 | 6 | 93 | 203 | 153 | 39 | 1 | 0 | 0 | 496 |
| Total Precipitation (") | 0.98 | 0.89 | 1.06 | 0.52 | 0.83 | 0.41 | 1.12 | 1.44 | 1.01 | 1.05 | 0.80 | 0.80 | 10.91 |
| Days ≥ 0.1" Precip | 3 | 3 | 3 | 1 | 2 | 1 | 3 | 4 | 3 | 3 | 2 | 3 | 31 |
| Total Snowfall (") | na | na | na | *0.2* | 0.0 | 0.0 | 0.0 | 0.0 | 0.0 | 0.3 | *2.0* | na | na |
| Days ≥ 1" Snow Depth | *18* | 10 | 2 | 0 | 0 | 0 | 0 | 0 | 0 | 0 | 1 | 9 | 40 |

## BRYCE CANYON N P HQ *Garfield County* ELEVATION 7907 ft LAT/LONG 37° 39' N / 112° 10' W

| | JAN | FEB | MAR | APR | MAY | JUN | JUL | AUG | SEP | OCT | NOV | DEC | YEAR |
|---|---|---|---|---|---|---|---|---|---|---|---|---|---|
| Maximum Temp °F | 36.0 | 38.6 | 44.4 | 53.1 | 63.1 | 73.8 | 79.2 | 76.8 | 69.6 | 58.7 | 44.1 | 36.5 | 56.2 |
| Minimum Temp °F | 9.4 | 11.6 | 18.2 | 23.8 | 31.3 | 39.2 | 46.2 | 44.8 | 36.7 | 27.3 | 18.0 | 10.2 | 26.4 |
| Mean Temp °F | 22.7 | 25.1 | 31.3 | 38.5 | 47.2 | 56.6 | 62.7 | 60.8 | 53.1 | 43.0 | 31.1 | 23.4 | 41.3 |
| Days Max Temp ≥ 90 °F | 0 | 0 | 0 | 0 | 0 | 0 | 0 | 0 | 0 | 0 | 0 | 0 | 0 |
| Days Max Temp ≤ 32 °F | 11 | 7 | 3 | 1 | 0 | 0 | 0 | 0 | 0 | 0 | 4 | 9 | 35 |
| Days Min Temp ≤ 32 °F | 31 | 28 | 30 | 27 | 17 | 5 | 0 | 1 | 8 | 24 | 29 | 31 | 231 |
| Days Min Temp ≤ 0 °F | 6 | 4 | 1 | 0 | 0 | 0 | 0 | 0 | 0 | 0 | 1 | 5 | 17 |
| Heating Degree Days | 1303 | 1119 | 1036 | 790 | 544 | 255 | 85 | 135 | 349 | 675 | 1011 | 1283 | 8585 |
| Cooling Degree Days | 0 | 0 | 0 | 0 | 0 | 12 | 22 | 12 | 0 | 0 | 0 | 0 | 46 |
| Total Precipitation (") | 1.43 | 1.60 | 1.61 | 0.88 | 1.16 | 0.56 | 1.47 | 2.11 | 1.51 | 1.34 | 1.28 | 1.24 | 16.19 |
| Days ≥ 0.1" Precip | 3 | 4 | 4 | 2 | 3 | 2 | 5 | 5 | 3 | 3 | 3 | 3 | 40 |
| Total Snowfall (") | 17.2 | 18.5 | 16.7 | *7.8* | *2.0* | 0.0 | 0.0 | 0.0 | 0.1 | 1.4 | 12.2 | 17.1 | 93.0 |
| Days ≥ 1" Snow Depth | 27 | 26 | 23 | 8 | 0 | 0 | 0 | 0 | 0 | 1 | 10 | 21 | 116 |

## CALLAO *Juab County* ELEVATION 4344 ft LAT/LONG 39° 54' N / 113° 43' W

| | JAN | FEB | MAR | APR | MAY | JUN | JUL | AUG | SEP | OCT | NOV | DEC | YEAR |
|---|---|---|---|---|---|---|---|---|---|---|---|---|---|
| Maximum Temp °F | 38.3 | 45.9 | 55.6 | 64.0 | 72.7 | 81.9 | 89.8 | 88.2 | 78.7 | 65.7 | 51.0 | 39.6 | 64.3 |
| Minimum Temp °F | 14.3 | 19.9 | 27.6 | 33.5 | 42.1 | 50.2 | 56.9 | 55.0 | 44.5 | 33.4 | 23.7 | 15.1 | 34.7 |
| Mean Temp °F | 26.3 | 32.9 | 41.6 | 48.8 | 57.4 | 66.1 | 73.4 | 71.6 | 61.7 | 49.5 | 37.4 | 27.3 | 49.5 |
| Days Max Temp ≥ 90 °F | 0 | 0 | 0 | 0 | 0 | 6 | 18 | *15* | 3 | 0 | 0 | 0 | 42 |
| Days Max Temp ≤ 32 °F | 10 | 3 | 0 | 0 | 0 | 0 | 0 | 0 | 0 | 0 | 1 | 7 | 21 |
| Days Min Temp ≤ 32 °F | 30 | 26 | 23 | 14 | 3 | 0 | 0 | 0 | 2 | 14 | 25 | 30 | 167 |
| Days Min Temp ≤ 0 °F | 3 | 1 | 0 | 0 | 0 | 0 | 0 | 0 | 0 | 0 | 0 | 2 | 6 |
| Heating Degree Days | 1194 | 899 | 718 | 483 | 249 | 74 | 5 | 11 | 146 | 475 | 822 | 1161 | 6237 |
| Cooling Degree Days | 0 | 0 | 0 | 3 | 25 | 123 | 269 | 244 | 51 | 1 | 0 | 0 | 716 |
| Total Precipitation (") | 0.34 | 0.33 | 0.39 | 0.41 | 0.78 | 0.62 | 0.60 | 0.67 | 0.57 | 0.70 | 0.35 | 0.27 | 6.03 |
| Days ≥ 0.1" Precip | 1 | 1 | 1 | 1 | 2 | 2 | 2 | 2 | 2 | 2 | 1 | 1 | 18 |
| Total Snowfall (") | *4.6* | *2.3* | 0.9 | 0.6 | 0.0 | 0.0 | 0.0 | 0.0 | 0.0 | 0.0 | 1.5 | *2.4* | 12.3 |
| Days ≥ 1" Snow Depth | na | *4* | *1* | 0 | 0 | 0 | 0 | 0 | 0 | 0 | *1* | na | na |

## CANYONLANDS T NEEDLE *San Juan County* ELEVATION 5043 ft LAT/LONG 38° 9' N / 109° 45' W

| | JAN | FEB | MAR | APR | MAY | JUN | JUL | AUG | SEP | OCT | NOV | DEC | YEAR |
|---|---|---|---|---|---|---|---|---|---|---|---|---|---|
| Maximum Temp °F | 39.4 | 48.3 | 58.5 | 67.5 | 78.1 | 89.3 | 95.0 | 92.5 | 83.5 | 69.8 | 53.7 | 41.9 | 68.1 |
| Minimum Temp °F | 14.5 | 22.0 | 29.7 | 36.3 | 45.0 | 54.2 | 61.2 | 59.4 | 49.4 | 37.3 | 26.3 | 17.3 | 37.7 |
| Mean Temp °F | 26.9 | 35.2 | 44.1 | 51.9 | 61.6 | 71.7 | 78.1 | 76.0 | 66.5 | 53.6 | 40.1 | 29.6 | 52.9 |
| Days Max Temp ≥ 90 °F | 0 | 0 | 0 | 0 | 2 | 16 | 27 | 23 | 7 | 0 | 0 | 0 | 75 |
| Days Max Temp ≤ 32 °F | 7 | 1 | 0 | 0 | 0 | 0 | 0 | 0 | 0 | 0 | 0 | 4 | 12 |
| Days Min Temp ≤ 32 °F | 30 | 25 | *21* | 11 | 2 | 0 | 0 | 0 | 0 | 9 | 23 | 30 | 151 |
| Days Min Temp ≤ 0 °F | 3 | 1 | 0 | 0 | 0 | 0 | 0 | 0 | 0 | 0 | 0 | 1 | 5 |
| Heating Degree Days | 1173 | 836 | 641 | 390 | 144 | 20 | 0 | 1 | 57 | 352 | 741 | 1091 | 5446 |
| Cooling Degree Days | 0 | 0 | 0 | 5 | 44 | 236 | 396 | 343 | 104 | 3 | 0 | 0 | 1131 |
| Total Precipitation (") | 0.57 | 0.46 | 0.72 | 0.67 | 0.57 | 0.40 | 0.93 | 0.99 | 0.76 | 1.15 | 0.68 | 0.57 | 8.47 |
| Days ≥ 0.1" Precip | 2 | 2 | *2* | 2 | 2 | 1 | 3 | 3 | 3 | 3 | 2 | 2 | 27 |
| Total Snowfall (") | *4.6* | *2.9* | na | 0.6 | 0.0 | 0.0 | 0.0 | 0.0 | 0.0 | 0.3 | *1.0* | *4.3* | na |
| Days ≥ 1" Snow Depth | *12* | 3 | *1* | 0 | 0 | 0 | 0 | 0 | 0 | 0 | *1* | 6 | 23 |

## CANYONLANDS THE NECK *San Juan County* ELEVATION 5905 ft LAT/LONG 38° 27' N / 109° 50' W

| | JAN | FEB | MAR | APR | MAY | JUN | JUL | AUG | SEP | OCT | NOV | DEC | YEAR |
|---|---|---|---|---|---|---|---|---|---|---|---|---|---|
| Maximum Temp °F | 36.0 | 43.5 | 53.0 | 62.4 | 73.1 | 84.5 | 90.3 | 88.0 | 78.7 | 65.2 | 48.4 | 37.7 | 63.4 |
| Minimum Temp °F | 19.0 | 25.3 | 32.1 | 38.8 | 48.3 | 59.0 | 64.8 | 62.8 | 54.5 | 43.1 | 30.6 | 21.3 | 41.6 |
| Mean Temp °F | 27.6 | 34.4 | 42.5 | 50.6 | 60.7 | 71.7 | 77.6 | 75.5 | 66.6 | 54.2 | 39.5 | 29.7 | 52.6 |
| Days Max Temp ≥ 90 °F | 0 | 0 | 0 | 0 | 0 | 8 | 20 | 13 | 2 | 0 | 0 | 0 | 43 |
| Days Max Temp ≤ 32 °F | 10 | 2 | 0 | 0 | 0 | 0 | 0 | 0 | 0 | 0 | 1 | 8 | 21 |
| Days Min Temp ≤ 32 °F | 30 | 23 | 16 | 8 | 1 | 0 | 0 | 0 | 0 | 4 | 18 | 29 | 129 |
| Days Min Temp ≤ 0 °F | 0 | 0 | 0 | 0 | 0 | 0 | 0 | 0 | 0 | 0 | 0 | 0 | 0 |
| Heating Degree Days | 1154 | 858 | 690 | 428 | 174 | 28 | 1 | 2 | 64 | 338 | 759 | 1088 | 5584 |
| Cooling Degree Days | 0 | 0 | 0 | 6 | 43 | 238 | 380 | 328 | 117 | 8 | 0 | 0 | 1120 |
| Total Precipitation (") | 0.53 | 0.36 | 0.81 | 0.76 | 0.78 | 0.47 | 1.07 | 0.78 | 0.80 | 1.23 | 0.81 | 0.64 | 9.04 |
| Days ≥ 0.1" Precip | 2 | 1 | 3 | 2 | 2 | 1 | 3 | 2 | 3 | 3 | 2 | 2 | 26 |
| Total Snowfall (") | 5.7 | ***2.9*** | 3.8 | ***1.7*** | 0.1 | 0.0 | 0.0 | 0.0 | 0.0 | 1.1 | 2.3 | 6.2 | 23.8 |
| Days ≥ 1" Snow Depth | 18 | 10 | 3 | 1 | 0 | 0 | 0 | 0 | 0 | 0 | 2 | ***10*** | 44 |

## CASTLE DALE *Emery County* ELEVATION 5604 ft LAT/LONG 39° 13' N / 111° 1' W

| | JAN | FEB | MAR | APR | MAY | JUN | JUL | AUG | SEP | OCT | NOV | DEC | YEAR |
|---|---|---|---|---|---|---|---|---|---|---|---|---|---|
| Maximum Temp °F | 35.1 | 42.4 | 53.5 | 62.9 | 72.7 | 83.3 | 89.3 | 87.1 | 78.6 | 66.1 | 49.9 | 38.1 | 63.3 |
| Minimum Temp °F | 7.6 | 14.3 | 23.9 | 30.3 | 38.9 | 46.8 | 53.4 | 51.4 | 42.3 | 31.9 | 21.2 | 11.1 | 31.1 |
| Mean Temp °F | 21.3 | 28.4 | 38.7 | 46.6 | 55.8 | 65.1 | 71.4 | 69.3 | 60.5 | 49.0 | 35.6 | 24.6 | 47.2 |
| Days Max Temp ≥ 90 °F | 0 | 0 | 0 | 0 | 0 | 7 | 17 | 11 | 2 | 0 | 0 | 0 | 37 |
| Days Max Temp ≤ 32 °F | 12 | 3 | 0 | 0 | 0 | 0 | 0 | 0 | 0 | 0 | 1 | 7 | 23 |
| Days Min Temp ≤ 32 °F | 31 | 28 | 28 | 19 | 6 | 0 | 0 | 0 | 3 | 17 | 28 | 31 | 191 |
| Days Min Temp ≤ 0 °F | 8 | 3 | 0 | 0 | 0 | 0 | 0 | 0 | 0 | 0 | 0 | 4 | 15 |
| Heating Degree Days | 1346 | 1029 | 808 | 545 | 282 | 74 | 3 | 13 | 153 | 488 | 875 | 1245 | 6861 |
| Cooling Degree Days | 0 | 0 | 0 | 0 | 4 | 83 | 187 | 144 | 24 | 0 | 0 | 0 | 442 |
| Total Precipitation (") | 0.61 | 0.53 | 0.64 | 0.48 | 0.71 | 0.50 | 0.83 | 0.93 | 0.72 | 0.79 | 0.50 | 0.51 | 7.75 |
| Days ≥ 0.1" Precip | 2 | 2 | 2 | 2 | 2 | 2 | 2 | 3 | 2 | 2 | 1 | 2 | 24 |
| Total Snowfall (") | ***8.1*** | ***4.6*** | ***1.7*** | 0.5 | 0.0 | 0.0 | 0.0 | 0.0 | 0.0 | 0.3 | 1.4 | 4.3 | 20.9 |
| Days ≥ 1" Snow Depth | na | ***8*** | ***1*** | 0 | 0 | 0 | 0 | 0 | 0 | 0 | 1 | ***6*** | na |

## CEDAR CITY MUNI AP *Iron County* ELEVATION 5620 ft LAT/LONG 37° 42' N / 113° 6' W

| | JAN | FEB | MAR | APR | MAY | JUN | JUL | AUG | SEP | OCT | NOV | DEC | YEAR |
|---|---|---|---|---|---|---|---|---|---|---|---|---|---|
| Maximum Temp °F | 42.0 | 46.9 | 53.7 | 61.8 | 72.0 | 83.8 | 90.1 | 87.7 | 79.3 | 66.8 | 52.2 | 42.7 | 64.9 |
| Minimum Temp °F | 18.0 | 22.6 | 28.2 | 33.7 | 41.3 | 49.8 | 57.9 | 56.6 | 47.2 | 36.1 | 26.4 | 18.5 | 36.4 |
| Mean Temp °F | 30.0 | 34.8 | 40.9 | 47.8 | 56.7 | 66.8 | 74.0 | 72.2 | 63.3 | 51.5 | 39.3 | 30.6 | 50.7 |
| Days Max Temp ≥ 90 °F | 0 | 0 | 0 | 0 | 0 | 8 | 18 | 13 | 2 | 0 | 0 | 0 | 41 |
| Days Max Temp ≤ 32 °F | 5 | 2 | 1 | 0 | 0 | 0 | 0 | 0 | 0 | 0 | 1 | 5 | 14 |
| Days Min Temp ≤ 32 °F | 29 | 24 | 22 | 13 | 3 | 0 | 0 | 0 | 1 | 9 | 23 | 29 | 153 |
| Days Min Temp ≤ 0 °F | 2 | 1 | 0 | 0 | 0 | 0 | 0 | 0 | 0 | 0 | 0 | 2 | 5 |
| Heating Degree Days | 1077 | 847 | 739 | 509 | 259 | 57 | 2 | 5 | 104 | 414 | 764 | 1059 | 5836 |
| Cooling Degree Days | 0 | 0 | 0 | 0 | 10 | 131 | 290 | 241 | 63 | 0 | 0 | 0 | 735 |
| Total Precipitation (") | 0.78 | 0.94 | 1.34 | 1.04 | 0.87 | 0.42 | 1.06 | 1.33 | 0.90 | 1.12 | 1.06 | 0.71 | 11.57 |
| Days ≥ 0.1" Precip | 3 | 3 | 4 | 3 | 3 | 1 | 3 | 3 | 2 | 3 | 3 | 2 | 33 |
| Total Snowfall (") | 8.2 | 8.5 | 8.7 | 5.0 | 1.4 | 0.1 | 0.0 | 0.0 | 0.1 | 2.1 | 5.9 | 7.9 | 47.9 |
| Days ≥ 1" Snow Depth | 14 | 9 | 4 | 2 | 0 | 0 | 0 | 0 | 0 | 1 | 4 | 10 | 44 |

## CEDAR POINT *San Juan County* ELEVATION 6785 ft LAT/LONG 37° 43' N / 109° 5' W

| | JAN | FEB | MAR | APR | MAY | JUN | JUL | AUG | SEP | OCT | NOV | DEC | YEAR |
|---|---|---|---|---|---|---|---|---|---|---|---|---|---|
| Maximum Temp °F | 36.2 | 40.8 | 47.5 | 57.5 | 68.4 | 80.1 | 85.7 | 82.9 | 74.4 | 62.0 | 46.9 | 37.7 | 60.0 |
| Minimum Temp °F | 14.1 | 17.9 | 24.9 | 31.0 | 38.7 | 47.1 | 53.8 | 52.3 | 44.9 | 35.2 | 24.4 | 16.1 | 33.4 |
| Mean Temp °F | 25.1 | 29.4 | 36.2 | 44.3 | 53.6 | 63.7 | 69.8 | 67.6 | 59.7 | 48.6 | 35.6 | 26.9 | 46.7 |
| Days Max Temp ≥ 90 °F | 0 | 0 | 0 | 0 | 0 | 3 | 8 | 4 | 0 | 0 | 0 | 0 | 15 |
| Days Max Temp ≤ 32 °F | 10 | 5 | 1 | 0 | 0 | 0 | 0 | 0 | 0 | 0 | 2 | 8 | 26 |
| Days Min Temp ≤ 32 °F | 31 | 28 | 27 | 18 | 5 | 0 | 0 | 0 | 1 | 10 | 26 | 31 | 177 |
| Days Min Temp ≤ 0 °F | 3 | 1 | 0 | 0 | 0 | 0 | 0 | 0 | 0 | 0 | 0 | 2 | 6 |
| Heating Degree Days | 1229 | 999 | 885 | 613 | 349 | 94 | 6 | 21 | 169 | 501 | 874 | 1173 | 6913 |
| Cooling Degree Days | 0 | 0 | 0 | 0 | 2 | 61 | 149 | 107 | 14 | 0 | 0 | 0 | 333 |
| Total Precipitation (") | 1.29 | 1.22 | 1.30 | 0.95 | 0.95 | 0.46 | 1.42 | 1.38 | 1.58 | 1.88 | 1.51 | 1.41 | 15.35 |
| Days ≥ 0.1" Precip | 3 | 3 | 4 | 3 | 3 | 1 | 4 | 4 | 4 | 4 | 3 | 3 | 39 |
| Total Snowfall (") | ***18.7*** | 14.9 | 10.9 | 4.4 | 0.8 | 0.0 | 0.0 | 0.0 | 0.0 | 2.6 | 8.8 | 14.5 | 75.6 |
| Days ≥ 1" Snow Depth | 22 | ***18*** | 10 | 2 | 0 | 0 | 0 | 0 | 0 | 1 | 5 | ***15*** | 73 |

## CIRCLEVILLE *Piute County* ELEVATION 5623 ft LAT/LONG 38° 10' N / 112° 16' W

| | JAN | FEB | MAR | APR | MAY | JUN | JUL | AUG | SEP | OCT | NOV | DEC | YEAR |
|---|---|---|---|---|---|---|---|---|---|---|---|---|---|
| Maximum Temp °F | 41.3 | 46.2 | 52.6 | 60.9 | 70.6 | 81.6 | 87.9 | 85.6 | 77.4 | 65.7 | 51.9 | 42.5 | 63.7 |
| Minimum Temp °F | 13.1 | 18.1 | 23.2 | 28.7 | 36.6 | 44.6 | 52.1 | 50.1 | 40.9 | 30.4 | 21.3 | 14.0 | 31.1 |
| Mean Temp °F | 27.2 | 32.2 | 37.9 | 44.8 | 53.6 | 63.1 | 70.0 | 67.9 | 59.1 | 48.0 | 36.6 | 28.3 | 47.4 |
| Days Max Temp ≥ 90 °F | 0 | 0 | 0 | 0 | 0 | 5 | 12 | 9 | 1 | 0 | 0 | 0 | 27 |
| Days Max Temp ≤ 32 °F | 5 | 2 | 1 | 0 | 0 | 0 | 0 | 0 | 0 | 0 | 1 | 5 | 14 |
| Days Min Temp ≤ 32 °F | 30 | 27 | 27 | 20 | 9 | 1 | 0 | 0 | 4 | 19 | 26 | 30 | 193 |
| Days Min Temp ≤ 0 °F | 4 | 1 | 0 | 0 | 0 | 0 | 0 | 0 | 0 | 0 | 1 | 4 | 10 |
| Heating Degree Days | 1164 | 919 | 832 | 598 | 347 | 108 | 7 | 21 | 187 | 521 | 846 | 1132 | 6682 |
| Cooling Degree Days | 0 | 0 | 0 | 0 | 1 | 56 | 150 | 107 | 15 | 0 | 0 | 0 | 329 |
| Total Precipitation (") | 0.55 | 0.50 | 0.72 | 0.57 | 0.94 | 0.54 | 0.91 | 1.34 | 0.96 | 0.79 | 0.57 | 0.56 | 8.95 |
| Days ≥ 0.1" Precip | 2 | 2 | 2 | 2 | 3 | 2 | 3 | 4 | 3 | 3 | 2 | 2 | 30 |
| Total Snowfall (") | 5.6 | 3.5 | 3.6 | 0.7 | 0.2 | 0.0 | 0.0 | 0.0 | 0.3 | 0.7 | 2.8 | 5.2 | 22.6 |
| Days ≥ 1" Snow Depth | 8 | 4 | 1 | 0 | 0 | 0 | 0 | 0 | 0 | 0 | 2 | 5 | 20 |

## COALVILLE *Summit County* ELEVATION 5551 ft LAT/LONG 40° 55' N / 111° 24' W

| | JAN | FEB | MAR | APR | MAY | JUN | JUL | AUG | SEP | OCT | NOV | DEC | YEAR |
|---|---|---|---|---|---|---|---|---|---|---|---|---|---|
| Maximum Temp °F | 36.4 | 41.7 | 50.4 | 60.1 | 69.6 | 78.7 | 86.0 | 84.5 | ***75.8*** | 65.1 | 49.3 | 38.1 | 61.3 |
| Minimum Temp °F | 11.2 | 14.0 | 22.4 | 28.3 | 34.5 | 40.5 | 46.2 | 44.7 | ***36.8*** | 27.9 | 20.8 | 12.6 | 28.3 |
| Mean Temp °F | 23.8 | 27.8 | 36.5 | 44.2 | 52.1 | 59.6 | 66.1 | 64.6 | ***56.3*** | 46.5 | 35.1 | ***25.6*** | 44.9 |
| Days Max Temp ≥ 90 °F | 0 | 0 | 0 | 0 | 0 | 3 | 8 | 6 | 0 | 0 | 0 | 0 | 17 |
| Days Max Temp ≤ 32 °F | 10 | 4 | 1 | 0 | 0 | 0 | 0 | 0 | 0 | 0 | ***2*** | ***7*** | 24 |
| Days Min Temp ≤ 32 °F | 30 | 27 | ***28*** | ***22*** | 12 | 2 | 0 | 1 | ***8*** | ***24*** | ***27*** | ***30*** | 211 |
| Days Min Temp ≤ 0 °F | 6 | 4 | 1 | 0 | 0 | 0 | 0 | 0 | 0 | 0 | 1 | ***4*** | 16 |
| Heating Degree Days | 1270 | 1042 | 871 | 617 | 394 | 169 | 28 | 52 | ***259*** | ***569*** | ***901*** | ***1214*** | 7386 |
| Cooling Degree Days | 0 | 0 | 0 | 0 | 1 | 19 | 75 | 56 | 5 | 0 | 0 | 0 | 156 |
| Total Precipitation (") | 1.12 | 1.11 | 1.49 | 1.91 | 1.94 | 1.17 | 0.96 | 1.00 | 1.39 | 1.68 | 1.61 | 1.20 | 16.58 |
| Days ≥ 0.1" Precip | 4 | 4 | 5 | 6 | 6 | 3 | 3 | 3 | 3 | 4 | ***5*** | 4 | 50 |
| Total Snowfall (") | na | ***15.2*** | na | na | ***2.1*** | 0.0 | 0.0 | 0.0 | 0.1 | 2.0 | na | na | na |
| Days ≥ 1" Snow Depth | na | na | na | na | ***0*** | 0 | 0 | 0 | 0 | ***0*** | na | na | na |

## CORINNE *Box Elder County* ELEVATION 4232 ft LAT/LONG 41° 34' N / 112° 7' W

| | JAN | FEB | MAR | APR | MAY | JUN | JUL | AUG | SEP | OCT | NOV | DEC | YEAR |
|---|---|---|---|---|---|---|---|---|---|---|---|---|---|
| Maximum Temp °F | 33.9 | 40.7 | 51.1 | 60.9 | 71.3 | 81.5 | 90.1 | 88.5 | 77.7 | 64.5 | 46.9 | 36.1 | 61.9 |
| Minimum Temp °F | 15.2 | 19.6 | 29.0 | 35.2 | 43.2 | 50.4 | 56.6 | 55.2 | 45.6 | 35.3 | 26.1 | 17.8 | 35.8 |
| Mean Temp °F | 24.6 | 30.2 | 40.1 | 48.1 | 57.3 | 66.0 | 73.4 | 71.9 | 61.7 | 49.9 | 36.5 | 27.0 | 48.9 |
| Days Max Temp ≥ 90 °F | 0 | 0 | 0 | 0 | 0 | 6 | 19 | 15 | 2 | 0 | 0 | 0 | 42 |
| Days Max Temp ≤ 32 °F | 13 | 5 | 1 | 0 | 0 | 0 | 0 | 0 | 0 | 0 | 1 | 9 | 29 |
| Days Min Temp ≤ 32 °F | 29 | 26 | 22 | 11 | 2 | 0 | 0 | 0 | 1 | 10 | ***24*** | 30 | 155 |
| Days Min Temp ≤ 0 °F | 5 | 2 | 0 | 0 | 0 | 0 | 0 | 0 | 0 | 0 | 0 | 2 | 9 |
| Heating Degree Days | 1246 | 978 | 765 | 503 | 251 | 73 | 7 | 11 | 140 | 461 | 848 | 1173 | 6456 |
| Cooling Degree Days | 0 | 0 | 0 | 1 | 22 | 105 | 252 | 229 | 52 | 0 | 0 | 0 | 661 |
| Total Precipitation (") | 1.45 | 1.56 | 1.65 | 1.75 | 1.83 | 1.29 | 0.79 | 0.90 | 1.53 | 1.76 | 1.47 | 1.45 | 17.43 |
| Days ≥ 0.1" Precip | 5 | 5 | ***4*** | 5 | 5 | 3 | 2 | 2 | 3 | 4 | 4 | 5 | 47 |
| Total Snowfall (") | na | na | na | ***2.6*** | 0.3 | 0.0 | 0.0 | 0.0 | 0.0 | 0.5 | na | na | na |
| Days ≥ 1" Snow Depth | na | ***14*** | ***5*** | 0 | 0 | 0 | 0 | 0 | 0 | 0 | na | na | na |

## COTTONWOOD WEIR *Salt Lake County* ELEVATION 4964 ft LAT/LONG 40° 37' N / 111° 47' W

| | JAN | FEB | MAR | APR | MAY | JUN | JUL | AUG | SEP | OCT | NOV | DEC | YEAR |
|---|---|---|---|---|---|---|---|---|---|---|---|---|---|
| Maximum Temp °F | 38.7 | 45.2 | 53.2 | 61.5 | 72.0 | 82.6 | 90.9 | 89.0 | 78.4 | 65.4 | 49.7 | 39.7 | 63.9 |
| Minimum Temp °F | 21.4 | 26.0 | 32.9 | 39.8 | 48.1 | 57.3 | 65.7 | 64.2 | 54.4 | 43.2 | 31.5 | 22.4 | 42.2 |
| Mean Temp °F | 30.1 | 35.6 | 43.2 | 50.7 | 60.1 | 70.0 | 78.3 | 76.6 | 66.4 | 54.3 | 40.6 | 31.1 | 53.1 |
| Days Max Temp ≥ 90 °F | 0 | 0 | 0 | 0 | 0 | 7 | 21 | 16 | 2 | 0 | 0 | 0 | 46 |
| Days Max Temp ≤ 32 °F | 8 | 2 | 0 | 0 | 0 | 0 | 0 | 0 | 0 | 0 | 1 | 7 | 18 |
| Days Min Temp ≤ 32 °F | 27 | 22 | 15 | 7 | 1 | 0 | 0 | 0 | 0 | 4 | 17 | 27 | 120 |
| Days Min Temp ≤ 0 °F | 0 | 0 | 0 | 0 | 0 | 0 | 0 | 0 | 0 | 0 | 0 | 1 | 1 |
| Heating Degree Days | 1076 | 824 | 670 | 433 | 201 | 49 | 2 | 4 | 84 | 337 | 724 | 1045 | 5449 |
| Cooling Degree Days | 0 | 0 | 0 | 16 | 61 | 218 | 404 | 381 | 141 | 12 | 0 | 0 | 1233 |
| Total Precipitation (") | 1.77 | 1.92 | 2.75 | 2.90 | 2.91 | 1.22 | 1.08 | 1.25 | 2.26 | 2.62 | 2.41 | 2.21 | 25.30 |
| Days ≥ 0.1" Precip | 5 | 5 | 7 | 6 | 5 | 3 | 3 | 3 | 4 | 5 | ***6*** | 5 | 57 |
| Total Snowfall (") | na | ***16.6*** | na | na | ***0.9*** | 0.0 | 0.0 | 0.0 | 0.2 | ***2.9*** | na | ***23.6*** | na |
| Days ≥ 1" Snow Depth | 25 | 18 | 7 | 2 | 0 | 0 | 0 | 0 | 0 | ***2*** | ***8*** | ***22*** | 84 |

## DEER CREEK DAM *Wasatch County* ELEVATION 5292 ft LAT/LONG 40° 24 ' N / 111° 32 ' W

| | JAN | FEB | MAR | APR | MAY | JUN | JUL | AUG | SEP | OCT | NOV | DEC | YEAR |
|---|---|---|---|---|---|---|---|---|---|---|---|---|---|
| Maximum Temp °F | 32.4 | 37.4 | 47.5 | 57.7 | 67.7 | 77.8 | 85.9 | 84.3 | 75.0 | 63.1 | 47.3 | 35.7 | 59.3 |
| Minimum Temp °F | 7.9 | 9.6 | 20.6 | 28.0 | 34.8 | 41.1 | 47.0 | 45.2 | 37.4 | 28.7 | 21.5 | 12.9 | 27.9 |
| Mean Temp °F | 20.1 | 23.5 | 34.1 | 42.9 | 51.3 | 59.5 | 66.5 | 64.8 | 56.2 | 45.9 | 34.4 | 24.3 | 43.6 |
| Days Max Temp ≥ 90 °F | 0 | 0 | 0 | 0 | 0 | 2 | 8 | 6 | 0 | 0 | 0 | 0 | 16 |
| Days Max Temp ≤ 32 °F | 15 | 7 | 2 | 0 | 0 | 0 | 0 | 0 | 0 | 0 | 3 | 11 | 38 |
| Days Min Temp ≤ 32 °F | 30 | 28 | 29 | 23 | 11 | 2 | 0 | 0 | 7 | 24 | 28 | 30 | 212 |
| Days Min Temp ≤ 0 °F | 9 | 7 | 1 | 0 | 0 | 0 | 0 | 0 | 0 | 0 | 0 | 3 | 20 |
| Heating Degree Days | 1384 | 1164 | 952 | 657 | 418 | 176 | 29 | 54 | 260 | 584 | 912 | 1254 | 7844 |
| Cooling Degree Days | 0 | 0 | 0 | 0 | 0 | 18 | 76 | 55 | 4 | 0 | 0 | 0 | 153 |
| Total Precipitation (") | 2.59 | 2.62 | 2.30 | 1.71 | 1.73 | 1.07 | 0.92 | 1.14 | 1.53 | 2.16 | 2.44 | 2.47 | 22.68 |
| Days ≥ 0.1" Precip | 6 | 5 | 5 | 5 | 5 | 3 | 2 | 3 | 4 | 4 | 6 | 5 | 53 |
| Total Snowfall (") | na | na | na | na | ***0.0*** | 0.0 | 0.0 | 0.0 | 0.0 | na | na | na | na |
| Days ≥ 1" Snow Depth | na | na | na | na | ***0*** | 0 | 0 | 0 | 0 | na | na | na | na |

## DELTA *Millard County* ELEVATION 4632 ft LAT/LONG 39° 21 ' N / 112° 34 ' W

| | JAN | FEB | MAR | APR | MAY | JUN | JUL | AUG | SEP | OCT | NOV | DEC | YEAR |
|---|---|---|---|---|---|---|---|---|---|---|---|---|---|
| Maximum Temp °F | 37.0 | 45.5 | 55.7 | 64.1 | ***73.9*** | 85.1 | 92.7 | 90.7 | 80.2 | 67.3 | 50.9 | 38.3 | 65.1 |
| Minimum Temp °F | 12.6 | 19.2 | 26.7 | 32.4 | ***41.1*** | 49.3 | 56.6 | 54.7 | 44.8 | 33.9 | 23.6 | 14.3 | 34.1 |
| Mean Temp °F | 24.8 | 32.4 | 41.2 | 48.3 | ***57.5*** | 67.2 | 74.6 | 72.7 | 62.5 | 50.6 | 37.2 | 26.3 | 49.6 |
| Days Max Temp ≥ 90 °F | 0 | 0 | 0 | 0 | 1 | 11 | 23 | 19 | 4 | 0 | 0 | 0 | 58 |
| Days Max Temp ≤ 32 °F | 10 | 3 | 0 | 0 | 0 | 0 | 0 | 0 | 0 | 0 | 2 | 8 | 23 |
| Days Min Temp ≤ 32 °F | 30 | 26 | 24 | 15 | 4 | 0 | 0 | 0 | 2 | 13 | 25 | 30 | 169 |
| Days Min Temp ≤ 0 °F | 6 | 2 | 0 | 0 | 0 | 0 | 0 | 0 | 0 | 0 | 0 | 3 | 11 |
| Heating Degree Days | 1239 | 916 | 731 | 496 | ***241*** | 56 | 3 | 6 | 123 | 440 | 826 | 1192 | 6269 |
| Cooling Degree Days | 0 | 0 | 0 | 1 | 15 | 130 | 298 | 263 | 60 | 1 | 0 | 0 | 768 |
| Total Precipitation (") | 0.57 | 0.56 | 0.82 | 0.74 | 0.94 | 0.44 | 0.54 | 0.62 | 0.85 | 0.91 | 0.66 | 0.58 | 8.23 |
| Days ≥ 0.1" Precip | 2 | 2 | 3 | 2 | ***3*** | 1 | 2 | 2 | 2 | 3 | 2 | 2 | 26 |
| Total Snowfall (") | 7.2 | 3.9 | ***4.3*** | 1.7 | ***0.8*** | 0.0 | 0.0 | 0.0 | 0.4 | 0.9 | 3.5 | ***5.3*** | 28.0 |
| Days ≥ 1" Snow Depth | 15 | 7 | ***2*** | ***0*** | 0 | 0 | 0 | 0 | 0 | 0 | 2 | 10 | 36 |

## DESERET *Millard County* ELEVATION 4541 ft LAT/LONG 39° 18 ' N / 112° 38 ' W

| | JAN | FEB | MAR | APR | MAY | JUN | JUL | AUG | SEP | OCT | NOV | DEC | YEAR |
|---|---|---|---|---|---|---|---|---|---|---|---|---|---|
| Maximum Temp °F | 37.2 | 45.5 | 55.8 | 64.7 | 75.0 | 86.1 | 93.9 | 91.2 | 80.4 | 66.7 | 51.0 | 38.4 | 65.5 |
| Minimum Temp °F | 12.6 | 18.9 | 25.7 | 31.3 | 39.8 | 47.9 | 55.7 | 53.3 | 43.0 | 32.0 | 22.9 | 14.0 | 33.1 |
| Mean Temp °F | 24.9 | 32.2 | 40.8 | 48.0 | 57.4 | 67.0 | 74.8 | 72.2 | 61.7 | 49.4 | 37.0 | 26.2 | 49.3 |
| Days Max Temp ≥ 90 °F | 0 | 0 | 0 | 0 | 1 | 12 | 25 | 21 | 5 | 0 | 0 | 0 | 64 |
| Days Max Temp ≤ 32 °F | 10 | 3 | 0 | 0 | 0 | 0 | 0 | 0 | 0 | 0 | 1 | 8 | 22 |
| Days Min Temp ≤ 32 °F | 30 | 26 | 25 | 17 | 5 | 0 | 0 | 0 | 4 | 17 | 26 | 30 | 180 |
| Days Min Temp ≤ 0 °F | 5 | 2 | 0 | 0 | 0 | 0 | 0 | 0 | 0 | 0 | 0 | 3 | 10 |
| Heating Degree Days | 1237 | 920 | 744 | 503 | 242 | 61 | 2 | 8 | 141 | 477 | 834 | 1194 | 6363 |
| Cooling Degree Days | 0 | 0 | 0 | 2 | 14 | 137 | 316 | 260 | 52 | 0 | 0 | 0 | 781 |
| Total Precipitation (") | 0.62 | 0.50 | 0.71 | 0.76 | 0.98 | 0.44 | 0.56 | 0.72 | 0.84 | 0.97 | 0.73 | 0.60 | 8.43 |
| Days ≥ 0.1" Precip | 2 | 2 | 2 | 2 | 3 | 1 | 2 | 2 | 2 | 3 | 2 | 2 | 25 |
| Total Snowfall (") | 5.1 | 2.4 | 1.8 | 1.2 | 0.6 | 0.0 | 0.0 | 0.0 | 0.3 | 0.5 | 1.9 | 4.2 | 18.0 |
| Days ≥ 1" Snow Depth | 13 | 6 | 1 | 0 | 0 | 0 | 0 | 0 | 0 | 0 | 2 | 8 | 30 |

## DINOSAUR QUARRY AREA *Uintah County* ELEVATION 4770 ft LAT/LONG 40° 26 ' N / 109° 18 ' W

| | JAN | FEB | MAR | APR | MAY | JUN | JUL | AUG | SEP | OCT | NOV | DEC | YEAR |
|---|---|---|---|---|---|---|---|---|---|---|---|---|---|
| Maximum Temp °F | 28.9 | 37.4 | 52.9 | 65.2 | 76.2 | 87.6 | 94.3 | 92.1 | 81.8 | 66.9 | 48.3 | 34.0 | 63.8 |
| Minimum Temp °F | 3.0 | 9.9 | 24.5 | 33.0 | 41.4 | 48.6 | 55.2 | 52.6 | 43.4 | 32.1 | 21.7 | 9.4 | 31.2 |
| Mean Temp °F | 15.9 | 23.7 | 38.7 | 49.1 | 58.8 | 68.1 | 74.8 | 72.4 | 62.6 | 49.5 | 35.0 | 22.0 | 47.6 |
| Days Max Temp ≥ 90 °F | 0 | 0 | 0 | 0 | 1 | 14 | 26 | 23 | 5 | 0 | 0 | 0 | 69 |
| Days Max Temp ≤ 32 °F | 20 | 9 | 1 | 0 | 0 | 0 | 0 | 0 | 0 | 0 | 2 | 13 | 45 |
| Days Min Temp ≤ 32 °F | 31 | 28 | 27 | 14 | 2 | 0 | 0 | 0 | 2 | 16 | 27 | 31 | 178 |
| Days Min Temp ≤ 0 °F | 13 | 7 | 0 | 0 | 0 | 0 | 0 | 0 | 0 | 0 | 0 | 6 | 26 |
| Heating Degree Days | 1515 | 1161 | 808 | 471 | 200 | 37 | 1 | 4 | 113 | 475 | 891 | 1327 | 7003 |
| Cooling Degree Days | 0 | 0 | 0 | 1 | 18 | 153 | 305 | 254 | 54 | 1 | 0 | 0 | 786 |
| Total Precipitation (") | 0.57 | 0.61 | 0.78 | 0.80 | 0.96 | 0.64 | 0.65 | 0.59 | 0.84 | 1.19 | 0.65 | 0.64 | 8.92 |
| Days ≥ 0.1" Precip | 2 | 2 | 3 | 3 | 3 | 2 | 2 | 2 | 3 | 3 | 2 | 2 | 29 |
| Total Snowfall (") | 6.2 | 4.5 | 1.7 | 0.5 | 0.1 | 0.0 | 0.0 | 0.0 | 0.4 | 0.4 | 2.4 | 7.6 | 23.8 |
| Days ≥ 1" Snow Depth | 25 | 21 | 6 | 0 | 0 | 0 | 0 | 0 | 0 | 0 | 2 | 16 | 70 |

## DUCHESNE *Duchesne County* ELEVATION 5525 ft LAT/LONG 40° 10 ' N / 110° 25 ' W

| | JAN | FEB | MAR | APR | MAY | JUN | JUL | AUG | SEP | OCT | NOV | DEC | YEAR |
|---|---|---|---|---|---|---|---|---|---|---|---|---|---|
| Maximum Temp °F | *30.7* | *37.4* | *51.1* | *62.3* | *72.6* | *81.4* | *87.2* | *85.6* | *76.3* | *62.9* | *46.2* | *33.1* | 60.6 |
| Minimum Temp °F | *5.9* | *11.8* | *24.0* | *32.0* | *39.8* | *47.8* | *53.9* | *52.8* | *43.1* | *32.8* | *21.5* | *9.4* | 31.2 |
| Mean Temp °F | *18.3* | *24.6* | *37.6* | *47.2* | *56.2* | *64.6* | *70.6* | *69.2* | *59.7* | *47.9* | *33.9* | *21.3* | 45.9 |
| Days Max Temp ≥ 90 °F | *0* | *0* | *0* | *0* | *0* | *5* | *11* | *8* | *0* | *0* | *0* | *0* | 24 |
| Days Max Temp ≤ 32 °F | *17* | *9* | *1* | *0* | *0* | *0* | *0* | *0* | *0* | *0* | *2* | *13* | 42 |
| Days Min Temp ≤ 32 °F | *31* | *28* | *28* | *16* | *4* | *0* | *0* | *0* | na | *15* | *28* | *31* | na |
| Days Min Temp ≤ 0 °F | *10* | *5* | *0* | *0* | *0* | *0* | *0* | *0* | *0* | *0* | *0* | *6* | 21 |
| Heating Degree Days | *1440* | *1133* | *843* | *528* | *270* | *79* | *4* | *12* | *168* | *524* | *927* | *1349* | 7277 |
| Cooling Degree Days | 0 | 0 | 0 | 0 | 4 | 85 | 174 | 140 | 19 | 0 | 0 | 0 | 422 |
| Total Precipitation (") | *0.45* | *0.52* | *0.65* | *0.84* | *1.04* | *0.88* | *0.97* | *1.04* | *1.11* | *1.06* | *0.47* | *0.68* | 9.71 |
| Days ≥ 0.1" Precip | *2* | *2* | *2* | *3* | *3* | *3* | *3* | *3* | *3* | *3* | *2* | *3* | 32 |
| Total Snowfall (") | na | na | na | na | na | na | *0.0* | na | na | na | na | na | na |
| Days ≥ 1" Snow Depth | na | na | na | na | na | na | *0* | na | na | na | na | na | na |

## DUGWAY *Tooele County* ELEVATION 4338 ft LAT/LONG 40° 12 ' N / 112° 57 ' W

| | JAN | FEB | MAR | APR | MAY | JUN | JUL | AUG | SEP | OCT | NOV | DEC | YEAR |
|---|---|---|---|---|---|---|---|---|---|---|---|---|---|
| Maximum Temp °F | 36.8 | 45.3 | 53.9 | 62.9 | 73.6 | 85.1 | 94.2 | 91.7 | 80.2 | 65.6 | 50.0 | 38.2 | 64.8 |
| Minimum Temp °F | 15.0 | 22.6 | 29.1 | 35.4 | 44.1 | 52.9 | 61.1 | 58.7 | 47.7 | 35.5 | 26.0 | 16.4 | 37.0 |
| Mean Temp °F | 25.9 | 33.9 | 41.5 | 49.2 | 58.8 | 69.0 | 77.7 | 75.2 | 63.9 | 50.6 | 38.0 | 27.3 | 50.9 |
| Days Max Temp ≥ 90 °F | 0 | 0 | 0 | 0 | 1 | 12 | 25 | 21 | 6 | 0 | 0 | 0 | 65 |
| Days Max Temp ≤ 32 °F | 11 | 3 | 0 | 0 | 0 | 0 | 0 | 0 | 0 | 0 | 1 | 9 | 24 |
| Days Min Temp ≤ 32 °F | 29 | 24 | 21 | 11 | 2 | 0 | 0 | 0 | 1 | 11 | 24 | 29 | 152 |
| Days Min Temp ≤ 0 °F | 4 | 1 | 0 | 0 | 0 | 0 | 0 | 0 | 0 | 0 | 0 | 3 | 8 |
| Heating Degree Days | 1204 | 871 | 721 | 471 | 217 | 48 | 2 | 5 | 115 | 441 | 803 | 1161 | 6059 |
| Cooling Degree Days | 0 | 0 | 0 | 4 | 34 | 187 | 388 | 341 | 95 | 1 | 0 | 0 | 1050 |
| Total Precipitation (") | 0.51 | 0.60 | 0.86 | 0.75 | 1.11 | 0.44 | 0.61 | 0.60 | 0.68 | 0.86 | 0.60 | 0.58 | 8.20 |
| Days ≥ 0.1" Precip | 2 | 2 | 3 | 2 | 3 | 2 | 2 | 2 | 2 | 3 | 2 | 2 | 27 |
| Total Snowfall (") | 3.9 | 2.3 | 1.6 | 0.8 | 0.3 | 0.0 | 0.0 | 0.0 | 0.0 | 0.1 | 1.5 | 4.6 | 15.1 |
| Days ≥ 1" Snow Depth | na | 3 | *1* | 0 | 0 | 0 | 0 | 0 | 0 | 0 | *0* | na | na |

## ECHO DAM *Summit County* ELEVATION 5472 ft LAT/LONG 40° 58 ' N / 111° 26 ' W

| | JAN | FEB | MAR | APR | MAY | JUN | JUL | AUG | SEP | OCT | NOV | DEC | YEAR |
|---|---|---|---|---|---|---|---|---|---|---|---|---|---|
| Maximum Temp °F | 33.9 | 39.3 | 48.8 | 58.0 | 68.2 | 78.6 | 87.4 | 86.0 | 76.3 | 63.8 | 46.3 | 35.4 | 60.2 |
| Minimum Temp °F | 10.6 | 13.1 | 23.0 | 29.9 | 36.7 | 43.1 | 49.4 | 47.7 | 38.7 | 29.6 | 21.4 | 12.6 | 29.7 |
| Mean Temp °F | 22.3 | 26.2 | 36.0 | 44.0 | 52.5 | 60.9 | 68.4 | 66.8 | 57.5 | 46.7 | 33.9 | 24.0 | 44.9 |
| Days Max Temp ≥ 90 °F | 0 | 0 | 0 | 0 | 0 | 3 | 12 | 9 | 1 | 0 | 0 | 0 | 25 |
| Days Max Temp ≤ 32 °F | 13 | 6 | 1 | 0 | 0 | 0 | 0 | 0 | 0 | 0 | 3 | 11 | 34 |
| Days Min Temp ≤ 32 °F | 30 | 27 | 28 | 20 | 7 | 1 | 0 | 0 | 6 | 21 | 27 | 30 | 197 |
| Days Min Temp ≤ 0 °F | 7 | 5 | 1 | 0 | 0 | 0 | 0 | 0 | 0 | 0 | 1 | 4 | 18 |
| Heating Degree Days | 1318 | 1089 | 891 | 624 | 380 | 146 | 15 | 32 | 229 | 560 | 925 | 1263 | 7472 |
| Cooling Degree Days | 0 | 0 | 0 | 0 | 0 | 32 | 122 | 104 | 12 | 0 | 0 | 0 | 270 |
| Total Precipitation (") | 0.94 | 0.96 | 1.31 | 1.63 | 1.76 | 1.12 | 0.84 | 0.80 | 1.37 | 1.65 | 1.51 | 1.13 | 15.02 |
| Days ≥ 0.1" Precip | 3 | 3 | 4 | 5 | 5 | 3 | 2 | 2 | 3 | 4 | 5 | 4 | 43 |
| Total Snowfall (") | 13.3 | 11.3 | 8.7 | 6.3 | 1.9 | 0.1 | 0.0 | 0.0 | 0.2 | 2.1 | 11.3 | 13.4 | 68.6 |
| Days ≥ 1" Snow Depth | 28 | 23 | 8 | 1 | 0 | 0 | 0 | 0 | 0 | 1 | 9 | 23 | 93 |

## ENTERPRISE BERYL JCT *Iron County* ELEVATION 5223 ft LAT/LONG 37° 43 ' N / 113° 39 ' W

| | JAN | FEB | MAR | APR | MAY | JUN | JUL | AUG | SEP | OCT | NOV | DEC | YEAR |
|---|---|---|---|---|---|---|---|---|---|---|---|---|---|
| Maximum Temp °F | 41.0 | 46.6 | 55.4 | 64.0 | 73.4 | 83.6 | 89.8 | 87.6 | 79.5 | 67.9 | 52.7 | 42.0 | 65.3 |
| Minimum Temp °F | 12.2 | 18.0 | 23.2 | 27.4 | 35.6 | 42.5 | 50.5 | 49.5 | 40.0 | 29.4 | 20.1 | 12.5 | 30.1 |
| Mean Temp °F | 26.6 | 32.3 | 39.3 | 45.7 | 54.5 | 63.1 | 70.1 | 68.6 | 59.8 | 48.7 | 36.4 | 27.3 | 47.7 |
| Days Max Temp ≥ 90 °F | 0 | 0 | 0 | 0 | 0 | 7 | 17 | 12 | 2 | 0 | 0 | 0 | 38 |
| Days Max Temp ≤ 32 °F | 6 | 2 | 0 | 0 | 0 | 0 | 0 | 0 | 0 | 0 | 1 | 5 | 14 |
| Days Min Temp ≤ 32 °F | 30 | 27 | 27 | 23 | 11 | 2 | 0 | 0 | 6 | 21 | 27 | 30 | 204 |
| Days Min Temp ≤ 0 °F | 5 | 2 | 0 | 0 | 0 | 0 | 0 | 0 | 0 | 0 | 1 | 4 | 12 |
| Heating Degree Days | 1183 | 916 | 790 | 572 | 321 | 104 | 8 | 22 | 169 | 501 | 850 | 1162 | 6598 |
| Cooling Degree Days | 0 | 0 | 0 | 0 | 4 | 61 | 172 | 144 | 21 | 0 | 0 | 0 | 402 |
| Total Precipitation (") | 0.75 | 0.90 | 1.24 | 0.84 | 0.69 | 0.46 | 1.16 | 1.14 | 0.78 | 0.97 | 0.86 | 0.66 | 10.45 |
| Days ≥ 0.1" Precip | 2 | 3 | 4 | 2 | 2 | 1 | 3 | 3 | 2 | 3 | 2 | 2 | 29 |
| Total Snowfall (") | 6.5 | 5.3 | 4.7 | 2.3 | 0.8 | 0.0 | 0.0 | 0.0 | 0.1 | 0.9 | 4.7 | 5.9 | 31.2 |
| Days ≥ 1" Snow Depth | 14 | 8 | 2 | 0 | 0 | 0 | 0 | 0 | 0 | 0 | 3 | *9* | 36 |

## EPHRAIM SORENSENS FD *Sanpete County* ELEVATION 5584 ft LAT/LONG 39° 21 ' N / 111° 35 ' W

| | JAN | FEB | MAR | APR | MAY | JUN | JUL | AUG | SEP | OCT | NOV | DEC | YEAR |
|---|---|---|---|---|---|---|---|---|---|---|---|---|---|
| Maximum Temp °F | 35.0 | 40.8 | 50.0 | 58.8 | 69.4 | 80.8 | 88.7 | 86.5 | 76.9 | 64.3 | 48.2 | 36.6 | 61.3 |
| Minimum Temp °F | 12.7 | 18.2 | 26.2 | 32.0 | 39.5 | 47.3 | 54.5 | 52.7 | 44.1 | 34.2 | 24.2 | 14.7 | 33.4 |
| Mean Temp °F | 23.9 | 29.6 | 38.1 | 45.4 | 54.5 | 64.1 | 71.6 | 69.6 | 60.5 | 49.3 | 36.3 | 25.7 | 47.4 |
| Days Max Temp ≥ 90 °F | 0 | 0 | 0 | 0 | 0 | 5 | 16 | 11 | 1 | 0 | 0 | 0 | 33 |
| Days Max Temp ≤ 32 °F | 11 | 4 | 1 | 0 | 0 | 0 | 0 | 0 | 0 | 0 | 2 | 10 | 28 |
| Days Min Temp ≤ 32 °F | 30 | 27 | 25 | 16 | 5 | 1 | 0 | 0 | 2 | 12 | 26 | 30 | 174 |
| Days Min Temp ≤ 0 °F | 5 | 1 | 0 | 0 | 0 | 0 | 0 | 0 | 0 | 0 | 0 | 3 | 9 |
| Heating Degree Days | 1269 | 995 | 826 | 581 | 322 | 95 | 5 | 15 | 158 | 480 | 855 | 1211 | 6812 |
| Cooling Degree Days | 0 | 0 | 0 | 0 | 3 | 85 | 218 | 175 | 36 | 0 | 0 | 0 | 517 |
| Total Precipitation (") | 0.90 | 0.96 | 1.31 | 1.07 | 1.10 | 0.71 | 0.75 | 0.84 | 1.09 | 1.32 | 1.05 | 0.94 | 12.04 |
| Days ≥ 0.1" Precip | 3 | 3 | 5 | 4 | 3 | 2 | 2 | 3 | 3 | 4 | 3 | 3 | 38 |
| Total Snowfall (") | na | na | na | na | ***0.0*** | 0.0 | 0.0 | 0.0 | 0.0 | ***0.0*** | na | na | na |
| Days ≥ 1" Snow Depth | na | na | na | na | ***0*** | 0 | 0 | 0 | 0 | ***0*** | na | na | na |

## ESCALANTE *Garfield County* ELEVATION 5764 ft LAT/LONG 37° 46 ' N / 111° 36 ' W

| | JAN | FEB | MAR | APR | MAY | JUN | JUL | AUG | SEP | OCT | NOV | DEC | YEAR |
|---|---|---|---|---|---|---|---|---|---|---|---|---|---|
| Maximum Temp °F | 40.9 | 46.8 | 55.1 | 64.0 | 73.8 | 84.7 | 90.0 | 86.9 | 78.8 | 67.1 | 52.7 | 42.5 | 65.3 |
| Minimum Temp °F | 14.6 | 20.9 | 27.1 | 32.6 | 40.3 | 47.8 | 54.6 | 53.0 | 44.7 | 35.3 | 25.2 | 16.7 | 34.4 |
| Mean Temp °F | 27.8 | 33.9 | 41.1 | 48.3 | 57.1 | 66.3 | 72.3 | 70.0 | 61.8 | 51.2 | 39.0 | 29.6 | 49.9 |
| Days Max Temp ≥ 90 °F | 0 | 0 | 0 | 0 | 0 | 8 | 19 | 12 | 2 | 0 | 0 | 0 | 41 |
| Days Max Temp ≤ 32 °F | 5 | 1 | 0 | 0 | 0 | 0 | 0 | 0 | 0 | 0 | 0 | 4 | 10 |
| Days Min Temp ≤ 32 °F | 30 | 27 | 25 | 15 | 3 | 0 | 0 | 0 | 1 | 10 | 26 | 30 | 167 |
| Days Min Temp ≤ 0 °F | 2 | 1 | 0 | 0 | 0 | 0 | 0 | 0 | 0 | 0 | 0 | 1 | 4 |
| Heating Degree Days | 1146 | 873 | 733 | 494 | 245 | 54 | 1 | 8 | 123 | 421 | 774 | 1090 | 5962 |
| Cooling Degree Days | 0 | 0 | 0 | 1 | 8 | 108 | 237 | 175 | 36 | 0 | 0 | 0 | 565 |
| Total Precipitation (") | 0.94 | 0.79 | 0.93 | 0.47 | 0.72 | 0.43 | 0.96 | 1.49 | 0.93 | 0.99 | 0.84 | 0.77 | 10.26 |
| Days ≥ 0.1" Precip | 3 | 2 | 3 | 1 | 2 | 1 | 3 | 4 | 3 | 3 | 2 | 2 | 29 |
| Total Snowfall (") | 9.5 | 5.5 | 4.5 | 1.4 | 0.1 | 0.0 | 0.0 | 0.0 | 0.0 | 0.3 | 3.1 | 7.1 | 31.5 |
| Days ≥ 1" Snow Depth | 16 | 8 | 1 | 0 | 0 | 0 | 0 | 0 | 0 | 0 | 2 | 7 | 34 |

## FAIRFIELD *Utah County* ELEVATION 4882 ft LAT/LONG 40° 16 ' N / 112° 5 ' W

| | JAN | FEB | MAR | APR | MAY | JUN | JUL | AUG | SEP | OCT | NOV | DEC | YEAR |
|---|---|---|---|---|---|---|---|---|---|---|---|---|---|
| Maximum Temp °F | 37.2 | 43.3 | 52.9 | 62.1 | 71.8 | 81.4 | 88.6 | 87.1 | 78.7 | 65.6 | 49.4 | 38.6 | 63.1 |
| Minimum Temp °F | 11.1 | 16.8 | 24.6 | 29.4 | 36.6 | 43.8 | 51.0 | 49.0 | 39.4 | 29.6 | 21.1 | 12.5 | 30.4 |
| Mean Temp °F | 24.2 | 30.1 | 38.7 | 45.8 | 54.2 | 62.6 | 69.8 | 68.1 | 59.1 | 47.7 | 35.3 | 25.6 | 46.8 |
| Days Max Temp ≥ 90 °F | 0 | 0 | 0 | 0 | 0 | 5 | ***14*** | ***11*** | 2 | 0 | 0 | 0 | 32 |
| Days Max Temp ≤ 32 °F | 9 | 3 | 0 | 0 | 0 | 0 | 0 | 0 | 0 | 0 | 1 | 7 | 20 |
| Days Min Temp ≤ 32 °F | 30 | 27 | 26 | 20 | 8 | 1 | 0 | 0 | 6 | ***22*** | ***27*** | 30 | 197 |
| Days Min Temp ≤ 0 °F | 7 | 3 | 0 | 0 | 0 | 0 | 0 | 0 | 0 | 0 | 1 | 5 | 16 |
| Heating Degree Days | 1258 | 978 | 805 | 569 | 330 | 111 | 8 | 21 | 190 | 529 | ***888*** | 1216 | 6903 |
| Cooling Degree Days | 0 | 0 | 0 | 0 | 3 | 54 | 164 | 137 | 21 | 0 | 0 | 0 | 379 |
| Total Precipitation (") | 1.03 | 1.02 | 1.12 | 1.00 | 1.21 | 0.78 | 1.12 | 1.18 | 1.06 | 1.23 | 1.06 | 0.85 | 12.66 |
| Days ≥ 0.1" Precip | 4 | 3 | 4 | 3 | 4 | 2 | 3 | 3 | 3 | 4 | 3 | 3 | 39 |
| Total Snowfall (") | 10.1 | 6.5 | 4.9 | 2.3 | 0.3 | 0.0 | 0.0 | 0.0 | 0.1 | 1.0 | 4.8 | 8.7 | 38.7 |
| Days ≥ 1" Snow Depth | 20 | 14 | 3 | ***0*** | 0 | 0 | 0 | 0 | 0 | 0 | 5 | 15 | 57 |

## FARMINGTON USU FIELD *Davis County* ELEVATION 4334 ft LAT/LONG 40° 59 ' N / 111° 53 ' W

| | JAN | FEB | MAR | APR | MAY | JUN | JUL | AUG | SEP | OCT | NOV | DEC | YEAR |
|---|---|---|---|---|---|---|---|---|---|---|---|---|---|
| Maximum Temp °F | 37.5 | 44.3 | 53.4 | 62.6 | 72.8 | 83.3 | 91.6 | 89.7 | 79.6 | 65.6 | 49.6 | 38.4 | 64.0 |
| Minimum Temp °F | 19.8 | 23.6 | 31.0 | 37.2 | 44.7 | 52.8 | 60.0 | 58.2 | 49.1 | 38.3 | 28.8 | 20.7 | 38.7 |
| Mean Temp °F | 28.7 | 33.9 | 42.2 | 49.9 | 58.8 | 68.1 | 75.8 | 74.0 | 64.4 | 52.0 | 39.2 | 29.6 | 51.4 |
| Days Max Temp ≥ 90 °F | 0 | 0 | 0 | 0 | 0 | 9 | 21 | 17 | 3 | 0 | 0 | 0 | 50 |
| Days Max Temp ≤ 32 °F | 9 | 2 | 0 | 0 | 0 | 0 | 0 | 0 | 0 | 0 | 1 | 7 | 19 |
| Days Min Temp ≤ 32 °F | 28 | 24 | 18 | 10 | 1 | 0 | 0 | 0 | 1 | 6 | 21 | 29 | 138 |
| Days Min Temp ≤ 0 °F | 1 | 0 | 0 | 0 | 0 | 0 | 0 | 0 | 0 | 0 | 0 | 1 | 2 |
| Heating Degree Days | 1119 | 870 | 699 | 450 | 217 | 52 | 2 | 5 | 94 | 397 | 766 | 1092 | 5763 |
| Cooling Degree Days | 0 | 0 | 0 | 7 | 40 | 184 | 359 | 325 | 92 | 1 | 0 | 0 | 1008 |
| Total Precipitation (") | 1.91 | 1.95 | 2.38 | 2.63 | 2.75 | 1.33 | 0.86 | 0.90 | 1.60 | 2.17 | 2.08 | 1.95 | 22.51 |
| Days ≥ 0.1" Precip | 5 | 5 | 6 | 6 | 5 | 3 | 2 | 2 | 3 | 4 | 5 | 5 | 51 |
| Total Snowfall (") | ***14.6*** | 7.9 | ***5.1*** | ***2.0*** | 0.4 | 0.0 | 0.0 | 0.0 | 0.2 | 0.5 | 5.9 | ***16.8*** | 53.4 |
| Days ≥ 1" Snow Depth | ***24*** | 14 | ***5*** | 1 | 0 | 0 | 0 | 0 | 0 | 0 | 5 | 17 | 66 |

## FERRON *Emery County* ELEVATION 5935 ft LAT/LONG 39° 5 ' N / 111° 8 ' W

| | JAN | FEB | MAR | APR | MAY | JUN | JUL | AUG | SEP | OCT | NOV | DEC | YEAR |
|---|---|---|---|---|---|---|---|---|---|---|---|---|---|
| Maximum Temp °F | 35.2 | 41.5 | 51.6 | 60.5 | 70.7 | 81.1 | 87.1 | 85.1 | 76.9 | 64.9 | 49.5 | 37.8 | 61.8 |
| Minimum Temp °F | 10.6 | 17.0 | 25.8 | 33.2 | 42.2 | 51.1 | 57.4 | 55.3 | 46.1 | 35.0 | 23.1 | 13.4 | 34.2 |
| Mean Temp °F | 22.9 | 29.3 | 38.7 | 46.9 | 56.4 | 66.1 | 72.2 | 70.2 | 61.5 | 50.0 | 36.3 | 25.6 | 48.0 |
| Days Max Temp ≥ 90 °F | 0 | 0 | 0 | 0 | 0 | 5 | 11 | 7 | 1 | 0 | 0 | 0 | 24 |
| Days Max Temp ≤ 32 °F | 12 | 4 | 1 | 0 | 0 | 0 | 0 | 0 | 0 | 0 | 1 | 8 | 26 |
| Days Min Temp ≤ 32 °F | 31 | 28 | 26 | 14 | 3 | 0 | 0 | 0 | 1 | 11 | 27 | 31 | 172 |
| Days Min Temp ≤ 0 °F | 4 | 1 | 0 | 0 | 0 | 0 | 0 | 0 | 0 | 0 | 0 | 2 | 7 |
| Heating Degree Days | 1298 | 1002 | 809 | 537 | 270 | 73 | 4 | 12 | 139 | 459 | 852 | 1214 | 6669 |
| Cooling Degree Days | 0 | 0 | 0 | 1 | 14 | 125 | 240 | 192 | 49 | 0 | 0 | 0 | 621 |
| Total Precipitation (") | 0.68 | 0.62 | 0.72 | 0.47 | 0.77 | 0.51 | 1.05 | 1.00 | 0.81 | 0.86 | 0.55 | 0.53 | 8.57 |
| Days ≥ 0.1" Precip | 2 | 2 | 2 | 2 | 2 | 2 | 3 | 3 | 2 | 2 | 2 | 2 | 26 |
| Total Snowfall (") | 9.6 | 7.2 | 3.8 | 1.0 | 0.1 | 0.0 | 0.0 | 0.0 | 0.1 | 0.8 | 2.6 | 5.8 | 31.0 |
| Days ≥ 1" Snow Depth | 18 | 13 | 3 | 0 | 0 | 0 | 0 | 0 | 0 | 0 | 2 | 10 | 46 |

## FILLMORE *Millard County* ELEVATION 5105 ft LAT/LONG 38° 58 ' N / 112° 20 ' W

| | JAN | FEB | MAR | APR | MAY | JUN | JUL | AUG | SEP | OCT | NOV | DEC | YEAR |
|---|---|---|---|---|---|---|---|---|---|---|---|---|---|
| Maximum Temp °F | 39.4 | 46.0 | 54.8 | 63.2 | 73.1 | 84.0 | 91.3 | 89.2 | 80.1 | 66.8 | 51.0 | 39.8 | 64.9 |
| Minimum Temp °F | 17.0 | 22.6 | 29.1 | 35.0 | 42.7 | 51.3 | 59.0 | 57.6 | 48.5 | 37.3 | 27.0 | 18.0 | 37.1 |
| Mean Temp °F | 28.2 | 34.3 | 42.0 | 49.2 | 57.9 | 67.7 | 75.2 | 73.4 | 64.3 | 52.1 | 39.0 | 28.9 | 51.0 |
| Days Max Temp ≥ 90 °F | 0 | 0 | 0 | 0 | 0 | 8 | 21 | 16 | 3 | 0 | 0 | 0 | 48 |
| Days Max Temp ≤ 32 °F | 8 | 2 | 0 | 0 | 0 | 0 | 0 | 0 | 0 | 0 | 1 | 7 | 18 |
| Days Min Temp ≤ 32 °F | 29 | 25 | 20 | 13 | 3 | 0 | 0 | 0 | 1 | 8 | 22 | 29 | 150 |
| Days Min Temp ≤ 0 °F | 2 | 1 | 0 | 0 | 0 | 0 | 0 | 0 | 0 | 0 | 0 | 2 | 5 |
| Heating Degree Days | 1133 | 860 | 707 | 469 | 236 | 56 | 2 | 5 | 99 | 398 | 772 | 1111 | 5848 |
| Cooling Degree Days | 0 | 0 | 0 | 4 | 27 | 152 | 310 | 279 | 88 | 3 | 0 | 0 | 863 |
| Total Precipitation (") | 1.33 | 1.21 | 1.96 | 1.78 | 1.51 | 0.80 | 0.78 | 0.85 | 1.13 | 1.60 | 1.48 | 1.47 | 15.90 |
| Days ≥ 0.1" Precip | 4 | 4 | 5 | 5 | 4 | 2 | 2 | 3 | 3 | 4 | 4 | 4 | 44 |
| Total Snowfall (") | 15.4 | 11.6 | 14.5 | 7.7 | 2.6 | 0.1 | 0.0 | 0.0 | 0.6 | 3.3 | 11.1 | 16.3 | 83.2 |
| Days ≥ 1" Snow Depth | 21 | 14 | 5 | 2 | 0 | 0 | 0 | 0 | 0 | 1 | 8 | 17 | 68 |

## FISH SPRINGS REFUGE *Juab County* ELEVATION 4354 ft LAT/LONG 39° 50 ' N / 113° 24 ' W

| | JAN | FEB | MAR | APR | MAY | JUN | JUL | AUG | SEP | OCT | NOV | DEC | YEAR |
|---|---|---|---|---|---|---|---|---|---|---|---|---|---|
| Maximum Temp °F | 38.5 | 46.1 | 55.7 | 64.4 | 74.6 | 85.6 | 94.3 | 92.3 | 81.3 | 67.0 | 51.3 | 39.6 | 65.9 |
| Minimum Temp °F | 17.1 | 23.0 | 31.3 | 37.7 | 47.2 | 56.4 | 63.9 | 61.7 | 50.6 | 37.9 | 27.5 | 18.5 | 39.4 |
| Mean Temp °F | 27.8 | 34.6 | 43.5 | 51.1 | 60.9 | 71.0 | 79.1 | 77.1 | 66.0 | 52.5 | 39.4 | 29.0 | 52.7 |
| Days Max Temp ≥ 90 °F | 0 | 0 | 0 | 0 | 1 | 12 | 26 | 23 | 6 | 0 | 0 | 0 | 68 |
| Days Max Temp ≤ 32 °F | 10 | 3 | 0 | 0 | 0 | 0 | 0 | 0 | 0 | 0 | 1 | 8 | 22 |
| Days Min Temp ≤ 32 °F | 29 | 24 | 18 | 8 | 1 | 0 | 0 | 0 | 1 | 7 | 22 | 29 | 139 |
| Days Min Temp ≤ 0 °F | 2 | 1 | 0 | 0 | 0 | 0 | 0 | 0 | 0 | 0 | 0 | 1 | 4 |
| Heating Degree Days | 1147 | 852 | 659 | 418 | 178 | 36 | 2 | 3 | 83 | 387 | 760 | 1108 | 5633 |
| Cooling Degree Days | 0 | 0 | 0 | 9 | 58 | 227 | 426 | 386 | 117 | 5 | 0 | 0 | 1228 |
| Total Precipitation (") | 0.41 | 0.50 | 0.76 | 0.86 | 1.03 | 0.61 | 0.60 | 0.63 | 0.74 | 0.91 | 0.53 | 0.40 | 7.98 |
| Days ≥ 0.1" Precip | 2 | 2 | 2 | 3 | 3 | 2 | 2 | 2 | 2 | 3 | 2 | 1 | 26 |
| Total Snowfall (") | ***3.0*** | ***2.0*** | 2.4 | 0.8 | 0.3 | 0.0 | 0.0 | 0.0 | 0.1 | 0.1 | 1.8 | ***2.9*** | 13.4 |
| Days ≥ 1" Snow Depth | ***8*** | ***4*** | ***0*** | 0 | 0 | 0 | 0 | 0 | 0 | 0 | 1 | ***3*** | 16 |

## FLAMING GORGE *Daggett County* ELEVATION 6270 ft LAT/LONG 40° 56 ' N / 109° 25 ' W

| | JAN | FEB | MAR | APR | MAY | JUN | JUL | AUG | SEP | OCT | NOV | DEC | YEAR |
|---|---|---|---|---|---|---|---|---|---|---|---|---|---|
| Maximum Temp °F | 35.0 | 39.5 | 47.1 | 56.6 | 67.5 | 77.7 | 85.1 | 83.3 | 74.1 | 61.3 | 44.6 | 35.3 | 58.9 |
| Minimum Temp °F | 9.1 | 12.0 | 20.7 | 27.9 | 35.5 | 42.9 | 49.7 | 47.7 | 39.2 | 29.7 | 20.2 | 11.1 | 28.8 |
| Mean Temp °F | 22.1 | 25.8 | 33.9 | 42.3 | 51.5 | 60.3 | 67.4 | 65.5 | 56.7 | 45.6 | 32.4 | 23.2 | 43.9 |
| Days Max Temp ≥ 90 °F | 0 | 0 | 0 | 0 | 0 | 2 | 6 | 4 | 0 | 0 | 0 | 0 | 12 |
| Days Max Temp ≤ 32 °F | 12 | 6 | 2 | 0 | 0 | 0 | 0 | 0 | 0 | 0 | 4 | 12 | 36 |
| Days Min Temp ≤ 32 °F | 31 | 28 | 29 | 22 | 11 | 2 | 0 | 0 | 6 | 20 | 27 | 30 | 206 |
| Days Min Temp ≤ 0 °F | 8 | 5 | 1 | 0 | 0 | 0 | 0 | 0 | 0 | 0 | 1 | 5 | 20 |
| Heating Degree Days | 1325 | 1102 | 957 | 675 | 413 | 162 | 21 | 46 | 249 | 596 | 970 | 1289 | 7805 |
| Cooling Degree Days | 0 | 0 | 0 | 0 | 1 | 30 | 102 | 68 | 8 | 0 | 0 | 0 | 209 |
| Total Precipitation (") | 0.44 | 0.52 | 1.09 | 1.59 | 1.64 | 1.27 | 1.13 | 1.10 | 1.06 | 1.35 | 0.85 | 0.63 | 12.67 |
| Days ≥ 0.1" Precip | 1 | 2 | 4 | 4 | 4 | 3 | 3 | 3 | 3 | 3 | 3 | 2 | 35 |
| Total Snowfall (") | 9.4 | 8.0 | 11.5 | 9.0 | 1.1 | 0.3 | 0.0 | 0.0 | 0.2 | 3.0 | ***9.5*** | ***10.2*** | 62.2 |
| Days ≥ 1" Snow Depth | na | ***18*** | 9 | ***2*** | 0 | 0 | 0 | 0 | 0 | 1 | ***9*** | ***18*** | na |

## FORT DUCHESNE *Uintah County* ELEVATION 4993 ft LAT/LONG 40° 17 ' N / 109° 52 ' W

| | JAN | FEB | MAR | APR | MAY | JUN | JUL | AUG | SEP | OCT | NOV | DEC | YEAR |
|---|---|---|---|---|---|---|---|---|---|---|---|---|---|
| Maximum Temp °F | 27.4 | 35.0 | 50.5 | 62.6 | 72.8 | 83.5 | 90.7 | 88.4 | 78.2 | 64.4 | 46.7 | 32.3 | 61.0 |
| Minimum Temp °F | 2.2 | 7.7 | 22.9 | 31.0 | 39.8 | 47.1 | 53.4 | 51.1 | 41.1 | 31.0 | 20.7 | 7.9 | 29.7 |
| Mean Temp °F | 14.8 | 21.2 | 36.7 | 46.7 | 56.3 | 65.3 | 72.1 | 69.8 | 59.7 | 47.7 | 33.7 | 20.1 | 45.3 |
| Days Max Temp ≥ 90 °F | 0 | 0 | 0 | 0 | 0 | 9 | *20* | 16 | 2 | 0 | 0 | 0 | 47 |
| Days Max Temp ≤ 32 °F | 20 | 11 | 2 | 0 | 0 | 0 | 0 | 0 | 0 | 0 | 3 | 14 | 50 |
| Days Min Temp ≤ 32 °F | 31 | 28 | 28 | 18 | 4 | 0 | 0 | 0 | 3 | 17 | 28 | 31 | 188 |
| Days Min Temp ≤ 0 °F | 13 | 7 | 0 | 0 | 0 | 0 | 0 | 0 | 0 | 0 | 0 | *7* | 27 |
| Heating Degree Days | 1550 | 1232 | 869 | 543 | 269 | 75 | 4 | 17 | 174 | 526 | 933 | 1381 | 7573 |
| Cooling Degree Days | 0 | 0 | 0 | 0 | 11 | 114 | 243 | 205 | 28 | 0 | 0 | 0 | 601 |
| Total Precipitation (") | 0.39 | 0.33 | 0.49 | 0.58 | 0.73 | 0.66 | 0.64 | 0.68 | 0.69 | 0.97 | 0.33 | 0.42 | 6.91 |
| Days ≥ 0.1" Precip | 1 | 1 | 1 | 2 | 2 | 2 | 2 | 2 | 2 | 3 | 1 | 2 | 21 |
| Total Snowfall (") | na | na | na | *0.2* | 0.0 | 0.0 | 0.0 | 0.0 | 0.3 | *0.0* | na | na | na |
| Days ≥ 1" Snow Depth | na | na | na | *0* | 0 | 0 | 0 | 0 | 0 | *0* | na | na | na |

## GARFIELD *Salt Lake County* ELEVATION 4314 ft LAT/LONG 40° 43 ' N / 112° 12 ' W

| | JAN | FEB | MAR | APR | MAY | JUN | JUL | AUG | SEP | OCT | NOV | DEC | YEAR |
|---|---|---|---|---|---|---|---|---|---|---|---|---|---|
| Maximum Temp °F | 36.1 | 41.4 | 50.9 | 59.9 | 70.3 | 81.2 | 90.5 | 88.2 | 76.9 | 63.0 | 48.2 | 38.2 | 62.1 |
| Minimum Temp °F | 22.8 | 26.9 | 34.6 | 42.0 | 50.7 | 60.5 | 69.0 | 66.3 | 55.9 | 44.5 | 33.4 | 24.3 | 44.2 |
| Mean Temp °F | 29.5 | 34.2 | 42.8 | 51.0 | 60.6 | 70.8 | 79.7 | 77.3 | 66.4 | 53.6 | 40.8 | 31.3 | 53.2 |
| Days Max Temp ≥ 90 °F | 0 | 0 | 0 | 0 | 0 | 6 | *19* | *15* | 2 | 0 | 0 | 0 | 42 |
| Days Max Temp ≤ 32 °F | 11 | 4 | 1 | 0 | 0 | 0 | 0 | 0 | 0 | 0 | 1 | *8* | 25 |
| Days Min Temp ≤ 32 °F | 27 | 22 | *11* | *3* | 0 | 0 | 0 | 0 | 0 | 1 | *13* | *26* | 103 |
| Days Min Temp ≤ 0 °F | 0 | 0 | 0 | 0 | 0 | 0 | 0 | 0 | 0 | 0 | 0 | 0 | 0 |
| Heating Degree Days | 1093 | 864 | 676 | 421 | 189 | 38 | 1 | 4 | 78 | 352 | 718 | *1043* | 5477 |
| Cooling Degree Days | 0 | 0 | 0 | 15 | *64* | 233 | 454 | 418 | 138 | 7 | *0* | *0* | 1329 |
| Total Precipitation (") | 1.20 | 1.23 | 1.81 | 2.15 | 2.15 | 0.98 | 0.96 | 0.93 | 1.71 | 2.00 | 1.70 | 1.30 | 18.12 |
| Days ≥ 0.1" Precip | 4 | 3 | *5* | *5* | 5 | 2 | 2 | 2 | 3 | 4 | *4* | *3* | 42 |
| Total Snowfall (") | na | *4.2* | *2.5* | 0.5 | 0.1 | 0.0 | 0.0 | 0.0 | 0.0 | 0.4 | 3.2 | *7.4* | na |
| Days ≥ 1" Snow Depth | na | *6* | *1* | 0 | 0 | 0 | 0 | 0 | 0 | 0 | 2 | na | na |

## GREEN RIVER AVIATION *Emery County* ELEVATION 4062 ft LAT/LONG 39° 0 ' N / 110° 9 ' W

| | JAN | FEB | MAR | APR | MAY | JUN | JUL | AUG | SEP | OCT | NOV | DEC | YEAR |
|---|---|---|---|---|---|---|---|---|---|---|---|---|---|
| Maximum Temp °F | 36.5 | 47.1 | 59.7 | 69.7 | 79.5 | 90.7 | 96.7 | 94.0 | *84.9* | 70.3 | 54.1 | 41.1 | 68.7 |
| Minimum Temp °F | 9.1 | 17.9 | 28.2 | 36.1 | 45.2 | 52.7 | 60.2 | 57.8 | *46.9* | 35.2 | 23.5 | 14.0 | 35.6 |
| Mean Temp °F | 22.8 | 32.5 | 44.0 | 52.9 | 62.4 | 71.7 | 78.5 | 75.9 | *66.0* | 52.8 | 38.8 | 27.6 | 52.2 |
| Days Max Temp ≥ 90 °F | 0 | 0 | 0 | 0 | 4 | 18 | 28 | 25 | *7* | 0 | 0 | 0 | 82 |
| Days Max Temp ≤ 32 °F | 10 | 1 | 0 | 0 | 0 | 0 | 0 | 0 | 0 | 0 | 0 | 4 | 15 |
| Days Min Temp ≤ 32 °F | 30 | 27 | 22 | 10 | 1 | 0 | 0 | 0 | 1 | *11* | *25* | 30 | 157 |
| Days Min Temp ≤ 0 °F | 8 | 2 | 0 | 0 | 0 | 0 | 0 | 0 | 0 | 0 | 0 | 2 | 12 |
| Heating Degree Days | 1302 | 909 | 646 | 360 | 124 | 15 | 0 | 1 | *60* | 374 | 778 | 1153 | 5722 |
| Cooling Degree Days | 0 | 0 | 0 | 7 | 61 | 260 | 432 | 366 | *107* | 1 | 0 | 0 | 1234 |
| Total Precipitation (") | 0.49 | 0.38 | 0.62 | 0.48 | 0.69 | 0.39 | 0.60 | 0.73 | 0.65 | 0.96 | 0.45 | 0.43 | 6.87 |
| Days ≥ 0.1" Precip | 2 | 2 | 2 | 2 | 2 | 1 | 2 | 2 | 2 | 3 | 1 | 1 | 22 |
| Total Snowfall (") | *5.5* | 0.7 | 0.5 | 0.0 | 0.0 | 0.0 | 0.0 | 0.0 | 0.0 | 0.1 | 0.4 | *3.9* | 11.1 |
| Days ≥ 1" Snow Depth | *14* | 5 | 0 | 0 | 0 | 0 | 0 | 0 | 0 | 0 | 0 | 4 | 23 |

## GROUSE CREEK *Box Elder County* ELEVATION 5299 ft LAT/LONG 41° 42 ' N / 113° 53 ' W

| | JAN | FEB | MAR | APR | MAY | JUN | JUL | AUG | SEP | OCT | NOV | DEC | YEAR |
|---|---|---|---|---|---|---|---|---|---|---|---|---|---|
| Maximum Temp °F | *35.4* | 40.8 | *48.4* | 58.1 | *67.8* | 77.0 | *87.4* | *86.2* | 76.0 | *62.9* | *46.5* | *35.5* | 60.2 |
| Minimum Temp °F | *9.6* | 15.1 | *23.0* | 27.7 | *36.0* | 42.8 | *49.4* | *47.1* | 38.2 | *28.1* | *19.9* | *10.0* | 28.9 |
| Mean Temp °F | *22.5* | 28.0 | *35.8* | 42.9 | *51.9* | 59.9 | *68.4* | *66.7* | 57.1 | *45.5* | *33.2* | *22.8* | 44.6 |
| Days Max Temp ≥ 90 °F | 0 | 0 | 0 | 0 | 0 | *3* | na | na | *1* | 0 | 0 | 0 | na |
| Days Max Temp ≤ 32 °F | *10* | 5 | 1 | 0 | 0 | 0 | 0 | 0 | 0 | 0 | 2 | na | na |
| Days Min Temp ≤ 32 °F | 30 | 28 | 28 | na | na | 2 | 0 | 0 | *5* | na | *28* | 30 | na |
| Days Min Temp ≤ 0 °F | 5 | 3 | 0 | 0 | 0 | 0 | 0 | 0 | 0 | 0 | 1 | 4 | 13 |
| Heating Degree Days | *1310* | 1040 | *898* | *653* | *393* | 182 | na | *51* | 239 | *593* | *938* | *1305* | na |
| Cooling Degree Days | 0 | 0 | 0 | 0 | 3 | 38 | *138* | 108 | 18 | 0 | 0 | 0 | 305 |
| Total Precipitation (") | 0.92 | 0.84 | 0.84 | 0.93 | 1.14 | 1.15 | 0.82 | 0.77 | 0.65 | 0.85 | 1.11 | 1.00 | 11.02 |
| Days ≥ 0.1" Precip | *3* | 3 | 3 | *2* | *3* | 3 | 2 | *2* | 2 | *2* | 3 | 4 | 32 |
| Total Snowfall (") | na | na | *2.5* | na | *0.0* | 0.0 | 0.0 | 0.0 | 0.0 | *0.2* | *5.6* | *10.4* | na |
| Days ≥ 1" Snow Depth | na | na | na | na | *0* | 0 | 0 | 0 | 0 | 0 | na | na | na |

## HANKSVILLE *Wayne County* ELEVATION 4308 ft LAT/LONG 38° 22 ' N / 110° 43 ' W

| | JAN | FEB | MAR | APR | MAY | JUN | JUL | AUG | SEP | OCT | NOV | DEC | YEAR |
|---|---|---|---|---|---|---|---|---|---|---|---|---|---|
| Maximum Temp °F | 40.2 | 49.6 | 60.9 | 70.7 | 81.5 | 93.2 | 98.7 | 95.8 | 86.0 | 71.6 | 54.8 | 42.4 | 70.5 |
| Minimum Temp °F | 10.3 | 18.6 | 28.2 | 36.1 | 44.8 | 53.1 | 59.7 | 57.7 | 47.4 | 35.0 | 22.9 | 13.5 | 35.6 |
| Mean Temp °F | 25.2 | 34.1 | 44.6 | 53.4 | 63.1 | 73.1 | 79.2 | 76.8 | 66.7 | 53.4 | 38.9 | 28.0 | 53.0 |
| Days Max Temp ≥ 90 °F | 0 | 0 | 0 | 1 | 7 | 21 | 29 | 26 | 12 | 1 | 0 | 0 | 97 |
| Days Max Temp ≤ 32 °F | 8 | 1 | 0 | 0 | 0 | 0 | 0 | 0 | 0 | 0 | 0 | 4 | 13 |
| Days Min Temp ≤ 32 °F | 30 | 26 | 22 | 11 | 2 | 0 | 0 | 0 | 1 | 12 | 26 | 30 | 160 |
| Days Min Temp ≤ 0 °F | 7 | 2 | 0 | 0 | 0 | 0 | 0 | 0 | 0 | 0 | 0 | 3 | 12 |
| Heating Degree Days | 1225 | 866 | 625 | 348 | 115 | 14 | 0 | 1 | 59 | 358 | 778 | 1141 | 5530 |
| Cooling Degree Days | 0 | 0 | 0 | 12 | 70 | 295 | 449 | 377 | 127 | 2 | 0 | 0 | 1332 |
| Total Precipitation (") | 0.43 | 0.25 | 0.55 | 0.39 | 0.54 | 0.30 | 0.55 | 0.64 | 0.67 | 0.71 | 0.39 | 0.35 | 5.77 |
| Days ≥ 0.1" Precip | 1 | 1 | 2 | 1 | 2 | 1 | 2 | 2 | 2 | 2 | 1 | 1 | 18 |
| Total Snowfall (") | *1.0* | 1.1 | *0.8* | 0.1 | 0.0 | 0.0 | 0.0 | 0.0 | 0.0 | 0.4 | 0.6 | *2.0* | 6.0 |
| Days ≥ 1" Snow Depth | 12 | 5 | 0 | 0 | 0 | 0 | 0 | 0 | 0 | 0 | 1 | 6 | 24 |

## HEBER *Wasatch County* ELEVATION 5594 ft LAT/LONG 40° 30 ' N / 111° 25 ' W

| | JAN | FEB | MAR | APR | MAY | JUN | JUL | AUG | SEP | OCT | NOV | DEC | YEAR |
|---|---|---|---|---|---|---|---|---|---|---|---|---|---|
| Maximum Temp °F | 34.3 | 39.6 | 48.9 | 59.1 | 69.1 | 78.9 | 86.6 | 85.1 | 76.1 | 64.4 | 47.6 | 36.2 | 60.5 |
| Minimum Temp °F | 9.3 | 13.1 | 23.0 | 29.1 | 35.6 | 42.0 | 48.4 | 46.8 | 38.9 | 29.5 | 21.5 | 11.8 | 29.1 |
| Mean Temp °F | 21.8 | 26.4 | 36.0 | 44.1 | 52.4 | 60.5 | 67.5 | 66.0 | 57.5 | 47.0 | 34.6 | 24.0 | 44.8 |
| Days Max Temp ≥ 90 °F | 0 | 0 | 0 | 0 | 0 | 3 | 9 | 7 | 1 | 0 | 0 | 0 | 20 |
| Days Max Temp ≤ 32 °F | 12 | 5 | 1 | 0 | 0 | 0 | 0 | 0 | 0 | 0 | 2 | 11 | 31 |
| Days Min Temp ≤ 32 °F | 30 | 28 | 28 | 21 | 10 | 2 | 0 | 0 | 6 | 21 | 27 | 30 | 203 |
| Days Min Temp ≤ 0 °F | 8 | 5 | 1 | 0 | 0 | 0 | 0 | 0 | 0 | 0 | 1 | 5 | 20 |
| Heating Degree Days | 1332 | 1083 | 893 | 620 | 385 | 152 | 17 | 39 | 227 | 553 | 906 | 1264 | 7471 |
| Cooling Degree Days | 0 | 0 | 0 | 0 | 0 | 28 | 103 | 86 | 12 | 0 | 0 | 0 | 229 |
| Total Precipitation (") | 1.71 | 1.61 | 1.37 | 1.28 | 1.34 | 0.85 | 0.86 | 1.00 | 1.22 | 1.69 | 1.65 | 1.52 | 16.10 |
| Days ≥ 0.1" Precip | 5 | 4 | 4 | 5 | 4 | 3 | 3 | 3 | 3 | 4 | 5 | 5 | 48 |
| Total Snowfall (") | 19.0 | 15.4 | 5.1 | 3.6 | 1.2 | 0.0 | 0.0 | 0.0 | 0.2 | 1.8 | 8.5 | 14.5 | 69.3 |
| Days ≥ 1" Snow Depth | 27 | 22 | 8 | 0 | 0 | 0 | 0 | 0 | 0 | 0 | 7 | 20 | 84 |

## HOVENWEEP NM *San Juan County* ELEVATION 5403 ft LAT/LONG 37° 23 ' N / 109° 4 ' W

| | JAN | FEB | MAR | APR | MAY | JUN | JUL | AUG | SEP | OCT | NOV | DEC | YEAR |
|---|---|---|---|---|---|---|---|---|---|---|---|---|---|
| Maximum Temp °F | 40.4 | 48.1 | 57.9 | 67.6 | 77.2 | 88.9 | 94.0 | 91.5 | 82.9 | 69.5 | 53.6 | 41.9 | 67.8 |
| Minimum Temp °F | 14.0 | 20.7 | 26.5 | 32.1 | 40.6 | 48.7 | 57.0 | 56.0 | 46.8 | 35.2 | 24.7 | 15.8 | 34.8 |
| Mean Temp °F | 27.3 | 34.4 | 42.2 | 49.9 | 59.0 | 68.8 | 75.5 | 73.8 | 64.9 | 52.4 | 39.2 | 28.9 | 51.4 |
| Days Max Temp ≥ 90 °F | 0 | 0 | 0 | 0 | 1 | 15 | 27 | 22 | 5 | 0 | 0 | 0 | 70 |
| Days Max Temp ≤ 32 °F | 5 | 1 | 0 | 0 | 0 | 0 | 0 | 0 | 0 | 0 | 0 | 4 | 10 |
| Days Min Temp ≤ 32 °F | 30 | 26 | 25 | 16 | 4 | 0 | 0 | 0 | 0 | 12 | 26 | 30 | 169 |
| Days Min Temp ≤ 0 °F | 4 | 1 | 0 | 0 | 0 | 0 | 0 | 0 | 0 | 0 | 0 | 2 | 7 |
| Heating Degree Days | 1164 | 857 | 700 | 447 | 195 | 27 | 0 | 1 | 67 | 384 | 767 | 1113 | 5722 |
| Cooling Degree Days | 0 | 0 | 0 | 0 | 13 | 155 | 319 | 281 | 68 | 0 | 0 | 0 | 836 |
| Total Precipitation (") | 1.04 | 1.04 | 1.17 | 0.80 | 0.78 | 0.36 | 0.96 | 1.09 | 1.01 | 1.39 | 1.18 | 1.13 | 11.95 |
| Days ≥ 0.1" Precip | 3 | 3 | 4 | 2 | 2 | 1 | 3 | 3 | 2 | 3 | 3 | 3 | 32 |
| Total Snowfall (") | 6.4 | 4.0 | 2.4 | 0.8 | 0.1 | 0.0 | 0.0 | 0.0 | 0.0 | 0.4 | 1.5 | 7.2 | 22.8 |
| Days ≥ 1" Snow Depth | 15 | 6 | 0 | 0 | 0 | 0 | 0 | 0 | 0 | 0 | 1 | 9 | 31 |

## IBAPAH *Tooele County* ELEVATION 5282 ft LAT/LONG 40° 0 ' N / 114° 0 ' W

| | JAN | FEB | MAR | APR | MAY | JUN | JUL | AUG | SEP | OCT | NOV | DEC | YEAR |
|---|---|---|---|---|---|---|---|---|---|---|---|---|---|
| Maximum Temp °F | 41.9 | 46.9 | 54.8 | *62.7* | 72.1 | 82.0 | 91.4 | 90.2 | 80.3 | 67.3 | 52.1 | 42.1 | 65.3 |
| Minimum Temp °F | 10.4 | 15.2 | 22.2 | *27.2* | 34.0 | 39.9 | 46.3 | 45.0 | 35.4 | 26.4 | 18.8 | 10.7 | 27.6 |
| Mean Temp °F | 26.1 | 31.0 | 38.5 | *45.0* | 53.1 | 60.9 | 68.9 | 67.7 | 57.9 | 46.9 | 35.5 | 26.4 | 46.5 |
| Days Max Temp ≥ 90 °F | 0 | 0 | 0 | 0 | 0 | *7* | *22* | *18* | 4 | 0 | 0 | 0 | 51 |
| Days Max Temp ≤ 32 °F | 6 | 2 | 0 | 0 | 0 | 0 | 0 | 0 | 0 | 0 | 0 | 5 | 13 |
| Days Min Temp ≤ 32 °F | 30 | 28 | 29 | *23* | *13* | 3 | 0 | 1 | 11 | 25 | 28 | 30 | 221 |
| Days Min Temp ≤ 0 °F | 6 | 3 | 0 | 0 | 0 | 0 | 0 | 0 | 0 | 0 | 1 | 5 | 15 |
| Heating Degree Days | 1198 | 952 | 815 | *595* | *359* | 150 | 20 | 31 | 222 | 555 | 878 | 1189 | 6964 |
| Cooling Degree Days | 0 | 0 | 0 | 0 | 3 | 38 | 138 | 114 | 17 | 0 | 0 | 0 | 310 |
| Total Precipitation (") | 0.61 | 0.58 | 0.97 | 0.98 | 1.20 | 0.98 | 0.85 | 0.93 | 0.92 | 0.95 | 0.56 | 0.50 | 10.03 |
| Days ≥ 0.1" Precip | *1* | 2 | 3 | *3* | 4 | 2 | 2 | 2 | 2 | 2 | *2* | 1 | 26 |
| Total Snowfall (") | *8.1* | *5.3* | *5.6* | *3.9* | 1.5 | 0.0 | 0.0 | 0.0 | 0.1 | 0.9 | *4.0* | na | na |
| Days ≥ 1" Snow Depth | *14* | na | *3* | *1* | 0 | 0 | 0 | 0 | 0 | 0 | *3* | na | na |

## JENSEN *Uintah County* ELEVATION 4744 ft LAT/LONG 40° 22 ' N / 109° 22 ' W

| | JAN | FEB | MAR | APR | MAY | JUN | JUL | AUG | SEP | OCT | NOV | DEC | YEAR |
|---|---|---|---|---|---|---|---|---|---|---|---|---|---|
| Maximum Temp °F | 27.9 | 36.4 | 52.0 | 64.0 | 74.2 | 84.0 | 90.6 | 88.6 | 79.0 | 65.5 | 47.4 | 32.4 | 61.8 |
| Minimum Temp °F | 1.5 | 8.1 | 21.9 | 30.4 | 39.4 | 46.4 | 52.6 | 49.9 | 40.8 | 30.2 | 19.6 | 7.1 | 29.0 |
| Mean Temp °F | 14.7 | 22.3 | 37.0 | 47.2 | 56.8 | 65.2 | 71.6 | 69.3 | 59.9 | 47.8 | 33.4 | 19.8 | 45.4 |
| Days Max Temp ≥ 90 °F | 0 | 0 | 0 | 0 | 0 | 9 | 20 | 15 | 2 | 0 | 0 | 0 | 46 |
| Days Max Temp ≤ 32 °F | 20 | 10 | 1 | 0 | 0 | 0 | 0 | 0 | 0 | 0 | 2 | 14 | 47 |
| Days Min Temp ≤ 32 °F | 31 | 28 | 29 | 19 | 4 | 0 | 0 | 0 | 4 | 20 | 29 | 31 | 195 |
| Days Min Temp ≤ 0 °F | 15 | 8 | 1 | 0 | 0 | 0 | 0 | 0 | 0 | 0 | 0 | 9 | 33 |
| Heating Degree Days | 1553 | 1201 | 861 | 527 | 252 | 63 | 2 | 11 | 164 | 525 | 940 | 1395 | 7494 |
| Cooling Degree Days | 0 | 0 | 0 | 0 | 5 | 89 | 213 | 168 | 25 | 0 | 0 | 0 | 500 |
| Total Precipitation (") | 0.49 | 0.54 | 0.67 | 0.73 | 0.85 | 0.64 | 0.72 | 0.56 | 0.83 | 1.15 | 0.60 | 0.60 | 8.38 |
| Days ≥ 0.1" Precip | 2 | 2 | 2 | 2 | 3 | 2 | 2 | 2 | 3 | 3 | 2 | 2 | 27 |
| Total Snowfall (") | 6.1 | 4.7 | 2.3 | 1.3 | 0.3 | 0.0 | 0.0 | 0.0 | 0.5 | 0.9 | 2.7 | 7.1 | 25.9 |
| Days ≥ 1" Snow Depth | na | na | na | na | 0 | 0 | 0 | 0 | 0 | *0* | na | na | na |

## KAMAS 3 NW *Summit County* ELEVATION 6410 ft LAT/LONG 40° 40 ' N / 111° 19 ' W

| | JAN | FEB | MAR | APR | MAY | JUN | JUL | AUG | SEP | OCT | NOV | DEC | YEAR |
|---|---|---|---|---|---|---|---|---|---|---|---|---|---|
| Maximum Temp °F | *35.3* | 39.8 | *46.5* | 55.6 | 65.9 | 75.8 | 84.3 | 83.0 | 73.7 | 61.8 | 45.4 | *36.6* | 58.6 |
| Minimum Temp °F | *11.1* | 14.0 | *21.6* | 27.2 | 34.4 | 40.7 | 47.6 | 46.3 | 38.2 | 29.6 | 20.4 | *12.3* | 28.6 |
| Mean Temp °F | *23.2* | 26.9 | *34.0* | 41.5 | 50.1 | 58.3 | 65.9 | 64.7 | 56.0 | 45.7 | 32.9 | *24.5* | 43.6 |
| Days Max Temp ≥ 90 °F | 0 | 0 | 0 | 0 | 0 | 1 | *4* | 3 | 0 | 0 | 0 | 0 | 8 |
| Days Max Temp ≤ 32 °F | *12* | *5* | *1* | 0 | 0 | 0 | 0 | 0 | 0 | 0 | *4* | na | na |
| Days Min Temp ≤ 32 °F | *31* | 27 | *28* | *23* | *12* | 2 | 0 | 0 | 7 | *21* | *28* | *30* | 209 |
| Days Min Temp ≤ 0 °F | *6* | 4 | *1* | 0 | 0 | 0 | 0 | 0 | 0 | 0 | 1 | *4* | 16 |
| Heating Degree Days | *1296* | 1073 | *949* | *701* | *456* | 208 | 38 | 62 | 270 | 592 | 953 | *1246* | 7844 |
| Cooling Degree Days | 0 | 0 | 0 | 0 | 0 | 12 | 67 | 58 | 5 | 0 | 0 | 0 | 142 |
| Total Precipitation (") | 1.36 | 1.79 | *1.61* | 1.67 | 1.63 | 1.04 | 1.28 | 1.20 | 1.45 | 1.84 | 1.67 | 1.41 | 17.95 |
| Days ≥ 0.1" Precip | *5* | *5* | na | na | *5* | 4 | 3 | 4 | 4 | *4* | *5* | na | na |
| Total Snowfall (") | na | na | na | na | na | *0.1* | 0.0 | 0.0 | 0.7 | *2.4* | na | na | na |
| Days ≥ 1" Snow Depth | na | na | na | na | *0* | 0 | 0 | 0 | 0 | *1* | na | na | na |

## KANAB *Kane County* ELEVATION 5013 ft LAT/LONG 37° 3 ' N / 112° 31 ' W

| | JAN | FEB | MAR | APR | MAY | JUN | JUL | AUG | SEP | OCT | NOV | DEC | YEAR |
|---|---|---|---|---|---|---|---|---|---|---|---|---|---|
| Maximum Temp °F | 48.1 | 53.2 | 59.1 | 67.4 | 76.9 | 87.5 | 92.5 | 90.0 | 83.0 | 72.4 | 58.4 | 49.1 | 69.8 |
| Minimum Temp °F | 22.8 | 26.7 | 31.1 | 35.9 | 43.4 | 51.2 | 58.3 | 57.1 | 49.7 | 39.9 | 30.1 | 23.6 | 39.2 |
| Mean Temp °F | 35.5 | 40.0 | 45.1 | 51.6 | 60.2 | 69.4 | 75.4 | 73.6 | 66.4 | 56.2 | 44.3 | 36.4 | 54.5 |
| Days Max Temp ≥ 90 °F | 0 | 0 | 0 | 0 | 1 | 13 | 23 | 17 | 5 | 0 | 0 | 0 | 59 |
| Days Max Temp ≤ 32 °F | 1 | 0 | 0 | 0 | 0 | 0 | 0 | 0 | 0 | 0 | 0 | 1 | 2 |
| Days Min Temp ≤ 32 °F | 28 | 23 | 18 | 9 | 2 | 0 | 0 | 0 | 0 | 4 | 19 | 28 | 131 |
| Days Min Temp ≤ 0 °F | 0 | 0 | 0 | 0 | 0 | 0 | 0 | 0 | 0 | 0 | 0 | 0 | 0 |
| Heating Degree Days | 909 | 699 | 608 | 395 | 165 | 23 | 0 | 1 | 43 | 272 | 614 | 881 | 4610 |
| Cooling Degree Days | 0 | 0 | 0 | 2 | 27 | 191 | 351 | 298 | 106 | 9 | 0 | 0 | 984 |
| Total Precipitation (") | 1.72 | 1.50 | 1.79 | 0.89 | 0.75 | 0.34 | 1.02 | 1.42 | 0.89 | 1.11 | 1.22 | 1.32 | 13.97 |
| Days ≥ 0.1" Precip | 4 | 4 | 5 | 2 | 2 | 1 | 3 | 4 | 2 | 3 | 3 | 3 | 36 |
| Total Snowfall (") | 7.8 | 4.9 | 2.7 | 1.6 | 0.0 | 0.0 | 0.0 | 0.0 | 0.0 | 0.2 | 1.9 | 5.9 | 25.0 |
| Days ≥ 1" Snow Depth | 7 | 3 | 0 | 0 | 0 | 0 | 0 | 0 | 0 | 0 | 1 | 4 | 15 |

## KANOSH *Millard County* ELEVATION 5023 ft LAT/LONG 38° 48 ' N / 112° 26 ' W

| | JAN | FEB | MAR | APR | MAY | JUN | JUL | AUG | SEP | OCT | NOV | DEC | YEAR |
|---|---|---|---|---|---|---|---|---|---|---|---|---|---|
| Maximum Temp °F | 39.8 | 46.4 | 54.9 | 63.0 | 72.7 | 84.2 | 92.0 | 89.8 | 80.8 | 67.2 | 52.0 | 40.7 | 65.3 |
| Minimum Temp °F | 17.7 | 23.6 | 30.5 | 36.2 | 44.8 | 54.2 | 62.8 | 60.9 | 51.4 | 40.3 | 28.4 | 19.2 | 39.2 |
| Mean Temp °F | 28.6 | 35.0 | 42.7 | 49.7 | 58.8 | 69.3 | 77.4 | 75.4 | 66.1 | 53.8 | 40.2 | 30.0 | 52.3 |
| Days Max Temp ≥ 90 °F | 0 | 0 | 0 | 0 | 0 | 9 | 22 | 18 | 4 | 0 | 0 | 0 | 53 |
| Days Max Temp ≤ 32 °F | 8 | 2 | 0 | 0 | 0 | 0 | 0 | 0 | 0 | 0 | 1 | 7 | 18 |
| Days Min Temp ≤ 32 °F | 29 | 23 | 18 | 12 | 2 | 0 | 0 | 0 | 1 | 6 | 20 | 28 | 139 |
| Days Min Temp ≤ 0 °F | 2 | 1 | 0 | 0 | 0 | 0 | 0 | 0 | 0 | 0 | 0 | 2 | 5 |
| Heating Degree Days | 1120 | 840 | 684 | 458 | 222 | 45 | 1 | 3 | 78 | 348 | 735 | 1079 | 5613 |
| Cooling Degree Days | 0 | 0 | 0 | 6 | 38 | 192 | 389 | 349 | 124 | 9 | 0 | 0 | 1107 |
| Total Precipitation (") | 1.17 | 1.07 | 1.92 | 1.73 | 1.43 | 0.71 | 0.88 | 1.02 | 1.07 | 1.48 | 1.42 | 1.33 | 15.23 |
| Days ≥ 0.1" Precip | 4 | 4 | 5 | 4 | 4 | 2 | 2 | 3 | 3 | 4 | 4 | 4 | 43 |
| Total Snowfall (") | 14.3 | 11.2 | *10.7* | *8.2* | 1.3 | 0.1 | 0.0 | 0.0 | 0.1 | 3.7 | 11.6 | 15.5 | 76.7 |
| Days ≥ 1" Snow Depth | 19 | 13 | 4 | 1 | 0 | 0 | 0 | 0 | 0 | 1 | 6 | 17 | 61 |

## KOOSHAREM *Sevier County* ELEVATION 6535 ft LAT/LONG 38° 31 ' N / 111° 53 ' W

| | JAN | FEB | MAR | APR | MAY | JUN | JUL | AUG | SEP | OCT | NOV | DEC | YEAR |
|---|---|---|---|---|---|---|---|---|---|---|---|---|---|
| Maximum Temp °F | 38.8 | 42.3 | 48.4 | 57.1 | 67.1 | *77.9* | 84.4 | 82.1 | 74.9 | 63.1 | 48.6 | 39.9 | 60.4 |
| Minimum Temp °F | 8.9 | 13.4 | 20.4 | 24.9 | 32.2 | 39.4 | 46.5 | 44.8 | 36.8 | 26.9 | 18.0 | 10.3 | 26.9 |
| Mean Temp °F | 23.9 | 27.8 | 34.4 | 41.1 | 49.6 | *58.7* | 65.5 | 63.5 | 55.9 | 45.0 | 33.4 | 25.2 | 43.7 |
| Days Max Temp ≥ 90 °F | 0 | 0 | 0 | 0 | 0 | 1 | 4 | 2 | 0 | 0 | 0 | 0 | 7 |
| Days Max Temp ≤ 32 °F | 7 | 3 | 1 | 0 | 0 | 0 | 0 | 0 | 0 | 0 | 2 | 7 | 20 |
| Days Min Temp ≤ 32 °F | 31 | 28 | 30 | 26 | 15 | 4 | 0 | 1 | 8 | 25 | 28 | 31 | 227 |
| Days Min Temp ≤ 0 °F | 7 | 4 | 1 | 0 | 0 | 0 | 0 | 0 | 0 | 0 | 1 | 5 | 18 |
| Heating Degree Days | 1268 | 1043 | 942 | 711 | 470 | *196* | 32 | 73 | 270 | 612 | 943 | 1229 | 7789 |
| Cooling Degree Days | 0 | 0 | 0 | 0 | *0* | na | 54 | 33 | 2 | 0 | 0 | 0 | na |
| Total Precipitation (") | 0.60 | 0.56 | 0.77 | 0.60 | 0.90 | 0.60 | 1.10 | 1.39 | 0.96 | 0.87 | 0.60 | 0.61 | 9.56 |
| Days ≥ 0.1" Precip | 2 | 2 | 3 | 2 | 3 | 2 | 4 | 4 | 2 | 3 | 2 | 2 | 31 |
| Total Snowfall (") | 8.6 | 6.6 | 6.1 | 2.8 | 1.2 | 0.1 | 0.0 | 0.0 | 0.2 | 1.5 | 4.1 | 6.3 | 37.5 |
| Days ≥ 1" Snow Depth | 20 | 16 | 5 | 0 | 0 | 0 | 0 | 0 | 0 | 0 | 4 | 11 | 56 |

## LA VERKIN *Washington County* ELEVATION 3202 ft LAT/LONG 37° 12 ' N / 113° 16 ' W

| | JAN | FEB | MAR | APR | MAY | JUN | JUL | AUG | SEP | OCT | NOV | DEC | YEAR |
|---|---|---|---|---|---|---|---|---|---|---|---|---|---|
| Maximum Temp °F | 53.0 | 59.2 | 65.8 | 74.0 | 83.5 | 93.9 | 98.4 | 96.0 | 89.0 | 77.4 | 62.8 | 53.0 | 75.5 |
| Minimum Temp °F | 26.1 | 31.2 | 37.1 | 42.1 | 49.6 | 58.1 | 64.8 | 63.7 | 55.3 | 43.8 | 33.1 | 26.1 | 44.3 |
| Mean Temp °F | 39.6 | 45.2 | 51.5 | 58.1 | 66.6 | 76.0 | 81.7 | 79.8 | 72.2 | 60.6 | 47.9 | 39.6 | 59.9 |
| Days Max Temp ≥ 90 °F | 0 | 0 | 0 | 1 | 8 | 22 | 29 | 27 | 15 | 2 | 0 | 0 | 104 |
| Days Max Temp ≤ 32 °F | 0 | 0 | 0 | 0 | 0 | 0 | 0 | 0 | 0 | 0 | 0 | 0 | 0 |
| Days Min Temp ≤ 32 °F | 25 | 16 | 7 | 2 | 0 | 0 | 0 | 0 | 0 | 2 | 14 | 26 | 92 |
| Days Min Temp ≤ 0 °F | 0 | 0 | 0 | 0 | 0 | 0 | 0 | 0 | 0 | 0 | 0 | 0 | 0 |
| Heating Degree Days | 781 | 552 | 415 | 224 | 67 | 3 | 0 | 0 | 15 | 165 | 505 | 780 | 3507 |
| Cooling Degree Days | 0 | 0 | 3 | 39 | 149 | 382 | 551 | 505 | 270 | 45 | 0 | 0 | 1944 |
| Total Precipitation (") | 1.48 | 1.44 | 1.84 | 0.81 | 0.52 | 0.28 | 0.73 | 1.02 | 0.76 | 0.82 | 1.12 | 1.00 | 11.82 |
| Days ≥ 0.1" Precip | 4 | 3 | 4 | 2 | 2 | 1 | 2 | 2 | 2 | 2 | 3 | 3 | 30 |
| Total Snowfall (") | 2.0 | 0.9 | 0.2 | 0.0 | 0.0 | 0.0 | 0.0 | 0.0 | 0.0 | 0.0 | 0.1 | 1.0 | 4.2 |
| Days ≥ 1" Snow Depth | na | 0 | 0 | 0 | 0 | 0 | 0 | 0 | 0 | 0 | 0 | *0* | na |

## LAKETOWN *Rich County* ELEVATION 5980 ft LAT/LONG 41° 49 ' N / 111° 19 ' W

| | JAN | FEB | MAR | APR | MAY | JUN | JUL | AUG | SEP | OCT | NOV | DEC | YEAR |
|---|---|---|---|---|---|---|---|---|---|---|---|---|---|
| Maximum Temp °F | 31.8 | 34.4 | 42.6 | 53.8 | 64.6 | 74.1 | 82.4 | 81.0 | 71.4 | 58.6 | 42.9 | 33.6 | 55.9 |
| Minimum Temp °F | 11.4 | 10.8 | 19.7 | 27.9 | 35.4 | 42.1 | 47.9 | 46.4 | 38.7 | 30.1 | 22.1 | 13.6 | 28.8 |
| Mean Temp °F | 21.6 | 22.6 | 31.1 | 40.8 | 50.0 | 58.1 | 65.2 | 63.7 | 55.1 | 44.4 | 32.5 | 23.6 | 42.4 |
| Days Max Temp ≥ 90 °F | 0 | 0 | 0 | 0 | 0 | 0 | 2 | 1 | 0 | 0 | 0 | 0 | 3 |
| Days Max Temp ≤ 32 °F | 16 | 10 | 3 | 0 | 0 | 0 | 0 | 0 | 0 | 0 | 4 | 14 | 47 |
| Days Min Temp ≤ 32 °F | 30 | 28 | 30 | 23 | 10 | 2 | 0 | 0 | 6 | 20 | 27 | 30 | 206 |
| Days Min Temp ≤ 0 °F | 6 | 6 | 2 | 0 | 0 | 0 | 0 | 0 | 0 | 0 | 1 | 3 | 18 |
| Heating Degree Days | 1339 | 1190 | 1043 | 718 | 457 | 212 | 48 | 73 | 295 | 632 | 968 | 1277 | 8252 |
| Cooling Degree Days | 0 | 0 | 0 | 0 | 0 | 16 | 60 | 44 | 3 | 0 | 0 | 0 | 123 |
| Total Precipitation (") | 0.96 | 0.80 | 1.00 | 1.14 | 1.28 | 1.06 | 0.79 | 0.88 | 1.16 | 1.20 | 1.15 | 1.01 | 12.43 |
| Days ≥ 0.1" Precip | 3 | 3 | 3 | 4 | 5 | 3 | 2 | 2 | 3 | 3 | 3 | 4 | 38 |
| Total Snowfall (") | *9.0* | 7.6 | *5.8* | 3.9 | 0.7 | 0.0 | 0.0 | 0.0 | 0.5 | 1.7 | 6.5 | 11.1 | 46.8 |
| Days ≥ 1" Snow Depth | *25* | *21* | *9* | 1 | 0 | 0 | 0 | 0 | 0 | 1 | *5* | 19 | 81 |

## LEVAN *Juab County* ELEVATION 5305 ft LAT/LONG 39° 33 ' N / 111° 52 ' W

| | JAN | FEB | MAR | APR | MAY | JUN | JUL | AUG | SEP | OCT | NOV | DEC | YEAR |
|---|---|---|---|---|---|---|---|---|---|---|---|---|---|
| Maximum Temp °F | 37.4 | 43.9 | 53.1 | 61.9 | 71.8 | 82.6 | 90.3 | 88.5 | 79.5 | 66.6 | 50.7 | 38.7 | 63.8 |
| Minimum Temp °F | 14.3 | 19.5 | 26.5 | 32.5 | 40.3 | 48.4 | 56.0 | 54.3 | 45.4 | 35.0 | 25.5 | 16.2 | 34.5 |
| Mean Temp °F | 25.8 | 31.7 | 39.8 | 47.2 | 56.1 | 65.5 | 73.1 | 71.4 | 62.5 | 50.8 | 38.1 | 27.5 | 49.1 |
| Days Max Temp ≥ 90 °F | 0 | 0 | 0 | 0 | 0 | 7 | 19 | 15 | 3 | 0 | 0 | 0 | 44 |
| Days Max Temp ≤ 32 °F | 9 | 3 | 1 | 0 | 0 | 0 | 0 | 0 | 0 | 0 | 1 | 8 | 22 |
| Days Min Temp ≤ 32 °F | 30 | 27 | 24 | 16 | 4 | 0 | 0 | 0 | 2 | 11 | 24 | 30 | 168 |
| Days Min Temp ≤ 0 °F | 4 | 1 | 0 | 0 | 0 | 0 | 0 | 0 | 0 | 0 | 0 | 2 | 7 |
| Heating Degree Days | 1207 | 933 | 773 | 527 | 279 | 77 | 3 | 8 | 122 | 434 | 800 | 1155 | 6318 |
| Cooling Degree Days | 0 | 0 | 0 | 1 | 10 | 111 | 261 | 235 | 65 | 3 | 0 | 0 | 686 |
| Total Precipitation (") | 1.25 | 1.24 | 1.60 | 1.39 | 1.46 | 0.82 | 0.82 | 0.94 | 1.31 | 1.46 | 1.22 | 1.34 | 14.85 |
| Days ≥ 0.1" Precip | 4 | 4 | 5 | 4 | 4 | 2 | 3 | 3 | 3 | 4 | 4 | 4 | 44 |
| Total Snowfall (") | 13.9 | 9.4 | 7.8 | 3.8 | 1.5 | 0.0 | 0.0 | 0.0 | 0.7 | 2.5 | 7.1 | 13.6 | 60.3 |
| Days ≥ 1" Snow Depth | na | na | *5* | *1* | 0 | 0 | 0 | 0 | 0 | 1 | *4* | na | na |

## LOA *Wayne County* ELEVATION 7054 ft LAT/LONG 38° 24' N / 111° 39' W

| | JAN | FEB | MAR | APR | MAY | JUN | JUL | AUG | SEP | OCT | NOV | DEC | YEAR |
|---|---|---|---|---|---|---|---|---|---|---|---|---|---|
| Maximum Temp °F | 39.0 | 43.0 | 49.9 | 58.4 | 67.2 | 77.2 | 82.7 | 80.1 | 72.9 | 62.4 | 48.7 | 39.9 | 60.1 |
| Minimum Temp °F | 8.0 | 13.0 | 19.8 | 25.6 | 33.4 | 40.3 | 47.1 | 45.7 | 37.4 | 27.4 | 17.5 | 9.5 | 27.1 |
| Mean Temp °F | 23.6 | 28.0 | 34.9 | 42.0 | 50.3 | 58.7 | 64.9 | 62.9 | 55.2 | 44.9 | 33.1 | 24.7 | 43.6 |
| Days Max Temp ≥ 90 °F | 0 | 0 | 0 | 0 | 0 | 1 | 2 | 1 | 0 | 0 | 0 | 0 | 4 |
| Days Max Temp ≤ 32 °F | 7 | 3 | 1 | 0 | 0 | 0 | 0 | 0 | 0 | 0 | 1 | 7 | 19 |
| Days Min Temp ≤ 32 °F | 31 | 28 | 30 | 25 | 13 | 3 | 0 | 0 | 7 | 23 | 28 | 30 | 218 |
| Days Min Temp ≤ 0 °F | 7 | 4 | 1 | 0 | 0 | 0 | 0 | 0 | 0 | 0 | 1 | 6 | 19 |
| Heating Degree Days | 1279 | 1038 | 928 | 682 | 447 | 192 | 39 | 81 | 289 | 616 | 949 | 1243 | 7783 |
| Cooling Degree Days | 0 | 0 | 0 | 0 | 0 | 13 | 43 | 26 | 2 | 0 | 0 | ***0*** | 84 |
| Total Precipitation (") | 0.40 | 0.32 | 0.56 | 0.40 | 0.81 | 0.54 | 1.08 | 1.42 | 0.91 | 0.68 | 0.41 | 0.35 | 7.88 |
| Days ≥ 0.1" Precip | 2 | 1 | 2 | 2 | 3 | 2 | 4 | 4 | 3 | 2 | 1 | 2 | 28 |
| Total Snowfall (") | 5.8 | 4.3 | 5.0 | ***1.3*** | 0.5 | 0.0 | 0.0 | 0.0 | 0.0 | 1.1 | 3.3 | ***5.4*** | 26.7 |
| Days ≥ 1" Snow Depth | ***8*** | na | ***1*** | ***0*** | 0 | 0 | 0 | 0 | 0 | 0 | ***2*** | ***7*** | na |

## LOGAN RADIO KVNU *Cache County* ELEVATION 4505 ft LAT/LONG 41° 46' N / 111° 50' W

| | JAN | FEB | MAR | APR | MAY | JUN | JUL | AUG | SEP | OCT | NOV | DEC | YEAR |
|---|---|---|---|---|---|---|---|---|---|---|---|---|---|
| Maximum Temp °F | 30.4 | 36.7 | 48.1 | 58.8 | 69.1 | 79.3 | 88.7 | 87.3 | 76.4 | 63.1 | 45.5 | 32.9 | 59.7 |
| Minimum Temp °F | 11.4 | 15.9 | 25.6 | 33.0 | 40.7 | 47.6 | 54.0 | 52.5 | 43.2 | 32.9 | 24.0 | 13.8 | 32.9 |
| Mean Temp °F | 20.9 | 26.3 | 36.9 | 45.9 | 55.0 | 63.5 | 71.4 | 69.9 | 59.8 | 48.0 | 34.8 | 23.4 | 46.3 |
| Days Max Temp ≥ 90 °F | 0 | 0 | 0 | 0 | 0 | 4 | 16 | 13 | 1 | 0 | 0 | 0 | 34 |
| Days Max Temp ≤ 32 °F | 16 | 9 | 2 | 0 | 0 | 0 | 0 | 0 | 0 | 0 | 3 | 14 | 44 |
| Days Min Temp ≤ 32 °F | 30 | 27 | 26 | 15 | 3 | 0 | 0 | 0 | 2 | 14 | 26 | 30 | 173 |
| Days Min Temp ≤ 0 °F | 8 | 4 | 1 | 0 | 0 | 0 | 0 | 0 | 0 | 0 | 1 | 4 | 18 |
| Heating Degree Days | 1360 | 1087 | 865 | 567 | 312 | 107 | 9 | 21 | 175 | 520 | 901 | 1282 | 7206 |
| Cooling Degree Days | 0 | 0 | 0 | 0 | 9 | 69 | 205 | 188 | 29 | 0 | 0 | 0 | 500 |
| Total Precipitation (") | 1.10 | 1.27 | 1.71 | 1.82 | 1.72 | 1.36 | 0.79 | 0.94 | 1.52 | 1.76 | 1.56 | 1.28 | 16.83 |
| Days ≥ 0.1" Precip | 3 | 4 | 5 | 6 | 5 | 3 | 2 | 3 | 3 | 4 | 5 | 4 | 47 |
| Total Snowfall (") | na | na | na | na | ***0.0*** | 0.0 | 0.0 | 0.0 | 0.0 | ***0.4*** | na | na | na |
| Days ≥ 1" Snow Depth | na | na | na | na | 0 | 0 | 0 | 0 | 0 | ***0*** | na | na | na |

## LOGAN UT STATE UNIV *Cache County* ELEVATION 4783 ft LAT/LONG 41° 44' N / 111° 49' W

| | JAN | FEB | MAR | APR | MAY | JUN | JUL | AUG | SEP | OCT | NOV | DEC | YEAR |
|---|---|---|---|---|---|---|---|---|---|---|---|---|---|
| Maximum Temp °F | 31.7 | 37.0 | 47.0 | 56.8 | 67.2 | 77.3 | 86.3 | 85.2 | 74.3 | 61.1 | 44.6 | 33.3 | 58.5 |
| Minimum Temp °F | 16.0 | 19.5 | 28.4 | 35.9 | 44.0 | 51.7 | 59.0 | 57.9 | 48.5 | 38.4 | 27.8 | 17.5 | 37.1 |
| Mean Temp °F | 23.9 | 28.3 | 37.7 | 46.4 | 55.7 | 64.5 | 72.7 | 71.6 | 61.4 | 49.8 | 36.2 | 25.4 | 47.8 |
| Days Max Temp ≥ 90 °F | 0 | 0 | 0 | 0 | 0 | 3 | 10 | 8 | 1 | 0 | 0 | 0 | 22 |
| Days Max Temp ≤ 32 °F | 16 | 8 | 2 | 0 | 0 | 0 | 0 | 0 | 0 | 0 | 3 | 14 | 43 |
| Days Min Temp ≤ 32 °F | 29 | 25 | 21 | 10 | 2 | 0 | 0 | 0 | 1 | 6 | 20 | 29 | 143 |
| Days Min Temp ≤ 0 °F | 4 | 1 | 0 | 0 | 0 | 0 | 0 | 0 | 0 | 0 | 0 | 2 | 7 |
| Heating Degree Days | 1268 | 1030 | 839 | 553 | 297 | 98 | 9 | 15 | 149 | 465 | 856 | 1221 | 6800 |
| Cooling Degree Days | 0 | 0 | 0 | 2 | 18 | 100 | 251 | 240 | 53 | 1 | 0 | 0 | 665 |
| Total Precipitation (") | 1.38 | 1.59 | 2.07 | 2.12 | 2.04 | 1.47 | 0.89 | 0.98 | 1.59 | 1.99 | 1.78 | 1.63 | 19.53 |
| Days ≥ 0.1" Precip | 5 | 5 | 6 | 6 | 6 | 4 | 2 | 2 | 4 | 4 | 5 | 5 | 54 |
| Total Snowfall (") | 12.8 | 12.1 | 10.3 | 5.8 | 0.9 | 0.0 | 0.0 | 0.0 | 0.1 | 1.9 | 8.3 | 16.5 | 68.7 |
| Days ≥ 1" Snow Depth | 25 | 19 | 8 | 2 | 0 | 0 | 0 | 0 | 0 | 1 | 6 | 22 | 83 |

## MANTI *Sanpete County* ELEVATION 5682 ft LAT/LONG 39° 16' N / 111° 39' W

| | JAN | FEB | MAR | APR | MAY | JUN | JUL | AUG | SEP | OCT | NOV | DEC | YEAR |
|---|---|---|---|---|---|---|---|---|---|---|---|---|---|
| Maximum Temp °F | 36.8 | 42.1 | 51.1 | 59.8 | 69.3 | 79.6 | 86.2 | 84.3 | 75.6 | 63.9 | 48.7 | 38.1 | 61.3 |
| Minimum Temp °F | 14.6 | 19.2 | 26.2 | 32.2 | 39.8 | 47.6 | 54.6 | 52.9 | 44.5 | 34.6 | 24.8 | 16.3 | 33.9 |
| Mean Temp °F | 25.8 | 30.7 | 38.7 | 46.1 | 54.6 | 63.6 | 70.4 | 68.6 | 60.1 | 49.3 | 36.8 | 27.2 | 47.7 |
| Days Max Temp ≥ 90 °F | 0 | 0 | 0 | 0 | 0 | 3 | 8 | 5 | 0 | 0 | 0 | 0 | 16 |
| Days Max Temp ≤ 32 °F | 9 | 4 | 1 | 0 | 0 | 0 | 0 | 0 | 0 | 0 | 2 | 8 | 24 |
| Days Min Temp ≤ 32 °F | 30 | 27 | 26 | 16 | 5 | 0 | 0 | 0 | 2 | 11 | 25 | 30 | 172 |
| Days Min Temp ≤ 0 °F | 3 | 1 | 0 | 0 | 0 | 0 | 0 | 0 | 0 | 0 | 0 | 2 | 6 |
| Heating Degree Days | 1210 | 962 | 810 | 562 | 319 | 96 | 5 | 16 | 164 | 481 | 840 | 1166 | 6631 |
| Cooling Degree Days | 0 | 0 | 0 | 0 | 3 | 68 | 182 | 146 | 26 | 0 | 0 | 0 | 425 |
| Total Precipitation (") | 1.00 | 0.98 | 1.46 | 1.33 | 1.34 | 0.80 | 0.80 | 0.96 | 1.25 | 1.45 | 1.14 | 1.00 | 13.51 |
| Days ≥ 0.1" Precip | 3 | 3 | 5 | 5 | 4 | 2 | 3 | 3 | 3 | 4 | 3 | 3 | 41 |
| Total Snowfall (") | 12.0 | 9.6 | 9.2 | 4.6 | 1.4 | 0.0 | 0.0 | 0.0 | 0.4 | 2.1 | 8.0 | 11.9 | 59.2 |
| Days ≥ 1" Snow Depth | 23 | 17 | 4 | 0 | 0 | 0 | 0 | 0 | 0 | 0 | 5 | 18 | 67 |

## MARYSVALE *Piute County* ELEVATION 5843 ft LAT/LONG 38° 27' N / 112° 14' W

| | JAN | FEB | MAR | APR | MAY | JUN | JUL | AUG | SEP | OCT | NOV | DEC | YEAR |
|---|---|---|---|---|---|---|---|---|---|---|---|---|---|
| Maximum Temp °F | *41.0* | *46.4* | *53.7* | *62.8* | *72.0* | *83.1* | *88.5* | 86.4 | 78.5 | 66.8 | 51.7 | 42.5 | 64.5 |
| Minimum Temp °F | *13.4* | *19.0* | *25.2* | *30.1* | *37.2* | *44.0* | *50.7* | 49.1 | 41.0 | 31.4 | 22.7 | 14.9 | 31.6 |
| Mean Temp °F | *27.2* | *32.7* | *39.5* | *46.5* | *54.6* | *63.6* | *69.6* | 67.8 | 59.7 | 49.1 | 37.3 | 28.7 | 48.0 |
| Days Max Temp ≥ 90 °F | *0* | *0* | *0* | *0* | *0* | *6* | *14* | 9 | 1 | 0 | 0 | 0 | 30 |
| Days Max Temp ≤ 32 °F | *6* | *2* | *1* | *0* | *0* | *0* | *0* | 0 | 0 | 0 | 1 | 5 | 15 |
| Days Min Temp ≤ 32 °F | *30* | *26* | *25* | *19* | *8* | *1* | *0* | 0 | *4* | 18 | *26* | *29* | 186 |
| Days Min Temp ≤ 0 °F | *4* | *1* | *0* | *0* | *0* | *0* | *0* | 0 | 0 | 0 | 0 | 3 | 8 |
| Heating Degree Days | *1164* | *906* | *785* | *550* | *317* | *89* | *5* | 16 | 168 | 485 | 825 | 1118 | 6428 |
| Cooling Degree Days | 0 | 0 | 0 | 0 | 3 | 57 | 148 | 107 | 17 | 0 | 0 | 0 | 332 |
| Total Precipitation (") | *0.53* | *0.49* | *0.80* | *0.59* | *0.72* | *0.39* | *0.87* | 1.14 | 0.72 | 0.84 | 0.61 | 0.54 | 8.24 |
| Days ≥ 0.1" Precip | *2* | *2* | *3* | *2* | *3* | *1* | *3* | 4 | 3 | 2 | *2* | 2 | 29 |
| Total Snowfall (") | na | na | na | na | *0.4* | *0.0* | *0.0* | 0.0 | *0.0* | *0.1* | *1.9* | na | na |
| Days ≥ 1" Snow Depth | na | *3* | *1* | *0* | *0* | *0* | *0* | 0 | *0* | *0* | na | *5* | na |

## MEXICAN HAT *San Juan County* ELEVATION 4252 ft LAT/LONG 37° 9' N / 109° 52' W

| | JAN | FEB | MAR | APR | MAY | JUN | JUL | AUG | SEP | OCT | NOV | DEC | YEAR |
|---|---|---|---|---|---|---|---|---|---|---|---|---|---|
| Maximum Temp °F | 43.5 | 52.0 | 61.3 | 70.8 | 80.9 | 92.5 | 97.7 | 94.9 | 86.5 | 73.5 | 57.6 | 45.0 | 71.4 |
| Minimum Temp °F | 19.2 | 25.2 | 31.7 | 38.8 | 48.2 | 57.2 | 65.2 | 63.2 | 52.9 | 39.9 | 29.2 | 20.4 | 40.9 |
| Mean Temp °F | 31.4 | 38.6 | 46.5 | 54.8 | 64.6 | 74.9 | 81.5 | 79.1 | 69.7 | 56.7 | 43.4 | 32.8 | 56.2 |
| Days Max Temp ≥ 90 °F | 0 | 0 | 0 | 0 | 5 | 21 | 29 | 26 | 11 | 0 | 0 | 0 | 92 |
| Days Max Temp ≤ 32 °F | 4 | 0 | 0 | 0 | 0 | 0 | 0 | 0 | 0 | 0 | 0 | 3 | 7 |
| Days Min Temp ≤ 32 °F | 30 | 24 | 17 | 6 | 0 | 0 | 0 | 0 | 0 | 4 | 22 | 29 | 132 |
| Days Min Temp ≤ 0 °F | 1 | 0 | 0 | 0 | 0 | 0 | 0 | 0 | 0 | 0 | 0 | 1 | 2 |
| Heating Degree Days | 1035 | 738 | 565 | 305 | 88 | 5 | 0 | 0 | 28 | 257 | 641 | 992 | 4654 |
| Cooling Degree Days | 0 | 0 | 0 | 9 | 84 | 325 | 515 | 449 | 185 | 7 | 0 | 0 | 1574 |
| Total Precipitation (") | 0.58 | 0.54 | 0.50 | 0.40 | 0.46 | 0.21 | 0.71 | 0.70 | 0.66 | 0.93 | 0.55 | 0.63 | 6.87 |
| Days ≥ 0.1" Precip | 2 | 2 | 1 | 2 | 1 | 1 | 2 | 2 | 2 | 2 | 2 | 2 | 21 |
| Total Snowfall (") | na | na | *0.0* | 0.0 | 0.0 | 0.0 | 0.0 | 0.0 | 0.0 | 0.0 | 0.1 | na | na |
| Days ≥ 1" Snow Depth | na | na | *0* | 0 | 0 | 0 | 0 | 0 | 0 | 0 | 0 | na | na |

## MOAB *Grand County* ELEVATION 3973 ft LAT/LONG 38° 36' N / 109° 36' W

| | JAN | FEB | MAR | APR | MAY | JUN | JUL | AUG | SEP | OCT | NOV | DEC | YEAR |
|---|---|---|---|---|---|---|---|---|---|---|---|---|---|
| Maximum Temp °F | 42.3 | 51.7 | 63.0 | 72.4 | 82.7 | 93.6 | 99.4 | 97.2 | 88.0 | 74.5 | 57.7 | 45.2 | 72.3 |
| Minimum Temp °F | 18.8 | 25.5 | 35.1 | 42.2 | 50.4 | 57.9 | 64.3 | 63.4 | 53.2 | 40.9 | 30.2 | 21.6 | 42.0 |
| Mean Temp °F | 30.6 | 38.6 | 49.1 | 57.3 | 66.6 | 75.8 | 81.9 | 80.3 | 70.6 | 57.8 | 44.0 | 33.4 | 57.2 |
| Days Max Temp ≥ 90 °F | 0 | 0 | 0 | 1 | 7 | 22 | 30 | 28 | 14 | 1 | 0 | 0 | 103 |
| Days Max Temp ≤ 32 °F | 5 | 1 | 0 | 0 | 0 | 0 | 0 | 0 | 0 | 0 | 0 | 2 | 8 |
| Days Min Temp ≤ 32 °F | 29 | 23 | 13 | 3 | 0 | 0 | 0 | 0 | 0 | 4 | 19 | 28 | 119 |
| Days Min Temp ≤ 0 °F | 1 | 0 | 0 | 0 | 0 | 0 | 0 | 0 | 0 | 0 | 0 | 0 | 1 |
| Heating Degree Days | 1059 | 739 | 488 | 245 | 65 | 3 | 0 | 0 | 22 | 235 | 624 | 971 | 4451 |
| Cooling Degree Days | 0 | 0 | 1 | 29 | 132 | 367 | 532 | 502 | 210 | 19 | 0 | 0 | 1792 |
| Total Precipitation (") | 0.62 | 0.45 | 0.84 | 0.93 | 0.79 | 0.42 | 0.87 | 0.83 | 0.65 | 1.18 | 0.76 | 0.68 | 9.02 |
| Days ≥ 0.1" Precip | 2 | 2 | 3 | 2 | 2 | 1 | 2 | 2 | 2 | 3 | 2 | 2 | 25 |
| Total Snowfall (") | na | na | 0.3 | 0.0 | 0.0 | 0.0 | 0.0 | 0.0 | 0.0 | 0.0 | 0.0 | *1.4* | na |
| Days ≥ 1" Snow Depth | na | na | 0 | 0 | 0 | 0 | 0 | 0 | 0 | 0 | 0 | na | na |

## MODENA *Iron County* ELEVATION 5476 ft LAT/LONG 37° 48' N / 113° 54' W

| | JAN | FEB | MAR | APR | MAY | JUN | JUL | AUG | SEP | OCT | NOV | DEC | YEAR |
|---|---|---|---|---|---|---|---|---|---|---|---|---|---|
| Maximum Temp °F | 42.1 | 47.8 | 55.5 | 64.4 | 74.0 | 85.0 | 91.1 | 88.7 | 80.9 | 68.9 | 53.5 | 43.0 | 66.2 |
| Minimum Temp °F | 14.6 | 19.7 | 24.6 | 29.5 | 37.4 | 45.2 | 52.8 | 51.7 | 43.0 | 31.8 | 22.2 | 14.6 | 32.3 |
| Mean Temp °F | 28.4 | 33.6 | 40.1 | 47.0 | 55.7 | 65.1 | 72.0 | 70.3 | 62.0 | 50.4 | 37.8 | 28.8 | 49.3 |
| Days Max Temp ≥ 90 °F | 0 | 0 | 0 | 0 | 1 | 9 | 21 | 14 | 4 | 0 | 0 | 0 | 49 |
| Days Max Temp ≤ 32 °F | 5 | 1 | 0 | 0 | 0 | 0 | 0 | 0 | 0 | 0 | 0 | 5 | 11 |
| Days Min Temp ≤ 32 °F | 30 | 27 | 26 | 20 | 8 | 1 | 0 | 0 | 3 | 16 | 27 | 30 | 188 |
| Days Min Temp ≤ 0 °F | 4 | 1 | 0 | 0 | 0 | 0 | 0 | 0 | 0 | 0 | 0 | 2 | 7 |
| Heating Degree Days | 1129 | 880 | 763 | 534 | 285 | 71 | 4 | 8 | 127 | 447 | 810 | 1126 | 6184 |
| Cooling Degree Days | 0 | 0 | *0* | 0 | 5 | 88 | 223 | 179 | 39 | 0 | 0 | 0 | 534 |
| Total Precipitation (") | 0.85 | 0.93 | 1.12 | 0.88 | 0.78 | 0.35 | 1.37 | 1.23 | 0.92 | 1.07 | 0.86 | 0.63 | 10.99 |
| Days ≥ 0.1" Precip | 3 | 3 | 3 | 3 | 2 | 1 | 4 | 3 | 2 | 3 | 2 | 2 | 31 |
| Total Snowfall (") | na | *5.2* | na | *2.8* | 0.9 | 0.0 | 0.0 | 0.0 | 0.0 | 0.9 | 2.6 | *5.8* | na |
| Days ≥ 1" Snow Depth | na | na | *1* | 0 | 0 | 0 | 0 | 0 | 0 | 0 | *2* | *8* | na |

## MONTICELLO *San Juan County* ELEVATION 7073 ft LAT/LONG 37° 52 ' N / 109° 20 ' W

| | JAN | FEB | MAR | APR | MAY | JUN | JUL | AUG | SEP | OCT | NOV | DEC | YEAR |
|---|---|---|---|---|---|---|---|---|---|---|---|---|---|
| Maximum Temp °F | 34.8 | 39.6 | 48.2 | 58.1 | 67.7 | 78.8 | 84.1 | 81.4 | 73.6 | 61.4 | 46.2 | 36.5 | 59.2 |
| Minimum Temp °F | 12.8 | 16.9 | 24.0 | 29.9 | 37.6 | 45.3 | 52.3 | 51.0 | 43.0 | 33.0 | 22.9 | 14.8 | 32.0 |
| Mean Temp °F | 23.8 | 28.3 | 36.1 | 44.0 | 52.7 | 62.1 | 68.2 | 66.2 | 58.4 | 47.2 | 34.6 | 25.7 | 45.6 |
| Days Max Temp ≥ 90 °F | 0 | 0 | 0 | 0 | 0 | 2 | 3 | 1 | 0 | 0 | 0 | 0 | 6 |
| Days Max Temp ≤ 32 °F | 12 | 5 | 1 | 0 | 0 | 0 | 0 | 0 | 0 | 0 | 3 | 10 | 31 |
| Days Min Temp ≤ 32 °F | 31 | 28 | 28 | 20 | 7 | 1 | 0 | 0 | 2 | 15 | 26 | 31 | 189 |
| Days Min Temp ≤ 0 °F | 4 | 1 | 0 | 0 | 0 | 0 | 0 | 0 | 0 | 0 | 0 | 2 | 7 |
| Heating Degree Days | 1269 | 1031 | 890 | 622 | 376 | 122 | 10 | 34 | 200 | 543 | 906 | 1211 | 7214 |
| Cooling Degree Days | 0 | 0 | 0 | 0 | 1 | 44 | 110 | 77 | 7 | 0 | 0 | 0 | 239 |
| Total Precipitation (") | 1.61 | 1.24 | 1.16 | 0.86 | 1.06 | 0.58 | 1.45 | 1.90 | 1.43 | 1.72 | 1.41 | 1.58 | 16.00 |
| Days ≥ 0.1" Precip | 4 | 3 | 3 | 3 | 3 | 2 | 4 | 5 | 4 | 4 | 3 | 3 | 41 |
| Total Snowfall (") | 18.4 | 12.7 | 9.1 | 2.9 | 0.7 | 0.0 | 0.0 | 0.0 | 0.0 | 1.1 | 7.3 | 15.8 | 68.0 |
| Days ≥ 1" Snow Depth | 24 | 20 | 12 | 2 | 0 | 0 | 0 | 0 | 0 | 1 | 5 | 17 | 81 |

## MORGAN COMO SPRINGS *Morgan County* ELEVATION 5072 ft LAT/LONG 41° 3 ' N / 111° 41 ' W

| | JAN | FEB | MAR | APR | MAY | JUN | JUL | AUG | SEP | OCT | NOV | DEC | YEAR |
|---|---|---|---|---|---|---|---|---|---|---|---|---|---|
| Maximum Temp °F | 34.9 | 40.5 | 50.1 | 60.0 | 70.4 | 81.0 | 89.0 | 87.4 | 77.6 | 65.0 | 47.5 | 36.4 | 61.7 |
| Minimum Temp °F | 11.2 | 14.9 | 24.3 | 30.6 | 37.2 | 43.6 | 50.2 | 48.1 | 39.4 | 30.1 | 22.0 | 12.7 | 30.4 |
| Mean Temp °F | 23.1 | 27.7 | 37.2 | 45.3 | 53.8 | 62.3 | 69.6 | 67.8 | 58.5 | 47.5 | 34.8 | 24.6 | 46.0 |
| Days Max Temp ≥ 90 °F | 0 | 0 | 0 | 0 | 0 | 5 | 17 | 13 | 2 | 0 | 0 | 0 | 37 |
| Days Max Temp ≤ 32 °F | 11 | 5 | 1 | 0 | 0 | 0 | 0 | 0 | 0 | 0 | 2 | 10 | 29 |
| Days Min Temp ≤ 32 °F | 30 | 27 | 26 | 19 | 7 | 1 | 0 | 0 | 6 | 21 | 26 | 30 | 193 |
| Days Min Temp ≤ 0 °F | 7 | 4 | 1 | 0 | 0 | 0 | 0 | 0 | 0 | 0 | 1 | 5 | 18 |
| Heating Degree Days | 1292 | 1046 | 854 | 584 | 340 | 117 | 9 | 27 | 203 | 535 | 898 | 1246 | 7151 |
| Cooling Degree Days | 0 | 0 | 0 | 0 | 1 | 47 | 157 | 131 | 16 | 0 | 0 | 0 | 352 |
| Total Precipitation (") | 1.75 | 1.89 | 1.78 | 2.26 | 1.95 | 1.25 | 0.72 | 0.94 | 1.49 | 1.89 | 1.97 | 1.75 | 19.64 |
| Days ≥ 0.1" Precip | 5 | 5 | 6 | 6 | 6 | 3 | 2 | 2 | 3 | 5 | 6 | 5 | 54 |
| Total Snowfall (") | ***16.1*** | ***15.5*** | ***6.6*** | ***6.1*** | 1.5 | 0.0 | 0.0 | 0.0 | 0.1 | 1.8 | 10.1 | 19.9 | 77.7 |
| Days ≥ 1" Snow Depth | na | na | na | na | 0 | 0 | 0 | 0 | 0 | 0 | ***5*** | na | na |

## MORONI *Sanpete County* ELEVATION 5535 ft LAT/LONG 39° 32 ' N / 111° 35 ' W

| | JAN | FEB | MAR | APR | MAY | JUN | JUL | AUG | SEP | OCT | NOV | DEC | YEAR |
|---|---|---|---|---|---|---|---|---|---|---|---|---|---|
| Maximum Temp °F | 35.6 | 41.7 | 51.6 | 61.1 | 70.8 | 81.7 | 88.9 | 86.9 | 78.8 | 65.2 | 48.8 | 37.5 | 62.4 |
| Minimum Temp °F | 10.5 | 15.8 | 23.4 | 28.5 | 35.9 | 42.3 | 49.2 | 48.0 | 39.7 | 30.4 | 21.5 | 12.3 | 29.8 |
| Mean Temp °F | 23.1 | 28.7 | 37.5 | 44.8 | 53.4 | 62.0 | 69.1 | 67.4 | 59.3 | 47.8 | 35.2 | 24.9 | 46.1 |
| Days Max Temp ≥ 90 °F | 0 | 0 | 0 | 0 | 0 | 6 | 15 | 11 | 2 | 0 | 0 | 0 | 34 |
| Days Max Temp ≤ 32 °F | 10 | 4 | 0 | 0 | 0 | 0 | 0 | 0 | 0 | 0 | 2 | 8 | 24 |
| Days Min Temp ≤ 32 °F | 31 | 28 | 28 | 22 | ***9*** | 1 | 0 | 0 | 5 | 20 | 28 | 31 | 203 |
| Days Min Temp ≤ 0 °F | 6 | 2 | 0 | 0 | 0 | 0 | 0 | 0 | 0 | 0 | 0 | 4 | 12 |
| Heating Degree Days | 1292 | 1017 | 844 | 599 | 353 | 120 | 9 | 24 | 179 | 527 | 889 | 1236 | 7089 |
| Cooling Degree Days | 0 | 0 | 0 | 0 | 1 | 43 | 138 | 108 | 14 | 0 | 0 | 0 | 304 |
| Total Precipitation (") | 0.86 | 0.79 | 0.91 | 0.60 | 0.82 | 0.56 | 0.71 | 0.76 | 0.96 | 1.05 | 0.86 | 0.96 | 9.84 |
| Days ≥ 0.1" Precip | na | ***2*** | 3 | 2 | 3 | 2 | 2 | 2 | 3 | 3 | 2 | 2 | na |
| Total Snowfall (") | 12.3 | 8.7 | 6.4 | 2.5 | 1.4 | 0.0 | 0.0 | 0.0 | 0.4 | 1.6 | 7.2 | 12.7 | 53.2 |
| Days ≥ 1" Snow Depth | na | na | na | na | 0 | 0 | 0 | 0 | 0 | ***0*** | na | na | na |

## MOUNTAIN DELL DAM *Salt Lake County* ELEVATION 5505 ft LAT/LONG 40° 45 ' N / 111° 44 ' W

| | JAN | FEB | MAR | APR | MAY | JUN | JUL | AUG | SEP | OCT | NOV | DEC | YEAR |
|---|---|---|---|---|---|---|---|---|---|---|---|---|---|
| Maximum Temp °F | 37.6 | 42.2 | 49.4 | 58.3 | 68.3 | 79.3 | 88.0 | 85.8 | 75.9 | ***62.7*** | ***47.6*** | ***38.1*** | 61.1 |
| Minimum Temp °F | 13.7 | 16.6 | 23.8 | 30.6 | 37.7 | 44.9 | 52.0 | 50.4 | 42.3 | ***32.7*** | ***24.3*** | ***15.7*** | 32.1 |
| Mean Temp °F | 25.7 | 29.5 | 36.6 | 44.4 | 53.0 | 62.1 | 70.0 | 68.1 | 59.1 | ***47.8*** | ***36.0*** | ***26.9*** | 46.6 |
| Days Max Temp ≥ 90 °F | 0 | 0 | 0 | 0 | 0 | 4 | ***14*** | 8 | 1 | 0 | 0 | 0 | 27 |
| Days Max Temp ≤ 32 °F | 9 | 4 | 1 | 0 | 0 | 0 | 0 | 0 | 0 | 0 | 2 | 8 | 24 |
| Days Min Temp ≤ 32 °F | 30 | 27 | 26 | 18 | 7 | 1 | 0 | 0 | 3 | 14 | 25 | 29 | 180 |
| Days Min Temp ≤ 0 °F | 5 | 2 | 1 | 0 | 0 | 0 | 0 | 0 | 0 | 0 | 0 | 3 | 11 |
| Heating Degree Days | 1213 | 997 | 872 | 612 | 367 | 130 | 8 | 22 | 191 | ***528*** | ***864*** | ***1175*** | 6979 |
| Cooling Degree Days | ***0*** | ***0*** | ***0*** | ***0*** | ***3*** | ***61*** | ***181*** | ***146*** | ***28*** | na | na | na | na |
| Total Precipitation (") | 1.86 | 1.81 | 2.39 | 2.45 | 2.47 | 1.47 | 1.19 | 1.30 | 2.17 | 2.54 | 2.19 | ***2.29*** | 24.13 |
| Days ≥ 0.1" Precip | 6 | ***5*** | 7 | 6 | 6 | 4 | 3 | 3 | 4 | 5 | ***5*** | ***6*** | 60 |
| Total Snowfall (") | ***24.0*** | ***17.6*** | ***15.0*** | na | ***0.8*** | ***0.2*** | 0.0 | 0.0 | 0.5 | ***2.3*** | ***11.7*** | ***23.4*** | na |
| Days ≥ 1" Snow Depth | ***30*** | ***25*** | ***12*** | ***3*** | ***0*** | 0 | 0 | 0 | 0 | ***1*** | ***8*** | ***23*** | 102 |

## MYTON *Duchesne County* ELEVATION 5082 ft LAT/LONG 40° 12' N / 110° 4' W

| | JAN | FEB | MAR | APR | MAY | JUN | JUL | AUG | SEP | OCT | NOV | DEC | YEAR |
|---|---|---|---|---|---|---|---|---|---|---|---|---|---|
| Maximum Temp °F | 28.9 | 36.3 | 51.3 | 63.1 | 72.4 | 83.0 | 89.2 | 87.1 | 78.0 | 64.2 | 46.6 | 33.3 | 61.1 |
| Minimum Temp °F | 2.1 | 8.4 | 22.6 | 31.3 | 39.9 | 47.9 | 54.0 | 52.3 | 42.9 | 31.5 | 19.6 | 7.0 | 30.0 |
| Mean Temp °F | 15.5 | 22.4 | 37.0 | 47.2 | 56.2 | 65.5 | 71.6 | 69.8 | 60.5 | 47.9 | 33.1 | 20.2 | 45.6 |
| Days Max Temp ≥ 90 °F | 0 | 0 | 0 | 0 | 0 | 8 | *17* | *12* | 2 | 0 | 0 | 0 | 39 |
| Days Max Temp ≤ 32 °F | 19 | 10 | 1 | 0 | 0 | 0 | 0 | 0 | 0 | 0 | 3 | *13* | 46 |
| Days Min Temp ≤ 32 °F | 31 | 28 | 28 | *17* | 4 | 0 | 0 | 0 | *2* | *17* | 29 | 31 | 187 |
| Days Min Temp ≤ 0 °F | *13* | *6* | 0 | 0 | 0 | 0 | 0 | 0 | 0 | 0 | 0 | *6* | 25 |
| Heating Degree Days | 1528 | 1199 | 861 | 526 | 273 | 72 | 5 | 14 | 157 | 524 | 950 | 1375 | 7484 |
| Cooling Degree Days | 0 | 0 | 0 | 0 | 7 | 97 | 219 | 180 | 28 | 0 | 0 | 0 | 531 |
| Total Precipitation (") | 0.38 | 0.35 | 0.55 | 0.62 | 0.81 | 0.63 | 0.60 | 0.67 | 0.66 | 0.92 | 0.37 | 0.29 | 6.85 |
| Days ≥ 0.1" Precip | 1 | 1 | *2* | 2 | 2 | 2 | 2 | 2 | 2 | 2 | *1* | 1 | 20 |
| Total Snowfall (") | *3.5* | *3.3* | na | *0.5* | 0.0 | 0.0 | 0.0 | 0.0 | 0.0 | *0.6* | na | 3.9 | na |
| Days ≥ 1" Snow Depth | na | na | na | *0* | 0 | 0 | 0 | 0 | 0 | *0* | na | na | na |

## NATURAL BRIDGES N M *San Juan County* ELEVATION 6506 ft LAT/LONG 37° 37' N / 109° 59' W

| | JAN | FEB | MAR | APR | MAY | JUN | JUL | AUG | SEP | OCT | NOV | DEC | YEAR |
|---|---|---|---|---|---|---|---|---|---|---|---|---|---|
| Maximum Temp °F | 39.5 | 44.7 | 51.5 | 61.7 | 72.5 | 84.1 | 89.4 | 86.4 | 77.6 | 64.6 | 49.5 | 40.4 | 63.5 |
| Minimum Temp °F | 17.4 | 22.2 | 27.8 | 33.8 | 42.8 | 52.4 | 58.8 | 56.9 | 49.4 | 38.6 | 27.7 | 19.1 | 37.2 |
| Mean Temp °F | 28.5 | 33.5 | 39.6 | 47.8 | 57.7 | 68.3 | 74.1 | 71.7 | 63.5 | 51.6 | 38.6 | 29.8 | 50.4 |
| Days Max Temp ≥ 90 °F | 0 | 0 | 0 | 0 | 0 | 7 | 17 | 10 | 1 | 0 | 0 | 0 | 35 |
| Days Max Temp ≤ 32 °F | 6 | 2 | 0 | 0 | 0 | 0 | 0 | 0 | 0 | 0 | 1 | 5 | 14 |
| Days Min Temp ≤ 32 °F | 30 | 26 | 23 | 13 | 3 | 0 | 0 | 0 | 0 | 7 | 21 | 30 | 153 |
| Days Min Temp ≤ 0 °F | 1 | 0 | 0 | 0 | 0 | 0 | 0 | 0 | 0 | 0 | 0 | 1 | 2 |
| Heating Degree Days | 1124 | 883 | 780 | 510 | 233 | 43 | 2 | 6 | 95 | 408 | 784 | 1085 | 5953 |
| Cooling Degree Days | 0 | 0 | 0 | 1 | 12 | 148 | 265 | 210 | 57 | 1 | 0 | 0 | 694 |
| Total Precipitation (") | 0.97 | 0.85 | 1.13 | 0.77 | 0.76 | 0.49 | 1.33 | 1.64 | 1.19 | 1.41 | 1.04 | 1.09 | 12.67 |
| Days ≥ 0.1" Precip | 3 | 3 | 4 | 2 | 2 | 1 | 4 | 4 | 3 | 3 | 3 | 3 | 35 |
| Total Snowfall (") | 12.1 | 7.5 | 7.1 | 3.1 | 0.2 | 0.0 | 0.0 | 0.0 | 0.0 | 0.9 | 4.9 | 10.9 | 46.7 |
| Days ≥ 1" Snow Depth | *22* | 14 | 5 | 1 | 0 | 0 | 0 | 0 | 0 | 0 | 3 | 16 | 61 |

## NEOLA *Duchesne County* ELEVATION 6004 ft LAT/LONG 40° 25' N / 110° 3' W

| | JAN | FEB | MAR | APR | MAY | JUN | JUL | AUG | SEP | OCT | NOV | DEC | YEAR |
|---|---|---|---|---|---|---|---|---|---|---|---|---|---|
| Maximum Temp °F | 30.2 | 36.4 | 48.3 | 59.3 | 68.2 | 77.9 | 84.4 | 82.2 | 73.6 | 60.6 | 43.6 | 32.2 | 58.1 |
| Minimum Temp °F | 7.1 | 11.2 | 22.9 | 30.8 | 39.7 | 47.7 | 54.0 | 52.5 | 43.5 | 33.1 | 20.9 | 10.3 | 31.1 |
| Mean Temp °F | 18.6 | 23.8 | 35.7 | 45.1 | 54.0 | 62.9 | 69.2 | 67.4 | 58.6 | 46.8 | 32.3 | 21.3 | 44.6 |
| Days Max Temp ≥ 90 °F | 0 | 0 | 0 | 0 | 0 | 1 | 3 | 1 | 0 | 0 | 0 | 0 | 5 |
| Days Max Temp ≤ 32 °F | 18 | 8 | 2 | 0 | 0 | 0 | 0 | 0 | 0 | 0 | 3 | 15 | 46 |
| Days Min Temp ≤ 32 °F | 31 | 28 | 28 | 18 | 5 | 0 | 0 | 0 | 2 | 13 | 28 | 31 | 184 |
| Days Min Temp ≤ 0 °F | 8 | 5 | 0 | 0 | 0 | 0 | 0 | 0 | 0 | 0 | 0 | 5 | 18 |
| Heating Degree Days | 1431 | 1156 | 903 | 591 | 337 | 109 | 10 | 25 | 199 | 556 | 976 | 1348 | 7641 |
| Cooling Degree Days | 0 | 0 | 0 | 0 | 2 | 63 | *160* | 119 | 15 | 0 | 0 | 0 | 359 |
| Total Precipitation (") | 0.53 | 0.47 | 0.64 | 0.70 | 1.13 | 0.77 | 0.70 | 0.79 | 0.90 | 1.19 | 0.59 | 0.50 | 8.91 |
| Days ≥ 0.1" Precip | 2 | 2 | 2 | 2 | 3 | 2 | 2 | 3 | 3 | 3 | 2 | 2 | 28 |
| Total Snowfall (") | na | na | na | *0.7* | *0.4* | 0.0 | 0.0 | 0.0 | 0.0 | 1.0 | na | na | na |
| Days ≥ 1" Snow Depth | na | na | na | *0* | *0* | 0 | 0 | 0 | 0 | 0 | *2* | na | na |

## NEPHI *Juab County* ELEVATION 5134 ft LAT/LONG 39° 42' N / 111° 50' W

| | JAN | FEB | MAR | APR | MAY | JUN | JUL | AUG | SEP | OCT | NOV | DEC | YEAR |
|---|---|---|---|---|---|---|---|---|---|---|---|---|---|
| Maximum Temp °F | 39.4 | 45.1 | 54.2 | 63.2 | 73.3 | 84.5 | 92.5 | 90.4 | 80.8 | 67.5 | 51.5 | 40.5 | 65.2 |
| Minimum Temp °F | 16.7 | 21.4 | 28.0 | 33.7 | 41.6 | 49.4 | 57.0 | 55.4 | 46.3 | 35.5 | 26.5 | 17.9 | 35.8 |
| Mean Temp °F | 28.1 | 33.3 | 41.1 | 48.5 | 57.5 | 67.0 | 74.7 | 72.9 | 63.6 | 51.6 | 39.0 | 29.2 | 50.5 |
| Days Max Temp ≥ 90 °F | 0 | 0 | 0 | 0 | 0 | 11 | 23 | 19 | 4 | 0 | 0 | 0 | 57 |
| Days Max Temp ≤ 32 °F | 7 | 2 | 1 | 0 | 0 | 0 | 0 | 0 | 0 | 0 | 1 | 6 | 17 |
| Days Min Temp ≤ 32 °F | 30 | 25 | 22 | 14 | 4 | 0 | 0 | 0 | 1 | 11 | 23 | 29 | 159 |
| Days Min Temp ≤ 0 °F | 3 | 1 | 0 | 0 | 0 | 0 | 0 | 0 | 0 | 0 | 0 | 1 | 5 |
| Heating Degree Days | 1139 | 888 | 734 | 490 | 245 | 63 | 2 | 7 | 106 | 412 | 774 | 1101 | 5961 |
| Cooling Degree Days | 0 | 0 | 0 | 2 | 19 | 137 | 292 | 264 | 68 | 2 | 0 | 0 | 784 |
| Total Precipitation (") | 1.27 | 1.23 | 1.70 | 1.46 | 1.39 | 0.79 | 0.85 | 1.05 | 1.18 | 1.46 | 1.40 | 1.32 | 15.10 |
| Days ≥ 0.1" Precip | 4 | 4 | 5 | 5 | 5 | 2 | 3 | 3 | 3 | 4 | 4 | 4 | 46 |
| Total Snowfall (") | *13.6* | na | *7.6* | *3.4* | 1.0 | 0.0 | 0.0 | 0.0 | 0.0 | 1.2 | *7.2* | na | na |
| Days ≥ 1" Snow Depth | na | na | na | *0* | 0 | 0 | 0 | 0 | 0 | *0* | *4* | na | na |

## NEW HARMONY *Washington County* ELEVATION 5282 ft LAT/LONG 37° 29' N / 113° 18' W

| | JAN | FEB | MAR | APR | MAY | JUN | JUL | AUG | SEP | OCT | NOV | DEC | YEAR |
|---|---|---|---|---|---|---|---|---|---|---|---|---|---|
| Maximum Temp °F | 44.5 | 49.0 | 55.1 | 63.4 | 73.0 | 83.4 | 88.6 | 86.4 | 79.1 | 68.0 | 54.4 | 45.4 | 65.9 |
| Minimum Temp °F | 20.5 | 24.2 | 29.2 | 34.3 | 42.0 | 50.9 | 58.1 | 56.9 | 48.8 | 38.5 | 28.3 | 21.2 | 37.7 |
| Mean Temp °F | 32.6 | 36.6 | 42.1 | 48.9 | 57.5 | 67.2 | 73.4 | 71.7 | 64.0 | 53.3 | 41.4 | 33.3 | 51.8 |
| Days Max Temp ≥ 90 °F | 0 | 0 | 0 | 0 | 0 | 6 | 14 | 9 | 1 | 0 | 0 | 0 | 30 |
| Days Max Temp ≤ 32 °F | 2 | 1 | 0 | 0 | 0 | 0 | 0 | 0 | 0 | 0 | 0 | 3 | 6 |
| Days Min Temp ≤ 32 °F | 30 | 25 | 22 | 12 | 3 | 0 | 0 | 0 | 1 | 6 | 21 | 29 | 149 |
| Days Min Temp ≤ 0 °F | 1 | 0 | 0 | 0 | 0 | 0 | 0 | 0 | 0 | 0 | 0 | 1 | 2 |
| Heating Degree Days | 999 | 794 | 702 | 477 | 238 | 47 | 1 | 4 | 87 | 358 | 702 | 975 | 5384 |
| Cooling Degree Days | 0 | 0 | 0 | 0 | 17 | 134 | 273 | 231 | 69 | 2 | 0 | 0 | 726 |
| Total Precipitation (") | 2.14 | 2.43 | 2.43 | 1.24 | 0.90 | 0.46 | 1.42 | 1.61 | 1.41 | 1.31 | 1.80 | 1.80 | 18.95 |
| Days ≥ 0.1" Precip | 4 | 4 | 5 | 3 | 3 | 1 | 3 | 4 | 3 | 3 | 3 | 3 | 39 |
| Total Snowfall (") | na | na | *5.4* | *1.5* | 0.0 | 0.0 | 0.0 | 0.0 | 0.0 | *0.0* | na | na | na |
| Days ≥ 1" Snow Depth | na | na | *2* | *0* | 0 | 0 | 0 | 0 | 0 | *0* | na | na | na |

## OAK CITY *Millard County* ELEVATION 4905 ft LAT/LONG 39° 23' N / 112° 20' W

| | JAN | FEB | MAR | APR | MAY | JUN | JUL | AUG | SEP | OCT | NOV | DEC | YEAR |
|---|---|---|---|---|---|---|---|---|---|---|---|---|---|
| Maximum Temp °F | 39.1 | 46.0 | 55.0 | 63.4 | 74.4 | 85.9 | 94.1 | 92.2 | 82.3 | 68.5 | 51.6 | 40.2 | 66.1 |
| Minimum Temp °F | 18.5 | 23.9 | 30.6 | 36.2 | 45.0 | 54.3 | 62.8 | 60.8 | 51.2 | 39.7 | 28.6 | 19.4 | 39.3 |
| Mean Temp °F | 28.8 | 35.0 | 42.8 | 49.8 | 59.7 | 70.1 | 78.5 | 76.5 | 66.8 | 54.1 | 40.1 | 29.8 | 52.7 |
| Days Max Temp ≥ 90 °F | 0 | 0 | 0 | 0 | 1 | 11 | 26 | 22 | 6 | 0 | 0 | 0 | 66 |
| Days Max Temp ≤ 32 °F | 8 | 2 | 0 | 0 | 0 | 0 | 0 | 0 | 0 | 0 | 1 | 7 | 18 |
| Days Min Temp ≤ 32 °F | 29 | 24 | 19 | 11 | 2 | 0 | 0 | 0 | 1 | 6 | 21 | 29 | 142 |
| Days Min Temp ≤ 0 °F | 2 | 0 | 0 | 0 | 0 | 0 | 0 | 0 | 0 | 0 | 0 | 1 | 3 |
| Heating Degree Days | 1115 | 842 | 680 | 454 | 197 | 42 | 1 | 2 | 69 | 339 | 739 | 1084 | 5564 |
| Cooling Degree Days | 0 | 0 | 0 | 4 | 45 | 213 | 419 | 382 | 135 | 7 | 0 | 0 | 1205 |
| Total Precipitation (") | 1.14 | 1.04 | 1.42 | 1.39 | 1.41 | 0.69 | 0.51 | 0.81 | 1.08 | 1.51 | 1.28 | 1.23 | 13.51 |
| Days ≥ 0.1" Precip | 3 | 3 | 4 | 4 | 4 | 2 | 2 | 3 | 3 | 4 | 4 | 3 | 39 |
| Total Snowfall (") | 11.2 | 7.5 | 7.6 | 4.9 | 1.6 | 0.0 | 0.0 | 0.0 | 0.4 | 1.9 | 7.0 | 11.3 | 53.4 |
| Days ≥ 1" Snow Depth | na | *10* | *2* | *1* | 0 | 0 | 0 | 0 | 0 | 0 | 4 | na | na |

## OGDEN PIONEER P H *Weber County* ELEVATION 4354 ft LAT/LONG 41° 15' N / 111° 57' W

| | JAN | FEB | MAR | APR | MAY | JUN | JUL | AUG | SEP | OCT | NOV | DEC | YEAR |
|---|---|---|---|---|---|---|---|---|---|---|---|---|---|
| Maximum Temp °F | 36.5 | 43.1 | 52.4 | 61.5 | 71.6 | 82.0 | 90.5 | 88.7 | 78.0 | 65.0 | 49.0 | 37.9 | 63.0 |
| Minimum Temp °F | 19.9 | 23.9 | 32.0 | 38.8 | 47.1 | 55.6 | 62.8 | 61.1 | 51.5 | 40.6 | 30.3 | 21.6 | 40.4 |
| Mean Temp °F | 28.2 | 33.5 | 42.2 | 50.2 | 59.4 | 68.8 | 76.7 | 74.9 | 64.8 | 52.8 | 39.7 | 29.8 | 51.8 |
| Days Max Temp ≥ 90 °F | 0 | 0 | 0 | 0 | 0 | 7 | 20 | 16 | 3 | 0 | 0 | 0 | 46 |
| Days Max Temp ≤ 32 °F | 10 | 3 | 0 | 0 | 0 | 0 | 0 | 0 | 0 | 0 | 1 | 8 | 22 |
| Days Min Temp ≤ 32 °F | 28 | 24 | 16 | 7 | 1 | 0 | 0 | 0 | 0 | 4 | 18 | 28 | 126 |
| Days Min Temp ≤ 0 °F | 1 | 0 | 0 | 0 | 0 | 0 | 0 | 0 | 0 | 0 | 0 | 1 | 2 |
| Heating Degree Days | 1132 | 882 | 700 | 444 | 210 | 51 | 2 | 7 | 96 | 374 | 753 | 1085 | 5736 |
| Cooling Degree Days | 0 | 0 | 0 | 9 | 49 | 190 | 361 | 342 | 106 | 5 | 0 | 0 | 1062 |
| Total Precipitation (") | 2.07 | 2.08 | 2.26 | 2.49 | 2.57 | 1.52 | 0.91 | 1.03 | 1.78 | 2.21 | 2.15 | 2.04 | 23.11 |
| Days ≥ 0.1" Precip | 6 | 5 | 6 | 6 | 6 | 3 | 2 | 2 | 3 | 4 | 6 | 6 | 55 |
| Total Snowfall (") | na | na | na | *0.6* | 0.0 | 0.0 | 0.0 | 0.0 | 0.0 | 0.3 | *3.1* | na | na |
| Days ≥ 1" Snow Depth | 21 | 13 | *3* | 0 | 0 | 0 | 0 | 0 | 0 | 0 | *4* | 12 | 53 |

## OGDEN SUGAR FACTORY *Weber County* ELEVATION 4281 ft LAT/LONG 41° 14' N / 112° 2' W

| | JAN | FEB | MAR | APR | MAY | JUN | JUL | AUG | SEP | OCT | NOV | DEC | YEAR |
|---|---|---|---|---|---|---|---|---|---|---|---|---|---|
| Maximum Temp °F | 35.8 | 42.8 | 52.6 | 62.2 | 72.3 | 82.9 | 91.6 | 89.8 | 79.0 | 65.6 | 49.1 | 37.7 | 63.5 |
| Minimum Temp °F | 17.8 | 22.3 | 30.5 | 37.0 | 44.9 | 53.1 | 59.6 | 57.6 | 47.9 | 37.7 | 28.2 | 20.0 | 38.1 |
| Mean Temp °F | 26.9 | 32.6 | 41.6 | 49.6 | 58.6 | 68.0 | 75.6 | 73.7 | 63.5 | 51.6 | 38.7 | 28.9 | 50.8 |
| Days Max Temp ≥ 90 °F | 0 | 0 | 0 | 0 | 0 | 8 | 22 | 18 | 3 | 0 | 0 | 0 | 51 |
| Days Max Temp ≤ 32 °F | 11 | 4 | 1 | 0 | 0 | 0 | 0 | 0 | 0 | 0 | 1 | 8 | 25 |
| Days Min Temp ≤ 32 °F | 29 | 25 | 19 | 9 | 1 | 0 | 0 | 0 | 1 | 7 | 21 | 29 | 141 |
| Days Min Temp ≤ 0 °F | 2 | 1 | 0 | 0 | 0 | 0 | 0 | 0 | 0 | 0 | 0 | 1 | 4 |
| Heating Degree Days | 1175 | 911 | 720 | 453 | 221 | 51 | 3 | 6 | 111 | 409 | 782 | 1112 | 5954 |
| Cooling Degree Days | 0 | 0 | 0 | 5 | 35 | 165 | 335 | 309 | 80 | 2 | 0 | 0 | 931 |
| Total Precipitation (") | 1.36 | 1.43 | 1.74 | 1.96 | 1.95 | 1.27 | 0.59 | 0.76 | 1.49 | 1.88 | 1.64 | 1.33 | 17.40 |
| Days ≥ 0.1" Precip | 4 | 4 | 4 | 5 | 5 | 3 | 2 | 2 | 3 | 4 | 4 | *4* | 44 |
| Total Snowfall (") | na | na | na | na | 0.0 | 0.0 | 0.0 | 0.0 | 0.0 | 0.0 | na | na | na |
| Days ≥ 1" Snow Depth | na | na | na | *0* | 0 | 0 | 0 | 0 | 0 | 0 | na | na | na |

## ORDERVILLE *Kane County* ELEVATION 5463 ft LAT/LONG 37° 16 ' N / 112° 38 ' W

| | JAN | FEB | MAR | APR | MAY | JUN | JUL | AUG | SEP | OCT | NOV | DEC | YEAR |
|---|---|---|---|---|---|---|---|---|---|---|---|---|---|
| Maximum Temp °F | 47.1 | 51.1 | 56.6 | 65.3 | 74.6 | 85.4 | 90.7 | 88.5 | 81.1 | 70.4 | 56.9 | 48.1 | 68.0 |
| Minimum Temp °F | 17.5 | 22.5 | 26.7 | 32.0 | 40.1 | 48.2 | 55.1 | 54.1 | 46.7 | 35.6 | 25.4 | 18.9 | 35.2 |
| Mean Temp °F | 32.3 | 36.8 | 41.8 | 48.7 | 57.4 | 66.8 | 72.9 | 71.3 | 63.9 | 53.0 | 41.2 | 33.5 | 51.6 |
| Days Max Temp ≥ 90 °F | 0 | 0 | 0 | 0 | 1 | 10 | 20 | 15 | 4 | 0 | 0 | 0 | 50 |
| Days Max Temp ≤ 32 °F | 2 | 1 | 0 | 0 | 0 | 0 | 0 | 0 | 0 | 0 | 0 | 2 | 5 |
| Days Min Temp ≤ 32 °F | 30 | 26 | 25 | 17 | 4 | 0 | 0 | 0 | 0 | 11 | 25 | 29 | 167 |
| Days Min Temp ≤ 0 °F | 2 | 1 | 0 | 0 | 0 | 0 | 0 | 0 | 0 | 0 | 0 | 2 | 5 |
| Heating Degree Days | 1006 | 789 | 712 | 484 | 239 | 47 | 1 | 4 | 80 | 369 | 706 | 970 | 5407 |
| Cooling Degree Days | 0 | 0 | 0 | 1 | 13 | 121 | 255 | 214 | 59 | 4 | 0 | 0 | 667 |
| Total Precipitation (") | 1.86 | 1.81 | 1.97 | 0.92 | 0.84 | 0.49 | 1.25 | 1.71 | 1.05 | 1.18 | 1.40 | 1.47 | 15.95 |
| Days ≥ 0.1" Precip | 4 | 4 | 5 | 2 | 3 | 2 | 3 | 4 | 3 | 3 | 3 | 4 | 40 |
| Total Snowfall (") | 12.1 | 8.5 | 4.8 | 1.7 | 0.1 | 0.0 | 0.0 | 0.0 | 0.0 | 0.5 | 3.4 | *7.5* | 38.6 |
| Days ≥ 1" Snow Depth | 12 | na | *2* | *0* | 0 | 0 | 0 | 0 | 0 | 0 | 2 | na | na |

## OURAY 4 NE *Uintah County* ELEVATION 4652 ft LAT/LONG 40° 5 ' N / 109° 41 ' W

| | JAN | FEB | MAR | APR | MAY | JUN | JUL | AUG | SEP | OCT | NOV | DEC | YEAR |
|---|---|---|---|---|---|---|---|---|---|---|---|---|---|
| Maximum Temp °F | 27.7 | 36.2 | 53.2 | 66.0 | 76.5 | 87.4 | 93.9 | 91.5 | 81.4 | 66.8 | 47.9 | *32.2* | 63.4 |
| Minimum Temp °F | 1.4 | 7.9 | 23.9 | 33.4 | 42.3 | 50.0 | 56.2 | 53.9 | 44.0 | 32.1 | 20.7 | *7.1* | 31.1 |
| Mean Temp °F | 14.6 | 22.1 | 38.6 | 49.7 | 59.5 | 68.7 | 75.1 | 72.8 | 62.7 | 49.5 | 34.3 | *19.6* | 47.3 |
| Days Max Temp ≥ 90 °F | 0 | 0 | 0 | 0 | 1 | *13* | 26 | *22* | 5 | 0 | 0 | 0 | 67 |
| Days Max Temp ≤ 32 °F | 20 | 11 | 1 | 0 | 0 | 0 | 0 | 0 | 0 | 0 | 2 | 14 | 48 |
| Days Min Temp ≤ 32 °F | 31 | 28 | 27 | *13* | 2 | 0 | 0 | 0 | 2 | *16* | 28 | 30 | 177 |
| Days Min Temp ≤ 0 °F | 15 | 8 | 1 | 0 | 0 | 0 | 0 | 0 | 0 | 0 | 0 | 7 | 31 |
| Heating Degree Days | 1573 | 1205 | 810 | 453 | 177 | 34 | 1 | 4 | 111 | *472* | 916 | *1402* | 7158 |
| Cooling Degree Days | 0 | 0 | 0 | 1 | 24 | 166 | 324 | 269 | 57 | 0 | 0 | *0* | 841 |
| Total Precipitation (") | 0.40 | 0.32 | 0.56 | 0.69 | 0.76 | 0.59 | 0.68 | 0.68 | 0.73 | 1.00 | 0.47 | 0.44 | 7.32 |
| Days ≥ 0.1" Precip | 1 | 1 | 2 | 2 | 3 | 2 | 2 | 2 | 2 | 3 | 1 | 1 | 22 |
| Total Snowfall (") | *4.9* | *3.0* | *1.6* | 0.6 | 0.0 | 0.0 | 0.0 | 0.0 | 0.0 | 0.5 | *1.8* | *4.6* | 17.0 |
| Days ≥ 1" Snow Depth | 16 | 12 | 2 | 0 | 0 | 0 | 0 | 0 | 0 | 0 | *2* | *7* | 39 |

## PANGUITCH *Garfield County* ELEVATION 6653 ft LAT/LONG 37° 49 ' N / 112° 27 ' W

| | JAN | FEB | MAR | APR | MAY | JUN | JUL | AUG | SEP | OCT | NOV | DEC | YEAR |
|---|---|---|---|---|---|---|---|---|---|---|---|---|---|
| Maximum Temp °F | 40.0 | 44.4 | 51.6 | 60.4 | 69.7 | 80.1 | 85.4 | 82.9 | 76.3 | 65.7 | 51.1 | 41.4 | 62.4 |
| Minimum Temp °F | 8.4 | 13.4 | 20.2 | 24.7 | 32.0 | 38.6 | 46.1 | 44.7 | 36.0 | 26.2 | 17.5 | 9.7 | 26.5 |
| Mean Temp °F | 24.2 | 29.0 | 35.9 | 42.5 | 50.8 | 59.4 | 65.8 | 63.8 | 56.2 | 46.0 | 34.3 | 25.5 | 44.4 |
| Days Max Temp ≥ 90 °F | 0 | 0 | 0 | 0 | 0 | 2 | 7 | 4 | 0 | 0 | 0 | 0 | 13 |
| Days Max Temp ≤ 32 °F | 5 | 3 | 1 | 0 | 0 | 0 | 0 | 0 | 0 | 0 | 1 | 5 | 15 |
| Days Min Temp ≤ 32 °F | 31 | 28 | 30 | 26 | 17 | 4 | 0 | 1 | 10 | 25 | 28 | 31 | 231 |
| Days Min Temp ≤ 0 °F | 8 | 3 | 0 | 0 | 0 | 0 | 0 | 0 | 0 | 0 | 1 | 6 | 18 |
| Heating Degree Days | 1258 | 1011 | 894 | 666 | 433 | 178 | 34 | 70 | 260 | 582 | 914 | 1216 | 7516 |
| Cooling Degree Days | 0 | 0 | 0 | 0 | 1 | 20 | 70 | 44 | 4 | 0 | 0 | 0 | 139 |
| Total Precipitation (") | 0.54 | 0.68 | 0.74 | 0.64 | 0.90 | 0.58 | 1.37 | 1.77 | 0.97 | 0.88 | 0.81 | 0.54 | 10.42 |
| Days ≥ 0.1" Precip | 2 | 2 | 3 | 2 | 3 | 2 | 4 | 5 | 3 | 2 | 2 | 2 | 32 |
| Total Snowfall (") | na | na | na | na | *0.4* | 0.0 | 0.0 | 0.0 | 0.0 | *0.2* | na | na | na |
| Days ≥ 1" Snow Depth | na | na | na | na | *0* | 0 | 0 | 0 | 0 | *0* | na | na | na |

## PAROWAN POWER PLANT *Iron County* ELEVATION 6033 ft LAT/LONG 37° 50 ' N / 112° 50 ' W

| | JAN | FEB | MAR | APR | MAY | JUN | JUL | AUG | SEP | OCT | NOV | DEC | YEAR |
|---|---|---|---|---|---|---|---|---|---|---|---|---|---|
| Maximum Temp °F | 41.9 | 46.0 | 52.2 | 60.4 | 70.2 | 81.5 | 87.4 | 85.2 | 77.9 | 66.3 | 51.6 | 42.8 | 63.6 |
| Minimum Temp °F | 14.3 | 18.8 | 25.4 | 31.6 | 39.4 | 47.5 | 54.8 | 52.9 | 44.2 | 33.6 | 23.4 | 15.7 | 33.5 |
| Mean Temp °F | 28.1 | 32.4 | 38.8 | 46.0 | 54.8 | 64.5 | 71.1 | 69.1 | 61.1 | 50.0 | 37.5 | 29.2 | 48.6 |
| Days Max Temp ≥ 90 °F | 0 | 0 | 0 | 0 | 0 | 5 | 12 | 7 | 1 | 0 | 0 | 0 | 25 |
| Days Max Temp ≤ 32 °F | 5 | 2 | 0 | 0 | 0 | 0 | 0 | 0 | 0 | 0 | 1 | 5 | 13 |
| Days Min Temp ≤ 32 °F | 30 | 27 | 27 | 17 | 5 | 0 | 0 | 0 | 2 | 13 | 26 | 30 | 177 |
| Days Min Temp ≤ 0 °F | 3 | 1 | 0 | 0 | 0 | 0 | 0 | 0 | 0 | 0 | 0 | 2 | 6 |
| Heating Degree Days | 1136 | 913 | 805 | 563 | 313 | 82 | 5 | 14 | 142 | 459 | 817 | 1102 | 6351 |
| Cooling Degree Days | 0 | 0 | 0 | 0 | 5 | 82 | 205 | 150 | 33 | 0 | 0 | 0 | 475 |
| Total Precipitation (") | 0.87 | 1.12 | 1.43 | 1.27 | 0.89 | 0.52 | 1.22 | 1.49 | 0.93 | 1.27 | 1.19 | 0.93 | 13.13 |
| Days ≥ 0.1" Precip | 3 | 3 | 4 | 4 | 3 | 2 | 3 | 4 | 2 | 3 | 3 | 3 | 37 |
| Total Snowfall (") | 12.6 | 13.1 | 11.9 | 6.6 | 1.6 | 0.1 | 0.0 | 0.0 | 0.1 | 2.5 | 10.0 | 12.8 | 71.3 |
| Days ≥ 1" Snow Depth | 18 | 12 | 5 | 2 | 0 | 0 | 0 | 0 | 0 | 1 | 6 | 14 | 58 |

## PARTOUN *Juab County* ELEVATION 4774 ft LAT/LONG 39° 37 ' N / 113° 52 ' W

| | JAN | FEB | MAR | APR | MAY | JUN | JUL | AUG | SEP | OCT | NOV | DEC | YEAR |
|---|---|---|---|---|---|---|---|---|---|---|---|---|---|
| Maximum Temp °F | 40.1 | 47.1 | 56.2 | 64.4 | 74.7 | 86.1 | 94.4 | 91.8 | 81.9 | 67.6 | 52.0 | 41.1 | 66.5 |
| Minimum Temp °F | 13.0 | 18.5 | 25.7 | 31.8 | 40.2 | 48.8 | 55.7 | 53.7 | 43.8 | 32.8 | 23.1 | 13.8 | 33.4 |
| Mean Temp °F | 26.6 | 32.9 | 41.0 | 48.1 | 57.5 | 67.5 | 75.1 | 72.8 | 62.9 | 50.3 | 37.7 | 27.5 | 50.0 |
| Days Max Temp ≥ 90 °F | 0 | 0 | 0 | 0 | 2 | 12 | 26 | 22 | 6 | 0 | 0 | 0 | 68 |
| Days Max Temp ≤ 32 °F | 7 | 2 | 0 | 0 | 0 | 0 | 0 | 0 | 0 | 0 | 1 | 6 | 16 |
| Days Min Temp ≤ 32 °F | 30 | 26 | 25 | 16 | 5 | 0 | 0 | 0 | 3 | 15 | 26 | 30 | 176 |
| Days Min Temp ≤ 0 °F | 5 | 1 | 0 | 0 | 0 | 0 | 0 | 0 | 0 | 0 | 0 | 3 | 9 |
| Heating Degree Days | 1183 | 900 | 738 | 500 | 244 | 53 | 2 | 6 | 121 | 452 | 814 | 1156 | 6169 |
| Cooling Degree Days | 0 | 0 | 0 | 0 | 18 | 148 | 309 | 256 | 63 | 0 | 0 | 0 | 794 |
| Total Precipitation (") | 0.38 | 0.42 | 0.53 | 0.69 | 0.97 | 0.56 | 0.65 | 0.62 | 0.71 | 0.74 | 0.45 | 0.34 | 7.06 |
| Days ≥ 0.1" Precip | *1* | 1 | 2 | 2 | 3 | 2 | 2 | 2 | 2 | 2 | 1 | 1 | 21 |
| Total Snowfall (") | na | na | na | na | *0.0* | 0.0 | 0.0 | 0.0 | 0.0 | *0.0* | na | na | na |
| Days ≥ 1" Snow Depth | na | na | na | na | *0* | 0 | 0 | 0 | 0 | *0* | na | na | na |

## PINE VIEW DAM *Weber County* ELEVATION 4823 ft LAT/LONG 41° 15 ' N / 111° 51 ' W

| | JAN | FEB | MAR | APR | MAY | JUN | JUL | AUG | SEP | OCT | NOV | DEC | YEAR |
|---|---|---|---|---|---|---|---|---|---|---|---|---|---|
| Maximum Temp °F | 30.6 | 35.8 | 46.0 | 56.2 | 66.7 | 77.0 | 86.0 | 84.3 | 73.9 | 61.0 | 44.0 | 31.9 | 57.8 |
| Minimum Temp °F | 7.4 | 9.8 | 21.7 | 30.2 | 37.2 | 43.6 | 50.0 | 48.7 | 40.1 | 30.6 | 22.0 | 11.3 | 29.4 |
| Mean Temp °F | 19.0 | 22.8 | 33.8 | 43.2 | 52.0 | 60.4 | 68.0 | 66.5 | 57.0 | 45.8 | 33.0 | 21.7 | 43.6 |
| Days Max Temp ≥ 90 °F | 0 | 0 | 0 | 0 | 0 | 2 | 10 | 7 | 0 | 0 | 0 | 0 | 19 |
| Days Max Temp ≤ 32 °F | 16 | 9 | 2 | 0 | 0 | 0 | 0 | 0 | 0 | 0 | 4 | 15 | 46 |
| Days Min Temp ≤ 32 °F | 31 | 28 | 29 | 20 | 7 | 1 | 0 | 0 | 4 | 20 | 28 | 31 | 199 |
| Days Min Temp ≤ 0 °F | 10 | 7 | 1 | 0 | 0 | 0 | 0 | 0 | 0 | 0 | 0 | 5 | 23 |
| Heating Degree Days | 1419 | 1184 | 959 | 646 | 398 | 159 | 22 | 39 | 241 | 589 | 953 | 1336 | 7945 |
| Cooling Degree Days | 0 | 0 | 0 | 0 | 1 | 31 | 117 | 98 | 8 | 0 | 0 | 0 | 255 |
| Total Precipitation (") | 3.27 | 3.09 | 3.16 | 2.96 | 3.08 | 1.67 | 1.07 | 1.12 | 2.03 | 2.93 | 3.36 | 3.29 | 31.03 |
| Days ≥ 0.1" Precip | 7 | 7 | 7 | 7 | 6 | 3 | 2 | 2 | 4 | 5 | 7 | 7 | 64 |
| Total Snowfall (") | 28.1 | *24.0* | 13.6 | 6.6 | 1.3 | 0.0 | 0.0 | 0.0 | 0.1 | *2.2* | 15.6 | 28.8 | 120.3 |
| Days ≥ 1" Snow Depth | 30 | 28 | 21 | 6 | 1 | 0 | 0 | 0 | 0 | 1 | 11 | 25 | 123 |

## PLEASANT GROVE *Utah County* ELEVATION 4662 ft LAT/LONG 40° 22 ' N / 111° 44 ' W

| | JAN | FEB | MAR | APR | MAY | JUN | JUL | AUG | SEP | OCT | NOV | DEC | YEAR |
|---|---|---|---|---|---|---|---|---|---|---|---|---|---|
| Maximum Temp °F | 38.0 | 44.9 | 54.2 | 62.7 | 72.7 | 82.6 | 89.9 | 88.1 | 78.9 | 66.1 | 50.5 | 39.5 | 64.0 |
| Minimum Temp °F | 19.1 | 23.4 | 30.3 | 36.3 | 43.7 | 51.7 | 59.1 | 57.0 | 48.2 | 37.9 | 29.1 | 21.1 | 38.1 |
| Mean Temp °F | 28.6 | 34.2 | 42.3 | 49.5 | 58.2 | 67.2 | 74.5 | 72.6 | 63.6 | 52.0 | 39.8 | 30.3 | 51.1 |
| Days Max Temp ≥ 90 °F | 0 | 0 | 0 | 0 | 0 | 7 | 19 | 14 | 2 | 0 | 0 | 0 | 42 |
| Days Max Temp ≤ 32 °F | 8 | 3 | 0 | 0 | 0 | 0 | 0 | 0 | 0 | 0 | 1 | 7 | 19 |
| Days Min Temp ≤ 32 °F | 29 | 24 | 20 | 10 | 2 | 0 | 0 | 0 | 1 | 7 | 21 | 29 | 143 |
| Days Min Temp ≤ 0 °F | 1 | 1 | 0 | 0 | 0 | 0 | 0 | 0 | 0 | 0 | 0 | 1 | 3 |
| Heating Degree Days | 1122 | 865 | 698 | 460 | 224 | 56 | 2 | 6 | 103 | 397 | 750 | 1069 | 5752 |
| Cooling Degree Days | 0 | 0 | 0 | 5 | 27 | 150 | 311 | 284 | 83 | 2 | 0 | 0 | 862 |
| Total Precipitation (") | 1.60 | 1.65 | 1.75 | 1.64 | 1.67 | 0.86 | 0.80 | 0.89 | 1.31 | 1.77 | 1.50 | 1.55 | 16.99 |
| Days ≥ 0.1" Precip | 5 | 5 | 5 | 5 | 4 | 2 | 2 | 3 | 3 | 4 | 4 | 5 | 47 |
| Total Snowfall (") | 10.9 | 7.5 | 3.9 | 3.8 | 0.2 | 0.0 | 0.0 | 0.0 | 0.1 | 0.9 | 5.6 | 10.7 | 43.6 |
| Days ≥ 1" Snow Depth | 17 | 7 | 1 | 0 | 0 | 0 | 0 | 0 | 0 | 0 | 3 | 11 | 39 |

## RICHFIELD RADIO KSVC *Sevier County* ELEVATION 5299 ft LAT/LONG 38° 46 ' N / 112° 5 ' W

| | JAN | FEB | MAR | APR | MAY | JUN | JUL | AUG | SEP | OCT | NOV | DEC | YEAR |
|---|---|---|---|---|---|---|---|---|---|---|---|---|---|
| Maximum Temp °F | 40.4 | 46.8 | 55.3 | 63.6 | 72.6 | 82.9 | 89.0 | 87.1 | 79.4 | 67.7 | 52.3 | 41.9 | 64.9 |
| Minimum Temp °F | 14.4 | 19.4 | 25.7 | 30.8 | 38.2 | 45.5 | 52.4 | 50.7 | 41.7 | 31.4 | 22.9 | 15.3 | 32.4 |
| Mean Temp °F | 27.4 | 33.1 | 40.5 | 47.2 | 55.4 | 64.2 | 70.8 | 69.0 | 60.6 | 49.6 | 37.6 | 28.7 | 48.7 |
| Days Max Temp ≥ 90 °F | 0 | 0 | 0 | 0 | 0 | 6 | 16 | 11 | 2 | 0 | 0 | 0 | 35 |
| Days Max Temp ≤ 32 °F | 7 | 2 | 0 | 0 | 0 | 0 | 0 | 0 | 0 | 0 | 1 | 6 | 16 |
| Days Min Temp ≤ 32 °F | 30 | 26 | 25 | 18 | 6 | 1 | 0 | 0 | 4 | 18 | 26 | 30 | 184 |
| Days Min Temp ≤ 0 °F | 4 | 1 | 0 | 0 | 0 | 0 | 0 | 0 | 0 | 0 | 0 | 3 | 8 |
| Heating Degree Days | 1158 | 892 | 751 | 526 | 293 | 82 | 4 | 13 | 152 | 471 | 815 | 1120 | 6277 |
| Cooling Degree Days | 0 | 0 | 0 | 0 | 4 | 74 | 184 | 144 | 29 | 0 | 0 | 0 | 435 |
| Total Precipitation (") | 0.57 | 0.52 | 0.78 | 0.66 | 0.98 | 0.57 | 0.78 | 0.67 | 0.86 | 1.01 | 0.67 | 0.55 | 8.62 |
| Days ≥ 0.1" Precip | 2 | 2 | 3 | 2 | 3 | 2 | 2 | 2 | 2 | 3 | 2 | 2 | 27 |
| Total Snowfall (") | na | na | *2.1* | *0.8* | 0.4 | 0.0 | 0.0 | 0.0 | 0.2 | 0.8 | *1.8* | na | na |
| Days ≥ 1" Snow Depth | na | *3* | *1* | *0* | 0 | 0 | 0 | 0 | 0 | 0 | *2* | *6* | na |

## RICHMOND *Cache County* ELEVATION 4524 ft LAT/LONG 41° 56 ' N / 111° 49 ' W

| | JAN | FEB | MAR | APR | MAY | JUN | JUL | AUG | SEP | OCT | NOV | DEC | YEAR |
|---|---|---|---|---|---|---|---|---|---|---|---|---|---|
| Maximum Temp °F | 31.4 | 37.8 | 48.7 | 59.2 | 69.8 | 80.6 | 90.1 | 88.5 | 77.2 | 62.9 | 45.2 | 33.0 | 60.4 |
| Minimum Temp °F | 13.5 | 17.4 | 25.6 | 32.4 | 39.4 | 46.5 | 52.5 | 51.9 | 43.2 | 33.8 | 24.9 | 15.2 | 33.0 |
| Mean Temp °F | 22.5 | 27.6 | 37.2 | 45.8 | 54.6 | 63.6 | 71.3 | 70.2 | 60.2 | 48.4 | 35.1 | 24.1 | 46.7 |
| Days Max Temp ≥ 90 °F | 0 | 0 | 0 | 0 | 0 | 5 | 19 | 15 | 2 | 0 | 0 | 0 | 41 |
| Days Max Temp ≤ 32 °F | 15 | 8 | 1 | 0 | 0 | 0 | 0 | 0 | 0 | 0 | 3 | 14 | 41 |
| Days Min Temp ≤ 32 °F | 30 | 27 | 25 | 16 | 5 | 0 | 0 | 0 | 3 | 13 | 25 | 30 | 174 |
| Days Min Temp ≤ 0 °F | 6 | 3 | 0 | 0 | 0 | 0 | 0 | 0 | 0 | 0 | 1 | 3 | 13 |
| Heating Degree Days | 1312 | 1048 | 855 | 569 | 320 | 107 | 10 | 19 | 170 | 508 | 891 | 1260 | 7069 |
| Cooling Degree Days | 0 | 0 | 0 | 1 | 5 | 79 | 218 | 211 | 39 | 0 | 0 | 0 | 553 |
| Total Precipitation (") | 1.48 | 1.49 | 2.01 | 2.20 | 2.21 | 1.36 | 0.93 | 1.09 | 1.48 | 1.94 | 1.77 | 1.56 | 19.52 |
| Days ≥ 0.1" Precip | 5 | 5 | 6 | 6 | 6 | 4 | 2 | 2 | 3 | 4 | 5 | 5 | 53 |
| Total Snowfall (") | 12.6 | 11.3 | 10.5 | 6.3 | 1.0 | 0.0 | 0.0 | 0.0 | 0.2 | 2.0 | 9.8 | 16.0 | 69.7 |
| Days ≥ 1" Snow Depth | 27 | 21 | 9 | 1 | 0 | 0 | 0 | 0 | 0 | 1 | 8 | 22 | 89 |

## ROOSEVELT RADIO *Uintah County* ELEVATION 5010 ft LAT/LONG 40° 17 ' N / 109° 58 ' W

| | JAN | FEB | MAR | APR | MAY | JUN | JUL | AUG | SEP | OCT | NOV | DEC | YEAR |
|---|---|---|---|---|---|---|---|---|---|---|---|---|---|
| Maximum Temp °F | 28.5 | 36.6 | 52.7 | 64.4 | 74.7 | 84.6 | 90.8 | 88.9 | 79.4 | 65.6 | 47.7 | 33.0 | 62.2 |
| Minimum Temp °F | 3.0 | 9.1 | 23.1 | 31.4 | 40.4 | 47.9 | 54.2 | 52.4 | 42.8 | 32.1 | 20.6 | 8.0 | 30.4 |
| Mean Temp °F | 15.8 | 22.8 | 37.9 | 47.9 | 57.5 | 66.3 | 72.5 | 70.7 | 61.1 | 48.9 | 34.2 | 20.5 | 46.3 |
| Days Max Temp ≥ 90 °F | 0 | 0 | 0 | 0 | 1 | 10 | 20 | 16 | 3 | 0 | 0 | 0 | 50 |
| Days Max Temp ≤ 32 °F | 19 | 10 | 1 | 0 | 0 | 0 | 0 | 0 | 0 | 0 | 3 | 14 | 47 |
| Days Min Temp ≤ 32 °F | 31 | 28 | 28 | 17 | 4 | 0 | 0 | 0 | 3 | 17 | 28 | 31 | 187 |
| Days Min Temp ≤ 0 °F | 13 | 7 | 0 | 0 | 0 | 0 | 0 | 0 | 0 | 0 | 0 | 7 | 27 |
| Heating Degree Days | 1521 | 1184 | 832 | 506 | 233 | 57 | 3 | 8 | 142 | 493 | 918 | 1372 | 7269 |
| Cooling Degree Days | 0 | 0 | 0 | 1 | 8 | 106 | 227 | 191 | 31 | 0 | 0 | 0 | 564 |
| Total Precipitation (") | 0.54 | 0.41 | 0.55 | 0.58 | 0.82 | 0.65 | 0.50 | 0.57 | 0.63 | 1.00 | 0.43 | 0.48 | 7.16 |
| Days ≥ 0.1" Precip | 2 | 2 | 2 | 2 | 2 | 2 | 2 | 2 | 2 | 3 | 2 | 2 | 25 |
| Total Snowfall (") | na | na | na | *0.5* | 0.0 | 0.0 | 0.0 | 0.0 | 0.0 | *0.2* | *2.3* | na | na |
| Days ≥ 1" Snow Depth | na | na | *3* | *0* | 0 | 0 | 0 | 0 | 0 | 0 | *2* | na | na |

## SALINA *Sevier County* ELEVATION 5203 ft LAT/LONG 38° 57 ' N / 111° 52 ' W

| | JAN | FEB | MAR | APR | MAY | JUN | JUL | AUG | SEP | OCT | NOV | DEC | YEAR |
|---|---|---|---|---|---|---|---|---|---|---|---|---|---|
| Maximum Temp °F | 39.9 | 46.5 | 55.9 | 64.7 | 74.5 | 85.0 | 91.7 | 89.6 | 80.7 | 68.1 | 52.5 | 41.3 | 65.9 |
| Minimum Temp °F | 13.1 | 18.7 | 25.6 | 31.0 | 38.7 | 46.5 | 53.8 | 51.7 | 42.4 | 31.6 | 23.0 | 14.3 | 32.5 |
| Mean Temp °F | 26.5 | 32.6 | 40.8 | 47.9 | 56.6 | 65.8 | 72.8 | 70.7 | 61.6 | 49.9 | 37.8 | 27.9 | 49.2 |
| Days Max Temp ≥ 90 °F | 0 | 0 | 0 | 0 | 1 | 10 | 21 | 17 | 4 | 0 | 0 | 0 | 53 |
| Days Max Temp ≤ 32 °F | 7 | 2 | 0 | 0 | 0 | 0 | 0 | 0 | 0 | 0 | 1 | 6 | 16 |
| Days Min Temp ≤ 32 °F | 30 | 27 | 26 | 19 | 6 | 1 | 0 | 0 | 4 | 18 | 26 | 30 | 187 |
| Days Min Temp ≤ 0 °F | 5 | 1 | 0 | 0 | 0 | 0 | 0 | 0 | 0 | 0 | 0 | 3 | 9 |
| Heating Degree Days | 1187 | 907 | 745 | 508 | 261 | 71 | 3 | 9 | 137 | 462 | 810 | 1146 | 6246 |
| Cooling Degree Days | 0 | 0 | 0 | 0 | 5 | 94 | 210 | 172 | 33 | 0 | 0 | 0 | 514 |
| Total Precipitation (") | 0.65 | 0.62 | 1.10 | 0.98 | 1.05 | 0.60 | 0.70 | 0.79 | 1.00 | 0.99 | 0.78 | 0.72 | 9.98 |
| Days ≥ 0.1" Precip | 2 | 2 | 3 | 3 | 3 | 2 | 2 | 3 | 3 | 3 | 3 | 2 | 31 |
| Total Snowfall (") | na | na | na | na | 0.1 | 0.0 | 0.0 | 0.0 | 0.3 | 0.2 | na | na | na |
| Days ≥ 1" Snow Depth | na | na | na | na | 0 | 0 | 0 | 0 | 0 | 0 | na | na | na |

## SALT LK CITY INTL AP *Salt Lake County* ELEVATION 4252 ft LAT/LONG 40° 46 ' N / 111° 58 ' W

| | JAN | FEB | MAR | APR | MAY | JUN | JUL | AUG | SEP | OCT | NOV | DEC | YEAR |
|---|---|---|---|---|---|---|---|---|---|---|---|---|---|
| Maximum Temp °F | 36.9 | 43.8 | 53.3 | 61.7 | 72.0 | 83.4 | 92.3 | 90.2 | 79.0 | 65.4 | 49.9 | 38.0 | 63.8 |
| Minimum Temp °F | 20.3 | 24.8 | 32.6 | 38.5 | 46.3 | 55.0 | 62.9 | 61.4 | 51.5 | 40.2 | 30.4 | 21.8 | 40.5 |
| Mean Temp °F | 28.6 | 34.3 | 43.0 | 50.1 | 59.2 | 69.3 | 77.6 | 75.8 | 65.3 | 52.8 | 40.2 | 29.9 | 52.2 |
| Days Max Temp ≥ 90 °F | 0 | 0 | 0 | 0 | 0 | 9 | 23 | 19 | 4 | 0 | 0 | 0 | 55 |
| Days Max Temp ≤ 32 °F | 10 | 3 | 0 | 0 | 0 | 0 | 0 | 0 | 0 | 0 | 1 | 8 | 22 |
| Days Min Temp ≤ 32 °F | 27 | 23 | 14 | 6 | 1 | 0 | 0 | 0 | 0 | 4 | 18 | 28 | 121 |
| Days Min Temp ≤ 0 °F | 1 | 0 | 0 | 0 | 0 | 0 | 0 | 0 | 0 | 0 | 0 | 1 | 2 |
| Heating Degree Days | 1121 | 859 | 675 | 445 | 210 | 46 | 2 | 5 | 92 | 374 | 738 | 1080 | 5647 |
| Cooling Degree Days | 0 | 0 | 0 | 7 | 45 | 200 | 398 | 375 | 118 | 4 | 0 | 0 | 1147 |
| Total Precipitation (") | 1.23 | 1.24 | 1.79 | 2.04 | 1.95 | 0.85 | 0.77 | 0.86 | 1.31 | 1.58 | 1.41 | 1.33 | 16.36 |
| Days ≥ 0.1" Precip | 4 | 4 | 6 | 5 | 5 | 2 | 2 | 2 | 3 | 4 | 4 | 4 | 45 |
| Total Snowfall (") | 13.6 | 9.8 | 9.2 | 6.8 | 0.9 | 0.0 | 0.0 | 0.0 | 0.2 | 2.0 | 7.7 | 14.2 | 64.4 |
| Days ≥ 1" Snow Depth | 19 | 11 | 3 | 1 | 0 | 0 | 0 | 0 | 0 | 0 | 4 | 14 | 52 |

## SANTAQUIN CHLOR *Utah County* ELEVATION 5253 ft LAT/LONG 39° 57 ' N / 111° 48 ' W

| | JAN | FEB | MAR | APR | MAY | JUN | JUL | AUG | SEP | OCT | NOV | DEC | YEAR |
|---|---|---|---|---|---|---|---|---|---|---|---|---|---|
| Maximum Temp °F | 37.9 | 43.4 | 52.0 | 60.5 | 71.0 | 81.8 | 90.0 | 88.3 | 78.2 | 64.9 | 49.2 | 38.6 | 63.0 |
| Minimum Temp °F | 16.6 | 21.1 | 28.3 | 34.4 | 43.1 | 51.6 | 59.1 | 57.0 | 46.8 | 36.6 | 26.4 | 18.0 | 36.6 |
| Mean Temp °F | 27.3 | 32.3 | 40.1 | 47.4 | 57.2 | 66.7 | 74.6 | 72.7 | 62.5 | 50.8 | 37.9 | 28.3 | 49.8 |
| Days Max Temp ≥ 90 °F | 0 | 0 | 0 | 0 | 0 | 6 | 19 | 14 | 2 | 0 | 0 | 0 | 41 |
| Days Max Temp ≤ 32 °F | 9 | 3 | 0 | 0 | 0 | 0 | 0 | 0 | 0 | 0 | 2 | 8 | 22 |
| Days Min Temp ≤ 32 °F | 29 | 25 | ***22*** | 13 | 3 | 0 | 0 | 0 | 1 | 9 | 24 | 29 | 155 |
| Days Min Temp ≤ 0 °F | 2 | 1 | 0 | 0 | 0 | 0 | 0 | 0 | 0 | 0 | 0 | 1 | 4 |
| Heating Degree Days | 1162 | 917 | 766 | 519 | 257 | 75 | 6 | 10 | 132 | 435 | 810 | 1129 | 6218 |
| Cooling Degree Days | 0 | 0 | 0 | 4 | 25 | 125 | 288 | 271 | 66 | 2 | 0 | 0 | 781 |
| Total Precipitation (") | 1.40 | 1.42 | 1.86 | 1.91 | 1.88 | 0.86 | 0.81 | 1.20 | 1.45 | 2.09 | 1.81 | 1.52 | 18.21 |
| Days ≥ 0.1" Precip | 4 | 4 | 5 | 5 | 4 | 2 | 2 | 3 | 3 | 4 | 4 | 4 | 44 |
| Total Snowfall (") | ***12.6*** | ***10.5*** | ***7.4*** | 4.6 | 1.4 | 0.0 | 0.0 | 0.0 | 0.2 | 1.9 | 8.7 | ***13.0*** | 60.3 |
| Days ≥ 1" Snow Depth | na | na | na | ***1*** | 0 | 0 | 0 | 0 | 0 | ***0*** | na | na | na |

## SCIPIO *Millard County* ELEVATION 5315 ft LAT/LONG 39° 15 ' N / 112° 6 ' W

| | JAN | FEB | MAR | APR | MAY | JUN | JUL | AUG | SEP | OCT | NOV | DEC | YEAR |
|---|---|---|---|---|---|---|---|---|---|---|---|---|---|
| Maximum Temp °F | 38.3 | 44.7 | 53.5 | 63.2 | 72.6 | 82.3 | 89.5 | 87.4 | 79.0 | 66.7 | 50.6 | 39.6 | 64.0 |
| Minimum Temp °F | 11.4 | 17.8 | 24.9 | 29.9 | 37.6 | 45.0 | 53.6 | 51.8 | 41.4 | 30.6 | 22.1 | 12.7 | 31.6 |
| Mean Temp °F | 24.9 | 31.3 | 39.2 | 46.6 | 55.1 | 63.7 | 71.6 | 69.6 | 60.2 | 48.7 | 36.4 | 26.2 | 47.8 |
| Days Max Temp ≥ 90 °F | 0 | 0 | 0 | 0 | 0 | 6 | 17 | 12 | 2 | 0 | 0 | 0 | 37 |
| Days Max Temp ≤ 32 °F | 8 | 2 | 0 | 0 | 0 | 0 | 0 | 0 | 0 | 0 | 2 | 7 | 19 |
| Days Min Temp ≤ 32 °F | 30 | 26 | 25 | 19 | 8 | 1 | 0 | 0 | 5 | 20 | 26 | 30 | 190 |
| Days Min Temp ≤ 0 °F | 7 | 2 | 0 | 0 | 0 | 0 | 0 | 0 | 0 | 0 | 1 | 5 | 15 |
| Heating Degree Days | 1236 | 946 | 792 | 547 | 307 | 105 | 5 | 15 | 166 | 501 | 852 | 1198 | 6670 |
| Cooling Degree Days | 0 | 0 | 0 | 1 | 7 | 87 | 234 | 198 | 35 | 0 | 0 | 0 | 562 |
| Total Precipitation (") | 1.15 | 1.12 | 1.25 | 1.02 | 1.33 | 0.79 | 0.85 | 1.06 | 1.14 | 1.41 | 1.20 | 1.17 | 13.49 |
| Days ≥ 0.1" Precip | 4 | 4 | ***5*** | 4 | 4 | 2 | 3 | 3 | 3 | 4 | 4 | 4 | 44 |
| Total Snowfall (") | na | na | na | na | 0.0 | 0.0 | 0.0 | 0.0 | 0.1 | ***0.1*** | na | na | na |
| Days ≥ 1" Snow Depth | na | na | na | na | 0 | 0 | 0 | 0 | 0 | ***0*** | na | na | na |

## SCOFIELD DAM *Carbon County* ELEVATION 7634 ft LAT/LONG 39° 47 ' N / 111° 8 ' W

| | JAN | FEB | MAR | APR | MAY | JUN | JUL | AUG | SEP | OCT | NOV | DEC | YEAR |
|---|---|---|---|---|---|---|---|---|---|---|---|---|---|
| Maximum Temp °F | 27.3 | 32.1 | 38.5 | 48.0 | 58.9 | ***69.8*** | ***77.3*** | 75.1 | 66.2 | ***54.9*** | 39.5 | 29.2 | 51.4 |
| Minimum Temp °F | -1.5 | 0.6 | 12.0 | 21.6 | 30.8 | ***38.0*** | ***44.7*** | 42.6 | 34.6 | ***25.8*** | 16.3 | 2.7 | 22.3 |
| Mean Temp °F | 12.9 | 16.4 | 25.3 | 34.8 | 44.8 | ***53.9*** | ***61.0*** | 58.9 | 50.4 | ***40.4*** | 27.9 | 16.0 | 36.9 |
| Days Max Temp ≥ 90 °F | 0 | 0 | 0 | 0 | 0 | ***0*** | ***0*** | 0 | 0 | 0 | 0 | 0 | 0 |
| Days Max Temp ≤ 32 °F | 22 | 14 | 7 | 2 | 0 | ***0*** | ***0*** | 0 | 0 | 1 | 8 | 19 | 73 |
| Days Min Temp ≤ 32 °F | 31 | 28 | 31 | 28 | 19 | ***5*** | ***0*** | 1 | 10 | 27 | 30 | 31 | 241 |
| Days Min Temp ≤ 0 °F | 17 | 14 | 5 | 1 | 0 | ***0*** | ***0*** | 0 | 0 | 0 | 2 | 13 | 52 |
| Heating Degree Days | 1610 | 1365 | 1225 | 898 | 618 | ***328*** | ***126*** | 184 | 431 | ***757*** | 1106 | 1515 | 10163 |
| Cooling Degree Days | ***0*** | ***0*** | ***0*** | ***0*** | ***0*** | na | na | ***3*** | ***0*** | na | ***0*** | ***0*** | na |
| Total Precipitation (") | 1.26 | 1.11 | 1.03 | 0.83 | 1.03 | ***1.00*** | ***1.24*** | 1.46 | 1.26 | 1.26 | 1.04 | 1.17 | 13.69 |
| Days ≥ 0.1" Precip | 3 | 3 | 4 | 3 | 4 | ***3*** | ***4*** | 4 | 4 | 4 | 3 | 3 | 42 |
| Total Snowfall (") | na | ***22.1*** | ***19.7*** | na | ***2.3*** | ***0.0*** | ***0.0*** | 0.0 | ***0.5*** | ***4.0*** | na | na | na |
| Days ≥ 1" Snow Depth | 30 | 27 | 26 | ***9*** | ***2*** | ***0*** | ***0*** | 0 | 0 | ***3*** | ***12*** | 25 | 134 |

## SILVER LAKE BRIGHTON *Salt Lake County* ELEVATION 8707 ft LAT/LONG 40° 36 ' N / 111° 35 ' W

| | JAN | FEB | MAR | APR | MAY | JUN | JUL | AUG | SEP | OCT | NOV | DEC | YEAR |
|---|---|---|---|---|---|---|---|---|---|---|---|---|---|
| Maximum Temp °F | 31.1 | 33.4 | 37.5 | 44.1 | 53.1 | 63.5 | 71.7 | 70.1 | 61.5 | 50.1 | 36.6 | 31.0 | 48.6 |
| Minimum Temp °F | 8.7 | 9.4 | 14.1 | 20.2 | 28.6 | 36.6 | 44.2 | 42.7 | 35.5 | 26.1 | 15.4 | 8.9 | 24.2 |
| Mean Temp °F | 19.9 | 21.4 | 25.8 | 32.2 | 40.9 | 50.1 | 58.0 | 56.4 | 48.5 | 38.1 | 26.0 | 20.0 | 36.4 |
| Days Max Temp ≥ 90 °F | 0 | 0 | 0 | 0 | 0 | 0 | 0 | 0 | 0 | 0 | 0 | 0 | 0 |
| Days Max Temp ≤ 32 °F | 18 | 13 | 9 | 3 | 0 | 0 | 0 | 0 | 0 | 2 | 11 | 17 | 73 |
| Days Min Temp ≤ 32 °F | 31 | 28 | 31 | 28 | 21 | 8 | 1 | 1 | 9 | 24 | 29 | 31 | 242 |
| Days Min Temp ≤ 0 °F | 7 | 5 | 3 | 1 | 0 | 0 | 0 | 0 | 0 | 0 | 3 | 6 | 25 |
| Heating Degree Days | 1389 | 1225 | 1208 | 978 | 742 | 440 | 213 | 260 | 487 | 827 | 1163 | 1391 | 10323 |
| Cooling Degree Days | 0 | 0 | 0 | 0 | 0 | 0 | 3 | 2 | 0 | 0 | 0 | 0 | 5 |
| Total Precipitation (") | 4.80 | 4.79 | 5.10 | 4.26 | 3.17 | 1.62 | 1.77 | 1.96 | 2.48 | 3.69 | 5.01 | 4.62 | 43.27 |
| Days ≥ 0.1" Precip | 10 | 9 | 11 | 9 | 7 | 4 | 5 | 5 | 5 | 6 | 9 | 10 | 90 |
| Total Snowfall (") | 60.6 | 62.6 | 65.6 | 49.1 | 18.7 | 1.9 | 0.0 | 0.1 | 3.0 | 25.5 | 61.3 | 64.0 | 412.4 |
| Days ≥ 1" Snow Depth | 31 | 28 | 31 | 30 | 24 | 4 | 0 | 0 | 1 | 12 | 28 | 31 | 220 |

## SNAKE CREEK POWERHOU *Wasatch County* ELEVATION 5955 ft LAT/LONG 40° 33' N / 111° 30' W

| | JAN | FEB | MAR | APR | MAY | JUN | JUL | AUG | SEP | OCT | NOV | DEC | YEAR |
|---|---|---|---|---|---|---|---|---|---|---|---|---|---|
| Maximum Temp °F | 33.7 | 38.3 | 46.5 | 57.0 | 67.2 | 76.8 | 84.3 | 82.7 | 73.9 | 62.1 | 45.6 | 35.2 | 58.6 |
| Minimum Temp °F | 10.3 | 12.8 | 21.6 | 27.7 | 34.0 | 39.8 | 46.0 | 44.9 | 37.4 | 28.9 | 20.9 | 11.9 | 28.0 |
| Mean Temp °F | 22.0 | 25.6 | 34.1 | 42.4 | 50.6 | 58.3 | 65.2 | 63.8 | 55.7 | 45.5 | 33.3 | 23.6 | 43.3 |
| Days Max Temp ≥ 90 °F | 0 | 0 | 0 | 0 | 0 | 1 | 4 | 3 | 0 | 0 | 0 | 0 | 8 |
| Days Max Temp ≤ 32 °F | 13 | 6 | 1 | 0 | 0 | 0 | 0 | 0 | 0 | 0 | 3 | 12 | 35 |
| Days Min Temp ≤ 32 °F | 31 | 28 | 29 | 24 | 12 | 3 | 0 | 1 | 7 | 22 | 28 | 31 | 216 |
| Days Min Temp ≤ 0 °F | 7 | 4 | 1 | 0 | 0 | 0 | 0 | 0 | 0 | 0 | 1 | 5 | 18 |
| Heating Degree Days | 1326 | 1106 | 952 | 671 | 439 | 204 | 40 | 69 | 277 | 598 | 946 | 1278 | 7906 |
| Cooling Degree Days | 0 | 0 | 0 | 0 | 0 | 17 | 64 | 53 | 6 | 0 | 0 | 0 | 140 |
| Total Precipitation (") | 2.74 | 2.69 | 1.87 | 1.66 | 1.63 | 0.96 | 0.99 | 1.23 | 1.45 | 2.06 | 2.43 | 2.46 | 22.17 |
| Days ≥ 0.1" Precip | 7 | 6 | 6 | 5 | 5 | 3 | 3 | 4 | 4 | 5 | 6 | 6 | 60 |
| Total Snowfall (") | na | na | na | na | na | 0.0 | 0.0 | 0.0 | 0.1 | *1.9* | na | na | na |
| Days ≥ 1" Snow Depth | na | na | na | na | *0* | 0 | 0 | 0 | 0 | *1* | na | na | na |

## SPANISH FORK P H *Utah County* ELEVATION 4715 ft LAT/LONG 40° 5' N / 111° 36' W

| | JAN | FEB | MAR | APR | MAY | JUN | JUL | AUG | SEP | OCT | NOV | DEC | YEAR |
|---|---|---|---|---|---|---|---|---|---|---|---|---|---|
| Maximum Temp °F | 36.3 | 43.2 | 53.6 | 63.1 | 73.8 | 84.5 | 92.2 | 89.7 | 79.7 | 65.8 | 49.1 | 37.5 | 64.0 |
| Minimum Temp °F | 19.5 | 23.8 | 30.9 | 37.1 | 45.0 | 52.3 | 59.4 | 57.7 | 49.3 | 39.8 | 30.0 | 21.3 | 38.8 |
| Mean Temp °F | 27.9 | 33.6 | 42.3 | 50.2 | 59.4 | 68.4 | 75.8 | 73.7 | 64.6 | 52.8 | 39.6 | 29.4 | 51.5 |
| Days Max Temp ≥ 90 °F | 0 | 0 | 0 | 0 | 1 | 10 | 23 | 17 | 3 | 0 | 0 | 0 | 54 |
| Days Max Temp ≤ 32 °F | 11 | 4 | 0 | 0 | 0 | 0 | 0 | 0 | 0 | 0 | 2 | 9 | 26 |
| Days Min Temp ≤ 32 °F | 28 | 24 | 18 | 8 | 1 | 0 | 0 | 0 | 0 | 5 | 18 | 28 | 130 |
| Days Min Temp ≤ 0 °F | 1 | 0 | 0 | 0 | 0 | 0 | 0 | 0 | 0 | 0 | 0 | 1 | 2 |
| Heating Degree Days | 1143 | 882 | 697 | 441 | 199 | 48 | 1 | 5 | 87 | 372 | 757 | 1096 | 5728 |
| Cooling Degree Days | 0 | 0 | 0 | 4 | 37 | 172 | 337 | 297 | 86 | 2 | 0 | 0 | 935 |
| Total Precipitation (") | 1.62 | 1.83 | 2.17 | 2.07 | 2.04 | 1.14 | 0.97 | 1.19 | 1.51 | 2.16 | 2.24 | 1.94 | 20.88 |
| Days ≥ 0.1" Precip | 5 | 5 | 6 | 6 | 5 | 3 | 3 | 3 | 4 | 5 | 5 | 5 | 55 |
| Total Snowfall (") | na | na | na | *3.2* | 0.2 | 0.0 | 0.0 | 0.0 | 0.0 | 0.6 | *8.7* | na | na |
| Days ≥ 1" Snow Depth | na | na | na | *1* | 0 | 0 | 0 | 0 | 0 | 0 | *5* | na | na |

## ST GEORGE *Washington County* ELEVATION 2760 ft LAT/LONG 37° 6' N / 113° 35' W

| | JAN | FEB | MAR | APR | MAY | JUN | JUL | AUG | SEP | OCT | NOV | DEC | YEAR |
|---|---|---|---|---|---|---|---|---|---|---|---|---|---|
| Maximum Temp °F | 53.9 | 60.6 | 68.0 | 76.8 | 86.7 | 97.0 | 102.2 | 100.0 | 92.8 | 80.5 | 64.5 | 53.9 | 78.1 |
| Minimum Temp °F | 28.1 | 32.4 | 38.6 | 44.9 | 54.1 | 62.5 | 69.4 | 67.4 | 58.4 | 46.0 | 35.1 | 28.0 | 47.1 |
| Mean Temp °F | 40.9 | 46.5 | 53.3 | 60.9 | 70.4 | 79.8 | 85.9 | 83.7 | 75.6 | 63.3 | 49.8 | 41.0 | 62.6 |
| Days Max Temp ≥ 90 °F | 0 | 0 | 0 | 2 | *13* | 25 | 31 | 30 | 21 | 4 | 0 | 0 | 126 |
| Days Max Temp ≤ 32 °F | 0 | 0 | 0 | 0 | 0 | 0 | 0 | 0 | 0 | 0 | 0 | 0 | 0 |
| Days Min Temp ≤ 32 °F | 23 | 13 | 5 | 1 | 0 | 0 | 0 | 0 | 0 | 1 | *11* | 24 | 78 |
| Days Min Temp ≤ 0 °F | 0 | 0 | 0 | 0 | 0 | 0 | 0 | 0 | 0 | 0 | 0 | 0 | 0 |
| Heating Degree Days | 741 | 514 | 356 | 155 | 32 | 0 | 0 | 0 | 7 | 106 | 449 | 744 | 3104 |
| Cooling Degree Days | 0 | 0 | 3 | 58 | 238 | 488 | 673 | 609 | 352 | 59 | *0* | 0 | 2480 |
| Total Precipitation (") | 1.19 | 1.04 | 1.13 | 0.53 | 0.40 | 0.17 | 0.56 | 0.72 | 0.52 | 0.63 | 0.87 | 0.80 | 8.56 |
| Days ≥ 0.1" Precip | 3 | 3 | 4 | 2 | 1 | 1 | 2 | 2 | 1 | 2 | 2 | 3 | 26 |
| Total Snowfall (") | *1.3* | *0.3* | 0.0 | 0.0 | 0.0 | 0.0 | 0.0 | 0.0 | 0.0 | 0.0 | 0.1 | 0.8 | 2.5 |
| Days ≥ 1" Snow Depth | *0* | 0 | 0 | 0 | 0 | 0 | 0 | 0 | 0 | 0 | 0 | 0 | 0 |

## THIOKOL PLANT 78 *Box Elder County* ELEVATION 4603 ft LAT/LONG 41° 43' N / 112° 26' W

| | JAN | FEB | MAR | APR | MAY | JUN | JUL | AUG | SEP | OCT | NOV | DEC | YEAR |
|---|---|---|---|---|---|---|---|---|---|---|---|---|---|
| Maximum Temp °F | 32.9 | 39.5 | 49.9 | 60.5 | 71.3 | 82.0 | 91.4 | 89.8 | 78.7 | 64.3 | 46.9 | 34.9 | 61.8 |
| Minimum Temp °F | 11.4 | 15.9 | 25.0 | 31.1 | 39.0 | 46.8 | 53.7 | 52.2 | 42.0 | 31.3 | 22.8 | 13.5 | 32.1 |
| Mean Temp °F | 22.2 | 27.7 | 37.5 | 45.9 | 55.2 | 64.4 | 72.6 | 71.0 | 60.4 | 47.8 | 34.9 | 24.2 | 47.0 |
| Days Max Temp ≥ 90 °F | 0 | 0 | 0 | 0 | 0 | 6 | 21 | 18 | 3 | 0 | 0 | 0 | 48 |
| Days Max Temp ≤ 32 °F | 14 | 6 | 1 | 0 | 0 | 0 | 0 | 0 | 0 | 0 | 2 | 11 | 34 |
| Days Min Temp ≤ 32 °F | 30 | 27 | 26 | 18 | 6 | 0 | 0 | 0 | 3 | 18 | 27 | 30 | 185 |
| Days Min Temp ≤ 0 °F | 7 | 4 | 0 | 0 | 0 | 0 | 0 | 0 | 0 | 0 | 1 | 4 | 16 |
| Heating Degree Days | 1321 | 1046 | 846 | 568 | 306 | 94 | 7 | 12 | 166 | 526 | 896 | 1258 | 7046 |
| Cooling Degree Days | 0 | 0 | 0 | 1 | 9 | 89 | 245 | 222 | 41 | 0 | 0 | 0 | 607 |
| Total Precipitation (") | 1.02 | 1.06 | 1.21 | 1.42 | 1.60 | 1.25 | 0.77 | 0.94 | 1.18 | 1.34 | 1.30 | 0.96 | 14.05 |
| Days ≥ 0.1" Precip | 3 | 3 | 4 | 5 | 4 | 3 | 2 | 2 | 3 | 3 | 4 | 3 | 39 |
| Total Snowfall (") | *7.3* | 5.3 | 1.0 | *0.1* | 0.0 | 0.0 | 0.0 | 0.0 | 0.0 | 0.0 | *3.6* | *8.2* | 25.5 |
| Days ≥ 1" Snow Depth | 24 | 15 | 5 | 0 | 0 | 0 | 0 | 0 | 0 | 0 | 4 | 16 | 64 |

## THOMPSON *Grand County* ELEVATION 5154 ft LAT/LONG 38° 58 ' N / 109° 43 ' W

| | JAN | FEB | MAR | APR | MAY | JUN | JUL | AUG | SEP | OCT | NOV | DEC | YEAR |
|---|---|---|---|---|---|---|---|---|---|---|---|---|---|
| Maximum Temp °F | 36.5 | 45.2 | 55.6 | 65.7 | 75.5 | 86.6 | 92.8 | 90.5 | 81.3 | 68.6 | 51.4 | ***39.8*** | 65.8 |
| Minimum Temp °F | 14.9 | 22.7 | 31.1 | 38.0 | 47.4 | 57.0 | 64.0 | 62.1 | 52.4 | 40.6 | 28.6 | ***18.0*** | 39.7 |
| Mean Temp °F | 25.7 | 34.0 | 43.3 | 51.9 | 61.4 | 71.8 | 78.5 | 76.3 | 66.9 | 54.7 | 40.0 | ***28.9*** | 52.8 |
| Days Max Temp ≥ 90 °F | 0 | 0 | 0 | 0 | 1 | 12 | 24 | 18 | 4 | 0 | 0 | 0 | 59 |
| Days Max Temp ≤ 32 °F | 11 | 2 | 0 | 0 | 0 | 0 | 0 | 0 | 0 | 0 | 0 | 6 | 19 |
| Days Min Temp ≤ 32 °F | 30 | 24 | 17 | 7 | 1 | 0 | 0 | 0 | 0 | 4 | 19 | 29 | 131 |
| Days Min Temp ≤ 0 °F | 3 | 1 | 0 | 0 | 0 | 0 | 0 | 0 | 0 | 0 | 0 | 1 | 5 |
| Heating Degree Days | 1211 | 869 | 665 | 388 | 156 | 25 | 1 | 1 | 56 | 320 | 743 | ***1112*** | 5547 |
| Cooling Degree Days | 0 | 0 | 0 | 5 | 48 | 251 | 404 | 351 | 109 | 3 | 0 | ***0*** | 1171 |
| Total Precipitation (") | 0.91 | 0.58 | 1.01 | 0.76 | 1.01 | 0.53 | 0.80 | 0.89 | 0.86 | 1.19 | 0.73 | 0.66 | 9.93 |
| Days ≥ 0.1" Precip | 3 | 2 | 3 | 2 | 3 | 2 | 2 | 2 | 2 | 3 | 2 | 2 | 28 |
| Total Snowfall (") | ***6.5*** | na | ***1.1*** | 0.0 | 0.0 | 0.0 | 0.0 | 0.0 | 0.0 | 0.2 | na | na | na |
| Days ≥ 1" Snow Depth | na | na | ***1*** | 0 | 0 | 0 | 0 | 0 | 0 | 0 | na | na | na |

## TIMPANOGOS CAVE *Utah County* ELEVATION 5525 ft LAT/LONG 40° 27 ' N / 111° 43 ' W

| | JAN | FEB | MAR | APR | MAY | JUN | JUL | AUG | SEP | OCT | NOV | DEC | YEAR |
|---|---|---|---|---|---|---|---|---|---|---|---|---|---|
| Maximum Temp °F | 33.3 | 40.1 | 49.8 | 60.1 | 70.9 | 81.8 | 90.9 | 89.8 | 78.5 | 62.0 | 42.9 | 33.8 | 61.2 |
| Minimum Temp °F | 19.5 | 22.6 | 28.1 | 34.0 | 41.7 | 49.3 | 56.8 | 55.4 | 47.3 | 38.0 | 28.1 | 20.5 | 36.8 |
| Mean Temp °F | 26.4 | 31.4 | 38.9 | 47.1 | 56.3 | 65.6 | 73.9 | 72.6 | 62.9 | 50.0 | 35.5 | 27.2 | 49.0 |
| Days Max Temp ≥ 90 °F | 0 | 0 | 0 | 0 | 0 | 6 | 20 | 17 | 3 | 0 | 0 | 0 | 46 |
| Days Max Temp ≤ 32 °F | 14 | 5 | 1 | 0 | 0 | 0 | 0 | 0 | 0 | 0 | 4 | 14 | 38 |
| Days Min Temp ≤ 32 °F | 29 | 25 | 21 | 12 | 3 | 0 | 0 | 0 | 1 | 6 | 21 | 29 | 147 |
| Days Min Temp ≤ 0 °F | 1 | 0 | 0 | 0 | 0 | 0 | 0 | 0 | 0 | 0 | 0 | 1 | 2 |
| Heating Degree Days | 1188 | 943 | 802 | 532 | 273 | 80 | 4 | 9 | 118 | 458 | 879 | 1167 | 6453 |
| Cooling Degree Days | 0 | 0 | 0 | 1 | 14 | 111 | 277 | 267 | 66 | 1 | 0 | 0 | 737 |
| Total Precipitation (") | 2.53 | 2.37 | 2.74 | 2.39 | 2.79 | 1.47 | 1.24 | 1.49 | 1.94 | 2.59 | 2.16 | 2.18 | 25.89 |
| Days ≥ 0.1" Precip | 6 | 5 | 6 | 6 | 6 | 3 | 3 | 4 | 4 | 5 | 5 | 5 | 58 |
| Total Snowfall (") | 22.7 | 17.7 | ***11.0*** | 6.2 | 0.9 | 0.0 | 0.0 | 0.0 | 0.2 | 2.0 | 10.7 | 19.4 | 90.8 |
| Days ≥ 1" Snow Depth | 31 | 26 | 19 | 3 | 0 | 0 | 0 | 0 | 0 | 1 | 10 | 25 | 115 |

## TOOELE *Tooele County* ELEVATION 4823 ft LAT/LONG 40° 32 ' N / 112° 18 ' W

| | JAN | FEB | MAR | APR | MAY | JUN | JUL | AUG | SEP | OCT | NOV | DEC | YEAR |
|---|---|---|---|---|---|---|---|---|---|---|---|---|---|
| Maximum Temp °F | 37.9 | 43.3 | 51.5 | 60.3 | 70.2 | 80.5 | 88.5 | 86.6 | 76.5 | 63.3 | 48.3 | 38.4 | 62.1 |
| Minimum Temp °F | 19.5 | 23.6 | 30.5 | 37.2 | 46.0 | 54.8 | 62.2 | 60.3 | 50.5 | 39.1 | 28.5 | 20.2 | 39.4 |
| Mean Temp °F | 28.7 | 33.5 | 41.0 | 48.8 | 58.1 | 67.7 | 75.4 | 73.5 | 63.5 | 51.2 | 38.4 | 29.3 | 50.8 |
| Days Max Temp ≥ 90 °F | 0 | 0 | 0 | 0 | 0 | 5 | 15 | 11 | 2 | 0 | 0 | 0 | 33 |
| Days Max Temp ≤ 32 °F | 9 | 3 | 1 | 0 | 0 | 0 | 0 | 0 | 0 | 0 | 1 | 8 | 22 |
| Days Min Temp ≤ 32 °F | 28 | 24 | 19 | 9 | 2 | 0 | 0 | 0 | 0 | 6 | 21 | 29 | 138 |
| Days Min Temp ≤ 0 °F | 1 | 0 | 0 | 0 | 0 | 0 | 0 | 0 | 0 | 0 | 0 | 1 | 2 |
| Heating Degree Days | 1119 | 883 | 735 | 485 | 242 | 66 | 4 | 8 | 116 | 424 | 792 | 1099 | 5973 |
| Cooling Degree Days | 0 | 0 | 0 | 9 | 39 | 170 | 339 | 308 | 89 | 3 | 0 | 0 | 957 |
| Total Precipitation (") | 1.19 | 1.37 | 2.25 | 2.44 | 2.04 | 0.99 | 0.91 | 0.94 | 1.41 | 2.01 | 1.87 | 1.47 | 18.89 |
| Days ≥ 0.1" Precip | 4 | 4 | 6 | 6 | 5 | 3 | 2 | 3 | 3 | 4 | 5 | 5 | 50 |
| Total Snowfall (") | 13.8 | 13.3 | 13.7 | 8.2 | 1.7 | 0.0 | 0.0 | 0.0 | 0.1 | 3.9 | 11.8 | 17.5 | 84.0 |
| Days ≥ 1" Snow Depth | 23 | 18 | 8 | 2 | 0 | 0 | 0 | 0 | 0 | 1 | 10 | 23 | 85 |

## TROPIC *Garfield County* ELEVATION 6306 ft LAT/LONG 37° 37 ' N / 112° 6 ' W

| | JAN | FEB | MAR | APR | MAY | JUN | JUL | AUG | SEP | OCT | NOV | DEC | YEAR |
|---|---|---|---|---|---|---|---|---|---|---|---|---|---|
| Maximum Temp °F | 40.7 | 44.4 | 51.4 | 59.6 | 69.2 | 79.5 | 84.8 | 81.8 | 74.1 | 64.0 | 50.3 | 42.0 | 61.8 |
| Minimum Temp °F | 15.1 | 18.9 | 24.4 | 29.3 | 36.4 | 44.4 | 51.0 | 49.8 | 42.0 | 33.8 | 23.3 | 15.8 | 32.0 |
| Mean Temp °F | 27.9 | 31.7 | 37.9 | 44.5 | 52.8 | 61.9 | 67.9 | 65.8 | 58.1 | 48.9 | 36.8 | 28.9 | 46.9 |
| Days Max Temp ≥ 90 °F | 0 | 0 | 0 | 0 | 0 | 2 | 5 | 3 | 0 | 0 | 0 | 0 | 10 |
| Days Max Temp ≤ 32 °F | 5 | 3 | 0 | 0 | 0 | 0 | 0 | 0 | 0 | 0 | 1 | 4 | 13 |
| Days Min Temp ≤ 32 °F | 31 | 28 | 28 | 21 | 9 | 1 | 0 | 0 | 2 | 13 | 27 | 30 | 190 |
| Days Min Temp ≤ 0 °F | 2 | 1 | 0 | 0 | 0 | 0 | 0 | 0 | 0 | 0 | 0 | 1 | 4 |
| Heating Degree Days | 1143 | 935 | 832 | 609 | 372 | 126 | 20 | 52 | 210 | 492 | 838 | 1111 | 6740 |
| Cooling Degree Days | 0 | 0 | 0 | 0 | 2 | 41 | 115 | 85 | 11 | 0 | 0 | 0 | 254 |
| Total Precipitation (") | 0.96 | 1.19 | 1.17 | 0.69 | 0.74 | 0.37 | 1.09 | 1.66 | 1.18 | 1.21 | 1.07 | 0.99 | 12.32 |
| Days ≥ 0.1" Precip | 3 | 3 | 3 | 2 | 3 | 1 | 4 | 5 | 3 | 3 | 3 | 3 | 36 |
| Total Snowfall (") | ***6.8*** | na | ***4.8*** | 1.3 | 0.0 | 0.0 | 0.0 | 0.0 | 0.0 | 0.1 | ***2.0*** | ***4.9*** | na |
| Days ≥ 1" Snow Depth | ***12*** | ***9*** | ***4*** | ***0*** | 0 | 0 | 0 | 0 | 0 | 0 | ***3*** | ***6*** | 34 |

## UTAH LAKE LEHI *Utah County* ELEVATION 4505 ft LAT/LONG 40° 22 ' N / 111° 54 ' W

| | JAN | FEB | MAR | APR | MAY | JUN | JUL | AUG | SEP | OCT | NOV | DEC | YEAR |
|---|---|---|---|---|---|---|---|---|---|---|---|---|---|
| Maximum Temp °F | 35.8 | 41.8 | 51.0 | 61.0 | 71.6 | 82.0 | 89.7 | 87.5 | 77.7 | 64.2 | 48.3 | 37.9 | 62.4 |
| Minimum Temp °F | 14.9 | 19.6 | 27.1 | 33.7 | 41.0 | 48.6 | 56.0 | 53.8 | 44.7 | 34.1 | 25.6 | 17.4 | 34.7 |
| Mean Temp °F | 25.4 | 30.7 | 39.1 | 47.4 | 56.3 | 65.3 | 72.9 | 70.7 | 61.2 | 49.2 | 37.0 | 27.7 | 48.6 |
| Days Max Temp ≥ 90 °F | 0 | 0 | 0 | 0 | 0 | 5 | 17 | 12 | 1 | 0 | 0 | 0 | 35 |
| Days Max Temp ≤ 32 °F | 10 | 3 | 0 | 0 | 0 | 0 | 0 | 0 | 0 | 0 | 1 | 7 | 21 |
| Days Min Temp ≤ 32 °F | 30 | 27 | 25 | 13 | 3 | 0 | 0 | 0 | 2 | 13 | 25 | 30 | 168 |
| Days Min Temp ≤ 0 °F | 3 | 1 | 0 | 0 | 0 | 0 | 0 | 0 | 0 | 0 | 0 | 2 | 6 |
| Heating Degree Days | 1221 | 961 | 797 | 523 | 269 | 68 | 3 | 9 | 139 | 484 | 835 | 1151 | 6460 |
| Cooling Degree Days | 0 | 0 | 0 | 1 | 8 | 91 | 258 | 216 | 34 | 0 | 0 | 0 | 608 |
| Total Precipitation (") | 0.93 | 0.92 | 1.09 | 1.20 | 1.23 | 0.66 | 0.64 | 0.98 | 1.16 | 1.29 | 1.15 | 0.69 | 11.94 |
| Days ≥ 0.1" Precip | *3* | 3 | 4 | 4 | 4 | 2 | 2 | 3 | 3 | 4 | 4 | 3 | 39 |
| Total Snowfall (") | *9.0* | *5.2* | *1.8* | *0.8* | 0.0 | 0.0 | 0.0 | 0.0 | 0.0 | 0.7 | 3.9 | na | na |
| Days ≥ 1" Snow Depth | na | na | na | *0* | 0 | 0 | 0 | 0 | 0 | 0 | 3 | na | na |

## VERNAL MUNICIPAL AP *Uintah County* ELEVATION 5259 ft LAT/LONG 40° 27 ' N / 109° 31 ' W

| | JAN | FEB | MAR | APR | MAY | JUN | JUL | AUG | SEP | OCT | NOV | DEC | YEAR |
|---|---|---|---|---|---|---|---|---|---|---|---|---|---|
| Maximum Temp °F | 27.8 | 35.8 | 50.5 | 62.5 | 72.9 | 83.3 | 89.7 | 87.6 | 77.3 | 62.7 | 45.4 | 31.4 | 60.6 |
| Minimum Temp °F | 5.3 | 10.9 | 23.4 | 31.1 | 39.4 | 47.2 | 53.1 | 51.0 | 42.1 | 31.8 | 20.8 | 9.4 | 30.5 |
| Mean Temp °F | 16.6 | 23.4 | 37.0 | 46.8 | 56.2 | 65.3 | 71.5 | 69.3 | 59.7 | 47.2 | 33.1 | 20.4 | 45.5 |
| Days Max Temp ≥ 90 °F | 0 | 0 | 0 | 0 | 0 | 8 | 18 | 13 | 1 | 0 | 0 | 0 | 40 |
| Days Max Temp ≤ 32 °F | 21 | 10 | 2 | 0 | 0 | 0 | 0 | 0 | 0 | 0 | 3 | 16 | 52 |
| Days Min Temp ≤ 32 °F | 31 | 28 | 28 | 17 | 5 | 0 | 0 | 0 | 3 | 17 | 28 | 31 | 188 |
| Days Min Temp ≤ 0 °F | 11 | 5 | 0 | 0 | 0 | 0 | 0 | 0 | 0 | 0 | 0 | 6 | 22 |
| Heating Degree Days | 1495 | 1168 | 861 | 539 | 271 | 70 | 4 | 13 | 172 | 543 | 949 | 1376 | 7461 |
| Cooling Degree Days | 0 | 0 | 0 | 0 | 5 | 89 | 202 | 158 | 20 | 0 | 0 | 0 | 474 |
| Total Precipitation (") | 0.43 | 0.43 | 0.69 | 0.81 | 0.95 | 0.81 | 0.56 | 0.62 | 0.83 | 1.21 | 0.60 | 0.59 | 8.53 |
| Days ≥ 0.1" Precip | 2 | 2 | 2 | 2 | 3 | 2 | 2 | 2 | 3 | 3 | 2 | 2 | 27 |
| Total Snowfall (") | na | na | na | na | 0.0 | 0.0 | 0.0 | 0.0 | 0.0 | *0.2* | na | na | na |
| Days ≥ 1" Snow Depth | na | na | na | *0* | 0 | 0 | 0 | 0 | 0 | 0 | na | na | na |

## VERNON *Tooele County* ELEVATION 5492 ft LAT/LONG 40° 5 ' N / 112° 27 ' W

| | JAN | FEB | MAR | APR | MAY | JUN | JUL | AUG | SEP | OCT | NOV | DEC | YEAR |
|---|---|---|---|---|---|---|---|---|---|---|---|---|---|
| Maximum Temp °F | 38.7 | 43.8 | *50.3* | 59.1 | 69.2 | 80.4 | 89.3 | 86.9 | 77.4 | 64.0 | 48.6 | 39.9 | 62.3 |
| Minimum Temp °F | 12.4 | 17.6 | *24.8* | 29.9 | 37.5 | 45.7 | 53.7 | 52.3 | 42.6 | 32.3 | 22.1 | 13.2 | 32.0 |
| Mean Temp °F | 25.6 | 30.7 | *37.6* | 44.5 | 53.4 | 63.1 | 71.5 | 69.6 | 60.0 | 48.2 | 35.4 | 26.5 | 47.2 |
| Days Max Temp ≥ 90 °F | 0 | 0 | 0 | 0 | 0 | 5 | 16 | 11 | 2 | 0 | 0 | 0 | 34 |
| Days Max Temp ≤ 32 °F | 8 | 3 | 1 | 0 | 0 | 0 | 0 | 0 | 0 | 0 | 2 | 7 | 21 |
| Days Min Temp ≤ 32 °F | 30 | 26 | *26* | 20 | 8 | 1 | 0 | 0 | 3 | 16 | 27 | 30 | 187 |
| Days Min Temp ≤ 0 °F | 5 | 2 | 0 | 0 | 0 | 0 | 0 | 0 | 0 | 0 | 1 | 4 | 12 |
| Heating Degree Days | 1216 | 961 | *843* | 608 | 358 | 117 | 7 | 17 | 175 | 513 | 882 | 1185 | 6882 |
| Cooling Degree Days | 0 | 0 | 0 | 0 | 4 | 78 | 212 | 180 | 33 | 0 | 0 | 0 | 507 |
| Total Precipitation (") | 0.62 | 0.63 | 0.99 | 0.79 | 1.11 | 0.65 | 0.90 | 0.99 | 0.84 | 1.11 | 0.90 | 0.70 | 10.23 |
| Days ≥ 0.1" Precip | 2 | 2 | 3 | 3 | 3 | 2 | 3 | 3 | 2 | 3 | 3 | 2 | 31 |
| Total Snowfall (") | na | *6.2* | *5.5* | *2.1* | 0.2 | 0.2 | 0.0 | 0.0 | 0.2 | 1.1 | *5.4* | na | na |
| Days ≥ 1" Snow Depth | na | na | na | *1* | 0 | 0 | 0 | 0 | 0 | *0* | *4* | na | na |

## VEYO POWERHOUSE *Washington County* ELEVATION 4600 ft LAT/LONG 37° 21 ' N / 113° 40 ' W

| | JAN | FEB | MAR | APR | MAY | JUN | JUL | AUG | SEP | OCT | NOV | DEC | YEAR |
|---|---|---|---|---|---|---|---|---|---|---|---|---|---|
| Maximum Temp °F | 46.8 | 51.4 | 57.1 | 65.2 | 75.0 | 86.1 | 92.2 | 90.0 | 82.4 | 70.9 | 56.7 | 48.0 | 68.5 |
| Minimum Temp °F | 24.6 | 28.2 | 32.6 | 38.0 | 45.6 | 54.2 | 59.8 | 58.2 | 51.4 | 42.0 | 31.2 | 25.0 | 40.9 |
| Mean Temp °F | 35.7 | 39.8 | 44.9 | 51.6 | 60.3 | 70.1 | 76.0 | 74.1 | 66.9 | 56.5 | 44.0 | 36.5 | 54.7 |
| Days Max Temp ≥ 90 °F | 0 | 0 | 0 | 0 | 1 | 11 | *23* | 18 | 4 | 0 | 0 | 0 | 57 |
| Days Max Temp ≤ 32 °F | 1 | 1 | 0 | 0 | 0 | 0 | 0 | 0 | 0 | 0 | 0 | 2 | 4 |
| Days Min Temp ≤ 32 °F | 28 | 21 | *14* | 6 | 1 | 0 | 0 | 0 | 0 | 3 | *17* | 27 | 117 |
| Days Min Temp ≤ 0 °F | 0 | 0 | 0 | 0 | 0 | 0 | 0 | 0 | 0 | 0 | 0 | 0 | 0 |
| Heating Degree Days | 901 | 704 | 617 | 397 | 172 | 25 | 1 | 2 | 50 | 272 | 624 | 877 | 4642 |
| Cooling Degree Days | 0 | 0 | 0 | 3 | 38 | 200 | 351 | 297 | 124 | 18 | 0 | 0 | 1031 |
| Total Precipitation (") | 1.40 | 1.68 | 2.17 | 1.10 | 0.78 | 0.33 | 1.00 | 1.18 | 1.07 | 1.09 | 1.32 | 1.29 | 14.41 |
| Days ≥ 0.1" Precip | 3 | 3 | 5 | 3 | 2 | 1 | 3 | 3 | 2 | 2 | 3 | 3 | 33 |
| Total Snowfall (") | na | *1.3* | *1.0* | 0.0 | 0.0 | 0.0 | 0.0 | 0.0 | 0.0 | 0.0 | *0.4* | *1.3* | na |
| Days ≥ 1" Snow Depth | na | na | *0* | 0 | 0 | 0 | 0 | 0 | 0 | 0 | *0* | na | na |

## WAH WAH RANCH *Beaver County* ELEVATION 4964 ft LAT/LONG 38° 29 ' N / 113° 25 ' W

| | JAN | FEB | MAR | APR | MAY | JUN | JUL | AUG | SEP | OCT | NOV | DEC | YEAR |
|---|---|---|---|---|---|---|---|---|---|---|---|---|---|
| Maximum Temp °F | 42.6 | 49.2 | 57.7 | 66.0 | 76.0 | 87.3 | 94.4 | 91.9 | 83.0 | 70.0 | 54.5 | 43.5 | 68.0 |
| Minimum Temp °F | 14.2 | 19.8 | 26.5 | 31.9 | 40.3 | 49.3 | 57.8 | 55.7 | 45.2 | 33.5 | 23.4 | 14.6 | 34.4 |
| Mean Temp °F | 28.4 | 34.5 | 42.1 | 49.0 | 58.2 | 68.3 | 76.1 | 73.8 | 64.1 | 51.8 | 38.9 | 29.1 | 51.2 |
| Days Max Temp ≥ 90 °F | 0 | 0 | 0 | 0 | 1 | 13 | 26 | 21 | 7 | 0 | 0 | 0 | 68 |
| Days Max Temp ≤ 32 °F | 5 | 1 | 0 | 0 | 0 | 0 | 0 | 0 | 0 | 0 | 1 | 5 | 12 |
| Days Min Temp ≤ 32 °F | 29 | 25 | 23 | 17 | 6 | 1 | 0 | 0 | 3 | 14 | 25 | 29 | 172 |
| Days Min Temp ≤ 0 °F | 4 | 2 | 0 | 0 | 0 | 0 | 0 | 0 | 0 | 0 | 0 | 3 | 9 |
| Heating Degree Days | 1128 | 853 | 703 | 474 | 225 | 49 | 2 | 4 | 100 | 405 | 776 | 1107 | 5826 |
| Cooling Degree Days | 0 | 0 | 0 | 2 | 20 | 162 | 347 | 296 | 88 | 1 | 0 | 0 | 916 |
| Total Precipitation (") | 0.35 | 0.30 | 0.66 | 0.61 | 0.67 | 0.34 | 0.71 | 1.08 | 0.76 | 0.72 | 0.45 | 0.27 | 6.92 |
| Days ≥ 0.1" Precip | 1 | 1 | 2 | 2 | 2 | 1 | 3 | 3 | 2 | 3 | 2 | 1 | 23 |
| Total Snowfall (") | na | *1.5* | na | *0.1* | 0.0 | 0.0 | 0.0 | 0.0 | 0.2 | 0.0 | *0.5* | *1.4* | na |
| Days ≥ 1" Snow Depth | *4* | *3* | 1 | 0 | 0 | 0 | 0 | 0 | 0 | 0 | 2 | 3 | 13 |

## WANSHIP DAM *Summit County* ELEVATION 5905 ft LAT/LONG 40° 48 ' N / 111° 24 ' W

| | JAN | FEB | MAR | APR | MAY | JUN | JUL | AUG | SEP | OCT | NOV | DEC | YEAR |
|---|---|---|---|---|---|---|---|---|---|---|---|---|---|
| Maximum Temp °F | 35.3 | 39.7 | 47.6 | 57.7 | 67.9 | 77.9 | 85.8 | 84.9 | 75.6 | 63.1 | 46.5 | 36.8 | 59.9 |
| Minimum Temp °F | 11.1 | 13.7 | 21.8 | 27.9 | 34.7 | 40.8 | 46.8 | 45.0 | 36.9 | 27.9 | 20.7 | 12.4 | 28.3 |
| Mean Temp °F | 23.2 | 26.7 | 34.7 | 42.8 | 51.3 | 59.4 | 66.3 | 65.0 | 56.3 | 45.4 | 33.6 | 24.7 | 44.1 |
| Days Max Temp ≥ 90 °F | 0 | 0 | 0 | 0 | 0 | 2 | 7 | 7 | 0 | 0 | 0 | 0 | 16 |
| Days Max Temp ≤ 32 °F | 11 | 5 | 1 | 0 | 0 | 0 | 0 | 0 | 0 | 0 | 3 | 9 | 29 |
| Days Min Temp ≤ 32 °F | 30 | 27 | 28 | 22 | 12 | 3 | 0 | 1 | 9 | 23 | 26 | 30 | 211 |
| Days Min Temp ≤ 0 °F | 7 | 4 | 1 | 0 | 0 | 0 | 0 | 0 | 0 | 0 | 1 | 4 | 17 |
| Heating Degree Days | 1289 | 1074 | 933 | 660 | 417 | 180 | 34 | 54 | 261 | 601 | 935 | 1245 | 7683 |
| Cooling Degree Days | 0 | 0 | 0 | 0 | 0 | 25 | 89 | 72 | 6 | 0 | 0 | 0 | 192 |
| Total Precipitation (") | 1.18 | 1.07 | 1.41 | 1.74 | 1.86 | 1.05 | 1.16 | 1.06 | 1.44 | 1.73 | 1.64 | 1.15 | 16.49 |
| Days ≥ 0.1" Precip | 4 | 4 | 5 | 5 | 6 | 3 | 3 | 3 | 4 | 5 | 5 | 4 | 51 |
| Total Snowfall (") | na | na | na | na | *0.5* | 0.0 | 0.0 | 0.0 | 0.1 | na | na | na | na |
| Days ≥ 1" Snow Depth | na | na | na | *1* | 0 | 0 | 0 | 0 | 0 | na | na | na | na |

## WENDOVER AF AUX FLD *Tooele County* ELEVATION 4239 ft LAT/LONG 40° 44 ' N / 114° 2 ' W

| | JAN | FEB | MAR | APR | MAY | JUN | JUL | AUG | SEP | OCT | NOV | DEC | YEAR |
|---|---|---|---|---|---|---|---|---|---|---|---|---|---|
| Maximum Temp °F | 34.6 | 42.2 | 52.5 | 61.3 | 71.7 | 82.1 | 91.1 | 88.4 | 77.2 | 62.0 | 46.2 | 35.0 | 62.0 |
| Minimum Temp °F | 18.9 | 24.5 | 33.1 | 40.4 | 50.3 | 59.6 | 67.6 | 64.8 | 54.1 | 41.4 | 29.6 | 19.7 | 42.0 |
| Mean Temp °F | 26.8 | 33.4 | 42.8 | 50.9 | 61.0 | 70.8 | 79.4 | 76.6 | 65.7 | 51.7 | 37.9 | 27.4 | 52.0 |
| Days Max Temp ≥ 90 °F | 0 | 0 | 0 | 0 | 1 | 7 | 20 | 15 | 2 | 0 | 0 | 0 | 45 |
| Days Max Temp ≤ 32 °F | 14 | 4 | 0 | 0 | 0 | 0 | 0 | 0 | 0 | 0 | 1 | 11 | 30 |
| Days Min Temp ≤ 32 °F | 29 | 24 | 14 | 4 | 0 | 0 | 0 | 0 | 0 | 3 | 19 | 30 | 123 |
| Days Min Temp ≤ 0 °F | 1 | 0 | 0 | 0 | 0 | 0 | 0 | 0 | 0 | 0 | 0 | 1 | 2 |
| Heating Degree Days | 1177 | 888 | 680 | 421 | 171 | 36 | 1 | 4 | 80 | 407 | 806 | 1160 | 5831 |
| Cooling Degree Days | 0 | 0 | 0 | 8 | 60 | 232 | 452 | 383 | 116 | 2 | 0 | 0 | 1253 |
| Total Precipitation (") | 0.26 | 0.28 | 0.45 | 0.53 | 0.83 | 0.49 | 0.32 | 0.44 | 0.40 | 0.52 | 0.34 | 0.24 | 5.10 |
| Days ≥ 0.1" Precip | 1 | 1 | 1 | 2 | 2 | 2 | 1 | 1 | 1 | 2 | 1 | 1 | 16 |
| Total Snowfall (") | na | na | *0.6* | 0.3 | 0.0 | 0.0 | 0.0 | 0.0 | 0.0 | 0.1 | 0.3 | *1.2* | na |
| Days ≥ 1" Snow Depth | na | na | 0 | 0 | 0 | 0 | 0 | 0 | 0 | 0 | 0 | *2* | na |

## WOODRUFF *Rich County* ELEVATION 6345 ft LAT/LONG 41° 32 ' N / 111° 9 ' W

| | JAN | FEB | MAR | APR | MAY | JUN | JUL | AUG | SEP | OCT | NOV | DEC | YEAR |
|---|---|---|---|---|---|---|---|---|---|---|---|---|---|
| Maximum Temp °F | 28.0 | 32.3 | 42.7 | 54.0 | 64.1 | 73.0 | 81.3 | 80.2 | 71.1 | 59.2 | 41.4 | 29.5 | 54.7 |
| Minimum Temp °F | 2.6 | 4.7 | 16.6 | 24.1 | 31.4 | 38.9 | 43.8 | 41.2 | 32.4 | 23.1 | 14.3 | 3.6 | 23.1 |
| Mean Temp °F | 15.3 | 18.5 | 29.7 | 39.1 | 47.8 | 56.0 | 62.6 | 60.7 | 51.8 | 41.1 | 27.9 | 16.6 | 38.9 |
| Days Max Temp ≥ 90 °F | 0 | 0 | 0 | 0 | 0 | 0 | 1 | 1 | 0 | 0 | 0 | 0 | 2 |
| Days Max Temp ≤ 32 °F | 18 | 13 | 4 | 0 | 0 | 0 | 0 | 0 | 0 | 1 | 7 | 18 | 61 |
| Days Min Temp ≤ 32 °F | 31 | 28 | 31 | 27 | 17 | 4 | 0 | 3 | 15 | 27 | 29 | 31 | 243 |
| Days Min Temp ≤ 0 °F | 14 | 11 | 3 | 0 | 0 | 0 | 0 | 0 | 0 | 0 | 4 | 13 | 45 |
| Heating Degree Days | 1536 | 1308 | 1088 | 772 | 526 | 268 | 88 | 136 | 390 | 733 | 1106 | 1495 | 9446 |
| Cooling Degree Days | 0 | 0 | 0 | 0 | 0 | 5 | 22 | 15 | 0 | 0 | 0 | 0 | 42 |
| Total Precipitation (") | 0.47 | 0.44 | 0.56 | 0.93 | 1.01 | 1.01 | 0.78 | 0.71 | 1.13 | 1.07 | 0.69 | 0.51 | 9.31 |
| Days ≥ 0.1" Precip | 2 | 2 | 2 | 3 | 3 | 3 | 3 | 2 | 3 | 3 | 2 | 2 | 30 |
| Total Snowfall (") | *7.8* | 7.8 | 5.7 | 4.5 | 1.0 | 0.1 | 0.0 | 0.0 | 0.6 | 3.0 | 6.4 | 8.0 | 44.9 |
| Days ≥ 1" Snow Depth | na | na | na | 1 | 0 | 0 | 0 | 0 | 0 | 1 | 5 | *12* | na |

## ZION NATIONAL PARK *Washington County* ELEVATION 4052 ft LAT/LONG 37° 13 ' N / 112° 59 ' W

| | JAN | FEB | MAR | APR | MAY | JUN | JUL | AUG | SEP | OCT | NOV | DEC | YEAR |
|---|---|---|---|---|---|---|---|---|---|---|---|---|---|
| Maximum Temp °F | 52.1 | 57.3 | 63.4 | 72.2 | 82.6 | 93.7 | 99.1 | 96.6 | 89.4 | 77.5 | 61.9 | 52.5 | 74.9 |
| Minimum Temp °F | 28.9 | 33.2 | 37.3 | 43.0 | 51.8 | 61.3 | 68.3 | 66.9 | 59.5 | 48.9 | 36.6 | 29.6 | 47.1 |
| Mean Temp °F | 40.5 | 45.3 | 50.4 | 57.6 | 67.2 | 77.6 | 83.7 | 81.8 | 74.5 | 63.2 | 49.3 | 41.1 | 61.0 |
| Days Max Temp ≥ 90 °F | 0 | 0 | 0 | 1 | 7 | 23 | 30 | 28 | 16 | 3 | 0 | 0 | 108 |
| Days Max Temp ≤ 32 °F | 0 | 0 | 0 | 0 | 0 | 0 | 0 | 0 | 0 | 0 | 0 | 0 | 0 |
| Days Min Temp ≤ 32 °F | 20 | 12 | 8 | 4 | 0 | 0 | 0 | 0 | 0 | 1 | 10 | 19 | 74 |
| Days Min Temp ≤ 0 °F | 0 | 0 | 0 | 0 | 0 | 0 | 0 | 0 | 0 | 0 | 0 | 0 | 0 |
| Heating Degree Days | 753 | 551 | 448 | 236 | 68 | 4 | 0 | 0 | 10 | 129 | 466 | 734 | 3399 |
| Cooling Degree Days | 0 | 0 | 1 | 32 | 146 | 405 | 582 | 522 | 304 | 79 | 1 | 0 | 2072 |
| Total Precipitation (") | 1.78 | 1.80 | 2.26 | 1.13 | 0.88 | 0.44 | 1.26 | 1.61 | 0.91 | 1.05 | 1.48 | 1.39 | 15.99 |
| Days ≥ 0.1" Precip | 4 | 4 | 5 | 3 | 3 | 1 | 3 | 4 | 2 | 3 | 3 | 4 | 39 |
| Total Snowfall (") | 3.2 | *1.4* | 0.5 | 0.2 | 0.0 | 0.0 | 0.0 | 0.0 | 0.0 | 0.1 | 0.8 | 2.5 | 8.7 |
| Days ≥ 1" Snow Depth | 2 | 1 | 0 | 0 | 0 | 0 | 0 | 0 | 0 | 0 | 0 | 1 | 4 |

## JANUARY MINIMUM TEMPERATURE °F

### LOWEST

| | | |
|---|---|---|
| 1 | Scofield Dam | -1.5 |
| 2 | Ouray | 1.4 |
| 3 | Jensen | 1.5 |
| 4 | Myton | 2.1 |
| 5 | Fort Duchesne | 2.2 |
| 6 | Woodruff | 2.6 |
| 7 | Dinosaur | 3.0 |
| | Roosevelt | 3.0 |
| 9 | Vernal | 5.3 |
| 10 | Altamont | 5.8 |
| 11 | Duchesne | 5.9 |
| 12 | Neola | 7.1 |
| 13 | Pine View Dam | 7.4 |
| 14 | Castle Dale | 7.6 |
| 15 | Deer Creek Dam | 7.9 |
| 16 | Loa | 8.0 |
| 17 | Panguitch | 8.4 |
| 18 | Silver Lake | 8.7 |
| 19 | Koosharem | 8.9 |
| 20 | Flaming Gorge | 9.1 |
| | Green River | 9.1 |
| 22 | Heber | 9.3 |
| 23 | Bryce Canyon | 9.4 |
| 24 | Grouse Creek | 9.6 |
| 25 | Hanksville | 10.3 |

### HIGHEST

| | | |
|---|---|---|
| 1 | Zion | 28.9 |
| 2 | St. George | 28.1 |
| 3 | La Verkin | 26.1 |
| 4 | Veyo | 24.6 |
| 5 | Garfield | 22.8 |
| | Kanab | 22.8 |
| 7 | Cottonwood | 21.4 |
| 8 | New Harmony | 20.5 |
| 9 | Salt Lake City | 20.3 |
| 10 | Ogden-Pioneer | 19.9 |
| 11 | Farmington | 19.8 |
| 12 | Spanish Fork | 19.5 |
| | Timpanogos | 19.5 |
| | Tooele | 19.5 |
| 15 | Mexican Hat | 19.2 |
| 16 | Pleasant Grove | 19.1 |
| 17 | Canyonlands-Nck | 19.0 |
| 18 | Wendover | 18.9 |
| 19 | Moab | 18.8 |
| 20 | Aneth | 18.7 |
| 21 | Oak City | 18.5 |
| 22 | Cedar City | 18.0 |
| 23 | Ogden-Sugar | 17.8 |
| 24 | Kanosh | 17.7 |
| 25 | Orderville | 17.5 |

## JULY MAXIMUM TEMPERATURE °F

### HIGHEST

| | | |
|---|---|---|
| 1 | St. George | 102.2 |
| 2 | Moab | 99.4 |
| 3 | Zion | 99.1 |
| 4 | Hanksville | 98.7 |
| 5 | La Verkin | 98.4 |
| 6 | Mexican Hat | 97.7 |
| 7 | Green River | 96.7 |
| 8 | Aneth | 96.0 |
| 9 | Bluff | 95.1 |
| 10 | Canyonlands-Ndle | 95.0 |
| 11 | Partoun | 94.4 |
| | Wah Wah | 94.4 |
| 13 | Dinosaur | 94.3 |
| | Fish Springs | 94.3 |
| 15 | Dugway | 94.2 |
| 16 | Oak City | 94.1 |
| 17 | Hovenweep | 94.0 |
| 18 | Deseret | 93.9 |
| | Ouray | 93.9 |
| 20 | Thompson | 92.8 |
| 21 | Delta | 92.7 |
| 22 | Black Rock | 92.5 |
| | Kanab | 92.5 |
| | Nephi | 92.5 |
| 25 | Salt Lake City | 92.3 |

### LOWEST

| | | |
|---|---|---|
| 1 | Blowhard Mntain | 62.3 |
| 2 | Alta | 71.0 |
| 3 | Silver Lake | 71.7 |
| 4 | Scofield Dam | 77.3 |
| 5 | Bryce Canyon | 79.2 |
| 6 | Woodruff | 81.3 |
| 7 | Laketown | 82.4 |
| 8 | Loa | 82.7 |
| 9 | Altamont | 83.1 |
| | Alton | 83.1 |
| 11 | Monticello | 84.1 |
| 12 | Kamas | 84.3 |
| | Snake Creek | 84.3 |
| 14 | Boulder | 84.4 |
| | Koosharem | 84.4 |
| | Neola | 84.4 |
| 17 | Tropic | 84.8 |
| 18 | Flaming Gorge | 85.1 |
| 19 | Panguitch | 85.4 |
| 20 | Cedar Point | 85.7 |
| 21 | Wanship Dam | 85.8 |
| 22 | Deer Creek Dam | 85.9 |
| 23 | Coalville | 86.0 |
| | Pine View Dam | 86.0 |
| 25 | Manti | 86.2 |

## ANNUAL PRECIPITATION (")

### HIGHEST

| | | |
|---|---|---|
| 1 | Alta | 56.75 |
| 2 | Silver Lake | 43.27 |
| 3 | Pine View Dam | 31.03 |
| 4 | Blowhard Mntain | 29.81 |
| 5 | Timpanogos | 25.89 |
| 6 | Cottonwood | 25.30 |
| 7 | Mountain Dll Dm | 24.13 |
| 8 | Ogden-Pioneer | 23.11 |
| 9 | Deer Creek Dam | 22.68 |
| 10 | Farmington | 22.51 |
| 11 | Snake Creek | 22.17 |
| 12 | Spanish Fork | 20.88 |
| 13 | Morgan | 19.64 |
| 14 | Logan-U St Univ | 19.53 |
| 15 | Richmond | 19.52 |
| 16 | New Harmony | 18.95 |
| 17 | Tooele | 18.89 |
| 18 | Santaquin | 18.21 |
| 19 | Garfield | 18.12 |
| 20 | Kamas | 17.95 |
| 21 | Corinne | 17.43 |
| 22 | Ogden-Sugar | 17.40 |
| 23 | Alton | 17.31 |
| 24 | Pleasant Grove | 16.99 |
| 25 | Logan-KVNU | 16.83 |

### LOWEST

| | | |
|---|---|---|
| 1 | Wendover | 5.10 |
| 2 | Hanksville | 5.77 |
| 3 | Callao | 6.03 |
| 4 | Myton | 6.85 |
| 5 | Green River | 6.87 |
| | Mexican Hat | 6.87 |
| 7 | Fort Duchesne | 6.91 |
| 8 | Wah Wah | 6.92 |
| 9 | Partoun | 7.06 |
| 10 | Roosevelt | 7.16 |
| 11 | Ouray | 7.32 |
| 12 | Castle Dale | 7.75 |
| 13 | Loa | 7.88 |
| 14 | Fish Springs | 7.98 |
| 15 | Dugway | 8.20 |
| 16 | Delta | 8.23 |
| 17 | Marysvale | 8.24 |
| 18 | Jensen | 8.38 |
| 19 | Deseret | 8.43 |
| 20 | Bluff | 8.44 |
| 21 | Canyonlands-Ndle | 8.47 |
| 22 | Vernal | 8.53 |
| 23 | St. George | 8.56 |
| 24 | Ferron | 8.57 |
| 25 | Richfield | 8.62 |

## ANNUAL SNOWFALL (")

### HIGHEST

| | | |
|---|---|---|
| 1 | Alta | 524.1 |
| 2 | Silver Lake | 412.4 |
| 3 | Blowhard Mntain | 239.0 |
| 4 | Pine View Dam | 120.3 |
| 5 | Bryce Canyon | 93.0 |
| 6 | Timpanogos | 90.8 |
| 7 | Alton | 89.8 |
| 8 | Tooele | 84.0 |
| 9 | Fillmore | 83.2 |
| 10 | Morgan | 77.7 |
| 11 | Kanosh | 76.7 |
| 12 | Cedar Point | 75.6 |
| 13 | Parowan | 71.3 |
| 14 | Richmond | 69.7 |
| 15 | Heber | 69.3 |
| 16 | Logan-U St Univ | 68.7 |
| 17 | Echo Dam | 68.6 |
| 18 | Monticello | 68.0 |
| 19 | Salt Lake City | 64.4 |
| 20 | Flaming Gorge | 62.2 |
| 21 | Levan | 60.3 |
| | Santaquin | 60.3 |
| 23 | Manti | 59.2 |
| 24 | Farmington | 53.4 |
| | Oak City | 53.4 |

### LOWEST

| | | |
|---|---|---|
| 1 | St. George | 2.5 |
| 2 | La Verkin | 4.2 |
| 3 | Hanksville | 6.0 |
| 4 | Zion | 8.7 |
| 5 | Green River | 11.1 |
| 6 | Bluff | 11.2 |
| 7 | Callao | 12.3 |
| 8 | Fish Springs | 13.4 |
| 9 | Dugway | 15.1 |
| 10 | Ouray | 17.0 |
| 11 | Deseret | 18.0 |
| 12 | Castle Dale | 20.9 |
| 13 | Circleville | 22.6 |
| 14 | Hovenweep | 22.8 |
| 15 | Canyonlands-Nck | 23.8 |
| | Dinosaur | 23.8 |
| 17 | Kanab | 25.0 |
| 18 | Thiokol | 25.5 |
| 19 | Jensen | 25.9 |
| 20 | Loa | 26.7 |
| 21 | Delta | 28.0 |
| 22 | Ferron | 31.0 |
| 23 | Enterprise | 31.2 |
| 24 | Escalante | 31.5 |
| 25 | Koosharem | 37.5 |

# VERMONT

PHYSICAL FEATURES. The Green Mountain State occupies 9,609 square miles. Though Vermont is the only New England state without a coastline on the Atlantic Ocean, most of its boundary is water. The Connecticut River forms the entire eastern border. Lake Champlain marks over 100 miles of the western boundary. Vermont extends southward from near the 45° parallel of latitude almost 160 miles to about 20 miles south of the 43d parallel. Vermont widens northward from about 40 to 90 miles across.

The terrain is hilly to mountainous. The Green Mountains extend the length of the State. They rise to their highest elevation at Mt. Mansfield, 4,393 feet above sea level. Many peaks in this range rise to over 3,000 feet, as do several others in eastern Vermont. Elevations of less than 500 feet above sea level are mostly confined to the lowlands paralleling Lake Champlain in the west and to the central and southern portions of the Connecticut Valley in the east. Much of the State ranges from 500 to 2,000 feet in elevation. The glacier of the great Ice Age accounts for many topographical features, lakes, and soils. Inland waters cover more than 300 square miles.

GENERAL CLIMATE. Vermont shares with the other New England states in the chief climatic characteristics. These include: (1) changeableness of the weather, (2) large range of temperature, both daily and annual, (3) great differences between the same seasons in different years, (4) equable distribution of precipitation, and (5) considerable diversity from place to place. The regional climatic influences are modified in Vermont by varying elevations, types of terrain, and distances from the Atlantic Ocean and from Lake Champlain. The State has been divided into three climatological divisions (Western, Northeastern, and Southeastern).

Vermont lies in the "prevailing westerlies", the belt of generally eastward air movement which encircles the globe in middle latitudes. Embedded in this circulation are extensive masses of air originating in higher or lower latitudes and interacting to produce low-pressure storm systems. Relative to most other sections of the country, a large number of such storms pass over or near Vermont. The majority of air masses affecting this State belong to three types: (1) cold, dry air pouring down from subarctic North America, (2) warm, moist air streaming up on a long overland journey from the Gulf of Mexico and other subtropical waters, and (3) cool, damp air moving in from the North Atlantic. Because the atmospheric flow is usually from a westerly direction, Vermont is more influenced by the first two types than it is by the third.

The procession of contrasting air masses and the relatively frequent passage of "Lows" bring about on the average a twice-weekly alternation from fair to cloudy or stormy conditions, attended by often abrupt changes in temperature, moisture, sunshine, wind direction and speed. There is no regular or persistent rhythm to this sequence, and it is interrupted by periods during which the weather patterns continue the same for several days, infrequently for several weeks.

TEMPERATURE. The annual mean temperature is near 43° F. in the Northeastern Division, 44° F. in the Southeastern, and 46° F. in the Western. Summer temperatures are delightfully comfortable as a rule. They are also reasonably uniform over the State, excepting topographical extremes. Long-period means for July average near 70° F. in the Western Division and near 68° F. in the other Divisions. Average daily minima in July are in the 50's over nearly the entire State. The average daily maxima reach only near 80° F. Hot days with maxima of 90° F. or higher average less than 10 per year at most stations. Even after one of these hot days the temperature is likely to fall to 60°F. or lower during the night.

Temperatures from place to place vary more in winter than in summer. The Northeastern Division average in January is near 17° F. The Southeastern Division average is nearly 19° F. and the Western Division, 21° F. Days with subzero readings are common at most stations in winter. They number from 10 to 40 per year in the southern portion and from 20 to 50 in the north. The growing season for vegetation subject to injury from freezing temperature averages 130 to 150 days in much of the Western Division and along the Connecticut River in the Southeastern Division. Elsewhere, and including the extreme southern portion of the Western Division, the season varies from 100 to 130 days.

PRECIPITATION. Vermont's precipitation, fortunately, is well distributed through the year. Winter precipitation is noticeably less than summer rainfall in the northern and western portions of the State. This difference is greater in those areas than in any other part of New England. New England as a whole is noted for the even distribution of its precipitation throughout the year, an effect due to the influence of the Atlantic Ocean. This ocean influence is still strongly felt in southeastern Vermont, but it becomes weaker with increasing distance from the ocean. Low-pressure, or frontal, storm systems are the principal year-round moisture producers. When this activity ebbs somewhat in summer, bands or patches of thunderstorms increase in activity, more than making up the difference. Though brief and often of small extent, the thunderstorms produce the heaviest local rainfall intensities.

Floods occur most often in the spring when they are caused by rainfall and melting snow. Stages of spring over-bank flooding are frequently increased by ice jams. Local flash floods result on occasions from short period summer storms between May and November. The Connecticut River and its tributaries drain the major portion of Vermont. In the northwest portion, rivers drain into Lake Champlain or directly to the St. Lawrence. A small area in southwest Vermont drains to the Hudson River.

Occasionally freezing rain occurs, coating exposed surfaces with troublesome ice. Most areas can expect at least one such occurrence in a winter. Frequency of days with measurable precipitation is between 120 and 160 days per year.

SNOWFALL. Average annual total snowfall is from 55 to 65 inches in much of the Western Division and also in parts of the Connecticut River Valley. Elsewhere the annual averages vary greatly. They range upward to as much as 100 inches. Topographical differences cause large variations in a short distance. The average number of days with 1 inch or more of snowfall in a season varies from near 20 to 40. The frequency increases with elevation. Most winters have several snowstorms of 5 inches or more per year. Snow cover is continuous throughout the winter season as a rule. Depth of snow on the ground reaches its maximum for much of the State in the latter part of February. At the highest elevations, however, the date falls in the middle of March. Water stored in the snow is an important contribution to the water supply. Spring melting is usually too gradual to produce serious flooding.

OTHER CLIMATIC ELEMENTS. Sunshine averages near 50 percent of possible on a year-round basis, but varies with topography. Higher elevations and peaks are much more cloudy, especially in winter. Sunshine is most abundant during the summer season. Persistent fogs are sometimes experienced on the higher elevations. The duration of fog diminishes over flatland valley locations. But the shorter duration heavy ground fogs of early morning occur frequently at susceptible places in these areas. The number of days with fog probably varies from 10 to 60 per year over the State.

WINDS AND STORMS. Vermont lies in the region of prevailing westerlies -- wind from the northwest in winter, and from the southwest in the warmer part of the year. But because the rugged topography has a strong influence on the direction of the wind, many areas have prevailing winds paralleling a valley. The major valleys tend to lie in a north-south direction. Thus prevailing winds may be from the north in winter and from the south in the warmer seasons in those areas.

Coastal storms, or "northeasters", are well known to New England. Their influence on Vermont is minimized by its inland location. They remain a factor, however, especially in the Southeastern Division. They generate very strong winds and heavy rain or snow. Storms of tropical origin may occasionally affect Vermont in summer or fall, but only rarely contain destructive winds. Vermont is far enough inland so that, usually, winds are considerably weakened by the time tropical storms reach the State. Rainfall associated with these storms may, however, remain heavy.

Tornadoes are not common phenomena, yet one or more of these most violent storms may occur in a year. The peak months are June and July. Thunder and hailstorms have a frequency maximum from midspring to early fall. Thunderstorms occur on 20 to 30 days per year. The most severe are attended by hail. The size of an area struck by a hailstorm, however, is usually small.

# COUNTY INDEX

***Addison County***
CORNWALL

***Bennington County***
READSBORO 1 SE

***Caledonia County***
SAINT JOHNSBURY
WEST BURKE

***Chittenden County***
BURLINGTON INTL AP

***Franklin County***
ENOSBURG FALLS

***Lamoille County***
MOUNT MANSFIELD

***Orange County***
CHELSEA

***Orleans County***
NEWPORT

***Rutland County***
RUTLAND

***Washington County***
MONTPELIER AP

***Windham County***
BELLOWS FALLS
VERNON

***Windsor County***
CAVENDISH
ROCHESTER

# ELEVATION INDEX

| FEET | STATION NAME |
|---|---|
| 226 | VERNON |
| 302 | BELLOWS FALLS |
| 335 | BURLINGTON INTL AP |
| 420 | ENOSBURG FALLS |
| 502 | CORNWALL |
| 620 | RUTLAND |
| 702 | SAINT JOHNSBURY |
| 751 | CHELSEA |
| 771 | NEWPORT |
| 801 | CAVENDISH |
| 860 | ROCHESTER |
| 902 | WEST BURKE |
| 1122 | READSBORO 1 SE |
| 1129 | MONTPELIER AP |
| 3963 | MOUNT MANSFIELD |

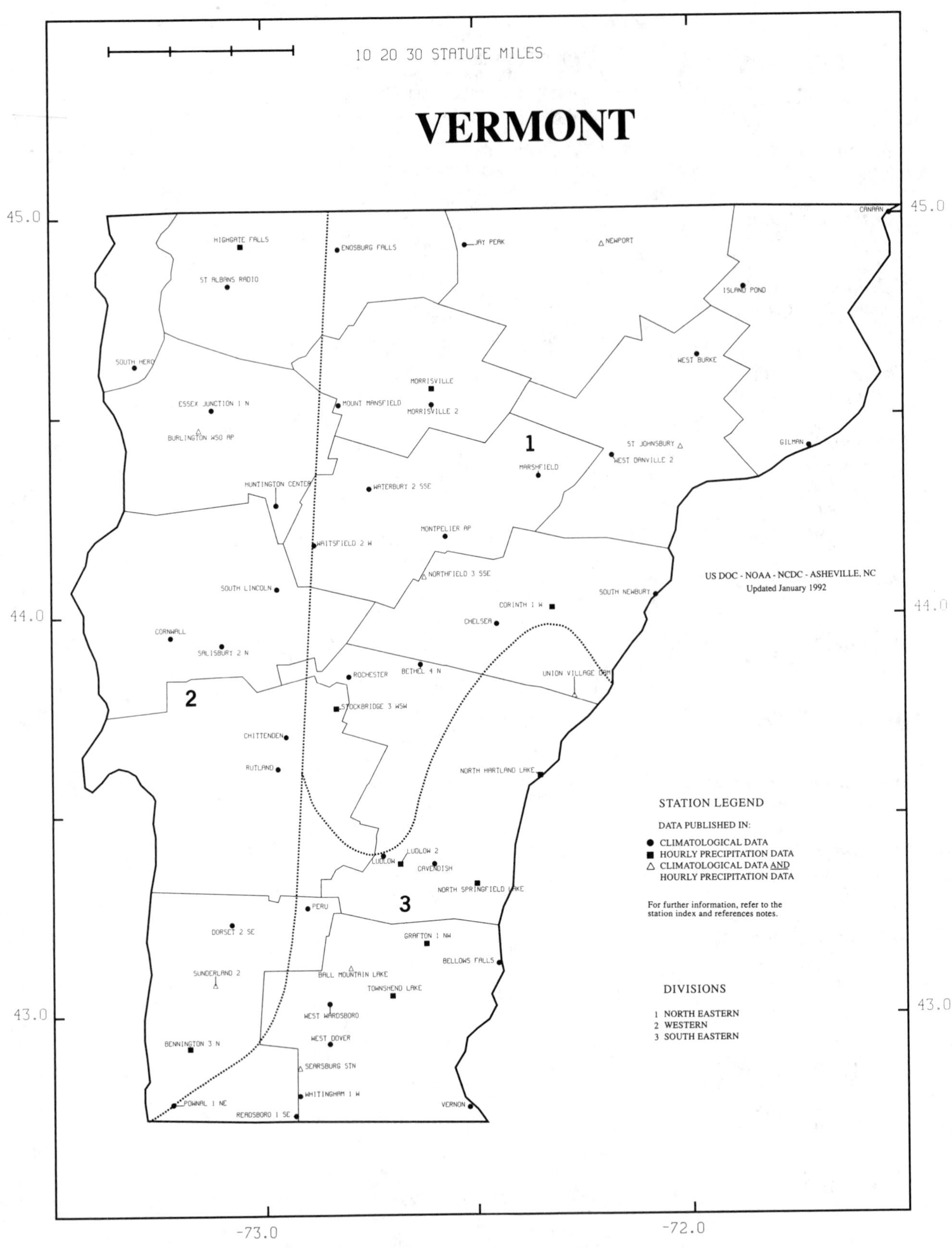
10 20 30 STATUTE MILES
VERMONT
45.0
44.0
43.0
-73.0
-72.0
HIGHGATE FALLS
ENOSBURG FALLS
JAY PEAK
NEWPORT
CANAAN
ST ALBANS RADIO
ISLAND POND
SOUTH HERO
WEST BURKE
MORRISVILLE
MOUNT MANSFIELD
MORRISVILLE 2
ESSEX JUNCTION 1 N
BURLINGTON WSO AP
1
ST JOHNSBURY
WEST DANVILLE 2
GILMAN
MARSHFIELD
HUNTINGTON CENTER
WATERBURY 2 SSE
MONTPELIER AP
WAITSFIELD 2 W
NORTHFIELD 3 SSE
US DOC - NOAA - NCDC - ASHEVILLE, NC
Updated January 1992
SOUTH LINCOLN
SOUTH NEWBURY
CORINTH 1 W
CHELSEA
CORNWALL
SALISBURY 2 N
ROCHESTER
BETHEL 4 N
UNION VILLAGE DAM
2
STOCKBRIDGE 3 WSW
CHITTENDEN
RUTLAND
NORTH HARTLAND LAKE
STATION LEGEND
DATA PUBLISHED IN:
● CLIMATOLOGICAL DATA
■ HOURLY PRECIPITATION DATA
△ CLIMATOLOGICAL DATA AND HOURLY PRECIPITATION DATA
LUDLOW 2
LUDLOW
CAVENDISH
NORTH SPRINGFIELD LAKE
PERU
3
For further information, refer to the station index and references notes.
DORSET 2 SE
GRAFTON 1 NW
BELLOWS FALLS
SUNDERLAND 2
BALL MOUNTAIN LAKE
TOWNSHEND LAKE
DIVISIONS
1 NORTH EASTERN
2 WESTERN
3 SOUTH EASTERN
WEST WARDSBORO
WEST DOVER
BENNINGTON 3 N
SEARSBURG STN
WHITINGHAM 1 W
POWNAL 1 NE
READSBORO 1 SE
VERNON

## BELLOWS FALLS *Windham County* ELEVATION 302 ft LAT/LONG 43° 8' N / 72° 27' W

| | JAN | FEB | MAR | APR | MAY | JUN | JUL | AUG | SEP | OCT | NOV | DEC | YEAR |
|---|---|---|---|---|---|---|---|---|---|---|---|---|---|
| Maximum Temp °F | 29.4 | 32.8 | 42.3 | 55.1 | 68.3 | 76.8 | 81.8 | 79.7 | 70.9 | 59.3 | 46.6 | 33.8 | 56.4 |
| Minimum Temp °F | 8.5 | 10.5 | 21.9 | 33.5 | 43.9 | 53.3 | 58.3 | 56.7 | 48.4 | 37.0 | 29.3 | 16.7 | 34.8 |
| Mean Temp °F | 19.0 | 21.7 | 32.1 | 44.3 | 56.1 | 65.1 | 70.1 | 68.2 | 59.6 | 48.2 | 38.0 | 25.3 | 45.6 |
| Days Max Temp ≥ 90 °F | 0 | 0 | 0 | 0 | 0 | 2 | 4 | 2 | 0 | 0 | 0 | 0 | 8 |
| Days Max Temp ≤ 32 °F | 18 | 13 | 4 | 0 | 0 | 0 | 0 | 0 | 0 | 0 | 1 | 13 | 49 |
| Days Min Temp ≤ 32 °F | 30 | 27 | 27 | 15 | 2 | 0 | 0 | 0 | 1 | 11 | 20 | 29 | 162 |
| Days Min Temp ≤ 0 °F | 8 | 6 | 1 | 0 | 0 | 0 | 0 | 0 | 0 | 0 | 0 | 3 | 18 |
| Heating Degree Days | 1420 | 1217 | 1012 | 615 | 287 | 78 | 15 | 33 | 189 | 515 | 804 | 1224 | 7409 |
| Cooling Degree Days | 0 | 0 | 0 | 1 | 19 | 80 | 186 | 145 | 34 | 2 | 0 | 0 | 467 |
| Total Precipitation (") | 2.82 | 2.67 | 3.21 | 3.20 | 3.51 | 3.32 | 3.49 | 3.87 | 3.51 | 3.24 | 3.59 | 3.52 | 39.95 |
| Days ≥ 0.1" Precip | 6 | 5 | 7 | 7 | 8 | 7 | 7 | 7 | 7 | 6 | 7 | 7 | 81 |
| Total Snowfall (") | 19.3 | 16.1 | 10.7 | 2.4 | 0.0 | 0.0 | 0.0 | 0.0 | 0.0 | 0.0 | 4.4 | 17.8 | 70.7 |
| Days ≥ 1" Snow Depth | 27 | 25 | 16 | 1 | 0 | 0 | 0 | 0 | 0 | 0 | 3 | 18 | 90 |

## BURLINGTON INTL AP *Chittenden County* ELEVATION 335 ft LAT/LONG 44° 28' N / 73° 9' W

| | JAN | FEB | MAR | APR | MAY | JUN | JUL | AUG | SEP | OCT | NOV | DEC | YEAR |
|---|---|---|---|---|---|---|---|---|---|---|---|---|---|
| Maximum Temp °F | 25.5 | 27.9 | 38.8 | 53.3 | 66.8 | 75.6 | 80.9 | 78.0 | 68.6 | 56.4 | 44.0 | 31.4 | 53.9 |
| Minimum Temp °F | 7.6 | 9.2 | 21.1 | 33.5 | 44.6 | 54.3 | 59.4 | 57.8 | 49.4 | 38.8 | 29.9 | 16.1 | 35.1 |
| Mean Temp °F | 16.6 | 18.6 | 29.9 | 43.5 | 55.7 | 65.0 | 70.2 | 68.0 | 59.0 | 47.6 | 36.9 | 23.8 | 44.6 |
| Days Max Temp ≥ 90 °F | 0 | 0 | 0 | 0 | 1 | 1 | 3 | 1 | 0 | 0 | 0 | 0 | 6 |
| Days Max Temp ≤ 32 °F | 21 | 18 | 9 | 0 | 0 | 0 | 0 | 0 | 0 | 0 | 4 | 16 | 68 |
| Days Min Temp ≤ 32 °F | 30 | 26 | 26 | 15 | 3 | 0 | 0 | 0 | 1 | 9 | 19 | 28 | 157 |
| Days Min Temp ≤ 0 °F | 10 | 8 | 2 | 0 | 0 | 0 | 0 | 0 | 0 | 0 | 0 | 5 | 25 |
| Heating Degree Days | 1495 | 1306 | 1080 | 642 | 303 | 84 | 19 | 42 | 206 | 533 | 835 | 1272 | 7817 |
| Cooling Degree Days | 0 | 0 | 0 | 2 | 19 | 85 | 200 | 151 | 37 | 2 | 0 | 0 | 496 |
| Total Precipitation (") | 1.89 | 1.64 | 2.25 | 2.83 | 3.13 | 3.43 | 3.63 | 4.02 | 3.49 | 3.01 | 3.10 | 2.40 | 34.82 |
| Days ≥ 0.1" Precip | 5 | 4 | 6 | 7 | 8 | 8 | 8 | 9 | 7 | 7 | 8 | 6 | 83 |
| Total Snowfall (") | 20.7 | 16.1 | 13.5 | 5.2 | 0.3 | 0.0 | 0.0 | 0.0 | 0.0 | 0.3 | 7.5 | 19.8 | 83.4 |
| Days ≥ 1" Snow Depth | 26 | 24 | 18 | 3 | 0 | 0 | 0 | 0 | 0 | 0 | 5 | 18 | 94 |

## CAVENDISH *Windsor County* ELEVATION 801 ft LAT/LONG 43° 23' N / 72° 36' W

| | JAN | FEB | MAR | APR | MAY | JUN | JUL | AUG | SEP | OCT | NOV | DEC | YEAR |
|---|---|---|---|---|---|---|---|---|---|---|---|---|---|
| Maximum Temp °F | 28.4 | 31.8 | 41.0 | 54.2 | 68.2 | 76.1 | 81.0 | 78.7 | 69.6 | 58.3 | 44.9 | 32.5 | 55.4 |
| Minimum Temp °F | 5.2 | 7.0 | 18.8 | 30.0 | 40.7 | 49.7 | 54.5 | 53.1 | 44.9 | 33.7 | 26.6 | 13.4 | 31.5 |
| Mean Temp °F | 16.8 | 19.4 | 30.0 | 42.1 | 54.5 | 63.0 | 67.7 | 65.9 | 57.3 | 46.0 | 35.8 | 23.0 | 43.5 |
| Days Max Temp ≥ 90 °F | 0 | 0 | 0 | 0 | 0 | 1 | 3 | 1 | 0 | 0 | 0 | 0 | 5 |
| Days Max Temp ≤ 32 °F | 19 | 14 | 5 | 0 | 0 | 0 | 0 | 0 | 0 | 0 | 2 | 14 | 54 |
| Days Min Temp ≤ 32 °F | 31 | 28 | 28 | 19 | 6 | 1 | 0 | 0 | 3 | 15 | 23 | 30 | 184 |
| Days Min Temp ≤ 0 °F | 12 | 10 | 3 | 0 | 0 | 0 | 0 | 0 | 0 | 0 | 0 | 5 | 30 |
| Heating Degree Days | 1488 | 1281 | 1079 | 680 | 331 | 114 | 37 | 61 | 245 | 581 | 870 | 1296 | 8063 |
| Cooling Degree Days | 0 | 0 | 0 | 0 | 12 | 51 | 128 | 93 | 18 | 1 | 0 | 0 | 303 |
| Total Precipitation (") | 3.02 | 2.81 | 3.53 | 3.84 | 4.09 | 4.16 | 3.58 | 4.07 | 3.88 | 3.58 | 4.07 | 3.77 | 44.40 |
| Days ≥ 0.1" Precip | 6 | 6 | 7 | 7 | 8 | 8 | 7 | 7 | 7 | 6 | 8 | 7 | 84 |
| Total Snowfall (") | 22.3 | 19.3 | 16.3 | 5.4 | 0.2 | 0.0 | 0.0 | 0.0 | 0.0 | 0.4 | 7.0 | 20.5 | 91.4 |
| Days ≥ 1" Snow Depth | 30 | 28 | 30 | 11 | 0 | 0 | 0 | 0 | 0 | 0 | 6 | 26 | 131 |

## CHELSEA *Orange County* ELEVATION 751 ft LAT/LONG 43° 57' N / 72° 27' W

| | JAN | FEB | MAR | APR | MAY | JUN | JUL | AUG | SEP | OCT | NOV | DEC | YEAR |
|---|---|---|---|---|---|---|---|---|---|---|---|---|---|
| Maximum Temp °F | 26.7 | 30.1 | 39.5 | 52.6 | 66.4 | 75.0 | 79.9 | 77.8 | 68.9 | 57.4 | 44.0 | 31.2 | 54.1 |
| Minimum Temp °F | -0.4 | 1.1 | 14.1 | 27.3 | 37.6 | 46.9 | 51.6 | 50.2 | 41.9 | 31.3 | 23.6 | 9.3 | 27.9 |
| Mean Temp °F | 13.2 | 15.6 | 26.8 | 40.0 | 52.0 | 60.9 | 65.8 | 64.0 | 55.4 | 44.4 | 33.8 | 20.3 | 41.0 |
| Days Max Temp ≥ 90 °F | 0 | 0 | 0 | 0 | 0 | 1 | 2 | 1 | 0 | 0 | 0 | 0 | 4 |
| Days Max Temp ≤ 32 °F | 21 | 16 | 7 | 0 | 0 | 0 | 0 | 0 | 0 | 0 | 3 | 16 | 63 |
| Days Min Temp ≤ 32 °F | 31 | 28 | 29 | 22 | 10 | 1 | 0 | 0 | 5 | 18 | 24 | 30 | 198 |
| Days Min Temp ≤ 0 °F | 17 | 15 | 5 | 0 | 0 | 0 | 0 | 0 | 0 | 0 | 0 | 8 | 45 |
| Heating Degree Days | 1602 | 1389 | 1176 | 744 | 402 | 156 | 61 | 96 | 294 | 633 | 929 | 1380 | 8862 |
| Cooling Degree Days | 0 | 0 | 0 | 0 | 4 | 29 | 84 | 66 | 10 | 0 | 0 | 0 | 193 |
| Total Precipitation (") | 2.50 | 2.13 | 2.64 | 2.88 | 3.39 | 3.27 | 3.51 | 4.02 | 3.44 | 3.20 | 3.37 | 3.01 | 37.36 |
| Days ≥ 0.1" Precip | 6 | 5 | 6 | 7 | 8 | 8 | 7 | 8 | 7 | 7 | 8 | 7 | 84 |
| Total Snowfall (") | 19.3 | 16.2 | 13.9 | 4.9 | 0.2 | 0.0 | 0.0 | 0.0 | 0.0 | 0.3 | 6.4 | 19.7 | 80.9 |
| Days ≥ 1" Snow Depth | 30 | 28 | 27 | 6 | 0 | 0 | 0 | 0 | 0 | 0 | 7 | 25 | 123 |

## CORNWALL *Addison County* ELEVATION 502 ft LAT/LONG 43° 58' N / 73° 12' W

| | JAN | FEB | MAR | APR | MAY | JUN | JUL | AUG | SEP | OCT | NOV | DEC | YEAR |
|---|---|---|---|---|---|---|---|---|---|---|---|---|---|
| Maximum Temp °F | 27.8 | 30.9 | 41.5 | 55.9 | 68.9 | 76.9 | 81.7 | 79.3 | 70.4 | 58.4 | 45.5 | 32.8 | 55.8 |
| Minimum Temp °F | 9.2 | 11.0 | 22.0 | 33.5 | 44.7 | 53.5 | 58.3 | 56.6 | 48.7 | 38.4 | 29.5 | 16.7 | 35.2 |
| Mean Temp °F | 18.5 | 20.9 | 31.8 | 44.7 | 56.8 | 65.3 | 70.0 | 68.0 | 59.6 | 48.4 | 37.5 | 24.8 | 45.5 |
| Days Max Temp ≥ 90 °F | 0 | 0 | 0 | 0 | 0 | 1 | 2 | 1 | 0 | 0 | 0 | 0 | 4 |
| Days Max Temp ≤ 32 °F | 20 | 15 | 5 | 0 | 0 | 0 | 0 | 0 | 0 | 0 | 2 | 14 | 56 |
| Days Min Temp ≤ 32 °F | 30 | 26 | 26 | 15 | 2 | 0 | 0 | 0 | 1 | 8 | 19 | 29 | 156 |
| Days Min Temp ≤ 0 °F | 9 | 7 | 1 | 0 | 0 | 0 | 0 | 0 | 0 | 0 | 0 | 4 | 21 |
| Heating Degree Days | 1436 | 1238 | 1023 | 605 | 268 | 74 | 17 | 37 | 190 | 509 | 818 | 1240 | 7455 |
| Cooling Degree Days | 0 | 0 | 0 | 3 | 19 | 76 | 179 | 131 | 32 | 1 | 0 | 0 | 441 |
| Total Precipitation (") | 2.06 | 1.81 | 2.34 | 2.60 | 3.33 | 3.14 | 3.36 | 4.27 | 3.52 | 3.00 | 3.16 | 2.72 | 35.31 |
| Days ≥ 0.1" Precip | 5 | 4 | 6 | 6 | 8 | 8 | 7 | 8 | 7 | 6 | 7 | 6 | 78 |
| Total Snowfall (") | 17.6 | 14.3 | 10.7 | 3.4 | 0.1 | 0.0 | 0.0 | 0.0 | 0.0 | 0.4 | 5.1 | 16.6 | 68.2 |
| Days ≥ 1" Snow Depth | na | na | na | na | 0 | 0 | 0 | 0 | 0 | 0 | na | na | na |

## ENOSBURG FALLS *Franklin County* ELEVATION 420 ft LAT/LONG 44° 55' N / 72° 48' W

| | JAN | FEB | MAR | APR | MAY | JUN | JUL | AUG | SEP | OCT | NOV | DEC | YEAR |
|---|---|---|---|---|---|---|---|---|---|---|---|---|---|
| Maximum Temp °F | 27.0 | 29.8 | 41.0 | 55.2 | 68.5 | 76.1 | 80.4 | 77.9 | 69.4 | 58.3 | 44.5 | 32.2 | 55.0 |
| Minimum Temp °F | 4.1 | 5.3 | 18.2 | 30.9 | 41.9 | 51.2 | 55.9 | 54.4 | 46.7 | 36.5 | 27.0 | 12.6 | 32.1 |
| Mean Temp °F | 15.6 | 17.6 | 29.6 | 43.1 | 55.2 | 63.7 | 68.2 | 66.2 | 58.1 | 47.5 | 35.8 | 22.5 | 43.6 |
| Days Max Temp ≥ 90 °F | 0 | 0 | 0 | 0 | 0 | 1 | 2 | 1 | 0 | 0 | 0 | 0 | 4 |
| Days Max Temp ≤ 32 °F | 20 | 16 | 7 | 1 | 0 | 0 | 0 | 0 | 0 | 0 | 4 | 15 | 63 |
| Days Min Temp ≤ 32 °F | 30 | 27 | 27 | 18 | 6 | 0 | 0 | 0 | 2 | 11 | 21 | 29 | 171 |
| Days Min Temp ≤ 0 °F | 13 | 11 | 4 | 0 | 0 | 0 | 0 | 0 | 0 | 0 | 0 | 7 | 35 |
| Heating Degree Days | 1527 | 1333 | 1089 | 653 | 313 | 103 | 36 | 61 | 227 | 539 | 870 | 1313 | 8064 |
| Cooling Degree Days | 0 | 0 | 0 | 1 | 15 | 63 | 146 | 107 | 26 | 2 | 0 | 0 | 360 |
| Total Precipitation (") | 2.56 | 2.04 | 2.80 | 3.33 | 3.79 | 4.15 | 4.17 | 4.93 | 4.22 | 3.97 | 4.40 | 3.24 | 43.60 |
| Days ≥ 0.1" Precip | *7* | *6* | 8 | 9 | 10 | 9 | 9 | 10 | 8 | 10 | 10 | *9* | 105 |
| Total Snowfall (") | 23.1 | 19.8 | 14.4 | 5.2 | 0.2 | 0.0 | 0.0 | 0.0 | 0.0 | 0.6 | 9.2 | 20.6 | 93.1 |
| Days ≥ 1" Snow Depth | na | na | na | *1* | 0 | 0 | 0 | 0 | 0 | 0 | na | na | na |

## MONTPELIER AP *Washington County* ELEVATION 1129 ft LAT/LONG 44° 12' N / 72° 34' W

| | JAN | FEB | MAR | APR | MAY | JUN | JUL | AUG | SEP | OCT | NOV | DEC | YEAR |
|---|---|---|---|---|---|---|---|---|---|---|---|---|---|
| Maximum Temp °F | 24.4 | 27.3 | 37.3 | 51.1 | 64.9 | 72.9 | 78.1 | 75.6 | 66.8 | 55.3 | 42.1 | 29.6 | 52.1 |
| Minimum Temp °F | 6.1 | 8.2 | 19.4 | 31.7 | 42.4 | 51.1 | 55.9 | 54.1 | 46.1 | 36.0 | 27.3 | 13.5 | 32.7 |
| Mean Temp °F | 15.4 | 17.8 | 28.4 | 41.5 | 53.7 | 62.0 | 67.0 | 64.9 | 56.4 | 45.7 | 34.7 | 21.6 | 42.4 |
| Days Max Temp ≥ 90 °F | 0 | 0 | 0 | 0 | 0 | 0 | 1 | 1 | 0 | 0 | 0 | 0 | 2 |
| Days Max Temp ≤ 32 °F | 23 | 19 | 10 | 1 | 0 | 0 | 0 | 0 | 0 | 0 | 6 | 18 | 77 |
| Days Min Temp ≤ 32 °F | 30 | 27 | 27 | 17 | 4 | 0 | 0 | 0 | 1 | 11 | 21 | 29 | 167 |
| Days Min Temp ≤ 0 °F | 11 | 8 | 2 | 0 | 0 | 0 | 0 | 0 | 0 | 0 | 0 | 6 | 27 |
| Heating Degree Days | 1534 | 1328 | 1128 | 701 | 357 | 131 | 45 | 81 | 267 | 592 | 903 | 1340 | 8407 |
| Cooling Degree Days | 0 | 0 | 0 | 1 | 10 | 47 | 118 | 84 | 15 | 1 | 0 | 0 | 276 |
| Total Precipitation (") | 2.32 | 1.96 | 2.32 | 2.45 | 3.24 | 3.49 | 2.99 | 4.01 | 3.14 | 2.85 | 3.10 | 2.81 | 34.68 |
| Days ≥ 0.1" Precip | 5 | 5 | 6 | 6 | 7 | 8 | 7 | 7 | 7 | 6 | 7 | 7 | 78 |
| Total Snowfall (") | 21.7 | 18.2 | 15.7 | 5.8 | 0.4 | 0.0 | 0.0 | 0.0 | 0.0 | 1.0 | 10.5 | 23.7 | 97.0 |
| Days ≥ 1" Snow Depth | 29 | 26 | 24 | 7 | 0 | 0 | 0 | 0 | 0 | 0 | 8 | 24 | 118 |

## MOUNT MANSFIELD *Lamoille County* ELEVATION 3963 ft LAT/LONG 44° 32' N / 72° 49' W

| | JAN | FEB | MAR | APR | MAY | JUN | JUL | AUG | SEP | OCT | NOV | DEC | YEAR |
|---|---|---|---|---|---|---|---|---|---|---|---|---|---|
| Maximum Temp °F | 16.9 | 18.4 | 27.1 | 38.9 | 53.2 | 61.4 | 65.5 | 63.2 | 55.3 | 44.2 | 32.2 | 21.9 | 41.5 |
| Minimum Temp °F | 0.6 | 2.7 | 12.3 | 24.4 | 36.9 | 46.1 | 50.9 | 49.6 | 41.6 | 30.5 | 19.6 | 6.8 | 26.8 |
| Mean Temp °F | 8.7 | 10.5 | 19.8 | 31.6 | 45.1 | 53.8 | 58.2 | 56.4 | 48.5 | 37.4 | 26.0 | 14.4 | 34.2 |
| Days Max Temp ≥ 90 °F | 0 | 0 | 0 | 0 | 0 | 0 | 0 | 0 | 0 | 0 | 0 | 0 | 0 |
| Days Max Temp ≤ 32 °F | 29 | 25 | 22 | 10 | 1 | 0 | 0 | 0 | 0 | 5 | 17 | 26 | 135 |
| Days Min Temp ≤ 32 °F | 31 | 28 | 30 | 24 | 10 | 2 | 0 | 0 | 5 | 19 | 27 | 31 | 207 |
| Days Min Temp ≤ 0 °F | 15 | 12 | 6 | 0 | 0 | 0 | 0 | 0 | 0 | 0 | 1 | 10 | 44 |
| Heating Degree Days | 1741 | 1534 | 1394 | 993 | 607 | 335 | 214 | 263 | 489 | 852 | 1164 | 1564 | 11150 |
| Cooling Degree Days | 0 | 0 | 0 | 0 | 0 | 7 | 12 | 5 | 0 | 0 | 0 | 0 | 24 |
| Total Precipitation (") | 5.14 | 4.01 | 5.50 | 5.74 | 5.59 | 6.85 | 6.57 | 7.79 | 6.84 | 5.86 | 7.27 | 6.29 | 73.45 |
| Days ≥ 0.1" Precip | 15 | 11 | 13 | 11 | 11 | 11 | 11 | 11 | 11 | 11 | 15 | 16 | 147 |
| Total Snowfall (") | 43.1 | 31.7 | 35.3 | 21.5 | 3.8 | 0.2 | 0.0 | 0.0 | 0.4 | 7.6 | 32.4 | 43.1 | 219.1 |
| Days ≥ 1" Snow Depth | 31 | 28 | 31 | 30 | 23 | 1 | 0 | 0 | 0 | 7 | 21 | 30 | 202 |

## NEWPORT *Orleans County* ELEVATION 771 ft LAT/LONG 44° 56 ' N / 72° 12 ' W

| | JAN | FEB | MAR | APR | MAY | JUN | JUL | AUG | SEP | OCT | NOV | DEC | YEAR |
|---|---|---|---|---|---|---|---|---|---|---|---|---|---|
| Maximum Temp °F | 24.3 | 28.4 | 39.0 | 52.6 | 66.7 | 74.9 | 79.5 | 77.1 | 68.1 | 55.8 | 42.0 | 29.2 | 53.1 |
| Minimum Temp °F | 3.3 | 5.1 | 17.2 | 30.4 | 41.8 | 51.4 | 56.1 | 54.3 | 46.0 | 36.4 | 26.8 | 11.7 | 31.7 |
| Mean Temp °F | 13.9 | 16.8 | 28.1 | 41.5 | 54.3 | 63.2 | 67.8 | 65.7 | 57.1 | 46.1 | 34.4 | 20.5 | 42.5 |
| Days Max Temp ≥ 90 °F | 0 | 0 | 0 | 0 | 0 | 1 | 2 | 1 | 0 | 0 | 0 | 0 | 4 |
| Days Max Temp ≤ 32 °F | 22 | 18 | 8 | 1 | 0 | 0 | 0 | 0 | 0 | 0 | 5 | 18 | 72 |
| Days Min Temp ≤ 32 °F | 31 | 27 | 28 | 19 | 5 | 0 | 0 | 0 | 2 | 11 | 22 | 29 | 174 |
| Days Min Temp ≤ 0 °F | 13 | 12 | 4 | 0 | 0 | 0 | 0 | 0 | 0 | 0 | 0 | 7 | 36 |
| Heating Degree Days | 1581 | 1356 | 1137 | 699 | 340 | 110 | 34 | 67 | 251 | 580 | 912 | 1374 | 8441 |
| Cooling Degree Days | 0 | 0 | 0 | 2 | 13 | 59 | 137 | 100 | 20 | 1 | 0 | 0 | 332 |
| Total Precipitation (") | 2.57 | 2.26 | 2.81 | 2.90 | 3.50 | 4.01 | 3.92 | 4.47 | 3.67 | 3.35 | 3.58 | 3.27 | 40.31 |
| Days ≥ 0.1" Precip | 7 | 7 | 8 | 8 | 9 | 8 | 8 | 9 | 8 | 8 | 9 | 8 | 97 |
| Total Snowfall (") | 22.6 | 18.3 | 17.3 | 6.2 | 0.4 | 0.0 | 0.0 | 0.0 | 0.0 | 1.0 | 10.9 | 23.2 | 99.9 |
| Days ≥ 1" Snow Depth | 30 | 28 | 28 | 8 | 0 | 0 | 0 | 0 | 0 | 0 | 10 | 25 | 129 |

## READSBORO 1 SE *Bennington County* ELEVATION 1122 ft LAT/LONG 42° 45 ' N / 72° 56 ' W

| | JAN | FEB | MAR | APR | MAY | JUN | JUL | AUG | SEP | OCT | NOV | DEC | YEAR |
|---|---|---|---|---|---|---|---|---|---|---|---|---|---|
| Maximum Temp °F | 28.9 | 31.5 | 40.2 | 53.1 | 66.1 | 74.5 | 79.7 | 77.7 | 69.1 | 58.2 | 45.6 | 33.6 | 54.9 |
| Minimum Temp °F | 8.2 | 9.2 | 19.7 | 30.9 | 41.1 | 49.9 | 54.7 | 53.2 | 45.4 | 34.7 | 27.6 | 15.6 | 32.5 |
| Mean Temp °F | 18.6 | 20.4 | 30.0 | 42.0 | 53.6 | 62.2 | 67.2 | 65.5 | 57.3 | 46.5 | 36.6 | 24.7 | 43.7 |
| Days Max Temp ≥ 90 °F | 0 | 0 | 0 | 0 | 0 | 1 | 2 | 1 | 0 | 0 | 0 | 0 | 4 |
| Days Max Temp ≤ 32 °F | 19 | 15 | 7 | 0 | 0 | 0 | 0 | 0 | 0 | 0 | 2 | 14 | 57 |
| Days Min Temp ≤ 32 °F | 31 | 28 | 28 | 18 | 4 | 0 | 0 | 0 | 2 | 14 | 22 | 29 | 176 |
| Days Min Temp ≤ 0 °F | 9 | 7 | 2 | 0 | 0 | 0 | 0 | 0 | 0 | 0 | 0 | 4 | 22 |
| Heating Degree Days | 1433 | 1253 | 1079 | 683 | 356 | 125 | 38 | 63 | 246 | 568 | 844 | 1244 | 7932 |
| Cooling Degree Days | 0 | 0 | 0 | 1 | 10 | 44 | 117 | 84 | 19 | 1 | 0 | 0 | 276 |
| Total Precipitation (") | 3.47 | 3.29 | 4.09 | 4.31 | 4.66 | 4.71 | 4.17 | 4.39 | 4.01 | 3.93 | 4.68 | 4.37 | 50.08 |
| Days ≥ 0.1" Precip | 7 | 6 | 8 | 8 | 8 | 7 | 7 | 8 | 7 | 7 | 9 | 8 | 90 |
| Total Snowfall (") | 21.9 | 16.7 | 15.5 | 3.9 | 0.2 | 0.0 | 0.0 | 0.0 | 0.0 | 0.1 | 6.2 | 19.3 | 83.8 |
| Days ≥ 1" Snow Depth | 26 | 25 | 24 | 8 | 0 | 0 | 0 | 0 | 0 | 0 | 4 | 19 | 106 |

## ROCHESTER *Windsor County* ELEVATION 860 ft LAT/LONG 43° 53 ' N / 72° 49 ' W

| | JAN | FEB | MAR | APR | MAY | JUN | JUL | AUG | SEP | OCT | NOV | DEC | YEAR |
|---|---|---|---|---|---|---|---|---|---|---|---|---|---|
| Maximum Temp °F | 28.1 | 31.3 | 40.5 | 53.1 | 66.6 | 74.9 | 79.7 | 77.5 | 69.4 | 58.4 | 45.3 | 32.7 | 54.8 |
| Minimum Temp °F | 4.1 | 5.0 | 16.6 | 28.9 | 39.2 | 48.5 | 53.2 | 52.2 | 44.2 | 33.7 | 25.9 | 12.5 | 30.3 |
| Mean Temp °F | 16.1 | 18.2 | 28.6 | 41.0 | 52.9 | 61.7 | 66.5 | 64.9 | 56.8 | 46.1 | 35.6 | 22.6 | 42.6 |
| Days Max Temp ≥ 90 °F | 0 | 0 | 0 | 0 | 0 | 1 | 1 | 1 | 0 | 0 | 0 | 0 | 3 |
| Days Max Temp ≤ 32 °F | 20 | 15 | 6 | 0 | 0 | 0 | 0 | 0 | 0 | 0 | 2 | 14 | 57 |
| Days Min Temp ≤ 32 °F | 31 | 28 | 29 | 21 | 7 | 0 | 0 | 0 | 2 | 15 | 23 | 30 | 186 |
| Days Min Temp ≤ 0 °F | 13 | 11 | 4 | 0 | 0 | 0 | 0 | 0 | 0 | 0 | 0 | 7 | 35 |
| Heating Degree Days | 1510 | 1315 | 1123 | 713 | 377 | 139 | 50 | 76 | 254 | 580 | 875 | 1308 | 8320 |
| Cooling Degree Days | 0 | 0 | 0 | 0 | 7 | 35 | 102 | 79 | 15 | 1 | 0 | 0 | 239 |
| Total Precipitation (") | 3.26 | 2.82 | 3.34 | 3.79 | 4.27 | 3.89 | 3.86 | 4.90 | 4.20 | 3.99 | 4.28 | 3.94 | 46.54 |
| Days ≥ 0.1" Precip | 6 | 6 | 7 | 8 | 9 | 8 | 8 | 9 | 7 | 8 | 8 | 8 | 92 |
| Total Snowfall (") | 22.6 | 19.1 | 14.9 | 4.8 | 0.1 | 0.0 | 0.0 | 0.0 | 0.0 | 0.7 | 6.8 | 21.4 | 90.4 |
| Days ≥ 1" Snow Depth | 28 | 26 | 21 | 5 | 0 | 0 | 0 | 0 | 0 | 0 | 6 | 22 | 108 |

## RUTLAND *Rutland County* ELEVATION 620 ft LAT/LONG 43° 36 ' N / 72° 58 ' W

| | JAN | FEB | MAR | APR | MAY | JUN | JUL | AUG | SEP | OCT | NOV | DEC | YEAR |
|---|---|---|---|---|---|---|---|---|---|---|---|---|---|
| Maximum Temp °F | 30.0 | 32.8 | 43.3 | 57.1 | 69.8 | 77.9 | 82.2 | 79.8 | 71.0 | 59.7 | 46.9 | 34.4 | 57.1 |
| Minimum Temp °F | 10.4 | 12.0 | 22.6 | 33.8 | 44.2 | 53.0 | 57.9 | 56.7 | 48.5 | 38.2 | 29.9 | 17.3 | 35.4 |
| Mean Temp °F | 20.2 | 22.4 | 33.0 | 45.4 | 57.0 | 65.5 | 70.1 | 68.2 | 59.8 | 49.0 | 38.4 | 25.9 | 46.2 |
| Days Max Temp ≥ 90 °F | 0 | 0 | 0 | 0 | 0 | 1 | 3 | 1 | 0 | 0 | 0 | 0 | 5 |
| Days Max Temp ≤ 32 °F | 18 | 14 | 5 | 0 | 0 | 0 | 0 | 0 | 0 | 0 | 2 | 13 | 52 |
| Days Min Temp ≤ 32 °F | 30 | 27 | 25 | 14 | 3 | 0 | 0 | 0 | 1 | 9 | 19 | 28 | 156 |
| Days Min Temp ≤ 0 °F | 8 | 6 | 1 | 0 | 0 | 0 | 0 | 0 | 0 | 0 | 0 | 3 | 18 |
| Heating Degree Days | 1382 | 1196 | 986 | 582 | 262 | 69 | 15 | 33 | 183 | 492 | 792 | 1205 | 7197 |
| Cooling Degree Days | 0 | 0 | 0 | 1 | 19 | 77 | 176 | 130 | 29 | 2 | 0 | 0 | 434 |
| Total Precipitation (") | 2.18 | 1.86 | 2.46 | 2.81 | 3.53 | 3.64 | 4.03 | 4.16 | 3.90 | 3.03 | 3.12 | 2.78 | 37.50 |
| Days ≥ 0.1" Precip | 5 | 5 | *6* | 7 | 8 | 8 | 7 | 8 | 6 | 6 | *7* | na | na |
| Total Snowfall (") | 15.2 | 14.9 | 10.8 | 3.2 | 0.1 | 0.0 | 0.0 | 0.0 | 0.0 | 0.4 | 4.8 | 14.4 | 63.8 |
| Days ≥ 1" Snow Depth | *25* | *22* | 12 | 1 | 0 | 0 | 0 | 0 | 0 | 0 | 3 | 17 | 80 |

## SAINT JOHNSBURY *Caledonia County* ELEVATION 702 ft LAT/LONG 44° 25 ' N / 72° 1 ' W

| | JAN | FEB | MAR | APR | MAY | JUN | JUL | AUG | SEP | OCT | NOV | DEC | YEAR |
|---|---|---|---|---|---|---|---|---|---|---|---|---|---|
| Maximum Temp °F | 26.9 | 31.0 | 41.6 | 55.9 | 70.0 | 78.0 | 82.3 | 79.8 | 71.0 | 58.8 | 43.9 | 31.1 | 55.9 |
| Minimum Temp °F | 5.6 | 7.5 | 19.2 | 31.0 | 42.2 | 51.7 | 56.6 | 55.2 | 47.2 | 36.4 | 27.5 | 13.2 | 32.8 |
| Mean Temp °F | 16.3 | 19.3 | 30.4 | 43.5 | 56.2 | 64.9 | 69.5 | 67.5 | 59.1 | 47.6 | 35.7 | 22.2 | 44.4 |
| Days Max Temp ≥ 90 °F | 0 | 0 | 0 | 0 | 1 | 2 | 3 | 1 | 0 | 0 | 0 | 0 | 7 |
| Days Max Temp ≤ 32 °F | 21 | 16 | 6 | 0 | 0 | 0 | 0 | 0 | 0 | 0 | 3 | 16 | 62 |
| Days Min Temp ≤ 32 °F | 30 | 27 | 27 | 17 | 5 | 0 | 0 | 0 | 2 | 12 | 21 | 29 | 170 |
| Days Min Temp ≤ 0 °F | 12 | 10 | 3 | 0 | 0 | 0 | 0 | 0 | 0 | 0 | 0 | 6 | 31 |
| Heating Degree Days | 1505 | 1285 | 1065 | 640 | 286 | 80 | 21 | 40 | 201 | 535 | 872 | 1320 | 7850 |
| Cooling Degree Days | 0 | 0 | 0 | 1 | 19 | 77 | 174 | 127 | 29 | 2 | 0 | 0 | 429 |
| Total Precipitation (") | 2.39 | 2.09 | 2.51 | 2.63 | 3.25 | 3.95 | 3.67 | 4.00 | 3.26 | 3.06 | 3.33 | 3.26 | 37.40 |
| Days ≥ 0.1" Precip | 6 | 5 | 6 | 7 | 8 | 8 | 8 | 8 | 7 | 7 | 7 | 8 | 85 |
| Total Snowfall (") | 22.0 | 18.2 | 14.0 | 4.8 | 0.0 | 0.0 | 0.0 | 0.0 | 0.0 | 0.4 | 6.7 | 22.5 | 88.6 |
| Days ≥ 1" Snow Depth | 30 | 28 | 22 | 4 | 0 | 0 | 0 | 0 | 0 | 0 | 7 | 25 | 116 |

## VERNON *Windham County* ELEVATION 226 ft LAT/LONG 42° 46 ' N / 72° 31 ' W

| | JAN | FEB | MAR | APR | MAY | JUN | JUL | AUG | SEP | OCT | NOV | DEC | YEAR |
|---|---|---|---|---|---|---|---|---|---|---|---|---|---|
| Maximum Temp °F | 31.4 | 34.8 | 44.3 | 57.3 | 70.0 | 78.4 | 83.5 | 81.6 | 73.0 | 61.4 | 48.4 | 35.9 | 58.3 |
| Minimum Temp °F | 9.0 | 11.4 | 22.9 | 33.3 | 43.9 | 53.2 | 58.4 | 56.9 | 48.7 | 37.0 | 29.3 | 17.0 | 35.1 |
| Mean Temp °F | 20.2 | 23.1 | 33.7 | 45.3 | 57.0 | 65.8 | 71.0 | 69.2 | 60.9 | 49.2 | 38.9 | 26.4 | 46.7 |
| Days Max Temp ≥ 90 °F | 0 | 0 | 0 | 0 | 1 | 3 | 5 | 4 | 1 | 0 | 0 | 0 | 14 |
| Days Max Temp ≤ 32 °F | 16 | 11 | 3 | 0 | 0 | 0 | 0 | 0 | 0 | 0 | 1 | 10 | 41 |
| Days Min Temp ≤ 32 °F | 31 | 27 | 27 | 15 | 3 | 0 | 0 | 0 | 1 | 11 | 21 | 29 | 165 |
| Days Min Temp ≤ 0 °F | 8 | 6 | 1 | 0 | 0 | 0 | 0 | 0 | 0 | 0 | 0 | 3 | 18 |
| Heating Degree Days | 1383 | 1177 | 965 | 586 | 266 | 68 | 12 | 26 | 162 | 484 | 776 | 1188 | 7093 |
| Cooling Degree Days | 0 | 0 | 0 | 2 | 29 | 103 | 220 | 176 | 46 | 4 | 0 | 0 | 580 |
| Total Precipitation (") | 3.28 | 3.13 | 3.83 | 3.82 | 4.20 | 4.09 | 3.69 | 4.08 | 3.72 | 3.66 | 4.20 | 3.91 | 45.61 |
| Days ≥ 0.1" Precip | 6 | 6 | 7 | 7 | 8 | 7 | 6 | 7 | 6 | 6 | 8 | 7 | 81 |
| Total Snowfall (") | 15.4 | 14.0 | 9.9 | 2.1 | 0.0 | 0.0 | 0.0 | 0.0 | 0.0 | 0.0 | 3.5 | 13.6 | 58.5 |
| Days ≥ 1" Snow Depth | 26 | 24 | 17 | 1 | 0 | 0 | 0 | 0 | 0 | 0 | 3 | 18 | 89 |

## WEST BURKE *Caledonia County* ELEVATION 902 ft LAT/LONG 44° 39 ' N / 71° 59 ' W

| | JAN | FEB | MAR | APR | MAY | JUN | JUL | AUG | SEP | OCT | NOV | DEC | YEAR |
|---|---|---|---|---|---|---|---|---|---|---|---|---|---|
| Maximum Temp °F | 23.2 | 26.7 | 37.2 | 50.8 | 65.5 | 74.2 | 79.0 | 76.8 | 67.2 | 54.3 | 40.9 | 28.1 | 52.0 |
| Minimum Temp °F | -2.8 | -2.2 | 10.8 | 25.5 | 36.0 | 45.7 | 50.4 | 48.7 | 40.4 | 30.5 | 22.0 | 6.7 | 26.0 |
| Mean Temp °F | 10.2 | 12.3 | 24.0 | 38.2 | 50.8 | 60.0 | 64.7 | 62.7 | 53.8 | 42.4 | 31.5 | 17.4 | 39.0 |
| Days Max Temp ≥ 90 °F | 0 | 0 | 0 | 0 | 0 | 1 | 2 | 1 | 0 | 0 | 0 | 0 | 4 |
| Days Max Temp ≤ 32 °F | 23 | 19 | 10 | 1 | 0 | 0 | 0 | 0 | 0 | 0 | 6 | 19 | 78 |
| Days Min Temp ≤ 32 °F | 31 | 28 | 30 | 24 | 13 | 2 | 0 | 1 | 7 | 19 | 26 | 30 | 211 |
| Days Min Temp ≤ 0 °F | 19 | 17 | 8 | 0 | 0 | 0 | 0 | 0 | 0 | 0 | 1 | 10 | 55 |
| Heating Degree Days | 1695 | 1484 | 1264 | 797 | 440 | 178 | 81 | 122 | 339 | 693 | 999 | 1471 | 9563 |
| Cooling Degree Days | 0 | 0 | 0 | 0 | 4 | 31 | 84 | 64 | 9 | 0 | 0 | 0 | 192 |
| Total Precipitation (") | 2.69 | 2.26 | 2.80 | 2.78 | 3.58 | 4.26 | 4.07 | 4.66 | 3.68 | 3.46 | 3.62 | 3.45 | 41.31 |
| Days ≥ 0.1" Precip | 7 | 6 | 8 | 7 | 9 | 9 | 9 | 9 | 8 | 8 | 9 | 9 | 98 |
| Total Snowfall (") | 22.7 | 18.9 | 15.8 | 5.2 | 0.0 | 0.0 | 0.0 | 0.0 | 0.0 | 0.5 | 8.7 | 24.1 | 95.9 |
| Days ≥ 1" Snow Depth | 30 | 28 | 29 | 10 | 0 | 0 | 0 | 0 | 0 | 0 | 10 | 28 | 135 |

## JANUARY MINIMUM TEMPERATURE °F

### LOWEST

| | | |
|---|---|---|
| 1 | West Burke | -2.8 |
| 2 | Chelsea | -0.4 |
| 3 | Mount Mansfield | 0.6 |
| 4 | Newport | 3.3 |
| 5 | Enosburg Falls | 4.1 |
| | Rochester | 4.1 |
| 7 | Cavendish | 5.2 |
| 8 | St. Johnsbury | 5.6 |
| 9 | Montpelier | 6.1 |
| 10 | Burlington | 7.6 |
| 11 | Readsboro | 8.2 |
| 12 | Bellows Falls | 8.5 |
| 13 | Vernon | 9.0 |
| 14 | Cornwall | 9.2 |
| 15 | Rutland | 10.4 |

### HIGHEST

| | | |
|---|---|---|
| 1 | Rutland | 10.4 |
| 2 | Cornwall | 9.2 |
| 3 | Vernon | 9.0 |
| 4 | Bellows Falls | 8.5 |
| 5 | Readsboro | 8.2 |
| 6 | Burlington | 7.6 |
| 7 | Montpelier | 6.1 |
| 8 | St. Johnsbury | 5.6 |
| 9 | Cavendish | 5.2 |
| 10 | Enosburg Falls | 4.1 |
| | Rochester | 4.1 |
| 12 | Newport | 3.3 |
| 13 | Mount Mansfield | 0.6 |
| 14 | Chelsea | -0.4 |
| 15 | West Burke | -2.8 |

## JULY MAXIMUM TEMPERATURE °F

### HIGHEST

| | | |
|---|---|---|
| 1 | Vernon | 83.5 |
| 2 | St. Johnsbury | 82.3 |
| 3 | Rutland | 82.2 |
| 4 | Bellows Falls | 81.8 |
| 5 | Cornwall | 81.7 |
| 6 | Cavendish | 81.0 |
| 7 | Burlington | 80.9 |
| 8 | Enosburg Falls | 80.4 |
| 9 | Chelsea | 79.9 |
| 10 | Readsboro | 79.7 |
| | Rochester | 79.7 |
| 12 | Newport | 79.5 |
| 13 | West Burke | 79.0 |
| 14 | Montpelier | 78.1 |
| 15 | Mount Mansfield | 65.5 |

### LOWEST

| | | |
|---|---|---|
| 1 | Mount Mansfield | 65.5 |
| 2 | Montpelier | 78.1 |
| 3 | West Burke | 79.0 |
| 4 | Newport | 79.5 |
| 5 | Readsboro | 79.7 |
| | Rochester | 79.7 |
| 7 | Chelsea | 79.9 |
| 8 | Enosburg Falls | 80.4 |
| 9 | Burlington | 80.9 |
| 10 | Cavendish | 81.0 |
| 11 | Cornwall | 81.7 |
| 12 | Bellows Falls | 81.8 |
| 13 | Rutland | 82.2 |
| 14 | St. Johnsbury | 82.3 |
| 15 | Vernon | 83.5 |

## ANNUAL PRECIPITATION (")

### HIGHEST

| | | |
|---|---|---|
| 1 | Mount Mansfield | 73.45 |
| 2 | Readsboro | 50.08 |
| 3 | Rochester | 46.54 |
| 4 | Vernon | 45.61 |
| 5 | Cavendish | 44.40 |
| 6 | Enosburg Falls | 43.60 |
| 7 | West Burke | 41.31 |
| 8 | Newport | 40.31 |
| 9 | Bellows Falls | 39.95 |
| 10 | Rutland | 37.50 |
| 11 | St. Johnsbury | 37.40 |
| 12 | Chelsea | 37.36 |
| 13 | Cornwall | 35.31 |
| 14 | Burlington | 34.82 |
| 15 | Montpelier | 34.68 |

### LOWEST

| | | |
|---|---|---|
| 1 | Montpelier | 34.68 |
| 2 | Burlington | 34.82 |
| 3 | Cornwall | 35.31 |
| 4 | Chelsea | 37.36 |
| 5 | St. Johnsbury | 37.40 |
| 6 | Rutland | 37.50 |
| 7 | Bellows Falls | 39.95 |
| 8 | Newport | 40.31 |
| 9 | West Burke | 41.31 |
| 10 | Enosburg Falls | 43.60 |
| 11 | Cavendish | 44.40 |
| 12 | Vernon | 45.61 |
| 13 | Rochester | 46.54 |
| 14 | Readsboro | 50.08 |
| 15 | Mount Mansfield | 73.45 |

## ANNUAL SNOWFALL (")

### HIGHEST

| | | |
|---|---|---|
| 1 | Mount Mansfield | 219.1 |
| 2 | Newport | 99.9 |
| 3 | Montpelier | 97.0 |
| 4 | West Burke | 95.9 |
| 5 | Enosburg Falls | 93.1 |
| 6 | Cavendish | 91.4 |
| 7 | Rochester | 90.4 |
| 8 | St. Johnsbury | 88.6 |
| 9 | Readsboro | 83.8 |
| 10 | Burlington | 83.4 |
| 11 | Chelsea | 80.9 |
| 12 | Bellows Falls | 70.7 |
| 13 | Cornwall | 68.2 |
| 14 | Rutland | 63.8 |
| 15 | Vernon | 58.5 |

### LOWEST

| | | |
|---|---|---|
| 1 | Vernon | 58.5 |
| 2 | Rutland | 63.8 |
| 3 | Cornwall | 68.2 |
| 4 | Bellows Falls | 70.7 |
| 5 | Chelsea | 80.9 |
| 6 | Burlington | 83.4 |
| 7 | Readsboro | 83.8 |
| 8 | St. Johnsbury | 88.6 |
| 9 | Rochester | 90.4 |
| 10 | Cavendish | 91.4 |
| 11 | Enosburg Falls | 93.1 |
| 12 | West Burke | 95.9 |
| 13 | Montpelier | 97.0 |
| 14 | Newport | 99.9 |
| 15 | Mount Mansfield | 219.1 |

# VIRGINIA

PHYSICAL FEATURES. Virginia is located on the east coast of the North American continent between latitudes 36½° and 39½° N. The State is triangular in shape with the longest north-south distance of about 200 miles and the longest east-west distance more than 400 miles. There are 40,815 square miles of area within the State of which 1,200 square miles are inland waters.

The State is composed of 3 natural topographic regions, namely: the Tidewater or coastal plains area, the Piedmont plateau or middle Virginia, and the western mountain region. Tidewater Virginia extends westward from the Atlantic Coast and west shore of the Chesapeake Bay to the "Fall Line." The "Fall Line" extends from Great Falls in the north, southward through Richmond to Emporia. It is divided into necks or peninsulas by 4 principal rivers and by numerous estuaries that open into the Chesapeake Bay. The principal rivers include the Potomac, Rappahannock, York, and the James. The Piedmont region is more than 200 miles wide in southern Virginia, but the Virginia section becomes quite narrow in the north. This region from east to west becomes more rolling and hilly with a few isolated mountains and ridges appearing a few miles east of the Blue Ridge. Elevations in general range from about 300 feet above sea level in the east to about 1,000 feet in the west. The James, the largest river crossing this region, divides it into two parts.

West of the Piedmont, the Blue Ridge Mountains traverse the State from southwest to northeast. They range from narrow ridges in the north to a high, wide plateau southwest from Roanoke. Elevations range generally from 1,500 to 3,500 feet. Mt. Rogers, in western Grayson County, towers to 5,719 feet, the highest point in the State. A great valley west of the Blue Ridge extends from Tennessee through Scott and Washington Counties in the south, northeastward to the northern-most point of the State. This great valley of Virginia is well drained. The north is drained by the north and south forks of the Shenandoah River, thence into the Potomac; the central portion by the Cow Pasture and Jackson Rivers flowing southeastward into the James; and the southwestern half of the valley is drained by the Roanoke River, the New River, and three forks of the Holston River.

GENERAL CLIMATE. The climate of Virginia is determined by its proximity to the Atlantic Ocean, latitude, and topography. The State is in the zone of prevailing westerly movement of the earth's atmosphere, in or near the mean path of winter storm tracks, and in the mean path of tropical, moist air from the southwest Atlantic and Gulf of Mexico much of the summer and early fall seasons. The mountains provide the usual elevation effects on temperatures, which are distinctly lower in this section, and there are wide variations over short distances as elevations change. Summers in the mountains are comparatively cool, and winters are more severe. In addition, these mountains produce various steering, blocking, and modifying effects on storms and general air movements in their vicinity. Temperature variations within the State due to latitude alone are very small. The longitudinal variations, however, show a sharper contrast, from the mountain extremes in the west toward an ocean influence in the east. The prevalence of winds with a westerly component prevents the extension of ocean influences very far westward from the coast.

Virginia lies in the zone of prevailing westerlies where the general motion is from west to east. Southerly and northerly winds are about equally frequent, reflecting the progression of weather systems over the State. The Appalachian mountains, however, act to deflect these winds to some extent with northeasterly and southwesterly directions occurring frequently. During the cold season a more intense circulation is present with frequent storms and outbreaks of cold polar air. Northerly winds are most common during this season. The storm track is well north of the State during the warm season and southerly winds with light speeds prevail.

Summers in Virginia are usually warm and humid, and several hot and humid periods usually occur each year. Principal sources of moisture are the Gulf of Mexico and the Atlantic Ocean. Relative humidity varies inversely with temperature -- high in the morning and low in the afternoon. Tidewater locations have a much higher frequency of humidity and temperature values in the range where human discomfort occurs.

PRECIPITATION. The annual precipitation ranges from about 35 to 50 inches. The heaviest amounts occur in the extreme southwest, the southeast, and the south-central areas. Minimum amounts are found in the sheltered valleys west of the Blue Ridge Mountains. Precipitation is well distributed throughout the year without distinct wet and dry periods. Maximum rainfall occurs in the summer months and minimum in the fall months. Precipitation during the cold season is associated with migratory low-pressure storms. The amounts are quite evenly distributed during this season in comparison to the warm season when showers and thundershowers account for most of the rainfall. Excessive rainfall usually occurs in the fall season with the passage of hurricanes. Snow is common in winter. Average seasonal amounts range from less than 10 inches in Tidewater Virginia to around 20 inches west of the Blue Ridge, and up to 30 inches on the mountains. Occasionally, a major snowstorm will occur with snow depths greater than 10 inches.

The greater portion of the State lies in the Atlantic drainage. The extreme southwestern portion drains to the Ohio Basin. Floods occur in all months of the year. The greatest frequency occurs in late winter and early spring; snowmelt occasionally is a factor. July is the month of least flooding. A second period of high water shows up in late summer and fall in the Piedmont and Tidewater sections associated mainly with tropical storms. Intense convectional storms in summer occasionally cause local flash floods.

Virginia is also subject to drought periods. Drought may be defined broadly as a prolonged and abnormal moisture deficiency. Some portions of Virginia sustain real damage from drought on the average of 1 year out of 3. Equitable distribution of ample precipitation has seldom, if ever, occurred over all of Virginia for an entire season. Almost every year some sections undergo periods of insufficient rainfall.

STORMS. Thunderstorms occur on the average of 32 to 50 days each year, the greater number occurring in the mountains of extreme southwestern Virginia, decreasing in number toward the northeastern part of the State. About 85 percent of the annual total occur during the period May to September. Only a small percentage of these can be classed as severe, however. Thunderstorms exact a sizable annual toll of damage when accompanied by severe lightning, wind, or hail. Tornadoes are local storms of short duration and usually small dimensions, formed of winds rotating at high speeds. Approximately 4 tornadoes are reported in Virginia each year. These tornadoes occur mainly east of the Blue Ridge, but a few have been observed in the mountains.

A hurricane is a tropical storm with winds of at least 74 m.p.h., which blow in a large spiral around a relatively calm center. Virginia has been affected by hurricanes since the early settlement days, but most have decreased in intensity before entering the State. Even though a hurricane may not enter the State, it can be much more destructive by passing closely offshore and maintaining its intense circulation. High tides along with waves and currents, and flooding from the torrential rains, cause damage. About 80 percent of the hurricanes occur during August, September, and October. An average of about 2 hurricanes each year come close enough to affect Virginia, but less than one enters the State.

Middle latitude storms sometimes develop south of Virginia and move northward along the Virginia coast. These storms, although usually weaker than hurricanes, produce similar type of damage. This type of storm, often referred to as a "Northeaster," generally occurs from late fall through the spring months. They often account for considerable damage from high tides, strong east or northeast winds, and heavy rain, mainly in Tidewater Virginia.

## COUNTY INDEX

***Accomack County***
PAINTER 2 W
TANGIER ISLAND

***Albemarle County***
CHARLOTTESVILLE 2 W

***Alleghany County***
COVINGTON FILTER PLT

***Appomattox County***
APPOMATTOX

***Arlington County***
WASHINGTN DC NATL AP

***Augusta County***
STAUNTON SEWAGE PLAN

***Botetourt County***
BUCHANAN

***Brunswick County***
LAWRENCEVILLE 3 E

***Buchanan County***
GRUNDY

***Campbell County***
LYNCHBURG MUNI AP

***Caroline County***
CORBIN

***Carroll County***
GALAX RADIO WBRF

***Charlotte County***
CHARLOTTE COURT HSE

***Cumberland County***
FARMVILLE 2 N

***Fauquier County***
WARRENTON 3 SE

***Floyd County***
FLOYD 2 NE

***Franklin County***
ROCKY MOUNT

***Frederick County***
WINCHESTER 3 ESE

***Giles County***
STAFFORDSVILLE 3 ENE

***Hanover County***
ASHLAND

***Henrico County***
RICHMOND BYRD AP

***Henry County***
MARTINSVILLE FLT PLT
PHILPOTT DAM 2

***King and Queen County***
WALKERTON 2 NW

***Lee County***
PENNINGTON GAP

***Loudoun County***
LINCOLN
MOUNT WEATHER
WASHINGTON DC DULLES

***Louisa County***
LOUISA

***Madison County***
BIG MEADOWS

***Mecklenburg County***
CHASE CITY
JOHN H KERR DAM

***Montgomery County***
BLACKSBURG 3 SE

***Nelson County***
TYE RIVER 1 SE

***New Kent County***
WEST POINT 2 SW

***Orange County***
PIEDMONT RESEARCH ST

***Page County***
LURAY 5 E

***Pittsylvania County***
CHATHAM
DANVILLE

***Prince George County***
HOPEWELL

***Pulaski County***
PULASKI

***Richmond County***
WARSAW 2 NW

***Rockbridge County***
LEXINGTON

***Rockingham County***
DALE ENTERPRISE

***Shenandoah County***
WOODSTOCK 2 NE

***Spotsylvania County***
FREDERICKSBURG N P

***Sussex County***
STONY CREEK 3 ESE

***Tazewell County***
BURKES GARDEN

***Westmoreland County***
COLONIAL BEACH

***Wise County***
WISE 2 E

***Wythe County***
WYTHEVILLE 1 S

***York County***
WILLIAMSBURG 2 N

***Bedford Independent City***
BEDFORD

### *Hampton Independent City*

LANGLEY AFB

### *Norfolk Independent City*

NORFOLK INTL ARPT

### *Roanoke Independent City*

ROANOKE WOODRUM AP

### *Staunton Independent City*

BACK BAY WILDLF REF

### *Suffolk Independent City*

HOLLAND 1 E
SUFFOLK LAKE KILBY

## ELEVATION INDEX

| FEET | STATION NAME |
|---|---|
| 10 | BACK BAY WILDLF REF |
| 10 | COLONIAL BEACH |
| 10 | LANGLEY AFB |
| 10 | TANGIER ISLAND |
| 20 | SUFFOLK LAKE KILBY |
| 20 | WEST POINT 2 SW |
| 30 | PAINTER 2 W |
| 39 | HOPEWELL |
| 39 | NORFOLK INTL ARPT |
| 49 | WALKERTON 2 NW |
| 66 | WASHINGTN DC NATL AP |
| 69 | STONY CREEK 3 ESE |
| 69 | WILLIAMSBURG 2 N |
| 79 | HOLLAND 1 E |
| 90 | FREDERICKSBURG N P |
| 141 | WARSAW 2 NW |
| 167 | RICHMOND BYRD AP |
| 220 | ASHLAND |
| 220 | CORBIN |
| 290 | WASHINGTON DC DULLES |
| 323 | JOHN H KERR DAM |
| 325 | LAWRENCEVILLE 3 E |
| 430 | FARMVILLE 2 N |
| 440 | LOUISA |
| 502 | LINCOLN |
| 502 | PIEDMONT RESEARCH ST |
| 502 | WARRENTON 3 SE |
| 541 | CHASE CITY |
| 551 | TYE RIVER 1 SE |
| 571 | DANVILLE |
| 590 | CHARLOTTE COURT HSE |
| 660 | WOODSTOCK 2 NE |
| 679 | WINCHESTER 3 ESE |
| 702 | CHATHAM |
| 761 | MARTINSVILLE FLT PLT |
| 830 | APPOMATTOX |
| 869 | CHARLOTTESVILLE 2 W |
| 879 | BUCHANAN |
| 951 | BEDFORD |
| 955 | LYNCHBURG MUNI AP |
| 1040 | GRUNDY |
| 1060 | LEXINGTON |
| 1123 | PHILPOTT DAM 2 |
| 1132 | ROCKY MOUNT |
| 1201 | LURAY 5 E |
| 1201 | ROANOKE WOODRUM AP |
| 1230 | COVINGTON FILTER PLT |
| 1352 | DALE ENTERPRISE |
| 1480 | STAUNTON SEWAGE PLAN |
| 1591 | PENNINGTON GAP |
| 1732 | MOUNT WEATHER |
| 1831 | STAFFORDSVILLE 3 ENE |
| 1850 | PULASKI |
| 2041 | BLACKSBURG 3 SE |
| 2402 | GALAX RADIO WBRF |
| 2451 | WYTHEVILLE 1 S |
| 2500 | WISE 2 E |
| 2602 | FLOYD 2 NE |
| 3251 | BURKES GARDEN |
| 3422 | BIG MEADOWS |

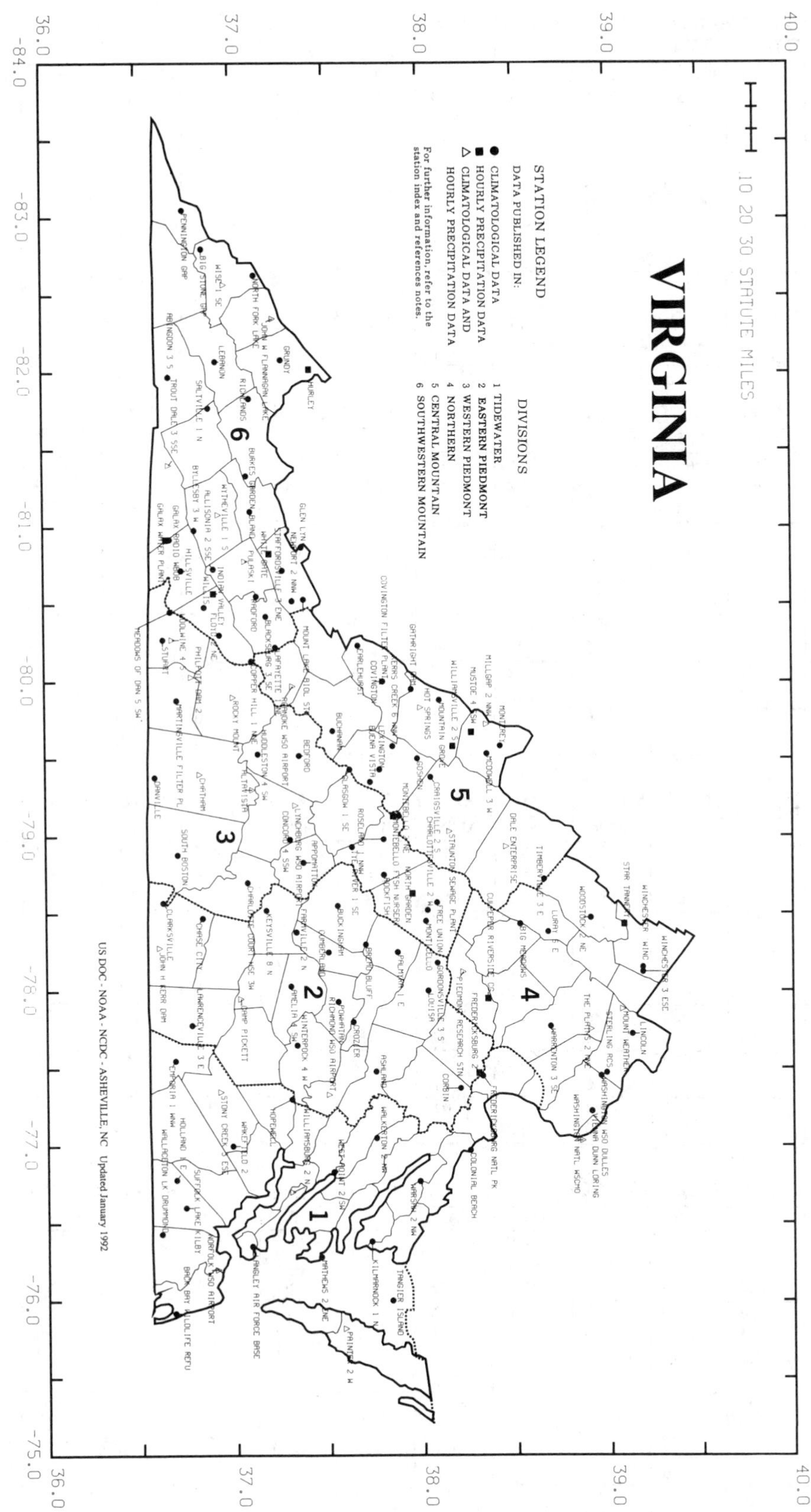
VIRGINIA
10 20 30 STATUTE MILES
STATION LEGEND
DATA PUBLISHED IN:
● CLIMATOLOGICAL DATA
■ HOURLY PRECIPITATION DATA
△ CLIMATOLOGICAL DATA AND HOURLY PRECIPITATION DATA
For further information, refer to the station index and references notes.
DIVISIONS
1 TIDEWATER
2 EASTERN PIEDMONT
3 WESTERN PIEDMONT
4 NORTHERN
5 CENTRAL MOUNTAIN
6 SOUTHWESTERN MOUNTAIN
US DOC - NOAA - NCDC - ASHEVILLE, NC Updated January 1992

## APPOMATTOX *Appomattox County* ELEVATION 830 ft LAT/LONG 37° 21 ' N / 78° 50 ' W

| | JAN | FEB | MAR | APR | MAY | JUN | JUL | AUG | SEP | OCT | NOV | DEC | YEAR |
|---|---|---|---|---|---|---|---|---|---|---|---|---|---|
| Maximum Temp °F | 44.1 | 47.4 | 56.8 | 67.5 | 75.3 | 82.8 | 87.0 | 85.2 | 78.9 | 68.0 | 58.5 | 48.8 | 66.7 |
| Minimum Temp °F | 23.6 | 25.6 | 33.5 | 42.2 | 51.1 | 59.8 | 64.6 | 63.1 | 56.1 | 42.8 | 35.5 | 27.8 | 43.8 |
| Mean Temp °F | 33.9 | 36.5 | 45.1 | 54.9 | 63.2 | 71.3 | 75.8 | 74.2 | 67.5 | 55.4 | 47.1 | 38.3 | 55.3 |
| Days Max Temp ≥ 90 °F | 0 | 0 | 0 | 1 | 1 | 5 | 11 | 8 | 2 | 0 | 0 | 0 | 28 |
| Days Max Temp ≤ 32 °F | 4 | 2 | 0 | 0 | 0 | 0 | 0 | 0 | 0 | 0 | 0 | 2 | 8 |
| Days Min Temp ≤ 32 °F | 26 | 22 | 15 | 4 | 0 | 0 | 0 | 0 | 0 | 4 | 12 | 22 | 105 |
| Days Min Temp ≤ 0 °F | 1 | 0 | 0 | 0 | 0 | 0 | 0 | 0 | 0 | 0 | 0 | 0 | 1 |
| Heating Degree Days | 959 | 798 | 614 | 313 | 115 | 16 | 1 | 3 | 50 | 303 | 533 | 822 | 4527 |
| Cooling Degree Days | 0 | 0 | 3 | 16 | 69 | 233 | 384 | 310 | 143 | 18 | 1 | 0 | 1177 |
| Total Precipitation (") | 3.06 | 2.95 | 4.23 | 3.38 | 4.37 | 3.17 | 4.59 | 4.05 | 3.56 | 3.65 | 3.37 | 3.23 | 43.61 |
| Days ≥ 0.1" Precip | 6 | 6 | 7 | 7 | 8 | 6 | 7 | 7 | 5 | 6 | 6 | 6 | 77 |
| Total Snowfall (") | 6.0 | 5.5 | 2.0 | 0.1 | 0.0 | 0.0 | 0.0 | 0.0 | 0.0 | 0.0 | 0.6 | 1.6 | 15.8 |
| Days ≥ 1" Snow Depth | ***5*** | 3 | 1 | 0 | 0 | 0 | 0 | 0 | 0 | 0 | 0 | 1 | 10 |

## ASHLAND *Hanover County* ELEVATION 220 ft LAT/LONG 37° 46 ' N / 77° 29 ' W

| | JAN | FEB | MAR | APR | MAY | JUN | JUL | AUG | SEP | OCT | NOV | DEC | YEAR |
|---|---|---|---|---|---|---|---|---|---|---|---|---|---|
| Maximum Temp °F | 45.1 | 49.1 | 59.2 | 69.6 | 76.5 | 83.6 | 87.5 | 85.6 | 79.7 | 68.6 | 59.7 | 49.5 | 67.8 |
| Minimum Temp °F | 24.4 | 26.9 | 34.7 | 43.2 | 52.6 | 61.3 | 66.1 | 64.9 | 57.7 | 45.3 | 36.4 | 28.8 | 45.2 |
| Mean Temp °F | 34.8 | 38.0 | 47.0 | 56.4 | 64.6 | 72.5 | 76.8 | 75.3 | 68.7 | 57.0 | 48.1 | 39.2 | 56.5 |
| Days Max Temp ≥ 90 °F | 0 | 0 | 0 | 1 | 1 | 5 | 12 | 8 | 3 | 0 | 0 | 0 | 30 |
| Days Max Temp ≤ 32 °F | 4 | 2 | 0 | 0 | 0 | 0 | 0 | 0 | 0 | 0 | 0 | 1 | 7 |
| Days Min Temp ≤ 32 °F | 24 | 21 | 14 | 4 | 0 | 0 | 0 | 0 | 0 | 3 | 12 | 21 | 99 |
| Days Min Temp ≤ 0 °F | 1 | 0 | 0 | 0 | 0 | 0 | 0 | 0 | 0 | 0 | 0 | 0 | 1 |
| Heating Degree Days | 929 | 755 | 557 | 275 | 94 | 12 | 0 | 3 | 40 | 264 | 505 | 795 | 4229 |
| Cooling Degree Days | 0 | 0 | 6 | 26 | 96 | 260 | 399 | 326 | 164 | 27 | 2 | 0 | 1306 |
| Total Precipitation (") | 3.37 | 3.02 | 4.21 | 3.02 | 4.15 | 3.24 | 4.11 | 4.30 | 3.70 | 3.24 | 3.09 | 3.44 | 42.89 |
| Days ≥ 0.1" Precip | 7 | 6 | 7 | 6 | 7 | 6 | 7 | 6 | 5 | 5 | 6 | 6 | 74 |
| Total Snowfall (") | 5.7 | 6.2 | 1.6 | 0.2 | 0.0 | 0.0 | 0.0 | 0.0 | 0.0 | 0.0 | 0.7 | 2.5 | 16.9 |
| Days ≥ 1" Snow Depth | ***5*** | 4 | 1 | 0 | 0 | 0 | 0 | 0 | 0 | 0 | 0 | 2 | 12 |

## BACK BAY WILDLF REF *Staunton Independent City* ELEVATION 10 ft LAT/LONG 36° 40 ' N / 75° 55 ' W

| | JAN | FEB | MAR | APR | MAY | JUN | JUL | AUG | SEP | OCT | NOV | DEC | YEAR |
|---|---|---|---|---|---|---|---|---|---|---|---|---|---|
| Maximum Temp °F | ***48.5*** | 51.2 | 57.9 | 66.9 | 74.4 | 82.4 | 86.4 | 85.2 | 80.8 | ***71.3*** | ***62.8*** | ***54.6*** | 68.5 |
| Minimum Temp °F | ***31.4*** | 33.3 | 39.6 | 47.5 | 56.9 | 65.7 | ***70.5*** | 69.8 | 65.7 | ***55.4*** | ***45.3*** | ***37.7*** | 51.6 |
| Mean Temp °F | ***39.9*** | 42.2 | 48.8 | 57.1 | 65.7 | 74.1 | ***78.5*** | 77.6 | 73.3 | ***63.4*** | ***54.0*** | ***46.2*** | 60.1 |
| Days Max Temp ≥ 90 °F | 0 | 0 | 0 | 0 | 1 | 5 | 11 | 7 | 2 | 0 | 0 | 0 | 26 |
| Days Max Temp ≤ 32 °F | 2 | 0 | 0 | 0 | 0 | 0 | 0 | 0 | 0 | 0 | 0 | 0 | 2 |
| Days Min Temp ≤ 32 °F | 14 | 13 | ***5*** | 0 | 0 | 0 | 0 | 0 | 0 | 0 | 2 | ***8*** | 42 |
| Days Min Temp ≤ 0 °F | 0 | 0 | 0 | 0 | 0 | 0 | 0 | 0 | 0 | 0 | 0 | 0 | 0 |
| Heating Degree Days | ***773*** | 638 | 498 | 252 | 69 | 5 | 0 | 0 | 4 | ***102*** | ***329*** | ***579*** | 3249 |
| Cooling Degree Days | na | 1 | ***2*** | 22 | ***118*** | 314 | ***457*** | 399 | ***262*** | na | na | na | na |
| Total Precipitation (") | 4.20 | 3.30 | 3.98 | 2.72 | 3.91 | 3.32 | 4.74 | 5.68 | 4.13 | 3.14 | 3.07 | 3.05 | 45.24 |
| Days ≥ 0.1" Precip | 6 | 5 | 6 | 5 | 6 | 5 | 7 | 6 | 5 | 4 | 5 | ***5*** | 65 |
| Total Snowfall (") | ***1.7*** | ***1.8*** | 0.4 | 0.0 | 0.0 | 0.0 | 0.0 | 0.0 | 0.0 | 0.0 | 0.0 | 0.1 | 4.0 |
| Days ≥ 1" Snow Depth | ***0*** | ***0*** | 0 | 0 | 0 | 0 | 0 | 0 | 0 | 0 | 0 | 0 | 0 |

## BEDFORD *Bedford Independent City* ELEVATION 951 ft LAT/LONG 37° 20 ' N / 79° 32 ' W

| | JAN | FEB | MAR | APR | MAY | JUN | JUL | AUG | SEP | OCT | NOV | DEC | YEAR |
|---|---|---|---|---|---|---|---|---|---|---|---|---|---|
| Maximum Temp °F | 45.0 | 48.7 | 59.6 | 70.0 | 75.9 | 82.6 | 85.9 | 84.1 | 77.8 | 67.6 | 58.7 | 49.0 | 67.1 |
| Minimum Temp °F | 26.3 | 27.8 | 35.8 | 44.2 | 52.6 | 60.6 | 65.0 | 63.5 | 57.3 | 46.0 | 37.7 | 30.5 | 45.6 |
| Mean Temp °F | 35.7 | 38.3 | 47.7 | 57.1 | 64.3 | 71.6 | 75.5 | 73.8 | 67.6 | 56.8 | 47.7 | 39.8 | 56.3 |
| Days Max Temp ≥ 90 °F | 0 | 0 | 0 | 0 | 0 | 3 | 8 | 5 | 1 | 0 | 0 | 0 | 17 |
| Days Max Temp ≤ 32 °F | 4 | 2 | 0 | 0 | 0 | 0 | 0 | 0 | 0 | 0 | 0 | 2 | 8 |
| Days Min Temp ≤ 32 °F | 23 | ***20*** | 12 | 3 | 0 | 0 | 0 | 0 | 0 | 2 | 10 | 18 | 88 |
| Days Min Temp ≤ 0 °F | 0 | 0 | 0 | 0 | 0 | 0 | 0 | 0 | 0 | 0 | 0 | 0 | 0 |
| Heating Degree Days | 902 | 749 | 532 | 253 | 91 | 10 | 1 | 3 | 48 | 261 | 498 | 774 | 4122 |
| Cooling Degree Days | 0 | 0 | 5 | 26 | 70 | 228 | 356 | 283 | 131 | 17 | 1 | 0 | 1117 |
| Total Precipitation (") | 3.02 | 3.12 | 3.73 | 3.61 | 4.43 | 3.60 | 4.43 | 4.00 | 3.30 | 3.96 | 3.14 | 3.11 | 43.45 |
| Days ≥ 0.1" Precip | 7 | 6 | 7 | 6 | 8 | 6 | 8 | 7 | 6 | 6 | 6 | 6 | 79 |
| Total Snowfall (") | 6.0 | 5.7 | 2.2 | 0.0 | 0.0 | 0.0 | 0.0 | 0.0 | 0.0 | 0.0 | ***0.4*** | 2.2 | 16.5 |
| Days ≥ 1" Snow Depth | na | na | ***0*** | 0 | 0 | 0 | 0 | 0 | 0 | 0 | 0 | 1 | na |

## BIG MEADOWS *Madison County* ELEVATION 3422 ft LAT/LONG 38° 31 ' N / 78° 26 ' W

| | JAN | FEB | MAR | APR | MAY | JUN | JUL | AUG | SEP | OCT | NOV | DEC | YEAR |
|---|---|---|---|---|---|---|---|---|---|---|---|---|---|
| Maximum Temp °F | 34.4 | 37.0 | 45.4 | 56.0 | 64.5 | 71.9 | 75.5 | 73.8 | 67.8 | 57.7 | 48.2 | 39.1 | 55.9 |
| Minimum Temp °F | 16.2 | 18.5 | 25.9 | 35.4 | 44.4 | 52.3 | 56.5 | 55.2 | 49.3 | 38.1 | 30.2 | 21.6 | 37.0 |
| Mean Temp °F | 25.4 | 27.8 | 35.7 | 45.7 | 54.4 | 62.2 | 66.0 | 64.5 | 58.6 | 48.0 | 39.2 | 30.4 | 46.5 |
| Days Max Temp ≥ 90 °F | 0 | 0 | 0 | 0 | 0 | 0 | 0 | 0 | 0 | 0 | 0 | 0 | 0 |
| Days Max Temp ≤ 32 °F | 14 | 10 | 4 | 0 | 0 | 0 | 0 | 0 | 0 | 0 | 2 | 9 | 39 |
| Days Min Temp ≤ 32 °F | 28 | 24 | 22 | 13 | 3 | 0 | 0 | 0 | 1 | 10 | 18 | 26 | 145 |
| Days Min Temp ≤ 0 °F | 3 | 2 | 0 | 0 | 0 | 0 | 0 | 0 | 0 | 0 | 0 | 1 | 6 |
| Heating Degree Days | 1222 | 1044 | 902 | 572 | 329 | 117 | 44 | 67 | 207 | 522 | 767 | 1066 | 6859 |
| Cooling Degree Days | 0 | 0 | 0 | 1 | 8 | 39 | 89 | 50 | 19 | 1 | 0 | 0 | 207 |
| Total Precipitation (") | 3.32 | 3.21 | 4.00 | 3.94 | 4.86 | 4.48 | 4.92 | 4.37 | 5.08 | 5.22 | 4.82 | 3.89 | 52.11 |
| Days ≥ 0.1" Precip | 6 | 6 | 7 | 7 | 8 | 7 | 8 | 7 | 6 | 6 | 7 | 6 | 81 |
| Total Snowfall (") | 12.0 | 10.2 | 7.0 | 2.0 | 0.0 | 0.0 | 0.0 | 0.0 | 0.0 | 0.9 | 3.7 | 6.4 | 42.2 |
| Days ≥ 1" Snow Depth | 14 | 16 | 9 | 2 | 0 | 0 | 0 | 0 | 0 | 0 | 2 | 8 | 51 |

## BLACKSBURG 3 SE *Montgomery County* ELEVATION 2041 ft LAT/LONG 37° 14 ' N / 80° 25 ' W

| | JAN | FEB | MAR | APR | MAY | JUN | JUL | AUG | SEP | OCT | NOV | DEC | YEAR |
|---|---|---|---|---|---|---|---|---|---|---|---|---|---|
| Maximum Temp °F | 40.6 | 44.2 | 53.7 | 63.6 | 71.7 | 78.9 | 82.8 | 81.4 | 75.6 | 65.2 | 55.2 | 44.5 | 63.1 |
| Minimum Temp °F | 19.7 | 21.7 | 29.7 | 38.1 | 46.6 | 54.8 | 59.5 | 58.0 | 50.9 | 38.5 | 31.1 | 23.5 | 39.3 |
| Mean Temp °F | 30.2 | 32.9 | 41.7 | 50.9 | 59.2 | 66.9 | 71.2 | 69.7 | 63.3 | 51.9 | 43.1 | 34.0 | 51.3 |
| Days Max Temp ≥ 90 °F | 0 | 0 | 0 | 0 | 0 | 1 | 3 | 2 | 0 | 0 | 0 | 0 | 6 |
| Days Max Temp ≤ 32 °F | 8 | 5 | 1 | 0 | 0 | 0 | 0 | 0 | 0 | 0 | 1 | 5 | 20 |
| Days Min Temp ≤ 32 °F | 28 | 24 | 19 | 9 | 2 | 0 | 0 | 0 | 0 | 10 | 18 | 26 | 136 |
| Days Min Temp ≤ 0 °F | 2 | 1 | 0 | 0 | 0 | 0 | 0 | 0 | 0 | 0 | 0 | 1 | 4 |
| Heating Degree Days | 1074 | 900 | 716 | 421 | 198 | 46 | 7 | 15 | 110 | 404 | 650 | 954 | 5495 |
| Cooling Degree Days | 0 | 0 | 1 | 5 | 28 | 127 | 239 | 175 | 70 | 5 | 0 | 0 | 650 |
| Total Precipitation (") | 2.87 | 2.92 | 3.79 | 3.77 | 4.14 | 3.49 | 4.17 | 3.83 | 3.40 | 3.57 | 2.94 | 2.87 | 41.76 |
| Days ≥ 0.1" Precip | 7 | 7 | 8 | 8 | 9 | 8 | 9 | 7 | 6 | 6 | 7 | 7 | 89 |
| Total Snowfall (") | 6.5 | 6.1 | 3.5 | 1.1 | 0.0 | 0.0 | 0.0 | 0.0 | 0.0 | 0.0 | 1.2 | 3.0 | 21.4 |
| Days ≥ 1" Snow Depth | 8 | 7 | 2 | 0 | 0 | 0 | 0 | 0 | 0 | 0 | 1 | 4 | 22 |

## BUCHANAN *Botetourt County* ELEVATION 879 ft LAT/LONG 37° 32 ' N / 79° 41 ' W

| | JAN | FEB | MAR | APR | MAY | JUN | JUL | AUG | SEP | OCT | NOV | DEC | YEAR |
|---|---|---|---|---|---|---|---|---|---|---|---|---|---|
| Maximum Temp °F | *45.3* | 50.0 | 59.9 | *70.9* | *78.0* | *84.2* | 87.6 | 86.3 | 80.0 | 70.1 | 59.0 | *49.4* | 68.4 |
| Minimum Temp °F | *25.1* | 26.7 | 34.5 | *42.2* | *51.2* | *59.3* | 63.9 | 62.8 | 56.2 | 44.3 | 35.0 | *28.6* | 44.2 |
| Mean Temp °F | *35.2* | 38.3 | 47.2 | *56.6* | *64.6* | *71.8* | 75.8 | 74.6 | 68.1 | 57.2 | 47.0 | *39.0* | 56.3 |
| Days Max Temp ≥ 90 °F | *0* | 0 | 0 | *1* | *2* | *6* | 11 | 9 | 3 | 0 | 0 | 0 | 32 |
| Days Max Temp ≤ 32 °F | *4* | 2 | 0 | *0* | *0* | *0* | 0 | 0 | 0 | 0 | 0 | *1* | 7 |
| Days Min Temp ≤ 32 °F | *24* | 21 | 14 | *5* | *1* | *0* | 0 | 0 | 0 | 4 | 13 | *20* | 102 |
| Days Min Temp ≤ 0 °F | *0* | 0 | 0 | *0* | *0* | *0* | 0 | 0 | 0 | 0 | 0 | 0 | 0 |
| Heating Degree Days | *917* | 747 | 546 | *264* | *87* | *10* | 0 | 1 | 43 | 252 | 534 | *799* | 4200 |
| Cooling Degree Days | *0* | *0* | *3* | na | na | na | *372* | *313* | *156* | *22* | *1* | *0* | na |
| Total Precipitation (") | 3.07 | 2.90 | 3.94 | *3.21* | *3.93* | *3.40* | 4.90 | 4.00 | 3.25 | 3.98 | 2.74 | 3.06 | 42.38 |
| Days ≥ 0.1" Precip | *6* | 6 | 8 | *7* | *8* | *6* | 8 | 7 | 6 | 6 | 6 | *6* | 80 |
| Total Snowfall (") | *5.5* | *5.2* | 2.8 | *0.3* | *0.0* | *0.0* | 0.0 | 0.0 | 0.0 | 0.0 | 1.2 | *3.5* | 18.5 |
| Days ≥ 1" Snow Depth | na | na | *1* | *0* | *0* | *0* | 0 | 0 | 0 | 0 | *0* | *2* | na |

## BURKES GARDEN *Tazewell County* ELEVATION 3251 ft LAT/LONG 37° 6 ' N / 81° 20 ' W

| | JAN | FEB | MAR | APR | MAY | JUN | JUL | AUG | SEP | OCT | NOV | DEC | YEAR |
|---|---|---|---|---|---|---|---|---|---|---|---|---|---|
| Maximum Temp °F | 39.2 | 42.4 | 51.9 | 61.6 | 69.0 | 75.6 | 79.0 | 77.6 | 72.0 | 62.3 | 52.3 | 43.4 | 60.5 |
| Minimum Temp °F | 19.6 | 21.3 | 29.4 | 36.0 | 44.6 | 52.0 | 56.6 | 54.8 | 48.4 | 37.0 | 29.9 | 23.6 | 37.8 |
| Mean Temp °F | 29.4 | 31.9 | 40.7 | 48.8 | 56.8 | 63.8 | 67.8 | 66.2 | 60.2 | 49.7 | 41.1 | 33.5 | 49.2 |
| Days Max Temp ≥ 90 °F | 0 | 0 | 0 | 0 | 0 | 0 | 0 | 0 | 0 | 0 | 0 | 0 | 0 |
| Days Max Temp ≤ 32 °F | 9 | 6 | 2 | 0 | 0 | 0 | 0 | 0 | 0 | 1 | 2 | 5 | 25 |
| Days Min Temp ≤ 32 °F | 27 | 24 | 20 | 11 | 3 | 0 | 0 | 0 | 1 | 11 | *18* | *25* | 140 |
| Days Min Temp ≤ 0 °F | 3 | 1 | 0 | 0 | 0 | 0 | 0 | 0 | 0 | 0 | 0 | 1 | 5 |
| Heating Degree Days | 1096 | 929 | 746 | 480 | 253 | 75 | 20 | 35 | 159 | 469 | 709 | 966 | 5937 |
| Cooling Degree Days | 0 | 0 | 0 | 1 | 8 | 53 | 130 | 82 | 26 | 2 | 0 | 0 | 302 |
| Total Precipitation (") | 3.32 | 3.39 | 3.94 | 3.76 | 4.67 | 4.06 | 4.57 | 3.98 | 3.34 | 3.31 | 3.14 | 3.44 | 44.92 |
| Days ≥ 0.1" Precip | 8 | 7 | 8 | 8 | 10 | 9 | 8 | 8 | 7 | 6 | 7 | 7 | 93 |
| Total Snowfall (") | 14.4 | 13.6 | 7.2 | 2.8 | 0.1 | 0.0 | 0.0 | 0.0 | 0.0 | 0.4 | 3.1 | 8.5 | 50.1 |
| Days ≥ 1" Snow Depth | 11 | 9 | 3 | 1 | 0 | 0 | 0 | 0 | 0 | 0 | 1 | 6 | 31 |

## CHARLOTTE COURT HSE *Charlotte County* ELEVATION 590 ft LAT/LONG 37° 4 ' N / 78° 42 ' W

| | JAN | FEB | MAR | APR | MAY | JUN | JUL | AUG | SEP | OCT | NOV | DEC | YEAR |
|---|---|---|---|---|---|---|---|---|---|---|---|---|---|
| Maximum Temp °F | 45.6 | 49.2 | 58.1 | 69.0 | 76.3 | 83.7 | 87.5 | 86.1 | 80.4 | 69.6 | 60.3 | 50.0 | 68.0 |
| Minimum Temp °F | 24.4 | 26.2 | 34.1 | 42.9 | 51.7 | 60.3 | 64.9 | 63.4 | 56.7 | 43.8 | 36.0 | 28.5 | 44.4 |
| Mean Temp °F | 35.0 | 37.7 | 46.1 | 56.0 | 64.0 | 72.0 | 76.3 | 74.7 | 68.6 | 56.7 | 48.3 | 39.3 | 56.2 |
| Days Max Temp ≥ 90 °F | 0 | 0 | 0 | 1 | 1 | 5 | 11 | 9 | 4 | 0 | 0 | 0 | 31 |
| Days Max Temp ≤ 32 °F | 4 | 2 | 0 | 0 | 0 | 0 | 0 | 0 | 0 | 0 | 0 | 2 | 8 |
| Days Min Temp ≤ 32 °F | 24 | 21 | 14 | 4 | 0 | 0 | 0 | 0 | 0 | 4 | 12 | 21 | 100 |
| Days Min Temp ≤ 0 °F | 0 | 0 | 0 | 0 | 0 | 0 | 0 | 0 | 0 | 0 | 0 | 0 | 0 |
| Heating Degree Days | 922 | 764 | 580 | 286 | 103 | 14 | 1 | 4 | 42 | 272 | 498 | 790 | 4276 |
| Cooling Degree Days | 0 | 0 | 5 | 24 | 85 | 249 | 383 | *304* | 163 | 29 | 1 | 0 | 1243 |
| Total Precipitation (") | 3.56 | 3.24 | 4.05 | 3.07 | 3.92 | 3.35 | 4.28 | 4.09 | 3.67 | 3.92 | 3.24 | 3.26 | 43.65 |
| Days ≥ 0.1" Precip | 6 | 6 | 7 | 6 | 7 | 6 | 7 | 6 | 5 | 5 | 5 | 6 | 72 |
| Total Snowfall (") | 4.4 | 2.8 | 2.4 | 0.0 | 0.0 | 0.0 | 0.0 | 0.0 | 0.0 | 0.0 | 0.3 | 1.6 | 11.5 |
| Days ≥ 1" Snow Depth | na | *1* | 1 | 0 | 0 | 0 | 0 | 0 | 0 | 0 | 0 | *1* | na |

## CHARLOTTESVILLE 2 W *Albemarle County* ELEVATION 869 ft LAT/LONG 38° 2 ' N / 78° 31 ' W

| | JAN | FEB | MAR | APR | MAY | JUN | JUL | AUG | SEP | OCT | NOV | DEC | YEAR |
|---|---|---|---|---|---|---|---|---|---|---|---|---|---|
| Maximum Temp °F | 43.6 | 47.2 | 56.8 | 68.2 | 75.5 | 83.2 | 87.0 | 85.1 | 79.0 | 68.3 | 58.4 | 47.7 | 66.7 |
| Minimum Temp °F | 25.6 | 27.8 | 35.7 | 45.2 | 54.1 | 61.8 | 66.0 | 64.3 | 58.6 | 47.2 | 38.9 | 29.9 | 46.3 |
| Mean Temp °F | 34.6 | 37.5 | 46.3 | 56.7 | 64.8 | 72.5 | 76.5 | 74.7 | 68.8 | 57.8 | 48.7 | 38.8 | 56.5 |
| Days Max Temp ≥ 90 °F | 0 | 0 | 0 | 1 | 1 | 6 | 10 | 8 | 3 | 0 | 0 | 0 | 29 |
| Days Max Temp ≤ 32 °F | 5 | 3 | 0 | 0 | 0 | 0 | 0 | 0 | 0 | 0 | 0 | 2 | 10 |
| Days Min Temp ≤ 32 °F | 24 | 20 | 12 | 2 | 0 | 0 | 0 | 0 | 0 | 1 | 8 | 19 | 86 |
| Days Min Temp ≤ 0 °F | 0 | 0 | 0 | 0 | 0 | 0 | 0 | 0 | 0 | 0 | 0 | 0 | 0 |
| Heating Degree Days | 935 | 772 | 580 | 273 | 93 | 11 | 1 | 2 | 38 | 242 | 490 | 806 | 4243 |
| Cooling Degree Days | 0 | 0 | 7 | 32 | 93 | 253 | 379 | 293 | 151 | 26 | 2 | 0 | 1236 |
| Total Precipitation (") | 3.30 | 3.27 | 4.09 | 3.23 | 4.77 | 3.76 | 5.49 | 4.51 | 4.15 | 4.49 | 3.59 | 3.31 | 47.96 |
| Days ≥ 0.1" Precip | 6 | 6 | 7 | 6 | 8 | 6 | 8 | 8 | 6 | 5 | 5 | 6 | 77 |
| Total Snowfall (") | 6.2 | 6.4 | 3.3 | 0.3 | 0.0 | 0.0 | 0.0 | 0.0 | 0.0 | 0.1 | 1.0 | 3.6 | 20.9 |
| Days ≥ 1" Snow Depth | 7 | 6 | 2 | 0 | 0 | 0 | 0 | 0 | 0 | 0 | 1 | 2 | 18 |

## CHASE CITY *Mecklenburg County* ELEVATION 541 ft LAT/LONG 36° 48 ' N / 78° 28 ' W

| | JAN | FEB | MAR | APR | MAY | JUN | JUL | AUG | SEP | OCT | NOV | DEC | YEAR |
|---|---|---|---|---|---|---|---|---|---|---|---|---|---|
| Maximum Temp °F | 48.2 | 52.6 | 61.0 | 71.9 | 78.4 | 85.9 | 90.1 | 88.4 | 82.6 | 71.3 | 62.3 | 52.4 | 70.4 |
| Minimum Temp °F | 26.0 | 28.3 | 35.4 | 44.0 | 53.4 | 61.9 | 66.8 | 65.2 | 58.6 | 45.9 | 38.1 | 29.9 | 46.1 |
| Mean Temp °F | 37.1 | 40.6 | 48.3 | 57.9 | 65.9 | 73.9 | 78.5 | 76.8 | 70.6 | 58.6 | 50.2 | 41.2 | 58.3 |
| Days Max Temp ≥ 90 °F | 0 | 0 | 0 | 1 | 2 | 10 | 18 | 14 | 6 | 0 | 0 | 0 | 51 |
| Days Max Temp ≤ 32 °F | 2 | 1 | 0 | 0 | 0 | 0 | 0 | 0 | 0 | 0 | 0 | 1 | 4 |
| Days Min Temp ≤ 32 °F | 23 | 20 | 13 | 3 | 0 | 0 | 0 | 0 | 0 | 2 | 10 | 19 | 90 |
| Days Min Temp ≤ 0 °F | 0 | 0 | 0 | 0 | 0 | 0 | 0 | 0 | 0 | 0 | 0 | 0 | 0 |
| Heating Degree Days | 863 | 684 | 530 | 235 | 74 | 6 | 0 | 1 | 24 | 213 | 440 | 731 | 3801 |
| Cooling Degree Days | 0 | 0 | 5 | 24 | 106 | 288 | 438 | 365 | 186 | 24 | 2 | 1 | 1439 |
| Total Precipitation (") | 3.69 | 3.19 | 3.95 | 3.19 | 4.01 | 3.29 | 3.88 | 3.56 | 3.68 | 3.92 | 3.42 | 3.23 | 43.01 |
| Days ≥ 0.1" Precip | *6* | *6* | 7 | 6 | 7 | 5 | 7 | 6 | 5 | 5 | *6* | 6 | 72 |
| Total Snowfall (") | *4.1* | 2.2 | 1.0 | 0.0 | 0.0 | 0.0 | 0.0 | 0.0 | 0.0 | 0.0 | 0.2 | 1.2 | 8.7 |
| Days ≥ 1" Snow Depth | *1* | *1* | *0* | 0 | 0 | 0 | 0 | 0 | 0 | 0 | 0 | *0* | 2 |

## CHATHAM *Pittsylvania County* ELEVATION 702 ft LAT/LONG 36° 50 ' N / 79° 22 ' W

| | JAN | FEB | MAR | APR | MAY | JUN | JUL | AUG | SEP | OCT | NOV | DEC | YEAR |
|---|---|---|---|---|---|---|---|---|---|---|---|---|---|
| Maximum Temp °F | 45.9 | 49.8 | 58.7 | 69.1 | 76.8 | 83.8 | 87.6 | 86.0 | 80.1 | 69.6 | 60.3 | 50.4 | 68.2 |
| Minimum Temp °F | 22.6 | 23.9 | 31.3 | 39.2 | 48.4 | 57.4 | 62.2 | 60.4 | *53.3* | 40.4 | 32.6 | 26.0 | 41.5 |
| Mean Temp °F | 34.4 | 36.9 | 45.1 | 54.2 | 62.6 | 70.7 | 74.9 | 73.1 | *66.8* | 55.0 | 46.4 | 38.2 | 54.9 |
| Days Max Temp ≥ 90 °F | 0 | 0 | 0 | 0 | 1 | 6 | 12 | 9 | 3 | 0 | 0 | 0 | 31 |
| Days Max Temp ≤ 32 °F | 3 | 2 | 0 | 0 | 0 | 0 | 0 | 0 | 0 | 0 | 0 | 1 | 6 |
| Days Min Temp ≤ 32 °F | 26 | 23 | 18 | 9 | 1 | 0 | 0 | 0 | 0 | 9 | 16 | 23 | 125 |
| Days Min Temp ≤ 0 °F | 1 | 0 | 0 | 0 | 0 | 0 | 0 | 0 | 0 | 0 | 0 | 0 | 1 |
| Heating Degree Days | 942 | 789 | 614 | 332 | 128 | 20 | 1 | 6 | *60* | 313 | 553 | 823 | 4581 |
| Cooling Degree Days | 0 | 0 | 2 | 9 | 54 | 193 | 320 | 248 | 104 | 13 | 0 | 0 | 943 |
| Total Precipitation (") | 3.44 | 3.23 | 4.34 | 3.54 | 4.22 | 3.45 | 4.36 | 3.52 | 3.58 | 3.84 | 3.26 | 3.38 | 44.16 |
| Days ≥ 0.1" Precip | 7 | 7 | 8 | 7 | 8 | 6 | 7 | 7 | 5 | 5 | 6 | 6 | 79 |
| Total Snowfall (") | *5.5* | *2.6* | 1.6 | 0.0 | 0.0 | 0.0 | 0.0 | 0.0 | 0.0 | 0.0 | 0.3 | 0.9 | 10.9 |
| Days ≥ 1" Snow Depth | *3* | *1* | 0 | 0 | 0 | 0 | 0 | 0 | 0 | 0 | 0 | 1 | 5 |

## COLONIAL BEACH *Westmoreland County* ELEVATION 10 ft LAT/LONG 38° 15' N / 76° 58' W

| | JAN | FEB | MAR | APR | MAY | JUN | JUL | AUG | SEP | OCT | NOV | DEC | YEAR |
|---|---|---|---|---|---|---|---|---|---|---|---|---|---|
| Maximum Temp °F | *45.9* | *49.9* | 58.2 | *69.0* | 78.2 | 86.0 | 90.1 | 88.3 | *82.3* | *70.8* | 60.9 | *51.5* | 69.3 |
| Minimum Temp °F | *25.8* | *28.1* | 35.5 | *44.8* | *54.9* | *63.6* | *68.6* | *66.7* | *61.0* | *47.8* | *39.7* | na | na |
| Mean Temp °F | *35.9* | *39.0* | 47.0 | *56.9* | *66.5* | *74.9* | *79.5* | *77.6* | *71.7* | *59.3* | *50.3* | na | na |
| Days Max Temp ≥ 90 °F | 0 | *0* | 0 | *0* | 2 | *10* | *18* | 14 | *5* | *0* | 0 | *0* | 49 |
| Days Max Temp ≤ 32 °F | *3* | *1* | 0 | *0* | 0 | 0 | 0 | 0 | 0 | *0* | 0 | *1* | 5 |
| Days Min Temp ≤ 32 °F | na | na | *11* | *2* | *0* | 0 | *0* | *0* | 0 | *1* | *7* | na | na |
| Days Min Temp ≤ 0 °F | *0* | *0* | 0 | *0* | 0 | 0 | *0* | *0* | 0 | *0* | 0 | *0* | 0 |
| Heating Degree Days | *896* | *728* | 558 | *257* | *63* | *5* | *0* | *1* | *18* | *202* | *443* | na | na |
| Cooling Degree Days | 0 | *0* | 5 | *25* | 118 | 325 | 476 | 402 | 226 | 37 | 6 | *1* | 1621 |
| Total Precipitation (") | 3.12 | 2.94 | 3.83 | 3.00 | 4.33 | 3.28 | 4.39 | 3.80 | 3.57 | 2.88 | 2.98 | 3.47 | 41.59 |
| Days ≥ 0.1" Precip | *6* | *6* | *8* | 7 | 8 | 6 | 7 | 6 | 6 | 5 | 6 | *7* | 78 |
| Total Snowfall (") | na | na | *0.1* | 0.0 | 0.0 | 0.0 | 0.0 | 0.0 | 0.0 | 0.0 | 0.0 | na | na |
| Days ≥ 1" Snow Depth | na | na | *0* | 0 | 0 | 0 | 0 | 0 | 0 | 0 | 0 | na | na |

## CORBIN *Caroline County* ELEVATION 220 ft LAT/LONG 38° 12' N / 77° 22' W

| | JAN | FEB | MAR | APR | MAY | JUN | JUL | AUG | SEP | OCT | NOV | DEC | YEAR |
|---|---|---|---|---|---|---|---|---|---|---|---|---|---|
| Maximum Temp °F | 43.2 | 46.8 | 56.0 | 66.6 | 75.0 | 82.8 | 87.0 | 85.5 | 79.4 | 68.3 | 58.7 | 48.1 | 66.5 |
| Minimum Temp °F | 23.2 | 25.8 | 33.8 | 42.4 | 51.8 | 60.7 | 65.4 | 64.1 | 57.1 | 44.6 | 36.4 | 28.2 | 44.5 |
| Mean Temp °F | 33.2 | 36.3 | 44.9 | 54.5 | 63.4 | 71.8 | 76.2 | 74.8 | 68.3 | 56.5 | 47.6 | 38.2 | 55.5 |
| Days Max Temp ≥ 90 °F | 0 | 0 | 0 | 0 | 1 | 5 | 11 | 9 | 3 | 0 | 0 | 0 | 29 |
| Days Max Temp ≤ 32 °F | 6 | 3 | 0 | 0 | 0 | 0 | 0 | 0 | 0 | 0 | 0 | 2 | 11 |
| Days Min Temp ≤ 32 °F | 26 | 21 | 15 | 5 | 0 | 0 | 0 | 0 | 0 | 4 | 12 | 21 | 104 |
| Days Min Temp ≤ 0 °F | 1 | 0 | 0 | 0 | 0 | 0 | 0 | 0 | 0 | 0 | 0 | 0 | 1 |
| Heating Degree Days | 978 | 803 | 620 | 328 | 120 | 18 | 1 | 4 | 49 | 279 | 520 | 825 | 4545 |
| Cooling Degree Days | 0 | 0 | 6 | 21 | 84 | 239 | 377 | 307 | 152 | 26 | 3 | 1 | 1216 |
| Total Precipitation (") | 3.35 | 2.91 | 4.02 | 3.07 | 4.06 | 3.48 | 4.24 | 4.08 | 3.66 | 3.69 | 3.24 | 3.49 | 43.29 |
| Days ≥ 0.1" Precip | 6 | 6 | 8 | 6 | 7 | 6 | 7 | 6 | 5 | 5 | 6 | 6 | 74 |
| Total Snowfall (") | 4.9 | 4.3 | 1.9 | 0.2 | 0.0 | 0.0 | 0.0 | 0.0 | 0.0 | 0.0 | 0.5 | 2.4 | 14.2 |
| Days ≥ 1" Snow Depth | 5 | 4 | 1 | 0 | 0 | 0 | 0 | 0 | 0 | 0 | 0 | 2 | 12 |

## COVINGTON FILTER PLT *Alleghany County* ELEVATION 1230 ft LAT/LONG 37° 48' N / 80° 0' W

| | JAN | FEB | MAR | APR | MAY | JUN | JUL | AUG | SEP | OCT | NOV | DEC | YEAR |
|---|---|---|---|---|---|---|---|---|---|---|---|---|---|
| Maximum Temp °F | 44.9 | 49.7 | 60.0 | 70.7 | 78.4 | 84.4 | 87.9 | 86.4 | 80.2 | 70.5 | 59.6 | 49.4 | 68.5 |
| Minimum Temp °F | 23.2 | 25.2 | 32.6 | 40.1 | 48.7 | 56.6 | 61.3 | 60.0 | 53.7 | 41.5 | 33.6 | 26.9 | 41.9 |
| Mean Temp °F | 34.1 | 37.5 | 46.3 | 55.5 | 63.6 | 70.5 | 74.6 | 73.2 | 67.0 | 56.0 | 46.6 | 38.2 | 55.3 |
| Days Max Temp ≥ 90 °F | 0 | 0 | 0 | 1 | 2 | 6 | 13 | 9 | 3 | 0 | 0 | 0 | 34 |
| Days Max Temp ≤ 32 °F | 4 | 2 | 0 | 0 | 0 | 0 | 0 | 0 | 0 | 0 | 0 | 2 | 8 |
| Days Min Temp ≤ 32 °F | 25 | 21 | 16 | 7 | 1 | 0 | 0 | 0 | 0 | 6 | 15 | 22 | 113 |
| Days Min Temp ≤ 0 °F | 1 | 0 | 0 | 0 | 0 | 0 | 0 | 0 | 0 | 0 | 0 | 0 | 1 |
| Heating Degree Days | 951 | 772 | 574 | 292 | 104 | 16 | 1 | 3 | 56 | 284 | 546 | 825 | 4424 |
| Cooling Degree Days | *0* | 0 | 2 | 14 | 75 | 209 | 340 | 280 | 131 | 15 | 1 | 0 | 1067 |
| Total Precipitation (") | 2.21 | 2.27 | 3.19 | 2.96 | 3.77 | 3.09 | 3.86 | 3.48 | 2.69 | 3.18 | 2.87 | 2.54 | 36.11 |
| Days ≥ 0.1" Precip | 6 | 5 | 7 | 7 | 8 | 7 | 8 | 6 | 5 | 5 | 6 | 5 | 75 |
| Total Snowfall (") | *4.4* | na | *0.6* | 0.3 | 0.0 | 0.0 | 0.0 | 0.0 | 0.0 | 0.4 | 0.5 | *2.4* | na |
| Days ≥ 1" Snow Depth | *5* | *2* | 1 | 0 | 0 | 0 | 0 | 0 | 0 | 0 | 0 | *1* | 9 |

## DALE ENTERPRISE *Rockingham County* ELEVATION 1352 ft LAT/LONG 38° 27' N / 78° 56' W

| | JAN | FEB | MAR | APR | MAY | JUN | JUL | AUG | SEP | OCT | NOV | DEC | YEAR |
|---|---|---|---|---|---|---|---|---|---|---|---|---|---|
| Maximum Temp °F | 41.2 | 45.3 | 55.4 | 65.8 | 74.5 | 82.6 | 86.0 | 84.2 | 77.7 | 66.5 | 55.6 | 45.4 | 65.0 |
| Minimum Temp °F | 21.6 | 24.1 | 32.1 | 40.2 | 49.3 | 57.5 | 61.9 | 60.1 | 53.7 | 42.2 | 34.5 | 26.3 | 42.0 |
| Mean Temp °F | 31.5 | 34.7 | 43.8 | 53.0 | 61.9 | 70.1 | 74.0 | 72.2 | 65.7 | 54.4 | 45.1 | 35.9 | 53.5 |
| Days Max Temp ≥ 90 °F | 0 | 0 | 0 | 0 | 0 | 5 | 9 | 6 | 2 | 0 | 0 | 0 | 22 |
| Days Max Temp ≤ 32 °F | 7 | 4 | 0 | 0 | 0 | 0 | 0 | 0 | 0 | 0 | 0 | 3 | 14 |
| Days Min Temp ≤ 32 °F | 26 | 22 | 17 | 7 | 1 | 0 | 0 | 0 | 0 | 5 | 13 | 23 | 114 |
| Days Min Temp ≤ 0 °F | 1 | 0 | 0 | 0 | 0 | 0 | 0 | 0 | 0 | 0 | 0 | 1 | 2 |
| Heating Degree Days | 1034 | 848 | 652 | 361 | 140 | 22 | 2 | 7 | 75 | 331 | 592 | 895 | 4959 |
| Cooling Degree Days | 0 | 0 | 2 | 9 | 56 | 194 | 315 | 249 | 112 | 14 | 0 | 0 | 951 |
| Total Precipitation (") | 2.02 | 2.05 | 2.67 | 2.65 | 3.43 | 2.99 | 3.87 | 3.60 | 3.27 | 2.88 | 2.67 | 2.32 | 34.42 |
| Days ≥ 0.1" Precip | 5 | 5 | 6 | 6 | 8 | 7 | 8 | 6 | 5 | 4 | 6 | 5 | 71 |
| Total Snowfall (") | 6.9 | 6.7 | 3.9 | 0.7 | 0.0 | 0.0 | 0.0 | 0.0 | 0.0 | 0.3 | 1.7 | 4.8 | 25.0 |
| Days ≥ 1" Snow Depth | 9 | 6 | 2 | 0 | 0 | 0 | 0 | 0 | 0 | 0 | 0 | 4 | 21 |

## DANVILLE *Pittsylvania County* ELEVATION 571 ft LAT/LONG 36° 36 ' N / 79° 23 ' W

| | JAN | FEB | MAR | APR | MAY | JUN | JUL | AUG | SEP | OCT | NOV | DEC | YEAR |
|---|---|---|---|---|---|---|---|---|---|---|---|---|---|
| Maximum Temp °F | 46.9 | 51.0 | 60.7 | 71.4 | 79.0 | 86.3 | 90.0 | 88.4 | 82.3 | 71.2 | 61.5 | 51.4 | 70.0 |
| Minimum Temp °F | 26.0 | 28.0 | 35.3 | 43.7 | 53.4 | 62.3 | 67.2 | 65.7 | 58.7 | 45.3 | 36.5 | 29.9 | 46.0 |
| Mean Temp °F | 36.4 | 39.5 | 48.0 | 57.6 | 66.2 | 74.3 | 78.6 | 77.1 | 70.5 | 58.3 | 49.1 | 40.7 | 58.0 |
| Days Max Temp ≥ 90 °F | 0 | 0 | 0 | 1 | 2 | 11 | 17 | 15 | 6 | 0 | 0 | 0 | 52 |
| Days Max Temp ≤ 32 °F | 2 | 1 | 0 | 0 | 0 | 0 | 0 | 0 | 0 | 0 | 0 | 1 | 4 |
| Days Min Temp ≤ 32 °F | 23 | 20 | 13 | 3 | 0 | 0 | 0 | 0 | 0 | 2 | 12 | 20 | 93 |
| Days Min Temp ≤ 0 °F | 0 | 0 | 0 | 0 | 0 | 0 | 0 | 0 | 0 | 0 | 0 | 0 | 0 |
| Heating Degree Days | 878 | 715 | 525 | 245 | 71 | 6 | 0 | 1 | 27 | 228 | 476 | 747 | 3919 |
| Cooling Degree Days | 0 | 0 | 5 | 30 | 121 | 318 | 461 | 385 | 204 | 33 | 2 | 0 | 1559 |
| Total Precipitation (") | 3.52 | 3.31 | 4.21 | 3.37 | 3.89 | 3.33 | 4.46 | 3.49 | 3.25 | 3.65 | 3.03 | 3.24 | 42.75 |
| Days ≥ 0.1" Precip | 7 | 7 | 8 | 6 | 7 | 6 | 7 | 7 | 5 | 5 | 6 | 6 | 77 |
| Total Snowfall (") | na | *1.4* | *0.9* | 0.1 | 0.0 | 0.0 | 0.0 | 0.0 | 0.0 | 0.0 | 0.1 | 0.7 | na |
| Days ≥ 1" Snow Depth | na | *1* | 0 | 0 | 0 | 0 | 0 | 0 | 0 | 0 | 0 | 0 | na |

## FARMVILLE 2 N *Cumberland County* ELEVATION 430 ft LAT/LONG 37° 18 ' N / 78° 23 ' W

| | JAN | FEB | MAR | APR | MAY | JUN | JUL | AUG | SEP | OCT | NOV | DEC | YEAR |
|---|---|---|---|---|---|---|---|---|---|---|---|---|---|
| Maximum Temp °F | 47.2 | 51.1 | 60.7 | 71.4 | 78.2 | 85.3 | 88.9 | 86.9 | 80.9 | 70.9 | 61.2 | 51.0 | 69.5 |
| Minimum Temp °F | 24.8 | 26.9 | 34.5 | 42.8 | 52.2 | 60.6 | 65.2 | 63.5 | 56.5 | 44.1 | 35.9 | 29.0 | 44.7 |
| Mean Temp °F | 36.0 | 39.0 | 47.6 | 57.1 | 65.2 | 73.0 | 77.1 | 75.2 | 68.7 | 57.5 | 48.6 | 40.0 | 57.1 |
| Days Max Temp ≥ 90 °F | 0 | 0 | 0 | 1 | 1 | 8 | 14 | 10 | 4 | 0 | 0 | 0 | 38 |
| Days Max Temp ≤ 32 °F | 3 | 1 | 0 | 0 | 0 | 0 | 0 | 0 | 0 | 0 | 0 | 1 | 5 |
| Days Min Temp ≤ 32 °F | 24 | 21 | 14 | 5 | 0 | 0 | 0 | 0 | 0 | 4 | 13 | 20 | 101 |
| Days Min Temp ≤ 0 °F | 0 | 0 | 0 | 0 | 0 | 0 | 0 | 0 | 0 | 0 | 0 | 0 | 0 |
| Heating Degree Days | 892 | 726 | 537 | 258 | 81 | 9 | 0 | 2 | 40 | 250 | 490 | 767 | 4052 |
| Cooling Degree Days | 0 | 0 | 6 | 26 | 97 | 272 | 409 | 327 | 163 | 29 | 2 | 0 | 1331 |
| Total Precipitation (") | 3.53 | 3.28 | 4.08 | 3.10 | 4.11 | 3.34 | 4.33 | 3.93 | 3.45 | 3.74 | 3.31 | 3.28 | 43.48 |
| Days ≥ 0.1" Precip | 7 | 6 | 7 | 6 | 7 | 5 | 7 | 6 | 5 | 6 | 6 | 6 | 74 |
| Total Snowfall (") | 5.5 | 5.4 | 2.2 | 0.0 | 0.0 | 0.0 | 0.0 | 0.0 | 0.0 | 0.0 | 0.8 | 2.8 | 16.7 |
| Days ≥ 1" Snow Depth | 5 | 4 | 1 | 0 | 0 | 0 | 0 | 0 | 0 | 0 | 0 | 2 | 12 |

## FLOYD 2 NE *Floyd County* ELEVATION 2602 ft LAT/LONG 36° 55 ' N / 80° 19 ' W

| | JAN | FEB | MAR | APR | MAY | JUN | JUL | AUG | SEP | OCT | NOV | DEC | YEAR |
|---|---|---|---|---|---|---|---|---|---|---|---|---|---|
| Maximum Temp °F | 42.5 | 46.4 | 55.6 | 65.1 | 72.2 | 78.6 | 81.9 | 80.9 | 75.3 | 65.8 | 55.4 | 46.3 | 63.8 |
| Minimum Temp °F | 20.7 | 22.4 | 29.8 | 37.1 | 45.7 | 53.3 | 58.1 | 56.3 | 49.8 | 38.3 | 30.9 | 24.3 | 38.9 |
| Mean Temp °F | 31.6 | 34.4 | 42.7 | 51.1 | 59.0 | 65.9 | 70.0 | 68.6 | 62.6 | 52.1 | 43.2 | 35.3 | 51.4 |
| Days Max Temp ≥ 90 °F | 0 | 0 | 0 | 0 | 0 | 0 | 2 | 1 | 0 | 0 | 0 | 0 | 3 |
| Days Max Temp ≤ 32 °F | 6 | 3 | 1 | 0 | 0 | 0 | 0 | 0 | 0 | 0 | 0 | 3 | 13 |
| Days Min Temp ≤ 32 °F | 27 | 23 | 19 | 11 | 3 | 0 | 0 | 0 | 1 | 10 | 18 | 24 | 136 |
| Days Min Temp ≤ 0 °F | 2 | 1 | 0 | 0 | 0 | 0 | 0 | 0 | 0 | 0 | 0 | 1 | 4 |
| Heating Degree Days | 1028 | 858 | 683 | 411 | 195 | 48 | 9 | 18 | 116 | 397 | 648 | 913 | 5324 |
| Cooling Degree Days | 0 | 0 | 0 | 1 | 16 | 96 | 197 | 144 | 55 | 4 | 0 | 0 | 513 |
| Total Precipitation (") | 2.76 | 2.77 | 3.86 | 3.70 | 3.93 | 3.66 | 3.78 | 3.35 | 3.63 | 3.91 | 3.35 | 2.65 | 41.35 |
| Days ≥ 0.1" Precip | 6 | 5 | 7 | 7 | 8 | 7 | 7 | 6 | 5 | 5 | 6 | 5 | 74 |
| Total Snowfall (") | *5.3* | 6.3 | 2.8 | 1.1 | 0.0 | 0.0 | 0.0 | 0.0 | 0.0 | 0.0 | 1.3 | 3.1 | 19.9 |
| Days ≥ 1" Snow Depth | na | na | 0 | 0 | 0 | 0 | 0 | 0 | 0 | 0 | 0 | *0* | na |

## FREDERICKSBURG N P *Spotsylvania County* ELEVATION 90 ft LAT/LONG 38° 19 ' N / 77° 27 ' W

| | JAN | FEB | MAR | APR | MAY | JUN | JUL | AUG | SEP | OCT | NOV | DEC | YEAR |
|---|---|---|---|---|---|---|---|---|---|---|---|---|---|
| Maximum Temp °F | 44.4 | 49.0 | 58.6 | 69.4 | 78.2 | 86.0 | 89.9 | 88.0 | 81.9 | 70.5 | 60.0 | 49.5 | 68.8 |
| Minimum Temp °F | 22.4 | 25.3 | 33.1 | 41.2 | 51.4 | 60.0 | 65.2 | 63.1 | 56.1 | 43.6 | 35.2 | 27.3 | 43.7 |
| Mean Temp °F | 33.4 | 37.2 | 45.9 | 55.3 | 64.8 | 73.0 | 77.6 | 75.6 | 69.1 | 57.1 | 47.7 | 38.4 | 56.3 |
| Days Max Temp ≥ 90 °F | 0 | 0 | 0 | 1 | 2 | 9 | 17 | 13 | 5 | 0 | 0 | 0 | 47 |
| Days Max Temp ≤ 32 °F | 4 | 1 | 0 | 0 | 0 | 0 | 0 | 0 | 0 | 0 | 0 | 1 | 6 |
| Days Min Temp ≤ 32 °F | 26 | 21 | 16 | 5 | 0 | 0 | 0 | 0 | 0 | 5 | 13 | 22 | 108 |
| Days Min Temp ≤ 0 °F | 1 | 0 | 0 | 0 | 0 | 0 | 0 | 0 | 0 | 0 | 0 | 0 | 1 |
| Heating Degree Days | 975 | 781 | 592 | 302 | 92 | 11 | 0 | 3 | 39 | 260 | 514 | 818 | 4387 |
| Cooling Degree Days | 0 | 0 | 5 | 16 | 87 | 261 | 415 | 323 | 158 | 21 | 2 | 0 | 1288 |
| Total Precipitation (") | 3.34 | 2.87 | 3.86 | 3.08 | 3.99 | 3.17 | 4.13 | 3.70 | 3.78 | 3.55 | 3.37 | 3.32 | 42.16 |
| Days ≥ 0.1" Precip | 6 | 6 | 7 | 6 | 8 | 6 | 7 | 6 | 5 | 5 | 6 | 6 | 74 |
| Total Snowfall (") | *6.4* | *2.8* | 2.0 | 0.1 | 0.0 | 0.0 | 0.0 | 0.0 | 0.0 | 0.0 | 0.6 | 1.9 | 13.8 |
| Days ≥ 1" Snow Depth | *4* | na | 1 | 0 | 0 | 0 | 0 | 0 | 0 | 0 | 0 | 1 | na |

## GALAX RADIO WBRF *Carroll County* ELEVATION 2402 ft LAT/LONG 36° 40 ' N / 80° 55 ' W

| | JAN | FEB | MAR | APR | MAY | JUN | JUL | AUG | SEP | OCT | NOV | DEC | YEAR |
|---|---|---|---|---|---|---|---|---|---|---|---|---|---|
| Maximum Temp °F | 42.5 | 46.8 | 56.4 | 66.1 | 73.3 | 79.6 | 82.8 | 81.2 | 75.3 | 65.6 | 55.2 | 46.3 | 64.3 |
| Minimum Temp °F | 22.0 | 24.1 | 31.3 | 38.7 | 47.4 | 54.8 | 59.2 | 58.2 | 51.8 | 40.2 | 32.5 | 25.7 | 40.5 |
| Mean Temp °F | 32.3 | 35.5 | 43.9 | 52.4 | 60.4 | 67.2 | 71.0 | 69.7 | 63.6 | 52.9 | 43.9 | 36.0 | 52.4 |
| Days Max Temp ≥ 90 °F | 0 | 0 | 0 | 0 | 0 | 0 | 3 | 1 | 0 | 0 | 0 | 0 | 4 |
| Days Max Temp ≤ 32 °F | 6 | 3 | 0 | 0 | 0 | 0 | 0 | 0 | 0 | 0 | 0 | 3 | 12 |
| Days Min Temp ≤ 32 °F | 26 | 22 | 18 | 8 | 2 | 0 | 0 | 0 | 0 | 8 | 16 | 23 | 123 |
| Days Min Temp ≤ 0 °F | 1 | 0 | 0 | 0 | 0 | 0 | 0 | 0 | 0 | 0 | 0 | 0 | 1 |
| Heating Degree Days | 1008 | 827 | 648 | 372 | 161 | 32 | 4 | 11 | 97 | 370 | 628 | 891 | 5049 |
| Cooling Degree Days | 0 | 0 | 0 | 2 | 26 | 118 | 217 | 166 | 63 | 4 | 0 | 0 | 596 |
| Total Precipitation (") | 2.70 | 3.24 | 3.86 | 3.68 | 4.42 | 3.85 | 4.15 | 3.34 | 3.59 | 3.93 | 3.36 | 2.99 | 43.11 |
| Days ≥ 0.1" Precip | 6 | 5 | 6 | 7 | 8 | 7 | 8 | 7 | 5 | 6 | 6 | 5 | 76 |
| Total Snowfall (") | 7.1 | 5.7 | 2.4 | 0.8 | 0.0 | 0.0 | 0.0 | 0.0 | 0.0 | 0.0 | 0.7 | 3.4 | 20.1 |
| Days ≥ 1" Snow Depth | na | na | *0* | 0 | 0 | 0 | 0 | 0 | 0 | 0 | 0 | *1* | na |

## GRUNDY *Buchanan County* ELEVATION 1040 ft LAT/LONG 37° 18 ' N / 82° 8 ' W

| | JAN | FEB | MAR | APR | MAY | JUN | JUL | AUG | SEP | OCT | NOV | DEC | YEAR |
|---|---|---|---|---|---|---|---|---|---|---|---|---|---|
| Maximum Temp °F | 45.0 | 50.0 | 59.8 | 70.1 | 76.8 | 84.0 | 87.4 | 85.9 | 80.5 | 70.3 | *60.4* | 49.9 | 68.3 |
| Minimum Temp °F | 22.7 | 24.7 | 32.3 | 39.9 | *49.3* | 58.0 | 63.1 | 62.2 | 55.9 | 41.4 | *33.3* | 26.3 | 42.4 |
| Mean Temp °F | 33.9 | 37.4 | 46.1 | 55.0 | *63.1* | 71.0 | 75.3 | 74.1 | 68.3 | 55.8 | *46.9* | 38.1 | 55.4 |
| Days Max Temp ≥ 90 °F | 0 | 0 | 0 | 1 | 1 | 6 | 12 | 8 | 3 | 0 | 0 | 0 | 31 |
| Days Max Temp ≤ 32 °F | *5* | *3* | 0 | 0 | 0 | 0 | 0 | 0 | 0 | 0 | 0 | 2 | 10 |
| Days Min Temp ≤ 32 °F | *26* | *22* | 17 | *8* | *0* | 0 | 0 | 0 | 0 | 6 | 15 | 23 | 117 |
| Days Min Temp ≤ 0 °F | 1 | *0* | 0 | 0 | 0 | 0 | 0 | 0 | 0 | 0 | 0 | 0 | 1 |
| Heating Degree Days | *964* | 776 | 580 | 307 | *118* | 15 | 1 | 2 | 42 | 294 | *538* | 828 | 4465 |
| Cooling Degree Days | 0 | 0 | 1 | 13 | 67 | 203 | 336 | 283 | 137 | 16 | 0 | 0 | 1056 |
| Total Precipitation (") | 3.14 | 3.01 | 3.88 | 3.75 | 4.24 | 4.09 | 4.60 | 3.95 | 3.42 | 3.17 | 3.01 | 3.19 | 43.45 |
| Days ≥ 0.1" Precip | *7* | 7 | 8 | 8 | 9 | 8 | *8* | 7 | 7 | 6 | *7* | 7 | 89 |
| Total Snowfall (") | *6.7* | *5.3* | *2.5* | 1.2 | 0.0 | 0.0 | 0.0 | 0.0 | 0.0 | 0.0 | 0.7 | *2.6* | 19.0 |
| Days ≥ 1" Snow Depth | *7* | *5* | 1 | 0 | 0 | 0 | 0 | 0 | 0 | 0 | *0* | *3* | 16 |

## HOLLAND 1 E *Suffolk Independent City* ELEVATION 79 ft LAT/LONG 36° 41 ' N / 76° 47 ' W

| | JAN | FEB | MAR | APR | MAY | JUN | JUL | AUG | SEP | OCT | NOV | DEC | YEAR |
|---|---|---|---|---|---|---|---|---|---|---|---|---|---|
| Maximum Temp °F | 48.2 | 51.3 | 60.0 | 69.9 | 77.6 | 84.9 | 88.4 | 86.9 | 81.8 | 71.7 | 63.2 | 53.3 | 69.8 |
| Minimum Temp °F | 27.7 | 29.7 | 36.9 | 44.9 | 54.3 | 62.6 | 67.4 | 65.8 | 59.3 | 47.1 | 39.4 | 31.5 | 47.2 |
| Mean Temp °F | 38.0 | 40.5 | 48.5 | 57.4 | 66.0 | 73.7 | 77.9 | 76.4 | 70.6 | 59.4 | 51.3 | 42.4 | 58.5 |
| Days Max Temp ≥ 90 °F | 0 | 0 | 0 | 1 | 2 | 7 | 14 | 11 | 4 | 0 | 0 | 0 | 39 |
| Days Max Temp ≤ 32 °F | 3 | 1 | 0 | 0 | 0 | 0 | 0 | 0 | 0 | 0 | 0 | 1 | 5 |
| Days Min Temp ≤ 32 °F | 22 | 18 | 11 | 3 | 0 | 0 | 0 | 0 | 0 | 2 | 9 | 18 | 83 |
| Days Min Temp ≤ 0 °F | 0 | 0 | 0 | 0 | 0 | 0 | 0 | 0 | 0 | 0 | 0 | 0 | 0 |
| Heating Degree Days | 831 | 687 | 511 | 251 | 75 | 9 | 0 | 1 | 24 | 204 | 412 | 694 | 3699 |
| Cooling Degree Days | 0 | 1 | 7 | 29 | 117 | 293 | 432 | 353 | 193 | 40 | 7 | 1 | 1473 |
| Total Precipitation (") | 4.07 | 3.33 | 4.39 | 3.28 | 4.04 | 3.83 | 5.35 | 5.36 | 3.79 | 3.56 | 2.92 | 3.37 | 47.29 |
| Days ≥ 0.1" Precip | 8 | 7 | 7 | 7 | 7 | 6 | 8 | 7 | 5 | 5 | 5 | 7 | 79 |
| Total Snowfall (") | 2.5 | 2.9 | 1.6 | 0.0 | 0.0 | 0.0 | 0.0 | 0.0 | 0.0 | 0.0 | 0.0 | 0.9 | 7.9 |
| Days ≥ 1" Snow Depth | 2 | 2 | 1 | 0 | 0 | 0 | 0 | 0 | 0 | 0 | 0 | 1 | 6 |

## HOPEWELL *Prince George County* ELEVATION 39 ft LAT/LONG 37° 18 ' N / 77° 18 ' W

| | JAN | FEB | MAR | APR | MAY | JUN | JUL | AUG | SEP | OCT | NOV | DEC | YEAR |
|---|---|---|---|---|---|---|---|---|---|---|---|---|---|
| Maximum Temp °F | 49.3 | 52.8 | 62.9 | 73.3 | 80.4 | 87.4 | 90.8 | 89.1 | 83.6 | 73.4 | 64.2 | 54.1 | 71.8 |
| Minimum Temp °F | 28.2 | 30.4 | 38.2 | 46.2 | 55.2 | 63.3 | 67.9 | 66.6 | 60.0 | 48.4 | 40.0 | 32.7 | 48.1 |
| Mean Temp °F | 38.8 | 41.6 | 50.5 | 59.8 | 67.8 | 75.4 | 79.4 | 77.9 | 71.8 | 60.9 | 52.1 | 43.4 | 59.9 |
| Days Max Temp ≥ 90 °F | 0 | 0 | 0 | 1 | 3 | 12 | 19 | 15 | 6 | 0 | 0 | 0 | 56 |
| Days Max Temp ≤ 32 °F | 2 | 0 | 0 | 0 | 0 | 0 | 0 | 0 | 0 | 0 | 0 | 1 | 3 |
| Days Min Temp ≤ 32 °F | 21 | 17 | 10 | 2 | 0 | 0 | 0 | 0 | 0 | 2 | 8 | 16 | 76 |
| Days Min Temp ≤ 0 °F | 0 | 0 | 0 | 0 | 0 | 0 | 0 | 0 | 0 | 0 | 0 | 0 | 0 |
| Heating Degree Days | 806 | 653 | 452 | 195 | 49 | 4 | 0 | 0 | 18 | 169 | 389 | 664 | 3399 |
| Cooling Degree Days | 0 | 1 | 11 | 46 | 144 | 337 | 476 | 392 | 229 | 50 | 9 | 1 | 1696 |
| Total Precipitation (") | 3.75 | 3.16 | 4.07 | 3.07 | 3.94 | 2.94 | 4.31 | 4.31 | 3.84 | 3.41 | 3.06 | 3.14 | 43.00 |
| Days ≥ 0.1" Precip | 7 | 6 | 8 | 6 | 7 | 5 | 8 | 7 | 5 | 5 | 5 | 6 | 75 |
| Total Snowfall (") | *3.8* | *2.4* | 0.8 | 0.0 | 0.0 | 0.0 | 0.0 | 0.0 | 0.0 | 0.0 | 0.1 | 0.4 | 7.5 |
| Days ≥ 1" Snow Depth | na | *1* | 0 | 0 | 0 | 0 | 0 | 0 | 0 | 0 | 0 | 0 | na |

## JOHN H KERR DAM *Mecklenburg County* ELEVATION 323 ft LAT/LONG 36° 36' N / 78° 18' W

| | JAN | FEB | MAR | APR | MAY | JUN | JUL | AUG | SEP | OCT | NOV | DEC | YEAR |
|---|---|---|---|---|---|---|---|---|---|---|---|---|---|
| Maximum Temp °F | 47.2 | 50.7 | 59.7 | 70.2 | 77.6 | 85.2 | 89.1 | 87.6 | 81.9 | 71.3 | 62.0 | 52.0 | 69.5 |
| Minimum Temp °F | 26.2 | 27.9 | 35.5 | 44.1 | 53.6 | 62.2 | 67.1 | 65.7 | 58.4 | 45.7 | 37.7 | 30.4 | 46.2 |
| Mean Temp °F | 36.7 | 39.4 | 47.6 | 57.2 | 65.6 | 73.7 | 78.1 | 76.7 | 70.2 | 58.5 | 49.9 | 41.3 | 57.9 |
| Days Max Temp ≥ 90 °F | 0 | 0 | 0 | 1 | 2 | 8 | 15 | 12 | 4 | 0 | 0 | 0 | 42 |
| Days Max Temp ≤ 32 °F | 3 | 1 | 0 | 0 | 0 | 0 | 0 | 0 | 0 | 0 | 0 | 1 | 5 |
| Days Min Temp ≤ 32 °F | 23 | 20 | 13 | 3 | 0 | 0 | 0 | 0 | 0 | 3 | 11 | 19 | 92 |
| Days Min Temp ≤ 0 °F | 0 | 0 | 0 | 0 | 0 | 0 | 0 | 0 | 0 | 0 | 0 | 0 | 0 |
| Heating Degree Days | 870 | 718 | 537 | 253 | 82 | 9 | 0 | 2 | 30 | 224 | 451 | 729 | 3905 |
| Cooling Degree Days | 0 | 0 | 6 | 26 | 113 | 294 | 442 | 378 | 206 | 39 | 3 | 0 | 1507 |
| Total Precipitation (") | 3.54 | 3.03 | 4.11 | 3.04 | 4.00 | 3.38 | 4.51 | 4.04 | 3.33 | 3.63 | 3.04 | 3.07 | 42.72 |
| Days ≥ 0.1" Precip | 7 | 6 | 8 | 6 | 7 | 6 | 7 | 7 | 5 | 5 | 5 | 6 | 75 |
| Total Snowfall (") | na | na | 0.9 | 0.0 | 0.0 | 0.0 | 0.0 | 0.0 | 0.0 | 0.0 | 0.1 | 0.3 | na |
| Days ≥ 1" Snow Depth | na | na | 0 | 0 | 0 | 0 | 0 | 0 | 0 | 0 | 0 | 0 | na |

## LANGLEY AFB *Hampton Independent City* ELEVATION 10 ft LAT/LONG 37° 5' N / 76° 21' W

| | JAN | FEB | MAR | APR | MAY | JUN | JUL | AUG | SEP | OCT | NOV | DEC | YEAR |
|---|---|---|---|---|---|---|---|---|---|---|---|---|---|
| Maximum Temp °F | 46.6 | 49.2 | 57.2 | 66.5 | 74.3 | 82.3 | 86.3 | 84.9 | 78.9 | 68.9 | 60.4 | 51.6 | 67.3 |
| Minimum Temp °F | 31.3 | 33.0 | 40.1 | 48.5 | 57.9 | 66.6 | 71.7 | 70.7 | 64.7 | 52.8 | 43.7 | 35.7 | 51.4 |
| Mean Temp °F | 39.0 | 41.1 | 48.7 | 57.6 | 66.1 | 74.5 | 79.1 | 77.8 | 71.8 | 60.8 | 52.1 | 43.7 | 59.4 |
| Days Max Temp ≥ 90 °F | 0 | 0 | 0 | 0 | 1 | 4 | 10 | 7 | 2 | 0 | 0 | 0 | 24 |
| Days Max Temp ≤ 32 °F | 3 | 2 | 0 | 0 | 0 | 0 | 0 | 0 | 0 | 0 | 0 | 1 | 6 |
| Days Min Temp ≤ 32 °F | 17 | 15 | 6 | 0 | 0 | 0 | 0 | 0 | 0 | 0 | 4 | 12 | 54 |
| Days Min Temp ≤ 0 °F | 0 | 0 | 0 | 0 | 0 | 0 | 0 | 0 | 0 | 0 | 0 | 0 | 0 |
| Heating Degree Days | 800 | 669 | 506 | 246 | 70 | 5 | 0 | 0 | 13 | 168 | 389 | 656 | 3522 |
| Cooling Degree Days | 0 | 1 | 8 | 36 | 128 | 323 | 484 | 414 | 236 | 56 | 8 | 1 | 1695 |
| Total Precipitation (") | 3.79 | 3.47 | 4.55 | 3.04 | 4.02 | 3.55 | 4.90 | 4.79 | 4.20 | 2.97 | 3.12 | 3.44 | 45.84 |
| Days ≥ 0.1" Precip | 7 | 7 | 7 | 6 | 7 | 6 | 8 | 6 | 5 | 5 | 5 | 6 | 75 |
| Total Snowfall (") | 3.2 | 3.5 | 1.4 | 0.2 | 0.0 | 0.0 | 0.0 | 0.0 | 0.0 | 0.0 | 0.1 | 0.7 | 9.1 |
| Days ≥ 1" Snow Depth | 2 | 2 | 0 | 0 | 0 | 0 | 0 | 0 | 0 | 0 | 0 | 1 | 5 |

## LAWRENCEVILLE 3 E *Brunswick County* ELEVATION 325 ft LAT/LONG 36° 46' N / 77° 47' W

| | JAN | FEB | MAR | APR | MAY | JUN | JUL | AUG | SEP | OCT | NOV | DEC | YEAR |
|---|---|---|---|---|---|---|---|---|---|---|---|---|---|
| Maximum Temp °F | 48.7 | 52.7 | 61.6 | 71.7 | 78.6 | 85.1 | 88.9 | 87.2 | 81.8 | 71.7 | 62.8 | 52.8 | 70.3 |
| Minimum Temp °F | 26.4 | 28.2 | 35.3 | 43.0 | 51.7 | 59.8 | 64.6 | 63.3 | 56.4 | 44.9 | 37.0 | 30.0 | 45.0 |
| Mean Temp °F | 37.6 | 40.5 | 48.5 | 57.4 | 65.2 | 72.5 | 76.8 | 75.3 | 69.1 | 58.3 | 50.0 | 41.4 | 57.7 |
| Days Max Temp ≥ 90 °F | 0 | 0 | 0 | 0 | 1 | 7 | 14 | 11 | 3 | 0 | 0 | 0 | 36 |
| Days Max Temp ≤ 32 °F | 2 | 1 | 0 | 0 | 0 | 0 | 0 | 0 | 0 | 0 | 0 | 1 | 4 |
| Days Min Temp ≤ 32 °F | 23 | 19 | 13 | 5 | 0 | 0 | 0 | 0 | 0 | 4 | 12 | 19 | 95 |
| Days Min Temp ≤ 0 °F | 0 | 0 | 0 | 0 | 0 | 0 | 0 | 0 | 0 | 0 | 0 | 0 | 0 |
| Heating Degree Days | 844 | 686 | 513 | 248 | 82 | 9 | 0 | 2 | 33 | 230 | 450 | 724 | 3821 |
| Cooling Degree Days | 0 | 0 | 6 | 25 | 95 | 261 | 400 | 320 | 166 | 32 | 4 | 1 | 1310 |
| Total Precipitation (") | 3.59 | 3.43 | 4.21 | 2.95 | 3.81 | 3.64 | 4.14 | 4.43 | 3.72 | 3.48 | 3.30 | 3.05 | 43.75 |
| Days ≥ 0.1" Precip | 7 | 6 | 7 | 6 | 6 | 6 | 7 | 7 | 5 | 5 | 5 | 6 | 73 |
| Total Snowfall (") | 4.0 | 4.0 | 1.9 | 0.1 | 0.0 | 0.0 | 0.0 | 0.0 | 0.0 | 0.0 | 0.1 | 1.5 | 11.6 |
| Days ≥ 1" Snow Depth | 3 | 2 | 1 | 0 | 0 | 0 | 0 | 0 | 0 | 0 | 0 | 1 | 7 |

## LEXINGTON *Rockbridge County* ELEVATION 1060 ft LAT/LONG 37° 47' N / 79° 26' W

| | JAN | FEB | MAR | APR | MAY | JUN | JUL | AUG | SEP | OCT | NOV | DEC | YEAR |
|---|---|---|---|---|---|---|---|---|---|---|---|---|---|
| Maximum Temp °F | 44.4 | 48.6 | 59.0 | 69.8 | 76.9 | 83.5 | 86.8 | 85.1 | 78.7 | 68.5 | 58.0 | 48.4 | 67.3 |
| Minimum Temp °F | 23.3 | 25.2 | 33.0 | 41.1 | 50.2 | 58.3 | 63.1 | 62.0 | 55.3 | 43.0 | 34.3 | 27.2 | 43.0 |
| Mean Temp °F | 33.9 | 36.9 | 46.0 | 55.4 | 63.6 | 70.9 | 75.0 | 73.6 | 67.0 | 55.7 | 46.2 | 37.9 | 55.2 |
| Days Max Temp ≥ 90 °F | 0 | 0 | 0 | 1 | 1 | 4 | 10 | 7 | 2 | 0 | 0 | 0 | 25 |
| Days Max Temp ≤ 32 °F | 4 | 2 | 0 | 0 | 0 | 0 | 0 | 0 | 0 | 0 | 0 | 1 | 7 |
| Days Min Temp ≤ 32 °F | 25 | 22 | 15 | 6 | 1 | 0 | 0 | 0 | 0 | 5 | 14 | 21 | 109 |
| Days Min Temp ≤ 0 °F | 1 | 0 | 0 | 0 | 0 | 0 | 0 | 0 | 0 | 0 | 0 | 0 | 1 |
| Heating Degree Days | 958 | 787 | 583 | 292 | 100 | 12 | 1 | 2 | 54 | 293 | 559 | 834 | 4475 |
| Cooling Degree Days | 0 | 0 | 2 | 14 | 64 | 203 | 332 | 276 | 122 | 16 | 0 | 0 | 1029 |
| Total Precipitation (") | 2.60 | 2.73 | 3.52 | 2.99 | 3.72 | 3.29 | 4.02 | 3.24 | 2.90 | 3.46 | 2.87 | 2.93 | 38.27 |
| Days ≥ 0.1" Precip | 6 | 5 | 7 | 6 | 7 | 6 | 7 | 6 | 5 | 5 | 5 | 6 | 71 |
| Total Snowfall (") | 6.0 | 5.2 | 2.2 | 0.2 | 0.0 | 0.0 | 0.0 | 0.0 | 0.0 | 0.0 | 0.7 | 3.2 | 17.5 |
| Days ≥ 1" Snow Depth | 7 | 5 | 1 | 0 | 0 | 0 | 0 | 0 | 0 | 0 | 0 | 3 | 16 |

## LINCOLN *Loudoun County* ELEVATION 502 ft LAT/LONG 39° 7 ' N / 77° 43 ' W

| | JAN | FEB | MAR | APR | MAY | JUN | JUL | AUG | SEP | OCT | NOV | DEC | YEAR |
|---|---|---|---|---|---|---|---|---|---|---|---|---|---|
| Maximum Temp °F | 43.1 | 47.5 | 57.1 | 68.1 | 76.6 | 84.6 | 88.3 | 86.9 | 80.7 | 69.8 | 58.7 | 47.7 | 67.4 |
| Minimum Temp °F | 23.0 | 25.1 | 33.1 | 42.2 | 51.5 | 60.6 | 65.0 | 63.2 | 56.4 | 44.7 | 36.6 | 27.9 | 44.1 |
| Mean Temp °F | 33.1 | 36.3 | 45.1 | 55.1 | 64.1 | 72.6 | 76.7 | 75.1 | 68.7 | 57.3 | 47.7 | 37.8 | 55.8 |
| Days Max Temp ≥ 90 °F | 0 | 0 | 0 | 0 | 2 | 7 | 13 | 10 | 3 | 0 | 0 | 0 | 35 |
| Days Max Temp ≤ 32 °F | 4 | 2 | 0 | 0 | 0 | 0 | 0 | 0 | 0 | 0 | 0 | 2 | 8 |
| Days Min Temp ≤ 32 °F | 26 | 23 | 16 | 4 | 0 | 0 | 0 | 0 | 0 | 3 | 11 | 22 | 105 |
| Days Min Temp ≤ 0 °F | 1 | 0 | 0 | 0 | 0 | 0 | 0 | 0 | 0 | 0 | 0 | 0 | 1 |
| Heating Degree Days | 982 | 814 | 615 | 305 | 102 | 9 | 0 | 2 | 40 | 249 | 517 | 836 | 4471 |
| Cooling Degree Days | 0 | 0 | 5 | 19 | 90 | 254 | 395 | 316 | 151 | 21 | 2 | 0 | 1253 |
| Total Precipitation (") | 2.85 | 2.45 | 3.49 | 3.47 | 4.49 | 3.98 | 3.77 | 4.09 | 3.35 | 3.41 | 3.49 | 3.28 | 42.12 |
| Days ≥ 0.1" Precip | 6 | 5 | 6 | 6 | 8 | 6 | 6 | 6 | 5 | 5 | 6 | 5 | 70 |
| Total Snowfall (") | 8.0 | 6.7 | 4.6 | 0.2 | 0.0 | 0.0 | 0.0 | 0.0 | 0.0 | 0.2 | 0.9 | 1.9 | 22.5 |
| Days ≥ 1" Snow Depth | 8 | 8 | 3 | 0 | 0 | 0 | 0 | 0 | 0 | 0 | 0 | 2 | 21 |

## LOUISA *Louisa County* ELEVATION 440 ft LAT/LONG 38° 2 ' N / 78° 0 ' W

| | JAN | FEB | MAR | APR | MAY | JUN | JUL | AUG | SEP | OCT | NOV | DEC | YEAR |
|---|---|---|---|---|---|---|---|---|---|---|---|---|---|
| Maximum Temp °F | 44.9 | 49.1 | 58.9 | 69.6 | 76.7 | 83.8 | 87.0 | 85.3 | 79.5 | 69.3 | 59.4 | 49.0 | 67.7 |
| Minimum Temp °F | 24.1 | 26.3 | 33.9 | 42.1 | 51.5 | 60.0 | 65.0 | 63.5 | 56.5 | 44.2 | 36.1 | 28.3 | 44.3 |
| Mean Temp °F | 34.5 | 37.8 | 46.4 | 55.9 | 64.1 | 71.9 | 76.0 | 74.4 | 68.0 | 56.8 | 47.8 | 38.7 | 56.0 |
| Days Max Temp ≥ 90 °F | 0 | 0 | 0 | 0 | 1 | 5 | 11 | 7 | 3 | 0 | 0 | 0 | 27 |
| Days Max Temp ≤ 32 °F | 4 | 2 | 0 | 0 | 0 | 0 | 0 | 0 | 0 | 0 | 0 | 2 | 8 |
| Days Min Temp ≤ 32 °F | 24 | 21 | 15 | 6 | 0 | 0 | 0 | 0 | 0 | 5 | 12 | 21 | 104 |
| Days Min Temp ≤ 0 °F | 1 | 0 | 0 | 0 | 0 | 0 | 0 | 0 | 0 | 0 | 0 | 0 | 1 |
| Heating Degree Days | 938 | 763 | 572 | 288 | 100 | 13 | 1 | 3 | 49 | 269 | 513 | 809 | 4318 |
| Cooling Degree Days | 0 | 0 | 5 | 20 | 85 | 238 | 373 | 297 | 148 | 25 | 2 | 0 | 1193 |
| Total Precipitation (") | 3.10 | 2.88 | 3.77 | 2.94 | 3.88 | 3.71 | 4.39 | 4.28 | 3.54 | 3.58 | 3.72 | 3.33 | 43.12 |
| Days ≥ 0.1" Precip | 6 | 6 | 7 | 6 | 7 | 6 | 7 | 6 | 5 | 5 | 6 | 6 | 73 |
| Total Snowfall (") | 6.3 | 5.0 | 2.5 | 0.4 | 0.0 | 0.0 | 0.0 | 0.0 | 0.0 | 0.1 | 0.8 | 3.4 | 18.5 |
| Days ≥ 1" Snow Depth | 7 | 4 | 1 | 0 | 0 | 0 | 0 | 0 | 0 | 0 | 0 | 2 | 14 |

## LURAY 5 E *Page County* ELEVATION 1201 ft LAT/LONG 38° 40 ' N / 78° 23 ' W

| | JAN | FEB | MAR | APR | MAY | JUN | JUL | AUG | SEP | OCT | NOV | DEC | YEAR |
|---|---|---|---|---|---|---|---|---|---|---|---|---|---|
| Maximum Temp °F | 42.1 | 45.8 | 54.9 | 66.5 | 74.9 | 82.1 | 85.8 | 84.0 | 78.0 | 67.5 | 57.9 | 47.0 | 65.5 |
| Minimum Temp °F | 18.7 | 20.9 | 29.2 | 38.3 | 47.3 | 55.1 | 59.8 | 58.3 | 51.4 | 39.7 | 31.9 | 23.8 | 39.5 |
| Mean Temp °F | 30.4 | 33.4 | 42.1 | 52.4 | 61.1 | 68.6 | 72.8 | 71.2 | 64.7 | 53.6 | 44.9 | 35.4 | 52.5 |
| Days Max Temp ≥ 90 °F | 0 | 0 | 0 | 0 | 1 | 4 | 8 | 5 | 2 | 0 | 0 | 0 | 20 |
| Days Max Temp ≤ 32 °F | 7 | 4 | 1 | 0 | 0 | 0 | 0 | 0 | 0 | 0 | 0 | 3 | 15 |
| Days Min Temp ≤ 32 °F | 28 | 24 | 20 | 9 | 2 | 0 | 0 | 0 | 0 | 8 | 17 | 26 | 134 |
| Days Min Temp ≤ 0 °F | 2 | 0 | 0 | 0 | 0 | 0 | 0 | 0 | 0 | 0 | 0 | 0 | 2 |
| Heating Degree Days | 1067 | 888 | 706 | 382 | 162 | 35 | 4 | 10 | 88 | 355 | 598 | 909 | 5204 |
| Cooling Degree Days | 0 | 0 | 2 | 12 | 52 | 158 | 274 | 209 | 88 | 15 | 0 | 0 | 810 |
| Total Precipitation (") | 2.86 | 2.47 | 3.32 | 3.11 | 3.74 | 3.48 | 3.94 | 3.51 | 3.58 | 3.44 | 3.40 | 2.99 | 39.84 |
| Days ≥ 0.1" Precip | 5 | 5 | 7 | 6 | 7 | 6 | 8 | 7 | 5 | 5 | 6 | 5 | 72 |
| Total Snowfall (") | 7.7 | 5.9 | 4.7 | 0.6 | 0.0 | 0.0 | 0.0 | 0.0 | 0.0 | 0.2 | 1.7 | 4.2 | 25.0 |
| Days ≥ 1" Snow Depth | 9 | 6 | 2 | 0 | 0 | 0 | 0 | 0 | 0 | 0 | 1 | 2 | 20 |

## LYNCHBURG MUNI AP *Campbell County* ELEVATION 955 ft LAT/LONG 37° 20 ' N / 79° 12 ' W

| | JAN | FEB | MAR | APR | MAY | JUN | JUL | AUG | SEP | OCT | NOV | DEC | YEAR |
|---|---|---|---|---|---|---|---|---|---|---|---|---|---|
| Maximum Temp °F | 44.0 | 47.7 | 57.3 | 68.1 | 75.6 | 82.7 | 86.5 | 85.0 | 78.6 | 68.2 | 58.1 | 48.4 | 66.7 |
| Minimum Temp °F | 25.4 | 27.7 | 35.6 | 44.2 | 52.8 | 61.1 | 65.7 | 64.6 | 58.1 | 45.7 | 37.3 | 29.8 | 45.7 |
| Mean Temp °F | 34.7 | 37.7 | 46.4 | 56.2 | 64.2 | 71.9 | 76.1 | 74.8 | 68.4 | 57.0 | 47.7 | 39.1 | 56.2 |
| Days Max Temp ≥ 90 °F | 0 | 0 | 0 | 0 | 1 | 4 | 9 | 7 | 2 | 0 | 0 | 0 | 23 |
| Days Max Temp ≤ 32 °F | 5 | 2 | 0 | 0 | 0 | 0 | 0 | 0 | 0 | 0 | 0 | 2 | 9 |
| Days Min Temp ≤ 32 °F | 23 | 20 | 12 | 3 | 0 | 0 | 0 | 0 | 0 | 2 | 10 | 19 | 89 |
| Days Min Temp ≤ 0 °F | 1 | 0 | 0 | 0 | 0 | 0 | 0 | 0 | 0 | 0 | 0 | 0 | 1 |
| Heating Degree Days | 932 | 763 | 571 | 279 | 95 | 11 | 0 | 1 | 40 | 262 | 513 | 797 | 4264 |
| Cooling Degree Days | 0 | 0 | 4 | 23 | 84 | 248 | 384 | 322 | 155 | 25 | 1 | 0 | 1246 |
| Total Precipitation (") | 2.96 | 3.01 | 3.75 | 3.22 | 4.06 | 3.34 | 4.32 | 3.53 | 3.15 | 3.58 | 3.08 | 3.34 | 41.34 |
| Days ≥ 0.1" Precip | 6 | 6 | 7 | 6 | 7 | 6 | 7 | 6 | 5 | 5 | 6 | 6 | 73 |
| Total Snowfall (") | 5.7 | 5.8 | 2.9 | 0.3 | 0.0 | 0.0 | 0.0 | 0.0 | 0.0 | 0.1 | 0.9 | 3.1 | 18.8 |
| Days ≥ 1" Snow Depth | 6 | 5 | 1 | 0 | 0 | 0 | 0 | 0 | 0 | 0 | 0 | 2 | 14 |

## MARTINSVILLE FLT PLT *Henry County* ELEVATION 761 ft LAT/LONG 36° 42 ' N / 79° 53 ' W

| | JAN | FEB | MAR | APR | MAY | JUN | JUL | AUG | SEP | OCT | NOV | DEC | YEAR |
|---|---|---|---|---|---|---|---|---|---|---|---|---|---|
| Maximum Temp °F | 46.6 | 50.6 | 60.1 | *70.7* | 77.2 | *84.2* | 87.5 | *86.1* | *80.0* | 69.7 | 60.2 | 50.5 | 68.6 |
| Minimum Temp °F | 23.2 | 24.3 | 32.3 | *40.2* | 49.2 | *58.0* | 63.1 | *61.7* | *54.7* | *41.0* | 32.8 | 26.5 | 42.3 |
| Mean Temp °F | 34.9 | 37.5 | 46.2 | *55.5* | 63.2 | *71.2* | 75.3 | *73.9* | *67.4* | *55.3* | 46.5 | 38.5 | 55.5 |
| Days Max Temp ≥ 90 °F | 0 | 0 | 0 | 0 | 1 | *6* | 11 | *9* | *3* | 0 | 0 | 0 | 30 |
| Days Max Temp ≤ 32 °F | *3* | 1 | 0 | *0* | 0 | 0 | 0 | *0* | *0* | 0 | 0 | 1 | 5 |
| Days Min Temp ≤ 32 °F | *26* | 23 | 16 | *8* | 1 | 0 | 0 | *0* | *0* | *7* | 16 | 23 | 120 |
| Days Min Temp ≤ 0 °F | 1 | 0 | 0 | 0 | 0 | 0 | 0 | *0* | *0* | 0 | 0 | 0 | 1 |
| Heating Degree Days | 924 | 771 | 575 | *293* | 112 | *14* | 0 | *2* | *52* | *304* | 548 | 814 | 4409 |
| Cooling Degree Days | *0* | *0* | *2* | *15* | *63* | *221* | *349* | *292* | *132* | *11* | 0 | *0* | 1085 |
| Total Precipitation (") | 3.46 | 3.47 | 4.63 | 3.42 | 4.40 | 3.88 | 4.76 | *3.59* | *3.50* | 3.63 | 2.84 | 3.33 | 44.91 |
| Days ≥ 0.1" Precip | 7 | 6 | 8 | *7* | 8 | *7* | 7 | *6* | *5* | 5 | 6 | 6 | 78 |
| Total Snowfall (") | na | na | *2.1* | *0.0* | 0.0 | 0.0 | 0.0 | *0.0* | *0.0* | 0.0 | *0.0* | *1.6* | na |
| Days ≥ 1" Snow Depth | na | na | *0* | *0* | 0 | 0 | 0 | *0* | *0* | 0 | 0 | *0* | na |

## MOUNT WEATHER *Loudoun County* ELEVATION 1732 ft LAT/LONG 39° 4 ' N / 77° 53 ' W

| | JAN | FEB | MAR | APR | MAY | JUN | JUL | AUG | SEP | OCT | NOV | DEC | YEAR |
|---|---|---|---|---|---|---|---|---|---|---|---|---|---|
| Maximum Temp °F | 35.3 | 38.2 | 47.5 | 59.3 | 68.6 | 76.8 | 80.6 | 79.0 | 72.0 | 60.4 | 50.4 | 40.3 | 59.0 |
| Minimum Temp °F | 20.4 | 22.4 | 30.3 | 40.8 | 50.7 | 59.7 | 64.2 | 62.9 | 56.3 | 44.7 | 36.0 | 26.0 | 42.9 |
| Mean Temp °F | 27.9 | 30.3 | 38.9 | 50.0 | 59.7 | 68.3 | 72.4 | 71.0 | 64.1 | 52.6 | 43.2 | 33.2 | 51.0 |
| Days Max Temp ≥ 90 °F | 0 | 0 | 0 | 0 | 0 | 1 | 2 | 1 | 0 | 0 | 0 | 0 | 4 |
| Days Max Temp ≤ 32 °F | 13 | 10 | 3 | 0 | 0 | 0 | 0 | 0 | 0 | 0 | 1 | 8 | 35 |
| Days Min Temp ≤ 32 °F | 27 | 23 | 19 | 6 | 0 | 0 | 0 | 0 | 0 | 2 | 12 | 24 | 113 |
| Days Min Temp ≤ 0 °F | 1 | 0 | 0 | 0 | 0 | 0 | 0 | 0 | 0 | 0 | 0 | 0 | 1 |
| Heating Degree Days | 1145 | 972 | 803 | 452 | 199 | 41 | 6 | 12 | 99 | 386 | 648 | 980 | 5743 |
| Cooling Degree Days | 0 | 0 | 2 | 10 | 49 | 157 | 269 | 207 | 84 | 10 | 1 | 0 | 789 |
| Total Precipitation (") | 2.52 | 2.31 | 3.30 | 3.47 | 4.35 | 3.60 | 3.92 | 3.74 | 3.63 | 3.79 | 3.47 | 2.98 | 41.08 |
| Days ≥ 0.1" Precip | 6 | 5 | 6 | 7 | 8 | 7 | 7 | 6 | 6 | 5 | 6 | 6 | 75 |
| Total Snowfall (") | *7.3* | *3.5* | *3.8* | 0.7 | 0.0 | 0.0 | 0.0 | 0.0 | 0.0 | 0.3 | *2.4* | *3.6* | 21.6 |
| Days ≥ 1" Snow Depth | 11 | 9 | 4 | 0 | 0 | 0 | 0 | 0 | 0 | 0 | 1 | 5 | 30 |

## NORFOLK INTL ARPT *Norfolk Independent City* ELEVATION 39 ft LAT/LONG 36° 53 ' N / 76° 12 ' W

| | JAN | FEB | MAR | APR | MAY | JUN | JUL | AUG | SEP | OCT | NOV | DEC | YEAR |
|---|---|---|---|---|---|---|---|---|---|---|---|---|---|
| Maximum Temp °F | 47.7 | 50.2 | 58.3 | 67.8 | 75.8 | 83.6 | 87.5 | 85.8 | 80.5 | 70.1 | 61.9 | 53.0 | 68.5 |
| Minimum Temp °F | 31.4 | 32.7 | 39.7 | 47.9 | 57.4 | 65.9 | 71.0 | 69.9 | 64.4 | 52.8 | 43.9 | 36.1 | 51.1 |
| Mean Temp °F | 39.6 | 41.5 | 49.0 | 57.9 | 66.6 | 74.8 | 79.2 | 77.8 | 72.5 | 61.5 | 53.0 | 44.6 | 59.8 |
| Days Max Temp ≥ 90 °F | 0 | 0 | 0 | 0 | 2 | 6 | 12 | 9 | 3 | 0 | 0 | 0 | 32 |
| Days Max Temp ≤ 32 °F | 3 | 1 | 0 | 0 | 0 | 0 | 0 | 0 | 0 | 0 | 0 | 1 | 5 |
| Days Min Temp ≤ 32 °F | 17 | 15 | 6 | 0 | 0 | 0 | 0 | 0 | 0 | 0 | 3 | 12 | 53 |
| Days Min Temp ≤ 0 °F | 0 | 0 | 0 | 0 | 0 | 0 | 0 | 0 | 0 | 0 | 0 | 0 | 0 |
| Heating Degree Days | 781 | 660 | 496 | 242 | 65 | 4 | 0 | 0 | 9 | 150 | 366 | 627 | 3400 |
| Cooling Degree Days | 0 | 1 | 9 | 37 | 135 | 327 | 491 | 411 | 245 | 56 | 12 | 2 | 1726 |
| Total Precipitation (") | 3.82 | 3.26 | 4.05 | 3.09 | 3.67 | 3.45 | 5.25 | 4.76 | 3.45 | 3.08 | 2.86 | 3.12 | 43.86 |
| Days ≥ 0.1" Precip | 7 | 7 | 7 | 6 | 7 | 6 | 7 | 7 | 5 | 5 | 5 | 6 | 75 |
| Total Snowfall (") | 2.8 | 3.5 | 1.2 | 0.0 | 0.0 | 0.0 | 0.0 | 0.0 | 0.0 | 0.0 | 0.0 | 0.5 | 8.0 |
| Days ≥ 1" Snow Depth | 2 | 2 | 0 | 0 | 0 | 0 | 0 | 0 | 0 | 0 | 0 | 0 | 4 |

## PAINTER 2 W *Accomack County* ELEVATION 30 ft LAT/LONG 37° 35 ' N / 75° 49 ' W

| | JAN | FEB | MAR | APR | MAY | JUN | JUL | AUG | SEP | OCT | NOV | DEC | YEAR |
|---|---|---|---|---|---|---|---|---|---|---|---|---|---|
| Maximum Temp °F | 46.1 | 48.4 | 56.5 | 66.2 | 74.7 | 82.5 | 86.8 | 85.2 | 79.9 | 69.5 | 60.4 | 51.2 | 67.3 |
| Minimum Temp °F | 28.7 | 30.1 | 37.2 | 45.0 | 54.6 | 63.3 | 68.7 | 67.1 | 60.7 | 49.7 | 41.6 | 33.6 | 48.4 |
| Mean Temp °F | 37.4 | 39.3 | 46.8 | 55.6 | 64.7 | 72.9 | 77.8 | 76.1 | 70.3 | 59.6 | 51.1 | 42.4 | 57.8 |
| Days Max Temp ≥ 90 °F | 0 | 0 | 0 | 0 | 1 | 4 | 10 | 6 | 2 | 0 | 0 | 0 | 23 |
| Days Max Temp ≤ 32 °F | 3 | 2 | 0 | 0 | 0 | 0 | 0 | 0 | 0 | 0 | 0 | 1 | 6 |
| Days Min Temp ≤ 32 °F | 20 | 18 | 11 | 3 | 0 | 0 | 0 | 0 | 0 | 1 | 6 | 15 | 74 |
| Days Min Temp ≤ 0 °F | 0 | 0 | 0 | 0 | 0 | 0 | 0 | 0 | 0 | 0 | 0 | 0 | 0 |
| Heating Degree Days | 847 | 720 | 560 | 293 | 92 | 11 | 0 | 2 | 22 | 196 | 416 | 693 | 3852 |
| Cooling Degree Days | 0 | 0 | 5 | 20 | 99 | 282 | 443 | 358 | 198 | 42 | 4 | 0 | 1451 |
| Total Precipitation (") | 3.74 | 3.23 | 4.41 | 2.88 | 3.33 | 3.06 | 4.46 | 3.88 | 3.20 | 3.11 | 2.91 | 3.42 | 41.63 |
| Days ≥ 0.1" Precip | 7 | 6 | 8 | 5 | 6 | 5 | 6 | 6 | 5 | 5 | 5 | 6 | 70 |
| Total Snowfall (") | 3.5 | 3.5 | 1.3 | 0.2 | 0.0 | 0.0 | 0.0 | 0.0 | 0.0 | 0.0 | 0.0 | 1.0 | 9.5 |
| Days ≥ 1" Snow Depth | 2 | 2 | 1 | 0 | 0 | 0 | 0 | 0 | 0 | 0 | 0 | 1 | 6 |

## PENNINGTON GAP *Lee County* ELEVATION 1591 ft LAT/LONG 36° 45' N / 83° 3' W

| | JAN | FEB | MAR | APR | MAY | JUN | JUL | AUG | SEP | OCT | NOV | DEC | YEAR |
|---|---|---|---|---|---|---|---|---|---|---|---|---|---|
| Maximum Temp °F | 44.9 | 49.7 | 59.9 | 69.8 | 76.5 | 83.1 | 86.3 | 85.2 | 80.3 | 70.6 | 59.3 | 49.4 | 67.9 |
| Minimum Temp °F | 23.0 | 24.8 | 32.8 | 40.1 | 48.8 | 57.1 | 61.7 | 60.4 | 53.9 | 40.9 | 33.0 | 26.7 | 41.9 |
| Mean Temp °F | 34.0 | 37.2 | 46.4 | 55.0 | 62.7 | 70.1 | 74.0 | 72.9 | 67.1 | 55.8 | 46.2 | 38.1 | 55.0 |
| Days Max Temp ≥ 90 °F | 0 | 0 | 0 | 0 | 0 | 3 | 8 | 5 | 2 | 0 | 0 | 0 | 18 |
| Days Max Temp ≤ 32 °F | 4 | 2 | 0 | 0 | 0 | 0 | 0 | 0 | 0 | 0 | 0 | 2 | 8 |
| Days Min Temp ≤ 32 °F | 24 | 21 | 16 | 7 | 1 | 0 | 0 | 0 | 0 | 7 | 16 | 22 | 114 |
| Days Min Temp ≤ 0 °F | 1 | 1 | 0 | 0 | 0 | 0 | 0 | 0 | 0 | 0 | 0 | 0 | 2 |
| Heating Degree Days | 956 | 778 | 571 | 303 | 119 | 16 | 1 | 3 | 51 | 290 | 558 | 828 | 4474 |
| Cooling Degree Days | 0 | 0 | 1 | 8 | 50 | 189 | 304 | 258 | 118 | 11 | 0 | 0 | 939 |
| Total Precipitation (") | 3.93 | 4.22 | 4.49 | 4.15 | 4.80 | 4.08 | 4.73 | 3.72 | 3.51 | 3.13 | 3.95 | 4.54 | 49.25 |
| Days ≥ 0.1" Precip | 8 | 8 | 8 | 7 | 9 | 7 | 8 | 7 | 7 | 6 | 7 | 7 | 89 |
| Total Snowfall (") | 6.9 | 4.7 | 1.7 | 0.6 | 0.0 | 0.0 | 0.0 | 0.0 | 0.0 | 0.0 | 0.8 | 2.6 | 17.3 |
| Days ≥ 1" Snow Depth | *4* | 2 | 0 | 0 | 0 | 0 | 0 | 0 | 0 | 0 | 0 | 1 | 7 |

## PHILPOTT DAM 2 *Henry County* ELEVATION 1123 ft LAT/LONG 36° 47' N / 80° 2' W

| | JAN | FEB | MAR | APR | MAY | JUN | JUL | AUG | SEP | OCT | NOV | DEC | YEAR |
|---|---|---|---|---|---|---|---|---|---|---|---|---|---|
| Maximum Temp °F | 45.4 | 49.3 | 58.0 | 68.5 | 75.9 | 83.0 | 86.8 | 85.2 | 79.5 | 69.6 | 59.8 | 49.8 | 67.6 |
| Minimum Temp °F | 24.3 | 25.9 | 33.0 | 41.7 | 50.7 | 59.2 | 64.1 | 62.9 | 56.3 | 43.7 | 35.9 | 28.4 | 43.8 |
| Mean Temp °F | 34.9 | 37.7 | 45.5 | 55.1 | 63.3 | 71.1 | 75.5 | 74.0 | 67.9 | 56.7 | 47.9 | 39.1 | 55.7 |
| Days Max Temp ≥ 90 °F | 0 | 0 | 0 | 0 | 1 | 5 | 10 | 7 | 2 | 0 | 0 | 0 | 25 |
| Days Max Temp ≤ 32 °F | 4 | 2 | 0 | 0 | 0 | 0 | 0 | 0 | 0 | 0 | 0 | 2 | 8 |
| Days Min Temp ≤ 32 °F | 26 | 22 | 16 | 5 | 0 | 0 | 0 | 0 | 0 | 3 | 12 | 21 | 105 |
| Days Min Temp ≤ 0 °F | 1 | 0 | 0 | 0 | 0 | 0 | 0 | 0 | 0 | 0 | 0 | 0 | 1 |
| Heating Degree Days | 927 | 763 | 598 | 304 | 112 | 14 | 1 | 3 | 44 | 267 | 509 | 795 | 4337 |
| Cooling Degree Days | 0 | 0 | 3 | 18 | 71 | 230 | 364 | 300 | 146 | 20 | 1 | 0 | 1153 |
| Total Precipitation (") | 3.44 | 3.35 | 4.50 | 3.72 | 4.91 | 4.12 | 5.35 | 4.74 | 4.13 | 4.38 | 3.17 | 3.23 | 49.04 |
| Days ≥ 0.1" Precip | 6 | 6 | 7 | 7 | 8 | 7 | 8 | 7 | 5 | 5 | 5 | 6 | 77 |
| Total Snowfall (") | *2.2* | 2.7 | 1.3 | 0.1 | 0.0 | 0.0 | 0.0 | 0.0 | 0.0 | 0.0 | 0.0 | *1.1* | 7.4 |
| Days ≥ 1" Snow Depth | 4 | 3 | 1 | 0 | 0 | 0 | 0 | 0 | 0 | 0 | 0 | 2 | 10 |

## PIEDMONT RESEARCH ST *Orange County* ELEVATION 502 ft LAT/LONG 38° 13' N / 78° 6' W

| | JAN | FEB | MAR | APR | MAY | JUN | JUL | AUG | SEP | OCT | NOV | DEC | YEAR |
|---|---|---|---|---|---|---|---|---|---|---|---|---|---|
| Maximum Temp °F | 42.6 | 46.2 | 55.5 | 66.5 | 74.7 | 82.7 | 86.4 | 84.7 | 78.6 | 67.8 | 57.8 | 47.2 | 65.9 |
| Minimum Temp °F | 22.7 | 25.2 | 33.2 | 42.8 | 52.2 | 61.0 | 65.6 | 64.0 | 56.9 | 44.3 | 36.1 | 27.6 | 44.3 |
| Mean Temp °F | 32.7 | 35.7 | 44.4 | 54.7 | 63.5 | 71.9 | 76.0 | 74.4 | 67.8 | 56.1 | 47.0 | 37.4 | 55.1 |
| Days Max Temp ≥ 90 °F | 0 | 0 | 0 | 0 | 1 | 5 | 10 | 7 | 3 | 0 | 0 | 0 | 26 |
| Days Max Temp ≤ 32 °F | 6 | 3 | 0 | 0 | 0 | 0 | 0 | 0 | 0 | 0 | 0 | 2 | 11 |
| Days Min Temp ≤ 32 °F | 26 | 23 | 16 | 3 | 0 | 0 | 0 | 0 | 0 | 2 | 12 | 22 | 104 |
| Days Min Temp ≤ 0 °F | 1 | 0 | 0 | 0 | 0 | 0 | 0 | 0 | 0 | 0 | 0 | 0 | 1 |
| Heating Degree Days | 994 | 821 | 635 | 322 | 114 | 14 | 1 | 2 | 46 | 285 | 536 | 847 | 4617 |
| Cooling Degree Days | 0 | 0 | 5 | 19 | 81 | 241 | 372 | 293 | 140 | 19 | 1 | 0 | 1171 |
| Total Precipitation (") | 2.92 | 2.64 | 3.71 | 3.13 | 4.44 | 3.51 | 4.68 | 4.14 | 3.55 | 3.93 | 3.60 | 3.18 | 43.43 |
| Days ≥ 0.1" Precip | 6 | 5 | 7 | 6 | 8 | 6 | 8 | 6 | 6 | 5 | 6 | 6 | 75 |
| Total Snowfall (") | 6.7 | 5.3 | 3.0 | 0.3 | 0.0 | 0.0 | 0.0 | 0.0 | 0.0 | 0.1 | 0.9 | 3.7 | 20.0 |
| Days ≥ 1" Snow Depth | 7 | 4 | 1 | 0 | 0 | 0 | 0 | 0 | 0 | 0 | 1 | 3 | 16 |

## PULASKI *Pulaski County* ELEVATION 1850 ft LAT/LONG 37° 3' N / 80° 46' W

| | JAN | FEB | MAR | APR | MAY | JUN | JUL | AUG | SEP | OCT | NOV | DEC | YEAR |
|---|---|---|---|---|---|---|---|---|---|---|---|---|---|
| Maximum Temp °F | 42.9 | 46.9 | 56.4 | 66.7 | 74.0 | 80.7 | 84.3 | 82.9 | 76.9 | 66.8 | 56.7 | 47.3 | 65.2 |
| Minimum Temp °F | 22.5 | 24.1 | 31.6 | 38.9 | 47.4 | 55.3 | 59.8 | 58.5 | 51.7 | 40.0 | 32.5 | 26.1 | 40.7 |
| Mean Temp °F | 32.7 | 35.5 | 44.1 | 52.8 | 60.7 | 68.0 | 72.1 | 70.7 | 64.3 | 53.4 | 44.6 | 36.7 | 53.0 |
| Days Max Temp ≥ 90 °F | 0 | 0 | 0 | 0 | 0 | 1 | 4 | 3 | 0 | 0 | 0 | 0 | 8 |
| Days Max Temp ≤ 32 °F | 5 | 3 | 1 | 0 | 0 | 0 | 0 | 0 | 0 | 0 | 0 | 3 | 12 |
| Days Min Temp ≤ 32 °F | 26 | 23 | 17 | 8 | 2 | 0 | 0 | 0 | 0 | 8 | 16 | 23 | 123 |
| Days Min Temp ≤ 0 °F | 1 | 0 | 0 | 0 | 0 | 0 | 0 | 0 | 0 | 0 | 0 | 0 | 1 |
| Heating Degree Days | 993 | 825 | 642 | 363 | 158 | 33 | 4 | 9 | 89 | 357 | 605 | 870 | 4948 |
| Cooling Degree Days | 0 | 0 | 1 | 6 | 35 | 142 | 246 | 195 | 74 | 6 | 0 | 0 | 705 |
| Total Precipitation (") | 2.43 | 2.73 | 3.29 | 3.04 | 3.92 | 3.43 | 3.93 | 3.16 | 2.93 | 3.26 | 2.46 | 2.69 | 37.27 |
| Days ≥ 0.1" Precip | 6 | 5 | 6 | 6 | 8 | 7 | 7 | 6 | 6 | 5 | 5 | 6 | 73 |
| Total Snowfall (") | na | na | *1.2* | 0.5 | 0.0 | 0.0 | 0.0 | 0.0 | 0.0 | 0.1 | 0.3 | *2.5* | na |
| Days ≥ 1" Snow Depth | na | na | *0* | 0 | 0 | 0 | 0 | 0 | 0 | 0 | 0 | *1* | na |

## RICHMOND BYRD AP *Henrico County* ELEVATION 167 ft LAT/LONG 37° 30' N / 77° 20' W

| | JAN | FEB | MAR | APR | MAY | JUN | JUL | AUG | SEP | OCT | NOV | DEC | YEAR |
|---|---|---|---|---|---|---|---|---|---|---|---|---|---|
| Maximum Temp °F | 46.0 | 49.8 | 59.4 | 70.3 | 77.8 | 85.4 | 88.9 | 87.1 | 81.2 | 70.5 | 61.3 | 50.9 | 69.1 |
| Minimum Temp °F | 26.4 | 28.6 | 36.5 | 45.1 | 54.3 | 62.9 | 68.2 | 66.6 | 59.5 | 46.9 | 38.4 | 30.8 | 47.0 |
| Mean Temp °F | 36.2 | 39.2 | 48.0 | 57.7 | 66.1 | 74.2 | 78.6 | 76.9 | 70.4 | 58.8 | 49.9 | 40.9 | 58.1 |
| Days Max Temp ≥ 90 °F | 0 | 0 | 0 | 1 | 2 | 9 | 15 | 11 | 4 | 0 | 0 | 0 | 42 |
| Days Max Temp ≤ 32 °F | 4 | 2 | 0 | 0 | 0 | 0 | 0 | 0 | 0 | 0 | 0 | 1 | 7 |
| Days Min Temp ≤ 32 °F | 22 | 19 | 11 | 2 | 0 | 0 | 0 | 0 | 0 | 2 | 9 | 19 | 84 |
| Days Min Temp ≤ 0 °F | 0 | 0 | 0 | 0 | 0 | 0 | 0 | 0 | 0 | 0 | 0 | 0 | 0 |
| Heating Degree Days | 886 | 723 | 528 | 246 | 73 | 7 | 0 | 1 | 28 | 221 | 453 | 742 | 3908 |
| Cooling Degree Days | 0 | 1 | 8 | 35 | 122 | 314 | 463 | 377 | 201 | 38 | 4 | 1 | 1564 |
| Total Precipitation (") | 3.20 | 3.00 | 3.94 | 2.96 | 3.86 | 3.30 | 5.13 | 4.27 | 3.38 | 3.40 | 2.94 | 3.21 | 42.59 |
| Days ≥ 0.1" Precip | 6 | 6 | 7 | 6 | 7 | 6 | 8 | 6 | 5 | 5 | 5 | 6 | 73 |
| Total Snowfall (") | 4.7 | 5.1 | 1.7 | 0.0 | 0.0 | 0.0 | 0.0 | 0.0 | 0.0 | 0.0 | 0.3 | 2.1 | 13.9 |
| Days ≥ 1" Snow Depth | 4 | 3 | 1 | 0 | 0 | 0 | 0 | 0 | 0 | 0 | 0 | 2 | 10 |

## ROANOKE WOODRUM AP *Roanoke Independent City* ELEVATION 1201 ft LAT/LONG 37° 19' N / 79° 58' W

| | JAN | FEB | MAR | APR | MAY | JUN | JUL | AUG | SEP | OCT | NOV | DEC | YEAR |
|---|---|---|---|---|---|---|---|---|---|---|---|---|---|
| Maximum Temp °F | 44.1 | 47.9 | 57.5 | 67.7 | 75.5 | 83.0 | 86.9 | 85.3 | 78.6 | 68.0 | 58.0 | 48.3 | 66.7 |
| Minimum Temp °F | 25.6 | 27.5 | 35.7 | 44.2 | 52.5 | 60.4 | 65.4 | 63.9 | 57.0 | 44.9 | 36.9 | 29.5 | 45.3 |
| Mean Temp °F | 34.9 | 37.7 | 46.6 | 56.0 | 64.0 | 71.7 | 76.2 | 74.6 | 67.8 | 56.5 | 47.5 | 38.9 | 56.0 |
| Days Max Temp ≥ 90 °F | 0 | 0 | 0 | 0 | 1 | 5 | 11 | 8 | 2 | 0 | 0 | 0 | 27 |
| Days Max Temp ≤ 32 °F | 5 | 3 | 0 | 0 | 0 | 0 | 0 | 0 | 0 | 0 | 0 | 2 | 10 |
| Days Min Temp ≤ 32 °F | 23 | 20 | 12 | 3 | 0 | 0 | 0 | 0 | 0 | 3 | 10 | 19 | 90 |
| Days Min Temp ≤ 0 °F | 0 | 0 | 0 | 0 | 0 | 0 | 0 | 0 | 0 | 0 | 0 | 0 | 0 |
| Heating Degree Days | 926 | 763 | 567 | 284 | 103 | 12 | 1 | 2 | 47 | 275 | 521 | 802 | 4303 |
| Cooling Degree Days | 0 | 0 | 5 | 23 | 81 | 239 | 384 | 315 | 143 | 22 | 1 | 0 | 1213 |
| Total Precipitation (") | 2.75 | 3.03 | 3.78 | 3.44 | 4.13 | 3.26 | 4.06 | 4.01 | 3.39 | 3.76 | 3.08 | 3.01 | 41.70 |
| Days ≥ 0.1" Precip | 6 | 6 | 7 | 6 | 8 | 6 | 7 | 7 | 6 | 5 | 5 | 6 | 75 |
| Total Snowfall (") | 6.4 | 6.9 | 3.3 | 0.6 | 0.0 | 0.0 | 0.0 | 0.0 | 0.0 | 0.0 | 1.5 | 4.1 | 22.8 |
| Days ≥ 1" Snow Depth | 5 | 4 | 1 | 0 | 0 | 0 | 0 | 0 | 0 | 0 | 0 | 3 | 13 |

## ROCKY MOUNT *Franklin County* ELEVATION 1132 ft LAT/LONG 37° 0' N / 79° 54' W

| | JAN | FEB | MAR | APR | MAY | JUN | JUL | AUG | SEP | OCT | NOV | DEC | YEAR |
|---|---|---|---|---|---|---|---|---|---|---|---|---|---|
| Maximum Temp °F | 44.0 | 48.1 | 56.9 | 67.7 | 75.3 | 82.8 | 86.7 | 85.1 | 78.6 | 68.3 | 58.5 | 48.4 | 66.7 |
| Minimum Temp °F | 24.5 | 26.3 | 33.5 | 41.9 | 50.4 | 59.1 | 64.0 | 62.4 | 55.7 | 43.5 | 35.9 | 28.5 | 43.8 |
| Mean Temp °F | 34.3 | 37.2 | 45.2 | 54.8 | 62.9 | 71.0 | 75.4 | 73.8 | 67.2 | 55.9 | 47.2 | 38.5 | 55.3 |
| Days Max Temp ≥ 90 °F | 0 | 0 | 0 | 0 | 1 | 5 | 9 | 8 | 2 | 0 | 0 | 0 | 25 |
| Days Max Temp ≤ 32 °F | 4 | 3 | 0 | 0 | 0 | 0 | 0 | 0 | 0 | 0 | 0 | 2 | 9 |
| Days Min Temp ≤ 32 °F | 24 | 21 | 15 | 5 | 1 | 0 | 0 | 0 | 0 | 5 | 12 | 21 | 104 |
| Days Min Temp ≤ 0 °F | 1 | 0 | 0 | 0 | 0 | 0 | 0 | 0 | 0 | 0 | 0 | 0 | 1 |
| Heating Degree Days | 944 | 778 | 608 | 315 | 122 | 14 | 1 | 3 | 52 | 289 | 529 | 815 | 4470 |
| Cooling Degree Days | 0 | 0 | 3 | 18 | 68 | 224 | 360 | 289 | 131 | 22 | 1 | 0 | 1116 |
| Total Precipitation (") | 3.26 | 3.08 | 4.22 | 3.82 | 4.30 | 3.61 | 5.04 | 4.15 | 3.82 | 4.01 | 3.13 | 3.10 | 45.54 |
| Days ≥ 0.1" Precip | 7 | 6 | 8 | 7 | 8 | 7 | 7 | 7 | 6 | 5 | 6 | 6 | 80 |
| Total Snowfall (") | 6.0 | 5.3 | 3.1 | 0.3 | 0.0 | 0.0 | 0.0 | 0.0 | 0.0 | 0.0 | 0.9 | 2.9 | 18.5 |
| Days ≥ 1" Snow Depth | 6 | 4 | 1 | 0 | 0 | 0 | 0 | 0 | 0 | 0 | 0 | 2 | 13 |

## STAFFORDSVILLE 3 ENE *Giles County* ELEVATION 1831 ft LAT/LONG 37° 16' N / 80° 44' W

| | JAN | FEB | MAR | APR | MAY | JUN | JUL | AUG | SEP | OCT | NOV | DEC | YEAR |
|---|---|---|---|---|---|---|---|---|---|---|---|---|---|
| Maximum Temp °F | *42.6* | *47.1* | *57.1* | *66.5* | *73.7* | *80.5* | *84.2* | *83.0* | *76.9* | *67.5* | *57.5* | *47.5* | 65.3 |
| Minimum Temp °F | *23.0* | *25.6* | *33.8* | *40.3* | *48.1* | *55.8* | *60.6* | *59.1* | *53.1* | *40.6* | *33.3* | *27.1* | 41.7 |
| Mean Temp °F | *32.8* | *36.4* | *45.5* | *53.4* | *60.9* | *68.2* | *72.4* | *71.1* | *65.1* | *54.1* | *45.4* | *37.3* | 53.6 |
| Days Max Temp ≥ 90 °F | *0* | *0* | *0* | *0* | *0* | *1* | *4* | *3* | *1* | *0* | *0* | *0* | 9 |
| Days Max Temp ≤ 32 °F | *6* | *3* | *1* | *0* | *0* | *0* | *0* | *0* | *0* | *0* | *0* | *3* | 13 |
| Days Min Temp ≤ 32 °F | *25* | *21* | na | *6* | *1* | *0* | *0* | *0* | *0* | *8* | *15* | *21* | na |
| Days Min Temp ≤ 0 °F | *1* | *0* | *0* | *0* | *0* | *0* | *0* | *0* | *0* | *0* | *0* | *1* | 2 |
| Heating Degree Days | *991* | *802* | *599* | *349* | *156* | *27* | *4* | *6* | *77* | *338* | *582* | *851* | 4782 |
| Cooling Degree Days | 0 | 0 | 1 | 7 | 40 | 144 | 273 | 213 | 95 | 9 | 1 | 0 | 783 |
| Total Precipitation (") | 2.66 | 2.80 | 3.39 | 3.54 | 3.81 | 3.51 | 3.90 | 3.36 | 3.10 | 2.99 | 2.60 | 2.61 | 38.27 |
| Days ≥ 0.1" Precip | 6 | 6 | 7 | 8 | 8 | 7 | 8 | 7 | 6 | 5 | 6 | 6 | 80 |
| Total Snowfall (") | 6.1 | 6.0 | 3.6 | 1.0 | 0.0 | 0.0 | 0.0 | 0.0 | 0.0 | 0.0 | 1.0 | 3.5 | 21.2 |
| Days ≥ 1" Snow Depth | 6 | 5 | 2 | 0 | 0 | 0 | 0 | 0 | 0 | 0 | 1 | 3 | 17 |

## STAUNTON SEWAGE PLAN *Augusta County* ELEVATION 1480 ft LAT/LONG 38° 9 ' N / 79° 4 ' W

| | JAN | FEB | MAR | APR | MAY | JUN | JUL | AUG | SEP | OCT | NOV | DEC | YEAR |
|---|---|---|---|---|---|---|---|---|---|---|---|---|---|
| Maximum Temp °F | 42.3 | 45.3 | 54.2 | 64.7 | 73.2 | 80.9 | 84.8 | 83.5 | 77.4 | 66.3 | 56.4 | 46.9 | 64.7 |
| Minimum Temp °F | 19.8 | 21.6 | 29.7 | 38.2 | 47.6 | 55.8 | 60.4 | 58.8 | 51.3 | 38.0 | 30.9 | 23.8 | 39.7 |
| Mean Temp °F | 31.0 | 33.5 | 42.0 | 51.5 | 60.4 | 68.4 | 72.6 | 71.2 | 64.4 | 52.2 | 43.7 | 35.3 | 52.2 |
| Days Max Temp ≥ 90 °F | 0 | 0 | 0 | 0 | 0 | 2 | 6 | 5 | 2 | 0 | 0 | 0 | 15 |
| Days Max Temp ≤ 32 °F | 6 | 4 | 1 | 0 | 0 | 0 | 0 | 0 | 0 | 0 | 0 | 3 | 14 |
| Days Min Temp ≤ 32 °F | 28 | 24 | 20 | 9 | 2 | 0 | 0 | 0 | 0 | 10 | 17 | 25 | 135 |
| Days Min Temp ≤ 0 °F | 2 | 1 | 0 | 0 | 0 | 0 | 0 | 0 | 0 | 0 | 0 | 1 | 4 |
| Heating Degree Days | 1046 | 882 | 707 | 405 | 174 | 36 | 4 | 11 | 93 | 394 | 633 | 913 | 5298 |
| Cooling Degree Days | 0 | 0 | 1 | 4 | 37 | 143 | *256* | 202 | 77 | 4 | 0 | 0 | 724 |
| Total Precipitation (") | 2.43 | 2.28 | 3.27 | 2.85 | 3.69 | 2.86 | 3.70 | 3.55 | 3.41 | 3.49 | 2.80 | 2.51 | 36.84 |
| Days ≥ 0.1" Precip | 6 | 5 | 7 | 6 | 8 | 6 | 7 | 6 | 6 | 5 | 5 | 5 | 72 |
| Total Snowfall (") | 7.3 | 6.9 | 3.7 | 0.3 | 0.0 | 0.0 | 0.0 | 0.0 | 0.0 | 0.3 | 1.2 | 3.8 | 23.5 |
| Days ≥ 1" Snow Depth | na | *5* | *2* | 0 | 0 | 0 | 0 | 0 | 0 | 0 | 1 | 2 | na |

## STONY CREEK 3 ESE *Sussex County* ELEVATION 69 ft LAT/LONG 36° 55 ' N / 77° 21 ' W

| | JAN | FEB | MAR | APR | MAY | JUN | JUL | AUG | SEP | OCT | NOV | DEC | YEAR |
|---|---|---|---|---|---|---|---|---|---|---|---|---|---|
| Maximum Temp °F | *47.2* | 51.4 | *60.5* | 72.3 | *79.4* | *87.6* | *90.8* | *88.0* | *82.3* | *71.2* | 61.8 | *50.5* | 70.3 |
| Minimum Temp °F | *26.3* | 28.8 | *35.1* | 43.9 | *52.7* | *61.7* | *65.9* | *63.9* | *57.1* | *44.3* | 36.5 | *28.3* | 45.4 |
| Mean Temp °F | *36.8* | 40.1 | *47.8* | 58.1 | *66.0* | *74.6* | *78.3* | *75.9* | *69.7* | *57.8* | 49.2 | *39.4* | 57.8 |
| Days Max Temp ≥ 90 °F | *0* | 0 | *0* | 1 | 4 | na | *18* | na | *5* | 0 | 0 | *0* | na |
| Days Max Temp ≤ 32 °F | *3* | *1* | *0* | 0 | 0 | *0* | *0* | 0 | 0 | 0 | 0 | *1* | 5 |
| Days Min Temp ≤ 32 °F | *23* | *19* | *13* | 4 | 0 | *0* | *0* | 0 | 0 | 4 | *11* | na | na |
| Days Min Temp ≤ 0 °F | 0 | 0 | *0* | 0 | 0 | *0* | *0* | 0 | 0 | 0 | 0 | *0* | 0 |
| Heating Degree Days | *867* | 695 | *530* | 236 | *80* | *7* | *0* | *3* | *40* | *247* | 471 | na | na |
| Cooling Degree Days | 0 | 1 | *5* | 42 | 135 | *330* | *468* | 357 | *192* | *33* | 4 | *0* | 1567 |
| Total Precipitation (") | 3.87 | *3.30* | *4.58* | 3.05 | 4.29 | *3.59* | *4.67* | 4.69 | 4.02 | *3.22* | 2.90 | *3.35* | 45.53 |
| Days ≥ 0.1" Precip | *7* | *6* | *7* | 5 | *6* | *5* | *6* | 6 | *5* | 4 | *5* | na | na |
| Total Snowfall (") | *2.9* | *4.2* | *1.8* | 0.2 | 0.0 | 0.0 | 0.0 | 0.0 | 0.0 | 0.0 | *0.1* | na | na |
| Days ≥ 1" Snow Depth | *2* | *2* | na | 0 | 0 | 0 | 0 | 0 | 0 | 0 | *0* | na | na |

## SUFFOLK LAKE KILBY *Suffolk Independent City* ELEVATION 20 ft LAT/LONG 36° 44 ' N / 76° 36 ' W

| | JAN | FEB | MAR | APR | MAY | JUN | JUL | AUG | SEP | OCT | NOV | DEC | YEAR |
|---|---|---|---|---|---|---|---|---|---|---|---|---|---|
| Maximum Temp °F | 47.8 | 51.2 | 59.8 | 69.3 | 77.0 | 84.3 | 87.9 | 86.1 | 80.5 | 70.4 | 61.9 | 52.9 | 69.1 |
| Minimum Temp °F | 29.2 | 31.1 | 38.2 | 46.3 | 55.6 | 63.7 | 68.6 | 67.3 | 61.5 | 49.8 | 41.0 | 33.3 | 48.8 |
| Mean Temp °F | 38.5 | 41.2 | 49.0 | 57.8 | 66.3 | 74.0 | 78.2 | 76.7 | 71.1 | 60.1 | 51.5 | 43.1 | 59.0 |
| Days Max Temp ≥ 90 °F | 0 | 0 | 0 | 0 | 1 | 6 | 12 | 9 | 3 | 0 | 0 | 0 | 31 |
| Days Max Temp ≤ 32 °F | 3 | 1 | 0 | 0 | 0 | 0 | 0 | 0 | 0 | 0 | 0 | 1 | 5 |
| Days Min Temp ≤ 32 °F | 20 | 17 | 9 | 1 | 0 | 0 | 0 | 0 | 0 | 1 | 6 | 16 | 70 |
| Days Min Temp ≤ 0 °F | 0 | 0 | 0 | 0 | 0 | 0 | 0 | 0 | 0 | 0 | 0 | 0 | 0 |
| Heating Degree Days | 814 | 668 | 495 | 239 | 66 | 5 | 0 | 1 | 16 | 183 | 406 | 673 | 3566 |
| Cooling Degree Days | 0 | 1 | 6 | 28 | 120 | 291 | 440 | 353 | 193 | 39 | 5 | 1 | 1477 |
| Total Precipitation (") | 3.80 | 3.50 | 4.37 | 3.16 | 3.77 | 3.83 | 5.06 | 5.41 | 4.06 | 3.13 | 2.82 | 3.32 | 46.23 |
| Days ≥ 0.1" Precip | 7 | 6 | 7 | 6 | 7 | 6 | 8 | 7 | 5 | 5 | 5 | 6 | 75 |
| Total Snowfall (") | 2.7 | 3.2 | 1.3 | 0.0 | 0.0 | 0.0 | 0.0 | 0.0 | 0.0 | 0.0 | 0.1 | 0.6 | 7.9 |
| Days ≥ 1" Snow Depth | 2 | 2 | 0 | 0 | 0 | 0 | 0 | 0 | 0 | 0 | 0 | 0 | 4 |

## TANGIER ISLAND *Accomack County* ELEVATION 10 ft LAT/LONG 37° 49 ' N / 75° 59 ' W

| | JAN | FEB | MAR | APR | MAY | JUN | JUL | AUG | SEP | OCT | NOV | DEC | YEAR |
|---|---|---|---|---|---|---|---|---|---|---|---|---|---|
| Maximum Temp °F | 43.9 | *46.1* | 54.3 | 65.1 | 74.8 | 82.7 | 87.3 | 85.9 | 80.9 | 70.0 | 59.9 | 50.7 | 66.8 |
| Minimum Temp °F | 30.4 | *31.6* | 38.0 | 47.5 | 56.7 | 65.4 | 71.1 | 70.0 | 64.4 | 53.6 | 44.5 | 36.2 | 50.8 |
| Mean Temp °F | 37.2 | *38.9* | 46.2 | 56.3 | 65.8 | 74.1 | 79.2 | 78.0 | 72.6 | 61.8 | 52.2 | 43.5 | 58.8 |
| Days Max Temp ≥ 90 °F | 0 | 0 | 0 | 0 | 0 | 2 | 10 | 7 | 1 | 0 | 0 | 0 | 20 |
| Days Max Temp ≤ 32 °F | 3 | *1* | 0 | 0 | 0 | 0 | 0 | 0 | 0 | 0 | 0 | 1 | 5 |
| Days Min Temp ≤ 32 °F | *18* | *15* | *6* | 0 | 0 | 0 | 0 | 0 | 0 | 0 | 1 | 9 | 49 |
| Days Min Temp ≤ 0 °F | 0 | 0 | 0 | 0 | 0 | 0 | 0 | 0 | 0 | 0 | 0 | 0 | 0 |
| Heating Degree Days | 856 | *732* | 577 | 266 | 59 | 2 | 0 | 0 | 6 | 135 | 379 | 660 | 3672 |
| Cooling Degree Days | *0* | *0* | 1 | 13 | 103 | 343 | 515 | 446 | 271 | 49 | 2 | *0* | 1743 |
| Total Precipitation (") | 2.74 | 2.74 | 4.17 | 2.42 | 2.98 | 2.60 | 3.42 | 3.61 | 3.11 | 2.58 | 2.67 | 2.92 | 35.96 |
| Days ≥ 0.1" Precip | *5* | *4* | *6* | 4 | 5 | 5 | 6 | 5 | 3 | *3* | *4* | 5 | 55 |
| Total Snowfall (") | *3.2* | *2.3* | *1.0* | 0.0 | 0.0 | 0.0 | 0.0 | 0.0 | 0.0 | 0.0 | *0.0* | *0.7* | 7.2 |
| Days ≥ 1" Snow Depth | na | na | na | 0 | 0 | 0 | 0 | 0 | 0 | 0 | *0* | *0* | na |

## TYE RIVER 1 SE *Nelson County* ELEVATION 551 ft LAT/LONG 37° 40' N / 78° 57' W

| | JAN | FEB | MAR | APR | MAY | JUN | JUL | AUG | SEP | OCT | NOV | DEC | YEAR |
|---|---|---|---|---|---|---|---|---|---|---|---|---|---|
| Maximum Temp °F | 45.3 | 49.0 | 58.2 | 69.2 | 76.1 | 83.5 | 87.2 | 85.9 | 79.9 | 70.0 | 60.3 | 50.0 | 67.9 |
| Minimum Temp °F | 24.3 | 26.0 | 33.6 | 42.5 | 51.4 | 60.0 | 65.1 | 63.4 | 57.0 | 44.0 | 35.7 | 28.3 | 44.3 |
| Mean Temp °F | 34.8 | 37.5 | 46.0 | 55.9 | 63.8 | 71.8 | 76.2 | 74.7 | 68.5 | 57.1 | 48.0 | 39.2 | 56.1 |
| Days Max Temp ≥ 90 °F | 0 | 0 | 0 | 1 | 1 | 6 | 11 | 9 | 4 | 0 | 0 | 0 | 32 |
| Days Max Temp ≤ 32 °F | 4 | 2 | 0 | 0 | 0 | 0 | 0 | 0 | 0 | 0 | 0 | 1 | 7 |
| Days Min Temp ≤ 32 °F | 24 | 22 | 15 | 4 | 0 | 0 | 0 | 0 | 0 | 4 | 12 | 21 | 102 |
| Days Min Temp ≤ 0 °F | 0 | 0 | 0 | 0 | 0 | 0 | 0 | 0 | 0 | 0 | 0 | 0 | 0 |
| Heating Degree Days | 929 | 769 | 585 | 289 | 109 | 13 | 0 | 2 | 42 | 260 | 505 | 794 | 4297 |
| Cooling Degree Days | 0 | 0 | 5 | 23 | 78 | 235 | 376 | 305 | 149 | 25 | 1 | 1 | 1198 |
| Total Precipitation (") | 3.32 | 3.26 | 4.13 | 3.32 | 4.72 | 3.34 | 4.65 | 3.65 | 3.71 | 4.05 | 3.35 | 3.41 | 44.91 |
| Days ≥ 0.1" Precip | 6 | 6 | 8 | 6 | 8 | 6 | 8 | 6 | 5 | 5 | 6 | 6 | 76 |
| Total Snowfall (") | na | na | *1.0* | 0.1 | 0.0 | 0.0 | 0.0 | 0.0 | 0.0 | 0.0 | 0.4 | *0.3* | na |
| Days ≥ 1" Snow Depth | na | 2 | 1 | 0 | 0 | 0 | 0 | 0 | 0 | 0 | 0 | 1 | na |

## WALKERTON 2 NW *King and Queen County* ELEVATION 49 ft LAT/LONG 37° 45' N / 77° 3' W

| | JAN | FEB | MAR | APR | MAY | JUN | JUL | AUG | SEP | OCT | NOV | DEC | YEAR |
|---|---|---|---|---|---|---|---|---|---|---|---|---|---|
| Maximum Temp °F | 46.5 | 50.5 | 60.0 | 70.8 | 78.1 | 85.2 | 88.5 | 87.1 | 81.3 | 70.7 | 61.2 | 51.2 | 69.3 |
| Minimum Temp °F | 25.2 | 27.3 | 34.8 | 43.3 | 53.4 | 62.1 | 66.8 | 65.2 | 58.0 | 45.7 | 37.1 | 29.4 | 45.7 |
| Mean Temp °F | 35.9 | 38.9 | 47.4 | 57.1 | 65.8 | 73.7 | 77.7 | 76.2 | 69.7 | 58.2 | 49.2 | 40.3 | 57.5 |
| Days Max Temp ≥ 90 °F | 0 | 0 | 0 | 1 | 2 | 7 | 14 | 11 | 4 | 0 | 0 | 0 | 39 |
| Days Max Temp ≤ 32 °F | 3 | 1 | 0 | 0 | 0 | 0 | 0 | 0 | 0 | 0 | 0 | 1 | 5 |
| Days Min Temp ≤ 32 °F | 23 | 20 | 14 | 5 | 0 | 0 | 0 | 0 | 0 | 4 | 11 | 20 | 97 |
| Days Min Temp ≤ 0 °F | 1 | 0 | 0 | 0 | 0 | 0 | 0 | 0 | 0 | 0 | 0 | 0 | 1 |
| Heating Degree Days | 895 | 731 | 543 | 259 | 74 | 7 | 0 | 1 | 31 | 233 | 473 | 759 | 4006 |
| Cooling Degree Days | 0 | 1 | 7 | 30 | 112 | 290 | 428 | 354 | 189 | 32 | 4 | 0 | 1447 |
| Total Precipitation (") | 3.40 | 3.14 | 4.04 | 2.86 | 4.08 | 3.50 | 4.43 | 3.78 | 3.79 | 3.24 | 3.08 | 3.33 | 42.67 |
| Days ≥ 0.1" Precip | 7 | 6 | 8 | 6 | 7 | 6 | 7 | 6 | 6 | 5 | 6 | 7 | 77 |
| Total Snowfall (") | 4.3 | 4.4 | 1.0 | 0.0 | 0.0 | 0.0 | 0.0 | 0.0 | 0.0 | 0.0 | 0.2 | 1.9 | 11.8 |
| Days ≥ 1" Snow Depth | 4 | 3 | 0 | 0 | 0 | 0 | 0 | 0 | 0 | 0 | 0 | 2 | 9 |

## WARRENTON 3 SE *Fauquier County* ELEVATION 502 ft LAT/LONG 38° 41' N / 77° 46' W

| | JAN | FEB | MAR | APR | MAY | JUN | JUL | AUG | SEP | OCT | NOV | DEC | YEAR |
|---|---|---|---|---|---|---|---|---|---|---|---|---|---|
| Maximum Temp °F | 41.6 | 45.2 | 54.5 | 65.5 | 74.1 | 82.1 | 86.2 | 85.1 | 79.1 | 67.8 | 57.2 | 46.3 | 65.4 |
| Minimum Temp °F | 22.1 | 24.2 | 32.2 | 41.7 | 50.9 | 59.5 | 64.2 | 62.7 | 55.5 | 43.2 | 36.0 | 27.1 | 43.3 |
| Mean Temp °F | 31.9 | 34.7 | 43.4 | 53.7 | 62.5 | 70.8 | 75.3 | 73.9 | 67.3 | 55.5 | 46.6 | 36.8 | 54.4 |
| Days Max Temp ≥ 90 °F | 0 | 0 | 0 | 0 | 1 | 4 | 10 | 8 | 3 | 0 | 0 | 0 | 26 |
| Days Max Temp ≤ 32 °F | 6 | 3 | 1 | 0 | 0 | 0 | 0 | 0 | 0 | 0 | 0 | 3 | 13 |
| Days Min Temp ≤ 32 °F | 27 | 23 | 17 | 4 | 0 | 0 | 0 | 0 | 0 | 3 | 11 | 23 | 108 |
| Days Min Temp ≤ 0 °F | 1 | 0 | 0 | 0 | 0 | 0 | 0 | 0 | 0 | 0 | 0 | 0 | 1 |
| Heating Degree Days | 1019 | 848 | 668 | 349 | 134 | 19 | 1 | 4 | 57 | 300 | 548 | 868 | 4815 |
| Cooling Degree Days | 0 | 0 | 4 | 14 | 70 | 212 | 350 | 275 | 132 | 18 | 1 | 0 | 1076 |
| Total Precipitation (") | 3.04 | 2.67 | 3.56 | 3.14 | 4.12 | 3.66 | 4.00 | 4.01 | 3.49 | 3.55 | 3.43 | 3.33 | 42.00 |
| Days ≥ 0.1" Precip | 6 | 5 | 7 | 6 | 8 | 6 | 7 | 6 | 5 | 5 | 6 | 6 | 73 |
| Total Snowfall (") | 6.7 | 5.3 | 3.1 | 0.3 | 0.0 | 0.0 | 0.0 | 0.0 | 0.0 | 0.1 | 1.1 | 2.6 | 19.2 |
| Days ≥ 1" Snow Depth | 10 | 8 | 2 | 0 | 0 | 0 | 0 | 0 | 0 | 0 | 1 | 3 | 24 |

## WARSAW 2 NW *Richmond County* ELEVATION 141 ft LAT/LONG 37° 59' N / 76° 46' W

| | JAN | FEB | MAR | APR | MAY | JUN | JUL | AUG | SEP | OCT | NOV | DEC | YEAR |
|---|---|---|---|---|---|---|---|---|---|---|---|---|---|
| Maximum Temp °F | 45.5 | 49.3 | 58.5 | 68.9 | 77.2 | 84.9 | 88.4 | 86.8 | 81.0 | 70.3 | 60.5 | 50.4 | 68.5 |
| Minimum Temp °F | 26.7 | 28.8 | 36.0 | 44.4 | 53.9 | 62.4 | 66.9 | 65.4 | 58.6 | 47.1 | 39.3 | 31.3 | 46.7 |
| Mean Temp °F | 36.1 | 39.0 | 47.3 | 56.7 | 65.6 | 73.7 | 77.7 | 76.1 | 69.9 | 58.7 | 49.9 | 40.9 | 57.6 |
| Days Max Temp ≥ 90 °F | 0 | 0 | 0 | 0 | 1 | 7 | 13 | 10 | 3 | 0 | 0 | 0 | 34 |
| Days Max Temp ≤ 32 °F | 4 | 2 | 0 | 0 | 0 | 0 | 0 | 0 | 0 | 0 | 0 | 1 | 7 |
| Days Min Temp ≤ 32 °F | 22 | 19 | 12 | 3 | 0 | 0 | 0 | 0 | 0 | 2 | 9 | 18 | 85 |
| Days Min Temp ≤ 0 °F | 0 | 0 | 0 | 0 | 0 | 0 | 0 | 0 | 0 | 0 | 0 | 0 | 0 |
| Heating Degree Days | 888 | 727 | 548 | 269 | 81 | 7 | 0 | 1 | 31 | 220 | 452 | 741 | 3965 |
| Cooling Degree Days | 0 | 0 | 7 | 28 | 117 | 293 | 432 | 349 | 188 | 38 | 5 | 1 | 1458 |
| Total Precipitation (") | 3.13 | 2.67 | 3.78 | 2.77 | 4.56 | 3.29 | 4.47 | 4.38 | 4.23 | 3.28 | 2.90 | 3.08 | 42.54 |
| Days ≥ 0.1" Precip | 6 | 6 | 7 | 6 | 8 | 6 | 7 | 6 | 5 | 5 | 5 | 6 | 73 |
| Total Snowfall (") | 5.8 | 5.5 | 2.2 | 0.1 | 0.0 | 0.0 | 0.0 | 0.0 | 0.0 | 0.0 | 0.4 | 2.7 | 16.7 |
| Days ≥ 1" Snow Depth | 6 | 4 | 1 | 0 | 0 | 0 | 0 | 0 | 0 | 0 | 0 | 2 | 13 |

## WASHINGTN DC NATL AP *Arlington County* ELEVATION 66 ft LAT/LONG 38° 51 ' N / 77° 2 ' W

| | JAN | FEB | MAR | APR | MAY | JUN | JUL | AUG | SEP | OCT | NOV | DEC | YEAR |
|---|---|---|---|---|---|---|---|---|---|---|---|---|---|
| Maximum Temp °F | 42.6 | 46.5 | 56.0 | 66.8 | 76.1 | 84.6 | 88.6 | 86.7 | 80.0 | 68.8 | 58.4 | 47.7 | 66.9 |
| Minimum Temp °F | 27.3 | 29.4 | 37.3 | 46.4 | 56.3 | 65.7 | 70.9 | 69.3 | 62.5 | 50.1 | 41.0 | 32.6 | 49.1 |
| Mean Temp °F | 35.0 | 38.0 | 46.7 | 56.6 | 66.2 | 75.2 | 79.8 | 78.0 | 71.3 | 59.5 | 49.7 | 40.2 | 58.0 |
| Days Max Temp ≥ 90 °F | 0 | 0 | 0 | 0 | 2 | 8 | 15 | 10 | 4 | 0 | 0 | 0 | 39 |
| Days Max Temp ≤ 32 °F | 5 | 2 | 0 | 0 | 0 | 0 | 0 | 0 | 0 | 0 | 0 | 2 | 9 |
| Days Min Temp ≤ 32 °F | 22 | 18 | 9 | 1 | 0 | 0 | 0 | 0 | 0 | 0 | 4 | 15 | 69 |
| Days Min Temp ≤ 0 °F | 0 | 0 | 0 | 0 | 0 | 0 | 0 | 0 | 0 | 0 | 0 | 0 | 0 |
| Heating Degree Days | 923 | 757 | 564 | 267 | 70 | 4 | 0 | 1 | 19 | 199 | 455 | 763 | 4022 |
| Cooling Degree Days | 0 | 0 | 5 | 24 | 123 | 326 | 490 | 406 | 212 | 35 | 3 | 0 | 1624 |
| Total Precipitation (") | 2.79 | 2.54 | 3.41 | 2.63 | 3.70 | 3.10 | 3.96 | 3.80 | 3.41 | 3.09 | 2.98 | 3.25 | 38.66 |
| Days ≥ 0.1" Precip | 6 | 5 | 7 | 6 | 7 | 6 | 6 | 6 | 5 | 5 | 5 | 6 | 70 |
| Total Snowfall (") | 5.7 | 5.4 | 1.6 | 0.0 | 0.0 | 0.0 | 0.0 | 0.0 | 0.0 | 0.0 | 0.9 | 2.4 | 16.0 |
| Days ≥ 1" Snow Depth | 6 | 4 | 1 | 0 | 0 | 0 | 0 | 0 | 0 | 0 | 0 | 2 | 13 |

## WASHINGTON DC DULLES *Loudoun County* ELEVATION 290 ft LAT/LONG 38° 57 ' N / 77° 27 ' W

| | JAN | FEB | MAR | APR | MAY | JUN | JUL | AUG | SEP | OCT | NOV | DEC | YEAR |
|---|---|---|---|---|---|---|---|---|---|---|---|---|---|
| Maximum Temp °F | 40.5 | 44.5 | 54.4 | 65.7 | 74.6 | 83.0 | 87.4 | 85.7 | 78.9 | 67.4 | 56.7 | 45.7 | 65.4 |
| Minimum Temp °F | 21.4 | 23.6 | 31.8 | 40.7 | 50.1 | 59.2 | 64.5 | 63.0 | 55.4 | 42.2 | 34.2 | 26.3 | 42.7 |
| Mean Temp °F | 31.0 | 34.1 | 43.1 | 53.2 | 62.4 | 71.1 | 76.0 | 74.4 | 67.2 | 54.9 | 45.5 | 36.0 | 54.1 |
| Days Max Temp ≥ 90 °F | 0 | 0 | 0 | 0 | 1 | 5 | 12 | 8 | 3 | 0 | 0 | 0 | 29 |
| Days Max Temp ≤ 32 °F | 7 | 4 | 1 | 0 | 0 | 0 | 0 | 0 | 0 | 0 | 0 | 3 | 15 |
| Days Min Temp ≤ 32 °F | 26 | 23 | 17 | 6 | 0 | 0 | 0 | 0 | 0 | 5 | 15 | 23 | 115 |
| Days Min Temp ≤ 0 °F | 1 | 0 | 0 | 0 | 0 | 0 | 0 | 0 | 0 | 0 | 0 | 0 | 1 |
| Heating Degree Days | 1048 | 866 | 674 | 357 | 137 | 19 | 1 | 4 | 60 | 322 | 581 | 891 | 4960 |
| Cooling Degree Days | 0 | 0 | 4 | 13 | 73 | 226 | 379 | 307 | 136 | 19 | 2 | 0 | 1159 |
| Total Precipitation (") | 2.69 | 2.66 | 3.43 | 3.11 | 4.06 | 3.88 | 3.73 | 3.92 | 3.55 | 3.26 | 3.24 | 3.31 | 40.84 |
| Days ≥ 0.1" Precip | 6 | 5 | 7 | 6 | 8 | 6 | 7 | 7 | 5 | 5 | 6 | 6 | 74 |
| Total Snowfall (") | 7.1 | 6.5 | 3.3 | 0.3 | 0.0 | 0.0 | 0.0 | 0.0 | 0.0 | 0.0 | 1.2 | 3.7 | 22.1 |
| Days ≥ 1" Snow Depth | 8 | 5 | 2 | 0 | 0 | 0 | 0 | 0 | 0 | 0 | 0 | 3 | 18 |

## WEST POINT 2 SW *New Kent County* ELEVATION 20 ft LAT/LONG 37° 31 ' N / 76° 49 ' W

| | JAN | FEB | MAR | APR | MAY | JUN | JUL | AUG | SEP | OCT | NOV | DEC | YEAR |
|---|---|---|---|---|---|---|---|---|---|---|---|---|---|
| Maximum Temp °F | 46.8 | 50.4 | 59.9 | 70.6 | 78.6 | 85.7 | 89.1 | 87.1 | 81.3 | 70.7 | 61.2 | 51.4 | 69.4 |
| Minimum Temp °F | 26.8 | 28.8 | 36.2 | 44.6 | 54.1 | 62.4 | 66.7 | 65.2 | 58.6 | 47.2 | 38.9 | 31.1 | 46.7 |
| Mean Temp °F | 36.8 | 39.6 | 48.1 | 57.7 | 66.4 | 74.1 | 77.9 | 76.1 | 70.0 | 58.9 | 50.1 | 41.3 | 58.1 |
| Days Max Temp ≥ 90 °F | 0 | 0 | 0 | 0 | 2 | 9 | 15 | 10 | 4 | 0 | 0 | 0 | 40 |
| Days Max Temp ≤ 32 °F | 3 | 1 | 0 | 0 | 0 | 0 | 0 | 0 | 0 | 0 | 0 | 1 | 5 |
| Days Min Temp ≤ 32 °F | 22 | 19 | 13 | 3 | 0 | 0 | 0 | 0 | 0 | 2 | 9 | 18 | 86 |
| Days Min Temp ≤ 0 °F | 0 | 0 | 0 | 0 | 0 | 0 | 0 | 0 | 0 | 0 | 0 | 0 | 0 |
| Heating Degree Days | 866 | 711 | 525 | 246 | 68 | 7 | 0 | 1 | 28 | 215 | 447 | 730 | 3844 |
| Cooling Degree Days | 0 | 1 | 8 | 36 | 133 | 312 | 443 | 351 | 190 | 40 | 6 | 1 | 1521 |
| Total Precipitation (") | 3.59 | 3.12 | 4.32 | 2.89 | 3.79 | 3.46 | 4.66 | 4.15 | 3.90 | 3.34 | 3.05 | 3.14 | 43.41 |
| Days ≥ 0.1" Precip | 7 | 6 | 8 | 5 | 7 | 6 | 7 | 7 | 5 | 5 | 5 | 6 | 74 |
| Total Snowfall (") | 4.7 | 3.8 | 1.4 | 0.0 | 0.0 | 0.0 | 0.0 | 0.0 | 0.0 | 0.0 | 0.2 | 1.7 | 11.8 |
| Days ≥ 1" Snow Depth | 4 | 3 | 0 | 0 | 0 | 0 | 0 | 0 | 0 | 0 | 0 | 2 | 9 |

## WILLIAMSBURG 2 N *York County* ELEVATION 69 ft LAT/LONG 37° 16 ' N / 76° 42 ' W

| | JAN | FEB | MAR | APR | MAY | JUN | JUL | AUG | SEP | OCT | NOV | DEC | YEAR |
|---|---|---|---|---|---|---|---|---|---|---|---|---|---|
| Maximum Temp °F | 48.1 | 51.4 | 60.4 | 70.7 | 78.1 | 85.1 | 88.6 | 87.0 | 81.5 | 71.2 | 62.7 | 52.9 | 69.8 |
| Minimum Temp °F | 27.2 | 28.9 | 36.0 | 44.2 | 53.8 | 62.0 | 67.1 | 65.9 | 59.8 | 48.0 | 39.3 | 31.6 | 47.0 |
| Mean Temp °F | 37.7 | 40.2 | 48.3 | 57.5 | 66.0 | 73.6 | 77.9 | 76.5 | 70.7 | 59.6 | 51.0 | 42.3 | 58.4 |
| Days Max Temp ≥ 90 °F | 0 | 0 | 0 | 0 | 2 | 7 | 14 | 10 | 3 | 0 | 0 | 0 | 36 |
| Days Max Temp ≤ 32 °F | 3 | 1 | 0 | 0 | 0 | 0 | 0 | 0 | 0 | 0 | 0 | 1 | 5 |
| Days Min Temp ≤ 32 °F | 22 | 19 | 13 | 3 | 0 | 0 | 0 | 0 | 0 | 1 | 8 | 17 | 83 |
| Days Min Temp ≤ 0 °F | 0 | 0 | 0 | 0 | 0 | 0 | 0 | 0 | 0 | 0 | 0 | 0 | 0 |
| Heating Degree Days | 841 | 694 | 519 | 247 | 69 | 6 | 0 | 1 | 20 | 196 | 419 | 699 | 3711 |
| Cooling Degree Days | 0 | 0 | 7 | 29 | 110 | 282 | 430 | 355 | 195 | 35 | 4 | 1 | 1448 |
| Total Precipitation (") | 3.83 | 3.33 | 4.52 | 2.99 | 4.41 | 3.76 | 5.18 | 5.07 | 4.12 | 3.19 | 3.27 | 3.24 | 46.91 |
| Days ≥ 0.1" Precip | 8 | 6 | 8 | 6 | 7 | 6 | 8 | 7 | 5 | 5 | 6 | 6 | 78 |
| Total Snowfall (") | 3.4 | 2.9 | 1.2 | 0.0 | 0.0 | 0.0 | 0.0 | 0.0 | 0.0 | 0.0 | 0.3 | 0.8 | 8.6 |
| Days ≥ 1" Snow Depth | 3 | 2 | 0 | 0 | 0 | 0 | 0 | 0 | 0 | 0 | 0 | 1 | 6 |

## WINCHESTER 3 ESE *Frederick County* ELEVATION 679 ft LAT/LONG 39° 11 ' N / 78° 7 ' W

| | JAN | FEB | MAR | APR | MAY | JUN | JUL | AUG | SEP | OCT | NOV | DEC | YEAR |
|---|---|---|---|---|---|---|---|---|---|---|---|---|---|
| Maximum Temp °F | 39.5 | 43.9 | 53.9 | 65.4 | 74.7 | 82.6 | 86.6 | 85.0 | 77.9 | 66.1 | 55.5 | 44.7 | 64.7 |
| Minimum Temp °F | 20.6 | 23.0 | 30.9 | 39.8 | 49.5 | 57.6 | 62.5 | 60.5 | 53.6 | 41.9 | 34.2 | 25.8 | 41.7 |
| Mean Temp °F | 30.1 | 33.5 | 42.4 | 52.6 | 62.1 | 70.2 | 74.6 | 72.8 | 65.8 | 54.0 | 44.9 | 35.2 | 53.2 |
| Days Max Temp ≥ 90 °F | 0 | 0 | 0 | 0 | 1 | 5 | 10 | 7 | 2 | 0 | 0 | 0 | 25 |
| Days Max Temp ≤ 32 °F | 8 | 5 | 1 | 0 | 0 | 0 | 0 | 0 | 0 | 0 | 0 | 3 | 17 |
| Days Min Temp ≤ 32 °F | 27 | 23 | 18 | 7 | 1 | 0 | 0 | 0 | 0 | 5 | 14 | 24 | 119 |
| Days Min Temp ≤ 0 °F | 1 | 0 | 0 | 0 | 0 | 0 | 0 | 0 | 0 | 0 | 0 | 0 | 1 |
| Heating Degree Days | 1075 | 884 | 695 | 376 | 146 | 24 | 2 | 7 | 77 | 343 | 599 | 916 | 5144 |
| Cooling Degree Days | 0 | 0 | 2 | 8 | 57 | 170 | 314 | 236 | 92 | *12* | *1* | 0 | 892 |
| Total Precipitation (") | 2.39 | 2.34 | 3.25 | 3.10 | 3.76 | 3.83 | 3.76 | 3.65 | 3.15 | 3.17 | 3.07 | 2.68 | 38.15 |
| Days ≥ 0.1" Precip | 6 | 5 | 7 | 6 | 7 | 6 | 7 | 6 | 5 | 5 | 6 | 5 | 71 |
| Total Snowfall (") | 6.4 | 6.4 | 4.2 | 0.3 | 0.0 | 0.0 | 0.0 | 0.0 | 0.0 | 0.0 | 1.3 | 3.0 | 21.6 |
| Days ≥ 1" Snow Depth | 9 | 8 | 2 | 0 | 0 | 0 | 0 | 0 | 0 | 0 | 1 | 3 | 23 |

## WISE 2 E *Wise County* ELEVATION 2500 ft LAT/LONG 36° 58 ' N / 82° 34 ' W

| | JAN | FEB | MAR | APR | MAY | JUN | JUL | AUG | SEP | OCT | NOV | DEC | YEAR |
|---|---|---|---|---|---|---|---|---|---|---|---|---|---|
| Maximum Temp °F | 41.5 | 45.6 | 55.5 | 65.3 | 71.9 | 78.2 | 81.1 | 79.8 | 74.7 | 65.9 | 55.8 | 45.8 | 63.4 |
| Minimum Temp °F | 23.4 | 25.9 | 34.2 | 42.5 | 50.0 | 57.2 | 61.1 | 60.1 | 54.3 | 43.4 | 36.1 | 28.1 | 43.0 |
| Mean Temp °F | 32.4 | 35.8 | 44.9 | 53.9 | 61.0 | 67.7 | 71.1 | 70.0 | 64.5 | 54.7 | 46.0 | 37.0 | 53.3 |
| Days Max Temp ≥ 90 °F | 0 | 0 | 0 | 0 | 0 | 0 | 1 | 0 | 0 | 0 | 0 | 0 | 1 |
| Days Max Temp ≤ 32 °F | 7 | 4 | 1 | 0 | 0 | 0 | 0 | 0 | 0 | 0 | 1 | 4 | 17 |
| Days Min Temp ≤ 32 °F | 24 | 20 | 14 | 6 | 1 | 0 | 0 | 0 | 0 | 4 | 12 | 20 | 101 |
| Days Min Temp ≤ 0 °F | 1 | 1 | 0 | 0 | 0 | 0 | 0 | 0 | 0 | 0 | 0 | 1 | 3 |
| Heating Degree Days | 1004 | 819 | 617 | 338 | 151 | 28 | 4 | 7 | 80 | 320 | 565 | 863 | 4796 |
| Cooling Degree Days | 0 | 0 | 1 | 12 | 38 | 136 | 235 | 186 | 78 | 9 | 0 | 0 | 695 |
| Total Precipitation (") | 3.35 | 3.54 | 4.35 | 3.89 | 4.49 | 3.85 | 5.22 | 4.12 | 3.55 | 3.08 | 3.53 | 3.63 | 46.60 |
| Days ≥ 0.1" Precip | 8 | 7 | 9 | 8 | 10 | 8 | 10 | 7 | 6 | 6 | 8 | 7 | 94 |
| Total Snowfall (") | 12.9 | 11.9 | 7.0 | 2.6 | 0.1 | 0.0 | 0.0 | 0.0 | 0.0 | 0.3 | 3.0 | 7.7 | 45.5 |
| Days ≥ 1" Snow Depth | 10 | 7 | 2 | 0 | 0 | 0 | 0 | 0 | 0 | 0 | 1 | 5 | 25 |

## WOODSTOCK 2 NE *Shenandoah County* ELEVATION 660 ft LAT/LONG 38° 54 ' N / 78° 28 ' W

| | JAN | FEB | MAR | APR | MAY | JUN | JUL | AUG | SEP | OCT | NOV | DEC | YEAR |
|---|---|---|---|---|---|---|---|---|---|---|---|---|---|
| Maximum Temp °F | 43.2 | 47.3 | 57.3 | 68.1 | 76.6 | 84.4 | 87.9 | 86.5 | 80.2 | 69.4 | 58.2 | 47.5 | 67.2 |
| Minimum Temp °F | 21.6 | 23.9 | 31.7 | 40.3 | 49.4 | 57.4 | 62.2 | 60.4 | 53.7 | 41.7 | 34.0 | 26.4 | 41.9 |
| Mean Temp °F | 32.4 | 35.7 | 44.5 | 54.3 | 63.1 | 70.9 | 75.1 | 73.5 | 67.0 | 55.5 | 46.2 | 37.0 | 54.6 |
| Days Max Temp ≥ 90 °F | 0 | 0 | 0 | 0 | 1 | 7 | 12 | 10 | 4 | 0 | 0 | 0 | 34 |
| Days Max Temp ≤ 32 °F | 5 | 2 | 0 | 0 | 0 | 0 | 0 | 0 | 0 | 0 | 0 | 2 | 9 |
| Days Min Temp ≤ 32 °F | 27 | 23 | 17 | 7 | 1 | 0 | 0 | 0 | 0 | 6 | 14 | 23 | 118 |
| Days Min Temp ≤ 0 °F | 1 | 0 | 0 | 0 | 0 | 0 | 0 | 0 | 0 | 0 | 0 | 0 | 1 |
| Heating Degree Days | 1003 | 822 | 631 | 327 | 120 | 16 | 1 | 5 | 59 | 298 | 560 | 862 | 4704 |
| Cooling Degree Days | 0 | 0 | 1 | 11 | 67 | 200 | 335 | 269 | 121 | 15 | 0 | 0 | 1019 |
| Total Precipitation (") | 2.47 | 2.25 | 3.04 | 2.74 | 3.59 | 3.14 | 3.99 | 3.21 | 3.15 | 3.06 | 2.78 | 2.55 | 35.97 |
| Days ≥ 0.1" Precip | 5 | 5 | 7 | 7 | 8 | 7 | 8 | 6 | 6 | 5 | 6 | 5 | 75 |
| Total Snowfall (") | 8.2 | 6.3 | 4.1 | 0.4 | 0.0 | 0.0 | 0.0 | 0.0 | 0.0 | 0.1 | 1.1 | 3.5 | 23.7 |
| Days ≥ 1" Snow Depth | 8 | 6 | 2 | 0 | 0 | 0 | 0 | 0 | 0 | 0 | 1 | 3 | 20 |

## WYTHEVILLE 1 S *Wythe County* ELEVATION 2451 ft LAT/LONG 36° 56 ' N / 81° 5 ' W

| | JAN | FEB | MAR | APR | MAY | JUN | JUL | AUG | SEP | OCT | NOV | DEC | YEAR |
|---|---|---|---|---|---|---|---|---|---|---|---|---|---|
| Maximum Temp °F | 42.7 | 47.1 | 57.0 | 66.7 | 74.2 | 80.7 | 84.3 | 83.0 | 77.6 | 68.1 | 56.8 | 46.8 | 65.4 |
| Minimum Temp °F | 21.3 | 23.0 | 30.1 | 37.4 | 46.5 | 54.2 | 59.0 | 57.7 | 51.3 | 38.9 | 31.4 | 25.0 | 39.7 |
| Mean Temp °F | 32.1 | 35.1 | 43.6 | 52.1 | 60.4 | 67.5 | 71.7 | 70.4 | 64.5 | 53.5 | 44.1 | 35.9 | 52.6 |
| Days Max Temp ≥ 90 °F | 0 | 0 | 0 | 0 | 0 | 1 | 4 | 3 | 1 | 0 | 0 | 0 | 9 |
| Days Max Temp ≤ 32 °F | 5 | 3 | 0 | 0 | 0 | 0 | 0 | 0 | 0 | 0 | 0 | 2 | 10 |
| Days Min Temp ≤ 32 °F | 27 | 23 | 19 | 10 | 3 | 0 | 0 | 0 | 1 | 9 | 17 | 24 | 133 |
| Days Min Temp ≤ 0 °F | 2 | 0 | 0 | 0 | 0 | 0 | 0 | 0 | 0 | 0 | 0 | 1 | 3 |
| Heating Degree Days | 1015 | 839 | 658 | 383 | 165 | 33 | 3 | 10 | 85 | 354 | 620 | 895 | 5060 |
| Cooling Degree Days | 0 | 0 | 1 | 3 | 29 | 123 | 240 | 189 | 77 | 5 | 0 | 0 | 667 |
| Total Precipitation (") | 2.50 | 2.76 | 3.18 | 3.23 | 3.99 | 3.16 | 4.05 | 3.36 | 3.01 | 3.25 | 2.62 | 2.55 | 37.66 |
| Days ≥ 0.1" Precip | 6 | 5 | 7 | 7 | 8 | 7 | 8 | 7 | 6 | 5 | 6 | 6 | 78 |
| Total Snowfall (") | 7.2 | 5.8 | 3.1 | 1.1 | 0.0 | 0.0 | 0.0 | 0.0 | 0.0 | 0.0 | 0.7 | 3.7 | 21.6 |
| Days ≥ 1" Snow Depth | 7 | 6 | 2 | 0 | 0 | 0 | 0 | 0 | 0 | 0 | 0 | 4 | 19 |

## JANUARY MINIMUM TEMPERATURE °F

### LOWEST

| | | |
|---|---|---|
| 1 | Big Meadows | 16.2 |
| 2 | Luray | 18.7 |
| 3 | Burkes Garden | 19.6 |
| 4 | Blacksburg | 19.7 |
| 5 | Staunton | 19.8 |
| 6 | Mount Weather | 20.4 |
| 7 | Winchester | 20.6 |
| 8 | Floyd | 20.7 |
| 9 | Wytheville | 21.3 |
| 10 | Washington-Dlles | 21.4 |
| 11 | Dale | 21.6 |
| | Woodstock | 21.6 |
| 13 | Galax | 22.0 |
| 14 | Warrenton | 22.1 |
| 15 | Fredericksburg | 22.4 |
| 16 | Pulaski | 22.5 |
| 17 | Chatham | 22.6 |
| 18 | Grundy | 22.7 |
| | Piedmont | 22.7 |
| 20 | Lincoln | 23.0 |
| | Pennington Gap | 23.0 |
| | Staffordsville | 23.0 |
| 23 | Corbin | 23.2 |
| | Covington | 23.2 |
| | Martinsville | 23.2 |

### HIGHEST

| | | |
|---|---|---|
| 1 | Back Bay wildlf | 31.4 |
| | Norfolk | 31.4 |
| 3 | Langley | 31.3 |
| 4 | Tangier Island | 30.4 |
| 5 | Suffolk | 29.2 |
| 6 | Painter | 28.7 |
| 7 | Hopewell | 28.2 |
| 8 | Holland | 27.7 |
| 9 | Washington-Natl | 27.3 |
| 10 | Williamsburg | 27.2 |
| 11 | West Point | 26.8 |
| 12 | Warsaw | 26.7 |
| 13 | Lawrenceville | 26.4 |
| | Richmond | 26.4 |
| 15 | Bedford | 26.3 |
| | Stony Creek | 26.3 |
| 17 | John H Kerr Dam | 26.2 |
| 18 | Chase City | 26.0 |
| | Danville | 26.0 |
| 20 | Colonial Beach | 25.8 |
| 21 | Charlottesville | 25.6 |
| | Roanoke | 25.6 |
| 23 | Lynchburg | 25.4 |
| 24 | Walkerton | 25.2 |
| 25 | Buchanan | 25.1 |

## JULY MAXIMUM TEMPERATURE °F

### HIGHEST

| | | |
|---|---|---|
| 1 | Hopewell | 90.8 |
| | Stony Creek | 90.8 |
| 3 | Chase City | 90.1 |
| | Colonial Beach | 90.1 |
| 5 | Danville | 90.0 |
| 6 | Fredericksburg | 89.9 |
| 7 | John H Kerr Dam | 89.1 |
| | West Point | 89.1 |
| 9 | Farmville | 88.9 |
| | Lawrenceville | 88.9 |
| | Richmond | 88.9 |
| 12 | Washington-Natl | 88.6 |
| | Williamsburg | 88.6 |
| 14 | Walkerton | 88.5 |
| 15 | Holland | 88.4 |
| | Warsaw | 88.4 |
| 17 | Lincoln | 88.3 |
| 18 | Covington | 87.9 |
| | Suffolk | 87.9 |
| | Woodstock | 87.9 |
| 21 | Buchanan | 87.6 |
| | Chatham | 87.6 |
| 23 | Ashland | 87.5 |
| | Charlotte Crt Hse | 87.5 |
| | Martinsville | 87.5 |

### LOWEST

| | | |
|---|---|---|
| 1 | Big Meadows | 75.5 |
| 2 | Burkes Garden | 79.0 |
| 3 | Mount Weather | 80.6 |
| 4 | Wise | 81.1 |
| 5 | Floyd | 81.9 |
| 6 | Blacksburg | 82.8 |
| | Galax | 82.8 |
| 8 | Staffordsville | 84.2 |
| 9 | Pulaski | 84.3 |
| | Wytheville | 84.3 |
| 11 | Staunton | 84.8 |
| 12 | Luray | 85.8 |
| 13 | Bedford | 85.9 |
| 14 | Dale | 86.0 |
| 15 | Warrenton | 86.2 |
| 16 | Langley | 86.3 |
| | Pennington Gap | 86.3 |
| 18 | Back Bay wildlf | 86.4 |
| | Piedmont | 86.4 |
| 20 | Lynchburg | 86.5 |
| 21 | Winchester | 86.6 |
| 22 | Rocky Mount | 86.7 |
| 23 | Lexington | 86.8 |
| | Painter | 86.8 |
| | Philpott Dam | 86.8 |

## ANNUAL PRECIPITATION (")

### HIGHEST

| | | |
|---|---|---|
| 1 | Big Meadows | 52.11 |
| 2 | Pennington Gap | 49.25 |
| 3 | Philpott Dam | 49.04 |
| 4 | Charlottesville | 47.96 |
| 5 | Holland | 47.29 |
| 6 | Williamsburg | 46.91 |
| 7 | Wise | 46.60 |
| 8 | Suffolk | 46.23 |
| 9 | Langley | 45.84 |
| 10 | Rocky Mount | 45.54 |
| 11 | Stony Creek | 45.53 |
| 12 | Back Bay wildlf | 45.24 |
| 13 | Burkes Garden | 44.92 |
| 14 | Martinsville | 44.91 |
| | Tye River | 44.91 |
| 16 | Chatham | 44.16 |
| 17 | Norfolk | 43.86 |
| 18 | Lawrenceville | 43.75 |
| 19 | Charlotte Crt Hse | 43.65 |
| 20 | Appomattox | 43.61 |
| 21 | Farmville | 43.48 |
| 22 | Bedford | 43.45 |
| | Grundy | 43.45 |
| 24 | Piedmont | 43.43 |
| 25 | West Point | 43.41 |

### LOWEST

| | | |
|---|---|---|
| 1 | Dale | 34.42 |
| 2 | Tangier Island | 35.96 |
| 3 | Woodstock | 35.97 |
| 4 | Covington | 36.11 |
| 5 | Staunton | 36.84 |
| 6 | Pulaski | 37.27 |
| 7 | Wytheville | 37.66 |
| 8 | Winchester | 38.15 |
| 9 | Lexington | 38.27 |
| | Staffordsville | 38.27 |
| 11 | Washington-Natl | 38.66 |
| 12 | Luray | 39.84 |
| 13 | Washington-Dlles | 40.84 |
| 14 | Mount Weather | 41.08 |
| 15 | Lynchburg | 41.34 |
| 16 | Floyd | 41.35 |
| 17 | Colonial Beach | 41.59 |
| 18 | Painter | 41.63 |
| 19 | Roanoke | 41.70 |
| 20 | Blacksburg | 41.76 |
| 21 | Warrenton | 42.00 |
| 22 | Lincoln | 42.12 |
| 23 | Fredericksburg | 42.16 |
| 24 | Buchanan | 42.38 |
| 25 | Warsaw | 42.54 |

## ANNUAL SNOWFALL (")

### HIGHEST

| | | |
|---|---|---|
| 1 | Burkes Garden | 50.1 |
| 2 | Wise | 45.5 |
| 3 | Big Meadows | 42.2 |
| 4 | Dale | 25.0 |
| | Luray | 25.0 |
| 6 | Woodstock | 23.7 |
| 7 | Staunton | 23.5 |
| 8 | Roanoke | 22.8 |
| 9 | Lincoln | 22.5 |
| 10 | Washington-Dlles | 22.1 |
| 11 | Mount Weather | 21.6 |
| | Winchester | 21.6 |
| | Wytheville | 21.6 |
| 14 | Blacksburg | 21.4 |
| 15 | Staffordsville | 21.2 |
| 16 | Charlottesville | 20.9 |
| 17 | Galax | 20.1 |
| 18 | Piedmont | 20.0 |
| 19 | Floyd | 19.9 |
| 20 | Warrenton | 19.2 |
| 21 | Grundy | 19.0 |
| 22 | Lynchburg | 18.8 |
| 23 | Buchanan | 18.5 |
| | Louisa | 18.5 |
| | Rocky Mount | 18.5 |

### LOWEST

| | | |
|---|---|---|
| 1 | Back Bay wildlf | 4.0 |
| 2 | Tangier Island | 7.2 |
| 3 | Philpott Dam | 7.4 |
| 4 | Hopewell | 7.5 |
| 5 | Holland | 7.9 |
| | Suffolk | 7.9 |
| 7 | Norfolk | 8.0 |
| 8 | Williamsburg | 8.6 |
| 9 | Chase City | 8.7 |
| 10 | Langley | 9.1 |
| 11 | Painter | 9.5 |
| 12 | Chatham | 10.9 |
| 13 | Charlotte Crt Hse | 11.5 |
| 14 | Lawrenceville | 11.6 |
| 15 | Walkerton | 11.8 |
| | West Point | 11.8 |
| 17 | Fredericksburg | 13.8 |
| 18 | Richmond | 13.9 |
| 19 | Corbin | 14.2 |
| 20 | Appomattox | 15.8 |
| 21 | Washington-Natl | 16.0 |
| 22 | Bedford | 16.5 |
| 23 | Farmville | 16.7 |
| | Warsaw | 16.7 |
| 25 | Ashland | 16.9 |

# WASHINGTON

PHYSICAL FEATURES. Washington's western boundary is formed by the Pacific Ocean. There are two ranges of mountains parallel to the coast. The Cascade Mountains, 90 to 125 miles inland and 4,000 to 10,000 feet in elevation, are a topographic and climatic barrier separating the State into eastern and western Washington. The higher, wider and more rugged sections are in the northern part of the State. Some of the highest isolated volcanic peaks are Mt. Rainier (14,408 ft.), Mt. Adams (12,307 ft.) and Mt. Baker (10,730 ft.). These and other high peaks are snowcapped throughout the year. The only break in the Cascade Range is the narrow Columbia River gorge.

GENERAL CLIMATE. The location of the State of Washington on the windward coast in mid-latitudes is such that the climatic elements combine to produce a predominantly marine-type climate west of the Cascade Mountains, while east of the Cascades, the climate possesses both continental and marine characteristics. Considering its northerly latitude, 46° to 49°, Washington's climate is mild. There are several climatic controls which have a definite influence on the climate: (a) terrain, (b) Pacific Ocean, and (c) semi-permanent high and low pressure regions located over the north Pacific Ocean. The effect of these various controls combine to produce entirely different conditions within short distances.

The seasonal change in the temperature of the ocean is less than the seasonal change in the temperature of the land, thus the ocean is warmer in winter and cooler in summer than the adjoining land surfaces. The average temperature of the water along the coast and in the Strait of Juan de Fuca ranges from 45° in January to 53° in July; however, during the summer, some of the shallow bays and protected coves are 5 to 10 degrees warmer.

The first orographic lifting and major release of moisture occurs along the western slope of the Coastal Range. The second area of heavy orographic precipitation is along the windward slopes of the Cascade Range. Warming and drying of air as it descends along the lee (eastern) slopes of the Cascade Range results in near desert conditions in the lowest section of the Columbia Basin. Another orographic lifting of the air occurs as it flows eastward from the lowest elevations of the Inland Basin toward the Rocky Mountains. This lifting of air results in a gradual increase in precipitation from the lowest section of the basin to the higher elevations along the eastern border of the State.

The location and intensity of the semi-permanent high and low pressure areas over the north Pacific Ocean have a definite influence on the climate. Air circulates in a clockwise direction around the semi-permanent high pressure cell and in a counter-clockwise direction around the semi-permanent low pressure cell. During the spring and summer, the low pressure cell becomes weak and moves north of the Aleutian Islands. At the same time, the high pressure area spreads over most of the north Pacific Ocean. A circulation of air around the high pressure center brings a prevailing westerly and northwesterly flow of comparatively dry, cool and stable air into the Pacific Northwest. As the air moves inland, it becomes warmer and drier which results in a dry season beginning in the late spring and reaching a peak in midsummer.

In the fall and winter, the Aleutian low pressure center intensifies and moves southward reaching a maximum intensity in midwinter. At the same time, the high pressure area becomes weaker and moves southward. A circulation of air around these two pressure centers over the ocean brings a prevailing southwesterly and westerly flow of air into the Pacific Northwest. This air from over the ocean is moist and near the temperature of the water. Condensation occurs as the air moves inland over the cooler land and rises along the windward slopes of the mountains. This results in a wet season beginning in October, reaching a peak in winter, then gradually decreasing in the spring.

WESTERN WASHINGTON. West of the Cascade mountains, summers are cool and comparatively dry and winters are mild, wet and cloudy. The average number of clear or only partly cloudy days each month varies from 4 to 8 in winter, 8 to 15 in spring and fall, and 15 to 20 in summer. The percent of possible sunshine received each month ranges from approximately 25 percent in winter to 60 percent in summer. In the interior valleys, measurable rainfall is recorded on 150 days each year, and on 190 days in the mountains and along the coast. Thunderstorms over the lower elevations occur on 4 to 8 days each year and over the mountains on 7 to 15 days. Damaging hail storms rarely, if ever, occur in most localities of western Washington. During July and August, the driest months, it is not unusual for 2 to 4 weeks to pass with only a few showers; however, in December and January, the wettest months, precipitation is frequently recorded on 20 to 25 days or more each month. The range in annual precipitation is from approximately 20 inches in an area northeast of the Olympic Mountains to 150 inches along the southwestern slopes of these mountains. Snowfall is light in the lower elevations and heavy in the mountains.

During the wet season, rainfall is usually of light to moderate intensity and continuous over a period of time rather than heavy downpours for brief periods. The heavier intensities occur along the windward slopes of the mountains. During the latter half of the summer and early fall, the lower valleys are sometimes filled with fog or low clouds until noon, while at the same time, the higher elevations are sunny. The strongest winds are generally from the south or southwest and occur during the late fall and winter. In the interior valleys, wind velocities can be expected to reach 40 to 50 m.p.h. each winter. The daily variation in relative humidity in January is from approximately 87 percent at 4 a.m. to 78 percent at 4 p.m., and in July, from 85 percent at 4 a.m. to 47 percent at 4 p.m. During periods of easterly winds, the relative humidity occasionally drops to 25 percent or lower. The highest summer and lowest winter temperatures are usually recorded during periods of easterly winds.

The Olympic Mountains, located on the northern section of the Olympic Peninsula, tower to nearly 8,000 feet. This area receives the full force of storms moving inland from over the ocean, thus heavy precipitation and winds of gale force occur frequently during the winter season. The "rainforest" area along the southwestern and western slopes of the Olympic Mountains receives the heaviest precipitation in the continental United States. A belt on the northeastern slope of the Olympic Mountains in the "rain shadow" of the Olympic Mountains is the driest area in western Washington. This area frequently receives drizzle or light rain while other localities are experiencing light to moderate rainfall, and has slightly more sunshine and slightly less cloudiness than other localities in Puget Sound..

EASTERN WASHINGTON. This section of the State is part of the large inland basin between the Cascade and Rocky Mountains. In an easterly and northerly direction, the Rocky Mountains shield the inland basin from the winter season's cold air masses traveling southward across Canada. In a westerly direction, the Cascade range forms a barrier to the easterly movement of moist and comparatively mild air in winter and cool air in summer. Some of the air from each of these source regions reaches this section of the State and produces a climate which has some of the characteristics of both continental and marine types. Most of the air masses and weather systems crossing eastern Washington are traveling under the influence of the prevailing westerly winds. Infrequently, dry continental air masses enter the inland basin from the north or east. In the summer season, this air from over the continent results in low relative humidity and high temperatures while in winter, clear cold weather prevails. Extremes in both summer and winter temperatures generally occur when the inland basin is under the influence of air from over the continent.

East of the Cascades, summers are warmer, winters are colder, and precipitation is less than in western Washington. The average number of clear or only partly cloudy days each month varies from 5 to 10 in winter, 12 to 18 in spring and fall, and 20 to 28 in summer. The percent of possible sunshine received each month is from 20 to 30 percent in winter, 50 to 60 percent in spring and fall, and 80 to 85 percent in summer. The number of hours of sunshine possible on a clear day ranges from approximately 8 in December to 16 in June. In the driest areas, rainfall is recorded on 70 days each year .

Annual precipitation ranges from 7 to 9 inches near the confluence of the Snake and Columbia Rivers, 15 to 30 inches along the eastern border, and 75 to 90 inches near the summit of the Cascade Mountains. During July and August, it is not unusual for 4 to 8 weeks to pass with only a few scattered showers. Thunderstorms can be expected on 1 to 3 days each month from April through September. Most thunderstorms in the warmest months occur as isolated cells covering only a few square miles. A few damaging hail storms are reported each summer.

During the coldest months, a loss of heat by radiation at night and moist air crossing the Cascades and mixing with the colder air in the inland basin results in cloudiness, fog, and occasional freezing drizzle. A "chinook" wind which produces a rapid rise in temperature occurs a few times each winter. During most of the year, the prevailing direction of the wind is from the southwest or west. The frequency of northeasterly winds is greatest in the fall and winter.

## COUNTY INDEX

### *Adams County*
LIND 3 NE
OTHELLO 6 ESE
RITZVILLE 1 SSE

### *Benton County*
KENNEWICK
MCNARY DAM
PROSSER 4 NE
RICHLAND

### *Chelan County*
CHELAN
HOLDEN VILLAGE
LEAVENWORTH 3 S
PLAIN
STEHEKIN 4 NW
WENATCHEE
WENATCHEE EXP STN
WENATCHEE PANGBORN

### *Clallam County*
ELWHA RS
FORKS 1 E
PORT ANGELES
SAPPHO 8 E

### *Clark County*
BATTLE GROUND
VANCOUVER 4 NNE

### *Columbia County*
DAYTON 1 WSW

### *Cowlitz County*
LONGVIEW

### *Douglas County*
CHIEF JOSEPH DAM
WATERVILLE

### *Ferry County*
REPUBLIC

### *Franklin County*
CONNELL 1 W
HATTON 9 SE

### *Garfield County*
POMEROY

### *Grant County*
COULEE DAM 1 SW
EPHRATA AP FCWOS
HARTLINE
PRIEST RAPIDS DAM
QUINCY 1 S
SMYRNA

### *Grays Harbor County*
ABERDEEN
ABERDEEN 20 NNE
ELMA
GRAYLAND
HOQUIAM AP
OAKVILLE

### *Island County*
COUPEVILLE 1 S

### *Jefferson County*
CLEARWATER
PORT TOWNSEND
QUILCENE 2 SW

### *King County*
CEDAR LAKE
KENT
LANDSBURG
MUD MOUNTAIN DAM
PALMER 3 ESE
SEATTLE-TACOMA AP
SNOQUALMIE FALLS
STEVENS PASS

### *Kitsap County*
BREMERTON

### *Kittitas County*
CLE ELUM
ELLENSBURG
STAMPEDE PASS

### *Klickitat County*
APPLETON
BICKLETON 3 ESE
DALLESPORT FCWOS AP
MOUNT ADAMS RS

### *Lewis County*
CENTRALIA
GLENOMA 1 W
PACKWOOD
TOLEDO

### *Lincoln County*
DAVENPORT
HARRINGTON 1 NW
ODESSA
WILBUR

### *Mason County*
GRAPEVIEW 3 SW
SHELTON

### *Okanogan County*
CONCONULLY
MAZAMA
WINTHROP 1 WSW

### *Pacific County*
GRAYS RIVER HATCHERY
LONG BEACH EXP STN

### *Pend Oreille County*
BOUNDARY DAM
NEWPORT

### *Pierce County*
BUCKLEY 1 NE
MC MILLIN RESERVOIR
PUYALLUP 2 W EXP STN
RAINIER PARADISE RNG

### *San Juan County*
OLGA 2 SE

### *Skagit County*
ANACORTES
CONCRETE PPL FISH ST
MOUNT VERNON 3 WNW
SEDRO WOOLLEY

### *Skamania County*
COUGAR 6 E
SKAMANIA FISH HATCHE

### *Snohomish County*
EVERETT
MONROE
STARTUP 1 E

### *Spokane County*
SPOKANE INTL ARPT

### *Stevens County*
CHEWELAH
NORTHPORT
WELLPINIT

### *Thurston County*
OLYMPIA AP

### *Walla Walla County*
ICE HARBOR DAM
WALLA WALLA CITY-COU
WHITMAN MISSION

### *Whatcom County*
BELLINGHAM INTL AP
BLAINE
CLEARBROOK
DIABLO DAM
NEWHALEM
ROSS DAM
UPPER BAKER DAM

### *Whitman County*
COLFAX
LACROSSE
PULLMAN 2 NW
ROSALIA
SAINT JOHN

### *Yakima County*
MOXEE CITY 10 E
SUNNYSIDE
WAPATO
YAKIMA AIR TERMINAL

## ELEVATION INDEX

| FEET | STATION NAME |
|---|---|
| 10 | ABERDEEN |
| 10 | LONGVIEW |
| 10 | MOUNT VERNON 3 WNW |
| 11 | HOQUIAM AP |
| 20 | GRAPEVIEW 3 SW |
| 20 | GRAYLAND |
| 20 | SHELTON |
| 30 | ANACORTES |
| 30 | KENT |
| 30 | LONG BEACH EXP STN |
| 39 | BLAINE |
| 40 | PORT ANGELES |
| 49 | COUPEVILLE 1 S |
| 49 | PUYALLUP 2 W EXP STN |
| 59 | CLEARBROOK |
| 59 | SEDRO WOOLLEY |
| 66 | PORT TOWNSEND |
| 69 | ELMA |
| 79 | OLGA 2 SE |
| 80 | CLEARWATER |
| 89 | OAKVILLE |
| 102 | GRAYS RIVER HATCHERY |
| 102 | VANCOUVER 4 NNE |
| 120 | EVERETT |
| 121 | MONROE |
| 121 | QUILCENE 2 SW |
| 154 | BELLINGHAM INTL AP |
| 161 | BREMERTON |
| 171 | STARTUP 1 E |
| 190 | CENTRALIA |
| 217 | OLYMPIA AP |
| 236 | DALLESPORT FCWOS AP |
| 269 | CONCRETE PPL FISH ST |
| 302 | BATTLE GROUND |
| 325 | TOLEDO |
| 351 | FORKS 1 E |
| 351 | MCNARY DAM |
| 361 | ELWHA RS |
| 384 | SEATTLE-TACOMA AP |
| 390 | KENNEWICK |
| 400 | RICHLAND |
| 430 | SNOQUALMIE FALLS |
| 440 | ABERDEEN 20 NNE |
| 440 | SKAMANIA FISH HATCHE |
| 469 | PRIEST RAPIDS DAM |
| 479 | ICE HARBOR DAM |
| 531 | NEWHALEM |
| 541 | LANDSBURG |
| 561 | SMYRNA |
| 581 | MC MILLIN RESERVOIR |
| 630 | COUGAR 6 E |
| 630 | WENATCHEE |
| 630 | WHITMAN MISSION |
| 689 | BUCKLEY 1 NE |
| 689 | UPPER BAKER DAM |
| 747 | SUNNYSIDE |
| 760 | SAPPHO 8 E |
| 820 | CHIEF JOSEPH DAM |
| 840 | GLENOMA 1 W |
| 840 | PROSSER 4 NE |
| 850 | WAPATO |
| 869 | WENATCHEE EXP STN |
| 889 | DIABLO DAM |
| 902 | PALMER 3 ESE |
| 1020 | CONNELL 1 W |
| 1060 | PACKWOOD |
| 1063 | YAKIMA AIR TERMINAL |
| 1120 | CHELAN |
| 1152 | STEHEKIN 4 NW |
| 1161 | LEAVENWORTH 3 S |
| 1191 | OTHELLO 6 ESE |
| 1207 | WALLA WALLA CITY-COU |
| 1229 | WENATCHEE PANGBORN |
| 1240 | ROSS DAM |
| 1270 | QUINCY 1 S |
| 1273 | EPHRATA AP FCWOS |
| 1312 | MUD MOUNTAIN DAM |
| 1332 | NORTHPORT |
| 1371 | HATTON 9 SE |
| 1480 | LACROSSE |
| 1522 | ELLENSBURG |
| 1542 | ODESSA |
| 1552 | MOXEE CITY 10 E |
| 1562 | CEDAR LAKE |
| 1631 | LIND 3 NE |
| 1670 | CHEWELAH |
| 1703 | COULEE DAM 1 SW |
| 1752 | DAYTON 1 WSW |
| 1762 | WINTHROP 1 WSW |
| 1801 | PLAIN |
| 1831 | BOUNDARY DAM |
| 1831 | RITZVILLE 1 SSE |
| 1900 | POMEROY |
| 1903 | CLE ELUM |
| 1913 | HARTLINE |
| 1952 | SAINT JOHN |
| 1962 | MOUNT ADAMS RS |
| 1972 | COLFAX |
| 2142 | NEWPORT |
| 2162 | WILBUR |
| 2182 | MAZAMA |
| 2260 | HARRINGTON 1 NW |
| 2270 | CONCONULLY |
| 2362 | APPLETON |
| 2365 | SPOKANE INTL ARPT |
| 2402 | ROSALIA |
| 2451 | DAVENPORT |
| 2451 | WELLPINIT |
| 2552 | PULLMAN 2 NW |
| 2621 | WATERVILLE |
| 2651 | REPUBLIC |
| 3002 | BICKLETON 3 ESE |
| 3222 | HOLDEN VILLAGE |
| 3958 | STAMPEDE PASS |
| 4062 | STEVENS PASS |
| 5554 | RAINIER PARADISE RNG |

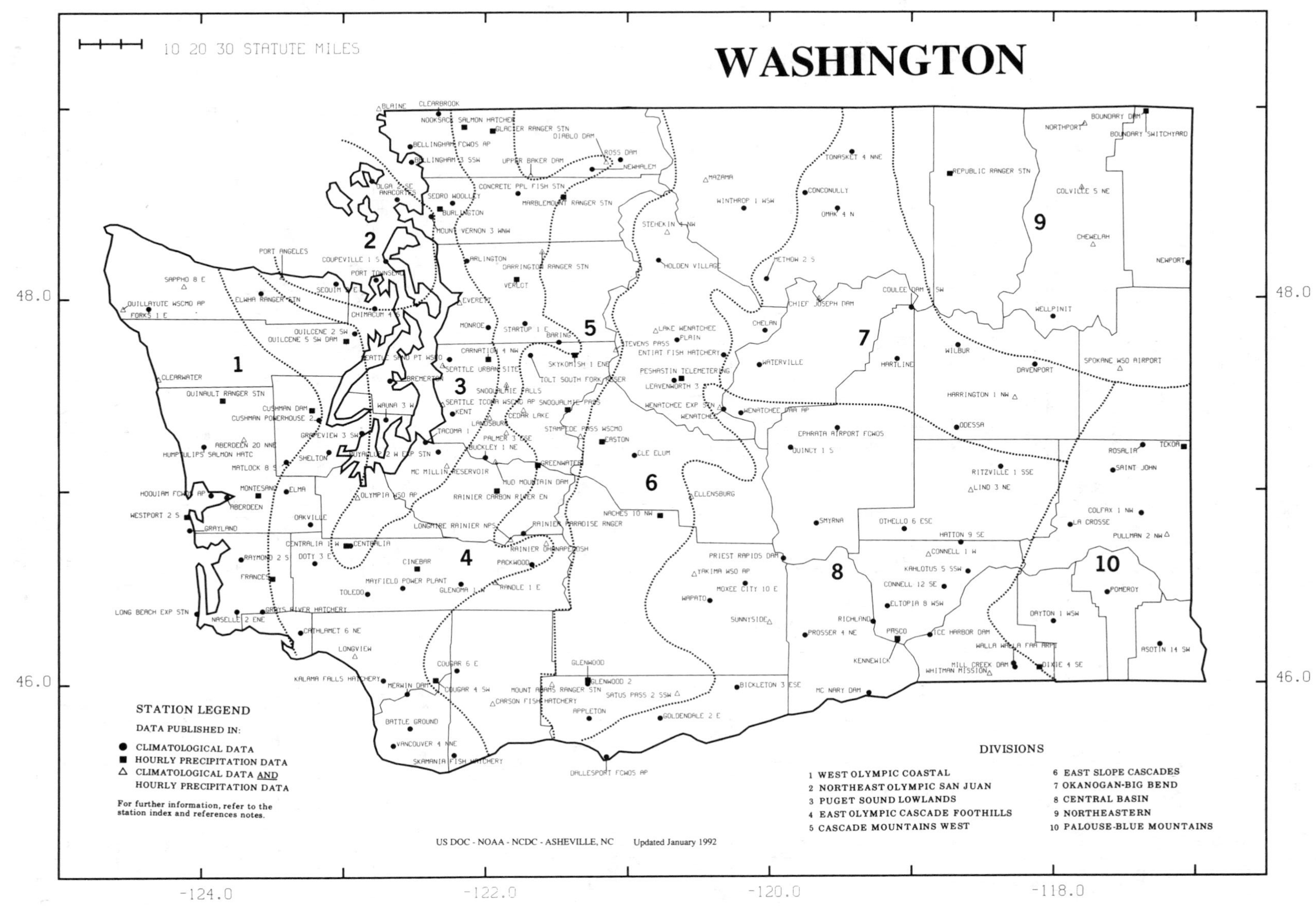
WASHINGTON
10 20 30 STATUTE MILES
STATION LEGEND
DATA PUBLISHED IN:
● CLIMATOLOGICAL DATA
■ HOURLY PRECIPITATION DATA
△ CLIMATOLOGICAL DATA AND HOURLY PRECIPITATION DATA
For further information, refer to the station index and references notes.
DIVISIONS
1 WEST OLYMPIC COASTAL
2 NORTHEAST OLYMPIC SAN JUAN
3 PUGET SOUND LOWLANDS
4 EAST OLYMPIC CASCADE FOOTHILLS
5 CASCADE MOUNTAINS WEST
6 EAST SLOPE CASCADES
7 OKANOGAN-BIG BEND
8 CENTRAL BASIN
9 NORTHEASTERN
10 PALOUSE-BLUE MOUNTAINS
US DOC - NOAA - NCDC - ASHEVILLE, NC Updated January 1992

## ABERDEEN *Grays Harbor County* ELEVATION 10 ft LAT/LONG 46° 59 ' N / 123° 49 ' W

| | JAN | FEB | MAR | APR | MAY | JUN | JUL | AUG | SEP | OCT | NOV | DEC | YEAR |
|---|---|---|---|---|---|---|---|---|---|---|---|---|---|
| Maximum Temp °F | 46.2 | 49.6 | 53.4 | 56.2 | 61.2 | 64.6 | 68.1 | 69.3 | 68.8 | 61.0 | 51.9 | 46.3 | 58.1 |
| Minimum Temp °F | 35.0 | 35.8 | 37.5 | 40.3 | 45.4 | 49.8 | 52.7 | 53.2 | 49.8 | 44.0 | 39.3 | 35.7 | 43.2 |
| Mean Temp °F | 40.6 | 42.8 | 45.5 | 48.3 | 53.3 | 57.2 | 60.4 | 61.2 | 59.3 | 52.5 | 45.6 | 41.0 | 50.6 |
| Days Max Temp ≥ 90 °F | 0 | 0 | 0 | 0 | 0 | 0 | 0 | 1 | 0 | 0 | 0 | 0 | 1 |
| Days Max Temp ≤ 32 °F | 1 | 0 | 0 | 0 | 0 | 0 | 0 | 0 | 0 | 0 | 0 | 1 | 2 |
| Days Min Temp ≤ 32 °F | 11 | 9 | 6 | 2 | 0 | 0 | 0 | 0 | 0 | 1 | 4 | 10 | 43 |
| Days Min Temp ≤ 0 °F | 0 | 0 | 0 | 0 | 0 | 0 | 0 | 0 | 0 | 0 | 0 | 0 | 0 |
| Heating Degree Days | 749 | 622 | 599 | 496 | 361 | 233 | 146 | 124 | 177 | 381 | 575 | 736 | 5199 |
| Cooling Degree Days | 0 | 0 | 0 | 0 | 5 | 6 | 11 | 18 | 15 | 1 | 0 | 0 | 56 |
| Total Precipitation (") | 12.86 | 10.13 | 8.63 | 6.28 | 3.49 | 2.58 | 1.37 | 1.56 | 3.04 | 7.15 | 11.65 | 13.00 | 81.74 |
| Days ≥ 0.1" Precip | 17 | 15 | 15 | 12 | 9 | 6 | 3 | 4 | 6 | 11 | 18 | 18 | 134 |
| Total Snowfall (") | 2.9 | 0.4 | 0.1 | 0.0 | 0.0 | 0.0 | 0.0 | 0.0 | 0.0 | 0.0 | 0.0 | *0.7* | 4.1 |
| Days ≥ 1" Snow Depth | *1* | 1 | 0 | 0 | 0 | 0 | 0 | 0 | 0 | 0 | 0 | *0* | 2 |

## ABERDEEN 20 NNE *Grays Harbor County* ELEVATION 440 ft LAT/LONG 47° 16 ' N / 123° 42 ' W

| | JAN | FEB | MAR | APR | MAY | JUN | JUL | AUG | SEP | OCT | NOV | DEC | YEAR |
|---|---|---|---|---|---|---|---|---|---|---|---|---|---|
| Maximum Temp °F | 42.5 | 46.5 | 51.3 | 55.3 | 62.6 | 66.9 | 71.8 | 72.4 | 67.5 | 58.1 | 48.1 | 42.3 | 57.1 |
| Minimum Temp °F | 32.2 | 32.9 | 34.2 | 36.7 | 41.4 | 46.2 | 49.0 | 50.0 | 46.2 | 40.5 | 36.3 | 32.9 | 39.9 |
| Mean Temp °F | 37.4 | 39.8 | 42.8 | 46.0 | 52.0 | 56.6 | 60.4 | 61.2 | 56.9 | 49.3 | 42.2 | 37.6 | 48.5 |
| Days Max Temp ≥ 90 °F | 0 | 0 | 0 | 0 | 0 | 0 | 1 | 1 | 0 | 0 | 0 | 0 | 2 |
| Days Max Temp ≤ 32 °F | 1 | 0 | 0 | 0 | 0 | 0 | 0 | 0 | 0 | 0 | 0 | 1 | 2 |
| Days Min Temp ≤ 32 °F | 18 | 15 | 14 | 8 | 1 | 0 | 0 | 0 | 0 | 3 | 10 | 16 | 85 |
| Days Min Temp ≤ 0 °F | 0 | 0 | 0 | 0 | 0 | 0 | 0 | 0 | 0 | 0 | 0 | 0 | 0 |
| Heating Degree Days | 849 | 706 | 682 | 563 | 398 | 254 | 155 | 132 | 239 | 479 | 677 | 842 | 5976 |
| Cooling Degree Days | 0 | 0 | 0 | 0 | 3 | 9 | 20 | 25 | 4 | 0 | 0 | 0 | 61 |
| Total Precipitation (") | 19.87 | 16.44 | 13.91 | 9.80 | 5.77 | 3.83 | 2.42 | 2.77 | 5.65 | 11.07 | 18.49 | 21.31 | 131.33 |
| Days ≥ 0.1" Precip | 19 | 16 | 17 | 15 | 10 | 8 | 4 | 5 | 7 | 12 | 19 | 19 | 151 |
| Total Snowfall (") | 8.6 | 4.7 | 2.2 | 0.4 | 0.0 | 0.0 | 0.0 | 0.0 | 0.0 | 0.0 | 1.2 | 3.7 | 20.8 |
| Days ≥ 1" Snow Depth | 8 | 5 | 3 | 0 | 0 | 0 | 0 | 0 | 0 | 0 | 1 | 5 | 22 |

## ANACORTES *Skagit County* ELEVATION 30 ft LAT/LONG 48° 31 ' N / 122° 37 ' W

| | JAN | FEB | MAR | APR | MAY | JUN | JUL | AUG | SEP | OCT | NOV | DEC | YEAR |
|---|---|---|---|---|---|---|---|---|---|---|---|---|---|
| Maximum Temp °F | 44.9 | 48.5 | 52.2 | 56.8 | 62.6 | 67.5 | 71.4 | 71.9 | 66.9 | 58.6 | 50.4 | 45.6 | 58.1 |
| Minimum Temp °F | 34.0 | 35.9 | 38.0 | 41.4 | 45.7 | 49.9 | 52.1 | 52.3 | 49.4 | 43.7 | 38.9 | 35.0 | 43.0 |
| Mean Temp °F | 39.5 | 42.2 | 45.1 | 49.1 | 54.2 | 58.7 | 61.8 | 62.1 | 58.2 | 51.2 | 44.7 | 40.3 | 50.6 |
| Days Max Temp ≥ 90 °F | 0 | 0 | 0 | 0 | 0 | 0 | 0 | 0 | 0 | 0 | 0 | 0 | 0 |
| Days Max Temp ≤ 32 °F | 1 | 0 | 0 | 0 | 0 | 0 | 0 | 0 | 0 | 0 | 0 | 1 | 2 |
| Days Min Temp ≤ 32 °F | 12 | 7 | 5 | 0 | 0 | 0 | 0 | 0 | 0 | 1 | 5 | 11 | 41 |
| Days Min Temp ≤ 0 °F | 0 | 0 | 0 | 0 | 0 | 0 | 0 | 0 | 0 | 0 | 0 | 0 | 0 |
| Heating Degree Days | 785 | 637 | 610 | 470 | 330 | 187 | 107 | 99 | 201 | 422 | 603 | 758 | 5209 |
| Cooling Degree Days | 0 | 0 | 0 | 0 | 1 | 8 | 16 | 22 | 3 | 0 | 0 | 0 | 50 |
| Total Precipitation (") | 3.76 | 2.40 | 2.02 | 1.87 | 1.54 | 1.34 | 1.01 | 0.99 | 1.45 | 2.20 | 3.78 | 3.75 | 26.11 |
| Days ≥ 0.1" Precip | 10 | 8 | 8 | 6 | 5 | 4 | 3 | 2 | 4 | 7 | 11 | 11 | 79 |
| Total Snowfall (") | 2.1 | 0.4 | 0.0 | 0.0 | 0.0 | 0.0 | 0.0 | 0.0 | 0.0 | 0.0 | 0.7 | *1.3* | 4.5 |
| Days ≥ 1" Snow Depth | *1* | 0 | 0 | 0 | 0 | 0 | 0 | 0 | 0 | 0 | 0 | *0* | 1 |

## APPLETON *Klickitat County* ELEVATION 2362 ft LAT/LONG 45° 49 ' N / 121° 16 ' W

| | JAN | FEB | MAR | APR | MAY | JUN | JUL | AUG | SEP | OCT | NOV | DEC | YEAR |
|---|---|---|---|---|---|---|---|---|---|---|---|---|---|
| Maximum Temp °F | 34.9 | 39.9 | 47.3 | 54.4 | 63.7 | 70.8 | 78.6 | 79.0 | 70.6 | 58.2 | 42.5 | 35.4 | 56.3 |
| Minimum Temp °F | 23.6 | 26.8 | 30.0 | 33.5 | 39.1 | 45.2 | 49.2 | 49.4 | 43.2 | 35.7 | 29.9 | 24.6 | 35.8 |
| Mean Temp °F | 29.3 | 33.4 | 38.7 | 44.0 | 51.4 | 58.0 | 63.9 | 64.3 | 57.0 | 47.0 | 36.2 | 30.0 | 46.1 |
| Days Max Temp ≥ 90 °F | 0 | 0 | 0 | 0 | 0 | 1 | 5 | 5 | 1 | 0 | 0 | 0 | 12 |
| Days Max Temp ≤ 32 °F | 10 | 4 | 0 | 0 | 0 | 0 | 0 | 0 | 0 | 0 | 3 | 9 | 26 |
| Days Min Temp ≤ 32 °F | 26 | 22 | 21 | 14 | 4 | 0 | 0 | 0 | 2 | 10 | 19 | 26 | 144 |
| Days Min Temp ≤ 0 °F | 1 | 1 | 0 | 0 | 0 | 0 | 0 | 0 | 0 | 0 | 0 | 1 | 3 |
| Heating Degree Days | 1099 | 886 | 808 | 625 | 419 | 227 | 106 | 98 | 250 | 553 | 857 | 1077 | 7005 |
| Cooling Degree Days | 0 | 0 | 0 | 0 | 7 | 27 | 76 | 85 | 22 | 0 | 0 | 0 | 217 |
| Total Precipitation (") | 6.02 | 3.86 | 3.09 | 1.80 | 1.03 | 0.84 | 0.32 | 0.56 | 1.03 | 2.33 | 4.90 | 5.98 | 31.76 |
| Days ≥ 0.1" Precip | 11 | 9 | 8 | 6 | 3 | 2 | 1 | 1 | 3 | 5 | 11 | 12 | 72 |
| Total Snowfall (") | 25.8 | 13.1 | 7.9 | 1.2 | 0.1 | 0.0 | 0.0 | 0.0 | 0.0 | 0.4 | 9.3 | 25.6 | 83.4 |
| Days ≥ 1" Snow Depth | 22 | 16 | 8 | 1 | 0 | 0 | 0 | 0 | 0 | 0 | 5 | 17 | 69 |

## BATTLE GROUND *Clark County* ELEVATION 302 ft LAT/LONG 45° 46 ' N / 122° 32 ' W

| | JAN | FEB | MAR | APR | MAY | JUN | JUL | AUG | SEP | OCT | NOV | DEC | YEAR |
|---|---|---|---|---|---|---|---|---|---|---|---|---|---|
| Maximum Temp °F | 45.4 | 50.6 | 55.4 | 59.3 | 66.1 | 71.6 | 77.6 | 78.8 | 74.2 | 63.8 | 51.9 | 45.1 | 61.7 |
| Minimum Temp °F | 31.5 | 33.6 | 35.9 | 38.7 | 43.1 | 48.0 | 50.2 | 50.0 | 46.3 | 40.1 | 36.4 | 32.8 | 40.6 |
| Mean Temp °F | 38.5 | 42.2 | 45.7 | 49.0 | 54.6 | 59.8 | 63.8 | 64.4 | 60.2 | 52.0 | 44.2 | 39.0 | 51.1 |
| Days Max Temp ≥ 90 °F | 0 | 0 | 0 | 0 | 0 | 1 | 3 | 4 | 2 | 0 | 0 | 0 | 10 |
| Days Max Temp ≤ 32 °F | 1 | 0 | 0 | 0 | 0 | 0 | 0 | 0 | 0 | 0 | 0 | 1 | 2 |
| Days Min Temp ≤ 32 °F | 17 | 12 | 10 | 5 | 1 | 0 | 0 | 0 | 0 | 4 | 9 | 14 | 72 |
| Days Min Temp ≤ 0 °F | 0 | 0 | 0 | 0 | 0 | 0 | 0 | 0 | 0 | 0 | 0 | 0 | 0 |
| Heating Degree Days | 815 | 636 | 593 | 473 | 322 | 172 | 84 | 74 | 159 | 397 | 617 | 799 | 5141 |
| Cooling Degree Days | 0 | 0 | 0 | 0 | 9 | 20 | 54 | 66 | 21 | 1 | 0 | 0 | 171 |
| Total Precipitation (") | 6.98 | 5.33 | 5.16 | 4.15 | 2.94 | 2.40 | 0.95 | 1.37 | 2.43 | 4.11 | 7.45 | 8.34 | 51.61 |
| Days ≥ 0.1" Precip | *14* | 12 | 13 | 11 | 7 | 6 | 2 | 2 | 5 | 9 | 15 | *15* | 111 |
| Total Snowfall (") | 2.9 | 1.5 | 0.2 | 0.0 | 0.0 | 0.0 | 0.0 | 0.0 | 0.0 | 0.0 | *0.0* | *2.0* | 6.6 |
| Days ≥ 1" Snow Depth | *1* | 0 | 0 | 0 | 0 | 0 | 0 | 0 | 0 | 0 | 0 | *1* | 2 |

## BELLINGHAM INTL AP *Whatcom County* ELEVATION 154 ft LAT/LONG 48° 48 ' N / 122° 32 ' W

| | JAN | FEB | MAR | APR | MAY | JUN | JUL | AUG | SEP | OCT | NOV | DEC | YEAR |
|---|---|---|---|---|---|---|---|---|---|---|---|---|---|
| Maximum Temp °F | 43.4 | 47.8 | 51.7 | 56.2 | 62.1 | 66.6 | 70.9 | 71.4 | 66.7 | 58.2 | 49.5 | 44.0 | 57.4 |
| Minimum Temp °F | 31.9 | 34.4 | 36.7 | 40.4 | 45.8 | 50.9 | 53.7 | 53.9 | 48.7 | 42.3 | 37.2 | 33.0 | 42.4 |
| Mean Temp °F | 37.7 | 41.1 | 44.2 | 48.3 | 54.0 | 58.8 | 62.3 | 62.7 | 57.7 | 50.3 | 43.4 | 38.5 | 49.9 |
| Days Max Temp ≥ 90 °F | 0 | 0 | 0 | 0 | 0 | 0 | 0 | 0 | 0 | 0 | 0 | 0 | 0 |
| Days Max Temp ≤ 32 °F | 3 | 0 | 0 | 0 | 0 | 0 | 0 | 0 | 0 | 0 | 1 | 2 | 6 |
| Days Min Temp ≤ 32 °F | 16 | 11 | 8 | 2 | 0 | 0 | 0 | 0 | 0 | 2 | 8 | 14 | 61 |
| Days Min Temp ≤ 0 °F | 0 | 0 | 0 | 0 | 0 | 0 | 0 | 0 | 0 | 0 | 0 | 0 | 0 |
| Heating Degree Days | 839 | 668 | 638 | 493 | 335 | 186 | 96 | 88 | 214 | 449 | 642 | 814 | 5462 |
| Cooling Degree Days | 0 | 0 | 0 | 0 | 1 | 7 | 21 | 26 | 3 | 0 | 0 | 0 | 58 |
| Total Precipitation (") | 4.81 | 3.53 | 2.88 | 2.79 | 2.22 | 1.87 | 1.29 | 1.33 | 1.81 | 3.24 | 5.00 | 4.78 | 35.55 |
| Days ≥ 0.1" Precip | 11 | 9 | 9 | 8 | 6 | 5 | 3 | 3 | 5 | 8 | 12 | 12 | 91 |
| Total Snowfall (") | 5.2 | 2.0 | 0.9 | 0.2 | 0.0 | 0.0 | 0.0 | 0.0 | 0.0 | 0.1 | 1.2 | 4.5 | 14.1 |
| Days ≥ 1" Snow Depth | 4 | 1 | 0 | 0 | 0 | 0 | 0 | 0 | 0 | 0 | 0 | 2 | 7 |

## BICKLETON 3 ESE *Klickitat County* ELEVATION 3002 ft LAT/LONG 46° 0 ' N / 120° 18 ' W

| | JAN | FEB | MAR | APR | MAY | JUN | JUL | AUG | SEP | OCT | NOV | DEC | YEAR |
|---|---|---|---|---|---|---|---|---|---|---|---|---|---|
| Maximum Temp °F | 35.8 | 40.3 | 47.8 | 54.9 | 64.1 | 72.5 | 80.4 | 81.0 | 71.3 | 59.9 | 43.9 | 36.2 | 57.3 |
| Minimum Temp °F | 22.7 | 26.2 | 30.4 | 34.1 | 40.2 | 46.6 | 52.3 | 53.1 | 46.3 | 37.9 | 29.0 | 22.9 | 36.8 |
| Mean Temp °F | 29.3 | 33.2 | 39.1 | 44.5 | 52.1 | 59.6 | 66.4 | 67.1 | 58.8 | 48.9 | 36.5 | 29.6 | 47.1 |
| Days Max Temp ≥ 90 °F | 0 | 0 | 0 | 0 | 0 | 1 | 5 | 5 | 1 | 0 | 0 | 0 | 12 |
| Days Max Temp ≤ 32 °F | 9 | 4 | 1 | 0 | 0 | 0 | 0 | 0 | 0 | 0 | 3 | 9 | 26 |
| Days Min Temp ≤ 32 °F | 26 | *23* | 19 | 13 | 5 | 1 | 0 | 0 | 1 | 7 | 19 | 27 | 141 |
| Days Min Temp ≤ 0 °F | 1 | 0 | 0 | 0 | 0 | 0 | 0 | 0 | 0 | 0 | 0 | 1 | 2 |
| Heating Degree Days | 1101 | 892 | 795 | 608 | 400 | 199 | 74 | 62 | 211 | 496 | 849 | 1091 | 6778 |
| Cooling Degree Days | 0 | 0 | 0 | 1 | 12 | 43 | 109 | 126 | 39 | 4 | 0 | 0 | 334 |
| Total Precipitation (") | 2.19 | 1.42 | 1.12 | 0.75 | 0.76 | 0.70 | 0.37 | 0.45 | 0.53 | 0.78 | 2.09 | 2.29 | 13.45 |
| Days ≥ 0.1" Precip | 7 | 5 | 4 | 3 | 2 | 2 | 1 | 1 | 2 | 2 | 6 | 7 | 42 |
| Total Snowfall (") | 9.4 | 4.8 | 2.1 | 0.8 | 0.1 | 0.0 | 0.0 | 0.0 | 0.0 | 0.2 | 3.9 | 10.6 | 31.9 |
| Days ≥ 1" Snow Depth | *16* | *8* | *2* | *0* | 0 | *0* | *0* | *0* | *0* | *0* | *4* | *13* | 43 |

## BLAINE *Whatcom County* ELEVATION 39 ft LAT/LONG 48° 59 ' N / 122° 45 ' W

| | JAN | FEB | MAR | APR | MAY | JUN | JUL | AUG | SEP | OCT | NOV | DEC | YEAR |
|---|---|---|---|---|---|---|---|---|---|---|---|---|---|
| Maximum Temp °F | 42.3 | 47.3 | 51.8 | 57.0 | 63.4 | 68.1 | 72.0 | 71.7 | 66.5 | 57.3 | 48.4 | 42.8 | 57.4 |
| Minimum Temp °F | 31.3 | 33.6 | 35.6 | 39.4 | 44.4 | 49.4 | 51.7 | 51.6 | 47.3 | 41.4 | 36.4 | 32.4 | 41.2 |
| Mean Temp °F | 36.8 | 40.5 | 43.7 | 48.2 | 53.9 | 58.8 | 61.8 | 61.7 | 56.9 | 49.4 | 42.4 | 37.6 | 49.3 |
| Days Max Temp ≥ 90 °F | 0 | 0 | 0 | 0 | 0 | 0 | 0 | 0 | 0 | 0 | 0 | 0 | 0 |
| Days Max Temp ≤ 32 °F | 2 | 1 | 0 | 0 | 0 | 0 | 0 | 0 | 0 | 0 | 1 | 3 | 7 |
| Days Min Temp ≤ 32 °F | 16 | 12 | 10 | 4 | 0 | 0 | 0 | 0 | 0 | 3 | 9 | 15 | 69 |
| Days Min Temp ≤ 0 °F | 0 | 0 | 0 | 0 | 0 | 0 | 0 | 0 | 0 | 0 | 0 | 0 | 0 |
| Heating Degree Days | 867 | 687 | 653 | 498 | 336 | 182 | 100 | 106 | 236 | 476 | 671 | 842 | 5654 |
| Cooling Degree Days | 0 | 0 | 0 | 0 | 1 | 3 | 10 | 12 | 1 | 0 | 0 | 0 | 27 |
| Total Precipitation (") | 5.43 | 4.22 | 3.49 | 2.91 | 2.27 | 1.96 | 1.36 | 1.50 | 2.08 | 3.83 | 5.95 | 5.76 | 40.76 |
| Days ≥ 0.1" Precip | 13 | 10 | 10 | 9 | 7 | 5 | 4 | 4 | 5 | 9 | 13 | 13 | 102 |
| Total Snowfall (") | 5.3 | 2.4 | 0.6 | 0.0 | 0.0 | 0.0 | 0.0 | 0.0 | 0.0 | 0.1 | 1.0 | 4.8 | 14.2 |
| Days ≥ 1" Snow Depth | 4 | 1 | 0 | 0 | 0 | 0 | 0 | 0 | 0 | 0 | 1 | 3 | 9 |

## BOUNDARY DAM *Pend Oreille County* ELEVATION 1831 ft LAT/LONG 48° 59 ' N / 117° 21 ' W

| | JAN | FEB | MAR | APR | MAY | JUN | JUL | AUG | SEP | OCT | NOV | DEC | YEAR |
|---|---|---|---|---|---|---|---|---|---|---|---|---|---|
| Maximum Temp °F | 31.3 | ***37.8*** | ***47.5*** | 57.2 | 65.1 | 71.3 | 80.0 | 81.8 | 71.6 | 56.3 | 40.1 | ***31.8*** | 56.0 |
| Minimum Temp °F | 17.2 | ***21.0*** | 26.3 | 33.0 | 40.4 | 47.2 | 50.5 | 49.0 | 41.9 | 34.1 | 27.5 | ***19.7*** | 34.0 |
| Mean Temp °F | 24.3 | ***29.3*** | ***37.0*** | 45.2 | 52.8 | 59.2 | 65.2 | 65.4 | 56.8 | 45.2 | 33.8 | ***25.8*** | 45.0 |
| Days Max Temp ≥ 90 °F | 0 | ***0*** | 0 | 0 | 0 | 1 | 6 | 7 | 1 | 0 | 0 | 0 | 15 |
| Days Max Temp ≤ 32 °F | 15 | ***5*** | 0 | 0 | 0 | 0 | 0 | 0 | 0 | 0 | 3 | ***15*** | 38 |
| Days Min Temp ≤ 32 °F | 30 | ***27*** | 27 | ***15*** | 2 | 0 | 0 | 0 | 2 | 12 | 23 | ***29*** | 167 |
| Days Min Temp ≤ 0 °F | 3 | ***2*** | 0 | 0 | 0 | 0 | 0 | 0 | 0 | 0 | 0 | 2 | 7 |
| Heating Degree Days | 1256 | ***1002*** | ***862*** | 589 | 377 | 187 | 70 | 65 | 248 | 605 | 929 | ***1208*** | 7398 |
| Cooling Degree Days | 0 | 0 | 0 | 0 | 7 | 27 | 86 | 80 | 7 | 0 | 0 | 0 | 207 |
| Total Precipitation (") | 2.75 | 2.09 | 1.96 | 2.03 | 2.48 | 2.66 | 1.63 | 1.66 | 1.39 | 1.74 | 3.06 | ***3.26*** | 26.71 |
| Days ≥ 0.1" Precip | 9 | 7 | ***6*** | 7 | 8 | 7 | 5 | 5 | 4 | ***6*** | 9 | ***9*** | 82 |
| Total Snowfall (") | ***21.6*** | ***10.2*** | ***2.4*** | 0.1 | 0.0 | 0.0 | 0.0 | 0.0 | 0.0 | 0.0 | ***6.5*** | ***20.1*** | 60.9 |
| Days ≥ 1" Snow Depth | ***30*** | ***27*** | ***16*** | 1 | 0 | 0 | 0 | 0 | 0 | 0 | ***6*** | na | na |

## BREMERTON *Kitsap County* ELEVATION 161 ft LAT/LONG 47° 34 ' N / 122° 40 ' W

| | JAN | FEB | MAR | APR | MAY | JUN | JUL | AUG | SEP | OCT | NOV | DEC | YEAR |
|---|---|---|---|---|---|---|---|---|---|---|---|---|---|
| Maximum Temp °F | 44.5 | 48.9 | 53.6 | 58.6 | 65.1 | 70.2 | 74.8 | 75.6 | 70.2 | 60.5 | 50.2 | 44.6 | 59.7 |
| Minimum Temp °F | 33.8 | 35.0 | 37.2 | 40.5 | 45.6 | 50.3 | 53.4 | 54.0 | 50.1 | 43.9 | 38.5 | 34.5 | 43.1 |
| Mean Temp °F | 39.2 | 42.0 | 45.4 | 49.6 | 55.4 | 60.3 | 64.1 | 64.8 | 60.2 | 52.2 | 44.3 | 39.6 | 51.4 |
| Days Max Temp ≥ 90 °F | 0 | 0 | 0 | 0 | 0 | 0 | 1 | 1 | 0 | 0 | 0 | 0 | 2 |
| Days Max Temp ≤ 32 °F | 1 | 0 | 0 | 0 | 0 | 0 | 0 | 0 | 0 | 0 | 0 | 1 | 2 |
| Days Min Temp ≤ 32 °F | 13 | 9 | 6 | 1 | 0 | 0 | 0 | 0 | 0 | 0 | 4 | 12 | 45 |
| Days Min Temp ≤ 0 °F | 0 | 0 | 0 | 0 | 0 | 0 | 0 | 0 | 0 | 0 | 0 | 0 | 0 |
| Heating Degree Days | 792 | 642 | 600 | 457 | 297 | 156 | 70 | 57 | 154 | 389 | 613 | 781 | 5008 |
| Cooling Degree Days | 0 | 0 | 0 | 0 | 7 | 22 | 45 | 65 | 19 | 1 | 0 | 0 | 159 |
| Total Precipitation (") | 7.44 | 6.26 | 5.65 | 3.25 | 2.09 | 1.48 | 0.88 | 1.02 | 2.02 | 4.05 | 7.83 | 9.13 | 51.10 |
| Days ≥ 0.1" Precip | 13 | 11 | 11 | 8 | 6 | 4 | 3 | 3 | 5 | 8 | 14 | 14 | 100 |
| Total Snowfall (") | ***3.2*** | 1.1 | 0.5 | 0.0 | 0.0 | 0.0 | 0.0 | 0.0 | 0.0 | 0.0 | 1.0 | 2.3 | 8.1 |
| Days ≥ 1" Snow Depth | na | ***0*** | ***0*** | 0 | 0 | 0 | 0 | 0 | 0 | 0 | 1 | ***1*** | na |

## BUCKLEY 1 NE *Pierce County* ELEVATION 689 ft LAT/LONG 47° 10 ' N / 122° 0 ' W

| | JAN | FEB | MAR | APR | MAY | JUN | JUL | AUG | SEP | OCT | NOV | DEC | YEAR |
|---|---|---|---|---|---|---|---|---|---|---|---|---|---|
| Maximum Temp °F | 45.0 | 49.7 | 54.1 | 59.0 | 66.0 | 71.2 | 76.6 | 77.6 | 71.4 | 60.6 | 50.0 | 44.6 | 60.5 |
| Minimum Temp °F | 32.5 | 34.4 | 35.8 | 38.4 | 43.2 | 47.9 | 50.3 | 50.5 | 46.8 | 41.3 | 36.7 | 32.6 | 40.9 |
| Mean Temp °F | 38.8 | 42.0 | 45.0 | 48.7 | 54.6 | 59.6 | 63.5 | 64.0 | 59.1 | 51.0 | 43.4 | 38.6 | 50.7 |
| Days Max Temp ≥ 90 °F | 0 | 0 | 0 | 0 | 0 | 1 | 2 | 2 | 1 | 0 | 0 | 0 | 6 |
| Days Max Temp ≤ 32 °F | 1 | 0 | 0 | 0 | 0 | 0 | 0 | 0 | 0 | 0 | 0 | 1 | 2 |
| Days Min Temp ≤ 32 °F | 15 | 11 | 9 | 4 | 0 | 0 | 0 | 0 | 0 | 2 | 7 | 15 | 63 |
| Days Min Temp ≤ 0 °F | 0 | 0 | 0 | 0 | 0 | 0 | 0 | 0 | 0 | 0 | 0 | 0 | 0 |
| Heating Degree Days | 805 | 642 | 614 | 483 | 321 | 177 | 86 | 74 | 181 | 428 | 642 | 811 | 5264 |
| Cooling Degree Days | 0 | 0 | 0 | 1 | 6 | 20 | 45 | 54 | 15 | 1 | 0 | 0 | 142 |
| Total Precipitation (") | 6.19 | 4.61 | 4.40 | 4.27 | 3.16 | 2.84 | 1.50 | 1.88 | 2.64 | 3.91 | 6.45 | 6.02 | 47.87 |
| Days ≥ 0.1" Precip | 13 | 11 | 12 | 12 | 9 | 7 | 4 | 4 | 7 | 10 | 14 | 14 | 117 |
| Total Snowfall (") | 4.6 | 1.9 | 1.0 | 0.1 | 0.0 | 0.0 | 0.0 | 0.0 | 0.0 | 0.1 | 0.8 | 3.1 | 11.6 |
| Days ≥ 1" Snow Depth | 2 | 1 | 0 | 0 | 0 | 0 | 0 | 0 | 0 | 0 | 0 | 2 | 5 |

## CEDAR LAKE *King County* ELEVATION 1562 ft LAT/LONG 47° 25 ' N / 121° 44 ' W

| | JAN | FEB | MAR | APR | MAY | JUN | JUL | AUG | SEP | OCT | NOV | DEC | YEAR |
|---|---|---|---|---|---|---|---|---|---|---|---|---|---|
| Maximum Temp °F | 40.2 | 44.6 | 48.7 | 53.3 | 60.4 | 65.7 | 71.6 | 72.7 | 66.8 | 57.8 | 45.9 | 40.2 | 55.7 |
| Minimum Temp °F | 30.3 | 32.0 | 33.2 | 35.9 | 41.2 | 46.0 | 49.5 | 50.3 | 46.3 | 40.3 | 34.9 | 30.7 | 39.2 |
| Mean Temp °F | 35.3 | 38.3 | 41.0 | 44.6 | 50.8 | 55.9 | 60.6 | 61.5 | 56.6 | 49.1 | 40.4 | 35.4 | 47.5 |
| Days Max Temp ≥ 90 °F | 0 | 0 | 0 | 0 | 0 | 0 | 1 | 1 | 0 | 0 | 0 | 0 | 2 |
| Days Max Temp ≤ 32 °F | 4 | 1 | 0 | 0 | 0 | 0 | 0 | 0 | 0 | 0 | 1 | 3 | 9 |
| Days Min Temp ≤ 32 °F | 19 | 15 | 15 | 8 | 1 | 0 | 0 | 0 | 0 | 2 | 11 | 20 | 91 |
| Days Min Temp ≤ 0 °F | 0 | 0 | 0 | 0 | 0 | 0 | 0 | 0 | 0 | 0 | 0 | 0 | 0 |
| Heating Degree Days | 914 | 747 | 738 | 605 | 437 | 278 | 161 | 139 | 257 | 489 | 731 | 909 | 6405 |
| Cooling Degree Days | 0 | 0 | 0 | 0 | 4 | 11 | 32 | 39 | 10 | 1 | 0 | 0 | 97 |
| Total Precipitation (") | 13.67 | 9.95 | 9.75 | 8.46 | 6.02 | 5.21 | 2.73 | 2.65 | 5.09 | 7.82 | 13.53 | 13.00 | 97.88 |
| Days ≥ 0.1" Precip | 17 | 14 | 16 | 15 | 12 | 9 | 6 | 5 | 8 | 12 | 17 | 17 | 148 |
| Total Snowfall (") | 14.1 | 12.5 | 8.7 | 3.8 | 0.3 | 0.0 | 0.0 | 0.0 | 0.0 | 0.3 | 5.5 | 14.2 | 59.4 |
| Days ≥ 1" Snow Depth | 15 | 10 | 8 | 2 | 0 | 0 | 0 | 0 | 0 | 0 | 4 | 13 | 52 |

## CENTRALIA *Lewis County* ELEVATION 190 ft LAT/LONG 46° 43' N / 122° 57' W

| | JAN | FEB | MAR | APR | MAY | JUN | JUL | AUG | SEP | OCT | NOV | DEC | YEAR |
|---|---|---|---|---|---|---|---|---|---|---|---|---|---|
| Maximum Temp °F | 45.6 | 50.4 | 55.4 | 60.6 | 67.5 | 72.8 | 77.9 | 78.5 | 72.9 | 62.1 | 51.2 | 45.3 | 61.7 |
| Minimum Temp °F | 34.0 | 35.1 | 36.7 | 39.7 | 44.7 | 49.5 | 52.6 | 53.0 | 48.9 | 42.9 | 38.6 | 34.7 | 42.5 |
| Mean Temp °F | 39.9 | 42.8 | 46.1 | 50.2 | 56.1 | 61.2 | 65.3 | 65.8 | 60.9 | 52.5 | 44.9 | 40.0 | 52.1 |
| Days Max Temp ≥ 90 °F | 0 | 0 | 0 | 0 | 0 | 1 | 3 | 3 | 1 | 0 | 0 | 0 | 8 |
| Days Max Temp ≤ 32 °F | 1 | 0 | 0 | 0 | 0 | 0 | 0 | 0 | 0 | 0 | 0 | 1 | 2 |
| Days Min Temp ≤ 32 °F | 12 | 10 | 9 | 4 | 0 | 0 | 0 | 0 | 0 | 2 | 6 | 11 | 54 |
| Days Min Temp ≤ 0 °F | 0 | 0 | 0 | 0 | 0 | 0 | 0 | 0 | 0 | 0 | 0 | 0 | 0 |
| Heating Degree Days | 773 | 620 | 579 | 438 | 276 | 135 | 52 | 45 | 135 | 379 | 595 | 767 | 4794 |
| Cooling Degree Days | 0 | 0 | 0 | 0 | 11 | 28 | 69 | 88 | 24 | 1 | 0 | 0 | 221 |
| Total Precipitation (") | 6.81 | 5.02 | 4.50 | 3.35 | 2.27 | 1.90 | 0.76 | 1.17 | 2.04 | 3.94 | 6.63 | 7.28 | 45.67 |
| Days ≥ 0.1" Precip | 14 | 12 | 12 | 9 | 7 | 5 | 2 | 3 | 6 | 9 | 14 | 14 | 107 |
| Total Snowfall (") | 2.5 | 0.8 | 0.4 | 0.0 | 0.0 | 0.0 | 0.0 | 0.0 | 0.0 | 0.0 | 0.2 | 1.1 | 5.0 |
| Days ≥ 1" Snow Depth | 1 | 0 | 0 | 0 | 0 | 0 | 0 | 0 | 0 | 0 | 0 | *0* | 1 |

## CHELAN *Chelan County* ELEVATION 1120 ft LAT/LONG 47° 50' N / 120° 2' W

| | JAN | FEB | MAR | APR | MAY | JUN | JUL | AUG | SEP | OCT | NOV | DEC | YEAR |
|---|---|---|---|---|---|---|---|---|---|---|---|---|---|
| Maximum Temp °F | 32.5 | 40.0 | 51.2 | 61.3 | 70.8 | 77.9 | 84.7 | 85.0 | 74.8 | 61.2 | 44.1 | 33.5 | 59.8 |
| Minimum Temp °F | 22.0 | 26.1 | 32.3 | 39.5 | 47.5 | 54.8 | 59.5 | 59.1 | 49.9 | 39.4 | 31.6 | 24.2 | 40.5 |
| Mean Temp °F | 27.3 | 33.1 | 41.8 | 50.4 | 59.2 | 66.4 | 72.1 | 72.1 | 62.4 | 50.3 | 37.9 | 28.9 | 50.2 |
| Days Max Temp ≥ 90 °F | 0 | 0 | 0 | 0 | 0 | 3 | 9 | 10 | 1 | 0 | 0 | 0 | 23 |
| Days Max Temp ≤ 32 °F | 15 | 5 | 0 | 0 | 0 | 0 | 0 | 0 | 0 | 0 | 2 | 13 | 35 |
| Days Min Temp ≤ 32 °F | 29 | 23 | 16 | 3 | 0 | 0 | 0 | 0 | 0 | 3 | 16 | 26 | 116 |
| Days Min Temp ≤ 0 °F | 1 | 0 | 0 | 0 | 0 | 0 | 0 | 0 | 0 | 0 | 0 | 1 | 2 |
| Heating Degree Days | 1163 | 895 | 713 | 432 | 194 | 59 | 13 | 11 | 117 | 448 | 808 | 1113 | 5966 |
| Cooling Degree Days | 0 | 0 | 0 | 0 | 24 | 102 | 226 | 233 | 49 | 0 | 0 | 0 | 634 |
| Total Precipitation (") | 1.40 | 1.00 | 0.82 | 0.78 | 0.64 | 0.75 | 0.34 | 0.50 | 0.48 | 0.57 | 1.51 | 1.75 | 10.54 |
| Days ≥ 0.1" Precip | 4 | 3 | 3 | 2 | 2 | 2 | 1 | 1 | 1 | 2 | 5 | 6 | 32 |
| Total Snowfall (") | 11.0 | 5.3 | 1.4 | 0.0 | 0.0 | 0.0 | 0.0 | 0.0 | 0.0 | 0.0 | 2.6 | 13.1 | 33.4 |
| Days ≥ 1" Snow Depth | 19 | 12 | 3 | 0 | 0 | 0 | 0 | 0 | 0 | 0 | 3 | *15* | 52 |

## CHEWELAH *Stevens County* ELEVATION 1670 ft LAT/LONG 48° 17' N / 117° 43' W

| | JAN | FEB | MAR | APR | MAY | JUN | JUL | AUG | SEP | OCT | NOV | DEC | YEAR |
|---|---|---|---|---|---|---|---|---|---|---|---|---|---|
| Maximum Temp °F | 33.1 | 40.5 | 51.0 | 60.4 | 70.0 | 77.2 | 85.0 | 86.0 | 75.7 | 61.3 | 42.5 | 33.7 | 59.7 |
| Minimum Temp °F | 17.5 | 21.9 | 27.2 | 32.9 | 39.5 | 45.5 | 48.2 | 46.9 | 38.8 | 30.4 | 26.8 | 20.0 | 33.0 |
| Mean Temp °F | 25.3 | 31.2 | 39.1 | 46.7 | 54.7 | 61.4 | 66.6 | 66.5 | 57.3 | 45.9 | 34.7 | 26.8 | 46.3 |
| Days Max Temp ≥ 90 °F | 0 | 0 | 0 | 0 | 1 | 4 | 12 | 12 | 3 | 0 | 0 | 0 | 32 |
| Days Max Temp ≤ 32 °F | 12 | 4 | 0 | 0 | 0 | 0 | 0 | 0 | 0 | 0 | 2 | 12 | 30 |
| Days Min Temp ≤ 32 °F | 29 | 25 | 24 | 15 | 5 | 0 | 0 | 0 | 5 | 20 | 23 | 28 | 174 |
| Days Min Temp ≤ 0 °F | 4 | 1 | 0 | 0 | 0 | 0 | 0 | 0 | 0 | 0 | 0 | 2 | 7 |
| Heating Degree Days | 1224 | 947 | 796 | 544 | 321 | 147 | 53 | 53 | 239 | 587 | 904 | 1176 | 6991 |
| Cooling Degree Days | 0 | 0 | 0 | 0 | 16 | 53 | 127 | 122 | 16 | 0 | 0 | 0 | 334 |
| Total Precipitation (") | 2.43 | 1.78 | 1.85 | 1.60 | 1.91 | 1.67 | 1.10 | 1.07 | 1.05 | 1.21 | 2.62 | 3.01 | 21.30 |
| Days ≥ 0.1" Precip | 7 | 5 | 6 | 5 | 5 | 5 | 3 | 3 | 3 | 4 | 7 | 8 | 61 |
| Total Snowfall (") | 14.4 | 5.9 | 2.0 | 0.1 | 0.0 | 0.0 | 0.0 | 0.0 | 0.0 | 0.1 | 5.1 | 14.0 | 41.6 |
| Days ≥ 1" Snow Depth | 25 | 17 | 5 | 0 | 0 | 0 | 0 | 0 | 0 | 0 | 4 | 18 | 69 |

## CHIEF JOSEPH DAM *Douglas County* ELEVATION 820 ft LAT/LONG 48° 0' N / 119° 39' W

| | JAN | FEB | MAR | APR | MAY | JUN | JUL | AUG | SEP | OCT | NOV | DEC | YEAR |
|---|---|---|---|---|---|---|---|---|---|---|---|---|---|
| Maximum Temp °F | 33.2 | 40.6 | 53.4 | 64.2 | 73.5 | 80.9 | 88.3 | 88.7 | 78.9 | 63.7 | 45.5 | 34.2 | 62.1 |
| Minimum Temp °F | 19.8 | 24.8 | 31.8 | 38.2 | 45.7 | 52.8 | 57.9 | 57.4 | 48.0 | 37.3 | 30.2 | 22.0 | 38.8 |
| Mean Temp °F | 26.5 | 32.7 | 42.6 | 51.2 | 59.6 | 66.9 | 73.1 | 73.1 | 63.5 | 50.5 | 37.9 | 28.1 | 50.5 |
| Days Max Temp ≥ 90 °F | 0 | 0 | 0 | 0 | 1 | 6 | 15 | 15 | 4 | 0 | 0 | 0 | 41 |
| Days Max Temp ≤ 32 °F | 14 | 4 | 0 | 0 | 0 | 0 | 0 | 0 | 0 | 0 | 2 | 12 | 32 |
| Days Min Temp ≤ 32 °F | 28 | 23 | 17 | 6 | 0 | 0 | 0 | 0 | 0 | 7 | 17 | 26 | 124 |
| Days Min Temp ≤ 0 °F | 2 | 1 | 0 | 0 | 0 | 0 | 0 | 0 | 0 | 0 | 0 | 1 | 4 |
| Heating Degree Days | 1187 | 904 | 687 | 408 | 186 | 53 | 8 | 8 | 106 | 442 | 807 | 1136 | 5932 |
| Cooling Degree Days | 0 | 0 | 0 | 2 | 37 | 124 | 266 | 272 | 74 | 1 | 0 | 0 | 776 |
| Total Precipitation (") | 1.13 | 0.94 | 0.80 | 0.68 | 0.81 | 0.83 | 0.52 | 0.52 | 0.47 | 0.53 | 1.40 | 1.69 | 10.32 |
| Days ≥ 0.1" Precip | 4 | 3 | 3 | 2 | 2 | 2 | 1 | 2 | 1 | 2 | 5 | 5 | 32 |
| Total Snowfall (") | na | na | na | *0.0* | *0.0* | *0.0* | *0.0* | *0.0* | *0.0* | *0.2* | na | na | na |
| Days ≥ 1" Snow Depth | na | na | na | *0* | *0* | *0* | *0* | *0* | *0* | *0* | na | na | na |

## CLE ELUM *Kittitas County* ELEVATION 1903 ft LAT/LONG 47° 11 ' N / 120° 55 ' W

| | JAN | FEB | MAR | APR | MAY | JUN | JUL | AUG | SEP | OCT | NOV | DEC | YEAR |
|---|---|---|---|---|---|---|---|---|---|---|---|---|---|
| Maximum Temp °F | 35.0 | 41.2 | 49.7 | 56.9 | 65.8 | 72.5 | 79.6 | 80.1 | 72.5 | 60.1 | 43.2 | 35.2 | 57.7 |
| Minimum Temp °F | 20.6 | 24.3 | 28.3 | 33.3 | 39.9 | 47.0 | 51.4 | 50.7 | 41.5 | 32.9 | 28.5 | 22.1 | 35.0 |
| Mean Temp °F | 27.8 | 32.8 | 39.0 | 45.1 | 52.9 | 59.8 | 65.5 | 65.4 | 57.0 | 46.5 | 35.8 | 28.7 | 46.4 |
| Days Max Temp ≥ 90 °F | 0 | 0 | 0 | 0 | 0 | 2 | 6 | 6 | 1 | 0 | 0 | 0 | 15 |
| Days Max Temp ≤ 32 °F | 10 | 3 | 0 | 0 | 0 | 0 | 0 | 0 | 0 | 0 | 2 | 9 | 24 |
| Days Min Temp ≤ 32 °F | 28 | 24 | 23 | 14 | 3 | 0 | 0 | 0 | 4 | 15 | 20 | 27 | 158 |
| Days Min Temp ≤ 0 °F | 2 | 1 | 0 | 0 | 0 | 0 | 0 | 0 | 0 | 0 | 0 | 1 | 4 |
| Heating Degree Days | 1146 | 904 | 799 | 590 | 373 | 182 | 71 | 72 | 244 | 566 | 868 | 1119 | 6934 |
| Cooling Degree Days | 0 | 0 | 0 | 0 | 7 | 30 | 87 | 90 | 14 | 0 | 0 | 0 | 228 |
| Total Precipitation (") | 4.09 | 2.36 | 1.57 | 1.22 | 0.83 | 0.86 | 0.42 | 0.60 | 0.82 | 1.57 | 3.65 | 4.31 | 22.30 |
| Days ≥ 0.1" Precip | 10 | 7 | 5 | 4 | 3 | 2 | 1 | 2 | 3 | 4 | 10 | 11 | 62 |
| Total Snowfall (") | 23.4 | 13.2 | 5.3 | 0.8 | 0.2 | 0.0 | 0.0 | 0.0 | 0.0 | 0.3 | 10.5 | 23.3 | 77.0 |
| Days ≥ 1" Snow Depth | 26 | 16 | 7 | 0 | 0 | 0 | 0 | 0 | 0 | 0 | 8 | 22 | 79 |

## CLEARBROOK *Whatcom County* ELEVATION 59 ft LAT/LONG 48° 58 ' N / 122° 20 ' W

| | JAN | FEB | MAR | APR | MAY | JUN | JUL | AUG | SEP | OCT | NOV | DEC | YEAR |
|---|---|---|---|---|---|---|---|---|---|---|---|---|---|
| Maximum Temp °F | 41.6 | 47.1 | 53.0 | 58.7 | 65.5 | 70.1 | 75.0 | 75.5 | 69.8 | 59.3 | 48.3 | 42.3 | 58.8 |
| Minimum Temp °F | 30.8 | 33.5 | 35.5 | 39.0 | 44.0 | 48.6 | 50.5 | 50.1 | 46.1 | 41.1 | 36.2 | 32.0 | 40.6 |
| Mean Temp °F | 36.2 | 40.3 | 44.3 | 48.9 | 54.8 | 59.3 | 62.8 | 62.8 | 58.0 | 50.2 | 42.3 | 37.2 | 49.8 |
| Days Max Temp ≥ 90 °F | 0 | 0 | 0 | 0 | 0 | 0 | 1 | 1 | 0 | 0 | 0 | 0 | 2 |
| Days Max Temp ≤ 32 °F | 4 | 1 | 0 | 0 | 0 | 0 | 0 | 0 | 0 | 0 | 1 | 3 | 9 |
| Days Min Temp ≤ 32 °F | 17 | 11 | 9 | 4 | 1 | 0 | 0 | 0 | 0 | 4 | 8 | 15 | 69 |
| Days Min Temp ≤ 0 °F | 0 | 0 | 0 | 0 | 0 | 0 | 0 | 0 | 0 | 0 | 0 | 0 | 0 |
| Heating Degree Days | 886 | 690 | 636 | 478 | 314 | 173 | 90 | 89 | 208 | 451 | 676 | 857 | 5548 |
| Cooling Degree Days | 0 | 0 | 0 | 0 | 5 | 11 | 30 | 34 | 6 | 0 | 0 | 0 | 86 |
| Total Precipitation (") | 5.29 | 4.24 | 3.75 | 3.63 | 2.96 | 2.60 | 1.73 | 1.79 | 2.74 | 4.33 | 5.74 | 5.63 | 44.43 |
| Days ≥ 0.1" Precip | 13 | 9 | 10 | 10 | 7 | 6 | 4 | 4 | 6 | 10 | 13 | 13 | 105 |
| Total Snowfall (") | 6.7 | 2.3 | 0.6 | 0.0 | 0.0 | 0.0 | 0.0 | 0.0 | 0.0 | 0.0 | 0.6 | 2.8 | 13.0 |
| Days ≥ 1" Snow Depth | na | ***0*** | 0 | 0 | 0 | 0 | 0 | 0 | 0 | 0 | 0 | na | na |

## CLEARWATER *Jefferson County* ELEVATION 80 ft LAT/LONG 47° 35 ' N / 124° 18 ' W

| | JAN | FEB | MAR | APR | MAY | JUN | JUL | AUG | SEP | OCT | NOV | DEC | YEAR |
|---|---|---|---|---|---|---|---|---|---|---|---|---|---|
| Maximum Temp °F | 46.0 | 50.3 | 53.6 | 57.2 | 62.6 | 65.9 | 70.0 | 71.2 | 69.1 | 61.0 | 51.2 | 45.4 | 58.6 |
| Minimum Temp °F | 34.0 | 34.8 | 35.8 | 38.3 | 42.2 | 46.5 | 49.0 | 49.0 | 46.2 | 41.7 | 37.3 | 34.5 | 40.8 |
| Mean Temp °F | 40.0 | 42.6 | 44.7 | 47.8 | 52.5 | 56.2 | 59.5 | 60.2 | 57.7 | 51.4 | 44.3 | 39.9 | 49.7 |
| Days Max Temp ≥ 90 °F | 0 | 0 | 0 | 0 | 0 | 0 | 0 | 1 | 0 | 0 | 0 | 0 | 1 |
| Days Max Temp ≤ 32 °F | 0 | 0 | 0 | 0 | 0 | 0 | 0 | 0 | 0 | 0 | 0 | 1 | 1 |
| Days Min Temp ≤ 32 °F | 14 | 12 | 11 | 6 | 1 | 0 | 0 | 0 | 0 | 4 | 8 | 14 | 70 |
| Days Min Temp ≤ 0 °F | 0 | 0 | 0 | 0 | 0 | 0 | 0 | 0 | 0 | 0 | 0 | 0 | 0 |
| Heating Degree Days | 768 | 626 | 622 | 511 | 383 | 260 | 170 | 152 | 217 | 416 | 616 | 773 | 5514 |
| Cooling Degree Days | 0 | 0 | 0 | 0 | 1 | 3 | 7 | 11 | 4 | 0 | 0 | 0 | 26 |
| Total Precipitation (") | 16.46 | 13.75 | 12.39 | 8.75 | 6.00 | 3.37 | 2.65 | 3.03 | 5.57 | 10.98 | 15.89 | 17.53 | 116.37 |
| Days ≥ 0.1" Precip | 18 | 16 | 16 | 14 | 10 | 8 | 4 | 5 | 7 | 13 | 19 | 19 | 149 |
| Total Snowfall (") | na | na | ***0.6*** | ***0.0*** | ***0.0*** | ***0.0*** | ***0.0*** | ***0.0*** | ***0.0*** | ***0.0*** | ***0.0*** | ***1.1*** | na |
| Days ≥ 1" Snow Depth | ***1*** | na | ***0*** | ***0*** | ***0*** | ***0*** | ***0*** | ***0*** | ***0*** | ***0*** | ***0*** | na | na |

## COLFAX *Whitman County* ELEVATION 1972 ft LAT/LONG 46° 53 ' N / 117° 22 ' W

| | JAN | FEB | MAR | APR | MAY | JUN | JUL | AUG | SEP | OCT | NOV | DEC | YEAR |
|---|---|---|---|---|---|---|---|---|---|---|---|---|---|
| Maximum Temp °F | 36.9 | 43.3 | 50.8 | 58.6 | 67.1 | 74.4 | 83.0 | 83.4 | 74.2 | 62.4 | 46.1 | 37.5 | 59.8 |
| Minimum Temp °F | 23.7 | 27.8 | 31.0 | 35.7 | 41.3 | 47.2 | 50.4 | 49.6 | 41.9 | 33.9 | 30.1 | 24.9 | 36.5 |
| Mean Temp °F | 30.3 | 35.6 | 40.9 | 47.2 | 54.2 | 60.8 | 66.7 | 66.5 | 58.1 | 48.1 | 38.1 | 31.3 | 48.2 |
| Days Max Temp ≥ 90 °F | 0 | 0 | 0 | 0 | 1 | 2 | 8 | 8 | 2 | 0 | 0 | 0 | 21 |
| Days Max Temp ≤ 32 °F | 9 | 3 | 0 | 0 | 0 | 0 | 0 | 0 | 0 | 0 | 2 | 8 | 22 |
| Days Min Temp ≤ 32 °F | 24 | 21 | 19 | 9 | 2 | 0 | 0 | 0 | 2 | 13 | 18 | 24 | 132 |
| Days Min Temp ≤ 0 °F | 2 | 0 | 0 | 0 | 0 | 0 | 0 | 0 | 0 | 0 | 0 | 1 | 3 |
| Heating Degree Days | 1069 | 824 | 739 | 531 | 337 | 159 | 54 | 56 | 221 | 517 | 799 | 1040 | 6346 |
| Cooling Degree Days | 0 | 0 | 0 | 2 | 13 | 37 | 101 | 103 | ***19*** | 1 | 0 | 0 | 276 |
| Total Precipitation (") | 2.50 | 1.73 | 1.92 | 1.66 | 1.64 | 1.32 | 0.74 | 0.76 | 0.80 | 1.07 | 2.31 | 2.75 | 19.20 |
| Days ≥ 0.1" Precip | 7 | 5 | 7 | 5 | 5 | 4 | 2 | 2 | 2 | 4 | 8 | 8 | 59 |
| Total Snowfall (") | 10.0 | 5.0 | 2.0 | 0.2 | 0.0 | 0.0 | 0.0 | 0.0 | 0.0 | 0.1 | 2.8 | 8.9 | 29.0 |
| Days ≥ 1" Snow Depth | 13 | 6 | 1 | 0 | 0 | 0 | 0 | 0 | 0 | 0 | 2 | 10 | 32 |

## CONCONULLY *Okanogan County* ELEVATION 2270 ft LAT/LONG 48° 34 ' N / 119° 45 ' W

| | JAN | FEB | MAR | APR | MAY | JUN | JUL | AUG | SEP | OCT | NOV | DEC | YEAR |
|---|---|---|---|---|---|---|---|---|---|---|---|---|---|
| Maximum Temp °F | 31.8 | 39.3 | 48.7 | 58.0 | 67.9 | 74.8 | ***81.3*** | 81.7 | 72.6 | 58.2 | 41.1 | 31.3 | 57.2 |
| Minimum Temp °F | 14.4 | 19.9 | 26.6 | 34.4 | 42.1 | 48.6 | ***53.4*** | 52.9 | 44.9 | 35.2 | 25.5 | 15.8 | 34.5 |
| Mean Temp °F | 23.1 | 29.6 | 37.7 | 46.2 | 55.0 | 61.7 | ***67.4*** | 67.3 | 58.8 | 46.7 | 33.4 | 23.6 | 45.9 |
| Days Max Temp ≥ 90 °F | 0 | 0 | 0 | 0 | 0 | 2 | ***6*** | 6 | 1 | 0 | 0 | 0 | 15 |
| Days Max Temp ≤ 32 °F | 15 | 4 | 1 | 0 | 0 | 0 | 0 | 0 | 0 | 0 | 4 | 14 | 38 |
| Days Min Temp ≤ 32 °F | 30 | 26 | 26 | 12 | 1 | 0 | ***0*** | 0 | 1 | 11 | 24 | 30 | 161 |
| Days Min Temp ≤ 0 °F | 5 | 1 | 0 | 0 | 0 | 0 | 0 | 0 | 0 | 0 | 1 | 3 | 10 |
| Heating Degree Days | 1292 | 993 | 840 | 558 | 313 | 141 | ***53*** | 51 | 202 | 559 | 943 | 1279 | 7224 |
| Cooling Degree Days | 0 | 0 | 0 | 0 | 17 | 44 | 129 | 133 | 30 | 0 | 0 | 0 | 353 |
| Total Precipitation (") | 1.43 | 1.34 | 1.15 | 1.05 | 1.55 | 1.55 | 0.87 | 1.01 | 0.79 | 0.90 | 1.61 | 1.81 | 15.06 |
| Days ≥ 0.1" Precip | 4 | 4 | 4 | 3 | 4 | 4 | 3 | 3 | 2 | 2 | 5 | 5 | 43 |
| Total Snowfall (") | na | na | ***2.2*** | ***0.2*** | 0.0 | 0.0 | 0.0 | 0.0 | 0.0 | 0.0 | na | na | na |
| Days ≥ 1" Snow Depth | na | na | na | ***0*** | 0 | 0 | 0 | 0 | 0 | 0 | na | na | na |

## CONCRETE PPL FISH ST *Skagit County* ELEVATION 269 ft LAT/LONG 48° 32 ' N / 121° 45 ' W

| | JAN | FEB | MAR | APR | MAY | JUN | JUL | AUG | SEP | OCT | NOV | DEC | YEAR |
|---|---|---|---|---|---|---|---|---|---|---|---|---|---|
| Maximum Temp °F | 41.6 | 46.8 | 52.6 | 58.3 | 65.4 | 69.7 | 75.0 | 75.8 | 70.4 | 60.4 | 48.1 | 42.0 | 58.8 |
| Minimum Temp °F | 31.3 | 32.8 | 35.2 | 39.0 | 44.5 | 49.5 | 52.1 | 52.7 | 48.9 | 42.9 | 36.9 | 32.5 | 41.5 |
| Mean Temp °F | 36.5 | 39.8 | 43.9 | 48.7 | 55.0 | 59.6 | 63.5 | 64.3 | 59.7 | 51.7 | 42.6 | 37.3 | 50.2 |
| Days Max Temp ≥ 90 °F | 0 | 0 | 0 | 0 | 0 | 0 | 1 | 1 | 0 | 0 | 0 | 0 | 2 |
| Days Max Temp ≤ 32 °F | 1 | 0 | 0 | 0 | 0 | 0 | 0 | 0 | 0 | 0 | 0 | 2 | 3 |
| Days Min Temp ≤ 32 °F | 16 | 12 | 9 | 2 | 0 | 0 | 0 | 0 | 0 | 1 | 6 | 15 | 61 |
| Days Min Temp ≤ 0 °F | 0 | 0 | 0 | 0 | 0 | 0 | 0 | 0 | 0 | 0 | 0 | 0 | 0 |
| Heating Degree Days | 878 | 704 | 646 | 483 | 310 | 173 | 83 | 67 | 167 | 406 | 667 | 852 | 5436 |
| Cooling Degree Days | 0 | 0 | 0 | 0 | 6 | 16 | 40 | 50 | 16 | 1 | 0 | 0 | 129 |
| Total Precipitation (") | 10.16 | 7.32 | 6.67 | 4.86 | 3.41 | 2.84 | 1.79 | 1.72 | 3.44 | 6.27 | 10.72 | 10.86 | 70.06 |
| Days ≥ 0.1" Precip | 16 | 13 | 14 | 12 | 9 | 7 | 4 | 4 | 7 | 11 | 17 | 17 | 131 |
| Total Snowfall (") | 8.1 | 4.6 | 1.5 | 0.3 | 0.0 | 0.0 | 0.0 | 0.0 | 0.0 | 0.0 | 1.5 | 6.3 | 22.3 |
| Days ≥ 1" Snow Depth | ***8*** | 4 | 1 | 0 | 0 | 0 | 0 | 0 | 0 | 0 | 1 | 5 | 19 |

## CONNELL 1 W *Franklin County* ELEVATION 1020 ft LAT/LONG 46° 40 ' N / 118° 53 ' W

| | JAN | FEB | MAR | APR | MAY | JUN | JUL | AUG | SEP | OCT | NOV | DEC | YEAR |
|---|---|---|---|---|---|---|---|---|---|---|---|---|---|
| Maximum Temp °F | 37.0 | 46.0 | 56.1 | 64.8 | 74.6 | 81.8 | 89.8 | 88.8 | 78.8 | ***65.5*** | 47.4 | 37.7 | 64.0 |
| Minimum Temp °F | 24.1 | 28.5 | 32.5 | 37.1 | 43.7 | 50.1 | ***55.4*** | 54.4 | 46.4 | 37.6 | 31.4 | 25.3 | 38.9 |
| Mean Temp °F | 30.6 | 37.3 | 44.3 | 51.0 | 59.2 | 66.0 | ***72.6*** | 71.6 | 62.6 | ***51.6*** | 39.4 | 31.5 | 51.5 |
| Days Max Temp ≥ 90 °F | 0 | 0 | 0 | 0 | 2 | 6 | 16 | 14 | 3 | 0 | 0 | 0 | 41 |
| Days Max Temp ≤ 32 °F | 10 | 3 | 0 | 0 | 0 | 0 | 0 | 0 | 0 | 0 | 2 | 9 | 24 |
| Days Min Temp ≤ 32 °F | 25 | 20 | 16 | 8 | 1 | 0 | 0 | 0 | 1 | ***8*** | ***16*** | 24 | 119 |
| Days Min Temp ≤ 0 °F | 1 | 0 | 0 | 0 | 0 | 0 | 0 | 0 | 0 | 0 | 0 | 1 | 2 |
| Heating Degree Days | 1061 | 776 | 636 | 414 | 204 | 64 | ***9*** | 13 | 115 | ***408*** | 760 | 1031 | 5491 |
| Cooling Degree Days | 0 | 0 | 0 | 3 | 37 | 87 | ***230*** | ***233*** | 50 | 2 | 0 | 0 | 642 |
| Total Precipitation (") | 0.91 | 0.70 | 0.80 | 0.67 | 0.74 | 0.43 | 0.30 | 0.33 | 0.35 | 0.65 | 1.17 | 1.09 | 8.14 |
| Days ≥ 0.1" Precip | 3 | 3 | 3 | 2 | 2 | 1 | 1 | 1 | 1 | 2 | 4 | 4 | 27 |
| Total Snowfall (") | na | na | ***0.2*** | ***0.0*** | ***0.0*** | ***0.0*** | ***0.0*** | ***0.0*** | ***0.0*** | ***0.0*** | ***1.0*** | na | na |
| Days ≥ 1" Snow Depth | na | na | ***0*** | ***0*** | ***0*** | ***0*** | ***0*** | ***0*** | 0 | ***0*** | ***0*** | ***2*** | na |

## COUGAR 6 E *Skamania County* ELEVATION 630 ft LAT/LONG 46° 4 ' N / 122° 12 ' W

| | JAN | FEB | MAR | APR | MAY | JUN | JUL | AUG | SEP | OCT | NOV | DEC | YEAR |
|---|---|---|---|---|---|---|---|---|---|---|---|---|---|
| Maximum Temp °F | 42.0 | 46.5 | 52.1 | 57.2 | 65.6 | 71.4 | 77.9 | 78.3 | 72.5 | 62.0 | 49.0 | 42.5 | 59.8 |
| Minimum Temp °F | 32.5 | 34.2 | 36.1 | 39.1 | 44.4 | 49.4 | 52.6 | 52.9 | 50.1 | 44.3 | 38.1 | 33.5 | 42.3 |
| Mean Temp °F | 37.3 | 40.4 | 44.1 | 48.2 | 55.0 | 60.4 | 65.3 | 65.6 | 61.3 | 53.2 | 43.6 | 38.0 | 51.0 |
| Days Max Temp ≥ 90 °F | 0 | 0 | 0 | 0 | 1 | 2 | 4 | 4 | 2 | 0 | 0 | 0 | 13 |
| Days Max Temp ≤ 32 °F | 2 | 0 | 0 | 0 | 0 | 0 | 0 | 0 | 0 | 0 | 0 | 1 | 3 |
| Days Min Temp ≤ 32 °F | 14 | 10 | 6 | 2 | 0 | 0 | 0 | 0 | 0 | 0 | 4 | 12 | 48 |
| Days Min Temp ≤ 0 °F | 0 | 0 | 0 | 0 | 0 | 0 | 0 | 0 | 0 | 0 | 0 | 0 | 0 |
| Heating Degree Days | 853 | 689 | 641 | 499 | 319 | 173 | 76 | 71 | 149 | 363 | 636 | 830 | 5299 |
| Cooling Degree Days | 0 | 0 | 0 | 1 | 23 | 43 | 95 | 108 | 55 | 6 | 0 | 0 | 331 |
| Total Precipitation (") | 17.90 | 13.86 | 11.72 | 8.31 | 4.86 | 3.78 | 1.49 | 2.09 | 4.28 | 8.27 | 16.70 | 18.96 | 112.22 |
| Days ≥ 0.1" Precip | 18 | 15 | 16 | 13 | 9 | 7 | 3 | 4 | 7 | 11 | 17 | 18 | 138 |
| Total Snowfall (") | na | ***5.0*** | 1.3 | 0.2 | 0.0 | 0.0 | 0.0 | 0.0 | 0.0 | 0.0 | 2.0 | ***4.5*** | na |
| Days ≥ 1" Snow Depth | na | ***4*** | ***2*** | 0 | 0 | 0 | 0 | 0 | 0 | 0 | 1 | ***4*** | na |

## COULEE DAM 1 SW *Grant County* ELEVATION 1703 ft LAT/LONG 47° 57' N / 119° 0' W

| | JAN | FEB | MAR | APR | MAY | JUN | JUL | AUG | SEP | OCT | NOV | DEC | YEAR |
|---|---|---|---|---|---|---|---|---|---|---|---|---|---|
| Maximum Temp °F | 32.6 | 40.1 | 51.2 | 61.2 | 70.7 | 79.0 | 86.4 | 86.2 | 76.0 | 61.4 | 43.4 | 33.9 | 60.2 |
| Minimum Temp °F | 21.5 | 26.4 | 31.9 | 38.4 | 45.9 | 53.3 | 58.5 | 58.3 | 49.9 | 40.2 | 31.6 | 23.8 | 40.0 |
| Mean Temp °F | 27.1 | 33.3 | 41.6 | 49.9 | 58.3 | 66.2 | 72.5 | 72.3 | 63.0 | 50.9 | 37.5 | 28.9 | 50.1 |
| Days Max Temp ≥ 90 °F | 0 | 0 | 0 | 0 | 1 | 4 | 12 | 11 | 2 | 0 | 0 | 0 | 30 |
| Days Max Temp ≤ 32 °F | 14 | 5 | 0 | 0 | 0 | 0 | 0 | 0 | 0 | 0 | 3 | 13 | 35 |
| Days Min Temp ≤ 32 °F | 28 | 22 | 17 | 6 | 0 | 0 | 0 | 0 | 0 | 3 | 15 | 26 | 117 |
| Days Min Temp ≤ 0 °F | 1 | 1 | 0 | 0 | 0 | 0 | 0 | 0 | 0 | 0 | 0 | 1 | 3 |
| Heating Degree Days | 1170 | 890 | 719 | 449 | 223 | 67 | 14 | 14 | 118 | 433 | 818 | 1113 | 6028 |
| Cooling Degree Days | 0 | 0 | 0 | 1 | 29 | 106 | 247 | 250 | 72 | 3 | 0 | 0 | 708 |
| Total Precipitation (") | 0.99 | 0.84 | 0.81 | 0.84 | 1.14 | 0.85 | 0.64 | 0.49 | 0.51 | 0.56 | 1.33 | 1.46 | 10.46 |
| Days ≥ 0.1" Precip | 4 | 3 | 3 | 3 | 3 | 3 | 2 | 2 | 2 | 2 | 5 | 5 | 37 |
| Total Snowfall (") | na | *1.1* | 0.3 | 0.0 | 0.0 | 0.0 | 0.0 | 0.0 | 0.0 | 0.0 | *0.5* | na | na |
| Days ≥ 1" Snow Depth | na | na | 0 | 0 | 0 | 0 | 0 | 0 | 0 | 0 | 1 | na | na |

## COUPEVILLE 1 S *Island County* ELEVATION 49 ft LAT/LONG 48° 12' N / 122° 42' W

| | JAN | FEB | MAR | APR | MAY | JUN | JUL | AUG | SEP | OCT | NOV | DEC | YEAR |
|---|---|---|---|---|---|---|---|---|---|---|---|---|---|
| Maximum Temp °F | 44.9 | 48.7 | 52.6 | 56.6 | 62.5 | 67.0 | 70.9 | 72.0 | 67.4 | 58.5 | 49.9 | 45.1 | 58.0 |
| Minimum Temp °F | 34.3 | 35.2 | 37.0 | 39.7 | 44.2 | 47.9 | 50.1 | 50.5 | 46.6 | 41.7 | 37.9 | 35.0 | 41.7 |
| Mean Temp °F | 39.6 | 42.0 | 44.8 | 48.2 | 53.4 | 57.5 | 60.5 | 61.3 | 57.0 | 50.1 | 43.9 | 40.1 | 49.9 |
| Days Max Temp ≥ 90 °F | 0 | 0 | 0 | 0 | 0 | 0 | 0 | 0 | 0 | 0 | 0 | 0 | 0 |
| Days Max Temp ≤ 32 °F | 1 | 0 | 0 | 0 | 0 | 0 | 0 | 0 | 0 | 0 | 0 | 1 | 2 |
| Days Min Temp ≤ 32 °F | 11 | 10 | 7 | 2 | 0 | 0 | 0 | 0 | 0 | 2 | 6 | 11 | 49 |
| Days Min Temp ≤ 0 °F | 0 | 0 | 0 | 0 | 0 | 0 | 0 | 0 | 0 | 0 | 0 | 0 | 0 |
| Heating Degree Days | 780 | 643 | 618 | 497 | 353 | 219 | 137 | 118 | 234 | 454 | 626 | 766 | 5445 |
| Cooling Degree Days | 0 | 0 | 0 | 0 | 0 | 1 | 4 | 9 | 1 | 0 | 0 | 0 | 15 |
| Total Precipitation (") | 2.59 | 1.70 | 1.81 | 1.75 | 1.61 | 1.24 | 0.87 | 1.00 | 1.27 | 1.58 | 2.61 | 2.80 | 20.83 |
| Days ≥ 0.1" Precip | 9 | 6 | 6 | 5 | 4 | 4 | 3 | 3 | 4 | 5 | 8 | 9 | 66 |
| Total Snowfall (") | *3.4* | 0.8 | 0.3 | 0.1 | 0.0 | 0.0 | 0.0 | 0.0 | 0.0 | 0.0 | 0.8 | *0.9* | 6.3 |
| Days ≥ 1" Snow Depth | *2* | 1 | 0 | 0 | 0 | 0 | 0 | 0 | 0 | 0 | 0 | *1* | 4 |

## DALLESPORT FCWOS AP *Klickitat County* ELEVATION 236 ft LAT/LONG 45° 37' N / 121° 9' W

| | JAN | FEB | MAR | APR | MAY | JUN | JUL | AUG | SEP | OCT | NOV | DEC | YEAR |
|---|---|---|---|---|---|---|---|---|---|---|---|---|---|
| Maximum Temp °F | 40.7 | 47.8 | 57.1 | 64.6 | 74.1 | 80.6 | 87.8 | 87.5 | 80.1 | 66.8 | 49.9 | 41.4 | 64.9 |
| Minimum Temp °F | 29.0 | 32.1 | 36.1 | 41.3 | 48.2 | 54.8 | 59.9 | 59.5 | 51.4 | 42.4 | 35.5 | 30.1 | 43.4 |
| Mean Temp °F | 34.9 | 39.9 | 46.6 | 52.9 | 61.2 | 67.7 | 73.8 | 73.5 | 65.8 | 54.6 | 42.8 | 35.8 | 54.1 |
| Days Max Temp ≥ 90 °F | 0 | 0 | 0 | 0 | 2 | 6 | 14 | 13 | 6 | 0 | 0 | 0 | 41 |
| Days Max Temp ≤ 32 °F | 6 | 1 | 0 | 0 | 0 | 0 | 0 | 0 | 0 | 0 | 1 | 4 | 12 |
| Days Min Temp ≤ 32 °F | 19 | 14 | 9 | 3 | 0 | 0 | 0 | 0 | 0 | 3 | 9 | 18 | 75 |
| Days Min Temp ≤ 0 °F | 0 | 0 | 0 | 0 | 0 | 0 | 0 | 0 | 0 | 0 | 0 | 1 | 1 |
| Heating Degree Days | 927 | 701 | 562 | 358 | 152 | 41 | 5 | 4 | 63 | 321 | 661 | 899 | 4694 |
| Cooling Degree Days | 0 | 0 | 0 | 4 | 46 | 119 | 273 | 273 | 97 | 7 | 0 | 0 | 819 |
| Total Precipitation (") | 2.54 | 1.52 | 1.06 | 0.65 | 0.44 | 0.36 | 0.14 | 0.35 | 0.46 | 0.95 | 2.00 | 2.62 | 13.09 |
| Days ≥ 0.1" Precip | 7 | 4 | 3 | 2 | 2 | 1 | 1 | 1 | 2 | 3 | 6 | 7 | 39 |
| Total Snowfall (") | 5.9 | 2.7 | 0.5 | 0.0 | 0.0 | 0.0 | 0.0 | 0.0 | 0.0 | 0.1 | 2.3 | 5.8 | 17.3 |
| Days ≥ 1" Snow Depth | 10 | 3 | 0 | 0 | 0 | 0 | 0 | 0 | 0 | 0 | 1 | 5 | 19 |

## DAVENPORT *Lincoln County* ELEVATION 2451 ft LAT/LONG 47° 39' N / 118° 9' W

| | JAN | FEB | MAR | APR | MAY | JUN | JUL | AUG | SEP | OCT | NOV | DEC | YEAR |
|---|---|---|---|---|---|---|---|---|---|---|---|---|---|
| Maximum Temp °F | 30.6 | 37.2 | 47.3 | 56.9 | 66.4 | 74.3 | 82.3 | 83.1 | 73.1 | 59.6 | 40.8 | 31.7 | 56.9 |
| Minimum Temp °F | 17.9 | 22.7 | 28.1 | 32.7 | 39.2 | 45.4 | 49.7 | 49.6 | 42.1 | 32.6 | 26.5 | 19.1 | 33.8 |
| Mean Temp °F | 24.3 | 30.0 | 37.8 | 44.8 | 52.8 | 59.8 | 66.0 | 66.4 | 57.6 | 46.1 | 33.7 | 25.5 | 45.4 |
| Days Max Temp ≥ 90 °F | 0 | 0 | 0 | 0 | 0 | 2 | 8 | 8 | 1 | 0 | 0 | 0 | 19 |
| Days Max Temp ≤ 32 °F | 16 | 7 | 1 | 0 | 0 | 0 | 0 | 0 | 0 | 0 | 4 | 15 | 43 |
| Days Min Temp ≤ 32 °F | 30 | 26 | 24 | 16 | 6 | 1 | 0 | 0 | 3 | 16 | 23 | 29 | 174 |
| Days Min Temp ≤ 0 °F | 3 | 1 | 0 | 0 | 0 | 0 | 0 | 0 | 0 | 0 | 1 | 2 | 7 |
| Heating Degree Days | 1256 | 982 | 838 | 600 | 378 | 189 | 73 | 69 | 233 | 579 | 933 | 1219 | 7349 |
| Cooling Degree Days | 0 | 0 | 0 | 1 | 10 | 35 | 92 | 105 | 19 | 0 | 0 | 0 | 262 |
| Total Precipitation (") | 1.53 | 1.15 | 1.25 | 1.03 | 1.23 | 0.98 | 0.80 | 0.55 | 0.55 | 0.85 | 1.84 | 1.96 | 13.72 |
| Days ≥ 0.1" Precip | 5 | 4 | 5 | 4 | 4 | 3 | 2 | 2 | 2 | 3 | 6 | 6 | 46 |
| Total Snowfall (") | 10.8 | 6.1 | 2.2 | 0.2 | 0.0 | 0.0 | 0.0 | 0.0 | 0.0 | 0.1 | 4.9 | 11.0 | 35.3 |
| Days ≥ 1" Snow Depth | *14* | *9* | *2* | 0 | 0 | 0 | 0 | 0 | 0 | 0 | 3 | na | na |

## DAYTON 1 WSW *Columbia County* ELEVATION 1752 ft LAT/LONG 46° 18' N / 117° 57' W

| | JAN | FEB | MAR | APR | MAY | JUN | JUL | AUG | SEP | OCT | NOV | DEC | YEAR |
|---|---|---|---|---|---|---|---|---|---|---|---|---|---|
| Maximum Temp °F | 39.9 | 46.0 | 53.7 | 60.7 | 69.5 | 77.6 | 86.5 | 86.4 | 76.7 | 64.6 | 49.1 | 41.0 | 62.6 |
| Minimum Temp °F | 25.7 | 29.3 | 33.6 | 38.2 | 44.5 | 50.5 | 55.1 | 54.7 | 47.0 | 38.1 | 32.3 | 26.6 | 39.6 |
| Mean Temp °F | 32.8 | 37.7 | 43.7 | 49.5 | 57.0 | 64.0 | 70.8 | 70.6 | 61.9 | 51.4 | 40.7 | 33.8 | 51.2 |
| Days Max Temp ≥ 90 °F | 0 | 0 | 0 | 0 | 1 | 5 | 13 | 12 | 3 | 0 | 0 | 0 | 34 |
| Days Max Temp ≤ 32 °F | 8 | 3 | 0 | 0 | 0 | 0 | 0 | 0 | 0 | 0 | 2 | 7 | 20 |
| Days Min Temp ≤ 32 °F | 22 | 18 | 13 | 5 | 0 | 0 | 0 | 0 | 1 | 6 | 14 | 23 | 102 |
| Days Min Temp ≤ 0 °F | 2 | 0 | 0 | 0 | 0 | 0 | 0 | 0 | 0 | 0 | 0 | 1 | 3 |
| Heating Degree Days | 988 | 764 | 654 | 460 | 264 | 108 | 25 | 25 | 141 | 418 | 721 | 959 | 5527 |
| Cooling Degree Days | 0 | 0 | 0 | 2 | 29 | 83 | 194 | 202 | 54 | 2 | 0 | 0 | 566 |
| Total Precipitation (") | 2.39 | 1.62 | 1.87 | 1.60 | 1.41 | 1.15 | 0.58 | 0.73 | 0.89 | 1.40 | 2.55 | 2.36 | 18.55 |
| Days ≥ 0.1" Precip | 7 | 5 | 6 | 5 | 4 | 3 | 2 | 2 | 3 | 4 | 8 | 7 | 56 |
| Total Snowfall (") | na | na | *1.2* | *0.0* | *0.0* | *0.0* | *0.0* | *0.0* | *0.0* | *0.1* | *1.5* | *6.1* | na |
| Days ≥ 1" Snow Depth | 8 | 4 | 1 | 0 | 0 | 0 | 0 | 0 | 0 | 0 | 2 | *5* | 20 |

## DIABLO DAM *Whatcom County* ELEVATION 889 ft LAT/LONG 48° 43' N / 121° 9' W

| | JAN | FEB | MAR | APR | MAY | JUN | JUL | AUG | SEP | OCT | NOV | DEC | YEAR |
|---|---|---|---|---|---|---|---|---|---|---|---|---|---|
| Maximum Temp °F | 38.0 | 43.3 | 50.1 | 56.8 | 65.3 | 70.8 | 76.7 | 77.8 | 70.6 | 58.2 | 44.7 | 38.5 | 57.6 |
| Minimum Temp °F | 28.1 | 30.1 | 33.0 | 37.0 | 43.0 | 48.5 | 51.9 | 52.9 | 47.6 | 40.7 | 34.3 | 29.7 | 39.7 |
| Mean Temp °F | 33.1 | 36.7 | 41.6 | 47.0 | 54.2 | 59.7 | 64.3 | 65.4 | 59.1 | 49.5 | 39.6 | 34.2 | 48.7 |
| Days Max Temp ≥ 90 °F | 0 | 0 | 0 | 0 | 1 | 2 | 4 | 4 | 1 | 0 | 0 | 0 | 12 |
| Days Max Temp ≤ 32 °F | 5 | 1 | 0 | 0 | 0 | 0 | 0 | 0 | 0 | 0 | 1 | 3 | 10 |
| Days Min Temp ≤ 32 °F | 21 | 18 | 13 | 4 | 0 | 0 | 0 | 0 | 0 | 1 | 10 | 19 | 86 |
| Days Min Temp ≤ 0 °F | 0 | 0 | 0 | 0 | 0 | 0 | 0 | 0 | 0 | 0 | 0 | 0 | 0 |
| Heating Degree Days | 982 | 792 | 719 | 536 | 338 | 182 | 85 | 68 | 189 | 473 | 755 | 950 | 6069 |
| Cooling Degree Days | 0 | 0 | 0 | 0 | 12 | 27 | 69 | 88 | 21 | 0 | 0 | 0 | 217 |
| Total Precipitation (") | 12.01 | 8.06 | 6.94 | 4.94 | 3.11 | 2.39 | 1.75 | 1.74 | 3.51 | 7.34 | 13.05 | 12.67 | 77.51 |
| Days ≥ 0.1" Precip | 15 | 12 | 13 | 11 | 7 | 6 | 4 | 4 | 6 | 11 | 17 | 16 | 122 |
| Total Snowfall (") | na | 10.8 | *2.2* | 0.0 | 0.0 | 0.0 | 0.0 | 0.0 | 0.0 | 0.0 | 2.7 | *11.3* | na |
| Days ≥ 1" Snow Depth | *18* | 12 | 3 | 0 | 0 | 0 | 0 | 0 | 0 | 0 | 2 | *11* | 46 |

## ELLENSBURG *Kittitas County* ELEVATION 1522 ft LAT/LONG 46° 59' N / 120° 32' W

| | JAN | FEB | MAR | APR | MAY | JUN | JUL | AUG | SEP | OCT | NOV | DEC | YEAR |
|---|---|---|---|---|---|---|---|---|---|---|---|---|---|
| Maximum Temp °F | 34.1 | 41.9 | 52.8 | 60.5 | 69.2 | 76.3 | 83.3 | 83.6 | 74.9 | 62.3 | 44.9 | 34.5 | 59.9 |
| Minimum Temp °F | 19.0 | 23.5 | 28.4 | 33.9 | 41.7 | 48.7 | 52.7 | 51.9 | 42.9 | 32.8 | 26.9 | 19.8 | 35.2 |
| Mean Temp °F | 26.7 | 32.7 | 40.6 | 47.2 | 55.5 | 62.5 | 68.0 | 67.8 | 58.9 | 47.5 | 35.9 | 27.2 | 47.5 |
| Days Max Temp ≥ 90 °F | 0 | 0 | 0 | 0 | 1 | 3 | 9 | 8 | 1 | 0 | 0 | 0 | 22 |
| Days Max Temp ≤ 32 °F | 13 | 4 | 0 | 0 | 0 | 0 | 0 | 0 | 0 | 0 | 2 | 12 | 31 |
| Days Min Temp ≤ 32 °F | 29 | 25 | 23 | 12 | 2 | 0 | 0 | 0 | 2 | 15 | 22 | 29 | 159 |
| Days Min Temp ≤ 0 °F | 2 | 1 | 0 | 0 | 0 | 0 | 0 | 0 | 0 | 0 | 0 | 2 | 5 |
| Heating Degree Days | 1180 | 905 | 749 | 528 | 298 | 127 | 41 | 42 | 196 | 534 | 866 | 1166 | 6632 |
| Cooling Degree Days | 0 | 0 | 0 | 1 | 14 | 53 | 125 | 125 | 19 | 0 | 0 | 0 | 337 |
| Total Precipitation (") | 1.27 | 0.78 | 0.69 | 0.56 | 0.51 | 0.64 | 0.33 | 0.40 | 0.47 | 0.55 | 1.14 | 1.45 | 8.79 |
| Days ≥ 0.1" Precip | 4 | 3 | 2 | 2 | 2 | 2 | 1 | 1 | 1 | 2 | 4 | 5 | 29 |
| Total Snowfall (") | 8.2 | 3.2 | 1.2 | 0.1 | 0.0 | 0.0 | 0.0 | 0.0 | 0.0 | 0.1 | 2.6 | 8.8 | 24.2 |
| Days ≥ 1" Snow Depth | *19* | *10* | 1 | 0 | 0 | 0 | 0 | 0 | 0 | 0 | *4* | 14 | 48 |

## ELMA *Grays Harbor County* ELEVATION 69 ft LAT/LONG 47° 0' N / 123° 24' W

| | JAN | FEB | MAR | APR | MAY | JUN | JUL | AUG | SEP | OCT | NOV | DEC | YEAR |
|---|---|---|---|---|---|---|---|---|---|---|---|---|---|
| Maximum Temp °F | 46.3 | 50.9 | 55.7 | 60.5 | 67.4 | 72.2 | 76.9 | 78.2 | 73.5 | 63.4 | 51.6 | 45.7 | 61.9 |
| Minimum Temp °F | 33.6 | 34.6 | 35.8 | 38.4 | 43.0 | 47.6 | 50.5 | 51.1 | 47.2 | 41.6 | 37.6 | 34.2 | 41.3 |
| Mean Temp °F | 40.0 | 42.8 | 45.8 | 49.4 | 55.2 | 59.9 | 63.7 | 64.7 | 60.4 | 52.5 | 44.7 | 40.0 | 51.6 |
| Days Max Temp ≥ 90 °F | 0 | 0 | 0 | 0 | 0 | 1 | 3 | 3 | 1 | 0 | 0 | 0 | 8 |
| Days Max Temp ≤ 32 °F | 1 | 0 | 0 | 0 | 0 | 0 | 0 | 0 | 0 | 0 | 0 | 1 | 2 |
| Days Min Temp ≤ 32 °F | 14 | 11 | 11 | 6 | 1 | 0 | 0 | 0 | 0 | 3 | 8 | 13 | 67 |
| Days Min Temp ≤ 0 °F | 0 | 0 | 0 | 0 | 0 | 0 | 0 | 0 | 0 | 0 | 0 | 0 | 0 |
| Heating Degree Days | 768 | 621 | 589 | 460 | 302 | 165 | 78 | 58 | 150 | 382 | 604 | 768 | 4945 |
| Cooling Degree Days | 0 | 0 | 0 | 0 | 7 | 18 | 41 | 60 | 19 | 1 | 0 | 0 | 146 |
| Total Precipitation (") | 10.38 | 7.89 | 6.87 | 4.83 | 2.88 | 2.12 | 1.12 | 1.42 | 2.92 | 5.59 | 9.68 | 10.88 | 66.58 |
| Days ≥ 0.1" Precip | 16 | 14 | 14 | 11 | 8 | 6 | 3 | 4 | 6 | 10 | 17 | 17 | 126 |
| Total Snowfall (") | *3.1* | *1.2* | *0.5* | *0.0* | *0.0* | *0.0* | *0.0* | *0.0* | *0.0* | *0.0* | *0.8* | *1.5* | 7.1 |
| Days ≥ 1" Snow Depth | *2* | *1* | *0* | *0* | *0* | *0* | *0* | *0* | *0* | *0* | *1* | *1* | 5 |

## ELWHA RS *Clallam County* ELEVATION 361 ft LAT/LONG 48° 1' N / 123° 35' W

| | JAN | FEB | MAR | APR | MAY | JUN | JUL | AUG | SEP | OCT | NOV | DEC | YEAR |
|---|---|---|---|---|---|---|---|---|---|---|---|---|---|
| Maximum Temp °F | 40.9 | 45.2 | 50.8 | 56.3 | 63.3 | 68.0 | 73.2 | 74.4 | 68.0 | 57.5 | 46.4 | 41.8 | 57.2 |
| Minimum Temp °F | 31.3 | 32.4 | 34.0 | 37.2 | 41.8 | 46.6 | 49.8 | 51.1 | 47.2 | 40.9 | 35.5 | 32.3 | 40.0 |
| Mean Temp °F | 36.2 | 38.8 | 42.4 | 46.8 | 52.6 | 57.3 | 61.5 | 62.8 | 57.6 | 49.2 | 41.0 | 37.0 | 48.6 |
| Days Max Temp ≥ 90 °F | 0 | 0 | 0 | 0 | 0 | 0 | 1 | 1 | 0 | 0 | 0 | 0 | 2 |
| Days Max Temp ≤ 32 °F | 2 | 1 | 0 | 0 | 0 | 0 | 0 | 0 | 0 | 0 | 1 | 2 | 6 |
| Days Min Temp ≤ 32 °F | *17* | 14 | 11 | 4 | 0 | 0 | 0 | 0 | 0 | 1 | *8* | *16* | 71 |
| Days Min Temp ≤ 0 °F | 0 | 0 | 0 | 0 | 0 | 0 | 0 | 0 | 0 | 0 | 0 | 0 | 0 |
| Heating Degree Days | 888 | 733 | 692 | 541 | 380 | 235 | 133 | 104 | 220 | 482 | 715 | 859 | 5982 |
| Cooling Degree Days | 0 | 0 | 0 | 0 | 1 | 10 | 28 | 40 | 6 | 0 | 0 | 0 | 85 |
| Total Precipitation (") | 8.69 | 6.54 | 5.76 | 3.37 | 1.84 | 1.16 | 0.64 | 1.13 | 1.78 | 4.65 | 8.54 | 9.82 | 53.92 |
| Days ≥ 0.1" Precip | 12 | 10 | 11 | 7 | 5 | 4 | 2 | 3 | 4 | 8 | 13 | *12* | 91 |
| Total Snowfall (") | *7.0* | 2.3 | 1.6 | 0.0 | 0.0 | 0.0 | 0.0 | 0.0 | 0.0 | 0.0 | 1.9 | *3.3* | 16.1 |
| Days ≥ 1" Snow Depth | na | *2* | *1* | *0* | *0* | *0* | 0 | *0* | 0 | 0 | *1* | *2* | na |

## EPHRATA AP FCWOS *Grant County* ELEVATION 1273 ft LAT/LONG 47° 18' N / 119° 32' W

| | JAN | FEB | MAR | APR | MAY | JUN | JUL | AUG | SEP | OCT | NOV | DEC | YEAR |
|---|---|---|---|---|---|---|---|---|---|---|---|---|---|
| Maximum Temp °F | 33.5 | 41.8 | 53.5 | 63.0 | 73.2 | 81.0 | 88.2 | 87.5 | 77.8 | 63.2 | 44.9 | 34.2 | 61.8 |
| Minimum Temp °F | 20.9 | 26.5 | 32.8 | 39.3 | 47.9 | 55.4 | 61.2 | 60.3 | 51.3 | 40.0 | 30.4 | 22.3 | 40.7 |
| Mean Temp °F | 27.2 | 34.2 | 43.2 | 51.2 | 60.6 | 68.3 | 74.7 | 73.9 | 64.6 | 51.6 | 37.7 | 28.2 | 51.3 |
| Days Max Temp ≥ 90 °F | 0 | 0 | 0 | 0 | 2 | 6 | 14 | 13 | 3 | 0 | 0 | 0 | 38 |
| Days Max Temp ≤ 32 °F | 14 | 5 | 0 | 0 | 0 | 0 | 0 | 0 | 0 | 0 | 3 | 13 | 35 |
| Days Min Temp ≤ 32 °F | 28 | 22 | 15 | 5 | 0 | 0 | 0 | 0 | 0 | 4 | 17 | 27 | 118 |
| Days Min Temp ≤ 0 °F | 2 | 1 | 0 | 0 | 0 | 0 | 0 | 0 | 0 | 0 | 0 | 1 | 4 |
| Heating Degree Days | 1164 | 863 | 669 | 411 | 174 | 45 | 9 | 8 | 94 | 412 | 814 | 1133 | 5796 |
| Cooling Degree Days | 0 | 0 | 0 | 4 | 46 | 128 | 284 | 275 | 86 | 5 | 0 | 0 | 828 |
| Total Precipitation (") | 0.83 | 0.66 | 0.64 | 0.46 | 0.50 | 0.52 | 0.39 | 0.25 | 0.34 | 0.44 | 1.02 | 1.23 | 7.28 |
| Days ≥ 0.1" Precip | 3 | 2 | 2 | 1 | 2 | 2 | 1 | 1 | 1 | 2 | 4 | 4 | 25 |
| Total Snowfall (") | 5.3 | 2.6 | 0.8 | 0.0 | 0.0 | 0.0 | 0.0 | 0.0 | 0.0 | 0.0 | 2.1 | 7.5 | 18.3 |
| Days ≥ 1" Snow Depth | 18 | 8 | 1 | 0 | 0 | 0 | 0 | 0 | 0 | 0 | 2 | 12 | 41 |

## EVERETT *Snohomish County* ELEVATION 120 ft LAT/LONG 48° 0' N / 122° 12' W

| | JAN | FEB | MAR | APR | MAY | JUN | JUL | AUG | SEP | OCT | NOV | DEC | YEAR |
|---|---|---|---|---|---|---|---|---|---|---|---|---|---|
| Maximum Temp °F | 44.9 | 49.0 | 53.0 | 57.5 | 63.5 | 68.4 | 72.5 | 73.5 | 68.1 | 59.6 | 50.3 | 44.9 | 58.8 |
| Minimum Temp °F | 33.3 | 34.8 | 37.2 | 40.8 | 46.0 | 51.2 | 53.8 | 53.6 | 48.6 | 42.4 | 37.5 | 34.1 | 42.8 |
| Mean Temp °F | 39.2 | 41.9 | 45.1 | 49.2 | 54.8 | 59.8 | 63.1 | 63.6 | 58.4 | 51.0 | 43.9 | 39.5 | 50.8 |
| Days Max Temp ≥ 90 °F | 0 | 0 | 0 | 0 | 0 | 0 | 0 | 0 | 0 | 0 | 0 | 0 | 0 |
| Days Max Temp ≤ 32 °F | 1 | 0 | 0 | 0 | 0 | 0 | 0 | 0 | 0 | 0 | 0 | 1 | 2 |
| Days Min Temp ≤ 32 °F | 13 | 10 | 7 | 2 | 0 | 0 | 0 | 0 | 0 | 2 | 8 | 12 | 54 |
| Days Min Temp ≤ 0 °F | 0 | 0 | 0 | 0 | 0 | 0 | 0 | 0 | 0 | 0 | 0 | 0 | 0 |
| Heating Degree Days | 794 | 645 | 609 | 469 | 313 | 161 | 83 | 75 | 197 | 426 | 627 | 783 | 5182 |
| Cooling Degree Days | 0 | 0 | 0 | 0 | 2 | 11 | 34 | 47 | 6 | 0 | 0 | 0 | 100 |
| Total Precipitation (") | 4.48 | 3.20 | 3.56 | 2.88 | 2.39 | 2.06 | 1.32 | 1.34 | 2.15 | 2.96 | 4.93 | 5.19 | 36.46 |
| Days ≥ 0.1" Precip | 11 | 10 | 11 | 9 | 7 | 6 | 3 | 3 | 5 | 8 | 12 | 13 | 98 |
| Total Snowfall (") | 2.1 | 0.6 | 0.0 | 0.3 | 0.0 | 0.0 | 0.0 | 0.0 | 0.0 | 0.0 | 0.3 | 0.9 | 4.2 |
| Days ≥ 1" Snow Depth | na | 0 | 0 | 0 | 0 | 0 | 0 | 0 | 0 | 0 | 0 | *0* | na |

## FORKS 1 E *Clallam County* ELEVATION 351 ft LAT/LONG 47° 57' N / 124° 22' W

| | JAN | FEB | MAR | APR | MAY | JUN | JUL | AUG | SEP | OCT | NOV | DEC | YEAR |
|---|---|---|---|---|---|---|---|---|---|---|---|---|---|
| Maximum Temp °F | 45.5 | 49.8 | 53.3 | 57.5 | 63.5 | 67.8 | 72.5 | 73.8 | 70.7 | 61.2 | 50.6 | 45.1 | 59.3 |
| Minimum Temp °F | 33.6 | 34.8 | 35.5 | 37.8 | 42.4 | 46.6 | 49.2 | 49.9 | 46.9 | 42.0 | 37.3 | 34.2 | 40.8 |
| Mean Temp °F | 39.6 | 42.3 | 44.4 | 47.7 | 52.9 | 57.2 | 60.9 | 61.9 | 58.8 | 51.6 | 43.9 | 39.7 | 50.1 |
| Days Max Temp ≥ 90 °F | 0 | 0 | 0 | 0 | 0 | 1 | 1 | 2 | 1 | 0 | 0 | 0 | 5 |
| Days Max Temp ≤ 32 °F | 0 | 0 | 0 | 0 | 0 | 0 | 0 | 0 | 0 | 0 | 0 | 1 | 1 |
| Days Min Temp ≤ 32 °F | 14 | 11 | 10 | 6 | 1 | 0 | 0 | 0 | 0 | 2 | 8 | 13 | 65 |
| Days Min Temp ≤ 0 °F | 0 | 0 | 0 | 0 | 0 | 0 | 0 | 0 | 0 | 0 | 0 | 0 | 0 |
| Heating Degree Days | 782 | 633 | 632 | 513 | 369 | 234 | 139 | 114 | 190 | 408 | 624 | 778 | 5416 |
| Cooling Degree Days | 0 | 0 | 0 | 0 | 4 | 10 | 22 | 30 | 12 | 1 | 0 | 0 | 79 |
| Total Precipitation (") | 16.63 | 14.77 | 12.83 | 8.93 | 5.75 | 3.30 | 2.42 | 2.73 | 4.80 | 11.11 | 16.41 | 18.82 | 118.50 |
| Days ≥ 0.1" Precip | 17 | 16 | 16 | 14 | 11 | 8 | 5 | 5 | 7 | 13 | 19 | 19 | 150 |
| Total Snowfall (") | 4.7 | 1.8 | 1.3 | 0.1 | 0.0 | 0.0 | 0.0 | 0.0 | 0.0 | 0.0 | 0.6 | 2.6 | 11.1 |
| Days ≥ 1" Snow Depth | *2* | 1 | 1 | 0 | 0 | 0 | 0 | 0 | 0 | 0 | 0 | *2* | 6 |

## GLENOMA 1 W *Lewis County* ELEVATION 840 ft LAT/LONG 46° 31 ' N / 122° 9 ' W

| | JAN | FEB | MAR | APR | MAY | JUN | JUL | AUG | SEP | OCT | NOV | DEC | YEAR |
|---|---|---|---|---|---|---|---|---|---|---|---|---|---|
| Maximum Temp °F | 45.1 | 50.3 | 54.2 | 58.7 | 65.5 | 70.8 | 76.5 | 77.7 | 72.5 | 63.0 | 51.0 | 45.3 | 60.9 |
| Minimum Temp °F | 30.6 | 32.3 | 34.3 | 37.1 | 42.4 | 47.1 | 50.2 | 49.4 | 45.2 | 38.4 | 34.9 | 31.3 | 39.4 |
| Mean Temp °F | 37.9 | 41.3 | 44.3 | 47.9 | 54.0 | 59.0 | 63.3 | 63.6 | 58.8 | 50.7 | 43.0 | 38.3 | 50.2 |
| Days Max Temp ≥ 90 °F | 0 | 0 | 0 | 0 | 0 | 1 | 2 | 3 | 1 | 0 | 0 | 0 | 7 |
| Days Max Temp ≤ 32 °F | 1 | 0 | 0 | 0 | 0 | 0 | 0 | 0 | 0 | 0 | 0 | 1 | 2 |
| Days Min Temp ≤ 32 °F | 18 | 14 | *12* | 6 | 1 | 0 | 0 | 0 | 1 | 6 | 10 | 18 | 86 |
| Days Min Temp ≤ 0 °F | 0 | 0 | 0 | 0 | 0 | 0 | 0 | 0 | 0 | 0 | 0 | 0 | 0 |
| Heating Degree Days | 833 | 662 | 636 | 505 | 344 | 196 | 95 | 88 | 194 | 438 | 654 | 820 | 5465 |
| Cooling Degree Days | 0 | 0 | 0 | 1 | 10 | 21 | 49 | 56 | 14 | 3 | 0 | 0 | 154 |
| Total Precipitation (") | 9.74 | 6.80 | 6.11 | 5.42 | 3.55 | 3.04 | 1.23 | 1.46 | 3.15 | 4.76 | 9.13 | 9.79 | 64.18 |
| Days ≥ 0.1" Precip | 16 | 13 | 14 | 13 | 10 | 8 | 3 | 3 | 7 | 10 | 16 | 16 | 129 |
| Total Snowfall (") | *7.5* | 4.7 | na | 0.3 | 0.0 | 0.0 | 0.0 | 0.0 | 0.0 | 0.0 | 1.6 | 3.6 | na |
| Days ≥ 1" Snow Depth | *4* | 3 | *1* | 0 | 0 | 0 | 0 | 0 | 0 | 0 | 1 | 3 | 12 |

## GRAPEVIEW 3 SW *Mason County* ELEVATION 20 ft LAT/LONG 47° 20 ' N / 122° 49 ' W

| | JAN | FEB | MAR | APR | MAY | JUN | JUL | AUG | SEP | OCT | NOV | DEC | YEAR |
|---|---|---|---|---|---|---|---|---|---|---|---|---|---|
| Maximum Temp °F | 44.9 | 49.0 | 54.0 | 58.9 | 66.0 | 70.9 | 75.7 | 76.1 | 70.2 | 59.8 | 50.3 | 45.2 | 60.1 |
| Minimum Temp °F | 34.4 | 35.8 | 37.2 | 39.6 | 44.4 | 49.2 | 52.2 | 53.0 | 49.1 | 43.5 | 39.1 | 35.4 | 42.7 |
| Mean Temp °F | 39.7 | 42.4 | 45.6 | 49.3 | 55.3 | 60.1 | 63.9 | 64.5 | 59.7 | 51.7 | 44.7 | 40.3 | 51.4 |
| Days Max Temp ≥ 90 °F | 0 | 0 | 0 | 0 | 0 | 0 | 1 | 1 | 0 | 0 | 0 | 0 | 2 |
| Days Max Temp ≤ 32 °F | 1 | 0 | 0 | 0 | 0 | 0 | 0 | 0 | 0 | 0 | 0 | 1 | 2 |
| Days Min Temp ≤ 32 °F | 13 | 9 | 6 | 3 | 0 | 0 | 0 | 0 | 0 | 1 | 5 | 10 | 47 |
| Days Min Temp ≤ 0 °F | 0 | 0 | 0 | 0 | 0 | 0 | 0 | 0 | 0 | 0 | 0 | 0 | 0 |
| Heating Degree Days | 779 | 631 | 593 | 464 | 298 | 156 | 69 | 55 | 162 | 406 | 601 | 759 | 4973 |
| Cooling Degree Days | 0 | 0 | 0 | 0 | 4 | 14 | 38 | 47 | 8 | 0 | 0 | 0 | 111 |
| Total Precipitation (") | 7.82 | 6.20 | 5.49 | 3.61 | 2.05 | 1.66 | 0.88 | 1.20 | 2.23 | 4.40 | 7.82 | 8.53 | 51.89 |
| Days ≥ 0.1" Precip | 14 | 12 | 11 | 9 | 5 | 4 | 3 | 3 | 5 | 9 | 14 | 14 | 103 |
| Total Snowfall (") | *1.7* | na | *0.4* | *0.0* | *0.0* | *0.0* | *0.0* | *0.0* | *0.0* | *0.0* | *1.0* | *1.1* | na |
| Days ≥ 1" Snow Depth | na | na | *0* | *0* | *0* | *0* | *0* | *0* | *0* | *0* | *1* | *1* | na |

## GRAYLAND *Grays Harbor County* ELEVATION 20 ft LAT/LONG 46° 46 ' N / 124° 5 ' W

| | JAN | FEB | MAR | APR | MAY | JUN | JUL | AUG | SEP | OCT | NOV | DEC | YEAR |
|---|---|---|---|---|---|---|---|---|---|---|---|---|---|
| Maximum Temp °F | *48.4* | 51.1 | 53.5 | 56.0 | 59.7 | 63.0 | 65.6 | 67.0 | 66.8 | 60.7 | 52.9 | *47.4* | 57.7 |
| Minimum Temp °F | *36.1* | 36.7 | 38.1 | 40.3 | 44.1 | 47.9 | 49.9 | 50.1 | 47.6 | 42.9 | 39.4 | *35.5* | 42.4 |
| Mean Temp °F | *42.3* | 44.0 | 45.8 | 48.2 | 51.9 | 55.5 | 57.8 | 58.6 | 57.2 | 51.9 | 46.2 | *41.5* | 50.1 |
| Days Max Temp ≥ 90 °F | 0 | 0 | 0 | 0 | 0 | 0 | 0 | 0 | 0 | 0 | 0 | 0 | 0 |
| Days Max Temp ≤ 32 °F | 0 | 0 | 0 | 0 | 0 | 0 | 0 | 0 | 0 | 0 | 0 | 1 | 1 |
| Days Min Temp ≤ 32 °F | *10* | 8 | 6 | 2 | 0 | 0 | 0 | 0 | 0 | 1 | *5* | *11* | 43 |
| Days Min Temp ≤ 0 °F | 0 | 0 | 0 | 0 | 0 | 0 | 0 | 0 | 0 | 0 | 0 | 0 | 0 |
| Heating Degree Days | *699* | 586 | 589 | 499 | 399 | 280 | 219 | 196 | 232 | 401 | 558 | *721* | 5379 |
| Cooling Degree Days | 0 | 0 | 0 | 1 | 1 | 3 | 3 | 6 | 7 | 0 | 0 | 0 | 21 |
| Total Precipitation (") | 10.08 | 8.14 | 7.86 | 5.39 | 3.46 | 2.29 | 1.33 | 1.72 | 3.09 | 6.87 | 10.23 | 10.98 | 71.44 |
| Days ≥ 0.1" Precip | *17* | 14 | 15 | 12 | 9 | 6 | 3 | 4 | 6 | 11 | na | *16* | na |
| Total Snowfall (") | na | na | *0.0* | *0.0* | *0.0* | *0.0* | *0.0* | *0.0* | *0.0* | *0.0* | na | na | na |
| Days ≥ 1" Snow Depth | na | na | *0* | *0* | *0* | *0* | *0* | *0* | *0* | *0* | na | na | na |

## GRAYS RIVER HATCHERY *Pacific County* ELEVATION 102 ft LAT/LONG 46° 23 ' N / 123° 34 ' W

| | JAN | FEB | MAR | APR | MAY | JUN | JUL | AUG | SEP | OCT | NOV | DEC | YEAR |
|---|---|---|---|---|---|---|---|---|---|---|---|---|---|
| Maximum Temp °F | 47.0 | 50.6 | 54.3 | 57.6 | 63.6 | 67.7 | 72.6 | 73.7 | 71.2 | 62.6 | 52.4 | 46.7 | 60.0 |
| Minimum Temp °F | 32.4 | 33.5 | 34.8 | 37.3 | 41.8 | 46.4 | 49.2 | 49.0 | 45.3 | 39.7 | 36.2 | 33.4 | 39.9 |
| Mean Temp °F | 39.7 | 42.1 | 44.6 | 47.5 | 52.6 | 57.1 | 60.9 | 61.4 | 58.3 | 51.2 | 44.3 | 40.0 | 50.0 |
| Days Max Temp ≥ 90 °F | 0 | 0 | 0 | 0 | 0 | 1 | 1 | 2 | 1 | 0 | 0 | 0 | 5 |
| Days Max Temp ≤ 32 °F | 0 | 0 | 0 | 0 | 0 | 0 | 0 | 0 | 0 | 0 | 0 | 1 | 1 |
| Days Min Temp ≤ 32 °F | 16 | 13 | 11 | 6 | 1 | 0 | 0 | 0 | 0 | 4 | 10 | 14 | 75 |
| Days Min Temp ≤ 0 °F | 0 | 0 | 0 | 0 | 0 | 0 | 0 | 0 | 0 | 0 | 0 | 0 | 0 |
| Heating Degree Days | 777 | 641 | 626 | 519 | 379 | 239 | 142 | 129 | 206 | 422 | 613 | 767 | 5460 |
| Cooling Degree Days | 0 | 0 | 0 | 0 | 4 | 9 | 20 | 25 | 11 | 0 | 0 | 0 | 69 |
| Total Precipitation (") | 16.41 | 12.74 | 11.32 | 7.82 | 4.93 | 3.67 | 1.95 | 2.26 | 4.47 | 9.15 | 15.23 | 17.52 | 107.47 |
| Days ≥ 0.1" Precip | 17 | 15 | 17 | 14 | 10 | 8 | 4 | 5 | 7 | 12 | 18 | 18 | 145 |
| Total Snowfall (") | 2.9 | 1.3 | 0.3 | 0.0 | 0.0 | 0.0 | 0.0 | 0.0 | 0.0 | 0.0 | 0.8 | 1.7 | 7.0 |
| Days ≥ 1" Snow Depth | *1* | *0* | 0 | 0 | 0 | 0 | 0 | 0 | 0 | 0 | 0 | *1* | 2 |

## HARRINGTON 1 NW *Lincoln County* ELEVATION 2260 ft LAT/LONG 47° 29 ' N / 118° 11 ' W

| | JAN | FEB | MAR | APR | MAY | JUN | JUL | AUG | SEP | OCT | NOV | DEC | YEAR |
|---|---|---|---|---|---|---|---|---|---|---|---|---|---|
| Maximum Temp °F | 31.6 | 38.8 | 49.0 | 58.6 | 68.2 | 76.1 | 83.9 | 83.3 | 73.6 | 59.6 | 41.4 | 32.4 | 58.0 |
| Minimum Temp °F | 19.9 | 25.1 | 29.5 | 33.0 | 38.6 | 44.4 | 48.5 | 48.4 | 41.3 | 32.9 | 27.5 | 21.0 | 34.2 |
| Mean Temp °F | 25.8 | 32.0 | 39.3 | 45.8 | 53.4 | 60.3 | 66.2 | 65.9 | 57.5 | 46.3 | 34.5 | 26.7 | 46.1 |
| Days Max Temp ≥ 90 °F | 0 | 0 | 0 | 0 | 0 | 2 | 9 | 8 | 1 | 0 | 0 | 0 | 20 |
| Days Max Temp ≤ 32 °F | 15 | 5 | 1 | 0 | 0 | 0 | 0 | 0 | 0 | 0 | 4 | 14 | 39 |
| Days Min Temp ≤ 32 °F | 28 | 24 | 22 | 15 | 7 | 1 | 0 | 0 | 4 | 15 | 21 | 28 | 165 |
| Days Min Temp ≤ 0 °F | 2 | 1 | 0 | 0 | 0 | 0 | 0 | 0 | 0 | 0 | 0 | 2 | 5 |
| Heating Degree Days | 1209 | 925 | 790 | 569 | 359 | 170 | 63 | 68 | 235 | 574 | 909 | 1179 | 7050 |
| Cooling Degree Days | 0 | 0 | 0 | 1 | 10 | 29 | 90 | 94 | 17 | 1 | 0 | 0 | 242 |
| Total Precipitation (") | 1.37 | 1.09 | 1.14 | 1.00 | 1.20 | 0.88 | 0.67 | 0.51 | 0.54 | 0.86 | 1.61 | 1.78 | 12.65 |
| Days ≥ 0.1" Precip | 5 | 4 | 4 | 4 | 4 | 3 | 2 | 2 | 2 | 3 | 5 | 6 | 44 |
| Total Snowfall (") | 9.5 | 4.3 | 1.4 | 0.2 | 0.0 | 0.0 | 0.0 | 0.0 | 0.0 | 0.1 | 3.6 | 9.5 | 28.6 |
| Days ≥ 1" Snow Depth | 19 | 9 | 2 | 0 | 0 | 0 | 0 | 0 | 0 | 0 | 3 | 14 | 47 |

## HARTLINE *Grant County* ELEVATION 1913 ft LAT/LONG 47° 41 ' N / 119° 6 ' W

| | JAN | FEB | MAR | APR | MAY | JUN | JUL | AUG | SEP | OCT | NOV | DEC | YEAR |
|---|---|---|---|---|---|---|---|---|---|---|---|---|---|
| Maximum Temp °F | 32.5 | 40.2 | 51.4 | 61.4 | 71.2 | 79.7 | 87.3 | 86.6 | 77.2 | 62.1 | 43.4 | 33.5 | 60.5 |
| Minimum Temp °F | 19.6 | 25.1 | 30.7 | 35.6 | 42.6 | 49.3 | 54.9 | 54.7 | 46.9 | 36.8 | 28.2 | 21.0 | 37.1 |
| Mean Temp °F | 26.1 | 32.7 | 41.1 | 48.5 | 56.9 | 64.5 | 71.1 | 70.7 | 62.1 | 49.5 | 35.8 | 27.3 | 48.9 |
| Days Max Temp ≥ 90 °F | 0 | 0 | 0 | 0 | 1 | 5 | 14 | 11 | 3 | 0 | 0 | 0 | 34 |
| Days Max Temp ≤ 32 °F | 14 | 4 | 0 | 0 | 0 | 0 | 0 | 0 | 0 | 0 | 3 | 13 | 34 |
| Days Min Temp ≤ 32 °F | 29 | 24 | 20 | 10 | 2 | 0 | 0 | 0 | 1 | 8 | 21 | 28 | 143 |
| Days Min Temp ≤ 0 °F | 2 | 1 | 0 | 0 | 0 | 0 | 0 | 0 | 0 | 0 | 0 | 1 | 4 |
| Heating Degree Days | 1199 | 907 | 734 | 489 | 262 | 95 | 21 | 24 | 139 | 477 | 869 | 1163 | 6379 |
| Cooling Degree Days | 0 | 0 | 0 | 2 | 26 | 77 | 202 | 203 | 64 | 3 | 0 | 0 | 577 |
| Total Precipitation (") | 1.05 | 0.88 | 0.83 | 0.82 | 0.94 | 0.80 | 0.68 | 0.45 | 0.48 | 0.58 | 1.43 | ***1.54*** | 10.48 |
| Days ≥ 0.1" Precip | ***3*** | 3 | 3 | 3 | 3 | 3 | 1 | 1 | 1 | 2 | 4 | ***4*** | 31 |
| Total Snowfall (") | ***6.7*** | 3.6 | 0.5 | 0.2 | 0.0 | 0.0 | 0.0 | 0.0 | 0.0 | 0.2 | 2.9 | ***9.9*** | 24.0 |
| Days ≥ 1" Snow Depth | na | na | 0 | 0 | 0 | 0 | 0 | 0 | 0 | 0 | na | na | na |

## HATTON 9 SE *Franklin County* ELEVATION 1371 ft LAT/LONG 46° 45 ' N / 118° 37 ' W

| | JAN | FEB | MAR | APR | MAY | JUN | JUL | AUG | SEP | OCT | NOV | DEC | YEAR |
|---|---|---|---|---|---|---|---|---|---|---|---|---|---|
| Maximum Temp °F | 36.2 | 44.3 | 54.2 | 63.1 | 73.0 | 81.5 | 88.8 | 88.1 | 78.1 | 63.9 | 46.0 | 36.7 | 62.8 |
| Minimum Temp °F | 24.0 | 28.9 | 32.6 | 36.7 | 43.1 | 50.1 | 55.4 | 55.0 | 47.1 | 37.5 | 31.3 | 25.0 | 38.9 |
| Mean Temp °F | 30.1 | 36.7 | 43.4 | 49.9 | 58.1 | 65.8 | 72.2 | 71.6 | 62.7 | 50.7 | 38.7 | 30.9 | 50.9 |
| Days Max Temp ≥ 90 °F | 0 | 0 | 0 | 0 | 1 | 7 | 15 | 13 | 3 | 0 | 0 | 0 | 39 |
| Days Max Temp ≤ 32 °F | 11 | 3 | 0 | 0 | 0 | 0 | 0 | 0 | 0 | 0 | 2 | 9 | 25 |
| Days Min Temp ≤ 32 °F | 24 | ***19*** | 15 | 9 | ***2*** | 0 | 0 | 0 | 1 | 8 | 16 | 24 | 118 |
| Days Min Temp ≤ 0 °F | 2 | 0 | 0 | 0 | 0 | 0 | 0 | 0 | 0 | 0 | 0 | 1 | 3 |
| Heating Degree Days | 1075 | 794 | 662 | 447 | 234 | 76 | 15 | 15 | 125 | 439 | 782 | 1051 | 5715 |
| Cooling Degree Days | 0 | 0 | 0 | 3 | 36 | 106 | 238 | 231 | 72 | 3 | 0 | 0 | 689 |
| Total Precipitation (") | 1.08 | 0.84 | 0.99 | 0.83 | 0.84 | 0.54 | 0.36 | 0.43 | 0.47 | 0.76 | 1.38 | 1.38 | 9.90 |
| Days ≥ 0.1" Precip | 4 | 3 | 3 | 3 | 3 | 2 | 1 | 1 | 2 | 2 | 5 | 5 | 34 |
| Total Snowfall (") | ***4.9*** | 2.0 | 0.4 | 0.0 | 0.0 | 0.0 | 0.0 | 0.0 | 0.0 | 0.0 | 1.0 | 4.5 | 12.8 |
| Days ≥ 1" Snow Depth | ***8*** | 2 | 0 | 0 | 0 | 0 | 0 | 0 | 0 | 0 | 1 | 6 | 17 |

## HOLDEN VILLAGE *Chelan County* ELEVATION 3222 ft LAT/LONG 48° 12 ' N / 120° 47 ' W

| | JAN | FEB | MAR | APR | MAY | JUN | JUL | AUG | SEP | OCT | NOV | DEC | YEAR |
|---|---|---|---|---|---|---|---|---|---|---|---|---|---|
| Maximum Temp °F | 30.3 | 36.6 | 44.2 | 51.9 | 62.2 | 69.2 | 76.9 | 77.2 | 68.6 | 54.8 | 37.2 | 29.5 | 53.2 |
| Minimum Temp °F | 15.2 | 18.0 | 22.5 | 28.1 | 34.2 | 40.3 | 43.4 | 43.9 | 37.4 | 30.2 | 23.1 | 15.2 | 29.3 |
| Mean Temp °F | 22.7 | 27.3 | 33.3 | 40.0 | 48.2 | 54.8 | 60.2 | 60.6 | 53.0 | 42.6 | 30.2 | 22.2 | 41.3 |
| Days Max Temp ≥ 90 °F | 0 | 0 | 0 | 0 | 0 | 0 | 2 | 3 | 0 | 0 | 0 | 0 | 5 |
| Days Max Temp ≤ 32 °F | 17 | 7 | 1 | 0 | 0 | 0 | 0 | 0 | 0 | 0 | 7 | 20 | 52 |
| Days Min Temp ≤ 32 °F | 30 | 28 | 30 | 25 | 13 | 3 | 0 | 1 | 7 | 21 | 28 | 30 | 216 |
| Days Min Temp ≤ 0 °F | 3 | 2 | 0 | 0 | 0 | 0 | 0 | 0 | 0 | 0 | 0 | 3 | 8 |
| Heating Degree Days | 1303 | 1057 | 975 | 743 | 514 | 304 | 168 | 157 | 355 | 688 | 1038 | 1318 | 8620 |
| Cooling Degree Days | 0 | 0 | 0 | 0 | 1 | 6 | 22 | 22 | 3 | 0 | 0 | 0 | 54 |
| Total Precipitation (") | 6.50 | 4.43 | 2.84 | 1.48 | 0.95 | 1.14 | 0.74 | 1.08 | 1.67 | 3.24 | 6.31 | 7.33 | 37.71 |
| Days ≥ 0.1" Precip | 11 | 9 | 6 | 4 | 3 | 3 | 2 | 3 | 4 | 7 | ***13*** | 14 | 79 |
| Total Snowfall (") | 62.1 | 40.0 | 21.3 | 6.3 | 0.5 | 0.0 | 0.0 | 0.0 | 0.0 | 5.3 | ***40.1*** | 72.4 | 248.0 |
| Days ≥ 1" Snow Depth | 31 | 28 | 31 | 24 | 3 | 0 | 0 | 0 | 0 | 2 | ***20*** | 30 | 169 |

## HOQUIAM AP *Grays Harbor County* ELEVATION 11 ft LAT/LONG 46° 58' N / 123° 56' W

| | JAN | FEB | MAR | APR | MAY | JUN | JUL | AUG | SEP | OCT | NOV | DEC | YEAR |
|---|---|---|---|---|---|---|---|---|---|---|---|---|---|
| Maximum Temp °F | 46.4 | 49.6 | 52.8 | 55.8 | 60.4 | 63.9 | 67.3 | 68.5 | 67.6 | 60.1 | 51.8 | 46.4 | 57.5 |
| Minimum Temp °F | 36.8 | 37.9 | 39.2 | 41.7 | 46.0 | 49.9 | 52.4 | 53.4 | 51.0 | 45.7 | 40.7 | 37.1 | 44.3 |
| Mean Temp °F | 41.6 | 43.8 | 46.0 | 48.8 | 53.2 | 57.0 | 59.9 | 60.9 | 59.3 | 52.9 | 46.3 | 41.8 | 51.0 |
| Days Max Temp ≥ 90 °F | 0 | 0 | 0 | 0 | 0 | 0 | 0 | 0 | 0 | 0 | 0 | 0 | 0 |
| Days Max Temp ≤ 32 °F | 1 | 0 | 0 | 0 | 0 | 0 | 0 | 0 | 0 | 0 | 0 | 1 | 2 |
| Days Min Temp ≤ 32 °F | 7 | 6 | 4 | 1 | 0 | 0 | 0 | 0 | 0 | 0 | 3 | 8 | 29 |
| Days Min Temp ≤ 0 °F | 0 | 0 | 0 | 0 | 0 | 0 | 0 | 0 | 0 | 0 | 0 | 0 | 0 |
| Heating Degree Days | 718 | 593 | 581 | 480 | 360 | 238 | 159 | 130 | 175 | 368 | 556 | 713 | 5071 |
| Cooling Degree Days | 0 | 0 | 0 | 0 | 2 | 4 | 6 | 13 | 11 | 0 | 0 | 0 | 36 |
| Total Precipitation (") | 9.76 | 7.97 | 6.97 | 4.86 | 3.21 | 2.15 | 1.16 | 1.48 | 3.02 | 6.08 | 9.73 | 10.55 | 66.94 |
| Days ≥ 0.1" Precip | 16 | 14 | 14 | 12 | 8 | 6 | 3 | 3 | 6 | 11 | 16 | 17 | 126 |
| Total Snowfall (") | 2.8 | 0.5 | 0.2 | 0.0 | 0.0 | 0.0 | 0.0 | 0.0 | 0.0 | 0.0 | 0.2 | 1.1 | 4.8 |
| Days ≥ 1" Snow Depth | 2 | 0 | 0 | 0 | 0 | 0 | 0 | 0 | 0 | 0 | 0 | 0 | 2 |

## ICE HARBOR DAM *Walla Walla County* ELEVATION 479 ft LAT/LONG 46° 15' N / 118° 52' W

| | JAN | FEB | MAR | APR | MAY | JUN | JUL | AUG | SEP | OCT | NOV | DEC | YEAR |
|---|---|---|---|---|---|---|---|---|---|---|---|---|---|
| Maximum Temp °F | 40.5 | 48.0 | 57.6 | 65.4 | 73.8 | 81.1 | 88.5 | 88.2 | 79.5 | 66.8 | 51.3 | 41.6 | 65.2 |
| Minimum Temp °F | 26.8 | 30.0 | 34.5 | 40.4 | 47.2 | 53.9 | 59.2 | 59.4 | 50.5 | 40.9 | 34.7 | 28.2 | 42.1 |
| Mean Temp °F | 33.7 | 39.0 | 46.1 | 52.9 | 60.6 | 67.5 | 73.9 | 73.8 | 65.0 | 53.9 | 43.0 | 35.0 | 53.7 |
| Days Max Temp ≥ 90 °F | 0 | 0 | 0 | 0 | 2 | 6 | 15 | 14 | 3 | 0 | 0 | 0 | 40 |
| Days Max Temp ≤ 32 °F | 7 | 2 | 0 | 0 | 0 | 0 | 0 | 0 | 0 | 0 | 1 | 5 | 15 |
| Days Min Temp ≤ 32 °F | 21 | 17 | 10 | 2 | 0 | 0 | 0 | 0 | 0 | 3 | 11 | 21 | 85 |
| Days Min Temp ≤ 0 °F | 1 | 0 | 0 | 0 | 0 | 0 | 0 | 0 | 0 | 0 | 0 | 1 | 2 |
| Heating Degree Days | 965 | 729 | 580 | 360 | 165 | 44 | 4 | 4 | 76 | 343 | 654 | 925 | 4849 |
| Cooling Degree Days | 0 | 0 | *0* | 3 | 40 | 120 | 269 | 279 | 85 | 4 | 0 | 0 | 800 |
| Total Precipitation (") | 1.22 | 0.92 | 0.96 | 0.73 | 0.83 | 0.65 | 0.27 | 0.56 | 0.48 | 0.67 | 1.47 | 1.36 | 10.12 |
| Days ≥ 0.1" Precip | 4 | 3 | 3 | 2 | 2 | 2 | 1 | 1 | 2 | 2 | 5 | 5 | 32 |
| Total Snowfall (") | na | *1.3* | 0.1 | 0.0 | 0.0 | 0.0 | 0.0 | 0.0 | 0.0 | 0.0 | *0.1* | *1.4* | na |
| Days ≥ 1" Snow Depth | na | *2* | 0 | 0 | 0 | 0 | 0 | 0 | 0 | 0 | 0 | *1* | na |

## KENNEWICK *Benton County* ELEVATION 390 ft LAT/LONG 46° 13' N / 119° 8' W

| | JAN | FEB | MAR | APR | MAY | JUN | JUL | AUG | SEP | OCT | NOV | DEC | YEAR |
|---|---|---|---|---|---|---|---|---|---|---|---|---|---|
| Maximum Temp °F | 40.3 | 48.1 | 58.6 | 66.9 | 75.6 | 83.2 | 90.2 | 89.3 | 79.5 | 66.3 | 50.1 | 41.1 | 65.8 |
| Minimum Temp °F | 27.1 | 30.4 | 35.1 | 41.1 | 47.9 | 54.7 | 59.3 | 58.8 | 50.2 | 40.6 | 34.5 | 28.1 | 42.3 |
| Mean Temp °F | 33.7 | 39.3 | 46.9 | 54.0 | 61.8 | 68.9 | 74.7 | 74.1 | 64.8 | 53.5 | 42.4 | 34.6 | 54.1 |
| Days Max Temp ≥ 90 °F | 0 | 0 | 0 | 0 | 3 | 8 | 18 | 16 | 4 | 0 | 0 | 0 | 49 |
| Days Max Temp ≤ 32 °F | 7 | 2 | 0 | 0 | 0 | 0 | 0 | 0 | 0 | 0 | 1 | 7 | 17 |
| Days Min Temp ≤ 32 °F | 21 | 16 | 11 | 3 | 0 | 0 | 0 | 0 | 0 | 4 | 11 | 21 | 87 |
| Days Min Temp ≤ 0 °F | 0 | 0 | 0 | 0 | 0 | 0 | 0 | 0 | 0 | 0 | 0 | 1 | 1 |
| Heating Degree Days | 963 | 718 | 555 | 325 | 141 | 29 | 3 | 4 | 76 | 354 | 673 | 935 | 4776 |
| Cooling Degree Days | 0 | 0 | 0 | 5 | 54 | 146 | 305 | 298 | 85 | 4 | 0 | 0 | 897 |
| Total Precipitation (") | 1.10 | 0.69 | 0.66 | 0.53 | 0.58 | 0.39 | 0.22 | 0.42 | 0.35 | 0.53 | 1.09 | 1.07 | 7.63 |
| Days ≥ 0.1" Precip | 4 | 3 | 2 | 2 | 2 | 1 | 1 | 1 | 1 | 2 | 4 | 4 | 27 |
| Total Snowfall (") | na | *0.5* | *0.0* | 0.0 | 0.0 | 0.0 | 0.0 | 0.0 | 0.0 | 0.0 | 0.2 | na | na |
| Days ≥ 1" Snow Depth | na | *0* | 0 | 0 | 0 | 0 | 0 | 0 | 0 | 0 | *0* | na | na |

## KENT *King County* ELEVATION 30 ft LAT/LONG 47° 23' N / 122° 14' W

| | JAN | FEB | MAR | APR | MAY | JUN | JUL | AUG | SEP | OCT | NOV | DEC | YEAR |
|---|---|---|---|---|---|---|---|---|---|---|---|---|---|
| Maximum Temp °F | 46.3 | 51.2 | 55.6 | 60.6 | 67.6 | 72.7 | 77.8 | 78.2 | 72.2 | *62.1* | 51.6 | 45.8 | 61.8 |
| Minimum Temp °F | 33.6 | 35.0 | 36.8 | 39.6 | *44.2* | 49.5 | 51.8 | 51.9 | 47.9 | *42.0* | 37.2 | 33.6 | 41.9 |
| Mean Temp °F | 40.0 | 43.1 | 46.2 | 50.1 | *55.9* | 61.2 | 64.8 | 65.0 | 60.1 | *52.1* | 44.4 | 39.7 | 51.9 |
| Days Max Temp ≥ 90 °F | 0 | 0 | 0 | 0 | 0 | 1 | 2 | 2 | 1 | 0 | 0 | 0 | 6 |
| Days Max Temp ≤ 32 °F | 1 | 0 | 0 | 0 | 0 | 0 | 0 | 0 | 0 | 0 | 0 | 1 | 2 |
| Days Min Temp ≤ 32 °F | 13 | 10 | 8 | 4 | 0 | 0 | 0 | 0 | 0 | *3* | 8 | 13 | 59 |
| Days Min Temp ≤ 0 °F | 0 | 0 | 0 | 0 | 0 | 0 | 0 | 0 | 0 | 0 | 0 | 0 | 0 |
| Heating Degree Days | 769 | 611 | 576 | 440 | *279* | 128 | 53 | 49 | 154 | *394* | 611 | 775 | 4839 |
| Cooling Degree Days | 0 | 0 | 0 | 0 | na | *18* | *49* | *62* | 20 | *0* | 0 | 0 | na |
| Total Precipitation (") | 5.43 | 4.07 | 3.91 | 2.92 | 1.82 | 1.68 | 0.83 | 1.25 | 1.98 | 3.06 | 5.46 | 5.98 | 38.39 |
| Days ≥ 0.1" Precip | 12 | 10 | 11 | 8 | *5* | 4 | 2 | 3 | 5 | *8* | 12 | 12 | 92 |
| Total Snowfall (") | *1.6* | 0.1 | 0.4 | 0.0 | 0.0 | 0.0 | 0.0 | 0.0 | 0.0 | 0.0 | 0.1 | *1.0* | 3.2 |
| Days ≥ 1" Snow Depth | *1* | 0 | 0 | 0 | 0 | 0 | 0 | 0 | 0 | *0* | 0 | *1* | 2 |

## LACROSSE *Whitman County* ELEVATION 1480 ft LAT/LONG 46° 49 ' N / 117° 53 ' W

| | JAN | FEB | MAR | APR | MAY | JUN | JUL | AUG | SEP | OCT | NOV | DEC | YEAR |
|---|---|---|---|---|---|---|---|---|---|---|---|---|---|
| Maximum Temp °F | 37.2 | 44.9 | 53.7 | 62.0 | 71.4 | 79.6 | 88.0 | 87.3 | 77.3 | 64.2 | 46.7 | 37.9 | 62.5 |
| Minimum Temp °F | 24.1 | 28.3 | 31.2 | 35.8 | 40.9 | 47.4 | 51.6 | 51.2 | 43.7 | 34.9 | 30.3 | 25.2 | 37.1 |
| Mean Temp °F | 30.7 | 36.6 | 42.5 | 48.9 | 56.2 | 63.5 | 69.9 | 69.3 | 60.5 | 49.6 | 38.5 | 31.6 | 49.8 |
| Days Max Temp ≥ 90 °F | 0 | 0 | 0 | 0 | 1 | 5 | 15 | 13 | 3 | 0 | 0 | 0 | 37 |
| Days Max Temp ≤ 32 °F | 9 | 3 | 0 | 0 | 0 | 0 | 0 | 0 | 0 | 0 | 2 | 8 | 22 |
| Days Min Temp ≤ 32 °F | 24 | 19 | 18 | 10 | 4 | 0 | 0 | 0 | 2 | 11 | 17 | 24 | 129 |
| Days Min Temp ≤ 0 °F | 2 | 1 | 0 | 0 | 0 | 0 | 0 | 0 | 0 | 0 | 0 | 1 | 4 |
| Heating Degree Days | 1057 | 795 | 690 | 478 | 280 | 105 | 25 | 28 | 163 | 472 | 787 | 1030 | 5910 |
| Cooling Degree Days | 0 | 0 | 0 | 2 | 20 | 64 | 168 | 168 | 35 | *1* | 0 | 0 | 458 |
| Total Precipitation (") | 1.82 | 1.32 | 1.38 | 1.18 | 1.11 | 0.88 | 0.48 | 0.48 | 0.60 | 0.89 | 1.87 | 2.01 | 14.02 |
| Days ≥ 0.1" Precip | 6 | 4 | 5 | 4 | 3 | 3 | 1 | 2 | 2 | 3 | 6 | 7 | 46 |
| Total Snowfall (") | 5.8 | 2.9 | 0.8 | 0.1 | 0.0 | 0.0 | 0.0 | 0.0 | 0.0 | 0.0 | 1.7 | 5.1 | 16.4 |
| Days ≥ 1" Snow Depth | 10 | 4 | 0 | 0 | 0 | 0 | 0 | 0 | 0 | 0 | 1 | 7 | 22 |

## LANDSBURG *King County* ELEVATION 541 ft LAT/LONG 47° 23 ' N / 121° 58 ' W

| | JAN | FEB | MAR | APR | MAY | JUN | JUL | AUG | SEP | OCT | NOV | DEC | YEAR |
|---|---|---|---|---|---|---|---|---|---|---|---|---|---|
| Maximum Temp °F | 43.3 | 48.3 | 52.7 | 57.4 | 64.5 | 69.6 | 75.2 | 75.5 | 69.6 | 59.6 | 48.8 | 42.7 | 58.9 |
| Minimum Temp °F | 31.7 | 32.8 | 34.1 | 37.2 | 42.1 | 47.2 | 50.3 | 49.8 | 45.2 | 39.6 | 35.4 | 31.6 | 39.8 |
| Mean Temp °F | 37.5 | 40.6 | 43.4 | 47.3 | 53.3 | 58.5 | 62.8 | 62.7 | 57.4 | 49.6 | 42.1 | 37.2 | 49.4 |
| Days Max Temp ≥ 90 °F | 0 | 0 | 0 | 0 | 0 | 1 | 1 | 2 | 0 | 0 | 0 | 0 | 4 |
| Days Max Temp ≤ 32 °F | 1 | 0 | 0 | 0 | 0 | 0 | 0 | 0 | 0 | 0 | 0 | 2 | 3 |
| Days Min Temp ≤ 32 °F | 16 | 14 | 12 | 6 | 1 | 0 | 0 | 0 | 0 | 4 | 10 | 16 | 79 |
| Days Min Temp ≤ 0 °F | 0 | 0 | 0 | 0 | 0 | 0 | 0 | 0 | 0 | 0 | 0 | 0 | 0 |
| Heating Degree Days | 845 | 683 | 662 | 523 | 359 | 203 | 99 | 102 | 227 | 469 | 679 | 855 | 5706 |
| Cooling Degree Days | 0 | 0 | 0 | 0 | 5 | 12 | 40 | 45 | 6 | 0 | 0 | 0 | 108 |
| Total Precipitation (") | 7.80 | 5.87 | 5.33 | 4.61 | 3.37 | 3.01 | 1.56 | 1.72 | 2.93 | 4.52 | 7.59 | 7.88 | 56.19 |
| Days ≥ 0.1" Precip | 15 | 13 | 13 | 12 | 9 | 7 | 4 | 4 | 7 | 10 | 15 | 15 | 124 |
| Total Snowfall (") | *2.7* | *1.7* | *0.3* | 0.0 | 0.0 | 0.0 | 0.0 | 0.0 | 0.0 | 0.0 | 1.2 | *2.8* | 8.7 |
| Days ≥ 1" Snow Depth | 6 | 2 | 1 | 0 | 0 | 0 | 0 | 0 | 0 | 0 | 1 | 4 | 14 |

## LEAVENWORTH 3 S *Chelan County* ELEVATION 1161 ft LAT/LONG 47° 36 ' N / 120° 39 ' W

| | JAN | FEB | MAR | APR | MAY | JUN | JUL | AUG | SEP | OCT | NOV | DEC | YEAR |
|---|---|---|---|---|---|---|---|---|---|---|---|---|---|
| Maximum Temp °F | 33.6 | 42.4 | 52.7 | 62.6 | 72.5 | 79.9 | 87.5 | 87.7 | 78.2 | 64.1 | 43.7 | 33.4 | 61.5 |
| Minimum Temp °F | 17.2 | 21.8 | 27.0 | 33.3 | 39.7 | 46.3 | 50.3 | 49.9 | 41.8 | 32.8 | 27.8 | 19.7 | 34.0 |
| Mean Temp °F | 25.5 | 32.1 | 39.8 | 48.0 | 56.1 | 63.1 | 68.9 | 68.8 | 60.0 | 48.4 | 35.7 | 26.5 | 47.7 |
| Days Max Temp ≥ 90 °F | 0 | 0 | 0 | 0 | 1 | 5 | 14 | 14 | 4 | 0 | 0 | 0 | 38 |
| Days Max Temp ≤ 32 °F | 12 | 3 | 0 | 0 | 0 | 0 | 0 | 0 | 0 | 0 | 2 | 13 | 30 |
| Days Min Temp ≤ 32 °F | 30 | 27 | 25 | 14 | 4 | 0 | 0 | 0 | 2 | 15 | 22 | 30 | 169 |
| Days Min Temp ≤ 0 °F | 3 | 1 | 0 | 0 | 0 | 0 | 0 | 0 | 0 | 0 | 0 | 1 | 5 |
| Heating Degree Days | 1220 | 922 | 773 | 504 | 280 | 113 | 35 | 32 | 173 | 507 | 871 | 1186 | 6616 |
| Cooling Degree Days | 0 | 0 | 0 | 0 | 16 | 64 | 155 | 156 | 33 | 0 | 0 | 0 | 424 |
| Total Precipitation (") | 4.78 | 2.90 | 1.88 | 1.11 | 0.67 | 0.74 | 0.34 | 0.61 | 0.71 | 1.66 | 3.80 | 5.06 | 24.26 |
| Days ≥ 0.1" Precip | 10 | 7 | 5 | 3 | 2 | 2 | 1 | 2 | 2 | 4 | 9 | 10 | 57 |
| Total Snowfall (") | 31.6 | 15.1 | *3.8* | 0.2 | 0.0 | 0.0 | 0.0 | 0.0 | 0.0 | 0.1 | 10.6 | 32.3 | 93.7 |
| Days ≥ 1" Snow Depth | 28 | 24 | *13* | 0 | 0 | 0 | 0 | 0 | 0 | 0 | 6 | *23* | 94 |

## LIND 3 NE *Adams County* ELEVATION 1631 ft LAT/LONG 47° 0 ' N / 118° 35 ' W

| | JAN | FEB | MAR | APR | MAY | JUN | JUL | AUG | SEP | OCT | NOV | DEC | YEAR |
|---|---|---|---|---|---|---|---|---|---|---|---|---|---|
| Maximum Temp °F | 35.7 | 43.9 | 53.8 | 62.6 | 72.4 | 80.5 | 88.3 | 87.6 | 77.9 | 64.0 | 45.6 | 36.2 | 62.4 |
| Minimum Temp °F | 23.0 | 27.9 | 31.5 | 35.3 | 41.3 | 47.7 | 52.6 | 52.3 | 44.8 | 35.9 | 30.1 | 24.0 | 37.2 |
| Mean Temp °F | 29.4 | 35.9 | 42.7 | 48.9 | 56.8 | 64.1 | 70.5 | 70.0 | 61.4 | 50.0 | 37.9 | 30.1 | 49.8 |
| Days Max Temp ≥ 90 °F | 0 | 0 | 0 | 0 | 1 | 5 | 15 | 13 | 3 | 0 | 0 | 0 | 37 |
| Days Max Temp ≤ 32 °F | 11 | 3 | 0 | 0 | 0 | 0 | 0 | 0 | 0 | 0 | 2 | 10 | 26 |
| Days Min Temp ≤ 32 °F | 25 | 21 | 18 | 11 | 3 | 0 | 0 | 0 | 1 | 10 | 18 | 26 | 133 |
| Days Min Temp ≤ 0 °F | 2 | 1 | 0 | 0 | 0 | 0 | 0 | 0 | 0 | 0 | 0 | 1 | 4 |
| Heating Degree Days | 1098 | 815 | 686 | 477 | 266 | 99 | 24 | 27 | 148 | 460 | 807 | 1075 | 5982 |
| Cooling Degree Days | 0 | 0 | 0 | 2 | 27 | 73 | 186 | 182 | 48 | 2 | 0 | 0 | 520 |
| Total Precipitation (") | 0.97 | 0.72 | 0.91 | 0.84 | 0.79 | 0.58 | 0.40 | 0.40 | 0.48 | 0.73 | 1.32 | 1.22 | 9.36 |
| Days ≥ 0.1" Precip | 4 | 3 | 3 | 3 | 2 | 2 | 1 | 1 | 2 | 2 | 5 | 5 | 33 |
| Total Snowfall (") | 5.3 | 2.5 | 0.6 | 0.1 | 0.0 | 0.0 | 0.0 | 0.0 | 0.0 | 0.1 | 1.8 | 5.2 | 15.6 |
| Days ≥ 1" Snow Depth | 10 | 3 | 0 | 0 | 0 | 0 | 0 | 0 | 0 | 0 | 1 | 7 | 21 |

## LONG BEACH EXP STN *Pacific County* ELEVATION 30 ft LAT/LONG 46° 22 ' N / 124° 2 ' W

| | JAN | FEB | MAR | APR | MAY | JUN | JUL | AUG | SEP | OCT | NOV | DEC | YEAR |
|---|---|---|---|---|---|---|---|---|---|---|---|---|---|
| Maximum Temp °F | 48.4 | 51.2 | 53.3 | 55.3 | 59.5 | 62.5 | 65.4 | 66.6 | 66.9 | 60.9 | 53.3 | 48.7 | 57.7 |
| Minimum Temp °F | 35.1 | 36.5 | 38.4 | 40.5 | 44.6 | 48.8 | 51.3 | 51.2 | 47.3 | 42.0 | 38.9 | 35.8 | 42.5 |
| Mean Temp °F | 41.8 | 43.9 | 45.9 | 47.9 | 52.1 | 55.7 | 58.4 | 58.9 | 57.1 | 51.5 | 46.1 | 42.3 | 50.1 |
| Days Max Temp ≥ 90 °F | 0 | 0 | 0 | 0 | 0 | 0 | 0 | 0 | 0 | 0 | 0 | 0 | 0 |
| Days Max Temp ≤ 32 °F | 0 | 0 | 0 | 0 | 0 | 0 | 0 | 0 | 0 | 0 | 0 | 1 | 1 |
| Days Min Temp ≤ 32 °F | 12 | 9 | 6 | 2 | 0 | 0 | 0 | 0 | 0 | 2 | 7 | 11 | 49 |
| Days Min Temp ≤ 0 °F | 0 | 0 | 0 | 0 | 0 | 0 | 0 | 0 | 0 | 0 | 0 | 0 | 0 |
| Heating Degree Days | 714 | 590 | 586 | 506 | 394 | 274 | 201 | 185 | 235 | 413 | 559 | 698 | 5355 |
| Cooling Degree Days | 0 | 0 | 0 | 0 | 1 | 2 | 2 | 4 | 5 | 1 | 0 | 0 | 15 |
| Total Precipitation (") | 11.03 | 9.49 | 8.76 | 5.98 | 3.91 | 2.92 | 1.70 | 1.85 | 3.82 | 6.73 | 11.34 | 12.82 | 80.35 |
| Days ≥ 0.1" Precip | 17 | 15 | 16 | 13 | 9 | 7 | 4 | 4 | 7 | 11 | 17 | 18 | 138 |
| Total Snowfall (") | *0.7* | 0.5 | 0.1 | 0.0 | 0.0 | 0.0 | 0.0 | 0.0 | 0.0 | 0.0 | 0.0 | 0.5 | 1.8 |
| Days ≥ 1" Snow Depth | *0* | *0* | 0 | 0 | 0 | 0 | 0 | 0 | 0 | 0 | 0 | 1 | 1 |

## LONGVIEW *Cowlitz County* ELEVATION 10 ft LAT/LONG 46° 10 ' N / 122° 55 ' W

| | JAN | FEB | MAR | APR | MAY | JUN | JUL | AUG | SEP | OCT | NOV | DEC | YEAR |
|---|---|---|---|---|---|---|---|---|---|---|---|---|---|
| Maximum Temp °F | 45.4 | 50.6 | 55.8 | 60.2 | 66.7 | 71.6 | 76.8 | 78.0 | 73.2 | 63.2 | 51.8 | 45.4 | 61.6 |
| Minimum Temp °F | 33.3 | 34.6 | 36.6 | 39.0 | 43.6 | 48.6 | 51.6 | 51.8 | 48.7 | 42.9 | 38.2 | 34.5 | 41.9 |
| Mean Temp °F | 39.4 | 42.6 | 46.2 | 49.6 | 55.2 | 60.1 | 64.2 | 64.9 | 61.0 | 53.1 | 45.0 | 40.0 | 51.8 |
| Days Max Temp ≥ 90 °F | 0 | 0 | 0 | 0 | 0 | 1 | 2 | 3 | 1 | 0 | 0 | 0 | 7 |
| Days Max Temp ≤ 32 °F | 1 | 0 | 0 | 0 | 0 | 0 | 0 | 0 | 0 | 0 | 0 | 1 | 2 |
| Days Min Temp ≤ 32 °F | 13 | 11 | 8 | 4 | 0 | 0 | 0 | 0 | 0 | 1 | 6 | 12 | 55 |
| Days Min Temp ≤ 0 °F | 0 | 0 | 0 | 0 | 0 | 0 | 0 | 0 | 0 | 0 | 0 | 0 | 0 |
| Heating Degree Days | 787 | 626 | 575 | 454 | 304 | 160 | 68 | 57 | 137 | 363 | 592 | 770 | 4893 |
| Cooling Degree Days | 0 | 0 | 0 | 1 | 10 | 19 | 52 | 65 | 28 | 1 | 0 | 0 | 176 |
| Total Precipitation (") | 6.61 | 4.72 | 4.23 | 3.57 | 2.55 | 2.19 | 1.01 | 1.27 | 2.25 | 3.65 | 6.50 | 7.08 | 45.63 |
| Days ≥ 0.1" Precip | 14 | 11 | 12 | 10 | 7 | 6 | 3 | 3 | 6 | 9 | 15 | 14 | 110 |
| Total Snowfall (") | 2.0 | 0.7 | 0.0 | 0.0 | 0.0 | 0.0 | 0.0 | 0.0 | 0.0 | 0.0 | 0.3 | 1.1 | 4.1 |
| Days ≥ 1" Snow Depth | 1 | 1 | 0 | 0 | 0 | 0 | 0 | 0 | 0 | 0 | 0 | 1 | 3 |

## MAZAMA *Okanogan County* ELEVATION 2182 ft LAT/LONG 48° 36 ' N / 120° 26 ' W

| | JAN | FEB | MAR | APR | MAY | JUN | JUL | AUG | SEP | OCT | NOV | DEC | YEAR |
|---|---|---|---|---|---|---|---|---|---|---|---|---|---|
| Maximum Temp °F | 28.2 | 36.2 | 46.6 | 57.3 | 67.0 | 74.3 | 81.9 | 82.1 | 72.4 | 56.9 | 37.7 | 27.2 | 55.7 |
| Minimum Temp °F | 11.5 | 16.7 | 24.3 | 31.1 | 39.5 | 46.4 | 50.8 | 50.0 | 40.6 | 30.7 | 22.9 | 12.6 | 31.4 |
| Mean Temp °F | 19.8 | 26.4 | 35.5 | 44.2 | 53.3 | 60.4 | 66.4 | 66.1 | 56.5 | 43.8 | 30.3 | 19.9 | 43.6 |
| Days Max Temp ≥ 90 °F | 0 | 0 | 0 | 0 | 0 | 2 | *8* | 7 | 1 | 0 | 0 | 0 | 18 |
| Days Max Temp ≤ 32 °F | 20 | 8 | 1 | 0 | 0 | 0 | 0 | 0 | 0 | 0 | 6 | *22* | 57 |
| Days Min Temp ≤ 32 °F | 31 | 28 | 28 | 18 | 5 | 1 | 0 | 0 | 4 | 20 | *27* | *31* | 193 |
| Days Min Temp ≤ 0 °F | 7 | 3 | 0 | 0 | 0 | 0 | 0 | 0 | 0 | 0 | 1 | 5 | 16 |
| Heating Degree Days | 1394 | 1084 | 908 | 618 | 361 | 168 | 64 | 62 | 260 | 651 | 1034 | 1390 | 7994 |
| Cooling Degree Days | 0 | 0 | 0 | 0 | 6 | 33 | 103 | 106 | 18 | 0 | 0 | 0 | 266 |
| Total Precipitation (") | 3.69 | 2.29 | 1.49 | 1.00 | 0.96 | 1.00 | 0.76 | 0.80 | 0.91 | 1.31 | 3.07 | 4.14 | 21.42 |
| Days ≥ 0.1" Precip | 8 | 7 | 4 | 3 | 3 | 3 | 2 | 2 | 3 | 4 | *8* | 9 | 56 |
| Total Snowfall (") | 34.6 | 18.8 | 6.3 | 0.4 | 0.0 | 0.0 | 0.0 | 0.0 | 0.0 | 1.7 | *15.6* | 38.6 | 116.0 |
| Days ≥ 1" Snow Depth | 31 | 28 | *27* | *5* | 0 | 0 | 0 | 0 | 0 | 0 | *13* | 30 | 134 |

## MC MILLIN RESERVOIR *Pierce County* ELEVATION 581 ft LAT/LONG 47° 8 ' N / 122° 16 ' W

| | JAN | FEB | MAR | APR | MAY | JUN | JUL | AUG | SEP | OCT | NOV | DEC | YEAR |
|---|---|---|---|---|---|---|---|---|---|---|---|---|---|
| Maximum Temp °F | 44.3 | 48.5 | 52.9 | 57.5 | 64.1 | 69.3 | 74.7 | 75.6 | 69.9 | 60.0 | 49.9 | 44.4 | 59.3 |
| Minimum Temp °F | 31.4 | 32.9 | 35.0 | 38.0 | 42.9 | 47.9 | 50.7 | 51.0 | 46.7 | 40.5 | 35.8 | 31.9 | 40.4 |
| Mean Temp °F | 37.8 | 40.7 | 44.0 | 47.7 | 53.5 | 58.6 | 62.7 | 63.3 | 58.4 | 50.3 | 42.9 | 38.2 | 49.8 |
| Days Max Temp ≥ 90 °F | 0 | 0 | 0 | 0 | 0 | 0 | 1 | 1 | 0 | 0 | 0 | 0 | 2 |
| Days Max Temp ≤ 32 °F | 1 | 0 | 0 | 0 | 0 | 0 | 0 | 0 | 0 | 0 | 0 | 1 | 2 |
| Days Min Temp ≤ 32 °F | 16 | 13 | 11 | 5 | 1 | 0 | 0 | 0 | 0 | 3 | 10 | 16 | 75 |
| Days Min Temp ≤ 0 °F | 0 | 0 | 0 | 0 | 0 | 0 | 0 | 0 | 0 | 0 | 0 | 0 | 0 |
| Heating Degree Days | 835 | 678 | 645 | 511 | 351 | 198 | 100 | 85 | 200 | 449 | 657 | 824 | 5533 |
| Cooling Degree Days | 0 | 0 | 0 | 0 | 3 | 14 | 34 | 43 | 9 | 0 | 0 | 0 | 103 |
| Total Precipitation (") | 5.59 | 4.22 | 4.07 | 3.42 | 2.36 | 2.22 | 1.07 | 1.38 | 2.13 | 3.29 | 5.78 | 5.83 | 41.36 |
| Days ≥ 0.1" Precip | 13 | 10 | 12 | 10 | 7 | 6 | 3 | 3 | 6 | 9 | 13 | 13 | 105 |
| Total Snowfall (") | 2.8 | 1.2 | 1.3 | 0.0 | 0.0 | 0.0 | 0.0 | 0.0 | 0.0 | 0.1 | 0.7 | 2.3 | 8.4 |
| Days ≥ 1" Snow Depth | 3 | 1 | 1 | 0 | 0 | 0 | 0 | 0 | 0 | 0 | 1 | 2 | 8 |

## MCNARY DAM *Benton County* ELEVATION 351 ft LAT/LONG 45° 57' N / 119° 18' W

| | JAN | FEB | MAR | APR | MAY | JUN | JUL | AUG | SEP | OCT | NOV | DEC | YEAR |
|---|---|---|---|---|---|---|---|---|---|---|---|---|---|
| Maximum Temp °F | 39.9 | 46.5 | 56.1 | 64.4 | 73.3 | 80.8 | 88.5 | 87.9 | 78.0 | 65.5 | 50.2 | 41.5 | 64.4 |
| Minimum Temp °F | 28.4 | 31.1 | 35.7 | 41.4 | 48.3 | 55.0 | 60.5 | 60.8 | 52.5 | 43.1 | 35.9 | 29.9 | 43.6 |
| Mean Temp °F | 34.1 | 38.8 | 45.9 | 53.0 | 61.0 | 68.0 | 74.6 | 74.4 | 65.3 | 54.4 | 43.0 | 35.7 | 54.0 |
| Days Max Temp ≥ 90 °F | 0 | 0 | 0 | 0 | 1 | 5 | 15 | 13 | 2 | 0 | 0 | 0 | 36 |
| Days Max Temp ≤ 32 °F | 7 | 2 | 0 | 0 | 0 | 0 | 0 | 0 | 0 | 0 | 1 | 5 | 15 |
| Days Min Temp ≤ 32 °F | 19 | 15 | 8 | 1 | 0 | 0 | 0 | 0 | 0 | 1 | 8 | 18 | 70 |
| Days Min Temp ≤ 0 °F | 0 | 0 | 0 | 0 | 0 | 0 | 0 | 0 | 0 | 0 | 0 | 1 | 1 |
| Heating Degree Days | 952 | 733 | 585 | 354 | 155 | 40 | 3 | 3 | 67 | 326 | 652 | 902 | 4772 |
| Cooling Degree Days | 0 | 0 | 0 | 3 | 41 | 124 | 288 | 290 | 83 | 5 | 0 | 0 | 834 |
| Total Precipitation (") | 1.06 | 0.67 | 0.68 | 0.57 | 0.55 | 0.41 | 0.24 | 0.40 | 0.38 | 0.49 | 1.11 | 1.10 | 7.66 |
| Days ≥ 0.1" Precip | 4 | 2 | 2 | 2 | 2 | 1 | 1 | 1 | 1 | 2 | 4 | 4 | 26 |
| Total Snowfall (") | na | *0.9* | 0.0 | 0.0 | 0.0 | 0.0 | 0.0 | 0.0 | 0.0 | 0.0 | 0.3 | na | na |
| Days ≥ 1" Snow Depth | na | *0* | 0 | 0 | 0 | 0 | 0 | 0 | 0 | 0 | 0 | na | na |

## MONROE *Snohomish County* ELEVATION 121 ft LAT/LONG 47° 51' N / 121° 59' W

| | JAN | FEB | MAR | APR | MAY | JUN | JUL | AUG | SEP | OCT | NOV | DEC | YEAR |
|---|---|---|---|---|---|---|---|---|---|---|---|---|---|
| Maximum Temp °F | 45.2 | 50.2 | 54.4 | 59.3 | 66.1 | 71.3 | 76.0 | 76.6 | 70.9 | 60.9 | 50.5 | 44.5 | 60.5 |
| Minimum Temp °F | 32.8 | 34.1 | 36.4 | 39.8 | 44.9 | 49.6 | 52.0 | 52.5 | 48.4 | 42.6 | 37.4 | 33.4 | 42.0 |
| Mean Temp °F | 39.0 | 42.2 | 45.5 | 49.6 | 55.6 | 60.5 | 64.0 | 64.6 | 59.7 | 51.8 | 44.0 | 39.0 | 51.3 |
| Days Max Temp ≥ 90 °F | 0 | 0 | 0 | 0 | 0 | 0 | 1 | 1 | 0 | 0 | 0 | 0 | 2 |
| Days Max Temp ≤ 32 °F | 1 | 0 | 0 | 0 | 0 | 0 | 0 | 0 | 0 | 0 | 0 | 1 | 2 |
| Days Min Temp ≤ 32 °F | 14 | 11 | 9 | 3 | 0 | 0 | 0 | 0 | 0 | 2 | 8 | 14 | 61 |
| Days Min Temp ≤ 0 °F | 0 | 0 | 0 | 0 | 0 | 0 | 0 | 0 | 0 | 0 | 0 | 0 | 0 |
| Heating Degree Days | 798 | 638 | 599 | 456 | 290 | 144 | 65 | 55 | 162 | 403 | 624 | 801 | 5035 |
| Cooling Degree Days | 0 | 0 | 0 | 0 | 4 | 15 | 43 | 56 | 11 | 0 | 0 | 0 | 129 |
| Total Precipitation (") | 6.25 | 4.43 | 4.75 | 3.75 | 2.90 | 2.37 | 1.44 | 1.79 | 2.84 | 4.15 | 6.50 | 6.85 | 48.02 |
| Days ≥ 0.1" Precip | 14 | 12 | 12 | 10 | 8 | 6 | 4 | 4 | 6 | 10 | 14 | 15 | 115 |
| Total Snowfall (") | 3.6 | 0.9 | 0.2 | 0.2 | 0.0 | 0.0 | 0.0 | 0.0 | 0.0 | 0.0 | 0.6 | 2.6 | 8.1 |
| Days ≥ 1" Snow Depth | 2 | 1 | 0 | 0 | 0 | 0 | 0 | 0 | 0 | 0 | 1 | 2 | 6 |

## MOUNT ADAMS RS *Klickitat County* ELEVATION 1962 ft LAT/LONG 46° 0' N / 121° 32' W

| | JAN | FEB | MAR | APR | MAY | JUN | JUL | AUG | SEP | OCT | NOV | DEC | YEAR |
|---|---|---|---|---|---|---|---|---|---|---|---|---|---|
| Maximum Temp °F | 36.9 | 42.6 | 50.0 | 57.8 | 67.4 | 75.0 | 82.5 | 82.3 | 73.2 | 61.6 | 44.2 | 36.5 | 59.2 |
| Minimum Temp °F | 22.7 | 26.3 | 29.1 | 32.8 | 38.3 | 44.6 | 48.2 | 47.3 | 41.0 | 34.2 | 29.7 | 24.5 | 34.9 |
| Mean Temp °F | 29.8 | 34.5 | 39.6 | 45.4 | 52.9 | 59.9 | 65.3 | 64.9 | 57.1 | 47.9 | 37.0 | 30.6 | 47.1 |
| Days Max Temp ≥ 90 °F | 0 | 0 | 0 | 0 | 0 | 2 | 8 | 7 | 1 | 0 | 0 | 0 | 18 |
| Days Max Temp ≤ 32 °F | 7 | 2 | 0 | 0 | 0 | 0 | 0 | 0 | 0 | 0 | 2 | 7 | 18 |
| Days Min Temp ≤ 32 °F | *26* | 22 | 21 | 15 | *6* | 1 | 0 | 0 | 4 | 13 | *20* | *26* | 154 |
| Days Min Temp ≤ 0 °F | 1 | 1 | 0 | 0 | 0 | 0 | 0 | 0 | 0 | 0 | 0 | 1 | 3 |
| Heating Degree Days | *1083* | 856 | 782 | 584 | 371 | 177 | 71 | 76 | 243 | 525 | 834 | 1062 | 6664 |
| Cooling Degree Days | 0 | 0 | 0 | 0 | 8 | 26 | 88 | 81 | 15 | 0 | 0 | 0 | 218 |
| Total Precipitation (") | 8.06 | 5.68 | 4.33 | 2.30 | 1.31 | 1.05 | 0.43 | 0.73 | 1.34 | 2.95 | 6.67 | 7.81 | 42.66 |
| Days ≥ 0.1" Precip | na | na | *8* | 6 | *3* | 2 | 1 | 2 | 3 | na | na | na | na |
| Total Snowfall (") | *34.8* | na | 6.7 | 1.1 | 0.0 | 0.0 | 0.0 | 0.0 | 0.0 | 0.2 | na | *28.3* | na |
| Days ≥ 1" Snow Depth | na | na | na | 1 | 0 | 0 | 0 | 0 | 0 | 0 | na | na | na |

## MOUNT VERNON 3 WNW *Skagit County* ELEVATION 10 ft LAT/LONG 48° 26' N / 122° 23' W

| | JAN | FEB | MAR | APR | MAY | JUN | JUL | AUG | SEP | OCT | NOV | DEC | YEAR |
|---|---|---|---|---|---|---|---|---|---|---|---|---|---|
| Maximum Temp °F | 44.9 | 49.0 | 52.9 | 57.1 | 63.6 | 68.2 | 72.6 | 73.6 | 68.3 | 59.5 | 50.3 | 45.3 | 58.8 |
| Minimum Temp °F | 33.2 | 34.9 | 37.0 | 39.8 | 44.5 | 48.6 | 50.5 | 50.7 | 46.8 | 41.5 | 37.7 | 34.2 | 41.6 |
| Mean Temp °F | 39.1 | 42.0 | 45.0 | 48.5 | 54.1 | 58.4 | 61.6 | 62.2 | 57.6 | 50.5 | 44.0 | 39.8 | 50.2 |
| Days Max Temp ≥ 90 °F | 0 | 0 | 0 | 0 | 0 | 0 | 0 | 0 | 0 | 0 | 0 | 0 | 0 |
| Days Max Temp ≤ 32 °F | 1 | 0 | 0 | 0 | 0 | 0 | 0 | 0 | 0 | 0 | 0 | 2 | 3 |
| Days Min Temp ≤ 32 °F | 14 | 11 | 8 | 2 | 0 | 0 | 0 | 0 | 0 | 3 | 8 | 13 | 59 |
| Days Min Temp ≤ 0 °F | 0 | 0 | 0 | 0 | 0 | 0 | 0 | 0 | 0 | 0 | 0 | 0 | 0 |
| Heating Degree Days | 797 | 643 | 615 | 489 | 332 | 195 | 111 | 97 | 219 | 443 | 623 | 776 | 5340 |
| Cooling Degree Days | 0 | 0 | 0 | 0 | 1 | 4 | 13 | 17 | 2 | 0 | 0 | 0 | 37 |
| Total Precipitation (") | 4.22 | 2.83 | 2.58 | 2.50 | 2.12 | 1.86 | 1.27 | 1.35 | 1.86 | 2.86 | 4.36 | 4.06 | 31.87 |
| Days ≥ 0.1" Precip | 11 | 9 | 9 | 8 | 6 | 5 | 4 | 3 | 5 | 8 | 12 | 12 | 92 |
| Total Snowfall (") | 1.9 | 0.4 | 0.1 | 0.0 | 0.0 | 0.0 | 0.0 | 0.0 | 0.0 | 0.0 | 0.2 | 1.4 | 4.0 |
| Days ≥ 1" Snow Depth | 2 | 0 | 0 | 0 | 0 | 0 | 0 | 0 | 0 | 0 | 0 | *1* | 3 |

## MOXEE CITY 10 E *Yakima County* ELEVATION 1552 ft LAT/LONG 46° 31 ' N / 120° 10 ' W

| | JAN | FEB | MAR | APR | MAY | JUN | JUL | AUG | SEP | OCT | NOV | DEC | YEAR |
|---|---|---|---|---|---|---|---|---|---|---|---|---|---|
| Maximum Temp °F | 36.1 | 43.8 | 53.6 | 61.6 | 70.6 | 78.4 | 85.3 | 84.4 | 75.7 | 62.7 | 45.9 | 36.1 | 61.2 |
| Minimum Temp °F | 22.6 | 26.8 | 30.7 | 35.2 | 41.0 | 48.0 | 52.2 | 51.9 | 45.4 | 36.7 | 29.7 | 23.1 | 36.9 |
| Mean Temp °F | 29.4 | 35.4 | 42.2 | 48.4 | 55.8 | 63.2 | 68.8 | 68.2 | 60.6 | 49.7 | 37.8 | 29.6 | 49.1 |
| Days Max Temp ≥ 90 °F | 0 | 0 | 0 | 0 | 0 | 3 | 11 | 9 | 1 | 0 | 0 | 0 | 24 |
| Days Max Temp ≤ 32 °F | 11 | 4 | 0 | 0 | 0 | 0 | 0 | 0 | 0 | 0 | 2 | 11 | 28 |
| Days Min Temp ≤ 32 °F | 27 | 23 | 20 | 12 | 4 | 0 | 0 | 0 | 1 | 9 | 19 | 27 | 142 |
| Days Min Temp ≤ 0 °F | 1 | 0 | 0 | 0 | 0 | 0 | 0 | 0 | 0 | 0 | 0 | 1 | 2 |
| Heating Degree Days | 1097 | 831 | 702 | 491 | 289 | 111 | 33 | 37 | 158 | 467 | 809 | 1089 | 6114 |
| Cooling Degree Days | 0 | 0 | *0* | 1 | *20* | *70* | *165* | *147* | *40* | 1 | 0 | 0 | 444 |
| Total Precipitation (") | 0.88 | 0.56 | 0.65 | 0.61 | 0.58 | 0.66 | 0.26 | 0.52 | 0.39 | 0.52 | 0.99 | 0.94 | 7.56 |
| Days ≥ 0.1" Precip | 3 | 2 | 2 | 2 | 2 | 2 | 1 | 1 | 1 | 2 | 4 | 3 | 25 |
| Total Snowfall (") | 5.0 | 1.9 | 0.8 | 0.1 | 0.0 | 0.0 | 0.0 | 0.0 | 0.0 | 0.1 | 1.3 | 4.8 | 14.0 |
| Days ≥ 1" Snow Depth | na | *2* | 0 | 0 | 0 | 0 | 0 | 0 | 0 | 0 | *0* | na | na |

## MUD MOUNTAIN DAM *King County* ELEVATION 1312 ft LAT/LONG 47° 9 ' N / 121° 56 ' W

| | JAN | FEB | MAR | APR | MAY | JUN | JUL | AUG | SEP | OCT | NOV | DEC | YEAR |
|---|---|---|---|---|---|---|---|---|---|---|---|---|---|
| Maximum Temp °F | 44.1 | 47.5 | 50.7 | 54.5 | 61.0 | 66.3 | 72.0 | 73.1 | 68.0 | 59.2 | 49.2 | 44.2 | 57.5 |
| Minimum Temp °F | 31.4 | 32.9 | 34.5 | 37.2 | 42.4 | 47.3 | 50.7 | 51.1 | 46.9 | 40.5 | 35.3 | 31.7 | 40.2 |
| Mean Temp °F | 37.8 | 40.2 | 42.6 | 45.9 | 51.7 | 56.9 | 61.4 | 62.1 | 57.5 | 49.9 | 42.3 | 37.9 | 48.8 |
| Days Max Temp ≥ 90 °F | 0 | 0 | 0 | 0 | 0 | 0 | 1 | 1 | 0 | 0 | 0 | 0 | 2 |
| Days Max Temp ≤ 32 °F | 2 | 0 | 0 | 0 | 0 | 0 | 0 | 0 | 0 | 0 | 1 | 2 | 5 |
| Days Min Temp ≤ 32 °F | 17 | 14 | 11 | 5 | 0 | 0 | 0 | 0 | 0 | 2 | 10 | 17 | 76 |
| Days Min Temp ≤ 0 °F | 0 | 0 | 0 | 0 | 0 | 0 | 0 | 0 | 0 | 0 | 0 | 0 | 0 |
| Heating Degree Days | 838 | 693 | 686 | 567 | 409 | 252 | 140 | 123 | 232 | 463 | 676 | 832 | 5911 |
| Cooling Degree Days | 0 | 0 | 0 | 0 | 5 | 13 | 35 | 42 | 13 | 1 | 0 | 0 | 109 |
| Total Precipitation (") | 7.01 | 5.02 | 5.11 | 5.23 | 4.25 | 3.82 | 2.03 | 2.30 | 3.24 | 4.51 | 7.65 | 6.69 | 56.86 |
| Days ≥ 0.1" Precip | 13 | 11 | 13 | 13 | 11 | 9 | 5 | 5 | 7 | 10 | 14 | 14 | 125 |
| Total Snowfall (") | *6.0* | 2.9 | 1.9 | 0.5 | 0.0 | 0.0 | 0.0 | 0.0 | 0.0 | 0.1 | 0.5 | *6.6* | 18.5 |
| Days ≥ 1" Snow Depth | *5* | 2 | 1 | 0 | 0 | 0 | 0 | 0 | 0 | 0 | 1 | 4 | 13 |

## NEWHALEM *Whatcom County* ELEVATION 531 ft LAT/LONG 48° 41 ' N / 121° 15 ' W

| | JAN | FEB | MAR | APR | MAY | JUN | JUL | AUG | SEP | OCT | NOV | DEC | YEAR |
|---|---|---|---|---|---|---|---|---|---|---|---|---|---|
| Maximum Temp °F | 38.9 | 43.4 | 50.3 | 57.1 | 65.3 | 70.6 | 76.2 | 76.9 | 69.8 | 57.6 | 45.6 | 39.4 | 57.6 |
| Minimum Temp °F | 30.6 | 32.3 | 35.0 | 38.7 | 44.4 | 49.3 | 52.5 | 53.4 | 49.4 | 42.8 | 36.2 | 31.6 | 41.4 |
| Mean Temp °F | 34.8 | 37.9 | 42.7 | 48.0 | 54.9 | 60.0 | 64.4 | 65.1 | 59.6 | 50.2 | 40.9 | 35.5 | 49.5 |
| Days Max Temp ≥ 90 °F | 0 | 0 | 0 | 0 | 0 | 1 | 3 | 3 | 1 | 0 | 0 | 0 | 8 |
| Days Max Temp ≤ 32 °F | 4 | 1 | 0 | 0 | 0 | 0 | 0 | 0 | 0 | 0 | 1 | 2 | 8 |
| Days Min Temp ≤ 32 °F | 18 | *13* | 8 | 2 | 0 | 0 | 0 | 0 | 0 | 1 | 7 | *15* | 64 |
| Days Min Temp ≤ 0 °F | 0 | 0 | 0 | 0 | 0 | 0 | 0 | 0 | 0 | 0 | 0 | 0 | 0 |
| Heating Degree Days | 928 | 759 | 686 | 505 | 316 | 175 | 83 | 68 | 180 | 451 | 716 | 906 | 5773 |
| Cooling Degree Days | 0 | 0 | 0 | 0 | 13 | 27 | 65 | 80 | 24 | 0 | 0 | 0 | 209 |
| Total Precipitation (") | 12.11 | 8.97 | 6.80 | 4.97 | 3.30 | 2.79 | 1.97 | 1.84 | 3.49 | 7.30 | 12.25 | 12.96 | 78.75 |
| Days ≥ 0.1" Precip | 15 | *12* | 12 | 11 | *7* | 7 | 4 | 4 | 7 | *12* | *15* | *15* | 121 |
| Total Snowfall (") | 19.5 | 9.6 | 1.8 | 0.1 | 0.0 | 0.0 | 0.0 | 0.0 | 0.0 | 0.1 | *2.7* | *10.9* | 44.7 |
| Days ≥ 1" Snow Depth | 15 | *10* | 3 | 0 | 0 | 0 | 0 | 0 | 0 | 0 | 1 | *8* | 37 |

## NEWPORT *Pend Oreille County* ELEVATION 2142 ft LAT/LONG 48° 11 ' N / 117° 3 ' W

| | JAN | FEB | MAR | APR | MAY | JUN | JUL | AUG | SEP | OCT | NOV | DEC | YEAR |
|---|---|---|---|---|---|---|---|---|---|---|---|---|---|
| Maximum Temp °F | 31.8 | 38.6 | 48.7 | 59.1 | 69.0 | 76.4 | 84.1 | 84.2 | 73.3 | 57.8 | 39.5 | 32.0 | 57.9 |
| Minimum Temp °F | 19.6 | 22.9 | 26.0 | 32.0 | 38.7 | 45.2 | 47.9 | 46.7 | 39.1 | 32.1 | 27.7 | 21.2 | 33.3 |
| Mean Temp °F | 25.7 | 30.8 | 37.4 | 45.6 | 53.9 | 60.9 | 66.0 | 65.5 | 56.2 | 44.9 | 33.6 | 26.6 | 45.6 |
| Days Max Temp ≥ 90 °F | 0 | 0 | 0 | 0 | 0 | 2 | 10 | 9 | 1 | 0 | 0 | 0 | 22 |
| Days Max Temp ≤ 32 °F | 14 | 4 | 1 | 0 | 0 | 0 | 0 | 0 | 0 | 0 | 4 | 15 | 38 |
| Days Min Temp ≤ 32 °F | 29 | 25 | 25 | 17 | 6 | 1 | 0 | 0 | 5 | 16 | 22 | 28 | 174 |
| Days Min Temp ≤ 0 °F | 3 | 1 | 0 | 0 | 0 | 0 | 0 | 0 | 0 | 0 | 0 | 2 | 6 |
| Heating Degree Days | 1211 | 960 | 848 | 577 | 344 | 151 | 56 | 62 | 264 | 615 | 937 | 1184 | 7209 |
| Cooling Degree Days | 0 | 0 | 0 | 0 | 9 | 35 | 95 | 88 | 8 | 0 | 0 | 0 | 235 |
| Total Precipitation (") | 3.29 | 2.36 | 2.10 | 1.98 | 2.12 | 1.81 | 1.31 | 1.27 | 1.21 | 1.54 | 3.51 | 3.76 | 26.26 |
| Days ≥ 0.1" Precip | 9 | 7 | 7 | 6 | 6 | 5 | 4 | 3 | 4 | 5 | 9 | 10 | 75 |
| Total Snowfall (") | 20.5 | 11.0 | 3.1 | 0.1 | 0.0 | 0.0 | 0.0 | 0.0 | 0.0 | 0.1 | 6.2 | 20.2 | 61.2 |
| Days ≥ 1" Snow Depth | *27* | 21 | 13 | 1 | 0 | 0 | 0 | 0 | 0 | 0 | 5 | 21 | 88 |

## NORTHPORT *Stevens County* ELEVATION 1332 ft LAT/LONG 48° 55' N / 117° 47' W

| | JAN | FEB | MAR | APR | MAY | JUN | JUL | AUG | SEP | OCT | NOV | DEC | YEAR |
|---|---|---|---|---|---|---|---|---|---|---|---|---|---|
| Maximum Temp °F | 32.3 | 39.7 | 52.0 | 63.6 | 73.6 | 80.5 | 87.4 | 87.1 | 76.1 | 59.5 | 41.7 | 32.8 | 60.5 |
| Minimum Temp °F | 20.5 | 24.4 | 28.6 | 34.8 | 41.4 | 47.8 | 51.4 | 50.9 | 43.5 | 35.3 | 28.9 | 22.5 | 35.8 |
| Mean Temp °F | 26.4 | 32.1 | 40.3 | 49.2 | 57.5 | 64.2 | 69.6 | 69.0 | 59.8 | 47.4 | 35.3 | 27.6 | 48.2 |
| Days Max Temp ≥ 90 °F | 0 | 0 | 0 | 0 | 2 | 6 | 14 | 14 | 2 | 0 | 0 | 0 | 38 |
| Days Max Temp ≤ 32 °F | 13 | 4 | 0 | 0 | 0 | 0 | 0 | 0 | 0 | 0 | 3 | 13 | 33 |
| Days Min Temp ≤ 32 °F | 29 | 24 | 23 | 11 | 3 | 0 | 0 | 0 | 2 | 11 | 20 | 27 | 150 |
| Days Min Temp ≤ 0 °F | 2 | 1 | 0 | 0 | 0 | 0 | 0 | 0 | 0 | 0 | 0 | 1 | 4 |
| Heating Degree Days | 1189 | 923 | 759 | 468 | 239 | 83 | 19 | 24 | 170 | 537 | 885 | 1151 | 6447 |
| Cooling Degree Days | 0 | 0 | 0 | 0 | 17 | 58 | 155 | 144 | 20 | 0 | 0 | 0 | 394 |
| Total Precipitation (") | 1.96 | 1.48 | 1.43 | 1.49 | 1.89 | 2.16 | 1.42 | 1.46 | 1.14 | 1.22 | 2.16 | 2.59 | 20.40 |
| Days ≥ 0.1" Precip | 7 | 5 | 5 | 5 | 6 | 5 | 4 | 4 | 3 | 4 | 7 | 8 | 63 |
| Total Snowfall (") | 17.7 | 8.9 | 2.0 | 0.3 | 0.0 | 0.0 | 0.0 | 0.0 | 0.0 | 0.1 | 7.4 | 19.6 | 56.0 |
| Days ≥ 1" Snow Depth | 28 | 22 | 7 | 0 | 0 | 0 | 0 | 0 | 0 | 0 | 6 | 21 | 84 |

## OAKVILLE *Grays Harbor County* ELEVATION 89 ft LAT/LONG 46° 51' N / 123° 14' W

| | JAN | FEB | MAR | APR | MAY | JUN | JUL | AUG | SEP | OCT | NOV | DEC | YEAR |
|---|---|---|---|---|---|---|---|---|---|---|---|---|---|
| Maximum Temp °F | 46.1 | 50.6 | 55.6 | 60.4 | 66.8 | 71.5 | 76.5 | 77.7 | 72.5 | 62.2 | 51.7 | 45.6 | 61.4 |
| Minimum Temp °F | 32.3 | 33.3 | 34.8 | 37.0 | 42.0 | 47.2 | 50.0 | 50.2 | 45.9 | 40.0 | 36.3 | 33.2 | 40.2 |
| Mean Temp °F | 39.2 | 41.9 | 45.2 | 48.7 | 54.4 | 59.3 | 63.3 | 63.9 | 59.2 | 51.1 | 44.0 | 39.4 | 50.8 |
| Days Max Temp ≥ 90 °F | 0 | 0 | 0 | 0 | 0 | 1 | 2 | 3 | 0 | 0 | 0 | 0 | 6 |
| Days Max Temp ≤ 32 °F | 1 | 0 | 0 | 0 | 0 | 0 | 0 | 0 | 0 | 0 | 0 | 1 | 2 |
| Days Min Temp ≤ 32 °F | 15 | 12 | 12 | 8 | 2 | 0 | 0 | 0 | 1 | 5 | 9 | 14 | 78 |
| Days Min Temp ≤ 0 °F | 0 | 0 | 0 | 0 | 0 | 0 | 0 | 0 | 0 | 0 | 0 | 0 | 0 |
| Heating Degree Days | 792 | 645 | 607 | 483 | 326 | 178 | 89 | 77 | 178 | 423 | 623 | 786 | 5207 |
| Cooling Degree Days | 0 | 0 | 0 | 0 | 6 | 16 | 51 | 60 | 13 | 0 | 0 | 0 | 146 |
| Total Precipitation (") | 8.92 | 6.57 | 5.98 | 3.91 | 2.49 | 2.17 | 0.84 | 1.26 | 2.51 | 4.88 | 8.28 | 9.49 | 57.30 |
| Days ≥ 0.1" Precip | 15 | 12 | 13 | 11 | 7 | 6 | 3 | 3 | 6 | 9 | 16 | 15 | 116 |
| Total Snowfall (") | 3.1 | 0.6 | 0.7 | 0.0 | 0.0 | 0.0 | 0.0 | 0.0 | 0.0 | 0.0 | 0.1 | 1.1 | 5.6 |
| Days ≥ 1" Snow Depth | na | 0 | 0 | 0 | 0 | 0 | 0 | 0 | 0 | 0 | 0 | na | na |

## ODESSA *Lincoln County* ELEVATION 1542 ft LAT/LONG 47° 20' N / 118° 41' W

| | JAN | FEB | MAR | APR | MAY | JUN | JUL | AUG | SEP | OCT | NOV | DEC | YEAR |
|---|---|---|---|---|---|---|---|---|---|---|---|---|---|
| Maximum Temp °F | 35.3 | 43.2 | 53.4 | 62.7 | 71.8 | 79.4 | 86.6 | 86.2 | 77.2 | 63.8 | 45.9 | 36.1 | 61.8 |
| Minimum Temp °F | 19.2 | 24.0 | 28.3 | 33.5 | 39.9 | 46.9 | 51.5 | 50.4 | 41.9 | 32.4 | 26.5 | 20.0 | 34.5 |
| Mean Temp °F | 27.2 | 33.6 | 40.9 | 48.1 | 55.9 | 63.2 | 69.0 | 68.3 | 59.5 | 48.0 | 36.2 | 28.1 | 48.2 |
| Days Max Temp ≥ 90 °F | 0 | 0 | 0 | 0 | 1 | 4 | 12 | 12 | 3 | 0 | 0 | 0 | 32 |
| Days Max Temp ≤ 32 °F | 11 | 4 | 0 | 0 | 0 | 0 | 0 | 0 | 0 | 0 | 2 | 10 | 27 |
| Days Min Temp ≤ 32 °F | 29 | 25 | 23 | 13 | 5 | 0 | 0 | 0 | 3 | 15 | 22 | 28 | 163 |
| Days Min Temp ≤ 0 °F | 2 | 1 | 0 | 0 | 0 | 0 | 0 | 0 | 0 | 0 | 0 | 2 | 5 |
| Heating Degree Days | 1164 | 879 | 741 | 501 | 291 | 113 | 34 | 40 | 186 | 520 | 857 | 1137 | 6463 |
| Cooling Degree Days | 0 | 0 | 0 | 2 | 18 | 50 | 132 | 125 | 23 | 1 | 0 | 0 | 351 |
| Total Precipitation (") | 1.11 | 0.89 | 0.91 | 0.78 | 0.80 | 0.54 | 0.48 | 0.35 | 0.47 | 0.62 | 1.39 | 1.58 | 9.92 |
| Days ≥ 0.1" Precip | 4 | 3 | 4 | 3 | 3 | 2 | 2 | 1 | 2 | 2 | 4 | 6 | 36 |
| Total Snowfall (") | 5.3 | 2.0 | 0.3 | 0.0 | 0.0 | 0.0 | 0.0 | 0.0 | 0.0 | 0.0 | 1.1 | 6.0 | 14.7 |
| Days ≥ 1" Snow Depth | na | na | *0* | *0* | *0* | *0* | *0* | *0* | *0* | *0* | na | na | na |

## OLGA 2 SE *San Juan County* ELEVATION 79 ft LAT/LONG 48° 37' N / 122° 48' W

| | JAN | FEB | MAR | APR | MAY | JUN | JUL | AUG | SEP | OCT | NOV | DEC | YEAR |
|---|---|---|---|---|---|---|---|---|---|---|---|---|---|
| Maximum Temp °F | 44.5 | 47.9 | 52.0 | 56.6 | 62.7 | 66.7 | 69.9 | 70.0 | 65.8 | 57.6 | 49.3 | 45.0 | 57.3 |
| Minimum Temp °F | 34.9 | 36.4 | 38.3 | 40.9 | 44.7 | 48.0 | 50.1 | 50.7 | 48.4 | 44.3 | 38.9 | 35.7 | 42.6 |
| Mean Temp °F | 39.8 | 42.1 | 45.2 | 48.8 | 53.7 | 57.4 | 60.1 | 60.4 | 57.1 | 51.0 | 44.2 | 40.4 | 50.0 |
| Days Max Temp ≥ 90 °F | 0 | 0 | 0 | 0 | 0 | 0 | 0 | 0 | 0 | 0 | 0 | 0 | 0 |
| Days Max Temp ≤ 32 °F | 1 | 0 | 0 | 0 | 0 | 0 | 0 | 0 | 0 | 0 | 1 | 1 | 3 |
| Days Min Temp ≤ 32 °F | 10 | 6 | 3 | 0 | 0 | 0 | 0 | 0 | 0 | 0 | 4 | 9 | 32 |
| Days Min Temp ≤ 0 °F | 0 | 0 | 0 | 0 | 0 | 0 | 0 | 0 | 0 | 0 | 0 | 0 | 0 |
| Heating Degree Days | 776 | 639 | 607 | 480 | 345 | 224 | 150 | 142 | 231 | 428 | 619 | 757 | 5398 |
| Cooling Degree Days | 0 | 0 | 0 | 0 | 2 | 2 | 4 | 7 | 2 | 0 | 0 | 0 | 17 |
| Total Precipitation (") | 4.01 | 2.76 | 2.21 | 2.04 | 1.60 | 1.32 | 0.97 | 1.02 | 1.49 | 2.59 | 3.91 | 4.08 | 28.00 |
| Days ≥ 0.1" Precip | 11 | 8 | 8 | 7 | 5 | 4 | 3 | 3 | 4 | 7 | 11 | 11 | 82 |
| Total Snowfall (") | 2.3 | 1.0 | 0.6 | 0.0 | 0.0 | 0.0 | 0.0 | 0.0 | 0.0 | 0.1 | 0.5 | 1.6 | 6.1 |
| Days ≥ 1" Snow Depth | *1* | 0 | 0 | 0 | 0 | 0 | 0 | 0 | 0 | 0 | 0 | 1 | 2 |

## OLYMPIA AP *Thurston County* ELEVATION 217 ft LAT/LONG 46° 58' N / 122° 54' W

| | JAN | FEB | MAR | APR | MAY | JUN | JUL | AUG | SEP | OCT | NOV | DEC | YEAR |
|---|---|---|---|---|---|---|---|---|---|---|---|---|---|
| Maximum Temp °F | 44.7 | 49.5 | 54.5 | 59.1 | 66.0 | 71.3 | 76.9 | 77.7 | 71.8 | 61.0 | 50.1 | 44.3 | 60.6 |
| Minimum Temp °F | 31.7 | 32.5 | 33.8 | 36.5 | 41.6 | 46.7 | 49.5 | 49.6 | 45.2 | 39.1 | 35.3 | 32.1 | 39.5 |
| Mean Temp °F | 38.2 | 41.0 | 44.2 | 47.8 | 53.8 | 59.0 | 63.2 | 63.7 | 58.5 | 50.1 | 42.7 | 38.2 | 50.0 |
| Days Max Temp ≥ 90 °F | 0 | 0 | 0 | 0 | 0 | 1 | 2 | 3 | 1 | 0 | 0 | 0 | 7 |
| Days Max Temp ≤ 32 °F | 1 | 0 | 0 | 0 | 0 | 0 | 0 | 0 | 0 | 0 | 0 | 1 | 2 |
| Days Min Temp ≤ 32 °F | 16 | 14 | 14 | 8 | 1 | 0 | 0 | 0 | 1 | 6 | 11 | 15 | 86 |
| Days Min Temp ≤ 0 °F | 0 | 0 | 0 | 0 | 0 | 0 | 0 | 0 | 0 | 0 | 0 | 0 | 0 |
| Heating Degree Days | 824 | 670 | 639 | 509 | 343 | 186 | 86 | 76 | 195 | 456 | 662 | 823 | 5469 |
| Cooling Degree Days | 0 | 0 | 0 | 0 | 4 | 12 | 36 | 48 | 10 | 0 | 0 | 0 | 110 |
| Total Precipitation (") | 7.79 | 5.47 | 4.81 | 3.56 | 2.10 | 1.67 | 0.78 | 1.21 | 2.16 | 4.16 | 7.47 | 8.07 | 49.25 |
| Days ≥ 0.1" Precip | 14 | 12 | 12 | 10 | 6 | 5 | 2 | 3 | 5 | 8 | 14 | 14 | 105 |
| Total Snowfall (") | 7.2 | 3.4 | 1.4 | 0.1 | 0.0 | 0.0 | 0.0 | 0.0 | 0.0 | 0.0 | 1.3 | 4.7 | 18.1 |
| Days ≥ 1" Snow Depth | 3 | 2 | 0 | 0 | 0 | 0 | 0 | 0 | 0 | 0 | 1 | 2 | 8 |

## OTHELLO 6 ESE *Adams County* ELEVATION 1191 ft LAT/LONG 46° 48' N / 119° 3' W

| | JAN | FEB | MAR | APR | MAY | JUN | JUL | AUG | SEP | OCT | NOV | DEC | YEAR |
|---|---|---|---|---|---|---|---|---|---|---|---|---|---|
| Maximum Temp °F | 35.5 | 44.1 | 54.6 | 62.4 | 71.1 | 78.0 | 84.8 | 84.3 | 75.6 | 63.3 | 46.6 | 36.8 | 61.4 |
| Minimum Temp °F | 22.1 | 27.3 | 31.7 | 36.4 | 43.2 | 49.9 | 53.6 | 53.4 | 45.7 | 36.4 | 30.1 | 23.9 | 37.8 |
| Mean Temp °F | 28.9 | 35.7 | 43.2 | 49.4 | 57.1 | 63.9 | 69.2 | 68.9 | 60.7 | 49.9 | 38.4 | 30.4 | 49.6 |
| Days Max Temp ≥ 90 °F | 0 | 0 | 0 | 0 | 1 | 4 | 10 | 9 | 1 | 0 | 0 | 0 | 25 |
| Days Max Temp ≤ 32 °F | 11 | 3 | 0 | 0 | 0 | 0 | 0 | 0 | 0 | 0 | 2 | 10 | 26 |
| Days Min Temp ≤ 32 °F | 26 | 22 | 18 | 9 | 1 | 0 | 0 | 0 | 1 | 9 | 19 | 27 | 132 |
| Days Min Temp ≤ 0 °F | 2 | 1 | 0 | 0 | 0 | 0 | 0 | 0 | 0 | 0 | 0 | 1 | 4 |
| Heating Degree Days | 1115 | 821 | 669 | 463 | 255 | 100 | 31 | 28 | 157 | 464 | 793 | 1067 | 5963 |
| Cooling Degree Days | 0 | *0* | 0 | 2 | 22 | 66 | *160* | 155 | 39 | 1 | 0 | 0 | 445 |
| Total Precipitation (") | 0.97 | 0.73 | 0.78 | 0.63 | 0.66 | 0.51 | 0.32 | 0.33 | 0.40 | 0.55 | 1.09 | 1.18 | 8.15 |
| Days ≥ 0.1" Precip | *3* | *3* | 2 | 2 | 2 | 2 | 1 | 1 | *1* | 2 | 4 | 4 | 27 |
| Total Snowfall (") | na | na | 0.0 | *0.0* | 0.0 | 0.0 | 0.0 | 0.0 | 0.0 | 0.0 | *0.0* | na | na |
| Days ≥ 1" Snow Depth | na | na | 0 | 0 | 0 | 0 | 0 | 0 | 0 | 0 | 0 | na | na |

## PACKWOOD *Lewis County* ELEVATION 1060 ft LAT/LONG 46° 37' N / 121° 40' W

| | JAN | FEB | MAR | APR | MAY | JUN | JUL | AUG | SEP | OCT | NOV | DEC | YEAR |
|---|---|---|---|---|---|---|---|---|---|---|---|---|---|
| Maximum Temp °F | 42.6 | 48.2 | 53.4 | 59.5 | 67.2 | 72.3 | 78.5 | 79.2 | 73.7 | 63.2 | *47.8* | *42.2* | 60.7 |
| Minimum Temp °F | 28.7 | 29.7 | 32.4 | 36.1 | 41.4 | 47.2 | 50.7 | 50.3 | 44.0 | 37.4 | *33.1* | *29.2* | 38.4 |
| Mean Temp °F | 35.7 | 38.9 | 43.1 | 47.8 | 54.3 | 59.7 | 64.6 | 64.8 | 58.9 | 50.4 | *40.5* | *35.7* | 49.5 |
| Days Max Temp ≥ 90 °F | 0 | 0 | 0 | 0 | 1 | 2 | 4 | 4 | 2 | 0 | 0 | 0 | 13 |
| Days Max Temp ≤ 32 °F | 2 | 0 | 0 | 0 | 0 | 0 | 0 | 0 | 0 | 0 | 0 | 2 | 4 |
| Days Min Temp ≤ 32 °F | *20* | *18* | *15* | *7* | 2 | 0 | 0 | 0 | 2 | *9* | *12* | 19 | 104 |
| Days Min Temp ≤ 0 °F | 0 | 0 | 0 | 0 | 0 | 0 | 0 | 0 | 0 | 0 | 0 | 0 | 0 |
| Heating Degree Days | *904* | 728 | 671 | 510 | 335 | 177 | 74 | 69 | 192 | 448 | *729* | *901* | 5738 |
| Cooling Degree Days | 0 | 0 | 0 | 1 | 10 | 28 | 76 | 84 | 18 | 1 | 0 | 0 | 218 |
| Total Precipitation (") | 9.80 | 6.35 | 4.88 | 3.56 | 2.38 | 1.99 | 0.88 | 1.19 | 2.32 | 4.55 | 8.24 | 9.04 | 55.18 |
| Days ≥ 0.1" Precip | *12* | *11* | *11* | *9* | 7 | 5 | 3 | 3 | 5 | *8* | *14* | na | na |
| Total Snowfall (") | *12.2* | 6.6 | *1.9* | *0.2* | 0.0 | 0.0 | 0.0 | 0.0 | 0.0 | 0.0 | *2.5* | *6.1* | 29.5 |
| Days ≥ 1" Snow Depth | *12* | *6* | *1* | 0 | 0 | 0 | 0 | 0 | 0 | 0 | *1* | na | na |

## PALMER 3 ESE *King County* ELEVATION 902 ft LAT/LONG 47° 18' N / 121° 50' W

| | JAN | FEB | MAR | APR | MAY | JUN | JUL | AUG | SEP | OCT | NOV | DEC | YEAR |
|---|---|---|---|---|---|---|---|---|---|---|---|---|---|
| Maximum Temp °F | 42.6 | 46.8 | 51.4 | 56.0 | 62.8 | 67.8 | 73.2 | 73.9 | 68.6 | 59.4 | 47.9 | 42.5 | 57.7 |
| Minimum Temp °F | 31.9 | 33.7 | 35.0 | 37.8 | 42.7 | 47.8 | 51.0 | 51.5 | 47.8 | 42.0 | 36.3 | 32.2 | 40.8 |
| Mean Temp °F | 37.2 | 40.3 | 43.2 | 46.9 | 52.8 | 57.8 | 62.2 | 62.7 | 58.2 | 50.7 | 42.2 | 37.4 | 49.3 |
| Days Max Temp ≥ 90 °F | 0 | 0 | 0 | 0 | 0 | 0 | 1 | 1 | 0 | 0 | 0 | 0 | 2 |
| Days Max Temp ≤ 32 °F | 2 | 1 | 0 | 0 | 0 | 0 | 0 | 0 | 0 | 0 | 1 | 2 | 6 |
| Days Min Temp ≤ 32 °F | 16 | 11 | 9 | 3 | 0 | 0 | 0 | 0 | 0 | 1 | 7 | 15 | 62 |
| Days Min Temp ≤ 0 °F | 0 | 0 | 0 | 0 | 0 | 0 | 0 | 0 | 0 | 0 | 0 | 0 | 0 |
| Heating Degree Days | 854 | 692 | 669 | 536 | 378 | 225 | 120 | 106 | 210 | 438 | 679 | 850 | 5757 |
| Cooling Degree Days | 0 | 0 | 0 | 0 | 7 | 15 | 35 | 41 | 12 | 2 | 0 | 0 | 112 |
| Total Precipitation (") | 11.35 | 8.57 | 8.39 | 7.56 | 5.61 | 5.05 | 2.84 | 2.87 | 4.66 | 6.59 | 10.93 | 11.16 | 85.58 |
| Days ≥ 0.1" Precip | 16 | 14 | 15 | 15 | 12 | 10 | 6 | 5 | 8 | 11 | 17 | 17 | 146 |
| Total Snowfall (") | 10.4 | 5.7 | 3.5 | 0.6 | 0.0 | 0.0 | 0.0 | 0.0 | 0.0 | 0.2 | 1.8 | 9.5 | 31.7 |
| Days ≥ 1" Snow Depth | 8 | 4 | 2 | 0 | 0 | 0 | 0 | 0 | 0 | 0 | 2 | 6 | 22 |

## PLAIN *Chelan County* ELEVATION 1801 ft LAT/LONG 47° 46 ' N / 120° 40 ' W

| | JAN | FEB | MAR | APR | MAY | JUN | JUL | AUG | SEP | OCT | NOV | DEC | YEAR |
|---|---|---|---|---|---|---|---|---|---|---|---|---|---|
| Maximum Temp °F | 34.1 | 41.1 | 49.8 | 58.3 | 67.6 | 74.4 | 81.4 | 81.7 | 73.7 | 60.2 | 41.8 | 33.2 | 58.1 |
| Minimum Temp °F | 19.6 | 23.2 | 26.4 | 31.3 | 37.6 | 44.1 | 47.6 | 47.7 | 40.2 | 32.6 | 27.8 | 21.0 | 33.3 |
| Mean Temp °F | 26.9 | 32.2 | 38.2 | 44.8 | 52.6 | 59.3 | 64.5 | 64.7 | 56.9 | 46.4 | 34.8 | 27.1 | 45.7 |
| Days Max Temp ≥ 90 °F | 0 | 0 | 0 | 0 | 0 | 2 | 7 | 6 | 1 | 0 | 0 | 0 | 16 |
| Days Max Temp ≤ 32 °F | 12 | 3 | 0 | 0 | 0 | 0 | 0 | 0 | 0 | 0 | 3 | 13 | 31 |
| Days Min Temp ≤ 32 °F | 30 | 27 | 27 | 19 | 6 | 1 | 0 | 0 | 4 | 16 | 24 | 30 | 184 |
| Days Min Temp ≤ 0 °F | 3 | 1 | 0 | 0 | 0 | 0 | 0 | 0 | 0 | 0 | 0 | 1 | 5 |
| Heating Degree Days | 1176 | 920 | 826 | 598 | 380 | 185 | 78 | 74 | 243 | 568 | 899 | 1167 | 7114 |
| Cooling Degree Days | 0 | 0 | 0 | 0 | 5 | 18 | 65 | 70 | 8 | 0 | 0 | 0 | 166 |
| Total Precipitation (") | 5.20 | 3.36 | 2.06 | 1.32 | 0.89 | 0.93 | 0.46 | 0.65 | 0.76 | 1.85 | 4.23 | 5.31 | 27.02 |
| Days ≥ 0.1" Precip | 11 | 8 | 6 | 5 | 3 | 2 | 2 | 2 | 3 | 5 | 11 | 12 | 70 |
| Total Snowfall (") | 42.1 | 22.2 | 9.5 | 1.4 | 0.1 | 0.0 | 0.0 | 0.0 | 0.0 | 1.2 | 19.2 | 42.4 | 138.1 |
| Days ≥ 1" Snow Depth | na | *14* | na | *1* | 0 | 0 | 0 | 0 | 0 | 0 | *6* | *15* | na |

## POMEROY *Garfield County* ELEVATION 1900 ft LAT/LONG 46° 29 ' N / 117° 35 ' W

| | JAN | FEB | MAR | APR | MAY | JUN | JUL | AUG | SEP | OCT | NOV | DEC | YEAR |
|---|---|---|---|---|---|---|---|---|---|---|---|---|---|
| Maximum Temp °F | 40.0 | 46.4 | 54.8 | 62.3 | 70.9 | 79.6 | 88.2 | 87.6 | 78.2 | 64.8 | *48.7* | 40.4 | 63.5 |
| Minimum Temp °F | 26.0 | 29.6 | 33.2 | 37.7 | 43.8 | 50.5 | 54.7 | 54.2 | 46.7 | 38.7 | *31.9* | 26.5 | 39.5 |
| Mean Temp °F | 33.1 | 38.1 | 44.0 | 50.0 | 57.4 | 65.1 | 71.5 | 70.9 | 62.5 | 51.8 | *40.3* | 33.5 | 51.5 |
| Days Max Temp ≥ 90 °F | 0 | 0 | 0 | 0 | 1 | 5 | 15 | 13 | 4 | 0 | 0 | 0 | 38 |
| Days Max Temp ≤ 32 °F | 7 | 2 | 0 | 0 | 0 | 0 | 0 | 0 | 0 | 0 | 1 | 6 | 16 |
| Days Min Temp ≤ 32 °F | 20 | 16 | 14 | 7 | 2 | 0 | 0 | 0 | 1 | *6* | *15* | 22 | 103 |
| Days Min Temp ≤ 0 °F | 1 | 0 | 0 | 0 | 0 | 0 | 0 | 0 | 0 | 0 | 0 | 1 | 2 |
| Heating Degree Days | 983 | 754 | 644 | 446 | 252 | 84 | 17 | 21 | 126 | 408 | *733* | 971 | 5439 |
| Cooling Degree Days | *0* | 0 | 0 | 3 | 29 | 92 | 205 | 210 | 57 | 4 | *0* | *0* | 600 |
| Total Precipitation (") | 2.28 | 1.48 | 1.52 | 1.26 | 1.33 | 1.07 | 0.64 | 0.84 | 0.81 | 1.08 | 1.90 | 1.89 | 16.10 |
| Days ≥ 0.1" Precip | *6* | 4 | 5 | 4 | 4 | 4 | 2 | 2 | 2 | 3 | 6 | 6 | 48 |
| Total Snowfall (") | na | *1.3* | 0.4 | 0.0 | 0.0 | 0.0 | 0.0 | 0.0 | 0.0 | 0.1 | 0.8 | na | na |
| Days ≥ 1" Snow Depth | na | na | 0 | 0 | 0 | 0 | 0 | 0 | 0 | 0 | *1* | na | na |

## PORT ANGELES *Clallam County* ELEVATION 40 ft LAT/LONG 48° 7 ' N / 123° 24 ' W

| | JAN | FEB | MAR | APR | MAY | JUN | JUL | AUG | SEP | OCT | NOV | DEC | YEAR |
|---|---|---|---|---|---|---|---|---|---|---|---|---|---|
| Maximum Temp °F | 45.2 | 47.8 | 51.2 | 54.9 | 60.5 | 64.6 | 68.2 | 68.8 | 65.8 | 57.6 | 49.8 | 45.5 | 56.7 |
| Minimum Temp °F | 33.8 | 35.2 | 36.9 | 39.6 | 44.4 | 48.5 | 51.0 | 51.4 | 48.3 | 42.8 | 37.9 | 34.7 | 42.0 |
| Mean Temp °F | 39.5 | 41.5 | 44.1 | 47.3 | 52.5 | 56.6 | 59.7 | 60.1 | 57.1 | 50.2 | 43.8 | 40.1 | 49.4 |
| Days Max Temp ≥ 90 °F | 0 | 0 | 0 | 0 | 0 | 0 | 0 | 0 | 0 | 0 | 0 | 0 | 0 |
| Days Max Temp ≤ 32 °F | 1 | 0 | 0 | 0 | 0 | 0 | 0 | 0 | 0 | 0 | 0 | 1 | 2 |
| Days Min Temp ≤ 32 °F | 12 | 8 | 6 | 2 | 0 | 0 | 0 | 0 | 0 | 1 | 5 | 11 | 45 |
| Days Min Temp ≤ 0 °F | 0 | 0 | 0 | 0 | 0 | 0 | 0 | 0 | 0 | 0 | 0 | 0 | 0 |
| Heating Degree Days | 784 | 656 | 642 | 525 | 382 | 249 | 170 | 155 | 234 | 451 | 628 | 765 | 5641 |
| Cooling Degree Days | 0 | 0 | 0 | 0 | 1 | 4 | 8 | 12 | 2 | 0 | 0 | 0 | 27 |
| Total Precipitation (") | 3.85 | 2.59 | 1.94 | 1.43 | 0.99 | 0.77 | 0.49 | 0.76 | 1.16 | 2.24 | 3.95 | 4.37 | 24.54 |
| Days ≥ 0.1" Precip | 10 | 7 | 6 | 4 | 3 | 3 | 2 | 2 | 3 | 7 | 9 | 10 | 66 |
| Total Snowfall (") | *0.8* | 0.0 | 0.0 | 0.0 | 0.0 | 0.0 | 0.0 | 0.0 | 0.0 | 0.0 | 0.2 | 0.8 | 1.8 |
| Days ≥ 1" Snow Depth | na | 0 | 0 | 0 | 0 | 0 | 0 | 0 | 0 | 0 | 0 | 0 | na |

## PORT TOWNSEND *Jefferson County* ELEVATION 66 ft LAT/LONG 48° 6 ' N / 122° 46 ' W

| | JAN | FEB | MAR | APR | MAY | JUN | JUL | AUG | SEP | OCT | NOV | DEC | YEAR |
|---|---|---|---|---|---|---|---|---|---|---|---|---|---|
| Maximum Temp °F | 45.7 | 48.6 | 52.6 | 57.0 | 62.9 | 67.5 | 71.8 | 72.5 | 67.7 | 58.6 | *49.8* | 45.7 | 58.4 |
| Minimum Temp °F | 36.7 | 37.9 | 39.3 | 42.2 | 46.3 | 49.6 | 52.1 | 52.6 | 50.0 | 45.3 | *40.5* | 37.3 | 44.2 |
| Mean Temp °F | 41.2 | 43.3 | 46.0 | 49.6 | 54.6 | 58.6 | 62.0 | 62.6 | 58.9 | 52.0 | *45.2* | 41.5 | 51.3 |
| Days Max Temp ≥ 90 °F | 0 | 0 | 0 | 0 | 0 | 0 | 0 | 0 | 0 | 0 | *0* | 0 | 0 |
| Days Max Temp ≤ 32 °F | 1 | 0 | 0 | 0 | 0 | 0 | 0 | 0 | 0 | 0 | *0* | *1* | 2 |
| Days Min Temp ≤ 32 °F | 6 | 3 | 2 | 0 | 0 | 0 | 0 | 0 | 0 | 0 | *2* | *6* | 19 |
| Days Min Temp ≤ 0 °F | 0 | 0 | 0 | 0 | 0 | 0 | 0 | 0 | 0 | 0 | *0* | 0 | 0 |
| Heating Degree Days | 731 | 606 | 583 | 455 | 314 | 191 | 105 | 89 | 182 | 397 | *588* | 720 | 4961 |
| Cooling Degree Days | 0 | 0 | 0 | 0 | 1 | 7 | 19 | 27 | 6 | 0 | *0* | 0 | 60 |
| Total Precipitation (") | 2.16 | 1.54 | 1.66 | 1.58 | 1.50 | 1.24 | 0.94 | 0.93 | 1.12 | 1.26 | 2.48 | 2.77 | 19.18 |
| Days ≥ 0.1" Precip | 7 | 5 | 6 | 5 | 4 | 4 | 2 | 3 | 3 | 5 | *7* | *8* | 59 |
| Total Snowfall (") | *1.4* | *0.6* | *0.1* | *0.0* | *0.0* | *0.0* | *0.0* | *0.0* | *0.0* | *0.0* | na | *0.7* | na |
| Days ≥ 1" Snow Depth | na | *0* | *0* | *0* | *0* | *0* | 0 | *0* | *0* | *0* | na | *0* | na |

## PRIEST RAPIDS DAM *Grant County* ELEVATION 469 ft LAT/LONG 46° 39 ' N / 119° 54 ' W

| | JAN | FEB | MAR | APR | MAY | JUN | JUL | AUG | SEP | OCT | NOV | DEC | YEAR |
|---|---|---|---|---|---|---|---|---|---|---|---|---|---|
| Maximum Temp °F | 40.0 | 47.0 | 57.4 | 66.0 | 75.7 | 83.1 | 90.8 | 90.1 | 79.9 | 65.9 | 49.7 | 40.4 | 65.5 |
| Minimum Temp °F | 26.3 | 30.6 | 36.9 | 43.4 | 51.6 | 58.6 | 64.3 | 64.3 | 55.9 | 43.9 | 34.5 | 27.2 | 44.8 |
| Mean Temp °F | 33.2 | 38.9 | 47.2 | 54.7 | 63.6 | 70.9 | 77.6 | 77.2 | 68.0 | 54.9 | 42.1 | 33.8 | 55.2 |
| Days Max Temp ≥ 90 °F | 0 | 0 | 0 | 0 | 2 | 8 | 18 | 16 | 4 | 0 | 0 | 0 | 48 |
| Days Max Temp ≤ 32 °F | 7 | 2 | 0 | 0 | 0 | 0 | 0 | 0 | 0 | 0 | 1 | 6 | 16 |
| Days Min Temp ≤ 32 °F | 23 | 15 | 7 | 2 | 0 | 0 | 0 | 0 | 0 | *2* | 12 | 23 | 84 |
| Days Min Temp ≤ 0 °F | 1 | 0 | 0 | 0 | 0 | 0 | 0 | 0 | 0 | 0 | 0 | 0 | 1 |
| Heating Degree Days | 981 | 733 | 547 | 307 | 103 | 19 | 2 | 1 | 43 | 314 | 681 | 960 | 4691 |
| Cooling Degree Days | 0 | 0 | 0 | 6 | 72 | 190 | 380 | 376 | 123 | 10 | 0 | 0 | 1157 |
| Total Precipitation (") | 0.79 | 0.53 | 0.59 | 0.42 | 0.38 | 0.43 | 0.21 | 0.30 | 0.37 | 0.45 | 0.99 | 1.10 | 6.56 |
| Days ≥ 0.1" Precip | 3 | 2 | 2 | 1 | 1 | 1 | 0 | 1 | 1 | 2 | 3 | 4 | 21 |
| Total Snowfall (") | na | na | *0.0* | *0.0* | *0.0* | *0.0* | *0.0* | *0.0* | *0.0* | *0.0* | na | na | na |
| Days ≥ 1" Snow Depth | na | na | *0* | *0* | *0* | *0* | *0* | *0* | *0* | *0* | na | na | na |

## PROSSER 4 NE *Benton County* ELEVATION 840 ft LAT/LONG 46° 15 ' N / 119° 45 ' W

| | JAN | FEB | MAR | APR | MAY | JUN | JUL | AUG | SEP | OCT | NOV | DEC | YEAR |
|---|---|---|---|---|---|---|---|---|---|---|---|---|---|
| Maximum Temp °F | 38.2 | 45.9 | 56.6 | 64.1 | 72.9 | 79.9 | 86.9 | 86.4 | 76.7 | 64.8 | 48.6 | 39.0 | 63.3 |
| Minimum Temp °F | 24.3 | 28.1 | 33.1 | 37.6 | 44.3 | 50.2 | 53.5 | 53.1 | 46.1 | 37.6 | 31.2 | 25.1 | 38.7 |
| Mean Temp °F | 31.3 | 37.0 | 44.9 | 50.9 | 58.6 | 65.1 | 70.2 | 69.8 | 61.4 | 51.2 | 39.9 | 32.1 | 51.0 |
| Days Max Temp ≥ 90 °F | 0 | 0 | 0 | 0 | 1 | 5 | 13 | 12 | 2 | 0 | 0 | 0 | 33 |
| Days Max Temp ≤ 32 °F | 9 | 3 | 0 | 0 | 0 | 0 | 0 | 0 | 0 | 0 | 2 | 9 | 23 |
| Days Min Temp ≤ 32 °F | 25 | 21 | 15 | 8 | 1 | 0 | 0 | 0 | 1 | 7 | 17 | 26 | 121 |
| Days Min Temp ≤ 0 °F | 1 | 0 | 0 | 0 | 0 | 0 | 0 | 0 | 0 | 0 | 0 | 1 | 2 |
| Heating Degree Days | 1039 | 782 | 617 | 419 | 217 | 82 | 22 | 21 | 142 | 421 | 745 | 1014 | 5521 |
| Cooling Degree Days | 0 | *0* | 0 | *3* | 31 | 82 | 185 | 182 | *45* | 2 | 0 | 0 | 530 |
| Total Precipitation (") | 0.98 | 0.57 | 0.57 | 0.62 | 0.58 | 0.58 | 0.24 | 0.36 | 0.48 | 0.56 | 1.03 | 1.12 | 7.69 |
| Days ≥ 0.1" Precip | 3 | 2 | 2 | 2 | 2 | 2 | 1 | 1 | 1 | 2 | 3 | 4 | 25 |
| Total Snowfall (") | 3.0 | *0.8* | 0.0 | 0.0 | 0.0 | 0.0 | 0.0 | 0.0 | 0.0 | 0.0 | 0.9 | *3.3* | 8.0 |
| Days ≥ 1" Snow Depth | *6* | *1* | 0 | 0 | 0 | 0 | 0 | 0 | 0 | 0 | 1 | na | na |

## PULLMAN 2 NW *Whitman County* ELEVATION 2552 ft LAT/LONG 46° 46 ' N / 117° 12 ' W

| | JAN | FEB | MAR | APR | MAY | JUN | JUL | AUG | SEP | OCT | NOV | DEC | YEAR |
|---|---|---|---|---|---|---|---|---|---|---|---|---|---|
| Maximum Temp °F | 35.0 | 40.7 | 47.8 | 55.7 | 64.5 | 71.8 | 80.9 | 82.1 | 72.4 | 60.0 | 43.3 | 35.7 | 57.5 |
| Minimum Temp °F | 23.4 | 27.5 | 31.2 | 35.7 | 41.4 | 46.8 | 49.9 | 50.2 | 44.1 | 36.8 | 30.4 | 24.4 | 36.8 |
| Mean Temp °F | 29.2 | 34.1 | 39.5 | 45.7 | 53.0 | 59.3 | 65.4 | 66.2 | 58.3 | 48.4 | 36.9 | 30.1 | 47.2 |
| Days Max Temp ≥ 90 °F | 0 | 0 | 0 | 0 | 0 | 1 | 6 | 7 | 1 | 0 | 0 | 0 | 15 |
| Days Max Temp ≤ 32 °F | 11 | 4 | 1 | 0 | 0 | 0 | 0 | 0 | 0 | 0 | 3 | 10 | 29 |
| Days Min Temp ≤ 32 °F | 25 | 21 | 18 | 10 | 3 | 0 | 0 | 0 | 2 | 8 | 18 | 25 | 130 |
| Days Min Temp ≤ 0 °F | 2 | 0 | 0 | 0 | 0 | 0 | 0 | 0 | 0 | 0 | 0 | 1 | 3 |
| Heating Degree Days | 1103 | 866 | 782 | 574 | 374 | 193 | 74 | 68 | 220 | 509 | 837 | 1076 | 6676 |
| Cooling Degree Days | 0 | 0 | 0 | 1 | 12 | 29 | 89 | 111 | 29 | 3 | 0 | 0 | 274 |
| Total Precipitation (") | 2.68 | 1.92 | 1.92 | 1.64 | 1.64 | 1.36 | 0.75 | 0.88 | 0.88 | 1.36 | 2.67 | 2.59 | 20.29 |
| Days ≥ 0.1" Precip | 8 | 6 | 7 | 5 | 5 | 4 | 2 | 2 | 3 | 4 | 9 | 8 | 63 |
| Total Snowfall (") | 11.5 | 5.4 | 2.6 | 0.6 | 0.0 | 0.0 | 0.0 | 0.0 | 0.0 | 0.4 | 4.6 | 9.3 | 34.4 |
| Days ≥ 1" Snow Depth | *15* | 7 | 3 | 0 | 0 | 0 | 0 | 0 | 0 | 0 | 4 | *12* | 41 |

## PUYALLUP 2 W EXP STN *Pierce County* ELEVATION 49 ft LAT/LONG 47° 12 ' N / 122° 20 ' W

| | JAN | FEB | MAR | APR | MAY | JUN | JUL | AUG | SEP | OCT | NOV | DEC | YEAR |
|---|---|---|---|---|---|---|---|---|---|---|---|---|---|
| Maximum Temp °F | 46.5 | 51.1 | 55.6 | 60.6 | 67.8 | 72.8 | 78.1 | 78.5 | 72.6 | 62.3 | 51.8 | 46.2 | 62.0 |
| Minimum Temp °F | 32.7 | 33.9 | 35.7 | 38.7 | 43.6 | 48.5 | 51.1 | 51.2 | 47.0 | 41.4 | 36.7 | 32.9 | 41.1 |
| Mean Temp °F | 39.6 | 42.5 | 45.7 | 49.7 | 55.7 | 60.7 | 64.6 | 64.9 | 59.8 | 51.9 | 44.2 | 39.6 | 51.6 |
| Days Max Temp ≥ 90 °F | 0 | 0 | 0 | 0 | 0 | 1 | 2 | 2 | 0 | 0 | 0 | 0 | 5 |
| Days Max Temp ≤ 32 °F | 1 | 0 | 0 | 0 | 0 | 0 | 0 | 0 | 0 | 0 | 0 | 1 | 2 |
| Days Min Temp ≤ 32 °F | 14 | 12 | 11 | 5 | 1 | 0 | 0 | 0 | 0 | 3 | 9 | 14 | 69 |
| Days Min Temp ≤ 0 °F | 0 | 0 | 0 | 0 | 0 | 0 | 0 | 0 | 0 | 0 | 0 | 0 | 0 |
| Heating Degree Days | 781 | 628 | 593 | 453 | 284 | 140 | 55 | 51 | 159 | 401 | 616 | 781 | 4942 |
| Cooling Degree Days | 0 | 0 | 0 | 0 | 6 | 16 | 51 | 64 | 14 | 0 | 0 | 0 | 151 |
| Total Precipitation (") | 5.67 | 4.36 | 3.89 | 3.07 | 1.95 | 1.81 | 0.84 | 1.22 | 1.86 | 3.19 | 5.58 | 6.00 | 39.44 |
| Days ≥ 0.1" Precip | 13 | 11 | 11 | 9 | 6 | 5 | 2 | 3 | 5 | 8 | 13 | 13 | 99 |
| Total Snowfall (") | 2.9 | 0.8 | 0.3 | 0.0 | 0.0 | 0.0 | 0.0 | 0.0 | 0.0 | 0.1 | 0.8 | 2.0 | 6.9 |
| Days ≥ 1" Snow Depth | 2 | 1 | 0 | 0 | 0 | 0 | 0 | 0 | 0 | 0 | 1 | 1 | 5 |

## QUILCENE 2 SW *Jefferson County* ELEVATION 121 ft LAT/LONG 47° 49 ' N / 122° 55 ' W

| | JAN | FEB | MAR | APR | MAY | JUN | JUL | AUG | SEP | OCT | NOV | DEC | YEAR |
|---|---|---|---|---|---|---|---|---|---|---|---|---|---|
| Maximum Temp °F | 44.1 | 49.5 | 55.1 | 60.2 | 66.8 | 71.9 | 76.9 | 78.0 | 72.1 | 61.5 | 49.9 | 43.5 | 60.8 |
| Minimum Temp °F | 30.6 | 32.2 | 34.2 | 37.7 | 42.8 | 47.6 | 50.3 | 50.2 | 45.7 | 39.4 | 35.2 | 31.5 | 39.8 |
| Mean Temp °F | 37.3 | 40.9 | 44.7 | 49.0 | 54.8 | 59.8 | 63.6 | 64.1 | 58.9 | 50.5 | 42.7 | 37.6 | 50.3 |
| Days Max Temp ≥ 90 °F | 0 | 0 | 0 | 0 | 0 | 1 | 2 | 3 | 0 | 0 | 0 | 0 | 6 |
| Days Max Temp ≤ 32 °F | 1 | 0 | 0 | 0 | 0 | 0 | 0 | 0 | 0 | 0 | 0 | 1 | 2 |
| Days Min Temp ≤ 32 °F | *19* | *16* | 13 | 6 | 1 | 0 | 0 | 0 | 0 | 4 | 10 | *18* | 87 |
| Days Min Temp ≤ 0 °F | 0 | 0 | 0 | 0 | 0 | 0 | 0 | 0 | 0 | 0 | 0 | 0 | 0 |
| Heating Degree Days | 852 | 675 | 623 | 473 | 312 | 167 | 81 | 71 | 185 | 439 | 661 | 846 | 5385 |
| Cooling Degree Days | 0 | 0 | 0 | 0 | 4 | 19 | 45 | 53 | 12 | 0 | 0 | 0 | 133 |
| Total Precipitation (") | 6.99 | 6.76 | 5.95 | 3.69 | 2.59 | 2.00 | 1.19 | 1.18 | 1.64 | 3.96 | 7.54 | 9.89 | 53.38 |
| Days ≥ 0.1" Precip | 10 | *10* | 9 | 8 | 7 | 5 | 4 | 3 | 4 | 6 | 10 | 10 | 86 |
| Total Snowfall (") | *3.6* | *1.1* | *0.1* | 0.0 | 0.0 | 0.0 | 0.0 | 0.0 | 0.0 | 0.0 | 0.6 | *2.8* | 8.2 |
| Days ≥ 1" Snow Depth | *2* | 1 | 0 | 0 | 0 | 0 | 0 | 0 | 0 | 0 | 0 | na | na |

## QUINCY 1 S *Grant County* ELEVATION 1270 ft LAT/LONG 47° 13 ' N / 119° 51 ' W

| | JAN | FEB | MAR | APR | MAY | JUN | JUL | AUG | SEP | OCT | NOV | DEC | YEAR |
|---|---|---|---|---|---|---|---|---|---|---|---|---|---|
| Maximum Temp °F | 35.0 | 43.3 | 55.2 | 63.9 | 73.0 | 79.9 | 86.4 | 85.7 | 77.2 | 63.4 | 46.0 | 35.3 | 62.0 |
| Minimum Temp °F | 19.3 | 25.1 | 30.9 | 37.3 | 45.2 | 51.7 | 55.6 | 55.1 | 46.9 | 36.3 | 28.6 | 20.7 | 37.7 |
| Mean Temp °F | 27.2 | 34.2 | 43.1 | 50.6 | 59.1 | 65.8 | 71.0 | 70.4 | 62.0 | 49.9 | 37.3 | 28.0 | 49.9 |
| Days Max Temp ≥ 90 °F | 0 | 0 | 0 | 0 | 1 | 4 | 12 | 10 | 2 | 0 | 0 | 0 | 29 |
| Days Max Temp ≤ 32 °F | 12 | 4 | 0 | 0 | 0 | 0 | 0 | 0 | 0 | 0 | 2 | 11 | 29 |
| Days Min Temp ≤ 32 °F | 29 | 24 | 19 | 8 | 1 | 0 | 0 | 0 | 1 | 10 | 20 | 28 | 140 |
| Days Min Temp ≤ 0 °F | 2 | 1 | 0 | 0 | 0 | 0 | 0 | 0 | 0 | 0 | 0 | 2 | 5 |
| Heating Degree Days | 1165 | 863 | 673 | 427 | 204 | 68 | 17 | 19 | 130 | 462 | 823 | 1139 | 5990 |
| Cooling Degree Days | 0 | 0 | 0 | 2 | 36 | 91 | 208 | 196 | 55 | 2 | 0 | 0 | 590 |
| Total Precipitation (") | 0.90 | 0.61 | 0.63 | 0.51 | 0.57 | 0.54 | 0.34 | 0.38 | 0.42 | 0.50 | 1.24 | 1.39 | 8.03 |
| Days ≥ 0.1" Precip | 3 | 2 | 2 | 2 | 2 | 2 | 1 | 1 | 1 | 2 | 4 | 5 | 27 |
| Total Snowfall (") | na | *1.0* | 0.2 | 0.0 | 0.0 | 0.0 | 0.0 | 0.0 | 0.0 | 0.0 | 1.1 | na | na |
| Days ≥ 1" Snow Depth | na | na | 0 | 0 | 0 | 0 | 0 | 0 | 0 | 0 | *0* | na | na |

## RAINIER PARADISE RNG *Pierce County* ELEVATION 5554 ft LAT/LONG 46° 47 ' N / 121° 44 ' W

| | JAN | FEB | MAR | APR | MAY | JUN | JUL | AUG | SEP | OCT | NOV | DEC | YEAR |
|---|---|---|---|---|---|---|---|---|---|---|---|---|---|
| Maximum Temp °F | 31.8 | 34.0 | 36.4 | 40.4 | 47.8 | 53.6 | 60.5 | 61.7 | 56.0 | 47.4 | 35.5 | 32.0 | 44.8 |
| Minimum Temp °F | 20.2 | 21.6 | 22.7 | 25.1 | 30.8 | 36.6 | 42.2 | 43.6 | 39.2 | 32.7 | 24.0 | 20.4 | 29.9 |
| Mean Temp °F | 26.0 | 27.8 | 29.6 | 32.8 | 39.4 | 45.1 | 51.4 | 52.7 | 47.6 | 40.1 | 29.7 | 26.2 | 37.4 |
| Days Max Temp ≥ 90 °F | 0 | 0 | 0 | 0 | 0 | 0 | 0 | 0 | 0 | 0 | 0 | 0 | 0 |
| Days Max Temp ≤ 32 °F | 18 | 13 | 11 | 7 | 1 | 0 | 0 | 0 | 0 | 3 | 12 | 16 | 81 |
| Days Min Temp ≤ 32 °F | 29 | 26 | 28 | 25 | 20 | 11 | 3 | 1 | 7 | 17 | 26 | 29 | 222 |
| Days Min Temp ≤ 0 °F | 1 | 0 | 0 | 0 | 0 | 0 | 0 | 0 | 0 | 0 | 0 | 1 | 2 |
| Heating Degree Days | 1202 | 1044 | 1092 | 961 | 788 | 591 | 424 | 387 | 518 | 766 | 1051 | 1196 | 10020 |
| Cooling Degree Days | 0 | 0 | 0 | 0 | 0 | 2 | 6 | 11 | 4 | 0 | 0 | 0 | 23 |
| Total Precipitation (") | 18.16 | 13.24 | 11.64 | 8.67 | 5.09 | 3.83 | 1.94 | 2.40 | 4.75 | 8.70 | 17.41 | 18.38 | 114.21 |
| Days ≥ 0.1" Precip | 18 | 16 | 17 | 15 | 11 | 9 | 5 | 5 | 8 | 12 | 19 | 20 | 155 |
| Total Snowfall (") | 131.6 | 101.3 | 98.7 | 72.1 | 25.2 | 6.2 | 0.6 | 0.0 | 3.4 | 28.8 | 97.2 | 121.0 | 686.1 |
| Days ≥ 1" Snow Depth | 31 | 28 | 31 | 30 | 31 | 29 | 12 | 1 | 1 | 11 | 27 | 31 | 263 |

## REPUBLIC *Ferry County* ELEVATION 2651 ft LAT/LONG 48° 39 ' N / 118° 44 ' W

| | JAN | FEB | MAR | APR | MAY | JUN | JUL | AUG | SEP | OCT | NOV | DEC | YEAR |
|---|---|---|---|---|---|---|---|---|---|---|---|---|---|
| Maximum Temp °F | 29.4 | 37.7 | 47.2 | 57.0 | 65.9 | 73.1 | 80.5 | 80.9 | 71.1 | 57.0 | 38.4 | 29.2 | 55.6 |
| Minimum Temp °F | 14.4 | 19.3 | 24.4 | 30.1 | 37.1 | 43.3 | 46.7 | 46.2 | 38.5 | 30.1 | 23.9 | 15.8 | 30.8 |
| Mean Temp °F | 21.9 | 28.5 | 35.9 | 43.6 | 51.5 | 58.2 | 63.6 | 63.5 | 54.9 | 43.6 | 31.1 | 22.5 | 43.2 |
| Days Max Temp ≥ 90 °F | 0 | 0 | 0 | 0 | 0 | 1 | 6 | 6 | 0 | 0 | 0 | 0 | 13 |
| Days Max Temp ≤ 32 °F | 18 | 6 | 1 | 0 | 0 | 0 | 0 | 0 | 0 | 0 | 6 | 19 | 50 |
| Days Min Temp ≤ 32 °F | 30 | 26 | 28 | 20 | 8 | 1 | 0 | 0 | 6 | 19 | 25 | 30 | 193 |
| Days Min Temp ≤ 0 °F | 5 | 2 | 0 | 0 | 0 | 0 | 0 | 0 | 0 | 0 | 1 | 4 | 12 |
| Heating Degree Days | 1330 | 1023 | 896 | 636 | 413 | 214 | 96 | 96 | 302 | 658 | 1009 | 1311 | 7984 |
| Cooling Degree Days | 0 | 0 | 0 | 0 | 4 | 15 | 55 | 55 | 4 | 0 | 0 | 0 | 133 |
| Total Precipitation (") | 1.50 | 1.20 | 1.22 | 1.22 | 1.64 | 1.62 | 1.17 | 1.20 | 0.88 | 0.86 | 1.60 | 1.98 | 16.09 |
| Days ≥ 0.1" Precip | 5 | 4 | 4 | 4 | 5 | 5 | 3 | 3 | 2 | 3 | 5 | 7 | 50 |
| Total Snowfall (") | 14.3 | 7.0 | 3.0 | 0.6 | 0.1 | 0.0 | 0.0 | 0.0 | 0.0 | 0.5 | 6.7 | 16.5 | 48.7 |
| Days ≥ 1" Snow Depth | 29 | 25 | 11 | 1 | 0 | 0 | 0 | 0 | 0 | 0 | 7 | 24 | 97 |

## RICHLAND *Benton County* ELEVATION 400 ft LAT/LONG 46° 18 ' N / 119° 18 ' W

| | JAN | FEB | MAR | APR | MAY | JUN | JUL | AUG | SEP | OCT | NOV | DEC | YEAR |
|---|---|---|---|---|---|---|---|---|---|---|---|---|---|
| Maximum Temp °F | 40.4 | 48.3 | 58.4 | 66.2 | 75.1 | 82.3 | 89.4 | 88.9 | 79.4 | 66.8 | 50.5 | 41.1 | 65.6 |
| Minimum Temp °F | 26.4 | 30.4 | 35.2 | 41.0 | 48.0 | 54.9 | 59.2 | 58.5 | 50.3 | 40.3 | 34.0 | 27.7 | 42.2 |
| Mean Temp °F | 33.4 | 39.4 | 46.8 | 53.6 | 61.6 | 68.6 | 74.3 | 73.7 | 64.9 | 53.6 | 42.3 | 34.4 | 53.9 |
| Days Max Temp ≥ 90 °F | 0 | 0 | 0 | 0 | 2 | 7 | 16 | 15 | 4 | 0 | 0 | 0 | 44 |
| Days Max Temp ≤ 32 °F | 8 | 2 | 0 | 0 | 0 | 0 | 0 | 0 | 0 | 0 | 1 | 6 | 17 |
| Days Min Temp ≤ 32 °F | 22 | 16 | 10 | 2 | 0 | 0 | 0 | 0 | 0 | 4 | 12 | 22 | 88 |
| Days Min Temp ≤ 0 °F | 1 | 0 | 0 | 0 | 0 | 0 | 0 | 0 | 0 | 0 | 0 | 1 | 2 |
| Heating Degree Days | 973 | 717 | 557 | 339 | 143 | 33 | 5 | 4 | 79 | 349 | 676 | 942 | 4817 |
| Cooling Degree Days | 0 | 0 | 0 | 4 | 51 | 135 | 283 | 276 | 75 | 2 | 0 | 0 | 826 |
| Total Precipitation (") | 0.97 | 0.67 | 0.60 | 0.52 | 0.55 | 0.41 | 0.20 | 0.33 | 0.32 | 0.46 | 1.07 | 1.03 | 7.13 |
| Days ≥ 0.1" Precip | 4 | 3 | 2 | 2 | 2 | 1 | 1 | 1 | 1 | 1 | 4 | 4 | 26 |
| Total Snowfall (") | 3.5 | 2.3 | 0.3 | 0.0 | 0.0 | 0.0 | 0.0 | 0.0 | 0.0 | 0.0 | 1.2 | 3.4 | 10.7 |
| Days ≥ 1" Snow Depth | 7 | 3 | 0 | 0 | 0 | 0 | 0 | 0 | 0 | 0 | 1 | 5 | 16 |

## RITZVILLE 1 SSE *Adams County* ELEVATION 1831 ft LAT/LONG 47° 7 ' N / 118° 22 ' W

| | JAN | FEB | MAR | APR | MAY | JUN | JUL | AUG | SEP | OCT | NOV | DEC | YEAR |
|---|---|---|---|---|---|---|---|---|---|---|---|---|---|
| Maximum Temp °F | 33.9 | 41.6 | 51.3 | 60.0 | 69.0 | 77.3 | 85.5 | 85.3 | 75.7 | 62.0 | 44.3 | 35.2 | 60.1 |
| Minimum Temp °F | 21.2 | 25.7 | 29.6 | 33.5 | 39.9 | 46.7 | 52.0 | 51.9 | 44.3 | 35.0 | 28.3 | 22.3 | 35.9 |
| Mean Temp °F | 27.6 | 33.7 | 40.4 | 46.7 | 54.5 | 62.0 | 68.7 | 68.6 | 60.0 | 48.5 | 36.3 | 28.8 | 48.0 |
| Days Max Temp ≥ 90 °F | 0 | 0 | 0 | 0 | 1 | 4 | 11 | 11 | 2 | 0 | 0 | 0 | 29 |
| Days Max Temp ≤ 32 °F | 12 | 4 | 0 | 0 | 0 | 0 | 0 | 0 | 0 | 0 | 2 | 10 | 28 |
| Days Min Temp ≤ 32 °F | 28 | 24 | 22 | 14 | 4 | 0 | 0 | 0 | 1 | 11 | 22 | 27 | 153 |
| Days Min Temp ≤ 0 °F | 2 | 0 | 0 | 0 | 0 | 0 | 0 | 0 | 0 | 0 | 0 | 1 | 3 |
| Heating Degree Days | 1153 | 879 | 755 | 542 | 332 | 144 | 44 | 42 | 183 | 506 | 855 | 1116 | 6551 |
| Cooling Degree Days | 0 | 0 | 0 | 1 | 18 | 60 | 152 | 155 | 42 | 3 | 0 | 0 | 431 |
| Total Precipitation (") | 1.27 | 0.99 | 1.09 | 0.91 | 0.89 | 0.69 | 0.52 | 0.42 | 0.54 | 0.82 | 1.63 | 1.64 | 11.41 |
| Days ≥ 0.1" Precip | 5 | 3 | 4 | 3 | 3 | 2 | 1 | 1 | 2 | 3 | 5 | 6 | 38 |
| Total Snowfall (") | 6.8 | 2.9 | 0.8 | 0.1 | 0.0 | 0.0 | 0.0 | 0.0 | 0.0 | 0.1 | 2.3 | 7.0 | 20.0 |
| Days ≥ 1" Snow Depth | 14 | 5 | 1 | 0 | 0 | 0 | 0 | 0 | 0 | 0 | 2 | 11 | 33 |

## ROSALIA *Whitman County* ELEVATION 2402 ft LAT/LONG 47° 14 ' N / 117° 22 ' W

| | JAN | FEB | MAR | APR | MAY | JUN | JUL | AUG | SEP | OCT | NOV | DEC | YEAR |
|---|---|---|---|---|---|---|---|---|---|---|---|---|---|
| Maximum Temp °F | 34.4 | 40.7 | 48.6 | 56.9 | 65.6 | 72.8 | 81.5 | 82.5 | 73.0 | 60.9 | 43.5 | 35.4 | 58.0 |
| Minimum Temp °F | 22.0 | 26.3 | 30.5 | 34.5 | 40.6 | 46.4 | 50.1 | 49.9 | 42.4 | 34.1 | 28.7 | 22.9 | 35.7 |
| Mean Temp °F | 28.2 | 33.5 | 39.6 | 45.7 | 53.2 | 59.6 | 65.8 | 66.3 | 57.8 | 47.5 | 36.1 | 29.2 | 46.9 |
| Days Max Temp ≥ 90 °F | 0 | 0 | 0 | 0 | 0 | 2 | 7 | 8 | 1 | 0 | 0 | 0 | 18 |
| Days Max Temp ≤ 32 °F | 11 | 4 | 1 | 0 | 0 | 0 | 0 | 0 | 0 | 0 | 2 | 9 | 27 |
| Days Min Temp ≤ 32 °F | 26 | 23 | 20 | 12 | 3 | 0 | 0 | 0 | 2 | 12 | 20 | 26 | 144 |
| Days Min Temp ≤ 0 °F | 2 | 1 | 0 | 0 | 0 | 0 | 0 | 0 | 0 | 0 | 0 | 1 | 4 |
| Heating Degree Days | 1134 | 883 | 782 | 573 | 369 | 192 | 71 | 65 | 233 | 536 | 860 | 1105 | 6803 |
| Cooling Degree Days | 0 | 0 | 0 | 1 | 13 | 38 | 103 | 112 | 26 | 2 | 0 | 0 | 295 |
| Total Precipitation (") | 2.26 | 1.52 | 1.62 | 1.47 | 1.58 | 1.33 | 0.76 | 0.73 | 0.84 | 1.13 | 2.25 | 2.33 | 17.82 |
| Days ≥ 0.1" Precip | 8 | 6 | 6 | 5 | 5 | 4 | 2 | 2 | 3 | 4 | 8 | 8 | 61 |
| Total Snowfall (") | 7.5 | 2.9 | 1.2 | 0.2 | 0.1 | 0.0 | 0.0 | 0.0 | 0.0 | 0.1 | 2.9 | 7.2 | 22.1 |
| Days ≥ 1" Snow Depth | *13* | *5* | 2 | 0 | 0 | 0 | 0 | 0 | 0 | 0 | *2* | 8 | 30 |

## ROSS DAM *Whatcom County* ELEVATION 1240 ft LAT/LONG 48° 44 ' N / 121° 3 ' W

| | JAN | FEB | MAR | APR | MAY | JUN | JUL | AUG | SEP | OCT | NOV | DEC | YEAR |
|---|---|---|---|---|---|---|---|---|---|---|---|---|---|
| Maximum Temp °F | 37.2 | 41.7 | 48.2 | 55.3 | 63.8 | 69.9 | 76.2 | 77.1 | 69.2 | 57.2 | 44.1 | 38.1 | 56.5 |
| Minimum Temp °F | 28.3 | 30.3 | 33.0 | 37.3 | 43.5 | 49.3 | 53.4 | 54.7 | 49.2 | 42.0 | 34.9 | 29.9 | 40.5 |
| Mean Temp °F | 32.7 | 36.0 | 40.6 | 46.3 | 53.7 | 59.6 | 64.8 | 65.9 | 59.2 | 49.6 | 39.5 | 34.0 | 48.5 |
| Days Max Temp ≥ 90 °F | 0 | 0 | 0 | 0 | 0 | 1 | 3 | 4 | 0 | 0 | 0 | 0 | 8 |
| Days Max Temp ≤ 32 °F | 6 | 2 | 0 | 0 | 0 | 0 | 0 | 0 | 0 | 0 | 1 | 4 | 13 |
| Days Min Temp ≤ 32 °F | 20 | 17 | 13 | 4 | 0 | 0 | 0 | 0 | 0 | 1 | 8 | 19 | 82 |
| Days Min Temp ≤ 0 °F | 0 | 0 | 0 | 0 | 0 | 0 | 0 | 0 | 0 | 0 | 0 | 0 | 0 |
| Heating Degree Days | 993 | 812 | 749 | 554 | 350 | 185 | 82 | 65 | 189 | 470 | 758 | 953 | 6160 |
| Cooling Degree Days | 0 | 0 | 0 | 0 | 9 | 33 | 85 | 100 | 26 | 1 | 0 | 0 | 254 |
| Total Precipitation (") | 8.99 | 6.38 | 4.97 | 3.19 | 2.00 | 1.64 | 1.27 | 1.19 | 2.33 | 5.19 | 9.32 | 9.32 | 55.79 |
| Days ≥ 0.1" Precip | 13 | 11 | 11 | 8 | 5 | 4 | 3 | 4 | 5 | 9 | 14 | 14 | 101 |
| Total Snowfall (") | na | na | na | ***0.3*** | ***0.0*** | ***0.0*** | ***0.0*** | ***0.0*** | ***0.0*** | ***0.1*** | na | na | na |
| Days ≥ 1" Snow Depth | 19 | 14 | 5 | 0 | 0 | 0 | 0 | 0 | 0 | 0 | *2* | *13* | 53 |

## SAINT JOHN *Whitman County* ELEVATION 1952 ft LAT/LONG 47° 6' N / 117° 35' W

| | JAN | FEB | MAR | APR | MAY | JUN | JUL | AUG | SEP | OCT | NOV | DEC | YEAR |
|---|---|---|---|---|---|---|---|---|---|---|---|---|---|
| Maximum Temp °F | 36.7 | 43.8 | 52.1 | 60.1 | 69.2 | 76.7 | 84.8 | 84.9 | 75.8 | 63.2 | 45.7 | 37.4 | 60.9 |
| Minimum Temp °F | 22.9 | 27.2 | 30.2 | 34.2 | 39.7 | 45.5 | 49.0 | 48.8 | 41.8 | 33.0 | 29.3 | 23.9 | 35.5 |
| Mean Temp °F | 29.9 | 35.5 | 41.2 | 47.2 | 54.5 | 61.1 | 66.9 | 66.9 | 59.0 | 48.0 | 37.5 | 30.6 | 48.2 |
| Days Max Temp ≥ 90 °F | 0 | 0 | 0 | 0 | 1 | 3 | 10 | 10 | 2 | 0 | 0 | 0 | 26 |
| Days Max Temp ≤ 32 °F | 9 | 3 | 0 | 0 | 0 | 0 | 0 | 0 | 0 | 0 | 2 | 8 | 22 |
| Days Min Temp ≤ 32 °F | 25 | 21 | 20 | 12 | 5 | 1 | 0 | 0 | 3 | 14 | 18 | 25 | 144 |
| Days Min Temp ≤ 0 °F | 2 | 0 | 0 | 0 | 0 | 0 | 0 | 0 | 0 | 0 | 0 | 1 | 3 |
| Heating Degree Days | 1082 | 826 | 732 | 529 | 329 | 151 | 49 | 51 | 197 | 520 | 820 | 1059 | 6345 |
| Cooling Degree Days | 0 | 0 | 0 | 1 | 13 | 39 | 106 | 116 | 23 | 1 | 0 | 0 | 299 |
| Total Precipitation (") | 2.00 | 1.30 | 1.50 | 1.42 | 1.50 | 1.16 | 0.75 | 0.70 | 0.80 | 1.11 | 2.12 | 2.19 | 16.55 |
| Days ≥ 0.1" Precip | 6 | 4 | 5 | 5 | 4 | 3 | 2 | 2 | 3 | 3 | 7 | 7 | 51 |
| Total Snowfall (") | ***8.2*** | 3.8 | 1.3 | 0.4 | 0.0 | 0.0 | 0.0 | 0.0 | 0.0 | 0.1 | 2.6 | 7.3 | 23.7 |
| Days ≥ 1" Snow Depth | na | ***2*** | 0 | 0 | 0 | 0 | 0 | 0 | 0 | 0 | 1 | ***3*** | na |

## SAPPHO 8 E *Clallam County* ELEVATION 760 ft LAT/LONG 48° 4' N / 124° 7' W

| | JAN | FEB | MAR | APR | MAY | JUN | JUL | AUG | SEP | OCT | NOV | DEC | YEAR |
|---|---|---|---|---|---|---|---|---|---|---|---|---|---|
| Maximum Temp °F | 42.7 | 46.7 | 51.1 | 55.1 | 62.1 | 67.6 | 72.6 | 73.9 | 69.7 | 59.3 | 47.7 | ***42.9*** | 57.6 |
| Minimum Temp °F | 33.0 | 34.1 | 35.0 | 37.3 | 42.0 | 46.3 | 48.9 | 49.4 | 46.7 | 41.7 | 36.5 | ***33.6*** | 40.4 |
| Mean Temp °F | 37.9 | 40.4 | 43.1 | 46.2 | 52.4 | 56.9 | 60.8 | 61.6 | 58.2 | 50.6 | 42.3 | ***38.3*** | 49.1 |
| Days Max Temp ≥ 90 °F | 0 | 0 | 0 | 0 | 0 | 0 | 1 | 2 | 0 | 0 | 0 | 0 | 3 |
| Days Max Temp ≤ 32 °F | 1 | 0 | 0 | 0 | 0 | 0 | 0 | 0 | 0 | 0 | 0 | 1 | 2 |
| Days Min Temp ≤ 32 °F | ***13*** | 10 | 10 | 4 | 1 | 0 | 0 | 0 | 0 | 2 | ***7*** | na | na |
| Days Min Temp ≤ 0 °F | 0 | 0 | 0 | 0 | 0 | 0 | 0 | 0 | 0 | 0 | 0 | 0 | 0 |
| Heating Degree Days | 834 | 688 | 673 | 557 | 397 | 247 | 150 | 129 | 208 | 441 | 676 | ***821*** | 5821 |
| Cooling Degree Days | 0 | 0 | 0 | 0 | 4 | 13 | 24 | 38 | 13 | 1 | 0 | ***0*** | 93 |
| Total Precipitation (") | 13.50 | 12.19 | 9.60 | 6.97 | 4.29 | 2.74 | 1.87 | 2.35 | 4.15 | 9.08 | 14.20 | 15.27 | 96.21 |
| Days ≥ 0.1" Precip | na | na | na | na | na | 6 | 5 | 5 | 7 | na | na | na | na |
| Total Snowfall (") | na | ***2.6*** | ***1.1*** | 0.1 | 0.0 | 0.0 | 0.0 | 0.0 | 0.0 | 0.0 | ***0.6*** | na | na |
| Days ≥ 1" Snow Depth | na | ***1*** | ***1*** | 0 | 0 | 0 | 0 | 0 | 0 | 0 | 0 | na | na |

## SEATTLE-TACOMA AP *King County* ELEVATION 384 ft LAT/LONG 47° 26' N / 122° 18' W

| | JAN | FEB | MAR | APR | MAY | JUN | JUL | AUG | SEP | OCT | NOV | DEC | YEAR |
|---|---|---|---|---|---|---|---|---|---|---|---|---|---|
| Maximum Temp °F | 45.3 | 49.7 | 53.3 | 57.7 | 64.5 | 70.0 | 75.4 | 75.5 | 69.7 | 59.8 | 50.5 | 45.3 | 59.7 |
| Minimum Temp °F | 35.6 | 37.3 | 39.0 | 41.8 | 46.9 | 51.8 | 55.1 | 55.7 | 52.0 | 45.8 | 40.1 | 36.0 | 44.8 |
| Mean Temp °F | 40.5 | 43.5 | 46.2 | 49.7 | 55.7 | 60.9 | 65.3 | 65.6 | 60.9 | 52.9 | 45.3 | 40.7 | 52.3 |
| Days Max Temp ≥ 90 °F | 0 | 0 | 0 | 0 | 0 | 0 | 1 | 1 | 0 | 0 | 0 | 0 | 2 |
| Days Max Temp ≤ 32 °F | 1 | 0 | 0 | 0 | 0 | 0 | 0 | 0 | 0 | 0 | 0 | 1 | 2 |
| Days Min Temp ≤ 32 °F | 9 | 5 | 3 | 0 | 0 | 0 | 0 | 0 | 0 | 0 | 3 | 9 | 29 |
| Days Min Temp ≤ 0 °F | 0 | 0 | 0 | 0 | 0 | 0 | 0 | 0 | 0 | 0 | 0 | 0 | 0 |
| Heating Degree Days | 753 | 601 | 576 | 451 | 286 | 138 | 51 | 43 | 138 | 370 | 584 | 746 | 4737 |
| Cooling Degree Days | 0 | 0 | 0 | 0 | 7 | 20 | 59 | 73 | 21 | 1 | 0 | 0 | 181 |
| Total Precipitation (") | 5.27 | 3.83 | 3.56 | 2.61 | 1.68 | 1.48 | 0.73 | 1.09 | 1.79 | 3.07 | 5.35 | 5.85 | 36.31 |
| Days ≥ 0.1" Precip | 12 | 10 | 10 | 7 | 5 | 4 | 2 | 3 | 5 | 8 | 12 | 13 | 91 |
| Total Snowfall (") | 4.0 | 1.1 | 0.7 | 0.1 | 0.0 | 0.0 | 0.0 | 0.0 | 0.0 | 0.1 | 1.0 | 3.1 | 10.1 |
| Days ≥ 1" Snow Depth | 2 | 1 | 0 | 0 | 0 | 0 | 0 | 0 | 0 | 0 | 1 | 1 | 5 |

## SEDRO WOOLLEY *Skagit County* ELEVATION 59 ft LAT/LONG 48° 30' N / 122° 13' W

| | JAN | FEB | MAR | APR | MAY | JUN | JUL | AUG | SEP | OCT | NOV | DEC | YEAR |
|---|---|---|---|---|---|---|---|---|---|---|---|---|---|
| Maximum Temp °F | 44.9 | 49.2 | 53.5 | 58.0 | 64.2 | 68.9 | 73.5 | 74.4 | 69.1 | 60.3 | 50.4 | 45.1 | 59.3 |
| Minimum Temp °F | 32.7 | 34.4 | 37.0 | 40.5 | 45.1 | 49.5 | 51.4 | 51.8 | 48.0 | 42.2 | 37.6 | 34.0 | 42.0 |
| Mean Temp °F | 38.9 | 41.8 | 45.3 | 49.3 | 54.7 | 59.2 | 62.4 | 63.1 | 58.6 | 51.3 | 44.0 | 39.5 | 50.7 |
| Days Max Temp ≥ 90 °F | 0 | 0 | 0 | 0 | 0 | 0 | 0 | 0 | 0 | 0 | 0 | 0 | 0 |
| Days Max Temp ≤ 32 °F | 2 | 0 | 0 | 0 | 0 | 0 | 0 | 0 | 0 | 0 | 0 | 2 | 4 |
| Days Min Temp ≤ 32 °F | 14 | 10 | 6 | 1 | 0 | 0 | 0 | 0 | 0 | 2 | 7 | 12 | 52 |
| Days Min Temp ≤ 0 °F | 0 | 0 | 0 | 0 | 0 | 0 | 0 | 0 | 0 | 0 | 0 | 0 | 0 |
| Heating Degree Days | 804 | 647 | 604 | 465 | 316 | 176 | 97 | 81 | 192 | 419 | 622 | 782 | 5205 |
| Cooling Degree Days | 0 | 0 | 0 | 0 | 4 | 10 | 26 | 33 | 6 | 1 | 0 | 0 | 80 |
| Total Precipitation (") | 5.80 | 4.15 | 3.91 | 3.77 | 2.73 | 2.55 | 1.75 | 1.70 | 2.88 | 3.94 | 6.31 | 5.80 | 45.29 |
| Days ≥ 0.1" Precip | 12 | 10 | 11 | 10 | 7 | 6 | 4 | 4 | 6 | 9 | 14 | 13 | 106 |
| Total Snowfall (") | 2.5 | 1.2 | 0.6 | 0.0 | 0.0 | 0.0 | 0.0 | 0.0 | 0.0 | 0.1 | 0.7 | 2.3 | 7.4 |
| Days ≥ 1" Snow Depth | 3 | 1 | 0 | 0 | 0 | 0 | 0 | 0 | 0 | 0 | 1 | 2 | 7 |

## SHELTON *Mason County* ELEVATION 20 ft LAT/LONG 47° 12 ' N / 123° 6 ' W

| | JAN | FEB | MAR | APR | MAY | JUN | JUL | AUG | SEP | OCT | NOV | DEC | YEAR |
|---|---|---|---|---|---|---|---|---|---|---|---|---|---|
| Maximum Temp °F | 44.7 | 49.1 | 54.3 | 59.4 | 66.8 | 71.9 | 77.0 | 77.5 | 71.7 | 60.9 | 50.4 | 44.6 | 60.7 |
| Minimum Temp °F | 33.8 | 34.8 | 36.5 | 39.5 | 44.9 | 49.8 | 52.9 | 53.3 | 48.8 | 42.5 | 38.0 | 34.5 | 42.4 |
| Mean Temp °F | 39.3 | 42.0 | 45.4 | 49.5 | 55.9 | 60.9 | 65.0 | 65.4 | 60.3 | 51.7 | 44.2 | 39.6 | 51.6 |
| Days Max Temp ≥ 90 °F | 0 | 0 | 0 | 0 | 0 | 1 | 2 | 3 | 0 | 0 | 0 | 0 | 6 |
| Days Max Temp ≤ 32 °F | 1 | 0 | 0 | 0 | 0 | 0 | 0 | 0 | 0 | 0 | 0 | 1 | 2 |
| Days Min Temp ≤ 32 °F | 12 | 11 | 9 | 4 | 0 | 0 | 0 | 0 | 0 | 2 | 6 | 12 | 56 |
| Days Min Temp ≤ 0 °F | 0 | 0 | 0 | 0 | 0 | 0 | 0 | 0 | 0 | 0 | 0 | 0 | 0 |
| Heating Degree Days | 790 | 644 | 600 | 458 | 284 | 143 | 56 | 48 | 150 | 405 | 617 | 781 | 4976 |
| Cooling Degree Days | 0 | 0 | 0 | 0 | 10 | 24 | 62 | 73 | 18 | 1 | 0 | 0 | 188 |
| Total Precipitation (") | 10.23 | 7.76 | 6.46 | 4.42 | 2.41 | 1.73 | 0.90 | 1.21 | 2.53 | 5.24 | 9.61 | 10.94 | 63.44 |
| Days ≥ 0.1" Precip | 15 | 13 | 12 | 10 | 6 | 5 | 3 | 3 | 5 | 9 | 15 | 16 | 112 |
| Total Snowfall (") | 3.5 | 0.3 | 0.1 | 0.0 | 0.0 | 0.0 | 0.0 | 0.0 | 0.0 | 0.0 | 0.8 | 2.1 | 6.8 |
| Days ≥ 1" Snow Depth | 3 | 1 | 0 | 0 | 0 | 0 | 0 | 0 | 0 | 0 | 0 | 1 | 5 |

## SKAMANIA FISH HATCHE *Skamania County* ELEVATION 440 ft LAT/LONG 45° 38 ' N / 122° 13 ' W

| | JAN | FEB | MAR | APR | MAY | JUN | JUL | AUG | SEP | OCT | NOV | DEC | YEAR |
|---|---|---|---|---|---|---|---|---|---|---|---|---|---|
| Maximum Temp °F | 44.8 | 49.4 | 55.0 | 59.4 | 66.7 | 72.3 | 78.5 | 79.5 | 74.2 | 64.0 | 51.1 | 44.3 | 61.6 |
| Minimum Temp °F | 30.0 | 31.9 | 33.3 | 35.5 | 39.6 | 44.3 | 46.3 | 45.7 | 42.0 | 37.2 | 34.4 | 30.7 | 37.6 |
| Mean Temp °F | 37.4 | 40.7 | 44.2 | 47.5 | 53.1 | 58.3 | 62.4 | 62.6 | 58.1 | 50.6 | 42.8 | 37.5 | 49.6 |
| Days Max Temp ≥ 90 °F | 0 | 0 | 0 | 0 | 1 | 1 | 4 | 5 | 2 | 0 | 0 | 0 | 13 |
| Days Max Temp ≤ 32 °F | 2 | 0 | 0 | 0 | 0 | 0 | 0 | 0 | 0 | 0 | 0 | 1 | 3 |
| Days Min Temp ≤ 32 °F | 19 | 15 | 15 | 9 | 4 | 0 | 0 | 0 | 2 | 7 | 11 | 18 | 100 |
| Days Min Temp ≤ 0 °F | 0 | 0 | 0 | 0 | 0 | 0 | 0 | 0 | 0 | 0 | 0 | 0 | 0 |
| Heating Degree Days | 848 | 681 | 637 | 520 | 366 | 211 | 113 | 107 | 212 | 441 | 660 | 845 | 5641 |
| Cooling Degree Days | 0 | 0 | 0 | 0 | 7 | 17 | 36 | 40 | 11 | 2 | 0 | 0 | 113 |
| Total Precipitation (") | 12.49 | 9.33 | 8.16 | 6.75 | 4.74 | 3.97 | 1.42 | 2.02 | 3.86 | 6.34 | 11.82 | 13.08 | 83.98 |
| Days ≥ 0.1" Precip | 16 | 14 | 15 | 13 | 10 | 7 | 3 | 4 | 6 | 11 | 17 | 17 | 133 |
| Total Snowfall (") | *2.6* | 1.9 | 0.6 | 0.0 | 0.0 | 0.0 | 0.0 | 0.0 | 0.0 | 0.0 | 1.2 | 2.7 | 9.0 |
| Days ≥ 1" Snow Depth | *4* | 3 | 1 | 0 | 0 | 0 | 0 | 0 | 0 | 0 | 1 | 4 | 13 |

## SMYRNA *Grant County* ELEVATION 561 ft LAT/LONG 46° 50 ' N / 119° 40 ' W

| | JAN | FEB | MAR | APR | MAY | JUN | JUL | AUG | SEP | OCT | NOV | DEC | YEAR |
|---|---|---|---|---|---|---|---|---|---|---|---|---|---|
| Maximum Temp °F | 38.7 | 46.3 | 57.7 | *66.7* | 75.6 | *82.9* | *90.5* | *90.4* | 80.1 | 67.2 | 49.0 | 38.5 | 65.3 |
| Minimum Temp °F | 22.7 | 26.6 | 32.1 | *38.7* | 46.1 | *53.3* | *57.9* | *56.7* | 46.5 | 36.2 | 29.5 | 22.9 | 39.1 |
| Mean Temp °F | 30.7 | 36.5 | 44.9 | *52.8* | 60.9 | *68.1* | *74.2* | *73.6* | 63.3 | 51.7 | 39.3 | 30.7 | 52.2 |
| Days Max Temp ≥ 90 °F | 0 | 0 | 0 | *0* | 3 | *7* | *17* | *17* | 4 | 0 | 0 | 0 | 48 |
| Days Max Temp ≤ 32 °F | 9 | 2 | 0 | *0* | 0 | 0 | *0* | *0* | 0 | 0 | 1 | 8 | 20 |
| Days Min Temp ≤ 32 °F | 27 | 22 | 17 | *6* | 0 | 0 | *0* | *0* | 1 | 10 | 18 | 27 | 128 |
| Days Min Temp ≤ 0 °F | 2 | 1 | 0 | *0* | 0 | 0 | *0* | *0* | 0 | 0 | 0 | 1 | 4 |
| Heating Degree Days | 1056 | 799 | 616 | *363* | 162 | *36* | *6* | *6* | 104 | 407 | 765 | 1057 | 5377 |
| Cooling Degree Days | 0 | 0 | 0 | 4 | 41 | 137 | 285 | 268 | 65 | 1 | 0 | 0 | 801 |
| Total Precipitation (") | 0.90 | 0.73 | 0.71 | *0.40* | 0.50 | 0.51 | *0.37* | *0.38* | 0.38 | 0.44 | 1.16 | 1.25 | 7.73 |
| Days ≥ 0.1" Precip | *3* | 3 | 3 | *1* | *1* | 2 | *1* | *1* | *1* | 2 | 4 | *4* | 26 |
| Total Snowfall (") | na | na | *0.0* | *0.0* | *0.0* | *0.0* | *0.0* | *0.0* | *0.0* | *0.0* | *0.3* | na | na |
| Days ≥ 1" Snow Depth | na | na | *0* | *0* | 0 | 0 | *0* | *0* | 0 | 0 | *1* | na | na |

## SNOQUALMIE FALLS *King County* ELEVATION 430 ft LAT/LONG 47° 33 ' N / 121° 51 ' W

| | JAN | FEB | MAR | APR | MAY | JUN | JUL | AUG | SEP | OCT | NOV | DEC | YEAR |
|---|---|---|---|---|---|---|---|---|---|---|---|---|---|
| Maximum Temp °F | 45.4 | 49.7 | 53.6 | 57.9 | 64.1 | 68.9 | 74.3 | 75.2 | 69.6 | 60.2 | 50.4 | 44.9 | 59.5 |
| Minimum Temp °F | 32.9 | 34.4 | 35.5 | 38.5 | 43.6 | 48.6 | 51.4 | 51.4 | 46.4 | 40.7 | 36.8 | 33.1 | 41.1 |
| Mean Temp °F | 39.2 | 42.1 | 44.5 | 48.2 | 53.9 | 58.8 | 62.9 | 63.3 | 58.1 | 50.5 | 43.6 | 39.1 | 50.4 |
| Days Max Temp ≥ 90 °F | 0 | 0 | 0 | 0 | 0 | 0 | 1 | 1 | 0 | 0 | 0 | 0 | 2 |
| Days Max Temp ≤ 32 °F | 1 | 0 | 0 | 0 | 0 | 0 | 0 | 0 | 0 | 0 | 0 | 1 | 2 |
| Days Min Temp ≤ 32 °F | 14 | 11 | 10 | 4 | 0 | 0 | 0 | 0 | 0 | 3 | 8 | 15 | 65 |
| Days Min Temp ≤ 0 °F | 0 | 0 | 0 | 0 | 0 | 0 | 0 | 0 | 0 | 0 | 0 | 0 | 0 |
| Heating Degree Days | 794 | 640 | 627 | 497 | 343 | 194 | 95 | 85 | 210 | 444 | 636 | 798 | 5363 |
| Cooling Degree Days | 0 | 0 | 0 | 1 | 5 | 14 | 38 | 45 | 9 | 1 | 0 | 0 | 113 |
| Total Precipitation (") | 8.69 | 5.96 | 5.91 | 4.84 | 3.44 | 2.74 | 1.51 | 1.60 | 2.82 | 4.84 | 8.61 | 9.18 | 60.14 |
| Days ≥ 0.1" Precip | 14 | 12 | 13 | 12 | 9 | 7 | 4 | 4 | 7 | 10 | 15 | 15 | 122 |
| Total Snowfall (") | 3.1 | 1.9 | 0.9 | 0.3 | 0.0 | 0.0 | 0.0 | 0.0 | 0.0 | 0.0 | 1.2 | 2.9 | 10.3 |
| Days ≥ 1" Snow Depth | *1* | 1 | 0 | 0 | 0 | 0 | 0 | 0 | 0 | 0 | 0 | *2* | 4 |

## SPOKANE INTL ARPT *Spokane County* ELEVATION 2365 ft LAT/LONG 47° 37' N / 117° 31' W

| | JAN | FEB | MAR | APR | MAY | JUN | JUL | AUG | SEP | OCT | NOV | DEC | YEAR |
|---|---|---|---|---|---|---|---|---|---|---|---|---|---|
| Maximum Temp °F | 32.5 | 39.3 | 48.3 | 57.1 | 66.6 | 74.5 | 82.8 | 82.6 | 72.4 | 58.7 | 41.0 | 33.0 | 57.4 |
| Minimum Temp °F | 21.3 | 25.6 | 30.0 | 35.3 | 42.5 | 49.4 | 54.7 | 54.6 | 45.9 | 35.9 | 28.4 | 21.8 | 37.1 |
| Mean Temp °F | 26.9 | 32.5 | 39.2 | 46.2 | 54.6 | 62.0 | 68.8 | 68.6 | 59.2 | 47.3 | 34.8 | 27.4 | 47.3 |
| Days Max Temp ≥ 90 °F | 0 | 0 | 0 | 0 | 0 | 2 | 8 | 7 | 1 | 0 | 0 | 0 | 18 |
| Days Max Temp ≤ 32 °F | 14 | 5 | 1 | 0 | 0 | 0 | 0 | 0 | 0 | 0 | 4 | 14 | 38 |
| Days Min Temp ≤ 32 °F | 26 | 23 | 21 | 10 | 2 | 0 | 0 | 0 | 1 | 9 | 20 | 27 | 139 |
| Days Min Temp ≤ 0 °F | 2 | 1 | 0 | 0 | 0 | 0 | 0 | 0 | 0 | 0 | 0 | 2 | 5 |
| Heating Degree Days | 1175 | 911 | 793 | 557 | 327 | 139 | 39 | 40 | 196 | 541 | 901 | 1158 | 6777 |
| Cooling Degree Days | 0 | 0 | 0 | 1 | 16 | 54 | 146 | 150 | 29 | 1 | 0 | 0 | 397 |
| Total Precipitation (") | 1.97 | 1.34 | 1.38 | 1.22 | 1.40 | 1.18 | 0.75 | 0.69 | 0.69 | 1.00 | 2.10 | 2.22 | 15.94 |
| Days ≥ 0.1" Precip | 7 | 5 | 5 | 4 | 4 | 3 | 2 | 2 | 2 | 3 | 6 | 7 | 50 |
| Total Snowfall (") | 13.8 | 6.9 | 3.1 | 0.7 | 0.2 | 0.0 | 0.0 | 0.0 | 0.0 | 0.4 | 6.5 | 14.0 | 45.6 |
| Days ≥ 1" Snow Depth | 18 | 9 | 3 | 0 | 0 | 0 | 0 | 0 | 0 | 0 | 5 | 16 | 51 |

## STAMPEDE PASS *Kittitas County* ELEVATION 3958 ft LAT/LONG 47° 17' N / 121° 20' W

| | JAN | FEB | MAR | APR | MAY | JUN | JUL | AUG | SEP | OCT | NOV | DEC | YEAR |
|---|---|---|---|---|---|---|---|---|---|---|---|---|---|
| Maximum Temp °F | 29.2 | 32.4 | 36.9 | 41.5 | 49.9 | 57.3 | 64.2 | 65.1 | 58.2 | 48.3 | 34.2 | 29.5 | 45.6 |
| Minimum Temp °F | 21.2 | 24.0 | 27.1 | 29.9 | 35.8 | 41.7 | 47.0 | 48.1 | 43.4 | 36.5 | 27.1 | 21.8 | 33.6 |
| Mean Temp °F | 25.2 | 28.2 | 32.0 | 35.7 | 42.9 | 49.6 | 55.6 | 56.6 | 50.8 | 42.5 | 30.7 | 25.7 | 39.6 |
| Days Max Temp ≥ 90 °F | 0 | 0 | 0 | 0 | 0 | 0 | 0 | 0 | 0 | 0 | 0 | 0 | 0 |
| Days Max Temp ≤ 32 °F | 20 | 15 | 10 | 4 | 0 | 0 | 0 | 0 | 0 | 1 | 12 | 21 | 83 |
| Days Min Temp ≤ 32 °F | 29 | 26 | 26 | 22 | 11 | 1 | 0 | 0 | 1 | 9 | 25 | 29 | 179 |
| Days Min Temp ≤ 0 °F | 1 | 0 | 0 | 0 | 0 | 0 | 0 | 0 | 0 | 0 | 0 | 1 | 2 |
| Heating Degree Days | 1226 | 1033 | 1015 | 871 | 680 | 462 | 305 | 280 | 424 | 692 | 1023 | 1213 | 9224 |
| Cooling Degree Days | 0 | 0 | 0 | *0* | *2* | *6* | *16* | *23* | *8* | *1* | *0* | *0* | 56 |
| Total Precipitation (") | 13.49 | 8.82 | 7.20 | 5.96 | 4.18 | 3.81 | 1.83 | 2.42 | 4.39 | 6.67 | 11.66 | 13.53 | 83.96 |
| Days ≥ 0.1" Precip | 17 | 14 | 15 | 13 | 10 | 9 | 5 | 5 | 8 | 11 | 17 | 18 | 142 |
| Total Snowfall (") | 85.8 | 57.2 | 54.7 | 43.2 | 16.5 | 1.2 | 0.5 | *0.0* | *2.3* | 18.3 | 67.9 | 79.4 | 427.0 |
| Days ≥ 1" Snow Depth | 30 | 28 | 31 | 30 | 25 | 6 | 0 | 0 | 1 | 5 | 23 | 31 | 210 |

## STARTUP 1 E *Snohomish County* ELEVATION 171 ft LAT/LONG 47° 52' N / 121° 43' W

| | JAN | FEB | MAR | APR | MAY | JUN | JUL | AUG | SEP | OCT | NOV | DEC | YEAR |
|---|---|---|---|---|---|---|---|---|---|---|---|---|---|
| Maximum Temp °F | 45.8 | 50.2 | 54.3 | 58.9 | 65.8 | 70.6 | 75.7 | 76.7 | 71.2 | 61.9 | 50.9 | 45.2 | 60.6 |
| Minimum Temp °F | 33.0 | 34.5 | 36.0 | 38.8 | 43.8 | 48.7 | 51.1 | 51.0 | 46.8 | 40.8 | 37.4 | 33.6 | 41.3 |
| Mean Temp °F | 39.4 | 42.4 | 45.1 | 48.9 | 54.8 | 59.7 | 63.4 | 63.9 | 59.1 | 51.4 | 44.2 | 39.4 | 51.0 |
| Days Max Temp ≥ 90 °F | 0 | 0 | 0 | 0 | 0 | 1 | 2 | 2 | 1 | 0 | 0 | 0 | 6 |
| Days Max Temp ≤ 32 °F | 1 | 0 | 0 | 0 | 0 | 0 | 0 | 0 | 0 | 0 | 0 | 1 | 2 |
| Days Min Temp ≤ 32 °F | 14 | 11 | 9 | 5 | 1 | 0 | 0 | 0 | 0 | 5 | 7 | 13 | 65 |
| Days Min Temp ≤ 0 °F | 0 | 0 | 0 | 0 | 0 | 0 | 0 | 0 | 0 | 0 | 0 | 0 | 0 |
| Heating Degree Days | 787 | 634 | 610 | 477 | 316 | 172 | 84 | 74 | 182 | 417 | 619 | 787 | 5159 |
| Cooling Degree Days | 0 | 0 | 0 | 1 | 9 | 19 | 38 | 51 | 10 | 1 | 0 | 0 | 129 |
| Total Precipitation (") | 8.44 | 5.69 | 5.89 | 5.54 | 4.43 | 3.57 | 2.21 | 2.20 | 3.58 | 5.11 | 8.80 | 8.68 | 64.14 |
| Days ≥ 0.1" Precip | 15 | 13 | 14 | 13 | 10 | 8 | 5 | 5 | 7 | 10 | 16 | 15 | 131 |
| Total Snowfall (") | 3.9 | 2.2 | 0.6 | 0.2 | 0.0 | 0.0 | 0.0 | 0.0 | 0.0 | 0.0 | 0.8 | *2.2* | 9.9 |
| Days ≥ 1" Snow Depth | 3 | 1 | 0 | 0 | 0 | 0 | 0 | 0 | 0 | 0 | 0 | *2* | 6 |

## STEHEKIN 4 NW *Chelan County* ELEVATION 1152 ft LAT/LONG 48° 20' N / 120° 42' W

| | JAN | FEB | MAR | APR | MAY | JUN | JUL | AUG | SEP | OCT | NOV | DEC | YEAR |
|---|---|---|---|---|---|---|---|---|---|---|---|---|---|
| Maximum Temp °F | 33.2 | 38.9 | 48.2 | 58.9 | 69.2 | 76.4 | 83.3 | 82.9 | 72.5 | 57.5 | 41.5 | 33.3 | 58.0 |
| Minimum Temp °F | 23.6 | 25.9 | 29.6 | 35.9 | 42.8 | 49.2 | 53.6 | 53.7 | 45.7 | 37.0 | 30.6 | 24.9 | 37.7 |
| Mean Temp °F | 28.4 | 32.4 | 38.9 | 47.4 | 56.0 | 62.8 | 68.5 | 68.4 | 59.2 | 47.3 | 36.1 | 29.1 | 47.9 |
| Days Max Temp ≥ 90 °F | 0 | 0 | 0 | 0 | 0 | 3 | 8 | 7 | 1 | 0 | 0 | 0 | 19 |
| Days Max Temp ≤ 32 °F | 12 | 3 | 0 | 0 | 0 | 0 | 0 | 0 | 0 | 0 | 2 | 11 | 28 |
| Days Min Temp ≤ 32 °F | 30 | 26 | 23 | 8 | 1 | 0 | 0 | 0 | 1 | 8 | 19 | 29 | 145 |
| Days Min Temp ≤ 0 °F | 0 | 0 | 0 | 0 | 0 | 0 | 0 | 0 | 0 | 0 | 0 | 1 | 1 |
| Heating Degree Days | 1127 | 913 | 801 | 520 | 281 | 112 | 34 | 35 | 190 | 543 | 861 | 1106 | 6523 |
| Cooling Degree Days | 0 | 0 | 0 | 0 | 17 | 64 | 159 | 159 | 30 | 0 | 0 | 0 | 429 |
| Total Precipitation (") | 6.51 | 3.94 | 2.73 | 1.57 | 0.90 | 0.98 | 0.59 | 0.83 | 1.24 | 2.87 | 6.10 | 7.19 | 35.45 |
| Days ≥ 0.1" Precip | 12 | 9 | 7 | 4 | 3 | 3 | 2 | 2 | 4 | 7 | 13 | 13 | 79 |
| Total Snowfall (") | 48.2 | 21.5 | 6.9 | 0.4 | 0.0 | 0.0 | 0.0 | 0.0 | 0.0 | 0.3 | 12.4 | 42.7 | 132.4 |
| Days ≥ 1" Snow Depth | 31 | 27 | 23 | 5 | 0 | 0 | 0 | 0 | 0 | 0 | 7 | 28 | 121 |

## STEVENS PASS *King County* ELEVATION 4062 ft LAT/LONG 47° 45' N / 121° 5' W

| | JAN | FEB | MAR | APR | MAY | JUN | JUL | AUG | SEP | OCT | NOV | DEC | YEAR |
|---|---|---|---|---|---|---|---|---|---|---|---|---|---|
| Maximum Temp °F | 28.7 | 31.9 | 36.7 | 41.6 | 49.8 | *56.8* | na | na | *58.3* | *48.1* | na | 29.1 | na |
| Minimum Temp °F | 19.1 | 21.8 | 24.7 | 28.5 | 34.0 | *40.0* | na | *46.7* | *41.0* | *34.4* | na | 19.7 | na |
| Mean Temp °F | 23.9 | 26.9 | 30.7 | 35.1 | 42.0 | *48.4* | na | na | *49.7* | *41.3* | na | 24.4 | na |
| Days Max Temp ≥ 90 °F | 0 | 0 | 0 | 0 | 0 | 0 | 0 | 0 | 0 | *0* | 0 | 0 | 0 |
| Days Max Temp ≤ 32 °F | 20 | 14 | 8 | 2 | 0 | 0 | 0 | 0 | 0 | *1* | *10* | 20 | 75 |
| Days Min Temp ≤ 32 °F | 30 | 28 | 29 | 23 | *13* | 2 | 0 | *0* | *2* | *13* | *23* | 29 | 192 |
| Days Min Temp ≤ 0 °F | 1 | 0 | 0 | 0 | 0 | 0 | 0 | 0 | 0 | 0 | 0 | 1 | 2 |
| Heating Degree Days | 1267 | 1070 | 1056 | 890 | 708 | *493* | na | na | *454* | *729* | na | 1252 | na |
| Cooling Degree Days | 0 | 0 | 0 | 0 | 0 | 1 | na | na | na | 0 | na | 0 | na |
| Total Precipitation (") | 13.36 | 9.10 | 6.98 | 4.99 | 3.57 | 3.19 | 1.53 | 1.94 | 3.51 | 6.50 | 12.55 | 13.65 | 80.87 |
| Days ≥ 0.1" Precip | 16 | 14 | 15 | 11 | *9* | 8 | 4 | *3* | *5* | *10* | *16* | 17 | 128 |
| Total Snowfall (") | 106.7 | 76.9 | 62.4 | 32.7 | 9.5 | 0.7 | 0.1 | *0.0* | *0.5* | *12.8* | *67.8* | 95.1 | 465.2 |
| Days ≥ 1" Snow Depth | 31 | 28 | 31 | 29 | *28* | 11 | 1 | 0 | 0 | *6* | 23 | 30 | 218 |

## SUNNYSIDE *Yakima County* ELEVATION 747 ft LAT/LONG 46° 19' N / 120° 0' W

| | JAN | FEB | MAR | APR | MAY | JUN | JUL | AUG | SEP | OCT | NOV | DEC | YEAR |
|---|---|---|---|---|---|---|---|---|---|---|---|---|---|
| Maximum Temp °F | 39.7 | 48.1 | 58.5 | 66.2 | 75.2 | 82.4 | 89.1 | 88.7 | 79.6 | 67.3 | 50.4 | 39.9 | 65.4 |
| Minimum Temp °F | 24.4 | 28.4 | 33.4 | 38.8 | 45.9 | 52.7 | 55.8 | 54.7 | 46.9 | 37.0 | 31.0 | 25.1 | 39.5 |
| Mean Temp °F | 32.1 | 38.2 | 46.0 | 52.6 | 60.5 | 67.6 | 72.5 | 71.7 | 63.3 | 52.2 | 40.7 | 32.5 | 52.5 |
| Days Max Temp ≥ 90 °F | 0 | 0 | 0 | 0 | 2 | 7 | 16 | 15 | 4 | 0 | 0 | 0 | 44 |
| Days Max Temp ≤ 32 °F | 8 | 2 | 0 | 0 | 0 | 0 | 0 | 0 | 0 | 0 | 1 | 8 | 19 |
| Days Min Temp ≤ 32 °F | 26 | 21 | 15 | 6 | 0 | 0 | 0 | 0 | 0 | 8 | 18 | 26 | 120 |
| Days Min Temp ≤ 0 °F | 1 | 0 | 0 | 0 | 0 | 0 | 0 | 0 | 0 | 0 | 0 | 1 | 2 |
| Heating Degree Days | 1015 | 749 | 582 | 370 | 173 | 50 | 11 | 10 | 103 | 392 | 722 | 1000 | 5177 |
| Cooling Degree Days | 0 | 0 | 0 | 7 | 52 | 133 | 250 | 235 | 65 | 3 | 0 | 0 | 745 |
| Total Precipitation (") | 0.88 | 0.52 | 0.50 | 0.48 | 0.45 | 0.45 | 0.18 | 0.32 | 0.43 | 0.47 | 0.96 | 1.12 | 6.76 |
| Days ≥ 0.1" Precip | 3 | 2 | 2 | 2 | 1 | 1 | 1 | 1 | 1 | 2 | 3 | 4 | 23 |
| Total Snowfall (") | 3.5 | 1.3 | 0.2 | 0.0 | 0.0 | 0.0 | 0.0 | 0.0 | 0.0 | 0.0 | 1.2 | 4.9 | 11.1 |
| Days ≥ 1" Snow Depth | *5* | 1 | 0 | 0 | 0 | 0 | 0 | 0 | 0 | 0 | 1 | 4 | 11 |

## TOLEDO *Lewis County* ELEVATION 325 ft LAT/LONG 46° 28' N / 122° 50' W

| | JAN | FEB | MAR | APR | MAY | JUN | JUL | AUG | SEP | OCT | NOV | DEC | YEAR |
|---|---|---|---|---|---|---|---|---|---|---|---|---|---|
| Maximum Temp °F | 45.0 | 50.7 | 56.0 | 60.6 | 67.5 | 72.7 | 78.5 | 79.4 | 74.3 | 63.4 | 50.9 | 44.7 | 62.0 |
| Minimum Temp °F | 32.6 | 33.8 | 35.8 | 38.2 | 42.9 | 47.5 | 50.0 | 49.7 | 45.8 | 40.7 | 37.2 | 33.8 | 40.7 |
| Mean Temp °F | 38.9 | 42.3 | 45.9 | 49.4 | 55.2 | 60.2 | 64.3 | 64.6 | 60.1 | 52.1 | 44.1 | 39.3 | 51.4 |
| Days Max Temp ≥ 90 °F | 0 | 0 | 0 | 0 | 0 | 1 | 3 | 4 | 1 | 0 | 0 | 0 | 9 |
| Days Max Temp ≤ 32 °F | 1 | 0 | 0 | 0 | 0 | 0 | 0 | 0 | 0 | 0 | 0 | 1 | 2 |
| Days Min Temp ≤ 32 °F | 15 | 12 | 11 | 6 | 1 | 0 | 0 | 0 | 1 | 4 | 8 | 13 | 71 |
| Days Min Temp ≤ 0 °F | 0 | 0 | 0 | 0 | 0 | 0 | 0 | 0 | 0 | 0 | 0 | 0 | 0 |
| Heating Degree Days | 803 | 635 | 584 | 461 | 302 | 159 | 70 | 65 | 154 | 394 | 621 | 790 | 5038 |
| Cooling Degree Days | 0 | 0 | 0 | *0* | 8 | 21 | *50* | *70* | *17* | 1 | 0 | 0 | 167 |
| Total Precipitation (") | *6.73* | *4.85* | *4.81* | *3.28* | na | na | na | na | na | *3.42* | 5.66 | *6.92* | na |
| Days ≥ 0.1" Precip | 14 | 11 | 12 | 9 | 6 | 6 | 2 | 3 | 5 | 8 | 13 | 14 | 103 |
| Total Snowfall (") | na | na | na | na | na | na | na | na | na | na | na | na | na |
| Days ≥ 1" Snow Depth | na | na | *0* | *0* | na | na | na | na | na | *0* | na | na | na |

## UPPER BAKER DAM *Whatcom County* ELEVATION 689 ft LAT/LONG 48° 39' N / 121° 41' W

| | JAN | FEB | MAR | APR | MAY | JUN | JUL | AUG | SEP | OCT | NOV | DEC | YEAR |
|---|---|---|---|---|---|---|---|---|---|---|---|---|---|
| Maximum Temp °F | 38.9 | 43.7 | 49.5 | 55.6 | 63.2 | 67.7 | 73.0 | 74.1 | 68.5 | 58.1 | 45.5 | 39.2 | 56.4 |
| Minimum Temp °F | 28.1 | 29.2 | 31.9 | 35.8 | 41.8 | 47.3 | 50.5 | 51.2 | 47.0 | 40.7 | 34.6 | 30.0 | 39.0 |
| Mean Temp °F | 33.5 | 36.5 | 40.7 | 45.7 | 52.5 | 57.5 | 61.8 | 62.7 | 57.8 | 49.4 | 40.1 | 34.6 | 47.7 |
| Days Max Temp ≥ 90 °F | 0 | 0 | 0 | 0 | 0 | 0 | 1 | 1 | 0 | 0 | 0 | 0 | 2 |
| Days Max Temp ≤ 32 °F | 3 | 1 | 0 | 0 | 0 | 0 | 0 | 0 | 0 | 0 | 1 | 2 | 7 |
| Days Min Temp ≤ 32 °F | 22 | 20 | 18 | 8 | 0 | 0 | 0 | 0 | 0 | 2 | 10 | 21 | 101 |
| Days Min Temp ≤ 0 °F | 0 | 0 | 0 | 0 | 0 | 0 | 0 | 0 | 0 | 0 | 0 | 0 | 0 |
| Heating Degree Days | 969 | 798 | 746 | 572 | 382 | 228 | 121 | 99 | 218 | 476 | 742 | 935 | 6286 |
| Cooling Degree Days | 0 | 0 | 0 | 0 | 5 | 11 | 32 | 39 | 9 | 0 | 0 | 0 | 96 |
| Total Precipitation (") | 14.59 | 10.58 | 9.44 | 6.41 | 4.60 | 3.57 | 2.55 | 2.12 | 4.37 | 9.03 | 15.20 | 15.61 | 98.07 |
| Days ≥ 0.1" Precip | 17 | 14 | 15 | 13 | 9 | 8 | 5 | 4 | 7 | 12 | 18 | 18 | 140 |
| Total Snowfall (") | 20.6 | 11.8 | 4.5 | 0.8 | 0.0 | 0.0 | 0.0 | 0.0 | 0.0 | 0.0 | 3.5 | 15.2 | 56.4 |
| Days ≥ 1" Snow Depth | 20 | 14 | 7 | 1 | 0 | 0 | 0 | 0 | 0 | 0 | 3 | 12 | 57 |

## VANCOUVER 4 NNE *Clark County* ELEVATION 102 ft LAT/LONG 45° 38 ' N / 122° 41 ' W

| | JAN | FEB | MAR | APR | MAY | JUN | JUL | AUG | SEP | OCT | NOV | DEC | YEAR |
|---|---|---|---|---|---|---|---|---|---|---|---|---|---|
| Maximum Temp °F | 45.3 | 50.0 | 55.3 | 59.4 | 66.2 | 71.9 | 77.2 | 78.1 | 73.5 | 63.4 | 51.9 | 45.2 | 61.5 |
| Minimum Temp °F | 31.3 | 33.2 | 36.2 | 39.0 | 44.0 | 49.2 | 52.0 | 51.5 | 46.8 | 40.5 | 36.5 | 32.5 | 41.1 |
| Mean Temp °F | 38.3 | 41.6 | 45.7 | 49.2 | 55.1 | 60.6 | 64.6 | 64.8 | 60.1 | 51.9 | 44.2 | 38.9 | 51.3 |
| Days Max Temp ≥ 90 °F | 0 | 0 | 0 | 0 | 0 | 1 | 2 | 3 | 1 | 0 | 0 | 0 | 7 |
| Days Max Temp ≤ 32 °F | 1 | 0 | 0 | 0 | 0 | 0 | 0 | 0 | 0 | 0 | 0 | 1 | 2 |
| Days Min Temp ≤ 32 °F | 17 | 13 | 10 | 4 | 0 | 0 | 0 | 0 | 0 | 4 | 9 | 14 | 71 |
| Days Min Temp ≤ 0 °F | 0 | 0 | 0 | 0 | 0 | 0 | 0 | 0 | 0 | 0 | 0 | 0 | 0 |
| Heating Degree Days | 821 | 653 | 591 | 467 | 307 | 153 | 69 | 69 | 160 | 398 | 616 | 803 | 5107 |
| Cooling Degree Days | 0 | 0 | 0 | 0 | 11 | 29 | 64 | 74 | 24 | 1 | 0 | 0 | 203 |
| Total Precipitation (") | 6.01 | 4.33 | 3.90 | 2.92 | 2.29 | 1.66 | 0.78 | 1.14 | 1.91 | 3.16 | 5.70 | 6.64 | 40.44 |
| Days ≥ 0.1" Precip | 13 | 11 | 11 | 10 | 6 | 5 | 2 | 3 | 5 | 8 | 13 | *13* | 100 |
| Total Snowfall (") | na | na | na | *0.0* | *0.0* | *0.0* | *0.0* | *0.0* | *0.0* | *0.0* | *0.3* | na | na |
| Days ≥ 1" Snow Depth | na | na | na | *0* | *0* | *0* | *0* | *0* | *0* | *0* | *0* | na | na |

## WALLA WALLA CITY-COU *Walla Walla County* ELEVATION 1207 ft LAT/LONG 46° 6 ' N / 118° 17 ' W

| | JAN | FEB | MAR | APR | MAY | JUN | JUL | AUG | SEP | OCT | NOV | DEC | YEAR |
|---|---|---|---|---|---|---|---|---|---|---|---|---|---|
| Maximum Temp °F | 39.6 | 46.2 | 55.4 | 62.8 | 71.5 | 80.3 | 89.3 | 88.2 | 78.1 | 65.0 | 49.0 | 40.4 | 63.8 |
| Minimum Temp °F | 28.6 | 32.8 | 37.2 | 41.8 | 48.1 | 54.8 | 60.7 | 61.1 | 52.8 | 43.7 | 35.8 | 29.5 | 43.9 |
| Mean Temp °F | 34.1 | 39.5 | 46.3 | 52.3 | 59.8 | 67.6 | 75.1 | 74.7 | 65.5 | 54.4 | 42.5 | 35.0 | 53.9 |
| Days Max Temp ≥ 90 °F | 0 | 0 | 0 | 0 | 1 | 6 | 17 | 14 | 4 | 0 | 0 | 0 | 42 |
| Days Max Temp ≤ 32 °F | 8 | 3 | 0 | 0 | 0 | 0 | 0 | 0 | 0 | 0 | 2 | 8 | 21 |
| Days Min Temp ≤ 32 °F | 18 | 12 | 5 | 1 | 0 | 0 | 0 | 0 | 0 | 2 | 8 | 18 | 64 |
| Days Min Temp ≤ 0 °F | 1 | 0 | 0 | 0 | 0 | 0 | 0 | 0 | 0 | 0 | 0 | 1 | 2 |
| Heating Degree Days | 951 | 713 | 573 | 376 | 190 | 52 | 7 | 7 | 79 | 329 | 670 | 925 | 4872 |
| Cooling Degree Days | 0 | 0 | 0 | 5 | 44 | 131 | 310 | 316 | 104 | 7 | 0 | 0 | 917 |
| Total Precipitation (") | 2.35 | 1.76 | 2.02 | 1.75 | 1.75 | 1.16 | 0.73 | 0.89 | 0.87 | 1.58 | 2.67 | 2.38 | 19.91 |
| Days ≥ 0.1" Precip | 7 | 6 | 6 | 5 | 4 | 3 | 2 | 2 | 3 | 4 | 8 | 7 | 57 |
| Total Snowfall (") | 6.3 | 3.2 | 1.2 | 0.0 | 0.0 | 0.0 | 0.0 | 0.0 | 0.0 | 0.2 | 2.0 | 5.8 | 18.7 |
| Days ≥ 1" Snow Depth | 9 | 4 | 1 | 0 | 0 | 0 | 0 | 0 | 0 | 0 | 2 | 6 | 22 |

## WAPATO *Yakima County* ELEVATION 850 ft LAT/LONG 46° 26 ' N / 120° 25 ' W

| | JAN | FEB | MAR | APR | MAY | JUN | JUL | AUG | SEP | OCT | NOV | DEC | YEAR |
|---|---|---|---|---|---|---|---|---|---|---|---|---|---|
| Maximum Temp °F | 38.9 | 47.7 | 57.9 | 66.0 | 75.5 | 82.6 | 89.4 | 88.8 | 79.8 | 66.9 | 49.9 | 39.0 | 65.2 |
| Minimum Temp °F | 23.3 | 27.8 | 33.2 | 39.2 | 46.8 | 53.8 | 58.8 | 57.5 | 48.3 | 37.5 | 30.4 | 23.5 | 40.0 |
| Mean Temp °F | 31.2 | 37.8 | 45.6 | 52.6 | 61.2 | 68.3 | 74.1 | 73.2 | 64.1 | 52.2 | 40.1 | 31.3 | 52.6 |
| Days Max Temp ≥ 90 °F | 0 | 0 | 0 | 0 | 2 | 8 | 16 | 15 | 4 | 0 | 0 | 0 | 45 |
| Days Max Temp ≤ 32 °F | 9 | 2 | 0 | 0 | 0 | 0 | 0 | 0 | 0 | 0 | 1 | 8 | 20 |
| Days Min Temp ≤ 32 °F | 26 | 21 | 15 | 5 | 1 | 0 | 0 | 0 | 0 | 7 | 18 | 27 | 120 |
| Days Min Temp ≤ 0 °F | 1 | 0 | 0 | 0 | 0 | 0 | 0 | 0 | 0 | 0 | 0 | 1 | 2 |
| Heating Degree Days | 1042 | 762 | 595 | 368 | 163 | 49 | 11 | 10 | 93 | 392 | 738 | 1038 | 5261 |
| Cooling Degree Days | 0 | 0 | 0 | 5 | 53 | 131 | 269 | 248 | 68 | 3 | 0 | 0 | 777 |
| Total Precipitation (") | 1.05 | 0.61 | 0.61 | 0.49 | 0.47 | 0.55 | 0.28 | 0.35 | 0.40 | 0.49 | 1.02 | 1.19 | 7.51 |
| Days ≥ 0.1" Precip | 3 | 2 | 2 | 2 | 2 | 1 | 1 | 1 | 1 | 1 | 3 | 4 | 23 |
| Total Snowfall (") | 5.3 | 2.0 | 0.5 | 0.0 | 0.0 | 0.0 | 0.0 | 0.0 | 0.0 | 0.1 | 1.4 | 6.0 | 15.3 |
| Days ≥ 1" Snow Depth | 12 | 4 | 0 | 0 | 0 | 0 | 0 | 0 | 0 | 0 | 2 | 10 | 28 |

## WATERVILLE *Douglas County* ELEVATION 2621 ft LAT/LONG 47° 39 ' N / 120° 4 ' W

| | JAN | FEB | MAR | APR | MAY | JUN | JUL | AUG | SEP | OCT | NOV | DEC | YEAR |
|---|---|---|---|---|---|---|---|---|---|---|---|---|---|
| Maximum Temp °F | 31.4 | 37.3 | 47.5 | 57.8 | 67.6 | 75.1 | 83.3 | 83.1 | 72.7 | 59.0 | 40.9 | 31.5 | 57.3 |
| Minimum Temp °F | 15.7 | 21.0 | 27.5 | 34.4 | 41.6 | 48.5 | 53.1 | 53.1 | 44.8 | 34.1 | 25.3 | 17.1 | 34.7 |
| Mean Temp °F | 23.6 | 29.2 | 37.5 | 46.1 | 54.6 | 61.8 | 68.2 | 68.1 | 58.8 | 46.5 | 33.1 | 24.3 | 46.0 |
| Days Max Temp ≥ 90 °F | 0 | 0 | 0 | 0 | 0 | 2 | 9 | 8 | 1 | 0 | 0 | 0 | 20 |
| Days Max Temp ≤ 32 °F | 15 | 7 | 1 | 0 | 0 | 0 | 0 | 0 | 0 | 0 | 4 | 15 | 42 |
| Days Min Temp ≤ 32 °F | 29 | 26 | 24 | 12 | 3 | 0 | 0 | 0 | 1 | 13 | 25 | 30 | 163 |
| Days Min Temp ≤ 0 °F | 4 | 1 | 0 | 0 | 0 | 0 | 0 | 0 | 0 | 0 | 1 | 2 | 8 |
| Heating Degree Days | 1278 | 1004 | 846 | 561 | 327 | 143 | 43 | 44 | 211 | 565 | 950 | 1255 | 7227 |
| Cooling Degree Days | *0* | 0 | 0 | 1 | 21 | 54 | 155 | 158 | 42 | 1 | 0 | 0 | 432 |
| Total Precipitation (") | 1.51 | 0.94 | 0.78 | 0.80 | 0.81 | 0.86 | 0.43 | 0.73 | 0.57 | 0.64 | 1.58 | 1.85 | 11.50 |
| Days ≥ 0.1" Precip | 5 | 3 | 3 | 3 | 2 | 2 | 2 | 2 | 2 | 2 | 5 | 6 | 37 |
| Total Snowfall (") | 12.4 | 5.7 | *3.2* | 0.2 | 0.0 | 0.0 | 0.0 | 0.0 | 0.0 | 0.4 | 7.2 | 16.0 | 45.1 |
| Days ≥ 1" Snow Depth | na | *8* | *2* | 0 | 0 | 0 | 0 | 0 | 0 | 0 | na | na | na |

## WELLPINIT *Stevens County* ELEVATION 2451 ft LAT/LONG 47° 53' N / 117° 59' W

| | JAN | FEB | MAR | APR | MAY | JUN | JUL | AUG | SEP | OCT | NOV | DEC | YEAR |
|---|---|---|---|---|---|---|---|---|---|---|---|---|---|
| Maximum Temp °F | 31.4 | 37.6 | 48.0 | 57.5 | 67.0 | 74.9 | 83.8 | 83.8 | 72.5 | 58.9 | ***40.3*** | 31.8 | 57.3 |
| Minimum Temp °F | 20.0 | 24.0 | 29.5 | 34.9 | 41.9 | 48.0 | 53.4 | 52.8 | 44.8 | 35.8 | ***28.8*** | 21.8 | 36.3 |
| Mean Temp °F | 25.7 | 30.8 | 38.7 | 46.2 | 54.5 | 61.6 | 68.7 | 68.3 | 58.7 | 47.4 | ***34.6*** | 26.9 | 46.8 |
| Days Max Temp ≥ 90 °F | 0 | 0 | 0 | 0 | 0 | 2 | 10 | 10 | 1 | 0 | 0 | 0 | 23 |
| Days Max Temp ≤ 32 °F | 15 | 7 | 1 | 0 | 0 | 0 | 0 | 0 | 0 | 0 | 5 | ***17*** | 45 |
| Days Min Temp ≤ 32 °F | ***30*** | 24 | ***20*** | 12 | 3 | 0 | 0 | 0 | 1 | ***10*** | ***21*** | ***28*** | 149 |
| Days Min Temp ≤ 0 °F | 2 | 1 | 0 | 0 | 0 | 0 | 0 | 0 | 0 | 0 | 0 | 1 | 4 |
| Heating Degree Days | 1213 | 961 | 809 | 560 | 331 | 149 | 42 | 44 | 213 | 542 | ***908*** | 1178 | 6950 |
| Cooling Degree Days | 0 | 0 | 0 | 3 | 18 | 52 | 164 | ***158*** | ***38*** | ***0*** | ***0*** | ***0*** | 433 |
| Total Precipitation (") | 1.96 | 1.62 | 1.60 | 1.42 | 1.61 | 1.29 | 0.92 | 0.76 | 0.84 | 0.98 | 2.73 | 2.72 | 18.45 |
| Days ≥ 0.1" Precip | 5 | ***3*** | 4 | 4 | ***4*** | 3 | 2 | 2 | 2 | ***3*** | ***6*** | na | na |
| Total Snowfall (") | na | ***3.9*** | 2.5 | 0.3 | 0.0 | 0.0 | 0.0 | 0.0 | 0.0 | ***0.4*** | na | na | na |
| Days ≥ 1" Snow Depth | na | na | ***5*** | 0 | 0 | 0 | 0 | 0 | 0 | 0 | na | na | na |

## WENATCHEE *Chelan County* ELEVATION 630 ft LAT/LONG 47° 25' N / 120° 19' W

| | JAN | FEB | MAR | APR | MAY | JUN | JUL | AUG | SEP | OCT | NOV | DEC | YEAR |
|---|---|---|---|---|---|---|---|---|---|---|---|---|---|
| Maximum Temp °F | 35.7 | 43.8 | 55.8 | 64.9 | 74.0 | 81.4 | 88.5 | 87.8 | 78.2 | 64.2 | 46.9 | 36.2 | 63.1 |
| Minimum Temp °F | 23.4 | 27.7 | 33.7 | 40.4 | 48.1 | 55.1 | 60.0 | 59.2 | 50.5 | 40.1 | 32.3 | 25.3 | 41.3 |
| Mean Temp °F | 29.6 | 35.8 | 44.7 | 52.6 | 61.1 | 68.3 | 74.3 | 73.6 | 64.4 | 52.2 | 39.6 | 30.8 | 52.3 |
| Days Max Temp ≥ 90 °F | 0 | 0 | 0 | 0 | 1 | 6 | 15 | 13 | 2 | 0 | 0 | 0 | 37 |
| Days Max Temp ≤ 32 °F | 10 | 3 | 0 | 0 | 0 | 0 | 0 | 0 | 0 | 0 | 1 | 9 | 23 |
| Days Min Temp ≤ 32 °F | 27 | 22 | 13 | 2 | 0 | 0 | 0 | 0 | 0 | 3 | 15 | 25 | 107 |
| Days Min Temp ≤ 0 °F | 1 | 0 | 0 | 0 | 0 | 0 | 0 | 0 | 0 | 0 | 0 | 0 | 1 |
| Heating Degree Days | 1091 | 818 | 621 | 367 | 152 | 37 | 5 | 7 | 85 | 393 | 755 | 1054 | 5385 |
| Cooling Degree Days | 0 | 0 | 0 | 3 | 47 | 141 | 294 | 286 | 85 | 3 | 0 | 0 | 859 |
| Total Precipitation (") | 1.33 | 0.78 | 0.59 | 0.60 | 0.45 | 0.65 | 0.30 | 0.48 | 0.40 | 0.49 | 1.28 | 1.54 | 8.89 |
| Days ≥ 0.1" Precip | 5 | 3 | 2 | 2 | 1 | 2 | 1 | 1 | 1 | 1 | 4 | 5 | 28 |
| Total Snowfall (") | 9.5 | 4.2 | 0.6 | 0.0 | 0.0 | 0.0 | 0.0 | 0.0 | 0.0 | 0.0 | 2.0 | 10.3 | 26.6 |
| Days ≥ 1" Snow Depth | 21 | 7 | 1 | 0 | 0 | 0 | 0 | 0 | 0 | 0 | 1 | 12 | 42 |

## WENATCHEE EXP STN *Chelan County* ELEVATION 869 ft LAT/LONG 47° 26' N / 120° 21' W

| | JAN | FEB | MAR | APR | MAY | JUN | JUL | AUG | SEP | OCT | NOV | DEC | YEAR |
|---|---|---|---|---|---|---|---|---|---|---|---|---|---|
| Maximum Temp °F | ***36.1*** | ***43.4*** | ***55.6*** | ***64.4*** | ***72.9*** | ***79.5*** | ***87.1*** | ***87.0*** | ***77.7*** | ***64.0*** | ***45.5*** | ***35.4*** | 62.4 |
| Minimum Temp °F | ***21.3*** | ***25.4*** | ***31.3*** | ***37.6*** | ***44.8*** | ***51.9*** | ***56.1*** | ***55.2*** | ***45.4*** | ***35.4*** | ***28.5*** | ***21.8*** | 37.9 |
| Mean Temp °F | ***28.7*** | ***34.4*** | ***43.5*** | ***51.0*** | ***58.9*** | ***65.7*** | ***71.6*** | ***71.1*** | ***61.5*** | ***49.7*** | ***37.1*** | ***28.6*** | 50.2 |
| Days Max Temp ≥ 90 °F | ***0*** | ***0*** | ***0*** | ***0*** | ***1*** | ***5*** | ***13*** | ***12*** | ***3*** | ***0*** | ***0*** | ***0*** | 34 |
| Days Max Temp ≤ 32 °F | ***10*** | ***3*** | ***0*** | ***0*** | ***0*** | ***0*** | ***0*** | ***0*** | ***0*** | ***0*** | ***2*** | ***11*** | 26 |
| Days Min Temp ≤ 32 °F | ***28*** | ***24*** | ***18*** | ***8*** | ***1*** | ***0*** | ***0*** | ***0*** | ***1*** | ***11*** | ***21*** | ***28*** | 140 |
| Days Min Temp ≤ 0 °F | ***2*** | ***0*** | ***0*** | ***0*** | ***0*** | ***0*** | ***0*** | ***0*** | ***0*** | ***0*** | ***0*** | ***1*** | 3 |
| Heating Degree Days | ***1118*** | ***858*** | ***659*** | ***414*** | ***206*** | ***66*** | ***13*** | ***13*** | ***136*** | ***467*** | ***832*** | ***1121*** | 5903 |
| Cooling Degree Days | 0 | 0 | 0 | 2 | 26 | 91 | 213 | 198 | 41 | 1 | 0 | 0 | 572 |
| Total Precipitation (") | ***1.53*** | ***1.04*** | ***0.71*** | ***0.56*** | ***0.51*** | ***0.70*** | ***0.35*** | ***0.43*** | ***0.51*** | ***0.59*** | ***1.50*** | ***1.83*** | 10.26 |
| Days ≥ 0.1" Precip | ***5*** | ***4*** | ***2*** | ***2*** | ***2*** | ***2*** | ***1*** | ***1*** | ***1*** | ***2*** | ***5*** | ***6*** | 33 |
| Total Snowfall (") | ***10.3*** | ***5.7*** | ***1.0*** | ***0.0*** | ***0.0*** | ***0.0*** | ***0.0*** | ***0.0*** | ***0.0*** | ***0.1*** | ***4.4*** | ***13.0*** | 34.5 |
| Days ≥ 1" Snow Depth | ***22*** | ***12*** | ***1*** | ***0*** | ***0*** | ***0*** | ***0*** | ***0*** | ***0*** | ***0*** | ***4*** | ***17*** | 56 |

## WENATCHEE PANGBORN *Chelan County* ELEVATION 1229 ft LAT/LONG 47° 24' N / 120° 12' W

| | JAN | FEB | MAR | APR | MAY | JUN | JUL | AUG | SEP | OCT | NOV | DEC | YEAR |
|---|---|---|---|---|---|---|---|---|---|---|---|---|---|
| Maximum Temp °F | 33.5 | 41.4 | 53.3 | 62.2 | 71.7 | 78.9 | 86.6 | 85.8 | 76.0 | 61.6 | 43.9 | 33.3 | 60.7 |
| Minimum Temp °F | 21.3 | 26.6 | 33.2 | 39.8 | 47.3 | 54.4 | 59.9 | 59.8 | 51.2 | 40.2 | 30.6 | 22.8 | 40.6 |
| Mean Temp °F | 27.4 | 34.0 | 43.3 | 51.0 | 59.5 | 66.7 | 73.3 | 72.9 | 63.6 | 50.9 | 37.3 | 28.1 | 50.7 |
| Days Max Temp ≥ 90 °F | 0 | 0 | 0 | 0 | 1 | 4 | 12 | 11 | 2 | 0 | 0 | 0 | 30 |
| Days Max Temp ≤ 32 °F | 14 | 5 | 0 | 0 | 0 | 0 | 0 | 0 | 0 | 0 | 3 | 14 | 36 |
| Days Min Temp ≤ 32 °F | 28 | 22 | 14 | 3 | 0 | 0 | 0 | 0 | 0 | 3 | 17 | 27 | 114 |
| Days Min Temp ≤ 0 °F | 1 | 0 | 0 | 0 | 0 | 0 | 0 | 0 | 0 | 0 | 0 | 1 | 2 |
| Heating Degree Days | 1157 | 867 | 666 | 414 | 195 | 60 | 11 | 10 | 105 | 431 | 824 | 1139 | 5879 |
| Cooling Degree Days | 0 | 0 | 0 | 2 | 39 | 111 | 255 | 254 | 75 | 2 | 0 | 0 | 738 |
| Total Precipitation (") | 1.13 | 0.71 | 0.60 | 0.58 | 0.53 | 0.63 | 0.31 | 0.42 | 0.40 | 0.43 | 1.14 | 1.41 | 8.29 |
| Days ≥ 0.1" Precip | 4 | 3 | 2 | 2 | 2 | 2 | 1 | 1 | 1 | 1 | 3 | 4 | 26 |
| Total Snowfall (") | 10.8 | 4.3 | 1.3 | 0.0 | 0.0 | 0.0 | 0.0 | 0.0 | 0.0 | 0.2 | 4.0 | 12.3 | 32.9 |
| Days ≥ 1" Snow Depth | 22 | 12 | 2 | 0 | 0 | 0 | 0 | 0 | 0 | 0 | 4 | 16 | 56 |

## WHITMAN MISSION *Walla Walla County* ELEVATION 630 ft LAT/LONG 46° 3 ' N / 118° 27 ' W

| | JAN | FEB | MAR | APR | MAY | JUN | JUL | AUG | SEP | OCT | NOV | DEC | YEAR |
|---|---|---|---|---|---|---|---|---|---|---|---|---|---|
| Maximum Temp °F | 40.3 | 47.3 | 56.7 | 64.0 | 72.7 | 80.9 | 89.1 | 88.5 | 79.0 | 66.1 | 50.2 | 41.7 | 64.7 |
| Minimum Temp °F | 25.7 | 29.3 | 33.1 | 37.7 | 43.5 | 49.5 | 52.8 | 51.4 | 43.4 | 35.0 | 32.0 | 26.7 | 38.3 |
| Mean Temp °F | 33.1 | 38.3 | 44.9 | 50.9 | 58.1 | 65.2 | 71.0 | 70.0 | 61.2 | 50.6 | 41.1 | 34.2 | 51.6 |
| Days Max Temp ≥ 90 °F | 0 | 0 | 0 | 0 | 1 | 6 | 16 | 15 | 4 | 0 | 0 | 0 | 42 |
| Days Max Temp ≤ 32 °F | 8 | 2 | 0 | 0 | 0 | 0 | 0 | 0 | 0 | 0 | 2 | 7 | 19 |
| Days Min Temp ≤ 32 °F | 23 | 19 | 15 | 7 | 1 | 0 | 0 | 0 | 3 | 13 | 16 | 23 | 120 |
| Days Min Temp ≤ 0 °F | 1 | 1 | 0 | 0 | 0 | 0 | 0 | 0 | 0 | 0 | 0 | 1 | 3 |
| Heating Degree Days | 984 | 748 | 615 | 418 | 227 | 78 | 17 | 23 | 146 | 442 | 709 | 949 | 5356 |
| Cooling Degree Days | 0 | 0 | 0 | 2 | 25 | 86 | 194 | 172 | 35 | 1 | 0 | 0 | 515 |
| Total Precipitation (") | 1.67 | 1.18 | 1.33 | 1.19 | 1.21 | 0.92 | 0.51 | 0.70 | 0.65 | 1.01 | 1.82 | 1.63 | 13.82 |
| Days ≥ 0.1" Precip | 5 | 4 | 5 | 4 | 3 | 3 | 1 | 2 | 2 | 3 | 6 | 6 | 44 |
| Total Snowfall (") | *4.3* | *1.9* | 0.4 | 0.0 | 0.0 | 0.0 | 0.0 | 0.0 | 0.0 | 0.0 | *0.3* | na | na |
| Days ≥ 1" Snow Depth | 8 | 3 | 1 | 0 | 0 | 0 | 0 | 0 | 0 | 0 | 1 | *4* | 17 |

## WILBUR *Lincoln County* ELEVATION 2162 ft LAT/LONG 47° 45 ' N / 118° 42 ' W

| | JAN | FEB | MAR | APR | MAY | JUN | JUL | AUG | SEP | OCT | NOV | DEC | YEAR |
|---|---|---|---|---|---|---|---|---|---|---|---|---|---|
| Maximum Temp °F | 31.9 | 39.0 | 49.3 | 59.0 | 68.5 | 76.3 | 83.9 | 84.0 | 74.2 | 60.4 | 42.1 | 32.7 | 58.4 |
| Minimum Temp °F | 18.3 | 23.2 | 28.1 | 32.6 | 39.0 | 45.1 | 49.6 | 49.6 | 42.0 | 33.0 | 26.5 | 19.7 | 33.9 |
| Mean Temp °F | 25.1 | 31.1 | 38.8 | 45.8 | 53.8 | 60.7 | 66.8 | 66.9 | 58.1 | 46.7 | 34.3 | 26.2 | 46.2 |
| Days Max Temp ≥ 90 °F | 0 | 0 | 0 | 0 | 0 | 3 | 9 | 9 | 1 | 0 | 0 | 0 | 22 |
| Days Max Temp ≤ 32 °F | 14 | 5 | 0 | 0 | 0 | 0 | 0 | 0 | 0 | 0 | 3 | 13 | 35 |
| Days Min Temp ≤ 32 °F | 29 | 25 | 24 | 16 | 6 | 1 | 0 | 0 | 2 | 15 | 23 | 29 | 170 |
| Days Min Temp ≤ 0 °F | 2 | 1 | 0 | 0 | 0 | 0 | 0 | 0 | 0 | 0 | 0 | 1 | 4 |
| Heating Degree Days | 1229 | 948 | 807 | 569 | 350 | 162 | 59 | 58 | 218 | 560 | 915 | 1197 | 7072 |
| Cooling Degree Days | 0 | 0 | 0 | 1 | 11 | 38 | 118 | 121 | 23 | 0 | 0 | 0 | 312 |
| Total Precipitation (") | 1.21 | 1.08 | 0.98 | 0.92 | 1.18 | 0.81 | 0.82 | 0.57 | 0.57 | 0.74 | 1.59 | 1.67 | 12.14 |
| Days ≥ 0.1" Precip | 4 | 4 | 4 | 3 | 4 | 3 | 2 | 2 | 2 | 2 | 5 | 6 | 41 |
| Total Snowfall (") | 6.9 | *3.9* | 1.4 | 0.2 | 0.0 | 0.0 | 0.0 | 0.0 | 0.0 | 0.1 | 2.8 | 9.3 | 24.6 |
| Days ≥ 1" Snow Depth | na | na | 1 | 0 | 0 | 0 | 0 | 0 | 0 | 0 | na | na | na |

## WINTHROP 1 WSW *Okanogan County* ELEVATION 1762 ft LAT/LONG 48° 28 ' N / 120° 11 ' W

| | JAN | FEB | MAR | APR | MAY | JUN | JUL | AUG | SEP | OCT | NOV | DEC | YEAR |
|---|---|---|---|---|---|---|---|---|---|---|---|---|---|
| Maximum Temp °F | 29.1 | 38.5 | 50.6 | 62.0 | 71.6 | 79.0 | 85.9 | 86.1 | 77.0 | 62.6 | 41.2 | 29.1 | 59.4 |
| Minimum Temp °F | 11.2 | 16.8 | 24.1 | 30.6 | 38.0 | 44.7 | 48.1 | 47.6 | 38.9 | 29.9 | 23.2 | 13.0 | 30.5 |
| Mean Temp °F | 20.2 | 27.7 | 37.4 | 46.3 | 54.8 | 61.9 | 67.0 | 66.9 | 57.9 | 46.3 | 32.2 | 21.1 | 45.0 |
| Days Max Temp ≥ 90 °F | 0 | 0 | 0 | 0 | 1 | 4 | 12 | 11 | 3 | 0 | 0 | 0 | 31 |
| Days Max Temp ≤ 32 °F | 18 | 5 | 0 | 0 | 0 | 0 | 0 | 0 | 0 | 0 | 4 | 19 | 46 |
| Days Min Temp ≤ 32 °F | 31 | 27 | 28 | 19 | 7 | 1 | 0 | 0 | 5 | 20 | 26 | 31 | 195 |
| Days Min Temp ≤ 0 °F | 7 | 3 | 0 | 0 | 0 | 0 | 0 | 0 | 0 | 0 | 1 | 5 | 16 |
| Heating Degree Days | 1383 | 1047 | 849 | 554 | 314 | 130 | 43 | 44 | 218 | 573 | 976 | 1355 | 7486 |
| Cooling Degree Days | 0 | 0 | 0 | 0 | 8 | 39 | 104 | 110 | 17 | 0 | 0 | 0 | 278 |
| Total Precipitation (") | 1.99 | 1.27 | 0.87 | 0.81 | 0.88 | 1.01 | 0.68 | 0.77 | 0.61 | 0.85 | 1.85 | 2.77 | 14.36 |
| Days ≥ 0.1" Precip | 6 | 4 | 3 | 3 | 3 | 2 | 2 | 2 | 2 | 2 | 6 | 8 | 43 |
| Total Snowfall (") | 18.8 | 8.3 | 3.3 | 0.1 | 0.0 | 0.0 | 0.0 | 0.0 | 0.0 | 0.6 | 8.1 | 25.7 | 64.9 |
| Days ≥ 1" Snow Depth | 31 | 27 | *17* | 1 | 0 | 0 | 0 | 0 | 0 | 0 | *9* | 29 | 114 |

## YAKIMA AIR TERMINAL *Yakima County* ELEVATION 1063 ft LAT/LONG 46° 34 ' N / 120° 32 ' W

| | JAN | FEB | MAR | APR | MAY | JUN | JUL | AUG | SEP | OCT | NOV | DEC | YEAR |
|---|---|---|---|---|---|---|---|---|---|---|---|---|---|
| Maximum Temp °F | 37.5 | 45.9 | 56.1 | 63.9 | 72.8 | 80.0 | 87.0 | 86.4 | 77.3 | 64.6 | 48.1 | 37.5 | 63.1 |
| Minimum Temp °F | 21.7 | 26.3 | 30.6 | 35.2 | 42.2 | 49.0 | 53.0 | 52.2 | 44.3 | 34.8 | 28.2 | 22.1 | 36.6 |
| Mean Temp °F | 29.7 | 36.1 | 43.4 | 49.6 | 57.5 | 64.5 | 70.0 | 69.3 | 60.8 | 49.7 | 38.2 | 29.8 | 49.9 |
| Days Max Temp ≥ 90 °F | 0 | 0 | 0 | 0 | 1 | 5 | 13 | 11 | 2 | 0 | 0 | 0 | 32 |
| Days Max Temp ≤ 32 °F | 10 | 3 | 0 | 0 | 0 | 0 | 0 | 0 | 0 | 0 | 2 | 9 | 24 |
| Days Min Temp ≤ 32 °F | 27 | 23 | 20 | 12 | 3 | 0 | 0 | 0 | 1 | 11 | 21 | 28 | 146 |
| Days Min Temp ≤ 0 °F | 1 | 0 | 0 | 0 | 0 | 0 | 0 | 0 | 0 | 0 | 0 | 1 | 2 |
| Heating Degree Days | 1089 | 809 | 663 | 457 | 243 | 87 | 25 | 25 | 149 | 467 | 798 | 1083 | 5895 |
| Cooling Degree Days | 0 | 0 | 0 | 2 | 23 | 71 | 174 | 156 | 32 | 1 | 0 | 0 | 459 |
| Total Precipitation (") | 1.18 | 0.68 | 0.62 | 0.49 | 0.42 | 0.63 | 0.17 | 0.39 | 0.40 | 0.50 | 1.03 | 1.35 | 7.86 |
| Days ≥ 0.1" Precip | 4 | 3 | 2 | 2 | 1 | 2 | 1 | 1 | 1 | 2 | 4 | 4 | 27 |
| Total Snowfall (") | 7.4 | 3.2 | 1.2 | 0.0 | 0.0 | 0.0 | 0.0 | 0.0 | 0.0 | 0.2 | 2.3 | 9.0 | 23.3 |
| Days ≥ 1" Snow Depth | 17 | 6 | 1 | 0 | 0 | 0 | 0 | 0 | 0 | 0 | 2 | 12 | 38 |

## JANUARY MINIMUM TEMPERATURE °F

### LOWEST

| | | |
|---|---|---|
| 1 | Winthrop | 11.2 |
| 2 | Mazama | 11.5 |
| 3 | Conconully | 14.4 |
| | Republic | 14.4 |
| 5 | Holden | 15.2 |
| 6 | Waterville | 15.7 |
| 7 | Boundary Dam | 17.2 |
| | Leavenworth | 17.2 |
| 9 | Chewelah | 17.5 |
| 10 | Davenport | 17.9 |
| 11 | Wilbur | 18.3 |
| 12 | Ellensburg | 19.0 |
| 13 | Stevens Pass | 19.1 |
| 14 | Odessa | 19.2 |
| 15 | Quincy | 19.3 |
| 16 | Hartline | 19.6 |
| | Newport | 19.6 |
| | Plain | 19.6 |
| 19 | Chief Joseph Dam | 19.8 |
| 20 | Harrington | 19.9 |
| 21 | Wellpinit | 20.0 |
| 22 | Rainier | 20.2 |
| 23 | Northport | 20.5 |
| 24 | Cle Elum | 20.6 |
| 25 | Ephrata | 20.9 |

### HIGHEST

| | | |
|---|---|---|
| 1 | Hoquiam | 36.8 |
| 2 | Port Townsend | 36.7 |
| 3 | Grayland | 36.1 |
| 4 | Seattle | 35.6 |
| 5 | Long Beach | 35.1 |
| 6 | Aberdeen | 35.0 |
| 7 | Olga | 34.9 |
| 8 | Grapeview | 34.4 |
| 9 | Coupeville | 34.3 |
| 10 | Anacortes | 34.0 |
| | Centralia | 34.0 |
| | Clearwater | 34.0 |
| 13 | Bremerton | 33.8 |
| | Port Angeles | 33.8 |
| | Shelton | 33.8 |
| 16 | Elma | 33.6 |
| | Forks | 33.6 |
| | Kent | 33.6 |
| 19 | Everett | 33.3 |
| | Longview | 33.3 |
| 21 | Mount Vernon | 33.2 |
| 22 | Sappho | 33.0 |
| | Startup | 33.0 |
| 24 | Snoqualmie Falls | 32.9 |
| 25 | Monroe | 32.8 |

## JULY MAXIMUM TEMPERATURE °F

### HIGHEST

| | | |
|---|---|---|
| 1 | Priest Rapids Dam | 90.8 |
| 2 | Smyrna | 90.5 |
| 3 | Kennewick | 90.2 |
| 4 | Connell | 89.8 |
| 5 | Richland | 89.4 |
| | Wapato | 89.4 |
| 7 | Walla Walla | 89.3 |
| 8 | Sunnyside | 89.1 |
| | Whitman | 89.1 |
| 10 | Hatton | 88.8 |
| 11 | Ice Harbor Dam | 88.5 |
| | McNary Dam | 88.5 |
| | Wenatchee | 88.5 |
| 14 | Chief Joseph Dam | 88.3 |
| | Lind | 88.3 |
| 16 | Ephrata | 88.2 |
| | Pomeroy | 88.2 |
| 18 | Lacrosse | 88.0 |
| 19 | Dallesport | 87.8 |
| 20 | Leavenworth | 87.5 |
| 21 | Northport | 87.4 |
| 22 | Hartline | 87.3 |
| 23 | Wenatchee-Exp | 87.1 |
| 24 | Yakima | 87.0 |
| 25 | Prosser | 86.9 |

### LOWEST

| | | |
|---|---|---|
| 1 | Rainier | 60.5 |
| 2 | Stampede Pass | 64.2 |
| 3 | Long Beach | 65.4 |
| 4 | Grayland | 65.6 |
| 5 | Hoquiam | 67.3 |
| 6 | Aberdeen | 68.1 |
| 7 | Port Angeles | 68.2 |
| 8 | Olga | 69.9 |
| 9 | Clearwater | 70.0 |
| 10 | Bellingham | 70.9 |
| | Coupeville | 70.9 |
| 12 | Anacortes | 71.4 |
| 13 | Cedar Lake | 71.6 |
| 14 | Aberdeen-20 NNE | 71.8 |
| | Port Townsend | 71.8 |
| 16 | Blaine | 72.0 |
| | Mud Mountn Dm | 72.0 |
| 18 | Everett | 72.5 |
| | Forks | 72.5 |
| 20 | Grays River | 72.6 |
| | Mount Vernon | 72.6 |
| | Sappho | 72.6 |
| 23 | Upper Baker Dam | 73.0 |
| 24 | Elwha | 73.2 |
| | Palmer | 73.2 |

## ANNUAL PRECIPITATION (")

### HIGHEST

| | | |
|---|---|---|
| 1 | Aberdeen-20 NNE | 131.33 |
| 2 | Forks | 118.50 |
| 3 | Clearwater | 116.37 |
| 4 | Rainier | 114.21 |
| 5 | Cougar | 112.22 |
| 6 | Grays River | 107.47 |
| 7 | Upper Baker Dam | 98.07 |
| 8 | Cedar Lake | 97.88 |
| 9 | Sappho | 96.21 |
| 10 | Palmer | 85.58 |
| 11 | Skamania | 83.98 |
| 12 | Stampede Pass | 83.96 |
| 13 | Aberdeen | 81.74 |
| 14 | Stevens Pass | 80.87 |
| 15 | Long Beach | 80.35 |
| 16 | Newhalem | 78.75 |
| 17 | Diablo Dam | 77.51 |
| 18 | Grayland | 71.44 |
| 19 | Concrete | 70.06 |
| 20 | Hoquiam | 66.94 |
| 21 | Elma | 66.58 |
| 22 | Glenoma | 64.18 |
| 23 | Startup | 64.14 |
| 24 | Shelton | 63.44 |
| 25 | Snoqualmie Falls | 60.14 |

### LOWEST

| | | |
|---|---|---|
| 1 | Priest Rapids Dam | 6.56 |
| 2 | Sunnyside | 6.76 |
| 3 | Richland | 7.13 |
| 4 | Ephrata | 7.28 |
| 5 | Wapato | 7.51 |
| 6 | Moxee City | 7.56 |
| 7 | Kennewick | 7.63 |
| 8 | McNary Dam | 7.66 |
| 9 | Prosser | 7.69 |
| 10 | Smyrna | 7.73 |
| 11 | Yakima | 7.86 |
| 12 | Quincy | 8.03 |
| 13 | Connell | 8.14 |
| 14 | Othello | 8.15 |
| 15 | Wnatchee-Pngbrn | 8.29 |
| 16 | Ellensburg | 8.79 |
| 17 | Wenatchee | 8.89 |
| 18 | Lind | 9.36 |
| 19 | Hatton | 9.90 |
| 20 | Odessa | 9.92 |
| 21 | Ice Harbor Dam | 10.12 |
| 22 | Wenatchee-Exp | 10.26 |
| 23 | Chief Joseph Dam | 10.32 |
| 24 | Coulee Dam | 10.46 |
| 25 | Hartline | 10.48 |

## ANNUAL SNOWFALL (")

### HIGHEST

| | | |
|---|---|---|
| 1 | Rainier | 686.1 |
| 2 | Stevens Pass | 465.2 |
| 3 | Stampede Pass | 427.0 |
| 4 | Holden | 248.0 |
| 5 | Plain | 138.1 |
| 6 | Stehekin | 132.4 |
| 7 | Mazama | 116.0 |
| 8 | Leavenworth | 93.7 |
| 9 | Appleton | 83.4 |
| 10 | Cle Elum | 77.0 |
| 11 | Winthrop | 64.9 |
| 12 | Newport | 61.2 |
| 13 | Boundary Dam | 60.9 |
| 14 | Cedar Lake | 59.4 |
| 15 | Upper Baker Dam | 56.4 |
| 16 | Northport | 56.0 |
| 17 | Republic | 48.7 |
| 18 | Spokane | 45.6 |
| 19 | Waterville | 45.1 |
| 20 | Newhalem | 44.7 |
| 21 | Chewelah | 41.6 |
| 22 | Davenport | 35.3 |
| 23 | Wenatchee-Exp | 34.5 |
| 24 | Pullman | 34.4 |
| 25 | Chelan | 33.4 |

### LOWEST

| | | |
|---|---|---|
| 1 | Long Beach | 1.8 |
| | Port Angeles | 1.8 |
| 3 | Kent | 3.2 |
| 4 | Mount Vernon | 4.0 |
| 5 | Aberdeen | 4.1 |
| | Longview | 4.1 |
| 7 | Everett | 4.2 |
| 8 | Anacortes | 4.5 |
| 9 | Hoquiam | 4.8 |
| 10 | Centralia | 5.0 |
| 11 | Oakville | 5.6 |
| 12 | Olga | 6.1 |
| 13 | Coupeville | 6.3 |
| 14 | Battle Ground | 6.6 |
| 15 | Shelton | 6.8 |
| 16 | Puyallup | 6.9 |
| 17 | Grays River | 7.0 |
| 18 | Elma | 7.1 |
| 19 | Sedro Woolley | 7.4 |
| 20 | Prosser | 8.0 |
| 21 | Bremerton | 8.1 |
| | Monroe | 8.1 |
| 23 | Quilcene | 8.2 |
| 24 | McMillin Rservoir | 8.4 |
| 25 | Landsburg | 8.7 |

# WEST VIRGINIA

PHYSICAL FEATURES. West Virginia has an area of over 24,000 square miles. From southwest to northeast, the State is about 200 miles in length; width averages a little over one-half the length. There are two projections: one, the Northeastern Panhandle, juts eastward between Maryland and Virginia; the other, the Northern Panhandle, is a narrow strip stretching northward along the Ohio River between Ohio and Pennsylvania. The easternmost extremity of the State is about 150 miles from the Atlantic Ocean and the southwestern corner adjacent to Kentucky is nearly 400 miles away from the ocean. As a result, West Virginia lies beyond the immediate climatic effect of the Atlantic, and its climate is much more of the continental than it is of the maritime type. The most important aspect of this type of climate is the marked temperature contrast between summer and winter.

The physical configuration of the State accentuates its interior location. Excluding the Northeastern Panhandle, the State lies in the Allegheny Plateau. The eastern third of the plateau is part of the Appalachian Mountain chain and contains the highest land in the State. Peak elevations in this area range from about 2,500 feet to 4,860 feet (above sea level) at Spruce Knob, the highest point in West Virginia. The central and western thirds of the plateau slope generally westward to the Ohio River which lies at about 550 to 650 feet above sea level.

The Northeastern Panhandle is marked by long ridges and valleys, oriented southwest-northeast, intersected by the winding courses of the Potomac River and its tributaries. The main stream of the Potomac with its North Branch forms the northern border of this part of the State. Summit elevations exceed 4,000 feet (above sea level), but the land in general slopes eastward away from the main ridgeline to the west and finally reaches the lowest elevation in the State of 274 feet at Harpers Ferry. This section lies in the Atlantic Ocean drainage and is drained by the Potomac River. The remainder of the State drains into the Ohio River.

GENERAL CLIMATE. Physical features considerably modify the effects of the major climatic controls. The State's latitudinal position (from about 37°15' N. latitude in the south to 40° in the north) places it in the zone of prevailing westerly winds, which are frequently interrupted by northward and southward surges of relatively warm and cold air, respectively. These atmospheric movements are accompanied by the passage of high and low-pressure areas; the latter are the large-dimension storms, known as extratropical cyclones, which are most common in the United States in the colder half-year. West Virginia lies near the average path of the extratropical cyclones that move in a general easterly direction across the United States. In the warmer half-year, the State is affected by the showers and thunderstorms that occur in the broad current of air that tends to sweep northeastward from the Gulf of Mexico.

The State has a moderately severe winter climate, accentuated and prolonged in the mountains, with frequent alternations of fair and stormy weather. Summer is marked by hot and showery weather; the heat is less pronounced in the mountains, but they are more subject to thunderstorms and have fewer clear days the year-round. There are marked variations in temperature, precipitation, and the other weather elements, due to the rugged topography occurring not only between the mountains and plateau areas but even between different parts of the same county.

TEMPERATURE. Locations in the mountainous belt, regardless of their latitude, tend to have lower temperatures than those in the rest of the State. Average winter minimum temperatures range from the low 20s in the mountains of the Central and Northeastern Divisions and in the Northern Panhandle, to near 30° F. in the extreme southern and southwestern corners of the State. Average winter maximum readings are in the middle and upper 40s, except in the mountains and in the Northern Panhandle where they are close to 40°F. In summer, maximum temperatures average over 85° F. everywhere except in the mountains, where they are 5° to 10° cooler; average minimum temperatures during this season range from the middle 50s in the mountains to the middle 60s elsewhere. Spring and autumn mean temperatures average in the 50s, with similar geographical variations. The average date of the last freezing temperature in spring ranges from mid-April in the southwest to mid-May in the mountains; the average first occurrence of 32° F. in the fall similarly varies from late October to late September.

Despite what has been said about the coolness of the mountains, they can on occasion be as hot as any other part of West Virginia. Temperatures near or over 100° F. have been recorded at all stations in the State. On the other hand, very low temperatures (below -30°) have been observed only in the mountains and in the North Central Division. These are extremes, and do not represent usual winter conditions. Cold waves, with near or subzero temperatures, come on an average of three times a winter, but as a rule do not last more than 2 or 3 days.

Fog conditions over the State are complicated as to their causes and distribution. The valley fogs are usually of the radiation type, and occur characteristically when a high-pressure area is centered over or near the State. This situation is most common in late summer and fall. Low cloudiness and fog in the mountains are generally orographic in nature, the result of moist winds moving upslope, so that there is usually a great difference in cloud and fog conditions on opposite sides of a ridge.

PRECIPITATION. The precipitation pattern can be directly related to the fact that the rain and snow-producing atmospheric currents generally move across West Virginia on an eastward course. As they approach the mountains, these air currents are subject to orographic lifting, which acts to trigger potential precipitation or to intensify the rain or snow that may already be falling. As a result, average annual precipitation increases from the Ohio eastward to the Appalachians. On the other side of the mountains, there is the well-marked "rain shadow" where the air currents descend the leeward slopes and precipitation is correspondingly reduced, to increase only when more favorable topographic influences are encountered farther eastward and where the influence of the ocean and coastal storms is more pronounced.

Mean annual snowfall exhibits the same features, but to a more remarkable degree. The mountain belt receives over 60 inches of snow a year, on the average. Amounts over 20 inches have been experienced everywhere else, except in that part of the State west of longitude 81°30' W. which usually receives about 15 inches. It is very unusual for a relatively small and compact area the size of West Virginia to exhibit such great differences in snowfall. Furthermore, the heavy snowfall at elevations under 5,000 feet (above sea level) is unusual here in the East, for an area located south of 40° N. latitude. The snow, as a general rule, does not remain on the ground for extended periods over most of the State. Except in the higher portions of the plateau and in the mountains themselves, the snow cover does not persist for anything like the duration of the winter. Snowstorms are usually followed by thawing periods and there is no large-scale melting in the spring of a seasonally accumulated snowpack.

SUNSHINE AND CLOUDINESS. West Virginia lies in a cloudy belt. Percentage of possible sunshine is only about 40 in winter, increasing to somewhat over 60 percent in early autumn. Cloudiness is most pronounced over the mountains. The average annual number of clear days ranges from about 80 in the mountains to about 120 in the western portion. In addition to cloudiness, the hours of sunshine are reduced by fog, particularly in the river valleys.

WINDS AND STORMS. The prevailing winds blow from westerly directions. There is a tendency outside of the mountain belt for southerly or southwesterly winds during summer and fall. Thunderstorms occur on an average of 40 to 50 days per year, being more frequent in the mountains. June and July are the months of most frequent occurrence. Violent local winds accompanying thunderstorms are experienced every year in some part of the State, but tornadoes are rare. Destructive hailstorms occur on an average of about three per year. Though hurricanes have damaged the State, principally as a result of heavy rains, it is uncommon for this type of storm to strike West Virginia with full force. The remnants of the hurricanes which have affected the State have been more noted for their accompanying heavy rainfalls than for any high winds produced. Much more frequent and costly is the damage from intense large-area storms -- that is, from exceptionally strong specimens of the ordinary lows that affect the State quite frequently during the colder half of the year. Such storms produce high winds and heavy rain or snow.

Warm-season thunderstorms, mostly those of June and July, often yield intense local rainfall and cause flash flooding in the narrow valleys that cut through the plateau and mountain districts. This kind of severe local flood is likely to occur in some part of the State every year. In contrast to flash flooding on the smaller streams, flooding in the larger streams is almost exclusively a cold season phenomenon. The ideal setup for the cold season floods requires the soil to be well saturated from previous rains, a good snow cover, and a more-or-less stationary front lying northeast-southwest across the State. Along this front separating two contrasting air masses, a succession of "waves" may move northeastward, resulting in copious warm rains for a period of at least several days and a rapid melting of the snow cover.

## COUNTY INDEX

***Barbour County***
BELINGTON

***Berkeley County***
MARTINSBURG E WV AP

***Boone County***
MADISON

***Braxton County***
GASSAWAY

***Brooke County***
WELLSBURG WTR TRMT P

***Fayette County***
OAK HILL

***Gilmer County***
GLENVILLE 1 ENE

***Grant County***
BAYARD

***Greenbrier County***
LEWISBURG 3 N
WHITE SULPHUR SPRNGS

***Hampshire County***
ROMNEY 1 SW

***Hardy County***
MATHIAS
MOOREFIELD 2 SSE
WARDENSVILLE RM FARM

***Harrison County***
CLARKSBURG 1

***Jackson County***
RIPLEY

***Jefferson County***
KEARNEYSVILLE

***Kanawha County***
CHARLESTON KNWA AP
CORTON
LONDON LOCKS

***Lewis County***
WESTON

***Lincoln County***
HAMLIN

***Logan County***
LOGAN

***Marion County***
FAIRMONT

***Marshall County***
MOUNDSVILLE

***Mason County***
HOGSETT GALLIPOLS DM

***Mercer County***
ATHENS CONCORD COLLE
BLUEFIELD MERCER AP

***Monongalia County***
MORGANTOWN L & D
MORGANTOWN MUNI AP

***Monroe County***
UNION 3 SSE

***Pendleton County***
FRANKLIN 2 NE

***Pocahontas County***
BUCKEYE 1 SE
SENECA STATE FOREST

***Preston County***
ROWLESBURG 1
TERRA ALTA NO 1

***Putnam County***
WINFIELD LOCKS

***Raleigh County***
BECKLEY RALEIGH AP
BECKLEY VA HOSPITAL

***Randolph County***
ELKINS RNDLPH CO AP

***Roane County***
SPENCER 1 SE

***Summers County***
BLUESTONE LAKE
FLAT TOP

***Tucker County***
CANAAN VALLEY
PARSONS 1 NE

***Tyler County***
MIDDLEBOURNE 3 ESE

***Upshur County***
BUCKHANNON

***Wayne County***
HNTNGTN TRI-STATE

***Wirt County***
CRESTON

***Wood County***
PARKERSBURG 1 NE
PARKERSBURG-WOOD AP

***Wyoming County***
KOPPERSTON
PINEVILLE

## ELEVATION INDEX

| FEET | STATION NAME |
|---|---|
| 540 | MARTINSBURG E WV AP |
| 550 | KEARNEYSVILLE |
| 561 | HOGSETT GALLIPOLS DM |
| 571 | WINFIELD LOCKS |
| 590 | RIPLEY |
| 620 | LONDON LOCKS |
| 620 | MOUNDSVILLE |
| 640 | CORTON |
| 640 | CRESTON |
| 640 | HAMLIN |
| 640 | PARKERSBURG 1 NE |

| FEET | STATION NAME |
|---|---|
| 664 | LOGAN |
| 669 | WELLSBURG WTR TRMT P |
| 670 | ROMNEY 1 SW |
| 679 | MADISON |
| 751 | MIDDLEBOURNE 3 ESE |
| 781 | GLENVILLE 1 ENE |
| 791 | SPENCER 1 SE |
| 820 | MOOREFIELD 2 SSE |
| 827 | HNTNGTN TRI-STATE |
| 830 | MORGANTOWN L & D |
| 840 | GASSAWAY |
| 840 | PARKERSBURG-WOOD AP |
| 951 | CHARLESTON KNWA AP |
| 961 | WARDENSVILLE RM FARM |
| 981 | CLARKSBURG 1 |
| 1020 | FAIRMONT |
| 1030 | WESTON |
| 1247 | MORGANTOWN MUNI AP |
| 1352 | PINEVILLE |
| 1381 | ROWLESBURG 1 |
| 1401 | BLUESTONE LAKE |
| 1450 | BUCKHANNON |
| 1552 | MATHIAS |
| 1650 | PARSONS 1 NE |
| 1680 | BELINGTON |
| 1703 | FRANKLIN 2 NE |
| 1732 | KOPPERSTON |
| 1913 | WHITE SULPHUR SPRNGS |
| 1948 | ELKINS RNDLPH CO AP |
| 1990 | OAK HILL |
| 2103 | BUCKEYE 1 SE |
| 2113 | UNION 3 SSE |
| 2303 | LEWISBURG 3 N |
| 2330 | BECKLEY VA HOSPITAL |
| 2382 | BAYARD |
| 2504 | BECKLEY RALEIGH AP |
| 2543 | TERRA ALTA NO 1 |
| 2602 | ATHENS CONCORD COLLE |
| 2651 | SENECA STATE FOREST |
| 2892 | BLUEFIELD MERCER AP |
| 3235 | FLAT TOP |
| 3250 | CANAAN VALLEY |

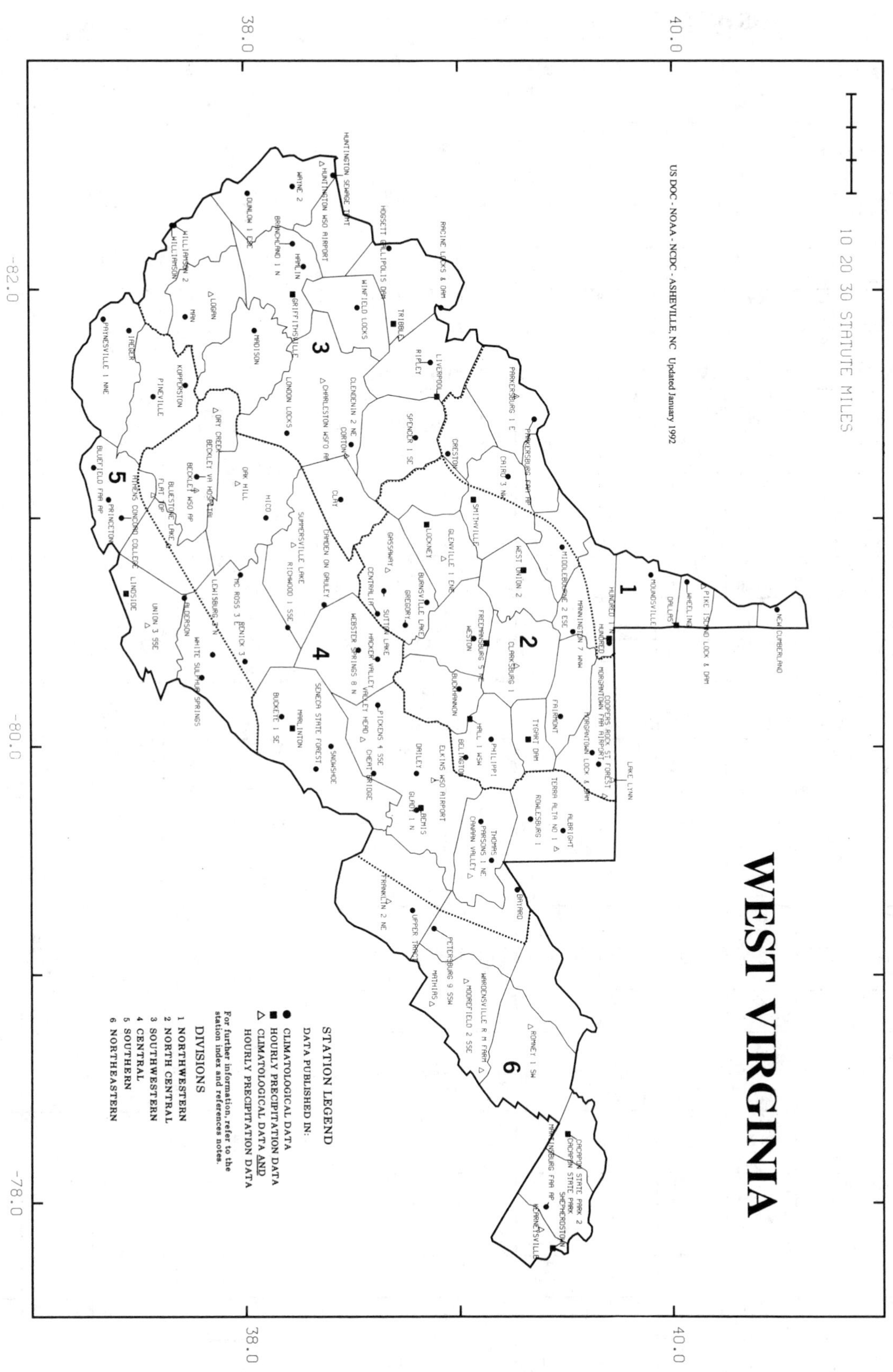
WEST VIRGINIA
US DOC - NOAA - NCDC - ASHEVILLE, NC Updated January 1992
10 20 30 STATUTE MILES
STATION LEGEND
DATA PUBLISHED IN:
● CLIMATOLOGICAL DATA
■ HOURLY PRECIPITATION DATA
△ CLIMATOLOGICAL DATA AND HOURLY PRECIPITATION DATA
For further information, refer to the station index and reference notes.
DIVISIONS
1 NORTHWESTERN
2 NORTH CENTRAL
3 SOUTHWESTERN
4 CENTRAL
5 SOUTHERN
6 NORTHEASTERN

## ATHENS CONCORD COLLE *Mercer County* ELEVATION 2602 ft LAT/LONG 37° 25 ' N / 81° 1 ' W

| | JAN | FEB | MAR | APR | MAY | JUN | JUL | AUG | SEP | OCT | NOV | DEC | YEAR |
|---|---|---|---|---|---|---|---|---|---|---|---|---|---|
| Maximum Temp °F | 41.7 | 45.3 | 55.7 | 65.6 | 72.7 | 78.7 | 81.9 | 80.7 | 75.0 | 65.8 | 55.5 | 46.1 | 63.7 |
| Minimum Temp °F | 21.5 | 23.5 | 31.3 | 39.6 | 47.7 | 55.1 | 59.4 | 57.9 | 51.7 | 40.7 | 32.9 | 26.4 | 40.6 |
| Mean Temp °F | 31.6 | 34.4 | 43.5 | 52.6 | 60.2 | 66.9 | 70.7 | 69.3 | 63.4 | 53.3 | 44.2 | 36.2 | 52.2 |
| Days Max Temp ≥ 90 °F | 0 | 0 | 0 | 0 | 0 | 0 | 1 | 1 | 0 | 0 | 0 | 0 | 2 |
| Days Max Temp ≤ 32 °F | 7 | 4 | 1 | 0 | 0 | 0 | 0 | 0 | 0 | 0 | 1 | 3 | 16 |
| Days Min Temp ≤ 32 °F | 26 | 22 | 17 | 8 | 2 | 0 | 0 | 0 | 1 | 7 | 15 | 23 | 121 |
| Days Min Temp ≤ 0 °F | 2 | 1 | 0 | 0 | 0 | 0 | 0 | 0 | 0 | 0 | 0 | 1 | 4 |
| Heating Degree Days | 1029 | 856 | 660 | 371 | 168 | 38 | 6 | 12 | 104 | 360 | 617 | 883 | 5104 |
| Cooling Degree Days | 0 | 0 | 0 | 8 | 31 | 111 | 210 | 163 | 64 | 6 | 0 | 0 | 593 |
| Total Precipitation (") | 2.69 | 2.62 | 3.29 | 3.28 | 3.73 | 3.18 | 3.82 | 2.97 | 2.97 | 2.70 | 2.79 | 2.75 | 36.79 |
| Days ≥ 0.1" Precip | 7 | 6 | 8 | 7 | 8 | 7 | 8 | 6 | 6 | 5 | 6 | 6 | 80 |
| Total Snowfall (") | ***10.6*** | 10.0 | 3.8 | 0.9 | 0.0 | 0.0 | 0.0 | 0.0 | 0.0 | 0.1 | 1.5 | 5.4 | 32.3 |
| Days ≥ 1" Snow Depth | na | ***8*** | 2 | 0 | 0 | 0 | 0 | 0 | 0 | 0 | 1 | ***3*** | na |

## BAYARD *Grant County* ELEVATION 2382 ft LAT/LONG 39° 16 ' N / 79° 22 ' W

| | JAN | FEB | MAR | APR | MAY | JUN | JUL | AUG | SEP | OCT | NOV | DEC | YEAR |
|---|---|---|---|---|---|---|---|---|---|---|---|---|---|
| Maximum Temp °F | 35.3 | 38.3 | 48.8 | 60.6 | 69.6 | 76.4 | 79.2 | 77.4 | 70.4 | 60.4 | 49.7 | 39.9 | 58.8 |
| Minimum Temp °F | 16.3 | 18.0 | 25.7 | 34.2 | 43.0 | 50.8 | 56.0 | 54.2 | 48.0 | 36.5 | 29.4 | 21.4 | 36.1 |
| Mean Temp °F | 25.8 | 28.2 | 37.3 | 47.4 | 56.3 | 63.6 | 67.6 | 65.8 | 59.2 | 48.5 | 39.5 | 30.7 | 47.5 |
| Days Max Temp ≥ 90 °F | 0 | 0 | 0 | 0 | 0 | 0 | 1 | 1 | 0 | 0 | 0 | 0 | 2 |
| Days Max Temp ≤ 32 °F | 12 | 9 | 3 | 0 | 0 | 0 | 0 | 0 | 0 | 0 | 2 | 8 | 34 |
| Days Min Temp ≤ 32 °F | 29 | 25 | 24 | 14 | 5 | 0 | 0 | 0 | 2 | 11 | 19 | 27 | 156 |
| Days Min Temp ≤ 0 °F | 4 | 2 | 0 | 0 | 0 | 0 | 0 | 0 | 0 | 0 | 0 | 1 | 7 |
| Heating Degree Days | 1208 | 1033 | 852 | 522 | 272 | 88 | 28 | 50 | 189 | 505 | 756 | 1058 | 6561 |
| Cooling Degree Days | 0 | 0 | 0 | 1 | 13 | 61 | 147 | 98 | 27 | 1 | 0 | 0 | 348 |
| Total Precipitation (") | 3.54 | 3.26 | 4.19 | 4.26 | 4.61 | 4.29 | 5.17 | 4.41 | 3.51 | 3.27 | 3.56 | 3.90 | 47.97 |
| Days ≥ 0.1" Precip | 10 | 9 | 11 | 10 | 10 | 9 | 9 | 8 | 8 | 8 | 9 | 10 | 111 |
| Total Snowfall (") | 26.0 | 22.6 | 17.0 | 4.2 | 0.0 | 0.0 | 0.0 | 0.0 | 0.0 | 0.3 | 6.7 | 19.2 | 96.0 |
| Days ≥ 1" Snow Depth | 23 | 22 | 14 | 2 | 0 | 0 | 0 | 0 | 0 | 0 | 5 | 16 | 82 |

## BECKLEY RALEIGH AP *Raleigh County* ELEVATION 2504 ft LAT/LONG 37° 47 ' N / 81° 7 ' W

| | JAN | FEB | MAR | APR | MAY | JUN | JUL | AUG | SEP | OCT | NOV | DEC | YEAR |
|---|---|---|---|---|---|---|---|---|---|---|---|---|---|
| Maximum Temp °F | 38.2 | 41.9 | 51.6 | 62.4 | 70.1 | 76.6 | 79.8 | 78.5 | 72.4 | 62.5 | 52.4 | 43.1 | 60.8 |
| Minimum Temp °F | 21.2 | 23.5 | 32.1 | 40.9 | 49.0 | 56.4 | 60.9 | 59.5 | 53.3 | 42.3 | 34.6 | 26.5 | 41.7 |
| Mean Temp °F | 29.8 | 32.7 | 41.9 | 51.7 | 59.5 | 66.5 | 70.4 | 69.0 | 62.9 | 52.4 | 43.6 | 34.8 | 51.3 |
| Days Max Temp ≥ 90 °F | 0 | 0 | 0 | 0 | 0 | 0 | 1 | 1 | 0 | 0 | 0 | 0 | 2 |
| Days Max Temp ≤ 32 °F | 10 | 7 | 2 | 0 | 0 | 0 | 0 | 0 | 0 | 0 | 2 | 6 | 27 |
| Days Min Temp ≤ 32 °F | 25 | 22 | 17 | 7 | 1 | 0 | 0 | 0 | 0 | 5 | 14 | 22 | 113 |
| Days Min Temp ≤ 0 °F | 2 | 1 | 0 | 0 | 0 | 0 | 0 | 0 | 0 | 0 | 0 | 1 | 4 |
| Heating Degree Days | 1086 | 905 | 710 | 402 | 192 | 48 | 10 | 18 | 116 | 387 | 637 | 929 | 5440 |
| Cooling Degree Days | 0 | 0 | 1 | 10 | 34 | 112 | 212 | 162 | 65 | 7 | 0 | 0 | 603 |
| Total Precipitation (") | 2.96 | 2.83 | 3.59 | 3.45 | 4.01 | 3.71 | 4.70 | 3.56 | 3.13 | 2.75 | 2.82 | 3.28 | 40.79 |
| Days ≥ 0.1" Precip | 7 | 7 | 9 | 8 | 9 | 7 | 9 | 7 | 6 | 6 | 7 | 8 | 90 |
| Total Snowfall (") | 17.1 | 15.5 | 9.1 | 2.7 | 0.0 | 0.0 | 0.0 | 0.0 | 0.0 | 0.5 | 3.6 | 11.4 | 59.9 |
| Days ≥ 1" Snow Depth | 14 | 12 | 5 | 1 | 0 | 0 | 0 | 0 | 0 | 0 | 2 | 8 | 42 |

## BECKLEY VA HOSPITAL *Raleigh County* ELEVATION 2330 ft LAT/LONG 37° 46 ' N / 81° 12 ' W

| | JAN | FEB | MAR | APR | MAY | JUN | JUL | AUG | SEP | OCT | NOV | DEC | YEAR |
|---|---|---|---|---|---|---|---|---|---|---|---|---|---|
| Maximum Temp °F | 39.8 | 44.0 | 54.1 | 65.0 | 72.4 | 77.2 | 79.9 | 78.6 | 72.6 | 63.4 | 53.7 | 44.3 | 62.1 |
| Minimum Temp °F | 19.5 | 21.4 | 29.4 | 37.2 | 45.5 | 53.4 | 58.1 | 56.8 | 50.7 | 39.2 | 31.3 | 24.2 | 38.9 |
| Mean Temp °F | 29.7 | 32.7 | 41.8 | 51.2 | 58.9 | 65.4 | 69.0 | 67.7 | 61.7 | 51.3 | 42.5 | 34.3 | 50.5 |
| Days Max Temp ≥ 90 °F | 0 | 0 | 0 | 0 | 0 | 0 | 1 | 1 | 0 | 0 | 0 | 0 | 2 |
| Days Max Temp ≤ 32 °F | 9 | 6 | 2 | 0 | 0 | 0 | 0 | 0 | 0 | 0 | 1 | 5 | 23 |
| Days Min Temp ≤ 32 °F | 26 | 24 | 20 | 10 | 3 | 0 | 0 | 0 | 1 | 9 | 17 | 24 | 134 |
| Days Min Temp ≤ 0 °F | 3 | 2 | 0 | 0 | 0 | 0 | 0 | 0 | 0 | 0 | 0 | 1 | 6 |
| Heating Degree Days | 1088 | 905 | 713 | 412 | 199 | 56 | 13 | 25 | 130 | 420 | 667 | 945 | 5573 |
| Cooling Degree Days | 0 | 0 | 0 | 4 | 19 | 83 | 166 | 125 | 40 | 4 | 0 | 0 | 441 |
| Total Precipitation (") | 2.98 | 2.47 | 3.14 | 3.38 | 3.92 | 3.33 | 4.60 | 3.75 | 3.28 | 2.72 | 2.80 | 3.01 | 39.38 |
| Days ≥ 0.1" Precip | 8 | 7 | 8 | 8 | 9 | 8 | 9 | 7 | 6 | 6 | 7 | 7 | 90 |
| Total Snowfall (") | 11.5 | ***9.6*** | 5.6 | 0.8 | 0.0 | 0.0 | 0.0 | 0.0 | 0.0 | 0.0 | 2.1 | 7.7 | 37.3 |
| Days ≥ 1" Snow Depth | 13 | 9 | 4 | 1 | 0 | 0 | 0 | 0 | 0 | 0 | 1 | ***7*** | 35 |

## BELINGTON *Barbour County* ELEVATION 1680 ft LAT/LONG 39° 2 ' N / 79° 56 ' W

| | JAN | FEB | MAR | APR | MAY | JUN | JUL | AUG | SEP | OCT | NOV | DEC | YEAR |
|---|---|---|---|---|---|---|---|---|---|---|---|---|---|
| Maximum Temp °F | 38.2 | 41.4 | 51.4 | 62.1 | 71.4 | 78.6 | 81.6 | 80.1 | 74.5 | 63.5 | 53.2 | 42.6 | 61.6 |
| Minimum Temp °F | 17.3 | 18.6 | 26.6 | 34.8 | 44.2 | 53.3 | 58.4 | 57.0 | 50.5 | 37.6 | 30.0 | 22.3 | 37.6 |
| Mean Temp °F | 27.8 | 30.2 | 39.0 | 48.5 | 57.8 | 66.0 | 70.0 | 68.6 | 62.5 | 50.6 | 41.6 | 32.5 | 49.6 |
| Days Max Temp ≥ 90 °F | 0 | 0 | 0 | 0 | 0 | 1 | 1 | 1 | 0 | 0 | 0 | 0 | 3 |
| Days Max Temp ≤ 32 °F | 11 | 8 | 3 | 0 | 0 | 0 | 0 | 0 | 0 | 0 | 1 | 7 | 30 |
| Days Min Temp ≤ 32 °F | 27 | 24 | 23 | 14 | 3 | 0 | 0 | 0 | 0 | 11 | 19 | 25 | 146 |
| Days Min Temp ≤ 0 °F | 3 | 2 | 1 | 0 | 0 | 0 | 0 | 0 | 0 | 0 | 0 | 1 | 7 |
| Heating Degree Days | 1144 | 979 | 798 | 492 | 240 | 63 | 13 | 24 | 126 | 444 | 695 | 988 | 6006 |
| Cooling Degree Days | 0 | 0 | 0 | *3* | 26 | 99 | 194 | 137 | 55 | 4 | 0 | 0 | 518 |
| Total Precipitation (") | 3.55 | 3.33 | 4.20 | 4.17 | 4.45 | 4.49 | 5.11 | 4.17 | 4.40 | 3.05 | 3.51 | 3.99 | 48.42 |
| Days ≥ 0.1" Precip | 10 | 9 | 10 | 10 | 10 | 9 | 10 | 8 | 8 | 7 | 9 | 10 | 110 |
| Total Snowfall (") | 15.6 | 12.9 | 8.4 | 0.6 | 0.0 | 0.0 | 0.0 | 0.0 | 0.0 | 0.1 | 3.0 | *7.9* | 48.5 |
| Days ≥ 1" Snow Depth | *16* | 12 | 5 | 1 | 0 | 0 | 0 | 0 | 0 | 0 | 2 | 9 | 45 |

## BLUEFIELD MERCER AP *Mercer County* ELEVATION 2892 ft LAT/LONG 37° 18 ' N / 81° 13 ' W

| | JAN | FEB | MAR | APR | MAY | JUN | JUL | AUG | SEP | OCT | NOV | DEC | YEAR |
|---|---|---|---|---|---|---|---|---|---|---|---|---|---|
| Maximum Temp °F | 39.1 | 42.6 | 52.5 | 62.9 | 69.9 | 75.9 | 79.0 | 77.8 | 72.0 | 62.7 | 53.2 | 44.1 | 61.0 |
| Minimum Temp °F | 23.3 | 25.4 | 34.2 | 43.3 | 51.3 | 58.6 | 62.5 | 61.4 | 55.7 | 44.9 | 36.8 | 28.5 | 43.8 |
| Mean Temp °F | 31.2 | 33.9 | 43.4 | 53.1 | 60.6 | 67.3 | 70.8 | 69.6 | 63.9 | 53.8 | 45.0 | 36.3 | 52.4 |
| Days Max Temp ≥ 90 °F | 0 | 0 | 0 | 0 | 0 | 0 | 1 | 1 | 0 | 0 | 0 | 0 | 2 |
| Days Max Temp ≤ 32 °F | 10 | 7 | 3 | 0 | 0 | 0 | 0 | 0 | 0 | 0 | 2 | 6 | 28 |
| Days Min Temp ≤ 32 °F | 24 | 20 | 14 | 5 | 0 | 0 | 0 | 0 | 0 | 3 | 11 | 20 | 97 |
| Days Min Temp ≤ 0 °F | 1 | 1 | 0 | 0 | 0 | 0 | 0 | 0 | 0 | 0 | 0 | 0 | 2 |
| Heating Degree Days | 1040 | 872 | 664 | 365 | 168 | 38 | 7 | 13 | 97 | 347 | 595 | 884 | 5090 |
| Cooling Degree Days | 0 | 0 | 3 | 17 | 39 | 119 | 216 | 175 | 75 | 11 | 1 | 0 | 656 |
| Total Precipitation (") | 2.61 | 2.77 | 3.45 | 3.51 | 4.00 | 3.45 | 4.08 | 3.30 | 3.17 | 2.75 | 2.59 | 2.88 | 38.56 |
| Days ≥ 0.1" Precip | 7 | 7 | 8 | 8 | 9 | 7 | 8 | 7 | 6 | 5 | 6 | 7 | 85 |
| Total Snowfall (") | 8.9 | 8.2 | 3.6 | 1.7 | 0.0 | 0.0 | 0.0 | 0.0 | 0.0 | 0.2 | 2.1 | 5.9 | 30.6 |
| Days ≥ 1" Snow Depth | 11 | 10 | 3 | 0 | 0 | 0 | 0 | 0 | 0 | 0 | 1 | 6 | 31 |

## BLUESTONE LAKE *Summers County* ELEVATION 1401 ft LAT/LONG 37° 38 ' N / 80° 53 ' W

| | JAN | FEB | MAR | APR | MAY | JUN | JUL | AUG | SEP | OCT | NOV | DEC | YEAR |
|---|---|---|---|---|---|---|---|---|---|---|---|---|---|
| Maximum Temp °F | 40.0 | 44.4 | 54.6 | 65.7 | 73.9 | 80.6 | 84.4 | 82.9 | 76.7 | 66.2 | 55.3 | 44.3 | 64.1 |
| Minimum Temp °F | 21.6 | 23.2 | 31.0 | 39.2 | 48.1 | 57.0 | 62.1 | 61.6 | 55.3 | 42.3 | 33.4 | 25.8 | 41.7 |
| Mean Temp °F | 30.8 | 33.9 | 42.8 | 52.5 | 61.0 | 68.8 | 73.3 | 72.3 | 66.0 | 54.3 | 44.4 | 35.1 | 52.9 |
| Days Max Temp ≥ 90 °F | 0 | 0 | 0 | 0 | 0 | 1 | 5 | 3 | 0 | 0 | 0 | 0 | 9 |
| Days Max Temp ≤ 32 °F | 7 | 4 | 1 | 0 | 0 | 0 | 0 | 0 | 0 | 0 | 0 | 4 | 16 |
| Days Min Temp ≤ 32 °F | 26 | 24 | 19 | 7 | 1 | 0 | 0 | 0 | 0 | 4 | 15 | 23 | 119 |
| Days Min Temp ≤ 0 °F | 2 | 0 | 0 | 0 | 0 | 0 | 0 | 0 | 0 | 0 | 0 | 0 | 2 |
| Heating Degree Days | 1053 | 874 | 681 | 374 | 157 | 27 | 2 | 4 | 65 | 334 | 613 | 921 | 5105 |
| Cooling Degree Days | 0 | 0 | 0 | 7 | 44 | 157 | 292 | 236 | 102 | 9 | 0 | 0 | 847 |
| Total Precipitation (") | 2.52 | 2.54 | 3.32 | 3.29 | 3.59 | 3.11 | 4.34 | 3.37 | 2.84 | 2.81 | 2.43 | 2.63 | 36.79 |
| Days ≥ 0.1" Precip | 6 | 6 | 7 | 8 | 8 | 7 | 8 | 7 | 6 | 6 | 6 | 6 | 81 |
| Total Snowfall (") | 7.3 | 6.2 | 3.0 | 0.4 | 0.0 | 0.0 | 0.0 | 0.0 | 0.0 | 0.0 | 0.8 | 4.0 | 21.7 |
| Days ≥ 1" Snow Depth | 8 | 6 | 2 | 0 | 0 | 0 | 0 | 0 | 0 | 0 | 0 | 4 | 20 |

## BUCKEYE 1 SE *Pocahontas County* ELEVATION 2103 ft LAT/LONG 38° 11 ' N / 80° 8 ' W

| | JAN | FEB | MAR | APR | MAY | JUN | JUL | AUG | SEP | OCT | NOV | DEC | YEAR |
|---|---|---|---|---|---|---|---|---|---|---|---|---|---|
| Maximum Temp °F | 37.3 | 42.2 | 53.1 | 63.7 | 72.2 | 79.2 | 82.3 | 80.9 | 75.1 | 64.7 | 52.5 | 41.8 | 62.1 |
| Minimum Temp °F | 15.3 | 18.0 | 25.8 | 34.1 | 43.7 | 52.4 | 57.5 | 56.3 | 50.0 | 36.8 | 28.1 | 20.5 | 36.5 |
| Mean Temp °F | 26.3 | 30.2 | 39.5 | 48.9 | 58.0 | 65.8 | 70.0 | 68.6 | 62.5 | 50.8 | 40.3 | 31.1 | 49.3 |
| Days Max Temp ≥ 90 °F | 0 | 0 | 0 | 0 | 0 | 1 | 2 | 1 | 0 | 0 | 0 | 0 | 4 |
| Days Max Temp ≤ 32 °F | 10 | 5 | 1 | 0 | 0 | 0 | 0 | 0 | 0 | 0 | 1 | 6 | 23 |
| Days Min Temp ≤ 32 °F | 29 | 25 | 24 | 13 | 4 | 0 | 0 | 0 | 1 | 11 | 20 | 27 | 154 |
| Days Min Temp ≤ 0 °F | 4 | 2 | 0 | 0 | 0 | 0 | 0 | 0 | 0 | 0 | 0 | 2 | 8 |
| Heating Degree Days | 1194 | 978 | 785 | 476 | 229 | 55 | 10 | 19 | 117 | 437 | 735 | 1042 | 6077 |
| Cooling Degree Days | 0 | 0 | 0 | 1 | 18 | 87 | 183 | 141 | 51 | 4 | 0 | 0 | 485 |
| Total Precipitation (") | 3.52 | 3.21 | 4.33 | 3.75 | 4.39 | 3.40 | 4.56 | 3.98 | 3.38 | 3.55 | 3.43 | 3.95 | 45.45 |
| Days ≥ 0.1" Precip | 8 | 7 | 9 | 8 | 8 | 7 | 9 | 7 | 6 | 6 | 7 | 8 | 90 |
| Total Snowfall (") | 11.8 | 9.3 | 7.1 | 2.0 | 0.0 | 0.0 | 0.0 | 0.0 | 0.0 | 0.2 | 2.4 | 7.8 | 40.6 |
| Days ≥ 1" Snow Depth | 15 | 14 | 5 | 1 | 0 | 0 | 0 | 0 | 0 | 0 | 2 | 8 | 45 |

## BUCKHANNON *Upshur County* ELEVATION 1450 ft LAT/LONG 39° 0 ' N / 80° 16 ' W

| | JAN | FEB | MAR | APR | MAY | JUN | JUL | AUG | SEP | OCT | NOV | DEC | YEAR |
|---|---|---|---|---|---|---|---|---|---|---|---|---|---|
| Maximum Temp °F | 39.2 | 43.5 | 54.2 | 65.1 | 74.0 | 80.8 | 83.6 | 82.3 | 76.7 | 66.2 | 54.9 | 45.5 | 63.8 |
| Minimum Temp °F | 19.1 | 20.8 | 29.4 | 37.5 | 46.2 | 54.4 | 59.8 | 57.8 | 51.7 | 39.5 | 32.3 | 25.7 | 39.5 |
| Mean Temp °F | 29.2 | 32.2 | 41.8 | 51.3 | 60.1 | 67.6 | 71.7 | 70.1 | 64.3 | 52.9 | 43.6 | 35.6 | 51.7 |
| Days Max Temp ≥ 90 °F | 0 | 0 | 0 | 0 | 0 | 1 | 3 | 2 | 1 | 0 | 0 | 0 | 7 |
| Days Max Temp ≤ 32 °F | 10 | 6 | 2 | 0 | 0 | 0 | 0 | 0 | 0 | 0 | 1 | 5 | 24 |
| Days Min Temp ≤ 32 °F | 27 | 23 | 19 | 11 | 2 | 0 | 0 | 0 | 0 | 9 | 16 | 23 | 130 |
| Days Min Temp ≤ 0 °F | 3 | 2 | 0 | 0 | 0 | 0 | 0 | 0 | 0 | 0 | 0 | 1 | 6 |
| Heating Degree Days | 1103 | 922 | 711 | 410 | 178 | 38 | 6 | 12 | 93 | 373 | 634 | 904 | 5384 |
| Cooling Degree Days | 0 | 0 | 0 | 6 | 37 | *127* | *241* | *176* | *77* | 5 | 0 | 0 | 669 |
| Total Precipitation (") | 3.59 | 3.18 | 4.26 | 4.06 | 4.14 | 4.22 | 5.04 | 4.42 | 3.87 | 3.15 | 3.73 | 4.25 | 47.91 |
| Days ≥ 0.1" Precip | 9 | 8 | 9 | 9 | 9 | 8 | 9 | 8 | 7 | 7 | 8 | 10 | 101 |
| Total Snowfall (") | 18.1 | 14.5 | 8.1 | 1.3 | 0.0 | 0.0 | 0.0 | 0.0 | 0.0 | 0.1 | 3.6 | 9.6 | 55.3 |
| Days ≥ 1" Snow Depth | 15 | 11 | 4 | 0 | 0 | 0 | 0 | 0 | 0 | 0 | 1 | 6 | 37 |

## CANAAN VALLEY *Tucker County* ELEVATION 3250 ft LAT/LONG 39° 3 ' N / 79° 26 ' W

| | JAN | FEB | MAR | APR | MAY | JUN | JUL | AUG | SEP | OCT | NOV | DEC | YEAR |
|---|---|---|---|---|---|---|---|---|---|---|---|---|---|
| Maximum Temp °F | 34.9 | 37.7 | 47.3 | 58.5 | 67.5 | 74.4 | 77.6 | 76.4 | 70.4 | 60.1 | 49.2 | 39.6 | 57.8 |
| Minimum Temp °F | 15.6 | 17.5 | 25.0 | 33.7 | 42.2 | 49.1 | 53.9 | 52.1 | 46.2 | 36.0 | 28.7 | 20.6 | 35.1 |
| Mean Temp °F | 25.3 | 27.6 | 36.2 | 46.1 | 54.9 | 61.8 | 65.8 | 64.2 | 58.3 | 48.1 | 39.0 | 30.1 | 46.5 |
| Days Max Temp ≥ 90 °F | 0 | 0 | 0 | 0 | 0 | 0 | 0 | 0 | 0 | 0 | 0 | 0 | 0 |
| Days Max Temp ≤ 32 °F | 13 | 9 | 5 | 0 | 0 | 0 | 0 | 0 | 0 | 0 | 3 | 9 | 39 |
| Days Min Temp ≤ 32 °F | 28 | 25 | 23 | 15 | 5 | 1 | 0 | 0 | 3 | 12 | 20 | 26 | 158 |
| Days Min Temp ≤ 0 °F | 5 | 3 | 1 | 0 | 0 | 0 | 0 | 0 | 0 | 0 | 0 | 2 | 11 |
| Heating Degree Days | 1226 | 1048 | 887 | 561 | 313 | 122 | 49 | 72 | 211 | 517 | 774 | 1074 | 6854 |
| Cooling Degree Days | 0 | 0 | 0 | 1 | 7 | 37 | 94 | 60 | 18 | 1 | 0 | 0 | 218 |
| Total Precipitation (") | 4.17 | 3.70 | 4.64 | 4.22 | 4.95 | 4.63 | 4.96 | 4.57 | 3.86 | 3.56 | 4.09 | 4.20 | 51.55 |
| Days ≥ 0.1" Precip | 11 | 10 | 11 | 11 | 11 | 10 | 10 | 9 | 8 | 8 | 10 | 11 | 120 |
| Total Snowfall (") | 32.6 | 27.9 | 23.4 | 9.4 | 0.2 | 0.0 | 0.0 | 0.0 | 0.0 | 2.4 | 11.7 | 24.0 | 131.6 |
| Days ≥ 1" Snow Depth | 22 | 20 | *11* | 3 | 0 | 0 | 0 | 0 | 0 | 1 | 6 | *16* | 79 |

## CHARLESTON KNWA AP *Kanawha County* ELEVATION 951 ft LAT/LONG 38° 22 ' N / 81° 36 ' W

| | JAN | FEB | MAR | APR | MAY | JUN | JUL | AUG | SEP | OCT | NOV | DEC | YEAR |
|---|---|---|---|---|---|---|---|---|---|---|---|---|---|
| Maximum Temp °F | 41.6 | 45.8 | 56.7 | 67.7 | 75.6 | 82.9 | 85.9 | 84.2 | 77.9 | 67.2 | 56.8 | 46.8 | 65.8 |
| Minimum Temp °F | 23.7 | 26.0 | 34.7 | 43.2 | 51.7 | 59.7 | 64.7 | 63.3 | 56.6 | 44.3 | 36.5 | 28.8 | 44.4 |
| Mean Temp °F | 32.6 | 35.9 | 45.7 | 55.5 | 63.6 | 71.3 | 75.3 | 73.8 | 67.3 | 55.8 | 46.7 | 37.8 | 55.1 |
| Days Max Temp ≥ 90 °F | 0 | 0 | 0 | 0 | 1 | 5 | 8 | 5 | 1 | 0 | 0 | 0 | 20 |
| Days Max Temp ≤ 32 °F | 8 | 5 | 1 | 0 | 0 | 0 | 0 | 0 | 0 | 0 | 0 | 4 | 18 |
| Days Min Temp ≤ 32 °F | 23 | 21 | 14 | 5 | 0 | 0 | 0 | 0 | 0 | 4 | 12 | 20 | 99 |
| Days Min Temp ≤ 0 °F | 1 | 0 | 0 | 0 | 0 | 0 | 0 | 0 | 0 | 0 | 0 | 0 | 1 |
| Heating Degree Days | 996 | 815 | 596 | 306 | 118 | 16 | 1 | 3 | 56 | 297 | 546 | 837 | 4587 |
| Cooling Degree Days | 0 | 0 | 6 | 29 | 89 | 224 | 365 | 300 | 139 | 24 | 4 | 1 | 1181 |
| Total Precipitation (") | 2.97 | 3.01 | 3.85 | 3.32 | 3.98 | 3.54 | 4.57 | 4.25 | 3.43 | 2.80 | 3.55 | 3.47 | 42.74 |
| Days ≥ 0.1" Precip | 8 | 8 | 9 | 8 | 9 | 7 | 8 | 7 | 7 | 6 | 8 | 8 | 93 |
| Total Snowfall (") | 12.3 | 9.6 | 5.5 | 1.1 | 0.0 | 0.0 | 0.0 | 0.0 | 0.0 | 0.1 | 1.7 | 5.4 | 35.7 |
| Days ≥ 1" Snow Depth | 11 | 8 | 3 | 0 | 0 | 0 | 0 | 0 | 0 | 0 | 1 | 4 | 27 |

## CLARKSBURG 1 *Harrison County* ELEVATION 981 ft LAT/LONG 39° 16 ' N / 80° 21 ' W

| | JAN | FEB | MAR | APR | MAY | JUN | JUL | AUG | SEP | OCT | NOV | DEC | YEAR |
|---|---|---|---|---|---|---|---|---|---|---|---|---|---|
| Maximum Temp °F | 38.6 | 42.4 | 53.4 | 64.6 | 73.8 | 81.8 | 84.7 | 82.9 | 76.7 | 65.4 | 54.2 | 43.7 | 63.5 |
| Minimum Temp °F | 19.9 | 21.2 | 29.2 | 37.7 | 47.1 | 56.2 | 61.4 | 60.1 | 53.5 | 40.3 | 32.7 | 25.6 | 40.4 |
| Mean Temp °F | 29.3 | 31.8 | 41.3 | 51.2 | 60.5 | 69.0 | 73.0 | 71.5 | 65.1 | 52.9 | 43.5 | 34.7 | 52.0 |
| Days Max Temp ≥ 90 °F | 0 | 0 | 0 | 0 | 0 | 3 | 6 | 3 | 1 | 0 | 0 | 0 | 13 |
| Days Max Temp ≤ 32 °F | 10 | 6 | 1 | 0 | 0 | 0 | 0 | 0 | 0 | 0 | 1 | 5 | 23 |
| Days Min Temp ≤ 32 °F | 27 | 23 | 21 | 9 | 1 | 0 | 0 | 0 | 0 | 7 | 16 | 23 | 127 |
| Days Min Temp ≤ 0 °F | 2 | 1 | 0 | 0 | 0 | 0 | 0 | 0 | 0 | 0 | 0 | 0 | 3 |
| Heating Degree Days | 1100 | 931 | 727 | 415 | 178 | 31 | 3 | 8 | 84 | 375 | 639 | 932 | 5423 |
| Cooling Degree Days | 0 | 0 | 1 | 9 | 46 | 162 | 284 | 221 | 94 | 7 | 0 | 0 | 824 |
| Total Precipitation (") | 3.11 | 2.77 | 3.93 | 3.57 | 4.09 | 3.86 | 4.09 | 4.44 | 3.21 | 2.95 | 3.57 | 3.49 | 43.08 |
| Days ≥ 0.1" Precip | 8 | 8 | 9 | 9 | 9 | 8 | 8 | 8 | 7 | 7 | 8 | 8 | 97 |
| Total Snowfall (") | *12.0* | *7.9* | 3.4 | 0.3 | 0.0 | 0.0 | 0.0 | 0.0 | 0.0 | 0.0 | *0.9* | *4.2* | 28.7 |
| Days ≥ 1" Snow Depth | *14* | 9 | 3 | 0 | 0 | 0 | 0 | 0 | 0 | 0 | 1 | 5 | 32 |

## CORTON *Kanawha County* ELEVATION 640 ft LAT/LONG 38° 29 ' N / 81° 16 ' W

| | JAN | FEB | MAR | APR | MAY | JUN | JUL | AUG | SEP | OCT | NOV | DEC | YEAR |
|---|---|---|---|---|---|---|---|---|---|---|---|---|---|
| Maximum Temp °F | 43.0 | ***47.8*** | 58.9 | 69.8 | 77.5 | 83.2 | 86.4 | 84.7 | ***79.1*** | ***69.1*** | 58.4 | 48.7 | 67.2 |
| Minimum Temp °F | 22.7 | 23.8 | 31.5 | 39.7 | 49.0 | 57.4 | 62.6 | 61.6 | 55.4 | ***43.8*** | 35.0 | 27.6 | 42.5 |
| Mean Temp °F | 32.9 | ***35.9*** | 45.2 | 54.8 | 63.4 | 70.4 | 74.6 | 73.2 | ***67.4*** | ***56.5*** | 46.7 | ***38.3*** | 54.9 |
| Days Max Temp ≥ 90 °F | 0 | 0 | 0 | 0 | 1 | 3 | 8 | 6 | 2 | 0 | 0 | 0 | 20 |
| Days Max Temp ≤ 32 °F | 6 | 3 | 0 | 0 | 0 | 0 | 0 | 0 | 0 | 0 | 0 | 2 | 11 |
| Days Min Temp ≤ 32 °F | 24 | 21 | 17 | 8 | 1 | 0 | 0 | 0 | 0 | 4 | 12 | 20 | 107 |
| Days Min Temp ≤ 0 °F | 1 | 1 | 0 | 0 | 0 | 0 | 0 | 0 | 0 | 0 | 0 | 0 | 2 |
| Heating Degree Days | 989 | ***815*** | 610 | 313 | 113 | 18 | 1 | 4 | ***50*** | ***274*** | 543 | ***820*** | 4550 |
| Cooling Degree Days | 0 | 0 | 2 | 17 | 83 | 205 | 343 | 278 | ***143*** | na | 0 | ***0*** | na |
| Total Precipitation (") | 3.12 | 2.66 | 4.07 | 3.60 | 3.88 | 3.89 | 4.61 | 4.52 | 3.75 | 2.94 | 3.47 | 3.68 | 44.19 |
| Days ≥ 0.1" Precip | na | ***7*** | 9 | 8 | 9 | 8 | 8 | 7 | 7 | 6 | 7 | 8 | na |
| Total Snowfall (") | na | na | na | 0.0 | 0.0 | 0.0 | 0.0 | 0.0 | 0.0 | 0.0 | 0.3 | na | na |
| Days ≥ 1" Snow Depth | na | na | na | 0 | 0 | 0 | 0 | 0 | 0 | 0 | 0 | na | na |

## CRESTON *Wirt County* ELEVATION 640 ft LAT/LONG 38° 57 ' N / 81° 16 ' W

| | JAN | FEB | MAR | APR | MAY | JUN | JUL | AUG | SEP | OCT | NOV | DEC | YEAR |
|---|---|---|---|---|---|---|---|---|---|---|---|---|---|
| Maximum Temp °F | 40.0 | 44.4 | 55.5 | 66.8 | 76.1 | 83.2 | 86.4 | 84.9 | 78.9 | 67.9 | 56.6 | 45.1 | 65.5 |
| Minimum Temp °F | 17.8 | 19.5 | 27.9 | 36.2 | 46.1 | 54.9 | 61.0 | 59.5 | 52.6 | 38.7 | 31.1 | 23.4 | 39.1 |
| Mean Temp °F | 28.9 | 32.0 | 41.7 | 51.5 | 61.1 | 69.1 | 73.7 | 72.2 | 65.8 | 53.3 | 43.9 | 34.3 | 52.3 |
| Days Max Temp ≥ 90 °F | 0 | 0 | 0 | 0 | 2 | 6 | 10 | 7 | 2 | 0 | 0 | 0 | 27 |
| Days Max Temp ≤ 32 °F | 9 | 6 | 1 | 0 | 0 | 0 | 0 | 0 | 0 | 0 | 0 | 4 | 20 |
| Days Min Temp ≤ 32 °F | 28 | 24 | 23 | 12 | 2 | 0 | 0 | 0 | 0 | 9 | 18 | 25 | 141 |
| Days Min Temp ≤ 0 °F | 4 | 3 | 0 | 0 | 0 | 0 | 0 | 0 | 0 | 0 | 0 | 1 | 8 |
| Heating Degree Days | 1111 | 926 | 716 | 407 | 169 | 33 | 4 | 8 | 78 | 366 | 629 | 947 | 5394 |
| Cooling Degree Days | 0 | 0 | 1 | 12 | 65 | 188 | 331 | 267 | 124 | 16 | 1 | 0 | 1005 |
| Total Precipitation (") | 3.31 | 3.03 | 3.91 | 3.57 | 4.16 | 3.73 | 4.64 | 4.22 | 3.45 | 3.25 | 3.55 | 3.76 | 44.58 |
| Days ≥ 0.1" Precip | 8 | 8 | 9 | 8 | 8 | 7 | 8 | 7 | 6 | 7 | 8 | ***8*** | 92 |
| Total Snowfall (") | ***8.6*** | ***5.4*** | 3.2 | 0.4 | 0.0 | 0.0 | 0.0 | 0.0 | 0.0 | 0.0 | 0.5 | na | na |
| Days ≥ 1" Snow Depth | ***11*** | ***7*** | ***1*** | 0 | 0 | 0 | 0 | 0 | 0 | 0 | ***0*** | ***3*** | 22 |

## ELKINS RNDLPH CO AP *Randolph County* ELEVATION 1948 ft LAT/LONG 38° 53 ' N / 79° 51 ' W

| | JAN | FEB | MAR | APR | MAY | JUN | JUL | AUG | SEP | OCT | NOV | DEC | YEAR |
|---|---|---|---|---|---|---|---|---|---|---|---|---|---|
| Maximum Temp °F | 38.1 | 41.7 | 51.8 | 62.2 | 70.9 | 77.9 | 81.0 | 79.6 | 73.6 | 63.5 | 52.9 | 43.3 | 61.4 |
| Minimum Temp °F | 16.9 | 18.6 | 27.1 | 35.1 | 44.1 | 52.2 | 57.6 | 56.5 | 50.2 | 37.0 | 29.7 | 22.0 | 37.3 |
| Mean Temp °F | 27.5 | 30.2 | 39.5 | 48.7 | 57.5 | 65.0 | 69.3 | 68.1 | 61.9 | 50.3 | 41.3 | 32.7 | 49.3 |
| Days Max Temp ≥ 90 °F | 0 | 0 | 0 | 0 | 0 | 0 | 1 | 1 | 0 | 0 | 0 | 0 | 2 |
| Days Max Temp ≤ 32 °F | 10 | 7 | 2 | 0 | 0 | 0 | 0 | 0 | 0 | 0 | 2 | 6 | 27 |
| Days Min Temp ≤ 32 °F | 27 | 24 | 22 | 13 | 3 | 0 | 0 | 0 | 1 | 11 | 19 | 25 | 145 |
| Days Min Temp ≤ 0 °F | 4 | 3 | 0 | 0 | 0 | 0 | 0 | 0 | 0 | 0 | 0 | 2 | 9 |
| Heating Degree Days | 1155 | 976 | 785 | 483 | 242 | 69 | 15 | 24 | 132 | 451 | 704 | 996 | 6032 |
| Cooling Degree Days | 0 | 0 | 0 | 2 | 19 | 79 | 178 | 135 | 47 | 3 | 0 | 0 | 463 |
| Total Precipitation (") | 3.17 | 3.04 | 3.86 | 3.71 | 4.21 | 4.21 | 4.79 | 4.38 | 3.70 | 2.94 | 3.34 | 3.64 | 44.99 |
| Days ≥ 0.1" Precip | 9 | 8 | 10 | 9 | 9 | 9 | 9 | 9 | 8 | 7 | 8 | 9 | 104 |
| Total Snowfall (") | 21.8 | 17.4 | 12.4 | 4.6 | 0.0 | 0.0 | 0.0 | 0.0 | 0.0 | 0.6 | 5.9 | 14.9 | 77.6 |
| Days ≥ 1" Snow Depth | 15 | 13 | 6 | 1 | 0 | 0 | 0 | 0 | 0 | 0 | 3 | 10 | 48 |

## FAIRMONT *Marion County* ELEVATION 1020 ft LAT/LONG 39° 28 ' N / 80° 9 ' W

| | JAN | FEB | MAR | APR | MAY | JUN | JUL | AUG | SEP | OCT | NOV | DEC | YEAR |
|---|---|---|---|---|---|---|---|---|---|---|---|---|---|
| Maximum Temp °F | 37.6 | 41.7 | 52.9 | 64.2 | 72.9 | 80.5 | 83.7 | 82.2 | 75.8 | 64.6 | 53.3 | 42.6 | 62.7 |
| Minimum Temp °F | 20.1 | 22.0 | 30.7 | 39.6 | 48.6 | 56.6 | 61.4 | 59.9 | 53.5 | 41.6 | 34.1 | 25.7 | 41.2 |
| Mean Temp °F | 28.9 | 31.9 | 41.8 | 51.9 | 60.8 | 68.5 | 72.6 | 71.1 | 64.7 | 53.1 | 43.7 | 34.2 | 51.9 |
| Days Max Temp ≥ 90 °F | 0 | 0 | 0 | 0 | 0 | 2 | 5 | 3 | 1 | 0 | 0 | 0 | 11 |
| Days Max Temp ≤ 32 °F | 11 | 7 | 2 | 0 | 0 | 0 | 0 | 0 | 0 | 0 | 1 | 7 | 28 |
| Days Min Temp ≤ 32 °F | 26 | 23 | 19 | 8 | 1 | 0 | 0 | 0 | 0 | 5 | 15 | 24 | 121 |
| Days Min Temp ≤ 0 °F | 2 | 1 | 0 | 0 | 0 | 0 | 0 | 0 | 0 | 0 | 0 | 1 | 4 |
| Heating Degree Days | 1113 | 929 | 714 | 397 | 173 | 35 | 4 | 10 | 90 | 369 | 633 | 948 | 5415 |
| Cooling Degree Days | 0 | 0 | 3 | 12 | 47 | 138 | 255 | 199 | 81 | 7 | 1 | 0 | 743 |
| Total Precipitation (") | 3.07 | 2.71 | 4.02 | 3.70 | 4.56 | 3.74 | 4.83 | 4.07 | 3.36 | 3.00 | 3.55 | 3.43 | 44.04 |
| Days ≥ 0.1" Precip | 8 | 7 | 8 | 9 | 9 | 8 | 9 | 8 | 7 | 7 | 8 | 9 | 97 |
| Total Snowfall (") | 13.7 | 9.6 | 5.7 | 1.3 | 0.0 | 0.0 | 0.0 | 0.0 | 0.0 | 0.0 | 2.2 | 6.1 | 38.6 |
| Days ≥ 1" Snow Depth | 14 | 10 | 4 | 0 | 0 | 0 | 0 | 0 | 0 | 0 | 1 | 6 | 35 |

## FLAT TOP *Summers County* ELEVATION 3235 ft LAT/LONG 37° 35 ' N / 81° 7 ' W

| | JAN | FEB | MAR | APR | MAY | JUN | JUL | AUG | SEP | OCT | NOV | DEC | YEAR |
|---|---|---|---|---|---|---|---|---|---|---|---|---|---|
| Maximum Temp °F | 34.7 | 38.0 | 47.3 | 58.3 | 66.0 | 72.8 | 76.4 | 75.3 | 69.4 | 59.3 | 49.2 | 39.3 | 57.2 |
| Minimum Temp °F | 18.2 | 20.3 | 28.4 | 37.8 | 47.0 | 55.3 | 59.7 | 58.4 | 52.2 | 40.7 | 32.1 | 23.2 | 39.4 |
| Mean Temp °F | 26.5 | 29.2 | 38.0 | 48.1 | 56.5 | 64.1 | 68.1 | 66.9 | 60.8 | 50.0 | 40.7 | 31.3 | 48.3 |
| Days Max Temp ≥ 90 °F | 0 | 0 | 0 | 0 | 0 | 0 | 0 | 0 | 0 | 0 | 0 | 0 | 0 |
| Days Max Temp ≤ 32 °F | 14 | 10 | 5 | 1 | 0 | 0 | 0 | 0 | 0 | 0 | 3 | 9 | 42 |
| Days Min Temp ≤ 32 °F | 27 | 24 | 21 | 10 | 2 | 0 | 0 | 0 | 0 | 7 | 16 | 25 | 132 |
| Days Min Temp ≤ 0 °F | 3 | 2 | 0 | 0 | 0 | 0 | 0 | 0 | 0 | 0 | 0 | 1 | 6 |
| Heating Degree Days | 1188 | 1006 | 832 | 504 | 267 | 79 | 23 | 33 | 152 | 459 | 723 | 1038 | 6304 |
| Cooling Degree Days | 0 | 0 | 0 | 4 | 12 | 69 | 151 | 109 | 36 | 2 | 0 | 0 | 383 |
| Total Precipitation (") | 3.44 | 3.27 | 4.06 | 4.10 | 4.51 | 3.68 | 4.87 | 3.82 | 3.47 | 3.45 | 3.11 | 3.44 | 45.22 |
| Days ≥ 0.1" Precip | 8 | 8 | 9 | 9 | 10 | 8 | 9 | 7 | 7 | 7 | 7 | 8 | 97 |
| Total Snowfall (") | 16.3 | *16.9* | *9.5* | 2.8 | 0.0 | 0.0 | 0.0 | 0.0 | 0.0 | 0.7 | 3.3 | 10.5 | 60.0 |
| Days ≥ 1" Snow Depth | 16 | 15 | 8 | 2 | 0 | 0 | 0 | 0 | 0 | 0 | 3 | 11 | 55 |

## FRANKLIN 2 NE *Pendleton County* ELEVATION 1703 ft LAT/LONG 38° 42 ' N / 79° 20 ' W

| | JAN | FEB | MAR | APR | MAY | JUN | JUL | AUG | SEP | OCT | NOV | DEC | YEAR |
|---|---|---|---|---|---|---|---|---|---|---|---|---|---|
| Maximum Temp °F | 42.1 | 45.2 | 54.8 | 65.0 | 73.4 | 80.1 | 83.5 | 82.1 | 76.0 | 66.4 | 55.9 | 46.1 | 64.2 |
| Minimum Temp °F | 19.3 | 21.4 | 29.0 | 37.1 | 45.9 | 53.1 | 57.9 | 56.3 | 49.9 | 38.5 | 31.1 | 24.1 | 38.6 |
| Mean Temp °F | 30.7 | 33.4 | 41.9 | 51.1 | 59.7 | 66.6 | 70.7 | 69.2 | 62.9 | 52.5 | 43.5 | 35.1 | 51.4 |
| Days Max Temp ≥ 90 °F | 0 | 0 | 0 | 0 | 0 | 1 | 4 | 3 | 1 | 0 | 0 | 0 | 9 |
| Days Max Temp ≤ 32 °F | 6 | 4 | 1 | 0 | 0 | 0 | 0 | 0 | 0 | 0 | 0 | 3 | 14 |
| Days Min Temp ≤ 32 °F | 28 | 24 | 21 | 10 | 2 | 0 | 0 | 0 | 1 | 9 | 17 | 25 | 137 |
| Days Min Temp ≤ 0 °F | 2 | 1 | 0 | 0 | 0 | 0 | 0 | 0 | 0 | 0 | 0 | 1 | 4 |
| Heating Degree Days | 1056 | 887 | 708 | 414 | 184 | 46 | 8 | 18 | 114 | 383 | 638 | 919 | 5375 |
| Cooling Degree Days | 0 | 0 | 0 | 3 | 26 | 107 | 211 | 155 | 58 | 4 | 0 | 0 | 564 |
| Total Precipitation (") | 1.97 | 1.91 | 2.89 | 2.76 | 3.46 | 2.73 | 3.86 | 3.34 | 2.85 | 2.93 | 2.62 | 2.25 | 33.57 |
| Days ≥ 0.1" Precip | 5 | 4 | 6 | 6 | 7 | 6 | 8 | 7 | 5 | 5 | 5 | 5 | 69 |
| Total Snowfall (") | 7.9 | 7.0 | 4.1 | 1.1 | 0.0 | 0.0 | 0.0 | 0.0 | 0.0 | 0.3 | 2.1 | 5.9 | 28.4 |
| Days ≥ 1" Snow Depth | na | na | na | 0 | 0 | 0 | 0 | 0 | 0 | 0 | 0 | na | na |

## GASSAWAY *Braxton County* ELEVATION 840 ft LAT/LONG 38° 41 ' N / 80° 46 ' W

| | JAN | FEB | MAR | APR | MAY | JUN | JUL | AUG | SEP | OCT | NOV | DEC | YEAR |
|---|---|---|---|---|---|---|---|---|---|---|---|---|---|
| Maximum Temp °F | 43.2 | 47.6 | 59.1 | 69.8 | 77.6 | 83.4 | 86.3 | 84.7 | 78.9 | 69.3 | 58.7 | 47.8 | 67.2 |
| Minimum Temp °F | 21.8 | 23.0 | 31.3 | 39.2 | 48.3 | 56.8 | 62.1 | 60.8 | 54.5 | 42.0 | 33.7 | 26.4 | 41.7 |
| Mean Temp °F | 32.5 | 35.3 | 45.1 | 54.5 | 63.0 | 70.1 | 74.2 | 72.8 | 66.7 | 55.7 | 46.2 | 37.1 | 54.4 |
| Days Max Temp ≥ 90 °F | 0 | 0 | 0 | 0 | 1 | 4 | 9 | 5 | 1 | 0 | 0 | 0 | 20 |
| Days Max Temp ≤ 32 °F | 6 | 4 | 0 | 0 | 0 | 0 | 0 | 0 | 0 | 0 | 0 | 3 | 13 |
| Days Min Temp ≤ 32 °F | 25 | 23 | 17 | 8 | 1 | 0 | 0 | 0 | 0 | 6 | 14 | 22 | 116 |
| Days Min Temp ≤ 0 °F | 2 | 1 | 0 | 0 | 0 | 0 | 0 | 0 | 0 | 0 | 0 | 0 | 3 |
| Heating Degree Days | 999 | 831 | 610 | 320 | 119 | 17 | 1 | 3 | 54 | 295 | 556 | 859 | 4664 |
| Cooling Degree Days | 0 | 0 | 2 | 15 | 71 | 183 | 323 | 262 | 115 | 17 | 0 | 0 | 988 |
| Total Precipitation (") | 3.12 | 3.18 | 4.06 | 4.04 | 4.34 | 4.18 | 5.14 | 4.54 | 3.86 | 3.16 | 3.83 | 3.89 | 47.34 |
| Days ≥ 0.1" Precip | 8 | 8 | 9 | 10 | 9 | 8 | 9 | 8 | 7 | 7 | 8 | 9 | 100 |
| Total Snowfall (") | 8.9 | *7.3* | 3.1 | 0.6 | 0.0 | 0.0 | 0.0 | 0.0 | 0.0 | 0.0 | 0.9 | 3.3 | 24.1 |
| Days ≥ 1" Snow Depth | 9 | 7 | 1 | 0 | 0 | 0 | 0 | 0 | 0 | 0 | 0 | *4* | 21 |

## GLENVILLE 1 ENE *Gilmer County* ELEVATION 781 ft LAT/LONG 38° 56 ' N / 80° 50 ' W

| | JAN | FEB | MAR | APR | MAY | JUN | JUL | AUG | SEP | OCT | NOV | DEC | YEAR |
|---|---|---|---|---|---|---|---|---|---|---|---|---|---|
| Maximum Temp °F | 40.4 | 44.5 | 55.1 | 66.0 | 75.0 | 82.2 | 85.5 | 84.1 | 78.2 | 67.3 | 56.2 | 45.4 | 65.0 |
| Minimum Temp °F | 18.4 | 20.1 | 28.3 | 37.2 | 46.6 | 55.7 | 61.4 | 60.0 | 52.8 | 38.9 | 30.9 | 24.2 | 39.5 |
| Mean Temp °F | 29.4 | 32.3 | 41.7 | 51.7 | 60.8 | 69.0 | 73.5 | 72.1 | 65.5 | 53.1 | 43.5 | 34.8 | 52.3 |
| Days Max Temp ≥ 90 °F | 0 | 0 | 0 | 0 | 1 | 3 | 7 | 5 | 1 | 0 | 0 | 0 | 17 |
| Days Max Temp ≤ 32 °F | 8 | 5 | 1 | 0 | 0 | 0 | 0 | 0 | 0 | 0 | 0 | 4 | 18 |
| Days Min Temp ≤ 32 °F | 27 | 24 | 21 | 10 | 2 | 0 | 0 | 0 | 0 | 9 | 18 | 24 | 135 |
| Days Min Temp ≤ 0 °F | 3 | 2 | 0 | 0 | 0 | 0 | 0 | 0 | 0 | 0 | 0 | 1 | 6 |
| Heating Degree Days | 1096 | 917 | 716 | 402 | 171 | 33 | 3 | 7 | 79 | 369 | 637 | 929 | 5359 |
| Cooling Degree Days | 0 | 0 | 1 | 10 | 55 | 172 | 308 | 250 | 107 | 11 | 0 | 0 | 914 |
| Total Precipitation (") | 3.01 | 2.82 | 4.04 | 3.83 | 4.10 | 3.37 | 5.23 | 4.24 | 3.41 | 3.12 | 3.68 | 3.53 | 44.38 |
| Days ≥ 0.1" Precip | 8 | 7 | 9 | 9 | 8 | 7 | 9 | 7 | 7 | 7 | 8 | 8 | 94 |
| Total Snowfall (") | 9.1 | 6.2 | 4.1 | 0.6 | 0.0 | 0.0 | 0.0 | 0.0 | 0.0 | 0.0 | 0.9 | 3.5 | 24.4 |
| Days ≥ 1" Snow Depth | 10 | 7 | 3 | 0 | 0 | 0 | 0 | 0 | 0 | 0 | 1 | 4 | 25 |

## HAMLIN *Lincoln County* ELEVATION 640 ft LAT/LONG 38° 17' N / 82° 6' W

| | JAN | FEB | MAR | APR | MAY | JUN | JUL | AUG | SEP | OCT | NOV | DEC | YEAR |
|---|---|---|---|---|---|---|---|---|---|---|---|---|---|
| Maximum Temp °F | 41.8 | 45.8 | 56.6 | 67.2 | 75.9 | 82.9 | 86.5 | 84.9 | 78.9 | 68.2 | 57.6 | 47.3 | 66.1 |
| Minimum Temp °F | 19.6 | 21.5 | 30.1 | 38.8 | 47.9 | 56.9 | 62.2 | 60.6 | 53.0 | 39.1 | 31.5 | 24.6 | 40.5 |
| Mean Temp °F | 30.8 | 33.7 | 43.4 | 53.0 | 61.9 | 69.9 | 74.4 | 72.8 | 66.1 | 53.8 | 44.7 | 36.1 | 53.4 |
| Days Max Temp ≥ 90 °F | 0 | 0 | 0 | 0 | 1 | 5 | 10 | 7 | 2 | 0 | 0 | 0 | 25 |
| Days Max Temp ≤ 32 °F | 7 | 5 | 1 | 0 | 0 | 0 | 0 | 0 | 0 | 0 | 0 | 3 | 16 |
| Days Min Temp ≤ 32 °F | 26 | 23 | 19 | 9 | 1 | 0 | 0 | 0 | 0 | 10 | 17 | 24 | 129 |
| Days Min Temp ≤ 0 °F | 3 | 1 | 0 | 0 | 0 | 0 | 0 | 0 | 0 | 0 | 0 | 1 | 5 |
| Heating Degree Days | 1052 | 879 | 666 | 367 | 152 | 27 | 2 | 6 | 74 | 351 | 603 | 890 | 5069 |
| Cooling Degree Days | 0 | 0 | 2 | 13 | 70 | 187 | 327 | 262 | 113 | 12 | 0 | 0 | 986 |
| Total Precipitation (") | 2.95 | 3.01 | 3.73 | 3.56 | 4.25 | 3.53 | 4.76 | 4.33 | 3.36 | 2.97 | 3.54 | 3.63 | 43.62 |
| Days ≥ 0.1" Precip | 7 | 7 | 8 | 8 | 9 | 7 | 8 | 7 | 6 | 6 | 7 | 8 | 88 |
| Total Snowfall (") | 8.6 | 7.2 | 3.4 | 0.7 | 0.0 | 0.0 | 0.0 | 0.0 | 0.0 | 0.0 | 0.7 | 3.6 | 24.2 |
| Days ≥ 1" Snow Depth | 9 | 7 | 2 | 0 | 0 | 0 | 0 | 0 | 0 | 0 | 0 | 3 | 21 |

## HNTNGTN TRI-STATE *Wayne County* ELEVATION 827 ft LAT/LONG 38° 22' N / 82° 33' W

| | JAN | FEB | MAR | APR | MAY | JUN | JUL | AUG | SEP | OCT | NOV | DEC | YEAR |
|---|---|---|---|---|---|---|---|---|---|---|---|---|---|
| Maximum Temp °F | 40.5 | 45.1 | 56.3 | 67.2 | 75.1 | 82.7 | 85.9 | 84.4 | 78.0 | 67.0 | 56.1 | 45.9 | 65.4 |
| Minimum Temp °F | 23.9 | 26.5 | 35.3 | 44.1 | 52.5 | 60.8 | 65.6 | 64.2 | 57.3 | 45.0 | 37.3 | 29.3 | 45.1 |
| Mean Temp °F | 32.3 | 35.8 | 45.8 | 55.7 | 63.8 | 71.8 | 75.7 | 74.3 | 67.7 | 56.0 | 46.7 | 37.6 | 55.3 |
| Days Max Temp ≥ 90 °F | 0 | 0 | 0 | 0 | 1 | 4 | 8 | 6 | 1 | 0 | 0 | 0 | 20 |
| Days Max Temp ≤ 32 °F | 9 | 5 | 1 | 0 | 0 | 0 | 0 | 0 | 0 | 0 | 0 | 4 | 19 |
| Days Min Temp ≤ 32 °F | 23 | 20 | 14 | 3 | 0 | 0 | 0 | 0 | 0 | 3 | 11 | 19 | 93 |
| Days Min Temp ≤ 0 °F | 1 | 0 | 0 | 0 | 0 | 0 | 0 | 0 | 0 | 0 | 0 | 0 | 1 |
| Heating Degree Days | 1008 | 819 | 596 | 298 | 115 | 13 | 1 | 3 | 54 | 292 | 545 | 844 | 4588 |
| Cooling Degree Days | 0 | 0 | 6 | 28 | 91 | 231 | 365 | 311 | 145 | 23 | 4 | 1 | 1205 |
| Total Precipitation (") | 2.88 | 2.91 | 3.83 | 3.45 | 4.23 | 3.40 | 4.55 | 3.92 | 2.93 | 2.86 | 3.27 | 3.44 | 41.67 |
| Days ≥ 0.1" Precip | 7 | 7 | 8 | 8 | 8 | 7 | 8 | 6 | 6 | 6 | 7 | 7 | 85 |
| Total Snowfall (") | 8.9 | 7.0 | 4.6 | 0.7 | 0.0 | 0.0 | 0.0 | 0.0 | 0.0 | 0.1 | 1.2 | 3.6 | 26.1 |
| Days ≥ 1" Snow Depth | 9 | 7 | 2 | 0 | 0 | 0 | 0 | 0 | 0 | 0 | 0 | 3 | 21 |

## HOGSETT GALLIPOLS DM *Mason County* ELEVATION 561 ft LAT/LONG 38° 41' N / 82° 11' W

| | JAN | FEB | MAR | APR | MAY | JUN | JUL | AUG | SEP | OCT | NOV | DEC | YEAR |
|---|---|---|---|---|---|---|---|---|---|---|---|---|---|
| Maximum Temp °F | 40.0 | 43.9 | 54.8 | 66.2 | 74.9 | 82.6 | 86.4 | 84.7 | 78.7 | 67.6 | 56.5 | 45.6 | 65.2 |
| Minimum Temp °F | 19.8 | 21.7 | 30.2 | 38.8 | 48.5 | 57.2 | 62.4 | 61.1 | 54.1 | 41.8 | 33.3 | 25.4 | 41.2 |
| Mean Temp °F | 30.0 | 32.8 | 42.6 | 52.5 | 61.7 | 69.9 | 74.4 | 72.9 | 66.4 | 54.7 | 45.1 | 35.5 | 53.2 |
| Days Max Temp ≥ 90 °F | 0 | 0 | 0 | 0 | 1 | 4 | 9 | 6 | 2 | 0 | 0 | 0 | 22 |
| Days Max Temp ≤ 32 °F | 9 | 6 | 1 | 0 | 0 | 0 | 0 | 0 | 0 | 0 | 0 | 4 | 20 |
| Days Min Temp ≤ 32 °F | 27 | 23 | 20 | 8 | 1 | 0 | 0 | 0 | 0 | 5 | 15 | 24 | 123 |
| Days Min Temp ≤ 0 °F | 2 | 1 | 0 | 0 | 0 | 0 | 0 | 0 | 0 | 0 | 0 | 0 | 3 |
| Heating Degree Days | 1080 | 903 | 690 | 377 | 152 | 24 | 2 | 4 | 66 | 325 | 591 | 908 | 5122 |
| Cooling Degree Days | 0 | 0 | 1 | 12 | 61 | 185 | 329 | 268 | 117 | 15 | 1 | 0 | 989 |
| Total Precipitation (") | 2.80 | 2.73 | 3.65 | 3.41 | 3.85 | 3.12 | 4.36 | 3.89 | 3.09 | 2.87 | 3.13 | 3.17 | 40.07 |
| Days ≥ 0.1" Precip | 7 | 6 | 8 | 8 | 8 | 7 | 8 | 7 | 6 | 6 | 7 | 7 | 85 |
| Total Snowfall (") | na | na | na | 0.0 | 0.0 | 0.0 | 0.0 | 0.0 | 0.0 | 0.0 | *0.1* | na | na |
| Days ≥ 1" Snow Depth | *11* | 7 | 2 | 0 | 0 | 0 | 0 | 0 | 0 | 0 | 0 | *1* | 21 |

## KEARNEYSVILLE *Jefferson County* ELEVATION 550 ft LAT/LONG 39° 23' N / 77° 53' W

| | JAN | FEB | MAR | APR | MAY | JUN | JUL | AUG | SEP | OCT | NOV | DEC | YEAR |
|---|---|---|---|---|---|---|---|---|---|---|---|---|---|
| Maximum Temp °F | 38.6 | 42.3 | 52.0 | 63.4 | 72.9 | 81.4 | 85.7 | 84.1 | 77.4 | 65.6 | 54.9 | 43.5 | 63.5 |
| Minimum Temp °F | 19.5 | 22.0 | 30.0 | 39.2 | 48.2 | 56.7 | 61.7 | 59.5 | 52.0 | 39.6 | 32.9 | 24.7 | 40.5 |
| Mean Temp °F | 29.1 | 32.1 | 41.0 | 51.3 | 60.6 | 69.1 | 73.7 | 71.8 | 64.7 | 52.6 | 43.9 | 34.2 | 52.0 |
| Days Max Temp ≥ 90 °F | 0 | 0 | 0 | 0 | 1 | 4 | 8 | 6 | 2 | 0 | 0 | 0 | 21 |
| Days Max Temp ≤ 32 °F | 9 | 6 | 1 | 0 | 0 | 0 | 0 | 0 | 0 | 0 | 0 | 4 | 20 |
| Days Min Temp ≤ 32 °F | 28 | 24 | 20 | 8 | 1 | 0 | 0 | 0 | 0 | 8 | 15 | 25 | 129 |
| Days Min Temp ≤ 0 °F | 2 | 0 | 0 | 0 | 0 | 0 | 0 | 0 | 0 | 0 | 0 | 0 | 2 |
| Heating Degree Days | 1108 | 921 | 739 | 412 | 179 | 35 | 4 | 12 | 96 | 384 | 627 | 949 | 5466 |
| Cooling Degree Days | 0 | 0 | 2 | 10 | 55 | 172 | 306 | 223 | 93 | 9 | 1 | 0 | 871 |
| Total Precipitation (") | 2.59 | 2.46 | 3.31 | 3.30 | 4.00 | 3.53 | 3.76 | 3.61 | 3.29 | 3.44 | 3.02 | 3.02 | 39.33 |
| Days ≥ 0.1" Precip | 6 | 5 | 7 | 7 | 7 | 6 | 7 | 6 | 5 | 5 | 6 | 6 | 73 |
| Total Snowfall (") | 7.6 | 6.3 | 3.9 | 0.4 | 0.0 | 0.0 | 0.0 | 0.0 | 0.0 | 0.0 | 0.8 | 3.6 | 22.6 |
| Days ≥ 1" Snow Depth | 11 | 7 | 3 | 0 | 0 | 0 | 0 | 0 | 0 | 0 | 1 | 3 | 25 |

## KOPPERSTON *Wyoming County* ELEVATION 1732 ft LAT/LONG 37° 44 ' N / 81° 35 ' W

| | JAN | FEB | MAR | APR | MAY | JUN | JUL | AUG | SEP | OCT | NOV | DEC | YEAR |
|---|---|---|---|---|---|---|---|---|---|---|---|---|---|
| Maximum Temp °F | 40.2 | 44.6 | 56.1 | 67.0 | 74.7 | 80.5 | 83.3 | 81.7 | 74.8 | 64.2 | 54.2 | 44.3 | 63.8 |
| Minimum Temp °F | 22.1 | 23.7 | 32.0 | 39.6 | 47.8 | 55.1 | 60.0 | 58.8 | 52.8 | 40.9 | 33.4 | 26.3 | 41.0 |
| Mean Temp °F | 31.2 | 34.2 | 44.0 | 53.3 | 61.3 | 67.8 | 71.7 | 70.2 | 63.9 | 52.6 | 43.9 | 35.2 | 52.4 |
| Days Max Temp ≥ 90 °F | 0 | 0 | 0 | 0 | 0 | 1 | 2 | 2 | 0 | 0 | 0 | 0 | 5 |
| Days Max Temp ≤ 32 °F | 7 | 5 | 1 | 0 | 0 | 0 | 0 | 0 | 0 | 0 | 1 | 4 | 18 |
| Days Min Temp ≤ 32 °F | 25 | 23 | 17 | 8 | 1 | 0 | 0 | 0 | 0 | 7 | 15 | 22 | 118 |
| Days Min Temp ≤ 0 °F | 2 | 1 | 0 | 0 | 0 | 0 | 0 | 0 | 0 | 0 | 0 | 0 | 3 |
| Heating Degree Days | 1042 | 863 | 643 | 352 | 145 | 30 | 3 | 8 | 96 | 382 | 627 | 916 | 5107 |
| Cooling Degree Days | 0 | 0 | 1 | 7 | 34 | 116 | 222 | 173 | 64 | 4 | 0 | 0 | 621 |
| Total Precipitation (") | 3.76 | 3.45 | 4.57 | 4.40 | 4.82 | 4.31 | 5.05 | 4.65 | 3.63 | 3.37 | 3.71 | 4.36 | 50.08 |
| Days ≥ 0.1" Precip | 9 | 8 | 10 | 10 | 10 | 8 | 9 | 9 | 7 | 7 | 8 | 9 | 104 |
| Total Snowfall (") | *8.4* | *7.7* | *3.6* | 0.1 | 0.0 | 0.0 | 0.0 | 0.0 | 0.0 | 0.0 | 1.4 | na | na |
| Days ≥ 1" Snow Depth | 11 | 9 | 2 | 0 | 0 | 0 | 0 | 0 | 0 | 0 | 1 | 6 | 29 |

## LEWISBURG 3 N *Greenbrier County* ELEVATION 2303 ft LAT/LONG 37° 51 ' N / 80° 24 ' W

| | JAN | FEB | MAR | APR | MAY | JUN | JUL | AUG | SEP | OCT | NOV | DEC | YEAR |
|---|---|---|---|---|---|---|---|---|---|---|---|---|---|
| Maximum Temp °F | 40.3 | 44.6 | 55.0 | 65.1 | 73.6 | 80.5 | 83.8 | 82.4 | 76.4 | 66.2 | 54.5 | 44.6 | 63.9 |
| Minimum Temp °F | 20.7 | 23.1 | 30.5 | 38.5 | 47.2 | 55.4 | 60.1 | 58.6 | 52.5 | 40.1 | 32.4 | 25.4 | 40.4 |
| Mean Temp °F | 30.5 | 33.9 | 42.7 | 51.8 | 60.4 | 68.0 | 72.0 | 70.5 | 64.5 | 53.2 | 43.5 | 35.0 | 52.2 |
| Days Max Temp ≥ 90 °F | 0 | 0 | 0 | 0 | 0 | 1 | 4 | 2 | 0 | 0 | 0 | 0 | 7 |
| Days Max Temp ≤ 32 °F | 7 | 4 | 1 | 0 | 0 | 0 | 0 | 0 | 0 | 0 | 1 | 4 | 17 |
| Days Min Temp ≤ 32 °F | 27 | 23 | 18 | 9 | 2 | 0 | 0 | 0 | 0 | 8 | 16 | 23 | 126 |
| Days Min Temp ≤ 0 °F | 1 | 1 | 0 | 0 | 0 | 0 | 0 | 0 | 0 | 0 | 0 | 1 | 3 |
| Heating Degree Days | 1061 | 872 | 683 | 394 | 167 | 34 | 4 | 9 | 88 | 367 | 639 | 922 | 5240 |
| Cooling Degree Days | 0 | 0 | 0 | 6 | 34 | 135 | 248 | 189 | 78 | 12 | 0 | 0 | 702 |
| Total Precipitation (") | 2.81 | 2.83 | 3.49 | 3.42 | 3.71 | 3.22 | 4.38 | 3.66 | 2.76 | 3.02 | 2.90 | 3.12 | 39.32 |
| Days ≥ 0.1" Precip | 7 | 6 | 8 | 7 | 8 | 7 | 8 | 7 | 5 | 5 | 6 | *7* | 81 |
| Total Snowfall (") | *7.8* | *6.6* | na | 0.7 | 0.0 | 0.0 | 0.0 | 0.0 | 0.0 | 0.2 | 1.4 | *5.3* | na |
| Days ≥ 1" Snow Depth | na | na | na | 0 | 0 | 0 | 0 | 0 | 0 | 0 | *1* | *5* | na |

## LOGAN *Logan County* ELEVATION 664 ft LAT/LONG 37° 51 ' N / 82° 0 ' W

| | JAN | FEB | MAR | APR | MAY | JUN | JUL | AUG | SEP | OCT | NOV | DEC | YEAR |
|---|---|---|---|---|---|---|---|---|---|---|---|---|---|
| Maximum Temp °F | 43.4 | 47.6 | 57.9 | 69.1 | 77.2 | 84.0 | 87.3 | 86.0 | 80.0 | 68.9 | 57.9 | 47.1 | 67.2 |
| Minimum Temp °F | 24.7 | 25.7 | 33.4 | 41.7 | 50.6 | 59.8 | 65.2 | 64.0 | 57.5 | 44.2 | 35.4 | 28.8 | 44.2 |
| Mean Temp °F | 34.1 | 36.7 | 45.7 | 55.4 | 63.9 | 71.9 | 76.3 | 75.0 | 68.8 | 56.5 | 46.6 | 38.0 | 55.7 |
| Days Max Temp ≥ 90 °F | 0 | 0 | 0 | 1 | 2 | 7 | 12 | 8 | 3 | 0 | 0 | 0 | 33 |
| Days Max Temp ≤ 32 °F | 6 | 4 | 1 | 0 | 0 | 0 | 0 | 0 | 0 | 0 | 0 | 3 | 14 |
| Days Min Temp ≤ 32 °F | 24 | 21 | 15 | 5 | 0 | 0 | 0 | 0 | 0 | 3 | 13 | 20 | 101 |
| Days Min Temp ≤ 0 °F | 1 | 0 | 0 | 0 | 0 | 0 | 0 | 0 | 0 | 0 | 0 | 0 | 1 |
| Heating Degree Days | 950 | 795 | 595 | 303 | 108 | 16 | 1 | 2 | 43 | 276 | 545 | 830 | 4464 |
| Cooling Degree Days | 0 | 0 | 4 | 27 | 90 | 241 | 367 | 326 | 163 | 20 | 1 | 0 | 1239 |
| Total Precipitation (") | 3.14 | 3.44 | 4.24 | 3.75 | 4.87 | 4.13 | 5.36 | 4.31 | 3.41 | 3.01 | 3.34 | 3.93 | 46.93 |
| Days ≥ 0.1" Precip | 8 | 8 | 10 | 9 | 9 | 8 | 9 | 7 | 7 | 6 | 7 | 8 | 96 |
| Total Snowfall (") | *6.7* | *6.0* | *2.3* | 0.0 | 0.0 | 0.0 | 0.0 | 0.0 | 0.0 | 0.0 | 0.5 | *2.6* | 18.1 |
| Days ≥ 1" Snow Depth | *8* | 6 | *1* | 0 | 0 | 0 | 0 | 0 | 0 | 0 | 0 | *3* | 18 |

## LONDON LOCKS *Kanawha County* ELEVATION 620 ft LAT/LONG 38° 12 ' N / 81° 22 ' W

| | JAN | FEB | MAR | APR | MAY | JUN | JUL | AUG | SEP | OCT | NOV | DEC | YEAR |
|---|---|---|---|---|---|---|---|---|---|---|---|---|---|
| Maximum Temp °F | 42.0 | 45.9 | 55.9 | 67.1 | 75.0 | 82.3 | 85.6 | 84.2 | 78.1 | 67.3 | 57.1 | 46.8 | 65.6 |
| Minimum Temp °F | 23.9 | 25.5 | 33.2 | 41.5 | 50.9 | 59.8 | 65.0 | 64.3 | 58.3 | 45.5 | 36.2 | 28.7 | 44.4 |
| Mean Temp °F | 32.9 | 35.7 | 44.6 | 54.3 | 63.0 | 71.1 | 75.3 | 74.2 | 68.3 | 56.4 | 46.7 | 37.8 | 55.0 |
| Days Max Temp ≥ 90 °F | 0 | 0 | 0 | 0 | 1 | 5 | 8 | 6 | 2 | 0 | 0 | 0 | 22 |
| Days Max Temp ≤ 32 °F | 7 | 4 | 1 | 0 | 0 | 0 | 0 | 0 | 0 | 0 | 0 | 3 | 15 |
| Days Min Temp ≤ 32 °F | 24 | 21 | 16 | 5 | 0 | 0 | 0 | 0 | 0 | 2 | 12 | 20 | 100 |
| Days Min Temp ≤ 0 °F | 1 | 0 | 0 | 0 | 0 | 0 | 0 | 0 | 0 | 0 | 0 | 0 | 1 |
| Heating Degree Days | 987 | 820 | 627 | 328 | 127 | 16 | 1 | 2 | 46 | 278 | 545 | 836 | 4613 |
| Cooling Degree Days | 0 | 0 | 2 | 17 | 74 | 219 | 357 | 297 | 152 | 22 | 0 | 0 | 1140 |
| Total Precipitation (") | 3.04 | 2.87 | 3.89 | 3.85 | 4.41 | 3.80 | 4.87 | 4.51 | 3.52 | 2.76 | 3.44 | 3.69 | 44.65 |
| Days ≥ 0.1" Precip | 8 | 7 | 9 | 9 | 10 | 8 | 8 | 8 | 7 | 6 | 8 | 8 | 96 |
| Total Snowfall (") | na | na | na | 0.0 | 0.0 | 0.0 | 0.0 | 0.0 | 0.0 | 0.0 | 0.3 | na | na |
| Days ≥ 1" Snow Depth | na | *7* | *1* | 0 | 0 | 0 | 0 | 0 | 0 | 0 | 0 | na | na |

## MADISON *Boone County* ELEVATION 679 ft LAT/LONG 38° 3 ' N / 81° 49 ' W

| | JAN | FEB | MAR | APR | MAY | JUN | JUL | AUG | SEP | OCT | NOV | DEC | YEAR |
|---|---|---|---|---|---|---|---|---|---|---|---|---|---|
| Maximum Temp °F | 41.7 | 46.8 | 58.0 | 69.1 | 76.7 | 83.6 | 86.9 | 85.4 | 79.5 | 68.7 | 57.6 | 46.8 | 66.7 |
| Minimum Temp °F | 20.8 | 22.7 | 30.8 | 39.1 | 48.3 | 57.3 | 62.5 | 61.2 | 54.4 | 40.7 | 32.2 | 25.4 | 41.3 |
| Mean Temp °F | 31.3 | 34.8 | 44.4 | 54.1 | 62.6 | 70.4 | 74.7 | 73.3 | 67.0 | 54.7 | 44.9 | 36.1 | 54.0 |
| Days Max Temp ≥ 90 °F | 0 | 0 | 0 | 1 | 1 | 5 | 11 | 7 | 2 | 0 | 0 | 0 | 27 |
| Days Max Temp ≤ 32 °F | 7 | 4 | 1 | 0 | 0 | 0 | 0 | 0 | 0 | 0 | 0 | 3 | 15 |
| Days Min Temp ≤ 32 °F | 26 | 23 | 19 | 8 | 1 | 0 | 0 | 0 | 0 | 7 | 17 | 23 | 124 |
| Days Min Temp ≤ 0 °F | 2 | 1 | 0 | 0 | 0 | 0 | 0 | 0 | 0 | 0 | 0 | 0 | 3 |
| Heating Degree Days | 1038 | 847 | 632 | 337 | 137 | 21 | 1 | 3 | 57 | 326 | 597 | 891 | 4887 |
| Cooling Degree Days | 0 | 0 | 2 | 17 | 70 | 201 | 333 | 270 | 121 | 16 | 0 | 0 | 1030 |
| Total Precipitation (") | 3.13 | 3.05 | 3.83 | 3.92 | 4.70 | 4.19 | 4.99 | 4.47 | 3.87 | 3.04 | 3.44 | 3.69 | 46.32 |
| Days ≥ 0.1" Precip | 8 | 7 | 9 | 9 | 9 | 8 | 9 | 7 | 7 | 6 | 7 | 7 | 93 |
| Total Snowfall (") | 8.5 | 5.6 | 2.7 | 1.0 | 0.0 | 0.0 | 0.0 | 0.0 | 0.0 | 0.0 | 0.7 | 2.7 | 21.2 |
| Days ≥ 1" Snow Depth | 6 | 5 | 2 | 0 | 0 | 0 | 0 | 0 | 0 | 0 | 0 | 2 | 15 |

## MARTINSBURG E WV AP *Berkeley County* ELEVATION 540 ft LAT/LONG 39° 24 ' N / 77° 59 ' W

| | JAN | FEB | MAR | APR | MAY | JUN | JUL | AUG | SEP | OCT | NOV | DEC | YEAR |
|---|---|---|---|---|---|---|---|---|---|---|---|---|---|
| Maximum Temp °F | 38.8 | 42.8 | 53.0 | 64.6 | 74.1 | 83.0 | 87.0 | 85.1 | 77.8 | 65.8 | 54.8 | 43.9 | 64.2 |
| Minimum Temp °F | 21.2 | 23.6 | 31.9 | 41.1 | 50.4 | 58.9 | 64.3 | 62.3 | 54.9 | 42.5 | 34.6 | 26.5 | 42.7 |
| Mean Temp °F | 30.0 | 33.2 | 42.5 | 52.9 | 62.3 | 71.0 | 75.6 | 73.7 | 66.3 | 54.2 | 44.7 | 35.2 | 53.5 |
| Days Max Temp ≥ 90 °F | 0 | 0 | 0 | 0 | 1 | 6 | 11 | 8 | 3 | 0 | 0 | 0 | 29 |
| Days Max Temp ≤ 32 °F | 8 | 5 | 1 | 0 | 0 | 0 | 0 | 0 | 0 | 0 | 0 | 4 | 18 |
| Days Min Temp ≤ 32 °F | 27 | 23 | 17 | 5 | 0 | 0 | 0 | 0 | 0 | 4 | 13 | 23 | 112 |
| Days Min Temp ≤ 0 °F | 1 | 0 | 0 | 0 | 0 | 0 | 0 | 0 | 0 | 0 | 0 | 0 | 1 |
| Heating Degree Days | 1077 | 890 | 692 | 368 | 142 | 18 | 1 | 5 | 70 | 339 | 603 | 916 | 5121 |
| Cooling Degree Days | 0 | 0 | 3 | 11 | 68 | 211 | 363 | 277 | 115 | 13 | 0 | 0 | 1061 |
| Total Precipitation (") | 2.34 | 2.34 | 3.36 | 3.17 | 4.08 | 3.28 | 3.49 | 3.50 | 3.10 | 3.50 | 3.01 | 2.83 | 38.00 |
| Days ≥ 0.1" Precip | 6 | 5 | 7 | 6 | 7 | 6 | 7 | 6 | 5 | 5 | 6 | 6 | 72 |
| Total Snowfall (") | 7.3 | 7.7 | 4.6 | 0.4 | 0.0 | 0.0 | 0.0 | 0.0 | 0.0 | 0.0 | 1.5 | 4.3 | 25.8 |
| Days ≥ 1" Snow Depth | 11 | 8 | 3 | 0 | 0 | 0 | 0 | 0 | 0 | 0 | 1 | 3 | 26 |

## MATHIAS *Hardy County* ELEVATION 1552 ft LAT/LONG 38° 52 ' N / 78° 52 ' W

| | JAN | FEB | MAR | APR | MAY | JUN | JUL | AUG | SEP | OCT | NOV | DEC | YEAR |
|---|---|---|---|---|---|---|---|---|---|---|---|---|---|
| Maximum Temp °F | 39.9 | 43.2 | 53.4 | 64.5 | 73.3 | 80.6 | 83.9 | 82.3 | 75.8 | 65.0 | 53.8 | 44.0 | 63.3 |
| Minimum Temp °F | 19.5 | 21.5 | 29.1 | 37.3 | 46.5 | 53.7 | 58.4 | 56.9 | 50.6 | 39.5 | 32.0 | 24.4 | 39.1 |
| Mean Temp °F | 29.7 | 32.4 | 41.3 | 50.9 | 59.9 | 67.2 | 71.2 | 69.6 | 63.2 | 52.3 | 42.9 | 34.2 | 51.2 |
| Days Max Temp ≥ 90 °F | 0 | 0 | 0 | 0 | 0 | 2 | 5 | 3 | 1 | 0 | 0 | 0 | 11 |
| Days Max Temp ≤ 32 °F | 8 | 5 | 1 | 0 | 0 | 0 | 0 | 0 | 0 | 0 | 0 | 5 | 19 |
| Days Min Temp ≤ 32 °F | 28 | 24 | 20 | 10 | 2 | 0 | 0 | 0 | 1 | 9 | 17 | 25 | 136 |
| Days Min Temp ≤ 0 °F | 2 | 1 | 0 | 0 | 0 | 0 | 0 | 0 | 0 | 0 | 0 | 1 | 4 |
| Heating Degree Days | 1086 | 915 | 729 | 419 | 185 | 43 | 7 | 18 | 112 | 393 | 655 | 948 | 5510 |
| Cooling Degree Days | 0 | *0* | 0 | 3 | 41 | *121* | 223 | 167 | 68 | 8 | 0 | 0 | 631 |
| Total Precipitation (") | 1.97 | 2.09 | 2.88 | 2.84 | 3.94 | 3.15 | 3.79 | 3.66 | 3.15 | 3.06 | 2.88 | 2.37 | 35.78 |
| Days ≥ 0.1" Precip | 5 | 5 | 6 | 6 | 9 | 7 | 8 | 7 | 5 | 5 | 6 | 5 | 74 |
| Total Snowfall (") | 9.1 | 9.0 | 6.9 | 1.8 | 0.0 | 0.0 | 0.0 | 0.0 | 0.0 | 0.4 | 3.0 | 6.9 | 37.1 |
| Days ≥ 1" Snow Depth | 11 | 10 | 3 | 0 | 0 | 0 | 0 | 0 | 0 | 0 | 1 | 5 | 30 |

## MIDDLEBOURNE 3 ESE *Tyler County* ELEVATION 751 ft LAT/LONG 39° 29 ' N / 80° 52 ' W

| | JAN | FEB | MAR | APR | MAY | JUN | JUL | AUG | SEP | OCT | NOV | DEC | YEAR |
|---|---|---|---|---|---|---|---|---|---|---|---|---|---|
| Maximum Temp °F | 39.0 | 43.4 | 54.6 | 65.7 | 74.2 | 81.6 | 84.8 | 83.1 | 77.2 | 66.2 | 54.9 | 44.1 | 64.1 |
| Minimum Temp °F | 19.6 | 21.5 | 29.9 | 38.4 | 47.8 | 56.7 | 62.2 | 60.7 | 54.2 | 40.8 | 33.0 | 25.5 | 40.9 |
| Mean Temp °F | 29.3 | 32.5 | 42.3 | 52.1 | 61.0 | 69.2 | 73.5 | 71.9 | 65.7 | 53.5 | 44.0 | 34.8 | 52.5 |
| Days Max Temp ≥ 90 °F | 0 | 0 | 0 | 0 | 0 | 2 | 6 | 3 | 1 | 0 | 0 | 0 | 12 |
| Days Max Temp ≤ 32 °F | 10 | 6 | 1 | 0 | 0 | 0 | 0 | 0 | 0 | 0 | 0 | 5 | 22 |
| Days Min Temp ≤ 32 °F | 26 | 23 | 19 | 10 | 2 | 0 | 0 | 0 | 0 | 7 | 15 | 23 | 125 |
| Days Min Temp ≤ 0 °F | 3 | 2 | 0 | 0 | 0 | 0 | 0 | 0 | 0 | 0 | 0 | 1 | 6 |
| Heating Degree Days | 1100 | 912 | 698 | 390 | 164 | 28 | 2 | 8 | 76 | 359 | 625 | 929 | 5291 |
| Cooling Degree Days | 0 | 0 | 1 | 11 | 49 | 160 | 294 | 230 | 105 | 12 | 0 | 0 | 862 |
| Total Precipitation (") | 3.06 | 2.73 | 3.69 | 3.68 | 4.46 | 3.74 | 4.74 | 4.22 | 3.61 | 3.13 | 3.61 | 3.24 | 43.91 |
| Days ≥ 0.1" Precip | 8 | 7 | 9 | 9 | 9 | 7 | 8 | 8 | 7 | 7 | 8 | 9 | 96 |
| Total Snowfall (") | 10.6 | 6.1 | 3.5 | 0.6 | 0.0 | 0.0 | 0.0 | 0.0 | 0.0 | 0.0 | 1.0 | 2.8 | 24.6 |
| Days ≥ 1" Snow Depth | 11 | 8 | 2 | 0 | 0 | 0 | 0 | 0 | 0 | 0 | 1 | 3 | 25 |

## MOOREFIELD 2 SSE *Hardy County* ELEVATION 820 ft LAT/LONG 39° 3 ' N / 78° 58 ' W

| | JAN | FEB | MAR | APR | MAY | JUN | JUL | AUG | SEP | OCT | NOV | DEC | YEAR |
|---|---|---|---|---|---|---|---|---|---|---|---|---|---|
| Maximum Temp °F | 41.7 | 46.2 | 56.5 | 67.9 | 76.8 | 84.3 | 87.4 | 86.1 | 80.2 | 69.8 | 57.3 | 46.3 | 66.7 |
| Minimum Temp °F | 20.3 | 23.3 | 31.2 | 39.7 | 49.6 | 58.0 | 62.0 | 60.3 | 53.9 | 42.0 | 33.6 | 25.7 | 41.6 |
| Mean Temp °F | 31.0 | 34.8 | 43.9 | 53.8 | 63.2 | 71.2 | 74.7 | 73.2 | 67.0 | 55.9 | 45.5 | 36.0 | 54.2 |
| Days Max Temp ≥ 90 °F | 0 | 0 | 0 | 0 | 1 | 6 | 12 | 9 | 3 | 0 | 0 | 0 | 31 |
| Days Max Temp ≤ 32 °F | 7 | 3 | 0 | 0 | 0 | 0 | 0 | 0 | 0 | 0 | 0 | 3 | 13 |
| Days Min Temp ≤ 32 °F | 27 | 23 | 18 | 8 | 1 | 0 | 0 | 0 | 0 | 6 | 14 | 24 | 121 |
| Days Min Temp ≤ 0 °F | 2 | 1 | 0 | 0 | 0 | 0 | 0 | 0 | 0 | 0 | 0 | 1 | 4 |
| Heating Degree Days | 1046 | 848 | 650 | 343 | 121 | 14 | 2 | 5 | 59 | 292 | 580 | 892 | 4852 |
| Cooling Degree Days | 0 | 0 | 2 | 14 | 76 | 208 | 324 | 268 | 129 | 19 | 1 | 0 | 1041 |
| Total Precipitation (") | 1.64 | 1.64 | 2.47 | 2.30 | 3.52 | 3.31 | 3.80 | 3.32 | 2.64 | 2.83 | 2.38 | 2.15 | 32.00 |
| Days ≥ 0.1" Precip | 4 | 4 | 6 | 6 | 8 | 7 | 8 | 7 | 6 | *5* | 5 | 4 | 70 |
| Total Snowfall (") | *8.4* | *5.0* | *4.3* | 0.5 | 0.0 | 0.0 | 0.0 | 0.0 | 0.0 | 0.0 | 1.3 | *4.3* | 23.8 |
| Days ≥ 1" Snow Depth | na | na | *2* | *0* | 0 | 0 | 0 | 0 | 0 | 0 | *0* | *2* | na |

## MORGANTOWN L & D *Monongalia County* ELEVATION 830 ft LAT/LONG 39° 38 ' N / 79° 58 ' W

| | JAN | FEB | MAR | APR | MAY | JUN | JUL | AUG | SEP | OCT | NOV | DEC | YEAR |
|---|---|---|---|---|---|---|---|---|---|---|---|---|---|
| Maximum Temp °F | 39.0 | 42.6 | 53.5 | 64.7 | 73.4 | 80.7 | 83.9 | 82.5 | 76.4 | 65.3 | 54.5 | 44.1 | 63.4 |
| Minimum Temp °F | 20.4 | 22.1 | 30.3 | 39.1 | 48.1 | 56.5 | 61.4 | 59.9 | 53.8 | 41.8 | 33.8 | 25.9 | 41.1 |
| Mean Temp °F | 29.7 | 32.4 | 41.9 | 51.9 | 60.8 | 68.7 | 72.7 | 71.2 | 65.1 | 53.6 | 44.2 | 35.0 | 52.3 |
| Days Max Temp ≥ 90 °F | 0 | 0 | 0 | 0 | 0 | 2 | 5 | 3 | 1 | 0 | 0 | 0 | 11 |
| Days Max Temp ≤ 32 °F | 9 | 6 | 1 | 0 | 0 | 0 | 0 | 0 | 0 | 0 | 1 | 5 | 22 |
| Days Min Temp ≤ 32 °F | 26 | 23 | 19 | 8 | 1 | 0 | 0 | 0 | 0 | 4 | 15 | 23 | 119 |
| Days Min Temp ≤ 0 °F | 2 | 1 | 0 | 0 | 0 | 0 | 0 | 0 | 0 | 0 | 0 | 0 | 3 |
| Heating Degree Days | 1088 | 916 | 709 | 394 | 168 | 32 | 4 | 8 | 80 | 354 | 619 | 922 | 5294 |
| Cooling Degree Days | 0 | 0 | 0 | 8 | 43 | 148 | 272 | 216 | 93 | 8 | 0 | 0 | 788 |
| Total Precipitation (") | 2.98 | 2.56 | 3.80 | 3.69 | 4.08 | 3.78 | 4.31 | 3.99 | 3.38 | 2.84 | 3.43 | 3.33 | 42.17 |
| Days ≥ 0.1" Precip | 8 | 7 | 9 | 9 | 9 | 8 | 8 | 8 | 7 | 7 | 8 | 8 | 96 |
| Total Snowfall (") | na | na | *2.4* | 0.1 | 0.0 | 0.0 | 0.0 | 0.0 | 0.0 | 0.0 | 0.4 | *2.1* | na |
| Days ≥ 1" Snow Depth | 10 | 7 | 3 | 0 | 0 | 0 | 0 | 0 | 0 | 0 | 0 | 4 | 24 |

## MORGANTOWN MUNI AP *Monongalia County* ELEVATION 1247 ft LAT/LONG 39° 38 ' N / 79° 55 ' W

| | JAN | FEB | MAR | APR | MAY | JUN | JUL | AUG | SEP | OCT | NOV | DEC | YEAR |
|---|---|---|---|---|---|---|---|---|---|---|---|---|---|
| Maximum Temp °F | 37.5 | 41.2 | 52.2 | 63.5 | 72.8 | 80.4 | 83.7 | 82.1 | 75.6 | 64.4 | 53.3 | 43.0 | 62.5 |
| Minimum Temp °F | 21.3 | 23.5 | 32.1 | 41.0 | 50.1 | 58.1 | 62.9 | 61.6 | 55.3 | 43.5 | 35.7 | 27.2 | 42.7 |
| Mean Temp °F | 29.4 | 32.3 | 42.2 | 52.3 | 61.4 | 69.3 | 73.3 | 71.9 | 65.5 | 54.0 | 44.5 | 35.1 | 52.6 |
| Days Max Temp ≥ 90 °F | 0 | 0 | 0 | 0 | 0 | 3 | 5 | 3 | 1 | 0 | 0 | 0 | 12 |
| Days Max Temp ≤ 32 °F | 11 | 8 | 2 | 0 | 0 | 0 | 0 | 0 | 0 | 0 | 1 | 7 | 29 |
| Days Min Temp ≤ 32 °F | 25 | 22 | 18 | 6 | 1 | 0 | 0 | 0 | 0 | 4 | 13 | 22 | 111 |
| Days Min Temp ≤ 0 °F | 2 | 1 | 0 | 0 | 0 | 0 | 0 | 0 | 0 | 0 | 0 | 1 | 4 |
| Heating Degree Days | 1094 | 916 | 704 | 392 | 167 | 40 | 4 | 9 | 85 | 348 | 610 | 921 | 5290 |
| Cooling Degree Days | 0 | 0 | 3 | 16 | 61 | 161 | 283 | 225 | 98 | 11 | 2 | 0 | 860 |
| Total Precipitation (") | 2.69 | 2.44 | 3.79 | 3.49 | 4.06 | 3.79 | 4.28 | 4.06 | 3.48 | 2.74 | 3.28 | 3.23 | 41.33 |
| Days ≥ 0.1" Precip | 7 | 6 | 9 | 9 | 8 | 8 | 8 | 7 | 7 | 6 | 8 | 8 | 91 |
| Total Snowfall (") | 11.1 | 7.4 | 5.8 | 0.9 | 0.0 | 0.0 | 0.0 | 0.0 | 0.0 | 0.1 | 2.0 | 4.9 | 32.2 |
| Days ≥ 1" Snow Depth | 12 | 9 | 4 | 0 | 0 | 0 | 0 | 0 | 0 | 0 | 1 | 6 | 32 |

## MOUNDSVILLE *Marshall County* ELEVATION 620 ft LAT/LONG 39° 54 ' N / 80° 45 ' W

| | JAN | FEB | MAR | APR | MAY | JUN | JUL | AUG | SEP | OCT | NOV | DEC | YEAR |
|---|---|---|---|---|---|---|---|---|---|---|---|---|---|
| Maximum Temp °F | *38.0* | 41.9 | 53.2 | 65.6 | 75.0 | 82.7 | 85.9 | 84.5 | 78.3 | 66.5 | 54.2 | 43.6 | 64.1 |
| Minimum Temp °F | *18.9* | 20.6 | 29.4 | 38.8 | 47.9 | 56.8 | 61.7 | 60.5 | 53.9 | 41.2 | 33.7 | 26.0 | 40.8 |
| Mean Temp °F | *28.5* | 31.3 | 41.3 | 52.2 | 61.5 | 69.7 | 73.8 | 72.5 | 66.2 | 53.9 | 44.0 | 34.8 | 52.5 |
| Days Max Temp ≥ 90 °F | 0 | 0 | 0 | 0 | 1 | 5 | 8 | 6 | 2 | 0 | 0 | 0 | 22 |
| Days Max Temp ≤ 32 °F | 10 | 6 | 1 | 0 | 0 | 0 | 0 | 0 | 0 | 0 | 0 | 4 | 21 |
| Days Min Temp ≤ 32 °F | 26 | 24 | 20 | *9* | 1 | 0 | 0 | 0 | 0 | 5 | 15 | 23 | 123 |
| Days Min Temp ≤ 0 °F | 3 | 2 | 0 | 0 | 0 | 0 | 0 | 0 | 0 | 0 | 0 | 0 | 5 |
| Heating Degree Days | *1124* | 945 | 728 | 386 | 160 | 26 | 3 | 7 | 75 | 348 | 624 | 929 | 5355 |
| Cooling Degree Days | na | *0* | *0* | *11* | *60* | *173* | *297* | *246* | *116* | *10* | *1* | *0* | na |
| Total Precipitation (") | 2.67 | 2.11 | 3.31 | 3.60 | 3.94 | 3.92 | 4.41 | 4.11 | 3.28 | 2.67 | 3.63 | 2.96 | 40.61 |
| Days ≥ 0.1" Precip | *7* | 6 | 8 | 9 | 8 | 7 | 8 | 7 | 6 | 6 | 8 | 8 | 88 |
| Total Snowfall (") | na | na | na | *0.0* | 0.0 | 0.0 | 0.0 | 0.0 | 0.0 | 0.0 | *0.7* | *2.1* | na |
| Days ≥ 1" Snow Depth | na | na | na | *0* | 0 | 0 | 0 | 0 | 0 | 0 | *0* | na | na |

## OAK HILL *Fayette County* ELEVATION 1990 ft LAT/LONG 37° 58 ' N / 81° 9 ' W

| | JAN | FEB | MAR | APR | MAY | JUN | JUL | AUG | SEP | OCT | NOV | DEC | YEAR |
|---|---|---|---|---|---|---|---|---|---|---|---|---|---|
| Maximum Temp °F | 39.2 | 43.1 | 53.1 | 63.9 | 71.9 | 79.1 | 82.5 | 81.2 | 75.2 | 64.6 | 54.5 | 44.3 | 62.7 |
| Minimum Temp °F | 19.7 | 22.0 | 30.4 | 39.3 | 47.2 | 55.0 | 59.6 | 58.5 | 52.4 | 40.7 | 33.3 | 25.1 | 40.3 |
| Mean Temp °F | 29.5 | 32.6 | 41.8 | 51.6 | 59.6 | 67.0 | 71.1 | 69.9 | 63.9 | 52.7 | 43.8 | 34.8 | 51.5 |
| Days Max Temp ≥ 90 °F | 0 | 0 | 0 | 0 | 0 | 1 | 3 | 2 | 0 | 0 | 0 | 0 | 6 |
| Days Max Temp ≤ 32 °F | 9 | 6 | 2 | 0 | 0 | 0 | 0 | 0 | 0 | 0 | 1 | 6 | 24 |
| Days Min Temp ≤ 32 °F | 27 | 23 | 19 | 8 | 1 | 0 | 0 | 0 | 0 | 7 | 15 | 24 | 124 |
| Days Min Temp ≤ 0 °F | 2 | 1 | 0 | 0 | 0 | 0 | 0 | 0 | 0 | 0 | 0 | 0 | 3 |
| Heating Degree Days | 1095 | 910 | 714 | 402 | 195 | 47 | 8 | 14 | 103 | 381 | 630 | 930 | 5429 |
| Cooling Degree Days | 0 | 0 | 1 | 8 | 33 | 114 | 220 | 171 | 70 | 6 | 1 | 0 | 624 |
| Total Precipitation (") | 3.31 | 3.02 | 3.99 | 3.91 | 4.21 | 3.76 | 5.31 | 4.28 | 3.57 | 3.13 | 3.31 | 3.64 | 45.44 |
| Days ≥ 0.1" Precip | 9 | 8 | 10 | 10 | 9 | 8 | 9 | 8 | 7 | 7 | 8 | 9 | 102 |
| Total Snowfall (") | 15.2 | 10.5 | 6.3 | 1.2 | 0.0 | 0.0 | 0.0 | 0.0 | 0.0 | 0.1 | 2.2 | 7.6 | 43.1 |
| Days ≥ 1" Snow Depth | 14 | 11 | 4 | 1 | 0 | 0 | 0 | 0 | 0 | 0 | 2 | 8 | 40 |

## PARKERSBURG 1 NE *Wood County* ELEVATION 640 ft LAT/LONG 39° 16 ' N / 81° 32 ' W

| | JAN | FEB | MAR | APR | MAY | JUN | JUL | AUG | SEP | OCT | NOV | DEC | YEAR |
|---|---|---|---|---|---|---|---|---|---|---|---|---|---|
| Maximum Temp °F | 38.4 | 42.6 | 53.7 | 65.1 | 74.5 | 82.3 | 85.5 | 84.2 | 77.5 | 66.0 | 54.5 | 43.9 | 64.0 |
| Minimum Temp °F | 21.8 | 23.9 | 32.4 | 41.5 | 51.1 | 59.5 | 64.4 | 62.8 | 56.0 | 43.7 | 35.7 | 27.6 | 43.4 |
| Mean Temp °F | 30.1 | 33.3 | 43.1 | 53.3 | 62.9 | 71.0 | 75.0 | 73.5 | 66.8 | 54.9 | 45.1 | 35.8 | 53.7 |
| Days Max Temp ≥ 90 °F | 0 | 0 | 0 | 0 | 1 | 4 | 8 | 6 | 2 | 0 | 0 | 0 | 21 |
| Days Max Temp ≤ 32 °F | 10 | 7 | 2 | 0 | 0 | 0 | 0 | 0 | 0 | 0 | 1 | 5 | 25 |
| Days Min Temp ≤ 32 °F | 26 | 22 | 17 | 5 | 0 | 0 | 0 | 0 | 0 | 4 | 13 | 22 | 109 |
| Days Min Temp ≤ 0 °F | 1 | 0 | 0 | 0 | 0 | 0 | 0 | 0 | 0 | 0 | 0 | 0 | 1 |
| Heating Degree Days | 1074 | 888 | 676 | 361 | 137 | 19 | 2 | 4 | 65 | 324 | 591 | 900 | 5041 |
| Cooling Degree Days | 0 | 0 | 2 | 17 | 78 | 208 | 341 | 281 | 125 | 19 | 2 | 0 | 1073 |
| Total Precipitation (") | 2.46 | 2.41 | 3.41 | 3.28 | 3.86 | 3.38 | 4.58 | 3.88 | 3.16 | 2.63 | 3.02 | 2.94 | 39.01 |
| Days ≥ 0.1" Precip | 6 | 6 | 8 | 8 | 8 | 6 | 8 | 7 | 6 | 6 | 7 | 7 | 83 |
| Total Snowfall (") | 8.3 | 4.9 | 3.0 | 0.9 | 0.0 | 0.0 | 0.0 | 0.0 | 0.0 | 0.0 | 0.9 | ***2.4*** | 20.4 |
| Days ≥ 1" Snow Depth | 8 | 7 | 1 | 0 | 0 | 0 | 0 | 0 | 0 | 0 | 0 | 2 | 18 |

## PARKERSBURG-WOOD AP *Wood County* ELEVATION 840 ft LAT/LONG 39° 21 ' N / 81° 26 ' W

| | JAN | FEB | MAR | APR | MAY | JUN | JUL | AUG | SEP | OCT | NOV | DEC | YEAR |
|---|---|---|---|---|---|---|---|---|---|---|---|---|---|
| Maximum Temp °F | 38.4 | 42.4 | 53.8 | 64.9 | 73.9 | 81.6 | 84.5 | 82.9 | 76.6 | 65.6 | 54.3 | 43.8 | 63.6 |
| Minimum Temp °F | 22.2 | 24.3 | 33.4 | 42.4 | 51.4 | 59.9 | 64.9 | 63.5 | 56.7 | 44.6 | 36.6 | 27.9 | 44.0 |
| Mean Temp °F | 30.3 | 33.4 | 43.6 | 53.6 | 62.6 | 70.8 | 74.7 | 73.2 | 66.7 | 55.1 | 45.5 | 35.9 | 53.8 |
| Days Max Temp ≥ 90 °F | 0 | 0 | 0 | 0 | 1 | 3 | 6 | 4 | 1 | 0 | 0 | 0 | 15 |
| Days Max Temp ≤ 32 °F | 10 | 7 | 1 | 0 | 0 | 0 | 0 | 0 | 0 | 0 | 0 | 5 | 23 |
| Days Min Temp ≤ 32 °F | 25 | 21 | 16 | 5 | 0 | 0 | 0 | 0 | 0 | 3 | 11 | 20 | 101 |
| Days Min Temp ≤ 0 °F | 2 | 1 | 0 | 0 | 0 | 0 | 0 | 0 | 0 | 0 | 0 | 0 | 3 |
| Heating Degree Days | 1067 | 886 | 659 | 351 | 141 | 19 | 2 | 5 | 65 | 318 | 581 | 894 | 4988 |
| Cooling Degree Days | 0 | 0 | 4 | 19 | 76 | 200 | 326 | 270 | 121 | 20 | 3 | 1 | 1040 |
| Total Precipitation (") | 2.81 | 2.66 | 3.76 | 3.47 | 4.01 | 3.62 | 4.30 | 4.35 | 3.17 | 2.98 | 3.30 | 3.29 | 41.72 |
| Days ≥ 0.1" Precip | 7 | 7 | 8 | 8 | 8 | 7 | 8 | 7 | 6 | 6 | 8 | 8 | 88 |
| Total Snowfall (") | 9.7 | 5.7 | 3.7 | 0.7 | 0.0 | 0.0 | 0.0 | 0.0 | 0.0 | 0.0 | 1.2 | 3.1 | 24.1 |
| Days ≥ 1" Snow Depth | 11 | 8 | 2 | 0 | 0 | 0 | 0 | 0 | 0 | 0 | 1 | 3 | 25 |

## PARSONS 1 NE *Tucker County* ELEVATION 1650 ft LAT/LONG 39° 6 ' N / 79° 40 ' W

| | JAN | FEB | MAR | APR | MAY | JUN | JUL | AUG | SEP | OCT | NOV | DEC | YEAR |
|---|---|---|---|---|---|---|---|---|---|---|---|---|---|
| Maximum Temp °F | 37.5 | 40.8 | 50.9 | 61.7 | 71.1 | 78.4 | 81.7 | 80.5 | 74.4 | 63.6 | 52.6 | 42.2 | 61.3 |
| Minimum Temp °F | 17.5 | 19.2 | 27.6 | 36.4 | 45.7 | 54.2 | 59.4 | 58.0 | 51.4 | 38.3 | 30.9 | 23.1 | 38.5 |
| Mean Temp °F | 27.6 | 30.0 | 39.2 | 49.1 | 58.4 | 66.3 | 70.6 | 69.3 | 62.9 | 51.0 | 41.8 | 32.7 | 49.9 |
| Days Max Temp ≥ 90 °F | 0 | 0 | 0 | 0 | 0 | 1 | 2 | 1 | 0 | 0 | 0 | 0 | 4 |
| Days Max Temp ≤ 32 °F | 10 | 8 | 3 | 0 | 0 | 0 | 0 | 0 | 0 | 0 | 2 | 7 | 30 |
| Days Min Temp ≤ 32 °F | 28 | 25 | 22 | 12 | 2 | 0 | 0 | 0 | 0 | 9 | 18 | 25 | 141 |
| Days Min Temp ≤ 0 °F | 3 | 2 | 0 | 0 | 0 | 0 | 0 | 0 | 0 | 0 | 0 | 1 | 6 |
| Heating Degree Days | 1154 | 981 | 792 | 475 | 225 | 60 | 11 | 19 | 118 | 431 | 690 | 994 | 5950 |
| Cooling Degree Days | 0 | 0 | 0 | 4 | 29 | 106 | 212 | 163 | 61 | 3 | 0 | 0 | 578 |
| Total Precipitation (") | 3.57 | 3.26 | 4.15 | 4.57 | 4.88 | 5.03 | 5.43 | 4.85 | 4.21 | 3.45 | 3.79 | 4.12 | 51.31 |
| Days ≥ 0.1" Precip | 9 | 8 | 10 | 10 | 10 | 9 | 10 | 9 | 7 | 7 | 8 | 10 | 107 |
| Total Snowfall (") | ***17.0*** | ***13.7*** | ***9.0*** | 1.9 | 0.0 | 0.0 | 0.0 | 0.0 | 0.0 | 0.1 | 3.5 | ***10.4*** | 55.6 |
| Days ≥ 1" Snow Depth | 17 | ***14*** | 6 | 1 | 0 | 0 | 0 | 0 | 0 | 0 | 3 | ***10*** | 51 |

## PINEVILLE *Wyoming County* ELEVATION 1352 ft LAT/LONG 37° 35 ' N / 81° 32 ' W

| | JAN | FEB | MAR | APR | MAY | JUN | JUL | AUG | SEP | OCT | NOV | DEC | YEAR |
|---|---|---|---|---|---|---|---|---|---|---|---|---|---|
| Maximum Temp °F | 41.0 | 45.5 | 56.0 | 67.3 | 75.5 | 82.4 | 85.3 | 84.2 | 78.2 | 67.4 | 56.7 | 45.7 | 65.4 |
| Minimum Temp °F | 20.9 | 22.1 | 30.0 | 38.1 | 47.5 | 56.4 | 61.9 | 60.9 | 54.4 | 40.6 | 32.1 | 25.1 | 40.8 |
| Mean Temp °F | 31.0 | 33.8 | 43.0 | 52.7 | 61.5 | 69.4 | 73.6 | 72.6 | 66.3 | 54.0 | 44.4 | 35.4 | 53.1 |
| Days Max Temp ≥ 90 °F | 0 | 0 | 0 | 0 | 0 | 3 | 7 | 4 | 1 | 0 | 0 | 0 | 15 |
| Days Max Temp ≤ 32 °F | 7 | 4 | 1 | 0 | 0 | 0 | 0 | 0 | 0 | 0 | 0 | 4 | 16 |
| Days Min Temp ≤ 32 °F | 26 | 24 | 20 | 9 | 2 | 0 | 0 | 0 | 0 | 6 | 16 | 24 | 127 |
| Days Min Temp ≤ 0 °F | 2 | 1 | 0 | 0 | 0 | 0 | 0 | 0 | 0 | 0 | 0 | 0 | 3 |
| Heating Degree Days | 1049 | 875 | 676 | 369 | 149 | 24 | 2 | 4 | 62 | 343 | 611 | 910 | 5074 |
| Cooling Degree Days | 0 | 0 | 1 | 9 | 53 | 178 | 310 | 255 | 112 | 11 | 0 | 0 | 929 |
| Total Precipitation (") | 3.42 | 3.06 | 3.85 | 3.96 | 4.53 | 3.78 | 4.79 | 3.96 | 3.63 | 3.31 | 3.17 | 3.55 | 45.01 |
| Days ≥ 0.1" Precip | 8 | 7 | 9 | 9 | 9 | 8 | 9 | 8 | 7 | 6 | 8 | 8 | 96 |
| Total Snowfall (") | 8.3 | 5.7 | 3.0 | 0.8 | 0.0 | 0.0 | 0.0 | 0.0 | 0.0 | 0.0 | 0.8 | 4.2 | 22.8 |
| Days ≥ 1" Snow Depth | 9 | 6 | 2 | 0 | 0 | 0 | 0 | 0 | 0 | 0 | 1 | 4 | 22 |

## RIPLEY *Jackson County* ELEVATION 590 ft LAT/LONG 38° 49 ' N / 81° 43 ' W

| | JAN | FEB | MAR | APR | MAY | JUN | JUL | AUG | SEP | OCT | NOV | DEC | YEAR |
|---|---|---|---|---|---|---|---|---|---|---|---|---|---|
| Maximum Temp °F | 42.1 | 46.6 | 57.7 | 68.6 | 77.6 | 84.5 | 87.7 | 85.7 | 80.1 | 69.1 | 57.5 | 46.9 | 67.0 |
| Minimum Temp °F | 20.9 | 23.2 | 31.5 | 39.9 | 49.2 | 57.5 | 62.5 | 60.6 | 53.6 | 41.2 | 33.3 | 26.4 | 41.7 |
| Mean Temp °F | 31.5 | 34.9 | 44.6 | 54.3 | 63.4 | 71.0 | 75.1 | 73.2 | 66.9 | 55.2 | 45.5 | 36.6 | 54.4 |
| Days Max Temp ≥ 90 °F | 0 | 0 | 0 | 1 | 2 | 7 | 11 | 8 | 3 | 0 | 0 | 0 | 32 |
| Days Max Temp ≤ 32 °F | 7 | 3 | 0 | 0 | 0 | 0 | 0 | 0 | 0 | 0 | 0 | 3 | 13 |
| Days Min Temp ≤ 32 °F | 26 | 22 | 17 | 8 | 1 | 0 | 0 | 0 | 0 | 7 | 15 | 22 | 118 |
| Days Min Temp ≤ 0 °F | 2 | 1 | 0 | 0 | 0 | 0 | 0 | 0 | 0 | 0 | 0 | 1 | 4 |
| Heating Degree Days | 1032 | 844 | 627 | 330 | 118 | 17 | 1 | 5 | 63 | 311 | 578 | 873 | 4799 |
| Cooling Degree Days | 0 | 0 | 2 | 15 | 72 | 197 | 333 | 263 | 119 | 14 | 0 | 0 | 1015 |
| Total Precipitation (") | 3.21 | 3.10 | 4.00 | 3.83 | 4.25 | 3.63 | 4.83 | 3.80 | 3.31 | 3.36 | 3.62 | 3.61 | 44.55 |
| Days ≥ 0.1" Precip | 9 | 8 | 9 | 9 | 9 | 8 | 8 | 7 | 7 | 7 | 8 | 9 | 98 |
| Total Snowfall (") | 8.5 | 6.3 | 2.5 | 0.7 | 0.0 | 0.0 | 0.0 | 0.0 | 0.0 | 0.1 | 0.8 | 2.9 | 21.8 |
| Days ≥ 1" Snow Depth | 10 | 7 | 1 | 0 | 0 | 0 | 0 | 0 | 0 | 0 | 0 | 3 | 21 |

## ROMNEY 1 SW *Hampshire County* ELEVATION 670 ft LAT/LONG 39° 20 ' N / 78° 46 ' W

| | JAN | FEB | MAR | APR | MAY | JUN | JUL | AUG | SEP | OCT | NOV | DEC | YEAR |
|---|---|---|---|---|---|---|---|---|---|---|---|---|---|
| Maximum Temp °F | 39.3 | 43.9 | 53.8 | 65.7 | 75.2 | 83.4 | 86.8 | 85.1 | 78.5 | 67.3 | 55.6 | 44.1 | 64.9 |
| Minimum Temp °F | 19.3 | 21.3 | 29.3 | 38.2 | 47.5 | 56.0 | 61.1 | 59.5 | 52.5 | 39.8 | 31.9 | 24.7 | 40.1 |
| Mean Temp °F | 29.3 | 32.6 | 41.6 | 52.0 | 61.4 | 69.7 | 74.0 | 72.3 | 65.5 | 53.6 | 43.8 | 34.4 | 52.5 |
| Days Max Temp ≥ 90 °F | 0 | 0 | 0 | 0 | 1 | 7 | 10 | 8 | 3 | 0 | 0 | 0 | 29 |
| Days Max Temp ≤ 32 °F | 8 | 4 | 1 | 0 | 0 | 0 | 0 | 0 | 0 | 0 | 0 | 4 | 17 |
| Days Min Temp ≤ 32 °F | 28 | 24 | 20 | 9 | 1 | 0 | 0 | 0 | 0 | 8 | 16 | 25 | 131 |
| Days Min Temp ≤ 0 °F | 2 | 1 | 0 | 0 | 0 | 0 | 0 | 0 | 0 | 0 | 0 | 0 | 3 |
| Heating Degree Days | 1099 | 909 | 721 | 394 | 161 | 26 | 3 | 9 | 82 | 355 | 628 | 942 | 5329 |
| Cooling Degree Days | 0 | 0 | 2 | 10 | 61 | 177 | 308 | 240 | 103 | 10 | 0 | 0 | 911 |
| Total Precipitation (") | 2.30 | 1.93 | 3.08 | 2.89 | 3.49 | 3.12 | 3.90 | 3.17 | 2.94 | 2.69 | 2.50 | 2.48 | 34.49 |
| Days ≥ 0.1" Precip | 6 | 5 | 6 | 7 | 8 | 7 | 8 | 7 | 6 | 5 | 6 | 6 | 77 |
| Total Snowfall (") | 9.2 | 7.4 | 5.9 | 0.3 | 0.0 | 0.0 | 0.0 | 0.0 | 0.0 | 0.0 | *1.2* | 4.6 | 28.6 |
| Days ≥ 1" Snow Depth | 12 | 7 | 4 | 0 | 0 | 0 | 0 | 0 | 0 | 0 | 1 | 5 | 29 |

## ROWLESBURG 1 *Preston County* ELEVATION 1381 ft LAT/LONG 39° 21 ' N / 79° 40 ' W

| | JAN | FEB | MAR | APR | MAY | JUN | JUL | AUG | SEP | OCT | NOV | DEC | YEAR |
|---|---|---|---|---|---|---|---|---|---|---|---|---|---|
| Maximum Temp °F | 37.5 | 41.0 | 51.5 | 62.6 | 71.9 | 79.0 | 82.2 | 81.0 | 75.1 | 64.1 | 53.2 | 42.1 | 61.8 |
| Minimum Temp °F | 18.8 | 20.2 | 28.2 | 36.9 | 46.1 | 54.6 | 60.1 | 59.1 | 52.6 | 40.0 | 32.2 | 24.3 | 39.4 |
| Mean Temp °F | 28.2 | 30.6 | 39.9 | 49.7 | 59.0 | 66.9 | 71.1 | 70.1 | 63.9 | 52.1 | 42.7 | 33.2 | 50.6 |
| Days Max Temp ≥ 90 °F | 0 | 0 | 0 | 0 | 0 | 1 | 2 | 2 | 1 | 0 | 0 | 0 | 6 |
| Days Max Temp ≤ 32 °F | 11 | 7 | 2 | 0 | 0 | 0 | 0 | 0 | 0 | 0 | 1 | 6 | 27 |
| Days Min Temp ≤ 32 °F | 28 | 24 | 22 | 10 | 2 | 0 | 0 | 0 | 0 | 6 | 16 | 25 | 133 |
| Days Min Temp ≤ 0 °F | 3 | 2 | 0 | 0 | 0 | 0 | 0 | 0 | 0 | 0 | 0 | 1 | 6 |
| Heating Degree Days | 1136 | 964 | 772 | 455 | 209 | 46 | 7 | 14 | 98 | 398 | 663 | 978 | 5740 |
| Cooling Degree Days | 0 | 0 | 0 | 4 | 31 | 98 | 213 | 173 | 67 | 3 | 0 | 0 | 589 |
| Total Precipitation (") | 4.18 | 3.83 | 4.75 | 4.80 | 5.02 | 5.05 | 5.63 | 4.64 | 4.03 | 3.76 | 4.41 | 4.73 | 54.83 |
| Days ≥ 0.1" Precip | 11 | 10 | 11 | 11 | 10 | 10 | 9 | 9 | 8 | 8 | 10 | 11 | 118 |
| Total Snowfall (") | 19.2 | 15.1 | 8.7 | 1.3 | 0.0 | 0.0 | 0.0 | 0.0 | 0.0 | 0.0 | 3.3 | 11.2 | 58.8 |
| Days ≥ 1" Snow Depth | 16 | 14 | 6 | 1 | 0 | 0 | 0 | 0 | 0 | 0 | 3 | 10 | 50 |

## SENECA STATE FOREST *Pocahontas County* ELEVATION 2651 ft LAT/LONG 38° 18' N / 79° 56' W

| | JAN | FEB | MAR | APR | MAY | JUN | JUL | AUG | SEP | OCT | NOV | DEC | YEAR |
|---|---|---|---|---|---|---|---|---|---|---|---|---|---|
| Maximum Temp °F | 38.2 | 42.8 | 51.8 | 62.7 | 71.6 | 78.7 | 82.1 | 80.5 | 74.4 | 63.9 | 52.5 | 42.3 | 61.8 |
| Minimum Temp °F | 13.9 | 16.4 | 24.2 | 31.3 | 40.8 | 49.2 | 54.2 | 53.2 | 46.4 | 34.0 | 26.1 | 18.7 | 34.0 |
| Mean Temp °F | 26.0 | 29.8 | 38.0 | 47.0 | 56.2 | 63.9 | 68.2 | 66.9 | 60.4 | 49.0 | 39.3 | 30.5 | 47.9 |
| Days Max Temp ≥ 90 °F | 0 | 0 | 0 | 0 | 0 | 1 | 2 | 2 | 0 | 0 | 0 | 0 | 5 |
| Days Max Temp ≤ 32 °F | 9 | 6 | 2 | 0 | 0 | 0 | 0 | 0 | 0 | 0 | 1 | 5 | 23 |
| Days Min Temp ≤ 32 °F | 29 | 27 | 25 | 17 | 7 | 1 | 0 | 0 | 2 | 15 | 22 | 28 | 173 |
| Days Min Temp ≤ 0 °F | 5 | 3 | 0 | 0 | 0 | 0 | 0 | 0 | 0 | 0 | 0 | 2 | 10 |
| Heating Degree Days | 1201 | 989 | 829 | 533 | 276 | 83 | 21 | 33 | 162 | 490 | 763 | 1062 | 6442 |
| Cooling Degree Days | 0 | 0 | 0 | 1 | 16 | 65 | 148 | 102 | 31 | 1 | 0 | 0 | 364 |
| Total Precipitation (") | 3.41 | 3.41 | 4.44 | 3.80 | 4.59 | 4.02 | 5.05 | 4.38 | 3.53 | 3.51 | 3.85 | 4.31 | 48.30 |
| Days ≥ 0.1" Precip | 8 | 7 | 9 | 8 | 9 | 9 | 9 | 8 | 6 | 6 | 7 | 8 | 94 |
| Total Snowfall (") | 17.2 | 14.8 | 12.3 | 2.8 | 0.1 | 0.0 | 0.0 | 0.0 | 0.0 | 0.7 | 5.5 | 12.2 | 65.6 |
| Days ≥ 1" Snow Depth | 19 | 17 | 8 | 1 | 0 | 0 | 0 | 0 | 0 | 0 | 3 | 11 | 59 |

## SPENCER 1 SE *Roane County* ELEVATION 791 ft LAT/LONG 38° 48' N / 81° 21' W

| | JAN | FEB | MAR | APR | MAY | JUN | JUL | AUG | SEP | OCT | NOV | DEC | YEAR |
|---|---|---|---|---|---|---|---|---|---|---|---|---|---|
| Maximum Temp °F | 41.2 | 45.0 | 56.1 | 67.0 | 75.6 | 82.7 | 85.8 | 84.4 | 78.4 | 67.8 | 56.2 | 45.3 | 65.5 |
| Minimum Temp °F | 20.2 | 21.5 | 29.7 | 38.7 | 47.7 | 56.6 | 61.6 | 59.7 | 52.7 | 40.5 | 32.7 | 25.1 | 40.6 |
| Mean Temp °F | 30.7 | 33.3 | 42.9 | 52.9 | 61.8 | 69.7 | 73.7 | 72.1 | 65.5 | 54.1 | 44.5 | 35.2 | 53.0 |
| Days Max Temp ≥ 90 °F | 0 | 0 | 0 | 0 | 1 | 4 | 7 | 5 | 1 | 0 | 0 | 0 | 18 |
| Days Max Temp ≤ 32 °F | 8 | 5 | 1 | 0 | 0 | 0 | 0 | 0 | 0 | 0 | 0 | 4 | 18 |
| Days Min Temp ≤ 32 °F | 26 | 23 | 19 | 9 | 2 | 0 | 0 | 0 | 1 | 8 | 15 | 23 | 126 |
| Days Min Temp ≤ 0 °F | 3 | 2 | 0 | 0 | 0 | 0 | 0 | 0 | 0 | 0 | 0 | 1 | 6 |
| Heating Degree Days | 1057 | 889 | 681 | 372 | 154 | 27 | 3 | 8 | 82 | 342 | 611 | 917 | 5143 |
| Cooling Degree Days | 0 | 0 | 2 | 13 | 64 | 175 | 308 | 234 | 100 | 12 | 1 | 1 | 910 |
| Total Precipitation (") | 3.27 | 2.95 | 3.71 | 3.48 | 4.02 | 3.35 | 4.83 | 3.91 | 3.80 | 3.13 | 3.65 | 3.44 | 43.54 |
| Days ≥ 0.1" Precip | 8 | 7 | 8 | 8 | 8 | 7 | 8 | 7 | 7 | 6 | 8 | 8 | 90 |
| Total Snowfall (") | ***10.6*** | na | ***3.9*** | 0.6 | 0.0 | 0.0 | 0.0 | 0.0 | 0.0 | 0.0 | 0.6 | ***3.1*** | na |
| Days ≥ 1" Snow Depth | ***10*** | na | ***3*** | 0 | 0 | 0 | 0 | 0 | 0 | 0 | 0 | ***4*** | na |

## TERRA ALTA NO 1 *Preston County* ELEVATION 2543 ft LAT/LONG 39° 26' N / 79° 31' W

| | JAN | FEB | MAR | APR | MAY | JUN | JUL | AUG | SEP | OCT | NOV | DEC | YEAR |
|---|---|---|---|---|---|---|---|---|---|---|---|---|---|
| Maximum Temp °F | 33.3 | 36.7 | 46.8 | 58.1 | 67.3 | 74.5 | 77.8 | 76.5 | 70.5 | 59.9 | 49.2 | 38.6 | 57.4 |
| Minimum Temp °F | 16.4 | 18.3 | 26.9 | 36.8 | 46.1 | 53.7 | 58.6 | 57.1 | 51.0 | 39.9 | 31.8 | 22.4 | 38.3 |
| Mean Temp °F | 24.9 | 27.5 | 36.8 | 47.5 | 56.7 | 64.1 | 68.2 | 66.8 | 60.8 | 49.9 | 40.5 | 30.6 | 47.9 |
| Days Max Temp ≥ 90 °F | 0 | 0 | 0 | 0 | 0 | 0 | 0 | 1 | 0 | 0 | 0 | 0 | 1 |
| Days Max Temp ≤ 32 °F | 15 | 10 | 5 | 1 | 0 | 0 | 0 | 0 | 0 | 0 | 3 | 11 | 45 |
| Days Min Temp ≤ 32 °F | 28 | 25 | 22 | 11 | 2 | 0 | 0 | 0 | 1 | 7 | 17 | 26 | 139 |
| Days Min Temp ≤ 0 °F | 4 | 2 | 0 | 0 | 0 | 0 | 0 | 0 | 0 | 0 | 0 | 1 | 7 |
| Heating Degree Days | 1238 | 1054 | 867 | 524 | 269 | 84 | 22 | 37 | 156 | 462 | 728 | 1061 | 6502 |
| Cooling Degree Days | 0 | 0 | 0 | 4 | 19 | 67 | 147 | 108 | 37 | 1 | 0 | 0 | 383 |
| Total Precipitation (") | 4.24 | 3.74 | 4.75 | 4.77 | 4.73 | 4.97 | 5.92 | 4.56 | 3.84 | 3.46 | 4.19 | 4.77 | 53.94 |
| Days ≥ 0.1" Precip | 11 | 10 | 11 | 11 | 10 | 9 | 10 | 9 | 8 | 8 | 10 | 11 | 118 |
| Total Snowfall (") | 39.5 | 31.3 | 24.6 | 10.7 | 0.2 | 0.0 | 0.0 | 0.0 | 0.0 | 1.4 | 12.0 | 26.5 | 146.2 |
| Days ≥ 1" Snow Depth | 22 | 19 | 12 | 3 | 0 | 0 | 0 | 0 | 0 | 0 | 5 | 16 | 77 |

## UNION 3 SSE *Monroe County* ELEVATION 2113 ft LAT/LONG 37° 33' N / 80° 32' W

| | JAN | FEB | MAR | APR | MAY | JUN | JUL | AUG | SEP | OCT | NOV | DEC | YEAR |
|---|---|---|---|---|---|---|---|---|---|---|---|---|---|
| Maximum Temp °F | 40.8 | 44.7 | 54.3 | 64.2 | 72.4 | 79.3 | 82.8 | 81.5 | 75.6 | 65.5 | 54.9 | 45.0 | 63.4 |
| Minimum Temp °F | 20.1 | 22.4 | 30.0 | 37.7 | 46.2 | 54.1 | 58.9 | 57.4 | 50.8 | 38.7 | 31.0 | 24.1 | 39.3 |
| Mean Temp °F | 30.4 | 33.5 | 42.2 | 51.0 | 59.3 | 66.7 | 70.9 | 69.5 | 63.2 | 52.1 | 43.0 | 34.6 | 51.4 |
| Days Max Temp ≥ 90 °F | 0 | 0 | 0 | 0 | 0 | 0 | 2 | 1 | 0 | 0 | 0 | 0 | 3 |
| Days Max Temp ≤ 32 °F | 7 | 5 | 1 | 0 | 0 | 0 | 0 | 0 | 0 | 0 | 1 | 4 | 18 |
| Days Min Temp ≤ 32 °F | 27 | 23 | 19 | 9 | 2 | 0 | 0 | 0 | 1 | 9 | 17 | 24 | 131 |
| Days Min Temp ≤ 0 °F | 2 | 1 | 0 | 0 | 0 | 0 | 0 | 0 | 0 | 0 | 0 | 1 | 4 |
| Heating Degree Days | 1050 | 882 | 701 | 416 | 191 | 45 | 8 | 14 | 109 | 395 | 653 | 936 | 5400 |
| Cooling Degree Days | 0 | 0 | 1 | 4 | 26 | 115 | 225 | 172 | 67 | 5 | 0 | 0 | 615 |
| Total Precipitation (") | 2.30 | 2.52 | 2.99 | 3.20 | 3.59 | 3.12 | 3.63 | 3.23 | 3.07 | 2.72 | 2.39 | 2.39 | 35.15 |
| Days ≥ 0.1" Precip | 6 | 6 | 7 | 7 | 8 | 6 | 8 | 7 | 6 | 5 | 5 | 5 | 76 |
| Total Snowfall (") | ***8.2*** | ***6.6*** | 3.6 | 1.1 | 0.0 | 0.0 | 0.0 | 0.0 | 0.0 | 0.1 | 1.1 | 4.2 | 24.9 |
| Days ≥ 1" Snow Depth | ***10*** | ***6*** | 2 | 0 | 0 | 0 | 0 | 0 | 0 | 0 | 1 | 4 | 23 |

## WARDENSVILLE RM FARM *Hardy County* ELEVATION 961 ft LAT/LONG 39° 6 ' N / 78° 35 ' W

| | JAN | FEB | MAR | APR | MAY | JUN | JUL | AUG | SEP | OCT | NOV | DEC | YEAR |
|---|---|---|---|---|---|---|---|---|---|---|---|---|---|
| Maximum Temp °F | 40.2 | 43.4 | 52.8 | 63.9 | 73.5 | 81.8 | 85.7 | 84.2 | 77.8 | 66.6 | 56.0 | 45.1 | 64.3 |
| Minimum Temp °F | 17.4 | 20.0 | 28.2 | 37.2 | 46.6 | 55.1 | 60.0 | 58.1 | 50.6 | 37.9 | 30.9 | 23.2 | 38.8 |
| Mean Temp °F | 28.9 | 31.7 | 40.5 | 50.6 | 60.0 | 68.5 | 72.9 | 71.2 | 64.3 | 52.3 | 43.5 | 34.1 | 51.5 |
| Days Max Temp ≥ 90 °F | 0 | 0 | 0 | 0 | 0 | 4 | 9 | 7 | 2 | 0 | 0 | 0 | 22 |
| Days Max Temp ≤ 32 °F | 8 | 5 | 1 | 0 | 0 | 0 | 0 | 0 | 0 | 0 | 0 | 4 | 18 |
| Days Min Temp ≤ 32 °F | 29 | 25 | 22 | 9 | 2 | 0 | 0 | 0 | 1 | 10 | 17 | 26 | 141 |
| Days Min Temp ≤ 0 °F | 2 | 1 | 0 | 0 | 0 | 0 | 0 | 0 | 0 | 0 | 0 | 1 | 4 |
| Heating Degree Days | 1113 | 933 | 752 | 433 | 190 | 39 | 5 | 13 | 100 | 393 | 640 | 950 | 5561 |
| Cooling Degree Days | 0 | 0 | 1 | 6 | 47 | 154 | 279 | 213 | 91 | 8 | 0 | 0 | 799 |
| Total Precipitation (") | 2.00 | 1.89 | 2.90 | 2.84 | 3.41 | 3.25 | 3.73 | 3.37 | 2.94 | 2.98 | 2.77 | 2.28 | 34.36 |
| Days ≥ 0.1" Precip | 5 | 4 | 6 | 7 | 7 | 7 | 7 | 6 | 5 | 6 | 5 | 5 | 70 |
| Total Snowfall (") | 9.3 | 6.2 | 4.9 | 0.6 | 0.0 | 0.0 | 0.0 | 0.0 | 0.0 | 0.1 | 1.6 | 4.9 | 27.6 |
| Days ≥ 1" Snow Depth | 9 | 5 | 2 | 0 | 0 | 0 | 0 | 0 | 0 | 0 | 1 | 2 | 19 |

## WELLSBURG WTR TRMT P *Brooke County* ELEVATION 669 ft LAT/LONG 40° 18 ' N / 80° 35 ' W

| | JAN | FEB | MAR | APR | MAY | JUN | JUL | AUG | SEP | OCT | NOV | DEC | YEAR |
|---|---|---|---|---|---|---|---|---|---|---|---|---|---|
| Maximum Temp °F | 36.7 | 40.8 | 52.3 | 64.5 | 74.6 | 81.7 | 85.0 | 83.7 | 77.2 | 65.5 | 53.6 | 42.8 | 63.2 |
| Minimum Temp °F | 19.3 | 21.5 | 29.9 | 38.7 | 48.4 | 56.3 | 61.1 | 60.0 | 53.7 | 42.2 | 34.7 | 26.3 | 41.0 |
| Mean Temp °F | 28.0 | 31.2 | 41.1 | 51.6 | 61.5 | 69.1 | 73.1 | 71.9 | 65.5 | 53.9 | 44.3 | 34.6 | 52.2 |
| Days Max Temp ≥ 90 °F | 0 | 0 | 0 | 0 | 0 | 3 | 7 | 3 | 1 | 0 | 0 | 0 | 14 |
| Days Max Temp ≤ 32 °F | 12 | 7 | 1 | 0 | 0 | 0 | 0 | 0 | 0 | 0 | 1 | 5 | 26 |
| Days Min Temp ≤ 32 °F | 26 | 23 | 19 | 8 | 1 | 0 | 0 | 0 | 0 | 5 | 13 | 23 | 118 |
| Days Min Temp ≤ 0 °F | 2 | 1 | 0 | 0 | 0 | 0 | 0 | 0 | 0 | 0 | 0 | 0 | 3 |
| Heating Degree Days | 1139 | 949 | 734 | 401 | 157 | 29 | 4 | 9 | 76 | 347 | 616 | 936 | 5397 |
| Cooling Degree Days | na | na | na | na | na | na | na | na | na | na | na | na | na |
| Total Precipitation (") | 2.28 | 2.11 | 3.04 | 3.20 | 4.60 | 3.67 | 4.41 | 3.71 | 2.95 | 2.86 | 3.26 | 2.74 | 38.83 |
| Days ≥ 0.1" Precip | 6 | 5 | 8 | 7 | 9 | 8 | 8 | 7 | 6 | 7 | 8 | 7 | 86 |
| Total Snowfall (") | 7.7 | na | 3.2 | 0.4 | 0.0 | 0.0 | 0.0 | 0.0 | 0.0 | 0.0 | 1.6 | 2.8 | na |
| Days ≥ 1" Snow Depth | na | na | 3 | 0 | 0 | 0 | 0 | 0 | 0 | 0 | 1 | 4 | na |

## WESTON *Lewis County* ELEVATION 1030 ft LAT/LONG 39° 2 ' N / 80° 28 ' W

| | JAN | FEB | MAR | APR | MAY | JUN | JUL | AUG | SEP | OCT | NOV | DEC | YEAR |
|---|---|---|---|---|---|---|---|---|---|---|---|---|---|
| Maximum Temp °F | 39.7 | 43.3 | 54.1 | 65.4 | 74.6 | 82.3 | 85.5 | 84.0 | 78.0 | 66.5 | 55.8 | 44.9 | 64.5 |
| Minimum Temp °F | 18.8 | 19.5 | 27.9 | 36.5 | 45.6 | 54.7 | 60.2 | 58.8 | 52.0 | 38.7 | 31.5 | 24.4 | 39.1 |
| Mean Temp °F | 29.3 | 31.5 | 41.0 | 51.0 | 60.1 | 68.5 | 72.9 | 71.4 | 65.0 | 52.6 | 43.7 | 34.7 | 51.8 |
| Days Max Temp ≥ 90 °F | 0 | 0 | 0 | 0 | 1 | 5 | 8 | 5 | 2 | 0 | 0 | 0 | 21 |
| Days Max Temp ≤ 32 °F | 10 | 6 | 2 | 0 | 0 | 0 | 0 | 0 | 0 | 0 | 0 | 5 | 23 |
| Days Min Temp ≤ 32 °F | 27 | 24 | 21 | 11 | 2 | 0 | 0 | 0 | 0 | 9 | 17 | 24 | 135 |
| Days Min Temp ≤ 0 °F | 3 | 2 | 0 | 0 | 0 | 0 | 0 | 0 | 0 | 0 | 0 | 1 | 6 |
| Heating Degree Days | 1100 | 941 | 737 | 421 | 186 | 38 | 5 | 10 | 83 | 382 | 634 | 932 | 5469 |
| Cooling Degree Days | 0 | 0 | 1 | 8 | 40 | 148 | 269 | 213 | 84 | 6 | 1 | 0 | 770 |
| Total Precipitation (") | 3.90 | 3.31 | 4.63 | 3.91 | 4.39 | 4.22 | 5.14 | 4.45 | 4.06 | 3.25 | 3.96 | 4.17 | 49.39 |
| Days ≥ 0.1" Precip | 10 | 9 | 10 | 10 | 9 | 8 | 10 | 8 | 7 | 7 | 9 | 10 | 107 |
| Total Snowfall (") | 15.3 | 10.8 | 7.5 | 1.2 | 0.1 | 0.0 | 0.0 | 0.0 | 0.0 | 0.1 | 2.9 | 7.8 | 45.7 |
| Days ≥ 1" Snow Depth | 13 | 10 | 4 | 0 | 0 | 0 | 0 | 0 | 0 | 0 | 1 | 7 | 35 |

## WHITE SULPHUR SPRNGS *Greenbrier County* ELEVATION 1913 ft LAT/LONG 37° 48 ' N / 80° 18 ' W

| | JAN | FEB | MAR | APR | MAY | JUN | JUL | AUG | SEP | OCT | NOV | DEC | YEAR |
|---|---|---|---|---|---|---|---|---|---|---|---|---|---|
| Maximum Temp °F | 41.5 | 46.3 | 56.8 | 67.2 | 75.2 | 81.4 | 84.8 | 83.6 | 77.8 | 68.0 | 56.1 | 45.6 | 65.4 |
| Minimum Temp °F | 20.5 | 22.3 | 29.8 | 37.2 | 46.4 | 54.4 | 59.9 | 58.8 | 52.2 | 39.3 | 31.6 | 24.4 | 39.7 |
| Mean Temp °F | 31.0 | 34.3 | 43.4 | 52.2 | 60.8 | 67.9 | 72.4 | 71.3 | 65.0 | 53.7 | 43.9 | 35.0 | 52.6 |
| Days Max Temp ≥ 90 °F | 0 | 0 | 0 | 0 | 0 | 2 | 5 | 4 | 2 | 0 | 0 | 0 | 13 |
| Days Max Temp ≤ 32 °F | 6 | 4 | 1 | 0 | 0 | 0 | 0 | 0 | 0 | 0 | 0 | 4 | 15 |
| Days Min Temp ≤ 32 °F | 27 | 23 | 19 | 10 | 2 | 0 | 0 | 0 | 1 | 9 | 17 | 24 | 132 |
| Days Min Temp ≤ 0 °F | 2 | 1 | 0 | 0 | 0 | 0 | 0 | 0 | 0 | 0 | 0 | 1 | 4 |
| Heating Degree Days | 1047 | 860 | 662 | 382 | 160 | 34 | 3 | 8 | 82 | 353 | 627 | 923 | 5141 |
| Cooling Degree Days | 0 | 0 | 1 | 4 | 35 | 120 | 239 | 190 | 75 | 9 | 0 | 0 | 673 |
| Total Precipitation (") | 2.92 | 2.84 | 3.54 | 3.31 | 4.02 | 3.10 | 4.19 | 3.50 | 3.12 | 3.19 | 2.76 | 3.03 | 39.52 |
| Days ≥ 0.1" Precip | 7 | 7 | 7 | 8 | 8 | 7 | 8 | 7 | 5 | 6 | 6 | 7 | 83 |
| Total Snowfall (") | 7.7 | 5.9 | 3.4 | 0.6 | 0.0 | 0.0 | 0.0 | 0.0 | 0.0 | 0.0 | 1.3 | 3.9 | 22.8 |
| Days ≥ 1" Snow Depth | na | 6 | 2 | 0 | 0 | 0 | 0 | 0 | 0 | 0 | 0 | na | na |

## WINFIELD LOCKS *Putnam County* ELEVATION 571 ft LAT/LONG 38° 32' N / 81° 55' W

| | JAN | FEB | MAR | APR | MAY | JUN | JUL | AUG | SEP | OCT | NOV | DEC | YEAR |
|---|---|---|---|---|---|---|---|---|---|---|---|---|---|
| Maximum Temp °F | 41.2 | 44.8 | 55.5 | 66.8 | 75.1 | 82.4 | 86.1 | 84.7 | 78.6 | 67.9 | 57.1 | 46.5 | 65.6 |
| Minimum Temp °F | 22.1 | 23.8 | 31.7 | 40.0 | 49.6 | 58.7 | 64.3 | 63.3 | 56.8 | 43.9 | 35.2 | 27.6 | 43.1 |
| Mean Temp °F | 31.7 | 34.3 | 43.6 | 53.3 | 62.4 | 70.6 | 75.2 | 74.0 | 67.7 | 56.0 | 46.2 | 37.1 | 54.3 |
| Days Max Temp ≥ 90 °F | 0 | 0 | 0 | 0 | 0 | 4 | 8 | 6 | 2 | 0 | 0 | 0 | 20 |
| Days Max Temp ≤ 32 °F | 8 | 5 | 1 | 0 | 0 | 0 | 0 | 0 | 0 | 0 | 0 | 3 | 17 |
| Days Min Temp ≤ 32 °F | 26 | 22 | 18 | 6 | 0 | 0 | 0 | 0 | 0 | 3 | 13 | 21 | 109 |
| Days Min Temp ≤ 0 °F | 2 | 0 | 0 | 0 | 0 | 0 | 0 | 0 | 0 | 0 | 0 | 0 | 2 |
| Heating Degree Days | 1026 | 861 | 657 | 355 | 136 | 19 | 1 | 3 | 50 | 289 | 557 | 857 | 4811 |
| Cooling Degree Days | 0 | 0 | 1 | 12 | 63 | 195 | 350 | 298 | 136 | 16 | 1 | 0 | 1072 |
| Total Precipitation (") | 2.78 | 2.64 | 3.70 | 3.42 | 3.85 | 3.16 | 4.65 | 4.22 | 3.66 | 2.85 | 3.26 | 3.24 | 41.43 |
| Days ≥ 0.1" Precip | 7 | 6 | 8 | 8 | 8 | 7 | 8 | 7 | 6 | 6 | 7 | 7 | 85 |
| Total Snowfall (") | *4.7* | 4.0 | 2.5 | 0.9 | 0.0 | 0.0 | 0.0 | 0.0 | 0.0 | 0.0 | 0.3 | 1.8 | 14.2 |
| Days ≥ 1" Snow Depth | 9 | 6 | 1 | 0 | 0 | 0 | 0 | 0 | 0 | 0 | 0 | 2 | 18 |

## JANUARY MINIMUM TEMPERATURE °F

| LOWEST | | | HIGHEST | | |
|---|---|---|---|---|---|
| 1 | Seneca | 13.9 | 1 | Logan | 24.7 |
| 2 | Buckeye | 15.3 | 2 | Huntington | 23.9 |
| 3 | Canaan Valley | 15.6 | | London Locks | 23.9 |
| 4 | Bayard | 16.3 | 4 | Charleston | 23.7 |
| 5 | Terra Alta | 16.4 | 5 | Bluefield | 23.3 |
| 6 | Elkins | 16.9 | 6 | Corton | 22.7 |
| 7 | Belington | 17.3 | 7 | Parkersburg-Wd | 22.2 |
| 8 | Wardensville | 17.4 | 8 | Kopperston | 22.1 |
| 9 | Parsons | 17.5 | | Winfield Locks | 22.1 |
| 10 | Creston | 17.8 | 10 | Gassaway | 21.8 |
| 11 | Flat Top | 18.2 | | Parkersburg-1 NE | 21.8 |
| 12 | Glenville | 18.4 | 12 | Bluestone Lake | 21.6 |
| 13 | Rowlesburg | 18.8 | 13 | Athens | 21.5 |
| | Weston | 18.8 | 14 | Morgantown | 21.3 |
| 15 | Moundsville | 18.9 | 15 | Beckley-Raleigh | 21.2 |
| 16 | Buckhannon | 19.1 | | Martinsburg | 21.2 |
| 17 | Franklin | 19.3 | 17 | Pineville | 20.9 |
| | Romney | 19.3 | | Ripley | 20.9 |
| | Wellsburg | 19.3 | 19 | Madison | 20.8 |
| 20 | Beckley-VA | 19.5 | 20 | Lewisburg | 20.7 |
| | Kearneysville | 19.5 | 21 | White Slphr Sprgs | 20.5 |
| | Mathias | 19.5 | 22 | Morgantown L&D | 20.4 |
| 23 | Hamlin | 19.6 | 23 | Moorefield | 20.3 |
| | Middlebourne | 19.6 | 24 | Spencer | 20.2 |
| 25 | Oak Hill | 19.7 | 25 | Fairmont | 20.1 |

## JULY MAXIMUM TEMPERATURE °F

| HIGHEST | | | LOWEST | | |
|---|---|---|---|---|---|
| 1 | Ripley | 87.7 | 1 | Flat Top | 76.4 |
| 2 | Moorefield | 87.4 | 2 | Canaan Valley | 77.6 |
| 3 | Logan | 87.3 | 3 | Terra Alta | 77.8 |
| 4 | Martinsburg | 87.0 | 4 | Bluefield | 79.0 |
| 5 | Madison | 86.9 | 5 | Bayard | 79.2 |
| 6 | Romney | 86.8 | 6 | Beckley-Raleigh | 79.8 |
| 7 | Hamlin | 86.5 | 7 | Beckley-VA | 79.9 |
| 8 | Corton | 86.4 | 8 | Elkins | 81.0 |
| | Creston | 86.4 | 9 | Belington | 81.6 |
| | Hogstt Gllipls Dm | 86.4 | 10 | Parsons | 81.7 |
| 11 | Gassaway | 86.3 | 11 | Athens | 81.9 |
| 12 | Winfield Locks | 86.1 | 12 | Seneca | 82.1 |
| 13 | Charleston | 85.9 | 13 | Rowlesburg | 82.2 |
| | Huntington | 85.9 | 14 | Buckeye | 82.3 |
| | Moundsville | 85.9 | 15 | Oak Hill | 82.5 |
| 16 | Spencer | 85.8 | 16 | Union | 82.8 |
| 17 | Kearneysville | 85.7 | 17 | Kopperston | 83.3 |
| | Wardensville | 85.7 | 18 | Franklin | 83.5 |
| 19 | London Locks | 85.6 | 19 | Buckhannon | 83.6 |
| 20 | Glenville | 85.5 | 20 | Fairmont | 83.7 |
| | Parkersburg-1 NE | 85.5 | | Morgantown | 83.7 |
| | Weston | 85.5 | 22 | Lewisburg | 83.8 |
| 23 | Pineville | 85.3 | 23 | Mathias | 83.9 |
| 24 | Wellsburg | 85.0 | | Morgantown L&D | 83.9 |
| 25 | Middlebourne | 84.8 | 25 | Bluestone Lake | 84.4 |

## ANNUAL PRECIPITATION (")

| HIGHEST | | | LOWEST | | |
|---|---|---|---|---|---|
| 1 | Rowlesburg | 54.83 | 1 | Moorefield | 32.00 |
| 2 | Terra Alta | 53.94 | 2 | Franklin | 33.57 |
| 3 | Canaan Valley | 51.55 | 3 | Wardensville | 34.36 |
| 4 | Parsons | 51.31 | 4 | Romney | 34.49 |
| 5 | Kopperston | 50.08 | 5 | Union | 35.15 |
| 6 | Weston | 49.39 | 6 | Mathias | 35.78 |
| 7 | Belington | 48.42 | 7 | Athens | 36.79 |
| 8 | Seneca | 48.30 | | Bluestone Lake | 36.79 |
| 9 | Bayard | 47.97 | 9 | Martinsburg | 38.00 |
| 10 | Buckhannon | 47.91 | 10 | Bluefield | 38.56 |
| 11 | Gassaway | 47.34 | 11 | Wellsburg | 38.83 |
| 12 | Logan | 46.93 | 12 | Parkersburg-1 NE | 39.01 |
| 13 | Madison | 46.32 | 13 | Lewisburg | 39.32 |
| 14 | Buckeye | 45.45 | 14 | Kearneysville | 39.33 |
| 15 | Oak Hill | 45.44 | 15 | Beckley-VA | 39.38 |
| 16 | Flat Top | 45.22 | 16 | White Slphr Sprgs | 39.52 |
| 17 | Pineville | 45.01 | 17 | Hogstt Gllipls Dm | 40.07 |
| 18 | Elkins | 44.99 | 18 | Moundsville | 40.61 |
| 19 | London Locks | 44.65 | 19 | Beckley-Raleigh | 40.79 |
| 20 | Creston | 44.58 | 20 | Morgantown | 41.33 |
| 21 | Ripley | 44.55 | 21 | Winfield Locks | 41.43 |
| 22 | Glenville | 44.38 | 22 | Huntington | 41.67 |
| 23 | Corton | 44.19 | 23 | Parkersburg-Wd | 41.72 |
| 24 | Fairmont | 44.04 | 24 | Morgantown L&D | 42.17 |
| 25 | Middlebourne | 43.91 | 25 | Charleston | 42.74 |

## ANNUAL SNOWFALL (")

| HIGHEST | | | LOWEST | | |
|---|---|---|---|---|---|
| 1 | Terra Alta | 146.2 | 1 | Winfield Locks | 14.2 |
| 2 | Canaan Valley | 131.6 | 2 | Logan | 18.1 |
| 3 | Bayard | 96.0 | 3 | Parkersburg-1 NE | 20.4 |
| 4 | Elkins | 77.6 | 4 | Madison | 21.2 |
| 5 | Seneca | 65.6 | 5 | Bluestone Lake | 21.7 |
| 6 | Flat Top | 60.0 | 6 | Ripley | 21.8 |
| 7 | Beckley-Raleigh | 59.9 | 7 | Kearneysville | 22.6 |
| 8 | Rowlesburg | 58.8 | 8 | Pineville | 22.8 |
| 9 | Parsons | 55.6 | | White Slphr Sprgs | 22.8 |
| 10 | Buckhannon | 55.3 | 10 | Moorefield | 23.8 |
| 11 | Belington | 48.5 | 11 | Gassaway | 24.1 |
| 12 | Weston | 45.7 | | Parkersburg-Wd | 24.1 |
| 13 | Oak Hill | 43.1 | 13 | Hamlin | 24.2 |
| 14 | Buckeye | 40.6 | 14 | Glenville | 24.4 |
| 15 | Fairmont | 38.6 | 15 | Middlebourne | 24.6 |
| 16 | Beckley-VA | 37.3 | 16 | Union | 24.9 |
| 17 | Mathias | 37.1 | 17 | Martinsburg | 25.8 |
| 18 | Charleston | 35.7 | 18 | Huntington | 26.1 |
| 19 | Athens | 32.3 | 19 | Wardensville | 27.6 |
| 20 | Morgantown | 32.2 | 20 | Franklin | 28.4 |
| 21 | Bluefield | 30.6 | 21 | Romney | 28.6 |
| 22 | Clarksburg | 28.7 | 22 | Clarksburg | 28.7 |
| 23 | Romney | 28.6 | 23 | Bluefield | 30.6 |
| 24 | Franklin | 28.4 | 24 | Morgantown | 32.2 |
| 25 | Wardensville | 27.6 | 25 | Athens | 32.3 |

# WISCONSIN

PHYSICAL FEATURES. Wisconsin lies in the upper Midwest between Lake Superior, Upper Michigan, Lake Michigan, and the Mississippi and St. Croix Rivers. Its greatest length is 320 miles, greatest width 295 miles, and total area 56,066 square miles. Glaciation has largely determined the topography and soils of the State. The various glaciations created a rolling terrain with nearly 9,000 lakes and several areas of marshes and swamps. Elevations range from about 600 feet above sea level along the Lake Superior and Lake Michigan shores and in the Mississippi flood plain in southwestern Wisconsin, to nearly 1,950 feet above sea level at Rib and Strawberry Hills.

The Northern Highlands, a plateau extending across northern Wisconsin, is an area of about 15,000 square miles with elevations from 1,000 to 1,800 feet above sea level. This area is the location of many lakes and the origin of most of the major streams in the State. The slope down to the narrow Lake Superior plain is quite steep. A comparatively flat, crescent shaped lowland lies immediately south of the Northern Highlands, embodying nearly one-fourth of Wisconsin. The eastern ridges and lowlands lie to the southeast of the central plains. The western uplands of southwestern Wisconsin west of the ridges and lowlands and south of the central plains contains approximately one-fourth of the State. This area is the roughest section of the State, rising 200 to 350 feet above the central plains and 100 to 200 feet above the Eastern Ridges and Lowlands. The Mississippi River bluffs rise 230 to 650 feet.

GENERAL CLIMATE. The Wisconsin climate is typically continental with some modification by Lakes Michigan and Superior. The winters are cold and snowy, and the summers are warm. About two-thirds of the annual precipitation falls during the growing season (freeze-free period). The rapid succession of storms moving from west to east and southwest to northeast account for the stimulating climate.

TEMPERATURE. The average annual temperature varies from 39° F. to 49.5° F. During more than one-half of the winters temperatures fall to -40° F. or lower, and almost every winter -30° F. or colder is reported from northern Wisconsin. Summer temperatures above 90° F. or higher average 2 to 4 days in northern counties to about 14 days in southern districts. During marked cool outbreaks in the summer months, the central lowlands occasionally report freezing temperatures. The freeze-free season averages around 80 days per year in the upper northeast and north-central lowlands, to about 180 days in the Milwaukee area. The pronounced moderating effect of Lake Michigan is well illustrated by the fact that the growing season of 140 to 150 days along the east-central coastal area is of the same duration as in the southwestern Wisconsin valleys. The short growing season in the central portion of the State is attributed to a number of factors, among them being an inward cold air drainage and the low heat capacities of the peat and sandy soils. The average date of last spring freeze ranges from early May along the Lake Michigan coastal area and southern counties to early June in the northernmost counties. The first autumn freezes occur in late August and early September in northern and central lowlands to mid-October along the Lake Michigan coast line. However, July freeze is not unusual in the north and central Wisconsin lowlands.

PRECIPITATION. The mean annual precipitation totals 30 to 34 inches over most of the Western Uplands and Northern Highlands, diminishing to about 28 inches along most of the Wisconsin coastal area bordering Lake Michigan and 28 to 30 inches over most of the Wisconsin Central Plain and Lake Superior Coastal area. The higher average annual precipitation coincides generally with the highest elevations, particularly to the windward slopes of the Western Uplands and Northern Highlands. Thunderstorms average about 30 per year in northern Wisconsin to about 40 per year in southern counties, occurring mostly in the summer. Occasional hail, wind, and lightning damage are reported.

The average seasonal snowfall varies from about 30 inches to well over 100 inches along the steep western slope of the Gogebic Range. The heavy snowfall along the Gogebic Range is a result of the prevailing cold northerly winter winds blowing across the relatively warm Lake Superior. Relatively greater average snowfall is recorded over the Western Uplands and Eastern Ridges than in adjacent lowland areas. The mean dates of the first snowfall of consequence, an inch or more, varies from early November in northern localities to around December 1 in southern Wisconsin counties. Average annual duration of snowcover ranges from 85 days in southern-most Wisconsin to more than 140 days along Lake Superior.

The drainage of Wisconsin is into Lake Superior, Lake Michigan, and the Mississippi River. The Mississippi and St. Croix Rivers form most of the western boundary. About one-half of the northwestern portion of the State is drained through the Chippewa River, while the remainder of this region drains directly into the Mississippi or the St. Croix and into Lake Superior. The Wisconsin River has its source at a small lake nearly 1,600 feet above mean sea level on the Upper Michigan boundary and drains most of central Wisconsin. Most of the Wisconsin River tributaries also spring from the many lakes in the north. Except for the Rock River, a Mississippi River tributary which flows through northern Illinois, eastern Wisconsin drains into Lake Michigan, a large part through Green Bay.

Most of the streams and lakes in Wisconsin are ice-covered from late November to late March. Snow covers the ground in practically all the winter months, except in the extreme southern areas. Flooding is most frequent, and most serious, during April due to the melting of snow associated with spring rains. During this period, flood conditions are often aggravated by ice jams which back up the flood waters. Excessive rains of the thunderstorm type sometimes produce tributary flooding or flash flooding along the smaller streams and creeks. Major flooding occurs on the Mississippi River, on the average, about 3 years in 10.

Tornadoes occur in Wisconsin. Most of the very destructive tornadoes occur in the northwestern quarter of the State. Wisconsin tornado frequency is highest in June and July, followed in order by April, May, and September.

## COUNTY INDEX

***Oneida County***
LONG LAKE DAM
MINOCQUA DAM
NORTH PELICAN
RAINBOW RSVR LAKE
RHINELANDER
WILLOW RESERVOIR

***Outagamie County***
APPLETON

***Ozaukee County***
PORT WASHINGTON

***Pierce County***
ELLSWORTH 1 E
RIVER FALLS

***Polk County***
AMERY
ST CROIX FALLS

***Portage County***
STEVENS POINT

***Price County***
PARK FALLS DNR HQ
PRENTICE NO. 2

***Racine County***
BURLINGTON
RACINE

***Richland County***
RICHLAND CENTER

***Rock County***
BELOIT

***Rusk County***
WEYERHAUSER

***Sauk County***
BARABOO
PRAIRIE DU SAC 2 N

***Sawyer County***
COUDERAY 7 W
WINTER 5 NW

***Shawano County***
BOWLER
SHAWANO 2 SSW

***Sheboygan County***
PLYMOUTH
SHEBOYGAN

***Taylor County***
JUMP RIVER 3 E
MEDFORD

***Trempealeau County***
BLAIR
TREMPEALEAU DAM 6

***Vernon County***
GENOA DAM 8
HILLSBORO
VIROQUA 2 NW

***Vilas County***
REST LAKE

***Walworth County***
LAKE GENEVA
WHITEWATER

***Washburn County***
SPOONER EXP FARM

***Washington County***
GERMANTOWN
HARTFORD 2 W
WEST BEND

***Waukesha County***
OCONOMOWOC
WAUKESHA

***Waupaca County***
CLINTONVILLE
NEW LONDON
WAUPACA

***Waushara County***
HANCOCK EXP FARM

***Winnebago County***
OSHKOSH

***Wood County***
MARSHFIELD EXP FARM
WISCONSIN RAPIDS

## ELEVATION INDEX

| FEET | STATION NAME |
|---|---|
| 600 | MARINETTE |
| 600 | PORT WASHINGTON |
| 600 | TWO RIVERS |
| 610 | MANITOWOC |
| 610 | WASHINGTON ISLAND |
| 620 | KENOSHA |
| 620 | MADELINE ISLAND |
| 630 | PRAIRIE DU CHIEN |
| 630 | RACINE |
| 630 | SUPERIOR |
| 633 | LYNXVILLE DAM 9 |
| 639 | GENOA DAM 8 |
| 650 | ASHLAND EXP FARM |
| 659 | OCONTO 4 W |
| 659 | STURGEON BAY EXP FAR |
| 660 | TREMPEALEAU DAM 6 |
| 669 | LA CROSSE MUNI AP |
| 670 | ALMA DAM 4 |
| 672 | MILWAUKEE MTCHLL FLD |
| 679 | SHEBOYGAN |
| 689 | KEWAUNEE 5 S |
| 699 | GREEN BAY AUSTIN STR |
| 732 | MILWAUKEE MT MARY CO |
| 732 | RICHLAND CENTER |
| 741 | APPLETON |
| 741 | MONDOVI |
| 751 | BURLINGTON |
| 751 | OSHKOSH |
| 751 | PRAIRIE DU SAC 2 N |
| 761 | FOND DU LAC |
| 761 | ST CROIX FALLS |
| 781 | BELOIT |
| 781 | NEW LONDON |
| 781 | SPARTA |
| 801 | BRODHEAD |
| 801 | FORT ATKINSON |
| 801 | SHAWANO 2 SSW |
| 810 | PORTAGE |
| 820 | BAYFIELD 6 N |
| 820 | CLINTONVILLE |
| 820 | DARLINGTON |
| 820 | MONTELLO |
| 820 | WATERTOWN |
| 820 | WHITEWATER |
| 840 | PLYMOUTH |
| 850 | WAUPACA |

| FEET | STATION NAME |
|---|---|
| 860 | BLAIR |
| 860 | CHILTON |
| 860 | DALTON |
| 860 | LAKE MILLS |
| 860 | STOUGHTON |
| 860 | WAUKESHA |
| 869 | BEAVER DAM |
| 873 | MADISON DANE CNTY AP |
| 879 | BREED 6 SSE |
| 879 | CRIVITZ HIGH FALLS |
| 879 | LAKE GENEVA |
| 879 | MAUSTON 1 SE |
| 879 | OCONOMOWOC |
| 889 | GERMANTOWN |
| 892 | EAU CLAIRE COUNTY AP |
| 902 | BARABOO |
| 902 | DANBURY |
| 902 | MENOMONIE |
| 902 | RIVER FALLS |
| 902 | WISCONSIN DELLS |
| 932 | FOXBORO |
| 932 | GRANTSBURG |
| 942 | NECEDAH |
| 942 | WEST BEND |
| 951 | RIDGELAND 1 NNE |
| 961 | MATHER 3 NW |
| 981 | HARTFORD 2 W |
| 991 | PLATTEVILLE |
| 1001 | BLOOMER |
| 1001 | GURNEY |
| 1001 | HILLSBORO |
| 1030 | ELLSWORTH 1 E |
| 1040 | GORDON |
| 1040 | LANCASTER 4 WSW |
| 1040 | WISCONSIN RAPIDS |
| 1050 | CHARMANY FARM |
| 1050 | HOLCOMBE |
| 1060 | NEILLSVILLE 3 SW |
| 1070 | AMERY |
| 1079 | ARLINGTON UNIV FARM |
| 1079 | FAIRCHILD RANGER STA |
| 1079 | SOLON SPRINGS |
| 1079 | STANLEY |
| 1080 | BOWLER |
| 1089 | HANCOCK EXP FARM |
| 1100 | SPOONER EXP FARM |
| 1112 | STEVENS POINT |
| 1138 | RICE LAKE |
| 1161 | ROSHOLT 9 NNE |
| 1191 | WAUSAU MUNICIPAL AP |

| FEET | STATION NAME |
|---|---|
| 1220 | DODGEVILLE |
| 1240 | CUMBERLAND |
| 1240 | OWEN |
| 1250 | MARSHFIELD EXP FARM |
| 1260 | JUMP RIVER 3 E |
| 1270 | MERRILL |
| 1280 | VIROQUA 2 NW |
| 1302 | COUDERAY 7 W |
| 1302 | MELLEN 4 NE |
| 1312 | WINTER 5 NW |
| 1342 | WEYERHAUSER |
| 1391 | GOODMAN |
| 1421 | MEDFORD |
| 1480 | ANTIGO |
| 1503 | LAONA 6 SW |
| 1525 | PARK FALLS DNR HQ |
| 1552 | NORTH PELICAN |
| 1552 | RHINELANDER |
| 1562 | WILLOW RESERVOIR |
| 1572 | NEWALD 4 N |
| 1572 | PRENTICE NO. 2 |
| 1591 | MINOCQUA DAM |
| 1600 | RAINBOW RSVR LAKE |
| 1601 | REST LAKE |
| 1631 | LONG LAKE DAM |

10 20 30 STATUTE MILES
WISCONSIN
STATION LEGEND
DATA PUBLISHED IN:
CLIMATOLOGICAL DATA
HOURLY PRECIPITATION DATA
CLIMATOLOGICAL DATA AND HOURLY PRECIPITATION DATA
For further information, refer to the station index and references notes.
DIVISIONS
1 NORTHWEST
2 NORTH CENTRAL
3 NORTHEAST
4 WEST CENTRAL
5 CENTRAL
6 EAST CENTRAL
7 SOUTHWEST
8 SOUTH CENTRAL
9 SOUTHEAST
US DOC - NOAA - NCDC - ASHEVILLE, NC
Updated January 1992
46.0
44.0
42.0
-92.0
-90.0
-88.0
SUPERIOR
ASHLAND EXP FARM
GURNEY
MERCER RANGER STN
LAC VIEUX DESERT
PHELPS DEERSKIN DAM
GRANTSBURG
PARK FALLS
RHINELANDER
THREE LAKES 10 SE
GOODMAN
ST CROIX FALLS
PRENTICE
SPIRIT FALLS
WASHINGTON ISLAND
MARINETTE
MERRILL
ANTIGO 1 SSW
MEDFORD
RIB FALLS
WAUSAU FAA AIRPORT
CHIPPEWA FALLS
EAU CLAIRE FAA AIRPORT
STURGEON BAY EXP FARM
STEVENS POINT
WISCONSIN RAPIDS
GREEN BAY WSO AIRPORT
APPLETON
TWO RIVERS
MANITOWOC
HANCOCK EXP FARM
LA CROSSE FAA AIRPORT
SHEBOYGAN
FOND DU LAC
PORTAGE
BEAVER DAM
MADISON WSO AIRPORT
PRAIRIE DU CHIEN
MILWAUKEE WSO AIRPORT
PLATTEVILLE
BURLINGTON
RACINE
KENOSHA
LAKE GENEVA
BELOIT
MONROE 1 W

## ALMA DAM 4 *Buffalo County* ELEVATION 670 ft LAT/LONG 44° 20' N / 91° 55' W

| | JAN | FEB | MAR | APR | MAY | JUN | JUL | AUG | SEP | OCT | NOV | DEC | YEAR |
|---|---|---|---|---|---|---|---|---|---|---|---|---|---|
| Maximum Temp °F | 23.9 | 29.2 | 41.3 | 57.4 | 70.2 | 78.9 | 82.9 | 80.5 | 71.4 | 59.6 | 42.4 | 28.9 | 55.5 |
| Minimum Temp °F | 6.3 | 11.0 | 23.8 | 37.2 | 48.4 | 57.6 | 62.7 | 60.7 | 51.8 | 40.7 | 27.2 | 14.2 | 36.8 |
| Mean Temp °F | 15.1 | 20.1 | 32.6 | 47.3 | 59.3 | 68.3 | 72.8 | 70.7 | 61.6 | 50.2 | 34.8 | 21.6 | 46.2 |
| Days Max Temp ≥ 90 °F | 0 | 0 | 0 | 0 | 0 | 2 | 4 | 2 | 0 | 0 | 0 | 0 | 8 |
| Days Max Temp ≤ 32 °F | 22 | 15 | 6 | 0 | 0 | 0 | 0 | 0 | 0 | 0 | 5 | 18 | 66 |
| Days Min Temp ≤ 32 °F | 31 | 27 | 24 | 9 | 1 | 0 | 0 | 0 | 0 | 6 | 21 | 30 | 149 |
| Days Min Temp ≤ 0 °F | 11 | 7 | 1 | 0 | 0 | 0 | 0 | 0 | 0 | 0 | 0 | 5 | 24 |
| Heating Degree Days | 1542 | 1263 | 998 | 529 | 212 | 39 | 5 | 17 | 152 | 459 | 900 | 1340 | 7456 |
| Cooling Degree Days | 0 | 0 | 0 | 8 | 47 | 149 | 260 | 203 | 60 | 3 | 0 | 0 | 730 |
| Total Precipitation (") | 1.00 | 0.64 | 1.75 | 3.32 | 3.81 | 4.38 | 5.13 | 4.02 | 4.24 | 2.55 | 2.12 | 1.08 | 34.04 |
| Days ≥ 0.1" Precip | 3 | 2 | 4 | 7 | 8 | 8 | 8 | 7 | 7 | 5 | 4 | 3 | 66 |
| Total Snowfall (") | ***10.4*** | 6.8 | 6.8 | 1.4 | 0.0 | 0.0 | 0.0 | 0.0 | 0.0 | 0.0 | 3.4 | ***8.5*** | 37.3 |
| Days ≥ 1" Snow Depth | 28 | 23 | 13 | 1 | 0 | 0 | 0 | 0 | 0 | 0 | 4 | 20 | 89 |

## AMERY *Polk County* ELEVATION 1070 ft LAT/LONG 45° 19' N / 92° 21' W

| | JAN | FEB | MAR | APR | MAY | JUN | JUL | AUG | SEP | OCT | NOV | DEC | YEAR |
|---|---|---|---|---|---|---|---|---|---|---|---|---|---|
| Maximum Temp °F | 19.0 | 25.3 | 37.9 | 54.3 | 67.7 | 76.2 | 81.1 | 78.7 | 68.8 | 56.6 | 38.9 | 24.7 | 52.4 |
| Minimum Temp °F | -2.2 | 3.1 | 17.9 | 32.6 | 44.1 | 53.3 | 58.3 | 55.5 | 46.2 | 35.0 | 22.1 | 6.8 | 31.1 |
| Mean Temp °F | 8.4 | 14.2 | 27.9 | 43.5 | 55.9 | 64.8 | 69.7 | 67.1 | 57.6 | 45.8 | 30.5 | 15.8 | 41.8 |
| Days Max Temp ≥ 90 °F | 0 | 0 | 0 | 0 | 0 | 1 | 3 | 2 | 0 | 0 | 0 | 0 | 6 |
| Days Max Temp ≤ 32 °F | 26 | 19 | 9 | 1 | 0 | 0 | 0 | 0 | 0 | 0 | 8 | 23 | 86 |
| Days Min Temp ≤ 32 °F | 31 | 28 | 27 | 15 | 3 | 0 | 0 | 0 | 2 | 13 | 26 | 31 | 176 |
| Days Min Temp ≤ 0 °F | 17 | 13 | 4 | 0 | 0 | 0 | 0 | 0 | 0 | 0 | 1 | 10 | 45 |
| Heating Degree Days | 1751 | 1430 | 1143 | 642 | 300 | 86 | 22 | 52 | 247 | 590 | 1027 | 1522 | 8812 |
| Cooling Degree Days | 0 | 0 | 0 | 4 | 25 | 93 | 180 | 131 | 30 | 1 | 0 | 0 | 464 |
| Total Precipitation (") | 1.02 | 0.77 | 1.64 | 2.72 | 3.05 | 4.75 | 3.92 | 4.32 | 3.95 | 2.51 | 1.89 | 1.23 | 31.77 |
| Days ≥ 0.1" Precip | 3 | 3 | 4 | 6 | 7 | 8 | 7 | 7 | 7 | 5 | 4 | 4 | 65 |
| Total Snowfall (") | 12.7 | 7.5 | 8.0 | ***2.4*** | 0.0 | 0.0 | 0.0 | 0.0 | 0.0 | 0.5 | ***7.4*** | ***10.9*** | 49.4 |
| Days ≥ 1" Snow Depth | 30 | 27 | 21 | 2 | 0 | 0 | 0 | 0 | 0 | 0 | ***8*** | 26 | 114 |

## ANTIGO *Langlade County* ELEVATION 1480 ft LAT/LONG 45° 9' N / 89° 9' W

| | JAN | FEB | MAR | APR | MAY | JUN | JUL | AUG | SEP | OCT | NOV | DEC | YEAR |
|---|---|---|---|---|---|---|---|---|---|---|---|---|---|
| Maximum Temp °F | 20.9 | 26.0 | 37.2 | 53.1 | 67.0 | 74.8 | 79.5 | 77.0 | 67.5 | 56.0 | 39.1 | 25.3 | 52.0 |
| Minimum Temp °F | 0.5 | 3.6 | 16.4 | 30.8 | 41.0 | 50.0 | 54.9 | 53.1 | 44.7 | 34.9 | 22.9 | 8.1 | 30.1 |
| Mean Temp °F | 10.7 | 14.8 | 26.8 | 42.0 | 54.0 | 62.4 | 67.2 | 65.1 | 56.1 | 45.5 | 31.0 | 16.7 | 41.0 |
| Days Max Temp ≥ 90 °F | 0 | 0 | 0 | 0 | 0 | 1 | 2 | 1 | 0 | 0 | 0 | 0 | 4 |
| Days Max Temp ≤ 32 °F | 27 | 20 | 10 | 1 | 0 | 0 | 0 | 0 | 0 | 0 | 8 | 23 | 89 |
| Days Min Temp ≤ 32 °F | 31 | 28 | 28 | 19 | 6 | 0 | 0 | 0 | 3 | 13 | 25 | 30 | 183 |
| Days Min Temp ≤ 0 °F | 15 | 12 | 5 | 0 | 0 | 0 | 0 | 0 | 0 | 0 | 1 | 9 | 42 |
| Heating Degree Days | 1680 | 1412 | 1177 | 688 | 349 | 126 | 44 | 76 | 280 | 599 | 1013 | 1491 | 8935 |
| Cooling Degree Days | 0 | 0 | 0 | 3 | 12 | 49 | 105 | 81 | ***16*** | 0 | 0 | ***0*** | 266 |
| Total Precipitation (") | 0.90 | 0.80 | 1.80 | 2.81 | 3.14 | 3.88 | 3.61 | 4.08 | 4.45 | 2.62 | 2.21 | 1.39 | 31.69 |
| Days ≥ 0.1" Precip | 3 | 2 | 5 | 6 | 6 | 8 | 7 | 7 | 7 | 6 | 5 | 4 | 66 |
| Total Snowfall (") | 13.4 | 9.2 | 9.1 | 3.7 | 0.5 | 0.0 | 0.0 | 0.0 | 0.0 | 1.1 | 7.9 | 14.8 | 59.7 |
| Days ≥ 1" Snow Depth | 31 | 28 | 24 | 6 | 0 | 0 | 0 | 0 | 0 | 1 | 9 | 27 | 126 |

## APPLETON *Outagamie County* ELEVATION 741 ft LAT/LONG 44° 15' N / 88° 23' W

| | JAN | FEB | MAR | APR | MAY | JUN | JUL | AUG | SEP | OCT | NOV | DEC | YEAR |
|---|---|---|---|---|---|---|---|---|---|---|---|---|---|
| Maximum Temp °F | 24.0 | 28.5 | 39.8 | 54.4 | 68.3 | 76.8 | 81.6 | 78.8 | 70.1 | 57.3 | 42.0 | 29.0 | 54.2 |
| Minimum Temp °F | 7.3 | 11.3 | 22.6 | 35.1 | 46.5 | 56.1 | 61.9 | 59.7 | 51.4 | 39.9 | 27.7 | 14.4 | 36.2 |
| Mean Temp °F | 15.7 | 19.9 | 31.2 | 44.8 | 57.4 | 66.4 | 71.8 | 69.3 | 60.8 | 48.7 | 34.9 | 21.7 | 45.2 |
| Days Max Temp ≥ 90 °F | 0 | 0 | 0 | 0 | 0 | 2 | 3 | 2 | 0 | 0 | 0 | 0 | 7 |
| Days Max Temp ≤ 32 °F | 22 | 17 | 7 | 0 | 0 | 0 | 0 | 0 | 0 | 0 | 5 | 18 | 69 |
| Days Min Temp ≤ 32 °F | 31 | 28 | 26 | 12 | 1 | 0 | 0 | 0 | 0 | 6 | 21 | 30 | 155 |
| Days Min Temp ≤ 0 °F | 10 | 6 | 1 | 0 | 0 | 0 | 0 | 0 | 0 | 0 | 0 | 5 | 22 |
| Heating Degree Days | 1524 | 1267 | 1041 | 603 | 265 | 64 | 11 | 26 | 169 | 502 | 896 | 1336 | 7704 |
| Cooling Degree Days | 0 | 0 | 0 | 7 | 45 | 125 | 245 | 177 | 56 | 1 | 0 | 0 | 656 |
| Total Precipitation (") | 1.18 | 1.02 | 2.10 | 2.97 | 3.08 | 3.51 | 3.31 | 3.69 | 3.55 | 2.40 | 2.23 | 1.49 | 30.53 |
| Days ≥ 0.1" Precip | 3 | 3 | 5 | 7 | 6 | 7 | 6 | 7 | 7 | 5 | 5 | 4 | 65 |
| Total Snowfall (") | 11.4 | 7.9 | 7.5 | 2.5 | 0.2 | 0.0 | 0.0 | 0.0 | 0.0 | 0.3 | 4.1 | 11.2 | 45.1 |
| Days ≥ 1" Snow Depth | 26 | 22 | 13 | 2 | 0 | 0 | 0 | 0 | 0 | 0 | 4 | 19 | 86 |

## ARLINGTON UNIV FARM *Columbia County* ELEVATION 1079 ft LAT/LONG 43° 18 ' N / 89° 21 ' W

| | JAN | FEB | MAR | APR | MAY | JUN | JUL | AUG | SEP | OCT | NOV | DEC | YEAR |
|---|---|---|---|---|---|---|---|---|---|---|---|---|---|
| Maximum Temp °F | 24.4 | 29.4 | 41.8 | 57.5 | 70.4 | 78.9 | 82.6 | 80.3 | 72.2 | 59.9 | 43.4 | 30.1 | 55.9 |
| Minimum Temp °F | 7.1 | 11.5 | 23.4 | 34.9 | 45.4 | 54.5 | 59.5 | 57.4 | 49.5 | 38.7 | 27.1 | 14.4 | 35.3 |
| Mean Temp °F | 15.8 | 20.5 | 32.6 | 46.2 | 57.9 | 66.7 | 71.1 | 68.8 | 60.9 | 49.4 | 35.3 | 22.3 | 45.6 |
| Days Max Temp ≥ 90 °F | 0 | 0 | 0 | 0 | 0 | 2 | 4 | 2 | 0 | 0 | 0 | 0 | 8 |
| Days Max Temp ≤ 32 °F | 22 | 16 | 6 | 0 | 0 | 0 | 0 | 0 | 0 | 0 | 4 | 17 | 65 |
| Days Min Temp ≤ 32 °F | 31 | 27 | 26 | 13 | 2 | 0 | 0 | 0 | 1 | 9 | 22 | 30 | 161 |
| Days Min Temp ≤ 0 °F | 11 | 7 | 1 | 0 | 0 | 0 | 0 | 0 | 0 | 0 | 0 | 5 | 24 |
| Heating Degree Days | 1520 | 1251 | 998 | 561 | 247 | 56 | 12 | 30 | 168 | 483 | 885 | 1318 | 7529 |
| Cooling Degree Days | 0 | 0 | 0 | 6 | 32 | 119 | 209 | 161 | 55 | 3 | 0 | 0 | 585 |
| Total Precipitation (") | 1.02 | 0.99 | 1.88 | 3.14 | 3.05 | 3.94 | 3.81 | 3.93 | 4.48 | 2.49 | 2.31 | 1.52 | 32.56 |
| Days ≥ 0.1" Precip | 3 | 3 | 5 | 7 | 7 | 7 | 7 | 7 | 7 | 6 | 5 | 4 | 68 |
| Total Snowfall (") | 7.9 | 6.7 | 5.3 | 1.9 | 0.1 | 0.0 | 0.0 | 0.0 | 0.0 | 0.1 | *3.3* | 8.8 | 34.1 |
| Days ≥ 1" Snow Depth | 23 | 22 | 10 | 1 | 0 | 0 | 0 | 0 | 0 | 0 | 3 | 17 | 76 |

## ASHLAND EXP FARM *Bayfield County* ELEVATION 650 ft LAT/LONG 46° 34 ' N / 90° 58 ' W

| | JAN | FEB | MAR | APR | MAY | JUN | JUL | AUG | SEP | OCT | NOV | DEC | YEAR |
|---|---|---|---|---|---|---|---|---|---|---|---|---|---|
| Maximum Temp °F | 22.1 | 27.6 | 38.2 | 51.9 | 65.2 | 74.8 | 80.5 | 78.2 | 68.7 | 56.8 | 39.4 | 26.7 | 52.5 |
| Minimum Temp °F | 0.6 | 4.2 | 16.5 | 28.4 | 38.6 | 48.2 | 54.8 | 53.3 | 45.2 | 34.6 | 23.2 | 9.0 | 29.7 |
| Mean Temp °F | 11.3 | 15.9 | 27.3 | 40.2 | 51.9 | 61.5 | 67.7 | 65.8 | 56.9 | 45.7 | 31.2 | 17.9 | 41.1 |
| Days Max Temp ≥ 90 °F | 0 | 0 | 0 | 0 | 0 | 1 | 3 | 2 | 0 | 0 | 0 | 0 | 6 |
| Days Max Temp ≤ 32 °F | 25 | 18 | 9 | 1 | 0 | 0 | 0 | 0 | 0 | 0 | 7 | 21 | 81 |
| Days Min Temp ≤ 32 °F | 31 | 28 | 29 | 20 | 8 | 1 | 0 | 0 | 3 | 13 | 25 | 30 | 188 |
| Days Min Temp ≤ 0 °F | 15 | 12 | 4 | 0 | 0 | 0 | 0 | 0 | 0 | 0 | 1 | 9 | 41 |
| Heating Degree Days | 1660 | 1381 | 1161 | 740 | 411 | 146 | 41 | 68 | 258 | 593 | 1006 | 1455 | 8920 |
| Cooling Degree Days | 0 | 0 | 0 | 3 | 15 | 53 | 136 | 105 | 21 | 1 | 0 | 0 | 334 |
| Total Precipitation (") | 1.13 | 0.71 | 1.84 | 2.24 | 3.05 | 3.44 | 3.92 | 4.18 | 3.89 | 2.57 | 2.34 | 1.31 | 30.62 |
| Days ≥ 0.1" Precip | 3 | 2 | 4 | 6 | 6 | 8 | 7 | 7 | 8 | 6 | 5 | 4 | 66 |
| Total Snowfall (") | 14.3 | 7.8 | 9.3 | 3.9 | 0.0 | 0.0 | 0.0 | 0.0 | 0.0 | 0.3 | 8.6 | 13.4 | 57.6 |
| Days ≥ 1" Snow Depth | 29 | 27 | 20 | 4 | 0 | 0 | 0 | 0 | 0 | 0 | 9 | 24 | 113 |

## BARABOO *Sauk County* ELEVATION 902 ft LAT/LONG 43° 28 ' N / 89° 44 ' W

| | JAN | FEB | MAR | APR | MAY | JUN | JUL | AUG | SEP | OCT | NOV | DEC | YEAR |
|---|---|---|---|---|---|---|---|---|---|---|---|---|---|
| Maximum Temp °F | 25.4 | 30.6 | 42.6 | 56.9 | 69.5 | 78.1 | 82.3 | 79.8 | 71.2 | 59.3 | 43.7 | 30.9 | 55.9 |
| Minimum Temp °F | 4.0 | 8.2 | 21.1 | 33.1 | 44.1 | 53.1 | 58.0 | 54.8 | 46.6 | 35.5 | 24.5 | 11.9 | 32.9 |
| Mean Temp °F | 14.7 | 19.4 | 31.9 | 45.0 | 56.8 | 65.6 | 70.2 | 67.3 | 59.0 | 47.3 | 34.1 | 21.4 | 44.4 |
| Days Max Temp ≥ 90 °F | 0 | 0 | 0 | 0 | 0 | 2 | 4 | 2 | 0 | 0 | 0 | 0 | 8 |
| Days Max Temp ≤ 32 °F | 21 | 15 | 6 | 0 | 0 | 0 | 0 | 0 | 0 | 0 | 4 | 16 | 62 |
| Days Min Temp ≤ 32 °F | 31 | 28 | 27 | 15 | 4 | 0 | 0 | 0 | 2 | 13 | 24 | 30 | 174 |
| Days Min Temp ≤ 0 °F | 13 | 9 | 2 | 0 | 0 | 0 | 0 | 0 | 0 | 0 | 1 | 7 | 32 |
| Heating Degree Days | 1554 | 1281 | 1021 | 598 | 280 | 76 | 20 | 49 | 216 | 547 | 921 | 1344 | 7907 |
| Cooling Degree Days | 0 | 0 | 0 | 9 | 32 | 108 | 201 | 137 | 43 | 1 | 0 | 0 | 531 |
| Total Precipitation (") | 0.98 | 0.91 | 1.94 | 3.37 | 3.07 | 3.96 | 4.01 | 4.29 | 4.09 | 2.43 | 2.31 | 1.39 | 32.75 |
| Days ≥ 0.1" Precip | 3 | 3 | 4 | 7 | 6 | 7 | 6 | 7 | 7 | 6 | 5 | 4 | 65 |
| Total Snowfall (") | 10.2 | 8.0 | 6.4 | 2.2 | 0.0 | 0.0 | 0.0 | 0.0 | 0.0 | 0.3 | 4.0 | 10.4 | 41.5 |
| Days ≥ 1" Snow Depth | 26 | 24 | 13 | 2 | 0 | 0 | 0 | 0 | 0 | 0 | 4 | 18 | 87 |

## BAYFIELD 6 N *Bayfield County* ELEVATION 820 ft LAT/LONG 46° 53 ' N / 90° 49 ' W

| | JAN | FEB | MAR | APR | MAY | JUN | JUL | AUG | SEP | OCT | NOV | DEC | YEAR |
|---|---|---|---|---|---|---|---|---|---|---|---|---|---|
| Maximum Temp °F | 21.9 | 26.6 | 37.4 | 50.8 | 63.9 | 72.7 | 78.0 | 75.8 | 66.2 | 54.9 | 39.1 | 26.7 | 51.2 |
| Minimum Temp °F | 4.1 | 6.9 | 18.1 | 30.1 | 39.3 | 48.2 | 55.5 | 54.8 | 47.1 | 36.7 | 24.9 | 12.0 | 31.5 |
| Mean Temp °F | 13.0 | 16.8 | 27.7 | 40.5 | 51.6 | 60.5 | 66.7 | 65.3 | 56.7 | 45.9 | 32.0 | 19.3 | 41.3 |
| Days Max Temp ≥ 90 °F | 0 | 0 | 0 | 0 | 0 | 0 | 2 | 1 | 0 | 0 | 0 | 0 | 3 |
| Days Max Temp ≤ 32 °F | 25 | 19 | 10 | 1 | 0 | 0 | 0 | 0 | 0 | 0 | 7 | 21 | 83 |
| Days Min Temp ≤ 32 °F | 31 | 28 | 29 | 19 | 6 | 0 | 0 | 0 | 1 | 9 | 25 | 30 | 178 |
| Days Min Temp ≤ 0 °F | 13 | 10 | 3 | 0 | 0 | 0 | 0 | 0 | 0 | 0 | 0 | 5 | 31 |
| Heating Degree Days | 1607 | 1357 | 1148 | 730 | 417 | 166 | 49 | 72 | 261 | 587 | 982 | 1412 | 8788 |
| Cooling Degree Days | 0 | 0 | 0 | 1 | 12 | 34 | 102 | 89 | 14 | 0 | 0 | 0 | 252 |
| Total Precipitation (") | 1.69 | 0.83 | 2.31 | 2.22 | 3.22 | 3.60 | 3.91 | 4.27 | 3.86 | 2.89 | 2.74 | 1.79 | 33.33 |
| Days ≥ 0.1" Precip | 5 | 3 | 5 | 5 | 7 | 8 | 7 | 7 | 7 | 7 | 5 | 6 | 72 |
| Total Snowfall (") | 25.2 | 10.5 | 13.1 | 5.0 | 0.3 | 0.0 | 0.0 | 0.0 | 0.0 | 0.7 | 13.9 | 24.6 | 93.3 |
| Days ≥ 1" Snow Depth | 31 | 28 | 28 | 8 | 0 | 0 | 0 | 0 | 0 | 0 | 11 | 28 | 134 |

## BEAVER DAM *Dodge County* ELEVATION 869 ft LAT/LONG 43° 28 ' N / 88° 50 ' W

| | JAN | FEB | MAR | APR | MAY | JUN | JUL | AUG | SEP | OCT | NOV | DEC | YEAR |
|---|---|---|---|---|---|---|---|---|---|---|---|---|---|
| Maximum Temp °F | 25.7 | 30.4 | 42.6 | 57.7 | 70.7 | 79.2 | 83.2 | 80.8 | 72.6 | 60.6 | 44.3 | 31.0 | 56.6 |
| Minimum Temp °F | 8.5 | 12.7 | 24.5 | 35.9 | 46.0 | 55.2 | 59.9 | 57.8 | 50.3 | 39.3 | 28.3 | 15.8 | 36.2 |
| Mean Temp °F | 17.1 | 21.6 | 33.6 | 46.8 | 58.4 | 67.2 | 71.5 | 69.3 | 61.5 | 50.0 | 36.3 | 23.4 | 46.4 |
| Days Max Temp ≥ 90 °F | 0 | 0 | 0 | 0 | 0 | 3 | 5 | 3 | 0 | 0 | 0 | 0 | 11 |
| Days Max Temp ≤ 32 °F | 21 | 15 | 5 | 0 | 0 | 0 | 0 | 0 | 0 | 0 | 3 | 16 | 60 |
| Days Min Temp ≤ 32 °F | 30 | 27 | 24 | 11 | 2 | 0 | 0 | 0 | 0 | 7 | 21 | 29 | 151 |
| Days Min Temp ≤ 0 °F | 10 | 6 | 1 | 0 | 0 | 0 | 0 | 0 | 0 | 0 | 0 | 4 | 21 |
| Heating Degree Days | 1483 | 1224 | 972 | 548 | 238 | 50 | 10 | 24 | 156 | 468 | 858 | 1286 | 7317 |
| Cooling Degree Days | 0 | 0 | 0 | 7 | 35 | 126 | 221 | 167 | 57 | 3 | 0 | 0 | 616 |
| Total Precipitation (") | 1.21 | 1.20 | 2.10 | 3.29 | 3.05 | 3.80 | 3.79 | 3.40 | 4.23 | 2.54 | 2.31 | 1.86 | 32.78 |
| Days ≥ 0.1" Precip | 4 | 3 | 5 | 7 | 7 | 7 | 7 | 6 | 6 | 6 | 5 | 5 | 68 |
| Total Snowfall (") | 9.1 | 7.7 | 5.1 | 1.3 | 0.1 | 0.0 | 0.0 | 0.0 | 0.0 | 0.1 | 2.0 | 10.0 | 35.4 |
| Days ≥ 1" Snow Depth | 24 | 22 | 10 | 1 | 0 | 0 | 0 | 0 | 0 | 0 | 2 | 17 | 76 |

## BELOIT *Rock County* ELEVATION 781 ft LAT/LONG 42° 30 ' N / 89° 2 ' W

| | JAN | FEB | MAR | APR | MAY | JUN | JUL | AUG | SEP | OCT | NOV | DEC | YEAR |
|---|---|---|---|---|---|---|---|---|---|---|---|---|---|
| Maximum Temp °F | 27.4 | 31.6 | 44.4 | 59.5 | 71.8 | 80.8 | 84.1 | 81.5 | 74.3 | 62.7 | 46.4 | 33.0 | 58.1 |
| Minimum Temp °F | 11.0 | 15.1 | 26.2 | 37.9 | 48.2 | 57.6 | 62.4 | 60.0 | 52.6 | 41.1 | 30.2 | 17.8 | 38.3 |
| Mean Temp °F | 19.2 | 23.3 | 35.4 | 48.7 | 60.0 | 69.2 | 73.3 | 70.8 | 63.4 | 51.7 | 38.4 | 25.4 | 48.2 |
| Days Max Temp ≥ 90 °F | 0 | 0 | 0 | 0 | 0 | 4 | 7 | 4 | 1 | 0 | 0 | 0 | 16 |
| Days Max Temp ≤ 32 °F | 19 | 14 | 4 | 0 | 0 | 0 | 0 | 0 | 0 | 0 | 2 | 13 | 52 |
| Days Min Temp ≤ 32 °F | 30 | 26 | 23 | 8 | 1 | 0 | 0 | 0 | 0 | 6 | 18 | 28 | 140 |
| Days Min Temp ≤ 0 °F | 7 | 5 | 0 | 0 | 0 | 0 | 0 | 0 | 0 | 0 | 0 | 3 | 15 |
| Heating Degree Days | 1414 | 1172 | 912 | 492 | 203 | 35 | 6 | 16 | 120 | 414 | 793 | 1223 | 6800 |
| Cooling Degree Days | 0 | 0 | 0 | 10 | 49 | 161 | 265 | 203 | 77 | 5 | 0 | 0 | 770 |
| Total Precipitation (") | 1.27 | 1.13 | 2.16 | 3.34 | 3.06 | 4.17 | 3.82 | 4.12 | 4.03 | 2.40 | 2.83 | 2.01 | 34.34 |
| Days ≥ 0.1" Precip | 4 | 3 | 5 | 7 | 7 | 7 | 6 | 6 | 6 | *5* | *6* | 5 | 67 |
| Total Snowfall (") | *7.5* | *7.2* | *4.3* | 0.8 | 0.0 | 0.0 | 0.0 | 0.0 | 0.0 | 0.0 | *1.0* | na | na |
| Days ≥ 1" Snow Depth | na | *17* | *5* | 0 | 0 | 0 | 0 | 0 | 0 | 0 | *1* | na | na |

## BLAIR *Trempealeau County* ELEVATION 860 ft LAT/LONG 44° 18 ' N / 91° 14 ' W

| | JAN | FEB | MAR | APR | MAY | JUN | JUL | AUG | SEP | OCT | NOV | DEC | YEAR |
|---|---|---|---|---|---|---|---|---|---|---|---|---|---|
| Maximum Temp °F | 22.7 | 28.8 | 41.1 | 56.8 | 69.7 | 78.0 | 82.2 | 79.9 | 70.7 | 58.7 | 42.0 | 28.2 | 54.9 |
| Minimum Temp °F | -0.1 | 4.9 | 19.5 | 32.9 | 43.6 | 52.8 | 57.5 | 55.1 | 46.0 | 34.9 | 23.3 | 8.7 | 31.6 |
| Mean Temp °F | 11.3 | 16.9 | 30.3 | 44.9 | 56.7 | 65.4 | 69.9 | 67.5 | 58.4 | 46.8 | 32.6 | 18.5 | 43.3 |
| Days Max Temp ≥ 90 °F | 0 | 0 | 0 | 0 | 0 | 2 | 4 | 3 | 0 | 0 | 0 | 0 | 9 |
| Days Max Temp ≤ 32 °F | 24 | 16 | 7 | 0 | 0 | 0 | 0 | 0 | 0 | 0 | 5 | 19 | 71 |
| Days Min Temp ≤ 32 °F | 31 | 28 | 27 | 16 | 4 | 0 | 0 | 0 | 2 | 14 | 25 | 30 | 177 |
| Days Min Temp ≤ 0 °F | 15 | 11 | 3 | 0 | 0 | 0 | 0 | 0 | 0 | 0 | 1 | 9 | 39 |
| Heating Degree Days | 1661 | 1354 | 1067 | 601 | 281 | 79 | 19 | 46 | 226 | 561 | 964 | 1437 | 8296 |
| Cooling Degree Days | 0 | 0 | 0 | 6 | 28 | 101 | 177 | 134 | 34 | 1 | 0 | 0 | 481 |
| Total Precipitation (") | 0.96 | 0.79 | 1.87 | 3.32 | 3.89 | 4.09 | 4.58 | 4.36 | 4.53 | 2.63 | 2.10 | 1.30 | 34.42 |
| Days ≥ 0.1" Precip | 3 | 2 | 5 | 7 | 7 | 8 | 7 | 7 | 7 | 5 | 4 | 4 | 66 |
| Total Snowfall (") | 10.7 | 7.0 | 6.8 | 1.7 | 0.0 | 0.0 | 0.0 | 0.0 | 0.0 | 0.2 | 4.0 | 10.0 | 40.4 |
| Days ≥ 1" Snow Depth | 31 | 26 | 18 | 2 | 0 | 0 | 0 | 0 | 0 | 0 | 5 | 24 | 106 |

## BLOOMER *Chippewa County* ELEVATION 1001 ft LAT/LONG 45° 6 ' N / 91° 29 ' W

| | JAN | FEB | MAR | APR | MAY | JUN | JUL | AUG | SEP | OCT | NOV | DEC | YEAR |
|---|---|---|---|---|---|---|---|---|---|---|---|---|---|
| Maximum Temp °F | 21.3 | 27.6 | 40.3 | 57.0 | 70.2 | 78.1 | 82.8 | 80.3 | 70.6 | 58.1 | 40.3 | 26.2 | 54.4 |
| Minimum Temp °F | 1.2 | 6.4 | 19.9 | 33.3 | 44.4 | 53.5 | 58.5 | 56.2 | 47.4 | 36.3 | 23.7 | 9.4 | 32.5 |
| Mean Temp °F | 11.3 | 17.0 | 30.1 | 45.2 | 57.3 | 65.8 | 70.7 | 68.3 | 59.0 | 47.2 | 32.0 | 17.8 | 43.5 |
| Days Max Temp ≥ 90 °F | 0 | 0 | 0 | 0 | 0 | 2 | 5 | 3 | 0 | 0 | 0 | 0 | 10 |
| Days Max Temp ≤ 32 °F | 25 | 17 | 7 | 0 | 0 | 0 | 0 | 0 | 0 | 0 | 7 | 22 | 78 |
| Days Min Temp ≤ 32 °F | 31 | 28 | 26 | 15 | 3 | 0 | 0 | 0 | 2 | 12 | 25 | 30 | 172 |
| Days Min Temp ≤ 0 °F | 15 | 10 | 3 | 0 | 0 | 0 | 0 | 0 | 0 | 0 | 1 | 9 | 38 |
| Heating Degree Days | 1662 | 1349 | 1075 | 592 | 264 | 69 | 16 | 38 | 211 | 547 | 982 | 1457 | 8262 |
| Cooling Degree Days | 0 | 0 | 0 | 7 | 37 | 114 | 211 | 162 | 43 | 0 | 0 | 0 | 574 |
| Total Precipitation (") | 0.98 | 0.69 | 1.72 | 2.91 | 3.57 | 4.62 | 3.80 | 4.84 | 3.94 | 2.47 | 2.00 | 1.16 | 32.70 |
| Days ≥ 0.1" Precip | 3 | 2 | 4 | 6 | 7 | 8 | 7 | 7 | 7 | 5 | 4 | 4 | 64 |
| Total Snowfall (") | 11.3 | 6.5 | 6.8 | 1.1 | 0.0 | 0.0 | 0.0 | 0.0 | 0.0 | 0.3 | 5.9 | 9.9 | 41.8 |
| Days ≥ 1" Snow Depth | 29 | 27 | 18 | 2 | 0 | 0 | 0 | 0 | 0 | 0 | 6 | 24 | 106 |

## BOWLER *Shawano County* ELEVATION 1080 ft LAT/LONG 44° 52 ' N / 88° 59 ' W

| | JAN | FEB | MAR | APR | MAY | JUN | JUL | AUG | SEP | OCT | NOV | DEC | YEAR |
|---|---|---|---|---|---|---|---|---|---|---|---|---|---|
| Maximum Temp °F | 22.0 | 27.3 | 38.5 | 53.9 | 67.4 | 75.6 | *80.4* | 77.6 | 68.0 | 56.5 | 40.8 | *26.8* | 52.9 |
| Minimum Temp °F | 0.2 | 3.9 | 16.9 | 30.7 | 40.9 | 49.8 | 55.2 | 52.8 | 43.8 | 33.1 | 22.7 | *9.0* | 29.9 |
| Mean Temp °F | 11.1 | 15.6 | 27.7 | 42.4 | 54.2 | 62.8 | *67.8* | 65.3 | 55.9 | 44.8 | 31.8 | *17.9* | 41.4 |
| Days Max Temp ≥ 90 °F | 0 | 0 | 0 | 0 | 0 | 1 | 2 | 1 | 0 | 0 | 0 | 0 | 4 |
| Days Max Temp ≤ 32 °F | 25 | 18 | 9 | 0 | 0 | 0 | 0 | 0 | 0 | 0 | 6 | 20 | 78 |
| Days Min Temp ≤ 32 °F | 30 | 28 | 28 | 18 | 7 | 0 | 0 | 0 | 3 | *15* | 24 | 29 | 182 |
| Days Min Temp ≤ 0 °F | 15 | 12 | 4 | 0 | 0 | 0 | 0 | 0 | 0 | 0 | 1 | 8 | 40 |
| Heating Degree Days | 1667 | 1388 | 1149 | 676 | 346 | 122 | *39* | 73 | 280 | 614 | 990 | *1453* | 8797 |
| Cooling Degree Days | 0 | 0 | 0 | 4 | 15 | 67 | *143* | 97 | 16 | 0 | 0 | *0* | 342 |
| Total Precipitation (") | 0.97 | 0.90 | 1.81 | 3.00 | 3.68 | 3.74 | 3.64 | 3.78 | 4.10 | 2.62 | 2.12 | 1.49 | 31.85 |
| Days ≥ 0.1" Precip | 3 | 3 | 4 | 7 | 6 | 7 | 6 | 7 | 7 | 5 | 4 | 4 | 63 |
| Total Snowfall (") | na | *6.2* | *7.9* | *2.2* | 0.0 | 0.0 | 0.0 | 0.0 | 0.0 | 0.4 | *4.8* | *10.5* | na |
| Days ≥ 1" Snow Depth | na | na | na | *2* | 0 | 0 | 0 | 0 | 0 | 0 | na | na | na |

## BREED 6 SSE *Oconto County* ELEVATION 879 ft LAT/LONG 45° 4 ' N / 88° 26 ' W

| | JAN | FEB | MAR | APR | MAY | JUN | JUL | AUG | SEP | OCT | NOV | DEC | YEAR |
|---|---|---|---|---|---|---|---|---|---|---|---|---|---|
| Maximum Temp °F | 24.5 | 29.5 | 40.4 | 55.7 | 69.4 | 77.7 | 82.2 | 79.7 | 70.3 | 58.3 | 41.7 | 29.1 | 54.9 |
| Minimum Temp °F | 1.9 | 5.4 | 17.6 | 30.0 | 40.2 | 49.7 | 55.0 | 52.8 | 44.7 | 34.7 | 23.8 | 11.0 | 30.6 |
| Mean Temp °F | 13.2 | 17.5 | 29.0 | 42.9 | 54.8 | 63.7 | 68.6 | 66.2 | 57.5 | 46.5 | 32.8 | 20.1 | 42.7 |
| Days Max Temp ≥ 90 °F | 0 | 0 | 0 | 0 | 0 | 2 | 4 | 2 | 0 | 0 | 0 | 0 | 8 |
| Days Max Temp ≤ 32 °F | 24 | 17 | 6 | 0 | 0 | 0 | 0 | 0 | 0 | 0 | 5 | 19 | 71 |
| Days Min Temp ≤ 32 °F | 31 | 28 | 28 | 19 | 7 | 1 | 0 | 0 | 4 | 14 | 25 | 30 | 187 |
| Days Min Temp ≤ 0 °F | 14 | 11 | 4 | 0 | 0 | 0 | 0 | 0 | 0 | 0 | 1 | 7 | 37 |
| Heating Degree Days | 1601 | 1338 | 1109 | 659 | 329 | 104 | 32 | 61 | 243 | 568 | 960 | 1386 | 8390 |
| Cooling Degree Days | 0 | 0 | 0 | 3 | 22 | 73 | 151 | 114 | 23 | 1 | 0 | 0 | 387 |
| Total Precipitation (") | 1.30 | 0.95 | 2.04 | 3.07 | 3.75 | 3.93 | 3.40 | 3.76 | 3.81 | 2.57 | 2.54 | 1.71 | 32.83 |
| Days ≥ 0.1" Precip | 4 | 3 | 5 | 7 | 7 | 7 | 7 | 7 | 7 | 5 | 6 | 5 | 70 |
| Total Snowfall (") | 13.2 | 8.5 | 9.2 | 3.9 | 0.4 | 0.0 | 0.0 | 0.0 | 0.0 | 0.3 | 5.7 | 13.4 | 54.6 |
| Days ≥ 1" Snow Depth | 24 | 22 | 17 | 4 | 0 | 0 | 0 | 0 | 0 | 0 | 4 | 18 | 89 |

## BRODHEAD *Green County* ELEVATION 801 ft LAT/LONG 42° 37 ' N / 89° 22 ' W

| | JAN | FEB | MAR | APR | MAY | JUN | JUL | AUG | SEP | OCT | NOV | DEC | YEAR |
|---|---|---|---|---|---|---|---|---|---|---|---|---|---|
| Maximum Temp °F | 26.3 | 30.8 | 43.4 | 58.2 | 70.7 | 79.9 | 83.6 | 81.1 | 73.3 | 61.3 | 45.5 | 32.0 | 57.2 |
| Minimum Temp °F | 6.3 | 10.6 | 23.1 | 35.0 | 45.7 | 54.8 | 59.4 | 56.7 | 48.0 | 36.4 | 26.3 | 13.7 | 34.7 |
| Mean Temp °F | 16.3 | 20.8 | 33.3 | 46.6 | 58.2 | 67.4 | 71.5 | 68.9 | 60.7 | 48.8 | 35.9 | 22.8 | 45.9 |
| Days Max Temp ≥ 90 °F | 0 | 0 | 0 | 0 | 1 | 3 | 6 | 4 | 1 | 0 | 0 | 0 | 15 |
| Days Max Temp ≤ 32 °F | 20 | 14 | 5 | 0 | 0 | 0 | 0 | 0 | 0 | 0 | 3 | 15 | 57 |
| Days Min Temp ≤ 32 °F | 31 | 28 | 26 | 12 | 2 | 0 | 0 | 0 | 2 | 12 | 23 | 29 | 165 |
| Days Min Temp ≤ 0 °F | 11 | 8 | 1 | 0 | 0 | 0 | 0 | 0 | 0 | 0 | 0 | 6 | 26 |
| Heating Degree Days | 1504 | 1241 | 976 | 554 | 246 | 52 | 12 | 31 | 173 | 501 | 865 | 1300 | 7455 |
| Cooling Degree Days | 0 | 0 | 0 | 7 | 35 | 124 | 214 | 152 | 46 | 2 | 0 | 0 | 580 |
| Total Precipitation (") | 1.22 | 1.21 | 2.28 | 3.36 | 3.30 | 4.47 | 3.78 | 4.35 | 3.94 | 2.66 | 2.34 | 1.92 | 34.83 |
| Days ≥ 0.1" Precip | 3 | 3 | 6 | 7 | 7 | 7 | 6 | 6 | 6 | 6 | 5 | 5 | 67 |
| Total Snowfall (") | *8.5* | 6.2 | *5.6* | 1.3 | 0.1 | 0.0 | 0.0 | 0.0 | 0.0 | 0.1 | *1.4* | *8.3* | 31.5 |
| Days ≥ 1" Snow Depth | na | na | na | 1 | 0 | 0 | 0 | 0 | 0 | 0 | *1* | na | na |

## BURLINGTON *Racine County* ELEVATION 751 ft LAT/LONG 42° 40 ' N / 88° 16 ' W

| | JAN | FEB | MAR | APR | MAY | JUN | JUL | AUG | SEP | OCT | NOV | DEC | YEAR |
|---|---|---|---|---|---|---|---|---|---|---|---|---|---|
| Maximum Temp °F | 26.2 | 30.6 | 42.1 | 55.9 | 68.5 | 78.1 | 82.2 | 80.0 | 72.3 | 60.3 | 45.6 | 32.8 | 56.2 |
| Minimum Temp °F | 9.0 | 12.8 | 24.0 | 35.5 | 45.1 | 54.7 | 60.0 | 58.0 | 50.0 | 38.2 | 28.5 | 16.9 | 36.1 |
| Mean Temp °F | 17.6 | 21.7 | 33.0 | 45.7 | 56.8 | 66.4 | 71.1 | 69.0 | 61.2 | 49.3 | 37.1 | 24.8 | 46.1 |
| Days Max Temp ≥ 90 °F | 0 | 0 | 0 | 0 | 0 | 3 | 5 | 3 | 1 | 0 | 0 | 0 | 12 |
| Days Max Temp ≤ 32 °F | 20 | 15 | 6 | 0 | 0 | 0 | 0 | 0 | 0 | 0 | 3 | 14 | 58 |
| Days Min Temp ≤ 32 °F | 30 | 27 | 25 | 12 | 2 | 0 | 0 | 0 | 0 | 9 | 20 | 29 | 154 |
| Days Min Temp ≤ 0 °F | 10 | 6 | 1 | 0 | 0 | 0 | 0 | 0 | 0 | 0 | 0 | 4 | 21 |
| Heating Degree Days | 1464 | 1213 | 985 | 578 | 281 | 71 | 16 | 31 | 163 | 486 | 829 | 1239 | 7356 |
| Cooling Degree Days | 0 | 0 | 0 | 9 | 37 | 126 | 216 | 172 | 58 | 2 | 0 | 0 | 620 |
| Total Precipitation (") | 1.47 | 1.19 | 2.34 | 3.48 | 2.77 | 4.15 | 4.14 | 4.03 | 3.44 | 2.50 | 2.72 | 2.02 | 34.25 |
| Days ≥ 0.1" Precip | 4 | 3 | 6 | 7 | 6 | 7 | 6 | 7 | 7 | 6 | 6 | 5 | 70 |
| Total Snowfall (") | 11.0 | 7.8 | 6.1 | 1.0 | 0.0 | 0.0 | 0.0 | 0.0 | 0.0 | 0.1 | 1.3 | 9.2 | 36.5 |
| Days ≥ 1" Snow Depth | na | na | na | 0 | 0 | 0 | 0 | 0 | 0 | 0 | *1* | na | na |

## CHARMANY FARM *Dane County* ELEVATION 1050 ft LAT/LONG 43° 3' N / 89° 28' W

| | JAN | FEB | MAR | APR | MAY | JUN | JUL | AUG | SEP | OCT | NOV | DEC | YEAR |
|---|---|---|---|---|---|---|---|---|---|---|---|---|---|
| Maximum Temp °F | 25.0 | 29.9 | 42.1 | 56.6 | 69.2 | 78.2 | 82.2 | 79.8 | 71.4 | 59.6 | 44.1 | 30.8 | 55.7 |
| Minimum Temp °F | 7.4 | 12.0 | 23.5 | 35.3 | 46.0 | 55.2 | 59.9 | 57.5 | 49.8 | 38.2 | 26.7 | 14.3 | 35.5 |
| Mean Temp °F | 16.2 | 21.0 | 32.8 | 46.0 | 57.6 | 66.7 | 71.1 | 68.6 | 60.6 | 49.0 | 35.4 | 22.6 | 45.6 |
| Days Max Temp ≥ 90 °F | 0 | 0 | 0 | 0 | 0 | 2 | 4 | 2 | 0 | 0 | 0 | 0 | 8 |
| Days Max Temp ≤ 32 °F | 22 | 16 | 6 | 0 | 0 | 0 | 0 | 0 | 0 | 0 | 4 | 16 | 64 |
| Days Min Temp ≤ 32 °F | 31 | 28 | 25 | 11 | 2 | 0 | 0 | 0 | 1 | 9 | 22 | 30 | 159 |
| Days Min Temp ≤ 0 °F | 10 | 6 | 1 | 0 | 0 | 0 | 0 | 0 | 0 | 0 | 0 | 5 | 22 |
| Heating Degree Days | 1509 | 1236 | 984 | 569 | 255 | 56 | 12 | 30 | 175 | 495 | 880 | 1310 | 7511 |
| Cooling Degree Days | 0 | 0 | 0 | 5 | 29 | 105 | 196 | 131 | 42 | 1 | 0 | 0 | 509 |
| Total Precipitation (") | 0.96 | 0.98 | 2.00 | 3.36 | 3.10 | 4.28 | 3.79 | 3.91 | 3.82 | 2.35 | 2.31 | 1.54 | 32.40 |
| Days ≥ 0.1" Precip | 3 | 3 | 5 | 7 | 7 | 7 | 6 | 7 | 6 | 5 | 5 | 4 | 65 |
| Total Snowfall (") | 8.9 | 6.9 | 4.8 | 1.3 | 0.0 | 0.0 | 0.0 | 0.0 | 0.0 | 0.1 | 1.8 | 9.3 | 33.1 |
| Days ≥ 1" Snow Depth | *16* | *14* | *4* | 1 | 0 | 0 | 0 | 0 | 0 | 0 | *1* | *10* | 46 |

## CHILTON *Calumet County* ELEVATION 860 ft LAT/LONG 44° 2' N / 88° 9' W

| | JAN | FEB | MAR | APR | MAY | JUN | JUL | AUG | SEP | OCT | NOV | DEC | YEAR |
|---|---|---|---|---|---|---|---|---|---|---|---|---|---|
| Maximum Temp °F | 24.5 | 28.8 | 40.4 | 55.7 | 69.3 | 78.2 | 82.4 | 80.3 | 71.8 | 59.3 | 43.2 | 30.0 | 55.3 |
| Minimum Temp °F | 7.9 | 11.8 | 22.8 | 34.5 | 44.6 | 54.2 | 59.3 | 57.7 | 49.9 | 39.3 | 28.1 | 15.3 | 35.5 |
| Mean Temp °F | 16.2 | 20.2 | 31.5 | 45.1 | 56.8 | 66.3 | 70.9 | 69.0 | 60.9 | 49.4 | 35.7 | 22.7 | 45.4 |
| Days Max Temp ≥ 90 °F | 0 | 0 | 0 | 0 | 0 | 2 | 4 | 3 | 0 | 0 | 0 | 0 | 9 |
| Days Max Temp ≤ 32 °F | 22 | 17 | 7 | 0 | 0 | 0 | 0 | 0 | 0 | 0 | 4 | 17 | 67 |
| Days Min Temp ≤ 32 °F | 31 | 28 | 26 | 14 | 3 | 0 | 0 | 0 | 1 | 7 | 21 | 29 | 160 |
| Days Min Temp ≤ 0 °F | 10 | 6 | 1 | 0 | 0 | 0 | 0 | 0 | 0 | 0 | 0 | 4 | 21 |
| Heating Degree Days | 1507 | 1259 | 1031 | 596 | 279 | 66 | 16 | 29 | 167 | 481 | 874 | 1306 | 7611 |
| Cooling Degree Days | 0 | 0 | 0 | 6 | 33 | 116 | 198 | 162 | 51 | 2 | 0 | 0 | 568 |
| Total Precipitation (") | 1.33 | 1.15 | 2.12 | 2.96 | 3.10 | 3.63 | 3.51 | 3.56 | 3.90 | 2.58 | 2.33 | 1.72 | 31.89 |
| Days ≥ 0.1" Precip | 4 | 3 | 6 | 7 | 6 | 7 | 7 | 6 | 7 | 6 | 6 | 5 | 70 |
| Total Snowfall (") | 11.7 | 10.1 | 7.6 | 2.2 | 0.3 | 0.0 | 0.0 | 0.0 | 0.0 | 0.4 | 4.0 | 10.2 | 46.5 |
| Days ≥ 1" Snow Depth | 26 | 21 | 10 | 1 | 0 | 0 | 0 | 0 | 0 | 0 | 3 | 17 | 78 |

## CLINTONVILLE *Waupaca County* ELEVATION 820 ft LAT/LONG 44° 37' N / 88° 45' W

| | JAN | FEB | MAR | APR | MAY | JUN | JUL | AUG | SEP | OCT | NOV | DEC | YEAR |
|---|---|---|---|---|---|---|---|---|---|---|---|---|---|
| Maximum Temp °F | 24.1 | 29.1 | 40.1 | 55.2 | 68.7 | 77.2 | 81.7 | 79.1 | 70.3 | 58.5 | 42.3 | 29.0 | 54.6 |
| Minimum Temp °F | 4.1 | 8.2 | 20.0 | 32.7 | 43.3 | 52.5 | 57.6 | 54.8 | 46.4 | 35.9 | 24.8 | 11.9 | 32.7 |
| Mean Temp °F | 14.2 | 18.7 | 30.1 | 44.0 | 56.0 | 64.9 | 69.7 | 67.0 | 58.3 | 47.2 | 33.6 | 20.5 | 43.7 |
| Days Max Temp ≥ 90 °F | 0 | 0 | 0 | 0 | 0 | 2 | 4 | 2 | 0 | 0 | 0 | 0 | 8 |
| Days Max Temp ≤ 32 °F | 23 | 17 | 6 | 0 | 0 | 0 | 0 | 0 | 0 | 0 | 5 | 19 | 70 |
| Days Min Temp ≤ 32 °F | 31 | 28 | 28 | 16 | 3 | 0 | 0 | 0 | 2 | 12 | 25 | 30 | 175 |
| Days Min Temp ≤ 0 °F | 13 | 9 | 2 | 0 | 0 | 0 | 0 | 0 | 0 | 0 | 0 | 6 | 30 |
| Heating Degree Days | 1572 | 1304 | 1075 | 627 | 296 | 86 | 20 | 46 | 222 | 547 | 935 | 1374 | 8104 |
| Cooling Degree Days | 0 | 0 | 0 | 4 | 24 | 89 | 163 | 111 | 25 | 0 | 0 | 0 | 416 |
| Total Precipitation (") | 1.25 | 0.96 | 2.15 | 2.72 | 3.58 | 3.49 | 3.86 | 3.82 | 3.66 | 2.49 | 2.32 | 1.63 | 31.93 |
| Days ≥ 0.1" Precip | 4 | 3 | 5 | 7 | 7 | 7 | 7 | 7 | 7 | 5 | 5 | 5 | 69 |
| Total Snowfall (") | 12.2 | 7.9 | 7.3 | 2.2 | 0.0 | 0.0 | 0.0 | 0.0 | 0.1 | 0.2 | 3.9 | 11.8 | 45.6 |
| Days ≥ 1" Snow Depth | 28 | 23 | 14 | 2 | 0 | 0 | 0 | 0 | 0 | 0 | 4 | 18 | 89 |

## COUDERAY 7 W *Sawyer County* ELEVATION 1302 ft LAT/LONG 45° 52' N / 91° 27' W

| | JAN | FEB | MAR | APR | MAY | JUN | JUL | AUG | SEP | OCT | NOV | DEC | YEAR |
|---|---|---|---|---|---|---|---|---|---|---|---|---|---|
| Maximum Temp °F | *20.8* | *26.6* | 38.7 | *52.7* | 68.0 | 76.1 | 81.5 | 78.1 | 68.1 | 56.6 | 38.6 | 25.8 | 52.6 |
| Minimum Temp °F | *-2.0* | *1.8* | *17.9* | 30.9 | *40.9* | *51.0* | *57.9* | *55.0* | *43.9* | *33.3* | 21.4 | *7.0* | 29.9 |
| Mean Temp °F | *9.9* | *14.3* | *28.4* | *41.9* | *54.8* | *63.6* | *69.7* | *66.5* | *56.0* | *45.1* | 30.1 | *16.4* | 41.4 |
| Days Max Temp ≥ 90 °F | 0 | 0 | 0 | 0 | 0 | 1 | 4 | 2 | 0 | 0 | 0 | 0 | 7 |
| Days Max Temp ≤ 32 °F | *25* | *19* | *8* | *1* | 0 | 0 | 0 | 0 | 0 | 0 | 8 | 22 | 83 |
| Days Min Temp ≤ 32 °F | *30* | *27* | *27* | *18* | *7* | *1* | 0 | 0 | *4* | *15* | *26* | 30 | 185 |
| Days Min Temp ≤ 0 °F | *16* | na | *4* | 0 | 0 | *0* | 0 | 0 | 0 | 0 | 2 | *11* | na |
| Heating Degree Days | *1715* | *1430* | *1127* | *691* | *340* | *115* | *29* | *69* | *287* | *612* | 1042 | *1499* | 8956 |
| Cooling Degree Days | *0* | 0 | 0 | 2 | 27 | 94 | 213 | 159 | 32 | 0 | 0 | 0 | 527 |
| Total Precipitation (") | 1.10 | 0.89 | 1.76 | 2.70 | 3.20 | 4.67 | 4.46 | 4.62 | 4.81 | 3.42 | 1.82 | 1.18 | 34.63 |
| Days ≥ 0.1" Precip | *4* | na | *4* | *7* | 7 | *9* | *8* | 8 | 9 | 7 | *5* | *4* | na |
| Total Snowfall (") | na | na | na | na | 0.2 | 0.0 | 0.0 | 0.0 | 0.0 | *0.8* | na | na | na |
| Days ≥ 1" Snow Depth | na | na | na | na | 0 | 0 | 0 | 0 | 0 | 0 | na | na | na |

## CRIVITZ HIGH FALLS *Marinette County* ELEVATION 879 ft LAT/LONG 45° 17' N / 88° 12' W

| | JAN | FEB | MAR | APR | MAY | JUN | JUL | AUG | SEP | OCT | NOV | DEC | YEAR |
|---|---|---|---|---|---|---|---|---|---|---|---|---|---|
| Maximum Temp °F | 24.2 | 28.4 | 39.1 | 53.5 | 67.8 | 76.1 | 80.5 | ***78.4*** | 68.4 | 56.7 | 41.0 | 29.0 | 53.6 |
| Minimum Temp °F | 1.1 | 4.5 | 16.9 | 29.6 | 40.3 | 50.0 | 55.0 | ***53.9*** | 45.1 | 34.4 | 23.3 | 10.7 | 30.4 |
| Mean Temp °F | 12.7 | 16.5 | 28.0 | 41.6 | 54.0 | 63.1 | 67.8 | ***66.2*** | 56.8 | 45.5 | 32.2 | 19.9 | 42.0 |
| Days Max Temp ≥ 90 °F | 0 | 0 | 0 | 0 | 0 | 2 | 3 | 2 | 0 | 0 | 0 | 0 | 7 |
| Days Max Temp ≤ 32 °F | 24 | 18 | 8 | 0 | 0 | 0 | 0 | 0 | 0 | 0 | 6 | 20 | 76 |
| Days Min Temp ≤ 32 °F | 31 | 28 | 28 | 19 | 7 | 0 | 0 | 0 | 2 | 14 | 25 | 30 | 184 |
| Days Min Temp ≤ 0 °F | 15 | 11 | 4 | 0 | 0 | 0 | 0 | 0 | 0 | 0 | 0 | 7 | 37 |
| Heating Degree Days | 1617 | 1365 | 1139 | 700 | 349 | 116 | 43 | ***62*** | 261 | 599 | 979 | 1393 | 8623 |
| Cooling Degree Days | 0 | 0 | 0 | ***3*** | ***12*** | 51 | 115 | na | ***14*** | ***0*** | ***0*** | 0 | na |
| Total Precipitation (") | 1.07 | 0.83 | 1.83 | 2.69 | 3.48 | 3.96 | 3.00 | 3.71 | 4.13 | 2.38 | 2.19 | 1.54 | 30.81 |
| Days ≥ 0.1" Precip | 3 | 3 | 5 | 6 | 7 | 7 | 6 | 6 | 8 | 5 | 4 | 5 | 65 |
| Total Snowfall (") | ***14.7*** | ***9.0*** | ***9.6*** | ***3.0*** | 0.4 | 0.0 | 0.0 | 0.0 | 0.0 | 0.3 | na | ***13.8*** | na |
| Days ≥ 1" Snow Depth | ***23*** | ***21*** | ***14*** | ***4*** | 0 | 0 | 0 | 0 | 0 | 0 | na | na | na |

## CUMBERLAND *Barron County* ELEVATION 1240 ft LAT/LONG 45° 32' N / 92° 1' W

| | JAN | FEB | MAR | APR | MAY | JUN | JUL | AUG | SEP | OCT | NOV | DEC | YEAR |
|---|---|---|---|---|---|---|---|---|---|---|---|---|---|
| Maximum Temp °F | 20.7 | 27.2 | 39.6 | 56.4 | 70.2 | 77.9 | 82.3 | 79.7 | 69.6 | 57.3 | 39.0 | 25.1 | 53.8 |
| Minimum Temp °F | -0.4 | 4.3 | 17.9 | 32.6 | 44.8 | 54.1 | 59.4 | 56.7 | 47.5 | 36.2 | 22.8 | 7.9 | 32.0 |
| Mean Temp °F | 10.1 | 15.7 | 28.8 | 44.5 | 57.5 | 66.1 | 70.9 | 68.2 | 58.6 | 46.8 | 30.9 | 16.5 | 42.9 |
| Days Max Temp ≥ 90 °F | 0 | 0 | 0 | 0 | 0 | 2 | 4 | 2 | 0 | 0 | 0 | 0 | 8 |
| Days Max Temp ≤ 32 °F | 25 | 17 | 8 | 0 | 0 | 0 | 0 | 0 | 0 | 0 | 8 | 23 | 81 |
| Days Min Temp ≤ 32 °F | 31 | 27 | 28 | 16 | 2 | 0 | 0 | 0 | 1 | 11 | 25 | 31 | 172 |
| Days Min Temp ≤ 0 °F | 16 | 11 | 4 | 0 | 0 | 0 | 0 | 0 | 0 | 0 | 1 | 9 | 41 |
| Heating Degree Days | 1697 | 1387 | 1117 | 610 | 257 | 60 | 12 | 35 | 216 | 561 | 1016 | 1497 | 8465 |
| Cooling Degree Days | 0 | 0 | 0 | 5 | 33 | 100 | 204 | 146 | 29 | 1 | 0 | 0 | 518 |
| Total Precipitation (") | 1.19 | 0.86 | 1.76 | 2.89 | 3.19 | 4.80 | 4.25 | 4.18 | 4.26 | 2.73 | 2.03 | 1.33 | 33.47 |
| Days ≥ 0.1" Precip | 4 | 3 | 4 | 6 | 7 | 8 | 7 | 7 | 7 | 6 | 5 | 4 | 68 |
| Total Snowfall (") | 13.4 | 8.7 | 9.5 | 3.9 | 0.0 | 0.0 | 0.0 | 0.0 | 0.0 | 0.6 | 7.9 | 11.9 | 55.9 |
| Days ≥ 1" Snow Depth | 24 | 21 | ***17*** | 4 | 0 | 0 | 0 | 0 | 0 | 0 | 6 | 20 | 92 |

## DALTON *Green Lake County* ELEVATION 860 ft LAT/LONG 43° 39' N / 89° 12' W

| | JAN | FEB | MAR | APR | MAY | JUN | JUL | AUG | SEP | OCT | NOV | DEC | YEAR |
|---|---|---|---|---|---|---|---|---|---|---|---|---|---|
| Maximum Temp °F | 25.4 | 30.5 | 42.3 | 57.5 | 70.5 | 79.1 | 83.0 | 80.7 | 71.9 | 60.0 | 43.7 | 30.6 | 56.3 |
| Minimum Temp °F | 7.1 | 11.3 | 22.9 | 34.7 | 45.3 | 54.5 | 59.4 | 57.3 | 49.6 | 38.9 | 27.1 | 14.3 | 35.2 |
| Mean Temp °F | 16.3 | 20.9 | 32.7 | 46.2 | 57.9 | 66.8 | 71.3 | 69.0 | 60.8 | 49.5 | 35.4 | 22.5 | 45.8 |
| Days Max Temp ≥ 90 °F | 0 | 0 | 0 | 0 | 0 | 2 | 4 | 2 | 0 | 0 | 0 | 0 | 8 |
| Days Max Temp ≤ 32 °F | 21 | 15 | 6 | 0 | 0 | 0 | 0 | 0 | 0 | 0 | 4 | 17 | 63 |
| Days Min Temp ≤ 32 °F | 31 | 27 | 26 | 13 | 3 | 0 | 0 | 0 | 1 | 8 | 21 | 30 | 160 |
| Days Min Temp ≤ 0 °F | 11 | 7 | 1 | 0 | 0 | 0 | 0 | 0 | 0 | 0 | 0 | 5 | 24 |
| Heating Degree Days | 1506 | 1240 | 996 | 565 | 251 | 58 | 12 | 30 | 172 | 480 | 881 | 1312 | 7503 |
| Cooling Degree Days | 0 | 0 | 0 | 7 | 37 | 126 | 222 | 172 | 56 | 3 | 0 | 0 | 623 |
| Total Precipitation (") | 1.16 | 1.05 | 2.11 | 3.05 | 3.27 | 3.79 | 3.71 | 3.64 | 4.20 | 2.47 | 2.32 | 1.61 | 32.38 |
| Days ≥ 0.1" Precip | 4 | 3 | 5 | 7 | 7 | 7 | 6 | 6 | 7 | 6 | 5 | 4 | 67 |
| Total Snowfall (") | 10.3 | 7.9 | 7.6 | 2.6 | 0.1 | 0.0 | 0.0 | 0.0 | 0.0 | 0.4 | 3.9 | 10.9 | 43.7 |
| Days ≥ 1" Snow Depth | 25 | 23 | 12 | 1 | 0 | 0 | 0 | 0 | 0 | 0 | 3 | 17 | 81 |

## DANBURY *Burnett County* ELEVATION 902 ft LAT/LONG 46° 0' N / 92° 22' W

| | JAN | FEB | MAR | APR | MAY | JUN | JUL | AUG | SEP | OCT | NOV | DEC | YEAR |
|---|---|---|---|---|---|---|---|---|---|---|---|---|---|
| Maximum Temp °F | 20.1 | 26.6 | 39.0 | 55.3 | 68.7 | 75.9 | 80.3 | 77.6 | 67.8 | 56.2 | 38.2 | 24.8 | 52.5 |
| Minimum Temp °F | -2.3 | 3.3 | 17.3 | 30.7 | 41.7 | 50.7 | 56.6 | 54.2 | 45.4 | 34.6 | 21.5 | 6.2 | 30.0 |
| Mean Temp °F | 8.9 | 14.9 | 28.2 | 43.0 | 55.2 | 63.3 | 68.4 | 65.9 | 56.6 | 45.4 | 29.9 | 15.5 | 41.3 |
| Days Max Temp ≥ 90 °F | 0 | 0 | 0 | 0 | 0 | 1 | 2 | 1 | 0 | 0 | 0 | 0 | 4 |
| Days Max Temp ≤ 32 °F | 26 | 19 | 8 | 1 | 0 | 0 | 0 | 0 | 0 | 0 | 9 | 23 | 86 |
| Days Min Temp ≤ 32 °F | 31 | 28 | 28 | 17 | 5 | 0 | 0 | 0 | 3 | 14 | 26 | 31 | 183 |
| Days Min Temp ≤ 0 °F | 17 | 12 | 4 | 0 | 0 | 0 | 0 | 0 | 0 | 0 | 2 | 11 | 46 |
| Heating Degree Days | 1736 | 1409 | 1135 | 654 | 317 | 106 | 32 | 68 | 269 | 602 | 1046 | 1528 | 8902 |
| Cooling Degree Days | 0 | 0 | 0 | 3 | 23 | 70 | 151 | 112 | 25 | 1 | 0 | 0 | 385 |
| Total Precipitation (") | 1.07 | 0.77 | 1.71 | 2.29 | 3.24 | 4.22 | 4.18 | 4.09 | 3.42 | 2.45 | 1.87 | 1.22 | 30.53 |
| Days ≥ 0.1" Precip | 3 | 2 | 5 | 6 | 7 | 8 | 7 | 7 | 7 | 6 | 4 | 4 | 66 |
| Total Snowfall (") | 12.2 | 7.4 | 8.7 | 3.5 | 0.2 | 0.0 | 0.0 | 0.0 | 0.0 | 0.7 | 8.5 | 11.0 | 52.2 |
| Days ≥ 1" Snow Depth | 25 | 22 | 19 | 4 | 0 | 0 | 0 | 0 | 0 | 0 | 7 | 21 | 98 |

## DARLINGTON *Lafayette County* ELEVATION 820 ft LAT/LONG 42° 41 ' N / 90° 7 ' W

| | JAN | FEB | MAR | APR | MAY | JUN | JUL | AUG | SEP | OCT | NOV | DEC | YEAR |
|---|---|---|---|---|---|---|---|---|---|---|---|---|---|
| Maximum Temp °F | 26.1 | 31.5 | 44.1 | 58.4 | 70.5 | 79.1 | 83.2 | 81.0 | 72.7 | 61.1 | 44.6 | 31.2 | 57.0 |
| Minimum Temp °F | 7.1 | 11.6 | 23.9 | 35.4 | 46.0 | 55.4 | 60.5 | 58.0 | 49.4 | 38.0 | 26.6 | 14.3 | 35.5 |
| Mean Temp °F | 16.6 | 21.6 | 34.0 | 47.0 | 58.3 | 67.2 | 71.9 | 69.5 | 61.0 | 49.5 | 35.6 | 22.8 | 46.3 |
| Days Max Temp ≥ 90 °F | 0 | 0 | 0 | 0 | 0 | 2 | 5 | 3 | 1 | 0 | 0 | 0 | 11 |
| Days Max Temp ≤ 32 °F | 21 | 14 | 4 | 0 | 0 | 0 | 0 | 0 | 0 | 0 | 4 | 15 | 58 |
| Days Min Temp ≤ 32 °F | 31 | 27 | 25 | 12 | 2 | 0 | 0 | 0 | 1 | 10 | 22 | 29 | 159 |
| Days Min Temp ≤ 0 °F | 11 | 7 | 1 | 0 | 0 | 0 | 0 | 0 | 0 | 0 | 0 | 5 | 24 |
| Heating Degree Days | 1495 | 1220 | 955 | 541 | 241 | 52 | 10 | 25 | 170 | 479 | 876 | 1303 | 7367 |
| Cooling Degree Days | 0 | 0 | 0 | 9 | 37 | 127 | 242 | 177 | 59 | 2 | 0 | 0 | 653 |
| Total Precipitation (") | 1.22 | 1.20 | 2.30 | 3.31 | 3.34 | 4.72 | 3.98 | 4.25 | 4.15 | 2.41 | 2.39 | 1.83 | 35.10 |
| Days ≥ 0.1" Precip | 4 | 3 | 6 | 7 | 7 | 7 | 6 | 7 | 6 | 5 | 5 | 4 | 67 |
| Total Snowfall (") | 9.3 | 7.6 | 6.7 | 2.7 | 0.3 | 0.0 | 0.0 | 0.0 | 0.0 | 0.3 | 2.9 | 9.1 | 38.9 |
| Days ≥ 1" Snow Depth | 23 | 21 | 9 | 1 | 0 | 0 | 0 | 0 | 0 | 0 | 3 | 16 | 73 |

## DODGEVILLE *Iowa County* ELEVATION 1220 ft LAT/LONG 42° 58 ' N / 90° 7 ' W

| | JAN | FEB | MAR | APR | MAY | JUN | JUL | AUG | SEP | OCT | NOV | DEC | YEAR |
|---|---|---|---|---|---|---|---|---|---|---|---|---|---|
| Maximum Temp °F | 25.0 | 29.3 | 42.6 | 57.5 | 69.3 | 78.0 | 82.0 | 79.5 | 71.3 | 59.2 | 43.3 | 30.4 | 55.6 |
| Minimum Temp °F | 7.0 | 10.6 | 23.6 | 35.4 | 45.7 | 55.0 | 59.9 | 57.4 | 49.5 | 37.9 | 26.3 | 14.0 | 35.2 |
| Mean Temp °F | 16.0 | 19.9 | 33.1 | 46.5 | 57.5 | 66.6 | 71.0 | 68.5 | 60.5 | 48.6 | 34.8 | 22.2 | 45.4 |
| Days Max Temp ≥ 90 °F | 0 | 0 | 0 | 0 | 0 | 1 | 4 | 2 | 0 | 0 | 0 | 0 | 7 |
| Days Max Temp ≤ 32 °F | 22 | 16 | 5 | 0 | 0 | 0 | 0 | 0 | 0 | 0 | 5 | 17 | 65 |
| Days Min Temp ≤ 32 °F | 31 | 28 | 25 | 12 | 2 | 0 | 0 | 0 | 1 | 9 | 22 | 30 | 160 |
| Days Min Temp ≤ 0 °F | 11 | 7 | 1 | 0 | 0 | 0 | 0 | 0 | 0 | 0 | 0 | 5 | 24 |
| Heating Degree Days | 1514 | 1265 | 982 | 555 | 258 | 59 | 13 | 32 | 176 | 508 | 898 | 1320 | 7580 |
| Cooling Degree Days | ***0*** | ***0*** | ***0*** | ***8*** | ***29*** | ***111*** | ***194*** | ***135*** | ***45*** | ***1*** | ***0*** | 0 | 523 |
| Total Precipitation (") | 1.33 | 1.27 | 2.65 | 3.43 | 3.24 | 4.23 | 4.14 | 4.68 | 3.99 | 2.33 | 2.36 | 1.77 | 35.42 |
| Days ≥ 0.1" Precip | 4 | 3 | 5 | 8 | 7 | 7 | 6 | 8 | 7 | 5 | 5 | 5 | 70 |
| Total Snowfall (") | 9.0 | 7.5 | 6.4 | 1.8 | 0.1 | 0.0 | 0.0 | 0.0 | 0.0 | 0.3 | 4.0 | 10.3 | 39.4 |
| Days ≥ 1" Snow Depth | 23 | 21 | 10 | 1 | 0 | 0 | 0 | 0 | 0 | 0 | 4 | 16 | 75 |

## EAU CLAIRE COUNTY AP *Chippewa County* ELEVATION 892 ft LAT/LONG 44° 52 ' N / 91° 29 ' W

| | JAN | FEB | MAR | APR | MAY | JUN | JUL | AUG | SEP | OCT | NOV | DEC | YEAR |
|---|---|---|---|---|---|---|---|---|---|---|---|---|---|
| Maximum Temp °F | 20.5 | 26.8 | 39.7 | 56.4 | 69.6 | 78.0 | 82.5 | 79.7 | 69.9 | 57.5 | 40.1 | 26.1 | 53.9 |
| Minimum Temp °F | 1.1 | 6.3 | 20.0 | 33.6 | 45.3 | 54.7 | 59.9 | 57.4 | 48.1 | 36.6 | 23.6 | 9.5 | 33.0 |
| Mean Temp °F | 10.8 | 16.6 | 29.9 | 45.0 | 57.5 | 66.4 | 71.2 | 68.6 | 59.0 | 47.1 | 31.9 | 17.8 | 43.5 |
| Days Max Temp ≥ 90 °F | 0 | 0 | 0 | 0 | 1 | 3 | 5 | 3 | 0 | 0 | 0 | 0 | 12 |
| Days Max Temp ≤ 32 °F | 24 | 18 | 8 | 1 | 0 | 0 | 0 | 0 | 0 | 0 | 7 | 21 | 79 |
| Days Min Temp ≤ 32 °F | 30 | 28 | 27 | 15 | 3 | 0 | 0 | 0 | 1 | 11 | 25 | 30 | 170 |
| Days Min Temp ≤ 0 °F | 15 | 10 | 3 | 0 | 0 | 0 | 0 | 0 | 0 | 0 | 1 | 9 | 38 |
| Heating Degree Days | 1677 | 1362 | 1082 | 598 | 264 | 63 | 12 | 35 | 214 | 552 | 988 | 1457 | 8304 |
| Cooling Degree Days | 0 | 0 | 0 | 7 | 40 | 119 | 222 | 163 | 44 | 1 | 0 | 0 | 596 |
| Total Precipitation (") | 1.06 | 0.73 | 1.72 | 3.03 | 3.67 | 4.29 | 4.08 | 4.38 | 3.81 | 2.44 | 1.82 | 1.17 | 32.20 |
| Days ≥ 0.1" Precip | 3 | 2 | 5 | 6 | 7 | 8 | 7 | 7 | 7 | 5 | 4 | 4 | 65 |
| Total Snowfall (") | 12.5 | 7.5 | 8.9 | 2.2 | 0.0 | 0.0 | 0.0 | 0.0 | 0.0 | 0.3 | 6.0 | 10.9 | 48.3 |
| Days ≥ 1" Snow Depth | 30 | 26 | 18 | 2 | 0 | 0 | 0 | 0 | 0 | 0 | 6 | 24 | 106 |

## ELLSWORTH 1 E *Pierce County* ELEVATION 1030 ft LAT/LONG 44° 44 ' N / 92° 28 ' W

| | JAN | FEB | MAR | APR | MAY | JUN | JUL | AUG | SEP | OCT | NOV | DEC | YEAR |
|---|---|---|---|---|---|---|---|---|---|---|---|---|---|
| Maximum Temp °F | 21.4 | 28.1 | 40.7 | 57.3 | 70.0 | 78.5 | 82.8 | ***80.1*** | 70.7 | 59.0 | 40.4 | 26.1 | 54.6 |
| Minimum Temp °F | 1.8 | 8.0 | 20.9 | 34.0 | 45.9 | 55.1 | 59.5 | ***56.8*** | 47.7 | 36.9 | 23.6 | 9.7 | 33.3 |
| Mean Temp °F | 11.6 | 18.1 | 30.9 | 45.7 | 58.0 | 66.9 | 71.2 | ***68.5*** | 59.2 | 48.0 | 32.0 | 18.0 | 44.0 |
| Days Max Temp ≥ 90 °F | 0 | 0 | 0 | 0 | 0 | 2 | 4 | ***2*** | 0 | 0 | 0 | 0 | 8 |
| Days Max Temp ≤ 32 °F | 24 | 17 | 7 | 0 | 0 | 0 | 0 | ***0*** | 0 | 0 | 7 | 22 | 77 |
| Days Min Temp ≤ 32 °F | 31 | 28 | 26 | 14 | 2 | 0 | 0 | ***0*** | 1 | 11 | 24 | 31 | 168 |
| Days Min Temp ≤ 0 °F | 14 | 9 | 2 | 0 | 0 | 0 | 0 | ***0*** | 0 | 0 | 1 | ***8*** | 34 |
| Heating Degree Days | 1651 | 1319 | 1050 | 577 | 245 | 55 | 12 | ***37*** | 202 | 525 | 982 | 1453 | 8108 |
| Cooling Degree Days | 0 | 0 | 0 | 4 | 25 | 112 | 187 | ***125*** | 38 | 2 | 0 | 0 | 493 |
| Total Precipitation (") | 1.05 | 0.79 | 1.91 | 3.31 | 3.94 | 4.68 | 4.59 | ***3.99*** | 4.21 | 2.66 | 2.25 | 1.32 | 34.70 |
| Days ≥ 0.1" Precip | 3 | 3 | 5 | 7 | 8 | 8 | 7 | ***7*** | 7 | 5 | 5 | 4 | 69 |
| Total Snowfall (") | 13.8 | 8.8 | 10.7 | 2.4 | 0.0 | 0.0 | 0.0 | ***0.0*** | 0.0 | 0.4 | 7.7 | 12.4 | 56.2 |
| Days ≥ 1" Snow Depth | na | na | na | 1 | 0 | 0 | 0 | ***0*** | 0 | 0 | na | na | na |

## FAIRCHILD RANGER STA *Eau Claire County* ELEVATION 1079 ft LAT/LONG 44° 36' N / 90° 58' W

| | JAN | FEB | MAR | APR | MAY | JUN | JUL | AUG | SEP | OCT | NOV | DEC | YEAR |
|---|---|---|---|---|---|---|---|---|---|---|---|---|---|
| Maximum Temp °F | ***18.9*** | 26.7 | 39.2 | 55.1 | 68.3 | 76.4 | 81.0 | 78.9 | 69.7 | 57.1 | 40.5 | ***25.5*** | 53.1 |
| Minimum Temp °F | ***-0.6*** | 6.0 | 19.6 | 33.8 | 45.0 | 53.8 | 58.5 | 56.3 | 47.0 | 35.8 | 24.0 | ***8.7*** | 32.3 |
| Mean Temp °F | ***9.2*** | 16.7 | 29.4 | 44.5 | 56.6 | 65.2 | 69.8 | 67.7 | 58.3 | 46.4 | 32.2 | ***17.1*** | 42.8 |
| Days Max Temp ≥ 90 °F | 0 | 0 | 0 | 0 | 0 | 1 | 3 | 2 | 0 | 0 | 0 | 0 | 6 |
| Days Max Temp ≤ 32 °F | 24 | 17 | 9 | 0 | 0 | 0 | 0 | 0 | 0 | 0 | 6 | 20 | 76 |
| Days Min Temp ≤ 32 °F | 29 | 27 | 27 | 15 | 3 | 0 | 0 | 0 | 1 | 12 | 23 | 28 | 165 |
| Days Min Temp ≤ 0 °F | 15 | 10 | 3 | 0 | 0 | 0 | 0 | 0 | 0 | 0 | 1 | 8 | 37 |
| Heating Degree Days | ***1728*** | 1362 | 1096 | 613 | 284 | 80 | 24 | 45 | 228 | 574 | ***968*** | ***1480*** | 8482 |
| Cooling Degree Days | na | ***0*** | 0 | 6 | 30 | 99 | ***186*** | 144 | ***39*** | 1 | na | na | na |
| Total Precipitation (") | 1.11 | 0.81 | 2.02 | 3.08 | 3.68 | 4.35 | 4.44 | 4.09 | 4.62 | 2.66 | 2.11 | 1.42 | 34.39 |
| Days ≥ 0.1" Precip | 3 | 3 | 4 | 7 | 7 | 8 | 7 | 6 | 7 | 5 | 4 | 4 | 65 |
| Total Snowfall (") | 10.2 | 7.1 | 8.2 | 2.8 | 0.0 | 0.0 | 0.0 | 0.0 | 0.0 | 0.5 | ***5.1*** | 11.1 | 45.0 |
| Days ≥ 1" Snow Depth | 28 | 26 | 21 | 4 | 0 | 0 | 0 | 0 | 0 | 0 | 5 | 22 | 106 |

## FOND DU LAC *Fond du Lac County* ELEVATION 761 ft LAT/LONG 43° 47' N / 88° 27' W

| | JAN | FEB | MAR | APR | MAY | JUN | JUL | AUG | SEP | OCT | NOV | DEC | YEAR |
|---|---|---|---|---|---|---|---|---|---|---|---|---|---|
| Maximum Temp °F | 24.4 | 28.7 | 40.3 | 54.6 | 68.1 | 77.1 | 81.6 | 79.0 | 70.5 | 58.4 | 43.0 | 30.1 | 54.7 |
| Minimum Temp °F | 8.1 | 12.1 | 23.5 | 36.1 | 47.1 | 56.6 | 61.9 | 59.9 | 51.9 | 40.7 | 28.6 | 15.5 | 36.8 |
| Mean Temp °F | 16.3 | 20.5 | 31.9 | 45.3 | 57.6 | 66.9 | 71.8 | 69.5 | 61.2 | 49.5 | 35.8 | 22.8 | 45.8 |
| Days Max Temp ≥ 90 °F | 0 | 0 | 0 | 0 | 0 | 2 | 4 | 2 | 0 | 0 | 0 | 0 | 8 |
| Days Max Temp ≤ 32 °F | 22 | 17 | 7 | 1 | 0 | 0 | 0 | 0 | 0 | 0 | 4 | 16 | 67 |
| Days Min Temp ≤ 32 °F | 30 | 27 | 26 | 11 | 1 | 0 | 0 | 0 | 0 | 5 | 20 | 29 | 149 |
| Days Min Temp ≤ 0 °F | 10 | 6 | 1 | 0 | 0 | 0 | 0 | 0 | 0 | 0 | 0 | 4 | 21 |
| Heating Degree Days | 1505 | 1250 | 1018 | 590 | 262 | 60 | 9 | 23 | 161 | 478 | 869 | 1300 | 7525 |
| Cooling Degree Days | 0 | 0 | 0 | 9 | 42 | 126 | 230 | 171 | 58 | 2 | 0 | 0 | 638 |
| Total Precipitation (") | 1.01 | 0.91 | 1.76 | 2.77 | 2.95 | 3.32 | 3.36 | 3.84 | 3.73 | 2.48 | 1.95 | 1.52 | 29.60 |
| Days ≥ 0.1" Precip | 3 | 3 | 5 | 6 | 6 | 6 | 6 | 6 | 6 | 6 | 5 | 4 | 62 |
| Total Snowfall (") | 9.8 | 7.4 | 6.3 | 1.4 | 0.1 | 0.0 | 0.0 | 0.0 | 0.0 | 0.1 | 2.7 | 10.0 | 37.8 |
| Days ≥ 1" Snow Depth | 24 | 22 | 10 | 1 | 0 | 0 | 0 | 0 | 0 | 0 | 3 | 17 | 77 |

## FORT ATKINSON *Jefferson County* ELEVATION 801 ft LAT/LONG 42° 55' N / 88° 51' W

| | JAN | FEB | MAR | APR | MAY | JUN | JUL | AUG | SEP | OCT | NOV | DEC | YEAR |
|---|---|---|---|---|---|---|---|---|---|---|---|---|---|
| Maximum Temp °F | 25.9 | 30.8 | 43.4 | 58.0 | 70.3 | 79.6 | 83.7 | 81.1 | 73.2 | 61.1 | 45.3 | 31.9 | 57.0 |
| Minimum Temp °F | 7.7 | 12.0 | 24.7 | 36.5 | 46.2 | 55.6 | 60.4 | 57.8 | 49.9 | 38.8 | 28.2 | 15.6 | 36.1 |
| Mean Temp °F | 16.8 | 21.4 | 34.1 | 47.3 | 58.3 | 67.7 | 72.0 | 69.5 | 61.5 | 49.9 | 36.7 | 23.8 | 46.6 |
| Days Max Temp ≥ 90 °F | 0 | 0 | 0 | 0 | 1 | 3 | 6 | 3 | 1 | 0 | 0 | 0 | 14 |
| Days Max Temp ≤ 32 °F | 21 | 15 | 5 | 0 | 0 | 0 | 0 | 0 | 0 | 0 | 3 | 14 | 58 |
| Days Min Temp ≤ 32 °F | 30 | 27 | 24 | 10 | 2 | 0 | 0 | 0 | 1 | 9 | 21 | 29 | 153 |
| Days Min Temp ≤ 0 °F | 11 | 7 | 1 | 0 | 0 | 0 | 0 | 0 | 0 | 0 | 0 | 5 | 24 |
| Heating Degree Days | 1489 | 1225 | 952 | 532 | 244 | 52 | 10 | 28 | 157 | 467 | 842 | 1271 | 7269 |
| Cooling Degree Days | 0 | 0 | 0 | 9 | 40 | 130 | 222 | 162 | 52 | 2 | 0 | 0 | 617 |
| Total Precipitation (") | 1.33 | 1.20 | 2.15 | 3.30 | 3.06 | 3.89 | 3.97 | 4.05 | 3.81 | 2.56 | 2.51 | 1.87 | 33.70 |
| Days ≥ 0.1" Precip | 4 | 4 | 5 | 7 | 7 | 7 | 6 | 7 | 7 | 5 | 6 | 5 | 70 |
| Total Snowfall (") | 11.1 | 7.4 | 6.0 | 1.5 | 0.2 | 0.0 | 0.0 | 0.0 | 0.0 | 0.1 | 2.2 | 9.5 | 38.0 |
| Days ≥ 1" Snow Depth | 23 | 19 | ***8*** | 1 | 0 | 0 | 0 | 0 | 0 | 0 | 2 | 15 | 68 |

## FOXBORO *Douglas County* ELEVATION 932 ft LAT/LONG 46° 30' N / 92° 17' W

| | JAN | FEB | MAR | APR | MAY | JUN | JUL | AUG | SEP | OCT | NOV | DEC | YEAR |
|---|---|---|---|---|---|---|---|---|---|---|---|---|---|
| Maximum Temp °F | 20.2 | 26.3 | 37.4 | 52.4 | 66.1 | 74.4 | 80.0 | 77.6 | 68.1 | 56.0 | 38.2 | 25.2 | 51.8 |
| Minimum Temp °F | -3.7 | 1.4 | 14.7 | 27.2 | 36.1 | 45.1 | 52.6 | 51.4 | 43.7 | 33.0 | 20.0 | 4.7 | 27.2 |
| Mean Temp °F | 8.2 | 13.9 | 26.1 | 39.8 | 51.1 | 59.8 | 66.3 | 64.5 | 55.9 | 44.5 | 29.1 | 15.0 | 39.5 |
| Days Max Temp ≥ 90 °F | 0 | 0 | 0 | 0 | 0 | 1 | 3 | 2 | 0 | 0 | 0 | 0 | 6 |
| Days Max Temp ≤ 32 °F | 26 | 20 | 10 | 1 | 0 | 0 | 0 | 0 | 0 | 0 | 9 | 22 | 88 |
| Days Min Temp ≤ 32 °F | 31 | 28 | 29 | 22 | 10 | 2 | 0 | 0 | 4 | 15 | 27 | 31 | 199 |
| Days Min Temp ≤ 0 °F | 18 | 13 | 5 | 0 | 0 | 0 | 0 | 0 | 0 | 0 | 2 | 12 | 50 |
| Heating Degree Days | 1756 | 1438 | 1202 | 750 | 431 | 181 | 57 | 89 | 285 | 628 | 1070 | 1545 | 9432 |
| Cooling Degree Days | 0 | 0 | 0 | 0 | 8 | 27 | 89 | 77 | 15 | 1 | 0 | 0 | 217 |
| Total Precipitation (") | 0.94 | 0.64 | 1.61 | 2.36 | 3.24 | 4.28 | 4.13 | 4.34 | 3.99 | 2.61 | 2.01 | 1.10 | 31.25 |
| Days ≥ 0.1" Precip | 3 | 2 | 4 | 5 | 7 | 8 | 7 | 7 | 7 | 5 | 5 | 4 | 64 |
| Total Snowfall (") | na | na | na | na | na | na | na | na | na | na | na | na | na |
| Days ≥ 1" Snow Depth | na | na | na | na | na | na | na | na | na | na | na | na | na |

## GENOA DAM 8 *Vernon County* ELEVATION 639 ft LAT/LONG 43° 35' N / 91° 13' W

| | JAN | FEB | MAR | APR | MAY | JUN | JUL | AUG | SEP | OCT | NOV | DEC | YEAR |
|---|---|---|---|---|---|---|---|---|---|---|---|---|---|
| Maximum Temp °F | 24.6 | 30.1 | 42.6 | 57.9 | 70.6 | 79.1 | 83.0 | 80.7 | 71.9 | 60.2 | 43.3 | 30.1 | 56.2 |
| Minimum Temp °F | 7.0 | 11.5 | 24.7 | 37.9 | 48.7 | 58.1 | 62.8 | 60.6 | 52.2 | 41.2 | 28.2 | 15.5 | 37.4 |
| Mean Temp °F | 15.8 | 20.8 | 33.7 | 47.9 | 59.7 | 68.6 | 72.9 | 70.7 | 62.1 | 50.7 | 35.8 | 22.8 | 46.8 |
| Days Max Temp ≥ 90 °F | 0 | 0 | 0 | 0 | 0 | 2 | 4 | 3 | 1 | 0 | 0 | 0 | 10 |
| Days Max Temp ≤ 32 °F | 21 | 15 | 5 | 0 | 0 | 0 | 0 | 0 | 0 | 0 | 4 | 17 | 62 |
| Days Min Temp ≤ 32 °F | 31 | 27 | 24 | 8 | 1 | 0 | 0 | 0 | 0 | 6 | 20 | 29 | 146 |
| Days Min Temp ≤ 0 °F | 11 | 7 | 1 | 0 | 0 | 0 | 0 | 0 | 0 | 0 | 0 | 4 | 23 |
| Heating Degree Days | 1520 | 1243 | 963 | 514 | 203 | 33 | 5 | 16 | 142 | 441 | 870 | 1302 | 7252 |
| Cooling Degree Days | 0 | 0 | 0 | 10 | 48 | 161 | 270 | 216 | 70 | 2 | 0 | 0 | 777 |
| Total Precipitation (") | 0.86 | 0.74 | 1.76 | 3.60 | 3.36 | 3.98 | 4.28 | 4.11 | 4.09 | 2.34 | 2.09 | 1.21 | 32.42 |
| Days ≥ 0.1" Precip | 3 | 2 | 4 | 7 | 7 | 7 | 7 | 6 | 7 | 5 | 5 | 3 | 63 |
| Total Snowfall (") | 8.5 | 5.9 | 4.3 | 0.9 | 0.0 | 0.0 | 0.0 | 0.0 | 0.0 | 0.0 | 3.3 | 7.9 | 30.8 |
| Days ≥ 1" Snow Depth | 25 | 21 | 11 | 1 | 0 | 0 | 0 | 0 | 0 | 0 | 3 | 19 | 80 |

## GERMANTOWN *Washington County* ELEVATION 889 ft LAT/LONG 43° 14' N / 88° 6' W

| | JAN | FEB | MAR | APR | MAY | JUN | JUL | AUG | SEP | OCT | NOV | DEC | YEAR |
|---|---|---|---|---|---|---|---|---|---|---|---|---|---|
| Maximum Temp °F | 25.5 | 29.8 | 40.9 | 54.5 | 67.2 | 76.6 | 81.3 | 79.1 | 71.2 | 59.1 | 44.6 | 31.6 | 55.1 |
| Minimum Temp °F | 7.6 | 11.8 | 22.9 | 33.8 | 42.9 | 52.1 | 57.6 | 55.8 | 48.5 | 37.4 | 27.2 | 14.9 | 34.4 |
| Mean Temp °F | 16.6 | 20.8 | 31.9 | 44.1 | 55.1 | 64.4 | 69.5 | 67.5 | 59.9 | 48.3 | 35.9 | 23.3 | 44.8 |
| Days Max Temp ≥ 90 °F | 0 | 0 | 0 | 0 | 0 | 2 | 4 | 2 | 1 | 0 | 0 | 0 | 9 |
| Days Max Temp ≤ 32 °F | 21 | 16 | 7 | 0 | 0 | 0 | 0 | 0 | 0 | 0 | 3 | 15 | 62 |
| Days Min Temp ≤ 32 °F | 30 | 27 | 26 | 14 | 4 | 0 | 0 | 0 | 1 | 10 | 21 | 29 | 162 |
| Days Min Temp ≤ 0 °F | 11 | 6 | 1 | 0 | 0 | 0 | 0 | 0 | 0 | 0 | 0 | 5 | 23 |
| Heating Degree Days | 1497 | 1239 | 1019 | 624 | 326 | 98 | 29 | 47 | 191 | 516 | 866 | 1289 | 7741 |
| Cooling Degree Days | 0 | 0 | 0 | 7 | 27 | 81 | 172 | 129 | 42 | 1 | 0 | 0 | 459 |
| Total Precipitation (") | 1.11 | 1.03 | 1.98 | 3.06 | 2.75 | 3.44 | 3.70 | 3.98 | 3.83 | 2.52 | 2.57 | 1.83 | 31.80 |
| Days ≥ 0.1" Precip | 4 | 3 | 5 | 6 | 6 | 6 | 7 | 6 | 7 | 5 | 6 | 5 | 66 |
| Total Snowfall (") | 11.9 | 8.7 | 6.9 | 1.9 | 0.3 | 0.0 | 0.0 | 0.0 | 0.0 | 0.2 | 3.0 | 10.0 | 42.9 |
| Days ≥ 1" Snow Depth | 21 | 18 | 11 | 2 | 0 | 0 | 0 | 0 | 0 | 0 | 3 | 15 | 70 |

## GOODMAN *Marinette County* ELEVATION 1391 ft LAT/LONG 45° 37' N / 88° 20' W

| | JAN | FEB | MAR | APR | MAY | JUN | JUL | AUG | SEP | OCT | NOV | DEC | YEAR |
|---|---|---|---|---|---|---|---|---|---|---|---|---|---|
| Maximum Temp °F | *20.8* | 26.7 | 37.4 | 52.1 | 66.2 | *73.4* | *78.5* | 75.6 | 65.8 | 54.6 | *38.9* | *26.4* | 51.4 |
| Minimum Temp °F | *-0.2* | 4.8 | 15.7 | 29.1 | 39.7 | 48.6 | *54.3* | *52.3* | 43.7 | 33.9 | *22.8* | *9.5* | 29.5 |
| Mean Temp °F | *10.1* | 15.8 | 26.6 | 40.7 | 53.0 | *61.1* | *66.4* | *63.9* | 54.8 | 44.2 | *30.9* | *18.0* | 40.5 |
| Days Max Temp ≥ 90 °F | 0 | 0 | 0 | 0 | 0 | 1 | 1 | 1 | 0 | 0 | 0 | 0 | 3 |
| Days Max Temp ≤ 32 °F | 24 | 19 | 10 | 1 | 0 | 0 | 0 | 0 | 0 | 0 | 8 | 20 | 82 |
| Days Min Temp ≤ 32 °F | 28 | 26 | 28 | 21 | 8 | 0 | 0 | 0 | 3 | 14 | 24 | 27 | 179 |
| Days Min Temp ≤ 0 °F | 14 | 11 | 4 | 0 | 0 | 0 | 0 | 0 | 0 | 0 | 1 | 7 | 37 |
| Heating Degree Days | *1699* | *1391* | 1183 | 726 | 382 | *156* | *55* | *95* | 312 | 638 | *1017* | *1452* | 9106 |
| Cooling Degree Days | na | na | *0* | 3 | 14 | na | na | na | *10* | *1* | na | na | na |
| Total Precipitation (") | 1.11 | 0.86 | 1.90 | 2.41 | 3.35 | 3.76 | 3.52 | 3.33 | 4.00 | 2.70 | 2.11 | 1.52 | 30.57 |
| Days ≥ 0.1" Precip | 4 | 3 | 4 | 6 | 6 | 7 | 6 | 6 | 7 | 5 | 5 | 5 | 64 |
| Total Snowfall (") | 12.1 | 7.4 | 9.5 | 3.9 | 0.6 | 0.0 | 0.0 | 0.0 | 0.0 | 1.2 | *4.9* | *12.3* | 51.9 |
| Days ≥ 1" Snow Depth | 28 | 26 | 25 | 9 | 0 | 0 | 0 | 0 | 0 | 1 | 8 | 25 | 122 |

## GORDON *Douglas County* ELEVATION 1040 ft LAT/LONG 46° 15' N / 91° 48' W

| | JAN | FEB | MAR | APR | MAY | JUN | JUL | AUG | SEP | OCT | NOV | DEC | YEAR |
|---|---|---|---|---|---|---|---|---|---|---|---|---|---|
| Maximum Temp °F | 19.5 | 26.5 | 38.6 | 54.2 | 68.1 | 76.4 | 81.1 | 77.8 | 67.7 | 55.3 | 37.6 | 24.4 | 52.3 |
| Minimum Temp °F | -6.4 | -1.3 | 13.5 | 27.5 | 38.9 | 47.7 | 53.3 | 50.8 | 42.1 | 31.4 | 18.6 | 3.2 | 26.6 |
| Mean Temp °F | 6.5 | 12.6 | 26.1 | 40.9 | 53.5 | 62.0 | 67.1 | 64.3 | 54.9 | 43.5 | 28.1 | 13.8 | 39.4 |
| Days Max Temp ≥ 90 °F | 0 | 0 | 0 | 0 | 0 | 2 | 3 | 2 | 0 | 0 | 0 | 0 | 7 |
| Days Max Temp ≤ 32 °F | 26 | 19 | 9 | 1 | 0 | 0 | 0 | 0 | 0 | 0 | 9 | 23 | 87 |
| Days Min Temp ≤ 32 °F | 31 | 28 | 30 | 22 | 9 | 1 | 0 | 1 | 6 | 18 | 27 | 31 | 204 |
| Days Min Temp ≤ 0 °F | 19 | 15 | 6 | 0 | 0 | 0 | 0 | 0 | 0 | 0 | 2 | 13 | 55 |
| Heating Degree Days | 1811 | 1478 | 1200 | 718 | 378 | 136 | 51 | 95 | 313 | 663 | 1092 | 1581 | 9516 |
| Cooling Degree Days | 0 | 0 | 0 | 2 | 17 | 47 | 113 | 78 | 9 | 0 | 0 | 0 | 266 |
| Total Precipitation (") | 1.07 | 0.77 | 1.70 | 2.27 | 3.36 | 3.86 | 4.87 | 4.49 | 3.94 | 2.64 | 1.86 | 1.12 | 31.95 |
| Days ≥ 0.1" Precip | 4 | *2* | 4 | *6* | 7 | 8 | 8 | 7 | 8 | 6 | 5 | 4 | 69 |
| Total Snowfall (") | 13.6 | 7.5 | 9.3 | 3.6 | 0.2 | 0.0 | 0.0 | 0.0 | 0.0 | 0.7 | 9.5 | 11.7 | 56.1 |
| Days ≥ 1" Snow Depth | 31 | 28 | 26 | 7 | 0 | 0 | 0 | 0 | 0 | 1 | *12* | 29 | 134 |

## GRANTSBURG *Burnett County* ELEVATION 932 ft LAT/LONG 45° 48' N / 92° 40' W

| | JAN | FEB | MAR | APR | MAY | JUN | JUL | AUG | SEP | OCT | NOV | DEC | YEAR |
|---|---|---|---|---|---|---|---|---|---|---|---|---|---|
| Maximum Temp °F | 19.0 | 25.9 | 38.5 | 55.6 | 68.7 | 76.1 | 80.7 | 78.1 | 67.7 | 56.2 | 38.8 | 23.4 | 52.4 |
| Minimum Temp °F | -3.2 | 2.6 | 16.9 | 31.6 | 43.1 | 52.1 | 57.2 | 54.9 | 45.4 | 34.4 | 21.4 | 5.1 | 30.1 |
| Mean Temp °F | 7.9 | 14.3 | 27.7 | 43.6 | 55.9 | 64.1 | 69.0 | 66.5 | 56.6 | 45.3 | 30.1 | 15.0 | 41.3 |
| Days Max Temp ≥ 90 °F | 0 | 0 | 0 | 0 | 0 | 1 | 3 | 2 | 0 | 0 | 0 | 0 | 6 |
| Days Max Temp ≤ 32 °F | 26 | 19 | 9 | 1 | 0 | 0 | 0 | 0 | 0 | 0 | *9* | *23* | 87 |
| Days Min Temp ≤ 32 °F | 31 | 28 | 28 | 17 | 4 | 0 | 0 | 0 | 2 | 14 | *27* | 31 | 182 |
| Days Min Temp ≤ 0 °F | 18 | 13 | 4 | 0 | 0 | 0 | 0 | 0 | 0 | 0 | *2* | *12* | 49 |
| Heating Degree Days | 1766 | 1429 | 1149 | 637 | 301 | 89 | 25 | 55 | 267 | 604 | 1046 | 1545 | 8913 |
| Cooling Degree Days | 0 | 0 | 0 | 4 | 28 | 70 | 157 | 111 | 16 | 1 | 0 | 0 | 387 |
| Total Precipitation (") | 1.20 | 0.83 | 1.71 | 2.28 | 3.42 | 4.81 | 4.14 | 4.27 | 3.54 | 2.61 | 1.89 | 1.30 | 32.00 |
| Days ≥ 0.1" Precip | 4 | 3 | 5 | 6 | 7 | 8 | 7 | 7 | 7 | 5 | 4 | *4* | 67 |
| Total Snowfall (") | 14.3 | 7.7 | *8.2* | *2.2* | 0.0 | 0.0 | 0.0 | 0.0 | 0.0 | 0.4 | na | *12.0* | na |
| Days ≥ 1" Snow Depth | *20* | *18* | na | *2* | 0 | 0 | 0 | 0 | 0 | 0 | na | na | na |

## GREEN BAY AUSTIN STR *Brown County* ELEVATION 699 ft LAT/LONG 44° 29' N / 88° 8' W

| | JAN | FEB | MAR | APR | MAY | JUN | JUL | AUG | SEP | OCT | NOV | DEC | YEAR |
|---|---|---|---|---|---|---|---|---|---|---|---|---|---|
| Maximum Temp °F | 23.0 | 27.4 | 38.8 | 53.9 | 67.3 | 76.3 | 81.3 | 78.5 | 69.6 | 57.2 | 41.8 | 28.6 | 53.6 |
| Minimum Temp °F | 6.3 | 10.3 | 21.9 | 34.0 | 44.1 | 53.6 | 58.9 | 56.8 | 48.6 | 38.1 | 26.6 | 13.8 | 34.4 |
| Mean Temp °F | 14.7 | 18.9 | 30.4 | 44.0 | 55.7 | 65.0 | 70.1 | 67.7 | 59.1 | 47.7 | 34.2 | 21.2 | 44.1 |
| Days Max Temp ≥ 90 °F | 0 | 0 | 0 | 0 | 0 | 2 | 3 | 2 | 0 | 0 | 0 | 0 | 7 |
| Days Max Temp ≤ 32 °F | 23 | 18 | 8 | 0 | 0 | 0 | 0 | 0 | 0 | 0 | 5 | 18 | 72 |
| Days Min Temp ≤ 32 °F | 31 | 27 | 26 | 14 | 3 | 0 | 0 | 0 | 1 | 9 | 23 | 29 | 163 |
| Days Min Temp ≤ 0 °F | 11 | 7 | 1 | 0 | 0 | 0 | 0 | 0 | 0 | 0 | 0 | 5 | 24 |
| Heating Degree Days | 1555 | 1297 | 1067 | 628 | 306 | 85 | 19 | 40 | 207 | 533 | 918 | 1351 | 8006 |
| Cooling Degree Days | 0 | 0 | 0 | 5 | 26 | 89 | 184 | 134 | 38 | 1 | 0 | 0 | 477 |
| Total Precipitation (") | 1.18 | 0.98 | 2.04 | 2.67 | 2.82 | 3.39 | 3.34 | 3.49 | 3.28 | 2.27 | 2.29 | 1.56 | 29.31 |
| Days ≥ 0.1" Precip | 4 | 3 | 5 | 7 | 6 | 6 | 6 | 7 | 6 | 5 | 6 | 5 | 66 |
| Total Snowfall (") | 12.5 | 8.1 | 8.8 | 2.6 | 0.1 | 0.0 | 0.0 | 0.0 | 0.0 | 0.2 | 5.0 | 12.4 | 49.7 |
| Days ≥ 1" Snow Depth | 27 | 22 | 15 | 2 | 0 | 0 | 0 | 0 | 0 | 0 | 4 | 19 | 89 |

## GURNEY *Iron County* ELEVATION 1001 ft LAT/LONG 46° 28' N / 90° 30' W

| | JAN | FEB | MAR | APR | MAY | JUN | JUL | AUG | SEP | OCT | NOV | DEC | YEAR |
|---|---|---|---|---|---|---|---|---|---|---|---|---|---|
| Maximum Temp °F | 22.2 | 27.6 | 38.8 | 52.2 | 65.4 | 73.5 | 78.5 | 76.1 | 66.9 | 56.3 | 39.8 | 27.1 | 52.0 |
| Minimum Temp °F | 1.9 | 5.0 | 17.2 | 29.8 | 40.1 | 48.9 | 54.9 | 53.6 | 45.7 | 36.4 | 24.1 | 10.0 | 30.6 |
| Mean Temp °F | 12.1 | 16.3 | 28.2 | 41.0 | 52.8 | 61.2 | 66.7 | 64.9 | 56.3 | 46.4 | 32.0 | 18.6 | 41.4 |
| Days Max Temp ≥ 90 °F | 0 | 0 | 0 | 0 | 0 | 1 | 2 | 1 | 0 | 0 | 0 | 0 | 4 |
| Days Max Temp ≤ 32 °F | 25 | 18 | 9 | 1 | 0 | 0 | 0 | 0 | 0 | 0 | 7 | 21 | 81 |
| Days Min Temp ≤ 32 °F | 31 | 28 | 28 | 20 | 7 | 1 | 0 | 0 | 2 | 11 | 25 | 30 | 183 |
| Days Min Temp ≤ 0 °F | 14 | 11 | 4 | 0 | 0 | 0 | 0 | 0 | 0 | 0 | 1 | 8 | 38 |
| Heating Degree Days | 1638 | 1371 | 1134 | 715 | 390 | 158 | 57 | 85 | 277 | 574 | 985 | 1434 | 8818 |
| Cooling Degree Days | 0 | 0 | 0 | 4 | 19 | 51 | 113 | 88 | 19 | 1 | 0 | 0 | 295 |
| Total Precipitation (") | 1.80 | 1.10 | 2.00 | 2.21 | 3.00 | 4.09 | 4.07 | 4.29 | 4.34 | 3.48 | 2.82 | 1.79 | 34.99 |
| Days ≥ 0.1" Precip | 6 | 3 | 5 | 7 | 7 | 8 | 7 | 8 | 9 | 8 | 7 | 5 | 80 |
| Total Snowfall (") | 34.1 | 21.3 | 21.6 | 8.1 | 0.8 | 0.0 | 0.0 | 0.0 | 0.0 | 2.2 | 21.8 | 30.7 | 140.6 |
| Days ≥ 1" Snow Depth | 31 | 28 | 30 | 15 | 0 | 0 | 0 | 0 | 0 | 1 | 16 | *30* | 151 |

## HANCOCK EXP FARM *Waushara County* ELEVATION 1089 ft LAT/LONG 44° 7' N / 89° 31' W

| | JAN | FEB | MAR | APR | MAY | JUN | JUL | AUG | SEP | OCT | NOV | DEC | YEAR |
|---|---|---|---|---|---|---|---|---|---|---|---|---|---|
| Maximum Temp °F | 24.2 | 29.7 | 41.6 | 57.8 | 71.2 | 79.2 | 82.7 | 80.4 | 71.6 | 59.8 | 42.8 | 29.1 | 55.8 |
| Minimum Temp °F | 4.2 | 8.4 | 20.7 | 33.6 | 44.8 | 54.1 | 58.7 | 56.6 | 48.5 | 38.1 | 25.5 | 11.8 | 33.8 |
| Mean Temp °F | 14.2 | 19.1 | 31.2 | 45.8 | 58.0 | 66.6 | 70.8 | 68.5 | 60.1 | 49.0 | 34.2 | 20.5 | 44.8 |
| Days Max Temp ≥ 90 °F | 0 | 0 | 0 | 0 | 0 | 3 | 5 | 3 | 0 | 0 | 0 | 0 | 11 |
| Days Max Temp ≤ 32 °F | 23 | 16 | 6 | 0 | 0 | 0 | 0 | 0 | 0 | 0 | 5 | 19 | 69 |
| Days Min Temp ≤ 32 °F | 31 | 28 | 26 | 14 | 3 | 0 | 0 | 0 | 1 | 9 | 23 | 30 | 165 |
| Days Min Temp ≤ 0 °F | 12 | 9 | 3 | 0 | 0 | 0 | 0 | 0 | 0 | 0 | 0 | 7 | 31 |
| Heating Degree Days | 1570 | 1290 | 1043 | 577 | 251 | 62 | 16 | 35 | 188 | 495 | 917 | 1375 | 7819 |
| Cooling Degree Days | 0 | 0 | 0 | 8 | 39 | 124 | 207 | 161 | 52 | 3 | 0 | 0 | 594 |
| Total Precipitation (") | 0.87 | 0.88 | 1.95 | 2.85 | 3.44 | 3.66 | 3.82 | 3.87 | 4.19 | 2.31 | 2.10 | 1.20 | 31.14 |
| Days ≥ 0.1" Precip | 3 | 3 | 5 | 6 | 7 | 7 | 7 | 7 | 7 | 6 | 5 | 3 | 66 |
| Total Snowfall (") | 11.7 | 9.3 | 8.9 | 3.0 | 0.1 | 0.0 | 0.0 | 0.0 | 0.0 | 0.6 | 5.4 | 11.9 | 50.9 |
| Days ≥ 1" Snow Depth | 29 | 26 | 17 | 2 | 0 | 0 | 0 | 0 | 0 | 0 | 5 | 22 | 101 |

## HARTFORD 2 W *Washington County* ELEVATION 981 ft LAT/LONG 43° 19 ' N / 88° 23 ' W

| | JAN | FEB | MAR | APR | MAY | JUN | JUL | AUG | SEP | OCT | NOV | DEC | YEAR |
|---|---|---|---|---|---|---|---|---|---|---|---|---|---|
| Maximum Temp °F | 25.3 | 29.9 | 41.8 | 56.2 | 69.2 | 78.0 | 82.4 | 80.0 | 71.8 | 59.7 | 44.2 | 31.1 | 55.8 |
| Minimum Temp °F | 7.1 | 11.1 | 23.4 | 34.2 | 44.2 | 53.1 | 58.0 | 56.0 | 48.4 | 37.9 | 27.4 | 14.6 | 34.6 |
| Mean Temp °F | 16.2 | 20.5 | 32.7 | 45.2 | 56.7 | 65.6 | 70.2 | 68.0 | 60.1 | 48.8 | 35.8 | 22.9 | 45.2 |
| Days Max Temp ≥ 90 °F | 0 | 0 | 0 | 0 | 0 | 2 | 4 | 2 | 0 | 0 | 0 | 0 | 8 |
| Days Max Temp ≤ 32 °F | 21 | 16 | 6 | 0 | 0 | 0 | 0 | 0 | 0 | 0 | 3 | 16 | 62 |
| Days Min Temp ≤ 32 °F | 30 | 27 | 25 | 13 | 3 | 0 | 0 | 0 | 1 | 9 | 21 | 29 | 158 |
| Days Min Temp ≤ 0 °F | 11 | 7 | 1 | 0 | 0 | 0 | 0 | 0 | 0 | 0 | 0 | 5 | 24 |
| Heating Degree Days | 1507 | 1249 | 994 | 592 | 281 | 77 | 19 | 42 | 185 | 499 | 869 | 1299 | 7613 |
| Cooling Degree Days | 0 | 0 | 0 | 7 | 28 | 94 | 179 | 137 | 42 | 1 | 0 | 0 | 488 |
| Total Precipitation (") | 1.18 | 1.02 | 1.95 | 2.89 | 2.97 | 3.80 | 3.85 | 3.83 | 4.16 | 2.70 | 2.28 | 1.78 | 32.41 |
| Days ≥ 0.1" Precip | 4 | 3 | 5 | 7 | 6 | 6 | 7 | 7 | 7 | 6 | 5 | 5 | 68 |
| Total Snowfall (") | 10.0 | 7.1 | 6.0 | 1.8 | 0.4 | 0.0 | 0.0 | 0.0 | 0.0 | 0.1 | 2.5 | 8.4 | 36.3 |
| Days ≥ 1" Snow Depth | 22 | 22 | 9 | 1 | 0 | 0 | 0 | 0 | 0 | 0 | 2 | 15 | 71 |

## HILLSBORO *Vernon County* ELEVATION 1001 ft LAT/LONG 43° 39 ' N / 90° 20 ' W

| | JAN | FEB | MAR | APR | MAY | JUN | JUL | AUG | SEP | OCT | NOV | DEC | YEAR |
|---|---|---|---|---|---|---|---|---|---|---|---|---|---|
| Maximum Temp °F | 25.0 | 30.6 | 43.0 | 57.6 | 70.3 | 78.9 | 83.4 | 80.8 | 72.0 | 60.5 | ***44.0*** | 30.6 | 56.4 |
| Minimum Temp °F | 3.5 | 7.4 | 21.6 | 33.1 | 43.9 | 53.4 | 58.6 | 55.8 | 47.2 | 35.9 | ***24.8*** | 11.2 | 33.0 |
| Mean Temp °F | 14.1 | 19.0 | 32.3 | 45.4 | 57.1 | 66.2 | 70.9 | 68.2 | 59.6 | 48.2 | ***34.4*** | 21.0 | 44.7 |
| Days Max Temp ≥ 90 °F | 0 | 0 | 0 | 0 | 0 | 2 | 6 | 3 | 1 | 0 | 0 | 0 | 12 |
| Days Max Temp ≤ 32 °F | 21 | 14 | 5 | 0 | 0 | 0 | 0 | 0 | 0 | 0 | ***4*** | 16 | 60 |
| Days Min Temp ≤ 32 °F | 30 | 27 | 26 | 14 | 3 | 0 | 0 | 0 | 2 | ***12*** | ***24*** | 30 | 168 |
| Days Min Temp ≤ 0 °F | 12 | 9 | 2 | 0 | 0 | 0 | 0 | 0 | 0 | 0 | ***0*** | 7 | 30 |
| Heating Degree Days | 1558 | 1292 | 1008 | 587 | 266 | 66 | 13 | 40 | 195 | 515 | ***911*** | 1354 | 7805 |
| Cooling Degree Days | 0 | 0 | 0 | 6 | 24 | 101 | 193 | 143 | 39 | 0 | 0 | 0 | 506 |
| Total Precipitation (") | 0.92 | 1.02 | 1.94 | 3.28 | 3.38 | 3.91 | 3.82 | 4.10 | 4.43 | 2.37 | 2.15 | 1.29 | 32.61 |
| Days ≥ 0.1" Precip | 3 | ***3*** | 5 | 6 | 7 | 7 | 6 | 6 | ***6*** | 5 | 4 | 4 | 62 |
| Total Snowfall (") | ***9.0*** | ***8.0*** | na | ***2.2*** | 0.0 | 0.0 | 0.0 | 0.0 | 0.0 | 0.1 | na | na | na |
| Days ≥ 1" Snow Depth | ***23*** | na | na | 2 | 0 | 0 | 0 | 0 | 0 | 0 | na | na | na |

## HOLCOMBE *Chippewa County* ELEVATION 1050 ft LAT/LONG 45° 14 ' N / 91° 7 ' W

| | JAN | FEB | MAR | APR | MAY | JUN | JUL | AUG | SEP | OCT | NOV | DEC | YEAR |
|---|---|---|---|---|---|---|---|---|---|---|---|---|---|
| Maximum Temp °F | 22.2 | 28.3 | 40.3 | 56.6 | 70.1 | 77.5 | 82.1 | 79.4 | 70.4 | 58.4 | 40.7 | 27.0 | 54.4 |
| Minimum Temp °F | -0.1 | 5.1 | 18.7 | 32.9 | 44.3 | 53.0 | 58.1 | 55.6 | 47.2 | 36.8 | 24.0 | 9.1 | 32.1 |
| Mean Temp °F | 11.1 | 16.8 | 29.5 | 44.8 | 57.2 | 65.2 | 70.1 | 67.6 | 58.8 | 47.6 | 32.4 | 18.1 | 43.3 |
| Days Max Temp ≥ 90 °F | 0 | 0 | 0 | 0 | 0 | 1 | 3 | 2 | 0 | 0 | 0 | 0 | 6 |
| Days Max Temp ≤ 32 °F | 24 | 17 | 7 | 0 | 0 | 0 | 0 | 0 | 0 | 0 | 5 | 21 | 74 |
| Days Min Temp ≤ 32 °F | 31 | 28 | 27 | 15 | 3 | 0 | 0 | 0 | 1 | 11 | 24 | 31 | 171 |
| Days Min Temp ≤ 0 °F | 15 | 11 | 4 | 0 | 0 | 0 | 0 | 0 | 0 | 0 | 1 | 9 | 40 |
| Heating Degree Days | 1669 | 1357 | 1092 | 603 | 264 | 70 | 17 | 42 | 215 | 534 | 970 | 1448 | 8281 |
| Cooling Degree Days | 0 | 0 | 0 | 4 | 27 | 86 | 187 | 132 | 35 | 1 | 0 | 0 | 472 |
| Total Precipitation (") | 1.04 | 0.75 | 1.66 | 2.75 | 3.49 | 4.63 | 3.66 | 4.38 | 4.18 | 2.53 | 2.08 | 1.22 | 32.37 |
| Days ≥ 0.1" Precip | 3 | 3 | 4 | 5 | 7 | 8 | 7 | 7 | 7 | 5 | 4 | ***3*** | 63 |
| Total Snowfall (") | na | na | ***5.6*** | ***1.1*** | 0.0 | 0.0 | 0.0 | 0.0 | 0.0 | 0.0 | na | na | na |
| Days ≥ 1" Snow Depth | na | na | ***16*** | 3 | 0 | 0 | 0 | 0 | 0 | 0 | na | na | na |

## JUMP RIVER 3 E *Taylor County* ELEVATION 1260 ft LAT/LONG 45° 21 ' N / 90° 47 ' W

| | JAN | FEB | MAR | APR | MAY | JUN | JUL | AUG | SEP | OCT | NOV | DEC | YEAR |
|---|---|---|---|---|---|---|---|---|---|---|---|---|---|
| Maximum Temp °F | 21.0 | 27.0 | 39.1 | 55.0 | 68.4 | 75.5 | 80.0 | 77.4 | 68.2 | 56.5 | 38.8 | 25.6 | 52.7 |
| Minimum Temp °F | -1.5 | 3.5 | 16.5 | 30.7 | 40.8 | 49.2 | 53.8 | 51.8 | 43.8 | 34.1 | 21.6 | 7.0 | 29.3 |
| Mean Temp °F | 9.8 | 15.2 | 27.8 | 42.8 | 54.7 | 62.3 | 66.9 | 64.6 | 56.0 | 45.3 | 30.2 | 16.3 | 41.0 |
| Days Max Temp ≥ 90 °F | 0 | 0 | 0 | 0 | 0 | 0 | 1 | 1 | 0 | 0 | 0 | 0 | 2 |
| Days Max Temp ≤ 32 °F | 25 | 18 | 8 | 0 | 0 | 0 | 0 | 0 | 0 | 0 | 8 | 23 | 82 |
| Days Min Temp ≤ 32 °F | 31 | 28 | 28 | 18 | 7 | 1 | 0 | 1 | 4 | 14 | 26 | 31 | 189 |
| Days Min Temp ≤ 0 °F | 16 | 11 | 5 | 0 | 0 | 0 | 0 | 0 | 0 | 0 | 1 | 10 | 43 |
| Heating Degree Days | 1710 | 1401 | 1146 | 661 | 331 | 127 | 50 | 89 | 286 | 606 | 1036 | 1503 | 8946 |
| Cooling Degree Days | 0 | 0 | 0 | 5 | 18 | 60 | 122 | 92 | 24 | 1 | 0 | 0 | 322 |
| Total Precipitation (") | 0.93 | 0.69 | 1.64 | 2.65 | 3.44 | 4.26 | 4.19 | 4.12 | 4.02 | 2.69 | 2.08 | 1.27 | 31.98 |
| Days ≥ 0.1" Precip | 3 | 2 | 4 | 6 | 7 | 7 | 8 | 7 | 8 | 6 | 4 | 4 | 66 |
| Total Snowfall (") | 13.1 | 8.9 | 8.9 | 2.3 | 0.1 | 0.0 | 0.0 | 0.0 | 0.0 | 0.9 | 6.5 | 12.4 | 53.1 |
| Days ≥ 1" Snow Depth | 31 | 27 | 23 | 3 | 0 | 0 | 0 | 0 | 0 | 0 | 8 | 26 | 118 |

## KENOSHA *Kenosha County* ELEVATION 620 ft LAT/LONG 42° 35 ' N / 87° 50 ' W

| | JAN | FEB | MAR | APR | MAY | JUN | JUL | AUG | SEP | OCT | NOV | DEC | YEAR |
|---|---|---|---|---|---|---|---|---|---|---|---|---|---|
| Maximum Temp °F | 28.4 | 32.1 | 41.7 | 52.0 | 62.3 | 73.1 | 78.7 | 77.8 | 70.4 | 59.2 | 46.2 | 34.2 | 54.7 |
| Minimum Temp °F | 12.3 | 16.6 | 26.1 | 35.7 | 44.7 | 54.2 | 60.9 | 60.3 | 53.0 | 41.8 | 30.8 | 19.2 | 38.0 |
| Mean Temp °F | 20.3 | 24.4 | 33.9 | 43.9 | 53.5 | 63.7 | 69.8 | 69.0 | 61.8 | 50.5 | 38.5 | 26.7 | 46.3 |
| Days Max Temp ≥ 90 °F | 0 | 0 | 0 | 0 | 0 | 1 | 3 | 2 | 0 | 0 | 0 | 0 | 6 |
| Days Max Temp ≤ 32 °F | 18 | 13 | 5 | 0 | 0 | 0 | 0 | 0 | 0 | 0 | 2 | 10 | 48 |
| Days Min Temp ≤ 32 °F | 30 | 26 | 24 | 9 | 1 | 0 | 0 | 0 | 0 | 4 | 17 | 27 | 138 |
| Days Min Temp ≤ 0 °F | 7 | 3 | 0 | 0 | 0 | 0 | 0 | 0 | 0 | 0 | 0 | 2 | 12 |
| Heating Degree Days | 1379 | 1141 | 957 | 628 | 363 | 111 | 19 | 23 | 138 | 445 | 787 | 1180 | 7171 |
| Cooling Degree Days | 0 | 0 | 0 | 2 | 18 | 80 | 188 | 162 | 51 | 2 | 0 | 0 | 503 |
| Total Precipitation (") | 1.40 | 1.15 | 2.27 | 3.52 | 2.95 | 3.68 | 3.69 | 4.05 | 3.70 | 2.63 | 2.75 | 2.11 | 33.90 |
| Days ≥ 0.1" Precip | 3 | 3 | 5 | 7 | 6 | 6 | 6 | 6 | 7 | 5 | 6 | 5 | 65 |
| Total Snowfall (") | 12.1 | 10.3 | 6.4 | 1.4 | 0.0 | 0.0 | 0.0 | 0.0 | 0.0 | 0.1 | 1.2 | 7.9 | 39.4 |
| Days ≥ 1" Snow Depth | 20 | 15 | 6 | 1 | 0 | 0 | 0 | 0 | 0 | 0 | 1 | 10 | 53 |

## KEWAUNEE 5 S *Kewaunee County* ELEVATION 689 ft LAT/LONG 44° 27 ' N / 87° 31 ' W

| | JAN | FEB | MAR | APR | MAY | JUN | JUL | AUG | SEP | OCT | NOV | DEC | YEAR |
|---|---|---|---|---|---|---|---|---|---|---|---|---|---|
| Maximum Temp °F | 25.7 | 29.3 | 38.7 | 50.0 | 61.8 | 71.6 | 77.6 | 76.2 | 68.5 | 56.4 | 43.2 | 31.2 | 52.5 |
| Minimum Temp °F | 10.9 | 13.8 | 23.8 | 33.9 | 42.2 | 51.3 | 58.2 | 58.0 | 50.7 | 40.1 | 29.5 | 18.0 | 35.9 |
| Mean Temp °F | 18.4 | 21.6 | 31.3 | 42.0 | 52.0 | 61.5 | 67.9 | 67.2 | 59.7 | 48.3 | 36.4 | 24.6 | 44.2 |
| Days Max Temp ≥ 90 °F | 0 | 0 | 0 | 0 | 0 | 1 | 2 | 1 | 0 | 0 | 0 | 0 | 4 |
| Days Max Temp ≤ 32 °F | 21 | 17 | 6 | 0 | 0 | 0 | 0 | 0 | 0 | 0 | 3 | 15 | 62 |
| Days Min Temp ≤ 32 °F | 30 | 27 | 25 | 11 | 2 | 0 | 0 | 0 | 0 | 5 | 19 | 28 | 147 |
| Days Min Temp ≤ 0 °F | 7 | 4 | 1 | 0 | 0 | 0 | 0 | 0 | 0 | 0 | 0 | 3 | 15 |
| Heating Degree Days | 1439 | 1221 | 1037 | 684 | 400 | 142 | 32 | 37 | 180 | 513 | 852 | 1246 | 7783 |
| Cooling Degree Days | 0 | 0 | 0 | 0 | 7 | 44 | 127 | 115 | 27 | 0 | 0 | 0 | 320 |
| Total Precipitation (") | 1.50 | 1.24 | 2.01 | 2.95 | 2.86 | 3.37 | 3.20 | 3.51 | 3.55 | 2.38 | 2.50 | 1.80 | 30.87 |
| Days ≥ 0.1" Precip | 5 | 3 | 5 | 7 | 6 | 6 | 6 | 6 | 6 | 6 | 6 | 5 | 67 |
| Total Snowfall (") | 12.1 | 9.0 | 6.5 | 2.2 | 0.1 | 0.0 | 0.0 | 0.0 | 0.0 | 0.0 | 2.5 | 9.5 | 41.9 |
| Days ≥ 1" Snow Depth | 26 | 25 | 16 | 2 | 0 | 0 | 0 | 0 | 0 | 0 | 3 | 17 | 89 |

## LA CROSSE MUNI AP *La Crosse County* ELEVATION 669 ft LAT/LONG 43° 56 ' N / 91° 17 ' W

| | JAN | FEB | MAR | APR | MAY | JUN | JUL | AUG | SEP | OCT | NOV | DEC | YEAR |
|---|---|---|---|---|---|---|---|---|---|---|---|---|---|
| Maximum Temp °F | 23.5 | 29.6 | 42.4 | 58.3 | 71.0 | 79.8 | 84.1 | 81.4 | 72.2 | 59.8 | 43.0 | 29.2 | 56.2 |
| Minimum Temp °F | 5.2 | 10.4 | 23.7 | 37.1 | 48.2 | 57.4 | 62.3 | 60.0 | 51.2 | 39.9 | 27.5 | 14.1 | 36.4 |
| Mean Temp °F | 14.4 | 20.1 | 33.1 | 47.7 | 59.6 | 68.6 | 73.2 | 70.7 | 61.7 | 49.9 | 35.3 | 21.7 | 46.3 |
| Days Max Temp ≥ 90 °F | 0 | 0 | 0 | 0 | 1 | 4 | 7 | 4 | 1 | 0 | 0 | 0 | 17 |
| Days Max Temp ≤ 32 °F | 22 | 15 | 6 | 0 | 0 | 0 | 0 | 0 | 0 | 0 | 5 | 18 | 66 |
| Days Min Temp ≤ 32 °F | 31 | 27 | 25 | 10 | 1 | 0 | 0 | 0 | 0 | 7 | 21 | 29 | 151 |
| Days Min Temp ≤ 0 °F | 12 | 8 | 1 | 0 | 0 | 0 | 0 | 0 | 0 | 0 | 0 | 5 | 26 |
| Heating Degree Days | 1565 | 1264 | 983 | 520 | 211 | 38 | 6 | 19 | 157 | 469 | 885 | 1336 | 7453 |
| Cooling Degree Days | 0 | 0 | 0 | 11 | 55 | 166 | 284 | 217 | 71 | 3 | 0 | 0 | 807 |
| Total Precipitation (") | 1.08 | 0.91 | 1.97 | 3.38 | 3.36 | 4.16 | 4.03 | 3.94 | 3.87 | 2.20 | 1.92 | 1.33 | 32.15 |
| Days ≥ 0.1" Precip | 3 | 3 | 5 | 7 | 7 | 7 | 7 | 7 | 7 | 5 | 4 | 4 | 66 |
| Total Snowfall (") | 11.7 | 8.1 | 6.4 | 1.7 | 0.0 | 0.0 | 0.0 | 0.0 | 0.0 | 0.1 | 3.9 | 10.0 | 41.9 |
| Days ≥ 1" Snow Depth | 26 | 22 | 12 | 1 | 0 | 0 | 0 | 0 | 0 | 0 | 3 | 18 | 82 |

## LAKE GENEVA *Walworth County* ELEVATION 879 ft LAT/LONG 42° 36 ' N / 88° 26 ' W

| | JAN | FEB | MAR | APR | MAY | JUN | JUL | AUG | SEP | OCT | NOV | DEC | YEAR |
|---|---|---|---|---|---|---|---|---|---|---|---|---|---|
| Maximum Temp °F | 27.6 | 32.6 | 44.2 | 58.4 | 71.3 | 80.9 | 85.0 | 82.3 | 74.3 | 61.7 | 45.8 | 33.0 | 58.1 |
| Minimum Temp °F | 11.3 | 15.3 | 25.8 | 36.8 | 47.0 | 56.7 | 62.0 | 60.1 | 52.4 | 41.2 | 30.0 | 18.3 | 38.1 |
| Mean Temp °F | 19.5 | 24.0 | 35.0 | 47.7 | 59.2 | 68.8 | 73.5 | 71.3 | 63.4 | 51.5 | 38.0 | 25.7 | 48.1 |
| Days Max Temp ≥ 90 °F | 0 | 0 | 0 | 0 | 1 | 4 | 8 | 4 | 1 | 0 | 0 | 0 | 18 |
| Days Max Temp ≤ 32 °F | 20 | 13 | 4 | 0 | 0 | 0 | 0 | 0 | 0 | 0 | 3 | 13 | 53 |
| Days Min Temp ≤ 32 °F | 30 | 26 | 24 | 10 | 1 | 0 | 0 | 0 | 0 | 6 | 19 | 28 | 144 |
| Days Min Temp ≤ 0 °F | 8 | 4 | 0 | 0 | 0 | 0 | 0 | 0 | 0 | 0 | 0 | 3 | 15 |
| Heating Degree Days | 1407 | 1153 | 924 | 521 | 223 | 42 | 7 | 15 | 122 | 421 | 805 | 1212 | 6852 |
| Cooling Degree Days | 0 | 0 | 0 | 9 | 52 | 172 | 292 | 228 | 84 | 4 | 0 | 0 | 841 |
| Total Precipitation (") | 1.87 | 1.57 | 2.75 | 3.70 | 3.22 | 3.99 | 4.18 | 3.99 | 3.90 | 2.85 | 2.85 | 2.47 | 37.34 |
| Days ≥ 0.1" Precip | 5 | 4 | 7 | 7 | 7 | 7 | 7 | 6 | 7 | 6 | 6 | 6 | 75 |
| Total Snowfall (") | 13.0 | 9.9 | 7.9 | 2.2 | 0.1 | 0.0 | 0.0 | 0.0 | 0.0 | 0.2 | 2.9 | 11.4 | 47.6 |
| Days ≥ 1" Snow Depth | 22 | 19 | 8 | 1 | 0 | 0 | 0 | 0 | 0 | 0 | 2 | 14 | 66 |

## LAKE MILLS *Jefferson County* ELEVATION 860 ft LAT/LONG 43° 5' N / 88° 55' W

| | JAN | FEB | MAR | APR | MAY | JUN | JUL | AUG | SEP | OCT | NOV | DEC | YEAR |
|---|---|---|---|---|---|---|---|---|---|---|---|---|---|
| Maximum Temp °F | 26.5 | 31.7 | 44.0 | 59.1 | 71.6 | 80.4 | 84.4 | 81.9 | 73.9 | 61.6 | 45.2 | 31.9 | 57.7 |
| Minimum Temp °F | 8.3 | 12.2 | 24.1 | 36.3 | 46.5 | 56.0 | 61.1 | 58.5 | 50.8 | 40.1 | 28.4 | 15.8 | 36.5 |
| Mean Temp °F | 17.4 | 22.0 | 34.1 | 47.7 | 59.0 | 68.3 | 72.7 | 70.3 | 62.4 | 50.9 | 36.8 | 23.9 | 47.1 |
| Days Max Temp ≥ 90 °F | 0 | 0 | 0 | 0 | 1 | 3 | 7 | 4 | 1 | 0 | 0 | 0 | 16 |
| Days Max Temp ≤ 32 °F | 20 | 14 | 4 | 0 | 0 | 0 | 0 | 0 | 0 | 0 | 3 | 15 | 56 |
| Days Min Temp ≤ 32 °F | 30 | 28 | 25 | 11 | 1 | 0 | 0 | 0 | 0 | 6 | 20 | 29 | 150 |
| Days Min Temp ≤ 0 °F | 10 | 6 | 1 | 0 | 0 | 0 | 0 | 0 | 0 | 0 | 0 | 4 | 21 |
| Heating Degree Days | 1469 | 1208 | 952 | 519 | 222 | 41 | 5 | 18 | 135 | 438 | 839 | 1269 | 7115 |
| Cooling Degree Days | 0 | 0 | 0 | 8 | 47 | 153 | 266 | 202 | 71 | 4 | 0 | 0 | 751 |
| Total Precipitation (") | 1.37 | 1.18 | 2.25 | 3.16 | 3.06 | 3.68 | 4.06 | 4.21 | 3.98 | 2.48 | 2.42 | 1.95 | 33.80 |
| Days ≥ 0.1" Precip | 4 | 3 | 5 | 7 | 7 | 6 | 7 | 6 | 6 | 5 | 6 | 5 | 67 |
| Total Snowfall (") | 10.0 | 7.7 | 5.3 | 1.4 | 0.2 | 0.0 | 0.0 | 0.0 | 0.0 | 0.1 | 2.5 | 10.5 | 37.7 |
| Days ≥ 1" Snow Depth | 24 | 22 | 8 | 1 | 0 | 0 | 0 | 0 | 0 | 0 | 3 | 17 | 75 |

## LANCASTER 4 WSW *Grant County* ELEVATION 1040 ft LAT/LONG 42° 50' N / 90° 47' W

| | JAN | FEB | MAR | APR | MAY | JUN | JUL | AUG | SEP | OCT | NOV | DEC | YEAR |
|---|---|---|---|---|---|---|---|---|---|---|---|---|---|
| Maximum Temp °F | 24.3 | 30.1 | 42.6 | 57.8 | 69.9 | 78.7 | 82.6 | 80.3 | 71.9 | 60.1 | 43.4 | 30.1 | 56.0 |
| Minimum Temp °F | 6.9 | 12.1 | 24.5 | 36.3 | 47.2 | 56.4 | 60.8 | 58.8 | 50.8 | 39.4 | 27.0 | 14.3 | 36.2 |
| Mean Temp °F | 15.7 | 21.1 | 33.6 | 47.0 | 58.6 | 67.6 | 71.7 | 69.6 | 61.3 | 49.8 | 35.3 | 22.2 | 46.1 |
| Days Max Temp ≥ 90 °F | 0 | 0 | 0 | 0 | 0 | 2 | 4 | 2 | 0 | 0 | 0 | 0 | 8 |
| Days Max Temp ≤ 32 °F | 22 | 16 | 6 | 0 | 0 | 0 | 0 | 0 | 0 | 0 | 5 | 17 | 66 |
| Days Min Temp ≤ 32 °F | 31 | 27 | 25 | 10 | 1 | 0 | 0 | 0 | 1 | 8 | 22 | 30 | 155 |
| Days Min Temp ≤ 0 °F | 11 | 7 | 1 | 0 | 0 | 0 | 0 | 0 | 0 | 0 | 0 | 5 | 24 |
| Heating Degree Days | 1524 | 1232 | 967 | 538 | 229 | 47 | 10 | 24 | 160 | 471 | 885 | 1320 | 7407 |
| Cooling Degree Days | 0 | 0 | 0 | 7 | 31 | 124 | 224 | 170 | 59 | 3 | 0 | 0 | 618 |
| Total Precipitation (") | 0.92 | 0.92 | 2.18 | 3.15 | 3.44 | 4.51 | 4.16 | 4.62 | 3.70 | 2.41 | 2.40 | 1.40 | 33.81 |
| Days ≥ 0.1" Precip | 3 | 3 | 5 | 7 | 7 | 7 | 7 | 7 | 6 | 5 | 5 | 4 | 66 |
| Total Snowfall (") | 10.4 | 7.7 | 6.3 | 2.2 | 0.1 | 0.0 | 0.0 | 0.0 | 0.0 | 0.1 | 3.4 | 9.8 | 40.0 |
| Days ≥ 1" Snow Depth | 23 | 19 | 10 | 1 | 0 | 0 | 0 | 0 | 0 | 0 | 4 | *14* | 71 |

## LAONA 6 SW *Forest County* ELEVATION 1503 ft LAT/LONG 45° 31' N / 88° 46' W

| | JAN | FEB | MAR | APR | MAY | JUN | JUL | AUG | SEP | OCT | NOV | DEC | YEAR |
|---|---|---|---|---|---|---|---|---|---|---|---|---|---|
| Maximum Temp °F | 21.0 | 26.5 | 37.8 | 53.0 | 66.4 | 72.7 | 76.3 | 73.6 | 64.6 | 54.0 | 37.2 | 25.0 | 50.7 |
| Minimum Temp °F | 2.0 | 5.4 | 16.2 | 29.1 | 39.7 | 48.7 | 54.0 | 52.2 | 44.6 | 34.4 | 22.3 | 9.3 | 29.8 |
| Mean Temp °F | 11.5 | 16.0 | 27.0 | 41.0 | 53.0 | 60.7 | 65.2 | 62.9 | 54.7 | 44.3 | 29.7 | 17.2 | 40.3 |
| Days Max Temp ≥ 90 °F | 0 | 0 | 0 | 0 | 0 | 0 | 0 | 0 | 0 | 0 | 0 | 0 | 0 |
| Days Max Temp ≤ 32 °F | 27 | 20 | 10 | 1 | 0 | 0 | 0 | 0 | 0 | 0 | 11 | 25 | 94 |
| Days Min Temp ≤ 32 °F | 31 | 28 | 29 | 20 | 8 | 1 | 0 | 0 | 3 | 14 | 26 | 31 | 191 |
| Days Min Temp ≤ 0 °F | 14 | 11 | 5 | 0 | 0 | 0 | 0 | 0 | 0 | 0 | 1 | 8 | 39 |
| Heating Degree Days | 1654 | 1380 | 1170 | 714 | 375 | 158 | 70 | 111 | 316 | 637 | 1051 | 1477 | 9113 |
| Cooling Degree Days | 0 | 0 | 0 | 3 | 11 | 34 | 74 | 53 | 11 | 0 | 0 | 0 | 186 |
| Total Precipitation (") | 1.11 | 0.85 | 1.95 | 2.72 | 3.37 | 3.87 | 3.30 | 3.53 | 4.10 | 2.68 | 2.25 | 1.47 | 31.20 |
| Days ≥ 0.1" Precip | 4 | 3 | 5 | 6 | 7 | 8 | 8 | 7 | 8 | 6 | 5 | 5 | 72 |
| Total Snowfall (") | 15.5 | 9.8 | 12.1 | 6.2 | 0.8 | 0.0 | 0.0 | 0.0 | 0.0 | 1.9 | 9.2 | 15.6 | 71.1 |
| Days ≥ 1" Snow Depth | 31 | 28 | 28 | 10 | 0 | 0 | 0 | 0 | 0 | 1 | 11 | 29 | 138 |

## LONG LAKE DAM *Oneida County* ELEVATION 1631 ft LAT/LONG 45° 54' N / 89° 8' W

| | JAN | FEB | MAR | APR | MAY | JUN | JUL | AUG | SEP | OCT | NOV | DEC | YEAR |
|---|---|---|---|---|---|---|---|---|---|---|---|---|---|
| Maximum Temp °F | 21.2 | 26.7 | 38.0 | 52.2 | 66.3 | 73.7 | 78.3 | 75.9 | 66.2 | 54.1 | 37.7 | 25.5 | 51.3 |
| Minimum Temp °F | -2.3 | 0.0 | 11.8 | 26.1 | 37.7 | 47.4 | 52.6 | 50.5 | 42.3 | 32.1 | 20.3 | 6.1 | 27.1 |
| Mean Temp °F | 9.5 | 13.4 | 25.0 | 39.2 | 52.0 | 60.6 | 65.5 | 63.3 | 54.3 | 43.1 | 29.0 | 15.8 | 39.2 |
| Days Max Temp ≥ 90 °F | 0 | 0 | 0 | 0 | 0 | 1 | 1 | 1 | 0 | 0 | 0 | 0 | 3 |
| Days Max Temp ≤ 32 °F | 26 | 20 | 9 | 1 | 0 | 0 | 0 | 0 | 0 | 0 | 10 | 24 | 90 |
| Days Min Temp ≤ 32 °F | 31 | 28 | 30 | 23 | 11 | 1 | 0 | 0 | 5 | 18 | 28 | 31 | 206 |
| Days Min Temp ≤ 0 °F | 17 | 14 | 7 | 0 | 0 | 0 | 0 | 0 | 0 | 0 | 2 | 11 | 51 |
| Heating Degree Days | 1719 | 1454 | 1235 | 769 | 408 | 167 | 69 | 111 | 327 | 672 | 1073 | 1520 | 9524 |
| Cooling Degree Days | 0 | 0 | 0 | 2 | 11 | 41 | 88 | 67 | 12 | 0 | 0 | 0 | 221 |
| Total Precipitation (") | 1.30 | 0.90 | 1.82 | 2.62 | 3.27 | 3.99 | 3.50 | 4.09 | 4.39 | 2.77 | 2.38 | 1.57 | 32.60 |
| Days ≥ 0.1" Precip | 4 | 3 | 5 | 7 | 7 | 8 | 8 | 7 | 8 | 7 | 6 | 5 | 75 |
| Total Snowfall (") | 16.0 | 10.1 | *11.8* | *5.9* | 0.6 | 0.0 | 0.0 | 0.0 | 0.0 | *1.3* | *10.7* | 16.6 | 73.0 |
| Days ≥ 1" Snow Depth | *25* | *20* | *20* | *8* | 0 | 0 | 0 | 0 | 0 | 1 | *9* | *24* | 107 |

## LYNXVILLE DAM 9 *Crawford County* ELEVATION 633 ft LAT/LONG 43° 13 ' N / 91° 6 ' W

| | JAN | FEB | MAR | APR | MAY | JUN | JUL | AUG | SEP | OCT | NOV | DEC | YEAR |
|---|---|---|---|---|---|---|---|---|---|---|---|---|---|
| Maximum Temp °F | 25.8 | 31.4 | 43.7 | 58.9 | 71.5 | 80.2 | 84.2 | 81.6 | 72.8 | 60.9 | 44.2 | 31.0 | 57.2 |
| Minimum Temp °F | 7.6 | 12.1 | 25.3 | 38.4 | 49.4 | 58.7 | 63.6 | 61.5 | 53.2 | 41.9 | 28.9 | 16.0 | 38.1 |
| Mean Temp °F | 16.7 | 21.8 | 34.5 | 48.7 | 60.5 | 69.5 | 73.9 | 71.6 | 63.0 | 51.4 | 36.6 | 23.5 | 47.6 |
| Days Max Temp ≥ 90 °F | 0 | 0 | 0 | 0 | 0 | 3 | 6 | 4 | 1 | 0 | 0 | 0 | 14 |
| Days Max Temp ≤ 32 °F | 20 | 14 | 4 | 0 | 0 | 0 | 0 | 0 | 0 | 0 | 4 | 16 | 58 |
| Days Min Temp ≤ 32 °F | 31 | 27 | 24 | 8 | 0 | 0 | 0 | 0 | 0 | 5 | 19 | 29 | 143 |
| Days Min Temp ≤ 0 °F | 10 | 7 | 1 | 0 | 0 | 0 | 0 | 0 | 0 | 0 | 0 | 4 | 22 |
| Heating Degree Days | 1492 | 1215 | 938 | 491 | 184 | 26 | 3 | 10 | 124 | 421 | 845 | 1278 | 7027 |
| Cooling Degree Days | 0 | 0 | 0 | 10 | 52 | 176 | 298 | 231 | 76 | 3 | 0 | 0 | 846 |
| Total Precipitation (") | 0.97 | 0.91 | 1.82 | 3.53 | 3.72 | 3.98 | 3.89 | 4.17 | 3.56 | 2.24 | 2.20 | 1.38 | 32.37 |
| Days ≥ 0.1" Precip | 3 | 3 | 5 | 7 | 7 | 7 | 6 | 7 | 6 | 5 | 5 | 4 | 65 |
| Total Snowfall (") | 8.5 | 7.3 | 4.1 | 1.0 | 0.0 | 0.0 | 0.0 | 0.0 | 0.0 | 0.0 | ***3.0*** | ***9.1*** | 33.0 |
| Days ≥ 1" Snow Depth | 27 | 24 | 13 | 1 | 0 | 0 | 0 | 0 | 0 | 0 | 3 | 20 | 88 |

## MADELINE ISLAND *Ashland County* ELEVATION 620 ft LAT/LONG 46° 50 ' N / 90° 38 ' W

| | JAN | FEB | MAR | APR | MAY | JUN | JUL | AUG | SEP | OCT | NOV | DEC | YEAR |
|---|---|---|---|---|---|---|---|---|---|---|---|---|---|
| Maximum Temp °F | 21.3 | ***25.2*** | ***35.8*** | 48.8 | 60.7 | 70.4 | 76.9 | 75.4 | 66.6 | 54.7 | 39.3 | ***27.4*** | 50.2 |
| Minimum Temp °F | 4.0 | ***5.0*** | ***16.3*** | 28.8 | 37.3 | 46.4 | 54.2 | 54.0 | 46.9 | 36.9 | 26.0 | ***13.8*** | 30.8 |
| Mean Temp °F | 12.7 | ***15.1*** | ***26.1*** | 38.8 | 49.0 | 58.4 | 65.6 | 64.7 | 56.8 | 45.8 | 32.6 | ***20.6*** | 40.5 |
| Days Max Temp ≥ 90 °F | 0 | 0 | 0 | 0 | 0 | 0 | 1 | 1 | 0 | 0 | 0 | 0 | 2 |
| Days Max Temp ≤ 32 °F | 26 | 20 | 10 | 1 | 0 | 0 | 0 | 0 | 0 | 0 | 7 | 19 | 83 |
| Days Min Temp ≤ 32 °F | 30 | 27 | 28 | 20 | 7 | 0 | 0 | 0 | 1 | 9 | 24 | 28 | 174 |
| Days Min Temp ≤ 0 °F | 13 | ***11*** | 3 | 0 | 0 | 0 | 0 | 0 | 0 | 0 | 0 | 4 | 31 |
| Heating Degree Days | 1619 | ***1403*** | ***1199*** | 779 | 491 | 211 | 60 | 72 | 252 | 588 | 965 | ***1370*** | 9009 |
| Cooling Degree Days | 0 | ***0*** | na | ***1*** | 7 | 30 | ***102*** | 80 | 12 | 0 | ***0*** | na | na |
| Total Precipitation (") | ***1.62*** | ***1.11*** | ***2.07*** | 2.57 | 3.42 | 4.23 | 3.72 | 4.46 | 3.54 | 2.76 | 2.42 | ***1.50*** | 33.42 |
| Days ≥ 0.1" Precip | ***5*** | na | 4 | 6 | 7 | 8 | 7 | 7 | 7 | 6 | 6 | ***4*** | na |
| Total Snowfall (") | 19.5 | na | ***10.0*** | 3.3 | 0.1 | 0.0 | 0.0 | 0.0 | 0.0 | 0.4 | 8.4 | ***15.3*** | na |
| Days ≥ 1" Snow Depth | ***26*** | ***24*** | 21 | 6 | 0 | 0 | 0 | 0 | 0 | 0 | 8 | ***20*** | 105 |

## MADISON DANE CNTY AP *Dane County* ELEVATION 873 ft LAT/LONG 43° 8 ' N / 89° 20 ' W

| | JAN | FEB | MAR | APR | MAY | JUN | JUL | AUG | SEP | OCT | NOV | DEC | YEAR |
|---|---|---|---|---|---|---|---|---|---|---|---|---|---|
| Maximum Temp °F | 25.2 | 30.3 | 43.3 | 57.9 | 70.0 | 78.9 | 82.9 | 80.2 | 71.7 | 59.8 | 43.9 | 30.9 | 56.2 |
| Minimum Temp °F | 7.6 | 11.7 | 23.6 | 34.4 | 44.3 | 54.0 | 59.2 | 56.7 | 48.5 | 37.6 | 26.7 | 14.9 | 34.9 |
| Mean Temp °F | 16.4 | 21.0 | 33.5 | 46.2 | 57.2 | 66.5 | 71.1 | 68.4 | 60.1 | 48.7 | 35.3 | 22.9 | 45.6 |
| Days Max Temp ≥ 90 °F | 0 | 0 | 0 | 0 | 0 | 3 | 5 | 3 | 0 | 0 | 0 | 0 | 11 |
| Days Max Temp ≤ 32 °F | 21 | 15 | 5 | 0 | 0 | 0 | 0 | 0 | 0 | 0 | 4 | 16 | 61 |
| Days Min Temp ≤ 32 °F | 30 | 27 | 25 | 13 | 3 | 0 | 0 | 0 | 1 | 10 | 22 | 29 | 160 |
| Days Min Temp ≤ 0 °F | 11 | 7 | 1 | 0 | 0 | 0 | 0 | 0 | 0 | 0 | 0 | 5 | 24 |
| Heating Degree Days | 1501 | 1237 | 970 | 564 | 270 | 67 | 14 | 38 | 188 | 503 | 884 | 1298 | 7534 |
| Cooling Degree Days | 0 | 0 | 0 | 8 | 34 | 124 | 226 | 164 | 52 | 2 | 0 | 0 | 610 |
| Total Precipitation (") | 1.12 | 1.18 | 2.19 | 3.12 | 3.11 | 3.83 | 3.71 | 4.22 | 3.59 | 2.23 | 2.25 | 1.93 | 32.48 |
| Days ≥ 0.1" Precip | 3 | 3 | 5 | 7 | 7 | 7 | 6 | 7 | 6 | 5 | 5 | 4 | 65 |
| Total Snowfall (") | 10.6 | 8.6 | 6.9 | 3.1 | 0.1 | 0.0 | 0.0 | 0.0 | 0.0 | 0.3 | 3.7 | 12.4 | 45.7 |
| Days ≥ 1" Snow Depth | 23 | 21 | 9 | 2 | 0 | 0 | 0 | 0 | 0 | 0 | 3 | 18 | 76 |

## MANITOWOC *Manitowoc County* ELEVATION 610 ft LAT/LONG 44° 6 ' N / 87° 40 ' W

| | JAN | FEB | MAR | APR | MAY | JUN | JUL | AUG | SEP | OCT | NOV | DEC | YEAR |
|---|---|---|---|---|---|---|---|---|---|---|---|---|---|
| Maximum Temp °F | 26.2 | 29.8 | 39.5 | 52.3 | 65.1 | 74.6 | 79.6 | 77.9 | 69.7 | 57.4 | 43.6 | 31.8 | 54.0 |
| Minimum Temp °F | 10.0 | 13.7 | 23.8 | 34.1 | 43.6 | 52.8 | 59.5 | 58.6 | 51.2 | 40.6 | 29.3 | 17.0 | 36.2 |
| Mean Temp °F | 18.1 | 21.7 | 31.6 | 43.3 | 54.4 | 63.7 | 69.6 | 68.3 | 60.5 | 49.0 | 36.5 | 24.4 | 45.1 |
| Days Max Temp ≥ 90 °F | 0 | 0 | 0 | 0 | 0 | 1 | 2 | 2 | 0 | 0 | 0 | 0 | 5 |
| Days Max Temp ≤ 32 °F | 20 | 15 | 6 | 0 | 0 | 0 | 0 | 0 | 0 | 0 | 3 | 14 | 58 |
| Days Min Temp ≤ 32 °F | 30 | 27 | 26 | 12 | 2 | 0 | 0 | 0 | 0 | 5 | 20 | 29 | 151 |
| Days Min Temp ≤ 0 °F | 8 | 4 | 0 | 0 | 0 | 0 | 0 | 0 | 0 | 0 | 0 | 3 | 15 |
| Heating Degree Days | 1448 | 1216 | 1027 | 645 | 332 | 99 | 19 | 29 | 164 | 491 | 850 | 1250 | 7570 |
| Cooling Degree Days | 0 | 0 | 0 | 1 | 12 | 70 | 164 | 143 | 37 | 0 | 0 | 0 | 427 |
| Total Precipitation (") | 1.33 | 1.22 | 1.93 | 2.94 | 2.73 | 3.06 | 3.37 | 3.23 | 3.37 | 2.33 | 2.40 | 1.95 | 29.86 |
| Days ≥ 0.1" Precip | 4 | 4 | 5 | 7 | 5 | 6 | 6 | 6 | 7 | 5 | 6 | 5 | 66 |
| Total Snowfall (") | ***5.9*** | na | na | ***0.8*** | 0.0 | 0.0 | 0.0 | 0.0 | 0.0 | 0.0 | ***1.1*** | na | na |
| Days ≥ 1" Snow Depth | ***21*** | ***18*** | ***9*** | ***1*** | 0 | 0 | 0 | 0 | 0 | 0 | 1 | na | na |

## MARINETTE *Marinette County* ELEVATION 600 ft LAT/LONG 45° 6' N / 87° 38' W

| | JAN | FEB | MAR | APR | MAY | JUN | JUL | AUG | SEP | OCT | NOV | DEC | YEAR |
|---|---|---|---|---|---|---|---|---|---|---|---|---|---|
| Maximum Temp °F | 26.1 | 30.3 | 40.9 | 54.8 | 68.0 | 77.6 | 83.0 | 80.3 | 70.8 | 58.2 | 43.5 | 31.1 | 55.4 |
| Minimum Temp °F | 8.3 | 11.5 | 21.9 | 33.2 | 43.9 | 53.4 | 59.0 | 57.2 | 49.7 | 39.4 | 27.9 | 15.8 | 35.1 |
| Mean Temp °F | 17.2 | 20.9 | 31.4 | 44.1 | 56.0 | 65.6 | 71.0 | 68.8 | 60.3 | 48.8 | 35.7 | 23.5 | 45.3 |
| Days Max Temp ≥ 90 °F | 0 | 0 | 0 | 0 | 1 | 3 | 5 | 3 | 0 | 0 | 0 | 0 | 12 |
| Days Max Temp ≤ 32 °F | 22 | 16 | 5 | 0 | 0 | 0 | 0 | 0 | 0 | 0 | 3 | 16 | 62 |
| Days Min Temp ≤ 32 °F | 31 | 28 | 27 | 14 | 3 | 0 | 0 | 0 | 1 | 7 | 21 | 29 | 161 |
| Days Min Temp ≤ 0 °F | 9 | 6 | 1 | 0 | 0 | 0 | 0 | 0 | 0 | 0 | 0 | 4 | 20 |
| Heating Degree Days | 1475 | 1239 | 1034 | 623 | 299 | 75 | 14 | 31 | 177 | 497 | 872 | 1281 | 7617 |
| Cooling Degree Days | 0 | 0 | 0 | 2 | 25 | 88 | 194 | 150 | 36 | 1 | 0 | 0 | 496 |
| Total Precipitation (") | 1.70 | 1.29 | 2.24 | 3.08 | 3.46 | 3.52 | 3.35 | 3.11 | 3.67 | 2.55 | 2.76 | 1.92 | 32.65 |
| Days ≥ 0.1" Precip | 5 | 3 | 5 | 7 | 7 | 7 | 7 | 7 | 7 | 6 | 6 | 5 | 72 |
| Total Snowfall (") | 16.0 | 10.4 | 8.3 | 2.6 | 0.1 | 0.0 | 0.0 | 0.0 | 0.0 | 0.1 | *3.1* | 12.9 | 53.5 |
| Days ≥ 1" Snow Depth | na | na | *12* | 2 | 0 | 0 | 0 | 0 | 0 | 0 | *1* | na | na |

## MARSHFIELD EXP FARM *Wood County* ELEVATION 1250 ft LAT/LONG 44° 39' N / 90° 8' W

| | JAN | FEB | MAR | APR | MAY | JUN | JUL | AUG | SEP | OCT | NOV | DEC | YEAR |
|---|---|---|---|---|---|---|---|---|---|---|---|---|---|
| Maximum Temp °F | 22.4 | 28.0 | 39.8 | 56.2 | 69.3 | 77.5 | 81.8 | 79.3 | 70.2 | 58.3 | 40.9 | 27.4 | 54.3 |
| Minimum Temp °F | 2.9 | 7.9 | 19.9 | 32.9 | 43.3 | 52.2 | 56.9 | 54.7 | 46.5 | 36.1 | 24.1 | 10.5 | 32.3 |
| Mean Temp °F | 12.7 | 18.0 | 29.9 | 44.5 | 56.3 | 64.9 | 69.4 | 67.1 | 58.4 | 47.2 | 32.6 | 19.0 | 43.3 |
| Days Max Temp ≥ 90 °F | 0 | 0 | 0 | 0 | 0 | 2 | 3 | 2 | 0 | 0 | 0 | 0 | 7 |
| Days Max Temp ≤ 32 °F | 24 | 18 | 8 | 0 | 0 | 0 | 0 | 0 | 0 | 0 | 6 | 21 | 77 |
| Days Min Temp ≤ 32 °F | 31 | 28 | 27 | 15 | 4 | 0 | 0 | 0 | 2 | 12 | 24 | 31 | 174 |
| Days Min Temp ≤ 0 °F | 13 | 9 | 3 | 0 | 0 | 0 | 0 | 0 | 0 | 0 | 1 | 7 | 33 |
| Heating Degree Days | 1619 | 1322 | 1082 | 610 | 287 | 81 | 23 | 50 | 225 | 548 | 967 | 1420 | 8234 |
| Cooling Degree Days | 0 | 0 | 0 | 3 | 27 | 97 | 179 | 146 | 38 | 1 | 0 | 0 | 491 |
| Total Precipitation (") | 0.94 | 0.85 | 1.95 | 3.10 | 4.14 | 4.00 | 4.23 | 4.01 | 4.44 | 2.64 | 2.28 | 1.46 | 34.04 |
| Days ≥ 0.1" Precip | 3 | 3 | 4 | 7 | 7 | 8 | 8 | 7 | 8 | 5 | 5 | 4 | 69 |
| Total Snowfall (") | 11.0 | 8.3 | 8.5 | 3.1 | 0.1 | 0.0 | 0.0 | 0.0 | 0.0 | 0.5 | 5.8 | 12.1 | 49.4 |
| Days ≥ 1" Snow Depth | 22 | *16* | *12* | *1* | 0 | 0 | 0 | 0 | 0 | 0 | *3* | *18* | 72 |

## MATHER 3 NW *Jackson County* ELEVATION 961 ft LAT/LONG 44° 9' N / 90° 21' W

| | JAN | FEB | MAR | APR | MAY | JUN | JUL | AUG | SEP | OCT | NOV | DEC | YEAR |
|---|---|---|---|---|---|---|---|---|---|---|---|---|---|
| Maximum Temp °F | 22.5 | 28.6 | 40.3 | 55.5 | 68.5 | 77.1 | 81.5 | 79.2 | 70.0 | 58.2 | 41.8 | 27.9 | 54.3 |
| Minimum Temp °F | 1.9 | 6.5 | 19.5 | 32.9 | 43.0 | 52.3 | 57.0 | 54.4 | 45.8 | 35.0 | 24.0 | 9.8 | 31.8 |
| Mean Temp °F | 12.3 | 17.5 | 29.9 | 44.2 | 55.8 | 64.7 | 69.3 | 66.8 | 57.9 | 46.6 | 32.9 | 18.9 | 43.1 |
| Days Max Temp ≥ 90 °F | 0 | 0 | 0 | 0 | 0 | 2 | 4 | 3 | 0 | 0 | 0 | 0 | 9 |
| Days Max Temp ≤ 32 °F | 24 | 17 | 7 | 1 | 0 | 0 | 0 | 0 | 0 | 0 | 6 | 20 | 75 |
| Days Min Temp ≤ 32 °F | 31 | 27 | 28 | 16 | 4 | 0 | 0 | 0 | 2 | 13 | 24 | 31 | 176 |
| Days Min Temp ≤ 0 °F | 15 | 10 | 3 | 0 | 0 | 0 | 0 | 0 | 0 | 0 | 1 | 8 | 37 |
| Heating Degree Days | 1631 | 1336 | 1080 | 621 | 305 | 89 | 24 | 52 | 237 | 565 | 955 | 1423 | 8318 |
| Cooling Degree Days | 0 | 0 | 0 | 5 | 23 | 85 | 158 | 116 | 31 | 1 | 0 | 0 | 419 |
| Total Precipitation (") | 1.16 | 0.86 | 2.05 | 3.18 | 3.58 | 3.94 | 4.43 | 4.04 | 4.28 | 2.34 | 2.32 | 1.59 | 33.77 |
| Days ≥ 0.1" Precip | 4 | 3 | 5 | 7 | 7 | 8 | 7 | 7 | 7 | 5 | 5 | 5 | 70 |
| Total Snowfall (") | 13.3 | 8.9 | 7.7 | 2.2 | 0.0 | 0.0 | 0.0 | 0.0 | 0.0 | 0.4 | 5.4 | 12.3 | 50.2 |
| Days ≥ 1" Snow Depth | 19 | *20* | 14 | 2 | 0 | 0 | 0 | 0 | 0 | 0 | 3 | *15* | 73 |

## MAUSTON 1 SE *Juneau County* ELEVATION 879 ft LAT/LONG 43° 48' N / 90° 5' W

| | JAN | FEB | MAR | APR | MAY | JUN | JUL | AUG | SEP | OCT | NOV | DEC | YEAR |
|---|---|---|---|---|---|---|---|---|---|---|---|---|---|
| Maximum Temp °F | 25.7 | 31.2 | 43.1 | 58.4 | 70.9 | 79.1 | 83.1 | 80.4 | 71.7 | 60.4 | 43.9 | 31.1 | 56.6 |
| Minimum Temp °F | 4.0 | 8.2 | 20.9 | 33.7 | 44.1 | 53.0 | 58.1 | 55.5 | 47.2 | 36.4 | 24.9 | 12.6 | 33.2 |
| Mean Temp °F | 14.9 | 19.7 | 32.0 | 46.0 | 57.5 | 66.1 | 70.6 | 68.0 | 59.5 | 48.4 | 34.4 | 21.9 | 44.9 |
| Days Max Temp ≥ 90 °F | 0 | 0 | 0 | 0 | 0 | 2 | 5 | 3 | 0 | 0 | 0 | 0 | 10 |
| Days Max Temp ≤ 32 °F | 22 | 14 | 4 | 0 | 0 | 0 | 0 | 0 | 0 | 0 | 4 | 17 | 61 |
| Days Min Temp ≤ 32 °F | 31 | 28 | 27 | 14 | 3 | 0 | 0 | 0 | 2 | 12 | 23 | 30 | 170 |
| Days Min Temp ≤ 0 °F | 13 | 9 | 2 | 0 | 0 | 0 | 0 | 0 | 0 | 0 | 1 | 6 | 31 |
| Heating Degree Days | 1547 | 1272 | 1018 | 568 | 257 | 67 | 14 | 38 | 198 | 511 | 913 | 1330 | 7733 |
| Cooling Degree Days | 0 | 0 | 0 | 8 | 32 | 115 | 207 | 142 | 39 | 0 | 0 | 0 | 543 |
| Total Precipitation (") | 1.07 | 1.00 | 2.06 | 3.48 | 3.46 | 3.95 | 3.91 | 4.13 | 4.69 | 2.44 | 2.37 | 1.50 | 34.06 |
| Days ≥ 0.1" Precip | 3 | 3 | 5 | 7 | 7 | 7 | 7 | 7 | 7 | 5 | 5 | 4 | 67 |
| Total Snowfall (") | 12.3 | 10.1 | 7.7 | 2.6 | 0.0 | 0.0 | 0.0 | 0.0 | 0.0 | 0.4 | 4.7 | 12.1 | 49.9 |
| Days ≥ 1" Snow Depth | 26 | 23 | 12 | 1 | 0 | 0 | 0 | 0 | 0 | 0 | 4 | 18 | 84 |

## MEDFORD *Taylor County* ELEVATION 1421 ft LAT/LONG 45° 8 ' N / 90° 20 ' W

| | JAN | FEB | MAR | APR | MAY | JUN | JUL | AUG | SEP | OCT | NOV | DEC | YEAR |
|---|---|---|---|---|---|---|---|---|---|---|---|---|---|
| Maximum Temp °F | 19.5 | 25.2 | 36.5 | 52.5 | 66.3 | 74.2 | 78.9 | 76.6 | 66.9 | 54.9 | 38.1 | 24.7 | 51.2 |
| Minimum Temp °F | -0.7 | 3.8 | 17.1 | 31.5 | 42.6 | 51.6 | 56.6 | 54.5 | 45.5 | 34.6 | 22.5 | 7.7 | 30.6 |
| Mean Temp °F | 9.4 | 14.5 | 26.9 | 42.0 | 54.5 | 62.9 | 67.8 | 65.6 | 56.2 | 44.8 | 30.3 | 16.3 | 40.9 |
| Days Max Temp ≥ 90 °F | 0 | 0 | 0 | 0 | 0 | 0 | 1 | 1 | 0 | 0 | 0 | 0 | 2 |
| Days Max Temp ≤ 32 °F | 27 | 20 | 11 | 1 | 0 | 0 | 0 | 0 | 0 | 0 | 9 | 24 | 92 |
| Days Min Temp ≤ 32 °F | 31 | 28 | 28 | 17 | 5 | 0 | 0 | 0 | 2 | 14 | 26 | 31 | 182 |
| Days Min Temp ≤ 0 °F | 16 | 12 | 4 | 0 | 0 | 0 | 0 | 0 | 0 | 0 | 1 | 10 | 43 |
| Heating Degree Days | 1719 | 1421 | 1176 | 685 | 337 | 118 | 36 | 68 | 279 | 622 | 1035 | 1506 | 9002 |
| Cooling Degree Days | 0 | 0 | 0 | 3 | 17 | 64 | 129 | 97 | 22 | 0 | 0 | 0 | 332 |
| Total Precipitation (") | 1.13 | 0.86 | 1.88 | 2.61 | 3.39 | 4.59 | 3.79 | 4.25 | 4.52 | 2.47 | 2.04 | 1.42 | 32.95 |
| Days ≥ 0.1" Precip | 4 | 3 | 5 | 6 | 7 | 8 | 8 | 7 | 8 | 6 | 5 | 5 | 72 |
| Total Snowfall (") | 11.6 | 6.8 | 8.6 | 2.2 | 0.0 | 0.0 | 0.0 | 0.0 | 0.0 | 0.5 | ***4.8*** | ***11.4*** | 45.9 |
| Days ≥ 1" Snow Depth | 30 | 27 | 21 | 3 | 0 | 0 | 0 | 0 | 0 | 0 | 6 | ***24*** | 111 |

## MELLEN 4 NE *Ashland County* ELEVATION 1302 ft LAT/LONG 46° 26 ' N / 90° 37 ' W

| | JAN | FEB | MAR | APR | MAY | JUN | JUL | AUG | SEP | OCT | NOV | DEC | YEAR |
|---|---|---|---|---|---|---|---|---|---|---|---|---|---|
| Maximum Temp °F | 20.4 | 26.0 | 37.4 | 52.0 | 65.8 | 73.9 | 78.9 | 76.4 | 66.5 | 54.7 | 37.8 | 24.8 | 51.2 |
| Minimum Temp °F | -2.3 | 1.1 | 13.9 | 28.4 | 39.7 | 48.8 | 54.6 | 52.6 | 43.9 | 33.9 | 21.6 | 6.1 | 28.5 |
| Mean Temp °F | 9.1 | 13.6 | 25.7 | 40.2 | 52.8 | 61.4 | 66.8 | 64.5 | 55.3 | 44.3 | 29.7 | 15.5 | 39.9 |
| Days Max Temp ≥ 90 °F | 0 | 0 | 0 | 0 | 0 | 1 | 2 | 1 | 0 | 0 | 0 | 0 | 4 |
| Days Max Temp ≤ 32 °F | 26 | 19 | 10 | 1 | 0 | 0 | 0 | 0 | 0 | 0 | 10 | 23 | 89 |
| Days Min Temp ≤ 32 °F | 31 | 28 | 29 | 21 | 8 | 1 | 0 | 0 | 3 | 15 | 26 | 31 | 193 |
| Days Min Temp ≤ 0 °F | 17 | 14 | 7 | 0 | 0 | 0 | 0 | 0 | 0 | 0 | 1 | 11 | 50 |
| Heating Degree Days | 1732 | 1447 | 1212 | 738 | 393 | 155 | 55 | 92 | 304 | 636 | 1052 | 1530 | 9346 |
| Cooling Degree Days | 0 | 0 | 0 | 3 | 20 | 54 | 110 | 88 | 15 | 0 | 0 | 0 | 290 |
| Total Precipitation (") | 1.50 | 0.90 | 1.99 | 2.36 | 3.28 | 3.85 | 4.07 | 3.94 | 4.02 | 3.19 | 2.79 | 1.65 | 33.54 |
| Days ≥ 0.1" Precip | 5 | 3 | 5 | 6 | 7 | 8 | 8 | 7 | 8 | 7 | 6 | 5 | 75 |
| Total Snowfall (") | 23.3 | 13.2 | 16.4 | 6.3 | 0.5 | 0.0 | 0.0 | 0.0 | 0.0 | 2.2 | 16.1 | 23.0 | 101.0 |
| Days ≥ 1" Snow Depth | ***31*** | 28 | ***29*** | 13 | 0 | 0 | 0 | 0 | 0 | 1 | 15 | ***30*** | 147 |

## MENOMONIE *Dunn County* ELEVATION 902 ft LAT/LONG 44° 54 ' N / 91° 56 ' W

| | JAN | FEB | MAR | APR | MAY | JUN | JUL | AUG | SEP | OCT | NOV | DEC | YEAR |
|---|---|---|---|---|---|---|---|---|---|---|---|---|---|
| Maximum Temp °F | 23.5 | 29.6 | 42.2 | 58.9 | 71.4 | 79.4 | 83.8 | 81.3 | 72.1 | 60.4 | 42.4 | 28.4 | 56.1 |
| Minimum Temp °F | 2.6 | 7.7 | 20.9 | 34.0 | 45.0 | 54.3 | 59.4 | 56.9 | 48.4 | 37.3 | 24.8 | 11.0 | 33.5 |
| Mean Temp °F | 13.1 | 18.7 | 31.6 | 46.5 | 58.2 | 66.9 | 71.7 | 69.1 | 60.3 | 48.9 | 33.6 | 19.7 | 44.9 |
| Days Max Temp ≥ 90 °F | 0 | 0 | 0 | 0 | 0 | 2 | 6 | 4 | 1 | 0 | 0 | 0 | 13 |
| Days Max Temp ≤ 32 °F | 23 | 16 | 6 | 0 | 0 | 0 | 0 | 0 | 0 | 0 | 5 | 19 | 69 |
| Days Min Temp ≤ 32 °F | 31 | 28 | 26 | 14 | 3 | 0 | 0 | 0 | 1 | 11 | 24 | 30 | 168 |
| Days Min Temp ≤ 0 °F | 13 | 10 | 3 | 0 | 0 | 0 | 0 | 0 | 0 | 0 | 1 | 8 | 35 |
| Heating Degree Days | 1607 | 1303 | 1029 | 553 | 240 | 54 | 11 | 30 | 183 | 497 | 935 | 1398 | 7840 |
| Cooling Degree Days | 0 | 0 | 0 | 6 | 34 | 116 | 216 | 161 | 48 | 1 | 0 | 0 | 582 |
| Total Precipitation (") | 0.86 | 0.59 | 1.48 | 2.79 | 3.38 | 4.46 | 4.02 | 3.81 | 3.76 | 2.32 | 1.77 | 1.02 | 30.26 |
| Days ≥ 0.1" Precip | 3 | 2 | 4 | 6 | 7 | 8 | 7 | 6 | 6 | 5 | 4 | 3 | 61 |
| Total Snowfall (") | ***12.4*** | 7.5 | 7.9 | 1.6 | 0.0 | 0.0 | 0.0 | 0.0 | 0.0 | 0.1 | ***5.0*** | ***9.3*** | 43.8 |
| Days ≥ 1" Snow Depth | 29 | 26 | 18 | 2 | 0 | 0 | 0 | 0 | 0 | 0 | 5 | 23 | 103 |

## MERRILL *Lincoln County* ELEVATION 1270 ft LAT/LONG 45° 11 ' N / 89° 41 ' W

| | JAN | FEB | MAR | APR | MAY | JUN | JUL | AUG | SEP | OCT | NOV | DEC | YEAR |
|---|---|---|---|---|---|---|---|---|---|---|---|---|---|
| Maximum Temp °F | 21.4 | 27.0 | 38.6 | 54.0 | 68.1 | 75.9 | 80.3 | 77.8 | 67.9 | 55.7 | 39.3 | 26.1 | 52.7 |
| Minimum Temp °F | -0.7 | 3.0 | 16.4 | 30.3 | 41.2 | 50.5 | 55.1 | 52.6 | 43.7 | 33.6 | 22.0 | 7.9 | 29.6 |
| Mean Temp °F | 10.4 | 15.1 | 27.5 | 42.2 | 54.7 | 63.3 | 67.7 | 65.2 | 55.9 | 44.7 | 30.7 | 17.0 | 41.2 |
| Days Max Temp ≥ 90 °F | 0 | 0 | 0 | 0 | 0 | 1 | 2 | 1 | 0 | 0 | 0 | 0 | 4 |
| Days Max Temp ≤ 32 °F | 26 | 18 | 9 | 1 | 0 | 0 | 0 | 0 | 0 | 0 | 7 | 22 | 83 |
| Days Min Temp ≤ 32 °F | 31 | 28 | 28 | 19 | 6 | 1 | 0 | 0 | 4 | 15 | 26 | 30 | 188 |
| Days Min Temp ≤ 0 °F | 16 | 12 | 4 | 0 | 0 | 0 | 0 | 0 | 0 | 0 | 1 | 9 | 42 |
| Heating Degree Days | 1690 | 1402 | 1154 | 680 | 332 | 114 | 39 | 75 | 288 | 624 | 1023 | 1484 | 8905 |
| Cooling Degree Days | 0 | 0 | 0 | 3 | 17 | 63 | 119 | 87 | 17 | 0 | 0 | 0 | 306 |
| Total Precipitation (") | 1.07 | 0.84 | 1.85 | 2.90 | 3.39 | 3.98 | 3.80 | 3.92 | 4.53 | 2.64 | 2.31 | 1.50 | 32.73 |
| Days ≥ 0.1" Precip | 4 | 3 | 4 | 6 | 7 | 8 | 8 | 7 | 8 | 6 | 5 | 4 | 70 |
| Total Snowfall (") | na | na | na | ***2.0*** | 0.0 | 0.0 | 0.0 | 0.0 | 0.0 | 0.5 | ***6.5*** | na | na |
| Days ≥ 1" Snow Depth | na | ***27*** | na | ***2*** | 0 | 0 | 0 | 0 | 0 | 0 | na | na | na |

## MILWAUKEE MT MARY CO *Milwaukee County* ELEVATION 732 ft LAT/LONG 43° 4 ' N / 88° 2 ' W

| | JAN | FEB | MAR | APR | MAY | JUN | JUL | AUG | SEP | OCT | NOV | DEC | YEAR |
|---|---|---|---|---|---|---|---|---|---|---|---|---|---|
| Maximum Temp °F | 27.7 | 32.2 | 43.6 | 57.2 | 70.1 | 79.7 | 84.3 | 81.9 | 73.8 | 60.7 | 46.0 | 33.5 | 57.6 |
| Minimum Temp °F | 13.0 | 17.0 | 26.9 | 37.4 | 47.3 | 57.0 | 63.1 | 61.6 | 54.0 | 42.8 | 31.5 | 20.0 | 39.3 |
| Mean Temp °F | 20.3 | 24.6 | 35.3 | 47.4 | 58.7 | 68.4 | 73.7 | 71.8 | 63.9 | 51.8 | 38.8 | 26.8 | 48.5 |
| Days Max Temp ≥ 90 °F | 0 | 0 | 0 | 0 | 1 | 4 | 8 | 4 | 1 | 0 | 0 | 0 | 18 |
| Days Max Temp ≤ 32 °F | 19 | 14 | 4 | 0 | 0 | 0 | 0 | 0 | 0 | 0 | 2 | 13 | 52 |
| Days Min Temp ≤ 32 °F | 29 | 26 | 22 | 9 | 1 | 0 | 0 | 0 | 0 | 4 | 17 | 27 | 135 |
| Days Min Temp ≤ 0 °F | 7 | 3 | 0 | 0 | 0 | 0 | 0 | 0 | 0 | 0 | 0 | 2 | 12 |
| Heating Degree Days | 1378 | 1134 | 915 | 531 | 236 | 47 | 7 | 12 | 109 | 410 | 779 | 1178 | 6736 |
| Cooling Degree Days | 0 | 0 | 1 | 10 | 55 | 166 | 299 | 245 | 91 | 5 | 0 | 0 | 872 |
| Total Precipitation (") | 1.51 | 1.28 | 2.06 | 3.23 | 2.68 | 3.38 | 3.32 | 3.66 | 3.51 | 2.40 | 2.38 | 2.06 | 31.47 |
| Days ≥ 0.1" Precip | 4 | 3 | 5 | 6 | 6 | 6 | 6 | 6 | 6 | 5 | 5 | 5 | 63 |
| Total Snowfall (") | 12.0 | 9.6 | 5.8 | 1.7 | 0.1 | 0.0 | 0.0 | 0.0 | 0.0 | 0.1 | 2.0 | 9.1 | 40.4 |
| Days ≥ 1" Snow Depth | 22 | 17 | 5 | 1 | 0 | 0 | 0 | 0 | 0 | 0 | 1 | 11 | 57 |

## MILWAUKEE MTCHLL FLD *Milwaukee County* ELEVATION 672 ft LAT/LONG 42° 57 ' N / 87° 54 ' W

| | JAN | FEB | MAR | APR | MAY | JUN | JUL | AUG | SEP | OCT | NOV | DEC | YEAR |
|---|---|---|---|---|---|---|---|---|---|---|---|---|---|
| Maximum Temp °F | 26.8 | 30.7 | 41.3 | 53.5 | 65.2 | 75.3 | 80.4 | 78.3 | 70.9 | 59.0 | 45.0 | 32.7 | 54.9 |
| Minimum Temp °F | 12.4 | 16.6 | 26.5 | 36.3 | 45.4 | 55.6 | 62.4 | 61.3 | 53.7 | 42.0 | 30.9 | 19.4 | 38.5 |
| Mean Temp °F | 19.6 | 23.7 | 33.9 | 44.9 | 55.3 | 65.5 | 71.4 | 69.8 | 62.3 | 50.6 | 38.0 | 26.1 | 46.8 |
| Days Max Temp ≥ 90 °F | 0 | 0 | 0 | 0 | 0 | 2 | 4 | 3 | 1 | 0 | 0 | 0 | 10 |
| Days Max Temp ≤ 32 °F | 20 | 15 | 6 | 1 | 0 | 0 | 0 | 0 | 0 | 0 | 3 | 14 | 59 |
| Days Min Temp ≤ 32 °F | 30 | 26 | 23 | 9 | 1 | 0 | 0 | 0 | 0 | 4 | 17 | 27 | 137 |
| Days Min Temp ≤ 0 °F | 7 | 3 | 0 | 0 | 0 | 0 | 0 | 0 | 0 | 0 | 0 | 3 | 13 |
| Heating Degree Days | 1400 | 1160 | 957 | 601 | 320 | 88 | 14 | 21 | 136 | 447 | 803 | 1201 | 7148 |
| Cooling Degree Days | 0 | 0 | 0 | 8 | 34 | 120 | 241 | 196 | 69 | 3 | 0 | 0 | 671 |
| Total Precipitation (") | 1.70 | 1.53 | 2.68 | 3.61 | 2.81 | 3.53 | 3.48 | 3.71 | 3.37 | 2.55 | 2.73 | 2.43 | 34.13 |
| Days ≥ 0.1" Precip | 5 | 4 | 6 | 7 | 6 | 6 | 6 | 7 | 6 | 5 | 6 | 6 | 70 |
| Total Snowfall (") | 13.6 | 12.0 | 8.0 | 2.4 | 0.1 | 0.0 | 0.0 | 0.0 | 0.0 | 0.4 | 3.0 | 11.3 | 50.8 |
| Days ≥ 1" Snow Depth | 22 | 18 | 9 | 1 | 0 | 0 | 0 | 0 | 0 | 0 | 1 | 13 | 64 |

## MINOCQUA DAM *Oneida County* ELEVATION 1591 ft LAT/LONG 45° 53 ' N / 89° 44 ' W

| | JAN | FEB | MAR | APR | MAY | JUN | JUL | AUG | SEP | OCT | NOV | DEC | YEAR |
|---|---|---|---|---|---|---|---|---|---|---|---|---|---|
| Maximum Temp °F | 20.0 | 25.9 | 37.1 | 51.6 | 65.9 | 73.2 | 77.7 | 75.4 | 65.7 | 53.7 | 37.0 | 24.1 | 50.6 |
| Minimum Temp °F | -1.9 | 0.8 | 13.0 | 27.6 | 39.8 | 49.2 | 54.1 | 51.8 | 43.7 | 33.1 | 20.5 | 5.7 | 28.1 |
| Mean Temp °F | 9.2 | 13.4 | 25.1 | 39.6 | 52.9 | 61.2 | 66.0 | 63.6 | 54.7 | 43.4 | 28.8 | 14.9 | 39.4 |
| Days Max Temp ≥ 90 °F | 0 | 0 | 0 | 0 | 0 | 0 | 1 | 1 | 0 | 0 | 0 | 0 | 2 |
| Days Max Temp ≤ 32 °F | 27 | 20 | 10 | 1 | 0 | 0 | 0 | 0 | 0 | 0 | 11 | 25 | 94 |
| Days Min Temp ≤ 32 °F | 31 | 28 | 29 | 22 | 8 | 0 | 0 | 0 | 3 | 16 | 27 | 31 | 195 |
| Days Min Temp ≤ 0 °F | 16 | 14 | 7 | 0 | 0 | 0 | 0 | 0 | 0 | 0 | 1 | 11 | 49 |
| Heating Degree Days | 1727 | 1454 | 1231 | 757 | 385 | 151 | 58 | 101 | 316 | 663 | 1080 | 1547 | 9470 |
| Cooling Degree Days | 0 | 0 | 0 | 2 | 14 | 41 | 88 | 63 | 8 | 0 | 0 | 0 | 216 |
| Total Precipitation (") | 1.08 | 0.79 | 1.75 | 2.43 | 3.37 | 4.03 | 3.78 | 4.32 | 4.24 | 2.65 | 2.24 | 1.29 | 31.97 |
| Days ≥ 0.1" Precip | 3 | 2 | 4 | 6 | 7 | 8 | 8 | 7 | 8 | 6 | 6 | 4 | 69 |
| Total Snowfall (") | 23.3 | 15.8 | 16.1 | 9.8 | 1.0 | 0.0 | 0.0 | 0.0 | 0.0 | 3.0 | 14.1 | 24.4 | 107.5 |
| Days ≥ 1" Snow Depth | 31 | 28 | 29 | 11 | 0 | 0 | 0 | 0 | 0 | 1 | 14 | 30 | 144 |

## MONDOVI *Buffalo County* ELEVATION 741 ft LAT/LONG 44° 34 ' N / 91° 40 ' W

| | JAN | FEB | MAR | APR | MAY | JUN | JUL | AUG | SEP | OCT | NOV | DEC | YEAR |
|---|---|---|---|---|---|---|---|---|---|---|---|---|---|
| Maximum Temp °F | 23.1 | 29.7 | 42.0 | 58.9 | 71.3 | 78.8 | 82.8 | 80.3 | 71.0 | 59.9 | 41.8 | 28.5 | 55.7 |
| Minimum Temp °F | 1.8 | 7.4 | 20.8 | 34.4 | 45.4 | 54.6 | 59.5 | 57.1 | 48.5 | 37.3 | 24.1 | 10.6 | 33.5 |
| Mean Temp °F | 12.4 | 18.6 | 31.4 | 46.7 | 58.4 | 66.7 | 71.2 | 68.8 | 59.8 | 48.6 | 33.0 | 19.5 | 44.6 |
| Days Max Temp ≥ 90 °F | 0 | 0 | 0 | 0 | 0 | 2 | 4 | 2 | 0 | 0 | 0 | 0 | 8 |
| Days Max Temp ≤ 32 °F | 22 | 15 | 6 | 0 | 0 | 0 | 0 | 0 | 0 | 0 | 6 | 18 | 67 |
| Days Min Temp ≤ 32 °F | 31 | 28 | 26 | 13 | 3 | 0 | 0 | 0 | 1 | 11 | 23 | 30 | 166 |
| Days Min Temp ≤ 0 °F | 14 | 10 | 3 | 0 | 0 | 0 | 0 | 0 | 0 | 0 | 1 | 8 | 36 |
| Heating Degree Days | 1627 | 1303 | 1035 | 549 | 237 | 55 | 12 | 33 | 196 | 504 | 955 | 1405 | 7911 |
| Cooling Degree Days | 0 | 0 | 0 | 9 | 38 | 128 | 218 | 170 | *51* | 2 | 0 | 0 | 616 |
| Total Precipitation (") | 1.01 | 0.80 | 1.93 | 3.27 | 4.00 | 4.08 | 4.50 | 3.64 | 4.33 | 2.49 | 2.17 | 1.29 | 33.51 |
| Days ≥ 0.1" Precip | 3 | 3 | 5 | 7 | *8* | 8 | 7 | 6 | 6 | 5 | 5 | 4 | 67 |
| Total Snowfall (") | 12.1 | 7.9 | 8.5 | 1.8 | 0.0 | 0.0 | 0.0 | 0.0 | 0.0 | 0.1 | 6.0 | 10.3 | 46.7 |
| Days ≥ 1" Snow Depth | *27* | *23* | 15 | 1 | 0 | 0 | 0 | 0 | 0 | 0 | *5* | *21* | 92 |

## MONTELLO *Marquette County* ELEVATION 820 ft LAT/LONG 43° 48 ' N / 89° 19 ' W

| | JAN | FEB | MAR | APR | MAY | JUN | JUL | AUG | SEP | OCT | NOV | DEC | YEAR |
|---|---|---|---|---|---|---|---|---|---|---|---|---|---|
| Maximum Temp °F | 24.4 | 29.2 | 41.5 | 56.9 | 69.8 | 78.9 | 82.6 | 80.0 | 71.3 | 59.5 | 43.0 | 29.7 | 55.6 |
| Minimum Temp °F | 5.4 | 9.4 | 22.0 | 34.7 | 45.6 | 54.8 | 59.6 | 56.6 | 48.6 | 37.2 | 25.4 | 12.7 | 34.3 |
| Mean Temp °F | 14.8 | 19.2 | 31.7 | 45.9 | 57.8 | 66.9 | 71.1 | 68.3 | 60.0 | 48.4 | 34.3 | 21.1 | 45.0 |
| Days Max Temp ≥ 90 °F | 0 | 0 | 0 | 0 | 0 | 3 | 5 | 2 | 0 | 0 | 0 | 0 | 10 |
| Days Max Temp ≤ 32 °F | 22 | 17 | 6 | 0 | 0 | 0 | 0 | 0 | 0 | 0 | 5 | 17 | 67 |
| Days Min Temp ≤ 32 °F | 31 | 28 | 26 | 14 | 3 | 0 | 0 | 0 | 1 | 11 | 23 | 29 | 166 |
| Days Min Temp ≤ 0 °F | 12 | 8 | 2 | 0 | 0 | 0 | 0 | 0 | 0 | 0 | 0 | 7 | 29 |
| Heating Degree Days | 1551 | 1290 | 1024 | 575 | 255 | 56 | 13 | 36 | 184 | 510 | 914 | 1354 | 7762 |
| Cooling Degree Days | *0* | *0* | *0* | *11* | *41* | *123* | 201 | 146 | *45* | *1* | *0* | *0* | 568 |
| Total Precipitation (") | 1.23 | 1.04 | 2.26 | 3.28 | 3.26 | 3.86 | 3.68 | 3.86 | 4.32 | 2.35 | 2.40 | 1.65 | 33.19 |
| Days ≥ 0.1" Precip | 4 | 3 | 5 | 7 | 7 | 7 | 6 | 7 | 7 | 6 | 6 | 5 | 70 |
| Total Snowfall (") | *10.5* | *8.3* | *6.7* | *2.1* | 0.1 | 0.0 | 0.0 | 0.0 | 0.0 | 0.1 | na | *10.5* | na |
| Days ≥ 1" Snow Depth | 28 | 25 | 17 | 2 | 0 | 0 | 0 | 0 | 0 | 0 | na | 21 | na |

## NECEDAH *Juneau County* ELEVATION 942 ft LAT/LONG 44° 3 ' N / 90° 6 ' W

| | JAN | FEB | MAR | APR | MAY | JUN | JUL | AUG | SEP | OCT | NOV | DEC | YEAR |
|---|---|---|---|---|---|---|---|---|---|---|---|---|---|
| Maximum Temp °F | 25.2 | 31.0 | 42.9 | 58.4 | 71.5 | 79.9 | 84.3 | 81.5 | 72.5 | 60.6 | 43.3 | 30.0 | 56.8 |
| Minimum Temp °F | 3.2 | 7.6 | 20.0 | 33.0 | 43.9 | 53.0 | 58.3 | 55.4 | 47.3 | 36.6 | 24.7 | 11.2 | 32.9 |
| Mean Temp °F | 14.2 | 19.4 | 31.5 | 45.7 | 57.7 | 66.5 | 71.3 | 68.5 | 59.9 | 48.6 | 34.0 | 20.6 | 44.8 |
| Days Max Temp ≥ 90 °F | 0 | 0 | 0 | 0 | 1 | 3 | 7 | 4 | 1 | 0 | 0 | 0 | 16 |
| Days Max Temp ≤ 32 °F | 22 | 15 | 5 | 0 | 0 | 0 | 0 | 0 | 0 | 0 | 4 | 17 | 63 |
| Days Min Temp ≤ 32 °F | 31 | 27 | 26 | 14 | 4 | 0 | 0 | 0 | 2 | 12 | 23 | 30 | 169 |
| Days Min Temp ≤ 0 °F | 13 | 10 | 3 | 0 | 0 | 0 | 0 | 0 | 0 | 0 | 1 | 7 | 34 |
| Heating Degree Days | 1571 | 1283 | 1033 | 578 | 256 | 66 | 14 | 38 | 194 | 506 | 923 | 1371 | 7833 |
| Cooling Degree Days | 0 | 0 | 0 | 9 | 43 | 134 | 232 | 173 | 58 | 3 | 0 | 0 | 652 |
| Total Precipitation (") | 0.93 | 0.88 | 1.91 | 3.02 | 3.58 | 3.55 | 4.15 | 4.13 | 4.41 | 2.28 | 2.12 | 1.31 | 32.27 |
| Days ≥ 0.1" Precip | 3 | 3 | 5 | 7 | 6 | 7 | 7 | 6 | 7 | 5 | 5 | 4 | 65 |
| Total Snowfall (") | 8.3 | 6.0 | 5.1 | 1.7 | 0.0 | 0.0 | 0.0 | 0.0 | 0.0 | 0.3 | 3.3 | 8.6 | 33.3 |
| Days ≥ 1" Snow Depth | 27 | 24 | 15 | 2 | 0 | 0 | 0 | 0 | 0 | 0 | 4 | 20 | 92 |

## NEILLSVILLE 3 SW *Clark County* ELEVATION 1060 ft LAT/LONG 44° 34 ' N / 90° 36 ' W

| | JAN | FEB | MAR | APR | MAY | JUN | JUL | AUG | SEP | OCT | NOV | DEC | YEAR |
|---|---|---|---|---|---|---|---|---|---|---|---|---|---|
| Maximum Temp °F | 22.6 | 28.6 | 40.5 | 56.4 | 69.7 | 77.4 | 81.6 | 79.3 | 70.0 | 58.3 | 40.8 | 27.5 | 54.4 |
| Minimum Temp °F | 1.4 | 6.7 | 19.3 | 32.5 | 43.1 | 51.5 | 56.4 | 54.5 | 46.2 | 35.7 | 23.1 | 9.3 | 31.6 |
| Mean Temp °F | 12.0 | 17.7 | 29.9 | 44.5 | 56.4 | 64.5 | 69.0 | 67.0 | 58.1 | 47.0 | 31.9 | 18.4 | 43.0 |
| Days Max Temp ≥ 90 °F | 0 | 0 | 0 | 0 | 0 | 1 | 3 | 2 | 0 | 0 | 0 | 0 | 6 |
| Days Max Temp ≤ 32 °F | 24 | 17 | 7 | 0 | 0 | 0 | 0 | 0 | 0 | 0 | 6 | 20 | 74 |
| Days Min Temp ≤ 32 °F | 31 | 28 | 27 | 16 | 4 | 0 | 0 | 0 | 3 | 12 | 26 | 31 | 178 |
| Days Min Temp ≤ 0 °F | 14 | 10 | 3 | 0 | 0 | 0 | 0 | 0 | 0 | 0 | 1 | 8 | 36 |
| Heating Degree Days | 1638 | 1332 | 1081 | 612 | 282 | 85 | 22 | 49 | 230 | 552 | 985 | 1439 | 8307 |
| Cooling Degree Days | 0 | 0 | 0 | 5 | 22 | 80 | 154 | 118 | 32 | 1 | 0 | 0 | 412 |
| Total Precipitation (") | 0.83 | 0.75 | 1.69 | 2.90 | 3.39 | 4.14 | 4.59 | 4.16 | 4.35 | 2.28 | 1.91 | 1.14 | 32.13 |
| Days ≥ 0.1" Precip | 3 | 2 | 4 | 7 | 7 | 7 | 7 | 7 | 7 | 5 | 5 | 3 | 64 |
| Total Snowfall (") | 10.7 | 7.4 | 7.5 | 1.9 | 0.0 | 0.0 | 0.0 | 0.0 | 0.0 | 0.3 | 4.5 | 10.1 | 42.4 |
| Days ≥ 1" Snow Depth | 29 | 25 | 17 | 2 | 0 | 0 | 0 | 0 | 0 | 0 | 5 | 22 | 100 |

## NEW LONDON *Waupaca County* ELEVATION 781 ft LAT/LONG 44° 23 ' N / 88° 44 ' W

| | JAN | FEB | MAR | APR | MAY | JUN | JUL | AUG | SEP | OCT | NOV | DEC | YEAR |
|---|---|---|---|---|---|---|---|---|---|---|---|---|---|
| Maximum Temp °F | 24.8 | 29.9 | 41.3 | 56.8 | 70.2 | 78.9 | 83.0 | 80.2 | 71.9 | 59.5 | 43.1 | 29.8 | 55.8 |
| Minimum Temp °F | 4.6 | 8.7 | 20.6 | 34.0 | 44.5 | 53.8 | 58.5 | 56.1 | 47.7 | 37.1 | 25.6 | 12.5 | 33.6 |
| Mean Temp °F | 14.8 | 19.3 | 30.9 | 45.4 | 57.4 | 66.3 | 70.8 | 68.2 | 59.8 | 48.4 | 34.4 | 21.2 | 44.7 |
| Days Max Temp ≥ 90 °F | 0 | 0 | 0 | 0 | 0 | 3 | 5 | 3 | 1 | 0 | 0 | 0 | 12 |
| Days Max Temp ≤ 32 °F | 22 | 16 | 6 | 0 | 0 | 0 | 0 | 0 | 0 | 0 | 4 | 17 | 65 |
| Days Min Temp ≤ 32 °F | 30 | 28 | 27 | 14 | 3 | 0 | 0 | 0 | 1 | 10 | 23 | 30 | 166 |
| Days Min Temp ≤ 0 °F | 12 | 9 | 2 | 0 | 0 | 0 | 0 | 0 | 0 | 0 | 0 | 6 | 29 |
| Heating Degree Days | 1553 | 1283 | 1050 | 585 | 265 | 63 | 16 | 36 | 192 | 514 | 912 | 1352 | 7821 |
| Cooling Degree Days | 0 | 0 | *0* | *7* | *32* | 101 | 182 | 130 | 39 | 1 | 0 | 0 | 492 |
| Total Precipitation (") | 1.40 | 1.12 | 1.97 | 2.76 | 3.38 | 3.55 | 3.71 | 3.77 | 3.44 | 2.32 | 2.13 | 1.69 | 31.24 |
| Days ≥ 0.1" Precip | 4 | 3 | 5 | 6 | 7 | 6 | 7 | 6 | 6 | 5 | 5 | 5 | 65 |
| Total Snowfall (") | 10.3 | *7.4* | *7.5* | *2.0* | 0.0 | 0.0 | 0.0 | 0.0 | 0.0 | 0.1 | 3.6 | *9.3* | 40.2 |
| Days ≥ 1" Snow Depth | 26 | *23* | 14 | 2 | 0 | 0 | 0 | 0 | 0 | 0 | 3 | 17 | 85 |

## NEWALD 4 N *Forest County* ELEVATION 1572 ft LAT/LONG 45° 44 ' N / 88° 41 ' W

| | JAN | FEB | MAR | APR | MAY | JUN | JUL | AUG | SEP | OCT | NOV | DEC | YEAR |
|---|---|---|---|---|---|---|---|---|---|---|---|---|---|
| Maximum Temp °F | 22.3 | 27.5 | 38.9 | 53.8 | 67.8 | 74.6 | 78.8 | ***76.1*** | 67.2 | 55.3 | 38.3 | 26.4 | 52.2 |
| Minimum Temp °F | -0.5 | 2.0 | 13.8 | 26.9 | 37.7 | 46.5 | 51.2 | ***49.8*** | 42.4 | 32.7 | 21.0 | 7.6 | 27.6 |
| Mean Temp °F | 10.9 | 14.8 | 26.4 | 40.3 | 52.8 | 60.6 | 65.0 | ***63.0*** | 54.8 | 44.0 | 29.7 | 17.0 | 39.9 |
| Days Max Temp ≥ 90 °F | 0 | 0 | 0 | 0 | 0 | 0 | 1 | 0 | 0 | 0 | 0 | 0 | 1 |
| Days Max Temp ≤ 32 °F | 26 | 19 | 9 | 1 | 0 | 0 | 0 | 0 | 0 | 0 | 9 | 23 | 87 |
| Days Min Temp ≤ 32 °F | 31 | 28 | 29 | 22 | 10 | 2 | 0 | 1 | 6 | 16 | 27 | 31 | 203 |
| Days Min Temp ≤ 0 °F | 15 | 13 | 6 | 0 | 0 | 0 | 0 | 0 | 0 | 0 | 2 | 9 | 45 |
| Heating Degree Days | 1673 | 1412 | 1191 | 735 | 384 | 164 | 75 | ***113*** | 313 | 646 | 1054 | 1481 | 9241 |
| Cooling Degree Days | 0 | 0 | 0 | 2 | 9 | 38 | 74 | 61 | 13 | 0 | 0 | 0 | 197 |
| Total Precipitation (") | 1.19 | 0.87 | 1.77 | 2.50 | 3.38 | 3.90 | 3.27 | 3.74 | 4.20 | 2.74 | 2.11 | 1.51 | 31.18 |
| Days ≥ 0.1" Precip | 4 | 3 | 4 | 6 | 7 | 8 | 7 | ***8*** | 7 | 6 | 5 | 5 | 70 |
| Total Snowfall (") | ***11.6*** | ***7.2*** | 8.3 | 4.7 | 0.7 | 0.0 | 0.0 | 0.0 | 0.1 | 1.5 | ***7.1*** | ***12.8*** | 54.0 |
| Days ≥ 1" Snow Depth | 26 | ***23*** | 18 | 5 | 0 | 0 | 0 | 0 | 0 | 1 | 9 | ***24*** | 106 |

## NORTH PELICAN *Oneida County* ELEVATION 1552 ft LAT/LONG 45° 38 ' N / 89° 15 ' W

| | JAN | FEB | MAR | APR | MAY | JUN | JUL | AUG | SEP | OCT | NOV | DEC | YEAR |
|---|---|---|---|---|---|---|---|---|---|---|---|---|---|
| Maximum Temp °F | 20.0 | 25.9 | 36.8 | 51.7 | 66.1 | 72.8 | 76.6 | 73.8 | 64.2 | 53.5 | 37.1 | 24.7 | 50.3 |
| Minimum Temp °F | -1.3 | 1.4 | 13.4 | 28.2 | 39.9 | 49.3 | 54.3 | 52.3 | 44.1 | 34.2 | 21.6 | 6.8 | 28.7 |
| Mean Temp °F | 9.4 | 13.7 | 25.1 | 40.0 | 53.0 | 61.1 | 65.5 | 63.1 | 54.2 | 43.9 | 29.4 | 15.8 | 39.5 |
| Days Max Temp ≥ 90 °F | 0 | 0 | 0 | 0 | 0 | 0 | 1 | 0 | 0 | 0 | 0 | 0 | 1 |
| Days Max Temp ≤ 32 °F | 26 | 20 | 10 | 1 | 0 | 0 | 0 | 0 | 0 | 0 | 10 | 24 | 91 |
| Days Min Temp ≤ 32 °F | 31 | 28 | 29 | 21 | 8 | 1 | 0 | 0 | 3 | 15 | 27 | 31 | 194 |
| Days Min Temp ≤ 0 °F | 16 | 14 | 7 | 0 | 0 | 0 | 0 | 0 | 0 | 0 | 1 | 10 | 48 |
| Heating Degree Days | 1723 | 1443 | 1229 | 746 | 379 | 151 | 62 | 108 | 329 | 649 | 1062 | 1521 | 9402 |
| Cooling Degree Days | 0 | 0 | 0 | 3 | 13 | 38 | 78 | 58 | 11 | 0 | 0 | 0 | 201 |
| Total Precipitation (") | 1.13 | 0.79 | 1.71 | 2.54 | 3.34 | 3.83 | 3.65 | 4.11 | 4.15 | 2.54 | 2.04 | 1.35 | 31.18 |
| Days ≥ 0.1" Precip | 4 | 3 | 5 | 6 | 7 | 8 | 7 | 7 | 8 | 6 | 6 | 5 | 72 |
| Total Snowfall (") | 15.2 | 8.9 | 10.3 | 4.6 | 0.4 | 0.0 | 0.0 | 0.0 | 0.0 | 1.1 | 8.0 | ***15.2*** | 63.7 |
| Days ≥ 1" Snow Depth | 31 | 28 | 27 | 9 | 0 | 0 | 0 | 0 | 0 | 1 | 10 | 29 | 135 |

## OCONOMOWOC *Waukesha County* ELEVATION 879 ft LAT/LONG 43° 6 ' N / 88° 31 ' W

| | JAN | FEB | MAR | APR | MAY | JUN | JUL | AUG | SEP | OCT | NOV | DEC | YEAR |
|---|---|---|---|---|---|---|---|---|---|---|---|---|---|
| Maximum Temp °F | 25.7 | 30.3 | 42.3 | 56.8 | 69.5 | 78.3 | 82.8 | 80.1 | 71.8 | 59.6 | 44.6 | 31.8 | 56.1 |
| Minimum Temp °F | 8.0 | 12.4 | 24.4 | 36.2 | 46.4 | 55.5 | 60.9 | 58.7 | 50.2 | 38.8 | 28.4 | 16.0 | 36.3 |
| Mean Temp °F | 16.9 | 21.4 | 33.4 | 46.5 | 57.9 | 66.9 | 71.9 | 69.4 | 61.0 | 49.3 | 36.5 | 24.0 | 46.3 |
| Days Max Temp ≥ 90 °F | 0 | 0 | 0 | 0 | 0 | 2 | 5 | 2 | 0 | 0 | 0 | 0 | 9 |
| Days Max Temp ≤ 32 °F | 21 | 15 | 5 | 0 | 0 | 0 | 0 | 0 | 0 | 0 | 3 | 15 | 59 |
| Days Min Temp ≤ 32 °F | 30 | 27 | 25 | 11 | 2 | 0 | 0 | 0 | 1 | 8 | 20 | 29 | 153 |
| Days Min Temp ≤ 0 °F | 10 | 6 | 1 | 0 | 0 | 0 | 0 | 0 | 0 | 0 | 0 | 4 | 21 |
| Heating Degree Days | 1486 | 1225 | 973 | 556 | 254 | 58 | 11 | 26 | 166 | 486 | 849 | 1265 | 7355 |
| Cooling Degree Days | 0 | 0 | 0 | 10 | 38 | 123 | 228 | 174 | 55 | 2 | 0 | 0 | 630 |
| Total Precipitation (") | 1.09 | 1.05 | 1.89 | 2.98 | 2.84 | 3.81 | 4.03 | 4.06 | 4.16 | 2.60 | 2.29 | 1.75 | 32.55 |
| Days ≥ 0.1" Precip | 4 | 3 | 5 | 7 | 7 | 7 | 6 | 6 | 7 | 6 | 5 | 4 | 67 |
| Total Snowfall (") | 9.7 | 8.0 | 6.1 | 1.6 | 0.3 | 0.0 | 0.0 | 0.0 | 0.0 | 0.1 | 2.8 | 10.5 | 39.1 |
| Days ≥ 1" Snow Depth | 24 | 20 | 8 | 1 | 0 | 0 | 0 | 0 | 0 | 0 | 2 | 16 | 71 |

## OCONTO 4 W *Oconto County* ELEVATION 659 ft LAT/LONG 44° 54 ' N / 87° 57 ' W

| | JAN | FEB | MAR | APR | MAY | JUN | JUL | AUG | SEP | OCT | NOV | DEC | YEAR |
|---|---|---|---|---|---|---|---|---|---|---|---|---|---|
| Maximum Temp °F | 24.0 | 28.2 | 38.9 | 53.0 | 66.5 | 75.4 | 80.7 | 78.1 | 69.4 | 57.6 | 42.2 | 29.4 | 53.6 |
| Minimum Temp °F | 4.3 | 7.1 | 18.7 | 31.6 | 42.1 | 51.8 | 57.1 | 54.5 | 46.2 | 35.9 | 24.8 | 12.2 | 32.2 |
| Mean Temp °F | 14.1 | 17.7 | 28.8 | 42.3 | 54.4 | 63.6 | 68.9 | 66.3 | 57.8 | 46.8 | 33.5 | 20.8 | 42.9 |
| Days Max Temp ≥ 90 °F | 0 | 0 | 0 | 0 | 0 | 1 | 3 | 1 | 0 | 0 | 0 | 0 | 5 |
| Days Max Temp ≤ 32 °F | 24 | 18 | 8 | 0 | 0 | 0 | 0 | 0 | 0 | 0 | 5 | 18 | 73 |
| Days Min Temp ≤ 32 °F | 31 | 28 | 28 | 18 | 4 | 0 | 0 | 0 | 2 | 12 | 24 | 30 | 177 |
| Days Min Temp ≤ 0 °F | 13 | 10 | 2 | 0 | 0 | 0 | 0 | 0 | 0 | 0 | 0 | 6 | 31 |
| Heating Degree Days | 1574 | 1331 | 1115 | 676 | 342 | 104 | 28 | 54 | 234 | 560 | 938 | 1363 | 8319 |
| Cooling Degree Days | 0 | 0 | 0 | 3 | 20 | 64 | 147 | 99 | 22 | 1 | 0 | 0 | 356 |
| Total Precipitation (") | 1.60 | 1.16 | 2.14 | 2.80 | 3.29 | 3.58 | 3.83 | 3.34 | 3.50 | 2.37 | 2.49 | 1.82 | 31.92 |
| Days ≥ 0.1" Precip | 4 | 4 | 5 | 6 | 7 | 7 | 7 | 6 | 7 | 5 | 5 | 5 | 68 |
| Total Snowfall (") | 12.7 | 8.6 | 8.5 | 3.0 | 0.0 | 0.0 | 0.0 | 0.0 | 0.0 | 0.3 | 3.4 | 12.3 | 48.8 |
| Days ≥ 1" Snow Depth | 28 | ***22*** | 13 | 2 | 0 | 0 | 0 | 0 | 0 | 0 | 3 | ***19*** | 87 |

## OSHKOSH *Winnebago County* ELEVATION 751 ft LAT/LONG 44° 2' N / 88° 32' W

| | JAN | FEB | MAR | APR | MAY | JUN | JUL | AUG | SEP | OCT | NOV | DEC | YEAR |
|---|---|---|---|---|---|---|---|---|---|---|---|---|---|
| Maximum Temp °F | 24.5 | 29.1 | 40.2 | 54.6 | 68.4 | 77.4 | 82.1 | 79.5 | 70.9 | 58.4 | 42.9 | 29.9 | 54.8 |
| Minimum Temp °F | 7.0 | 10.8 | 22.1 | 34.7 | 46.3 | 56.0 | 60.9 | 58.6 | 50.5 | 39.5 | 27.2 | 14.2 | 35.7 |
| Mean Temp °F | 15.7 | 19.9 | 31.2 | 44.7 | 57.4 | 66.7 | 71.6 | 69.1 | 60.7 | 49.0 | 35.1 | 22.1 | 45.3 |
| Days Max Temp ≥ 90 °F | 0 | 0 | 0 | 0 | 0 | 2 | 4 | 2 | 0 | 0 | 0 | 0 | 8 |
| Days Max Temp ≤ 32 °F | 22 | 17 | 7 | 0 | 0 | 0 | 0 | 0 | 0 | 0 | 4 | 17 | 67 |
| Days Min Temp ≤ 32 °F | 31 | 28 | 26 | 13 | 2 | 0 | 0 | 0 | 1 | 7 | 22 | 30 | 160 |
| Days Min Temp ≤ 0 °F | 11 | 7 | 1 | 0 | 0 | 0 | 0 | 0 | 0 | 0 | 0 | 5 | 24 |
| Heating Degree Days | 1522 | 1265 | 1041 | 608 | 265 | 63 | 11 | 28 | 172 | 495 | 891 | 1323 | 7684 |
| Cooling Degree Days | 0 | 0 | 0 | 7 | 36 | 122 | 221 | 158 | 50 | 1 | 0 | 0 | 595 |
| Total Precipitation (") | 1.28 | 1.05 | 2.14 | 3.04 | 2.97 | 3.43 | 3.50 | 3.65 | 3.73 | 2.38 | 2.45 | 1.71 | 31.33 |
| Days ≥ 0.1" Precip | 4 | 3 | 5 | 7 | 6 | 6 | 7 | 6 | 7 | 5 | 5 | 4 | 65 |
| Total Snowfall (") | 11.8 | 7.6 | 7.9 | 1.7 | 0.0 | 0.0 | 0.0 | 0.0 | 0.0 | 0.1 | 3.6 | 11.4 | 44.1 |
| Days ≥ 1" Snow Depth | na | na | *10* | 1 | 0 | 0 | 0 | 0 | 0 | 0 | *1* | na | na |

## OWEN *Clark County* ELEVATION 1240 ft LAT/LONG 44° 57' N / 90° 34' W

| | JAN | FEB | MAR | APR | MAY | JUN | JUL | AUG | SEP | OCT | NOV | DEC | YEAR |
|---|---|---|---|---|---|---|---|---|---|---|---|---|---|
| Maximum Temp °F | 18.7 | 24.2 | 36.3 | 52.3 | 66.4 | 74.7 | 79.5 | 77.1 | 67.8 | 55.4 | 38.7 | 24.6 | 51.3 |
| Minimum Temp °F | -1.6 | 2.9 | 17.0 | 32.0 | 42.7 | 51.7 | 56.3 | 53.9 | 44.9 | 34.2 | 22.3 | 7.0 | 30.3 |
| Mean Temp °F | 8.5 | 13.6 | 26.7 | 42.2 | 54.5 | 63.2 | 67.9 | 65.5 | 56.3 | 44.8 | 30.5 | 15.8 | 40.8 |
| Days Max Temp ≥ 90 °F | 0 | 0 | 0 | 0 | 0 | 1 | 2 | 1 | 0 | 0 | 0 | 0 | 4 |
| Days Max Temp ≤ 32 °F | 26 | 20 | 11 | 1 | 0 | 0 | 0 | 0 | 0 | 0 | 9 | 23 | 90 |
| Days Min Temp ≤ 32 °F | 31 | 28 | 28 | 16 | 4 | 0 | 0 | 0 | 2 | 14 | 26 | 31 | 180 |
| Days Min Temp ≤ 0 °F | 17 | 13 | 5 | 0 | 0 | 0 | 0 | 0 | 0 | 0 | 1 | 10 | 46 |
| Heating Degree Days | 1748 | 1448 | 1180 | 680 | 339 | 115 | 37 | 72 | 276 | 621 | 1029 | 1519 | 9064 |
| Cooling Degree Days | 0 | 0 | 0 | 5 | 22 | 73 | 141 | 104 | 25 | 1 | 0 | 0 | 371 |
| Total Precipitation (") | 0.96 | 0.81 | 1.81 | 2.60 | 3.68 | 4.53 | 4.30 | 4.31 | 4.28 | 2.52 | 2.06 | 1.36 | 33.22 |
| Days ≥ 0.1" Precip | 3 | 3 | 4 | 6 | 7 | 8 | 8 | 7 | 8 | 6 | 5 | 4 | 69 |
| Total Snowfall (") | 11.4 | 8.1 | 8.3 | 3.2 | 0.0 | 0.0 | 0.0 | 0.0 | 0.0 | 0.5 | 5.4 | *11.6* | 48.5 |
| Days ≥ 1" Snow Depth | 31 | 28 | 24 | 5 | 0 | 0 | 0 | 0 | 0 | 0 | 7 | 27 | 122 |

## PARK FALLS DNR HQ *Price County* ELEVATION 1525 ft LAT/LONG 45° 56' N / 90° 27' W

| | JAN | FEB | MAR | APR | MAY | JUN | JUL | AUG | SEP | OCT | NOV | DEC | YEAR |
|---|---|---|---|---|---|---|---|---|---|---|---|---|---|
| Maximum Temp °F | 19.8 | 25.7 | 37.1 | 52.0 | 65.7 | *73.2* | 77.3 | *74.7* | 65.4 | 53.5 | 36.8 | *23.5* | 50.4 |
| Minimum Temp °F | 1.0 | 5.4 | 17.3 | 30.9 | 42.8 | *51.9* | 56.7 | *54.4* | 45.7 | 35.4 | 22.6 | *8.2* | 31.0 |
| Mean Temp °F | 10.4 | 15.6 | 27.2 | 41.5 | 54.3 | *62.5* | 67.0 | *64.6* | 55.6 | 44.5 | 29.7 | *15.9* | 40.7 |
| Days Max Temp ≥ 90 °F | 0 | 0 | 0 | 0 | 0 | 0 | 1 | 0 | 0 | 0 | 0 | 0 | 1 |
| Days Max Temp ≤ 32 °F | 26 | 20 | 11 | 1 | 0 | 0 | 0 | 0 | 0 | 1 | 11 | 24 | 94 |
| Days Min Temp ≤ 32 °F | 30 | 28 | 28 | 18 | 4 | 0 | 0 | 0 | 2 | 13 | 26 | 30 | 179 |
| Days Min Temp ≤ 0 °F | 15 | 11 | 4 | 0 | 0 | 0 | 0 | 0 | 0 | 0 | 1 | 9 | 40 |
| Heating Degree Days | 1690 | 1390 | 1164 | 701 | 347 | *122* | 45 | *84* | 295 | 630 | 1051 | *1518* | 9037 |
| Cooling Degree Days | 0 | 0 | 0 | 3 | 19 | *52* | 113 | na | 16 | 0 | 0 | *0* | na |
| Total Precipitation (") | 1.18 | 0.77 | 1.85 | 2.44 | 3.35 | 4.41 | 3.65 | *4.47* | 4.04 | 2.93 | 1.98 | 1.35 | 32.42 |
| Days ≥ 0.1" Precip | 4 | 3 | 5 | 7 | 7 | 9 | 7 | 7 | 8 | 7 | 5 | 4 | 73 |
| Total Snowfall (") | na | na | na | na | 0.1 | 0.0 | 0.0 | 0.0 | 0.0 | *0.5* | na | na | na |
| Days ≥ 1" Snow Depth | 29 | 26 | 23 | *6* | 0 | 0 | 0 | 0 | 0 | *1* | na | na | na |

## PLATTEVILLE *Grant County* ELEVATION 991 ft LAT/LONG 42° 44' N / 90° 29' W

| | JAN | FEB | MAR | APR | MAY | JUN | JUL | AUG | SEP | OCT | NOV | DEC | YEAR |
|---|---|---|---|---|---|---|---|---|---|---|---|---|---|
| Maximum Temp °F | 25.5 | 31.0 | 44.0 | 59.1 | 70.9 | 79.5 | 83.5 | 81.2 | 73.0 | 61.1 | 44.3 | 31.1 | 57.0 |
| Minimum Temp °F | 8.2 | 13.3 | 25.1 | 36.5 | 47.4 | 56.6 | 61.3 | 59.0 | 50.8 | 39.8 | 27.8 | 15.5 | 36.8 |
| Mean Temp °F | 16.9 | 22.2 | 34.6 | 47.8 | 59.2 | 68.1 | 72.4 | 70.2 | 61.9 | 50.4 | 36.1 | 23.3 | 46.9 |
| Days Max Temp ≥ 90 °F | 0 | 0 | 0 | 0 | 0 | 2 | 5 | 3 | 1 | 0 | 0 | 0 | 11 |
| Days Max Temp ≤ 32 °F | 21 | 15 | 5 | 0 | 0 | 0 | 0 | 0 | 0 | 0 | 4 | 16 | 61 |
| Days Min Temp ≤ 32 °F | 31 | 27 | 24 | 11 | 2 | 0 | 0 | 0 | 1 | 8 | 21 | 29 | 154 |
| Days Min Temp ≤ 0 °F | 10 | 6 | 1 | 0 | 0 | 0 | 0 | 0 | 0 | 0 | 0 | 5 | 22 |
| Heating Degree Days | 1486 | 1204 | 936 | 516 | 218 | 43 | 9 | 21 | 151 | 452 | 861 | 1286 | 7183 |
| Cooling Degree Days | 0 | 0 | 0 | 10 | 43 | 148 | 256 | 198 | 71 | 4 | 0 | 0 | 730 |
| Total Precipitation (") | 1.07 | 1.14 | 2.23 | 3.37 | 3.64 | 4.85 | 4.45 | 4.36 | 4.02 | 2.60 | 2.38 | 1.65 | 35.76 |
| Days ≥ 0.1" Precip | 3 | 3 | 5 | 7 | 7 | 7 | 7 | 7 | 7 | 6 | 5 | 4 | 68 |
| Total Snowfall (") | 9.9 | 7.7 | 5.7 | 1.9 | 0.1 | 0.0 | 0.0 | 0.0 | 0.0 | 0.1 | 2.8 | 10.3 | 38.5 |
| Days ≥ 1" Snow Depth | 22 | 20 | 7 | 1 | 0 | 0 | 0 | 0 | 0 | 0 | 3 | 16 | 69 |

## PLYMOUTH *Sheboygan County* ELEVATION 840 ft LAT/LONG 43° 45 ' N / 87° 59 ' W

| | JAN | FEB | MAR | APR | MAY | JUN | JUL | AUG | SEP | OCT | NOV | DEC | YEAR |
|---|---|---|---|---|---|---|---|---|---|---|---|---|---|
| Maximum Temp °F | 25.8 | 30.1 | 41.0 | 55.1 | 68.0 | 77.3 | 81.9 | 79.8 | 71.3 | 58.8 | 43.7 | 31.2 | 55.3 |
| Minimum Temp °F | 9.4 | 13.1 | 23.6 | 34.5 | 44.2 | 54.1 | 59.9 | 58.3 | 50.5 | 39.7 | 28.5 | 16.5 | 36.0 |
| Mean Temp °F | 17.6 | 21.6 | 32.3 | 44.8 | 56.1 | 65.7 | 70.9 | 69.1 | 61.0 | 49.2 | 36.1 | 23.9 | 45.7 |
| Days Max Temp ≥ 90 °F | 0 | 0 | 0 | 0 | 0 | 2 | 4 | 2 | 0 | 0 | 0 | 0 | 8 |
| Days Max Temp ≤ 32 °F | 21 | 16 | 6 | 0 | 0 | 0 | 0 | 0 | 0 | 0 | 3 | 16 | 62 |
| Days Min Temp ≤ 32 °F | 30 | 27 | 26 | 13 | 3 | 0 | 0 | 0 | 0 | 7 | 20 | 29 | 155 |
| Days Min Temp ≤ 0 °F | 9 | 5 | 1 | 0 | 0 | 0 | 0 | 0 | 0 | 0 | 0 | 4 | 19 |
| Heating Degree Days | 1461 | 1218 | 1006 | 603 | 295 | 79 | 18 | 30 | 165 | 486 | 859 | 1269 | 7489 |
| Cooling Degree Days | 0 | 0 | 0 | 6 | 32 | 110 | 212 | 170 | 52 | 2 | 0 | 0 | 584 |
| Total Precipitation (") | 1.32 | 1.20 | 2.39 | 3.64 | 3.75 | 3.68 | 3.87 | 4.23 | 4.48 | 3.10 | 2.95 | 2.21 | 36.82 |
| Days ≥ 0.1" Precip | 3 | 3 | 5 | 7 | 7 | 7 | 7 | 7 | 7 | 7 | 6 | 5 | 71 |
| Total Snowfall (") | 14.6 | 12.9 | 10.3 | 3.4 | 0.2 | 0.0 | 0.0 | 0.0 | 0.0 | 0.2 | 5.1 | 15.3 | 62.0 |
| Days ≥ 1" Snow Depth | 27 | 25 | 15 | 2 | 0 | 0 | 0 | 0 | 0 | 0 | 4 | 18 | 91 |

## PORT WASHINGTON *Ozaukee County* ELEVATION 600 ft LAT/LONG 43° 23 ' N / 87° 52 ' W

| | JAN | FEB | MAR | APR | MAY | JUN | JUL | AUG | SEP | OCT | NOV | DEC | YEAR |
|---|---|---|---|---|---|---|---|---|---|---|---|---|---|
| Maximum Temp °F | 27.3 | 31.0 | 40.1 | 50.7 | 61.7 | 72.0 | 78.2 | 77.1 | 69.6 | 58.0 | 44.8 | 33.0 | 53.6 |
| Minimum Temp °F | 11.6 | 15.4 | 25.3 | 35.1 | 44.2 | 53.5 | 60.5 | 59.9 | 52.5 | 41.3 | 30.3 | 18.5 | 37.3 |
| Mean Temp °F | 19.5 | 23.2 | 32.8 | 43.0 | 53.0 | 62.9 | 69.4 | 68.5 | 61.1 | 49.7 | 37.6 | 25.7 | 45.5 |
| Days Max Temp ≥ 90 °F | 0 | 0 | 0 | 0 | 0 | 1 | 3 | 2 | 0 | 0 | 0 | 0 | 6 |
| Days Max Temp ≤ 32 °F | 19 | 14 | 6 | 0 | 0 | 0 | 0 | 0 | 0 | 0 | 2 | 12 | 53 |
| Days Min Temp ≤ 32 °F | 30 | 27 | 24 | 10 | 1 | 0 | 0 | 0 | 0 | 4 | 18 | 28 | 142 |
| Days Min Temp ≤ 0 °F | 7 | 3 | 0 | 0 | 0 | 0 | 0 | 0 | 0 | 0 | 0 | 3 | 13 |
| Heating Degree Days | 1406 | 1175 | 994 | 654 | 375 | 122 | 24 | 27 | 150 | 469 | 816 | 1211 | 7423 |
| Cooling Degree Days | 0 | 0 | 0 | 1 | 11 | 74 | 174 | 159 | 45 | 1 | 0 | 0 | 465 |
| Total Precipitation (") | 1.30 | 1.04 | 1.84 | 2.97 | 2.72 | 3.16 | 3.47 | 3.88 | 3.67 | 2.45 | 2.30 | 1.98 | 30.78 |
| Days ≥ 0.1" Precip | 3 | 3 | 5 | 6 | 6 | 6 | 6 | 6 | 6 | 5 | 5 | 4 | 61 |
| Total Snowfall (") | 10.7 | 9.1 | 6.1 | 1.6 | 0.0 | 0.0 | 0.0 | 0.0 | 0.0 | 0.2 | 1.5 | 8.2 | 37.4 |
| Days ≥ 1" Snow Depth | 22 | 18 | 8 | 1 | 0 | 0 | 0 | 0 | 0 | 0 | 1 | 13 | 63 |

## PORTAGE *Columbia County* ELEVATION 810 ft LAT/LONG 43° 32 ' N / 89° 28 ' W

| | JAN | FEB | MAR | APR | MAY | JUN | JUL | AUG | SEP | OCT | NOV | DEC | YEAR |
|---|---|---|---|---|---|---|---|---|---|---|---|---|---|
| Maximum Temp °F | 25.9 | 31.4 | 43.2 | 58.4 | 71.0 | 79.5 | 83.5 | 81.2 | 72.7 | 60.9 | 44.8 | 31.2 | 57.0 |
| Minimum Temp °F | 6.4 | 10.7 | 22.8 | 34.6 | 44.7 | 53.5 | 58.3 | 55.6 | 47.7 | 37.2 | 26.4 | 14.0 | 34.3 |
| Mean Temp °F | 16.2 | 21.1 | 33.1 | 46.5 | 57.9 | 66.5 | 70.9 | 68.4 | 60.3 | 48.9 | 35.5 | 22.6 | 45.7 |
| Days Max Temp ≥ 90 °F | 0 | 0 | 0 | 0 | 0 | 3 | 6 | 3 | 0 | 0 | 0 | 0 | 12 |
| Days Max Temp ≤ 32 °F | 21 | 15 | 5 | 0 | 0 | 0 | 0 | 0 | 0 | 0 | 4 | 16 | 61 |
| Days Min Temp ≤ 32 °F | 31 | 27 | 25 | 13 | 4 | 0 | 0 | 0 | 2 | 11 | 22 | 29 | 164 |
| Days Min Temp ≤ 0 °F | 11 | 7 | 1 | 0 | 0 | 0 | 0 | 0 | 0 | 0 | 0 | 6 | 25 |
| Heating Degree Days | 1509 | 1232 | 983 | 554 | 251 | 65 | 16 | 39 | 184 | 495 | 878 | 1307 | 7513 |
| Cooling Degree Days | 0 | 0 | 0 | 10 | 39 | 130 | 219 | 171 | 52 | 2 | 0 | 0 | 623 |
| Total Precipitation (") | 1.13 | 1.11 | 2.09 | 3.29 | 3.17 | 3.95 | 4.14 | 4.08 | 4.21 | 2.49 | 2.39 | 1.59 | 33.64 |
| Days ≥ 0.1" Precip | 4 | 3 | 5 | 7 | 6 | 7 | 6 | 7 | 7 | 5 | 5 | 4 | 66 |
| Total Snowfall (") | 9.4 | *8.7* | 5.3 | 1.5 | 0.0 | 0.0 | 0.0 | 0.0 | 0.0 | 0.2 | 2.8 | *8.3* | 36.2 |
| Days ≥ 1" Snow Depth | na | na | na | 1 | 0 | 0 | 0 | 0 | 0 | 0 | 3 | na | na |

## PRAIRIE DU CHIEN *Crawford County* ELEVATION 630 ft LAT/LONG 43° 3 ' N / 91° 9 ' W

| | JAN | FEB | MAR | APR | MAY | JUN | JUL | AUG | SEP | OCT | NOV | DEC | YEAR |
|---|---|---|---|---|---|---|---|---|---|---|---|---|---|
| Maximum Temp °F | 27.1 | 33.1 | 45.8 | 61.2 | 73.1 | 81.8 | 85.8 | 83.5 | 75.1 | 63.4 | 46.1 | 32.5 | 59.0 |
| Minimum Temp °F | 7.7 | 12.7 | 25.3 | 37.5 | 48.2 | 57.7 | 62.1 | 60.0 | 51.7 | 40.8 | 28.5 | 15.8 | 37.3 |
| Mean Temp °F | 17.4 | 22.9 | 35.6 | 49.3 | 60.7 | 69.8 | 74.0 | 71.8 | 63.4 | 52.1 | 37.3 | 24.2 | 48.2 |
| Days Max Temp ≥ 90 °F | 0 | 0 | 0 | 0 | 1 | 4 | 9 | 6 | 2 | 0 | 0 | 0 | 22 |
| Days Max Temp ≤ 32 °F | 19 | 13 | 4 | 0 | 0 | 0 | 0 | 0 | 0 | 0 | 3 | 14 | 53 |
| Days Min Temp ≤ 32 °F | 30 | 27 | 23 | 10 | 1 | 0 | 0 | 0 | 1 | 7 | 19 | 29 | 147 |
| Days Min Temp ≤ 0 °F | 10 | 7 | 1 | 0 | 0 | 0 | 0 | 0 | 0 | 0 | 0 | 5 | 23 |
| Heating Degree Days | 1470 | 1182 | 904 | 475 | 185 | 27 | 5 | 13 | 125 | 404 | 824 | 1257 | 6871 |
| Cooling Degree Days | 0 | 0 | 0 | 14 | 57 | 182 | 301 | 239 | 92 | 7 | 0 | 0 | 892 |
| Total Precipitation (") | 1.02 | 1.13 | 1.93 | 3.36 | 3.70 | 4.09 | 3.53 | 4.19 | 3.47 | 2.29 | 2.25 | 1.46 | 32.42 |
| Days ≥ 0.1" Precip | 3 | 3 | 4 | 7 | 8 | 7 | 6 | 6 | 6 | 5 | 5 | 4 | 64 |
| Total Snowfall (") | 11.0 | *7.9* | *4.7* | 1.6 | 0.0 | 0.0 | 0.0 | 0.0 | 0.0 | 0.0 | 3.1 | 9.9 | 38.2 |
| Days ≥ 1" Snow Depth | na | na | na | 1 | 0 | 0 | 0 | 0 | 0 | 0 | 1 | na | na |

## PRAIRIE DU SAC 2 N *Sauk County* ELEVATION 751 ft LAT/LONG 43° 17 ' N / 89° 44 ' W

| | JAN | FEB | MAR | APR | MAY | JUN | JUL | AUG | SEP | OCT | NOV | DEC | YEAR |
|---|---|---|---|---|---|---|---|---|---|---|---|---|---|
| Maximum Temp °F | 25.3 | 30.5 | 42.0 | 56.6 | 69.6 | 78.6 | 82.8 | 79.9 | 71.3 | 59.3 | 43.3 | 30.6 | 55.8 |
| Minimum Temp °F | 7.0 | 11.3 | 24.0 | 36.7 | 47.9 | 57.1 | 62.0 | 59.4 | 51.3 | 39.8 | 27.5 | 14.9 | 36.6 |
| Mean Temp °F | 16.1 | 20.9 | 33.0 | 46.7 | 58.8 | 67.9 | 72.4 | 69.7 | 61.3 | 49.5 | 35.5 | 22.8 | 46.2 |
| Days Max Temp ≥ 90 °F | 0 | 0 | 0 | 0 | 0 | 2 | 5 | 2 | 0 | 0 | 0 | 0 | 9 |
| Days Max Temp ≤ 32 °F | 21 | 15 | 6 | 0 | 0 | 0 | 0 | 0 | 0 | 0 | 4 | 17 | 63 |
| Days Min Temp ≤ 32 °F | 31 | 27 | 25 | 10 | 1 | 0 | 0 | 0 | 1 | 7 | 21 | 30 | 153 |
| Days Min Temp ≤ 0 °F | 11 | 7 | 1 | 0 | 0 | 0 | 0 | 0 | 0 | 0 | 0 | 5 | 24 |
| Heating Degree Days | 1509 | 1239 | 985 | 549 | 228 | 43 | 7 | 23 | 158 | 478 | 879 | 1304 | 7402 |
| Cooling Degree Days | 0 | 0 | 0 | 9 | 42 | 143 | 253 | 181 | 59 | 2 | 0 | 0 | 689 |
| Total Precipitation (") | 1.05 | 0.99 | 1.88 | 2.91 | 2.92 | 3.75 | 3.69 | 4.02 | 3.94 | 2.27 | 2.09 | 1.51 | 31.02 |
| Days ≥ 0.1" Precip | 3 | 3 | 5 | 6 | 6 | 7 | 6 | 7 | 7 | 5 | 5 | 4 | 64 |
| Total Snowfall (") | 8.5 | 6.3 | 4.1 | 1.1 | 0.0 | 0.0 | 0.0 | 0.0 | 0.0 | 0.0 | 2.0 | 7.3 | 29.3 |
| Days ≥ 1" Snow Depth | 20 | 18 | 8 | 1 | 0 | 0 | 0 | 0 | 0 | 0 | 2 | 12 | 61 |

## PRENTICE NO. 2 *Price County* ELEVATION 1572 ft LAT/LONG 45° 33 ' N / 90° 17 ' W

| | JAN | FEB | MAR | APR | MAY | JUN | JUL | AUG | SEP | OCT | NOV | DEC | YEAR |
|---|---|---|---|---|---|---|---|---|---|---|---|---|---|
| Maximum Temp °F | 19.6 | 25.5 | 37.3 | 52.5 | 66.1 | 73.4 | 77.8 | 75.5 | 66.3 | 54.6 | 37.5 | 24.2 | 50.9 |
| Minimum Temp °F | -3.2 | 0.7 | 15.0 | 29.4 | 40.0 | 48.5 | 53.1 | 51.4 | 43.0 | 33.2 | 20.8 | 5.4 | 28.1 |
| Mean Temp °F | 8.2 | 13.1 | 26.3 | 41.0 | 53.1 | 61.0 | 65.5 | 63.5 | 54.7 | 43.9 | 29.2 | 14.8 | 39.5 |
| Days Max Temp ≥ 90 °F | 0 | 0 | 0 | 0 | 0 | 0 | 1 | 0 | 0 | 0 | 0 | 0 | 1 |
| Days Max Temp ≤ 32 °F | 27 | 20 | 11 | 1 | 0 | 0 | 0 | 0 | 0 | 0 | 10 | 24 | 93 |
| Days Min Temp ≤ 32 °F | 31 | 28 | 28 | 19 | 8 | 1 | 0 | 1 | 5 | 16 | 26 | 31 | 194 |
| Days Min Temp ≤ 0 °F | 17 | 13 | 6 | 0 | 0 | 0 | 0 | 0 | 0 | 0 | 1 | 11 | 48 |
| Heating Degree Days | 1758 | 1460 | 1194 | 717 | 375 | 159 | 69 | 106 | 320 | 647 | 1069 | 1550 | 9424 |
| Cooling Degree Days | 0 | 0 | 0 | 3 | 13 | 44 | 90 | 67 | 14 | 0 | 0 | 0 | 231 |
| Total Precipitation (") | 0.88 | 0.59 | 1.48 | 2.38 | 3.47 | 4.12 | 3.77 | 3.94 | 4.51 | 2.71 | 1.95 | 1.14 | 30.94 |
| Days ≥ 0.1" Precip | 3 | 2 | 4 | 6 | 7 | 8 | 7 | 7 | 8 | 6 | 5 | 4 | 67 |
| Total Snowfall (") | 12.9 | 8.5 | 9.9 | 3.6 | 0.3 | 0.0 | 0.0 | 0.0 | 0.0 | 1.1 | 7.1 | 12.0 | 55.4 |
| Days ≥ 1" Snow Depth | na | na | *16* | 3 | 0 | 0 | 0 | 0 | 0 | 0 | 7 | 21 | na |

## RACINE *Racine County* ELEVATION 630 ft LAT/LONG 42° 43 ' N / 87° 47 ' W

| | JAN | FEB | MAR | APR | MAY | JUN | JUL | AUG | SEP | OCT | NOV | DEC | YEAR |
|---|---|---|---|---|---|---|---|---|---|---|---|---|---|
| Maximum Temp °F | 27.7 | 31.6 | 41.4 | 52.7 | 63.5 | 74.5 | 79.5 | 78.5 | 70.8 | 58.9 | 45.8 | 33.8 | 54.9 |
| Minimum Temp °F | 12.0 | 16.5 | 26.7 | 36.6 | 45.6 | 55.6 | 62.4 | 61.8 | 54.3 | 42.7 | 31.7 | 19.5 | 38.8 |
| Mean Temp °F | 19.9 | 24.1 | 34.1 | 44.7 | 54.5 | 65.1 | 71.0 | 70.2 | 62.5 | 50.8 | 38.8 | 26.7 | 46.9 |
| Days Max Temp ≥ 90 °F | 0 | 0 | 0 | 0 | 0 | 2 | 4 | 3 | 0 | 0 | 0 | 0 | 9 |
| Days Max Temp ≤ 32 °F | 19 | 14 | 6 | 0 | 0 | 0 | 0 | 0 | 0 | 0 | 2 | 12 | 53 |
| Days Min Temp ≤ 32 °F | 30 | 26 | 23 | 8 | 1 | 0 | 0 | 0 | 0 | 4 | 16 | 27 | 135 |
| Days Min Temp ≤ 0 °F | 8 | 3 | 0 | 0 | 0 | 0 | 0 | 0 | 0 | 0 | 0 | 3 | 14 |
| Heating Degree Days | 1391 | 1150 | 953 | 605 | 334 | 87 | 14 | 18 | 126 | 438 | 781 | 1183 | 7080 |
| Cooling Degree Days | 0 | 0 | 0 | 2 | 14 | 74 | 186 | 176 | 54 | 1 | 0 | 0 | 507 |
| Total Precipitation (") | 1.56 | 1.35 | 2.47 | 3.77 | 2.89 | 3.84 | 3.60 | 4.00 | 3.90 | 2.53 | 2.93 | 2.21 | 35.05 |
| Days ≥ 0.1" Precip | 4 | 4 | 6 | 7 | 6 | 7 | 7 | 6 | 7 | 6 | 6 | 5 | 71 |
| Total Snowfall (") | 13.9 | 10.3 | 6.4 | 1.3 | 0.1 | 0.0 | 0.0 | 0.0 | 0.0 | 0.3 | 1.7 | 8.8 | 42.8 |
| Days ≥ 1" Snow Depth | *19* | *15* | *5* | 0 | 0 | 0 | 0 | 0 | 0 | 0 | *1* | na | na |

## RAINBOW RSVR LAKE *Oneida County* ELEVATION 1600 ft LAT/LONG 45° 50 ' N / 89° 32 ' W

| | JAN | FEB | MAR | APR | MAY | JUN | JUL | AUG | SEP | OCT | NOV | DEC | YEAR |
|---|---|---|---|---|---|---|---|---|---|---|---|---|---|
| Maximum Temp °F | 20.2 | 25.6 | 36.9 | 51.5 | 65.5 | 73.0 | 77.5 | 75.1 | 65.5 | 53.7 | 37.2 | 24.7 | 50.5 |
| Minimum Temp °F | -2.1 | 0.2 | 12.0 | 26.5 | 38.7 | 48.3 | 53.4 | 51.6 | 43.4 | 33.1 | 20.7 | 6.3 | 27.7 |
| Mean Temp °F | 9.1 | 12.9 | 24.5 | 39.0 | 52.1 | 60.7 | 65.5 | 63.4 | 54.5 | 43.4 | 29.0 | 15.5 | 39.1 |
| Days Max Temp ≥ 90 °F | 0 | 0 | 0 | 0 | 0 | 0 | 1 | 0 | 0 | 0 | 0 | 0 | 1 |
| Days Max Temp ≤ 32 °F | 27 | 20 | 11 | 1 | 0 | 0 | 0 | 0 | 0 | 0 | 11 | 24 | 94 |
| Days Min Temp ≤ 32 °F | 31 | 28 | 30 | 23 | 10 | 1 | 0 | 0 | 3 | 17 | 27 | 31 | 201 |
| Days Min Temp ≤ 0 °F | 17 | 15 | 7 | 0 | 0 | 0 | 0 | 0 | 0 | 0 | 1 | 11 | 51 |
| Heating Degree Days | 1732 | 1466 | 1250 | 775 | 404 | 164 | 65 | 104 | 324 | 663 | 1073 | 1529 | 9549 |
| Cooling Degree Days | 0 | 0 | 0 | 2 | 9 | 35 | 76 | 60 | 11 | 0 | 0 | 0 | 193 |
| Total Precipitation (") | 1.08 | 0.83 | 1.73 | 2.38 | 3.31 | 4.11 | 3.72 | 4.27 | 4.44 | 2.85 | 2.24 | 1.38 | 32.34 |
| Days ≥ 0.1" Precip | 3 | 3 | 5 | 6 | 7 | 8 | 8 | 8 | 9 | 7 | 6 | 4 | 74 |
| Total Snowfall (") | *13.1* | *9.2* | 8.7 | 3.8 | 0.5 | 0.0 | 0.0 | 0.0 | 0.0 | 1.1 | *7.3* | 13.7 | 57.4 |
| Days ≥ 1" Snow Depth | 31 | 28 | *29* | 11 | 0 | 0 | 0 | 0 | 0 | 1 | *12* | *30* | 142 |

## REST LAKE *Vilas County* ELEVATION 1601 ft LAT/LONG 46° 7 ' N / 89° 53 ' W

| | JAN | FEB | MAR | APR | MAY | JUN | JUL | AUG | SEP | OCT | NOV | DEC | YEAR |
|---|---|---|---|---|---|---|---|---|---|---|---|---|---|
| Maximum Temp °F | 19.6 | 25.1 | 36.9 | 52.3 | 66.6 | 73.0 | 77.1 | 74.5 | 65.0 | 53.6 | 36.5 | 24.0 | 50.4 |
| Minimum Temp °F | -0.4 | 2.7 | 14.9 | 28.8 | 41.1 | 50.3 | 55.5 | 53.5 | 45.3 | 34.8 | 22.2 | 7.9 | 29.7 |
| Mean Temp °F | 9.6 | 13.9 | 25.9 | 40.6 | 53.9 | 61.7 | 66.3 | 64.0 | 55.2 | 44.2 | 29.4 | 15.9 | 40.1 |
| Days Max Temp ≥ 90 °F | 0 | 0 | 0 | 0 | 0 | 0 | 1 | 0 | 0 | 0 | 0 | 0 | 1 |
| Days Max Temp ≤ 32 °F | 27 | 21 | 11 | 1 | 0 | 0 | 0 | 0 | 0 | 0 | 12 | 25 | 97 |
| Days Min Temp ≤ 32 °F | 31 | 28 | 28 | 20 | 7 | 0 | 0 | 0 | 2 | 13 | 26 | 31 | 186 |
| Days Min Temp ≤ 0 °F | 15 | 12 | 6 | 0 | 0 | 0 | 0 | 0 | 0 | 0 | 1 | 9 | 43 |
| Heating Degree Days | 1712 | 1438 | 1204 | 727 | 356 | 139 | 55 | 89 | 303 | 639 | 1061 | 1515 | 9238 |
| Cooling Degree Days | 0 | 0 | 0 | 2 | 14 | 44 | 98 | 62 | 12 | 0 | 0 | 0 | 232 |
| Total Precipitation (") | 1.20 | 0.78 | 1.61 | 2.29 | 3.73 | 4.24 | 4.01 | 4.37 | 4.23 | 3.00 | 2.08 | 1.48 | 33.02 |
| Days ≥ 0.1" Precip | 4 | 2 | 4 | 6 | 7 | 8 | 8 | 7 | 9 | 7 | 5 | 5 | 72 |
| Total Snowfall (") | 19.3 | 12.3 | 12.0 | 5.3 | 0.4 | 0.0 | 0.0 | 0.0 | 0.0 | 1.5 | 12.8 | 20.9 | 84.5 |
| Days ≥ 1" Snow Depth | 30 | 27 | 26 | 11 | 0 | 0 | 0 | 0 | 0 | 1 | 13 | 28 | 136 |

## RHINELANDER *Oneida County* ELEVATION 1552 ft LAT/LONG 45° 38 ' N / 89° 24 ' W

| | JAN | FEB | MAR | APR | MAY | JUN | JUL | AUG | SEP | OCT | NOV | DEC | YEAR |
|---|---|---|---|---|---|---|---|---|---|---|---|---|---|
| Maximum Temp °F | 21.2 | 26.6 | 38.1 | 52.9 | 67.0 | 74.6 | 79.3 | 76.5 | 66.8 | 54.6 | 38.0 | 25.2 | 51.7 |
| Minimum Temp °F | -1.1 | 2.6 | 15.4 | 29.7 | 41.4 | 50.5 | 55.7 | 53.7 | 44.9 | 34.5 | 21.7 | 6.9 | 29.7 |
| Mean Temp °F | 10.1 | 14.6 | 26.8 | 41.3 | 54.2 | 62.6 | 67.5 | 65.1 | 55.9 | 44.6 | 29.9 | 16.1 | 40.7 |
| Days Max Temp ≥ 90 °F | 0 | 0 | 0 | 0 | 0 | 1 | 2 | 1 | 0 | 0 | 0 | 0 | 4 |
| Days Max Temp ≤ 32 °F | 26 | 20 | 10 | 1 | 0 | 0 | 0 | 0 | 0 | 0 | 10 | 24 | 91 |
| Days Min Temp ≤ 32 °F | 31 | 28 | 29 | 20 | 6 | 0 | 0 | 0 | 3 | 15 | 27 | 31 | 190 |
| Days Min Temp ≤ 0 °F | 16 | 13 | 5 | 0 | 0 | 0 | 0 | 0 | 0 | 0 | 1 | 10 | 45 |
| Heating Degree Days | 1700 | 1417 | 1179 | 705 | 347 | 124 | 40 | 76 | 286 | 627 | 1046 | 1510 | 9057 |
| Cooling Degree Days | 0 | 0 | 0 | 3 | 17 | 55 | 118 | 87 | 16 | 0 | 0 | 0 | 296 |
| Total Precipitation (") | 1.14 | 0.77 | 1.52 | 2.49 | 3.31 | 4.00 | 3.49 | 4.21 | 4.49 | 2.66 | 2.03 | 1.42 | 31.53 |
| Days ≥ 0.1" Precip | 4 | 2 | 4 | 6 | 7 | 8 | 7 | 7 | 8 | 6 | 5 | 5 | 69 |
| Total Snowfall (") | *11.8* | *6.0* | *6.7* | *2.1* | 0.2 | 0.0 | 0.0 | 0.0 | 0.0 | *0.1* | na | na | na |
| Days ≥ 1" Snow Depth | 31 | 27 | 19 | 4 | 0 | 0 | 0 | 0 | 0 | 0 | *9* | 26 | 116 |

## RICE LAKE *Barron County* ELEVATION 1138 ft LAT/LONG 45° 30 ' N / 91° 44 ' W

| | JAN | FEB | MAR | APR | MAY | JUN | JUL | AUG | SEP | OCT | NOV | DEC | YEAR |
|---|---|---|---|---|---|---|---|---|---|---|---|---|---|
| Maximum Temp °F | 20.2 | 26.6 | 39.1 | 56.0 | 69.1 | 76.9 | 81.2 | 78.7 | 68.9 | 57.3 | 38.8 | 24.9 | 53.1 |
| Minimum Temp °F | -1.7 | 4.2 | 18.1 | 31.9 | 43.5 | 52.3 | 57.5 | 55.1 | 46.3 | 35.4 | 22.2 | 7.3 | 31.0 |
| Mean Temp °F | 9.3 | 15.4 | 28.6 | 44.0 | 56.3 | 64.6 | 69.4 | 66.9 | 57.6 | 46.4 | 30.5 | 16.1 | 42.1 |
| Days Max Temp ≥ 90 °F | 0 | 0 | 0 | 0 | 0 | 1 | 3 | 2 | 0 | 0 | 0 | 0 | 6 |
| Days Max Temp ≤ 32 °F | 26 | 18 | 8 | 0 | 0 | 0 | 0 | 0 | 0 | 0 | 8 | 23 | 83 |
| Days Min Temp ≤ 32 °F | 31 | 28 | 27 | 17 | 4 | 0 | 0 | 0 | 2 | 12 | 26 | 31 | 178 |
| Days Min Temp ≤ 0 °F | 16 | 11 | 4 | 0 | 0 | 0 | 0 | 0 | 0 | 0 | 1 | 10 | 42 |
| Heating Degree Days | 1725 | 1396 | 1120 | 628 | 290 | 86 | 24 | 52 | 243 | 571 | 1028 | 1512 | 8675 |
| Cooling Degree Days | 0 | 0 | 0 | 5 | 31 | 93 | 177 | 133 | 33 | 1 | 0 | 0 | 473 |
| Total Precipitation (") | 1.04 | 0.79 | 1.78 | 2.74 | 3.12 | 4.35 | 3.95 | 4.21 | 4.44 | 2.65 | 1.92 | 1.24 | 32.23 |
| Days ≥ 0.1" Precip | 3 | 2 | 4 | 6 | 7 | 8 | 6 | 7 | 8 | 6 | 4 | 3 | 64 |
| Total Snowfall (") | 12.3 | 7.6 | 8.1 | 2.6 | 0.0 | 0.0 | 0.0 | 0.0 | 0.0 | 0.7 | 7.3 | 11.1 | 49.7 |
| Days ≥ 1" Snow Depth | 30 | 27 | 19 | 2 | 0 | 0 | 0 | 0 | 0 | 0 | *8* | *26* | 112 |

## RICHLAND CENTER *Richland County* ELEVATION 732 ft LAT/LONG 43° 20 ' N / 90° 23 ' W

| | JAN | FEB | MAR | APR | MAY | JUN | JUL | AUG | SEP | OCT | NOV | DEC | YEAR |
|---|---|---|---|---|---|---|---|---|---|---|---|---|---|
| Maximum Temp °F | 26.4 | 31.7 | 44.0 | 58.6 | 71.5 | 80.0 | 84.7 | 81.9 | 73.0 | 61.3 | 44.5 | 31.4 | 57.4 |
| Minimum Temp °F | 5.1 | 9.6 | 23.1 | 34.5 | 44.5 | 53.4 | 58.5 | 56.0 | 48.2 | 36.6 | 25.1 | 12.8 | 34.0 |
| Mean Temp °F | 15.7 | 20.7 | 33.5 | 46.6 | 58.0 | 66.8 | 71.6 | 69.0 | 60.6 | 49.0 | 34.8 | 22.1 | 45.7 |
| Days Max Temp ≥ 90 °F | 0 | 0 | 0 | 0 | 1 | 3 | 7 | 4 | 1 | 0 | 0 | 0 | 16 |
| Days Max Temp ≤ 32 °F | 21 | 14 | 5 | 0 | 0 | 0 | 0 | 0 | 0 | 0 | 4 | 16 | 60 |
| Days Min Temp ≤ 32 °F | 31 | 27 | 25 | 13 | 3 | 0 | 0 | 0 | 1 | 12 | 23 | 30 | 165 |
| Days Min Temp ≤ 0 °F | 12 | 8 | 1 | 0 | 0 | 0 | 0 | 0 | 0 | 0 | 1 | 6 | 28 |
| Heating Degree Days | 1523 | 1245 | 969 | 552 | 244 | 59 | 10 | 31 | 174 | 496 | 899 | 1323 | 7525 |
| Cooling Degree Days | 0 | 0 | 0 | 6 | 23 | 109 | 206 | 150 | 47 | 2 | 0 | 0 | 543 |
| Total Precipitation (") | 1.15 | 1.13 | 2.15 | 3.65 | 3.64 | 4.21 | 4.31 | 4.03 | 4.26 | 2.25 | 2.48 | 1.59 | 34.85 |
| Days ≥ 0.1" Precip | 4 | 3 | 5 | 7 | 7 | 7 | 6 | 6 | 7 | 5 | 5 | 4 | 66 |
| Total Snowfall (") | 10.2 | 8.1 | 5.8 | 2.2 | 0.0 | 0.0 | 0.0 | 0.0 | 0.0 | 0.1 | 4.4 | 10.3 | 41.1 |
| Days ≥ 1" Snow Depth | 25 | 22 | 11 | 1 | 0 | 0 | 0 | 0 | 0 | 0 | 4 | 18 | 81 |

## RIDGELAND 1 NNE *Barron County* ELEVATION 951 ft LAT/LONG 45° 12 ' N / 91° 54 ' W

| | JAN | FEB | MAR | APR | MAY | JUN | JUL | AUG | SEP | OCT | NOV | DEC | YEAR |
|---|---|---|---|---|---|---|---|---|---|---|---|---|---|
| Maximum Temp °F | 21.3 | 27.1 | 39.8 | 56.8 | 69.9 | 77.4 | 81.8 | 79.2 | 69.6 | 58.0 | 40.2 | 26.2 | 53.9 |
| Minimum Temp °F | -1.3 | 3.6 | 17.8 | 31.4 | 42.9 | 52.1 | 57.0 | 54.6 | 46.1 | 34.9 | 22.5 | 7.8 | 30.8 |
| Mean Temp °F | 10.0 | 15.4 | 28.8 | 44.1 | 56.5 | 64.7 | 69.4 | 66.9 | 57.9 | 46.5 | 31.3 | 17.0 | 42.4 |
| Days Max Temp ≥ 90 °F | 0 | 0 | 0 | 0 | 0 | 1 | 3 | 2 | 0 | 0 | 0 | 0 | 6 |
| Days Max Temp ≤ 32 °F | 25 | 18 | 8 | 0 | 0 | 0 | 0 | 0 | 0 | 0 | 7 | 21 | 79 |
| Days Min Temp ≤ 32 °F | 31 | 28 | 28 | 17 | 5 | 0 | 0 | 0 | 2 | 13 | 25 | 31 | 180 |
| Days Min Temp ≤ 0 °F | 16 | 12 | 4 | 0 | 0 | 0 | 0 | 0 | 0 | 0 | 1 | 10 | 43 |
| Heating Degree Days | 1702 | 1395 | 1116 | 626 | 283 | 84 | 23 | 53 | 238 | 570 | 1003 | 1482 | 8575 |
| Cooling Degree Days | 0 | 0 | 0 | 6 | 27 | 88 | 167 | 126 | 31 | 1 | 0 | 0 | 446 |
| Total Precipitation (") | 1.10 | 0.80 | 1.89 | 2.77 | 3.37 | 4.67 | 3.94 | 4.76 | 4.02 | 2.61 | 1.76 | 1.23 | 32.92 |
| Days ≥ 0.1" Precip | 3 | 2 | 4 | 6 | 7 | 8 | 7 | 7 | 7 | 6 | 4 | 4 | 65 |
| Total Snowfall (") | 11.0 | 6.8 | 8.2 | 1.9 | 0.0 | 0.0 | 0.0 | 0.0 | 0.0 | 0.3 | 5.6 | 9.2 | 43.0 |
| Days ≥ 1" Snow Depth | 29 | 27 | 21 | 3 | 0 | 0 | 0 | 0 | 0 | 0 | 7 | 23 | 110 |

## RIVER FALLS *Pierce County* ELEVATION 902 ft LAT/LONG 44° 52 ' N / 92° 38 ' W

| | JAN | FEB | MAR | APR | MAY | JUN | JUL | AUG | SEP | OCT | NOV | DEC | YEAR |
|---|---|---|---|---|---|---|---|---|---|---|---|---|---|
| Maximum Temp °F | 22.0 | 27.8 | 40.7 | 57.9 | 70.7 | 79.0 | 83.0 | 80.3 | 70.5 | 58.9 | 40.8 | 27.1 | 54.9 |
| Minimum Temp °F | 2.7 | 8.0 | 21.6 | 34.9 | 46.5 | 55.8 | 60.6 | 58.4 | 49.6 | 38.5 | 24.8 | 11.0 | 34.4 |
| Mean Temp °F | 12.3 | 17.9 | 31.2 | 46.4 | 58.6 | 67.4 | 71.8 | 69.3 | 60.1 | 48.7 | 32.8 | 19.1 | 44.6 |
| Days Max Temp ≥ 90 °F | 0 | 0 | 0 | 0 | 0 | 2 | 4 | 3 | 0 | 0 | 0 | 0 | 9 |
| Days Max Temp ≤ 32 °F | 23 | 17 | 7 | 0 | 0 | 0 | 0 | 0 | 0 | 0 | 6 | 20 | 73 |
| Days Min Temp ≤ 32 °F | 31 | 27 | 25 | 13 | 2 | 0 | 0 | 0 | 1 | 9 | 23 | 30 | 161 |
| Days Min Temp ≤ 0 °F | 14 | 10 | 2 | 0 | 0 | 0 | 0 | 0 | 0 | 0 | 1 | 8 | 35 |
| Heating Degree Days | 1629 | 1323 | 1040 | 555 | 235 | 49 | 10 | 28 | 188 | 501 | 960 | 1419 | 7937 |
| Cooling Degree Days | 0 | 0 | 0 | 7 | 44 | 139 | 232 | 178 | 53 | 2 | 0 | 0 | 655 |
| Total Precipitation (") | 0.82 | 0.69 | 1.51 | 2.60 | 3.51 | 4.58 | 4.09 | 4.21 | 3.67 | 2.43 | 1.53 | 0.98 | 30.62 |
| Days ≥ 0.1" Precip | 3 | 3 | 4 | 6 | 7 | 8 | 7 | 6 | 7 | 5 | 4 | 3 | 63 |
| Total Snowfall (") | 11.3 | 7.6 | 9.4 | 2.7 | 0.1 | 0.0 | 0.0 | 0.0 | 0.0 | 0.5 | 7.3 | 9.2 | 48.1 |
| Days ≥ 1" Snow Depth | 29 | 26 | 18 | 2 | 0 | 0 | 0 | 0 | 0 | 0 | 7 | 23 | 105 |

## ROSHOLT 9 NNE *Marathon County* ELEVATION 1161 ft LAT/LONG 44° 46 ' N / 89° 15 ' W

| | JAN | FEB | MAR | APR | MAY | JUN | JUL | AUG | SEP | OCT | NOV | DEC | YEAR |
|---|---|---|---|---|---|---|---|---|---|---|---|---|---|
| Maximum Temp °F | 23.3 | 28.4 | 39.9 | 55.6 | 69.3 | 77.1 | 81.2 | ***79.5*** | ***69.9*** | ***58.4*** | 41.0 | 27.9 | 54.3 |
| Minimum Temp °F | 2.4 | 7.1 | 18.2 | 31.5 | 42.5 | 51.3 | 56.2 | 54.2 | 45.6 | 35.5 | 23.4 | 10.2 | 31.5 |
| Mean Temp °F | 12.8 | 17.8 | 29.1 | 43.6 | 56.0 | 64.2 | 68.7 | ***66.9*** | ***57.7*** | ***47.0*** | 32.2 | 19.1 | 42.9 |
| Days Max Temp ≥ 90 °F | 0 | 0 | 0 | 0 | 0 | 2 | 3 | 2 | 0 | 0 | 0 | 0 | 7 |
| Days Max Temp ≤ 32 °F | 24 | 17 | 7 | 0 | 0 | 0 | 0 | 0 | 0 | 0 | 6 | 20 | 74 |
| Days Min Temp ≤ 32 °F | 31 | 28 | 27 | 17 | 5 | 0 | 0 | 0 | 2 | 12 | 26 | 30 | 178 |
| Days Min Temp ≤ 0 °F | 14 | 10 | 4 | 0 | 0 | 0 | 0 | 0 | 0 | 0 | 1 | 7 | 36 |
| Heating Degree Days | 1614 | 1326 | 1107 | 641 | 299 | 94 | 29 | ***52*** | ***240*** | ***554*** | 977 | 1407 | 8340 |
| Cooling Degree Days | 0 | 0 | 0 | 3 | 19 | 66 | 123 | 102 | 24 | 1 | 0 | 0 | 338 |
| Total Precipitation (") | 1.10 | 0.91 | 1.72 | 3.18 | 3.87 | 3.76 | 3.79 | 3.96 | 4.19 | 2.60 | 2.42 | 1.56 | 33.06 |
| Days ≥ 0.1" Precip | 4 | 3 | 5 | 7 | 7 | 7 | 8 | 6 | 8 | 6 | 5 | 5 | 71 |
| Total Snowfall (") | 10.4 | 8.5 | 9.0 | 2.8 | 0.0 | 0.0 | 0.0 | 0.0 | 0.1 | 0.3 | 6.6 | 12.0 | 49.7 |
| Days ≥ 1" Snow Depth | 30 | 27 | 20 | 3 | 0 | 0 | 0 | 0 | 0 | 0 | 6 | 25 | 111 |

## SHAWANO 2 SSW *Shawano County* ELEVATION 801 ft LAT/LONG 44° 47 ' N / 88° 37 ' W

| | JAN | FEB | MAR | APR | MAY | JUN | JUL | AUG | SEP | OCT | NOV | DEC | YEAR |
|---|---|---|---|---|---|---|---|---|---|---|---|---|---|
| Maximum Temp °F | 23.7 | 28.8 | 40.3 | 56.1 | 69.3 | 77.8 | 82.1 | 79.5 | 70.4 | 58.4 | 41.8 | 28.7 | 54.7 |
| Minimum Temp °F | 3.8 | 7.4 | 20.2 | 32.7 | 42.9 | 52.1 | 57.2 | 54.9 | 46.5 | 36.3 | 24.9 | 11.8 | 32.6 |
| Mean Temp °F | 13.7 | 18.1 | 30.3 | 44.5 | 56.1 | 65.0 | 69.7 | 67.2 | 58.4 | 47.4 | 33.4 | 20.3 | 43.7 |
| Days Max Temp ≥ 90 °F | 0 | 0 | 0 | 0 | 0 | 2 | 4 | 2 | 0 | 0 | 0 | 0 | 8 |
| Days Max Temp ≤ 32 °F | 23 | 17 | 7 | 0 | 0 | 0 | 0 | 0 | 0 | 0 | 5 | 19 | 71 |
| Days Min Temp ≤ 32 °F | 31 | 28 | 27 | 16 | 4 | 0 | 0 | 0 | 2 | 12 | 24 | 30 | 174 |
| Days Min Temp ≤ 0 °F | 13 | 10 | 2 | 0 | 0 | 0 | 0 | 0 | 0 | 0 | 0 | 7 | 32 |
| Heating Degree Days | 1586 | 1318 | 1069 | 613 | 296 | 84 | 22 | 49 | 222 | 544 | 943 | 1380 | 8126 |
| Cooling Degree Days | 0 | 0 | 0 | 6 | 29 | 95 | 181 | 135 | 34 | 1 | 0 | 0 | 481 |
| Total Precipitation (") | 1.29 | 0.95 | 1.84 | 2.97 | 3.73 | 3.53 | 3.68 | 3.70 | 3.94 | 2.53 | 2.46 | 1.51 | 32.13 |
| Days ≥ 0.1" Precip | 4 | 3 | 5 | 7 | 7 | 7 | 7 | 7 | 7 | 5 | 5 | 5 | 69 |
| Total Snowfall (") | 13.0 | 8.0 | 7.8 | 2.4 | 0.2 | 0.0 | 0.0 | 0.0 | 0.0 | 0.2 | 4.7 | 12.7 | 49.0 |
| Days ≥ 1" Snow Depth | 31 | 26 | ***15*** | 3 | 0 | 0 | 0 | 0 | 0 | 0 | 5 | 21 | 101 |

## SHEBOYGAN *Sheboygan County* ELEVATION 679 ft LAT/LONG 43° 45 ' N / 87° 43 ' W

| | JAN | FEB | MAR | APR | MAY | JUN | JUL | AUG | SEP | OCT | NOV | DEC | YEAR |
|---|---|---|---|---|---|---|---|---|---|---|---|---|---|
| Maximum Temp °F | 28.0 | 32.0 | 41.3 | 52.2 | 63.9 | 74.4 | 80.4 | 79.2 | 71.3 | 59.0 | 45.0 | 33.1 | 55.0 |
| Minimum Temp °F | 13.0 | 16.9 | 26.3 | 35.9 | 44.8 | 54.2 | 61.3 | 60.9 | 53.5 | 42.8 | 31.8 | 19.9 | 38.4 |
| Mean Temp °F | 20.5 | 24.5 | 33.8 | 44.1 | 54.4 | 64.3 | 70.9 | 70.0 | 62.5 | 50.9 | 38.4 | 26.5 | 46.7 |
| Days Max Temp ≥ 90 °F | 0 | 0 | 0 | 0 | 0 | 2 | 4 | 3 | 1 | 0 | 0 | 0 | 10 |
| Days Max Temp ≤ 32 °F | 19 | 14 | 4 | 0 | 0 | 0 | 0 | 0 | 0 | 0 | 2 | 12 | 51 |
| Days Min Temp ≤ 32 °F | 29 | 26 | 23 | 8 | 0 | 0 | 0 | 0 | 0 | 3 | 15 | 27 | 131 |
| Days Min Temp ≤ 0 °F | 6 | 3 | 0 | 0 | 0 | 0 | 0 | 0 | 0 | 0 | 0 | 2 | 11 |
| Heating Degree Days | 1373 | 1138 | 960 | 622 | 334 | 91 | 13 | 14 | 122 | 432 | 791 | 1186 | 7076 |
| Cooling Degree Days | 0 | 0 | 0 | 2 | 16 | 85 | 206 | 193 | 60 | 2 | 0 | 0 | 564 |
| Total Precipitation (") | 1.61 | 1.30 | 2.20 | 2.98 | 2.72 | 3.27 | 3.02 | 3.46 | 3.63 | 2.61 | 2.46 | 2.06 | 31.32 |
| Days ≥ 0.1" Precip | 5 | 4 | 6 | 6 | 6 | 7 | 6 | 6 | 7 | 6 | 6 | 5 | 70 |
| Total Snowfall (") | 14.3 | 10.8 | 7.5 | 1.7 | 0.0 | 0.0 | 0.0 | 0.0 | 0.0 | 0.2 | 2.4 | 10.3 | 47.2 |
| Days ≥ 1" Snow Depth | 23 | 21 | 10 | 1 | 0 | 0 | 0 | 0 | 0 | 0 | 1 | 15 | 71 |

## SOLON SPRINGS *Douglas County* ELEVATION 1079 ft LAT/LONG 46° 21 ' N / 91° 49 ' W

| | JAN | FEB | MAR | APR | MAY | JUN | JUL | AUG | SEP | OCT | NOV | DEC | YEAR |
|---|---|---|---|---|---|---|---|---|---|---|---|---|---|
| Maximum Temp °F | 20.0 | 26.8 | 38.8 | 54.8 | 69.1 | 77.3 | 82.2 | 79.6 | 68.8 | 56.4 | 37.8 | 24.2 | 53.0 |
| Minimum Temp °F | -4.0 | 1.1 | 14.9 | 28.0 | 39.0 | 48.1 | 54.2 | 52.0 | 43.3 | 32.7 | 19.9 | 4.1 | 27.8 |
| Mean Temp °F | 8.0 | 14.0 | 26.9 | 41.5 | 54.1 | 62.7 | 68.2 | 65.8 | 56.0 | 44.6 | 28.9 | 13.9 | 40.4 |
| Days Max Temp ≥ 90 °F | 0 | 0 | 0 | 0 | 1 | 2 | 4 | 3 | 0 | 0 | 0 | 0 | 10 |
| Days Max Temp ≤ 32 °F | 27 | 19 | 9 | 1 | 0 | 0 | 0 | 0 | 0 | 0 | 10 | 24 | 90 |
| Days Min Temp ≤ 32 °F | 31 | 28 | 29 | 21 | 8 | 1 | 0 | 0 | 4 | 16 | 27 | 31 | 196 |
| Days Min Temp ≤ 0 °F | 18 | 13 | 6 | 0 | 0 | 0 | 0 | 0 | 0 | 0 | 2 | 12 | 51 |
| Heating Degree Days | 1763 | 1437 | 1175 | 701 | 351 | 118 | 33 | 69 | 283 | 629 | 1078 | 1579 | 9216 |
| Cooling Degree Days | 0 | 0 | 0 | 2 | 18 | 56 | 135 | 95 | 14 | 0 | 0 | 0 | 320 |
| Total Precipitation (") | 1.11 | 0.75 | 1.72 | 2.31 | 3.20 | 3.97 | 4.53 | 4.26 | 3.70 | 2.68 | 2.00 | 1.11 | 31.34 |
| Days ≥ 0.1" Precip | 3 | 2 | 4 | ***6*** | ***7*** | na | 7 | ***7*** | 7 | ***6*** | 4 | 3 | na |
| Total Snowfall (") | 13.3 | 8.1 | 9.4 | 4.0 | 0.2 | 0.0 | 0.0 | 0.0 | 0.0 | 0.6 | 8.8 | 12.1 | 56.5 |
| Days ≥ 1" Snow Depth | 30 | 28 | 26 | 7 | 0 | 0 | 0 | 0 | 0 | 0 | 11 | 29 | 131 |

## SPARTA *Monroe County* ELEVATION 781 ft LAT/LONG 43° 57 ' N / 90° 48 ' W

| | JAN | FEB | MAR | APR | MAY | JUN | JUL | AUG | SEP | OCT | NOV | DEC | YEAR |
|---|---|---|---|---|---|---|---|---|---|---|---|---|---|
| Maximum Temp °F | 23.1 | 29.8 | 42.3 | 57.7 | 70.6 | 79.1 | 83.5 | 81.2 | ***71.8*** | ***59.8*** | 42.6 | 28.9 | 55.9 |
| Minimum Temp °F | 2.4 | 7.9 | 21.5 | 34.1 | 44.4 | 53.9 | 58.5 | 56.1 | ***47.2*** | 36.6 | 25.1 | 11.0 | 33.2 |
| Mean Temp °F | 12.9 | 18.8 | 31.9 | 45.9 | 57.5 | 66.5 | 71.0 | 68.7 | ***59.5*** | ***48.3*** | 33.9 | 20.0 | 44.6 |
| Days Max Temp ≥ 90 °F | 0 | 0 | 0 | 0 | 1 | 3 | 6 | 4 | ***1*** | 0 | 0 | 0 | 15 |
| Days Max Temp ≤ 32 °F | 23 | 16 | 6 | 0 | 0 | 0 | 0 | 0 | ***0*** | 0 | 5 | 18 | 68 |
| Days Min Temp ≤ 32 °F | 31 | 28 | 26 | ***14*** | 3 | 0 | 0 | 0 | ***2*** | 11 | 23 | 30 | 168 |
| Days Min Temp ≤ 0 °F | 14 | 9 | 2 | 0 | 0 | 0 | 0 | 0 | ***0*** | 0 | 1 | 7 | 33 |
| Heating Degree Days | 1612 | 1298 | 1019 | 571 | 261 | 65 | 15 | 37 | ***203*** | ***516*** | 928 | 1391 | 7916 |
| Cooling Degree Days | 0 | 0 | 0 | ***9*** | 29 | 125 | ***212*** | 168 | na | ***2*** | 0 | 0 | na |
| Total Precipitation (") | ***0.91*** | 0.83 | 1.91 | 3.29 | 3.93 | 4.24 | 4.23 | 4.10 | ***4.78*** | 2.34 | 2.20 | ***1.31*** | 34.07 |
| Days ≥ 0.1" Precip | ***3*** | 2 | 5 | 7 | 7 | 8 | 7 | 7 | ***7*** | 5 | 5 | ***3*** | 66 |
| Total Snowfall (") | 10.9 | 7.9 | 6.1 | 1.7 | 0.0 | 0.0 | 0.0 | 0.0 | ***0.0*** | 0.2 | 3.7 | 10.1 | 40.6 |
| Days ≥ 1" Snow Depth | 26 | 22 | 12 | 1 | 0 | 0 | 0 | 0 | ***0*** | 0 | 3 | 18 | 82 |

## SPOONER EXP FARM *Washburn County* ELEVATION 1100 ft LAT/LONG 45° 49 ' N / 91° 53 ' W

| | JAN | FEB | MAR | APR | MAY | JUN | JUL | AUG | SEP | OCT | NOV | DEC | YEAR |
|---|---|---|---|---|---|---|---|---|---|---|---|---|---|
| Maximum Temp °F | 20.9 | 27.5 | 40.0 | 56.4 | 69.6 | 77.2 | 81.5 | 78.8 | 69.2 | 57.5 | 39.1 | 25.3 | 53.6 |
| Minimum Temp °F | -1.4 | 3.8 | 17.7 | 31.6 | 42.8 | 51.7 | 56.9 | 54.3 | 46.0 | 35.4 | 22.4 | 7.2 | 30.7 |
| Mean Temp °F | 9.7 | 15.7 | 28.9 | 44.0 | 56.2 | 64.5 | 69.2 | 66.6 | 57.6 | 46.5 | 30.8 | 16.2 | 42.2 |
| Days Max Temp ≥ 90 °F | 0 | 0 | 0 | 0 | 0 | 1 | 3 | 2 | 0 | 0 | 0 | 0 | 6 |
| Days Max Temp ≤ 32 °F | 26 | 18 | 8 | 0 | 0 | 0 | 0 | 0 | 0 | 0 | 9 | 23 | 84 |
| Days Min Temp ≤ 32 °F | 31 | 28 | 27 | 17 | 5 | 0 | 0 | 0 | 3 | 13 | 26 | 31 | 181 |
| Days Min Temp ≤ 0 °F | 16 | 12 | 4 | 0 | 0 | 0 | 0 | 0 | 0 | 0 | 1 | 10 | 43 |
| Heating Degree Days | 1711 | 1386 | 1113 | 625 | 293 | 87 | 26 | 58 | 244 | 570 | 1019 | 1507 | 8639 |
| Cooling Degree Days | 0 | 0 | 0 | 4 | 27 | 84 | 169 | 123 | 29 | 1 | 0 | 0 | 437 |
| Total Precipitation (") | 0.82 | 0.65 | 1.36 | 2.26 | 3.11 | 4.07 | 4.11 | 4.29 | 3.90 | 2.65 | 1.75 | 0.93 | 29.90 |
| Days ≥ 0.1" Precip | 3 | 2 | 4 | 6 | 7 | 8 | 7 | 7 | 7 | 6 | 4 | 3 | 64 |
| Total Snowfall (") | 14.4 | 7.5 | 8.5 | 2.7 | 0.0 | 0.0 | 0.0 | 0.0 | 0.0 | 0.7 | 8.3 | 11.7 | 53.8 |
| Days ≥ 1" Snow Depth | 30 | 26 | 18 | 4 | 0 | 0 | 0 | 0 | 0 | 0 | 9 | 25 | 112 |

## ST CROIX FALLS *Polk County* ELEVATION 761 ft LAT/LONG 45° 25 ' N / 92° 39 ' W

| | JAN | FEB | MAR | APR | MAY | JUN | JUL | AUG | SEP | OCT | NOV | DEC | YEAR |
|---|---|---|---|---|---|---|---|---|---|---|---|---|---|
| Maximum Temp °F | 21.1 | 27.6 | 40.3 | 56.9 | 70.1 | 78.4 | 83.0 | 80.2 | 70.4 | 58.2 | 40.2 | 26.3 | 54.4 |
| Minimum Temp °F | -1.9 | 4.0 | 18.6 | 33.2 | 44.9 | 54.3 | 59.4 | 57.3 | 47.9 | 36.3 | 22.8 | 7.6 | 32.0 |
| Mean Temp °F | 9.6 | 15.8 | 29.5 | 45.1 | 57.5 | 66.4 | 71.2 | 68.8 | 59.2 | 47.3 | 31.5 | 16.9 | 43.2 |
| Days Max Temp ≥ 90 °F | 0 | 0 | 0 | 0 | 1 | 2 | 5 | 3 | 0 | 0 | 0 | 0 | 11 |
| Days Max Temp ≤ 32 °F | 24 | 17 | 7 | 0 | 0 | 0 | 0 | 0 | 0 | 0 | 7 | 21 | 76 |
| Days Min Temp ≤ 32 °F | 31 | 28 | 27 | 15 | 3 | 0 | 0 | 0 | 1 | 11 | 25 | 31 | 172 |
| Days Min Temp ≤ 0 °F | 17 | 12 | 3 | 0 | 0 | 0 | 0 | 0 | 0 | 0 | 1 | 10 | 43 |
| Heating Degree Days | 1714 | 1385 | 1095 | 595 | 261 | 62 | 14 | 34 | 209 | 546 | 999 | 1485 | 8399 |
| Cooling Degree Days | 0 | 0 | 0 | 5 | 43 | 121 | 225 | 171 | 44 | 2 | 0 | 0 | 611 |
| Total Precipitation (") | 0.95 | 0.69 | 1.56 | 2.74 | 3.27 | 4.62 | 3.93 | 4.36 | 3.75 | 2.64 | 1.57 | 0.98 | 31.06 |
| Days ≥ 0.1" Precip | 3 | 2 | 4 | 6 | 7 | 8 | 7 | 6 | 7 | 5 | 4 | 3 | 62 |
| Total Snowfall (") | 12.2 | 6.6 | 7.5 | 1.4 | 0.0 | 0.0 | 0.0 | 0.0 | 0.0 | 0.4 | 7.2 | 8.5 | 43.8 |
| Days ≥ 1" Snow Depth | 30 | 27 | 20 | 2 | 0 | 0 | 0 | 0 | 0 | 0 | 7 | 25 | 111 |

## STANLEY *Chippewa County* ELEVATION 1079 ft LAT/LONG 44° 58 ' N / 90° 54 ' W

| | JAN | FEB | MAR | APR | MAY | JUN | JUL | AUG | SEP | OCT | NOV | DEC | YEAR |
|---|---|---|---|---|---|---|---|---|---|---|---|---|---|
| Maximum Temp °F | 21.4 | 26.7 | 39.7 | 55.5 | 68.8 | 76.5 | 81.1 | 78.9 | 69.5 | 57.6 | 40.0 | 26.0 | 53.5 |
| Minimum Temp °F | 1.0 | 5.3 | 19.4 | 32.4 | 43.2 | 52.2 | 57.3 | 54.8 | 46.1 | 35.5 | 23.0 | 8.6 | 31.6 |
| Mean Temp °F | 11.2 | 16.0 | 29.6 | 44.0 | 56.0 | 64.3 | 69.2 | 66.9 | 57.8 | 46.6 | 31.6 | 17.3 | 42.5 |
| Days Max Temp ≥ 90 °F | 0 | 0 | 0 | 0 | 0 | 1 | 3 | 2 | 0 | 0 | 0 | 0 | 6 |
| Days Max Temp ≤ 32 °F | 24 | 18 | *8* | 0 | 0 | 0 | 0 | 0 | 0 | 0 | 7 | 22 | 79 |
| Days Min Temp ≤ 32 °F | 30 | 27 | *27* | 16 | 4 | 0 | 0 | 0 | 2 | 13 | 25 | 30 | 174 |
| Days Min Temp ≤ 0 °F | 15 | 11 | 3 | 0 | 0 | 0 | 0 | 0 | 0 | 0 | 1 | 9 | 39 |
| Heating Degree Days | 1662 | 1378 | 1091 | 627 | 296 | 88 | 24 | 49 | 238 | 567 | 997 | 1472 | 8489 |
| Cooling Degree Days | 0 | 0 | 0 | 4 | 21 | 77 | 154 | 113 | 27 | 1 | 0 | 0 | 397 |
| Total Precipitation (") | 1.03 | 0.74 | 1.56 | 2.76 | 3.69 | 4.37 | 4.07 | 4.14 | 3.81 | 2.32 | 2.08 | 1.32 | 31.89 |
| Days ≥ 0.1" Precip | 3 | *2* | *4* | 7 | 7 | 8 | 7 | 7 | 7 | 5 | 5 | 4 | 66 |
| Total Snowfall (") | 13.2 | *7.5* | *7.3* | 2.7 | 0.0 | 0.0 | 0.0 | 0.0 | 0.0 | 0.5 | 6.7 | 12.0 | 49.9 |
| Days ≥ 1" Snow Depth | na | na | *15* | 2 | 0 | 0 | 0 | 0 | 0 | 0 | 4 | *18* | na |

## STEVENS POINT *Portage County* ELEVATION 1112 ft LAT/LONG 44° 31 ' N / 89° 34 ' W

| | JAN | FEB | MAR | APR | MAY | JUN | JUL | AUG | SEP | OCT | NOV | DEC | YEAR |
|---|---|---|---|---|---|---|---|---|---|---|---|---|---|
| Maximum Temp °F | 23.0 | 28.2 | 39.8 | 55.3 | 68.3 | 76.8 | 81.0 | 78.6 | 69.5 | 57.7 | 41.7 | 28.2 | 54.0 |
| Minimum Temp °F | 4.5 | 8.7 | 20.9 | 34.1 | 45.2 | 54.1 | 59.2 | 57.0 | 48.6 | 38.0 | 25.7 | 12.4 | 34.0 |
| Mean Temp °F | 13.8 | 18.5 | 30.4 | 44.7 | 56.8 | 65.5 | 70.2 | 67.9 | 59.1 | 47.9 | 33.7 | 20.4 | 44.1 |
| Days Max Temp ≥ 90 °F | 0 | 0 | 0 | 0 | 0 | 1 | 3 | 2 | 0 | 0 | 0 | 0 | 6 |
| Days Max Temp ≤ 32 °F | 23 | 17 | 8 | 0 | 0 | 0 | 0 | 0 | 0 | 0 | 6 | 19 | 73 |
| Days Min Temp ≤ 32 °F | 31 | 28 | 26 | 14 | 2 | 0 | 0 | 0 | 1 | 9 | 23 | 30 | 164 |
| Days Min Temp ≤ 0 °F | 12 | 9 | 2 | 0 | 0 | 0 | 0 | 0 | 0 | 0 | 0 | 6 | 29 |
| Heating Degree Days | 1583 | 1309 | 1066 | 607 | 282 | 77 | 17 | 41 | 209 | 527 | 933 | 1378 | 8029 |
| Cooling Degree Days | 0 | 0 | 0 | 7 | 33 | 102 | 186 | 144 | 41 | 2 | 0 | 0 | 515 |
| Total Precipitation (") | 1.06 | 0.96 | 2.02 | 2.94 | 3.86 | 3.47 | 4.00 | 3.86 | 4.17 | 2.40 | 2.30 | 1.51 | 32.55 |
| Days ≥ 0.1" Precip | 3 | 3 | 5 | 7 | 7 | 8 | 7 | 7 | 8 | 6 | 5 | 4 | 70 |
| Total Snowfall (") | 10.4 | 7.9 | 6.7 | 1.9 | 0.0 | 0.0 | 0.0 | 0.0 | 0.0 | 0.1 | 4.4 | 11.2 | 42.6 |
| Days ≥ 1" Snow Depth | 30 | 27 | 19 | 3 | 0 | 0 | 0 | 0 | 0 | 0 | 5 | 24 | 108 |

## STOUGHTON *Dane County* ELEVATION 860 ft LAT/LONG 42° 55 ' N / 89° 13 ' W

| | JAN | FEB | MAR | APR | MAY | JUN | JUL | AUG | SEP | OCT | NOV | DEC | YEAR |
|---|---|---|---|---|---|---|---|---|---|---|---|---|---|
| Maximum Temp °F | 26.6 | 30.7 | 42.6 | 57.2 | 69.9 | 79.1 | 82.7 | 80.4 | 72.5 | 61.2 | *45.2* | *31.8* | 56.7 |
| Minimum Temp °F | *8.5* | 11.6 | 23.3 | 35.0 | 46.0 | 55.4 | 60.2 | 57.8 | 49.1 | 37.9 | 27.2 | 14.9 | 35.6 |
| Mean Temp °F | *17.6* | 21.2 | 33.0 | 46.1 | 58.0 | 67.3 | 71.5 | 69.1 | 60.9 | 49.6 | *36.4* | *23.3* | 46.2 |
| Days Max Temp ≥ 90 °F | 0 | 0 | 0 | 0 | 0 | 2 | 4 | 3 | 0 | 0 | 0 | 0 | 9 |
| Days Max Temp ≤ 32 °F | 20 | 15 | 5 | 0 | 0 | 0 | 0 | 0 | 0 | 0 | *3* | *15* | 58 |
| Days Min Temp ≤ 32 °F | 30 | 27 | 26 | 12 | 2 | 0 | 0 | 0 | 1 | 10 | *22* | 29 | 159 |
| Days Min Temp ≤ 0 °F | *9* | 7 | 1 | 0 | 0 | 0 | 0 | 0 | 0 | 0 | 0 | 4 | 21 |
| Heating Degree Days | *1459* | 1232 | 986 | 564 | 245 | 51 | 10 | 29 | 167 | 476 | *854* | *1287* | 7360 |
| Cooling Degree Days | 0 | 0 | 0 | 7 | 31 | *120* | 209 | 153 | 43 | 2 | *0* | 0 | 565 |
| Total Precipitation (") | 1.21 | 1.18 | 1.92 | 3.25 | 3.23 | 3.94 | 3.83 | 3.86 | 3.89 | 2.46 | 2.47 | 1.75 | 32.99 |
| Days ≥ 0.1" Precip | 4 | 3 | 5 | 7 | 7 | 7 | 6 | 6 | 6 | 5 | *5* | 4 | 65 |
| Total Snowfall (") | *8.5* | 7.4 | *5.6* | 1.4 | 0.0 | 0.0 | 0.0 | 0.0 | 0.0 | 0.1 | *1.3* | 9.1 | 33.4 |
| Days ≥ 1" Snow Depth | 23 | *20* | *10* | 1 | 0 | 0 | 0 | 0 | 0 | 0 | *2* | 15 | 71 |

## STURGEON BAY EXP FAR *Door County* ELEVATION 659 ft LAT/LONG 44° 52' N / 87° 20' W

| | JAN | FEB | MAR | APR | MAY | JUN | JUL | AUG | SEP | OCT | NOV | DEC | YEAR |
|---|---|---|---|---|---|---|---|---|---|---|---|---|---|
| Maximum Temp °F | 25.0 | 28.6 | 38.5 | 51.7 | 64.7 | 74.2 | 79.3 | 77.7 | 69.0 | 56.7 | 42.4 | 30.6 | 53.2 |
| Minimum Temp °F | 8.8 | 11.6 | 22.1 | 33.0 | 42.2 | 51.7 | 58.0 | 57.1 | 50.1 | 40.0 | 29.4 | 17.6 | 35.1 |
| Mean Temp °F | 16.9 | 20.1 | 30.3 | 42.4 | 53.5 | 63.0 | 68.7 | 67.4 | 59.6 | 48.4 | 35.9 | 24.1 | 44.2 |
| Days Max Temp ≥ 90 °F | 0 | 0 | 0 | 0 | 0 | 1 | 2 | 1 | 0 | 0 | 0 | 0 | 4 |
| Days Max Temp ≤ 32 °F | 22 | 17 | 7 | 0 | 0 | 0 | 0 | 0 | 0 | 0 | 3 | 16 | 65 |
| Days Min Temp ≤ 32 °F | 30 | 28 | 27 | 14 | 3 | 0 | 0 | 0 | 0 | 5 | 19 | 29 | 155 |
| Days Min Temp ≤ 0 °F | 9 | 6 | 1 | 0 | 0 | 0 | 0 | 0 | 0 | 0 | 0 | 3 | 19 |
| Heating Degree Days | 1484 | 1261 | 1067 | 673 | 358 | 110 | 25 | 41 | 186 | 509 | 866 | 1260 | 7840 |
| Cooling Degree Days | 0 | 0 | 0 | 0 | 11 | 59 | 145 | 126 | 31 | 0 | 0 | 0 | 372 |
| Total Precipitation (") | 1.60 | 1.09 | 2.09 | 2.90 | 3.05 | 3.33 | 3.47 | 3.34 | 3.85 | 2.71 | 2.64 | 1.88 | 31.95 |
| Days ≥ 0.1" Precip | 5 | 4 | 5 | 7 | 6 | 7 | 7 | 7 | 7 | 6 | 6 | 6 | 73 |
| Total Snowfall (") | 13.7 | 7.9 | 7.2 | 2.4 | 0.1 | 0.0 | 0.0 | 0.0 | 0.0 | 0.1 | 2.9 | 11.1 | 45.4 |
| Days ≥ 1" Snow Depth | 28 | 26 | 22 | 5 | 0 | 0 | 0 | 0 | 0 | 0 | 3 | 18 | 102 |

## SUPERIOR *Douglas County* ELEVATION 630 ft LAT/LONG 46° 42' N / 92° 1' W

| | JAN | FEB | MAR | APR | MAY | JUN | JUL | AUG | SEP | OCT | NOV | DEC | YEAR |
|---|---|---|---|---|---|---|---|---|---|---|---|---|---|
| Maximum Temp °F | 20.1 | 25.3 | 35.3 | 47.3 | 58.8 | 68.2 | 76.7 | 74.4 | 65.8 | 54.0 | 38.4 | 25.3 | 49.1 |
| Minimum Temp °F | -0.1 | 5.6 | 17.9 | 30.0 | 38.9 | 47.5 | 55.4 | 55.2 | 46.5 | 35.9 | 23.2 | 8.2 | 30.4 |
| Mean Temp °F | 10.0 | 15.5 | 26.6 | 38.7 | 48.9 | 57.9 | 66.1 | 64.8 | 56.2 | 45.0 | 30.8 | 16.8 | 39.8 |
| Days Max Temp ≥ 90 °F | 0 | 0 | 0 | 0 | 0 | 1 | 2 | 1 | 0 | 0 | 0 | 0 | 4 |
| Days Max Temp ≤ 32 °F | 27 | 20 | 12 | 1 | 0 | 0 | 0 | 0 | 0 | 0 | 8 | 22 | 90 |
| Days Min Temp ≤ 32 °F | 31 | 28 | 29 | 18 | 4 | 0 | 0 | 0 | 1 | 10 | 25 | 30 | 176 |
| Days Min Temp ≤ 0 °F | 16 | 11 | 3 | 0 | 0 | 0 | 0 | 0 | 0 | 0 | 1 | 10 | 41 |
| Heating Degree Days | 1701 | 1394 | 1183 | 784 | 497 | 232 | 66 | 85 | 272 | 614 | 1019 | 1490 | 9337 |
| Cooling Degree Days | 0 | 0 | 0 | 0 | 5 | 27 | 113 | 96 | 15 | 0 | 0 | 0 | 256 |
| Total Precipitation (") | 1.00 | 0.67 | 1.97 | 2.33 | 3.15 | 3.93 | 4.00 | 4.01 | 4.11 | 2.40 | 1.91 | 1.04 | 30.52 |
| Days ≥ 0.1" Precip | 3 | 2 | 5 | 5 | 7 | 8 | 7 | 7 | 7 | 5 | 4 | 4 | 64 |
| Total Snowfall (") | 12.9 | 7.1 | 8.5 | 2.4 | 0.1 | 0.0 | 0.0 | 0.0 | 0.0 | 0.3 | 7.9 | 11.0 | 50.2 |
| Days ≥ 1" Snow Depth | 30 | 25 | 22 | 4 | 0 | 0 | 0 | 0 | 0 | 0 | 8 | 23 | 112 |

## TREMPEALEAU DAM 6 *Trempealeau County* ELEVATION 660 ft LAT/LONG 44° 0' N / 91° 26' W

| | JAN | FEB | MAR | APR | MAY | JUN | JUL | AUG | SEP | OCT | NOV | DEC | YEAR |
|---|---|---|---|---|---|---|---|---|---|---|---|---|---|
| Maximum Temp °F | 23.7 | 29.7 | 42.0 | 57.9 | 70.4 | 79.1 | 83.3 | 80.7 | 71.6 | 59.7 | 42.3 | 29.0 | 55.8 |
| Minimum Temp °F | 4.4 | 9.2 | 23.0 | 36.9 | 48.1 | 57.3 | 61.8 | 59.3 | 50.8 | 39.7 | 26.6 | 12.9 | 35.8 |
| Mean Temp °F | 14.1 | 19.5 | 32.6 | 47.4 | 59.3 | 68.2 | 72.6 | 70.0 | 61.2 | 49.7 | 34.5 | 21.0 | 45.8 |
| Days Max Temp ≥ 90 °F | 0 | 0 | 0 | 0 | 0 | 2 | 5 | 3 | 0 | 0 | 0 | 0 | 10 |
| Days Max Temp ≤ 32 °F | 22 | 15 | 6 | 0 | 0 | 0 | 0 | 0 | 0 | 0 | 5 | 17 | 65 |
| Days Min Temp ≤ 32 °F | 31 | 28 | 25 | 10 | 1 | 0 | 0 | 0 | 1 | 8 | 22 | 30 | 156 |
| Days Min Temp ≤ 0 °F | 13 | 8 | 2 | 0 | 0 | 0 | 0 | 0 | 0 | 0 | 0 | 6 | 29 |
| Heating Degree Days | 1574 | 1280 | 999 | 528 | 212 | 37 | 7 | 21 | 158 | 471 | 909 | 1359 | 7555 |
| Cooling Degree Days | 0 | 0 | 0 | 9 | 40 | 145 | 248 | 182 | 53 | 1 | 0 | 0 | 678 |
| Total Precipitation (") | 1.03 | 0.83 | 1.98 | 3.35 | 3.76 | 4.08 | 4.24 | 4.17 | 4.32 | 2.39 | 2.05 | 1.28 | 33.48 |
| Days ≥ 0.1" Precip | 3 | 2 | 4 | 7 | 7 | 7 | 7 | 7 | 7 | 5 | 5 | 4 | 65 |
| Total Snowfall (") | na | na | na | *0.9* | 0.0 | 0.0 | 0.0 | 0.0 | 0.0 | 0.0 | *2.7* | na | na |
| Days ≥ 1" Snow Depth | na | na | *11* | 1 | 0 | 0 | 0 | 0 | 0 | 0 | *2* | *13* | na |

## TWO RIVERS *Manitowoc County* ELEVATION 600 ft LAT/LONG 44° 9' N / 87° 34' W

| | JAN | FEB | MAR | APR | MAY | JUN | JUL | AUG | SEP | OCT | NOV | DEC | YEAR |
|---|---|---|---|---|---|---|---|---|---|---|---|---|---|
| Maximum Temp °F | 26.2 | 29.6 | 38.7 | 49.3 | 60.0 | 69.3 | 75.1 | 74.8 | 67.3 | 55.6 | 42.8 | 31.6 | 51.7 |
| Minimum Temp °F | 11.0 | 14.5 | 24.3 | 34.2 | 42.6 | 51.1 | 57.8 | 58.0 | 51.2 | 41.0 | 30.0 | 18.1 | 36.2 |
| Mean Temp °F | 18.6 | 22.1 | 31.5 | 41.8 | 51.3 | 60.2 | 66.5 | 66.5 | 59.3 | 48.3 | 36.4 | 24.9 | 43.9 |
| Days Max Temp ≥ 90 °F | 0 | 0 | 0 | 0 | 0 | 0 | 1 | 1 | 0 | 0 | 0 | 0 | 2 |
| Days Max Temp ≤ 32 °F | 20 | 16 | 6 | 0 | 0 | 0 | 0 | 0 | 0 | 0 | 3 | 14 | 59 |
| Days Min Temp ≤ 32 °F | 30 | 27 | 25 | 10 | 1 | 0 | 0 | 0 | 0 | 4 | 18 | 28 | 143 |
| Days Min Temp ≤ 0 °F | 7 | 4 | 0 | 0 | 0 | 0 | 0 | 0 | 0 | 0 | 0 | 3 | 14 |
| Heating Degree Days | 1432 | 1206 | 1031 | 689 | 419 | 162 | 37 | 39 | 183 | 512 | 852 | 1238 | 7800 |
| Cooling Degree Days | 0 | 0 | 0 | 0 | 2 | 32 | 93 | 105 | 19 | 0 | 0 | 0 | 251 |
| Total Precipitation (") | 1.52 | 1.33 | 2.32 | 2.93 | 2.73 | 3.00 | 2.77 | 3.25 | 3.56 | 2.39 | 2.33 | 1.93 | 30.06 |
| Days ≥ 0.1" Precip | 4 | 4 | 5 | 7 | 6 | 6 | 5 | 6 | 7 | 6 | 6 | 5 | 67 |
| Total Snowfall (") | 9.8 | 7.2 | *6.0* | *1.2* | 0.0 | 0.0 | 0.0 | 0.0 | 0.0 | 0.0 | *1.4* | 7.9 | 33.5 |
| Days ≥ 1" Snow Depth | 24 | 24 | 12 | 1 | 0 | 0 | 0 | 0 | 0 | 0 | *2* | 15 | 78 |

## VIROQUA 2 NW *Vernon County* ELEVATION 1280 ft LAT/LONG 43° 33 ' N / 90° 53 ' W

| | JAN | FEB | MAR | APR | MAY | JUN | JUL | AUG | SEP | OCT | NOV | DEC | YEAR |
|---|---|---|---|---|---|---|---|---|---|---|---|---|---|
| Maximum Temp °F | 23.0 | 28.9 | 41.3 | 57.3 | 69.1 | 77.4 | 81.7 | 79.3 | 70.3 | 58.9 | 42.2 | 28.2 | 54.8 |
| Minimum Temp °F | 3.9 | 8.7 | 21.6 | 34.1 | 44.8 | 54.0 | 58.6 | 56.2 | 47.6 | 36.9 | 24.9 | 11.6 | 33.6 |
| Mean Temp °F | 13.4 | 18.8 | 31.5 | 45.7 | 57.0 | 65.7 | 70.2 | 67.7 | 59.0 | 48.0 | 33.5 | 19.8 | 44.2 |
| Days Max Temp ≥ 90 °F | 0 | 0 | 0 | 0 | 0 | 1 | 3 | 2 | 0 | 0 | 0 | 0 | 6 |
| Days Max Temp ≤ 32 °F | 23 | 16 | 7 | 0 | 0 | 0 | 0 | 0 | 0 | 0 | 6 | 20 | 72 |
| Days Min Temp ≤ 32 °F | 31 | 28 | 26 | 14 | 3 | 0 | 0 | 0 | 2 | 11 | 24 | 30 | 169 |
| Days Min Temp ≤ 0 °F | 13 | 9 | 2 | 0 | 0 | 0 | 0 | 0 | 0 | 0 | 1 | 7 | 32 |
| Heating Degree Days | 1600 | 1300 | 1033 | 580 | 271 | 71 | 19 | 42 | 214 | 527 | 936 | 1396 | 7989 |
| Cooling Degree Days | 0 | 0 | 0 | 7 | 23 | 90 | *169* | 126 | 33 | 1 | 0 | 0 | 449 |
| Total Precipitation (") | 0.88 | 0.81 | 1.60 | 3.53 | 3.49 | 3.86 | 4.55 | 4.38 | 4.31 | 2.27 | 1.93 | 17.86 | 49.47 |
| Days ≥ 0.1" Precip | 3 | 2 | 4 | 7 | 7 | 7 | 6 | 7 | 7 | 5 | 5 | 4 | 64 |
| Total Snowfall (") | 9.9 | 7.2 | 5.6 | 2.1 | 0.0 | 0.0 | 0.0 | 0.0 | 0.0 | 0.4 | 4.1 | 10.2 | 39.5 |
| Days ≥ 1" Snow Depth | 26 | 22 | 12 | 2 | 0 | 0 | 0 | 0 | 0 | 0 | 5 | 21 | 88 |

## WASHINGTON ISLAND *Door County* ELEVATION 610 ft LAT/LONG 45° 21 ' N / 86° 57 ' W

| | JAN | FEB | MAR | APR | MAY | JUN | JUL | AUG | SEP | OCT | NOV | DEC | YEAR |
|---|---|---|---|---|---|---|---|---|---|---|---|---|---|
| Maximum Temp °F | 25.7 | 28.2 | 37.2 | 48.4 | 60.2 | 69.3 | 76.2 | 75.1 | 66.9 | 55.3 | 42.4 | 31.0 | 51.3 |
| Minimum Temp °F | 9.6 | 10.4 | 20.1 | 30.5 | 39.4 | 49.1 | 56.4 | 56.0 | 49.3 | 39.9 | 29.5 | 18.6 | 34.1 |
| Mean Temp °F | 17.8 | 19.3 | 28.7 | 39.5 | 49.9 | 59.2 | 66.3 | 65.6 | 58.1 | 47.7 | 35.9 | 24.8 | 42.7 |
| Days Max Temp ≥ 90 °F | 0 | 0 | 0 | 0 | 0 | 0 | 0 | 0 | 0 | 0 | 0 | 0 | 0 |
| Days Max Temp ≤ 32 °F | 22 | 18 | 8 | 1 | 0 | 0 | 0 | 0 | 0 | 0 | 3 | 16 | 68 |
| Days Min Temp ≤ 32 °F | 31 | 28 | 28 | 17 | 5 | 0 | 0 | 0 | 1 | 5 | 20 | 29 | 164 |
| Days Min Temp ≤ 0 °F | 8 | 7 | 1 | 0 | 0 | 0 | 0 | 0 | 0 | 0 | 0 | 2 | 18 |
| Heating Degree Days | 1458 | 1284 | 1119 | 759 | 464 | 184 | 42 | 56 | 214 | 531 | 865 | 1239 | 8215 |
| Cooling Degree Days | 0 | 0 | 0 | 0 | 1 | 21 | 97 | 91 | 16 | 0 | 0 | 0 | 226 |
| Total Precipitation (") | 1.42 | 0.80 | 1.61 | 2.28 | 2.80 | 3.41 | 2.87 | 3.21 | 3.45 | 2.85 | 2.53 | 1.57 | 28.80 |
| Days ≥ 0.1" Precip | 5 | 3 | 4 | 6 | 6 | 7 | 6 | 6 | 7 | 6 | 6 | 5 | 67 |
| Total Snowfall (") | 14.5 | 9.1 | 9.1 | 2.9 | 0.2 | 0.0 | 0.0 | 0.0 | 0.0 | 0.1 | 3.5 | 11.4 | 50.8 |
| Days ≥ 1" Snow Depth | 22 | *17* | *15* | 3 | 0 | 0 | 0 | 0 | 0 | 0 | *3* | na | na |

## WATERTOWN *Jefferson County* ELEVATION 820 ft LAT/LONG 43° 12 ' N / 88° 43 ' W

| | JAN | FEB | MAR | APR | MAY | JUN | JUL | AUG | SEP | OCT | NOV | DEC | YEAR |
|---|---|---|---|---|---|---|---|---|---|---|---|---|---|
| Maximum Temp °F | 25.7 | 30.2 | 42.5 | 57.2 | 70.4 | 79.6 | 83.5 | 81.1 | 72.8 | 60.4 | 44.6 | 31.4 | 56.6 |
| Minimum Temp °F | 8.3 | 12.5 | 24.8 | 36.3 | 46.7 | 56.0 | 61.0 | 58.7 | 50.8 | 39.9 | 29.1 | 15.9 | 36.7 |
| Mean Temp °F | 17.0 | 21.4 | 33.6 | 46.8 | 58.5 | 67.8 | 72.3 | 69.9 | 61.8 | 50.2 | 36.9 | 23.7 | 46.7 |
| Days Max Temp ≥ 90 °F | 0 | 0 | 0 | 0 | 1 | 3 | 6 | 3 | 1 | 0 | 0 | 0 | 14 |
| Days Max Temp ≤ 32 °F | 21 | 15 | 5 | 0 | 0 | 0 | 0 | 0 | 0 | 0 | 3 | 15 | 59 |
| Days Min Temp ≤ 32 °F | 30 | 27 | 24 | 10 | 2 | 0 | 0 | 0 | 0 | 7 | 20 | 29 | 149 |
| Days Min Temp ≤ 0 °F | 10 | 6 | 1 | 0 | 0 | 0 | 0 | 0 | 0 | 0 | 0 | 4 | 21 |
| Heating Degree Days | 1483 | 1226 | 967 | 547 | 238 | 48 | 8 | 23 | 150 | 458 | 838 | 1273 | 7259 |
| Cooling Degree Days | 0 | 0 | 0 | 9 | 40 | 135 | 237 | 178 | 60 | 3 | 0 | 0 | 662 |
| Total Precipitation (") | 1.25 | 1.19 | 2.19 | 3.06 | 2.81 | 4.13 | 4.24 | 4.15 | 4.22 | 2.55 | 2.41 | 1.87 | 34.07 |
| Days ≥ 0.1" Precip | 4 | 3 | 5 | 7 | 6 | 7 | 6 | 7 | 7 | 5 | 5 | 5 | 67 |
| Total Snowfall (") | 11.2 | 8.2 | 6.2 | 1.5 | 0.2 | 0.0 | 0.0 | 0.0 | 0.0 | 0.1 | 2.6 | 9.7 | 39.7 |
| Days ≥ 1" Snow Depth | 21 | 20 | 8 | 1 | 0 | 0 | 0 | 0 | 0 | 0 | 2 | 15 | 67 |

## WAUKESHA *Waukesha County* ELEVATION 860 ft LAT/LONG 43° 1 ' N / 88° 14 ' W

| | JAN | FEB | MAR | APR | MAY | JUN | JUL | AUG | SEP | OCT | NOV | DEC | YEAR |
|---|---|---|---|---|---|---|---|---|---|---|---|---|---|
| Maximum Temp °F | *25.6* | 31.1 | 42.2 | 56.0 | 68.8 | 78.1 | 82.8 | 80.2 | 72.1 | 60.1 | 44.6 | 32.4 | 56.2 |
| Minimum Temp °F | *10.1* | 15.4 | 25.5 | 36.5 | 46.7 | 56.4 | 62.0 | 60.1 | 52.4 | 41.4 | 29.8 | 18.3 | 37.9 |
| Mean Temp °F | *17.8* | 23.3 | 33.9 | 46.3 | 57.8 | 67.3 | 72.4 | 70.2 | 62.3 | 50.8 | 37.2 | 25.4 | 47.1 |
| Days Max Temp ≥ 90 °F | *0* | 0 | 0 | 0 | 1 | 2 | 5 | 3 | 1 | 0 | 0 | 0 | 12 |
| Days Max Temp ≤ 32 °F | *21* | 15 | 6 | 0 | 0 | 0 | 0 | 0 | 0 | 0 | 3 | 15 | 60 |
| Days Min Temp ≤ 32 °F | *30* | 27 | 24 | 10 | 1 | 0 | 0 | 0 | 0 | 5 | 19 | 28 | 144 |
| Days Min Temp ≤ 0 °F | *9* | 4 | 0 | 0 | 0 | 0 | 0 | 0 | 0 | 0 | 0 | 3 | 16 |
| Heating Degree Days | *1456* | 1172 | 957 | *565* | 255 | 58 | 9 | 19 | 141 | 441 | 828 | 1222 | 7123 |
| Cooling Degree Days | na | *0* | *0* | na | *38* | *139* | *268* | *207* | *74* | *4* | *0* | *0* | na |
| Total Precipitation (") | *1.39* | 1.23 | 2.28 | 3.56 | 2.84 | 3.75 | 3.65 | 4.25 | 3.64 | 2.64 | 2.59 | 2.02 | 33.84 |
| Days ≥ 0.1" Precip | *4* | 3 | 5 | 7 | 6 | 6 | 7 | 7 | 6 | 5 | 6 | 5 | 67 |
| Total Snowfall (") | *11.3* | 8.3 | 7.9 | 2.3 | 0.0 | 0.0 | 0.0 | 0.0 | 0.0 | 0.1 | 2.4 | 10.0 | 42.3 |
| Days ≥ 1" Snow Depth | *23* | *20* | *7* | *1* | 0 | 0 | 0 | 0 | 0 | 0 | 2 | 15 | 68 |

## WAUPACA *Waupaca County* ELEVATION 850 ft LAT/LONG 44° 22 ' N / 89° 5 ' W

| | JAN | FEB | MAR | APR | MAY | JUN | JUL | AUG | SEP | OCT | NOV | DEC | YEAR |
|---|---|---|---|---|---|---|---|---|---|---|---|---|---|
| Maximum Temp °F | 24.8 | 29.9 | 41.2 | 56.5 | 70.0 | 78.3 | 82.7 | 80.0 | 71.2 | 59.3 | 43.1 | 29.8 | 55.6 |
| Minimum Temp °F | 5.9 | 10.0 | 22.0 | 34.3 | 45.0 | 54.1 | 59.3 | 57.0 | 48.9 | 38.1 | 26.5 | 13.5 | 34.6 |
| Mean Temp °F | 15.4 | 20.0 | 31.6 | 45.4 | 57.5 | 66.2 | 71.0 | 68.5 | 60.1 | 48.7 | 34.8 | 21.7 | 45.1 |
| Days Max Temp ≥ 90 °F | 0 | 0 | 0 | 0 | 0 | 3 | 5 | 3 | 0 | 0 | 0 | 0 | 11 |
| Days Max Temp ≤ 32 °F | 22 | 16 | 6 | 0 | 0 | 0 | 0 | 0 | 0 | 0 | 4 | 18 | 66 |
| Days Min Temp ≤ 32 °F | 31 | 28 | 26 | 13 | 3 | 0 | 0 | 0 | 1 | 9 | 23 | 30 | 164 |
| Days Min Temp ≤ 0 °F | 11 | 7 | 1 | 0 | 0 | 0 | 0 | 0 | 0 | 0 | 0 | 5 | 24 |
| Heating Degree Days | 1535 | 1266 | 1028 | 587 | 262 | 68 | 14 | 34 | 185 | 502 | 899 | 1336 | 7716 |
| Cooling Degree Days | 0 | 0 | 0 | 8 | 36 | 109 | 205 | 157 | 44 | 1 | 0 | 0 | 560 |
| Total Precipitation (") | 1.24 | 0.97 | 2.14 | 3.16 | 3.88 | 3.85 | 4.22 | 3.80 | 4.07 | 2.44 | 2.41 | 1.55 | 33.73 |
| Days ≥ 0.1" Precip | 4 | 3 | 5 | 7 | 7 | 7 | 7 | 7 | 7 | 5 | 5 | 5 | 69 |
| Total Snowfall (") | 11.4 | 7.0 | 8.6 | 2.2 | 0.1 | 0.0 | 0.0 | 0.0 | 0.0 | 0.2 | 4.4 | 11.7 | 45.6 |
| Days ≥ 1" Snow Depth | 29 | 26 | 17 | 2 | 0 | 0 | 0 | 0 | 0 | 0 | 4 | 21 | 99 |

## WAUSAU MUNICIPAL AP *Marathon County* ELEVATION 1191 ft LAT/LONG 44° 55 ' N / 89° 37 ' W

| | JAN | FEB | MAR | APR | MAY | JUN | JUL | AUG | SEP | OCT | NOV | DEC | YEAR |
|---|---|---|---|---|---|---|---|---|---|---|---|---|---|
| Maximum Temp °F | 21.5 | 27.2 | 38.6 | 54.0 | 67.5 | 75.7 | 80.3 | 77.5 | 67.9 | 55.5 | 39.3 | 26.3 | 52.6 |
| Minimum Temp °F | 2.9 | 7.3 | 19.6 | 33.3 | 44.6 | 53.7 | 59.0 | 56.9 | 48.0 | 37.3 | 24.7 | 11.0 | 33.2 |
| Mean Temp °F | 12.2 | 17.2 | 29.1 | 43.7 | 56.1 | 64.7 | 69.7 | 67.2 | 57.9 | 46.4 | 32.0 | 18.7 | 42.9 |
| Days Max Temp ≥ 90 °F | 0 | 0 | 0 | 0 | 0 | 1 | 2 | 2 | 0 | 0 | 0 | 0 | 5 |
| Days Max Temp ≤ 32 °F | 25 | 18 | 8 | 1 | 0 | 0 | 0 | 0 | 0 | 0 | 8 | 22 | 82 |
| Days Min Temp ≤ 32 °F | 30 | 28 | 27 | 14 | 3 | 0 | 0 | 0 | 1 | 10 | 24 | 30 | 167 |
| Days Min Temp ≤ 0 °F | 13 | 9 | 3 | 0 | 0 | 0 | 0 | 0 | 0 | 0 | 0 | 7 | 32 |
| Heating Degree Days | 1634 | 1343 | 1104 | 636 | 297 | 85 | 21 | 46 | 235 | 571 | 984 | 1430 | 8386 |
| Cooling Degree Days | 0 | 0 | 0 | 6 | 27 | 88 | 173 | 128 | 29 | 0 | 0 | 0 | 451 |
| Total Precipitation (") | 1.01 | 0.83 | 1.94 | 3.00 | 3.76 | 4.15 | 4.04 | 4.08 | 4.27 | 2.71 | 2.20 | 1.47 | 33.46 |
| Days ≥ 0.1" Precip | 3 | 2 | 5 | 7 | 7 | 8 | 7 | 7 | 8 | 6 | 5 | 4 | 69 |
| Total Snowfall (") | 12.4 | 8.3 | 9.8 | 3.1 | 0.1 | 0.0 | 0.0 | 0.0 | 0.0 | 0.9 | 6.8 | 12.8 | 54.2 |
| Days ≥ 1" Snow Depth | 31 | 27 | 20 | 3 | 0 | 0 | 0 | 0 | 0 | 0 | 6 | 25 | 112 |

## WEST BEND *Washington County* ELEVATION 942 ft LAT/LONG 43° 25 ' N / 88° 11 ' W

| | JAN | FEB | MAR | APR | MAY | JUN | JUL | AUG | SEP | OCT | NOV | DEC | YEAR |
|---|---|---|---|---|---|---|---|---|---|---|---|---|---|
| Maximum Temp °F | 25.5 | 29.9 | 40.9 | 54.7 | 67.7 | 76.9 | 81.4 | 78.9 | 71.2 | 58.8 | 43.9 | 31.3 | 55.1 |
| Minimum Temp °F | 9.4 | 13.6 | 24.1 | 34.7 | 43.8 | 53.4 | 59.2 | 57.4 | 49.9 | 39.5 | 28.6 | 16.8 | 35.9 |
| Mean Temp °F | 17.5 | 21.8 | 32.6 | 44.7 | 55.8 | 65.2 | 70.3 | 68.2 | 60.6 | 49.2 | 36.3 | 24.1 | 45.5 |
| Days Max Temp ≥ 90 °F | 0 | 0 | 0 | 0 | 0 | 2 | 3 | 2 | 0 | 0 | 0 | 0 | 7 |
| Days Max Temp ≤ 32 °F | 21 | 16 | 7 | 0 | 0 | 0 | 0 | 0 | 0 | 0 | 4 | 15 | 63 |
| Days Min Temp ≤ 32 °F | 30 | 27 | 26 | 13 | 3 | 0 | 0 | 0 | 1 | 7 | 20 | 29 | 156 |
| Days Min Temp ≤ 0 °F | 9 | 5 | 1 | 0 | 0 | 0 | 0 | 0 | 0 | 0 | 0 | 3 | 18 |
| Heating Degree Days | 1467 | 1215 | 999 | 606 | 306 | 82 | 19 | 36 | 171 | 488 | 855 | 1262 | 7506 |
| Cooling Degree Days | 0 | 0 | 0 | 7 | 29 | 96 | 189 | 143 | 42 | 2 | 0 | 0 | 508 |
| Total Precipitation (") | 1.32 | 1.01 | 1.92 | 3.08 | 2.88 | 3.60 | 3.74 | 3.74 | 3.89 | 2.68 | 2.60 | 1.82 | 32.28 |
| Days ≥ 0.1" Precip | 4 | 3 | 5 | 7 | 6 | 7 | 7 | 7 | 7 | 6 | 6 | 5 | 70 |
| Total Snowfall (") | 13.2 | 10.8 | 8.4 | *3.1* | 0.1 | 0.0 | 0.0 | 0.0 | 0.0 | 0.3 | 3.6 | 13.2 | 52.7 |
| Days ≥ 1" Snow Depth | 25 | 23 | 12 | 1 | 0 | 0 | 0 | 0 | 0 | 0 | 3 | 17 | 81 |

## WEYERHAUSER *Rusk County* ELEVATION 1342 ft LAT/LONG 45° 26 ' N / 91° 24 ' W

| | JAN | FEB | MAR | APR | MAY | JUN | JUL | AUG | SEP | OCT | NOV | DEC | YEAR |
|---|---|---|---|---|---|---|---|---|---|---|---|---|---|
| Maximum Temp °F | 21.1 | 27.6 | 40.6 | 56.9 | 69.9 | 76.9 | 81.0 | 78.7 | 69.3 | 57.3 | 39.6 | 25.9 | 53.7 |
| Minimum Temp °F | -1.9 | 3.4 | 17.2 | 30.6 | 41.2 | 50.1 | 54.8 | 52.3 | 44.3 | 33.1 | 20.9 | 6.3 | 29.4 |
| Mean Temp °F | 9.6 | 15.5 | 28.9 | 43.8 | 55.6 | 63.5 | 67.9 | 65.5 | 56.8 | 45.3 | 30.2 | 16.1 | 41.6 |
| Days Max Temp ≥ 90 °F | 0 | 0 | 0 | 0 | 0 | 1 | 3 | 2 | 0 | 0 | 0 | 0 | 6 |
| Days Max Temp ≤ 32 °F | 25 | 18 | 7 | 0 | 0 | 0 | 0 | 0 | 0 | 0 | *8* | *21* | 79 |
| Days Min Temp ≤ 32 °F | 31 | 28 | 28 | 18 | 6 | 1 | 0 | 0 | 3 | *15* | *27* | 31 | 188 |
| Days Min Temp ≤ 0 °F | 17 | 12 | 4 | 0 | 0 | 0 | 0 | 0 | 0 | 0 | *2* | *10* | 45 |
| Heating Degree Days | 1714 | 1395 | 1110 | 634 | 303 | 98 | 34 | 70 | 258 | 607 | 1035 | 1511 | 8769 |
| Cooling Degree Days | 0 | 0 | 0 | 3 | 17 | 68 | 135 | 99 | 20 | 0 | 0 | 0 | 342 |
| Total Precipitation (") | 1.19 | 0.84 | 1.99 | 2.61 | 3.46 | 4.43 | 4.19 | 3.94 | 4.58 | 2.81 | 1.97 | 1.26 | 33.27 |
| Days ≥ 0.1" Precip | 4 | 3 | 5 | 7 | 7 | *8* | 7 | 7 | 8 | 6 | 5 | 4 | 71 |
| Total Snowfall (") | *13.2* | *7.8* | 8.5 | 2.1 | 0.0 | 0.0 | 0.0 | 0.0 | 0.0 | 0.6 | *7.3* | *10.6* | 50.1 |
| Days ≥ 1" Snow Depth | 30 | 27 | 21 | 3 | 0 | 0 | 0 | 0 | 0 | 0 | *7* | *25* | 113 |

## WHITEWATER *Walworth County* ELEVATION 820 ft LAT/LONG 42° 50' N / 88° 44' W

| | JAN | FEB | MAR | APR | MAY | JUN | JUL | AUG | SEP | OCT | NOV | DEC | YEAR |
|---|---|---|---|---|---|---|---|---|---|---|---|---|---|
| Maximum Temp °F | 26.8 | 31.5 | 43.8 | 58.4 | 71.2 | 79.9 | 83.8 | 81.3 | 73.8 | 61.8 | 46.0 | 32.6 | 57.6 |
| Minimum Temp °F | 9.3 | 13.5 | 24.8 | 35.6 | 45.8 | 55.4 | 60.0 | 57.6 | 49.9 | 39.1 | 28.7 | 16.2 | 36.3 |
| Mean Temp °F | 18.0 | 22.5 | 34.3 | 47.0 | 58.5 | 67.7 | 71.9 | 69.5 | 61.9 | 50.5 | 37.3 | 24.3 | 46.9 |
| Days Max Temp ≥ 90 °F | 0 | 0 | 0 | 0 | 1 | 3 | 6 | 3 | 1 | 0 | 0 | 0 | 14 |
| Days Max Temp ≤ 32 °F | 20 | 14 | 5 | 0 | 0 | 0 | 0 | 0 | 0 | 0 | 3 | 14 | 56 |
| Days Min Temp ≤ 32 °F | 30 | 27 | 24 | 12 | 3 | 0 | 0 | 0 | 1 | 8 | 20 | 29 | 154 |
| Days Min Temp ≤ 0 °F | 10 | 6 | 1 | 0 | 0 | 0 | 0 | 0 | 0 | 0 | 0 | 4 | 21 |
| Heating Degree Days | 1450 | 1194 | 944 | 540 | 239 | 50 | 11 | 29 | 151 | 450 | 823 | 1256 | 7137 |
| Cooling Degree Days | 0 | 0 | 0 | 9 | 41 | 132 | 226 | 176 | 60 | 2 | 0 | 0 | 646 |
| Total Precipitation (") | 1.17 | 1.02 | 2.11 | 3.24 | 3.07 | 3.82 | 4.01 | 4.11 | 3.71 | 2.60 | 2.55 | 1.73 | 33.14 |
| Days ≥ 0.1" Precip | 3 | 3 | 5 | 7 | 7 | 6 | 6 | 7 | 6 | 5 | 5 | 4 | 64 |
| Total Snowfall (") | 8.7 | 6.7 | 4.8 | 1.3 | 0.1 | 0.0 | 0.0 | 0.0 | 0.0 | 0.0 | 1.8 | 8.0 | 31.4 |
| Days ≥ 1" Snow Depth | 22 | 18 | 7 | 1 | 0 | 0 | 0 | 0 | 0 | 0 | 1 | 13 | 62 |

## WILLOW RESERVOIR *Oneida County* ELEVATION 1562 ft LAT/LONG 45° 43' N / 89° 51' W

| | JAN | FEB | MAR | APR | MAY | JUN | JUL | AUG | SEP | OCT | NOV | DEC | YEAR |
|---|---|---|---|---|---|---|---|---|---|---|---|---|---|
| Maximum Temp °F | 18.8 | 24.6 | 35.8 | 50.9 | 65.1 | 73.0 | 77.7 | 75.3 | 65.7 | 53.3 | 36.8 | 24.2 | 50.1 |
| Minimum Temp °F | -3.1 | 0.2 | 12.9 | 27.9 | 39.4 | 49.1 | 53.9 | 51.3 | 43.2 | 33.4 | 21.0 | 6.4 | 28.0 |
| Mean Temp °F | 8.0 | 12.4 | 24.4 | 39.5 | 52.2 | 61.1 | 65.8 | 63.3 | 54.5 | 43.4 | 28.9 | 15.3 | 39.1 |
| Days Max Temp ≥ 90 °F | 0 | 0 | 0 | 0 | 0 | 0 | 1 | 0 | 0 | 0 | 0 | 0 | 1 |
| Days Max Temp ≤ 32 °F | 27 | 21 | 12 | 1 | 0 | 0 | 0 | 0 | 0 | 1 | 10 | 24 | 96 |
| Days Min Temp ≤ 32 °F | 31 | 28 | 28 | 22 | 8 | 1 | 0 | 0 | 4 | 15 | 27 | 30 | 194 |
| Days Min Temp ≤ 0 °F | 17 | 15 | 7 | 0 | 0 | 0 | 0 | 0 | 0 | 0 | 1 | 10 | 50 |
| Heating Degree Days | 1764 | 1480 | 1253 | 761 | 400 | 152 | 59 | 106 | 324 | 663 | 1076 | 1535 | 9573 |
| Cooling Degree Days | 0 | 0 | 0 | 1 | 8 | 35 | 78 | 60 | 11 | 0 | 0 | 0 | 193 |
| Total Precipitation (") | 1.02 | 0.75 | 1.65 | 2.36 | 3.24 | 4.02 | 3.58 | 4.07 | 4.38 | 2.53 | 1.99 | 1.23 | 30.82 |
| Days ≥ 0.1" Precip | 3 | 3 | 4 | 6 | 7 | 8 | 8 | 7 | 8 | 6 | 5 | 4 | 69 |
| Total Snowfall (") | ***12.5*** | ***7.6*** | ***8.8*** | ***3.6*** | 0.3 | 0.0 | 0.0 | 0.0 | 0.0 | 0.6 | na | ***10.2*** | na |
| Days ≥ 1" Snow Depth | na | na | na | ***7*** | 0 | 0 | 0 | 0 | 0 | 0 | na | na | na |

## WINTER 5 NW *Sawyer County* ELEVATION 1312 ft LAT/LONG 45° 53' N / 91° 4' W

| | JAN | FEB | MAR | APR | MAY | JUN | JUL | AUG | SEP | OCT | NOV | DEC | YEAR |
|---|---|---|---|---|---|---|---|---|---|---|---|---|---|
| Maximum Temp °F | 18.2 | 24.3 | 36.5 | 51.3 | 65.4 | 73.5 | 77.8 | 75.0 | 65.2 | 53.8 | 37.6 | 24.1 | 50.2 |
| Minimum Temp °F | -5.2 | -1.4 | 13.4 | 27.8 | 38.9 | 48.4 | 53.6 | 51.2 | 42.6 | 31.7 | 20.8 | 4.9 | 27.2 |
| Mean Temp °F | 6.5 | 11.5 | 24.9 | 39.6 | 52.2 | 61.0 | 65.7 | 63.1 | 53.9 | 42.8 | 29.2 | 14.5 | 38.7 |
| Days Max Temp ≥ 90 °F | 0 | 0 | 0 | 0 | 0 | 0 | 1 | 0 | 0 | 0 | 0 | 0 | 1 |
| Days Max Temp ≤ 32 °F | 27 | 20 | 11 | 2 | 0 | 0 | 0 | 0 | 0 | 0 | ***10*** | 24 | 94 |
| Days Min Temp ≤ 32 °F | 31 | 28 | 29 | 22 | 9 | 1 | 0 | 0 | 4 | 18 | 27 | 31 | 200 |
| Days Min Temp ≤ 0 °F | 19 | 15 | 7 | 0 | 0 | 0 | 0 | 0 | 0 | 0 | 1 | 12 | 54 |
| Heating Degree Days | 1812 | 1508 | 1236 | 757 | 403 | 155 | 65 | 108 | 336 | 683 | 1067 | 1560 | 9690 |
| Cooling Degree Days | ***0*** | ***0*** | ***0*** | ***2*** | ***12*** | ***46*** | ***99*** | 69 | 10 | ***0*** | 0 | 0 | 238 |
| Total Precipitation (") | 1.12 | 0.69 | 1.74 | 2.31 | 3.22 | 4.31 | 4.94 | 4.26 | 4.71 | 2.92 | 2.04 | 1.20 | 33.46 |
| Days ≥ 0.1" Precip | 4 | 2 | 4 | 6 | 7 | 8 | 8 | 7 | 8 | 6 | 5 | 4 | 69 |
| Total Snowfall (") | 16.4 | 8.8 | 8.8 | 3.6 | 0.2 | 0.0 | 0.0 | 0.0 | 0.0 | 0.9 | 8.3 | 12.8 | 59.8 |
| Days ≥ 1" Snow Depth | na | na | na | ***7*** | 0 | 0 | 0 | 0 | 0 | 1 | ***7*** | na | na |

## WISCONSIN DELLS *Columbia County* ELEVATION 902 ft LAT/LONG 43° 38' N / 89° 46' W

| | JAN | FEB | MAR | APR | MAY | JUN | JUL | AUG | SEP | OCT | NOV | DEC | YEAR |
|---|---|---|---|---|---|---|---|---|---|---|---|---|---|
| Maximum Temp °F | 24.5 | 29.9 | 42.2 | 56.8 | 69.5 | 77.5 | 81.5 | 79.2 | 70.4 | 58.9 | 42.8 | 30.2 | 55.3 |
| Minimum Temp °F | 5.3 | 8.7 | 22.1 | 33.9 | 45.1 | 54.3 | 59.4 | 56.9 | 48.8 | 37.1 | 25.5 | 13.2 | 34.2 |
| Mean Temp °F | 14.9 | 19.3 | 32.1 | 45.3 | 57.3 | 65.9 | 70.5 | 68.0 | 59.6 | 48.0 | 34.2 | 21.7 | 44.7 |
| Days Max Temp ≥ 90 °F | 0 | 0 | 0 | 0 | 0 | 1 | 3 | 2 | 0 | 0 | 0 | 0 | 6 |
| Days Max Temp ≤ 32 °F | 22 | 16 | 6 | 0 | 0 | 0 | 0 | 0 | 0 | 0 | 5 | 17 | 66 |
| Days Min Temp ≤ 32 °F | 30 | 27 | 26 | 14 | 3 | 0 | 0 | 0 | 1 | 11 | 23 | 30 | 165 |
| Days Min Temp ≤ 0 °F | 12 | 9 | 1 | 0 | 0 | 0 | 0 | 0 | 0 | 0 | 0 | 6 | 28 |
| Heating Degree Days | 1548 | 1277 | 1012 | 589 | 264 | 66 | 15 | 39 | 198 | 523 | 919 | 1336 | 7786 |
| Cooling Degree Days | 0 | 0 | 0 | 8 | 34 | 104 | 200 | 147 | 47 | 2 | 0 | 0 | 542 |
| Total Precipitation (") | 1.00 | 0.90 | 2.03 | 3.33 | 3.33 | 4.11 | 3.96 | 4.42 | 4.15 | 2.34 | 1.97 | 1.42 | 32.96 |
| Days ≥ 0.1" Precip | 3 | 3 | 5 | 7 | 7 | 7 | 6 | 7 | 7 | 5 | 4 | 4 | 65 |
| Total Snowfall (") | 10.2 | 7.1 | 5.9 | 2.0 | 0.0 | 0.0 | 0.0 | 0.0 | 0.0 | 0.3 | 3.7 | 10.3 | 39.5 |
| Days ≥ 1" Snow Depth | 23 | 23 | 12 | 1 | 0 | 0 | 0 | 0 | 0 | 0 | 3 | 17 | 79 |

## WISCONSIN RAPIDS *Wood County* ELEVATION 1040 ft LAT/LONG 44° 23 ' N / 89° 51 ' W

| | JAN | FEB | MAR | APR | MAY | JUN | JUL | AUG | SEP | OCT | NOV | DEC | YEAR |
|---|---|---|---|---|---|---|---|---|---|---|---|---|---|
| Maximum Temp °F | 22.9 | 29.0 | 40.6 | 56.1 | 69.5 | 77.8 | 82.2 | 79.6 | 69.9 | 57.6 | 41.4 | 28.2 | 54.6 |
| Minimum Temp °F | 2.9 | 7.1 | 19.4 | 32.5 | 43.6 | 52.8 | 58.1 | 55.5 | 46.8 | 35.9 | 23.8 | 11.0 | 32.4 |
| Mean Temp °F | 12.9 | 18.1 | 30.0 | 44.3 | 56.6 | 65.4 | 70.2 | 67.6 | 58.4 | 46.7 | 32.6 | 19.6 | 43.5 |
| Days Max Temp ≥ 90 °F | 0 | 0 | 0 | 0 | 1 | 3 | 5 | 3 | 0 | 0 | 0 | 0 | 12 |
| Days Max Temp ≤ 32 °F | 24 | 17 | 7 | 0 | 0 | 0 | 0 | 0 | 0 | 0 | 6 | 20 | 74 |
| Days Min Temp ≤ 32 °F | 31 | 28 | 28 | 16 | 4 | 0 | 0 | 0 | 2 | 12 | 25 | 30 | 176 |
| Days Min Temp ≤ 0 °F | 14 | 10 | 3 | 0 | 0 | 0 | 0 | 0 | 0 | 0 | 1 | 8 | 36 |
| Heating Degree Days | 1610 | 1319 | 1076 | 618 | 287 | 78 | 20 | 45 | 226 | 562 | 965 | 1400 | 8206 |
| Cooling Degree Days | 0 | 0 | 0 | 7 | 30 | 96 | 190 | 136 | 34 | 0 | 0 | 0 | 493 |
| Total Precipitation (") | 1.13 | 0.98 | 2.06 | 3.00 | 3.68 | 3.68 | 4.21 | 3.82 | 4.01 | 2.56 | 2.11 | 1.50 | 32.74 |
| Days ≥ 0.1" Precip | 4 | 3 | 5 | 7 | 7 | 7 | 7 | 7 | 7 | 6 | 5 | 5 | 70 |
| Total Snowfall (") | 11.2 | 8.4 | 9.6 | 2.7 | 0.0 | 0.0 | 0.0 | 0.0 | 0.0 | 0.3 | 3.4 | ***10.0*** | 45.6 |
| Days ≥ 1" Snow Depth | 28 | 25 | 15 | 1 | 0 | 0 | 0 | 0 | 0 | 0 | 5 | ***23*** | 97 |

## JANUARY MINIMUM TEMPERATURE °F

### LOWEST

| | | |
|---|---|---|
| 1 | Gordon | -6.4 |
| 2 | Winter | -5.2 |
| 3 | Solon Springs | -4.0 |
| 4 | Foxboro | -3.7 |
| 5 | Grantsburg | -3.2 |
| | Prentice | -3.2 |
| 7 | Willow Reservoir | -3.1 |
| 8 | Danbury | -2.3 |
| | Long Lake Dam | -2.3 |
| | Mellen | -2.3 |
| 11 | Amery | -2.2 |
| 12 | Rainbow Rsvr Lk | -2.1 |
| 13 | Couderay | -2.0 |
| 14 | Minocqua Dam | -1.9 |
| | St. Croix Falls | -1.9 |
| | Weyerhauser | -1.9 |
| 17 | Rice Lake | -1.7 |
| 18 | Owen | -1.6 |
| 19 | Jump River | -1.5 |
| 20 | Spooner | -1.4 |
| 21 | North Pelican | -1.3 |
| | Ridgeland | -1.3 |
| 23 | Rhinelander | -1.1 |
| 24 | Medford | -0.7 |
| | Merrill | -0.7 |

### HIGHEST

| | | |
|---|---|---|
| 1 | Milwkee-Mt. Mry | 13.0 |
| | Sheboygan | 13.0 |
| 3 | Milwaukee-Mtchll | 12.4 |
| 4 | Kenosha | 12.3 |
| 5 | Racine | 12.0 |
| 6 | Port Washington | 11.6 |
| 7 | Lake Geneva | 11.3 |
| 8 | Beloit | 11.0 |
| | Two Rivers | 11.0 |
| 10 | Kewaunee | 10.9 |
| 11 | Waukesha | 10.1 |
| 12 | Manitowoc | 10.0 |
| 13 | Washington Island | 9.6 |
| 14 | Plymouth | 9.4 |
| | West Bend | 9.4 |
| 16 | Whitewater | 9.3 |
| 17 | Burlington | 9.0 |
| 18 | Sturgeon Bay | 8.8 |
| 19 | Beaver Dam | 8.5 |
| | Stoughton | 8.5 |
| 21 | Lake Mills | 8.3 |
| | Marinette | 8.3 |
| | Watertown | 8.3 |
| 24 | Platteville | 8.2 |
| 25 | Fond du Lac | 8.1 |

## JULY MAXIMUM TEMPERATURE °F

### HIGHEST

| | | |
|---|---|---|
| 1 | Prairie du Chien | 85.8 |
| 2 | Lake Geneva | 85.0 |
| 3 | Richland Center | 84.7 |
| 4 | Lake Mills | 84.4 |
| 5 | Milwkee-Mt. Mry | 84.3 |
| | Necedah | 84.3 |
| 7 | Lynxville Dam | 84.2 |
| 8 | Beloit | 84.1 |
| | La Crosse | 84.1 |
| 10 | Menomonie | 83.8 |
| | Whitewater | 83.8 |
| 12 | Fort Atkinson | 83.7 |
| 13 | Brodhead | 83.6 |
| 14 | Platteville | 83.5 |
| | Portage | 83.5 |
| | Sparta | 83.5 |
| | Watertown | 83.5 |
| 18 | Hillsboro | 83.4 |
| 19 | Trempealeau Dam | 83.3 |
| 20 | Beaver Dam | 83.2 |
| | Darlington | 83.2 |
| 22 | Mauston | 83.1 |
| 23 | Dalton | 83.0 |
| | Genoa Dam | 83.0 |
| | Marinette | 83.0 |

### LOWEST

| | | |
|---|---|---|
| 1 | Two Rivers | 75.1 |
| 2 | Washington Island | 76.2 |
| 3 | Laona | 76.3 |
| 4 | North Pelican | 76.6 |
| 5 | Superior | 76.7 |
| 6 | Madeline Island | 76.9 |
| 7 | Rest Lake | 77.1 |
| 8 | Park Falls | 77.3 |
| 9 | Rainbow Rsvr Lk | 77.5 |
| 10 | Kewaunee | 77.6 |
| 11 | Minocqua Dam | 77.7 |
| | Willow Reservoir | 77.7 |
| 13 | Prentice | 77.8 |
| | Winter | 77.8 |
| 15 | Bayfield | 78.0 |
| 16 | Port Washington | 78.2 |
| 17 | Long Lake Dam | 78.3 |
| 18 | Goodman | 78.5 |
| | Gurney | 78.5 |
| 20 | Kenosha | 78.7 |
| 21 | Newald | 78.8 |
| 22 | Medford | 78.9 |
| | Mellen | 78.9 |
| 24 | Rhinelander | 79.3 |
| | Sturgeon Bay | 79.3 |

## ANNUAL PRECIPITATION (")

### HIGHEST

| | | |
|---|---|---|
| 1 | Viroqua | 49.47 |
| 2 | Lake Geneva | 37.34 |
| 3 | Plymouth | 36.82 |
| 4 | Platteville | 35.76 |
| 5 | Dodgeville | 35.42 |
| 6 | Darlington | 35.10 |
| 7 | Racine | 35.05 |
| 8 | Gurney | 34.99 |
| 9 | Richland Center | 34.85 |
| 10 | Brodhead | 34.83 |
| 11 | Ellsworth | 34.70 |
| 12 | Couderay | 34.63 |
| 13 | Blair | 34.42 |
| 14 | Fairchild | 34.39 |
| 15 | Beloit | 34.34 |
| 16 | Burlington | 34.25 |
| 17 | Milwaukee-Mtchll | 34.13 |
| 18 | Sparta | 34.07 |
| | Watertown | 34.07 |
| 20 | Mauston | 34.06 |
| 21 | Alma Dam | 34.04 |
| | Marshfield | 34.04 |
| 23 | Kenosha | 33.90 |
| 24 | Waukesha | 33.84 |
| 25 | Lancaster | 33.81 |

### LOWEST

| | | |
|---|---|---|
| 1 | Washington Island | 28.80 |
| 2 | Green Bay | 29.31 |
| 3 | Fond du Lac | 29.60 |
| 4 | Manitowoc | 29.86 |
| 5 | Spooner | 29.90 |
| 6 | Two Rivers | 30.06 |
| 7 | Menomonie | 30.26 |
| 8 | Superior | 30.52 |
| 9 | Appleton | 30.53 |
| | Danbury | 30.53 |
| 11 | Goodman | 30.57 |
| 12 | Ashland | 30.62 |
| | River Falls | 30.62 |
| 14 | Port Washington | 30.78 |
| 15 | Crivitz High Falls | 30.81 |
| 16 | Willow Reservoir | 30.82 |
| 17 | Kewaunee | 30.87 |
| 18 | Prentice | 30.94 |
| 19 | Prairie du Sac | 31.02 |
| 20 | St. Croix Falls | 31.06 |
| 21 | Hancock | 31.14 |
| 22 | Newald | 31.18 |
| | North Pelican | 31.18 |
| 24 | Laona | 31.20 |
| 25 | New London | 31.24 |

## ANNUAL SNOWFALL (")

### HIGHEST

| | | |
|---|---|---|
| 1 | Gurney | 140.6 |
| 2 | Minocqua Dam | 107.5 |
| 3 | Mellen | 101.0 |
| 4 | Bayfield | 93.3 |
| 5 | Rest Lake | 84.5 |
| 6 | Long Lake Dam | 73.0 |
| 7 | Laona | 71.1 |
| 8 | North Pelican | 63.7 |
| 9 | Plymouth | 62.0 |
| 10 | Winter | 59.8 |
| 11 | Antigo | 59.7 |
| 12 | Ashland | 57.6 |
| 13 | Rainbow Rsvr Lk | 57.4 |
| 14 | Solon Springs | 56.5 |
| 15 | Ellsworth | 56.2 |
| 16 | Gordon | 56.1 |
| 17 | Cumberland | 55.9 |
| 18 | Prentice | 55.4 |
| 19 | Breed | 54.6 |
| 20 | Wausau | 54.2 |
| 21 | Newald | 54.0 |
| 22 | Spooner | 53.8 |
| 23 | Marinette | 53.5 |
| 24 | Jump River | 53.1 |
| 25 | West Bend | 52.7 |

### LOWEST

| | | |
|---|---|---|
| 1 | Prairie du Sac | 29.3 |
| 2 | Genoa Dam | 30.8 |
| 3 | Whitewater | 31.4 |
| 4 | Brodhead | 31.5 |
| 5 | Lynxville Dam | 33.0 |
| 6 | Charmany | 33.1 |
| 7 | Necedah | 33.3 |
| 8 | Stoughton | 33.4 |
| 9 | Two Rivers | 33.5 |
| 10 | Arlington | 34.1 |
| 11 | Beaver Dam | 35.4 |
| 12 | Portage | 36.2 |
| 13 | Hartford | 36.3 |
| 14 | Burlington | 36.5 |
| 15 | Alma Dam | 37.3 |
| 16 | Port Washington | 37.4 |
| 17 | Lake Mills | 37.7 |
| 18 | Fond du Lac | 37.8 |
| 19 | Fort Atkinson | 38.0 |
| 20 | Prairie du Chien | 38.2 |
| 21 | Platteville | 38.5 |
| 22 | Darlington | 38.9 |
| 23 | Oconomowoc | 39.1 |
| 24 | Dodgeville | 39.4 |
| | Kenosha | 39.4 |

# WYOMING

PHYSICAL FEATURES AND GENERAL CLIMATE. Wyoming is a name of Delaware Indian origin and is variously interpreted as "large plains" or "end of the plains." Thus, the name describes the State's outstanding topographic feature. There are, of course, several mountain ranges, but the mountains themselves cover less area than the high plains. The topography and variations in elevation make it difficult to divide the State into homogeneous, climatological areas.

The mean elevation is given as about 6,700 feet above sea level. Even excluding the mountain ranges, the average elevation over the southern portion is well over 6,000 feet, while much of the northern portion is some 2,500 feet lower. The lowest point, 3,125 feet, is near the northeast corner where the Belle Fourche River crosses the State line to South Dakota. The highest point is Gannett Peak at 13,785 feet, which is part of the Wind River Range in the west-central portion. Since the mountain ranges lie in a general north to south direction, they are perpendicular to the prevailing westerlies. Therefore, the mountain ranges provide effective barriers which force the air currents moving in from the Pacific Ocean to drop much of their moisture. It naturally follows that the mountain ranges and the western slopes receive the greatest amount of precipitation. Outside of the mountains, the State is considered semiarid.

The Continental Divide splits the State from near the northwest corner to a point along the southern border about midway. This leaves most of the drainage areas to the east. Precipitation drains into three great river systems: the Columbia, the Colorado, and the Missouri. The Snake with its tributaries in the northwest portion flows to the Columbia; the Green River draining most of the Southwest portion joins the Colorado; the Yellowstone, Wind River, Big Horn, Tongue, and Powder drainage areas cover most of the north portion and flow northward into Montana, entering the Missouri there; the Belle Fourche, Cheyenne, and Niobrara covering the east-central portion flow eastward; and the Platte (mostly North Platte), draining all of the southeast, flows eastward over Nebraska. There is a relatively small area along the southwest border that is drained by the Bear going to the Great Salt Lake. In the south-central portion west of Rawlins, there is an area called the Great Divide Basin which extends from near Rawlins to nearly 100 miles westward and about 50 miles in a north to south direction. Part of this is often referred to as the Red Desert. There is no drainage from the Great Divide Basin. Precipitation here, which averages only 7 to 10 inches annually, follows usually dry creekbeds to ponds or small lakes, also often dry.

Rapid runoff from heavy thunderstorm rains causes flash flooding on the headwater streams of the State, and when the time of these storms coincides with the melting of the snowpack, the flooding is intensified.

PRECIPITATION. Like the other states in the western part of the country, precipitation varies considerably from one location to another. The period of maximum precipitation occurs in the spring and early summer for most of the State. It is greater over the mountain ranges and usually at the higher elevations, although elevation alone is not the only influence. Mountain ranges block the flow of moisture laden air from the east as well as from the west. During the summer months showers are quite frequent but light and often amount to only a few hundredths of an inch. Occasionally there will be some very heavy rain associated with thunderstorms covering a few square miles. There are usually several local storms each year with from 1 to 2 inches of rain in a 24-hour period. On rarer occasions 24-hour amounts range from 3 to 5 inches.

SNOWFALL. Snow falls frequently from November through May. Generally snowfall at lower elevations is light to moderate. About five times a year on the average, stations at the lower elevations will have snowfall exceeding 5 inches. Of course, wind will frequently accompany or follow a snowstorm and pile the snow into drifts several feet deep. Wind with the snow will quite often cause blizzard or near blizzard conditions in parts of the State for a few hours; however, it is uncommon for a severe blizzard to last long.

The total annual amount of snow varies considerably over the State as does the rainfall. At the lower elevations of the east portion, the range is mostly from 60 to 70 inches annually. Over the drier southwest portion, amounts vary from 45 to 55 inches at most places. Snow is very light in the Big Horn Basin with annual averages from 15 to 20 inches over the lower portion to 30 to 40 inches on the sides of the Basin where elevations range from 5,000 to 6,000 feet. Of course the mountains receive a great deal more and over the higher ranges annual amounts are well over 200 inches.

TEMPERATURES. Because of the elevation, Wyoming has a relatively cool climate. Above the 6,000-foot level the temperature rarely exceeds 100° F. The warmest parts of the State are the lower portion of the Big Horn Basin, the lower elevations of the central and northeast portions, and along the east border. With increasing elevation, average values drop rapidly. A few places in the mountains at about the 9,000-foot level show an average high in July close to 70° F. Summer nights are almost invariably cool, even though daytime readings may be quite high at times. For most places outside of the mountains, the mean low temperature in July is in the range from 50° to 60° F. Of course, the mountains and high valleys are much cooler with average lows in the middle of the summer in the 30's and low 40's with occasional drops below freezing.

In the wintertime it is characteristic to have rapid and frequent changes between mild and cold spells. Usually there are less than 10 cold waves during a winter, and frequently less than half that number for most of the State. The majority of cold waves move southward on the east side of the Divide, with only an occasional cold wave for the west side. Sometimes only the northeast portion will be affected by the cold air as it slides on to the east over the plains. Many of the cold waves are not accompanied by enough snow to cause severe conditions. Numerous valleys provide ideal pockets for the collection of cold air drainage at night. Protecting mountain ranges prevent the wind from stirring the air, and the colder heavier air continues to deepen in valleys, often sending readings well below zero. It is common to have temperatures in the valleys considerably lower than on the nearby mountain sides. Except for the occasional cold waves and an infrequent blizzard, the winters are not severe. Even January, the coldest month, has occasional mild periods when maximum readings will reach the 50's and 60's.

Early freezes in the fall and late in the spring are characteristic of the Wyoming climate. This has the effect of seemingly long winters and short growing seasons.

SUNSHINE. For most of the State, sunshine ranges from approximately 60 percent of the possible amount during the winter to about 75 percent during the summer. Mountain areas receive less, and in the wintertime the estimated amount over the northwestern mountains is about 45 percent. Although the average amount of sunshine is less in winter, the low point on the annual variations is not during the coldest month (January or February). One low period of sunshine comes in November or December, and another in April or May. These periods of low sunshine correspond fairly closely to the periods of greatest temperature changes, i.e., in the late fall when average temperatures are dropping rapidly and in the spring when the average is climbing rapidly. To be sure, sunshine will not be much higher during the coldest months, but cold air masses are apt to be more stable at that time, and frontal activity is followed by a slightly longer period of sunshine. In the summertime when sunshine is greatest--not only in time but also intensity--it is characteristic for the forenoons to be mostly clear. Cumulus clouds develop nearly every day and frequently blot out the sun for a portion of the afternoons. Because of the altitude--providing less atmosphere for the sun's rays to penetrate--and because of the very small amount of fog, haze, and smoke, the intensity of sunshine is unusually high.

OTHER CLIMATIC ELEMENTS. Hailstorms are the most destructive type of local storm for this State. Tornadoes occur over Wyoming, but they are small and have a short duration. Many of them touch the ground for only a few minutes before receding into the clouds. The season for these local storms extends from April through September. June has the greatest number on the average, with May next.

Wind is an important factor of the Wyoming climate. This is largely due to the high elevation and the enormous stretches of rolling plains. Most of Wyoming is quite windy, and during the colder months from November through March there are frequent periods when the wind reaches 30 to 40 m.p.h., with occasional gusts much higher. Prevailing directions in the different localities vary from west-southwest through west to northwest. In many localities winds are so strong and constant from those directions that trees show a definite lean toward the east or southeast.

The average relative humidity is quite low and, while this has a distinct advantage in providing delightful summer weather, it is related to the rather low amount of moisture. During the warmer part of the summer days, the average drops to about 25 to 30 percent. Late at night when the temperature is lowest the humidity will generally be up to 65 to 75 percent. Low relative humidity, high percentage of sunshine, and rather high average winds add their influence in causing a large amount of evaporation.

## COUNTY INDEX



## ELEVATION INDEX

| FEET | STATION NAME |
|---|---|
| 4291 | WHALEN DAM (USBR) |
| 4304 | DILLINGER |
| 4320 | YODER 2 WSW |
| 4354 | DULL CENTER 1 SE |
| 4380 | NEWCASTLE |
| 4442 | EMBLEM |
| 4491 | ROCHELLE 3 E |
| 4593 | LA GRANGE |
| 4632 | WHEATLAND 4 N |
| 4640 | GILLETTE 9 ESE |
| 4662 | KAYCEE |
| 4682 | TEN SLEEP 16 SSE |
| 4692 | BUFFALO |
| 4754 | SUNDANCE |
| 4790 | HEART MOUNTAIN |
| 4803 | TEN SLEEP 4 NE |
| 4813 | BOYSEN DAM |
| 4852 | MIDWEST |
| 4905 | PHILLIPS |
| 4954 | BILLY CREEK |
| 4954 | GLENROCK 5 ESE |
| 4954 | RIVERTON |
| 5002 | CODY |
| 5013 | LUSK 2 SW |
| 5164 | BUFFALO BILL DAM |
| 5282 | CHUGWATER |
| 5321 | CASPER NATRONA CO AP |
| 5325 | CODY 12 SE |
| 5344 | ALBIN |
| 5390 | CARPENTER |
| 5556 | LANDER HUNT FIELD |
| 5564 | THERMOPOLIS 25 WNW |
| 5574 | DIVERSION DAM |
| 5643 | BLACK MOUNTAIN |
| 5882 | CODY 21 SW |
| 5964 | POWDER RIVER 2 |
| 6014 | ARCHER |
| 6083 | GREEN RIVER |
| 6106 | SYBILLE RESEARCH UNI |
| 6120 | BURRIS |
| 6125 | AFTON |
| 6158 | CHEYENNE MUNI AP |
| 6204 | DOUBLE FOUR RANCH |
| 6243 | YELLOWSTONE MAMMOTH |
| 6247 | JACKSON |
| 6273 | TOWER FALLS |
| 6316 | MUDDY GAP |
| 6325 | SAGE 4 NNW |
| 6414 | ALTA 1 NNW |
| 6443 | SUNSHINE 2 ENE |

| FEET | STATION NAME |
|---|---|
| 6473 | GAS HILLS 4 E |
| 6473 | MOOSE |
| 6483 | FONTENELLE DAM |
| 6506 | BONDURANT |
| 6565 | MEDICINE BOW |
| 6726 | BITTER CREEK 4 NE |
| 6735 | LA BARGE |
| 6745 | MORAN 5 WNW |
| 6745 | ROCK SPRINGS AP |
| 6781 | RAWLINS MUNI AP |
| 6795 | SARATOGA |
| 6824 | WAMSUTTER |
| 6863 | EVANSTON 1 E |
| 6883 | SNAKE RIVER |
| 6926 | DUBOIS |
| 6926 | KEMMERER 2 N |
| 7054 | CHURCH BUTTES GAS PL |
| 7116 | ELK MOUNTAIN |
| 7185 | PINEDALE |
| 7274 | LARAMIE GEN BREES FD |
| 7395 | ENCAMPMENT 11 ESE |
| 7736 | LAKE YELLOWSTONE |
| 8205 | BURGESS JUNCTION |

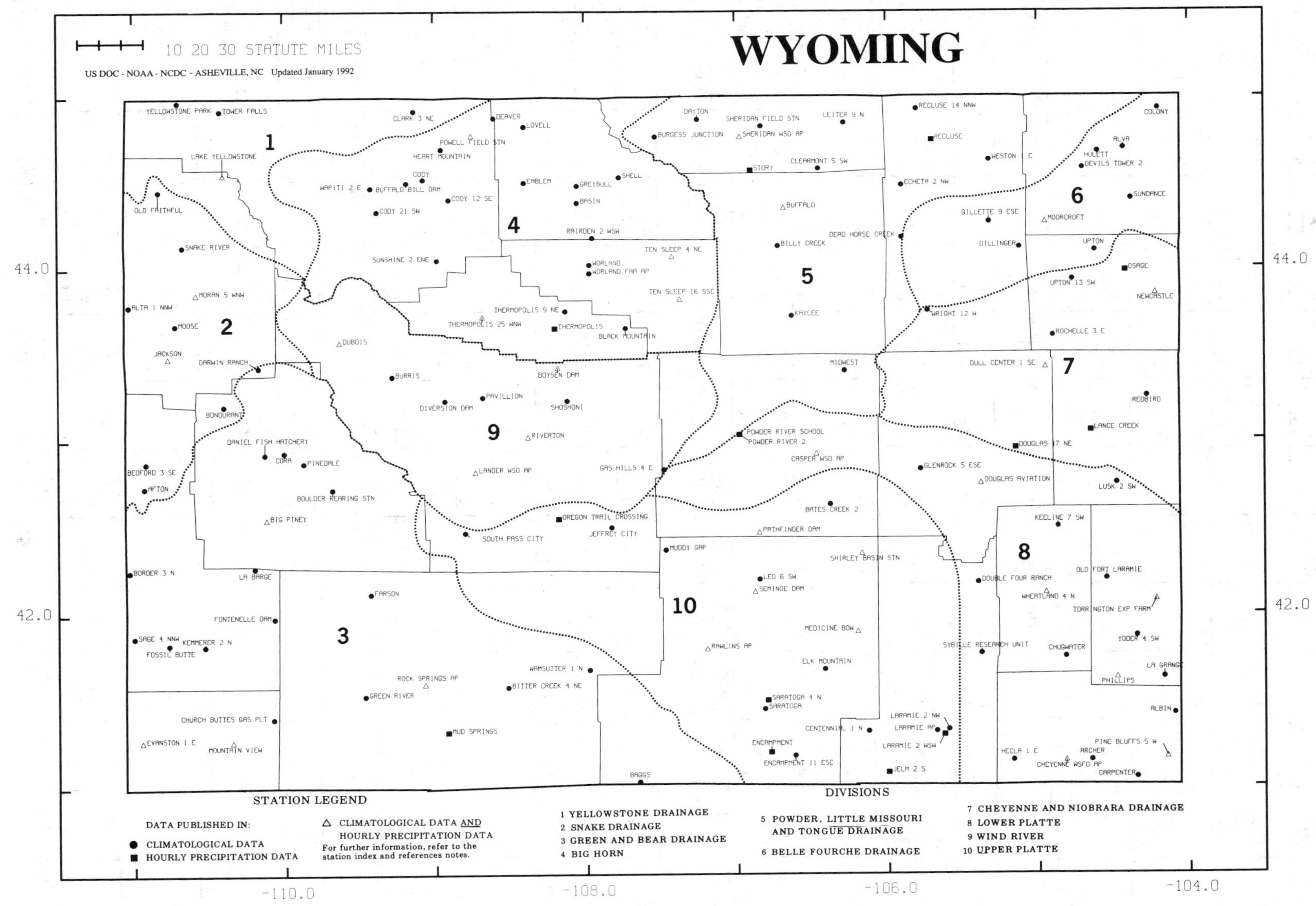
WYOMING
10 20 30 STATUTE MILES
US DOC - NOAA - NCDC - ASHEVILLE, NC Updated January 1992
44.0
42.0
-110.0
-108.0
-106.0
-104.0
YELLOWSTONE PARK
TOWER FALLS
LAKE YELLOWSTONE
OLD FAITHFUL
SNAKE RIVER
ALTA 1 NNW
MOOSE
JACKSON
MORAN 5 WNW
DARWIN RANCH
BONDURANT
DANIEL FISH HATCHERY
CORA
PINEDALE
BIG PINEY
BOULDER REARING STN
AFTON
BEDFORD 3 SE
BORDER 3 N
LA BARGE
FONTENELLE DAM
SAGE 4 NNW
KEMMERER 2 N
FOSSIL BUTTE
EVANSTON 1 E
CHURCH BUTTES GAS PLT
MOUNTAIN VIEW
FARSON
GREEN RIVER
ROCK SPRINGS AP
MUD SPRINGS
BITTER CREEK 4 NE
WAMSUTTER 1 N
BAGGS
CLARK 3 NE
POWELL FIELD STN
HEART MOUNTAIN
CODY
WAPITI 2 E
BUFFALO BILL DAM
CODY 12 SE
CODY 21 SW
SUNSHINE 2 ENE
DEAVER
LOVELL
EMBLEM
GREYBULL
SHELL
BASIN
RAIRDEN 2 WSW
TEN SLEEP 4 NE
WORLAND
WORLAND FAA AP
TEN SLEEP 16 SSE
THERMOPOLIS 9 NE
THERMOPOLIS 25 WNW
THERMOPOLIS
BLACK MOUNTAIN
DUBOIS
BURRIS
BOYSEN DAM
PAVILLION
DIVERSION DAM
SHOSHONI
RIVERTON
LANDER WSO AP
GAS HILLS 4 E
OREGON TRAIL CROSSING
SOUTH PASS CITY
JEFFREY CITY
DAYTON
SHERIDAN FIELD STN
SHERIDAN WSO AP
LEITER 9 N
BURGESS JUNCTION
STORY
CLEARMONT 5 SW
BUFFALO
DEAD HORSE CREEK
BILLY CREEK
KAYCEE
MIDWEST
POWDER RIVER SCHOOL
POWDER RIVER 2
CASPER WSO AP
BATES CREEK 2
PATHFINDER DAM
MUDDY GAP
SHIRLEY BASIN STN
LEO 6 SW
SEMINOE DAM
MEDICINE BOW
RAWLINS AP
ELK MOUNTAIN
SARATOGA 4 N
SARATOGA
CENTENNIAL 1 N
LARAMIE 2 NW
LARAMIE AP
LARAMIE 2 WSW
ENCAMPMENT
ENCAMPMENT 11 ESE
JELM 2 S
RECLUSE 14 NNW
RECLUSE
WESTON 1 E
ECHETA 2 NW
GILLETTE 9 ESE
DILLINGER
WRIGHT 12 W
COLONY
ALVA
HULETT
DEVILS TOWER 2
SUNDANCE
MOORCROFT
UPTON
OSAGE
UPTON 13 SW
NEWCASTLE
ROCHELLE 3 E
DULL CENTER 1 SE
REDBIRD
LANCE CREEK
DOUGLAS 7 NE
GLENROCK 5 ESE
DOUGLAS AVIATION
LUSK 2 SW
KEELINE 7 SW
OLD FORT LARAMIE
DOUBLE FOUR RANCH
WHEATLAND 4 N
TORRINGTON EXP FARM
SYBILLE RESEARCH UNIT
CHUGWATER
YODER 4 SW
LA GRANGE
PHILLIPS
ALBIN
PINE BLUFFS 5 W
HECLA 1 E
ARCHER
CHEYENNE WSFO AP
CARPENTER
1
2
3
4
5
6
7
8
9
10
STATION LEGEND
DATA PUBLISHED IN:
● CLIMATOLOGICAL DATA
■ HOURLY PRECIPITATION DATA
△ CLIMATOLOGICAL DATA AND HOURLY PRECIPITATION DATA
For further information, refer to the station index and references notes.
DIVISIONS
1 YELLOWSTONE DRAINAGE
2 SNAKE DRAINAGE
3 GREEN AND BEAR DRAINAGE
4 BIG HORN
5 POWDER, LITTLE MISSOURI AND TONGUE DRAINAGE
6 BELLE FOURCHE DRAINAGE
7 CHEYENNE AND NIOBRARA DRAINAGE
8 LOWER PLATTE
9 WIND RIVER
10 UPPER PLATTE

## AFTON *Lincoln County* ELEVATION 6125 ft LAT/LONG 42° 45' N / 110° 55' W

| | JAN | FEB | MAR | APR | MAY | JUN | JUL | AUG | SEP | OCT | NOV | DEC | YEAR |
|---|---|---|---|---|---|---|---|---|---|---|---|---|---|
| Maximum Temp °F | 28.7 | 33.6 | 41.7 | 52.0 | 63.8 | 72.8 | 81.1 | 80.2 | 71.0 | 59.1 | 40.1 | 29.2 | 54.4 |
| Minimum Temp °F | 5.0 | 7.2 | 15.6 | 23.8 | 31.4 | 37.4 | 42.2 | 40.7 | 32.9 | 24.7 | 15.8 | 5.4 | 23.5 |
| Mean Temp °F | 16.9 | 20.4 | 28.7 | 37.9 | 47.6 | 55.1 | 61.7 | 60.5 | 52.0 | 41.9 | 28.1 | 17.3 | 39.0 |
| Days Max Temp ≥ 90 °F | 0 | 0 | 0 | 0 | 0 | 0 | 2 | 1 | 0 | 0 | 0 | 0 | 3 |
| Days Max Temp ≤ 32 °F | 18 | 11 | 3 | 0 | 0 | 0 | 0 | 0 | 0 | 0 | 7 | 19 | 58 |
| Days Min Temp ≤ 32 °F | 30 | 28 | 30 | 27 | 18 | 6 | 1 | 3 | 14 | 27 | 29 | 31 | 244 |
| Days Min Temp ≤ 0 °F | 12 | 9 | 3 | 0 | 0 | 0 | 0 | 0 | 0 | 0 | 3 | 12 | 39 |
| Heating Degree Days | 1480 | 1254 | 1119 | 806 | 533 | 296 | 116 | 147 | 386 | 709 | 1100 | 1471 | 9417 |
| Cooling Degree Days | 0 | 0 | 0 | 0 | 0 | 5 | 23 | 16 | 2 | 0 | 0 | 0 | 46 |
| Total Precipitation (") | 1.49 | 1.23 | 1.35 | 1.82 | 2.06 | 1.75 | 1.35 | 1.29 | 1.69 | 1.57 | 1.53 | 1.41 | 18.54 |
| Days ≥ 0.1" Precip | 6 | 4 | 4 | 6 | 6 | 5 | 4 | 4 | 5 | 5 | 6 | 5 | 60 |
| Total Snowfall (") | *16.6* | 12.6 | *12.2* | na | *1.2* | 0.0 | 0.0 | 0.0 | 0.3 | *2.5* | na | *15.8* | na |
| Days ≥ 1" Snow Depth | *30* | 26 | *24* | na | *0* | 0 | 0 | 0 | 0 | *1* | na | *25* | na |

## ALBIN *Laramie County* ELEVATION 5344 ft LAT/LONG 41° 25' N / 104° 6' W

| | JAN | FEB | MAR | APR | MAY | JUN | JUL | AUG | SEP | OCT | NOV | DEC | YEAR |
|---|---|---|---|---|---|---|---|---|---|---|---|---|---|
| Maximum Temp °F | 38.4 | 42.2 | 48.8 | 58.3 | 68.0 | 78.5 | 85.2 | 83.9 | 75.1 | 63.0 | 47.1 | 39.3 | 60.7 |
| Minimum Temp °F | 14.6 | 17.8 | 23.0 | 30.3 | 39.6 | 48.8 | 54.8 | 53.4 | 44.2 | 33.6 | 22.7 | 16.0 | 33.2 |
| Mean Temp °F | 26.5 | 30.0 | 35.9 | 44.3 | 53.8 | 63.7 | 70.0 | 68.7 | 59.7 | 48.3 | 34.9 | 27.7 | 47.0 |
| Days Max Temp ≥ 90 °F | 0 | 0 | 0 | 0 | 0 | 3 | 8 | 5 | 1 | 0 | 0 | 0 | 17 |
| Days Max Temp ≤ 32 °F | 9 | 6 | 3 | 1 | 0 | 0 | 0 | 0 | 0 | 1 | 4 | 9 | 33 |
| Days Min Temp ≤ 32 °F | 30 | 27 | 26 | 18 | 5 | 0 | 0 | 0 | 3 | 13 | 25 | 29 | 176 |
| Days Min Temp ≤ 0 °F | 5 | 3 | 1 | 0 | 0 | 0 | 0 | 0 | 0 | 0 | 1 | 3 | 13 |
| Heating Degree Days | 1187 | 982 | 894 | 614 | 344 | 100 | 14 | 22 | 185 | 511 | 897 | 1151 | 6901 |
| Cooling Degree Days | 0 | 0 | 0 | 0 | 5 | 77 | 174 | 149 | 35 | 1 | 0 | 0 | 441 |
| Total Precipitation (") | 0.75 | 0.60 | 1.57 | 1.87 | 2.84 | 2.74 | 2.21 | 1.58 | 1.38 | 1.10 | 0.98 | 0.76 | 18.38 |
| Days ≥ 0.1" Precip | 3 | 2 | 4 | 5 | 7 | 5 | 5 | 4 | 4 | 3 | 3 | 3 | 48 |
| Total Snowfall (") | 12.1 | 8.7 | 17.1 | 10.3 | 1.4 | 0.0 | 0.0 | 0.0 | 0.8 | 4.1 | 12.3 | 13.3 | 80.1 |
| Days ≥ 1" Snow Depth | 16 | 11 | 9 | 4 | 0 | 0 | 0 | 0 | 0 | 2 | 9 | *15* | 66 |

## ALTA 1 NNW *Teton County* ELEVATION 6414 ft LAT/LONG 43° 47' N / 111° 2' W

| | JAN | FEB | MAR | APR | MAY | JUN | JUL | AUG | SEP | OCT | NOV | DEC | YEAR |
|---|---|---|---|---|---|---|---|---|---|---|---|---|---|
| Maximum Temp °F | 29.1 | 34.2 | 40.0 | 48.5 | 60.3 | 69.8 | 78.7 | 77.3 | 67.7 | 55.1 | 38.1 | 29.8 | 52.4 |
| Minimum Temp °F | 8.8 | 11.6 | 17.3 | 25.0 | 33.2 | 41.0 | *47.0* | 45.0 | 37.6 | 28.4 | 17.4 | 9.0 | 26.8 |
| Mean Temp °F | 19.0 | 22.9 | 28.7 | 36.8 | 46.7 | 55.4 | *62.7* | 61.1 | 52.7 | 41.7 | 27.8 | 19.4 | 39.6 |
| Days Max Temp ≥ 90 °F | 0 | 0 | 0 | 0 | 0 | 0 | 1 | 1 | 0 | 0 | 0 | 0 | 2 |
| Days Max Temp ≤ 32 °F | 20 | 12 | 5 | 1 | 0 | 0 | 0 | 0 | 0 | 1 | 9 | 19 | 67 |
| Days Min Temp ≤ 32 °F | 31 | 28 | 30 | 25 | 15 | 4 | 1 | 1 | 9 | 21 | 28 | 31 | 224 |
| Days Min Temp ≤ 0 °F | 9 | 5 | 2 | 0 | 0 | 0 | 0 | 0 | 0 | 0 | 2 | 8 | 26 |
| Heating Degree Days | 1421 | 1183 | 1119 | 841 | 561 | 291 | *104* | 137 | 365 | 719 | 1109 | 1400 | 9250 |
| Cooling Degree Days | 0 | 0 | 0 | 0 | 0 | 13 | 41 | 21 | 2 | 0 | 0 | 0 | 77 |
| Total Precipitation (") | 2.05 | 1.62 | 1.69 | 2.13 | 3.21 | 2.26 | 1.69 | 1.58 | 1.89 | 2.01 | 2.09 | 1.96 | 24.18 |
| Days ≥ 0.1" Precip | 6 | 5 | 6 | 6 | 8 | 7 | 5 | 5 | 5 | 5 | 7 | 6 | 71 |
| Total Snowfall (") | 22.5 | 15.4 | 13.2 | 11.0 | 4.9 | 0.3 | 0.0 | 0.0 | 0.9 | 6.2 | 16.4 | 19.2 | 110.0 |
| Days ≥ 1" Snow Depth | 29 | 26 | 29 | 21 | 4 | 0 | 0 | 0 | 0 | 4 | 17 | 30 | 160 |

## ARCHER *Laramie County* ELEVATION 6014 ft LAT/LONG 41° 9' N / 104° 39' W

| | JAN | FEB | MAR | APR | MAY | JUN | JUL | AUG | SEP | OCT | NOV | DEC | YEAR |
|---|---|---|---|---|---|---|---|---|---|---|---|---|---|
| Maximum Temp °F | 39.3 | 42.2 | 47.7 | 56.6 | 66.6 | 76.7 | 84.4 | 82.8 | 74.1 | 62.7 | 47.4 | 40.3 | 60.1 |
| Minimum Temp °F | 13.9 | 16.3 | 21.3 | 28.2 | 37.3 | 46.5 | 52.3 | 50.8 | 41.7 | 31.5 | 21.1 | 14.8 | 31.3 |
| Mean Temp °F | 26.6 | 29.3 | 34.5 | 42.5 | 52.0 | 61.6 | 68.4 | 66.8 | 57.9 | 47.2 | 34.2 | 27.5 | 45.7 |
| Days Max Temp ≥ 90 °F | 0 | 0 | 0 | 0 | 0 | 3 | 8 | 5 | 1 | 0 | 0 | 0 | 17 |
| Days Max Temp ≤ 32 °F | 8 | 6 | 4 | 1 | 0 | 0 | 0 | 0 | 0 | 0 | 4 | 8 | 31 |
| Days Min Temp ≤ 32 °F | 30 | 27 | 28 | 21 | 7 | 1 | 0 | 0 | 3 | 16 | 27 | 30 | 190 |
| Days Min Temp ≤ 0 °F | 5 | 3 | 1 | 0 | 0 | 0 | 0 | 0 | 0 | 0 | 1 | 4 | 14 |
| Heating Degree Days | 1180 | 1002 | 938 | 670 | 398 | 138 | 26 | 36 | 225 | 543 | 916 | 1156 | 7228 |
| Cooling Degree Days | 0 | 0 | 0 | 0 | 2 | 47 | 129 | 103 | 21 | 0 | 0 | 0 | 302 |
| Total Precipitation (") | 0.40 | 0.43 | 1.07 | 1.53 | 2.57 | 2.54 | 2.08 | 1.71 | 1.41 | 0.94 | 0.64 | 0.42 | 15.74 |
| Days ≥ 0.1" Precip | 1 | 1 | 3 | 4 | 6 | 5 | 5 | 5 | 4 | 3 | 2 | 2 | 41 |
| Total Snowfall (") | 6.6 | *6.2* | *9.3* | *6.1* | 1.3 | 0.0 | 0.0 | 0.0 | 0.8 | *3.0* | *7.8* | *6.8* | 47.9 |
| Days ≥ 1" Snow Depth | 13 | *11* | *9* | *3* | 0 | 0 | 0 | 0 | 0 | 2 | 8 | 13 | 59 |

## BASIN *Big Horn County* ELEVATION 3832 ft LAT/LONG 44° 23 ' N / 108° 3 ' W

| | JAN | FEB | MAR | APR | MAY | JUN | JUL | AUG | SEP | OCT | NOV | DEC | YEAR |
|---|---|---|---|---|---|---|---|---|---|---|---|---|---|
| Maximum Temp °F | 28.9 | 38.3 | 51.5 | 62.4 | 72.6 | 82.7 | 90.4 | 88.8 | 76.7 | 63.1 | 44.6 | 32.3 | 61.0 |
| Minimum Temp °F | 2.2 | 10.1 | 22.4 | 32.0 | 41.4 | 49.9 | 55.2 | 52.3 | 41.4 | 30.3 | 17.7 | 6.1 | 30.1 |
| Mean Temp °F | 15.6 | 24.2 | 37.0 | 47.2 | 57.0 | 66.3 | 72.8 | 70.6 | 59.0 | 46.7 | 31.2 | 19.2 | 45.6 |
| Days Max Temp ≥ 90 °F | 0 | 0 | 0 | 0 | 1 | 8 | 19 | 16 | 3 | 0 | 0 | 0 | 47 |
| Days Max Temp ≤ 32 °F | 17 | 8 | 2 | 0 | 0 | 0 | 0 | 0 | 0 | 0 | 5 | 14 | 46 |
| Days Min Temp ≤ 32 °F | 31 | 28 | 28 | 15 | 4 | 0 | 0 | 0 | 4 | 19 | 29 | 31 | 189 |
| Days Min Temp ≤ 0 °F | 14 | 6 | 1 | 0 | 0 | 0 | 0 | 0 | 0 | 0 | 2 | 9 | 32 |
| Heating Degree Days | 1527 | 1145 | 861 | 527 | 259 | 69 | 9 | 19 | 203 | 560 | 1009 | 1414 | 7602 |
| Cooling Degree Days | 0 | 0 | 0 | 2 | 24 | 124 | 242 | 204 | 30 | 0 | 0 | 0 | 626 |
| Total Precipitation (") | 0.22 | 0.15 | 0.32 | 0.74 | 1.21 | 1.07 | 0.60 | 0.53 | 0.80 | 0.54 | 0.26 | 0.28 | 6.72 |
| Days ≥ 0.1" Precip | 1 | 1 | 1 | 2 | 4 | 3 | 2 | 2 | 2 | 2 | 1 | 1 | 22 |
| Total Snowfall (") | 3.7 | 2.8 | 2.8 | 2.2 | 0.2 | 0.0 | 0.0 | 0.0 | 0.9 | 0.5 | 2.6 | 4.4 | 20.1 |
| Days ≥ 1" Snow Depth | *23* | *13* | *2* | 0 | 0 | 0 | 0 | 0 | 0 | 0 | *3* | 13 | 54 |

## BILLY CREEK *Johnson County* ELEVATION 4954 ft LAT/LONG 44° 8 ' N / 106° 44 ' W

| | JAN | FEB | MAR | APR | MAY | JUN | JUL | AUG | SEP | OCT | NOV | DEC | YEAR |
|---|---|---|---|---|---|---|---|---|---|---|---|---|---|
| Maximum Temp °F | 35.8 | 40.0 | 46.9 | 56.2 | 65.6 | 75.6 | 83.8 | 82.9 | 71.9 | 59.8 | 45.1 | 37.9 | 58.5 |
| Minimum Temp °F | 9.1 | 14.0 | 21.3 | 29.3 | 38.4 | 47.0 | 52.6 | 51.1 | 41.3 | 31.2 | 19.4 | 11.4 | 30.5 |
| Mean Temp °F | 22.5 | 27.0 | 34.1 | 42.8 | 52.0 | 61.3 | 68.2 | 67.0 | 56.6 | 45.5 | 32.3 | 24.7 | 44.5 |
| Days Max Temp ≥ 90 °F | 0 | 0 | 0 | 0 | 0 | 3 | 8 | 7 | 1 | 0 | 0 | 0 | 19 |
| Days Max Temp ≤ 32 °F | 12 | 7 | 4 | 1 | 0 | 0 | 0 | 0 | 0 | 1 | 5 | 10 | 40 |
| Days Min Temp ≤ 32 °F | 30 | 27 | 28 | 20 | 8 | 0 | 0 | 0 | 5 | 17 | 27 | 30 | 192 |
| Days Min Temp ≤ 0 °F | 8 | 4 | 1 | 0 | 0 | 0 | 0 | 0 | 0 | 0 | 2 | 6 | 21 |
| Heating Degree Days | 1312 | 1065 | 951 | 660 | 403 | 156 | 37 | 53 | 268 | 598 | 976 | 1245 | 7724 |
| Cooling Degree Days | 0 | 0 | 0 | 1 | 7 | 62 | 149 | 132 | 25 | 0 | 0 | 0 | 376 |
| Total Precipitation (") | 0.28 | 0.31 | 0.63 | 1.35 | 2.24 | 2.30 | 1.42 | 1.15 | 1.07 | 0.97 | 0.43 | 0.35 | 12.50 |
| Days ≥ 0.1" Precip | 1 | 1 | 2 | 3 | 6 | 6 | 4 | 3 | 3 | 3 | 1 | 1 | 34 |
| Total Snowfall (") | 5.7 | 5.7 | 8.2 | 6.6 | 2.0 | 0.0 | 0.0 | 0.0 | 1.0 | 3.5 | 5.9 | 5.8 | 44.4 |
| Days ≥ 1" Snow Depth | 18 | 12 | 8 | 2 | 0 | 0 | 0 | 0 | 0 | 1 | 6 | 12 | 59 |

## BITTER CREEK 4 NE *Sweetwater County* ELEVATION 6726 ft LAT/LONG 41° 35 ' N / 108° 31 ' W

| | JAN | FEB | MAR | APR | MAY | JUN | JUL | AUG | SEP | OCT | NOV | DEC | YEAR |
|---|---|---|---|---|---|---|---|---|---|---|---|---|---|
| Maximum Temp °F | 31.7 | 35.4 | 43.9 | 54.7 | 65.6 | 76.1 | 83.6 | 82.2 | 72.3 | 59.0 | 42.2 | 32.9 | 56.6 |
| Minimum Temp °F | 7.1 | 10.0 | 18.7 | 25.0 | 32.6 | 40.2 | 46.4 | 44.3 | 35.4 | 26.1 | 16.6 | 7.9 | 25.9 |
| Mean Temp °F | 19.4 | 22.8 | 31.4 | 39.9 | 49.1 | 58.1 | 65.0 | 63.3 | 53.8 | 42.6 | 29.4 | 20.4 | 41.3 |
| Days Max Temp ≥ 90 °F | 0 | 0 | 0 | 0 | 0 | 1 | 3 | 2 | 0 | 0 | 0 | 0 | 6 |
| Days Max Temp ≤ 32 °F | 15 | 10 | 3 | 1 | 0 | 0 | 0 | 0 | 0 | 0 | 6 | 14 | 49 |
| Days Min Temp ≤ 32 °F | 31 | 28 | 30 | 25 | 16 | 4 | 0 | 2 | 11 | 24 | 29 | 31 | 231 |
| Days Min Temp ≤ 0 °F | 8 | 6 | 2 | 0 | 0 | 0 | 0 | 0 | 0 | 0 | 3 | 8 | 27 |
| Heating Degree Days | 1406 | 1186 | 1036 | 747 | 485 | 213 | 50 | 84 | 332 | 687 | 1062 | 1375 | 8663 |
| Cooling Degree Days | 0 | 0 | 0 | 0 | 0 | 19 | 66 | 45 | 4 | 0 | 0 | 0 | 134 |
| Total Precipitation (") | 0.34 | *0.35* | 0.38 | 0.53 | 1.21 | 0.81 | 0.73 | 0.54 | 0.67 | 0.67 | *0.43* | *0.32* | 6.98 |
| Days ≥ 0.1" Precip | na | na | *1* | *2* | *3* | 2 | *2* | 2 | *2* | 2 | *2* | na | na |
| Total Snowfall (") | na | na | na | na | *0.0* | 0.0 | 0.0 | 0.0 | *0.0* | *0.9* | na | na | na |
| Days ≥ 1" Snow Depth | na | na | na | na | 0 | 0 | 0 | 0 | *0* | *0* | na | na | na |

## BLACK MOUNTAIN *Hot Springs County* ELEVATION 5643 ft LAT/LONG 43° 39 ' N / 107° 44 ' W

| | JAN | FEB | MAR | APR | MAY | JUN | JUL | AUG | SEP | OCT | NOV | DEC | YEAR |
|---|---|---|---|---|---|---|---|---|---|---|---|---|---|
| Maximum Temp °F | 34.3 | 38.1 | 44.3 | 53.6 | 63.9 | 75.8 | 85.2 | 83.6 | 71.0 | 57.9 | 42.7 | 35.0 | 57.1 |
| Minimum Temp °F | 14.7 | 18.1 | 23.8 | 31.9 | 40.7 | 49.8 | 56.9 | 55.4 | 44.3 | 34.0 | 23.2 | 16.5 | 34.1 |
| Mean Temp °F | 24.3 | 28.2 | 34.1 | 42.7 | 52.3 | 62.9 | 70.9 | 69.7 | 57.7 | 46.0 | 33.0 | 25.8 | 45.6 |
| Days Max Temp ≥ 90 °F | 0 | 0 | 0 | 0 | 0 | 3 | 9 | 7 | 1 | 0 | 0 | 0 | 20 |
| Days Max Temp ≤ 32 °F | 11 | 7 | 5 | 1 | 0 | 0 | 0 | 0 | 0 | 1 | 5 | 11 | 41 |
| Days Min Temp ≤ 32 °F | 29 | 26 | 26 | 17 | 6 | 0 | 0 | 0 | 4 | 14 | 24 | 29 | 175 |
| Days Min Temp ≤ 0 °F | 5 | 3 | 1 | 0 | 0 | 0 | 0 | 0 | 0 | 0 | 1 | 3 | 13 |
| Heating Degree Days | 1255 | 1033 | 952 | 664 | 396 | 138 | 25 | 32 | 253 | 584 | 953 | 1210 | 7495 |
| Cooling Degree Days | 0 | 0 | 0 | 1 | 9 | 83 | 195 | 179 | 31 | 1 | 0 | 0 | 499 |
| Total Precipitation (") | 0.50 | 0.49 | 1.02 | 1.73 | 2.19 | 1.82 | 0.94 | 0.84 | 1.15 | 1.32 | 0.78 | 0.60 | 13.38 |
| Days ≥ 0.1" Precip | 2 | 2 | 3 | 5 | 5 | 4 | 3 | 2 | 3 | 3 | 3 | 2 | 37 |
| Total Snowfall (") | *8.8* | 6.5 | 11.2 | 10.8 | 4.5 | 0.4 | 0.0 | 0.0 | 1.8 | 6.3 | *8.8* | *9.9* | 69.0 |
| Days ≥ 1" Snow Depth | 24 | 20 | 14 | 7 | 2 | 0 | 0 | 0 | 1 | 2 | 13 | 22 | 105 |

## BONDURANT *Sublette County* ELEVATION 6506 ft LAT/LONG 43° 14' N / 110° 26' W

| | JAN | FEB | MAR | APR | MAY | JUN | JUL | AUG | SEP | OCT | NOV | DEC | YEAR |
|---|---|---|---|---|---|---|---|---|---|---|---|---|---|
| Maximum Temp °F | 23.7 | 28.9 | 38.2 | 47.9 | 61.4 | 70.6 | 78.3 | 77.1 | 67.6 | 55.1 | 35.8 | ***22.4*** | 50.6 |
| Minimum Temp °F | -4.2 | -3.4 | 6.5 | 18.4 | 27.6 | 32.0 | 34.8 | 33.1 | 25.7 | 17.8 | 8.2 | ***-4.8*** | 16.0 |
| Mean Temp °F | 9.8 | 12.8 | 22.4 | 33.2 | 44.5 | 51.3 | 56.7 | 55.1 | 46.7 | 36.4 | 22.0 | ***8.8*** | 33.3 |
| Days Max Temp ≥ 90 °F | 0 | 0 | 0 | 0 | 0 | 0 | 0 | 0 | 0 | 0 | 0 | 0 | 0 |
| Days Max Temp ≤ 32 °F | ***24*** | 18 | 6 | 1 | 0 | 0 | 0 | 0 | 0 | 1 | 11 | ***25*** | 86 |
| Days Min Temp ≤ 32 °F | 31 | 28 | 31 | 29 | 24 | 17 | 13 | 16 | 24 | 29 | 30 | 30 | 302 |
| Days Min Temp ≤ 0 °F | 19 | 17 | 10 | 2 | 0 | 0 | 0 | 0 | 0 | 1 | 9 | 19 | 77 |
| Heating Degree Days | 1708 | 1472 | 1315 | 946 | 628 | 404 | 252 | 300 | 542 | 878 | 1283 | ***1739*** | 11467 |
| Cooling Degree Days | 0 | 0 | 0 | 0 | 0 | 0 | 0 | 1 | 0 | 0 | 0 | 0 | 1 |
| Total Precipitation (") | 2.36 | 1.71 | 1.58 | 1.22 | 1.51 | 1.48 | 1.35 | 1.34 | 1.47 | 1.38 | 2.05 | 2.17 | 19.62 |
| Days ≥ 0.1" Precip | 7 | 5 | 4 | 4 | 5 | 5 | 4 | 4 | 4 | 3 | 6 | 6 | 57 |
| Total Snowfall (") | 28.4 | 18.5 | 14.7 | 5.2 | 1.3 | 0.1 | 0.0 | 0.0 | 0.4 | 3.2 | 21.0 | 28.6 | 121.4 |
| Days ≥ 1" Snow Depth | 31 | 28 | 29 | 21 | 1 | 0 | 0 | 0 | 0 | 1 | 18 | 29 | 158 |

## BOYSEN DAM *Fremont County* ELEVATION 4813 ft LAT/LONG 43° 25' N / 108° 12' W

| | JAN | FEB | MAR | APR | MAY | JUN | JUL | AUG | SEP | OCT | NOV | DEC | YEAR |
|---|---|---|---|---|---|---|---|---|---|---|---|---|---|
| Maximum Temp °F | 29.2 | 36.9 | 49.6 | 60.4 | 70.6 | 81.5 | 89.6 | 87.8 | 75.8 | 61.9 | 43.0 | 30.1 | 59.7 |
| Minimum Temp °F | 6.9 | 12.5 | 24.2 | 34.1 | 43.7 | 53.0 | 59.7 | 58.4 | 47.5 | 36.5 | 23.3 | 10.2 | 34.2 |
| Mean Temp °F | 18.1 | 24.7 | 36.9 | 47.2 | 57.2 | 67.3 | 74.7 | 73.1 | 61.7 | 49.2 | 33.3 | 20.1 | 47.0 |
| Days Max Temp ≥ 90 °F | 0 | 0 | 0 | 0 | 0 | 7 | 17 | 14 | 2 | 0 | 0 | 0 | 40 |
| Days Max Temp ≤ 32 °F | 17 | 9 | 2 | 0 | 0 | 0 | 0 | 0 | 0 | 0 | 5 | 16 | 49 |
| Days Min Temp ≤ 32 °F | 31 | 28 | 26 | 12 | 2 | 0 | 0 | 0 | 1 | 8 | 25 | 30 | 163 |
| Days Min Temp ≤ 0 °F | 9 | 5 | 1 | 0 | 0 | 0 | 0 | 0 | 0 | 0 | 1 | 6 | 22 |
| Heating Degree Days | 1449 | 1131 | 865 | 527 | 256 | 60 | 5 | 7 | 149 | 486 | 944 | 1383 | 7262 |
| Cooling Degree Days | 0 | 0 | 0 | 1 | 21 | 145 | 307 | 273 | 59 | 1 | 0 | 0 | 807 |
| Total Precipitation (") | 0.25 | 0.29 | 0.58 | 1.20 | 1.76 | 1.33 | 0.76 | 0.58 | 0.83 | 0.84 | 0.41 | 0.32 | 9.15 |
| Days ≥ 0.1" Precip | 1 | 1 | 2 | 3 | 5 | 3 | 2 | 2 | 2 | 2 | 1 | 1 | 25 |
| Total Snowfall (") | na | na | na | na | 0.0 | 0.0 | 0.0 | 0.0 | 0.0 | ***0.0*** | na | na | na |
| Days ≥ 1" Snow Depth | na | na | na | na | 0 | 0 | 0 | 0 | 0 | ***0*** | na | na | na |

## BUFFALO *Johnson County* ELEVATION 4692 ft LAT/LONG 44° 21' N / 106° 41' W

| | JAN | FEB | MAR | APR | MAY | JUN | JUL | AUG | SEP | OCT | NOV | DEC | YEAR |
|---|---|---|---|---|---|---|---|---|---|---|---|---|---|
| Maximum Temp °F | 35.8 | 39.8 | 47.7 | 57.7 | 67.5 | 77.4 | 85.6 | 84.6 | 73.6 | 61.2 | 45.3 | 37.5 | 59.5 |
| Minimum Temp °F | 9.2 | 13.5 | 21.4 | 30.8 | 39.5 | 48.6 | 54.4 | 52.5 | 42.1 | 31.6 | 19.6 | 11.5 | 31.2 |
| Mean Temp °F | 22.5 | 26.7 | 34.6 | 44.3 | 53.5 | 63.0 | 70.1 | 68.6 | 57.9 | 46.5 | 32.5 | 24.5 | 45.4 |
| Days Max Temp ≥ 90 °F | 0 | 0 | 0 | 0 | 0 | 4 | 11 | 10 | 1 | 0 | 0 | 0 | 26 |
| Days Max Temp ≤ 32 °F | 10 | 7 | 4 | 1 | 0 | 0 | 0 | 0 | 0 | 1 | 5 | 10 | 38 |
| Days Min Temp ≤ 32 °F | 31 | 28 | 29 | 18 | 6 | 0 | 0 | 0 | 4 | 17 | 27 | 30 | 190 |
| Days Min Temp ≤ 0 °F | 8 | 4 | 1 | 0 | 0 | 0 | 0 | 0 | 0 | 0 | 1 | 5 | 19 |
| Heating Degree Days | 1312 | 1076 | 938 | 616 | 358 | 125 | 24 | 34 | 235 | 569 | 971 | 1250 | 7508 |
| Cooling Degree Days | 0 | 0 | 0 | 2 | 13 | 92 | 208 | 175 | 30 | 1 | 0 | 0 | 521 |
| Total Precipitation (") | 0.47 | 0.43 | 0.75 | 1.46 | 2.42 | 2.31 | 1.32 | 0.86 | 1.38 | 0.97 | 0.54 | 0.46 | 13.37 |
| Days ≥ 0.1" Precip | 2 | 2 | 2 | 4 | 5 | 5 | 4 | 3 | 4 | 3 | 2 | 2 | 38 |
| Total Snowfall (") | na | na | na | na | 0.6 | ***0.0*** | 0.0 | 0.0 | ***0.0*** | ***1.9*** | na | na | na |
| Days ≥ 1" Snow Depth | na | na | na | na | ***0*** | ***0*** | 0 | 0 | 0 | ***1*** | na | na | na |

## BUFFALO BILL DAM *Park County* ELEVATION 5164 ft LAT/LONG 44° 30' N / 109° 11' W

| | JAN | FEB | MAR | APR | MAY | JUN | JUL | AUG | SEP | OCT | NOV | DEC | YEAR |
|---|---|---|---|---|---|---|---|---|---|---|---|---|---|
| Maximum Temp °F | 36.0 | 39.4 | 45.3 | 52.7 | 60.8 | 68.5 | 74.9 | 75.1 | 68.4 | 58.6 | ***44.8*** | ***37.0*** | 55.1 |
| Minimum Temp °F | 19.8 | 23.2 | 28.1 | 35.0 | 42.9 | 50.0 | 56.2 | 56.0 | 49.1 | 40.3 | 29.6 | ***22.2*** | 37.7 |
| Mean Temp °F | 27.7 | 31.2 | 36.6 | 43.9 | 51.9 | 59.3 | 65.6 | 65.6 | 58.8 | 49.5 | ***37.6*** | ***29.4*** | 46.4 |
| Days Max Temp ≥ 90 °F | 0 | 0 | 0 | 0 | 0 | 0 | 0 | 0 | 0 | 0 | 0 | 0 | 0 |
| Days Max Temp ≤ 32 °F | 10 | 6 | 3 | 1 | 0 | 0 | 0 | 0 | 0 | 1 | 4 | 9 | 34 |
| Days Min Temp ≤ 32 °F | 24 | 21 | 20 | 11 | 2 | 0 | 0 | 0 | 1 | 6 | 17 | 23 | 125 |
| Days Min Temp ≤ 0 °F | 4 | 2 | 1 | 0 | 0 | 0 | 0 | 0 | 0 | 0 | 1 | 2 | 10 |
| Heating Degree Days | 1148 | 947 | 872 | 625 | 399 | 182 | 45 | 43 | 207 | 475 | ***817*** | ***1097*** | 6857 |
| Cooling Degree Days | 0 | 0 | 0 | 0 | 1 | ***19*** | 61 | 67 | 22 | 1 | ***0*** | ***0*** | 171 |
| Total Precipitation (") | 0.37 | 0.35 | 0.69 | 1.38 | 2.18 | 1.79 | 0.88 | 0.77 | 1.09 | 0.88 | 0.51 | 0.28 | 11.17 |
| Days ≥ 0.1" Precip | 1 | 1 | 2 | 3 | 5 | 4 | 3 | 3 | 3 | 2 | 2 | 1 | 30 |
| Total Snowfall (") | ***4.6*** | na | na | ***4.3*** | 0.0 | 0.0 | 0.0 | 0.0 | 0.5 | 1.0 | na | na | na |
| Days ≥ 1" Snow Depth | ***10*** | na | na | 2 | 0 | 0 | 0 | 0 | 0 | 0 | na | ***6*** | na |

## BURGESS JUNCTION *Sheridan County* ELEVATION 8205 ft LAT/LONG 44° 46 ' N / 107° 32 ' W

| | JAN | FEB | MAR | APR | MAY | JUN | JUL | AUG | SEP | OCT | NOV | DEC | YEAR |
|---|---|---|---|---|---|---|---|---|---|---|---|---|---|
| Maximum Temp °F | 27.1 | 29.5 | 33.7 | 40.5 | 50.4 | 61.5 | na | na | ***57.5*** | 46.6 | 33.9 | 28.1 | na |
| Minimum Temp °F | 4.8 | 5.3 | 9.3 | 17.4 | 27.4 | 35.1 | na | na | ***30.3*** | 21.7 | 11.4 | 5.7 | na |
| Mean Temp °F | 16.0 | 17.4 | 21.6 | 29.0 | 38.9 | 48.3 | na | na | ***43.9*** | 34.2 | 22.7 | 16.9 | na |
| Days Max Temp ≥ 90 °F | 0 | 0 | 0 | 0 | 0 | 0 | 0 | 0 | 0 | 0 | 0 | 0 | 0 |
| Days Max Temp ≤ 32 °F | 21 | 16 | 13 | 7 | 1 | 0 | 0 | 0 | 1 | 4 | 13 | 19 | 95 |
| Days Min Temp ≤ 32 °F | 31 | 28 | 31 | 28 | ***21*** | ***10*** | ***3*** | ***4*** | na | 28 | 30 | 31 | na |
| Days Min Temp ≤ 0 °F | 11 | 9 | 6 | 2 | 0 | 0 | 0 | 0 | 0 | 1 | 5 | 10 | 44 |
| Heating Degree Days | 1515 | 1337 | 1341 | 1075 | 802 | 493 | na | na | na | 949 | 1263 | 1484 | na |
| Cooling Degree Days | 0 | 0 | 0 | 0 | 0 | 1 | na | na | ***0*** | 0 | 0 | 0 | na |
| Total Precipitation (") | 1.32 | 1.17 | 2.07 | 2.57 | 2.24 | 2.46 | 1.54 | 1.38 | 1.87 | 1.83 | 1.49 | 1.30 | 21.24 |
| Days ≥ 0.1" Precip | 5 | 4 | 7 | 7 | 6 | 6 | 4 | 3 | 4 | 5 | 5 | 5 | 61 |
| Total Snowfall (") | 28.7 | 24.5 | 39.4 | 34.5 | 13.2 | 4.3 | 0.1 | 0.0 | 7.6 | 21.0 | 27.6 | 28.8 | 229.7 |
| Days ≥ 1" Snow Depth | 31 | 28 | 31 | 30 | 20 | 3 | 0 | 0 | 4 | 14 | 27 | 31 | 219 |

## BURRIS *Fremont County* ELEVATION 6120 ft LAT/LONG 43° 22 ' N / 109° 16 ' W

| | JAN | FEB | MAR | APR | MAY | JUN | JUL | AUG | SEP | OCT | NOV | DEC | YEAR |
|---|---|---|---|---|---|---|---|---|---|---|---|---|---|
| Maximum Temp °F | 36.9 | 40.8 | 46.8 | 55.6 | 64.6 | 73.8 | 80.7 | 79.4 | 69.7 | 60.0 | 44.3 | 37.0 | 57.5 |
| Minimum Temp °F | 10.9 | 13.7 | 20.0 | 27.8 | 36.3 | 44.0 | 49.0 | 47.6 | 38.8 | 30.9 | 19.4 | 11.2 | 29.1 |
| Mean Temp °F | 23.9 | 27.3 | 33.5 | 41.7 | 50.5 | 58.9 | 64.8 | 63.5 | 54.3 | 45.5 | 31.9 | 24.1 | 43.3 |
| Days Max Temp ≥ 90 °F | 0 | 0 | 0 | 0 | 0 | 1 | 1 | 0 | 0 | 0 | 0 | 0 | 2 |
| Days Max Temp ≤ 32 °F | 10 | 5 | 3 | 0 | 0 | 0 | 0 | 0 | 0 | 0 | 5 | 9 | 32 |
| Days Min Temp ≤ 32 °F | 30 | 27 | 29 | 21 | 10 | 1 | 0 | 0 | 6 | 17 | 27 | 29 | 197 |
| Days Min Temp ≤ 0 °F | 7 | 4 | 2 | 0 | 0 | 0 | 0 | 0 | 0 | 0 | 2 | 6 | 21 |
| Heating Degree Days | 1267 | 1059 | 971 | 693 | 445 | 197 | 60 | 83 | 315 | 601 | 984 | 1258 | 7933 |
| Cooling Degree Days | 0 | 0 | 0 | 0 | 1 | 24 | 67 | 42 | 4 | 0 | 0 | 0 | 138 |
| Total Precipitation (") | 0.15 | 0.16 | 0.54 | 1.00 | 1.68 | 1.55 | 0.98 | 0.76 | 0.96 | 0.67 | 0.37 | 0.19 | 9.01 |
| Days ≥ 0.1" Precip | 1 | 1 | 2 | 3 | 5 | 4 | 3 | 2 | 2 | 2 | 1 | 1 | 27 |
| Total Snowfall (") | ***3.3*** | ***2.0*** | na | na | 1.8 | 0.0 | 0.0 | 0.0 | 1.3 | 1.6 | ***3.9*** | na | na |
| Days ≥ 1" Snow Depth | na | na | na | na | ***0*** | 0 | 0 | 0 | 0 | ***1*** | na | na | na |

## CARPENTER *Laramie County* ELEVATION 5390 ft LAT/LONG 41° 3 ' N / 104° 21 ' W

| | JAN | FEB | MAR | APR | MAY | JUN | JUL | AUG | SEP | OCT | NOV | DEC | YEAR |
|---|---|---|---|---|---|---|---|---|---|---|---|---|---|
| Maximum Temp °F | 40.3 | 44.2 | 50.0 | 59.7 | 69.2 | 79.4 | 86.7 | 84.9 | 76.0 | 64.0 | 48.6 | 41.0 | 62.0 |
| Minimum Temp °F | 13.2 | 16.3 | 21.9 | 29.6 | 38.7 | 48.0 | 53.9 | 52.0 | 43.1 | 32.2 | 21.2 | 14.1 | 32.0 |
| Mean Temp °F | 26.8 | 30.3 | 36.0 | 44.7 | 54.0 | 63.7 | 70.3 | 68.5 | 59.6 | 48.1 | 34.9 | 27.6 | 47.0 |
| Days Max Temp ≥ 90 °F | 0 | 0 | 0 | 0 | 0 | 4 | 12 | 9 | 2 | 0 | 0 | 0 | 27 |
| Days Max Temp ≤ 32 °F | 7 | 5 | 3 | 0 | 0 | 0 | 0 | 0 | 0 | 0 | 3 | 7 | 25 |
| Days Min Temp ≤ 32 °F | 30 | 27 | 28 | 19 | 6 | 0 | 0 | 0 | 3 | 15 | 27 | 30 | 185 |
| Days Min Temp ≤ 0 °F | 5 | 3 | 1 | 0 | 0 | 0 | 0 | 0 | 0 | 0 | 1 | 4 | 14 |
| Heating Degree Days | 1179 | 974 | 892 | 603 | 339 | 97 | 12 | 22 | 183 | 516 | 895 | 1156 | 6868 |
| Cooling Degree Days | 0 | 0 | 0 | 0 | 6 | 74 | 179 | 148 | 30 | 0 | 0 | 0 | 437 |
| Total Precipitation (") | 0.32 | 0.20 | 0.86 | 1.23 | 2.13 | 2.48 | 2.41 | 1.44 | 1.37 | 0.70 | 0.43 | 0.33 | 13.90 |
| Days ≥ 0.1" Precip | 1 | 1 | 2 | 3 | 5 | 6 | 5 | 4 | 3 | 2 | 1 | 1 | 34 |
| Total Snowfall (") | 5.1 | 2.9 | 7.0 | 4.1 | 0.6 | 0.0 | 0.0 | 0.0 | 0.5 | 2.5 | 4.6 | 5.1 | 32.4 |
| Days ≥ 1" Snow Depth | 6 | 4 | 4 | 1 | 0 | 0 | 0 | 0 | 0 | 1 | 3 | 6 | 25 |

## CASPER NATRONA CO AP *Natrona County* ELEVATION 5321 ft LAT/LONG 42° 55 ' N / 106° 28 ' W

| | JAN | FEB | MAR | APR | MAY | JUN | JUL | AUG | SEP | OCT | NOV | DEC | YEAR |
|---|---|---|---|---|---|---|---|---|---|---|---|---|---|
| Maximum Temp °F | 33.3 | 37.3 | 46.4 | 56.6 | 66.9 | 78.9 | 87.3 | 85.8 | 74.3 | 60.3 | 43.6 | 34.3 | 58.7 |
| Minimum Temp °F | 12.6 | 16.1 | 22.6 | 29.8 | 38.2 | 47.2 | 53.9 | 52.1 | 41.9 | 32.2 | 21.8 | 14.2 | 31.9 |
| Mean Temp °F | 23.0 | 26.7 | 34.5 | 43.3 | 52.6 | 63.1 | 70.6 | 69.0 | 58.1 | 46.3 | 32.7 | 24.2 | 45.3 |
| Days Max Temp ≥ 90 °F | 0 | 0 | 0 | 0 | 0 | 5 | 13 | 11 | 1 | 0 | 0 | 0 | 30 |
| Days Max Temp ≤ 32 °F | 13 | 8 | 4 | 1 | 0 | 0 | 0 | 0 | 0 | 1 | 6 | 13 | 46 |
| Days Min Temp ≤ 32 °F | 29 | 26 | 27 | 19 | 7 | 0 | 0 | 0 | 4 | 15 | 25 | 29 | 181 |
| Days Min Temp ≤ 0 °F | 7 | 4 | 1 | 0 | 0 | 0 | 0 | 0 | 0 | 0 | 2 | 5 | 19 |
| Heating Degree Days | 1296 | 1073 | 937 | 646 | 382 | 117 | 13 | 27 | 226 | 574 | 962 | 1257 | 7510 |
| Cooling Degree Days | 0 | 0 | 0 | 0 | 6 | 79 | 197 | 176 | 28 | 1 | 0 | 0 | 487 |
| Total Precipitation (") | 0.55 | 0.62 | 0.95 | 1.48 | 2.21 | 1.52 | 1.28 | 0.74 | 0.88 | 1.04 | 0.86 | 0.69 | 12.82 |
| Days ≥ 0.1" Precip | 2 | 2 | 3 | 4 | 6 | 4 | 3 | 2 | 2 | 3 | 3 | 2 | 36 |
| Total Snowfall (") | 10.5 | 10.2 | 13.4 | 13.3 | 4.9 | 0.2 | 0.0 | 0.0 | 1.5 | 6.6 | 11.7 | 12.2 | 84.5 |
| Days ≥ 1" Snow Depth | 17 | 12 | 8 | 4 | 1 | 0 | 0 | 0 | 0 | 2 | 9 | 14 | 67 |

## CHEYENNE MUNI AP *Laramie County* ELEVATION 6158 ft LAT/LONG 41° 9 ' N / 104° 49 ' W

| | JAN | FEB | MAR | APR | MAY | JUN | JUL | AUG | SEP | OCT | NOV | DEC | YEAR |
|---|---|---|---|---|---|---|---|---|---|---|---|---|---|
| Maximum Temp °F | 38.2 | 40.8 | 45.9 | 54.9 | 64.5 | 75.0 | 82.3 | 80.4 | 71.2 | 59.4 | 45.8 | 39.0 | 58.1 |
| Minimum Temp °F | 16.0 | 18.3 | 23.0 | 30.4 | 39.5 | 48.4 | 54.4 | 52.9 | 43.7 | 33.5 | 23.2 | 17.0 | 33.4 |
| Mean Temp °F | 27.1 | 29.6 | 34.5 | 42.7 | 52.0 | 61.7 | 68.4 | 66.6 | 57.5 | 46.5 | 34.5 | 28.0 | 45.8 |
| Days Max Temp ≥ 90 °F | 0 | 0 | 0 | 0 | 0 | 1 | 5 | 2 | 0 | 0 | 0 | 0 | 8 |
| Days Max Temp ≤ 32 °F | 9 | 7 | 5 | 1 | 0 | 0 | 0 | 0 | 0 | 1 | 5 | 9 | 37 |
| Days Min Temp ≤ 32 °F | 29 | 27 | 27 | 18 | 4 | 0 | 0 | 0 | 2 | 13 | 25 | 29 | 174 |
| Days Min Temp ≤ 0 °F | 4 | 2 | 1 | 0 | 0 | 0 | 0 | 0 | 0 | 0 | 1 | 3 | 11 |
| Heating Degree Days | 1168 | 994 | 939 | 664 | 398 | 136 | 25 | 39 | 235 | 568 | 908 | 1140 | 7214 |
| Cooling Degree Days | 0 | 0 | 0 | 0 | 1 | 49 | 132 | 100 | 18 | 0 | 0 | 0 | 300 |
| Total Precipitation (") | 0.44 | 0.41 | 1.01 | 1.41 | 2.46 | 2.14 | 2.11 | 1.72 | 1.21 | 0.80 | 0.64 | 0.44 | 14.79 |
| Days ≥ 0.1" Precip | 1 | 1 | 3 | 4 | 6 | 5 | 5 | 4 | 3 | 3 | 2 | 1 | 38 |
| Total Snowfall (") | 6.7 | 5.9 | 11.7 | 8.2 | 2.7 | 0.0 | 0.0 | 0.0 | 1.0 | 4.5 | 8.0 | 7.0 | 55.7 |
| Days ≥ 1" Snow Depth | 12 | 9 | 8 | 4 | 1 | 0 | 0 | 0 | 0 | 2 | 8 | 12 | 56 |

## CHUGWATER *Platte County* ELEVATION 5282 ft LAT/LONG 41° 45 ' N / 104° 49 ' W

| | JAN | FEB | MAR | APR | MAY | JUN | JUL | AUG | SEP | OCT | NOV | DEC | YEAR |
|---|---|---|---|---|---|---|---|---|---|---|---|---|---|
| Maximum Temp °F | 39.5 | 42.6 | 49.7 | 59.0 | 68.7 | 79.1 | 86.6 | 85.0 | 76.1 | 63.8 | 48.2 | 40.7 | 61.6 |
| Minimum Temp °F | 16.1 | 17.9 | 22.6 | 29.2 | 37.7 | 46.3 | 51.7 | 49.7 | 40.1 | 31.0 | 22.4 | 16.6 | 31.8 |
| Mean Temp °F | 27.8 | 30.3 | 36.2 | 44.1 | 53.3 | 62.7 | 69.2 | 67.4 | 58.1 | 47.4 | 35.3 | 28.7 | 46.7 |
| Days Max Temp ≥ 90 °F | 0 | 0 | 0 | 0 | 0 | 4 | 12 | 8 | 2 | 0 | 0 | 0 | 26 |
| Days Max Temp ≤ 32 °F | 8 | 5 | 3 | 1 | 0 | 0 | 0 | 0 | 0 | 1 | 4 | 8 | 30 |
| Days Min Temp ≤ 32 °F | 28 | 25 | 26 | 20 | 8 | 1 | 0 | 0 | 5 | 17 | 25 | 28 | 183 |
| Days Min Temp ≤ 0 °F | 5 | 3 | 1 | 0 | 0 | 0 | 0 | 0 | 0 | 0 | 1 | 4 | 14 |
| Heating Degree Days | 1146 | 974 | 886 | 620 | 361 | 114 | 19 | 31 | 218 | 539 | 883 | 1121 | 6912 |
| Cooling Degree Days | 0 | 0 | 0 | 0 | 3 | 55 | 144 | 114 | 18 | 0 | 0 | 0 | 334 |
| Total Precipitation (") | 0.60 | 0.56 | 1.03 | 1.50 | 2.77 | 2.18 | 1.97 | 1.41 | 1.14 | 0.95 | 0.70 | 0.60 | 15.41 |
| Days ≥ 0.1" Precip | 2 | 2 | 3 | 4 | 7 | 5 | 5 | 4 | 3 | 3 | 2 | 2 | 42 |
| Total Snowfall (") | 9.3 | 8.2 | 11.5 | 9.3 | 3.8 | 0.0 | 0.0 | 0.0 | 1.8 | 5.3 | 10.6 | 9.9 | 69.7 |
| Days ≥ 1" Snow Depth | 15 | 12 | 9 | 3 | 0 | 0 | 0 | 0 | 1 | 2 | 9 | 14 | 65 |

## CHURCH BUTTES GAS PL *Uinta County* ELEVATION 7054 ft LAT/LONG 41° 26 ' N / 110° 2 ' W

| | JAN | FEB | MAR | APR | MAY | JUN | JUL | AUG | SEP | OCT | NOV | DEC | YEAR |
|---|---|---|---|---|---|---|---|---|---|---|---|---|---|
| Maximum Temp °F | 29.6 | 33.6 | 41.7 | 51.5 | 63.1 | 73.9 | 82.5 | 80.6 | *70.9* | 57.7 | 41.0 | 31.4 | 54.8 |
| Minimum Temp °F | 8.8 | 11.8 | 18.9 | 26.4 | 35.0 | *43.5* | 49.7 | *48.6* | *39.8* | *30.6* | 18.5 | 10.5 | 28.5 |
| Mean Temp °F | 19.2 | 22.7 | 30.4 | 39.0 | 49.1 | *58.8* | 66.1 | *64.7* | *55.4* | *44.1* | 29.7 | 21.1 | 41.7 |
| Days Max Temp ≥ 90 °F | 0 | 0 | 0 | 0 | 0 | 1 | 3 | 1 | 0 | 0 | 0 | 0 | 5 |
| Days Max Temp ≤ 32 °F | 18 | 12 | 5 | 1 | 0 | 0 | 0 | 0 | 0 | 1 | 6 | 15 | 58 |
| Days Min Temp ≤ 32 °F | 31 | 27 | 29 | 24 | 11 | *1* | 0 | 0 | 5 | 17 | 27 | 29 | 201 |
| Days Min Temp ≤ 0 °F | 8 | 4 | 1 | 0 | 0 | 0 | 0 | 0 | 0 | 0 | 1 | 6 | 20 |
| Heating Degree Days | 1414 | 1188 | 1068 | 775 | 488 | *204* | 41 | *63* | *288* | *641* | 1051 | 1356 | 8577 |
| Cooling Degree Days | 0 | 0 | 0 | 0 | *0* | *31* | *83* | 67 | *6* | *0* | 0 | 0 | 187 |
| Total Precipitation (") | 0.54 | 0.33 | 0.65 | 0.92 | 1.13 | 0.87 | 0.63 | 0.66 | 0.69 | 0.74 | 0.65 | 0.48 | 8.29 |
| Days ≥ 0.1" Precip | *2* | 1 | 2 | 3 | 3 | 2 | 2 | 2 | 2 | 2 | 2 | 1 | 24 |
| Total Snowfall (") | na | na | na | na | *0.5* | 0.0 | 0.0 | 0.0 | 0.2 | *2.3* | na | na | na |
| Days ≥ 1" Snow Depth | na | na | na | *3* | 0 | 0 | 0 | 0 | 0 | *1* | na | na | na |

## CLARK 3 NE *Park County* ELEVATION 4090 ft LAT/LONG 44° 55 ' N / 109° 8 ' W

| | JAN | FEB | MAR | APR | MAY | JUN | JUL | AUG | SEP | OCT | NOV | DEC | YEAR |
|---|---|---|---|---|---|---|---|---|---|---|---|---|---|
| Maximum Temp °F | 35.5 | 43.4 | 51.2 | 61.1 | 70.0 | 79.0 | 85.6 | 84.7 | 74.2 | 62.8 | 47.0 | 37.3 | 61.0 |
| Minimum Temp °F | 7.3 | 13.1 | 20.4 | 29.1 | 38.5 | 47.3 | 52.3 | 50.2 | 39.9 | 29.4 | 18.4 | 9.0 | 29.6 |
| Mean Temp °F | 21.5 | 28.3 | 35.9 | 45.0 | 54.3 | 63.2 | 69.0 | 67.5 | 57.1 | 46.1 | 32.7 | 23.2 | 45.3 |
| Days Max Temp ≥ 90 °F | 0 | 0 | 0 | 0 | 1 | 4 | 11 | 9 | 2 | 0 | 0 | 0 | 27 |
| Days Max Temp ≤ 32 °F | 11 | 5 | 2 | 0 | 0 | 0 | 0 | 0 | 0 | 0 | 4 | 11 | 33 |
| Days Min Temp ≤ 32 °F | 30 | 27 | 28 | 20 | 7 | 0 | 0 | 0 | 6 | 20 | 27 | 30 | 195 |
| Days Min Temp ≤ 0 °F | 9 | 4 | 1 | 0 | 0 | 0 | 0 | 0 | 0 | 0 | 2 | 7 | 23 |
| Heating Degree Days | 1347 | 1031 | 897 | 595 | 334 | 113 | 26 | 36 | 248 | 580 | 962 | 1291 | 7460 |
| Cooling Degree Days | 0 | 0 | 0 | 2 | *11* | 74 | 147 | 135 | 21 | 1 | 0 | 0 | 391 |
| Total Precipitation (") | 0.27 | 0.16 | 0.26 | 0.54 | 1.22 | 1.51 | 0.94 | 0.78 | 0.82 | 0.62 | 0.31 | 0.19 | 7.62 |
| Days ≥ 0.1" Precip | *1* | 1 | 1 | 2 | 4 | 4 | 3 | 2 | 2 | 2 | 1 | 1 | 24 |
| Total Snowfall (") | *6.5* | *2.5* | 4.1 | 2.4 | 0.3 | 0.0 | 0.0 | 0.0 | 0.8 | 1.5 | *3.5* | 4.5 | 26.1 |
| Days ≥ 1" Snow Depth | *7* | na | *2* | 1 | 0 | 0 | 0 | 0 | 0 | 1 | *3* | *7* | na |

## CLEARMONT 5 SW *Sheridan County* ELEVATION 4062 ft LAT/LONG 44° 36' N / 106° 27' W

| | JAN | FEB | MAR | APR | MAY | JUN | JUL | AUG | SEP | OCT | NOV | DEC | YEAR |
|---|---|---|---|---|---|---|---|---|---|---|---|---|---|
| Maximum Temp °F | 30.8 | 36.4 | 46.6 | 57.9 | 67.8 | 77.0 | 84.8 | 83.1 | 71.7 | 60.0 | 42.9 | 32.8 | 57.7 |
| Minimum Temp °F | 4.7 | 10.3 | 20.3 | 28.8 | 37.8 | 46.8 | 51.7 | 49.2 | 39.2 | 28.6 | 16.9 | 7.0 | 28.4 |
| Mean Temp °F | 17.8 | 23.4 | 33.5 | 43.4 | 52.9 | 61.9 | 68.3 | 66.2 | 55.5 | 44.3 | 29.9 | 19.9 | 43.1 |
| Days Max Temp ≥ 90 °F | 0 | 0 | 0 | 0 | 0 | 3 | 9 | 7 | 1 | 0 | 0 | 0 | 20 |
| Days Max Temp ≤ 32 °F | 14 | 9 | 4 | 1 | 0 | 0 | 0 | 0 | 0 | 1 | 5 | 14 | 48 |
| Days Min Temp ≤ 32 °F | 31 | 28 | 29 | 20 | 7 | 1 | 0 | 0 | 6 | 21 | 29 | 31 | 203 |
| Days Min Temp ≤ 0 °F | 11 | 6 | 2 | 0 | 0 | 0 | 0 | 0 | 0 | 0 | 2 | 9 | 30 |
| Heating Degree Days | 1459 | 1170 | 970 | 642 | 374 | 135 | 28 | 53 | 290 | 635 | 1046 | 1392 | 8194 |
| Cooling Degree Days | 0 | 0 | 0 | 0 | 4 | 56 | 127 | 96 | 11 | 0 | 0 | 0 | 294 |
| Total Precipitation (") | 0.67 | 0.50 | 0.89 | 1.70 | 2.24 | 2.29 | 1.44 | 1.12 | 1.42 | 1.11 | 0.70 | 0.54 | 14.62 |
| Days ≥ 0.1" Precip | 2 | 2 | 3 | 4 | 6 | 6 | 3 | 3 | 4 | 4 | 3 | 2 | 42 |
| Total Snowfall (") | 9.7 | 6.5 | 7.5 | 6.7 | 1.1 | 0.1 | 0.0 | 0.0 | 1.1 | 2.3 | 7.1 | 9.2 | 51.3 |
| Days ≥ 1" Snow Depth | 24 | 19 | 10 | 2 | 1 | 0 | 0 | 0 | 0 | 1 | 7 | 18 | 82 |

## CODY *Park County* ELEVATION 5002 ft LAT/LONG 44° 32' N / 109° 4' W

| | JAN | FEB | MAR | APR | MAY | JUN | JUL | AUG | SEP | OCT | NOV | DEC | YEAR |
|---|---|---|---|---|---|---|---|---|---|---|---|---|---|
| Maximum Temp °F | 35.5 | 40.6 | 48.1 | 56.4 | 66.2 | 76.3 | 84.0 | 82.0 | 71.5 | 60.8 | 44.9 | 36.9 | 58.6 |
| Minimum Temp °F | 13.2 | 17.2 | 24.3 | 31.7 | 40.4 | 48.8 | 54.3 | 52.7 | 43.3 | 34.8 | 23.6 | 15.2 | 33.3 |
| Mean Temp °F | 24.4 | 28.9 | 36.2 | 44.1 | 53.3 | 62.6 | 69.2 | 67.4 | 57.5 | 47.8 | 34.3 | 26.1 | 46.0 |
| Days Max Temp ≥ 90 °F | 0 | 0 | 0 | 0 | 0 | 2 | 7 | 5 | 1 | 0 | 0 | 0 | 15 |
| Days Max Temp ≤ 32 °F | 11 | 6 | 3 | 1 | 0 | 0 | 0 | 0 | 0 | 1 | 4 | 10 | 36 |
| Days Min Temp ≤ 32 °F | 28 | 25 | 25 | 17 | 5 | 0 | 0 | 0 | 3 | 12 | 23 | 28 | 166 |
| Days Min Temp ≤ 0 °F | 7 | 3 | 1 | 0 | 0 | 0 | 0 | 0 | 0 | 0 | 1 | 4 | 16 |
| Heating Degree Days | 1253 | 1013 | 885 | 622 | 364 | 126 | 25 | 42 | 248 | 526 | 916 | 1200 | 7220 |
| Cooling Degree Days | 0 | 0 | 0 | 0 | 8 | 60 | 144 | 120 | 27 | 1 | 0 | 0 | 360 |
| Total Precipitation (") | 0.35 | 0.23 | 0.49 | 0.99 | 1.93 | 1.61 | 1.11 | 0.87 | 1.12 | 0.82 | 0.44 | 0.27 | 10.23 |
| Days ≥ 0.1" Precip | 1 | 1 | 2 | 3 | 6 | 5 | 3 | 3 | 3 | 2 | 1 | 1 | 31 |
| Total Snowfall (") | 8.5 | *5.5* | *6.9* | *4.8* | 0.7 | 0.0 | 0.0 | 0.0 | 0.5 | 4.0 | *6.2* | 7.8 | 44.9 |
| Days ≥ 1" Snow Depth | na | na | na | na | 0 | 0 | 0 | 0 | 0 | *1* | na | na | na |

## CODY 12 SE *Park County* ELEVATION 5325 ft LAT/LONG 44° 23' N / 108° 56' W

| | JAN | FEB | MAR | APR | MAY | JUN | JUL | AUG | SEP | OCT | NOV | DEC | YEAR |
|---|---|---|---|---|---|---|---|---|---|---|---|---|---|
| Maximum Temp °F | 36.1 | 39.6 | 46.0 | 55.0 | 64.7 | 75.1 | 83.2 | 82.1 | 70.4 | 58.5 | 43.8 | 37.6 | 57.7 |
| Minimum Temp °F | 9.8 | 14.2 | 21.3 | 29.6 | 38.5 | 46.7 | 52.6 | 50.8 | 40.0 | 29.9 | 18.6 | 11.1 | 30.3 |
| Mean Temp °F | 23.2 | 27.0 | 33.7 | 42.3 | 51.6 | 60.9 | 67.9 | 66.4 | 55.2 | 44.2 | 31.2 | 24.4 | 44.0 |
| Days Max Temp ≥ 90 °F | 0 | 0 | 0 | 0 | 0 | 2 | 7 | 6 | 1 | 0 | 0 | 0 | 16 |
| Days Max Temp ≤ 32 °F | 11 | 7 | 4 | 1 | 0 | 0 | 0 | 0 | 0 | 1 | 5 | 9 | 38 |
| Days Min Temp ≤ 32 °F | 29 | 27 | 27 | 20 | 7 | 0 | 0 | 0 | 6 | 19 | 27 | 29 | 191 |
| Days Min Temp ≤ 0 °F | 8 | 5 | 2 | 0 | 0 | 0 | 0 | 0 | 0 | 0 | 3 | 7 | 25 |
| Heating Degree Days | 1289 | 1063 | 963 | 674 | 413 | 164 | 39 | 60 | 302 | 638 | 1008 | 1254 | 7867 |
| Cooling Degree Days | 0 | 0 | 0 | 0 | 4 | 52 | 126 | 111 | 15 | 0 | 0 | 0 | 308 |
| Total Precipitation (") | 0.38 | 0.25 | 0.53 | 0.91 | 2.00 | 1.97 | 1.02 | 0.94 | 0.94 | 0.63 | 0.43 | 0.31 | 10.31 |
| Days ≥ 0.1" Precip | *1* | 1 | 2 | 3 | 5 | 5 | 3 | 3 | 3 | 2 | 2 | 1 | 31 |
| Total Snowfall (") | na | na | na | na | *0.2* | 0.0 | 0.0 | 0.0 | 0.1 | *0.4* | na | na | na |
| Days ≥ 1" Snow Depth | na | na | na | na | *0* | 0 | 0 | 0 | 0 | *0* | na | na | na |

## CODY 21 SW *Park County* ELEVATION 5882 ft LAT/LONG 44° 20' N / 109° 22' W

| | JAN | FEB | MAR | APR | MAY | JUN | JUL | AUG | SEP | OCT | NOV | DEC | YEAR |
|---|---|---|---|---|---|---|---|---|---|---|---|---|---|
| Maximum Temp °F | *38.0* | *40.8* | *46.3* | *54.9* | *63.3* | *74.0* | 81.4 | 80.4 | 69.5 | *59.4* | *44.1* | *38.1* | 57.5 |
| Minimum Temp °F | *14.4* | *16.4* | *21.6* | *28.5* | *35.9* | *43.1* | 47.8 | 46.7 | 38.1 | *30.8* | *21.0* | *15.2* | 30.0 |
| Mean Temp °F | *26.2* | *28.6* | *34.0* | *41.7* | *49.6* | *58.6* | 64.6 | 63.6 | 53.8 | *45.1* | *32.6* | *26.7* | 43.8 |
| Days Max Temp ≥ 90 °F | *0* | *0* | *0* | *0* | *0* | *1* | 4 | 3 | 0 | *0* | *0* | *0* | 8 |
| Days Max Temp ≤ 32 °F | *8* | *6* | *3* | *1* | *0* | *0* | 0 | 0 | 0 | *1* | *4* | *9* | 32 |
| Days Min Temp ≤ 32 °F | *27* | *26* | *27* | *21* | *10* | *2* | 0 | 0 | 7 | *18* | *25* | *28* | 191 |
| Days Min Temp ≤ 0 °F | *6* | *4* | *2* | *0* | *0* | *0* | 0 | 0 | 0 | *0* | *2* | *5* | 19 |
| Heating Degree Days | *1195* | *1021* | *954* | *691* | *471* | *208* | 69 | 89 | 337 | *610* | *966* | *1181* | 7792 |
| Cooling Degree Days | 0 | 0 | 0 | 0 | 1 | 24 | 66 | 52 | 9 | 0 | 0 | 0 | 152 |
| Total Precipitation (") | *0.42* | *0.45* | *0.89* | *1.44* | *2.11* | *1.45* | 1.37 | 0.99 | 1.32 | *1.26* | *0.75* | *0.55* | 13.00 |
| Days ≥ 0.1" Precip | *2* | *1* | *3* | *4* | *6* | *4* | 4 | 3 | 3 | *3* | *3* | *2* | 38 |
| Total Snowfall (") | *4.9* | *5.1* | *7.9* | *7.3* | *1.0* | *0.1* | 0.0 | 0.0 | *2.3* | *5.8* | *7.8* | *7.1* | 49.3 |
| Days ≥ 1" Snow Depth | *15* | *10* | *6* | *4* | *0* | *0* | 0 | 0 | 1 | *2* | *10* | *14* | 62 |

## COLONY *Crook County* ELEVATION 3524 ft LAT/LONG 44° 54' N / 104° 10' W

| | JAN | FEB | MAR | APR | MAY | JUN | JUL | AUG | SEP | OCT | NOV | DEC | YEAR |
|---|---|---|---|---|---|---|---|---|---|---|---|---|---|
| Maximum Temp °F | 31.9 | 36.7 | 45.5 | 58.0 | 69.0 | 79.0 | 87.7 | 86.7 | 74.8 | 60.9 | 44.0 | 34.2 | 59.0 |
| Minimum Temp °F | 10.8 | 15.4 | 23.0 | 32.6 | 42.5 | 51.7 | 58.0 | 56.4 | 46.2 | 35.6 | 23.4 | 14.1 | 34.1 |
| Mean Temp °F | 21.5 | 26.0 | 34.3 | 45.3 | 55.8 | 65.4 | 72.9 | 71.6 | 60.5 | 48.3 | 33.7 | 24.2 | 46.6 |
| Days Max Temp ≥ 90 °F | 0 | 0 | 0 | 0 | 1 | 5 | 14 | 13 | 3 | 0 | 0 | 0 | 36 |
| Days Max Temp ≤ 32 °F | 14 | 10 | 6 | 1 | 0 | 0 | 0 | 0 | 0 | 0 | 6 | 13 | 50 |
| Days Min Temp ≤ 32 °F | 29 | 26 | 26 | 16 | 3 | 0 | 0 | 0 | 2 | 11 | 24 | 29 | 166 |
| Days Min Temp ≤ 0 °F | 9 | 5 | 1 | 0 | 0 | 0 | 0 | 0 | 0 | 0 | 1 | 5 | 21 |
| Heating Degree Days | 1341 | 1093 | 946 | 587 | 301 | 83 | 11 | 20 | 189 | 515 | 933 | 1259 | 7278 |
| Cooling Degree Days | 0 | 0 | 0 | 5 | 28 | 115 | 278 | 241 | 64 | 4 | 0 | 0 | 735 |
| Total Precipitation (") | 0.33 | 0.40 | 0.76 | 1.57 | 2.27 | 2.83 | 1.80 | 1.18 | 1.33 | 1.10 | 0.55 | 0.49 | 14.61 |
| Days ≥ 0.1" Precip | 1 | 2 | 2 | 4 | 5 | 6 | 4 | 3 | 3 | 3 | 2 | 1 | 36 |
| Total Snowfall (") | 5.0 | 4.7 | 7.5 | 5.7 | 1.0 | 0.2 | 0.0 | 0.0 | 0.1 | 1.5 | 5.9 | 7.4 | 39.0 |
| Days ≥ 1" Snow Depth | 24 | 17 | 13 | 3 | 1 | 0 | 0 | 0 | 0 | 1 | 9 | 20 | 88 |

## DEAVER *Big Horn County* ELEVATION 4104 ft LAT/LONG 44° 53' N / 108° 36' W

| | JAN | FEB | MAR | APR | MAY | JUN | JUL | AUG | SEP | OCT | NOV | DEC | YEAR |
|---|---|---|---|---|---|---|---|---|---|---|---|---|---|
| Maximum Temp °F | 30.6 | 38.8 | 50.9 | 61.5 | 71.5 | 80.4 | 88.1 | 86.7 | 75.1 | 62.4 | 44.2 | 32.8 | 60.3 |
| Minimum Temp °F | 4.9 | 10.7 | 20.4 | 29.0 | 39.4 | 47.6 | 53.0 | 50.6 | 40.2 | 29.6 | 17.6 | 7.0 | 29.2 |
| Mean Temp °F | 17.6 | 24.8 | 35.7 | 45.3 | 55.5 | 64.0 | 70.6 | 68.7 | 57.7 | 46.0 | 30.9 | 19.9 | 44.7 |
| Days Max Temp ≥ 90 °F | 0 | 0 | 0 | 0 | 1 | 5 | 14 | 13 | 2 | 0 | 0 | 0 | 35 |
| Days Max Temp ≤ 32 °F | 16 | 7 | 2 | 0 | 0 | 0 | 0 | 0 | 0 | 0 | 4 | 14 | 43 |
| Days Min Temp ≤ 32 °F | 31 | 28 | 29 | 21 | 5 | 0 | 0 | 0 | 5 | 20 | 29 | 31 | 199 |
| Days Min Temp ≤ 0 °F | 11 | 5 | 1 | 0 | 0 | 0 | 0 | 0 | 0 | 0 | 2 | 7 | 26 |
| Heating Degree Days | 1466 | 1129 | 901 | 585 | 299 | 96 | 14 | 28 | 236 | 582 | 1016 | 1393 | 7745 |
| Cooling Degree Days | 0 | 0 | 0 | 0 | 12 | 80 | 178 | 156 | 18 | 1 | 0 | 0 | 445 |
| Total Precipitation (") | 0.17 | 0.09 | 0.19 | 0.36 | 0.99 | 1.12 | 0.64 | 0.62 | 0.62 | 0.33 | 0.17 | 0.15 | 5.45 |
| Days ≥ 0.1" Precip | 1 | 0 | 1 | 1 | 3 | 3 | 2 | 2 | 2 | 1 | 1 | 1 | 18 |
| Total Snowfall (") | *3.3* | na | na | *1.2* | 0.1 | 0.0 | 0.0 | 0.0 | 0.6 | *0.3* | *1.5* | na | na |
| Days ≥ 1" Snow Depth | na | na | *1* | *0* | 0 | 0 | 0 | 0 | 0 | *0* | *2* | na | na |

## DEVILS TOWER 2 *Crook County* ELEVATION 3862 ft LAT/LONG 44° 35' N / 104° 42' W

| | JAN | FEB | MAR | APR | MAY | JUN | JUL | AUG | SEP | OCT | NOV | DEC | YEAR |
|---|---|---|---|---|---|---|---|---|---|---|---|---|---|
| Maximum Temp °F | 34.1 | 39.5 | 48.9 | 60.0 | 70.2 | 79.3 | 87.3 | 86.5 | 75.8 | 62.9 | 45.4 | 35.7 | 60.5 |
| Minimum Temp °F | 4.5 | 9.6 | 18.7 | 28.2 | 37.6 | 46.9 | 52.5 | 49.6 | 39.3 | 27.9 | 16.9 | 7.4 | 28.3 |
| Mean Temp °F | 19.3 | 24.5 | 33.8 | 44.1 | 53.9 | 63.1 | 69.9 | 68.1 | 57.6 | 45.5 | 31.2 | 21.6 | 44.4 |
| Days Max Temp ≥ 90 °F | 0 | 0 | 0 | 0 | 1 | 4 | 13 | 12 | 3 | 0 | 0 | 0 | 33 |
| Days Max Temp ≤ 32 °F | 11 | 7 | 3 | 0 | 0 | 0 | 0 | 0 | 0 | 0 | 4 | 11 | 36 |
| Days Min Temp ≤ 32 °F | 31 | 28 | 30 | 21 | 9 | 1 | 0 | 0 | 7 | 23 | 29 | 31 | 210 |
| Days Min Temp ≤ 0 °F | 11 | 6 | 2 | 0 | 0 | 0 | 0 | 0 | 0 | 0 | 2 | 8 | 29 |
| Heating Degree Days | 1410 | 1135 | 959 | 620 | 346 | 116 | 24 | 38 | 246 | 599 | 1008 | 1339 | 7840 |
| Cooling Degree Days | 0 | 0 | 0 | 1 | 12 | 76 | 190 | 150 | 28 | 0 | 0 | 0 | 457 |
| Total Precipitation (") | 0.60 | 0.52 | 0.90 | 1.90 | 2.53 | 3.09 | 1.95 | 1.41 | 1.49 | 1.32 | 0.67 | 0.74 | 17.12 |
| Days ≥ 0.1" Precip | 2 | 2 | 2 | 5 | 6 | 6 | 5 | 4 | 4 | 3 | 2 | 3 | 44 |
| Total Snowfall (") | 9.3 | 7.6 | 8.9 | 7.0 | 0.6 | 0.0 | 0.0 | 0.0 | 0.5 | 2.4 | 6.6 | 11.6 | 54.5 |
| Days ≥ 1" Snow Depth | 29 | 25 | 15 | 3 | 0 | 0 | 0 | 0 | 0 | 2 | 8 | 22 | 104 |

## DILLINGER *Campbell County* ELEVATION 4304 ft LAT/LONG 44° 7' N / 105° 7' W

| | JAN | FEB | MAR | APR | MAY | JUN | JUL | AUG | SEP | OCT | NOV | DEC | YEAR |
|---|---|---|---|---|---|---|---|---|---|---|---|---|---|
| Maximum Temp °F | 32.6 | 37.7 | 46.0 | 56.8 | 67.2 | 78.0 | 87.0 | 85.5 | 74.1 | 60.4 | 44.3 | 34.5 | 58.7 |
| Minimum Temp °F | 6.7 | 11.6 | 20.1 | 28.8 | 38.5 | 47.3 | 53.3 | 51.4 | 40.6 | 29.4 | 17.5 | 8.1 | 29.4 |
| Mean Temp °F | 19.7 | 24.7 | 33.0 | 42.8 | 53.0 | 62.7 | 70.2 | 68.5 | 57.3 | 44.9 | 30.9 | 21.3 | 44.1 |
| Days Max Temp ≥ 90 °F | 0 | 0 | 0 | 0 | 0 | 4 | 13 | 12 | 2 | 0 | 0 | 0 | 31 |
| Days Max Temp ≤ 32 °F | 14 | 8 | 5 | 1 | 0 | 0 | 0 | 0 | 0 | 1 | 6 | 13 | 48 |
| Days Min Temp ≤ 32 °F | 31 | 28 | 29 | 21 | 7 | 1 | 0 | 0 | 5 | 20 | 28 | 31 | 201 |
| Days Min Temp ≤ 0 °F | 10 | 6 | 2 | 0 | 0 | 0 | 0 | 0 | 0 | 0 | 2 | 8 | 28 |
| Heating Degree Days | 1400 | 1133 | 984 | 659 | 373 | 127 | 23 | 37 | 254 | 617 | 1016 | 1348 | 7971 |
| Cooling Degree Days | 0 | 0 | 0 | 1 | 8 | 76 | 197 | 169 | 30 | 0 | 0 | 0 | 481 |
| Total Precipitation (") | 0.38 | 0.51 | 0.72 | 1.69 | 2.42 | 2.17 | 1.84 | 1.05 | 1.18 | 1.02 | 0.54 | 0.53 | 14.05 |
| Days ≥ 0.1" Precip | 1 | 2 | 3 | 4 | 6 | 5 | 4 | 3 | 3 | 3 | 2 | 2 | 38 |
| Total Snowfall (") | 6.0 | 6.9 | *6.5* | *6.1* | 0.6 | 0.1 | 0.0 | 0.0 | 0.5 | *2.3* | 6.0 | 8.4 | 43.4 |
| Days ≥ 1" Snow Depth | na | na | na | na | 0 | 0 | 0 | 0 | 0 | *0* | *4* | na | na |

## DIVERSION DAM *Fremont County* ELEVATION 5574 ft LAT/LONG 43° 14 ' N / 108° 56 ' W

| | JAN | FEB | MAR | APR | MAY | JUN | JUL | AUG | SEP | OCT | NOV | DEC | YEAR |
|---|---|---|---|---|---|---|---|---|---|---|---|---|---|
| Maximum Temp °F | 35.2 | 41.4 | 49.9 | 58.4 | 67.2 | 77.5 | 85.4 | 83.3 | 73.5 | 62.2 | 44.3 | 36.3 | 59.5 |
| Minimum Temp °F | 8.1 | 12.5 | 21.4 | 29.8 | 38.7 | 46.8 | 51.8 | 50.0 | 40.8 | 31.3 | 17.4 | 8.3 | 29.7 |
| Mean Temp °F | 21.6 | 27.2 | 35.7 | 44.2 | 52.9 | 62.1 | 68.6 | 66.7 | 57.1 | 46.7 | 30.8 | 22.4 | 44.7 |
| Days Max Temp ≥ 90 °F | 0 | 0 | 0 | 0 | 0 | 3 | 9 | 5 | 1 | 0 | 0 | 0 | 18 |
| Days Max Temp ≤ 32 °F | 11 | 6 | 2 | 0 | 0 | 0 | 0 | 0 | 0 | 0 | 5 | 11 | 35 |
| Days Min Temp ≤ 32 °F | 30 | 27 | 28 | 20 | 7 | 0 | 0 | 0 | 4 | 17 | 27 | 30 | 190 |
| Days Min Temp ≤ 0 °F | 9 | 5 | 1 | 0 | 0 | 0 | 0 | 0 | 0 | 0 | 3 | 8 | 26 |
| Heating Degree Days | 1341 | 1059 | 903 | 619 | 374 | 133 | 28 | 41 | 244 | 561 | 1019 | 1314 | 7636 |
| Cooling Degree Days | 0 | 0 | 0 | 1 | 10 | 75 | 164 | 126 | 20 | 3 | *0* | 0 | 399 |
| Total Precipitation (") | 0.15 | 0.18 | 0.48 | 1.06 | 1.91 | 1.52 | 0.93 | 0.64 | 1.00 | 0.67 | 0.37 | 0.20 | 9.11 |
| Days ≥ 0.1" Precip | 1 | 1 | 2 | 3 | 4 | 3 | 3 | 2 | 2 | 2 | 1 | 1 | 25 |
| Total Snowfall (") | na | *1.4* | *3.6* | na | 0.9 | 0.0 | 0.0 | 0.0 | 0.3 | *0.7* | na | na | na |
| Days ≥ 1" Snow Depth | na | na | na | *0* | 0 | 0 | 0 | 0 | 0 | *0* | *4* | na | na |

## DOUBLE FOUR RANCH *Albany County* ELEVATION 6204 ft LAT/LONG 42° 11 ' N / 105° 24 ' W

| | JAN | FEB | MAR | APR | MAY | JUN | JUL | AUG | SEP | OCT | NOV | DEC | YEAR |
|---|---|---|---|---|---|---|---|---|---|---|---|---|---|
| Maximum Temp °F | 34.2 | 37.4 | 44.6 | 54.1 | 63.7 | 74.5 | 81.6 | 80.1 | 70.7 | 58.7 | 43.6 | 35.0 | 56.5 |
| Minimum Temp °F | 13.3 | 15.5 | 20.6 | 26.8 | 34.7 | 41.8 | 47.4 | 45.1 | 36.6 | 28.6 | 21.0 | 14.2 | 28.8 |
| Mean Temp °F | 23.8 | 26.5 | 32.6 | 40.5 | 49.3 | 58.2 | 64.5 | 62.6 | 53.6 | 43.7 | 32.3 | 24.6 | 42.7 |
| Days Max Temp ≥ 90 °F | 0 | 0 | 0 | 0 | 0 | 1 | 3 | 2 | 0 | 0 | 0 | 0 | 6 |
| Days Max Temp ≤ 32 °F | 12 | 8 | 4 | 1 | 0 | 0 | 0 | 0 | 0 | 1 | 5 | 11 | 42 |
| Days Min Temp ≤ 32 °F | 30 | 27 | 28 | 23 | 12 | 2 | 0 | 1 | 9 | 22 | 26 | 29 | 209 |
| Days Min Temp ≤ 0 °F | 5 | 3 | 2 | 0 | 0 | 0 | 0 | 0 | 0 | 0 | 1 | 4 | 15 |
| Heating Degree Days | 1271 | 1079 | 998 | 728 | 482 | 210 | 61 | 100 | 337 | 654 | 975 | 1245 | 8140 |
| Cooling Degree Days | 0 | 0 | 0 | 0 | 0 | 11 | 41 | 28 | 2 | 0 | 0 | 0 | 82 |
| Total Precipitation (") | 0.38 | 0.47 | 0.93 | 1.73 | 2.76 | 2.42 | 1.73 | 1.37 | 1.18 | 0.83 | 0.79 | 0.53 | 15.12 |
| Days ≥ 0.1" Precip | 1 | 2 | 3 | 4 | 7 | 5 | 5 | 4 | 3 | 3 | 3 | 2 | 42 |
| Total Snowfall (") | 8.0 | 8.1 | 11.9 | 10.4 | 3.7 | 0.3 | 0.0 | 0.0 | 1.4 | 3.2 | 10.5 | 10.8 | 68.3 |
| Days ≥ 1" Snow Depth | 18 | 15 | 11 | 4 | 1 | 0 | 0 | 0 | 0 | 1 | 9 | 17 | 76 |

## DUBOIS *Fremont County* ELEVATION 6926 ft LAT/LONG 43° 33 ' N / 109° 37 ' W

| | JAN | FEB | MAR | APR | MAY | JUN | JUL | AUG | SEP | OCT | NOV | DEC | YEAR |
|---|---|---|---|---|---|---|---|---|---|---|---|---|---|
| Maximum Temp °F | *33.8* | *36.7* | *41.8* | *49.7* | *60.5* | *70.2* | *78.4* | 77.9 | 66.8 | *55.5* | *41.3* | *33.9* | 53.9 |
| Minimum Temp °F | *12.8* | *12.0* | *17.1* | *23.7* | *31.2* | *38.9* | *42.8* | 41.2 | 33.4 | *26.8* | *18.3* | *12.6* | 25.9 |
| Mean Temp °F | *23.3* | *24.4* | *29.5* | *36.7* | *45.9* | *54.6* | *60.6* | 59.6 | 50.2 | *41.2* | *29.8* | *23.3* | 39.9 |
| Days Max Temp ≥ 90 °F | 0 | 0 | 0 | 0 | 0 | 0 | *1* | *0* | 0 | 0 | 0 | 0 | 1 |
| Days Max Temp ≤ 32 °F | na | *8* | *4* | *1* | 0 | 0 | 0 | 0 | 0 | 1 | na | na | na |
| Days Min Temp ≤ 32 °F | *30* | na | na | na | na | na | *1* | 2 | na | na | *28* | *30* | na |
| Days Min Temp ≤ 0 °F | na | na | *3* | 0 | *0* | 0 | 0 | 0 | 0 | 0 | *2* | *4* | na |
| Heating Degree Days | *1287* | *1136* | *1094* | *842* | *586* | *312* | *149* | 168 | 437 | *730* | *1049* | *1293* | 9083 |
| Cooling Degree Days | *0* | na | *0* | na | *0* | *10* | *22* | 8 | 0 | 0 | 0 | 0 | na |
| Total Precipitation (") | 0.21 | 0.24 | 0.40 | 0.98 | 1.20 | 1.45 | 0.96 | 0.87 | 1.08 | 0.65 | 0.46 | 0.25 | 8.75 |
| Days ≥ 0.1" Precip | *1* | 1 | 2 | *3* | *4* | 4 | 3 | 3 | 3 | 2 | *2* | *1* | 29 |
| Total Snowfall (") | na | na | na | *5.0* | *2.0* | 0.2 | 0.0 | 0.0 | 2.3 | *1.3* | na | na | na |
| Days ≥ 1" Snow Depth | na | na | na | na | *1* | 0 | 0 | 0 | 1 | *0* | na | na | na |

## DULL CENTER 1 SE *Converse County* ELEVATION 4354 ft LAT/LONG 43° 25 ' N / 104° 59 ' W

| | JAN | FEB | MAR | APR | MAY | JUN | JUL | AUG | SEP | OCT | NOV | DEC | YEAR |
|---|---|---|---|---|---|---|---|---|---|---|---|---|---|
| Maximum Temp °F | 34.7 | 40.0 | 48.6 | 59.1 | 69.3 | 80.6 | 89.5 | 88.2 | 76.8 | 63.4 | 45.8 | 36.4 | 61.0 |
| Minimum Temp °F | 10.9 | 15.2 | 22.3 | 31.0 | 40.4 | 49.2 | 55.0 | 53.4 | 43.1 | 32.3 | 21.5 | 12.6 | 32.2 |
| Mean Temp °F | 22.8 | 27.6 | 35.5 | 45.0 | 54.9 | 64.9 | 72.2 | 70.8 | 59.9 | 47.9 | 33.7 | 24.5 | 46.6 |
| Days Max Temp ≥ 90 °F | 0 | 0 | 0 | 0 | 0 | 6 | 15 | *14* | 4 | 0 | 0 | 0 | 39 |
| Days Max Temp ≤ 32 °F | 12 | 7 | 3 | 0 | 0 | 0 | 0 | 0 | 0 | 0 | 5 | 11 | 38 |
| Days Min Temp ≤ 32 °F | 29 | 26 | 27 | 17 | 5 | 0 | 0 | 0 | 3 | 15 | 25 | 29 | 176 |
| Days Min Temp ≤ 0 °F | 8 | 4 | 1 | 0 | 0 | 0 | 0 | 0 | 0 | 0 | 2 | 6 | 21 |
| Heating Degree Days | 1301 | 1050 | 909 | 593 | 318 | 89 | 10 | 17 | 188 | 524 | 934 | 1250 | 7183 |
| Cooling Degree Days | 0 | 0 | 0 | 1 | 12 | 100 | 232 | 212 | 42 | 0 | 0 | 0 | 599 |
| Total Precipitation (") | 0.25 | 0.37 | 0.70 | 1.44 | 2.28 | 2.09 | 1.73 | 1.28 | 0.92 | 0.80 | 0.56 | 0.38 | 12.80 |
| Days ≥ 0.1" Precip | 1 | 1 | 2 | 4 | 5 | 5 | 4 | 3 | 3 | 2 | 2 | 2 | 34 |
| Total Snowfall (") | 6.2 | 6.6 | 9.5 | 8.0 | 1.8 | 0.2 | 0.0 | 0.0 | 0.7 | 3.0 | 7.3 | 8.2 | 51.5 |
| Days ≥ 1" Snow Depth | 18 | 11 | 7 | 2 | 0 | 0 | 0 | 0 | 0 | 1 | 7 | 14 | 60 |

## ECHETA 2 NW *Campbell County* ELEVATION 4081 ft LAT/LONG 44° 28 ' N / 105° 52 ' W

| | JAN | FEB | MAR | APR | MAY | JUN | JUL | AUG | SEP | OCT | NOV | DEC | YEAR |
|---|---|---|---|---|---|---|---|---|---|---|---|---|---|
| Maximum Temp °F | 33.1 | 38.9 | 48.1 | 59.4 | 69.2 | 80.0 | 89.2 | 87.6 | 75.8 | 62.8 | 45.2 | 35.5 | 60.4 |
| Minimum Temp °F | 6.5 | 12.3 | 20.9 | 29.9 | 38.7 | 47.8 | 54.0 | 51.7 | 41.0 | 30.5 | 18.8 | 8.9 | 30.1 |
| Mean Temp °F | 19.9 | 25.6 | 34.5 | 44.7 | 54.0 | 63.9 | 71.7 | 69.7 | 58.5 | 46.7 | 32.0 | 22.2 | 45.3 |
| Days Max Temp ≥ 90 °F | 0 | 0 | 0 | 0 | 1 | 5 | *15* | 13 | 3 | 0 | 0 | 0 | 37 |
| Days Max Temp ≤ 32 °F | 13 | 8 | 3 | 1 | 0 | 0 | 0 | 0 | 0 | 1 | 5 | 12 | 43 |
| Days Min Temp ≤ 32 °F | 30 | 27 | 28 | 19 | 6 | 0 | 0 | 0 | 4 | 18 | 28 | 30 | 190 |
| Days Min Temp ≤ 0 °F | 10 | 6 | 2 | 0 | 0 | 0 | 0 | 0 | 0 | 0 | 3 | 8 | 29 |
| Heating Degree Days | 1395 | 1103 | 939 | 605 | 344 | 107 | 18 | 31 | 226 | 560 | 981 | 1320 | 7629 |
| Cooling Degree Days | 0 | 0 | 0 | 1 | 13 | 106 | 238 | *185* | 33 | 1 | 0 | 0 | 577 |
| Total Precipitation (") | 0.55 | 0.52 | 0.85 | 2.19 | 2.72 | 2.48 | 1.31 | 1.08 | 1.39 | 1.34 | 0.81 | 0.60 | 15.84 |
| Days ≥ 0.1" Precip | 3 | 2 | 3 | 5 | 7 | 6 | 4 | 3 | 3 | 4 | 3 | 3 | 46 |
| Total Snowfall (") | *9.5* | *8.0* | 10.4 | *6.6* | 1.5 | 0.0 | 0.0 | 0.0 | 0.9 | *3.0* | *8.5* | *11.0* | 59.4 |
| Days ≥ 1" Snow Depth | na | na | na | *2* | 0 | 0 | 0 | 0 | 0 | *1* | na | na | na |

## ELK MOUNTAIN *Carbon County* ELEVATION 7116 ft LAT/LONG 41° 42 ' N / 106° 25 ' W

| | JAN | FEB | MAR | APR | MAY | JUN | JUL | AUG | SEP | OCT | NOV | DEC | YEAR |
|---|---|---|---|---|---|---|---|---|---|---|---|---|---|
| Maximum Temp °F | 32.6 | 34.8 | 40.9 | 50.2 | 61.1 | 72.1 | 79.4 | 77.1 | 68.1 | 55.9 | 40.5 | 33.5 | 53.9 |
| Minimum Temp °F | 14.1 | 15.1 | 20.3 | 26.7 | 34.4 | 42.3 | 47.4 | 45.8 | 37.8 | 30.3 | 20.7 | 14.6 | 29.1 |
| Mean Temp °F | 23.4 | 25.0 | 30.6 | 38.4 | 47.8 | 57.2 | 63.4 | 61.4 | 53.0 | 43.1 | 30.6 | 24.1 | 41.5 |
| Days Max Temp ≥ 90 °F | 0 | 0 | 0 | 0 | 0 | 0 | 0 | 0 | 0 | 0 | 0 | 0 | 0 |
| Days Max Temp ≤ 32 °F | 14 | 10 | 5 | 2 | 0 | 0 | 0 | 0 | 0 | 1 | 7 | 13 | 52 |
| Days Min Temp ≤ 32 °F | 30 | 27 | 28 | 22 | 11 | 3 | 0 | 0 | 7 | 18 | 26 | 29 | 201 |
| Days Min Temp ≤ 0 °F | 4 | 3 | 2 | 0 | 0 | 0 | 0 | 0 | 0 | 0 | 1 | 3 | 13 |
| Heating Degree Days | 1284 | 1124 | 1059 | 790 | 527 | 237 | 76 | 120 | 355 | 672 | 1024 | 1262 | 8530 |
| Cooling Degree Days | 0 | 0 | 0 | 0 | 0 | 11 | 29 | 16 | 1 | 0 | 0 | 0 | 57 |
| Total Precipitation (") | 0.72 | 0.69 | 1.03 | 1.40 | 1.77 | 1.27 | 0.99 | 0.99 | 1.03 | 1.02 | 1.04 | 0.79 | 12.74 |
| Days ≥ 0.1" Precip | 3 | 2 | 3 | 4 | 4 | 4 | 2 | 3 | 3 | 3 | 3 | 3 | 37 |
| Total Snowfall (") | 12.2 | 10.5 | 13.2 | 11.6 | 3.8 | 0.5 | 0.0 | 0.0 | 1.6 | 5.1 | 13.3 | 13.5 | 85.3 |
| Days ≥ 1" Snow Depth | 26 | 25 | 20 | 7 | 1 | 0 | 0 | 0 | 1 | 3 | 13 | 20 | 116 |

## EMBLEM *Big Horn County* ELEVATION 4442 ft LAT/LONG 44° 31 ' N / 108° 22 ' W

| | JAN | FEB | MAR | APR | MAY | JUN | JUL | AUG | SEP | OCT | NOV | DEC | YEAR |
|---|---|---|---|---|---|---|---|---|---|---|---|---|---|
| Maximum Temp °F | 29.5 | 37.1 | 48.3 | 59.1 | 69.3 | 79.1 | 86.8 | 85.0 | 73.2 | 59.8 | 42.6 | 31.5 | 58.4 |
| Minimum Temp °F | 6.4 | 12.8 | 22.6 | 31.4 | 41.3 | 49.1 | 54.2 | 51.9 | 42.3 | 32.0 | 19.7 | 8.9 | 31.1 |
| Mean Temp °F | 18.0 | 24.9 | 35.5 | 45.3 | 55.3 | 64.1 | 70.5 | 68.5 | 57.8 | 45.9 | 31.2 | 20.2 | 44.8 |
| Days Max Temp ≥ 90 °F | 0 | 0 | 0 | 0 | 0 | 5 | 12 | 9 | 1 | 0 | 0 | 0 | 27 |
| Days Max Temp ≤ 32 °F | 16 | 8 | 3 | 0 | 0 | 0 | 0 | 0 | 0 | 0 | 5 | 14 | 46 |
| Days Min Temp ≤ 32 °F | 31 | 28 | 28 | 17 | 4 | 0 | 0 | 0 | 3 | 16 | 28 | 31 | 186 |
| Days Min Temp ≤ 0 °F | 9 | 4 | 1 | 0 | 0 | 0 | 0 | 0 | 0 | 0 | 2 | 7 | 23 |
| Heating Degree Days | 1454 | 1125 | 909 | 585 | 305 | 98 | 16 | 32 | 233 | 585 | 1009 | 1381 | 7732 |
| Cooling Degree Days | 0 | 0 | 0 | 1 | 14 | 84 | 192 | 160 | 23 | 0 | 0 | 0 | 474 |
| Total Precipitation (") | 0.26 | 0.19 | 0.45 | 0.74 | 1.33 | 1.39 | 0.71 | 0.60 | 0.88 | 0.65 | 0.32 | 0.25 | 7.77 |
| Days ≥ 0.1" Precip | 1 | 1 | 2 | 2 | 4 | 4 | 2 | 2 | 2 | 2 | 1 | 1 | 24 |
| Total Snowfall (") | *3.3* | *1.9* | na | *1.7* | 0.3 | 0.1 | 0.0 | 0.0 | 0.3 | 1.7 | *2.0* | *3.2* | na |
| Days ≥ 1" Snow Depth | na | na | na | *0* | 0 | 0 | 0 | 0 | 0 | 0 | na | na | na |

## ENCAMPMENT 11 ESE *Carbon County* ELEVATION 7395 ft LAT/LONG 41° 11 ' N / 106° 37 ' W

| | JAN | FEB | MAR | APR | MAY | JUN | JUL | AUG | SEP | OCT | NOV | DEC | YEAR |
|---|---|---|---|---|---|---|---|---|---|---|---|---|---|
| Maximum Temp °F | 32.7 | 35.7 | 41.9 | 52.2 | 62.6 | 72.1 | 78.5 | 77.5 | 69.0 | 57.7 | 42.4 | 34.2 | 54.7 |
| Minimum Temp °F | 10.7 | 12.0 | 17.6 | 24.6 | 33.0 | 40.2 | 46.4 | 44.8 | 36.5 | 27.3 | 17.6 | 10.8 | 26.8 |
| Mean Temp °F | 21.7 | 23.9 | 29.8 | 38.4 | 47.9 | 56.1 | 62.4 | 61.2 | 52.8 | 42.5 | 30.0 | 22.6 | 40.8 |
| Days Max Temp ≥ 90 °F | 0 | 0 | 0 | 0 | 0 | 0 | 0 | 0 | 0 | 0 | 0 | 0 | 0 |
| Days Max Temp ≤ 32 °F | 14 | 9 | 5 | 1 | 0 | 0 | 0 | 0 | 0 | 1 | 6 | 13 | 49 |
| Days Min Temp ≤ 32 °F | 31 | 28 | 30 | 25 | 14 | 3 | 0 | 1 | 8 | 23 | 28 | 31 | 222 |
| Days Min Temp ≤ 0 °F | 6 | 4 | 2 | 0 | 0 | 0 | 0 | 0 | 0 | 0 | 2 | 6 | 20 |
| Heating Degree Days | 1336 | 1154 | 1085 | 791 | 524 | 263 | 95 | 124 | 360 | 689 | 1044 | 1309 | 8774 |
| Cooling Degree Days | 0 | 0 | 0 | 0 | 0 | 4 | 16 | 8 | 1 | 0 | 0 | 0 | 29 |
| Total Precipitation (") | 0.68 | 0.86 | 1.23 | 1.54 | 1.65 | 1.31 | 1.33 | 1.29 | 1.17 | 1.39 | 1.10 | 0.89 | 14.44 |
| Days ≥ 0.1" Precip | 3 | 3 | 5 | 5 | 5 | 4 | 4 | 4 | 3 | 4 | 3 | 3 | 46 |
| Total Snowfall (") | 9.6 | 12.0 | 13.2 | 9.9 | 2.3 | 0.6 | 0.0 | 0.0 | 1.3 | 4.6 | *12.0* | 11.5 | 77.0 |
| Days ≥ 1" Snow Depth | na | na | na | na | 0 | 0 | 0 | 0 | 0 | *1* | na | na | na |

## EVANSTON 1 E *Uinta County* ELEVATION 6863 ft LAT/LONG 41° 16 ' N / 110° 57 ' W

| | JAN | FEB | MAR | APR | MAY | JUN | JUL | AUG | SEP | OCT | NOV | DEC | YEAR |
|---|---|---|---|---|---|---|---|---|---|---|---|---|---|
| Maximum Temp °F | 31.7 | 35.3 | 42.3 | 53.0 | 63.2 | 73.5 | 82.2 | 80.5 | 71.0 | 58.4 | 41.8 | 33.2 | 55.5 |
| Minimum Temp °F | 9.2 | 10.7 | 18.1 | 26.0 | 32.9 | 39.6 | 45.4 | 44.1 | 36.2 | 27.4 | 17.6 | 9.5 | 26.4 |
| Mean Temp °F | 20.5 | 23.0 | 30.3 | 39.6 | 48.1 | 56.6 | 63.8 | 62.3 | 53.6 | 42.9 | 29.7 | 21.4 | 41.0 |
| Days Max Temp ≥ 90 °F | 0 | 0 | 0 | 0 | 0 | 0 | 2 | 1 | 0 | 0 | 0 | 0 | 3 |
| Days Max Temp ≤ 32 °F | 15 | 10 | 4 | 0 | 0 | 0 | 0 | 0 | 0 | 1 | 6 | 15 | 51 |
| Days Min Temp ≤ 32 °F | 30 | 28 | 30 | 25 | 15 | 4 | 0 | 1 | 9 | 24 | 29 | 31 | 226 |
| Days Min Temp ≤ 0 °F | 8 | 5 | 1 | 0 | 0 | 0 | 0 | 0 | 0 | 0 | 2 | 7 | 23 |
| Heating Degree Days | 1373 | 1177 | 1070 | 754 | 517 | 253 | 67 | 99 | 337 | 678 | 1052 | 1347 | 8724 |
| Cooling Degree Days | 0 | 0 | 0 | *1* | 0 | 10 | 41 | 27 | *2* | 0 | 0 | 0 | 81 |
| Total Precipitation (") | 0.59 | 0.60 | 0.87 | 1.12 | 1.31 | 0.95 | 0.94 | 0.94 | 1.14 | 1.19 | 0.79 | 0.64 | 11.08 |
| Days ≥ 0.1" Precip | 2 | 2 | 3 | 3 | 4 | 3 | 3 | 3 | 3 | 4 | 3 | 2 | 35 |
| Total Snowfall (") | na | na | na | na | na | *0.0* | *0.0* | *0.0* | na | na | na | na | na |
| Days ≥ 1" Snow Depth | na | na | na | na | *0* | *0* | *0* | *0* | na | na | na | na | na |

## FONTENELLE DAM *Lincoln County* ELEVATION 6483 ft LAT/LONG 41° 59 ' N / 110° 4 ' W

| | JAN | FEB | MAR | APR | MAY | JUN | JUL | AUG | SEP | OCT | NOV | DEC | YEAR |
|---|---|---|---|---|---|---|---|---|---|---|---|---|---|
| Maximum Temp °F | 26.5 | 31.8 | 42.0 | 53.3 | 63.3 | 74.3 | 82.8 | 81.1 | 70.3 | 57.7 | 38.6 | 27.7 | 54.1 |
| Minimum Temp °F | -2.1 | 1.9 | 14.8 | 22.7 | 31.6 | 39.5 | 45.1 | 42.5 | 32.3 | 21.6 | 10.6 | -0.7 | 21.7 |
| Mean Temp °F | 12.1 | 16.8 | 28.4 | 38.0 | 47.5 | 56.9 | 63.9 | 61.8 | 51.3 | 39.7 | 24.6 | 13.5 | 37.9 |
| Days Max Temp ≥ 90 °F | 0 | 0 | 0 | 0 | 0 | 1 | 3 | 1 | 0 | 0 | 0 | 0 | 5 |
| Days Max Temp ≤ 32 °F | 20 | 13 | 4 | 0 | 0 | 0 | 0 | 0 | 0 | 1 | 8 | 20 | 66 |
| Days Min Temp ≤ 32 °F | 30 | 28 | 31 | 28 | 17 | 4 | 1 | 2 | 15 | 29 | 29 | 31 | 245 |
| Days Min Temp ≤ 0 °F | 16 | 12 | 3 | 0 | 0 | 0 | 0 | 0 | 0 | 0 | 6 | 17 | 54 |
| Heating Degree Days | 1634 | 1356 | 1127 | 802 | 537 | 248 | 76 | 118 | 405 | 778 | 1204 | 1590 | 9875 |
| Cooling Degree Days | 0 | 0 | 0 | 0 | 0 | 12 | 49 | 23 | 0 | 0 | 0 | 0 | 84 |
| Total Precipitation (") | 0.24 | 0.24 | 0.33 | 0.64 | 1.08 | 0.94 | 0.77 | 0.67 | 0.87 | 0.67 | 0.38 | 0.25 | 7.08 |
| Days ≥ 0.1" Precip | 1 | 1 | 1 | 2 | 4 | 3 | 3 | 2 | 2 | 2 | 1 | 1 | 23 |
| Total Snowfall (") | na | na | na | na | *0.7* | 0.0 | 0.0 | 0.0 | 0.1 | *1.4* | na | na | na |
| Days ≥ 1" Snow Depth | na | na | na | *0* | 0 | 0 | 0 | 0 | 0 | *1* | na | na | na |

## GAS HILLS 4 E *Natrona County* ELEVATION 6473 ft LAT/LONG 42° 50 ' N / 107° 29 ' W

| | JAN | FEB | MAR | APR | MAY | JUN | JUL | AUG | SEP | OCT | NOV | DEC | YEAR |
|---|---|---|---|---|---|---|---|---|---|---|---|---|---|
| Maximum Temp °F | *29.2* | *32.0* | 41.2 | *51.8* | *62.9* | *73.9* | na | *80.4* | *69.7* | *56.7* | na | na | na |
| Minimum Temp °F | *11.8* | *13.4* | *20.0* | *28.7* | *37.6* | *46.8* | na | *52.3* | *42.4* | *32.6* | na | na | na |
| Mean Temp °F | na | *22.7* | *30.5* | *40.3* | *50.3* | *60.4* | na | *66.5* | *56.2* | *44.7* | na | na | na |
| Days Max Temp ≥ 90 °F | 0 | 0 | 0 | 0 | 0 | 1 | *3* | 1 | 0 | 0 | 0 | 0 | 5 |
| Days Max Temp ≤ 32 °F | 16 | 11 | 5 | 1 | 0 | 0 | 0 | 0 | 0 | 1 | 7 | 15 | 56 |
| Days Min Temp ≤ 32 °F | 29 | 26 | 27 | *19* | 8 | 1 | 0 | 0 | 4 | 13 | *24* | 27 | 178 |
| Days Min Temp ≤ 0 °F | 4 | 3 | 1 | 0 | 0 | 0 | 0 | 0 | 0 | 0 | 1 | 3 | 12 |
| Heating Degree Days | na | *1188* | *1062* | *734* | *443* | *178* | na | *49* | *268* | *623* | na | na | na |
| Cooling Degree Days | na | na | *0* | na | na | na | na | na | na | na | na | na | na |
| Total Precipitation (") | 0.36 | 0.39 | 0.80 | 1.26 | 1.59 | 1.26 | 0.96 | 0.65 | 0.68 | 0.61 | 0.46 | 0.40 | 9.42 |
| Days ≥ 0.1" Precip | 1 | 1 | 2 | 3 | 3 | 3 | 2 | 2 | 2 | 2 | 1 | 2 | 24 |
| Total Snowfall (") | 7.4 | 8.3 | 11.0 | 12.5 | 2.5 | 0.3 | 0.0 | 0.0 | 1.1 | 5.4 | 7.9 | 8.2 | 64.6 |
| Days ≥ 1" Snow Depth | 20 | 17 | 12 | 8 | 1 | 0 | 0 | 0 | 0 | 3 | 10 | 17 | 88 |

## GILLETTE 9 ESE *Campbell County* ELEVATION 4640 ft LAT/LONG 44° 16 ' N / 105° 19 ' W

| | JAN | FEB | MAR | APR | MAY | JUN | JUL | AUG | SEP | OCT | NOV | DEC | YEAR |
|---|---|---|---|---|---|---|---|---|---|---|---|---|---|
| Maximum Temp °F | 31.8 | 36.7 | 45.2 | 55.7 | 65.7 | 76.1 | 85.3 | 84.2 | 72.8 | 59.5 | 43.0 | 33.7 | 57.5 |
| Minimum Temp °F | 10.8 | 15.4 | 22.4 | 30.8 | 40.0 | 48.8 | 54.8 | 53.5 | 43.5 | 33.2 | 21.5 | 13.0 | 32.3 |
| Mean Temp °F | 21.3 | 26.0 | 33.8 | 43.3 | 52.8 | 62.5 | 70.1 | 68.8 | 58.2 | 46.4 | 32.3 | 23.4 | 44.9 |
| Days Max Temp ≥ 90 °F | 0 | 0 | 0 | 0 | 0 | 3 | 10 | 9 | 2 | 0 | 0 | 0 | 24 |
| Days Max Temp ≤ 32 °F | 14 | 9 | 5 | 1 | 0 | 0 | 0 | 0 | 0 | 1 | 6 | 13 | 49 |
| Days Min Temp ≤ 32 °F | 30 | 27 | 27 | 18 | 6 | 0 | 0 | 0 | 3 | 14 | 26 | 29 | 180 |
| Days Min Temp ≤ 0 °F | 8 | 4 | 1 | 0 | 0 | 0 | 0 | 0 | 0 | 0 | 2 | 6 | 21 |
| Heating Degree Days | 1347 | 1094 | 959 | 645 | 377 | 132 | 24 | 36 | 234 | 571 | 974 | 1284 | 7677 |
| Cooling Degree Days | 0 | 0 | 0 | 1 | 9 | 75 | 194 | 171 | 34 | 1 | 0 | 0 | 485 |
| Total Precipitation (") | 0.50 | 0.48 | 0.88 | 1.89 | 2.90 | 2.87 | 1.78 | 1.26 | 1.43 | 1.38 | 0.65 | 0.61 | 16.63 |
| Days ≥ 0.1" Precip | 2 | 2 | 3 | 5 | 6 | 6 | 4 | 3 | 3 | 3 | 2 | 2 | 41 |
| Total Snowfall (") | 8.9 | 7.7 | 10.9 | 10.6 | 3.1 | 0.1 | 0.0 | 0.0 | 1.1 | 4.3 | 8.3 | 10.5 | 65.5 |
| Days ≥ 1" Snow Depth | 21 | 14 | 9 | 3 | 1 | 0 | 0 | 0 | 0 | 1 | 8 | 18 | 75 |

## GLENROCK 5 ESE *Converse County* ELEVATION 4954 ft LAT/LONG 42° 50 ' N / 105° 47 ' W

| | JAN | FEB | MAR | APR | MAY | JUN | JUL | AUG | SEP | OCT | NOV | DEC | YEAR |
|---|---|---|---|---|---|---|---|---|---|---|---|---|---|
| Maximum Temp °F | 37.8 | 41.4 | 48.8 | 58.9 | 69.2 | 80.7 | 89.0 | 86.8 | 76.2 | 62.6 | 45.9 | 38.2 | 61.3 |
| Minimum Temp °F | 16.0 | 19.0 | 25.1 | 32.2 | 40.8 | 49.7 | 55.7 | 53.8 | 43.7 | 33.7 | 24.0 | 17.4 | 34.3 |
| Mean Temp °F | 26.9 | 30.2 | 37.0 | 45.6 | 55.0 | 65.2 | 72.4 | 70.3 | 59.9 | 48.2 | 34.9 | 27.8 | 47.8 |
| Days Max Temp ≥ 90 °F | 0 | 0 | 0 | 0 | 0 | 7 | 16 | 12 | 2 | 0 | 0 | 0 | 37 |
| Days Max Temp ≤ 32 °F | 9 | 6 | 3 | 1 | 0 | 0 | 0 | 0 | 0 | 1 | 4 | 9 | 33 |
| Days Min Temp ≤ 32 °F | 27 | 24 | 24 | 16 | 4 | 0 | 0 | 0 | 3 | 14 | 23 | 27 | 162 |
| Days Min Temp ≤ 0 °F | 5 | 3 | 1 | 0 | 0 | 0 | 0 | 0 | 0 | 0 | 1 | 4 | 14 |
| Heating Degree Days | 1174 | 977 | 863 | 577 | 312 | 82 | 9 | 16 | 186 | 516 | 897 | 1148 | 6757 |
| Cooling Degree Days | 0 | 0 | 0 | 1 | 9 | 102 | 229 | 184 | 34 | 0 | 0 | 0 | 559 |
| Total Precipitation (") | 0.38 | 0.44 | 0.83 | 1.66 | 2.30 | 1.78 | 1.25 | 0.74 | 0.98 | 1.00 | 0.69 | 0.46 | 12.51 |
| Days ≥ 0.1" Precip | 1 | 2 | 3 | 4 | 5 | 4 | 3 | 2 | 3 | 3 | 2 | 1 | 33 |
| Total Snowfall (") | ***3.8*** | na | na | na | ***0.1*** | 0.0 | 0.0 | 0.0 | 0.0 | 1.4 | na | na | na |
| Days ≥ 1" Snow Depth | na | na | na | ***0*** | ***0*** | 0 | 0 | 0 | 0 | 1 | na | na | na |

## GREEN RIVER *Sweetwater County* ELEVATION 6083 ft LAT/LONG 41° 32 ' N / 109° 28 ' W

| | JAN | FEB | MAR | APR | MAY | JUN | JUL | AUG | SEP | OCT | NOV | DEC | YEAR |
|---|---|---|---|---|---|---|---|---|---|---|---|---|---|
| Maximum Temp °F | 31.4 | 36.1 | 46.1 | 56.2 | 67.1 | 77.4 | 85.6 | 83.9 | 73.8 | 61.8 | 43.9 | 32.8 | 58.0 |
| Minimum Temp °F | 3.2 | 6.5 | 18.6 | 26.7 | 35.2 | 42.9 | 48.6 | 46.1 | 35.7 | 25.5 | 14.9 | 3.7 | 25.6 |
| Mean Temp °F | 17.3 | 21.3 | 32.4 | 41.5 | 51.2 | 60.2 | 67.1 | 64.8 | 54.8 | 43.8 | 29.4 | 18.3 | 41.8 |
| Days Max Temp ≥ 90 °F | 0 | 0 | 0 | 0 | 0 | 2 | 7 | 5 | 0 | 0 | 0 | 0 | 14 |
| Days Max Temp ≤ 32 °F | 16 | 10 | 3 | 0 | 0 | 0 | 0 | 0 | 0 | 0 | 5 | 15 | 49 |
| Days Min Temp ≤ 32 °F | 31 | 28 | 30 | 24 | 11 | 1 | 0 | 1 | 10 | 26 | 30 | 31 | 223 |
| Days Min Temp ≤ 0 °F | 13 | 9 | 1 | 0 | 0 | 0 | 0 | 0 | 0 | 0 | 3 | 12 | 38 |
| Heating Degree Days | 1473 | 1227 | 1004 | 699 | 423 | 168 | 30 | 60 | 304 | 651 | 1062 | 1442 | 8543 |
| Cooling Degree Days | 0 | 0 | 0 | 0 | 0 | 30 | 99 | 57 | 3 | 0 | 0 | 0 | 189 |
| Total Precipitation (") | 0.34 | 0.28 | 0.60 | 0.83 | 1.09 | 0.94 | 0.77 | 0.65 | 0.76 | 0.74 | 0.48 | 0.37 | 7.85 |
| Days ≥ 0.1" Precip | 1 | 1 | 2 | 2 | 3 | 3 | 2 | 2 | 2 | 3 | 2 | 1 | 24 |
| Total Snowfall (") | ***5.1*** | ***4.1*** | na | ***2.9*** | ***0.6*** | 0.0 | 0.0 | 0.0 | 0.4 | ***0.8*** | na | na | na |
| Days ≥ 1" Snow Depth | na | na | na | na | ***0*** | 0 | 0 | 0 | ***0*** | 1 | na | na | na |

## HEART MOUNTAIN *Park County* ELEVATION 4790 ft LAT/LONG 44° 42 ' N / 108° 57 ' W

| | JAN | FEB | MAR | APR | MAY | JUN | JUL | AUG | SEP | OCT | NOV | DEC | YEAR |
|---|---|---|---|---|---|---|---|---|---|---|---|---|---|
| Maximum Temp °F | 32.7 | 39.9 | 49.2 | 59.2 | 68.7 | 78.0 | 84.4 | 82.7 | 72.6 | 60.7 | 43.8 | 34.3 | 58.9 |
| Minimum Temp °F | 8.5 | 14.0 | 21.6 | 30.2 | 39.4 | 46.7 | 51.3 | 49.5 | 40.5 | 31.2 | 19.5 | 10.5 | 30.2 |
| Mean Temp °F | 20.6 | 27.0 | 35.4 | 44.8 | 54.1 | 62.4 | 67.8 | 66.2 | 56.6 | 46.0 | 31.6 | 22.4 | 44.6 |
| Days Max Temp ≥ 90 °F | 0 | 0 | 0 | 0 | 0 | 3 | 7 | 5 | 1 | 0 | 0 | 0 | 16 |
| Days Max Temp ≤ 32 °F | 13 | 7 | 3 | 0 | 0 | 0 | 0 | 0 | 0 | 1 | 5 | 13 | 42 |
| Days Min Temp ≤ 32 °F | 30 | 27 | 28 | 19 | 6 | 0 | 0 | 0 | 5 | 18 | 27 | 30 | 190 |
| Days Min Temp ≤ 0 °F | 9 | 4 | 1 | 0 | 0 | 0 | 0 | 0 | 0 | 0 | 2 | 6 | 22 |
| Heating Degree Days | 1370 | 1067 | 912 | 601 | 340 | 126 | 28 | 49 | 261 | 583 | 994 | 1314 | 7645 |
| Cooling Degree Days | 0 | 0 | 0 | 1 | 10 | 58 | 122 | 97 | 15 | 0 | 0 | 0 | 303 |
| Total Precipitation (") | 0.24 | 0.21 | 0.60 | 0.89 | 1.90 | 1.39 | 1.00 | 0.67 | 0.88 | 0.69 | 0.37 | 0.22 | 9.06 |
| Days ≥ 0.1" Precip | 1 | 1 | 2 | 3 | 5 | 4 | 3 | 2 | 3 | 2 | 1 | 1 | 28 |
| Total Snowfall (") | na | na | na | ***3.3*** | ***0.1*** | 0.0 | 0.0 | 0.0 | 0.7 | ***3.3*** | na | na | na |
| Days ≥ 1" Snow Depth | na | na | na | ***1*** | ***0*** | 0 | 0 | 0 | 0 | ***1*** | ***6*** | na | na |

## JACKSON *Teton County* ELEVATION 6247 ft LAT/LONG 43° 28 ' N / 110° 46 ' W

| | JAN | FEB | MAR | APR | MAY | JUN | JUL | AUG | SEP | OCT | NOV | DEC | YEAR |
|---|---|---|---|---|---|---|---|---|---|---|---|---|---|
| Maximum Temp °F | 27.3 | 32.2 | 41.6 | 52.1 | 62.7 | 72.8 | 81.6 | 80.2 | 70.2 | 57.6 | 38.8 | 27.1 | 53.7 |
| Minimum Temp °F | 4.9 | 7.2 | 16.2 | 24.1 | 30.1 | 36.6 | 40.5 | 38.6 | 30.8 | 22.9 | 15.9 | 5.0 | 22.7 |
| Mean Temp °F | 16.1 | 19.8 | 29.0 | 38.1 | 46.4 | 54.7 | 61.1 | 59.4 | 50.5 | 40.2 | 27.4 | 16.1 | 38.2 |
| Days Max Temp ≥ 90 °F | 0 | 0 | 0 | 0 | 0 | 1 | 2 | 1 | 0 | 0 | 0 | 0 | 4 |
| Days Max Temp ≤ 32 °F | 20 | 13 | 4 | 0 | 0 | 0 | 0 | 0 | 0 | 1 | 8 | 21 | 67 |
| Days Min Temp ≤ 32 °F | 30 | 27 | 30 | 27 | 20 | 8 | 3 | 6 | 19 | 26 | 28 | 30 | 254 |
| Days Min Temp ≤ 0 °F | 13 | 10 | 4 | 0 | 0 | 0 | 0 | 0 | 0 | 0 | 4 | 13 | 44 |
| Heating Degree Days | 1509 | 1271 | 1111 | 801 | 570 | 305 | 130 | 176 | 428 | 761 | 1120 | 1509 | 9691 |
| Cooling Degree Days | 0 | 0 | 0 | 0 | 0 | 6 | 18 | 14 | 1 | 0 | 0 | 0 | 39 |
| Total Precipitation (") | 1.44 | 0.98 | 1.03 | 1.18 | 2.01 | 1.67 | 1.19 | 1.32 | 1.35 | 1.21 | 1.54 | 1.38 | 16.30 |
| Days ≥ 0.1" Precip | 5 | 4 | 4 | 4 | 6 | 5 | 4 | 5 | 4 | 4 | 5 | 5 | 55 |
| Total Snowfall (") | 21.0 | 12.6 | 9.1 | ***5.0*** | 1.1 | 0.2 | 0.0 | 0.0 | 0.1 | 1.2 | ***12.6*** | 18.8 | 81.7 |
| Days ≥ 1" Snow Depth | na | ***24*** | ***22*** | na | 0 | 0 | 0 | 0 | 0 | ***1*** | na | na | na |

## KAYCEE *Johnson County* ELEVATION 4662 ft LAT/LONG 43° 44' N / 106° 38' W

| | JAN | FEB | MAR | APR | MAY | JUN | JUL | AUG | SEP | OCT | NOV | DEC | YEAR |
|---|---|---|---|---|---|---|---|---|---|---|---|---|---|
| Maximum Temp °F | 37.6 | 42.1 | 49.7 | 58.7 | 68.2 | 79.1 | 87.4 | 86.3 | 75.3 | 62.9 | 47.1 | 39.0 | 61.1 |
| Minimum Temp °F | 6.9 | 12.8 | 21.1 | 29.0 | 38.2 | 47.0 | 52.6 | 50.3 | 39.8 | 29.4 | 17.6 | 9.2 | 29.5 |
| Mean Temp °F | 22.3 | 27.5 | 35.4 | 43.9 | 53.2 | 63.1 | 70.0 | 68.3 | 57.6 | 46.2 | 32.4 | 24.1 | 45.3 |
| Days Max Temp ≥ 90 °F | 0 | 0 | 0 | 0 | 0 | 5 | 13 | 12 | 2 | 0 | 0 | 0 | 32 |
| Days Max Temp ≤ 32 °F | 9 | 6 | 3 | 0 | 0 | 0 | 0 | 0 | 0 | 0 | 4 | 8 | 30 |
| Days Min Temp ≤ 32 °F | 31 | 27 | 29 | 20 | 7 | 0 | 0 | 0 | 5 | 20 | 28 | 30 | 197 |
| Days Min Temp ≤ 0 °F | 10 | 5 | 1 | 0 | 0 | 0 | 0 | 0 | 0 | 0 | 3 | 7 | 26 |
| Heating Degree Days | 1319 | 1054 | 910 | 627 | 362 | 115 | 17 | 28 | 234 | 577 | 971 | 1260 | 7474 |
| Cooling Degree Days | 0 | 0 | 0 | 0 | 6 | 84 | 196 | 164 | 20 | 0 | 0 | 0 | 470 |
| Total Precipitation (") | 0.44 | 0.38 | 0.79 | 1.48 | 2.52 | 2.11 | 1.21 | 0.86 | 1.01 | 1.15 | 0.54 | 0.44 | 12.93 |
| Days ≥ 0.1" Precip | 2 | 1 | 3 | 4 | 6 | 5 | 3 | 2 | 3 | 3 | 2 | 2 | 36 |
| Total Snowfall (") | 6.0 | 5.0 | 7.6 | 6.6 | 1.3 | 0.1 | 0.0 | 0.0 | 0.5 | 2.1 | 5.4 | 7.1 | 41.7 |
| Days ≥ 1" Snow Depth | 16 | 10 | 4 | 2 | 0 | 0 | 0 | 0 | 0 | 1 | 7 | 13 | 53 |

## KEMMERER 2 N *Lincoln County* ELEVATION 6926 ft LAT/LONG 41° 49' N / 110° 32' W

| | JAN | FEB | MAR | APR | MAY | JUN | JUL | AUG | SEP | OCT | NOV | DEC | YEAR |
|---|---|---|---|---|---|---|---|---|---|---|---|---|---|
| Maximum Temp °F | 29.2 | 31.8 | 40.1 | 51.0 | 63.0 | 72.1 | 81.0 | 79.3 | 69.0 | 55.8 | 39.4 | 29.4 | 53.4 |
| Minimum Temp °F | 5.2 | 5.5 | 14.2 | 22.6 | 32.0 | 38.7 | 44.6 | 42.8 | 34.1 | 24.7 | 14.5 | 4.4 | 23.6 |
| Mean Temp °F | 17.2 | 18.7 | 27.2 | 36.8 | 47.5 | 55.4 | 62.8 | 61.1 | 51.6 | 40.2 | 27.0 | 16.9 | 38.5 |
| Days Max Temp ≥ 90 °F | 0 | 0 | 0 | 0 | 0 | 0 | 1 | 0 | 0 | 0 | 0 | 0 | 1 |
| Days Max Temp ≤ 32 °F | 18 | 14 | 6 | 1 | 0 | 0 | 0 | 0 | 0 | 1 | 8 | 19 | 67 |
| Days Min Temp ≤ 32 °F | 31 | 28 | 31 | 28 | 16 | 5 | 1 | 2 | 12 | 27 | 30 | 31 | 242 |
| Days Min Temp ≤ 0 °F | 11 | 9 | 3 | 0 | 0 | 0 | 0 | 0 | 0 | 0 | 3 | 11 | 37 |
| Heating Degree Days | 1476 | 1303 | 1167 | 840 | 534 | 284 | 94 | 132 | 396 | 761 | 1135 | 1485 | 9607 |
| Cooling Degree Days | 0 | 0 | 0 | 0 | 0 | 6 | 20 | 8 | 0 | 0 | 0 | 0 | 34 |
| Total Precipitation (") | 0.66 | 0.59 | 0.70 | 0.92 | 1.20 | 1.16 | 0.81 | 0.94 | 1.17 | 0.91 | 0.91 | 0.64 | 10.61 |
| Days ≥ 0.1" Precip | 2 | 2 | 3 | 3 | 4 | 3 | 3 | 3 | 3 | 3 | 3 | 2 | 34 |
| Total Snowfall (") | na | na | *7.8* | 5.8 | 2.4 | 0.1 | 0.0 | 0.0 | 0.5 | *3.1* | 10.4 | na | na |
| Days ≥ 1" Snow Depth | *29* | *27* | na | *4* | 1 | 0 | 0 | 0 | 0 | *2* | 13 | *26* | na |

## LA BARGE *Lincoln County* ELEVATION 6735 ft LAT/LONG 42° 14' N / 110° 16' W

| | JAN | FEB | MAR | APR | MAY | JUN | JUL | AUG | SEP | OCT | NOV | DEC | YEAR |
|---|---|---|---|---|---|---|---|---|---|---|---|---|---|
| Maximum Temp °F | *30.7* | *34.8* | *43.1* | *54.2* | 64.7 | *73.8* | *82.9* | *81.0* | *70.7* | 58.8 | *40.4* | 30.8 | 55.5 |
| Minimum Temp °F | -2.6 | 0.3 | 13.3 | *23.1* | 31.6 | *38.8* | *43.5* | *41.6* | 32.6 | 22.2 | *9.6* | -1.4 | 21.0 |
| Mean Temp °F | *14.0* | *17.6* | *28.2* | *38.6* | 48.2 | *56.4* | *63.3* | *61.4* | *51.6* | 40.5 | *25.0* | 14.7 | 38.3 |
| Days Max Temp ≥ 90 °F | 0 | 0 | 0 | 0 | 0 | 0 | *3* | *1* | 0 | 0 | *0* | 0 | 4 |
| Days Max Temp ≤ 32 °F | *15* | *10* | 3 | *0* | 0 | 0 | *0* | *0* | 0 | 1 | *7* | 17 | 53 |
| Days Min Temp ≤ 32 °F | 29 | *27* | 29 | 27 | *16* | *5* | *1* | na | *15* | 28 | *30* | 31 | na |
| Days Min Temp ≤ 0 °F | 17 | *13* | 3 | 0 | 0 | 0 | *0* | *0* | 0 | 0 | *7* | 18 | 58 |
| Heating Degree Days | *1576* | *1334* | *1134* | *786* | 515 | *258* | *77* | *123* | *396* | 741 | *1192* | 1554 | 9686 |
| Cooling Degree Days | *0* | *0* | *0* | *0* | *0* | *7* | na | *9* | 0 | 0 | 0 | 0 | na |
| Total Precipitation (") | 0.37 | *0.39* | 0.41 | 0.79 | 1.27 | 1.02 | *0.69* | *0.92* | 0.73 | 0.59 | 0.56 | 0.38 | 8.12 |
| Days ≥ 0.1" Precip | *1* | *1* | *1* | *2* | 3 | *3* | *2* | *2* | 2 | 2 | *2* | *1* | 22 |
| Total Snowfall (") | na | na | na | na | *1.7* | *0.0* | *0.0* | *0.0* | *0.1* | *2.3* | na | *6.5* | na |
| Days ≥ 1" Snow Depth | na | na | na | na | *0* | *0* | *0* | *0* | *0* | *1* | na | na | na |

## LA GRANGE *Goshen County* ELEVATION 4593 ft LAT/LONG 41° 38' N / 104° 10' W

| | JAN | FEB | MAR | APR | MAY | JUN | JUL | AUG | SEP | OCT | NOV | DEC | YEAR |
|---|---|---|---|---|---|---|---|---|---|---|---|---|---|
| Maximum Temp °F | 39.6 | 43.5 | 50.3 | 60.2 | 70.3 | 80.8 | 88.4 | 86.8 | 76.7 | 64.3 | 48.6 | 40.5 | 62.5 |
| Minimum Temp °F | 12.8 | 16.4 | 22.3 | 30.2 | 39.0 | 48.2 | 54.0 | 51.6 | 41.5 | 30.9 | 20.7 | 13.8 | 31.8 |
| Mean Temp °F | 26.2 | 30.0 | 36.3 | 45.2 | 54.7 | 64.4 | 71.2 | 69.3 | 59.1 | 47.6 | 34.7 | 27.2 | 47.2 |
| Days Max Temp ≥ 90 °F | 0 | 0 | 0 | 0 | 1 | 6 | 14 | 12 | 3 | 0 | 0 | 0 | 36 |
| Days Max Temp ≤ 32 °F | 9 | 5 | 3 | 0 | 0 | 0 | 0 | 0 | 0 | 1 | 3 | 8 | 29 |
| Days Min Temp ≤ 32 °F | 30 | 26 | 27 | 18 | 6 | 0 | 0 | 0 | 4 | 17 | 26 | 29 | 183 |
| Days Min Temp ≤ 0 °F | 6 | 3 | 1 | 0 | 0 | 0 | 0 | 0 | 0 | 0 | 1 | 4 | 15 |
| Heating Degree Days | 1197 | 984 | 881 | 587 | 323 | 94 | 12 | 21 | 201 | 532 | 904 | 1166 | 6902 |
| Cooling Degree Days | 0 | 0 | 0 | 0 | 10 | 92 | 199 | 163 | 33 | 1 | 0 | 0 | 498 |
| Total Precipitation (") | 0.56 | 0.51 | 1.28 | 1.86 | 2.80 | 2.39 | 2.13 | 1.39 | 1.23 | 1.04 | 0.79 | 0.63 | 16.61 |
| Days ≥ 0.1" Precip | 2 | 2 | 3 | 4 | 6 | 5 | 5 | 4 | 3 | 3 | 2 | 2 | 41 |
| Total Snowfall (") | 8.3 | *7.0* | 13.3 | 8.9 | 0.7 | 0.0 | 0.0 | 0.0 | 0.8 | 4.3 | 9.4 | 10.0 | 62.7 |
| Days ≥ 1" Snow Depth | 12 | 8 | 7 | 3 | 0 | 0 | 0 | 0 | 0 | 2 | 6 | 10 | 48 |

## LAKE YELLOWSTONE *Teton County* ELEVATION 7736 ft LAT/LONG 44° 34 ' N / 110° 24 ' W

| | JAN | FEB | MAR | APR | MAY | JUN | JUL | AUG | SEP | OCT | NOV | DEC | YEAR |
|---|---|---|---|---|---|---|---|---|---|---|---|---|---|
| Maximum Temp °F | 22.6 | 28.2 | 34.9 | 42.5 | ***52.5*** | ***61.3*** | 71.0 | 70.9 | 61.1 | ***49.0*** | 33.2 | 24.4 | 46.0 |
| Minimum Temp °F | -2.4 | -1.7 | 3.3 | 14.4 | 24.6 | ***32.6*** | 37.9 | 36.6 | 28.8 | ***21.3*** | 10.9 | 1.1 | 17.3 |
| Mean Temp °F | 10.1 | 13.2 | 19.1 | 28.5 | ***38.5*** | ***46.9*** | 54.5 | 53.8 | 45.0 | ***35.2*** | 22.0 | 12.7 | 31.6 |
| Days Max Temp ≥ 90 °F | 0 | 0 | 0 | 0 | 0 | 0 | 0 | 0 | 0 | 0 | 0 | 0 | 0 |
| Days Max Temp ≤ 32 °F | 27 | 19 | 10 | 3 | 0 | 0 | 0 | 0 | 0 | 2 | 14 | 26 | 101 |
| Days Min Temp ≤ 32 °F | 31 | 28 | 30 | 30 | 28 | ***15*** | 5 | 7 | ***23*** | ***30*** | 30 | 31 | 288 |
| Days Min Temp ≤ 0 °F | 17 | 15 | 12 | 4 | 0 | 0 | 0 | 0 | 0 | 1 | ***6*** | 15 | 70 |
| Heating Degree Days | 1697 | 1457 | 1416 | 1089 | ***814*** | ***535*** | 319 | 342 | 594 | ***919*** | 1283 | 1616 | 12081 |
| Cooling Degree Days | 0 | 0 | 0 | 0 | 0 | 0 | 1 | 0 | 0 | 0 | 0 | 0 | 1 |
| Total Precipitation (") | 2.00 | 1.32 | 1.62 | 1.44 | 1.83 | 2.18 | 1.64 | 1.66 | 1.73 | 1.37 | 1.66 | 1.59 | 20.04 |
| Days ≥ 0.1" Precip | 7 | 5 | 6 | 5 | 6 | ***6*** | 5 | 5 | 5 | 4 | 6 | 6 | 66 |
| Total Snowfall (") | 33.9 | 23.4 | 26.1 | 16.8 | 6.5 | ***0.8*** | 0.0 | 0.0 | 1.7 | ***6.6*** | 24.5 | 29.0 | 169.3 |
| Days ≥ 1" Snow Depth | 31 | 28 | 30 | 29 | ***15*** | ***0*** | 0 | 0 | 1 | ***6*** | 25 | 31 | 196 |

## LANDER HUNT FIELD *Fremont County* ELEVATION 5556 ft LAT/LONG 42° 49 ' N / 108° 44 ' W

| | JAN | FEB | MAR | APR | MAY | JUN | JUL | AUG | SEP | OCT | NOV | DEC | YEAR |
|---|---|---|---|---|---|---|---|---|---|---|---|---|---|
| Maximum Temp °F | 31.8 | 36.8 | 46.5 | 56.1 | 66.2 | 77.6 | 85.6 | 84.0 | 72.4 | 59.4 | 41.6 | 32.0 | 57.5 |
| Minimum Temp °F | 8.7 | 13.3 | 22.9 | 31.1 | 40.2 | 48.9 | 55.5 | 53.9 | 44.0 | 33.1 | 18.8 | 9.4 | 31.7 |
| Mean Temp °F | 20.2 | 25.1 | 34.7 | 43.7 | 53.3 | 63.3 | 70.6 | 69.0 | 58.2 | 46.3 | 30.2 | 20.7 | 44.6 |
| Days Max Temp ≥ 90 °F | 0 | 0 | 0 | 0 | 0 | 3 | 10 | 6 | 1 | 0 | 0 | 0 | 20 |
| Days Max Temp ≤ 32 °F | 15 | 9 | 3 | 1 | 0 | 0 | 0 | 0 | 0 | 1 | 7 | 15 | 51 |
| Days Min Temp ≤ 32 °F | 31 | 28 | 28 | 17 | 5 | 0 | 0 | 0 | 3 | 14 | 27 | 31 | 184 |
| Days Min Temp ≤ 0 °F | 8 | 4 | 1 | 0 | 0 | 0 | 0 | 0 | 0 | 0 | 2 | 6 | 21 |
| Heating Degree Days | 1382 | 1121 | 933 | 634 | 361 | 118 | 14 | 26 | 221 | 576 | 1033 | 1362 | 7781 |
| Cooling Degree Days | 0 | 0 | 0 | 0 | 6 | 87 | 207 | 173 | 29 | 0 | 0 | 0 | 502 |
| Total Precipitation (") | 0.48 | 0.52 | 1.20 | 2.00 | 2.31 | 1.44 | 0.87 | 0.54 | 1.03 | 1.26 | 0.97 | 0.62 | 13.24 |
| Days ≥ 0.1" Precip | 2 | 2 | 3 | 5 | 5 | 3 | 2 | 2 | 2 | 3 | 3 | 2 | 34 |
| Total Snowfall (") | 9.5 | 9.6 | 18.2 | 22.2 | 7.7 | 0.2 | 0.0 | 0.0 | 3.3 | 9.9 | 15.3 | 12.3 | 108.2 |
| Days ≥ 1" Snow Depth | 25 | 22 | 12 | 6 | 1 | 0 | 0 | 0 | 1 | 2 | 13 | 23 | 105 |

## LARAMIE GEN BREES FD *Albany County* ELEVATION 7274 ft LAT/LONG 41° 19 ' N / 105° 41 ' W

| | JAN | FEB | MAR | APR | MAY | JUN | JUL | AUG | SEP | OCT | NOV | DEC | YEAR |
|---|---|---|---|---|---|---|---|---|---|---|---|---|---|
| Maximum Temp °F | 32.4 | 35.0 | 41.3 | 50.8 | 61.4 | 72.8 | 79.6 | 77.7 | 68.2 | 55.9 | 40.7 | 33.3 | 54.1 |
| Minimum Temp °F | 9.1 | 10.8 | 17.3 | 24.2 | 33.1 | 41.8 | 47.6 | 45.7 | 37.1 | 27.2 | 16.2 | 10.0 | 26.7 |
| Mean Temp °F | 20.8 | 22.9 | 29.3 | 37.5 | 47.3 | 57.3 | 63.6 | 61.7 | 52.7 | 41.6 | 28.5 | 21.7 | 40.4 |
| Days Max Temp ≥ 90 °F | 0 | 0 | 0 | 0 | 0 | 0 | 1 | 0 | 0 | 0 | 0 | 0 | 1 |
| Days Max Temp ≤ 32 °F | 14 | 11 | 7 | 2 | 0 | 0 | 0 | 0 | 0 | 1 | 8 | 14 | 57 |
| Days Min Temp ≤ 32 °F | 31 | 28 | 30 | 26 | 14 | 2 | 0 | 0 | 7 | 23 | 29 | 30 | 220 |
| Days Min Temp ≤ 0 °F | 8 | 6 | 3 | 0 | 0 | 0 | 0 | 0 | 0 | 0 | 4 | 7 | 28 |
| Heating Degree Days | 1364 | 1182 | 1099 | 818 | 542 | 233 | 68 | 109 | 362 | 718 | 1084 | 1334 | 8913 |
| Cooling Degree Days | 0 | 0 | 0 | 0 | 0 | 11 | 27 | 13 | 1 | 0 | 0 | 0 | 52 |
| Total Precipitation (") | 0.40 | 0.49 | 0.83 | 1.09 | 1.45 | 1.36 | 1.56 | 1.14 | 0.94 | 0.82 | 0.72 | 0.48 | 11.28 |
| Days ≥ 0.1" Precip | 1 | 1 | 2 | 3 | 4 | 4 | 4 | 3 | 2 | 3 | 2 | 1 | 30 |
| Total Snowfall (") | 5.1 | 6.2 | 9.3 | 8.1 | 3.1 | 0.2 | 0.0 | 0.0 | 1.5 | 4.7 | 8.3 | 7.2 | 53.7 |
| Days ≥ 1" Snow Depth | 17 | 15 | 11 | 5 | 1 | 0 | 0 | 0 | 1 | 3 | 11 | 15 | 79 |

## LEITER 9 N *Sheridan County* ELEVATION 4203 ft LAT/LONG 44° 51 ' N / 106° 17 ' W

| | JAN | FEB | MAR | APR | MAY | JUN | JUL | AUG | SEP | OCT | NOV | DEC | YEAR |
|---|---|---|---|---|---|---|---|---|---|---|---|---|---|
| Maximum Temp °F | 32.6 | 38.2 | 47.3 | 58.0 | 67.9 | 78.0 | 87.7 | 87.1 | 74.3 | 61.3 | 44.2 | 34.7 | 59.3 |
| Minimum Temp °F | 10.7 | 15.8 | 23.8 | 32.1 | 41.2 | 50.1 | 56.9 | 55.5 | 45.0 | 34.0 | 22.2 | 13.5 | 33.4 |
| Mean Temp °F | 21.7 | 27.0 | 35.5 | 45.1 | 54.6 | 64.1 | 72.3 | 71.4 | 59.7 | 47.7 | 33.2 | 24.1 | 46.4 |
| Days Max Temp ≥ 90 °F | 0 | 0 | 0 | 0 | 0 | 4 | 14 | 13 | 3 | 0 | 0 | 0 | 34 |
| Days Max Temp ≤ 32 °F | 13 | 8 | 4 | 1 | 0 | 0 | 0 | 0 | 0 | 1 | 5 | 12 | 44 |
| Days Min Temp ≤ 32 °F | 30 | 27 | 27 | 16 | 4 | 0 | 0 | 0 | 3 | 13 | 26 | 30 | 176 |
| Days Min Temp ≤ 0 °F | 8 | 4 | 1 | 0 | 0 | 0 | 0 | 0 | 0 | 0 | 1 | 5 | 19 |
| Heating Degree Days | 1337 | 1065 | 907 | 592 | 328 | 103 | 15 | 20 | 205 | 531 | 947 | 1261 | 7311 |
| Cooling Degree Days | 0 | 0 | 0 | 1 | 16 | 99 | 253 | 240 | 52 | 1 | 0 | 0 | 662 |
| Total Precipitation (") | 0.58 | 0.42 | 0.84 | 1.77 | 2.34 | 2.55 | 1.38 | 1.11 | 1.44 | 1.29 | 0.75 | 0.55 | 15.02 |
| Days ≥ 0.1" Precip | 2 | 1 | 3 | 5 | 6 | 6 | 4 | 3 | 4 | 4 | 3 | 2 | 43 |
| Total Snowfall (") | 8.7 | 6.6 | 10.2 | 8.9 | 1.6 | 0.2 | 0.0 | 0.0 | 1.6 | 3.6 | 7.9 | 8.6 | 57.9 |
| Days ≥ 1" Snow Depth | 24 | 16 | 8 | 2 | 0 | 0 | 0 | 0 | 0 | 1 | 8 | 20 | 79 |

## LOVELL *Big Horn County* ELEVATION 3832 ft LAT/LONG 44° 51 ' N / 108° 23 ' W

| | JAN | FEB | MAR | APR | MAY | JUN | JUL | AUG | SEP | OCT | NOV | DEC | YEAR |
|---|---|---|---|---|---|---|---|---|---|---|---|---|---|
| Maximum Temp °F | 29.4 | 37.8 | 48.9 | 59.2 | 69.0 | 78.9 | 87.4 | 86.0 | 73.7 | 61.0 | 44.0 | 32.6 | 59.0 |
| Minimum Temp °F | 4.6 | 11.2 | 21.8 | 31.2 | 41.2 | 49.5 | 54.3 | 51.4 | 40.3 | 29.9 | 18.5 | 8.0 | 30.2 |
| Mean Temp °F | 17.1 | 24.5 | 35.4 | 45.2 | 55.2 | 64.3 | 70.8 | 68.7 | 57.0 | 45.5 | 31.2 | 20.4 | 44.6 |
| Days Max Temp ≥ 90 °F | 0 | 0 | 0 | 0 | 1 | 5 | 14 | 12 | 2 | 0 | 0 | 0 | 34 |
| Days Max Temp ≤ 32 °F | 16 | 8 | 3 | 0 | 0 | 0 | 0 | 0 | 0 | 0 | 5 | 14 | 46 |
| Days Min Temp ≤ 32 °F | 31 | 28 | 28 | 17 | 4 | 0 | 0 | 0 | 4 | 20 | 28 | 31 | 191 |
| Days Min Temp ≤ 0 °F | 11 | 5 | 1 | 0 | 0 | 0 | 0 | 0 | 0 | 0 | 2 | 7 | 26 |
| Heating Degree Days | 1481 | 1136 | 912 | 587 | 312 | 101 | 20 | 33 | 254 | 599 | 1005 | 1377 | 7817 |
| Cooling Degree Days | 0 | 0 | 0 | 1 | 15 | 86 | 188 | 152 | 17 | 0 | 0 | 0 | 459 |
| Total Precipitation (") | 0.21 | 0.15 | 0.28 | 0.61 | 1.20 | 1.22 | 0.67 | 0.67 | 0.77 | 0.58 | 0.24 | 0.21 | 6.81 |
| Days ≥ 0.1" Precip | 1 | 1 | 1 | 2 | 4 | 4 | 2 | 2 | 2 | 2 | 1 | 1 | 23 |
| Total Snowfall (") | 5.0 | 3.1 | 2.8 | 1.2 | 0.1 | 0.0 | 0.0 | 0.0 | 0.7 | 0.5 | 2.0 | 4.0 | 19.4 |
| Days ≥ 1" Snow Depth | na | na | na | 0 | 0 | 0 | 0 | 0 | 0 | 0 | 3 | na | na |

## LUSK 2 SW *Niobrara County* ELEVATION 5013 ft LAT/LONG 42° 46 ' N / 104° 26 ' W

| | JAN | FEB | MAR | APR | MAY | JUN | JUL | AUG | SEP | OCT | NOV | DEC | YEAR |
|---|---|---|---|---|---|---|---|---|---|---|---|---|---|
| Maximum Temp °F | 36.4 | 41.2 | 47.4 | 57.3 | 67.5 | 77.8 | 85.2 | 84.2 | 74.8 | 62.1 | 46.6 | 37.3 | 59.8 |
| Minimum Temp °F | 10.9 | 15.4 | 20.2 | 28.7 | 37.9 | 47.0 | 52.6 | 50.1 | 40.3 | 29.8 | 20.2 | 11.5 | 30.4 |
| Mean Temp °F | 23.7 | 28.3 | 33.8 | 43.0 | 52.8 | 62.4 | 68.9 | 67.2 | 57.6 | 46.0 | 33.4 | 24.4 | 45.1 |
| Days Max Temp ≥ 90 °F | 0 | 0 | 0 | 0 | 0 | 3 | 9 | 7 | 1 | 0 | 0 | 0 | 20 |
| Days Max Temp ≤ 32 °F | 10 | 6 | 3 | 1 | 0 | 0 | 0 | 0 | 0 | 1 | 4 | 10 | 35 |
| Days Min Temp ≤ 32 °F | 30 | 27 | 28 | 21 | 7 | 1 | 0 | 0 | 5 | 19 | 26 | 30 | 194 |
| Days Min Temp ≤ 0 °F | 7 | 4 | 1 | 0 | 0 | 0 | 0 | 0 | 0 | 0 | 1 | 6 | 19 |
| Heating Degree Days | 1275 | 1026 | 959 | 653 | 376 | 124 | 27 | 37 | 234 | 582 | 941 | 1251 | 7485 |
| Cooling Degree Days | na | 0 | 0 | 0 | 4 | na | na | na | 19 | 0 | 0 | 0 | na |
| Total Precipitation (") | 0.62 | 0.57 | 1.34 | 2.35 | 2.82 | 2.31 | 2.04 | 1.09 | 1.15 | 1.09 | 0.69 | 0.72 | 16.79 |
| Days ≥ 0.1" Precip | 2 | 2 | 3 | 5 | 6 | 5 | 5 | 3 | 3 | 2 | 2 | 2 | 40 |
| Total Snowfall (") | na | 7.5 | na | na | 1.9 | 0.1 | 0.0 | 0.0 | 0.0 | na | na | na | na |
| Days ≥ 1" Snow Depth | na | na | na | na | 0 | 0 | 0 | 0 | 0 | na | na | na | na |

## MEDICINE BOW *Carbon County* ELEVATION 6565 ft LAT/LONG 41° 54 ' N / 106° 12 ' W

| | JAN | FEB | MAR | APR | MAY | JUN | JUL | AUG | SEP | OCT | NOV | DEC | YEAR |
|---|---|---|---|---|---|---|---|---|---|---|---|---|---|
| Maximum Temp °F | 32.1 | 35.2 | 43.4 | 55.6 | 65.5 | 76.5 | 83.2 | 81.2 | 72.1 | 59.6 | 41.5 | 33.4 | 56.6 |
| Minimum Temp °F | 11.6 | 12.6 | 19.1 | 26.0 | 34.0 | 42.5 | 47.1 | 44.7 | 35.6 | 26.8 | 18.0 | 11.7 | 27.5 |
| Mean Temp °F | 21.9 | 23.9 | 31.3 | 40.8 | 49.8 | 59.5 | 65.2 | 63.0 | 53.9 | 43.2 | 29.7 | 22.6 | 42.1 |
| Days Max Temp ≥ 90 °F | 0 | 0 | 0 | 0 | 0 | 1 | 3 | 1 | 0 | 0 | 0 | 0 | 5 |
| Days Max Temp ≤ 32 °F | 15 | 10 | 5 | 1 | 0 | 0 | 0 | 0 | 0 | 1 | 7 | 13 | 52 |
| Days Min Temp ≤ 32 °F | 30 | 28 | 30 | 25 | 13 | 2 | 0 | 1 | 11 | 23 | 28 | 30 | 221 |
| Days Min Temp ≤ 0 °F | 6 | 5 | 2 | 0 | 0 | 0 | 0 | 0 | 0 | 0 | 2 | 5 | 20 |
| Heating Degree Days | 1330 | 1154 | 1039 | 719 | 466 | 176 | 40 | 87 | 329 | 669 | 1052 | 1307 | 8368 |
| Cooling Degree Days | na | 0 | 0 | na | na | na | na | 33 | na | na | na | na | na |
| Total Precipitation (") | 0.43 | 0.52 | 0.66 | 1.04 | 1.56 | 1.23 | 1.21 | 0.88 | 0.95 | 0.95 | 0.56 | 0.58 | 10.57 |
| Days ≥ 0.1" Precip | 1 | 2 | 2 | 3 | 5 | 3 | 3 | 2 | 2 | 3 | 2 | na | na |
| Total Snowfall (") | na | 6.6 | 4.6 | 5.3 | 1.3 | 0.1 | 0.0 | 0.0 | 1.8 | 3.3 | na | 5.2 | na |
| Days ≥ 1" Snow Depth | na | na | na | na | 0 | 0 | 0 | 0 | 0 | 1 | na | na | na |

## MIDWEST *Natrona County* ELEVATION 4852 ft LAT/LONG 43° 25 ' N / 106° 17 ' W

| | JAN | FEB | MAR | APR | MAY | JUN | JUL | AUG | SEP | OCT | NOV | DEC | YEAR |
|---|---|---|---|---|---|---|---|---|---|---|---|---|---|
| Maximum Temp °F | 35.9 | 40.8 | 50.0 | 59.5 | 70.2 | 81.1 | 89.2 | 88.1 | 77.3 | 64.3 | 46.1 | 37.0 | 61.6 |
| Minimum Temp °F | 11.4 | 15.6 | 23.4 | 30.8 | 39.9 | 48.4 | 54.0 | 52.0 | 41.6 | 32.5 | 21.0 | 13.6 | 32.0 |
| Mean Temp °F | 23.7 | 28.2 | 36.7 | 45.2 | 55.1 | 64.8 | 71.6 | 70.1 | 59.5 | 48.4 | 33.6 | 25.3 | 46.8 |
| Days Max Temp ≥ 90 °F | 0 | 0 | 0 | 0 | 0 | 6 | 17 | 15 | 4 | 0 | 0 | 0 | 42 |
| Days Max Temp ≤ 32 °F | 10 | 6 | 2 | 1 | 0 | 0 | 0 | 0 | 0 | 1 | 4 | 9 | 33 |
| Days Min Temp ≤ 32 °F | 30 | 27 | 27 | 17 | 5 | 0 | 0 | 0 | 4 | 16 | 25 | 29 | 180 |
| Days Min Temp ≤ 0 °F | 7 | 4 | 1 | 0 | 0 | 0 | 0 | 0 | 0 | 0 | 2 | 5 | 19 |
| Heating Degree Days | 1274 | 1032 | 869 | 588 | 311 | 87 | 12 | 18 | 197 | 509 | 936 | 1224 | 7057 |
| Cooling Degree Days | 0 | 0 | 0 | 1 | 14 | 100 | 213 | 183 | 32 | 0 | 0 | 0 | 543 |
| Total Precipitation (") | 0.63 | 0.62 | 0.90 | 1.72 | 2.34 | 2.03 | 1.38 | 0.75 | 0.94 | 1.06 | 0.67 | 0.77 | 13.81 |
| Days ≥ 0.1" Precip | 2 | 2 | 3 | 4 | 6 | 4 | 3 | 2 | 2 | 2 | 2 | 2 | 34 |
| Total Snowfall (") | 8.4 | 8.9 | 8.2 | 7.7 | 2.1 | 0.1 | 0.0 | 0.0 | 0.3 | 3.4 | 7.8 | 9.5 | 56.4 |
| Days ≥ 1" Snow Depth | 16 | 10 | 6 | 3 | 0 | 0 | 0 | 0 | 0 | 1 | 7 | 14 | 57 |

## MOORCROFT *Crook County* ELEVATION 4209 ft LAT/LONG 44° 16 ' N / 104° 57 ' W

| | JAN | FEB | MAR | APR | MAY | JUN | JUL | AUG | SEP | OCT | NOV | DEC | YEAR |
|---|---|---|---|---|---|---|---|---|---|---|---|---|---|
| Maximum Temp °F | 32.0 | 36.1 | 45.4 | 56.0 | ***66.6*** | 77.3 | ***86.1*** | 85.3 | 73.9 | 59.8 | ***44.3*** | 33.8 | 58.0 |
| Minimum Temp °F | 7.7 | 12.3 | 21.0 | 29.9 | ***39.3*** | 48.5 | ***54.2*** | 52.5 | 41.7 | 30.5 | ***19.9*** | 9.4 | 30.6 |
| Mean Temp °F | 19.9 | 24.2 | 33.2 | 43.0 | ***53.0*** | 62.9 | ***70.2*** | 68.9 | 57.8 | 45.2 | ***32.1*** | 21.6 | 44.3 |
| Days Max Temp ≥ 90 °F | 0 | 0 | 0 | 0 | ***0*** | 4 | ***11*** | 12 | 2 | 0 | ***0*** | 0 | 29 |
| Days Max Temp ≤ 32 °F | 14 | 10 | 5 | 1 | ***0*** | 0 | 0 | 0 | 0 | 1 | ***5*** | 13 | 49 |
| Days Min Temp ≤ 32 °F | 30 | 28 | 28 | 18 | ***6*** | 0 | 0 | 0 | 4 | 18 | ***28*** | 30 | 190 |
| Days Min Temp ≤ 0 °F | 9 | 5 | 2 | 0 | ***0*** | 0 | 0 | 0 | 0 | 0 | ***1*** | 7 | 24 |
| Heating Degree Days | 1393 | 1145 | 979 | 654 | ***373*** | 126 | ***24*** | 37 | 242 | 609 | ***966*** | 1339 | 7887 |
| Cooling Degree Days | 0 | 0 | 0 | 1 | ***10*** | 82 | ***195*** | 171 | 29 | 1 | 0 | 0 | 489 |
| Total Precipitation (") | 0.41 | 0.37 | 0.65 | 1.33 | 2.38 | 2.40 | 1.80 | 1.27 | 1.12 | 0.99 | 0.53 | 0.44 | 13.69 |
| Days ≥ 0.1" Precip | ***1*** | 1 | 2 | 3 | 5 | 6 | ***4*** | 3 | 3 | 2 | 2 | 2 | 34 |
| Total Snowfall (") | na | na | na | na | ***4.3*** | 0.1 | 0.0 | 0.0 | 0.3 | 2.1 | na | ***7.8*** | na |
| Days ≥ 1" Snow Depth | na | na | ***4*** | ***2*** | 2 | 0 | 0 | 0 | 0 | ***1*** | na | na | na |

## MOOSE *Teton County* ELEVATION 6473 ft LAT/LONG 43° 40 ' N / 110° 43 ' W

| | JAN | FEB | MAR | APR | MAY | JUN | JUL | AUG | SEP | OCT | NOV | DEC | YEAR |
|---|---|---|---|---|---|---|---|---|---|---|---|---|---|
| Maximum Temp °F | 25.5 | 30.6 | 39.2 | 48.9 | 60.8 | 70.4 | 79.2 | 78.4 | 68.7 | 55.3 | 37.1 | 25.5 | 51.6 |
| Minimum Temp °F | 1.0 | 2.5 | 12.2 | 21.6 | 30.4 | 36.8 | 40.9 | 39.5 | 31.8 | 22.9 | 13.2 | 0.6 | 21.1 |
| Mean Temp °F | 13.3 | 16.6 | 25.7 | 35.3 | 45.6 | 53.6 | 60.1 | 59.0 | 50.3 | 39.1 | 25.2 | 13.1 | 36.4 |
| Days Max Temp ≥ 90 °F | 0 | 0 | 0 | 0 | 0 | 0 | 0 | 1 | 0 | 0 | 0 | 0 | 1 |
| Days Max Temp ≤ 32 °F | 23 | 15 | 5 | 0 | 0 | 0 | 0 | 0 | 0 | 1 | 9 | 23 | 76 |
| Days Min Temp ≤ 32 °F | 31 | 28 | 30 | 28 | 19 | 7 | 2 | 3 | 16 | 28 | 29 | 31 | 252 |
| Days Min Temp ≤ 0 °F | 15 | 13 | 6 | 0 | 0 | 0 | 0 | 0 | 0 | 0 | 5 | 17 | 56 |
| Heating Degree Days | 1599 | 1362 | 1209 | 884 | 593 | 335 | 154 | 189 | 435 | 796 | 1187 | 1605 | 10348 |
| Cooling Degree Days | 0 | 0 | 0 | 0 | 0 | 3 | 10 | 8 | 0 | 0 | 0 | 0 | 21 |
| Total Precipitation (") | 2.45 | 1.81 | 1.48 | 1.39 | 1.81 | 1.71 | 1.28 | 1.35 | 1.35 | 1.26 | 2.14 | 2.28 | 20.31 |
| Days ≥ 0.1" Precip | 8 | 6 | 5 | 5 | 6 | 6 | 4 | 4 | 4 | 4 | 7 | 7 | 66 |
| Total Snowfall (") | 41.3 | 28.0 | 18.5 | 10.0 | 3.2 | 0.1 | 0.0 | 0.0 | 0.4 | 4.4 | 25.8 | 38.0 | 169.7 |
| Days ≥ 1" Snow Depth | 31 | 28 | 31 | 22 | 3 | 0 | 0 | 0 | 0 | 3 | 20 | 30 | 168 |

## MORAN 5 WNW *Teton County* ELEVATION 6745 ft LAT/LONG 43° 52 ' N / 110° 35 ' W

| | JAN | FEB | MAR | APR | MAY | JUN | JUL | AUG | SEP | OCT | NOV | DEC | YEAR |
|---|---|---|---|---|---|---|---|---|---|---|---|---|---|
| Maximum Temp °F | 25.5 | 31.6 | 39.2 | 47.6 | 58.1 | 68.7 | 77.6 | 76.5 | 66.4 | 53.6 | 35.8 | 25.7 | 50.5 |
| Minimum Temp °F | 1.9 | 3.4 | 11.6 | 21.2 | 30.1 | 37.3 | 41.8 | 40.4 | 32.7 | 24.5 | 14.8 | 3.3 | 21.9 |
| Mean Temp °F | 13.7 | 17.5 | 25.4 | 34.4 | 44.2 | 53.1 | 59.7 | 58.5 | 49.6 | 39.1 | 25.4 | 14.5 | 36.3 |
| Days Max Temp ≥ 90 °F | 0 | 0 | 0 | 0 | 0 | 0 | 0 | 0 | 0 | 0 | 0 | 0 | 0 |
| Days Max Temp ≤ 32 °F | 25 | 14 | 5 | 1 | 0 | 0 | 0 | 0 | 0 | 1 | 11 | 24 | 81 |
| Days Min Temp ≤ 32 °F | 31 | 28 | 31 | 28 | 20 | 6 | 1 | 2 | 15 | 27 | 29 | 31 | 249 |
| Days Min Temp ≤ 0 °F | 15 | 12 | 6 | 1 | 0 | 0 | 0 | 0 | 0 | 0 | 4 | 14 | 52 |
| Heating Degree Days | 1585 | 1336 | 1220 | 911 | 639 | 353 | 163 | 201 | 456 | 798 | 1184 | 1559 | 10405 |
| Cooling Degree Days | 0 | 0 | 0 | 0 | 0 | 2 | 7 | 7 | 0 | 0 | 0 | 0 | 16 |
| Total Precipitation (") | 3.15 | 2.19 | 2.00 | 1.93 | 2.04 | 1.57 | 1.29 | 1.24 | 1.49 | 1.55 | 2.72 | 2.77 | 23.94 |
| Days ≥ 0.1" Precip | 10 | 7 | 7 | 6 | 7 | 5 | 4 | 4 | 4 | 4 | 9 | 8 | 75 |
| Total Snowfall (") | na | na | na | ***8.9*** | 1.7 | 0.1 | 0.0 | 0.0 | 0.7 | 2.8 | na | na | na |
| Days ≥ 1" Snow Depth | 31 | 28 | 31 | 26 | 5 | 0 | 0 | 0 | 0 | 1 | 18 | 31 | 171 |

## MUDDY GAP *Carbon County* ELEVATION 6316 ft LAT/LONG 42° 21 ' N / 107° 28 ' W

| | JAN | FEB | MAR | APR | MAY | JUN | JUL | AUG | SEP | OCT | NOV | DEC | YEAR |
|---|---|---|---|---|---|---|---|---|---|---|---|---|---|
| Maximum Temp °F | 30.9 | 34.6 | 44.5 | 55.0 | 65.9 | 76.9 | ***84.3*** | ***83.2*** | 72.4 | ***59.4*** | 41.4 | ***31.5*** | 56.7 |
| Minimum Temp °F | 13.8 | 15.6 | 22.7 | 28.9 | 37.5 | 46.5 | ***52.9*** | 51.5 | 41.6 | ***32.0*** | 21.9 | ***13.7*** | 31.6 |
| Mean Temp °F | 22.4 | 25.2 | 33.6 | 42.0 | 51.7 | 61.7 | ***68.7*** | ***67.3*** | 57.0 | ***45.8*** | 31.7 | ***22.7*** | 44.2 |
| Days Max Temp ≥ 90 °F | 0 | 0 | 0 | 0 | 0 | 2 | 5 | ***4*** | 0 | ***0*** | 0 | ***0*** | 11 |
| Days Max Temp ≤ 32 °F | ***15*** | 10 | 4 | 1 | 0 | 0 | 0 | 0 | 0 | ***1*** | 7 | ***16*** | 54 |
| Days Min Temp ≤ 32 °F | 30 | 27 | 26 | 20 | ***8*** | 1 | 0 | 0 | 4 | ***15*** | 24 | ***29*** | 184 |
| Days Min Temp ≤ 0 °F | ***4*** | 3 | 1 | 0 | 0 | 0 | 0 | 0 | 0 | ***0*** | 1 | ***4*** | 13 |
| Heating Degree Days | 1315 | 1111 | 967 | 685 | 406 | 141 | ***21*** | 35 | 247 | 591 | 992 | ***1306*** | 7817 |
| Cooling Degree Days | 0 | 0 | 0 | 0 | 2 | 59 | 145 | 128 | 18 | 0 | 0 | 0 | 352 |
| Total Precipitation (") | 0.39 | 0.43 | 0.75 | 1.31 | 1.90 | 1.02 | 1.02 | 0.61 | 0.94 | 0.90 | 0.82 | ***0.60*** | 10.69 |
| Days ≥ 0.1" Precip | 2 | 1 | 2 | 4 | 5 | 2 | 3 | 2 | 3 | 3 | 3 | ***2*** | 32 |
| Total Snowfall (") | ***5.8*** | ***6.9*** | ***8.5*** | ***9.9*** | 2.0 | 0.0 | 0.0 | 0.0 | ***1.0*** | ***3.6*** | ***9.5*** | ***7.8*** | 55.0 |
| Days ≥ 1" Snow Depth | na | na | na | ***2*** | 1 | 0 | 0 | 0 | 0 | 1 | na | na | na |

## NEWCASTLE *Weston County* ELEVATION 4380 ft LAT/LONG 43° 51' N / 104° 12' W

| | JAN | FEB | MAR | APR | MAY | JUN | JUL | AUG | SEP | OCT | NOV | DEC | YEAR |
|---|---|---|---|---|---|---|---|---|---|---|---|---|---|
| Maximum Temp °F | 33.6 | 38.2 | 47.3 | 58.7 | 69.7 | 79.9 | 87.7 | 85.2 | 74.2 | 60.8 | 44.2 | 35.0 | 59.5 |
| Minimum Temp °F | 11.2 | 14.7 | 23.3 | 32.8 | 42.5 | 51.6 | 58.2 | 55.9 | 45.6 | 34.2 | 22.3 | 12.9 | 33.8 |
| Mean Temp °F | 22.4 | 26.5 | 35.3 | 45.7 | 56.1 | 65.7 | 73.0 | 70.6 | 59.9 | 47.5 | 33.3 | 24.0 | 46.7 |
| Days Max Temp ≥ 90 °F | 0 | 0 | 0 | 0 | 0 | 5 | 13 | 10 | 2 | 0 | 0 | 0 | 30 |
| Days Max Temp ≤ 32 °F | 12 | 8 | 3 | 0 | 0 | 0 | 0 | 0 | 0 | 0 | 5 | 12 | 40 |
| Days Min Temp ≤ 32 °F | 31 | 28 | 26 | 15 | 3 | 0 | 0 | 0 | 2 | 13 | 25 | 31 | 174 |
| Days Min Temp ≤ 0 °F | 7 | 4 | 1 | 0 | 0 | 0 | 0 | 0 | 0 | 0 | 1 | 5 | 18 |
| Heating Degree Days | 1315 | 1080 | 914 | 572 | 288 | 82 | 10 | 22 | 192 | 536 | 944 | 1266 | 7221 |
| Cooling Degree Days | 0 | 0 | 0 | 2 | 21 | 121 | 259 | 219 | 53 | 2 | 0 | 0 | 677 |
| Total Precipitation (") | 0.40 | 0.53 | 0.65 | 1.60 | 2.35 | 2.37 | 2.05 | 1.64 | 1.15 | 1.08 | 0.65 | 0.57 | 15.04 |
| Days ≥ 0.1" Precip | 1 | 2 | 2 | 4 | 6 | 5 | 5 | 4 | 3 | 3 | 2 | 2 | 39 |
| Total Snowfall (") | 6.5 | 6.9 | *6.0* | *4.3* | 0.6 | 0.1 | 0.0 | 0.0 | 0.2 | 1.1 | 5.9 | 8.1 | 39.7 |
| Days ≥ 1" Snow Depth | *18* | *12* | *5* | na | 0 | 0 | 0 | 0 | 0 | *0* | *6* | *16* | na |

## PHILLIPS *Goshen County* ELEVATION 4905 ft LAT/LONG 41° 38' N / 104° 29' W

| | JAN | FEB | MAR | APR | MAY | JUN | JUL | AUG | SEP | OCT | NOV | DEC | YEAR |
|---|---|---|---|---|---|---|---|---|---|---|---|---|---|
| Maximum Temp °F | 40.5 | 43.7 | 50.2 | 59.4 | 68.8 | 78.9 | 86.3 | 84.6 | 75.8 | 64.1 | 49.2 | 41.8 | 61.9 |
| Minimum Temp °F | 16.3 | 18.7 | 24.1 | 31.1 | 40.1 | 48.6 | 54.6 | 52.3 | 43.0 | 33.1 | 23.3 | 17.2 | 33.5 |
| Mean Temp °F | 28.4 | 31.2 | 37.2 | 45.3 | 54.4 | 63.8 | 70.4 | 68.5 | 59.4 | 48.7 | 36.3 | 29.5 | 47.8 |
| Days Max Temp ≥ 90 °F | 0 | 0 | 0 | 0 | 0 | 4 | 11 | 7 | 2 | 0 | 0 | 0 | 24 |
| Days Max Temp ≤ 32 °F | 7 | 5 | 3 | 0 | 0 | 0 | 0 | 0 | 0 | 0 | 3 | 7 | 25 |
| Days Min Temp ≤ 32 °F | 28 | 26 | 26 | 17 | 5 | 0 | 0 | 0 | 3 | 14 | 25 | 28 | 172 |
| Days Min Temp ≤ 0 °F | 5 | 3 | 1 | 0 | 0 | 0 | 0 | 0 | 0 | 0 | 1 | 3 | 13 |
| Heating Degree Days | 1127 | 947 | 855 | 585 | 327 | 101 | 11 | 26 | 193 | 501 | 855 | 1093 | 6621 |
| Cooling Degree Days | 0 | 0 | 0 | 0 | 9 | 82 | 179 | 145 | 32 | 1 | 0 | 0 | 448 |
| Total Precipitation (") | 0.42 | 0.39 | 0.86 | 1.49 | 2.72 | 2.43 | 2.25 | 1.64 | 1.09 | 0.86 | 0.66 | 0.41 | 15.22 |
| Days ≥ 0.1" Precip | 2 | 1 | 3 | 4 | 6 | 5 | 5 | 4 | 3 | 3 | 2 | 1 | 39 |
| Total Snowfall (") | 8.3 | 6.2 | 10.0 | 6.5 | 1.4 | 0.0 | 0.0 | 0.0 | 0.6 | 4.4 | 8.3 | 8.1 | 53.8 |
| Days ≥ 1" Snow Depth | 11 | 8 | 6 | 3 | 0 | 0 | 0 | 0 | 0 | 2 | 7 | 11 | 48 |

## PINEDALE *Sublette County* ELEVATION 7185 ft LAT/LONG 42° 52' N / 109° 51' W

| | JAN | FEB | MAR | APR | MAY | JUN | JUL | AUG | SEP | OCT | NOV | DEC | YEAR |
|---|---|---|---|---|---|---|---|---|---|---|---|---|---|
| Maximum Temp °F | 26.5 | 30.4 | 38.4 | 49.7 | 61.2 | 70.1 | 77.7 | 76.5 | 67.4 | 55.3 | 37.2 | 27.4 | 51.5 |
| Minimum Temp °F | -0.7 | 1.3 | 10.4 | 20.8 | 28.4 | 36.6 | 41.4 | 37.8 | 30.0 | 21.8 | 10.9 | 0.7 | 20.0 |
| Mean Temp °F | 12.9 | 15.9 | 24.4 | 35.3 | 44.8 | 53.4 | 59.5 | 57.2 | 48.8 | 38.6 | 24.1 | 14.1 | 35.8 |
| Days Max Temp ≥ 90 °F | 0 | 0 | 0 | 0 | 0 | 0 | 0 | 0 | 0 | 0 | 0 | 0 | 0 |
| Days Max Temp ≤ 32 °F | 22 | 15 | 6 | 1 | 0 | 0 | 0 | 0 | 0 | 1 | 10 | 21 | 76 |
| Days Min Temp ≤ 32 °F | 31 | 28 | 31 | 28 | 23 | 7 | 1 | 6 | 20 | 29 | 30 | 31 | 265 |
| Days Min Temp ≤ 0 °F | 17 | 13 | 6 | 1 | 0 | 0 | 0 | 0 | 0 | 0 | 6 | 15 | 58 |
| Heating Degree Days | 1610 | 1382 | 1250 | 886 | 620 | 343 | 166 | 236 | 480 | 811 | 1222 | 1574 | 10580 |
| Cooling Degree Days | 0 | 0 | 0 | 0 | 0 | 1 | 4 | 2 | 0 | 0 | 0 | 0 | 7 |
| Total Precipitation (") | 0.67 | 0.51 | 0.64 | 0.86 | 1.71 | 1.24 | 1.17 | 1.13 | 1.19 | 0.83 | 0.79 | 0.59 | 11.33 |
| Days ≥ 0.1" Precip | 3 | 2 | 2 | 3 | 5 | 3 | 4 | 4 | 3 | 3 | 3 | 2 | 37 |
| Total Snowfall (") | 13.1 | 10.7 | 8.9 | 6.5 | 3.0 | 0.2 | 0.0 | 0.0 | 1.0 | 3.6 | 11.0 | 11.0 | 69.0 |
| Days ≥ 1" Snow Depth | *30* | 27 | *25* | 7 | 1 | 0 | 0 | 0 | 0 | 1 | *14* | *26* | 131 |

## POWDER RIVER 2 *Natrona County* ELEVATION 5964 ft LAT/LONG 43° 1' N / 107° 0' W

| | JAN | FEB | MAR | APR | MAY | JUN | JUL | AUG | SEP | OCT | NOV | DEC | YEAR |
|---|---|---|---|---|---|---|---|---|---|---|---|---|---|
| Maximum Temp °F | *33.5* | *37.9* | 46.9 | 56.2 | *66.7* | *78.0* | *87.6* | 86.0 | *74.8* | *60.5* | *44.0* | *35.8* | 59.0 |
| Minimum Temp °F | *7.9* | 13.0 | 21.0 | 28.4 | *36.5* | *44.7* | *51.7* | 49.2 | 40.2 | *30.6* | *18.8* | *10.0* | 29.3 |
| Mean Temp °F | *20.8* | *25.5* | 34.0 | 42.3 | *51.6* | *61.3* | *69.7* | 67.6 | *57.7* | *45.6* | *31.4* | *23.0* | 44.2 |
| Days Max Temp ≥ 90 °F | *0* | 0 | 0 | 0 | 0 | 4 | 13 | 10 | 2 | 0 | *0* | 0 | 29 |
| Days Max Temp ≤ 32 °F | *12* | 7 | 3 | 1 | 0 | 0 | 0 | 0 | 0 | 1 | *5* | 10 | 39 |
| Days Min Temp ≤ 32 °F | 29 | 26 | 27 | 20 | 9 | 1 | 0 | 0 | 5 | 17 | 26 | 28 | 188 |
| Days Min Temp ≤ 0 °F | *7* | 5 | 1 | 0 | 0 | 0 | 0 | 0 | 0 | 0 | 2 | 6 | 21 |
| Heating Degree Days | *1367* | *1107* | 956 | 676 | *411* | *148* | *24* | 42 | *240* | *596* | *1002* | *1297* | 7866 |
| Cooling Degree Days | na | *0* | *0* | *0* | *3* | *40* | *176* | 125 | *23* | *0* | *0* | na | na |
| Total Precipitation (") | *0.43* | 0.50 | *0.97* | 1.35 | 2.05 | 1.87 | 1.40 | 0.74 | 0.79 | 0.85 | 0.67 | 0.54 | 12.16 |
| Days ≥ 0.1" Precip | *2* | 2 | *3* | *3* | 5 | *4* | 3 | 2 | 2 | *2* | *2* | 2 | 32 |
| Total Snowfall (") | na | na | na | na | *1.6* | 0.0 | 0.0 | 0.0 | 0.4 | *5.2* | *7.7* | na | na |
| Days ≥ 1" Snow Depth | na | na | na | na | 1 | 0 | 0 | 0 | 0 | na | na | na | na |

## RAWLINS MUNI AP *Carbon County* ELEVATION 6781 ft LAT/LONG 41° 48 ' N / 107° 12 ' W

| | JAN | FEB | MAR | APR | MAY | JUN | JUL | AUG | SEP | OCT | NOV | DEC | YEAR |
|---|---|---|---|---|---|---|---|---|---|---|---|---|---|
| Maximum Temp °F | 30.2 | 33.5 | 41.4 | 52.1 | 63.2 | 74.8 | 82.5 | 80.4 | 69.4 | 56.0 | 39.8 | 31.2 | 54.5 |
| Minimum Temp °F | 12.4 | 14.3 | 21.3 | 27.8 | 36.3 | 45.1 | 51.5 | 49.8 | 40.6 | 31.2 | 20.2 | 13.3 | 30.3 |
| Mean Temp °F | 21.3 | 23.9 | 31.3 | 40.0 | 49.8 | 60.0 | 67.0 | 65.2 | 55.0 | 43.7 | 30.1 | 22.3 | 42.5 |
| Days Max Temp ≥ 90 °F | 0 | 0 | 0 | 0 | 0 | 1 | 3 | 1 | 0 | 0 | 0 | 0 | 5 |
| Days Max Temp ≤ 32 °F | 17 | 12 | 6 | 1 | 0 | 0 | 0 | 0 | 0 | 1 | 8 | 16 | 61 |
| Days Min Temp ≤ 32 °F | 30 | 28 | 28 | 21 | 9 | 1 | 0 | 0 | 5 | 17 | 26 | 30 | 195 |
| Days Min Temp ≤ 0 °F | 5 | 4 | 1 | 0 | 0 | 0 | 0 | 0 | 0 | 0 | 2 | 5 | 17 |
| Heating Degree Days | 1348 | 1153 | 1036 | 743 | 463 | 171 | 26 | 52 | 296 | 656 | 1041 | 1318 | 8303 |
| Cooling Degree Days | 0 | 0 | 0 | 0 | 0 | 33 | 98 | 69 | 7 | 0 | 0 | 0 | 207 |
| Total Precipitation (") | 0.52 | 0.47 | 0.63 | 1.04 | 1.32 | 0.87 | 0.90 | 0.83 | 0.87 | 0.91 | 0.68 | 0.55 | 9.59 |
| Days ≥ 0.1" Precip | 2 | 1 | 2 | 3 | 4 | 3 | 3 | 2 | 2 | 3 | 3 | 2 | 30 |
| Total Snowfall (") | 8.8 | 7.1 | 7.6 | 8.0 | 2.0 | 0.3 | 0.0 | 0.0 | 1.6 | 4.6 | 9.2 | 8.3 | 57.5 |
| Days ≥ 1" Snow Depth | 20 | 17 | 10 | 5 | 1 | 0 | 0 | 0 | 0 | 2 | 11 | 19 | 85 |

## REDBIRD *Niobrara County* ELEVATION 3891 ft LAT/LONG 43° 15 ' N / 104° 17 ' W

| | JAN | FEB | MAR | APR | MAY | JUN | JUL | AUG | SEP | OCT | NOV | DEC | YEAR |
|---|---|---|---|---|---|---|---|---|---|---|---|---|---|
| Maximum Temp °F | 34.3 | 40.3 | 50.0 | 60.8 | 70.7 | 81.1 | 89.5 | 87.9 | 77.1 | 64.1 | 46.2 | 36.2 | 61.5 |
| Minimum Temp °F | 7.4 | 12.7 | 21.5 | 31.2 | 40.9 | 50.0 | 56.2 | 53.3 | 41.8 | 30.2 | 18.4 | 8.4 | 31.0 |
| Mean Temp °F | 20.8 | 26.7 | 35.8 | 46.0 | 55.8 | 65.5 | 72.8 | 70.6 | 59.5 | 47.2 | 32.4 | 22.3 | 46.3 |
| Days Max Temp ≥ 90 °F | 0 | 0 | 0 | 0 | 1 | 6 | 16 | 14 | 4 | 0 | 0 | 0 | 41 |
| Days Max Temp ≤ 32 °F | 12 | 7 | 3 | 0 | 0 | 0 | 0 | 0 | 0 | 0 | 5 | 11 | 38 |
| Days Min Temp ≤ 32 °F | 30 | 27 | 27 | 16 | 4 | 0 | 0 | 0 | 5 | 19 | 28 | 30 | 186 |
| Days Min Temp ≤ 0 °F | 10 | 5 | 2 | 0 | 0 | 0 | 0 | 0 | 0 | 0 | 2 | 8 | 27 |
| Heating Degree Days | 1364 | 1077 | 899 | 563 | 292 | 78 | 9 | 19 | 200 | 548 | 973 | 1318 | 7340 |
| Cooling Degree Days | 0 | 0 | 0 | 1 | 16 | 109 | 263 | 222 | 42 | 2 | 0 | 0 | 655 |
| Total Precipitation (") | 0.28 | 0.37 | 0.70 | 1.65 | 2.44 | 2.42 | 2.04 | 1.60 | 1.21 | 0.91 | 0.63 | 0.32 | 14.57 |
| Days ≥ 0.1" Precip | 1 | 1 | 2 | 4 | 6 | 5 | 4 | 4 | 3 | 3 | 2 | 1 | 36 |
| Total Snowfall (") | 6.4 | 6.7 | 9.9 | 5.8 | 0.7 | 0.0 | 0.0 | 0.0 | 0.1 | 2.3 | 7.4 | 7.6 | 46.9 |
| Days ≥ 1" Snow Depth | 17 | 11 | 7 | 2 | 0 | 0 | 0 | 0 | 0 | 1 | 7 | 14 | 59 |

## RIVERTON *Fremont County* ELEVATION 4954 ft LAT/LONG 43° 1 ' N / 108° 23 ' W

| | JAN | FEB | MAR | APR | MAY | JUN | JUL | AUG | SEP | OCT | NOV | DEC | YEAR |
|---|---|---|---|---|---|---|---|---|---|---|---|---|---|
| Maximum Temp °F | 29.4 | 36.6 | 48.7 | 58.9 | 69.2 | 80.1 | 88.3 | 86.9 | 74.8 | 61.1 | 42.0 | 29.3 | 58.8 |
| Minimum Temp °F | -0.6 | 5.7 | 18.6 | 28.2 | 37.8 | 45.9 | 50.9 | 48.1 | 37.7 | 27.0 | 12.9 | 0.2 | 26.0 |
| Mean Temp °F | 14.4 | 21.2 | 33.7 | 43.6 | 53.6 | 63.1 | 69.6 | 67.5 | 56.3 | 44.1 | 27.5 | 14.8 | 42.5 |
| Days Max Temp ≥ 90 °F | 0 | 0 | 0 | 0 | 0 | 6 | 15 | 13 | 2 | 0 | 0 | 0 | 36 |
| Days Max Temp ≤ 32 °F | 17 | 10 | 3 | 1 | 0 | 0 | 0 | 0 | 0 | 1 | 6 | 17 | 55 |
| Days Min Temp ≤ 32 °F | 30 | 28 | 30 | 21 | 7 | 1 | 0 | 0 | 7 | 24 | 29 | 30 | 207 |
| Days Min Temp ≤ 0 °F | 16 | 9 | 2 | 0 | 0 | 0 | 0 | 0 | 0 | 0 | 4 | 14 | 45 |
| Heating Degree Days | 1563 | 1233 | 963 | 635 | 353 | 119 | 21 | 33 | 264 | 645 | 1116 | 1547 | 8492 |
| Cooling Degree Days | 0 | 0 | 0 | 0 | 7 | 74 | 163 | 125 | 18 | 0 | 0 | 0 | 387 |
| Total Precipitation (") | 0.19 | 0.21 | 0.38 | 0.94 | 1.66 | 1.22 | 0.74 | 0.49 | 0.77 | 0.76 | 0.43 | 0.29 | 8.08 |
| Days ≥ 0.1" Precip | 1 | 1 | 1 | 3 | 4 | 3 | 2 | 1 | 2 | 2 | 1 | 1 | 22 |
| Total Snowfall (") | *4.0* | *4.4* | *4.9* | *3.7* | 0.5 | 0.0 | 0.0 | 0.0 | 0.7 | *2.3* | *5.8* | *6.9* | 33.2 |
| Days ≥ 1" Snow Depth | na | na | *4* | *2* | 0 | 0 | 0 | 0 | 0 | *1* | na | na | na |

## ROCHELLE 3 E *Weston County* ELEVATION 4491 ft LAT/LONG 43° 36 ' N / 104° 56 ' W

| | JAN | FEB | MAR | APR | MAY | JUN | JUL | AUG | SEP | OCT | NOV | DEC | YEAR |
|---|---|---|---|---|---|---|---|---|---|---|---|---|---|
| Maximum Temp °F | 32.6 | 37.9 | 46.4 | 56.6 | 68.1 | 79.5 | 88.1 | 86.4 | 75.7 | 63.8 | 45.2 | 33.9 | 59.5 |
| Minimum Temp °F | 8.1 | 12.9 | 21.2 | 29.0 | 38.0 | 47.8 | 54.7 | 52.8 | 39.7 | 28.9 | 19.4 | 9.0 | 30.1 |
| Mean Temp °F | 20.4 | 25.4 | 33.9 | 42.8 | 53.0 | 63.7 | 71.4 | 69.6 | 57.7 | 46.4 | 32.3 | 21.5 | 44.8 |
| Days Max Temp ≥ 90 °F | 0 | 0 | 0 | 0 | 0 | 5 | 14 | 12 | 2 | 0 | 0 | 0 | 33 |
| Days Max Temp ≤ 32 °F | 14 | 8 | 3 | 0 | 0 | 0 | 0 | 0 | 0 | 0 | 4 | 13 | 42 |
| Days Min Temp ≤ 32 °F | 31 | 28 | 28 | 20 | 7 | 0 | 0 | 0 | 6 | 21 | 29 | 31 | 201 |
| Days Min Temp ≤ 0 °F | 10 | 6 | 2 | 0 | 0 | 0 | 0 | 0 | 0 | 0 | 3 | 9 | 30 |
| Heating Degree Days | 1378 | 1111 | 959 | 660 | 368 | 100 | 13 | 28 | 237 | 570 | 976 | 1344 | 7744 |
| Cooling Degree Days | 0 | 0 | 0 | 1 | 4 | 66 | 210 | 183 | 24 | 0 | 0 | 0 | 488 |
| Total Precipitation (") | 0.25 | 0.33 | 0.73 | 1.55 | 2.47 | 1.90 | 1.68 | 1.27 | 0.87 | 0.82 | 0.53 | 0.39 | 12.79 |
| Days ≥ 0.1" Precip | 1 | 1 | 3 | 4 | 6 | 5 | 4 | 3 | 3 | 3 | 2 | 2 | 37 |
| Total Snowfall (") | 4.8 | 5.7 | 7.0 | 6.4 | 1.4 | 0.2 | 0.0 | 0.0 | 0.5 | 2.3 | 6.1 | 7.4 | 41.8 |
| Days ≥ 1" Snow Depth | 19 | 13 | 10 | 3 | 1 | 0 | 0 | 0 | 0 | 1 | 8 | 18 | 73 |

## ROCK SPRINGS AP *Sweetwater County* ELEVATION 6745 ft LAT/LONG 41° 36 ' N / 109° 4 ' W

| | JAN | FEB | MAR | APR | MAY | JUN | JUL | AUG | SEP | OCT | NOV | DEC | YEAR |
|---|---|---|---|---|---|---|---|---|---|---|---|---|---|
| Maximum Temp °F | 29.3 | 34.1 | 43.0 | 53.7 | 64.4 | 75.1 | 83.2 | 81.0 | 70.5 | 57.1 | 40.1 | 30.6 | 55.2 |
| Minimum Temp °F | 11.4 | 14.0 | 21.7 | 28.9 | 37.1 | 45.8 | 52.3 | 50.3 | 41.0 | 31.2 | 19.7 | 11.6 | 30.4 |
| Mean Temp °F | 20.4 | 24.0 | 32.3 | 41.3 | 50.8 | 60.5 | 67.8 | 65.7 | 55.7 | 44.2 | 29.9 | 21.1 | 42.8 |
| Days Max Temp ≥ 90 °F | 0 | 0 | 0 | 0 | 0 | 1 | 3 | 2 | 0 | 0 | 0 | 0 | 6 |
| Days Max Temp ≤ 32 °F | 18 | 12 | 5 | 1 | 0 | 0 | 0 | 0 | 0 | 1 | 8 | 17 | 62 |
| Days Min Temp ≤ 32 °F | 30 | 28 | 29 | 22 | 8 | 1 | 0 | 0 | 4 | 18 | 28 | 31 | 199 |
| Days Min Temp ≤ 0 °F | 5 | 3 | 1 | 0 | 0 | 0 | 0 | 0 | 0 | 0 | 1 | 5 | 15 |
| Heating Degree Days | 1375 | 1151 | 1005 | 703 | 434 | 166 | 22 | 48 | 278 | 639 | 1046 | 1354 | 8221 |
| Cooling Degree Days | 0 | 0 | 0 | 0 | 0 | 43 | 118 | *72* | 8 | 0 | 0 | 0 | 241 |
| Total Precipitation (") | 0.55 | 0.50 | 0.75 | 1.06 | 1.21 | 0.96 | 0.94 | 0.59 | 0.79 | 0.85 | 0.68 | 0.63 | 9.51 |
| Days ≥ 0.1" Precip | 2 | 2 | 2 | 3 | 4 | 3 | 2 | 2 | 2 | 3 | 2 | 2 | 29 |
| Total Snowfall (") | 6.4 | 5.0 | 7.2 | 7.0 | 2.2 | 0.1 | 0.0 | 0.0 | 1.2 | 3.7 | 6.4 | 6.8 | 46.0 |
| Days ≥ 1" Snow Depth | 22 | 16 | 9 | 3 | 1 | 0 | 0 | 0 | 0 | 1 | 8 | 17 | 77 |

## SAGE 4 NNW *Lincoln County* ELEVATION 6325 ft LAT/LONG 41° 52 ' N / 111° 1 ' W

| | JAN | FEB | MAR | APR | MAY | JUN | JUL | AUG | SEP | OCT | NOV | DEC | YEAR |
|---|---|---|---|---|---|---|---|---|---|---|---|---|---|
| Maximum Temp °F | 26.5 | 31.1 | 42.1 | *54.4* | 64.8 | 73.9 | 82.0 | 81.2 | 71.4 | 59.4 | 40.9 | 27.4 | 54.6 |
| Minimum Temp °F | -1.9 | 0.5 | 14.3 | *22.9* | 30.5 | 37.5 | 41.6 | 38.2 | 29.5 | 20.1 | 11.8 | -1.6 | 20.3 |
| Mean Temp °F | 12.3 | 15.8 | 28.2 | *38.7* | 47.7 | 55.7 | 61.9 | 59.7 | 50.5 | 39.8 | 26.4 | 12.9 | 37.5 |
| Days Max Temp ≥ 90 °F | 0 | 0 | 0 | 0 | 0 | 1 | 2 | 2 | 0 | 0 | 0 | 0 | 5 |
| Days Max Temp ≤ 32 °F | 20 | 13 | 4 | *0* | 0 | 0 | 0 | 0 | 0 | 1 | 7 | 20 | 65 |
| Days Min Temp ≤ 32 °F | 31 | 28 | *31* | *28* | 19 | 6 | 3 | 6 | 20 | 29 | 29 | 31 | 261 |
| Days Min Temp ≤ 0 °F | 17 | 14 | 4 | *0* | 0 | 0 | 0 | 0 | 0 | 0 | 5 | 18 | 58 |
| Heating Degree Days | 1631 | 1383 | 1133 | *784* | 530 | 275 | 110 | 164 | 430 | 776 | 1152 | 1610 | 9978 |
| Cooling Degree Days | 0 | 0 | 0 | 0 | 0 | 4 | 17 | 7 | 0 | 0 | 0 | 0 | 28 |
| Total Precipitation (") | *0.54* | *0.37* | *0.71* | *1.07* | 1.16 | 1.07 | 0.87 | 0.75 | 1.25 | 1.02 | *0.81* | *0.50* | 10.12 |
| Days ≥ 0.1" Precip | na | na | na | *3* | 4 | 3 | 3 | 2 | 3 | 3 | *2* | *2* | na |
| Total Snowfall (") | na | na | na | na | *1.1* | 0.1 | 0.0 | 0.0 | *0.2* | *1.8* | *4.0* | na | na |
| Days ≥ 1" Snow Depth | na | na | na | na | *0* | 0 | 0 | 0 | *0* | *0* | na | na | na |

## SARATOGA *Carbon County* ELEVATION 6795 ft LAT/LONG 41° 27 ' N / 106° 48 ' W

| | JAN | FEB | MAR | APR | MAY | JUN | JUL | AUG | SEP | OCT | NOV | DEC | YEAR |
|---|---|---|---|---|---|---|---|---|---|---|---|---|---|
| Maximum Temp °F | 33.1 | 36.9 | 44.3 | 54.9 | 65.6 | 75.9 | 82.5 | 80.8 | 71.6 | 59.1 | 42.5 | 35.0 | 56.9 |
| Minimum Temp °F | 10.6 | 12.8 | 20.3 | 27.4 | 35.5 | 43.7 | 49.5 | 47.2 | 38.3 | 29.0 | 19.1 | 11.3 | 28.7 |
| Mean Temp °F | 21.9 | 24.9 | 32.3 | 41.2 | 50.6 | 59.8 | 66.0 | 64.0 | 55.0 | 44.1 | 30.9 | 23.1 | 42.8 |
| Days Max Temp ≥ 90 °F | 0 | 0 | 0 | 0 | 0 | 1 | 2 | 1 | 0 | 0 | 0 | 0 | 4 |
| Days Max Temp ≤ 32 °F | 13 | 8 | 3 | 1 | 0 | 0 | 0 | 0 | 0 | 1 | 6 | 12 | 44 |
| Days Min Temp ≤ 32 °F | 30 | 28 | 29 | 22 | 10 | 1 | 0 | 0 | 7 | 21 | 28 | 30 | 206 |
| Days Min Temp ≤ 0 °F | 6 | 4 | 1 | 0 | 0 | 0 | 0 | 0 | 0 | 0 | 2 | 6 | 19 |
| Heating Degree Days | 1330 | 1125 | 1007 | 708 | 440 | 174 | 34 | 65 | 296 | 641 | 1018 | 1291 | 8129 |
| Cooling Degree Days | 0 | 0 | 0 | 0 | 0 | 25 | 67 | 45 | 4 | 0 | 0 | 0 | 141 |
| Total Precipitation (") | 0.40 | 0.46 | 0.79 | 0.91 | 1.39 | 0.94 | 1.17 | 0.85 | 0.89 | 1.05 | 0.66 | 0.49 | 10.00 |
| Days ≥ 0.1" Precip | 1 | 2 | 3 | 3 | 4 | 3 | 3 | 3 | 3 | 3 | 2 | 2 | 32 |
| Total Snowfall (") | 8.3 | 8.7 | 10.1 | 5.8 | 1.7 | 0.2 | 0.0 | 0.0 | 1.4 | 3.9 | 10.5 | 9.8 | 60.4 |
| Days ≥ 1" Snow Depth | 21 | 19 | 10 | 3 | 0 | 0 | 0 | 0 | 0 | 2 | 14 | 21 | 90 |

## SHERIDAN COUNTY AP *Sheridan County* ELEVATION 3942 ft LAT/LONG 44° 46 ' N / 106° 58 ' W

| | JAN | FEB | MAR | APR | MAY | JUN | JUL | AUG | SEP | OCT | NOV | DEC | YEAR |
|---|---|---|---|---|---|---|---|---|---|---|---|---|---|
| Maximum Temp °F | 33.2 | 38.2 | 47.1 | 57.3 | 67.0 | 76.8 | 85.6 | 85.1 | 73.4 | 61.0 | 44.8 | 35.3 | 58.7 |
| Minimum Temp °F | 9.1 | 14.2 | 22.3 | 30.8 | 39.3 | 47.4 | 53.0 | 51.9 | 41.8 | 31.6 | 19.7 | 10.7 | 31.0 |
| Mean Temp °F | 21.2 | 26.2 | 34.7 | 44.1 | 53.2 | 62.1 | 69.3 | 68.5 | 57.6 | 46.3 | 32.3 | 23.0 | 44.9 |
| Days Max Temp ≥ 90 °F | 0 | 0 | 0 | 0 | 0 | 3 | 10 | 11 | 2 | 0 | 0 | 0 | 26 |
| Days Max Temp ≤ 32 °F | 13 | 8 | 4 | 1 | 0 | 0 | 0 | 0 | 0 | 1 | 5 | 12 | 44 |
| Days Min Temp ≤ 32 °F | 30 | 27 | 27 | 18 | 5 | 0 | 0 | 0 | 3 | 16 | 28 | 30 | 184 |
| Days Min Temp ≤ 0 °F | 9 | 5 | 1 | 0 | 0 | 0 | 0 | 0 | 0 | 0 | 2 | 7 | 24 |
| Heating Degree Days | 1353 | 1089 | 933 | 622 | 365 | 132 | 24 | 35 | 242 | 574 | 975 | 1296 | 7640 |
| Cooling Degree Days | 0 | 0 | 0 | 1 | 7 | 67 | 185 | 177 | 33 | 2 | 0 | 0 | 472 |
| Total Precipitation (") | 0.77 | 0.61 | 1.03 | 1.71 | 2.37 | 2.13 | 1.02 | 0.79 | 1.30 | 1.28 | 0.86 | 0.69 | 14.56 |
| Days ≥ 0.1" Precip | 3 | 2 | 4 | 5 | 6 | 5 | 3 | 2 | 4 | 3 | 3 | 2 | 42 |
| Total Snowfall (") | 12.3 | 9.9 | 12.8 | 10.8 | 2.4 | 0.1 | 0.0 | 0.0 | 2.0 | 5.0 | 9.8 | 12.1 | 77.2 |
| Days ≥ 1" Snow Depth | 24 | 20 | 11 | 3 | 1 | 0 | 0 | 0 | 1 | 2 | 10 | 20 | 92 |

## SHERIDAN FIELD STN *Sheridan County* ELEVATION 3802 ft LAT/LONG 44° 51 ' N / 106° 52 ' W

| | JAN | FEB | MAR | APR | MAY | JUN | JUL | AUG | SEP | OCT | NOV | DEC | YEAR |
|---|---|---|---|---|---|---|---|---|---|---|---|---|---|
| Maximum Temp °F | 32.4 | 37.3 | 46.6 | 57.6 | 67.5 | 77.4 | 86.6 | 86.4 | 74.4 | 61.4 | 44.8 | 35.0 | 58.9 |
| Minimum Temp °F | 5.2 | 10.1 | 20.1 | 29.4 | 38.5 | 47.3 | 52.0 | 49.3 | 39.1 | 28.7 | 17.3 | 7.3 | 28.7 |
| Mean Temp °F | 18.8 | 23.7 | 33.3 | 43.5 | 53.1 | 62.4 | 69.3 | 67.9 | 56.7 | 45.1 | 31.1 | 21.2 | 43.8 |
| Days Max Temp ≥ 90 °F | 0 | 0 | 0 | 0 | 1 | 3 | 13 | 13 | 3 | 0 | 0 | 0 | 33 |
| Days Max Temp ≤ 32 °F | 13 | 9 | 4 | 1 | 0 | 0 | 0 | 0 | 0 | 1 | 4 | 12 | 44 |
| Days Min Temp ≤ 32 °F | 31 | 28 | 29 | 20 | 7 | 1 | 0 | 0 | 6 | 21 | 29 | 31 | 203 |
| Days Min Temp ≤ 0 °F | 12 | 6 | 2 | 0 | 0 | 0 | 0 | 0 | 0 | 0 | 3 | 9 | 32 |
| Heating Degree Days | 1426 | 1159 | 975 | 638 | 371 | 131 | 31 | 44 | 266 | 611 | 1009 | 1351 | 8012 |
| Cooling Degree Days | 0 | 0 | 0 | 1 | 9 | 69 | 172 | 148 | 23 | 1 | 0 | 0 | 423 |
| Total Precipitation (") | 0.54 | 0.43 | 0.86 | 1.56 | 2.58 | 2.51 | 1.16 | 0.89 | 1.42 | 1.18 | 0.75 | 0.52 | 14.40 |
| Days ≥ 0.1" Precip | 2 | 2 | 3 | 5 | 6 | 6 | 3 | 2 | 4 | 3 | 2 | 2 | 40 |
| Total Snowfall (") | na | na | na | *5.2* | 0.6 | 0.1 | 0.0 | 0.0 | 0.5 | *1.1* | na | na | na |
| Days ≥ 1" Snow Depth | na | na | na | *2* | 0 | 0 | 0 | 0 | 0 | *1* | na | na | na |

## SNAKE RIVER *Teton County* ELEVATION 6883 ft LAT/LONG 44° 8 ' N / 110° 40 ' W

| | JAN | FEB | MAR | APR | MAY | JUN | JUL | AUG | SEP | OCT | NOV | DEC | YEAR |
|---|---|---|---|---|---|---|---|---|---|---|---|---|---|
| Maximum Temp °F | 26.0 | 31.4 | *37.8* | *46.8* | *57.5* | 68.3 | 76.9 | 76.6 | *66.3* | *52.4* | *35.1* | *26.3* | 50.1 |
| Minimum Temp °F | -1.5 | 0.1 | *8.2* | *18.0* | *26.6* | 33.4 | 37.1 | 35.0 | *26.5* | *19.0* | *8.5* | *-1.7* | 17.4 |
| Mean Temp °F | 12.3 | 15.8 | *23.0* | *32.4* | *42.1* | 50.9 | 57.0 | 55.8 | *46.4* | *35.7* | *21.8* | *12.3* | 33.8 |
| Days Max Temp ≥ 90 °F | 0 | 0 | *0* | *0* | 0 | 0 | 0 | 0 | *0* | *0* | *0* | *0* | 0 |
| Days Max Temp ≤ 32 °F | 24 | 15 | *7* | *1* | 0 | 0 | 0 | 0 | *0* | *1* | *12* | *24* | 84 |
| Days Min Temp ≤ 32 °F | 31 | 28 | *31* | na | *26* | *15* | 7 | *10* | *25* | *30* | *30* | *31* | na |
| Days Min Temp ≤ 0 °F | 17 | 14 | *8* | *2* | 0 | 0 | 0 | 0 | *0* | *1* | *8* | *17* | 67 |
| Heating Degree Days | 1630 | 1385 | *1296* | *971* | *704* | 419 | 241 | 282 | *552* | *900* | *1290* | *1628* | 11298 |
| Cooling Degree Days | 0 | 0 | 0 | 0 | 0 | 1 | 1 | 1 | 0 | 0 | 0 | 0 | 3 |
| Total Precipitation (") | 3.75 | 2.85 | *2.50* | *2.16* | *2.42* | 2.37 | 1.63 | 1.58 | *1.82* | *2.10* | *3.12* | *4.01* | 30.31 |
| Days ≥ 0.1" Precip | 11 | 9 | *8* | *6* | 7 | 6 | 5 | 5 | *5* | *6* | *9* | *11* | 88 |
| Total Snowfall (") | *49.6* | *42.0* | *35.0* | *18.0* | *7.6* | 0.5 | 0.0 | 0.0 | *0.5* | *8.8* | *41.4* | *58.6* | 262.0 |
| Days ≥ 1" Snow Depth | 31 | 28 | *31* | *28* | *16* | 0 | 0 | 0 | *0* | *6* | *25* | *31* | 196 |

## SUNDANCE *Crook County* ELEVATION 4754 ft LAT/LONG 44° 24 ' N / 104° 21 ' W

| | JAN | FEB | MAR | APR | MAY | JUN | JUL | AUG | SEP | OCT | NOV | DEC | YEAR |
|---|---|---|---|---|---|---|---|---|---|---|---|---|---|
| Maximum Temp °F | 31.5 | 35.6 | 44.1 | 54.9 | 65.1 | 74.7 | 82.5 | 81.4 | 71.1 | 58.4 | 42.1 | 32.8 | 56.2 |
| Minimum Temp °F | 10.0 | 13.6 | 21.0 | 30.0 | 39.5 | 48.5 | 54.6 | 53.0 | 43.3 | 32.7 | 21.1 | 12.6 | 31.7 |
| Mean Temp °F | 20.8 | 24.6 | 32.6 | 42.4 | 52.3 | 61.7 | 68.6 | 67.2 | 57.2 | 45.6 | 31.6 | 22.8 | 43.9 |
| Days Max Temp ≥ 90 °F | 0 | 0 | 0 | 0 | 0 | 1 | 5 | 3 | 1 | 0 | 0 | 0 | 10 |
| Days Max Temp ≤ 32 °F | 14 | 9 | 5 | 1 | 0 | 0 | 0 | 0 | 0 | 1 | 6 | 14 | 50 |
| Days Min Temp ≤ 32 °F | 30 | 28 | 27 | 19 | 6 | 0 | 0 | 0 | 4 | 14 | 26 | 30 | 184 |
| Days Min Temp ≤ 0 °F | 8 | 5 | 2 | 0 | 0 | 0 | 0 | 0 | 0 | 0 | 1 | 5 | 21 |
| Heating Degree Days | 1365 | 1133 | 999 | 671 | 392 | 147 | 33 | 48 | 253 | 595 | 995 | 1304 | 7935 |
| Cooling Degree Days | 0 | 0 | 0 | 0 | 7 | 64 | 164 | 133 | 28 | 0 | 0 | 0 | 396 |
| Total Precipitation (") | 0.66 | 0.61 | 0.89 | 2.09 | 2.73 | 3.12 | 1.91 | 1.57 | 1.41 | 1.27 | 0.78 | 0.71 | 17.75 |
| Days ≥ 0.1" Precip | 2 | 2 | 3 | 6 | 6 | 7 | 5 | 4 | 3 | 3 | 3 | 3 | 47 |
| Total Snowfall (") | 12.0 | 10.9 | 12.5 | 14.6 | 3.0 | 0.1 | 0.0 | 0.0 | 0.8 | 5.3 | 10.7 | 13.5 | 83.4 |
| Days ≥ 1" Snow Depth | 28 | 24 | 14 | 4 | 1 | 0 | 0 | 0 | 0 | 2 | 11 | 23 | 107 |

## SUNSHINE 2 ENE *Park County* ELEVATION 6443 ft LAT/LONG 44° 3 ' N / 108° 59 ' W

| | JAN | FEB | MAR | APR | MAY | JUN | JUL | AUG | SEP | OCT | NOV | DEC | YEAR |
|---|---|---|---|---|---|---|---|---|---|---|---|---|---|
| Maximum Temp °F | 36.0 | 39.1 | 44.3 | 52.1 | 62.0 | 70.9 | 78.5 | 77.6 | 67.8 | 57.7 | 42.7 | 35.7 | 55.4 |
| Minimum Temp °F | 6.1 | 8.9 | 15.7 | 23.4 | 31.9 | 38.8 | 43.4 | 42.7 | 33.8 | 25.2 | 13.8 | 6.7 | 24.2 |
| Mean Temp °F | 21.1 | 24.0 | 30.1 | 37.9 | 46.9 | 54.9 | 61.0 | 60.2 | 50.8 | 41.5 | 28.3 | 21.2 | 39.8 |
| Days Max Temp ≥ 90 °F | 0 | 0 | 0 | 0 | 0 | 0 | 0 | 0 | 0 | 0 | 0 | 0 | 0 |
| Days Max Temp ≤ 32 °F | 10 | 6 | 4 | 1 | 0 | 0 | 0 | 0 | 0 | 1 | 6 | 10 | 38 |
| Days Min Temp ≤ 32 °F | 29 | 27 | *30* | 27 | *17* | *5* | 1 | 1 | 11 | 26 | *29* | 31 | 234 |
| Days Min Temp ≤ 0 °F | 10 | 7 | 2 | 0 | 0 | 0 | 0 | 0 | 0 | 0 | 3 | 8 | 30 |
| Heating Degree Days | 1355 | 1151 | 1077 | 807 | 553 | 303 | 132 | 154 | 419 | 723 | 1095 | 1351 | 9120 |
| Cooling Degree Days | 0 | *0* | 0 | 0 | 0 | 5 | 16 | 10 | 0 | 0 | 0 | 0 | 31 |
| Total Precipitation (") | 0.37 | 0.34 | 0.98 | 1.64 | 2.90 | 2.36 | 1.55 | 1.36 | 1.37 | 0.99 | 0.65 | 0.49 | 15.00 |
| Days ≥ 0.1" Precip | 2 | 1 | 3 | 5 | 7 | 6 | 5 | 4 | 3 | 3 | 2 | 2 | 43 |
| Total Snowfall (") | *6.1* | na | 11.5 | na | *6.9* | 0.3 | 0.0 | 0.0 | 3.3 | 8.2 | *8.0* | na | na |
| Days ≥ 1" Snow Depth | na | na | na | na | *1* | 0 | 0 | 0 | 0 | 2 | na | na | na |

## SYBILLE RESEARCH UNI *Albany County* ELEVATION 6106 ft LAT/LONG 41° 46 ' N / 105° 23 ' W

| | JAN | FEB | MAR | APR | MAY | JUN | JUL | AUG | SEP | OCT | NOV | DEC | YEAR |
|---|---|---|---|---|---|---|---|---|---|---|---|---|---|
| Maximum Temp °F | 35.9 | 39.0 | 46.7 | 56.5 | 66.5 | 77.4 | 84.4 | 83.1 | 74.0 | 61.8 | 45.1 | 36.9 | 58.9 |
| Minimum Temp °F | 16.0 | 18.2 | 23.2 | 30.5 | 38.5 | 47.1 | 53.7 | 52.4 | 43.6 | 34.2 | 24.0 | 17.0 | 33.2 |
| Mean Temp °F | 26.0 | 28.6 | 35.0 | 43.5 | 52.5 | 62.3 | 69.1 | 67.7 | 58.9 | 48.0 | 34.5 | 27.0 | 46.1 |
| Days Max Temp ≥ 90 °F | 0 | 0 | 0 | 0 | 0 | 2 | 7 | 3 | 0 | 0 | 0 | 0 | 12 |
| Days Max Temp ≤ 32 °F | 10 | 7 | 3 | 1 | 0 | 0 | 0 | 0 | 0 | 0 | 5 | 10 | 36 |
| Days Min Temp ≤ 32 °F | 29 | 26 | 26 | 18 | 6 | 1 | 0 | 0 | 3 | 13 | 24 | 29 | 175 |
| Days Min Temp ≤ 0 °F | 4 | 2 | 1 | 0 | 0 | 0 | 0 | 0 | 0 | 0 | 1 | 3 | 11 |
| Heating Degree Days | 1203 | 1021 | 924 | 638 | 382 | 129 | 19 | 29 | 204 | 519 | 907 | 1173 | 7148 |
| Cooling Degree Days | 0 | 0 | 0 | 0 | 2 | 61 | 149 | 128 | 27 | 1 | 0 | 0 | 368 |
| Total Precipitation (") | 0.64 | 0.62 | 1.08 | 1.80 | 2.43 | 2.33 | 1.68 | 1.36 | 1.30 | 1.17 | 0.95 | 0.64 | 16.00 |
| Days ≥ 0.1" Precip | 3 | 2 | 3 | 4 | 6 | 5 | 4 | 3 | 3 | 3 | 3 | 2 | 41 |
| Total Snowfall (") | 11.5 | 9.4 | 14.0 | 11.5 | 3.6 | 0.1 | 0.0 | 0.0 | 2.1 | 7.2 | 12.4 | 11.6 | 83.4 |
| Days ≥ 1" Snow Depth | 15 | 11 | 9 | 4 | 1 | 0 | 0 | 0 | 1 | 2 | 8 | 12 | 63 |

## TEN SLEEP 16 SSE *Washakie County* ELEVATION 4682 ft LAT/LONG 43° 49 ' N / 107° 22 ' W

| | JAN | FEB | MAR | APR | MAY | JUN | JUL | AUG | SEP | OCT | NOV | DEC | YEAR |
|---|---|---|---|---|---|---|---|---|---|---|---|---|---|
| Maximum Temp °F | 31.6 | 38.6 | 47.8 | 57.4 | 67.5 | 78.4 | 87.5 | 86.0 | 73.8 | 60.5 | 44.1 | 33.8 | 58.9 |
| Minimum Temp °F | -2.8 | 5.2 | 19.1 | 27.4 | 36.7 | 45.3 | 50.3 | 47.3 | 36.2 | 25.5 | 13.4 | 0.9 | 25.4 |
| Mean Temp °F | 14.5 | 21.9 | 33.5 | 42.4 | 52.1 | 61.9 | 68.9 | 66.7 | 55.0 | 43.1 | 28.8 | 17.4 | 42.2 |
| Days Max Temp ≥ 90 °F | 0 | 0 | 0 | 0 | 0 | 5 | 15 | 11 | 2 | 0 | 0 | 0 | 33 |
| Days Max Temp ≤ 32 °F | 14 | 7 | 3 | 0 | 0 | 0 | 0 | 0 | 0 | 1 | 5 | 13 | 43 |
| Days Min Temp ≤ 32 °F | 31 | 28 | 30 | 23 | 8 | 1 | 0 | 0 | 9 | 27 | 30 | 31 | 218 |
| Days Min Temp ≤ 0 °F | 17 | 10 | 2 | 0 | 0 | 0 | 0 | 0 | 0 | 0 | 5 | 14 | 48 |
| Heating Degree Days | 1562 | 1211 | 970 | 672 | 397 | 140 | 28 | 43 | 303 | 673 | 1081 | 1471 | 8551 |
| Cooling Degree Days | 0 | 0 | 0 | 0 | 2 | 47 | 132 | 89 | 6 | 0 | 0 | 0 | 276 |
| Total Precipitation (") | 0.52 | 0.51 | 1.02 | 1.49 | 2.20 | 1.84 | 0.87 | 0.76 | 1.13 | 1.20 | 0.75 | 0.66 | 12.95 |
| Days ≥ 0.1" Precip | 2 | 2 | 3 | 4 | 5 | 5 | 3 | 2 | 3 | 4 | 3 | 2 | 38 |
| Total Snowfall (") | 9.5 | 8.6 | 10.2 | 6.1 | 1.8 | 0.1 | 0.0 | 0.0 | 0.5 | 3.9 | 9.1 | 11.2 | 61.0 |
| Days ≥ 1" Snow Depth | 29 | 27 | 15 | 3 | 1 | 0 | 0 | 0 | 0 | 2 | 13 | 24 | 114 |

## TEN SLEEP 4 NE *Washakie County* ELEVATION 4803 ft LAT/LONG 44° 4 ' N / 107° 25 ' W

| | JAN | FEB | MAR | APR | MAY | JUN | JUL | AUG | SEP | OCT | NOV | DEC | YEAR |
|---|---|---|---|---|---|---|---|---|---|---|---|---|---|
| Maximum Temp °F | 36.5 | 40.9 | 49.7 | 59.6 | 69.5 | 79.2 | 87.1 | 85.7 | 75.0 | 62.3 | 45.7 | 38.1 | 60.8 |
| Minimum Temp °F | 12.6 | 17.4 | 25.8 | 33.9 | 42.2 | 50.3 | 56.3 | 54.7 | 44.7 | 34.6 | 23.5 | 15.0 | 34.3 |
| Mean Temp °F | 24.6 | 29.2 | 37.8 | 46.8 | 55.9 | 64.8 | 71.7 | 70.2 | 59.9 | 48.5 | 34.7 | 26.6 | 47.6 |
| Days Max Temp ≥ 90 °F | 0 | 0 | 0 | 0 | 0 | 3 | 12 | 9 | 1 | 0 | 0 | 0 | 25 |
| Days Max Temp ≤ 32 °F | 10 | 5 | 2 | 0 | 0 | 0 | 0 | 0 | 0 | 0 | 4 | 8 | 29 |
| Days Min Temp ≤ 32 °F | 30 | 27 | 25 | 14 | 4 | 0 | 0 | 0 | 2 | 12 | 25 | 29 | 168 |
| Days Min Temp ≤ 0 °F | 6 | 3 | 1 | 0 | 0 | 0 | 0 | 0 | 0 | 0 | 1 | 3 | 14 |
| Heating Degree Days | 1247 | 1005 | 837 | 542 | 290 | 89 | 11 | 20 | 190 | 507 | 903 | 1185 | 6826 |
| Cooling Degree Days | 0 | 0 | 0 | 3 | 17 | 95 | 207 | 186 | 39 | 1 | 0 | 0 | 548 |
| Total Precipitation (") | 0.53 | 0.36 | 0.84 | 1.45 | 2.24 | 2.13 | 0.87 | 0.72 | 1.36 | 1.17 | 0.81 | 0.61 | 13.09 |
| Days ≥ 0.1" Precip | 2 | 2 | 3 | 4 | 5 | 5 | 2 | 2 | 4 | 4 | 3 | 2 | 38 |
| Total Snowfall (") | 9.7 | 6.7 | 8.1 | 5.7 | 2.0 | 0.0 | 0.0 | 0.0 | 0.9 | 3.3 | 7.9 | 10.2 | 54.5 |
| Days ≥ 1" Snow Depth | 24 | 19 | 7 | 2 | 0 | 0 | 0 | 0 | 0 | 1 | 10 | 19 | 82 |

## THERMOPOLIS 25 WNW *Hot Springs County* ELEVATION 5564 ft LAT/LONG 43° 43 ' N / 108° 40 ' W

| | JAN | FEB | MAR | APR | MAY | JUN | JUL | AUG | SEP | OCT | NOV | DEC | YEAR |
|---|---|---|---|---|---|---|---|---|---|---|---|---|---|
| Maximum Temp °F | 36.8 | 40.2 | 47.2 | 56.0 | 65.0 | 74.6 | 81.5 | ***80.0*** | 70.6 | ***59.4*** | 44.6 | 37.2 | 57.8 |
| Minimum Temp °F | 7.1 | 11.9 | 19.3 | 27.2 | 36.5 | 43.3 | 48.6 | ***46.8*** | 37.5 | ***28.2*** | 16.5 | 8.1 | 27.6 |
| Mean Temp °F | 22.0 | 26.1 | 33.3 | 41.6 | 50.8 | 58.9 | 65.0 | ***63.4*** | 54.1 | ***43.9*** | 30.6 | 22.7 | 42.7 |
| Days Max Temp ≥ 90 °F | 0 | 0 | 0 | 0 | 0 | 1 | 3 | 1 | 0 | 0 | 0 | 0 | 5 |
| Days Max Temp ≤ 32 °F | 9 | 6 | 3 | 0 | 0 | 0 | 0 | 0 | 0 | 0 | ***5*** | ***9*** | 32 |
| Days Min Temp ≤ 32 °F | 30 | 27 | 28 | na | ***9*** | 2 | 0 | 1 | 8 | na | ***28*** | ***30*** | na |
| Days Min Temp ≤ 0 °F | 8 | 4 | 1 | 0 | 0 | 0 | 0 | 0 | 0 | 0 | ***2*** | ***6*** | 21 |
| Heating Degree Days | 1328 | 1093 | 973 | 696 | 434 | 197 | 62 | ***89*** | 327 | ***650*** | 1021 | 1304 | 8174 |
| Cooling Degree Days | 0 | 0 | 0 | 0 | 1 | 25 | 69 | na | ***3*** | ***0*** | ***0*** | 0 | na |
| Total Precipitation (") | 0.35 | 0.29 | 0.75 | 1.23 | 2.38 | 1.92 | 1.47 | 1.12 | 1.09 | 0.83 | 0.52 | 0.36 | 12.31 |
| Days ≥ 0.1" Precip | 1 | 1 | 3 | ***4*** | 6 | 4 | 3 | 3 | 2 | 2 | 2 | 1 | 32 |
| Total Snowfall (") | 6.9 | 5.7 | 10.5 | 11.0 | 3.7 | 0.1 | 0.0 | 0.0 | 2.3 | ***5.4*** | ***8.7*** | ***6.3*** | 60.6 |
| Days ≥ 1" Snow Depth | 15 | 9 | 6 | ***3*** | 1 | 0 | 0 | 0 | 0 | ***2*** | ***9*** | ***11*** | 56 |

## TORRINGTON EXP FARM *Goshen County* ELEVATION 4104 ft LAT/LONG 42° 5 ' N / 104° 13 ' W

| | JAN | FEB | MAR | APR | MAY | JUN | JUL | AUG | SEP | OCT | NOV | DEC | YEAR |
|---|---|---|---|---|---|---|---|---|---|---|---|---|---|
| Maximum Temp °F | 40.0 | 44.9 | 52.3 | 62.4 | 72.0 | 82.5 | 89.3 | 87.4 | 78.1 | 66.1 | 50.2 | 40.8 | 63.8 |
| Minimum Temp °F | 10.6 | 14.8 | 22.4 | 31.0 | 40.9 | 50.0 | 55.2 | 52.2 | 41.3 | 29.4 | 19.5 | 11.1 | 31.5 |
| Mean Temp °F | 25.3 | 29.9 | 37.4 | 46.7 | 56.5 | 66.2 | 72.3 | 69.9 | 59.7 | 47.8 | 34.8 | 26.0 | 47.7 |
| Days Max Temp ≥ 90 °F | 0 | 0 | 0 | 0 | 1 | 8 | 16 | 14 | 4 | 0 | 0 | 0 | 43 |
| Days Max Temp ≤ 32 °F | 8 | 5 | 2 | 0 | 0 | 0 | 0 | 0 | 0 | 0 | 3 | 7 | 25 |
| Days Min Temp ≤ 32 °F | 30 | 27 | 27 | 17 | 4 | 0 | 0 | 0 | 4 | 20 | 28 | 30 | 187 |
| Days Min Temp ≤ 0 °F | 7 | 4 | 1 | 0 | 0 | 0 | 0 | 0 | 0 | 0 | 2 | 6 | 20 |
| Heating Degree Days | 1224 | 984 | 850 | 541 | 272 | 67 | 9 | 20 | 184 | 528 | 898 | 1205 | 6782 |
| Cooling Degree Days | 0 | 0 | 0 | 1 | 18 | 113 | 218 | 177 | 31 | 0 | 0 | 0 | 558 |
| Total Precipitation (") | 0.30 | 0.38 | 0.69 | 1.61 | 2.27 | 2.33 | 1.86 | 1.15 | 1.12 | 0.96 | 0.48 | 0.37 | 13.52 |
| Days ≥ 0.1" Precip | 1 | 1 | 2 | 4 | 5 | 5 | 5 | 3 | 3 | 2 | 2 | 1 | 34 |
| Total Snowfall (") | 4.8 | 5.0 | 5.7 | 2.7 | 0.2 | 0.0 | 0.0 | 0.0 | 0.5 | 1.8 | 5.2 | 6.9 | 32.8 |
| Days ≥ 1" Snow Depth | 12 | 8 | 4 | 1 | 0 | 0 | 0 | 0 | 0 | 1 | 5 | 11 | 42 |

## TOWER FALLS *Park County* ELEVATION 6273 ft LAT/LONG 44° 55 ' N / 110° 25 ' W

| | JAN | FEB | MAR | APR | MAY | JUN | JUL | AUG | SEP | OCT | NOV | DEC | YEAR |
|---|---|---|---|---|---|---|---|---|---|---|---|---|---|
| Maximum Temp °F | 27.0 | 33.4 | 42.3 | 51.0 | 61.8 | 70.4 | 78.3 | 78.2 | 68.1 | 55.7 | 36.5 | 26.6 | 52.4 |
| Minimum Temp °F | 0.6 | 3.4 | 11.6 | 20.2 | 28.2 | 33.9 | 37.6 | 35.6 | 27.7 | 20.2 | 10.0 | 0.7 | 19.1 |
| Mean Temp °F | 13.8 | 18.4 | 27.0 | 35.6 | 45.0 | 52.2 | 58.0 | 57.1 | 47.9 | 38.0 | 23.3 | 13.6 | 35.8 |
| Days Max Temp ≥ 90 °F | 0 | 0 | 0 | 0 | 0 | 0 | 1 | 1 | 0 | 0 | 0 | 0 | 2 |
| Days Max Temp ≤ 32 °F | 20 | na | na | 0 | 0 | 0 | 0 | 0 | 0 | 1 | 11 | 22 | na |
| Days Min Temp ≤ 32 °F | 30 | 28 | na | 29 | 24 | 12 | 5 | 8 | 23 | 30 | 30 | 31 | na |
| Days Min Temp ≤ 0 °F | 15 | na | 4 | 0 | 0 | 0 | 0 | 0 | 0 | 0 | 7 | 15 | na |
| Heating Degree Days | 1581 | 1308 | 1169 | 876 | 612 | 381 | 215 | 242 | 506 | 829 | 1243 | 1588 | 10550 |
| Cooling Degree Days | 0 | 0 | 0 | 0 | 0 | 1 | 3 | 2 | 0 | 0 | 0 | 0 | 6 |
| Total Precipitation (") | 1.25 | 0.77 | 1.01 | 1.07 | 1.84 | 2.24 | 1.85 | 1.51 | 1.49 | 1.11 | 1.21 | 1.20 | 16.55 |
| Days ≥ 0.1" Precip | 4 | 3 | 3 | 3 | 5 | 7 | 5 | 5 | 4 | 3 | 4 | 4 | 50 |
| Total Snowfall (") | 21.1 | 12.2 | na | 8.3 | 2.4 | 0.1 | 0.0 | 0.0 | 1.0 | 2.5 | na | 19.6 | na |
| Days ≥ 1" Snow Depth | 31 | 28 | 31 | 16 | 2 | 0 | 0 | 0 | 0 | 2 | na | 30 | na |

## UPTON *Weston County* ELEVATION 4232 ft LAT/LONG 44° 6 ' N / 104° 37 ' W

| | JAN | FEB | MAR | APR | MAY | JUN | JUL | AUG | SEP | OCT | NOV | DEC | YEAR |
|---|---|---|---|---|---|---|---|---|---|---|---|---|---|
| Maximum Temp °F | 30.3 | 35.3 | 45.7 | 57.4 | 68.2 | 78.4 | 86.5 | 84.8 | 73.9 | 60.1 | 42.4 | 31.9 | 57.9 |
| Minimum Temp °F | 5.0 | 10.2 | 19.9 | 29.1 | 39.1 | 48.7 | 54.9 | 52.8 | 41.6 | 30.3 | 18.2 | 7.6 | 29.8 |
| Mean Temp °F | 17.7 | 22.8 | 32.8 | 43.3 | 53.7 | 63.6 | 70.7 | 68.8 | 57.8 | 45.2 | 30.4 | 19.7 | 43.9 |
| Days Max Temp ≥ 90 °F | 0 | 0 | 0 | 0 | 0 | 4 | 12 | 10 | 2 | 0 | 0 | 0 | 28 |
| Days Max Temp ≤ 32 °F | 16 | 10 | 5 | 1 | 0 | 0 | 0 | 0 | 0 | 1 | 6 | 15 | 54 |
| Days Min Temp ≤ 32 °F | 31 | 27 | 29 | 21 | 6 | 1 | 0 | 0 | 5 | 19 | 27 | 31 | 197 |
| Days Min Temp ≤ 0 °F | 11 | 6 | 2 | 0 | 0 | 0 | 0 | 0 | 0 | 0 | 2 | 8 | 29 |
| Heating Degree Days | 1463 | 1186 | 990 | 647 | 351 | 113 | 17 | 33 | 240 | 613 | 1032 | 1397 | 8082 |
| Cooling Degree Days | 0 | 0 | 0 | 1 | 12 | 89 | 219 | 187 | 33 | 0 | 0 | 0 | 541 |
| Total Precipitation (") | 0.39 | 0.46 | 0.57 | 1.53 | 2.45 | 2.31 | 1.97 | 1.41 | 1.14 | 1.07 | 0.59 | 0.55 | 14.44 |
| Days ≥ 0.1" Precip | 1 | 2 | 2 | 4 | 6 | 6 | 4 | 3 | 3 | 3 | 2 | 2 | 38 |
| Total Snowfall (") | 7.3 | 6.8 | 7.3 | 5.5 | 0.8 | 0.2 | 0.0 | 0.0 | 0.5 | 1.9 | 7.7 | 10.1 | 48.1 |
| Days ≥ 1" Snow Depth | 23 | 19 | 8 | 2 | 0 | 0 | 0 | 0 | 0 | 1 | na | na | na |

## WAMSUTTER *Sweetwater County* ELEVATION 6824 ft LAT/LONG 41° 41 ' N / 107° 59 ' W

| | JAN | FEB | MAR | APR | MAY | JUN | JUL | AUG | SEP | OCT | NOV | DEC | YEAR |
|---|---|---|---|---|---|---|---|---|---|---|---|---|---|
| Maximum Temp °F | 27.4 | 32.7 | na | 54.0 | 65.1 | 76.1 | 83.5 | 81.4 | 71.9 | 59.0 | 40.6 | 29.3 | na |
| Minimum Temp °F | 6.9 | 9.8 | 18.8 | 26.0 | 33.6 | 42.5 | 48.6 | 46.6 | 38.2 | 28.5 | 16.7 | 8.3 | 27.0 |
| Mean Temp °F | 17.1 | 21.3 | na | 40.0 | 49.4 | 59.3 | 66.1 | 64.0 | 55.1 | 43.8 | 28.8 | 18.6 | na |
| Days Max Temp ≥ 90 °F | 0 | 0 | 0 | 0 | 0 | 1 | 4 | 1 | 0 | 0 | 0 | 0 | 6 |
| Days Max Temp ≤ 32 °F | 21 | 12 | 4 | 1 | 0 | 0 | 0 | 0 | 0 | 0 | 8 | 19 | 65 |
| Days Min Temp ≤ 32 °F | 30 | 28 | 29 | 23 | 12 | 2 | 0 | 0 | 7 | 21 | 28 | 30 | 210 |
| Days Min Temp ≤ 0 °F | 8 | 6 | 2 | 0 | 0 | 0 | 0 | 0 | 0 | 0 | 2 | 7 | 25 |
| Heating Degree Days | 1479 | 1228 | na | 742 | 478 | 188 | 34 | 68 | 296 | 651 | 1080 | 1432 | na |
| Cooling Degree Days | 0 | 0 | na | 0 | 0 | 31 | 73 | 52 | 5 | na | 0 | 0 | na |
| Total Precipitation (") | 0.28 | 0.22 | 0.35 | 0.69 | 0.98 | 0.85 | 0.98 | 0.88 | 0.75 | 0.59 | 0.41 | 0.23 | 7.21 |
| Days ≥ 0.1" Precip | 1 | 1 | 1 | 3 | 3 | 3 | 3 | 3 | 2 | 2 | 2 | 1 | 25 |
| Total Snowfall (") | na | na | na | na | na | 0.1 | 0.0 | 0.0 | 0.2 | na | na | na | na |
| Days ≥ 1" Snow Depth | na | na | na | na | 0 | 0 | 0 | 0 | 0 | na | na | na | na |

## WESTON 1 E *Campbell County* ELEVATION 3525 ft LAT/LONG 44° 41 ' N / 105° 20 ' W

| | JAN | FEB | MAR | APR | MAY | JUN | JUL | AUG | SEP | OCT | NOV | DEC | YEAR |
|---|---|---|---|---|---|---|---|---|---|---|---|---|---|
| Maximum Temp °F | 33.5 | 38.5 | 47.6 | 58.5 | 68.8 | 78.7 | 87.3 | 86.6 | 75.3 | 62.4 | 46.0 | 35.4 | 59.9 |
| Minimum Temp °F | 7.0 | 12.3 | 21.4 | 30.4 | 39.6 | 48.3 | 53.5 | 51.5 | 41.0 | 30.2 | 19.1 | 9.2 | 30.3 |
| Mean Temp °F | 20.3 | 25.4 | 34.5 | 44.5 | 54.2 | 63.5 | 70.5 | 69.1 | 58.2 | 46.3 | 32.6 | 22.3 | 45.1 |
| Days Max Temp ≥ 90 °F | 0 | 0 | 0 | 0 | 1 | 4 | 13 | 13 | 3 | 0 | 0 | 0 | 34 |
| Days Max Temp ≤ 32 °F | 12 | 8 | 4 | 1 | 0 | 0 | 0 | 0 | 0 | 0 | 4 | 11 | 40 |
| Days Min Temp ≤ 32 °F | 31 | 28 | 28 | 19 | 6 | 0 | 0 | 0 | 5 | 18 | 27 | 30 | 192 |
| Days Min Temp ≤ 0 °F | 10 | 5 | 2 | 0 | 0 | 0 | 0 | 0 | 0 | 0 | 2 | 8 | 27 |
| Heating Degree Days | 1382 | 1112 | 939 | 609 | 335 | 108 | 22 | 34 | 229 | 571 | 967 | 1317 | 7625 |
| Cooling Degree Days | 0 | 0 | 0 | 1 | 9 | 81 | 200 | 172 | 30 | 0 | 0 | 0 | 493 |
| Total Precipitation (") | 0.30 | 0.33 | 0.53 | 1.28 | 2.47 | 2.48 | 1.40 | 1.02 | 1.25 | 1.03 | 0.45 | 0.36 | 12.90 |
| Days ≥ 0.1" Precip | 1 | 1 | 2 | 4 | 6 | 6 | 4 | 3 | 3 | 3 | 2 | 1 | 36 |
| Total Snowfall (") | 6.8 | 5.7 | 7.7 | 5.9 | 1.5 | 0.1 | 0.0 | 0.0 | 0.6 | 2.1 | 5.0 | 7.9 | 43.3 |
| Days ≥ 1" Snow Depth | 23 | 16 | 10 | 2 | 0 | 0 | 0 | 0 | 0 | 1 | 6 | 18 | 76 |

## WHALEN DAM (USBR) *Goshen County* ELEVATION 4291 ft LAT/LONG 42° 14 ' N / 104° 38 ' W

| | JAN | FEB | MAR | APR | MAY | JUN | JUL | AUG | SEP | OCT | NOV | DEC | YEAR |
|---|---|---|---|---|---|---|---|---|---|---|---|---|---|
| Maximum Temp °F | 40.3 | 45.0 | 52.1 | 61.6 | 71.5 | 82.6 | 90.3 | 88.6 | 78.9 | 66.6 | 51.0 | 41.8 | 64.2 |
| Minimum Temp °F | 10.7 | 16.1 | 22.9 | 31.8 | 40.4 | 50.0 | 55.6 | 53.2 | 42.5 | 31.6 | 20.8 | 12.3 | 32.3 |
| Mean Temp °F | 25.5 | 30.6 | 37.6 | 46.6 | *55.9* | 66.3 | 72.9 | 70.9 | 60.7 | 49.1 | 35.9 | 27.0 | 48.2 |
| Days Max Temp ≥ 90 °F | 0 | 0 | 0 | 0 | 1 | 8 | 19 | 16 | 5 | 0 | 0 | 0 | 49 |
| Days Max Temp ≤ 32 °F | 8 | 5 | 3 | 0 | 0 | 0 | 0 | 0 | 0 | 0 | 3 | 7 | 26 |
| Days Min Temp ≤ 32 °F | 29 | 26 | 27 | 16 | 4 | 0 | 0 | 0 | 4 | 18 | 27 | *30* | 181 |
| Days Min Temp ≤ 0 °F | 8 | 4 | 1 | 0 | 0 | 0 | 0 | 0 | 0 | 0 | 1 | *6* | 20 |
| Heating Degree Days | 1218 | 965 | 843 | 547 | *287* | 67 | 7 | 14 | 171 | 487 | 866 | 1171 | 6643 |
| Cooling Degree Days | *0* | *0* | *0* | *1* | *12* | *132* | *266* | *226* | *58* | *4* | *0* | *0* | 699 |
| Total Precipitation (") | 0.31 | 0.30 | 0.66 | 1.55 | 2.31 | 2.18 | 1.96 | 1.01 | 1.16 | 0.79 | 0.45 | 0.31 | 12.99 |
| Days ≥ 0.1" Precip | 1 | 1 | 2 | 4 | 5 | 5 | 4 | 3 | 3 | 2 | *1* | 1 | 32 |
| Total Snowfall (") | na | na | na | na | *0.0* | *0.0* | *0.0* | *0.0* | *0.0* | na | na | na | na |
| Days ≥ 1" Snow Depth | na | na | na | na | *0* | *0* | *0* | *0* | *0* | na | na | na | na |

## WHEATLAND 4 N *Platte County* ELEVATION 4632 ft LAT/LONG 42° 7 ' N / 104° 57 ' W

| | JAN | FEB | MAR | APR | MAY | JUN | JUL | AUG | SEP | OCT | NOV | DEC | YEAR |
|---|---|---|---|---|---|---|---|---|---|---|---|---|---|
| Maximum Temp °F | 41.3 | 44.2 | 52.2 | 61.5 | 70.9 | 81.4 | 88.5 | 86.7 | 77.4 | 65.7 | 50.0 | 41.9 | 63.5 |
| Minimum Temp °F | 16.0 | 18.8 | 24.9 | 31.6 | 40.8 | 49.4 | 55.0 | 52.5 | 42.8 | 33.5 | 24.4 | 17.4 | 33.9 |
| Mean Temp °F | 29.0 | 31.4 | 38.5 | 46.6 | 55.9 | 65.4 | 71.8 | 69.6 | 60.1 | 49.6 | 37.2 | 29.7 | 48.7 |
| Days Max Temp ≥ 90 °F | 0 | 0 | 0 | 0 | 0 | 6 | 14 | 12 | 3 | 0 | 0 | 0 | 35 |
| Days Max Temp ≤ 32 °F | 7 | 5 | 2 | 0 | 0 | 0 | 0 | 0 | 0 | 0 | 3 | 6 | 23 |
| Days Min Temp ≤ 32 °F | 27 | 24 | 24 | 15 | 4 | 0 | 0 | 0 | 4 | 14 | 23 | 26 | 161 |
| Days Min Temp ≤ 0 °F | 5 | 3 | 1 | 0 | 0 | 0 | 0 | 0 | 0 | 0 | 1 | 4 | 14 |
| Heating Degree Days | 1111 | 942 | 814 | 548 | 287 | 74 | 8 | 15 | 174 | 473 | 826 | 1086 | 6358 |
| Cooling Degree Days | 0 | 0 | 0 | 1 | 13 | 108 | 226 | 189 | 41 | 2 | 0 | 0 | 580 |
| Total Precipitation (") | 0.22 | 0.29 | 0.67 | 1.38 | 2.21 | 2.19 | 1.71 | 1.10 | 1.16 | 0.76 | 0.47 | 0.25 | 12.41 |
| Days ≥ 0.1" Precip | 1 | 1 | 2 | 4 | 5 | 4 | 4 | 3 | 3 | 2 | 1 | 1 | 31 |
| Total Snowfall (") | 6.0 | 5.4 | 7.2 | 4.5 | 1.3 | 0.0 | 0.0 | 0.0 | 1.1 | 2.7 | 6.2 | 7.0 | 41.4 |
| Days ≥ 1" Snow Depth | 7 | 5 | 4 | 2 | 0 | 0 | 0 | 0 | 0 | 1 | 4 | 7 | 30 |

## WORLAND *Washakie County* ELEVATION 4062 ft LAT/LONG 44° 1 ' N / 107° 58 ' W

| | JAN | FEB | MAR | APR | MAY | JUN | JUL | AUG | SEP | OCT | NOV | DEC | YEAR |
|---|---|---|---|---|---|---|---|---|---|---|---|---|---|
| Maximum Temp °F | 28.8 | 38.0 | 49.9 | 60.6 | 70.7 | 81.2 | 89.6 | 87.6 | 75.4 | 61.8 | 44.2 | 31.2 | 59.9 |
| Minimum Temp °F | 2.1 | 9.7 | 22.2 | 31.6 | 41.8 | 50.3 | 54.7 | 51.7 | 40.8 | 30.5 | 17.7 | 5.7 | 29.9 |
| Mean Temp °F | 15.5 | 23.9 | 36.1 | 46.1 | 56.3 | 65.7 | 72.1 | 69.7 | 58.1 | 46.1 | 30.9 | 18.4 | 44.9 |
| Days Max Temp ≥ 90 °F | 0 | 0 | 0 | 0 | 1 | 7 | 18 | 15 | 3 | 0 | 0 | 0 | 44 |
| Days Max Temp ≤ 32 °F | 17 | 7 | 3 | 0 | 0 | 0 | 0 | 0 | 0 | 0 | 5 | 15 | 47 |
| Days Min Temp ≤ 32 °F | 31 | 28 | 29 | 17 | 3 | 0 | 0 | 0 | 4 | 19 | 29 | 31 | 191 |
| Days Min Temp ≤ 0 °F | 13 | 6 | 1 | 0 | 0 | 0 | 0 | 0 | 0 | 0 | 2 | 9 | 31 |
| Heating Degree Days | 1531 | 1154 | 890 | 561 | 280 | 81 | 14 | 26 | 226 | 579 | 1015 | 1440 | 7797 |
| Cooling Degree Days | 0 | 0 | 0 | 2 | 23 | 121 | 242 | 200 | 32 | 0 | 0 | 0 | 620 |
| Total Precipitation (") | 0.21 | 0.18 | 0.36 | 0.82 | 1.49 | 1.13 | 0.62 | 0.63 | 0.80 | 0.67 | 0.35 | 0.27 | 7.53 |
| Days ≥ 0.1" Precip | 1 | 1 | 1 | 2 | 4 | 4 | 2 | 2 | 2 | 2 | 1 | 1 | 23 |
| Total Snowfall (") | na | na | na | na | *0.0* | *0.0* | 0.0 | 0.0 | *0.0* | na | na | na | na |
| Days ≥ 1" Snow Depth | na | na | na | na | *0* | *0* | 0 | 0 | *0* | na | na | na | na |

## WORLAND MUNI AP *Washakie County* ELEVATION 4176 ft LAT/LONG 43° 58 ' N / 107° 58 ' W

| | JAN | FEB | MAR | APR | MAY | JUN | JUL | AUG | SEP | OCT | NOV | DEC | YEAR |
|---|---|---|---|---|---|---|---|---|---|---|---|---|---|
| Maximum Temp °F | 29.8 | 38.7 | 50.5 | 61.1 | 71.3 | 82.1 | 89.8 | 87.8 | 75.5 | 62.3 | 44.5 | 32.0 | 60.5 |
| Minimum Temp °F | 3.4 | 11.3 | 23.3 | 32.6 | 42.3 | 50.8 | 56.4 | 54.3 | 43.5 | 32.1 | 18.3 | 5.8 | 31.2 |
| Mean Temp °F | 16.6 | 25.1 | 36.9 | 46.9 | 56.8 | 66.5 | 73.1 | 71.1 | 59.5 | 47.2 | 31.4 | 18.9 | 45.8 |
| Days Max Temp ≥ 90 °F | 0 | 0 | 0 | 0 | 1 | 8 | 18 | 15 | 3 | 0 | 0 | 0 | 45 |
| Days Max Temp ≤ 32 °F | 17 | 7 | 2 | 0 | 0 | 0 | 0 | 0 | 0 | 0 | 5 | 14 | 45 |
| Days Min Temp ≤ 32 °F | 31 | 28 | 27 | 15 | 3 | 0 | 0 | 0 | 2 | 15 | 29 | 31 | 181 |
| Days Min Temp ≤ 0 °F | 11 | 5 | 1 | 0 | 0 | 0 | 0 | 0 | 0 | 0 | 2 | 9 | 28 |
| Heating Degree Days | 1493 | 1122 | 863 | 538 | 266 | 69 | 9 | 19 | 196 | 545 | 1001 | 1422 | 7543 |
| Cooling Degree Days | 0 | 0 | 0 | 2 | 24 | 128 | 257 | 224 | 37 | 0 | 0 | 0 | 672 |
| Total Precipitation (") | 0.25 | 0.19 | 0.39 | 0.81 | 1.49 | 1.12 | 0.63 | 0.61 | 0.78 | 0.64 | 0.37 | 0.26 | 7.54 |
| Days ≥ 0.1" Precip | 1 | 1 | 1 | 2 | 4 | 3 | 2 | 2 | 2 | 2 | 1 | 1 | 22 |
| Total Snowfall (") | 5.1 | 3.4 | 3.8 | 3.1 | 0.6 | 0.0 | 0.0 | 0.0 | 0.9 | 2.2 | 4.7 | 5.8 | 29.6 |
| Days ≥ 1" Snow Depth | 24 | 16 | 5 | 1 | 0 | 0 | 0 | 0 | 0 | 1 | 7 | 16 | 70 |

## YELLOWSTONE MAMMOTH *Teton County* ELEVATION 6243 ft LAT/LONG 44° 58 ' N / 110° 42 ' W

| | JAN | FEB | MAR | APR | MAY | JUN | JUL | AUG | SEP | OCT | NOV | DEC | YEAR |
|---|---|---|---|---|---|---|---|---|---|---|---|---|---|
| Maximum Temp °F | 29.3 | 34.1 | 40.4 | 49.4 | 60.7 | 70.1 | 79.4 | 78.4 | *67.5* | *55.1* | 38.1 | 29.9 | 52.7 |
| Minimum Temp °F | 10.5 | 13.1 | 18.2 | 26.3 | 34.7 | 41.9 | 47.3 | 46.0 | *37.4* | *29.6* | 19.1 | 11.1 | 27.9 |
| Mean Temp °F | 19.9 | 23.7 | 29.3 | 37.9 | 47.8 | 56.0 | 63.4 | 62.2 | *52.5* | *42.4* | 28.8 | 20.8 | 40.4 |
| Days Max Temp ≥ 90 °F | 0 | 0 | 0 | 0 | 0 | 0 | 1 | 1 | 0 | 0 | 0 | 0 | 2 |
| Days Max Temp ≤ 32 °F | 17 | 10 | *5* | 1 | 0 | 0 | 0 | 0 | 0 | 1 | *8* | na | na |
| Days Min Temp ≤ 32 °F | 31 | 28 | 29 | 23 | *11* | *2* | 0 | 1 | *7* | *19* | *27* | 31 | 209 |
| Days Min Temp ≤ 0 °F | 6 | 4 | 2 | 0 | 0 | 0 | 0 | 0 | 0 | 0 | 1 | *5* | 18 |
| Heating Degree Days | 1391 | 1161 | 1098 | 806 | 526 | 275 | 92 | 117 | *371* | *695* | 1080 | 1365 | 8977 |
| Cooling Degree Days | 0 | 0 | 0 | 0 | 0 | 13 | 50 | *36* | *3* | *0* | 0 | 0 | 102 |
| Total Precipitation (") | 1.05 | 0.65 | 0.99 | 1.07 | 2.01 | 2.02 | 1.58 | 1.43 | 1.30 | 1.03 | 0.98 | 0.84 | 14.95 |
| Days ≥ 0.1" Precip | 3 | 2 | 3 | 3 | 5 | 6 | 5 | *4* | 4 | 3 | 3 | 3 | 44 |
| Total Snowfall (") | *12.5* | 8.9 | 11.1 | 5.9 | *2.3* | 0.0 | 0.0 | 0.0 | *0.3* | *3.6* | 9.7 | 11.9 | 66.2 |
| Days ≥ 1" Snow Depth | 30 | 26 | *20* | 5 | *1* | 0 | 0 | 0 | *0* | *2* | *15* | *28* | 127 |

## YODER 2 WSW *Goshen County* ELEVATION 4320 ft LAT/LONG 41° 52 ' N / 104° 21 ' W

| | JAN | FEB | MAR | APR | MAY | JUN | JUL | AUG | SEP | OCT | NOV | DEC | YEAR |
|---|---|---|---|---|---|---|---|---|---|---|---|---|---|
| Maximum Temp °F | 41.0 | 45.5 | 53.0 | 62.9 | 72.1 | 82.7 | 89.8 | 87.9 | 78.6 | 66.5 | 50.2 | 41.9 | 64.3 |
| Minimum Temp °F | 13.8 | 16.6 | 22.9 | 30.5 | 39.7 | 48.8 | 54.4 | 51.7 | 41.4 | 31.4 | 21.5 | 14.7 | 32.3 |
| Mean Temp °F | 27.5 | 31.1 | 38.0 | 46.7 | 55.9 | 65.8 | 72.1 | 69.8 | 60.0 | 48.9 | 35.8 | 28.3 | 48.3 |
| Days Max Temp ≥ 90 °F | 0 | 0 | 0 | 0 | 1 | 8 | 18 | 15 | 3 | 0 | 0 | 0 | 45 |
| Days Max Temp ≤ 32 °F | 7 | 4 | 2 | 0 | 0 | 0 | 0 | 0 | 0 | 0 | 3 | 7 | 23 |
| Days Min Temp ≤ 32 °F | 29 | 26 | 26 | 18 | 5 | 0 | 0 | 0 | 4 | 17 | 26 | 29 | 180 |
| Days Min Temp ≤ 0 °F | 6 | 3 | 1 | 0 | 0 | 0 | 0 | 0 | 0 | 0 | 1 | 5 | 16 |
| Heating Degree Days | 1158 | 952 | 831 | 543 | 285 | 71 | 7 | 15 | 175 | 492 | 868 | 1132 | 6529 |
| Cooling Degree Days | 0 | 0 | 0 | 1 | 12 | 109 | 219 | 175 | 34 | 1 | 0 | 0 | 551 |
| Total Precipitation (") | 0.29 | 0.30 | 0.87 | 1.69 | 2.65 | 2.42 | 1.93 | 1.12 | 1.27 | 0.92 | 0.54 | 0.32 | 14.32 |
| Days ≥ 0.1" Precip | 1 | 1 | 2 | 4 | 6 | 5 | 5 | 3 | 3 | 3 | 2 | 1 | 36 |
| Total Snowfall (") | 7.1 | 5.7 | 8.2 | 5.7 | 0.4 | 0.0 | 0.0 | 0.0 | 0.6 | 3.1 | 7.6 | 8.1 | 46.5 |
| Days ≥ 1" Snow Depth | 12 | 7 | 5 | 2 | 0 | 0 | 0 | 0 | 0 | 1 | 6 | 11 | 44 |

## JANUARY MINIMUM TEMPERATURE °F

### LOWEST

| | | |
|---|---|---|
| 1 | Bondurant | -4.2 |
| 2 | Ten Sleep-16 SSE | -2.8 |
| 3 | La Barge | -2.6 |
| 4 | Lake Yellowstone | -2.4 |
| 5 | Fontenelle Dam | -2.1 |
| 6 | Sage | -1.9 |
| 7 | Snake River | -1.5 |
| 8 | Pinedale | -0.7 |
| 9 | Riverton | -0.6 |
| 10 | Tower Falls | 0.6 |
| 11 | Moose | 1.0 |
| 12 | Moran | 1.9 |
| 13 | Worland | 2.1 |
| 14 | Basin | 2.2 |
| 15 | Green River | 3.2 |
| 16 | Worland-Muni | 3.4 |
| 17 | Devils Tower | 4.5 |
| 18 | Lovell | 4.6 |
| 19 | Clearmont | 4.7 |
| 20 | Burgess Junction | 4.8 |
| 21 | Deaver | 4.9 |
| | Jackson | 4.9 |
| 23 | Afton | 5.0 |
| | Upton | 5.0 |
| 25 | Kemmerer | 5.2 |

### HIGHEST

| | | |
|---|---|---|
| 1 | Buffalo Bill Dam | 19.8 |
| 2 | Phillips | 16.3 |
| 3 | Chugwater | 16.1 |
| 4 | Cheyenne | 16.0 |
| | Glenrock | 16.0 |
| | Sybille | 16.0 |
| | Wheatland | 16.0 |
| 8 | Black Mountain | 14.7 |
| 9 | Albin | 14.6 |
| 10 | Cody-21 SW | 14.4 |
| 11 | Elk Mountain | 14.1 |
| 12 | Archer | 13.9 |
| 13 | Muddy Gap | 13.8 |
| | Yoder | 13.8 |
| 15 | Double Four | 13.3 |
| 16 | Carpenter | 13.2 |
| | Cody | 13.2 |
| 18 | Dubois | 12.8 |
| | La Grange | 12.8 |
| 20 | Casper | 12.6 |
| | Ten Sleep-4 NE | 12.6 |
| 22 | Rawlins | 12.4 |
| 23 | Gas Hills | 11.8 |
| 24 | Medicine Bow | 11.6 |
| 25 | Midwest | 11.4 |

## JULY MAXIMUM TEMPERATURE °F

### HIGHEST

| | | |
|---|---|---|
| 1 | Basin | 90.4 |
| 2 | Whalen Dam | 90.3 |
| 3 | Worland-Muni | 89.8 |
| | Yoder | 89.8 |
| 5 | Boysen Dam | 89.6 |
| | Worland | 89.6 |
| 7 | Dull Center | 89.5 |
| | Redbird | 89.5 |
| 9 | Torrington | 89.3 |
| 10 | Echeta | 89.2 |
| | Midwest | 89.2 |
| 12 | Glenrock | 89.0 |
| 13 | Wheatland | 88.5 |
| 14 | La Grange | 88.4 |
| 15 | Riverton | 88.3 |
| 16 | Deaver | 88.1 |
| | Rochelle | 88.1 |
| 18 | Colony | 87.7 |
| | Leiter | 87.7 |
| | Newcastle | 87.7 |
| 21 | Powder River | 87.6 |
| 22 | Ten Sleep-16 SSE | 87.5 |
| 23 | Kaycee | 87.4 |
| | Lovell | 87.4 |
| 25 | Casper | 87.3 |

### LOWEST

| | | |
|---|---|---|
| 1 | Lake Yellowstone | 71.0 |
| 2 | Buffalo Bill Dam | 74.9 |
| 3 | Snake River | 76.9 |
| 4 | Moran | 77.6 |
| 5 | Pinedale | 77.7 |
| 6 | Bondurant | 78.3 |
| | Tower Falls | 78.3 |
| 8 | Dubois | 78.4 |
| 9 | Encampment | 78.5 |
| | Sunshine | 78.5 |
| 11 | Alta | 78.7 |
| 12 | Moose | 79.2 |
| 13 | Elk Mountain | 79.4 |
| | Yellowstone | 79.4 |
| 15 | Laramie | 79.6 |
| 16 | Burris | 80.7 |
| 17 | Kemmerer | 81.0 |
| 18 | Afton | 81.1 |
| 19 | Cody-21 SW | 81.4 |
| 20 | Thermopolis | 81.5 |
| 21 | Double Four | 81.6 |
| | Jackson | 81.6 |
| 23 | Sage | 82.0 |
| 24 | Evanston | 82.2 |
| 25 | Cheyenne | 82.3 |

## ANNUAL PRECIPITATION (")

### HIGHEST

| | | |
|---|---|---|
| 1 | Snake River | 30.31 |
| 2 | Alta | 24.18 |
| 3 | Moran | 23.94 |
| 4 | Burgess Junction | 21.24 |
| 5 | Moose | 20.31 |
| 6 | Lake Yellowstone | 20.04 |
| 7 | Bondurant | 19.62 |
| 8 | Afton | 18.54 |
| 9 | Albin | 18.38 |
| 10 | Sundance | 17.75 |
| 11 | Devils Tower | 17.12 |
| 12 | Lusk | 16.79 |
| 13 | Gillette | 16.63 |
| 14 | La Grange | 16.61 |
| 15 | Tower Falls | 16.55 |
| 16 | Jackson | 16.30 |
| 17 | Sybille | 16.00 |
| 18 | Echeta | 15.84 |
| 19 | Archer | 15.74 |
| 20 | Chugwater | 15.41 |
| 21 | Phillips | 15.22 |
| 22 | Double Four | 15.12 |
| 23 | Newcastle | 15.04 |
| 24 | Leiter | 15.02 |
| 25 | Sunshine | 15.00 |

### LOWEST

| | | |
|---|---|---|
| 1 | Deaver | 5.45 |
| 2 | Basin | 6.72 |
| 3 | Lovell | 6.81 |
| 4 | Bitter Creek | 6.98 |
| 5 | Fontenelle Dam | 7.08 |
| 6 | Wamsutter | 7.21 |
| 7 | Worland | 7.53 |
| 8 | Worland-Muni | 7.54 |
| 9 | Clark | 7.62 |
| 10 | Emblem | 7.77 |
| 11 | Green River | 7.85 |
| 12 | Riverton | 8.08 |
| 13 | La Barge | 8.12 |
| 14 | Church Buttes | 8.29 |
| 15 | Dubois | 8.75 |
| 16 | Burris | 9.01 |
| 17 | Heart Mountain | 9.06 |
| 18 | Diversion Dam | 9.11 |
| 19 | Boysen Dam | 9.15 |
| 20 | Gas Hills | 9.42 |
| 21 | Rock Springs | 9.51 |
| 22 | Rawlins | 9.59 |
| 23 | Saratoga | 10.00 |
| 24 | Sage | 10.12 |
| 25 | Cody | 10.23 |

## ANNUAL SNOWFALL (")

### HIGHEST

| | | |
|---|---|---|
| 1 | Snake River | 262.0 |
| 2 | Burgess Junction | 229.7 |
| 3 | Moose | 169.7 |
| 4 | Lake Yellowstone | 169.3 |
| 5 | Bondurant | 121.4 |
| 6 | Alta | 110.0 |
| 7 | Lander | 108.2 |
| 8 | Elk Mountain | 85.3 |
| 9 | Casper | 84.5 |
| 10 | Sundance | 83.4 |
| | Sybille | 83.4 |
| 12 | Jackson | 81.7 |
| 13 | Albin | 80.1 |
| 14 | Sheridan-County | 77.2 |
| 15 | Encampment | 77.0 |
| 16 | Chugwater | 69.7 |
| 17 | Black Mountain | 69.0 |
| | Pinedale | 69.0 |
| 19 | Double Four | 68.3 |
| 20 | Yellowstone | 66.2 |
| 21 | Gillette | 65.5 |
| 22 | Gas Hills | 64.6 |
| 23 | La Grange | 62.7 |
| 24 | Ten Sleep-16 SSE | 61.0 |
| 25 | Thermopolis | 60.6 |

### LOWEST

| | | |
|---|---|---|
| 1 | Lovell | 19.4 |
| 2 | Basin | 20.1 |
| 3 | Clark | 26.1 |
| 4 | Worland-Muni | 29.6 |
| 5 | Carpenter | 32.4 |
| 6 | Torrington | 32.8 |
| 7 | Riverton | 33.2 |
| 8 | Colony | 39.0 |
| 9 | Newcastle | 39.7 |
| 10 | Wheatland | 41.4 |
| 11 | Kaycee | 41.7 |
| 12 | Rochelle | 41.8 |
| 13 | Weston | 43.3 |
| 14 | Dillinger | 43.4 |
| 15 | Billy Creek | 44.4 |
| 16 | Cody | 44.9 |
| 17 | Rock Springs | 46.0 |
| 18 | Yoder | 46.5 |
| 19 | Redbird | 46.9 |
| 20 | Archer | 47.9 |
| 21 | Upton | 48.1 |
| 22 | Cody-21 SW | 49.3 |
| 23 | Clearmont | 51.3 |
| 24 | Dull Center | 51.5 |
| 25 | Laramie | 53.7 |

# UNITED STATES RANKINGS

## JANUARY MINIMUM TEMPERATURE °F

### LOWEST

| | | | | | | | | | | |
|---|---|---|---|---|---|---|---|---|---|---|
| 1 | Northway, AK | -27.3 | 26 | Upham, ND | -9.5 | 51 | Cook, MN | -7.1 | | McVille, ND | -5.4 |
| 2 | Tok, AK | -25.7 | 27 | Argyle, MN | -9.4 | | Westhope, ND | -7.1 | | Mohall, ND | -5.4 |
| 3 | Eagle, AK | -21.0 | | Hallock, MN | -9.4 | 53 | Gunnison, CO | -7.0 | | Tioga, ND | -5.4 |
| 4 | Bettles, AK | -20.1 | 29 | Langdon, ND | -9.1 | 54 | Gunflint Lake, MN | -6.9 | 79 | Grand Rapids, MN | -5.3 |
| 5 | Barrow, AK | -19.7 | | Thorhult, MN | -9.1 | | Hansboro, ND | -6.9 | 80 | Isle, MN | -5.2 |
| 6 | Snowshoe Lake, AK | -18.8 | 31 | Baudette, MN | -9.0 | 56 | Leeds, ND | -6.7 | | Sharon, ND | -5.2 |
| 7 | Eielson, AK | -18.7 | 32 | Big Falls, MN | -8.8 | | Red Lake Falls, MN | -6.7 | | Valley City, ND | -5.2 |
| 8 | Clearwater, AK | -18.5 | 33 | Petersburg, ND | -8.6 | | Squa Pan Dam, ME | -6.7 | | Winter, WI | -5.2 |
| 9 | Fairbanks, AK | -18.2 | 34 | Agassiz, MN | -8.5 | 59 | Hibbing, MN | -6.6 | 84 | Bowbells, ND | -5.1 |
| 10 | Tanana, AK | -18.1 | 35 | Oklee, MN | -8.4 | 60 | Powers Lake, ND | -6.5 | | Grafton, ND | -5.1 |
| 11 | McGrath, AK | -17.3 | 36 | McKinley, AK | -8.3 | 61 | Gordon, WI | -6.4 | | Leech Lake Dm, MN | -5.1 |
| 12 | Glennallen, AK | -17.0 | | Red Lake, MN | -8.3 | 62 | Detroit Lakes, MN | -6.3 | | Oakes, ND | -5.1 |
| 13 | Tonsina, AK | -14.4 | 38 | Hermit, CO | -8.2 | 63 | Crookston, MN | -6.2 | 88 | Rugby, ND | -5.0 |
| 14 | Gulkana, AK | -13.5 | | Kotzebue, AK | -8.2 | | Mahnomen, MN | -6.2 | | Silverton, CO | -5.0 |
| 15 | College, AK | -12.9 | | Van Buren, ME | -8.2 | | Park Rapids, MN | -6.2 | 90 | Westby, MT | -4.9 |
| 16 | Taylor Park, CO | -12.7 | 41 | Bottineau, ND | -8.1 | | Wales, AK | -6.2 | 91 | Colgate, ND | -4.8 |
| 17 | Big Delta, AK | -11.1 | 42 | Fosston, MN | -8.0 | 67 | Cochetopa Creek, CO | -6.1 | | Fortuna, ND | -4.8 |
| | Tower, MN | -11.1 | 43 | Cotton, MN | -7.9 | 68 | Puntilla, AK | -6.0 | | Pine River Dam, MN | -4.8 |
| 19 | Cass Lake, MN | -11.0 | 44 | Bemidji, MN | -7.8 | 69 | Crested Butte, CO | -5.9 | 94 | Hurdsfield, ND | -4.7 |
| 20 | Roseau, MN | -10.8 | 45 | Brainerd, MN | -7.6 | | Mora, MN | -5.9 | | Stanley, ND | -4.7 |
| 21 | Pembina, ND | -10.5 | | Itasca, MN | -7.6 | 71 | Rolla, ND | -5.8 | | Wright, MN | -4.7 |
| 22 | Warroad, MN | -10.1 | 47 | Winton, MN | -7.5 | 72 | Grand Forks-Intl, ND | -5.7 | 97 | Mt Washington, NH | -4.6 |
| 23 | Internat'l Falls, MN | -9.7 | 48 | McHenry, ND | -7.3 | | Winnibigoshish MN | -5.7 | 98 | Granville, ND | -4.5 |
| | Willow City, ND | -9.7 | 49 | Cavalier, ND | -7.2 | 74 | Wadena, MN | -5.6 | | Kenmare, ND | -4.5 |
| 25 | Belcourt, ND | -9.6 | | Towner, ND | -7.2 | 75 | Fort Kent, ME | -5.4 | 100 | Drake, ND | -4.4 |

### HIGHEST

| | | | | | | | | | | | |
|---|---|---|---|---|---|---|---|---|---|---|---|
| 1 | Kaneohe Mauka, HI | 65.5 | 26 | Fort Myers, FL | 53.9 | 51 | Bartow, FL | 49.8 | 76 | Chapman, TX | 46.7 |
| 2 | Key West, FL | 65.3 | 27 | Everglades, FL | 53.7 | 52 | Hawaii Volcano, HI | 49.5 | | Kanalohuluhulu, HI | 46.7 |
| 3 | Honolulu, HI | 65.2 | | Naples, FL | 53.7 | | Kissimmee, FL | 49.5 | | Mission, TX | 46.7 |
| 4 | Lihue, HI | 65.1 | 29 | Canal Point, FL | 52.6 | | La Belle, FL | 49.5 | 79 | Lisbon, FL | 46.6 |
| 5 | Waikiki, HI | 64.2 | 30 | Immokalee, FL | 52.0 | 55 | Clermont, FL | 49.4 | 80 | Corpus Christi, TX | 46.2 |
| 6 | Hana, HI | 63.9 | 31 | Fort Pierce, FL | 51.8 | 56 | Weslaco, TX | 49.3 | | Rockport, TX | 46.2 |
| | Lahaina, HI | 63.9 | 32 | Belle Glade, FL | 51.6 | 57 | Arcadia, FL | 49.0 | 82 | San Francisco, CA | 46.1 |
| 8 | Molokai, HI | 63.8 | 33 | Moore Haven Lk, FL | 51.5 | | Los Angeles-CC, CA | 49.0 | | Santa Ana, CA | 46.1 |
| 9 | Hilo, HI | 63.6 | 34 | Punta Gorda, FL | 51.4 | | Plant City, FL | 49.0 | | Zapata, TX | 46.1 |
| 10 | Tavernier, FL | 63.5 | 35 | Melbourne, FL | 51.2 | | San Diego, CA | 49.0 | 85 | Gold Rock, CA | 45.9 |
| 11 | Kahului, HI | 63.3 | 36 | Okeechobee, FL | 51.1 | 61 | Myakka River, FL | 48.8 | 86 | Federal Point, FL | 45.8 |
| 12 | Miami Beach, FL | 62.8 | 37 | Venice, FL | 51.0 | 62 | Brooksville, FL | 48.6 | 87 | Jacksonville Bch, FL | 45.7 |
| 13 | Naalehu, HI | 62.4 | 38 | Vero Beach, FL | 50.6 | | Mountain Lake, FL | 48.6 | | Matagorda, TX | 45.7 |
| 14 | Miami, FL | 59.9 | 39 | Brownsville, TX | 50.5 | 64 | St. Leo, FL | 48.5 | | Torrance, CA | 45.7 |
| 15 | Kainaliu, HI | 58.9 | | Lakeland, FL | 50.5 | 65 | Los Angeles-Intl, CA | 48.2 | 90 | Kofa, AZ | 45.6 |
| 16 | Pompano Beach, FL | 58.5 | 41 | Ls Angeles-Ucla, CA | 50.3 | 66 | Port Mansfield, TX | 48.1 | | Ocala, FL | 45.6 |
| 17 | Fort Lauderdale, FL | 58.0 | 42 | Winter Haven, FL | 50.2 | 67 | Avon Park, FL | 48.0 | 92 | Bushnell, FL | 45.4 |
| 18 | Hialeah, FL | 57.6 | 43 | Devils Garden, FL | 50.1 | | Titusville, FL | 48.0 | | McCook, TX | 45.4 |
| 19 | Opihihale, HI | 57.3 | | Fort Drum, FL | 50.1 | 69 | Galveston, TX | 47.9 | 94 | Long Beach, CA | 45.3 |
| 20 | West Palm Beach, FL | 56.8 | | Parrish, FL | 50.1 | 70 | Daytona Beach, FL | 47.5 | 95 | Chula Vista, CA | 45.2 |
| 21 | Tamiami Trail, FL | 56.6 | 46 | Bradenton, FL | 50.0 | 71 | Harlingen, TX | 47.4 | | Point Comfort, TX | 45.2 |
| 22 | Flamingo, FL | 56.2 | | Tampa, FL | 50.0 | 72 | Newport Beach, CA | 47.3 | 97 | Falcon Dam, TX | 45.1 |
| 23 | Stuart, FL | 54.9 | | Tarpon Springs, FL | 50.0 | 73 | Lake Alfred, FL | 47.1 | 98 | Deland, FL | 45.0 |
| 24 | Clewiston, FL | 54.4 | 49 | Santa Monica, CA | 49.9 | 74 | Culver City, CA | 46.8 | | Inverness, FL | 45.0 |
| 25 | St. Petersburg, FL | 54.3 | | Wauchula, FL | 49.9 | | Raymondville, TX | 46.8 | | Port O'Connor, TX | 45.0 |

## JULY MAXIMUM TEMPERATURE °F

### HIGHEST

| 1 | Death Valley, CA | 114.9 | 26 | Casa Grande, AZ | 105.8 | 51 | Castle Hot Sprgs, AZ | 102.7 | | Three Rivers, CA | 99.3 |
|---|---|---|---|---|---|---|---|---|---|---|---|
| 2 | Buckeye, AZ | 109.0 | | Phoenix-Sky H, AZ | 105.8 | 52 | Inyokern, CA | 102.6 | | Tucson-Intl, AZ | 99.3 |
| | Gila Bend, AZ | 109.0 | | Tempe, AZ | 105.8 | 53 | South Phoenix, AZ | 102.2 | 78 | Zion, UT | 99.1 |
| 4 | Palm Springs, CA | 108.5 | 29 | Phoenix, AZ | 105.7 | | St. George, UT | 102.2 | 79 | Coalinga, CA | 99.0 |
| 5 | Blythe-Riverside, CA | 108.4 | 30 | Yuma-Citrus, AZ | 105.6 | 55 | Ajo, AZ | 102.1 | | Falcon Dam, TX | 99.0 |
| | Needles, CA | 108.4 | 31 | Twntynine Plms, CA | 105.4 | | Childs, AZ | 102.1 | | Fowlerton, TX | 99.0 |
| | Parker, AZ | 108.4 | | Youngtown, AZ | 105.4 | 57 | Roosevelt, AZ | 101.8 | 82 | Carrizo Springs, TX | 98.9 |
| 8 | Blythe, CA | 108.2 | 33 | Florence, AZ | 105.3 | 58 | Boulder City, NV | 101.7 | | Wasco, CA | 98.9 |
| 9 | Iron Mountain, CA | 107.9 | | Trona, CA | 105.3 | | Desert, NV | 101.7 | | Zapata, TX | 98.9 |
| | Mecca, CA | 107.9 | 35 | Laveen, AZ | 104.9 | 60 | Yucca, AZ | 101.6 | 85 | Hanksville, UT | 98.7 |
| 11 | Litchfield Park, AZ | 107.8 | | Mesa, AZ | 104.9 | 61 | Aguila, AZ | 101.5 | | Mangum, OK | 98.7 |
| 12 | Bouse, AZ | 107.6 | 37 | Bartlett Dam, AZ | 104.7 | 62 | Montezuma Cstl, AZ | 100.9 | 87 | Bakersfield, CA | 98.6 |
| 13 | Parker Reservoir, CA | 107.4 | 38 | Wikieup, AZ | 104.6 | 63 | Presidio, TX | 100.7 | 88 | Pecos, TX | 98.5 |
| 14 | El Centro, CA | 107.2 | 39 | Eloy, AZ | 104.5 | | Tucson-Magnetic, AZ | 100.7 | 89 | Corcoran, CA | 98.4 |
| 15 | Borrego Desert, CA | 107.1 | 40 | Eagle Mountain, CA | 104.3 | | Tucson-U of A, AZ | 100.7 | | Encinal, TX | 98.4 |
| | Casa Grande-NM, AZ | 107.1 | 41 | Las Vegas, NV | 104.1 | 66 | Pahrump, NV | 100.6 | | La Verkin, UT | 98.4 |
| 17 | Gold Rock, CA | 106.9 | 42 | Chandler Heights, AZ | 104.0 | 67 | Tucson-CP Ave, AZ | 100.5 | 92 | Fresno, CA | 98.3 |
| 18 | Maricopa, AZ | 106.8 | | Wickenburg, AZ | 104.0 | 68 | Candelaria, TX | 99.8 | 93 | Porterville, CA | 98.2 |
| 19 | Indio, CA | 106.7 | 44 | Daggett, CA | 103.9 | | San Carlos Rsvr, AZ | 99.8 | | Rio Grande City, TX | 98.2 |
| 20 | Yuma-Intl, AZ | 106.6 | 45 | Hayfield, CA | 103.8 | 70 | Laredo, TX | 99.7 | 95 | Chattanooga, OK | 98.1 |
| 21 | Brawley, CA | 106.5 | 46 | Stewart Mount, AZ | 103.6 | 71 | Clifton, AZ | 99.6 | | Elsinore, CA | 98.1 |
| | Sacaton, AZ | 106.5 | 47 | Mormon Flat, AZ | 103.4 | 72 | Catarina, TX | 99.5 | | Ysleta, TX | 98.1 |
| 23 | Imperial, CA | 106.2 | 48 | Kofa, AZ | 103.3 | 73 | Anvil, AZ | 99.4 | 98 | Grandfalls, TX | 98.0 |
| | Palm Sprgs-Ther, CA | 106.2 | 49 | Organ Pipe Cact, AZ | 103.2 | | Moab, UT | 99.4 | | Independence, CA | 98.0 |
| 25 | Yuma-Proving, AZ | 106.1 | 50 | Lees Ferry, AZ | 103.1 | 75 | Friant, CA | 99.3 | | Randsburg, CA | 98.0 |

### LOWEST

| 1 | Barrow, AK | 45.5 | 26 | King Salmon, AK | 63.2 | 51 | Wolf Creek Pass, CO | 65.9 | | Scotia, CA | 68.6 |
|---|---|---|---|---|---|---|---|---|---|---|---|
| 2 | Shemya, AK | 49.2 | 27 | Puntilla, AK | 63.6 | 52 | Fort Ross, CA | 66.1 | 77 | Orick Prairie Crk, CA | 68.7 |
| 3 | St. Paul Island, AK | 50.0 | 28 | Sn Frncsco-Ocns, CA | 63.9 | 53 | North Bend, OR | 66.3 | 78 | Eklutna, AK | 69.1 |
| 4 | Wales, AK | 51.8 | 29 | Half Moon Bay, CA | 64.0 | 54 | Bandon, OR | 66.6 | 79 | Grand Marais, MN | 69.3 |
| 5 | Mt Washington, NH | 54.2 | | Juneau, AK | 64.0 | | Klamath, CA | 66.6 | 80 | San Gregorio, CA | 69.4 |
| 6 | Cold Bay, AK | 55.2 | 31 | Stampede Pass, WA | 64.2 | | Palmer, AK | 66.6 | | Santa Monica, CA | 69.4 |
| 7 | Mauna Loa, HI | 56.2 | 32 | Beaver Falls, AK | 64.3 | | San Francisco, CA | 66.6 | 82 | Tonsina, AK | 69.5 |
| 8 | Nome, AK | 58.3 | 33 | Annette Island, AK | 64.5 | 58 | Snowshoe Lake, AK | 66.8 | 83 | Grandfather Mt, NC | 69.6 |
| 9 | Kotzebue, AK | 59.7 | | Climax, CO | 64.5 | 59 | McKinley, AK | 66.9 | | Otis, OR | 69.6 |
| | Yakutat, AK | 59.7 | | Wrangell, AK | 64.5 | 60 | Hoquiam, WA | 67.3 | | Wasilla, AK | 69.6 |
| 11 | Kitoi Bay, AK | 60.5 | 36 | Newport, OR | 64.6 | | Matanuska, AK | 67.3 | 86 | Northway, AK | 69.8 |
| | Rainier, WA | 60.5 | 37 | Anchorage-Elmn, AK | 64.7 | | Tillamook, OR | 67.3 | 87 | Berkeley, CA | 69.9 |
| 13 | Homer, AK | 60.6 | 38 | Ketchikan, AK | 65.0 | 63 | Governmnt Cmp, OR | 67.5 | | Olga, WA | 69.9 |
| 14 | Sitka, AK | 61.0 | 39 | Intricate Bay, AK | 65.1 | 64 | Port Alsworth, AK | 67.6 | 89 | Clearwater, WA | 70.0 |
| 15 | Cordova, AK | 61.7 | 40 | Anchorage-Intl, AK | 65.3 | 65 | Astoria, OR | 67.7 | 90 | Bettles, AK | 70.2 |
| 16 | Dillingham, AK | 61.8 | 41 | Auke Bay, AK | 65.4 | | Crater Lake, OR | 67.7 | | Big Delta, AK | 70.2 |
| 17 | Kenai, AK | 61.9 | | Kulani Camp, HI | 65.4 | 67 | Seaside, OR | 67.8 | | Cloverdale, OR | 70.2 |
| | Little Prt Walter, AK | 61.9 | | Long Beach, WA | 65.4 | | Talkeetna, AK | 67.8 | 93 | Glennallen, AK | 70.7 |
| 19 | Seward, AK | 62.0 | 44 | Crescent City, CA | 65.5 | 69 | Brookings, OR | 67.9 | 94 | Twin Lakes, CA | 70.8 |
| 20 | Eureka, CA | 62.1 | | Fort Bragg, CA | 65.5 | 70 | Aberdeen, WA | 68.1 | 95 | Bellingham, WA | 70.9 |
| 21 | Blowhard Mt, UT | 62.3 | | Haleakala, HI | 65.5 | | Gold Beach, OR | 68.1 | | Coupeville, WA | 70.9 |
| 22 | Bethel, AK | 62.5 | | Mount Mansfield, VT | 65.5 | | Monterey, CA | 68.1 | | Eielson, AK | 70.9 |
| | Valdez, AK | 62.5 | 48 | Grayland, WA | 65.6 | 73 | Port Angeles, WA | 68.2 | | Pismo Beach, CA | 70.9 |
| 24 | Kasilof, AK | 62.7 | | Morro Bay, CA | 65.6 | 74 | Gulkana, AK | 68.5 | | Ruxton Park, CO | 70.9 |
| 25 | Iliamna, AK | 63.0 | 50 | Cooper Lake, AK | 65.7 | 75 | McGrath, AK | 68.6 | 100 | Alta, UT | 71.0 |

## ANNUAL PRECIPITATION (")

### HIGHEST

| | | | | | | | | | | | |
|---|---|---|---|---|---|---|---|---|---|---|---|
| 1 | Little Prt Waltr, AK | 225.62 | 26 | Detroit Dam, OR | 85.51 | 51 | Belknap Springs, OR | 72.25 | 76 | Hialeah, FL | 66.08 |
| 2 | Yakutat, AK | 155.43 | 27 | Governmnt Cmp, OR | 85.06 | 52 | Coweeta, NC | 72.16 | | Tallahassee, FL | 66.08 |
| 3 | Beaver Falls, AK | 148.39 | 28 | Sitka, AK | 84.91 | 53 | Helen, GA | 71.74 | 78 | Oberlin, LA | 66.01 |
| 4 | Ketchikan, AK | 144.75 | 29 | Hana, HI | 84.36 | 54 | Grayland, WA | 71.44 | 79 | Falls City, OR | 65.79 |
| 5 | Aberdn-20 nne, WA | 131.33 | 30 | Skamania, WA | 83.98 | 55 | Brookings, OR | 71.38 | 80 | Bay Minette, AL | 65.78 |
| 6 | Hilo, HI | 131.23 | 31 | Stampede Pass, WA | 83.96 | 56 | Three Lynx, OR | 70.54 | | Fairhope, AL | 65.78 |
| 7 | Kailua, HI | 128.02 | 32 | Cloverdale, OR | 82.10 | 57 | Concrete, WA | 70.06 | 82 | Sierra City, CA | 65.77 |
| 8 | Forks, WA | 118.50 | 33 | Aberdeen, WA | 81.74 | 58 | Lake Spaulding, CA | 69.88 | 83 | Franklin, LA | 65.76 |
| 9 | Clearwater, WA | 116.37 | 34 | Stevens Pass, WA | 80.87 | 59 | Seward, AK | 69.45 | 84 | Hammond, LA | 65.66 |
| 10 | Rainier, WA | 114.21 | 35 | Scotts Mills, OR | 80.42 | 60 | Wewahitchka, FL | 69.26 | 85 | Kitoi Bay, AK | 65.50 |
| 11 | Kulani Camp, HI | 112.39 | 36 | Illahe, OR | 80.40 | 61 | Newport, OR | 69.23 | 86 | Amite, LA | 65.38 |
| 12 | Cougar, WA | 112.22 | 37 | Long Beach, WA | 80.35 | 62 | Saucier, MS | 68.67 | 87 | Pascagoula, MS | 65.35 |
| 13 | Hawaii Volcano, HI | 110.20 | 38 | Wrangell, AK | 79.25 | 63 | Milton, FL | 68.09 | 88 | Paradis, LA | 65.31 |
| 14 | Grays River, WA | 107.47 | 39 | Kaneohe Mauka, HI | 78.94 | 64 | Blowing Rock, NC | 67.87 | 89 | Brewton, AL | 65.26 |
| 15 | Annette Island, AK | 102.89 | 40 | Newhalem, WA | 78.75 | 65 | Woodville, MS | 67.43 | 90 | Pisgah Forest, NC | 65.21 |
| 16 | Mt Washington, NH | 99.95 | 41 | Portland-Hdwrks, OR | 78.39 | 66 | Blue Canyon, CA | 67.37 | 91 | Dahlonega, GA | 64.99 |
| 17 | Cordova, AK | 99.56 | 42 | Gold Beach, OR | 77.94 | 67 | Niceville, FL | 67.30 | 92 | Tryon, NC | 64.87 |
| 18 | Upper Bkr Dm, WA | 98.07 | 43 | Strawberry Vlly, CA | 77.67 | 68 | Kanalohuluhulu, HI | 66.96 | 93 | Astoria, OR | 64.86 |
| 19 | Cedar Lake, WA | 97.88 | 44 | Diablo Dam, WA | 77.51 | 69 | Hoquiam, WA | 66.94 | 94 | Houma, LA | 64.74 |
| 20 | Sappho, WA | 96.21 | 45 | Silver Crk Falls, OR | 77.03 | 70 | Valdez, AK | 66.70 | 95 | Coden, AL | 64.62 |
| 21 | Otis, OR | 96.13 | 46 | Klamath, CA | 76.33 | 71 | Elma, WA | 66.58 | 96 | Corvallis-Water, OR | 64.54 |
| 22 | Tidewater, OR | 88.62 | 47 | Bonneville Dam, OR | 74.00 | 72 | Richardson Grve, CA | 66.38 | 97 | Mobile, AL | 64.53 |
| 23 | Tillamook, OR | 87.19 | 48 | Seaside, OR | 73.50 | 73 | Marion, OR | 66.29 | 98 | De Funiak Spris, FL | 64.51 |
| 24 | Highlands, NC | 85.59 | 49 | Mount Mansfield, VT | 73.45 | 74 | Robertsdale, AL | 66.15 | 99 | Bowman Dam, CA | 64.40 |
| 25 | Palmer, WA | 85.58 | 50 | Clayton, GA | 72.95 | 75 | Morgan City, LA | 66.10 | 100 | Gulfport, MS | 64.39 |

### LOWEST

| | | | | | | | | | | | |
|---|---|---|---|---|---|---|---|---|---|---|---|
| 1 | Death Valley, CA | 2.30 | 26 | Dyer, NV | 5.18 | | Priest Rapids, WA | 6.56 | 76 | Wamsutter, WY | 7.21 |
| 2 | El Centro, CA | 3.13 | 27 | Fallon, NV | 5.20 | 52 | Wahweap, AZ | 6.57 | 77 | Center, CO | 7.26 |
| | Imperial, CA | 3.13 | 28 | Lovelock-Derby, NV | 5.22 | 53 | Basin, WY | 6.72 | 78 | Ephrata, WA | 7.28 |
| 4 | Yuma-Intl, AZ | 3.38 | 29 | Parker, AZ | 5.25 | 54 | Pahranagat, NV | 6.73 | 79 | Corcoran, CA | 7.29 |
| 5 | Brawley, CA | 3.45 | 30 | Needles, CA | 5.30 | 55 | Mojave, CA | 6.74 | 80 | Ouray, UT | 7.32 |
| | Mecca, CA | 3.45 | 31 | Lahontan Dam, NV | 5.35 | 56 | Sunnyside, WA | 6.76 | 81 | Smokey Valley, NV | 7.35 |
| 7 | Plm Spngs-Thr, CA | 3.60 | 32 | Bishop, CA | 5.38 | 57 | Lovell, WY | 6.81 | 82 | Golconda, NV | 7.37 |
| 8 | Indio, CA | 3.68 | 33 | Independence, CA | 5.43 | | Randsburg, CA | 6.81 | 83 | Bruneau, ID | 7.44 |
| 9 | Yuma-Citrus, AZ | 3.90 | 34 | Deaver, WY | 5.45 | 59 | Goldfield, NV | 6.85 | 84 | Haiwee, CA | 7.49 |
| 10 | Iron Mountain, CA | 3.91 | 35 | Lovelock, NV | 5.46 | | Myton, UT | 6.85 | 85 | Wapato, WA | 7.51 |
| 11 | Gold Rock, CA | 3.95 | 36 | Yerington, NV | 5.49 | 61 | Green River, UT | 6.87 | 86 | Worland, WY | 7.53 |
| 12 | Blythe, CA | 3.98 | 37 | Palm Springs, CA | 5.64 | | Mexican Hat, UT | 6.87 | 87 | Worland-Muni, WY | 7.54 |
| 13 | Blythe-Rvrside, CA | 4.02 | 38 | Tonopah, NV | 5.67 | | Wasco, CA | 6.87 | 88 | Palmdale, CA | 7.55 |
| 14 | Yuma-Proving, AZ | 4.10 | 39 | Mina, NV | 5.72 | 64 | Fort Duchesne, UT | 6.91 | 89 | Moxee City, WA | 7.56 |
| 15 | Barrow, AK | 4.14 | 40 | Hanksville, UT | 5.77 | 65 | Wah Wah, UT | 6.92 | 90 | Challis, ID | 7.62 |
| 16 | Daggett, CA | 4.28 | 41 | Buttonwillow, CA | 5.81 | 66 | Bitter Creek, WY | 6.98 | | Clark, WY | 7.62 |
| 17 | Eagle Mountain, CA | 4.46 | 42 | Bouse, AZ | 6.03 | 67 | Grand View, ID | 7.02 | 92 | Delta, CO | 7.63 |
| 18 | Twntynine Pms, CA | 4.47 | | Callao, UT | 6.03 | 68 | Partoun, UT | 7.06 | | Kennewick, WA | 7.63 |
| 19 | Trona, CA | 4.48 | 44 | Bakersfield, CA | 6.07 | 69 | Five Points, CA | 7.07 | 94 | Kofa, AZ | 7.65 |
| 20 | Las Vegas, NV | 4.53 | 45 | Parker Reservoir, CA | 6.17 | 70 | Fontenelle Dam, WY | 7.08 | 95 | McNary Dam, WA | 7.66 |
| 21 | Hayfield, CA | 4.56 | 46 | Victorville, CA | 6.18 | 71 | Borrego Desert, CA | 7.11 | 96 | Prosser, WA | 7.69 |
| 22 | Desert, NV | 4.75 | 47 | Boulder City, NV | 6.44 | 72 | Richland, WA | 7.13 | 97 | Imlay, NV | 7.72 |
| 23 | Inyokern, CA | 4.96 | 48 | Lees Ferry, AZ | 6.53 | 73 | Gila Bend, AZ | 7.14 | 98 | Smyrna, WA | 7.73 |
| 24 | Wendover, UT | 5.10 | 49 | Page, AZ | 6.54 | 74 | Reno, NV | 7.15 | 99 | Alamosa, CO | 7.75 |
| 25 | Pahrump, NV | 5.16 | 50 | Deep Springs, CA | 6.56 | 75 | Roosevelt, UT | 7.16 | | Castle Dale, UT | 7.75 |

## ANNUAL SNOWFALL (")

### HIGHEST

| 1 | Rainier, WA | 686.1 | 26 | Hebgen Dam, MT | 208.0 | 51 | Oswego, NY | 153.9 | 76 | Swan Lake, MT | 133.0 |
|---|---|---|---|---|---|---|---|---|---|---|---|
| 2 | Alta, UT | 524.1 | 27 | Telluride, CO | 205.9 | 52 | Lindbergh Lake, MT | 153.7 | 77 | Stehekin, WA | 132.4 |
| 3 | Crater Lake, OR | 475.3 | 28 | Truckee, CA | 201.1 | 53 | Mineral, CA | 152.5 | 78 | Talkeetna, AK | 131.8 |
| 4 | Stevens Pass, WA | 465.2 | 29 | Dixie, ID | 199.0 | 54 | Maple City, MI | 150.3 | 79 | Chester, CA | 131.7 |
| 5 | Wolf Crk Pass, CO | 452.1 | 30 | Yakutat, AK | 193.4 | 55 | Grand Marais, MI | 149.9 | 80 | Canaan Valley, WV | 131.6 |
| 6 | Stampede Pass, WA | 427.0 | 31 | Stillwater Rsrvr, NY | 192.2 | 56 | Spicer, CO | 149.6 | 81 | Corry, PA | 131.3 |
| 7 | Silver Lake, UT | 412.4 | 32 | Grant Grove, CA | 184.8 | 57 | Gaylord, MI | 149.0 | 82 | Elk City, ID | 130.2 |
| 8 | Mt Washington, NH | 316.3 | 33 | Donner Memrial, CA | 183.7 | 58 | McCall, ID | 148.8 | | West Glacier, MT | 130.2 |
| 9 | Valdez, AK | 309.9 | 34 | Manzanita Lake, CA | 181.4 | 59 | Silverton, CO | 148.4 | 84 | Wolf Canyon, NM | 128.1 |
| 10 | Gvernmnt Cmp, OR | 266.5 | 35 | Lead, SD | 180.8 | 60 | Terra Alta, WV | 146.2 | 85 | Camden, NY | 127.3 |
| 11 | Climax, CO | 263.8 | | Rico, CO | 180.8 | 61 | Alberta, MI | 145.7 | 86 | Sault Ste. Marie, MI | 126.5 |
| 12 | Snake River, WY | 262.0 | 37 | Red Lodge, MT | 179.2 | 62 | Munising, MI | 145.4 | 87 | Vallecito Dam, CO | 126.2 |
| 13 | Lake Spaulding, CA | 253.0 | 38 | Mystic Lake, MT | 176.7 | 63 | Ruxton Park, CO | 144.0 | 88 | Calaveras, CA | 123.9 |
| 14 | Holden, WA | 248.0 | 39 | Powell, ID | 175.9 | 64 | Shonkin, MT | 143.3 | 89 | Chasm Falls, NY | 122.6 |
| 15 | Bowman Dam, CA | 245.6 | 40 | Bergland Dam, MI | 175.6 | 65 | Ouray, CO | 142.9 | 90 | Bondurant, WY | 121.4 |
| 16 | Blowhard Mt, UT | 239.0 | 41 | Ironwood, MI | 174.7 | 66 | Grand Lk-1 NW, CO | 142.8 | 91 | Lowville, NY | 121.0 |
| 17 | Blue Canyon, CA | 236.9 | 42 | Tahoe City, CA | 172.9 | 67 | Pinkham Notch, NH | 142.4 | 92 | Cordova, AK | 120.5 |
| 18 | Bozeman-12 ne, MT | 232.4 | 43 | Steamboat Sprgs, CO | 170.2 | 68 | Gurney, WI | 140.6 | 93 | Pine View Dam, UT | 120.3 |
| 19 | Boonville, NY | 229.8 | 44 | Moose, WY | 169.7 | 69 | Little Valley, NY | 140.5 | 94 | Cherry Valley, NY | 120.0 |
| 20 | Burgess Jnctn, WY | 229.7 | 45 | Lk Yellowstone, WY | 169.3 | 70 | Plain, WA | 138.1 | 95 | Hayden, CO | 119.9 |
| 21 | Old Forge, NY | 225.5 | 46 | Warren, ID | 169.2 | 71 | Champion, MI | 137.6 | 96 | Seney, MI | 119.6 |
| 22 | Crested Butte, CO | 224.1 | 47 | Marquette-Co, MI | 166.4 | 72 | Gascon, NM | 136.2 | 97 | Boyne Falls, MI | 118.4 |
| 23 | Mt Mansfield, VT | 219.1 | 48 | Colden, NY | 162.6 | 73 | Howard Prairie, OR | 135.8 | 98 | Canyon Dam, CA | 117.8 |
| 24 | Hancock, MI | 216.1 | 49 | Red River, NM | 160.5 | 74 | Wanakena, NY | 134.8 | 99 | Chemult, OR | 117.2 |
| 25 | Island Park, ID | 210.3 | 50 | First Conn Lake, NH | 155.0 | 75 | Whitefish Point, MI | 133.5 | 100 | Syracuse, NY | 116.9 |

### LOWEST*

| 1 | Ajo, AZ | 0.1 | | Edisto Island, SC | 0.1 | | La Tuna, TX | 0.1 | | Sacramento-Exc, CA | 0.1 |
|---|---|---|---|---|---|---|---|---|---|---|---|
| | Albany, GA | 0.1 | | Elberton, GA | 0.1 | | Live Oak, FL | 0.1 | | Sinton, TX | 0.1 |
| | Amistad Dam, TX | 0.1 | | Eudora, AR | 0.1 | | Los Gatos, CA | 0.1 | | Sugar Land, TX | 0.1 |
| | Bakersfield, CA | 0.1 | | Fairfield, CA | 0.1 | | Lumpkin, GA | 0.1 | | Sullivans Island, SC | 0.1 |
| | Belzoni, MS | 0.1 | | Falcon Dam, TX | 0.1 | | Marion Junction, AL | 0.1 | | Thompsons, TX | 0.1 |
| | Biloxi, MS | 0.1 | | Falfurrias, TX | 0.1 | | Matagorda, TX | 0.1 | | Tilden, TX | 0.1 |
| | Brunswck-McK, GA | 0.1 | | Fitzgerald, GA | 0.1 | | Mathis, TX | 0.1 | | Ukiah, CA | 0.1 |
| | Cameron, TX | 0.1 | | Five Points, CA | 0.1 | | New Gulf, TX | 0.1 | | Victoria, TX | 0.1 |
| | Camilla, GA | 0.1 | | Fort Stewart, GA | 0.1 | | Pahrump, NV | 0.1 | | Vista, CA | 0.1 |
| | Campo, CA | 0.1 | | Franklin, LA | 0.1 | | Palacios, TX | 0.1 | | Washington, TX | 0.1 |
| | Candelaria, TX | 0.1 | | Gadsden, AL | 0.1 | | Palm Springs, CA | 0.1 | | West Point, GA | 0.1 |
| | Carville, LA | 0.1 | | Gila Bend, AZ | 0.1 | | Paradis, LA | 0.1 | | Wiggins, MS | 0.1 |
| | Chipley, FL | 0.1 | | Grand Coteau, LA | 0.1 | | Paso Robles, CA | 0.1 | | Wikieup, AZ | 0.1 |
| | Chula Vista, CA | 0.1 | | Gulfport, MS | 0.1 | | Paso Robles-Mni, CA | 0.1 | | Winnsboro, LA | 0.1 |
| | Clifton, AZ | 0.1 | | Hackberry, LA | 0.1 | | Pinopolis Dam, SC | 0.1 | | Winters, CA | 0.1 |
| | Coden, AL | 0.1 | | Hallettsville, TX | 0.1 | | Port Arthur, TX | 0.1 | | Woodland, CA | 0.1 |
| | Colusa, CA | 0.1 | | High Springs, FL | 0.1 | | Port O'Connor, TX | 0.1 | 92 | Andalusia, AL | 0.2 |
| | Conroe, TX | 0.1 | | Hilton Head, SC | 0.1 | | Poteet, TX | 0.1 | | Angleton, TX | 0.2 |
| | Cordele, GA | 0.1 | | Homerville, GA | 0.1 | | Presidio, TX | 0.1 | | Anvil, AZ | 0.2 |
| | Crowley, LA | 0.1 | | Hords Creek Da, TX | 0.1 | | Quincy, FL | 0.1 | | Ashburn, GA | 0.2 |
| | Cuero, TX | 0.1 | | Houma, LA | 0.1 | | Quitman, GA | 0.1 | | Bankhead L&D, AL | 0.2 |
| | Davis, CA | 0.1 | | Jacksonville Bch, FL | 0.1 | | Reserve, LA | 0.1 | | Bardwell Dam, TX | 0.2 |
| | DeFuniak Sprgs, FL | 0.1 | | Jasper, FL | 0.1 | | Rio Grande City, TX | 0.1 | | Baton Rge-Ryn, LA | 0.2 |
| | Donaldsonville, LA | 0.1 | | Jeanerette, LA | 0.1 | | Rockefeller, LA | 0.1 | | Borrego Desert, CA | 0.2 |
| | Eastman, GA | 0.1 | | La Grange, TX | 0.1 | | Roosevelt, AZ | 0.1 | | Carrollton, GA | 0.2 |

* Does not include stations which receive no snowfall. 38 additional stations receive 0.2" snowfall.